KU-208-393

McGraw-Hill Encyclopedia of the Geological Sciences

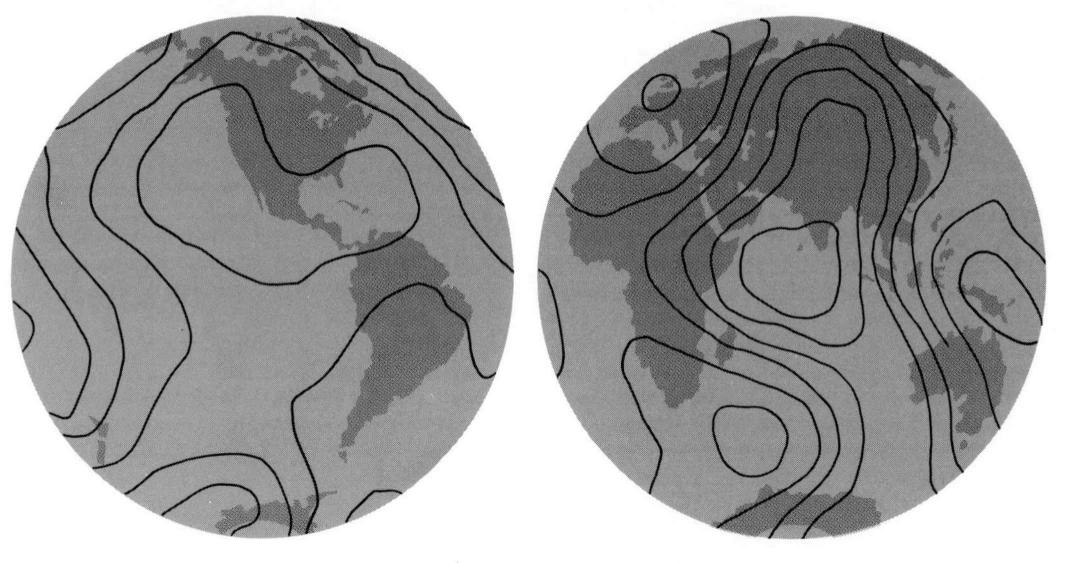

McGraw-Hill Encyclopedia of the Geological Sciences

Daniel N. Lapedes
Editor in Chief

McGraw-Hill Book Company

New York
St. Louis
San Francisco

Auckland
Bogotá
Düsseldorf
Johannesburg
London
Madrid
Mexico
Montreal
New Delhi
Panama
Paris
São Paulo
Singapore
Sydney
Tokyo
Toronto

Library of Congress Cataloging in Publication Data

McGraw-Hill encyclopedia of the geological sciences.

Includes index.
1. Geology—Dictionaries. I. Lapedes, Daniel N.
QE5.M29 550′.3 78-18425
ISBN 0-07-045265-2

Contents

Editorial Staff

Daniel N. Lapedes, Editor in Chief

Sybil P. Parker, Senior editor

Jonathan Weil, Staff editor

Edward J. Fox, Art director

Joe Faulk, Editing manager

Patricia Albers, Senior editing assistant
Judith Alberts, Editing assistant
Dolores Messina, Editing assistant

Ellen Okin Powers, Editing supervisor

Ann D. Bonardi, Art production supervisor
Richard A. Roth, Art editor
Eda Grilli, Art/traffic

Consulting Editors

Dr. John C. Drake. *Associate Professor, Department of Geology, University of Vermont.* GEOLOGY (MINERALOGY AND PETROLOGY).

Prof. Frederic L. Schwab. *Department of Geology, Washington and Lee University.* GEOLOGY (SURFICIAL AND HISTORICAL); PHYSICAL GEOGRAPHY.

Dr. Donald J. P. Swift. *Research Oceanographer, Marine Geology & Geophysics Laboratory, National Oceanic and Atmospheric Administration, Miami, FL.* OCEANOGRAPHY.

Preface

Public interest in the Earth until recently has been transitory, being limited to natural disasters such as earthquakes and erupting volcanoes. However, the environmental movement of the 1960s and the energy problems of the 1970s have changed that attitude considerably. As a result, there has been a great increase in people's awareness of the fragility of the earth system and of the finite limit of its nonrenewable natural resources.

We who live on planet Earth have come to realize that this platform spinning through space is still undergoing fundamental change. Although the Earth with its present mass and density is approximately 4.6 billion years old, many of the mechanisms that produced the features we see are still operating at an undiminished pace. One of the most exciting ideas in the geological sciences today is plate tectonics, which theorizes that the surface of the Earth is broken up into a series of plates which move with respect to each other. Interactions between the plates at their boundaries are responsible for earthquakes, volcanism, and mountain building. A logical extension of plate tectonics is the theory of continental drift. The most recent episode of such drift has given the continents their present form and distribution.

Scientists studying the Earth as a whole consider it to be a dynamic system with solid, liquid, and gaseous phases. These are known to us as the solid earth, the oceans, and the atmosphere. The dynamics of this system have been of interest mainly to the geological scientist, oceanographer, and meteorologist.

The geological scientist is interested in the solid earth and may follow that interest in the areas of geology, geophysics, or geochemistry. Geology is the study of the nature of earth materials and processes and how they have interacted through time to leave a record of past events in existing earth features and materials. Geochemistry is the study of the earth system and the way matter has interacted in the system through time. Geophysics deals with the physical characteristics and dynamic behavior of the earth system and thus concerns itself with the diversity of complex problems involving natural phenomena.

While the primary interest of the geological scientist is the solid earth, that interest is not isolated from considering the effects of the oceans and the atmosphere upon the Earth. This encyclopedia provides a comprehensive treatment of the geological sciences, including geology, geochemistry, geophysics, and those aspects of oceanography and meteorology which are essential to understanding the materials, processes, composition, and physical characteristics of the solid part of the Earth.

The articles included in the Encyclopedia were selected by a Board of Consulting Editors and either have been taken from the *McGraw-Hill Encyclopedia of Science and Technology* (4th ed., 1977) or have been written especially for this volume. The broad scope of geological sciences is covered in great depth in the 560 alphabetically arranged articles on subjects such as plate tectonics, oceans, geological time scale, paleontology, mineralogy, physiographic provinces, orogeny, high-pressure phenomena, geophysical explorations, geochemistry, and atmospheric chemistry.

The volume contains over 700 photographs, maps, tables, drawings, graphs, and diagrams. Cross-references appear within each article to guide the reader to articles on related subjects. All information in the Encyclopedia is easily accessible through an extensive analytical index. An additional useful feature is a table in the Appendix which lists the properties of 1500 mineral species.

DANIEL N. LAPEDES
EDITOR IN CHIEF

McGraw-Hill Encyclopedia of the Geological Sciences

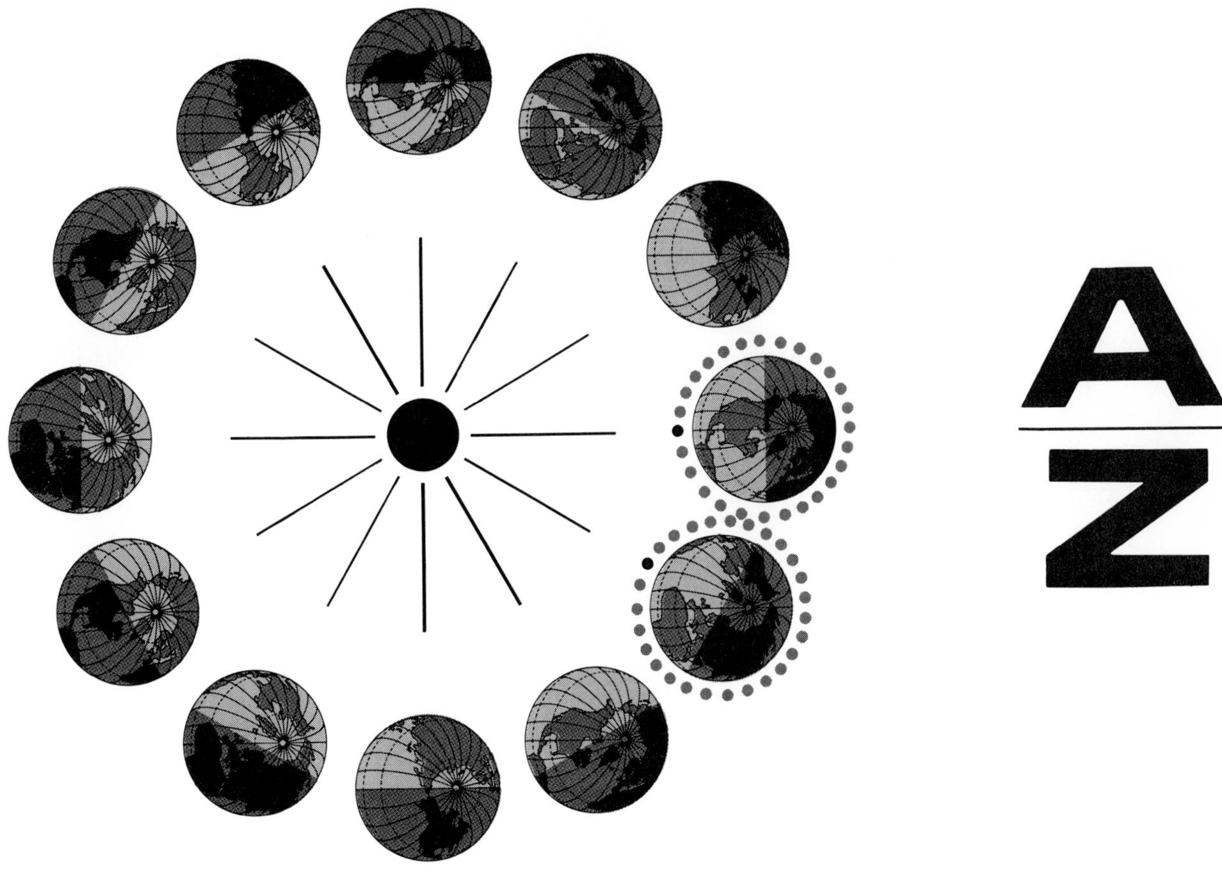

Aeolian (desert) landforms

Topographic features generated by the wind. The most commonly seen aeolian landforms are sand dunes created by transportation and accumulation of windblown sand. Blankets of wind-deposited loess, consisting of fine-grained silt, are less obvious than dunes, but cover extensive areas in some parts of the world.

Movement of particles by wind. Although wind is not capable of moving large particles, it can transport large amounts of sand, silt, and clay, especially in desert or coastal areas where loose sand and finer particles are exposed at the surface as a result of the scarcity of vegetation or the continuous supply of loose material.

The size of particles which can be moved by wind depends primarily on wind velocity. Swirls and eddies associated with wind turbulence have upward components of movement which pick up loose material. The velocity of upward gusts is usually not very constant, but generally averages about 20% of mean wind velocity. Where turbulence is strong enough to overcome the force of gravity, the particles remain suspended in the air and are carried downwind. Because the maximum size of a particle suspended in the air varies with the square of its radius, wind is generally limited to the movement of material of sand size or smaller, and suspended particles are quite sensitive to changes in wind velocity. Thus, the wind is an effective winnowing agent, separating finer from coarser material, and grain size of aeolian deposits is typically quite uniform.

Windblown sand seldom rises more than a few feet above the ground. However, fine silt and dust may rise to altitudes of hundreds or thousands of feet during desert windstorms. Individual sand grains rise and fall as they are blown downwind and travel in a bouncing fashion known as saltation. Sand grains bouncing along the ground within a few feet of the surface abrade materials with which they collide. Such natural sand blasting produces polished, pitted, grooved, and faceted rocks known as ventifacts, and mutual abrasion of the sand results in highly rounded, spherical grains.

Silt- and clay-sized particles can be held in suspension much longer than sand grains, and thus may travel long distances before settling out.

Fig. 1. Barchan dunes at Moses Lake in Washington.

Sand dunes. Where abundant loose sand is available for the wind to carry, sand dunes develop. As soon as enough sand accumulates in one place, it interferes with the movement of air and a wind shadow is produced which contributes to the shaping of the pile of sand. Sand grains bounce up the windward side of the sand pile until they reach the crest, then tumble down the lee side in the wind shadow behind the crest. Sand trapped in the wind shadow accumulates until the slope reaches the angle of repose for loose sand, where any additional increase in slope causes sliding of the sand and development of a slip face. Dunes advance downwind by erosion of sand on the windward side and redeposition on the slip face. Dunes may have a variety of shapes, depending on wind conditions, vegetation, and sand supply.

Barchan dunes. These are crescent-shaped forms in which the ends of the crescent point downwind and the steep slip face of the dune is concave downwind (Fig. 1). The crescent shape is maintained as the dune advances downwind because the rate of movement of sand is somewhat slower in the central part of the crescent where dune height is greatest. Barchans are commonly found on barren desert floors, where they may occur singly or in clusters.

Parabolic dunes. Parabolic dunes also have crescent-shaped forms, except that the crescent faces the opposite direction and the steep slip face is on the convex rather than the concave side of the dune. They typically occur where vegetation impedes the advance of the points of the crescent, allowing the higher vegetation-free central part of the dune to move at a faster rate, leaving the ends of the dune trailing behind.

Transverse dunes. These are elongate forms whose long axes are at right angles to the prevailing wind direction. They frequently develop as a result of coalescence of other dune types.

Longitudinal dunes. These consist of long ridges parallel to the prevailing wind direction, often found where sand is blown through a gap in a high ridge. However, very large elongate dunes up to 700 ft (210 m) high and 50 mi (80 km) long also occur in regular rows in the desert regions of Africa and the Middle East.

Star-shaped dunes. Dunes of this type are found in some areas where the direction of prevailing winds shifts from season to season, piling up sand in forms with long radial arms extending from a central high point (Fig. 2).

Fig. 2. Star-shaped dune at Death Valley in California.

Fine-grained deposits. The fine silt and clay winnowed out from coarser sand is often blown longer distances before coming to rest as a blanket of loess mantling the preexisting topography. Thick deposits of loess are most often found in regions downwind from glacial outwash plains or alluvial valleys such as the Mississippi Valley, southeastern Washington, and portions of Europe, China, and the Soviet Union. *See* LOESS; SAND.

[DON J. EASTERBROOK]

Bibliography: R. A. Bagnold, *The Physics of Blown Sand and Desert Dunes*, 1941; W. S. Cooper, *Coastal Sand Dunes of Oregon and Washington*, Geol. Soc. Amer. Mem. 72, 1958; D. J. Easterbrook, *Principles of Geomorphology*, pp. 288–303, 1969.

Agate

A variety of chalcedonic quartz that is distinguished by color banding in curved or irregular patterns (see illustration). Most agate used for ornamental purposes is composed of two or more tones or intensities of brownish-red, often interlayered with white, but is also commonly composed of various shades of gray and white. Since agate is relatively porous, it can be dyed permanently in red, green, blue, and a variety of other colors. The difference in porosity of the adjacent layers permits the dye to penetrate unevenly and preserves marked differences in appearance between layers.

The term agate is also used with prefixes to de-

Section of polished agate showing the characteristic color banding. (*Field Museum of Natural History, Chicago*)

scribe certain types of chalcedony in which banding is not evident. Moss agate is a milky or almost transparent chalcedony which contains dark inclusions in a dendritic pattern. Iris agate exhibits an iridescent color effect. Fortification, or landscape, agate is translucent and contains inclusions that give it an appearance reminiscent of familiar natural scenes. Banded agate is distinguished from onyx by the fact that its banding is curved or irregular, in contrast to the straight, parallel layers of onyx. The properties of agate are those of chalcedony: refractive indices of 1.535 and 1.539, a hardness of 6 1/2 to 7, and a specific gravity of about 2.60. *See* CHALCEDONY; GEM; QUARTZ.

[RICHARD T. LIDDICOAT, JR.]

Albertite

A naturally occurring, brown to black, carbonaceous substance associated with oil shales. It has a specific gravity of 1.07–1.10, is infusible, insoluble in carbon disulfide, and consists of approximately 79–89% carbon, 7–13% hydrogen, 0–5% sulfur, some oxygen, and traces of nitrogen. The material derives its name from the Albert Mines, Albert County, New Brunswick, Canada, where veins up to 17 ft wide have been traced for over 1/2 mi. Albertite also occurs in Pictou County, Nova Scotia; Uinta County, Utah; the Falkland Islands; Germany; Tasmania; and Portuguese West Africa.

Albertite is classified on the basis of its physical properties. Different deposits may have had variable contributions from different sources. Low oxygen contents of various samples of albertite indicate, however, that the material is not derived from coal in spite of its occasional occurrence near coal seams.

Albertite was mined for many years in New Brunswick and Nova Scotia for use in enriching bituminous coal in the manufacture of illuminating gas. *See* ASPHALT AND ASPHALTITE; ELATERITE; IMPSONITE; WURTZILITE.

[IRVING A. BREGER]

Albite

A sodium feldspar, $NaAlSi_3O_8$, with an Al-Si distribution producing triclinic symmetry (see illustration). Albite is usually found in nature with the highest state of Al-Si order which appears to be possible at low temperature. As a result of other Al-Si distributions of a less ordered character, several other states of $NaAlSi_3O_8$ exist; for example, monalbite with monoclinic symmetry is stable under equilibrium conditions at temperatures (≈ 1000°C) near the melting point. Monalbite inverts upon cooling under nonequilibrium conditions into analbite that, like albite, is triclinic but has a somewhat different lattice geometry. The transformation temperature depends upon the degree of Al-Si disorder and the amount of $KAlSi_3O_8$ or $CaAl_2Si_2O_8$ present. Albite in its most ordered form cannot take more than about 2 mole % of $KAlSi_3O_8$ or $CaAl_2Si_2O_8$ as solid solution in its structure. *See* FELDSPAR; IGNEOUS ROCKS.

[FRITZ H. LAVES]

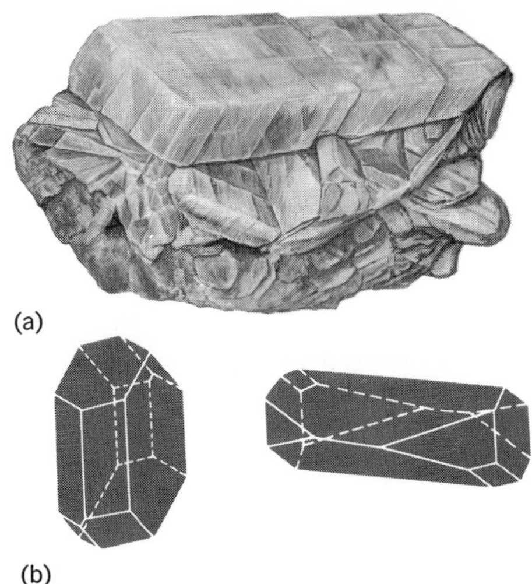

Albite. (*a*) Crystals, Amelia Court House, Va. (*specimen from Department of Geology, Bryn Mawr College*). (*b*) Crystal habits (*from C. S. Hurlbut, Jr., Dana's Manual of Mineralogy, 17th ed., copyright © 1959 by John Wiley & Sons, Inc.; reprinted by permission*)

Allanite

A mineral, also known as orthite, distinguished from other members of the epidote group of silicates by a relatively high content of rare earths, chiefly cerium, lanthanum, and yttrium. The chemical composition is $(Ca, Ce, La, Y)_2(Al, Fe)_3Si_3O_{12}(OH)$. Other lanthanide elements are generally present, as is Be in some varieties. Small amounts of thorium and uranium are often present, and the mineral then may be metamict. Allan-

Allanite from Goiaz, Brazil. (*Specimen courtesy of Department of Geology, Bryn Mawr College*)

ite is monoclinic in crystallization. The color is black or brownish-black, and the specific gravity ranges from 3.4 to 4.2 (see illustration). Allanite is widespread as an accessory constituent in acid to intermediate igneous rocks and in gneissic metamorphic rocks. It is typical pegmatite mineral. *See* EPIDOTE; METAMICT STATE; RADIOACTIVE MINERALS.

[CLIFFORD FRONDEL]

Amber

A fossil resin derived from a coniferous tree. It has been used for ornamental purposes since prehistoric times, for it is mentioned among the first recorded references to beads and other ornamental objects. When used for jewelry purposes, it is usually a transparent yellow, orange, or reddish-brown color. It frequently contains insects that were entrapped when the resinous tree exudation was still in a semiliquid state (see illustration). Amber of a translucent or semitranslucent type is used for decorating small boxes, for pipe stems, and for a variety of ornamental purposes.

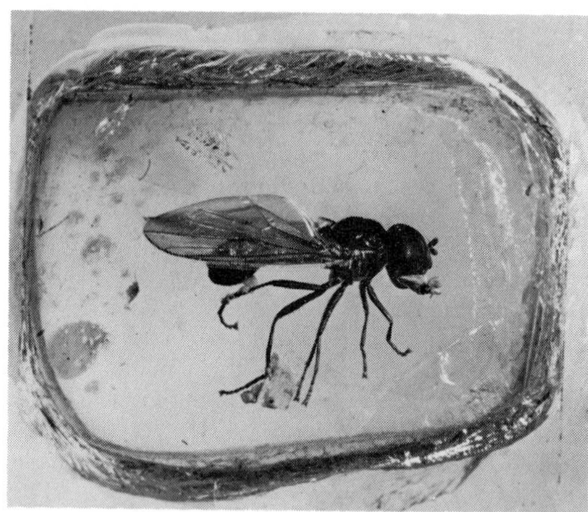

Pseudosphegina carpenteri Hull (Diptera), syrphid fly in Baltic amber, Oligocene. (*Courtesy of F. M. Carpenter*)

For centuries, the most important source of amber has been along the south shore of the Baltic Sea, particularly in the section of Poland that was Germany before World War II (East Prussia). Other sources are Burma, Sicily, and Rumania. There are only minor differences in composition and properties of material from the various sources. Amber is amorphous, has a refractive index of about 1.54, a specific gravity of 1.05–1.10, and a hardness of 2–2.5 on Mohs scale. In polarized light, irregular interference colors are exhibited as a result of pronounced internal strain. *See* GEM; MINERALOGY.

[RICHARD T. LIDDICOAT, JR.]

Amblygonite

A mineral consisting of lithium, sodium, and aluminum phosphate (Li, Na)Al(PO_4)(F,OH). Lithium and sodium substitute for each other but generally Li is in excess of Na; F and OH also substitute mutually. These substitutions give rise to an amblygonite series of minerals. Amblygonite crystallizes in the triclinic system. Crystals are short and prismatic. Colors range from white through shades of gray, yellow, green, and brown.

Amblygonite, found at many places in the world, occurs mostly in granitic pegmatites. It is mined in the Black Hills of South Dakota, in southwestern Africa, and elsewhere for its lithium content.

[WAYNE R. LOWELL]

Amethyst

The transparent purple to violet variety of the mineral quartz. Although quartz is perhaps the commonest gem mineral known, amethyst is rare in the deep colors that characterize fine quality. Amethyst is usually colored unevenly and is often heated slightly in an effort to distribute the color more evenly. Heating at higher temperatures usually changes it to yellow or brown (rarely green), and further heating removes all color. The principal sources are Brazil, Arizona, Uruguay, and the Soviet Union. Amethyst is often cut in step or brilliant shapes, and drilled or carved for beads. Carvings are made both from transparent and nontransparent material. *See* GEM; QUARTZ.

[RICHARD T. LIDDICOAT, JR.]

Amphibole

A large group of common rock-forming inosilicate (metasilicate) minerals. The amphiboles exhibit a wide range of compositional variation, as indicated by the generalized formulas

$$(Na,Ca)_{2-3}(Mg,Fe^{2+},Fe^{3+},Al)_5(Si,Al)_8O_{22}(OH,O,F)_2$$

$$(Mg,Fe^{2+},Fe^{3+},Al)_7(Si,Al)_8O_{22}(OH,O,F)_2$$

The amphiboles represent a complex series of solid solutions between a variety of idealized end members. The species names of these end members are usually the best known; the more important ones are listed below. *See* SILICATE MINERALS.

Orthorhombic amphiboles	
Anthophyllite	$(Mg,Fe^{2+})_7Si_8O_{22}(OH)_2$
Gedrite	$(Mg,Fe^{2+})_5Al_2(Si_6Al_2)O_{22}(OH)_2$
Monoclinic amphiboles	
Cummingtonite	$(Fe^{2+},Mg)_7Si_8O_{22}(OH)_2$
Tremolite	$Ca_2Mg_5Si_8O_{22}(OH)_2$
Tschermakite	$Ca_2Mg_3Al_2(Si_6Al_2)(OH)_2$
Edenite	$NaCa_2Mg_5(Si_7Al)O_{22}(OH)_2$
Glaucophane	$Na_2(Mg_3Al_2)Si_8O_{22}(OH)_2$
Riebeckite	$Na_2(Fe_3{}^{2+}Fe_2{}^{3+})Si_8O_{22}(OH)_2$
Arfvedsonite	$Na_3Fe_4{}^{2+}Fe^{3+}Si_8O_{22}(OH)_2$
Eckermannite	$Na_3Mg_4(Fe^{3+},Al)Si_8O_{22}(OH)_2$

Composition. Each of the listed calcium (Ca) amphiboles has a ferrous iron equivalent and would be indicated by the prefix fero-, for example, ferrotremolite. Also, some ferric iron and oxygen in the place of the OH groups can occur in the ferroamphiboles and is indicated by the prefix ferri-, for example, ferritremolite; the term oxyhornblende is commonly applied to the ferriamphiboles. A high content of fluorine is indicated by the prefix fluoro-, for example, fluorocummingtonite. The iron content in the anthophyllite series seems to be limited to about 60 mole % of the iron end member; a higher concentration of iron results in the monoclinic cummingtonite. Cummingtonite with less then 25 mole % of the iron end member is unknown. The overlap, between 25–60 mole % iron, between the two series is probably the result of different amounts of aluminum, particularly in the anthophyllite. For the Na-Ca amphiboles, solid solution between any of the end members is possible so that a single crystal could be composed of contributions from some 10 different end members. The general term hornblende is usually applied to these solid solutions of the Na-Ca amphiboles. Obviously, a chemical analysis is needed to determine the composition. However, certain physical features suggest the dominance of certain compositions. *See* ANTHOPHYLLITE; CUMMINGTONITE; HORNBLENDE.

Physical and optical properties. All amphiboles are characterized by two directions of well-developed prismatic cleavages which intersect at approximately 124°. In detail, the magnesium-rich amphiboles are generally light colored (white, gray, light green) with the color darkening to dark green, dark brown, or black with increasing iron content. The presence of sodium is often indicated by a bluish color, especially in thin sections. The Mg-rich amphiboles commonly develop long needle-like or fibrous crystals, and there is a rough correlation such that, as the iron and aluminum content increases, the individual crystals become progressively more short and stubby.

Occurrence. The amphiboles as a group are present as minor constituents in many volcanic, igneous, and metamorphic rocks and are thus able to form over most of the temperature range observed in the Earth's crust. Amphiboles are present as major constituents in many metamorphic schists and gneisses, for example cummingtonite schist, glaucophane schist, and hornblende schist. These amphibole schists are most common in the middle grades of metamorphism. Amphiboles occur as gangue minerals in certain ore deposits and are found in many skarn deposits. Ferriamphiboles or oxyhornblendes are usually the product of oxidation of ordinary hornblendes in lava flows. Intertwinned fibrous crystals of tremolite are known as nephrite and classed as one of the jades. Fibrous riebeckite is known as the semiprecious stone crocidolite or tiger's eye. Fibrous tremolite, anthopyllite, and riebeckite are used in commercial asbestos. Amphiboles can result from the alteration of pyroxenes. Amphiboles of this type are occasionally referred to as uralite and the process as uralization. The amphiboles can be altered to a variety of decomposition products with talc, antigorite, chlorite, and epidote the most commonly observed. *See* ASBESTOS; GLAUCOPHANE; JADE; TREMOLITE.

Crystal structure. The compositional variation is best explained from the viewpoint of crystal structure. In silicates four oxygens arranged at the corners of a tetrahedron surround each 4-valent silicon atom. These SiO_4 tetrahedrons can share one, two, three, or four oxygen anions with one, two, three, or four neighboring tetrahedrons (one oxygen to each neighboring tetrahedron) and thereby reduce in steps the excess charges on the oxygen atoms. The amphiboles are characterized by the polymerization of the SiO_4 tetrahedrons into endless double chains or ribbons. These double chains with the associated (OH,F,O) atoms are bonded to adjacent parallel double chains by the other metal atoms of the crystal. Extensive miscibility between atoms of similar size occupying essentially equivalent atomic sites (within charge limitations) is very common in silicates. The ionic sizes of Mg, Fe^{2+}, Fe^{3+}, and Al are similar enough that these atoms occupy the same type of atomic site in the mineral; Na and Ca occupy equivalent atomic sites, as do OH, O, and F. Limited substitution of the silicon atoms by aluminum atoms also can occur, which extends the range of compositional variation. Very commonly one-fourth of the silicon atoms are replaced by aluminum, but greater replacement than this by aluminum is rare. Solid solution between the cummingtonite series and the calcium amphibole series is not present even though the calcium sites can be completely replaced by the iron-magnesium atoms in cummingtonite.

Genetic relations. Because of the lack of solid solution between them, cummingtonite and the calcium amphiboles can exist together at equilibrium and they often occur intimately intergrown. Anthophyllite and cummingtonite can exist together, apparently at equilibrium, but anthophyllite coexisting with the calcium amphiboles seems to be rare.

The temperature of formation of the mineral assemblage is not particularly indicated by the amphiboles. However, the aluminum- and iron-rich hornblendes tend to occur in the higher grades of metamorphism and the aluminum-poor hornblendes in the lower grades of metamorphism. The sodium- and aluminum-rich hornblende hastingsite is often used as an index mineral for the highest grade of metamorphism (granulite facies).

[GEORGE W. DE VORE]

Amphibolite

A class of metamorphic rocks. Amphibolites are crystalline schists (recrystallized, foliated metamorphic rocks) of greenish-black color composed of amphibole (hornblende) and plagioclase feldspar as essential minerals. Prisms of hornblende show a preferred orientation defining a lineation and, particularly if biotite is present, also a schistosity. Amphibolites are formed by regional metamorphism under the conditions of the amphibolite facies (about 500°C). *See* METAMORPHIC ROCKS; SCHIST.

Among the metamorphic rocks, the amphibolites occupy a position similar to that of the basalt-gabbro rocks of the igneous suite. However, whereas basalts and gabbros have been extensively investigated, and well-known chemical and mineralogical criteria exist to distinguish the various types, such as tholeiites, tephrites, and basanites, no corresponding characterization has been made of amphibolites. A wide and virgin field is here open, and calculation of the newly proposed mesonorms will provide a means of standard comparison and thus serve as a convenient way of studying amphibolites chemically. A great number of chemical elements are able to enter into the crystal structure of hornblende [iron (II) for magnesium, sodium for calcium, iron (III) for aluminum, and others], whereby the number of constituent mineral phases in the rock is greatly reduced (the principle of the paucity of mineral phases). Thus amphibolites of very complicated chemical compositions are usually bimineralic, containing just hornblende and plagioclase. Three or four chief minerals are not unusual, but to encounter five or six chief minerals is rare.

Amphibolites may be formed from rocks of rather diverse kinds, igneous and sedimentary. Frequently the nature of the original rock cannot be determined.

An igneous parentage (derived from gabbros, diabases, basalts, or basic tuffs) may reveal itself, for example, in weak foliation, but this is not a rule. Quartz and biotite are often present and in some rocks garnets may be present as conspicuously large, round porphyroblasts. The presence of epidote usually indicates a slightly lower temperature of formation. Minor amounts of sphene, apatite, and opaque ores are almost always present. Amphibolites derived from ultrabasic igneous rocks are very dark (black) with a high content of hornblende; biotite may be present, and sometimes anthophyllite. *See* PORPHYROBLAST.

Amphibolites of sedimentary parentage exhibit a much wider range of composition. The lime content especially is subject to large variations which are reflected in the mineral contents: diopside amphibolites, epidote (zoisite) amphibolites, and sphene may be abundant. In other types cordierite is present, often associated with anthophyllite.

Some amphibolites are truly metasomatic and may derive from impure limestone or dolomite which has reacted with "emanations" containing silica, magnesium, and iron.

Amphibolite has a wide distribution in areas of Precambrian rocks and in areas of deeply eroded younger mountain ranges. Next to granitic gneiss, amphibolite probably is the most abundant crystalline schist.

[T. F. W. BARTH]

Amygdule

A mineral filling formed in vesicles (cavities) of lava flows. The lava rock containing amygdules is called an amygdaloid, or amygdaloidal lava.

Gases dissolved in liquid lava at depth come out of solution and form bubbles as the rising liquid reaches the surface of the earth. When the lava congeals to solid rock, the trapped bubbles are preserved as holes, or vesicles (see illustration). In pahoehoe lava the vesicles are rather regular spheroids or groups of coalescing spheroids—sometimes nearly spherical, but generally elongated and flattened because of flowage of the enclosing lava. The vesicles of aa and block lava are very irregular. *See* VOLCANO.

Amygdules may consist of mineral matter deposited by gases or liquids released during the consolidation of the surrounding rock. Generally, however, they are deposited by fluids of extraneous origin moving through the rocks. The depositing agents may be hot gases and hydrous solutions rising in volcanic vent areas or in mineralized areas unrelated to volcanic vents; or they may be cold solutions in which the water is of meteoric (atmospheric) origin. The material deposited may be brought in from remote sources or derived by alteration of the adjacent rocks. The common minerals of amygdules are chalcedony, opal, calcite, chlorite, prehnite, pectolite, apophyllite, datolite, and various zeolites. More rarely, native copper and silver may be found. Well-formed amygdules of gibbsite are found, sometimes abundantly, in areas of lateritic alteration of basaltic lavas.

[GORDON A. MACDONALD]

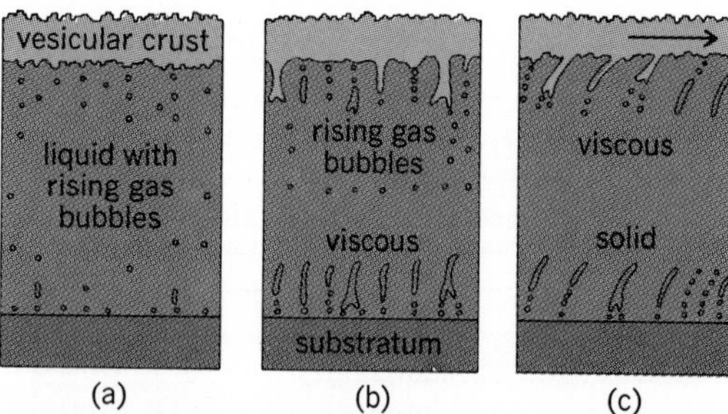

Diagram of spheroidal and pipe vesicles in lava flows. (*a*) Liquid flow with vesicular crust. (*b*) Congealing flow. (c) Nearly solidified flow. (*From R. R. Shrock, Sequence in Layered Rocks, McGraw-Hill, 1948*)

Analcime

A mineral belonging to the zeolite family of silicates, with composition $Na(AlSi_2O_6)\cdot H_2O$. It crystallizes in the isometric system and is usually found in trapezohedral crystals (see illustration).

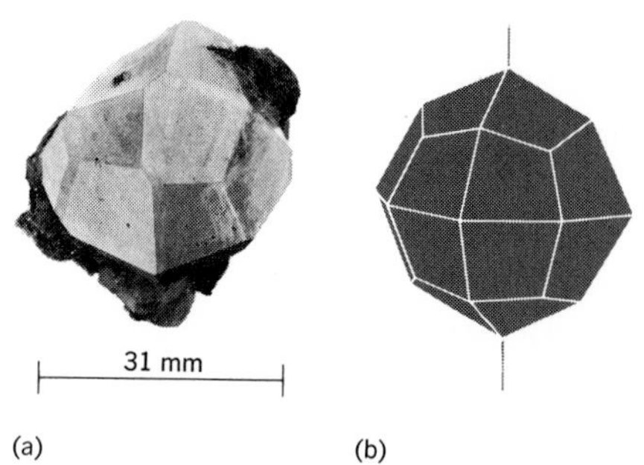

Analcime. (*a*) Specimen from Keweenaw County, Mich. (*American Museum of Natural History*). (*b*) Trapezohedral crystal typical of analcime (*from C. S. Hurlbut, Jr., Dana's Manual of Mineralogy, 17th ed., copyright © 1959 by John Wiley & Sons, Inc.; reprinted by permission*)

The crystals are characterized by a complex internal structure composed of birefringent lamellae or sectors. Only rarely is it massive or granular. The hardness is 5–5 1/2 on Mohs scale; specific gravity is 2.27. Crystals are white or colorless with a vitreous luster. In color and crystal form analcime resembles leucite but can usually be distinguished from it by its free-growing crystals. Leucite is always embedded in a rock matrix. *See* LEUCITE ROCK; ZEOLITE.

Analcime is most commonly a secondary mineral found in veins and cavities of basic igneous rocks (usually basalts), where it is associated with other zeolites, datolite, prehnite, and calcite. Although rare, analcime has been found as a primary constituent of certain igneous rocks. It occurs in the copper deposits of Lake Superior and has been found in hydrothermal sulfide vein deposits. It also occurs as small bedded deposits in saline lakes in arid regions.

[CORNELIUS S. HURLBUT, JR.]

Andalusite

A nesosilicate mineral, composition Al_2SiO_5, crystallizing in the orthorhombic system. It occurs commonly in large, nearly square prismatic crys-

Fig. 1. Andalusite, variety chiastolite. Prismatic crystal specimens from Worchester County, Mass. *(American Museum of Natural History specimens)*

tals. The variety chiastolite has inclusions of dark-colored carbonaceous material arranged in a regular manner. When these crystals are cut at right angles to the *c* axis, the inclusions form a cruciform pattern (Fig. 1). There is poor prismatic cleavage; the luster is vitreous and the color red, reddish-brown, olive-green, or bluish. Transparent crystals may show strong dichroism, appearing red in one direction and green in another in transmitted light. The specific gravity is 3.1–3.2; hardness is 7 1/2 on Mohs scale, but may be less on the surface because of alteration. *See* SILICATE MINERALS.

Occurrence and use. Andalusite was first described in Andalusia, Spain, and was named after this locality. It is found abundantly in the White Mountains near Laws, Calif., where for many years it was mined for manufacture of spark plugs and other highly refractive porcelain. Chiastolite, in crystals largely altered to mica, is found in Lancaster and Sterling, Mass. Water-worn pebbles of gem quality are found at Minas Gerais, Brazil.

[CORNELIUS S. HURLBUT, JR.]

Aluminosilicate phase relations. The three polymorphic forms of Al_2SiO_5 are andalusite (Fig. 1), kyanite, and sillimanite. These minerals occur in metamorphic rocks. Experimental determination of the stability fields of the three minerals, in terms of pressure and temperature, has proved to be very difficult, but the phase diagram shown in Fig. 2 is a version now generally accepted; the experimental results are consistent with available thermochemical data. The three univariant reaction lines separating the stability fields of the minerals meet at an invariant point located near 4 kilobars pressure and 520°C.

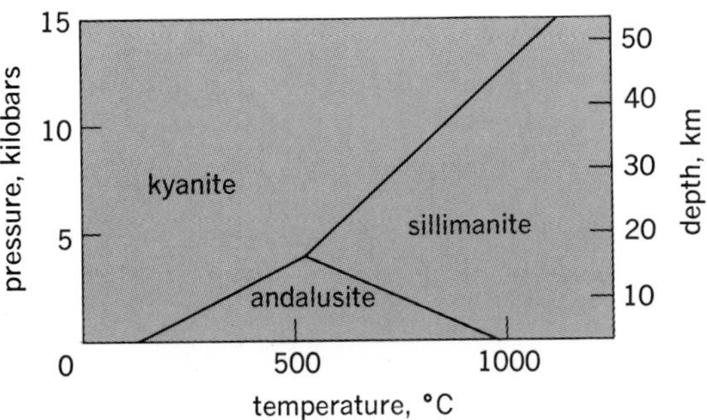

Fig. 2. Phase diagram for the polymorphs of Al_2SiO_5. *(After R. C. Newton, Kyanite-andalusite equilibrium from 700° to 800°C, Science, 153:170–172, copyright 1966 by the American Association for the Advancement of Science)*

The degree of metamorphism in rocks is indicated by index minerals. In many terranes of regional metamorphism, kyanite is succeeded by sillimanite, with increasing intensity of metamorphism. Figure 2 shows that for this sequence to develop the rocks must have been subjected to pressures greater than 4 kilobars, which corresponds to depths greater than 16 km in the Earth's crust. In other metamorphic terranes andalusite is developed on a regional scale, without kyanite or sillimanite, which indicates that the rocks were metamorphosed at higher levels in the Earth's crust. Thus, knowledge of the phase relationships of the polymorphic forms of Al_2SiO_5, combined with geological studies of rocks containing these minerals, is a useful aid in deducing the extent of vertical movements occurring in the Earth's crust. *See* KYANITE; SILLIMANITE.

[PETER J. WYLLIE]

Andesine

A plagioclase feldspar with a composition ranging from $Ab_{70}An_{30}$ to $Ab_{50}An_{50}$, where Ab = $NaAlSi_3O_8$ and An = $CaAl_2Si_2O_8$ (see illustration). In the high-temperature state, andesine has albite-type structure. In the course of cooling, natural material develops a peculiar structural state which, investigated by x-rays, shows reflections that indicate the beginning of an exsolution process, sometimes accompanied by a beautiful variously colored luster (labradorizing). If Fe_2O_3 is present as thin flakes and oriented parallel to certain structurally defined planes, such andesine is called aventurine or sunstone. *See* FELDSPAR; GEM; IGNEOUS ROCKS.

Andesine grains with biotite, in a specimen from Zoutpansberg, Transvaal, South Africa. Andesine is rarely found except as grains in igneous rocks.

[FRITZ H. LAVES]

Andesite

A volcanic rock intermediate in composition and color between basalt and rhyolite and characterized by excess SiO_2 (that is, presence of silica minerals in completely crystalline examples). Most andesite is porous tephra (tuff) erupted explosively from steep-sided volcanoes (Fig. 1). Many of the most explosive volcanic eruptions yielded andesite (Krakatoa, 1883; Sakurajima, 1914; and Bezymianny, 1956). Such eruptions awe people by devastating wide areas, turning day into night, and reddening sunsets around the Earth for several years after the explosion. Powerful eruptions throw enough dust into the stratosphere to have a possible effect on the climate and even to influence the coming of ice ages.

Together with basalt, andesite is a principal volcanic rock of island arcs and continental margins. Andesite is generally rich in large crystals of plagioclase feldspar and has a partly glassy, finely crystalline groundmass. The average composition of andesite closely resembles that of continental crust. It appears likely that the formation of andesite and continental crust is a single complex response to the thrusting of oceanic lithosphere into the Earth.

Definition. Andesite is difficult to define because of usage which varied as techniques available to geologists changed from microscopic to analytical. The development of the meaning of the word reveals significant aspects of the rock type and of the interplay between words and understanding in scientific progress. The rock is named for examples in the Andes Mountains, where widespread occurrences of the rock were first found, and dates to times (early 1800s) when igneous rocks were classified primarily according to their mineralogical makeup and grain size. Andesites proved difficult to identify owing to their fine grain size. Indeed, many andesites contain substantial proportions of glass. When chemical analyses became more readily available in the 20th century, rocks called andesite were found by F. Chayes in 1969 to have a broad range of compositions not intended by systematic schemes of classification. Since about 1960 andesite has been applied to rocks like those in the original region, that is, volcanic rocks of intermediate composition with an excess of SiO_2. Intermediate rocks deficient in SiO_2 are called hawaiite and other names indicative of occurrence in oceanic regions, in contrast to andesites which are characteristic of mountainous continental margins and island arcs.

The distinction between andesite and basalt and dacite or rhyolite is variously defined according to proportions of light and dark minerals, composition of feldspar, or compositional relations between SiO_2 and alkalies. Chayes noted that 1749 out of 1775 andesites have excess SiO_2 and suggested that excess SiO_2 be part of the definition of andesite. As basalts almost never have excess SiO_2, this is a natural dividing line between basalt and andesite. There is no comparable basis to distinguish between andesite and dacite, and most geologists term andesite all rocks with excess silica

Fig. 1. Two andesite volcanoes (Pavlof and Pavlof Sister, in front) in Alaska. *(Photograph by Finley C. Bishop, 1974)*

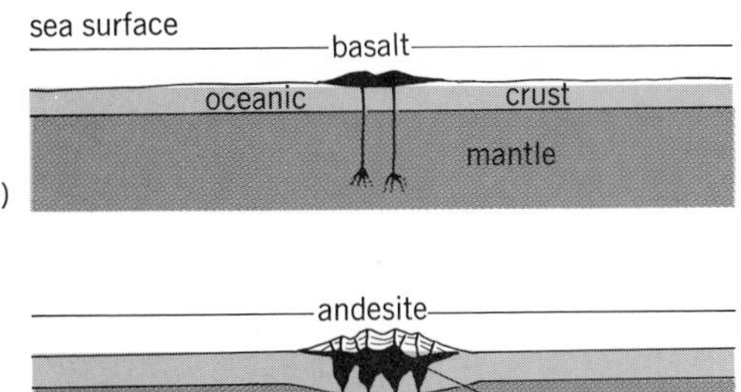

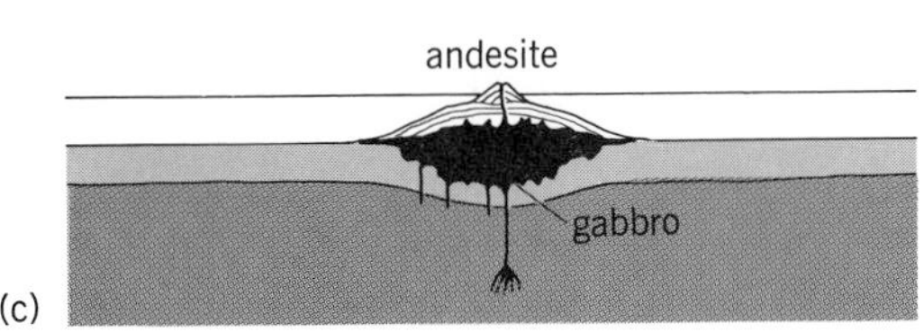

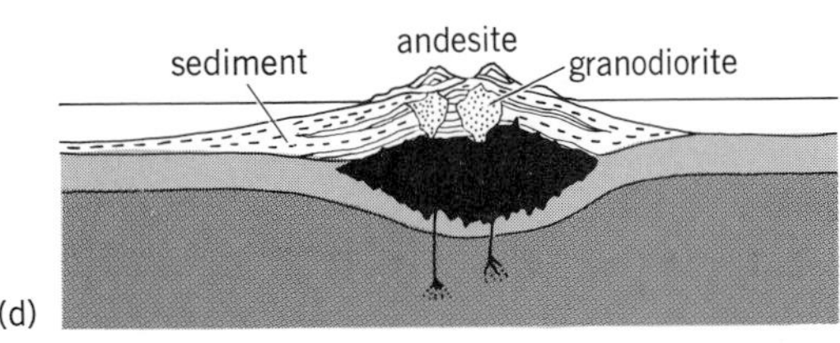

Fig. 2. Successive stages of development of island arc structure. (*a*) Initial formation of basaltic material on the sea floor in an island arc. (*b*) Development of reservoirs of gabbroic magma within the oceanic crust and submarine extrusion of andesite. (*c*) Coalescence of magma reservoirs, continued andesite extrusion, and crustal thickening adequate to support volcanoes above sea level. (*d*) Continued andesite extrusion from subaerial volcanoes and growth of silicic magma reservoirs (granodiorite) within the upper crust of volcanic and sedimentary materials, with further crust thickening. (*From H. Kuno, Origin of andesite and its bearing on the island arc structure, Bull. Volcanol., 37:141–176, 1968*)

and more than about 10% of dark minerals (rhyolites have less).

Structure, texture, and mineralogy. Most andesite is erupted explosively and consists of porous fragments of tephra. Andesite tephra contains up to about 30 vol % of crystals larger than 1 mm set in a groundmass of tiny crystals and glass. The big crystals are generally well formed and dominated by plagioclase and pyroxene, both of which are commonly compositionally zoned. Big crystals of olivine and magnetite occur in many andesites; amphibole, biotite, and quartz are common only in the more siliceous andesites. The groundmass minerals are largely plagioclase with minor pyroxene, magnetite, and apatite. The groundmass glass is commonly rhyolitic in composition, but may be andesitic in rapidly cooled, glass-rich varieties. Crystals of cristobalite, tridymite, and hematite are common in the vesicles.

Flows and domes of andesite have craggy to blocky surfaces and platy, slabby, or massive interiors. The texture of flows and domes is similar to that of tephra, except that tephra is more porous and generally contains higher proportions of glass than do flows and domes. Some andesites rich in big crystals have bulk compositions outside the range found for andesites poor in big crystals.

Composition, occurrence, and origin. The chemical and mineralogical composition of andesite varies in space and time in a broadly regular way, but with many exceptions. Andesites on continental margins (such as the Alaskan peninsula) are more silicic, in general, than andesites on oceanic islands (such as the Aleutian Islands). Andesites near the ocean and above shallow zones of earthquakes tend to be poorer in alkalies and richer in lime than those farther inland associated with deeper seismicity. Volcanoes formed in the early stage of development of island arcs contain the most iron-rich andesites, except for analogs on truly oceanic islands.

Geological thinking during the 1970s is dominated by plate tectonic theory. Accordingly, the space-time variations of andesites are explained according to a hypothesis of partial melting of lithosphere in subduction zones. It appears unlikely that andesites are generated as such in subduction zones, although many scientists favor this possibility. In 1969 H. Kuno extended the hypothesis of N. Bowen that basaltic magmas feed volcanoes with andesite from large bodies of gabbroic magma near the base of the crust (Fig. 2). However, Kuno's hypothesis fails to give subduction an essential role in the origin of andesite. This was first done by R. Coats, in 1962, who proposed that water-rich sedimentary rocks were thrust deep into the Earth beneath island arcs and partially melted together with overlying mantle to yield basalt and related andesite. Geochemical studies showed that less than about 2% of the radiogenic lead and strontium in typical andesites could be derived from the kind of subducted rock proposed by Coats. C. B. Rayleigh and W. Lee proposed that the material of basalts and related andesites might be generated by partial melting of hydrous oceanic crust in subduction zones, and A. McBirney suggested that water derived from the subducted slab of dehydrating lithosphere would flux melting in the overlying mantle. The last two ideas can help explain some of the space-time variations of andesites mentioned above: the nature of melting may change with progressive subduction, and water-fluxed melting of the overlying mantle may lead to variation in iron and alkalies with time and depth (extent) of dehydration of the subducted lithosphere. *See* BASALT; IGNEOUS ROCKS; OCEANIC ISLANDS; TECTONIC PATTERNS; VOLCANO.

[ALFRED T. ANDERSON, JR.]

Bibliography: I. S. E. Carmichael, F. J. Turner, and J. Verhoogen, *Igneous Petrology*, 1974; H. Kuno, Origin of andesite and its bearing on the island arc structure, *Bull. Volcanol.*, 37:141–176, 1968; A. R. McBirney (ed.), *Proceedings of the Andesite Conference*, Oreg. Dep. Geol. Miner. Ind. Bull. 65, 1969; G. A. Macdonald, *Volcanoes*, 1972; H. W. Menard, *Geology, Resources and Society*, 1974; W. F. Whittard and K. Bradshaw (eds.), *Submarine Geology and Geophysics*, 1965.

Anglesite

A mineral with the chemical composition $PbSO_4$. Anglesite occurs in white or gray, orthorhombic, tabular or prismatic crystals or compact masses (see illustration). It is a common secondary mineral, usually formed by the oxidation of galena. Fracture is conchoidal and luster is adamantine. Hard-

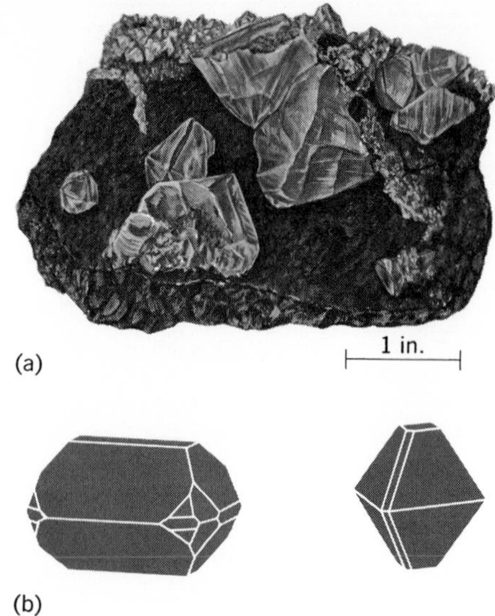

Anglesite. (*a*) Crystals on galena from Phoenixville, Pa. (*Bryn Mawr College specimen*). (*b*) Crystal habits (*from L. G. Berry and B. Mason, Mineralogy, copyright © 1959 by W. H. Freeman and Co.*)

ness is 2.5–3 on Mohs scale and specific gravity is 6.38. Anglesite fuses readily in a candle flame. It is soluble with difficulty in nitric acid. The mineral does not occur in large enough quantity to be mined as an ore of lead, and is therefore of no particular commercial value. Fine exceptional crystals of anglesite have been found throughout the world. In the United States good crystals of anglesite have been found at the Wheatley Mine, Phoenixville, Chester County, Pa., and in the Coeur d'Alene district of Shoshone County, Idaho. *See* GALENA. [EDWARD C. T. CHAO]

Anhydrite

A mineral with the chemical composition $CaSO_4$. Anhydrite occurs commonly in white and grayish granular masses, rarely in large, orthorhombic crystals (see illustration). Fracture is uneven and luster is pearly to vitreous. Hardness is 3–3.5 on Mohs scale and specific gravity is 2.98. It fuses readily to a white enamel. It is soluble in acids and slightly soluble in water. Anhydrite is an important rock-forming mineral and occurs in association with gypsum, limestone, dolomite, and salt beds. It is deposited directly by evaporation of sea water of high salinity at or above 42 °C. Anhydrite can be produced artificially by dehydration of gypsum at about 200°C. Under natural conditions anhydrite hydrates slowly, but readily, to gypsum. It is not used as widely as gypsum. Anhydrite is of worldwide distribution. Large deposits occur in the Carlsbad district, Eddy County, N.Mex., and in salt-dome areas in Texas and Louisiana. *See* EVAPORITE, SALINE; GYPSUM. [EDWARD C. T. CHAO]

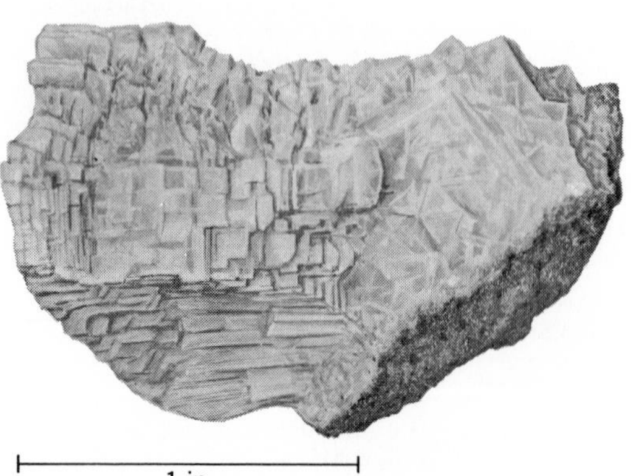

Anhydrite from Montanzas, Cuba. (*Specimen from Department of Geology, Bryn Mawr College*)

Animal evolution

The theory that modern animals are the modified descendants of animals that formerly existed and that these earlier organisms descended from still earlier and different forms. Evidence for animal evolution should, where possible, be drawn from fossils representing actual ancestors. However, even for groups with hard skeletal parts the fossil record is far from complete. In addition, many types are soft-bodied and hence not preserved as fossils. As a result, much of the history of evolution must be deduced from study of existing forms. An increasing amount of biochemical evidence of animal relationship is accumulating, but reliance is placed mainly on comparison of adult structures and on embryonic and larval development. It was once assumed that embryos recapitulated the phylogeny of the race; this theory is largely abandoned today, but comparison of patterns of development and of larval types is of great value. The evidence, drawn from the various sources cited above, indicates that animals have evolved according to the evolutionary scheme shown in Fig. 1. *See* EVOLUTION, ORGANIC; FOSSIL; PALEONTOLOGY.

Lower animals. There are no definitive characters which apply to all supposed animals other than the fact that they are cellular or multicellular organisms which, unlike plants, do not manufacture their own food materials. It is often assumed that all such animals form a single evolutionary stock, advancing from single-celled forms, the Protozoa, through a colonial stage from which the sponges, the Parazoa, are an offshoot, to the level of the Metazoa. It is reasonable to believe that these all derive ultimately from single-celled flagellated plants, but there is no guarantee that they form an evolutionary unit. The Protozoa include the flagellates, the Mastigophora; the amebas and their relatives, the Sarcodina (Rhizopoda); the parasitic Sporozoa; and the complexly built Ciliophora. No one of the last three appears to be particularly related to any other. The Mastigophora seem to be a miscellaneous collection of flagellates which may have independently of one another, lost their food-manufacturing potentialities. The sponges seem to be derived from one peculiar type of collar-bearing flagellates, quite unrelated to typical protozoans. Metazoans show no indication of relationship to sponges or to protozoans in general.

Early metazoan history. In contrast with doubts regarding lower forms, there is essential unanimity

Fig. 1. (*Opposite page*) Evolution of animal groups. (*From G. B. Moment, General Zoology, 2d ed., Houghton Mifflin, 1967; used by permission*)

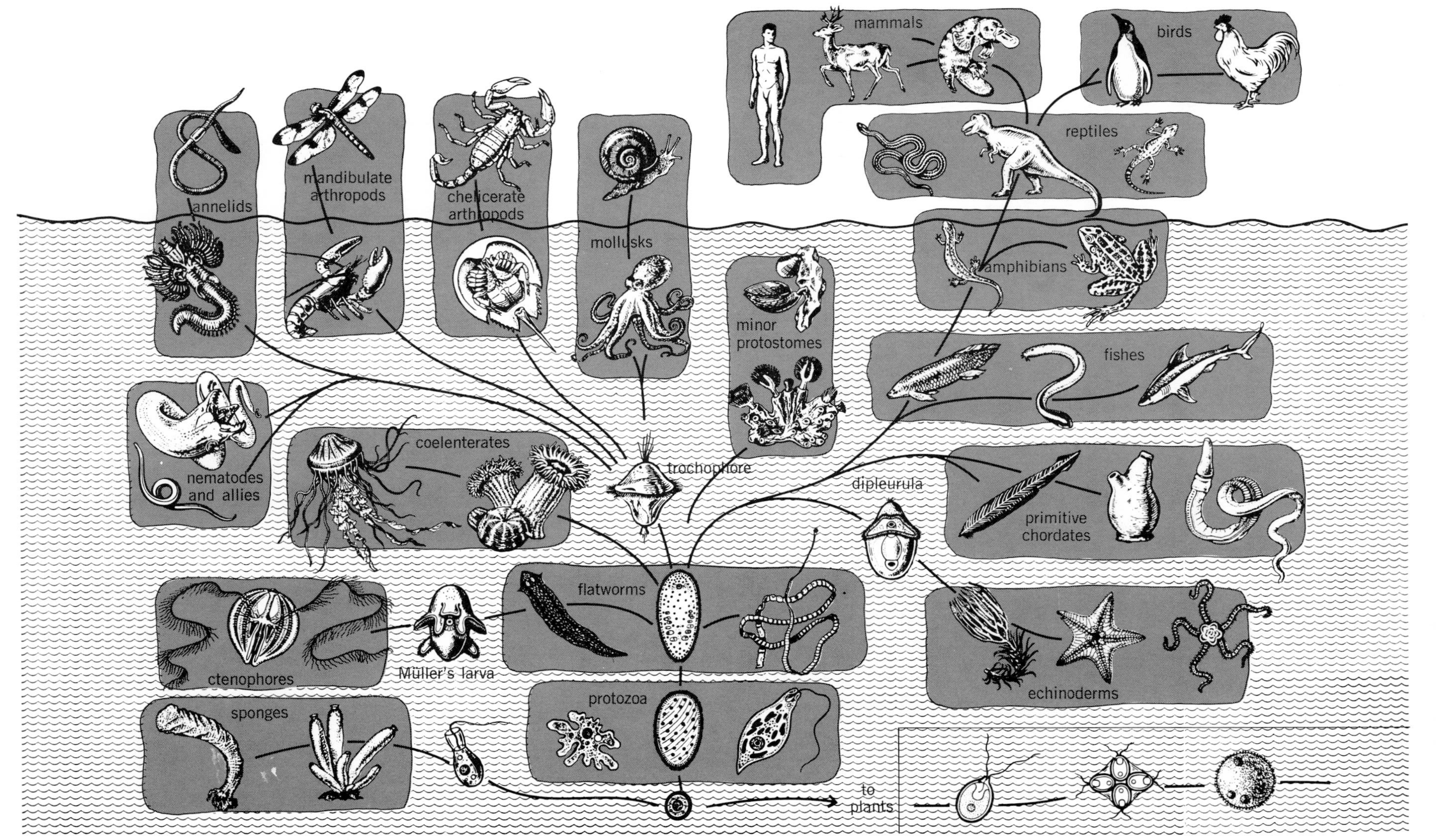
mammals
birds
reptiles
annelids
mandibulate arthropods
chelicerate arthropods
mollusks
amphibians
minor protostomes
fishes
coelenterates
nematodes and allies
trochophore
dipleurula
primitive chordates
flatworms
ctenophores
Müller's larva
echinoderms
protozoa
sponges
to plants

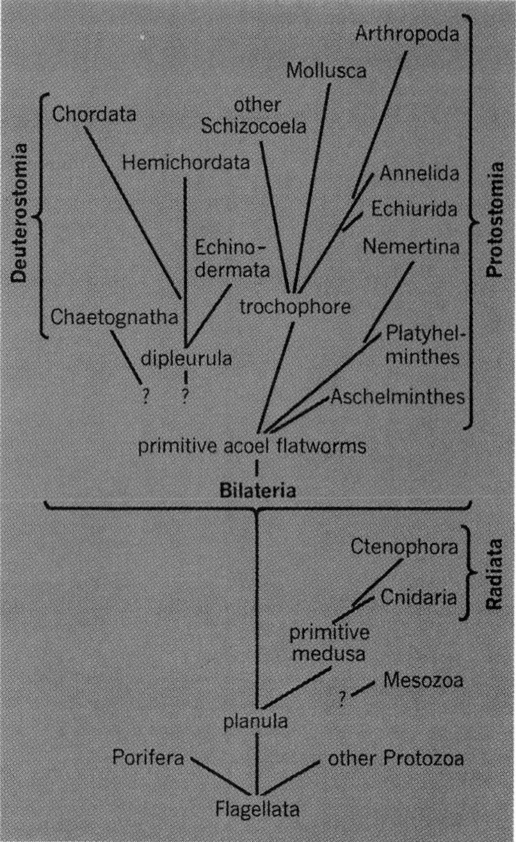

Fig. 2. Hypothetical diagram of the relationships of the phyla of the animal kingdom. (*L. H. Hyman, The Invertebrates, vol. 1, McGraw-Hill, 1940*)

of opinion that the Metazoa, including all multicellular forms with differentiated tissues, are a true phyletic unit; it is also common opinion that the phylum Coelenterata (to which the ctenophores, or comb jellies, are often appended) lies close to the base of this stock. The coelenterates proper (Cnidaria) include such sessile forms as the corals and hydroids, and free-floating medusae, such as the jellyfishes. Their bodies consist essentially of two simple epithelia – an external, protective ectoderm, and an internal, essentially digestive endoderm, lining a gut cavity which has but a single opening. Typical living coelenterates are specialized in that they have peculiar stinging cells which enable them to catch and eat relatively large prey; truly ancestral metazoans presumably lacked these structures and subsisted on smaller organisms or organic debris.

Beyond this lowly metazoan level there developed a vast radiation of animals making up, according to some systems, a score or more of phyla. Some became degenerate parasites; most advanced in many regards. A second gut opening often developed, giving a separate mouth and anus. Most had a third, intermediate body layer, the mesoderm, between ectoderm and endoderm, in which there often developed the coelom, a cavity surrounding the viscera. Body shape and mode of life became highly variable.

In a number of phyla, typical extant forms are sessile (like coelenterate polyps), gathering food particles by means of cilia-created currents along outstretched arms or tentacles. In others food is actively sought, and the body assumes an elongate, bilaterally symmetrical, wormlike form.

The interrelations of the various phyla are obscure in many cases. There is, however, general agreement that a number can be grouped in two major series (Fig. 2). One line, including the wormlike forms (to which the arthropods and mollusks are allied), is sometimes termed the Protostomia or Ecterocoelia. In them the original gut opening becomes the mouth, the secondary opening the anus; the mesoderm forms from a solid mass of cells; there is often a characteristic larva, termed the trochophore, with a horizontal circular band of cilia above the mouth. Among the sedentary cilia feeders, the primitive echinoderms are outstanding, and are considered typical of the Deuterostomia or Enterocoelia. Here the new gut opening is the mouth, the mesoderm forms from pouches derived from the gut; the larva is of a different sort, with complex longitudinal bands of cilia.

Worm phyla. Most primitive of the worm types included in the Protostomia are the flatworms, the Platyhelminthes, including free-living planarians and a host of parasitic forms, such as flukes and tapeworms. The digestive tract, which is lost in some parasites, has but one opening; the mesoderm is poorly developed. Nearly as simply built, except for the presence of an anus, but specialized in the presence of an evertible prey-catching head structure are the proboscis worms, the Nemertea. Grouped by some writers as the phylum Aschelminthes is a grab bag of half a dozen or so minor groups, including, as well as more obscure types, the rotifers and the exceedingly abundant roundworms (nematodes), of which most are minute and harmless, but a few, such as hookworm and trichina worm, are dangerous parasites. Minor phyla usually bracketed with the worms are the Mesozoa, Sipunculida, Echinoidea, and Priapulida.

Annelids, arthropods, and mollusks. The phyla Annelida, Arthropoda, and Mollusca exhibit highly diverse structures and modes of life, but are definitely related to one another. The annelids, including numerous marine types and leeches, as well as the familiar earthworms, show a division of the body into metameric segments, each of which repeats, with regional variations, all the structures found in its neighbors. The arthropods, including crustaceans, arachnids, millipedes, centipedes, and insects, in many ways most successful of animal phyla, are considered to be of annelid origin, but are characterized by an external armor of chitin and jointed appendages. Seemingly quite unlike either annelids or arthropods are the mollusks, most of which are sluggish shell-bearers, without a trace of segmentation. However, it has long been suspected, on the basis of similarity of the larvae of annelids and mollusks, that the two phyla are related, and this has been proved by the discovery in the depths of the Pacific Ocean of a primitive mollusk, *Neopilina*, which is distinctly segmented.

Sessile arm feeders. In a series of phyla which is, on the whole, much less progressive, the typical members are persistently sessile, stalked forms. Above a simple body armlike structures stretch out, bearing ciliated bands whose function is to direct food particles to the mouth. The tiny but abundant Bryozoa (Polyzoa), or moss animalcules, which are sometimes divided into two phyla as the Ectoprocta and Entoprocta, are forms of this nature, as are the tube-dwelling members of the minor phylum Phoronida. Unimportant today but abundant in earlier geologic periods, members of the phylum Brachiopoda (lampshells) bear coiled ciliated arms within their paired shells. The relationships of these phyla are uncertain. They are sometimes appended, on feeble evidence, to the proterostome group; but equally likely is relationship to the typical deuterostomes, the phylum Echinodermata. Echinoderms are known for such specializations as a calcareous armor and a unique water-vascular system. Most modern echinoderms are free-moving, if sluggish, forms; the older fossil echinoderm groups and the crinoids today resemble the series of phyla listed above as stalked sessile forms, feeding on food particles brought to the mouth by ciliated bands on outstretched arms.

Chordate origins. The phylum Chordata includes the vertebrates as a major component and in addition certain lower types, such as the lancelets, tunicates, acorn worms, and pterobranchs. The last two forms are sometimes regarded as constituting a separate but related phylum, Hemichordata. Basic chordate characters seen in the lancelet (*Branchiostoma*) include a dorsal hollow nerve cord; a longitudinal supporting structure, the notochord, running the length of the body beneath the nerve cord; and a series of pharyngeal gill slits which primitively function as a food-filtering device. The origin of vertebrates has been sought among the annelids and arachnids, but there are few real resemblances. In recent decades increasing evidence has accumulated which indicates that the closest chordate relatives are the echinoderms, whose mesoderm develops in similar fashion; acorn worms have a larva similar to that of some echinoderms; and there are biochemical similarities. Vertebrates are typically active forms, and therefore it has been argued that

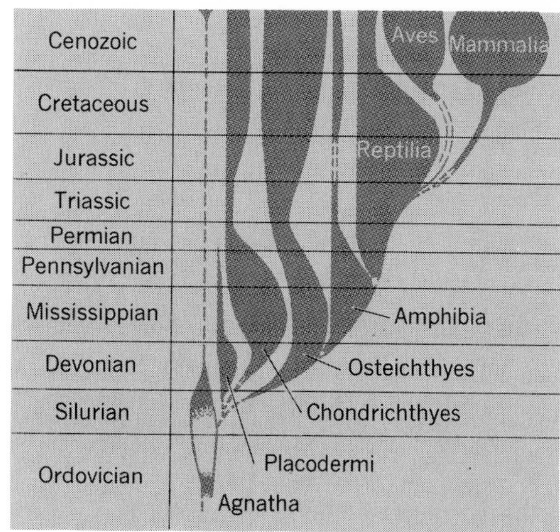

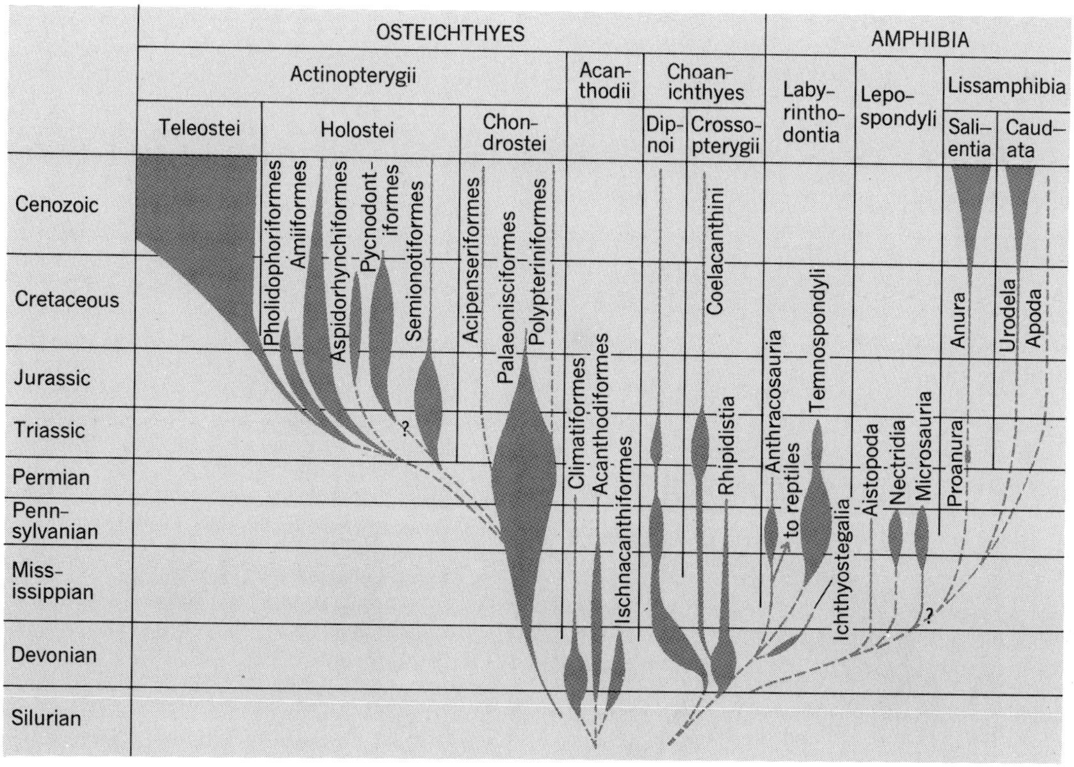

Fig. 3. Family tree of vertebrate classes. The thickness of various branches is an approximation of comparative abundance of various groups during the geologic eras. (*A. S. Romer, Vertebrate Paleontology, University of Chicago Press, 3d ed., 1966*)

Fig. 4. Geologic timetable of the development of the various groups of the higher bony fishes and the amphibians. (*A. S. Romer, Vertebrate Paleontology, University of Chicago Press, 3d ed., 1966*)

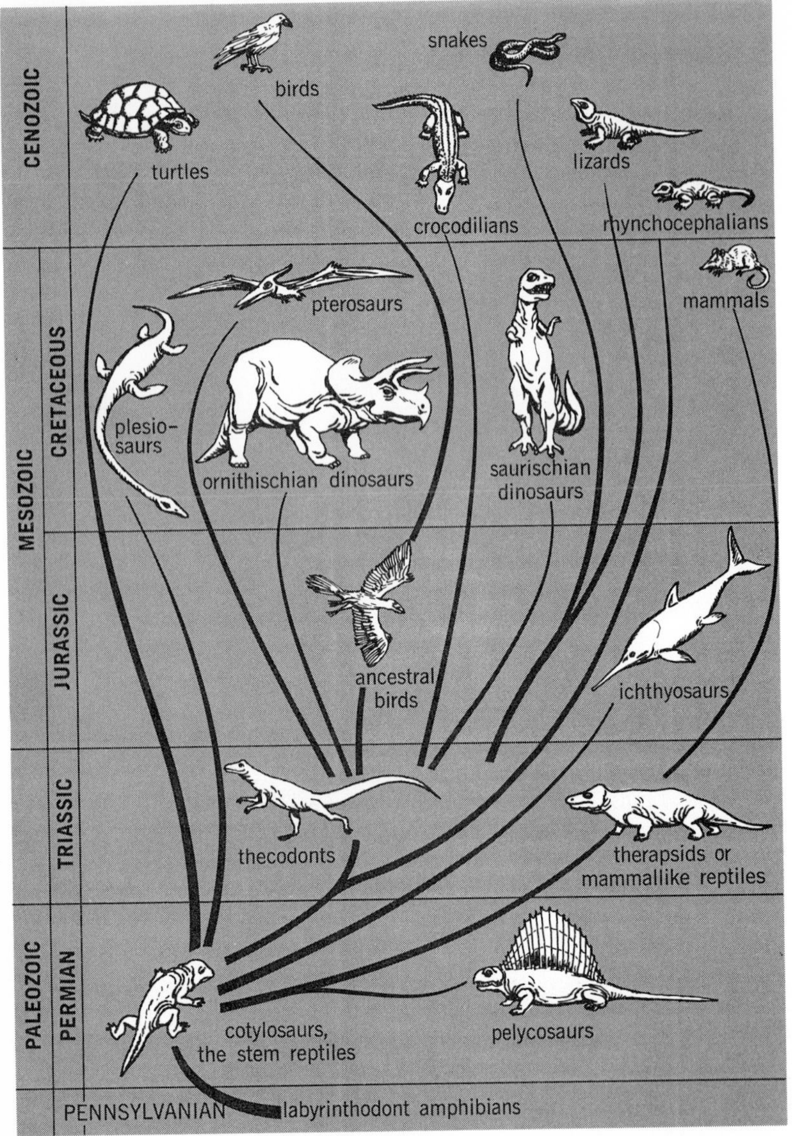

Fig. 5. The family tree of the reptiles, showing evolution through geologic time. (*E. H. Colbert, The Dinosaur Book, American Museum of Natural History, 1951*)

the common ancestor of vertebrates and echinoderms was a simply built, active, bilaterally symmetrical animal, from which the echinoderms branched off to become sessile, assuming a superficially radial symmetry. But the lower chordates are essentially sessile types, living on food particles brought to them by ciliary currents; the most primitive chordates, the pterobranchs, are small stalked forms with extended arms with ciliated food-gathering bands. They are thus much like the cilia-feeding phyla discussed above, and could be close to the ancestry of echinoderms as well as higher chordates, in which gill filtering replaces arms in feeding. In the tunicates the typical adults are sessile, but there is in some a tadpole larva, which has not only gills but nerve cord and notochord. It is highly probable that the evolution of higher members of the phylum took place by pedomorphosis – the retention of the tadpole body in the adult condition and consequent introduction of an active swimming life into the vertebrate ancestry. The evolution of the vertebrate classes is shown in Fig. 3.

Fish evolution. Study of the fossil record indicates that the oldest fishes, of the class Agnatha, lived much like tunicates and the lancelet in that they filtered food particles from a water current which entered the mouth and passed out through a complex gill apparatus. These were small armored types termed ostracoderms. The lampreys and hagfishes of today are degenerate in the loss of bone from the skeleton, but although they lack jaws, they have become predaceous in habit through the development of a rasping tongue structure.

A class entirely extinct is the Placodermi, armored Paleozoic fishes in which jaws and paired appendages (as fins) were appearing. Beyond this stage fish evolution progressed along two paths. One led to the class Chondrichthyes, including the sharks, skates, rays, and chimaeras, in which (as the lampreys) bone is lost, although there is a highly developed skeleton of cartilage. A second, more progressive group is the Osteichthyes, the higher bony fishes which dominate the life of modern waters (Fig. 4). *See* PALEOZOIC.

Conquest of land. The evolutionary sequence Osteichthyes – Amphibia – Reptilia indicates stages in a major development in vertebrate history, the evolution of four-footed (tetrapod) vertebrates which freed themselves from an aquatic existence to conquer the land. The fish ancestors of land vertebrates were the Crossopterygii, a group of higher bony fishes of which only one specialized form (*Latimeria*) survives. Studies of amphibians of the great extinct group Labyrinthodontia, including Late Devonian forerunners, have in considerable measure bridged the gap between the fish stage on the one hand and reptiles on the other, but the early evolution of the existing amphibian orders is still for the most part a closed book. The major steps in tetrapod evolution appear to be closely correlated with adaptations countering seasonal drought conditions which were widespread in late Paleozoic days. *See* DEVONIAN.

Reptile radiation. Toward the close of the Carboniferous there began a great radiation of reptiles, resulting in the establishment of a score or more of orders, of which only four survive. Much of the phyletic pattern is known, but considerable gaps remain in knowledge of the earliest stages in the origin of various lines. Prominent in reptile evolutionary history was the rise of marine reptiles, such as the plesiosaurs and ichthyosaurs, and on land of the subclass Archosauria, the ruling reptiles, to which the dinosaurs belonged. From Archosauria, with *Archaeopteryx* as a transitional form, came the class Aves, the birds (Fig. 5). *See* CARBONIFEROUS.

Mammal origins. The line leading to the class Mammalia diverged almost at the beginning of reptile history. From Late Carboniferous times on into the Triassic, mammal-like reptiles, first the Pelycosauria and then the advanced mammal-like forms, the Therapsida, dominated the terrestrial scene. Before the end of the Triassic, however, the therapsids waned with the advent of the dinosaurs and soon disappeared, leaving descendants behind them as the first true mammals. The egg-laying monotremes, the Prototheria, must have diverged from the main mammalian line almost at its inception, but the history of these living Australian forms is almost unknown. Remains of Jurassic and Cretaceous mammals are scanty; under dinosaurian dominance they formed an inconspicuous element in the fauna. By the Late Cretaceous both marsupials (the Metatheria) and the Eutheria, the more highly developed placental mammals, had come into existence. With the extinction of the dinosaurs, mammals came into their own. The marsupials flourished in Australia, where eutherians were long absent. In other continents the placental mammals rapidly expanded, in the early Tertiary, into a host of types, most of which are still represented in the existing fauna. *See* CRETACEOUS; JURASSIC; TRIASSIC.

Primates and man. Man is a member of the eutherian order Primates, to which also belong the lemurs, monkeys, and great apes. Many of the ordinal characters and trends, such as an agile and flexible locomotor apparatus, good hands, high development of vision, and a trend toward large brain size, are to be correlated with the arboreal life characteristic of the order as a whole. The lemurs of the Old World tropics are little-changed representatives of an early stage in primate evolution; *Tarsius* of the Oriental region shows advanced features leading toward the monkey level. Monkeys developed in two independent lines, one in South America, the second in the Old World. Allied to the latter were the ancestors of still higher primates, the great apes. By Miocene times there was present in Eurasia and Africa a number of such apes, from which may have descended the living gibbons, orangutans, chimpanzees, gorillas, and possibly man as well. *Australopithecus* and related types of the Pleistocene of South Africa are small-brained and apelike in certain features, but are morphologically transitional to later men in such characters as dentition and upright gait. Definitely human types appear in the middle Pleistocene, in *Pithecanthropus* and *Sinanthropus* of the Far East and fragmentary remains from the Occident (the Piltdown skull and jaw, long a stumbling block in interpretation of the record, is now known to be a hoax). Of later date, Neanderthal and other types show a closer approach to modern man, and toward the close of the Pleistocene, forms appear which are definitely identifiable as belonging to our own species. *See* FOSSIL MAN; MIOCENE; PLEISTOCENE.

[ALFRED S. ROMER]

Ankerite

The carbonate mineral $Ca(Fe,Mg)(CO_3)_2$, also commonly containing some manganese. The mineral has hexagonal (rhombohedral) symmetry (see illustration) and has the cation-ordered structure of dolomite. The name is applied only to those species in which at least 20% of the magnesium positions are occupied by iron or manganese; species containing less iron are termed ferroan dolomites. The pure compound, $CaFe(CO_3)_2$, has never been found in nature and has never been synthesized as an ordered compound. *See* DOLOMITE.

Ankerite typically occurs in sedimentary rocks as a result of low-temperature metasomatism. It is commonly white to light brown, its specific gravity is about 3, and its hardness is about 4 on Mohs scale. *See* CARBONATE MINERALS; METASOMATISM.

[ALAN M. GAINES]

Ankerite. (*a*) Specimen of the mineral (*specimen from Department of Geology, Bryn Mawr College*). (*b*) Crystal habits (*from C. S. Hurlbut, Jr., Dana's Manual of Mineralogy, 17th ed., copyright © 1959 by John Wiley & Sons, Inc.; reprinted by permission*)

Anorthite

The calcium-rich end member of the plagioclase feldspar series. Its composition ranges from $Ab_{10}An_{90}$ to An_{100} ($Ab = NaAlSi_3O_8$ and $An = CaAl_2Si_2O_8$). Its crystal structure is shown by x-ray investigations to possess a highly ordered Al-Si distribution within the $(Al,Si)_4O_8$-framework, differing from that in microcline and albite (see illustration). In the pure $CaAl_2Si_2O_8$ compound, each AlO_4 tetrahedron is apparently surrounded by four SiO_4 tetrahedrons, and vice versa. Whereas this kind of ordering is, under equilibrium conditions, always present up to the melting point (1544°C), structural differences occur as a result of different Ca positioning and as a function of temperature above 1000°C. The temperature at which the Ca disorder sets in is a function of the Ab content. The larger the Ab content, the lower is this temperature. In consequence, the following phase relations exist in the An-rich plagioclases anorthite-bytownite: Anorthite (low) is the highest-ordered Ca-rich structure as far as the Al-Si distribution and the Ca positioning are concerned; anorthite (high) is similarly ordered with respect to the Al-Si distribution as anorthite (low), but is disor-

Anorthite crystals, Miakejima, Japan. (*Specimen from Department of Geology, Bryn Mawr College*)

dered with respect to the Ca positioning; bytownite (low) corresponds to anorthite (high), but with part of the Ca randomly replaced by Na; and bytownite (high) corresponds to the albite-type structure forming a continuous series of solid solutions with albite ($NaAlSi_3O_8$) at high temperatures. *See* FELDSPAR; IGNEOUS ROCKS. [FRITZ H. LAVES]

Anorthoclase

The name usually given to alkali feldspars having a chemical composition ranging from $Or_{40}Ab_{60}$ to $Or_{10}Ab_{90} \pm$ up to approximately 20 mole % An (Or, Ab, An = $KAlSi_3O_8$, $NaAlSi_3O_8$, $CaAl_2Si_2O_8$) and which deviate in one way or another from monoclinic symmetry tending toward triclinic symmetry. When found in nature, they usually do not consist of a single phase but are composed of two or more kinds of K- and Na-rich domains mostly of submicroscopic size. In addition, they are frequently polysynthetically twinned after either or both of the albite and pericline laws. It appears that they originally grew as the monoclinic monalbite phase inverting and unmixing in the course of cooling during geological times. They are typically found in lavas or high-temperature rocks. *See* FELDSPAR; IGNEOUS ROCKS.

[FRITZ H. LAVES]

Anorthosite

A rock composed of 90 vol % or more of plagioclase feldspar. Strictly, the rock is composed entirely of crystals discernible with the eye, but some finely crystalline examples from the Moon have been called anorthosite or anorthositic breccia. Two principal types of anorthosite are based on field occurrence: (1) layers in stratified complexes of igneous rock, and (2) large massifs of rock up to 30,000 km^2 in area. Scientists have been fascinated with anorthosites because they are spectacular rocks (dark varieties are quarried and polished for ornamental use); valuable deposits of iron and titanium ore are associated with anorthosites; and the massif anorthosites appear to have been produced during a unique episode of ancient Earth history (about $1-2 \times 10^9$ years ago).

Definition, occurrence, and structure. Pure anorthosite has less than 10% of dark minerals—generally some combination of pyroxene, olivine, and oxides of iron and titanium; amphibole and biotite are rare, as are the light minerals apatite, zircon, scapolite, and calcite. Rocks with less than 90% but more than 78% of plagioclase are modified anorthosites (such as gabbroic anorthosite), and rocks with 78–65% of plagioclase are anorthositic (such as anorthositic gabbro). *See* GABBRO.

The structure, texture, and mineralogy vary with type of occurrence. One type of occurrence is as layers (up to several meters thick) interstratified with layers rich in pyroxene or olivine. The bulk composition of the layered rock samples containing anorthosite layers is gabbroic. The Stillwater igneous complex in Montana is an example where the lower part of the complex is rich in olivine, pyroxene, and chromite, and the upper part is rich in plagioclase. The second type of occurrence is the massifs type. Commonly, the massifs are domical in shape and weakly layered. There appear to be two kinds of massif anorthosites: irregular massifs typified by the Lake St. John body in Quebec, and domical massifs similar to the Adirondack (NY) massif made classic through the studies of A. Buddington. The irregular massifs are older than the domical massifs and commonly contain olivine. The domical massifs are associated with silicic rocks of uncertain origin.

Possibly there is a third group of anorthosite occurrences: extremely ancient bodies of layered rock in which the layers of anorthosite contain calcium-rich plagioclase and the adjacent layers are rich in chromite and amphibole in addition to pyroxene. There are only a few examples of these apparently igneous complexes, in Greenland, southern Africa, and India. However, they appear to be terrestrial counterparts of lunar anorthosites. Recrystallization and deformation have pervasively affected these rocks, and it is difficult to discover their original attributes.

Layered anorthosites. Zones of anorthosite in layered complexes are composed mostly of well-formed, moderately zoned crystals of plagioclase (labradorite to bytownite) up to several millimeters long. The tabular crystals of plagioclase commonly are aligned subparallel to the layering. Irregular crystals of pyroxene, olivine, and rarely magnetite are interstitial to the plagioclase crystals. These features are best explained by the hypothesis of accumulation of plagioclase crystals onto the floor of the reservoir of melt, where gravity and perhaps motion of melt caused them to lie with their centers of gravity in the lowest position (like pennies spilled on the floor). After accumulation, the intergranular melt gradually crystallized as overgrowths of plagioclase and interstitial pyroxene. The separation of accumulated plagioclase crystals in anorthosite layers from pyroxene or olivine in pryoxenite and peridotite layers is a puzzle to scientists, for it is expected that plagioclase and pyroxene crystallize together in subequal proportions from basaltic magmas, not first one mineral, then the other and back again as suggested by the

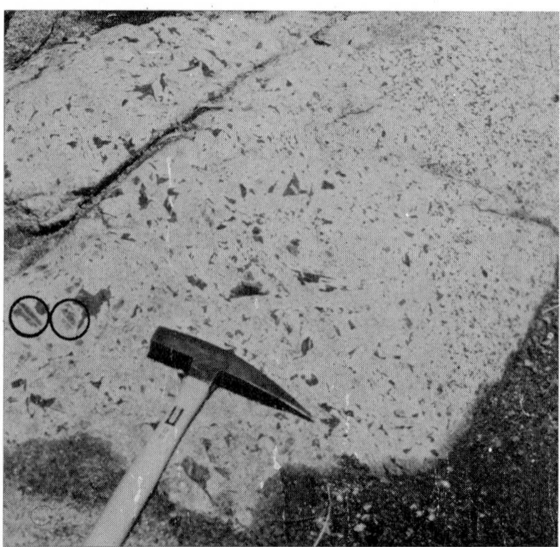

Fig. 1. Outcrop of massif anorthosite on the shore of Pipmuacan Reservoir, Quebec. Exceptionally well-preserved texture shows well-formed relict plagioclase crystals (now recrystallized) outlined by irregular black pyroxene grains. Two dark cores of twinned plagioclase are shown (circles). The longest plagioclase relic is about 12 cm.

layering. The faster rate of fall through the liquid of the denser pyroxene crystals and a possible finite yield strength of the liquid may explain the layering. *See* MAGMA.

Massif anorthosites. In irregular massifs the plagioclase has an intermediate composition, mostly labradorite. The plagioclase crystals are poorly formed and practically unzoned and may be up to a few meters long, but crystals about a centimeter long constitute the bulk of most massifs. Rare dark varieties have moderately well-formed plagioclase crystals (Fig. 2) filled with tiny specks of dark minerals. In many specimens the large labradorite crystals display a play of colors (blue and green are common) in a particular orientation, an attractive quality related to very slow cooling and sought after in ornamental stone. Lighter varieties have smaller crystals of sugary plagioclase lacking inclusions of dark minerals, which instead occur as millimeter-sized interstitial grains. (Thus the various shades of anorthosite colors result from variations in the size or granularity of the pigmenting dark minerals.) In transitional varieties large dark crystals of plagioclase are set in a fine-grained sugary white matrix. Many of the plagioclase grains are bent and broken. The associated dark minerals are pyroxene, olivine, magnetite, and ilmenite. Deposits of almost pure magnetite occur and have been used for iron ore, although most deposits have too much titanium to be highly desired for iron ore. Other deposits are rich in olivine, spinel, and apatite as well as magnetite. The irregular massifs are surrounded by gneissic rocks, and the original age relationships between anorthosite and gneiss are commonly uncertain. The textural and mineralogical fractures of irregular massifs are indicative mainly of extensive recrystallization and deformation at high temperature. The dark varieties have textures similar to those of layered anorthosites and presumably have a similar igneous origin, but no one can be certain.

Domical massifs (Fig. 1) associated with silicic rocks consist predominantly of irregular crystals of plagioclase (andesine), generally a few millimeters to a few centimeters long. Dark varieties are rare, although large individual crystals may be dark and filled with tiny inclusions of dark minerals. Chatoyancy (play of colors) is rare, but many plagioclase crystals have minute inclusions of potassium feldspar which probably segregated out of the plagioclase during very slow cooling. Olivine is rare or absent in domical massifs, and pyroxenes and ilmenite (rather than magnetite) are the principal dark minerals. Many domical anorthosites have zones or layers rich in ilmenite and pyroxene, as well as massive deposits of ilmenite and apatite which are locally quarried as titanium ore. Commonly, domical anorthosites have a border zone of gabbroic anorthosite, in both of which the plagioclase grains are well formed as in typical igneous rocks. In many massifs of domical anorthosite there are angular inclusions of anorthosite made of plagioclase richer in calcium and similar to the anorthosite of the irregular massifs. These features have convinced scientists that domical anorthosites are igneous rocks, but the relationship to the associated silicic rocks is uncertain in most cases.

The domical anorthosite massifs, and possibly the irregular massifs as well, formed about 1.1–1.7 $\times 10^9$ years ago. They possibly mark a unique episode in the history of the differentiation of the Earth. N. Herz suggested this might be related to the capture of the Moon. A. T. Anderson and M. Morin suggested an episode of unusual liberation of heat and volatile material from deep within the Earth, but they could find no special cause for such an event. The restricted age of massif anorthosite still awaits a convincing explanation.

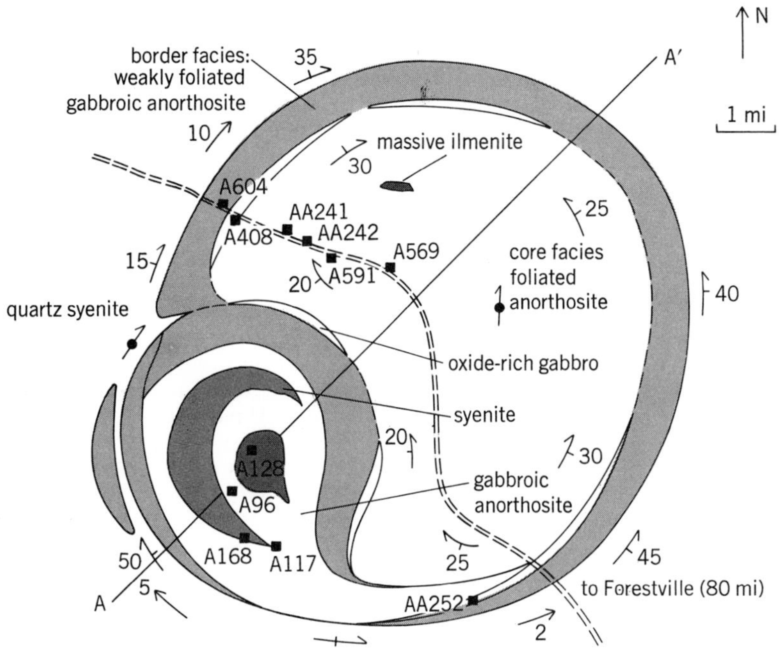

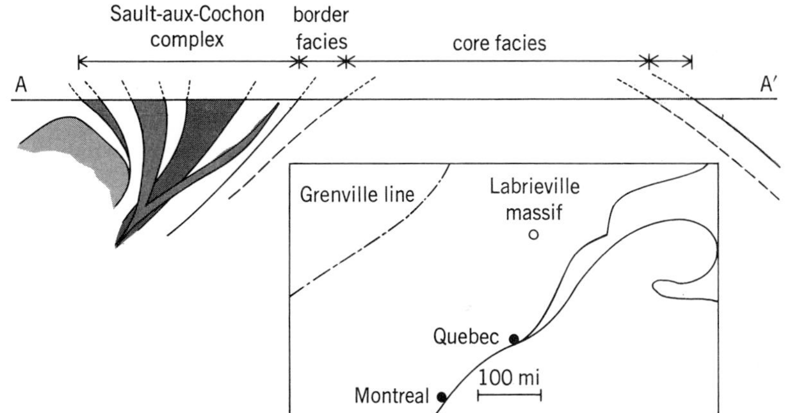

Fig. 2. Diagrammatic plan and cross section of the Labrieville anorthosite massif, showing sample localities; 1 mi = 1.6 km. (*From A. T. Anderson, Mineralogy of the Labrieville anorthosite, Amer. Mineral., 51:1671–1711, copyright 1966 by Mineralogical Society of America*)

Lunar anorthosites. By comparison with terrestrial occurrences, most lunar anorthosites are very fine grained, although one rock has crystals up to a centimeter long. Much of the fine grain size results from comminution by meteorite impact, and some of it probably results from rapid crystallization of impact melts. The unbrecciated lunar anorthosites have well-formed crystals of calcium-rich plagioclase (anorthite) intergrown with traces of olivine, spinel, pyroxene, ilmenite, metallic iron, and a few other exotic minerals. Associated rocks and minor minerals suggest that lunar anorthosites are of the layered type formed by accumulation of crystals of plagioclase from basaltic melts. It is uncertain whether lunar plagioclases sank or floated in the iron-rich lunar basaltic melts, but most scientists agree that rocks rich in plagioclase and including anorthosite compose a major part of the skin of the Moon. *See* ANDESINE; IGNEOUS ROCKS; LABRADORITE; METAMORPHISM; MOON.

[ALFRED T. ANDERSON, JR.]

Bibliography: I. S. E. Carmichael, F. J. Turner, and J. Verhoogen, *Igneous Petrology*, 1974; Y. Isachsen (ed.), *Origin of Anorthosite and Related Rocks*, New York State Museum and Science Service, Mem. no. 18, 1969; B. Mason and W. G. Melson, *The Lunar Rocks*, 1970.

Anthophyllite

The name given to the orthorhombic amphiboles. The anthophyllite amphiboles form a limited solid-solution series between the magnesian end member $Mg_7Si_8O_{22}(OH)_2$ and an approximately 50 mole % ferric iron member. Minerals with more iron than this apparently form the monoclinic mineral cummingtonite, thereby indicating a limited tolerance of the orthorhombic form for iron. Anthophyllite can accept various amounts of aluminum up to the approximate composition $(Mg,Fe)_5Al_2(Al_2Si_6)O_{22}(OH)_2$. Depending on the iron content, it can be white, gray, green, or brown, but usually is colorless in thin sections. Anthophyllite commonly occurs in fibrous masses, sometimes as asbestiform masses and in prismatic needles (see illustration).

Anthophyllite from East Goshen, Pa. (*Specimen from Department of Geology, Bryn Mawr College*)

Anthophyllite (110) cleavages of 54°30′ are approximately 1° smaller than the monoclinic amphiboles (except cummingtonite). In optical examination the parallel extinction of anthophyllite distinguishes it from the other amphibole varieties. *See* AMPHIBOLE.

Anthophyllite is a metamorphic mineral occurring in many schists and gneisses in association with such minerals as tremolite, chlorite, cordierite, garnet, talc, serpentine, spinel, and quartz. Asbestiform varieties are sometimes used as asbestos. *See* CUMMINGTONITE; TREMOLITE.

[GEORGE W. DE VORE]

Anticline

A fold in layered rocks in which the strata are inclined down and away from the axes. The simplest anticlines (see illustration) are symmetrical, but in more highly deformed regions they may be asymmetrical, overturned, or recumbent. Most anticlines are elongate with axes that plunge toward the extremities of the fold, but some have no distinct trend; the latter are called domes. Generally, the stratigraphically older rocks are found toward the center of curvature of an anticline, but in more complex structures these simple relations need not hold. Under such circumstances, it is sometimes convenient to recognize two types of anticlines. Stratigraphic anticlines are those folds, regardless of their observed forms, that are inferred from stratigraphic information to have been anticlines originally. Structural anticlines are those that have forms of anticlines, regardless of their original form. *See* FOLD AND FOLD SYSTEMS; SYNCLINE.

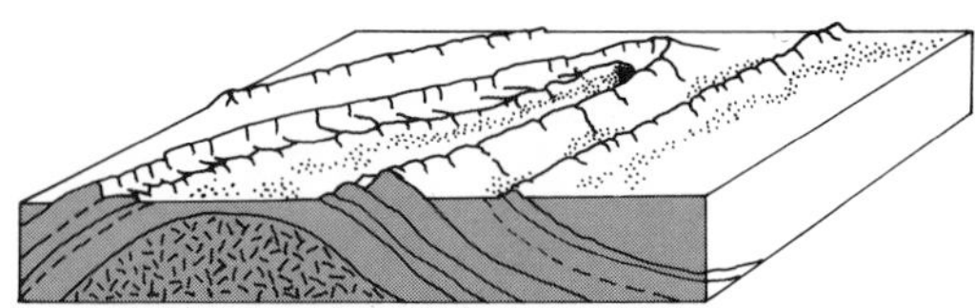

Diagram relating anticlinal structure to topography.

[PHILIP H. OSBERG]

Apatite

The most abundant and widespread of the phosphate minerals, crystallizing in the hexagonal system, space group $P6_3/m$. The apatite structure type includes no less than 10 mineral species and has the general formula $X_5(YO_4)_3Z$, where X is usually Ca^{2+} or Pb^{2+}, Y is P^{5+} or As^{5+}, and Z is F^-, Cl^-, or $(OH)^-$. The apatite series takes X = Ca, whereas the pyromorphite series includes those members with X = Pb. Three end members form a complete solid-solution series involving the halide and hydroxyl anions and these are fluorapatite, $Ca_5(PO_4)_3F$; chlorapatite, $Ca_5(PO_4)_3Cl$; and hydroxyapatite, $Ca_5(PO_4)_3(OH)$. Thus, the general series can be written $Ca_5(PO_4)_3(F,Cl,OH)$, the fluoride member being the most frequent and often simply called apatite.

The apatite series is further complicated with other substitutions. Carbonate-apatites involve the coupled substitution $(PO_4)^{3-} \rightleftharpoons (H_4O_4),(CO_3)^{2-}$; $Ca^{2+} \rightleftharpoons H_2O$; but the exact nature of this substitution is not yet known. The fluoride- and chloride-bearing carbonate-apatites are called francolite and dahllite, respectively. They are of considerable biological interest since they occur in human calculi, bones, and teeth. Dental enamels are composed in part of dahllite, and the efficacy of fluoride-bearing toothpaste partly concerns apatite crystal chemistry. Other less frequent substitutions occur in nature, involving $(AsO_4)^{3-} \rightleftharpoons (PO_4)^{3-}$ and the arsenate end members are known as svabite, $Ca_5(AsO_4)_3F$, and hedyphane, $Ca_5(AsO_4)_3Cl$. More limited substitutions can include many other ions, such as Mn^{2+} for Ca^{2+} (mangan-apatite), and Si^{4+} and S^{6+} for P^{5+}.

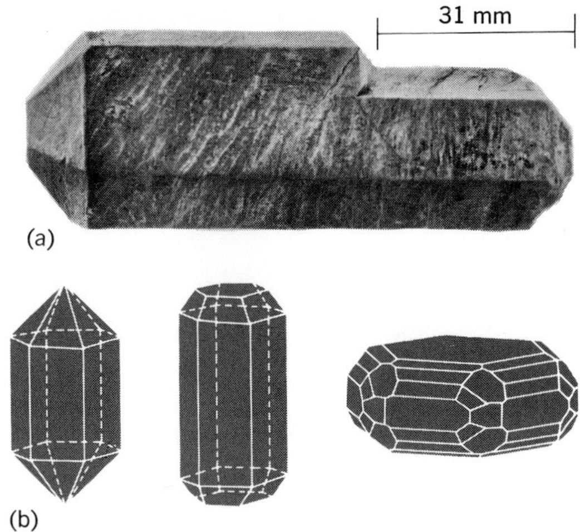

Apatite. (*a*) Specimen from Eganville, Ontario, Canada (*American Museum of Natural History specimens*). (*b*) Typical crystal habits (*from C. S. Hurlbut, Jr., Dana's Manual of Mineralogy, 17th ed., copyright © 1959 by John Wiley & Sons, Inc.; reprinted by permission*)

Apatites have been synthesized hydrothermally by the hydrolysis of monetite, $CaH(PO_4)$, or by direct fusion of the components, such as $Ca_3(PO_4)_2$ + CaF_2, which yields fluorapatite. Other apatite structures have been synthesized, usually by fusion of the components, involving the following elements and ions: Ba, Pb, Mg, Ni, Sr, Zn, Cr (3+ and 6+), Al, Fe, Na, Ce, Y, O^{2-}, $(OH)^-$, F^-, Cl^-,$(CO_3)^{2-}$, and $(SO_4)^{2-}$.

Diagnostic features. The apatite isomorphous series of minerals occur as grains, blebs, or short to long hexagonal prisms terminated by pyramids, dipyramids, and the basal pinacoid (see illustration). The minerals are transparent to opaque, and can be asparagus-green (asparagus stone), greyish-green, greenish-yellow, gray, brown, brownish-red, and more rarely violet, pink, or colorless. Mangan-apatite is usually blue or blue-gray to blue-green. Some apatites fluoresce in ultraviolet radiation, in particular the mangan-apatite variety. The cleavage is poor parallel to the base and fracture uneven. Apatites are brittle, with hardness 5 on Mohs scale, and specific gravity 3.1–3.2; they are also botryoidal, fibrous, and earthy.

Occurrence. Apatite occurs in nearly every rock type as an accessory mineral. In pegmatites it occurs as a late-stage fluid segregate, sometimes in large crystals and commonly containing Mn^{2+}. It often crystallizes in regional and contact metamorphic rocks, especially in limestone and associated with chondrodite and phlogopite. It is very common in basic to ultrabasic rocks; enormous masses occur associated with nepheline-syenites in the Kola Peninsula, Soviet Union, and constitute valuable ores which also contain rare-earth elements. Apatite is a common accessory with magnetite ores associated with norites and anorthosites and is detrimental because it renders a high phosphorus content to the ores. Apatite occurs infrequently in meteorites along with other phosphates.

Large beds of oolitic, pulverulent, and compact fine-grained carbonate-apatites occur as phosphate rock, phosphorites, or collophanes. They have arisen by the precipitation of small concretions formed by organisms and by the action of phosphatic water on bone material or corals. Extensive deposits of this kind occur in the United States in Montana and Florida and in North Africa. The material is mined for fertilizer and for the manufacture of elemental phosphorus. *See* PHOSPHATE MINERALS.

[PAUL B. MOORE]

Aphanite

An igneous rock in which the constituents cannot be distinguished by the unaided eye. Aphanitic material may be minutely crystalline, or it may be glassy. Most lavas solidify so rapidly that coarse mineral grains cannot form; thus, volcanic rocks are mostly aphanites. Under conditions of slow cooling, however, such as exist deep below the Earth's surface, molten rock material (magma) solidifies to coarse-grained plutonic rocks. Such visibly crystalline rocks are called phanerites. *See* IGNEOUS ROCKS.

[CARLETON A. CHAPMAN]

Aplite

A fine-grained, sugary-textured rock, generally of granitic composition; also any body composed of such rock.

This light-colored rock consists chiefly of quartz, microcline, or orthoclase perthite and sodic plagioclase, with small amounts of muscovite, biotite, or hornblende and traces of tourmaline, garnet, fluorite, and topaz. Much quartz and potash feldspar may be micrographically intergrown in cuneiform fashion. *See* GRANITE; IGNEOUS ROCKS.

Aplites may form dikes, veins, or stringers, generally not more than a few feet thick, with sharp or gradational walls. Some show banding parallel to their margins. Aplites usually occur within bodies of granite and more rarely in the country rock surrounding granite. They are commonly associated with pegmatites and may cut or be cut by pegmatites. Aplite and pegmatite may be gradational or interlayered, or one may occur as patches within the other. *See* PEGMATITE.

Aplites form in different ways. Some represent granite or pegmatite which recrystallized along fractures and zones of shearing. Others are of metasomatic (replacement) origin. Many form from residual solutions derived from crystallizing granitic magma (rock melt). If these fluids retain their volatiles, pegmatites may form. If the volatiles escape, a more viscous fluid may be created, and a fine-grained (aplitic) texture may be developed. *See* MAGMA; METASOMATISM.

[CARLETON A. CHAPMAN]

Apophyllite

A hydrous calcium potassium silicate containing fluorine. The composition is variable but approximates to $KFCa_4(Si_2O_5)_4 \cdot 8H_2O$. It resembles the zeolites, with which it is sometimes classified, but differs from most zeolites in having no aluminum. It exfoliates (swells) when heated, losing water, and is named from this characteristic; the water can be reabsorbed. The mineral decomposes in hydrochloric acid, with separation of silica. It is essentially white, with a vitreous luster, but may show shades of green, yellow, or red. The symme-

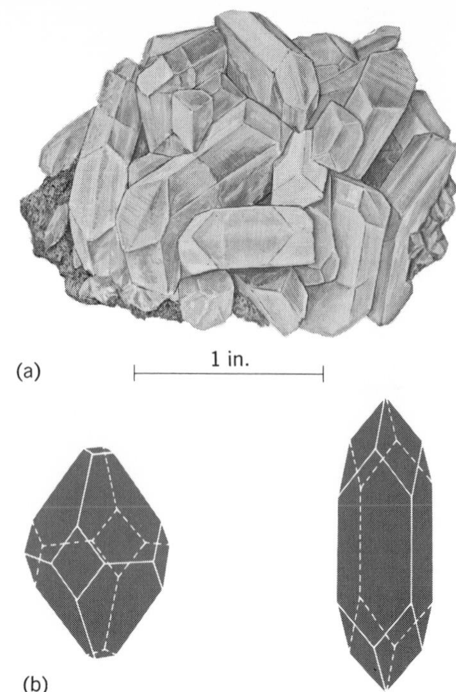

Apophyllite. (*a*) Crystals with basalt from French Creek, Pa. (*specimen from Department of Geology, Bryn Mawr College*) (*b*) Crystal habits (*from C. S. Hurlbut, Jr., Dana's Manual of Mineralogy, 17th ed., copyright © 1959 by John Wiley & Sons, Inc.; reprinted by permission*)

Aragonite. (*a*) Specimen from Girgenti, Sicily (*American Museum of Natural History specimens*). (*b*) Pseudohexagonal twinned crystal (*from C. S. Hurlbut, Jr., Dana's Manual of Mineralogy, 17th ed., copyright © 1959 by John Wiley & Sons, Inc.; reprinted by permission*)

try is tetragonal and the crystal structure contains sheets of linked SiO_4 groups, and this accounts for the perfect basal cleavage of the mineral (see illustration).

Apophyllite occurs as a secondary mineral in cavities in basic igneous rocks, commonly in association with zeolites. The specific gravity is about 2.3–2.4, the hardness 4.5–5 on Mohs scale, the mean refractive index about 1.535, and the birefringence 0.002. *See* SILICATE MINERALS; ZEOLITE. [GEORGE W. BRINDLEY]

Aragonite

A mineral species of calcium carbonate, but with a crystal structure different from that of vaterite and calcite, the other two polymorphs of the same composition. Strontium and lead are sometimes substituted for some of the calcium in aragonite. Aragonite is much less common than calcite and under atmospheric conditions is metastable with respect to calcite, to which it is frequently partially inverted. Complete inversion can be effected by boiling powdered aragonite in water. In all of its natural occurrences, aragonite is a low-temperature, near-surface deposit. It commonly occurs as stalactites and stalagmites in limestone caverns and as secondary growths in cavities in rocks containing calcium-bearing minerals. Aragonite with some calcite constitutes the substance of pearls. It is also present in the hard parts (shells) of some fossils, pelecypods, and gastropods. *See* CALCITE; PALEOECOLOGY, GEOCHEMICAL ASPECTS OF.

Aragonite is found in many localities throughout the world. In the United States fine crystals associated with calcite and cerussite are found in the Magdalena district, Socorro County, N.Mex.

Aragonite has orthorhombic symmetry and a structure in which the calcium atoms are approximately arranged as in hexagonal close packing. The CO_3 groups fall into two sublayers between the layers of calcium atoms. Single crystals are often prismatic, short, or elongated, and frequently twinned. They commonly have a hexagonal outline (see illustration).

When pure, aragonite is white, but impurities may induce a variety of tints such as grays, blues, greens, and pinks.

Hardness is 3.5–4 on Mohs scale and specific gravity is 2.947. Aragonite may be precipitated at 80°C by adding a solution of calcium chloride to a solution of sodium carbonate. *See* CARBONATE MINERALS.

[ROBERT I. HARKER]

Archeozoic

A term for the earlier interval in the twofold subdivision of Precambrian time. It has been more or less synonymous with Archean, which has been more widely used, especially outside the United States. Most students of the Precambrian find it very difficult to apply the old twofold subdivision, however, and many have a strong tendency to discard it and to build a new time scale more in keeping with the numerous events of Precambrian time. *See* PRECAMBRIAN.

[J. PAUL FITZSIMMONS]

Arenaceous rocks

The arenaceous rocks (arenites) include all those clastic rocks whose particle sizes range from 2 to 1/16 mm, or if silt is included, to 1/256 mm. Some arenites are composed primarily of carbonate particles, in which case they are called calcarenites and grouped with the limestones. Some oolitic iron ores and glauconite beds are properly classified as arenites. But the vast majority of arenites are commonly called sandstones, and the two words are almost synonymous. *See* CALCARENITE; GRAYWACKE; OOLITE AND PISOLITE; ORTHOQUARTZITE; SANDSTONE; SEDIMENTARY ROCKS; SUBGRAYWACKE.

[RAYMOND SIEVER]

Argentite

A mineral having composition Ag_2S. Argentite crystals are rare and are cubes or octahedrons. These are paramorphs of the isometric form of silver sulfide that is stable only above 179°C. Below this temperature the orthorhombic modification, acanthite, is stable. Argentite most commonly occurs in massive form or as coatings with a lead-gray color (see illustration). Its hardness is 2.5

Argentite with quartz crystals from Sarrabus, Sardinia. (*Specimen from Department of Geology, Bryn Mawr College*)

(Mohs scale) and specific gravity 7.3; it is very sectile and is bright on the fresh surface but becomes dull black on exposure. It occurs in veins associated with other silver minerals and at some places is a major silver ore. Important localities are in Mexico, Peru, Chile, and Bolivia. In the United States it has been extensively mined in Nevada at the Comstock Lode and at Tonopah.

[CORNELIUS S. HURLBUT, JR.]

Argillaceous rocks

The argillaceous rocks (lutites) include shales, argillites, siltstones, and mudstones; they are clastic sediments whose constituent particles are less than 1/16 mm (if siltstones are included) or less than 1/256 mm (if siltstones are excluded). They are the most abundant sedimentary rock type, varying according to different estimates from 44 to 56% of the total sedimentary rock column. Claystone is indurated clay, which consists dominantly of fine material of which at least a major proportion is clay mineral (hydrous aluminum silicates). Shale is a laminated or fissile claystone or siltstone, in general more consolidated than claystone. Mudstone is a claystone that is blocky and massive.

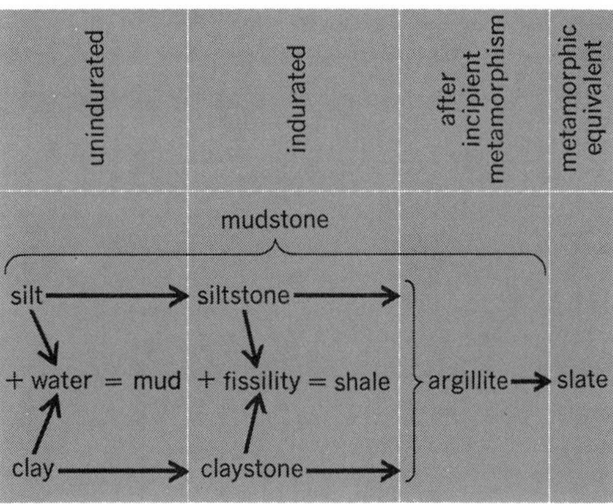

Classification of fine-grained mechanical sediments. (*After W. H. Twenhofel, in F. J. Pettijohn, Sedimentary Rocks, 2d ed., copyright © 1949, 1957 by Harper & Row, Publishers, Inc.*)

The term argillite is used for rocks which are more indurated than claystone or shale but not metamorphosed to slate. All these argillaceous rocks are consolidated equivalents of muds, oozes, silts, and clays (see illustration). Loess is a fine-grained, unconsolidated, windblown deposit. The term shale has been used by many authors generically to denote all of these types of rock. *See* ARGILLITE; BENTONITE; CLAY; CLAY MINERALS; LOESS; SEDIMENTARY ROCKS; SHALE.

[RAYMOND SIEVER]

Argillite

A nonfissile argillaceous rock that seems to have formed as the result of incipient metamorphism. The degree of induration is greater than that of shales, and the rock is dense and hard. The term has been used to designate argillaceous rocks in low-grade metamorphic belts that do not have the parting or cleavage of slates. Argillites are normally finely laminated, but the term has also been applied to dense, hard, massive, nonbedded argillaceous rocks. Argillites seem to have a gross mineral and chemical composition similar to that of shales. The argillites have no fissility, for which there is no ready explanation. They may represent conditions under which original settling did not result in preferred orientation of platy minerals, or in which there has been no reorientation due to compaction and pressure. Alternatively, the clay mineral crystal habit in the argillites may be rod-shaped and fibrous rather than platy. *See* ARGILLACEOUS ROCKS; SEDIMENTARY ROCKS; SHALE.

[RAYMOND SIEVER]

Argyrodite

A mineral having composition Ag_8GeS_6 and crystallizing in the isometric system. Argyrodite crystals show the octahedron and dodecahedron, but the mineral is usually massive or in crusts with a crystalline surface. The hardness is 2.5 (Mohs scale) and the specific gravity is 6.3. The luster is metallic and the color black with a blue-to-purple tone. A complete series exists between argyrodite and canfieldite, Ag_8SnS_6, with substitution of tin for germanium. The principal occurrences of both minerals are at Freiberg, Germany, and Potosi,

Bolivia, associated with other silver minerals. Argyrodite is one of two germanium minerals and is thus a source of this rare but important element.

[CORNELIUS S. HURLBUT, JR.]

Arkose

Arenite (rock composed of sand-size fragments) that contains a high proportion of feldspar in addition to quartz and other detrital minerals. Arkose is also known as feldspathic sandstone. Although there is no universal agreement, many geologists consider a minimum of 25% feldspar a requisite for calling sandstone an arkose. Other geologists accept a lower value.

Composition. Arkoses may contain a high proportion of other nonquartz detritus, such as igneous and metamorphic rock fragments, micas, amphiboles, and pyroxenes. Frequently the accessory heavy mineral suite consists of a variety of species. Though the arkoses are rarely as well sorted as orthoquartzites, they may be moderately well sorted. The grains are angular or poorly rounded. Clay matrix is generally subordinate in arkoses, but there may be as much as 10–15%. If there is that much, the rocks have textural similarity to feldspathic graywackes, though the mineralogy may be appreciably different in the latter. The clay is dominantly kaolinite with smaller proportions of micaceous and montmorillonitic clay. Presumably much of the kaolinite has come from the weathering of feldspar. Conglomeratic zones are common in many arkoses. Feldspathic sandstones that contain less than 25% of feldspar have been termed subarkoses. They are in general similar to the more highly feldspathic true arkoses.

Structure. Sedimentary structures of arkoses are similar in kind to those of the orthoquartzites. Cross-bedding, the major feature, may be displayed on a huge scale, some cross-bedded units being many feet thick. Bedding is crude and many times not distinguishable; many beds are thick and massive. Ripple marks may be present but are not common. Some arkoses, such as those of the Newark Series along the Atlantic Coast of the United States, contain mud cracks, frost crystal impressions, and footprints of small dinosaurs. *See* BEDDING; SEDIMENTARY STRUCTURES.

Occurrence. Arkoses are associated with a variety of clastic rocks, dominantly conglomerates, and reddish-colored shales. Arkoses also are found with basic lava flows. The formations occur as thick, wedge-shaped deposits, the thick end of the wedge being in close proximity to the source area and sometimes separated from it by normal faults of large displacement. Other arkoses are relatively thin formations that overlie granitic terrain. These formations grade laterally into other kinds of sandstone away from the area of underlying granite. Most arkoses are found in geosynclinal areas, but the thin, reworked, granite-wash arkoses can be found on stable continental platforms. *See* GEOSYNCLINE.

Origin. The granite-wash arkoses appear to have formed as the result of a transgression of the sea over a land area underlain by granite. The fragmented granite in the soil and mantle rock is incorporated in the basal sediment. In some areas the original granite is changed so slightly that the arkose is called recomposed granite and may be almost indistinguishable from the original granite.

The origin of the arkoses is best understood in terms of the abundance of feldspar. Feldspar is unstable both chemically and mechanically as compared with quartz and, given sufficient rigors of chemical weathering at the source and abrasion during transportation, will disappear. The lack of appreciable chemical weathering at the source (which allows the contribution of much feldspar to the sediment) may be due to high topographic relief or to climatic extremes.

Since high relief and climatic extremes generally are associated with orogenic movements, arkoses are usually interpreted as sediments that result from tectonically active regions. Rift valleys formed by the divergence of two continental plates at a spreading center are now favored as the tectonic environment for arkose formation. This explains too the association with basaltic lavas. The abundant iron oxides present in most arkoses, the mud cracks, and the fanglomerates all point to a predominantly terrestrial mode of deposition. The thin granite washes are marine and represent different conditions. *See* FELDSPAR; GRAYWACKE; ORTHOQUARTZITE; SANDSTONE; SEDIMENTARY ROCKS.

[RAYMOND SIEVER]

Arsenolite

A mineral having composition As_2O_3 and crystallizing in the isometric system. Arsenolite forms small octahedral crystals but also is in botryoidal and stalactitic aggregates. There is octahedral cleavage. The hardness is 1.5 (Mohs scale) and the specific gravity is 3.87. The luster is vitreous to silky and the color white, occasionally tinged with blue, yellow, or red. The mineral is slightly soluble in water and has a sweetish taste. Arsenolite is a secondary mineral formed by the oxidation of arsenic minerals such as native arsenic, arsenopyrite, enargite, and tennantite. *See* ARSENOPYRITE; ENARGITE.

[CORNELIUS S. HURLBUT, JR.]

Arsenopyrite

A mineral having composition FeAsS. Although arsenopyrite crystallizes in the monoclinic system, the crystals have pseudoorthorhombic symmetry because of twinning (see illustration). The hardness is 5.5–6 (Mohs scale) and the specific gravity is 6.0. The luster is metallic and the color silver-white. Arsenopyrite is the commonest arsenic-bearing mineral. It is associated with ores of tin, tungsten, and gold formed at high temperatures and to a lesser extent is found in cobalt-nickel veins formed at lower temperatures. It is a widespread mineral found abundantly at many localities, notably at Freiberg, Germany; Cornwall, England; and Cobalt and Dolero, Ontario, Canada. It is associated with the gold ores at Lead, S.Dak.

[CORNELIUS S. HURLBUT, JR.]

Asbestos

General name for the useful, fibrous varieties of six rock-forming minerals. The value of asbestos derives from its strength as a reinforcing agent, its relative chemical inertness, and the incombustible nature of certain asbestos-containing products. Originally the term applied only to amphibole vari-

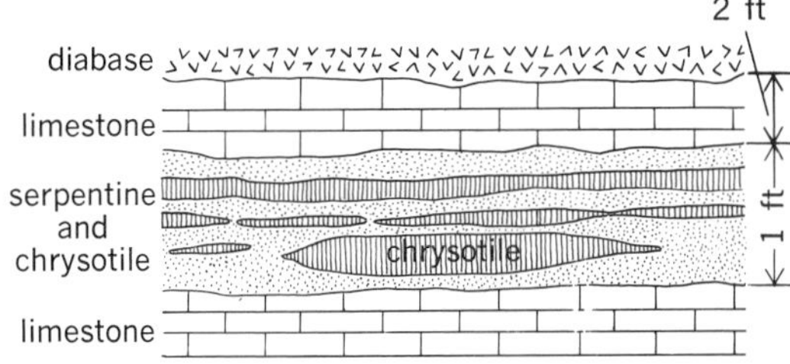

Fig. 1. Asbestos bands in serpentine bands in dolomite, Sierra Ancha. Ariz. (*From A. M. Bateman, An Arizona asbestos deposit, Econ. Geol., 18:663–683, 1923*)

eties, but it now applies to all types used in commerce. Over 95% of production is fibrous serpentine.

World production of asbestos in 1976 reached 5.6×10^6 tons (5×10^6 metric tons), a 70% increase over that in 1966. Production in the Soviet Union more than tripled during that period to 2.5×10^6 tons (2.25×10^6 metric tons). Canadian production increased 13% to 1.7×10^6 tons (1.53×10^6 metric tons). These two countries produce 75% of the world's supply of asbestos. United States production, 115,000 tons (103,500 metric tons) in 1976, was just 91% of 1966 production. United States imports, more than 90% from Canada, at 658,000 tons (592,200 metric tons), were also 91% of the 1966 figure. The Republic of South Africa, Italy, and Southern Rhodesia were other significant asbestos producers in 1976.

Environment. Prolonged exposure to high concentrations of airborne asbestos fibers have been demonstrated to be injurious to human health. However, there is still disagreement as to the possible hazard from ingestion of asbestos particles. The Environmental Protection Agency (EPA), Occupational Safety and Health Administration (OSHA), and Mine Safety and Health Administration (MSHA) are Federal agencies which conduct the enforcement programs in the United States. In their published regulations, the term "asbestos" includes only the fibrous form of one serpentine group and five amphibole-group minerals.

Amphibole asbestos. The fibrous amphiboles that are termed asbestos are generally more heat- and acid-resistant than serpentine asbestos, though not as strong, alkali-resistant, or available. These fibrous amphiboles include amosite, a trade name for a complex iron-magnesium silicate (mostly cummingtonite-grunerite), $(Mg,Fe^{2+})_7Si_8O_{22}(OH)_2$; and crocidolite, a sodium-iron hydrous silicate variety of riebeckite, $Na_2Fe_3{}^{2+}Fe_2{}^{3+}Si_8O_{22}(OH,F)_2$. Of minor importance are tremolite asbestos, $Ca_2(Mg,Fe^{2+})_5Si_8O_{22}(OH,F)_2$, anthophyllite asbestos, $(Mg,Fe^{2+})_7Si_8O_{22}(OH,F)_2$; and actinolite asbestos, a variety of tremolite containing 15–20% ferrous iron.

Serpentine asbestos. Fibrous serpentine (chrysotile), $Mg_3Si_2(OH)_4O_5$, is mined in the three principal producing areas. The eastern part of Quebec province in Canada and, so far as is known, the Asbest region in the Ural Mountains of the Soviet Union produce only chrysotile; in South Africa chrysotile production exceeds that of more specialized varieties. *See* SERPENTINE.

Chrysotile asbestos is usually found associated with altered massive serpentine bodies in olivine-rich ultrabasic rocks, such as dunite and peridotite. In such cases, the chrysotile characteristically occupies irregularly intersecting fracture systems

Fig. 2. Jeffrey Mine in Quebec province. (*Canadian Johns-Manville Co., Ltd., Asbestos, Quebec*)

in the host rock. These are presumed to be associated with more acid intrusions into the older, basic host. In other instances, such as in Arizona, chrysotile is found in serpentinized dolomitic limestone in association with diabase sills (Fig. 1). The more frequent occurrences in either type of host are of the cross-fiber orientation, in which fibers lie perpendicular to the vein walls. Slip-fiber occurrences, with fibers oriented along veins, are less perfect and are not suitable for spinning. *See* SERPENTINITE.

Mining. In Vermont and in the Copperopolis district of California, the fiber-bearing rock is mined from an open pit. In the Coalinga district of California, the highly sheared ore is simply "plowed" and allowed to air-dry, and the coarse fraction is then screened out from the mill feed. In Arizona, a heading is driven beneath the fiber zone; the zone is later blasted down and milled. The Canadian mines generally are open pits (Fig. 2). The chrysotile of Southern Rhodesia, the Republic of South Africa, and Swaziland is obtained from underground mines. Amosite is obtained chiefly from large underground workings, and blue asbestos (crocidolite) from small open pits and shallow mines. The Soviet deposits are worked both in open pits and underground. Cyprus has large open-pit workings.

The data analyzed by least-squares regression analysis (Fig. 3) show trends in asbestos recovery from established mines over a period of time. In 1951 in the Quebec asbestos mines, 75% of the rock mined was milled and 9.9% of that was recovered as fibers. About 20 years later (1970), only 32.2% of the rock mined was milled, and it yielded 6.1% fibers. Trend projection to the year 2000 indicates that only 24.4% of the mined rock will be milled and 2.9% of the milled material will be made into fibers.

Processing. Asbestos milling is a complex operation involving crushing, fiberizing, screening, vacuum separation of fiber from rock, and classification of fiber by length. Special milling techniques have been developed for the matted short-fiber chrysotile of the Coalinga district, including grinding and wet-milling. In Copperopolis, Vermont, and Canada, the mills are large and complex. The fiber is classified into many grades, such as spinning, cement stock, and paper stock.

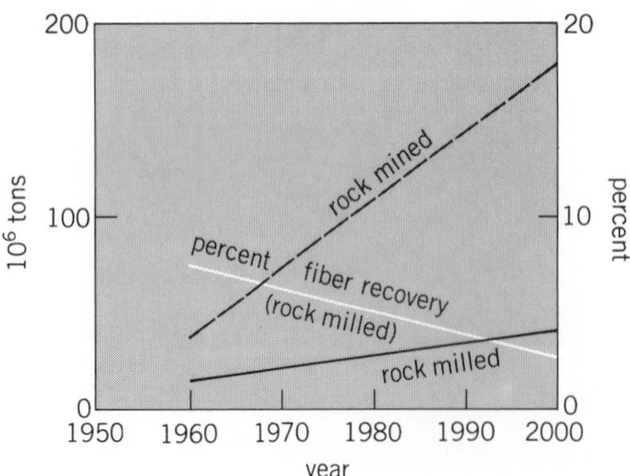

Fig. 3. Quebec production trends from analysis of 1950–1971 data. "Rock mined" refers to ore and waste rock exclusive of overburden. 10^6 tons = 0.9×10^6 metric tons. *(From R. A. Clifton, Asbestos—1977, U.S. Bur. Mines MCP-6, 1977)*

Uses. Long fibers are spun into yarn and woven into cloth, either with or without auxiliary fiber material, such as cotton, glass wool, or copper wire. Brake linings, heavy packings and gaskets, electrical insulating materials, and protective clothing require the best fibers. Medium and short fibers are used to make asbestos shingles, sheet siding, pipe, floor tile, less critical packings and gaskets, paper, binders for other heat insulators, and fillers for asphalts, plastics, paints, and greases.

Current research. The most significant area of asbestos research for the last few years has been concerned with the controversial health aspects of the fibers. The National Institute of Environmental Health Sciences, for example, started a multiyear animal feeding study in 1976 to determine the effects on health of ingested asbestos and asbestos-related minerals.

In late 1976, the U.S. Bureau of Mines established a Particulate Mineralogy Unit at College Park, MD, to provide technical information on particulate mineralogy, especially in reference to asbestos; to develop a solid scientific basis for research into particle-related pollution problems; and to facilitate decision making by regulatory bodies.

Efforts aimed at either synthesizing analogs of the natural fibers or finding substitutes have met with limited success. Most efforts seem to be aimed at thermal insulation. There are several commercially available synthetic inorganic fibers, and more are becoming available each year.

Glass-reinforced cement, now commercially available in the United States and Europe, could affect the future of the asbestos industry. The glass used is a high-zirconia, alkali-resistant fiber developed by the United Kingdom's Building Research Station.

[ROBERT A. CLIFTON]

Bibliography: H. Berger, *Asbestos Fundamentals: Origins, Properties, Mining, Processing, Utilization*, 1963; R. A. Clifton, *Asbestos—1977*, U.S. Bur. Mines MCP-6, 1977; S. J. Lefond et al. (eds.), *Industrial Minerals and Rocks*, 4th ed., 1975.

Asphalt and asphaltite

Varities of naturally occurring bitumen. Asphalt is also produced as a petroleum by-product. Both substances are black and largely soluble in carbon disulfide. Asphalts are of variable consistency, ranging from a highly viscous fluid to a solid, whereas asphaltites are all solid. Asphalts fuse readily, but asphaltites fuse only with difficulty. Asphalts may, moreover, occur with or without appreciable percentages of mineral matter, but asphaltites usually have little or no associated mineral matter. *See* BITUMEN.

Natural occurrence. Many asphalts occur as viscous impregnations in sandstones, siltstones, and limestones. Most such deposits are thought to be petroleum reservoirs from which volatile constituents have been stripped by exposure of the rock.

The asphaltites (gilsonite, grahamite, and glance pitch) were probably derived from a saline lacustrine sapropel and owe their variable properties to differences in environment of deposition. These substances occur on a large scale in the Uinta Basin of northeastern Utah, where they are derived from upper Eocene Green River sediments, most of which are oil shales high in carbonate content. *See* OIL SHALE; SAPROPEL.

Asphalt. Relatively pure asphalt occurs in Kern, San Luis Obispo, and Santa Barbara counties, Calif. The asphalt, or tar seep, may contain up to 30% mineral matter and occasionally, as in the La Brea tar pits, has remains of insects or animals that became entrapped in the viscous material. Many of these occurrences are associated with the Miocene Monterey shale, which is thought to have been a source bed for the local crude oil. Occurrences of asphalt are also known in Kentucky and Oklahoma. Although asphalt seeps have long been known in France, Greece, Russia, Cuba, and other countries, the best known and largest are those of Venezuela and Trinidad.

Venezuelan asphalt has been recovered from the Bermudez Pitch Lake covering over 1000 acres and averaging 5 ft in depth. The lake is thought to be fed by asphalt springs. The asphalt at the edges of the lake becomes hard enough to walk on. After treatment to remove water and volatile constituents, the asphalt fuses at 130–140°F, is 92–97% soluble in carbon disulfide, and contains approximately 83% carbon, 11% hydrogen, 6% sulfur, and 1% nitrogen.

The Trinidad Asphalt Lake covers 115 acres and is 135 ft deep at its center. The asphalt is softest at the center, where it is probably fed by underground springs, and hardest at the edges. Even at the center, however, the material is hard enough to support a man and can be broken into blocks. Gas is evolved from the asphalt, and rainwater tends to collect in depressions forming an emulsion of asphalt and water containing gas, sand, and clay. The crude material is 39% soluble in carbon disulfide, contains 27% mineral matter, and loses 29% of its weight as water and gas on heating to 100°C. After the refining process which drives off gas and water, the product fuses at about 190°F, is 56% soluble in carbon disulfide, contains 38% mineral matter, and consists of 80–82% carbon, 10–11% hydrogen, 6–8% sulfur, and less than 1% nitrogen, all on an ashfree basis. The Trinidad lake has been estimated to contain 10,000,000–15,000,000 tons of asphalt.

Sandstones impregnated with asphalt occur in Oklahoma, Kentucky, Arkansas, Alabama, Utah, California, and other states. Many of these deposits have been developed as a source of paving material. Similar deposits also occur in Canada, South America, Europe, Asia, and Africa. One of the most extensive occurrences is that of Asphalt Ridge near Vernal, Utah, where asphalt saturates sandstones of the Uinta formation of Eocene age and of the Mesaverde formation of Upper Cretaceous age. This deposit crops out over a distance of 11.5 mi, and the sandstone contains 8–15% asphalt. It has been estimated that over 1,000,000,000 tons of high-grade material is readily available. The asphaltic sandstone is scooped from the deposit and used directly in road construction. If necessary, it is mixed with sand to attain the desired consistency.

Table 1. Major physical differences of asphaltite groups

Group	Specific gravity at 77°F	Softening point, °F
Gilsonite	1.03–1.10	230–350
Glance pitch	1.10–1.15	230–350
Grahamite	1.15–1.20	350–600

Most asphalts are of marine origin and consist of the high-molecular-weight compounds normally present in petroleum residues. Asphalts and asphaltites frequently contain unusually high percentages of vanadium.

Asphaltites. These substances are divided into three groups: gilsonite, glance pitch, and grahamite. The major physical differences in these substances are in specific gravity and softening point, as shown in Table 1. All three substances are nearly completely soluble in carbon disulfide. *See* GILSONITE; IMPSONITE; WURTZILITE.

Differentiation of the asphaltites into three groups is based only on physical properties and not on a genetic basis. For this reason, similarly categorized substances may have somewhat different origins and variable compositions.

Glance pitch occurs on Barbados, and material from this deposit has been marketed as manjak. Other veins of glance pitch, some of which contain up to 27% mineral matter and up to 7.4% sulfur, also occur in Haiti, Cuba, Mexico, Argentina, Colombia, Chile, the Baltic states, and the Near East. Glance pitch has been used for the manufacture of lacquers.

Grahamite occurs in West Virginia, Texas, Oklahoma, and Colorado. It is also known in Mexico, Cuba, Trinidad, Argentina, and Peru. The Peruvian grahamite is particularly rich in vanadium, and some vanadium minerals are associated with it. In general, most deposits are relatively small and are no longer of commercial interest. *See* ALBERTITE; ELATERITE. [IRVING A BREGER]

Table 2. Asphalts and their uses

Asphalt type and % of production	Manufacturing process	Properties	Uses
Straight-run, 70–75%	Distillation or solvent precipitation	Nearly viscous flow	Roads, airport runways, hydraulic works
Air-blown, 25–30%	Reacting with air at 400–600°F	Resilient; viscosity less susceptible to temperature change than straight-run	Roofing, pipe coating, paints, underbody coatings, paper laminates
Cracked, less than 5%	Heating to 800–1000°F	Nearly viscous flow; viscosity more susceptible to temperature change than that of straight-run asphalt	Insulation board saturant, dust laying

Petroleum by-product. Asphalt is derived from petroleum in commercial quantities by removal of volatile components. It is an inexpensive construction material used primarily as a cementing and waterproofing agent. Over 27,000,000 tons of asphalt is used in the United States annually, of which more than 98% is derived from petroleum.

Asphalt is composed of hydrocarbons and heterocyclic compounds containing nitrogen, sulfur, and oxygen; its components vary in molecular weight from about 400 to 5000. It is thermoplastic and viscoelastic; at high temperatures or over long loading times it behaves as a viscous fluid, while at low temperatures or short loading times it behaves as an elastic body.

The three distinct types of asphalt made from petroleum residues are described in Table 2.

In the construction of pavement surface for major roads and airport runways, hot asphalt is mixed with hot graded-stone aggregate. The mixture is spread on a dense, compacted stone base and rolled while still hot to give a smooth surface.

Roads having only light traffic are often given a thin, inexpensive wearing surface by spraying fluid asphalt on the road base and covering it immediately with stone. The fluid asphalt may be hot paving asphalt or a liquid asphalt. Liquid asphalts are produced by blending asphalt with various petroleum distillates or by emulsifying hot asphalt with water containing a small amount of soap. The liquid asphalts are fluid at ambient temperatures but harden as the solvent or water evaporates.

Air-blown asphalt is used mainly for roofs. Hot asphalt may be mopped on the roof and covered with decorative gravel or prefabricated asphalt shingles may be nailed onto the roof.

Asphalt is increasingly used in hydraulic works to line canals and reservoirs, to face dams and dikes, and to bind together the rocks in breakwaters. [THOMAS K. MILES]

Bibliography: H. Abraham, *Asphalts and Allied Substances*, 6th ed., 5 vols., 1960–1963; A. J. Hoiberg (ed.), *Bituminous Materials: Asphalts, Tars, and Pitches*, vol. 1, 1964; E. M. Spieker, *Bituminous Sandstone near Vernal, Utah*, U.S. Geol. Surv. Bull. no. 822-C, 1930.

Asthenosphere

A region in the interior of the Earth characterized by less mechanical strength and less resistance to deforming stresses than regions above and below. The term "asthenosphere" was introduced by J. Barrell in 1914 as one of three zones of the Earth's interior differentiated on the basis of strength: (1) the lithosphere, comprising the outer 100 km or so of the Earth and characterized by rigidity, stiffness, and strength; the tectonic plates are made up of lithosphere; (2) the asthenosphere, a weaker layer lying below the lithosphere and much more susceptible to deformation by plastic or viscous flow; and (3) the centrosphere, another stronger region lying beneath the asthenosphere and, according to Barrell, extending to the center of the Earth. This zonation is in contrast to the more familiar subdivision of the interior into crust, mantle, and core, based on discontinuities in seismic-wave velocities. The asthenosphere lies within the seismically defined upper mantle. With the discovery that the Earth's core is primarily liquid, the term "centrosphere" has been superseded by "mesosphere," introduced by H. S. Washington to designate the region of the mantle between the asthenosphere and the core. *See* EARTH, INTERIOR OF; PLATE TECTONICS.

Geodynamics. The asthenosphere, because of its ability to flow, has figured prominently in geodynamic theories on the causes of vertical motions observed at the Earth's surface, such as postglacial rebound. The compensatory adjustments which take place in the interior in response to changing mass distributions at the surface arising from erosion, sedimentation, glaciation and deglaciation, and volcanism are thought to occur through flow in the asthenosphere. Likewise, the asthenosphere plays a prominent role in models of the large horizontal movements of the lithosphere as observed in continental drift and plate tectonics. The asthenosphere is envisioned as the lubricating layer over which the plates glide; moreover, the thermal convection thought to be the force driving the tectonic plates lies within and, in fact, probably defines the vertical extent of the asthenosphere. Some debate exists as to whether convection is confined to the upper mantle in a thin asthenosphere or occurs throughout the mantle, in which case the asthenosphere would occupy the entire mantle beneath the lithosphere. *See* ISOSTASY.

Viscosity. The reasons for the existence of the asthenosphere are principally thermal. It is well known that many physical properties of rocks vary greatly with temperature; in particular, viscosity, or the resistance of rock to creep and flowage, is reduced by higher temperatures. The relationship

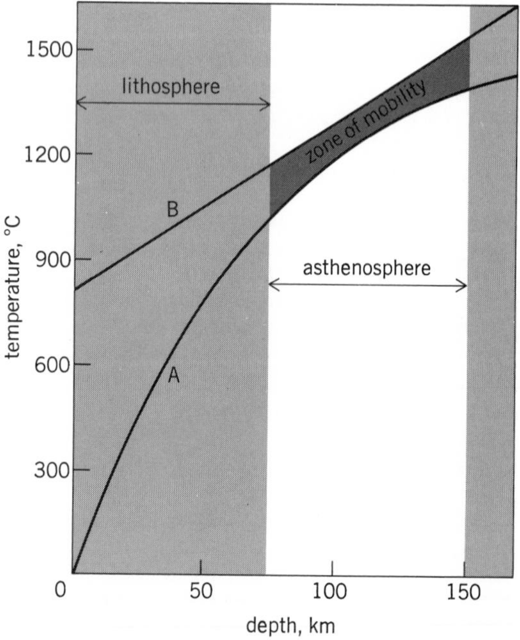

Fig. 1. Thermal condition of the Earth's crust and upper mantle, showing how and at what depth the asthenosphere develops. Curve A shows how at a typical continental site the temperature of rock increases with depth under the surface. Curve B represents the temperatures at which that rock would melt under the conditions of pressure at increasing depth. (*From Univ. Mich. Res. News, vol. 27, no. 8, August 1976*)

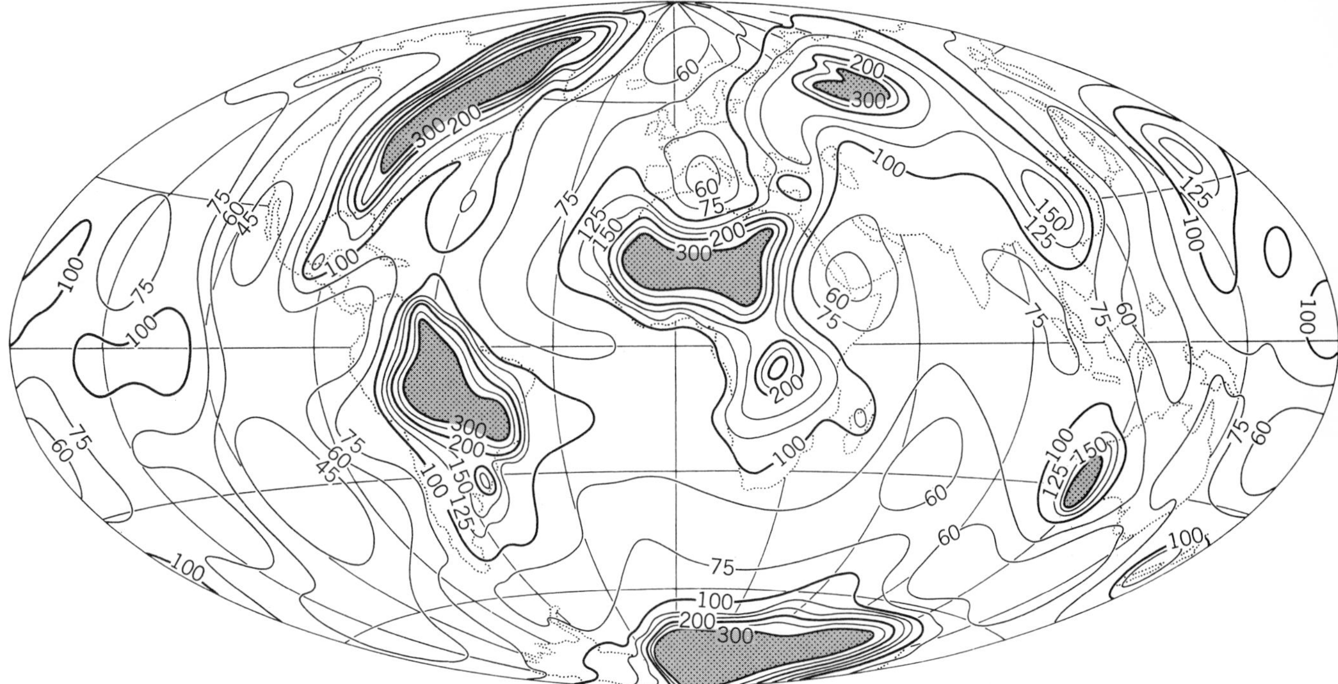

Fig. 2. Global map of the depth to the asthenosphere in kilometers. The asthenosphere occurs at shallow depths in areas of high heat flow from the Earth's interior, and at greater depths where heat flow is low, principally in the geologically older regions of the continents. *(From H. N. Pollack and D. S. Chapman, On the regional variation of heat flow, geotherms, and lithospheric thickness, Tectonophysics, 38:279–296, 1977)*

of temperature to rock viscosity is usually characterized by the quotient T/T_m, the ratio of ambient temperature to the temperature at which the rock begins to melt. Thus it is not the absolute temperature at a given depth that determines a rock's viscosity, but rather how close the temperature is to the melting temperature at that depth. The asthenosphere is thought to be that range of depth where the actual earth temperature is closest to the temperature of incipient melting of mantle rocks (Fig. 1). In some places the temperature may actually reach the rock-melting temperature and produce magma, thus giving the asthenosphere another important role in geodynamics—as the source region of many types of igneous rocks.

Heat flow. The temperature at a given depth within the Earth is dependent upon the amount of heat flowing out of the Earth at that location, a quantity which varies from place to place. Thus the depth at which the asthenosphere develops is regionally variable, reflecting the geographic pattern of terrestrial heat flow. The depth to the top of the asthenosphere (that is, the thickness of the lithosphere) is illustrated in the global map of Fig. 2. *See* EARTH, HEAT FLOW IN.

Earthquakes and seismic waves. The closeness to, and sometimes achievement of, melting conditions in the asthenosphere has, in addition to reducing the viscosity, other significant effects, such as slowing down seismic waves and reducing their amplitude as they traverse the region. Thus, by careful observations of the regional and depth variations of seismic-wave velocities, the position of the asthenosphere can be determined seismologically. Another consequence of the ability of the asthenosphere to relieve stress through creep and flow is that it appears to be largely earthquake-free. Those intermediate- and deep-focus earthquakes that do occur in the depth range of the asthenosphere are nearly all found in regions where rigid lithosphere has been subducted (recycled) into the Earth's interior and has penetrated the asthenosphere at subduction zones. Deep earthquake activity appears to occur within the subducted lithosphere and not in the surrounding asthenosphere. *See* EARTHQUAKE; SEISMOLOGY.

[HENRY N. POLLACK]

Bibliography: R. A. Daly, *Strength and Structure of the Earth*, 1940; H. N. Pollack and D. S. Chapman, *Tectonophysics*, 38:279–296, 1977; R. I. Walcott, *Annual Review of Earth and Planetary Science*, vol. 1, 1973; J. Weertman and J. R. Weertman, *Annual Review of Earth and Planetary Science*, vol. 3, 1975.

Atmosphere, evolution of

Variation with time of the chemical composition and total weight of the Earth's atmosphere. The atmosphere is a most tenuous envelope; its mass is less than one-millionth that of the solid Earth; its density even at sea level is less than one-thousandth that of rocks, and virtually all of the atmosphere is below a height only one-hundredth of an earth radius above the surface of the Earth. But the atmosphere is taken so much for granted that one tends to be surprised at the thought that it has a history, that its chemical composition and total weight have varied through time. Indeed, it would be odd to find that the atmosphere had not changed during the long years of the Earth's existence, and that its weight and composition had not responded in some way to the complicated series of events that have left their mark on the Earth's crust.

The study of these changes is difficult, because

Table 1. Nonvariable constituents of atmospheric air*

Constituent	Content
N_2	78.084%
O_2	20.946%
CO_2	0.033%
A	0.934%
Ne	18.18×10^{-6}
He^4	5.24×10^{-6}
He^3	6.55×10^{-12}
Kr	1.14×10^{-6}
Xe	0.087×10^{-6}
H_2	0.5×10^{-6}
N_2O	0.5×10^{-6}

*The classification of H_2 and N_2O as nonvariable constituents is uncertain.

there seems to be no way of obtaining reliable samples of the atmosphere in the past. But the quest is not altogether hopeless. The origin of life, the continued existence of animals since at least 600,000,000 years ago, the marks which interaction of the atmosphere with surface rocks have left on ancient sediments, and the nature of the input of volcanic gases through geologic time all give clues about the chemical evolution of the atmosphere. At present one cannot pretend to solve all the puzzles of the history of the Earth's atmosphere with these clues, but the broad pattern is emerging and a greater understanding is well within reach.

It may be wise to define the problem rather precisely before looking at the clues. Today the atmosphere contains a small number of major components and a very large number of minor components. Each component exerts a pressure, which is essentially constant for some components at sea level and which is variable in time and space for other components. The most important constituents of the atmosphere are shown in Tables 1 and 2. One could consider the problem of the chemical evolution of the Earth's atmosphere solved when an accurate plot can be made of the variation with time of the pressure of these and other components that may have existed in the atmosphere in the past. This is obviously a large undertaking, but it is made somewhat less so by the mutual exclusion of certain gases as major components of the atmosphere. At present, for instance, gases like ammonia, NH_3, methane, CH_4, and hydrogen can

Table 2. Variable constituents of dry air

Constituent	Content
O_3	0 to 0.07 ppm (summer)
	0 to 0.02 ppm (winter)
SO_2	0 to 1 ppm
NO_2	0 to 0.02 ppm
CH_4	0 to 2 ppm
CH_2O	Uncertain
I_2	0 to 10^{-10} g cm^{-3}
NaCl	Order of 10^{-10} g cm^{-3}
NH_3	0 to trace
CO	0 to trace
	(0.8 cm atm)

*Also from I_2 evaporation from the oceans following photooxidation of I^- in the ocean surface.

exist in the atmosphere as trace components only, because they are unstable in the presence of the large quantities of oxygen. Conversely, oxygen could not have been a major component of an atmosphere in which ammonia, methane, and hydrogen were abundant. *See* ATMOSPHERIC CHEMISTRY.

Origin of free oxygen. Free oxygen is somewhat of an anomaly on the Earth. Rocks more than a few feet below the Earth's surface are out of equilibrium with free oxygen and are oxidized in contact with the atmosphere. This is seen most easily in the development of red hydrous ferric oxide minerals in soil zones above many rock types in temperate and tropical areas and in the development of the extensive red to reddish-brown sediments of the western United States.

The origin of oxygen in the Earth's atmosphere has been a source of continuing controversy for many decades. Of the two theories that have been dominant, the first proposes that atmospheric oxygen has been produced through geologic time by the continuing effect of photosynthesis, during which carbon is effectively separated from oxygen in carbon dioxide. In a very rough manner this reaction can be written as shown in Eq. (1).

$$6CO_2 + 6H_2O \xrightarrow{\text{Photosynthesis}} \underset{\text{Carbohydrate}}{C_6H_{12}O_6} + 6O_2 \quad (1)$$

This reaction runs in the opposite direction during the decay of plant material and the breathing of animals. It can be shown that nearly all of the oxygen produced by photosynthesis during a given period of time is lost by plant decay, but that the small amount not lost could account for the present rather large quantity of atmospheric oxygen.

The second theory proposes an alternate way of producing free oxygen. Ultraviolet light from the Sun decomposes water molecules in the upper atmosphere. Most of these recombine, but there is a finite possibility that a given hydrogen atom will manage to escape from the Earth's atmosphere before recombination has taken place. Oxygen atoms, being 16 times as heavy as hydrogen atoms, escape very much more slowly or not at all. The decomposition of water vapor followed by hydrogen escape is therefore a distinctly plausible manner of generating free atmospheric oxygen. The escape rate of hydrogen from the atmosphere is almost certainly rapid. The critical factor in determining escape rates is the temperature in the upper atmosphere, and this is now well known from rocket measurements. But a strongly limiting factor for oxygen production by this mechanism is the formation of ozone, O_3, as a by-product of the photodissociation of water. Ozone absorbs ultraviolet light very readily and tends to form a screen preventing it from reaching the lower levels of the atmosphere, where water vapor is abundant. M. McElroy pointed out that most of the hydrogen which does manage to reach the upper atmosphere today is a constituent not of water but of biologically produced methane, and that the rate of hydrogen escape from the Earth's atmosphere is controlled largely by the rate of upward transport of methane rather than of water. Nevertheless, the hydrogen loss rate is currently a small fraction of the oxygen production rate by photosynthesis. The oxygen content of the atmosphere today is therefore only slightly influenced by hydrogen loss from the upper atmosphere; rather, it depends almost exclusively on the operation of a feedback system which links the rate of oxygen production during photosynthesis to the rate of oxygen use by weathering and the decay of organic matter.

Biologic evidence. The proposition that the present high concentration of oxygen in the atmosphere is largely due to oxygen production during photosynthesis implies that oxygen was much less abundant prior to the existence of photosynthesis. This view is in harmony with the requirement that free oxygen was absent from the atmosphere during the development of life. *See* LIFE, ORIGIN OF.

It is very likely that life evolved early in Earth history. The oldest known unmetamorphosed sedimentary rocks, those found in the Barberton Mountain area of South Africa, are about 3.3×10^9 years old and contain microscopic bits of carbon which, according to M. Muir, could well be of biologic origin. Limestones of similar age in South Africa contain stromatolitic structures, which are almost certainly the work of calcareous algae. There is little doubt, then, that life began more than 3×10^9 years ago. It is not yet clear when organisms developed which were able to produce free oxygen, but evidence from the nature of sedimentary rocks suggests that this event occurred well before 2×10^9 years ago. Biologic evidence for the rise of oxygen to levels approaching those of the present day is still scant until the close of the Precambrian era, some 600×10^6 years ago, despite the discovery of superb microfossils by J. W. Schopf and P. E. Cloud in earlier rocks.

D. C. Rhoads pointed out that the sequence of animals which developed during the latest part of the Precambrian era and the beginning of the Phanerozoic era parallels rather strikingly the zoning of animals in contemporary settings of progressively greater oxygen content. This suggests, but hardly proves, that evolutionary events were related to changes in the levels of atmospheric oxygen 600×10^6 years ago. A. G. Fischer suggested that, prior to the general rise in the level of atmospheric oxygen, animals were restricted to "oxygen oases" in the vicinity of photosynthetic organisms (Fig. 1), and that they were able to populate the oceans as a whole only after the level of oxygen in the atmosphere had risen above some threshold value.

L. V. Berkner and L. C. Marshall proposed that the invasion of the land by plants and animals about 400×10^6 years ago occurred when the oxygen pressure had risen to about one-tenth of its present value. At this oxygen pressure the intensity of ultraviolet radiation at the Earth's surface would have been so low that it no longer presented a health hazard. Although this is no more than an interesting speculation, the persistence of animals requiring oxygen in more or less large quantities indicates that oxygen levels in the atmosphere have probably never been lower than one-tenth of the present value during the past 400×10^6 years. There may well have been times when the oxygen pressure was greater than at present. *See* BIOSPHERE, GEOCHEMISTRY OF.

Evidence from sediments. Today oxidation of rocks at the Earth-atmosphere interface is widespread. If oxygen had been essentially absent from the atmosphere in times past, one might expect to see relatively less oxidation in the minerals of ancient sediments, and to see the formation of new minerals in these sediments which are less oxidized than their modern counterparts. Iron, manganese, uranium, and sulfur are among the elements which today respond most readily to oxidation at the Earth-atmosphere interface. Uraninite, UO_2, for instance, reacts rapidly with atmospheric oxygen to form a variety of higher oxides and hydrous oxides; concerted search for uraninite in black sands during World War II was quite unsuccessful. And yet, uraninite, which has apparently survived weathering and transport, occurs in the ores of the Dominion Reef series and the Witwatersrand series of South Africa (Figs. 2 and 3), as well as at Blind River, Canada, and at Serra de Jacobina in Brazil.

In these areas the sediments are more than 1.8×10^9 years old. The origin of the uraninite as a residue of weathering has been proposed by geologists who have closely studied these ore deposits,

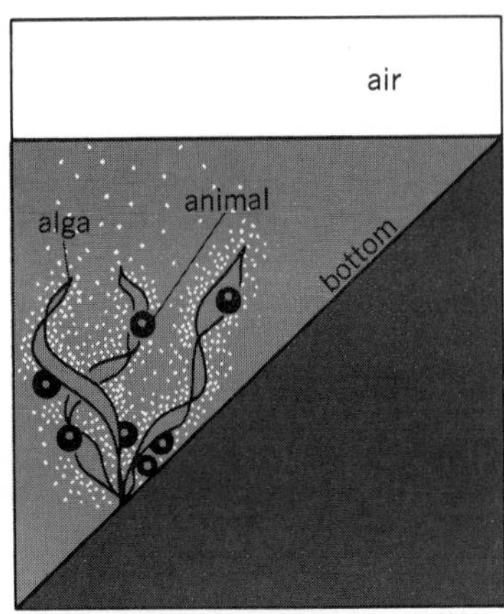

Precambrian

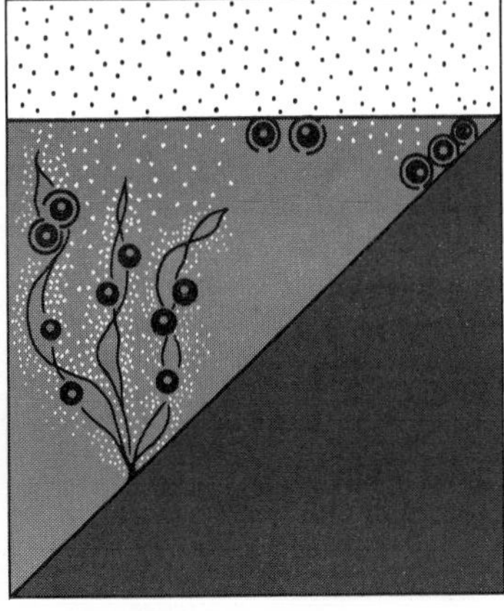
Early Paleozoic

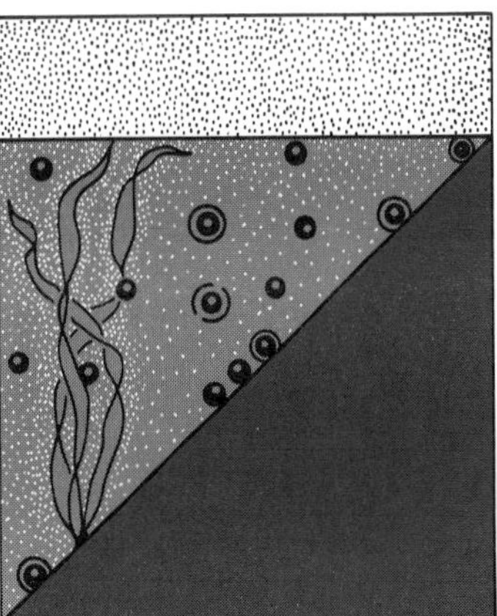
Present

Fig. 1. Relationship between atmospheric oxygen and animal evolution. Precambrian (stage 1): Atmosphere is essentially devoid of free oxygen; animals are living in "oxygen oases" in complete respiratory dependence on host plants. Early Paleozic (stage 2): Oxygen pressure has increased to level at which animals may leave plants but flock toward air-water interface. Present (stage 3): Atmosphere and water are highly oxygenated; animals are widely distributed. (*A. G. Fischer, Proc. Nat. Acad. Sci., 53:1205–1213, 1965*)

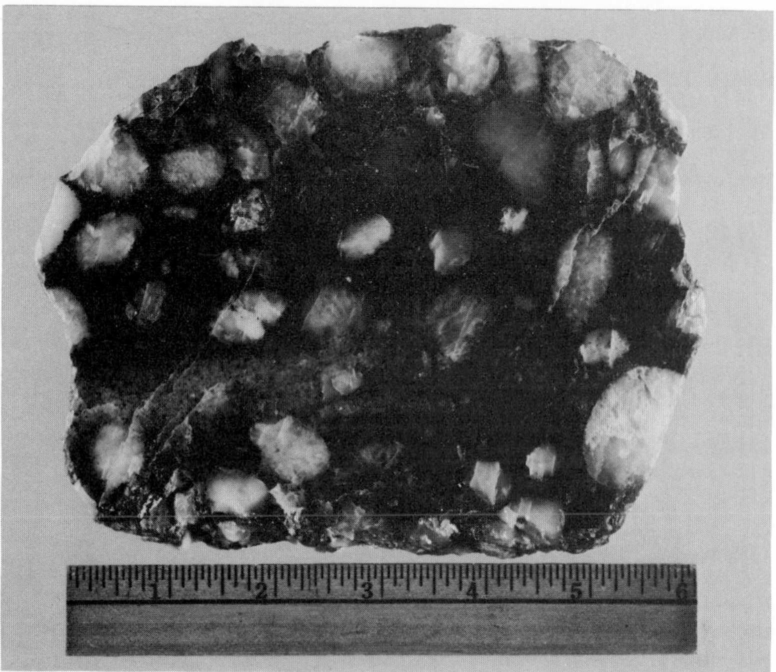

Fig. 2. Polished slab of gold ore, Modder Deep, South Africa. The white pebbles are quartz, SiO_2, the major metallic constituent is pyrite, FeS_2. Most of the gold ore also contains uraninite, UO_2, which apparently is detrital.

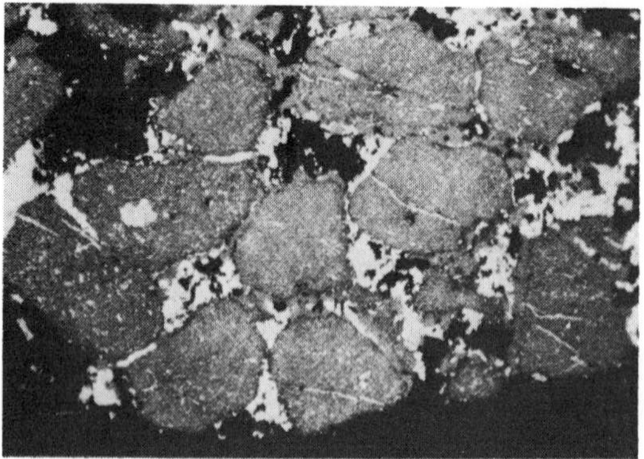

Fig. 3. Photomicrograph (× 280) of a polished section of uranium ore, West Rand Cons Mine, South Africa. Gray grains are uraninite, UO_2, and probably detrital. White material in cracks and between grains of uraninite is brannerite, an oxide of uranium, titanium, and calcium. (*Ramdohr, Abhandl. Deut. Akad. Wiss. Berlin, 1958*)

and has been corroborated by work on the age of the uraninite grains. These appear to be considerably older than the time of accumulation of the sediments. On the other hand, C. F. Davidson maintained that the uraninite is not a weathering residue but was introduced into the sediments after they were deposited. If these grains are indeed detrital, then the rate of oxidation of uraninite must have been much slower prior to 1.8×10^9 years ago than it is today. A careful study by D. Grandstaff of the rate of oxidation of uraninite has shown that the oxygen pressure was probably less than 4×10^{-3} atm (1 atm = 101.325 Pa) to permit the survival of uraninite during weathering and transport.

Macgregor's observation of the abnormal abundance of detrital pyrite in ancient sediments from South Africa is certainly also in accord with this conclusion. But at the same time large amounts of calcite, dolomite, biogenically precipitated pyrite, and iron ore, much of it apparently in the form of hematite or a hydrated ferric oxide, were being deposited in North America. The precipitation of calcite and dolomite demands that at least some of the atmospheric carbon was present as CO_2. Although hematite and its hydrated analogs are stable at extremely low oxygen pressures, the total amount of oxygen which must have been used up in oxidizing the predominantly ferrous iron of most igneous rocks to ferric iron was large. It therefore seems likely that the atmosphere between 2 and 3×10^9 years ago was sufficiently less oxidizing than at present to prevent the oxidation or uraninite, at least under some circumstances, yet was sufficiently oxidizing to permit the formation of widespread accumulations of ferric oxide and hydrous ferric oxide. *See* LITHOSPHERE, GEOCHEMISTRY OF.

Evidence from gaseous emissions. Fortunately, a third line of evidence is available to tell something about the oxidation state of the atmosphere in a nonbiotic state. W. W. Rubey showed that many of the volatile constituents of the atmosphere and oceans cannot have been derived from the weathering of igneous rocks. He has developed a very plausible argument for the concept that these "excess volatiles" have boiled out of the interior of the Earth during the course of its long history. Volcanoes are eloquent evidence for such boiling out today, and it seems likely that a portion of the discharge of at least some hot springs has a deep source. In a nonbiotic state the chemistry of the atmosphere would be controlled largely by the chemistry of such emanations and by their interaction with surface rocks (Fig. 4). Thus, if one knew something about the oxidation of volcanic gases in the past, one might be able to predict, at least within broad limits, the oxidation state of an atmosphere unaffected by biologic processes.

At present, volcanic gases consist mainly of water, carbon dioxide, sulfur dioxide, hydrogen, carbon monoxide, and nitrogen. Table 3 contains an average of representative analyses of Hawaiian

Fig. 4. Eruption of Mount Vesuvius in 1943. The gases emitted by volcanoes throughout geologic time have played a major role in the chemical evolution of the atmosphere of the Earth.

Table 3. Composition of typical Hawaiian volcanic gases*

Gas	Vol. %
H_2O	79.3
CO_2	11.6
SO_2	6.5
N_2	1.3
H_2	0.6
CO	0.4
Cl_2	0.05
Ar	0.04

*After Eaton and Murata, 1960.

volcanic gases. Free oxygen is almost completely absent: The oxygen pressure in these gases as they emerge is about 10^{-7} atm. If a gas is defined as being neutral from an oxidation-reduction point of view when it contains neither free hydrogen nor free oxygen, then these gases are slightly on the reduced side since they do contain a small amount of free hydrogen and CO. Today, these reduced gases react rapidly with atmospheric oxygen. In the absence of atmospheric oxygen a variety of reactions might take place. All of these would tend to produce an essentially neutral atmosphere which would be similar to the present atmosphere with the difference that oxygen would be virtually absent.

The oxidation state of volcanic gases today is controlled in large part by the oxidation state of the associated lavas. This in turn is reflected in the ratio of ferrous iron, Fe^{+2}, to ferric iron, Fe^{+3}, in lavas, a ratio which tends to be preserved on cooling. Studies by S. Steinthorssen of this ratio in basalts have shown that it has probably not changed significantly with time during at least the past 2×10^9 years. One can be fairly certain, then, that the oxidation state of volcanic gases has not varied greatly during the second half of Earth history. There are good theoretical reasons for believing that this same state of affairs prevailed at least back to 3×10^9 years ago, but it is quite possible that shortly after its birth the Earth vented volcanic gases which were more highly reducing. The cause for this state of affairs would have been the presence of metallic iron in the upper part of the Earth's mantle. Today this iron is largely concentrated in the Earth's core, far below the part of the mantle where lavas are generated.

Rise of oxygen pressure. It may be well to summarize the evidence regarding the history of the pressure of oxygen in the Earth's atmosphere before considering the pressure of the other constituents through geologic time. The largest area of ignorance surrounds the events prior to the accumulation of the earliest known sediments, some 3.3×10^9 years ago. It is possible that the atmosphere was quite reducing as a consequence of the injection of highly reduced volcanic gases into the atmosphere during the first few hundred years of Earth history.

Shortly after 3×10^9 years ago, the atmosphere contained less than 4×10^{-3} atm, but more than 0 atm, oxygen. That is indicated on the one hand by the presence of detrital uraninite and on the other hand by the presence of hematite, calcite, and dolomite in these early sediments. Photosynthesis was already under way, but presumably at a rate which was insufficient to support more than a small oxygen pressure in the atmosphere. This state of affairs probably prevailed until about 1.8×10^9 years ago. The first widespread red bed sequences were deposited about that time, and no uranium deposits of the Blind River type younger than 1.8×10^9 years have been discovered. Both observations are consistent with an increase in atmospheric oxygen at that time.

Another rise may well have taken place at the end of the Precambrian and may have been important in the development of animal life. Finally, the colonization of the land became possible about 400×10^6 years ago after a further rise in the oxygen pressure and a further drop in the intensity of ultraviolet radiation at sea level. Since that time the oxygen pressure has, presumably, continued to rise; whether a plateau has been reached or whether there have been fluctuations in the oxygen pressure during the last 400×10^6 years is uncertain. It is most unlikely, however, that large fluctuations have occurred recently. Figure 5 is an effort at illustrating the curve of oxygen pressure versus time; the assigned uncertainties are reasonably generous, but future work may well prove that they are not generous enough.

Rare gases other than helium. Of the atmospheric constituents other than oxygen, the rare gases, nitrogen, and carbon dioxide are probably the most interesting. H. Brown and H. E. Suess pointed out that the Earth is quite depleted in rare gases in comparison with the Sun. Even xenon, an element which certainly cannot escape from the Earth's atmosphere today at anything but a geologically insignificant rate, appears to be only about one-millionth as abundant in comparison with silicon as in the Sun. Brown concluded that the Earth was essentially devoid of an atmosphere after it had reached its present size and gravitational field. It is possible but unlikely that an original gaseous envelope containing the missing rare gases was swept away by strong magnetic fields which penetrated throughout the solar system, or during the violent Hayashi phase early in the history of the Sun. Whichever view turns out to be correct, the Earth seems to have been left nearly devoid of an atmospheric blanket at some

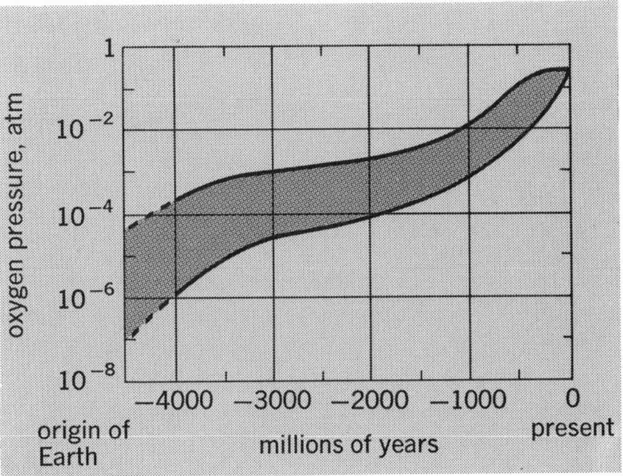

Fig. 5. The pressure of oxygen in the atmosphere during geologic time. Available evidence suggests that there has been a nearly continuous increase in the oxygen pressure since the Earth was formed.

time not long after the end of the accumulation phase.

A corollary of this conclusion is that the present atmosphere has been acquired since that time. Rubey's argument for the degassing of the Earth's interior has been given above, and this is strongly confirmed by the rather anomalous abundance and isotopic composition of atmospheric argon. This gas is much more abundant than the other rare gases, and consists almost entirely of the isotope of mass 40, which should, in the normal course of nuclear events, not be particularly more abundant than the argon isotopes of mass 36 and mass 38. The anomaly is readily explained by the degassing of argon-40 from the Earth after its production from the radioactive decay of potassium-40. Although the potassium content of the Earth is not very well known, there is little doubt that the available potassium could account for the abundance of atmospheric argon-40. It seems likely that degassing was most intense during the early history of the Earth. N. C. Craig's discovery of excess He^3 in the Pacific Ocean indicates that degassing of primary material is still continuing. The pressure of neon, argon, krypton, and xenon have probably been building up gradually in the atmosphere.

Other gases. The abundance of helium in the atmosphere is anomalously low. The quantity of helium which has been generated in the solid Earth by the decay of uranium and thorium series nuclides and has entered the atmosphere with other gases is much larger than the quantity of helium now present in the atmosphere. The inference that helium has escaped into interplanetary space is amply corroborated by analyses of the mechanics of the escape of helium from the Earth's atmosphere.

Nitrogen is a good deal more reactive than the rare gases. It is cycled through living organisms at a geologically rapid rate but seems to be accumulating mainly in the atmosphere. From a thermodynamic point of view nitrate should today be quite abundant in the oceans. Its concentration is, however, kept to very low levels by living organisms for whom nitrate is an important and frequently limiting nutrient.

Ammonia is thermodynamically unstable in the presence of free oxygen, and it is easy to show that a major fraction of the nitrogen now in the atmosphere would be converted to ammonia only in the presence of a hydrogen pressure in excess of 10^{-3} atm. Such a hydrogen pressure could have existed only very early in the Earth's history.

However, the value of the hydrogen pressure, and even the very existence of a highly reducing early atmosphere, is presently very uncertain. R. Brett pointed out that the relatively large nickel content of olivine in the upper mantle is inconsistent with the presence of large amounts of metallic iron in the upper mantle early in Earth history. If this is correct, volcanic gases may never have been much more reducing than at present. Interestingly, McElroy suggested that a sufficient concentration of H_2 could have been built up in the early atmosphere from the injection of volcanic gases no more reducing than those of the present day to permit the production of the organic compounds which were necessary for the development of life.

Methane could have been a major constituent of an early reducing atmosphere. It is likely that its pressure was already less than 10^{-3} atm during the deposition of the sediments of the Issua area in West Greenland more than 3.7×10^9 years ago. Methane must have been a very minor component ever since atmospheric oxygen became more than a trace constituent of the atmosphere. The CH_4 concentration since then has almost certainly been controlled by the balance between the rate of biologic production and the rate of its photochemical decomposition.

It now seems likely that the CO_2 pressure has been hovering near the value at which hydrous magnesium aluminum silicate minerals are in equilibrium with kaolinite, quartz, calcite, and dolomite, as shown in Eq. (2). All of these minerals

$$\underset{\text{Chlorite}}{Mg_5Al_2Si_3O_{10}(OH)_8} + \underset{\text{Calcite}}{5CaCO_3} + 5CO_2 \rightleftharpoons \underset{\text{Dolomite}}{5CaMg(CO_3)_2} + \underset{\text{Kaolinite}}{Al_2Si_2O_5(OH)_4} + \underset{\text{Quartz}}{SiO_2} + 2H_2O \quad (2)$$

are abundant in sediments and sedimentary rocks, and it is likely that such equilibria have helped to maintain the CO_2 pressure near its present value for much of geologic time. But it also seems likely that the CO_2 pressure in the atmosphere was somewhat greater than today, perhaps by as much as a factor of 5, before the invasion of the land by plants near the end of the Silurian, some 400×10^6 years ago.

The CO_2 content of the atmosphere has increased by about 10% since the turn of the century because of fossil fuel burning. If all of the current reserves of coal, oil, and gas were suddenly burned, the CO_2 content of the atmosphere would increase by about a factor of 10. Most projections suggest that the peak in CO_2 pressure will occur about the year 2100 at a level about five times the present CO_2 pressure. After that date the CO_2 pressure will decrease in response to the removal of CO_2 from the atmosphere, largely into the

Fig. 6. Sample of Tertiary limestone from Florida. Nearly all carbon dioxide injected into the atmosphere has been removed in limestones and dolomites. During the past 600×10^9 years much removal has taken place through the action of organisms with carbonate shells. Prior to that time, removal by inorganic precipitation and by stromatolites was probably dominant.

Table 4. Summary of data on the probable chemical composition of the atmosphere during stages 1, 2, and 3

Components	Stage 1	Stage 2	Stage 3
Major components: $P > 10^{-2}$ atm	CH_4 H_2 (?)	N_2	N_2 O_2
Minor components: $10^{-2} > P > 10^{-4}$ atm	H_2 (?) H_2O N_2 H_2S NH_3 Ar	 H_2O CO_2 A O_2 (?)	Ar H_2O CO_2
Trace components: $10^{-4} > P > 10^{-6}$ atm	He	Ne He CH_4 NH_3 (?) SO_2 (?) H_2S (?)	Ne He CH_4 Kr

oceans, and CO_2 pressures close to the present values should again prevail by the year 3000.

Today, carbon dioxide is second in abundance only to water in gases issuing from volcanoes. Yet the ratio of CO_2 to water in the atmosphere and oceans is minuscule. CO_2 has clearly been scavenged very thoroughly from the atmosphere-hydrosphere system. Its resting place is easily discovered in the elemental carbon in sedimentary rocks and in the carbonate rocks which have been deposited throughout the entire range of Earth history that is accessible through the study of sedimentary rocks (Fig. 6). The two dominant carbonate minerals, calcite, $CaCO_3$, and dolomite, $CaMg(CO_3)_2$, are the major components of limestones and dolomites. The removal of CO_2 from the atmosphere, its reaction with surface rocks, and its ultimate burial are processes which are chemically, mineralogically, and biologically complex.

Three stages of evolution. The discussion seems to lead quite naturally to the threefold division in Table 4 of the history of the Earth's atmosphere. During the first stage, very shortly after the accretion of the Earth, the atmosphere may have been quite reducing. Reduced gases issuing from volcanoes probably would have given rise to an atmosphere consisting of methane with minor quantities of hydrogen, nitrogen, and ammonia. After metallic iron was removed from the upper mantle, the oxidation state of volcanic gases probably approached its present value, methane was replaced by carbon dioxide, and ammonia was converted to nitrogen. In the atmosphere during this (its second) stage, nitrogen was its dominant component, and carbon dioxide and argon were its most important minor constituents.

The third stage opened when the rate of oxygen production by photosynthesis became sufficiently great so that oxygen became more than a trace component of the atmosphere. The transition from stage 2 to stage 3 may well have occurred about 1.8×10^9 years ago. Since the opening of stage 3, the oxygen pressure has climbed to its present value by a path which was probably simple but which could have been complex in detail. During this period the nitrogen, neon, argon, krypton, and xenon pressures have also gradually climbed to their present value, while the helium and CO_2 pressures have remained reasonably constant, the former suspended between the rate of input and the rate of escape, the latter controlled in large part by reactions which involve carbonate and silicate minerals.

This sequence seems a reasonable synthesis of the rather diverse sets of clues which have turned up in atmospheric research. But in conclusion, even if the analysis of the plot turns out to be correct in its main features, it would be too much to ask for an absence of new surprises as the search continues for a more complete understanding of the history of the atmosphere.

[HEINRICH D. HOLLAND]

Bibliography: P. J. Brancazio and A. G. W. Cameron (eds.), *The Origin and Evolution of Atmospheres and Oceans*, 1964; R. M. Garrels and F. T. Mackenzie, *Evolution of Sedimentary Rocks*, 1971; P. M. Hurley (ed.), *Advances of Earth Science*, 1966; K. B. Krauskopf, *Introduction to Geochemistry*, 1967; K. A. Kvenvolden (ed.), *Geochemistry and the Origin of Life*, 1974; B. Mason, *Principles of Geochemistry*, 3d ed., 1966; S. I. Rasool (ed.), *Chemistry of the Lower Atmosphere*, 1973.

Atmospheric chemistry

A subdivision of atmospheric science concerned with the chemistry and physics of atmospheric constituents, including studies of their sources, circulation, and sinks and their perturbations caused by anthropogenic activity.

Known gaseous constituents have mixing ratios with air by volume (or equivalently by number), f, ranging from 0.78 for N_2 to 6×10^{-20} for Rn. Known particulate constituents (solid or liquid) have mixing ratios with air by mass, χ, ranging from about 10^{-3} for liquid water in raining clouds to about 10^{-16} for large hydrated ions in otherwise

Composition of tropospheric air

Gas	Volume mixing ratio
Nitrogen, N_2	0.781 (in dry air)
Oxygen, O_2	0.209 (in dry air)
Argon, ^{40}Ar	9.34×10^{-3} (in dry air)
Water vapor, H_2O	Up to 4×10^{-2}
Carbon dioxide, CO_2	2 to 4×10^{-4}
Neon, Ne	1.82×10^{-5}
Helium, ^4He	5.24×10^{-6}
Methane, CH_4	1 to 2×10^{-6}
Krypton, Kr	1.14×10^{-6}
Hydrogen, H_2	4 to 10×10^{-7}
Nitrous oxide, N_2O	2 to 6×10^{-7}
Carbon monoxide, CO	1 to 20×10^{-8}
Xenon, Xe	8.7×10^{-8}
Ozone, O_3	Up to 5×10^{-8}
Nitrogen dioxide, NO_2	Up to 3×10^{-9}
Nitric oxide, NO	Up to 3×10^{-9}
Sulfur dioxide, SO_2	Up to 2×10^{-8}
Hydrogen sulfide, H_2S	2 to 20×10^{-9}
Ammonia, NH_3	Up to 2×10^{-8}
Formaldehyde, CH_2O	Up to 1×10^{-8}
Nitric acid, HNO_3	Up to 1×10^{-9}
Methyl chloride, CH_3Cl	Up to 3×10^{-9}
Hydrochloric acid, HCl	Up to 1.5×10^{-9}
Freon-11, $CFCl_3$	About 8×10^{-11}
Freon-12, CF_2Cl_2	About 10^{-10}
Carbon tetrachloride, CCl_4	About 10^{-10}

clear air. These constituents are involved in cyclic processes of varying complexity which, in addition to the atmosphere, may involve the hydrosphere, biosphere, lithosphere, and even the deep interior of the Earth. The subject of atmospheric chemistry has assumed considerable importance in recent years because a number of the natural chemical cycles in the atmosphere may be particularly sensitive to perturbation by the industrial and related activities of humans.

Atmospheric composition. A summary of the important gaseous constituents of tropospheric air is given in the table. The predominance of N_2 and O_2 and the presence of the inert gases ^{40}Ar, Ne, ^4He, Kr, and Xe are considered to be the result of a very-long-term evolutionary sequence in the atmosphere. These seven gases have extremely long atmospheric lifetimes, the shortest being 10^6 years for ^4He, which escapes from the top of the atmosphere. In contrast, all the other gases listed in the table participate in relatively rapid chemical cycles and have atmospheric residence times of a few decades or less. *See* ATMOSPHERE, EVOLUTION OF.

Particles in the atmosphere range in size from about 10^{-3} to more than 10^2 μm in radius. The term "aerosol" is usually reserved for particulate material other than water or ice. A summary of tropospheric aerosol size ranges and compositions is given in Fig. 1. Concentrations of the very smallest aerosol particles in the atmosphere are limited by coagulation to form larger particles, and the concentrations of the larger aerosols are restricted by sedimentation, the rate of which increases as the square of the aerosol radius. In addition, the size distributions of water droplets and ice crystals are affected by evaporation, condensation, and coalescence processes. Some dry aerosols are water-soluble and can also grow by condensation. Aerosols are important in the atmosphere as nuclei for the condensation of water droplets and ice crystals, as absorbers and scatterers of radiation, and as participants in various chemical cycles.

Above the tropopause, the composition of the atmosphere begins to change, primarily because of the decomposition of molecules by ultraviolet radiation and the subsequent chemistry. For example, decomposition of O_2 produces an O_3 layer in

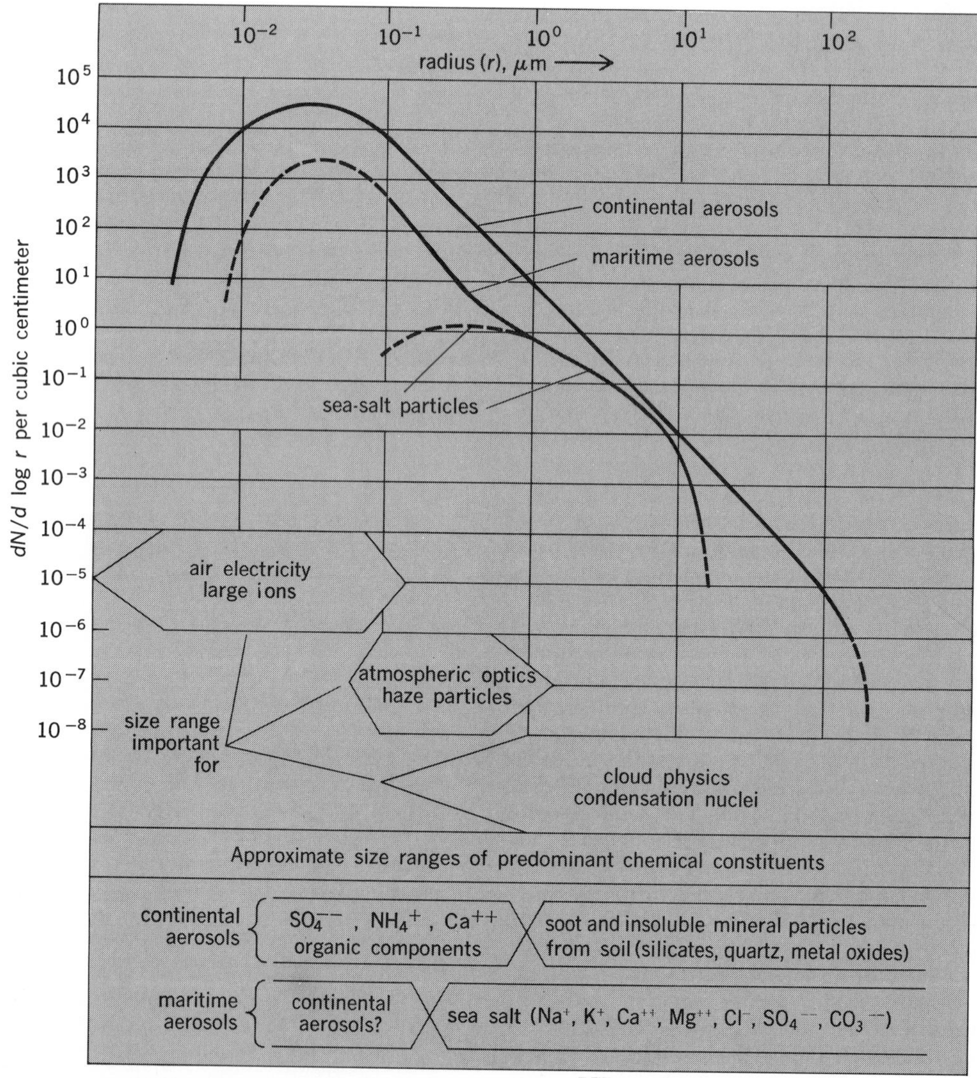

Fig. 1. Chart of the average size distributions and the predominant chemical constituents of some natural aerosols. The size ranges which are important for the various fields of meteorology are shown.

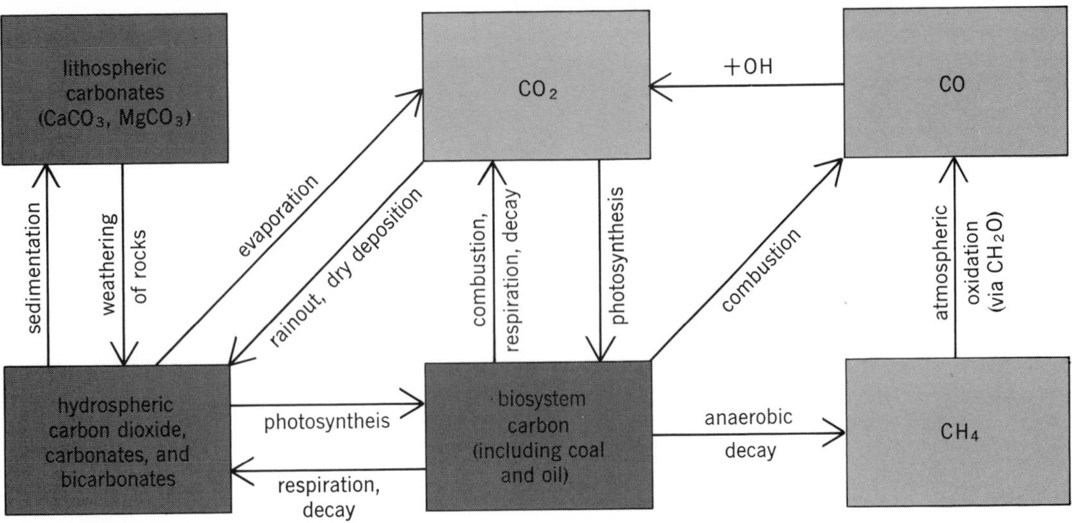

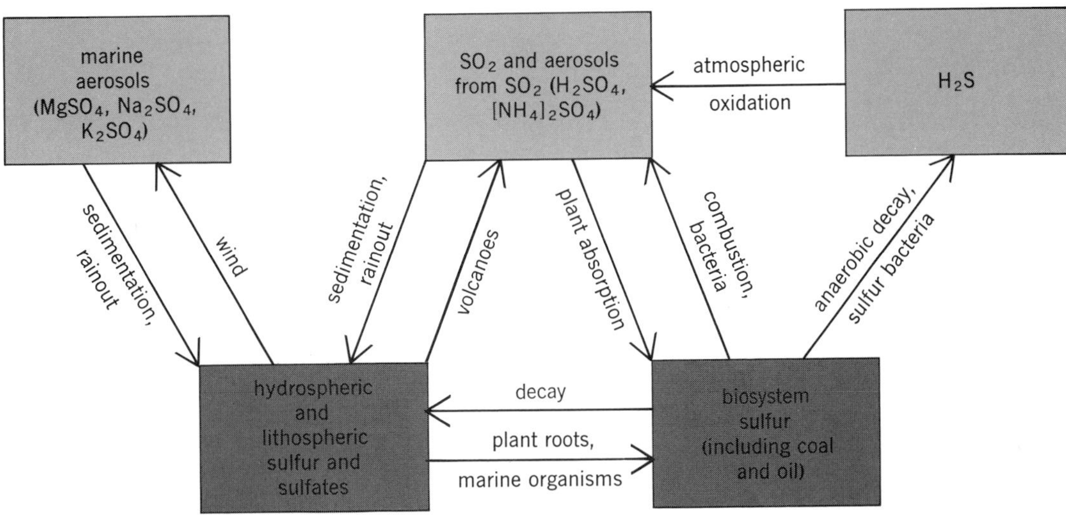

Fig. 2. The carbon and sulfur cycles.

the stratosphere. Although the peak value of f for O_3 in this layer is only about 10^{-5}, this amount of ozone is sufficient to shield the Earth's surface from biologically lethal ultraviolet radiation. About 100 km above the surface, ultraviolet dissociation of O_2 is so intense that the predominant atmospheric constituents become N_2 and O. There is also a layer of aerosols in the lower stratosphere composed primarily of sulfuric acid and dust particles. The sulfuric acid is probably produced by oxidation and hydration of sulfur gases. This same process, but strongly amplified, is the probable source of the much thicker clouds of sulfuric acid recently identified on the planet Venus.

A number of radioactive nuclides are formed naturally in the atmosphere by decay of Rn and by cosmic radiation. Radon, which is produced by decay of U and Th in the crust, enters the atmosphere, where it in turn decays to produce a number of radioactive heavy metals. These metal atoms become attached to aerosol particles and sediment out. Cosmic rays striking N_2, O_2, and Ar principally in the stratosphere give rise to a number of radioactive isotopes, including ^{14}C, ^{7}Be, ^{10}Be, and ^{3}H. The incorporation of ^{14}C into organic matter, where it decays with a half-life of about 5600 years, forms the basis of the radiocarbon dating method. In addition to naturally occurring radioactivity, large quantities of radioactive material have been injected into the atmosphere as a result of nuclear bomb tests. The most dangerous isotope is ^{90}Sr, which can be incorporated into human bones, where it radioactively decays with a half-life of about 28 years. Both natural and anthropogenic radioisotopes have been used as tracers for tropospheric and stratospheric motions. *See* RADIOCARBON DATING.

Atmospheric chemical models. In order both to adequately understand the present chemistry of the atmosphere and to predict the effects of anthropogenic perturbations on this chemistry, it has been necessary to construct quantitative chemical models of the atmosphere. In general, gas and particle mixing ratios show considerable variation with space and time. This variability can be quantitatively analyzed using the continuity equation

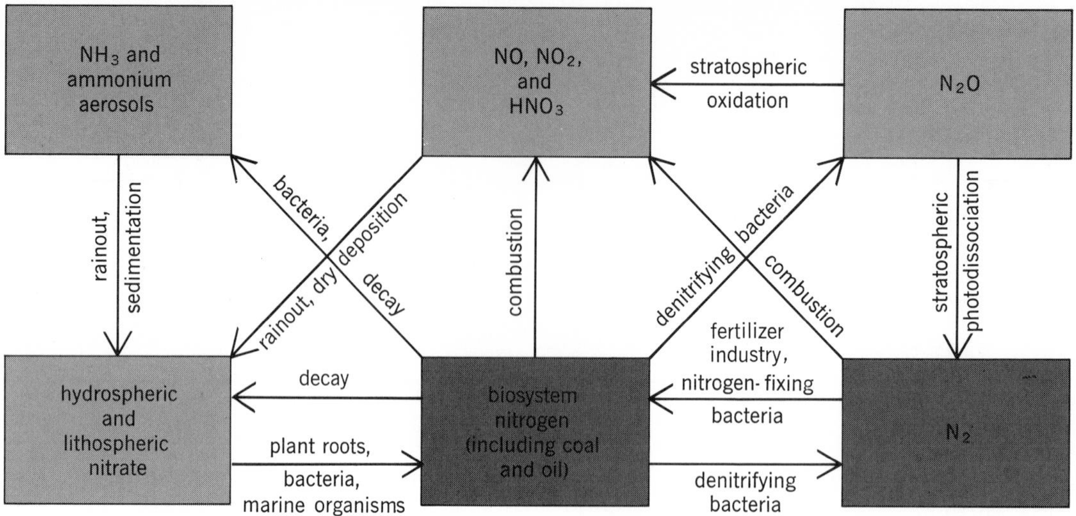

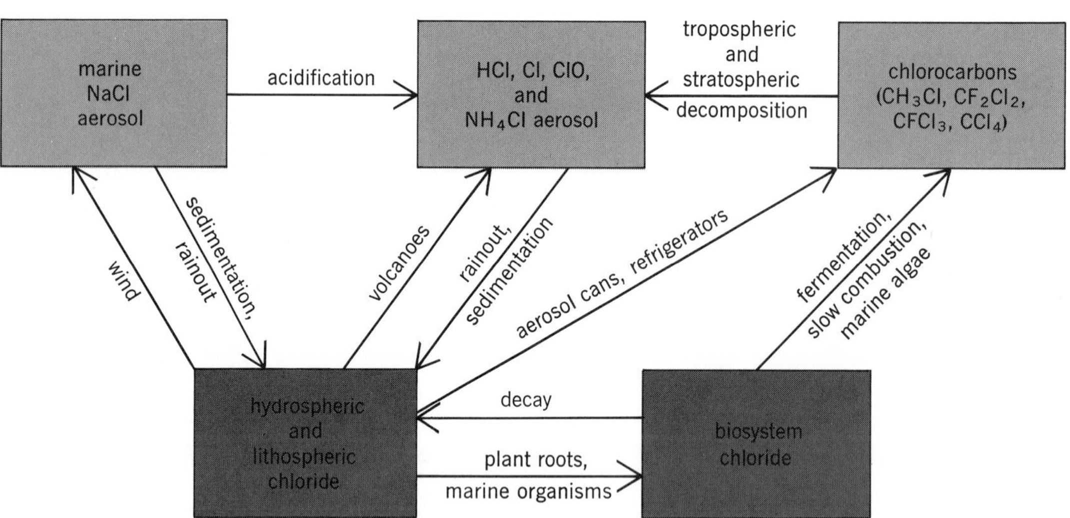

Fig. 3. The nitrogen and chlorine cycles.

for the particular species. In terms of χ, time t, wind velocity $\mathbf{v}$, average particle sedimentation velocity W_p, and air density ρ, this equation is conveniently written as shown below, where an over-

$$\frac{\partial \bar{\chi}}{\partial t} \simeq -\bar{\mathbf{v}} \cdot \nabla \bar{\chi} - W_p \frac{\partial \bar{\chi}}{\partial z} - \frac{1}{\bar{\rho}} \nabla \cdot (\overline{\rho \chi' \mathbf{v}'}) + \frac{d\bar{\chi}}{dt}$$

bar denotes an average over a time scale that is long compared to that associated with turbulence, and a prime denotes an instantaneous fluctuation from this average value. The first term on the right-hand side of the equation describes the changes in $\bar{\chi}$ due to the mean atmospheric circulation; the second term gives the $\bar{\chi}$ alternation due to sedimentation which is, of course, zero for gases; the third term denotes fluctuations in $\bar{\chi}$ due to turbulence or eddies, and in this term the eddy flux $\overline{\bar{\rho}\chi'\mathbf{v}'}$ is often roughly approximated by $-K\bar{\rho}\nabla\bar{\chi}$, where K is a three-dimensional matrix of eddy diffusion coefficients; and the last term describes changes in $\bar{\chi}$ caused by chemical production or destruction which may involve simple condensation or evaporation or more complex chemical reactions.

From the equation above, it is seen that the variability of gas or aerosol concentrations is generally due to a combination of transport and true production or destruction. If a constituent has a destruction time T_0, then $d\bar{\chi}/dt = -\bar{\chi}/T_0$, and the equation then implies that the larger the T_0 value the less variability one expects to see in the atmosphere. A constituent for which $\bar{\chi}$ is completely independent of space and time is said to be well mixed. For example, the very-long-lived gases N_2, O_2, ^{40}Ar, Ne, ^{4}He, Kr, and Xe are essentially well mixed in the lower atmosphere.

Atmospheric chemical models involve simultaneous solution of the above equation for each atmospheric constituent involved in a particular chemical cycle. Often, simplified versions of this equation can be utilized, for example, when chemical lifetimes are much shorter than typical atmospheric transport times, or vice versa. Models for the ozone layer have now progressed from historical one-dimensional models which neglected transport to sophisticated three-dimensional models which, in addition to solving the above equation, including transport, solve the equations of motion to obtain **v** as a function of position and time.

Chemical cycles. In studying the chemical cycles of atmospheric gases, it is important to consider both the overall budgets on a global scale and the kinetics of the elementary chemical reactions on a local scale. The study of chemical cycles is in its infancy. Some of the minor details in the cycles outlined below may be subject to change, but this present lack of definition does not detract from their importance.

Carbon cycle. The atmospheric cycle which is of primary significance to life on Earth is that of carbon, which is illustrated in Fig. 2. The CO_2 content of the oceans is about 60 times that of the atmosphere and is controlled by the temperature and acidity of sea water. Release of CO_2 into the atmosphere over tropical oceans and uptake by polar oceans result in a CO_2 residence time of about 5 years. On the other hand, the cycle of CO_2 through the biosphere has a turnover time of a few decades. The amount of CO_2 buried as carbonate in limestone, marble, chalk, dolomite, and related deposits is about 600 times that in the ocean-atmosphere system. If all this CO_2 were released to the atmosphere, a massive CO_2 atmosphere similar to that on the planet Venus would result. Because CO is a poisonous gas, its production from automobile engines in urban areas must be closely monitored. On a global scale, the principal sources of CO are combustion of oil and coal and oxidation of the CH_4 produced naturally during anaerobic decay.

Sulfur cycle. The important aspects of the sulfur cycle are also illustrated in Fig. 2. Sulfur dioxide is produced in the atmosphere by combustion of high-sulfur fuels, by plants and bacteria, and by oxidation of H_2S introduced by anaerobic decay. On a global scale, the SO_2 budget is perturbed significantly by the anthropogenic source, and this perturbation is even more pronounced in urban localities. Oxidation of SO_2 produces sulfuric acid, which is a particularly noxious pollutant; thus regulation of high-sulfur fuel combustion, at least on the local scale, is now required.

Nitrogen cycle. The nitrogen cycle is shown in Fig. 3. The nitrogen oxides (NO, NO_2) are important because they are presently the major compounds governing ozone concentrations in the stratosphere. The main natural source of stratospheric NO and NO_2 is decomposition of the N_2O produced from soil nitrate ions (NO_3^-) by denitrifying bacteria. There has been considerable concern about injection of nitrogen oxides directly into the stratosphere by supersonic aircraft. Projected fleet levels for the year 2000 suggest a future anthropogenic source for these oxides comparable to their natural source. This anthropogenic perturbation would cause a decrease of about 10% in total atmospheric ozone and significantly increase the ultraviolet radiation dosage at the Earth's surface.

Chlorine cycle. The main natural source for atmospheric chlorine is acidification of chloride aerosols producing HCl near the surface. The chlorine cycle (Fig. 3) has received considerable attention because any significant concentrations of Cl and ClO in the stratosphere will lead to depletion of ozone in a manner similar to that caused by NO and NO_2. Because the HCl produced at the ground is severely depleted by rain-out, it does not give rise to significant stratospheric chlorine concentrations. However, the chlorocarbons CCl_4, $CFCl_3$, CF_2Cl_2, and CH_3Cl are relatively insoluble and are not rained out. They appear to decompose in the stratosphere, releasing chlorine; and although the influence of this chlorine on the present stratospheric ozone budget is small, an unchecked build-up of these compounds could lead to significant ozone depletion. The compounds CF_2Cl_2 and $CFCl_3$ are manufactured for use as propellants in aerosol cans, for use in refrigerators and air conditioners, and for manufacture of plastic foams. Methyl chloride is produced naturally by microbial fermentation and by combustion of vegetation, and CCl_4 is probably derived from both natural and industrial sources. [RONALD G. PRINN]

Bibliography: S. S. Butcher and R. J. Charlson, *An Introduction to Air Chemistry*, 1972; C. E. Junge, *Air Chemistry and Radioactivity*, 1963; E. Robinson and R. E. Robbins, in W. Strauss (ed.), *Air Pollution Control*, pt. 2, 1972.

Atoll

An annular coral reef, with or without small islets, that surrounds a lagoon without projecting land area.

Physiography. Most atolls are isolated reefs rising from the deep sea, and vary considerably in size. Small rings, usually without islets, may be less than a mile in diameter, but many atolls have a diameter of about 20 mi and bear numerous islets. The largest atoll, Kwajalein in the Marshall Islands, has an irregular shape and covers 840 mi^2.

The reefs of the atoll ring are flat, pavementlike areas, large parts of which, particularly along the seaward margin, may be exposed at times of low tide (Fig. 1). The reefs vary in width from narrow ribbons to broad bulging areas more than a mile across. The most prominent feature may be a low, cuestalike, wave-resistant ridge whose steeper side faces the sea. Usually brownish or pinkish in color, this ridge is composed mainly of calcareous algae. To seaward of the ridge the gentle submarine slope is cut by a series of regularly spaced grooves. These may be slightly sinuous but in general they extend seaward at right angles to the reef front. They are separated by flattened spurs or buttresses that appear to have been developed mainly by coral and algal growth. The grooves and buttresses form the toothed edge that is so conspicuous when an atoll is viewed from the air. Beyond the toothed edge may be a shallow terrace, then a steep (35°) submarine slope that flattens progressively at deeper levels until it touches the ocean floor. Some grooves of the toothed margin extend through the marginal algal ridge as surge channels, ending on the reef flat in a series of "blow holes" due to constriction by algal growth. These structures taken together form a most effective baffle that robs the incoming waves of much of their destructive power, and at the same time bring a constant supply of refreshing sea water with oxygen, food, and nutrient salts to wide expanses of reef.

Inside the margin of the reef may be several zones that are rich in corals or other reef builders,

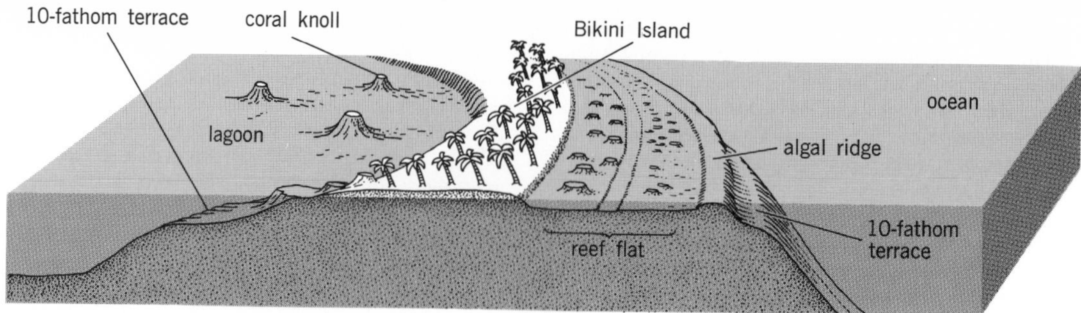

Fig. 1. Diagrammatic section showing features of ring of Bikini Atoll.

but large parts of the reef flat are composed of cemented reef rock, bare or thinly veneered with loose sand or coarser reef debris. When the reef flats of Eniwetok Atoll were drilled, a solid plate of hard limestone was found below that extends to depths of 10–15 ft. Growth of organisms and deposition of sediments on the reef are controlled by the prevailing winds and currents and other ecologic factors. The sediments, including cemented rock, that make up the reef surface and its islets are composed entirely of carbonates secreted by shallow-water organisms.

The reef rings of most atolls are broken, usually to leeward, by one or more passes. Some of these are as deep as the lagoon and give ready access to that body of water. In general, lagoon depths are proportional to size. A lagoon with a diameter of 20 mi may have a depth of 300 ft. Hundreds of coral knolls rise from the lagoon floor. The broad crests of these structures, covered by a growth of coral, lie at all depths, the highest approaching low-tide level.

On many atolls islets are spaced at irregular intervals along the reef. Most of these are small and rarely rise as much as 20 ft above the reef flat. They are composed of reef detritus (coral heads and sand) piled up by waves and winds. The sediments are unconsolidated except for layers of beach rock (cemented conglomerate and sandstone) at intertidal levels. Islets located on reefs that are subjected periodically to violent storms may exhibit one or more ramparts built of blocks of reef rock, coral heads, and finer material. These structures apparently can be built, moved, or destroyed by the waves of a single major storm. Inland from the beaches and ramparts, sand and gravel flats occur and large areas are covered with coconut palms, screw pines, and thickets of bay cedar and other plants. Low sand dunes may occur on the borders of islands inside sand beaches.

Many atolls exhibit terraces around their margins and inside their lagoons. These structures were probably developed during the Pleistocene glacial epochs when, at intervals, sea level stood appreciably lower than now. The most recent change in sea level may have been a negative shift of 6 ft, but convincing evidence for such a change has not been preserved on most atolls. Numerous investigations on reef islands other than atolls are now under way to determine—with the aid of age measurements such as the carbon-14 and uranium natural series techniques—if there was a eustatic shift of this magnitude in fairly recent times. If such a shift did occur, it would offer a plausible explanation for the origin of atoll islets: When the sea level dropped, the reefs were eroded by waves and the debris concentrated to form the existing islets.

Distribution. Existing atolls, like other types of coral reefs, require strong light and warm waters and are limited in the existing seas to tropical and near-tropical latitudes. A large percentage of the world's atolls are contained in an area known as the former Darwin Rise that covers much of the central and southwestern Pacific. Atolls are also numerous in parts of the Indian Ocean and a number are found, mostly on continental shelves, in the Caribbean area. *See* REEF.

Structures properly described as atolls existed in shallow Paleozoic seas. These ancient atolls were built by a variety of organisms; the types of corals and algae that are primarily responsible for today's reefs were not then in existence. Many Paleozoic atolls became reservoirs of petroleum, and are being studied and mapped in detail by deep drilling.

Origin and development. The many atolls that rise from the deep sea have probably been built on submerged volcanoes. The reef caps of two such atolls (Eniwetok and Midway) have been drilled and the existence of a foundation of volcanic flows confirmed. In both instances it was found that the volcanoes had been at least partly truncated by erosion prior to the initiation of reef growth. Many of the existing atolls probably originated as near-surface reefs in Tertiary time. The oldest limestones beneath Eniwetok are late Eocene, and

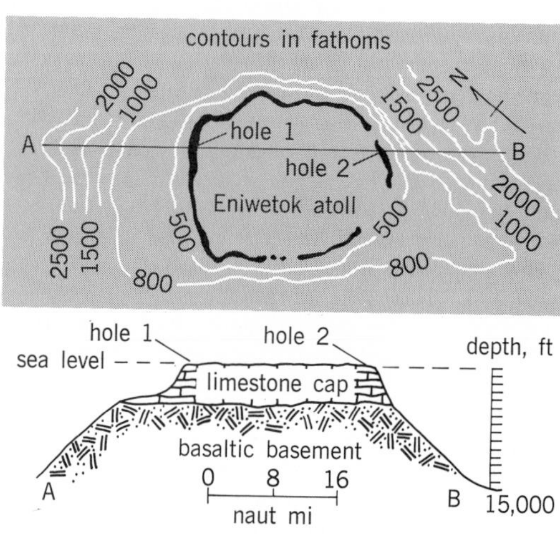

Fig. 2. Structure of Eniwetok Atoll in Marshall Islands.

Midway's oldest limestones are early Miocene. The bases of other atolls may prove to be appreciably younger.

Wherever extended three-dimensional studies have been carried out on atolls (Funafuti in the Ellice Islands, Kita-Daito-Jima, Bikini, and Eniwetok in the Marshall Islands, and Midway in Hawaii), they have proved the atolls to be complex structures; surface similarities may mask striking differences in underground structure. Wherever tested by the drill, the thicknesses of reef caps of atolls greatly exceed the depth limits of the reef builders, and these caps appear to have been developed during long periods of subsidence. *See* SEAMOUNT AND GUYOT.

The structure of Eniwetok Atoll in the Marshall Islands, as revealed by drilling, is shown in Fig. 2. The limestone cap, nearly a mile in thickness, rests on the summit of a basaltic volcano that rises 2 mi above the ocean floor. In hole 1 the lower limestones are late Eocene and formed as forereef deposits on an outer slope. Correlative limestones in hole 2 were formed in shallow water. The 4000-ft section of limestone that forms the cap of Eniwetok records an overall subsidence of that amount over a period exceeding 40,000,000 years. Subsidence of this magnitude even exceeds the amounts so brilliantly postulated by Charles Darwin. The sinking has not been steady or continuous, but was interrupted by periods of emergence. The emergent stages are recorded in zones of recrystallized and partially dolomitized sediments. Such zones are termed solution unconformities and were formed when the top of the atoll stood above the sea for appreciable time. During such periods land snails lived on the island and land plants left a record of spores and pollen; Eniwetok and perhaps many other atolls functioned as stepping stones in the distribution of shallow-water marine life in the Pacific. *See* OCEANIC ISLANDS. [HARRY S. LADD]

Bibliography: R. Berner, Dolomitization of the mid-Pacific atolls, *Science*, 147:1297–1299, 1965; K. Emery et al., *Bikini and Nearby Atolls, Marshall Islands*, USGS Prof. Pap. no. 260-A, 1954; F. R. Fosberg (ed.), *Atoll Res. Bull.*, nos. 1–117, Pacific Science Board, National Academy of Sciences, 1951–1966; and nos. 118–119, Smithsonian Institution, 1967; H. Ladd and S. Schlanger, *Drilling Operations on Eniwetok Atoll*, Geol. Soc. Amer. Prof. Pap. no. 260-Y, 1960; H. Ladd et al., Drilling on Midway Atoll, Hawaii, *Science*, 156(3778):1088–1094, 1967; H. Menard and H. Ladd, Oceanic islands, seamounts, guyots and atolls, in M. N. Hill (ed.), *The Sea*, vol. 3, 1963; H. Wiens, *Atoll Environment and Ecology*, 1962.

Augite

A group of monoclinic calcic pyroxenes which have the general chemical formula (Ca,Mg,Fe)(Mg,Fe)Si_2O_6, in which calcium is the dominant cation in the first cation position. Monoclinic pyroxene with substantial iron or magnesium in place of calcium is called pigeonite, and has a different crystal structure from augite. Augite is generally considered to be a combination of the four end members diopside ($CaMgSi_2O_6$), hedenbergite ($CaFe^{2+}Si_2O_6$), enstatite ($Mg_2Si_2O_6$), and ferrosilite ($Fe_2^{2+}Si_2O_6$), but it almost always has substantial aluminum and minor to substantial amounts of sodium, ferric iron, chromium, and titanium. These minor constituents enter as solid solutions of augite toward acmite ($NaFe^{3+}Si_2O_6$), jadeite ($NaAlSi_2O_6$), calcium Tschermak's molecule ($CaAl_2SiO_6$), ureyite ($NaCrSi_2O_6$), and $CaTiAl_2O_6$. The amount of substitution by any of these minor components depends upon the bulk composition of the rock as well as the conditions of formation. Four common varieties of augite are omphacite, which has about 50% substitution of jadeite; aegerine augite, which has substantial acmite substitution; fassaite, which has a considerable amount of calcium Tschermak's molecule in solution; and titanaugite, which is rich in titanium and has a characteristic purple color. *See* DIOPSIDE; ENSTATITE; PIGEONITE.

Augite in igneous and metamorphic rocks does not commonly occur as large well-formed crystals; when it does, the crystals are dark green to black, short and prismatic. Augite crystallizes in a *C*-centered monoclinic cell, space group $C2/c$, and is commonly recognized by its prominent (110) pyroxene cleavages which form 87° angles. It displays typical pyroxene crystal chemistry—single chains of tetrahedral (SiO_4) groups alternating with strips containing two distinct types of octahedral sites accommodating Ca, Fe, Mg, Mn, Na, Cr, and Ti. The smaller Al cation can enter both tetrahedral and octahedral sites.

Augite occurs in both igneous and metamorphic rocks. It is nearly universal in basalts and gabbros, and occurs somewhat less frequently in less mafic igneous rocks. Alkali olivine basalts and other alkali-rich volcanic rocks contain augite which is commonly enriched in Na, Ti, Al, and Fe^{3+}. Magnesium-rich augite is a characteristic mineral in many ultramafic rocks and in rocks of the Earth's mantle. Augite and pigeonite are also rather common constituents of lunar basalts and basaltic meteorites. In metamorphic rocks, augite of a composition close to diopside occurs in marbles and in calc-silicate rocks. More compositionally complex augite occurs in high-grade metamorphosed amphibolites and granulites where it typically coexists with hornblende or orthopyroxene. The augite variety omphacite is found with magnesium-rich garnet in eclogite which is of basaltic composition, and apparently forms at very high pressures. *See* ECLOGITE; PYROXENE.

[ROBERT J. TRACY]

Bibliography: W. A. Deer, R. A. Howie, and J. Zussman, *Rock-forming Minerals*, vol. 2: *Chain Silicates*, 1963; J. J. Papike (ed.), *Pyroxenes and Amphiboles*, Mineral. Soc. Amer. Spec. Pap. 2, 1969.

Aureole, contact

A halo or zone of alteration surrounding a body of igneous rock (such as granite or granodiorite) and presumably formed by heat and emanations from the igneous mass. Aureoles are usually not more than a few thousand feet wide. They may be detected by mineralogical or textural changes in the surrounding (country) rocks, which become progressively more intense as the igneous contact is approached. The aureole, thus, represents a shell of changed or metamorphosed rock (see illustration).

Aureoles are generally most pronounced where developed in previously unmetamorphosed or

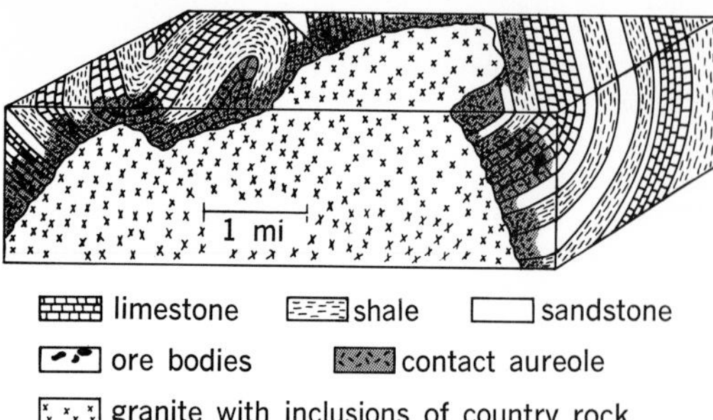

Contact aureole around granite.

weakly metamorphosed rocks. They may not be detected where impressed upon intensely metamorphosed rocks. Shale and limestone generally show a more conspicuous aureole than does sandstone. The metamorphic effect around small igneous bodies is generally less extensive and intensive than that around large ones. The country rock against small dikes and sills may only be baked or indurated, whereas against stocks, large sheets, and laccoliths it is usually recrystallized. *See* PLUTON.

In the outer part of the aureole, recrystallization may be highly sporadic and restricted to certain minerals. Further inward it may be more extensive, and small grains may integrate to form larger ones, giving rise to coarser textures. Reconstitution may entail the formation of new minerals (neomineralization) from old ones. Thus biotite, pyroxene, andalusite, or cordierite may develop at the expense of original constituents. As might be expected, the higher-temperature mineral assemblages appear nearest the igneous contact.

Reconstitution in the outer part of the aureole may be indicated by the presence of small spots or knotlike aggregates of new minerals, which have grown by accretion, or by the formation of large scattered crystals (porphyroblasts). In addition, thin flakes of mica may grow along the bedding planes or schistosity to accentuate the fissility of the rock. Rocks exhibiting these features are known as spotted and knotted slates and schists. *See* CLEAVAGE, ROCK; PORPHYROBLAST; SCHISTOSITY, ROCK.

Well within the aureole, where recrystallization and neomineralization are pronounced, old structures (bedding or schistosity) may be obliterated. Grains tend to become equidimensional with little or no preferred orientation. The pattern of grain outline is that of a mosaic, and the rock is called hornfels. A hornfels may carry numerous scattered prophyroblasts. *See* HORNFELS.

As a rule, reconstitution entails relatively little change in bulk composition within the rocks. There is, however, a tendency for water and carbon dioxide to be eliminated and, in many cases, for volatiles (boron, fluorine, and chlorine) driven off from the crystallizing magma to be taken up or fixed in new minerals (tourmaline, fluorite, topaz, and scapolite) in the country rock. Large quantities of iron may be introduced into certain limestones to form deposits of skarn (rocks rich in silicates of calcium, iron, and magnesium). Magnesium may be introduced to form cordierite-anthophyllite rocks, and some ore deposits may be formed in the aureole by addition of material. Aureoles developed about some diabase bodies are enriched in sodium. *See* SKARN.

Alkali feldspar appears to have developed in abundance near some granite contacts because of introduction of alkalies from the magma. Frequently this feldspar forms large crystals (porphyroblasts) which replace earlier minerals in the country rock and inclusions within the granite. The fact that these porphyroblasts appear identical with large feldspars in the granite suggests that the latter feldspars are also of replacement origin. This relation, furthermore, is believed by some petrologists to indicate a replacement (metasomatic) origin for the granite body as well. *See* METASOMATISM.

The inner boundary of the aureole (against the plutonic rock) may be sharp or transitional. Along some contacts abundant fragments (inclusions) of country rock appear to have been ripped off and enclosed in the granite, and apophyses (offshoots) of granite extend well into the country rock. Along other contacts, notably where granite comes against well-bedded or foliated rocks, migmatites (mixed rocks) have developed. Some of these may represent injection gneiss, produced by intercalation of granite magma between thin layers of country rock. Such rocks (arterites) are difficult to distinguish from veinites, in which the granitic layers have been sweated out of the adjacent rock, and from replacement migmatites, formed by introduction of granitic material metasomatically. *See* MIGMATITE; XENOLITH.

Changes are generally most pronounced within a very short distance of the contact or in inclusions in the igneous rock. Magma tends to react with this foreign material and to convert it into minerals which are stable in the magma itself. This conversion generally involves an exchange of constituent ions between fluid (magma) and solid (inclusions) and is simply expressed by what is known as the reaction principle. *See* MAGMA.

Incorporation of foreign material, therefore, may result in contamination of the magma and consequent crystallization of unusual minerals from the melt. Digestion and assimilation of great quantities of foreign rock may cause such a marked change that an entirely different rock type is created.

Along many igneous contacts, reaction between magma and country rock has produced hybrid rock types of varying composition. This phenomenon is strikingly exhibited where granite is separated from older gabbro by a zone of hybrid rock (diorite, quartz diorite, and granodiorite). Much material in this intervening zone may represent original magma, contaminated by iron, magnesium, and calcium; but a large proportion may represent solid rock converted by metasomatism to more granitic types.

Contact aureoles around granitic bodies have been variously interpreted. According to one interpretation (magmatic), they represent halos (essentially thermal) formed around intrusions of molten rock. In another (metasomatic), they are zones of changed rock formed by an advancing wave of granitization; the granite represents the final product of this transformation. From this viewpoint, as the requisite material takes its place in the solid rock to form granite, unessential constituents (usually calcium, iron, and magnesium) are expelled and may become fixed in the surrounding region or contact aureole. Heat, accompanying the granitization process, is held responsible for reconstitution within both the granite and adjacent aureole. *See* GRANITIZATION.

[CARLETON A. CHAPMAN]

Authigenic minerals

Minerals that are formed "on the spot" within sediments or sedimentary rocks during or long after their deposition. Their in-place origin distinguishes them from detrital (allogenic) minerals, which are formed elsewhere and transported to the site of deposition, and from biogenic minerals, which compose the hard parts of organisms or are indirectly precipitated by organisms at the site of deposition. Authigenic minerals are among the principal products of diagenetic processes and are frequently the only tangible evidence of these processes. *See* DIAGENESIS.

Examples. Authigenic minerals can occur in almost all types of sedimentary rock and can range in abundance from a trace to virtually the total rock. The more common authigenic minerals are shown in the table. Probably the most abundant are calcite and dolomite, the principal minerals composing limestones and dolostones, respectively. Much of the calcite in limestones was initially aragonite, whereas most dolomite has been formed by the chemical alteration of calcite. The relatively common sedimentary rock called chert (flint, jasper) often consists of microcrystalline quartz (chalcedony), and some recent chert also contains cristobalite; both types crystallized from amorphous opaline silica. The detrital minerals composing sandstones are often cemented by authigenic quartz, calcite, dolomite, or the iron oxides hematite or goethite. Overgrowths of authigenic quartz on detrital quartz and of authigenic feldspar on detrital feldspar are very common. Authigenic clay minerals—chlorite, illite, kaolinite, and smectite—are much more common in sandstones than once realized. Authigenic pyrite and marcasite are frequent constituents of shales and mudstones. Recent marine sediments often contain one or more of the zeolite minerals which have formed from the devitrification of volcanic glass. *See* CALCITE; CHERT; DOLOMITE; ZEOLITE.

Common authigenic minerals

Mineral	Formula
Albite	$NaAlSi_3O_8$
Anatase	TiO_2
Anhydrite	$CaSO_4$
Apatites*	$Ca_5(PO_4)_3(F,Cl,OH)_3$
Aragonite (orthorhombic)	$CaCO_3$
Barite	$BaSO_4$
Boehmite	$AlO(OH)$
Calcite (hexagonal)	$CaCO_3$
Celestite	$SrSO_4$
Chalcedony	SiO_2
Clay minerals	
Chlorites*	$(MgFe^{2+}Fe^{3+})_6(AlSi_3)O_{10}(OH)_8$
Illites*	$K(AlMgFe)_2(AlSi_3)O_{10}(OH)_2$
Koalinite	$Al_2Si_2O_5(OH)_4$
Sepiolite	$Mg_2Si_3O_8 \cdot 2H_2O$
Smectites*	$(NaCa)(AlMgFe)_2(AlSi_3)O_{10}(OH)_2 \cdot nH_2O$
Dolomite	$CaMg(CO_3)_2$
Gibbsite	$Al(OH)_3$
Glauconite*	$K(AlMgFe^{2+}Fe^{3+})_2(AlSi_3)O_{10}(OH)_2$
Goethite	$Fe_2O_3 \cdot n(H_2O)$
Gypsum	$CaSO_4 \cdot 2H_2O$
Halite	$NaCl$
Hematite	Fe_2O_3
Leucoxene	TiO_2
Limonite	$FeO(OH) \cdot n(H_2O)$
Marcasite (orthorhombic)	FeS_2
Microcline	$KAlSl_3O_8$
Muscovite	$KAl_2(AlSl_3)O_{10}(OH)_2$
Opal	SiO_2
Orthoclase	$KalSi_3O_8$
Psilomelane	$BMn^{2+}Mn_8{}^{4+}O_{16}(OH)_4$
Pyrite (isometric)	FeS_2
Pyrolusite	MnO_2
Quartz	SiO_2
Siderite	$FeCO_3$
Zeolites*	$X^{1+,2+}{}_yAl^{3+}{}_xSi^{4+}{}_{1-x}O_z \cdot nH_2O$

*Group of minerals characterized by considerable chemical variation.

Formation processes. Authigenic minerals may be formed by direct precipitation, by recrystallization or alteration of preexisting minerals, or by structural transformation of one mineral to another. Their formation indicates an attempt to establish an equilibrated mineral assemblage. Critical factors in the formation of authigenic minerals are temperature, pressure, ionic concentration, pH, and Eh (standard oxidation-reduction potential).

Fig. 1. Scanning electron micrograph showing pseudohexagonal authigenic kaolinite crystals in the pore space of the St. Peter (Ordovician) Sandstone. Quartz grains at upper left and lower right contain authigenic quartz overgrowths.

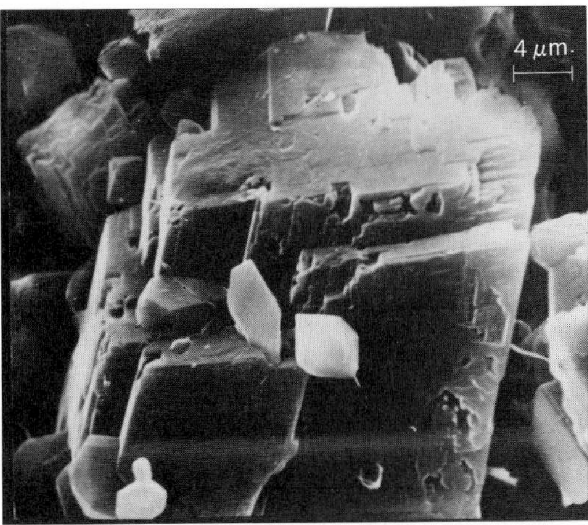

Fig. 2. Scanning electron micrograph showing the monoclinic form of an authigenic feldspar crystal in the St. Peter Sandstone. The smaller, light-colored crystal in the foreground is a doubly terminated authigenic quartz crystal.

Commonly, the authigenic clay minerals present in sandstones (Fig. 1) have formed by direct precipitation from pore fluids. Other authigenic clay minerals formed by direct precipitation are pyrite, formed by the reaction of iron and hydrogen sulfide, and quartz, as single crystals (Fig. 2) or as overgrowths, formed as a result of the precipitation of silica.

Recrystallization in authigenic mineral formation leads to an enlargement of crystal size through the process of solution followed by reprecipitation. Noteworthy examples are large halite, gypsum, and anhydrite crystals in evaporite deposits and calcite crystals in some limestones.

The conversion of feldspars to clay minerals is frequently observed in sandstones. This involves both structural reorganization of ions and hydration. The chemical composition of the feldspar and the pH and Eh conditions determine the type of clay mineral that is formed. Authigenic processes involving dehydration, on the other hand, may change gypsum to anhydrite, goethite to hematite, and opaline silica to chalcedony. It is commonly believed that glauconite forms authigenically by the chemical reconstitution of degraded three-layer clay minerals.

The substitution of magnesium for calcium is responsible for the conversion of calcite or aragonite to dolomite, and it has been shown that dedolomitization (replacement of magnesium by calcium) is also possible.

Structural inversion may change aragonite to calcite. Aragonite (orthorhombic) is a naturally unstable form of calcium carbonate. With the passage of geologic time, aragonite normally inverts to the more stable calcite (hexagonal) structure, either directly or indirectly, through the dissolution of aragonite and reprecipitation of calcite, unless the presence of foreign ions or hydrocarbons prevents this change. *See* ARAGONITE.

Distinguishing characteristics. Since many of the authigenic minerals shown in the table may also form in other ways and may be deposited in sediments as detrital minerals, it is important that their authigenic origin be clearly established, especially when these minerals are utilized to interpret diagenetic chemical, temperature, and pressure conditions. Although the authigenic origin of minerals formed by alteration or by precipitation in the pores of sandstones can often be established from the study of thin sections, the scanning electron microscope and electron microprobe are considerably more useful for their identification and chemical study. Because many authigenic minerals are small in size and scarce in amount, the high magnification and resolution attainable with the scanning electron microscope frequently make it possible to observe these minerals visually and to differentiate them from like species of detrital origin by their euhedral crystal form (Figs. 1 and 2). Conversely, the form of detrital minerals is irregular, roughed, or smoothed because they were formed in a weathering environment or were abraded during transportation.

[I. EDGAR ODOM]

Avalanche

A mass of snow or ice moving rapidly down a mountain slope or cliff. Avalanches, or snowslides, range from small movements on established avalanche tracks to large, sporadic, very rapid movements capable of taking a heavy toll of life. Avalanches are infrequent on slopes of less than 25°, especially numerous on those exceeding 35°, and most commonly start on convex slopes. Both old and new snow may avalanche, but serious avalanches are always possible when 12 in. or more of new snow is present. Movement may be set off by temperature, vibration, shearing, or other slope disturbance.

Dry snow avalanches usually occur during, or within several days after, snowfall. They may affect whole slopes, even if wooded, and may exceed 100 mph. Wet snow avalanches are formed during thaws or rainy weather. Their movement is less rapid but may be destructive. Slab avalanches of wind-packed snow are broader and deeper and move rapidly. This type of avalanche is an extreme hazard to life and property.

Avalanche prevention and prediction has become an important service. Reforestation, snowsheds, and avalanche breakers and barriers help to prevent or control movement. Careful coordination in planning land use, avalanche forecasting, and safety patrolling, with artificial release of slides and temporary closing of hazardous zones, help make mountain areas safe for human use.

[C. F. STEWART SHARPE]

Azurite

A basic carbonate of copper with the chemical formula $Cu_3(OH)_2(CO_3)_2$. Azurite is normally associated with copper ores and often occurs with malachite. Azurite is monoclinic. It may be massive or may occur in tabular, prismatic, or equant crystals (see illustration). Invariably blue, azurite was originally used extensively as a pigment. Hardness is 3 1/2–4 (Mohs scale) and specific gravity is 3.8. It can be synthesized by gentle heating of cupric

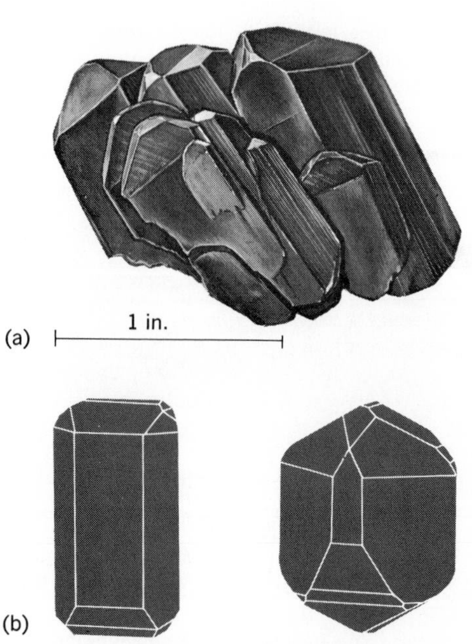

Azurite. (*a*) Crystals from Tsumeb, Southwest Africa (*specimen from Department of Geology, Bryn Mawr College*). (*b*) Crystal habits (*from C. S. Hurlbut, Jr., Dana's Manual of Mineralogy, 17th ed., copyright © 1959 by John Wiley & Sons, Inc.; reprinted by permission*)

nitrate or sulfate solutions with calcium carbonate in a closed tube. Notable localities for azurite are at Tsumeb, Southwest Africa, and Bisbee, Ariz. [ROBERT I. HARKER]

Barite

A mineral with the chemical composition $BaSO_4$. Barite occurs mainly in white and yellowish, orthorhombic, tabular crystals or masses (see illustration). Fracture is uneven and luster is vitreous to pearly. Hardness is 3–3.5 on Mohs scale and specific gravity is 4.5. It decrepitates and fuses with difficulty, and is insoluble in acids. The barium present in barite imparts a characteristic yellowish-green color to the flame. *See* BARIUM.

Barite. (*a*) Specimen from Dufton, Westmoreland, England (*American Museum of Natural History*). (*b*) Typical crystals. Crystals may be very complex (*from C. S. Hurlbut, Jr., Dana's Manual of Mineralogy, 17th ed., copyright © 1959 by John Wiley & Sons, Inc.; reprinted by permission*)

Because of the high specific gravity of barite, nearly three-quarters of the domestic ouput of the mineral is used as a weighing agent in rotary well-drilling fluids. Barite has numerous other uses, for example, in the manufacture of glass and white pigment, as a filler or extender in paint, inks, oilcloth, linoleum, and rubber, and in the manufacture of barium compounds for various industrial applications.

Barite will continue to be in great demand as well-drilling mud. It has also been used to weight concrete around pipelines in river crossings and swampy areas, and in the manufacture of barium titanate for use in ultrasonics. Large tonnages of barite might also be used as an aggregate in concrete shield for future atomic power plants.

Barite is commonly found in ore veins, frequently associated with fluorite, calcite, dolomite, and quartz. It is also widely distributed in limestone as veins, lenses, or cavity fillings. The principal domestic barite-producing areas are in Washington County, Mo., and the Ouachita Mountains of Arkansas. Although known reserves of commercial-grade material in excess of 40,000,000 tons are adequate for the near future, new deposits are being sought to meet the expanding demand for barite in drilling and in the metallurgical, chemical, and electronic fields. [EDWARD C. T. CHAO]

Basalt

An igneous rock characterized by small grain size (less than about 5 mm) and approximately equal proportions of calcium-rich plagioclase feldspar and calcium-rich pyroxene, with less than about 20% by volume of other minerals. Olivine, calcium-poor pyroxene, and iron-titanium oxide minerals are the most prevalent other minerals. Most basalts are dark gray or black, but some are light gray. Various structures and textures of basalts are useful in inferring both their igneous origin and their environment of emplacement. Basalts are the predominant surficial igneous rocks on the Earth, Moon, and probably other bodies in the solar system. Several chemical-mineralogical types of basalts are recognized. The nature of basaltic rocks provides helpful clues about the composition and temperature within the Earth and Moon. The magnetic properties of basalts are responsible in large part for present knowledge of the past behavior of the Earth's magnetic field and of the rate of sea-floor spreading. *See* IGNEOUS ROCKS.

Extrusion of basalt. Basalt flows out of fissures and cylindrical vents. Fissures commonly are a few meters wide. Cylindrical vents generally are several tens of meters in diameter and may pass downward into cylindrical shell-shaped or planar fissures. Repeated or continued extrusion of basalt from cylindrical vents generally builds up a volcano of accumulated lava and tephra around the vent. Fissure eruptions commonly do not build volcanoes, but small cones of tephra may accumulate along the fissure. *See* VOLCANO; VOLCANOLOGY.

Form and structure. Basalts display a variety of structures mostly related to their environments of consolidation. On land, basalt flows form pahoehoe, aa, and block lava, while under water, pillow lava is formed. Basaltic lava flows on land have rugged to smooth surfaces. The rugged flows are called aa and have top surfaces made of spines and ridges up to several meters high. They are typically composed of three layers: a massive interior overlain and underlain by rubbly zones. Flows with smooth surfaces are called pahoehoe. Pahoehoe has been seen to form on fluid lava flows as the incandescent surface rapidly advances (1–10 m/s), cools below red heat, and stiffens into a rigid crust a few centimeters thick in about a minute. There are transitional types of surfaces, as well as a separate kind called block lava, which is similar to aa lava but has a surface made up of equidimensional blocks up to about 1 m in diameter.

Pillow lava is characterized by lens-shaped bodies (pillows) of basalt up to a few meters in length which have upward convex tops and partially upward convex bottoms conforming to the convex tops of underlying pillows. Such structure is found in basalts dredged from the deep-sea floor, in basalts which extruded under ice caps, as in Iceland, and in basalts which have flowed into lakes, rivers, and the sea from vents on land.

Basalt occurs as pumice and bombs as well as lava flows. Commonly, basaltic pumice is called scoria to distinguish it from the lighter-colored, more siliceous rhyolitic pumice. Basaltic pumice can have more than 95% porosity. Bombs are relatively smooth, round bodies up to about 1 m in diameter whose surfaces commonly are cracked as a result of expansion of the effervescing interior which remains fluid longer than the quickly cooled surface. Bombs with cracked surfaces are called bread-crust bombs, the name attesting both to the fluid condition and to the gas content of the extruded material. Pele's hair and Pele's tears are terms applied to threads and drops of basaltic glass blown out of vents. Pasty material intermediate between pumice and bombs is called spatter and commonly forms agglutinate or vent breccia near vents of basalts. Accumulations of pumice, bombs, and spatter commonly develop around vents and form cinder cones, spatter cones, and spatter ramparts. Similar, but commonly much smaller, secondary vent deposits may develop on lava flows distant from the primary vent. Secondary vents are rooted within the flow and do not penetrate it; they are called hornitos. Littoral cones may form where a lava flow encounters abundant water and water vapor blows through the liquid lava.

Basalt commonly contains tiny round voids called vesicles. In pumice, vesicles generally make up most of the volume. The vesicle porosity in flows is high at the top and bottom of flows, but it decreases dramatically a few meters away from cooling surfaces. In general, vesicles are smallest (1–10 mm in diameter) in the top and bottom half-meter of flows, reach a maximum of a few centimeters in diameter a few meters from the cooling surface, and then decrease in size and abundance toward the center of the flow. Vesicles in basalts on the sea floor (about 5 km depth) are very tiny (about 10 μm diameter), and the total porosity is commonly less than 1%. Many pillow basalts now on land lack vesicles or have only a few tiny vesicles and presumably formed in deep water. However, there are basalts on land which are vesicle-poor, and some sea-floor basalts have significant vesicles. In sum, the presence of vesicles indicates that basalts evolve gas in low-pressure environments, and the distribution of vesicles within flows provides a means of identifying the tops and bottoms of flows. Flows of basalt develop a characteristic kind of fracture called columnar jointing whereby the rock splits into polygonal columns which are oriented perpendicular to the cooling surfaces.

Physical properties. Many physical properties of basalt have been measured in the field and laboratory, making it one of the best-characterized rock types. The physical properties of basalts vary greatly because of the range of textures and compositions. Temperature ranges from 1000 to 1220°C for flowing basalt. The viscosity for magma is $10^3 - 10^6$ poise (1 poise = 0.1 N · s/m^2), increasing as the temperature falls and the crystal content of the magma increases. The surface tension for magma is 270–350 dynes/cm (1 dyne = 0.00001 N). For melt the density is 2.6–2.7 g/cm^3 at 1200°C, while for crystalline rock at standard temperature and pressure (STP), void-free (calculated from simple mixtures of pyroxene, plagioclase, and iron oxides consistent with the definition of basalt), density is 2.8–3.1 g/km^3. The velocity of sound is 2.3 km/s for compressional wave in melt at 1200°C, and 5.7 km/s for rock at 700°C. Both isothermal and adiabatic compressibilities are closely equal to 7×10^{-12} cm^2/dyne for melt at 1200°C, and 2×10^{-12} for rock at 800°C. The crushing strength and cohesive strength of cold rock are 1700–2200 atm and 320–440 atm (1 atm = 101,325 N/m^2), respectively. The coefficient for volumetric thermal expansion of melt is about 2×10^{-5} per °C. The electrical conductivity is about 100 per ohm per centimeter for melt at 1200°C. Thermal conductivity is about $1-3 \times 10^{-3}$ cal/cm s °C for melt at 1200°C, and $4-6 \times 10^{-3}$ cal/cm s °C (1 cal = 4.19 J) for rock at STP. Heat capacity at constant pressure is about 0.2–0.3 cal per °C per g. About 100 cal/g is generally assumed for the heat of melting, but the figure is possibly uncertain by 30%. Theoretical and observed diffusion coefficients for Mg^{+2} in basaltic melt at 1170°C are 4×10^{-9} cm^2/s and 72×10^{-9} cm^2/s.

The magnetic susceptibility of basalts is commonly about 10^{-4} to 4×10^{-3} emu/g. The natural remanent magnetization intensity of basalts depends upon the amount of magnetite, the composition of the magnetite, the grain size, the cooling history, and the paleointensity of the Earth's field. Observed values of natural remanent magnetization intensity are commonly between 10^{-4} and 10^{-3} emu/g for subaerial basalts, but they increase to 10^{-2} to 10^{-1} emu/g for basalts from the sea floor. Thus, the remanent magnetization of submarine basalts is stronger than their induced magnetization—a relationship that has made it possible for the sea floor to preserve a "tape-recorded" history of the reversals of the Earth's magnetic field. The

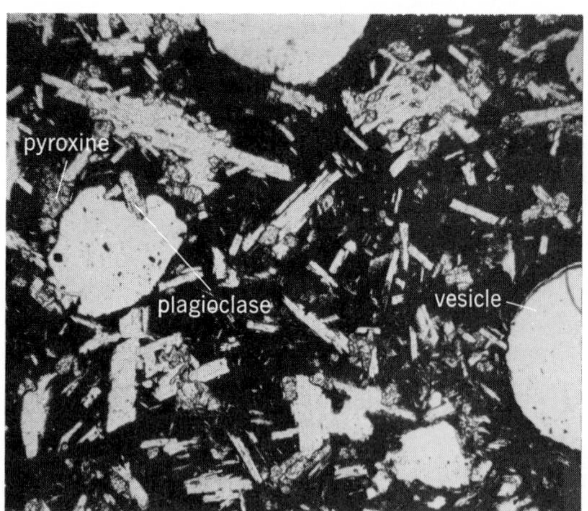

Fig. 1. Thin section of upper crust of lava about 3 cm from top; black is microcrystalline groundmass.

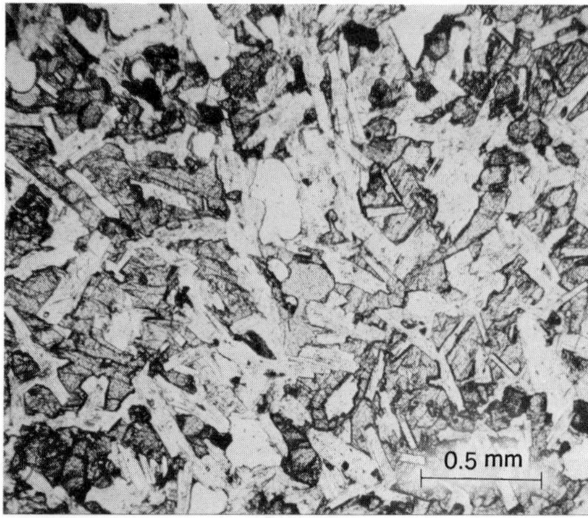

Fig. 2. Thin section of interior of lava shown in Fig. 1, but 7.9 m below the top of the flow. Clear plagioclase crystals are ophitically enclosed in large pyroxene grains up to 1 mm in diameter. Crystals of black magnetite are minor. A few irregular voids are present.

natural magnetization of many basalts is sufficiently strong to be measured with a homemade astatic magnetometer having a sensitivity of about 500 radians per oersted. *See* ROCK MAGNETISM.

Mineralogy and texture. The mineralogy and texture of basalts vary with cooling history and with chemical composition. With slow cooling (few °C per day 10 m deep in a flow), crystals grow to large sizes (millimeters) and acquire equant shapes. In basaltic melt, diffusion in the melt limits the rate at which a given crystal-liquid boundary can move because the crystals differ in composition from the melt. If crystallization is forced at a rapid rate by rapid cooling (about 100°C per minute 10 cm deep in the flow), either the crystals have high ratios of surface area to volume (needles, plates) or many seeds (nuclei) develop. Crystals which grow rapidly under strongly supercooled conditions have different compositions from those which form if slower cooling prevails. Consequently, the same basalt flow may contain different minerals in variously cooled parts of the flow.

The size of crystals in basalt reflects the rate of production of seed crystals (or nuclei) as well as the rate of growth of individual crystals. Crystals of pyroxene are able to grow at an average linear rate of about 2×10^{-5} μm/s, or 2 μm/day. At higher temperatures, in melts rich in Mg and Fe, crystals may grow as rapidly as 6×10^{-3} μm/s, or about 0.5 mm/day. The large crystals (phenocrysts) present in most basalts possibly formed in a few hours or days during the ascent of the magma through the crust. Alternatively, many large crystals possibly grew weeks or years before extrusion in a large reservoir of melt which crystallized slowly because of its large size and slow cooling due to a large ratio of volume to surface area.

As basalt crystallizes, both the minerals and the residual melt change in composition because of differences between the composition of the melt and the crystals forming from it. In basalts, because of the rapid cooling, there is little chance for crystals to react with the residual melt after the crystals have formed. Early-formed crystals tend to become armored with later overgrowths. Because diffusion is sluggish in crystals even at the high temperatures at which basalts crystallize, the minerals generally retain their compositional zoning during cooling. Completely solid basalts generally preserve a record of their crystallization in the zoned crystals and residual glass. Figures 1 and 2 illustrate the development of crystalline texture in basalt. Figure 3 shows the sequence of crystallization in basaltic lava on Hawaii.

Basalts may alternatively yield a silica-saturated residual melt (rhyolite) or one deficient in silica to form nepheline, depending upon the initial bulk composition of the basalt. Basalts yielding nepheline upon crystallization are termed alkaline, those yielding rhyolite are called subalkaline (or tholeiitic).

Most basalts contain minor amounts of chromite, magnetite, ilmenite, apatite, and sulfides in addition to the minerals mentioned above. Chromite and the sulfides commonly form early in the sequence of crystallization, whereas magnetite, ilmenite, and apatite form late. Magnetite in basalts contains a history of the strength and orientation of the Earth's magnetic field at the time of cooling. Therefore, although magnetite is minor in amount, it is probably the most important mineral in terrestrial basalts, because it enables earth scientists to infer both the magnetic history of the Earth and the rate of the production of basaltic ocean floor at the oceanic ridges.

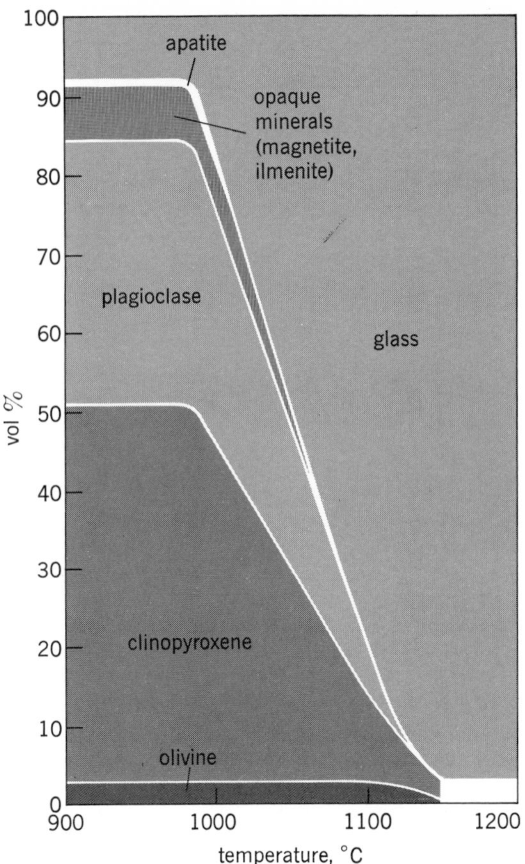

Fig. 3. Growth of crystals as a function of temperature in basalt in Hawaiian lava. (*From D. L. Peck, T. L. Wright, and J. G. Moore, Crystallization of tholeiitic basalt in Alae lava lake, Hawaii, Bull. Volcanol., 29:629–656, 1966*)

Chemical composition. Significant differences exist in the composition of basalt which relate to different tectonic environments. The table lists typical compositions of basalts from oceanic ridges, island arcs, continental platforms, and oceanic islands and of alkalic basalts which have no particular tectonic association. In addition to the common elements, the table gives the concentrations of H_2O, CO_2, F, Cl, and S typical of the various basalts. It is common practice to reduce chemical analyses of igneous rock to a volatile-free basis for comparison purposes, and this practice has been followed in the table.

Chemical analyses of basalts are now used instead of, or together with, the textural criteria as a basis of classification. Geologists customarily recast the chemical analysis into a set of ideal minerals according to a set of rules. The result is called the norm of the rock. The principal normative minerals of basalts are the feldspars; diopside, $Ca(MgFe)Si_2O_6$; olivine, $(MgFe)_2SiO_4$; hypersthene, $(MgFe)SiO_3$; nepheline, $NaAlSiO_4$; and quartz, SiO_2. Because quartz reacts with olivine to form hypersthene and with nepheline to form feldspar (albite), basalts can be classified in order of in-

Chemical compositions of basalts, in weight percent

	1*	2	3	4	5	6
SiO_2	49.92	49.20	49.56	51.5	45.90	45.5
TiO_2	1.51	2.32	1.53	1.1	1.80	2.97
Al_2O_3	17.24	11.45	17.88	17.1	15.36	9.69
Fe_2O_3	2.01	1.58	2.78	n.d.	1.22	0.00
Cr_2O_3	0.04	n.d.	n.d.	n.d.	n.d.	0.50
FeO	6.90	10.08	7.26	8.9	8.13	19.7
MnO	0.17	0.18	0.14	n.d.	.08	0.27
MgO	7.28	13.62	6.97	7.0	13.22	10.9
CaO	11.85	8.84	9.99	9.3	10.71	10.0
Na_2O	2.76	2.04	2.90	4.3	2.29	0.33
K_2O	0.16	0.46	0.73	0.80	0.67	0.06
P_2O_5	0.16	0.23	0.26	n.d.	0.62	0.10
Sum	100.00	100.00	100.00	100.00	100.00	100.02
H_2O	0.4	0.3	n.d.	2	n.d.	0.0
CO_2	0.02	0.01	n.d.	n.d.	n.d.	n.d.
F	0.02	0.03	n.d.	n.d.	n.d.	0.002
Cl	0.02	0.03	n.d.	0.09	0.01	0.0005
S	0.08	0.07	n.d.	0.19	n.d.	0.07

*(1) Average of 10 basalts from oceanic ridges; (2) submarine basalt, Eastern Rift Zone, Kilauea Volcano, Hawaii; (3) Average high-alumina basalt of Oregon Plateau; (4) Initial melt of Pacaya Volcano, Guatemala; (5) Alkali basalt, Hualalai Volcano, Hawaii; (6) Average *Apollo 12* lunar basalt.

creasing SiO_2 saturation by the presence of normative nepheline (lowest SiO_2), olivine, or quartz. Diagrams illustrating these relations and the particular names associated with individual combinations are shown in Fig. 4. In general, there is rather close correspondence between the normative minerals and the observed minerals in basaltic rocks.

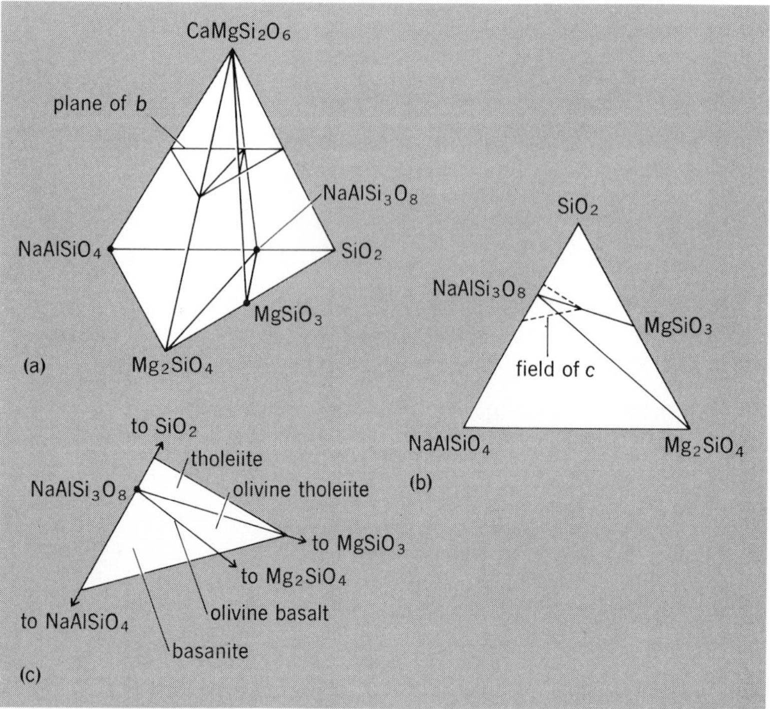

Fig. 4. Diagrams illustrating the compositions of basalts in terms of normative mineral formulas. Customarily, $CaAl_2Si_2O_8$ (anorthite) is grouped together with $NaAlSi_3O_8$ (albite), and Fe_2SiO_4, $FeSiO_3$, and $CaFeSi_2O_6$ are grouped with the corresponding Mg formulas. *(a)* The position of the basalt plane in a tetrahedron with $CaMgSi_2O_6$ (representing calcium-rich pyroxene) at the apex and plagioclase (represented by $NaAlSi_3O_8$) and other minerals indicates SiO_2 saturation on the base. *(b)* Diagram shows that natural basalts occupy only a small portion of the basalt plane. *(c)* Basalt plane from *b*.

Lunar basalts. Lunar basalts belong exclusively to the subalkaline class. They are similar to subalkaline basalts on Earth but have metallic iron rather than ferric-iron-bearing magnetite. Many lunar basalts are exceptionally rich in iron and titanium. They also differ compositionally from terrestrial basalts in being poor in alkali elements and siderophile elements such as Ni. *See* MOON.

Meteoritic basalts. Some meteorites are basaltic rocks. They differ significantly from lunar basalts and appear to have originated elsewhere in the solar system at a time close to the initial condensation of the solar nebula. Meteoritic basaltic rocks are called eucrites and howardites. *See* METEORITE.

Tectonic environment and origin. Basalts occur in all four major tectonic environments: ridges in the sea floor, islands in ocean basins, island arcs and mountainous continental margins, and interiors of continents. The principal environment is the deep sea floor, where the rate of production of basalt is about 1.5×10^{16} g/year. Exploration of the Mid-Atlantic Ridge revealed that the ridge trough contains longitudinal depressions and mounds. J. Moore and others suggest that the floor of the ridge crest cracks open and is subsequently filled with a mound of pillow basalt on a time scale of 10,000–20,000 years. The chemical composition of ridge basalts, their tectonic environment, and their melting behavior as a function of pressure suggest that ridge basalts originate by crystal-liquid separation from a primary magma generated at depths of a few tens of kilometers by melting 10–30% of the mantle rock.

Basaltic islands in the ocean basins contain diverse basalts. There is a general association of alkaline-basalts with volcanoes developed at significant distances from oceanic ridges. The alkalinity increases and silica decreases with distances from the ridge. Major oceanic centers of volcanism (hot spots) such as Hawaii, Galapagos, and Réunion generally extrude subalkaline basalts first and alkaline basalts later (after the site of principal activity has shifted or after the lithospheric plate on which the volcano ridges moves the volcano off the principal source or hot spot). Thus, both islands related to ridges and islands related to hot spots share the feature that near the place of the highest mass rate of extrusion the basalts are subalkaline, whereas farther away, the volcanoes and their basalts become increasingly alkaline.

Basalts in island arc–continental margin environments show similar variations: where volcanism is extensive near the ocean trench, basalts are subalkaline; where volcanism is sporadic closer to the continental interior, basalts become increasingly alkaline. However, the basalts in the island arc environments are significantly different in chemical composition from those of oceanic ridges and hot spot island (see table). The causes of the differences are not completely established, but the special environment created where oceanic lithosphere plunges back into the Earth probably includes volatile ingredients and preexisting basaltic rock which may help explain the nature of the basalts extruded.

Basalts in continental interiors fall into two classes: (1) major plateaus of lava flows extruded from giant fissures (Columbia River, Deccan Traps, Siberian Traps, Piranhas Basin, Karroo, Tasmania), and (2) volcanoes (mostly small) of alkaline basalt commonly containing nodules of olivine-rich rocks apparently derived from the mantle.

The basalts of the major plateaus vary in chemical composition but are, with rare exception, subalkaline. Some of the plateau basalts are strongly similar to basalts from oceanic ridges, but are slightly richer in some of the elements normally found in high concentration in the granitic continental crust. Other plateau basalts are similar to the subalkaline basalts of oceanic islands. Still others are rather unique (Tasmania). The plateau basalts are unquestionably derived from the mantle, because they are erupted at too high a temperature to have been generated within the crust. However, the particular conditions of initial generation in the mantle are uncertain.

Evidently, basalt is the most widespread material available at the surface of the Earth which is derived from the mantle. Study of the regional distribution of various basalts helps earth scientists formulate some details regarding the tectonic processes affecting their generation. After these factors are taken into account, the differences remain which may be understood in terms of the compositionally inhomogeneous mantle. Some inhomogeneities may relate to earlier episodes of mountain building and possibly to primal planetary differentiation. *See* MAGMA; PETROGRAPHIC PROVINCE.

[ALFRED T. ANDERSON, JR.]

Bibliography: I. S. E. Carmichael, F. J. Turner, and J. Verhoogen, *Igneous Petrology*, 1974; H. H. Hess (ed.), *Basalts*, 2 vols., 1968; G. A. Macdonald, *Volcanoes*, 1972; E. W. Spencer, *The Dynamics of the Earth*, 1972.

Basement rock

The more resistant, generally crystalline rock beneath layers or irregular deposits of younger, relatively deformed sedimentary rock. The term is frequently equated with "basement complex," but the latter term should be used only when the composition and structure are truly complex. In much of the world basement rock consists of Precambrian igneous and metamorphic rocks of various types, but locally the basement rock beneath layers of sedimentary strata may be Paleozoic, Mesozoic, or even Cenozoic. It may lie at great depths, beneath hundreds or thousands of feet of sediments, or it may occur at the surface, where it has been exposed by faulting or erosion or both. *See* PRECAMBRIAN.

[J. PAUL FITZSIMMONS]

Bibliography: J. Edwards, Jr., The petrology and structure of the buried Precambrian basement of Colorado, *Colo. Sch. Mines Quart.*, vol. 61, no. 4, 1966; P. T. Flawn et al., *Basement Map of North America between Latitudes 24° and 60°N*, U.S. Geological Survey and American Association of Petroleum Geologists, 1967; W. R. Muehlberger, R. E. Denison, and E. G. Lidiak, Basement rocks in continental interior of United States, *Amer. Ass. Petrol. Geol. Bull.*, 51:2351–2380, 1967.

Basin

A low-lying area, wholly or largely surrounded by higher land. Basins vary from small, nearly enclosed valleys to extensive mountain-rimmed depressions, such as the Bighorn Basin of north-central Wyoming and the larger Tarim (see illustration) and Dzungarian basins of central Asia, or to equally broad but shallower features, such as the Congo Basin of central Africa.

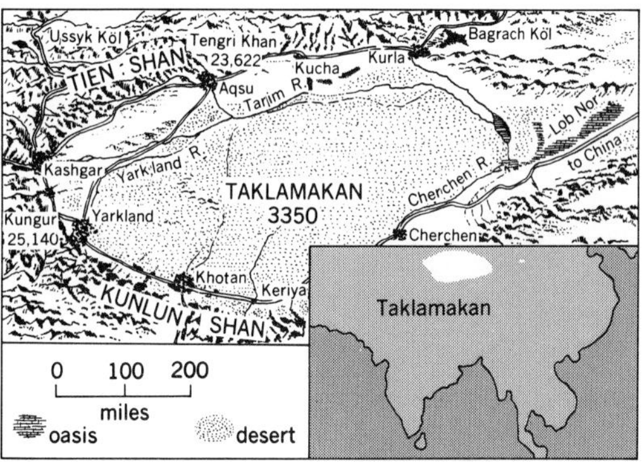

Tarim Basin in central Asia, a mountain-rimmed desert basin of interior drainage. (*From P. E. James, An Outline of Geography, copyright 1935 by Ginn and Co.*)

The term is also applied (sometimes as drainage basin) to the entire area drained by a given stream and its tributaries. In this sense one speaks of the Mississippi Basin, occupying most of the United States between the Rocky Mountains and the Appalachians, or of the Amazon Basin, an area of similar size in northern South America. Depressions that drain entirely inward, without outlet to the sea, are basins of interior drainage. Examples are the Aral and Caspian depressions (the Volga River drains into the latter) in Russian Turkistan, and the Great Basin of the United States, actually an area of many smaller mountain-ringed basins of interior drainage occupying most of Nevada and parts of the adjacent states.

A geologic basin is an area in which the rock strata are inclined downward from all sides toward the center. Because of erosion and deposition, the surface may not show the form of a depression. The Paris Basin, the London Basin, and the Illinois – Western Kentucky Coal Basin are examples of this type.

[EDWIN H. HAMMOND]

Batholith

A geologic name for a body of igneous rock of moderate or large size (40 mi^2 or more in cross-sectional diameter) emplaced at great or intermediate depth in the Earth's crust. Batholiths are commonly discordant with (cut across) the enclosing rocks, but on a grand scale most of the large batholiths are elongated parallel to the dominant structure of the enclosing crustal rocks. The shape of batholiths in depth is a point of debate. One school argues that batholiths continue downward with increasing size or at least with no reduction in size. The other school argues that large batholiths are composites of many smaller units, each of which maintains a nearly constant cross section, thus suggesting variation in shape with downward development. *See* PLUTON.

[JAMES A. NOBLE]

Bauxite

A rock name for weathering products composed largely of hydrous aluminum oxides and aluminous laterite. Bauxite is the ore of aluminum. Red, nonplastic, claylike material from Les Baux in southeastern France was first analyzed by P. Berthier in 1821, named beauxite by A. Dufrénoy about 1845–1847, and corrected to bauxite by H. Sainte-Claire Deville in 1861. The type material is a mixture and not a mineral species.

Bauxite occurs in a variety of structures, textures, and colors. The color depends on the content of iron oxides and ranges from white to dark red or brown. Commonly bauxite is pisolitic and concretionary, but it may be fine-grained, dense, or vermicular (see illustration). Earthy bauxite is difficult to identify in the field and may require chemical, x-ray, petrographic, or differential thermal analysis. Portable differential thermal analysis units and quick chemical methods commonly are used in exploration.

Commercial bauxite. These deposits consist largely of gibbsite, $Al(OH)_3$, or boehmite, $AlO(OH)$, or mixtures of these minerals. The chief impurities are clay minerals and iron oxides, and bauxite may grade into clay or into ferruginous laterite. Some bauxite derived from basaltic rocks in India and Hawaii contains large amounts of titania. Clayey bauxite is beneficiated by washing to remove the finer-grained clay minerals (Binton Island, Palau Islands, and Brazil). *See* LATERITE.

C. S. Fox in 1927 suggested two main groups of bauxite deposits: laterite type and terra rossa type. All deposits are not easily placed in this classification, but bauxite formed more or less in place from crystalline rocks is of the first type, whereas the deposits that overlie limestone or dolomite commonly are referred to as the terra rossa type.

Bauxite from crystalline rocks generally is gibbsitic (Arkansas, Surinam, British Guiana, Brazil, French West Africa, Gold Coast, India, and Australia). The terra rossa bauxite usually contains appreciable amounts of boehmite (France, Italy, Yugoslavia, Greece, Hungary, Jamaica, Haiti, and Dominican Republic). Diaspore, $HAlO_2$, in some Mediterranean deposits is attributed to metamorphism and is not a product of weathering.

Production. The world production of bauxite has steadily increased. The domestic production amounts to about one-fourth of the total bauxite used. Imports are mainly from Jamaica and Surinam. Ferruginous bauxite has been explored in Oregon and Hawaii. [SAMUEL S. GOLDICH]

Bibliography: E. C. Fischer, *Annotated Bibliography of the Bauxite Deposits of the World*, USGS Bull. no. 999, 1955; C. S. Fox, *Bauxite and Aluminous Laterite*, 2d ed., 1932; S. S. Goldich and H. R. Bergquist, *Aluminous Lateritic Soil of the Republic of Haiti, W. I.*, USGS Bull. no. 954-C, 1948.

Bedding

A layering of rocks, most commonly sedimentary rocks, resulting from the original processes of deposition. (The term may be used with reference to igneous and metamorphic rocks, but is applied only to sedimentary rocks in this article.) Bedding is commonly used as a synonym for stratification; some researchers use stratification as the general term and restrict bedding to stratification in which the layers (strata) are thicker than 1 cm. Stratification into layers thinner than 1 cm is known as lamination.

Bedding generally results from a combination of two main factors: (1) Processes acting during the original deposition of the sediment grains (mineral or rock particles) arrange the grains in layers that differ somewhat in structure, texture, or composition. (2) Consolidation and chemical transformations, taking place after deposition, accentuate original differences and produce a tendency for rocks to split along the planes of bedding. Bedding was generally, though not always, horizontal when first formed (the "law" of original horizontality of strata was first formally enunciated by Nicholas Steno in 1669). When observed in rocks, however,

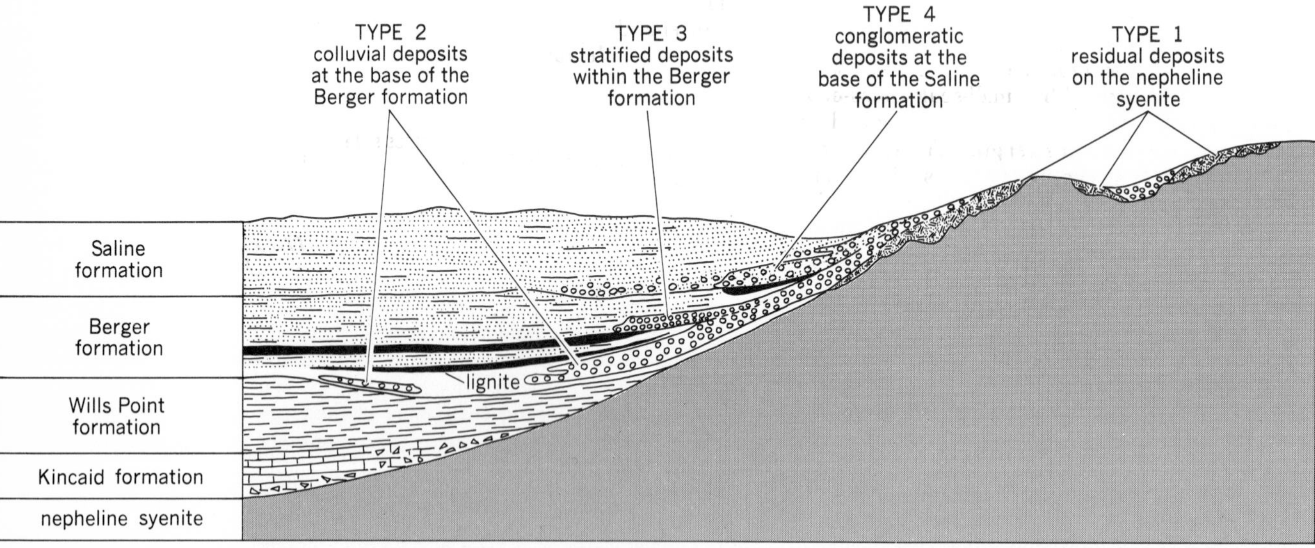

Diagrammatic section of the principal types of deposits in the Arkansas bauxite region. (*M. Gordon, J. I. Tracey, Jr., and M. W. Ellis, Geology of the Arkansas Bauxite Region, USGS Prof. Pap. no. 299, 1958*)

bedding is frequently inclined at an angle to the horizontal, due to subsequent tilting produced by folding or faulting of the originally horizontal strata. In some cases, bedding is even completely overturned. The original tops and bottoms of the beds may be determined from a careful study of bedding and the structures within beds. *See* SEDIMENTARY STRUCTURES; STRATIGRAPHY.

Not all parallel planar structures in rocks are termed bedding; layering produced by a thorough alteration of preexisting rocks by the action of heat and pressure (metamorphism) is called cleavage, schistosity, or gneissosity. Many rocks, whether or not they are bedded, also tend to break along several systems of parallel planes (called joints). In rocks that are bedded, one system of joints is generally parallel to the bedding (bedding plane joints) and the other systems are almost at right angles to the bedding. The tendency of bedded rocks to split and weather along bedding plane joints makes bedding a prominent structure on most outcrops of stratified rocks. *See* CLEAVAGE, ROCK; GNEISS; JOINT (GEOLOGY); SCHISTOSITY, ROCK.

Graded bedding. In some cases, bedding results from abrupt changes in the conditions of sedimentation. For example, silt may be carried over the bank of a river during floods and deposited as a layer several centimeters thick on the alluvial plain, with very little sediment accumulating between floods. Sand may be carried out from beaches and deposited in relatively deep water on the continental shelf during rare intense storms, such as hurricanes; between storms, only fine-grained muds accumulate. In the deep sea, fine clays or organic oozes accumulate at very slow rates (a few millimeters per 1000 years), but from time to time a layer of sand may suddenly be deposited by the action of a dense, sediment-laden current (turbidity current). In all the examples above (particularly the last), the deposited bed of sand tends to show a progressive grading, with the coarsest sediment near the base and the finest at the top; this structure is known as graded bedding (Fig. 1). Because graded beds are deposited from decelerating surges of water laden with suspended sediment, they tend to have sharp, erosional bases, and a regular sequence of internal structures (no internal bedding near the base, passing upward into plane lamination produced by traction of sand over a flat bed, followed by small-scale cross-bedding produced by migration of ripples formed as the current decreases in strength). The structures originally formed on the underlying sediment are preserved as casts (sole marks) on the base of the graded bed. Common types of sole marks are organic markings (burrows, trails), small scours elongated parallel to the current (flutes), and grooves cut by "tools" such as large fragments of mud dragged over the bottom by the current. *See* SEDIMENTATION (GEOLOGY); TURBIDITY CURRENT.

Cross-bedding. Not all bedding results from sudden changes in the type or rate of sedimentation. Migration of sand waves, produced by the action of waves or currents, takes place by deposition on the downstream (lee) side of the wave and by erosion on the upstream (stoss) side. If there is net accumulation of sediment, migration produces a series of composite beds (called sets), each of which is separated from the sets above and below by a prominent structural and textural discontinuity (bedding plane) that was originally nearly horizontal. Within each set, however, the bedding is inclined at the angle of the original lee side of the migrating sand wave. Such bedding is called cross-bedding. The scale (thickness) of cross-bedded sets varies from about 1 cm to more than 20 m. Small-scale cross-bedding (set thickness less than 5 cm) generally results from the migration of small sand ripples. Medium-scale cross-bedding (5 cm to

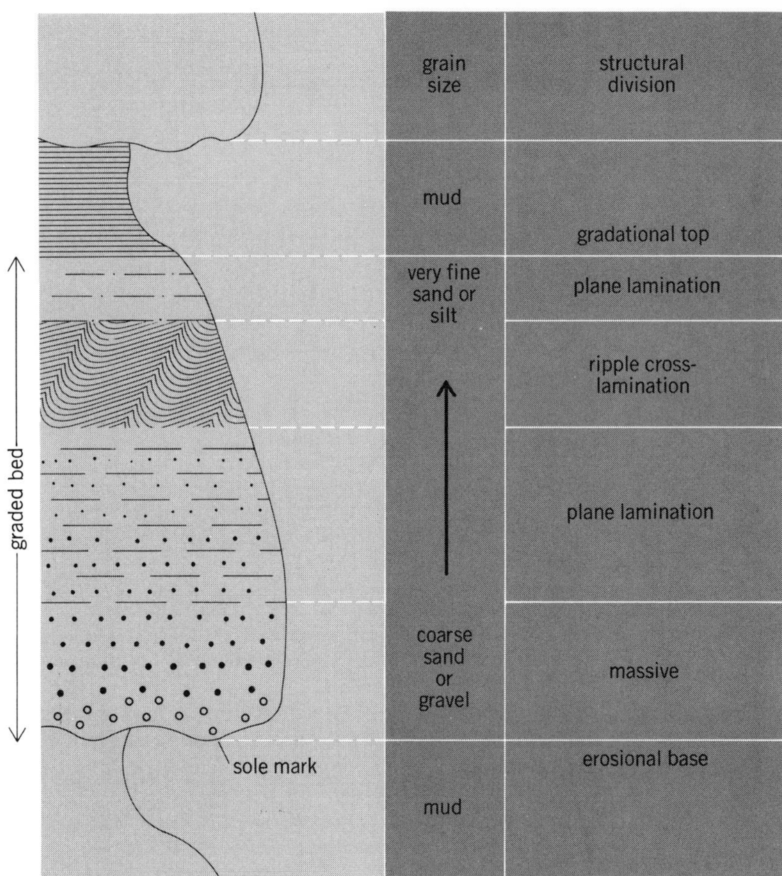

Fig. 1. A typical graded bed, showing erosional-base (with sole marks) size decline from gravel at base to fine sand at top, and a sequence of four main structural divisions within the bed.

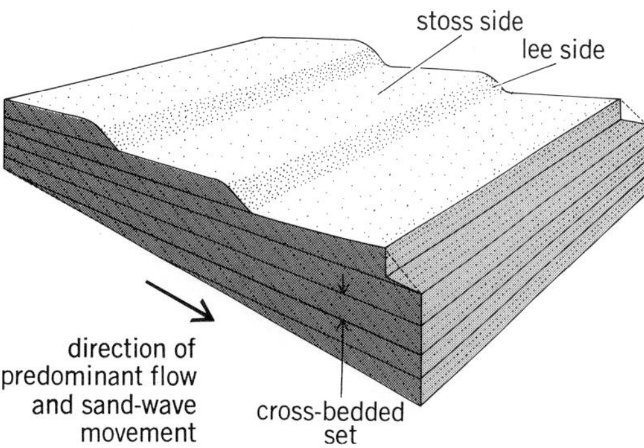

Fig. 2. Block diagram showing the formation of tabular sets of cross-bedding by migration of regular, straight-crested sand waves. (*From H. E. Reineck and I. B. Singh, Depositional Sedimentary Environments, Springer-Verlag, p. 30, 1973*)

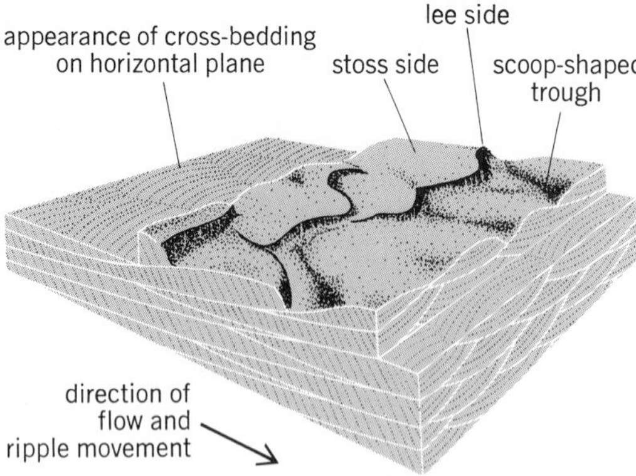

Fig. 3. Block diagram showing the formation of trough-shaped sets of cross-bedding by the migration of sinuous (linguoid) sand waves (ripples). Note the different appearance of both set boundaries and cross-beds in sections parallel and transverse to the flow. (*From H. E. Reineck and I. B. Singh, Depositional Sedimentary Environments, Springer-Verlag, p. 33, 1973*)

1 m) results from the migration of subaqueous sand dunes (also called megaripples) which are generally 0.2 to 2 m high and 1 to 10 m long, and larger-scale cross-beds result from the migration of larger sand waves, particularly subaerial sand dunes, or from the growth of deltas.

The exact nature and geometry of the surfaces bounding sets, and of the cross-beds within sets, depend on the geometry of the sand waves and the rate of net accumulation of sediment as compared with the rate of migration of the sand waves. Most ripples and megaripples produced by currents in rivers and tidal streams migrate by erosion and transportation of sediment on the stoss side and deposition at the crest; deposition at the crest produces oversteepening, followed by avalanching of sand down the lee side of the ripple. The ripple (or megaripple) is therefore strongly asymmetrical, with the lee slope (and therefore the angle of the cross-beds) at the angle of repose of sand (about 35°). Ripples produced by waves, however, may be either symmetrical or asymmetrical, with lee slopes less than the angle of repose. Migration of ripples with relatively straight, regular crests and troughs produces cross-bed sets with regular, tabular shapes, whereas migration of ripples with strongly curved crests and irregular scoop-shaped troughs produces trough-shaped sets (Figs. 2 and 3). Both tabular and trough cross-bedding are common.

The shape of the cross-bedding may be used to indicate the original tops and bottoms of beds. Most cross-beds approach the lower, near-horizontal set boundary asymptotically but are truncated at an acute angle by the upper set boundary. The original direction of dip of cross-beds may also be used to reconstruct the direction of the predominant sediment transport (paleocurrents).

Cross-bedding produced by the growth of deltas can be described as consisting of three types of beds: the bottom-set beds, deposited on the near-horizontal bottom some distance away from the delta; the inclined fore-set beds, deposited at the delta front; and the horizontal top-set beds, deposited on the upper surface of the delta. Cross-bedding produced by the growth of very small deltas (also variously called transverse bars, splays, or washover fans, depending on morphology and physiographic setting) may be very similar to that produced by migration of sand waves. In fact, before echo-sounding techniques revealed the size and abundance of megaripples in rivers and tidal environments, most cross-bedding was thought to result from growth of small deltas. In large deltas, however, the angle of the fore-set beds is only a few degrees and the structure of the delta is complex, so that large, simple, easily recognizable cross-beds are not produced. *See* DELTA.

Sedimentary rhythms. In sediment that is deposited slowly, bedding may result from progressive changes in the nature of sedimentation. Such changes may be related to the changing seasons, and give rise to yearly repeated beds, called varves. The most common types of varves are formed in lakes fed directly by the meltwaters of valley glaciers or ice sheets. In the spring, meltwaters carry much suspended sediment into the lakes, in some cases forming turbidity currents which spread out over the bottom of the lake. Sediment deposited during the spring and summer is relatively coarse and light in color. In the winter, little mud is carried into the lake, and suspended fine mud and organic matter slowly settle to the bottom to form the darker, finer-grained upper part of the varve. Varves may also be formed in nonglacial lakes by seasonal turnover of the waters and

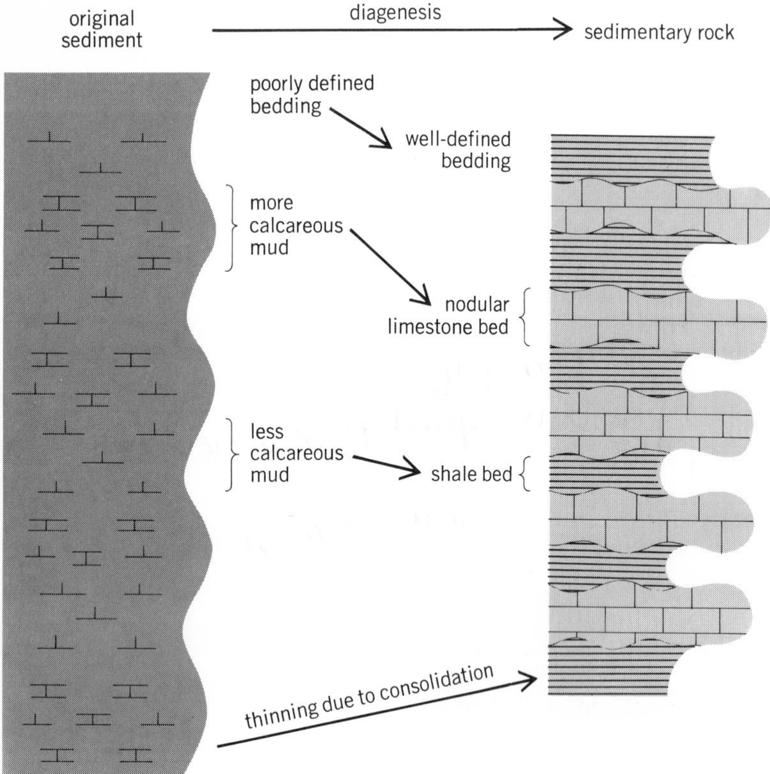

Fig. 4. The accentuation of bedding in an original calcareous mud produced by the chemical dissolution and reprecipitation of $CaCO_3$ (calcite) during the diagenetic change to a sedimentary rock.

the resulting regular changes in chemical conditions and biological productivity. Bedding may also result from changes in sedimentation related to much longer climatic rhythms, which may in turn be related to astronomical cycles such as the precession of the equinoxes. *See* VARVE.

Diagenesis. Slight changes in the mineralogy of sediments, produced by slight changes in sedimentation conditions, may be greatly accentuated by slow chemical and mineralogical changes taking place in the sediment after deposition (diagenesis; Fig. 4). Examples are regular alternations of beds of limestone and shale or of chert and shale. The original sediment probably consisted of a mixture of terrigenous mud and organically produced $CaCO_3$ (calcite and aragonite) or SiO_2 (opal). During diagenesis, interaction of mineral grains and pore waters leads to solution and reprecipitation of $CaCO_3$ (calcite) or SiO_2 (microcrystalline quartz) to produce the beds of limestone or chert. Such beds may vary in thickness or may break up laterally into a series of discrete nodules. *See* DIAGENESIS.

Bioturbation. In sedimentary environments populated by abundant bottom fauna or flora, the action of burrowing organisms or plant roots (bioturbation) may have the reverse effect: a partial to near-complete destruction of bedding. Rocks which lack bedding, or show only faint traces of it, are called massive.

Some sedimentary rocks appear massive only because the slight textural or mineralogical changes that indicate bedding have not been revealed by weathering or are masked, rather than accentuated, by later diagenetic changes such as formation of a mineral cement. In such cases the original bedding or cross-bedding can frequently be revealed by chemical etching or x-radiography. *See* SEDIMENTARY ROCKS.

[GERARD V. MIDDLETON]

Bibliography: H. Blatt, G. Middleton, and R. Murray, *Origin of Sedimentary Rocks*, 1972; F. J. Pettijohn, *Sedimentary Rocks*, 3d ed., 1975; H. E. Reineck and I. B. Singh, *Depositional Sedimentary Environments*, 1973.

Benioff zone

A zone of earthquake hypocenters distributed in well-defined planes that dip at an angle of 45° from a shallow depth into the Earth's mantle to depths as great as 700 km. Benioff zones are found in close association with deep-sea trenches, island arcs, young mountains, and volcanoes. Geographically, Benioff zones are well developed in Tonga-Kermadec, Izu-Bonin, the Marianas, Japan, the Kuriles, Peru-Chile, the Philippines, Sunda, the New Hebrides, New Zealand, Middle America, Ryukyu, and the Aleutians. In the framework of plate tectonics, the Benioff zone is considered to represent seismic activity associated with subducted plates which are plunging into the mantle beneath an overriding plate. *See* PLATE TECTONICS.

History. The discovery of deep-focus earthquakes dates back to 1922. Later works revealed the characteristic dipping geometry of intermediate and deep seismic zones at many island arcs. In 1954, H. Benioff interpreted the dipping seismic

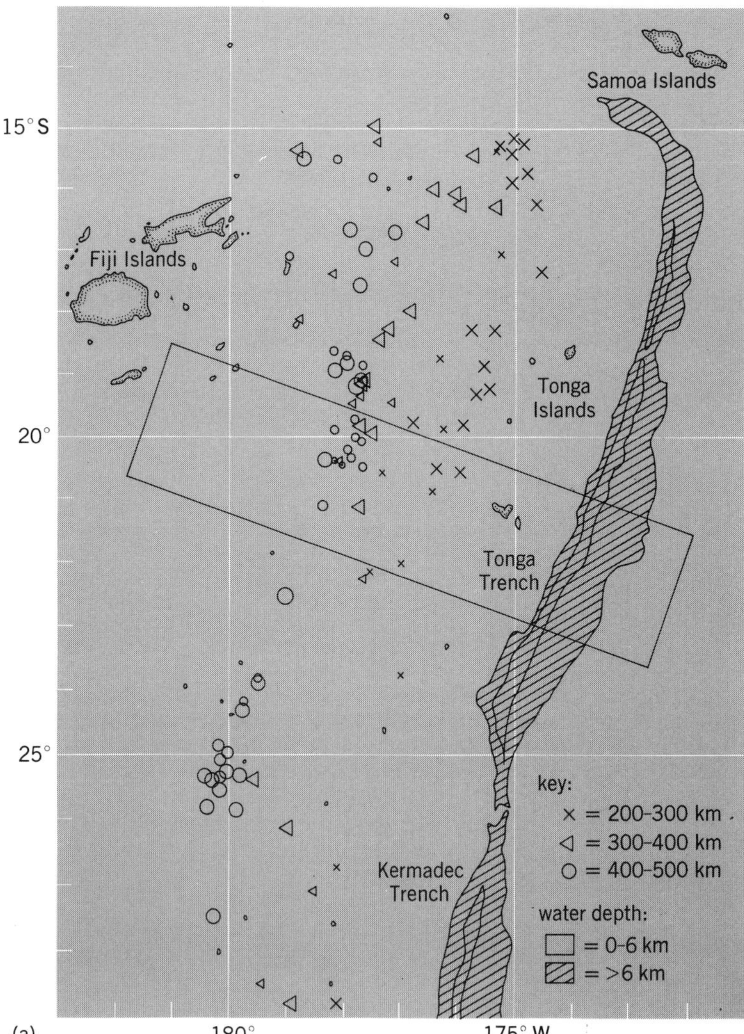

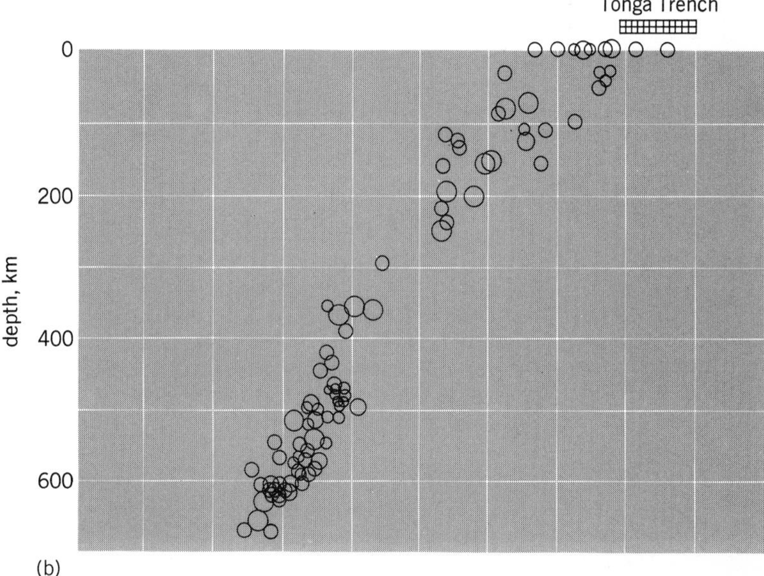

Fig. 1. Structure of a Benioff zone in which the distribution of earthquakes is nearly continuous. (*a*) Earthquakes in the Tonga-Kermadec arc with focal depths ranging from 200 to 500 km. (*b*) Vertical section of the rectangular region shown in *a*. (*From L. R. Sykes, The seismicity and deep structure of island arcs, J. Geophys. Res., 71:2981–3006, copyright 1966 by American Geophysical Union*)

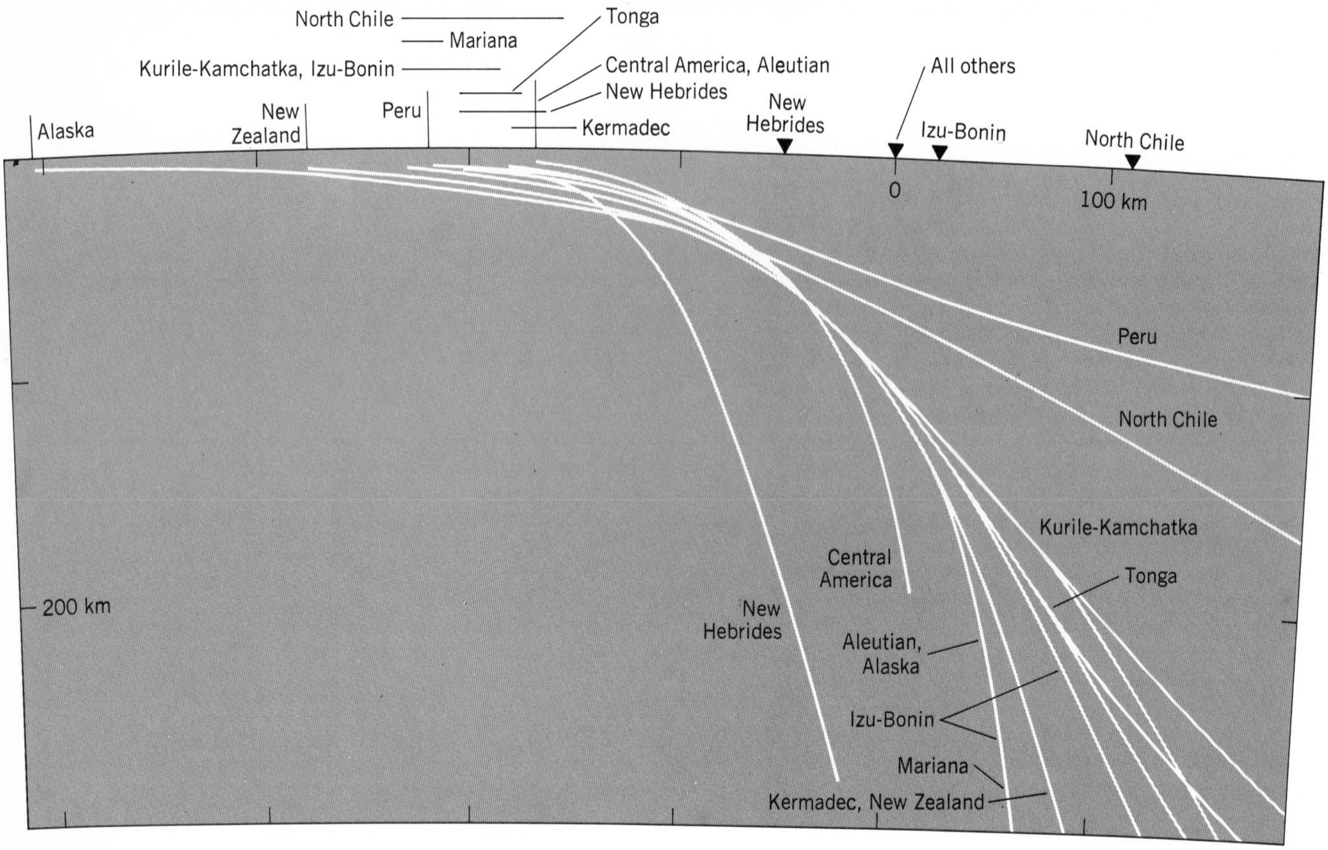

Fig. 2. Comparison of Benioff zones in various regions. The sections are offset in certain cases with respect to the volcanic lines (solid triangles) to emphasize similarities and differences in geometry. The central and southern Chile segments are essentially identical to the Peru and northern Chile segments, respectively, and are not shown. (*From B. L. Isacks and M. Barazangi, in M. Talwani and W. C. Pitman, III, eds., Island Arcs, Deep Sea Trenches and Back-Arc Basins, pp. 99–114, copyright 1977 by the American Geophysical Union*)

zones in the Pacific as a plane where the oceanic block is being thrust under the overlying continental block. This interpretation was extended to a more systematic concept of plate subduction in the later development of plate tectonics, and the dipping seismic zones became known as Benioff zones.

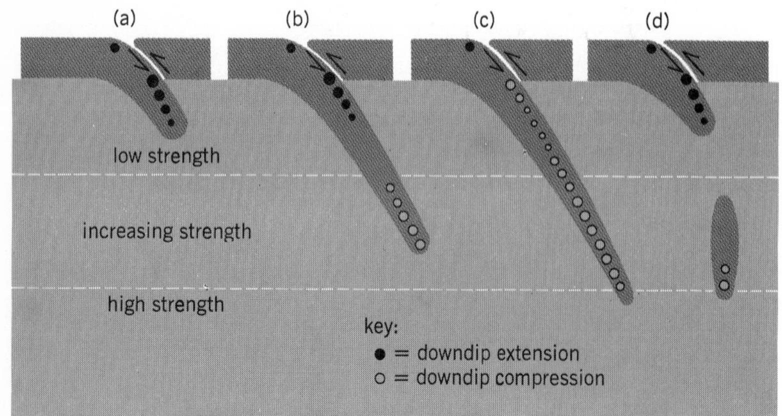

Fig. 3. Possible distribution of stresses in downgoing slabs of lithosphere that sink into the asthenosphere. (*a*) Extensional stress. (*b*) Compressional stress near leading edge; extensional stress in shallow part. (*c*) Compressional stress. (*d*) Broken piece of lithosphere sinks and comes under compression. (*From B. Isacks and P. Molnar, Mantle earthquake mechanisms and the sinking of the lithosphere, Nature, 223: 1121–1124, 1969*)

Structure. One of the best examples of the Benioff zone is found in the Tonga-Kermadec arc (Fig. 1). In this region, the Benioff zone dips about 45° and extends to a depth of 700 km. The distribution of earthquakes is nearly continuous. The dip angle, shape, and downdip length of the Benioff zone vary regionally as shown in Fig. 2. As the accuracy of establishing earthquake locations improved, it became clear that the thickness of the seismic zone is as small as 20 km. Studies of seismic-wave propagation through Benioff zones have revealed that the seismic-wave velocity is higher (typically by 10 to 15%) and the absorption of seismic waves is lower (high Q) in the mantle along Benioff zones than in the surrounding mantle at a comparable depth. This suggests that the Benioff zone delineates downgoing oceanic lithosphere (sometimes called a slab) which has not been heated to the ambient temperature in the mantle; seismic-wave velocity and Q in rocks are higher at lower temperatures, other conditions being equal. Recent studies of the structure of island arcs indicate that most earthquakes in Benioff zones occur inside the downgoing slab rather than along its boundary. A recent analysis with very high resolution revealed a double Benioff zone in northern Japan, with one zone located along the upper boundary of the slab and the other inside it. The distribution of earthquakes is not always continuous along the entire length of the

Benioff zone. At some subduction zones (for example, Chile, Sunda, the New Hebrides, Izu-Bonin, and the Kuriles), the spatial distribution of hypocenters is interrupted by aseismic gaps. *See* SEISMOLOGY.

Earthquake mechanism. Studies of the mechanism of earthquakes in Benioff zones indicate that either the pressure axes (often called the maximum compression axes) or the tension axes (often called the minimum compression axes) of the earthquakes are often aligned in the downdip direction of the slab. This observation suggests that these earthquakes occur within downgoing slabs in response to stresses built up in the slab during subduction. When the gravitational downward pull in the slab dominates, events with downdip extension occur, and when the resistance of the surrounding mantle against subduction is large, earthquakes with downdip compression occur.

The downdip stress types vary in a systematic manner from region to region, depending on the continuity of the distribution of earthquakes along the Benioff zone and its total downdip length. The schematic model shown in Fig. 3 explains this variation. When the downgoing slab extends only to a relatively shallow depth where the material strength is low, the stress in the slab is extensional because of its own weight (Fig. 3*a*). As the slab penetrates into a region of increasing strength, the stress near the leading edge of the slab becomes compressional, while the stress in the shallow part of the lithosphere remains extensional (Fig. 3*b*). When the slab reaches the rigid "bottom," the stress in the entire slab becomes compressional (Fig. 3*c*). In some cases, a piece of lithosphere breaks off, sinks independently, hits bottom, and comes under compression (Fig. 3*d*). *See* EARTHQUAKE. [HIROO KANAMORI]

Bibliography: H. Benioff, *Bull. Geol. Soc. Amer.*, 65:385–400, 1954; B. Isacks and P. Molnar, *Rev. Geophys. Space Phys.*, 9:103–174, 1971; B. Isacks, J. Oliver, and L. R. Sykes, *J. Geophys. Res.*, 73: 5855–5899, 1968; M. Talwani and W. C. Pitman, III (eds.), *Island Arcs, Deep Sea Trenches and Back-Arc Basins*, 1977.

Bentonite

The term first applied to a particular, highly colloidal plastic clay found near Fort Benton in the Cretaceous beds of Wyoming. This clay swells to several times its original volume when placed in water and forms thixotropic gels when small amounts are added to water. Later investigations showed that this clay was formed by the alteration of volcanic ash in place; thus, the term bentonite was redefined by geologists to limit it to highly colloidal and plastic clay materials composed largely of montmorillonite clay minerals, and produced by the alteration of volcanic ash in place. Many mineralogists and geologists now use the term bentonite without reference to the physical properties of the clay. On the other hand, the term has been used commercially for any plastic, colloidal, and swelling clays without reference to a particular mode of origin. *See* CLAY; MONTMORILLONITE.

Uses. Bentonites are of great commercial value. They are used in decolorizing oils, in bonding molding sands, in the manufacture of catalysts, in the preparation of oil well drilling muds, and in numerous other relatively minor ways. Since the properties of bentonites vary widely, they are not all suitable for commercial use. The properties of a particular bentonite determine its economic use. For example, the bentonite found in Wyoming is excellent for drilling muds and for foundry use, but it is not useful for the making of catalysts or for oil decolorizing.

Origin. The occurrence of shard structures of ash as pseudomorphs in the clay often indicates the volcanic ash parent material of bentonites. The presence of nonclay minerals characteristic of igneous material also provides evidence for the origin from ash.

Evidence strongly indicates that the transformation of volcanic glass to montmorillonite takes place either during or soon after accumulation of the ash. The alteration process is essentially a devitrification of the natural glass of the ash and the crystallization of montmorillonite. The ash often contains excess silica which may remain in the bentonite as cristobalite. *See* VOLCANIC GLASS.

Occurrence. Bentonites have been found in almost all countries and in rocks of a wide variety of ages. They appear to be most abundant in rocks of Cretaceous age and younger. In such material in older rocks the montmorillonite is often almost completely collapsed by compaction and metamorphism. This altered bentonite swells very little and does not have the usual high colloidal properties of bentonite; it is sometimes called metabentonite.

Beds of bentonite show a variable thickness and can range from a fraction of an inch up to as much as 50 ft thick. The color of bentonite is also variable, but perhaps most often it is yellow or yellowish-green. When observed in the weathered outcrop, bentonite tends to develop a characteristic cellular structure. It shrinks upon drying and often shows a characteristic jigsaw-puzzle set of fractures.

Not all beds of ash in the geologic column have altered to bentonite. When present, bentonite is most frequently found in marine beds; thus, it seems certain that alteration from ash to bentonite is favored in sea water.

In the United States, bentonites are mined extensively in Wyoming, Arizona, and Mississippi. England, Germany, Yugoslavia, the Soviet Union, Algeria, Japan, and Argentina also produce large tonnages of bentonite.

[RALPH E. GRIM; FLOYD M. WAHL]

Beryl

An aluminum beryllosilicate mineral, $Al_2[Be_3Si_6O_{18}]$, crystallizing in the hexagonal system with space group *P6/mcc*. The hardness is 8 on Mohs scale, cleavage distinct on {0001}, specific gravity 2.8, luster vitreous, and color bluish-green, light yellow, and less frequently green, pink, golden yellow, or colorless. It is easily confused with apatite. The crystal structure includes oxygen atoms octahedrally coordinated to aluminum and tetrahedrally coordinated to beryllium and silicon. The rare scandium analog, $Sc_2[Be_3Si_6O_{18}]$, is called bazzite. The Si—O tetrahedrons link to form six-membered rings by corner-sharing; further corner-sharing with Be—O

Beryl. (*a*) Specimen from Bikita, Southern Rhodesia (*American Museum of Natural History specimens*). (*b*) Crystal habits (*from C. S. Hurlbut, Jr., Dana's Manual of Mineralogy, 17th ed., copyright © 1959 by John Wiley & Sons, Inc.; reprinted by permission*)

tetrahedrons results in a three-dimensional honeycomb structure. Open channels run parallel to the hexad axis, admitting limited alkali ions, such as Li^+, Na^+, K^+, Rb^+, and Cs^+, and water molecules, and these may affect the color and crystal habit of the mineral (see illustration). The alkalies, symbolized as X^+, are believed to substitute according to the equation $3X^+ + 1Al^{3+} \rightleftharpoons 3Be^{2+}$.

Beryl and its gem varieties aquamarine (pale blue), goshenite (colorless), and morganite (pink) most frequently occur in granite pegmatite druses along with quartz, feldspar, lepidolite, topaz, and tourmaline, where the beryllium ions have concentrated in residual pegmatitic fluids. Beryl crystals up to many tons in weight have been recovered and the mineral constitutes an important ore of beryllium. Beryl occurs more rarely in mica schists and bituminous limestones (sometimes as emerald, the green gem variety of great value) and nepheline-syenites. The mineral sometimes is altered to kaolinite and muscovite; hydrothermal action may result in the formation of secondary beryllium minerals bertrandite and phenacite.

Beryl has been synthesized from a mixture of $BeCO_3 + Al_2O_3 + SiO_2$ over the pressure intervals of 400 to 1500 bars and at temperatures of 600°C, and it may be stable at much higher pressures and temperatures.

Beryl is frequent in minor amounts in granite pegmatites throughout the world. Important occurrences include certain pegmatites in New England and California and pegmatites in Brazil, Madagascar, and Siberia. Emeralds occur in certain mica schists in the Urals and in bituminous limestones in Colombia. *See* EMERALD.

[PAUL B. MOORE]

Bioherm

An organic reef composed of fragmental bioclastics, the major percentage of which are of marine origin and are not found in growth position. The concentration of these bioclastics is due to intense localized organic activity or to local aggregation by waves and currents, forming rigid mounds or submarine bars which are wave-resistant as a result of cementation in place by living calcereous algae and physicochemical precipitation of calcite. Their prime characteristics are that the structure is unstratified or poorly stratified, composed of bioclastics, and thicker than surrounding time-equivalent nonbioclastic or biostromal strata. Bioherms may be of any size and shape and in any stratigraphic position from Cambrian to Recent, but they are usually associated with topographic irregularities or the break in slope between the shelf and basin within the light zone of ancient seas (see illustration). *See* BIOSTROME; REEF.

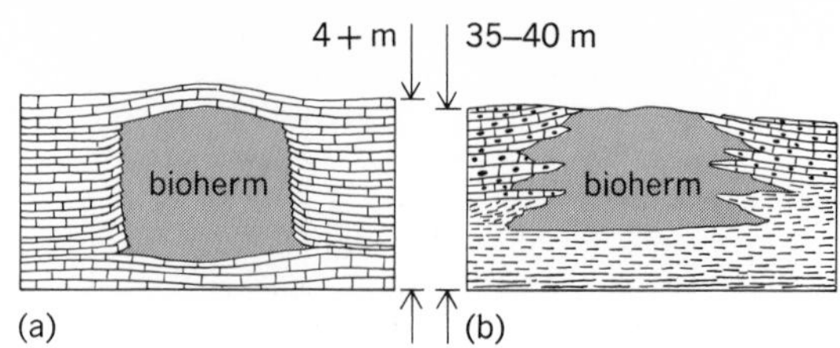

Bioherms. (*a*) A small stumplike mass of cavernous dolostone in midst of thin-bedded dolomitic limestone of Middle Silurian age near Lomira, Wis. There has been prominent steepening of initial dip around periphery of structure because of differential compaction in surrounding calcareous muds. (*b*) Coelenterate bioherm lying in midst of shale and thin-bedded cherty limestone of Middle Silurian age at Wabash, Ind. (*From R. R. Shrock, Sequence in Layered Rocks, McGraw-Hill, 1948*)

The term bioherm is applied by some geologists (E. R. Cummings and R. R. Shrock, 1928) to any domelike, moundlike, lenslike, or other circumscribed mass built exclusively or mainly by sedentary organisms and enclosed in a normal rock of different lithologic character, but it seems preferable to restrict its usage to masses of fragmented debris which were moved out of growth positions before the broken fragments were cemented together. This in fact seems to be the case for the greater part of rock bodies to which the term has been applied.

[SHERMAN A. WENGERD]

Bibliography: P. E. Cloud, Jr., Facies relations of organic reefs, *Bull. Amer. Ass. Petrol. Geol.*, 36:2125–2149, 1952; R. R. Compton, *Manual of Field Geology*, 1962; E. R. Cummings and R. R. Shrock, Niagaran coral reefs of Indiana and adjacent states and their stratigraphic relations, *Bull. Geol. Soc. Amer.*, 3:579–620, 1928; T. A. Link, Theory of transgressive and regressive reef (bioherm) development and origin of oil, *Bull. Amer. Ass. Petrol. Geol.*, 34:263–294, 1950.

Biosphere, geochemistry of

The term biosphere indicates the totality of organisms which are alive on Earth at a given time. Biogeochemistry is the description and the reconstruction of the genetical history of the biosphere. This article first discusses briefly the problem of the primitive genesis of the biosphere, and the possible evolutionary changes in the chemistry of the biosphere throughout the geological ages since the emergence of the protoorganism. Then the chemical properties and other characteristics of the biosphere at the present time are described. The article terminates with a brief description of the processes of degradation of deeply buried biological debris.

Chemical interaction exists between the atmosphere, hydrosphere, lithosphere, and biosphere. The biosphere is particularly closely interconnected with nonliving organic matter within the Earth's crust, which may be referred to by the term carbosphere. The carbosphere is the subject matter of organic geochemistry. Most of the interactions between the biosphere and its surroundings are, as a first approximation, of a reversible-fluxing character. The chemical elements enter the organism, are later ejected from the biosphere through the products of metabolism and death, and are thereafter incorporated into another organism, and so forth. These processes are termed the biogeochemical cycles. A relatively smaller portion of organic matter within certain localities, however, becomes so deeply buried within the Earth's crust that it would not be available for further biological recycling.

The importance of living organisms in geochemical processes was first appreciated by J. B. Lamarck, J. Liebig, E. Suess, and other investigators of the 19th century, but it was not until the beginning of the 20th century that biogeochemical data were systematically accumulated into a coherent body of information. The first comprehensive studies were undertaken by V. I. Verdansky (1863–1945) of the Soviet Union, who is regarded as the modern founder of the subject. Other leaders in the field have been A. P. Vinogradov (Soviet Union), G. Bertrand and D. Bertrand (France), and F. W. Clarke and G. E. Hutchinson (United States).

Genesis of biosphere. The spectral data relating to stars, interstellar space, and smaller celestial bodies indicate that simple gases such as CH_4, CO, CO_2, NH_3, and H_2O are abundant throughout the cosmos. Numerous experiments clearly demonstrate that the above mentioned gases can condense to complex organic molecules, including amino acids and so on, on irradiation with ionizing rays or on being subjected to electrical discharges; these conditions are presumed to occur in the course of the condensation of a given celestial body. Therefore, prebiogenic carbonaceous complexes would be expected to have preceded the emergence of the organism on any condensing celestial body, and the cosmochemical and geological evidences for such carbonaceous complexes are briefly outlined in the following. *See* LIFE, ORIGIN OF.

There are some 30 carbonaceous meteorites, which contain 0.3–4.0% carbon, corresponding to approximately twice as much organic complex. According to the generally accepted theory, the meteorites originate from asteroid-sized celestial bodies during the course of condensation, of which the conditions have been adequate for the genesis of a prebiogenical carbonaceous complex, but too variable and most likely too dry for the emergence of the organism. The abiogenic nature of the carbonaceous complex is partially supported by the following evidence: (1) There is absence or virtual absence of optical rotation, which is characteristic to all biological products which contain asymmetrical molecules; (2) paraffins and other hydrocarbons do not show a marked trend toward dominance of carbon chains of odd numbers, as is the case in most biogenic products; (3) C_{12}/C_{13} ratio is well within the range of juvenile carbons; and (4) the carbonaceous complex contains up to 5% organic chlorine, and chlorinated compounds are extremely rare within the organism of the terrestrial type of biochemistry.

Of the terrestrial geological settings, there are several candidates for organic substances which may have been preserved since times prior to the emergence of life, or they are condensation products from the interior of the Earth. Thus, small quantities of amino acids have been detected in impermeable chert from the 3,200,000,000-year-old Fig-tree formation of Transvaal, South Africa. It is possible that these sediments deposited prior to the genesis of life. Bitumenlike substances with a high percentage of organic acids, and ashes rich in uranium and thorium, termed thucholites, are found in numerous pegmatite veins, which traverse exclusively igneous rocks. It appears, therefore, that the hot vapors or solutions which produced such veins must have transported the organic matter from the deeper zones of the Earth. Distillation from surrounding organic sediments is unlikely, because they were not available within these areas. The same considerations may apply to the small quantities (1–2 ppm) of amino acids, which have been detected within some active fumaroles from the Caribbean, Hawaii, and Iceland.

The above-mentioned evidence does not establish the presence of prebiogenic carbonaceous complex either on the parent bodies of meteorites or on the Earth, but all the same are strongly indicative of its existence.

Evolution of biosphere. Little is known about possible chemical changes which may have occurred in the course of evolution from the protoorganism to the present-day biosphere. At the early Precambrian the presence of exclusively heterotrophic organisms in a reducing atmosphere is postulated. The presence of some three valent iron in the form of magnetite ($FeO \cdot Fe_2O_3$), however, does not seem to corroborate the hypothesis of a markedly reducing atmosphere even at that early age. The evolution of photosynthetizing organisms in the sea, possibly in later Precambrian, may have quite fundamentally changed the cycles of the diverse elements, resulting in the liberation of considerable quantities of oxygen into the atmosphere. This trend may have received yet another impetus on the spreading of photosynthetizing plants to the land surface during the Paleozoic and particularly during the Carboniferous age. It is likely that since the Paleozoic, the chemical composition and the geochemical cycles of the biosphere did not change in any fundamental manner.

Living matter. Living matter occurs in the environment wherever water exists in the liquid state in conjunction with a source of energy for metabolism. Some extreme conditions under which certain species maintain permanent populations are listed in Table 1.

The fundamental distinction between organisms with regard to geochemical activities and importance results in two classes: (1) those organisms that contain chlorophyll pigments and utilize the energy of visible light for the synthesis of energy-rich organic compounds (pigmented algae, photosynthetic bacteria, and higher plants); and (2) those lacking the photosynthetic apparatus (viruses, nonphotosynthetic bacteria, fungi, and animals). Those without chlorophyll derive their

Table 1. Conditions under which living organisms are known to reproduce in nature

Factor	Minimum	Maximum
Temperature	−7°C (*Pyramidomonas* in saline lakes)	95°C (blue-green algae in hot springs)
Hydrostatic pressure	Not determined	1070 atm (deep-sea forms)
pH	1.7 (the diatom *Pinnularia*)	12.0 (blue-green algae in Kenya lakes)
Oxidation potential, E_h	−0.35 volt (marine sedimentary bacteria)	0.83 volt (aerobic organisms)
Salinity	Very low as in rainwater	220 parts per thousand (halophytic bacteria and *Artemia salina*)

energy for metabolism from the respiratory oxidation of food materials in the form of organic and in some cases inorganic compounds.

In separate parts of the biosphere, fairly distinct communities of organisms can be recognized, for example, the plankton communities of natural waters, coral reefs, sphagnum bogs, spruce-fir forests, and steppes. Because there is an exchange of energy and matter between the living community and the nonliving environment, the complex is often studied as a unit, called an ecosystem. The physical dimensions of an ecosystem may range from a puddle of water to a maximum which is the entire biosphere. The passage of energy through an ecosystem can be conceived as proceeding through a series of trophic (feeding) levels. These trophic levels are recognized: photosynthetic plants, herbivorous animals and microorganisms of decay, primary carnivores, and higher carnivorous levels. Under steady-state conditions, the rate of energy production by one trophic level must necessarily exceed the rate of energy utilization by a succeeding trophic level. This is in accordance with the second law of thermodynamics as applied to steady-state systems. It follows that, for the biosphere as a whole, green plants tend to be of greater geochemical importance than consumer organisms, in terms of both biomass and metabolism. However, photosynthetic plants are unable to maintain permanent populations at depths greater than 100 m in the hydrosphere or greater than a few meters (in the case of plant roots) in soils, because of the lack of light. In such environments, all biochemical activity is the result of animals and nonphotosynthetic microorganisms. This situation exists in the more peripheral parts of the biosphere.

The linear dimensions attained by living organisms lie between the limits of 10^{-6} cm in the case of viruses and 10^{4} cm for sequoia trees. Within a given taxonomic group, the rate of respiration per unit weight of tissue tends to increase as the two-thirds power of the weight. There is, however, much variation from one taxonomic group to another, and many exceptions to this general rule are known. The overall result is that, per unit weight, smaller individuals tend to respire and also to increase in mass at faster rates than do larger individuals. Thus, under maximal growth conditions a virus population will double its weight in a matter of minutes, bacteria in less than 1 hr, small animals in days or weeks, and larger forms in months or years. As a consequence, small individuals tend to be more active geochemically than their biomass might suggest.

Although the total mass of living matter in the biosphere has never been accurately estimated, its probable value is $n \times 10^{17}$ g live weight (where n is any small integer), which corresponds to $20n$ mg per cm^2 of Earth surface. The amount of carbon dioxide annually fixed in photosynthesis in the entire biosphere is probably equivalent to 7×10^{16} g of carbon, with the true value not less than one-half nor greater than twice this amount. This rate approximately corresponds to a 0.1% efficiency in the utilization of the visible light energy that actually reaches the Earth's surface. Photosynthesis by marine plants accounts for at least two-thirds of the total photosynthesis in the biosphere. The greatest amount of photosynthesis on land occurs in forests, followed in decreasing order by cultivated land, steppes, and deserts.

Influence of living matter. As a result of enzymatic action, many chemical reactions occur at low temperatures within the bodies of living organisms that would otherwise proceed only at slow rates in the biosphere. Prime examples are photosynthesis and the formation of nitrates from molecular nitrogen by the combined action of nitrogen-fixing and nitrifying microorganisms. In spite of the small mass of living matter in the biosphere relative to that of the atmosphere, hydrosphere, and lithosphere, living organisms do have an appreciable geochemical influence. Terrestrial plants are known to regulate soil erosion rates, to transport certain elements from the lower soil layers to the surface, and to accelerate the decomposition of some minerals and rocks by the excretion of carbon dioxide and polyvalent organic compounds that act as ligands (chelating agents). In addition, many of the physical and chemical characteristics of soils and aquatic sediments are determined by the activities of microorganisms, such as pH, oxidation potential E_h, and texture.

The concentrations of certain elements in plants and animals may exceed those in the media of growth by factors up to several hundred thousand. One of the more marked cases is the accumulation of vanadium by some ascidians. The concentration of vanadium in some species of *Ascidia* may reach 0.01% of the live weight, whereas sea water contains 0.0000002% vanadium. The extensive deposition of foraminiferan and radiolarian oozes on the sea floor emphasizes the importance of these forms in the deposition of $CaCO_3$ and SiO_2, respectively. Evidence of the former activity or organisms in the biosphere is provided by acaustobioliths (inorganic bioherms) such as reef limestones, and by caustobioliths (organic accumulations) such as oil shales, and deposits of coal and petroleum. *See* COAL; MARINE SEDIMENTS; PETROLEUM, ORIGIN OF.

Considerable indirect geochemical evidence, as well as more direct indications from the occurrence of fossil microorganisms in Precambrian deposits, indicates that the biosphere has been in existence for some 2,000,000,000 years, and possibly for twice that length of time. As a result of evolutionary processes, both the numbers and kinds of species have been changing from time to time. New environments have been occupied and even biologically created in the process. The quantitative importance of living organisms in the biosphere is quickly appreciated when it is realized that a weight of water equivalent to that present in all the oceans (1.4×10^{24} g) is decomposed by photosynthetic plants once every 10,000,000 years. The supply of oxygen in the Earth's atmosphere, a characteristic that serves to differentiate the Earth from all other planets, with the possible exception of Mars, has been derived mostly from photosynthesis resulting from a lack of balance in the biospheric carbon cycle. The total amount of carbon photosynthetically incorporated into green plants during the history of Earth is approximately 10^{26} g, an amount which is equivalent to one-fiftieth of the weight of the globe. It is estimated that about 10^{21} g of this carbon still persists as fossil organic compounds in coal, petroleum, and organic shales.

Following the appearance of *Homo sapiens* in the biosphere, both direct and indirect geochemical changes have arisen as a result of land clearing, water conservation practices, road construction, combustion of fossil fuels, mining operations, and extinction and redistribution of species. The term anthroposphere (or noosphere) has been used to designate the modern biosphere in which human changes play a significant part. As a result of atomic fission processes, both in weapons testing and in nuclear reactors, new biogeochemical problems have arisen. The radioactive nuclides present in atomic fallout and in low-level reactor wastes are accumulated by organisms; the nuclide-enrichment factors sometimes amount to as much as 100,000 times the environmental concentrations. In general, the radioactive nuclides most concentrated by organisms are those of elements that occur in the environment in physiologically limiting concentrations. This is the case, for example, in the accumulation of P^{32} by phytoplankton and some vertebrates in the Columbia River (United States) below the level of the Hanford atomic energy plant.

Chemical composition of living matter. A living organism can perhaps be best described in general terms as a complex colloidal system, enclosed and separated from the environment by a semipermeable membrane, with enzyme-controlled feedback mechanisms that serve to maintain and reproduce the entire system. A characteristic which serves to distinguish the living or once living from the nonliving is the molecular asymmetry of biological matter; certain enantiomorphs (optically active isomers) predominate. Thus D-glucose is the only known biological enantiomorph of glucose, and in proteins all the amino acids appear to be of the L series. In chemical syntheses, a racemate (50:50 mixture of D- and L-isomers) is usually produced.

Elements of living matter. Excluding technetium, promethium, astatine, francium, and the transuranium elements, all of which are either very rare or nonexistent in nature, there are 88 elements that could conceivably enter into the composition of living matter. Of this number, 71 have been identified, but the data for three of these (niobium, zirconium, and tantalum) are questionable and need confirmation. Among the elements still undetected in living matter are four of the noble gases (helium, neon, krypton, and xenon) that undoubtedly occur there as atmospheric contaminants, all the platinum metals (ruthenium, rhodium, palladium, osmium, iridium, and platinum), and seven other elements of atomic numbers in the range 49–91 (indium, tellurium, hafnium, rhenium, polonium, actinium, and protactinium). It is improbable that any of the unidentified elements is completely excluded from protoplasm.

The major biological elements, hydrogen (H), carbon (C), nitrogen (N), and oxygen (O), compose 96–99% of the live weight of nonskeletal tissues. These elements enter into the composition of the four main groups of protoplasmic molecules, namely, water (89.4% O and 10.6% H by weight), proteins (51.3% C, 22.4% O, 17.8% N, and 6.9% H, by weight), carbohydrates (49.4% O, 44.4% C, and 6.2% H, by weight), and lipids (69.0% C, 17.9% O, and 10.0% H, by weight). In terms of the number of atoms, the abundance of elements in living matter decreases in the order hydrogen > oxygen > carbon > nitrogen. Skeletal or supporting structures may be composed of organic compounds such as cellulose, lignin, chitin, and sclero-proteins; or inorganic compounds such as calcium carbonate, silicon dioxide, and calcium phosphate.

Concentrations of elements. The average percentage composition of terrestrial vegetation is listed in Table 2 for those elements that have been studied in sufficient detail. The concentration of a minor element in living matter provides no indication of its physiological importance or necessity. Concentrations of essential elements may vary by a factor of 100,000 or higher. Some elements of unknown function occur in very small amounts. The cow liver cell, for example, contains only some 23 atoms of radium, probably in nonessential sites. The protistan *Euglena gracilis*, on the other hand, requires only 5000 molecules per cell of the essential cobalt-containing vitamin cyanocobalamin (vitamin B_{12}).

Several attempts have been made to construct a biological classification of the elements, but no single satisfactory tabulation has yet been discovered. The biological elements are generally those

Table 2. The average percentage by weight of elements in terrestrial vegetation*

Element	Percentage	Element	Percentage	Element	Percentage
O	70	Fe	2×10^{-2}	Br	1×10^{-4}
C	18	Mn	7×10^{-3}	Mo	5×10^{-5}
H	10.5	F	3×10^{-3}	Y†	4×10^{-5}
Ca	0.5	Ba	3×10^{-3}	Ni	2×10^{-5}
N	0.3	Al	2×10^{-3}	V	2×10^{-5}
K	0.3	Sr	2×10^{-3}	Pb	2×10^{-5}
Si	0.15	B	1×10^{-3}	Li	1×10^{-5}
Mg	7×10^{-2}	Zn	3×10^{-4}	U	1×10^{-5}
P	7×10^{-2}	Rb	2×10^{-4}	Ga	3×10^{-6}
S	5×10^{-2}	Cs	2×10^{-4}	Co	2×10^{-6}
Cl	4×10^{-2}	Ti	1×10^{-4}	I	1×10^{-6}
Na	2×10^{-2}	Cu	1×10^{-4}	Ra	2×10^{-12}

*Modified slightly from a tabulation by G. E. Hutchinson.

†Data are given for yttrium and the rare earths.

of low atomic weight; however, iodine, with an atomic weight of 127, is a notable exception. Partly as a result of the high percentage of water in living tissues, there is actually more resemblance between the cosmic abundance of elements and the abundance of elements in living matter than there is between that of living matter and the lithosphere. Perhaps the best-documented relationship is that discovered by G. E. Hutchinson for elements of constant valence, as shown in Fig. 1. The elements enriched by plants relative to the upper part of the lithosphere are those with either a high or a low ionic potential (ratio of ionic charge to radius). The basis of the rule is water solubility. The elements in the upper left group of Fig. 1 form soluble cations in water; those in the lower right group form anionic complexes with oxygen that are soluble in water, whereas the elements in the central group tend to form insoluble oxides.

The chemical composition of single-species populations cannot be taken as a completely invariable characteristic, although it is approximately so. Chemical composition changes with genetic constitution, growth, and environmental conditions. The availability of many soil elements for plant growth depends upon the pH of the soil, the state of the element, concentration, and many other factors. The ability of plants to reflect subsoil concentrations of certain elements has been used as a basis of biogeochemical prospecting for subsurface ore deposits. In the nickel-rich areas of the southern Ural Mountains, some abnormal forms of *Anemone patens* may contain up to 8 times more cobalt and 50 times more nickel than the same species growing on normal soils. Biogeochemical prospecting has been similarly used for copper, lead, selenium, and uranium deposits. *See* GEOCHEMICAL PROSPECTING.

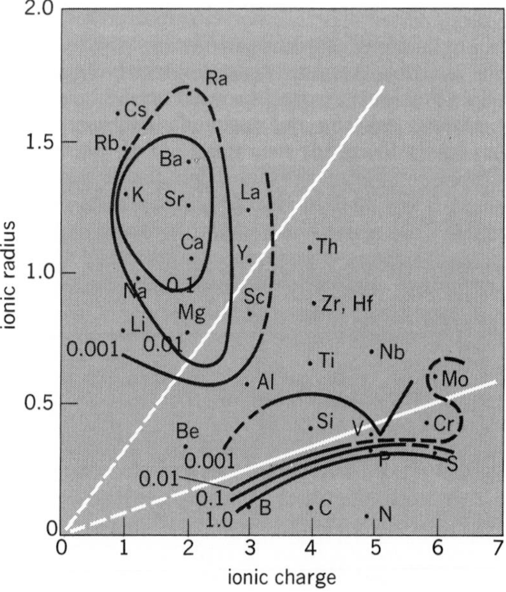

Fig. 1. Plot of ionic radius versus ionic charge for elements of constant valence. Curved lines show ratios of average concentration in terrestrial plants to concentrations in accessible lithosphere. (*From Goldschmidt, in G. E. Hutchinson, The biogeochemistry of aluminum and of certain related elements, Quart. Rev. Biol., 18(4):337, 1943*)

Trace elements and isotopes. Many of the trace metals are incorporated into protoplasm by forming complexes with ligands such as proteins, porphyrins, amino acids, mucopolysaccharides, and other polyvalent compounds. Among metals of the first transition series, there exists a stability sequence for the metal-ligand complexes that is largely independent of the nature of the ligand: $Mn^{++} < Fe^{++} < Co^{++} < Ni^{++} < Cu^{++} > Zn^{++}$.

It has occasionally been suggested that specific isotopes of some elements may be essential for living matter, but there is no evidence for such a concept. A fractionation of isotopes does occur in most biological processes, with a preferential accumulation of the lighter members in the pairs $C^{12}-C^{13}$ and $S^{32}-S^{34}$. Isotope ratios of natural materials provide useful data in determining the mode of origin of certain deposits, and in the case of $O^{16}-O^{18}$ in carbonate shells, in determining paleotemperatures. *See* GEOLOGIC THERMOMETRY; PALEOBIOCHEMISTRY.

Biogeochemical cycles of elements. Elements go through characteristic processes or cycles of entering an organism, returning to the organism's surroundings through the products of metabolism and death, entering another organism, and so forth.

Primary elements. Each of the primary elements of living matter (hydrogen, carbon, nitrogen, oxygen, and phosphorus) constitutes more than 0.5% of the weight of the living body.

Hydrogen occurs in nearly all biological compounds. Molecular hydrogen is both produced and oxidized by different kinds of bacteria. Hydrogen sulfide and methane are also metabolized under anaerobic conditions. The basis of many enzymatic oxidation reactions is a dehydrogenation.

Carbon is unique in that it forms more compounds than all the other elements combined. The cycle of carbon in the biosphere is shown in Fig. 2. Photosynthetic plants reduce carbon dioxide during photosynthesis to organic carbon compounds that are ultimately converted back to carbon dioxide in plant and animal respiration. Because some carbon compounds are not readily attacked by animals and microorganisms, there is a slight but steady loss of carbon to sediments, a process that has led to the accumulation of coal and petroleum deposits. *See* COAL; PETROLEUM, ORIGIN OF.

Much of the carbonate deposition in the oceans is mediated by calcareous organisms such as corals, mollusks, brachiopods, and some bacteria and algae. Carbonates may be indirectly precipitated from natural waters by a rise in pH following periods of intense phytoplankton photosynthesis. About 1 mg of carbon dioxide per cm^2 of Earth surface is added annually to the atmosphere as a result of the combustion of fossil fuels. This appears to have led to a 9% increase in atmospheric carbon dioxide in the period 1900–1960.

Oxygen is an essential element for all organisms; however, certain microorganisms are adapted to an anaerobic existence which in some cases is facultative and in others obligate. The oxygen produced by green plants during photosynthesis is used in plant and animal respiration, the overall reaction (proceeding to the right) being a reversal of photosynthesis, as shown in the equation be-

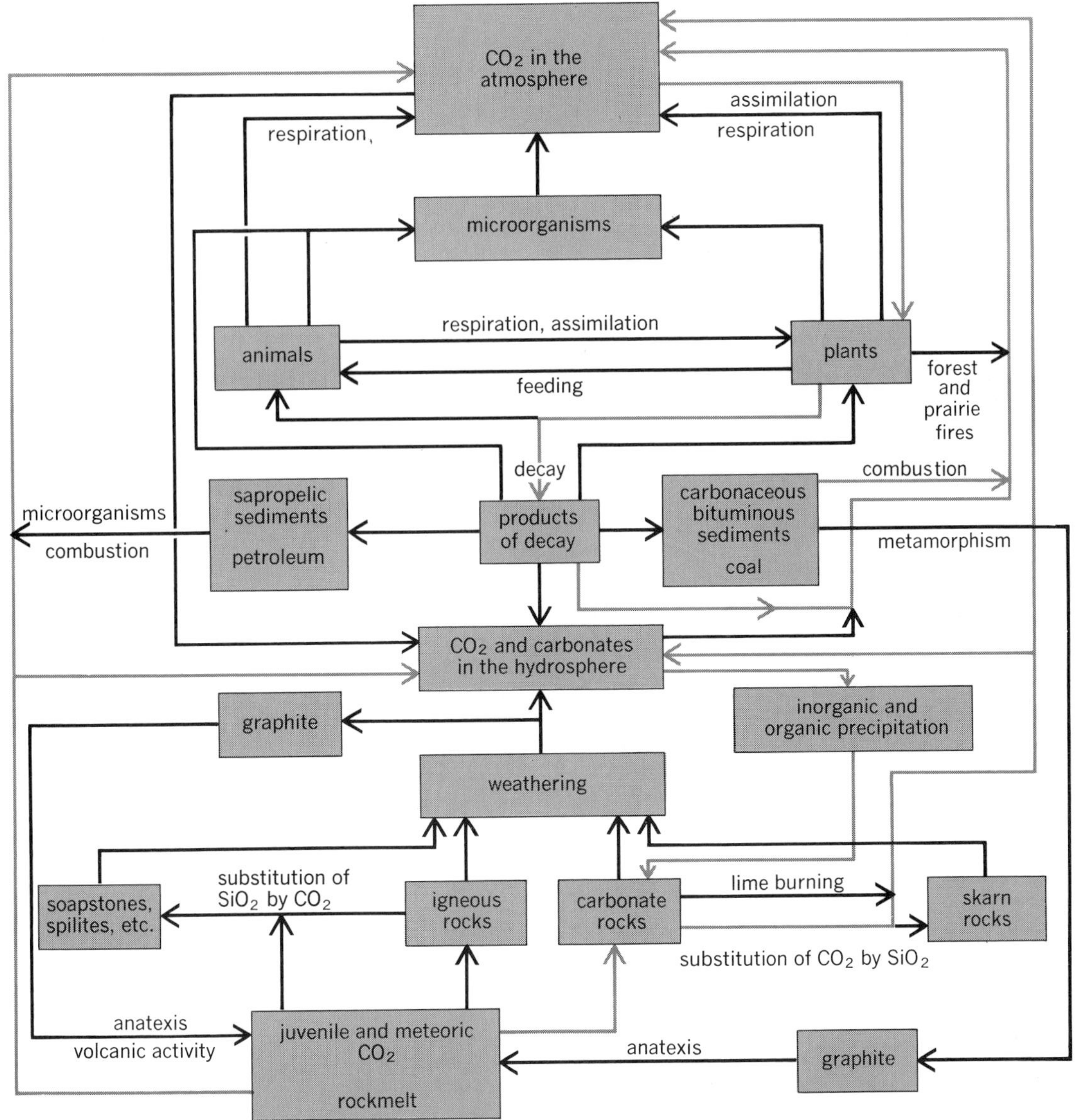

Fig. 2. Cycle of carbon. *(From K. Rankama and T. G. Sahama, Geochemistry, University of Chicago Press, 1950)*

low. The oxygen produced during photosynthesis arises from the decomposition of water.

$$(CH_2O)_n + nO_2 \rightleftharpoons nCO_2 + nH_2O + 106n \text{ kcal}$$

The geochemical cycle of nitrogen is outlined in Fig. 3. In most plant tissues, nitrogen averages 2–6% of the dry weight, whereas in animals and some bacteria the corresponding value is 6–13% for ash-free weights. *Azotobacter*, some blue-green algae, and *Rhizobium* in association with plant roots convert molecular nitrogen into organic compounds. Hutchinson estimates that the total annual biological fixation of nitrogen amounts to $n \times 10^{12}$ g of nitrogen, or approximately $n/5$ μg of nitrogen per cm^2 of Earth surface. The return of molecular nitrogen to the atmosphere is accomplished by a number of facultative anaerobic bacteria that convert nitrates and nitrites into molecular nitrogen. The most important biological forms of nitrogen are proteins and nucleic acids. The predominant excreted forms of nitrogen in terrestrial animals are uric acid and urea, whereas in aquatic forms the nitrogen is usually excreted as ammonia.

Phosphorus usually forms 0.2–1.0% of the dry weight of plant and animal tissues. Many organic phosphorus compounds occur biologically (for example, phosphoproteins, phospholipids, carbohydrate-phosphates, and nucleic acids). In all these compounds, the phosphorus occurs as phosphate esters. Phosphate is taken up from natural waters by algae and bacteria when the concentrations are only a few parts per 1,000,000,000 parts of water. The concentrations of phosphate in both waters and soils are low enough to be limiting to plant growth. Because of erosion and agricultural practices, about 10^{13} g of phosphorus is annually delivered to the sea. This is largely a unidirectional process because only $n \times 10^{10}$ g of phosphorus is annually removed from the sea, mostly in the form of guano deposited by marine birds.

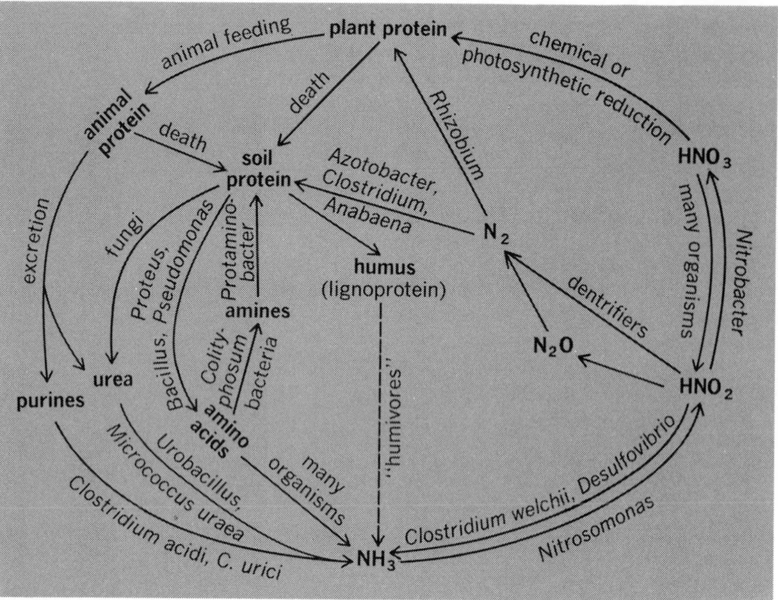

Fig. 3. Outline of the major processes involved in the nitrogen cycle in the biosphere. (*From K. V. Thimann, The Life of Bacteria, copyright 1950 by Macmillan Publishing Co., Inc.; reprinted by permission*)

Secondary elements. The other essential elements for living organisms in approximate order of decreasing concentration by weight are sodium, calcium, magnesium, sulfur, chlorine, potassium, iron, silicon, vanadium, zinc, boron, fluorine, manganese, cobalt, copper, and iodine.

Among the alkali-metal cations, sodium (Na) occurs primarily in circulating fluids, whereas potassium (K) is located intracellularly. The biological K/Na ratio is usually higher than in the growth medium. Lithium is a minor and variable constituent of living matter. Rubidium, on the other hand, is a normal constituent, behaving essentially like potassium; however, potassium cannot be completely replaced by rubidium. Chlorine occurs as chloride in biological fluids. The antibiotic chloromycetin is one of the few known organic chloride compounds. Iodine is concentrated from sea water by brown and red algae where it may occur in concentrations of 1% of the live weight. In sponges, corals, and the thyroid glands of vertebrates, iodine occurs in the form of iodinated amino acids. Bromine occurs biologically as bromide and also in the form of organic compounds. Fluorine enters the apatite lattice of bones and teeth.

Calcium is biologically the most important member of the alkaline-earth family, occurring as skeletal matter in the form of carbonates (aragonite and calcite) and as phosphate (hydroxyapatite). Calcite skeletons are generally rich (2–28%) in $MgCO_3$, whereas aragonite skeletons usually have less than 1%. Corals and some red and brown algae accumulate strontium (Sr) preferentially to calcium (Ca) from sea water, but most plants and animals are characterized by higher Sr/Ca ratios in their diets than in their bodies. Sr^{90} is the most serious of the radioactive fallout nuclides because it has a long biological residence time as well as a long physical half-life. Sr^{90} fallout is apt to be most serious in areas of calcium-deficient soils. Magnesium is an integral part of the chlorophyll molecule and is also required as a cofactor in enzymatic phosphate reactions. Radium is accumulated by some plants. Marine animals (especially fish) appear to contain less radium than marine plants.

Boron in the form of borates is essential for the growth of higher plants, but its function is unknown. Silicon is primarily a skeletal element. It is accumulated by diatoms, radiolarians, some sponges, and some terrestrial plants. Copper is accumulated from sea water by certain mollusks and crustaceans where it occurs in the form of the respiratory pigment hemocyanin. Copper is also known to participate in some enzyme reactions, but ionic copper is toxic to most forms of life when present in appreciable concentrations.

The geochemical cycle of sulfur is markedly affected by biological reactions. The photosynthetic sulfur bacteria have a photosynthesis based on the metabolism of sulfur compounds. Hydrogen sulfide is generated during the anaerobic decomposition of plankton. Selenium is selectively accumulated by some land plants such as *Astragalus*, but it is otherwise toxic to plants and animals.

Aluminum is the third most abundant element in the lithosphere, but it is scant in most living tissues. One exception is the club moss *Lycopodium alpinum*, which may contain up to 33% Al_2O_3 in its ash.

Vanadium, manganese, iron, and cobalt appear to be indispensable but only for some species in the cases of the ferrides vanadium and cobalt. Iron occurs biologically in a number of organic compounds, the most important of which are hemoglobin and the respiratory enzymes (cytochromes). The geochemical cycle of iron is influenced directly by the deposition of hydroxides by the iron bacteria and indirectly through changes in pH and E_h conditions. Manganese is a cofactor for some enzymes, and vanadium appears to play some role in photosynthesis. Vanadium also occurs in the form of vanadyl porphyrins in petroleum.

The other elements of the periodic table are, so far as is known, neither essential for organisms nor present therein in appreciable amounts. The geochemistry of such elements may be secondarily influenced by biological events, as in the precipitation of heavy metals by organic sediments rich in hydrogen sulfide.

Degradation of biogenic molecules. Under average conditions of burial and subsequent low-temperature maturing of biogenic sediments, the main chemical trend appears to be dehydration. As a result, all the molecules which have an H/O ratio close to that of water have a relatively low geological time-scale stability. Such types of mole-

Table 3. Comparison of Recent and Oligocene marine mud*

Content	Recent	Oligocene
Carbonate carbon, %	0.71	1.49
Organic carbon, %	0.53	0.27
Organic nitrogen, %	0.044	0.032
Amino acids, $\mu M/g$	3.0	0.51

*From J. G. Erdman, E. M. Marlett, and W. E. Hanson, Survival of amino acids in marine sediments, *Science*, 124:1026, 1956.

Table 4. Principal amino acids in Recent and Oligocene sediments*

Recent†	Oligocene†
Valine	Alanine
Leucines	Glutamic acid
Alanine	Glycine
Glutamic acid	Proline
Aspartic acid	Leucines
Glycine	Aspartic acid
Proline	
Tyrosine	
Phenylalanine	

*From J. G. Erdman, E. M. Marlett, and W. E. Hanson, Survival of amino acids in marine sediments, *Science*, 124:1026, 1956.

†Arranged in order of decreasing abundance.

cules are lignin, cellulose, proteins, and nucleic acids.

The higher H/O ratios of the lipids render these molecules of greater geological time-scale stability. The same applies to fossil resins, which can become preserved as ambers from the Mesozoic age onward, particularly in specimens from East Prussia; and occasionally insects, mosses, and even a small lizard, which lived some 150,000,000 years ago, are present in a state of perfect preservation. Some pigments of high H/O ratios are also well preserved. This applies in particular to the V, Ni, and Ca complexes of porphyrins, which can be detected in Paleozoic sediments. Hydroxyquinoline dyes have been identified from fossil sea lilies of Mesozoic age, which are closely comparable to those which exist in Recent species. *See* PETROLEUM GEOLOGY.

It appears from works on members of the coal series that diamino acids are preserved only in peats, lignite, and subbituminous coals, whereas the geologically more stable monoamino acids can be detected also in bituminous coals and anthracites. Table 3 indicates that the percentage of carbonate increases with the aging of marine muds and that amino acids are depleted at a considerably higher rate than either organic carbon or nitrogen. According to the data in Table 4, alanine is the amino acid of major geological stability, and it is interesting to note that the results of thermal decomposition experiments also demonstrated that alanine is the stablest of all the amino acids.

Little is presently known regarding the stabilities of the individual sugar and other carbohydrate molecules on a geological time scale. Analyses of samples from a test bore hole from the bottom of the Pacific, collected in the course of the Mohole project, demonstrated the reduction in the total organic C of sugars from 3.8 to 0.2% at a depth of 138 m. Galactose showed in this case at least a relatively lower stability than glucose. Depolymerization of cellulose in Miocene lignites has also been proven. [GEORGE MUELLER]

Bibliography: I. A. Breger (ed.), *Organic Geochemistry*, 1963; F. W. Clarke, *The Data of Geochemistry*, 5th ed., USGS Bull. no. 770, 1924; U. Colombo and G. D. Hobson (eds.), *Advances in Organic Geochemistry*, 1964; S. W. Fox (ed.), *The Origins of Prebiological Systems*, 1965; G. E. Hutchinson, *A Treatise on Limnology*, vol. 1, 1957; K. Rankama and T. G. Sahama, *Geochemistry*, 1950; H. U. Sverdrup, M. W. Johnson, and R. H. Fleming, *The Oceans*, 1946; W. Vernadsky, *La Biosphere*, 1929; A. P. Vinogradov, The elementary chemical composition of marine organisms, *J. Mar. Res.*, Sears Foundation for Marine Research, 1953.

Biostrome

A bedded structure, or layer (bioclastic stratum), composed of any calcite and dolomitized calcarenitic fossil fragments distributed over the sea bottom as finite lentils independent of, or in association with, bioherms or other areas of organic growth. Fragmental bioclastic layers may be of any thickness, are found in sedimentary rocks ranging from Cambrian to Recent, are typically developed on marine shelves and platforms within the zone of light penetration, and are deposited under sedimentary controls similar to those involving the distribution of quartzose sand on the sea floor. *See* BIOHERM; COQUINA; REEF.

[SHERMAN A. WENGERD]

Bibliography: P. E. Cloud, Jr., Facies relationships of organic reefs, *Bull. Amer. Ass. Petrol. Geol.*, 36:2125–2149, 1952; E. R. Cumings, Reefs or bioherms?, *Bull. Geol. Soc. Amer.*, 43:331–352, 1932; C. O. Dunbar and J. Rodgers, *Principles of Stratigraphy*, 1957.

Biotite

The most abundant and widely distributed species of the mica group of minerals, also called black mica or magnesium-iron mica. Its chemical composition is variable:

$$K_2[Fe(II),Mg]_{6-4}[Fe(III),Al,Ti]_{0-2}(Si_{6-5},Al_{2-3})O_{20-22}(OH,F)_{4-2}$$

Biotite itself is of no commercial importance, but vermiculite, which expands (exfoliates) upon heating, has widespread application as lightweight concrete and plaster aggregate, insulation, plant-growing medium, and lubrication.

Biotite grades into phlogopite, and no sharp division exists between the two. In a general way, biotite represents the iron-rich half of the biotite-phlogopite series. Some types contain little Mg; others are rich in Fe(III), Ti, or Mn; a few contain Li. The variety lepidomelane is a dark-colored iron-rich biotite, containing Fe(II), Fe(III), or both. It is common in syenites and nepheline syenites. The variety siderophyllite is rich in Fe(II) with little Fe(III); haughtonite, another varietal name, refers to biotite rich in both Fe(II) and Fe(III). Meroxene is that variety in which the optic plane is parallel with the *b* axis of the crystal, in contrast to anomite, which has the optic plane normal to the *b* axis.

The color ranges from black to dark brown or green. Unaltered basal cleavage sheets (easy and perfect) are both flexible and elastic (see illustration). The hardness is 2.5–3 (Mohs scale) and specific gravity 2.8–3.2. Polymorphism is extensive: 1-layer monoclinic, 2-layer monoclinic, 3-layer hexagonal, 4-, 10-, and 20-layer monoclinic, and 3-, 6-, 8-, 14-, 18-, 23-, 24-, and 25-layer triclinic. It is strongly pleochroic in brown, less commonly in green.

BIOTITE

Biotite specimen from Burgess, Ontario, Canada. (*American Museum of Natural History specimens*)

Biotite, an important rock-forming mineral, may be an essential constituent of both intrusive and extrusive igneous rocks, ranging in composition from gabbro to granite. It is especially common in granite, granodiorite, and syenite. In porphyritic extrusive rocks biotite appears chiefly as phenocrysts, very rarely in the matrix. Phenocrystic biotites commonly have increased amounts of Fe(III) (oxybiotite). Many pegmatites also contain biotite in large amounts in coarse flakes or rough crystals. It is also formed under moderate-grade metamorphic conditions, appearing commonly in schists and gneisses of both sedimentary and igneous parentage, in some cases associated with such species as muscovite, kyanite, sillimanite, garnet, potash feldspar, oligoclase, and quartz. Under weathering conditions, biotite alters more readily than muscovite; hence it is much less common in clastic sedimentary rocks. Biotite is readily transformed to chlorite, during which change by-product magnetite and sphene may also be produced. Much of this chloritization is the result of low-grade hydrothermal action. Some biotite is transformed, either under hydrothermal conditions, or as the result of groundwater action, first to hydrobiotite and then to vermiculite. *See* MICA; SILICATE MINERALS.

[E. WILLIAM HEINRICH]

Bitumen

A term used to designate naturally occurring or pyrolytically obtained substances of dark to black color consisting almost entirely of carbon and hydrogen with very little oxygen, nitrogen, and sulfur. Bitumen may be of variable hardness and volatility, ranging from crude oil to asphaltites, and is largely soluble in carbon disulfide. *See* ASPHALT AND ASPHALTITE. [IRVING A. BREGER]

Black shale

Shales, which are soft, fine-grained (silt- and clay-sized particles) detrital sedimentary rocks, can be subdivided into a series of varieties on the basis of color (red, gray, black, and so forth), with each color generally reflecting a predominant chemical constituent. Black shales are very fissile, almost papery, and usually thinly laminated. Their dark color is due primarily to a high content of carbonaceous organic matter (5% or more). Most black shales contain abundant sulfide sulfur (typically as the iron sulfide mineral pyrite), and many also contain abnormally high concentrations of trace elements such as uranium, copper, nickel, and vanadium. Generally, a prerequisite for deposition of shale, whether red (ferruginous), white (siliceous), gray (calcareous), or black (carbonaceous), is the comparative absence of deposition of both coarser clastic material such as sand and gravel, and chemical sediment such as limestone, evaporate, or chert. Deposition of black shale requires an additional precondition—the existence of an anaerobic (oxygen-poor) reducing environment within the depositional basin. The lack of oxygen retards or completely arrests the normal oxidation (decay) of any organic material buried with the sediment, preserving it instead as a compositional component. Anaerobic environments are produced in settings where clastic influx, normal water circulation, and bottom-surface interchanges are restricted by geography and climate.

Stagnant, quiet-water, "starved" reducing basins exist today; the Black Sea is a classic example. Similar settings existed at times in the geological past within the epicontinental seas which periodically covered the submerged continental blocks. The best-known ancient black shale deposit is the Upper Paleozoic Chattanooga Shale, which is widely exposed in the southeastern United States. Because they often contain unusually high concentrations of certain trace elements such as uranium, ancient black shale deposits such as the Chattanooga Shale are considered to be likely sources for nuclear fuels.

[FREDERIC L. SCHWAB]

Bibliography: American Geological Institute, *Glossary of Geology*, 1972; F. J. Pettijohn, *Sedimentary Rocks*, 3d ed., 1976.

Blueschist metamorphism

Regional metamorphism with the highest pressures and lowest temperatures, commonly above 5 kilobars (kb; 500 MPa) and below 400°C. Metamorphic rocks of the relatively uncommon blueschist facies contain assemblages of minerals that record these high pressures and low temperatures. The name "blueschist" derives from the fact that at this metamorphic grade, rocks of ordinary basaltic composition are often bluish because they contain the sodium-bearing blue amphiboles glaucophane or crossite rather than the calcium-bearing green or black amphiboles actinolite or hornblende, which are developed in the more common greenschist- or amphibolite-facies metamorphism. This difference in amphiboles in metamorphosed basalts led J. P. Smith in 1906 to conclude that glaucophane-bearing metamorphosed basalts had experienced different temperature and pressure conditions from those of other metamorphic rocks. In 1939 P. Eskola proposed a glaucophane-schist facies of regional metamorphism. More recently the term "blueschist facies" has been preferred because glaucophane is not always developed in many important rock types, for example, graywacke sandstones, which may contain other minerals that are characteristic of high pressure and low temperature such as aragonite, lawsonite, and jadeite + quartz. "Blueschist facies" is more generally applicable than "glaucophane-schist facies."

Blueschist metamorphism developed almost exclusively in young Mesozoic and Cenozoic mountain belts. As shown in Fig. 1, blueschist-facies metamorphism has taken place primarily during the last 5 to 10% of geologic time, an observation of considerable importance in understanding

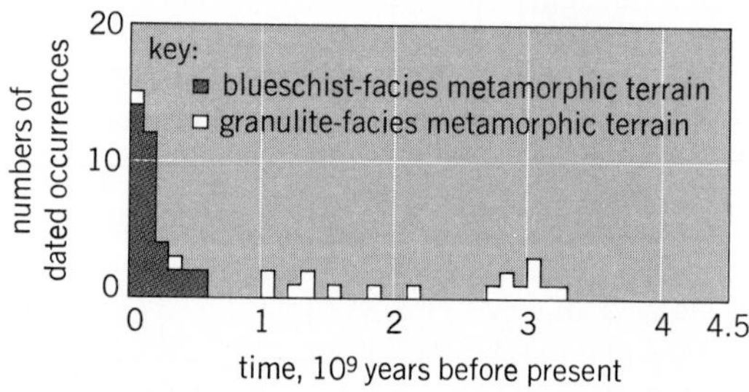

Fig. 1. Histogram showing the occurrence of blueschist-facies and granulite-facies metamorphism as a function of geologic time.

the evolution of the Earth. During most of geologic time, high-pressure metamorphism has been dominated by high-temperature amphibolite-facies metamorphism, as well as by very high-temperature granulite-facies metamorphism, the latter being uncommon during the last 5 to 10% of geologic time (Fig. 1). These phenomena are generally interpreted as being in some way an effect of the cooling of the Earth.

Mineral assemblages and physical conditions. A variety of mineral assemblages are characteristic of blueschist-facies metamorphism, depending upon the chemical composition of the rock and the actual temperature and pressure within the field of blueschist metamorphism. Detailed consideration of these assemblages and their mineral composition allows the determination of the physical conditions of metamorphism. Rocks of basaltic composition near the middle of the field of blueschist metamorphism commonly contain glaucophane + lawsonite + phengitic (silica-rich) white mica + sphene + aragonite or calcite; with increasing temperature, epidote, almandine garnet, actinolite, and rutile may appear in glaucophane-bearing assemblages; and at the very highest temperatures the sodic pyroxene omphacite or the sodic hornblende barrowisite may be present. At the lowest grades of blueschist metamorphism, blue amphibole may be missing entirely in metamorphosed basalts, with albite + pumpellyite + chlorite mineral assemblages being present. A common mineral assemblage in metamorphosed graywacke sandstone is albite + lawsonite + phengitic white mica + quartz + aragonite. At higher pressures jadeitic pyroxene or glaucophane crystallizes, and at lower pressures and temperatures pumpellyite may be present instead of lawsonite. Epidote may appear at higher temperatures.

The physical conditions of blueschist metamorphism are most easily determined from the phase relations of calcium carbonate ($CaCO_3$) and plagioclase feldspar (Fig. 2). With increasing pressure, calcite, the trigonal polymorph of $CaCO_3$, breaks down to form aragonite, the more dense orthorhombic polymorph. For example, at 200°C the breakdown reaction occurs at a pressure of about 6 kb (600 MPa; about 20 km depth). Aragonite rather than calcite is widespread in many but not all blueschist metamorphic terranes. Plagioclase feldspar also breaks down to denser phases in blueschist metamorphism. At very high pressures, for example, 8.5 kb (850 MPa) at 200°C, the albite ($NaAlSi_3O_8$) component of plagioclase feldspar breaks down to the sodic pyroxene jadeite ($NaAlSi_2O_6$) and quartz (SiO_2) (Fig. 2). Jadeite + quartz is present in metamorphosed graywacke sandstones of the highest-pressure blueschist metamorphic terranes (for example, in western California). The dense calcium-aluminum silicate lawsonite [$CaAl_2(OH)_2Si_2O_7H_2O$] is equivalent in chemical composition to anorthite feldspar ($CaAl_2Si_2O_8$) + water. In the presence of quartz, lawsonite is not stable below pressures of about 3 kb (300 MPa; about 10 km depth), and dehydrates to anorthite + water at temperatures above about 400°C (see Fig. 2).

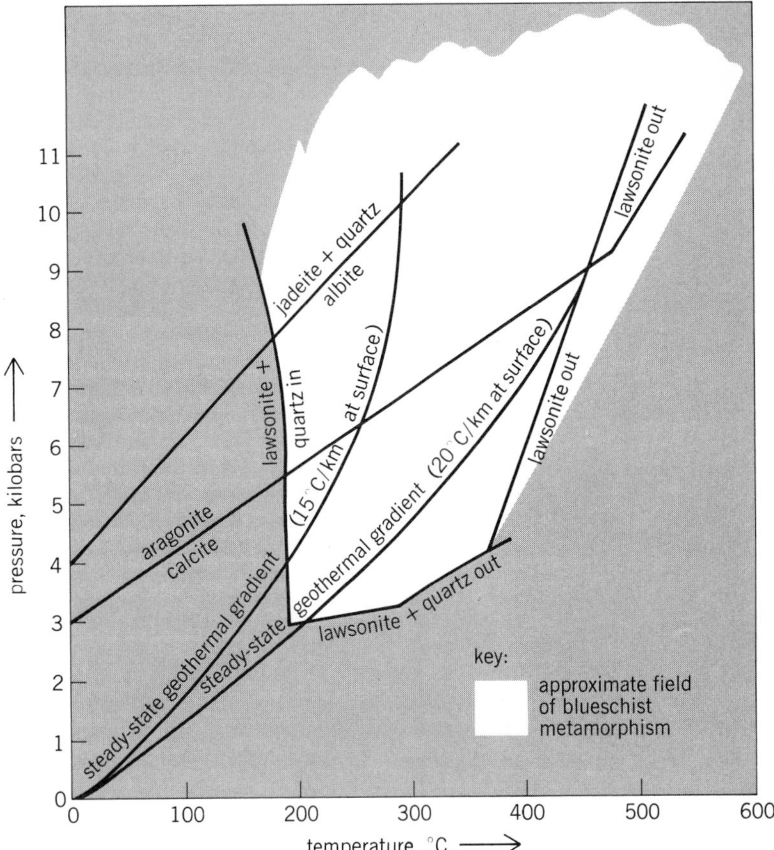

Fig. 2. Approximate temperature-pressure field of blueschist metamorphism.

Oxygen isotope geothermometry, based on the variation in coexisting minerals of $^{18}O/^{16}O$ ratios with temperature, yields temperatures of about 300°C for typical blueschists from western Califor-

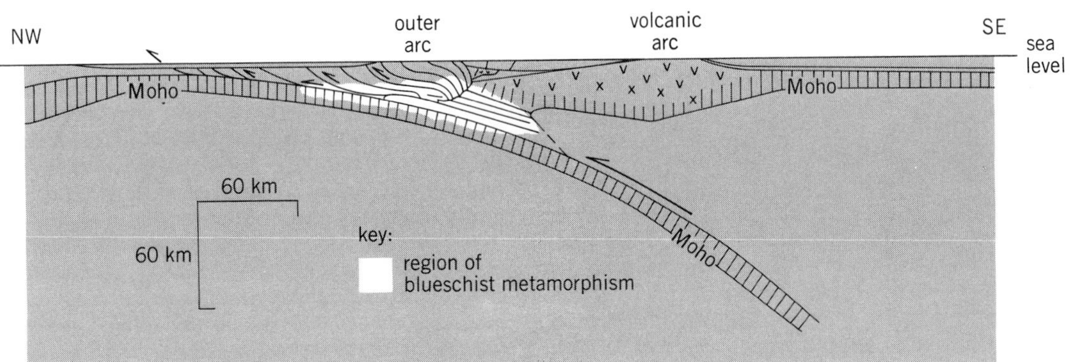

Fig. 3. Cross section of typical island arc and subduction zone with inferred region of blueschist metamorphism.

nia. Higher-temperature garnet- and epidote-bearing blueschists yield temperatures of 400–500°C. An approximate pressure-temperature field for blueschist metamorphism based on phase relations and oxygen isotopes is shown in Fig. 2.

Tectonics. Blueschist metamorphic rocks are found almost exclusively in the young mountain belts of the circum-Pacific and Alpine-Himalayan chains. The rocks are usually metamorphosed oceanic sediments and basaltic oceanic crust. Previously continental rocks rarely exhibit blueschist metamorphism. The tectonic mechanism for blueschist metamorphism must move the rocks to depths of more than 10 to 20 km while maintaining relatively cool temperatures (200–400°C). These temperatures are much cooler than for continental crust at those depths. For example, surface geothermal gradients of the order of 30°C per kilometer are common in continental crust and in thick sedimentary basins. In contrast, a steady-state surface gradient of about 15 to 20°C per kilometer would be required for typical blueschist metamorphism (Fig. 2). Heat flow measurements above long-lived subduction zones, together with thermal models, suggest that the conditions of blueschist metamorphism exist today above subduction zones just landward of deep-sea trenches (Fig. 3). This tectonic setting at the time of blueschist metamorphism is independently inferred for a number of metamorphic terranes. *See* TECTONIC PATTERNS.

What is not well understood is how the blueschist metamorphic rocks return to the surface; it is clear that the mechanism is not simple uplift and erosion of 20–30 km of the Earth's crust. Blueschist metamorphic rocks are usually in immediate fault contact with much less metamorphosed or unmetamorphosed sediments, indicating that they have been tectonically displaced relative to their surroundings since metamorphism. *See* METAMORPHIC ROCKS; METAMORPHISM. [JOHN SUPPE]

Bibliography: W. G. Ernst, Do mineral parageneses reflect unusually high-pressure conditions of Franciscan metamorphism?, *Amer. J. Sci.*, 270: 81–108, 1971; W. G. Ernst, Occurrence and mineralogic evolution of blueschist belts with time, *Amer. J. Sci.*, 272:657–668, 1972; A. Miyashiro, *Metamorphism and Metamorphic Belts*, 1973; J. Suppe, Interrelationships of high-pressure metamorphism, deformation, and sedimentation in Franciscan tectonics U.S.A., *24th International Geological Congress, Montreal, Reports Section 3 (Tectonics)*, pp. 552–559, 1972.

Boracite

A borate mineral with chemical composition $Mg_3B_7O_{13}Cl$. It occurs in Germany, England, and the United States, usually in bedded sedimentary deposits of anhydrite ($CaSO_4$), gypsum ($CaSO_4 \cdot 2H_2O$), and halite (NaCl), and in potash deposits of oceanic type. The chemical composition of natural boracites varies, with Fe^{++} or Mn^{++} replacing part of the Mg^{++} to yield ferroan boracite or manganoan boracite. *See* BORATE MINERALS.

Boracite occurs in crystals which appear to be isometric in external form, despite the fact that the arrangement of the atoms in the crystal structure has only orthorhombic symmetry (see illustration). When this natural low-temperature form is heated

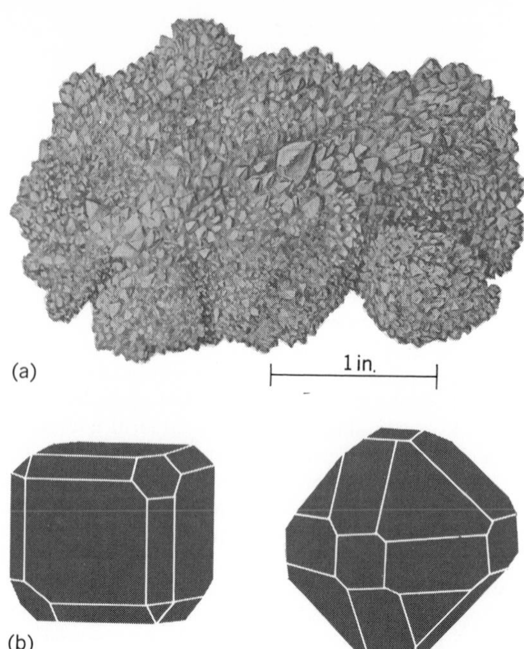

Boracite. (*a*) Specimen from Segeberg, Germany (*specimen from Department of Geology, Bryn Mawr College*). (*b*) Crystal habits (*from C. S. Hurlbut, Jr., Dana's Manual of Mineralogy, 17th ed., copyright © 1959 by John Wiley & Sons, Inc.; reprinted by permission*)

to 265°C, a slight readjustment of the atoms takes place internally without change in external form. The atomic symmetry of this resulting high-temperature form, which does not occur naturally, is then consistent with the external habit.

The hardness is 7–7½ on Mohs scale, and specific gravity is 2.91–2.97 for colorless crystals and 2.97–3.10 for green and ferroan types. Luster is vitreous, inclining toward adamantine. Boracite is colorless to white, inclining to gray, yellow, and green, and rarely pink (manganoan); its streak is white; and it is transparent to translucent. It is strongly piezoelectric and pyroelectric and does not cleave. [CHARLES L. CHRIST]

Borate minerals

A large and complex group of naturally occurring crystalline solids in which boron occurs in chemical combination with oxygen; hence the term borate. With the exception of avogadrite, $(K,Cs)BF_4$, and ferrucite, $NaBF_4$, all known boron minerals are borates. Boron, having an oxidation number (valence) of +3, does not act as a cation; for example, compounds of the type $B(NO_3)_3$ are not known to exist. Rather, boron combines with oxygen to form anions of varying degrees of complexity. Borate minerals may contain silicon, phosphorus, or arsenic. These, specifically, are the borosilicates, the borophosphates, and the boroarsenates. Of these three groups, the first is by far the largest.

Occurrence. The average amount of boron in the Earth's crust is estimated as 3 ppm. Fortunately, large quantities are concentrated in deposits of hydrated borate minerals that are easily worked, and boron is not, like some minor elements, so completely dispersed that it is hard to find and use. Deposits are found in the United States (Mojave

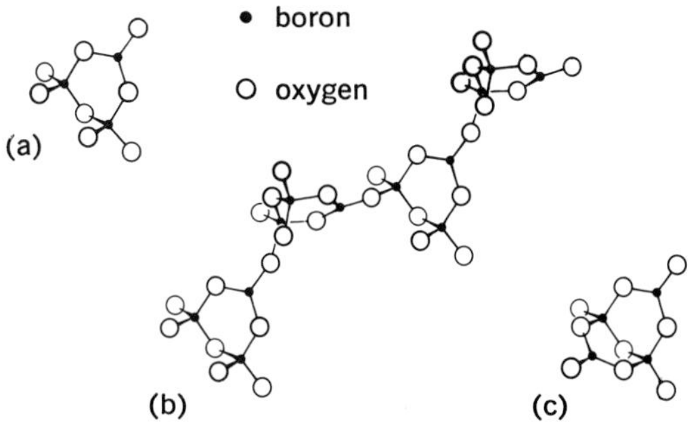

Some structural elements of borate minerals. (*a*) Meyerhofferite. (*b*) Colemanite. (*c*) Borax.

Desert), Turkey, Soviet Union (Inder district), Tibet, Italy, Germany, Argentina, Bolivia, Canada, China, India, Iran, New Zealand, New Guinea, and Syria. The minerals of chief commercial importance are borax, $Na_2B_4O_7{\cdot}10H_2O$, and kernite, $Na_2B_4O_7{\cdot}4H_2O$. Large quantities of these minerals are mined at the Kramer deposit, Boron, Calif. Borax is extracted in large amounts from brine at Searles Lake, Calif. *See* KERNITE.

Structural classification. Any rational classification of the borate minerals must rest directly on a knowledge of their crystal structures. Not all of the structures are known, but a large number have been worked out so that a satisfactory classification can be made. The crystal chemistry of the borates is similar to that of the silicates, with the additional complication that boron can occur in either triangular or tetrahedral coordination. Boron will combine with three oxygen atoms to form a planar triangular group BO_3, or with four oxygen atoms to form a tetrahedral group BO_4. These groups of threefold and fourfold coordination result in isolated ions such as BO_3^{-3}, or $B(OH)_4^{-1}$, or the molecule $B(OH)_3$, contained in sassolite. Alternatively, the groups may share corners in various ways to form polyanions of varying complexity. These polyanions may be relatively simple, as in the pyro ion $O_2B{-}O{-}BO_2$, or more complicated. For example, the mineral meyerhofferite, $Ca_2B_6O_{11}{\cdot}7H_2O$, contains insular polyions of composition $[B_3O_3(OH)_5]^{-2}$, formed by corner sharing of two $BO_2(OH)_2$ tetrahedra with a $BO_2(OH)$ triangle, as shown in the illustration. Other examples shown are the insular polyion $[B_4O_5(OH)_4]^{-2}$ found in borax, $Na_2B_4O_7{\cdot}10H_2O$, and the infinite chains of composition $[B_3O_4(OH)_3]_n^{-2n}$, found in colemanite, $Ca_2B_6O_{11}{\cdot}5H_2O$. Three rules have emerged for predicting the nature of the polyanions in borates (in the absence of Si, P, or As): (1) Polyanions are formed by the corner sharing only of triangles and tetrahedra in such manner that a compact insular group of low negative charge results; (2) these insular groups may polymerize to form infinite chains; an example is the equation below; (3) in hydrated borates those

$$\underset{\text{Meyerhofferite}}{n[B_3O_3(OH)_5]^{-2}} = \underset{\text{Colemanite}}{[B_3O_4(OH)_3]_n^{-2n}} + nH_2O$$

oxygens not shared between two borons or more always attach a hydrogen atom and become hydroxyl groups.

In the borosilicates there is much evidence for

Borate minerals

Mineral name	Empirical formula	Structural formula
Anhydrous borates		
Suanite	$Mg_2B_2O_5$	$Mg_2(B_2O_5)$
Kotoite	$Mg_3B_2O_6$	$Mg_3(BO_3)_2$
Nordenskiöldine	$CaSnB_2O_6$	$CaSn(BO_3)_2$
Ludwigite	$(Mg,Fe^{2+})_2Fe^{3+}BO_5$	$(Mg,Fe^{3+})_2Fe^{3+}BO_3O_2$
Boracite	$Mg_3B_7O_{13}Cl$	Complex 3-dimensional network
Hydrated borates		
Sassolite	H_3BO_3	$B(OH)_3$
Pinnoite	$Mg(BO_2)_2{\cdot}3H_2O$	$Mg[B_2O(OH)_6]$
Teepleite	$Na_2BO_2Cl{\cdot}2H_2O$	$Na_2[B(OH)_4]Cl$
Fluoborite	$Mg_3(BO_3)(OH,F)_3$	
Colemanite	$Ca_2B_6O_{11}{\cdot}5H_2O$	$Ca[B_3O_4(OH)_3]{\cdot}H_2O$
Meyerhofferite	$Ca_2B_6O_{11}{\cdot}7H_2O$	$Ca[B_3O_3(OH)_5]{\cdot}H_2O$
Inyoite	$Ca_2B_6O_{11}{\cdot}13H_2O$	$Ca[B_3O_3(OH)_5]{\cdot}4H_2O$
Inderite	$Mg_2B_6O_{11}{\cdot}15H_2O$	$Mg[B_3O_3(OH)_5]{\cdot}5H_2O$
Borax	$Na_2B_4O_7{\cdot}10H_2O$	$Na_2[B_4O_5(OH)_4]{\cdot}8H_2O$
Tincalconite	$Na_2B_4O_7{\cdot}5H_2O$	$Na_2[B_4O_5(OH)_4]{\cdot}3H_2O$
Kernite	$Na_2B_4O_7{\cdot}4H_2O$	$Na_2[B_4O_6(OH)_2]{\cdot}3H_2O$
Veatchite	$Sr_4B_{22}O_{37}{\cdot}7H_2O$	$Sr_2[B_5O_8(OH)]_2{\cdot}B(OH)_3{\cdot}H_2O$
Probertite	$NaCaB_5O_9{\cdot}5H_2O$	$NaCa[B_5O_7(OH)_4]{\cdot}3H_2O$
Ulexite	$NaCaB_5O_9{\cdot}8H_2O$	$NaCa[B_5O_6(OH)_6]{\cdot}5H_2O$
Borosilicates		
Datolite	$CaBsiO_5{\cdot}H_2O$	$Ca_4[BSiO_4(OH)]_4$
Reedmergnerite	$NaBSi_3O_8$	Feldspar structure
Danburite	$CaB_2Si_2O_8$	Feldspar structure
Tourmaline	$(Na,Ca)(Li,Al)_3Al_6(OH)_4(BO_3)_3Si_6O_{18}$	
Borophosphates and boroarsenates		
Lunebergite	$Mg_3B_2(OH)_6(PO_4)_2{\cdot}6H_2O$	
Seamanite	$Mn_3(PO_4)(BO_3){\cdot}3H_2O$	
Cahnite	$Ca_2B(OH)_4(AsO_4)$	

the linking of boron-oxygen polyanions with silicon-oxygen tetrahedra, but the details for most minerals are unknown. In borophosphates and boroarsenates there does not appear to be sharing between phosphate or arsenate tetrahedra and borate polyanions. In the table is a list of some of the commoner borate minerals. Structural formulas are given where it is possible to do so. The table also includes several minerals that serve as examples of structural types, even though they actually may be comparatively rare in occurrence. *See* BORACITE; DATOLITE; TOURMALINE.

[C. L. CHRIST]

Bibliography: R. M. Adams (ed.), *Boron, Metallo-Boron Compounds and Boranes*, 1964; C. Palache, H. Berman, and C. Frondel, *The System of Mineralogy*, vol. 2, 7th ed., 1951.

Borax

A hydrated borate mineral with chemical composition $Na_2B_4O_5(OH)_4 \cdot 8H_2O$. It is not uncommon in monoclinic prismatic crystals (see illustration) with symmetry $2/m$, but well-formed crystals are rare. Borax is also found in massive occurrences with conchoidal fracture and as encrustations. It has a hardness of 2–2½ and a specific gravity of approximately 1.7. It is colorless or white, exhibits perfect {100} cleavage, and has a sweetish alkaline taste. On exposure to the atmosphere, borax becomes chalky white due to dehydration to tincalconite, $Na_2B_4O_5(OH)_4 \cdot 3H_2O$. The structure of borax, with space group $C\ 2/c$, consists of $[BO_3OH]$ tetrahedra and $[BO_2OH]$ triangles which are joined parallel to the *c* crystallographic axis. The Na^+ is in six-coordination with H_2O molecules, producing $Na(H_2O)_6$ octahedra which have common edges forming ribbons parallel to the *c* crystallographic axis.

Borax is the most widespread of the borate mineral group. It is the result of evaporation of enclosed lakes in arid regions (playa deposits) and occurs as a surface efflorescence on the ground. It is the most abundant single mineral in the borate deposits of the Kramer borate district in the western part of the Mojave Desert in California. It occurs there as a massive, crystalline variety interbedded with clay-shale and siltstone. Other noteworthy borax-rich deposits are in Lake County, Inyo County, and Death Valley in California. Another great reserve of borax, also in California, comprises the brines of Searles Lake. Searles Lake is dry for most of the year with its surface covered by a white crust of salt from which borax can be extracted. Extensive borax deposits are present also in the Kirka area of Turkey, with estimated reserves several times greater than those of Boron, CA. South American regions with extensive borax and other associated borates are the contiguous parts of Argentina, Bolivia, and Chile. The three major producers of crude borates are the United States, Turkey, and Argentina.

Borax crystals, altered to white, chalky tincalconite, from Baker mine, Kramer borate district, in California. (*Courtesy of R. C. Erd, U.S. Geological Survey, Menlo Park, CA*)

Borates are used in glass manufacturing, especially in glass wool used for insulation. Borate compounds are also used in soap, in porcelain enamels for coating of metal surfaces, and in the preparation of fertilizers and herbicides. *See* BORATE MINERALS.

[CORNELIS KLEIN]

Bibliography: L. F. Aristarain and C. S. Hurlbut, Jr., Boron minerals and deposits, *Mineral. Rec.*, 3:165–172, 213–220, 1972; C. S. Hurlbut, Jr., and C. Klein, *Manual of Mineralogy*, 19th ed., 1977; V. Morgan and R. C. Erd, Minerals of the Kramer borate district, California, *Mineral Information Service (California Division of Mines)*, vol. 22, pp. 143–153, 165–172, 1969; N. Morimoto, The crystal structure of borax, *Mineral. J. Japan*, 2:1–18, 1956; C. Palache, H. Berman, and C. Frondel, *The System of Mineralogy*, vol. 2, 1951.

Bornite

A sulfide of composition Cu_5FeS_4, specific gravity 5.07, and hardness 3 (Mohs scale), commonly occurring as a primary mineral in many copper ore deposits. Crystals are rare; bornite is usually massive or granular (Fig. 1). The metallic and brassy color of a fresh surface rapidly tarnishes upon exposure to air to a characteristic iridescent purple, giving rise to the name "peacock ore." Bornite exhibits considerable solid solution at elevated temperatures. Upon cooling, other copper sulfides, such as chalcopyrite, chalcocite, and digenite, may

Fig. 1. Example of bornite from Messina, Transvaal, South Africa. (*Specimen from Department of Geology, Bryn Mawr College*)

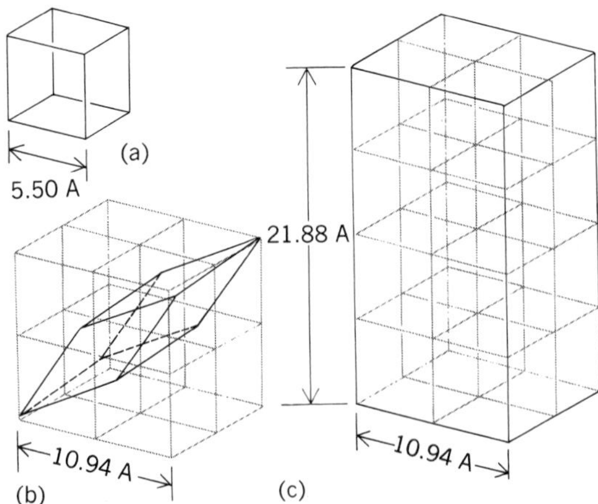

Fig. 2. Relation of the unit cells of the three polymorphic modifications of bornite to a face-centered cubic array of sulfur atoms. (*a*) High-temperature cubic form (one sulfur subcell). (*b*) Metastable rhombohedral form. (*c*) Low-temperature tetragonal form.

be exsolved. Association or intergrowths of bornite with these minerals is common. Though of lesser importance as an ore than chalcocite or chalcopyrite, masses of bornite have been mined in Chile, Peru, Bolivia, and Mexico and in the United States in Arizona and Montana.

Three polymorphic modifications of bornite are known. A cubic form, stable above 228 °C, cannot be quenched, and transforms into a metastable rhombohedral structure upon rapid cooling. This intermediate form slowly transforms into a tetragonal modification stable at room temperature. The atomic arrangement in all three phases is based upon a face-centered cubic arrangement of sulfur atoms (Fig. 2). The metal atoms are statistically distributed among the tetrahedral interstices available in this array: 3/4 atom is disordered among 24 equivalent sites within every available tetrahedron in the high-temperature phase; and 1 atom is distributed over 4 positions within 3 of the 4 tetrahedral sites available in the rhombohedral modification. The rhombohedral and tetragonal phases both mimic cubic symmetry because of fine-scale twinning. *See* CHALCOCITE; CHALCOPYRITE. [BERNHARDT J. WUENSCH]

Breccia

A rock composed of fragments, similar to a conglomerate in distribution of sizes of the materials (coarser than 2 mm in diameter). Breccias differ from conglomerates in the sharp, angular nature of the fragments. The sedimentary breccias are to be distinguished from volcanic breccias formed by the agglomeration of volcanic materials and tectonic breccias formed by fracturing and disruption due to faulting.

Breccias derived from erosion of cliffs or scarps with little or no transportation are called talus breccia or scree. The intraformational conglomerates are frequently called breccias if the pebbles are angular.

Some intraformational breccias may be called desiccation breccias, for they apparently resulted from the mud cracking and desiccation shrinking of a mud bottom after withdrawal of water. Later invasion of water disturbed the cracked pieces and arranged them in a thin layer of relatively flat pebbles. *See* TALUS.

A similar origin is inferred for flat limestone pebble conglomerates. Reef flank breccias are formed from the talus of coral and algal limestone reefs. Such breccias include fragments of reef materials eroded by wave action and mixed with calcareous fossil remains.

A type of breccia that has received increased attention is a soft rock deformation product. Such breccias are similar both to flat shale pebble conglomerates and to fault breccias. They develop as a result of gravitational slumping or sliding down a slope where a soft shale bed is interbedded with more competent beds, such as sandstone or siltstone. Very large pieces of the shale bed may become detached and deformed and even squeezed up through overlying layers, giving a picture of badly distorted and in some cases, almost unrecognizable bedding. A characteristic of these breccias is that they are intraformational; that is, higher beds carry across the deformed zone without disturbance. *See* CONGLOMERATE; GRAVEL; SEDIMENTARY ROCKS. [RAYMOND SIEVER]

Brochantite

A mineral with chemical composition $Cu_4(SO_4)(OH)_6$. Brochantite occurs in light- to dark-green, prismatic or loosely coherent aggregates of needlelike monoclinic crystals or granular masses (see illustration). It is a secondary mineral found in the oxidized zone of copper deposits, especially in arid regions. Fracture is conchoidal to uneven and luster is vitreous. Hardness is 3.5–4 on Mohs scale and specific gravity is 3.97. It fuses and is soluble in acids.

The occurrence of brochantite indicates the presence of other copper minerals, notably malachite and azurite, but brochantite itself has no

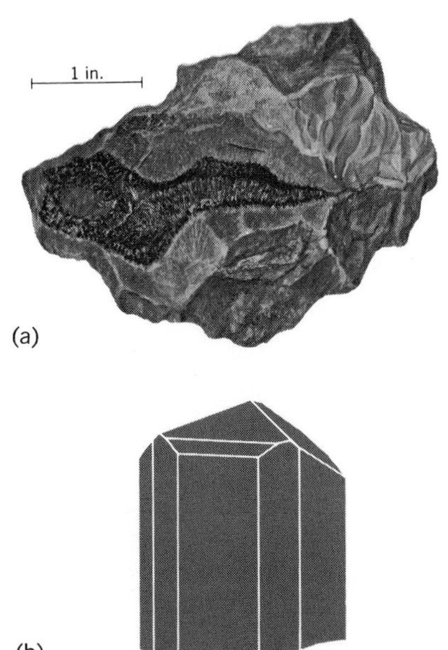

Brochantite. (*a*) Needlelike crystals from Tsumeb, Namibia (*specimen from Department of Geology, Bryn Mawr College*). (*b*) Crystal habit (*from L. G. Berry and B. Mason, Mineralogy, copyright © 1959 by W. H. Freeman and Co.*).

particular commercial value. It is found in many parts of Europe, Africa, South America, and Australia. In the United States it is found in the Bisbee district, Cochise County, Ariz., and in Colorado, Utah, Oregon, New Mexico, and California.

[EDWARD C. T. CHAO]

Brucite

A magnesium hydroxide mineral, $Mg(OH)_2$, crystallizing in the trigonal system with space group $P\bar{3}m$. It is a member of the important $Cd(OH)_2$ structure type, consisting of hexagonal close-packed oxygen atoms with alternate octahedral layers occupied by Mg. The "brucite layer" is an important structural component in the clay, mica, and chlorite mineral groups.

Brucite occurs as tabular crystals and as elongated fibers (as the variety nemalite), cleavage on {0001} perfect, hardness 2 1/2 (Mohs scale), color white to greenish, and specific gravity 2.4. Fe^{++} and Mn^{++} commonly substitute for Mg^{++} (see illustration).

Specimen of brucite found in Texas, Pa. (*American Museum of Natural History specimens*)

Brucite often occurs in a low-temperature vein paragenesis, usually with serpentine and accessory magnesite. It is also derived by the action of water on periclase, MgO, which results from the thermal metamorphism of dolomites and limestones. Carbonate rocks rich in periclase and brucite are called predazzites. *See* DOLOMITE; MAGNESITE; SERPENTINE.

[PAUL B. MOORE]

Calcarenite

A mechanically deposited (clastic) limestone in which the constituent particles of carbonate are sand-grain size. If the grains are larger than 2 mm, the rock is a calcirudite. If the grains are smaller than 1/16 mm, it is a calcilutite. The calcarenites are normally cemented by clear calcite. The detrital carbonate grains are fossil fragments, broken fragments of calcilutite, oolites, and fecal pellets. The detritus is fairly well sorted as to size, and many grains show slightly abraded edges, indicating a normal, current-deposited sediment. Where oolites are dominant, the rock is called an oolite or oolitic limestone. Calcarenites are often found with orthoquartzites, either interbedded or grading laterally into them. They have the same sedimentary structures as the orthoquartzites, dominantly cross-bedding.

The fine-grained equivalent of calcarenite, calcilutite, is called lithographic limestone if it is homogeneous, dense, and very-fine-grained and breaks with conchoidal or subconchoidal fracture; the name is derived from the use of the rock for lithography. The very-fine-grained particles may be the product of inorganic precipitation or of finely comminuted algal debris. *See* COQUINA; LIMESTONE; SEDIMENTARY ROCKS.

[RAYMOND SIEVER]

Calcite

One of the most common and widespread minerals in the Earth's crust. Calcite, $CaCO_3$, may be found in a variety of sedimentary, metamorphic, and igneous rocks. It is also an important rock-forming mineral in calcite marbles, carbonatites, and pure limestones.

Although it is commonly of high purity, calcite itself may contain other cations such as manganese, iron, magnesium, cobalt, barium, and strontium, substituting for the calcium in variable amounts. Well-developed crystals of calcite are common in cavities in limestones and basic igneous rocks. Of special interest are the large crystals of optical quality obtained from near the Eskefiord, Iceland. This material is known as Iceland spar and single crystals several feet across have been reported. In the United States very large crystals have been mined in Taos County, N.Mex.

Calcite in the form of pure limestones is the main source of the world's quicklime and hydrated or slaked lime. It is also used as a metallurgical flux and in the manufacture of cement and glass.

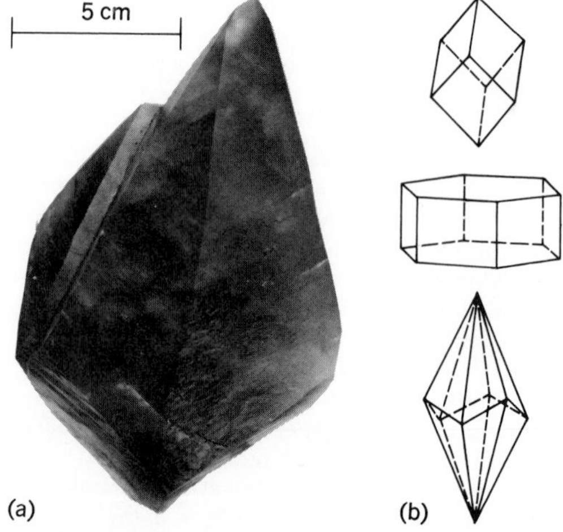

Fig. 1. Calcite. (*a*) Scalenohedral crystal, Joplin, Mo. (*American Museum of Natural History specimens*). (*b*) Common crystal forms (*C. S. Hurlbut, Jr., Dana's Manual of Mineralogy, 17th ed., copyright © 1959 by John Wiley & Sons, Inc.; reprinted by permission*).

Material which is less pure may be used as dimension stone, soil conditioners, industrial acid neutralizers, and as aggregate in concrete and road building. Calcite in transparent well-formed crystals is used in certain optical instruments. *See* LIMESTONE.

Crystallography. Calcite has hexagonal (rhombohedral) symmetry and a structure built up of alternate layers of Ca^{++} ions and CO_3^{--} groups stacked up so that the layers lie in planes perpendicular to the main axis of symmetry. Calcite may exhibit a wide variety of forms in addition to simple and complex rhombohedrons. Tabular, prismatic, and scalenohedral varieties are common and twinned crystals are often found (Fig. 1).

Some fibrous varieties of calcite have gem value as a kind of satin spar. Deposits in streams in calcareous areas and in limestone caves as stalactites and stalagmites occur as masses with overall rounded forms made up of very small crystals. Travertine and onyx are often of such material. *See* ONYX; TRAVERTINE.

When pure, calcite is either colorless or white but impurities can introduce a wide variety of colors; blues, pinks, yellow-browns, greens, and grays have all been reported. Hardness is 3 on Mohs scale. The specific gravity of pure calcite is 2.7102 ± 0.0002 at 20°C.

Calcite has a very low solubility in pure water but the solubility increases greatly with the pressure of carbon dioxide in the water, as a result of the formation of carbonic acid.

The equilibrium amounts of certain cations that replace calcium in the calcite structure have been determined experimentally for magnesium, iron, and manganese. The amounts of each have been found to increase with temperature. At a fixed temperature the equilibrium amounts for these three increase in the order Mg–Fe–Mn as the size of the substituting cation approaches that of calcium.

The temperature at which calcite dissociates to calcium oxide and carbon dioxide at atmospheric pressure depends on the heating rate, grain size, presence of moisture, and partial pressure of carbon dioxide. Dissociation temperatures range from 700 to 1000°C. The equilibrium thermal dissociation temperatures have been determined experimentally and nearly pure calcite has been melted at approximately 1340°C and 15,000 psi of carbon dioxide. Attempts to grow large single crystals of calcite of optical quality by high-temperature and high-pressure techniques have been encouraging, although the dimensions of synthetic crystals do not yet compete with those of natural crystals.

Partly because it is available in pure, clear, well-developed crystals, calcite has played an important role in the formation of certain fundamental concepts in crystallography and mineralogy. The phenomenon of double refraction was first observed in calcite and the discovery of the polarization of light resulted from studies of this mineral. Detailed work has shown anisotropism in most of its physical properties. Calcite is somewhat less hard on certain faces {0001} than on others {1010}. Single crystals of calcite have different coefficients of expansion and show variations in magnetic susceptibility and electrical conductivity which are dependent upon the crystallographic directions involved. *See* CARBONATE MINERALS.

[ROBERT I. HARKER]

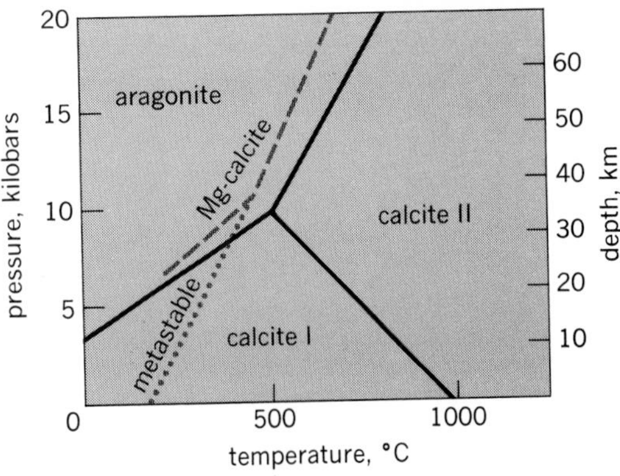

Fig. 2. Calcium carbonate. Heavy lines show the phase diagram for the system. Dashed lines indicate the coexistence of aragonite with magnesium-saturated calcite. Dotted line points out the metastable extension of transition of magnesian calcite to aragonite.

Calcium carbonate relations. Calcite, aragonite, and vaterite are three polymorphs of $CaCO_3$ occurring in rocks. Vaterite is rare and probably formed under metastable conditions. Although only one form of calcite, calcite I, is known in rocks, there is evidence for a rapid transition, calcite I $\rightleftharpoons$ calcite II. Figure 2 shows the probable phase relationships. Aragonite does not accept $MgCO_3$ in solid solution, whereas calcite does. The dashed lines show the position of the phase transition between aragonite and magnesian calcite. The magnesium content of the calcite increases along the transition boundary with increasing temperature.

Aragonite has been recognized in regionally metamorphosed rocks, and it appears from Fig. 2 that these rocks have been subjected to pressures corresponding to depths of the order of 30 kms. Such great depths are difficult to reconcile with some geological evidence, and the existence of tectonic overpressures during metamorphism is an alternative proposal. A more likely explanation for the formation of aragonite in metamorphic rocks at lower pressures is that metastable magnesian calcites in fossiliferous limestones have undergone reaction to aragonite at a metastable extension of the high-temperature portion of the transition boundary for magnesian calcites, as shown in Fig. 2. The utility of aragonite as a geobarometer in metamorphic rocks thus remains in question. *See* ARAGONITE.

[PETER J. WYLLIE]

Caliche

A term used to describe the crude soda niter of the Chilean nitrate deposits. It is a variable admixture of salts, including, in addition to the soda niter, bloedite, anhydrite, gypsum, polyhalite, halite, glauberite, darapskite, and minor amounts of various iodate, chromate, and borate minerals. *See* BORATE MINERALS; SODA NITER.

The term caliche has several other meanings. It is commonly used in Mexico and the southwestern United States to designate gravel, sand, or desert debris cemented by porous calcium carbonate. Other uses of the term include a thin layer of clayey soil capping auriferous veins; whitish clay in the selvage of veins; a white mineral vein recently discovered; and in placer mining, a bank composed of clay, sand, and gravel.

[GEORGE SWITZER]

Cambrian

An interval of time in Earth history (Cambrian Period) and its rock record (Cambrian System). The Cambrian Period spanned about 70,000,000 years and began with the first appearance of marine animals with calcareous shells. The Cambrian System includes many different kinds of marine sandstones, shales, limestones, dolomites, and volcanics. There is very little provable record of nonmarine Cambrian environments.

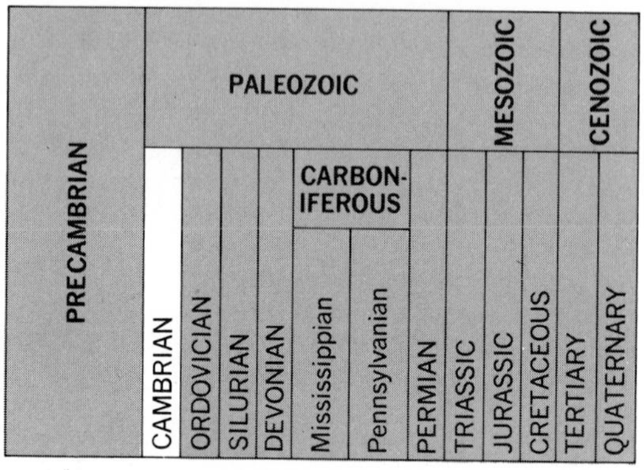

PRECAMBRIAN	PALEOZOIC							MESOZOIC			CENOZOIC	
	CAMBRIAN	ORDOVICIAN	SILURIAN	DEVONIAN	CARBONIFEROUS		PERMIAN	TRIASSIC	JURASSIC	CRETACEOUS	TERTIARY	QUATERNARY
					Mississippian	Pennsylvanian						

The concept that great systems of rocks recorded successive periods of Earth history was developed in England in the early 19th century. The Cambrian, which was one of the first systems to be formally named, was proposed by the Reverend Adam Sedgwick in 1835 for a series of sedimentary rocks in Wales that seemed to constitute the oldest sediments in the British Isles. At that time, there was no real idea of the antiquity of Cambrian rocks. They were recognized by distinctive fossils and by their geologic relations to other systems. In the early part of the 20th century, radiometric techniques for obtaining the ages of igneous and metamorphic rocks evolved. Because of the difficulty of finding rocks that can be dated radiometrically in association with rocks that can be dated empirically by fossils, the age in years of most Cambrian deposits is only approximate. The best present estimates suggest that Cambrian time began about 570,000,000 years ago and ended about 475,000,000 years ago. It is one of the longest Phanerozoic periods.

For most practical purposes, rocks of Cambrian age are recognized by their content of distinctive fossils. On the basis of the successive changes in the evolutionary record of Cambrian life that have been worked out during the past century, the Cambrian System has been divided globally into Lower, Middle, and Upper series, each of which has been further divided on each continent into stages, each stage consisting of several zones. The divisions of the Cambrian System presently recognized in North America are shown in Fig. 1. Despite the amount of work already done, precise intercontinental correlation of series and stage boundaries, and of zones, is still difficult, and refinement of intercontinental correlation of these ancient rocks is a topic of research.

Life. The record preserved in rocks indicates that essentially all Cambrian plants and animals lived in the sea. The few places where terrestrial sediments have been preserved suggest that the land was barren of major plant life, and there are no known records of Cambrian insects or of terrestrial vertebrate animals of any kind.

Plants. The plant record consists entirely of algae, preserved either as carbonized impressions in marine black shales or as filamentous or blotchy microstructures within marine buildups of calcium carbonate, called stromatolites, produced by the actions of these organisms. Cambrian algal stromatolites were generally low domal structures, rarely more than a few meters high or wide, which were built up by the trapping or precipitation of calcium carbonate by one or more species of algae. Such structures, often composed of upwardly arched laminae, were common in regions of carbonate sedimentation in the shallow Cambrian seas. *See* STROMATOLITE.

Animals. The animal record is composed almost entirely of invertebrates that had either calcereous or phosphatic shells (Fig. 2). A few rare occurrences of impressions or of carbonized remains of soft-bodied organisms indicate that the fossil record is incomplete and biased in favor of shell-bearing organisms. The fossils of shell-bearing organisms include representatives of several different classes of arthropods, mollusks, echinoderms, brachiopods, and poriferans. Coelenterates (other than jellyfish impressions), bryozoans, radiolarians, and foraminiferans are unknown from Cambrian rocks. Some fossil groups of widespread occurrence, such as Archaeocyatha, are known only from Cambrian rocks, and several extinct groups of Paleozoic organisms such as hyolithids and conodonts have their first appearance in Cambrian rocks.

Diversity. Although the record of marine life in the Cambrian seems rich, one of the dramatic differences between Cambrian marine rocks and those of younger periods is the low phyletic diversity of most fossiliferous localities. The most diverse faunas of Cambrian age have been found along the ocean-facing margins of the shallow seas that covered large areas of the Cambrian continents. Because these margins were often involved in later geologic upheavals, their rich record of Cambrian life has been largely destroyed. Only a few localities in the world remain to provide a more accurate picture of the diversity of organisms living in Cambrian time. In North America, the richest localities are in the Kinzers Formation of southeastern Pennsylvania, the Spence Shale of northern Utah, and the Burgess Shale of British Columbia.

Trilobites. The most abundant remains of organisms in Cambrian rocks are of trilobites (Fig. 2*a*–*c*). They are present in almost every fossiliferous Cambrian deposit and are the principal tools used to describe divisions of Cambrian time and to correlate Cambrian rocks. These marine arthropods ranged from a few millimeters to 50 cm in length, but most were less than 10 cm long. Although some groups of trilobites such as the Agnostida (Fig. 2*a*) were predominantly planktonic in habitat, most trilobites seem to have been benthic or nektobenthic and show a reasonably close correlation with bottom environments. For this reason, there are distinct regional differences in the Cambrian trilobite faunas of the shallow seas of different parts of the Cambrian world.

Brachiopods. The next most abundant Cambrian fossils are brachiopods (Fig. 2*d–f*). These bivalved animals were often gregarious and lived on the sediment surface or on the surfaces of other organisms. Brachiopods with phosphatic shells, referred to the Acrotretida (Fig. 2*f*), are particularly abundant in many limestones and can be recovered in nearly perfect condition by dissolving these limestones in acetic or formic acids. Upper Cambrian limestones from Texas, Oklahoma, and the Rocky Mountains yield excellent silicified shells of formerly calcareous brachiopods when they are dissolved in dilute hydrochloric acid.

Archaeocyathids. Limestones of Early Cambrian age may have large stomatolites formed by an association of algae and an extinct phylum of invertebrates called Archaeocyatha (Fig. 2*m* and *n*). Typical archeocyathids grew conical or cylindrical shells with two walls separated by elaborate radial partitions. The walls often have characteristic patterns of perforations.

Mollusks and echinoderms. The Cambrian record of mollusks and echinoderms is characterized by many strange-looking forms that lived for only short periods of time and left no clear descendants (Fig. 2*g–l*). Representatives of these phyla, such as cephalopods, clams, and true crinoids, which are abundant in younger rocks, are absent from Cambrian rocks or are found only in the very latest Cambrian deposits. Snails, however, are found throughout the Cambrian. Discoveries of primitive clams have been made in Early Cambrian beds, but they are apparently absent from the later record of life for tens of millions of years until post-Cambrian time.

Extinction. The stratigraphic record of Cambrian life in North America shows perhaps five major extinctions of most of the organisms living in the shallow seas. These extinction events form the boundaries of evolutionary units called biomeres (Fig. 1). Their cause, and their presence in the Cambrian records of other continents, is under investigation. However, perhaps it was these periodic disasters that prevented clear continuity in the evolutionary records of many groups and which led, particularly, to the records of the echinoderms and mollusks. *See* ANIMAL EVOLUTION; EXTINCTION (BIOLOGY).

Faunal origin. Another major unsolved problem is the origin of the entire Cambrian fauna. Animal life was already quite diverse before Cambrian time. The earliest Cambrian beds contain representatives of more than 20 distinctly different invertebrate groups. All of these have calcified shells, but none of the Precambrian organisms have any evidence of shells. There is still no clear evidence to determine whether shells evolved in response to predation or to environmental stress, or as the result of some change in oceanic or atmospheric chemistry. *See* PRECAMBRIAN.

Geography. Knowledge of Cambrian geography and of the dynamic aspects of evolution and history in Cambrian time is derived from rocks of this age that have been exposed by present-day erosion or penetrated by borings into the Earth's surface. Despite the antiquity of Cambrian time, a surprisingly good record of marine rocks of Cambrian age has been preserved at many localities throughout the world. Each of the different rock types contains clues about its environment of deposition that have been derived from analogy with modern marine environments. From this information, together with knowledge gained from fossils of about the same age within the Cambrian and information about the present geographic distribution of each Cambrian locality, a general picture of world geography and its changes through Cambrian time is available.

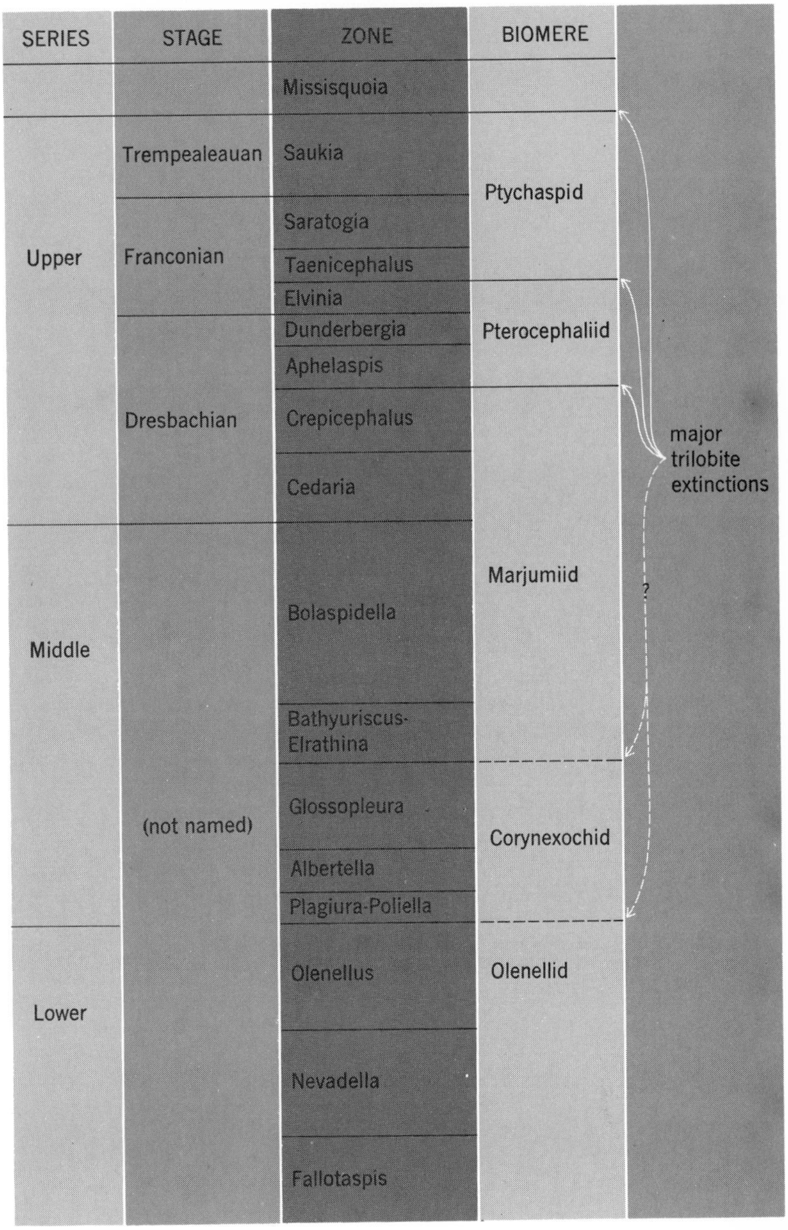

SERIES	STAGE	ZONE	BIOMERE
		Missisquoia	
Upper	Trempealeauan	Saukia	Ptychaspid
	Franconian	Saratogia	
		Taenicephalus	
		Elvinia	Pterocephaliid
	Dresbachian	Dunderbergia	
		Aphelaspis	
		Crepicephalus	Marjumiid
		Cedaria	
Middle	(not named)	Bolaspidella	
		Bathyuriscus-Elrathina	
		Glossopleura	Corynexochid
		Albertella	
		Plagiura-Poliella	
Lower		Olenellus	Olenellid
		Nevadella	
		Fallotaspis	

major trilobite extinctions

Fig. 1. North American divisions of the Cambrian System.

The developing theory of plate tectonics has provided criteria whereby ancient continental margins can be identified. By using these criteria and the spatial information about marine environments derived from study of the rocks, the Cambrian world can be resolved into at least five major continents. These were (1) North America, minus a narrow belt along the eastern coast from eastern Newfoundland to northern Georgia which is most

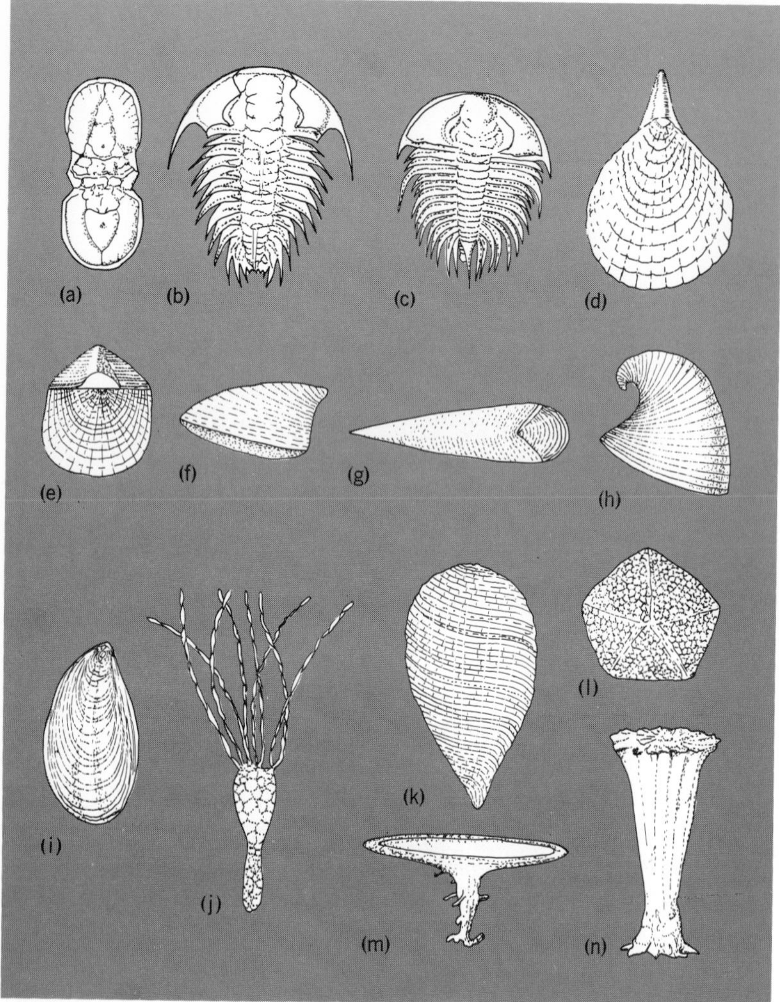

Fig. 2. Representative Cambrian fossils: (*a–c*) trilobites; (*d–f*) brachiopods; (*g*) hyolithid; (*h, i*) mollusks; (*j–l*) echinoderms; and (*m, n*) archaeocyathids.

closely related to Cambrian Gondwana; (2) Baltica, consisting of present-day northern Europe north of France and west of the Ural Mountains, but excluding most of Scotland and northern Ireland, which are remains of marginal Cambrian North America; (3) Gondwana, a giant continent whose present-day fragments are Africa, South America, India, Australia and Antarctica, parts of southern Europe, the Middle East, and southeast Asia; (4) Siberia, including much of the northeastern quarter of Asia; and (5) Mongolia, including present-day Mongolia and the northern half of China. Unfortunately, there is not enough reliable information to accurately locate these continents relative to one another on the Cambrian globe. *See* CONTINENT FORMATION; PLATE TECTONICS.

History. At the beginning of Cambrian time, the continents of the world were largely exposed, much as they are now. Following some still-unexplained event, the seas were suddenly populated by a rich fauna of shell-bearing invertebrates after 4,000,000,000 years of nearly barren existence. Belts of volcanic islands comparable to those of the western Pacific Ocean today fringed eastern North America, the Australian and western Antarctic margins of Gondwana, and southern Siberia. These belts suggest that crustal plates analogous to those of the present-day were in motion at that time. Thick evaporites in Siberia and the Middle Eastern and Indian parts of Gondwana suggest regions of warm temperature and high evaporation rate. Absence of significant development of limestones around Baltica suggests that it was a cool region, probably at high latitudes. Near the continental margins of eastern and western North America, Siberia, and Mongolia and the western Antarctic, eastern Australian, northwestern African, and southern European margins of Gondwana, archaeocyathid bioherms developed and flourished. By the end of Early Cambrian time, archaeocyathids had become extinct, and shell-bearing organisms capable of building bioherms did not reappear until Middle Ordovician time, at least 40,000,000 years later.

Volcanism and evaporitic conditions continued into the Middle Cambrian in Siberia and parts of Gondwana, and evaporites of this age are also known from northern Canada. However, a dramatic change took place in the southern European and northwestern African parts of Gondwana. Carbonate sedimentation virtually ceased throughout that region as those parts of Gondwana reached areas of cooler water and probably higher latitudes. Sea level was rising over much of the world throughout Middle Cambrian time, flooding the interiors of most continents.

In the Late Cambrian, parts of western Baltica and eastern North America began to show signs of crustal deformation suggesting that Iapetus, the ocean between North America, Gondwana, and Baltica, was beginning to close. Crustal deformation was also taking place in southern Siberia, eastern Australia, and western Antarctica. In the broad, shallow seas over all of the continents except Baltica and the southern European and northwestern African parts of Gondwana, extensive areas of carbonate sediments developed. At least five times in North America and three of the those five times in Australia, large parts of the animal populations of these seas became extinct and had to be replenished from the oceanic regions. The last of these extinction events marks the end of Cambrian time in North America.

Throughout Cambrian time, landscapes were stark and barren. Life in the sea was primitive and struggling for existence. Only in post-Cambrian time did the shallow marine environment stabilize and marine life really flourish. Only then did vertebrates evolve and plants and animals invade the land.

[ALLISON R. PALMER]

Bibliography: C. H. Holland (ed.), *Cambrian of the British Isles, Norden and Spitzbergen*, 1974; C. H. Holland (ed.), *Cambrian of the New World*, 1971; A. R. Palmer, Search for the Cambrian world, *Amer. Sci.*, 62:216–224, 1974.

Cameo

A type of carved gemstone in which the background is cut away to leave the subject in relief. Often cameos are cut from stones in which the coloring is layered, resulting in a figure of one color and a background of another. The term cameo, when used without qualification, is usually reserved for those cut from a gem mineral, although they are known also as stone cameos. The commonly encountered cameo cut from shell is properly called a shell cameo.

Most cameos are cut from onyx or agate, but many other varieties of quartz, such as tiger's-eye, bloodstone, sard, carnelian, and amethyst, are used; other materials used include beryl, malachite, hematite, labradorite, and moonstone. *See* GEM. [RICHARD T. LIDDICOAT, JR.]

Cancrinite

A mineral tectosilicate belonging to the feldspathoid group and crystallizing in the hexagonal system. The rare crystals are prismatic and have a perfect prismatic cleavage; the mineral is usually in compact or disseminated masses (see illustration). The hardness is 5–6 on Mohs scale, and the specific gravity is 2.45.

The luster is pearly on the cleavages and the color is usually brownish-yellow but may be reddish, green, white, or colorless. The composition has the formula $Na_3CaAl_3Si_3O_{12}CO_3(OH)_2$, but there is considerable variation with SO_4 substituting for CO_3. Cancrinite occurs in nepheline syenites where it may be a primary mineral, but more commonly it is formed by the alteration of nepheline. It is found at Miask in the Ural Mountains; in Finland, Sweden, and Ontario, Canada; and at Litchfield, Maine. *See* FELDSPATHOID; NEPHELINE SYENITE; SILICATE MINERALS.

[CORNELIUS S. HURLBUT, JR.]

Cancrinite with biotite in nepheline syenite rock, Bigwood Township, Ontario, Canada. (*Specimen from Department of Geology, Bryn Mawr College*)

Carat

The unit of weight now used for all gemstones except pearls. It is also called the metric carat (m.c.). By international agreement, the carat weight is set at 200 milligrams (mg). Pearls are weighed in grains, a unit of weight equal to 50 mg, or 1/4 carat.

Despite the great value of many of the gemstones weighed against a unit called the carat, the weight differed from country to country until early in the 20th century. Although most of the weights to which the name was applied were near 205 mg, the range was from less than 190 to over 210 mg. The 200-mg carat was proposed in 1907 in Paris. This weight slowly gained acceptance, and by 1914 most of the important nations in the gem trade had accepted the metric carat as the legal standard for gemstones.

The word carat comes from the Greek word for the locust tree that is common in the Mediterranean area. This tree produces seeds that are fairly uniform in weight, averaging about 205 mg each. As a result, locust seeds came to be used for gem weight comparisons.

The application of the term carat as a unit of weight must not be confused with the term karat used to indicate fineness or purity of the gold in which gems are mounted.

[RICHARD T. LIDDICOAT, JR.]

Carbonate minerals

Mineral species containing CO_3^{--} ion as the fundamental anionic unit. The carbonate minerals can be classified as (1) anhydrous normal carbonates, (2) hydrated normal carbonates, (3) acid carbonates (bicarbonates), and (4) compound carbonates containing hydroxide, halide, or other anions in addition to the carbonate.

Most of the common carbonate minerals belong to group (1), and can be further classified according to their structures. The rhombohedral carbonates are typified by calcite, $CaCO_3$, and by dolomite, $CaMg(CO_3)_2$. Minerals with the calcite structure include magnesite, $MgCO_3$; siderite, $FeCO_3$; rhodochrosite, $MnCO_3$; smithsonite, $ZnCO_3$; cobaltocalcite, $CoCO_3$; otavite, $CdCO_3$; and gaspeite, $NiCO_3$. Ankerite, $Ca(Fe, Mg)(CO_3)_2$, and kutnahorite, $CaMn(CO_3)_2$, have the dolomite structure. The other structural type within this group is that of aragonite, a metastable high-pressure polymorph of $CaCO_3$, which has orthorhombic symmetry. Minerals with the aragonite structure are strontianite, $SrCO_3$; witherite, $BaCO_3$; and cerussite, $PbCO_3$. Less common group (1) minerals which do not fall into these three structural types include vaterite, another metastable polymorph of $CaCO_3$; huntite, $Mg_3Ca(CO_3)_4$; and several other multication minerals. *See* ANKERITE; ARAGONITE; CALCITE; CERUSSITE; DOLOMITE; HUNTITE; KUTNAHORITE; MAGNESITE; RHODOCHROSITE; SIDERITE; SMITHSONITE; STRONTIANITE; WITHERITE.

The minerals in groups (2) and (3) all decompose at relatively low temperatures and therefore occur only in sedimentary deposits (typically evaporites) and as low-temperature hydrothermal alteration products. The only common mineral in these groups is trona, $Na_3H(CO_3)_2 \cdot 2H_2O$. *See* EVAPORITE, SALINE.

Similarly, the group (4) minerals are relatively rare and are characteristically low-temperature hydrothermal alteration products. The commonest members of this group are malachite, $Cu_2CO_3(OH)_2$, and azurite, $Cu_3(CO_3)_2(OH)_2$, which are often found in copper ore deposits. *See* AZURITE; MALACHITE.

Important occurrences of carbonates include ultrabasic igneous rocks such as carbonatites and serpentinites, and metamorphosed carbonate sediments, which may recrystallize to form marble. The major occurrences of carbonates, however, are in sedimentary deposits as limestone and dolomite rock. *See* DOLOMITE ROCK; LIMESTONE; MARBLE; ROCK; SEDIMENTARY ROCKS.

[ALAN M. GAINES]

Bibliography: G. V. Chilingar, H. J. Bissell, and R. W. Fairbridge (eds.), *Developments in Sedimentology*, vol. 9B: *Carbonate Rocks*, 1967; W. A. Deer, R. A. Howie, and J. Zussman, *Rock Forming Minerals*, vol. 5: *Non-silicates*, 1962; E. T. Degens, *Geochemistry of Sediments*, 1965; J. R. Goldsmith, Some aspects of the geochemistry of carbonates, in P. H. Abelson (ed.), *Researches in Geochemistry*, 1959; D. L. Graf and J. E. Lamar, Properties of calcium and magnesium carbonates and their bearing on some uses of carbonate rocks, *Econ. Geol.*, 50th anniversary vol., pp. 639–713, 1955; C. Palache, H. Berman, and C. Frondel, *Dana's System of Mineralogy*, vol. 2, 7th ed., 1951.

Carbonatites

Carbonate-rich rocks of magmatic derivation or descent. Carbonatites, among the rarest rocks of magmatic descent, are important scientifically because their study provides information (1) on the origin of their associated subsilicic alkalic rocks (nepheline syenites and so forth), (2) on the origin of low-silicate or nonsilicate magmas, (3) on explosive vulcanism, (4) on the origin of kimberlites and subcrustal processes, and (5) on the tectonics of cratonic areas. Their economic significance also is immense. In addition to being the major source of niobium (chiefly as pyrochlore), some carbonatites yield large amounts of rare-earth elements (for example, at Mountain Pass, Calif.) and others are sources or potential sources for phosphate (apatite), barite, thorium, vermiculite, magnetite, titanium, and carbonate rock. One of the world's largest copper deposits (at Palabora, South Africa) is in a carbonatite.

A carbonatite may have crystallized from a carbonatitic magma or have been deposited by CO_2-rich fluids fractionated from such magmas. The chief carbonate species are calcite (usually Sr and Ba rich), dolomite, and ankerite. Most of the features that characterize silicate igneous rocks also typify carbonatites. There are both intrusive and extrusive types. Intrusive types, the more common, have been emplaced at various depth levels (plutonic, hypabyssal, and subvolcanic). Extrusive varieties are known both as flows and pyroclastics. Carbonatites are spatially, chemically, mineralogically, and genetically related to alkalic subsilic rocks (syenites, foidal syenites, ijolites, and kimberlites), and the emplacement of such carbonatitic–alkalic complexes has been accompanied by distinctive metasomatic transformations of their wall rocks, a process called fenitization, in which, in a typical example, host granite gneiss is converted to metasomatic alkalic syenite (fenite) by the formation of aegirine and potash feldspar with or without albite and sodic amphibole, with quartz being removed.

Carbonatites, being usually the youngest rocks in their multi-intrusive complexes, are subject to manifold deuteric and autohydrothermal mineralogical changes. Many support cognate xenoliths that may be oriented, reacted upon, or partly assimilated. Textures are simple to heterogeneous. Some are granitoid; others show well-defined primary foliation; and a few are porphyritic with either carbonate or silicate phenocrysts. Many carbonatite bodies are multiphase, and these display abrupt and marked changes in texture, grain size, and mineralogy. Carbonatites occur as dikes (some with chilled margins), sills, sheets, pipes, stocks, and more irregular bodies, all of limited size. Those forming the more-or-less central pipes of ring dike–cone sheet complexes generally range in size from less than 1 to about 2–3 mi^2. The ratio of the area of carbonatite to that of the ring complex is from <0.01 to >0.5.

Carbonatites tend to occur in broad petrographic provinces characterized by a cratonic tectonic environment, closely associated major faults (rift valleys and lineaments), explosive (Mt. Elgon–type) vulcanism, associated lamprophyres, kimberlites, and miascitic alkalic rocks (not agpaitic types). They range in age from Precambrian to Recent. Oldonyo Lengai in Tanzania erupted Na-carbonatite in 1960.

By 1956 about 35 carbonatites had been recognized, and the known localities in 1968 numbered nearly 350. They have been found on all continents except Australia and Antarctica. Some of the best-known occurrences of carbonatites are in the following countries:

United States: Magnet Cove, Ark.; Iron Hill, Colo.; Mountain Pass, Calif.

Canada: Oka, Quebec; Lake Nipissing, Ontario.

Brazil: Araxá and Tapira, Minas Gerais.

Europe: Fen, Norway; Alnö, Sweden; Kaiserstuhl, Baden, Germany.

Soviet Union: Kola Peninsula; Urals; Tuva region, Meimecha–Kotui region, Siberia.

Africa: Sukulu and Tororo, Uganda; Mrima Hill and Homa Mountain, Kenya; Oldoinyo Lengai, Sangu, Ngualla, and Mbeya, Tanzania; Kaluwe, Zambia; Dorowa and Shawa, Rhodesia; Chilwa Island and Tundulu, Malawi; Muambe, Mozambique; Palabora and Spitskop, South Africa.

Asia: Amba Dongar in Gujarat, India.

Carbonatite occurrences may be grouped according to the following outline:

I. Associated with alkalic ring complexes.
   A. Core carbonatites (plugs or stocks), cone sheets, ring dikes, and other dikes of carbonatite.
   B. Dikes, ring dikes, cone sheets, or breccia zones of carbonatite without a core carbonatite.
   C. Ring complexes essentially of only a carbonatite stock plus fenite.
II. Associated with nonring-type alkalic complexes.
   A. Large, thick sheets.
   B. Dike swarms, sills, and stockworks within or satellite to irregular alkalic complexes.
III. Not directly associated with alkalic complexes.
   A. Large composite sheets.
   B. Dike and sill swarms.
IV. Extrusive carbonatites.
   A. Flows.
   B. Pyroclastic.

About 170 primary minerals have been recognized in carbonatites, plus about 20 others of supergene origin. Chief components, beyond the common carbonates, are rare-earth carbonates, apatite, monazite, barite, fluorite, pyrite, magnetite, hematite, perovskite, rutile, pyrochlore, fersmite, columbite, forsterite, monticellite, Ca-Ti-Zr garnets, diopside, aegirine, richterite, riebeckite, biotite, phlogopite, potash feldspar, albite, and quartz (in late veinlike carbonatites). The sequence of crystallization involves the following stages: (1) silicates, (2) carbonates in which the usual sequence is Sr-Ba calcite, dolomite, and ankerite, (3) pyrochlore (4) rare-earth carbonate (deuteric stage), and (5) veinlet.

Although they are characterized by a distinctive minor element assemblage (Ti, Nb, Zr, rare earths of the Ce subgroup, P, F, Ba, Sr, and Th), carbonatites differ in the relative concentrations of these various minor elements (that is, a niobium type versus a rare-earth type) and in the way these elements may be represented mineralogically. Minor elements may be camouflaged in early carbonate or silicate species or, in highly differentiated carbonatites, they may form their own minerals.

The strontium isotope ratios of carbonatites are useful for recognizing these rocks. These Sr^{87}/Sr^{86} ratios, distinctly lower than those of sedimentary (and metamorphic) carbonate rocks, are identical with those of their associated silicate rocks. Studies of oxygen and carbon isotope ratios in carbonatites support the thesis that some carbonatites are magmatic and others hydrothermal.

Simple carbonatite magmas have been synthesized in the range of 680–450°C and 1000–10 bars, with environments in good accord with those deduced from geological and mineralogical evidence.

Carbonatite magmas are secondary magmas derived by a complex series of fractional crystallizations from subcrustal (probably kimberlitic) material. Such development takes place in tectonically stable continental areas over a very long period of time. Carbonatitic–alkalic complexes are not unique merely because of their contents of juvenile carbon dioxide and juvenile carbonate species, but also in that they represent the end product of development within a magmatic–tectonic environment that has produced local, extraordinary concentrations of carbon dioxide and carbon dioxide–derivative minerals and phenomena. Other magmas have similar initial carbon dioxide concentrations, which are normally gradually dissipated. Only in carbonatites is the original carbon dioxide retained and accumulated and its concentration multiplied. *See* IGNEOUS ROCKS. [E. WILLIAM HEINRICH]

Carboniferous

A division of later Paleozoic rocks, economically important because of the fossil fuel (coal) deposits that its name implies. The Carboniferous represents a 65×10^6 year time span that began 345×10^6 years ago and ended 280×10^6 years ago.

PRECAMBRIAN	PALEOZOIC							MESOZOIC			CENOZOIC	
					CARBONIFEROUS							
	CAMBRIAN	ORDOVICIAN	SILURIAN	DEVONIAN	Mississippian	Pennsylvanian	PERMIAN	TRIASSIC	JURASSIC	CRETACEOUS	TERTIARY	QUATERNARY

The name Carboniferous was first applied, in 1822, to the coal-bearing rocks in England: it is now widely used in Europe, Asia, northern Africa, northern South America, and Greenland. In North America the term Carboniferous is not widely used; instead, the time interval and representative rock record is subdivided into two major units, the earlier Mississippian period (345 to 310×10^6 years) and the later Pennsylvanian period (310 to 280×10^6 years). In southern Africa, southern South America, and Antarctica, the areas referred to as Gondwanaland, the late Paleozoic and early Mesozoic Gondwana System, include Carboniferous age rocks. *See* MISSISSIPPIAN; PENNSYLVANIAN.

Although no type locality was designated, the Pennine uplands of Yorkshire and Derbyshire, England, were intended to be used as the typical area of the Carboniferous system of rocks and period of time and are limited below (earlier) by the Devonian and above (later) by the Permian. The Carboniferous rocks are separated from the overlying and underlying systems by unconformities in many places; however, where there are no physical interruptions, the positions of the lower and upper limits have been based on a comparison of fossil faunas and floras with those from the type Devonian (southwestern England) and the type Permian (European Soviet Union). A widespread and distinct unconformity separates the Lower Carboniferous (Mississippian) rocks from the Upper Carboniferous (Pennsylvanian) rocks throughout much of North America. The recognition of this unconformity, separating the largely non-coal-bearing Pennsylvanian strata, led to the early disuse of the term Carboniferous in North America. *See* DEVONIAN; PERMIAN.

Distribution. Carboniferous rocks are widespread on all continents, being absent or poorly known in areas of much older rocks, such as the Canadian or African Shield, or in areas where younger Mesozoic and Cenozoic sediments and volcanics constitute the bulk of the rocks. In the United States, Carboniferous rocks are widespread; they are absent only in northern New England, parts of the Atlantic Coastal Plain, Wisconsin, Minnesota, and Hawaii. In Canada, Carboniferous sequences are present in the Maritime Provinces, the Canadian Rockies, the Coast Ranges, and the Arctic Archipelago, including Ellesmere Island, 84°N. Little Carboniferous is found in Mexico, Central America, or the West Indies. In South America, the Carboniferous is present in Colombia, Peru, and Venezuela. In the British Isles, there are important Carboniferous basins in the Pennines and central England coalfields, southern Wales, the Scottish midlands, and Ireland. Well-known European Carboniferous basins are the Westphalian (Germany, Belgium, and Netherlands), Silesian (Poland), Moscow and Donets (Soviet Union), and smaller basins in France, Spain, Italy, Austria, Czechoslovakia, and Switzerland. In Asia, important Carboniferous areas are in Japan, northern China and Korea, southeastern Asia, and Turkey (Asia Minor). There are Carboniferous basins in Morocco, Algeria, Libya, and Egypt, and outcrop areas in eastern Greenland and Spitsbergen.

The differentiation of the Carboniferous in the Gondwana System is difficult because few diagnostic animal or plant fossils are common to Gondwanaland and the rest of the world. The most noteworthy marine Carboniferous rocks in the Gondwana area are in the Tasman geosyncline of eastern Australia, where total thickness exceeds 9000 m. Basal glacial sediments of the Gondwana succession are generally unfossiliferous and therefore difficult to date. The Carboniferous age of some sequences in Brazil, Paraguay, and Uruguay (South America), eastern Australia, southern Africa, and Antarctica is generally accepted. Glacial successions less certainly Carboniferous are found in Victoria, South Australia, Tasmania, and Western Australia; Madagascar; Argentina; Chile; the Falkland Islands; and India.

Classification. The importance of coal research, records of Carboniferous life, depositional and tectonic history, and intercontinental correlations prompted six International Congresses of Carboniferous Stratigraphy and Geology between 1927 and

Principal major series names applied to Carboniferous rocks in western Europe and in North America

Western Europe	United States	
Stephanian (A and B)	Virgilian	Pennsylvanian
Westphalian (Upper D)	Missourian	
Westphalian (C and lower D)	Desmoinesian	
Westphalian (B)	Atokan	
Westphalian (A)	Morrowan	
Namurian (B and C)	Springeran	
Namurian (A)	Upper Chesterian	Mississippian
Upper Visean	Lower Chesterian	
Lower Visean	Meramecian	
Upper Dinantian	Osagean	
Lower Dinantian	Kinderhookian	

1967. These congresses proposed that Carboniferous stratigraphic series successions are the result of different events and exhibit different facies (see table). European stage terminology is not extensively used in North America, largely because the European stage boundaries are based on plant index fossils, whereas in North America the Carboniferous successions are dominated by marine rocks in which plant fossils are not abundant. In Gondwanaland, where the oldest deposits are glacial, the series are referred to as Talchir in India, Dwyka in Africa, Kuttung in eastern Australia, and Itarari in Brazil. In most of these areas, strata overlying the glacial deposits are coal-bearing and are generally considered to be Permian in age because of a distinctive flora which is abundant in Gondwanaland but absent elsewhere.

Correlation. Interbasin correlation of Lower Carboniferous (Mississippian) sequences has been based largely on the use of marine index fossils, such as brachiopods, crinoids, blastoids, foraminifera, and conodonts (microfossils of unknown affinity). *See* INDEX FOSSIL.

Correlation between separated Late Carboniferous (Pennsylvanian) marine sequences of the world are based on fusulinids (large foraminifera), of which particular genera and species are limited to small spans of time and are excellent index fossils. Where the sequences are dominantly nonmarine or transitional and coal-bearing, plant spores extracted from coals and shales have been used as index fossils for short time intervals in western Europe, Maritime Canada, and the eastern United States. The spores, distributed by the wind, are also found in marine sediments. Generally, the spore assemblages reflect the local flora, which is controlled by environment, climate, and water table, as well as evolutionary sequence (time).

Gondwana areas have a limited number of useful spore associations, permitting correlations of deposits as distant as South America and Australia. There is little or no intermingling of spore associations of the Gondwana type with those of North America and Eurasia. For correlation of the Gondwana sequences with Carboniferous sequences of the northern areas, goniatites (ammonoid cephalopods) have been used. Impressions of foliage, roots, seeds, and sporangia of land plants have been widely used for interbasin and intercontinental correlation. A few species of fossil foliage appear to be common to Gondwana and northern areas, offering some prospect for more accurate correlation between these two diverse parts of the Carboniferous world.

Carboniferous sediments. Carboniferous successions include virtually all types of sedimentary rocks. Volcanic rocks are uncommon in North America, limited largely to the far northwestern United States, western British Columbia, and the Boston Basin in Massachusetts. Intrusions and interbedded volcanics are present in the Lower Carboniferous of central England, the Midland Valley of Scotland, and the Tasman geosyncline of eastern Australia.

The Lower Carboniferous has thick marine carbonates (limestones and dolomites) in the British Isles and borders of the Westphalian Basin. In North America there are carbonates in equivalent intervals (Mississippian) in the Rocky Mountains, the Colorado Plateau, and the Great Basin. Although interrupted by the Transcontinental Arch from South Dakota to New Mexico, Lower and Middle Mississippian (Dinantian and Visean) successions are predominantly marine carbonate rocks from Iowa southwest to Texas, and east to Ohio, West Virginia, Tennessee, and Alabama. The Mammoth Cave of Kentucky occurs in Middle Mississippian limestone.

In the Upper Visean and Lower Namurian (Millstone Grit) of northern England, the Midland Valley of Scotland, Belgium, and western Germany, a major change in sedimentary facies is apparent. Sandstones, shales, and occasional coalbeds are interbedded with limestones as repeated cyclic units 6 m thick or less. Similar cyclic successions, limestone at the base, thick shale in the middle, and sandstone at the top, are very well developed in the Chester Series (Upper Mississippian) of the central and east-central United States from northern Alabama and eastern Tennessee and Kentucky northwestward to western Illinois. These successions are interpreted as a record of the seaward progradation and abandonment of numerous deltas onto a widespread shallow marine shelf.

In the later Namurian and Westphalian (Pennsylvanian), cyclic sedimentation continued in the British Isles and western Europe, but marine limestones are much less frequent, and coal beds thicker and more important as fuel resources. Widespread marine transgressions are recorded as beds of shale only a few feet thick, but traceable as far as 1100 km between Great Britain and Germany. These key zones have been correlated by using species of goniatites (cephalopods). Correlations have also been made in Great Britain and on the European continent utilizing thin beds of hard kaolinitic clay (tonstein), each bed 25 cm or less in thickness and continuous for hundreds of kilometers. The origin of these thin widespread beds is unclear, but they may represent ashfalls from volcanoes. Because they are more persistent than beds of sandstone, mudstone, and coal, they are used extensively as boundaries between series and smaller stratigraphic subdivisions. For example, Tonstein 60 was officially adopted as the boundary between the Westphalian and Stephanian.

In the United States, there is a widespread unconformity between the Mississippian (Lower Carboniferous) and Pennsylvanian (Upper Carboniferous) systems. Strata equivalent to Middle and Upper Namurian are absent except in bordering geosynclines in the eastern (Appalachian), southern (Ouachita), and western (Cordilleran) United

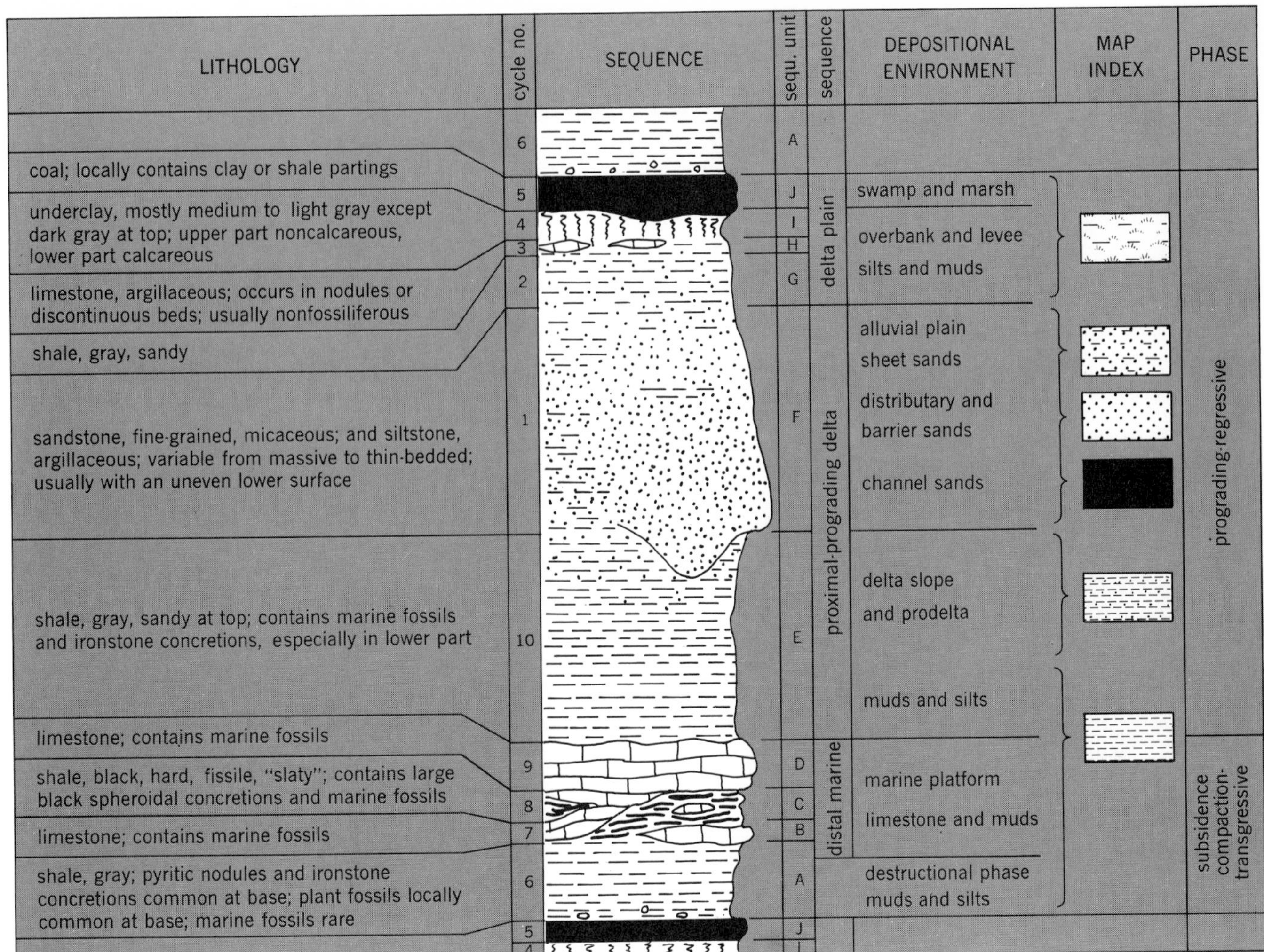

Fig. 1. Idealized Pennsylvanian cyclothem with depositional environments. Map index patterns refer to Fig. 2.

States. In the central and eastern United States, erosional valleys locally as deep as 75 m were cut into Mississippian strata and later filled by Early Pennsylvanian riverine sandstones. Where limestones and dolomites were exposed at this Late Mississippian erosional surface, sink holes and caves developed from Illinois west to Utah and from Montana south to Arizona. These solution cavities subsequently were filled with refractory clays, iron oxides, and iron sulfides, and, locally, with zinc, lead, and copper sulfides. Filled sink holes are well displayed in the Ozark uplands of southern Missouri, where clay and metals have been mined. *See* UNCONFORMITY.

During the Early Pennsylvanian Morrowan (Westphalian A), much of the continental interior of the United States was a low-relief, hilly, erosional topography, with the earliest Pennsylvanian sediments filling in the valleys, caves, and sink holes. Many rock units are local and are difficult to correlate from place to place. Cyclic sedimentation was less prominent than in the Chester Series and later intervals. There were highlands due to tectonic uplift after the Mississippian, such as the Nemaha Ridge of Nebraska and Oklahoma and the LaSalle anticline of Illinois. Adjacent areas became sites of earlier Pennsylvanian deposition, but by the Desmoinesian (Westphalian C and lower D) most of the topographic irregularities had disappeared, and a nearly featureless alluvial plain and shallow marine shelf covered vast areas between the Colorado Rockies and the Appalachian highlands, a distance of about 2000 km. To the south, a deep, narrow basin (the Ouachita trough), extending from Alabama through Arkansas and curving south into northern Mexico, continued to receive bathyl turbidite sediments from the north, east, and, possibly, source lands to the south. During the Morrowan the pre-Pennsylvanian hills were inundated, first in Alabama and Mississippi in the southeast; and during the Atokan, level plains of deltaic and riverine deposition became widespread in Tennessee, Kentucky, southern West Virginia, southern Illinois, and northern Arkansas.

The Pennsylvanian sequence in the eastern and central United States exhibits some of the best-developed cyclic sedimentation in the world. Fifty-one formally named cycles (cyclothems) have been recognized in the Pennsylvanian of the Eastern Interior coal basin, each cyclothem being composed of a maximum of 10 individual and distinct rock types in a distinct sequence (Figs. 1 and 2). Ideally, this cyclic sequence is the result of prograding delta masses, and each sequence repre-

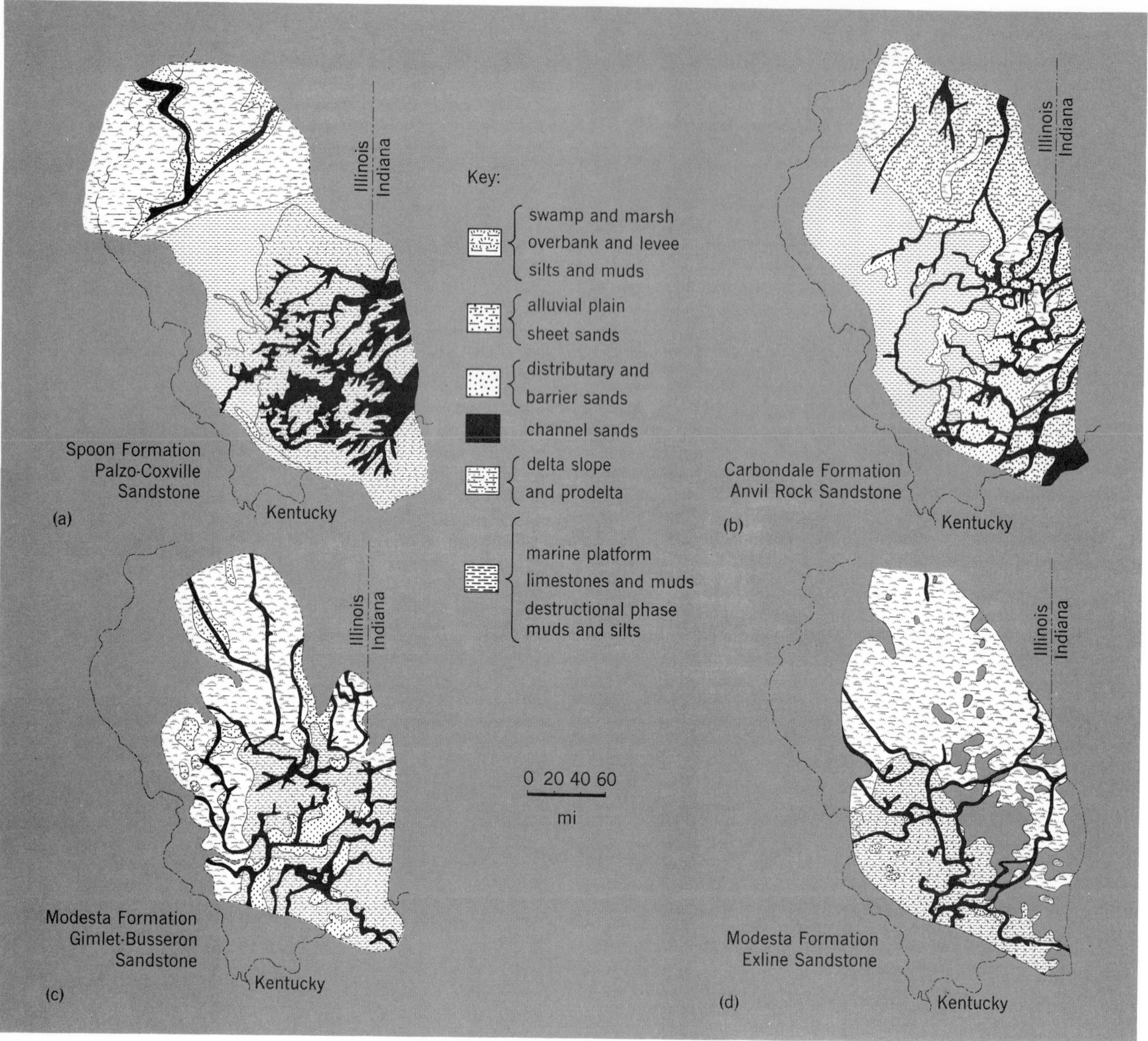

Fig. 2. Maps of several Pennsylvanian cyclothems of the Illinois Basin.

sents the initiation, growth, and abandonment of a delta mass on a shallow marine platform which, when repeated, resulted in cyclic alternations of detrital and nondetrital deposits. *See* CYCLOTHEM.

Cyclothemic sedimentation is especially well developed in the Desmoinesian Series. There are at least 20 marine limestones of Desmoinesian age in Missouri, and 13 in Illinois, and 6 marine shales or limestones in Ohio, and 1 in Pennsylvania. Most marine zones of Missouri are traceable westward across Kansas and eastern Colorado.

Tectonism during the Atokan or Early Desmoinesian produced several new uplands, including the Ouachita Mountains of Arkansas, Oklahoma, and Texas; the Wichita Mountains–Front Range uplift of Oklahoma, the Texas Panhandle, Colorado, and southern Wyoming; and the Uncompahgre uplift of Utah, Colorado, and New Mexico. Among the basins that developed or deepened at this time are the Paradox Basin of southwestern Colorado and southeastern Utah and the Eagle Valley of northwestern Colorado, west and east of the Uncompahgre uplift. In the Paradox Basin more than 1500 m of rock salt, potash salts, and anhydrite were deposited, and in the Eagle Valley basin more than 300 m of gypsum and anhydrite accumulated.

During the later Pennsylvanian, Missourian (Upper Westphalian), and Virgilian (Stephanian) epochs, cyclic sedimentation continued with the deposition of cyclothems that included 21 Missourian marine limestones in Missouri, one of which extended east to near Wilkes-Barre, PA, the most extensive marine transgression. During the Virgilian, 25 limestones were deposited in eastern Kansas, but marine conditions did not reach the Appalachian coalfield. In the northern Appalachians there are several thick beds of fresh-water

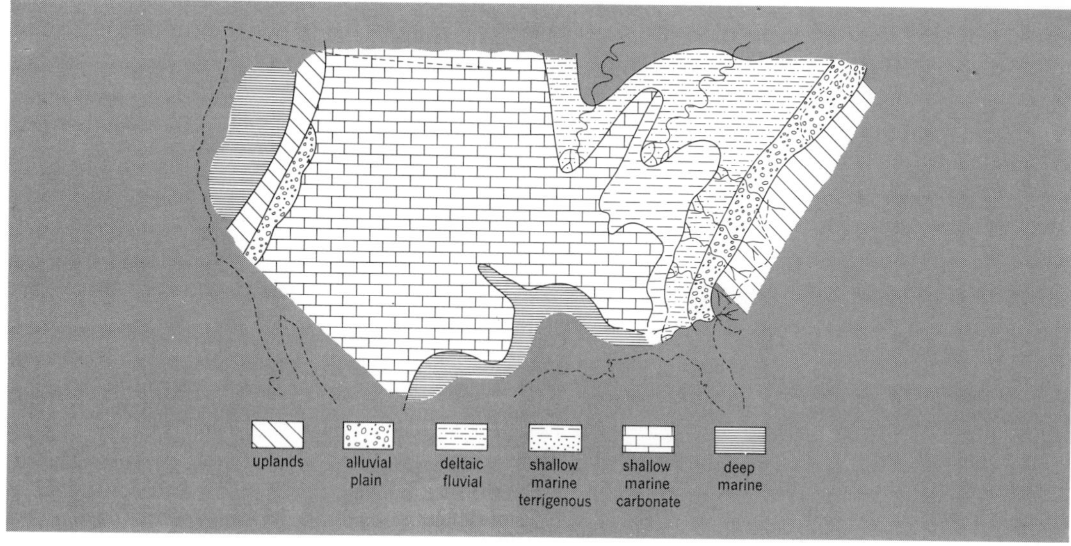

Fig. 3. Paleogeography of the United States during mid-Mississippian time.

(lacustrine) limestone, some of which have abundant invertebrate and fish remains. *See* SEDIMENTATION.

History of deposition and paleogeography. A useful approach to the reconstruction of the depositional history and understanding of the Mississippian and Pennsylvanian has been the preparation of paleoenvironmental maps, such as the one shown for the Pennsylvanian cyclic units in Fig. 2. Figures 3 and 4, which are summations of a series of such maps, show the average paleogeography for the Mississippian and Pennsylvanian time periods in the United States. These maps portray the succession of events, such as the advance and retreat of shorelines; the progradation of deltas; and the introduction of wedges of sand and mud into basins caused by the uplift of tectonic uplands or of increased upland erosion due to climatic change. *See* PALEOGEOGRAPHY.

Outside of the United States, cyclic sedimentation has been described in Morocco and Algeria, North Africa, the Silesian Basin in Poland, the Donets and Moscow basins in the Soviet Union, a large area in northern China and Korea, and Maritime Canada in rocks similar in age to those of the Middle and Upper Pennsylvanian of the United

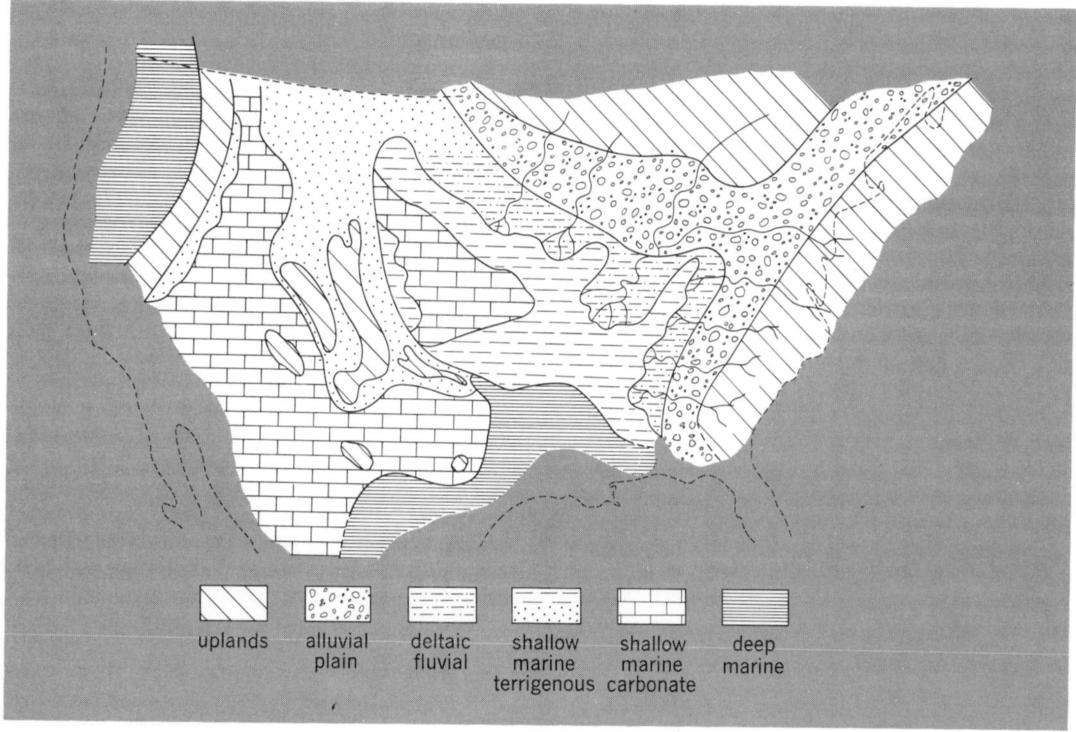

Fig. 4. Paleogeography of the United States during mid-Pennsylvanian time.

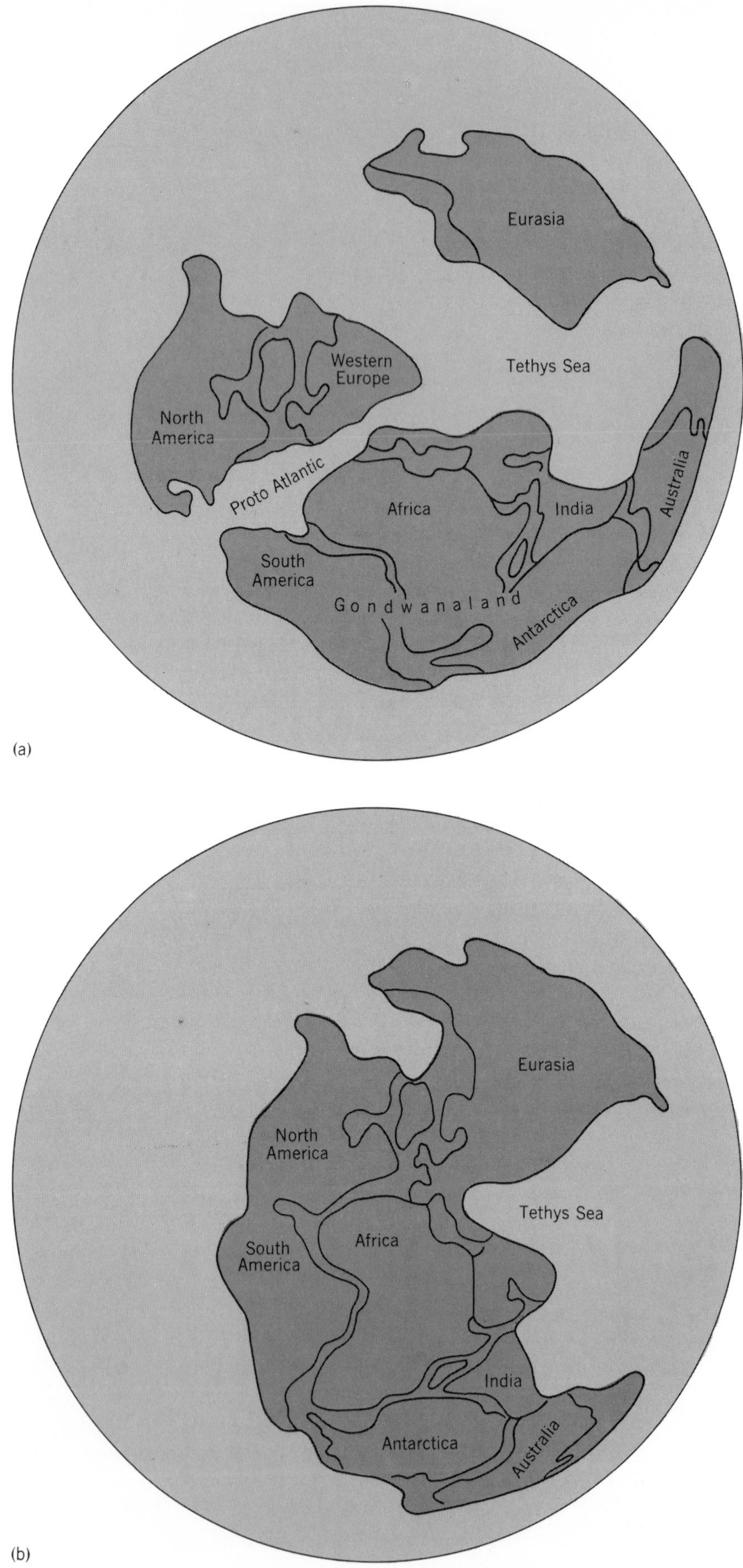

Fig. 5. Relative position of continents during (*a*) beginning and (*b*) end of Carboniferous time.

States, the British Isles, and western Europe.

Widespread development of cyclic sedimentation is not unique to the Carboniferous sediments of the world, but the phenomenon is most widely recorded and studied from these rocks. Some explanations for cyclic sedimentation on a broad scale are given below.

1. Repeated transgressions and regressions of the shoreline were the results of worldwide rising and lowering of sea level, from either upwarping or downwarping of the ocean floor, or from the withdrawal of water from the ocean to form Gondwana glaciers, followed by its return in interglacial times. Associated climatic changes played a part in geologic processes even on continents remote from glaciated areas.

2. Worldwide climatic cycles, not necessarily associated with glaciation, shifted the balance between erosion and less interrupted soil development.

3. Uplift of an area adjoining a sedimentary basin, increased erosion in the elevated area, and transportation of the detritus to the basin occurred. Then the upland was lowered, or at times reduced, by erosion, so that basin sediment spread widely. Then both basins and source uplands were depressed so that both areas became swampy and were later inundated. The cycle ended with renewed uplift, first in the source uplands and later in the basin.

4. A widespread shift in sites of successive deltas from the drainage system, as in the Mississippi subdeltas, took place. This theory assumes a constant progradation and abandonment of depositional systems in a regularly subsiding basin and noncontemporaneity of corresponding cyclic phases in different areas. This process is the most satisfactory, in that it can be observed in Recent sedimentary environments, and is becoming the accepted theory for the kinds of cyclic deposits so prominent in the Carboniferous. *See* FLUVIAL EROSION CYCLE.

Tectonics and continental drift. During the Carboniferous, Variscan mountain building in several stages affected the British Isles and western Europe, and the Kanimblan orogeny affected western Australia. Deformation also took place in the Appalachian region of the eastern United States; the Maritime Provinces of Canada; the Ouachita, Arbuckle, and Wichita mountain belts of Oklahoma, New Mexico, Colorado, and Utah; and the Antler orogenic belt in Nevada and Idaho. These and other uplifts produced highlands which yielded vast quantities of sediment to nearby and distant centers of deposition.

Theories of global tectonics and continental drift propose that at the beginning of Carboniferous time there were three major continental masses moving toward each other; they eventually came together at the end of the Carboniferous to form the supercontinent Pangaea. These three continents were North America–Greenland–western Europe; eastern Europe–northern and eastern Asia; and Gondwanaland (South America, Africa, India, Antarctica, and Australia). North America–western Europe was separated from northern Africa and eastern Europe by a narrow proto-Atlantic, and eastern Europe–northern Asia was separated from northern Africa and Gondwanaland by an eastern widening sea called Tethys (Fig. 5). It has been proposed that Gondwanaland was in the region of the South Pole during the Carboniferous and early Permian time. These interpretations are based on paleomagnetic studies of carboniferous polar positions and by the distribution of Carboniferous Gondwana glacial sediments. *See* CONTINENT FORMATION; TECTONIC PATTERNS.

Carboniferous life. During the Carboniferous the broad epeiric seas supported an abundant diversity of life of invertebrates, of which brachiopods, cephalopods, pelecypods (clams), gastropods (snails), crinoids, corals, and bryozoa, and, among the microfossils, fusulinids (large protozoans), ostracods (minute crustaceans), and conodonts, were abundant and diverse. Trilobites were present, but nearing the end of their long history. Corals formed reefs, and calcareous algae formed reeflike mounds in southern Kansas and western and central Texas. Although insect fossils are uncommon, insects were highly diverse and included large dragonflies and giant cockroaches.

Among vertebrates there were large sharks, some 6 m in length, adapted to fresh and marine waters, some with teeth for crushing shells. Amphibians, descendants of the first air-breathing vertebrates of the Devonian, were present. During the later Carboniferous the first reptiles appeared, but they were uncommon and small.

Fossil plants include large trees, some 30 m tall, in widespread swampy, coastal forests throughout many continents. Better-known varieties are relatives of the horsetail rushes, club mosses, and tree ferns of today. There were also primitive gymnosperms, related to conifers of today, which seem to have grown in somewhat less swampy and drier-climate areas. The first seed-bearing plants (pteridosperms), with foliage resembling ferns, appeared during the Middle Carboniferous. *See* PALEOBOTANY; PALEONTOLOGY.

Mineral resources. The Upper Carboniferous (Pennsylvanian) is the world's leading source of medium- and high-ranked coal. It is produced in the Appalachian coalfield, the Illinois Basin, and eastern Mid-Continent in the United States, the Maritime Provinces of eastern Canada, England, southern Wales, the Midland Valley of Scotland, the Westphalian coal basin of western Europe, the Silurian coal basin of Poland, the Donets Basin of the southern Soviet Union, and smaller basins in France, Spain, Morocco, Algeria, Norway, and northern China and Korea. *See* COAL.

Petroleum is produced from Carboniferous rocks in many parts of the central and western United States, western Canada, western Europe, and the North Sea. Oil shales occur in Carboniferous rocks in Scotland. Refractory clays (diaspore and kaolinite) are derived from Lower Pennsylvanian sediments in Pennsylvania, Ohio, eastern Kentucky, northern and western Illinois, eastern Missouri, and the Missouri Ozarks. Underclays or seat earths, associated with coals, are especially important sources of refractory clays. Many mudstones and shales in all continents are used for making brick and tile. Some sands in northern Tennessee are pure enough for glass sand. The Crab Orchard Sandstone (Pennsylvanian) of northern Tennessee and the Fountain-Lyons Sandstone (Pennsylvanian) of central Colorado are important building stones. Limestones are widely quarried for road material, fertilizer, lime, and cement. The Salem Limestone of southern Indiana (Middle Mississippian) is the leading building stone of the United States. Gypsum is produced from the Eagle Valley basin (Pennsylvanian) of Colorado, the Middle Mississippian rocks of western Indiana, and the Mississippian of Nova Scotia. Salt and potash salts are produced from the Paradox Basin of southwestern Colorado and southeastern Utah (Middle Pennsylvanian). Iron ores have been produced from sink fillings in the Missouri Ozarks, and lead, zinc, and copper ores are mined from solution cavities in Mississippian limestone in Missouri, Kansas, and Oklahoma. Lead and zinc are mined from veins associated with dolerite intrusives in Visean limestones of northern England. Fluorite is produced from veins in Mississippian limestones and sandstones in Illinois and western Kentucky, Flints from the Mississippian and Pennsylvanian of the Ohio Valley were important items of trade for aboriginal Americans. *See* PETROLEUM, ORIGIN OF.

[WAYNE A. PRYOR]

Bibliography: G. Briggs, *Carboniferous of the Southeastern United States,* Geol. Soc. Amer. Spec. Pap. no. 148, 1974; H. R. Wanless, Marine and non-marine facies of the Upper Carboniferous of North America, *International Congress of Carboniferous Stratigraphy and Geology*, 1(6):293–336, 1967.

Carnotite

A mineral which is a hydrous vanadate of potassium and uranium, $K_2(UO_2)_2(VO_4)_2 \cdot nH_2O$. The water content varies at ordinary temperatures from one to three molecules.

Carnotite generally occurs as a powder or as a slightly coherent microcrystalline aggregate. Its microscopic crystals are monoclinic. Color ranges from bright yellow to lemon- and greenish-yellow.

Carnotite is a secondary mineral resulting from the action of groundwater on preexisting uranium-bearing minerals. In the United States the principal region of carnotite mineralization is the Colorado Plateau and adjoining districts of Utah, New Mexico, and Arizona, where it occurs as relatively pure masses near fossilized tree trunks and other vegetal matter and as disseminated grains in cross-bedded sandstones of Triassic and Jurassic age. Carnotite is found also in Wyoming and in Carbon County, Pa. Deposits are located at Radium Hill near Olary, Australia, and in Katanga (in the former Belgian Congo). Carnotite is the chief source of uranium in the United States. It is also a source of radium and vanadium. *See* RADIOACTIVE MINERALS.

[WAYNE R. LOWELL]

Cassiterite

A mineral having the composition SnO_2. Cassiterite is the principal ore of tin. It crystallizes in the tetragonal system in prisms terminated by dipyramids. Twins with a characteristic notch—called visor tin are common (see illustration). Cassiterite is usually massive granular, but may be in radiating fibrous aggregates with reniform shapes (wood tin). The hardness is 6–7 (Mohs scale), and the specific gravity is 6.8–7.1 (unusually high for a nonmetallic mineral). The luster is adamantine to submetallic. Pure tin oxide is white, but cassiterite is usually yellow, brown, or black because of the presence of iron.

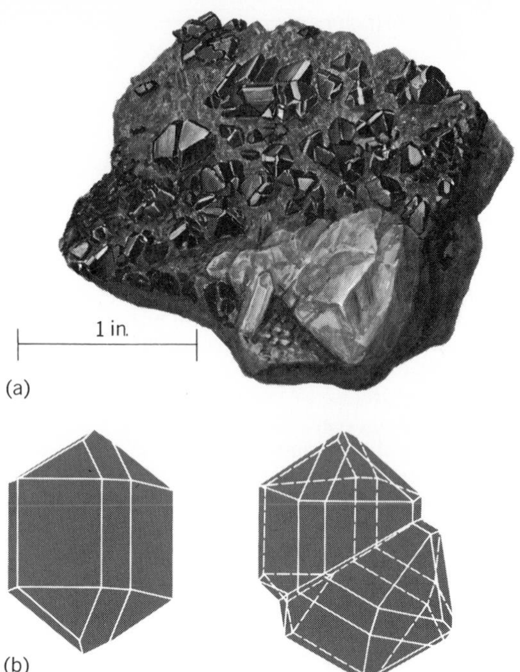

Cassiterite. (*a*) Crystals in granite-pegmatite rock from Temescol Mines, San Bernardino County, Calif. (*specimen from Department of Geology, Bryn Mawr College*). (*b*) Crystal habits (*from C. S. Hurlbut, Jr., Dana's Manual of Mineralogy, 17th ed., copyright © 1959 by John Wiley & Sons, Inc.; reprinted by permission*).

Cassiterite is genetically associated with granitic rocks and pegmatites and may be present as a primary constituent. More commonly it is found in quartz veins in or near such rocks associated with tourmaline, topaz, and wolframite. Cassiterite is most abundantly found as stream tin (rolled pebbles in placer deposits). Early production of tin came from cassiterite mined at Cornwall, England, but at present, the world's supply comes mostly from placer or residual deposits in the Malay States, Indonesia, the Belgian Congo, and Nigeria. It is also mined in Bolivia.

[CORNELIUS S. HURLBUT, JR.]

Cave

A natural cavity in rock beneath the surface of the earth. Most caves have been formed in calcareous rock (limestone and dolomite) by the dissolving action of groundwater circulating along the partings in the rocks. Such partings include existing joints and bedding (stratification) planes. Joints commonly occur in sets as semiparallel and nearly vertical planes, the sets intersecting each other at large angles. Bedding planes usually are horizontal. The intersection of these planes accounts for the three-dimensional pattern observed in caves, but in most caves the horizontal extent greatly exceeds vertical development.

Only beneath swamps and marshes does groundwater completely saturate subjacent soil and the deeper bedrock. Beneath uplands the saturated zone may be far below the surface of the land with a zone of aeration intervening. In this zone, mete-

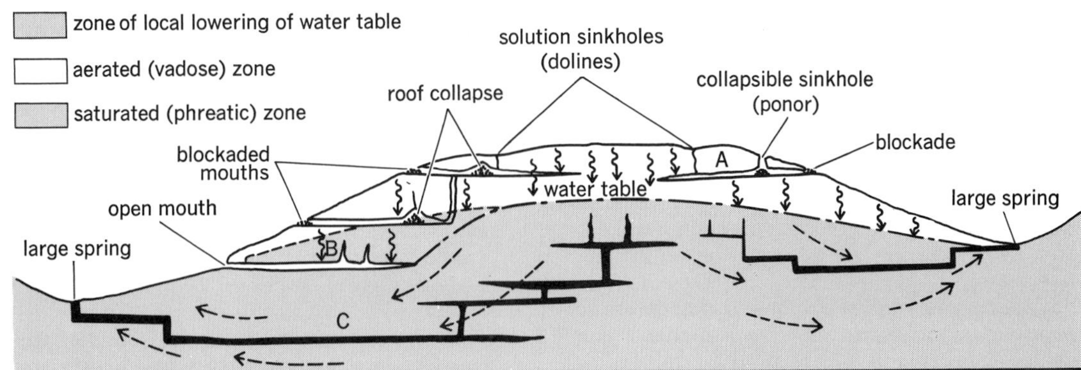

Fig. 1. Generalized cross section showing relations of caves to water table in horizontally bedded, vertically jointed rock. A indicates streamless caves; they may have wet-weather contributions from sinkholes. B indicates caves with streams on the floor; they may have perennial flow if they tap the saturated zone. C indicates tube-full caves; they are below the water table and are in the process of growing caves.

oric water (of atmospheric origin) moves downward through the partings and interstitial spaces until it reaches the top of the saturated zone, or water table. Below the water table, groundwater movement is largely lateral. Where the underground flow nears the surface of adjacent lowlands, it may emerge as springs and seepages. The zone of aeration is known as the vadose zone, and the saturated zone as the phreatic zone. Enterable caves must be in the vadose zone.

Many caves have streams with fairly low gradients. This has given rise to the idea that the cave is essentially a stream-made valley with a roof, the vertical dimensions of the cave being considered a measure of the stream's downcutting since its initiation as a subterranean watercourse. Dry caves, by this view, have simply lost their causal streams. In 1930 W. M. Davis challenged this once prevalent idea. Davis proposed that the cave-floor stream found its route already largely developed; that the cave had been made earlier by a hydraulic circulation below the water table, a circulation comparable to that in a water main under hydrostatic pressure. Under these conditions in the phreatic or saturated zone, tube-full flow attacked walls and ceilings as well as floors. Caves thus become enlarged upward and sidewise as readily as they do downward.

The Davis theory considered that, because of later deepening of adjacent surface river valleys, the water table became lowered to drain the cave and allow air to enter for the first time. Being thus shifted to the vadose zone, the cave collected drip water to constitute the free-surface stream on the floor (Fig. 1).

One of the most convincing evidences for the Davis concept is the ground plan of intersecting joint-controlled passages in many caves. Cameron Cave, near Hannibal, Mo., is an outstanding illustration of the network type of cave (Fig. 2). With an essentially level floor, the area involved is about 30 acres while the total of all passage lengths is almost 4 mi. Four sets of joints are recognizable in Cameron's labyrinthine maze, and widths of passages along them are nearly uniform. The pattern cannot be resolved into a main channel and its tributaries. Cameron Cave underlies the full width of a hill about 150 ft high and has about two dozen passage terminations in the hill's lower slopes, now blockaded by debris which has crept or slid down the slope. An impervious shale overlies the limestone formation which contains the cave and has denied any possible entrance of water from above. The many intersections with the hillside make it clear that the cave system is older than the stream-carved topography of the vicinity. This conclusion is confirmed by the existence of another and very similar cave, Mark Twain Cave, just across the local stream valley. The two are but surviving portions of a once larger cave, now largely destroyed by the severing valley. Both caves ceased to grow when that deepening valley lowered the water ta-

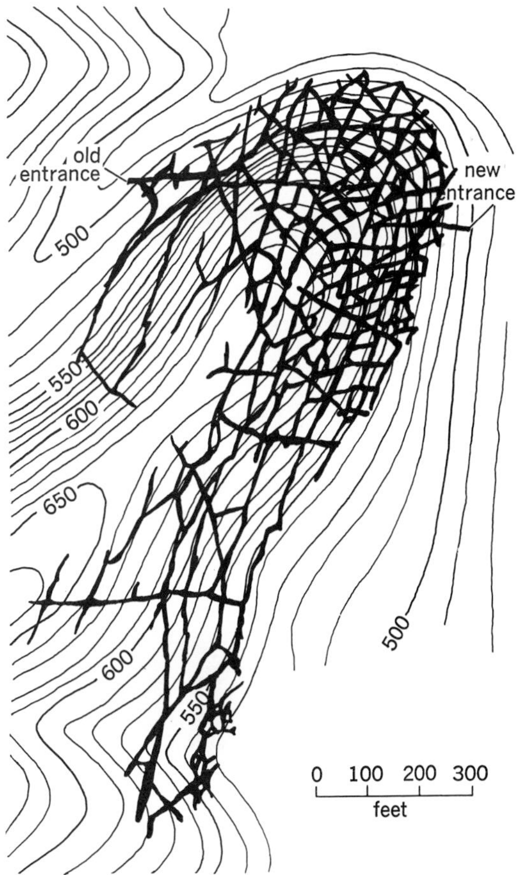

Fig. 2. Map of Cameron Cave, a network type of cave. (*Missouri Geological Survey*)

ble past their level and thus brought the phreatic circulation to an end. *See* JOINT.

Bedding planes instead of joints may be the dominant factor that determines the ground plan of many caves. A recurring feature in such caves is that of satellitic caves connected with the main cave. Such systems of repeatedly uniting and dividing passages may constitute very irregular anastomoses, chiefly in the horizontal. Again is indicated a hydraulic circulation before the water table was lowered.

Most enterable caves are undergoing deterioration from fall of ceiling rock, deposition of stream debris, and growth of stalactites, stalagmites, flowstone, and other forms of secondary limestone deposited by drip water. These alterations result from transfer of the cave to the vadose zone. *See* STALACTITES AND STALAGMITES.

Caves of no great length may be formed by mechanical erosion of waves against coastal cliffs and by weathering attack on cliffs of jointed rock. Lava flows may, on cooling, develop a stiff upper crust over still molten rock which, continuing to flow after supply at the source has ceased, may leave long caves. *See* KARST TOPOGRAPHY; WEATHERING PROCESSES. [J. HARLEN BRETZ]

Bibliography: J. H. Bretz, *Caves of Missouri*, Missouri Geological Survey, vol. 39, 2d ser., 1956; J. H. Bretz, Vadose and phreatic features of limestone caverns, *J. Geol.*, 50(6):675–811, 1942; W. M. Davis, Origin of limestone caverns, *Bull. Geol. Soc. Amer.*, 41:475, 1930; G. W. More and G. Nicholas, *Speleology: The Study of Caves*, 1964.

Celestite

A mineral with the chemical composition $SrSO_4$. Celestite occurs commonly in colorless to sky-blue, orthorhombic, tabular crystals. Fracture is uneven and luster is vitreous. Hardness is 3–3.5 on Mohs scale and specific gravity is 3.97. It fuses readily to a white pearl. It is only slowly soluble in hot concentrated acids or alkali carbonate solutions. The strontium present in celestite imparts a characteristic crimson color to the flame.

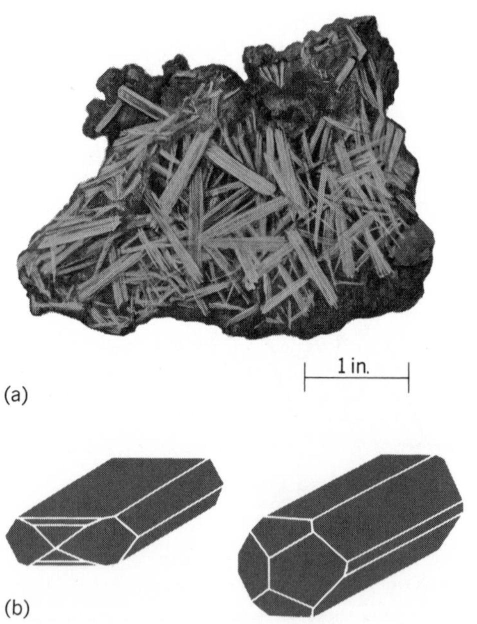

Celestite. (*a*) Bladed crystals in limestone from Clay Center, Ohio (*specimen from Department of Geology, Bryn Mawr College*). (*b*) Crystal habits (*from C. S. Hurlbut, Jr., Dana's Manual of Mineralogy. 17th ed., copyright © 1959 by John Wiley & Sons, Inc.; reprinted by permission*).

Celestite occurs in association with gypsum, anhydrite, salt beds, limestone, and dolomite. Large crystals are found in vugs or cavities of limestone (see illustration). It is deposited directly from sea water, by groundwater, or from hydrothermal solutions.

Celestite is the major source of strontium. The principal use of strontium is in tracer bullets and in various red flares used by the armed forces. Minor applications of strontium compounds are in ceramics, depilatories, and medicine.

Although celestite deposits occur in Arizona and California, domestic production of celestite has been small and sporadic. Much of the strontium demand is satisfied by imported ores from England and Mexico. [EDWARD C. T. CHAO]

Cenozoic

The youngest of the eras, or major subdivisions of geologic time, extending from the end of the Mesozoic Era to the present, or Recent. The term Cenozoic is also applied to all the rocks formed during this time and all the remains of past life which they contain. Although no unit term, such as era, has in the past been used for fossiliferous rock groups of era magnitude, the term erathem has recently been proposed for such major time-rock units. Cenozoic strata are characterized (1) by marked evolutionary changes, especially in stocks and faunas of marine shellfish such as the bivalved and the more snaillike mollusks, the sea urchins and sand dollars, the microscopic foraminifers, and to some extent corals and other invertebrate animals; (2) by a great expansion of grasses and the higher types of flowering plants; and (3) by the rapid development of birds and mammals.

PRECAMBRIAN	PALEOZOIC							MESOZOIC			CENOZOIC	
					CARBONIFEROUS							
	CAMBRIAN	ORDOVICIAN	SILURIAN	DEVONIAN	Mississippian	Pennsylvanian	PERMIAN	TRIASSIC	JURASSIC	CRETACEOUS	TERTIARY	QUATERNARY

Originally written Kainozoic and then Cainozoic (from the Greek *kainos*, "recent," plus *zoe*, "life") as first used by John Phillips in 1840 and 1841, it was so called to distinguish it from the much more ancient Paleozoic Era, so called by Adam Sedgwick, and the intermediate Mesozoic Era of Phillips. The distinction was made on the basis of the life forms, rather than the rock, mineral, or chemical qualities, that prevailed in the highest, the lowest, and the intermediate beds, respectively, of the geologic strata known at that time.

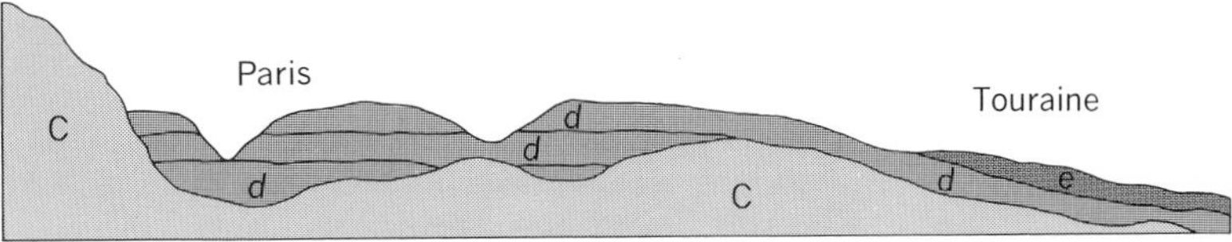

C = chalk and other secondary formations (Cretaceous)
d = Tertiary formations of Paris basin (Eocene)
e = superimposed marine Tertiary beds of the Loire (Miocene)

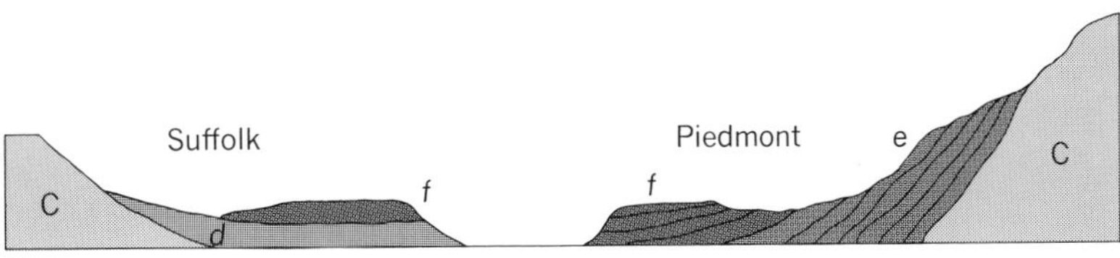

C = chalk and older formations (Cretaceous)
d = London clay (Eocene)
e = Tertiary strata of same age as beds of the Loire (Miocene)
f = crag and subapennine Tertiary deposits (Pliocene)

Diagrams of Cenozoic strata of Europe used by C. Lyell as a key to his three subdivisions. (*After C. Lyell, Principles of Geology, vol. 3, John Murray, 1833*)

Deposition and deformation. Rocks formed during Cenozoic time were deposited mechanically or chemically upon many kinds of older rocks, or in volcanic form extruded to rest also upon older rocks or intruded into or below such older rocks. The sedimentary as distinguished from the volcanic rocks include widespread limestones, mudstones, sandstones, and conglomerates (lithified gravels) of both marine and terrestrial origin. From time to time these Cenozoic rocks in certain areas of concentrated deposition have been elevated, locally broken, and widely crumpled by stresses operating in the crust of the Earth, to produce great mountain ranges such as those around the entire rim of the Pacific Ocean (the Andes, the West Coast ranges of North America, those of Kamchatka, Japan, the East Indies, New Zealand, and so on) and the east-west-trending Alpine-Himalayan chain. Geologically, these are the youngest mountains.

Marine transgressions. Distribution of land and sea was different at different times during the Cenozoic. Although this was a continuous process of change, with local exceptions to the general trend, three great expansions of the oceans over the continental shelf (with several lesser fluctuations) were interspersed between four extended intervals of great land emergence, uplift, and increased surface relief, all prior to the attainment of geologically modern times. Of these events the major marine transgressions of sea over land provided the most continuous depositional record of the prehistoric past. The intervals of greater crustal instability resulted in a greater diversity of natural phenomena from place to place, that is, greater diversities in aspect (facies) of both geological and life phenomena albeit of the same geologic age; and also more in the way of local gaps, or hiatuses, in the depositional record. It is thus not surprising that, historically, recognition of subdivisions within the Cenozoic first arrived at a threefold classification (Eocene, Miocene, Pliocene), with intervening gaps filled in only subsequently (Paleocene, Oligocene, "Mio-Pliocene," Pleistocene, Holocene).

Modern landforms and the present configuration of the continental land masses have come into being largely from this dynamic interplay of deposition, erosion, mountain-making (orogeny), and volcanism during Cenozoic time. Near its close continental glaciation became widespread, during a great Ice Age (Pleistocene) that finally began to wane some 25,000 years ago (beginning of the Holocene Epoch).

Divisions. The entire Cenozoic Era lasted on the order of 70,000,000 years. Such an estimate is based on the known deterioration rate of certain radioactive materials found in certain mineral or rock types. More numerous and widespread than such rock types suitable for geochronological dating, however—in the Cenozoic as in the Mesozoic and Paleozoic eras also (though not in the pre-Paleozoic rocks)—are fossilized forms of former life. On the basis of fossils and their positions relative to each other in the stratal sequence, the Cenozoic was at first, in 1833, divided by Sir Charles Lyell into three (subsequently four) periods now formally termed epochs (see illustration), to which E. von Beyrich, W. Schimper, and P. Gervais have added three more. Customarily, nevertheless, it is also formally divided into two larger subdivisions, or periods, of unequal length: the older, or Tertiary, period which is relatively long and in turn now subdivided into five epochs; and the younger, or Quaternary, period which is relatively short, including only two epochs (Pleistocene and Holocene). Rocks formed during such periods are classified as systems. The terms Tertiary and Quaternary are holdovers from earlier geological nomenclature in which, largely on the basis of physical evidence, rocks were classed as Primary (oldest), Secondary, Tertiary (from the proposals of Giovanni Arduino, an 18th-century naturalist), and Quaternary (from J. Desnoyers in 1829). Another though less formal subdivision of these Cenozoic rocks is also twofold: Paleogene for those of older Tertiary age, Neogene for the younger Tertiary and Quaternary combined. *See* EOCENE; MIOCENE; OLIGOCENE; PALEOCENE; PLEISTOCENE; PLIOCENE; RECENT.

As a whole, the Quaternary is distinguished from the Tertiary by the apparent emergence of men and relics of their presence. Yet the lesser subdivisions of the Quaternary are still based primarily on inorganic criteria, principally the waxing and waning of continental glaciation. Lyell's subdivisions of the Tertiary, on the other hand, stemmed from the observation that in ascending the Tertiary stratal column the percentage of living species present in the fossil faunas increased progressively from the bottom of the column to the top. *See* PALEOBOTANY; PALEONTOLOGY; QUATERNARY; TERTIARY.

[ROBERT M. KLEINPELL]

Bibliography: W. B. N. Berry, *Growth of a Prehistoric Time Scale Based on Organic Evolution*, 1968; A. M. Davies, *Tertiary Faunas*, 2 vols., 1934, 1935; M. Gignoux, *Stratigraphic Geology*, 1955; E. Neaverson, *Stratigraphical Paleontology*, 1955; R. A. Stirton, *Time, Life, and Man*, 1959; M. G. Wilmarth, *The Geologic Time Classification of the U.S. Geological Survey Compared with Other Classifications*, U.S. Geol. Surv. Bull. no. 769, 1925; A. O. Woodford, *Historical Geology*, 1965.

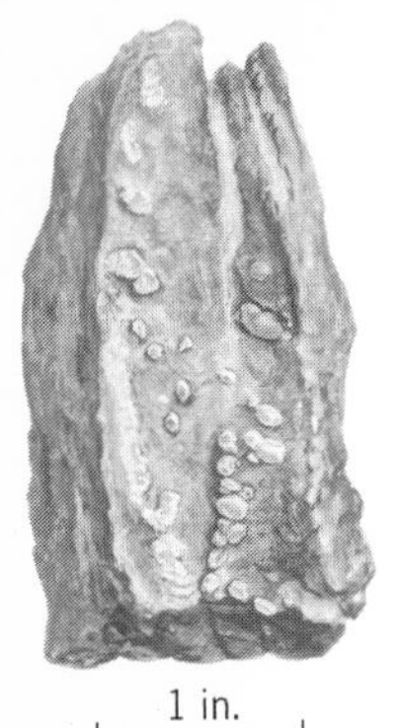

Cerargyrite crystals from Leadville, Colo. (*Specimen from Department of Geology, Bryn Mawr College*)

Cerargyrite

A mineral with composition AgCl. It crystallizes in the isometric system but crystals, usually cubic, are rare (see illustration). Most commonly it is in crusts and coatings resembling wax. The hardness is 2½ on Mohs scale and specific gravity 5.5. Cerargyrite is colorless to pearl-gray but darkens to violet-brown on exposure to light. It is perfectly sectile and can be cut with a knife like horn; hence the name horn silver. Bromyrite, AgBr, is physically indistinguishable from cerargyrite and the two minerals form a complete series. Both minerals are secondary ores of silver and occur in the oxidized zone of silver deposits.

[CORNELIUS S. HURLBUT, JR.]

Cerussite

The mineral form of lead carbonate, $PbCO_3$. Cerussite is common as a secondary mineral associated with lead ores. In the United States it occurs mostly in the central and far western regions. Cerussite is white when pure but is sometimes darkened by impurities. Hardness is 3¼ on Mohs scale and specific gravity is 6.5. Cerussite has orthorhombic symmetry and the same crystal structure as aragonite. Crystals may be tabular, elongated, or arranged in clusters (see illustration). *See* CARBONATE MINERALS. [ROBERT I. HARKER]

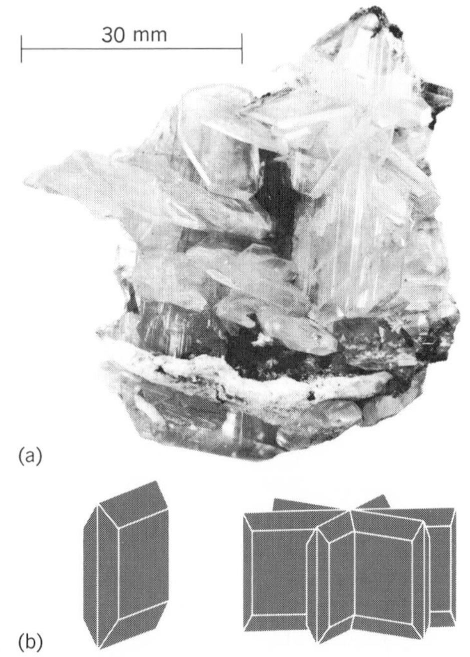

Cerussite crystals. (*a*) Specimen from Ems, Nassau (*American Museum of Natural History specimen*). (*b*) Common crystal habits (*from C. S. Hurlbut, Jr., Dana's Manual of Mineralogy, 17th ed., copyright © 1959 by John Wiley & Sons, Inc.; reprinted by permission*).

Chabazite

A mineral belonging to the zeolite family of silicates. It crystallizes in the hexagonal system and usually occurs in simple rhombohedral crystals with nearly cubic angles (see illustration). Penetration twins are common. There is a poor rhombohedral cleavage. Hardness is 4–5 on Mohs scale; specific gravity is 2.05–2.15. The luster is vitreous and the color usually white but may be yellow, red, or pink. Chabazite resembles calcite in its crystal form and cleavage but may be distinguished from it by the lack of effervescence in hydrochloric acid. *See* ZEOLITE.

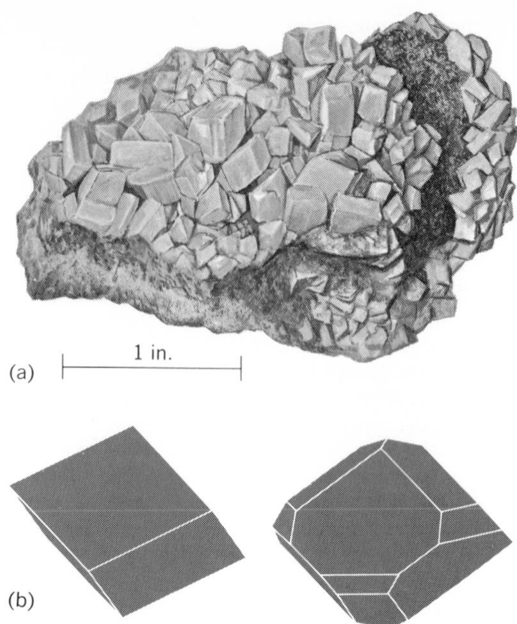

The mineral chabazite, in the zeolite family. (*a*) Crystalline specimen found at Two Islands, Nova Scotia (*specimen from Department of Geology, Bryn Mawr College*). (*b*) Crystal habits (*from C. S. Hurlbut, Jr., Dana's Manual of Mineralogy, 17th ed., copyright © 1959 by John Wiley & Sons, Inc.; reprinted by permission*).

Chabazite is essentially a hydrous calcium sodium aluminum silicate, $(CaNa_2)[Al_2Si_4O_{12}]\cdot 6H_2O$. The ratio of calcium to sodium may vary, and potassium is usually present in small amounts. Chabazite has been the most widely used natural zeolite for cation-exchange and gas-sorption purposes, but synthetic zeolites are now widely employed.

Chabazite is a secondary mineral usually found lining cavities in basalt and related rocks. It is associated with other zeolites and calcite. Notable localities are the Giant's Causeway, Ireland; Faeroe Islands; Trentino, Italy; and Oberstein, Germany. In the United States it is found at West Paterson and Bergen Hill, N.J.; and Table Mountain, Ore. [CLIFFORD FRONDEL; CORNELIUS S. HURLBUT, JR.]

Chalcanthite

A mineral with the chemical composition $CuSO_4\cdot 5H_2O$. Chalcanthite commonly occurs in blue to greenish-blue triclinic crystals or in massive fibrous veins or stalactites (see illustration). Fracture is conchoidal and luster is vitreous. Hardness is 2.5 on Mohs scale and specific gravity is 2.28. It has a nauseating taste and is readily soluble in water. It dehydrates in dry air to a greenish-white powder.

Chalcanthite is a secondary mineral associated with gypsum, melanterite, brochantite, and other sulfate minerals found in copper or iron sulfide

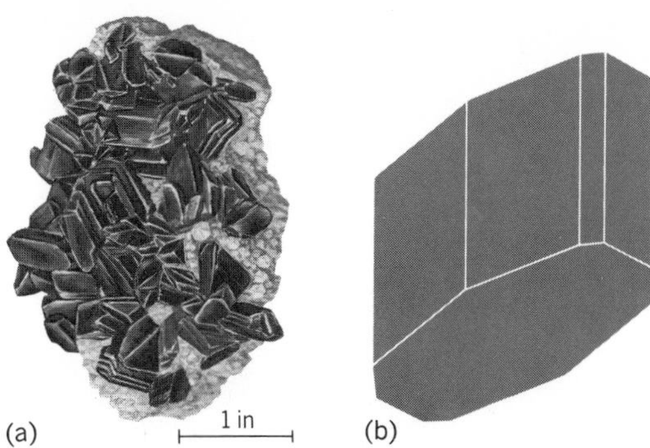

Chalcanthite. (*a*) Crystals associated with quartz, Clifton, Ariz. (*specimen from Department of Geology, Bryn Mawr College*). (*b*) Crystal habit (*from L. G. Berry and B. Mason, Mineralogy, copyright © 1959 by W. H. Freeman and Co.*).

deposits. It is also found in mine workings.

Although deposits of commercial size occur in arid areas, chalcanthite is generally not an important source of copper ore. Its occurrence is widespread in the western United States.

[EDWARD C. T. CHAO]

Chalcedony

A fine-grained fibrous variety of quartz, silicon dioxide. The individual fibers that compose the mineral aggregate usually are visible only under the microscope. Subvarieties of chalcedony recognized on the basis of color differences (induced by impurities), some valued since ancient times as semiprecious gem materials, include carnelian (translucent, deep flesh red to clear red in color), sard (orange-brown to reddish-brown), and chrysoprase (apple green). Chalcedony sometimes contains dendritic enclosures resembling plants or trees. Major kinds of impurities that give color to chalcedony are iron oxides (carnelian and sard), nickel (chrysoprase), and manganese. *See* GEM; QUARTZ.

Chalcedony occurs as crusts with a rounded, mammillary, or botryoidal surface and as a major constituent of nodular and bedded cherts. The hardness is 6.5–7 on Mohs scale. The specific gravity is 2.57–2.64. The ultrafine structure of chalcedony has been deduced from x-ray diffraction and electron microscopy to consist of a network of microcrystalline quartz with many micropores. The amount of amorphous silica, if any, is less than 10%. The yellowish color and anomalously low indices of refraction commonly observed under a transmitted light microscope result from scattering of light from the micropores. Paleozoic and older chalcedony is usually more coarsely crystallized than younger examples, grain growth of microcrystalline quartz being a result of time.

Crusts of chalcedony generally are composed of fairly distinct layers concentric to the surface. Agate is a common and important type of chalcedony in which successive layers differ markedly in color and degree of translucency. In the most common kind of agate the layers are curved and concentric to the shape of the cavity in which the material formed. The successive layers of chalcedony and agate usually differ in permeability to solutions, and much colored agate sold commercially is artificially pigmented by dyes or inorganic chemical compounds. *See* AGATE. [RAYMOND SIEVER]

Bibliography: C. Frondel, *Dana's System of Mineralogy*, vol. 3: *Silica Minerals*, 1962.

Chalcocite

A mineral having composition Cu_2S. Chalcocite crystallizes in the orthorhombic system (below 105°C) but crystals are rare and small, usually with hexagonal outline as a result of twinning (see illustration). Most commonly the mineral is fine-grained and massive with a metallic luster and a shining lead-gray color which tarnishes to dull black on exposure. It is imperfectly sectile; the hardness is 2.5 (Mohs scale) and specific gravity is 5.7. Chalcocite is an important copper ore formed principally as a supergene mineral in the enriched zones of sulfide deposits as at Miami, Morenci, and Bisbee, Ariz. *See* BORNITE.

[CORNELIUS S. HURLBUT, JR.]

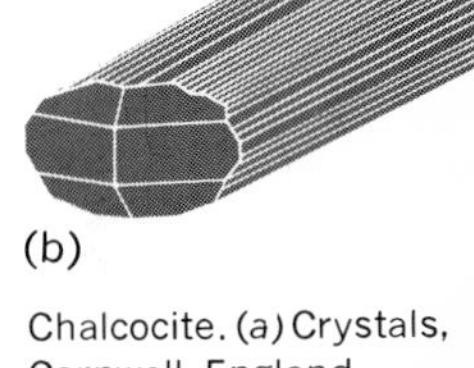

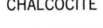

Chalcocite. (*a*) Crystals, Cornwall, England (*specimen from Department of Geology, Bryn Mawr College*). (*b*) A crystal habit (*from C. S. Hurlbut, Jr., Dana's Manual of Mineralogy, 17th ed., copyright © 1959 by John Wiley & Sons, Inc.; reprinted by permission*).

Chalcopyrite

A major ore mineral of copper with composition $CuFeS_2$. Chalcopyrite crystallizes in the tetragonal crystal system, but crystals are usually small with disphenoidal faces resembling the tetrahedron (see illustration). It is usually massive with metallic luster and a brass-yellow color, often with a bronze or iridescent tarnish. The hardness is 3.5 (Mohs scale) and the specific gravity is 4.2. Chalcopyrite is a so-called fool's gold, but it is brittle and thus may be distinguished from sectile gold. It is distinguished from pyrite, the most common and widespread fool's gold, by its lesser hardness; pyrite cannot be scratched by the knife.

Chalcopyrite is the most widely occurring of the primary copper ore minerals, and many of the secondary copper minerals have been derived by its alteration. It is an original constituent of igneous rocks and occurs in pegmatite dikes, contact metamorphic deposits, and metallic veins. It is often disseminated through large masses of pyrite and pyrrhotite, making them serve as low-grade copper ore. A few localities where chalcopyrite is the chief copper ore are Cornwall, England; Freiberg, Germany; and Rio Tinto, Spain. In the United States it is mined at Butte, Mont.; Jerome, Ariz.; and Ducktown, Tenn. *See* CHALCOCITE; PYRITE.

[CORNELIUS S. HURLBUT, JR.]

Chalcopyrite. (*a*) With galena and mica schist from Bottino, Tuscany, Italy (*American Museum of Natural History*). (*b*) Crystal habit (*from C. S. Hurlbut, Jr., Dana's Manual of Mineralogy, 17th ed., copyright © 1959 by John Wiley & Sons, Inc.; reprinted by permission*).

Chalk

A variety of limestone formed from pelagic, or floating, organisms that is very fine-grained, porous, and friable. It is white or very light-colored and consists almost entirely of calcite. The rock is made up of the calcite shells of microorganisms partially cemented by structureless calcite. The most conspicuous are the Foraminiferida, *Globigerina* and *Textularia.* Also present are remains of floating algae, rhabdoliths and coccoliths, some sponge spicules, and radiolaria. In some chalk, remains of microorganisms constitute more than one-third of the rock. *See* LIMESTONE.

The best-known chalks are those of the Cretaceous, exposed in cliffs on both sides of the English Channel. The Selma (Cretaceous) chalk of Alabama, Mississippi, and Tennessee and the Niobrara chalk of the same age in Nebraska are well-known deposits in the United States. Chalk is used for cements, powders (as soft abrasives and polishers), crayons, and fertilizers.

[RAYMOND SIEVER]

Chert

A dense, microcrystalline or cryptocrystalline rock composed of free silica, either in the form of chalcedony and microcrystalline quartz or in the form of opaline silica of various kinds. Chalcedony is a microcrystalline, fibrous form of silica that is now thought to consist mainly of quartz with only small amounts, if any, of amorphous silica or opal between the fibers. It may be almost any color, depending on impurities, but most cherts are light colored. It breaks with conchoidal to splintery fracture. Opaline silica ranges from amorphous varieties to opals that are disordered varieties of cristobalite. Flint is synonymous with chert. The latter term is preferred for the rock, although the former is still used for artifacts, such as arrowheads. Other names that have been used for chert are silexite, hornstone, and phthanite. Porcellanite is a dense, hard siliceous rock with the texture and fracture of unglazed porcelain. It usually is a chert with much included material, mainly argillaceous or calcareous. Novaculite is a bedded chert that consists dominantly of microcrystalline quartz; the term has been used chiefly in Arkansas and Oklahoma to describe some of the cherts of the Ouachita Mountains.

Composition. The composition of cherts varies with age. Opal occurs in Tertiary cherts, both on land and in cherts drilled in the Deep Sea Drilling Program, but not in older beds. The older cherts consist of microcrystalline quartz and chalcedony. Common nonsilica components of cherts are calcite and dolomite, detrital quartz, and clay minerals.

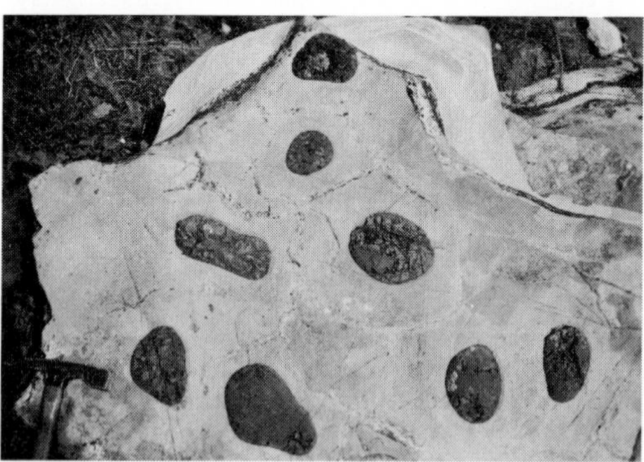

Fig. 1. Chert nodules in Hardy Creek limestone of southwest Virginia. (*Photograph by L. D. Harris, USGS*)

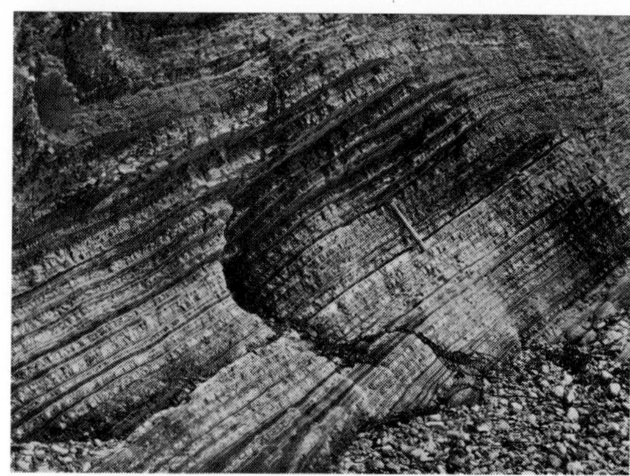

Fig. 2. Bedded chert, Monterey formation (Miocene), California. (*Photograph by M. N. Bramlette, USGS*)

Occurrence. Chert is found in two forms, as nodules in limestones and dolomites, and as bedded formations. The nodules are flattened parallel to the bedding and vary in size from a few inches to several feet in diameter (Fig. 1). Some nodules show traces of bedding that is continuous with the enclosing rock. Fossils are common in the nodules. The bedded cherts are tens of feet thick and extend in area for hundreds of square miles. Typically they are layered; the layers are a few inches thick and may be separated by thin beds of shale, siderite, or hematite (Fig. 2). The bedded cherts of the Precambrian are associated with extensive iron ore deposits. Early Precambrian cherts contain the remains of the earliest organisms to have evolved on Earth. Some cherts are oolitic and clearly show evidence of having formed by the replacement of original calcareous oolites. Carbonate inclusions in the chert are common, and many of the original concentric rings of the oolites have been obliterated. *See* OOLITE AND PISOLITE.

Much attention has been focused on deep-sea cherts, since the Deep Sea Drilling Program demonstrated the widespread occurrence of Tertiary-age lithified cherts in many areas of the ocean. At the same time plate tectonic theory has stimulated interest in the ophiolites, rock assemblages containing cherts and volcanic rocks that are found in strongly deformed mountain belts. Ophiolites are now interpreted as the remains of deep-sea cherts and clays which, together with volcanics, become caught up in subduction zones.

Origin. Two theories have been suggested for the origin of chert. One is that of replacement origin of originally calcareous materials. The other calls for primary precipitation of silica from the water of the depositional environment.

Replacement origin. In the case of the nodular cherts the evidence is good for replacement origin. The evidence includes the presence of silicified fossils, the irregular shape of the nodules, the inclusion of patches of limestone within the nodule, the preservation of original bedding and other textures in the nodules, and the undoubted replacement origin of the silicified oolites. The most likely sources of the silica are the siliceous skeletal materials of sponges, diatoms, and radiolaria, which are originally present and disseminated throughout the sediment. Early diagenetic processes lead to solution of the amorphous silica that composes all of the shells and spicules and reprecipitation in segregations.

The process of solution and reprecipitation may be linked to decay of organic matter and its effect on the pH (alkalinity-acidity) of the interstitial waters, which affects the solubility of silica. Evidence for this process in some limestones is the presence of abundant spicules and siliceous tests in nodules and complete lack of silica in the surrounding limestone. It is doubtful, however, whether this explanation will hold for Precambrian cherts, for silica-secreting species apparently did not evolve until the Cambrian. *See* DIAGENESIS.

Primary precipitation. The origin of bedded cherts has long been a problem. The theory that such cherts form from primary inorganic precipitation from sea water is difficult to accept because modern sea water is undersaturated with respect to amorphous silica (opal) and all modern marine siliceous deposits are biogenic. But in 1965 M. N. A. Peterson and C. C. von der Borch described modern deposits of nonbiogenic opal in a hypersaline lagoonal arm of the sea in South Australia, and in 1967 H. P. Eugster described deposits of inorganic sodium silicates and chert from Lake Magadi, Kenya. Thus there may be some possibility of primary precipitation in special environments, but it is doubtful that this was the condition under which the extensive bedded cherts formed. The weight of the evidence for their origin is on the side of the cherts being altered equivalents of widespread biogenic silica deposits, such as diatomaceous oozes found on the present-day sea floor. There is no doubt that some Tertiary deposits are the result of lithification and recrystallization of diatomaceous deposits. Detailed microscopic examination of Mesozoic and Paleozoic cherts reveals enough faint outlines of siliceous sponge spicules and radiolaria to point to the probability that these deposits also are of biogenic origin. Inorganic precipitation of silica from waters that were, at least locally, supersaturated with silica may have occurred in the Precambrian before the evolution of sponges and radiolaria.

The relationship of volcanic activity to chert deposition is not wholly clear, for there are many deposits with no apparent link to volcanism; but in those deposits where there is a clear association, high dissolved silica concentrations are thought to have come from hydrolysis of volcanic glass. *See* CHALCEDONY; DIATOMACEOUS EARTH; SEDIMENTARY ROCKS. [RAYMOND SIEVER]

Bibliography: H. A. Ireland (ed.), *Silica in Sediments*, Soc. Econ. Paleontol. Mineral. Spec. Publ. no. 7, 1959; F. J. Pettijohn, *Sedimentary Rocks*, 2d ed., 1957.

Chlorite

A group of hydrous silicates which are closely related to the micas. They are normally green and contain appreciable amounts of ferrous iron. *See* CLAY MINERALS; MICA.

Structurally, the chlorites are regular interstratifications of single biotite mica layers and brucite layers (see figure). The mica-like layers are trioctahedral and are unbalanced by substitution

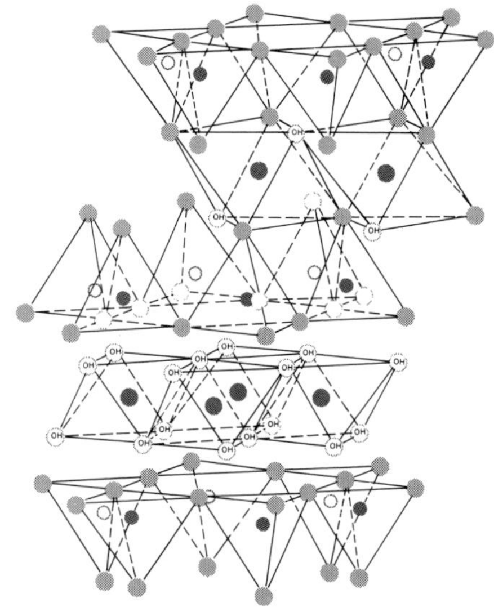

Diagram of the structure of chlorite. (*From R. E. Grim, Clay Mineralogy, McGraw-Hill, 1953*)

of aluminum for silicon. The brucite-like layer has the general composition $(Mg{\cdot}Al)_6(OH)_{12}$ and balances the charge deficiency of the mica layer. The chlorite unit cell is composed of two brucite-like and two mica-like layers.

Various members of the chlorite group differ from each other in the kind and amount of structural substitutions. They also exhibit polymorphic forms and show different stacking arrangements of the chlorite layers.

Chlorites can form from magnesium-rich minerals such as hornblende and biotite by hydrothermal or weathering processes. Poorly crystalline varieties of chlorite are frequent components of Recent marine sediments and soils. The mineral forms readily in the marine environment from degraded micas, montmorillonite, and possibly kaolinite by a process involving taking up magnesium from sea water. Chlorites, somewhat better crystallized, are also frequent clay-mineral constituents of ancient sediments. *See* AUTHIGENIC MINERALS; DIAGENESIS.

[FLOYD M. WAHL; RALPH E. GRIM]

Chloritoid

A member of the brittle mica (clintonite) group, which are hydrous aluminum silicates resembling both micas and chlorites. Although they also possess a perfect basal cleavage, cleavage flakes of the clintonites are distinctly more brittle than those of the micas. The composition is [Fe(II),Mg] [Al,Fe(III)] $(Si,Al)O_5(OH)_2$. The variety ottrelite also contains manganese. Margarite, prehnite, and stilpnomelane are other members of the group. Chloritoid has both monoclinic and triclinic modifications which may occur together. It forms rosettes, foliated aggregates, and plates, the last with pseudohexagonal, or rhombic, outlines. The color is gray to green, and crystals are weakly pleochroic. Face-loci zoning (hourglass structure) is distinctive. The hardness is 6.5, specific gravity 3.5. Inclusions of quartz and magnetite are common.

Chloritoid is principally a metamorphic mineral in schists, phyllites, and quartzites, and is associated with quartz, muscovite, and chlorite. Chloritoid rocks result from the low- to medium-grade metamorphism of pelitic sediments rich in aluminum and iron but relatively low in potassium, calcium, amd magnesium. Chloritoid also forms in some quartz-ankerite veins. It alters to pennine. *See* MICA; SILICATE MINERALS.

[E. WILLIAM HEINRICH]

Chromite

CHROMITE

Chromite veins in a peridotite sample found in Selukwe, Rhodesia. (*Specimen from Department of Geology, Bryn Mawr College*)

The only important ore mineral of chromium. Chromite crystallizes in the isometric system but crystals are rare and it usually is massive. The hardness is 5.5 (Mohs scale) and the specific gravity 4.6. The luster is submetallic and the color iron-black to brownish-black. Pure chromite with composition $FeCr_2O_4$ is rare, because magnesium usually substitutes for some ferrous iron, and aluminum and ferric iron substitute for chromium. Chromite is a common constituent of peridotites and serpentines and in some is present in large masses (see illustration). Major producing countries are Turkey, the Union of South Africa, the Soviet Union, the Philippines, and Rhodesia. *See* PERIDOTITE; SERPENTINE.

[CORNELIUS S. HURLBUT, JR.]

Chrysoberyl

A mineral having composition $BeAl_2O_4$ and crystallizing in the orthorhombic system. Chrysoberyl crystals are usually tabular parallel to the front pinacoid and frequently in pseudohexagonal twins (see illustration). There is good prismatic cleavage. The hardness is 8.5 (Mohs scale) and the specific gravity is 3.7–3.8. The luster is vitreous and the color various shades of green, yellow, and brown. There are two gem varieties of chrysoberyl. Alexandrite, one of the most prized of gemstones, is an emerald green but in transmitted or in artificial light is red. Cat's eye, or cymophane, is a green chatoyant variety with an opalescent luster. When cut en cabochon, it is crossed by a narrow beam of light. This property results from minute tabular cavities that are arranged in parallel position.

(a) 1 in (b)

Chrysoberyl. (*a*) Crystals in pegmatite, Greenwood, Maine (*specimen from Department of Geology, Bryn Mawr College*). (*b*) Crystal habit (*from C. S. Hurlbut, Jr., Dana's Manual of Mineralogy, 17th ed., copyright © 1959 by John Wiley & Sons, Inc.; reprinted by permission*).

Chrysoberyl is a rare mineral found most commonly in pegmatite dikes and occasionally in granitic rocks and mica schists. Gem material is found in stream gravels in Ceylon and Brazil. The alexandrite variety is found in the Ural Mountains. In the United States chrysoberyl is found in pegmatites in Maine, Connecticut, and Colorado. *See* GEM.

[CORNELIUS S. HURLBUT, JR.]

Chrysocolla

A silicate mineral, composition $CuSiO_3 \cdot 2H_2O$. Small acicular crystals have been observed, but it ordinarily occurs in impure cryptocrystalline crusts and masses with conchoidal fracture. The hardness varies from 2 to 4 on Mohs scale, and the specific gravity varies from 2.0 to 2.4. The luster is vitreous and it is normally green to greenish-blue, but may be brown to black when impure. Chrysocolla is a secondary mineral occurring in the oxidized zones of copper deposits, where it is associated with malachite, azurite, native copper, and cuprite. It is a minor ore of copper.

[CORNELIUS S. HURLBUT, JR.]

Cinnabar

A mineral of composition HgS. Cinnabar is the only important ore of mercury. It crystallizes in the hexagonal system but crystals are rare, usually of rhombohedral habit and often in penetration twins. It most commonly occurs in fine, granular, massive form. There is perfect prismatic cleavage; the hardness is 2.5 (Mohs scale) and the specific

(b)

Cinnabar. (*a*) On quartz crystals, Hunan Province, China (*American Museum of Natural History*). (*b*) Crystal habits (*from L. G. Berry and B. Mason, Mineralogy, copyright © 1959 by W. H. Freeman and Co.*).

gravity is 8.10. When pure, cinnabar has an adamantine luster and vermillion-red color; when impure, the luster may be dull and the color brownish-red (see illustration).

Metacinnabar is polymorphous with cinnabar and the two minerals occur together. It is isometric, with hardness 3, specific gravity 7.6, and color grayish-black. It inverts to cinnabar, the stable form, on heating to 400–500°C.

Cinnabar occurs in minable deposits at comparatively few localities. It is commonly found in veins and impregnations deposited near the surface and appears to be genetically associated with recent volcanic rocks and hot springs. Important localities for the occurrence of cinnabar are at Almaden, Spain; Idria, Italy; and in Kweichow and Hunan provinces, China. In the United States it was early found at New Almaden and New Idra, Calif. It also has been mined in Nevada, Utah, Oregon, Arkansas, Texas, and Idaho.

[CORNELIUS S. HURLBUT, JR.]

Cirque

A deep, steep-walled recess, roughly semicircular in plan, cut into the bedrock of a mountain or other highland. Ideally it has the form of half a bowl, and results from erosion beneath and around a snowbank or a glacier (see illustration). Many cirques occur at the heads of valleys.

Cirque in Yosemite area of Sierra Nevada, Calif. (*F. E. Matthes, U.S. Geological Survey*)

They range in diameter from a few tens of meters to much more than 1 km. Meltwater penetrates the bedrock beneath snow or ice and, expanding as it freezes, wedges out particles of rock, which are moved downslope by several processes, including glacial flow if a glacier is present. The floors of many glaciers are shallow basins cut into the bedrock or formed behind end moraines; most such basins contain small lakes. Many cirques are not now active, but are relict from the latest glacial age. *See* GLACIATED TERRAIN. [RICHARD F. FLINT]

Clay

The term clay has a dual meaning. It is used both as a rock term and as a particle-size term. As a rock term, clay has no genetic significance. It is used for material that has been deposited as a sediment, has formed by hydrothermal action, or is the product of weathering. Chemical analyses of clays show them to be composed primarily of silica, alumina, and water. Iron, alkalies, and alkaline earths are also often present in appreciable quantities. As a rock term clay implies a natural, earthy, fine-grained material which develops plasticity when mixed with a limited amount of water. Plasticity is that property of the moistened material which allows deformation of the material when pressure is applied. In a plastic substance the deformed shape is maintained when the deforming force is removed.

As a particle-size term, clay refers to that fraction of an earthy material containing the smallest particles. In the mechanical analysis of sedimentary rocks and soils, material is normally divided into three size grades, or fractions, on the basis of particle size. These are sand, silt, and clay fractions. The maximum size of the particles comprising the clay-size grade is defined differently by different observers. In geology the tendency has been to define the clay grade as any material finer than about 4 microns (μ) in diameter. In the investigation of soils there is a tendency to use 2 μ as the upper limit of the clay-size grade. There is a fundamental reason for choosing this value as the upper limit. A large number of analyses have shown that the clay minerals tend to be concentrated in a size less than 2 μ. It has also been observed that when naturally occurring larger flakes of clay-mineral particles are slaked in water, they break down more easily to this size. *See* CLAY MINERALS.

There is no sharp universal boundary between the particle size of the clay minerals and non-clay minerals in argillaceous sediments, but numerous investigations have shown that non-clay minerals usually are not present in abundance in particles much smaller than 1–2 μ. (For the best split of clay-mineral and non-clay-mineral components, a separation at 2 μ is now frequently used.) *See* ARGILLACEOUS ROCKS.

Composition. The clay-size-grade materials which make up the rock clay are always present in varying percentages. Coarser grades composed generally of the non-clay minerals are thus also present in varying relative amounts. In addition to clay minerals and non-clay minerals, clays may also contain organic material and water-soluble salts.

The designation of a material as clay is often based solely on its bulk properties and appearance. Unfortunately, in many such materials the clay-size grade and clay-mineral fraction comprise less than 50% of the total rock. A more complete study and analysis of the material often warrants reclassification of the rock as silt or even sand.

Some material called clay does not meet all the above specifications. So-called flint clay, a refractory ceramic clay, does not develop plasticity when mixed with water. It does, however, have the other attributes of clay.

Some types of sedimentary rocks are substantially of the same composition as clay, but differ from clays in their textural characteristics. The expression clay material is often used for any fine-grained, natural, earthy, argillaceous material, when a more precise designation is not possible or desirable. Clay material includes clays, shales, and argillites. Soils would also be included if such materials contained an appreciable quantity of clay-size-grade material and were argillaceous. *See* ARGILLITE; SHALE.

Properties. The properties of clay materials are controlled by at least five major factors. These attributes which characterize a clay material are clay-mineral composition, non-clay-mineral composition, organic material, soluble salts and exchangeable ions, and texture. Generally the clay-mineral composition is the most important factor, and sometimes as little as 5% of a particular clay mineral may largely determine the properties of the whole clay.

The economic use of a clay material is determined largely by its clay-mineral composition. In the ceramic industry only certain clays of a particular clay-mineral composition will withstand high temperatures and can be used for making refractories. The oil industry requires certain types of bentonite clay for the preparation of drilling muds. Other types of bentonite clay are used for the preparation of catalysts in the refining of petroleum products. *See* BENTONITE.

The paper industry utilizes certain types of kaolinite clays for the coating and filling of paper. In construction engineering a knowledge of the clay material on or through which a structure is to be built is essential. The presence of even small amounts (5%) of montmorillonite warrants additional testing of the material to predict how the soil material will act when it is placed under different conditions. *See* SOIL.

In the field of agriculture the tilth of a soil, its content of plant nutrients, and its treatment possibilities with fertilizers are all to a large extent contingent on the clay-mineral composition of the soil.

[RALPH E. GRIM; FLOYD M. WAHL]

Clay minerals

The clay minerals are the major components of clay materials. They occur in extremely small particles which are essentially crystalline and are limited in number. The clay minerals are essentially hydrous aluminum silicates. Alkalies or alkaline earths are present as principal constituents in some of them. Also, in some clay minerals magnesium or iron or both substitute wholly or in part for the aluminum. The ultimate chemical constituents of the clay minerals vary not only in amounts, but also in the way in which they are combined or are present in various clay minerals. *See* CLAY.

Clay-mineral concepts. The investigation of clay materials goes far back into antiquity because of their importance in industry, agriculture, and geology. A considerable number of concepts have been suggested to portray the essential components of all clay materials and to explain their variation in properties. Most of these concepts are concerned with the way in which the chemical components are put together in clays, that is, the portrayal of the fundamental building blocks of clay materials.

Before the advent of adequate analytical techniques and tools for the study of materials as fine-grained as clays, there was no general agreement as to the exact nature of these fundamental building blocks. A very old idea is that all clay materials are composed of kaolinite, frequently containing other materials considered to be impurities.

Another concept, particularly held by soil investigators, was that a colloidal complex was the essential component. This complex was thought to be amorphous and either inorganic or organic. Some investigators regarded it as a loose mixture of the oxides of silicon, aluminum, and iron. Another idea regarded this complex as a mixture of salts of weak ferroaluminosiliceous acids.

A different concept considered clay to have two essential components. One, called "clayite," was thought to be the true clay substance in kaolins. It was considered to be an amorphous substance with about the same chemical composition as the mineral kaolinite. The other essential component, called "pelinite," was suggested as the true clay substance in other clay materials. The latter was thought to be an amorphous material with a higher silica content and more alkalies or alkaline earths or both than clayite.

Still another concept was that fineness was the sole determining factor—thus clay materials could have any mineral composition.

For many years some students have suggested that clays are composed of extremely small particles of a limited number of crystalline minerals. This is the clay mineral concept which is now generally accepted. It is not new, but until the advent of x-ray diffraction analysis, the crystalline nature of the extremely small clay minerals could not be proved. Recent work indicates that some clay materials contain extremely poorly crystalline or amorphous components. Such material is not present in all clay materials, or even most of them, and thus cannot be considered a constituent generally responsible for clay properties.

Classification. No completely satisfactory classification of the clay minerals has yet been suggested. Classification is difficult because clays are often composed of complex mixtures of clay minerals of such poor crystallinity that the individual components cannot be adequately characterized. Also, there seems to be a continuous gradation between some of the types. The following classification has proved to be a workable one. A major subdivision into amorphous and crystalline groups is made even though the amorphous components are relatively rare and of little importance.

- I. Amorphous
  - Allophane group
- II. Crystalline
  - *A*. Two-layer type (sheet structures composed of units of one layer of silica tetrahedrons and one layer of alumina octahedrons)
    - 1. Equidimensional
      - Kaolinite group: kaolinite, nacrite, dickite, and so forth
    - 2. Elongate
      - Halloysite group
  - *B*. Three-layer types (sheet structures composed of two layers of silica tetrahedrons and one central dioctahedral or triotahedral layer)
    - 1. Expanding structure
      - *a*. Equidimensional
        - Montmorillonite group: montmorillonite, sauconite, and so forth
        - Vermiculite

b. Elongate

Montmorillonite group: nontronite, saponite, hectorite

2. Nonexpanding structure

Illite group

C. Regular mixed-layer types (ordered stacking of alternate structural types)

Chlorite group

D. Chain-structure types (similar to hornblende—chains of silica tetrahedrons linked together by octahedral groups of oxygens and hydroxyls containing Al and Mg atoms)

Attapulgite

Sepiolite

Palygorskite

The three-layer minerals are divided into the expanding and nonexpanding types in this classification. This subdivision is made even though there may be a continuous gradation between the two types. The expanding attribute is a readily determinable diagnostic property, since it imparts some unique physical properties to the clay materials composed of these minerals. The expanding minerals are divided into equidimensional and elongate divisions, because shape is a readily recognizable characteristic that probably reflects structural attributes.

Nomenclature. During the early work with clay minerals, the name allophane came to be associated with amorphous constituents of clay. The study of this material showed that some of it was actually crystalline, although much of it was amorphous to x-ray diffraction. It has been suggested that the term allophane be used for all such amorphous clay-mineral materials regardless of their composition.

Kaolinite group. Perhaps the best-known clay minerals are those of the kaolinite group. Three distinct mineral species are represented by this group. They are kaolinite, nacrite, and dickite. Anauxite has also been defined as a mineral with essentially the same attributes as kaolinite but with a higher silica-alumina molecular ratio. Early investigators believed anauxite to be an interlayer mixture of a double silica layer and a two-layer sheet structure. X-ray studies suggest that material called anauxite is simply a mixture of kaolinite and quartz. *See* KAOLINITE.

Halloysite group. Minerals of the halloysite group are somewhat similar chemically to kaolinite, but have different structural attributes. There are two forms of the mineral halloysite. One form has the same chemical composition as kaolinite; the other is more hydrous, with $4H_2O$ instead of $2H_2O$ in its composition. Unfortunately, there is no agreement among clay mineralogists regarding proper nomenclature for the halloysite minerals. Thus, the more hydrated form has been called endellite by some investigators, and simply halloysite by others. The lower hydration form has been called metahalloysite by some, and halloysite by other students. It has been suggested that halloysite be used as a general term for all naturally occurring specimens of the mineral regardless of their state of hydration. When it is possible and necessary to describe the state of hydration, additional self-explanatory qualifications should be used. *See* HALLOYSITE.

Montmorillonite group. Montmorillonite has as an essential characteristic its expanding structure, that is, the property of adsorbing variable amounts of water between individual unit layers. Members of this group show considerable variation in the ratio of SiO_2 to R_2O_3. There can also be replacement of aluminum by iron and magnesium. The iron-rich montmorillonite is called nontronite, and the magnesium-rich variety is known as saponite. *See* MONTMORILLONITE.

Vermiculite. Vermiculites also have an expanding structure. They differ from montmorillonite in that only a limited degree of expansion can take place, and their particle size is larger. Vermiculites, once thought to be closely related to the micas, are now recognized as clay mineral constituents of clay materials. *See* VERMICULITE.

Illite. The term illite was proposed not as a specific clay-mineral name, but as a general term for the mica-like clay minerals. It refers to any mica-type clay mineral with 10-A *c*-axis spacing which shows no expanding structure characteristics. Illites include both trioctahedral and dioctahedral types of clay-mineral micas. *See* ILLITE.

Chlorite group. The chlorites are a group of green hydrous silicates which contain ferrous iron and are closely related to the micas. Structurally, they are regular interstratifications of single biotite mica layers and brucite layers, but they have been recognized as important constituents of clay materials. The identification of chlorite is particularly difficult when kaolinite is present in the same clay material. *See* CHLORITE.

Chain-structure types. Attapulgite, sepiolite, and palygorskite are clay minerals with chain structures similar to hornblende. These minerals are not yet completely understood. They are easily missed in clay-mineral analyses, and may be much more abundant than is now considered to be the case.

Mixed-layer combinations. Many clay materials are composed of more than one clay mineral. These clay-mineral components are often random mixtures of discrete units so that there is no preferred geometric orientation. In another type of mixing, the layer-type clay minerals are interstratified. This is possible because of their structural similarity. There are two different types of such so-called mixed-layers structures. In one type there is a regular repetition of the different layers which are stacked along the *c* axis. The other kind of mixed-layer structure is a random, irregular interstratification of layers without a uniform repetition of layers. Mixed layers of illite and montmorillonite and of chlorite and vermiculite are particularly common in clay materials and can only be detected by careful x-ray diffraction techniques.

Structure. The atomic structures of the clay minerals consist of combinations of two basic structural units. One unit consists of silica tetrahedrons, in which each tetrahedron is made up of a silicon atom equidistant from four oxygen atoms or hydroxyl ions arranged in the form of a tetrahedron with the silicon atom at the center. The silica tetrahedral groups are arranged to form a hexagonal network, which is repeated in two directions to

CLAY MINERALS

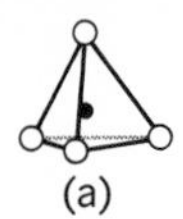
(a)

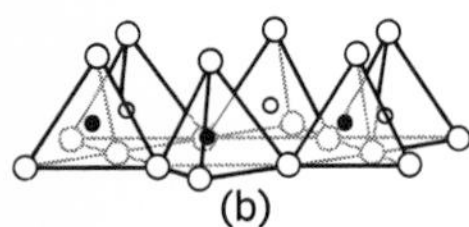
(b)

○ and ○ = oxygens
• and ∘ = silicons

Fig. 1. Diagrammatic sketch showing (*a*) single silica tetrahedron and (*b*) sheet structure of silica tetrahedrons arranged in a hexagonal network. (*From R. E. Grim, Clay Mineralogy, McGraw-Hill, 1953*)

form a sheet (Fig. 1), with the composition $Si_4O_6(OH)_4$. The tips of all tetrahedrons point in the same direction and their bases are all in the same plane. The composite structure of this unit may be viewed as a perforated basal plane of oxygen atoms, a plane of silicon atoms forming a hexagonal network, and a plane of hydroxyl atoms with each hydroxyl positioned at the tip of the tetrahedron, directly above the silicon. The thickness of this tetrahedrally coordinated unit is 4.93 A in clay-mineral structures.

The other structural unit consists of two sheets of closely packed oxygen atoms or hydroxyl ions in which aluminum, iron, or magnesium atoms are octahedrally coordinated (Fig. 2). When aluminum is present, only two-thirds of all the possible positions are filled to balance the structure. This is the dioctahedral gibbsite structure and has the formula $Al_2(OH)_6$. If all the positions are filled, as when magnesium is present, the trioctahedral brucite structure is formed, with the formula $Mg_3(OH)_6$. The thickness of this sheet is 5.05 A in clay-mineral structures.

Shape. Because of the sheetlike nature of their structural units, most of the clay minerals are flake-shaped. Some of the flakes have a distinct hexagonal outline. A few of the clay minerals are tubular or elongate or both. Some are fibrous and are composed of structural units different from those listed above. The sepiolite, palygorskite, and attapulgite minerals have structural characteristics similar to the amphiboles. Their basic structural unit is composed of silica tetrahedrons arranged in a double chain of composition Si_4O_{11}. The chains are bound together by aluminum or magnesium atoms or both. This structure is continuous in only one direction and it is restricted to a width of about 11.5 A.

The precise determination of the shape and size of the various clay-mineral particles has been made possible by the use of the electron microscope. With this instrument the clay-mineral particles can be magnified directly up to 15,000 times. Subsequent photographic enlargement can then increase this magnification several times.

Properties. The varied properties of the clay minerals are important in that they govern the economic use of clay materials.

Ion exchange. The ability of clay minerals to hold certain cations and anions which are readily exchangeable for other cations and anions is most significant. The commonest exchangeable cations in clay materials are Ca^{2+}, Mg^{2+}, H^+, K^+, NH_4^+, and Na^+. The more common exchangeable anions are SO_4^{2-}, Cl^-, PO_4^{3-}, and NO_3^-.

There are two major causes for the exchange capacity of the clay minerals. Broken bonds around the edges of silica-alumina units give rise to unsatisfied charges which are balanced by adsorbed cations. Substitutions within the structure result in unbalanced charges satisfied by exchangeable cations. In addition, some anions may be adsorbed by replacement of exposed hydroxyl ions, and because of the structural fit of some of them with tetrahedral units.

This property of ion exhange and the exchange reaction are of fundamental importance in all fields in which clay materials are used. For example, in soils the retention and availability of potash added in fertilizers depends on cation exchange between the potash salt and the clay mineral in the soil. *See* SOIL.

CLAY MINERALS

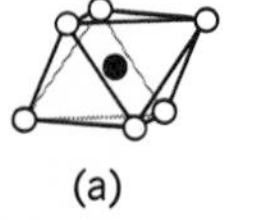
(a)

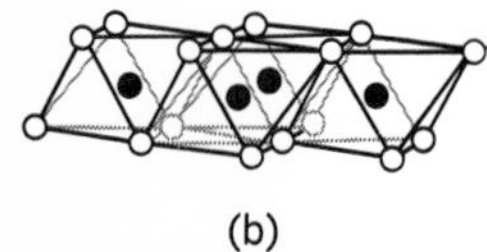
(b)

○ and ○ = hydroxyls
● = aluminums, magnesiums, etc.

Fig. 2. Diagrammatic sketch showing (*a*) single octahedral unit and (*b*) sheet structure of octahedral units. (*From R. E. Grim, Clay Mineralogy, McGraw-Hill, 1953*)

Clay-water system. Another important property of clay materials is their ability to hold water. This water is of two types, low-temperature water and OH structural water. The low-temperature water is driven off by heating to 100–150°C. This is the water in pores, on the surfaces, and around the edges of the minerals composing the material. It is also the interlayer water between the unit cell layers of minerals such as montmorillonite. The water which occupies the tubular openings between the elongate structural units of the sepiolite-attapulgite-palygorskite minerals is also of this type.

Abundant evidence shows that there is some sort of definite configuration of the water molecules initially adsorbed on the surfaces of clay materials; however, there is not agreement regarding the precise nature of the structure of this initially adsorbed water. The configuration of this low-temperature water and the factors controlling it largely determine the plastic, bonding, suspension, compaction, and other properties of clay materials, which in turn govern economic usage.

Dehydration. When clay materials are heated, dehydration occurs. Thus there is a loss of any water (adsorbed, interlayer, or structural OH water) held by the clay minerals. The heating of clay materials also causes changes in the clay-mineral structures. At relatively high temperatures these structural changes facilitate the formation of new mineral phases. These structural modifications are particularly important in the firing of clay materials.

Clay-mineral organic reactions. Another property of the clay minerals is their ability to react with organic materials. Clays with a high adsorbing capacity are used in decolorizing oils, while others provide catalysts in the cracking of organic compounds.

Some analytical techniques for the analysis of clay materials are based on clay-mineral organic reactions. For example, the identification of montmorillonite by x-ray diffraction is greatly simplified by treating the material with glycerol or ethylene glycol, which substitutes for water in the interlayer position and causes measurable expansion of the lattice along the *c* axis. Some organic materials are also used in the identification of clay minerals by staining techniques.

Optical properties. The optical properties of the clay minerals are usually difficult to determine because of their extremely small size, and because of inherent variations in the clay minerals themselves. Replacements within the structure and variations in the amount of interlayer water are reflected by changes in the optical properties.

Origin. Many of the clay minerals have been synthesized in the laboratory under conditions varying from room temperature and atmospheric pressure to elevated temperatures and pressures. From these experiments some general conclusions can be drawn regarding the environmental conditions favorable for the formation of the individual minerals. At low temperatures acid conditions favor the formation of the kaolinite type of mineral; whereas alkaline conditions favor the formation of montmorillonite, if magnesium is present, or mica, if sufficient potassium is present.

At elevated temperatures such generalities do not always hold, and the minerals which form depend upon ionic concentration, temperature, and the Al_2O_3/SiO_2 ratio.

Many of the clay minerals are of a hydrothermal origin. With the possible exceptions of attapulgite, palygorskite, and vermiculite, all have been reported in hydrothermal bodies, frequently with metalliferous ore deposits. Some hydrothermal clay deposits are monomineralic, but most consist of a mixture of clay minerals. *See* ORE AND MINERAL DEPOSITS.

The weathering of soils and of different rock types also leads to the formation of clay minerals. The type of clay mineral formed, however, depends upon a number of factors. These are parent rock, climate, topography, vegetation, and time. A complex interplay of these factors determines the weathering environment and its processes, and thus the type of clay mineral which will form. *See* WEATHERING PROCESSES.

Clay minerals are abundant in sediments, both recent and ancient. In some cases at least, the character of the clay minerals changes in passing from one environment to another, such as from fresh water to marine, so that the clay mineral composition reflects the geologic history of the sediment. *See* MARINE SEDIMENTS; SEDIMENTARY ROCKS; SEDIMENTATION.

[R. E. GRIM; F. M. WAHL]

Bibliography: G. W. Brindley, *X-Ray Identification and Crystal Structures of Clay Minerals*, 1951; R. E. Grim, *Clay Mineralogy*, 1953; C. I. Rich and G. W. Kunze; *Soil Clay Mineralogy*, 1964; H. Van Olphen, *Introduction to Clay Colloid Chemistry*, 1963.

Cleavage, rock

A secondary foliation produced in a rock by deformation. At present there exists no general agreement on the terminology or the mechanical significance of the various types of cleavage. As a compromise to existing opinions, cleavage is thus described as a special type of foliation. Foliation is a descriptive term that includes all types of parallel or subparallel primary and secondary planar structures that pervade a rock. Such surfaces usually exist as planes of weakness along which a rock may split.

Rock cleavage is usually associated with folding, but may also occur in response to faulting. It is generally best developed in fine-grained rocks and in rocks with a high content of platy minerals such as mica. Usually, cleavage planes either are parallel to the axial plane of a fold or exhibit a fan-shaped distribution around the hinge of a fold (Fig. 1). Cleavage may be parallel to the limbs of a fold, especially when the fold is very tight or isoclinal. Refraction of cleavage occurs on passing from one rock layer to another of different composition or texture.

Most rock cleavage is associated with dynamically metamorphosed rocks. The processes most commonly recognized in the generation of cleavage include shear, flow, extension and rotation, flattening, solution, and recrystallization. However, rock cleavage also occurs in deformed unmetamorphosed rocks. J. C. Maxwell proposed a dewatering process to explain the development of rock cleavage in slates that he believed had not involved metamorphism. This interpretation for the rocks that Maxwell studied has been challenged by a number of workers, such as J. B. Epstein and A. G. Epstein. However, cleavage has been recognized in other parts of the world in unconsolidated sediments, which may have formed by this process.

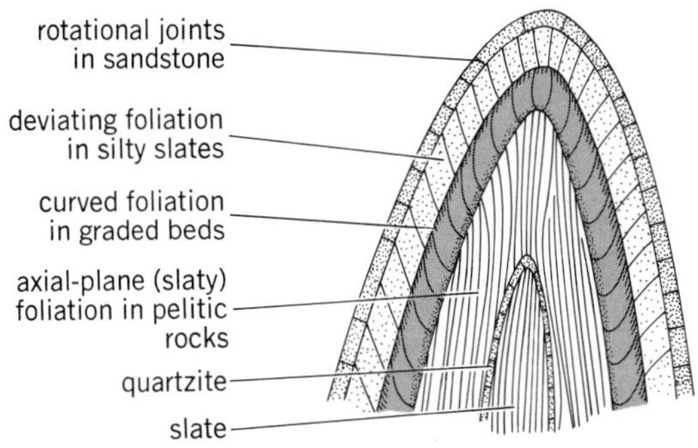

Fig. 1. Diagrammatic sketch of foliation attitudes in beds of different lithologies. (*From W. F. J. Kleinsmiede, Geology of the Valle de Aran (central Pyrenees), Leid. Geol. Meded., 25:127–244, 1960*)

Two classifications of cleavage are generally recognized. The first classification recognizes two types of cleavage:

1. Penetrative cleavage includes slaty or flow cleavage and schistosity (which is a coarse-grained variety of a slate texture). This type of cleavage is dependent on the parallel arrangement of minerals in a rock.
2. Nonpenetrative cleavage includes fracture cleavage or false cleavage and strain-slip or crenulation cleavage. It is independent of a parallel arrangement of minerals and is characterized by cemented or welded fractures.

Fig. 2. An example of cleavage; crenulation cleavage cuts across primary layering in Dartmouth slates, Devon. (*National Environment Research Council copyright; reproduced by permission of the Director, Institute of Geological Sciences, London*)

The second classification recognizes at least four categories of cleavage:

1. Axial-plane cleavage (also referred to as slaty and flow cleavage) is approximately parallel to the axial plane of the folds (Fig. 1).

2. Crenulation cleavage (Fig. 2; also called slip cleavage, shear cleavage, strain-slip cleavage, cataclastic cleavage, transposition cleavage, and kinked cleavage) is interpreted as indicating either a separate deformation or a late phase of the original movements that gave rise to the original foliation.

3. Secondary bedding or lithologic cleavage.

4. Fracture cleavage is a type of closely spaced jointing and is usually symmetrically oriented with respect to folded rock layers.

More than one type or generation and direction of cleavage may exist in a rock, and the chronological order in which these planes or S-surfaces have formed is indicated by the notation S_1, S_2, S_3, and so on. Cleavage can also be designated as pre-, syn-, and postkinematic, according to whether it was formed before, during, or after movement that deformed the rock. [PAUL J. ROPER]

Bibliography: J. B. Epstein and A. G. Epstein, in S. Subitsky (ed.), *Geology of the Selected Areas in New Jersey and Eastern Pennsylvania and Guidebook of Excursions*, Geological Society of America Annual Meeting, New Branswick, NJ, 1969; J. C. Maxwell, Origin of slaty and fracture cleavage in the Delaware Water Gap area, New Jersey and Pennsylvania, in *Petrologic Studies: A Volume to Honor A. F. Buddington*, Geological Society of America, Boulder, 1962; E. W. Spencer, *Introduction to the Structure of the Earth*, 1977.

Coal

The general name for the natural, rocklike, brown to black derivative of forest-type plant material, usually accumulated in peat beds. By burial and subsequent geological processes, coal is progressively compressed and indurated, finally altering into graphite or graphitelike material. In American terminology, the rank varieties of coal comprising this carbonification series consist of lignitic coal, represented by brown coal and lignite; bituminous coal, also including subbituminous coals; and anthracite coals, consisting of semianthracite, anthracite, and metaanthracite. *See* LIGNITE; PEAT.

Formation. Coal may originate from isolated fragments of vegetation, but most coal represents the carbonification of woody plants accumulated in peat beds. These are mainly of two kinds: autochthonous deposits representing accumulations at the place of plant growth, such as those found in the Great Dismal Swamp of Virginia, and allochthonous deposits accumulated elsewhere than at the place of growth by the drifting action of stream, lake, or sea currents, such as the Red River "rafts." Generally autochthonous coal deposits overlie seat rock, or underclay containing traces of plant roots called Stigmaria in the case of coals of Paleozoic age. *See* COAL PALEOBOTANY; SEDIMENTATION.

Biochemical activity modifies the character of the unsubmerged, lightly submerged, or lightly buried peat. This process consists in part of general oxidation, but mainly of attack by aerobic bacteria and fungi that can live only where oxygen is available and of anaerobic bacteria where water or thin sediments cover the peat. Fires set by lightning or other causes may consume part of the peat from time to time, leaving in places a residue of charcoal which may eventually be incorporated into the coal bed in the form of fusain, known to miners as mineral charcoal, mother-of-coal, and mothercoal. *See* DIAGENESIS.

The forest fire origin of fusain is disputed by many botanists who believe the presence of certain combustible components in fusain, such as resins, indicates that chemical causes operating under special conditions bring about the formation of fusain. No completely satisfactory explanation of the origin of fusain has been stated. It is found in all ranks of coal with relatively little difference in composition. There is also transitional material, between normally coalified wood or bark and fusain, called semifusain. Because of its porosity, fusain is commonly mineralized into a hard and heavy substance; unmineralized fusain is soft and light. Fusain occurs in all sizes from particles of microscopic dimension to aggregates forming fairly continuous thin sheets or lenses several feet across and several inches thick.

The material composing the peat which is finally transformed into coal varies with differences in source material, in conditions of accumulation and diagenesis, and in the length of time involved prior to burial. Figure 1 shows the relative resistance of the principal peat-forming plant substances to microbic decomposition. Based upon the peat-forming components, which are more or less segregated into bands, several common types or varieties of coal can be recognized.

Unbanded coal is represented by cannel, or sapropelic, coal and includes common cannel and algal cannel, variously designated boghead cannel and torbanite. The microscope is used to determine the presence of algae in the latter type. Cannel coals are also distinguished by their greasy luster and blocky, conchoidal fracture. *See* SAPROPEL; TORBANITE.

Banded coal may be either bright or dull. Banded bituminous coal is produced by thin lenses of highly lustrous coalified wood or bark called vitrain, if of megascopically distinguishable thick-

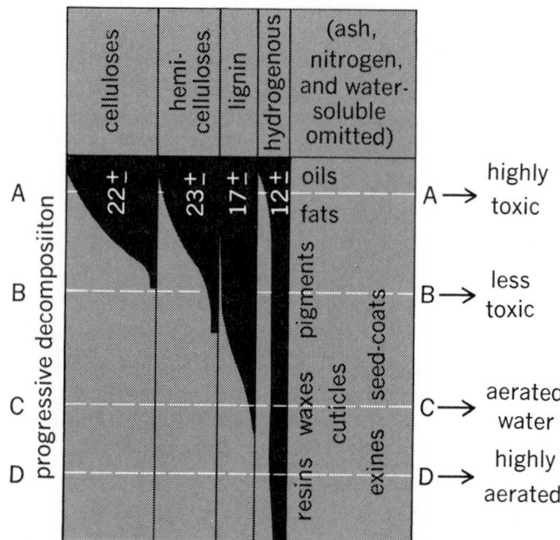

Fig. 1. Approximate proportions of principal peat-forming components in the dry-ingredient plant debris plotted according to relative resistance to microbic decomposition and order of disappearance. (*After D. White*)

ness (1/2 mm or 1/50 in.). Intervening between such black, lustrous bands are layers of more or less striated bright or dull coal (clarain or durain, respectively). The clarain contains a predominance of fine vitrainlike laminae or lenses (microvitrain); microvitrain is of minor importance in durain, and this material has a dull luster. Dullness may also result from a predominance of mineral matter or from a relatively high content of bituminous matter, such as spores, cuticles, resins, and waxes. Likewise the presence of opaque matter or fine fusain contributes to dullness in banded coal.

Carbonification. This term refers to the process of coal metamorphism brought about by increasing weight of overriding sediments, by tectonic movements, by an increase in temperature resulting from depth of burial, or from close approach to, or contact with, igneous intrusions or extrusions. Increase in pressure affects principally the physical properties of the coal, that is, hardness, strength, optical anisotropy, and porosity. Increase in temperature acts chiefly to modify the chemical composition by increasing the carbon content and decreasing the content of oxygen and hydrogen, decreasing the volatile matter and increasing the amount of fixed carbon, and increasing the calorific value to a maximum with about 20% volatile matter. Rapid metamorphism or carbonification of coal, effected by close approach to or contact with igneous intrustions or extrusions, may result in the formation of natural coke. Natural coke is found in some coal fields but is relatively rare. *See* METAMORPHISM.

Classification. The classification and description of coal depends largely upon information supplied by chemical analysis and the results of a number of empirical tests. Chemical analyses are of two kinds: the elementary or ultimate analysis and the proximate or commercial type of analysis. The ultimate analysis is usually limited to the percent of hydrogen, carbon, oxygen, nitrogen, and sulfur, exclusive of the mineral matter or ash. Phosphorus may be determined for those coals which are used for metallurgical purposes. The proximate analysis, which is empirical in character, is the prevailing form of analysis in North America. In this analysis, systematically standardized procedures are used to provide values for volatile matter, fixed carbon, moisture, and ash, all of which add up to 100%. Calorific and sulfur values are also determined, the latter sometimes in terms of forms of sulfur such as pyritic, organic, and sulfate sulfur.

It is common practice to use qualifying terms, such as "moisture-free" (mf), "moisture-and-ash-free" (maf), "pure coal," and "as received," in presenting the findings of proximate and ultimate analyses. Usually values are determined initially on a "moisture-free" or "dry" basis, the other forms of analysis being calculated with the use of determined moisture and ash values. The maf value is much used on the mistaken assumption that it represents the heat value or composition of the pure coal material, as though the formation of ash from the original mineral matter were without calorific or other effects. To make a rapid and convenient allowance for the errors inherent in the maf values, various procedures have been devised to arrive at a mineral-matter-free (mmf) basis of comparison and classification. The best known of these devices is that proposed by S. W. Parr whereby so-called unit coal, or the correction to be applied to obtain approximately correct values for mmf coal, is obtained:

$$\text{Unit coal} = 1.00 - (1.08 \text{ ash } \% + 0.55 \text{ sulfur } \%)$$

Using this formula,

$$\text{Moist mmf Btu} = \frac{\text{As rec'd Btu} - 5000 \text{ sulfur } \%}{1.00 - (1.08 \text{ ash } \% + 0.55 \text{ sulfur } \%)}$$

and

$$\text{Dry mmf fixed carbon} = \frac{\text{Fixed carbon} - 0.15 \text{ sulfur } \%}{1.00 - (\text{moisture } \% + 1.08 \text{ ash } \% + 0.55 \text{ sulfur } \%)}$$

Table 1. Classification of coals by rank[a]

Class	Group	Fixed carbon limits, % (dry, mineral-matter-free basis)		Volatile matter limits, % (dry, mineral-matter-free basis)		Calorific value limits, Btu/lb (moist,[b] mineral-matter-free basis)		Agglomerating character
		Equal or greater than	Less than	Greater than	Equal or less than	Equal or greater than	Less than	
I. Anthracitic	Metaanthracite	98	–	–	2	–	–	Nonagglomerating
	Anthracite	92	98	2	8	–	–	
	Semianthracite[c]	86	92	8	14	–	–	
II. Bituminous	Low-volatile bituminous coal	78	86	14	22	–	–	
	Medium-volatile bituminous coal	69	78	22	31	–	–	Commonly agglomerating[e]
	High-volatile A bituminous coal	–	69	31	–	14,000[d]	–	
	High-volatile B bituminous coal	–	–	–	–	13,000[d]	14,000	
	High-volatile C bituminous coal	–	–	–	–	11,500	13,000	
						10,500	11,500	Agglomerating
III. Subbituminous	Subbituminous A coal	–	–	–	–	10,500	11,500	Nonagglomerating
	Subbituminous B coal	–	–	–	–	9,500	10,500	
	Subbituminous C coal	–	–	–	–	8,300	9,500	
IV. Lignitic	Lignite A	–	–	–	–	6,300	8,300	
	Lignite B	–	–	–	–	–	6,300	

[a]This classification does not include a few coals, principally nonbanded varieties, which have unusual physical and chemical properties and which come within the limits of fixed carbon or calorific value of the high-volatile bituminous and subbituminous ranks. All these coals either contain less than 48% dry, mineral-matter-free fixed carbon or have more than 15,500 moist, mineral-matter-free Btu/lb.

[b]Moist refers to coal containing its natural inherent moisture but not including visible water on the surface of the coal.

[c]If agglomerating, classify in low-volatile group of the bituminous class.

[d]Coals having 69% or more fixed carbon on the dry, mineral-matter-free basis shall be classified according to fixed carbon, regardless of calorific value.

[e]There may be nonagglomerating varieties in these groups of the bituminous class, and there are notable exceptions in high volatile *C* bituminous group.

SOURCE: From *Book of ASTM Standards*, copyright 1967 by the American Society for Testing and Materials; reprinted by permission.

The calculations are all based upon the "as received" values, and both moist and dry mmf values are used in the standard classification of coal by rank (Table 1).

Mineral matter. The determination of mineral matter composition on an elementary basis requires a much more elaborate procedure than is necessary for the coal material itself, since the mineral substances found in coal show important local and regional variations. The following detrital minerals, other than clay minerals and quartz, are reported from an Illinois coal bed: feldspar, garnet, common hornblende, apatite, zircon, muscovite, epidote, biotite, augite, kyanite, rutile, staurolite, topaz, tourmaline, and chloritic material. Secondary minerals consist of kaolinite, calcite, pyrite, siderite, and ankerite. In specially mineralized areas, the range of secondary minerals in coal is almost unlimited. *See* MINERAL.

The character of the nonorganic matter in the coal has a considerable effect upon the fusion temperature of the ash, the determination of which is one of the common subsidiary tests in coal analysis procedure. In general, ash that fuses at high temperatures is preferred to low-fusion ash, the temperature being below 2000°F for low-fusion ash and above 2400°F for high-fusion ash; the range is about 1600–2800°F.

Rank. The term rank refers to the stage of carbonification reached in the course of metamorphism. (In international usage the term type is sometimes given the meaning commonly given to rank in North America and in much of the world. No equivalent of the American term type is given.) Classification of coal by rank is fundamental in coal description and generally is based upon the chemical composition as stated in the previous section on carbonification, and upon proximate values based upon mmf coal obtained from standard face samples. In the case of low-rank bituminous and subbituminous coals, rank also is based upon agglomerating characteristics and, in the case of brown coal and lignite, upon calorific value, separation being at 6300 Btu. In Europe, coking properties are a more common basis of classification than in North America. The American Society for Testing and Materials standards of rank classification of coals of North America are shown in Table 1. The higher-rank coals are classified with respect to fixed carbon on the dry basis (mmf) and the lower-rank coals according to Btu on the moist basis (mmf).

The difference between common banded and the sapropelic, or canneloid, types of coal of the same rank, which occur in the same general region, is mainly in the higher hydrogen and slightly higher calorific value of the latter type.

Calorific values. The calorific values of coal are expressed in British thermal units (Btu) per pound in English-speaking countries, and in calories per gram where the decimal system prevails. Heat values, determined by a calorimeter, vary for coal material from about 6300 Btu for California lignite to about 16,300 Btu for some maf cannel coal. Proximate and ultimate analyses of representative United States coals by rank are given in Table 2, the analyses being on the "as received" basis.

Physical characteristics. The physical characteristics of coal concern the structural aspects of the coal bed and texture as determined by the megascopic and microscopic physical constitution of the coal itself. Structurally the coal bed (also called seam and, less appropriately, vein) is a geological stratum characterized by the same irregularities in thickness, uniformity, and continuity as other strata of sedimentary origin. Thickness varies greatly. German brown coals occasionally approach 300 ft. A drill hole in the Lake De Smet area in Wyoming penetrated 223 ft of lignite or subbituminous coal essentially in one bed. In 1960 62% of the coal produced in the United States came from beds 3 to 6 ft thick, the average thickness that year being 61 in., with 1.5% produced from beds 6 ft or less thick.

Coals beds may consist of essentially uniform continuous strata or, like other sedimentary deposits, may be made up of distinctly different bands or benches of varying thickness. The benches may be separated by thin layers of clay, shale, fusain, pyrite, or other mineral matter, commonly called partings by the miner. Clay or shale bands may be called blue bands, as is the persistent clay parting in the Herrin (No. 6) coal bed of Illinois. Like other sedimentary strata, coal beds may be structurally

Table 2. Proximate and ultimate analyses of samples of each rank of common banded coal in the United States*

Rank	State	Proximate analysis, %				Ultimate analysis, %					Heating value, Btu/lb
		Moisture	Volatile matter	Fixed carbon	Ash	S	H	C	N	O	
Anthracite	Pa.	4.4	4.8	81.8	9.0	0.6	3.4	79.8	1.0	6.2	13,130
Semianthracite	Ark.	2.8	11.9	75.2	10.1	2.2	3.7	78.3	1.7	4.0	13,360
Bituminous coal											
Low-volatile	Md.	2.3	19.6	65.8	12.3	3.1	4.5	74.5	1.4	4.2	13,220
Medium-volatile	Ala.	3.1	23.4	63.6	9.9	0.8	4.9	76.7	1.5	6.2	13,530
High-volatile A	Ky.	3.2	36.8	56.4	3.6	0.6	5.6	79.4	1.6	9.2	14,090
High-volatile B	Ohio	5.9	43.8	46.5	3.8	3.0	5.7	72.2	1.3	14.0	13,150
High-volatile C	Ill.	14.8	33.3	39.9	12.0	2.5	5.8	58.8	1.0	19.9	10,550
Subbituminous coal											
Rank A	Wash.	13.9	34.2	41.0	10.9	0.6	6.2	57.5	1.4	23.4	10,330
Rank B	Wyo.	22.2	32.2	40.3	4.3	0.5	6.9	53.9	1.0	33.4	9,610
Rank C	Colo.	25.8	31.1	38.4	4.7	0.3	6.3	50.0	0.6	38.1	8,580
Lignite	N. Dak.	36.8	27.8	30.2	5.2	0.4	6.9	41.2	0.7	45.6	6,960

**Technology of Lignitic Coals*, U.S. Bur. Mines Inform. Circ. no. 769, 1954. Sources of information omitted.

disturbed by folding and faulting so that the originally approximate horizontality of position is lost to the extent that beds may become vertically oriented or even overturned, as in the anthracite fields of the eastern United States and in other places in the world where similar high-rank coals are found.

Texture. The texture of the coal itself is determined by the character, grain, and distribution of its megascopic and microscopic components. In general banded coals composed or relatively coarse highly lustrous vitrain lenses (1/4 in. or more in thickness) are considered coarsely textured. As the thickness of the vitrain bands progressively lessens, the texture becomes finebanded and then microbanded, with bright laminae composed of microvitrain. Coals with less than 5% vitrain or microvitrain are regarded as nonbanded or canneloid.

In general, the same textural units are observed in bituminous and anthracite coals. The names employed for the lithotypes, that is, vitrain, clarain, and durain, are somewhat less suitable for lignitic coals, because in these ranks of coal the bands of material that appear as vitrain in bituminous coals and anthracite commonly have the unmistakable appearance of tree trunks and pieces of wood or bark.

Microscopic texture is determined by the physical composition of the lithotypes as noted above, or of the anthraxylon and attritus, in terms of the microscopic constituents. The microscopy and petrology of coal are concerned very largely with these constituents. This textural aspect of coal is of imporatnce in accurate coal description and classification, and for an understanding of the behavior of coal in its preparation and utilization.

Botanical and petrologic entities. The fundamental physical constituents of coal have been investigated from two points of view. In North America the conventional point of view established by Reinhardt Thiessen and long followed by the U.S. Bureau of Mines is microscopic, regarding coal as an aggregate of original botanical entities identifiable only by miscroscopic means. This approach has commonly been referred to as coal microscopy, although the use of the phrase coal petrography has become more frequent in recent years. The other point of view, that assumed by Marie C. Stopes of England and accepted in much of the world, is based upon the concept of the four megascopic ingredients now called lithotypes: vitrain, clarain, durain, and fusain. Since these concern coal materials as rock substances, the concept provides a basis for the petrologic study of coal. See the section below on petrology and petrography.

Microscopy. The microscopic study of coal, as developed in North America, is primarily concerned with the botanical entities or phyterals of coal, with fusain regarded as a coal substance of unique character and with appropriate consideration being given to mineral matter. By 1930 Thiessen had established three categories of microscopic components, anthraxylon, attritus, and fusain, and had recognized and described most of the botanical constituents of coal. This established a system of nomenclature, description, and classification which has been followed essentially in publications of the U.S. Bureau of Mines concerned with coal microscopy since that date. Anthraxylon consists of coal occurring in bands in which wood or bark structure is microscopically evident. All vitrain of the European classification, if more than 14 microns (μ) thick, is regarded as anthraxylon; below this threshold it is classified as attritus. The attritus consists of finely textured coalified plant entities or phyterals not classified as anthraxylon (less than 14 μ) or fusain (less than 40 μ). Attritus, therefore, may contain very fine shreds of anthraxylonlike material, fine particles of fusain, disintegrated or macerated humic material, or "humic degradation matter" (HDM) in addition to the following constituents: resins, waxes, cuticles, spore and pollen exines, algae, opaque matter, fungal bodies such as sclerotia, and fine mineral matter of various kinds with clay minerals usually predominating. The subdivision of attritus into translucent and opaque attritus on the basis of its content of opaque matter is the only subdivision made of the attritus with respect to variations in its heterogeneous constitution. In North America banded bituminous coals with 30% or more of opaque matter are classified as splint coals; those with 20–30% opaque matter are classified as semisplint coals. Generally, no equivalent classification is recognized in other parts of the world.

Analysis and classification of coals on the microscopic basis have usually been made in terms of the major components anthraxylon, attritus, and fusain, with consideration given to the quantity of opaque attritus and mineral matter. These distinctions in regard to opaque attritus have been applied because splint coals are generally not amenable to hydrogenation and commonly not to carbonization. Coal microscopy technique depends almost entirely on the use of thin, translucent sections of coal, a technique which is not adapted for use with high rank low-volatile and anthracitic coals. Maceration has also been used as a means of breaking down the coal and isolating the more resistant constituents, thus incidentally providing the fossil spores that are the basis of the science of palynology.

Petrology and petrography. The use of the word petrology as applied to coal assumes that coal can correctly be regarded as a rock substance; its description is therefore consistently regarded as petrographic, that is, as a field of petrography.

The initial contribution of Marie C. Stopes (1919) to the field of coal microscopy included the adoption of a petrographic concept of coal as a rock substance composed of banded "ingredients" now called lithotypes—vitrain, clarain, durain, and fusain and also mineral matter. She also introduced, in 1935, the "maceral" concept (Table 3), macerals being the individual components of the lithotypes, comparable to minerals of nonorganic rocks.

A primary dissatisfaction with the treatment of macerals as the equivalent of minerals arose from the realization that macerals possess no fixed chemical composition, such as that possessed by minerals. Each maceral becomes progressively modified chemically and physically as the rank of the coal advances. Hence it has become the practice in coal petrology to indicate the rank position of individual macerals solely by reliance upon measurement of some physical attribute, power of

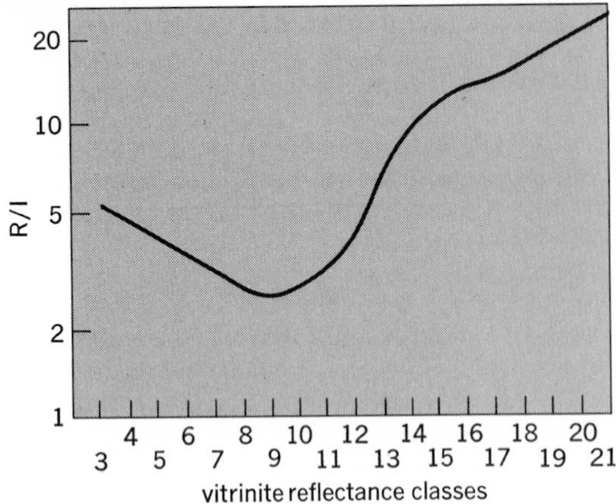

Fig. 2. Optimum ratio of the reactives to inerts (*R/I*) for each vitrinite reflectance class. (*Modification by J. A. Harrison, H. W. Jackman, and J. A. Simon, Predicting Coke Stability from Petrographic Analysis of Illinois Coals, Ill. State Geol. Surv. Circ. no. 366, 1964, of a figure by N. Schapiro, R. S. Gray, and G. R. Eusner, 1961*)

reflectance now being the most favored, because of the relative simplicity of its application.

Coal rank and coke. The usefulness of a bituminous coal for the production of metallurgical coke is primarily determined by its rank, that is, by its position in the lignite to anthracite series. The possibility of predicting the production of satisfactory metallurgical coke from coal on the basis of the rank of the coal determined by reflectance of certain macerals, particularly vitrinite, was introduced by I. Ammosov and associates in 1957. The usefulness of this method with respect to the coals of the United States has since been thoroughly investigated by N. Schapiro, R. J. Gray, and G. R. Eusner (1960–1967) and further tested by J. A. Harrison, H. W. Jackman, J. A. Simon (1964), and others. These investigations of many coals involved determination of reflectance, particularly of vitrinite; chemical analyses; and determination of stability or strength of laboratory cokes produced from the coal.

In the early years of activity in this field of investigation, emphasis was mainly on the determination of the reflectance of vitrinite because of its common occurrence and conspicuous reflectance. However, as investigations progressed, it was found that the presence of other macerals, and even of minerals, also affected coke strength even though some did not actually produce coke.

Reflectance procedure. Investigations concerned with the reflectance of coal between 1958 and 1968 consisted largely in the accumulation of petrographic and experimental data providing the basis for graphs used in predicting the suitability of a particular coal or blends of coal for the production of metallurgical coke suitable for use in the steel industry. Three such graphs (Figs. 2, 3, and 4) have come into frequent use in industrial and research laboratories concerned with the production of metallurgical coke.

For the preparation of these three graphs it was necessary to accumulate a large volume of laboratory data on reflectance, petrographic compositions, chemical analyses, and the stability of the coke made from many bituminous coals. Petrographic data were acquired by the point-count method (F. Chayes, 1956), using polished surfaces or broken coals (–8 mesh) mounted in a suitable

Table 3. Maceral reflectance classes and reactivity during carbonization*†

Reactives			Inerts		
Group macerals	Macerals	Reflectance class	Group macerals	Macerals	Reflectance class
Vitrinite		V0 to V21	Inert vitrinite		V22 to V80
	Collinite	C0 to C21		Inert resinite	R22 to R80
	Telinite	T0 to T21	Inertinite		I18 to I80
				Fusinite	F40 to F80
				Micrinite	M18 to M80
				Semifusinite‡	SF22 to SF80
Exinite		E0 to E15		Sclerotinite	Sc22 to Sc80
	Sporinite	St0 to St15			
	Cutinite	Ct0 to Ct15			
			Group minerals	Minerals	
	Alginite	At0 to At15	Sulfides	Pyrite, etc.	
	Resinite	R0 to R15	Carbonates	Calcite, etc.	
			Silicates	Illite, etc.	
Fusible inertinite	Semifusinite‡	SF0 to SF21			
	Micrinite	MO to M18			

*From *Ill. State Geol. Surv. Circ.*, no. 366, 1964.

†Nomenclature as defined in Glossary of International Committee for Coal Petrology and based primarily on Stopes-Heerlen system of classification. Range of reflectance values of macerals based on values of N. Schapiro and R. J. Gray, Petrographic classification applicable to coals of all ranks, *Proc. Ill. Min. Inst.*, pp. 83–97, 1960.

‡Estimated values; reactive group is about one-third and inert group about two-thirds of semifusinite total. From I. I. Ammosov et al., Calculation of coking charges on the basis of petrographic characteristics of coke, *Koks i Khimiya*, no. 12, pp. 9–12, 1957.

medium, forming cylindrical blocks about 1 in. in diameter and 1 in. in height, size not being a critical consideration.

Of the three graphs that have come into use for predicting a coke stability, Fig. 2 is of fundamental importance. The experimentally determined position of the optimum coke for each reflectance class of vitrinite (classes 3 to 21; see Table 3) and the best ratio of reactives in inerts (*R/I*) is determined by the position of the heavy curved line. For all practical purposes, *R/I* varies between 0 and 25. When a coal or blend of coals contains reactives of several reflectance classes, as is usually the case, the inert index is derived from the equation

$$N = \frac{Q}{P_1/M_1 + P_2/M_2 + \cdot\cdot\cdot + P_{21}/M_{21}} = \text{inert index}$$

where Q = total percent by volume of inerts in coal blend from analysis; P_1, P_2, etc. = percent of reactives (exinite, resinite, and 1/3 of total semifusinite) from analysis; and M_1, M_2, etc. = ratio of reactives to inerts to produce optimum coke. Inerts consist of fusinite, 2/3 of total semifusinite, micrinite, and ash. For the application of inert index see the explanation of Fig. 4.

Figure 3 consists of a set of curves, based upon experimental data, designed to show how the volume percent of the inerts (not the inert index) affects the strength of the coke (sometimes called the coking coefficient) made from coals of various reflectance classes. The position, spacing, and curvature of the curves for vitrinite reflectance classes are based upon experimental determinations with reference to an arbitrary scale of strength index (0–10) and to the volume of inerts (0–50% in steps of 5%). The strength index (coking coefficient) of a coal or blend of coals composed of reactives of various reflectance classes can be determined from the following equation in conjunction with Fig. 3:

K_T = strength index of a coal or blends

$$= \frac{(K_1 \times P_1) + (K_2 \times P_2) + \cdot\cdot\cdot + K_{21} \times P_{21}}{P_T}$$

Here P_T = total percentage of reactives; K_1, K_2, etc. = strength index of reactives in the reflectance classes present in the coal sample, obtained from the family of curves in Fig. 3; and P_1, P_2, etc. = percentage of reactives in reflectance classes present in the coal sample. The values K_1, K_2, etc. are obtained from Fig. 3 on the basis of the volume of inerts reported in the analytical data. A vertical line projected from the abscissa at this position will successively intersect the curves representing progressively higher reflectance classes present in the sample. By projecting lines horizontally from the points of intersection to the ordinate, what is designated as the strength index of the successive reflectance classes can be read, assuming the same percentage of inerts. The analytical data supply values for P_1, P_2, etc. The products indicated by the calculation K_1, P_1, etc. are the results obtained by multiplying the strength index of a particular reflectance class by the percentage of such class as provided by the petrographic analysis. Adding the resulting values of the various reflectance classes provides a figure representing the total strength index (K_T) of the reactives. This

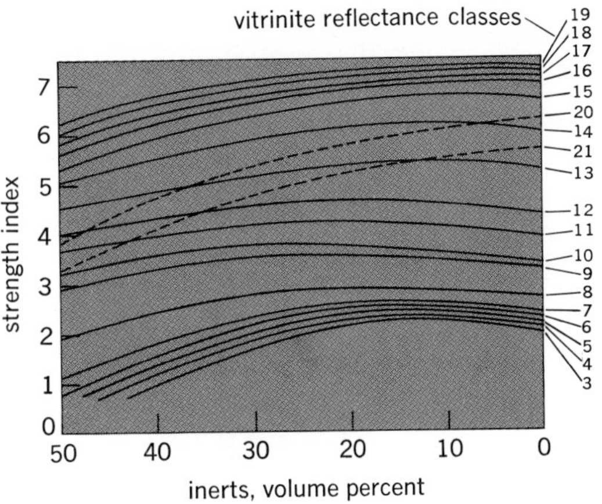

Fig. 3. Strength index of vitrinite reflectance classes depending on the amounts of inerts present. (*Modification by J. A. Harrison, H. W. Jackman, and J. A. Simon, Predicting Coke Stability from Petrographic Analysis of Illinois Coals, Ill. State Geol. Surv. Circ. no. 366, 1964, of curves by N. Schapiro, R. S. Gray, and G. R. Eusner, 1961*)

value divided by the total amount of reactives (P_T) provides the calculated strength index, which is not the equivalent of stability (Fig. 4) based upon tumbler tests.

Many tests by N. Schapiro and associates and by J. A. Harrison and associates and by others provide the basis for graphs similar to Fig. 4. The purpose of such graphs is to predict the stability factor of coke made from various coals and from blends of coals. The curves are based on the results obtained in laboratory practice in terms of stability factors varying between 10 and 65%, strength indexes between 2 and 10, and inert indexes between 0.2 and 10.0. In order to apply this chart, the coal or coals must be subjected to the various petrographic tests and analyses already cited whereby the strength and inert indexes are obtained. By the use of these data and Fig. 4, a close approach to the actual stability of the coke can be forecast by using relatively small samples of coal in the laboratory.

Rank. There has been wide acceptance of the Stopes-Heerlen nomenclature of coal petrography. But the realization, at least by American coal geol-

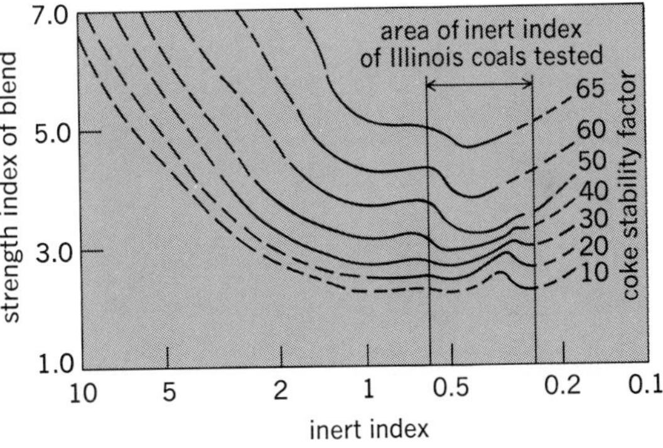

Fig. 4. Curves showing the relation between the strength index, inert index, and stability factor. (*Modification by J. A. Harrison, H. W. Jackman, and J. A. Simon, Predicting Coke Stability from Petrographic Analysis of Illinois Coals, Ill. State Geol. Surv. Circ. no. 366, 1964, of a figure by N. Schapiro, R. S. Gray, and G. R. Eusner, 1961*)

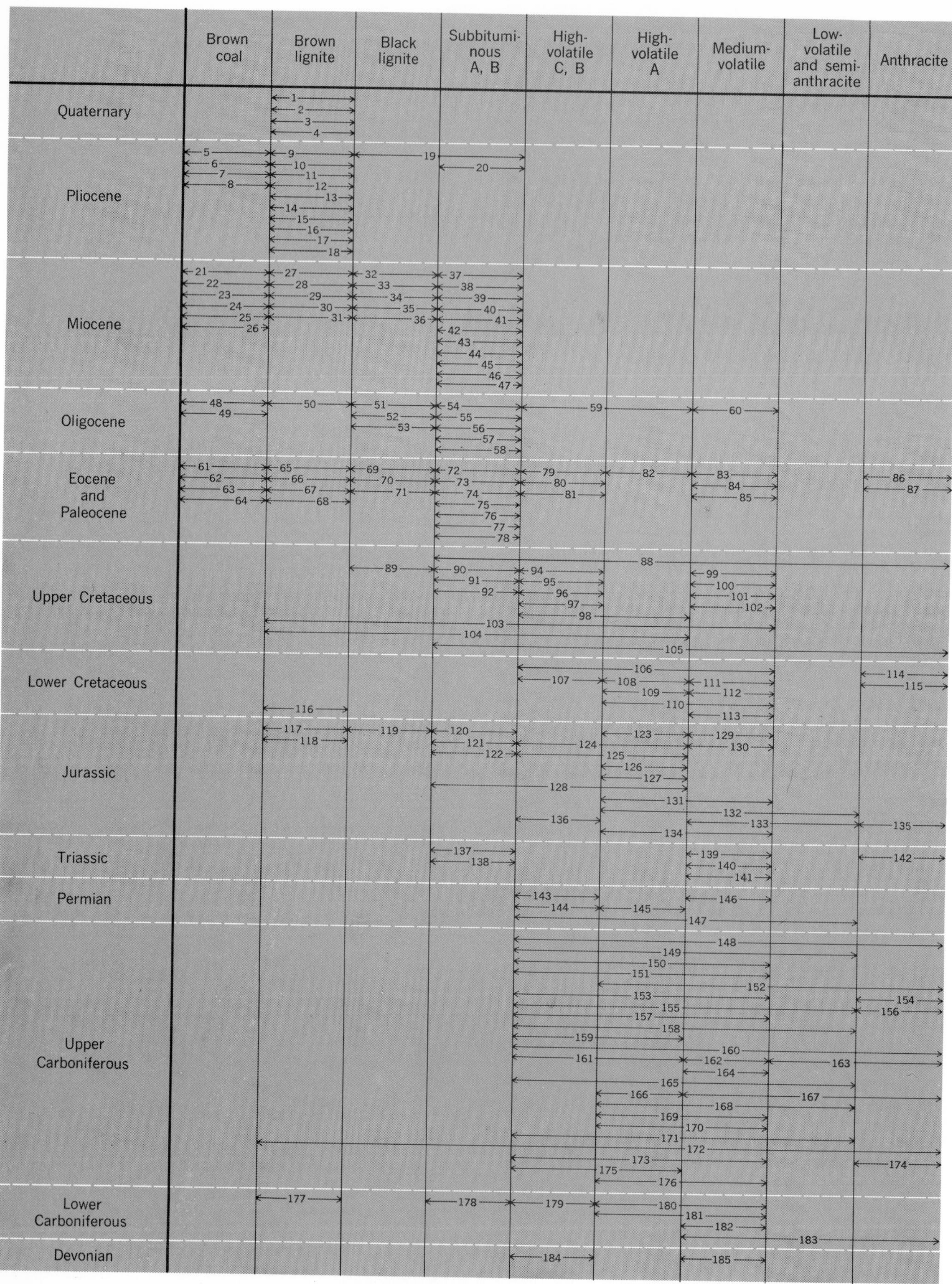

Fig. 5. Age and rank of coal in the better-known coal fields of the world. The index numbers are identified in the last paragraph of the article. (*After W. Petrascheck and W. E. Petrascheck, Lagerstättenlehre, Springer, 1950*)

ogists, that this nomenclature makes no provision for even the major subdivisions of coals by rank — anthracite, bituminous coal, and lignite — has led to at least one important series of proposals for modification of the terminology by W. Spackman. The Spackman system accepts the general validity of the three maceral groups of the Stopes-Heerlen system — vitrinite, exinite, and inertinite — but prefers to designate these as suites and to use the term liptinite rather than exinite, as the former term includes both resinite and exinite. The suite subdivision, at least with respect to the vitrinite suite, on the basis of rank is subdivided into anthrinoid, vitrinoid, and xylinoid groups for anthracite, bituminous coal, and lignite ranks. Within the three maceral groups are various maceral types identified not by name but by physical properties, and particularly by reflectivity (Table 1).

Genetic varieties. Soviet petrographers and coal geologists have proposed the recognition petrographically of varieties of coal macerals produced not by diastrophism, that is, by changes in rank after the coal bed is buried, but resulting from "genetic changes" in peat accumulations before burial, and the subsequent geological effects of such burial. Although the Soviet coal geologists accept the maceral concept, they believe important sources of variability of coal reside in the dissimilar conditions of decay that may affect the accumulated peat prior to its permanent burial. The terminology proposed by the Soviet coal petrographers to indicate the varieties to be expected in coal of a relatively high woody origin as a result of "genetic processes" seems to be restricted to two general categories of alteration, which grade into one another. These are fusinization and gelification. Those varieties of coal upon which fusinization of woody material had worked the most distinctive effect had been subjected as peat to strong atmospheric oxidation resulting in the formation of much fusain, whereas peat existing under conditions of prevailing but not too deep submergence would often be subjected to the process of gelification of the woody substances before burial. Types of coal are therefore recognized emphasizing either fusinization or gelification. These two types of coal, both more or less of the vitrinite class, are thought to provide six lithologic varieties depending mainly on the predominance of fusinite, on the one hand, and collinite, on the other hand, namely, fusinito-telinite, fusinito-precollinite, fusinito-post-telinite, fusinito-collinite, and gelito-telinite, gelito-precollinite, gelito-post-telinite, and gelito-collinite. Other varieties of coal derived from weathered peat in which the amount of woody matter was relatively small have names such as cutinite, resinite, or sporinite, depending upon the abundance (over 50%) of cuticular or resinous bodies or fossil spores.

Evaluation and uses. Although the utilization of coal varies widely with rank, three general fields may be distinguished. They are combustion (domestic, industrial, railroads, and public utilities), gasification, and carbonization (high-temperature coke for metallurgical uses and low-temperature coke for production of smokeless fuel). Anthracite is used mainly for domestic and related types of heating; some fine sizes are blended with bituminous coal to make metallurgical coke. Besides the main use of lignitic coal for combustion either in the raw state or in dehydrated and briqueted form, it is a source of industrial carbon, industrial gases, and montan wax. *See* LIGNITE.

The evaluation of coal for particular uses is usually based upon the information provided by the ultimate and proximate analyses, the values for sulfur, and the heat value. The proximate analysis is preferred in the United States. These criteria provide the means of classifying the coal by rank and, with the cost per ton known, the value of the coal in terms of 1,000,000 Btu. Ash fusion values are considered incidentally, and if the coal is to be used for making metallurgical coke, the information provided by certain coking tests is usually required. These tests consist of plasticity tests, swelling and swelling pressure tests, agglutinating tests, and agglomerating tests.

Occurrence. Coal is found on every continent, in the islands of Oceania, and in the West Indies. The amount of coal in the various countries varies greatly. The eight countries with the greatest reserves (in 10^9 short tons) of coal, including lignite, in 1977 were as follows: Soviet Union, 440; United States, 263; China, 60; West Germany, 8; United Kingdom, 6; Poland, 5; Republic of South Africa, 3.5; Australia, 3. Only two countries have data available on lignitic coal reserves (in 10^9 short tons): Soviet Union, 190; United States, 34.

The distribution of coal by rank, geologic age, and district is indicated in Fig. 5. The index numbers, which refer to coal districts, are grouped by continent and country in the following list:

Africa: Belgian Congo, 144; Natal, 147. *Asia:* China, 163, 132, 167, 135; India, 19; Siberia (U.S.S.R.), 118, 172, 119, 122, 128, 131, 133, 183, 171; Turkestan, 125. *Australia:* 8, 134; New Zealand, 31, 36, 89, 98. *Oceania:* Borneo, 81; Sumatra, 78. *Europe:* Austria, 1, 3, 14, 29, 33, 38, 72, 123, 139; Bavaria, 2, 24; Bulgaria, 7, 20, 100; Czechoslovakia, 16, 28, 32, 37, 55, 91, 161, 164; England, 48, 153, 155, 158, 157, 180, 182; France, 53, 165, 166, 168; East Germany, 5, 21, 22, 23, 90, 121, 137, 106, 143; West Germany, 24, 49, 61, 63, 62, 64, 2, 27, 57, 59, 127, 150; Greece, 75, 86; Hungary, 30, 35, 42, 44, 74, 95, 130; Italy, 18, 45, 83; Poland, 6, 138, 149, 151; Ruhr district, 148; Romania, 17, 129, 162; Russia (U.S.S.R.), 117, 177, 124, 126, 179, 184, 185; Scotland, 120, 181; Spain, 160; Spitzbergen, 76, 178, 79, 82, 84, 107, 108; Switzerland, 4, 87, 154; Turkey, 58; Yugoslavia, 10, 11, 12, 13, 15, 50, 34, 51, 52, 39, 40, 41, 47, 54, 56, 60, 85, 99, 142, 156. *North America:* Canada, 68, 116, 104, 110, 176, 112, 113, 115; Mexico, 141; United States, 43, 77, 92, 65, 66, 67, 69, 70, 71, 80, 96, 97, 103, 101, 102, 88, 105, 140, 173, 174, 175. *South America:* Brazil, 145; Peru, 46, 111, 109, 114.

[GILBERT H. CADY]

Bibliography: American Society for Testing and Materials, *1967 Book of ASTM Standards*, pt. 19, 1967; Felix Chayes, *Petrographic Model Analysis: An Elemental Statistical Appraisal*, 1956; A. C. Fieldner and W. A. Selvig, *Methods of Analyzing Coal and Coke*, U.S. Bur. Mines Bull. no. 492, 1951; J. A. Harrison, H. W. Jackman, and J. A. Simon, *Predicting Coke Stability from Petrographic Analysis of Illinois Coals*, Ill. State Geol. Surv. Circ.

no. 366, 1964; International Committee for Coal Petrology, *International Handbook of Coal Petrography*, 2d ed., 1963; D. W. van Krevelen and J. Schuyer, *Coal Science*, 1957; B. C. Parks and H. J. O'Donnell, *Petrography of American Coals*, U.S. Bur. Mines Bull. no. 550, 1956; N. Schapiro and R. J. Gray, *Petrographic Classification Applicable to Coals of All Ranks*, Ill. Mining Inst. 1960 Proc., 1960; N. Schapiro, R. J. Gray, and G. R. Eusner, Recent developments in coal petrography, in *Blast Furnace, Coke Oven and Raw Materials Conference*, Amer. Inst. Mining Engn. Proc., 20:89–112, 1961; W. H. Young and R. L. Anderson, *Thickness of Bituminous Coal and Lignite Seams Mined in 1960*, U.S. Bur. Mines Inform. Circ. no. 8118, 1962.

Coal paleobotany

A special branch of the paleobotanical sciences concerned with the origin, composition, mode of occurrence, and significance of the fossil plant materials that occur in, or are associated with, coal seams. Information developed in this field of science provides knowledge useful to the biologist in his efforts to describe the development of the plant world, aids the geologist in unravelling the complexities of coal measure stratigraphy in order to reconstruct the geography of past ages and to describe ancient climates, and has practical application in the coal, coke, and coal chemical industries.

Nature of coal seams. All coal seams consist of countless fragments of fossilized plant material admixed with varying percentages of mineral matter. The organic and inorganic materials initially accumulate in some type of swamp environment. Any chemical or physical alteration experienced by the organic fragments during the course of transportation and deposition is followed by another series of changes effected by the chemical, physical, and microbiological agents characterizing the environment in which the particle comes to rest. Subsequent burial beneath a thick cover of sediment induces further physical and chemical alteration of the particles comprising the coal seam. Usually a consolidated layer of well-bonded fragments results. In some instances, crustal deformation and even volcanism add their modifying effects. Thus, the chemical composition, size, shape, and orientation of the fossilized plant remains are influenced both before and after death by biological processes and by the chemical and physical processes attending their postdepositional history. Even in coal seams that have been metamorphosed to anthracitic rank, certain of the constituents remain recognizable as portions of particular plant organs, tissues, or cells. Such entities are classed as phyterals, and in paleobotanical descriptions of coal seams these are identified as specifically as possible as megaspores, cuticles, periderm, and so on. In seams of peat, lignite, and high-volatile bituminous coal, entities are more readily recognized as particular phyterals than they are in higher-rank deposits. This is partly because fewer are destroyed by the metamorphic processes and partly because of the distinctiveness of the substances composing the entity.

In some instances, the fossilized plant fragments can be recognized as remnants of a plant of some particular family, genus, or species. When this is possible, information can be obtained on the vegetation extant at the time the source peat was formed, and such data aid greatly in reconstructing paleogeographies and paleoclimatic patterns. Perhaps the most extensive studies of this type are those conducted on the Tertiary brown coals of the Rhine Valley in Germany. From these, detailed reconstructions have been prepared, describing and illustrating the three major swamp environments that gave rise to the sedimentary layers comprising these spectacular coal seams.

Plant fossils in coal seams. Pollen grains and spores are more adequately preserved in coals than are most other plant parts. Recognition of this fact, coupled with an appreciation of the high degree to which these fossils are diagnostic of floral composition, has led to the rapid development of the paleobotanical subscience of palynology. Fruits, seeds, and identifiable woods also occur as coalified fossils and are deserving of more attention than has been accorded them in the past. Often minute structural details are preserved (see illustration). Identifiable leaf fossils are comparatively uncommon although the coalified cuticles of leaves are frequently encountered and their botanical affinities determined. Occasionally, coal seams contain fossil-rich coal balls. Essentially all types of plant fossil, including entire leaves, cones, and seeds, are encountered in these discrete nodular masses. Within the coal ball, the altered but undistorted plant tissues are thoroughly impregnated with mineral matter, usually with calcite or dolomite. Comparatively few coal seams have been encountered with large concentrations of coal balls, but these seams have provided a wealth of detailed information on the nature of the plants which gave rise to the coals concerned. *See* FOSSIL SEEDS AND FRUITS.

Sedimentary units of coal seams. As implied previously, coal seams generally are composed of several superposed sedimentary layers, each having formed under somewhat different environmental conditions. The coal petrologist and paleobotanist recognize these layers because of their distinctive textural appearance and because each consists of a particular association of organic and inorganic materials. Accordingly, each coal seam usually contains several types of coal. These coal types, or lithotypes, possess characteristic suites of physical properties, and knowledge of these properties is very profitably employed in manipulating coal composition in coal preparation and beneficiation plants.

A given lithotype may form several of the constituent layers of lithobodies of a coal seam. Each lithobody often possesses a characteristic assemblage of botanically identifiable fossil plant fragments rendering the unit recognizable to the coal paleobotanist without reference to its textural or compositional features. Paleobotanical descriptions of coal beds may relate to the entire thickness of the seam without regard for the seam's composite nature, or the fossils may be described in relation to the particular lithobody sequence. The more detailed type of paleobotanical description is required if paleoecological interpretations are to be made or in instances where detailed stratigraphic work is involved.

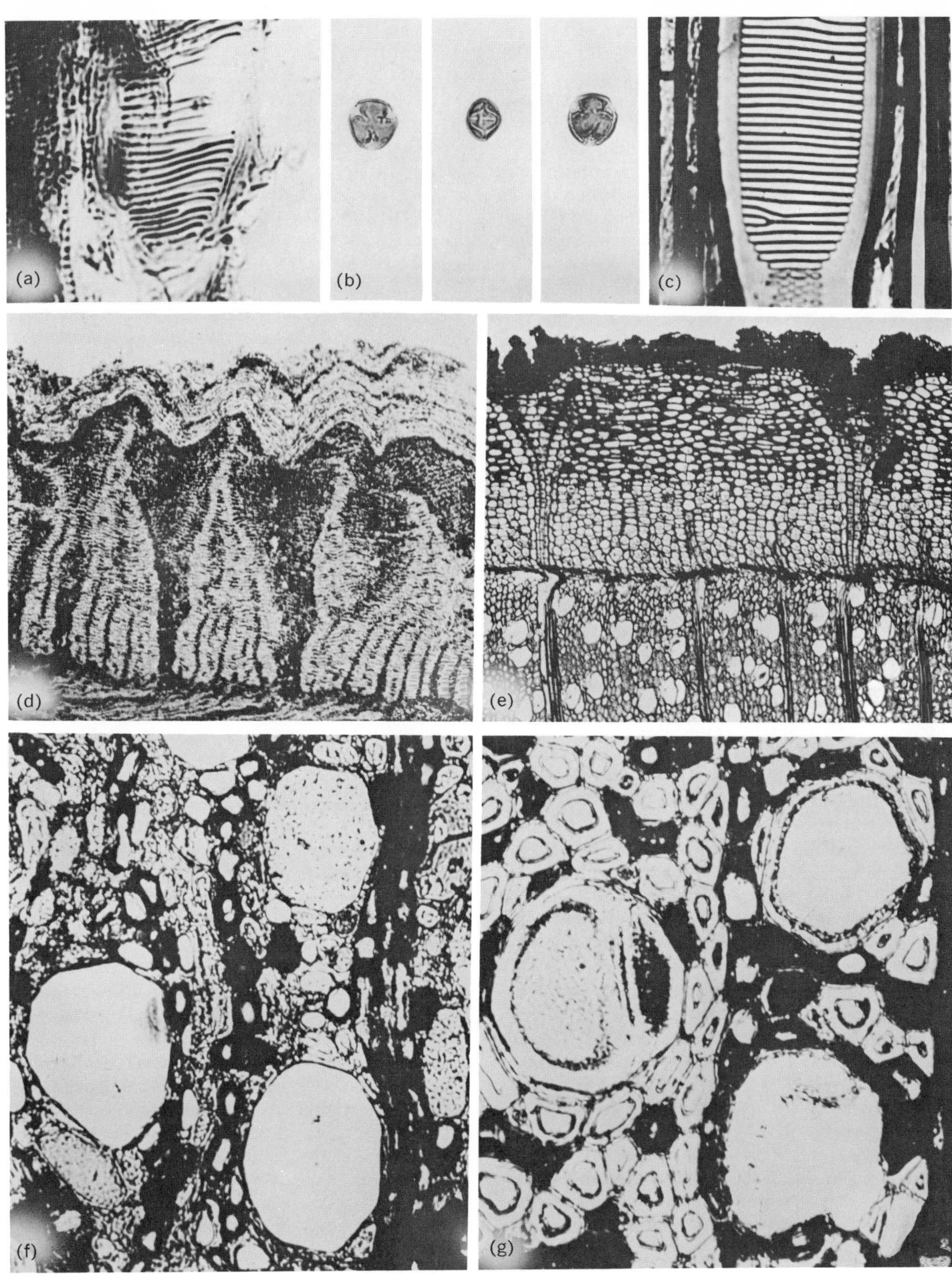

Plant fossils in coal. (*a*) Coalified remnants of scalariform perforation plate in vessel element of piece of lignitized *Cyrilla* wood from Oligocene lignite deposit near Brandon, Vt. (*b*) Coalified *Cyrilla* pollen grains extracted from Brandon lignite. (*c*) Scalariform perforation plate of vessel in wood of modern *Cyrilla racemiflora*. (*d*) Transverse section of lignitized bark of *Cyrilla* from Brandon lignite. (*e*) Transverse section of extant *Cyrilla racemiflora* bark with portion of secondary xylem shown in lower half of photo. (*f*) Transverse section of coalified secondary xylem of *Cyrilla* wood from Brandon lignite showing preservation of fine structural detail. (*g*) Transverse section of secondary xylem of *Cyrilla racemiflora* showing the structure of vessels, fiber tracheids, wood parenchyma, and rays. Note the coalification effects by making a comparison with *f*.

Interrelations with coal petrology. Another facet of coal paleobotany concerns the nature of the substances which compose the coalified plant fragments. When his attention is focused upon such matters, the coal paleobotanist is indistinguishable from the coal petrologist and both begin to encroach on the province of the coal chemist. It is noteworthy, however, that the paleobotanist tends to concentrate his attention on the genetic relationships of the various coal substances and hence upon the steps in the derivation and subsequent evolution of maceral materials. The initial stages of coalification tend to be ignored by the coal chemist and petrologist, and the study of these has, quite properly, been thought of as an integral part of coal paleobotany. Thus, through the study of the botanical character of coalified fossils and by means of investigating the substances produced by the coalification process, coal paleobotanists extend man's knowledge of the organically derived sediments of the Earth's crust.

[WILLIAM SPACKMAN]

Coastal landforms

The characteristic features and patterns of land in a coastal zone subject to marine and subaerial processes of erosion and deposition. Coastland lies between the shoreline, or mean low-water line, and the hinterland and includes many minor features of coastal relief as well as major features of shoreline topography. As treated here, emphasis is placed upon regional aspects of coastal morphology. For a discussion of beach and nearshore processes *see* NEARSHORE PROCESSES.

Vigorous wave action, a variety of currents, water of high saline content, and steady winds produce both constructional and destructional landforms as they act on land with whatever rock composition and structure, topographic characteristics, and vegetational cover it may possess. In addition to the agents working upon landforms, the changing position of shorelines with respect to sea level is an important factor in coastal evolution.

The attack of the sea has been concentrated at or near its present level only for some 5000–6000 years. During the preceding 130 centuries or so, the sea level rose some 400 ft as a result of an increase in ocean volume by meltwaters from the waning continental ice in late Pleistocene time. Coastal landforms which originated during low stands of the sea now lie submerged offshore.

Constructional landforms. Included among these forms are deltas, deltaic coastal plains, beaches, bars, spits, reefs, dunes, and many minor features. Also included in this group are structures of organic origin. Along the coasts of warm seas, algae, coral, and other organisms may accumulate in huge quantities to form reefs, which sometimes extend considerable distances offshore. These are constructional landforms, as are the low flats of sediment trapped around the roots of mangroves and other plants on some shores. The position and development of these features are related to the stand of the sea during their formation.

Fig. 1. Quaternary wave-cut terrace, north of Point Harford, Calif. Unreduced masses rise above it. (*USGS*)

Fig. 2. Caverns resulting from marine erosion in Eocene sandstone near La Jolla, Calif. (*USGS*)

Deltas grow seaward in estuaries and at river mouths. They may even extend outward into seas where rivers deposit their loads more rapidly than land subsides or currents carry sediment away. The location of a delta depends on the topography of the mainland, which determines the location of the river mouth. Where the land is flat, as in the case of a broad flood plain leading to the sea, the river mouth occurs at the lower end of a meander belt, the position of which depends on diversions that commonly occur as far as several hundred miles inland. If diversions of river courses into new meander belts occur frequently, delta growth may take place vigorously, first at one site, then at another which may be many miles away. Eventually the coastal flat becomes a deltaic coastal plain. Beaches fringe its front. Grassy marshes, lakes, and embayments of various kinds commonly form a low zone behind the beaches, across which the natural levees of branching distributary streams form relatively firm and dry strips of land leading to the sea. *See* DELTA; FLOODPLAINS; PLAINS.

Destructional landforms. Erosional landforms include sea cliffs, wave-cut beaches, arches and caves near sea level, detached masses of rock which form islands and sea stacks, and many other forms whose characteristic features have developed under the dominant influence of local topography, rock type, and rock structure (Figs. 1 and 2).

Erosional forms depend to a large extent upon the type of coast under attack. Thus, resistant rock may have changed but little during 50–60 centuries of exposure to wave attack, as in the case of massive crystalline bedrock along coasts in the vicinity of Rio de Janeiro and many rugged coasts elsewhere. Unconsolidated rock, such as most Tertiary and Quaternary sediments, commonly has been driven back to form smooth coasts with long, straight beaches which may border the mainland or lie some distance offshore. In these cases wave action and chemical weathering processes have proved capable of disintegrating the rock into fine debris of sizes small enough for currents to transport and deposit as beach. Contrasts in coastal types which are related to the kind of rock at the shoreline may be noted on opposite sides of the Baltic Sea, to the east and west of the Rhone Delta, or between the smooth coast of Texas and the irregular coast of Newfoundland. See WEATHERING PROCESSES.

Fiord coasts. A bedrock region that is deeply scoured by glaciers may become a fiord coast. Along such coasts deep estuaries may extend far inland, often between sheer cliffs over which high waterfalls plunge, as along the coasts of Norway, southern Alaska, and southern Chile. *See* FIORD.

Ria coasts. Ria coasts are characterized by irregular coastal indentations similar to fiorded coasts, such as the Dalmatian Coast of the Adriatic produced by drowning of nonglacial topography (Fig. 3).

Karst coasts. Karstic features are produced in strata of relatively soluble limestone along some coasts, such as the Dalmatian Coast of the Adriatic Sea. *See* KARST TOPOGRAPHY.

Emergent and submergent shorelines. Smooth coasts were regarded by many geomorphologists as exhibiting shorelines of emergence, based on the hypothesis that sea bottoms are flat, so that when uplifted to become coastal plains they formed relatively straight shores. More complicated shorelines were regarded as submergent, in the belief that land had lowered to some extent, permitting seas to drown complicated topography which had been produced by weathering and erosion of surfaces exposed to the atmosphere. However, with the realization that the Recent rise of sea level is of much greater magnitude than formerly supposed, there has come greater acceptance of the idea that oceanic shorelines are universally drowned.

Contrasts between smooth and irregular coasts have largely developed during the brief, latest stillstand of sea level. The classic examples of shorelines of emergence, in fact, exhibit abundant evidence of drowning. Smooth coasts from New Jersey southward demonstrate the ability of the Atlantic to smooth coastal outlines and produce long beaches in unconsolidated rock, whereas the irregular outlines of Chesapeake and other bays inland are products of flooding valleys which were cut during the earlier low stand of the sea.

New shorelines. Uplift of land occurs rapidly along many coasts, particularly those with mountainous shores. Thus beaches, sea cliffs, coastal flats, and other landforms are elevated well above sea level. The original flats when elevated form coastal terraces. A more gentle upwarping in flatter regions also tilts and elevates coastal plains so that some become coastwise terrraces. These ordinarily continue inland as river terraces, or flats which were once flood plains. Land may also be downfaulted so that escarpments form the shoreline locally. New shorelines form around volcanic islands or lava flows that reach the sea. In time these will be modified by erosional attack, which will produce sea cliffs and beaches similar to those which have developed rapidly along shores of submerged glacial deposits, such as drumlins on the Maine coast or huge masses of drift on the south coast of Long Island. [RICHARD J. RUSSELL]

Bibliography: A. Guilcher, *Coastal and Submarine Morphology*, 1958; C. A. M. King, *Beaches and Coasts*, 1959; P. H. Kuenen, *Marine Geology*, 1950; R. J. Russell, *River Plains and Sea Coasts*, 1967; F. P. Shepard, *Submarine Geology*, 2d ed., 1963; J. A. Steers, *The Coastline of England and Wales*, 1947.

Coastal plain

A lowland of slight relief, bordering the sea, and ideally, bordered on the landward side by noticeably higher land, either plateau or mountain. It is an emerged portion of the continental shelf along a coast composed of little-consolidated materials as compared to the rock base, that is, an emerged shelf with horizontal or nearly horizontal sedimentary layers of relatively recent age which dip gently seaward. Among the most extensive coastal plain areas are those of southeastern England, northern France, and the Atlantic and Gulf coasts of North America. *See* CONTINENTAL SHELF AND SLOPE.

The North American coastal plain extends from northern New Jersey along the Atlantic and Gulf coasts to southern Mexico (Fig. 1). It is commonly divided into the Atlantic coastal plain, the Florida peninsula, the East Gulf coastal plain, the Mississippi embayment including the Mississippi alluvial plain, the West Gulf coastal plain, and the Yucatàn Peninsula. The surface is low and has slight relief. The seaward portions are mostly lagoonal or swampy. At its inner border the Atlantic coastal plain is separated from the Piedmont plateau by the Fall Line. The surface of the Florida peninsula slopes gently east and west from a low central divide. The north border of the East Gulf and Mississippi embayment divisions is generally placed along the contact line between outcropping Paleozoic and Mesozoic formations. In southeastern Texas the coastal plain is separated from the Edwards Plateau by the Balcones escarpment. *See* FALL LINE.

The origin and history are not identical for all coastal plains. By uplift of land or depression of sea level, the shallow sea floor emerges. The sea may advance and retreat repeatedly, leaving marine and nonmarine sediments that outcrop in bands roughly parallel to the shoreline. Coastal areas may be downfaulted or downwarped relative to surfaces farther inland.

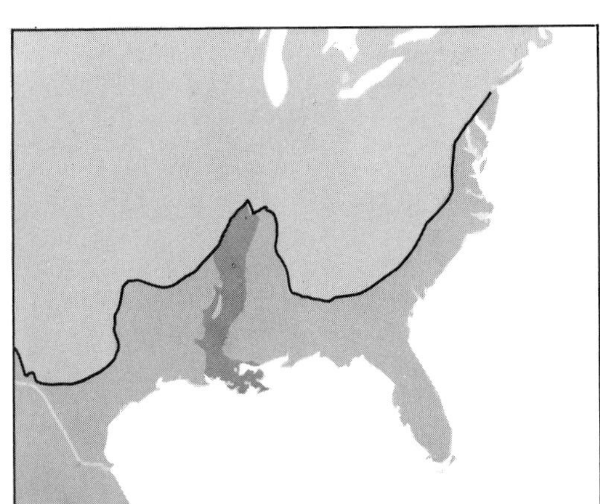

Fig. 1. Diagram showing coastal plains of United States.

Fig. 2 Block diagram of belted coastal plain, showing ridges and depressions. (*After A. K. Lobeck*)

COASTAL LANDFORMS

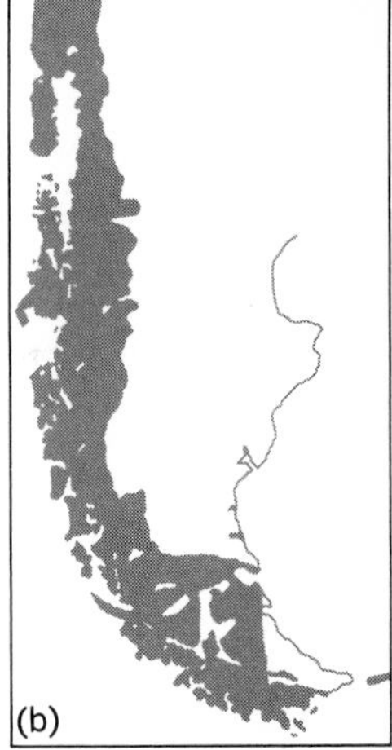

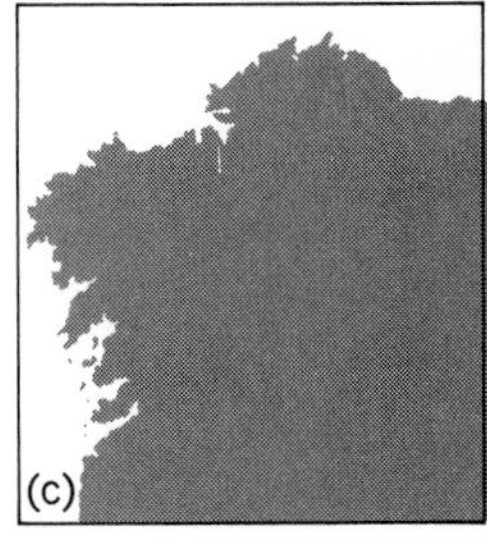

Fig. 3. Fiords of (*a*) Norway and (*b*) southern Chile. (*c*) Rias of northwestern Spain. (*From V. C. Finch et al., Elements of Geography, 4th ed., McGraw-Hill, 1957*)

Belted coastal plains result from the erosion of rock formations of varying resistance (Fig. 2). Resistant formations give rise to parallel asymmetric ridges or infacing escarpments called cuestas, and the nonresistant formations give rise to linear depressions known as vales. Such belted coastal plains occur in Texas, Alabama, and New Jersey, in southern England, in the Po valley of northeastern Italy, and in the Paris basin of northern France. *See* COASTAL LANDFORMS; GEOMORPHOLOGY.

[ARTHUR C. TROWBRIDGE]

Coesite

Naturally occurring coesite, a mineral of wide interest and the high-pressure polymorph of SiO_2, was first discovered and identified from shocked Coconino sandstone of the Meteor Crater in Arizona by E. C. T. Chao of the U.S. Geological Survey in June, 1960. Since then coesite has been identified from the Wabar (meteorite) Crater in Saudi Arabia, from the Ries Crater in Bavaria in southern Germany, from the Lake Bosumtwi Crater in Ashanti, Ghana, Africa, and from Lake Mien in Sweden. Coesite has also been identified from some Thailand tektites which are considered to have been formed also by an impact cratering process. Pertinent data on the craters are given in the table. The finding of natural coesite elevates it as a true mineral species. Because it requires a unique physical condition, extremely high pressure, for its formation, its occurrence is diagnostic of a special natural phenomenon, in this case, the hypervelocity impact of a meteorite. The fact that an instant shock can create material like coesite has opened a vast field of research on the phenomenon of shock and on the possible creation of high-pressure polymorphs with combined properties that are unfamiliar to daily life. This discovery has led to the study of the equation of state of mineral phases by shock-wave experiments, a study that is important to the understanding of the mineral assemblages in and below the mantle of the Earth at pressures in excess of 300 kilobars. Such regions of high pressures cannot be explored by current static high-pressure experimental techniques.

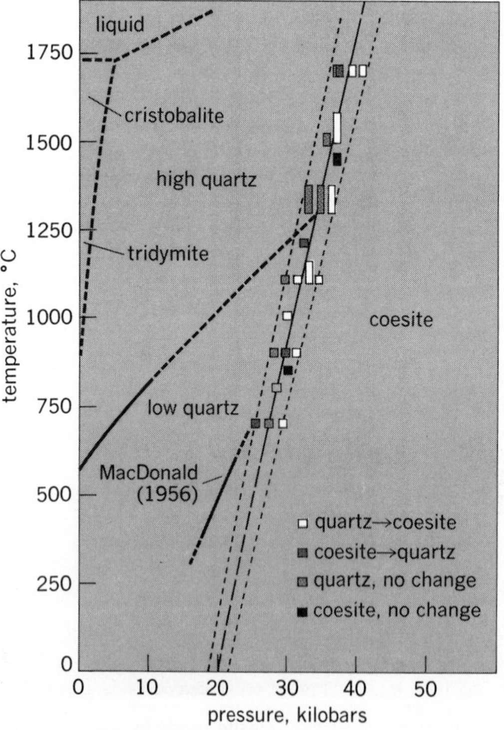

Fig. 1. Phase diagram showing the stability field of coesite. (*After F. R. Boyd and J. L. England*)

Synthesis. In 1953 coesite was first synthesized in the laboratory by L. Coes, Jr., as a chemical compound at pressures of about 35 kilobars, in the temperature range of 500–800°C. Later work by G. MacDonald in 1956, by F. Dachille and R. Roy in 1958, and by F. R. Boyd and J. L. England in 1960 established and defined part of the stability field of this high-pressure polymorph of silica, as shown in Fig. 1. It is clear from this phase diagram that coesite cannot form at pressures less than 20 kilobars (approximately 20,000 atmospheres). Coesite has also been found in synthetic diamonds and has been formed by transformation from alpha quartz by the application of shearing stress. *See* HIGH-PRESSURE PHENOMENA.

Since the synthesis of coesite, investigators have searched for it in its naturally occurring state. They have looked in kymberlites, a volcanic rock in which diamonds occur, and in eclogites, a garnet-pyroxene rock believed by geologists and petrologists to have been formed under great pressure, but have failed to find coesite. On the basis of the laboratory data it was previously concluded that coesite could not be formed naturally in the Earth's crust at depths of less than 60 mi.

Properties. Coesite occurs in grains that are usually less than 5 microns (μ) in size and are generally present in small amounts; therefore it was not identified by previous investigators, such as George Merrill, who first studied and described the petrography of the fractured Coconino sandstones as early as 1907. The properties of the mineral are known mainly from studies of synthesized crystals. It is colorless with vitreous luster and has no cleavage. It has a specific gravity of 2.915 ± 0.015 and a hardness of about 8. It is biaxial positive with $2V$ about 64°. Its indices of refraction are α 1.5940, β 1.5955, and γ 1.5970 ± 0.0005. Its dispersion is horizontal, with r less than v (weak). The optical orientation is $X = b$, $Z \angle C = 4-6°$, and $\beta = 120°$. Synthetic crystals occur as euhedral to subhedral hexagonal platelets, and laths with positive elongation. Simple contact twins occur with (021) as twin and composition plane. The mean index of the naturally occurring coesite extracted from the Meteor Crater in Arizona is 1.595 ± 0.002. Coesite is monoclinic, with $a = 7.16$ A, $b = 12.39$, $c = 7.16$, and $\beta = 120°$. It is nearly insoluble in 5% HF at room temperature but is readily soluble in concentrated HF at elevated temperatures.

Thermal expansion of the unit cell of coesite has been determined by B. J. Skinner. Coesite was shown to be unstable by F. Dachille, R. J. Zeto, and R. Roy when annealed at temperatures above 1300°C. Below such temperatures, however, coesite will remain indefinitely. This accounts for the finding of coesite from the Ries Crater in southern Germany; this crater is approximately 15,000,000 years old.

Natural coesite can be identified from thin sections of rocks by its peculiar habit and high index

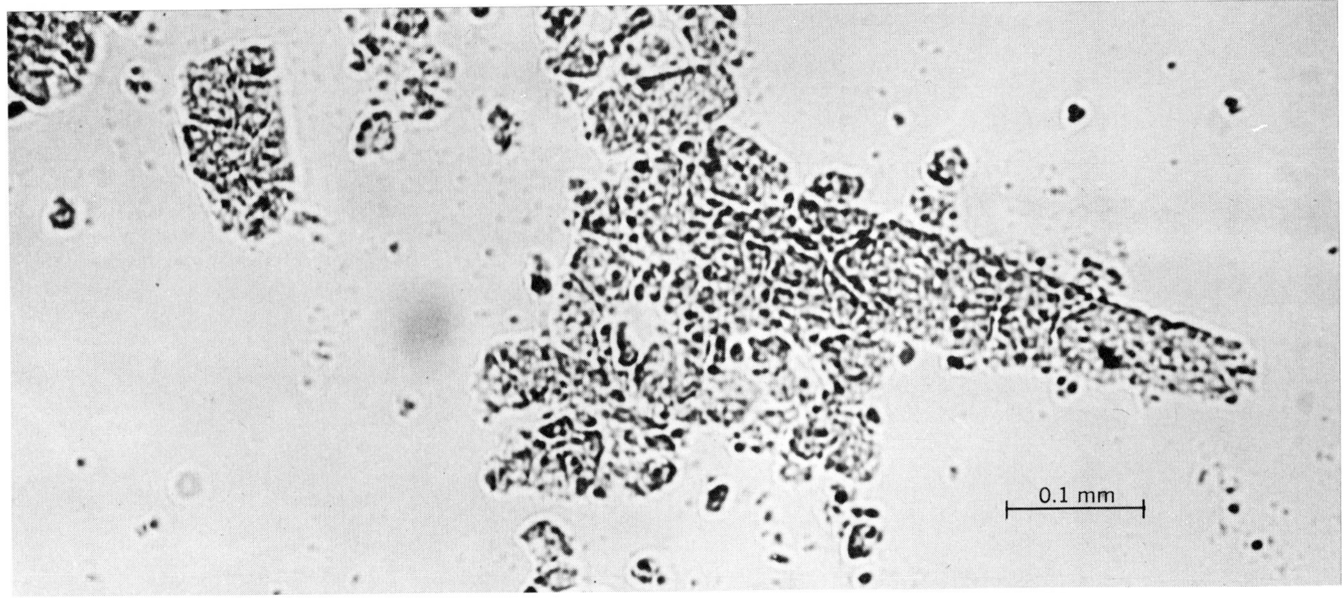

Fig. 2. Natural coesite with inclusions of quartz. Plain reflected light.

of refraction (Fig. 2). However, its positive identification must rest on the x-ray powder diffraction technique (Fig. 3).

Applications and origins. Coesite has as yet no evident commercial use and therefore has no obvious economic value. As a stepping-stone in scientific research, it serves in at least two ways: (1) Where it occurs naturally, coesite is diagnostic of a past history of high pressure; (2) the occurrence of coesite from Meteor Crater clearly suggests that shock as a process can transform a low-density ordinary substance to one of high density and unique properties.

Impact origin. Iron meteorites have been found both at the Meteor Crater in Arizona and at the Wabar Crater in Arabia; thus there is no doubt that the crater as well as the formation of coesite in siliceous rocks were the direct result of hypervelocity impact of the iron meteorites. If other energy sources, for example, volcanic explosion or tectonic movement, cannot produce a pressure high enough to convert quartz to coesite, then the presence of coesite, especially if associated with a crater, is diagnostic of a meteorite impact origin. The Ries Crater and the Lake Bosumtwi Crater, long of debatable origin, can now be said to be of meteoritic impact origin, although meteoritic remains have yet to be found from these two large craters.

Shock origin. The finding of natural coesite from meteorite impact craters opened up the field of experiment by shock loading and the possibility of the creation of other high-pressure synthetic compounds by shock. The finding of coesite from the Meteor Crater helps to explain the occurrence of diamond in the Canyon Diablo meteorite, the shock origin of which was proposed by M. E. Lipschutz and Edward Anders prior to the discovery of natural coesite. Subsequent experiment by P. S. DeCarli and J. C. Jamieson succeeded in producing diamonds by shock loading. It is clearly possible that high-pressure polymorphs of other substances can be formed by shock, which, like diamonds, may have extensive potential economic value. *See* DIAMOND; METEORITE.

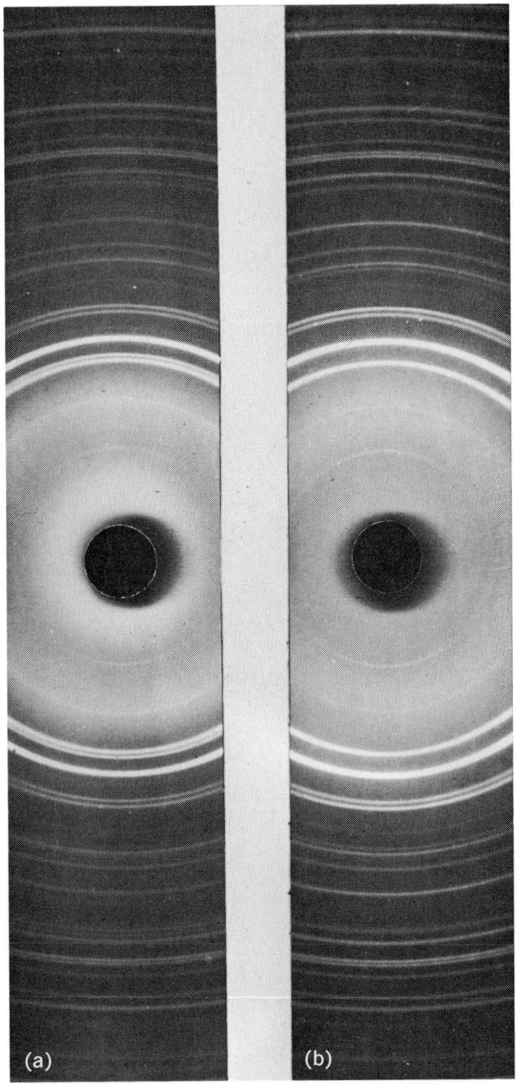

Fig. 3. Photographs of coesite made by x-ray powder diffraction technique. (*a*) Natural coesite with minor amount of quartz. (*b*) Synthetic coesite.

Meteorite craters in which coesite has been found

Crater	Location	Dimension	Bedrock geology	Remarks
Meteor Crater, Arizona	Near Canyon Diablo; south of Route 66, near Winslow, Ariz.	About 4000 ft in diameter, 600 ft deep; rim rises 100–200 ft above surrounding plain.	Crater is developed in Permian Coconino sandstone, Toroweap formation, and Kaibab limestone; and in the Triassic Moenkopi formation. Crater floor is underlain by Pleistocene and Recent talus, alluvium and lake beds, resting on 30-ft-thick layer of mixed debris and 600-ft-thick breccia lens. Upper part and rim of crater consist of an inverted section of the above formations.	Coesite occurs mostly in the intensely fractured and partly fused white Coconino sandstone. A trace of coesite was also present in the fractured sandy Kaibab limestone. The meteorite that created this crater is estimated to weigh about 60,000 tons. Crater is estimated to be about 30,000 years old.
Wabar Crater, Saudi Arabia	Near Al Hadida (21°30′N, 50°28′E) in the sand desert of Arabia	About 300 ft in diameter, 40 ft deep.	Crater is developed in late Tertiary soft sandstone formation and is partly covered by drift sand.	Coesite occurs in intensely fractured and partly fused sandstone. It is the smallest natural crater where coesite occurs, equivalent to that made in alluvium by the explosion of a 1-kiloton nuclear device at depth of 67 ft. Age is unknown.
Ries Crater, Bavaria, Germany	The town of Nördlingen located within crater, 50 mi east of Stuttgart, on Schwabian and Franken Elbe	About 17 mi in diameter; 600 ft deep as a result of filling of sediments and extensive erosion of rim; rim rises 100–200 ft above plateau.	Crater is developed in Jurassic limestone overlying Triassic sandstone and older crystalline basement rock. Numerous huge blocks of limestones were heaved out of the crater and scattered on the rim and for miles in the vicinity of the crater. Crater was filled by late upper Miocene freshwater deposits.	Coesite was found in a tufflike breccia known as suevite. Age of the crater is late upper Miocene, 15,000,000–20,000,000 years old.
Lake Bosumtwi Crater, Ghana	About 20 mi southeast of Kumasi, capital of Ashanti	About 6.5 mi in diameter; depth from lowest part of rim to bottom of lake is about 1150 ft; rim rises 300–600 ft above adjacent upland surface.	Crater is developed in Precambrian phyllite, graywacke, and schists. Scattered areas of breccia are found outside of crater rim. Crater is filled by Lake Bosumtwi.	Coesite was found in a suevitelike rock. Age of crater is unknown.

Coesite has since been found in materials ejected from man-made craters formed by the explosion of 500,000 tons of TNT. Research on the occurrence of coesite from rocks deformed by other energy sources, such as volcanic explosions and deep-seated tectonic movement, is continuing. The study of coesite and craters is essential to the understanding of the impact craters on the Earth as well as those on the Moon.

[EDWARD C. T. CHAO]

Bibliography: F. R. Boyd and J. L. England, The quartz-coesite transition, *J. Geophys. Res.*, 65(2): 749–756, 1960; E. C. T. Chao, Natural coesite: An unexpected geological discovery, *Foote Prints*, 32(1):25–32, 1960; E. C. T. Chao, E. M. Shoemaker, and B. M. Madsen, The first natural occurrence of coesite from Meteor Crater, Arizona, *Science*, 132(3421):220–222, 1960; A. J. Cohen, T. E. Bunch, and A. M. Reid, Coesite discoveries establish cryptovolcanics as fossil meteorite craters, *Science*, 134(3490):1624–1625, 1961; Cratering symposium, *J. Geophys. Res.*, 66(10):3371–3470,

1961; F. Dachille and R. Roy, High pressure region of the silica isotypes, *Z. Krist.*, 111(6):451–461, 1959; G. J. F. MacDonald, Quartz-coesite stability relations at high temperatures and pressures, *Amer. J. Sci.*, 254(12):713–721, 1956; C. B. Sclar, L. C. Carrison, and C. M. Schwartz, Optical crystallography of coesite, *Amer. Miner.*, 47:1292–1302, 1962; N. B. Svenson and F. E. Wickman, Coesite from Lake Mien, southern Sweden, *Nature*, 205:1202–1203, 1965; L. S. Walter, Coesite discovered in tektites, *Science*, 147:1029–1032, 1965; T. Zoltai and M. J. Buerger, The crystal structure of coesite, the dense, high pressure form of silica, *Z. Krist.*, 111:129–141, 1959.

Concretion

A loosely defined term used for a sedimentary mineral segregation that may range in size from inches to many feet. Concretions are usually distinguished from the sedimentary matrix enclosing them by a difference in mineralogy, color, hardness, and weathering characteristics. Some concretions show definite sharp boundaries with the matrix, while others have gradational boundaries. Most concretions are composed dominantly of calcium carbonate, with or without an admixture of various amounts of silt, clay, or organic material. Less common are the clay-ironstone concretions characteristic of the Carboniferous coal measures in many parts of the world. The latter are mixtures of iron carbonate minerals and iron silicate minerals. Coal balls are calcareous concretions, found in or immediately above coal beds, in which there may be a high percentage of original plant organic matter, showing wonderfully preserved plant fossils in a noncompressed condition. Concretions are normally spherical or ellipsoidal; some are flattened to disklike shapes. Frequently a concretion is dumbbell-shaped, indicating that two separate concretionary centers have grown together (see illustration). *See* SEDIMENTARY ROCKS.

[RAYMOND SIEVER]

Variously shaped calcareous concretions found in clay beds. (*From W. H. Emmons et al., Geology, Principles and Processes, McGraw-Hill, 1955*)

Lithified gravels. (*a*) Conglomerate, composed of rounded pebbles. (*b*) Breccia, containing many angular fragments. (*Specimens from Princeton University Museum of Natural History; photographs by Willard Starks*)

Conglomerate

The consolidated equivalent of gravel. Conglomerates are aggregates of more or less rounded particles greater than 2 mm in diameter. Frequently they are subdivided on the basis of size of particles into pebble (fine), cobble (medium), and boulder (coarse) conglomerates. The common admixture of sand-sized and gravel-sized particles in the same deposit leads to further subdivisions, into conglomerates (50% or more pebbles), sandy conglomerates (25–50% pebbles), and pebbly or conglomeratic sandstones (less than 25% pebbles). The pebbles of conglomerates are always somewhat rounded, giving evidence of abrasion during transportation; this distinguishes them from some tillites and from breccias, whose particles are sharp and angular (see illustration).

Conglomerates fall into two general classes, based on the range of lithologic types represented by the pebbles, and on the degree of sorting and amount of matrix present. The well-sorted, matrix-poor conglomerates with homogeneous pebble lithology are one type, and the poorly sorted, matrix-rich conglomerates with heterogeneous pebble lithology are the other. The well-sorted class includes quartz-pebbles, chert-pebble, and limestone-pebble conglomerates. The quartz pebbles were derived from long, continued erosion of the source-rock terrain that resulted in the disappearance of all unstable minerals, such as feldspars and ferromagnesian minerals, that make up most igneous and metamorphic rocks. Typically the quartz-pebble conglomerates are thin sheets that overlie an unconformity and are basal to a series of overlapping marine beds. Chert pebble conglomerates are derived from weathered limestone terrains. Limestone pebble conglomerates seemingly are the result of special conditions involving rapid mechanical erosion and short transport distances; otherwise the limestone would quickly become degraded to finer sizes and dissolve. The well-sorted conglomerates tend to be distributed in thin, widespread sheets, normally interbedded with well-sorted, quartzose sandstones. *See* UNCONFORMITY.

The poorly sorted, lithologically heterogeneous conglomerates include many different types, all related in having very large amounts of sandy or clayey matrix and pebbles of many different rock classes. The graywacke conglomerates are the outstanding representatives. They are composed of pebbles, many times only slightly rounded, of many different kinds of igneous and metamorphic rock as well as sedimentary rocks, bound together by a matrix that is a mixture of sand, silt, and clay. They seem to have been formed from the products of rigorous mechanical erosion of highlands and transported in large part by turbidity or density currents. Another representative of the poorly sorted class is fanglomerate, a conglomerate formed on an alluvial fan. Fanglomerates in many cases are much better sorted than the graywacke conglomerates but have heterogeneous pebble composition. A tillite is another representative of this class of conglomerate. All of these poorly sorted conglomerates tend to occur in fairly thick sequences, and some of them, typically the fanglomerates, are wedge-shaped accumulations. *See* SEDIMENTARY ROCKS; TURBIDITY CURRENT.

Special types of conglomerates, such as volcanic conglomerates and agglomerates and some intraformational conglomerates composed of shale pebbles or deformed limestone pebbles, do not seem to fall easily into either class. Some of the shale and limestone intraformational conglomerates have formed from the tearing up of previously deposited beds at the sea bottom while they were still relatively soft and only partially consolidated. Because of the angularity of some of these pebbles, they have often been called breccias. *See* BRECCIA; GRAVEL; GRAYWACKE; TILL.

[RAYMOND SIEVER]

Conservation of resources

Conservation is concerned with the utilization of resources—the rate, purpose, and efficiency of use. This article emphasizes integrated conservation trends and policies.

Nature of resources. Universal natural resources are the land and soil, water, forests, grassland and other vegetation, fish and wildlife, rocks and minerals, and solar and other forms of energy. Some natural resources, such as metallic ores, coal, petroleum, and stone, are called fund or stock resources. They usually are referred to as nonrenewable natural resources because extraction from the stock depletes the usable quantity remaining and, even if some is being formed, the rate of formation is too slow for practical meaning. Other natural resources, such as living organisms and their products and solar and atomic radiation, are called flow resources. They usually are referred to as renewable natural resources because they involve organic growth and reproduction or because they are relatively quickly recycled or renewed in nature, as in the case of water in the hydrologic cycle and certain atmospheric phenomena. Some natural resources are difficult to fit into such a simple system. Soil, for example, is commonly thought of as renewable, as erosion and nutrient depletion can in some cases be rather quickly corrected, but if the upper layers of the soil are removed or bedrock is exposed, renewal may take thousands of years. Water also is commonly renewable, but rapid extraction of water by wells from deep aquifers may be equivalent to mining minerals.

Human resources are of two types: the people themselves (their numbers, qualities, knowledge, and skills) and their culture (the tools and institutions of society). The three broad classes of resources correspond to the economic factors of production: land (natural resources), labor (personal human resources), and capital (cultural tools and institutions). The natural resources have meaning only as there is human ability to make use of them.

Nature of conservation. Conservation has received many definitions because it has many aspects, concerns issues arising between individuals and groups, and involves private and public enterprise. Conservation receives impetus from the social conscience aware of an obligation to future generations and is viewed differently according to one's social and economic philosophy. To some extent, the meaning of conservation changes with the time and place. It is understood differently when approached from the natural sciences and technologies than when it is approached from the social sciences. Conservation for the petroleum engineer is largely the avoidance of waste from incomplete extraction; for the forester it may be sustained yield of products; and for the economist it is a change in the intertemporal distribution of use toward the future. In all cases, conservation deals with the judicious development and manner of use of natural resources of all kinds.

No definition of conservation exists that is satisfactory to all elements of the public and applicable to all resources. In its absence, an operational or functional definition can be arrived at by considering a series of conservation measures.

1. Preservation is the protection of nature from commercial exploitation to prolong its use for recreation, watershed protection, and scientific study. It is familiar in the establishment and protection of parks and reserves of many kinds.

2. Restoration, another widely familiar conservation measure, is essentially the correction of past willful and inadvertent abuses that have impaired the productivity of the resources base. This measure is familiar in modern soil and water conservation practices applied to agricultural land.

3. Beneficiation is the upgrading of the usefulness or quality of something, for instance, the utilization of ores that were formerly of uneconomic grade. Modern technology has provided many examples of this type of conservation.

4. Maximization includes all measures to avoid waste and increase the quantity and quality of production from resources.

5. Reutilization, in industry commonly called recycling, is the reuse of waste materials, as in the use of scrap iron in steel manufacture or of industrial water after it has been cleaned and cooled.

6. Substitution, an important conservation measure, has two aspects: the use of a common

resource instead of a rare one when it serves the same end and the use of renewable rather than nonrenewable resources when conditions permit.

7. Allocation concerns the strategy of use—the best use of a resource. For many resources and products from them, the market price, as determined by supply and demand, establishes to what use a resource is put, but under certain circumstances the general welfare may dictate usage and resources may be controlled by government through the use of quotas, rationing, or outright ownership.

8. Integration in resources management is a conservation measure because it maximizes over a period of time the sum of goods and services that can be had from a resource or a resource complex, such as a river valley; this is preferable to maximizing certain benefits from a single resource at the expense of other benefits or other resources. This is one of the meanings of multiple use, and integration is a central objective of planning.

A generalized definition that fits many but not all meanings of conservation is "the maximization over time of the net social benefits in goods and services from resources." Although it is technologically based, conservation cannot escape socially determined values. There is an ethic involved in all aspects of conservation. Certain values are accepted in conservation, but they are the creation of society, not of conservation.

Conservation trends. There has been an important trend in conservation from an almost exclusive interest in production from individual natural resources to a balanced interest in that need and in the human resources and social goals for which resources are managed. The conservation movement originated with the realization that the economic doctrine of laissez-faire and quick profits—whether from forest, farm, or oil field—was resulting in tremendous waste that was socially harmful, even if it seemed to be good business. The beginning of the conservation movement stemmed from revulsion against destructive and wasteful lumbering. In time, the movement spread to farm and grazing lands, water, wildlife, and oil and gas. It was gradually learned that conservation management was good business in the long run.

The second trend in conservation was toward integrated management of resources. Students and administrators of resources—in colleges, business, and government— were discovering that the way one resource was handled affected the usability of others. Forestry broadened its interest to include forest influences on the watershed and the relations of the forest to wildlife and human uses for recreation. For many industries working directly with natural resources or their products, as in paper and chemical manufacturing and in coal mining, it was discovered that waste products produced costly and dangerous pollution of streams and air. Engineering on great river systems moved on from problems of flood hazard abatement and hydropower development to the design of structures with regard to fisheries and recreational values, and it was slowly realized that the way the land of a valley was managed affected erosion, siltation, water retention, and flooding.

Conservation in its third phase extends the ecological or integrated approach to resources management to include a more complete acceptance of the force of societal factors (such as economics, government, and social conditions) in determining resource management. There also is a closer examination of social costs and benefits and of human goals for which resources are employed.

Conservation of human resources. Many students of conservation prefer for practical reasons to limit conservation to the management of natural resources and to leave problems of human resources to other disciplines and fields of action. However, because of the inevitable interplay of natural resources and the resourcefulness of man in utilizing them, others emphasize the role of man himself as a resource. This leads to the application of conservation measures to man and his institutions. The many measures that tend to preserve, rehabilitate, renew, maximize, allocate, and integrate human abilities are coming to be referred to as conservation measures. These measures tend to be organized and institutionalized so that the institutions themselves become means to an end within the society and thus are considered to be resources as truly as water, petroleum, or the labor force.

Conservation of recreation resources. More than 30,000,000 acres of rural public land in the United States are managed for recreational use or to preserve scenic, historical, or scientific and natural history values, and additional space is allocated for more intensive recreational activities by units of local government. There has been in recent decades a tremendous growth in the use of such lands for all forms of recreation, and the forecast is that outdoor recreation will increase more rapidly than population growth because of the increase in urban living and personal incomes, more and better roads, the shorter workweek, and paid vacations. Rapidly growing new trends in recreation require special uses of space, such as bow and arrow shooting, skiing, and motorboating. With increasing numbers of persons camping, picnicking, and swimming, the provision of adequate facilities and physical maintenance of crowded sites have become problems.

Except for the more intensive recreational uses, rural and wildland areas serve multiple purposes. An abundance of game can be raised in agricultural regions. Watershed protection and the maintenance of scenic values go naturally together. Stream impoundments for multiple purposes create new recreational facilities. Yet some land uses are incompatible: municipal, manufacturing, and mining pollution destroys water recreation values; commercial developments destroy wilderness values.

Many trends indicate that there are not enough acres dedicated to recreation, especially smaller areas near strongly urbanized regions. There is a growing need for research on trends in wildland usage by people and for better use of space in large parks and forests so as to avoid wearing out the sites by the persistent concentration of people, as at campsites. It is clear that certain human facilities must be provided where many people use wildlands and that inappropriate ones, such as amusement facilities, must be kept separate from activities requiring special terrain. Some growing recreational demands are being met by private enterprise, such as in ski resorts and privately owned public hunting grounds. Although game is owned by the public, it lives and breeds largely on private land. In time more farm and ranch owners will be paid for hunting privileges. Further development in recreation is coming from increased knowledge of the biology of fish and wildlife and the consequent improvement of management arising from it and from a better understanding by the sportsman of the management problems.

Conservation policies. Individuals, corporations, and governmental entities at all levels have policies pertaining to resources. There could be a single national policy concerning a natural resource, such as water, only if the central government had complete authority over all governmental agencies and private enterprises. In the United States, however, many Federal and state departments, bureaus, agencies, and commissions have some authority over natural resources. There usually are several policies concerning the use and management of soil, water, forests, rangelands, fish, wildlife, minerals, and space, and they are not necessarily uniform as to objectives or program. One exception is the Atomic Energy Commission, which has centralized authority in the creation and execution of policy regarding radioactive resources. This authority is granted by Congress and can be modified by congressional action.

Because each resource is capable of being utilized for a variety of goods and services, and because individuals, enterprises, and regions tend to value certain uses more than others, conflicts arise in the allocation of a resource among competing uses when the supply is inadequate for all desirable uses. The demand for water, for example, may be for rural domestic needs, urban and industrial needs, irrigation, power development, and recreation in its many aspects.

Because situations such as this exist or are potential with respect to every resource, two outstanding needs arise in the conservation of natural resources. Detailed information is needed concerning the location, quantity, and quality of each resource and of the interrelations among the resources. As a result, each agency of government needs to strengthen its own fact-finding, analysis, and programming machinery. The second need is for coordinating machinery that will improve the efficiency of allocation of resources so as to maximize the net private and public benefits from them. This is the goal of conservation policy, and it must, in a democratic society, be approached as far as possible through the voluntary cooperation of government and private enterprise. However, as pressures on society increase, whether because of actual depletion of resources or because of an increase in critical demand, as in war, authority to allocate resources tends to be delegated by the citizenry to the central government.

Federal conservation legislation. National resources policy in the United States is framed by acts of Congress and in the states by acts of legislatures. The language of specific acts, however, usually permits some freedom for administrative decisions and also permits different interpretations that must be settled in the courts of law. As the country has developed and conditions have changed, a sequence of laws has been passed to deal with the exigencies of natural resource conditions. Also, there has occurred some evolution of political philosophy to meet the changing conditions. In addition to legislative and court actions, some natural resources are subject to treaties and other international agreements.

Water. In 1824 Congress assigned to the Army Corps of Engineers responsibility for improving rivers and harbors for navigation, and this assignment has become its principal civilian activity. The Inland Waterways Commission in 1907 and the National Waterways Commission in 1909 were created by Congress with broad responsibilities for planning the development and conservation of water and related resources. The Flood Control Act of 1927 provided for Federal surveys, and the act of 1936 provided for Federal construction of projects. Subsequent amendments led to the 1954 Watershed Protection and Flood Prevention Act, which clearly recognized the relation of upland management practices to erosion, siltation, streamflow, and other water-related resource problems. Interest in irrigation has resulted in a series of acts from 1866 onward (the Desert Land Act of 1877 being significant), while water facilities, drainage, hydroelectric power, and water pollution also have received repeated attention. The Water Power Act of 1920 set up the Federal Power Commission and aimed to safeguard the rights of government and hydroelectric companies. Water pollution control is largely state responsibility, but the Federal Water Pollution Control Act of 1948, amended in 1956, enables financial and other assistance to the states.

Soil. Although soil surveys had been carried out for a century, it was the Soil Conservation Act of 1935 that created the Soil Conservation Service and the autonomous Soil Conservation Districts. Soil conservation payments were started under the Agricultural Adjustment Act of 1933. Research, education, cooperation, and financial inducements have been stressed more than regulatory measures for soil and water conservation on the land.

Minerals. In 1866 Congress provided for the sale of mineral lands, but a general minerals policy did not exist until later. The Minerals Leasing Act of 1920 provided that nonmetalliferous minerals on public lands could be utilized only under lease. In 1953 Congress confirmed jurisdiction of the states over minerals under navigable inland waters and on the continental shelf to state boundaries.

Forests. A long series of acts has been concerned with the disposal of the public lands and the exploitation of timber and forage. An act in 1891 empowered the President to set aside forest reserves. In 1901 a Forestry Division was created in the General Land Office in the Department of Interior, and in 1905 forestry activities were transferred to the Department of Agriculture. An act in 1907 changed the reserves to the national forests. The Forest Service soon became recognized as one of the most efficient bureaus in Washington as it dealt, on a decentralized basis, with the conservational use of all resources, not just timber. The Weeks Act in 1911 inaugurated cooperation with the states and is best known because it started the policy of land acquisition. Today management of the nation's forest lands has become a balanced cooperation between government and private enterprise on a mosaic of ownership.

Wildlife. The first Federal recognition of responsibility for fisheries and wildlife was in regard to research, with the establishment in 1905 of the Bureau of Biological Survey. An international convention for protection of fur seals was signed in 1911, with protection of other marine animals coming later. The Migratory Bird Act was passed in 1913. Federal refuges were started in 1903, and sizable funds for refuges became available with the Migratory Bird Conservation Act of 1929 and at an accelerated rate from 1933 to 1953. State activities in the fields of wildlife and fishery management have been greatly facilitated by Federal aid under the Pittman-Robertson Act of 1937 and the Dingell-Johnson Act of 1950. [STANLEY A. CAIN]

Bibliography: S. W. Allen and J. W. Leonard, *Conserving Natural Resources: Principles and Practice in a Democracy*, 2d ed., 1966; G. Borgstrom, *The Hungry Planet: The Modern World at the Edge of Famine*, 1965; C. H. Callison (ed.), *America's Natural Rsources*, 1957; S. V. Ciriacy-Wantrup, *Resource Conservation: Economics and Policies*, 1952; D. C. Coyle, *Conservation: An American Story of Conflict and Accomplishment*, 1957; S. T. Dana, *Forest and Range Policy*, 1956; R. F. Dasman, *Environmental Conservation*, 1959; E. S. Helfman, *Rivers and Watersheds in America's Future*, 1965; G.-H. Smith, *Conservation of Natural Resources*, 3d ed., 1965; E. W. Zimmermann, *World Resources and Industries*, rev. ed., 1951.

Continent formation

The continents, or more accurately, the protocontinents, began to form in the final stages of the origin of the Earth. But because the precise processes contributing to the origin of the Earth are speculative, the initial stages of continent formation also remain enigmatic. It is known that the Earth accreted from clouds and chunks of solar matter about 4.6×10^9 years ago. What is not known in detail is the actual rate, or kind of accretionary process involved, or the precise composition of the particles. From about 1950 until the era of lunar exploration, the data available to earth scientists suggested that the Earth had accreted as a relatively cool conglomerate of solar debris, much like the more common stony meteorites in average composition. The Earth's accretion was assumed to have occurred over a period of perhaps 25 to 100×10^6 years. During and immediately after the accretionary process, heat produced within the Earth both from the gravitative accumulation of matter and from its compaction, coupled with radioactive heat, may have partially melted the Earth. The heavier elements and compounds, especially the native iron in meteorites, would have melted and sunk into the interior of the Earth, producing its dense core and a high internal temperature. These processes would have triggered a simultaneous migration of the more volatile, lighter elements such as hydrogen, nitrogen, and carbon, as well as potassium-rich silicates, to the Earth's surface. Their accumulation would have produced the Earth's primordial atmosphere, oceans, and the continents. In effect, it seemed that during the first 100 to perhaps 500×10^6 years of the Earth's history the Earth became sufficiently heated and molten to literally turn itself inside out. If this actually occurred, it was the first of a series of major refractionating events that have resulted in the growth of the contemporary outer terrestrial rind. This rind includes the several kinds of crusts and subjacent elastic rocks called the lithosphere. *See* EARTH; EARTH, AGE OF; LITHOSPHERE, GEOCHEMISTRY OF.

Some of these interpretations seem less probable in the light of lunar explorations. To the surprise of almost every scientist involved, lunar investigations indicate that the Moon not only was once largely molten but also has a composition quite unlike that of either the stony meteorites or the rocks of the outer, cool lithosphere of the Earth. The Moon's outer shell, and hence probably all of the Moon, not only is devoid of water but also is depleted in other more volatile elements, including potassium. These and other findings lead to the speculation that if the Earth and the Moon accreted as companion (binary) planets the accretionary process was less haphazard than previously assumed. *See* MOON.

Accretion of both Earth and Moon may well have occurred as the solar plasma and gases cooled and precipitated at successively lower temperatures. In this scenario, the higher-temperature, more refractory elements that compose the Earth's core and mantle would have precipitated and accreted first, followed sequentially by successively lower-temperature silicates, oxides, and other substances. This inhomogeneous or anisotropic accretion of the Earth would have resulted in the initial coating or encrusting of the Earth with the more volatile compounds. If so, the crust of the primordial Earth, and especially the protocontinents, evolved during the final stages of the Earth's origin, essentially as the last stage of its accretionary growth. The Moon, being by pure chance a much smaller planet evolving near the Earth, would have lacked the mass necessary to attract and hold the lighter and more volatile elements that uniquely distinguish the outer Earth, with its oceans and air, from the dry and sterile lunar surface.

Ages. If, indeed, the Earth accreted inhomogeneously, this merely indicates that the initial stages of continent formation were less cataclysmic than previously assumed. It does not greatly change subsequent stages in the evolution of the continents that are seen today, the growth of which during the last 3.7×10^9 years can be readily inferred. New insights regarding the growth of continents are in large part due to the advent of radiometric age studies of the continents. The dating of crustal rocks, and hence of continent-forming events, by radiometric techniques is now widely employed as an invaluable means of studying continental history. Studies of the nature, and the ages, of continental rocks clearly indicate that the rate of continent formation has varied with time. The oldest continental rocks discovered to date occur in Minnesota, in southwestern Greenland, and in southern Rhodesia. In these places granitic rocks, as well as metamorphosed sedimentary rocks formed in ancient seas, range from 3.5 to almost 4×10^9 years old. It was a surprising discovery that these ancient segments of the continents are almost as old as the oldest definitively dated lunar crust.

Continent changes. If the hypothesis of inhomogeneous accretion is correct, the early fragments of the continents were probably quite large, con-

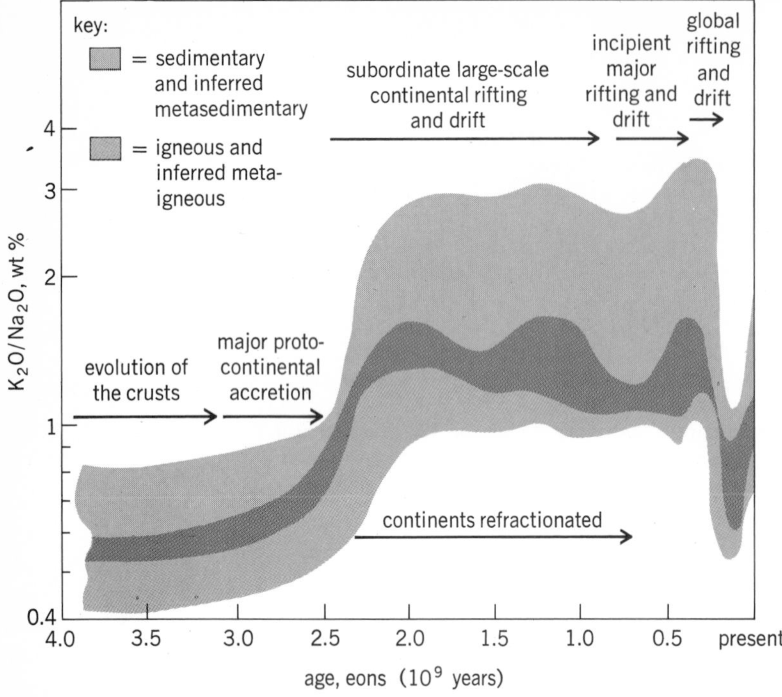

Fig. 1. Evolution of continents as interpreted from crustal studies. An early period of differentiation of the outer Earth into oceanic, arc, and continental crusts was followed, about 2.5×10^9 years ago, by continued refractionation, and by global rifting and drift of large continental fragments.

ceivably covering the entire Earth. If this conclusion is accepted, then the continents have decreased in area but have greatly increased in thickness and become more granitic as they have aged. Hence the more striking changes in the formation of the contemporary continents are the shape, area, and distribution over the Earth. Less dramatic but critical changes have occurred in the structure and composition of the continents, and in their relationships to the Earth's other crusts: the oceanic crust and the island arcs.

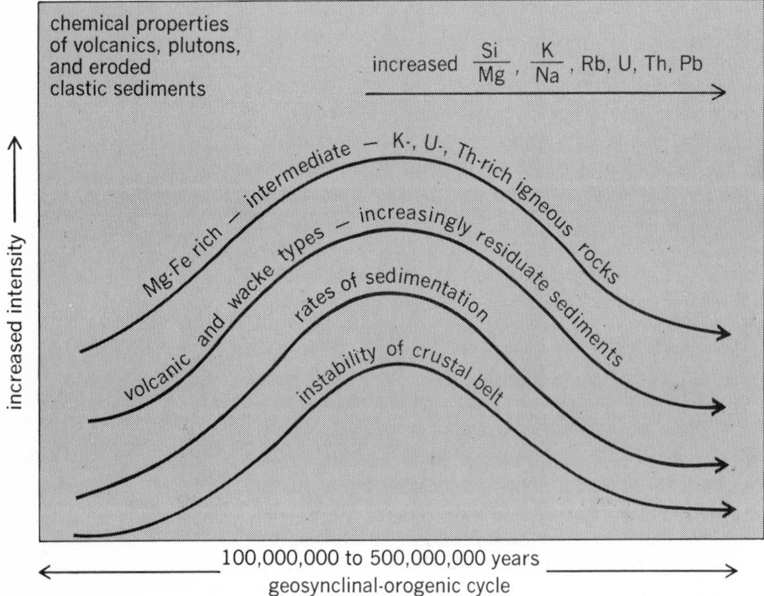

Fig. 2. Diagrammatic representation of the major mountain-building cycles that are associated with continent formation.

Episodic growth. All of the changes appear to have been episodic. At least six major episodes of continental growth, refractionation, and change are identifiable, as are many more subordinate episodes (Figs. 1 and 2). The major episodes in the Earth's evolution, which are quasi-cataclysmic, seem to have resulted from the buildup of radioactive heat within parts of the Earth, causing melting or partial melting of the Earth's interior. The molten rock consists largely of the lowest-melting fractions of the Earth. The melts (magmas) rise and are either erupted at the surface or emplaced within the crusts. On the continents, granitic magmas outnumber the more magnesium-iron–rich basaltic rocks by factors of 10 to 50. In the oceanic areas, the reverse is true. Why granitic melts tend to be emplaced along and within the continents, and the basalts within the oceans, is one of the fascinating puzzles now being unraveled by earth scientists. But equally important and exciting is the fact that the rise of magma and softened rock tends to cause the existing crusts to crack and move laterally over the Earth's surface. Consequently, continents, and fragments of continents, drift over the face of the Earth. *See* EARTH, INTERIOR OF; MAGMA.

Continental drift. Generally, the continents are coupled with large fragments of oceanic crust and ride piggyback on the entire outer, cooler lithosphere. Great fragments, or plates, of the lithosphere appear to jostle and move over the softer, perhaps incipiently melted upper mantle of the Earth. A crude analogy is that of the movements of great plates of Arctic and Antarctic ice in response to the currents in the underlying ocean.

The most recent major episode of continental drift that has given the continents their present form and distribution began about 200×10^6 years ago (Fig. 1). Immediately prior to that time, the Americas, India, Australia, and Antarctica were joined to Africa and Europe and formed one vast megacontinental cluster sometimes called Pangea. The reconstruction of events involved in the breakup of Pangea, and of one of its largest fragments, called Gondwana, which resulted in the opening of new oceans, especially the Atlantic and Indian oceans, has preoccupied earth scientists for decades. But actual proof of this majestic reorientation of the continents and oceans is the product of intensive studies since about 1960.

The most obvious evidence of continental fragmentation and drift is derived from the shapes of the continents and their submerged borderlands. Long ago, A. von Humboldt, and others, especially the Austrian A. Wegener, noted that the eastern margins of the Americas could be fitted rather neatly together with the western outlines of Africa, Europe, and the whole of Greenland (Fig. 3). Further study has shown that the outlines of Antarctica, Australia, and India may be grouped into a cluster that fits into the outlines of southeastern Africa. Moreover, this grouping aligns what seem to be quite independent mountain belts of these several continents into one very coherent crustal fabric (Fig. 4). These highly significant interrelationships are matched by equally, perhaps more, compelling evidence of crustal movements found on the sea floor. Indeed, much of the ultimate proof of more recent aspects of continental formation, fragmentation, and drift has come from studies of the oceans and island arcs.

Oceanic insights. Studies of the oceans that intensified several decades ago have revealed many of the major features of the Earth's crusts and their origins. One of the most imposing and significant features of the sea floor is the "ridge and rise complex." This is a vast, largely submarine mountain chain dotted with numerous volcanoes. The complex is world-girdling, extending some 60,000 km largely through the median parts of ocean basins (Fig. 5). In the Pacific Ocean, however, one great segment of the complex, the East Pacific Rise, converges with the western edge of Central and North America. Detailed photography, dredging, and related geophysical studies indicate that the ridge-rise complex is a tensional feature, with great central rifts from which emerge floods of a unique basaltic magma. This magma ponds or puddles along the rift zones and also intrudes the bounding zones of rifted oceanic crust. Consequently, these eruptions of basalt continuously add to the existing ocean crust, which is spreading away from the ridge-rise complex as it forms. *See* MARINE GEOLOGY.

Sea floor spreading. The evidence for sea floor spreading is severalfold. One indication of spreading lies in the patterns of the stripelike masses of basalt episodically welded to the crust. These basaltic stripes, which crudely parallel the ridges, have frozen into them the orientation of the Earth's magnetic field extant at the time of volcanic eruption. Studies of exposed basalts whose ages have been determined indicate that this magnetic field reversed itself at intervals of 0.5 to 2×10^6 years during recent geologic time. The ocean floor stripes are consequently the alternately reversed patterns imposed by reversals of the Earth's magnetic field during successive eruptions of basalt (Fig. 5). In places, the ocean floor is striped for about 1000 km or more from the zones of spreading and on both sides of this zone. The result is a series of symmetrical magnetic patterns, indicative of the rate of sea floor spreading. Spreading rates of 0.5 to 10 cm per year are commonly observed. *See* ROCK MAGNETISM.

Further corroboration of sea floor spreading and continental drift is found in the patterns of sediments formed on the sea floor, and in the trench–island arc systems that almost encircle the Pacific, offshore from the continents. Coring of the sedimentary cover in the sea by the drilling ship *Glomar Challenger* clearly indicates that the sea floor sediments are essentially absent at the spreading centers, but tend to thicken and increase in age away from them. The age of the oldest sediments lying at the base of the sedimentary column and just above the basaltic ocean crust is essentially that of the crust as indicated by the particular magnetic stripe mapped in the underlying basalt.

Plate tectonics. A tantalizing question is the fate of the striped ocean crust that is being conveyed away from the ridge-rise spreading centers, presumably by convection in the Earth's mantle. Clearly, this process is a dynamic one that forces the great platelike masses of the lithosphere, with its overlying continents and ocean crust, against other plates. To find solutions to the plate movements, that is, plate tectonics, it has been necessary to map the boundaries of the plates and to trace their motions through time. Fortunately, plate patterns can be defined by their pushing and frictional movements relative to one another, because the movements of adjoining or colliding plates produce most of the world's earthquakes. Hence, a world map of recorded earthquakes defines the outlines and patterns of the great lithospheric plates active today throughout the world. Plate outlines and relative motions are further defined both by the direction of slip during the earthquakes and by the relative displacements of the magnetic stripes frozen into the basalts of the sea floor. *See* SEISMOLOGY; TECTONIC PATTERNS.

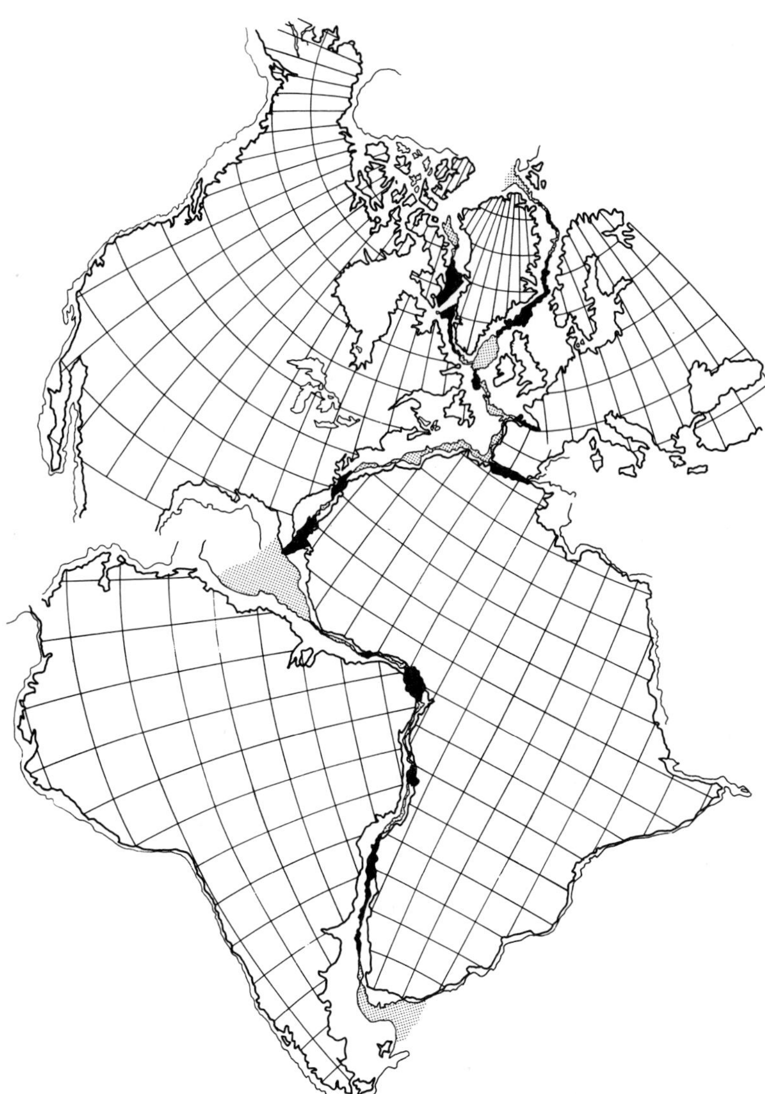

Fig. 3. Map showing how the Atlantic margins of the Americas, Greenland, Europe, and Africa can be fitted together. The fit is made along the continental slope at the 500-fathom (3000-ft) contour line. Regions where the continents overlap are shown in black, gaps in the fit in white.

One of the most fascinating relationships that emerges is the appearance of essentially all the world's recent mountain belts, the evolving island arcs. and the great trenches of the sea floor at sites where moving lithospheric plates are converging and either colliding or overriding one another (Fig.

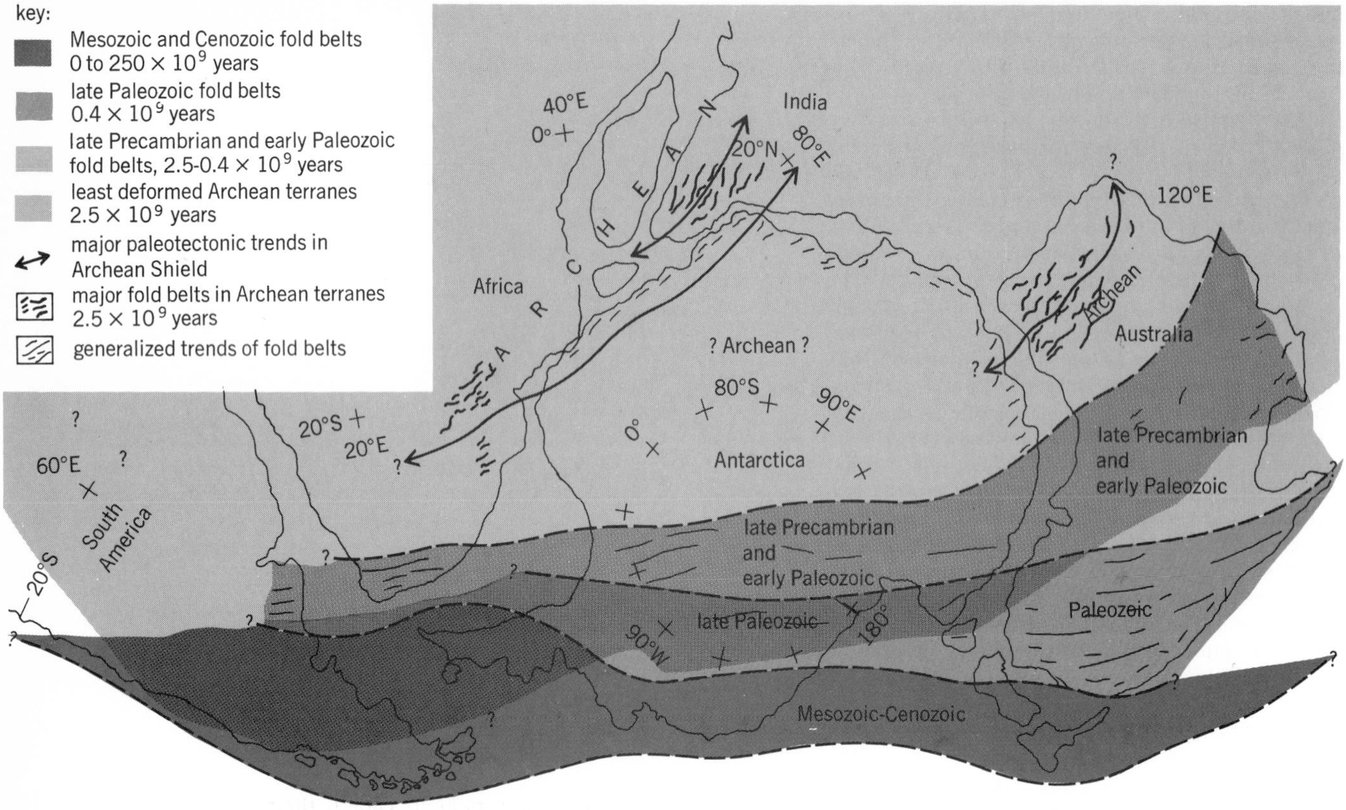

key:

- Mesozoic and Cenozoic fold belts 0 to 250 × 10^9 years
- late Paleozoic fold belts 0.4 × 10^9 years
- late Precambrian and early Paleozoic fold belts, 2.5-0.4 × 10^9 years
- least deformed Archean terranes 2.5 × 10^9 years
- major paleotectonic trends in Archean Shield
- major fold belts in Archean terranes 2.5 × 10^9 years
- generalized trends of fold belts

Fig. 4. Subparallel alignment of major fold belts ranging in age from Archean to Cenozoic that appear when these continents are clustered into Gondwana. The gross trend of progressively younger orogenies from north to south suggests the progressive secular thickening and refractionation of Gondwana in this direction.

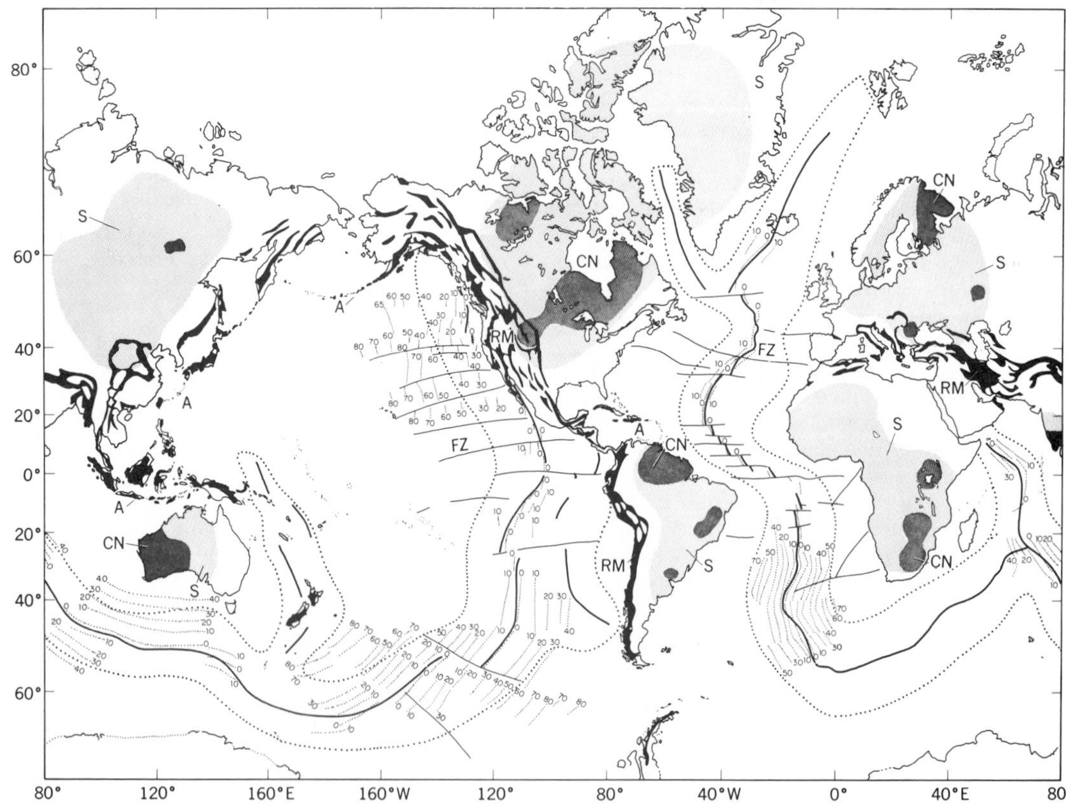

Fig. 5. Generalized map of some of the major crustal features, including oldest continental nuclei (CN), with probable extensions (S) now overprinted by younger orogenic-granite-forming events, recent to mature island arcs (A) and young (0.2 eons) mountain belts (RM), oceanic Ridge and Rise System, major striped magnetic patterns in the oceans (dotted) with ages in millions of years, and oceanic fracture zones (FZ).

5). The worldwide pattern is completely consistent with the patterns of rift and drift of the continents and with the emergence of the Atlantic, Indian, and smaller oceans. *See* PLATE TECTONICS.

Island arcs. The island arcs are of special interest because they exhibit features intermediate between those of the oceanic and continental crust (Fig. 6). The incipient arcs, such as the Tongas, Solomons, or the Kurils, north of Japan, and the more southwesterly Aleutians, southwest of Alaska, are chains of volcanoes built upon slightly to distinctly thickened oceanic crust which overrides a sinking lithospheric plate.

The distribution of the island arcs and associated deep ocean trenches indicates their origin and evolution. The Kurils and Aleutians merge laterally into the recently formed continental borderland of Asia and North America. Studies of these arcs and borderlands reveal essentially all the gradations between oceanic crust, arcs, and the mountain belts within and marginal to the continents. These relationships indicate that island arcs commonly represent steps in the growth of new continental areas at the expense of ocean basins. In fact, some of the oldest segments of continents in the Canadian, Australian, and South African shields contain fossilized remnants of island arcs and floors of ancient oceans formed more than 3×10^9 years ago.

Most arcs lie offshore of the continents, encircling the great Pacific Basin. The arcs commonly evolve on the continental side of the trenches. The trenches mark sites of lithospheric plate margins, where the oceanward plate is sinking under the evolving arc. Elsewhere on Earth, plates migrating from oceanic spreading are being crushed against or thrust under the continental borderlands (Fig. 7). Hence, the sites of plate convergence are commonly marked by either trenches or island arcs, or by the recent great mountain belts of the continents. *See* OCEANIC ISLANDS.

Mountain belts. All of the mountain ranges of the western Americas as well as those in eastern Australia and Asia define zones of recent and contemporary plate convergence and collision. Similarly, the great Alpine-Himalayan mountain chains are the products of collisions of Africa and India with Eurasia as these great continent-covered plates collided.

Studies of these recent and continuing geologic processes suggest that most of the older great mountain belts of the continents, now variously dissected or beveled by erosion, mark the sites of lithospheric plate convergence and collision. It follows that the major rifting, drift, and collisions of continents, and of continents with arcs or oceanic crust, are the major episodic processes reflected in the great granite-forming episodes on Earth. Undoubtedly, all Earth motions are perennial, but waxing and waning in intensity. *See* OROGENY.

Continent refractionation. Viewed in this light, the emplacement of granitic rocks on and marginal to the continents and the extrusion of basalts in the oceans are integral parts of plate tectonics. Under most circumstances, basalt is the lowest-melting fraction in the inner Earth that will form and rise in great volumes at the oceanic ridge-rise complex. This basalt, when combined with oceanic sediment and fragments of arcs and subducted into the inner Earth at sites of plate convergence, can produce a more granitic melt, especially the andesites characteristic of island arcs and continental borderlands. And if arcs and continents collide, their depressed and underthrust root zones will melt to form the far more fractionated granitic rocks common to the continental mountain belts.

These stepwise processes of refractionation of the Earth's several crusts are greatly aided by the weathering and erosion of evolving mountains, which contribute great volumes of even more uniquely fractionated sediments to adjacent ocean basins. The continental melting and refractionation of successively more granitic and sedimentary rocks through billions of years have produced the thick, granitic continents seen today, each made up largely of a complex of successively formed, overprinted, and variously obliterated mountain

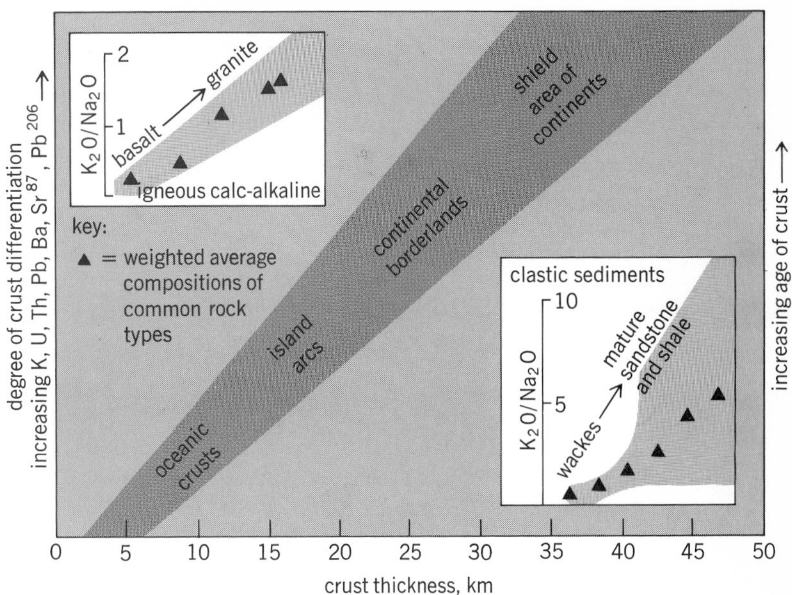

Fig. 6. Relationships of age, thickness, and composition of Earth's crusts, and complementary variations in the petrogenic index K_2O/Na_2O for major crustal rocks.

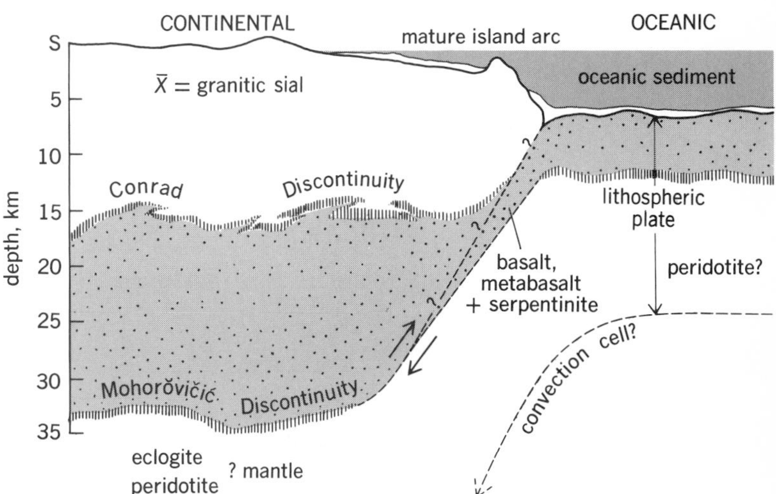

Fig. 7. Generalized vertical section through the Earth's crust, showing the inferred relations of the Americas or Asia to the Pacific Ocean, to the associated island arcs, to the trenches, and to the underlying mantle.

belts. The major mountain-building sedimentary cycles characteristic of continent formation and refractionation can be illustrated as in Fig. 8.

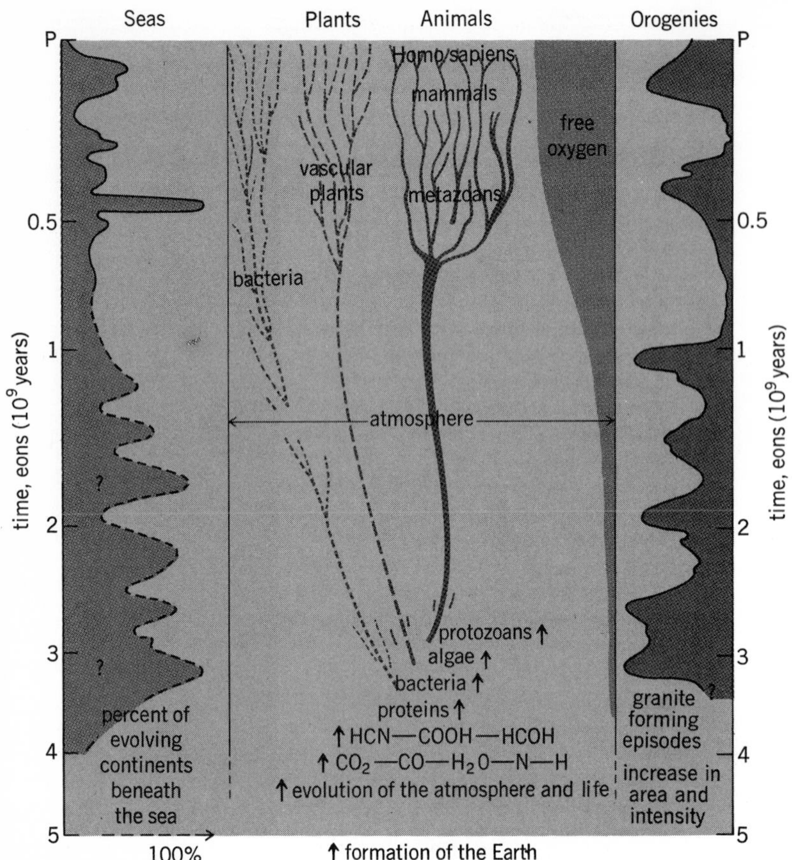

Fig. 8. Interrelationships of the fossil record, atmospheric evolution, mountain building, orogenies, and the advance and retreat of seas across the continents.

Continent-biosphere interrelations. Whereas these continent-forming processes have been cumulative throughout geologic time, sea floors are ephemeral, accreted and consumed in periods of several million to several hundred million years. Fortunately, the complex continents contain relics of various ages of all of the crusts—oceanic crusts, arcs, interarc basins, borderlands, and the ancient continental shields. Studies of the relative abundances, ages, and distribution of these continental components yield evidence not only that the major episodes of continent formation are intermittent but also that the early and intermediate episodes differed from the more recent ones. Figures 1 and 8 are attempts to recapitulate the history of continents and the related features of the outer Earth and its life. Note, for example, in Fig. 1 that the major accretion of primordial crust into the earliest continents was essentially complete at least 2.5×10^9 years ago. In the following 2×10^9 years, these continents were continuously refractionated by plate motions into more potassium-rich granitic masses of at least hemispheric dimensions. About 600×10^6 years ago continental rifting and drift increased in scale and frequency. About 250×10^6 years ago the most recent worldwide breakup of megacontinents took place, with the fragments approaching the forms of the contemporary continents. The last 200×10^6 years has been the era of worldwide drift and collision of these fragments, as briefly outlined in the preceding paragraphs.

Figure 8 is an attempt to interrelate major, presently decipherable, continent-forming events with (1) the origin and evolution of life as indicated by fossils found in continental rocks, (2) the evolution of the present atmosphere as suggested by both rock and fossil evidence, and (3) the periods when seas invaded the continents, and the extent of inundation of the continents. These events can be crudely reconstructed from thoughtful studies of

Compositions of important rock types in the Earth's crust and the average continental crust

Composition	Anorthosite	Peridotite	Oceanic basalt	Andesite	Dacite	Granodiorite	Granite	Graywacke	Sandy shale	Continental crust upper 15 km
Chemical					*Weight, %*					
SiO_2	54.0	44.0	50.0	60.0	66.5	66.0	70.5	64.0	65.5	66.0
TiO_2	.8	.2	1.5	.8	.3	0.5	.3	.5	.5	0.5
Al_2O_3	24.0	2.5	15.5	17.5	15.0	15.5	14.6	14.5	14.0	15.5
Fe_2O_3	.8	1.0	1.5	3.0	.8	2.0	1.6	1.5	3.5	2.0
FeO	2.5	8.0	8.0	3.2	2.5	2.6	1.8	3.5	2.0	3.0
MgO	1.5	40.0	7.0	2.8	2.0	2.0	.8	2.2	1.7	2.0
CaO	10.0	2.5	10.5	6.0	3.7	4.0	2.0	2.6	2.5	4.2
Na_2O	4.5	.1	2.9	3.5	3.8	3.6	3.5	3.2	1.5	3.5
K_2O	.1	.02	.25	3.0	2.4	2.8	4.3	2.0	4.0	3.0
Mineralogical					*Approximate volume*					
Olivine	—	*	†	—	—	—	—	—	—	—
Fe, T, Mg oxides	†	†	†	—	—	—	—	—	—	†
Pyroxene	†	*	*	*	—	—	—	—	—	—
Amphibole	—	—	—	*	†	†	—	—	—	†
Plagioclase	*	—	*	*	*	*	†	*	†	*
K-feldspar	—	—	—	†	†	*	*	†	*	*
Micas	—	—	—	—	*	†	*	*	*	*
Quartz	—	—	—	—	—	*	*	*	*	*
Chlorites	—	—	—	—	—	—	—	*	*	—
Clay minerals	—	—	—	—	—	—	—	†	*	—

*Major constituent. †Subordinate mineral.

continental rocks. For example, there are fossil algal reefs in South African rocks that exceed 3×10^9 years in age. In contrast, the oldest of the more complex invertebrate fossils, the metazoans, are only 600 to 700 $\times 10^6$ years old. Sedimentary rock complexes of the continents date back almost 4×10^9 years and were clearly formed in ancient seas. Some of these sediments, especially the iron formations and the carbonate- and sulfate-bearing sediments, indicate that some oxygen existed in the atmosphere 3.5×10^9 years ago. The metazoan fossil life, widely entombed in rocks 500×10^6 years old, suggests that the air and seas extant at that time were much like those of today in their composition. *See* ATMOSPHERE, EVOLUTION OF; LIFE, ORIGIN OF; PALEONTOLOGY.

The oldest vascular plants are found as fossils about 375 $\times$ 10^6 years old. This suggests that, throughout much of the preceding geologic time, the continents were barren and inhospitable sites. It also suggests that the biosphere as it is now known existed for at least 100 $\times$ 10^6 years before the present continents fragmented and began their continued migration over the Earth's surface. Curiously, humans have evolved and lived on Earth for only 2 or 3×10^6 years.

Summary. In summary, the continents are old, granitic, and relatively thick segments of the Earth's crust; but their form, composition, and geographic position have undergone constant change. Probably a thin outer layer of the Earth, having many constituents of the continents and biosphere, was accreted to the Earth in the final stages of its formation, 4.5 $\times$ 10^9 years ago. Since that time many oceanic crusts have formed and have been destroyed, each adding to the continents during their destruction those lowest-melting constituents such as the potassium-aluminum silicates common to the granitic continents.

Today, continents cover roughly one-third of the Earth's surface but constitute less than 0.3 of its mass. In most areas where erosion has stripped away the surficial blanket of shelf sediments, 30–90% of the underlying rock is granitic. Most of this granitic rock exists as a complex mosaic of faulted and folded mountain belts, the oldest more than 3.6×10^9 years old. The roots of these mountains contain remnants of altered sedimentary and igneous rocks formed in equally old oceans.

Existing continents, although generally granitic in their average composition, consist of a bewildering variety of igneous, metamorphic, and sedimentary rocks. The compositions of the more important and more common continental rocks are given in the table. At depths below 15 to 20 km continents probably become less granitic and more granodioritic in composition. The bases of continents lie at depths of 25 to 60 km, the thickness being crudely proportional to continental topography (Fig. 7). In regions of recently formed mountains such as the Coast Ranges of North America, the Andes of South America, and the Alpine-Himalayan chains, the continental crust is often thicker than in the lower, flatter areas of most continental interiors. These insights have been gained by means of remote sensing, based on the use of elastic waves generated by earthquakes and atomic explosions.

[A. E. J. ENGEL]

Bibliography: P. Cloud (ed.), *Adventures in Earth History*, 1970; *Geologic Map of North America*, U.S. Geological Survey, 1965; J. Gilluly, A. C. Waters, and A. O. Woodford, *Principles of Geology*, 3d rev. ed., 1968; M. N. Hill et al. (eds.), Ideas and observations on progress in the study of the seas, *The Sea*, vol. 3, 1963, vol. 4, 1969.

Continental shelf and slope

The continental shelf is the zone around the continent, extending from the low-water line to the depth at which there is a marked increase in slope to greater depth. The continental slope is the declivity from the edge of the shelf extending down to great ocean depths. The shelf and slope comprise the continental terrace, which is the submerged fringe of the continent, connecting the shoreline with the 2½-mi-deep (4-km) abyssal ocean floor (see illustration). About 18% of the world's oil comes from offshore, and this proportion is increasing yearly. The value of oil from the continental shelves around the United States exceeds that of fisheries.

Continental shelf. This comparatively featureless plain, with an average width of 45 mi (72 km), slopes gently seaward at about 10 ft/mi (1.9m/km). At a depth of about 70 fathoms (128 m) there generally is an abrupt increase in declivity called the shelf break, or the shelf edge. This break marks the limit of the shelf, the top of the continental slope, and the brink of the deep sea. However, some shelves are as deep as 200–300 fathoms (180–550 m), especially in past or presently glaciated regions. For some purposes, especially legal, the 100-fathom line or 200-m line is conventionally taken as the limit of the shelf. Characteristically, the shelves are thinly veneered with clastic sands, silts, and silty muds, which are patchily distributed. Geologically, the shelf is an extension of, and in unity with, the adjacent coastal plain. The position of the shoreline is geologically ephemeral, being subject to constant prograding and retrograding, so that its precise position at any particular time is not important. Genetically, the origin of the shelf seems to be primarily related to shallow wave cutting (waves cut effectively as breakers and surf only down to about 5 fathoms (9 m), the depth of vigorous abrasion), shoreline deposition, and oscillations of sea level, which have been especially strong during the Pleistocene and Recent. Although worldwide in distribution and comprising 5% of the area of the Earth, shelves differ considerably in width. Off the east coast of the United States the shelf is about 75 mi wide (120 km), while off the west coast it is about 20 mi wide (32 km). Especially broad shelves fringe northern Australia, Argentina, and the Arctic Ocean. As along the eastern United States, continental shelves commonly acquire a prism of sediments as the continental margin downflexes. Such capping prisms appear to be nascent miogeoclines.

Continental slope. The drowned edges of the low-density "granitic" or sialic continental masses are the continental slopes. The continental plateaus float like icebergs in the Earth's mantle with the slopes marking the transition between the low-density continents and the heavier oceanic segments of the Earth's crust. Averaging 2½ mi high (4 km) and in some places attaining 6 mi (10 km),

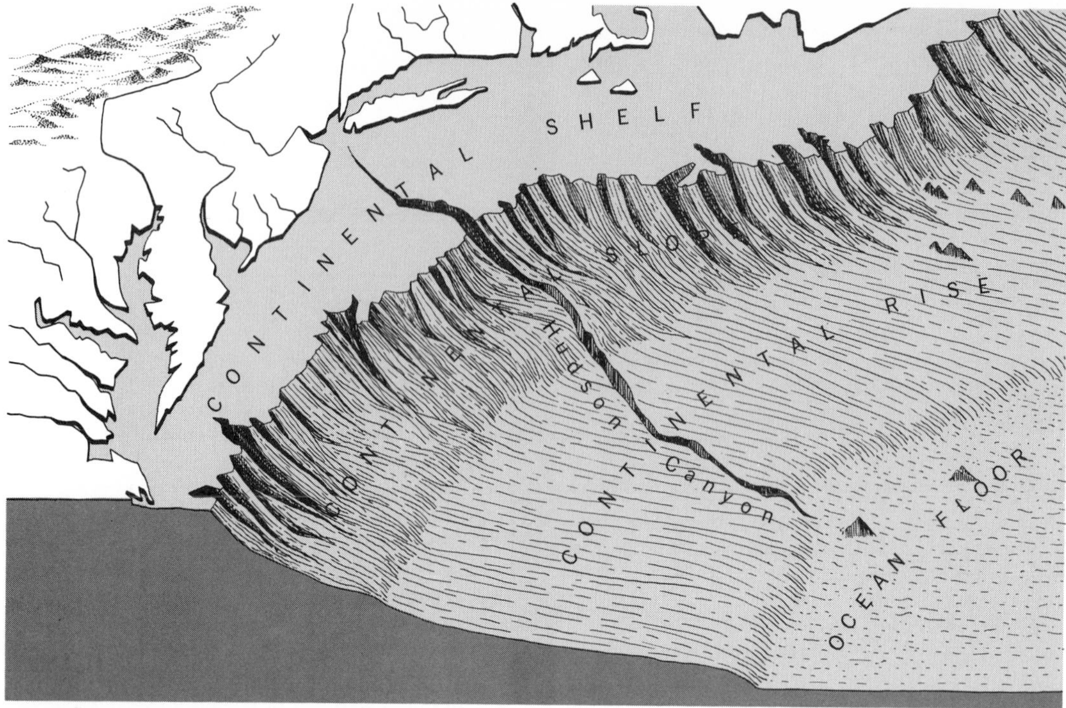

Continental margin off northeastern United States.

the continental slopes are the most imposing escarpments on the Earth. The slope is comparatively steep with an average declivity of 4.25° for the upper 1000 fathoms (1.8 km). Most slopes resemble a straight mountain front but are highly irregular in detail; in places they are deeply incised by submarine canyons, some of which cut deeply into the shelf. Usually the slope does not connect directly with the sea floor; instead there is a transitional area, the continental rise, or apron, built by the shedding of sediments from the continental block. *See* MARINE GEOLOGY; SUBMARINE CANYON.

[ROBERT S. DIETZ]

Bibliography: C. Burk and C. Drake, *The Geology of Continental Margins*, 1974; F. P. Shepard, *Submarine Geology*, 3d ed., 1973.

Coquina

A calcarenite or clastic limestone whose detrital particles are chiefly fossils, whole or fragmented. The term is most frequently used for an aggregate of large shells more or less cemented by calcite. If the rock consists of fine-sized shell debris, it is called a microcoquina. Encrinite is a microcoquina made up primarily of crinoid fragments (see illustration). Some coquinas show little evidence of any transportation by currents; articulated bivalves are preserved in entirety and the shells are not broken or abraded. *See* CALCARENITE; LIMESTONE.

[RAYMOND SIEVER]

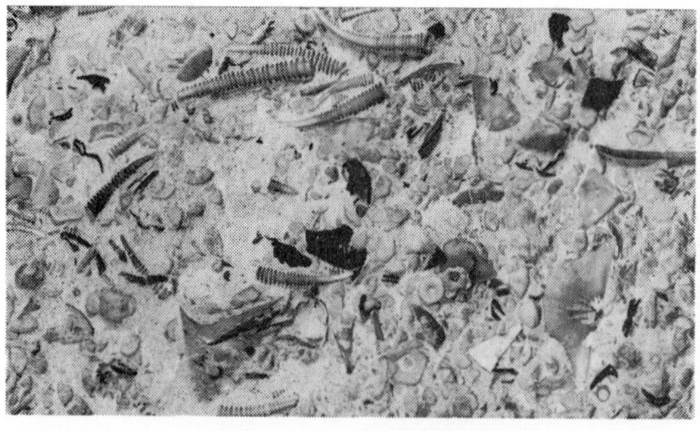

Coquinoid limestone, encrinite, showing many entire fossils in fine-grained matrix. Note lack of assortment and diversity of types. *(From F. J. Pettijohn, Sedimentary Rocks, 3d ed., copyright © 1949, 1959 by Harper & Row, Publishers, Inc.; copyright © 1975 by Francis J. Pettijohn)*

Cordierite

An orthorhombic magnesium aluminosilicate mineral of composition $Mg_2[Al_4Si_5O_{18}]$, space group Cccm. The crystal structure is related to beryl; magnesium is octahedrally coordinated by oxygen atoms, and the tetrahedrally coordinated aluminum and silicon atoms form a pseudohexagonal honeycomb, admitting limited amounts of K^+ and Na^+ ions and water molecules in the open channels. Limited amounts of Fe^{2+} may substitute for Mg^{2+}, and Fe^{3+} for Al^{3+}. The hardness is 7 (Mohs scale), specific gravity 2.6; luster vitreous; cleavage poor; and color greenish-blue, lilac blue, or dark blue, often strongly pleochroic colorless to deep blue (see illustration). Transparent pleochroic crystals are used as gem material. The disordered cordierite structure, $Mg_2[(Al,Si)_9O_{18}]$, is called indialite and is hexagonal, space group P6/m cc, isotypic with beryl. Osumilite, $KMg_2Al_3[(Al,Si)_{12}O_{30}] \cdot H_2O$, is also related but has a structure built of double six-membered rings of tetrahedrons. Cordierite, indialite, and osumilite are difficult to distinguish. Pale colored varieties are often misidentified as quartz, since these minerals have many physical properties in common. *See* BERYL; SILICATE MINERALS.

Cordierite is an important phase in the system MgO-Al_2O_3-SiO_2. Between 800 to 900°C, glasses of cordieritic composition crystallize. Below 500°C at pressures up to 5000 bars, cordierite in the pres-

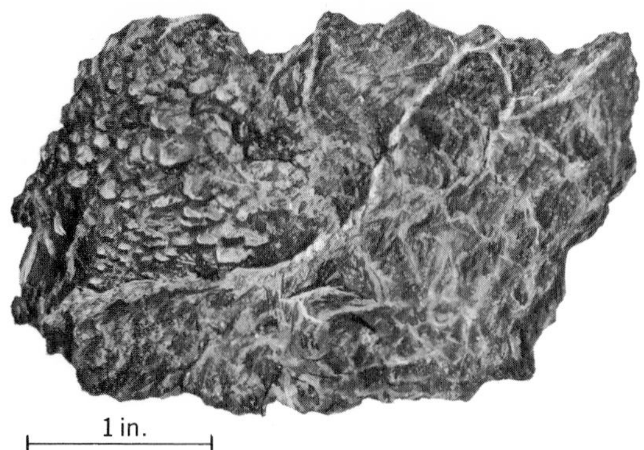

Cordierite, from Tsilaizina, Madagascar. (*Specimen from Department of Geology, Bryn Mawr College*)

ence of water breaks down to form chlorite + pyrophyllite. Indeed, it is often found naturally altered to these minerals. Cordierite melts incongruently at 2000 bars and 1125°C to mullite + spinel + liquid.

Cordierite frequently occurs associated with thermally metamorphosed rocks derived from argillaceous sediments. A common reaction in regional metamorphism resulting in gneisses is garnet + muscovite → cordierite + biotite. It may occur in aluminous schists, gneisses, and granulites; though usually appearing in minor amounts, cordierite occurs at many localities throughout the world. [PAUL B. MOORE]

Cordilleran belt

A mountain belt or chain which is an assemblage of individual mountain ranges and associated plateaus and intermontane lowlands. A cordillera is usually of continental extent and linear trend; component elements may trend at angles to its length or be nonlinear.

The term "cordillera" is most frequently used in reference to the mountainous regions of western South and North America, which lie between the Pacific Ocean and interior lowlands to the east (see illustration). The term was first applied to the Andes Mountains of South America in their entirety (Cordillera de los Andes), but individual mountain ranges within the Andean belt are now called cordilleras by some authorities. Farther north, the extensive and geologically diverse mountain terrane of western North America is formally known as the Cordilleran belt or orogen. This belt includes such contrasting elements within the United States as the Sierra Nevada, Central Valley of California, Cascade Range, Basin and Range Province, Colorado Plateau, and Rocky Mountains. *See* MOUNTAIN SYSTEMS.

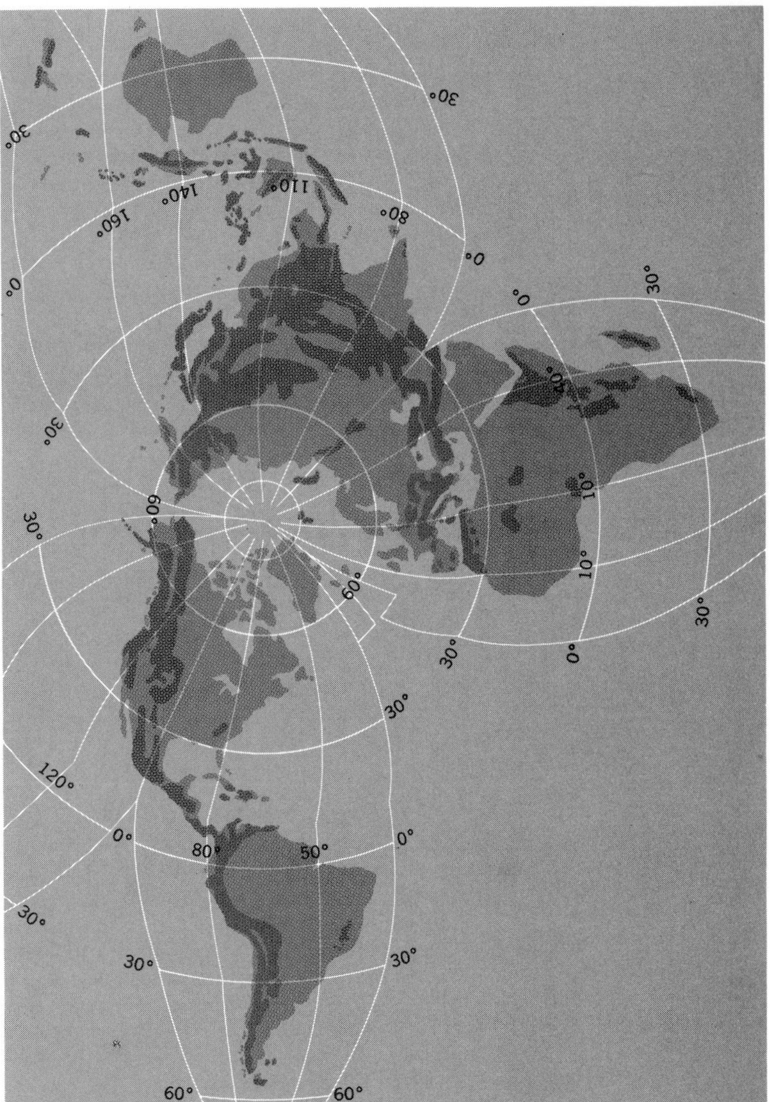

The arrangement of the mountains and the continents of the world. (*Modified from P. E. James, An Outline of Geography, copyright 1935 by Ginn and Co.*)

Cordilleras represent zones of intense deformation of the Earth's crust produced by the convergence and interaction of large, relatively stable areas known as plates. J. F. Dewey and J. M. Bird have analyzed mountain belts in terms of different modes of plate convergence. They contrast cordilleran-type mountain belts, such as the North American Cordillera, with collision-type belts, such as the Himalayas. The former develop during long-term convergence of an oceanic plate toward and beneath a continental plate, whereas the latter are produced by the convergence and collision of one continental plate with another or with an island arc. Characteristics of cordilleran-type mountain belts include their position along a continental margin, their widespread volcanic and plutonic igneous activity, and their tendency to be bordered on both sides by zones of low-angle thrust faulting directed away from the axis of the belt. The deformational, igneous, and metamorphic characteristics of cordilleran belts appear to be largely related to crustal compression, to melting of rocks at depth along the inclined plate boundary, and to attendant thermal effects in the overlying continental plate. *See* OROGENY. [GREGORY A. DAVIS]

Bibliography: B. C. Burchfiel and G. A. Davis, Nature and controls of Cordilleran orogenesis, western United States: Extensions of an earlier synthesis, *Amer. J. Sci.*, Rodgers Volume, vol. 275A, 1975; J. F. Dewey and J. M. Bird, Mountain belts and the new global tectonics, *J. Geophys. Res.*, 75(14), 1970.

Corundum

A mineral with the ideal composition Al_2O_3. It is one of a large group of isostructural compounds including hematite (Fe_2O_3) and ilmenite ($FeTiO_3$),

CORUNDUM

(a)

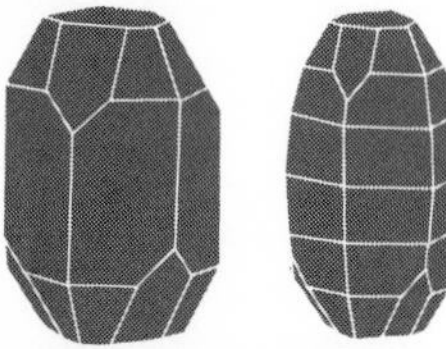

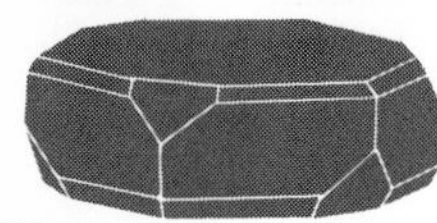

(b)

Corundum. (*a*) Specimen taken from Steinkopf, South Africa (*American Museum of Natural History*). (*b*) Crystal habits (*from C. S. Hurlbut, Jr., Dana's Manual of Mineralogy, 17th ed., copyright © 1959 by John Wiley & Sons, Inc.; reprinted by permission*).

all of which crystallize in the hexagonal crystal system, trigonal subsystem. The structure is based on hexagonal closest packing of oxygen ions with aluminum in octahedrally coordinated sites. Corundum has the high hardness of 9 on Mohs scale and is therefore commonly used as an abrasive, either alone or in the form of the rock called emery, which consists principally of the minerals corundum and magnetite. Crystals occurring in igneous rocks usually have an elongated barrellike shape, while crystals from metamorphic rocks are generally tabular (see illustration). It is commonly complexly twinned. Such specimens have a pronounced parting along twin planes and when broken may have blocky shapes. There is no discernible cleavage. The specific gravity is approximately 3.98.

Pure corundum is transparent and colorless, but most specimens contain some transition elements substituting for aluminum, resulting in the presence of color. Substitution of chromium results in a deep red color; such red corundum is known as ruby. The term "sapphire" is used in both a restricted sense for the "cornflower blue" variety containing iron and titanium, and in a general sense for gem-quality corundums of any color other than red. Green, yellow, gray, violet, and orange hues are not uncommon. Star ruby and star sapphire contain tiny needles of the mineral rutile. Reflection of light from these needles produces the six-rayed star in specimens cut and polished en cabochon, with the cabochon base perpendicular to the *c*-axis. *See* GEM; RUBY; SAPPHIRE.

Corundum occurs as a rock-forming mineral in both metamorphic and igneous rocks, but only in those which are relatively poor in silica, and never in association with free silica. Igneous rocks which most commonly contain corundum include syenites, nepheline syenites, and syenite pegmatites. It is also found as megacrysts in basaltic igneous rocks and in xenoliths contained in igneous rocks. Both contact and regionally metamorphosed silica-poor rocks may contain corundum. Original unmetamorphosed source rocks include bauxite and other aluminous sediments, as well as basic igneous rocks. Some metamorphic rocks with corundum are probably generated through the extraction of alkali and silica-rich magma. Corundum is chemically resistant to weathering processes and is therefore frequently found in alluvial deposits. Most of the gem-quality material is recovered from such sediments, as in the Mogok district of Burma (ruby), Thailand, Ceylon, and Australia. Deposits of gem-quality sapphire occur in Montana, U.S.A. *See* IGNEOUS ROCKS; METAMORPHIC ROCKS.

Corundum is synthesized by a variety of techniques for use as synthetic gems of a variety of colors and as a laser source (ruby). The most important technique is the Verneuil flame-fusion method, but others include the flux-fusion and Czochralski "crystal-pulling" techniques.

[DONALD R. PEACOR]

Covellite

A mineral having composition CuS and crystallizing in the hexagonal system. Tabular hexagonal crystals of covellite are rare (see illustration); it is usually massive or occurs in disseminations through other copper minerals. The luster is metallic and the color indigo blue. There is perfect basal cleavage yielding flexible plates; the hardness is 1.5 (Mohs scale), the specific gravity 4.7. Covellite is a common though not abundant mineral in most copper deposits. It is a supergene mineral and is thus found in the zone of sulfide enrichment associated with other copper minerals, principally chalcocite, chalcopyrite, bornite, and enargite, and is derived by their alteration. *See* BORNITE; CHALCOCITE; CHALCOPYRITE; ENARGITE.

[CORNELIUS S. HURLBUT, JR.]

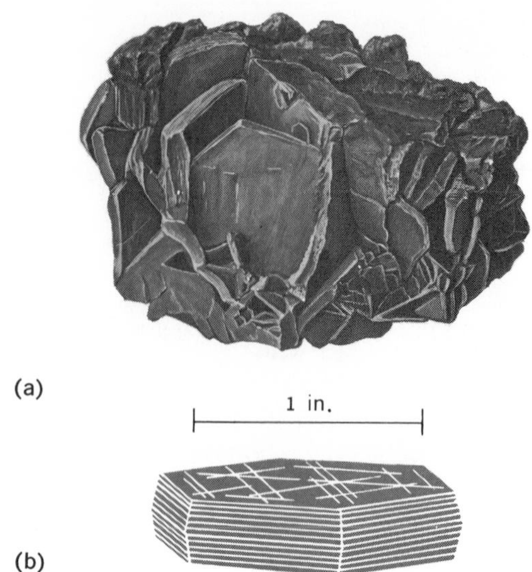

(a)

1 in.

(b)

Covellite. (*a*) Crystals that were found near Butte, Mont. (*Specimen courtesy of Department of Geology, Bryn Mawr College*). (*b*) Crystal habit (*from L. G. Berry and B. Mason, Mineralogy, copyright © 1959 by W. H. Freeman and Co.*).

Cretaceous

The latest system of rocks or period of the Mesozoic Era. It includes that part of the Earth's history from about 127,000,000 to 64,000,000 years ago—the time of the greatest flooding of the Earth's surface during the Mesozoic Era. It was also the time of extensive chalk deposition over much of the Northern Hemisphere; hence the origin of the term *terrains crétacés* by d'Omalius d'Halloy in 1822.

The Cretaceous, bounded below by the Jurassic and above by the Tertiary, marks the rapid rise and spread of deciduous trees over the landscape. Its close was marked by the withdrawal of the seas from the continents, and by the extinction of the dinosaurs, flying reptiles and huge marine reptiles, and the ammonites and certain other groups of marine mollusks that had been so conspicuous throughout the Mesozoic Era. *See* JURASSIC; TERTIARY.

Paleogeography. Much of the Earth's present land surface was submerged by shallow seas during the Cretaceous. Areas that were most extensively and frequently flooded include Europe, northernmost Africa, Madagascar, northern India, Japan, the western margin of North and South America, Mexico and the Gulf coastal plain of the United States, and the western interior of the United States and Canada. Two seas covered much of Europe most of the time. A Northern sea extended from the British Isles eastward across Germany and Poland to Russia. A southern sea, Tethys, covered most of the present Mediterranean Sea and the surrounding parts of southern Europe and northernmost Africa. The Tethys sea also crossed Asia as a belt through Turkey, the Persian Gulf area, West Pakistan, the Himalayan region, Assam, Burma, and Sumatra. The Tethys is believed to have been moderately deep, whereas the Northern sea and the vast Western Interior sea of North America were much shallower.

According to concepts of continental drift, at the beginning of the Cretaceous the Earth consisted of four closely spaced landmasses and a vast Pacific Ocean. The landmasses consisted of South America united to Africa, an India landmass lying to the east, North America united to Europe by way of Greenland, and Antarctica united to Australia. The northern landmass (North America–Eurasia) is known as Laurasia, and the rest of the landmasses are known collectively as Gondwana. Laurasia and Gondwana almost touched in one place, at what is now Spain and Morocco. Eastward from this point, a narrow seaway (Tethys Sea) extended to the Pacific Ocean and separated the Eurasian part of Laurasia from the north African part of Gondwana and India. Another and broader arm of the Pacific Ocean crossed present-day Central America and formed a North Atlantic Ocean that reached the present British Isles. *See* CONTINENT FORMATION.

A rift began splitting South America from Africa near the close of the Jurassic. This rift began in the south and, by the beginning of the Cretaceous, had worked its way north as far as present-day Nigeria. Lacustrine sediments formed at first in the rift, and later beds of salt, anhydrite, and black shales accumulated. By the end of the Early Cretaceous, the rift had widened and deepened enough to allow marine water to advance as far north as Nigeria. A rift had also developed trending northwest from Nigeria to the North Atlantic Ocean. Inasmuch as North America was drifting westward and Africa northward, the continents separated completely in the early part of the Late Cretaceous. A rift that had formed earlier in the middle of the North Atlantic Ocean eventually split Greenland from Eurasia as North America drifted westward. By the close of the Cretaceous, Madagascar had rifted from Africa, and India had drifted northward toward its present position.

Broad climatic belts were present in the Cretaceous. Most workers recognize two belts, a southern Mediterranean or Tethyan belt and a northern

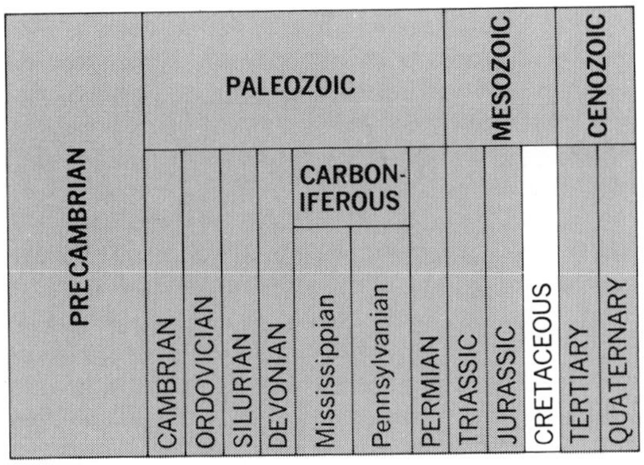

PRECAMBRIAN
PALEOZOIC
MESOZOIC
CENOZOIC
CARBONIFEROUS
CAMBRIAN
ORDOVICIAN
SILURIAN
DEVONIAN
Mississippian
Pennsylvanian
PERMIAN
TRIASSIC
JURASSIC
CRETACEOUS
TERTIARY
QUATERNARY

Stages of the Cretaceous

Epoch	Stage	Source of name
Upper Cretaceous	Maestrichtian	Maastricht, Netherlands
	Senonian	Sens, France
	Turonian	Touraine, France
	Cenomanian	Cenomanum, Latin for Mans, France
Lower Cretaceous	Albian	Aube, France
	Aptian	Apt, France
	Neocomian	Neocomum, Latin for Neuchâtel, Switzerland

or boreal belt. The Mediterranean belt, extending from Central America eastward across southern Europe and northern Africa to India and the East Indies, was characterized by much deposition of limestone and other calcareous rocks. Characteristic fossils include large thick-shelled gastropods, certain ammonites, and sessile bivalves known as rudistids which formed reefs. The boreal belt was characterized by an extinct group of squids known as belemnites, certain bivalves (such as aucellas), and numerous ammonites. Temperatures were warmer than at present and probably more uniform. Palms grew as far north as Alaska and Greenland.

The Cretaceous is generally divided into two epochs, Lower Cretaceous and Upper Cretaceous. These are subdivided into stages whose names derive from localities in western Europe, where the Cretaceous was first studied intensively (see table). The Senonian is commonly divided into three substages: Campanian (Champagne, France), Santonian (Saintonge, France), and Coniacian (Cognac, France). The Neocomian is often divided into four substages: Barremian (Barrème, France), Hauterivian (Hauterive, France), Valanginian (Valangin, France), and Berriasian (Berrias, France).

Lower Cretaceous. During Early Cretaceous time three seas—Tethys, Northern, and Russian—covered much of Europe. The Tethys submerged the northern part of Morocco, Algeria, and Tunisia, southern and eastern Spain, southern France, Italy, Albania, western Greece and Yugoslavia, and parts of Switzerland, Austria, Hungary, Romania, Bulgaria, and Turkey. The Northern sea, an arm of the Arctic or Boreal Ocean, extended through the present North Sea area and flooded the eastern part of England and much of northern Germany and Poland. The Russian sea was at first an arm of the Boreal Ocean and later an extension of the Tethys sea. Limestone and marl were the dominant sediments formed in the Tethys whereas marl, clay, shale, and sandstone are characteristic of sediments of the Northern sea and the Russian sea.

The fauna of the warm Tethys sea differed considerably from that of the cooler Northern and Russian seas. In southeastern France thick masses of limestone (Urgonian limestone) were formed by the accumulation of large foraminifers (orbitolines), corals, stromatoporids, bryozoans, rudistids, and large gastropods (nerineas). The cephalopods in the deeper parts of the Tethys were dominantly the rather smooth ammonites

Fig. 1. Land and sea in Europe and adjoining areas during the Neocomian. Darker shading over contemporary land and sea indicates Neocomian land.

(phylloceratids, lytoceratids) and flat belemnites. The shallow Northern sea contained a more varied pelecypod fauna and very ornate ammonites. Aucellas and distinctive forms of ammonites were featured in the cool Russian sea.

During the Neocomian, the Tethys and Northern seas may have been connected by a strait through Poland. West of this possible strait was a land mass covering the southern half of Germany, western Czechoslovakia, Belgium, northern and western France, northern Spain, and most of the eastern edge of the British Isles. In some areas along the margins of this land mass deltaic deposits of sandstone and clay accumulated. The largest area of deltaic sediments was in a lowland covering part of Belgium, northern France, and southern England. Here these nonmarine rocks, known as the Wealden Beds, contain fossil plants, freshwater gastropods and pelecypods, and dinosaurs known as iguanodons. During the Aptian, the Wealden area sank below sea level and the Tethys and Northern seas were connected by a strait through northern France and southeastern England. The land mass centering in southern Germany was reduced to an island known as the mid-European island, which persisted throughout Cretaceous time (Fig. 1).

The Russian sea was a south-trending gulf of the Boreal Ocean during the early part of the Neocomian. Later in the Neocomian the sea extended southward and merged with the Tethys. During the Aptian the sea was closed off from the Boreal Ocean and became a narrow northeastern arm of the Tethys sea.

The Albian seas were very extensive in Europe. The Northern sea spread eastward across southern Russia and merged with the Russian sea. The mid-European island was reduced in size. The Northern sea transgressed westward too and submerged most of England, where the Gault clay, famous for its abundant and exceptionally well-preserved fossils, was deposited. Glauconitic clay, marl, and sandstone characterize the rocks of Albian age.

Africa. In Africa marine waters invaded Angola and Nigeria while sandstones of nonmarine origin accumulated in northern Algeria. During the Early Cretaceous a seaway extended northeastward through the present Mozambique Channel to the Persian Gulf. This sea encroached upon the western part of Madagascar and the eastern margin of Africa. In southeastern Saudi Arabia this sea merged with the Tethys.

Asia. The Tethys seaway through the Himalayan area of northern India seems to have persisted throughout the Early Cretaceous. Nonmarine strata totaling as much as 12,000 ft thick were deposited in northwestern and west-central China during the Neocomian. Nonmarine rocks also characterize the older part of the Neocomian of Japan. The Aptian there is chiefly marine sandstone and shale, but includes some reef-limestone containing fossils reminiscent of the Urgonian reef-limestones of the European Tethys. The Albian is represented by conglomerates, sandstones, and shales and marks an extensive transgression of marine waters.

Australia. Australia was above the sea during the Neocomian. The east-central part was a lowland in which fluviatile and lacustrine deposition took place. In the Aptian and Albian this lowland area was invaded by shallow marine waters which submerged about a third of the continent and divided it into two or three large islands.

North and South America. Along the Pacific Coast of North America many thousands of feet of shale, sandstone, and conglomerate of marine origin reveal the encroachment of the Pacific Ocean into the continent. The rich ammonite faunas of California show the presence of all the stages of the Lower Cretaceous. Pacific waters also transgressed upon the western part of South America. Marine Aptian limestone, shale, and sandstone in northern Colombia and Venezuela suggest a strait connecting the Pacific and Atlantic Oceans. The area of the present Caribbean Sea was probably land that contained numerous volcanoes.

Mexico and much of the Gulf Coastal area of the United States were flooded by a gulf of the Atlantic Ocean. During the Neocomian a peninsula extended from Arizona south through Sonora and thence southeast along the western edge of Mexico as far as the latitude of Mexico City. A much smaller peninsula extended from Texas south into Coahuila. Central America and the southwestern part of Mexico were above sea level. The Pacific and Atlantic Oceans were united by means of the narrow passage west of Mexico City. Near the shores coarse clastic sediments were formed, whereas farther east away from the shore limestone and marl accumulated. During the Aptian and Albian the Mexican sea spread widely, crossed the Coahuila peninsula, and submerged the Gulf Coastal area of the United States as well as southwestern Texas and southern New Mexico and Arizona. Limestone was formed in central and eastern Mexico, but near the shore in Sonora a thick sequence of limestone, shale, agglomerate, and lava was deposited.

Neocomian rocks are not known with certainty

in the western interior of the United States. In Alberta and British Columbia nonmarine deposits with many important beds of coal formed during this time. Nonmarine rocks of Aptian and early Albian age are widely distributed over the western interior of the United States and Canada. During middle Albian time an arm of the Arctic Ocean covered northern Alaska and the Mackenzie River area and advanced southeastward across northeastern British Columbia, most of Alberta and Saskatchewan, and southwestern Manitoba. About the same time marine waters spread north from the Gulf Coastal area and inundated Kansas, Nebraska, and the Dakotas, and the eastern halves of New Mexico, Colorado, Wyoming, and Montana, to merge with the Canadian sea. This vast seaway thus split North America into two land masses. Sandstone and dark noncalcareous shale characterize the sediments.

Nonmarine deposition occurred in Virginia, Maryland, and Delaware during the Neocomian and Albian. In west Greenland nonmarine rocks also accumulated during the Early Cretaceous. East Greenland was partly submerged in the Neocomian and Albian.

Upper Cretaceous. The great transgression of the seas in Europe, which began in the Aptian and Albian, reached its peak during the Late Cretaceous. The mid-European island was further reduced in size and most of the British Isles was submerged. An arm of the Northern sea extended eastward across Poland to Russia. The Tethys sea still covered the Mediterranean area but spread widely and merged with the Northern sea through wide straits in France, Romania, and in the region of the Caspian Sea.

Cenomanian deposits of the Northern sea are typically glauconitic clay, sandstone, and sandy marl near the shores, and glauconitic chalk, marly chalk, and marl farther from the shores. Sediments formed in the Tethys sea area ranged from sandstones and conglomerates near the shores to rudistid limestones or shales and sandstones with oysters. During the Turonian, marly chalk was deposited in much of the Northern sea, and marls, sandstones, and rudistid limestones in the Tethys. The Senonian of the Northern sea area is characterized by white chalk with flint nodules in France, the British Isles, Denmark, and northernmost Germany. Farther south in Germany sandy marl and sandstone were deposited near the mid-European island. In the northern part of the Tethys sea the Senonian rocks display many facies. In southwestern France marly limestone, rudistid limestone, and sandy limestone are typical. In southeastern France the upper part of the Senonian consists of brackish-water, lacustrine, and fluviatile beds. In Italy, Yugoslavia, Albania, and Greece, rudistid limestones mark the Senonian. Rocks of Maestrichtian age are not as widespread as those of the Senonian. The beds are largely tuffs in the Netherlands and Belgium, chalks and greensands in southwestern France, chalks in northern Germany, and marly and sandy beds farther south near the mid-European island. In southeastern France the Maestrichtian is represented by a continental facies of lacustrine marls, lignites, and sandstones. The Danian also has a very restricted distribution. The type Danian of Denmark is limestone with abundant hydrocorals, bryozoans, and mollusks. The Danian is represented by marine limestone in southwestern France and by nonmarine rocks in southeastern France. Continental formations also feature the Danian of Spain and Portugal.

Africa. During the Cenomanian the Tethys sea spread southward over the Sahara and deposited marls and marly limestones with oysters, echinoids, and some ammonites. Sandstone, shale, and limestone of marine origin are present in Nigeria and north of there in the eastern part of French West Africa. This suggests the probability that the Tethys sea and the Atlantic Ocean were connected through the Sahara, thus dividing Africa into an eastern and a western land mass. A seaway still existed through the Mozambique Channel to West Pakistan, and Cenomanian sediments were deposited in western Madagascar and in places along the east coast of Africa. Limy sediments continued to form in North Africa during the Turonian. Nigeria, the eastern part of French West Africa, and the Atlantic coastal part of French Equatorial Africa were submerged, as well as the western part of Madagascar and the eastern part of South Africa. The Tethys sea continued to cover the northern edge of Africa from Morocco to Egypt during the Maestrichtian. Chalk and marl were deposited in Morocco and Algeria. Shale, sandstone, and coal beds formed in Nigeria and, in Angola, marls and limestones. In North Africa the Danian is represented by marine beds with numerous nautiloids.

Australia and Asia. Most of Australia was above the sea during Late Cretaceous time. Chalk, chalky clay, and greensand of Senonian age are present along the westernmost part of the continent.

The Himalayan seaway across the northern border of India persisted through the Late Cretaceous. In addition the east coastal area of India was submerged.

Cenomanian and Turonian times were marked in Japan by crustal unrest and by a regression of marine waters from parts of the area. Coarse sediments are common. The Senonian was a time of marine transgression, and shale and fine-grained sandstone were the dominant rocks formed. In the Maestrichtian a regression of the seas occurred and coarse clastic sediments were again common.

North and South America. North America was extensively flooded during the Late Cretaceous. The pattern of the seaways established by the great Albian transgression was followed throughout most of the later Cretaceous. During the Cenomanian, Turonian, and Senonian stages Pacific waters invaded much of Alaska, Oregon, and California, and deposited shales, sandstones, and conglomerates. Marine rocks of Senonian or Maestrichtian age are found along the west coast of parts of British Columbia. The Pacific Ocean also encroached upon the western margin of South America and laid down mostly shales in the Cenomanian, limy beds in the Turonian, and shale and sandstone in the Senonian. Part of Central America and much of the present area of the Caribbean Sea were probably land with many active volcanoes. An arm of the Pacific Ocean crossed northern Colombia and Venezuela and merged with the Atlantic. The Gulf Coastal region of the United States and most of Mexico were under water during the Cenomanian. Only the Sonoran peninsula and the westernmost edge of southern Mexico were emer-

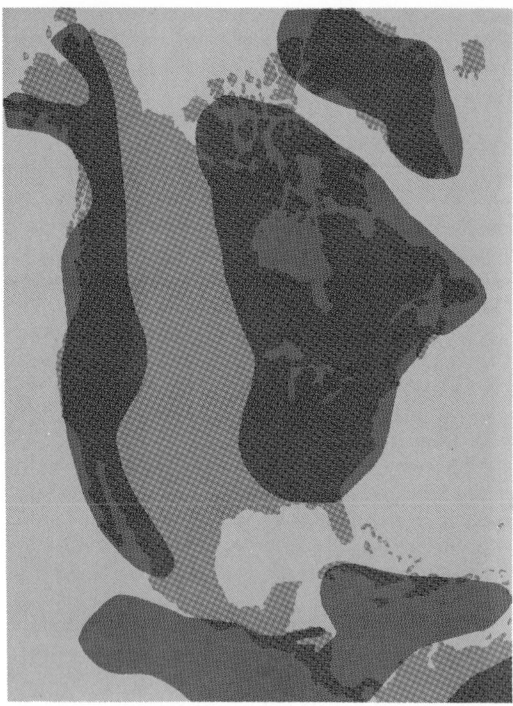

Fig. 2. Land and sea in North America and Central America during the Turonian. Darker shading over contemporary land and sea indicates Turonian land.

gent. Limestone and marl were the commonest rocks formed. Carbonate deposition continued through the Turonian.

In the early Cenomanian the sea seems to have withdrawn from the great seaway established during the Albian through the western interior of the United States and Canada. In the late Cenomanian, marine waters again filled this low area and united the Gulf of Mexico with the Arctic Ocean. The seaway spread somewhat during the early Turonian, and at one time its width reached from central Utah to western Iowa. Limestones and marls were formed over most of this region in the early Turonian, but sandstones and noncalcareous shales in the later Turonian reflected rising mountains in western Utah (Fig. 2).

During the early Senonian the Mexican–Western Interior sea was gradually reduced in size as mountains formed to the west. Near the western shore thick deposits of shale and sandstone accumulated with thick coal beds in some areas. Chalky beds were deposited farther east. Later in the Senonian, carbonate deposition almost ceased in the western interior of the United States and sandstone and shale was the rule. By the late Maestrichtian, the sea had completely withdrawn from the region and fluviatile deposition occurred to the end of the Cretaceous. These latest Cretaceous rocks, usually assigned to the Danian, have yielded remarkable horned dinosaurs.

Along the Atlantic Coastal Plain of the United States, nonmarine deposition occurred during the Cenomanian and in the early Senonian. Later in the Senonian and in the Maestrichtian, Atlantic waters transgressed upon the continent, depositing marl, clay, and sand from central South Carolina northeastward through New Jersey.

In Greenland the Cenomanian is known from a very small area near the center of the east coast, and the Turonian is known from small areas near the center of the east and west coasts. The Senonian rocks are more widespread; they are represented mainly by marine shales on both coasts. *See* MESOZOIC.

[WILLIAM A. COBBAN]

Bibliography: M. Gignoux, *Stratigraphic Geology*, 1955; W. B. Harland, A. G. Smith, and B. Wilcock (eds.), *The Phanerozoic Time-scale*, J. Geol. Soc. London, vol. 1205, 1964; B. Kummel, *History of the Earth*, 1961; R. C. Moore, *Introduction to Historical Geology*, 2d ed., 1958; J. B. Reeside, Jr., Paleoecology of the Cretaceous seas of the western interior of the United States, in H. S. Ladd (ed.), *Treatise on Marine Ecology and Paleoecology*, Geol. Soc. Amer. Memo. no. 67, vol. 2, 1957.

Cryolite

A mineral with chemical composition Na_3AlF_6. Although it crystallizes in the monoclinic system, cryolite has three directions of parting at nearly 90°, giving it a pseudocubic aspect. Hardness is $2\frac{1}{2}$ on Mohs scale and the specific gravity is 2.95. Crystals are usually snow-white but may be colorless and more rarely brownish, reddish, or even black (see illustration). The mean refraction index is 1.338, approximately that of water, and thus fragments become invisible when immersed in water. Cryolite, associated with siderite, galena, and chalcopyrite, was discovered at Ivigtut, Greenland, in 1794. This locality remains the only important occurrence.

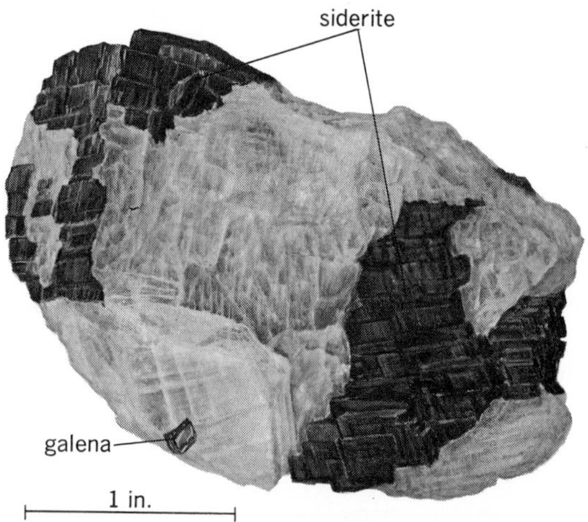

White translucent cryolite in association with siderite and galena, Ivigtut, Greenland. (*Specimen courtesy of Department of Geology, Bryn Mawr College*)

Cryolite was once used as a source of metallic sodium and aluminum, but now is used chiefly as a flux in the electrolytic process in the production of aluminum from bauxite.

[CORNELIUS S. HURLBUT, JR.]

Crystal

This term, as used in science and technology, usually denotes a single crystal. A single crystal is a solid throughout which the atoms or molecules are arranged in a regularly repeating pattern. In electronics the term crystal is usually restricted to mean a single crystal which is piezoelectric. Examples of single crystals are most gems, piezoelectric quartz crystals used in controlling the frequencies of radio transmitters, and single crystals of galena (lead sulfide) used in crystal radios.

Most crystalline solids are made up of millions of tiny single crystals called grains and are said to be polycrystalline. These grains are oriented randomly with respect to each other. Any single crystal, however, no matter how large, is a single grain. Single crystals of metals many cubic centimeters in volume are relatively easy to prepare in the laboratory.

Single crystals differ from polycrystalline and amorphous substances in that their properties are anisotropic. Young's modulus, for example, is different for different directions in the crystal. Anisotropy is responsible for the fact that crystals will cleave (split) along very flat planes which are characteristic of the atomic stacking pattern. *See* CRYSTAL GROWTH; CRYSTAL STRUCTURE; CRYSTALLOGRAPHY.

[HERMAN H. HOBBS]

Crystal growth

The growth of crystals, of which all crystalline solids are composed, generally occurs by means of the following sequence of processes: (1) diffusion of the molecules of the crystallizing substance through the surrounding environment (or solution) to the surface of the crystal, (2) diffusion of these molecules over the surface of the crystal to special sites on the surface, (3) incorporation of molecules into the crystal at these sites, and (4) diffusion of the heat of crystallization away from the crystal surface. The rate of crystal growth may be limited by any of these four steps. The initial formation of the centers from which crystal growth proceeds is known as nucleation.

Increasing the supersaturation of the crystallizing component or increasing the temperature independently increases the rate of crystal growth. However, in many physical situations the supersaturation is increased by decreasing the temperature. In these circumstances the rate of crystal growth increases with decreasing temperature at first, goes through a maximum, and then decreases. Often the growth is greatly retarded by traces of certain impurities.

After nucleation, the crystals in the medium grow isolated from one another for a time. However, if several differently oriented crystals are growing, they may finally impinge on one another, and intercrystalline (grain) boundaries will be formed. At relatively high temperatures, the average grain size in these polycrystalline aggregates increases with time by a process called grain growth, whereby the larger grains grow at the expense of the smaller.

Regular growth. During its growth into a fluid phase, a crystal often develops and maintains a definite polyhedral form which may reflect the characteristic symmetry of the molecular pattern of the crystal. The bounding faces of this form are those which are perpendicular to the directions of slowest growth. How this comes about is illustrated in Fig. 1, in which it is seen that the faces *b*, normal to the faster-growing direction, disappear, and the faces *a*, normal to the slower-growing directions, become predominant. Growth forms, like that shown in the figure, are not necessarily equilibrium forms, but they are likely to be most regular when the departures from equilibrium are not large. *See* CRYSTAL STRUCTURE.

J. W. Gibbs pointed out that the molecular binding sites on the surface of a crystal can be of several kinds. Thus a molecule must be more weakly bound on a perfectly developed plane of molecules at the crystal surface than at a ledge formed by an incomplete plane one molecule thick (Fig. 2). Therefore, the binding of molecules in an island monolayer on the crystal surface will be less per molecule than it would be within a completed surface layer.

The potential energy of a crystal is most likely to be minimum in forms containing the fewest possible ledge sites. This means that, in a regime of regular crystal growth, dilute fluid, and moderate departure from equilibrium, the crystal faces of the growth form are likely to be densely packed and molecularly smooth. There will be a critical size of monolayer, which will be a decreasing function of supersaturation, such that all monolayers smaller than the critical size will shrink out of existence, and those which are larger will grow to a complete layer. The critical monolayers form by a fluctuation process. Kinetic analyses along the lines initiated by M. Volmer and others indicate that the probability of critical fluctuations is so small that in finite systems perfect crystals will not grow, except at substantial departures from equilibrium. That, in ordinary experience, finite crystals do grow in a regular regime only at infinitesimal departures from equilibrium is explained by the screw dislocation theory of F. C. Frank. According to this theory, growth is sustained by indestructible surface ledges which result from the emergence of screw dislocations in the crystal face.

The theory of interface structure and regular growth is not sufficiently developed to predict definitively the growth behavior and morphology of crystals growing into concentrated fluids or their own melts. However, a highly successful correlation, due to K. A. Jackson, indicates that the condition for the occurrence of a regular growth regime is that the entropy of crystallization be greater than $2k$ per molecule (k is Boltzmann's constant).

Irregular growth. When the departures from equilibrium (supersaturation or undercooling) are sufficiently large, the more regular growth shapes become unstable and dendritic (treelike), or cellu-

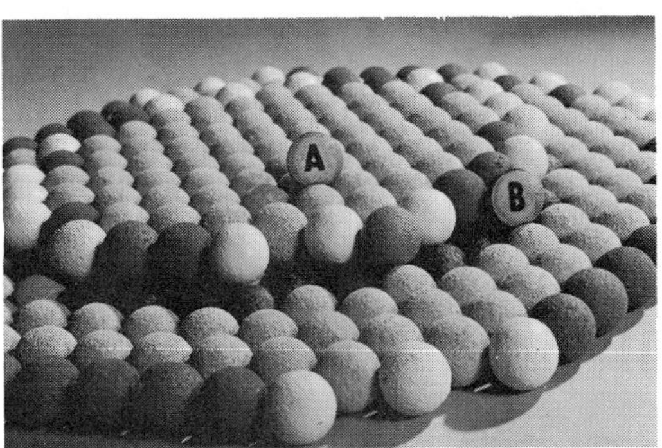

Fig. 2. A model of a crystal surface showing the inequivalent binding sites at which the atoms labeled A and B are located. (*General Electric Co.*)

CRYSTAL GROWTH

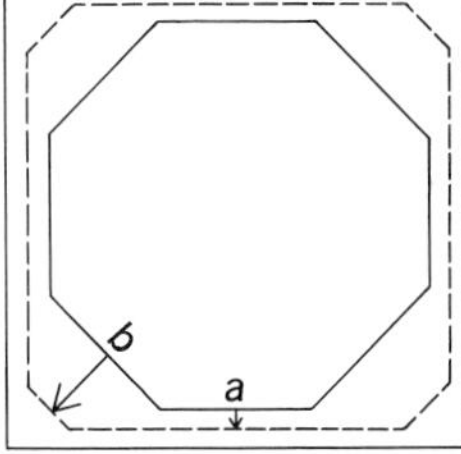

Fig. 1. Schematic representation of cross section of crystal at three stages of growth.

lar morphologies develop. Essentially, the development of protuberances on an initially regular crystal permits higher diffusive fluxes, but at the cost of higher interfacial area–crystal volume ratios. Theories which define the conditions for the onset of this morphological instability were developed by W. W. Mullins and R. F. Sekerka and by A. A. Chernov. These theories were generally confirmed by the studies of S. R. Coriell and coworkers, and revealed that a critical undercooling is indeed required for the development of protuberances on the edges of initially disk-shaped single ice crystals growing in water.

[DAVID TURNBULL]

Bibliography: W. K. Burton et al., *Phil. Roy. Soc. London*, 243A:299–358, 1951; A. A. Chernov, *Sov. Phys. Crystallogr.*, 16:734–753, 1972; R. H. Doremus et al. (eds.), *Growth and Perfection of Crystals*, 1958; R. L. Parker, in H. Ehrenreich (ed.), *Solid State Physics*, 1970.

Crystal structure

The arrangement of atoms or ions in a crystalline solid. Knowledge of the precise ways in which atoms and ions are distributed in crystals is of prime importance in solid-state physics, chemistry, metallurgy, mineralogy, geochemistry, and other fields. In 1912 M. von Laue, W. Friedrich, and P. Knipping first discovered that x-rays could be diffracted by crystals. Prior to this discovery, little was known about the arrangement of atoms and ions in solid materials. Knowledge of crystal structure is and has been obtained mainly from x-ray diffraction data, although electron diffraction and neutron diffraction have become important tools in crystal analysis.

This article discusses the important concepts and terminology involved in the structure of crystalline solids and describes the structure of various metals and some relatively simple crystalline compounds. For related information *see* CRYSTALLOGRAPHY; MINERALOGY; SILICATE MINERALS.

CONCEPTS AND TERMINOLOGY

In order to understand and describe crystal structures, several terms and concepts have been developed. Brief explanations of the most important are given in the following paragraphs.

Space lattices. A three-dimensional, indefinitely extended array of points, each of which is surrounded in an identical way by its neighbors, is known as a lattice or space lattice. The space lattice of a crystal is the representation of the periodicity with which matter is distributed in it. It is essential to distinguish a lattice from a crystal structure; a crystal structure is formed by associating with every lattice point an assembly of atoms identical in composition, arrangement, and orientation. The space-lattice concept was introduced by R. J. Haüy as an explanation for the special geometric properties of crystal polyhedrons. It was postulated that an elementary unit, having all the properties of the crystal, should exist, or conversely that a crystal was built up by the juxtaposition of such elementary units. If mathematical points forming the vertices of a parallelepiped $OABC$ (defined by three vectors $\overline{OA}, \overline{OB}, \overline{OC}$) are considered (Fig. 1), a space lattice is obtained by translations parallel to and equal to $\overline{OA}, \overline{OB}, \overline{OC}$.

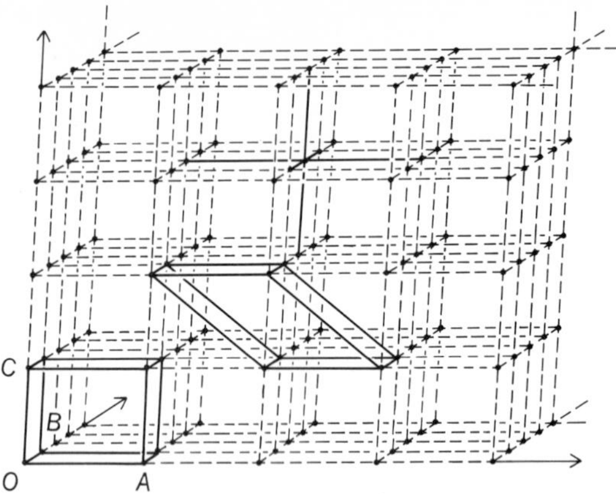

Fig. 1. A space lattice, two possible unit cells, and the environment of a point.

The parallelepiped is called the unit cell. The vector **r** joining O to any lattice point can be written as $\mathbf{r} = m\overline{OA} + n\overline{OB} + r\overline{OC}$, where m, n, and r are integers.

In a space lattice two points define a row, and three define a lattice plane. Taking the directions of OA, OB, OC as axes and ABC as the unit plane, all planes and rows can be expressed by Miller indices.

In a row $[uvw]$ two neighboring points are separated by a distance p_{uvw} which is characteristic for that row and all parallel ones. Two adjacent parallel rows of the family represented by $[uvw]$ are a distance r_{uvw} apart. The plane (hkl) defines a family of lattice planes in which the points are identically distributed and form a two-dimensional lattice. Figure 2 is an example.

This two-dimensional lattice can be deduced from two vectors such as $\overline{OA}, \overline{OB}$. The parallelogram OAB is the smallest unit from which the whole assembly can be obtained by parallel translation. This statement applies equally well to the other two parallelograms $OA'B'$ and $OA''B''$ in Fig. 2 and, in fact, to an infinite number of such parallelograms, all of which have the same surface area $S_{hkl} = p_{uvw} r_{uvw}$. Thus the density of points in a given row $1/p_{uvw}$ is proportional to the distance between rows.

In the same way an infinity of unit cells can be considered in the space lattice. These cells all have the same volume $V = S_{hkl} d_{hkl}$, if d_{hkl} is the distance between two adjacent planes (hkl). The distance d_{hkl} is proportional to $1/S_{hkl}$, the latter providing a measure for the density of points or the packing in (hkl). This remark, as well as the similar one made for the rows, is important when the properties of the crystals are considered on an atomic scale. The faces and the edges of crystals are, respectively, densely packed lattice planes and rows. It is clear from a mathematical standpoint that from all possible cells the most convenient one should be chosen and that, whenever possible, preference should be given to a parallelepiped with mutually perpendicular edges.

The volume V of the unit cell is given by Eq. (1),

$$V = abc\sqrt{1 - \cos^2\alpha - \cos^2\beta - \cos^2\gamma + 2\cos\alpha\cos\beta\cos\gamma} \qquad (1)$$

$$\frac{1}{d^2}=\frac{\frac{h}{a}\begin{vmatrix} h/a & \cos\gamma & \cos\beta \\ k/b & 1 & \cos\alpha \\ l/c & \cos\alpha & 1 \end{vmatrix}+\frac{k}{b}\begin{vmatrix} 1 & h/a & \cos\beta \\ \cos\gamma & k/b & \cos\alpha \\ \cos\beta & l/c & 1 \end{vmatrix}+\frac{l}{c}\begin{vmatrix} 1 & \cos\gamma & h/a \\ \cos\gamma & 1 & k/b \\ \cos\beta & \cos\alpha & l/c \end{vmatrix}}{\begin{vmatrix} 1 & \cos\gamma & \cos\beta \\ \cos\gamma & 1 & \cos\alpha \\ \cos\beta & \cos\alpha & 1 \end{vmatrix}} \qquad (2)$$

where α, β, and γ are the angles defined by the crystal axes, as shown in Fig. 3*a*. The general expression for d_{hkl} (called d for convenience) is given by Eq. (2).

If $\alpha=\beta=\gamma=90°$, $\cos\alpha=\cos\beta=\cos\gamma=0$, and Eq. (2) reduces to Eq. (3), and if furthermore $a=b=c$, as in cubic lattices, this becomes $1/d^2=(h^2+k^2+l^2)/a^2$. When $a=b\neq c$, $\alpha=\beta=90°$, and

$$\frac{1}{d^2}=\frac{h^2}{a^2}+\frac{k^2}{b^2}+\frac{l^2}{c^2} \qquad (3)$$

$\gamma=120°$ as in the hexagonal case, Eq. (4) holds.

$$1/d^2=(4/3a^2)(h^2+k^2+l^2)+l^2/c^2 \qquad (4)$$

Symmetry can also be considered in a lattice. The lattice being indefinitely extended, there is no longer a point group but an array of regularly repeating symmetry elements. Each lattice point is a center of symmetry. Symmetry planes are parallel to lattice planes. Symmetry axes must be perpendicular to a lattice plane and coincide with, or be parallel to, a row.

Bravais lattices. These are the 14 different possible space lattices obtained on the basis that two lattices are different when the environment of their points is different. The 14 unit cells are represented in Fig. 3. If considered as solids, the combination of symmetry elements they exhibit can be determined. Seven point groups which are respectively the most symmetrical of each system are found. Accordingly the unit cells, and by extension the lattices, can be divided into seven groups corresponding to the seven crystallographic systems. Among the cells shown in Fig. 3, some have points at places other than corners. These cells are not primitive but are multiple cells chosen for convenience. As an example, the three primitive cells of the cubic lattices are, respectively, a cube, a rhombohedron with a plane angle of 109°28′, and a rhombohedron with an angle of 60°. The two rhombohedrons are extremely inconvenient to handle; consequently, the body-centered and face-centered cubes are adopted in their stead. Figure 4 shows the three cubic systems and the relationship of the last two to their primitive rhombohedrons. The letters P, I, and F stand for primitive, body-centered, and face-centered, respectively. A P cell contains one point, since each corner is used eight times in the formation of the complete assembly. An I cell contains two points, and an F cell four (one at the corners and six halves, as each point in the center of the faces belongs to two cells).

The external symmetry of a crystal is due to the periodically repeated arrangement of its atoms or ions. The Bravais lattices give the possible periodicities. A further step must consist in finding the number of possible periodic arrangements, two such arrangements being considered as different when they give rise to a different symmetry.

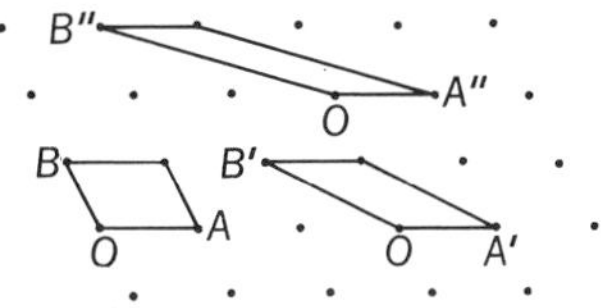

Fig. 2. Two-dimensional lattice, formed by the points in a lattice plane (*hkl*). All possible primitive parallelograms have the same surface area.

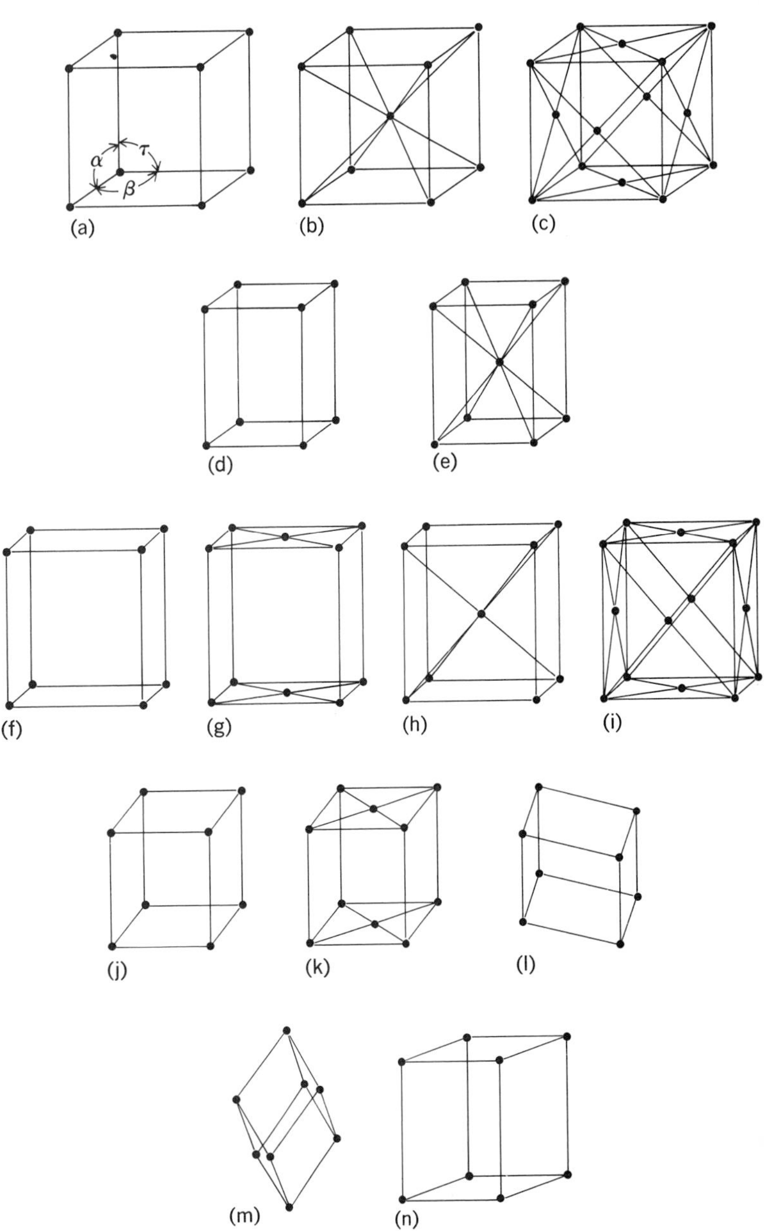

Fig. 3. The 14 Bravais space lattices. (*a*) Simple cubic. (*b*) Body-centered cubic. (*c*) Face-centered cubic. (*d*) Tetragonal. (*e*) Body-centered tetragonal. (*f*) Orthorhombic. (*g*) Base-centered orthorhombic. (*h*) Body-centered orthorhombic. (*i*) Face-centered orthorhombic. (*j*) Monoclinic. (*k*) Base-centered monoclinic. (*l*) Triclinic. (*m*) Trigonal. (*n*) Hexagonal.

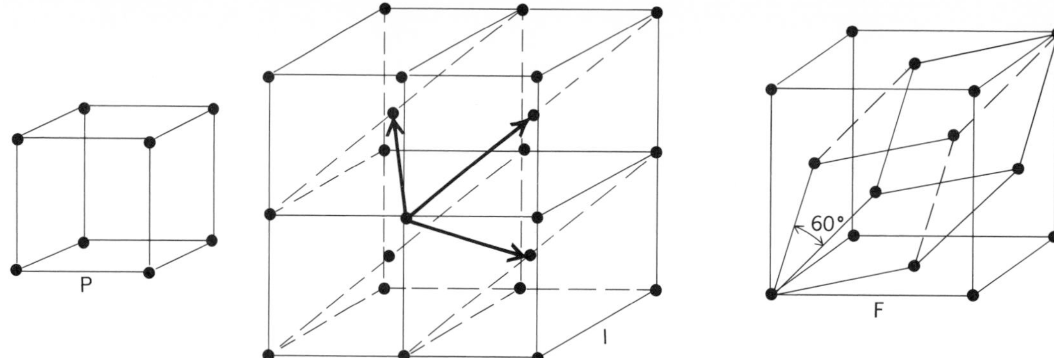

Fig. 4. The three cubic systems. The relationship of the F and I cells to their primitive rhombohedrons is demonstrated. (*After C. Kittel, Introduction to Solid State Physics, 2d ed., copyright © 1956 by John Wiley & Sons, Inc.; reprinted by permission*)

Two-dimensional structures are very well suited to illustrate this. In Figs. 5 and 6 a numeral 9 is chosen as the object; it is fully asymmetric and no false symmetry can therefore be introduced. It is multiplied by a periodic array of mirror planes and fourfold axes. Several interpenetrating identical lattices are formed. The two arrangements have the same Bravais lattice but not the same symmetry. In three-dimensional space the problem is similar, but other symmetry elements are considered.

Screw axes. These combine the rotation of an ordinary symmetry axis with a translation parallel to it and equal to a fraction of the unit distance in this direction. Figure 7 illustrates such an operation, which is symbolically denoted 3_1 and 3_2. The translation is respectively 1/3 and 2/3. The helices are added to help the visualization, and it is seen that they are respectively right- and left-handed. The projection on a plane perpendicular to the axis shows that the relationship about the axis remains in spite of the displacement. A similar type of arrangement can be considered around the other symmetry axes, and the following possibilities arise: 2_1, 3_1, 3_2, 4_1, 4_2, 4_3, 6_1, 6_2, 6_3, 6_4, and 6_5.

If screw axes are present in crystals, it is clear that the displacements involved are of the order of a few angstroms and that they cannot be distinguished macroscopically from ordinary symmetry axes. The same is true for glide mirror planes.

Glide mirror planes. These combine the mirror image with a translation parallel to the mirror plane over a distance which is half the unit distance in the glide direction. This is illustrated in Fig. 8. Axial glide planes, denoted by a, b, c, have translations which are equal to $a/2$, $b/2$, $c/2$, respectively, where a, b, c are the lattice vectors. Diagonal glide planes, denoted by n, have translations of $(a+b)/2$, $(b+c)/2$, or $(c+a)/2$.

These new symmetry elements must be taken into account when the number of possible periodic arrangements is considered. This problem was solved independently by R. S. Federow (1885), A. M. Schoenflies (1891), and W. Barlow (1894). A total of 230 arrangements or space groups is possible; of these, 32 can be distinguished macroscopically.

Space groups. These are indefinitely extended arrays of symmetry elements disposed on a space lattice. A space group acts as a three-dimensional kaleidoscope: An object submitted to its symmetry

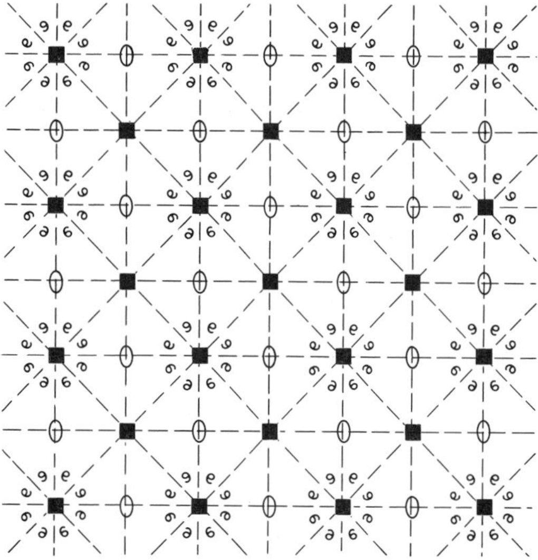

Fig. 5. Periodic pattern obtained by multiplication of asymmetric object by indicated array of symmetry elements. Both pattern and lattice have same symmetry elements. Mirror planes indicated by dotted lines.

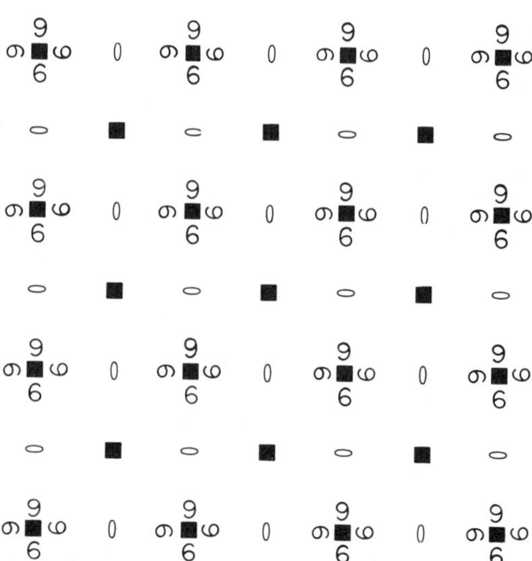

Fig. 6. Arrangement in which the periodic pattern has fewer symmetry elements than the lattice. This shows that the symmetry of the Bravais lattices can be lowered by the structure of the lattice points.

operations is multiplied and periodically repeated in such a way that it generates a number of interpenetrating identical space lattices. The fact that 230 space groups are possible means, of course, 230 kinds of periodic arrangement of objects in space. When only two dimensions are considered, 17 space groups are possible.

Space groups are denoted by the Hermann-Mauguin notation preceded by a letter indicating the Bravais lattice on which it is based. For example, P $2_12_12_1$ is an orthorhombic space group; the cell is primitive and three mutually perpendicular screw axes are the symmetry elements. All space groups are listed in advanced textbooks.

J. D. H. Donnay and D. Harker have shown that it is possible to deduce the space group from a detailed examination of the external morphology of crystals.

COMMON STRUCTURES

A discussion of the three structural systems found in metals and of the five systems found in crystalline compounds is given below.

Metals. In general, metallic structures are relatively simple, characterized by a dense packing and a high degree of symmetry. Manganese, gallium, mercury, and one form of tungsten are exceptions. Metallic elements situated in the subgroups of the periodic table gradually lose their metallic character and simple structure as the number of the subgroup increases. A characteristic of metallic structures is the frequent occurrence of allotropic forms; that is, the same metal can have two or more different structures which are most frequently stable in a different temperature range.

The forces which link the atoms together in metallic crystals are nondirectional. This means that each atom tends to surround itself by as many others as possible. This results in a dense packing, similar to that of spheres of equal radius, and yields three distinct systems: close-packed (face-centered) cubic, hexagonal close-packed, and body-centered cubic.

Close packing. For spheres of equal radius, close packing is interesting to consider in detail with respect to metal structures. Close packing is a way of arranging spheres of equal radius in such a manner that the volume of the interstices between the spheres is minimal. The problem has an infinity of solutions. The manner in which the spheres can be most closely packed in a plane *A* is shown in Fig. 9. Each sphere is in contact with six others; the centers form a regular pattern of equilateral triangles. The cavities between the spheres are numbered. A second, similar plane can be positioned in such a way that its spheres rest in the cavities 1, 3, 5 between those of the layer *A*. The new layer, *B*, has an arrangment similar to that of *A* but is shifted with respect to *A*. Two possibilities exist for adding a third layer. Its spheres can be put exactly above those of layer *A* (an assembly *ABA* is then formed), or they may come above the interstices 2, 4, 6. In the latter case the new layer is shifted with respect to *A* and *B* and is called *C*. For each further layer two possibilities exist and any sequence such as *ABCBABCACBA* . . . in which two successive layers have not the same denomination is a solution of the problem. They are all characterized by the fact that each sphere touches 12 others and all are equally densely packed. Such assemblies are rare in crystals. Complicated periodic sequences have, however, been observed, especially in carborundum. One structure is known where 89 layers form a sequence which is regularly repeated. In the vast majority of cases, periodic assemblies with a very short repeat distance occur. The cavities between the spheres, occupying 27% of the total volume, are of two types: tetrahedral cavities between four spheres, and octahedral ones between six spheres. For an assembly which consists of *N* spheres, 3*N* cavities exist; 2*N* of these are tetrahedral and *N* are octahedral.

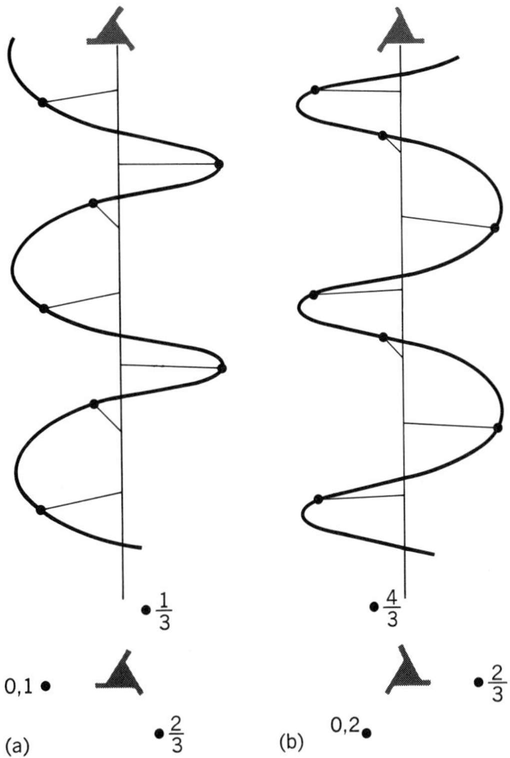

Fig. 7. Screw axes (*a*) 3_1 and (*b*) 3_2. Lower parts of figure are projections on a plane perpendicular to the axis. Numbers indicate heights of points above that plane.

Face-centered cubic structure. This utilizes close packing characterized by the regular repetition of the sequence *ABC*. The centers of the spheres form a cubic lattice *F*, as shown in Fig. 10*a*. The densely packed planes of the type *A*, *B*, *C* are per-

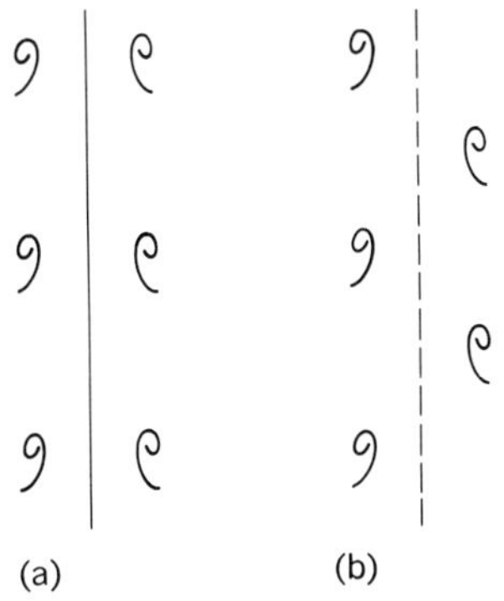

Fig. 8. (*a*) Mirror and (*b*) glide plane.

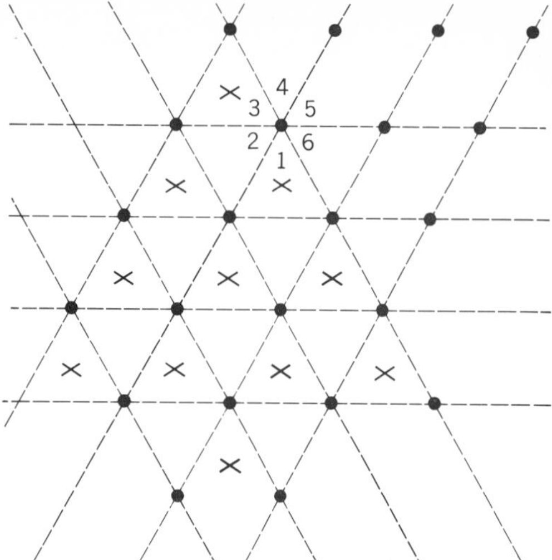

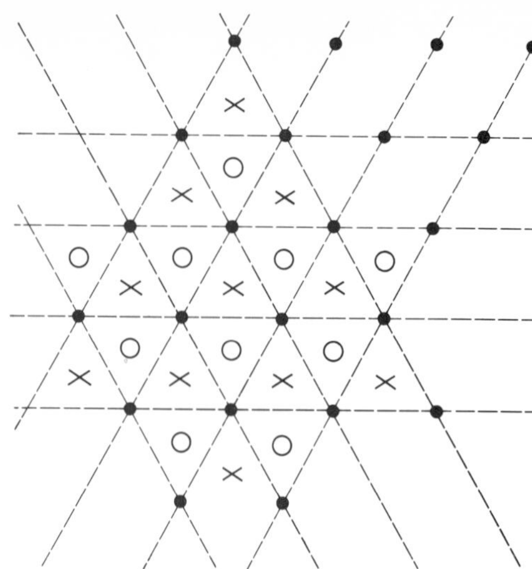

Fig. 9. Close packing of spheres of equal radius. The centers for the *A*, *B*, and *C* layers are indicated by a dot, a cross, and an open circle, respectively. Cavities between spheres are numbered.

pendicular to the threefold axis and can therefore be written as {111}. This form contains four sets of planes. These being the closest packed planes of the structure, d_{111} is greater than any other d_{hkl} of the lattice. The densest rows in these planes are <110>.

It is relatively easy to calculate the percentage of the volume occupied by the spheres. The unit cell has a volume a^3 and contains four spheres of radius R, their volume being $16\pi R^3/3$. The spheres touch each other along the face diagonal (Fig. 10*a*); $a\sqrt{2}$ is therefore equal to $4R$, or $R=a\sqrt{2}/4$. Substitution gives for the volume of the spheres $\pi\sqrt{2}a^3/6$, which is 73% of the volume of the cube. It is clear that the same percentage of the unit volume is filled in all other close-packed assemblies, that is, assemblies corresponding to other alternations of the planes *A*, *B*, *C*.

Hexagonal close-packed structure. This is a close packing characterized by the regular alternation of two layers, or *ABAB* . . . . The assembly has hexagonal symmetry (Fig. 10*b*); the unit cell is shown in Fig. 11. Six spheres are at the corners of an orthogonal parallelepiped having a parallelo-

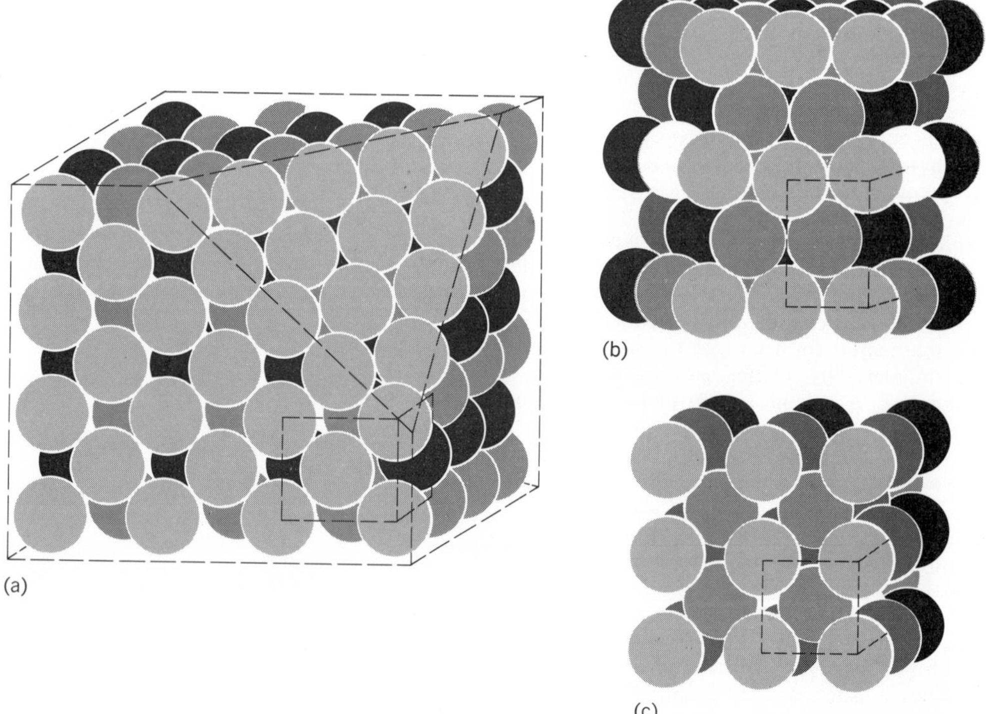

Fig. 10. (*a*) Cubic close packing of spheres (face-centered cubic). One set of close-packed {111} planes (*A*, *B*, or *C*) is shown. (*b*) Hexagonal close packing of spheres. (*c*) Body-centered cubic arrangement.

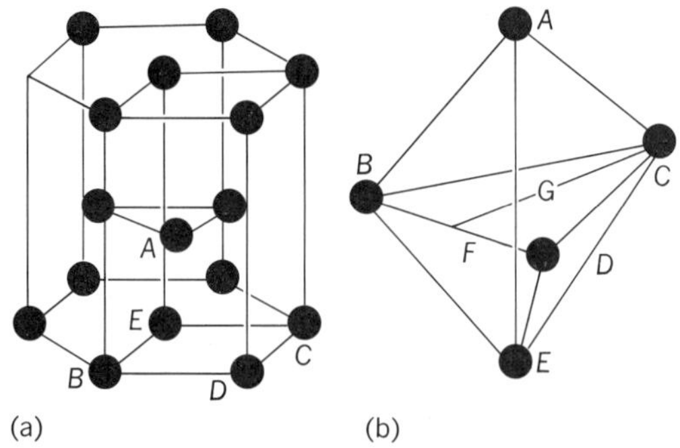

Fig. 11. (*a*) Three unit cells of the hexagonal close-packed structure, showing how the hexagonal axis results. One of the cells is fully outlined. (*b*) Calculation of the ratio *c*/*a*. Distance *AE* is equal to height of cell.

gram as its base; another atom has as coordinates (1/3, 1/3, 1/2). The ratio c/a is easily calculated. The length $BD=a$, the edge of the unit cell. The height of the cell is $AE=2AG=c$, defined in Eqs. (5) and (6).

$$AG=\sqrt{a^2-\left(\frac{2}{3}\frac{a\sqrt{3}}{2}\right)^2}=a\sqrt{2/3} \qquad (5)$$

$$c=2AG=a\sqrt{8/3} \qquad c/a=\sqrt{8/3}=1.633 \qquad (6)$$

This latter value is important, for it permits determination of how closely an actual hexagonal structure approaches ideal close packing.

Body-centered cubic structure. This is an assembly of spheres in which each one is in contact with eight others, as shown in Fig. 10*c*.

The spheres of radius R touch each other along the diagonal of the cube, so that measuring from the centers of the two corner cubes, the length of the cube diagonal is $4R$. The length of the diagonal is also equal to $a\sqrt{3}$, if a is the length of the cube edge, and thus $R=a\sqrt{3}/4$. The unit cell contains two spheres (one in the center and 1/8 in each corner) so that the total volume of the spheres in each cube is $8\pi R^3/3$. Substituting $R=a\sqrt{3}/4$ and dividing by a^3, the total volume of the cube, gives the percentage of filled space as 67%. Thus the structure is less dense than the two preceding cases.

The closest packed planes are {110}; this form contains six planes. They are, however, not as dense as the A, B, C planes considered in the preceding structures. The densest rows have the four <111> directions.

Tabulation of structures. The structures of various metals are listed in the table, in which the abbreviations fcc, hcp, and bcc, respectively, stand for face-centered cubic, hexagonal close-packed, and body-centered cubic. In this table the hexagonal structures classified as ? are still in doubt. The structures listed hcp are only roughly so; only magnesium has a c/a ratio (equal to 1.62) which is very nearly equal to the ratio (1.63) calculated earlier for the ideal hcp structure. For cadmium and zinc the ratios are respectively 1.89 and 1.86, which are significantly larger than 1.63. Strictly speaking, each zinc or cadmium atom has therefore not twelve nearest neighbors but only six. This departure from the ideal case for subgroup metals follows a general empirical rule, formulated by W. Hume-Rothery and known as the 8-N rule. It states that a subgroup metal has a structure in which each atom has 8-N nearest neighbors, N being the number of the subgroup.

Crystalline compounds. Simple crystal structures are usually named after the compounds in which they were first discovered (diamond or zinc sulfide, cesium chloride, sodium chloride, and calcium fluoride). Many compounds of the type A^+X^-, $A^{++}X_2^-$ have such structures. They are highly symmetrical, the unit cell is cubic, and the atoms or ions are disposed at the corners of the unit cell and at points having coordinates which are combinations of 0, 1, 1/2, or 1/4.

Sodium chloride structure. This is an arrangement in which each positive ion is surrounded by six negative ions and vice versa. The arrangement is expressed by stating that the coordination is 6/6. The centers of the positive and the negative ions each form a face-centered cubic lattice. They are shifted one with respect to the other over a distance $a/2$, where a is the repeat distance (Fig. 12*a*). Systematic study of the dimensions of the unit cells of compounds having this structure has revealed that:

1. Each ion can be assigned a definite radius. A positive ion is smaller than the corresponding atom and a negative ion is larger.
2. Each ion tends to surround itself by as many others as possible of the opposite sign because the binding forces are nondirectional.

On this basis the structure is determined by two factors, a geometrical factor involving the size of the two ions which behave in first approximation as hard spheres, and an energetical one involving electrical neutrality in the smallest possible volume. In the ideal case all ions will touch each other; therefore, if r_A and r_X are the radii of the ions, $4r_X=a\sqrt{2}$ and $2(r_A+r_X)=a$. Expressing a as a function of r_X gives $r_A/r_X=\sqrt{2}-1=0.41$. When r_A/r_X becomes smaller than 0.41, the positive and negative ions are no longer in contact and the structure becomes unstable. When r_A/r_X is greater than 0.41, the positive ions are no longer in con-

Metal structures

Metal	Modification	Stability range	Structure
Beryllium	α	To melting point	hcp
	?β	630°C to melting point	Hexagonal
Cadmium	α	To melting point	hcp
Calcium	α	To 450°C	fcc
	?β	300–450°C	Hexagonal
	γ	450°C to melting point	bcc
Cerium	α	To melting point	fcc
	?β	50°C	Hexagonal
Chromium	α	To melting point	bcc
	?β	Electrolytic form	Hexagonal
Cobalt	α	To 420°C	hcp mixed with fcc
	β	420°C to melting point	fcc
Gold	α	To melting point	fcc
Iron	α	To 909°C	bcc
	γ	909–1403°C	fcc
	δ	1403°C to melting point	bcc
Lead	α	To melting point	fcc
Magnesium	α	To melting point	Nearly hcp
Nickel	α	To melting point	fcc
	β	Electrolytic form	Hexagonal
Silver	α	To melting point	fcc
Tungsten	α	To melting point	bcc
Zirconium	α	To 862°C	hcp
	β	862°C to melting point	bcc
Zinc	α	To melting point	hcp

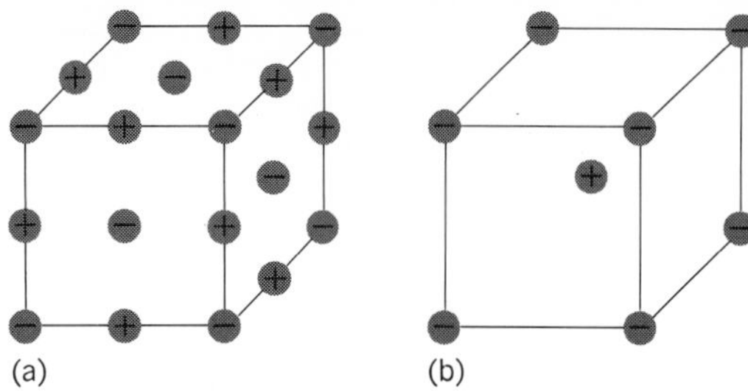

Fig. 12. (*a*) Sodium chloride structure. (*b*) Cesium chloride structure.

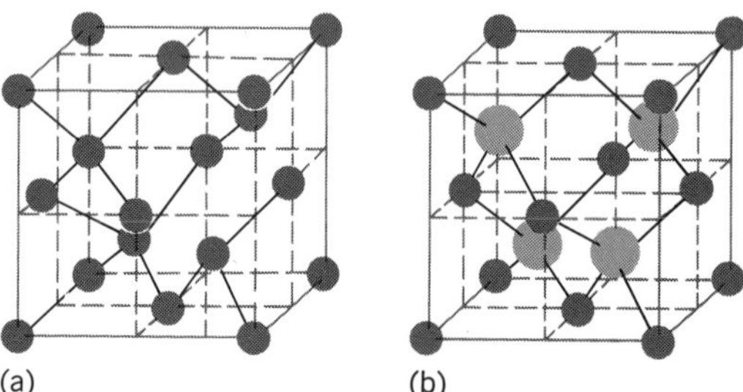

Fig. 13. (*a*) Diamond and (*b*) zinc blende structures.

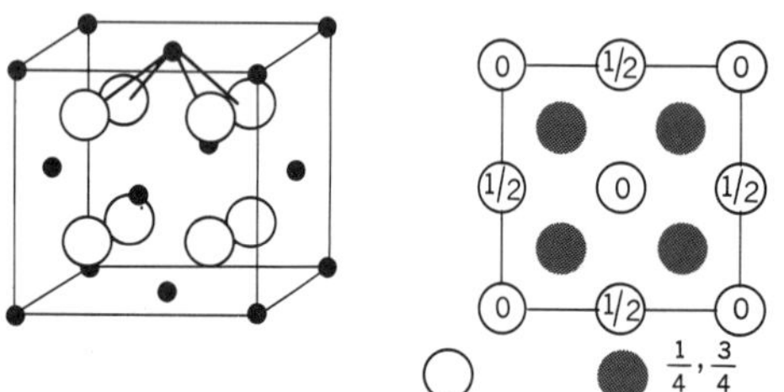

Fig. 14. Calcium fluoride structure.

tact, but ions of different sign still touch each other. The structure is stable up to $r_A/r_X = 0.73$, which occurs in the cesium chloride structure.

Cesium chloride structure. This is characterized by a coordination 8/8 (Fig. 12*b*). Each of the centers of the positive and negative ions forms a primitive cubic lattice; the centers are mutually shifted over a distance $a\sqrt{3}/2$. The stability condition for this structure can be calculated as in the preceding case. Contact of the ions of opposite sign here is along the cube diagonal.

Diamond structure. In this arrangement each atom is in the center of a tetrahedron formed by its nearest neighbors. The 4-coordination follows from the well-known bonds of the carbon atoms. This structure is illustrated in Fig. 13*a*. The atoms are at the corners of the unit cell, the centers of the faces, and at four points having as coordinates (1/4, 1/4, 1/4), (3/4, 3/4, 1/4), (3/4, 1/4, 3/4), (1/4, 3/4, 3/4). The atoms can be divided into two groups, each forming a face-centered cubic lattice; the mutual shift is $a\sqrt{3}/4$.

Zinc blende structure. This structure, shown in Fig. 13*b*, has coordination 4/4 and is basically similar to the diamond structure. Each zinc atom (small circles in Fig. 13*b*) is in the center of a tetrahedron formed by sulfur atoms (large circles) and vice versa. The zinc atoms form a face-centered cubic lattice, as do the sulfur atoms.

Calcium fluoride structure. Figure 14 shows the calcium fluoride structure. If the unit cell is divided into eight equal cubelets, calcium ions are situated at corners and centers of the faces of the cell. The fluorine ions are at the centers of the eight cubelets. There exist three interpenetrating face-centered cubic lattices, one formed by the Ca ion and two by the F ions, the mutual shifts being (0, 0, 0), (1/4, 1/4, 1/4), (3/4, 3/4, 3/4).

[WILLY C. DEKEYSER]

Bibliography: Akademische Verlagsgesell schaft, Strukturbericht, vols. 1–7, *Z. Krist.*, suppl. vols., 1913–1939, reprint, 1943, and continued by International Union of Crystallography, *Structure Reports*, 1945–; C. S. Barrett, *Structure of Metals*, 2d ed., 1952; M. J. Buerger, *Crystal-Structure Analysis*, 1960; J. D. H. Donnay and W. Nowacki, *Crystal Data*, Geol. Soc. Amer., Mem., no. 60, 1954; R. C. Evans, *Introduction to Crystal Chemistry*, 1939; N. F. M. Henry and K. Lonsdale, *International Tables for X-ray Crystallography*, vol. 1, 1952; F. M. Jaeger, *Le principe de la symétrie*, 1925.

Crystallography

The branch of science that deals with the geometric description of crystals, their internal arrangement, and their properties. Long a part of mineralogy, crystallography is now a vast no-man's-land between physics, chemistry, physical metallurgy, mineralogy, and even biology.

This article deals with the geometric description of crystals; for other aspects of crystallography *see* CRYSTAL STRUCTURE.

Formal description of crystals. The anisotropic nature of crystals results in a freely growing crystal, bounded by flat faces. The following discussion presents a formalism for describing crystals exactly in terms of these faces.

Faces which are parallel to the same edge are said to form a zone; the direction of the edge is the zone axis. The number and relative importance of the faces bounding crystals of the same chemical compound can be very different. The crystals are then said to have a different habit. Faces and edges which are present as well as those which are geometrically possible must therefore be taken into account, and in the following discussion the term face is extended to any plane determined by two edges. An edge is, in the same sense, the intersection of two faces.

N. Steno found that the angle between similar faces is constant for different-sized crystals of the same substance. This means that a crystal grows by the parallel displacement of its faces. Therefore, only the direction of a normal to a face is of importance. As a consequence, the crystal is considered to be in the center of a sphere (Fig. 1). The normals to its faces from the center cut the sphere in points or poles whose position can be fixed by two coordinates φ and ρ (Fig. 2*a*). The constellation of faces is in this way replaced by a constellation of points on a sphere. The zone axes are replaced by great circles or zone circles; they are the intersections of the sphere with planes passing through the center of the sphere, perpendicular to the zone axis. In practice, φ and ρ values are obtained by "measuring" the crystal; two-circle goniometers are used for the purpose. All problems

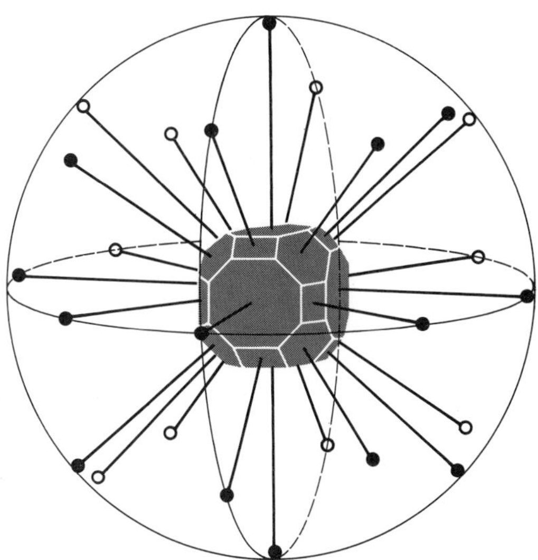

Fig. 1. Replacement of the constellation of faces by a constellation of poles on a sphere. Faces parallel to the same edge, that is, forming a zone, have their poles on a great circle called the zone circle.

which can arise are connected with angles between faces and between zone axes; they can in principle be solved by spherical trigonometry once the φ and ρ values are known. Graphical constructions are, however, most helpful, and gnomonic and stereographic projections are used in such constructions. These projections are shown in Fig. 2. In stereographic projection, circles are projected as circles, and the angles between them are not altered. In gnomonic projection, great circles are projected as straight lines.

Crystal symmetry. Symmetry is a basic property of crystals. A crystal polyhedron (or better, the constellation of the normals) can be brought into successive indistinguishable positions by nontrivial operations. These are inversion, rotation around an axis over a specific angle, reflection into a plane, or combinations of these (Fig. 3). It is, however, convenient to consider all operations as rotations, and two types of axis can be considered: (1) a rotation or symmetry axis involving a rotation over an angle of 360°/n (n is the multiplicity of the axis) and (2) an inversion axis which combines the preceding operation with an inversion with respect to a point of the axis. It is found that only multiplicities of 1, 2, 3, 4, and 6 occur. This has a deeper meaning and is connected with the homogeneous filling of space. It can be understood intuitively by considering the two-dimensional example in Fig. 4. An assembly of identical polygons can completely cover a surface only when the polygons possess one of the following symmetry elements: 2, 3, 4, 6. The symmetry axes are represented by their multiplicity; inversion axes are written $\bar{1}$, $\bar{2}$, $\bar{3}$, $\bar{4}$, $\bar{6}$. Axis $\bar{1}$ is in fact an inversion center; $\bar{2}$ is equivalent to a mirror plane and is therefore written m. Not all of the other inversion axes are specific symmetry elements; that is, $\bar{3}$ is a combination of the operations 3 and $\bar{1}$, and $\bar{6}$ a combination of a threefold axis and a mirror plane perpendicular to it. Finally, the distinct symmetry elements are 1, 2, 3, 4, 6, $\bar{1}$, m, $\bar{4}$.

It can be proved that a symmetry axis, the perpendicular to a symmetry plane, the plane perpendicular to a symmetry axis, and a symmetry plane are all possible edges or faces of the crystal.

A crystal can exhibit a number of these symmetry elements, all going through a common point. A collection of symmetry elements applied about such a point is called a point group.

Combinations of symmetry elements are ruled by the following theorems:

1. When two out of the three elements $\bar{1}$, m, 2 are present, the third follows automatically.
2. The intersection of two mirror planes forming an angle φ is a symmetry axis with multiplicity $n = 360/2\varphi$.
3. When a mirror plane contains a symmetry axis of multiplicity n, this axis is the intersection of n mirror planes forming angles of 360°/n.
4. When two twofold axes forming an angle φ intersect, the perpendicular at the intersection point on the plane they determine is a symmetry axis with multiplicity $n = 360/2\varphi$.
5. Any crystal polyhedron containing more than two axes with a multiplicity higher than two must necessarily present the combination of symmetry axes present in a cube or in a tetrahedron.

Taking these five theorems into account, it can be shown that 32 combinations or point groups are possible. These are conveniently represented by

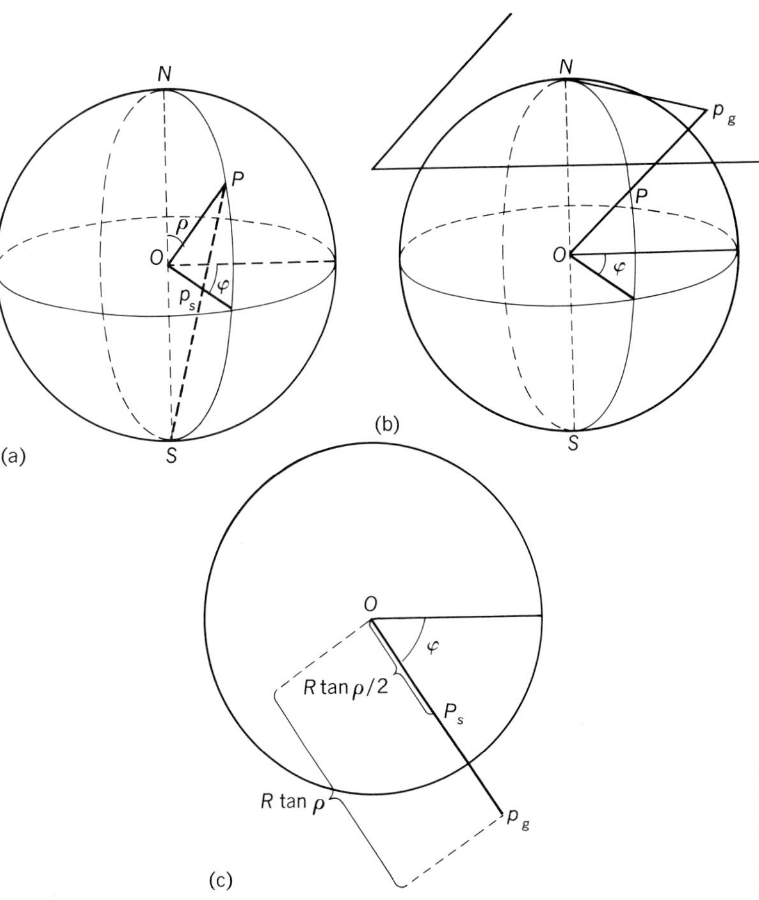

Fig. 2. (*a*) Stereographic projection of *P*. Line joining pole *P*, defined by φ and ρ, to south pole *S* cuts equator at p_s. (*b*) Gnomonic projection of *P*. Line *OP* cuts the plane tangent to the sphere at *N* at point p_g. (*c*) Stereognomogram. Both projection planes are superimposed; p_s and p_g are on a radius making an angle φ with the origin and at distances respectively equal to $R \tan \rho/2$ and $R \tan \rho$.

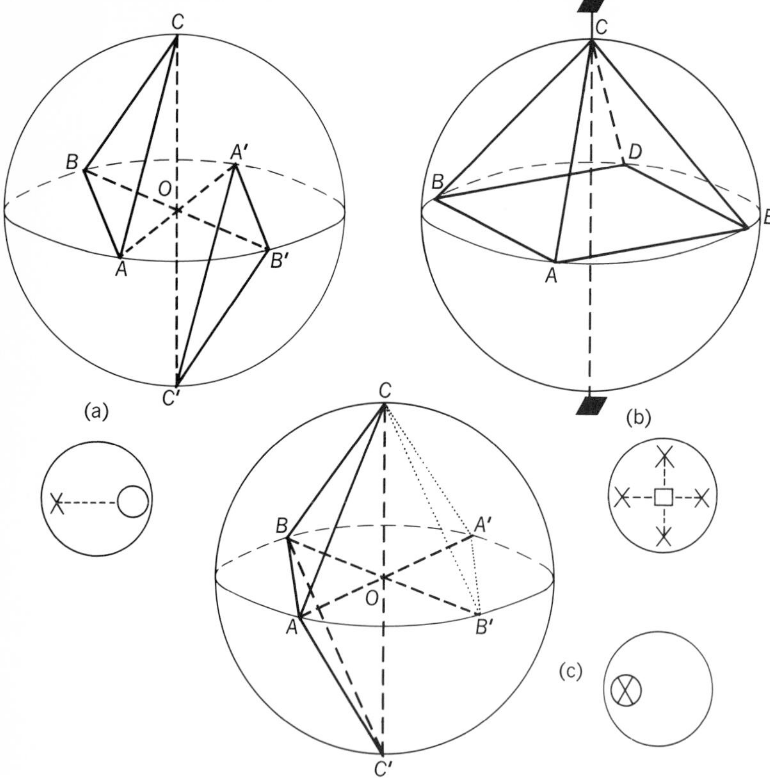

Fig. 3. Elementary symmetry operations in space and in projection. (*a*) *O* is in an inversion center. The pole of *ABC* (× in the small diagram) is projected stereographically from *C'*, that of *A'B'C'* from *C* (○ in the diagram). (*b*) Fourfold axis and stereographic projection of the four planes related by its symmetry operation. (*c*) *CC'* is a twofold inversion axis. *ABC* is first rotated over 180° into the intermediate position *CA'B'* and by inversion to *ABC'*. This operation $\bar{2}$ gives the same result as a reflection of *ABC* into the equatorial plane, thus showing that $\bar{2} = m$.

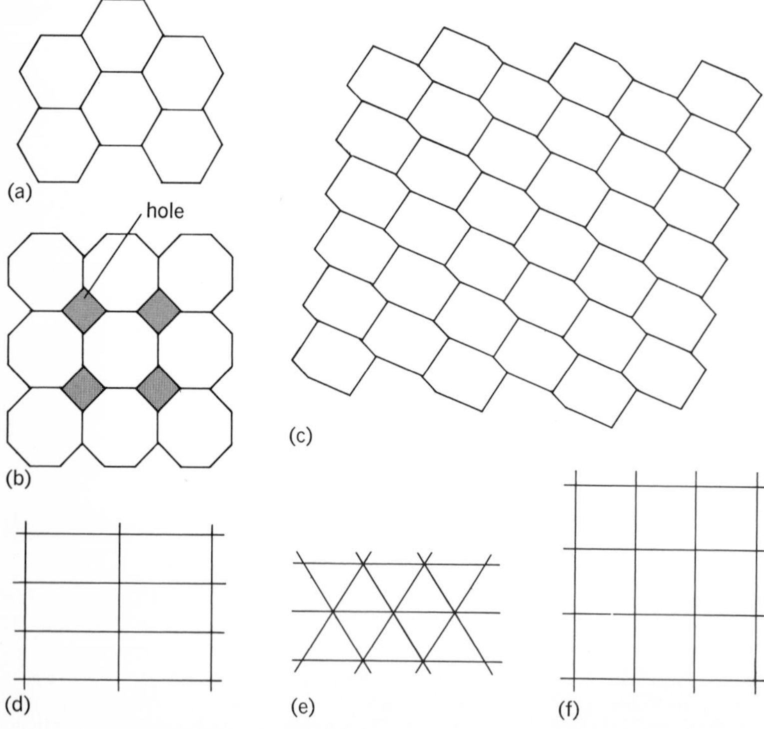

Fig. 4. Mosaics. Homogeneous filling of a surface with polygons is possible only when they have definite symmetry elements. Note the holes formed in *b* when octagons are used. Mosaics *a, c, d, e, f* fill space completely.

the Hermann-Mauguin notation, which provides the minimum information from which, with the aid of the five theorems, all the symmetry elements of a point group can be deduced. The code is as follows: If the axis of highest multiplicity is written first, the succeeding directions are perpendicular to it, and the succeeding planes are either perpendicular to it or pass through it. A mirror plane perpendicular to an axis is written $3/m$. The diagonal is not present when the plane passes through the axis; that is, one writes $3m$. The notation $m3$ means that the threefold axis is neither in the plane m nor perpendicular to it. A combination such as $3/mm$ represents a threefold axis, three planes passing through it and one perpendicular to it. The intersections of these mirror planes are twofold axes which are not written, but deduced.

All crystals can accordingly be divided into 32 classes if the point group is used as a criterion. Examples belonging to each class are known.

Crystal systems. The seven groups obtained when only the dominant symmetry elements of the classes are taken into account are called crystal systems. In order of decreasing symmetry, they are called cubic, hexagonal, tetragonal, trigonal, orthorhombic, monoclinic, and triclinic. Sometimes the hexagonal and trigonal systems are considered as a single system. The systems and classes are summarized in Table 1, in which the Hermann-Mauguin notation is used. The expression "a cubic crystal" means that the crystal belongs to the cubic system.

Miller indices. These are used for the analytical description of crystal faces, and are shown in Fig. 5. A specially well-suited reference system used for that purpose consists of three nonparallel possible edges of the crystal which are chosen as axes and a possible plane of the crystal which intersects all three axes respectively at *A*, *B*, C. It is defined by the three angles α, β, γ and the ratio $OA:OB:OC$. Another plane, HKL, of the same crystal is defined by the ratio $OH:OK:OL$. The lengths OH, OK, OL are respectively measured with $OA = a$, $OB = b$, $OC = c$ as units, that is, by the ratios a/OH, b/OK, c/OL. A measurement is a comparison (ratio) between what must be measured and a unit. The unit is generally written as the denominator in such ratios; the reverse is done here for a special reason which will be explained later. In Eq. (1), h, k, l are rational num-

$$(a/OH)/(b/OK)/(c/OL) = h/k/l \qquad (1)$$

bers which are small when the axes are well chosen (the law of rational indices). A different unit is used along each axis to take the anisotropy of the crystal into account. When, for instance, the temperature is raised, a, b, and c expand differently, but the ratios a/OH, b/OK, c/OL remain constant. The ratios $h:k:l$ define the plane HKL; h, k, l are called its Miller indices, and by convention, this is written (hkl), or ($\bar{h}\bar{k}\bar{l}$) when the indices are negative numbers. In geometrical crystallography, it is customary to precede the indices with a letter. The letters a, b, c are generally used for the axial planes, and the letter o is used for the plane ABC. The indices of these planes are respectively $a(100)$, $b(010)$, $c(001)$, $o(111)$. The last is therefore called unit plane. A plane (hkl) cuts from the axis segments OH, OK, OL whose ratios are, following the definition, equal to $(a/h)/(b/k)/(c/l)$. The equation of

Table 1. Crystal systems and classes

Cubic	Hexagonal	Tetragonal	Trigonal	Orthorhombic	Monoclinic	Triclinic
23	6	4	3	2*mm*	2	1
$(2/m)\bar{3}$	$\bar{6}$	$\bar{4}$	$\bar{3}$	222		$\bar{1}$
$\bar{4}3m$	$6/m$	$4/m$	$3m$	$(2/m)(2/m)(2/m)$	m	
432	$6mm$	$4mm$	$\bar{3}(2/m)$		$2/m$	
$(4/m)\bar{3}(2/m)$	$\bar{6}m2$	$\bar{4}2m$	32			
	622	422				
	$(6/m)(2/m)(2/m)$	$(4/m)(2/m)(2/m)$				

the plane in cartesian coordinates is therefore given by Eq. (2). The quantity d is a parameter, since

$$h\frac{x}{a}+k\frac{y}{b}+l\frac{z}{c}=d \tag{2}$$

the plane can be shifted parallel to itself. The indices are respectively the coefficients in x/a, y/b, z/c in the equation of the plane. This is the special reason mentioned earlier.

Zone indices. Two planes $(h_1k_1l_1)$ and $(h_2k_2l_2)$ define an edge, the direction coefficients of which are determined by simultaneous solution of Eqs. (3*a*) and (3*b*). The direction coefficients are given

$$h_1\,x/a+k_1\,y/b+l_1\,z/c=0 \tag{3a}$$

$$h_2\,x/a+k_2\,y/b+l_2\,z/c=0 \tag{3b}$$

by notation (4) or by ua, vb, wc, when the determi-

$$a\begin{vmatrix}k_1 & l_1\\ k_2 & l_2\end{vmatrix} \qquad b\begin{vmatrix}l_1 & h_1\\ l_2 & h_2\end{vmatrix} \qquad c\begin{vmatrix}h_1 & k_1\\ h_2 & k_2\end{vmatrix} \tag{4}$$

nants are respectively denoted by u, v, w. If a, b, c are taken as units along each of the axes, u, v, w are the coordinates of a point. The edge, or zone axis, determined by the two planes has the direction of the line joining the origin to the point with crystallographic coordinates u, v, w; it is represented by $[uvw]$.

Conversely, two zones $[u_1v_1w_1]$ $[u_2v_2w_2]$ define a plane, whose indices are found by the same procedure. The condition that a plane (hkl) belongs to a zone $[uvw]$ is found by requiring that this plane be parallel to $[uvw]$, that is, that the plane passing through the origin must contain the point ua, vb, wc. This gives $hu+kv+lw=0$.

Choice of reference system. It is useful to consider the stereographic and gnomonic projection of the reference system (Fig. 6). The c axis is vertical, that is, perpendicular to the projection plane. The poles $a(100)$, $b(010)$, $c(001)$, $o(111)$ define a number of zones, and the representative circles are easily constructed. Their indices, as well as those of the planes defined by their intersection, are found by the rules given earlier. In this way 26 planes having indices which are combinations of 0, 1, and $\bar{1}$ are obtained. The zone circles [100], [010], [001] divide the projection into four quadrants. Each intersection of one of these with the reference circle as well as an interior point having indices formed by 1 or $\bar{1}$ is the intersection of three zone circles. Such a pattern is referred to as the basic pattern in the discussion that follows. The angles α, β, γ and the ratios $a{:}b{:}c$ can be determined graphically.

With the method of gnomonic projection (as in Fig. 7), indices can be assigned to a plane such as e by the following construction. Zones [100] and [010] are considered as axes of a coordinate system. The coordinates of the gnomonic projection of the plane e are measured with the coordinates of (111) as units, that is, by the ratio OE/OA, OD/OB. It can be shown that, provided the c axis is vertical, the indices of this plane are (OE/OA, OD/OB, 1) or in the case under consideration (2, 1/2, 1) or (4, 1, 2). This graphical construction only works for simple indices; 1/3 can be easily distinguished from 1/4, but not from 1/7 and 1/8. In such cases, the indices must be calculated, but such small fractions are highly improbable provided the axes are well chosen. This choosing of the axes is a major problem. Symmetry elements and the related elements which are possible faces or edges are usually chosen as axis and unit plane, as explained in the earlier discussion of crystal symmetry. The axis of highest symmetry is chosen as the c axis and is oriented vertically except in the monoclinic system, where b is usually the vertical axis. Inspection of the symmetry elements of the different classes indicates plainly that it is only in the cubic system that enough symmetry elements are present to permit an unambiguous determination of the reference system, that is, three mutually perpendicular twofold axes and a plane perpendicular to a threefold axis as unit plane; further inspection shows that mutually perpendicular axes can be chosen except in the monoclinic ($\beta \neq 90°$) and triclinic systems ($\alpha \neq \beta \neq \gamma \neq 90°$).

This means that two different observers describing the same noncubic crystal will probably choose

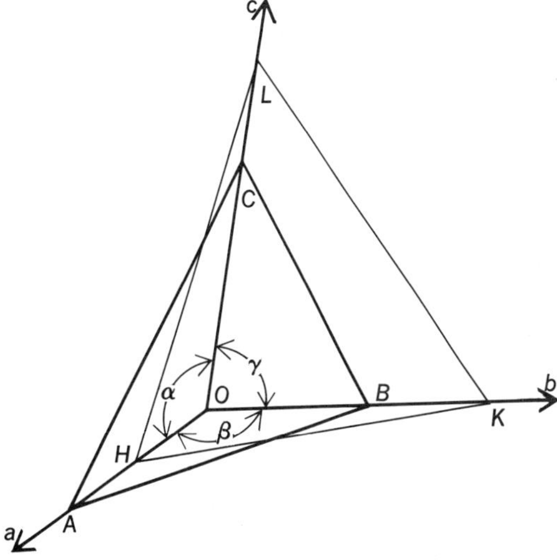

Fig. 5. The reference system used to define Miller indices; *ABC* is the unit plane.

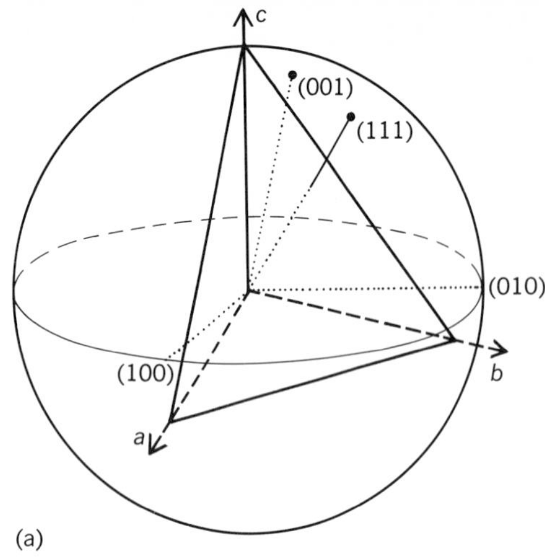

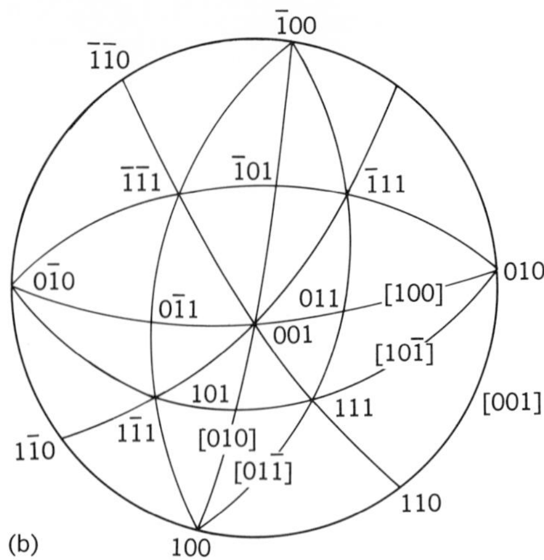

Fig. 6. Aspects of reference system. (*a*) Orientation of the reference system and poles of the axial and unit plane. (*b*) Projection. By the construction of great circles, four quadrants of type 001-010-100 are obtained.

CRYSTALLOGRAPHY

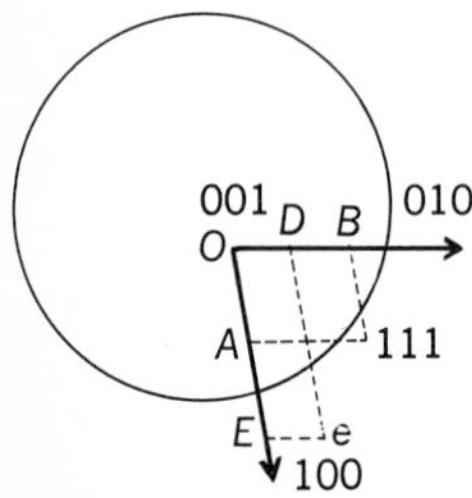

Fig. 7. Gnomonic projection of reference system, illustrating how indices can be graphically assigned to a plane *e*.

CRYSTALLOGRAPHY

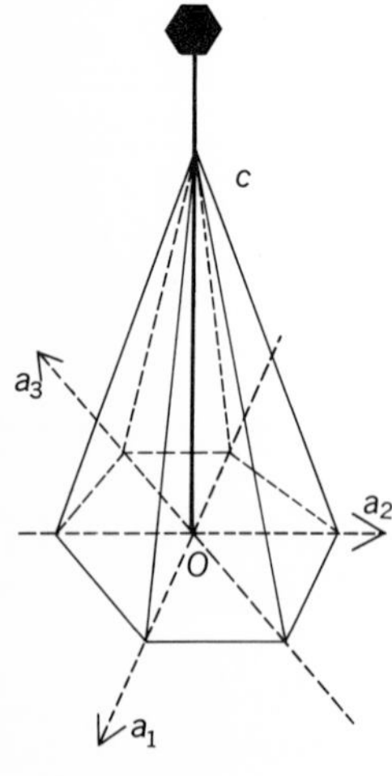

Fig. 8. Bravais indices in hexagonal system.

a different reference system, that is, orient or set the crystal differently. The same face of the crystal will consequently have different indices in the two settings. Different procedures have been worked out to obtain uniformity. The Barker method is the simplest and best known. It consists of a number of convenient rules which allow two observers to come independently to an identical choice. This is based on the consideration that the basic pattern considered earlier is the generalization of what exists in the cubic system. The main idea is therefore to find in a stereographic projection a quadrant as in a basic pattern and to choose the corners and an inside intersection of three zones as elements of the reference system. Rules are provided to distinguish between them, that is, to find out which pole will be denoted as *a*, *b*, *c*, or *o*, and what sense will be assigned to the axis. At the same time, one has also a maximum chance that a maximum number of planes will have simple indices. It is clear that the number of rules grows as the indeterminacy becomes greater. It is also clear that the rules have no physical meaning.

Bravais indices. These are frequently used in the hexagonal and trigonal systems because they reveal the symmetry more clearly than do other indices. In the hexagonal system, the axes a_1, a_2, a_3 are perpendicular to the *c* axis (Fig. 8) and make angles of 120° with each other. The four indices are denoted (*hkil*), but it is easy to prove that $h+k=\bar{i}$; therefore one may write (*hk.l*). Thus, *i* is a redundant index but is used because of the additional clarity which it yields in this system. The unit plane is $(1\bar{1}01)$.

The zone axes are found by considering the first two and the last indices as usual and inserting the sum of the first two with an opposite sign for the third one. As an example, the planes (0001) and $(1\bar{1}01)$ define the zone $[11\bar{2}0]$.

When Miller indices are used in the trigonal system, the three axes are chosen equally inclined around the threefold axis and making an equal angle α between them (Fig. 9). A plane normal to the symmetry axis is chosen as the (111) face. The system is therefore defined by α, $a{:}a{:}a$. Bravais indices (*hkil*) are easily transformed into Miller indices, here denoted (*pqr*), by the relations shown as Eq. (5).

$$\begin{aligned} p &= 2h + i - l \\ q &= -h + i + l \\ r &= -h - 2i + l \end{aligned} \tag{5}$$

The reference system used in the seven crystal systems is shown in Table 2.

Crystal form. The set of planes obtained by submitting a plane (*hkl*) to the symmetry operations of a point group is called a crystal form. All these planes together are written as {*hkl*}; they all have indices which are cyclic permutations of *h*, *k*, *l*, $\bar{h}$, $\bar{k}$, $\bar{l}$. The plane (*hkl*) can occupy a general or special position with respect to the symmetry elements; accordingly, one obtains a general or special form.

A crystal polyhedron can be described by using zone axes [*uvw*]; the group obtained is written <*uvw*>.

Practical description of crystals. In practice, a crystal description is made as follows. The crystal is measured; that is, the φ and ρ values of the faces are determined and are plotted in a stereognomogram. Zone circles are drawn, and the intersections give possible faces. The Barker rules are applied to find the reference systems. The indices are found graphically, as are all other needed elements. Some of them are calculated afterward if more accuracy is wanted. A crystal drawing deduced from the projection is also made. A description of the crystals measured is to be found in the

Table 2. Crystallographic reference systems

Crystal system	Angles formed by crystal axes	Defining ratio
Cubic	$\alpha=\beta=\gamma=90°$	$a{:}a{:}a$
Hexagonal	$\alpha=\beta=90°, \gamma=120°$	$a{:}a{:}c$
Tetragonal	$\alpha=\beta=\gamma=90°$	$a{:}a{:}c$
Trigonal	$\alpha=\beta=90°, \gamma=120°$	$a{:}a{:}c$
	$\alpha=\beta=\gamma\neq 90°$	$a{:}a{:}a$
Orthorhombic	$\alpha=\beta=\gamma=90°$	$a{:}b{:}c$
Monoclinic	$\alpha=\gamma=90°, \beta\neq 90°$	$a{:}b{:}c$
Triclinic	$\alpha\neq\beta\neq\gamma\neq 90°$	$a{:}b{:}c$

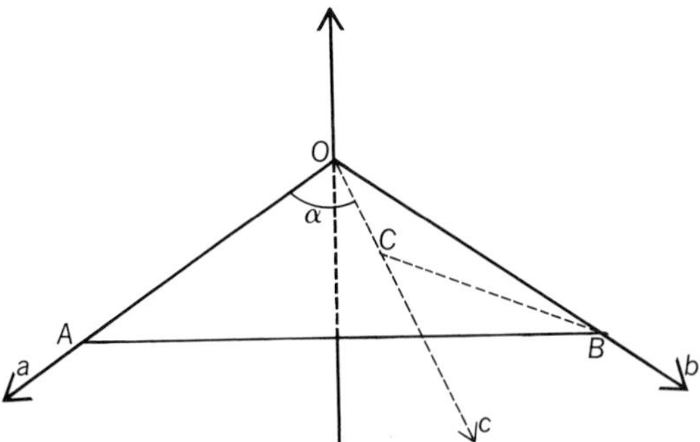

Fig. 9. Miller indices in trigonal system.

book by P. H. Groth listed in the bibliography. These have been revised and completed in the Barker index. [WILLY C. DEKEYSER]

Bibliography: L. V. Azároff, *Introduction to Solids*, 1960; T. V. Barker, *Graphical and Tabular Methods in Crystallography*, 1922; M. J. Buerger, *Elementary Crystallography*, 1956; P. H. Groth, *Chemische Krystallographie*, vols. 1–5, 1906–1919; F. C. Phillips, *Introduction to Crystallography*, 2d ed., 1956; M. W. Porter and R. C. Spiller, *The Barker Index of Crystals*, vols. 1 and 2, 1951, 1957; R. W. G. Wyckoff, *Crystal Structures*, 1948–1957.

Cummingtonite

The name given to the calcium-poor monoclinic amphiboles that form a limited solid-solution series between the iron end member grunerite, $Fe_7Si_8O_{22}(OH)_2$, and a magnesian end member of approximately $Mg_5Fe_2Si_8O_{22}(OH)_2$. More magnesium-rich minerals apparently form the closely related orthorhombic form, anthophyllite. The exact composition for the change of form and the equilibrium relations are unknown. The observed ranges of composition of cummingtonite and anthophyllite overlap considerably; the two minerals can occur in the same specimen. Cummingtonite is optically positive and thus distinct from the other monoclinic amphiboles; its inclined extinction distinguishes it from the very similar appearing anthophyllite. It is usually yellow-brown to brown and is slightly colored in thin sections. It forms elongate prismatic to fibrous crystals that exhibit the 54°30′ amphibole {110} cleavages, an angle approximately 1° smaller than that of the other monoclinic amphiboles. Grunerite commonly occurs as brown fibrous crystals with a silky luster and is usually found associated with iron formation.

Cummingtonite is a metamorphic mineral that occurs in various schists, in contact metamorphic rocks, and in some ore deposits. It is found associated with such minerals as hornblende, biotite, garnet, plagioclase, quartz, and anthophyllite. *See* AMPHIBOLE; ANTHOPHYLLITE.

[GEORGE W. DE VORE]

Cuprite

A mineral having composition Cu_2O and crystallizing in the isometric system. Cuprite is commonly in crystals showing the cube, octahedron, and dodecahedron (see illustration). More rarely, it is found in long capillary crystals known as chalcotrichite. It is also in fine-grained aggregates or massive.

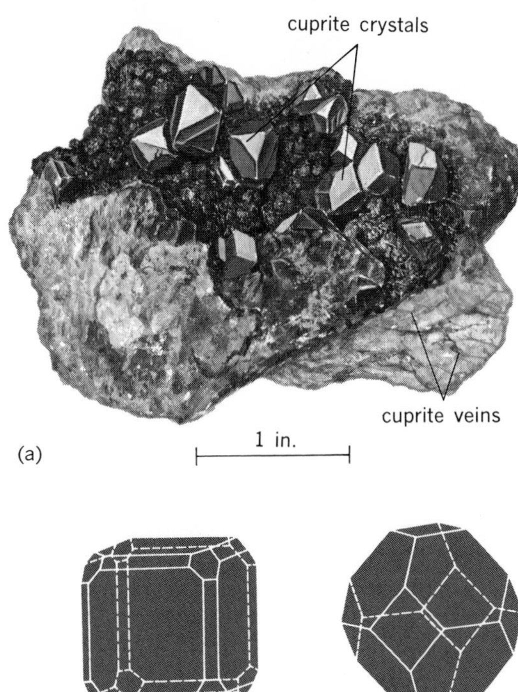

Cuprite. (*a*) Crystals and veins in limonite-bearing rock, Globe, Ariz. (*Specimen courtesy of Department of Geology, Bryn Mawr College*). (*b*) Crystal habits (*from C. S. Hurlbut, Jr., Dana's Manual of Mineralogy, 17th ed., copyright © 1959 by John Wiley & Sons, Inc.; reprinted by permission*)

Cuprite is various shades of red and a fine ruby-red in transparent crystals which have a metallic to adamantine luster. The hardness is 3.5–4 (Mohs scale) and the specific gravity is 6.1.

Cuprite is a widespread supergene copper ore found in the upper oxidized portions of copper deposits in which it is associated with limonite, malachite, azurite, native copper, and chrysocolla. Fine crystals have been found at Cornwall, England, and Chessy, France. In the early mining operations of many copper deposits cuprite has been an important ore mineral but gives way to the primary ores in depth. It has served as an ore in the Congo, Chile, Bolivia, and Australia. In the United States it has been found at Clifton, Morenci, Globe, and Bisbee, all in Arizona.

[CORNELIUS S. HURLBUT, JR.]

Cyclothem

A specific sequence of different kinds of rocks, one of which is generally coal. These sequences, each on the order of tens of feet in thickness, occur repeatedly, one above the other, in coal-bearing strata. Cyclothems have been recognized mainly in rocks of late Paleozoic age in the eastern United States but also have been described in late Mesozoic coal-bearing strata of the Rocky Mountain region. The illustration shows a diagram of a composite cyclothem of western Illinois. The diagram is idealized in the sense that not all of the different kinds of rocks shown occur in each cyclothem of that region. Cyclothems in other areas depart significantly from those in Illinois. In Iowa, Kansas,

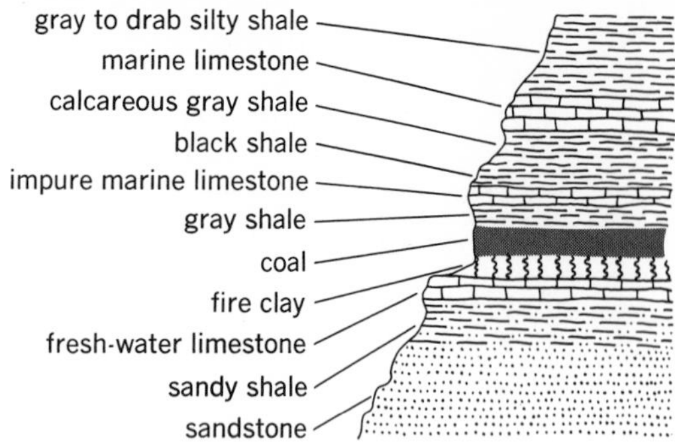

Cyclothem showing the complete succession of members that are recognized in Illinois.

and Nebraska, sandstones and coals are thinner and, in many places, absent, but limestones containing marine fossils are abundant. In the Appalachian region sandstones and coal beds are thicker and more numerous than in Illinois, whereas marine limestones or even shales containing marine fossils are relatively rare.

The lateral extent and boundaries of cyclothems are a subject of debate. Some geologists claim that cyclothems are very widespread, extending across the entire eastern half of the United States; others, while admitting lateral continuity of some rock units in smaller areas, believe that many cyclothems are mainly local deposits. Similarly, some authorities state that the boundary of the cyclothem should be placed at the base of the sandstone shown in the diagram because they believe that that sandstone rests on an eroded surface of great lateral extent. Others think that the eroded surface at the base of the sandstone is only a local phenomenon and that the boundary should be placed at the coal bed or at the top of the fireclay.

Because there has been no real agreement about factual knowledge concerning cyclothems, there is no agreement about their origin. Those who advocate the notion of widespread cyclothems limited at top and bottom by regional erosion surfaces are divided in opinion concerning their origin. Some believe that the erosion surface represents an episode of uplift of the land, and the succeeding deposits a period of subsidence, resulting first in nonmarine deposition and then marine deposition. Others believe that the land surface may have remained stable but that the sea level rose and fell in response to the growth and melting of continental ice sheets in the Southern Hemisphere. Some support for this notion is found in the late Paleozoic glacial deposits recognized in portions of South Africa and Australia. At the time when cyclothems were being formed, these now isolated areas are believed to have been joined in a single continent designated Gondwana.

Geologists who do not agree that cyclothems are deposits of regional extent are inclined to favor combinations of (1) local subsidence influenced by differential compaction of underlying sediments, (2) regional periodic subsidence, and (3) differential rates of local sediment influx due to shifting of stream patterns such as are observed on modern alluvial plains and coastal regions. *See* SEDIMENTATION; STRATIGRAPHY.

[JOHN C. FERM]

Bibliography: P. McL. Duff et al., *Cyclic Sedimentation*, 1967; J. C. Ferm, in E. D. McKee et al., (eds.), *Paleotectonic Investigations of the Pennsylvanian System in the United States*, U.S. Geol. Surv. Prof. Pap. no. 853, 1975; D. F. Merriam (ed.), *Symposium on Cyclic Sedimentation*, Kans. Geol. Surv. Bull. no. 169, 1964; H. R. Wanless, Late Paleozoic cycles of sedimentation in the United States, *18th International Geological Congress Report*, pt. 4, pp. 17–28, 1950; J. M. Weller, Cyclic sedimentation of the Pennsylvanian period and its significance, *J. Geol.*, 38:97–135, 1930.

Dacite

Aphanitic (very finely crystalline or glassy) rock of volcanic origin, composed chiefly of sodic plagioclase (oligoclase or andesine) and free silica (quartz or tridymite) with subordinate dark-colored (mafic) minerals (biotite, amphibole, or pyroxene). If alkali feldspar exceeds 5% of the total feldspar, the rock is a quartz latite. As quartz decreases in abundance, dacite passes into andesite. Thus, dacite is roughly intermediate between andesite and quartz latite. *See* ANDESITE.

[CARLETON A. CHAPMAN]

Dating methods

Methods and techniques used in archeology, biology, and geology to fix dates, assign periods of time, and determine age. The uranium-lead, rubidium-strontium, potassium-argon, and carbon-14 methods have yielded reliable results on suitable samples and have permitted the construction of absolute geologic history in many areas of the world. *See* EARTH, AGE OF; GEOLOGICAL TIME SCALE; LEAD ISOTOPES, GEOCHEMISTRY OF; METEORITE; RADIOCARBON DATING; ROCK, AGE DETERMINATION OF.

For descriptions of other methods and techniques used to establish the sequence of events, the succession of strata, and relative chronologies *see* GEOCHRONOMETRY; GEOLOGY; INDEX FOSSIL; MARINE SEDIMENTS; PALEOBIOCHEMISTRY; PALEOBOTANY; PALEONTOLOGY; ROCK MAGNETISM; STRATIGRAPHY; TREE-RING HYDROLOGY; UNCONFORMITY; VARVE.

[DAVID E. FOGARTY]

Datolite

A mineral nesosilicate, composition $CaBSiO_4(OH)$, crystallizing in the monoclinic system. It usually occurs in crystals showing many faces and having an equidimensional habit. It may also be fine granular or compact and massive. Hardness is 5–5½ on Mohs scale; specific gravity is 2.8–3.0. The luster is vitreous, the crystals colorless or white with a greenish tinge. Datolite is a secondary mineral found in cracks and cavities in basaltic lavas or similar rocks associated with zeolites, apophyllite, prehnite, and calcite. It is found in the Harz Mountains, Germany; Bologna, Italy; and Arendal, Norway. In the United States fine crystals have come from Westfield, Mass.; Bergen Hill, N.J.; and various places in Connecticut. In Michigan, in the Lake Superior copper district, datolite occurs in fine-grained porcelainlike masses which may be coppery red because of inclusions of native copper. *See* SILICATE MINERALS.

[CORNELIUS S. HURLBUT, JR.]

Delta

A deposit of sediment at the mouth of a river or tidal inlet. The name, from the Greek letter Δ, was first used by Herodotus (5th century B.C.) for the triangular delta of the Nile. It is also used for storm washovers of barrier islands and for sediment accumulations at the mouths of submarine canyons. *See* FLOODPLAINS; NEARSHORE PROCESSES.

Structure and growth. The shape and internal structure of a delta depend on the nature and interaction of two forces: the sediment-carrying stream from a river, tidal inlet, or submarine canyon, and the current and wave action of the water body in which the delta is building. This interaction ranges from complete dominance of the sediment-carrying stream (still-water deltas) to complete dominance of currents and waves, resulting in redistribution of the sediment over a wide area (no deltas). This interaction has a large effect on the shape and structure of the delta body.

Most of the sediment carried into the basin is deposited when the inflowing stream decelerates. If there is little density contrast, this deceleration is sudden and most sediment is deposited near the mouth of the river. If the inflowing water is much lighter than the basin water, for example, fresh water flowing into a colder sea, the outflow spreads at the surface over a large distance away from the outlet. If the inflow is very dense, for instance, cold muddy water in a warm lake, it may form a density flow on or near the bottom, and the principal deposition may occur at great distance from the outlet. A good example is Lake Mead, where 134 ft of sediment have been deposited against Hoover Dam at 75 mi from the inlet of the Colorado River. However, not all long-distance transport of river sediment in the sea can be attributed to deep density flow; almost half of the total load of the Amazon River is deposited more than 1000 mi away near the Orinoco delta and in the southeastern Caribbean because of longshore transport by currents and waves.

Three principal components make up the bodies of most deltas in varying proportions: topset, foreset, and bottomset beds. They were defined originally by G. K. Gilbert for lake and shallow marine deltas; there, sizable differences in the depositional slopes of the three units, which form part of the definition, are readily seen (Fig. 1). The concept has been redefined to apply to most deltas, and as now understood, the topset beds comprise the sediments formed on the subaerial delta: channel deposits, natural levees, floodplains, marshes, and swamp and bay sediments. The foreset beds are those formed in shallow water, mostly as a broad platform fronting the delta shore, and the bottomset beds are the deep-water deposits beyond the deltaic bulge. In marine deltas the fluviatile influence decreases and the marine influence increases from the topset to the bottomset beds. Concurrently, the mean grain size and lithological variability also decrease. The topset beds are variable in grain size and composition over short distances, ranging from coarse channel deposits through layered levee silts and clays to the fine clays of the floodplains and marshes. The foreset beds consist of a uniform blanket of laminated silts and silty clays reflecting the seasonal influence of currents and waves on the shallow platform, while the bottomset beds are homogeneous silty clays,

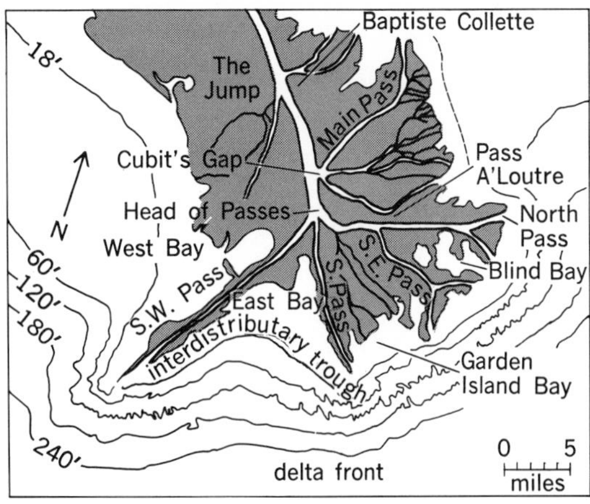

Fig. 2. Modern Mississippi bird-foot delta. (*From H. N. Fisk et al., Sedimentary framework of the modern Mississippi delta, J. Sediment. Petrol., 24:76–99, 1954*)

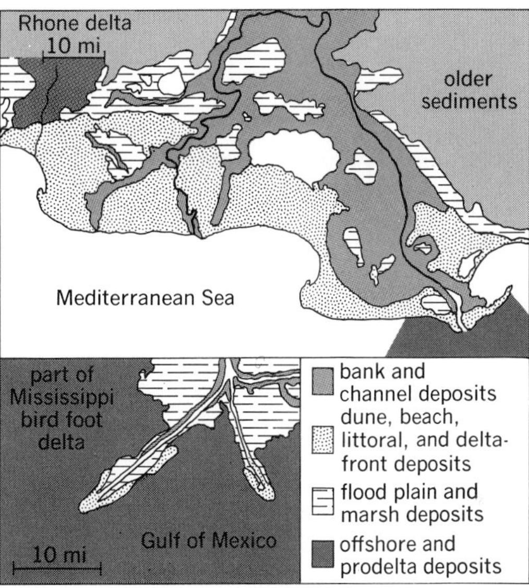

Fig. 3. Examples of deltaic sediment distribution.

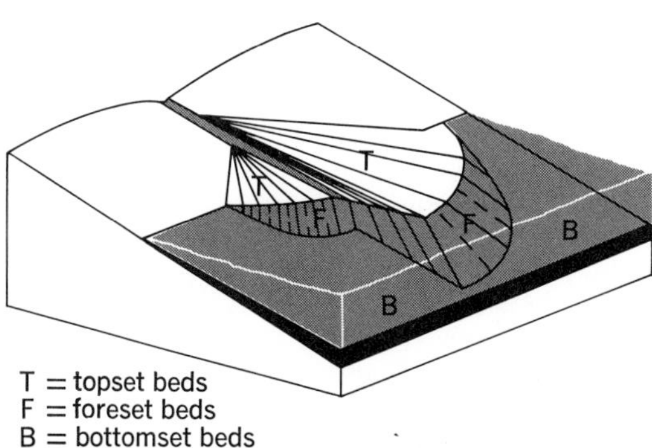

Fig. 1. Schematic diagram showing two stages of growth of a typical Gilbert-type delta. (*Adapted from P. H. Kuenen, Marine Geology, copyright © 1950 by John Wiley & Sons, Inc.; reprinted by permission*)

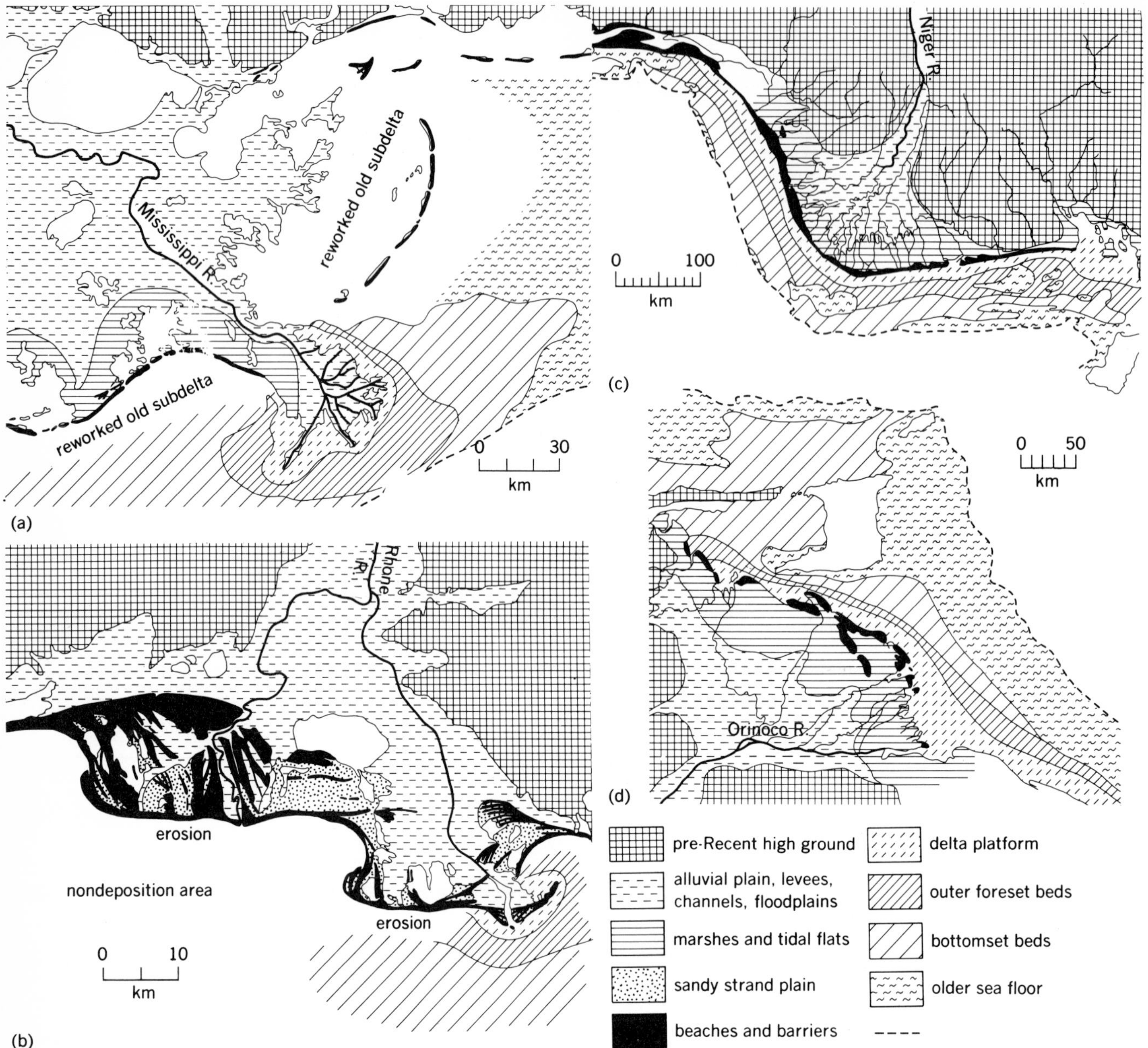

Fig. 4. Sketch maps of the sediments of the (*a*) Mississippi, (*b*) Rhone, (*c*) Niger, and (*d*) Orinoco deltas. (*After T. H. van Andel, The Orinoco delta, J. Sediment. Petrol., 37(2):297–310, 1967*)

often characterized by abundant plant fragments. The boundary between bottomset beds and neritic shelf clays is completely arbitrary. Similarly, the faunas incorporated in the sediments range from great local variability in the topset beds, with dominance of barren sediments and local pockets of rich brackish faunas, to the almost completely marine fossils uniformly distributed in the bottomset beds.

Through progressive outbuilding, the delta can become overextended with long river courses and very low gradients. Eventually, shorter and steeper paths to the sea will be developed and the existing subdelta will be abandoned in favor of a shorter course. The Mississippi delta shows good examples of such subdelta migration, and the present active delta would be abandoned for a new one off the Atchafalaya River, if artificial control did not keep the flow in check. The abandoned delta body gradually submerges and is eroded by waves and wind. The submergence results from two factors: temporary continuation of subsidence caused by loading of the deltaic sediments on the substrate, and compaction of the fine sediments. These are overcompensated during delta growth, but cause it to sink below sea level after abandonment. Winnowing of the sediments by waves produces a lag deposit of sand at the advancing edge of the sea, which ultimately results in a thin and discontinuous blanket of sand, often bordered by a string of sand bars and small islands near the previous delta shore. This sequence is called the destructive phase of delta formation; a good example is the Breton Sound-Chandeleur Island area of the Mississippi delta, which marks the site of the old St. Bernard subdelta.

In a different way, deltas can be viewed as being composed of three structural elements: (1) a framework of elongate coarse bodies (channels, river-mouth bars, levee deposits), which radiate from the apex to the distributary mouths (sand fingers); (2) a matrix of fine-grained floodplain, marsh, and bay sediments; and (3) a littoral zone, usually of beach and dune sands which result from sorting and longshore transport of river-mouth deposits by waves, currents, tides, and wind. The relative proportions of these components vary widely. The Mississippi delta consists almost entirely of framework and matrix (Fig. 2; Fig. 3, bottom); its rapid seaward growth is the result of deposition of river-mouth bars and extension of levees, and the areas in between are filled later with matrix. This gives the delta its characteristic bird-foot outline. A different makeup is presented by the Rhone delta (Fig. 3, top), where the supply of coarse material at the distributary mouths is slow, and dispersal by wave action and longshore drift fairly efficient, so that nearly all material is evenly redistributed as a series of coastal bars and dunes across a large part of the delta front. This delta advances as a broad lobate front, while the present Mississippi delta grows at several localized and sharply defined points.

Pattern of deposition. Many different types of deltas exist, few of which have been examined in detail. It appears that most of them can be explained by the interplay, in various proportions, of two major controlling forces: the coarseness and the rate of sediment supply by the river, and the rate of reworking and dispersal along the delta front by marine currents and waves. Other factors, such as the shape of the preexisting coastline and of the sea floor, seasonal fluctuations of marine and river forces, the tidal range, and so on, cannot as yet be evaluated, but appear to be secondary. The Mississippi delta (Fig. 4*a*) is dominantly fine-grained and consists in large part of thick sand fingers with intervening matrix clays. It builds rapidly outward, is almost entirely controlled by river action, and has a typical bird-foot shape.

In the Rhone delta (Fig. 4*b*) the rate of sediment supply is much lower, but the sediment is coarser, and the rate of reworking and dispersal by waves and currents is higher. Consequently, this delta is more sandy, and its principal growth takes place through the accretion over a broad front of bars and beaches, resulting from winnowing of sand at the river mouth and dispersal by wind and waves. The advance of the delta is slow, but occurs over a broadly lobate front, and the bulk of the delta body consists of the deposits of the sandy strand plain. Matrix and framework deposits are restricted to the topset beds in the upper delta. Littoral marine forces control to a large extent the shape of this delta.

A broad coastal plain is also found in the Orinoco delta (Fig. 4*d*). The large supply of very fine sediment of this river is augmented by Amazon mud brought along the shore from the east. It builds rapidly in a sheltered sea with little wave action, but longshore drift to the northwest is marked. Consequently, the coastal plain, although also littoral marine, is not wave-built and sandy, but formed by tidal depositon on mud flats. The delta, as a result of the longshore currents, advances over its entire perimeter, and the action of the river plays an insignificant part in its development.

Delta advance over the entire perimeter also takes place in the Niger delta (Fig. 4*c*), which resembles the Orinoco delta in shape and distribution of sediments. However, enough sediment is supplied by this river to form, under the influence of moderate wave action on an exposed shelf, a marginal zone of sand bars, beaches, and barrier islands. Behind this barrier, tidal sedimentation of a muddy coastal plain takes place, and framework-matrix units are restricted to the upper delta. The delta advances over a broad front as a result of redistribution of sediments by currents. In a sense it is intermediate between the Orinoco and Rhone deltas.

Characteristics of modern deltas. While over 150 major deltas are formed today, not all rivers, or even all major ones, have deltas. This is the result of a rise in sea level following the last glacial period, which produced deep estuaries in many parts of the world which have not yet been filled (for example, the Amazon estuary). Delta thick-

Statistics pertaining to modern deltas

River	Dimension of subaerial delta, statute mi		Amount of sediment discharged		Annual extension of subaerial delta	
	Length	Breadth	River water by weight (avg), ppm	Annual volume of sediment, mi³	Measurement period, years	Approximate distance, ft
Mississippi's present bird-foot delta	12	30	550	0.068	1838–1947	250
Hwang-Ho	300	470*	50,000†		1870–1937	950
Ganges-Brahmaputra	220	200	870	0.043 (Ganges only)		
Rhone into Mediterranean Sea	30	47	400–590	0.005	1737–1870	190
Danube	46	46	310	0.008		40
Nile (prior to barrages)	96	145	1600	0.001	1100–1870	45
Colorado above Hoover Dam	43	0.05–0.6	8300	0.032	1936–1948	3.6 mi (gorge)
Euphrates-Tigris	350	90			1793–1853	180

*Includes 100 mi of nondeltaic Shantung Peninsula.
†Maximum is 400,000 ppm.

nesses vary widely: The Nile is depositing a layer 50 ft thick in a shallow embayment, whereas the Mississippi, building out in deeper water, has constructed a delta body more than 850 ft thick. This thick wedge exerts pressure on the underlying beds, causing downbending of the shelf and producing mudlumps in areas where the underlying sediment is soft (see table).

Engineering problems. Despite difficult engineering problems, many cities, such as Calcutta, Shanghai, Venice, Alexandria (Egypt), and New Orleans, were constructed on deltas. These problems include shifting and extending shipping channels; lack of firm footing for construction except on levees; steady subsidence, which may reach a rate of 5 ft per century; poor drainage; and extensive flood danger. Submergence to a depth of 15 ft during hurricanes or typhoons is not uncommon. Moreover, in certain deltas the tendency of the main flow to shift away to entirely different areas, with resulting disappearance of the main channels for water traffic, is a constant problem that is difficult and costly to counter. *See* MARINE SEDIMENTS. [TJEERD H. VAN ANDEL]

Bibliography: C. C. Bates, Rational theory of delta formation, *Bull. Amer. Ass. Petrol. Geol.*, 37(9):2119–2162, 1953; H. N. Fisk et al., Sedimentary framework of the modern Mississippi delta, *J. Sediment. Petrol.*, 24:76–99, 1954; P. C. Scruton, Delta building and the deltaic sequence, in F. P. Shepard (ed.), *Recent Sediments, Northwestern Gulf of Mexico*, Amer. Ass. Petrol. Geol. Spec. Publ. 82–102, 1960; T. H. van Andel, The Orinoco delta, *J. Sediment. Petrol.*, 37(2):297–310, 1967.

Desert erosion features

A distinctive topography carved by erosion in regions of low rainfall and high evaporation where vegetation is scanty or absent. Although rainfall is low, it is the most important climatic factor in the formation of desert erosion features. Desert rains commonly occur as torrential downpours of short duration with a consequent high percentage of runoff. As a result of the dryness, wind and mechanical weathering also play an important part in desert erosion. *See* WEATHERING PROCESSES; *see also* SEDIMENTATION.

DESERT EROSION FEATURES

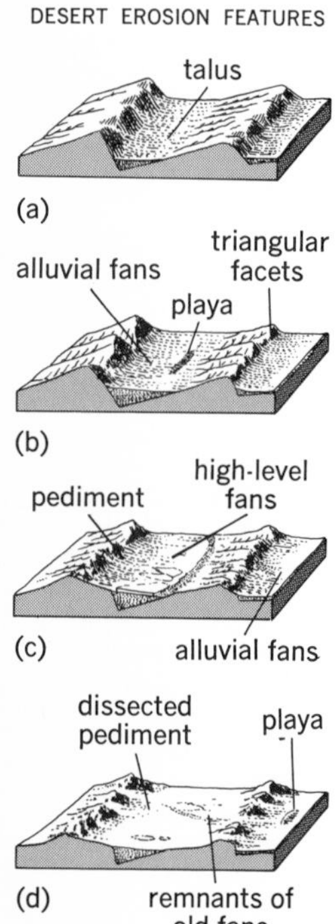

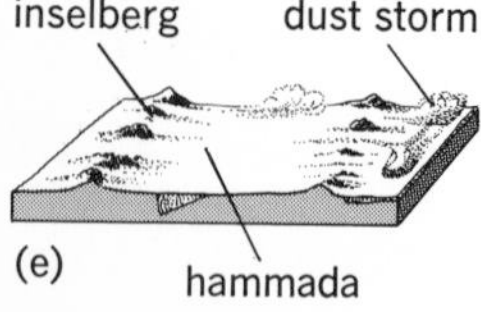

Fig. 1. (*a–e*). Series of block diagrams showing a sequence of landforms in an arid climate. (*From P. E. James, An Outline of Geography, copyright 1935 by Ginn and Co.*)

Erosion agents. The principal agents of erosion in deserts are the atmosphere, running water, and wind.

Weathering involves both mechanical and chemical processes. Since chemical processes require moisture, mechanical weathering predominates in the desert, although chemical action is not altogether lacking. Rocks are broken by unequal expansion and contraction of constituent minerals and by unequal heating and cooling of outer and inner layers. These processes are aided during the rare periods when moisture is available by the swelling of some minerals as they become hydrated or oxidized, and in some localities by crystallization of wind-blown salts in cracks. Frost-wedging may also take place in the winter months when there is a combination of rare rain with freezing temperatures.

When storms of the so-called cloudburst type occur in the desert, sudden rushes of water, or flash floods, sweep down the normally dry washes or the narrow canyons in the mountains bordering the basins. The comparatively large volume of water combined with a high velocity due to the steepness of the slopes give the short-lived streams power to carry large amounts of fine and coarse rock fragments. As a result, the streams have great erosive power.

When intermittent streams leave the canyons and spread out at the foot of a desert mountain, they lose velocity and quickly drop the coarsest of the transported material to build an alluvial fan. Some of the water sinks into the fan, and some evaporates, but whatever remains may follow one of the channels on the fan or spread out in the form of a sheetflood, in either case carrying coarse sand, silt, and clay, and perhaps rolling some larger rock fragments along.

When the water reaches the toe of the fan, it spreads still more, dropping all but the finest silt and clay. Any excess water follows shallow washes to the lowest part of the basin, where it may form a playa lake. This evaporates in a few hours or a few days, depositing the silt and clay, mixed perhaps with soluble salts. The flat-surfaced area resulting from the silt and clay deposition is a playa. *See* PLAYA.

A variation of the action of running water occurs if a large accumulation of completely disintegrated material becomes thoroughly water-soaked by a sudden hard rain and moves down a canyon and out on the fan as a mudflow. Because of the high viscosity and density of the mass of mud and water, large boulders may be carried or rolled considerable distances.

The lack of moisture during most of the year and the scanty vegetation make the wind a more potent agent of erosion in deserts than in humid lands. The finest material is blown high in the air and may be carried entirely out of the area, a process known as deflation. The larger sand grains are rolled along the surface, bouncing into the air when they strike an obstacle, knocking more grains into the air as they hit the ground again, until eventually a sheet of sand is moving along in the 3 or 4 ft above the surface. This moving sand abrades rocks and other objects with which it comes in contact; at the same time the grains themselves become rounded and frosted. If movement is impeded by vegetation or other obstacles, sand accumulates to form dunes.

Erosion cycle. Some knowledge of the erosion cycle in arid regions seems necessary to an understanding of the formation of the distinctive erosion features of deserts, as well as the relationships between them.

Youthful stage. During initial development the bold mountain ranges in or bordering desert areas become gashed by steep canyons and shed waste into the adjoining basins or lowlands as erosion is accelerated during the infrequent but violent rainstorms. Alluvial fans are built, washes develop, playas form, and the basins slowly fill with detritus (Fig. 1). As this stage progresses, some alluvial fans coalesce to form bajadas or piedmont alluvial plains along the mountain fronts, and individual basins may become deeply filled with waste to form bolsons. Desert flats develop between alluvial fans (or bajadas) and playas, and isolated dunes accumulate on the lee sides of the latter. If the original highlands are flat-topped rather than tilted mountain blocks, mesas develop (Fig. 2).

As the mountain fronts slowly retreat under the attack of the atmosphere and running water, small bare rock surfaces or pediments form at the canyon mouths, the result of lateral cutting by the intermittent streams. The pediments increase in size in the late part of the stage. The general tendency during youth is for relief to decrease.

Mature stage. The middle stage is initiated by the development of exterior drainage or the capture of higher basins by lower ones as drainage channels erode headward through low divides (Fig. 1*c* and *d*). The fill deposited during youth undergoes erosion, and pediments become more widely developed, cut not only on the bedrock of the original mountain blocks but also upon the deposits of the captured basin.

The mountains are worn still lower, and more and more channels extend completely through them, cut by the streams engaged in draining and dissecting the higher basins. Playa deposits or other easily eroded sediments are cut into badlands before being entirely removed, and mesas are reduced to buttes. Undissected remnants of older deposits become covered with desert pavement. Where winds are turbulent and large supplies of sand are available, complex dune areas or even great ergs develop. Relief shows some net increase during maturity.

Old-age stage. The original mountains are so reduced in elevation that the winds sweep over them with little or no condensation of moisture, and rains become still more infrequent. Great expanses of wind-scoured bare rock or hammada are exposed, with here and there a more resistant remnant standing above the general level as an inselberg (Fig. 1*e*). Buttes are reduced to bornhardts and finally disappear.

Those parts of the flat surface floored by earlier deposits are covered and protected by extensive areas of desert pavement. The rock fragments may be colored brown to black by desert varnish, a coating of manganese and iron oxides.

Sand blown from the bare rock surfaces and from the sediments may form large dune areas. If there are no obstacles to obstruct movement or cause wind turbulence, the sand may move as a sheet, forming large expanses of flat or gently undulating sand surfaces. Relief slowly decreases in old age.

The final result of desert erosion, as of erosion under humid conditions, is the peneplain. While such a surface is theoretically possible, it is doubtful that one has been attained anywhere during recent geologic time. *See* FLUVIAL EROSION CYCLE.

Physiographic features. There are a number of physiographic features characteristic of desert erosion and deposition.

Alluvial fan. Where intermittent streams flow down steep canyons in mountains bordering desert areas, alluvial fans are formed. As the streams suddenly lose velocity on emerging from the canyons at the mountain front, they drop most of their load, building a fan-shaped deposit. Such fans consist of a rudely cross-bedded mixture of coarse and fine rock fragments, largely subangular.

Badlands. The intricate dissection of relatively fine-grained, more or less horizontally bedded, poorly consolidated sediments results in badlands, which are characterized by sharp-edged, sinuous ridges separated by steep-sided, narrow, winding gullies.

Fig. 2. Present desert floor shown in the foreground with remnants of higher structures in the background. From left are a small mesa, a pinnacle remnant, and a butte. (*W. T. Lee, U.S. Geological Survey*)

Bajada. A bajada is formed as the result of lateral growth of adjacent alluvial fans until they finally coalesce to form a continuous deposit along a mountain front.

Bolson. A desert basin of interior drainage which is almost filled by waste from the surrounding mountains is called a bolson.

Bornhardt. The last remnant of a once-elevated area, a bornhardt is reduced to small dimensions by almost equal backweathering on all sides.

Butte. The erosion, under arid conditions, of a flat-topped surface of soft sediments protected by a resistant cap forms a relatively small remnant of a few acres called a butte. Its sides are steep, approaching the vertical, and may be some hundreds of feet high.

Deflation. Fine material is blown completely out of a desert region by wind, an erosive process called deflation, which results in lowering of the surface.

Desert flat. A large part of a desert area may consist of desert flats, which are essentially level surfaces extending from the edge of a playa to the alluvial fans or bajadas bordering a basin.

Desert pavement. When the finer particles have been removed by deflation, the coarser materials form a desert pavement. This mosaic of flat-lying, interlocking, angular-to-subrounded stones is left as a protective covering over the remainder of the fine material on the desert floor.

Dry wash. The bed of an intermittent stream in arid or semiarid regions, a dry wash (wadi) is generally flat-bottomed; its sides are usually vertical, ranging from a few feet in height to over 100 ft. Arroyo is the term used for a deeper wash.

Hammada. Ordinarily, a hammada is a bare rock surface composed of relatively flat-lying consolidated sedimentary rocks from which overlying softer sediments have been stripped, principally by wind erosion. Hammadas may also be extensive surfaces cut on bedrock by means of protracted desert erosion.

Inselberg. A resistant remnant of a former mountain mass rising above the general level of an almost flat, bare rock surface, an inselberg is formed by stream planation. An outlying peak almost buried by alluvial deposits is sometimes called an inselberg.

Mesa. A large flat-topped surface with an area of a few to many square miles, and bounded by steep to nearly vertical sides, is called a mesa.

Pediment. A pediment is a piedmont slope much like a bajada but formed from a combination of processes which are mainly erosional. The surface is chiefly bare rock but may have a covering veneer of alluvium or gravel. A pediment may be formed at the mouth of a canyon by lateral cutting of an intermittent stream as it swings back and forth seeking new channels at the head of an alluvial fan; the cutting may be aided by sheetwash. A pediment also may be formed when gradients have been reduced by the filling of a basin or when a higher basin is captured either by a lower basin or by exterior drainage. Deposits of the captured basin are beveled by headward erosion of the capturing drainage. One type of pediment may grade into the other.

[THOMAS CLEMENTS]

Bibliography: T. Clements et al., *A Study of Desert Surface Conditions*, U.S. Army Tech. Rept. no. EP-53, 1957; E. F. Gautier, *Sahara: The Great Desert*, 1935; E. C. Jaeger, *North American Deserts*, 1957; A. S. Leopold and *Life* (eds.), *The Desert*, 1961; G. Pickwell, *Deserts*, 1939; P. B. Sears, *Deserts on the March*, 1935.

Devonian

The fourth period of the Paleozoic Era, usually taken as the base of the upper Paleozoic. The period lasted about 50,000,000 years, approximately from 400,000,000 to 350,000,000 years ago, but limiting dates are not precisely agreed upon. The period is especially noted for the widespread occurrence of primitive fish, and it is sometimes called the Age of Fishes. During the Devonian, vascular plants became abundant for the first time and the earliest known forests appeared. The earliest known insects are Devonian in age, and the first tetrapods occur at about the close of the period.

The Devonian was first named by Adam Sedgwick and Roderick Murchison in 1839 for marine rocks in Devon and Cornwall in southwest England, and rocks of equivalent age were soon recognized throughout the world. In Europe there are excellent sequences in the Ardennes, Rhenish Schiefergebirge, and Harz Mountains. Devonian deposits occur widely in Asia. In the Soviet Union, they are known underground across the Russian platform, and they crop out in the Leningrad area and the Urals. The Devonian succession in New York is the finest in North America and was first described by James Hall in 1842 (Fig. 1). Devonian rocks are widespread in the Midwest, and occur in many areas of the Rocky Mountains and Cordilleran belt of western North America. Rocks of Devonian age are also found in regions of South America, Antarctica, Australia, and southern and northern Africa.

Rocks which formed in terrestrial conditions, including desert sandstones, lake deposits, and alluvial plain sediments, are common in some areas, and often have a distinctive red coloration. These were first described in detail and named by Hugh Miller in 1841, and Miller's term "Old Red Sandstone" is widely used for the rocks of continental facies such as are well developed in Scotland, Greenland, and parts of the Canadian arctic islands and northeastern North America.

Economic deposits. Since Devonian rocks comprise a wide range of rock types and chemical compositions, and since they crop out over wide areas, they are of considerable local economic importance. On several continents Devonian rocks provide building stones, cement rocks, brick clays, roofing slates, glass sands, salt, anhydrite, and abrasives. Quarries in Michigan, working Devonian limestones, are among the largest limestone quarries in the world. Sedimentary iron ore was formerly important in Germany, and lodes of metalliferous ores are common especially in altered Devonian rocks.

Oil and gas have been obtained from Devonian rocks in Pennsylvania and New York since the last century. In 1944 large discoveries of hydrocarbons were made in Devonian sandstones at depth in the Ural-Volga region of the Soviet Union, and other Soviet discoveries subsequently have been made in other areas. In 1947 oil was found in a buried Devonian carbonate reef at Leduc in Alberta, and other discoveries followed. Devonian evaporites, especially rock salt and anhydrite, are important in many areas; those discovered in recent years in Saskatchewan, for example, provide enormous potash reserves.

Paleogeography. In Devonian times the continents were quite differently arranged than they are today (Fig. 2). Paleomagnetic and geologic evidence indicates that Europe, Greenland, and North America were joined together as a single continent called Laurasia, which was tropical in position. Similarly, the present southern continents of India, Australia, Antarctica, Africa, and South America were united as a single huge continent called Gondwanaland. The southern pole apparently lay in the region of southern Africa or southern South America with Gondwanaland situated such that North Africa and Australia were closest to the Equator. Marine waters covered large areas of these continents and hence, unless the Earth's radius was smaller, a very much larger area of the Earth was covered by seas than at present. *See* CONTINENT FORMATION; PALEOMAGNETICS.

Certain belts of thick marine sedimentation occur in the Devonian, forming geosynclinal tracts. These were mostly established before Devonian times and continued afterward. Such belts include the Hercynian or Variscan geosyncline of Europe, which stretched from Ireland and France, through France, Belgium, and Germany, and continued at

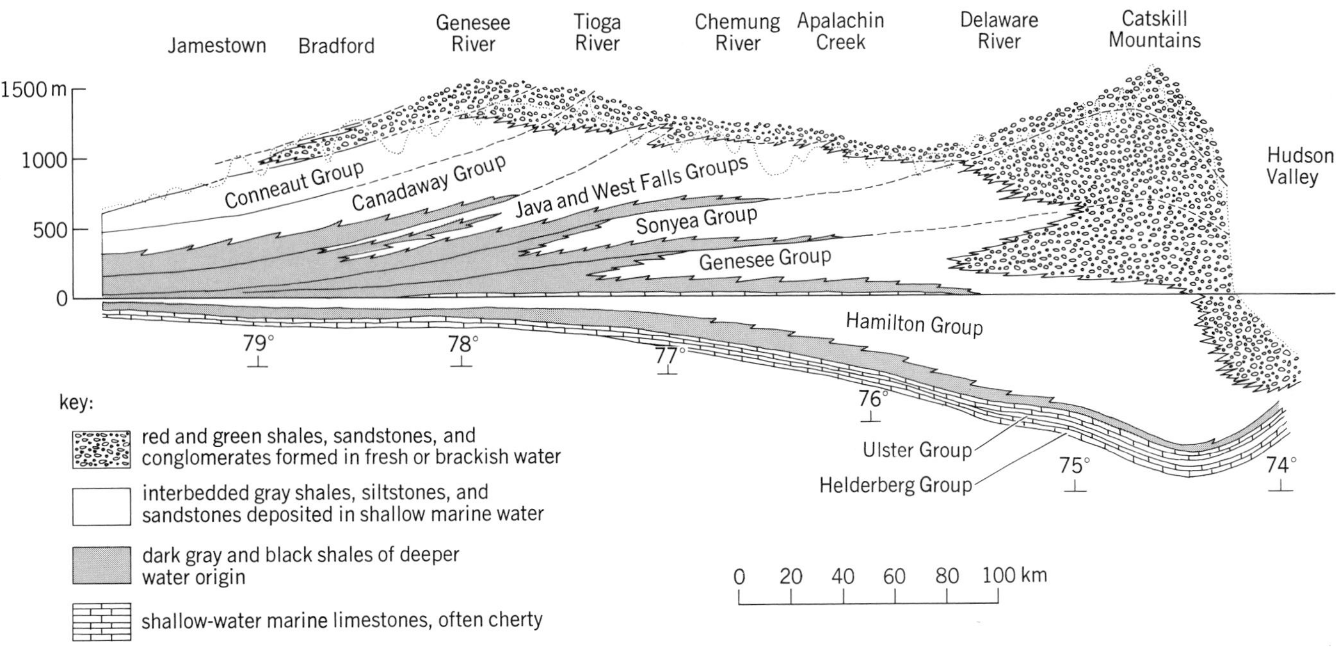

Fig. 1. East-west section of Devonian strata along the New York–Pennsylvania state line showing the marked easterly thickening of units toward the Catskill Delta and the progressive facies change from offshore sediments in the west to continental clastic deposits in the east. (*From M. R. House, Facies and time in Devonian tropical areas, Proc. Yorkshire Geol. Soc., 40:272, 1975*)

least to southern Poland. This belt was folded in late Carboniferous times. The Ural geosyncline stretched from Novaya Zemlya southward through the Urals to Kazakhstan. More extensive tracts in central Asia, south of the Siberian platform, form the Angara geosyncline, which passes from the Tien Shan mountains through Mongolia to the Pacific coast at the Sea of Okhotsk. In southern Europe, Devonian rocks crop out in Spain and in ancient massifs within the Alpine belt, which stretches eastward to Turkey and Afghanistan. Rocks of this age also occur within the Tertiary fold belts from the Himalayas to Malaysia. *See* GEOSYNCLINE.

In North America, great thicknesses, perhaps formerly reaching 25,000 ft (7500 m), were laid down in the Appalachian geosyncline, but the successions thin to the west, toward the North American craton (Fig. 1). The Devonian rocks of the eastern part of the geosyncline were substantially folded, thrust, and often metamorphosed in the late Paleozoic; those of the west (in the present Appalachian Plateau) are still almost completely undisturbed. A period of uplift in mid-Devonian time, the Acadian orogeny, led to the influx of considerable clastic sedimentation of terrestrial type which formed red sandstones and conglomerates in the area of the Catskill Mountains known as the Catskill Delta. In contrast, in the central areas of North America, Devonian deposits are relatively thin, are mainly carbonates, and rest almost undisturbed on older rocks with disconformity, with the Lower Devonian mostly missing.

In western North America, a broad geosynclinal tract usually referred to as the Cordilleran geosyncline stretched from Alaska to Mexico. Today this is structurally limited to the east by the Rocky Mountain front, and the whole area was to a large extent affected by faulting and folding during the Mesozoic and Tertiary. A westerly belt in the geosyncline can be distinguished, characterized by thick, deep-water shales with volcanic rocks; this contrasts with an easterly belt in which the sequence is thinner and is largely composed of limestones and dolomites. These are referred to as the eugeosynclinal and miogeosynclinal belts, respectively. A localized period of folding and eastward overthrusting characterizing the area of southern California, central Nevada, and Idaho, termed the Antler orogeny, began in the Late Devonian and continued through the early Mississippian. Devonian rocks occur widely in the Canadian arctic islands.

In South America, Devonian rocks occur primarily in Colombia, Brazil, and Argentina, and mostly consist of clastic rocks of shallow-water type. The fossils in northern areas show affinities with the Appalachians, but those of more southern rocks show affinities with South Africa and Antarctica; these compose areas of the Malvinokaffric province or realm which are characterized by a distinctive brachiopod fauna and probably occupied a polar position in Devonian times. In South Africa, faunas of this type occur in the Bokkeveld.

Devonian rocks in Australia are found in the eastern belt of the Tasman geosyncline, which is thought to have been contiguous with the Transantarctic geosyncline. Rocks of Old Red Sandstone type occur in the Amadeus basin of central Australia. In Western Australia, Devonian rocks lie unconformably on lower Paleozoic or Precambrian rocks in the coastal areas of the Canning and adjacent basins, where spectacular reef carbonates occur. *See* PALEOGEOGRAPHY.

Climate. In the Devonian the world climate was probably somewhat warmer than at present. Despite the fact that paleomagnetic evidence indicates that the southern continents were grouped together with a southern pole in the region of Argentina, there is no evidence for more than limited glacial deposits. The deduction seems inescapable that there were no large ice caps in the Devonian. The wide distribution, especially in the Middle and early Late Devonian, of extensive reef carbonate in equatorial areas supports this hypothesis, as does the high clastic sedimentation rate in the Catskill Delta, which suggests rapid erosion rates like those found in areas of tropical rain belts.

The Malvinokaffric province carries a fauna of marine animals quite distinct from that known in supposed lower latitudes. Certain brachiopods are endemic, and the limited fauna includes trilobites, bivalves, hyolithids, and conulariids. Corals, both tabulates and rugosans, and bryozoans and stromatoporoids — all of which are common in the inferred equatorial areas — are virtually absent from the province, the faunas of which are interpreted as those of a cold, high-latitude type.

Some interesting estimates have been made by using daily and annual banding on the epithecae of fossil Devonian corals which indicate that there were about 400 days in the Devonian year and about 30½ days in the lunar month. *See* PALEOCLIMATOLOGY.

Devonian life. The Devonian was a period of prolific marine life. A more significant development of the period, however, is that the continents were first colonized on a large scale by plants, invertebrates, and vertebrates. During the Early Devonian, vascular plants were abundant for the first time, and forests developed in many areas. Similarly, the development of fish in large numbers characterizes the period, and these lived in fresh as well as marine waters. The arid nature of much of the Old Red Sandstone continent (Laurasia), with desiccating pools and restricted water supply, is thought to be a major explanation for the evolution of the first tetrapods, the amphibians, at the close of the period.

Considering the Paleozoic as a whole, the Devonian seems to represent the period of greatest diversity of life. This is true in terms of the number of families known, and probably is also true of genera. However, in the Late Devonian large numbers of groups became extinct, and there is a particular time within the Upper Devonian when most were lost; other groups became extinct at the close of the period. The cause of these extinctions is difficult to determine precisely, but undoubtedly one significant factor was a series of sea-level rises

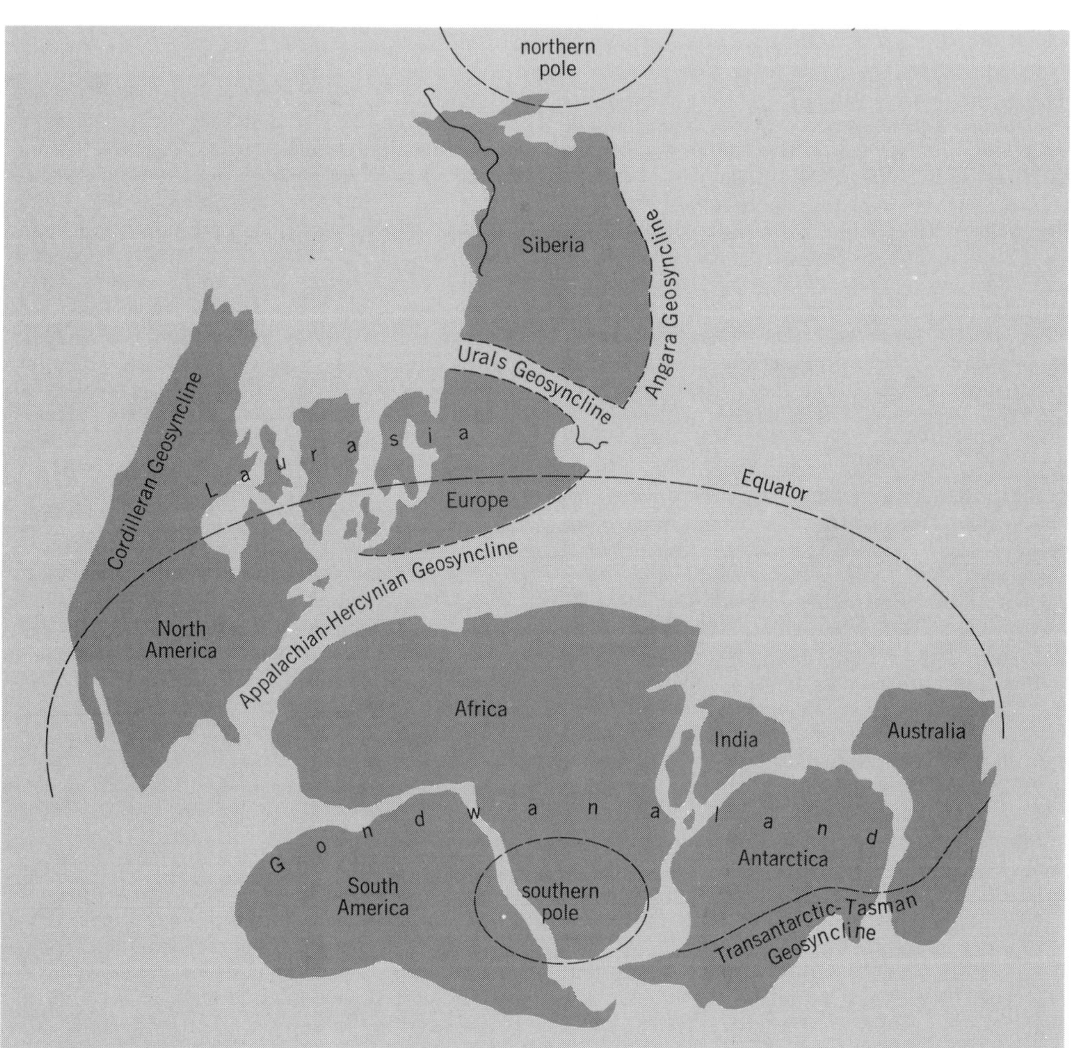

Fig. 2. Reconstruction of the globe in Devonian times with the continents placed in the position suggested by paleomagnetic and geological evidence. Positions of Peking platform and massifs of Southeast Asia are uncertain.

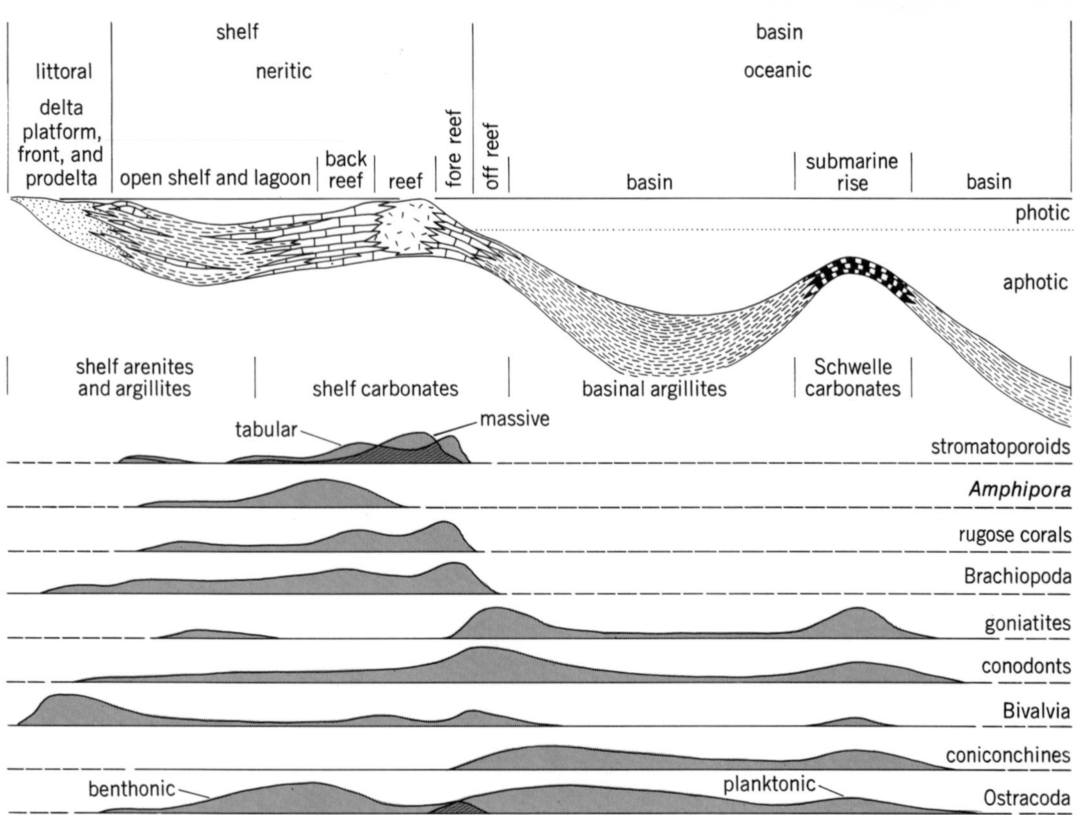

Fig. 3. The distribution of facies types and the faunal groups which are common in them, based largely on the Middle and Upper Devonian of Europe. (*From M. R. House, Faunas and time in the marine Devonian, Proc. Yorkshire Geol. Soc., 40:464, 1975*)

in the early Upper Devonian. It is thought that these led to the extinction of most of the reef complexes, thus destroying a large number of marine ecological niches. *See* EXTINCTION (BIOLOGY).

Plants. Algae had important and diverse roles in the period, but many algal groups do not preserve well in the fossil record. The stromatolitic algae, in which calcium carbonate is precipitated into a mucoid pad surrounding the algal filaments, are an important constituent of Devonian reefs; indeed, they form spectacular reefs in the latest Devonian in Western Australia after reefs are lost in other areas.

The origin of vascular plants took place before the Devonian Period began, and there are well-preserved floras of this type in the Late Silurian. In such plants, the presence of strengthened water-conducting tissue, which is often well preserved, shows that they had developed a rather erect habit with water transpiration into air from the upper surfaces. The earliest Devonian plants of this type, the Psiliphytopsida, comprise little more than a rhizoid base, a stem without leaves, and terminal spore cases or sporangia. Stomata on the stem surfaces indicate that these were mostly aerial. Such types include *Gosslingia* and *Rhynia*, the latter being one of several types known from the Rhynie Chert in Scotland in which most of the cell structure is well preserved. However, even in the Lower Devonian, forms such as *Asteroxylon* occurred which developed a dense clothing of the stems and branches with small, leaflike structures in which sporangia are borne laterally. These structures are characteristic of the Lycopsida. The Lycopsida are thought to have achieved tree habit in the Devonian and to have been a major constituent of forests such as that in Gilboa, NY. Another group, the Sphenopsida, appears in the Devonian; it includes present-day horsetails. They are distinguished by a characteristic pattern of jointed stems and branches with side branches and sporangia arising from the nodes.

In addition to the larger remains of plants, plant spores are common in many types of sediment, and since these differed from time to time and different types characterize different periods, they are valuable in the international correlation of rocks in the Devonian Period. *See* PALEOBOTANY.

Invertebrates. The invertebrate faunas, which are quite abundant as fossils in the Devonian Period, almost wholly represent groups which evolved much earlier, either at the beginning of the Cambrian or in the Ordovician. The Devonian representatives of these groups are mostly distinguished at the generic or family level. These faunas are best considered in relation to the environments which they inhabited (Fig. 3).

In the marine environment, the littoral or tidal range sediments are mostly delta sands and sand bars, and were characterized by bivalves and other organisms that burrowed in the sea-floor sediment. In the deeper, and generally finer-grained sediment of the neritic areas, brachiopods, bivalves, bryozoans, simple corals, and trilobites abounded. Such facies are well represented, for example, in the mid-Devonian rocks of upstate New York and the Eifel district of Germany. In offshore areas, which are freer of land detritus, carbonate reef complexes occur in many regions. The actual skeletal element of the reef usually consisted of large growths of stromatoporoids, finely layered and pore-covered calcareous organisms traditionally considered as coelenterates. These formed masses from a few inches to many feet across. Other coelenterates, notably the tabulate and rugosan corals, were abundant in the reef complexes, and these were usually of compound or massive type with many corallites; such corals were smaller and sparser in deeper-water facies. Today, reef growth is usually most active in wave agitation zones, and this is thought to have been true in the Devonian. Within the reef complex, other animal groups flourished, including gastropods, trilobites, sponges, cephalopods, brachiopods, echinoids, and especially crinoids. These were often of highly specialized and quite bizarre types.

The deeper-water, basinal facies deposits of the Devonian are well represented in many areas of the world. In environments of excessive depth, bottom-living organisms are rare, restricted mainly to specialized trilobites, arthropods, and echinoderms. The faunas found in basinal sediments are mostly those which lived in the upper parts of the seas, the remains of which dropped to the sea floor after death. Such types certainly included both microfauna and microflora. Since most of these were soft-bodied, very few are preserved as fossils. Well-preserved remains of larger groups occur which dominated in the photic zone during the Devonian. Thus, in the earliest Devonian, graptolites occur in slates called graptolitenschiefer; these graptolites represent the last survivors of a group abundant in the Silurian. These become replaced in the late Lower Devonian by abundant, small, conelike forms of Coniconchia, including *Styliolina* and *Tentaculites*, which produced the distinctive slates named the styliolinenschiefer and tentaculitenschiefer. During the early Upper Devonian, these occupants of the photic zone were replaced by distinctive, small, planktonic ostracods, and forms of these gave rise to the name cypridinenschiefer.

These various slate facies characterize the European basinal deposits, but on submarine rises thin carbonate sequences occur. In eastern North America a black shale facies occurs which seems comparable with the slate facies of Europe; the Cleveland and New Albany shales are of this type. In these shales bottom-living organisms are rare, but large numbers of small bivalves and brachiopods occur which are thought to have attached themselves to floating weeds during life.

Two groups of invertebrates are particularly useful in the Devonian for the international correlation of rocks: the goniatites and the conodonts. The goniatites are a group of cephalopods with coiled, chambered shells in which the partitions are simply folded: they have a marginal siphuncle which connects to the body chamber. These evolved so rapidly during the Devonian that their evolutionary stages can be used for time correlation. A special group, the clymenids, characterize the Late Devonian, and these have the siphuncle on the inner side of the coil. The conodonts, on the other hand, are an enigma. They are small toothlike structures usually less than 1 mm in size and are composed of calcium phosphate. Their affinities are uncertain. They are sometimes found in assemblages of different types, and are thought to comprise internal support structures within a soft-bodied organism which has not been found preserved. However, because they are resistant to acids, they can be readily extracted from rocks. Their rapid evolution makes them extremely useful for time discrimination and correlation during the Devonian.

Fish and tetrapods. Fish, although present earlier, first became abundant in the Late Silurian and Early Devonian. During the Devonian the major evolutionary radiation of the group was accomplished, and a range of different types were evolved. Many groups are associated with fresh-water or estuarine facies.

The earliest groups comprise the Agnatha, a group without true jaws which are thought to have been mud eaters and scavengers. Some of these developed extremely thick armor, such as *Pteraspis* and *Cephalaspis*. The true jawed fish, or Gnathostomata, appeared in the period. Some of these, too, were heavily armored, such as *Dunkleosteus*, a giant fish from the Late Devonian of Ohio. Shark-like fish, bony fish, lung fish, and coelacanths also evolved in this period. Rhiphiditid fish are thought to have given rise to the tetrapods by the close of the period. These appear to have developed their pectoral and pelvic fins for movement on land, and the fin lobes evolved to become the fore- and hind-feet of the amphibians and later tetrapods.

Global events. Following the fusion of North America and Europe at the close of the Silurian during the Caledonian orogeny, North America and Europe were united in Devonian time. The present southern continents were united as Gondwanaland (Fig. 2). Siberia, and probably other areas of present-day Europe and Asia, probably formed separate continental masses.

The geosynclinal tract that comprised the Appalachians and the European Hercynides was folded in the late Paleozoic, apparently as a result of a collision between these two supercontinents. However, details of the relationships are far from clear. Paleomagnetic evidence suggests a general lateral sinistral movement of each supercontinent during the Paleozoic, so this was not a simple convergent collision, but the eugeosynclinal deposits of the eastern Appalachians and along the Hercynian geosyncline suggest a subduction zone along that belt.

Similarly, the Ural geosynclinal belt was folded by the collision of Siberia and the North American–European supercontinent in the late Paleozoic. The easterly tract of eugeosynclinal deposits in the Urals suggests a subduction zone during the Devonian which dipped westward, with the whole belt approximately at right angles to the Hercynides.

The Devonian events are but a prelude to the major mountain building of the late Paleozoic. However, during the Devonian the series of transgressive events seems best explained in terms of changing sea-floor spreading rates which were part of the process. Certainly, with little evidence of polar ice, this is the best explanation for sea-level changes. The varied paleogeographic distribution of environments, both marine and nonmarine, which resulted seems, in turn, to be the cause of the extraordinary diversity of Devonian life. However, some of the changes led to the spectacular extinctions which also characterize the latter part of the period. [M. R. HOUSE]

Bibliography: M. A. Murphy, W. B. N. Berry, and C. A. Sandberg (eds.), *Western North America: Devonian*, Univ. Calif. Riverside Mus. Contrib. no. 4, 1977; D. H. Oswald (ed.), *International Symposium on the Devonian System*, Alberta Society of Petroleum Geologists, 1968; L. V. Rickard, *Correlation of the Silurian and Devonian Rocks in New York State*, N.Y. State Mus. Sci. Serv. Map and Chart Ser. no. 24, 1975.

Diabase

A fine-textured, dark-gray to black igneous rock composed mostly of plagioclase feldspar (labradorite) and pyroxene and exhibiting ophitic texture. It is commonly used for crushed stone. Its resistance to weathering and its general appearance make it a first-class material for monuments.

The most diagnostic feature is the ophitic texture, in which small rectangular plagioclase crystals are enclosed or partially wrapped by large crystals of pyroxene. As the quantity of pyroxene decreases, the mineral becomes more interstitial to feldspar. The rock is closely allied chemically and mineralogically with basalt and gabbro. As grain size increases, the rock passes into gabbro; as it decreases, diabase passes into basalt.

Diabase forms by relatively rapid crystallization of basaltic magma (rock melt). It is a common and extremely widespread rock type. It forms dikes, sills, sheets, and other small intrusive bodies. The Palisades of the Hudson, near New York City, are formed of a thick horizontal sheet of diabase. In the lower part of this sheet is a layer rich in the mineral olivine. This concentration is attributed by some investigators to settling of heavy olivine crystals through the molten diabase and by others to movement of early crystals away from the walls of the passageway along which the melt flowed upward from depth, before it spread horizontally to form the sill.

As defined, diabase is equivalent to the British term dolerite. The British term diabase is an altered diabase in the sense defined here. *See* BASALT; GABBRO; IGNEOUS ROCKS.

[CARLETON A. CHAPMAN]

Diagenesis

Those processes that alter the structure, texture, and mineralogy of a sediment during its deposition, lithification, and ultimate burial, but which exclude high-temperature and high-pressure modifications attributed to metamorphism. Diagenetic changes are intergradational with those of metamorphism at elevated temperatures and pressures, and with those of weathering at, or near, surface conditions.

Among the processes involved are purely physical ones such as compaction, desiccation, and soft-sediment deformation; physicochemical ones such as grain solution, corrosion, bleaching, oxidation-reduction, crystal inversion, recrystallization, intercrystalline bonding, cementation, decementation, authigenic mineral growth, and mineral replacement; also, such biochemical and organic processes as particle accretion, flocculation, sediment mixing, boring, and decomposition and synthesis of organic compounds. Of the chemical processes, some tend to establish chemical equilibria between certain minerals in the sediment, the incorporated biota and its shells, and interstitial fluids during three somewhat separate stages in the history of a sedimentary rock.

Stages. (1) In the initial, or depositional, stage modification in the raw detritus is controlled by the environment of the water-sediment interface. (2) The intermediate, or early burial, stage is confined to changes which occur primarily in the upper few feet of accumulated debris as compaction occurs. This stage is transitional with lithification, as well as the initial stage, but represents a time of major modification in bedding, texture, and to a lesser degree in mineralogy. Bonding of grains initiates cementation. (3) The late burial, or premetamorphic, stage involves lithification and is best developed by the environment of deep burial where temperatures above the boiling point of water and pressures of thousands of pounds per square inch promote important changes in bedding cleavage, porosity, clay-mineral dehydration, cementation, and mineral authigenesis. *See* AUTHIGENIC MINERALS.

Initial stage. An important control of the initial stage is the salinity and acidity of the fluid, although local chemical changes are influenced by infauna, depth of water, and intensity of currents. The condition is that of an open system, and chemical reactions between the sediment and water do not attain equilibrium. Reactions proceed toward simplification of products: reduction in the variety of minerals in the raw detritus, and concentration of resistates (quartz) and opaline biogenic remains, precipitates (carbonates such as the contribution of epifauna), and hydrolyzates (clays).

Certain hydrolyzates and other minerals of clay size are degraded or decomposed, whereas others are formed by authigenesis. Examples in the marine environment are devitrification of volcanic glass to montmorillonite; alteration of degraded illite to montmorillonite (high pH) and illite to glauconite; and authigenesis of certain zeolites, particularly where evaporation is pronounced.

Early burial stage. The environment of early burial is a system which is partially closed, and important chemical conditions are introduced by incorporated organic matter, sediment-dwelling organisms, and interstitial fluids. The latter are in slow motion, as they are squeezed from the sediment pores during compaction, and may remain in contact with individual minerals sufficiently long for the two to react. Primary effects are the accentuation of bedding by compaction, and oxidation-reduction reactions to crystallize pyrite and other sulfides, particularly in organic-rich sediments (−Eh), or iron oxides where oxidizing conditions prevail. Biogenic opal is reprecipitated as poorly ordered cristobalite; aragonite is recrystallized as low-Mg calcite; aggregates of clay minerals tend to become bonded to one another along surfaces with unsatisfied charges; uncommon clay minerals, such as palygorskite, crystallize from other clay minerals; and overgrowths may be crystallized on certain minerals, such as quartz. Collectively, all such manifestations can be considered initiation of lithification or cementation.

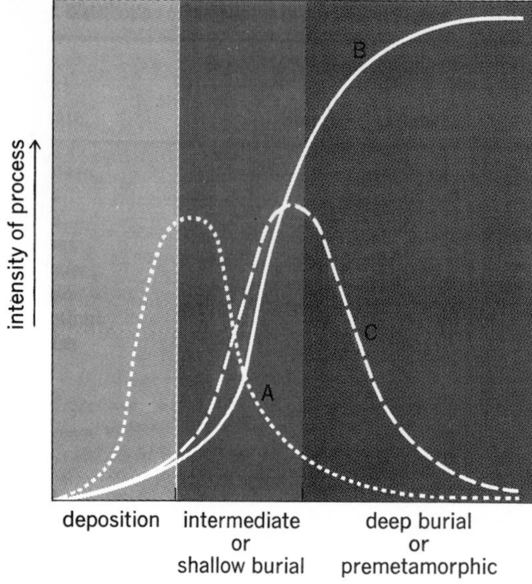

Diagram of intensity of major processes through three diagenetic stages. Curve A: rounding, pitting, and frosting of grains; overgrowths on quartz, solution of detrital chert; and reactions controlled by Eh and pH. Curve B: grain interpenetrations, mineral authigenesis, carbonate cementation, recrystallization, and development of fissile bedding. Curve C: development of concretions, compaction, silicification, and dolomitization.

Late burial stage. This stage reflects the modification brought about by increases in depth of burial and temperature, as well as the length of time since deposition; for example, crystallization of cristobalite into chert (quartz) is not observed in modern deep-sea sediments, but rather in sediments as old as Middle Tertiary. Aging, folding, and high temperatures advance progression in cementation, ranging in complexity from crystallization of incomplete overgrowths on quartz grains to well-developed, interlocked, planar grain boundaries. The increase in complexity of grain interpenetrations toward a sutured interlock is attributed to "pressure solution." Such a boundary is typical of quartz or calcite grains that are accompanied by considerable recrystallization. Each ultimate texture of low porosity is typical of those textures associated with metamorphic rocks.

Clay minerals tend to release water, increase in lattice order and crystallinity, and commonly are rearranged as aggregates of mica crystals. Some become unstable and alter to a variety more stable at the greater depths, for example, from montmorillonite to illite. All modifications listed above collectively constitute the basis for the concept of diagenetic grades representing a measure of the progress of diagenesis in the development of textures and mineralogy, each of which reflects increasingly high temperatures, pressures, and ion concentrations (see illustration).

In sandstones. Among quartzose sandstones the initial stage of diagenesis is characterized by rounding and solution pitting of quartz. Chert is unstable and is reduced in amount, and early glauconite is formed, transported, and redeposited locally. Early burial is characterized by precipitation of overgrowths on loose quartz grains; initially there are many small crystals producing the "frosted" grain, but actual cementation or interlocking of grains is not an important process. Commonly, precipitation of small amounts of carbonate as cement follows precipitation of quartz overgrowths. Where such sandstone passes laterally into limestone, increasing deposition of carbonate locally forces quartz grains apart and produces a texture recognized as "floating" sand grains. The late stage of diagenesis, associated with deep burial, is recognized either by strong addition of carbonate cement and concomitant solution of quartz grains or, in the absence of carbonate cement, intergranular penetration of grains to produce sutured or planar boundaries intersecting at a triple junction. *See* STYLOLITES.

Very fine-grained arkoses with little clay matrix tend to develop overgrowths on feldspar grains, such overgrowths being nearly pure potassium feldspar; albite overgrowths are rare. Argillaceous sandstones display a commonplace diagenetic sequence. During early burial silica is deposited on some of the large grains as overgrowths but is less prominent than in grains in which clay is absent. Some of the clay minerals become metastable; chert is precipitated locally and is welded to the grains of quartz. Siderite concretions are considered to develop during this stage in brackish water or marine sediments where the local pore water contains a high concentration of carbonate ions and incorporated organic matter. Late and deep burial are indicated principally by introduction of carbonates, which corrode quartz and replace the clay matrix, by chertification of the matrix detritus, principally clay, and by local grain interlock. Mixed-layer clay minerals and montmorillonites tend to be reconstituted as illites, chlorites, and well-crystallized biotite. The latter is favored where coaly fragments occur in the sandstone, whereas in the oxidized red sandstone, well-ordered illite and muscovite tend to dominate.

Typical graywacke sandstones show well-developed reaction effects between mineral grains and destruction of original boundaries. Such alteration suggests that at least some of the matrix is neoformed from feldspar grains, and the typical texture is not entirely of primary origin. Concentration of authigenic chlorite is typical. The principal diagenetic relationships, such as welding of grains and recrystallization of clays, are considered to be late in development and to be attendant upon deep burial and elevated temperature. The presence of certain zeolites, clinoptilolite in particular, derived from the alteration of volcanic debris and associated with thick accumulations of graywacke transitional into the chlorite grade, has led to the proposal that a zeolite metamorphic grade exists as transitional into truly metamorphosed strata.

In shales. Diagenesis in shales is manifested primarily by color changes, an increase in separation along the bedding, decomposition of unstable clay minerals and recrystallization of others, and development of micas of restricted composition. Parting along bedding is accentuated by carbonization of organic debris, particularly woody material, and is inhibited by precipitation of carbonate and silica. Although fissility is inherited from rhythmic pulsation in amounts of clay and silt particles to the suspension load, the interlamina separation is accentuated by loading and unloading of strata and by orientation of clay minerals and organic remains during compaction. As long as the rock is under significant load, bedding-plane cleavage is poorly developed, but upon unloading and during weathering, the incipient planes of failure become enlarged until parting occurs. Such parting is gradational with fracture cleavage, as not all of the separation planes are coincident with bedding. Certain aspects of this cleavage are produced during dewatering accompanying folding, and clay minerals are reoriented to parallel the direction of the escaping pore fluid.

Compaction is an important diagenetic process in all shales, particularly those which are deposited as hydrosols. Continued loading of such material causes destruction of the sol and ejection of water until silts and clays are brought into contact, after which reduction in pore water approaches the zero limit asymptotically.

Early burial is marked by reactions controlled by pH (alkalinity-acidity) and Eh (oxidation-reduction potential). Also observed are diminution of Mg ions of interstitial waters as this ion is utilized by incipient development of montmorillonites or dolomite. Such waters also tend to show increases in K ions and SiO_2, presumably the result of decomposition of unstable feldspars. Dark-gray colors and sulfide minerals signify the existence of reducing conditions; whereas light-gray and red colors and oxides and hydroxide minerals are associated with oxidizing environments. Calcite is the principal indicator of the pH (precipitated at about 7.8).

Early burial of deep-sea clays illustrates reactions of high-pH pore waters that result in decomposition of particles of volcanic glass into montmorillonites, opal-cristobalite, and certain zeolites as neoformed products. Similarly, degraded illite (land-derived) alters to more stable montmorillonite.

Principal modifications during late burial are increase in structural ordering of clay minerals to micas and chlorites, and recrystallization to produce a welded aggregate of crystals, particularly chert (quartz) and certain zeolites (clinoptilolite) in pore waters of low pH. Well-established mineral transformations with increased depth of burial are montmorillonite to illite, and reduction in *b*-axis—disordered kaolinite with increase in temperature. More than half of the illite becomes well crystallized, and in total may constitute more than 80% of the clay minerals. Correspondingly, there is a tendency for kaolinite to be absent in deeply buried shales. Shales associated with volcanic rocks often are very siliceous, the silica having been precipitated during late burial and probably having been derived from waters moving through the volcanic sequence.

In carbonates. Carbonates are recognized to be exceptionally influenced by diagenetic modifications which involve prominent changes in particle size, porosity, and authigenic mineral growth. Dominating such changes in the original carbonate are recrystallization and wholesale introduction of magnesia and silica. Early modifications accompanied by initial cementation are recrystallization of aragonite shell matter to acicular aragonite or spar calcite, of high-Mg calcite to low-Mg calcite, or of low-Mg to high-Mg calcite. Algal metabolism is capable of producing high-Mg calcite in laminated form as an initial step toward dolomite. Recrystallization of fossil-fragmental texture destroys original organic structures and substitutes an interlocking crystal meshwork held without cement. The course of recrystallization is progressive from stages in which fossil fragments are attached by a crystalline cement to complete alteration to a granoblastic texture.

Selective dolomitization of entire formations is known, in the majority of cases, to postdate precipitation of siliceous concretions, and is developed preferentially in sites of broad structural arches which were tectonically stable for long periods. A preferred mechanism is mixing of groundwater in low-lying carbonate islands with up to 30% sea water to dissolve calcite and crystallize dolomite.

Introduction of silica is principally in the form of chert as nodules and lenticular beds generally emplaced in stratigraphically favorable positions. Mineral paragenetic relationships favor the interpretation that such silica is introduced during early burial and preceding lithification. The initial source is biogenic opal which is reprecipitated as cristobalite. Later, the cristobalite is reorganized into nodules of chert (quartz). Other late-burial modifications include reversal in dolomitization, described as dedolomitization, and authigenesis of feldspars, in particular, low-temperature albite. *See* SEDIMENTARY ROCKS; SEDIMENTATION.

[EDWARD C. DAPPLES]

Bibliography: G. Larsen and G. V. Chilingar (eds.), *Diagensis in Sediments*, 2d ed., 1976; F. J. Pettijohn, *Sedimentary Rocks*, 2d ed., 1957.

Diamond

A mineral composed entirely of the element carbon crystallized in the isometric system.

Physical properties. Gem diamonds have a density of 3.53, but the tough, black, cokelike aggregates of microscopic crystals sold as carbons and known to the layman as black diamond may have a density as low as 3.15. It is the hardest known substance. Boron nitride, which has been synthesized by the General Electric Company and called borazon, is the second hardest material known. Boron carbide, silicon carbide, tungsten carbide, and aluminum oxide rank below boron nitride in hardness, in that order. *See* GEM; MINERALOGY.

Because these substances are all crystalline, the bonds between atoms are arranged in definite patterns; thus certain planes and directions across a crystal surface have greater concentrations of bonds than others. Therefore hardness varies with the direction of abrasion. Diamond crystals can be cut only by diamond dust on a lap, or rapidly rotating horizontal plate, when the softer directions of the diamond crystal are presented to the diamond particles that attack it. In the random,distribution of diamond dust on a lap, some particles will present their hardest directions to the diamond that is being cut.

Diamond slowly burns to carbon dioxide in oxygen at a temperature as low as 900°C and slowly inverts to graphite at temperatures as low as

1000°C. At higher temperatures, conversion of diamond to graphite can be exceedingly rapid. The rate of graphitization is greatly accelerated when the diamond is in contact with any of the group VIII elements or with alloys such as iron, nickel, or cobalt.

Diamond has the highest thermal conductivity of any known substance. The thermal conductivity of pure diamond at room temperature is approximately five times that of copper. The points of diamonds used as cutting tools do not become hot owing to this very high thermal conductivity. This quality contributes greatly to their usefulness in abrasion-resistant cutting tools.

All except a few diamonds are nonconductors of electricity, but all are excellent heat conductors, superior to iron and steel. Some diamonds will conduct electricity only when subjected to radioactive emanations. The current conducted is proportional to the intensity of radiation. Devices (proportional counters) using this principle have been developed for measuring radioactivity. Under intense radioactive bombardment, diamonds first become green, then brown, and finally black. If not carried too far, the process may be reversed by heating to a white heat.

DIAMOND

10 mm

(a)

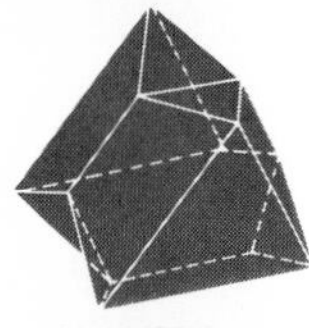

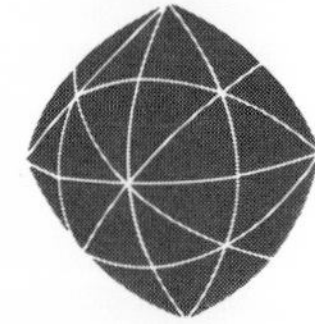

(b)

Diamond. (*a*) Crystals from Kimberley, South Africa (*American Museum of Natural History specimens*). (*b*) Crystal habits (*from C. S. Hurlbut, Jr., Dana's Manual of Mineralogy, 17th ed., copyright © 1959 by John Wiley & Sons, Inc; reprinted by permission*).

Crystal structure. Diamond crystallizes as octahedrons, dodecahedrons, and cubes, the first two forms being by far the most common. Overgrowths of dodecahedrons and cubes on octahedrons are not uncommon. Some crystals from Sierra Leone and the Republic of the Congo (formerly Belgian Congo) show all three forms nearly equally developed. The opaque yellow to brown crystals often display a concentric layering with clear, colorless octahedral diamond in the center. In the trade they are known as coated stones. The outer layers are slightly impure diamond, usually much twinned.

Some evidence of twinning parallel to the octahedron faces is present in nearly every crystal. The crystals of highest purity are often irregularly shaped. Those that are faintly yellow in color have the most perfect external forms and are less often twinned. The twinning in all types of crystals is often associated with minute or larger inclusions of dark foreign matter, generally called carbon spots. Twinning is so common that diamond cutters refer to the most obvious as macles. Large twinned areas are called blocks and the smaller ones knots or pings. *See* CRYSTAL STRUCTURE.

It is now generally accepted that diamond crystals have an octahedral structure ($m3m$). For many years the idea of a tetrahedral structure persisted, although it was contrary to x-ray and etch-figure evidence. The tetrahedral theory developed from the external morphology of diamonds from the Kimberley district, by acceptance of the unusual rather than the commonplace characteristics (see illustration).

Octahedral crystals are typical of the better qualities of diamond from Sierra Leone, Ghana, Angola, and the Republic of the Congo. The dodecahedral crystals are typical of Brazil. Octahedrons and dodecahedrons, with the latter predominating, are typical of the Kimberley district. Irregular shapes are characteristic of South-West Africa and Tanzania, although these usually reveal good external evidence of crystal forms.

Occurrence. Platelets of polycyanogen, C_3N, are dispersed throughout many naturally occurring diamonds. Natural diamonds without polycyanogen platelets have higher thermal conductivity and low optical absorption bands and are known as type 2 diamonds. The more common naturally occurring diamond, type 1, has varying amounts of polycyanogen platelets. Their thermal conductivities are lower by a factor of three than of those pure type 2 diamonds.

Diamonds crystallize directly from rock melts rich in magnesium, at depths of 150 km or more in the Earth. The melts from which they crystallize are essentially saturated in carbon dioxide gas at exceedingly high pressures; temperatures in excess of 1400°C are required. Thus rocks which contain diamonds as a natural component are samples of the deep mantle of the Earth and worldwide are remarkably similar in composition. Only garnet-bearing lherzolite and garnet-pyroxene eclogites are known to contain diamond. Explosive CO_2 gas – driven eruptions of frozen material bring these diamond-bearing rocks up through cracks to the surface of the Earth. In the near-surface environment, craters form. These craters, filled with deep-mantle rock as well as shallower-mantle and surface material, are called diamond pipes. This hot, magnesium-rich, deep-mantle material reacts with the groundwater and carbon dioxide, and a suite of hydrated magnesium minerals forms. This complex of hydrated, deep-mantle, magnesium-rich material, mixed with shallower-mantle and surface material, changes to form a rock type of highly variable chemistry called kimberlite. The variability in the chemistry of kimberlite results from varying ratios of deep-mantle, shallow-mantle, and surface rock chemically mixed into a clastic mass.

Diamond pipes are commonly called volcanic necks in the literature. However, this terminology is incorrect, for there is no evidence that the material in diamond pipes that has reached the surface is a liquid or is related to volcanoes. They are more properly called gas-driven diatremes. These dikes and pipes of diamond-bearing hydrated magnesium silicates form primary diamond deposits. Productive dikes have been mined in Sierra Leone, the Orange Free State, and the Transvaal. Pipes which are roughly elliptical in form have been mined in the Kimberley district of Cape Province and the Orange Free State; in the Transvaal at the Premier Mine; in Tanzania at Mwadui and Mabuki in the Shinyanga District; in the Republic of the Congo at Bakwanga; in Sierra Leone near Yengema; and in the United States at Murfreesboro, AR. In addition, a substantial number of these pipes have been found in Yakutsk, Siberia, and are now being actively mined.

Several hundred dikes and pipes of kimberlite that contained diamonds have been found. Only a few have been profitable to mine at depth. All pipes have an enriched surface layer which has been developed by erosional processes. Initially the upper few feet of many pipes may be profitably mined, although they may not warrant mining at depth. The pipe at Murfreesboro is of this type.

Most diamonds have been produced from secondary (placer) deposits, called alluvials. In Brazil and Angola conglomerates of an early geological age have been profitably mined. These are the exceptions, and most alluvial diamonds are recovered from modern gravels. All of these are stream gravels except the beach gravels of the Atlantic Ocean, extending from Conception Bay on the north to Buffels River on the south in South-West Africa and Namaqualand. In this area the deposits adjacent to both sides of the Orange River mouth are the richest. The beaches that have been mined are from a few feet to 500 ft above the present sea level. The productive gravels are usually covered with tens of feet of drifting dune sand. These rich Orange River beach gravels extend out to sea; vast quantities of diamonds are known to exist on the floor of the Atlantic Ocean off the southwest coast of Africa. Unfortunately, no profitable way of recovering these ocean-floor diamonds has been developed to date. They are distributed through a thick submarine gravel bed, and the cost of mining these submarine gravels and recovering the diamonds exceeds the value of the stones.

Diamonds also have been found as irregular microcrystalline clumps up to a millimeter or so in size and as tiny crystals in meteorites, notably the Canyon Diablo iron meteorite (a coarse octahedrite). These meteoritic diamonds were formed from graphite nodules as the iron meteorites were subjected to intense high pressures by the shock of the meteorite impacting on the surface of the Earth. Lonsdaleite, a hexagonal polymorph of diamond, has also been found in meteorites and has been synthesized by shock conversion of graphite. *See* METEORITE.

History. Diamonds were mined from stream gravels in India and Borneo in prehistoric times. Originally the words adamas, adamant, and diamant were given to the very hard, colorless, transparent minerals now known as diamond, corundum, spinel, topaz, and quartz. Pliny describes the geometrical shape of six varieties of adamas, one of which is obviously the mineral now known as diamond. The authentic history of diamond mining begins with Jean Baptiste Tavernier's visit to Golconda, India (1638–1668).

About 1720, diamonds were identified in the gold washings of the Jequitinhonha River near Diamantia, Minas Gerais, about 300 mi due north of Rio de Janeiro, Brazil. For a century and a half this district and the area near the headwaters of the Paraguay River in the State of Mato Grosso, Brazil, were the chief sources of the world's diamond supply. Diamonds have been found in every state in Brazil and along the northward flowing tributaries of the Orinoco River in Venezuela and Guyana. Dodecahedral crystals are characteristic of Brazil, and the name Brazilian diamonds, as now used, describes this shape and not the geographic origin of the diamonds thus designated. All the Brazilian production is from placers, either ancient or modern.

In 1866 the first South African diamond was identified. It was a 21½-carat stone among the playthings of a small boy living near the banks of the Orange River at Hopetown. In 1868 a small diamond was found 80 mi north of Hopetown at the German mission of Pneill on the banks of the Vaal River, a tributary of the Orange. The village of Klipdrift, now known as Barkly West, across the river has since become a center of alluvial diamond mines known as wet diggings. This designation distinguishes these secondary (placer) stream deposits from the dry diggings or pipe mines which were discovered shortly after 1868. In August, 1870, the discovery of a 50-carat diamond in an intermittent stream led to the discovery of the first kimberlite pipe. This, the Jagersfontein Mine, lies 80 mi east of Hopetown. It is a nearly circular pipe with a cross-sectional area of 25 acres, and is the erosional remnant of the feeder neck of an old volcano. The second pipe mine, the Dutoitspan (60 acres), was discovered in the following month 20 mi southeast of Pneill, where the city of Kimberley now stands. The Bulfontein (62 acres) was discovered early in 1871, the DeBeers (43 acres) in May, 1871, and the Kimberley (38 acres) the following month. These last four and the city of Kimberley all lie within an area 3 mi in diameter. The fifth member of the Kimberley group, the Wesselton (49 acres), discovered in September, 1890, originally called Premier, lies 1¾ mi east of the Dutoitspan. Of the many other pipes in this area only the Koffyfontein, between Kimberley and Jagersfontein, has been profitably operated. *See* CARAT.

The only other profitably operated pipe mines are the Premier (80 acres), 70 mi northeast of Johannesburg, discovered in 1903; the Williamson Mine (400 acres) at Mwadui, Shinyanga District, Tanzania (then Tanganyika), 1940; the Bakwanga (80 acres) in the Republic of the Congo, 1949; and the Koidu (1/2 acre) near Yengema, Sierra Leone, 1956.

Little accurate information is available on the kimberlite diamond pipes and associated stream deposits discovered in the Yakutia district of Siberia. They lie west of Yakutsk on the Lena River and north of Lake Baikal within the Arctic Circle. They are within the permafrost region, and the difficulties to be surmounted in successful mining are great.

Diamond recovery. Of the world's output of diamonds, 95% comes from Africa. Most of the production is by large mining companies, but a significant amount comes from individual operations from stream-bed deposits in Sierra Leone, Ghana, and the Union of South Africa. These small operations recover diamonds by a method similar to the panning of gold from placers, taking advantage of the fact that the density of diamonds is greater than that of most other minerals. Concentrates of the heavy minerals recovered by panning are handpicked for diamonds.

Separation of concentrates. The large mining companies, which operate both alluvial (stream) deposits and pipe mines, also use other methods for separating concentrates based on the higher density of diamonds. The majority use large circular pans up to 16 ft in diameter for the first stage of recovery.

A large horizontal wheel is supported above the pan by a shaft that passes up through an oversized cylinder in the center of the pan. The top of this cylinder is lower than the outside walls of the pan. Water and diamond-bearing earth are continually fed into the pan and stirred by spokes extending downward nearly to the bottom of the pan from the rotating wheel. The lighter material is carried by the water over the rim of the central cylinder. The heavier minerals work slowly to the bottom outside edge of the pan, where they are periodically removed. The heavy minerals are further separated according to relative density by jigs, which are known as pulsators in the diamond mining industry. A later innovation is known as the sink-or-float method, in which a separation according to density is made by mud of ferrosilicon flowing upward in a large cone.

Separation of the diamonds. Diamonds are separated from the concentrates by hand sorting, grease tables, electrostatic methods, fusion with alkalies, surface tension, or abrasion of the gangue minerals. Grease tables are of two types: (1) vibrating stationary tables coated with a layer of grease which is periodically removed and renewed by hand; and (2) an endless belt to which a layer of grease is automatically applied before it moves across the table and is continuously removed after it leaves the table. The concentrates fall onto the grease table and a stream of water carries the hydrophilic gangue across the table, while the hydrophobic diamonds adhere to the grease.

In electrostatic separation the concentrates are fed onto a grounded and rotating horizontal steel cylinder which lies beneath a strong electrostatic field. Diamond is a nonconductor and retains the induced charge, while the gangue minerals lose their charge to the grounded cylinders. When the concentrates fall from the rotating cylinder, they pass through a strong electric field of the same sign as the induced charge on the diamonds. Small diamonds are deflected away from this second electric field to the far side of an adjustable knife edge which separates the falling diamond from the gangue.

All the minerals in the concentrate have approximately the same density, and all of these except diamond dissolve in molten alkali. The chief drawback to this fusion process of separation is that the high temperature necessary may induce color changes in some diamonds.

Because diamond is hydrophobic, small crystals will float on water, supported by surface tension. An endless belt carrying the dry concentrates passes into a tank in which the water is slowly moving away from the point at which the belt enters. The diamonds float, while the hydrophilic gangue minerals sink.

Before the treated concentrates are finally discarded, they may be ground in a ball mill. The gangue minerals are reduced to a fine powder, but any diamond that has not previously been removed resists the abrasion in the mill and can be recovered by screening.

Diamond cutting. After recovery, diamonds are referred to as rough. Those of gem quality are called cuttable rough, and all others are classed as industrial rough. The poor grades of gem-quality diamonds and finer industrials are synonymous.

Inspection. In cutting diamonds, the objective is to obtain the maximum price for the finished gems, not the maximum weight of the finished stones. It may be desirable to cut away a portion containing flaws so that the stone can be marketed as a first-quality stone of lesser weight; or the flaws may be left in the stone if the sales price of a second-quality stone of greater weight will be above the price of the smaller stone of first quality. The rough crystal may be divided into two or more smaller stones whose total value will be greater than a single large stone. Flaws may often be eliminated by this subdivision. The most important step in diamond cutting is the decision as to how the stone will be cut.

In subdividing an irregularly shaped crystal or eliminating flaws, the stone may be either cleaved or sawed, but only in certain crystallographic directions. There are four directions in which a diamond may be cleaved parallel to any octahedron face, and nine directions in which it may be sawed—three parallel to any cube face and six parallel to any dodecahedron face. Some of the more expert cleavers can cleave parallel to the dodecahedron face, but sawing in this direction is generally preferred.

Cleaving. In order to cleave a diamond, a small groove is first cut in the edge of the diamond to be cleaved with the sharp edge of another diamond. The cleaving iron (blade) is inserted in this groove parallel to the cleavage to be made; it is struck a sharp blow, and the diamond breaks along smooth flat surfaces.

Sawing. Sawing is done with the edge of a rapidly rotating phosphor bronze disk that has been impregnated with diamond dust. The starting saw which makes the first cut is thick enough to be rigid, but after this initial groove a rapidly rotating, paper-thin, phosphor bronze disk is used. Initially the saw must be impregnated with diamond dust, but if properly started it continually recharges itself with material removed from the groove. If the plane of the saw departs from parallelism to the proper crystallographic direction (the sawing grain), progress is slower and ceases when the departure is 10–15°. The saw must also rotate in the proper direction or no progress is made.

Cutting. If the finished diamond is to be round or oval in shape, it goes to a man who is known in the trade as a cutter. He rounds up or "girdles" the stone, held in a rotating chuck, in a lathe much like a wood-cutting lathe. Another diamond is held against the rotating stone and "brutes" off the corners.

Polishing. The process of putting on the facets is known as polishing. The diamond is held in a dop (holder) against a cast-iron lap (skeif) that has been charged with oil and diamond dust. The lap must move across the diamond in the proper crystallographic direction or it will not cut. The term cutting grain is used by diamond cutters to indicate the proper crystallographic direction for polishing and, like the cleaving and sawing grains, is somewhat analogous to the grain in wood. Its direction may usually be determined from the external shape of the crystal or from markings on the crystal faces.

Weight. The weight of the finished gems is 50–60% of the weight of the rough stones if they are well-formed crystals with few flaws. The finished weight of irregularly shaped or badly flawed crystals is often very much less. The metric carat, 0.200 g, is the unit of weight by which both gem and industrial diamonds are sold. A point is 1/100 of a carat and is used only in reference to gem diamonds.

Industrial diamonds. Industrial diamonds vary from the better grades, which are identical with inferior gems, to crushing bort, which is suitable only for crushing to grit and powder sizes. The better qualities are made into shaped diamond-cutting tools or wire-drawing dies. Diamond-cutting tools break on ferrous alloys because of the chattering which results from lack of ductility of the metals.

Shaped diamond tools have been supplanted in many of their former uses by sintered tungsten carbide.

Tungsten carbide wire-drawing dies have also replaced diamond for the larger sizes of dies. Diamond is still used for drawing wire smaller than 0.0025 in. in diameter (average size of human hair). Tungsten filament wire for light bulbs and radio tubes is drawn at a low red heat, and diamond is especially desirable because its hardness is little affected by these temperatures.

For truing and shaping grinding wheels of alumina or silicon carbide, diamond crystals, in the shapes in which they are mined, are used. In most grinding and finishing operations it is only necessary to impart a smooth, even finish to the wheel. In form grinding the reverse of the shape to be ground is imparted to the grinding surface of the wheel. Diamonds used for truing are usually called dressers and range in size from a fraction of a carat to several carats. Those used for form grinding are referred to as thread grinders if in the original crystal form, but if they have been shaped they may be called phonopoints because of their similarity to diamond phonograph needles. Diamonds used for form grinding usually weigh less than 1/25 carat.

Drill bort consists of crystals in their original form as mined which are mounted in the end of a cylinder called a bit or crown. When the rotating cylinder is forced against a rock surface, it wears its way into the rock. That portion of the rock which extends through the cylinder up into the hollow drill rod which rotates the bit is called the core and is periodically removed. A fluid, usually water, is circulated down the hollow drill rod and up the outside of the rod. It cools and lubricates the drill bit and flushes out the rock particles abraded by the diamond. The mining industry is the largest user of drill bort, although this method of coring has been adapted to the testing of concrete and the foundations of buildings, dams, and bridges. For these and mining purposes, the average size of the diamonds used is less than 1/10 carat. Large bits up to a foot in diameter are used by the oil industry both for coring and "making hole." These large bits use diamonds from 1/4 to 1 carat in weight. Prior to World War II, drill bort was the most important use of industrial diamonds.

Beginning with World War II, the greatly expanded use of tungsten carbide tools made bonded diamond grinding wheels by far the largest market for industrial diamonds. A grinding wheel with a thin layer of embedded diamond grit was developed to shape and sharpen these ultrahard tools. Originally the manufacturer of these wheels crushed the bort and sized it, but after World War II Industrial Distributors Ltd. (1946), the sales outlet for industrial diamonds of the so-called diamond syndicate, processed the diamonds and sold the material as fragmented bort. The annual worldwide market before synthetic diamonds were developed totaled 12,000,000 carats, three-fourths of which was sold in the United States. The mines at Bakwanga in the Republic of the Congo produce nearly all of this, the lowest grade of diamond.

Synthetic or man-made diamonds. Many attempts had been made to manufacture diamonds prior to the announcement (Feb. 15, 1955) by the General Electric Company of their successful synthesis. All claims of success prior to 1955 have proved erroneous. The synthesis of diamond has been effected in various ways: by static crystallization from certain molten metals or alloys at pressures upward of about 50 kilobars and temperatures over 1500 K; by shock conversion from graphite at transient pressures of about 300 kilobars and temperatures of about 1300 K; and by static conversion from graphite at pressures more than 130 kilobars and transient temperatures more than about 3300 K.

Synthetic diamonds are identical with natural diamonds in fundamental properties but differ in those characteristics that depend on the process of manufacture, such as impurities, size, and shape. Synthetic diamonds are made in the grit sizes (approximately 0.1 mm). These sizes are in greatest demand for the manufacture of bonded diamond grinding wheels for shaping and sharpening tungsten carbide tools. This is the greatest single use of industrial diamonds. Synthetics are superior to natural diamond for this use because they are single crystals, roughly octahedral in shape, with many cutting edges. In crushing natural diamond, many elongated slivers and flats are produced which reduce its efficiency.

A few large synthetic diamonds of gem quality have also been made by the slow growth of diamonds from high-purity carbon dissolved in a molten iron-nickel alloy. Stones up to 1 carat in size have been made. They are essentially indistinguishable from natural stones. A few synthetic crystals even purer than any known naturally occurring stones have been made. Unfortunately, the cost of producing the synthetic gem-quality diamonds is greater than their value; thus they are not commercially available to date. All of the so-called synthetic diamonds in the trade are actually imitation diamonds and are not made of carbon crystallized in its densest form. Most of these imitation diamonds are yttrium-aluminum-garnet or strontium-titanate. The distinction between synthetic gems and imitation gems is not always made. Synthetic gems are chemically and crystallographically identical to the natural stone, whereas imitation gems are made of some other component, such as glass, or another crystal, to look like the natural stone.

[CLIFFORD FRONDEL; GEORGE C. KENNEDY]

Diapiric structures

Anticlines or domes in which a core of mobile rock has broken through the overlying strata. In the process of concentric folding, the rocks of the core occupy increasingly more limited space as the

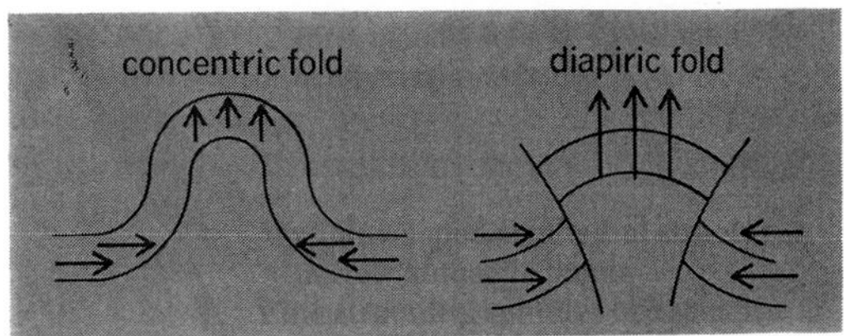

Fig. 1. Horizontal and vertical movement in a concentric fold, leading to diapirism. *(From L. U. de Sitter, Structural Geology, McGraw-Hill, 1956)*

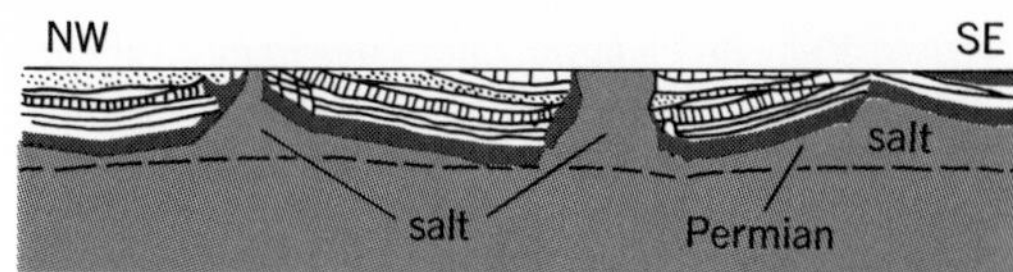

Fig. 2. Penetration of slightly deformed strata by salt in the Hanoverian salt dome region. *(After A. Roll, from L. U. de Sitter, Structural Geology, McGraw-Hill, 1956)*

curvature of the fold becomes smaller. If the rocks of the core are extremely plastic, they may exert an upward pressure sufficient to cause expulsion of the core through the crest (Fig. 1). Diapiric structures may also form as a result of unequal loading by the sediments overlying a plastic layer. The plastic rocks flow laterally away from places of high pressure and accumulate at places of low pressure. Where accumulations of plastic rock reach a critical size, the plastic material, if it has a sufficiently low specific gravity compared to that of the surrounding rocks, may penetrate the overlying rocks under the impetus of hydrostatic forces (Fig. 2). *See* FOLD AND FOLD SYSTEMS; SALT DOME.

[PHILIP H. OSBERG]

Diastem

A temporal break between adjacent geologic strata that represents nondeposition or local erosion but not a change in the general regimen of deposition (in contrast to unconformity). Diastems may be produced by the scouring action of shifting submarine currents which temporarily interrupt deposition on the continental shelf, or by a shifting river within the deposits of its floodplain. Or they may be produced simply by nondeposition in either environment of deposition where the absence of sediment reflects normal shifting of currents rather than an overall change in conditions. *See* UNCONFORMITY.

The existence of such breaks and their importance for interpreting the stratigraphic record was first pointed out by J. Barrell in 1917. Barrell showed that the time represented by the deposition of the beds actually observed in a stratigraphic sequence may be only a small fraction of the total time represented by the sequence as a whole, even if the entire sequence was deposited under an essentially uniform regimen; the rest of the time is represented by diastems. It is now generally accepted that deposition on a shallow sea floor or on a floodplain is a discontinuous process. *See* FLOODPLAINS; MARINE SEDIMENTS; SEDIMENTATION.

[JOHN RODGERS]

Diastrophism

The general process or combination of processes by which the Earth's crust is deformed; also, the results of this deforming action. The term diastrophism was first used by J. W. Powell when, in his study and discussion of major geologic features in the Cordilleran region of the United States, he felt the need of a single word equivalent to the somewhat cumbersome phrase "deformation of the Earth's crust." G. K. Gilbert, a coworker, adopted the new term as defined by Powell and suggested a dual subdivision of diastrophic processes and effects to distinguish the strong and comparatively localized deformation in mountain belts from the simpler structural patterns of broad plateaus and basins that are bounded by zones of faulting and warping. Gilbert approved the term orogeny (mountain making), already in general use, for the more intensive deformation, and proposed epeirogeny (continent making) as the kindred term applying to simpler uplifts and depressions that affect wide segments of the crust.

Diastrophism has operated continuously or repeatedly throughout geologic history. Modern movements in disturbed zones have caused major earthquakes and measurable displacements of land surfaces. Diastrophic effects in late geologic time are both topographic and structural. In the Alps, Andes, and other lofty mountains, layers of sedimentary rock that were formed on sea floors during the present geologic era are now at high altitudes and are much broken, tilted, and folded. Similar structure in older formations, now partly or completely leveled by erosion, marks locations of earlier mountain belts. Dissection of these deformed belts has revealed not only buckling and fracturing, but also large-scale metamorphism of the sedimentary rocks. The oldest deformed belts, widely exposed in areas of long-continued erosion, have bedrock consisting largely of crumpled and metamorphosed sedimentary strata, partly engulfed in large bodies of granitic rock. Thus intensive diastrophic action involves not only folding and fracturing, but also metamorphism and major igneous activity. The basic cause of diastrophism is unknown. *See* OROGENY; STRUCTURAL GEOLOGY; TECTONIC PATTERNS; WARPING OF EARTH'S CRUST.

[CHESTER R. LONGWELL]

Bibliography: M. P. Billings, Diastrophism and mountain building, *Bull. Geol. Soc. Amer.*, 71(4): 363–398, 1960; L. U. de Sitter, *Structural Geology*, 2d ed., 1964.

Diatomaceous earth

Earth consisting of a friable, porous silica deposit made up of the opaline silica tests (shells) of diatoms. It is the dry, relatively unconsolidated equivalent of diatom ooze found on some parts of the sea floor today. It is usually white or cream-colored.

Diatomite is the indurated equivalent of diatomaceous earth; the pores are partially or completely filled with silica. *See* CHERT.

[RAYMOND SIEVER]

Diopside

The monoclinic pyroxene mineral which in pure form has the formula $CaMgSi_2O_6$. Its space group is $C2/c$. Pure diopside melts congruently at 1391°C at atmospheric pressure; the effect of pressure up to 50 kilobars on its melting temperature is given by Eq. (1), where P is in kilobars and T is in degrees Celsius.

$$P = 23.3\left[\left(\frac{T}{1665}\right)^{4.46} - 1\right]. \qquad (1)$$

Diopside has no known polymorphs. Its structure consists of chains of SiO_4 tetrahedrons in which each silicon ion shares an oxygen with each of its two nearest silicon neighbors. These chains are linked together by Ca and Mg ions in octahedral coordination.

Diopside forms gray to white, short, stubby, pris-

Diopside crystals from St. Lawrence County, N.Y. (*Specimen from Department of Geology, Bryn Mawr College*)

matic, often equidimensional, crystals with (110) cleavages intersecting at 87° (see illustration). Small amounts of iron impart a greenish color to the mineral. The indices of refraction increase with increasing iron; for pure diopside they are $n\alpha = 1.664$, $n\beta = 1.671$, and $n\gamma = 1.694$. Pure diopside is common and occurs as a metamorphic alteration of impure dolomites in medium and high grades of metamorphism, as shown by Eq. (2).

$$CaMg(CO_3)_2 + 2SiO_2 = CaMg(SiO_3)_2 + 2CO_2 \quad (2)$$

Diopside shows extensive solid solution with a variety of other pyroxenes. The most important are $CaFeSi_2O_6$ (hedenbergite), $FeSiO_3$ (ferrosilite), $MgSiO_3$ (enstatite), $NaAlSi_2O_6$ (jadeite), $NaFeSi_2O_6$ (acmite), and $CaAl_2SiO_6$ (a synthetic pyroxene). Natural diopsidic pyroxenes also commonly contain significant amounts of chromium, titanium, and manganese. $CaMgSi_2O_6$ combined with variable amounts of $CaFeSi_2O_6$, $MgSiO_3$, and $FeSiO_3$ form the common Ca-rich pyroxenes of basaltic and gabbroic rocks. A few natural Fe-poor diopsides containing up to 40% $MgSiO_3$ have been found in nodules in kimberlite. Natural diopsidic pyroxenes which show extensive solid solution with jadeite and to a lesser extent with acmite are called omphacite. Omphacite is a principal constituent of eclogites, rocks of basaltic composition which have formed at high pressure. *See* DOLOMITE; ECLOGITE; PYROXENE. [F. R. BOYD]

Diorite

A phaneritic (visibly crystalline) plutonic rock with granular texture composed largely of plagioclase feldspar (oligoclase or andesine) with smaller amounts of dark-colored (mafic) minerals (hornblende, biotite, or pyroxene). This dark-gray rock is used occasionally as ornamental and building stone and is known commercially as black granite. For a general discussion of textural, structural, and compositional characteristics *see* IGNEOUS ROCKS.

Mineralogy. Gray or white feldspar grains commonly show fine, parallel striations or twin lines on broken surfaces. Under the microscope plagioclase feldspar exhibits striking examples of zonal structure in which individual grains are composed of concentric shells of rectangular outline and differing composition. Most commonly, internal shells are calcium-rich; more external shells are more sodic. Not uncommonly the change in composition of successive shells from center to exterior may be reversed, interrupted, oscillatory, or repetitive. As the plagioclase feldspar becomes more calcic than andesine, the rock passes into gabbro. With increase in plagioclase content and decrease in other constituents, diorite passes into anorthosite.

One or more mafic minerals may be present. Hornblende (microscopically green or brown), as irregular or elongate grains, is the most common. Black flaky biotite mica (microscopically brown) may be intergrown with hornblende. It is most abundant in quartz-rich diorites. Augite is uncommon and may be converted to hornblende. Olivine and orthopyroxene are rare.

Quartz, generally interstitial to plagioclase, may be present in small amounts. Where it constitutes over 5% of the minerals, the rock is called quartz diorite (tonalite). Small amounts of potassium feldspar closely associated or intergrown with quartz (micropegmatitic texture) are common; but where present in excess of 5%, the rock is a monzonite. Nepheline and other feldspathoids are rare constituents. Magnetite, ilmenite, and apatite are the most common accessory minerals; but zircon and sphene occur in certain varieties.

Textures and structures. Although generally equigranular, the texture may be porphyritic with large crystals (phenocrysts) of plagioclase or hornblende. Porphyritic types may grade into diorite porphyry. Some diorites and quartz diorites (tonalites) carry orbicular structures, and many show flow structures and banding.

Occurrence. Diorite occurs as isolated small bodies such as dikes, sills, and stocks. In more irregular forms it may be associated with granodiorite and granite or with gabbro. It occurs most abundantly in orogenic (fold-mountain) belts.

Origin. Diorite forms in many different ways. Some has crystallized directly from a dioritic magma (rock melt). Some is of hybrid origin and formed by reaction between a magma and contaminating, foreign rock fragments (xenoliths). Many diorites are products of solid-state transformation or metasomatism. Gabbro may be converted to diorite by a relative loss in calcium, iron, and magnesium and a gain in sodium and silicon. *See* MAGMA; METASOMATISM; PETROGRAPHIC PROVINCE. [CARLETON A. CHAPMAN]

Dolomite

The carbonate mineral $CaMg(CO_3)_2$. Often small amounts of iron, manganese, or excess calcium replace some of the magnesium; cobalt, zinc, lead, and barium are more rarely found. Dolomite merges into ankerite with increasing substitution of iron for magnesium, and into kutnahorite with increasing manganese substitution. *See* ANKERITE; KUTNAHORITE.

Dolomite has hexagonal (rhombohedral) symmetry and a structure similar to that of calcite, but with alternate layers of calcium ions being completely replaced by magnesium. Thus each layer of CO_3^{--} ions lies between a layer consisting entirely of Ca^{++} ions and one consisting entirely of Mg^{++} ions. This ordered arrangement of the cations distinguishes the structure of dolomite from that of a randomly substituted calcite. *See* CALCITE.

Dolomite is a very common mineral, occurring in a variety of geologic settings. It is often found in

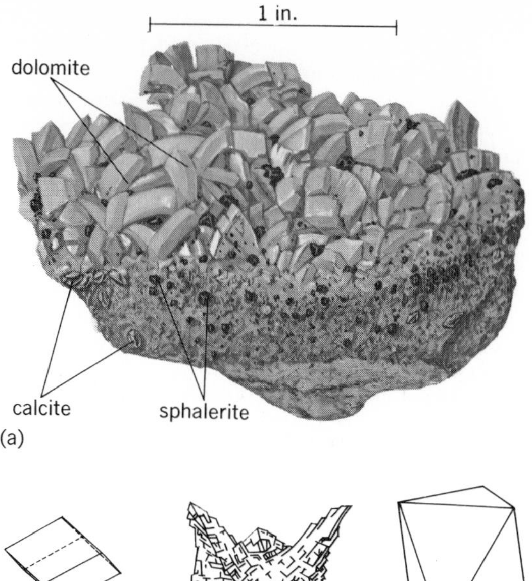

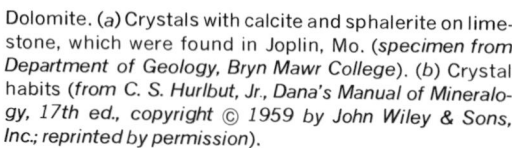

Dolomite. (*a*) Crystals with calcite and sphalerite on limestone, which were found in Joplin, Mo. (*specimen from Department of Geology, Bryn Mawr College*). (*b*) Crystal habits (*from C. S. Hurlbut, Jr., Dana's Manual of Mineralogy, 17th ed., copyright © 1959 by John Wiley & Sons, Inc.; reprinted by permission*).

ultrabasic igneous rocks, notably in carbonatites and serpentinites, in metamorphosed carbonate sediments, where it may recrystallize to form dolomite marbles, and in hydrothermal veins. The primary occurrence of dolomite is in sedimentary deposits, where it constitutes the major component of dolomite rock and is often present in limestones (see illustration). *See* DOLOMITE ROCK; LIMESTONE; SEDIMENTARY ROCKS.

Dolomite is normally white or colorless with a specific gravity of 2.9 and a hardness of 3.5–4 on Mohs scale. It can be distinguished from calcite by its extremely slow reaction with cold dilute acid. *See* CARBONATE MINERALS.

[ALAN M. GAINES]

Dolomite rock

A limestone whose carbonate fraction contains more than 50% of the mineral dolomite, $CaMg(CO_3)_2$. To avoid confusion between the mineral and the rock, R. R. Shrock recommended the name dolostone for the rock, but most geologists still use the name for both mineral and rock. F. J. Pettijohn followed common usage somewhat and suggested the names magnesian limestone for a dolomite content of 5–10%, dolomitic limestone for 10–15%, calcitic dolomite for 50–90%, and dolomite for over 90% (see illustration). Though this terminology is still followed by many, the term "magnesian limestone" has fallen into disuse, partly because of confusion with "magnesian calcite," a term used for an impure calcite. Most rocks called dolomite have more than 90% dolomite. Newer classifications of carbonate rocks based on textural elements introduce the amount of dolomite as a secondary criterion. *See* LIMESTONE.

Composition and texture. Dolomites show gross textures similar to those of limestones, but the finer details differ. Most coarsely crystalline dolomite appears to be secondary; rhombohedral crystals of dolomite that have grown in the rock after original limestone deposition usually have destroyed or obliterated many original structures and textures. An original clastic texture may be revealed only by scattered quartz grains floating in a mosaic of dolomite crystals. The original outlines of fossils or oolites may show as faint dust outlines in the dolomite, but the original structure has been lost. A sharply contrasting type of dolomite is finely crystalline, shows little or no evidence of replacement origin, and preserves many sedimentary structures such as weak current bedding. Such rocks have been interpreted as being the lithified equivalent of an original precipitate of dolomite mud. The fine-grained dolomites are typically associated with evaporite rocks.

The characteristic habit of dolomite crystals growing in a calcite matrix is rhombohedral. The rhombohedrons are clearly formed after the calcite and in many cases have replaced it, as is evidenced by the dolomite cutting across fossil shells and oolites. Complete dolomitization results in an interlocking mosaic of dolomite crystals, in which, because of interference between adjacent crystals, rhombohedral faces may be lacking. Partially dolomitized limestones tend to have a mottled appearance, which results from the uneven distribution of dolomite crystals in the calcite matrix.

In most of the rocks in which both chert and dolomite are found, chert appears to have replaced dolomite, that is, to have formed later. In most rocks there is no firm evidence that dolomite has replaced chert. Dolocasts, hollow cavities in chert that have the shape of dolomite rhombohedrons, are common in the insoluble residues left from the acid digestion of dolomites and limestones. These may be the result of dolomite replacing chert but may also represent selective replacement of a dolomitic limestone by chert.

Occurrence and origin. Dolomite and dolomitic limestone are known from rocks of all ages but are more common in older rocks, particularly the Paleozoic. Dolomite is most often found in association with limestone, with which it may be interbedded or laterally gradational. Some dolomitized zones do not follow bedding planes and are thought to be controlled by faults or folds. Dolomitization of limestones may be highly selective; for example,

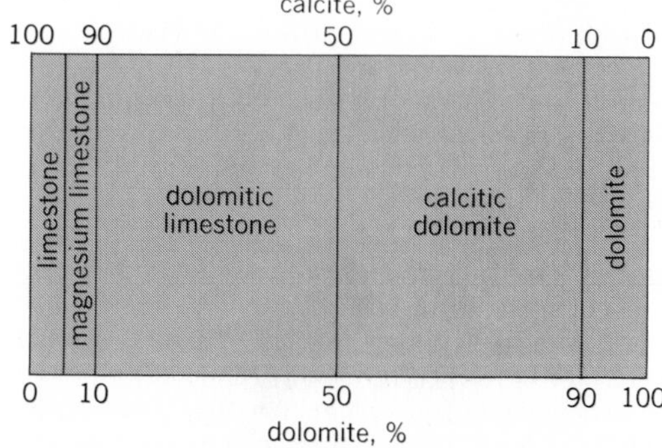

Classification of calcite-dolomite mixtures. (*From F. J. Pettijohn, Sedimentary Rocks, 3d ed., copyright © 1949, 1957 by Harper & Row, Publishers, Inc.; copyright © 1975 by Francis J. Pettijohn*)

the cores of the Silurian reefs of Illinois, Indiana, and Wisconsin are dolomite, whereas the reef-flank material may be only partially dolomitic. Modern dolomite has been found in supratidal flats in carbonate depositional areas such as the Bahamas and in a variety of hypersaline lagoons or arms of the sea in warm climates. The deposits are associated with precipitates of calcite, aragonite, and magnesian calcite as well as gypsum and anhydrite. The major condition necessary for this kind of precipitation is evaporation of sea water, which results in depletion of Ca^{++} ions by precipitation of calcium carbonate and calcium sulfate. The consequent buildup of Mg^{++} ion finally results in the production of dolomite. Analysis of O^{18}/O^{16} isotopes in the carbonate suggests a secondary origin for all low-temperature dolomites, but the conversion of a fine-grained calcite or aragonite precipitate to dolomite in a high Mg^{++} ion hypersaline environment must take place almost immediately.

Many, if not most, dolomites are replaced limestones, as is shown by crystals of dolomite cutting across original textures, by structural control of some dolomitization, and by dolomitization cutting across stratigraphic boundaries. The replacement is the product of the reaction between Mg^{++} ions in interstitial waters and calcite to form dolomite. An alternative explanation for some dolomite is that there has been a transformation from a magnesian calcite (calcite that has magnesium replacing some calcium in the calcite lattice) to a thermodynamically more stable association of magnesium-free calcite and dolomite. Many carbonate-secreting invertebrates incorporate magnesium in their calcite shells. These shells may be the source of some dolomite. But the reaction with magnesium in groundwaters must be of major importance in producing the pure dolomites, for there is not enough magnesium in the magnesian calcites to account for all dolomite. *See* LIMESTONE; SEDIMENTARY ROCKS. [RAYMOND SIEVER]

Bibliography: L. C. Pray and R. C. Murray (eds.), *Dolomitization and Limestone Diagenesis*, 1965.

Dopplerite

A naturally occurring gel of humic acids found in peat bogs or where an aqueous extract from a low-rank coal can collect. Dopplerite is soluble in alkali, contains organically bound calcium, iron, or magnesium, and has an ash content of about 5%. On an ash-free basis it consists, on the average, of 56.5% carbon, 5.5% hydrogen, 36.0% oxygen, and 2.0% nitrogen. The composition of the material usually is nearly identical to that of the coal from which it is derived. On dehydration dopplerite becomes a black, brittle solid with conchoidal fracture. *See* COAL; PEAT. [IRVING A. BREGER]

Drumlin

A hill of glacial drift or bedrock having a half-ellipsoidal streamline form like the inverted bowl of a spoon, with its long axis paralleling the direction of movement of the glacier that fashioned it. The "upstream" end is higher and steeper than the tail, which tapers in the "downstream" direction (see illustration). Most drumlins are 5–50 m high, 400–600 m wide, and 1–2 km long. Drumlins grade from 100% bedrock, through individuals consisting of a bedrock core covered with till, into 100% drift, without change in the typical streamline form. They are made by glacial abrasion, glacial accretion (a plastering-on process), or both. In the mainly accretional forms the constituent till is rich in clay. Drumlins occur in families or in larger fields. Conspicuous fields include east-central Wisconsin, central-western New York, south-central New England, southwestern Nova Scotia, and northern Ireland. *See* GLACIATED TERRAIN. [RICHARD F. FLINT]

Side view of a drumlin, in the background, located in east-central Massachusetts. The former glacier flowed in a right-to-left direction. (*W. C. Alde, Geological Survey*)

Dunite

An ultrabasic rock consisting almost solely of olivine. The type locality is Dun Mountain, situated in the so-called Mineral Belt in New Zealand. Dunites generally form sills and lenses in layered basic complexes such as the Bushveld lopolith in South Africa. Together with less pure olivine rocks, such as peridotites, dunites also occur as conformable lenticular bodies in high- and medium-grade gneisses, for example, in the gneiss complex in southwest Norway. Crosscutting dikes or pipes of dunite are rarer. *See* OLIVINE; PERIDOTITE; PLUTON.

Composition. Besides a magnesium-rich olivine, chromite and picotite are common constituents of dunites. Dunites are therefore important sources of chromium. Spinel, magnetite, ilmenite, pyrrhotite, and native platinum also occur in some dunites.

If the amount of ferromagnesian silicates other than olivine (wehrlite and harzburgite) exceeds a few percent, the rock is classified as peridotite.

Structure and texture. Most dunites show various kinds of deformation structure. The olivine in the dunite from the type locality usually shows undulose extinction. Even crushed structures are common. Most dunites studied show preferred lattice orientation of the olivine.

Alteration. Serpentinization of olivine in dunites is widespread. Some dunites have changed completely to serpentine bodies. *See* SERPENTINE; SERPENTINITE.

Origin. Dunites may form by fractional crystallization of basic magma or by replacement of rocks such as limestone and dolomites; metamorphic differentiation has also been suggested as a mechanism of formation. It is also possible that some dunites are directly derived from a deep-seated basic layer in the earth.

Whatever the complete story of the origin of these rocks, it is evident that much of their structure and mineral association developed by a metamorphic process. Some dunites have probably been emplaced as solids. *See* MAGMA; METAMORPHISM. [HANS RAMBERG]

Earth

The third planet in the solar system, lying between Venus and Mars, and the only part of the known universe to present the intermingling conditions of air, water, and land, making possible such a life zone as that at the face of this terrestrial globe. Various scientific investigations are aimed at increasing knowledge of the inner, surficial, and outer parts of the Earth and of their relationships with external parts of the universe. Most of the topics outlined in this article lie within the general field of geophysical investigation. *See* GEOPHYSICS.

Shape, gravitation, and density. The Earth is roughly elliptical in shape (Table 1), the equatorial bulge being approximately what would be expected for a rotating fluid. Surface elevations are referred to a theoretical surface called the geoid, the equipotential surface which most nearly approximates mean sea level. It can be visualized as the surface passing through the continents which the water table would assume if the rock were a porous, permeable mass which exerted no forces on the water other than purely gravitational forces. The geoid departs from the approximating spheroid probably by 100 m, at most.

The shape of the Earth is commonly specified in terms of its gravitational field as shown in the equation below, where g_0, a, b, c, and λ' are con-

$$g = g_0[1 + a\sin^2\varphi + b\sin^2 2\varphi + c\cos^2\varphi\cos 2(\lambda + \lambda')]$$

stants, φ is latitude, and λ is longitude measured westward from Greenwich (Table 2). The Earth closely approximates a spheroid of revolution, in which case c is zero. Computations based on the shapes of satellite orbits suggest that additional terms may be needed to describe accurately the Earth's shape. *See* GEODESY.

The average density of the Earth is 5.517 g/cm³. Because the density of surface rocks is only 1.6–3.4 g/cm³, the interior of the Earth must consist largely of rocks of greater density (Fig. 1). This is indicated also by the Earth's moment of inertia. A uniform sphere would have a moment of 0.4 MR^2, where M is its mass and R its equatorial radius. The Earth's moment is 0.3337 MR^2.

Water and land distribution. Water covers 70.8% of the Earth's surface (Table 3). The land is very unequally distributed on the globe. Over 80% lies in the hemisphere centered at 38°N lat 0° long. Furthermore, the land is predominantly in the Northern Hemisphere (Fig. 2), most of it lying in south-pointing wedges around the Arctic Basin. The proportion of land over sea surface is at a max-

Table 1. Dimensions of the Earth*

Equatorial radius	6,378,099 ± 116 m
Polar radius	6,356,631 m
Radius of sphere of equal volume	6,371,200 m
Ellipticity	0.0033659 ± 0.0000041
Volume	1.083×10^{27} cm³
Mass	5.975×10^{27} g
Average density	5.517 g/cm³
Moment about polar axis, A	8.05×10^{44} g/cm²
Moment about equatorial axis, C	8.08×10^{44} g/cm²
$(C-A)/A$	0.003273

*B. Gutenberg, *Internal Constitution of the Earth*, Dover, 1951; H. Jeffreys, *The Earth: Its Origin, History and Physical Constitution*, 3d ed., Cambridge, 1952.

Table 2. Values of constants in the gravity formula*

g_0	a	b	c	λ'	Source
978.0490	0.0052884	0.0000059	0		International Formula of 1930
978.0496	0.0052934	0.0000059	0		Uotila 1957
978.0516	0.0052910	0.0000059	0.0000106	6°	Uotila 1957

*W. A. Heiskanen and F. A. Vening Meinesz, *The Earth and Its Gravity Field*, McGraw-Hill, 1958.

Table 3. Area and elevation of the land and oceans

Earth divisions, land and water	Area, % of Earth	Area, km² × 10⁶	Average elevation, m	Source
Land	29.2	148.892	840	Kossinna (1921)*
Ocean	70.8	361.059	−3800	Kossinna (1921)
Continental shelves	5.4	27.5	−100	Kossinna (1921)
Continental slopes	9.8	50	−2200	Howell (1959)
Ocean basins	47.8	243.6	−4860	
Seas	7.8	39.928	−1210	Kossinna (1921)
Whole Earth	100.0	510.1	−2440	Kossinna (1921)

*E. Kossinna, Die Tiefen des Weltmeeres, *Veröff. Inst. Meereskunde Univ. Berlin*, Neue Folge A, *Geogr.-naturwiss.*, 9:1–70, 1921.

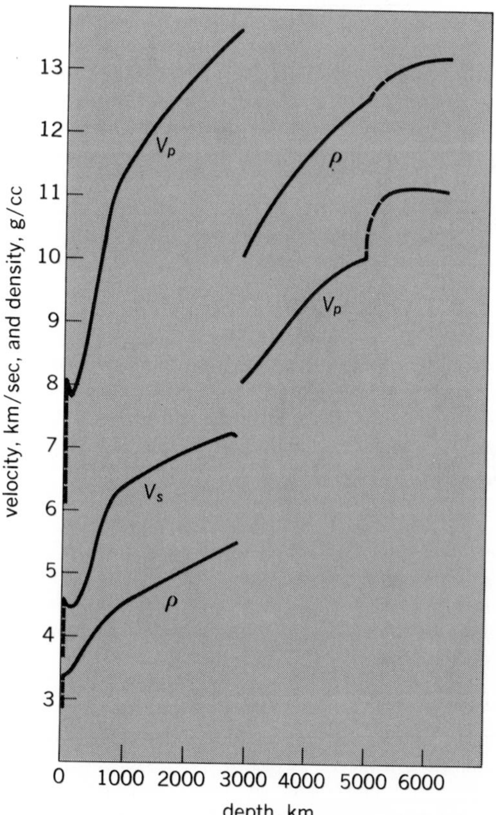

Fig. 1. Density (ρ), compressional wave (V_p), and shear wave (V_s) velocities. (*After B. Gutenberg, Trans. Amer. Geophys. Union, 39:488, 1958, and D. L. Anderson and R. L. Kovach, J. Geophys. Res., 68:3499, copyright 1963 by American Geophysical Union*)

imum at 66°N lat. The land itself has a generally concentric arrangement, with mountain ranges fringing the continents and old crystalline rocks outcropping in the centers. The Himalayan region of Asia is the highest land, Mount Everest (8840 m) being the highest peak.

The oceans may be visualized as a series of basins separated by ridges. The deepest parts of the oceans are not in their centers but in narrow troughs along their edges or adjoining submarine ridges and island arcs (Table 4). *See* MARINE GEOLOGY; OCEANS.

The ocean floors are usually covered by 1–3 km of sediments. From them rise mountains of volcanic rocks in whole ranges, lines of cones, or isolated peaks. These volcanic rocks appear to be different chemically from rocks of the continents (Table 5).

The oceanic volcanoes produce only basaltic lavas, whereas both basalts and more acidic extrusives are found on the continents. The acidic rocks are lighter in weight than the basaltic, and the height of the continents is thus explained. The surface of the Earth can be looked upon as being divided into two predominant levels (Fig. 3), the ocean basins at a depth of 4–6 km and the continental plateaus from −200 m to +1 km. Above these stand the mountain ridges; below them dip the ocean deeps. They are separated by the abrupt (2–3.5°) drop of the continental slopes.

Enveloping atmosphere. The Earth is surrounded by an envelope of gas, the atmosphere. This is arbitrarily divided into layers on the basis of temperature (Fig. 4). The lowermost layer, the troposphere, contains about three-fourths of the mass of the atmosphere and is the only layer capable of supporting life as known on Earth. It varies in thickness from about 8 to 17 km, bulging at the Equator. The tops of even the highest mountains are all within the troposphere, as are most of the clouds and all of the weather experienced on the ground.

At the critical level of 450–550 km the horizontal mean free path of the gas particles (distance between collisions) becomes equal to the elevation, and the gas is so tenuous that the term temperature ceases to have its conventional meaning. Above this point, light atoms such as hydrogen and even helium when given sufficient thermal energy can escape from the Earth altogether (escape velocity = 11.2 km/sec).

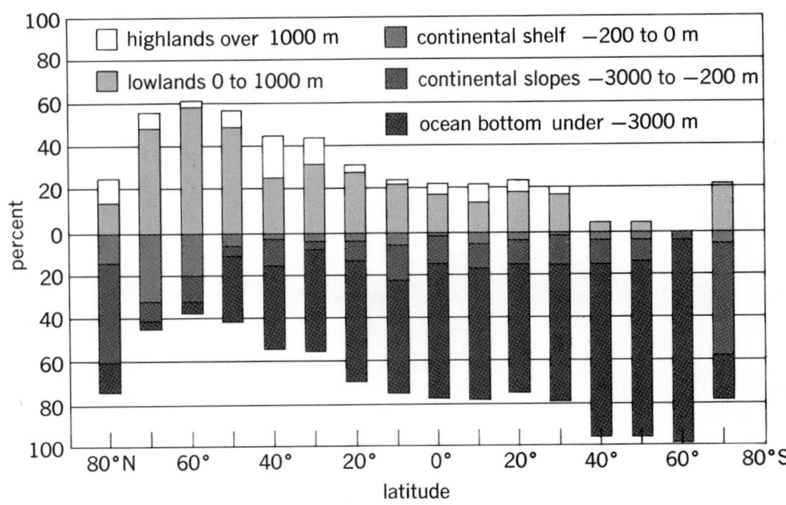

Fig. 2. Graph indicating the variation of elevation with latitude along a number of selected parallels. (*After Howell, 1959*)

The atmosphere becomes increasingly ionized with elevation. Because of the high concentration of electrically charged particles, the upper atmosphere (ionosphere) is a good conductor of electricity. The electrical charges move in loops in the Earth's magnetic field, and this motion tends to restrain particles in the Earth's neighborhood far beyond the critical level. Two regions of exceptionally high charge concentration, known as Van Allen radiation belts, have been recognized (Fig. 5).

Surface and interior temperatures. At the surface of the Earth the average annual temperature varies from around 90°F (32°C) to lower than −25°F (−32°C). Among the highest and lowest recorded temperatures (in shade) are −125°F (Vostok, Antarctica) and +136.4°F (Libyan Desert). The gradient with depth at the surface generally lies in the range 0.01–0.04°C/m. Figure 6 gives an estimate of the temperature in the interior. The average heat loss at the surface by conduction from below is about $1.2 \times 10^{-6} \pm 50\%$ g-cal/(cm^2)(sec). This is equivalent to 6×10^{12} cal/sec for the whole Earth. The additional heat loss through volcanoes and

Table 4. Greatest depths in the oceans*

Name	Depth in water, m	Ocean basin	Lying near
Challenger Deep	11,500	Pacific	Mariana Islands
Mindanao Deep	10,497	Pacific	Philippines
Ramapo Depth	10,374	Pacific	Honshu, Japan
Tonga-Kermadec Trench	10,035	Pacific	Tonga-Kermadec Islands
Planet Depth	9,410	Pacific	New Britain
Milwaukee Depth	8,750	Northwest Atlantic	Puerto Rico
Bonin Trench	8,660	Pacific	Bonin Islands
Byrd Deep	8,590	Pacific	Southeast of New Zealand
Tuscarora Depth	8,500	Pacific	Kurile Islands
South Sandwich Trench	8,264	Southwest Atlantic	South Sandwich Islands
Aleutian Trench	7,680	Pacific	Aleutian Islands
Atacama Trench	7,635	Pacific	Northern Chile
Ryukyu Trench	7,480	Pacific	Ryukyu Islands
Sunda Trench	7,455	East Indian	Java

*From G. P. Kuiper (ed.), *The Earth as a Planet*, University of Chicago Press, 1954; H. U. Sverdrup, M. W. Johnson, and R. H. Fleming, *The Oceans*, Prentice-Hall, 1942.

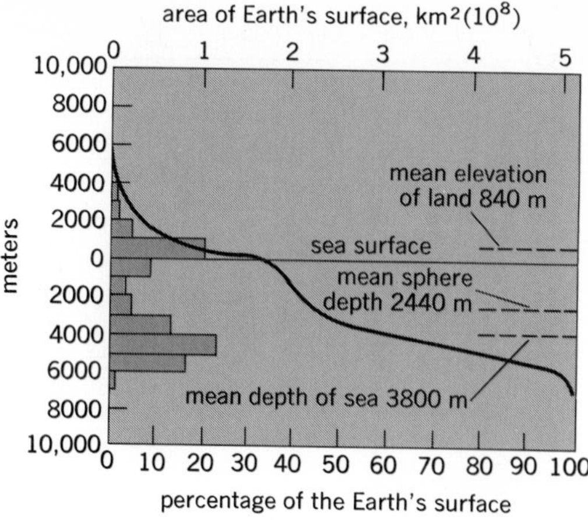

Fig. 3. Hypsographic curve of land surface elevation. (*After E. Kossinna, Die Tiefen des Weltmeeres, Veröff. Inst. Meereskunde Univ. Berlin, Neue Folge A, Geogr.-naturwiss., 9:1–70, 1921*)

hot springs is at least two orders of magnitude smaller. The estimated concentrations of radioactive elements in the Earth's crust are sufficient to provide all this heat by their disintegration. Therefore, it is believed that their concentrations decrease rapidly with depth, at least under the continents. *See* EARTH, HEAT FLOW IN.

Layering of the interior. Most information about the Earth's interior is provided by seismology. From a study of the arrival times of seismic pulses, it is known that the Earth contains three main layers: the crust, the mantle, and the core (Fig. 7). The composition of the crust beneath the continents is believed to vary with depth. Much of the surface is covered by a thin veneer of sedimentary rocks, 0–3 km thick. In a few places great thicknesses (up to 20 km) have accumulated in local basins. Beneath this are crystalline rocks which have been formed under conditions of high temperature and pressure. The rocks of the continents are largely silicates, aluminum-rich at the surface (sial) grading to iron- and magnesium-rich at depth (sima). The sial is missing beneath the oceans. It is uncertain how sharp a boundary exists between the sial and sima. There is probably an intermediate zone in which the velocities of earthquake waves are at a minimum. Seismic pulses are not returned directly to the surface from this zone; hence little is known of its nature.

At the lower boundary of the crust a sudden increase in seismic wave velocity occurs. This is known as the Mohorovičić discontinuity, after its discoverer. It may represent a change in composition or only a phase change from one suite of mineral species to another of identical chemical composition. No direct evidence of the composition of the mantle or core exists. The increase in velocity at the Mohorovičić discontinuity is consistent with what one would expect if the composition changed from a gabbro to a more ultrabasic rock such as dunite or eclogite. On the assumption that meteorites are a sample of a fragmented former planet like the Earth, it would be expected that the bulk of the Earth is composed of ultrabasic silicates and metallic iron (Table 5). According to this analogy, the outer mantle would be composed of silicates, similar to ultrabasic rocks found at the Earth's surface. Between 150 and 900 km depth, the zone of rapidly increasing seismic wave velocity (Fig. 1), the various minerals constituting the mantle rocks are thought to have phase transitions to denser forms. *See* MOHO (MOHOROVIČIĆ DISCONTINUITY).

According to the meteorite analogy, the core would be composed of nickel-iron, liquid at the top and probably solid at the center.

In the crust and the outermost part of the mantle most rocks are brittle and break before they will flow plastically. Studies of the variation of gravity over the Earth's surface in relation to its history of erosion and sedimentation indicate that the surface rises and falls to maintain a sort of hydrostatic

Table 5. Average composition of rocks believed typical of the major layers of the Earth, wt%

Element	Continental crust*	Plateau basalt†	Stone meteorites‡	Iron meteorites‡
O	46.59	44.67	36.15±0.89	
Si	27.72	22.81	18.12±0.22	
Al	8.13	7.40	1.53±0.13	
Fe	5.01	10.11	24.18±1.08	90.78±0.26
Ca	3.63	6.70	1.74±0.18	
Na	2.85	1.92	0.69±0.05	
K	2.60	0.57	0.18±0.02	
Mg	2.09	4.04	13.93±0.29	
Ti	0.63	1.31	0.08±0.03	
H	0.13	0.20	0.06±0.02	
P	0.13	0.14	0.14±0.01	
Mn	0.10	0.13	0.26±0.06	
S	0.052		1.79±0.08	
Cl	0.048			
Cr	0.037		0.30±0.05	
C	0.032			
Ni	0.020		1.53±0.16	8.59±0.24
Co	0.001		0.10±0.02	0.63±0.02
Other	0.200			

*B. Gutenberg (ed.), *Internal Constitution of the Earth*, Dover, 1951.
†R. A. Daly, *Igneous Rocks and the Depths of the Earth*, McGraw-Hill, 1933.
‡H. Brown and C. Patterson, The composition of meteoric matter II, *J. Geol.*, 56:85–111, 1948.

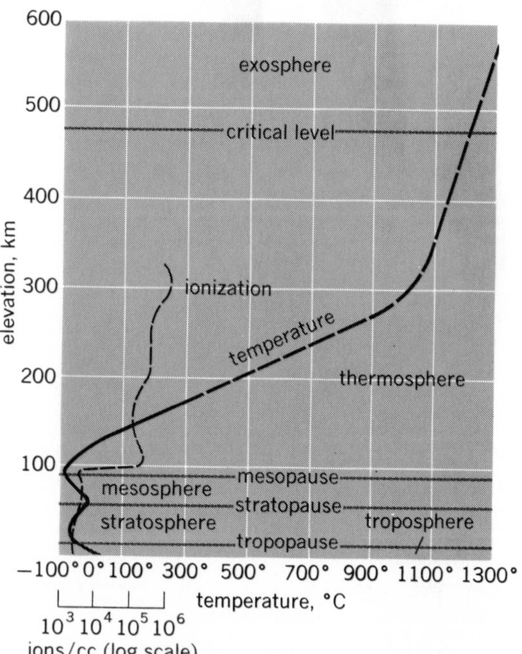

Fig. 4. Structure of the atmosphere.

balance called isostasy. This implies that the rocks beneath a certain depth (96 km), called the depth of compensation, are plastic and of low strength. However, there is a finite breaking strength at least to a depth of 700 km, because earthquakes occur to that depth. *See* ASTHENOSPHERE; ISOSTASY; TERRESTRIAL GRAVITATION.

The surface of the Earth is continually undergoing deformation. All coastlines show evidence of repeated rise and fall of the land relative to the sea. Some of this deformation, but not all, is due to changes in sea volume. Sea-level falls of 75–100 m have occurred when water was locked in continental glaciers. If all the present glaciers were melted, the sea level would rise roughly 60 m. The volume of surface water is presumably being increased slowly by the new water in volcanic gases. The amount is so small that sea level has not been increased noticeably in the last 5×10^8 years, as proved by the distribution of sediments. One might expect that in this length of time erosion would have worn away all the continents. Some process acts to preserve the continents, causing them to rise repeatedly over large areas, and to be thrown into mountains in long narrow belts. The forces causing these changes must come from the Earth's interior. One theory is that the interior of the Earth is shrinking in size, causing the crust to wrinkle as it adjusts. Another is that subcrustal convection currents sweep parts of the crust together, producing thick welts of mountain ranges. *See* CONTINENT FORMATION; GLACIOLOGY; OROGENY.

Both dilatational and shear waves are transmitted throughout the crust and mantle. In the outer part of the core, however, no shear waves are observed, suggesting that it is fluid. The decrease in seismic wave velocity at the core boundary (Fig. 1) is too large to be easily explained only by a change from the solid to the liquid state. The division of seismic energy between reflected and transmitted phases suggests a sharp increase in density, implying a change in composition.

It has been suggested that the core is a plasma, a material so compressed that the orbital electrons have been collapsed around the nuclei of the atoms. If this is the case, the composition may be largely hydrogen and helium as in stars. However, it is doubtful whether sufficient pressure exists to create a plasma of this sort within the Earth. Within the core is a central body of higher seismic velocity than that of the outer part. The size of the velocity increase at 5000 km depth (Fig. 1) is consistent with a change from liquid to solid with no change in chemical composition. *See* EARTH, INTERIOR OF; SEISMOLOGY.

Major planetary relationships. The Earth moves about the Sun in an elliptical orbit (eccentricity = 0.01674, average radius = 1.495×10^8 km) in a little over a year. The tropical year on which the Gregorian calendar is based is the period between vernal equinoxes, which precess in position with respect to the stars with a period of 25,800 years. The rate of precession, 50.2 sec/year, is not constant, but has a fluctuation, called nutation, of about 9.23 sec. Both precession and nutation are due to the gravitational attraction of the Sun and Moon on the Earth's equatorial bulge.

The Earth rotates 365.2422 times on its axis over a period of a year. This period is not constant, however, but is increasing at a rate of a little more than 1 microsecond (μsec) per 100 years. The cause of this change is a transfer of energy from the rotation of the Earth to the revolution of the moon about the Earth through the action of tidal friction. There are also irregular fluctuations of as much as 5 msec in the length of a day. The best explanation of these is that the Earth's core is in turbulent fluid motion. This motion is coupled to the mantle and crust of the Earth in such a way that changes in the surface motion compensate for changes in the core currents to keep the total moment of inertia constant.

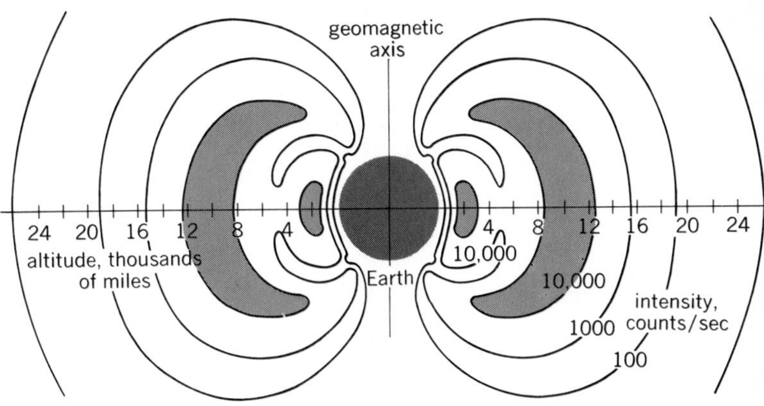

Fig. 5. The Van Allen radiation belts.

The axis of the Earth is tilted 23°26′59″ to the plane of its revolution. The axis of rotation does not coincide with the axis of figure, but circles about it in a counterclockwise direction with a maximum separation of about 0.4 sec. The period of this motion, called the Chandler (Eulerian) motion, is about 14 months. In addition, it is believed that the position of the axis of rotation of the Earth (or of the crust with respect to the interior) may have shifted greatly in geologic time. Evidence for such changes comes from studies of the direction of remnant magnetization of rocks. *See* ROCK MAGNETISM.

Satellites. The Earth has one natural satellite and a variable number of artificial ones.

The natural satellite is the Moon, which moves about the Earth in an elliptical orbit at a mean distance of 383,403 km with an eccentricity of 0.05490 and a period of 27 days 7 hr 43 min 11.5 sec. Its orbit has an average inclination to the orbit of the Earth of 5°8′33″. Actually, the Earth and Moon move about a common center of gravity, whose motion about the Sun is more regular than that of

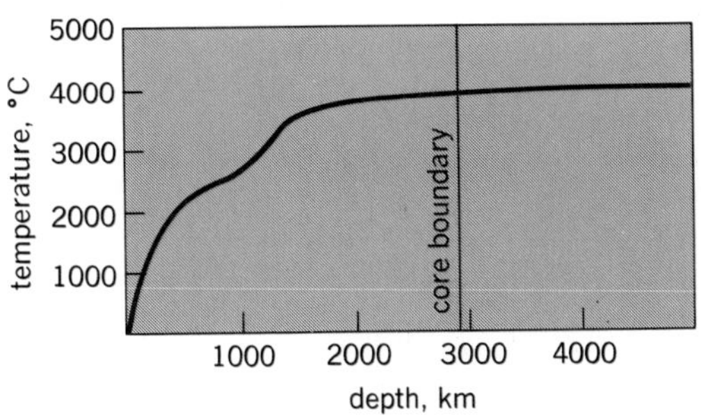

Fig. 6. Temperature in the interior of the Earth. (*After E. Lubimova, Geophys. J., 1:115, 1958*)

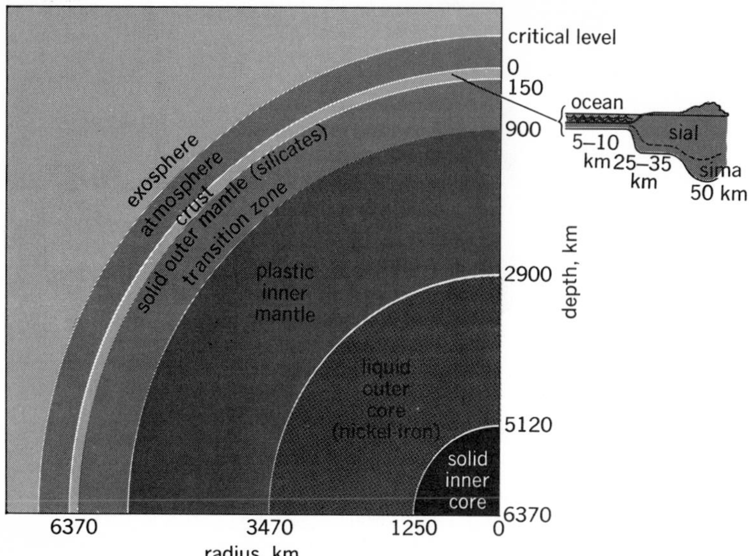

Fig. 7. The principal layers of the Earth.

the Earth. Because of the lesser mass of the Moon (1/81.3 of that of the Earth), this center is within the Earth at an average radius of 4645 km.

Theories on age and evolution. Many lines of evidence suggest an age for the Earth of $4-6 \times 10^9$ years. The presence of any radioactive elements in the Earth at all indicates that the material of which it is formed was created under entirely different conditions from any found in the Earth today. The only places known today where radioactive elements are likely to form are the centers of exceptionally hot, dense stars. Assuming all the heavy nuclides were formed in nearly equal quantities, as seems likely, their relative abundance suggests an age for these elements of about 6×10^9 years. The relative amounts of radioactive elements and their end products in the Earth suggest an age of $5-5.5 \times 10^9$ years for the Earth's crust. The oldest rocks found anywhere on the Earth appear to have been formed $2-3.5 \times 10^9$ years ago. Meteorites have ages (from studies of their radioactivity) of about 4.5×10^9 years. *See* DATING METHODS; GEOCHRONOMETRY.

The rate of recession of distant galaxies has been estimated to be such that about $6-8 \times 10^9$ years ago all were grouped tightly in a small volume of the universe. The origin of the Earth may thus have been a minor event in the origin of all known galaxies.

At one time it was thought that the Earth was a fragment of a larger star which had somehow been disrupted. A more widely held hypothesis is that the Sun and all bodies of the solar system condensed from a dispersed gas. The smaller planets lacked strong enough gravitational fields to retain their hydrogen and helium, and thus they differ in composition from the larger planets and the Sun. According to one variation of this theory, the Earth formed originally by coagulation of cold particles in a turbulent gas cloud. At first the Earth was uniform in composition, but as it grew in size the temperature of the interior increased through compression and radioactive decay, and iron separated from the silicate phase to form the core. The bulk of the elements, particularly the light and the radioactive elements, was concentrated upward. This process is presumed to be still going on. Radiogenic heat and gravitative forces are the two major sources called on to explain the deformations to which the crust is continually subject. The thermal conductivity of the Earth is so low that it is possible that the Earth's interior may still be warming. It is believed by some on geologic grounds that the rate of change of the Earth's surface features has been accelerating throughout its history. In any case the Earth is still undergoing change, and can be expected to evolve for some time to come. *See* EARTH, AGE OF. [BENJAMIN F. HOWELL, JR.]

Bibliography: K. E. Bullen, An earth model based on a compressibility-pressure hypothesis, *Mon. Notic. Roy. Astron. Soc.*, Geophys. Suppl., 6: 50–59, 1950; K. E. Bullen, *Introduction to the Theory of Seismology*, 2d ed., 1954; R. A. Daly, *Igneous Rocks and the Depths of the Earth*, 1933; T. F. Gaskell, *The Earth's Mantle*, 1967; B. Gutenberg, *Physics of the Earth's Interior*, 1959; W. A. Heiskenen, *Size and Shape of the Earth*, Inst. Geod. Photogramm. Cartogr. Publ. no. 7, 1957; B. F. Howell, Jr., *Introduction to Geophysics*, 1959; H. Jeffreys, *The Earth: Its Origin, History and Physical Constitution*, 5th ed., 1968; G. P. Kuiper (ed.), *Earth as a Planet*, 1954; A. N. Strahler, *The Earth Sciences*, 1963.

Earth, age of

The age of the Earth may be defined as the time that this planet has existed with approximately its present mass and density. Today the best estimate for this period is 4.6×10^9 years.

Once the general physical features of the Earth and its solar environment became known, the age of the Earth was determined by means of long-term and steady processes whose rates and extent of progress could be measured. Such processes as the cooling of the Earth, the accumulating of sediments, and the salting of the oceans were used. Today the process of radioactivity is utilized, where the time required to form a measured amount of decay product associated with a radioactive element is calculated from the known rate of disintegration of the parent element.

Minimum age. It is difficult to apply this method to the whole Earth, since any rock chosen as a sample for study is a secondary system younger than the Earth. For this reason the age of the oldest rocks that can be sampled will give only a minimum age for the Earth. In every continent there are limited regions which have somehow escaped pervasive destruction by weathering and igneous activity for very long periods of time. These oldest regions, called shield areas because of their stability, consist of rocks little different from any formed during subsequent times. The ages of minerals in some of these oldest available rocks have been determined by various radioactive decay systems, such as $K^{40} \rightarrow A^{40}$, $Rb^{87} \rightarrow Sr^{87}$, or $U^{238} \rightarrow PB^{206}$. In this method the concentrations of both the radioactive parent and the decay product elements are determined in a mineral, and the age is calculated from known physical relations. The case of a uranium-containing mineral is shown as an example by Eq. (1). The amount of Pb^{206} increases in the

$$Pb^{206} = U^{238}(e^{\lambda_8 T} - 1) \tag{1}$$

mineral as the U^{238} in it decays according to its characteristic constant λ_8. By measuring the amount of Pb^{206} formed and the amount of U^{238} remaining within the mineral, the equation can be solved for the only unknown, *T*, the age of the mineral. The oldest mineral ages on the Earth obtained in this manner are about 3.5×10^9 years. *See* ROCK, AGE DETERMINATION OF.

Maximum age. A maximum age for the Earth is given by theories for the cosmogenic origin of the elements. According to present theory, the nuclides U^{235} and U^{238} were formed in the ratio 1.64:1 within an ancient star just before it exploded and scattered debris in space. The Sun and Earth were subsequently formed of material which included this debris. U^{235} and U^{238} are both radioactive, but U^{235} decays much faster than U^{238} so that the ratio of U^{235} to U^{238} is constantly decreasing. The U^{235}/U^{238} ratio is 0.007 in the Earth today. If one calculates how far back in time he must go in order to increase this ratio to the maximum prescribed by cosmogenic theory, one finds that it takes 6.6×10^9 years. Since the Earth is younger than any element formed in this manner, the maximum age of the Earth from this point of view is 6.6×10^9 years. Corrections must be made for the production of uranium in stars exploding during a finite interval or at a constant rate, but a maximum age is not seriously increased by the choice of a reasonable cosmological model.

Ore-lead age. The ratio of radioactive parent and accumulated daughter of a decay scheme cannot be measured within the whole Earth, but it is possible to make a direct measurement of the age of the Earth by comparing the relative progress of the two radioactive decay schemes $U^{235} \rightarrow Pb^{207}$ and $U^{238} \rightarrow Pb^{206}$. This group of four nuclides has an important and singular property. The ratio of radiogenic Pb^{207} and Pb^{206} formed during a period of uranium decay is time-dependent, and the age of any closed system containing uranium is easily determined by measuring only the ratio of radiogenic Pb^{207} and PB^{206} contained in it, since the different decay rates of the uranium nuclides and their relative abundances are known. This is shown by Eq. (2). According to this relation,

$$\frac{Pb^{206}}{Pb^{207}} = \frac{U^{238}(e^{\lambda_8 T} - 1)}{U^{235}(e^{\lambda_5 T} - 1)} \tag{2}$$

it is not necessary to measure the amounts of lead or uranium in the mineral chemically, for only isotropic ratios are needed. Since the isotopic composition of uranium is everywhere the same, the Pb^{206}/Pb^{207} ratio is all that is required to measure *T*, the only unknown left in Eq. (2). *See* LEAD ISOTOPES, GEOCHEMISTRY OF.

Of all the lead in the Earth, only part of the Pb^{206} and Pb^{207} has been formed by uranium decay because there was some primordial load Pb^{206} and Pb^{207} in the Earth when it was formed. A very useful characteristic of primordial lead is that it contains Pb^{204}, whose abundance in the Earth does not change, for it does not have a long-lived radioactive parent. As a consequence, changes in the abundances of Pb^{206} and Pb^{207} may be determined simply by comparing their abundances against Pb^{204}. In primordial lead the Pb^{206}/Pb^{204} and Pb^{207}/Pb^{204} ratios have fixed values. When radiogenic Pb^{206} and Pb^{207} are added, these ratios increased to higher values.

Nearly all lead in the Earth is a mixture of primordial lead and radiogenic lead. At any given time there is only a small quantity of radiogenic lead which exists separately in uranium minerals. These minerals are continuously being formed and destroyed so that their radiogenic leads are continuously being mixed into the common pool of Earth lead. The radiogenic component of common Earth lead is therefore the sum of many small increments added during the Earth's lifetime.

If a uranium mineral contains primordial lead, then the radiogenic Pb^{206} which has accumulated may be found by subtracting the primordial Pb^{206}/Pb^{204} ratio from the total Pb^{206}/Pb^{204} ratio in a sample of lead from the mineral. The radiogenic Pb^{207} may be similarly found. The ratio of these two radiogenic uranium leads is the time-dependent Pb^{207}/Pb^{206} ratio useful for age calculations. This is shown by Eqs. (3) and (4), where the abbreviations indicate "primitive" and "radiogenic."

$$\frac{(Pb^{206})_{prim.} + (Pb^{206})_{rad.}}{(Pb^{204})_{prim.}} - \frac{(Pb^{206})_{prim.}}{(Pb^{204})_{prim.}} = \frac{(Pb^{206})_{rad.}}{(Pb^{204})_{prim.}} \tag{3}$$

$$\frac{(Pb^{206})_{rad.}/(Pb^{204})_{prim.}}{(Pb^{207})_{rad.}/(Pb^{204})_{prim.}} = \frac{(Pb^{206})_{rad.}}{(Pb^{207})_{rad.}} = \frac{U^{238}(e^{\lambda_8 T} - 1)}{U^{235}(e^{\lambda_5 T} - 1)} \tag{4}$$

In a practical case, if two minerals have the same age and both contain primordial lead but only one contains uranium, then the age can be readily calculated from the ratio of the differences between the uranium leads in the two lead samples. Similarly, if two minerals have the same age and both contain primordial lead and both contain uranium but in different proportions, then the age can be calculated as before from the ratio of the differences between the two leads without knowledge of the composition of primordial lead. This is shown by Eqs. (5) and (6).

$$\frac{(Pb^{206})_{prim.} + (Pb^{206})^{I}_{rad.}}{(Pb^{204})_{prim.}} - \frac{(Pb^{206})_{prim.} + (Pb^{206})^{II}_{rad.}}{(Pb^{204})_{prim.}} = \frac{\Delta(Pb^{206})_{rad.}}{(Pb^{204})_{prim.}} \tag{5}$$

$$\frac{\Delta(Pb^{206})_{rad.}/(Pb^{204})_{prim.}}{\Delta(Pb^{207})_{rad.}/(Pb^{204})_{prim.}} = \frac{(Pb^{206})_{rad.}}{(Pb^{207})_{rad.}} = \frac{U^{238}(e^{\lambda_8 T} - 1)}{U^{235}(e^{\lambda_5 T} - 1)} \tag{6}$$

The first truly significant measurement of the age of the Earth, the ore-lead method, was based on this principle. In order to measure the age of the Earth, it is not necessary to determine how much radiogenic Pb^{206} has accumulated in the Earth or how much U^{238} remains, according to Eqs. (5) and (6). Common Earth leads are obtained from lead ore minerals which exist separately in small amounts just as uranium minerals do. The Earth is such a vast object that the lead from one continent may never mix well with that from another, and if there were more uranium or less primordial lead in one continent than another, their common leads might contain different proportions of radiogenic lead. The age of the Earth might be calculated from the ratio of the differences between common ore leads in the two continents. Actual differences among common Earth leads are so small that many different lead samples and statistical approaches must be used in practical calculations. The age obtained in this manner is about 3×10^9 years. There are many theoretical difficulties involved in this method, but by using a more reliable age of the Earth (as given below) this same type of calculation can be used to study the mixing of rocks within the Earth.

Meteorite-lead age. The most acceptable way to measure the Earth's age is by the meteorite-lead method. This is simply an extension of the ore-lead method to include meteorites. When leads are isolated from various iron and stone meteorites and their different isotopic compositions are compared, a simple pattern emerges: The differences, shown in Eqs. (7) and (8), cover a wide range of

$$[Pb^{206}/Pb^{204}]_{\text{meteorite 1}} - [Pb^{206}/Pb^{204}]_{\text{meteorite 2}} = \Delta[Pb^{206}/Pb^{204}]_{1,2} \quad (7)$$

$$[Pb^{207}/Pb^{204}]_{\text{meteorite 1}} - [Pb^{207}/Pb^{204}]_{\text{meteorite 2}} = \Delta[Pb^{207}/Pb^{204}]_{1,2} \quad (8)$$

numerical values among different meteorites, but the ratios

$$\Delta[Pb^{206}/Pb^{207}]_{1,2} \text{ and } \Delta[Pb^{206}/Pb^{207}]_{2,3}$$

and so forth are found to equal a constant. The age calculated according to Eq. (2) from this constant is 4.6×10^9 years, and for meteorites it has been confirmed in a general way by other radioactive decay methods; that is, meteorites may be considered separate little planets, all formed at the same time, all containing the same kind of primordial lead, and all having varying proportions of radiogenic lead. Because some meteorites contain essentially no uranium, enough information is available not only to calculate an accurate age for meteorites but to describe and define completely all the possible meteoritic leads which can exist.

These meteoritic leads constitute only a very small fraction of the total of every possible kind of lead, so that if the lead from an object of unknown age fits the description of a meteoritic lead, it is probably safe to conclude that the object has the age calculated for meteorites. This is the case for the Earth. Samples of common lead are obtained from the oceans which contain mixtures of leads derived from all lands, and these leads fit the description of meteoritic lead. One therefore may conclude that the Earth has the same age as meteorites, namely, 4.6×10^9 years.

The isotopic composition of lead from those meteorites which contain virtually no uranium has at present acquired the status of a solar constant, and it may not be generally recognized that it appears in calculations of the age of the Earth, which use terrestrial lead isotopes, without explicit mention of the links which thus are necessarily involved between the genesis of the Earth and of meteorites. It is these latter relationships which are most obscure, and to this extent so is the age of the Earth. The age calculations found in the literature which involve the meteorite lead constant and terrestrial lead isotopes therefore clarify the evolution of uranium-lead systems within the Earth by using a language that is both self-consistent and intellectually satisfying but whose meaning regarding the absolute age of the Earth is still poorly understood. *See* GEOLOGICAL TIME SCALE; METEORITE.

[CLAIR C. PATTERSON]

Bibliography: A. Holmes, *Principles of Physical Geology*, rev. ed., 1965; F. Hoyle and W. A. Fowler, On the abundances of uranium and thorium in solar system material, in H. Craig, S. L. Miller, and G. J. Wasserburg (eds.), *Isotopic and Cosmic Chemistry*, 1964; R. Ostic, R. Russell, and P. Reynolds, A new calculation for the age of the Earth from abundances of lead isotopes, *Nature*, 199:1150, 1963; C. Patterson, Age of meteorites and the Earth, *Geochim. Cosmochim. Acta*, 10:230; 1956; R. Russell and R. Farquhar, *Lead Isotopes in Geology*, 1960; G. Tilton and R. Steiger, Lead isotopes and the age of the Earth, *Science*, 150:1805, 1965; T. Ulrych, Oceanic basalt leads: A new interpretation and an independent age for the Earth, *Science*, 158:252, 1967.

Earth, deformations and vibrations in

Alterations of form or shape of the Earth. These changes and vibrations range from minute to great in magnitude at intervals differing from regular to irregular and from extremely short to long portions of the geologic history of the Earth. Large displacements of rock masses along a geologic fault or the shock of an earthquake can be impressive to a casual observer, but many types of deformations are revealed only by instruments of the highest sensitivity, or by precise measurements extending over tens or hundreds of years. A systematic survey of these and related questions is briefly developed in this article and may be carried further by consulting the articles cited by cross reference.

Background considerations. The Earth is an engine powered by heat and at times by gravitational energy which has been operating vigorously for some 4,000,000,000 years. Some of the results to date of this activity may be seen in the present surface features of the Earth, for example, the distribution of continents, oceans, and mountain systems. More basic to the understanding of the energetics of the evolving Earth is the concealed structure of its interior, as it is revealed by the behavior of seismic waves. These indicate a solid inner core, a fluid outer core, a thick solid mantle, and thin superficial crust above a well-defined Mohorovičić discontinuity.

The subject "Earth deformations" has literally acquired new dimensions with the accumulation of striking evidence that continental drift has occurred. Thus the accepted magnitude of observable deformations has been multiplied by more than 10, from displacements of several hundred kilometers observed in major fault systems (such as the San Andreas in California) to distances of the order of 5000 km involved in the separation of the continents Africa and South America. The record of the sea floor spreading during the last 70,000,000 years has been recovered with especial clarity through the realization that the newly cooled rocks being injected at the crests of mid-ocean ridges receive a permanent imprint of the then existing Earth's magnetic field as these rocks cool below the Curie point. In particular, the pattern of reversals of the Earth's magnetic field leads to recognizable signatures in the rocks identifiable in both hemispheres. The upper layer of solid rock (the lithosphere) is being transported bodily away from the midoceanic ridge, as on a conveyor belt. The permanent magnetization acquired at the source serves as a magnetic tape record of the lateral motion. Thus it is found that in the last 70,000,000 years steady transport has existed at rates of 1–5 cm or more per year. The subject of Earth deformations and movements has therefore acquired a new degree of freedom. It is no longer reasonable to try to identify the large observed deformations with the thermal contraction or expansion of a cooling or heating Earth. The new evidence at sea fully supports the classical values of rates of movement obtained from studies on land of active faults. These rates of a few centimeters per year are great enough for direct measurement with refined geodetic instruments. The newly expanded ability to study geophysically the great ocean basins brings essentially the whole area of the Earth under observation. Thus the full integrated pattern of Earth deformations is emerging for the first time. *See* CONTINENT FORMATION; EARTH; MOHO (MOHOROVIČIĆ DISCONTINUITY).

Systematic geophysical studies. Much of the evidence concerning the present nature and past evolution of the Earth is obtained from observation of its motions and deformations. The kinds of motions available for study are conveniently classified into three groups in respect to their time scales: (1) major events of geological history, (2) dynamic behavior patterns of the Earth, and (3) elastic waves in the Earth.

Major events of geological history. These include major events throughout a long time scale of several billion years. Investigations of these events focus on the growth of continents, oceans, and mountains, and on the transport of large amounts of material by lava flows from depth. These subjects are the province of special geophysical studies. *See* OROGENY; TECTONOPHYSICS; VOLCANOLOGY.

The abundant evidence of large horizontal shearing displacements along extensive fault systems seems to be especially significant. The San Andreas Fault, extending northwest in a nearly straight line for 900 mi across California, shows evidence of accumulative clockwise shearing displacements of about 300 km during the last 100,000,000 years. Similarly, magnetic mapping at sea clearly reveals clockwise shearing displacements of 250 km along Pioneer Ridge (38°30′N, between 127° and 136°W) along an east-west fault. These are examples of a significant type of mobility of the crust. They represent large continuing shear strains and associated shearing couples. Shearing couples on this scale seem difficult to explain without postulating a mobile subcrust supporting slow convection on a large scale.

Dynamic behavior patterns of the Earth. During the span of historic time, observations of astronomy and geodesy have established significant features of the Earth's dynamic behavior. The rate of precession of the equinoxes provides the best value of the ratio H involving the moments of inertia of the Earth, $H = C - [(A+B)/2C]$, where A, B, and C are principal moments of inertia. The period of the 14-month Chandlerian wobble provides a value for the Love number k. The rate of transfer of angular momentum from the Earth to the Moon and Sun by virtue of tidal torques has been best determined by astronomical observations during the last several hundred years. Observations of Earth tides with gravimeters, horizontal pendulums, and strain gages reveal the amplitude of the elastic tidal deformation of the Earth. Significant local and regional differences in the tidal yielding appear to exist. The tidal periods are much longer than the longest free period of oscillation of the Earth (which is slightly less than 1 hr). Accordingly, the solid Earth probably responds to tidal forces in an essentially static manner. Unfortunately, the modifications in this response caused by ocean tides are still undetermined. *See* EARTH TIDES; GEODESY; TERRESTRIAL GRAVITATION.

Elastic waves in the Earth. Such waves traversing the Earth from earthquake foci or from man-made explosions have furnished all the precise data about the Earth's internal geometry and of the changing values of its elastic parameters with depth. The large atomic explosions whose time and place of origin are accurately known have significantly improved precision of seismic interpretations concerning the Earth's interior. Furthermore, the increasing availability of large digital computers has made feasible the heavy computational programs involved in power-spectra analyses of seismic records and of earth-tide records. Such analyses increase the amount of significant information deducible from the records. *See* SEISMOLOGY. [LOUIS B. SLICHTER]

Bibliography: J. C. Crowell, The San Andreas Fault in southern California, *Proc. 21st Int. Geol. Congr.*, Norden, Germany, 1960; H. H. Hess, *Petrologic Studies*, 1962; P. M. Hurley, Test of continental drift by comparison of radiometric ages, *Science*, vol. 157, no. 3788, 1967; H. Jeffreys, *The Earth: Its Origin, History, and Physical Constitution*, 4th ed., 1959; D. P. McKenzie and R. L. Parker, The North Pacific: An example of tectonics on a sphere, *Nature*, vol. 216, no. 5122, 1967; W. H. Munk and G. F. MacDonald, *The Rotation of the Earth: A Geophysical Study*, 1960; V. Vacquier, Measurement of horizontal displacement along faults in the ocean floor, *Nature*, 183(4659):452–453, 1959; F. J. Vine, Spreading of the ocean floor: New evidence, *Science*, vol. 154, no. 3755, 1966; F. J. Vine and D. H. Matthews, Magnetic anomalies over oceanic ridges, *Nature*, vol. 199, no. 4897, 1963; J. E. White, *Seismic Waves*, 1965.

Earth, heat flow in

Terrestrial heat flow, or the amount of thermal energy escaping from the Earth per unit area and unit time, is of fundamental importance to geology and geophysics. The heat leading to the fusion and metamorphism of rocks and to the forces of mountain building are thought to originate in deep-seated thermal processes. A most useful quantity both for the general theory of these processes and for the estimation of temperatures beyond accessible depths is the terrestrial heat flow. *See* GEOLOGY; GEOPHYSICS.

There are two main reasons for dealing with heat flow rather than the closely related thermal gradient. Changes in thermal conductivity with depth in the Earth affect the thermal gradient but do not affect the heat flow. Besides, heat flow is closely related to heat production, which in the Earth is mainly due to radioactive decay. Local variations in heat flow in many cases reflect locally heterogeneous distributions of radioactive sources.

Heat flow at the surface. This is determined by multiplying the rate of increase of temperature with depth (the geothermal gradient) by the local thermal conductivity. The latter quantity is usually measured in the laboratory on samples taken from the site of the temperature observations. The geothermal gradient can be determined on land wherever underground temperatures can be measured. It is usually necessary to have data extending to a depth of at least 1000 ft to avoid disturbances from diurnal and annual temperature fluctuations, climatic change, and circulation of groundwater. Suitable measurements have been made in boreholes, mines, and tunnels. Heat flow through the ocean floor is measured by dropping a metal probe or a coring tube equipped with temperature-sensing devices into the sea bottom. Samples for measurement of thermal conductivity are collected by conventional coring techniques.

An extensive discussion of heat flow at the surface, including numerical values, regional variations, and regional and worldwide averages, is given in the last section of this article. The average heat flow is about 1.5×10^{-6} cal/cm^2 sec. Mean heat flow in the continents is essentially the same as in the oceans. The total amount of heat lost by conduction is about 2×10^{20} cal/year for the whole Earth, a number which is apparently significantly greater than the losses due to other processes, such as volcanism. On the other hand, the conductive loss represents only about 0.01% of the energy received from the Sun, so that its effect on the temperature at the surface is negligible.

Radioactive heat production. The important heat-producing elements in the Earth are thorium, potassium, and the two natural isotopes of uranium. Heat production in igneous rocks ranges from about 600×10^{-15} cal/cm^3 sec in granites, through 70×10^{-15} cal/cm^3 sec in basalts, to $0.2-15 \times 10^{-15}$ cal/cm^3 sec in ultramafic rocks (dunites and peridotites). It is associated with the tenor of feldspar and silica, being highest in the light-colored, low-density rocks and low in the dense, dark-colored rocks.

It is immediately obvious that rocks having radioactivities in the range from basalts to granites, which constitute the majority of those exposed at the surface of the Earth, must be confined to a thin shell. Otherwise their heat production would cause a higher heat flow than is observed. This implies strong upward concentration of the radioactive elements.

Further considerations about the distribution of radioactivity with depth stem from the equality of mean continental and oceanic heat flow. The measurements of heat production imply that more than half of the continental heat flow originates in crustal rocks. Since the crust in the ocean basins is comparatively thin and apparently devoid of granitic rocks, its heat production must be correspondingly lower. The simplest explanation of the constancy of heat flow is that the total amount of radioactivity beneath unit area of the Earth's surface is roughly the same everywhere, but the degree of upward concentration is greater in continental regions. The mantle beneath the continents has presumably been impoverished in radioactive elements during the formation of the continental crust. *See* CONTINENT FORMATION.

A straightforward consequence of this simple model is that, over a considerable range of depth, temperature beneath continents are lower than beneath oceans. The mantle therefore should be denser under continents because of thermal expansion. The lack of correlation between the Earth's gravity field and the distribution of continents and oceans shows that no large density contrasts associated purely with land and sea exist deep in the Earth. It is plausible that the same process which impoverished the mantle beneath the continents in radioactivity swept out other elements as well, leaving an intrinsically lighter residue. Thus differences in composition may largely offset the effect of the different temperatures.

Furthermore, there is evidence that, even within the Earth's crust, the radioactive elements are segregated toward the top. The sampling, which perforce is confined to the outer surface of the Earth, may therefore be biased toward abnormally radioactive material not representative of the crust as a whole. If this is true, less of the continental heat flow originates in crustal radioactivity, more must be derived from beneath the crust, and temperature differences between continental and oceanic areas are reduced. Horizontal temperature differences may also be prevented from building up by processes other than heat conduction. An example is the movement of hot material.

Information about changes in the thermal properties of the Earth with depth can be deduced from the following argument. The mantle beneath the oceans is commonly thought to be ultramafic in composition. The heat flow changes slowly with depth in such material because of the low radioactivity, and it can be regarded as constant in the outermost 100 km of the Earth. If the thermal conductivity does not change with depth, the thermal gradient will likewise be essentially constant. Taking a heat flow of 1.2×10^{-6} cal/cm sec °C, a typical value in deep ocean basins, and a conductivity of 0.005 cal/cm sec °C, a value representative of rocks at moderately elevated temperatures, one finds that the thermal gradient is 24°C/km. The temperature at a depth of 100 km is thus predicted to be well in excess of 2000°C. Any reasonable mantle material would be wholly molten at such temperatures, even after allowance for the effect of pressure on melting relations. Seismic data show emphatically that the Earth is essentially solid to a depth of 2900 km, and a clear contradiction has appeared.

The radioactivity of basalts dredged from the deep sea is abnormally low, falling in the range of ultramafic rocks. This observation seems to preclude the possibility of escaping the difficulty by postulating much higher radioactivity in the mantle than given by the ultramafic model. Apparently the effective thermal conductivity must increase drastically, perhaps by as much as an order of magnitude, between the surface and a depth of 100 km.

This conclusion represents direct geophysical evidence that the effective thermal conductivity changes with depth. As already noted, the heat sources must do the same, but there is little quantitative information about how either property varies. Under these circumstances, it is all but impossible to estimate the temperatures at depths greater than 100 km or so with confidence; deeper in the Earth one can give little more than guided guesses.

Heat transfer in the Earth. All available evidence suggests that temperatures of a few thousand degrees must be expected in the Earth. Few materials remain solid at these temperatures unless they are subjected to high pressure, and mechanisms of heat transfer which are unimportant under common laboratory conditions may be very important in the Earth. Effects of pressure must also be considered.

Heat conduction. This takes place in solids in two basically different ways. Conduction by phonons (lattice vibrations) is the only important mechanism of heat transfer in dielectric solids at room temperature; in metals conduction by mobile charged particles dominates. Both types of conduction take place in the Earth.

Phonon conductivity in dielectric solids at high temperatures is usually observed to decrease with temperature according to the relation $K=(AT+B)^{-1}$, where K is conductivity, T is temperature, and A and B are adjustable constants. Its variation with pressure has been little studied experimentally, but a few theoretical estimates have been made. The results indicate that the effect of pressure may nullify or even reverse the decrease in conductivity resulting from increasing temperature in the Earth. Near the surface where the thermal gradient is steep, the temperature effect dominates, but at greater depths the effect of pressure may be the more important. There appears to be no hope of achieving a rapid rise in conductivity near 100 km depth by this mechanism, however. The conductivity due to phonons is likely to be about 0.005 cal/cm sec °C at shallow depths. It might be increased by an order of magnitude at most by the high pressures near the bottom of the Earth's mantle.

Additional heat transfer takes place in electrically conducting substances. In a metal the relation between electrical and thermal conductivity is given approximately by the Wiedemann-Franz law, which states that the ratio of the thermal conductivity to the electrical conductivity times the temperature is constant. Application of this relation to the Earth's core leads to a thermal conductivity of about 0.2 cal/cm sec °C, but this numerical value depends on the electrical conductivity assumed and is very uncertain.

In the semiconducting mantle the Wiedemann-Franz formula must be modified to take account of the transport of excitation energy as well as kinetic energy. The resulting thermal conductivity, however, is unlikely to exceed 0.001 cal/cm sec °C in the lower mantle if the electrical conductivity σ equals 1 ohm^{-1} cm^{-1}. This is at least an order of magnitude less than the value expected for the phonon conductivity, but if $\sigma=10$ ohm^{-1} cm^{-1}, which is probably within the uncertainty of the electrical data in the Earth, account should be taken of this effect.

The transfer of heat by excitons has also been considered. Excitons are electron-hole pairs bound together by their Coulomb attraction, and as such they cannot contribute to electrical conduction. They can, however, carry thermal energy. Evidence for their existence in oxides and silicates is scanty, and the importance of this process of heat transfer in the Earth is highly uncertain.

Radiative transfer. Such transfer may become very important in nonopaque solids at high temperatures. The contribution of radiation to the thermal conductivity is given to ample accuracy by the expression $K=16n^2sT^3/3\bar{\epsilon}$, where n is refractive index, s the Stefan-Boltzmann constant, and $\bar{\epsilon}$ the sum of the absorption and scattering coefficients averaged over all wavelengths. The other symbols have their previous meaning.

The mean refractive index in the mantle can be obtained from one of several relations between index and density. Well-known examples are the Gladstone-Dale law and the Clausius-Mosotti relation. Since the density is known to relatively high precision, the index is well determined. No such relatively direct procedure for estimating $\bar{\epsilon}$ is available, although it appears that absorption is more important than scattering in the mantle.

Three processes leading to absorption of light in semiconducting solids may be distinguished. Photon-phonon interactions produce strong absorption in the infrared, but this is unimportant in the present context because there is little radiant energy at these long wavelengths. Optical excitation of electrons to higher energy levels in the crystal produces absorption in the visible and ultraviolet, and such absorption may be important in the Earth.

In electrically conducting materials, absorption may take place at all wavelengths. The relation between absorption coefficient α (in cm^{-1}) and conductivity (in ohm^{-1} cm^{-1}) is $\alpha=120\pi\sigma/n$. All terms in this expression must be evaluated at the same wavelength, or frequency.

Many dielectric solids are almost perfectly transparent in visible light, but this is often not the case if the crystal contains transition elements. In the Earth the most important transition element is iron, which in oxide compounds produces a rather weak absorption peak at a wavelength of about 1 μ and intense absorption at wavelengths shorter than about 0.3 μ. Neither spectral feature is sharp. The tail of the 0.3-μ absorption overlaps a wing of the 1-μ peak in the visible, producing the green color characteristic of ferrous compounds. The other wing of the 1-μ peak extends further into the infrared. Other minor peaks are also usually identifiable in the spectra of iron compounds.

The absorption between peaks is of paramount importance to radiative transfer. Each spectral interval acts like an electrical resistor in a bank of resistances in parallel. One highly transparent "window" gives very high thermal conductivity, regardless of the absorption at other wavelengths. The exact value of the absorption in the comparatively opaque intervals is relatively unimportant. The limitation of the mean free path by processes which are relatively independent of frequency can be highly significant because they limit transmission through the transparent intervals.

Room-temperature measurements on a few ferromagnesian silicates have shown that the lowest absorption coefficients are about 1 cm^{-1}. These results are not directly applicable to the mantle because the absorption increases with temperature and possibly with pressure as well. The magnitudes of the temperature and pressure coefficients are known roughly at best.

High temperatures broaden the wings of the absorption peaks in ferromagnesian silicates, and at temperatures of 1000–1300°C $\bar{\epsilon}$ is of the order of 10 cm^{-1}. Thus the rise in temperature increases the absorption, and the radiative contribution to the thermal conductivity comes out to be about 0.008 cal/cm sec °C at 1300°C if $\bar{\epsilon}=10$ cm^{-1} and $n=1.7$. This figure is larger than the phonon conductivity by a factor of 1–2, implying that the total conductivity is increased by a factor of 2–3 over the value of 0.005 cal/cm sec °C quoted earlier. This total increase in thermal conductivity is inadequate to escape the catastrophe of total melting at depths of about 100 km and greater; lower values of $\bar{\epsilon}$ are required. In view of the various uncertainties, it is perhaps not clear that radiative transfer, in addition to phonon conduction, is incapable of accounting for the thermal properties of the outer mantle, but present evidence suggests that other mechanisms of heat transfer must constitute part of the picture.

The electrical conductivity in the outer mantle is too low to lead to appreciable absorption, but at a depth of about 600 km the conductivity rises sharply to a value of about 1 ohm^{-1} cm^{-1}. High absorption may be associated with this feature of the Earth because of the general absorption connected with electrical conductivity. Evaluation of these effects depends on interpretation of the rise in electrical conductivity.

The electrical conductivity of a semiconductor may increase for two important reasons other than because of an increase in temperature. The drift mobilities (mean velocities in unit electric field) of the charge carriers may increase, or the activation energy of conduction may decrease. If the drift mobility increases, the conductivity is likely to depend on frequency and to decrease at high frequency. The conductivity in the Earth is determined at essentially zero frequency, and the large frequency effect implied by this model suggests that the transparency may remain high. However, if rise in conductivity is due to a decrease in activation energy, the transparency is not as great. A smaller frequency effect is anticipated, and strong absorption may cut off much of the visible part of the spectrum. Experimentation indicates that the activation energy decreases with pressure and temperature, and this may be the most important factor contributing to the rise in electrical conductivity near 600-km depth. If this interpretation is correct, the lower mantle is fairly opaque and radiative transfer does not dominate over phonon conduction in this part of the Earth.

An alternative, rather unlikely, explanation of the rise in conductivity is that electricity is conducted by ions rather than electrons in the lower mantle. In this case the conductivity would have almost no effect on the transparency, since ions are too massive to interact with visible light.

Convective transport of heat. Circulation has been considered to take place in both the mantle and core. The fluid outer core is thought to be in convective equilibrium, that is, the thermal gradient is thought to be that required to start convective motion. The amount of heat actually transferred by fluid motion may not be large compared to conduction, however, because magnetically induced viscosity restricts the fluid motions. It is also possible that some process other than thermal convection is responsible for providing the energy to generate the Earth's magnetic field. In this case convection need not occur.

It is widely considered that convection may occur in the mantle despite seismic evidence for its rigidity. Although the mantle behaves as a solid for short-period stresses, it may be unable to support shearing stresses of long duration and thus may exhibit properties like those of a liquid over geologic times. If convective motions of 0.03 cm/year on the average occur, then motion of material is the dominant mechanism of heat transfer in the Earth. The strength of solids, which is a measure of their resistance to convective motions, is greatly reduced as temperatures approach those of melting. If temperatures rise near the surface of the Earth because other mechanisms of heat transfer are incapable of preventing it, weakening of solids or partial melting will enable motions of material to act as a highly efficient means of heat transfer which severely limits the rise in temperature. Such a mechanism may represent the Earth's safety valve against overheating. *See* EARTH; EARTH, INTERIOR OF. [SYDNEY P. CLARK, JR.]

Heat-flow analysis. In heat-flow analysis, measurements of outflow from the Earth's interior are cataloged, analyzed, and interpreted with respect to geological processes.

Basic concepts. All geological processes require an energy source to drive them. For example, heat is needed to melt rocks to produce lavas for volcanoes; some sources of energy must exist to spread sea floors and move continents; and the occurrence of earthquakes implies a previous accumulation of strain energy. Measurements of heat flow from the Earth's interior provide the basic constraint on the energy sources within the Earth.

A general increase of temperature with depth became known in the 16th century when mines were excavated to depths of a few hundred meters. The rise of temperature with depth in the Earth implies that the Earth is losing heat to the surface. The amount of heat flowing to the Earth's surface per unit of area per unit of time is called the surface heat flow or, simply, heat flow.

Heat transfer occurs from points at higher temperature to points at lower temperature in three distinct ways: conduction, convection, and radiation. In fluids, convection is of great importance; but in solids, convection is usually absent, and radiation is negligible at ordinary temperatures. As most of the Earth's upper crust is solid and at relatively low temperatures, outflow of heat from the Earth's interior occurs mostly by thermal conduction near the Earth's surface. But deeper in the

Earth, convection and radiation may be more important than conduction because of the higher temperatures and, probably, the fluidlike behavior of the Earth's interior. This is also true for geothermal areas on land and for sea-floor spreading centers (that is, places where new crust is created) because magmas occur near the Earth's surface.

When the steady state of temperature is reached, heat flow by thermal conduction $\vec{q}$ in a solid is found experimentally to be proportional to the temperature gradient ∇T, as shown in Eq. (1), where K is called the thermal conductivity.

$$\vec{q} = -K \cdot \nabla T \qquad (1)$$

The minus sign indicates that heat flows down the gradient, from a higher temperature to a lower one. The amount of heat conducted increases with increasing values of thermal conductivity. For example, silver, with a conductivity value of 1 cal/cm s °C, can conduct 200 times more heat per unit of time for a given temperature gradient than a typical rock, which has a conductivity value of 0.005 cal/cm s °C, can. Therefore, rocks such as granite, basalt, and sandstone are considered to be poor thermal conductors.

The speed with which heat can be transferred by conduction in a body with internal heat sources can be estimated from the heat-conduction equation, Eq. (2), where T is temperature; t, time; ρ,

$$\frac{\partial T}{\partial t} = \frac{K}{\rho c} \nabla^2 T + \frac{A}{\rho c} \qquad (2)$$

density; c, specific heat; and A, rate of heat production per unit of volume. The term $K/\rho c$ is known as the thermal diffusivity κ; it has the dimension of (cal/cm s °C)/(g/cm³)(cal/g °C) = cm²/s. Thus Eq. (2) implies that the heat-transfer time by conduction can be expressed as relation (3),

$$\tau \propto L^2/\kappa \qquad (3)$$

where L is the linear dimension of the body.

Thermal diffusivity of rocks at ordinary temperature and pressure is about 0.01 cm²/s. Therefore, relation (3) implies that the heat-transfer time by conduction for the Earth is very large. For example, conduction of heat from a depth of 100 km to the surface will require 10^{16} s, or 300,000,000 years. On the other hand, heat-transfer time by convection can be much shorter. For example, a convection current with a small velocity of 1 cm/yr will deliver heat from a depth of 100 km to the surface in 10,000,000 years.

Measurements. In geophysics, heat-flow measurements are usually made in a very small area near the Earth's surface and over a depth that is small compared with the Earth's radius. Therefore, the surface heat flow by conduction q can be obtained from the heat-flow equation, Eq. (1), as the product of thermal conductivity and temperature gradient with depth (that is, geothermal gradient), as shown in Eq. (4), where the flow is vertically outward.

$$q = K(dT/dZ) \qquad (4)$$

To determine surface heat flow by conduction at a location, geophysicists measure (1) the temperatures through some finite interval of depth to obtain the geothermal gradient and (2) the thermal conductivity of the same interval, either in place or in the laboratory, on an appropriate number of samples taken. In practice, great difficulties arise because the geothermal gradient is often disturbed, and the thermal conductivity measured may not be representative.

On land, the disturbances include the effect of drilling a borehole, variation of surface temperature, uplift and erosion, topographic and conductivity irregularities, water circulation, and volcanic activity. Most of these disturbances can be avoided by the use of deep boreholes made through hard rocks, but this solution is very expensive. Obtaining a representative conductivity is difficult because conductivity often varies greatly from rock to rock and even from sample to sample of an apparently uniform rock.

At sea, the bottom-water temperature in deep oceans is remarkably constant with time, so that the sea-floor sediment is at thermal equilibrium; also, the sea-floor sediment is easily penetrable by heat-flow probes. Moreover, the thermal conductivity of ocean sediments varies only slightly, usually less than 25%. The temperature gradients are measured with a probe that penetrates a few meters into the sediment, and thermal conductivities are measured from cores taken at or near the site of temperature measurements. The operation at sea is fast (a few hours for a measurement) and inexpensive, so that oceanic heat-flow measurements have been far more numerous than those on land.

The thermal conductivity of sediments and rocks depends on composition, porosity, water content, temperature, and pressure. Usual values are between 0.002 and 0.010 cal/cm s °C. The geothermal gradient varies from place to place; usual values are between 10 to 100°C/km. However, in most areas, a high geothermal gradient is usually associated with low thermal conductivity, so that most of the surface heat-flow values lie in the range of 1 to 3 μcal/cm² s (1 μcal = 10^{-6} cal). The traditional centimeter-gram-second (cgs) system of units is used in heat-flow measurements (length, cm; mass, g; and time, s); temperature is measured in degrees Celsius; and calorie (cal) is used as the unit quantity of heat. However, it may be preferable to use the International System of Units (SI), in which length is measured in meters, mass in kilograms, time in seconds, temperature in degrees kelvin, and heat in joules (1 cal = 4.1868 J). Thus, the unit of heat flow is expressed as joule per square meter per second or watt per square meter (1 μcal/cm² s = 4.1868×10^{-2} W · m^{-2}).

As more measurements were made, it became evident that heat flow often varies from place to place. Some heat-flow variations are due to local disturbances, such as groundwater motions and climatic changes. These variations should be eliminated from the measurements, but this is often difficult to do in practice. Most variations in heat flow, however, arise from the complexity in the structure, history, and composition of the Earth's crust and upper mantle. The aim of heat-flow data analysis is, therefore, to establish the pattern of heat-flow variations and relate it to the Earth's thermal evolution and internal structure.

The number of heat-flow measurements has in-

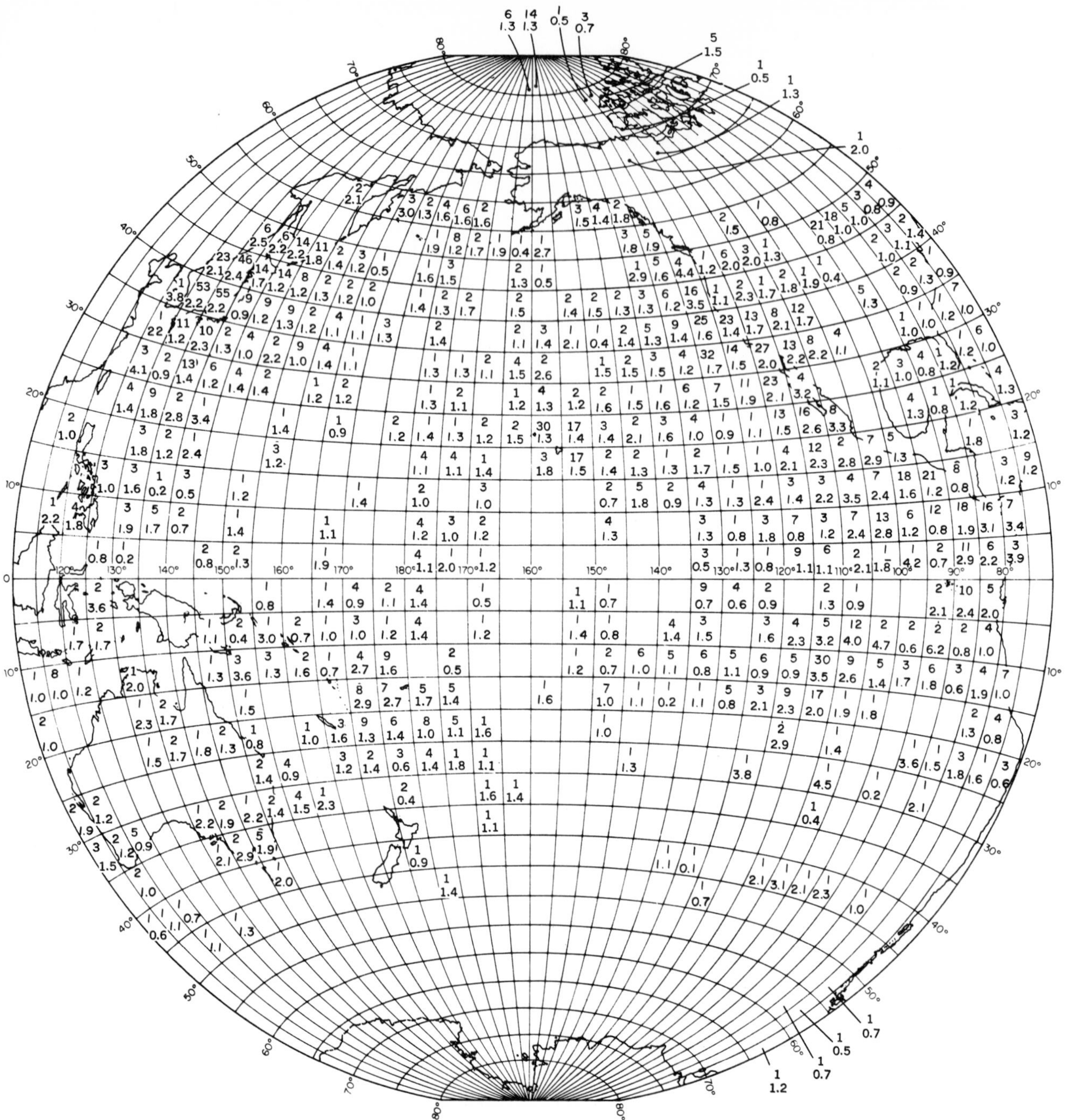

Fig. 1. Selected heat-flow data in 5 × 5° grid for the first hemisphere. In each grid element, the upper value is the number of selected data and the lower value the arithmetic mean in μcal/cm² s.

creased rapidly since 1960. About 5000 individual observations were made from 1939 to 1975, and about 500 new measurements are reported per year.

Data analysis. Because heat-flow data include measurements of different quality, only data of reasonable reliability are selected for analysis. A convenient way to summarize selected heat-flow data is by means of maps containing averages of measurements in areas of 5° latitude by 5° longitude, as shown in Figs. 1 and 2. Although measurements are fairly well distributed over the oceans, large gaps exist over continents and high-latitude regions.

Statistics of heat-flow data from individual measurements in various geographical regions are summarized in Table 1. Because of the uneven geographic distribution of heat-flow stations, these results contain significant sampling bias and must be interpreted with caution. To compensate for sampling bias, heat-flow averages for grid elements of equal areas (9×10^4 naut mi²; 1 naut mi = 1.852 km) are also shown in Table 1. Comparison of information in Table 1 reveals that the arithmetic mean and standard deviation of grid averages are less than those for the original heat-flow data without grid averaging. The plausible interpretations are that individual measurements are biased toward regions of high heat flow and that some local variations are removed by grid averaging.

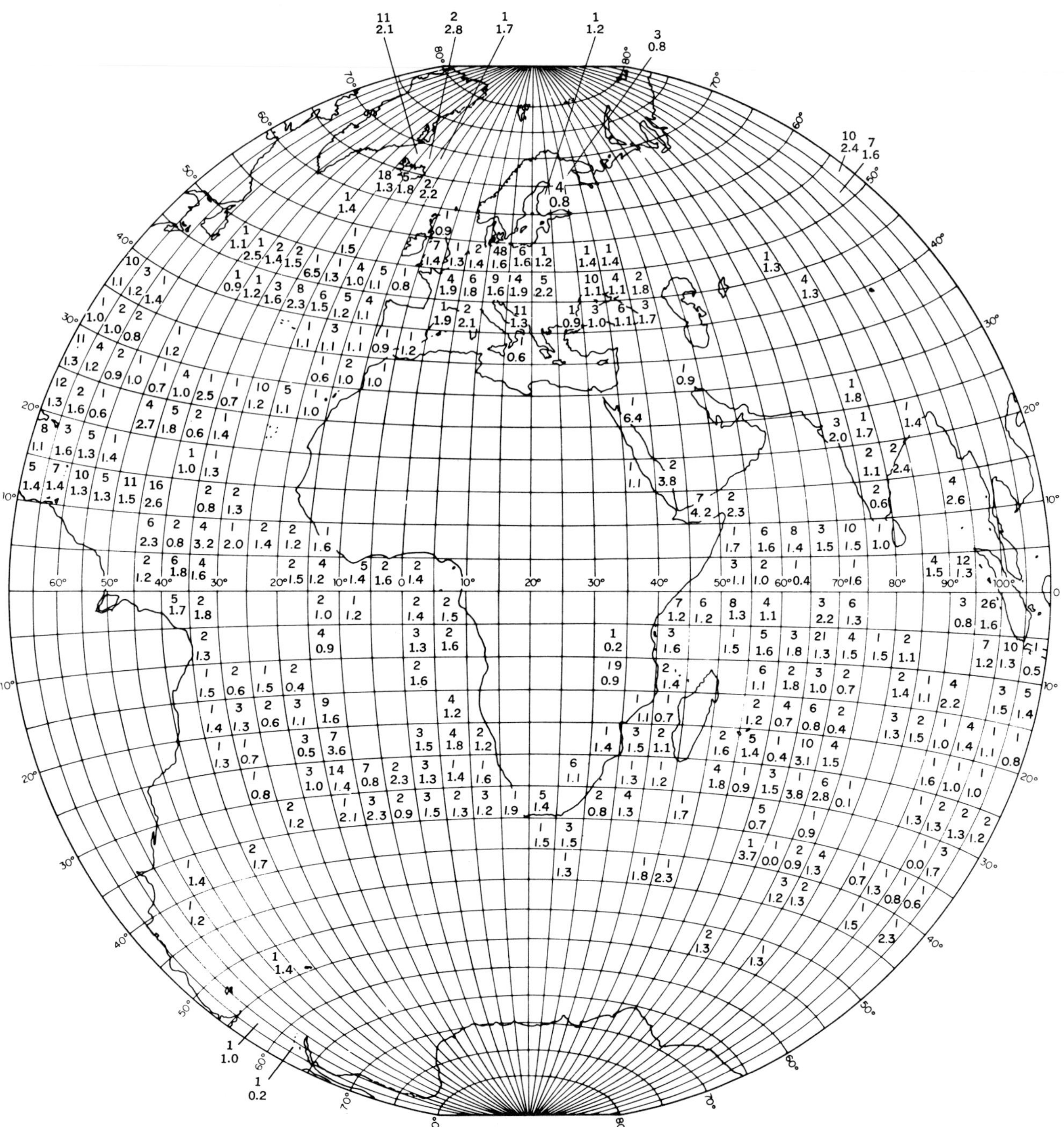

Fig. 2. Selected heat-flow data in 5×5° grid for the second hemisphere. In each grid element, the upper value is the number of selected data and the lower value the arithmetic mean in μcal/cm² s.

Histograms of equal-area grid averages of heat flow are shown in Fig. 3. The continental grid averages do not extend beyond 3 μcal/cm² s, because heat-flow data from geothermal areas on land (such as Yellowstone Park) have been excluded from the analysis. Such data have not been excluded from measurements made at sea because of a lack of information. A characteristic of these histograms is that most heat-flow grid averages fall between 1 and 2 μcal/cm² s.

Heat loss from Earth's interior. The global average of conduction heat flow from several thousand measurements is $1.5 \pm 10\%$ μcal/cm² s at a 95% confidence level. This amount of heat flux is exceedingly small; for example, a 75-W light bulb produces 100,000 times more energy at its surface. However, the Earth is very large, with a surface area of 5.1×10^{18} cm²; thus the present rate of heat loss from the Earth's interior by thermal conduction is (1.5 μcal/cm² s)$\times$(5.1×10^{18} cm²)$=2.4 \times 10^{20}$ cal/yr.

The heat loss of 2.4×10^{20} cal/yr is large in comparison to that created by human activities; for example, the world's total electric power output is about 1×10^{18} cal/yr, and the world's total food consumption is about 3×10^{18} cal/yr. However, the heat loss is very small compared to the solar energy received by the Earth. which is about 5.5×10^{24} cal/yr. As a result, heat loss from the Earth's interior has a negligible effect on the

Table 1. Statistics of heat-flow data and heat-flow grid averages*

Area	Individual values			Equal-area grid averages		
	N	$\bar{q}$	SD	N	$\bar{q}$	SD
World	3127	1.63	1.07	673	1.47	0.74
All continents	597	1.45	0.57	95	1.46	0.46
All oceans	2530	1.67	1.15	591	1.47	0.78
Atlantic Ocean	436	1.47	1.14	126	1.34	0.57
Indian Ocean	358	1.36	0.95	108	1.32	0.52
Pacific Ocean	1308	1.70	1.24	310	1.50	0.84

*N = number of data; $\bar{q}$ = arithmetic mean in μcal/cm² s; SD = standard deviation from the mean in μcal/cm² s; area per grid element $= 9 \times 10^4$ naut mi².

Earth's surface temperature, or climate.

In estimating the total heat loss from the Earth's interior, one must also consider heat loss transported by hot mobile constituents, such as volcanic ash and lavas, steam, and hot water. In geothermal areas on land and in sea-floor spreading centers at sea, hydrothermal convection may play an important role in transporting heat from the Earth's interior to its surface. No satisfactory methods have been developed to measure the heat flow in these anomalous areas, but crude measurements suggest heat-flow values in the tens or hundreds of μcal/cm² s. In 1974 D. L. Williams and R. P. Von Herzen estimated that the heat loss due to thermal convection may be as much as one-third of that due to thermal conduction.

If the present rate persisted throughout a geologic history of 4.5×10^9 years, as evidence suggests, then the total heat loss since the Earth was formed is about 10^{30} cal, which is equivalent to the energy released by 10^{18} kilotons of TNT, or by 10^{15} megaton bombs. It was first supposed that this heat came from the cooling of a molten Earth. However, simple calculation shows that this cannot be the case. Relation (3) implies that the depth from which heat can be conducted to the surface in 4,500,000,000 years is $\sqrt{(4.5 \times 10^9 \text{ yr}) \times (0.01 \text{ cm}^2/\text{s})} \approx 300$ km.

Cooling of 1 g of magma to 0°C yields about 400 cal. The outer 300 km of the Earth has a mass of about 5×10^{26}g; thus the total of 2×10^{29} cal available from cooling can account for only 20% of the total heat loss of about 10^{30} cal from the Earth's interior since the Earth was formed.

The discovery of radioactivity resolved this apparent dilemma. It was recognized that radioactive isotopes of potassium (^{40}K), uranium (^{235}U and ^{238}U), and thorium (^{232}Th) have half-lives comparable to the Earth's age, and also produce large quantities of heat. For example, granite contains enough potassium, uranium, and thorium to produce 2.6×10^{-13} cal/g s, or roughly 5×10^4 cal/g in 4,500,000,000 years. As a result, if the Earth contained a layer of granite about 10 km thick, all the heat loss from the Earth's interior since the Earth was formed could have been supplied by the granite.

Equality of heat flow. It was known by the late 1940s that the oceanic crust is much thinner than the continental crust (7 versus 33 km), and contains a much smaller amount of radioactive elements. When oceanic heat flow was first measured in 1952, geophysicists were greatly surprised to learn that this heat flow is approximately equal to that of the continents. E. C. Bullard suggested that (1) the total number of radioactive elements in the Earth's upper few hundred kilometers is the same beneath land and sea or (2) a large portion of the oceanic heat flow is due to thermal convection in the upper mantle.

Either one of these explanations for the observed equality of heat flow has great implications concerning the composition, structure, and history of the Earth's interior. In the late 1950s and early 1960s the first hypothesis was preferred. It was argued that if the radioactivity is the same beneath continents and oceans to a depth of several hundred kilometers, then a process of vertical differentiation must predominate. Since most radioactive material had already been concentrated in the continental crust, very little of it would remain in the continental mantle. Thus, continental heat flow originates mainly from the crust, and very little comes from the upper mantle. The case is just the opposite for the oceans. Since the oceanic crust contains insufficient radioactivity to produce the observed heat flow, most of the observed heat flux must derive from the oceanic mantle. Proponents of Bullard's first hypothesis then argued that differences between continents and oceans extend to a depth of several hundred kilometers, thus making continental drift unlikely.

A major flaw in the above argument is the assumption that the oceanic crust is as old as the continental crust. By the early 1960s it was clear that the oceanic crust is at least an order of magnitude younger than the continental crust. In

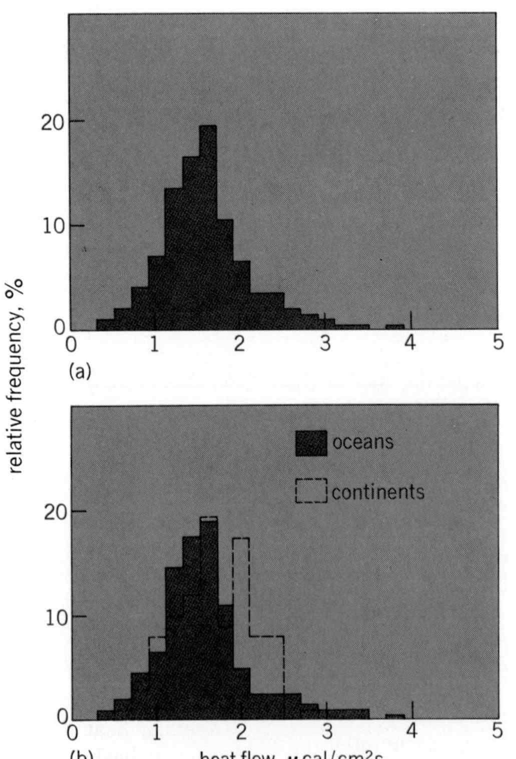

Fig. 3. Histograms of equal-area grid averages of heat-flow (*a*) for the world and (*b*) for the continents and oceans.

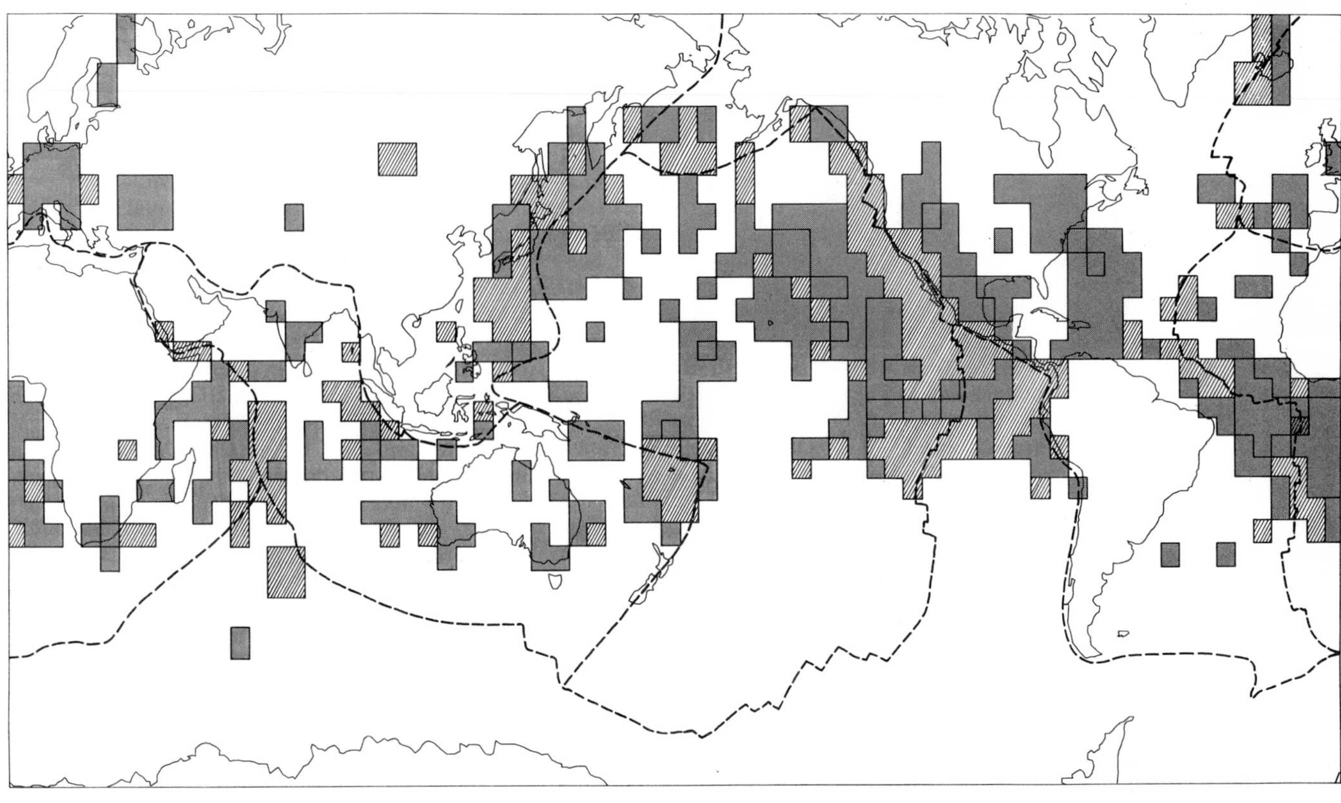

Fig. 4. Heat-flow standard deviations (σ) in 5° × 5° grid. Data are available only in enclosed rectangular areas.

order for the sea floor to be renewed every 100,000,000 years or so, H. H. Hess proposed in 1962 that new oceanic crust is created at the crest of oceanic ridges, spreads away from it, and eventually descends beneath the deep-sea trenches. This hypothesis of sea-floor spreading was quickly demonstrated in the magnetic patterns of the ocean floor. It was then evident that the lithospheric plates (crust and part of the upper mantle to a depth of about 100 km) are capable of large horizontal movements. There are three types of plate boundaries; in addition to the sea-floor spreading centers and deep-sea trenches mentioned above, there is the transform fault, along which plates slide by one another horizontally but without creating or consuming the crust. The driving force of plate motions may be provided by thermal convection in the asthenosphere, where partial melting occurs.

Continental heat flow derives mainly from the radioactive material in the thick continental plate, whereas oceanic heat flow is due to cooling of the thinner oceanic plate as well as transference of heat from the asthenosphere. The equality of heat flow in continents and oceans may be purely accidental.

Heat flow and plate tectonics. The concepts of sea-floor spreading and transform faults, and their tectonic implications, led to the formulation of the theory of plate tectonics by the late 1960s. To a first approximation, the Earth's lithosphere may be divided into six main plates, as shown in Fig. 4. The boundaries are ridge crests, trenches, or transform faults. Since each plate moves more or less as a single rigid unit, active tectonic processes occur mainly along its edges, as suggested by the distribution of earthquake foci.

Since tectonic processes involve transfer of energy, it can be expected that the heat flow will fluctuate more near the edge of each plate than at its interior. This assumption can be tested by plotting the heat-flow standard deviation in 5° × 5° squares (see Fig. 4). Because the standard deviation in each 5° × 5° square can be computed only if there are two or more values in it, there are fewer data to work with than there are in the averages in Figs. 1 and 2. For the purpose of discussion, variation of heat flow is considered low if the standard deviation is less than 0.5 $\mu cal/cm^2\ s$. Examination of Fig. 4 shows that the heat-flow data are generally consistent with plate tectonics: large heat-flow variations occur approximately along the edges of the tectonic plates, and small variations occur in the interior regions. There are, however, some exceptions which may be caused by insufficient numbers of measurements or by local conditions, or both.

Great interest has been focused on the island-arc trench regions where the lithospheric plates descend. A plot of heat flow versus the distance from the trench axis for the western Pacific Ocean is shown in Fig. 5. Average heat flow is about 1.3 $\mu cal/cm^2\ s$ in the ocean basin, but it dips below 1 $\mu cal/cm^2\ s$ at about 200 km behind the trench axis, and rises rapidly to 2 $\mu cal/cm^2\ s$ in the inland seas. This pattern of heat flow is

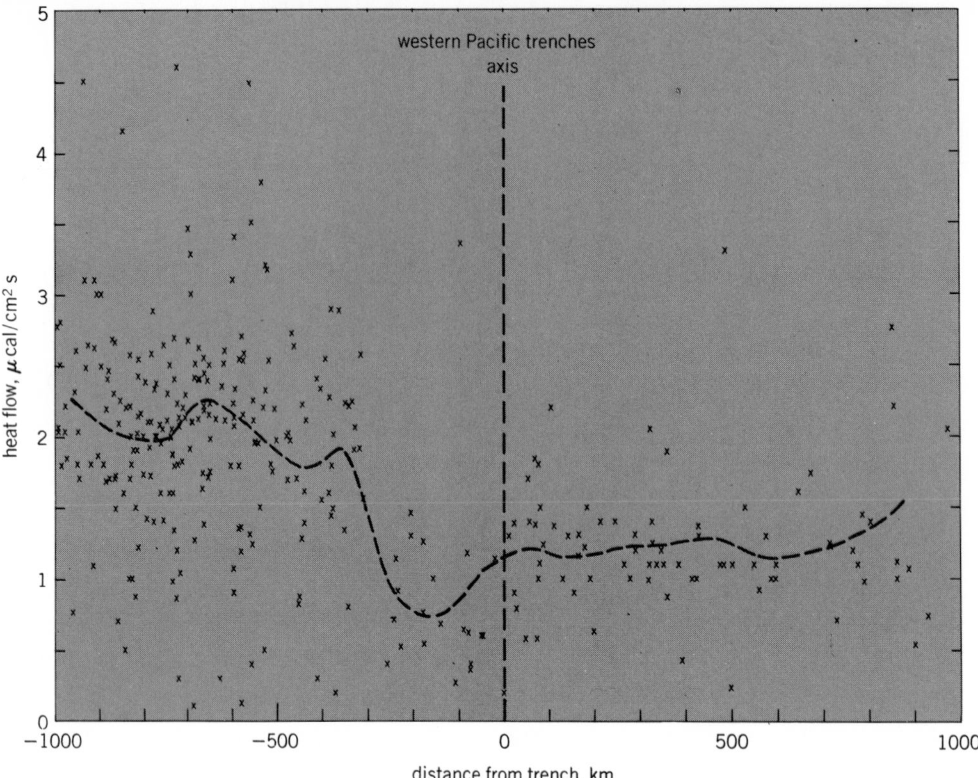

Fig. 5. Heat flow versus the distance from the trench axis for the western Pacific Ocean. Distance is positive toward the oceans and negative toward the continents. The curve is based on the arithmetic average in each 100-km distance interval.

consistent with the idea that the lithospheric plate is descending beneath the trench. The high heat flow in the inland seas may be due to various heating processes—for example, friction—near the boundary of the descending plate.

The theory of sea-floor spreading implies that oceanic heat flow is related to the age of sea floor. This is shown in Fig. 6*a*; the dashed curve is a least-squares fit of the data to fifth-degree polynomials, Eq. (5), where q is the heat flow, t the age of the sea floor, and C_i the coefficients obtained from the least-squares fit. Equation (5) has been used because the fit does not improve significantly beyond the fifth degree.

$$q(t) = \sum_{i=0}^{5} C_i t^i \qquad (5)$$

Equation (5) was used to fit data from the east side of the Mid-Atlantic Ridge, data from the west side of the Mid-Atlantic Ridge, and combined data without regard to the direction from the Mid-Atlantic Ridge. Since the resulting curves do not differ appreciably, the curve shown in Fig. 6*a* is based on the combined data and is therefore automatically symmetrical with respect to the ridge axis. From Fig. 6*a* an examination of how the data are fitted to a symmetric pattern can be made. For comparison, a magnetic profile is shown in Fig. 6*b*. One important feature of Fig. 6 is the change in character of both magnetic and heat-flow profiles at about 10,000,000 years.

Heat-flow profiles for the East Pacific Rise and the Mid-Indian Ocean Ridge are similar to that for the Mid-Atlantic Ridge. These observations can be accounted for by a simple thermal model of an oceanic plate (50–100 km thick) moving away from the ridge crest with a velocity of a few centimeters per year. The boundary conditions are 0°C at the surface and near melting temperature at the crest and at the bottom of the plate.

Heat-flow provinces. In the early 1960s it was recognized that distinct heat-flow provinces exist. Since different geologic provinces represent various stages of crustal development, it is natural to correlate heat flow and major geological features. The results are summarized in Table 2. Tectonically active areas are characterized by higher heat-flow averages and more scattered values, whereas stable areas are characterized by lower averages and more uniform values. For example, the Precambrian shields (such as the Canadian Shield in east-central Canada) have an average heat flow of only half that found in the Mesozoic and Cenozoic orogenic areas (such as the western United States). The standard deviation of heat-flow values in ocean basins (such as the central Pacific) is one-third that in oceanic ridges (such as the East Pacific Rise and the Mid-Atlantic Ridge).

Within the framework of sea-floor spreading and plate tectonics, several attempts have been made to construct models for different heat-flow provinces. Of particular interest are the models of J. G. Sclater and J. Francheteau for continental Precambrian shields and ocean basins. They showed that heat flow decreased systematically with age to about 1 μcal/cm² s. Since the age for Precambrian shields is an order of magnitude larger than that for the ocean basins (10^9 versus

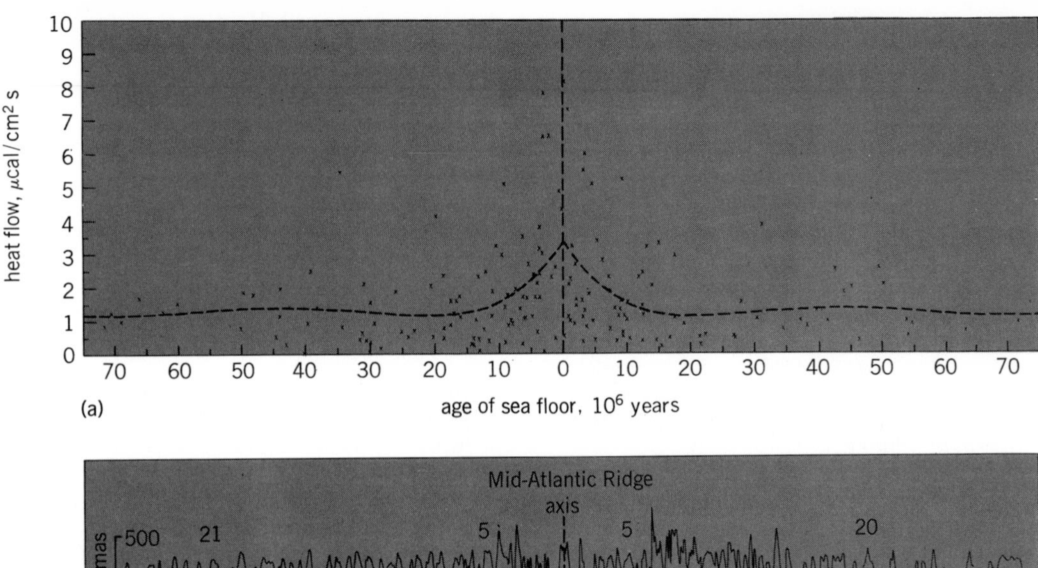

Fig. 6. Mid-Atlantic Ridge data. (*a*) Heat flow and (*b*) magnetic profile. Note the change in character of both the heat-flow and magnetic profiles which occurs at about 10,000,000 years.

10^8 years), very different processes are needed for explanation.

With an increasing number of heat-flow measurements, more detailed delineation of heat-flow provinces becomes evident. F. Birch first observed in the late 1960s that heat flow and near-surface heat production are linearly related for the granitic rocks of a given heat-flow province, as shown in Eq. (6), where q is the observed heat flow; A_0 is the heat production of the surface rocks; and q^* and D are constants within a given heat-flow province that have the dimensions of heat flow and depth respectively. A. H. Lachenbruch suggested that in order for Eq. (6) to remain valid as the surface erodes, the radioactive elements must be distributed as a function of depth Z, as expressed in Eq. (7). If heat flow and heat production in a heat-flow province are measured, an insight into how much heat is produced in the granitic rocks, and how much comes from below it (that is, q^*), can be gained.

$$q = q^* + DA_0 \qquad (6)$$

$$A(Z) = A_0 e^{-Z/D} \qquad (7)$$

Table 2. Statistics of heat-flow values for major geological features

Geological province	Surface area, %	Heat flow, μcal/cm² s: Number of values	Arithmetic mean	Standard deviation
Precambrian shields	9.2	214	0.98	0.24
Post-Precambrian nonorogenic areas	9.5	96	1.49	0.41
Paleozoic orogenic areas	2.1	88	1.43	0.40
Mesozoic-Cenozoic orogenic areas	8.3	159	1.76	0.58
Ocean basins	29.3	683	1.27	0.53
Mid-oceanic ridges	25.6	1065	1.90	1.48
Ocean trenches	0.2	78	1.16	0.70
Continental margins	15.8	642	1.80	0.93

The combined studies of heat flow and radioactivity provide geophysicists with a powerful method to map variations of temperature in the Earth's crust and upper mantle. Knowledge of the temperature within the Earth will lead to a better understanding of the geological processes that shape the Earth.

[W. H. K. LEE]

Bibliography: L. H. Ahrens, K. Rankama, and S. K. Runcorn (eds.), *Physics and Chemistry of the Earth*, vol. 1, 1956; G. O. Dickson et al., Magnetic anomalies in the South Atlantic and ocean floor spreading, *J. Geophys. Res.*, 73:2087–2100, 1968; T. F. Gaskell (ed.), *The Earth's Mantle*, 1967; P. M. Hurley (ed.), *Advances in Earth Science*, 1966; G. P. Kuiper (ed.), *The Earth as a Planet*, 1954; M. G. Langseth and R. P. Von Herzen, in A. E. Maxwell (ed.), *The Sea*, vol. 4, pt. 1, 1970; W. H. K. Lee, *Earth Planet. Sci. Lett.*, 4:270–276, 1968; W. H. K. Lee, On the global variations of terrestrial heat-flow, *Phys. Earth Planet. Interiors*, 2:332–341, 1970; W. H. K. Lee (ed.), *Terrestrial Heat Flow*, Amer. Geophys. Union, Geophys. Monogr. no. 8, 1965; X. Le Pichon et al., *Plate Tectonics*, 1973; R. F. Roy et al., in E. C. Robertson (ed.), *The Nature of the Solid Earth*, 1972; J. G. Sclater and J. Francheteau, The implications of terrestrial heat flow observations on current tectonic and geochemical models of the crust and upper mantle of the Earth, *Geophys. J. Roy. Astron. Soc.*, 20:509–542, 1970; J. Verhoogen et al., *The Earth*, 1970; D. L. Williams and R. P. Von Herzen, Heat loss from the Earth: New estimate, *Geology*, 2:327–328, 1974.

Earth, interior of

Features of the interior of the Earth are revealed through the application of the principles and techniques of physics and related sciences and subsequent mathematical analysis. Although the Earth's interior is not available to direct visual observation, many of its properties are nevertheless firmly established. This article discusses the following properties: spherical stratification, mean density and moment of inertia, seismic *P* and *S* and surface waves, density variation, pressure variation, gravitational intensity, incompressibility (bulk-modulus), compression, rigidity, effects of nuclear explosions, free Earth oscillations, composition, internal heat and temperature, and magnetism. *See* ASTHENOSPHERE.

Spherical stratification. The Earth is a planet of mean radius 6371 km. The structure and composition of the outermost 70 km depend upon the geographical region, the differences being specially marked between the regions below continental shield areas, certain mountain ranges, and the oceans. These lateral differences become comparatively less important as the depth increases, but are now detectable down to several hundred kilometers depth. For most of the Earth, however, the deviations from spherical symmetry are very slight, apart from the ellipticities (flattenings at the axial poles) of internal surfaces of constant density and composition. The ellipticity is the ratio of the excess of the equatorial diameter over the polar diameter to the equatorial diameter. Its value is only 1/298 at the Earth's surface, and diminishes with increasing depth below the surface, being about 1/400 near the Earth's center. Thus the Earth may, to a good first approximation, be treated as concentrically layered, the chief properties varying only with the depth.

The exceptional outermost 70 km include the outer layers, commonly called the crust, which is bounded below by the Mohorovičić discontinuity, so called because findings of the Balkan seismologist A. Mohorovičić in 1909 provided the first evidence on the point. Under continental shields the discontinuity is in general 30–40 km deep; it is deeper under some mountain ranges, but lies only about 5–10 km below ocean floors. *See* MOHO (MOHOROVIČIĆ DISCONTINUITY).

Mean density and moment of inertia. The mean density of the Earth compared with water is 5.517. Certain astronomical data, theory of the dynamics of the Earth-Moon system, and theory of the Earth's ellipticity enable the Earth's moment of inertia to be estimated. An early calculation gave the moment of inertia as 0.834 times that of a body of constant density with the same size and mass as the Earth. This factor was revised to 0.827 in 1963 as a result of important new information provided by analyses of orbits of artificial satellites which move through the Earth's gravitational field. The smaller that factor is, the greater the central concentration of matter in the body. It follows that the Earth's deep interior is appreciably denser than the rocks that are formed near the surface. Further, the results on the mean density and moment of inertia provide two of the most important numerical criteria which have to be fitted in determining the internal distribution of matter in the Earth.

Seismic P and S and surface waves. Evidence from seismology supplies much further detail that has to be fitted. A large earthquake sends out seismic waves, some of which (the "bodily" waves) penetrate the whole interior of the Earth. The bodily waves rise again to the surface, and are recorded on seismographs in the 2000 or so seismological observatories set up all over the world. Through solid parts of the Earth, both *P* (primary) and *S* (secondary) bodily waves are transmitted, *S* waves having about two-thirds (or less) of the speed of *P* waves. Through fluid zones, only *P* waves are transmitted. From the times of arrival of the various wave pulses at the Earth's surface, seismologists have derived values of the *P* and *S* velocities throughout much of the interior. *See* SEISMOGRAPH; SEISMOLOGY.

The results supply the chief direct evidence on the nearly spherical character of the Earth's subcrustal stratification. The variation of the velocities with depth then enables the interior to be charted out into regions whose approximately spherical boundaries are mostly characterized by fairly sudden changes either in the *P* or *S* velocity, or in the rate of change with respect to depth. The table, which emerged from work of H. Jeffreys and K. E. Bullen over the period 1932–1942, continues to serve as a useful basis for describing the broad internal layering of the Earth.

Mantle and core. The separation of the Earth into mantle and core was suggested by E. Wiechert in the 1890s. The existence of the core was established in 1906, when R. D. Oldham found that *P* waves from a given earthquake arrived so late at

The Earth's layers and their seismic velocities

Region	Name	Range of depth, km	*P*-wave velocity, km/s	*S*-wave velocity, km/s
A	Crust	Base varying from 10 to 70 km depth	Very variable	Very variable
B, *C*	Upper mantle	From crust to 1000 km depth	8–9	4–6
D′, *D*″	Lower mantle	1000–2700 2700–2900	9–13	6–7
E	Outer core	2900–4600	8–10	0
F	Transition region(s)	4600–5150	(10–11)	Uncertain
G	Inner core	5150–6371	11	3

observatories in the opposite hemisphere that there must be a central region, the core, in which the average speed of *P* waves is appreciably less than in the mantle. The core was also observed to cause a partial shadow in the recording of *P* waves between angular distances of about 100 and 142° from an earthquake source, the waves being refracted toward the Earth's center on passing from mantle to core. The boundary between mantle and core is more sharply defined than any other concentric boundary inside the Earth. In 1913 B. Gutenberg calculated its depth to be 2900 km (1800 mi); in 1939 Jeffreys arrived at a closely similar estimate after applying detailed theory of statistics to a large quantity of *P*- and *S*-wave data. The preferred valve now is a little less than 2890 km.

Upper mantle. The layering inside the upper mantle has proved specially difficult to unravel, and data from a number of sources additional to seismic bodily-wave data, including seismic surface-wave data, have been brought to bear. Surface waves travel over the surface of the Earth, their amplitudes in general decreasing strongly with depth below the surface. Surface waves are subject to dispersion (spreading out into long trains of waves), and details of the dispersion depend on the layering and physical properties in the vicinity of the surface. The longer the surface waves, the greater is the vertical extent of the region which can influence their dispersion and on which observations of the waves can provide useful structural information. The information is supplementary to the basic information provided by observations of *P* and *S* waves. Until the development of modern instruments, the existing surface-wave observations could supply useful evidence only about the Earth's crust. Modern instrumental developments have led, however, to a notable increase in the lengths of recordable surface waves, which can now supply useful evidence on the structure down to some hundreds of kilometers depth.

A first approximation to the structure of the upper mantle, as originally given by Jeffreys and Bullen, includes a layer *B* of fairly uniform composition nearly 400 km thick and a layer *C*, between 500 and 600 km thick, inside which the chemical composition or phase is variable. A section of the bodily-wave evidence had, further, been interpreted by Gutenberg as indicating the presence, inside the outermost 100–200 km of *B*, of a layer with abnormally low *P* and *S* velocities. Later results, for example, evidence by I. Lehmann from bodily waves and by J. Dorman from surface waves, have supported the presence in this part of the Earth of reduced *S* velocities, but probably not *P*. This is further discussed in the section on free Earth oscillations.

Lower mantle. The transition from upper to lower mantle is gradual, the depth of the separating surface being usually taken as between 900 and 1000 km. The lower mantle appears to have a less complex structure than the upper mantle, the *P* and *S* velocity gradients being apparently very steady throughout the region D', that is, between depths of 1000 and 2700 km. Variations of properties inside D' appear to be predominantly due to the effects of simple compression. However, inside the region D'', that is, from 2700 km depth to the core, the *P* and *S* velocity gradients fall continuously and may even become negative, indicating some continuous changes of property over and above those due to compression.

Layering in the core. The separation into the so-called outer core and inner core was made in 1936, when Lehmann showed, for analyses of European records of two New Zealand earthquakes, that descending *P* waves are strongly refracted away from the Earth's center at a distance later shown to be between 1200 and 1250 km from the center. In 1938–1939, Jeffreys and Gutenberg found suggestions that some form of transition region may lie between the outer and inner core. Subsequently, other investigators, including B. A. Bolt and Nguyen Hai, proposed a variety of structures for such a transition region, with the suggestion that more than a single transition layer may be involved. An interpretation by R. A. Haddon in 1972 of the relevant seismological data in terms of wave scattering in the neighborhood of the mantle-core boundary has, however, thrown the existence of the transition region *F* into some doubt. It is thus possible that the core consists essentially of only two regions, inner and outer.

Density variation. At any point inside the Earth, the values of the *P* and *S* velocities depend principally on the local density, incompressibility, and rigidity. The two latter moduli describe the Earth's elastic response to stresses which, as in earthquake waves, are of short duration compared with geological time. The incompressibility, or bulk modulus, measures the resistance to change of density under pressure; the rigidity measures the resistance to distortion of shape. A fluid has negligible rigidity and hence does not transmit *S* waves.

The *P* velocities are known directly from seismology to considerable accuracy throughout the Earth, and the *S* velocities throughout the mantle. The *S* velocities can also be taken as zero throughout the outer core because of its lack of rigidity. Knowledge of both the *P* and *S* velocities at any point of the Earth's interior provides two numerical relations which involve the values of the density and elastic moduli at that point. In this way, important information, though not sufficient in itself, is provided toward determining the various physical properties of the whole Earth interior, apart from the inner core.

The additional evidence needed to determine the values of the density, rigidity, and incompressibility separately inside the Earth includes the use of the known values of the mean density and moment of inertia of the Earth, and the use of the theory of gravitational attraction. Further use is made of the *P* and *S* velocity data through the fact that the changes of density due to the increasing pressure in the Earth are linked with the incompressibility. Matching laboratory experiments on rocks against *P* and *S* velocities also yields a useful partial correlation between velocities and densities; an empirical equation connecting *P* velocities and densities of rocks likely to predominate in the Earth's mantle, obtained by F. Birch in 1961, has been much used for this purpose.

From evidence of the type just listed, it was calculated, starting from work of Bullen in 1936, that the density ranges from 3.3 just below the Earth's crust to about 5.7 at the base of the mantle, jumps abruptly to 9.9 at the top of the core, and reaches about 12.7 at the boundary of the inner core and 13 at the Earth's center. These values were estimated to be reliable within about 0.2 units in the mantle and outer core, and about 0.5 in the inner core. The density value of 3.3 just below the crust rests on geological and related observational evidence initially provided by E. D. Williamson and L. H. Adams in 1923. F. Press suggested that the value may need to be increased to about 3.5, with a density inversion occurring some distance below the crust; the inversion would be associated with a negative density gradient in the outermost 100–200 km of the crust, presumably because of an abnormally high temperature gradient in this part of the Earth, the density falling to 3.3 before starting to rise again. However, the evidence for this density inversion is inconclusive.

In 1961 Birch assembled evidence from laboratory shock-wave experiments, in which transitory pressures exceeding a million atmospheres were realized, indicating that the central density of the Earth does not much exceed 13. The earlier evidence had established that the central density is at least 12.3. Thus the central density is fairly reliably known within close limits. Additional evidence on the Earth's density variation comes from observations of free Earth oscillations.

Pressure variation. The pressure distribution in the Earth was derived in the course of the calculations which gave the density distribution. The pressure (strictly speaking, minus the mean of the three principal stresses) is about 10,000 atmospheres (atm) at the base of the continental crust, 400,000 atm at 1000 km depth, about 1,350,000 and atm at the mantle-core boundary, and about 3,700,000 atm at the center of the Earth. These pressures can now be reached in transitory laboratory experiments.

Gravitational intensity. The internal distribution of the gravitational intensity *g*, which at the Earth's surface is approximately equal to the acceleration of a freely falling body, is also derived from the density calculations. The value of *g* remains within about 2% of 10 m/s^2 (32.8 ft/s^2) down to a depth of nearly 2400 km (1500 mi), rises to a maximum of about 10.7 m/s^2 at the bottom of the mantle, and then falls fairly steadily throughout the core to zero at the center. *See* GRAVITATION.

Incompressibility. On the whole, the Earth's incompressibility increases steadily with depth. Unlike density and rigidity, the variation of incompressibility with pressure in the lower mantle and outer core was shown by Bullen in 1949 to be remarkably smooth and little affected by variations of chemical composition such as may occur in the Earth's deeper interior. A compressibility-pressure theory developed from this result has proved to be of much value in investigating the properties below 1000 km depth. In the vicinity of the mantle-core boundary the materials of the lower mantle and outer core are about 4 times as incompressible as ordinary steel, and at the center of the Earth about 10 times.

Another striking property that emerged from work on the Earth's density distribution is that the gradient of the incompressibility with respect to pressure is fairly steady below 1000 km depth. The gradient has a value of 3 units at the mantle-core boundary and lies between 3 and 4 inside the core. These results have received confirmation in independent theoretical investigations by Birch on the finite strain of materials subjected to high pressures. The numerical values of the gradient have proved important, among other things, in enabling the rates of density increase to be estimated in regions of the Earth where chemical composition is not uniform.

Compression. Due to the increasing pressure, the materials of the interior are increasingly compressed as the depth inside the Earth increases. At the mantle-core boundary, the compression amounts to 30% of the volume the material would occupy at zero pressure, and to nearly 40% at the Earth's center. These percentages are minima and could need to be increased should sizable changes of chemical phase occur inside the Earth.

Rigidity. Using evidence from fortnightly tides, Lord Kelvin in 1863 showed the mean rigidity of the Earth to be appreciably greater than that of ordinary steel. The crust and mantle were later found to be solid throughout (apart from the oceans and limited pockets of magmatic material), since *S* as well as *P* waves are transmitted everywhere outside the core. Below 400 km depth the rigidity is in fact now known to increase steadily with depth in the mantle, its value reaching about 4 times that of ordinary steel at the bottom. The terms solid and rigid here relate to the elastic behavior under stresses of limited duration. Convection currents or other forms of solid flow in the mantle, taking place over geologically long periods of time, are not precluded by the use of these terms; conclusions are, however, less precisely established than conclusions on the properties chiefly discussed in this article.

The outer core is much less rigid than the mantle and is in an essentially fluid or molten state. Negative evidence of this is the failure to detect *S* waves below the mantle. The positive evidence lies in direct calculations, initiated by Jeffreys in 1926 and carried to greater detail by H. Takeuchi in 1951 and M. S. Molodenski in 1955, of the core rigidity based on the known rigidity distribution in the mantle and on measurements of the Earth's tidal straining. *See* EARTH TIDES.

Application of the compressibility-pressure theory by Bullen in 1946 showed that the inner core is probably solid. Lehmann had shown that the refraction of *P* waves at the boundary between the outer and inner core is associated with a sharp rise in the *P* velocity. Consequently, a sharp rise is required in either the incompressibility or the rigidity, and the smooth variation of the incompressibility elsewhere in the Earth makes a rise in rigidity much more probable. The available evidence from laboratory experiments and theoretical physics supported this conclusion. A tentative estimate gives the mean rigidity of the inner core as somewhat greater than that of ordinary steel. In 1964 Bullen showed that Birch's estimate of 13 for the Earth's central density leads to independent evidence that the inner core is solid. Other important evidence was provided by F. Gilbert and A. M. Dziewonski in 1973 by using analyses of observations of free Earth oscillations.

Effects of nuclear explosions. Nuclear explosions close to or below the Earth's surface, like earthquakes, generate P and S waves which can penetrate all parts of the Earth. In investigations of the Earth's interior, they have an advantage over natural earthquakes in that their source locations and origin times can be precisely known. In consequence, records of waves produced by nuclear explosions have increased the precision with which several of the Earth's internal physical properties have been determined.

Free Earth oscillations. The catastrophic Chilean earthquake of May, 1960, heralded another important development in the investigation of the Earth's interior. In P and S and ordinary seismic surface-wave transmission, the motion is looked upon as a series of traveling disturbances which affect only a relatively small part of the Earth at any given time, without reference to what is happening at that time to the whole Earth. In the Chilean earthquake, in addition to the regular P, S, and surface waves, there were also recorded, indisputably for the first time in history, a series of free oscillations of the whole Earth. This earthquake had set the Earth vibrating as a single unit, like a bell, so to speak. Starting from 1935, H. Benioff had developed a new type of seismograph, which measured changes of strain in the Earth and was capable of recording earth movements with periods of the order of an hour and more. In November, 1952, immediately following a very large earthquake in Kamchatka in eastern Siberia, one of the instruments showed a period of nearly an hour which Benioff associated with a fundamental free Earth oscillation. The observation stimulated Benioff and others to improve long-period recording still further, and still others, including N. Jobert, C. L. Pekeris, and Takeuchi, to calculate theoretical values of the periods of free Earth oscillations.

Just as vibrating drums and bells give out sounds that are the superposed result of various types of pure vibration, so the Earth will vibrate in a great many modes when suitably excited. The oscillations fall into two main classes, called spheroidal and torsional, each with a large number of fundamental modes and with numerous overtones corresponding to each fundamental. Each oscillation period observed supplies a value which must be fitted by acceptable distributions of the Earth's density, incompressibility, and rigidity. Moreover, whereas the immediate data from P and S bodily waves yield only the quotients of the elastic moduli and the density, the free oscillation data are not tied to these quotients (a property shared with seismic surface-wave observations). Observations, mainly from the 1960 Chilean earthquake and the Alaskan earthquake of March, 1946, have now supplied periods of a few hundred free Earth oscillation modes, of durations up to 54 min.

Observations from the Chilean earthquake gathered by Benioff, L. B. Slichter, F. Press, and B. P. Bogert, along with theoretical results derived by Pekeris, confirmed in 1961 that the earlier derived distributions of density and elastic moduli were reliable within 5% or less throughout most of the Earth.

Since 1961, numerous analyses of free Earth oscillations have led to considerable refinement in the numerical detail and have strengthened various conclusions relating to the Earth's interior. The long-standing estimate, 3473 ± 3 km, made by Jeffreys in 1939 of the radius of the Earth's core, has been increased by about 10 km, though this revised estimate may still need to be modified when allowances for possible damping of the oscillations, not yet finely calculated, have been made. The oscillation data have also contributed part of the evidence relating to the rigidity of the inner core and evidence leading to the questioning of the existence of transition region F.

The overall evidence now available, including the earlier evidence and the free Earth oscillation data, indicates that the most suitable value for the representative average crustal thickness (the region A) is of the order of 15 km, and that each of the regions B and C needs to be subdivided into at least two layers; a tentative division designates these layers as B' (15–60 km depth), B'' (60–400 km), C' (400–650 km), and C'' (650–950 km). The stated ranges of depth are by no means precisely determined, and some investigators would include one or two additional subdivisions. The overall evidence establishes fairly firmly that the average S velocity gradient and density gradient are abnormally small in the region B and (along with the P velocity gradient) abnormally large in the region C. In most of the lower mantle and outer core, the oscillation data are compatible with the density and P and S velocity gradients indicated in earlier work.

Composition. Information on the Earth's internal chemical composition is derived from the results of studies of the variation of the density, incompressibility, and compression with pressure, along with geophysical laboratory studies and studies in geochemistry. In certain regions of the Earth where the rate of density increase is found to be greater than normal, departures from chemical homogeneity, or possibly phase changes, are indicated. By comparing the dependence of density and incompressibility on pressure as derived for any internal region of the Earth with results from physical and chemical theory and experiment, an estimate can be made of the representative average atomic number for that region.

Just below the crust only a few known rocks, chiefly dunite, peridotite, pyroxenite, and eclogite, go close to fitting observed P and S velocities; evidence favors the presence of rocks of the type of dunite and peridotite, consisting mostly of magnesium-iron olivines and pyroxenes. There is general agreement with the conclusion of E. D. Williamson and L. G. Adams in 1923 that the upper mantle consists mostly of ultrabasic rock.

The region B (including B' and B'') is probably fairly uniform in chemical composition. It was, however, shown by Bullen in 1936 that significant changes of chemical composition or phase must occur over a range of depth inside the mantle, most probably the region C (including C' and C''). Jeffreys and J. D. Bernal interpreted this conclusion in terms of changes in crystal structure (phase changes), and the findings were later supported and amplified in experimental and theoretical work of Birch and A. E. Ringwood. Birch also found evidence of limited compositional changes inside C.

The region D' of the lower mantle probably consists mainly of ingredients chemically equivalent to those in the upper mantle, according to Birch possibly in the form of distinct phases such as silica, magnesia, and iron oxide; there may also be some free iron as well. In 1949, Bullen inferred a density gradient three times the normal inside D'', suggesting an accumulation of denser materials at the bottom of the mantle; J. Cleary proposed in 1969 that there may also be a rather sharp fall in the rigidity inside D''; Birch suggested some extra accumulation of free iron inside D'' to account for the increased density gradient; and Ringwood suggested that some core material may have diffused upward into D'' across the mantle-core boundary.

A long-standing theory, based originally on analyses of meteorites, regarded as samples of the interior of an Earthlike planet, and on the Earth's high core density, is that the core consists predominantly of iron and nickel. The currently favored estimate of the representative atomic number for the outer core is, however, only about 23 (a smaller value is possible). Birch, Ringwood, and others have suggested that the outer core consists of an alloy of iron with less dense material, possibly carbon, silicon, or sulfur. An alternative suggestion put forward by W. H. Ramsey and Bullen in 1948–1949 is that the outer core is a phase transformation of the lower mantle material, but there is some difficulty in reconciling this suggestion with the large density jump (by about 75%) from mantle to core. The representative atomic number estimated for the inner core is at least 26, which indicates a composition predominantly of iron and nickel.

A suggestion put forward and examined by O. G. Soroktin and Bullen in 1971–1973 is that the outer core may consist of the iron oxide Fe_2O. This oxide is unstable at ordinary pressures, but theoretical calculations have shown that it would be stable at the pressures reached in the Earth's core and, further, that its pressure-density relation matches conditions in the outer core.

Internal heat and temperature. Prior to around 1950 the Earth was assumed to have had a hot origin. In 1952 H. Urey and others, taking results from theoretical chemistry into account, favored a cold origin. In the latter case the gravitational energy of accumulation would raise the temperature so that, on either theory, most of the Earth's interior would have been molten at a fairly early stage. According to Jeffreys, the present mantle would then have formed by solidification from the bottom upward. Jeffreys attributed the formation of mountains principally to thermal contraction within the outermost 700 km of the Earth, a region inside which all recorded earthquakes have originated. Other theories postulate solid convection currents in the mantle, continental drift (separation would be a more appropriate term), and an expanding Earth. *See* CONTINENT FORMATION.

The fact that the Earth has not by now cooled to zero temperature shows that significant heat sources, almost certainly due to radioactivity, exist below the surface. The total radioactivity inside the Earth is, however, severely limited by the circumstance that the Earth is solid (in the ordinary sense) to nearly 2900 km depth. In fact, most investigators have concluded that the radioactivity is largely confined to the vicinity of the crust. It was earlier thought that the main heat flow from the Earth's interior comes from continental regions, but measurements at sea since 1950 indicate that the heat flow does not deviate very far from a uniform distribution over the whole of the Earth's surface. *See* EARTH, HEAT FLOW IN.

The distribution of temperature in the Earth is less precisely determined than the distributions of density, pressure, and the other physical properties discussed above, principally because direct information on the distribution of heat sources is lacking. The temperature gradient is high near the surface, and the average value down to a depth of order 350 km may be as much as 5°C/km. Thermodynamical studies and the abnormally low gradients of the density and *S* velocity indicate that the temperature gradients are abnormally high in this part of the Earth. The fact that the whole mantle is solid sets an upper limit to the temperature inside the mantle. In 1954 R. J. Uffen gave 5000°C as an upper limit to the temperature at the base of the mantle, and J. Verhoogen gave 2000°C as a lower limit. Through the core the temperature may rise by a further 500°C, but probably not much more.

In an extension of the work of Jeffreys, J. A. Jacobs suggested in 1953 that the Earth started to solidify from the center about the time the mantle started to form, the fluid outer core being trapped in between. Lord Simon showed in 1954 that a solid iron inner core would be compatible with an inner-core temperature of about 4000°C.

Most investigators estimate the temperature at the center of the Earth to be about 4000°C, though this figure is still uncertain by at least 1000°C.

Magnetism. Early theories attributed the Earth's magnetism to permanently magnetized iron in the deep interior. This is now considered improbable because of seismic and other evidence on the fluidity of the outer core. Nevertheless, it is well established that the Earth's magnetic field does originate predominantly in the deep interior. The only theory which does not appear to meet insuperable objections is the dynamo theory of W. M. Elasser and E. C. Bullard. This theory attributes the Earth's magnetism to currents flowing inside the fluid outer core and involves the conversion of mechanical energy into electromagnetic energy. The theory is compatible with either a molten-iron outer core or an electrically conducting outer core composed of the same ingredients as the lower mantle or of the oxide Fe_2O. It is also compatible with the presence of a solid inner core, the boundary of which would enable convection currents of the required modes to occur in the outer core. The theory requires a source of energy for the convection currents. Radioactivity has been suggested, but there is a problem of obtaining enough energy from this source. This is one of the chief difficulties the theory has to face. *See* EARTH; ROCK MAGNETISM; TERRESTRIAL MAGNETISM.

[KEITH E. BULLEN]

Bibliography: K. E. Bullen, *The Earth's Density*, 1975; K. E. Bullen, *Introduction to the Theory of Seismology*, 3d ed., 1965; T. F. Gaskell (ed.), *The Earth's Mantle*, 1967; J. A. Jacobs, *The Earth's Core*, 1976; H. Jeffreys, *The Earth: Its Origin, History and Physical Constitution*, 6th ed., 1970.

Earth, rotation and orbital motion of

The Earth rotates on its axis with respect to inertial axes, as demonstrated by the Foucault pendulum experiment, and revolves about the Sun, as demonstrated by the annual parallactic displacement of nearby stars against the background of distant stars. Because the Earth, including its oceans and atmosphere, is not a symmetric rigid body and because it is not the only body revolving about the Sun, the motions vary with time. *See* EARTH.

ROTATION ABOUT AXIS

Astronomical observations show that planet Earth undergoes three types of variations in speed of rotation: secular, irregular, and periodic. The secular increase in the length of the day, chiefly as a result of tidal friction due to the Moon, is about 0ˢ.0015 per century. Irregular changes in speed are random and persist for several years. The total range in the length of the day during the past 200 years is nearly 0ˢ.01. Periodic variations have periods of 1 year, 0.5 year, 27.55 days, and 13.66 days. These changes in the length of a day are on the order of 0ˢ.0005.

Causes of variations. The speed of rotation of Earth is measured with respect to stations fixed on land. However, the total angular momentum of planet Earth is the sum of the momenta of land, water, and air. An interchange of momentum between these three will change the length of a day.

The crust of the Earth is elastic. Tides generated by Moon and Sun change the shape, changing the moment of inertia. The Earth may have a fluid core with turbulent motion. Changes in coupling between such a core and the crust would change the speed of rotation. Ocean and bodily tides decrease the angular momentum of the Earth through friction. The Earth may shrink or expand as a whole, or landmasses may rise or fall with a change in moment of inertia.

Measurement of variation. Mean solar time is defined so as to be a strict measure of the angular rotation of the Earth. Hence, variations in speed of rotation are obtained by comparing mean solar time with some other form of time which is independent of the rotation of the Earth. In the past, ephemeris time based on the Moon's orbital motion was used chiefly. However, construction of the cesium-beam atomic clock in 1955 provided a much more precise reference, which permits details of the irregular variation to be observed.

Tidal friction. The Moon raises tides in the ocean. Friction carries the maximum tide ahead of the line joining the center of the Earth and Moon (Fig. 1). The resulting couple diminishes the speed of rotation of the Earth. The couple reacts on the Moon to increase its orbital momentum. The sum of the angular momentum of the Earth and the orbital momentum of the Moon remains constant. The effect is to increase the size of the orbit of the Moon and to diminish its angular speed about the Earth.

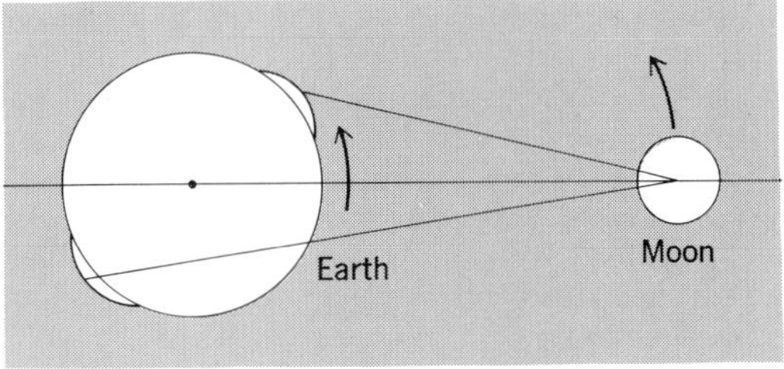

Fig. 1. Couple produced by tidal friction.

The effect of tidal friction is to increase both the distance D and the lunar month with respect to ephemeris time. Because of the change in the lunar month, the Moon has an orbital deceleration in terms of ephemeris time. The proportional change in D, however, is greater than in the month. Hence, in terms of mean solar time the Moon appears to have a secular orbital acceleration, an effect discovered by E. Halley in 1693 from a study of ancient eclipses. If the Earth has changed its speed of rotation since ancient times, the path of an eclipse which occurred 2000 years ago would be displaced in longitude with respect to the path that would have occurred if the speed had remained constant.

The secular acceleration, when discovered, was puzzling; it was not then attributed to tidal friction. In 1786 P. S. Laplace announced that the secular acceleration was due to a neglected effect in the gravitational theory of the motion of the Moon. However, in 1853 J. C. Adams found that gravitational theory could account for only about half the acceleration found by Halley. It has become clear since then that the rotation of the Earth is slowing down because of tidal friction.

Irregular variation. Researches by S. Newcomb, W. de Sitter, E. W. Brown, Harold Spencer Jones, and others have shown that the Earth has irregular variations in addition to the secular retardation. The orbital motions of Earth, Venus, and Mercury about the Sun show discordances similar to that shown by the Moon, but proportional to their mean speeds.

It was formerly thought that changes in speed of rotation occurred suddenly. However, D. Brouwer suggested in 1952 that sudden changes in acceleration occurred, but not sudden changes in speed. Comparisons of mean solar time with atomic time made since 1955 confirm this hypothesis. Changes in acceleration (or deceleration) occur about every 5 to 10 years.

Periodic variations. Seasonal variations were detected in 1937 by N. Stoyko, who used a number of pendulum and quartz-crystal clocks. The early determinations of the seasonal variation appear to be too large, judged by results obtained with improved quartz-crystal clocks since 1950 and with atomic clocks since 1956, by a factor of two. The Earth is behind its mean position by about 0ˢ.035 at the end of May and about 0ˢ.030 ahead at the beginning of October.

The seasonal variation is probably caused by an exchange of momentum between winds and the crust of the Earth.

Lunar tidal variations of periods 27.55 days and 13.66 days were predicted by Jeffreys in 1928. The amplitude of each term is about 0ˢ.001 in time. These terms have been determined by observation.

Energy dissipation. The energy lost by tidal friction must be dissipated somehow. G. I. Taylor and Harold Jeffreys have studied the effects in shallow waters. Jeffreys considered that the Bering Sea alone would account for 80% of the energy dissipation required. However, studies made by C. A. Murray and by W. H. Munk and G. J. F. McDonald indicate that only about 10% of the required effect can be accounted for by tidal currents in shallow seas. This question is, therefore, still open.

[WILLIAM MARKOWITZ]

REVOLUTION ALONG ORBIT

Motion of Earth about Sun is seen as an apparent annual motion of the Sun along the ecliptic. That the effect is caused by motion of Earth and not of the Sun is proved by the annual parallactic displacement of near stars and by the aberration of light, causing an apparent annual displacement of all stars on the celestial sphere.

Period of revolution. The true period of revolution of the Earth around the Sun is determined by the time interval between successive returns of the Sun to the direction of the same star. This interval is the sidereal year $T = 365$ days 6 hr 9 min 9.5 sec of mean solar time or 365.25636 mean solar days. The period between successive returns to the moving vernal equinox is the tropical year $T' = 365$ days 5 hr 48 min 46.0 sec or 365.24220 days. Chronology is based on the tropical year. The period of time between successive passages at perihelion is called the anomalistic year $T'' = 365$ days 6 hr 13 min 53.0 sec or 365.25964 days.

The length of the year varies slowly as a consequence of the long period perturbations of the Earth orbit by the other planets. The second of ephemeris time was defined in 1956 as the fraction 1/31,556,925.9747 of the tropical year 1900.0.

Mean radius of orbit. The mean distance from Earth to Sun or semimajor axis of the Earth's orbit is the astronomical unit of distances in the solar system. Its absolute value fixes the scale of the solar system and the whole universe in terms of terrestrial standards of length. This value can be determined by a variety of methods; the results are usually expressed in terms of solar parallax, or more precisely the Sun's mean equatorial horizontal parallax p, which is the angle subtended by the equatorial radius r of the Earth at the mean distance a of the Sun. The relation between r and p is $a = r \sin p = 206{,}265r/p$, if p is in seconds of arc. The equatorial radius r of the Earth is 6,378,156 m.

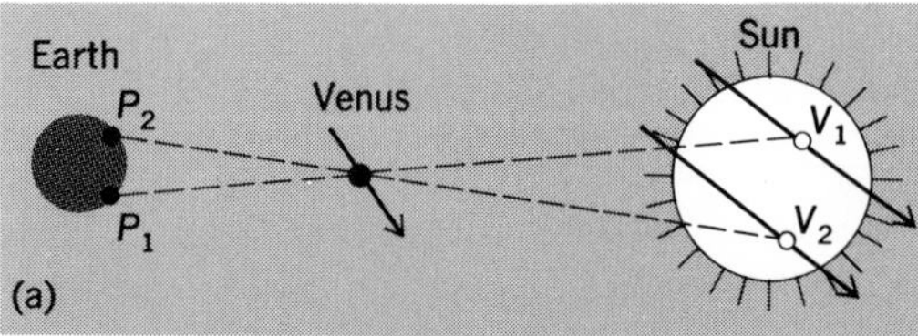

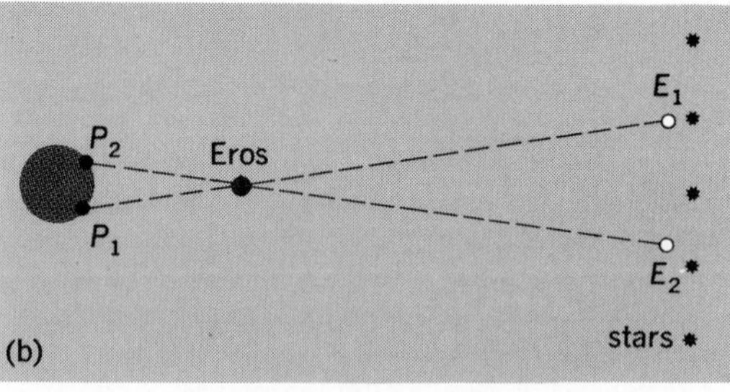

Fig. 2. Determination of solar parallax. (*a*) Observation of Venus in transit. (*b*) Observation of Eros in opposition.

Geometrical, gravitational, and physical methods have all been used to measure solar parallax.

Geometrical methods. The geometrical methods involve the direct measurement by optical triangulation of the parallax of a nearby planet (Mars or Venus) or asteroid (Eros) at its closest approach to Earth (opposition or transit). Because the relative distances in the solar system are accurately known in terms of the astronomical unit, the absolute measurement of one distance gives the scale of the system (Fig. 2).

The most accurate trigonometric determination, derived by H. Spencer Jones in 1941 from an international campaign of observation of the minor planet Eros near its opposition in 1930–1931, is $p = 8''.790 \pm 0''.001$, corresponding to a mean solar distance $a = 149{,}675{,}000 \pm 17{,}000$ km $= 93{,}004{,}000 \pm 11{,}000$ mi. This determination is now known to be in error by about ±0.05%.

Gravitational methods. The gravitational methods involve the determination of the ratio of the mass m of the Earth to the mass M of the Sun from the perturbations in the motion of a minor planet (Eros) caused by the Earth. The method rests essentially on a comparison between length $l = gt^2/4\pi^2$ of a pendulum of double oscillation period t (seconds of mean solar time) on Earth, which gives the acceleration of gravity $g = Gm/r^2$ in terms of gravitational constant G, and the length of the radius a of the Earth's orbit, related by Newton's law to the duration T of the sidereal year (in seconds of mean solar time), by the equation below.

$$4\pi^2 a^3/T^2 = G(M+m) = g(1 + M/m)$$

Because g, G, and T are known with high accuracy, a determination of M/m gives a directly in terms of terrestrial standards of length. Although it is customary to express the result as a parallax angle, the Earth's radius is not needed to compute a, but it is necessary to make allowance for the mass of the Moon relative to the mass of Earth.

Until 1960 this was the most accurate method, and the best result, derived by W. Rabe in 1948 from 50 years of observations of Eros at its nearest approaches to the Earth, was $p = 8''.7984 \pm 0''.0004$, corresponding to a mean solar distance $a = 149{,}532{,}000 \pm 7000$ km $= 92{,}915{,}000 \pm 4500$ mi. This result is now known to be in error by about −0.05%.

Physical methods. The older physical methods rest on a determination of the ratio of the mean orbital velocity $V = 2\pi a/T$ to the accurately known velocity of light. This ratio can be derived either from the annual variations of the radial velocities of ecliptic stars (or occasionally of planets) determined by observations of the Doppler shift of spectral lines, or with less accuracy from the constant of aberration.

The best results by spectroscopic methods were obtained by W. S. Adams in 1948, $p = 8''.805 \pm 0''.007$, from velocities of the star Arcturus; and by J. Guinot in 1958, $p = 8''.787 \pm 0''.008$, from observations of the planet Venus. The mean of the two determinations $8''.796$ was found to be in error by less than 0.02%.

The most precise physical method rests on measurements of the travel time of radar signals bounced off the planet Venus; the distance of the planet Δ is simply the ratio $t/2c$ of half the round trip travel of the signal to the velocity c of the radio waves. Since the distance of the planet is known precisely (in astronomical units) from basic orbital computations, measuring the actual distance of the planet in kilometers at a given time gives the scale factor of the astronomical unit. Observations made in the United States, United Kingdom, and Soviet Union since 1961 give a well-determined mean value of the astronomical unit, 1AU= 149,598,500 km, corresponding to a solar parallax $p = 8''.79412$. The estimated uncertainty is less than 5 parts in 1,000,000 and arises mainly from the uncertainty in the adopted value for the velocity of light ($c = 299792.5$ km/sec).

Orbital velocity. The mean orbital velocity of the Earth is 29.80 km/sec ≡ 18.5 mi/sec. The curvature of the orbit, as measured by the departure from its tangent at the end of the arc traveled in 1 sec of time is 0.296 cm ≡ 0.117 in.; the acceleration toward the Sun is numerically twice this quantity or 0.593 cm/sec^2; since the acceleration of gravity at the Earth's surface is $g = 880.66$ cm/sec^2, the attraction of the Sun at the mean distance of the Earth, GM/a^2, is about 0.605% of the attraction of the Earth on bodies at its surface, Gm/r^2. To produce this attraction at a distance equal to $a/r = 23{,}480$ earth radii, the mass of the Sun is therefore approximately $0.605\ (23{,}480)^2 =$ 333,000 times the mass of the Earth.

Eccentricity of orbit. The eccentricity of the Earth's orbit was initially determined by the variations of the apparent diameter of the Sun's disk; it is accurately determined by the variable speed of the Sun's apparent motion along the ecliptic and the laws of elliptic motion. The nonuniformity of the Sun's motion manifests itself in the equation of time, which is the difference between apparent (or true) and mean solar time. This difference arises in part from the obliquity of the ecliptic and in part from the eccentricity of the Earth's orbit. The present value of the eccentricity of the Earth's orbit is 0.01675; it decreases slowly under the effect of planetary perturbations, the present rate of decrease being −0.000042 per century counting from 1900. However, the eccentricity will not decrease to zero and will increase again after reaching a minimum. The Earth is at perihelion on January 2 and at aphelion on July 2; the corresponding variation of temperature tends to reduce the seasonal amplitude in the Northern Hemisphere and to increase it in the Southern Hemisphere.

[GERARD DE VAUCOULEURS]

Earth sciences

Sciences primarily concerned with the atmosphere, the oceans, and the solid Earth. They deal with the history, chemical composition, physical characteristics, and dynamic behavior of solid Earth, fluid streams and oceans, and gaseous atmosphere. Because of the three-phase nature of the Earth system, Earth scientists generally have to consider the interaction of all three phases – solid, liquid, and gaseous – in most problems that they investigate.

The geosciences (geology, geochemistry, and geophysics) are concerned with the solid part of the Earth system. Geology is largely a study of the nature of Earth materials and processes, and how these have interacted through time to leave a record of past events in existing Earth features and materials. Hence, geologists study minerals, rocks, ore deposits, mineral fuels and fossils, and the long-term effects of terrestrial and oceanic waters and of the atmosphere. They also investigate present processes in order to explain past events. *See* GEOLOGY.

Geochemistry involves the composition of the Earth system and the way that matter has interacted in the system through time. For example, by studying the behavior of radioactive substances it is possible to determine how old the substances are and how much energy has been released through time as a result of decay of radioactive compounds. *See* GEOCHEMISTRY.

Geophysics deals with the physical characteristics and dynamic behavior of the Earth system and thus concerns itself with a great diversity of complex problems involving natural phenomena. For example, earthquakes, vulcanism, and mountain building throw light on the structure and constitution of the Earth's interior and lead to consideration of the Earth as a great heat engine. Study of the magnetic field involves considering the Earth as a self-sustaining dynamo. *See* GEOPHYSICS.

The atmospheric sciences, commonly grouped together as meteorology, are concerned with all chemical, physical, and biological aspects of the Earth's atmosphere. Although the study of weather used to be the chief occupation of meteorologists, man's entry into the space age calls for a vast increase in knowledge of the environment through which vehicles and ultimately living things will go and return. Consequently, many aspects of the Earth's atmosphere are now being studied intensively for the first time. As an example, great planetary currents in the atmosphere, and also in the oceans, are now being investigated not only for the light they may shed on a better understanding of weather but also as a basis for understanding more fully the motion of the entire atmosphere and oceans.

Oceanography encompasses the study of all aspects of the oceans – their history, composition, physical behavior, and life content. Before World War II little was known about the oceans of the world. During that war many important characteristics of the ocean were discovered, and since then, with instruments and facilities developed during the war, oceanographic research has been going on at a quickened pace.

[ROBERT R. SHROCK]

Bibliography: *Investigating the Earth*, Earth Science Curriculum Project of the American Geological Institute, 1967; K. Krauskopf, *Introduction to Geochemistry*, 1967; H. Takeuchi, S. Uyeda, and H. Kanamori, *Debate About the Earth*, 1967.

Earth tides

Cyclic motions, sometimes over a foot in height, caused by the same lunar and solar forces which produce tides in the sea. These forces also react on the Moon and Sun, and thus are significant in astronomy in evaluations of the dynamics of the three bodies. For example, the secular spin-down of the Earth due to lunar tidal torques is best computed from the observed acceleration of the Moon's orbital velocity. In oceanography, earth tides and ocean tides are very closely related. *See* GEODESY; TERRESTRIAL GRAVITATION.

Efforts to measure earth tides date from the end of the 19th century; accounts are found in George Darwin's *Scientific Papers* and in other writings on natural philosophy about 1890. In 1919 A. A. Michelson and H. G. Gale measured the daily tidal tilt near Geneva, Wisconsin, by recording the changes in water level at the ends of 500-ft-long (152 m) horizontal pipes buried in the ground. Because of its precision and significance their work may be regarded as initiating modern earth tide measurements. By far the most widely used earth tide instruments are the tiltmeter and the gravimeter. Both instruments have the merits of portability, high potential precision, and low cost. Thus they are able to advance economically an important mission—the global mapping of earth tides and ocean tides, both still in a rudimentary stage.

Tide-producing potential. Tidal theory begins simply, by evaluating the tidal forces produced on a rigid, unyielding spherical Earth by the Moon and Sun. In this discussion, the Earth, Moon, and Sun will be regarded as mass points at their respective centers. The results of this idealized assumption will be modified later to relate earth tide observations to deformable Earth models.

In the expansion of the total gravitational potential of the satellite as a sum of solid spherical harmonics, the tide-producing potential consists only of those terms which vary within the Earth and are the source of Earth deformations. The omitted terms are the constant which is arbitrarily chosen to produce null potential at the Earth's center and the first-degree term expressing the uniform acceleration of every part of the Earth toward the satellite (thus also entailing no distortion). Accordingly, the tide-producing potential U is expressible as the sum of solid spherical harmonics of degree 2 and higher in the notation which follows.

Let the mass of the satellite be denoted by m its distance from the Earth's center by R, the distance of the observing point from the Earth's center by r, and the geocentric zenith angle measured from the line of centers by θ. Then for points $r<R$, Eq. (1) holds, where G is Newton's constant ($6.67 \times 10^{-8} c^3 g^{-1} s^{-2}$).

$$U=U_2+U_3+\cdots=GmR^{-1}\left[\left(\frac{r}{R}\right)^2 P_2(\cos\theta) + \left(\frac{r}{R}\right)^3 P_3(\cos\theta)+\cdots\right] \quad (1)$$

On an idealized oblate Earth of equatorial radius r_e and flattening constant f, a point on the surface at geocentric latitude ϕ has radial distance $r=r_eC(\phi)$, where $C(\phi)\equiv 1-f\sin^2\phi$. The vertical (upward) component of tidal gravity $\partial U/\partial r$ and the horizontal component $r^{-1}\partial U/\partial\theta$ in the azimuth direction away from the satellite, with $P_n{}^1(\cos\theta)=-\partial P_n(\cos\theta)/\partial\theta$, are given by Eqs. (2) and (3).

$$g_r=\frac{\partial U}{\partial r}=GmC\alpha^3 r_e^{-2}\sum_{n=2}^{\infty} n(\alpha C)^{n-2}P_n(\cos\theta) \quad (2)$$

$$g_\theta=r^{-1}\frac{\partial U}{\partial\theta}= -GmC\alpha^3 r_e^{-2}\sum_{n=2}^{\infty}(\alpha C)^{n-2}P_n{}^1(\cos\theta) \quad (3)$$

Here $\alpha_m\equiv r_e/R_m$ and $\alpha_s\equiv r_e/R_s$, for the Moon and Sun, respectively. R. A. Broucke and coworkers provided a computer program for computing g_r and g_θ in Eqs. (2) and (3) for the Moon within a tested precision of ±0.004 microgal (1 gal = 1 cm · s^{-2}); a similar precision is available for the solar tide.

Harmonic constituents. Equations (2) and (3) indicate the simple nature of the dependence of tidal gravity on the rigid Earth upon the distance R and the zenith angle θ of the satellite. But the time variations of R and θ are complex, because of the Earth's rotation and the complexity of the orbital motions. It is in the analysis of these complexities that earth tide studies show their inheritance from the extensive studies of ocean tides of the last century. For predictions at the world's harbors a year or more in advance, the ocean tides were analyzed in the Darwin-Doodson system as a sum of many harmonic constituents, —386 in A. T. Doodson's 1922 analysis.

However, the objective in earth tide observations is not prediction but assistance in determining present relevant properties of the Earth. For this purpose, only the major tidal constituents in each class provide the critical evidence. Table 1 lists the periods and amplitude factors b of the larger tidal harmonic constituents over the broad range of tidal periods. This discussion has been shortened by deemphasizing the purely geometric complexities essential in applied tidal theory and by adopting, when feasible, the simple zenith-centered satellite coordinate system used in Eqs. (1)–(3). Thus this discussion omits mention of the geographic variation with latitude, which differs for the three species in Table 1, and of the tides resolved into their important spectral constituents. Such resolution is especially fertile in the interpretation of earth tide residuals in terms of their ocean tide causes. It should be remembered that the importance of the subject of earth tides is greatly enhanced because of the content of the wide range of known periodicities and known excitation forces.

Tidal torques. The orbital acceleration $\dot{n}_m$ of the Moon is the critical quantity in Eq. (4), giving the

$$N_m=-\tfrac{1}{3}[M_eM_m/(M_e+M_m)]r_m{}^2\dot{n}_m \quad (4)$$

lunar torque N_m, which slows the Earth's spin.

Table 1. Parameters for some equilibrium tides

Species	Symbol	Period	b
Long period	Lunar	18.6 yr	0.066
	Sa	1 yr	0.012
	SSa	½ yr	0.073
	MSm	31.85 day	0.016
	Mm	27.55 day	0.083
	MSf	14.77 day	0.014
	Mf	13.66 day	0.156
	—	13.63 day	0.065
Diurnal	O_1	25.82 hr	0.377
	P_1	24.07 hr	0.176
	K_1	23.93 hr	0.531
Semidiurnal	N_2	12.66 hr	0.174
	M_2	12.42 hr	0.908
	S_2	12.00 hr	0.423
	K_2	11.97 hr	0.115

(Here M_e and M_m are masses of the Earth and Moon respectively, and r_m is the distance to the Moon.) In 1939 H. Spencer-Jones, by using observations made subsequent to 1680, inferred that $\dot{n}_m$ has been constant at 22″.4 ± 1″.0 century^{-2}. W. H. Munk and G. J. F. McDonald, in adopting this value, found $N_m = -3.9 \times 10^{23}$ dyne-cm, to which corresponds the energy loss rate $\dot{E}$, shown in Eq. (5), and the relative spin-down, $-\dot{\omega}/\omega = N/c\omega =$

$$\dot{E} = N(\omega - n_m) = 2.7 \times 10^{19} \text{ erg/s} \qquad (5)$$

0.21 eon $^{-1}$ (1 eon = 10^9 yr). (Here ω is Earth's rotational velocity, and c its principal moment of inertia.) The factor $\dot{n}_m$ is not as firm as might be desired. K. Lambeck reported values from several sources which are almost twice the Spencer-Jones value used by Munk. Concerning such differences in $\dot{n}_m$, W. M. Kaula proposed that the matter be ignored until the time base is long enough to get a good figure from laser ranging to the Moon.

Tidal loss in the solid earth. The gravest free mode of the Earth, period 0.9 hr, has the same external form and nearly the same internal geometry as the M_2 tidal bulge. Its observed $Q = 350 \pm 100$ was used by Munk to estimate the loss rate $\dot{E}_B$ in the bodily M_2 tide (Q^{-1} is the loss rate number, and M_2 is the principal semidiurnal tide). The result, $\dot{E}_B = 7 \times 10^{17}$ erg/s, is only 3% of the required total in Eq. (5). According to Lambeck, the rate at which the Earth dissipates the total lunar-solar tidal energy is $5.7 \pm 0.5 \times 10^{19}$ erg/s, which is about double Munk's estimate of loss rate due to the Moon alone. The share attributed by Lambeck to the oceans is $5.0 \pm 0.3 \times 10^{19}$ erg/s. Commenting about the small difference, 7×10^{18} erg/s, he says that "if not all, at least a very major part of the secular change in the Moon's mean longitude is caused by dissipation of tidal energy in oceans, and we do not have to invoke significant energy sinks in the Earth's mantle and core."

Symmetric standard earth. The precision in generating symmetrical oceanless Earth models based upon seismic travel times and the periods of the Earth's numerous observed free vibrations had led to remarkable uniformity in the Love numbers of these models. W. E. Farrell performed a test aimed at producing differences in the Love numbers by changing the upper 1000 km of the Earth. Farrell substituted in the Gutenberg-Bullen A Earth model first a 1000-km upper layer of oceanic crustal structure, then the same thickness of continental shield structure, and compared the three models in respect to the computed Love numbers. Table 2 shows the results for the second-degree Love numbers. Differences exist chiefly in the fourth decimal place. Any of several modern Earth models would serve as a suitable reference for earth tide reductions. Presumably such a standard will be named. For the present, the values $\delta = 1.160$, $\gamma = 0.690$, and $l = 0.0840$ will serve, with an

Table 2. Earth tide parameters for different upper mantles

Model	k	h	l	δ	γ
Gutenberg-Bullen Earth model	0.3040	0.6114	0.0832	1.1554	0.6926
Oceanic mantle	0.3055	0.6149	0.0840	1.1567	0.6906
Shield mantle	0.3062	0.6169	0.0842	1.1576	0.6893

assumed zero phase lag of the tidal bulge. (This lag certainly is small, probably less than 0°.1, but as noted present estimates of tidal energy dissipation in the solid Earth are imprecise.)

Instrumental dimensionless amplitude factors. The following paragraphs show how the reading of an earth tide meter is altered by the fact that it is anchored to a yielding Earth. The Earth chosen is assumed to be the symmetric standard of the previous paragraph.

Gravimeter. Basically, a gravimeter consists of a mass generally supported by a spring. (The superconducting model uses a magnetic field.) Variations in gravity are measured by the extension of the spring or in the null method by the small corrections required to restore the original configuration. A satellite affects the mean local value of gravity in three ways: (1) by its direct attraction, (2) by the tidal change in elevation of the observing station, and (3) by the redistribution of mass in the deformed Earth. Of these, the value of the first is expressed by Eq. (2), namely, $\Delta g_1 = \partial U/\partial r = 2U(r_0)\, r_0^{-1}$. An increase in elevation Δr diminishes the Earth's downward gravity field, that is, it increases the upward component (which is taken as positive in the present context). This is expressed as $\Delta g_2 = (\partial g/\partial r)\ \Delta r = +2g_0 r_0^{-1}\ \Delta r$, where g_0 is local gravity. For Δr the following expression defining the Love number h is substituted, $\Delta r = hg_0^{-1}U(r_0)$, thus obtaining $\Delta g_2 = 2hr_0^{-1}U(r_0)$. Finally, the change in gravity caused by the redistribution of mass is expressed by introducing another proportionality factor, the Love number k, defined by $V_0 = kU(r_0)$, where V_0 is the potential of the altered mass distribution at points on the Earth's surface. Outside the Earth, Eq. (6) holds.

$$V = kU(r_0)\, r_0^3 r^{-3} \qquad (6)$$

The contribution Δg_3 at the surface $r = r_0$ is $\Delta g_3 = -3kr_0^{-1}\ U(r_0)$. Total change is per Eq. (7).

$$\Delta g = \Delta g_1 + \Delta g_2 + \Delta g_e = 2U(r_0) r_0^{-1}(1 + h - \tfrac{3}{2}k) \qquad (7)$$

The factor $1 + h - \frac{3}{2}k$ is called the gravitational factor δ, but obviously the pertinent geophysical quantity is $\delta - 1 = h - \frac{3}{2}k$, which alone characterizes the tidal deformation. On a rigid Earth, $h = k = 0$. On an Earth covered with a fluid layer, the equilibrium displacement Δr is determined by the requirement that the deformed surface remain equipotential, or "level," at its original value W_0. Thus $U(r_0) + (\partial W_0/\partial r)\ \Delta r + kU(r_0) = 0$. Using the relations $\Delta r = hg_0^{-1}U$ and $\partial W_0/\partial r = -g_0$, one finds that Eq. (8) holds for such an Earth. Furthermore,

$$1 - h + k = 0 \qquad (8)$$

for a homogeneous incompressible Earth of density ρ, it is easy to compute the external potential due to the superficial mass distribution associated with the displacement $\Delta r = hg_0^{-1}U_2$. This potential is in fact given by Eq. (9), which at $r = r_0$ is $kU(r_0)$

$$V(r,\theta) = \tfrac{4}{5}\pi G\rho r_0^4 r^{-3} hg_0^{-1}U(r_0,\theta)$$
$$= \tfrac{3}{5}g_0 r_0 r^{-1} hg_0^{-1}U(r_0,\theta) \qquad (9)$$

by definition. Thus $k = \frac{3}{5}h$ for a homogeneous Earth, fluid or solid. For a homogeneous incompressible fluid Earth, in view of Eq. (8), $h = 5/2$ and $k = \frac{3}{2}$. For this case, $\delta = 1.25$. Values of $\delta > 1.25$ deduced from observations must be due to errors or to extreme local conditions.

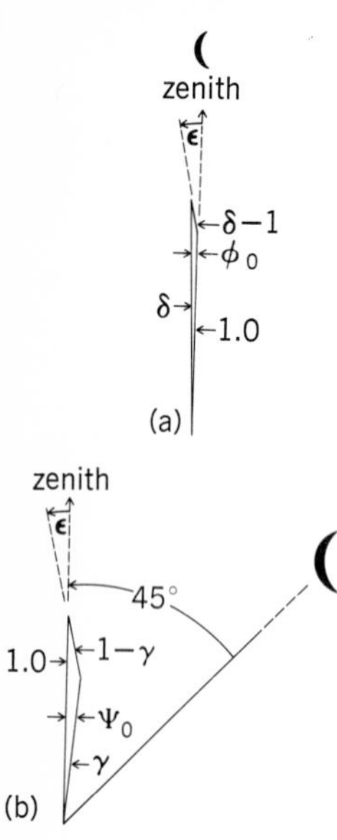

Fig. 1. Observed phase lags in relation to the true lag ϵ of the tidal bulge. (a) ϕ_0 observed by a gravimeter. (b) ψ_0 observed by a tiltmeter.

Tiltmeters. The tiltmeter measures changes in the angle of tilt between the tiltmeter's foundation and the local vertical, both subject to tidal variations. Several types of tiltmeters are in use. In the horizontal pendulum the rotation axis is fixed at a small angle with the vertical to produce high sensitivity to tilts. In the level tube, which may be several hundred meters or more long, the difference in elevation of a fluid is measured at the two ends. In this way a sample of the tilt of a large region is obtained, but the horizontal pendulum sometimes also samples a large volume, as in A. Marussi's installation in the Grotto Gigante, Trieste, where these pendulums are 75 m high. Generally two orthogonal tiltmeters are used to deduce the total tilt angle. In terms of the Love numbers h and k, the tilt observations have the following significance. The ground itself is elevated an amount $\Delta r = hg^{-1}U$ by the tidal potential, so that in the direction of maximum tilt the ground's tilt angle ψ is given by Eq. (10).

$$\tan \psi \doteq \psi = hg^{-1}r^{-1}\frac{\partial U}{\partial \theta} \tag{10}$$

This tilt of the solid Earth, for which h is about 0.61, is clearly less than that of the local level (that is, fluid) surface for which h exceeds unity (that is, $h=1+k$). Since the tilt of the level surface is $(1+k)g^{-1}r^{-1}(\partial U/\partial\theta)$, the net tilt angle observed (ψ) is given by Eq. (11). This quantity achieves its maximum value when the satellite's zenith angle is 45°, attains an equal minimum at $\theta = 135°$, and is zero when $\theta = 0$, 90°, or 180°. The factor $1+k-h$ is called the tilt number γ but, in analogy with the previous case, the significant part is only $1-\gamma = h-k$. In magnitude $h-k$ is about twice the corresponding gravitational quantity $\delta - 1 = h - \frac{3}{2}k$.

$$\psi = g^{-1}r^{-1}\frac{\partial U}{\partial \theta}(1+k-h) \tag{11}$$

Extensometer. The extensometer, or linear strainmeter, measures the change in distance between two reference points. In the Benioff design these positions are connected with a quartz tube to bring the points into juxtaposition for precise measurements of relative displacement. By using laser beams, the measurements may be made over distances of kilometers, but the small mechanical meters still have their use. They offer considerably greater utility per unit cost than other equivalent instruments.

The total horizontal component of displacement s toward the satellite serves to define the dimensionless proportionality constant, or Shida's number, l in accordance with Eq. (12), where g is the local value of gravity. However, the absolute displacement s is not measured, but instead the relative displacement between two points, namely a strain component. While the general specification of the strain in an isotropic solid requires six numbers, on the Earth's free surface the two components of shear stress and their associated strain components vanish. Further, a horizontal strainmeter is immune to vertical strains, so that, in all, only three constituents of strain remain effective. To determine the horizontal areal strain (ratio of change in area to original area), since this is independent of azimuth, only two rectangular strain components are needed. These two strain components are sufficient to determine the combination $h-3l$ of the Love and Shida numbers. With a third horizontal strainmeter, l and h may be independently obtained.

$$s = lg^{-1}\frac{\partial U}{\partial \theta} \tag{12}$$

Indicated and true phase lags. The phase indications of a gravimeter or tiltmeter are only a fraction of the true angular lag of the tidal bulge. The tidal bulge is assumed to retain its no-loss equilibrium form, but with axis carried forward from the Moon's zenith by the small angle ϵ required to produce the Moon's known orbital acceleration. In the expressions for the dimensionless amplitude factors, $\delta = 1 + (h - {}^{3}\!/_{2}k)$ and $\gamma = 1 + (k-h)$, the tidal losses are produced only by the terms in parentheses. On a rigid, lossless Earth, $\delta = \gamma = 1.0$. Figure 1*a* is a vector diagram illustrating the relation between a phase angle ϕ observed with a gravimeter and the bulge lag angle ϵ (geometrically a lead angle with respect to the Moon's zenith, but in time, a lag). For small angles, $\epsilon = \delta\phi/(\delta-1) = 7.25\phi$, if $\delta = 1.16$. The similar relation between true bulge lag ϵ and the phase lag ψ observed in tilt measurements is shown in Fig. 1*b*. Here $\epsilon = -\gamma\psi/(1-\gamma) = -2.23\psi$, if $\gamma = 0.69$. Strainmeters, of course, record strain-amplitude components directly, and the true phase lag ϵ of the tidal bulge on the idealized model.

Residuals. The observations of a gravimeter, tiltmeter, or strainmeter are vector quantities, often written in dimensionless form. $\vec{\delta}_0 = \delta_0 \exp(i\phi_0)$, $\vec{\gamma}_0 = \gamma_0 \exp(i\psi_0)$, $\vec{l}_0 = l_0 \exp(i\lambda_0)$, where δ_0, γ_0, and l_0 are the dimensionless amplitudes, and ϕ_0, ψ_0, and λ_0 are the observed phase angles. The residuals, $\vec{\delta} = \vec{\delta}_0 - 1.16$, $\vec{\gamma} = \vec{\gamma}_0 - 0.69$, and $\vec{l} = \vec{l}_0 - 0.084$, referred to the standard symmetrical Earth, contain all the nontrivial information. In practice they are resolved into their harmonic constituents and displayed as functions of these known frequencies. These resolved residuals are the essence of the observational subject.

Earth tides and ocean tides. A major part of research in earth tides relates to ocean tides. Where local ocean tides are large and well known, as in the Bay of Fundy, the Irish Sea, and the Alaskan area, the ocean loads afford a tempting opportunity to investigate the local crustal response to such loads and perhaps to revise ideas about its mechanical structure. Observations by gravimeters may be appreciably affected by distant ocean tides. Near the coastline, ocean loads are commonly responsible for 10% of the gravity tide, 25% of the strain tide, and 90% of the tilt tide. Studies of earth tides and ocean tides have become mutually supportive. Solutions for mean values of ocean tides which are free of the local distortions from effects of shallow water can be provided by earth tides. W. E. Farrell produced evidence of the precision of the transcontinental gravity tide profile of J. T. Kuo and coworkers by showing that agreement was produced only when the omitted condition of conservation of mass in the Atlantic ocean tides was introduced. Global coverage with precise earth tide measurements is a major continuing subject of study.

Nearly diurnal resonance. H. Jeffreys and R. O. Vicente, and M. S. Molodensky deduced theoretically the existence of a variation in the amplitudes of the nearly diurnal earth tides due to resonance in the coupling of the Earth's mantle and liquid core. M. G. Rochester noted that either the observations are too inaccurate to determine the location of the resonant frequency or the resonance-band structure is more complicated than expected. With this nearly diurnal resonance should also be observed a strangely neglected nutation, rediscovered by A. Toomre in 1974, of period 460 days, which thus far has equally strangely escaped observation. The continuation of the observations by earth tide meters at well-selected samples of different geographic locations should provide the needed additional evidence. Because the maximum tilt amplitudes in the diurnal constituents occur theoretically at the South Pole, tilt observations in bore holes in the ice have been proposed.

Dilatation. In some seismically active areas it has been observed that slow changes in the dilatation of crustal rocks and their associated absorption of additional interstitial water occur prior to earthquakes. Such variations have been detected by noting changes in the observed ratio of the seismic-wave velocities of the compressional and shear waves. C. Beaumont and J. Berger produced theoretical arguments which show that tidal strains and tidal tilts should be somewhat more strongly affected by the dilatations than are seismic waves. The records of strainmeters, and especially of tiltmeters in the preferred new sites in drill holes, should be monitored in the search for changes in dilatation, hithertofore detected only by seismic waves.

High-latitude stations. At Longyearbyen, Spitzbergen (78°20′N, 15°52′E), observations with three Askania gravimeters and 3 north-south and 3 east-west Verbaandert Melchior quartz pendulums were taken during June 1969 to July 1970. For the gravity factor, δ_{mf} of the fortnightly tide the weighted mean was 1.142 ± 0.014, but the spread in the values obtained by the three gravimeters seems large, 1.124 to 1.183. For the semiannual tide (SSa) the gravity factor $\delta_{ss} = 1.094 \pm 0.045$ was obtained, with a phase lag of 3 or 4 days.

At the South Pole, latitude −90°, elevation 2810.4 m, observations have been carried on since 1967. The observations during 107 days in 1970 gave the value $\delta = 1.153 \pm 0.002$ for the fortnightly and monthly tides (unresolved), and $\lambda = 0.°60 \pm 0.°40$ for the phase lag for the fortnightly period. B. V. Jackson and L. B. Slichter reported observations of the "forbidden" daily tides, with maximum amplitude in the diurnal band $= 0.606 \pm 0.003$ microgal (K_1), and in the semidiurnal band $= 0.341 \pm 0.0015$ (M_2), as shown in Fig. 2. Tests concerning the presence of a persistent inner-core oscillation showed that no such oscillation producing a signal of amplitude greater than 0.008 microgal could have been present.

The advantages of the South Pole location are the following: (1) The amplitudes of all the long-period gravity tides are theoretically maximal, whereas the short-period gravity tides, which theoretically vanish on a symmetric Earth, have small observed amplitudes in the fractional microgal range. (2) Concerning the observations of tilts, the long-period and semidiurnal amplitudes vanish in theory, but the diurnal constituents, significant in the display of nearly diurnal resonances of the liquid core, all have maximum amplitude. (3) In the observation of free modes, the splitting by rotation vanishes to first order. The records should be simpler. (4) The corrections for ocean tides are small and will be increasingly better known, aided by more abundant ocean tide measurements in southern seas. The location is good for operating earth tide meters designed for highest precision.

Instrumentation. Improvements in the sensitivity, portability, and economy of earth tide instruments are always in progress. The use of boreholes as stable sites for tilt pendulums has been pioneered. G. Cabanias has reported results on the M_2 tides from three Arthur D. Little borehole tiltmeters installed in holes 100 m apart at Bedford, MA. The three instruments agreed within 2% in amplitude and 1° in phase. Work by J. C. Harrison in the Poorman mine shows the importance there of the mine cavity and of surface topography upon measurements of tidal tilt and strain. A minimodel of the low-drift-rate superconducting gravimeter developed by J. Goodkind is designed to increase to several months the intervals between replenishing the helium supply. It is hoped that this meter will become a widely used standard. At the South Pole the plumb-bob suspensions of the two LaCoste-Romberg gravimeters preserve true levels conveniently. Careful corrections were introduced there in 1971 for the direct and load effects of barometric changes, and for changes in drift rate. This station produces readings of superior quality,

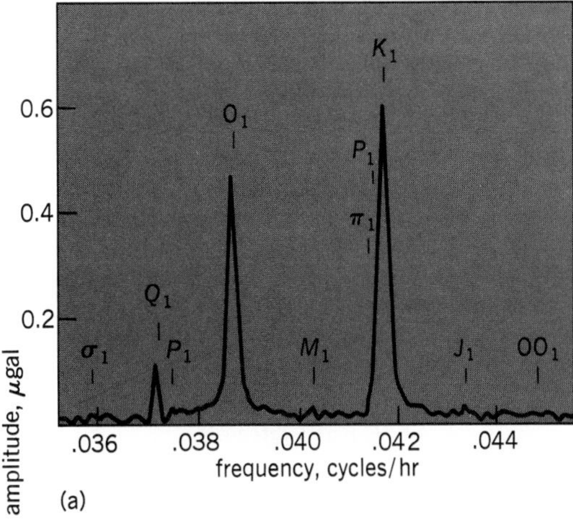

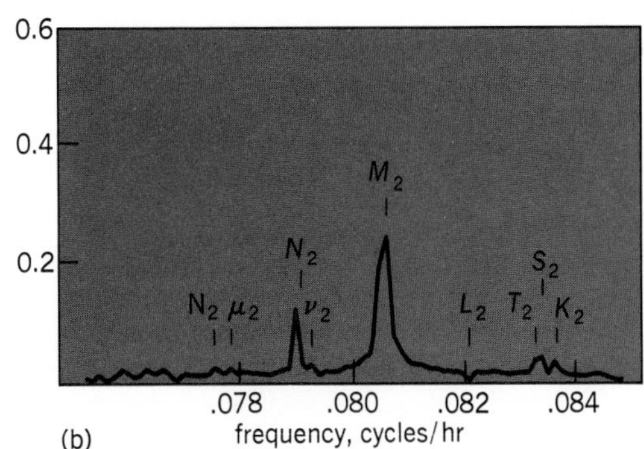

Fig. 2. Observed amplitude spectrum of (*a*) diurnal and (*b*) semidiurnal constituents at South Pole. (*From B. V. Jackson and L. B. Slichter, The residual daily earth tides at the South Pole, J. Geophys. Res., 79(11):1711–1715, copyright 1974 by American Geophysical Union*)

significant at the nanogal level in the diurnal and shorter-period range. [LOUIS B. SLICHTER]

Bibliography: C. Beaumont and J. Berger, Earthquake prediction: Modification of the earth tide tilts and strains by dilatancy, *Geophys. J. Roy. Astron. Soc.*, 39:111–122, 1974; M. Bonatz and T. Chojnicki, International Astro-Geo-Project Spitsbergen 1969/70, *Berechnung langperiodischer Gezeitenwellen fur die Gravimeterstation Longyearbyen, Mitteilungen ars dem Institut fur Theoretische Geodasie der Universitat Bonn*, vol. 7, 1972; R. A. Broucke, W. E. Zűrn, and L. B. Slichter, Lunar tidal acceleration on a rigid Earth, in H. C. Heard et al. (eds.), *Flow and Fracture of Rocks*, Geophys. Monogr. 16, American Geophysical Union, 1972; W. E. Farrell, Deformation of the Earth by surface loads, *Rev. Geophys. Space Phys.*, 10:761–797, 1972; J. Goodkind and W. Prothero, Tidal measurements with a superconducting gravimeter, *J. Geophys. Res.*, 77:926–937, 1972; J. C. Harrison, Cavity and topographic effects in tilt and strain measurements, *J. Geophys. Res.*, 81: 319–328, 1976; M. C. Hendershott, Ocean tides, *E. O. S.*, 54:76–86, 1973; B. V. Jackson and L. B. Slichter, The residual daily earth tides at South Pole, *J. Geophys. Res.*, 79:1711–1715, 1974; J. T. Kuo, Earth tides, *Rev. Geophys. Space Phys.*, 13:260–262, 1975; J. T. Kuo et al., Transcontinental gravity profile across the U. S., *Science*, 168: 968–971, 1970; P. L. Lagus and D. L. Anderson, Tidal dissipation in the Earth and planets, *Phys. Earth Planet Interiors*, 1:57–62, 1968; K. Lambeck, Effects of tidal dissipation in the oceans on the Moon's orbit and the Earth's rotation, *J. Geophys. Res.*, 80:2917–2925, 1975; P. Melchior, *Earth Tides*, 1966; P. Melchior, Earth tides and polar motions, *Technophysics*, 13:361–372, 1972; W. H. Munk, Once again—tidal friction, *Geophys. J. Roy. Astron. Soc.*, 9:352–375, 1968; W. H. Munk and G. J. F. MacDonald, *The Rotation of the Earth*, 1960; M. G. Rochester et al., A search for the Earth's "nearly diurnal free wobble," *Geophys. J. Roy. Astron. Soc.*, 38:349–363, 1974; P. H. Sydenham, 2000 hr Comparison of 10 m quartz-tube and quartz-catenary tidal strainmeters, *Geophys. J. Roy. Astron. Soc.*, 38:377–387, 1974; A. Toomre, On the "nearly diurnal wobble" of the Earth, *Geophys. J. Roy. Astron. Soc.*, 38:335–348, 1974.

Earthquake

A phenomenon during the occurrence of which the Earth's crust is set shaking for a period of time. The shaking is caused by the passage through the Earth of seismic waves—low-frequency sound waves that are emanated from a point in the Earth's interior where a sudden, rapid motion has taken place. It is more proper today to use the term earthquake to refer to the source of seismic waves, rather than the shaking phenomenon, which is an effect of the earthquake. Earthquakes vary immensely in size, from tiny events that can be detected only with the most sensitive seismographs, to great earthquakes that can cause extensive damage over widespread areas.

Although thousands of earthquakes occur every day, and have for billions of years, a truly great earthquake occurs somewhere in the world only once every 2 or 3 years. When a great earthquake occurs near a highly populated region, tremendous destruction can occur within a few seconds. In 1556, in the Shensi Province of China, 800,000 people were killed in a single earthquake. The city of Lisbon, one of the principal capitals of that day, was utterly destroyed, with high loss of life, in 1755. In the 20th century such cities as Tokyo and San Francisco have been leveled by earthquakes. In these more modern cases, much of the damage was not due to the shaking of the earthquake itself, but was caused by fires originating in the gas and electrical lines which interweave modern cities, and by damage to fire-fighting capability which rendered the cities helpless to fight the conflagrations.

Cause. The locations of earthquakes which occurred between 1961 and 1967 are shown on the map in Fig. 1. The map shows that earthquakes are not distributed randomly over the globe but tend to occur in narrow, continuous belts of activity. These earthquake belts link up so that they encircle large seismically quiet regions, which are known as plates. The plates are in continuous motion with respect to one another at rates on the order of centimeters per year; this plate motion is responsible for most geological activity, including earthquakes.

Plate motion occurs because the outer cold, hard skin of the Earth, the lithosphere, overlies a hotter, soft layer known as the asthenosphere. Heat from decay of radioactive minerals in the Earth's interior sets the asthenosphere into thermal convection. This convection has broken the lithosphere into plates which move about in response to the convective motion in a manner shown schematically in Fig. 2. The plates move apart at oceanic ridges. Magma wells up in the void created by this motion and solidifies to form new sea floor. This process, in which new sea floor is continually created at oceanic ridges, is called sea-floor spreading. Since new lithosphere is continually being created at the oceanic ridges by sea-floor spreading, a like amount of lithosphere must be destroyed somewhere. This occurs at the oceanic trenches, where plates converge and the oceanic lithosphere is thrust back down into the asthenosphere and remelted. The melting of the lithosphere in this way supplies the magma for the volcanic arcs which occur behind the trenches. Where two continents collide, however, the greater bouyancy of the less dense continental material prevents the lithosphere from being underthrust, and the lithosphere buckles under the force of the collision, forming great mountain ranges such as the Alps and Himalayas. Where the relative motion of the plates is parallel to their common boundary, slip occurs along great faults which form that boundary, such as the San Andreas fault in California. *See* EARTH, HEAT FLOW IN; MAGMA; VOLCANOLOGY.

According to the theory of plate tectonics, the motion of the plates is very similar to the movement of ice floes in arctic waters. Where floes diverge, leads form and water wells up, freezing to the floes and producing new floe ice. The formation of pressure ridges where floes converge is analogous to the development of mountain ranges where plates converge. *See* OROGENY; PLATE TECTONICS.

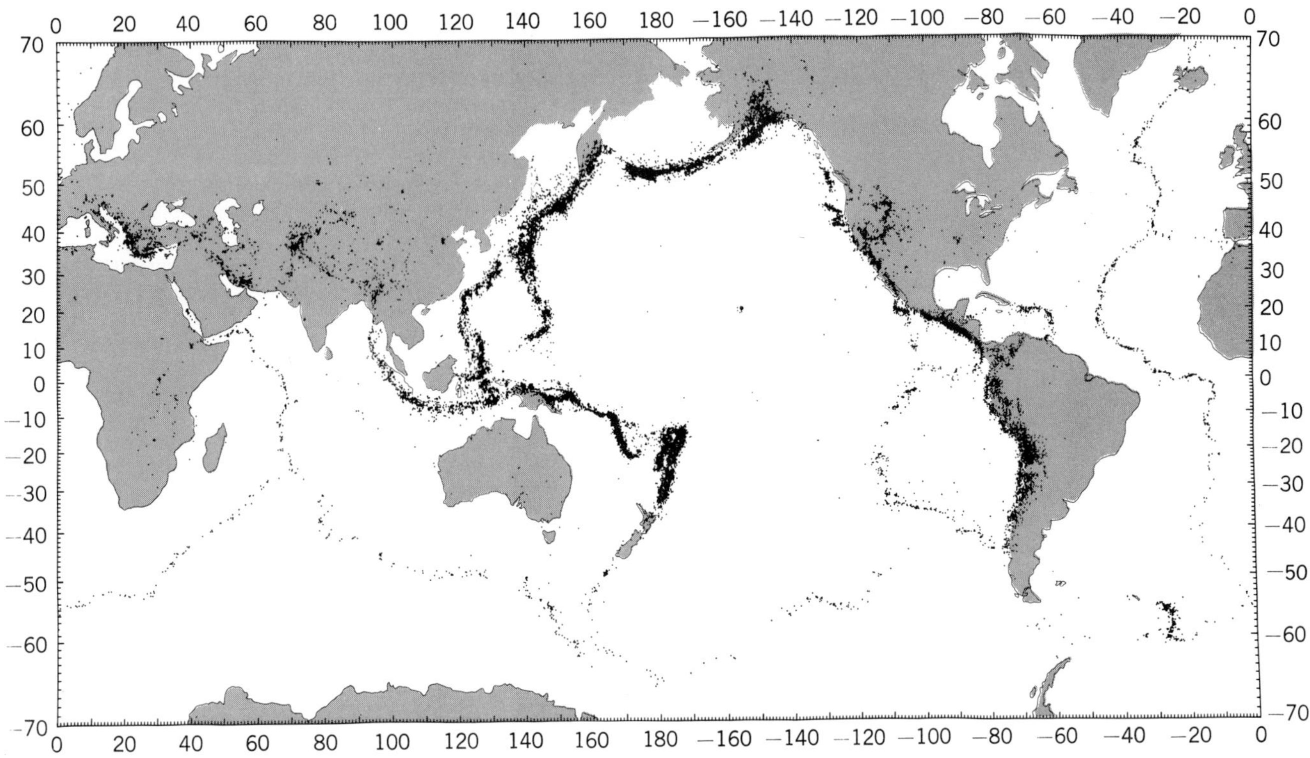

Fig. 1. Seismicity of the Earth from 1961 to 1967; depths to 700 km. The earthquake belts mark the plate boundaries. The number scales indicate latitude and longitude. *(From M. Barazangi and J. Dorman, World seismicity maps compiled from ESSA, Coast and Geodetic Survey, Epicenter Data, 1961–1967, Bull. Seis. Soc. Amer., 59:369–380, 1969)*

Stick-slip friction and elastic rebound. As the plates move past each other, the motion at their boundaries does not occur by continuous slippage but in a series of rapid jerks. Each jerk is an earthquake. This happens because, under the pressure and temperature conditions of the shallow part of the Earth's lithosphere, the frictional sliding of rock exhibits a property known as stick-slip, in which frictional sliding occurs in a series of jerky movements, interspersed with periods of no motion—or sticking. In the geologic time frame, then, the lithospheric plates chatter at their boundaries, and at any one place the time between chatters may be hundreds of years.

The peroids between major earthquakes is thus one during which strain slowly builds up near the plate boundary in response to the continuous movement of the plates. The strain is ultimately released by an earthquake when the frictional strength of the plate boundary is exceeded. This pattern of strain buildup and release was discovered by H. F. Reid in his study of the 1906 San Francisco earthquake. During that earthquake, a 250-mi-long (400 km) portion of the San Andreas fault, from Cape Mendicino to the town of Gilroy, south of San Francisco, slipped an average of 12 ft (3.6 m). Subsequently, the triangulation network in the San Francisco Bay area was resurveyed; it was found that the west side of the fault had moved northward with respect to the east side, but that these motions died out at distances of 20 mi (32 km) or more from the fault. Reid had noticed, however, that measurements made about 40 years prior to the 1906 earthquake had shown that points far to the west of the fault were moving northward at a slow rate. From these clues, he deduced his theory of elastic rebound, illustrated schematically in Fig. 3. The figure is a map view, the vertical line representing the fault separating two moving plates. The unstrained rocks in Fig. 3*a* are distorted by the slow movement of the plates in Fig. 3*b*. Slippage in an earthquake, returning the rocks to an unstrained state, occurs as in Fig. 3*c*. *See* FAULT AND FAULT STRUCTURES.

Classification. Most great earthquakes occur on the boundaries between lithospheric plates and arise directly from the motions between the plates. Although these may be called plate boundary earthquakes, there are many earthquakes, sometimes of substantial size, that can not be related so

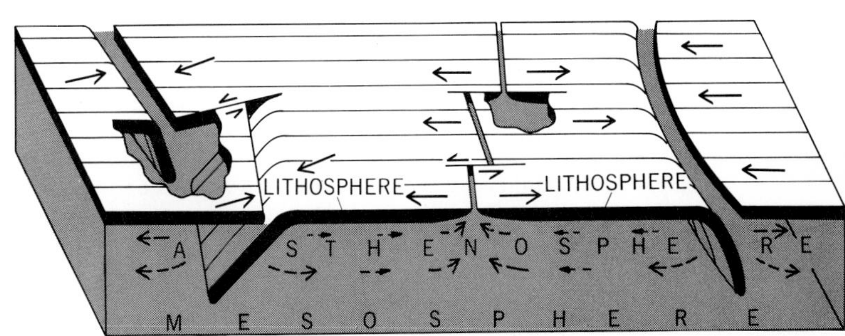

Fig. 2. Movement of the lithosphere over the more fluid asthenosphere. In the center, the lithosphere spreads away from the oceanic ridges. At the edges of the diagram, it descends again into the asthenosphere at the trenches. *(From B. Isacks, Oliver, and L. R. Sykes, Seismology and the new global tectonics, J. Geophys. Res., 73: 5855–5899, copyright 1968 by American Geophysical Union)*

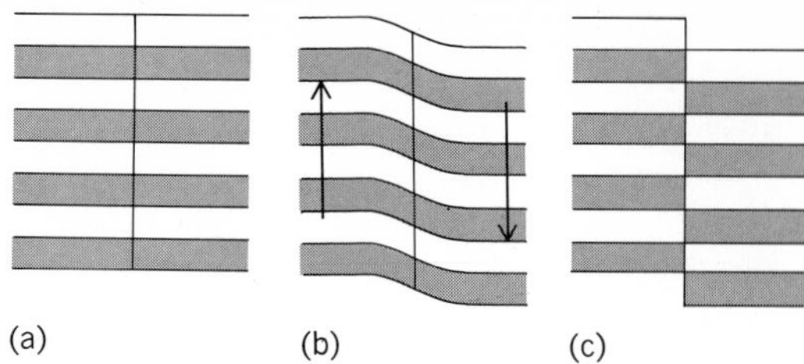

Fig. 3. Schematic of elastic rebound theory. (*a*) Unstrained rocks (*b*) are distorted by relative movement between the two plates, causing strains within the fault zone that finally become so great that (*c*) the rocks break and rebound to a new unstrained position. (*From C. R. Allen, The San Andreas Fault, Eng. Sci. Mag., Calif. Inst. of Technol., pp. 1–5, May 1957*)

simply to the movements of the plates.

Near many plate boundaries, earthquakes are not restricted to the plate boundary itself, but occur over a broad zone—often several hundred miles wide—adjacent to the plate boundary. These earthquakes, which may be called plate boundary-related earthquakes, do not reflect the plate motions directly, but are secondarily caused by the stresses set up at the plate boundary. In Japan, for example, the plate boundaries are in the deep ocean trenches offshore of the Japanese islands, and that is where the great plate boundary earthquakes occur. Many smaller events occur scattered throughout the Japanese islands, caused by the overall compression of the whole region. Although these small events are energetically minor when compared to the great offshore earthquakes, they are often more destructive, owing to their greater proximity to population centers.

Although most earthquakes occur on or near plate boundaries, some also occur, although infrequently, within plates. These earthquakes, which are not related to plate boundaries, are called intraplate earthquakes, and can sometimes be quite destructive. Although intraplate earthquakes are probably caused by the same convective forces which drive the plates, their immediate cause is not understood. Some of them can be quite large. One of the largest earthquakes known to have occurred in the United States was one of a series of intraplate earthquakes which took place in the Mississippi Valley, near New Madrid, MO, in 1811 and 1812. Another intraplate earthquake, in 1886, caused moderate damage to Charleston, SC.

In addition to the tectonic types of earthquakes described above, some earthquakes are directly associated with volcanic activity. These volcanic earthquakes result from the motion of undergound magma that leads to volcanic eruptions.

Sequences. Earthquakes often occur in well-defined sequences in time. Tectonic earthquakes are often preceded, by a few days to weeks, by several smaller shocks (foreshocks), and are nearly always followed by large numbers of aftershocks. Foreshocks and aftershocks are usually much smaller than the main shock. Volcanic earthquakes often occur in flurries of activity, with no discernible main shock. This type of sequence is called a swarm.

Size. Earthquakes range enormously in size, from tremors in which slippage of a few tenths of an inch occurs on a few feet of fault, to the greatest events, which may involve a rupture many hundreds of miles long, with tens of feet of slip. Accelerations as high as 1 *g* (acceleration due to gravity) can occur during an earthquake motion. The velocity at which the two sides of the fault move during an earthquake is only 1–10 mph, but the rupture front spreads along the fault at a velocity of nearly 5000 mph. The earthquake's primary damage is due to the generated seismic waves, or sound waves which travel through the Earth, excited by the rapid movement of the earthquake. The energy radiated as seismic waves during a large earthquake can be as great as 10^{12} cal, ($10^{12} \times 4.19$ J) and the power emitted during the few hundred seconds of movement as great as a billion megawatts.

The size of an earthquake is in terms of a scale of magnitude based on the amount of seismic waves generated. Magnitude 2.0 is about the smallest tremor that can be felt. Most destructive earthquakes are greater than magnitude 6; the largest shock known measured 8.9. The scale is logarithmic, so that a magnitude 7 shock is about 30 times more energetic than one of magnitude 6, and 30×30, or 900 times, more energetic than one of magnitude 5. Because of this great increase in size with magnitude, only the largest events (greater than magnitude 8) significantly contribute to plate movements. The smaller events occur much more often but are almost incidental to the process.

The intensity of an earthquake is a measure of the severity of shaking and its attendant damage at a point on the surface of the Earth. The same earthquake may therefore have different intensities at different places. The intensity usually decreases away from the epicenter (the point on the surface directly above the onset of the earthquake), but its value depends on many factors in addition to earthquake magnitude. Intensity is usually higher in areas with thick alluvial cover or landfill than in areas of shallow soil or bare rock. Poor building construction leads to high intensity ratings because the damage to structures is high. Intensity is therefore more a measure of the earthquake's effect on humans than an innate property of the earthquake.

Effects. Many different effects are produced by earthquake shaking. Although the fault motion that produced the earthquake is sometimes observed at the surface, often other earth movements, such as landslides, are triggered by earthquakes. On rare occasions the ground has been observed to undulate in a wavelike manner, and cracks and fissures often form in soil. The flow of springs and rivers may be altered, and the compression of aquifers sometimes causes water to spout from the ground in fountains. Undersea earthquakes often generate very-long-wavelength water waves, which are sometimes called tidal waves but are more properly called seismic sea waves, or tsunami. These waves, almost imperceptible in the open ocean, increase in height as they approach a coast and often inflict great damage to coastal cities and ports.

Prediction. The largest earthquakes occur along the boundaries between moving lithospheric plates, and are caused by a sudden slip of portions of these boundaries when the pent-up stresses exceed the frictional resistance to motion. Each segment of a plate boundary must therefore periodically experience great earthquakes, the frequency of which will depend on the relative rate of plate motion between the two plates in contact, and on the nature of the boundary itself. It is thus possible to state in general that, on a given plate boundary, the most likely sites of the next great earthquakes are those segments, called seismic gaps, in which no large earthquakes have occurred in the longest time. Known seismic gaps in the western Pacific are shown in Fig. 4.

Recognizing seismic gaps is a crude means of predicting earthquakes, although it really provides only a rough estimate of earthquake risk, since it does not permit a precise estimation of the time, place, and magnitude of future shocks. Furthermore, there are many smaller earthquakes, many of which are destructive, which occur on or adjacent to plate boundaries; they do not play such a direct role in plate motion, and hence cannot be anticipated in terms of seismic gaps. These earthquakes occur almost randomly in time, and the hazard that they present can be estimated only statistically. Thus it is known that areas near plate boundaries, such as the San Andreas Fault, are more prone to earthquakes than are other areas, say, Colorado, which are far from currently active plate margins; thus it is possible to statistically rate the difference in hazard between the two areas.

Precursory phenomena. Neither the seismic gap approach nor the statistical approach provides a means of actual prediction of individual earthquakes. Research on earthquake prediction has concentrated instead in discovering phenomena which are precursory to earthquakes, and which can therefore be used as warnings. A number of such phenomena have been discovered to precede earthquakes by days or months or even years. These precursors are usually changes in some property of the Earth in the vicinity of the coming earthquake. They include such phenomena as anomalous uplift of the ground, changes in electrical conductivity of rock, changes in the isotopic composition of deep well water, and changes in the nature of small earthquake activity.

The most often observed precursory event is the relative change in the velocity at which the two types of seismic waves, the *p* (or pressure) and the *s* (or shear) waves (also written as P waves and S waves) travel through the source region of an impending earthquake. The velocity of the *p* wave v_p is always greater than that of the *s* wave v_s. Usually it is about 1.75 times that of the *s* wave. However, when the *p* and *s* waves travel through a region in which an earthquake is about to occur, it has been observed that the *p* wave travels more slowly than it normally does, only about 1.5 times faster than the *s* wave. This anomalous behavior persists for some time, and then the velocities return to normal; shortly thereafter, the earthquake occurs. This

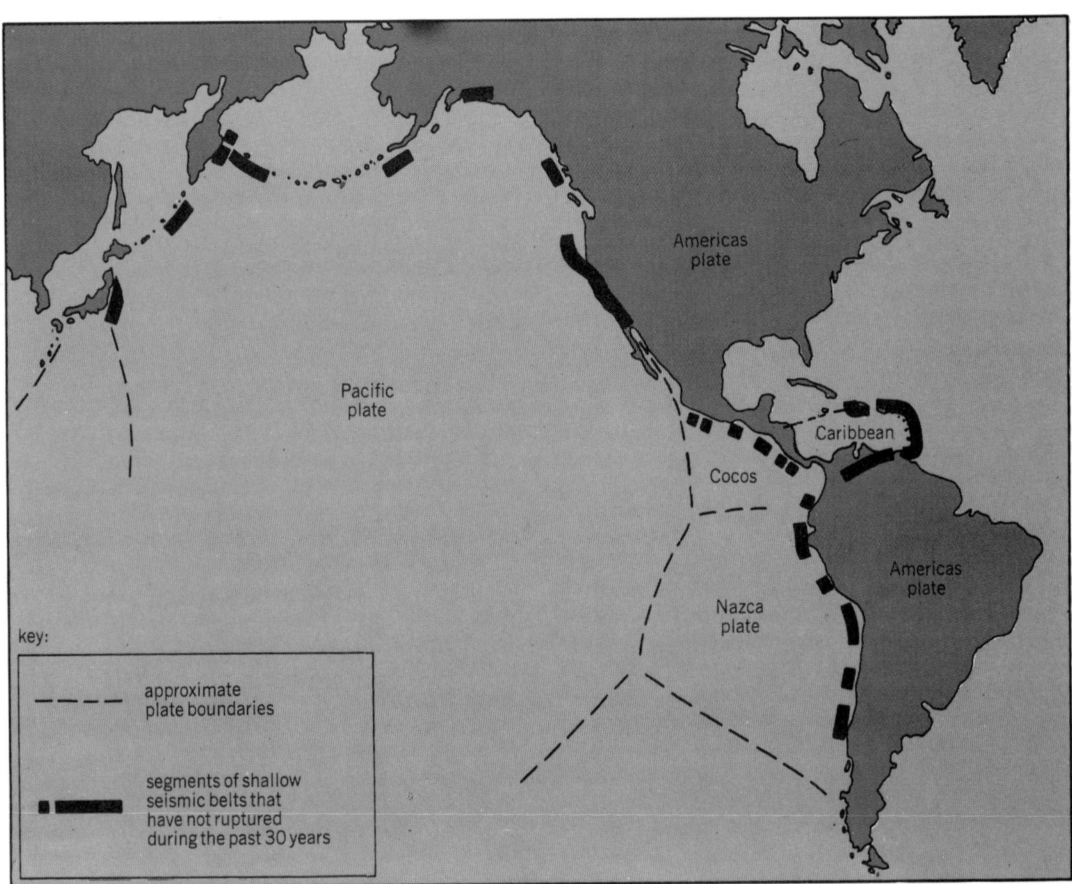

Fig. 4. Locations of major known seismic gaps in the western Pacific. (*From J. Kelleher et al., J. Geophys. Res., 78:2547–2585, copyright 1973 by American Geophysical Union*)

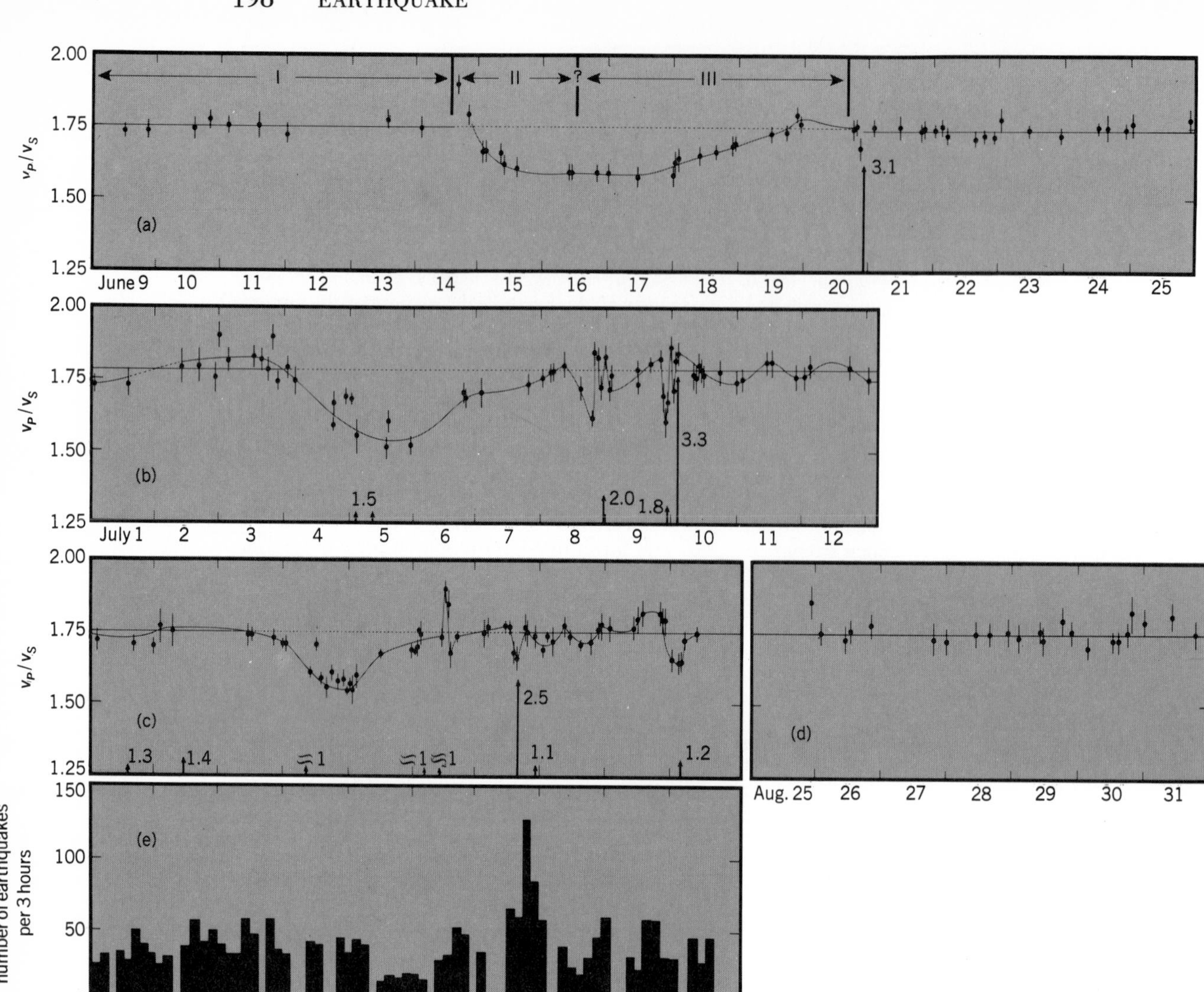

Fig. 5. (*a–c*) Diagrams of changes in the ratio v_p/v_s preceding three of the largest events in the Blue Mountain Lake, NY, earthquake swarm of 1971. All quakes of magnitude greater than 1 are shown as arrows. (*d*) A similar plot during a time when no large events took place. (*e*) A plot of the frequency of occurrence of small earthquakes in the swarm plotted on the same time scale as *c*. (*From C. H. Scholz et al., Earthquake prediction: A physical basis, Science, 181:803–810, copyright 1973 by the American Association for the Advancement of Science*)

phenomenon was found to occur repeatedly before small earthquakes took place in the Adirondack Mountains of New York (Fig. 5); one earthquake was successfully predicted in advance on the basis of this precursor. *See* SEISMOLOGY.

Dilatancy. These precursors indicate that some fundamental change takes place in the Earth's crust just before the occurrence there of an earthquake. Many scientists now believe that these precursory effects are caused by a property of rock called dilatancy, which is an increase in rock volume caused by an increase in distortional strain on the rock. When stress builds up prior to fracture, numerous cracks open up within a rock, causing it to swell.

According to the dilatancy theory, stress slowly builds up over many years in an earthquake-prone area. Eventually, in some part of that area, the stress becomes high enough to cause cracks to open; hence dilatancy occurs. Since the cracks and pores of the rock have been filled with water and other fluids at some pressure, known as the pore pressure, the increase in total crack volume reduces this pore pressure, thereby strengthening the rock. At the same time, the various premonitory phenomena begin to occur, because properties such as electrical conductivity and seismic wave velocity are very sensitive to the amount of open cracks in the rock. At a later time, these effects begin to disappear because water flows in from adjacent areas, filling the cracks and bringing the pore pressure back to normal, which weakens the rock again and triggers the earthquake.

An important consequence of this process is that the larger the upcoming earthquake, the larger the dilatant volume, and hence the longer the time required for the influx of water. Thus precursors last longer for larger earthquakes (Fig. 6). This means that the magnitude, as well as the time and place, of the earthquake can be predicted. *See*

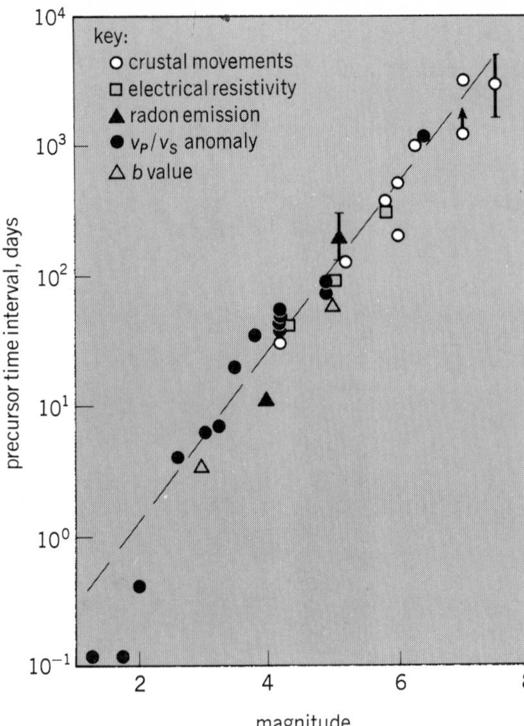

Fig. 6. Duration time of various precursory phenomena as a function of the magnitude of the following earthquake. Here *b* is a parameter that describes the ratio of small to large shocks. (*From C. H. Scholz et al., Earthquake prediction: A physical basis, Science, 181:803–810, copyright 1973 by the American Association for the Advancement of Science*)

ROCK MECHANICS.

These precursory phenomena have been found to precede about 50 earthquakes. It is not known whether they occur before all earthquakes, or whether the phenomena are always the same. Many aspects of the dilatancy theory remain to be proved. This line of research, however, holds out the greatest promise for providing a practical means of predicting earthquakes.

[CHRISTOPHER H. SCHOLZ]

Eclogite

A class of metamorphic rocks. Eclogites are different from all other rocks, essentially consisting of the typical eclogite minerals: a strongly green pyroxene (omphacite) and a Mg-rich, brilliant red-brown garnet (pyrope). Plagioclase feldspar has never been observed associated with the typical eclogite minerals, and the eclogite minerals are not known from any other rocks but are critical for eclogites. Other stable minerals occurring in eclogites are diopside, enstatite, olivine, kyanite, rutile, and rarely, diamond. Calcite is also observed, but may be back-reacted from aragonite, the high-pressure modification. The bulk composition corresponds to that of gabbro (or basalt). *See* CALCITE; GARNET; PYROXENE.

Hornblende occurs in many eclogites as a hysterogenic product; it is often alkali-bearing. Another characteristic hysterogenic product is kelyphite, a peripheral alteration of garnet into pyroxene or amphibole. Complete gradation is often traced from unaltered eclogite through eclogite-amphibolites containing relics of garnet and omphacite, together with newly generated plagioclase and hornblende, to amphibolites of normal composition.

Properties. The properties of some of the eclogite minerals are worthy of notice. Omphacite is a pyroxene high in Na and Al, representing a solid solution of mainly diopside, jadeite, and the Tschermak silicate: $Ca(Mg,Fe)Si_2O_6$, $NaAlSi_2O_6$, and $MgAl_2SiO_6$, respectively. Their content of calcium and sodium explains the absence of plagioclase in eclogites. The eclogite garnets are also unusual in that almandite forms extensive series of solid solutions with pyrope (up to 70 mole %, that is, more than in any garnet of other rocks) and with grossularite. Thus the garnets exhibit an exceptionally large variation in the Ca/Mg/Fe ratio, and a high magnesia content is typical.

The difference in mineral contents between a gabbro and an eclogite of the same bulk chemical composition is illustrated schematically by the reaction

$$\underbrace{\text{Olivine} + \text{pyroxene} + \text{plagioclase}}_{\text{gabbro}} \rightarrow \underbrace{\text{omphacite} + \text{garnet} + \text{quartz}}_{\text{eclogite}}$$

and partitioned into two simple reactions

$$\text{Olivine} + \text{anorthite} \rightarrow \text{garnet}$$
$$\text{Albite} \rightarrow \text{jadeite} + \text{quartz}$$

Both of these reactions proceed to the right under high pressure, about 10–20 kilobars (kb) corresponding to 40–60 km depth. *See* GABBRO.

Subgroups. Eclogites may be subdivided into three groups that are found in different geological settings and are associated with different rock complexes.

1. Garnet pyroxenites (griquaites) are associated with ultrabasites and occur as fragments in kimberlites and rounded inclusions in basalt together with olivine nodules; they probably derive from the upper mantle. The constituent pyroxene is chromiferous; the garnet is high in magnesium.

2. Ophiolitic eclogites (alpine type) are products of early orogenic volcanism that by later metamorphism transformed into rocks of the high-pressure facies series: eclogites with transitions into prasinites and glaucophane schists. The constituent pyroxene is rich in soda; the garnet is relatively poor in magnesium, and rich in iron and calcium.

3. Common eclogites occur as intercalations, lenses, or boudines in paragneisses and migmatities of the amphibolite or granulite facies. The bulk composition corresponds to that of high-alumina basalt manifested by the presence of abundant kyanite (or secondary zoisite). The pyroxene has a medium soda content; the garnet is usually rich in calcium. The origin of these eclogites is diversified, with basaltic sills and lava sheets, dolomitic marl, or carbonate-rich siltstones forming intercalations within geosynclinal graywacke and siltstone sediments.

The existence of a separate eclogite facies is under discussion. Following K. Smulikowski, eclogite is here used as the name of a rock type (divided into three subgroups) whose relation to

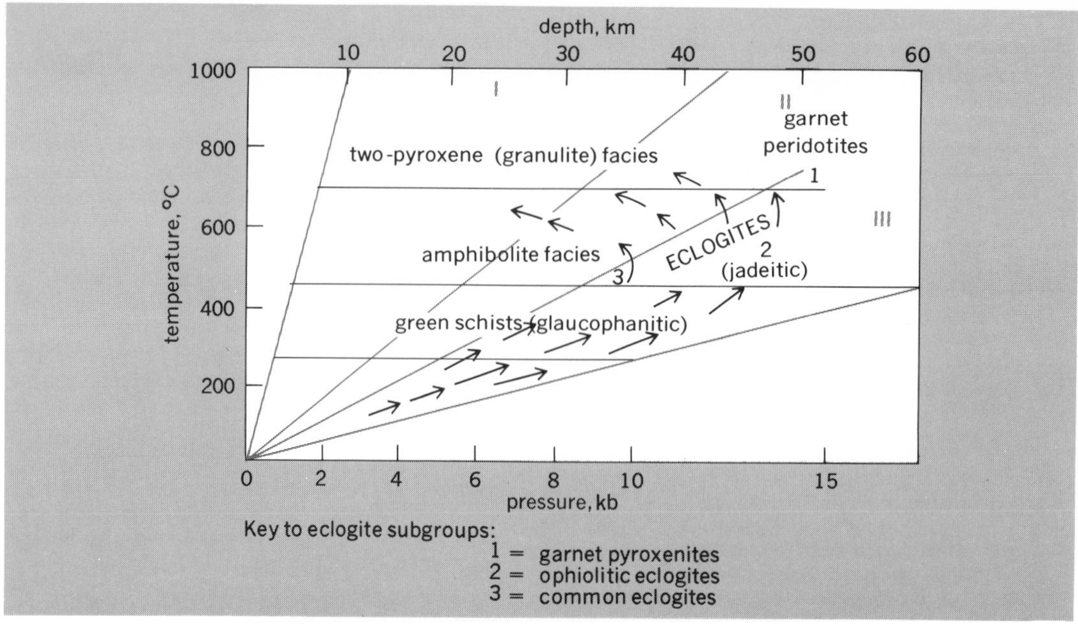

The relation of the eclogites to the mineral facies. Tracks of arrows mark the directions of later metamorphic evolution of eclogite into amphibolites and pyroxene granulites. Low-, intermediate-, and high-pressure metamorphic facies series are represented by the numerals I, II, and III, respectively.

the mineral facies appears from the illustration. For an explanation of mineral facies *see* METAMORPHIC ROCKS.

Much attention has been paid to eclogite, its origin, and geological relations. H. Yoder and C. Tilley have presented the following conclusions pertinent to its ultimate origin:

1. For every major basalt type there is a chemically equivalent eclogite.

2. At high confining pressure basalt magma crystallizes to eclogite.

3. Eclogite melts in a eutecticlike fashion, suggesting that eclogite itself is the partial melting product of a more primitive rock (for example, garnet peridotite). It is believed, therefore, that initial melting products in the region of magma generation in the mantle are eclogitic.

4. The Mohorovičić discontinuity under the continents (as distinct from the Moho discontinuity under the oceans) may be due to the phase change of basalt to eclogite. P. Eskola suggested as early as 1921 that there may exist a continuous shell of eclogite under the sial crust of the Earth. *See* BASALT; MOHO (MOHOROVIČIĆ DISCONTINUITY).

[T. F. W. BARTH]

Bibliography: K. Smulikowski, Differentiation of eclogites and its possible causes, *Lithos*, 1:89–101, 1968; H. S. Yoder and C. E. Tilley, Origin of basalt magmas, *J. Petrol.*, 3:342–532, 1962.

Elaterite

A light-brown to black, naturally occurring carbonaceous substance having specific gravity 0.90–1.05. Elaterite is insoluble in carbon disulfide and infusible. It is moderately soft and elastic, and on heating leaves only 2–5% fixed carbon. Elaterite is thought to be of algal origin and occurs in small quantities in Derbyshire County, England, in the Coorong District of South Australia, and in Turkestan, Soviet Union. *See* ALBERTITE; ASPHALT AND ASPHALTITE; IMPSONITE; WURTZILITE.

[IRVING A. BREGER]

Elements, geochemical distribution of

The distribution of chemical elements in major zones (crust, mantle, and core) of the Earth. This distribution is dependent upon the early history and subsequent evolution of both the Earth and solar system. Since these events occurred long ago and there is no direct evidence for what actually happened, considerable speculation is necessarily involved in explaining the postulated present-day distribution of elements among the major zones of the Earth. Some of the most pertinent questions are: (1) How were elements distributed during the initial accretion of the protoearth as a part of the early development of the solar system? (2) What processes were involved in the formation of the Earth's core and mantle, and how were elements distributed between these two main zones? (3) How did the Earth's crust form, and what processes controlled the distribution of elements between crust and mantle? (4) Inasmuch as the crust is probably the most chemically heterogeneous of the three zones, by what means did it achieve this heterogeneity, and how are the elements distributed among its major parts?

Initial accretion. The proto–solar system, before it evolved into the Sun and orbiting planets, was probably a rotating, swirling, lens-shaped cloud of gas, dust, and other matter. The interior of this cloud contracted and heated primarily through gravitational attraction, until finally temperatures and pressures were high enough to initiate nuclear reactions, giving off light and heat. Matter in swirling vortices within peripheral regions of the cloud eventually coalesced and formed individual planets. Portions of lighter, more volatile elements (such as N, C, O, and H) escaped outward from the hotter interior of the system and were preferentially enriched in the less dense, larger outer planets (Jupiter, Saturn, Uranus, and Neptune). Heavier, less volatile elements (such as Ca, Na, Mg, Al, Si, K, Fe, Ni, and S) tended to remain closer to the center of the system, eventually becoming preferentially enriched in the more dense, smaller inner planets (Mercury, Venus, Earth, and Mars).

The exact mechanism by which the planets accreted is not known. A question that is extremely critical to the origin of the Earth's megastructure is whether the process took place by "cold" or "hot" accretion. If the Earth formed primarily by cold accretion, temperatures would have been sufficiently low for much of the accretionary materials to have existed in a solid or liquid state, and some volatile elements would have been retained. The most significant aspect of this process is that it probably would have led to an initially homogeneous, unzoned Earth, that is, without core, mantle, or crust. Hot accretion, in which temperatures would have been hot enough that most of the accretionary material would have been in a gaseous state, would have produced a heterogeneous, zoned Earth, because of the differential condensation temperatures of the elements and compounds that would have formed the planets. In other words, substances with the highest condensation temperatures would have condensed first in the cooling gas clouds and eventually have formed the cores of the planets. For various reasons, cold accretion, forming a homogeneous, unzoned Earth, is at present the preferred theory. For example, only cold accretion can account for the high concentration of certain volatile elements on the Earth.

Accretion of the Earth is believed to have taken place from a cloud whose composition was quite similar to that of the type of stony meteorites known as chondrites (Table 1). The average composition of chondrites is thought to approximate at least the nonvolatile average composition of the Earth, thus the term "chondritic Earth model." The protoearth was probably a homogeneous, unzoned spheroid of approximately chondritic composition.

Distribution between core and mantle. Assuming the hypothesis of an initially unzoned Earth, how were the mantle and core formed? The core, with an approximate density of 11 g/cm^3, is thought to be composed primarily of Fe, about 6% Ni, and minor amounts of other elements such as Si and S. The mantle, with an average density of about 4.5 g/cm^3, is normally considered to consist largely of Fe-Mg silicates.

The average mineralogical composition of chondrites is 70% olivine and pyroxene (Fe-Mg silicates), 15% Ni-Fe alloy, 10% plagioclase ($CaAl_2Si_2O_8$-$NaAlSi_3O_8$), and 5% troilite (FeS). Many minor and trace elements can fit into the lattice positions of these minerals. Assuming the chondritic Earth model hypothesis, the Ni-Fe alloy

Table 1. Averages of major and selected trace metal concentrations for several rock types that approximate compositions of principal zones in the Earth*

Rock type:	Chondrites	Ultramafics	Oceanic basalts	Andesites	High-Ca granites
Zone:	Whole Earth	Mantle	Oceanic crust	Continental crust	Upper continental crust
		Major elements (in wt %)			
Al	1.3	2.0	8.1	9.1	8.2
Ca	1.4	2.5	8.1	5.0	2.5
Fe	25.1	9.4	8.5	4.7	3.0
K	0.1	0.01	0.2	1.3	2.5
Mg	14.4	20.4	4.7	2.1	0.9
Na	0.7	0.02	2.1	2.7	2.8
Si	17.8	20.5	23.2	27.8	31.4
Ti	0.1	0.03	0.8	0.4	0.3
		Trace elements (in parts per million)			
B	0.43	3	5	—	9
Ba	4.5	0.4	10	270	420
Be	0.04	0.5	1	—	2
Co	520	110	40	24	7
Cr	3,000	2,980	300	56	22
Cs	0.1	0.1	0.5	1	2
Cu	90	10	60	54	30
Ga	5.3	1.5	10	16	17
Li	2.5	0.5	10	10	24
Mn	2,600	1,040	1,200	1,200	540
Ni	13,500	2,000	150	18	15
P	1,100	220	440	—	920
Pb	0.18	1	5	67	15
Rb	3	0.13	1	31	110
Sc	8.5	16	40	30	14
Sr	11	5.8	100	385	440
Th	0.4	0.004	0.2	2.2	8.9
U	0.01	0.001	0.1	0.69	2.3
V	65	40	300	175	88
Y	2	5.0	20	21	44
Zn	54	50	50	—	60
Zr	12	45	100	110	140

*Number of significant figures for trace elements reflects reliability of data.

Table 2. Geochemical tendencies of the elements

Atomic number	Symbol	Tendency*	Group in periodic table	Electro-negativity†
1	H	L	IA	2.1
2	He	A	VIIIA	–
3	Li	L	IA	1.0
4	Be	L	IIA	1.5
5	B	L	IIIA	2.0
6	C	LSA	IVA	2.5
7	N	A	VA	3.0
8	O	LA	VIA	3.5
9	F	L	VIIA	4.0
10	Ne	A	VIIIA	–
11	Na	L	IA	0.9
12	Mg	L	IIA	1.2
13	Al	L	IIIA	1.5
14	Si	L	IVA	1.8
15	P	LS	VA	2.1
16	S	C	VIA	2.5
17	Cl	L	VIIA	3.0
18	Ar	A	VIIIA	–
19	K	L	IA	0.8
20	Ca	L	IIA	1.0
21	Sc	L	IIIB	1.3
22	Ti	L	IVB	1.5
23	V	L	VB	1.6
24	Cr	LC	VIB	1.6
25	Mn	L	VIIB	1.5
26	Fe	SCL	VIII	1.8
27	Co	S	VIII	1.8
28	Ni	S	VIII	1.8
29	Cu	C	IB	1.9
30	Zn	C	IIB	1.6
31	Ga	CL	IIIA	1.6
32	Ge	SCL	IVA	1.8
33	As	CS	VA	2.0
34	Se	C	VIA	2.4
35	Br	L	VIIA	2.8
36	Kr	A	VIIIA	–
37	Rb	L	IA	0.8
38	Sr	L	IIA	1.0
39	Y	L	IIIB	1.3
40	Zr	L	IVB	1.4
41	Nb	L	VB	1.6
42	Mo	SC	VIB	1.8
44	Ru	S	VIII	2.2
45	Rh	S	VIII	2.2
46	Pd	S	VIII	2.2
47	Ag	C	IB	1.9
48	Cd	C	IIB	1.7
49	In	C	IIIA	1.7
50	Sn	SC	IVA	1.8
51	Sb	C	VA	1.9
52	Te	C	VIA	2.1
53	I	L	VIIA	2.5
54	Xe	A	VIIIA	–
55	Cs	L	IA	0.7
56	Ba	L	IIA	0.9
57	La	L	IIIB	1.1
58–71	Ce–Lu	L	IIIB	1.1–1.2
72	Hf	L	IVB	1.3
73	Ta	L	VB	1.5
74	W	SL	VIB	1.7
75	Re	S	VIIB	1.9
76	Os	S	VIII	2.2
77	Ir	S	VIII	2.2
78	Pt	S	VIII	2.2
79	Au	S	IB	2.4
80	Hg	C	IIB	1.9
81	Tl	CL	IIIA	1.8
82	Pb	CS	IVA	1.8
83	Bi	C	VA	1.9
90	Th	L	IIIB	1.3
92	U	L	IIIB	1.7

*L = lithophile; C = chalcophile; A = atmophile; S = siderophile.

†After L. Pauling.

eventually formed the core, and the remaining phases formed the mantle.

At a very early stage in its history (4–5 × 10^9 years ago), the Earth probably was mostly in solid form. Most earth scientists now believe that subsequent heating, due to adiabatic contraction and radioactive decay, led to a widespread melting event. The lowest-melting component, the Ni-Fe alloy, is believed to have melted initially; because of its much greater density, this alloy would have settled and formed the core. This event has been referred to as the iron catastrophe. Continued melting would have created three immiscible liquids: silicate, sulfide, and alloy. The remaining silicates, sulfides, and other compounds then would have formed the surrounding mantle.

V. M. Goldschmidt, an early pioneer in geochemistry, used this theory as a partial basis for a geochemical classification of the elements, according to whether they had siderophile, chalcophile, lithophile, or atmophile tendencies (Table 2). Obviously, volatile elements such as Ar, O, C, and N are classified as atmophile elements, and are the major constituents of the atmosphere. Siderophile elements tend to form alloys rather than compounds, and are preferentially concentrated in the core. Chalcophile elements chemically combine with S to form sulfides and concentrate in the mantle (and eventually the crust), whereas lithophile elements combine with O to form oxides and with O plus Si to form silicates. Lithophile elements also tend to be concentrated in the mantle and crust. An analogous situation can be found in a blast furnace, which has three immiscible melts: slag is equivalent to lithophile elements, matte to chalcophile, and ore to siderophile. Goldschmidt's geochemical classification of the elements can be related to the groups in the periodic table (Table 2). It follows that strongly lithophile elements are generally either highly electropositive or electronegative, whereas siderophile and chalcophile elements have intermediate electronegativities. Some elements, such as iron, exhibit lithophile, chalcophile, and siderophile tendencies.

Distribution between crust and mantle. S. R. Taylor has shown that a relationship exists between a cation's ionic size (and, to a lesser degree, its ionic charge) and its distribution between the mantle and crust (Table 3). In general, cations with comparatively large or small ionic radii tend to be concentrated in the crust relative to the mantle. Cations with intermediate radii (such as Mg, Fe, Mn, Co, Ni, and Cr), which fit well into the lattice positions of minerals believed to be stable in the mantle, tend to be concentrated in the mantle relative to the crust. Thus if the Earth were initially homogeneous, comparatively large cations (such as Ba, Rb, U, Th, and Cs) and small lithophile cations (such as Li, Be, and B) would be concentrated upward through time. The exact mechanism by which this occurs is not known.

Whatever the mechanism for this upward migration of abnormally sized cations, it very likely involves modern concepts of plate tectonics and the origin of the crust. New oceanic crust, composed primarily of basaltic rocks, is thought to be forming at the mid-oceanic ridges (spreading centers) by means of partial melting of the underlying mantle. Relative to the mantle, the basaltic crust is comparatively enriched in Si, Al, Ca, Na, K, and large ionic lithophile elements, but is depleted in Mg, Fe, and certain transition metals (group VIII in particular). The enriched-in-crust group tends to

Table 3. Crustal enrichment in some common metallic elements

Cation	Radius, angstroms*	Percent in crust†
Cs^+	1.69	>50
Rb^+	1.48	>50
Ba^{2+}	1.35	>50
K^+	1.33	>50
Sr^{2+}	1.13	30–50
Ca^{2+}	0.99	1–3
U^{4+}	0.97	20–30
Th^{4+}	0.95	30–50
Na^+	0.95	1–3
Rare-earth elements	0.93–1.12	8–50
Zr^{4+}	0.80	8–20
Mn^{2+}	0.80	<1
Fe^{2+}	0.76	<1
Co^{2+}	0.74	<1
V^{3+}	0.74	1–3
Ni^{2+}	0.72	<1
Cr^{3+}	0.69	<1
Ti^{4+}	0.68	3–8
Mg^{2+}	0.65	<1
Li^+	0.60	1–3
Al^{3+}	0.50	3–8
Si^{4+}	0.41	1–3
Be^{2+}	0.31	3–8
B^{3+}	0.20	3–8

*10 A = 1 nm.

†Percentage of each element's total mass in the Earth that is found in the crust (assuming an andesitic crustal composition and a chondritic Earth model, after Taylor). Because the crust makes up approximately 1% of the Earth by volume, elements with percentages greater than 1% are enriched-in-crust elements; those with values less than 1%, enriched-in-mantle elements.

have lithophile or chalcophile characteristics, whereas the enriched-in-mantle group, with the exception of Mg, tends to be siderophile. Hence, the process of partial melting of upper mantle material and ascent of magma, eventually forming new crust, may be the dominant mechanism for concentrating the enriched-in-crust elements in the overlying crust at the expense of the mantle.

Heterogeneity within the crust. Inasmuch as the continental crust and oceanic crust are compositionally different, how is the continental crust formed? Its overall composition is reportedly similar to the average composition of volcanic rocks known as andesites (Table 1), the dominant volcanic rocks that compose most of the active volcanoes around the margins of the Pacific Ocean. Again plate tectonics is involved, for these andesite volcanoes are generally found in island arcs (such as Japan and the Aleutian Islands) and in young mobile belts (such as the Andes Mountains). These are associated with subduction zones, regions in which the downgoing oceanic lithospheric plate plunges beneath the less dense continental plate. Partial melting occurs in the downgoing plate or the mantle wedge above the plate; this results in primarily andesite volcanism. Andesites are still richer in enriched-in-crust elements and poorer in enriched-in-mantle elements than are basalts. These andesitic volcanoes add new material to the continental crust, material derived from the oceanic crust or directly from the mantle by partial melting.

This process of partial melting also goes on within the continental crust, leading to the formation and ascent of magmas comparatively rich in enriched-in-crust elements and poor in enriched-in-mantle elements relative to the rocks from which the magmas derive. These magmas tend to move upward through time, eventually solidifying and forming a zoned continental crust, with the upper zone (sial) having an overall granitic composition (Table 1) and the lower zone (sima) an unknown composition, probably similar to that of basalt. The upper granitic crust is even more abundant in enriched-in-crust elements. Further modifications of the upper continental crust can occur through processes such as weathering, sedimentation, metamorphism, and igneous differentiation. *See* EARTH, INTERIOR OF; GEOCHEMISTRY; PLANETOLOGY.

[PAUL C. RAGLAND]

Bibliography: L. H. Ahrens, *Distribution of the Elements in Our Planet*, 1965; B. Mason, *Principles of Geochemistry*, 3d ed., 1966; K. K. Turekian, *Chemistry of the Earth*, 1972; K. H. Wedepohl, *Geochemistry*, 1971.

Emerald

The medium- to dark-green gem variety of the mineral beryl, $Al_2[Be_3Si_6O_{18}]$, crystallizing in the hexagonal system. A flawless emerald with good color is one of the most sought after and highly prized of all precious gems; an exceptional stone may be worth up to $10,000 per carat. Emerald is restricted in its occurrence, and only infrequently are exceptional stones found; most emeralds are flawed and cloudy, and few stones command high prices. What constitutes a true emerald is well defined by the gem dealers, and the dividing line between emerald and mere green beryl is sharp.

Occurrence. In contradistinction to beryl and its other gem varieties, aquamarine, morganite, and goshenite, which almost exclusively occur in granite pegmatite druses, emeralds have only been found in mica schists or metasomatized limestones. The most outstanding occurrences include the Muzo and El Chivor mines in Colombia. Here emeralds occur in calcite veins associated with a dark limestone which has been metasomatized by pegmatitic solutions. Noteworthy occurrences in mica schists include Tokovoja in the Ural Mountains, where emerald occurs with the beryllium minerals chrysoberyl (and its gem variety alexandrite) and phenakite; Habachtal, Austria; Transvaal, South Africa; and Kaliguman, India. The ultimate source of an emerald can often be assessed by a study of its inclusions.

Synthesis. Exceptional emeralds have been synthesized that rival the finest natural stones in quality and color. Broadly speaking, there are two main synthesis techniques. The molten-flux technique involves the compound oxides or gels of approximate beryl composition mixed with fluxes such as lithium molybdate and vanadium pentoxide. Fusion followed by slow cooling results in the growth of small prismatic crystals. The hydrothermal synthesis technique is capable of producing exceptional stones; emerald is hydrothermally grown on the plate of cut aquamarine which is used as a seed. The grown emerald is then sawed and used in turn as a seed for other emeralds. In all techniques 0.05–1.4% Cr_2O_3 is added. As in the natural emeralds, the Cr^{3+} chromophore is responsible for the green color. Detailed studies have been undertaken to define means of distinguishing natural emeralds from synthetics. Careful study (involving determination of birefringence and infrared absorption spectrum) will not only reveal whether a stone is natural or synthetic but will also distinguish the mode of synthesis if the emerald proved to be a synthetic stone. *See* BERYL; GEM.

[PAUL B. MOORE]

Emery

A natural mixture of corundum with magnetite or with hematite and spinel. Emery has been used for centuries as an abrasive or polishing material. Because the mixture is very intimate and appears to be quite homogeneous, it was considered to be a single mineral species until the middle of the 19th century. The aggregate has a gray-to-black color and is extremely tough and difficult to break. The specific gravity varies from 3.7 to 4.3, depending upon the relative amounts of the constituent minerals. The hardness is about 8 (Mohs scale), less than that of pure corundum which is 9, and is more dependent upon the physical state of aggregation than on the percentage of corundum. *See* CORUNDUM.

Since early times emery has been recovered from Cape Emeri on the island of Naxos and from other islands in the Grecian archipelago. Here it occurs as irregular beds and lenses and in loose blocks associated with crystalline limestone and schists. It is also found at several localities in Asia Minor under similar conditions, notably at Gumach Dagh, east of Ephesus, and at Kula, near Alashehr. Emery was worked during the latter part of the 19th century at Chester, Mass., where it was associated with diaspore, margarite, and chloritoid. Because of its magnetic properties, resulting from the admixed magnetite, the Chester material was first worked unsuccessfully as an iron ore. Only after the similarity of the associated minerals with those of the Naxos emery was noted was its true nature determined.

Although synthetic abrasives have replaced emery in many of its earlier uses, it is still used as an abrasive and polishing material by lapidaries and in the manufacture of lenses, prisms, and other optical equipment. Emery wheels, emery paper, and emery cloth are used not only by lapidaries but also by machinists in the grinding and polishing of steel. *See* MAGNETITE.

[CORNELIUS S. HURLBUT, JR.]

Enargite

A mineral having composition Cu_3AsS_4. In some places enargite is a valuable ore of copper. It is found in orthorhombic crystals but is more commonly columnar, bladed, or massive (see illustration). The mineral has perfect prismatic cleavage, metallic luster, and grayish-black color. The hardness is 3 on Mohs scale, and the specific gravity is 4.44. Enargite is one of the rarer copper ore minerals and is found in vein and replacement deposits associated with pyrite, galena, sphalerite, tetrahedrite, and bornite. It has been mined in Yugoslavia, Peru, the Philippines, and the United States at Butte, Mont., and Bingham Canyon, Utah. Probably the largest deposit is at Chuquicamata, Chile, where enargite with other primary copper minerals has been altered to form the great deposit of copper sulfates. *See* COVELLITE.

[CORNELIUS S. HURLBUT, JR.]

ENARGITE

1 in.

Enargite crystals, Silver Bow County, Mont. (*Specimen from Department of Geology, Bryn Mawr College*)

Engineering geology

The application of education and experience in geology and other geosciences to solve geological problems posed by civil engineering structures. The branches of the geosciences most applicable are (1) surficial geology, (2) petrofabrics; (3) rock and soil mechanics; (4) geohydrology; and (5) geophysics, particularly exploration geophysics and earthquake seismology. This article discusses some of the practical aspects of engineering geology.

Geotechnics is the combination of pertinent geoscience elements with civil engineering elements to formulate the civil engineering system that has the optimal interaction with the natural environment.

Engineering properties of rock. The civil engineer and the engineering geologist consider most hard and compact natural materials of the earth crust as rock, and their derivatives, formed mostly by weathering processes, as soil. A number of useful soil classification systems exist. Because of the lack of a rock classification system suitable for civil engineering purposes, most engineering geology reports use generic classification systems modified by appropriate rock-property adjectives. *See* ROCK; ROCK MECHANICS.

Rock sampling. The properties of a rock element can be determined by tests on cores obtained from boreholes. These holes are made by one or a combination of the following basic types of drills: the rotary or core drill (Fig. 1), the cable-tool or churn drill, and the auger. The rotary type generally is used to obtain rock cores. The rotary rig has a motor or engine (gasoline, diesel, electric, or compressed air) that drives a drill head that rotates a drill rod (a thick-walled hollow pipe) fastened to a core barrel with a bit at its end. Downward pressure on the bit is created by hydraulic pressure in the drill head. Water or air is used to remove the rock that is comminuted (chipped or ground) by the diamonds or hard-metal alloy used to face the bit. The core barrel may be in one piece or have an inner metal tube to facilitate recovery of soft or badly broken rock ("double-tube" core barrel). The churn-type drill may be used to extend the hole through the soil overlying the rock, to chop through boulders, or occasionally to deepen a hole in rock where core is not desired. When the rock is too broken to support itself, casing (steel pipe) is

Fig. 1. Rotary or core drill on damsite investigation.

driven or drilled through the broken zone. Drill rigs range in size from those mounted on the rear of large multiwheel trucks to small, portable ones that can be packed to the investigation site on a man's back or parachuted from a small plane.

The rock properties most useful to the engineering geologist are compressive and triaxial shear strengths, permeability, Young's modulus of elasticity, erodability under water action, and density (in pounds per cubic foot, or pcf).

Compressive strength. The compressive (crushing) strength of rock generally is measured in pounds per square inch (psi) or kilograms per square centimeter (kg/cm^2). It is the amount of stress required to fracture a sample unconfined on the sides and loaded on the ends (Fig. 2). If the load P of 40,000 lb is applied to a sample with a diameter of 2 in. (3.14 in.^{2}), the compressive stress is $40{,}000 \div 3.14 = 12{,}738$ psi. If this load breaks the sample, the ultimate compressive strength equals the compressive stress acting at the moment of failure, in this case 12,738 psi. The test samples generally are cylindrical rock cores that have a length-to-diameter ratio (L/D) of about 2. The wide variety of classification systems used for rock results in a wide variation in compressive strengths for rocks having the same geologic name. The table gives a statistical evaluation of the compressive strengths of several rocks commonly encountered in engineering geology.

Most laboratory tests show that an increase in moisture in rock causes a decrease in its compressive strength and elastic modulus; what is not generally known, however, is that the reverse situation shown in Fig. 3 has been encountered in certain types of volcanic rocks. In sedimentary rock the compressive strength is strongly dependent upon the quality of the cement that bonds the mineral grains together (for example, clay cement gives low strength) and upon the quantity of cement (a rock may have only a small amount of cement, and despite a strong bond between the grains, the strength is directly related to the inherent strength of the grains). Strength test results are adversely affected by microfractures that may be present in the sample prior to testing, particularly if the microfractures are oriented parallel to the potential failure planes.

Compressive strength of rocks*

Type of rock	No. of tests	Weighted mean, psi	Minimum and maximum values, psi
Igneous			
Basalt	22	21,159	2,400 – 52,000
Diabase	14	31,680	6,000 – 46,600
Diorite	14	29,530	12,200 – 48,300
Granite	81	22,420	5,100 – 51,200
Porphyry	72	17,790	5,000 – 63,000
Sedimentary			
Dolomite	62	12,740	900 – 52,000
Limestone	211	10,760	200 – 37,800
Marlstone	14	15,620	8,100 – 28,200
Sandstone	257	9,200	300 – 47,600
Shale	67	9,660	100 – 33,500
Siltstone	14	15,740	500 – 45,800
Metamorphic			
Amphibolite	11	47,550	22,300 – 61,700
Gneiss	23	25,470	5,200 – 44,200
Greenstone	11	28,930	16,600 – 45,500
Marble	31	14,590	4,400 – 39,700
Quartzite	25	42,400	3,700 – 91,200
Schist	16	7,290	1,000 – 23,500
Mineral			
Hematite	18	35,050	16,200 – 99,900
Iron Ore	18	4,933	2,300 – 7,800

*Based on data used by W. R. Judd, *Correlating rock properties by statistical methods*, Purdue Research Foundation Report, 1969.

The value of compressive strength to be used in an engineering design must be related to the direction of the structure's load and the orientation of the bedding in the foundation rock. This relationship is important because the highest compressive strength usually is obtained when the compressive stress is normal to the bedding. Conversely, the highest Young's modulus of elasticity (E) usually results when the compressive stress parallels the bedding. When these strength and elastic properties apparently are not affected by the direction of applied load, the rock is described as isotropic; if load applied parallel to the bedding provides physical property data that are significantly different than those obtained when the load is applied normal (perpendicular) to the bedding, the rock is anisotropic or aeolotropic. If the physical components of the rock element or rock system have equal dimensions and equal fabric relationships, the rock is homogeneous; significant variance in these relationships results in a heterogeneous rock. Most rocks encountered in foundations are anisotropic and heterogeneous.

Shear in rocks. Shearing stresses tend to separate portions of the rock (or soil) mass. Faults and folds are examples of shear failures in nature. In engineering structures, every compression is accompanied by shear stresses. For example, an arch dam compresses the abutment rock and, if the latter is intersected by fissures or weak zones, it may fail in shear with a resulting tensile stress in the dam concrete that may rupture the concrete. The application of loads over long periods of time on most rocks will cause them to creep or even to flow like a dense fluid (plastic flow). *See* STRUCTURAL GEOLOGY.

Ambient stress. This type of stress in a rock system is actually potential energy, probably created by ancient natural forces, recent seismic activity, or nearby man-made disturbances. Ambient (residual, stored, or primary) stress may remain in rock long after the disturbance is removed. An excavation, such as a tunnel or quarry, will relieve the ambient stress by providing room for displacement of the rock, and thus the potential energy is converted to kinetic energy. In tunnels and quarries, the release of this energy can cause spalling, the slow outward separation of rock slabs from the rock massif; when this movement is rapid or explosive, a rock burst occurs. The latter is a different phenomenon from a rock bump, which is a rapid upward movement of a large portion of a rock system and, in a tunnel, can have sufficient force to flatten a steel mine car against the roof or break the legs of a man standing on the floor when the bump occurs.

One of two fundamental principles generally is used to predict the possible rock load on a tunnel roof, steel or timber supports, or a concrete or steel lining: (1) The weight of the burden (the rock and soil mass between the roof and the ground surface) and its shear strength control the load, and therefore the resultant stresses are depth-dependent, or (2) the shear strength of the rock system and the ambient stresses control the stress distribution, so the resultant loading is only indirectly

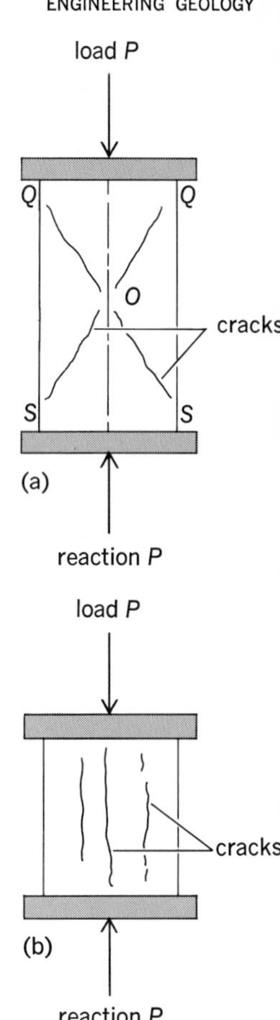

Fig. 2. Unconfined compression test. (*a*) Shear failure showing failure planes *QS*. (*b*) Tension failure. (*From D. P. Krynine and W. R. Judd, Principles of Engineering Geology and Geotechnics, McGraw-Hill, 1957*)

dependent upon depth. The excavation process can cause rapid redistribution of these stresses to produce high loads upon supports some distance from the newly excavated face in the tunnel. Lined tunnels can be designed so that the reinforced concrete or steel lining will have to carry only a portion of the ambient or burden stresses.

Construction material. Rock as a construction material is used in the form of dimension, crushed, or broken stone. Broken stone is placed as riprap on slopes of earth dams, canals, and riverbanks to protect them against water action. Also, it is used as the core and armor stone for breakwater structures. For all such uses, the stone should have high density (±165 pcf), be insoluble in water, and be relatively nonporous to resist cavitation. Dimension stone (granite, limestone, sandstone, and some basalts) is quarried and sawed into blocks of a shape and size suitable for facing buildings or for interior decorative panels. For exterior use dimension stone preferably should be isotropic (in physical properties), have a low coefficient of expansion when subjected to temperature changes, and be resistant to deleterious chemicals in the atmosphere (such as sulfuric acid). Crushed stone (primarily limestone but also some basalt, granite, sandstone, and quartzite) is used as aggregate in concrete and in bituminous surfaces for highways, as a base course or embankment material for highways, and for railroad ballast (to support the ties). When used in highway construction, the crushed stone should be resistant to abrasion as fine stone dust reduces the permeability of the stone layer; the roadway then is more susceptible to settling and heaving caused by freezing and thawing of water in the embankment. Concrete aggregate must be free of deleterious material such as opal and chalcedony; volcanic rocks containing glass, devitrified glass, and tridymite; quartz with microfractures; phyllites containing hydromica (illites); and other rocks containing free silica (SiO_2). These materials will react chemically with the cement in concrete and release sodium and potassium oxides (alkalies) or silica gels. Preliminary petrographic analyses of the aggregate and chemical analysis of the cement can indicate the possibility of alkali reactions and thus prevent construction difficulties such as expansion, cracking, or a strength decrease of the concrete. *See* PETROGRAPHY.

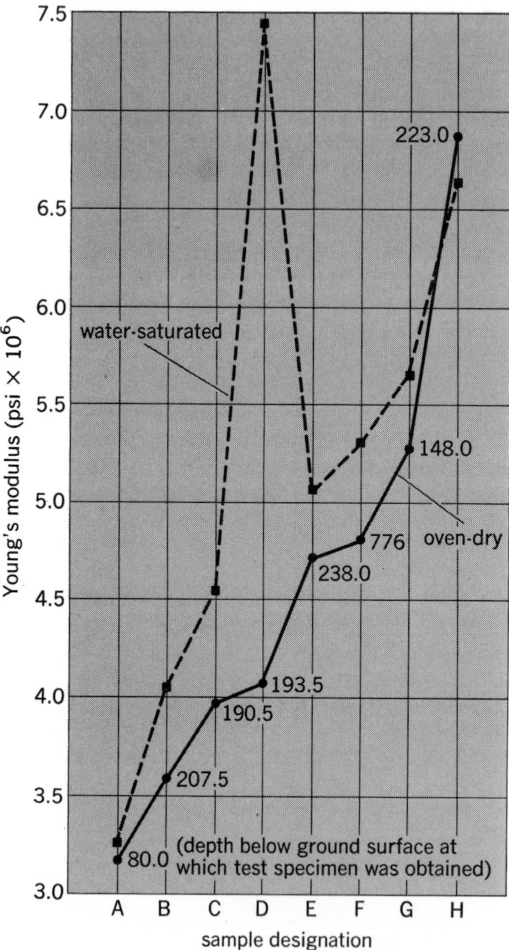

Fig. 3. Increase in Young's modulus caused by saturation of dacite porphyry. (*After J. R. Ege and R. B. Johnson, Consolidated tables of physical properties of rock samples from area 401, Nevada Test Site, U. S. Geol. Surv. Tech. Letter Pluto-21, 1962*)

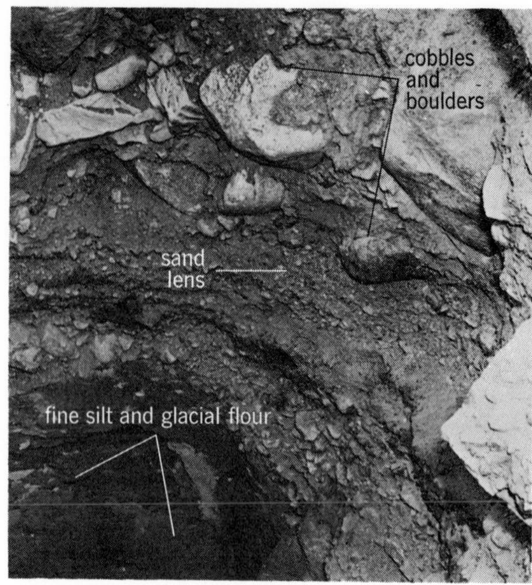

Fig. 4. Glacial deposit in test pit.

Geotechnical significance of soils. Glacial and alluvial deposits contain heterogeneous mixtures of pervious (sand and gravel) and impervious (clay, silt, and rock flour) soil materials (Fig. 4). The pervious materials can be used for highway subgrade, concrete aggregate, and filters and pervious zones in earth embankments. Dam reservoirs may be endangered by the presence of stratified or lenticular bodies of pervious materials or ancient buried river channels filled with pervious material. Deep alluvial deposits in or close to river deltas may contain very soft materials such as organic silt or mud. An unsuitable soil that has been found in dam foundations is open-work gravel. This material may have a good bearing strength because of a natural cement bond between grains, but it is highly pervious because of the almost complete lack of fine soil to fill the voids between the gravel pebbles. *See* DELTA; FLOODPLAINS.

Concrete or earth dams can be built safely on sand foundations if the latter receive special treatment. One requirement is to minimize seepage losses by the construction of cutoff walls (of concrete, compacted clay, or interlocking-steel-sheet piling) or by use of mixed-in-place piles 3 ft or more in diameter. The latter are constructed by augering to the required depth but not removing any of the sand. At the desired depth, cement grout is pumped through the hollow stem of the auger, which is slowly withdrawn while still rotating; this mixes the grout and the sand into a relatively impervious concrete pile. The cutoff is created by overlapping these augered holes. Some sand foundations may incur excessive consolidation when loaded and then saturated, particularly if there is a vibratory load from heavy machinery or high-velocity water in a spillway. This problem is minimized prior to loading by using a vibrating probe inserted into the sand or vibratory rollers on the sand surface or by removing the sand and then replacing it under vibratory compaction and water sluicing.

Aeolian (windblown) deposits. Loess is a relatively low-density soil composed primarily of silt grains cemented by clay or calcium carbonate. It has a vertical permeability considerably greater than the horizontal. When a loaded loess deposit is wetted, it rapidly consolidates, and the overlying structure settles. When permanent open excavations ("cuts") are required for highways or canals through loess, the sides of the cut should be as near vertical as possible: Sloping cuts in loesses will rapidly erode and slide because of the high vertical permeability. To avoid undesirable settlement of earth embankments, the loess is "prewetted" prior to construction by building ponds on the foundation surface (Fig. 5). Permanently dry loess is a relatively strong bearing material. Aeolian sand deposits present the problem of stabilization for the continually moving sand. This can be done by planting such vegetation as heather or young pine or by treating it with crude oil. Cuts are traps for moving sand and should be avoided.

Organic deposits. Excessive settlement will occur in structures founded on muskeg terrain. Embankments can be stabilized by good drainage, the avoidance of cuts, and the removal of the organic soil and replacement by sand and gravel or, when removal is uneconomical, displacement of it by the continuous dumping of embankment material upon it. Structures imposing concentrated loads are supported by piling driven through the soft layers into layers with sufficient bearing power.

Residual soils. These soils are derived from the in-place deterioration of the underlying bedrock. The granular material caused by the in-place disintegration of granite generally is sufficiently thin to cause only nominal problems. However, there are regions (such as California, Australia, and Brazil) where the disintegrated granite (locally termed "DG") may be hundreds of feet thick; although it may be competent to support moderate loads, it is unstable in open excavations and is pervious. A thickness of about 200 ft of DG and weathered gneiss on the sides of a narrow canyon was a major cause for construction of the Tumut-1 Power Plant (New South Wales) in hard rock some 1200 ft underground. Laterite (a red clayey soil) derived from the in-place disintegration of limestone in tropical to semitropical climates is another critical residual soil. It is unstable in open cuts on moderately steep slopes, compressible under load, and when wet produces a slick surface that is unsatisfactory for vehicular traffic. This soil frequently is encountered in the southeastern United States and southeastern Asia, including Indonesia and Vietnam.

Fig. 5. Prewetting loess foundation for earth dam. (*U.S. Bureau of Reclamation*)

Clays supporting structures may consolidate slowly over a long period of time and cause structural damage. When clay containing montmorillonite is constantly dried and rewetted by climatic or drainage processes, it alternately contracts and expands. During the drying cycle, extensive networks of fissures are formed that facilitate the rapid introduction of water during a rainfall. This cyclic volume change of the clay can produce uplift forces on structures placed upon the clay or compressive and uplift forces on walls of structures placed within the clay. These forces have been known to rupture concrete walls containing 3/4-in.-diameter steel reinforcement bars. A thixotropic or "quick" clay has a unique lattice structure that causes the clay to become fluid when subjected to vibratory forces. Various techniques are used to improve the foundation characteristics of critical types of clay: (1) electroosmosis that uses electricity to force redistribution of water molecules and subsequent hardening of the clay around the anodes inserted in the foundation; (2) provision of adequate space beneath a foundation slab or beam so the clay can expand upward and not lift the structure; (3) belling, or increasing in size, of the diameter of the lower end of concrete piling so the pile will withstand uplift forces imposed by clay layers around the upper part of the pile; (4) treatment of the pile surface with a frictionless coating (such as Teflon or a loose wrapping of asphalt-impregnated paper) so the upward-moving clay cannot adhere to the pile; (5) sufficient drainage around the structure to prevent moisture from contacting the clay; and (6) replacement of the clay by a satisfactory foundation material. Where none of these solutions are feasible, the structure then must be relocated to a satisfactory site or designed so it can withstand uplift or compressive forces without expensive damage. *See* CLAY.

Silt may settle rapidly under a load or offer a

"quick" condition when saturated. For supporting some structures (such as residences), the bearing capacity of silts and fine sands can be improved by intermixing them with certain chemicals that will cause the mixture to "set" or harden when exposed to air or moisture; some of the chemicals used are sodium silicate with the later application of calcium chloride, bituminous compounds, phenolic resins, or special cements (to form "soil cement"). The last mixture has been used for surfacing secondary roads, for jungle runways in Vietnam, and as a substitute for riprap of earth dams. Some types of silt foundations can be improved by pumping into them soil-cement or clay mixtures under sufficient pressures to create large bulbs of compacted silt around the pumped area.

Geotechnical investigation. For engineering projects these investigations may include preliminary studies, preconstruction or design investigations, consultation during construction, and the maintenance of the completed structure.

Preliminary studies. These are made to select the best location for a project and to aid in formulating the preliminary designs for the structures. The first step in the study is a search for pertinent published material in libraries, state and Federal agencies, private companies, and university theses. Regional, and occasionally detailed reports on local geology, including geologic maps, are available in publications of the U.S. Geological Survey; topographic maps are available from that agency and from the U.S. Army Map Service. Oil companies occasionally will release the geologic logs of any drilling they may have done in a project area. Sources of geologic information are listed in the *Directory of Geological Material in North America* by J. A. Howell and A. I. Levorsen (NAS-NRC Publ. 556, 1957). Air photos should be used to supplement map information (or may be the only surficial information readily available). The U.S. Geological Survey maintains a current index map of the air-photo coverage of the United States. The photos are available from that agency, the U.S. Forest Service, the Soil Conservation Service, and commercial air-photo companies; for some projects the military agencies will provide air-photo coverage. The topographic maps and air photos can be used to study rock outcrop and drainage patterns, landforms, geologic structures, the nature of soil and vegetation, moisture conditions, and land use by man (cultural features).

Fig. 6. Drill core obtained in dam foundation.

Field reconnaissance may include the collection of rock and soil specimens; inspection of road cuts and other excavations; inspection of the condition of nearby engineering structures such as bridges, pavements, and buildings; and location of sources of construction material. Aerial reconnaissance is essential at this stage and can be performed best in helicopters and second-best in slow-flying small planes.

Preconstruction. Surface and subsurface investigations are required prior to design and construction. Surface studies include the preparation of a detailed map of surficial geology, hydrologic features, and well-defined landforms. For dam projects, a small-scale geologic map (for example, 1 in. = 200 or 500 ft) is made of the reservoir area and any adjacent areas that may be directly influenced by the project; in addition, a large-scale geologic map (for example, 1 in. = 50 ft) is required of the specific sites of the main structures (the dam, spillway, power plant, tunnels, and so on). These maps can be compiled by a combination of field survey methods and aerial mapping procedures. They should have a grid system (coordinates) and show the proposed locations for subsurface investigations.

Subsurface investigations are required to confirm and amplify the surficial geologic data. These may include test pits, trenches, short tunnels (drifts or adits), and the drilling of vertical, horizontal, or oblique (angle) boreholes (Fig. 6). Geologic data obtained by these direct methods can be supplemented by indirect or interpreted data obtained by geophysical methods on the surface or in subsurface holes and by installation of special instruments to measure strain or deformation in a borehole or tunnel.

The geology disclosed by subsurface investigations is "logged" on appropriate forms. Tunnel logs display visual measurements of fractures and joint orientations (strike and dip); rock names and a description of their estimated engineering properties; alteration, layering, and other geologic defects; the location and amount of water or gas inflow; the size and shape of blocks caused by fracturing or jointing and the width of separation or the filling material between blocks; and the irregularities in the shape of the tunnel caused by the displacement of blocks during or after excavation (rock falls, rock bursts, chimneying, and overbreakage). Geophysical seismic methods may be used to define the thickness of loosened rock around the tunnel; geoacoustical techniques that detect increases in microseismic noise during tunneling may be used to determine if the excavation is causing excessive loosening in the tunnel rock. This detection of "subaudible rock noise" occasionally is used to detect the potential movement of rock slopes in open excavations.

The borehole data can be logged on a form such as shown in Fig. 7. These data can be obtained by direct examination of the core, by visual inspection of the interior of the borehole using a borehole camera (a specially made television camera) or a stratoscope (a periscopelike device), or by geophysical techniques. Direct viewing of the interior of the hole is the only positive method of determining the in-place orientation and characteristics of separations and of layering in the rock system. The

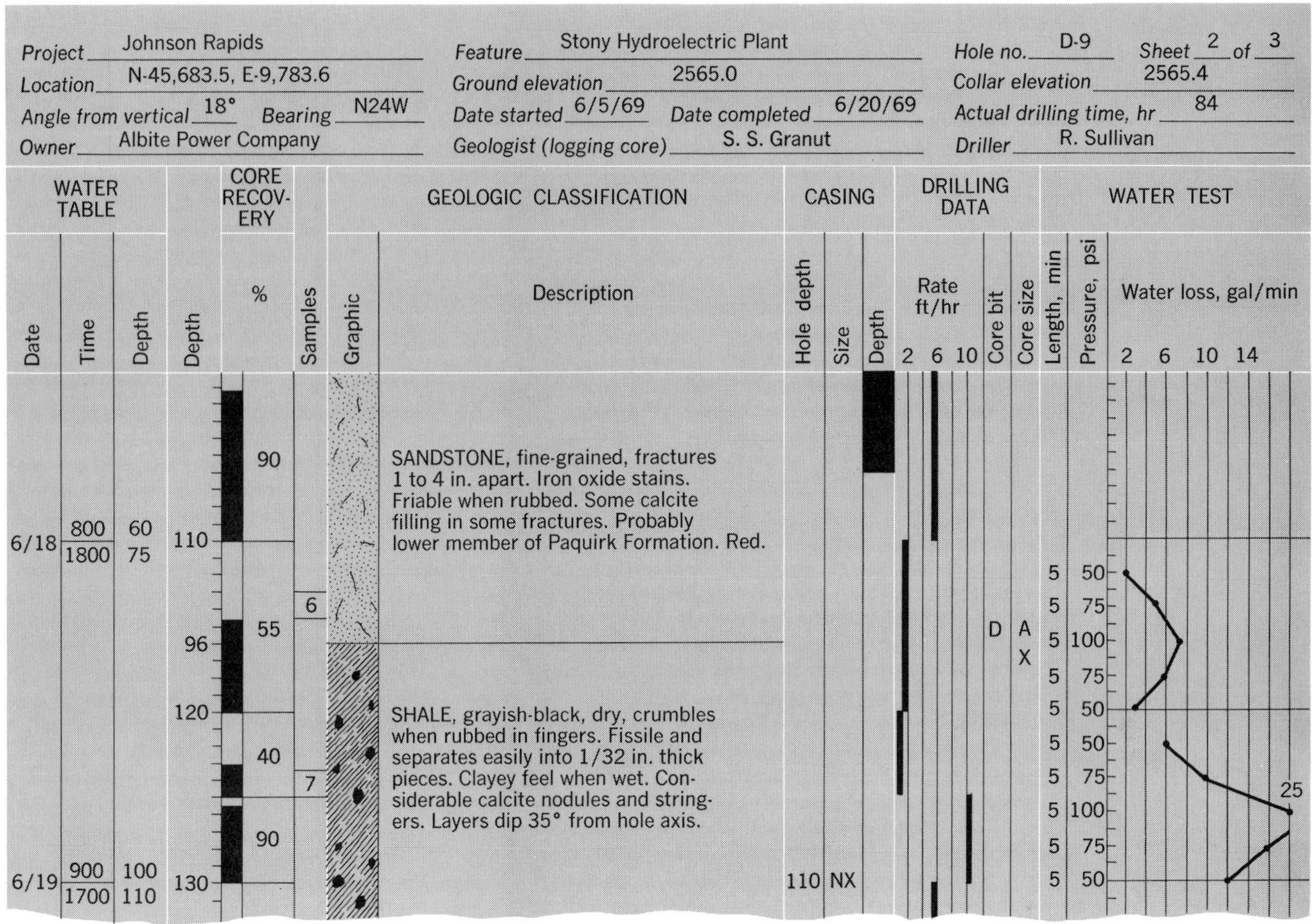

Project Johnson Rapids
Location N-45,683.5, E-9,783.6
Angle from vertical 18° Bearing N24W
Owner Albite Power Company
Feature Stony Hydroelectric Plant
Ground elevation 2565.0
Date started 6/5/69 Date completed 6/20/69
Geologist (logging core) S. S. Granut
Hole no. D-9 Sheet 2 of 3
Collar elevation 2565.4
Actual drilling time, hr 84
Driller R. Sullivan
WATER TABLE
Date
Time
Depth
Depth
CORE RECOVERY
%
Samples
GEOLOGIC CLASSIFICATION
Graphic
Description
CASING
Hole depth
Size
Depth
DRILLING DATA
Rate ft/hr
2 6 10
Core bit
Core size
WATER TEST
Length, min
Pressure, psi
Water loss, gal/min
2 6 10 14
6/18 800 60
1800 75
6/19 900 100
1700 110
110
96
120
130
90
55
40
90
6
7
SANDSTONE, fine-grained, fractures 1 to 4 in. apart. Iron oxide stains. Friable when rubbed. Some calcite filling in some fractures. Probably lower member of Paquirk Formation. Red.
SHALE, grayish-black, dry, crumbles when rubbed in fingers. Fissile and separates easily into 1/32 in. thick pieces. Clayey feel when wet. Considerable calcite nodules and stringers. Layers dip 35° from hole axis.
110 NX
D
AX
5 50
5 75
5 100
5 75
5 50
5 50
5 75
5 100
5 75
5 50
25

Fig. 7. Log recording of geological information from a borehole.

geophysical techniques include use of gamma-gamma logging that evaluates the density of the rock surrounding the borehole or at depths as great as 150 ft beneath the gamma probe; neutron logging to determine the moisture content of the rock system by measuring the depth of penetration of the neutrons; traversing the borehole with a sonic logger that, by calibration, measures differences in the velocity of wave propagation in different strata (and thus can determine in place Young's modulus of elasticity and the thickness of each stratum encountered by the borehole); and electric logging that uses differences in the electrical resistivity of different strata to define their porosity, moisture content, and thickness.

Occasionally a hole is drilled through a talus deposit containing the same type of rock as the underlying rock in place (bedrock). Because of the similarity in rock types, the talus-bedrock contact sometimes is best identified by determining the orientation of the remnant magnetism in the core: The magnetic lines in the core will have a regular orientation, but the talus magnetism will have random directions. This method is useful only in rocks that contain appreciable remnant magnetism such as some basalts.

Geophysical seismic or electrical resistivity methods also can be used on the ground surface to define the approximate depth of bedrock or various rock layers. The results require verification by occasional boreholes, but this is an inexpensive and satisfactory technique for planning and design investigations. The seismic methods are not useful when it is necessary to locate soft strata (wherein the seismic waves travel at relatively low velocity) that are overlain by hard strata (that have higher wave velocity); the latter conceal and block the signal from the soft strata. Also, difficulties may occur when the strata to be located are overlain by soil containing numerous large boulders composed of rock having higher velocities than the surrounding soil, or when the soil is very compact (such as glacial till) because its velocity characteristics may resemble those of the underlying bedrock. Another problem is that the seismic method seldom can identify narrow and steep declivities in the underlying hard rock (because of improper reflection of the waves).

Construction. Geotechnical supervision is desirable during construction in or on earth media. The engineering geologist must give advice and keep a record of all geotechnical difficulties encountered during the construction and of all geological features disclosed by excavations. During the operation and maintenance of a completed project the services of the engineering geologist often are required to determine causes and assist in the preparation of corrective measures for cracks in linings of water tunnels, excessive settlement of structures, undesirable seepage in the foundations of

Fig. 8. Kortes Dam, Wyo. Arrows show diverter walls protecting power plant against rock falls.

dams, slides in canal and other open excavations, overturning of steel transmission-line towers owing to a foundation failure, and rock falls onto hydroelectric power plants at the base of steep canyon walls (Fig. 8).

Legal aspects. An important consideration for the engineering geologist is the possibility of a contractor making legal claims for damages, purportedly because of unforeseen geologic conditions he encounters during construction. Legal support for such claims can be diminished if the engineering geologist supplies accurate and detailed geologic information in the specifications and drawings used for bidding purposes. These documents should not contain assumptions about the geological conditions (for a proposed structure), but they should show all tangible geologic data obtained during the investigation for the project: for example, an accurate log of all boreholes and drifts and a drawing showing the boundaries of the outcrops of all geological formations in the project area. The engineering geologist should have sufficient experience with design and construction procedures to formulate an investigation program that results in a minimum of subsequent uncertainties by a contractor. Numerous uncertainties about geologic conditions can result not only in increased claims but also may cause a contractor to submit a higher bid (to minimize his risks) than if detailed geologic information were available to him.

Special geotechnical problems. In arctic zones, structures built on permafrosted soils may be heaved or may cause thawing and subsequent disastrous settlement (Fig. 9). The growth of permafrost upward into earth dams seriously affects their stability and permeability characteristics. Obtaining natural construction materials in permafrosted areas requires thawing of the borrow area to permit efficient excavation; once excavated, the material must be protected against refreezing prior to placement in the structures. Permafrost in rock seldom will cause foundation difficulties. In planning reservoirs, it is essential to evaluate their watertightness, particularly in areas containing carbonate or sulfate rock formations or lava flows. These formations frequently contain extensive systems of caverns and channels that may or may not be filled with claylike material or water. Sedimentation studies are required for the design of efficient harbors or reservoirs because soil carried by the moving water will settle and block or fill these structures. In areas with known earthquake activity, aseismic (earthquakeproof) design requires knowledge of the intensity and magnitude of earthquake forces. The prevention and rehabilitation of slides (landslides) in steep natural slopes and in excavations are important considerations in many construction projects and are particularly important in planning reservoirs, as was disastrously proved by the Vaiont Dam catastrophe in 1963.

Geohydrologic problems. In the foundation material under a structure, water can occur in the form of pore water locked into the interstices or pores of the soil or rock, as free water that is moving through openings in the earth media, or as included water that is a constituent or chemically bound part of the soil or rock. When the structure load compresses the foundation material, the resulting compressive forces on the pore water can produce undesirable uplift pressures on the base of the structure. Free water is indicative of the permeability of the foundation material and possible excessive water loss (from a reservoir, canal, or tunnel); uplift on the structure because of an increase in hydrostatic head (caused by a reservoir or the like); or piping, which is the removal of particles of the natural material by flowing water with a consequent unfilled opening that weakens the foundation and increases seepage losses.

The possibility of excessive seepage or piping can be learned by appropriate tests during the boring program. For example, water pressure can be placed on each 5-ft section of a borehole, after the core is removed, and any resulting water loss can be measured. The water pressure is maintained within the 5-ft section by placing an expandable rubber ring (packer) around the drill pipe at the top of the test section and then sealing off the section by using mechanical or hydraulic pressure on the pipe to force expansion of the packer. When only one packer is used, because it is desired to test only the section of hole beneath it, it is a "single-packer" test. In a double-packer test a segment of hole is isolated for pressure testing by placing packers at the top and the bottom of the test section. The best information on the permeability

Fig. 9. Door-frame distortion and floor settlement which occurred as a result of permafrost thaw and heave. (*U.S. Geological Survey*)

characteristics of the rock can be obtained by the use of three or more increments of increasing and then decreasing water pressure for each tested length of hole. The permeability K of the rock in feet per minute is $(Q \ln_e L/r)/2\pi LH$, where Q is the rate of flow loss in gallons per minute (gpm), r is the radius of the test section, L is the length of hole being tested and should be $\geqq 10r$, and H (in feet) is the height of the water column between the ground surface and the center of the test section plus 2.3 times the psi pump pressure. If the water loss continues to increase when the pressure is decreased, piping of the rock or filling material in fractures may be occurring or fractures are widening or forming. The water-pressure test can be supplemented by a groutability test in the same borehole. This test is performed in the same way as the water test except, instead of water, a mixture of cement, sand, and water (cement grout) or a phenolic resin (chemical grout) is pumped under pressure into the test section. The resulting information is used to design cutoff walls and grout curtains for dams. The pressures used in water-pressure or grouting tests should not exceed the pressure exerted by the weight of the burden between the ground surface and the top of the test section. Excessive test pressure can cause uplift in the rock, and the resulting test data will be misleading.

Included or pore water generally is determined by laboratory tests on cores; these are shipped from the borehole to the laboratory in relatively impervious containers that resist loss of moisture from the core. The cores with their natural moisture content are weighed when received and then dried in a vacuum oven at about 110°F until their dry weight stabilizes. The percentage of pore water (by dry weight) is (wet weight − dry weight) × 100 ÷ dry weight. Temperatures up to 200°F can be used for more rapid drying, provided the dried specimens are not to be used for strength or elastic property determinations. (High temperatures can significantly affect the strength because the heat apparently causes internal stresses that disturb the rock fabric or change the chemical composition of the rock by evaporation of the included moisture.)

Protective construction. Civilian and military structures may be designed to minimize the effects of nuclear explosions. The most effective protection is to place the facility in a hardened underground excavation. A hardened facility, including the excavation and its contents, is able to withstand the effects generated by a specified size of nuclear weapon. The degree of hardening commonly is indicated by a numerical "psi" value; for example, a facility having 100-psi hardness is designed to maintain operations despite nearby nuclear explosions that generate 100-psi air-blast overpressure on the surface directly above the structure. However, this type of designation does not necessarily indicate the capability of the facility to withstand all direct-induced effects resulting from the nuclear blast, that is, the action on the structure that is induced by shock waves propagating through the earth media surrounding the structure. Direct-induced effects include the amount of acceleration, particle velocity, and displacement that can occur in earth media and in the structure.

Desirable depths and configurations for hardened facilities are highly dependent upon the shock-wave propagation characteristics of the surrounding earth media. These characteristics are influenced by the type of rock (Fig. 10), discontinuities in the rock system, free water, and geologic structure. Therefore, prior to the design and construction of such facilities, extensive geotechnical field and laboratory tests are performed. In addition to conventional physical properties, the laboratory tests also are directed toward determination of the appropriate Hugoniot curve for the media; this is done by exposing small elements of the rock to high pressures generated in a shock tube. The field tests may include small-scale nuclear explosions or chemical explosions designed to produce shock effects that simulate the nuclear explosion, and measurement of the shock effects by special instruments installed in the earth environment surrounding the blast point ("ground zero") and in any structures that are to be tested. Accurate interpretation of the explosion effects and adequate planning for the explosion require an accurate geologic map of the surface and of the underground environment that will be affected by the explosion. The map should show the precise location and orientation of all geologic defects that would influence the wave path, such as joints, fractures, and layers of alternately hard and soft rock. The importance of discontinuities was illustrated by the Rainier test (a 1.7-kiloton nuclear burst in tuff, 899 ft underground, in 1957); the shock wave caused a displacement of several inches along a fault that intersected the tunnel used for access to the shot point.

Application of nuclear energy. The use of nuclear energy for the efficient construction of civil engi-

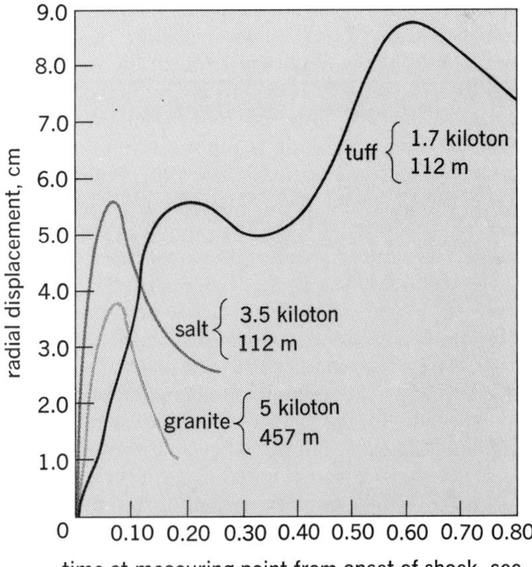

Fig. 10. Effect of rock type on the propagation of nuclear shock waves. The measuring point was 457 m from explosion source; all displacement measurements are extrapolated to the movement of the particle this distance from the source of a 5-kiloton explosion. (*Data from G. C. Werth and R. C. Herbst, Comparison of amplitudes of seismic waves from nuclear explosions in four media, Lawrence Berkeley Laboratory Report, UCRL-7350, University of California, Berkeley*)

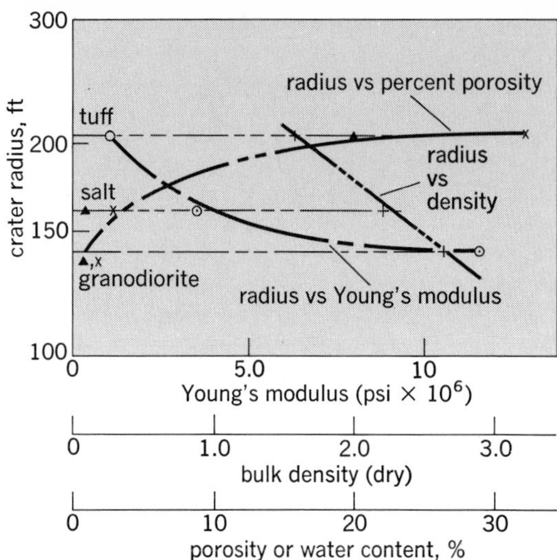

Fig. 11. Effect of rock properties and type on cavity radii for a 100-kiloton nuclear explosion buried 2000 ft. Solid triangles represent percentage moisture; curve not plotted. (*After C. R. Boardman et al., Characteristic effects of contained nuclear explosions for evaluation of mining applications, Lawrence Berkeley Laboratory Report UCRL-6862, University of California, Berkeley*)

neering projects is under investigation in the Plowshare Program. Examples are the proposed use of nuclear explosions for the rapid excavation of a new route for the Panama Canal; the Gas Buggy nuclear shot in 1968 to test the possibility of increasing the production of natural gas by opening fractures in the reservoir rock; and a proposed underground nuclear explosion that might expedite the production of low-grade copper ore by causing extensive fracturing and possible concentration of the ore. All such uses of nuclear explosives must be preceded by extensive geotechnical investigations of more than usual accuracy: Sufficient geologic detail must be obtained to minimize the possible venting of the underground explosion; otherwise, unpredicted continuous fractures or joints would allow the radiation products from the blast to be released into the atmosphere in the vicinity of the shot. Geotechnical conditions also will influence the size and shape of the cavity that will be caused by the explosion (Fig. 11).

Waste disposal. Another geotechnical nuclear problem occurs in the use of nuclear energy to generate power and to produce radioisotopes: safe disposal of the radioactive waste products. These products can be mixed with concrete and buried in the ground, but geohydrologic conditions or the chemical nature of the earth environment must not be allowed to cause excessive deterioration of the concrete. Dangerous waste products from nuclear reactors or from the production of agents for chemical or bacterial warfare also may be disposed of in deep wells, but then it is necessary to determine if the groundwater is likely to transport the waste products to areas where the same water may be used by man. Another potential problem from this method became known when disposal of chemical agents in a deep well in the Denver, Colo., area between 1962 and 1965 apparently disturbed the ambient stress regime sufficiently to trigger a succession of small earthquakes. Dangerous wastes also are disposed of by sealing them in large concrete blocks that are dropped into the deep parts of the ocean.

[WILLIAM R. JUDD]

Bibliography: E. C. Eckel (ed.), *Landslides and Engineering Practice*, National Academy of Sciences, Highway Research Board Special Report 29, 1958 (also HRB Reports 216, 1959, and 236, 1960); T. Fluhr and R. F. Legget (eds.), *Reviews in Engineering Geology*, vol. 1, Geological Society of America, 1962; Geological Society of America, *Engineering Geology Case Histories*, vols. 1–7, 1957–1968; N. V. Glazov and A. N. Glazov, *New Instruments and Methods of Engineering Geology*, 1959; M. E. Harr, *Groundwater and Seepage*, 1962; Idaho State University Department of Geology, *Proceedings of the 5th Annual Engineering Geology and Soil Engineering Symposium*, 1967; W. R. Judd, Geological factors in choosing underground sites, in J. J. O'Sullivan (ed.) *Protective Construction in a Nuclear Age*, 2 vols., 1961; D. P. Krynine and W. R. Judd, *Principles of Engineering Geology and Geotechnics*, 1957; U. Langefors and B. Kihlström, *Rock Blasting*, 1963; R. F. Legget, *Geology and Engineering*, 2d ed., 1962; D. S. Parasins, *Mining Geophysics*, 1966; U.S. Coast and Geodetic Survey, *Earthquake Investigations in the U.S.*, Special Publication No. 282, 1965; U.S. Department of the Army Office of the Chief of Research and Development, *Scientific and Technical Applications Forecast—1964—Excavation*, AD-607077 and AD-611555 (bibliography), 1963; R. C. S. Walters, *Dam Geology*, 1962.

Enstatite

The name given to the magnesian end member ($MgSiO_3$) of the orthorhombic pyroxene solid-solution series. The mineral is characterized by the (110) cleavages 87° apart. It is usually yellowish-gray, becoming greenish with a little iron present; transparent in thin sections; and optically positive with refractive indices $n\alpha = 1.650$, $n\beta = 1.653$, and $n\gamma = 1.658$. The calcium-free enstatite inverts at approximately 990°C to protoenstatite (also orthorhombic). Above 990°C protoenstatite is the stable form up to the decomposition temperature. On rapid cooling below 990°C protoenstatite inverts to a metastable monoclinic form called clinoenstatite. However, clinoenstatite becomes the stable high-temperature form if small amounts of calcium are in solid solution in the mineral. *See* PYROXENE.

Enstatite is a common constituent in many basalts, dunites, serpentinites, and peridotites and occurs in slags and meteorites. Olivine, diopside, calcium pyroxenes, and calcium-rich plagioclases are common associated minerals. Single crystals of enstatite altered to a single crystal of antigorite (serpentine), called bastite, that preserves the pyroxene crystal outline are frequent in certain serpentine masses. Enstatite alteration to amphibole is also common.

Enstatite is an important mineral in the upper mantle of the Earth, and coexists with clinopyroxene, garnet, olivine, and plagioclase. The solubility of aluminum in enstatite increases with increasing temperature. If temperature of formation can be determined, for example, from the enstatite content of the coexisting clinopyroxene, the aluminum content of the enstatite can be used to determine the depth of origin for the rock if garnet is present to establish aluminum saturation. Enstatite, clinopyroxene, and garnet rocks called lherzolites are common in certain alpine mountain terrains and as nodules in some diamond-bearing rocks called kimberlites. The lherzolites in alpine mountain terrains are interpreted as uplifted upper mantle materials now exposed at the surface. The aluminum contents of the lherzolite enstatites in the kimberlites suggest depths of origin for the nodules of 100 to 225 km and, thus, are samples of crystalline material from the deepest part of the Earth available for direct study.

[GEORGE W. DE VORE]

Eocene

The next to the oldest of the five major worldwide divisions (epochs) of the Tertiary Period (Cenozoic Era); the epoch of geologic time extending from the end of the Paleocene Epoch to the beginning of the Oligocene Epoch; the middle epoch of the older Tertiary (Paleogene or Nummulitic). *See* CENOZOIC; OLIGOCENE; PALEOCENE; TERTIARY.

Eocene strata include all the common sedimentary types, varying from marine through estuarine to terrestrial in origin. Igneous activity was still not as widespread as in the later Tertiary, though it was notable in some areas (Oregon, Washington, British Columbia).

Sir Charles Lyell in 1833 subdivided the Tertiary into Pliocene, Miocene, and Eocene (the last being the oldest, with about 3.5% living species). This original Eocene included those segments of the Tertiary which W. Schimper subsequently designated as Paleocene below, that is, older, and E. von Beyrich designated as Oligocene above, that is, younger. The sporadic occurrence of fossils in his lowermost and uppermost typical Eocene beds of the Paris and London basins led Lyell to propose that, when better sequences in these parts of his column were found, the middle beds of his original sequence should be retained as the type standard of comparison for his Eocene subdivision (Fig. 1). This restricted usage of Eocene, as originally recommended by Lyell, is the usage of today.

PRECAMBRIAN	PALEOZOIC							MESOZOIC			CENOZOIC	
					CARBONIFEROUS							
	CAMBRIAN	ORDOVICIAN	SILURIAN	DEVONIAN	Mississippian	Pennsylvanian	PERMIAN	TRIASSIC	JURASSIC	CRETACEOUS	TERTIARY	QUATERNARY

TERTIARY					QUATERNARY	
Paleocene	Eocene	Oligocene	Miocene	Pliocene	Pleistocene	Recent

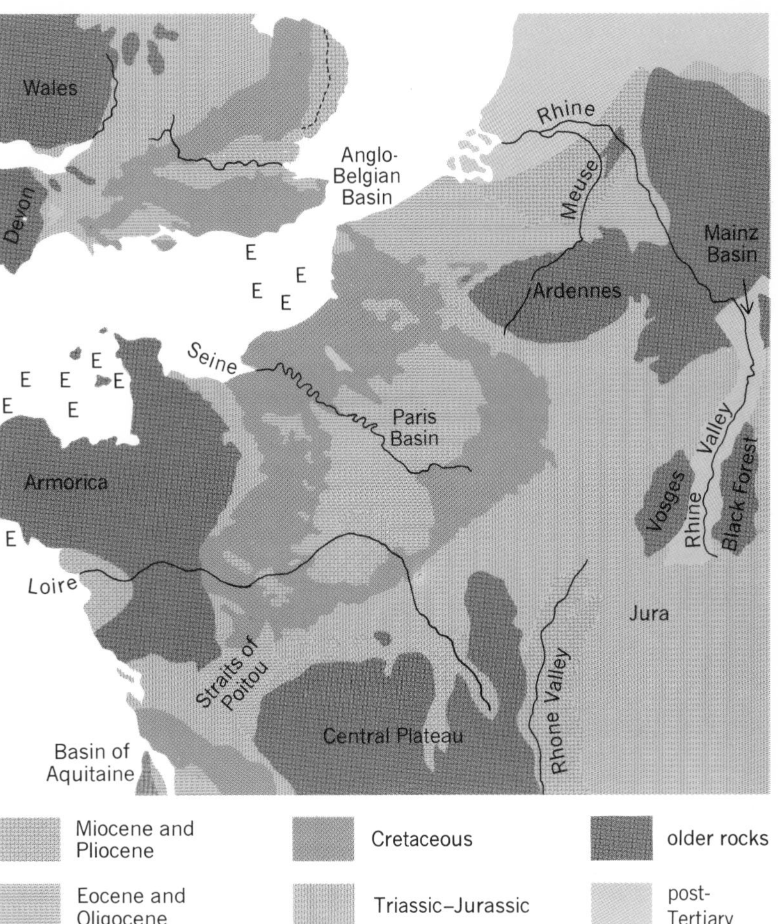

Fig. 1. The Paris and London (Anglo-Belgian) basins and their framework. E = Eocene outcrop. *(From A. M. Davies, Tertiary Faunas, vol. 2, Murby, 1934)*

Subdivisions. In its restricted sense, Eocene time encompasses two major depositional cycles. Major stage subdivisions in Europe are Ypresian (with a Cuisian substage) below, a Lutetian middle, and an English Bartonian (shallow-water) or French Auversian-Marinesian-Ludian or Italian Priabonian (deep-water) above (Fig. 2). Studies in the Paris Basin have suggested that some of Schimper's "Paleocene" fossils may have come from beds stratigraphically as high as those now assigned to the Eocene Cuisian stage, and that the Cuisian and Ypresian may be essentially facies of time-equivalent strata. On the American Gulf Coast a formational terminology (Wilcox, Claiborne, Jackson) prevails, as it has on the Pacific Coast (Capay, Domengine, Tejon-Cowlitz), where, however, a local stage classification based on foraminifers has also come into use: Penutian, Ulatisian, Narizian. Inland, Wasatchian, Bridgerian, Uintan, Duchesnean are the mammalian ages. Bortonian and Tahuian are local Eocene time subdivisions in New Zealand, and Tertiary a2, b1, b2 are used in the East Indies.

Eocene marine waters expanded far beyond those of the Paleocene. Generally, the early Eocene seas transgressed over land areas of low relief; limestones and mudstones of relatively uniform thickness are widespread, with basal beds of coarser texture most conspicuous in areas of marine transgression. Mid-Eocene diastrophic dis-

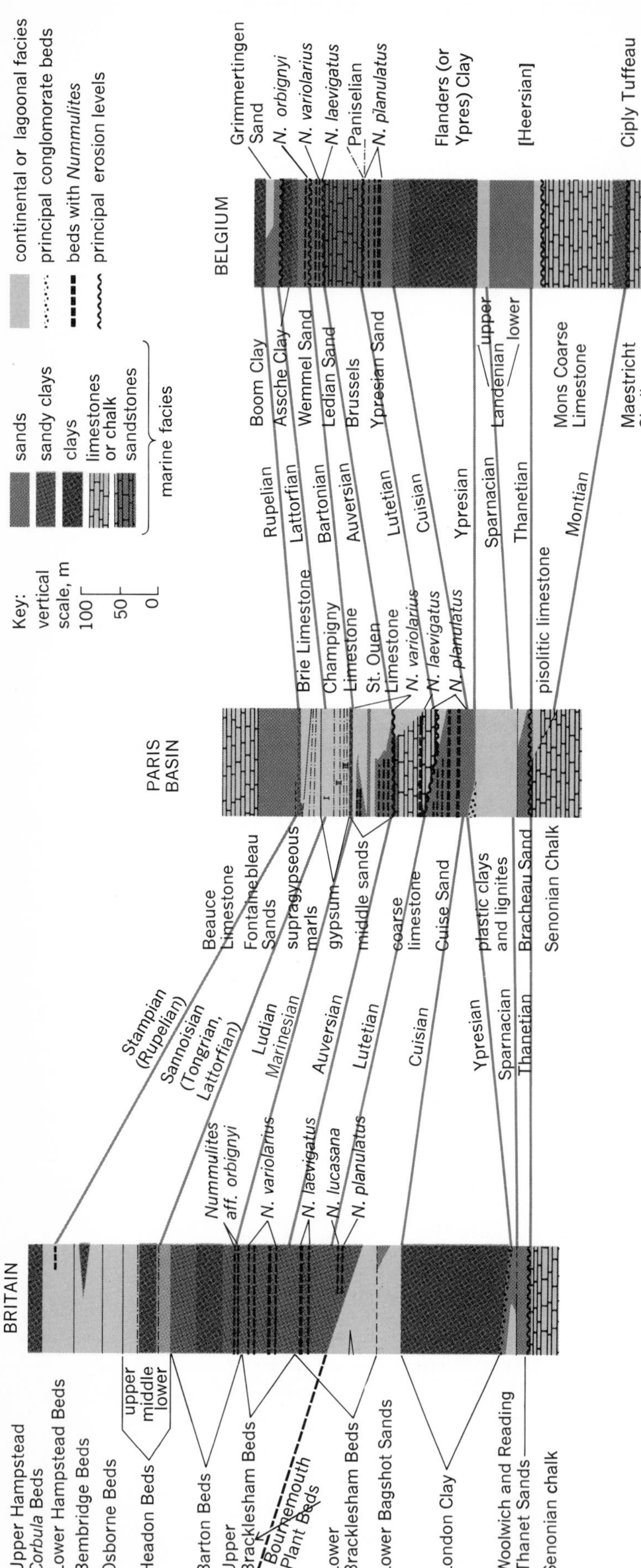

Fig. 2 (*at left*). Correlation of Paleocene (bottom of diagram), Eocene, and Oligocene (top of diagram) stages in Britain, the Paris Basin, and Belgium. (*After A. Wrigley and A. G. Davis, modified from M. Gignoux, Stratigraphic Geology, copyright © 1955 by W. H. Freeman and Co.*)

turbances modified somewhat the previous low-relief surfaces; breaks in the depositional record do not indicate intensive folding, but in many areas diastrophism is reflected by a coarsening in texture of the sedimentary deposits (including conspicuous reworked white sandstones in southeastern and western United States). Increasing tectonic instability, local hiatus (lower Indus), and increasing surface relief characterized late Eocene; sedimentary rock became more diversified laterally and, in some areas (Paris Basin, California), a widespread three-fold upper Eocene sequence resulted: strata of shallow-water origin below and above, and of deeper-water origin between.

Life. Eocene life was essentially an amplification of that of the Paleocene. Birds diversified and the modern families appeared. True bony fish and marine siphonate gastropods flourished. The pelecypod *Venericardia planicosta* characterized the world's last tropicopolitan faunas of bottom-living shellfish. Echinoids thrived. Nummulites multiplied and diversified, and the worldwide discocycline foraminifers made their last stand. Microscopic floating life (siliceous diatoms, calcareous coccoliths, and foraminifers) flowered in the widespread open oceans.

With the poleward expansions of tropical seas, the fragmented Paleocene faunal provinces tended to coalesce; by mid-Eocene time a Tethyan marine shellfish fauna had incorporated that of the previously Atlantic early Eocene Gulf of Aquitaine, and so forth. Nevertheless, Old World and New World retained some distinction, and restricted cool-water faunas persisted.

On land nipa palms grew where London is today, and the flora there was much as in Japan and Australia. On the North American Pacific Coast palms reached Puget Sound and southern Alaska, whereas today their farthest north is in the latitude of Point Conception; in the humid interior a huge lake deposited the Green River Shale. On the shrunken isolated Eocene land masses, increased competition and endemic animal evolution went hand in hand. Large flightless birds developed locally. Bats, flying lemurs, creodont carnivores, perissodactyls, artiodactyls, notoungulates (mostly South American), and edentates reflected the diversification of primitive placental mammals; uintatheres, rhinoceroslike titanotheres, and other huge grotesque browsers eventually appeared alongside the last multituberculate mammals, and Eohippus (*Hyracotherium*), the multitoed, little "dawn horse," was already present (Fig. 3). Toward the close of the Eocene, the aquatic ancestors of the proboscidians developed in Egypt, and mammals began their trek into the free-swimming life of the ocean: Zeuglodonts (the first whales) and the first sea cows appeared.

Australia, already separated from Paleocene southeastern Asia, was further isolated by the ex-

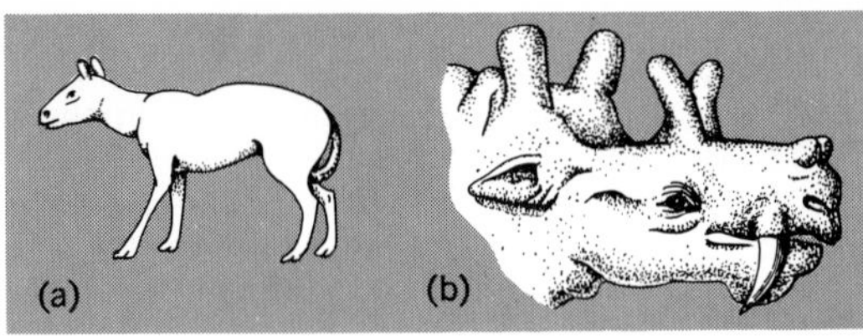

Fig. 3. Eocene life forms. (*a*) Eohippus (*Hyracotherium*), early Eocene. (*b*) *Eobasileus*, one of largest mammals from late Eocene in Rocky Mountain province. (*From R. A. Stirton, Time, Life, and Man, copyright © 1959 by John Wiley & Sons, Inc.; reprinted by permission*)

pansion of the warm-water Eocene seas over much of the East Indies, laying the foundations for the most sharply drawn of all modern terrestrial life zones, Wallace's Line. Eocene marine expansions in Middle America cut off South America from North, isolating its fauna of marsupials, ancestral anteaters, armadillos, ground sloths, oversized rodents, and other now-extinct placentals until toward the end of the Tertiary. *See* PALEOBOTANY; PALEONTOLOGY.

[ROBERT M. KLEINPELL]

Epidiorite

A rock name applied to an altered diorite or gabbro. Gabbroic rocks exposed to mechanical pressure and hydrothermal action easily alter into green rocks. Many greenstones are in reality epidiorites. The plagioclase decomposes and is replaced by an aggregate of epidote and albite (saussuritization) and the pyroxene becomes transformed to hornblende (uralitization). Patches of the original rock may still remain and be recognized here and there. The term epidiorite has been generally used to cover rocks of this category. In such rocks the hornblende may retain parting on the crystallographic plane (001) or on (100) inherited from the pyroxene it has replaced, and the saussuritized plagioclase may retain the idiomorphic, tabular habit and complex twinning (such as Carlsbad-albite and pericline) of the pristine feldspar. Minor recrystallized constituents may be epidote, garnet, sphene, and granular quartz. *See* METAMORPHIC ROCKS.

Epidiorites have a widespread distribution among low-metamorphic schists and phyllites of mountain ranges in Scotland, the Alps, the Appalachians, and the Rocky Mountains.

[T. F. W. BARTH]

Epidosite

A rare metamorphic rock composed of epidote and quartz. It may form by normal metamorphism of calcareous grits, but most epidosites are of metasomatic-hydrothermal origin. They correspond in a way to skarn rocks, but are formed at a lower temperature. *See* METAMORPHIC ROCKS; METASOMATISM; SKARN.

The original material may be basic lavas or sediments. The truly metasomatic epidosites are found in connection with limestones that have reacted with adjacent silicate rocks under the conditions of the epidote-amphibolite facies. Or they form bands, veins, streaks, or nodules in amphibolites, hornblende schists, and masses of epidiorite. Epidosite from such occurrences is sometimes cut and polished and used for ornamental purposes.

[T. F. W. BARTH]

Epidote

A group of minerals having monoclinic symmetry and the ideal composition $Ca_2Fe^{3+}Al_2O(OH)[Si_2O_7]$-$[SiO_4]$. The common rock-forming epidote group minerals are members of a ternary system with trivalent octahedral cations Al, Fe^{3+}, and Mn^{3+} as the primary substitutions. They include zoisite (gemstone variety, tanzanite), clinozoisite, epidote, and piemontite (manganiferous epidote). Substitution of lanthanides for calcium and concomitant replacement of Fe^{3+} by Fe^{2+} occur in the isostructural allanites. The ideal Al end member is dimorphous with a monoclinic form, clinozoisite, and an orthorhombic form, zoisite.

The general crystal structure contains chains of edge-sharing octahedra of two types (zoisite differs by having only the second type): a single chain of M(2) octahedra and a multiple or zigzag chain composed of central M(1) and peripheral M(3) octahedra. The chains are cross-linked by $[Si_2O_7]$ and $[SiO_4]$ groups. Finally, there remains between the chains and cross-links relatively large polyhedra, A(1) and A(2), which contain calcium [except in allanite where it is replaced by some trivalent lanthanides in site A(2)]. Electron microprobe analysis of epidote group minerals indicate $(Fe^{3+} + Mn^{3+} + Cr^{3+} + V^{3+}) \leq 1$ ion per formula unit and $(Fe^{3+} + Mn^{3+}) \leq 1.25$. These ions occur in the largest and most distorted octahedral site, namely M(3), except in the case of piemontite, where some Mn^{3+} is in M(1). The crystal structure refinements demonstrate that these compositional limits are expected.

Epidote crystals are elongated and striated parallel to the twofold symmetry axis (*b* crystallographic axis) and are often twinned on the *bc* plane (illustration *a*). Their color is usually a distinctive

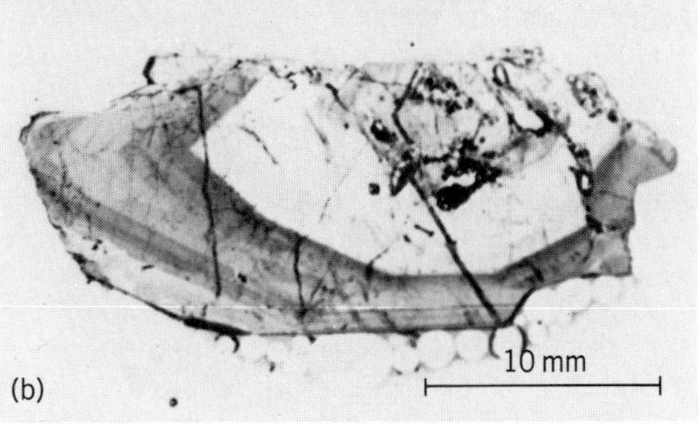

Epidote. (*a*) Crystals from Prince of Wales Island, AK (*Yale University*). (*b*) Oscillatory zoning of these crystals.

yellowish-green (pistachio), but it may be gray, brown, or blackish-green. Hardness is 6–7 and specific gravity is 3.2–3.5. Cleavage is perfect on *c* (001) and imperfect on *a*(100) with an angle of 115°. In petrographic thin section, epidote may have yellow-green pleochroism with spectacular color-zoning common (illustration *b*).

Clinozoisite has a composition $Fe^{3+} \leq 0.45$, due to Al^{3+} replacing Fe^{3+}, and is distinguished by being gray and optically positive in both optic axial and extinction angles. Clinozoisite and epidote form a continuous solid-solution series, as demonstrated by the smooth variation in x-ray diffraction parameters, band shifts in the infrared spectra, and optical properties. Each of these physical properties is correlated to substitution of Fe^{3+} for Al, and allows mean value compositional estimates on as little as 1 mg of sample; however, microprobe analysis is recommended.

Zoisite (orthorhombic) occurs in bladed to needle-shaped crystals elongated parallel to the *b* crystallographic axis. Brightly colored crystals are uncommon except for tanzanite, which is noted for its striking pleochroism and ranges from red-violet, blue, and yellow-green before heating to violet-red and sapphire blue after heating. Cleavage is good parallel to *a*(100) and *c*(001) with an angle of 90°. Zoisite has a composition that is restricted to $Fe^{3+} \leq 0.15$; even so, it exhibits a remarkable variation in optical properties. It occurs in two optically positive orientations, with the optic plane perpendicular to cleavage *a*(100) in the β form, which is very low in Fe^{3+}, and parallel to cleavage in the α form (ferrian zoisite). Both forms have straight extinction and noticeable anomalous interference colors of the first order. Zoisite can exhibit complex lamellar structures caused by an oriented (epitaxial) intergrowth of the α and β forms.

Clinozoisite can be distinguished from zoisite by its definitive x-ray diffraction pattern, infrared spectrum, inclined extinction, cleavage angle, and dispersion of the optic axial angle. Coexisting zoisite and clinozoisite minerals seldom overlap in composition, which suggests an unmixing relation between the two structural types as well as within the zoisite.

Epidote is a common rock-forming mineral in regionally metamorphosed mafic rocks from low- to medium-grade terranes. It is not characteristic of very-low-grade conditions, and in rocks of appropriate bulk compositions it replaces early-formed pumpellyite. Epidote usually occurs with quartz, sodium-rich plagioclase, chlorite, muscovite, and calcium-rich amphibole. Its composition field enlarges as grade of metamorphism increases. It is a common constituent of igneous rocks, crystallizing as a late-stage mineral in cavities or as an alteration of primary calcium-aluminum silicates such as plagioclase feldspar. Zoisite, clinozoisite, and epidote are common products of regional and thermal metamorphism of impure limestone, occurring with such skarn minerals as pyroxene, calcite, wollastonite, idocrase, garnet, and calcium-rich amphibole and often in association with iron ores. In the majority of occurrences of the epidote group minerals, predictable associations are found. *See* SILICATE MINERALS. [GEORGE H. MYER]

Bibliography: W. A. Dollase, *Zeitschrift für Kristallographie*, 938:41–63, 1973; K. Langer and M. Raith, *Amer. Mineral.*, pp. 1249–1258, 1974; G. H. Myer, *Amer. J. Sci.* 264:364–385, 1966; D. Shelley, *Manual of Optical Mineralogy*, pp. 116–119, 1975.

Erosion

The loosening and transporting of rock debris at the Earth's surface, aptly described as the wearing away of the land. Agents of erosion include surface, ground, and ocean water; ice (especially glaciers); wind; gravity; and organisms. *See* WEATHERING PROCESSES.

Erosion removes, on the average, 1 ft (vertically) of rock material in the order of thousands of years. From the time it begins until the lowest possible level has been reached, erosion progresses downward through topographic stages of gradually diminishing slope described by W. M. Davis as youth, maturity, and old age, where each stage is characterized by a distinctive group of landforms. According to an alternate description by W. Penck and L. King, erosion progresses by laterally directed planation and parallel retreat of slopes. *See* FLUVIAL EROSION CYCLE; GEOMORPHOLOGY; ROCK CONTROL (GEOMORPHOLOGY).

Erosion is of great concern to man because it may remove the fertile topsoil, change watercourses and landforms, and cause damage to valuable man-made structures. Erosion has removed valuable ore deposits and rendered some land uninhabitable, but has also stripped off worthless overburden, making some mineral deposits available, and smoothed wide areas, making land suitable for agricultural purposes. Quickening of the pace of erosion (accelerated erosion), brought about by man, has produced landforms and other abnormal conditions that are detrimental to the productivity of the land. *See* SOIL.

[WALTER D. KELLER]

Escarpment

A cliff or steep slope of some extent, generally the margin of a plateau or the steep face of an asymmetrical ridge. Escarpments, or scarps, may be produced by faulting, erosion, or sapping of less-resistant underlying strata to create cliffy rock faces in massive layers above, as on the walls of the Grand Canyon of the Colorado River. Spectacular waterfalls may plunge over scarps, as along the glacially steepened sides of Yosemite Valley, where jointing of granitic rock determines the locations of cliffs. The north face of the San Jacinto Range in southern California is the steepest and highest fault scarp in the United States. *See* FAULT AND FAULT STRUCTURES.

More commonly scarps are lower and result from differential rates of weathering and erosion of contrasted rock types. The denudation of inclined strata may produce cuestas (cuesta escarpments) or hogbacks with dip slopes just as steep as the escarpment. Both have backslopes which approximate the dip of their sedimentary layers and steeper scarps which truncate the bedding. Niagara Falls is supported by a cuesta of erosionally resistant limestone which crops out for several hundred miles along an arc from New York through Ontario, and which eventually swings southward through Wisconsin into Illinois. Low scarps may occur at the boundary between older, more resistant bedrock and younger, less resistant strata. A hidden scarp east of Ozark Plateau, which lies buried beneath alluvial fill, once attained heights of as much as 80 ft. Many coastal plains exhibit low cuestas with scarps directed inland for the reason that their sedimentary strata dip seaward. Some are called wolds in the South because they were originally densely wooded, as was the English weald, or the Schwarzwald of Germany. *See* COASTAL PLAIN; FALL LINE.

[RICHARD J. RUSSELL]

Esker

A sinuous ridge of constructional form, consisting of glacial sand and gravel. Most eskers are 3–50 m high, 10–200 m wide, and 100–500 km long. Side slopes are steep, approaching the angle of rest of the sediments (see illustration). Eskers may go up and down over hills through a vertical range of as much as 250 m; the trends of eskers parallel the regional directions of flow of former ice sheets. Most eskers are the deposits made by streams flowing through tunnels in or near the base of an ice sheet, late in its history when the ice had become nearly motionless. Regions in which numerous large eskers are present include southern Sweden, Finland, eastern Maine, and Ungava. *See* GLACIATED TERRAIN. [RICHARD F. FLINT]

View along the crest of an esker in southeastern Wisconsin. (*W. C. Alden, USGS*)

Evaporite, saline

A sedimentary deposit of soluble salts resulting from the evaporation of a standing body of water. Quantitatively the most important evaporites are anhydrite, $CaSO_4$; gypsum, $CaSO_4 \cdot 2H_2O$; and halite (rock salt), NaCl. Other evaporites, of much more limited distribution and volume but of economic significance, include potassium chlorides, sodium carbonates, borates, and nitrates. *See* ANHYDRITE; GYPSUM; ROCK SALT.

Evaporites, being soluble, are rarely exposed at the surface except in arid regions. By far the greater part of the data on evaporites are derived from deep borings, chiefly those drilled in the search for oil and gas, and from underground workings developed to exploit economically valuable deposits. Gypsum, the mineral used in plaster and cement manufacture, is common near the surface but is invariably replaced by anhydrite at greater depths, the latter mineral being vastly more abundant. Other evaporite minerals are not related to depth of burial, except for the effects of near-surface solution.

Anhydrite and halite, the dominant evaporite minerals, occur in bedded deposits ranging from thin laminae to massive beds several tens of feet in thickness. These may be present in vertical succession, separated by partings of shale or carbonates, to make up aggregate thicknesses of several hundreds, or even thousands, of feet. Such large accumulations are commonly interbedded with other strata bearing marine fossils which clearly demonstrate a relationship between major evaporite deposits and marine waters. Typically, the evaporites are found in repeated cycles which approximate this vertical order from base upward: (1) marine fossil-bearing limestone; (2) dolomitized marine limestone; (3) fine-grained, finely laminated, unfossiliferous dolomite; (4) anhydrite; and (5) halite. Many cycles lack the halite member, while in a few localities a sixth member bearing potash minerals is present.

Marine basin evaporites. The geographic distribution of major evaporite deposits and the thickness and character of the other rocks with which they are associated strongly indicate a relationship between evaporites and marine sedimentary basins which tended to subside during deposition of the sediments. It is estimated that more than 95% of the volume of known evaporite deposits occur in such sedimentary basins. Two distinct patterns of

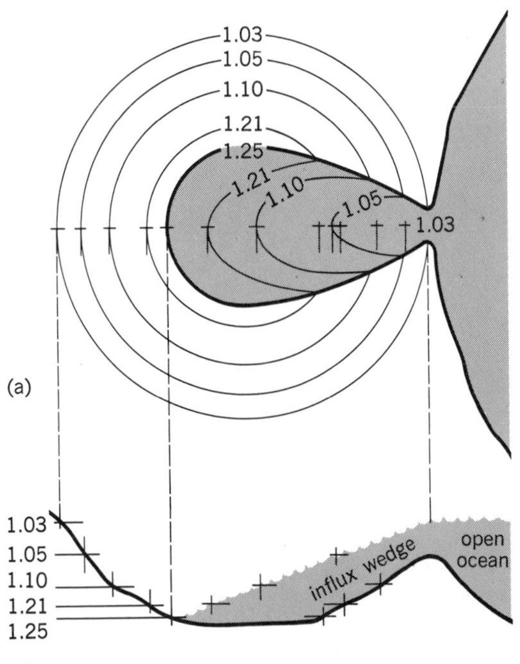

Fig. 1. Geometry of model evaporite basin, showing lines of equal brine density at stage of saline equilibrium. (*a*) Plan view. (*b*) Cross section. (*From L. I. Briggs, Evaporite facies, J. Sediment. Petrol., 28(1):46–56, 1958*)

basinal evaporite occurrence are noted: basin-center evaporites occupying the interior of sedimentary basins; and basin-margin evaporites localized in a ring at the margin of the basins.

The basin-center occurrences commonly exhibit a concentric pattern in plan view, duplicating in a lateral fashion the vertical succession of the evaporite cycle, with marine carbonates (limestone or dolomite) at the outside and halite at the center. The peripheral carbonate belt in many instances includes elongate trends developed parallel to the margin of the basin and forming an encircling barrier. Such a barrier may be either of two types: apparently wave-resistant trends of coral-algal reefs, or bar- or banklike trends of shell fragments, ooliths, and carbonate pellets which appear to have been built by wave action in shoaling water. Basins with extreme subsiding tendencies during evaporite deposition appear to have the greatest possibilities of containing potash salts. These are found with halite along the axis of maximum subsidence.

Basin-margin evaporites, usually anhydrite without large volumes of halite, occupy zones of varying width at the edge of a sedimentary basin. At greater distances from the basin, these evaporites are typically intercalated with, and pass into, red shales and siltstones. Toward the basin they tend to end relatively abruptly against trends of limestone or dolomite which may form an inner encirclement of the basin in the form of reefs, bars, or banks, as noted above. The interiors of such basins, within the encircling reefs, bars or banks, are filled with marine limestones and shales, commonly dark in color and bearing a fossil fauna lacking the remains of bottom-dwelling organisms.

Theory of origin. Both major types of evaporite occurrence give evidence of forming in partially isolated bodies of water separated from the open sea by barriers established by organic growths (reefs) or by wave action (banks and bars) which restrict and confine the circulation of the waters. In the case of basin-center evaporites, it is thought that evaporation of water in the basin produces an inflow of water from the surrounding or adjacent open sea. The inflowing water, of normal salinity and therefore of low density, flows readily as surface currents through passes in the barrier. Dense, highly saline water produced through evaporation sinks to the bottom and is unable to pass back through the barrier to mix with normal sea water (Fig. 1). Thus sea water can continue to enter; but heavy brines are confined in the basin, where they collect until the saturation points, first of the sulphates, then of the chlorides, are reached and precipitation of the salts takes place.

Basin-margin evaporites appear to represent a reversal of this pattern of openly circulating and restricted waters. In the basin-margin case the waters of the basin interior are in open communication with the sea and are of normal salinity. At the edge of the basin, confined between barriers of reefs, banks, or bars, and a surrounding land area, marginal lagoons form evaporating pans. Here water passing through the barrier from the basin is confined, and evaporation leads to the precipitation of salts, as noted before. It is apparent that basin-margin evaporites require an arid climate in the surrounding land area, since large streams of fresh water flowing into the lagoons would cancel the effects of evaporation.

Shelf evaporites. Although basinal evaporites include by far the greatest volume of these salts, there are extensive deposits which show no relationship to subsiding sedimentary basins. The lateral continuity in thickness and character of relatively thin evaporite strata and the other bedded sediments with which they occur indicates deposition in, or at the margin of, shallow shelf seas covering broad areas of slow, uniform subsidence. Shelf evaporites, because of their wide distribution and the shallow depths at which they may be encountered, include many of the commercially exploited deposits of gypsum.

Two types of shelf evaporite occurrence are recognized; both are known as extensions of basin-margin deposits and also occur at great distances from basinal influences. One type is characterized by repeated thin cycles involving interbedded marine carbonates and evaporites. The second type is intercalated with red silts and shales which may contain remains of land animals and plants. The carbonate association is believed to represent deposition in shallow, ephemeral, evaporating pans at a distance from land areas or adjoining land areas of very low relief. The redbed type is interpreted as an alternation of land-derived detritus and the deposits of a shallow bordering shelf sea. *See* REDBEDS.

Nonmarine evaporites. Deposits suggesting precipitation from nonmarine waters are extremely rare in ancient rocks. However, present-day lakes, many of them temporary or seasonal, in areas of arid or semiarid climate are sites of evaporite deposition.

Streams in areas such as these carry the soluble products of rock weathering to closed depressions occupied by ephemeral bodies of water, where the salts are concentrated by evaporation. Nonmarine

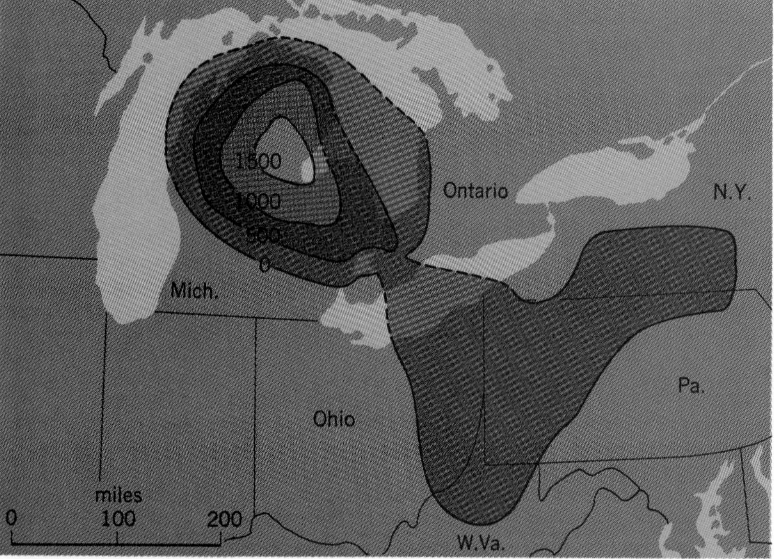

Fig. 2. Upper Silurian salt deposits, as shown by drilling. The salt deposits are known to occur throughout most of the southern peninsula of Michigan, with thickness ranging upward to 1800 ft. Silurian salt beds aggregating 325 ft or less are distributed as indicated by the map. (*Data from USGS and from Fettke, Martens, Pepper, and Alling, Baltimore and Ohio Railway Co.*)

evaporites include locally important halite deposits and the major occurrences of borates, nitrates, and sodium carbonate.

Geologic distribution. In North America Cambrian evaporites appear to be confined to the arctic areas of Canada. Ordovician strata are notably poor in thick and extensive evaporite deposits. Strata of all subsequent geologic ages include major accumulations in one or more areas of North America. Certain sedimentary basins of long-continued subsiding tendencies contain thick evaporites, representing several successive periods of geologic time (Fig. 2). Examples are the Michigan Basin (Silurian, Devonian, and Mississippian) and the Gulf Coast Basin (Permian ?, Jurassic, Cretaceous, and Tertiary). [L. L. SLOSS]

Bibliography: H. Borchert and R. C. Muir, *Salt Deposits: The Origin, Metamorphism and Deformation of Evaporites*, 1964; W. C. Krumbein, Occurrence and lithologic associations of evaporites in the United States, *J. Sediment. Petrol.*, 21(2): 63–81, 1951; P. C. Scruton, Deposition of evaporites, *Bull. Amer. Ass. Petrol. Geol.*, 37(11):2498–2512, 1953; L. L. Sloss, The significance of evaporites, *J. Sediment. Petrol.*, 23(3):143–161, 1953.

Evolution, organic

The processes of change in organisms by which descendants come to differ from their ancestors, and a history of the sequence of such changes. Formerly regarded as expressing simply a theory, organic evolution is now an integral part of any modern biological synthesis, as a "fact" in the main body of science. Discussion centers not on the validity of the theory but on the factors that determine the routes of evolution and on the detailed routes that evolution has followed.

On biochemical and biological evidence, evolution is all-inclusive, at least of cellular organisms. However diverse they may be, cellular organisms have a fundamental structure, the nucleated cell, that may imply a relationship backward in time to a common ancestor. Nevertheless, many links in the ancestor-descendant chain are lacking in most groups of organisms. The evidence of organic affinity consequently has to be sought in many sources, from which the course of evolution can in great part be reconstructed. The factors of evolution also prove to be multiple and varying in their influence. A present-day view of evolutionary process is synthetic in embracing both inorganic and organic components, of which an environmentally determined natural selection is only one factor. *See* PALEOBIOCHEMISTRY.

The problem of the origin of living organisms at the initiation of the evolutionary process is mainly biochemical. Primeval air and water can only be surmised as containing carbon, nitrogen, and hydrogen compounds responsive to electrical and radioactive discharge in ways that encouraged the production of such compounds as amino acids, hydrocarbons, and lipids—compounds that by combination and polymerization gave rise to macromolecules, including proteins and nucleic acids ultimately capable of catalytic ("instructional") self-assembly into still more elaborate structures. Conceptually (but with some basis in laboratory experiment), the association of colloids was congregated into coacervate droplets that became link-grouped in protocaryote nonnucleated unicells, each enclosed with a lipid- or protein-based cell wall. Fission or aggregation then produced multicellular strings like strings of bacterial growth. In support of the general hypothesis, *Eobacterium*, an apparent protocaryote organism, has been recognized in South African rocks of some 3.2 to 3.1×10^9 years ago—the oldest known relics of life yet discovered.

The nucleated eucaryote cell of more complex organisms, surmised to have arisen by a concentration of complex compounds (some of them to become chromosomes) in the nucleus of the cell unit within the cytoplasmic envelope and controlled in their growth mainly by DNA, was a much later development: recognizable bacterial and primitive plant growths were the only kinds of organisms for over 2×10^9 years, if the negative evidence of rocks is to be accepted; and the time and the process of evolutionary divergence of animal from plant protists is unknown. There is an enormous gap in the fossil record: the first animal fossils, in many and elaborately diversified kinds, suddenly and cryptically appeared in the rocks about 600×10^6 years ago.

Fossil series. The most direct evidence of evolution (as change with descent) is provided by fossil series. When difficulties arising from conditions of fossilization are overcome or allowed for—organisms without skeletons are rarely known as fossils—fossiliferous rocks of successive geological ages prove to be characterized by different kinds of fossils whose differences increase with age differences among the rocks. *See* FOSSIL.

In groups locally preserved as fossils in exceptional abundance (for example, oysters, horses, and sea urchins), linear series of specimens, collected sometimes inch by inch from successive strata, progressively younger in upward sequence, show gradual changes in skeletal form from ancestral to descendant types that are a visible witness to the reality of evolutionary change. Collecting may be so complete as to make the series virtually continuous in the near identity of one specimen to the next, and yet the end members of the series may be so different from those early in the series as to be placed in different species, genera, or even families. When fossils are scarce, there may be gaps in the graded series, but the generalized evidence is of the same relevance; thus, while the stages of evolution of present-day one-toed grazing horses from four-toed browsing ancestors of 50×10^6 years ago are well known, the stages of evolution of the amphibians from the fishes and of apes and men from more primitive primates are known only in broader, but only to a degree less convincing, terms. *See* MACROEVOLUTION.

Divergence. In further analytical refinement, linear fossil series (lineages), graded in evolutionary change, often radiate by divergence from a common ancestral form. Thus three-toed horses, occupants of woodland, were contemporaries in their slow evolution with the progressive one-toed grassland forms. This divergence from four-toed ancestors reflects a divergent adaptation to different kinds of environment. In the wider range of divergent (branching or cladogenic) adaptation, the many groups of mammals appear to have had a common ancestry in the therapsid reptiles of

Permian times. A corollary of adaptive divergence is the chance that environments may sooner or later become discouraging or inhospitable (as by the onset of an ice age). Some or all of the divergent evolutionary lines then are in danger of extinction, as illustrated by the disappearance of the once-abundant trilobites, graptolites, ammonites, dinosaurs, and pterodactyls.

Population clusters. In yet further refinement, the individual specimens of the fossil series, graded in geological time, are merely accidental samples of their contemporary populations (demes), which were to be counted in hundreds, thousands, or millions of individuals. It is not, however, the individual that forms the link in the evolutionary series, but the population it happens to represent. At any geological moment the population is an interbreeding association of variants statistically clustered about a mode of optimum biometric form. Evolution is thus more appropriately expressed as a progressive shift in the norms of population cluster, reflected in the survival of some kinds of variants and the extinction of others, until a later population may span a variation field not overlapping anywhere the variation field of an ancestral population.

Contemporary demes. Selection pressure, a term used to cover the many factors that influence the variant composition of a surviving population, is not likely to be uniform in any but the most narrowly contained deme. The external environment, one of the arbiters of selection and survival, changes over even short distances—from field to wood, from cliff to beach, from sandstone to granite substrate—the selective bias changing accordingly.

Graded topocline. Any kind of organism having a wide range, or ranging into different kinds of environment, is thus subjected to differential selection among its members. If there is no tendency through extensive and constant migration to counteract the local selective bias, a divergence between local races may become sufficient to warrant the creation of varietal names, sometimes specific names, as the signs of a graded topocline. Thus the great titmouse (*Parus*) has a virtually continuous range from Europe through southern Asia into China, with scarcely distinguishable merging demes along its length. The many factors of the accompanying transition in environmental circumstances are reflected, however, in a steady change in color and other characters that allows *Parus major* of Europe to be readily distinguished from *P. minor* of China. Among the distinguishing characters is sexual incompatibility, *P. major* reluctantly interbreeding with *P. minor* to produce fertile offspring, the two forms thus recognized by their behavior to be distinct biological species.

Geographic isolates. The chances of evolutionary diversification are increased when the continuous range of a widespread species is fragmented by partial extinction in intermediate ground, physical or ecological barriers, or changed migration habits. The graded topocline is then broken, and remnant populations, breeding in their isolation, are protected from gene flow along the cline and evolve along wholly independent routes often to highly differentiated extremes. Such geographical variation is almost universal. Initial stages of the process are illustrated by the varieties and subspecies of birds of paradise, each confined to its own mountain core in New Guinea; by the species of land snail, each in its own deep-cut valley in Oahu; or by the subspecies of kob, one on the right bank and the other on the left bank of the Nile. Darwin's finches in the stepped cline of the Galapagos are instances of the peculiar kinds of organisms that evolve in the isolation of oceanic islands; conversely, the Lake Baikal seal and the cichlids of Lake Tanganyika have evolved in similar independence in the confines of their water traps. On a world scale, the camels of Arabia and the llamas of Patagonia are the last survivors of a family that once ranged over most of the Americas and of the Old World; the dipnoan lungfishes, found only in the rivers of Australia, Africa, and South America, fall into different genera as the representatives, nearing extinction, of an ancient and widely dispersed family that at one time swam the seas.

Hybridization. Occasionally a new form occurs when varieties of a species, differentiated by distance but not yet sexually isolated, meet in their migrations and produce hybrid offspring. Man himself tended for many thousands of years to radiate outward in the Old World and to become diversified, notably in pigmentation. In recent centuries, however, the differentiation has been partly undone by cross migration and hybridization.

Topocline-chronocline complementation. The sequential variant populations of a fossil chronocline are thus precisely matched by the replacement populations of a geographical topocline. Both kinds reflect the influence of a sustained selective bias. In complementing each other, they combine the dimensions of time and space to provide a three-dimensional framework that incorporates both orthogenic and cladogenic processes in evolution.

Genetic evolution. Evolutionary possibilities reside in the universal occurrence of variation.

Phenotypic regulation. Variation in most animals and plants is mainly the product of diversity of genotypes due to different allelic combinations, chromosomal interchange, inversion, mutation, recombination, and polyploidy. Almost all individuals, except identical twins, are unique in the structure of their genotype. The physique of the individual organism as it grows (phenotype) is thus a reaction system whose factors are the genotype as the organizing instrument and the permissive external environment that allows growth to proceed within the limits of genotypic functioning. Moreover, the genotype is itself the "internal" environment of the individual gene, the information system that instructs the gene in its operations and that accommodates (selects) only those mutant genes which can function in harmony with their neighbors.

Demic gene balance. In a large deme there is thus an enormous number of genotypes with their corresponding phenotypes, not all of which find in their environment precisely the same encouragement to reach reproductive maturity. While many individuals die randomly "accidental" deaths before maturity, there is a statistical probability that some genotypes are environmentally preferred to others—and are thus more likely to survive—in the adult population, whose genic balance, in the proportions of different mutants in the demic gene pool, therefore shows systematic change through a sequence of generations linked by ancestor-descendant relationships. This changing balance in time is evolution.

Differential selection. Evolution as a product of competition between organisms or as adaptation to environment thus requires subtle definition. Competition between one genotype and another is measured statistically by the proportionate transfer of genes from one generation to the next. Natural selection of successful individuals is then expressed statistically as the differential selection of some genes in preference to others.

The differential, the effects of which are seen in the chronocline and topocline, defines selection pressure. When the environment is tolerant or static, a successful phenotypic form may survive over long geological periods, for example, the slowly evolving (bradytelic) brachiopod *Lingula*, which has survived for some 400×10^6 years—although in fossils there are, of course, no means of verifying whether constancy of phenotype also means constancy of genotype. When the environment changes and exerts a directed selection pressure (either actively, in geological time, or passively, as organisms expand their geographical range by migration), the preferred genotypes change correspondingly. Differential selection may then result in rapid (tachytelic) evolution, as in *Homo* which has evolved in Quaternary times at the rate of a new species perhaps every 100,000 years.

Similarly adaptation is obviously functional in the sense that every organism which survives must be able to live in the environment in which it finds itself. An evaluation of adaptation by relative survival designates the preferred genotypic pattern to be better adapted than any other patterns, which, conversely, in being selected against, are therefore ill-adapted—as is proved by their ultimate extinction. Consequently, it is not helpful to seek a teleological explanation for the origin or emergence of adaptive characters when they are defined as adaptive because they have emerged. Moreover, a genotype may be repeatedly discouraged by a continuing hostile environment, but when the environment changes it may find the opportunity of becoming adapted by being selectively approved. The genotype is thus preadapted when the circumstance arises.

Comparative morphology and taxonomy. At any moment in geological time, stocks revealed in cladogenic fossil sequences are represented by forms on the one hand comparable in being descended from a common ancestry and on the other hand different in being divergent. To a biologist comparing the divergent forms found in living organisms of present geological time, the similarity-with-a-difference manifest between tiger and lion, among Galapagos finches, or between primrose and cowslip is a sign of kinship. This kinship is expressed in taxonomic classification, the several kinds being placed in a single genus to indicate the unity of basic structural plan but in different species to indicate modified forms of the plan. Conversely, when fossil lineages are unknown and the cladogenic stages cannot be demonstrated, the significance of morphological similarities described in a hierarchical taxonomy resides, insofar as it can have any claim to be natural, in inferred phylogenetic sequence. C. Linnaeus, 100 years before Charles Darwin, unwittingly embodied the phylogenetic principle when he invented his classification; T. H. Huxley, in justification of Darwin, expressly invoked the principle when he identified man's place in nature by a morphological comparison of man with the great apes. When it has claims to a natural objectivity, a taxonomical hierarchy professes at each stage to define grades of cladogenic proliferation: The horse and the ass are species falling into the genus *Equus*; the genera *Equus* and (three-toed) *Pliohippus* fall into the family of equids; the equids and the rhinoceratids are families falling into the order of the (odd-toed) perissodactyls; the perissodactyls and the (cloven-hoofed) artiodactyls are orders falling into the class of mammals; and the mammals and the reptiles fall into the phylum of the vertebrates.

Indirect evidence of evolution, revealed in morphological comparison, is provided by obsolescent, vestigial, or infantile characters. Although the relative speeds of development (the allometric growth) of different organs or characters reflect the permutative instructions of changing genotypes, a persistent element in a genotypic frame may continue to promote the development of a character, even when the character no longer retains the ancestral function or has no function at all. Thus the fetal teeth of the platypus, the deeply buried hindlimbs of the whale, and the embryonic "gill" slits of man are signs in ontogeny of remotely functional stages in phylogeny (or would be inexplicable unless they were).

Evidence and process. Evolution is demonstrated in fossil series, in the geographical variation and distribution of organisms, in degrees of unity of plan between one organism and another, in the fine adaptation of organisms to an inconstant environment, and in the stages of individual growth. It is brought about by recurrent changes in the genotype due to mutation and recombination, and by differential environmental selection measured by relative survival. *See* ANIMAL EVOLUTION; PALEOBOTANY; PALEONTOLOGY. [T. NEVILLE GEORGE]

Bibliography: E. H. Colbert, *Evolution of the Vertebrates*, 1969; S. W. Fox and K. Dose, *Molecular Evolution and the Origin of Life*, 1971; J. Huxley, *Evolution: The Modern Synthesis*, 1974; A. Mayr, *Populations, Species, and Evolution*, 1970; G. G. Simpson, *The Major Features of Evolution*, 1953; B. J. Stahl, *Vertebrate History*, 1974; J. W. Valentine, *Evolutionary Palaeoecology of the Marine Biosphere*, 1973; G. C. Williams, *Adaptation and Natural Selection*, 1966.

Exfoliation, rock

The splitting off of concentric thin sheets or shells from the surface of massive rocks, a common weathering process in moist climates. Both above and beneath the ground surface, hydration of complex silicate minerals through the action of percolating water carrying minute amounts of carbon dioxide causes peripheral expansion in rocks. The process is particularly effective at edges and corners of joint blocks where penetration can be from several directions. Separation of successive curved shells tends to produce increasingly rounded forms. Chemical exfoliation is distinct from the spalling of rocks by fire or sheeting of rocks in response to release of load. *See* WEATHERING PROCESSES. [C. F. STEWART SHARPE]

Extinction (biology)

The death and disappearance throughout the world of diverse groups of organisms under circumstances that suggest common or related causes. The time span involved in mass extinctions varies greatly, but commonly it is short compared with evolutionary history. It may, however, seem to be long as compared with human records, and there is no scientific evidence that such mass extinctions were cataclysmic or involved extraordinary processes.

One of the greatest of all mass extinctions started a few hundred years ago and is still in progress. Approximately 500 species of animals and many plants have disappeared in that time or may soon become extinct as the result of ecological disturbances caused by man. This revolution in the organic world is expected to become climactic within a few decades because of the rapid spread of human population.

Evidence from fossils. Mass extinctions of past eras are recorded in hundreds of successive assemblage zones of fossil animals and plants, many of which contain widespread and abundant species and genera that first appear together at one level and disappear abruptly at a higher stratigraphic level without any, or only a few, known descendants. More than 160 years ago Georges Cuvier recognized revolutionary, apparently sudden, changes in the fossil content of successive strata in the Paris Basin. Without knowledge of the great span of geologic time, he had to attribute these revolutions to "catastrophes." He did, however, correctly explain the repopulations as owing to migrations from distant regions not affected by the environmental events.

The most critical episodes of faunal extinction are recorded at the close of the Cambrian (52% of all families of the time), Devonian (30% of families), Permian (50% of families), Triassic (35% of families), and Cretaceous (26% of families) (Fig. 1). These, and many lesser episodes of mass extinction, are used by biostratigraphers as a basis for delimiting many major and minor stratigraphic divisions.

Patterns. Mass extinctions are of two sorts. They may take place as a result of competition between native and immigrant faunas, or they may stem from external physical changes in environment in which competition is not a major factor. In the former case, the immediate cause of extinction stems from biological stresses between better- and poorer-adapted species that are using the same environmental facilities. Portions of an invading fauna replace the native animals in a sort of relay.

Within historical times there have been many well-documented examples of such competition and replacement of indigenous species by aggressive elements from other regions. For example, the terrestrial biotas of Australia, Madagascar, the

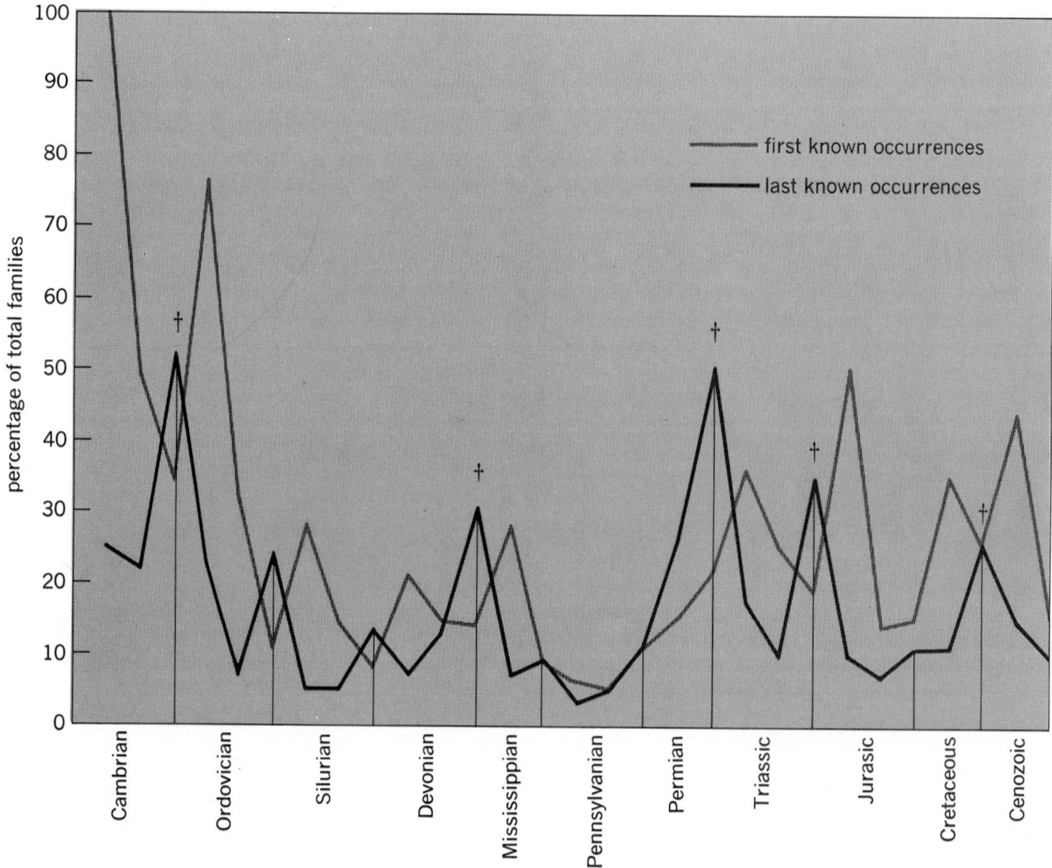

Fig. 1. Episodic fossil record of animals. (*N. D. Newell, Geol. Soc. Amer. Spec. Pap., no. 89, pp. 63–91, 1967*)

Antilles, New Zealand, innumerable smaller islands, and even the major continents have been greatly modified because of competition with species introduced by man and consequent extinctions of native species.

Natural migrations of animals in the past have also resulted in mass extinctions. The Isthmus of Panama rose out of the sea in the Pliocene Epoch, some 6,000,000 or 7,000,000 years ago, forming a land connection between the Americas and an intermingling of faunal elements of the previously separate continents. Many species, especially the marsupial mammals of South America, were decimated by unequal competition.

In a second pattern of mass extinction there is little or no intermingling of old and new faunas. In this case, many niches may be vacated for a time until new animals eventually replace those that had died out before. Slow recovery of benthonic invertebrates in the Triassic and, similarly, the slow radiation of the Paleocene mammals into habitats vacated by the dinosaurs are examples (Fig. 1). Evidently, invasion of a region by a new fauna is not the immediate cause of the elimination of the native species. By implication, physical factors of the environment are responsible.

The Paleocene Epoch (of some 11,000,000 years' duration) elapsed before many of the niches vacated by the vanished dinosaurs were reoccupied by the arising mammals. The pattern of such episodes is rapid extinction followed by an interval of recovery and subsequent replacement by new communities (Fig. 1) that were not in intimate contact with the vanished forms.

Paul S. Martin, of the University of Arizona, believes that the disappearance of many kinds of large herbivorous mammals in the Americas (Fig. 2) about 8000–14,000 years ago, shortly after the advent of man in the Western world, should be attributed to early human hunters.

His hypothesis stresses the fact that man frequently kills animals for motives other than hunger—overkill. Several authorities, however, question the premise that human populations were sufficiently dense and widely distributed or technically so advanced 10,000 years ago as to threaten the existence of widespread abundant animals such as the horses, which then ranged in great numbers over most of the Americas. These persons point to the recently increasing harshness and drying of climate, especially in intermediate latitudes, as a more likely cause of this extinction.

Mechanics. In simple terms, extinction results from an excess of deaths over births over much of the geographic range of a species. The equation $dn/dt = an - bn$ indicates the rate of change in population density through time, where n represents the number of individuals per unit area (that is, population density); t, time interval; a, birth rate; b, death rate. The values of a and b commonly depend on both population density and environmental conditions. When n falls below a critical value, population variability may be significantly reduced, breeding patterns disturbed, and a threshold passed below which the species can no longer cope with the ordinary vicissitudes of the environment, and extinction follows. H. Tappan, and Tappan and A. R. Loeblich have shown from the fossil record that whenever an expiring species

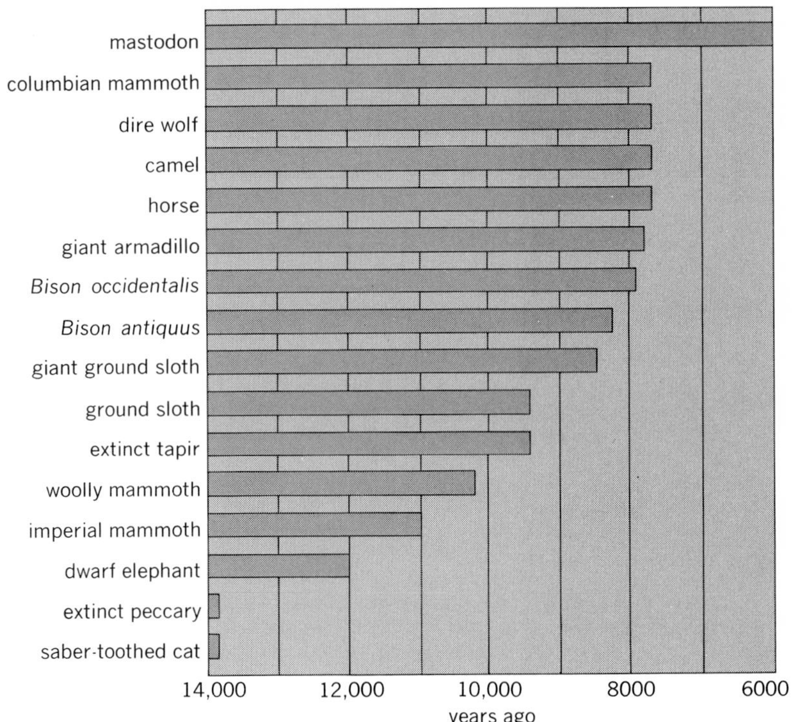

Fig. 2. Recent mass extinction of large mammals in North America. (*K. MacGowan and J. A. Hester, Early Man in the New World, copyright © 1950, 1962 by Kenneth MacGowan; reproduced by permission of Doubleday and Co., Inc.*)

was an important link near the base of the food web, all dependent species were vulnerable. On the other hand, if a population irrupts and the density exceeds a critical level, irreversible damage to the environment may occur, and intraspecific competition increases greatly. These changes may be followed by rapid reduction in population numbers, and extinction. Thus, there seems to be an optimal population density, or climax, which varies according to the carrying capacity of the environment.

Causes. Many ingenious hypotheses, ranging from unsupported speculation to sophisticated scientific arguments, have been advanced to explain past episodes of mass extinction. Commonly, these are specially designed for particular examples, and most are unsatisfactory because they do not suggest significant tests or corollaries needed to assess probabilities for a particular hypothesis of extinction.

Published hypotheses include meteorite storms, diseases, shifts of the Earth's axis or crust, variations in composition of the atmosphere, variations in supply of metallic trace elements, reduction of salinity in the ocean, racial old age, insufficient nutrient salts in the sea, excessive juvenile mortality, and climatic changes. All of these have been advanced as causes of mass extinctions, but mostly they are not supported by compelling evidence.

Ionizing radiation. An idea currently popular and repeated in variant forms since the 1930s is that mass extinctions and subsequent accelerated evolution of faunas might be attributed to bursts of ionizing radiation from the Sun or supernovae. K. Terry and W. Tucker estimated the frequency of such explosions to be about once in 60,000,000 years.

R. Uffen suggested, in 1963, that each of many reversals of polarity of the Earth's magnetic field recorded in remanent magnetism of stratified rocks must have been accompanied by temporary dissipation of the protective Van Allen belt, thus allowing floods of lethal and mutagenic radiation to reach the surface of the Earth and cause mass extinctions of the most sensitive animals. Most of the proponents of this hypothesis have postulated accelerated genetic mutations among the survivors to explain the appearance of mainly new faunas after the major episodes of extinction. *See* PALEOMAGNETICS.

C. Waddington argues that the shielding effect of the Earth's atmosphere is so great that a radiation dose at sea level during a magnetic reversal would be negligible.

Geologic factors. Historical geology has demonstrated that physical environments have been in flux since the origin of the Earth. Of these changes, no single set of factors has influenced the later history of life more than the continuous oscillations in the average level of the sea, and the drifting of the continents, with dependent climatic changes. The resulting disturbances to ecosystems at times must have been very great. During intervals of very low crustal relief, the changes probably were also rapid as compared with organic evolution. A. B. Ronov after measuring the distributions of marine strata on maps has estimated that at times the contenents, after being broadly inundated for long intervals by shallow seas, intermittently became almost completely emergent. N. D. Newell stressed these geographic changes as probable causes of climatic disruptions from maritime to continental conditions over large parts of the world. The extinction events apparently coincided with severe reductions in area of the adaptive zone populated by shallow marine invertebrates. By analogy with present-day island populations, the marine communities suffered widespread extinctions and loss of diversity, since diversity has been shown to be area-dependent. Axelrod and H. P. Bailey had already called attention to deleterious effects of heightened seasonality on breeding patterns of organisms adapted to equable climate.

T. J. M. Schopf and D. S. Simberloff reinforced the area-dependency argument by showing that dwindling diversity of marine organisms in the late Paleozoic apparently bears a simple mathematical relationship to the progressive emergence of the continents. The emergence they attribute to plate tectonics. *See* TECTONIC PATTERNS.

Scientific investigations of these relationships in the future may provide definite conclusions about the causes of mass extinctions. Work to date suggests that a single comprehensive theory cannot explain all mass extinctions, and it is clear that more observational field evidence is needed. *See* EVOLUTION, ORGANIC. [NORMAN D. NEWELL]

Bibliography: M. Buls, *The discovery of America, Science*, 179:969–974, Mar. 9, 1973; N. D. Newell, An outline of history of tropical organic reefs, *Amer. Mus. Novitates*, no. 2465, Sept. 21, 1971; N. D. Newell, *Geol. Soc. Amer. Spec. Pap.*, no. 89, 1967; T. J. M. Schopf, Permo-Triassic extinctions: Relation to sea-floor spreading, *J. Geol.*, 82:129–143, March 1974; D. S. Simberloff, Permo-Triassic extinctions: Effects of area on biotic equilibrium, *J. Geol.*, 82:267–274, March 1974; H. Tappan and A. R. Loeblich, Jr., Geobiologic implications of fossil phytoplankton evolution and space-time distribution, *Geol. Soc. Amer. Spec. Pap.*, no. 127, 1971; K. D. Terry and W. H. Tucker, Biological effects of supernovae, *Science*, 159: 421–423, 1968; C. J. Waddington, Paleomagnetic field reversals and cosmic radiation, *Science*, 158: 913–915, 1967.

Facies

Any observable attribute of rocks, such as overall appearance, composition, or conditions of formation, and changes that may occur in these attributes over a geographic area. The word facies has several meanings in common geological usage. It may refer to bodies of specific rock content or combinations of rock content, such as redbed facies and black shale facies; it may refer to some areally restricted and mappable part of a stratigraphic rock body; or it may be used as a generic term including all specialized or restricted types of facies. *See* STRATIGRAPHY.

The term facies is widely used in connection with sedimentary rock bodies, but is not restricted to these. In sedimentary rocks the changes in facies reflect changes in environments of deposition. Thus, fresh-water environments in rivers change to brackish-water environments in estuaries and to marine conditions in the ocean. In general, facies are not defined for sedimentary rocks by features produced during weathering, metamorphism, or structural disturbance. In metamorphic rocks specifically, however, facies may be identified by the presence of minerals that denote degrees of metamorphic change. *See* PETROLOGY.

Lithofacies. In geological studies the facies most commonly designated are lithofacies, defined as a lateral subdividing of a stratigraphic rock body differentiated from other adjacent subdivisions by its lithologic characteristics. Lithofacies may be separated laterally by vertical cutoff planes that divide the rock unit into sharply differentiated classes or by interfingering surfaces that produce intergrading types. This kind of facies is a mappable subdivision of the rocks.

Mappable lithofacies can be differentiated on any selected lithologic basis. One basis is gross lithologic composition, such as the percentages of sandstone, shale, and limestone, or the ratio of any of these rock types to the others. A second basis may be specific lithologic attributes such as colors of shale or varieties of limestone types. The differentiation may also be based on combinations of rock attributes that designate conditions of sediment deposition, such as terrestrial, marine, or geosynclinal facies. *See* SEDIMENTARY ROCKS; SEDIMENTATION.

Lithofacies maps show the areal distribution of the several differentiated aspects. Such maps find application in oil exploration, where they are used in conjunction with structure-contour maps and maps of the total thickness of the stratigraphic unit (isopach maps). The isopach and structure maps convey information on the overall geometry of the stratigraphic rock body. The isopach map shows its areal extent and thickness variations; the structure map shows the structural attitude of its upper surface. The lithofacies map gives information on the changing composition of the rock body throughout its geographic occurrence. This changing composition may indicate areas favorable for oil and gas exploration, inasmuch as certain combinations of sedimentary rocks tend to occur in known areas of oil production. In combination with structurally high areas, such as domes and anticlines, changes in facies may be important factors in localization of oil or gas. Stratigraphic oil fields, in which the controlling factors are the facies attributes themselves, may sometimes be located on a regional basis by means of lithofacies maps. Similarly, other kinds of mineral deposits associated with sedimentary rock bodies, such as uranium, lead, and zinc ores, commonly bear some relation to the facies patterns of the sedimentary rocks. *See* PETROLEUM GEOLOGY.

Lithofacies studies also shed light on the historical geology of sedimentary rocks, because they provide information which is useful in determining the conditions under which the sediments were laid down, such as shallow or deep marine waters, deltaic conditions, tidal lagoons, or other conditions of deposition. They also provide some indication of the broad features of relative subsidence in the Earth's crust during accumulation of the deposits.

Biofacies. Lithofacies studies contribute much to the knowledge of past environments; however, certain aspects of those environments can be unraveled only by studies of the fossil organisms contained in the sedimentary rock body. Lateral subdivision of a stratigraphic unit differentiated from other adjacent subdivisions by its biological characteristics constitutes a biofacies. Biofacies maps may be based on proportions or ratios among the fossil organisms, much as lithofacies maps are based on lithologic attributes. They are particularly useful in reconstructing the environment of deposition, inasmuch as the organisms commonly reflect conditions of water depth, salinity, temperature, or other conditions. *See* PALEOECOLOGY.

Facies maps. The importance of facies changes in rocks, in the reconstruction of geologic history and in the exploration of ore and mineral deposits, has given rise to a wide variety of methods for preparing facies maps. Most maps are prepared somewhat in the manner of a contour map by drawing lines of equal magnitude through a field of numbers representing the observed values of the measured rock attributes. The basic lithofacies data are obtained by making thickness measurements of the several kinds of rocks penetrated by wells in the mapped area or exposed in outcrops along valley walls. In the simplest kind of facies map (sometimes called an isolith map), the total thickness of a selected rock type in the stratigraphic unit, such as a sandstone, is plotted at each point of observation, and contours of net sand thickness are drawn through the numbers. Such a map shows the absolute quantity of sandstone in the stratigraphic unit, and may prove useful in study of the distribution of possible oil reservoirs in an area of interest. Maps of relative rock content, such as the percentage of sandstone in the unit or of the ratio of sandstone to shale, give other kinds of information about the same rock body. Commonly, several lithofacies maps are prepared for the unit under study, to provide information on more than one lithologic attribute.

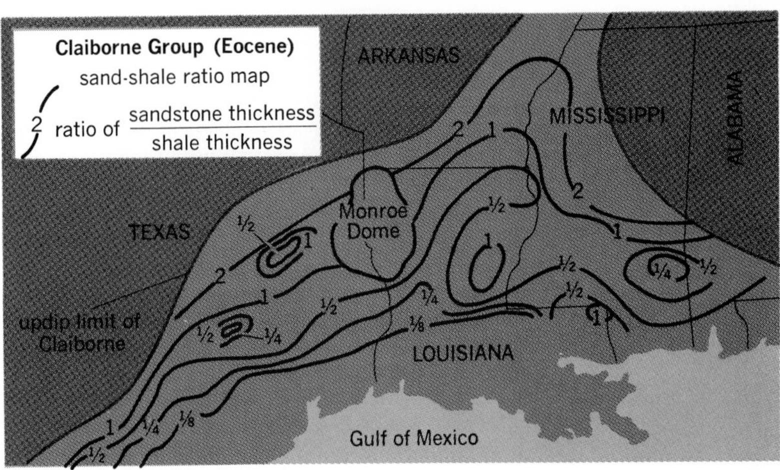

Lithofacies map showing the ratio of two rock components. (*After R. G. Strand*)

The illustration provides a lithofacies map of the Claiborne Group (Eocene) of the Gulf Coast region, showing the ratio of sandstone to shale in that body of rock. The ratios are contoured on a geometric basis and show a rapid decrease to the south. In the north there is more than 2 ft of sandstone per foot of shale, whereas near the southern limit of available data the amount of sandstone per foot of shale is less than 1/8 ft. The map reflects the commonly observed decrease in sand as mixtures of sand and mud are redistributed by waves and currents along ancient shorelines and into deeper water offshore. Irregularities in the subparallel facies lines indicate the influence of local factors, usually variations in water depth or structural growth in the depositional area.

Another application of ratio techniques in facies mapping is to contrast the detrital portions of a sedimentary body (the sand and mud carried into the depositional site by streams or other agencies) with nondetrital sediments, such as limestone, that may be formed directly within the environment by chemical or biological processes. This contrast may be expressed as the clastic ratio, the ratio of the combined thicknesses of sandstone and shale in the unit to the thickness of limestone. In this case the ratio indicates the number of feet of sand and mud deposited per foot of limestone. It is a measure of the relative contribution by materials carried into the environment compared to sediments formed locally. Such combination ratios provide a means for showing the mutual relations among several lithologic components by a single set of contour lines. For an illustration of lithofacies mapping techniques utilizing the clastic ratio *see* SILURIAN. [WILLIAM C. KRUMBEIN]

Bibliography: R. R. Compton, *Manual of Field Geology*, 1962; W. C. Krumbein and L. L. Sloss, *Stratigraphy and Sedimentation*, 1951; L. L. Sloss, E. C. Dapples, and W. C. Krumbein, *Lithofacies Maps*, 1960; J. M. Weller, Stratigraphic facies differentiation and nomenclature, *Bull. Amer. Ass. Petrol. Geol.*, 42:609–639, 1958.

Fall line

The zone or boundary between resistant rocks of older land and weaker strata of plains. Here the easily eroded strata of the plains lap onto the more resistant rocks of the old land (Fig. 1). Streams that cross this boundary have steeper gradients along the width of the zone and make sudden descents over falls and rapids. In geologic and geographic literature, a line connecting such falls on several main rivers is referred to as a fall line.

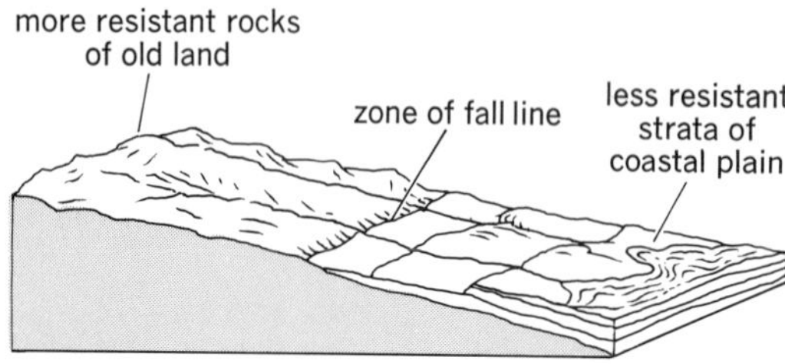

Fig. 1. Diagram showing fall-line zone.

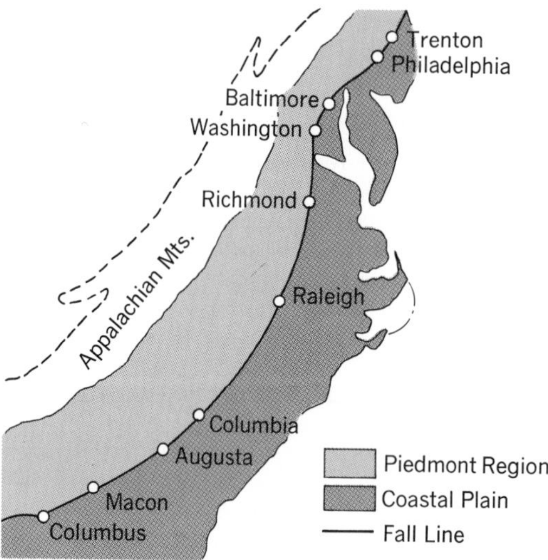

Fig. 2. Fall Line in eastern United States.

In the eastern United States, the boundary between the Coastal Plain and the Piedmont geographic section to its west (Fig. 2) has long been known as the Fall Line, and serves as the origin for the general term. Where such rivers as the Delaware, Potomac, James, Roanoke, and Savannah cross the boundary zone, there are falls or rapids that limit upstream navigation and provide a source of water power. Some geologists believe that this fall line represents the surface trace of an ancient peneplain (old erosion surface) of pre-Cretaceous age (the Fall Line Peneplain) now buried beneath younger strata of the coastal plain. A similar, buried fall line, with falls that once attained heights of 80 ft, lies beneath alluvial fill along the east scarp of the Ozark Plateau. *See* COASTAL PLAIN; FLUVIAL EROSION CYCLE.

[SHELDON JUDSON]

Bibliography: C. B. Hunt, *Physiography of the United States*, 1967.

Fault and fault structures

Fractures in rock along which the adjacent rock surfaces are differentially displaced. Some faults are only a few inches long and have displacements measured in fractions of 1 in.; others are miles long with displacements in thousands of feet.

The trace of a fault on the Earth's surface is a fault line. Its position may be indicated by an escarpment. If the escarpment is the direct result of movement along the fault, it is a fault scarp. On the other hand, if the escarpment is the result of differential erosion of the rocks brought together along the fault, it is a fault-line scarp. In practice the distinction between an escarpment and a scarp may be difficult to make. *See* ESCARPMENT.

The surface of a fault may be sinuous and its inclination may change with depth. Where the surface of the fault is other than vertical, the hanging wall is that face of rock which lies above the fault and the foot wall is that which lies beneath. The displacement is not necessarily confined to a single surface but may be distributed over many anastomosing faults.

Differential movement along a fault generally causes granulation and polishing of the adjacent rocks. Gouge is a claylike aggregate produced by the pulverization of the displaced rock. Along some faults the crushed material contains fragments of various sizes, and in this event it is called fault breccia. Mylonite is a microbreccia which is extremely fine-grained and has considerable coherence. It is characterized by a dark color and a streaked or platy structure. Slickensides are polished and striated surfaces that develop along a fault during its dislocation. *See* MYLONITE.

Nomenclature of displacements. In the absence of precise leveling surveys before and after faulting, the absolute movements on faults cannot be determined. The displacements along faults are generally examined in terms of apparent movement or relative movement.

Apparent movement describes the dislocation of beds or other contacts in a given section (Fig. 1). With reference to apparent movement, separation indicates the apparent displacement of two comparable parts of a bed measured in any indicated direction. The normal horizontal separation, or offset, is the separation measured perpendicularly to the disrupted horizon and in a horizontal plane. The vertical separation is the vertical distance between two comparable parts of the displaced bed. The dip separation is the separation measured in a direction directly down the dip of the

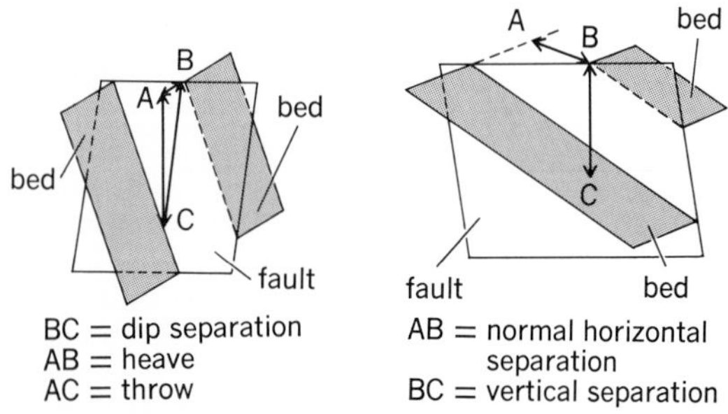

Fig. 1 Disruption of rock bed by fault.

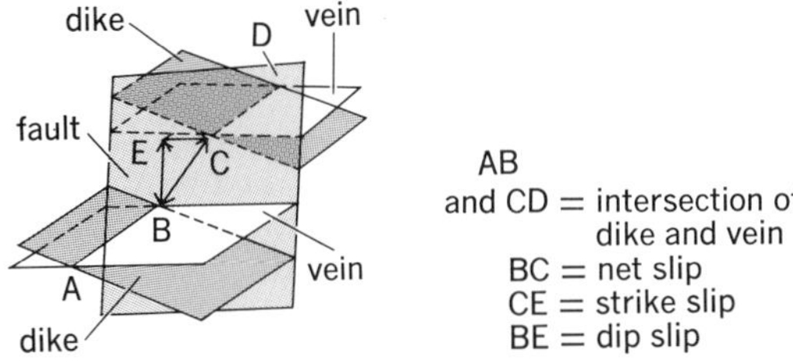

Fig. 2. Displacement of intersection of dike and vein by fault.

fault. The throw is the vertical component of the dip separation, and the heave is the horizontal component of the dip separation, both being measured in a section perpendicular to the strike of the fault. Dip is the angle between the maximum slope of a given surface and the horizontal. Strike is the bearing of any horizontal line along a given surface.

Relative movement may be ascertained by combining information concerning the direction of movement on the fault surface with information concerning the apparent movement. In exceptional cases, the amount of relative movement can be determined. For example, the intersection of a vein and a dike, if cut by a fault, would be broken in such a way that the points where the intersection pierced the fault surface would be displaced (Fig. 2). The distance between these points is the net slip. The strike slip is the component of the net slip parallel to the strike of the fault, and the dip slip is the component of the net slip parallel to the dip of the fault.

Classification. Faults may be classified with reference to either apparent or relative movement. With regard to apparent movements, dislocations in which the effects of disrupted contacts can be accounted for by a downward displacement of the hanging wall with respect to the foot wall are normal faults, and those in which the effects of disrupted contacts can be explained by an upward movement of the hanging wall with respect to the foot wall are reverse faults. If the relative movement is used as the basis of classification, a gravity fault is one along which the hanging wall has moved down relative to the foot wall, and a thrust fault is one along which the hanging wall has moved up relative to the foot wall. Along a strike-slip fault (wrench fault) the movement is essentially parallel to the strike of the fault.

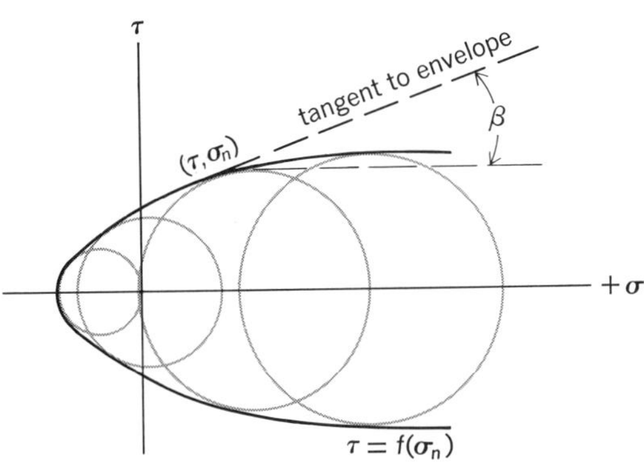

Fig. 3. Mohr's diagram showing the curve $\tau = f(\sigma_n)$ and the angle β between tangent and abscissa.

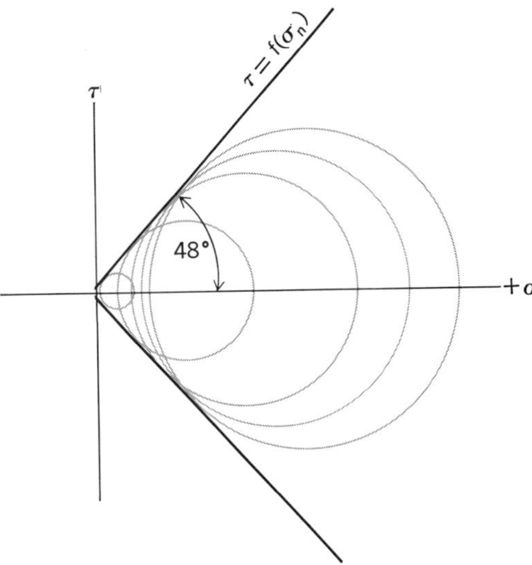

Fig. 4. Mohr's diagram, assuming a linear relationship between τ and σ_n and a stress difference of $\sigma_1 = 7\sigma_3$.

Mechanics of faulting. According to the Mohr criterion of failure, rupture occurs on planes that include the direction of the intermediate principal stress at values of shearing stress τ determined by some function of the normal stress σ_n, as in Eq. (1).

$$\tau = f(\sigma_n) \tag{1}$$

The shearing stress τ and the normal stress σ_n on these planes are given by Eqs. (2) and (3), where σ_1

$$\tau = \frac{\sigma_1 - \sigma_3}{2} \sin 2\alpha \tag{2}$$

$$\sigma_n = \frac{\sigma_1 + \sigma_3}{2} - \frac{\sigma_1 - \sigma_3}{2} \cos 2\alpha \tag{3}$$

is the greatest principal stress, σ_3 is the least principal stress, and α is the angle between the direction of shearing stress and the direction of the greatest principal stress.

The angle α between the direction of greatest principal stress and the plane on which shearing rupture occurs can be calculated by assuming that the value of the shearing stress τ_c necessary to cause rupture is that sufficient to overcome a resistance to the occurrence of shearing fracture. Equation (4) may be written, where R is resistance

$$R = c + \sigma_n \tan \beta \tag{4}$$

to the occurrence of shearing fracture, c is the threshold value of the shearing strength which is a constant for the material, and β is the angle of internal friction. Substituting for σ_n in terms of the principal stresses gives Eq. (5), and the difference

$$R = c + \left(\frac{\sigma_1 + \sigma_3}{2} - \frac{\sigma_1 - \sigma_3}{2} \cos 2\alpha\right) \tan \beta \tag{5}$$

between resistance to shear and the shearing stress is Eq. (6). Rupture occurs on those planes

$$R - \tau_c = c + \left(\frac{\sigma_1 + \sigma_3}{2} - \frac{\sigma_1 - \sigma_3}{2} \cos 2\alpha\right) \tan \beta - \frac{\sigma_1 - \sigma_3}{2} \sin 2\alpha \tag{6}$$

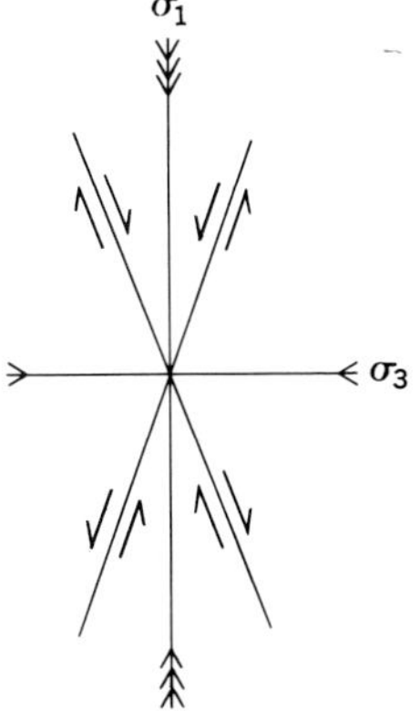

Fig. 5. Displacement on shear planes in relation to directions of principal stresses.

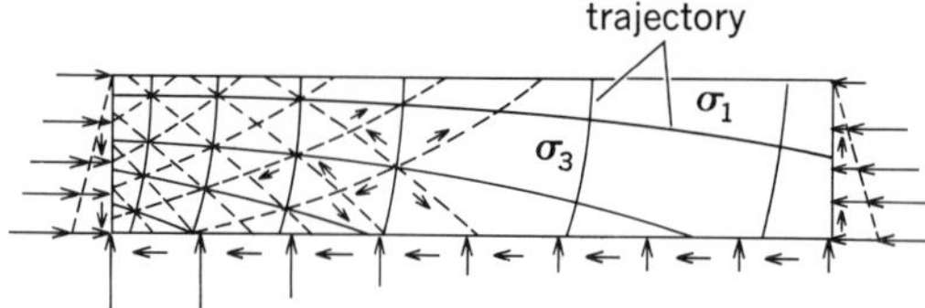

Fig. 6. Directions of principal stresses (solid lines) and of potential fault surfaces (broken lines) compatible with indicated boundary stresses. (*From M. K. Hubbert, Bull. Geol. Soc. Amer., 62:335–372, 1951*)

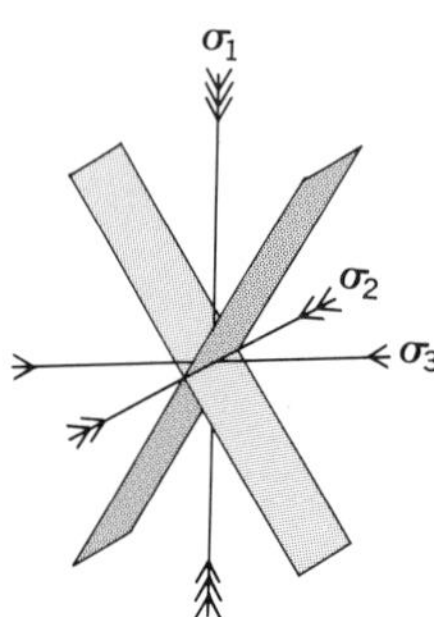

Fig. 7. Orientation of shear planes and principal stresses in normal faulting; σ_1 is nearly vertical. (*After E. M. Anderson*)

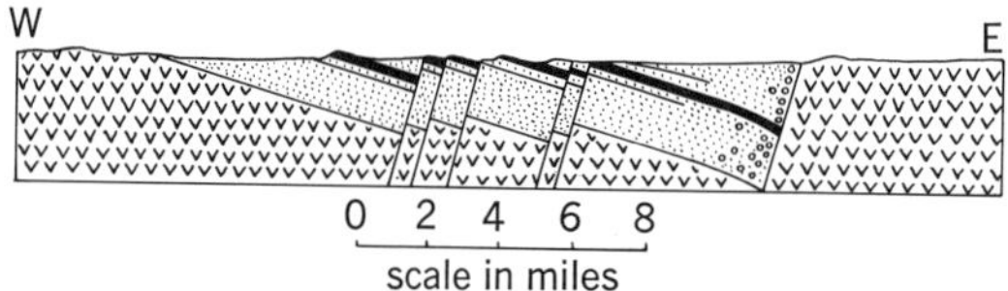

Fig. 8. Step faults and tilted fault blocks of Connecticut Valley, Conn. (*From M. P. Billings, Structural Geology, 2d ed., copyright, 1959 by Prentice-Hall, Inc.; reprinted by permission*)

for which the difference between the resistance to shear and the shearing stress is minimal, as shown in Eq. (7).

$$\frac{d(R-\tau_c)}{d\alpha}=\tan\beta\sin 2\alpha-\cos 2\alpha=0$$

$$\tan\beta=\cot 2\alpha, \beta=90°-2\alpha, \alpha=45°-\frac{\beta}{2} \qquad (7)$$

The value of β can be found by plotting the curve $\tau=f(\sigma_n)$ graphically, using Mohr's major stress circles. The normal stresses are plotted along the abscissa and the shearing stresses along the ordinate. The envelope of the major stress circles is the curve $\tau=f(\sigma_n)$, and β is the angle between the tangent to the envelope at any coordinates, τ, σ_n, and the abscissa (Fig. 3). Experiments determining failure in cylinders of silicate rocks and minerals under various confining pressures suggest that at failure a linear function exists between the calculated mean stress and the calculated maximum shearing stress. This relation implies that a stress difference of $\sigma_1=7\sigma_3$ exists at failure. By using this stress difference to find the envelope of Mohr's major stress circles, β is found to be about 48° and α is therefore about 21° (Fig. 4). Thus, faults in silicate rocks should make angles of about 21° to the maximum principal stress, and the direction of movement on these faults is such that the wedge receiving the greatest compressive force moves inward, whereas the wedge receiving the least compressive force moves outward (Fig. 5).

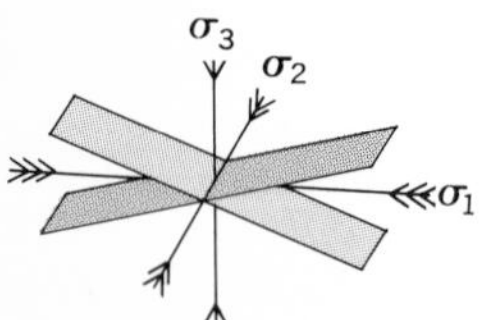

Fig. 9. Orientation of shear planes and principal stresses in thrust faulting; σ_3 is nearly vertical. (*After E. M. Anderson*)

The geometry of natural faults can be analyzed from the point of view of deriving the orientation of a stress field compatible with the fault pattern. In such a study, a hypothetical segment of the Earth's crust is isolated from its surroundings, and certain boundary stresses are imposed on its sides in accordance with Newton's laws of motion. By assuming various values for the boundary stresses, the pattern of the principal stress distribution in the interior of an isotropic block can be established. Possible surfaces of faulting contain the directions in which the intermediate principal stress operates and make an angle of about 21° with the direction in which the greatest principal stress acts (Fig. 6). The set of conditions producing a hypothetical fault pattern identical to that existing in nature serves to establish the stress field causing the natural faults. Analyses of this type assume that the rock is isotropic and that the fault pattern is the result of a single episode of deformation.

Gravity faults. The dips of gravity faults have a maximum frequency of about 60°, suggesting that at the time of faulting the greatest principal stress was oriented approximately vertically and the intermediate and the least principal stresses nearly horizontally (Fig. 7). In many cases, two or more faults parallel one another, commonly with an en echelon pattern. Step faults (Fig. 8) are parallel displacements on which the downthrown side is on the same side of each fault. Under certain circumstances, the blocks between two parallel faults become rotated, and in such cases they are called tilted fault blocks. Gravity faults may be genetically related to local structures, such as folds and domes, or they may be related to broad regional warpings.

Thrust faults. Thrusts have shallow inclinations, the majority having dips between 20 and 30°. Those with dips less than 10° are called overthrusts. If during the course of erosion a part of a thrust mass becomes separated from the main part, the separated remnant is called a klippe. If, on the other hand, erosion cuts a hole through the thrust sheet into the rocks beneath the thrust, the resulting erosional feature is called a fenster. Thrusts are almost entirely confined to regions of folded rocks, and this close association suggests that the two are genetically related. The orientations of these structures indicate that at the time of their development the greatest and intermediate principal stresses were nearly horizontal, and the

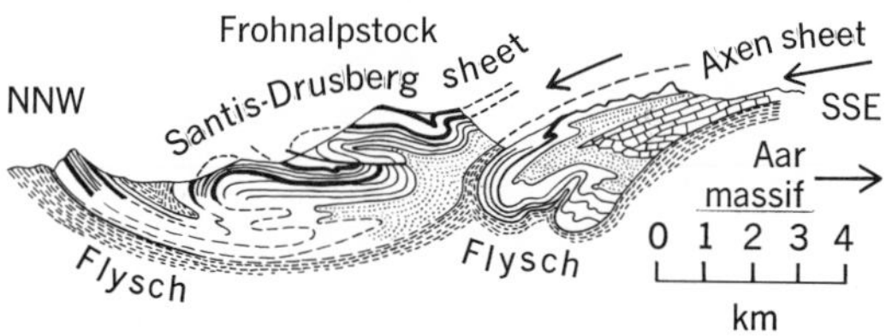

Fig. 10. The piling up of the Upper and Middle Helvetian sheets in the marginal trough of the Alps, north of the Aar massif. (*After Heim, as used in L. U. De Sitter, Structural Geology, McGraw-Hill, 1956*)

least principal stress was vertical (Fig. 9). The frictional drag beneath large thrust blocks is thought to be reduced by the buoyant effect of fluids compressed within the thrust surfaces.

Certain thrusts are abnormal in that their thrust sheets appear to have moved downhill and to have no visible source (Fig. 10). Some of these thrusts are attributed to gravitational gliding away from the crests of adjacent anticlinal structures.

Strike-slip faults. Strike-slip faults are nearly vertical and commonly are characterized by well-developed shear zones. The orientation of these faults suggests that during their formation the greatest and least principal stresses were approximately horizontal and the intermediate principal stress was vertical (Fig. 11). Although some strike-slip faults may be related to the structures they cut, others clearly postdate the associated structures. Some of the largest strike-slip faults are considered to belong to the major tectonic pattern of the Earth, and thus are thought to be the result of deforming pressures that are operative at great depths. *See* GEOLOGY, EVOLUTION OF; GRABEN; HORST; STRUCTURAL GEOLOGY; TECTONIC PATTERNS. [PHILIP H. OSBERG]

Bibliography: E. M. Anderson, *The Dynamics of Faulting and Dyke Formation with Applications to Britain*, 2d ed., 1951; P. C. Badgley, *Structural and Tectonic Principles*, 1965; W. Hainer, Stress distributions and faulting, *Bull. Geol. Soc. Amer.*, 62:373–398, 1951; M. K. Hubbert, Mechanical basis for certain familiar geologic structures, *Bull. Geol. Soc. Amer.*, 62:355–372, 1951; A. Nadai, *Theory of Flow and Fracture of Solids*, vol. 1, 2d ed., 1950; G. J. Ramsay, *Folding and Fracturing of Rocks*, 1967; E. C. Robertson, Experimental study of the strength of rocks, *Bull. Geol. Soc. Amer.*, 66: 1275–1314, 1955; F. J. Turner and L. E. Weiss, *Structural Analysis of Metamorphic Tectonites*, 1963.

FAULT AND FAULT STRUCTURES

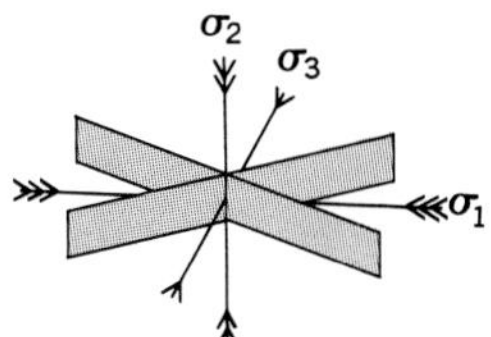

Fig. 11. Orientation of shear planes and principal stresses in strike-slip faulting; σ_2 is nearly vertical. (*After E. M. Anderson*)

Feldspar

A group of silicate minerals that make up about 60% of the outer 15 km of the Earth's crust. The feldspars constitute the most abundant group of minerals and are also important economic minerals, particularly in the ceramic and glass industries which utilize most of the output as raw materials. Feldspars are also used as gemstones if nicely colored or distinguished by an attractive luster. *See* SILICATE MINERALS.

Chemical composition. The feldspars are silicates of aluminum, Al, with the metals potassium, K, sodium, Na, and calcium, Ca, and, rarely, barium, Ba.

Their chemical composition can usually be expressed as a sum, $Or_xAb_yAn_z$, with $x+y+z=100$ and the symbols having the following significance: Or, $KAlSi_3O_8$; Ab, $NaAlSi_3O_8$; and An, $CaAl_2Si_2O_8$. The symbols Or, Ab, and An are derived from the special feldspar species: orthoclase, albite, and anorthite. Rubidium, Rb, and strontium, Sr, may enter the feldspars in amounts usually below 1 mole % of $RbAlSi_3O_8$ or $SrAl_2Si_2O_8$. Larger amounts of Ba may be found as $BaAl_2Si_2O_8$ in potassium feldspars (which are then called hyalophane). A pure barium feldspar, $BaAl_2Si_2O_8$ (celsian), is very rare. Al can be replaced by Fe (rarely more than 5%). All intermediate members, Or-Ab and Ab-An, are known. Whereas Or does not take much An into solid solution or vice versa, Ab can take up both Or and An (up to a maximum composition of about $Or_{10}Ab_{80}An_{10}$).

Occurrence. The feldspars occur as components of all kinds of rocks. In igneous and metamorphic rocks their particle size usually lies between 0.1 and 10 mm; in porphyritic rocks it frequently reaches 5–10 cm; in pegmatites, crystals of 10 m or larger are found. The size of authigenic (rather pure potassium or sodium) feldspars, developed in sedimentary rocks, is usually below 1 mm. Feldspars are also found beautifully crystallized, as fissure minerals in clefts or as druse minerals in cavities, for example, in granites.

Properties. The properties of the feldspars are variable and depend upon their chemical composition, upon the conditions under which they grew, and on their subsequent history (changes of the external conditions with time).

Optical properties. In agreement with their symmetry (triclinic or monoclinic) they are birefringent (relatively low, approximately 0.01 and smaller) with refractive indices about 1.52 (potassium feldspar), 1.53 (sodium feldspar), and 1.58 (calcium feldspar). They can be optically negative or positive corresponding to their chemical composition, crystal structure, and state of homogeneity.

Color. When feldspars appear in the state in which they grew, that is, as homogeneous crystals, they are usually clear and translucent. However, they have usually undergone changes in the course of geological times (exsolutions, changes in symmetry with subsequent small-scale twinning, and metasomatic alterations with or without deviation from a possible feldspar composition) as a result of which they have lost their translucency and assumed a milky or dirty appearance. Feldspars have no color of their own but are frequently colored (yellow, brown, reddish, or dirty green to black) by impurities. The cause of the beautiful bright green color of amazonite is still unknown. Crystals more or less exsolved on a submicroscopic scale may show a whitish-bluish luster (moonstones are alkali feldspars or cryptoperthites), a blue luster (peristerite is an oligoclase), or a bright luster of all colors (labradorites). Aventurine exhibits a shiny red or green color as a result of small but microscopically visible exsolved hematite or included mica particles.

Mechanical properties. The hardness (Mohs scale 6) is somewhat lower than that of quartz (Mohs scale 7). Feldspars are brittle but usually show an excellent cleavage parallel to the (001) plane and another, somewhat less distinct, parallel to the (010) plane. In addition (and sometimes even more pronounced than the cleavages), there are parting planes parallel to (110), $(1\bar{1}0)$, and other directions. These occur especially in clear, unmixed plagioclases that may not show the normal cleavage behavior at all.

Morphology. Feldspars are monoclinic, $\alpha=\gamma=90°$; $\beta\approx116°$, or triclinic (α and γ deviating from 90° by up to about 4°). The ratio a:b:c is, in angstroms, 8.6:13.0:7.2 for potassium feldspar and 8.1:12.8:7.2 for sodium and calcium feldspar. Depending upon chemical composition and growth conditions, their habit may be isometric, elongated

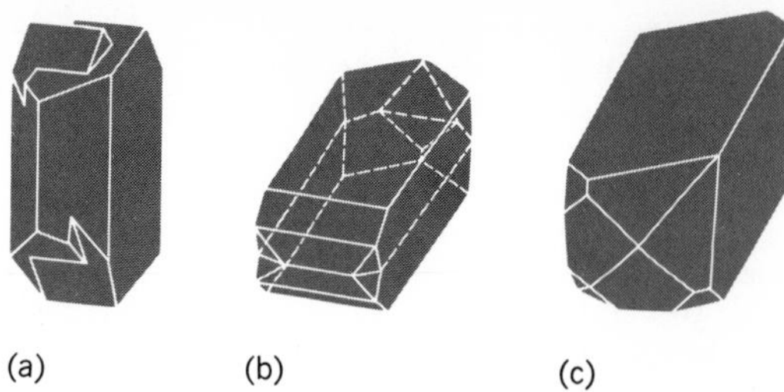

Twinning in feldspars. (*a*) Carlsbad twin. (*b*) Manebach twin. (*c*) Baveno twin. (*From C. S. Hurlbut, Jr., Dana's Manual of Mineralogy, 17th ed., copyright © 1959 by John Wiley & Sons, Inc.; reprinted by permission*)

after the *a*, *b*, or *c* axis or tabular after (010) or (001). The main limiting faces (110), $(1\bar{1}0)$ are always present in well-developed crystals, and (010), (001), $(\bar{1}01)$, $(\bar{2}01)$. If (110) and $(1\bar{1}0)$ are the dominating faces (as in adularia and in anorthoclase of the rhomb porphyries), a double-wedge-shaped morphology results with rhomblike cross sections.

Twinning. Twinning is frequently present in feldspars (see illustration). A distinction can be made between simple and multiple twinning. The former is characteristic of a few individuals related to each other by one or more of the following twin laws (only the most common are listed here): (*a*) after (100) or [001], Carlsbad law; (*b*) after (021), Baveno law; (*c*) after (001), Manebach law; (*d*) after (010), albite law; and (*e*) after [010], pericline law. The twin laws *d* and *e* cannot, for reasons of symmetry, occur in monoclinic feldspars. In addition, triclinic feldspars untwinned or twinned in the above simple sense frequently show multiple twinning, that is, many alternating lamellae whose orientation is governed throughout by the same twin law. The twin laws that most frequently lead to multiple twinning are the albite and the pericline laws. The width of the lamellae in multiple twins of plagioclases is usually about 1 mm and smaller; in the alkali feldspars anorthoclase and orthoclase, the width is frequently below the range of microscopic visibility. In such cases the existence of twinning can be determined only by x-rays. If the multiple twinning is governed by the albite law, the composition plane of the lamellae is (010). In the case of pericline law the composition plane is the so-called rhombic section, an irrational plane in the zone [010], the orientation of which depends on the lattice geometry. If the angle between the rhombic section and (001) is called $+s$ and conforms to the definition that the $+a$-axis direction corresponds to 0° and the $+c$-axis direction corresponds to 0° and the $+c$-axis direction to 116°, then this angle s is expressed by the relation $\tan s = \cos \gamma \cdot \tan \alpha^*$, where α^* is the angle (001) $\wedge$ (010). Frequently the orientation of the rhombic section is expressed by the value of σ, an angle between the trace of the rhombic section on (010) and the *a*-axis. σ is expressed by the relation $\cot \sigma = \tan \gamma \cdot \cos \alpha^*$. In feldspars the difference between the values s and σ is small (usually less than 4°). Because γ varies considerably in the plagioclases as a function of Al/Si distribution and chemical composition, s and σ are rather important as diagnostic magnitudes.

Microcline-type twinning. Because the alkali feldspars and the acid plagioclases can be monoclinic or triclinic, a transition from monoclinic to triclinic may be connected with simultaneous (sometimes submicroscopic) twinning after the albite and pericline laws with the twin plane (010) perpendicular to the twin axis [010]. Thus, a multiple twin system is formed, consisting of numerous interpenetrating domains divided into four groups of orientations: A_1 and A_2 (with albite-twin orientation) and P_1 and P_2 (with pericline-twin orientation).

Genetically, three main types of twinning can be recognized: (1) growth twins, usually observable as simple twinning according to all the laws mentioned; (2) transformation twins, usually observable as an interpenetrating system of multiple twinning after the albite and pericline laws; and (3) deformation twins (also called mechanical twins), usually observable as multiple twinning after the albite, or pericline laws, or both. Because, in principle, multiple twinning after a single law may be produced by growth, transformation, or deformation, it is frequently difficult to reach an unequivocal decision regarding the genetic interpretation of such twins. In the case of microcline-type twinning, however, it can be rigidly concluded that such material originally grew with monoclinic symmetry.

Melting point. Alkali feldspars melt (potassium feldspar incongruently, forming a mixture of leucite and glass) at about 1100°C. Plagioclases melt (as a function of An content) between 1100 and 1544°C.

Crystal structure. Although the crystal structure of the feldspars has been known in principle since 1933 (W. H. Taylor), details are still the object of investigation in several laboratories. Each Si^{4+} (radius ≈ 0.4 A) and Al^{3+} (radius ≈ 0.5 A) is surrounded in a nearly ideal tetrahedral coordination by four 0^{2-} (radius $\approx$ 1.3 A). Thus, each 0^{2-} belongs to two neighboring SiO_4 or AlO_4 tetrahedrons. Considering the Si^{4+} or Al^{3+} alone, they form a three-dimensional network with the coordination number 4 and the O^{2-} (nearly) centering the shortest distances between them.

This $(Si,Al)O_2$ framework has sizable voids in which the larger cations K^+ (radius ≈ 1.3), Na^+ (radius ≈ 1.0), and Ca^{2+} (radius ≈ 1.0 A) are located. In the solid solutions the K, Na, and Ca ions always appear to be randomly distributed. On the other hand, the distribution of the Si and Al ions may differ largely for feldspars of the same chemical composition. The ordered for disordered state of Si/Al is greatly responsible for the different feldspar varieties and their varying properties.

In discussing the various main possibilities connected with the Al/Si–order-disorder, it must be mentioned that the feldspar unit cell contains 4 molecules $(K,Na,Ca)(Si,Al)_4O_8$. There are, therefore, 16 small ions (Si or Al) per cell. In the case of monoclinic symmetry these 16 ions occupy two unequivalent 8-fold point positions that may be called *A* and *B*.

Restricting the discussion at the outset to the alkali feldspars, $KAlSi_3O_8$ and $NaAlSi_3O_8$, two extremes are theoretically possible: (1) 6Si + 2Al are situated in *A*, and 6Si + 2Al are situated in *B*; (2) 8Si are situated in *A*, and 4Si + 4Al are situated in *B*. In both cases some disorder or random distribution occurs, in case (1) more so than in (2). States intermediate between (1) and (2) are possible. A higher state of order can be reached only if the symmetry sinks from monoclinic (mcl) into triclinic (tr). Then the two 8-fold point positions A and B split up into four 4-fold point positions A_1, A_2, B_1, and B_2, and the 4 Al can completely fill one of these, for example B_2, whereas A_1, A_2, and B_1 are occupied solely by the 12 Si. Such an arrangement, case (3), would be the most ordered one compatible with the feldspar-type structure. Whereas case (3) must necessarily be triclinic, it must be realized that in cases (1) and (2) the symmetry may be monoclinic, but is not necessarily so. Summarizing this discussion, the table lists those cases of varying order-disorder and varying symmetry that appear to be pertinent for the understanding of the alkali feldspar varieties now to be discussed in detail. (In the plagioclases, presented below, additional cases have to be discussed, as additional order-disorder states are present because the Al/Si ratio varies from 1:3 in $NaAlSi_3O_8$ to 1:1 in $CaAl_2Si_2O_8$.) *See* CRYSTAL STRUCTURE; CRYSTALLOGRAPHY.

The existence of a high degree of disorder in sanidine and analbite and a high degree of order in microcline and albite has been proved by nuclear magnetic resonance experiments.

ALKALI FELDSPARS

Alkali feldspars is the name for feldspars composed of Or and Ab in any ratio. Usually some An is present as a minor constituent. The amount of An that can be taken up by the alkali feldspars increases with the Ab/Or ratio. The maximum value of An/(Or + Ab) that is found in natural crystals is approximately 1% in Or-rich feldspars and may increase to approximately 15% in Ab-rich feldspars. Because $NaAlSi_3O_8$ (Ab) is the Na-rich member of the alkali feldspars as well as of the plagioclases, Ab feldspars are sometimes considered to belong to the former, sometimes to the latter. It appears reasonable to separate the alkali feldspars from the plagioclases by the Or/An ratio (in alkali feldspars Or/An > 1; in plagioclases Or/An < 1). There are many names for the different varieties of alkali feldspars, most of which are listed below.

Sanidine. This is the monoclinic modification of $KAlSi_3O_8$. Its optical and structural properties vary with the Ab content and the temperature of equilibrium. At high temperatures the Al/Si distribution appears to be that of case (1) in the table. At lower temperature the Al/Si distribution changes in the direction of case (1–2) mcl. Corresponding to this change, a subdivision into sanidine (high, intermediate, and low) appears justified.

Connected with these structural changes are those of the optical properties. In Na-free sanidine (high) the optical plane is parallel to (010) with $2V \approx 60°$. With the change (high $\rightarrow$ intermediate $\rightarrow$ low), 2V becomes smaller, may pass through 0° and again increase, but now within an optical plane perpendicular (010) and inclined about 10° to (001). The maximum value of 2V in the plane perpendicular to (010) is not yet known, because there is a continuous transition of the properties of sanidine into those of orthoclase. Within the optical plane (010) 2V diminishes not only with increasing Al/Si order but also with increasing Ab content. A very useful parameter, easily measurable by an x-ray method, for determining the Ab content is the decrease with increasing Na content of the a-lattice parameter (~8.6 for potassium feldspar, ~8.1 for sodium feldspar). Sanidine appears to be stable under equilibrium conditions above approximately 500°C. Below this temperature microcline becomes stable (the temperature is not yet exactly known).

Microcline. Microcline is the triclinic modification of potassium feldspar. Its Al/Si distribution is ordered in such a way that monoclinic

Main Al/Si distributions under equilibrium conditions in monoclinic and triclinic alkali feldspars as function of temperature

	Point positions				Symmetry	
Temperature	A		B		Monoclinic	Triclinic
~1100°C	6Si + 2Al		6Si + 2Al		Case (1)mcl[a]	Case (1)tr[b]
↑	6–8Si + 2–0Al		6–4Si + 2–4Al		Case (1–2)mcl	Case (1–2)tr[b]
	8Si		4Si + 4Al		Case (2)mcl	Case (2)tr[c]
	A_1	A_2	B_1	B_2		
	3Si + 1Al	3Si + 1Al	3Si + 1Al	3Si + 1Al	Case (1)mcl[d]	Case (1)tr[e]
	3–4Si + 1–0Al		mSi + nAl ($m > n$)	mSi + nAl ($m < n$)		Case (1–3)tr
~0°C	4Si	4Si	4Si	4Al		Case (3)tr

[a]As the positions A and B possess different environments, it is improbable that the Al distribution can be exactly equal in both positions. Therefore, the case (1)mcl can only be a theoretical limit which may, however, be very nearly attained at temperatures near the melting point (about 1100°C).

[b]For reasons of structure theory these cases cannot have distributions exactly as listed. For the monoclinic positions A and B are split up into the triclinic positions A_1, A_2 and B_1, B_2, and it cannot be expected that the Al distribution be exactly the same in A_1 and A_2 as in B_1 and B_2, respectively.

[c]Case (2) is the theoretically possible case of monoclinic symmetry with highest possible order. It shows the direction in which a completely disordered feldspar can develop a more ordered (or less disordered) distribution, while retaining monoclinic symmetry. For thermodynamical reasons only the intermediate states of case (1–2)mcl can be expected to occur until about (7Si + 1Al) are in A and (5Si + 3Al) in B.

[d]This case is monoclinic in the exact sense only if both A_1 and A_2 and B_1 and B_2 are indistinguishable. If this is true, the case is identical with the one listed in the first row of the table.

[e]Considerations similar to those mentioned in footnote b hold for the case (1)tr.

symmetry is impossible from a structural point of view. The triclinic deviation from monoclinic symmetry may be small. In the most ordered state it approaches case (3) tr of the table, having values $\alpha^* = (010)/(001) \approx 90°25'$ and $\gamma^* = (010)/(100) \approx 92°16'$ (maximum microcline). All intermediate states between these values and $\alpha^* = \gamma^* = 90°$ are possible and l nown from natural occurrences. These intermediate states can be characterized as possessing Al/Si distributions lying between case (1–2)mcl ≈ case (1–2) tr and case (3) tr of the table. These intermediate structural states have optical properties that are also of intermediate character [that is, lying between the triclinic optics of maximum microcline with $2V \approx 80°$, in a plane not quite perpendicular to (010), and the monoclinic optics of sanidine]. Under equilibrium conditions maximum microcline does not take much Ab into solid solution (less than about 2 mole %). Most of the Ab content found in untreated microcline crystals (usually amounting to about 20 mole %) is present as perthitic, or cryptoperthitic, exsolution, or both. However, it is possible to prepare the whole series of solid solutions between microcline and albite (see below) under nonequilibrium conditions (superheating) by a heat treatment at approximately 1000°C, that is, above the transformation temperature microcline → sanidine. The time of heat treatment must be long enough to allow a K/Na exchange to take place, but short enough to prevent changes of the Al/Si distribution. Microcline usually shows multiple twinning of the microcline type. From this type of twinning it may be concluded that most microclines originally grew as sanidine, either stably at high temperature or metastably at low temperature. In the course of geological time they underwent a continuous change of diffusive transformation into the triclinic state now present. This transformation involving Al/Si exchange at fairly low temperatures proceeds very slowly and has not yet been reproduced in the laboratory. The slowness of this ordering process is the reason why all states of microcline are found in nature reaching from monoclinic properties to the triclinic ones of maximum microcline. The domain size of twinned microcline is variable, usually < 1 mm. The domains can be submicroscopic in size and then the material appears to be optically monoclinic. In these cases a distinction of microcline and orthoclase can be made only by means of x-rays. Whereas it appears to be impossible to produce microcline in the laboratory by synthesis or by heat treatment of sanidine below the sanidine microcline transformation temperature (assumed to be about 500°C), the reverse process can be effected within times available for extended experiments. It takes place at about 1000°C or higher in days, weeks, months, or years, depending on factors (impurities, imperfections) still unknown. Bright-green microclines (the cause of the color is still unknown) are used as gemstones and called amazonite or amazon stone. Such material is always maximum microcline. The reverse does not hold (most maximum microclines are not colored).

Pegmatites sometimes develop as graphic granite, that is, as an intergrowth of potassium feldspar and quartz having the appearance of an eutectic growth produced by simultaneous crystallization. Most potassium feldspars from such pegmatites have already changed into the microcline structure.

Orthoclase. The name orthoclase is used differently by different people. In the widest sense the name applies (in contrast to anorthoclase, plagioclase, and microcline) to those feldspars that are or appear to be monoclinic with $\alpha=\gamma=90°$. According to such a definition, sanidine, submicroscopically twinned microcline, adularia, and submicroscopically twinned analbite would be orthoclases or sodium orthoclases if a considerable amount of Na (more than 25 mole % Ab) is found by chemical analysis. Petrographers usually call a material orthoclase if it appears to be optically monoclinic with the optical plane ⊥ (010) and $2V > 30°$ (if $2V < 30°$, such material is frequently called sanidine). As stated above, there are true monoclinic and true triclinic states of potassium feldspar, sanidine and microcline. Therefore, it appears practical to include in the term normal orthoclase a material whose monoclinic appearance depends on the twinning of domains having an Al/Si distribution of triclinic symmetry. If these domains are submicroscopic in size or if their Al/Si order is small, such material shows optically perfect or nearly perfect monoclinic character. The same holds for normally exposed x-ray photographs which show reflections that can be given monoclinic indices. Long-exposed x-ray photographs, however, reveal additional diffuse streaks originating from the main reflections. The position of these streaks indicates the presence of triclinic domains within the normal orthoclase with a mutual orientation similar to that of microcline twinning. Thus, the normal orthoclases may display a broad variety of structural features with respect to the size of domains and the degree of Al/Si order within them. This concept also explains the variability of their optical behavior.

Adularia is another variety of potassium feldspar, grown as well-developed crystals in fissures, especially in the region of the Swiss Alps. The predominant forms are (110) and $(\bar{1}01)$. The form (001) is frequently very small or absent. The morphology is monoclinic and the structure is in principle similar to that of normal orthoclase. However, thin sections investigated with a polarizing microscope between crossed nicol prisms usually reveal a much more complex behavior than the normal orthoclases. Frequently, different parts of the same individual display varying optical properties ranging from those of typical orthoclase to those of typical sanidine. In places they deviate strongly from monoclinic symmetry and are usually most triclinic where the (110) and $(\bar{1}10)$ faces or the $(1\bar{1}0)$ and $(\bar{1}\bar{1}0)$ faces meet. The triclinicity of such parts can be measured by x-rays. Deviations from monoclinic symmetry of 30′ for γ^* are frequently met with in Swiss adularias. A variety from Mexico (valencianite) shows deviations in places extending to those of maximum microcline. The manner in which the optically different parts are related to each other indicates that adularia originally grew metastably as sanidine at low temperatures. In time the variety adularia changed to triclinic symmetry, a process taking place at different speed in the outer and inner parts of the crystals. Different conditions in different localities are the reason why the adularias display a fascinating variety of optical behavior when investigated between crossed nicols. Adularia frequently develops simple (growth) twinning after the Baveno and Manebach laws. The Carlsbad law is rare. The crystals are usually clear, sometimes milky, but colorless. They may reach sizes up to 10 cm (rarely even more) and are frequently covered by green chlorite.

Albite. $NaAlSi_3O_8$ is chemically the Na-rich end member of the alkali feldspars. If it is ordered with respect to the Al/Si distribution to the same extent as microcline, it is called albite and is the phase of $NaAlSi_3O_8$ most frequently found in nature. It usually occurs in the most highly ordered state, case (3) tr of the table (corresponding to maximum microcline), with a, b, c=8.16, 12.77, 7.17 and α^*, β, γ^*=86.3°, 116.7°, 90.5°. Albite occurs in all groups of rocks and is usually colorless or milky white. In fissures it frequently forms beautifully developed glassy-clear crystals, tabular after (010) and mostly simply twinned after the albite law. Sometimes fourlings are found and then two albite-twinned plates are, in addition, twinned after the Carlsbad law. A variety twinned after [010] (pericline law) and elongated along the same direction is called pericline. There are indications, such as the position of the rhombic section, that such material (which always appears milky as a consequence of numerous small voids) did not grow originally as a pure albite phase, but as oligoclase which subsequently underwent metasomatic changes in chemical composition. Albite crystals in rocks frequently exhibit polysynthetic twinning, predominantly after the albite law.

Under equilibrium conditions and at low temperatures albite does not take more than 1 or 2 mole % of Or or An into solid solution. However, with increasing Al/Si disorder and temperature the solubility increases, especially in the high-temperature modification monalbite.

Monalbite. Monalbite is the monoclinic modification of $NaAlSi_3O_8$ which under conditions of equilibrium is stable only at high temperature. Its existence and monoclinic symmetry can be proved by x-ray investigations at high temperature and by the twinning behavior. This albite material kept for long periods (weeks to months) at temperatures near the melting point (1000–1100°C) exhibits a displacive, unquenchable transformation into a triclinic form, analbite. If an untwinned single crystal of albite is treated in this way, submicroscopic twins of the microcline type fashioned after both the albite and the pericline laws appear in the course of cooling. Optically, such material seems to be monoclinic but its triclinic character can easily be proved by x-rays. Such triclinic $NaAlSi_3O_8$ that upon heating becomes monoclinic by a displacive transformation (that is, by a reversible transformation which takes place at once when the transformation temperature is passed) should be called analbite. The temperature at which the transformation monalbite $\leftrightarrow$ analbite takes place is a function of the Al/Si distribution. Highly disordered monalbite that approximates case (1) mcl of the table transforms into analbite at a lower temperature than does monalbite with less disorder, having an Al/Si distribution similar to case (1–2) mcl of the table. This influence of the Al/Si distribution on the transformation temperature of monalbite $\leftrightarrow$ analbite is very impressive. It appears that the highest transformation temperature (strictly speaking, the only one possible under equilibrium conditions if the influence of pressure is neglected) is about 900°C. Under nonequilibrium conditions, temperatures as low as −60°C are reported (W. L. Brown, 1959). Thus a monoclinic $NaAlSi_3O_8$ can exist metastably at room temperature, provided it has a very high degree of disorder resulting either from direct growth or assumed during prolonged heat treatment at very high temperatures near the melting point. Monalbite forms a complete series of solid solutions with sanidine. Thus the monoclinic high-temperature modifications of the alkali feldspars may be subdivided into sanidine, sodium sanidine, potassium monalbite, and monalbite. How high an An-content can be taken up by monalbite is still unknown. Present data indicate the existence of calcium-monalbites extending to about $Ab_{80}An_{20}$ at high temperatures. This fact is rather important for a genetic interpretation of the feldspars in rhomb porphyries which have compositions of about $Or_{10}Ab_{80}An_{10}$. The peculiar appearance of their twinning indicates that they originally grew as monoclinic potassium, calcium, or calcium-potassium monalbites that became polysynthetically twinned and unmixed during the course of cooling.

Analbite. Analbite, a triclinic $NaAlSi_3O_8$ that is not stable at any temperature, has an Al/Si distribution lying between cases (1) tr and (1–2) tr of the table, which according to structure theory cannot, strictly speaking, exist stably under any conditions. It can be produced by cooling monalbite and can be changed back to monalbite by heating, provided the Al/Si distribution does not change and assume the character of cases (1–3) tr or (3) tr given in the table. After having undergone such a change, the material could approach monoclinic symmetry only when reheated. It would actually be structurally intermediate between analbite and albite, if albite is defined as a series of materials having Al/Si distributions which necessitate triclinic symmetry and are capable of existing under equilibrium conditions. Lattice constants of most triclinic analbite are about a, b, $c \approx 8.12$, 12.87, 7.13 and α^*, β, $\gamma^* \approx 86.0$, 116.5, 88.3. All intermediate values between these and $\alpha^* = \gamma^* =$ 90° (monoclinic symmetry) may be met with as a function of temperature and Al/Si distribution (analbite-monalbite series). In addition, intermediate values between those given here and those given above for albite can also be expected (unstable analbite-albite series) as well as those of intermediate states that are stable as a function of temperature under equilibrium conditions forming a series albite (low, intermediate, high)—monalbite.

Perthite and antiperthite. Potassium- and sodium-rich areas forming parallel or subparallel intergrowths are called perthite if the K-rich phase appears to be the host from which the Na-rich phase exsolved; an intergrowth is called antiperthite if the Na-rich phase appears to be the host. Corresponding to the size of the exsolved areas, a distinction is made between perthites (visible to the naked eye), microperthites (visible in the microscope), and cryptoperthites (detectable only by x-rays). In perthites and microperthites, the exsolved Na-rich phase usually occurs as albite forming somewhat irregular bands that are more or less parallel to the plane near ($\bar{8}01$). Within these bands the albite usually shows multiple twinning after the albite law. The K-rich host has usually changed into microcline (mostly into maximum microcline). The cryptoperthites can be rather complex. In these cases the K-rich host can be sanidine, orthoclase, or microcline; the Na-rich exsolved domains can be material with properties near those of analbite (they are then usually oriented after pericline-twin law) or near those of albite (then usually in albite-twin law orientation). The cryptoperthites frequently display a bluish to whitish milky luster. If they are flawless and possess special qualities, they are also called moonstone and are used as gemstones (for example, moonstone from Ceylon).

Anorthoclase. The name anorthoclase is usually applied to Na-rich alkali feldspars ($Or_{40}AB_{60}$ to $Or_{10}Ab_{90}$±An content, up to approximately 10%). In one way or another they show deviations from monoclinic symmetry. As mentioned above, Or and Ab form a complete series of monoclinic solid solutions (sanidine-monalbite). When cooled, K-rich members change slowly by diffusive transformation into microcline. Na-rich members, however, change quickly into K-analbite by displacive transformation and in addition show a slow diffusive transformation trending to produce albite structure. The temperature of the displacive transformation monoclinic ↔ triclinic (monalbite ↔ analbite) is a function of the Ab/Or ratio and of the degree of Al/Si order. The larger these values, the higher the temperature required. Exact data are not yet known. Whereas no displacive transformation has yet been found to occur above room temperature in material with an Ab/(Ab + Or) ratio < 0.6, natural material with an higher Ab content has been described as showing this transformation. The relation between temperature and Ab/(Ab + Or) ratio shown by natural material has been reported to be approximately as follows: Ab/(Ab + Or) ratio = 0.7, 0.8, 0.9, at 250, 500, 750°C, respectively. During cooling both the displacive and the diffusive transformation processes can act concurrently in the formation of the final structural state of Na-rich alkali feldspars as found in nature. Therefore, the properties and state of twinning (visible lamellae or submicroscopical twinning) vary considerably from locality to locality. The situation is further complicated by the fact that under equilibrium conditions alkali feldspar of originally intermediate composition may have been exsolved into potassium and sodium feldspar, each of relatively pure composition (see discussion of perthite and antiperthite above). Because the relative speed of the three processes—displacive transformation, diffusive transformation, and exsolution—depends on several factors such as bulk chemical composition, rate of cooling, and others that are not yet known, the optical and structural properties of the anorthoclases vary strongly from crystal to crystal and can even within the same crystal be different from point to point. It must be emphasized, therefore, that anorthoclase is not a name for a particular phase with definite properties, but usually applies to a mixture of several phases which may not even have a stability field of their own at any temperature.

PLAGIOCLASE FELDSPARS

Plagioclase feldspars form at high temperature a complete series of solid solutions, ranging from $NaAlSi_3O_8$(Ab) to $CaAl_2Si_2O_8$(An). The Ab end may take up to approximately 10 mole % $KAlSi_3O_8$(Or) into solid solution. Under equilibrium conditions they have a rather disordered Al/Si distribution near their melting points (~1100 to 1544°C). At lower temperatures the distributions become more ordered. This is the reason for the existence of so-called high-temperature and low-temperature optics connected by intermediate optics. The differing optics met with in natural materials can also be produced by heat treatment of material that has low-temperature optics. Besides the Al/Si distribution the chemical composition has an essential influence on the optics. This fact has played a very important role in the development of microscopical methods for determining the chemical composition of plagioclases.

The plagioclase series is arbitrarily subdivided and named according to increasing An content as follows: $Ab_{100}An_0 - Ab_{90}An_{10}$ (albite); $Ab_{90}An_{10} - Ab_{70}An_{30}$ (oligoclase); $Ab_{70}An_{30} - Ab_{50}An_{50}$ (andesine); $Ab_{50}An_{50} - Ab_{30}An_{70}$ (labradorite); $Ab_{30}An_{70} - Ab_{10}An_{90}$ (bytownite); and $Ab_{10}An_{90} - Ab_0An_{100}$ (anorthite). Because the content of silicic acid (in its anhydrous form SiO_2) is relatively higher in $NaAlSi_3O_8$ than in $CaAl_2Si_2O_8$, the Ab-rich members albite-oligoclase are frequently called acid plagioclases and the An-rich members bytownite-anorthite basic plagioclases. It is important to note that in contrast to the alkali feldspars, in which the Al/Si ratio is always 1:3, this ratio varies in the plagioclases with increasing An content from 1:3 to 1:1. In consequence, solid-state reactions involving homogenization or unmixing as a function of temperature take place much more sluggishly than in the alkali feldspars. Whereas in the latter a compositional change involves only a Na/K exchange, a Na/Ca exchange in the plagioclases necessitates a change of the Al/Si ratio as well. The Al and Si atoms have to cover quite considerable distances by diffusion and require correspondingly long times for such processes. It is therefore not astonishing to find most natural plagioclases zoned; that is, great changes are often found in the Ab/An ratio between the core and the rim of crystals. So-called normal zoning obtains if the core is richer in An than the rim; reverse zoning if the opposite is the case. Another consequence is the absence in nature of exsolution structures on a visible scale between intergrown Ab- and An-rich parts, resembling those of the perthites. There are, however, many signs indicating that plagioclases of intermediate Ab/An ratio are not stable at low temperature, and, indeed, submicroscopic exsolutions have been detected by the application of x-rays as discussed below.

In the sections on the alkali feldspars an account was given of the structural variations which are caused by the various possibilities of Al/Si distribution. The same considerations hold true in principle for the plagioclases. With these the conditions are, however, even more complex to survey, as the varying Al/Si ratio now introduces a new factor. For this reason the phase relations within the plagioclase series are at present less well known than is the case of the alkali feldspars. This discussion is limited to listing the main structural varieties:

1. Phases with An content in states (*a*) similar to those of albite (low, intermediate, high); (*b*) similar to those of analbite; and (*c*) similar to those of monalbite.
2. Labradorite state (see below).
3. Anorthite state low.
4. Anorthite state high.

The states 1*a*, 1*b*, and 1*c* have been discussed above (albite, analbite, monalbite). How far they extend into the plagioclase region under equilibrium conditions while retaining their main features is still unknown. In all cases, however, the increase of the Al/Si ratio with increasing An content adds additional possibilities of order-disorder. These are indications that the states 1*b* and 1*c* may exist up to approximately 20% An content. State 1*a* may extend as calcium albite (intermediate to high) at intermediate and high temperatures up to approximately 80% An. State 2 appears to be a preliminary state of unmixing on a infrasubmicroscopic scale which is responsible for the labrador luster and develops from state 1*a* during the course of cooling. States 3 and 4 are characterized by a new type of Al/Si ordering, caused by the Al/Si ratio approaching the value 1:1, which is reached exactly in pure $CaAl_2Si_2O_8$ (anorthite). X-ray photographs of states 3 and 4 show superstructure reflections, the character of which (necessitating a doubling of the *c* axis) can be explained by the experimentally supported assumption that each SiO_4 tetrahedron is surrounded by four AlO_4 tetrahedrons and vice versa. States 3 and 4 are distinguished by positional differences of the Ca ions within the voids of the framework. In pure $CaAl_2Si_2O_8$ state 3 is stable up to approximately 1000°C and has fixed Ca positions. Above this temperature, state 3 changes continuously and reversibly into state 4, in which the Ca ions occupy several positions more or less randomly. The transformation temperature at which state 3 inverts to state 4 decreases rapidly with increasing Ab content of the anorthite.

Albite. The name usually given by petrographers and mineralogists to plagioclases ranging from $Ab_{100}An_0$ to $Ab_{90}An_{10}$ is albite. The structure of albite and its high-temperature polymorphs was discussed above. Whereas the pure $NaAlSi_3O_8$ can be ideally ordered as case (3) tr of the table, any An content (as $CaAl_2Si_2O_8$) necessarily introduces Al/Si disorder. Therefore the properties of the low- and high-temperature modifications approach each other with increasing An content. Natural albite with more than 2 mole % An is usually unmixed on a submicroscopic scale to be discussed below. In addition to the remarks on albite above, some varieties may also be discussed here.

Cleavelandite. This is structurally albite (low) and almost pure $NaAlSi_3O_8$. It has a tabular habit and is twinned after the albite law. The tabular individuals frequently show mosaic developments and present a bent appearance. They tend to occur as fan-shaped aggregates formed during the late pegmatitic stages of granite rock formation.

Peristerite. Natural plagioclases with an An content lying between $Ab_{98}An_2$ and $Ab_{85}An_{15}$ are usually submicroscopically unmixed into relatively pure albite Ab_mAn_n with $n<2$ when $m+n=100$, and into an oligoclase of the composition Ab_mAn_n when $n\approx 25-30$. This unmixing is easily detected by x-rays. If this unmixing is accompanied by a blue luster, such albite-oligoclase material is called peristerite and used as a gemstone, falsely called moonstone if of attractive appearance.

Adventurine or sunstone. This is an albite or oligoclase or andesine with a reddish luster caused by thin visible flakes oriented parallel to several structurally defined planes. They are probably Fe_2O_3 formed by exsolution.

Oligoclase. Oligoclase has the composition $Ab_{90}An_{10}$ to $Ab_{70}An_{30}$. For its phase relations to the other plagioclases see the introductory discussion of plagioclase feldspars; for its unmixing exsolution processes see the discussion of peristerite and sunstone.

Andesine. Andesine has the composition $Ab_{70}An_{30}$ to $Ab_{50}An_{50}$. For its phase relations to the other plagioclases see the discussion of plagioclase feldspars; exsolution and unmixing processes are discussed in the sections on sunstone and labradorite.

Labradorite. Labradorite has the composition $Ab_{50}An_{50}$ to $Ab_{30}An_{70}$. Its phase relations to the other plagioclases have been covered in the introductory discussion of plagioclase feldspars. Some additional remarks may be made here. X-ray photographs of albite differ from those of anorthite in that more reflections are present in the case of anorthite. In the literature these additional reflections are called *b* reflections (caused by Al/Si distribution) and *c* reflections (caused by the Ca position). These additional reflections indicate that the anorthite structure is a superstructure of the albite type. An explanation for this was given above. X-ray photographs of natural labradorite normally show *b*-type reflections also. However, in the range of approximately $Ab_{80}An_{20}$ up to $Ab_{25}An_{75}$ these *b* reflections are split into two, the degree of splitting increasing with increasing Ab content. Parallel to this increase of splitting goes a decrease of the relative intensities and an increase of the diffuseness of the split *b* reflections. This behavior indicates that labradorites and the contiguous plagioclases (ranging in composition from approximately $AB_{80}An_{20}$ up to $Ab_{25}An_{75}$) are not stable as solid solutions at low temperature. It appears that the labradorite state, state 2, is the beginning of an unmixing on a submicroscopical scale (in terms of angstrom units) into an oligoclase and a bytownite. Whether such a combination of oligoclase and bytownite is stable at low temperature under equilibrium conditions is not yet known. Obviously, such a mixture appears to be more stable at low temperature than the solid solution. But it might nevertheless be unstable with respect to other non-all-feldspar combinations within the quaternary system Na_2O-CaO-Al_2O_3-SiO_2. Numerous observations of natural alterations of the chemically intermediate plagioclases into products (saussurite) composed of zoisite-epidote, scapolite, and albite or into combinations of epidote, sericite, albite, calcite, and other minerals can be interpreted as confirming such a view, even when it is admitted the presence of water seems essential for these alterations.

Bytownite. This mineral has the composition $Ab_{30}An_{70}$ to $Ab_{10}An_{90}$. For its phase relations to the other plagioclases and for its structure see the general remarks under plagioclase feldspar and the discussion of labradorite above.

Anorthite. Anorthite is the most basic member of the plagioclases, its composition ranging from $Ab_{10}An_{90}$ to pure An ($CaAl_2Si_2O_8$). Its phase relations to the other plagioclases and its structure are discussed in the sections on plagioclase feldspar and labradorite. It occurs in basic rocks (gabbro, norite, anorthosite); rarely as well-developed druse mineral (for example, in ejectamenta of the Somma, Vesuvius); sometimes in tuffs (good crystals are known from Miyaka, Japan); and very rarely in metamorphic rocks. [FRITZ H. LAVES]

Bibliography: W. A. Deer, J. Zussman, and R. A. Howie, *Rock Forming Minerals*, 5 vols., 1962–1963; W. H. J. Eitel, *Structural conversions in crystalline systems and their importance for geological problems*, Geol. Soc. Amer. Spec. Pap. no. 66, 1958; E. W. Heinrich, *Microscopic Petrography*, 1956; E. H. Kraus, W. F. Hunt, and L. S. Ramsdell, *Mineralogy*, 5th ed., 1959; J. V. Smith, *Feldspar Minerals*, 3 vols., 1974; A. N. Winchell and H. Winchell, *Elements of Optical Mineralogy*, pt. 2, 4th ed., 1951.

Feldspathoid

A member of the feldspathoid group of minerals. Members of this group are characterized by the following related features: (1) All are aluminosilicates with one or more of the large alkali ions (for example, sodium, potassium) or alkaline-earth ions (for example, calcium, barium). (2) The proportion of aluminum relative to silicon, both of which are tetrahedrally coordinated by oxygen, is high. (3) Although the crystal structures of many members are different, they are all classed as tektosilicates; that is, they consist of a three-dimensional framework of aluminum and silicon tetrahedrons, each unit of which shares all four vertexes with other tetrahedrons, giving rise to a ratio of oxygen to aluminum plus silicon of 2:1. Ions of other elements occupy interframework sites. (4) They occur principally in igneous rocks, but only in silica-poor rocks, and do not coexist with quartz (SiO_2). Feldspathoids react with silica to yield feldspars, which also are alkali–alkaline-earth aluminosilicates. Feldspathoids commonly occur with feldspars. *See* SILICATE MINERALS.

The principal species of this group are the following:

Nepheline	$KNa_3[AlSiO_4]_4$
Leucite	$K[AlSi_2O_6]$
Cancrinite	$Na_6Ca[CO_3\vert(AlSiO_4)_6]\cdot 2H_2O$
Sodalite	$Na_8[Cl_2\vert(AlSiO_4)_6]$
Nosean	$Na_8[SO_4\vert(AlSiO_4)_6]$
Haüyne	$(Na,Ca)_{8-4}[(SO_4)_{2-1}\vert(AlSiO_4)_6]$
Lazurite	$(Na,Ca)_8[(SO_4,S,Cl)_2\vert(AlSiO_4)_6]$

The last four species (sodalite group) are isostructural, and extensive solid solution occurs between end members; but members of the sodalite group, cancrinite, leucite, and nepheline have different crystal structures.

Nepheline is an essential constituent of intrusive nepheline syenites and nepheline syenite pegmatites (as in the extensive deposits of the Kola Peninsula, in the Soviet Union) and of their extrusive equivalents, phonolites. It also occurs in rocks of metamorphic character, having originated by metamorphism of silica-poor source rocks or, more commonly, by metasomatic replacement of rocks such as marbles and gneisses (as near Bancroft, Ontario, Canada). Cancrinite occurs principally with, and as a replacement of, nepheline in plutonic rocks, having been derived by the reaction of preexisting nepheline with $CaCO_3$-containing solutions. Sodalite also occurs principally in association with nepheline in syenites and related silica-poor intrusive igneous rocks, but it may also occur in equivalent extrusive rocks. Both haüyne and nosean are found almost exclusively in extrusive, silica-deficient rocks such as phonolites. Leucite occurs in potassium-rich basic lavas, frequently in large phenocrysts having the form of trapezohedrons. Pseudoleucite is a mixture of feldspar and nepheline which has the crystal form of preexisting leucite; it occurs in both extrusive and intrusive rocks. Unlike most other members of the feldspathoid group, lazurite occurs in metamorphic rocks, specifically, metamorphosed impure limestones. Lapis lazuli is the name of a blue lazurite-rich rock used as a gem material. *See* CANCRINITE; LAZURITE; LEUCITE; LEUCITE ROCK; NEPHELINE SYENITE; NEPHELINITE. [DONALD R. PEACOR]

Felsite

An igneous rock with a felsitic or aphanitic (not visibly crystalline) texture, composed largely of light-colored (felsic) minerals (quartz and feldspar). The constituents may consist of glass, very fine crystalline material, or both. If relatively large crystals (generally visible) are abundant in the aphanitic portion, the rock becomes a felsite porphyry. Felsite is an aphanite in which felsic minerals predominate. *See* APHANITE; IGNEOUS ROCKS; PORPHYRY. [CARLETON A. CHAPMAN]

Ferberite

A mineral with chemical composition $FeWO_4$. Ferberite is the iron member of the wolframite solid-solution series. It generally contains some

Black, prismatic ferberite crystals on granite, found in Boulder County, Colo. (*Specimen from Department of Geology, Bryn Mawr College*)

manganese. It occurs in black, monoclinic, short, striated, prismatic crystals (see illustration). Fracture is uneven; luster is submetallic. Hardness is 4.5 on Mohs scale and specific gravity is 7.5. Streak is brownish-black. Ferberite fuses readily to a magnetic globule. It differs from the manganese member huebnerite in that it is slightly harder, heavier, darker, and nearly opaque. For occurrence, test for tungsten, and use *see* WOLFRAMITE.

[EDWARD C. T. CHAO]

Fiord

A segment of a troughlike glaciated valley partly filled by an arm of the sea. It differs from other glaciated valleys only in the fact of submergence. The floors of many fiords are elongate basins excavated in bedrock, and in consequence are shallower at the fiord mouths than in the inland direction. The seaward rims of such basins represent lessening of glacial erosion at the coastline, where the former glacier ceased to be confined by valley walls and could spread laterally. Some rims are heightened by glacial drift deposited upon them in the form of an end moraine.

Fiords occur conspicuously in British Columbia and southern Alaska (Figs. 1 and 2), Greenland, Arctic islands, Norway, Chile, New Zealand, and Antarctica—all of which are areas of rock resistant to erosion, with deep valleys, and with strong former glaciation. Depths of the floors of many fiords exceed 800 m; the deepest on record is 1933 m, in Skelton Inlet, Antarctica. Depths are the result of enhanced glacial erosion caused by the rapid flow due to confinement of the glacier between valley sides.

Fig. 1. Fiord in Alaska. *(W. W. Atwood, USGS)*

Fig. 2. Typical fiord-side scenery of southeastern Alaska in June. Steep forested slopes give way upward to rocky ice-scoured upland. *(R. M. Chapman, USGS)*

Some shallow fiords were glaciated while their floors stood above sea level, and were submerged later. Others were deepened below the level of the sea at the time of glaciation. A valley glacier flowing into sea water continues to exert stress on its floor until flotation occurs. A vigorous glacier of density 0.9 and thickness of 1000 m would continue to erode its floor even when submerged to a depth of nearly 900 m. As net submergence of coasts through postglacial rise of sea level is unlikely to have much exceeded 100 m, excavation below the sea level of the time probably occurred in most valleys that are now deep fiords. *See* GLACIATED TERRAIN.

[R. F. FLINT]

Bibliography: R. F. Flint, *Glacial and Quaternary Geology*, 1971; J. W. Gregory, *The Nature and Origin of Fiords*, 1913.

Fission track dating

A method of dating geological and archeological specimens by counting the radiation-damage tracks produced by spontaneous fission of uranium impurities in minerals and glasses. During fission two fragments of the uranium nucleus fly apart with high energy, traveling a total distance of ~25 μ (~0.001 in.) and creating a single, narrow but continuous, submicroscopic trail of altered material, where atoms have been ejected from their normal positions. Such a trail, or track, can be revealed by using a chemical reagent to dissolve the altered material, and the trail can then be seen in an ordinary microscope. The holes produced in this way can be enlarged by continued chemical attack until they are visible to the unaided eye.

Theory. Track dating is possible because most natural materials contain some uranium in trace amounts and because the most abundant isotope of uranium, U^{238}, fissions spontaneously. Over the lifetime of a rock substantial numbers of fissions occur; their tracks are stored and thus leave a record of the time elapsed since track preservation began. The illustration shows etched fission tracks in a mica crystal. The number of tracks produced in a given volume of material depends on the uranium content as well as the age, so that it is necessary to measure the uranium content before an age can be determined. This measurement is most conveniently made by exposing the sample to a known dose of thermal (slow) neutrons, which induce new fissions whose number then gives the uranium content.

The age A of a sample is given by the relation shown in the equation below, where ρ_s is the num-

$$\rho_s/\rho_i = [\exp(\lambda_D A) - 1]\,\lambda_F/\lambda_D f$$

ber of spontaneous fission tracks per unit area, ρ_i is the number induced per unit area, f is the fraction of the uranium caused to fission by the neutron irradiation, and λ_D and λ_F are the total decay rate and the spontaneous-fission decay rate (decays/atom per unit time) for U^{238}.

Many minerals and glasses are suitable for fission track dating. To be usable, a sample both must retain tracks over geological time and must contain sufficient uranium so that the track density is high enough to be countable in a reasonable pe-

FISSION TRACK DATING

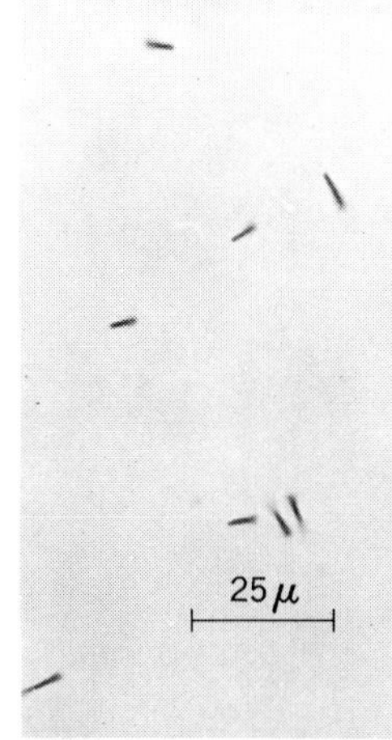

Etched natural fission tracks in muscovite mica crystal. The sample was etched in concentrated hydrofluoric acid.

riod of time. Thus quartz and olivine are highly track retentive but too low in uranium to be usable, whereas calcite may have enough uranium but appears to be inadequately track retentive. Minerals that have been successfully used for fission track dating include the following: zircon, apatite, allanite, epidote, sphene, hornblende, albite, diopside, enstatite, muscovite, phlogopite, and assorted natural glasses.

Applications. One feature unique to this dating technique is the time span to which it is applicable. It ranges from less than 100 years for certain man-made, decorative glasses to approximately 4,500,000,000 years, the age of the solar system. This great flexibility in age span allows fission track dating to fill the gap that previously existed between the upper limit of 50,000 years, in which carbon dating is applicable, and the lower limit of roughly 1,000,000 years, below which potassium-argon dating becomes excessively laborious.

A second useful feature is that measurements can sometimes be made on extremely minute specimens, such as chips of meteoritic minerals or fragments of glass from the ocean bottom.

A third useful feature is that each mineral dates the last cooling through the temperature below which tracks are retained permanently. Since this temperature is different for each mineral, it is possible to measure the cooling rate of a rock by dating several minerals—each with a different track-retention temperature. It is similarly possible to detect and date the reheating of an old rock (caused, for example, by a local igneous intrusion or regional metamorphism) by the finding of a low age for the less retentive minerals and a greater age for the more retentive ones. *See* EARTH, AGE OF; GEOLOGICAL TIME SCALE; MINERAL; ROCK, AGE DETERMINATION OF.

[R. L. FLEISCHER; P. B. PRICE; R. M. WALKER]

Bibliography: R. L. Fleischer, P. B. Price, and R. M. Walker, Nuclear tracks in solids, *Sci. Amer.*, 220(6):30–39, June, 1969; R. L. Fleischer, P. B. Price, and R. M. Walker, Solid state track detectors: Applications to nuclear science and geophysics, *Annu. Rev. Nucl. Sci.*, 1965; R. L. Fleischer, P. B. Price, and R. M. Walker, Tracks of charged particles in solids, *Science*, 149:383, 1965.

Floodplains

The valley floors formed by alluviating rivers which are subject to overflow. They may be narrow, between the valley walls of a small stream, or extremely wide, for example, 125 mi in the Lower Mississippi Valley near Memphis. Toward the sea, floodplains may grade imperceptibly into deltas or deltaic coastal plains. The floodplain of the Nile extends 700 mi below the first cataract at Aswan to the Nile delta. Though it is not much wider than 10 mi at any point, the floodplain and delta account for the support of practically all of Egypt's dense population. Ancient potamic civilizations in China, northern India, Mesopotamia, and Egypt arose along the floodplains of the Sian, Indus, Tigris-Euphrates, and Nile, where fertile alluvial soils and dependable water supply favored agriculture. Dikes and drainage canals now prevent or restrict natural flooding, but in protecting cities and habitations, they rob floodplains of the beneficial effects of overflow.

Valley floor deposits. The depth of alluvium in wide floodplains toward oceanic coasts commonly amounts to 400 ft or more (Fig. 1), a thickness which is a measure of the volume of water returned to the oceans by melting of continental ice masses during the last 12,000, or possibly 18,000, years. Gravels ordinarily occur abundantly toward the base of the alluvial section and are commonly overlain by finer materials such as sand or silt. Braided streams, with numerous branches around bars and lenticular islands, such as exist on floodplains where sand and coarser sediments predominate, deposited the coarser materials at depth. Later, when gradients and loads were re-

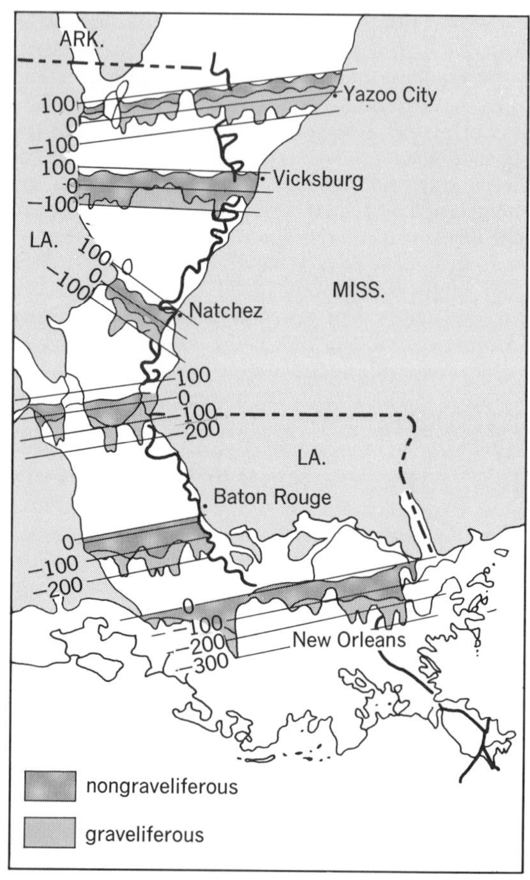

Fig. 1. Depth of sediments at various points in Mississippi alluvial valley. (*After H. N. Fisk, Mississippi River Commission, U.S. Army, Corps of Engineers, 1944*)

Fig. 2. Cutoff lakes, White River, Ark. 1949. (*USGS*)

duced, many floodplain rivers adopted sinuous, meandering courses through finer-grained alluvium.

Floodplain surface features. On a section across a typical wide floodplain, highest elevations commonly occur at the crests of natural levees which flank the immediate sides of channels. Levee backslopes flatten gradually toward backswamp basins, where network drainage, irregular lakes, and high water tables are characteristic and land vegetation may be grassy marsh or tree-covered swamp. Tongues of crevasse deposits, produced during one or more floods, extend natural levee widths irregularly toward the basins. Natural levees are ordinarily a mile or more in width along the Lower Mississippi, and crests rise above levels of backswamps by 5 ft toward the mouth of the river, over 20 ft at New Orleans, and more than twice that elevation at various points farther upstream. Crevasse deposits extend the useful, solid, fertile, and high land of natural levee belts laterally for 10 mi and more at various places. *See* PLAINS.

Meander-sediment relationships. Meandering rivers continually shift their courses as bars grow around the convex banks of points. These deposits, which accumulate mainly during floods, tend to narrow the channel, causing it to deepen and thus promote undermining of the concave cut banks across from the points. In this way the Lower Mississippi each decade abandons about half its channel area to occupy an equal amount of new space, mainly at the expense of cut-bank natural levees. This process also accentuates the size of meander loops, some of which are about 20 mi long. Cut banks, both up- and downstream, however, gradually approach each other, narrowing the neck of the loop, eventually creating a natural cutoff. The abandoned channel becomes a cutoff lake, or oxbow (Fig. 2). These lakes slowly fill with sediment containing considerable clay to become clay plugs—crescentic masses which erode much more slowly than other types of alluvium (Fig. 3). As they are aligned along both sides of the channel, with horns of the crescents toward the river, they serve effectively to restrict channel migrations to definite meander belts.

It is rare to find diversions from such meander belts along the Lower Mississippi and other rivers flanked by clay plugs. On many other rivers, however, floodplains lack fine materials and are incapable of trapping themselves in meander belts. Such is the case along the Great Meander River of Anatolia, where the floodplain is sandy and diversions from the main channel of the river have been numerous.

Other characteristics. Natural levees, definite meander belts, and numerous cutoff lakes are not characteristic of floodplains of rivers that transport little or no bed load, or which differ little in stage between low water and extreme flood. The smaller rivers of Belgium are examples. Many floodplains are modified by alluvial cones of tributary streams. The Great Meander above Nazilli is forced against its south valley wall by an overwhelming alluvial apron, which has been deposited by tributaries on its northern right side and has developed only a narrow floodplain. In a similar manner the San Joaquin has been forced westward by debris from the Sierra Nevada. Cones of tributary streams of the lower Rhone account for alternations between steeper gradients, where the Rhone braids, and flatter stretches upstream from each cone, where the Rhone meanders. *See* COASTAL LANDFORMS; DELTA; FLUVIAL EROSION CYCLE; STREAM TRANSPORT AND DEPOSITION.

[RICHARD J. RUSSELL]

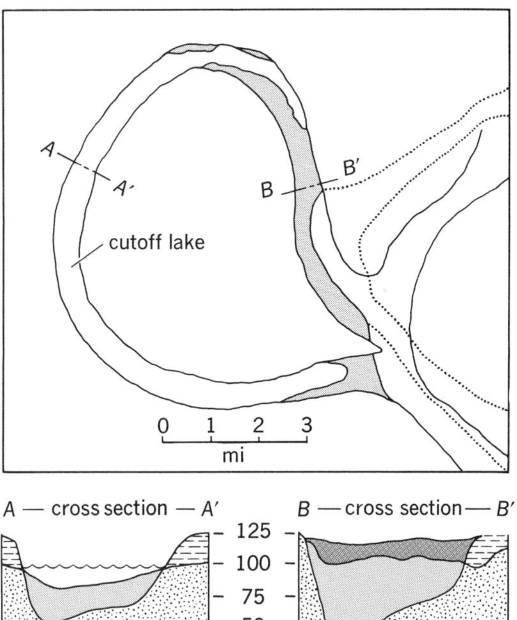

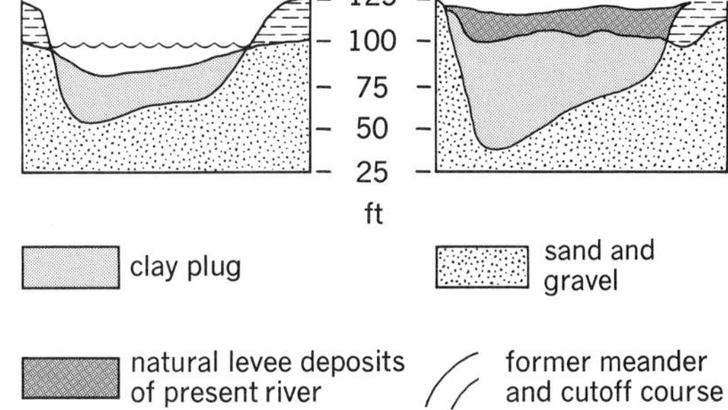

Fig. 3. Clay plugs in abandoned meander. Cross sections along lines *A-A'* and *B-B'* are for cutoff lake and one plug, respectively. *(After H. N. Fisk, Mississippi River Commission, U.S. Army, Corps of Engineers, 1944)*

Bibliography: H. N. Fisk, *Geological Investigation of the Alluvial Valley of the Lower Mississippi River*, Mississippi River Commission, U.S. Army, Corps of Engineers, 1944; R. J. Russell, Geological geomorphology, *Bull. Geol. Soc. Amer.*, 69(1):1–22, 1958; R. J. Russell, *River Plains and Sea Coasts*, 1967.

Fluorite

A mineral of composition CaF_2 and an important structure type with space group $Fm3m$. The fluorite structure type is written RX_2, where the R cations are in cubic (eightfold) coordination by the X anions. Alternatively, the X atoms are tetrahedrally (fourfold) coordinated by the R atoms. The fluorite arrangement may occur when the ratio of $r(R)$, the ionic radius of the R atoms, to $r(X)$, the ionic radius of the X atoms, is equal to or greater than 0.73. The arrangement XR_2 is termed the antifluorite arrangement. Here the anions are eight-coordinated and the cations are four-coordinated. Halides of larger divalent cations, oxides of larger quadrivalent (4+) cations, some intermetal-

FLUORITE

(a)

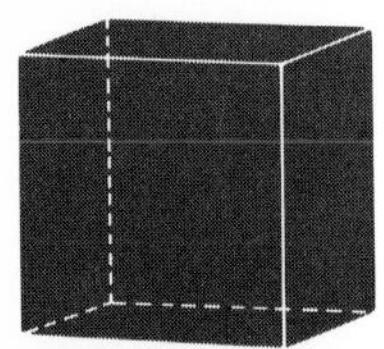

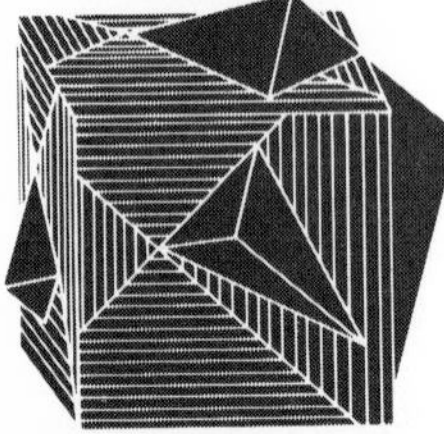

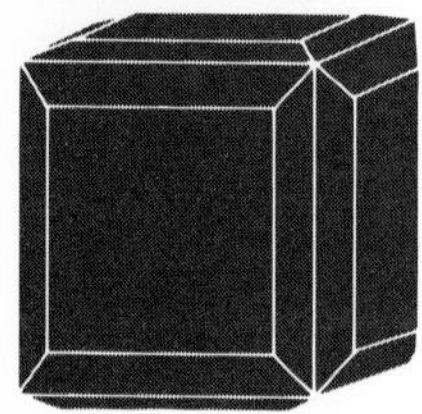

(b)

Fluorite. (*a*) Crystals from Elizabethtown, IL. (*American Museum of Natural History specimens*). (*b*) Crystal habits (*from C. S. Hurlbut, Jr., Dana's Manual of Mineralogy, 17th ed., copyright © 1959 by John Wiley & Sons, Inc.; reprinted by permission*).

lic compounds, and XR_2 alkali univalent oxides frequently belong to the fluorite structure type. Approximately 100 synthetic compounds are known to have this arrangement; distortions may result in similar arrangements but with lower symmetry, such as in ZrO_2, which is monoclinic.

The most abundant fluorine-bearing mineral, fluorite, occurs as cubes or compact masses and more rarely as octahedra with complex modifications (see illustration). The cube surfaces frequently show lineage features. Fluorite has a perfect octahedral cleavage, hardness 4 (Mohs scale), and specific gravity 3.18. The color is extremely variable, the most common being green and purple; but fluorite may also be colorless, white, yellow, blue, or brown. Colors may result from the presence of impurity ions such as rare earths or Mn^{4+}, hydrocarbon inclusions, or lattice defects. Fluorite frequently emits a blue-to-green fluorescence under ultraviolet radiation, especially if rare-earth or hydrocarbon material is present. Some fluorites are thermoluminescent; that is, they emit light when heated.

Fluorite is usually very pure, but some varieties, especially yttrofluorite, are known to carry up to 10 wt % CeF_3, the extra fluorine atoms occurring interstitially in the structure. Fluorite can be recrystallized from crushed material and used as optical glass; its more important uses are as flux material in smelting and as an ore of fluorine and hydrofluoric acid.

Fluorite occurs as a typical hydrothermal vein mineral with quartz, barite, calcite, sphalerite, and galena. Crystals of great beauty from Cumberland, England, and Rosiclare, Ill., are highly prized by mineral fanciers. It also occurs as a metasomatic replacement mineral in limestones and marbles. Fluorite is a minor accessory mineral in many other rocks, including granites, granite pegmatites, and nepheline syenites.

[PAUL B. MOORE]

Fluvial erosion cycle

That landscapes develop in a definite sequence of steps from an initial uplifted mountain mass to a nearly plane surface of low relief is a concept developed by William Morris Davis. This great physiographer made the concept so clear that the "geographical cycle," as he called it, dominated thinking on landform development for about half a century. However, objections were raised to this concept by proponents of the equilibrium theory of landscape development.

Stages in geographical cycle. Davis visualized the cycle as beginning with the uplift of a land mass into mountains. Erosion and transportation by running water are intensified by the topographic relief resulting from the uplift. In the first stage steep-sided valleys develop because of the large differences in elevation between mountaintop and sea level, which give the streams a large potential for erosive work. The first stage is characterized as the period of youth.

There follows a stage of maturity, during which downcutting of valleys essentially ceases because an equilibrium is established between the debris supplied from the eroding mountain mass and the ability of rivers to carry it away. Thus valleys widen while preserving the established river gradients. Hilltops gradually become rounded, and the sharp peaks and spurs give way to an undulating surface.

Through a much longer period of time the mature hills become ever more reduced in height; the river gradients progressively flatten as relief is reduced, and as a result the rivers become even less capable of carrying debris. Movement of weathered material in dissolved form assumes greater importance because of the rivers' lessened ability for transport of a clastic load.

Finally, as a result of a very long period of this condition, called old age, mountains are reduced to low and insignificant mounds on an extensive, nearly plane surface, or peneplain.

If a period of mountain building again uplifts the area, the peneplain is elevated and its dissection begins in a new period of youth. Thus the sequence from youth through maturity to old age can be viewed as a geographical cycle of erosion.

Interpretation of any landscape involves, according to the Davisian concept, the identification of the stage in the cycle presently represented and the recognition of earlier cycles, remnants of which would be preserved. Especially pertinent to such recognition is the occurrence of accordant summits of hill- or mountaintops, which indicate that they had been planed once to a nearly smooth surface which subsequently had been uplifted and then dissected. In the Appalachian Mountains, for example, remnants of at least two former peneplains have been described by various workers, and these examples are classic prototypes of cyclic landscape development.

Landform classification. A great strength of the idea lies not only in its simplicity, but in the possibility of elaboration of the details. Any particular landform or group of forms must be interpreted, as Davis lucidly explained, in terms of process, structure, and stage. Process, whether the operation of wind, ice, stream erosion, gravitational movement, or weathering, gives rise to specific detail of erosional and depositional forms. But the same process acting on rocks of different resistance leads to somewhat different results. Thus structure, meaning the sequence and attitude of rocks, alters the resulting forms. The erosion, for example, of a dome or an anticline consisting of layered rocks of different hardness provides characteristic patterns of stream networks, as well as ridges, cuestas, cliffs, and valleys. But the effects of a given set of processes acting on a particular rock structure or layered sequence may look different, depending on the time during which the processes have been acting; that is, the stage in the geographical cycle yields a different topography, even when given the same processes and structure.

These details of process, structure, and stage within the concept of the geographic cycle of sequential landscape development give rise to a special vocabulary to describe particular forms and their inferred derivations. Physiography, then, became increasingly the description of a landscape history utilizing this expanding nomenclature. *See* PHYSIOGRAPHIC PROVINCES.

Objections to cyclic concept. Although the concept of the cyclic sequence has been widely accepted, certain objections have been raised. Among the principal objections, three technical ones will be mentioned as well as a broad philosophical one.

The first objection is that there is nowhere in the world a land feature that can be called a peneplain exemplifying the final, or old-age, stage of the cycle. Those few large areas of low relief that cut across rocks of varied structure can be accounted for without involving the sequential stages leading to peneplanation. The Canadian Shield, for example, is one such large area of low relief, but its Recent glacial history accounts for many of its characteristics. Many small areas of low relief exist but do not have the attributes required of the concept. The principal examples cited as classic are either uplifted and dissected or buried. An extensive, nearly plane unconformity between beveled heterogeneous old rocks and the overlying sediments is exposed near the bottom of the Grand Canyon. This well-known example was first recognized by John Wesley Powell as an ancient plain of erosion. A similar extant situation does not exist.

The second technical objection is that weathering and erosion operate continuously. The Davisian simplification that mountain uplift is rapid relative to the downcutting process is often not tenable. Since erosion modifies the uplifted mass during the uplift process, the cyclic sequence is not as clear-cut as the cyclic concept envisions. Moreover, tectonic history makes it doubtful whether any large area of the Earth would in fact remain stable for a sufficiently long time to allow the complete development of a landscape through youth and maturity to peneplanation. The lack of stability might well account for the absence of any fully developed peneplain on the Earth today.

Third, Davis assumed that during the mature stage valley widening proceeds primarily from lateral corrosion by rivers, and that this process results in many of the broad and flat valley floors seen in nature. These valley floors were supposed to be underlain for the most part by beveled bedrock. In fact, a large percentage of flat-floored valleys, both in humid and arid areas, including valleys in Virginia where the Appalachian peneplain remnants were considered exemplary, are features of deposition; that is, many valleys are flat floored because of valley filling by stream deposition. *See* FLOODPLAINS.

Philosophically, there is another problem with the concept of sequential landscape development. Its inclusive nature discouraged the formulation of alternative hypotheses, especially those stressing uniformitarianism, a foundation block of the geological science. Further, the emphasis on fitting a given landscape into its inferred proper position in the cycle led to an increasing disregard for detailed field studies, both of process and mapping. However, there has been a surge of interest in quantitative methods, detailed fieldwork, and the application of principles of physics to geomorphology. These interests have led rapidly to the development of an alternative concept of the landscape as an open system in dynamic equilibrium.

Equilibrium concept. The equilibrium concept originated with G. K. Gilbert in the 1880s, and though discussed somewhat in the following half century, was revitalized and expanded by John T. Hack in the 1950s and 1960s. Hack studied in far more detail than had Davis the surficial geology of those portions of Virginia which had long served as the model of the cyclic theory. Hack's quantitative studies showed that the principal features of that part of the Appalachians including the Shenandoah Valley are better explained as the results of long-continued erosion of a thick sequence of rocks, during which an approximately balanced condition of the slopes was maintained throughout. The necessity to postulate a peneplain was eliminated.

The idea of a geographical cycle has been useful as a broad concept. Its application in detail to specific landscapes is no longer the main methodology in geomorphology. *See* DESERT EROSION FEATURES; FLUVIAL EROSION LANDFORMS; GEOMORPHOLOGY; ROCK CONTROL (GEOMORPHOLOGY).

[LUNA B. LEOPOLD]

Bibliography: W. M. Davis, *Geographical Essays*, reprint, 1954; W. M. Davis, The geographical cycle, *Geograph. J.*, 14(5):481–504, 1899; J. T. Hack, Geomorphology of the Shenandoah Valley, Virginia and West Virginia and origin of the residual ore deposits, *U.S. Geol. Surv. Profess. Papers*, 484:84, 1965; J. T. Hack, Interpretation of erosional topography in humid temperate regions, *Am. J. Sci.*, 258-A:80–97, 1960.

Fluvial erosion landforms

Landforms that result from erosion by water running on the Earth's surface. This water may concentrate in channels as streams and rivers or run

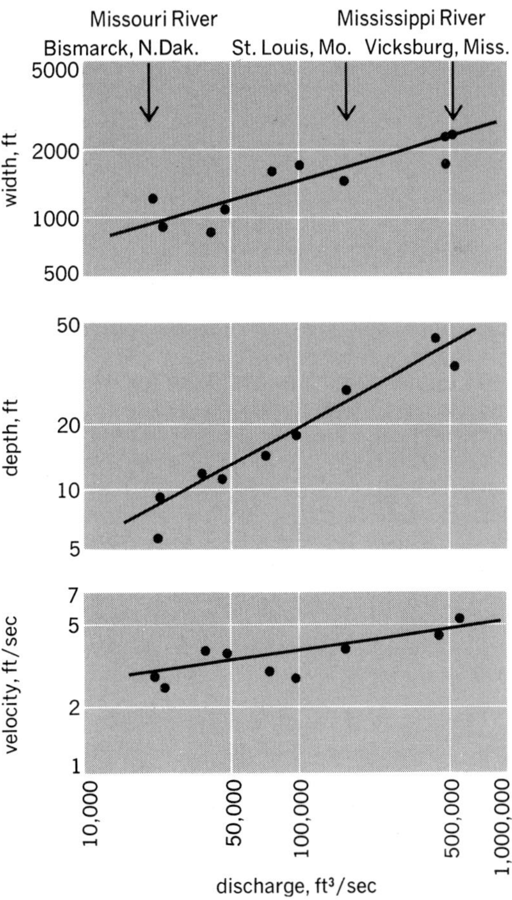

Fig. 1. Width, depth, and velocity of streams increase downstream with increasing discharge, as illustrated by the Mississippi-Missouri river system. (*From L B. Leopold and T. Maddock, Jr., Hydraulic Geometry of Stream Channels and Some Physiographic Implications, USGS Prof. Pap. no. 252, 1953*)

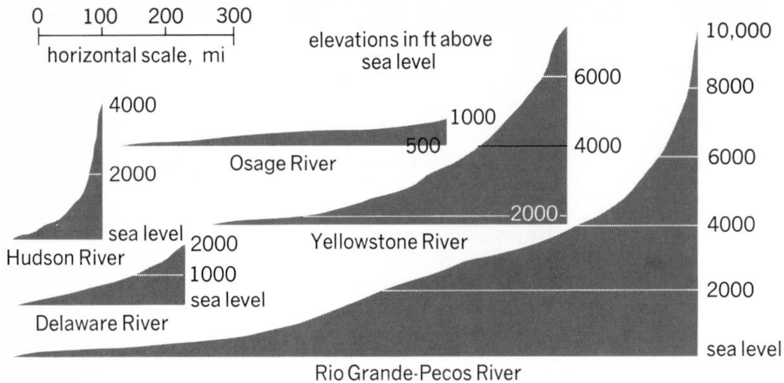

Fig. 2. The profiles of five different rivers. Each profile decreases in gradient from head to mouth and is concave upward. *(From H. Gannett, Profiles of Rivers in the United States, USGS Water Supply Pap. no. 44, 1901)*

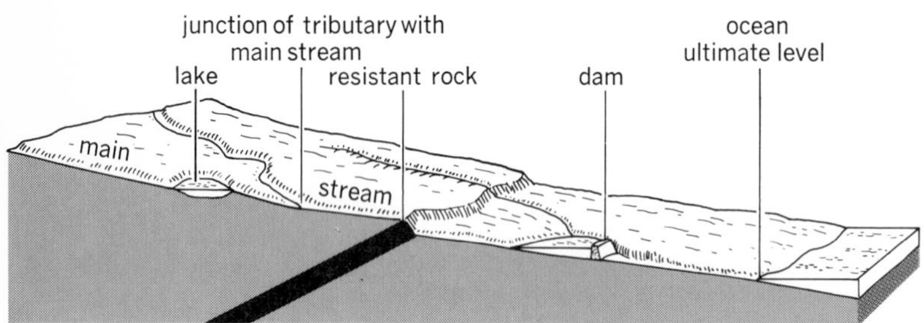

Fig. 3. Differing base levels for streams. *(From L. D. Leet and S. Judson, Physical Geology, 2d ed., copyright © 1958 by Prentice-Hall, Inc.; reprinted by permission)*

in thin sheets down slopes. All land surfaces, even those in the driest deserts, are subject to modification by running water.

Factors affecting fluvial erosion. Running water erodes in the following ways.

1. Direct lifting. Most stream flow is turbulent. The whirling eddies of turbulent water can dislodge material and carry it downstream.

2. Abrasion and impact. The solid particles carried by the stream can wear down bedrock portions of the stream channel or the large rock fragments that pave the bottom. The impact of rock particles thrown against rock may also knock loose fragments which in turn are carried off by the stream.

3. Cavitation. At stream velocities of 25–30 ft/sec cavitation can be effective. In this process vapor bubbles in the water suddenly collapse. This collapse theoretically generates sudden pressures up to 1500 or 2000 lb/in.^{2} Thus bubbles which collapse against a portion of the stream channel act as a sort of hammer blow and provide an effective erosive force.

4. Solution. Enormous amounts of rock material are carried annually to the oceans in dissolved form. Some solution may take place in the stream channel, but most dissolved material probably is contributed to streams by underground water.

5. Undercutting. A stream that erodes laterally by one or more of the above processes may undercut its channel bank or valley wall to the extent that gravity causes material to slump into the river, where it is moved downstream. Once in a stream, material may be carried along in solution, buoyed up in suspension by turbulent water, or rolled, slid, or bounced along the channel bottom.

FLUVIAL EROSION LANDFORMS

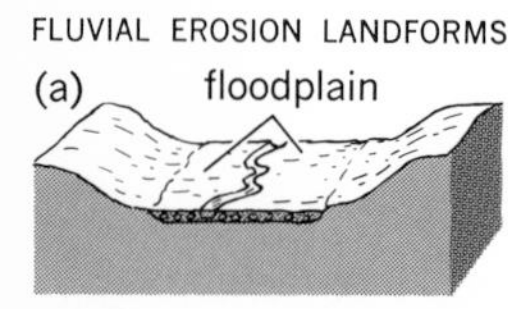

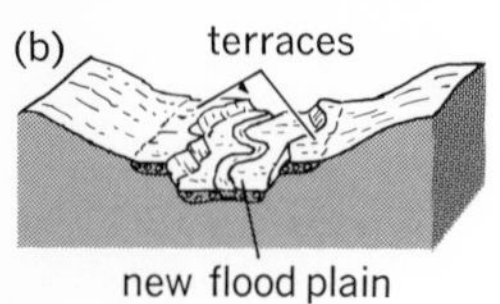

Fig. 4. Development of stream terrace. (*a*) Original floodplain. (*b*) Terrace marks level of abandoned floodplain above level of new floodplain.

Stream profile. Streams erode, transport, and deposit material in an orderly fashion. Each stream and its tributaries form a system that is in dynamic equilibrium. For instance, the width and depth of the channel, as well as the velocity, change in a predictable fashion with changes in the amount of discharge. This is true for both changes of discharge at a particular spot on a river and increases in discharge downstream.

Figure 1 shows that if a stream's depth, width, and velocity are measured at a standard flow frequency (here median flow) at several points along the stream, then as discharge increases downstream, the width and depth increase as expected. But surprisingly velocity also increases downstream. This happens to satisfy the relation that discharge must equal width times depth times velocity. Erosion increases the stream's depth and width but not rapidly enough to accommodate the discharge at a constant, much less a decreasing, velocity.

A corollary to the increase in width, depth, and velocity of a stream as discharge increases downstream is the overall shape of the longitudinal profile of a stream from its headwaters to the mouth. This profile is parabolic in shape and concave to the sky, and its gradient decreases rather than increases downstream (Fig. 2).

Base level. This general concept is useful in considering the erosional activity of a stream. A base level is the point below which a stream cannot erode. Thus the point at which a stream enters the ocean is the ultimate base level of the stream. If the ocean level is constant, the stream over the course of years will reduce its profile toward this base level, although it will reach it only at its junction with the ocean. As a result, although the profile will become gentler with time, it will always maintain its parabolic form, concave upward. If base level is lowered, either by falling sea level or rising land, then erosion will take place as the stream seeks to adjust itself to the new, lower base level. Conversely, a rise of base level, by whatever means, tends to cause stream deposition. Although the ocean is the ultimate base level, other local base levels may exist along a stream's course. Thus the presence of resistant rock will provide a local, although temporary, base level for all segments of the stream upstream from it. Other temporary base levels include lakes and swamps along the course. For a tributary base level is the elevation of the trunk stream it enters (Fig. 3).

Differential erosion. This aspect of erosion is critical in considering stream erosion. Earth materials erode at different rates. Thus quartzite is more resistant to erosion than shale. Climate may also affect the rate at which rocks are eroded. Limestone, for instance, is soluble in natural waters and therefore can be removed more rapidly in well-watered climates than in arid or semiarid regions. As a result limestones often underlie valleys in the eastern United States, whereas they form ridges in the drier, western states.

Features of fluvial erosion. Features of river erosion range in size from minute features of the channel to regional landscapes. The features that are most intimately related to fluvial erosion are undoubtedly those that are directly associated with the stream channel. Among these are included not only the cross section of a stream channel but also the long, concave-upward profile of a stream from head to mouth. Even when this shape is not directly related to a stream channel, it may still be due to running water which flows down slopes as sheets or in broad shallow rills during exceptional rains.

Stream terraces. A stream terrace is a relatively flat surface extending along a valley with a steep bank separating it from a lower terrace or from the floodplain of a river. The flat surface of the terrace represents a remnant of the stream's former floodplain. The terrace now stands above the general level of the stream's new floodplain because the stream has eroded downward and abandoned its earlier floodplain (Fig. 4).

Waterfalls and rapids. These are impressive, but geologically transitory, features of a stream. They are present because of a sudden drop in the stream course. This interruption in the profile of the stream is in many instances due to resistant rock which the stream cannot easily remove. Thus Niagara Falls is held up by a thick layer of resistant dolomite which overlies less resistant shale. Other falls may exist because a main valley has been deepened much more rapidly than a side valley, leaving tributary streams to plunge into the main valley over a falls. Several of the falls in the Yosemite National Park in California are of this type. The main valley of the Merced River has been deepened by now vanished glacier ice to a level much below less severely eroded side valleys. Whatever the particular cause of waterfalls, the water which moves over them works constantly toward their destruction. As the falls erode, they tend to migrate upstream as suggested in Fig. 5. Rapids may form as falls are destroyed or because of a rapid increase in the stream gradient.

Stream patterns. These are varied and each pattern indicates something about the underlying earth materials. A stream system may describe a branching, treelike (dendritic) pattern. Such an arrangement of a stream and its tributaries is actually a random orientation of channel ways and develops on earth materials that have a relatively uniform resistance to erosion, such as granite or flat-lying sedimentary rocks.

When the subsurface rocks are not uniform in resistance, streams tend to cut their valleys in the zones of least resistance. Therefore the resulting stream patterns tend to follow the underlying rock pattern. As an example, bedrock fractured into rectangular blocks tends to produce a rectangular drainage pattern reflecting the easily eroded fracture zones. A trellis pattern also develops because of varying resistance to erosion of the underlying rock. In this case, parallel bands of resistant and nonresistant rock are etched out by streams carving valleys in the least resistant bands. These streams often join a master stream which cuts across the resistant and nonresistant beds alike, giving the entire drainage the aspect of a garden trellis. Sometimes the stream pattern reflects the original topography of an area. Thus, on a domal area, streams tend to radiate outward from the high center; the term radial applies to this drainage.

Summary. Running water is the most important of all the processes which fashion the landscape. Directly or indirectly, water flowing in streams is responsible for carving most of the valleys of the continents. Streams cut into the land, eroding both downward and from side to side. These incisions allow gravity to operate along the valley walls. This motion of material under the influence of gravity may be either rapid, as in landslides, or slow, by almost imperceptible creep of the unconsolidated surficial cover derived by weathering from the firmer rock beneath. In addition to direct stream erosion and indirect contributions by gravity, material comes to the stream by sheets of water flowing down slopes. No matter how streams acquire their load of material, they serve as endless conveyor belts moving debris from the land to the oceans. As an area is worn lower and lower, its landscape passes through a series of stages known as the fluvial erosion cycle. *See* FLUVIAL EROSION CYCLE. [SHELDON JUDSON]

Bibliography: L. B. Leopold, M. G. Wolman, and J. P. Miller, *Fluvial Processes in Geomorphology*, 1964; W. D. Thornbury, *Principles of Geomorphology*, 1954.

FLUVIAL EROSION LANDFORMS

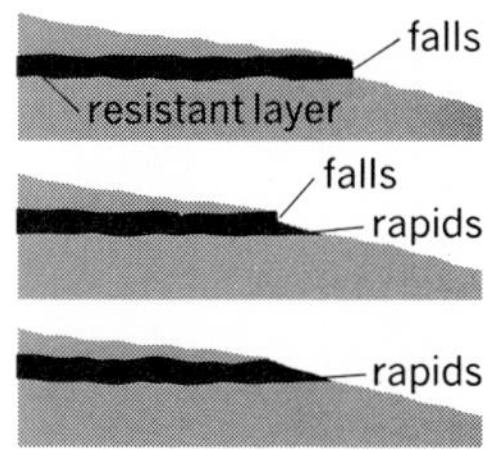

Fig. 5. Falls formed where stream crosses resistant rock. As falls retreat upstream, they may develop into rapids.

Fold and fold systems

Folds are recognized where layered rocks have been distorted into wavelike forms. Some folds are fractions of an inch across and have lengths measured in inches, whereas others are a few miles wide and tens of miles long.

Elements of folds. The axial surface (Fig. 1) divides a fold into two symmetrical parts, and the intersection of the axial surface with any bed is an axis. In general, an axis is undulatory, its height changing along the trend of the fold. Relatively high points on an axis are culminations; low points are depressions. The plunge of a fold is the angle beween an axis and its horizontal projection. The limbs or flanks are the sides. A limb extends from the axial surface of one fold to the axial surface of the adjacent fold. Generally, the radius of curvature of a fold is small compared to its wavelength and amplitude, so that much of the limbs is planar. The region of curvature is the hinge. *See* ANTICLINE; SYNCLINE.

Geometry of folds. The geometry of folds is described by the inclination of their axial surfaces and their plunges (Fig. 2). Upright folds have axial surfaces that dip from 81° to 90°; inclined folds have axial surfaces that dip from 10° to 80°; and recumbent folds have axial surfaces that dip less than 10°. Vertical folds plunge from 81° to 90°; plunging folds plunge from 10° to 80°; and horizontal folds plunge less than 10°. Auxiliary descriptive terms depend on the attitude or the

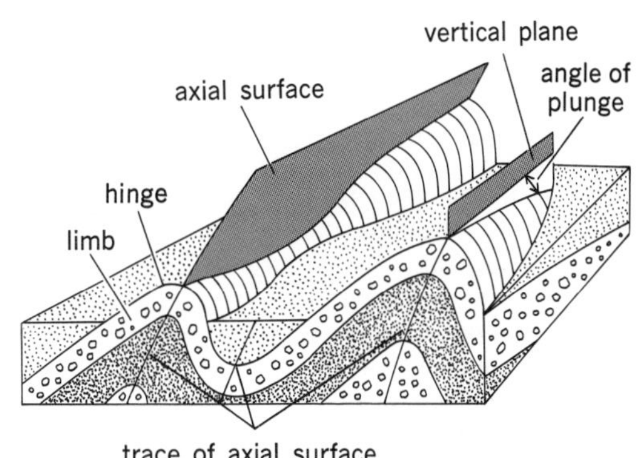

Fig. 1. Elements of folds.

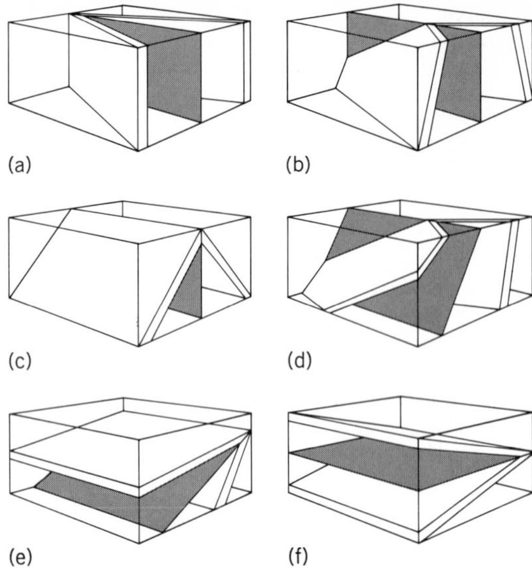

Fig. 2. Description of folds based on the attitudes of their axial surfaces and their plunges. (*a*) Vertical. (*b*) Upright plunging. (*c*) Upright horizontal. (*d*) Inclined plunging. (*e*) Inclined horizontal. (*f*) Recumbent. (*Modified from M. J. Rickard, A classification diagram for fold orientations, Geol. Mag., 188:23, 1971*)

relative lengths of the limbs. Overturned folds are inclined folds in which both limbs dip in the same direction; isoclinal folds are those in which both limbs are parallel; symmetrical folds have limbs of equal length; and asymmetrical folds have limbs of unequal length. The descriptions of folds consist of combinations of the above terms, for example, isoclinal upright horizontal fold, overturned plunging fold, asymmetrical inclined horizontal fold.

Character of folds. In a section of folded rocks the layers possess different rheological properties; some have apparent stiffness (competency), whereas others behave less stiffly (incompetency). The most competent layers or group of layers control the folding, and the less competent units tend to conform to the fold-form of the most competent units. The least competent units take up the strain between the adjacent fold-controlling units.

The following nomenclature has been devised. The most competent units in the folded sequence are the dominant structural members. Those layers slightly less competent that conform to the fold-form of the dominant structural member are competent conforming members, and the least competent units are passive structural members.

The dominant structural members deform through much of their history of folding so that their thicknesses measured perpendicularly to their boundary surfaces are nearly constant at all points in the fold-form (Fig. 3*a*). Such folds are said to be concentric. The passive structural members fold so that their thicknesses measured perpendicularly to their boundary surfaces vary from point-to-point in the fold-form, but thicknesses measured in the plane of their axial surfaces and perpendicularly to their axes are relatively constant (Fig. 3*b*). Such folds are called similar. The competent conforming members deform in a manner intermediate between that of concentric

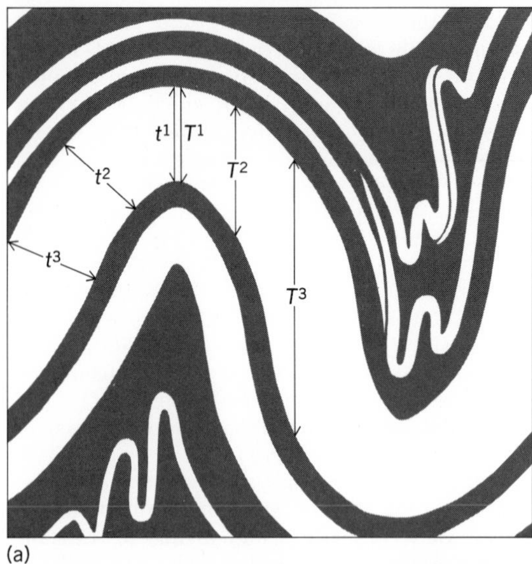

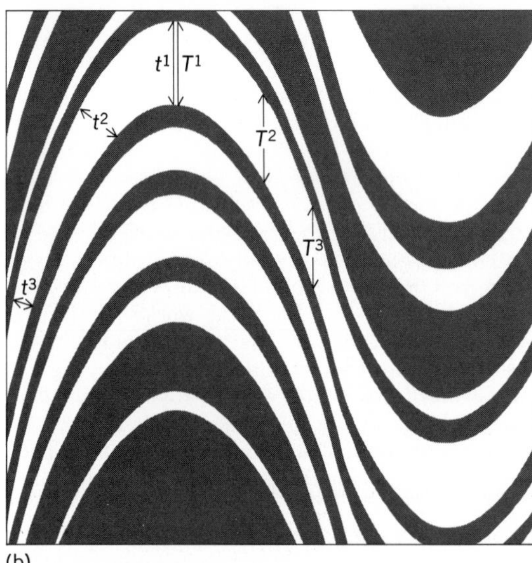

Fig. 3. Geometrical features of (*a*) concentric folds and (*b*) similar folds. Here t^1, t^2, and so on, are thicknesses measured perpendicular to boundary surface of folded layer. T^1, T^2, and so on, are thicknesses measured in axial surface and perpendicularly to fold axis. (*From J. G. Ramsay, The geometry and mechanics of "similar" type folds, J. Geol., 70:310, 1962*)

and that of similar folds. Under conditions of extreme strain all structural members may be flattened and take on the form of similar folds. Fold-forms when followed either upward or downward in a layered sequence tend to be disharmonic.

The wavelengths of concentric folds are related to the thickness of the dominant structural member. Studies of the relation between wavelength and thickness of the dominant structural member in folds of all sizes conclude $L \sim 27T$, where L = wavelength and T = thickness of dominant structural member. In a layered sequence, first-cycle folds are controlled by the dominant structural member. Thin conforming competent members have, in addition to first-cycle fold-forms, second-cycle folds (minor folds) developed in the limbs of the first-cycle fold. The wavelengths of these minor folds are related to the thickness of the conforming competent member, that is, the thinner the member the shorter the wavelength.

Strain-stress relationships. Strain in rocks can be measured by comparing the shape of objects of known original form (fossils, pebbles, ooids, reduction spots) to their form after strain. Most studies of strain in folds have dealt with rocks that must be considered to be passive structural members in regions of intense folding. Where the folding has not been complicated by more than a single episode of folding, the orientations of the strain axes are as follows: the greatest direction of shortening is perpendicular to the axial surface; the greatest direction of extension lies in the axial surface and is subperpendicular to the axis; and the intermediate strain direction is perpendicular to the other two directions. Strains as great as 60 to 120% are recorded. Doubtless, detailed studies of strain in competent units would have orientations that differ with position within the unit, but such observations are generally beyond the resolution of field studies.

The measured strain orientations in folds of many regions are consistent with folds developed by axial compression of the layered sequence. Theoretical models have been derived on this assumption. Such models also assume the following: the structural members are ideally elastic or viscous; strains are infinitesimal; the thickness of the individual structural members is small relative to the thickness of the layered sequence; the axial load is constant throughout the length of the structural member; vertical normal stresses vary continuously; the fiber strains are linearly related to the distance from the center of the dominant structural member; and the layered sequence is confined by rigid boundaries. The wavelength of the multilayered sequence is given by Eq. (1),

$$L = 2\pi\left(\frac{T}{n\pi}\right)^{1/2}\left(\frac{B_h I}{btB_v}\right)^{1/4} \qquad (1)$$

where L = wavelength; T = total thickness of multilayered sequence; t = thickness of dominant structural member and associated members; b = breadth of structural members; n = number of cycles of fold; B_h = horizontal modulus of elasticity; B_v = vertical modulus of elasticity; and I = moment of inertia. The critical load necessary for buckling with the wavelength given by Eq. (1) is expressed as Eq. (2), where P = critical axial load and G_a = modulus of shear.

$$P = 2\left(\frac{n\pi}{T}\right)(btB_vB_hI)^{1/2} + G_a tb \qquad (2)$$

Equations for viscous structural members are similar, but time becomes a significant parameter in them, and there is no critical axial load for buckling, although the amount of deflection is sensitive to the magnitude of the axial load. Equations based on both elastic and viscous structural members give wavelengths and numbers of dominant structural members per layered sequence in good agreement with observed characteristics of folded sequences.

Folding may be simulated with computers by programming suitable mathematical equations that relate the necessary boundary conditions, physical properties, and selected effects. These equations can be solved for any set of desired parameters. Two-dimensional maps of the distribution of stress and strain have been constructed for a single dominant structural member embedded in a viscous medium during folding (Fig. 4). The strain directions in simulated folds with finite amplitudes and steep limbs have a remarkable similarity to the distribution of strains measured in folded strata.

Minor folds and cleavage. Many first-cycle folds are accompanied by higher-cycle folds, called minor folds, and by cleavage. Empirical data indicate that, where these features formed in the same episode of deformation, the axes of minor folds are oriented approximately parallel to the axes of the first-cycle folds. Minor folds also have symmetries that vary with position in the first-cycle folds; they are asymmetrical in the limbs of the larger folds and symmetrical in their hinges.

The direction of deflection in the hinges of adjacent lower-cycle folds can be inferred from the minor folds (Fig. 5a). The short limbs of the minor folds face the hinge of the larger fold to be considered, and the direction of deflection in this hinge relative to the limb observed is that given by following the long limb into the short limb of the minor folds.

Investigations also show that cleavage, which is

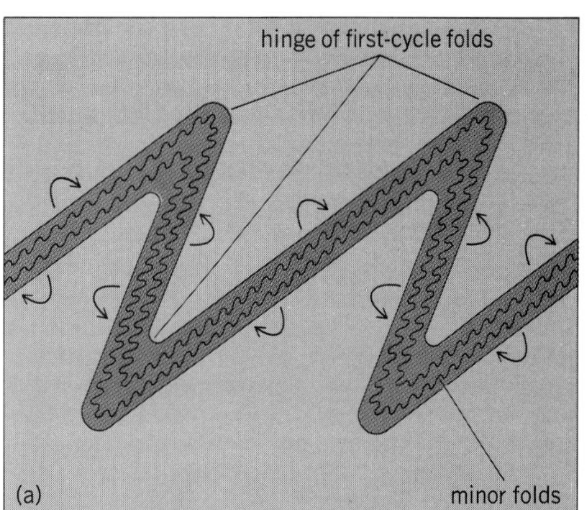

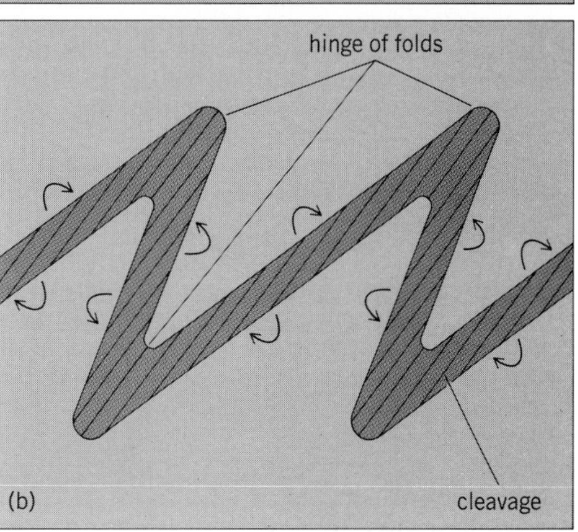

Fig. 5. Relationship between (*a*) minor folds and larger folds and (*b*) cleavage and folds; arrows indicate direction of deflection of adjacent hinge relative to limb observed.

FOLD AND FOLD SYSTEMS

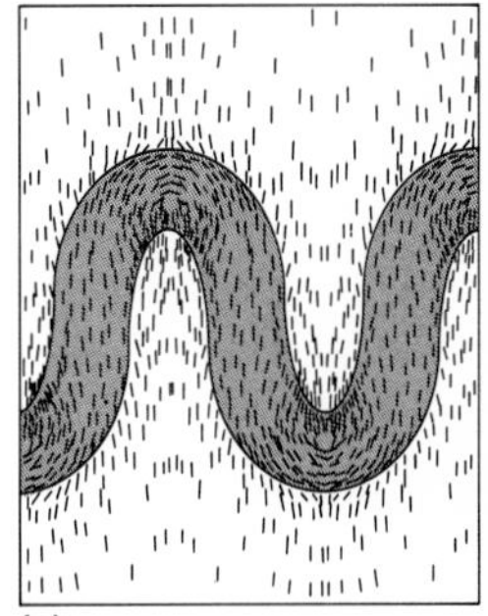

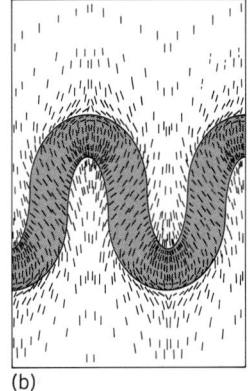

Fig. 4. Computed two-dimensional fold model after 100% shortening. (*a*) Stress relations; short lines are perpendicular to the maximum normal stress. (*b*) Strain relations; short lines are perpendicular to direction of maximum shortening. (*From J. H. Dieterich, Computer experiments on mechanics of finite amplitude folds, Can. J. Earth Sci., 7:472, 473, 1969*)

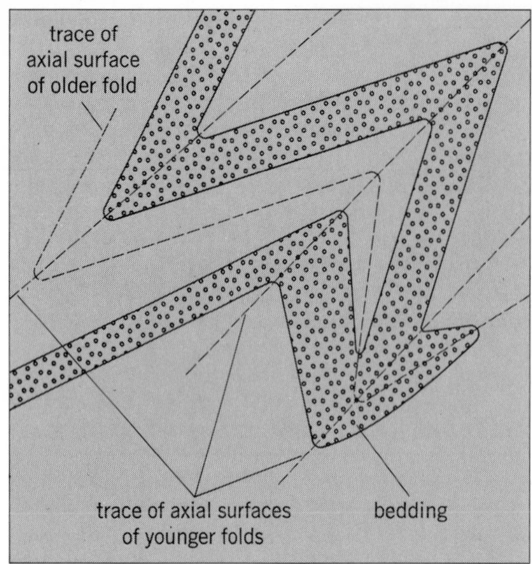

Fig. 6. Superimposed folds showing their relative ages.

manifest by closely spaced surfaces of rupture, approximately parallels the axial surfaces of folds, if these features were formed in the same episode of deformation. The geometric relation between cleavage and bedding is such that their intersection is approximately parallel to the axis of the fold. The direction of deflection in the adjacent hinges of folds can be inferred from the cleavage-bedding relationship (Fig. 5*b*). The acute angle between cleavage and bedding point to the hinge to be considered, and the direction of deflection in this hinge relative to the limb observed is that given by following the trace of bedding into the trace of cleavage around the acute angle between them. Generalizations of the relationships between minor folds and cleavage to larger folds allows the reconstruction of large folds from scattered observations.

Multiple folding. In many regions, layered rocks have had a long history of deformation including several episodes of folding. Because younger deformations do not completely obliterate earlier deformations, a complex array of folds develops. The relative timing of the various folds can be ascertained by their mutual relationships; the axial surfaces of older folds are folded by younger folds, and the axial surfaces of the youngest folds are not deflected (Fig. 6). Locally, folds of different ages may be directly superimposed, and it is at these localities that the relative ages are most readily determined. Once the relative timing of folding has been established at one or more localities, the relative ages can be applied to other localities on the basis of fold style. The orientations of superposed folds depend on the orientation of the forces producing them and on the attitude of the surfaces that are folded.

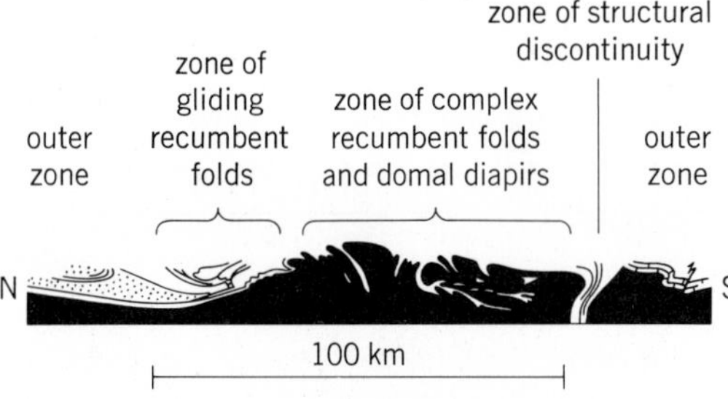

Fig. 7. Section through the Alpine Fold System. (*From J. F. Dewey and J. M. Bird, Mountain belts and the new global tectonics, J. Geophys. Res., 75:2639, 1970*)

Fold systems. Folds generally do not occur singly but are arranged in festoons in mobile belts with lengths of thousands of miles and widths of hundreds of miles. The folds of these belts, or fold systems, commonly consist of great complex structures composed of many smaller folds. In this regard, a large anticlinal structure consisting of numerous smaller folds is called an anticlinorium, and a large synclinal structure consisting of many smaller folds is termed a synclinorium. The trends of these smaller folds are more or less parallel, although the relationships may be complex. In some regions, folds belonging to several generations may be present, and locally these may have divergent trends.

Although comparisons of sections across fold systems are difficult to make because no two fold systems are identical and many are incompletely exposed for study, certain generalities emerge. Most fold systems are asymmetrical with respect to a zone of structural discontinuity (Fig. 7). The fold system is more extensively developed on one side of this zone than on the other, and the direction of structural transport on either side is directed toward the margins of the fold system. On the less-developed side of the zone of structural discontinuity, the folds are recumbent or asymmetrical and are accompanied by steeply dipping thrust faults. On the more extensively developed side, and adjacent to the zone of structural discontinuity, is a region of complex recumbent folds and domal massifs. On the outer side of these complex structures, recumbent folds that have moved toward the margin under the influence of gravity are extensively developed, and in the outer margin the folds are asymmetric and inclined and are associated with thrust faults having gentle dips.

The development of fold systems is closely tied to concepts of global tectonics. The favored hypothesis is that of plate tectonics, in which fold systems are formed at converging margins where continents are underridden by oceanic crust, or where collision occurs between a continent and an island arc or between two continents. Alternatively, the hypothesis of an expanding Earth, though less favored, is consistent with much of the geologic and geophysical data. According to this hypothesis, major zones of extension are the loci of gravity-driven diapirs whose upper parts spread laterally under their own weight to give rise to fold systems. *See* GEOLOGY, EVOLUTION OF; TECTONIC PATTERNS. [PHILIP H. OSBERG]

Bibliography: M. P. Billings, *Structural Geology*, 3d ed., 1972; S. W. Carey, The expanding Earth: An essay review, *Earth Sci. Rev.*, 1975; J. B. Currie et al., Development of folds in sedimentary strata, *Geol. Soc. Amer. Bull.*, 1962; J. F. Dewey and J. M. Bird, Mountain belts and the new global tectonics, *Geophys. Res.*, 1970; J. H. Dieterich, Computer experiments on mechanics of finite amplitude folds, *Can. J. Earth Sci.*, 1970; J. H. Dieterich and N. L.

Carter, Stress history of folding, *Amer. J. Sci.*, 1969; A. M. Johnson, *Physical Processes in Geology*, 1970; G. J. Ramsay, *Folding and Fracturing in Rocks*, 1967; M. J. Richard, A classification diagram for fold orientations, *Geol. Mag.*, 1971.

Fossil

The remains and traces, such as footprints and trails, of organisms preserved from the geologic past. The term fossil was originally used for a wide range of curios, both organic and mineral, found in rocks; but it came to be restricted to organic remains about the close of the 17th century. The term fossil is still used as an adjective and, in a figurative sense, to indicate that nonorganic structures in the rocks, such as fossil mudcracks, rain imprints, ripple marks, and even fuels, were formed in the geologic past.

Nature of fossils. Several different modes of preservation are known. Generally only the hard parts, such as bones and shells, have been preserved, but in some instances even the soft tissues are preserved almost intact.

Unaltered remains. Extinct mammals, such as the woolly mammoth, have been found in the ice or frozen ground of the Arctic, where they remained in cold storage for some thousands of years. The Beresovka mammoth, found in eastern Siberia in 1901, was virtually complete, with clotted blood in the chest and unswallowed food in the mouth. Part of the flesh was still red and was eagerly devoured by the collectors' dog team. In Poland, entire carcasses of the extinct woolly rhinoceros are preserved in oil seeps. Such preservation of animal tissues is limited to late glacial and postglacial deposits. Fossil wood is more durable, however, and is found virtually unaltered in rocks as old as the Cretaceous.

Petrified fossils. Some organic objects are turned to stone. This usually involves only the hard parts such as bones, shells, and wood. Petrifaction takes one of three forms:

1. Permineralization involves only the addition of mineral matter from underground solutions which fills the voids (mostly microscopic) in solid structures such as bone, shell, or wood.

2. Replacement involves the substitution of some other mineral, from underground solutions, for the original material. This is commonly accomplished molecule by molecule so that even microscopic structures are preserved while the object is literally turned to stone. The fossil wood shown in Fig. 1 illustrates both replacement and permineralization. The original cellulose of the cell walls has been replaced by quartz, and the voids of the empty cells have been filled by quartz.

Fig. 1. Petrified wood from the Fossil Forest of Yellowstone National Park. (*a*) Wood block. (*b*) Thin slice enlarged to show the cells in three growth rings.

3. Distillation concerns organic tissues, such as cellulose and flesh, that are largely made of compounds of carbon, hydrogen, and oxygen. After burial, and in the absence of oxygen, such compounds undergo destructive distillation that liberates carbon dioxide and water until only free carbon remains. This carbon forms a black film in the rock, a carbon copy of the animal or plant. Fossil leaves are commonly preserved in this way, and, more rarely, the outline of the fleshy body is preserved surrounding the skeleton of a fossil animal (Fig. 2). Even the remains of completely soft-bodied animals are thus preserved in the Burgess black shale of Middle Cambrian age near Field, British Columbia.

Mold, casts, and imprints. These form another category of fossils. They are formed when the sediment has solidified about an organic object and the latter is subsequently dissolved, leaving a hole in

Fig. 2. A marine reptile, *Ichthyosaurus,* from black shale of the Lower Jurassic period, which was discovered at Holzmaden, Germany. (*American Museum of Natural History*)

Fig. 3. Fern leaf from the Coal Measures.

the rock. This is a mold. By pressing a plastic substance such as dental wax into the mold an artificial cast of the object is obtained. Deposition of mineral matter from underground solutions may also fill the hole, producing a natural cast. Fossils which preserve the form but not the substance of the originals are called pseudomorphs. Molds of thin objects such as leaves are commonly called imprints. Figure 3 illustrates the imprint of a fern leaf which has undergone distillation. The carbon film faithfully records the shape of the leaf, but even if the carbon were removed it could be recognized from the imprint in the rock.

Tracks, trails, and burrows. These traces constitute a fourth category of fossils. Although involving no remnant of the organism, they are structures made by a living animal and, as such, may give evidence of its size, shape, and habits. They tell, for example, whether a dinosaur was bipedal or quadrupedal, whether it was slender and agile or ponderous and slow, and whether it ran like an ostrich or leapt like a kangaroo (Fig. 4).

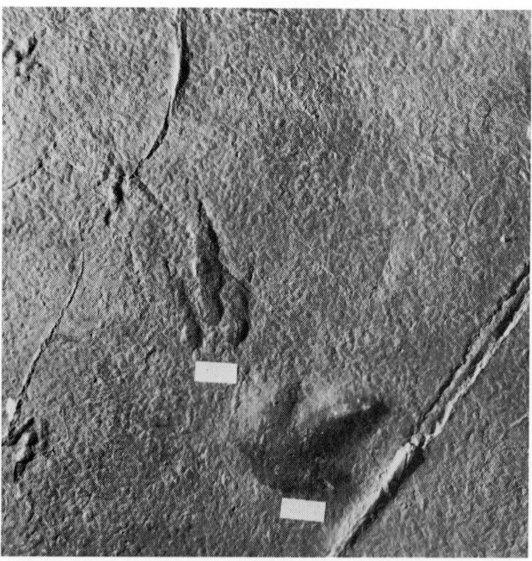

Fig. 4. Tracks of a small bipedal dinosaur in Triassic sandstone from Turners Falls, Mass.

Even worm tubes in an ancient sandstone prove that living organisms were present and may give clues as to the local environment under which the rocks were formed.

Coprolites. These fossils are petrified excrement. Undigested particles of food preserved in such fossils may indicate the feeding habits of animals long since extinct. Plant fragments in the dung associated with the extinct ground sloths, for example, have demonstrated the kind of vegetation they preferred.

Uses of fossils. The fossil record is of major importance to the geologist. Several aspects of this record are considered.

Correlation. Fossils permit correlation of sedimentary rocks of equivalent age the world over. By this means a synthesis of local details into broad regional interpretation is possible.

The use of fossils in correlation was discovered accidentally by W. Smith just before 1800, as a result of fieldwork in the richly fossiliferous Jurassic rocks of southern England. In that region a long sequence of formations (alternating limestones and shales) lies gently tilted and crops out in bands that are easily traced across the country. After extensive collecting, Smith perceived that each of these formations carries a unique assemblage of fossils by which it could be recognized in any outcrop. This insight was soon used by other geologists and found to be widely applicable. Fossils serve as a guide or index to any fossiliferous rock and characterize the deposits of a limited part of geologic time. *See* INDEX FOSSIL.

Smith had no idea why this should be so. It was an empirical discovery based on tedious and detailed observations and rested on no theory. Darwin's *Origin of Species*, published in 1859, provided the explanation. Life has been steadily evolving throughout geologic time and each genus and species had a relatively short duration; thus the faunas entombed in successive ages must necessarily differ each from the next. *See* STRATIGRAPHY.

Chronology. Fossils date the rocks in terms of geologic time. Once the sequence of faunas has been worked out in a long succession of undisturbed sedimentary deposits, it is possible to determine the relative ages of the fossils by the law of superposition, the oldest formation being the lowest and each in turn being younger than the one on which it rests.

Of course, no single region contains a complete record of all geologic time, and if it did the succession of strata would be so thick that the lower part would be deeply buried and inaccessible. The fact that part of the record is preserved in one region and other parts in other regions makes it necessary to correlate only enough partially overlapping sequences to establish a composite record for all of geologic time. In this way the geologic column has been built up by the collaboration of geologists throughout the world, and the corresponding geologic time scale has been established in terms of which events can be placed in proper chronologic order. If the stratified rocks are a manuscript in stone, fossils supply the pagination. No other principle has had such far-reaching influence in geology. *See* GEOLOGICAL TIME SCALE; GEOLOGY.

Past environments. Fossils indicate the local environment under which many of the stratified deposits were laid down. The presence of corals, echinoderms, brachiopods, cephalopods, and other such exclusively marine organisms proves that the enclosing sediments were laid down on the sea floor, even though they are now far inland and even high in mountain crests. On the other hand, the presence of abundant land animals or of land plants, and of stumps in their place of growth prove deposition to have taken place on land. More specific details about the environment may be indicated by certain fossils, especially when their study is combined with that of the sediments. Coral reefs, for example, will signify a warm and shallow sea floor. Other associations may indicate shallow mud flats, and still others broad shallow sand flats. The presence of exclusively pelagic organisms in dark shale or limestone will suggest deeper water and foul stagnant bottoms on which benthonic organisms could not survive. Certain types of vegetation will suggest a swampy lowland, whereas buried stumps and standing tree trunks in sandstone indicate rapid deposition and quick burial on a river flood plain. These are only samples of the use of fossils in judging the environment of deposition of rocks of the past. *See* PALEOECOLOGY; SEDIMENTATION.

Geography of past ages. When the local environment of a sufficient number of areas has been worked out and all the deposits of a given age are correlated, it becomes possible to construct a map showing the distribution of lands and seas of a particular geologic age for a limited region, a whole continent, or even for the whole Earth.

It is known, for example, that for a time during the middle of the Cretaceous Period, North America was divided into two land masses separated by a vast inland sea that stretched from the Gulf of Mexico to the Arctic Ocean. *See* PALEOGEOGRAPHY.

Record of life. The history of life on Earth and evidence of organic evolution are provided in the fossil record. Because soft-bodied organisms are rarely preserved there are numerous groups of animals and plants whose history is virtually unknown, but for those possessing hard parts (skeletons, shells, or woody tissue) a vast amount of knowledge is now at hand and is still growing as more and more fossils are discovered.

In regions where deposition was nearly continuous for considerable spans of time, fossils from successive horizons show an orderly and progressive change in the morphology of many different groups of animals or plants. The evolution of the horse from a tiny ancestor with four toes on the front feet and three on the hind feet during Cenozoic time is one example of the many phylogenies that are clearly documented by fossils.

Connecting links showing a common ancestry of stocks that have now diverged widely are not uncommon. A classic example is *Archeopteryx*, the Jurassic bird whose reptilelike teeth, claws, and tail clearly indicate that birds evolved from reptiles. The theriodont reptiles of Triassic time likewise record almost every step in the evolution from reptile to mammal. In smaller details the phylogeny of many smaller taxa is known from suites of fossils gathered from successive horizons in the rocks. *See* EVOLUTION, ORGANIC; PALEOBOTANY; PALEONTOLOGY.

Ancient climatic conditions. Many groups of modern animals and plants are adapted to rather restricted climatic environments: the musk ox and reindeer, for example, to the Arctic tundra; the crocodile, great tortoises, and palm trees to the tropics and the subtropics; and hardwood forests to the temperate zone. Wherever such organisms are preserved in sedimentary deposits far from their present range the indications are strong that the climate has changed.

Caution must be used, however, because some groups are highly adaptive and individual species may have adjusted themselves to conditions far from the norm. The woolly mammoth and the woolly rhinoceros, for example, were Arctic adaptations during Pleistocene time of groups otherwise limited to the tropics, but a whole fauna or flora is unlikely to show such abnormal specialization. The rich flora in the Cretaceous rocks of southwest Greenland, for example, records a forest of deciduous trees normal to the temperate zone with some species, such as the breadfruit tree, now confined to the tropics. Surely this barren, treeless land had a much milder climate in Cretaceous time. Fossil plants have also been found in the rocks jutting out of icebound Antarctica. During the ice ages of the Pleistocene Epoch the reindeer spread southward across Europe and was the chief game animal of prehistoric man. In America at the same time the musk ox migrated southward into New York and into Arkansas and Utah. For this period the fossil evidence of cold climate is abundantly confirmed by glacial deposits. During middle Cenozoic time Nevada was occupied by a flora composed of genera now restricted to the low, humid, coastal plain of Louisiana and eastern Texas; it could not therefore have been arid as it is today.

The evidence of past climates is clear for Cenozoic time, when the plants and invertebrate animals were mostly of genera and even of species which are still living; the evidence of climates is less and less obvious, however, in earlier periods, when the investigator must judge by extinct genera or larger taxa of organisms long since extinct. *See* PALEOCLIMATOLOGY. [CARL O. DUNBAR]

Bibliography: C. O. Dunbar, *Historical Geology*, 2d ed., 1960; C. L. Fenton and M. A. Fenton, *The Fossil Book*, 1959; J. F. Kirkaldy, *Study of Fossils*, 1963; G. G. Simpson, *Life of the Past*, 1953.

Fossil man

All prehistoric skeletal remains of humans which are archeologically earlier than Neolithic (necessarily an imprecise limit), regardless of degree of mineralization or fossilization of the bone, and regardless of whether the remains may be classed as *Homo sapiens sapiens*, modern man.

Discoveries began in the early years of the 19th century, although their meaning and antiquity were not recognized before the finding of Neanderthal man in 1856. Remains have come principally from Europe and adjacent portions of Asia, from Africa, from North China, and from Java. Because of the rather late entry of humans into the New World, American Indian remains are all of relatively recent origin and recognizable as *H. sapiens*.

Dating fossil man. The family of man (Hominidae) was once thought only to have come into existence at the beginning of the Pleistocene. It is now known to have been fully separate from

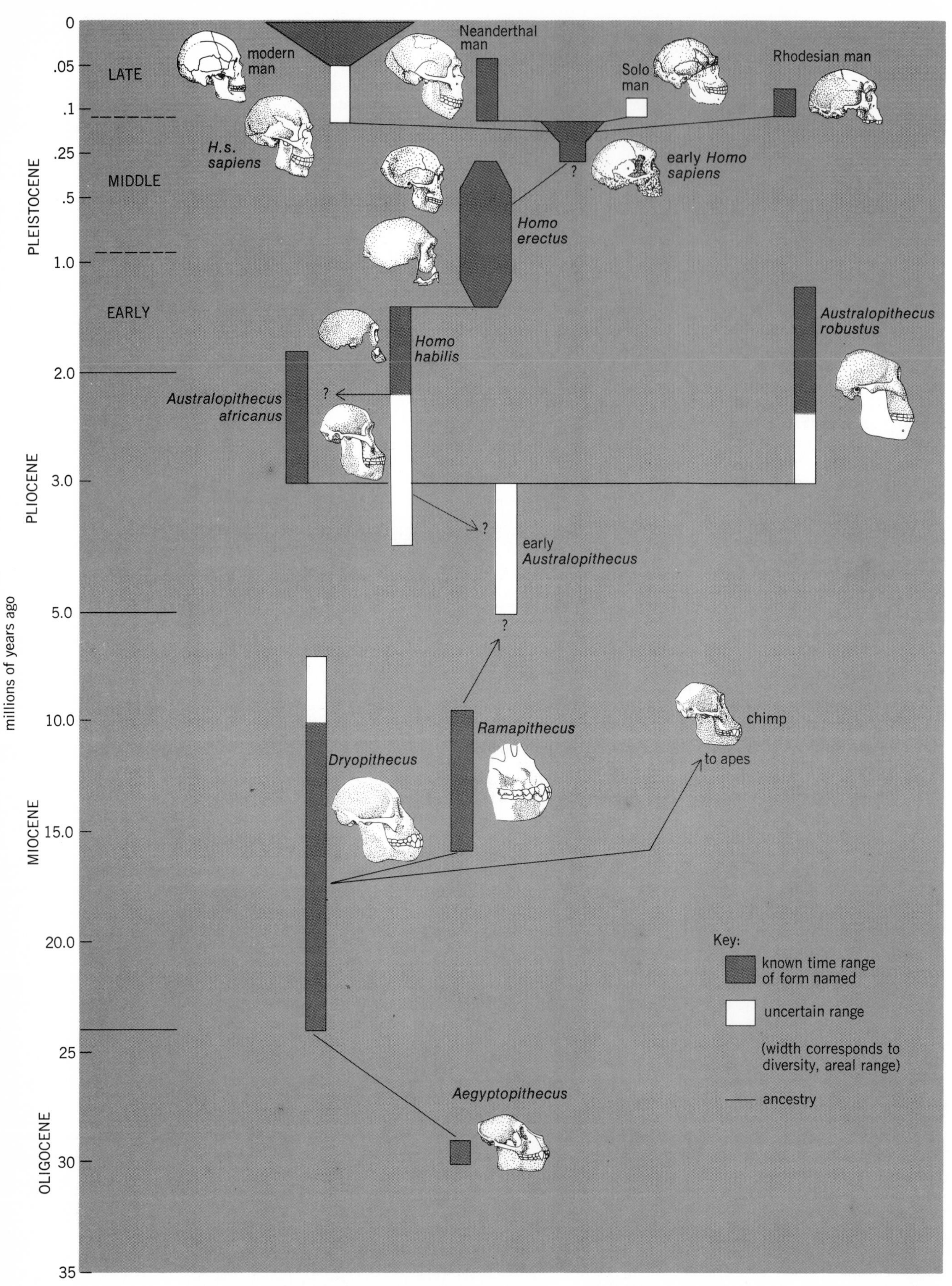

A diagram of human phylogeny from the Oligocene to the present time, showing the skulls of the major known fossil relatives and the possible ancestors of modern humans.

the family of the apes (Pongidae) at least as far back as the Miocene. While this view was widely held by anthropologists, on theoretical grounds, in the early part of the 20th century, it was abandoned in the 1940s because of lack of supporting evidence. Discoveries in Pliocene and Miocene deposits since then have led to a reappraisal of the evidence (see illustration).

Dating within the Pleistocene is accomplished by the methods of conventional paleontology and geology, by association with human implements, by several chemical and physical tests for relative age, such as the fluorine test for accumulation of fluorine in bone, and by chronometric tests, for age in years, involving radioactive substances such as radiocarbon and potassium-argon, as well as by paleomagnetic and amino acid racemization methods, among others. *See* PALEOMAGNETICS.

Carbon-14, a radioactive isotope with a half-life of 5760 years, forms a fairly constant percentage of the carbon atmosphere and, hence, of living matter. Its rate of decay allows an estimate in actual years of the time elapsed since death of an organism (charcoal, wood, shell, and vegetable matter are especially suitable for application of this method). The range of this method extends to approximately 50,000 years ago. *See* RADIOCARBON DATING.

Potassium-40 decays gradually to form calcium-40 and argon-40, argon-40 being a gas that may remain trapped in a mineral unless driven off at a critical heat, as in lava. After cooling, argon-40 begins to accumulate again in suitable rock and can be measured. Dates may be obtained ranging from approximately the upper limit of the carbon-14 method back to and beyond the origin of the Earth, some 4,500 m.y. (million years) ago.

Prehuman ancestry of man. Man is a catarrhine primate, part of a group including Old World monkeys, apes, and various extinct forms. The oldest known representatives of the Catarrhini are fossils from the Fayum beds of northern Egypt, dated around 30 m.y. years old. The best-known of these is *Aegyptopithecus zeuxis*, a possible ancestor for apes and humans. This animal was the size of a cat, with a monkeylike body, apelike teeth, and a distinctive skull: a mosaic of evolutionary features. Between about 23 and 15 m.y. ago, in the early Miocene, several species of *Dryopithecus* (= *Proconsul*) in East Africa represent relatives of the ancestors of humans. These "apes" ranged in size from small chimp to small gorilla, with a somewhat chimplike skull, large projecting canine teeth, and limb bones seemingly adapted to quadrupedal running, as well as some suspensory or supportive use of the arms, but not to the true brachiation or knuckle-walking of modern apes. Evidence from both comparative morphology and molecular studies of proteins shows that humans' closest living relatives are the African apes: the chimpanzee and the gorilla. This group probably had a most recent common ancestor about 17–15 m.y. ago, a form which would probably be considered a species of *Dryopithecus* if precise knowledge were obtainable.

The first evidence of a distinctive line of human evolution is the genus *Ramapithecus*, which comprises several species from East Africa (= *Kenyapithecus*), India, China, Turkey, and perhaps Europe in the middle and late Miocene (14–10 m.y. ago), contemporary with younger species of Eurasian *Dryopithecus*. *Ramapithecus* is recognized as being related to humans by the shape and wear pattern of its teeth, its apparently short face, and, especially, its short canines, projecting only slightly beyond the other teeth. Unfortunately, only teeth and partial jaws of *Ramapithecus* exist (no skull or limb bones have been recognized so far), but it probably had a rather apelike body form. No specimens of clearly human lineage between 10 and 6 m.y. ago have been found, but two teeth discovered in East Africa may represent this line.

Pliocene Hominidae. *Australopithecus*, the first truly humanlike beings, appear in the fossil record in quantity some 3 m.y. ago, during the Pliocene. One partial jaw with a single tooth has been found that may be as old as 6 m.y., and a few other fragments are a bit younger, but the majority of more complete finds date between 2 and 3 m.y. *Australopithecus* occurs in two main varieties, generally termed gracile and robust. These creatures share a number of basic characteristics distinguishing them from living and fossil apes and also from later humans, although clearly linking them to the latter. Such features include an apparently humanlike body form and upright posture, with long legs; a foramen magnum placed rather forward under the skull; a large brain relative to body size; a pelvis adapted to bipedalism, although perhaps not of a modern human type; and teeth of human form, especially with small canines in both sexes. The two varieties of australopith also broadly share a smaller body size than modern humans and have back teeth that are large for this body size, although the actual sizes involved are among the features distinguishing the two types from each other.

Specimens. The fossils of these animals (or humans, if one prefers) were first found in South Africa in 1924, but the most recent major discoveries and the best evidence of their age come from East Africa. The South African fossils come from five sites, which are the remains of ancient cave systems. These hominids did not live in caves, but their carcasses may have been dropped there by leopards or other carnivores. In two sites especially, many fossils are known from a relatively short span of time (less than 50,000 years), but the evidence for dating is not definite. Then, in 1959, Mary and L. S. B. Leakey discovered a nearly complete skull at Olduvai Gorge, Tanzania, and colleagues dated it at about 1.75 m.y. old, far older than previously had been thought. Since then, joint American-French expeditions have found hundreds of fossils in the Omo and Afar areas of Ethiopia, while a Kenyan team has worked on the east shore of Lake Turkana (formerly Lake Rudolf), Kenya, and studies have continued at Olduvai. These regions have yielded smaller numbers of specimens at many separate subsites, but the age of each site can usually be estimated closely by potassium-argon and paleomagnetic dating with subsequent faunal correlation. Also, many specimens are more complete and show less distortion than their contemporaries from South Africa. To date, no definite examples of *Australopithecus* are known outside these areas, although some claims have been made. Evidence of archeological activity has been found with these specimens, mostly choppers or "pebble tools" of the Oldowan culture, and also the remains of small animal prey. It is not possible to tell which type(s) of humans made the tools, but at present the oldest evidence is from the Omo valley, about 2.5–3 m.y. ago.

Gracile variety. In summary, it appears that between perhaps 6 and 3 m.y. ago (when few fossils are known), a generalized type of *Australopithecus* lived in eastern (and, perhaps, southern) Africa. By 2.5 to 3 m.y. ago, the two major varieties had diverged. *A. africanus*, the so-called gracile form, probably was little changed from its ancestor; it is known especially in South Africa, and less certainly in East Africa, between 3 and 2 m.y. ago. It may have stood 4 to 5 ft (120 to 150 cm) tall, weighed 50 to 90 lb (22.5–40.5 kg), and had a brain size of some 450 cm^3, compared to 350 cm^3 in a 100-lb (45 kg) chimp, 500 cm^3 in a 300-lb (135 kg) gorilla, and 1400 cm^3 in a 150-lb (67.5 kg) human. The skull is lightly built and rounded, with projecting face and teeth quite similar to those of modern humans. Although the absolute tooth size of *A. africanus* is nearly equal to that of small gorillas, the proportions are human, with a smooth decrease in size from molars through incisors. A partial skeleton found in the Afar region of Ethiopia probably represents *A. africanus*. It includes the mandible, but unfortunately not the skull.

Robust variety. The second form, known as *Australopithecus robustus* (or also *A. boisei*, and sometimes given generic or subgeneric rank as *Paranthropus*), lived in East and South Africa between roughly 2.25 and 1.25 m.y., and perhaps somewhat earlier. At some sites, bones of the two forms occur together. Specimens of *A.* (*P.*) *robustus* may be distinguished, however, by their larger size and craniodental specializations. It may have been 4½ to 5½ ft tall (135 to 165 cm), weighed 80 to 120 lb (36 to 55 kg), had a heavy, muscular body build, and a brain size of about 525 cm^3. The skull is more robust that *A. africanus*, with deep cheekbones and a lower-set jaw and often a slightly raised sagittal (midline) crest in the middle part of the skull roof from back to front. These features indicate strong chewing muscles and perhaps a diet of tough foods. The teeth themselves are distinctive: the back teeth (molars and premolars) are large to huge; the front teeth (incisors and canines) are quite small and run nearly straight across the front of the mouth. This difference from graciles and humans, and also from apes (which have generally large front teeth and small back teeth), combined with a low forehead and a flat, nearly upright face, further suggests adaptation to powerful chewing. The anterior teeth were probably used as much for grinding as for cutting. Postcranial evidence suggests to some that *A. robustus* was perhaps less adapted to bipedalism and more to tree climbing than the graciles.

Single-lineage hypothesis. These features combine to separate the robust australopiths from the main line of human evolution, in the view of most investigators, while the less specialized graciles are nearer to what may have actually become later humans. On the other hand, some workers argue that the variation found among all australopiths is not sufficient to justify more than one species (or lineage) at any given time. The idea of sexual dimorphism has been suggested to support this view further. Thus, the majority of so-called gracile specimens may be females, and the best-preserved robust skulls may be males of a single species. Two lines of evidence contradict this view: (1) both large (male?) and small (female?) robust skulls are now known from East Rudolf about 2 m.y. ago, and the presumed female is a smaller version of the male, not closely similar to South African gracile skulls; (2) the dental differences, especially those involving the canine and neighboring teeth, do not seem to fit any single-lineage hypothesis, sexual dimorphism notwithstanding. Nonetheless, the two types of this early hominid are generally similar, and part of their differences may be a result of allometry, and thus they may be placed in a single genus, *Australopithecus*.

Early Pleistocene. The only other genus of the family Hominidae is *Homo*, true humans, into which all later forms (including ourselves) are placed. The identification of the earliest specimens of *Homo* is a subject of debate among paleoanthropologists. This problem has been complicated by recent finds in Kenya, Ethiopia, and especially Tanzania which may document *Homo* as old as 3–3.75 m.y., thus older but more advanced than well-known *Australopithecus*. In late 1976 the scientific pendulum was swinging back to an idea proposed on less secure grounds by L. S. B. Leakey and colleagues in 1964. They named a species *Homo habilis*, based on several finds from Olduvai. Especially significant was the discovery of the 1.8-m.y.-old remains of a juvenile's lower jaw, with teeth much like those of *A. africanus*, and its partial skull, with an estimated cranial capacity of about 650 to 700 cm^3. Although most experts did not consider this find to be evidence of a new species, and showed that other specimens involved here might really be a later type of human, newer discoveries suggested that there was indeed an early form of *Homo* in East Africa, and possibly elsewhere, between 1.5 m.y. and more than 2 m.y. years ago. The so-called 1470 skull (named for its catalog number) found at East Rudolf by Richard Leakey may be a more complete adult form of this species. It has a brain size of some 750 to 800 cm^3, a high rounded skull vault, and probably large teeth (the crowns are broken off). Some thigh bones found in a neighboring locality are very modern indeed. This skull was originally dated at nearly 3 m.y., but studies revealed an age only slightly greater than 2 m.y., much like several Olduvai specimens of similar form. If all of these examples do represent a single species that may be called *Homo habilis*, it is a larger-brained and more bipedal animal than *A. africanus*, but one with essentially the same dental apparatus—another example of mosaic evolution. It is this combination of apparently significant change in the two major human adaptations of locomotion and "intelligence" that classifies this species as *Homo* but so far denies that placement to *A. africanus*. Tools from Olduvai are thought to have been made by *H. habilis*. It is not certain if *H. habilis* evolved from something like *A. africanus* about 2–3 m.y. or rather earlier. The importance of *Australopithecus* is that it preserves the morphology of an early evolutionary stage, even if known fossils are younger than the oldest *Homo*.

Homo erectus or "Pithecanthropus." While *H. habilis* is apparently a short-lived and rare East African species, its apparent descendant, *Homo erectus*, is common, widespread, and long-surviving. The first fossils were found in Java in 1893 and termed *Pithecanthropus erectus*. Later finds in China, North Africa, sub-Saharan Africa, and possibly Europe were each given distinctive generic and specific names, but are now all considered local variants or subspecies of the single species *H. erectus*. The major anatomical characteristics of this form are the following: a body of nearly modern proportions below the neck, topped by a low and slightly elongated skull with cranial capacity averaging 1100 cm^3 (range of about 800 to 1300 cm^3), smaller teeth in a less projecting face than *A. africanus* or *H. habilis*, large brow ridges, thick cranial bones, and, of course, no chin. Most finds are associated with examples of the Acheulean biface (or handaxe) culture, and there is evidence of group hunting of large animals (such as the mammoth), the use of fire, and of fabricated structures as well as caves for dwellings.

The earliest known representatives are once again from East Africa, dating from 1.5 to 1 m.y. ago. *Homo erectus* probably spread from Africa into Eurasia more than 1 m.y. ago, perhaps the first type of human to do so in large numbers. In addition, the eventual extinction of *A. robustus* in this time interval may have been caused by direct or indirect competition with the more advanced *H. erectus*. Fossils in Asia may date from over 1 m.y. through nearly 250,000 years ago, all associated with fauna from the warmer intervals in this time of alternating glacial climate. New finds in Africa are nearly as recent. In Europe, however, there is definite evidence of man-made tools between 1 m.y. and 500,000 years ago, but no human fossils are older than about 350,000 years. The earliest of these, in central and southeastern Europe, already seem to be more advanced in some ways than Afro-Asian *H. erectus*.

Pre-modern Homo sapiens. It is possible that the increased rigor of the glacial climate in Europe at this time was the impetus leading to the evolution of *H. sapiens*, modern-type humans. It is about equally likely that *H. sapiens* may have evolved elsewhere, but between about 500,000 and 150,000 years ago, the only known human fossils are those from Europe and those of typical *H. erectus* in China and North Africa. A number of partial skulls from this time period have been found in England, France (Arago), Germany, Hungary (Vertesszöllös), and Greece (Petralona). They share somewhat larger brains (for body size), smaller teeth, and less robustness than in *H. erectus*. All may be classified early *H. sapiens* and probably represent part of the spread of this species across the Old World, eventually replacing its ancestor by its superior efficiency. They still used Acheulean tools, but perhaps more effectively, and by 200,000 years ago may have been the only human type on Earth.

For the next 100,000 years there is still a dearth of evidence, but it may be suggested that local varieties of *H. sapiens* developed in southern Africa, southeastern (and perhaps northern) Asia, Europe, and elsewhere. The best-known of these varieties is Neanderthal (final syllable pronounced -tal) man, from Europe and the Near East. The most extreme (or classic) form of *H. sapiens neanderthalensis* is found in western Europe, with more modern-looking varieties farther east. Neanderthals peaked between 100,000 and 40,000 years ago and may represent an adaptation to the intensely cold, dry climate of northwestern Europe at that time. They were essentially stocky humans, but had long, low skulls with a projecting occipital region, large faces, teeth, brow ridges, and brains averaging 1500 cm^3. They made Mousterian tools (a variant of Middle Paleolithic flake-based tool kits), often lived in caves or wooden shelters, hunted big game, and had primitive religious beliefs. Fewer remains have been found further east and south, but at least two, and perhaps four or more, other varieties of humans may have existed in Asia and Africa. These populations were all distinct from *H. erectus*, also from each other to a greater degree than is true among living varieties or "races" of anatomically modern humans. They are each often classified as subspecies of *H. sapiens*. In southern Africa, one skull was found at Broken Hill, Zambia (previously Rhodesia, hence the name Rhodesian man), and similar specimens are known in South Africa (Saldanha) and Tanzania (Lake Eyasi). These men made Middle Paleolithic (Middle Stone Age) tools, and some may have lived around 100,000 years ago. From central Java, Indonesia, come a dozen partial skulls known as Solo man; all lack the face, and there is strong evidence of ritualized brain eating. A single Chinese skull fragment from Mapa is quite different from Solo man, but like Solo is not well dated. These last three types have often been called "Neanderthaloids" or "tropical Neanderthals," but as David Pilbeam has pointed out, that is like calling a modern Javanese a tropical European! They are best considered as independent developments paralleling modern subspecies *H. sapiens sapiens* in evolving from an early *H. sapiens* ancestry.

Spread of modern man. There is virtually no evidence of the area of origin and early history of anatomically modern *H. sapiens sapiens*, characterized by a small upright face, small teeth and brow ridges, chin, and high rounded skull. There are some possibly early, suggestive (but incomplete), fossils from East Africa, and around the southern and eastern shores of the Mediterranean are fossils which have previously been thought to represent hybrids of moderns and Neanderthals, but later findings place them as extinct races of modern humans. Wherever the source, sometime between 100,000 and 50,000 years ago, modern humans' direct ancestors began to spread across the Old World, so that by 35,000 to 40,000 years ago they were the sole form of humans to be found. It has been suggested that the Neanderthals were somehow directly ancestral to the first anatomically modern people, but most scientists today think that they were probably replaced, although some interbreeding possibly occurred. One reason for the success of *H. s. sapiens* may have been their still greater tool-making efficiency, as documented by the Late (or Upper) Paleolithic blade-and-burin industries. In many parts of the world, they also engaged in artistic pursuits, including carving small animal statues and perhaps calendars as well as painting on the walls of deep caves.

Many names have been given to early modern humans, especially in Europe, but these only indicate minor differences. Cro-Magnon man and his contemporaries were already essentially Europeans, while early Africans are known from Border

Cave in southern Africa. Australia was colonized by about 30,000 years ago, with important finds at Keilor and Lake Mungo. The place of origin of New World Indians was certainly Siberia, by means of crossing a land bridge over what is now the Bering Strait. Many human fossil remains are known in the Americas as far back as 20,000 years ago, but some reliable dates as old as 50,000 years have been obtained, indicating that perhaps several crossings of the land bridge occurred.

[ERIC DELSON]

Bibliography: Y. Coppens et al. (eds.), *Earliest Man and Environments in the Lake Rudolf Basin*, 1976; W. W. Howells, *Evolution of the Genus Homo*, 1973; W. W. Howells, Neanderthals, *Amer. Anthropol.*, 76:24–38; C. J. Jolly (ed.), *Early African Hominids*, in press; D. Pilbeam, Middle Pleistocene hominids, in K. Butzer and G. Isaac (eds.), *After the Australopithecines*, pp. 809–856, 1975; D. Pilbeam, *The Ascent of Man*, 1972.

Fossil seeds and fruits

Seeds, ovules containing a fertilized egg and ready to be shed from the plant, are reproductive organs characteristic of both gymnospermous and angiospermous plants. In angiosperms (Magnoliophyta) an additional structure, the matured ovary, encloses one or more seeds to form a fruit.

Seeds and fruits are less commonly found as fossils than are vegetative remains. They may be preserved structurally as casts, or as compressions which are sometimes found with leaf compressions. Seeds and fruits often occur in lignites. *See* LIGNITE.

The oldest known seed plants are of Mississippian age. Carboniferous seed plants include the extinct Cordaitales, probable conifer ancestors, and Pteridospermae, seed plants with fernlike foliage. Both groups had similar seeds, basically like those of modern cycads. No embryos have been found in Paleozoic seeds.

During the Mesozoic Era, all major modern groups of seed plants were represented, along with members of the declining cordaitalean and pteridospermous stocks. Among the most completely known Mesozoic seeds are those of the cycadeoids, extinct cycad relatives. Their silicified, beehive-shaped trunks may include shoots bearing numerous small seeds between thick scales. The Mesozoic Caytoniales, whose small seeds were borne in fleshy enclosures, were thought at one time to be ancestral angiosperms. Now they are recognized as gymnospermous forms, probably related to the pteridosperms.

Angiosperm fruits are rare in Lower Cretaceous beds; Upper Cretaceous fruits are known from northern Africa, Long Island, N.Y., and elsewhere. Tertiary fruits and seeds have been found in numbers in the United States in the Brandon lignite of Vermont (see illustration) and in the Clarno Formation of central Oregon. Silicified cones of araucarian conifers are known from Patagonia. The best known European Tertiary fruits and seeds are from the brown coals of Germany and the Eocene London Clay Formation of England.

Important paleobotanical findings resulting from the study of fossil seeds and fruits include the knowledge obtained of the independent evolution of the seed habit in unrelated groups; the discovery that much Carboniferous fernlike foliage was borne on seed plants rather than on ferns; and the discovery that *Glossopteris*, an important plant in widespread Permian floras of the Southern Hemisphere, was a seed plant. Pyritized fruits from the London Clay Formation reveal the presence of many extinct genera along with modern genera in early Tertiary time. Morphological changes in herbaceous angiosperm seeds from sequences of Tertiary beds furnish data on rates of evolution in plants. Because plant classification is based primarily upon reproductive structures, fossil seeds

Early Tertiary fruits and seeds which were found in Brandon lignite of Vermont. (*a*) Endocarps of *Symplocos*. (*b*) Seeds of grape (*Vitis*). (*c*) Seeds of *Rubus*. (*d*) Endocarp of extinct species of *Nyssa*. (*e*) Seeds of fossil genus of family Rutaceae, which are related to *Phellodendron*. (*f*) Endocarps of a species of *Alangium*. (*g*) Acorn cups, *Quercus* sp. (*Courtesy of Elso S. Barghoorn*).

and fruits provide highly reliable evidence for identification and interpretation of fossil plants. *See* PALEOBOTANY.

[RICHARD A. SCOTT]

Bibliography: H. N. Andrews, *Studies in Paleobotany*, 1961; C. A. Arnold, *An Introduction to Paleobotany*, 1947; F. Kirchheimer, *Die Laubgewachse der Braunkohlenzeit*, 1957; E. M. Reid and M. E. J. Chandler, *The London Clay Flora*, 1933; R. A. Scott, Fossil fruits and seeds from the Eocene Clarno Formation of Oregon, *Palaeontographica*, 96(B):66–97, 1954.

Franklinite

A natural member of the spinel structure type, with composition $Zn^{2+}Fe_2^{3+}O_4$, crystallizing with space group *Fd3m*. Franklinite usually possesses extensive substitution of Mn^{2+} and Fe^{2+} for the divalent Zn^{2+} cations, and limited Mn^{3+} for the trivalent Fe^{3+} cations. The habit is octahedral, often modified by the cube and dodecahedron, but the mineral usually occurs as bands of isolated rounded grains, blebs, or compact masses (see illustration). It is black with a metallic luster and red internal reflections; its hardness is 6 (Mohs scale); specific gravity, 5.3; streak, reddish-brown; and weakly magnetic.

Franklinite crystals in calcite, Franklin, N. J. (*Specimen from Department of Geology, Bryn Mawr College*)

Franklinite is confined in its occurrence to the unique ore bodies at Franklin and Sterling Hill (Ogdensburg), Sussex County, N.J. The highly unusual ore consists of intimate granular mixtures of franklinite; willemite, $(Zn, Mn)_2[SiO_4]$; tephroite, $(Mn,Mg)_2[SiO_4]$; and zincite, ZnO, in coarsely crystalline marble. No less than 200 mineral species have been recorded from these localities, at least 40 of which are found nowhere else in the world. Franklinite is the major ore mineral and, along with willemite and zincite, is still mined at Sterling Hill for spiegeleisen and zinc. *See* WILLEMITE; ZINCITE.

[PAUL B. MOORE]

Frost action

The weathering process caused by repeated cycles of freezing and thawing. This definition was adopted by the American Geological Institute in 1957. Most scientists use the term to include any effects of freezing and thawing, whether direct or indirect. The term also is used loosely to indicate any action in which frost is involved. Because of the multiple meanings of the term, many geologists avoid its usage except informally and have introduced more specific terms. Some common self-evident subdivisions are frost splitting, wedging, heaving, thrusting, and stirring. Frost action has been recognized for many decades as exceedingly important in weathering and erosion, in disturbing vegetation, and in engineering problems. Frost action is intensified in the zone of seasonal near-surface freezing and thawing above perennially frozen ground (permafrost), and has been more widespread in the geologic past than it is today. *See* PERMAFROST.

Frost action is initiated during growth of ice in the ground, from which may result either hydrostatic pressure or directed pressure by individual ice crystals, usually growing in the direction of heat loss. The transition of water to ice is accompanied by 9% volumetric expansion. In fine-grained soils, frost heaves are much greater and result from growth of ice lenses (see illustration). The ice grows slowly as water migrates, molecule by molecule, because of "tension" to the freezing zone. The process is complicated and not entirely understood. Many factors are involved, such as the surcharge above the growing ice lens that must be lifted, the physicochemical nature of the material, and its permeability, porosity, and rate of freezing. Plant roots are broken, soils are stirred and sorted, and many small related features are produced. In fine-grained unconsolidated materials, ice much in excess of the volume of pores is commonplace. Under these conditions, thawing yields suspensions of soil or high pore-water pressure that promotes loss of strength and subsequent slump and flow. *See* GEOLOGY; MASS WASTING; WEATHERING PROCESSES.

FROST ACTION

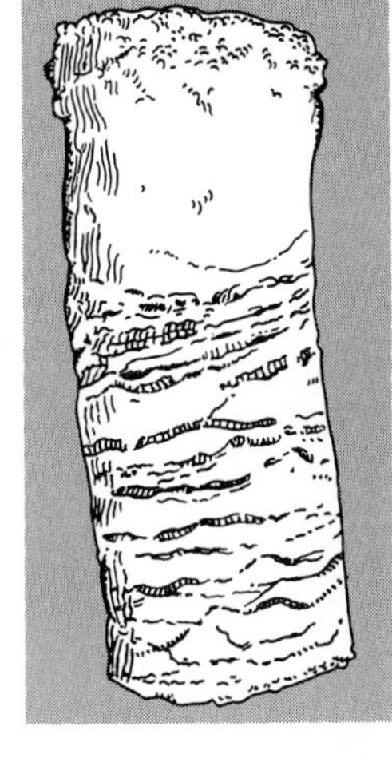
Ice lenses in frozen soil cylinder. (*After S. Taber, in D. P. Krynine and W. R. Judd, Principles of Engineering Geology and Geotechnics, McGraw-Hill, 1957*)

[ROBERT F. BLACK]

Bibliography: J. B. Bird, *The Physiography of Arctic Canada*, 1967; K. A. Jackson, D. R. Uhlmann, and B. Chalmers, Frost heave in soils, *J. Appl. Phys.*, 37:848–852, 1966; National Academy of Science–National Research Council, *Proceedings of the Permafrost International Conference*, 1966.

Fuller's earth

Any natural earthy material (such as clay materials) which decolorizes mineral or vegetable oils to a sufficient extent to be of economic importance. It has no mineralogic significance. *See* CLAY.

In ancient times raw wool was cleaned by kneading the wool in water with certain earths which adsorbed the dirt and oil. The process was known as fulling; the earth, as fuller's earth. The earth used by fullers was found to have oil-decolorizing properties.

Composition. Fuller's earths are not always composed entirely of clay. Some materials which have been used for decolorizing are primarily silts with a relatively low clay content. The decolorizing power of such earths is relatively low, and they are no longer used. The clay minerals present in fuller's earth may include montmorillonite, attapulgite, and kaolinite. However, some materials containing these clay minerals often have very low decolorizing power. *See* CLAY MINERALS.

Some montmorillonite clays with low decolorizing activity in their natural state develop very high activity following treatment with acids. Such clays are not usually defined as fuller's earths. Fuller's earths possess natural activity and do not ordinarily respond satisfactorily to acid activation treatment.

Fuller's earth from different parts of the country, or even from different parts of the same deposit, frequently shows a wide variation in properties. The fuller's earth type clay is often characterized by a high water content. In a dried state it often adheres strongly to the tongue. The chemical composition of fuller's earths is extremely variable, depending on its mineral composition.

Bleaching properties. The action of fuller's earth and other bleaching clays in decolorizing oils is not completely understood. This action involves a selective adsorption of color bodies and other impurities from oils. These color bodies and impurities are strongly held within the clay structure after adsorption. Because they can be removed from the clay only by drastic treatment, it is evident that bleaching clays operate mainly by chemical adsorption.

Large tonnages of fuller's earth are used to decolorize petroleum products, cottonseed oil, tallow, soy oils, and other products. For satisfactory use the earth must only decolorize—it must not change the character of the oil or impart an undesirable taste or odor to it. Also it must not retain too much oil which cannot be reclaimed, and in some cases it must be regenerated easily by solvent or heat-treatment.

Some fuller's earths are used for other purposes, for example, oil-well drilling muds, insecticide carriers, and fillers.

Fuller's earth deposits are distributed throughout the world, but the largest deposits have been found in the United States, England, and Japan. In the United States, principal deposits are found in Georgia, Florida, Illinois, Texas, Nevada, and California. [FLOYD M. WAHL; RALPH E. GRIM]

Gabbro

A rock made up of centimeter-sized crystals mostly of calcium-rich plagioclase (35–65 vol %) and pyroxene in subequal amounts. Minor amounts of calcium-poor pyroxene, olivine, hornblende, biotite, magnetite, apatite, and nepheline or quartz may occur, but not all together.

The textures of most gabbros are indicative of solidification from a melt. Gabbro occurs widely, but is not as common as basalt (its extrusive equivalent). *See* BASALT.

Definition. The terms gabbro, diabase (or dolerite), and basalt refer to mineralogically identical rocks which differ in grain size. There is a good correlation between grain size and mode of occurrence: Gabbro, with its centimeter-size crystals, generally forms kilometer-size bodies of intrusive rock, and diabase, with centimeter- to millimeter-size crystals, forms sills and dikes tens to hundreds of meters thick, whereas basalt forms flows and dikes commonly a few tens of meters thick. Because of the general correlation between grain size and mode of occurrence, it is standard practice to apply the terms to the respective occurrences regardless of the particular grain size.

Forms and structure. Gabbro is primarily an intrusive igneous rock. It forms irregular bodies as large as about 100 km² in area, as well as dikes, sills, and laccoliths. The contacts of bodies of gabbro with surrounding rocks commonly are sharp, and the gabbro generally cuts across trends in older adjacent rock. The contact rocks may be recrystallized at high temperature. Commonly, the gabbro is finer-grained adjacent to the country rock than within the interior of the body of gabbro. *See* IGNEOUS ROCKS.

Many gabbros are massive, but many have either a foliation or a layering or both. Crystals may be aligned parallel to the walls or to internal layers; such alignment imparts foliation to the rock. In some cases crystals are aligned perpendicular to walls and layers. Layers of such crystals give rise to Willow Lake layering. Layering may be mineralogical or textural. In mineralogical layering, the layers have different proportions of various minerals. There are about a dozen large gabbroic bodies of mineralogically layered rocks. Many of the layers are not gabbro, but are particularly rich in one mineral; the total assemblage of layers, however, is gabbroic. Mineralogical layering is thought to develop as a result of mechanical (gravitative) separation of dense crystals from less dense liquid. In textural layering, the proportions of the minerals remain approximately constant, but their size or the relationship of their arrangement varies. Textural layering is commonly expressed by variations in grain size where zones of pegmatitic gabbro with crystals up to 10 cm long alternate with finer-grained gabbro. Such textural layering is widespread in gabbros associated with ophiolite sequences made up of peridotite, gabbro, diabase, pillow basalt, and red chert, and is thought by many investigators to character-

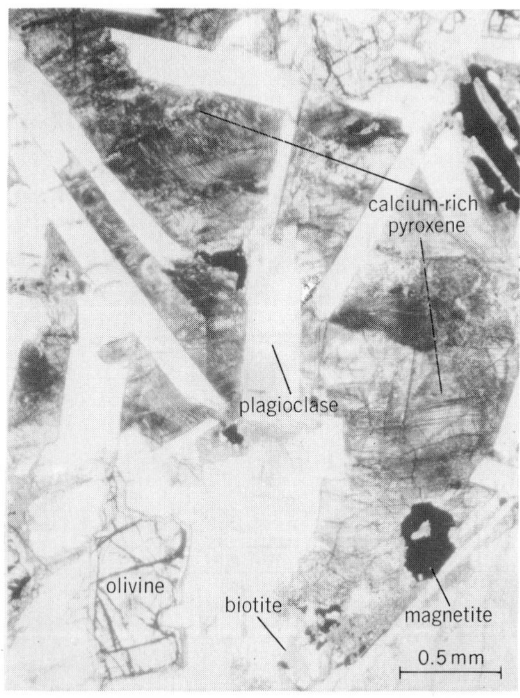

Fig. 1. Thin section of fine-grained gabbro. Plagioclase is subophitically enclosed in calcium-rich pyroxene which shows typical schiller or lamellae of dark material.

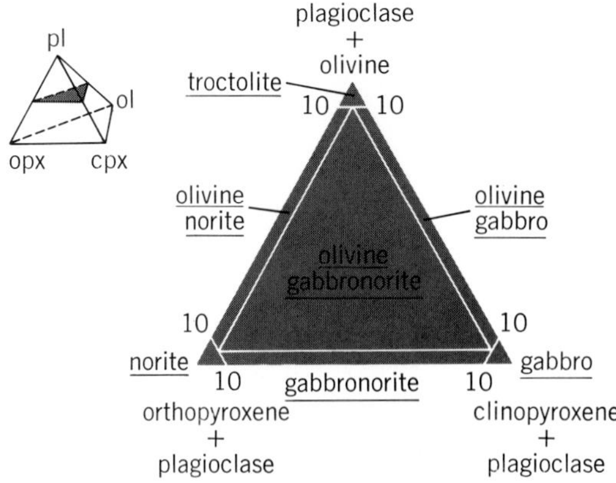

Fig. 2. Nomenclature of gabbroic rocks; rock names are underlined.

ize oceanic crust. Textural layering is not well understood, but it probably involves either deformation during consolidation or effervescence of gas.

Mineralogy and texture. Gabbros commonly have ophitic texture (Fig. 1), and normal gabbros have only a small percentage of minerals other than calcic plagioclase and augite (Ca-rich pyroxene). Alkali gabbros contain olivine or feldspathoids (minerals unsaturated with silica), or both; olivine gabbros have olivine but no feldspathoids; quartz gabbros have up to about 3% quartz. Significant varietal names are shown in Fig. 2. Varieties of alkali gabbros containing significant amounts of potassium feldspar are not shown. Amphibole and biotite are common in many gabbros.

The minerals in gabbros are less strongly zoned than their counterparts in basalts are. Evidently, the slower rate of cooling for gabbros allows greater equilibration to take place between early-formed crystals and leftover melt. In addition, the minerals in gabbros commonly are charged with numerous tiny inclusions of other minerals; plagioclases may be dusted with opaque inclusions or blebs of K feldspar (antiperthite); augites may have lamellae of orthopyroxene (and orthopyroxenes may have lamellae of augite); iron oxide minerals may have complex intergrowths. Most of these features are the result of mineralogical unmixing in response to temperature decreased below the igneous temperature of crystallization. Such unmixing, called exsolution, is promoted by slow cooling. Thus, the comparatively large grain size, minor zoning, and exsolution typical of gabbros result from slow cooling.

Because most gabbros crystallize at depths where the pressure is high, gases can remain in solution in the melt as it crystallizes. This leads to the crystallization of hydrous silicates such as amphibole and mica from the melt, and to the conversion of some of the anhydrous minerals to other hydrous silicates as the rock cools in the presence of gas. Such low-temperature hydration (and sometimes carbonation and oxidation) reactions are called deuteric alterations.

Properties and composition. The physical properties of gabbros are generally similar to those of basalt. Gabbros preserve a poor record of their initial temperatures of crystallization; consequently they are not well known. Because of the slow cooling of gabbros, the magnetic properties are characterized by a high Curie temperature and low ratio between natural remnant magnetization and induced magnetization.

The porosity of gabbros is less than a few percent. Consequently, gabbros are denser than basalts and may have greater cohesive strength. These qualities are of value in ornamental building stone.

The chemical compositions of gabbros are similar to those of basalts. Most gabbros have more H_2O, CO_2, and S than basalts do because of crystallization under considerable confining pressure. Some varieties of gabbros which occur as layers in gabbroic complexes have chemical compositions dominated by individual minerals, and therefore exceed the range of compositions found for basalts, in which rapid cooling prevents such accumulations from developing.

Lunar gabbros. On the Moon, as on Earth, gabbros are less common than basalts are. The lunar gabbros have textures similar to those of terrestrial gabbros. They are poor in alkalies, as are lunar basalts, and the plagioclases are richer in Ca than most terrestrial gabbros are. Minor exotic minerals rich in Ti and other refractory elements are present in lunar gabbros, as are olivine, metallic iron, and apatite.

Mineralogical banding is weakly evident in some lunar norites. Consequently, since the sorting of minerals according to density does take place in the weaker gravitational field of the Moon, the gravitational acceleration required for density sorting of crystals in gabbroic magmas apparently is less than that found on the Moon.

Lunar gabbros are practically free of deuteric alteration, and presumably formed and cooled in an environment poor in H_2O and CO_2. Among the secondary minerals observed in lunar gabbros are metal developed within iron-titanium oxide minerals and veinlets of sulfide minerals. *See* MOON.

Tectonic environment and origin. Gabbros occur in all major tectonic environments (oceanic ridge, continental margin, oceanic island, and continental interior). Many continental interior occurrences are in fossil environments of continental margins and, possibly, oceanic ridges. As with basalts, the alkalinity of gabbros in general increases with distance from major sites of igneous activity. Consequently, gabbroic magmas are inferred to originate from decreasing degrees of partial melting of mantle rock at increasing depths and with increasing alkalinity. *See* MAGMA; PETROGRAPHIC PROVINCE.

[ALFRED T. ANDERSON, JR.]

Bibliography: I. S. E. Carmichael, F. J. Turner, and J. Verhoogen, *Igneous Petrology*, 1974; E. W. Spencer, *The Dynamics of the Earth*, 1972; L. R. Wager and G. M. Brown, *Layered Igneous Rocks*, 1968.

Galena

A mineral with composition PbS (lead sulfide) and belonging to the rock salt (NaCl) structure type, crystallizing with space group Fm3m, with $a = 5.94$ A. Each lead atom is octahedrally surrounded by six sulfur atoms and vice versa, a common arrangement for binary 1:1 compounds.

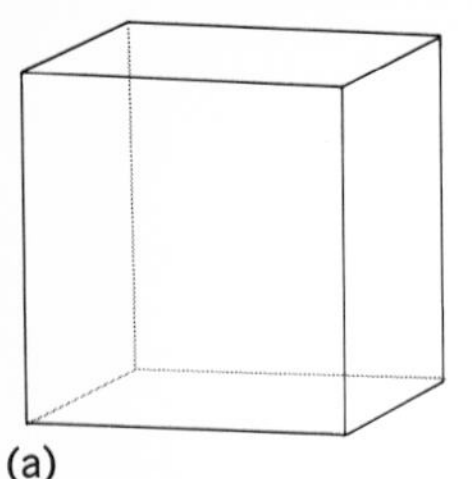

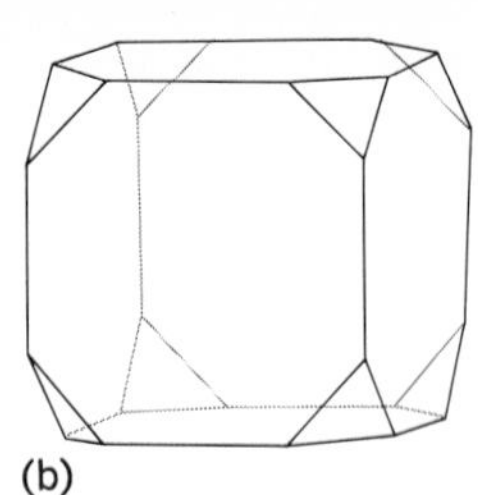

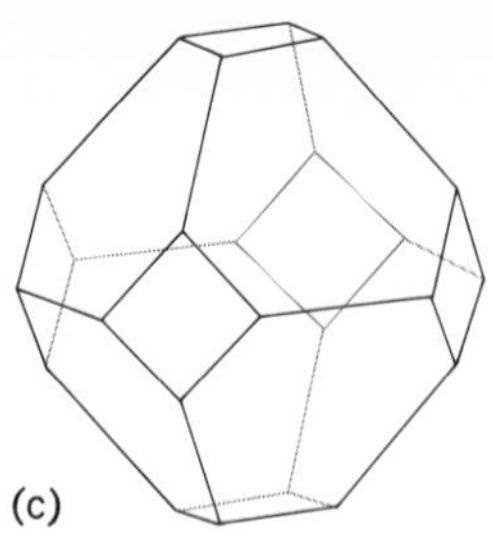

Galena crystals. (*a*) Cube. (b, c) Octahedral truncations of cube. (*From C. S. Hurlbut, Jr., Dana's Manual of Mineralogy, 17th ed., copyright © 1959 by John Wiley & Sons, Inc.; reprinted by permission*)

Blocks of distorted galena structure are often the basis of the atomic structures of more complex compounds, in particular the sulfosalts.

Galena usually occurs as cubes, sometimes modified by the octahedral form, with perfect cubic cleavage, brilliant metallic luster, color lead gray, specific gravity 7.5, and hardness 2½ on Mohs scale (see illustration). Galena is easily synthesized by heating the elements in a sealed capsule.

Galena is the most widely distributed of all sulfide minerals and constitutes by far the most important ore for lead. Silver, antimony, arsenic, copper, and zinc minerals often occur in intimate association with galena; consequently, galena ores mined for lead also include many valuable by-products. Galena occurs commonly with sphalerite, ZnS, in low-, intermediate-, and high-temperature veins, often in enormous amounts filling open cavities and fissures in limestone. Other common associates include pyrite, marcasite, chalcopyrite, barite, quartz, fluorite, and dolomite. Important localities include Broken Hill, Australia; the tri-state district of Missouri, Kansas, and Oklahoma; and numerous occurrences in Colorado, Montana, and Idaho.

Galena is a common, but minor, accessory associated with nearly every rock type; consequently it is a frequent mineral used for Pb^{206}, Pb^{207}, and Pb^{208} age dating. *See* ROCK, AGE DETERMINATION OF; SPHALERITE. [PAUL B. MOORE]

Garnet

A generic term applied to a group of mineral silicates that are isometric in crystallization and that conform to the general chemical formula $A_3B_2(SiO_4)_3$, where A is Fe^2, Mn^2, Mg, or Ca, and B is Al, Fe^3, or Cr^3.

Garnet group. The group is divided into a number of individual mineral species on the basis of chemical composition. The names of the more common of these species and the idealized compositions to which they refer are as follows:

Pyrope $Mg_3Al_2(SiO_4)_3$	Grossularite $Ca_3Al_2(SiO_4)_3$
Almandite $Fe_3Al_2(SiO_4)_3$	Andradite $Ca_3Fe_2(SiO_4)_3$
Spessartite $Mn_3Al_2(SiO_4)_3$	Uvarovite $Ca_3Cr_2(SiO_4)_3$

A wide range of solid solubility exists in the group by mutual substitution of the several cations of the A position or of the B position. Solid solubility is particularly marked between pyrope, almandite, and spessartite, comprising the so-called pyralspite series, and between grossularite and andradite in the grandite series. Since the chemical composition of garnets as found in nature is gradational, the species name is arbitrarily assigned on the basis of the A cation and of the B cation that is dominant in atomic percent. Many of the pure end compositions have been synthesized in the laboratory, including members of the group not found naturally. Among these are germanate garnets with the general composition $A_3B_2Ge_3O_{12}$, and iron garnets such as $Y_3Fe_2Fe_3O_{12}$.

A less common type of compositional variation found chiefly in grossularite and andradite involves the substitution of $(OH)_4$ groups for (SiO_4) tetrahedrons (hydrogrossular). In natural material this substitution is quite limited, but in synthetic material series have been obtained to the end composition $A_3B_2(OH)_{12}$. Titanium often is present in solid solution in andradite, where a complete series extends in synthetic material to the end composition $Ca_3Ti_2Fe_2SiO_{12}$ (schorlomite). Zirconium is a major constituent in kimzeyite, $Ca_3Zr_2(Al,Fe)_2SiO_{12}$.

Garnets usually are well crystallized. Typical forms are the dodecahedron and the trapezohedron (211), either alone or in combination (see illustration). Garnet also occurs as granular masses

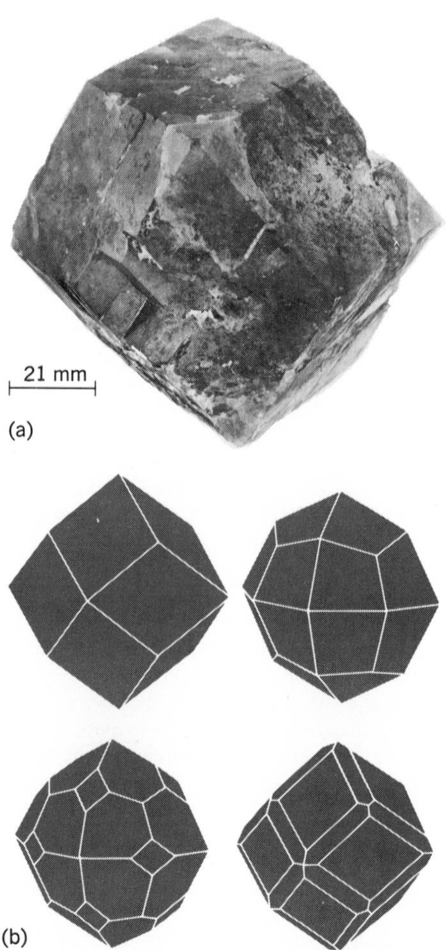

Garnet. (*a*) Subspecies grossularite, Lake Jaco, Mexico (*American Museum of Natural History specimens*). (*b*) Crystal habits (*from C. S. Hurlbut, Jr., Dana's Manual of Mineralogy, 17th ed., copyright © 1959 by John Wiley & Sons, Inc.; reprinted by permission*).

and as disseminated grains. Rarely, it is dense and microcrystalline, the green types then closely resembling nephrite.

Properties. Garnet is a nesosilicate based on isolated (SiO_4) tetrahedrons, with the A cations in 8-coordination and the B cations in 6-coordination with oxygen. The space group is *Ia3d*. The unit-cell dimension varies between about a_0 11.52 A and a_0 12.05 A, depending on the chemical composition. The index of refraction and the specific gravity also vary with composition between the approximate limits of 1.71–1.89 and 3.53–4.33, respectively. These properties, sometimes supplemented with qualitative chemical tests, can be used to identify individual garnet species. Graphical representations of the variation in unit-cell size, specific gravity, and index of refraction with variation in chemical composition are available. *See* SILICATE MINERALS.

The color of garnet is primarily determined by the chemical composition. Common garnet, chiefly comprising almandite and in part spessartite and andradite, is generally reddish-brown to black; grossularite usually is white, yellowish-brown to brown; uvarovite, emerald green; spessartite, dark hyacinth red, often with a tinge of violet; pyrope, deep red. When transparent and finely colored, garnet is prized as a gemstone. Gem varieties include chiefly pyrope, and some almandite and andradite. Demantoid is a rare grass-green to emerald-green gem variety of andradite. The Bohemian garnet jewelry is cut from pyrope. Asteriated varieties containing minute, oriented inclusions of rutile are known. The hardness of garnet ranges from 6½ to 7½ on Mohs scale. Cleavage is lacking, although a dodecahedral parting sometimes is present, and the fracture is subconchoidal to irregular. The relatively high hardness and the angular shape of crushed grains of garnet lead to its use as an abrasive. *See* GEM.

Normally, garnet is optically isotropic. Some species, particularly the grossularite and andradite of contact metamorphic deposits, show biaxial optical anomalies of a sectoral nature. It is relatively stable chemically but sometimes is found surficially or completely altered to chlorite, or to mixtures of hornblende or biotite with feldspar.

Occurrence. Garnet is a very common mineral. Almandite occurs widely in gneissic and schistose metamorphic rocks formed by the recrystallization of argillaceous sedimentary rocks at relatively elevated temperatures and pressures. The presence or absence of garnet and of certain associated minerals in metamorphic rocks of appropriate chemical composition is an index of the grade or intensity of the metamorphic process. Almandine also is typical of granulites, and solid solutions between almandine and pyrope occur in eclogites. Grossularite is a typical mineral of contact metamorphic zones in limestone. Andradite also occurs in such deposits, particularly in association with magnetite and iron-rich silicates, and it is found as an accessory mineral in igneous rock types containing feldspathoids. Uvarovite occurs chiefly in chromite deposits. Spessartite and almandite high in manganese occur in granite pegmatites and in acidic igneous rocks, and spessartite also forms during the thermal metamorphism of argillaceous sedimentary manganese deposits. Pyrope occurs in basic and ultrabasic igneous rocks, such as periodotite, and in serpentine rocks formed from them. Garnet is one of the most persistent and widespread minerals in detrital sediments. In beach and river sands it accumulates with zircon, ilmenite, monazite, and other heavy and resistant minerals in the so-called black sands. *See* METAMORPHIC ROCKS.

[CLIFFORD FRONDEL]

Garnierite

A mineral used as a minor ore of nickel. Garnierite is a monoclinic serpentine, but commonly occurs in reniform to earthy masses. The hardness is 2–3 on Mohs scale, and the specific gravity is 2.2–2.8. The luster is earthy and the color apple green to white. The composition may be expressed by the formula $(Ni,Mg)SiO_3 \cdot nH_2O$, but varies greatly in the relative amounts of nickel and magnesium and in the amount of water. Garnierite is a secondary mineral (a nickel-bearing serpentine) derived from the alteration of nickel-bearing olivine-rich rocks. It has been mined as an ore of nickel in New Caledonia, the Soviet Union, South Africa, and Madagascar. In the United States it is found at Riddle, Ore., and Webster, N.C.

[CORNELIUS S. HURLBUT, JR.]

Gem

A mineral or other material that has sufficient beauty for use as personal adornment and has the durability to make this feasible. With the exception of a few materials of organic origin, such as pearl, amber, coral, and jet, and inorganic substances of variable composition, such as natural glass, gems are lovely varieties of minerals.

A mineral is defined as an inorganic substance with a characteristic chemical composition and usually a characteristic crystal structure. Each distinct mineral is called a species by the gemologist. Two stones that have the same essential composition and crystal structure but that differ in color are considered varieties of the same species. Thus ruby and sapphire are distinct varieties of the mineral species corundum, and emerald and aquamarine are varieties of beryl. Two or more minerals that have the same structure and are related chemically are called groups. The garnet minerals form a group. *See* MINERAL; MINERALOGY.

Most gemstones are crystalline (that is, they have a definite atomic structure) and have characteristic properties, most of which are related directly to either beauty or durability.

Durability. Each mineral has a characteristic hardness (resistance to being scratched) and toughness (resistance to cleavage and fracture).

Cleavage. Cleavage is the term applied both to the tendency toward, and to the accomplished fact of, a separation within a gemstone parallel to a crystal face or a possible crystal face.

Fracture. Fracture is a break or a tendency to break in a direction that bears no relationship to the atomic structure of the mineral.

Hardness. The hardness of a gem material determines its resistance to being scratched and thus, in conjunction with resistance to cleavage and fracture, its practicability for use as a jewelry stone.

The hardness of minerals is measured with reference to an empirical scale known as Mohs scale, which consists of 10 minerals, ascending from talc (hardness 1) to diamond (hardness 10). The scale was formulated by determining which of the various well-known minerals scratched others. Gypsum (2) scratches talc, calcite (3) scratches gypsum, and so on up the scale. However, the hardness of diamond is much greater in relation to corundum (9) than is corundum in relation to topaz (8). With few exceptions, the most important gemstones are those at the top of the hardness ladder; for example, diamond is 10, ruby and sapphire are 9, chrysoberyl is 8½, and topaz, beryl (emerald and aquamarine), and spinel are 8.

Beauty. Optical properties are particularly important to the beauty of the various gem materials. The important optical properties include color, dispersion (or "fire"), refractive index, and pleochroism. Gemstones usually are cherished for their color, brilliancy, fire, or one of the several optical phenomena, such as the play of color of a fine opal or the star effect in a sapphire.

Brilliancy. Brilliancy depends on the refractive index, transparency, polish, and proportions of a cut stone. Refractive index is a measure of a gem's ability to reduce the velocity of light that passes through it. It is defined either as the ratio of the velocity of light in air to the velocity of light within a gemstone, or as the ratio of the sine of the angle of incidence to the sine of the angle of refraction. Although diamond does not have the highest refractive index known, it has one of the highest to be found among transparent materials. This quality, combined with its unsurpassed hardness (which permits it to be given a magnificent polish) and great transparency, gives diamond the highest brilliancy potential of any mineral or man-made material. Although synthetic rutile (titania, or titanium dioxide) has a higher refractive index, its lower transparency, relative softness (which makes a superior polish impossible), and other factors combine to reduce its brillancy well below that of diamond.

In order for a gemstone to display its latent brilliancy to best advantage, it must be cut to the proper angles for its refractive index. For example, the correct angle of the pavilion (bottom portion) of a brilliant-cut diamond is 41°, whereas the stones of lower refractive index must be cut to slightly greater angles (measured from the plane of the girdle).

Refractive indices. The refractive indices of the important gems vary from a low of approximately 1.45 for opal to a high of 2.6–2.9 for synthetic rutile. The value for diamond is 2.42. Refractive index is a rough measure of brilliancy; in other words, the brilliancy of two colorless stones of equal transparency and cutting quality is approximately in proportion to their refractive indices. The refractive indices of other important gems are as follows: ruby and sapphire, 1.76; spinel, 1.72; chrysoberyl, 1.74; topaz and tourmaline, 1.62; beryl, 1.58.

Dispersion. Dispersion is the breaking up of white light into its component colors. It is measured by the difference in the refractive indices of certain of the red and blue wavelengths of light as they pass through a gem. In order to give comparable figures for dispersion, two specified wavelengths in the red and blue ends of the Sun's spectrum are used. For example, the refractive indices for these two wavelengths in spinel are 1.710 and 1.730, respectively; therefore, the amount of dispersion is expressed as 0.020. A comparable figure for synthetic rutile is 0.330.

Selective absorption. Although diamond and colorless zircon are valued for their brilliancy and prismatic fire, gemstones are more often cherished for the loveliness of their colors. The color of most gemstones is caused by selective absorption. Selective absorption refers to a gem's ability to absorb or transmit certain wavelengths of light more readily than others. The wavelengths transmitted with least absorption are those that give a gem its color. Absorption is usually caused by very small percentages of metallic oxides that are present as impurities.

Pleochroism. Some gemstones exhibit pleochroism, the property of some doubly refractive materials of absorbing light unequally in the different directions of transmission, resulting in color differences. If two different colors are exhibited, the result is called dichroism (displayed by ruby, sapphire, and emerald); if three, the result is called trichroism (displayed by alexandrite).

Asterism. Some of the most highly prized gem stones are those that depend on unusual optical effects for their beauty. Perhaps the most important of these are the star stones, which display the phenomenon called asterism. In a star, the reflection of light from lustrous inclusions is reduced to sharp lines of light by a domed cabochon style of cutting. The usual star effect is the six-rayed star seen in star sapphires and rubies. Four-rayed stars occur in garnets and some spinels. Other six-rayed stars are occasionally seen in such stones as quartz and beryl. In corundum, the most important species in which asterism is a significant feature, the phenomenon is caused by reflection from three sets of needlelike inclusions arranged in planes perpendicular to the sides of the hexagonal (six-sided) crystal, with each of the three sets of inclusions in planes parallel to a pair of faces of the crystal. *See* CORUNDUM.

Chatoyancy. Another important optical phenomenon is chatoyancy, or a cat's-eye effect. The most important gem in which it is a prominent feature is the variety of the mineral chrysoberyl known as precious cat's-eye. This effect, when seen in chrysoberyl, is usually much more silkily lustrous than in any other gemstone. As a result, the term precious cat's-eye has come to be applied to the finer specimens of the chrysoberyl variety. Other gemstones in which cat's-eyes are sometimes encountered include tourmaline, quartz, beryl, scapolite, diopside, and some of the other rarer gem minerals.

Play of color. The gem mineral opal is one of the most attractive of the phenomenal gems. The colorful display called play of color results from light interference and diffraction effects from tiny spheres of material joined to make up the body of the opal.

Adularescence. Another optical phenomenon that produces an interesting gem variety is adularescence, the billowy light effect seen in adularia or moonstone varieties of orthoclase feldspar. In its most attractive form, it has a slightly bluish cast. Moonstone and opal are among the softer of the fairly common gem minerals and have hardnesses of approximately 6 on Mohs scale.

Precious and semiprecious stones. Gemstones are commonly designated as precious or semiprecious. This is a somewhat meaningless practice, however, and often misleading, since many of the so-called precious gem varieties are inexpensive and many of the more attractive varieties of the semiprecious stones are exceedingly expensive and valuable. For example, a piece of fine-quality jadeite may be valued at approximately $1000 per carat whereas a low-quality star ruby may be worth $1 per carat or less. Fine black opals, chrysoberyl cat's-eyes, and alexandrites are often much more expensive than many of the colors of sapphire. *See* CARAT.

Occurrence. Gemstones occur under a variety of conditions in nature. A number of the highly priced ones occur as primary constituents in igneous rocks (those that result from the cooling of molten rock masses) or in alluvial deposits derived from such deposits. The only known occurrences of diamonds in primary deposits are in the ultrabasic rock known as kimberlite, a type of peridotite. Although most of these are volcanic necks or conduits (so-called pipe deposits), dikes and sills of kimberlite also contain diamonds. Diamond-bearing pipes are known in South Africa, Tanzania, the two Congos, Sierra Leone, India, Lesotho, the Soviet Union, and the United States (Arkansas). Extensive alluvial deposits are mined in Ghana, Angola, Guinea, Ivory Coast, Namibia (South-West Africa), South Africa, Brazil, Guyana, and Venezuela.

Perhaps the most important types of primary deposits in which gemstones other than diamond are found are pegmatite dikes and contact metamorphic deposits. The intrusion of molten rock into impure limestone often brings about the crystallization of calcite to form a coarse-grained marble with a broad assemblage of other minerals. Frequently, some of the minerals produced in the contact zone include gem-quality crystals such as corundum and spinel. The Mogok area of Burma is the foremost supplier of both the finest rubies and the finest sapphires. The gems in this area are mined both from original contact metamorphic deposits and from gravels derived from those deposits. The deposits of Burma produce an amazing variety of gem-quality materials. In addition to both of the major corundum varieties, other colors of sapphire are also found in that country, as well as peridot, topaz, quartz, garnet, and a wide variety of rarely encountered gem minerals, such as scapolite and enstatite.

Pegmatite dikes are found in a number of places throughout the world, but only a few of them produce gem-quality crystals. The most important gem pegmatite areas include Brazil, Madagascar, and southern California. Minerals recovered from pegmatite dikes include most varieties of beryl other than fine emerald, the kunzite variety of spodumene, topaz, and tourmaline. Fine emeralds occur in calcite veins in a shale in the famous mine near Muzo, Colombia, which had been worked for an unknown period even before the Spanish conquerors stumbled upon it in the 16th century. Elsewhere most emeralds occur in mica schist, sometimes with other beryllium minerals, such as chrysoberyl.

Most of the important colored-stone mining is done in alluvial deposits. For example, the island of Ceylon, long known as the "island of gems," has produced a variety of gemstones from alluvial deposits for many years.

Identification. Since most gems are varieties of mineral species, and since mineral species have fairly constant chemical compositions and characteristic crystal structures, their physical and optical properties vary within rather narrow limits. Thus, identification is a matter of measuring these properties and other characteristics, the most important of which are refractive index, optic character, the nature of inclusions, and specific gravity. Because less than 10% of all known mineral species have varieties that are useful as gems, comparatively few tests are needed to separate them; only the first three tests are required in the majority of identifications.

Synthetic gem materials. The properties of synthetic gem materials are identical to those of their natural counterparts; therefore, it is usually necessary to study suspected stones under magnification to determine whether their growth characteristics are natural or the result of synthetic production. For example, the accumulation lines in the form of color banding in natural sapphires are arranged in a hexagonal pattern, whereas those in synthetic materials are curved. Spherical gas bubbles are characteristic of synthetics, but the inclusions in natural sapphires are angular. These same conditions also hold true for ruby and synthetic ruby. To make this determination, binocular magnification with dark-field illumination is the most effective (Fig. 1).

GEM

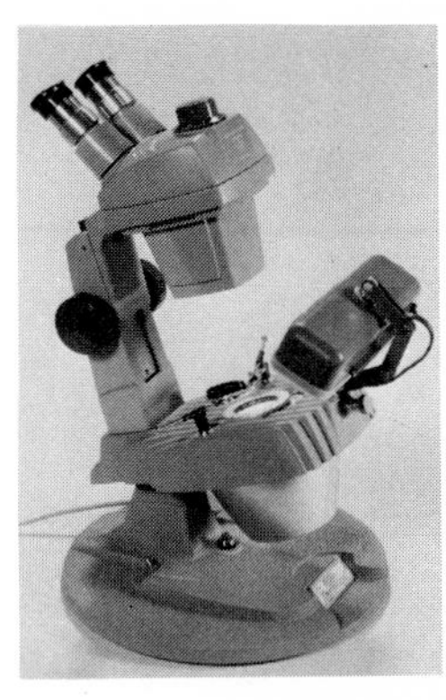

Fig. 1. The Gemolite, a binocular magnifier with dark-field illuminator. (*Gemological Institute of America*)

Refractive index. The refractive index is measured on a refractometer, an instrument that measures the critical angle between an optically dense hemisphere and the gem stone being tested. The instrument is calibrated directly into refractive indices and, as a result, readings from the scale are not in angles but in refractive indices. Doubly refractive minerals show two readings if the birefringence is sufficiently pronounced (birefringence is the difference between the two indices of a doubly refractive material). Thus, a characteristic birefringence may be read as well as a characteristic refractive index. For example, tourmaline shows two readings (one at 1.624, another at 1.644), which alone are sufficient to identify this gem mineral. If monochromatic sodium light is used instead of ordinary incandescent illumination, it is possible to read indices accurately to the third decimal place. When two readings are seen, the necessity for the use of the polariscope, the instrument used to distinguish between single and double infraction, is obviated. The polariscope (Fig. 2) utilizes two polaroid plates, set so that their vibration directions are at 90° to one another, and the gemstone is examined between the two plates. A singly refractive gemstone rotated between the plates remains uniformly dark, whereas a doubly refractive stone becomes alternately light and dark. Diamonds, garnets, and glass remain dark in the polariscope, whereas emerald, ruby, tourmaline, topaz, and zircon become alternately light and dark.

Specific gravity. The density of a gem material compared with an equal volume of water is measured either by heavy liquids or by weighing the

Fig. 2. Polariscope to distinguish between single and double refraction. (*Gemological Institute of America*)

stone in air and in water on a sensitive balance adapted for this purpose. Gem materials vary in specific gravity from amber, at about 1.05, to hematite, at 5.20.

Fluorescence characteristics in ultraviolet light, spectroscopy, x-ray diffraction, chemical tests, and hardness and other tests may be helpful in identification.

Hardness, specific gravity, and refractive indices of gem materials

Gem material	Hardness	Specific gravity	Refractive index
Amber	2–2½	1.05	1.54
Beryl	7½–8	2.67–2.85	1.57–1.58
Chrysoberyl	8½	3.73	1.746–1.755
Corundum	9	4.0	1.76–1.77
Diamond	10	3.52	2.42
Feldspar	6–6½	2.55–2.75	1.5–1.57
Garnet			
Almandite	7½	4.05	1.79
Pyrope	7–7½	3.78	1.745
Rhodolite	7–7½	3.84	1.76
Andradite	6½–7	3.84	1.875
Grossularite	7	3.61	1.735
Spessartite	7–7½	4.15	1.80
Hematite	5½–6½	5.20	
Jade			
Jadeite	6½–7	3.34	1.66–1.68
Nephrite	6–6½	2.95	1.61–1.63
Lapis lazuli	5–6	2.4–3.05	1.50
Malachite	3½–4	3.34–3.95	1.66–1.91
Opal	5–6½	2.15	1.45
Pearl	4	2.7	
Peridot	6½–7	3.34	1.654–1.690
Quartz			
Crystalline	7	2.65	1.54–1.55
Chalcedonic	6½–7	2.60	1.535–1.539
Spinel	8	3.60	1.72
Spodumene	6–7	3.18	1.66–1.676
Topaz	8	3.53	1.61–1.62
Tourmaline	7–7½	3.06	1.624–1.644
Turquois	5–6	2.76	1.61–1.65
Zircon			
Blue and colorless	7½	4.7	1.92–1.98
Green	6	4.0	1.81

Gem materials. More than 100 natural materials have been fashioned at one time or another for ornamental purposes. Of these, however, only a relatively small number are likely to be encountered in jewelry articles. The following paragraphs describe the important gem materials (see table for hardness, specific gravity, and refractive indices).

Amber. Amber is a yellow to brown fossil resin that is used mostly for beads, pipe stems, cigarette holders, and other items. *See* AMBER.

Beryl. Beryl is a transparent beryllium-aluminum silicate. When green, it is known as emerald; when light blue, as aquamarine; when pink to red, as morganite. It also occurs in yellow, light green, brown, orange, and colorless forms. *See* BERYL; EMERALD.

Chrysoberyl. Chrysoberyl is a transparent to translucent beryllium aluminate. Its varieties, known as cat's-eye and alexandrite, are among the important gemstones. Cat's-eye exhibits a band of light across the dome of a cabochon, which is caused by reflections from parallel silklike inclusions. The color is greenish yellow to yellowish green. Alexandrite changes color from green in daylight to garnet red in candlelight. *See* CHRYSOBERYL.

Coral. The gem material coral is an assemblage of colonies of the tiny marine animal coral. It is usually an orange to flesh color and semitranslucent to opaque.

Corundum. Corundum is transparent to translucent aluminum oxide. Ruby is the orange-red to violet-red variety; all other colors are called sapphire. The principal colors of sapphire other than blue are yellow, colorless, orange, pink, purple (amethystine), and green. Both ruby and sapphire occur with a beautiful six-rayed star effect. *See* CORUNDUM; RUBY; SAPPHIRE.

Diamond. Diamond, the transparent form of carbon, is the most important gemstone. In addition to the popular colorless form, it occurs in yellow, brown, pink, blue, and green. Bombardment by subatomic particles (followed by heat treatment for some colors) yields green, yellow, orange, and brown; blue is also possible. *See* DIAMOND.

Feldspar. The gemstone of importance in the feldspar group is moonstone. It is a semitransparent form of potassium feldspar that exhibits a floating light effect. Less important varieties include the amazonite variety of microcline and the iridescent, translucent gray labradorite. Since several minerals are included in the feldspar group, the properties are variable. *See* FELDSPAR; LABRADORITE; MICROCLINE; ORTHOCLASE.

Garnet. Within the garnet group are several distinct minerals that share the same crystal structure but differ in properties and chemical composition (all are silicates of aluminum and two other metals). Two of the garnet species, almandite and pyrope, have the appearance commonly ascribed to garnet. A mixture of the almandite and pyrope compositions in about a 3:1 ratio produces a distinctively colored stone that is given the separate name rhodolite. It is a lighter, more transparent gemstone, with a distinctive violet-red color. Other garnet species include andradite, of which the lovely green demantoid is a variety; grossularite, of which the orange-brown hessonite is a variety; and spessartite, which has varieties closely resembling some of the hessonites. *See* GARNET.

Hematite. The principal ore of iron, hematite, sometimes occurs in dense, hard masses that take a very high polish; these are often fashioned into intaglios. Hematite is metallic grayish black in color. *See* CAMEO.

Jade. The material popularly known as jade may be a variety of either jadeite or nephrite. They are characterized by exceptional toughness, which makes even delicate carvings durable, and by their attractive coloring and translucency. The lovelier and more valuable of the two is jadeite, a member of the pyroxene group of minerals; it is a sodium-aluminum silicate. It is semitransparent to semi-translucent and may be intense green, white with green streaks or patches, brown, yellow, orange, violet, or pink. Nephrite is translucent to opaque and is a darker, less intense green than jadeite; it may also be white, gray, or black. *See* JADE.

Lapis lazuli. Lapis lazuli is an opaque, intensely deep-blue gemstone flecked with golden-yellow pyrite. It has been used for many centuries and is undoubtedly the stone to which the term "sapphire" was first applied. Pliny, for example, described sapphire as an intensely deep-blue stone flecked with gold. *See* LAZURITE.

Malachite. The colorful, opaque mineral malachite is often banded in two or more tones of green and may show a radial fibrous structure. It is always green and is often accompanied by the deep violet-blue azurite, another copper mineral. Malachite is used mostly for cameos and intaglios and for inexpensive scarab bracelets. *See* AZURITE; MALACHITE.

Opal. The fabled beauty of opal is due to a form of light interference in which patches of intense colors are seen against either a white or a nearly black background. There is also a transparent orange to red variety called fire opal. Since it is a particularly fragile material, opal must be treated with care. *See* OPAL.

Pearl. Oriental pearls are those found as lustrous concretions in one of the three species of the salt-water mollusk genus *Pinctada*. Concretions in edible oysters are without pearly luster and are valueless. Pearls found in several genera of fresh-water clams are called fresh-water pearls, to distinguish them from Oriental pearls. Pearls usually occur in white, cream, or yellow colors with rose or other overtones, although black and gray are also very desirable.

Peridot. Peridot is a yellowish-green to green variety of the mineral group olivine; it is a magnesium-iron-aluminum silicate. In gem quality, it is always transparent and olive green. *See* OLIVINE.

Quartz. In its two major types, single crystal and cryptocrystalline (or chalcedonic), quartz, the most common mineral, has more gem varieties than any other mineral species. The important varieties of crystalline quartz are amethyst, citrine, aventurine, and tiger's-eye; the most important varieties of cryptocrystalline quartz are carnelian, sard, chrysoprase, bloodstone, agate, and onyx. Amethyst is purple to violet and transparent. Citrine is yellow to brown and also transparent; it is more commonly known as topaz-quartz and, unfortunately, is often sold as a topaz. Aventurine is usually green with lustrous or colored spangles; it is translucent. Tiger's-eye is a translucent, fibrous, broadly chatoyant, yellow-brown stone that may be dyed other colors. Carnelian is red to orange-red, and sard is a darker brownish red to red-brown; both are translucent. Chrysoprase is light yellowish green. Bloodstone is dark green with red spots. Agate is composed of curved bands in a variety of colors. Onyx is similar to agate, except that the bands are straight. *See* AGATE; AMETHYST; ONYX; QUARTZ.

Spinel. This gem mineral is of particular interest because of the strong resemblance of many of its varieties to comparable colors of corundum. In general, red spinel is less intense in color than ruby; similarly, blue spinel is less intensely colored than blue sapphire. On the other hand, some varieties of spinel are lovely in their own right. Flame spinel is an intense orange-red and is a very attractive gemstone; spinel also occurs in green and amethystine colors. All gem varieties of spinel are transparent, with the exception of the rare black star spinel. *See* SPINEL.

Spodumene. Spodumene is a member of the pyroxene group of minerals; in contrast to jadeite, however, it is fragile. The principal variety for gem purposes is kunzite, which is a lovely light-red to light-purple transparent stone. *See* SPODUMENE.

Topaz. Topaz is best known in its yellow to brown variety, but the red, pink, and blue varieties are also attractive and desirable. *See* TOPAZ.

Tourmaline. Although the color range of tourmaline is as wide as that of any gem material known in nature, its best known varieties are red (rubellite) and dark green; it may also be colorless, yellow, blue, black, brown, or other colors. *See* RUBELLITE; TOURMALINE.

Turquois. Turquois is an opaque gemstone with an intense light-blue color. Its intense color has attracted mankind from the earliest times. *See* TURQUOIS.

Zircon. Zircon is best known as a transparent colorless or blue gemstone. The colorless variety has been used principally as an inexpensive substitute for diamond. It also occurs in green, yellow, brown, red, and flame colors. The properties of zircon vary rather widely. *See* ZIRCON.

[RICHARD T. LIDDICOAT, JR.]

Bibliography: Gemological Institute of America, *Colored-Stone Course*, 1968; E. H. Kraus and C. B. Slawson, *Gems and Gem Materials*, 5th ed., 1947; R. T. Liddicoat, Jr., *Handbook of Gem Identification*, 8th ed., 1969; G. F. H. Smith, *Gemstones*, 13th ed., 1958; R. Webster, *Gems: Their Sources, Descriptions and Identification*, 2d ed., 1969.

Geobotanical indicators

The use of plant indicators in groundwater surveys, geologic mapping, and mineral prospecting. Species assemblages and plant density are used to map strata of different chemical composition and reservoir capacity. Plants are also used in prospecting by analyzing plant tissue for metal content and by observing the species association, toxicity effects, and morphological changes in plants whose roots are in contact with ore. A review of the use of vegetation in interpreting geologic phenomena has been published by H. L. Cannon.

Geologic mapping. Variations in plant associations are often useful in mapping the areal distribution of specific rock formations. A significant relationship between the vegetation and underlying rocks is commonly found in arid regions where the soils bear a chemical similarity to the rocks from which they have been derived. Plants are especially useful in geologic mapping wherever the dominant plant species differ on the formations being mapped. The outlines of the formations may be discernible on air photos. Plants may also be used in detailed, on-the-ground studies where the general floristic composition is the same but where there are minor species differences.

A geobotanist has been regularly assigned to each geologic field team in the Soviet Union since 1945. The floristic composition of the cover, the pattern of plant distribution, and the presence of deformities or growth anomalies resulting from soil conditions are observed both from the air and on the ground. Lithology and structure are also mapped by using plant associations. Similarly, changes in plant species and in growth habits have been used in the United States to mark the location of buried faults or formation contacts. Extensive use has been made of plants in geologic mapping in Texas, where abrupt and distinctive vegetational changes occur at contacts between marl, sandstone, and clay. Pure stands of endemic *Arctostaphylos myrtifolia* (manzanita) are reported to occur on laterites of the Sierra Nevada foothills.

Plants and groundwater. The dependence upon groundwater of certain plant species in contrast to the ability of others to survive on occasional rainfall was first pointed out by O. E. Meinzer in 1923. He gave the name "phreatophyte" to plants that use water from the zone of saturation, "xerophyte" to drought-resistant plants, and "halophyte" to salt-resistant plants. Typical species of phreatophytes are *Prosopis glandulosa* (mesquite), *Chrysothamnus nauseosus* (rabbit brush), *Atriplex canescens* (saltbush), *Acacia greggii* (cat's-claw), *Populus* sp. (cottonwood), *Tamarix aphylla* (tamarisk), *Salix* sp. (willow), *Elymus condensatus* (wild rye), and *Sarcobatus vermiculatus* (greasewood). Typical species of xerophytes are *Atriplex confertifolia* (shadscale), *Eurotia lanata* (winter fat), *Ephedra nevadensis* (Mormon tea), *Yucca baccata* (Spanish dagger), *Larrea divaricata* (creosote bush), and *Gutierrezia sarothrae* (snakeweed).

To absorb water from the soil, the plant sap must have a higher osmotic pressure and therefore a higher salt content than the soil water.

Plants are thus useful as indicators of the quality of water and depth of groundwater; the depth can be estimated from the known depth of root penetration of the species present. Halophytic plants have been especially useful in the Soviet Union in locating buried salt domes.

Indicator plants. The use of plant distribution, plant appearance, and growth anomalies in prospecting for ore deposits has been called geobotanical prospecting, in contrast to biogeochemical prospecting, in which common plant species are used as a sampling medium for chemical analysis. Plants have a third-dimensional advantage in prospecting because the roots may extend through many feet of surficial material for contact with mineralized water or rock. *See* GEOCHEMICAL PROSPECTING.

A plant species that grows more extensively near an ore body because it needs one or more of the ore elements or because an element on which this particular species depends is more available is called an indicator. Some useful indicator plants are given in the table. N. G. Nesvetailova has applied the term universal indicator to those plant species that are adapted to living exclusively on rocks and soils with a particular mineral content and are never found under any other living conditions; local-indicator plants are those species that have wide distribution, but under local situations may serve as indicators of certain properties of the soils and rocks.

In the late 1880s R. W. Raymond described a yellow violet first called *Viola calamineria.* This was later shown to be a variety of *V. lutea* which is restricted entirely to the zinc-rich soils of Aachen in Westphalia. The occurrence of the plant on zinc-bearing soils is so consistent that successful mining operations have been undertaken in areas of this indicator plant.

R. Palou and coworkers have made a study of two indicators of zinc in the Pyrenees. *Armeria halleri* Wallr. (statice) and *Hutchinsia alpina* R.B. occur either singly or together at 21 stations, all of which have indications of zinc mineralization. These indicators have been used to relocate lost mines and to discover new areas of mineralization. Palou and coworkers believe that the plants do not require zinc but rather are tolerant of zinc-rich soils because they reject it, while other species that absorb it are poisoned.

Indicator plants that are restricted in distribution to copper-bearing soils are used in many countries in the search for copper. As early as 1889, F. M. Bailey reported that *Polycarpea spyrostylis* has been used with great success in prospecting for copper in Australia. The plant associations that grow on ground mineralized with copper, lead, and zinc have since been carefully worked out and have been used successfully in prospecting in many areas of Australia.

Another member of the pink family, *Gypsophila patrinii*, has been used successfully in prospecting

Some useful indicator plants in mineral prospecting

Plant	Universal (U) or local (L)	Ore element	Area
Gypsophila patrinii	U	Cu	Soviet Union
Polycarpea spyrostylis	U	Cu	Australia
Tephrosia sp.	L	Cu	Australia
Elscholtzia haichowensis	U	Cu	China
Acrocephalus robertii	U	Cu	Katanga
Becium obovatum (Ocimum homblei)	U	Cu	Katanga
Merceya ligulata	U	Cu	Sweden
Merceya latifolia	U	Cu	Sweden
Mielichhoferia macrocarpa	U	Cu	Sweden, Alaska
Armeria maritima	L	Cu	Wales
Armeria maritima (vulgaris)	L	Zn	Aachen
Armeria halleri	L	Zn	Pyrenees
Hutchinsia alpina	L	Zn	Pyrenees
Thlaspi alpestre	L	Zn	Aachen
Stellaria verna	L	Zn	Aachen
Viola lutea var. *calamineria*	U	Zn	Aachen
Silene cobalticola	U	Co	Katanga
Crotaluria cobalticola	U	Co	Katanga
Astragalus bisulcatus	U	Se	United States (western)
Astragalus pattersoni	L	Se and U	United States (western)
Astragalus preussi	L	Se and U	United States (western)
Astragalus garbancillus	L	Se and U	Peru
Mechovia grandiflora	L	Mn	Katanga

for copper in the Soviet Union. According to Nesvetailova, up to 1% copper was found in rock underlying luxuriant stands of *Gypsophila*, whereas values of less than 10 ppm were found where no such plants grew.

Katanga geologists have long known that a small blue-flowered mint, *Acrocephalus robertii*, grows only on outcrops of copper-bearing rocks. G. W. Woodward of the Rhodesian Selection Trust has discovered a second indicator with mauve-colored flowers that does not require rock outcrop but grows on copper-bearing soils. The plant is also a mint and is now believed to be a variety of *Becium obovatum*. This particular variety consistently contains from 1000 to 4500 ppm copper, while the typical *Becium*, growing in a similar environment, never contains more than 250 ppm. This indicator plant has been used successfully by the Rhodesian Selection Trust since 1949 to locate large reserves of ore.

Another mint, *Elscholtzia haichowensis* Sun., was described by Se Syue-Tszin and Syuy Bay-Lyan as an indicator of copper deposits in China. The plant flourishes on soils containing as much as 4200 ppm copper, and it has been found to contain as much as 6580 ppm in the ash of leaves and 17,240 ppm in the roots. Known ore bodies have been extended in several districts by observing the distribution of *Elscholtzia*.

"Copper mosses" have long been known to occur in Sweden. These include several species of *Merceya* and *Mielichhoferia*. The affinity for copper does not extend to all species of the genera. Three new copper deposits were discovered in Sweden by investigating collection localities of "copper moss" specimens filed in Swedish herbariums. Several species of copper mosses have also been found by H. T. Shacklette in Alaska.

P. Duvigneaud found in Katanga a separate grouping of plants within the copper-loving plant association in areas where the cobalt values are extremely high (2000 ppm). Here two plants, *Silene cobalticola* Duvign. and *Crotalaria cobalticola* Duvign., act as indicators of cobalt. *Crotalaria* accumulates as much as 17,700 ppm cobalt in the ash. Several manganese indicators, including red-flowered *Mechovia grandiflora*, are also used in Katanga.

A large group of plants require selenium in such quantities that the plants are toxic to livestock. These plants are universal indicators of selenium and have been closely studied because of the agricultural implications. The most widely known indicators are species of *Astragalus* and *Stanleya*, although the selenium requirements of the various species of *Astragalus* vary markedly.

Selenium commonly accompanies uranium in the carnotite deposits of the Colorado Plateau, and for this reason selenium-indicator plants can be used in prospecting for uranium. The most useful species on the plateau are *Astragalus pattersoni* and *A. preussi* (see figure). A comparison of plant distribution with drilling data in the Yellow Cat district showed that indicator plants grew over 81% of the ore that occurred within 32 ft of the surface. *A. garbancillus* has also been used to locate uranium ore in Peru.

Selenium indicators used in prospecting for uranium. (*a*) *Astragalus pattersoni* and (*b*) *A. preussi.*

Physiological changes in plants. Toxicity symptoms and morphological changes in plants resulting from variations in the chemistry of the substrate may be useful in prospecting.

Several metals when absorbed by plants in greater than normal amounts cause a chlorosis, or yellowing, of the leaves because the metals in excess interfere with the normal iron metabolism. This chlorosis has been observed in many zinc and copper districts. A change in color of *Pulsatilla patens* and *Linosyris villosa* over nickel-cobalt deposits in the Urals has also been used in prospecting. Excesses of boron, on the other hand, darken the color of foliage and may in severe cases cause advanced necrosis.

Striking changes in plants exposed to bitumen in the soil have been noted in the Soviet Union. The toxicity results in gigantism and a two-cycle flowering habit. Morphological changes have been noted to result from radiation emanating from the Canadian uranium deposits. Fruits of *Vaccinium uliginosum* (blueberry) with unusual shapes and the production of pure white flowers in *Epilobium angustifolium* have been described by H. T. Shacklette.

Finally, plants are used as a sampling medium and are analyzed to detect differences in concentration of a particular element that occurs in the ore. *See* ORE AND MINERAL DEPOSITS.

[HELEN L. CANNON]

Bibliography: H. L. Cannon, The use of plant indicators in ground water surveys, geologic mapping and mineral prospecting, *Taxon*, 20:197–226, 1971; M. M. Cole, *Use of Vegetation in Mineral Exploration in Australia*, 8th Commonwealth Mineralogy and Metallurgy Congress, 6:1429–1458, 1965; G. B. Grigoryan, Plant accumulators and indicators of copper and molybdenum in the Vokhchi River Basin, *Biol. Zh. Arm.*, 22(12): 79–83, 1969; D. P. Malyuga, *Biochemical Methods of Prospecting*, 1964; H. T. Shacklette, *Copper Mosses as Indicators of Metal Concentrations*, U.S. Geol. Surv. Bull. 1198, 1967.

Geochemical prospecting

The use of geochemical and biogeochemical principles and data in the search for economic deposits of minerals, petroleum, and natural gases. In modern exploration programs, geochemical prospecting surveys are generally carried out in conjunction with geological and geophysical surveys. *See* GEOPHYSICAL EXPLORATION.

HISTORY

Both chemistry and geology stretch far back into antiquity and, as might be expected, suggestions as to how chemistry can be applied in the search for mineral deposits have been advanced since early times. Georgius Agricola in *De re metallica*, published in 1556, makes frequent reference to the use of springs, natural waters, and various chemical phenomena in prospecting for veins, and one can also find many interesting statements in the writings of the early Chinese and medieval European writers about the early lore of geochemical prospecting.

The techniques of modern geochemical prospecting had their origin mainly in the Soviet Union and in the Scandinavian countries, where much research on methods was carried out in the late 1930s. After World War II, the various methods were introduced into the United States, Canada, Great Britain, and other countries, where they have since been used extensively in mineral- and petroleum-exploration programs by both mining companies and government agencies.

The modern methods of geochemical prospecting owe their rapid development in the 20th century to the following:

1. Recognition of the nature of primary and secondary dispersion halos, trains, and fans that are associated with all mineral deposits and accumulations of hydrocarbons.
2. Development of accurate, rapid, inexpensive analytical methods utilizing the optical spectrograph, x-ray fluorescence spectrograph, atomic absorption spectrometer, and the various specific sensitive colorimetric reagents, especially dithizone (diphenylthiocarbazone).
3. Development of polyethylene laboratory ware of all types and the development of resins. These permit greater freedom of analysis in the field and reduce the incidence of contamination.
4. Development of rapid and precise methods of analyses of various volatile elements (such as mercury and radon) in rocks, soil gases, waters, and the atmosphere.
5. Development of airborne gamma-ray spectrometry for geochemical analysis of potassium, uranium, and thorium on a broad-scale reconnaissance basis.
6. Development of rapid and precise methods of analysis of various types of both organic and inorganic particulates in the atmosphere.
7. Development of gas chromatography and other precise methods of trace analysis of hydrocarbon compounds and various gaseous inorganic substances.
8. Refinement of field techniques for carrying out reconnaissance and detailed geochemical surveys of all types, especially those based on stream sediments, lake bottom sediments, heavy minerals, stream and lake waters, groundwaters, springs and their precipitates, and biological materials. The use of helicopters has revolutionized sample collection in reconnaissance surveys in practically all terrains.
9. Development and refinement of methods of detailed geochemical prospecting using overburden drilling techniques, especially in glacial terrains.
10. Development and refinement of methods using ore boulder and heavy mineral trains for prospecting in glacial terrains.
11. Research and development of efficient methods for the processing and assessment of geochemical prospecting data by statistical and computer techniques.

Geochemical prospecting is now being employed in all parts of the world, from the tundra to the tropical belts. Among its successes are a number of large low-grade deposits that yield gold, copper, nickel, lead, zinc, and other metals. Some oil fields have been discovered by detailed studies of hydrocarbons in groundwaters and overlying soils and glacial deposits. Even the hydrocarbon content of the sediments of the sea has been utilized in the search for submarine oil pools.

GENERAL PRINCIPLES

The Earth is divisible into five spheres: lithosphere (rocks), pedosphere (soils, glacial till, and so forth), hydrosphere (natural waters), atmosphere (gases), and biosphere (living organisms and their products). The chemistry of these spheres is termed, respectively, lithochemistry, pedochemistry, hydrochemistry, atmospheric chemistry, and biochemistry. Geochemical prospecting methods have been developed that utilize analyses of the materials from each of these spheres, namely: lithogeochemical methods, pedogeochemical methods, hydrogeochemical methods, and so on. *See* GEOCHEMISTRY.

Mineral deposits and accumulations of hydrocarbons represent anomalous concentrations of specific elements, usually within a relatively confined volume of the Earth's crust. Some of these deposits are syngenetic, that is, formed contemporaneously or nearly so with their enclosing rocks. Examples are diamond pipes, certain sedimentary copper and uranium ores, and gypsum and salt deposits. Other deposits are epigenetic; that is, they were introduced into fractures, faults,

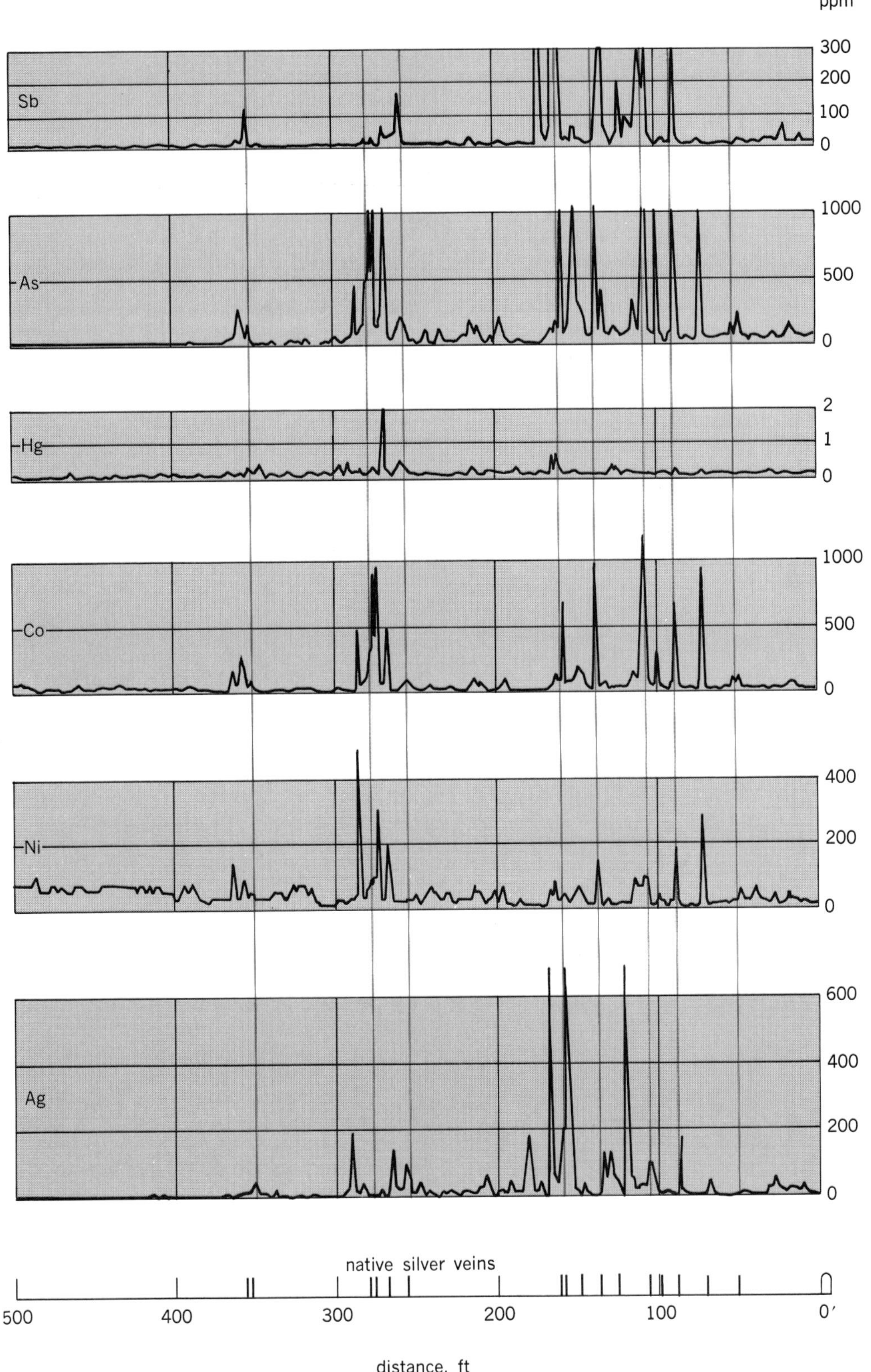

Fig. 1. Graph showing trace-element distribution in wall rocks of native silver veins, Silverfields Mine, Cobalt, Ontario. *(After Boyle et al., Geol. Surv. Can. Pap., 67-35, 1969)*

porous zones, and structural traps generally long after their host rocks were formed. Examples are veins and lodes of metallic minerals in fractures and faults and accumulations of petroleum and natural gas in anticlinal traps and porous reef structures. *See* ORE AND MINERAL DEPOSITS; PETROLEUM GEOLOGY.

Primary and leakage halos. Most mineral deposits and accumulations of hydrocarbons are characterized by a central core or layer in which the valuable elements or minerals are concentrated in percentage quantities. Surrounding this core or layer, the valuable elements generally progressively diminish in quantity to amounts measured in parts per million (ppm) or parts per billion (ppb) which constitute the normal content or background of the enclosing rocks. The region through which this diminution of valuable elements takes place is called the primary halo of the deposit. The term "primary" refers to the fact that the elements in the halo were dispersed into the enclosing rocks at the same time, or nearly so, as those in the central core or layer. Another term, "leakage halo," refers to the dispersion of elements along channels and paths followed by mineralizing solutions leading into and away from the central focus of mineralization. Primary halos show infinite variety. Some are recognizable only a few inches from the central focus of mineralization; others can be detected over distances of hundreds, and in places thousands, of feet from deposits. Most are controlled by microfractures in the rock and by porosity and permeability considerations. Leakage halos are also controlled by the geometry of the available channels, such as fractures, faults, and shear zones, and by the porosity and permeability of the rocks. Leakage halos have been identified in fault and fracture systems up to 500 ft (150 m) and more from mineral deposits. Gaseous leakage halos, such as those related to oil and gas pools, are often detected several thousands of feet above or lateral to the economic concentrations of hydrocarbons.

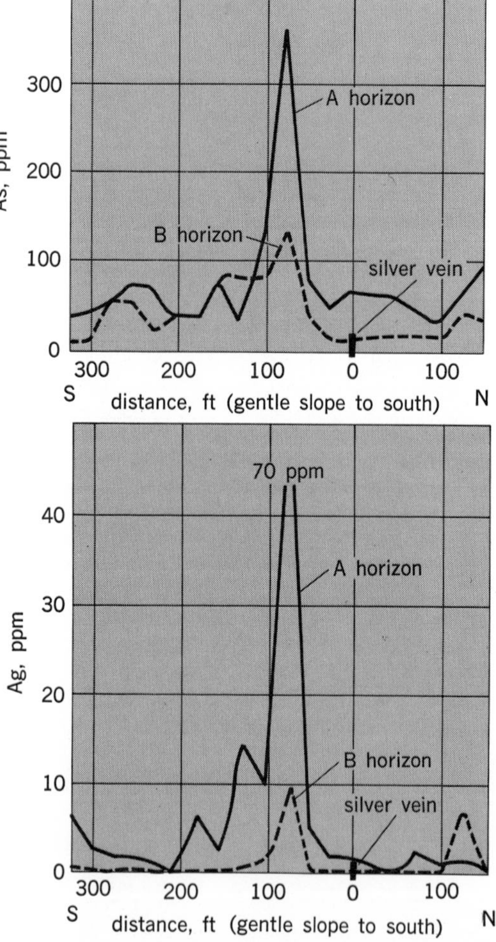

Fig. 2. Arsenic and silver contents on traverse across O'Brien No. 6 vein, Cobalt, Ontario. (*After Boyle and Dass, Econ. Geol., 62:274–276, 1967*)

Secondary halos or dispersion trains. Mineral deposits exposed at the surface undergo extensive oxidation, during which the ore and gangue minerals are disintegrated, and some of their elemental constituents go into solution in the groundwaters. As the deposits are weathered down, the disintegrated particles of the ore and gangue minerals, and some of the elements in solution in the ground waters, are dispersed into the soil and weathered debris overlying the deposits. This dispersion produces a secondary halo, or dispersion train, in the soil. Plants growing in this soil take up elements generally in excess of that required for their physiological processes, thus producing an anomalous elemental halo or dispersion in the vegetation. The shape of the dispersion trains in the soil and vegetation are variable, depending on the topography and a host of other factors. Some are essentially halos that surround the locus of mineralization below. Others are fan-shaped with their apexes at or near the mineral deposits. These are generally referred to as dispersion fans.

Additional dispersion trains are produced by groundwaters that have dissolved some of the ore and gangue minerals. These waters ultimately appear at the surface as springs that feed the streams and rivers of an area. Since these waters frequently have higher-than-normal contents of ore and gangue minerals, derived from the deposits, they produce a hydrogeochemical train whose elemental content is markedly high at the spring orifices and decreases in intensity downstream as dilution by surface waters and other factors come into play. Finally, a dispersion train in the stream and river sediments may be associated with mineral deposits. This train results principally from adsorption of the elemental ions in stream water on the fine stream silt or from complex coprecipitation processes in the stream bed. In addition, there may also be a contribution of fine fragmented particles of ore and gangue minerals which reach the streams by mechanical processes from the deposits. The anomalous values in the trains of ore and gangue minerals in stream sediments are generally highest near the source of the dispersion (that is, the deposits), and they fall off progressively with distance.

Summarizing briefly, most mineral deposits have associated geochemical halos and trains. Those associated with the primary mineralization processes are called primary halos and leakage halos; those associated with weathering processes are called secondary dispersion trains and include those developed in the soil, in the vegetation, in the groundwater system, in the springs, streams, and rivers, and in the stream, river, and lake sediments. All of these halos and trains provide means of tracing and locating the source from which ore and gangue minerals are dispersed, namely, economic deposits. They provide the geochemical anomalies for which all geochemical prospectors search.

Background. Before anomalous conditions can be recognized, however, it is necessary to establish a background against which the anomalies can be compared. In any given area this is generally done by analyzing a relatively large number of samples of the materials to be used in the geochemical survey (rock, soil, vegetation, stream sediment, and so forth), excluding as far as possible any mineralized material. The values obtained may show a fairly large scatter, but the most frequently recurring values tend to fall within a relatively restricted range about a mode. This modal value is generally considered to represent the normal abundance or background of the area for the particular material sampled. Samples which contain amounts of elements twice background or more are generally assumed to be anomalous. When extensive analytical data are available for sampling materials in a surveyed area, a variety of statistical methods are commonly employed to evaluate the background, threshold, and anomalous values.

Interpretation of anomalies. Where anomalous samples are geometrically grouped in a fairly definite pattern, as in a train or halo, a geochemical anomaly or dispersion pattern is present. Figures 1 to 4 represent such anomalies. Figure 1 shows the distribution of a number of elements, expressed as parts per million, in the wall rocks of some silver veins at Cobalt, Ontario. Figure 2 shows the arsenic and silver contents in the soil horizons over another silver vein at Cobalt, Ontario. Figure 3 shows anomalous conditions in the water and sediments of a stream in the Keno Hill area, Yukon. Figure 4 represents a geochemical anomaly in lead in the ash of the twigs of trees growing in soil overlying a lead deposit in Nova Scotia.

When a geochemical anomaly is discovered, a decision has to be made as to its origin. Three possibilities exist: that the anomaly is genetically related to a mineral deposit or concentration of hydrocarbons, that the anomaly is related to subeconomic deposits, or that the anomaly is due to a concentration of elements resulting from a combination of chemical, topographic, and other factors. This is the most difficult part of geochemical prospecting. Every available piece of information must be brought to bear on the problem, such as a knowledge of the climatic, topographic, and geologic conditions, groundwater and surface-water movement, glacial movement, type and distribution of vegetation and humic deposits, and certain physicochemical parameters, such as the Eh, pH, and organic activity, which control the mobility and migration of the elements. These physicochemical parameters affect each element differently, and hence a basic knowledge of the geochemistry of the elements is imperative if the geochemical prospector is to sort out those anomalies which are related to deposits from those that are not. After all interpretations have been made and ancillary data, such as those from geophysical surveys, have been integrated into the pattern, diamond drilling or other exploratory methods, such as trenching, are used to investigate those anomalies that are deemed worthy of detailed investigation.

Indicator elements. In geochemical surveys the key elements in the deposits may be used to trace out the anomalies, for example, lead and zinc for lead-zinc deposits and copper for copper deposits. It often happens, however, that certain economic elements are present in amounts too small for the analytical methods employed or do not give good dispersion patterns because of poor mobility or other factors, for example, gold. In such cases elements occurring in larger amounts or with more favorable dispersion characteristics are employed. Such elements are generally referred to as indicator or pathfinder elements. Examples are arsenic for gold deposits; cobalt, nickel, or arsenic

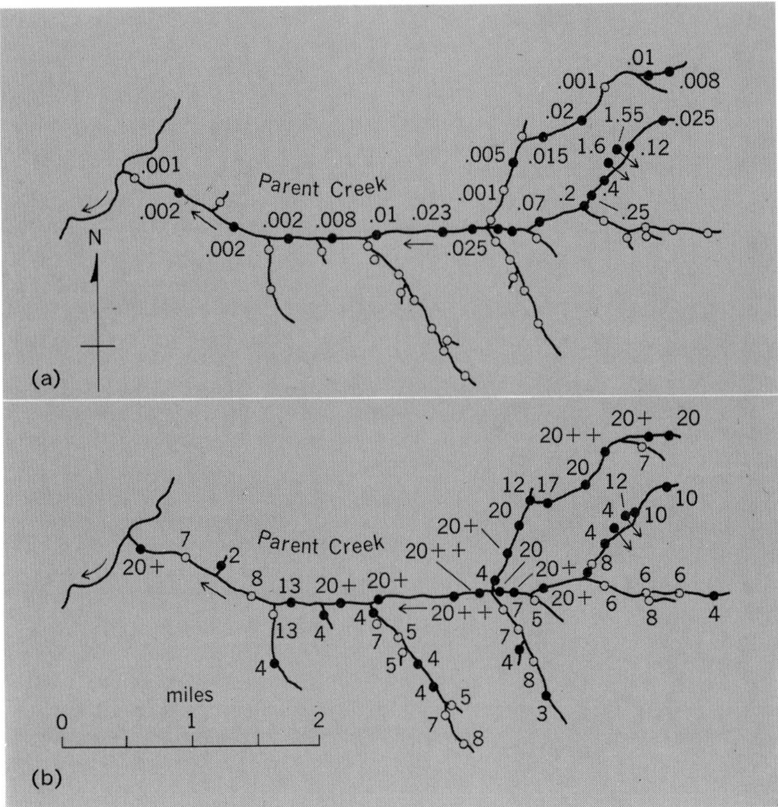

Fig. 3. Heavy-metal content (mainly zinc) in stream and spring (*a*) waters and (*b*) sediments, Parent Creek, Keno Hill area, Yukon Territory. All values in parts per million. (*After Gleeson et al., Geol. Surv. Can. Maps nos. 20–1964 and 21–1964*)

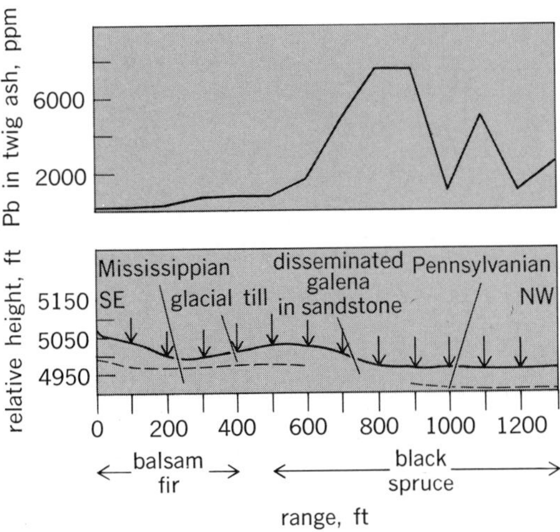

Fig. 4. Lead content in twigs, Silver Mine area, Cape Breton Island, Nova Scotia. (*After M. Carter, M.Sc. Thesis, Carleton University, Department of Geology, 1965*)

for certain types of native silver deposits; and molybdenum for certain types of copper deposits.

Sampling techniques. All geochemical prospecting surveys must be based on adequate sampling of the material or materials used in the survey. These materials may be rocks, waters, gases, soils, stream sediments, lake sediments, or vegetation. Where possible, sampling should be confined to one type of material for any particular survey, such as one rock type in lithogeochemical surveys, soil from one horizon in pedogeochemical surveys, or twigs from one species of tree. When diverse materials are sampled because of the heterogeneous nature of the geology or vegetation, correlative factors must be applied to the results before the geochemical patterns can be interpreted.

It should be emphasized that the sampling techniques used in geochemical surveys must be such that the sample obtained is representative of the rock, water, gas, soil, vegetation, or other types of organisms at the point of sampling. Failure to consider carefully this aspect of a survey leads only to spurious results and difficulties in interpretation. The spacing and distribution of the sampling points are controlled by the estimated size of the deposit, the mobility of the dispersed elements, and the type of dispersion pattern expected.

Chemical analysis. The analytical techniques used to determine the quantity of elements in geological and biological materials are varied. Some techniques have been devised for use at sampling sites in the field, whereas others have been developed for analysis in well-equipped mobile field laboratories or central laboratories operated by commercial firms, or government agencies. The analytical results are generally expressed in parts per million or parts per billion.

Analytical techniques for on-site sampling of water, soil, stream sediment, lake sediment and so on are usually based on colorimetry and use a specific selective organic reagent, such as dithizone. The loosely bonded or exchangeable metal in the geological material is extracted by means of a citrate, acetate, acid, or other extractant and then determined by the colorimetric reagent. When it is desirable to determine the total metal or other element content of a sample, various analytical methods are employed in mobile or central laboratories. These include optical spectrography, x-ray fluorescence spectrography, atomic absorption, and colorimetric, chromatographic, and polarographic methods. For hydrocarbons, gas chromatography and a variety of other methods employed in analytical organic chemistry have found wide usage.

Analytical techniques used for geochemical prospecting should be rapid, permitting a large number of samples to be done in a short time, and inexpensive, yet precise to within about 20%. Portability of equipment and reagents is the prime consideration in the case of methods used in the field at the sample sites.

GEOCHEMICAL PROSPECTING SURVEYS

Geochemical prospecting surveys are generally classified according to the type of materials sampled. These include lithogeochemical surveys (rocks), pedogeochemical surveys (soil and till), hydrogeochemical surveys (water and sediments), surveys based on volatiles and airborne particulates, biogeochemical surveys (plants and animals), remote-sensing surveys, and isotope surveys.

Geochemical prospecting surveys may be either reconnaissance surveys or detailed surveys. In reconnaissance surveys analyses of materials are carried out over a broad region, over a mineral or oil concession, or over a large number of claims with the express purpose of defining mineral belts, zones of mineralization, or sites of accumulation of hydrocarbons. Detailed surveys are carried out on a local basis with the purpose of locating individual deposits, oil pools, or favorable structures where these might occur. The amount of detail varies with the geological situation but may go down to sampling at 5-ft (1.5 m) intervals or less where primary halos associated with veins are sought.

Lithogeochemical surveys. These have not been used as extensively as other types of geochemical prospecting surveys, mainly because the sampling techniques and interpretation of the surveys have not been investigated sufficiently. Both reconnaissance and detailed lithogeochemical surveys may be carried out.

Most reconnaissance surveys are carried out on a grid or on traverses across a geological terrane, samples being taken of all available rock outcrops or at some specific interval. One or several rock types may be selected for sampling and analyzed for various elements. The distribution of the volatiles such as Cl, F, H_2O, S, and CO_2 in intrusives with associated mineralization has received some attention as indicators. Geochemical maps are compiled from the analyses, and contours of equal elemental values are drawn. These are then interpreted, often by using statistical methods, in the light of the geological and geochemical parameters. Under favorable conditions, mineralized zones or belts may be outlined in which more detailed work can be concentrated. If the survey is carried out over a large expanse of territory, geochemical provinces may be outlined.

Detailed lithogeochemical surveys are generally carried out on a local basis and have as their purpose the discovery or definition of primary or leakage halos associated with mineral deposits or petroleum accumulations. Chip samples of rocks on a definite grid are used in some cases, and samples from drill cores in others. All of the analytical data are plotted on plans and sections and compared with the geological situation. Frequently, primary halos can be discovered by this method (Fig. 1), and these and leakage halos can be traced to the focus of mineralization or to petroleum accumulations if sufficient work is done.

In detailed drilling and development work the use of ratios obtained from analyses of rocks along traverses or diamond drill holes can frequently be employed to estimate proximity to mineralized loci. Particularly useful ratios include K_2O/Na_2O, SiO_2/CO_2, and SiO_2/total volatiles; the volatiles commonly include H_2O, CO_2, S, As, and B. Many types of mineral deposits are characterized by a consistent increase in the ratio K_2O/Na_2O, which is essentially a manifestation of increasing potassic alteration. Similarly, a number of mineral deposits, particularly those enriched in gold and silver, are marked by a consistent decrease in the ratio

SiO_2/CO_2 as ore is approached. The ratio SiO_2/total volatiles exhibits considerable variation among the various types of mineral deposits; in most cases, skarns excepted, there is a consistent decrease in the ratio as mineralization is approached.

Pedogeochemical surveys. Soil and glacial-till surveys have been used extensively in geochemical prospecting and have resulted in the discovery of a number of orebodies. Similar surveys have been used in searching for petroleum pools. Generally, soil and glacial-till surveys are of a detailed nature and are run over a closely spaced grid. One of the soil horizons is chosen for sampling, generally the B horizon, and a plan of the metal or hydrocarbon contents is plotted and contoured. Under favorable conditions, the highest values are centered over a deposit, but more generally the dispersion pattern is a train or a fan that requires careful interpretation to locate its source.

Certain geological conditions may require that all horizons of the soil be sampled, including the A, B, and C horizons. Frequently, sampling of the A (organic) horizons is effective in some areas (Fig. 2), whereas in others the B or C horizons are more rewarding. In some places deep sampling of the C horizon by drilling is the only satisfactory method for soil and glacial-till surveys. This technique has proven most effective in many of the heavily overburdened, glaciated terrains of Canada, in the permafrost regions of Canada and the Soviet Union, and in the deeply weathered lateritic regions of the tropics.

Heavy- and resistate-mineral surveys of soil, till, and weathered debris have found increasing usefulness in recent years. In these surveys the geological materials are panned, and the heavy and resistate minerals obtained. These are then examined microscopically for ore minerals or analyzed for ore or indicator elements, and the results are plotted on maps. Heavy- and resistate-mineral maps, prepared on the same grid as those for soil surveys, provide valuable ancillary data and often aid in the interpretation of the elemental dispersion patterns.

In soil analyses the fine fraction (minus 80 mesh) is generally analyzed for the chemically dispersed elements, whereas for heavy and resistate minerals a coarser fraction is used. *See* SOIL.

In glaciated terrains, as well as in certain other terrains, heavy- and light-clast (mineral fragments, stones and boulders) tracing has proven effective in the discovery of certain types of mineral deposits. Examples are quartz boulders or fragments as indicators of gold-quartz veins, and galena boulders or fragments as indicators of lead-zinc deposits. Surveys of this type are generally carried out on grids, the abundance of light or heavy clasts being visually noted and plotted at each sampling point or wherever they occur along the grid lines. In other surveys, large samples of the till or overburden are obtained at each sampling point, and the light- and heavy-clast indicators are counted in the light and heavy concentrates obtained from the samples. When all of the data from clast surveys are plotted, fans or trains are commonly outlined whose apexes or starting points often mark the sites of underlying mineralization.

Hydrogeochemical surveys. These include water surveys, sediment surveys, and heavy-mineral surveys. The last two are not strictly hydrogeochemical in nature, although they are generally carried out along the drainage systems of an area.

The background metal content of most natural waters is only a few parts per billion, rising to a few parts per million for certain elements in the vicinity of mineral deposits. Seasonal and diurnal fluctuations often occur in these values, features that have to be considered in water surveys. Accurate and sensitive analytical methods are a requirement in water surveys, and on-site analysis is recommended if possible to avoid problems introduced by transportation and contamination of water samples. Most water analyses are done by sensitive colorimetric methods or by atomic absorption spectrometry.

Either surface waters or groundwaters may be tested in water surveys, depending on local conditions. Surface waters are sampled at regular intervals along the drainage net, and a map of the values is prepared (Fig. 3). An increase in the metal content of the water upstream may indicate approach to a mineralized zone. Surveys based on groundwaters can be done only where there is a good distribution of wells, springs, or diamond drill holes. The metal content of the water in these is plotted on a map and contoured. Higher-than-normal contents in the water system may indicate sites or zones of mineralization. Extensive surveys of this type are being carried out in the Canadian Shield, Turkey, and elsewhere in the search for uranium and other metallic deposits.

Sediment and heavy-mineral surveys are carried out to determine the migration path of dispersed elements and minerals along the surface-drainage channels of an area. Samples are collected from the fresh sediment in the bottoms of streams and also from old sediment on the terraces and floodplains. For chemically dispersed elements, the fine fraction (minus 80 mesh) is generally used for analysis; for mechanically dispersed heavy minerals, a coarser fraction is panned from the sediment. Sampling points are located at intervals along the length of the drainage system. The results of the chemical analyses of the stream sediment are plotted on a map of the drainage (Fig. 3). An increase in the metal content of the stream sediment upstream may indicate approach to a mineralized zone. The heavy-mineral fraction may be examined microscopically for ore minerals or accompanying gangue minerals, or the fraction may be analyzed for ore elements or indicator elements. Both results are plotted on the drainage map and interpreted in the same way as the stream-sediment data.

Volatiles and airborne particulates. These surveys are based on gases, such as hydrocarbons, H_2S, and SO_2, and on volatile elements, such as mercury. Airborne particulates, of both an organic and inorganic nature, can also form the basis of surveys. These are usually carried out by sophisticated airborne equipment.

Despite its obvious importance, geochemical prospecting for accumulations of petroleum and natural gas has not been extensively employed in North America, mainly because of a lack of research into methods. It is evident that many oil and gas fields are marked by macroseeps (leakage halos) along faults and porous zones. It is logical to suppose that there are also microseeps, which should be detectable by modern methods of hydrocarbon analysis, especially gas chromatography.

The methods investigated using hydrocarbon analysis have been based mainly on soils. The technique is similar to that used for metalliferous deposits. A grid of soil samples is analyzed for hydrocarbons over suspected areas; the results are plotted and contoured and then interpreted. Techniques similar to those using rocks and groundwaters for the discovery of mineral deposits, but employing hydrocarbons as the indicators, should also be effective in locating accumulations of petroleum and natural gas.

Gases, such as SO_2 from oxidizing sulfide orebodies, have been suggested as good indicators in soil and water analyses. Proof of their effectiveness, however, requires much more research. Mercury, a volatile element, has been used effectively as an indicator element in soil and rock surveys for locating sulfide deposits.

Surveys based on the analysis of organic and inorganic particulates in the near-surface atmosphere have been conducted in Canada, Australia, and South Africa. The methods used are essentially at the research stage and require much more work to prove their usefulness in localizing mineralization.

Biogeochemical surveys. These surveys are of two types. One type utilizes the trace-element content of plants to outline dispersion halos, trains, and fans related to mineralization; the other uses specific plants or the deleterious effects of an excess of elements in soils on plants as indicators of mineralization. The latter type of survey is often referred to as a geobotanical survey.

In the first type of survey the trace-element content of selected plant material is determined on a grid over an area. Generally, samples are collected from parts of individuals of the same species of vegetation, such as twigs, needles, leaves, and seeds. These are dry- or wet-ashed, the trace-element content is determined, and the results are plotted for interpretation with respect to mineralized zones or deposits (Fig. 4). Interpretation of vegetation surveys is frequently difficult, since a number of factors enter into the uptake of elements by plants, some of which are unrelated to mineralization. These include the content and nature of exchangeable elements in the soil, drainage conditions, soil pH, and growth factors in the vegetation.

Geobotanical surveys are carried out by mapping the distribution of indicator-plant species, or plant symptoms diagnostic of high metal-bearing soils, such as chlorosis and dwarfing. Where suitable indicator plants are present, this method is rapid and inexpensive, but unfortunately such plants are seldom consistent in distribution from one area to another. Geobotanical techniques require careful preliminary orientation surveys by trained personnel. *See* GEOBOTANICAL INDICATORS.

Some biogeochemical surveys of a research nature have been conducted in Canada and the Soviet Union utilizing various animals as the sampling media. The animals, or parts thereof, used have been fish (livers), mollusks (soft parts), and insects (whole organisms). The results of these surveys show that these animals commonly reflect the presence of mineralization in regions in which they occur by having higher-than-normal amounts of various elements.

An interesting development in petroleum prospecting is based on population counts of microflora and microfauna which oxidize hydrocarbons, particularly propane, during their metabolic processes. Where soils, rocks, and groundwaters are enriched in hydrocarbons, certain strains of bacteria and other similar forms of life flourish, and their density of population is apparently proportional to the content of hydrocarbons present. By utilizing specialized bacterial counting techniques, it is possible, as a number of geochemists in the Soviet Union have indicated, to plot contour maps showing the distribution of the bacteria and, hence, the hydrocarbons. Some of these contour maps show peaks and halos that mark accumulations of oil and gas at depth.

Dogs can locate mineral deposits by sniffing out boulders of ore occurring in the dispersion trains and fans of sulfide deposits. They can be trained to become quite sensitive to SO_2 and other gases associated with oxidizing sulfides and are said to be quite effective in the Scandinavian countries and the Soviet Union.

Remote sensing surveys. These surveys utilize various techniques such as airborne detection of radioactivity, infrared detection of anomalies (such as sulfide zones undergoing oxidation) in the geological terrane, and reflection studies. This is a large subject, and only a few examples will be given here.

Airborne gamma-ray spectrometers utilizing large detection crystals have been used extensively in many countries to outline positive radioactive belts and zones mineralized with uranium and thorium. Similar equipment can be utilized in demarcating zones rich in potassium (K^{40}) associated with gold and other types of mineralization. Negative radioactive zones, that is, those from which uranium and thorium have been leached during alteration processes often associated with certain types of gold, silver, lead, zinc, and copper mineralization, can also be located and outlined by these surveys.

Geochemists of the U.S. Geological Survey have observed that over certain types of ore deposits the reflectivity of leaves differs measurably from that of leaves on trees over barren ground. Efforts are being made to apply this method in low-lying jungle areas, where accessibility and deep weathering restricts the use of more conventional geochemical methods.

Isotope surveys. These surveys employ isotopic ratios such as those for lead (Pb^{204}, Pb^{206}, Pb^{207}, Pb^{208}) and sulfur (S^{32}, S^{34}). Only elements with two or more isotopes can be used in such surveys.

Lead isotopes. Most surveys use the isotopic ratios of minerals in "fingerprinting" or indicating certain types of deposits. For instance, lead minerals in uranium deposits tend to have a high proportion of (radiogenic) uranium-lead, that is, they are enriched in Pb^{206} and Pb^{207}, the derivatives of U^{238} and U^{235}; in thorium-rich deposits, a high enrichment of (radiogenic) thorium-lead, Pb^{208}, can be expected in the lead minerals. In ordinary lead-bearing deposits, the lead minerals have a component of original (primal) lead isotopes (Pb^{204}, Pb^{206}, Pb^{207}, Pb^{208}) plus varying amounts of radiogenic lead (Pb^{206}, Pb^{207}, Pb^{208}), depending, among other factors, on the age of the deposit. Furthermore, there are deposits in which the lead minerals have isotopic ratios that are unusual or anomalous (the J-lead of the Mississippi Valley deposits and also of such deposits as Keno Hill, Yukon). The reasons for the variable isotopic composition of lead in deposits are extremely complex, to say the least, and certainly not understood as yet. Scientists need to understand the processes involved in the migration and concentration of lead isotopes before their theories can be placed on a firm basis. These problems notwithstanding, it is possible in certain cases to fingerprint lead deposits and minerals derived from lead-bearing deposits by means of their isotopic ratios. Thus, during geochemical prospecting surveys where the lead in soils, stream sediments, or in particles of galena in heavy concentrates exhibits high concentrations of radiogenic lead, uranium or thorium deposits should be suspected in the area. Similarly, galena particles exhibiting the so-called J-lead characteristics may indicate the presence of Mississippi-Valley-type deposits. Finally, the lead in rocks (feldspars, apatite, and so on), when isotopically analyzed, may give a clue to the type of lead deposit to be expected or may indicate the presence of uranium or thorium deposits in the terrane.

Sulfur isotopes. Sulfur isotopes commonly fingerprint certain types of deposits in a region, and hence isotopic analyses of the sulfides in heavy concentrates from soils and stream sediments, of sulfur in the ground and surface waters, and of sulfur in stream and lake sediments can be used in a general way to decide what types of deposits occur within a terrane. Thus, the sulfur in waters and stream sediments in regions containing abundant barite or evaporites is generally greatly enriched in the heavy isotope S^{34}; on the other hand, waters leaching sulfide deposits in most regions often have sulfur that is relatively enriched in the lighter isotope S^{32}. Sulfur isotopic ratios also vary systematically with distance from deposits; in some cases, the lighter isotope is enriched as ore is approached; in others, the heavier isotope is enriched. This information may be useful in detailed drilling work and in the interpretation of the ratios K_2O/Na_2O, SiO_2/CO_2, and so forth, obtained during lithogeochemical work.

Sulfur isotopes, as well as those of hydrogen and carbon, should find greater use in hydrocarbon prospecting, particularly in diffusion studies and in determining the migration characteristics of hydrocarbons from source to reservoirs.

EXPLORATION PROGRAM

A full-scale geochemical prospecting program for metals would include the following stages.

1. Preliminary evaluation of areas, selected on the basis of available geological data, by sampling and testing intrusive, metamorphic, and sedimentary rocks and by noting the presence of mineralized zones, faults, fractures, layers, and so forth associated with these rocks. In this way, a metallogenetic province can be identified.

2. Primary reconnaissance and orientation surveys, based on sampling major drainage basins, using water, stream sediment, lake sediment, and heavy-mineral surveys.

3. Secondary reconnaissance surveys based on detailed testing of drainage basins containing anomalous values. Poorly drained areas can be tested by widely spaced sampling of soil and groundwaters.

4. Followup surveys along dispersion trains or fans to determine the cutoff points and the extent of dispersion patterns. These surveys are normally a combination of stream-sediment, heavy-mineral, water, and soil testing, but biogeochemical surveys may also be useful. Priority for followup surveys should be based on the presence of favorable rocks and geological structures, favorable geophysical indications, and intensity of the geochemical anomaly.

5. Detailed surveys carried out in the vicinity of the suspected metalliferous source by soil or vegetation sampling at closely spaced intervals. Interpretation of the results at this stage generally suggests sites for trenching, sinking of shallow shafts, or drilling to locate the precise source of the body which is giving rise to the geochemical anomaly.

For hydrocarbons the stages in the geochemical exploration program would include the following stages.

1. Preliminary evaluation of areas, selected on the basis of available geological data and known or suspected to contain hydrocarbons. Normally, these will be underlain by relatively unmetamorphosed sediments, younger than Precambrian, and usually containing a high organic (kerogen) content.

2. Primary reconnaissance surveys based on relatively widely spaced sampling of rocks, waters, soils, glacial materials, lake sediments, or oceanic sediments for hydrocarbons, particularly those with molecular weights higher than CH_4.

3. Secondary reconnaissance surveys based on detailed sampling of anomalous areas indicated in stage 2. More closely spaced sampling of the materials in stage 2 should be carried out in addition to sampling of groundwater and stratal water by means of test holes. The cores from the test holes should also be analyzed for hydrocarbons.

4. Interpretation of hydrocarbon anomalies in conjunction with analysis of favorable geophysical indications followed by drilling of the most favorable anomalies.

[R. W. BOYLE]

Bibliography: R. W. Boyle and J. I. McGerrigle (eds.), *Geochemical Exploration*, Can. Inst. Min. Met. Spec. Vol. II, 1971; R. R. Brooks, *Geobotany and Biogeochemistry in Mineral Exploration*, 1972; J. B. Davis, *Petroleum Microbiology*, 1967; I. L. Elliott and W. K. Fletcher (eds.), *Geochemical Exploration*, 1974; I. I. Ginzburg, *Principles of Geochemical Prospecting*, 1960; H. E. Hawkes and J. S. Webb, *Geochemistry in Mineral Exploration*, 1962; M. J. Jones (ed.), *Geochemical Exploration*, 1972, Inst. Min. Met. London, 1973; A. A. Kartsev et al., *Geochemical Methods of Prospecting and Exploration for Petroleum and Natural Gas*, 1959; A. A. Levinson, *Introduction to Exploration Geochemistry*, 1974.

Geochemistry

The science concerned with the chemical elements that make up the Earth and with the chemical reactions which produce and alter these elements. Therefore, a primary objective of geochemistry is to determine the chemical composition of the major divisions of the Earth: the mantle, oceanic crust, and continental crust, as well as the groundwater, surface water, and atmosphere. This can be achieved, however, only by determining the compositions of the rocks, minerals, and waters that constitute the Earth. Thus, this aspect of geochemistry is analytical, including methods and techniques ranging from the statistical problem of determining an average chemical composition for the mantle to the use of finely focused electron beams to analyze single micrometer-sized grains of individual minerals.

However, the distribution of elements in the Earth at present represents only a short moment of geologic time. Elements are constantly being rearranged via the processes of weathering, crystallizing of lavas and deep-seated magmas, ore deposition, and sediment diagenesis, and by the long-time-scale fractionation of the Earth itself. Experimental geochemistry, using thermodynamics, especially phase equilibria, translates these Earth processes into chemical processes that can be modeled in the laboratory. Experimental geochemistry can construct accurate models of the equilibrium processes, but chemical processes take place on a vast scale of space and time, and only recently has the subject of chemical kinetics been added to the repertory of geochemistry so that the time scale of chemical reactions can be considered as well as their final equilibrium state.

Distribution of elements. Determining the chemical composition of the Earth is perhaps the oldest problem in geochemistry. From thousands of individual analyses of rocks, glacial soils, and meteorites, tables of relative abundances of elements in the continental crust, the oceanic crust, and the mantle have been compiled, and even some hypotheses about the Earth's core have been advanced. These data have been continuously refined as more analyses become available. Knowledge of the composition of planetary crusts has now been extended to the Moon by the thousands of analyses on the rocks returned by the Apollo missions.

The Earth's elements are distributed quite unevenly. Eight elements—oxygen, silicon, aluminum, iron, magnesium, calcium, sodium, and potassium—account for 99% by weight of the Earth's crust. Many other elements occur only at concentrations of millionths of a percent. The mantle is believed to consist of ferromagnesian silicates, whereas the core is thought to be composed mainly of metals, the "siderophile" elements. Some elements, such as lead, are rare on the world average but are highly concentrated in ores, whereas other elements, such as gallium, although more common on the world average, are so widely dispersed that they form no primary minerals of their own. The wide range in average concentrations reflects the combination of processes by which the elements were originally created, as well as the processes of differentiation—first of the planets in the formation of the solar system and then of the fractionation of the Earth into its core-mantle-crust parts. *See* ELEMENTS, GEOCHEMICAL DISTRIBUTION OF; NATIVE METALS.

Two basic problems emerge from this large number of chemical analyses: the development of a plausible theory to explain the origin of the elements and the solar system that will reconcile the elemental abundances with known nuclear reactions and models of planetary evolution; and the determination of the geochemical processes that lead to the great enrichment of certain elements in minerals of limited occurrence and the wide dispersal of other elements throughout the terrestrial rocks.

Crystal chemistry. V. M. Goldschmidt long ago discovered that since crystals are held together by electrostatic forces, the size and charge of the constituent ions determine crystal structure. From this discovery was born crystal chemistry, which in its early form was mainly concerned with explaining the observed structures of crystals in terms of ionic radii, charges, and polarizabilities. The few common elements determine the crystal chemistry of the several hundred most common minerals. The dispersal of the less common elements can thus be partially explained in crystal chemical terms. Gallium readily substitutes for aluminum, which is a common constituent of many minerals in the crust. Because of the similarity in the ionic radii of gallium and aluminum, gallium is widely dispersed as a minor component in common rock-forming minerals. Lead, gold, and related elements with ionic radii or charge very different from those of rock-forming elements cannot readily substitute for them, and therefore do not occur as minor substituting constituents in common minerals but rather form minerals with their own unique crystal chemistry. Reactions by which these minerals are formed lead to the concentration of rare elements and thus to the formation of ore deposits.

Although considerations of size and charge are very useful for explaining the distribution of chemical elements in the Earth, there are many exceptions, and much of the thrust of solid-state physics and chemical bonding theory applied to geochemistry is to explain the exceptions. Crystal field theory accounts for the geochemical behavior of nickel and chromium and their unexpected enrichment in early-crystallizing minerals. The sulfide and pnictide minerals that make up important ore deposits do not conform to simple crystal chemical theory very well, and the covalent and metallic bonding of these compounds must be taken into account. *See* CRYSTAL STRUCTURE.

Geochemical prospecting. Concentrations of rare elements in a background of common elements from rock-forming minerals are the basis of geochemical prospecting. The elements in ore bodies are dispersed by weathering processes, and some migrate upward into overlying rocks and the soil. Unexpected increases in the concentrations of rare elements above background even when total concentrations are very low may indicate the presence of ore bodies at depth. Likewise, sensitive analysis of metals in stream sediments may permit the metals to be traced upstream to their source. *See* GEOCHEMICAL PROSPECTING.

Organic materials. Organic geochemistry deals primarily with two areas: (1) the chemistry of petroleum, asphalts, coal, humic acids, and other organic constituents found mostly in sedimentary rocks; and (2) the role of organic processes and living organisms in the dispersion of inorganic chemical elements (sometimes known as biogeochemistry).

Isotopes. Isotopes of the elements are fractionated during many geochemical processes. Fractionation is dependent on the kind of reaction and on the temperature. $^{16}O/^{18}O$ and $^{2}H/^{1}H$ ratios are used to provide information on paleotemperatures in marine sediments and in other low-temperature deposits. $^{34}S/^{32}S$ isotope ratios provide information about the sources of the constituents of mineral deposits. $^{12}C/^{13}C$ ratios relate to processes of sedimentation.

Age dating. Radioactive elements generate decay products which, if they remain with the parent element in the form of a recognizable isotope, can provide a method of dating. Rb/Sr, K/Ar, U/Th, and U/U measure times on the order of millions of years. Radioactive ^{14}C created in the upper atmosphere is incorporated into living matter, and its gradual decay after the organism dies is a dating method accurate to 50,000 years. *See* DATING METHODS.

Experimental geochemistry. Much of experimental geochemistry is the duplication in the laboratory of reactions believed to take place in the Earth. These experimental conditions include a range of pressures from atmospheric pressure to more than 1,000,000 atm (1.01325×10^{11} Pa) and a range of temperatures from Earth surface temperature to several thousand degrees Celsius. Very-high-pressure experiments, 20 kbars to 1 mbar (2×10^{9} Pa to 10^{11} Pa), attempt to duplicate the conditions in the mantle. Under these extreme pressures the minerals found in crustal rocks collapse into more densely packed structures.

Experiments at crustal pressures of a few hundred bars to 10 kbars (10^{9} Pa) are frequently carried out in the presence of water so that reaction kinetics are faster and the observed assemblage of phases more nearly matches natural conditions. Most of the important reactions describing the interactions of the rock-forming silicates have now been mapped, and most of the processes of magmatic crystallization, pegmatite genesis, and metamorphic reactions can be described. The results are presented as phase diagrams, with composition, temperature, and pressure as variables or as projections onto a pressure-temperature petrogenetic grid. The petrogenetic grid is particularly useful in describing the evolution of metamorphic rocks. *See* HIGH-PRESSURE PHENOMENA.

Fluids which transmit ore and vein-filling minerals have temperatures in the range of a few hundred degrees Celsius and pressures of a few hundred bars. However, many such systems are open to H^{+}, O, S, and other mobile species. The central problems concern the mechanisms by which heavy metals are transported and deposited in concentrated form as ores. *See* ORE AND MINERAL DEPOSITS; ORE DEPOSITS, GEOCHEMISTRY OF.

Low-temperature aqueous geochemistry is concerned with Earth surface processes: the chemistry of groundwater; water quality in surface streams; transport and deposition of minerals on the ocean floor and in lakes; diagenesis of sediments; and controlling factors on transport of agricultural runoff, sewage effluents, landfill leachates, and industrial wastes to ground and surface waters. Chemical reactions are complex but can be written specifically. Equilibrium states can be calculated from thermodynamic data on the coexisting solid phases and dissolved complexes as functions of temperature, Eh, pH, CO_2 activity, and other open-system variables. *See* HYDROSPHERE, GEOCHEMISTRY OF.

The rates of geochemical reactions can be regulated by intrinsic reaction rates or transport controls. Reactions between solid and liquid phases are often controlled by surface reactions which may involve inhibitors in addition to intrinsic reaction-rate constraints. Such processes are of greatest importance at low temperatures. Reaction-rate inhibitors can act at very low concentrations, and their identification is a time-consuming and difficult task.

Reaction rates depend on the transport of atoms to the reaction interface and the removal of reaction products. In solid/liquid reactions this involves the mechanics of fluid flow, diffusion of reactants across boundary layers, and buildup of insoluble boundary layers on the reaction interface. This buildup is a key process in mineral weathering. In the weathering of feldspars, the alkali ions are preferentially leached across a boundary of alkali-depleted silicates on the surface of the feldspar grains. *See* WEATHERING PROCESSES.

Many mineral reactions, particularly those taking place at high temperatures, depend on solid-state diffusion. In this process, ions must migrate directly through the solid at rates described by Fick's laws of diffusion. The theoretical basis of diffusion transport is more reliable than the theory of reaction rate and transport controls, but there are not yet sufficient diffusivity data for calculations of high temperature–high pressure reaction rates to be made reliably.

In general, rate theory in geochemical processes has undergone only rudimentary development and may be expected to be an important area of research in the immediate future.

Theory. Geochemistry, in common with all of the earth sciences, has become more quantitative in recent years, and for this reason students of geochemistry need to be familiar with aspects of statistics, thermodynamics, and quantum mechanics.

Statistics. Statistical methods include the analysis of variance by factor analysis, components analysis, and cluster analysis for the examination of rock and mineral compositions. The chemical composition of large units of the Earth from the continental crust to the individual stock or batholith can be described only in statistical terms. Most of these calculations also require the use of large computers.

Thermodynamics. Thermodynamics forms the backbone of experimental geochemistry and is a useful guideline to the limitations on reactions taking place in the Earth. Equilibrium thermodynamics is used to predict chemical reactions between minerals. Phase diagrams and associated laws of heterogeneous equilibria are used to describe the assemblages of minerals that are permitted to coexist at equilibrium under a stated regime of temperature and pressure. In recent years some use has been made of statistical thermodynamics, particularly in the construction of models for solid solutions. There is now quite a good data base of thermochemical data on the more important rock-forming minerals, and some progress has been made at using such data in large computer programs to generate theoretical phase diagrams.

Quantum mechanics. Formal quantum mechanics has had relatively little impact on geochemistry. However, the related theory from solid-state physics has become an important field of mineralogical investigation. "Mineral physics" has developed mainly in two areas: the use of crystal field theory to explain the colors, luminescence, and magnetic properties of transition-metal ions in minerals, and the use of molecular orbital theory to investigate chemical bonding in silicate and other minerals. Other applications of quantum mechanics, such as the application of the formally more difficult band theory to the geochemistry of the metallic chalcogenide and pnictide minerals, are only just under way.

Characterization. Geochemistry has benefited from many new methods for the characterization of solid materials. These may be divided broadly into methods of chemical analysis and structural analysis, the latter including techniques that give insight into chemical bonding.

Chemical analysis. The classical techniques for chemical analysis of rocks and minerals have been a group of wet-chemical methods for the determination of major elements, emission spectroscopy for rapid multielement analysis and for the analysis of minor and trace elements, and x-ray emission spectroscopy as an alternate method for the analysis of major and minor elements. To these have been added atomic absorption spectroscopy, which allows fairly precise analysis down to the trace level (a few parts per million), and neutron activation analysis, which extends the detection limits of certain elements to the part-per-billion level.

A complete revolution, however, was initiated by the introduction of focused-electron and ion beams to permit chemical analysis of very small volumes and of surfaces. The electron microprobe can determine the compositions of individual mineral grains in a rock or the compositional zonation in grains of minerals. Impacting ions penetrate surfaces for distances of only a few atomic layers. They sputter other ions from the surface whose composition can be determined by secondary ion mass spectrometry (SIMS) and generate secondary electrons which can be analyzed (by Auger spectroscopy). These methods give the chemical composition of the outer surface layers of the mineral rather than the bulk composition.

Structural analysis. Minerals are identified primarily by x-ray powder diffraction and optical microscopy. X-ray and electron diffraction from single crystals obtained by using elaborate diffractometers often interfaced with computers are used to determine crystal structures. To these methods—which are still the backbone of structural characterization—has been added a group of spectroscopies that yield either additional structural information or some insight into chemical bonding. Mössbauer, optical absorption, nuclear magnetic resonance (NMR), and electron spin resonance (ESR) spectroscopy depends on the presence of some target element in the structure, and the spectra reflect the immediate environment of the target. Infrared and Raman spectroscopy measures structural information on the scale of a unit cell. High-energy electron beams generate secondary electrons, and their spectral energy distribution can be used for chemical analysis (termed electron spectroscopy for chemical analysis, or ESCA) or to provide information about the orbital structure of atoms. The results of secondary electron and x-ray spectroscopy are the primary experimental checks for the various molecular orbital calculations of chemical bonding in minerals. *See* LITHOSPHERE, GEOCHEMISTRY OF.

[WILLIAM B. WHITE]

Bibliography: L. H. Ahrens, *Origin and Distribution of the Elements*, 1968; G. Faure, *Principles of Isotope Geology*, 1977; R. M. Garrels and C. L. Christ, *Solutions, Minerals and Equilibria*, 1965; A. W. Hofmann et al., *Geochemical Transport and Kinetics*, 1974; K. Krauskopf, *Introduction to Geochemistry*, 1967; R. F. Mueller and S. K. Saxena, *Chemical Petrology*, 1977; B. Nagy and U. Colombo, *Fundamental Aspects of Petroleum Geochemistry*, 1967; F. M. Swain, *Non-marine Organic Geochemistry*, 1970.

Geochronometry

The study of the absolute age of the rocks of the Earth using methods based on the radioactive decay of such isotopes as U^{238}, U^{235}, Th^{232}, Rb^{87}, K^{40}, and C^{14}. Measurements using these decay schemes permit definition of the age of the Earth and meteorites (4.5×10^9 years); the oldest available rocks (3×10^9 years); significant events in the geologic time scale; the rate of evolution of organisms; the time and duration of major events in crustal evolution; the advance and retreat of continental glaciers; and the evolution of human culture.

These isotopic chronometers have superseded many earlier qualitative methods for estimating geologic time, and their application introduces new quantitative data into Earth science. *See* EARTH, AGE OF; FISSION TRACK DATING; GEOLOGICAL TIME SCALE; LEAD ISOTOPES, GEOCHEMISTRY OF; RADIOCARBON DATING; ROCK, AGE DETERMINATION OF.

[J. LAURENCE KULP]

Bibliography: E. I. Hamilton, *Applied Geochronology*, 1965.

Geode

A roughly spheroidal hollow body, lined on the inside with inward-projecting small crystals (see illustration). Geodes are found most frequently in limestone beds but may occur in some shales. Typically, a geode consists of a thin outer shell of dense chalcedonic silica and an inner shell of quartz crystals, sometimes beautifully terminated, pointing toward the hollow interior. Many geodes are filled with water; others, having been exposed for some time at the surface, are dry. Calcite or dolomite crystals line the interior of some geodes, and a host of other minerals are less commonly found. In some geodes there is an alternation of layers of silica and calcite, but almost all geodes show some banding suggestive of rhythmic precipitation. *See* CHALCEDONY.

The origin of geodes lies in the presence of an original cavity, in many cases voids within fossil

Geode, lined with quartz crystals, Keokuk, Iowa. (*Brooks Museum, University of Virginia*)

shells, from which the geode originally grew. The geode grows by expansion, the layer of chalcedonic silica being the hardened equivalent of an original silica gel. The expansion is due to osmotic pressure from original sea water trapped inside the silica gel shell and fresh water on the outside of the gel. The projecting quartz crystals are precipitated from later groundwaters infiltrating the already hardened, hollow spheroid. *See* SEDIMENTARY ROCKS. [RAYMOND SIEVER]

Geodesy

A subdivision of geophysics which includes the determination of the size and shape of the Earth, the Earth's gravitational field, and the location of points fixed to the Earth's crust in an Earth-referred coordinate system. The shape of the Earth as defined in geodesy is the geoid: the equipotential, or surface of constant potential, of the Earth's gravity field which most closely approximates the mean sea level. Thus the location of the geoid and the accelerations of gravity are really two different modes—one geometrical, the other physical—of representing the same thing and require only a scale factor to relate them. *See* GEOPHYSICS; TERRESTRIAL GRAVITATION.

The purposes of geodesy as defined here include mapping of gravity variations in order to study the structure of the Earth's crust and mantle in combination with seismological, geological, and other data; the establishment of a control network of accurately located points for mapping, navigation, construction, and so on; the establishment of a reference network of gravity accelerations for geophysical prospecting by gravimetry; and the provision of an expression of the gravity field for the accurate determination of artificial satellite orbits.

Geodesy can be divided into the measurement of both constant and time-varying quantities. Because the temporal variations are all so small compared to spatial variations which are constant (in historical time), the measurement and analysis of the two divisions are normally treated separately. The major part of this description will be devoted to the constant component.

Mathematical expression. The geoid is irregular, and hence any mathematical representation is to some extent an approximation. The most important approximation is an oblate ellipsoid of revolution, which is customarily defined by its equatorial radius a and its flattening f, which is defined by Eq. (1), where b is the polar semidiameter (Fig. 1).

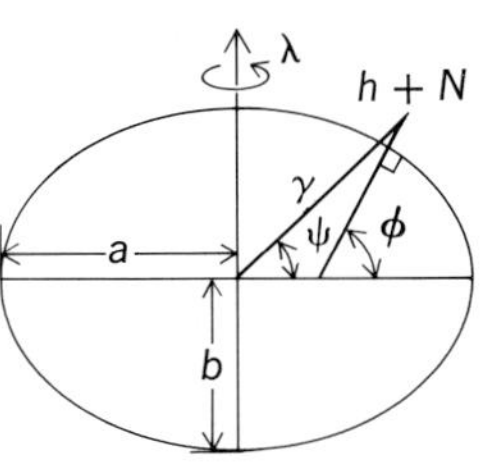

Fig. 1. Meridional section of reference ellipsoid.

$$f=\frac{a-b}{a} \qquad (1)$$

Geodetic location is conventionally expressed in coordinates referred to such an ellipsoid: the latitude φ, the angle between the normal to the ellipsoid and the equator; the longitude λ, the angle about the rotation axis measured from the Greenwich meridian; and the altitude h above or below the geoid, which in addition has a height N with respect to the ellipsoid.

If the ellipsoid is rotating with a rate ω, contains a total mass M, and is assumed to be an equipotential for the combined effects of centrifugal acceleration arising from the rotation ω and centripetal acceleration from the mass M, then the resulting net acceleration γ at latitude φ on the surface, called normal gravity, can be calculated from Eqs. (2) and (3), where G is the constant of gravitation

$$GM=a^2\gamma_e\left[1-f+\frac{3m}{2}-\frac{15mf}{14}+O(f^3)\right] \qquad (2)$$

$$\beta_2=\frac{5m}{2}-f-\frac{17mf}{14} \qquad \beta_4=\frac{f^2}{8}-\frac{5mf}{8}$$
$$\gamma=\gamma_e[1+\beta_2\sin^2\varphi+\beta_4\sin^2 2\varphi+O(f^3)] \qquad (3)$$

(6.67×10^{-8} cm^3 g^{-1} sec^{-2}), γ_e is the acceleration of gravity at the Equator, m is as defined in Eq. (4),

$$m=\frac{\omega^2 a}{\gamma_e} \qquad (4)$$

and $O(f^3)$ comprises terms of order of magnitude f^3.

The form shown in Eq. (3) is suitable for comparison with measurements of acceleration at the Earth's surface. For calculation of effects on orbits, there is required instead the potential of gravitation U in geocentric coordinates: the geocentric latitude ψ, the longitude λ, and the radial distance r (Fig. 1). This potential is conventionally expressed by means of Eq. (5), where $P_{20}(\sin\psi)$ and

$$U=\frac{GM}{r}\left[1-J_2\left(\frac{a_e}{r}\right)^2 P_{20}(\sin\psi) - J_4\left(\frac{a_e}{r}\right)^4 P_{40}(\sin\psi)-O(f^3)\right] \qquad (5)$$

$P_{40}(\sin\psi)$ are Legendre functions defined in Eq. (6),

$$P_{lm}(\sin\psi) = \cos^m\psi \sum_{t=0}^{k}\frac{(-1)^t(2l-2t)!}{2^l t!(l-t)!(l-m-2t)!}\sin^{l-m-2t}\psi \qquad (6)$$

and k is the integer part of $(l-m)/2$.

The ellipsoid parameters J_2 and J_4 are related to f and m by Eqs. (7) and (8). In Eq. (5) the conven-

$$J_2=\frac{2f(1-f/2)}{3}-\frac{m(1-3m/2-2f/7)}{3}+O(f^3) \qquad (7)$$

$$J_4=\frac{-4f(7f-5m)}{35}+O(f^3) \qquad (8)$$

tion of geodesy and astronomy is followed in making a gravitiational potential positive. The oblateness J_2 is also related to the moments of inertia about the polar (C) and equatorial (A) axes by means of Eq. (9).

$$J_2=\frac{(C-A)}{Ma^2} \tag{9}$$

The ellipsoid approximates the geoid within 2×10^{-5} of the radial coordinate. The physical basis of this approximation is that an ellipsoid is very close to the shape which would be assumed by a fluid body under the combined efforts of gravitation and rotation. Such a fluid body with the same mass M, radius a, rotation rate ω, and moment of inertia C as the actual Earth will have a flattening which is given approximately by Eq. (10).

$$f=\frac{5m}{2}\bigg/\left[1+\left(\frac{5}{2}-\frac{15C}{4Ma^2}\right)^2\right] \tag{10}$$

The difference between the actual potential of the Earth V and the reference ellipsoid potential U is known as the disturbing function T, which in turn is directly related to the geoid height N by Eq. (11).

$$N=\frac{T}{\gamma}=\frac{V-U}{\gamma} \tag{11}$$

The conventional manner of expression of the actual acceleration of gravity g at height h above the geoid is the gravity anomaly Δg, defined as the difference of g from normal gravity γ at the same height h above the ellipsoid. Hence, to relate Δg to the disturbing function T, an extra term is required to account for the change in gravity $N\partial\gamma/\partial r$ over the geoid height N, and Δg is given by Eq. (12).

$$\Delta g=\frac{-\partial T}{\partial r}-\frac{2T}{r} \tag{12}$$

The relationship between the geoid height N and the gravity anomaly Δg is expressed by an integral transform known as Stokes' theorem, which is shown in Eq. (13), where the integration is over the

$$N(\varphi_N,\lambda_N)=\frac{a}{4\pi\gamma}\int S(\theta)\,\Delta g(\varphi,\lambda)\,d\sigma \tag{13}$$

unit sphere; θ is the arc distance from (φ_N,λ_N) to (φ,λ); and $S(\theta)$ is a complicated sum of trigonometric functions of θ.

For satellite orbit perturbations, an expression of the gravitational potential V in terms of spherical harmonics is more convenient and can be written as in Eq. (14), where P_{lm} is defined by Eq. (6)

$$V=\frac{GM}{r}\left[1+\sum_{l=2}^{\infty}\left(\frac{a_e}{r}\right)^l\sum_{m=0}^{l}P_{lm}(\sin\psi)\cdot(C_{lm}\cos m\lambda+S_{lm}\sin m\lambda)\right] \tag{14}$$

and C_{lm},S_{lm} are arbitrary coefficients of order 10^{-6} which must be determined by observation.

The direction of the gravity vector is customarily known as the astronomical position, because it is determinable by observations of the stars. The difference between this astronomical position φ_a,λ_a and the direction implied by the geodetic coordinates φ,λ is known as the deflection of the vertical and has components given by Eq. (15). The

$$\xi=\varphi_a-\varphi \qquad \eta=(\lambda_a-\lambda)\cos\varphi \tag{15}$$

relationship of the deflection η and the difference between azimuths A_a,A as measured clockwise from north about the two axes φ_a,λ_a, and φ,λ is known in geodesy as Laplace's equation. Equation (16) demonstrates the relationship.

$$A=A_a-\eta\tan\varphi \tag{16}$$

A system of measurements of the gravity field is automatically referred to the Earth's center of mass by omitting the first degree harmonics $l=1$ from the mathematical expression of the field. A system of relative positions obtained by measurements of direction and distance will have correct orientation through observations of the stars and use of radio time signals, but in general it will have a difference of translation of its coordinate origin from the Earth's center of mass. This difference is usually expressed in terms of the differences in latitude, longitude, and height at an arbitrarily selected point on the Earth's surface called a datum origin.

Geodetic observing systems. Geodesy utilizes five distinct observing systems: horizontal control, vertical control, astronomical positions, gravimetry, and satellite tracking.

Horizontal control. Combinations of lengths, measured by invar tapes or electronic ranging devices, and angles about vertical axes, measured by theodolites in series of triangles or other plane geometric figures, are called triangulation, trilateration, or traverse, and are used to obtain geodetic position φ,λ dependent upon an assumed position φ_o,λ_o at a datum origin. The accuracy of such concatenations, as a ratio of error in position difference to distance between points in the same system, $\varepsilon(\Delta P)/D$, is on the order of $1/(17{,}000D^{1/3})$ for D in kilometers; for example, for points on opposite coasts of the United States, $\varepsilon(\Delta P)$ will be on the order of 15 m. For special surveys this accuracy can be increased to the order of 10^{-6} by conventional techniques.

Horizontal control has been extended to the sea-floor bottom for special purposes in limited areas by acoustic ranging to locate sea-floor transponders with respect to surface vessels located by other means. The accuracy of the acoustic ranging is about ±1 m.

Vertical control. Differences in elevation between points are determined by leveling, a series of readings on measured staffs made through carefully leveled telescopes. The extent of such systems approaches that of the horizontal control; the accuracy is such that the elevation above sea level of any point in a major leveling network should be known within a foot. Leveling tells nothing about heights above the mathematical fiction, the ellipsoid, since it is a mathematical artifact.

Astronomical positions. The direction φ_a,λ_a of the gravity vector and the azimuth A_a of a terrestrial point about this axis can be obtained within $\pm0''.2$ by a series of observations of the stars carefully distributed to minimize atmospheric refraction.

The deflections of the vertical ξ,η defined by Eq. (15) are the slopes of the geoid with respect to the ellipsoid. Starting with an arbitrary geoid height N_o and integrating these slopes, a geoid map can be constructed. Such astrogeodetic geoids have been calculated for most of the major triangulation nets and yield the difference in geoid height ΔN with an accuracy which is comparable to that of horizontal position ΔP.

Gravimetry. Absolute gravity can be measured in the laboratory within about 0.0005 cm/sec² by timing the fall of an object in a vacuum. Connected to the absolute measurements is a worldwide network of relative measurements, comprising pendulum measurements over large ranges (more than 1.0 cm/sec²), where scale accuracy is more important, and gravity meter (spring balance) measurements, where sensitivity is more important. The accuracy of reference of a typical field point with respect to the absolute datum is about ±0.001 cm/sec² on land and about ±0.005 cm/sec² at sea. The coverage of gravimetry over the land is fairly good, but considerable areas of the ocean in the Southern Hemisphere have not been measured.

Satellite tracking. Both artificial satellites and the Moon are used. Artificial satellites may be used for geodetic purposes in two ways: first, in a purely geometrical manner, as very high beacons observable simultaneously from widely separated points on the Earth's surface and, second, in a dynamical manner, as objects whose orbits are affected by the Earth's gravitational field. Both radio and optical tracking techniques are used. The principal radio technique is Doppler. Prior to about 1970 the main optical technique used cameras photographing the satellite against the star background; since then, laser ranging has become the more accurate method.

The most extensive geometrical application of satellites has been the use of camera observations of balloon satellites at altitudes of about 3500 km to establish a global network of control points at about the same interval. Satellites at lower levels, some with flashing lights, have also been used to connect stations at closer intervals, using radio and laser ranging as well as cameras. It is now planned to launch a high-altitude, very dense satellite with laser retroreflectors to obtain a more accurate global net.

The dynamical use of satellites depends on the orbits being different from the Keplerian ellipse which would exist in a purely central potential GM/r. The principal effect of the oblateness, J_2 in Eq. (5), is to cause steady motions $\dot{\omega}$ of the perigee (the point of closest approach) and $\dot{\Omega}$ of the node (the intersection of the orbit and the equator) at rates on the order of 0.001 times the mean motion n of the satellite. The other even-degree zonal harmonics, J_4, J_6, and so on, also make minor contributions to these steady motions. All the other spherical harmonics C_{lm},S_{lm} of the potential, Eq. (14), cause periodic perturbations which have rates that are various combinations of the three orbital rates n, $\dot{\omega}$, and $\dot{\Omega}$ and the rate of rotation of the Earth, ω. In particular, any tesseral harmonic ($m \neq 0$) perturbation will contain the rate $m(\dot{\Omega} - \omega)$; that is, it is an oscillation forced by the Earth's rotation.

To determine the harmonics C_{lm},S_{lm} of the gravitational potential from orbits requires, first, an extensive net of accurate tracking stations, 12 or more, well distributed around the Earth, and second, satellite orbits which are low enough to be appreciably perturbed but high enough to avoid drag and to be frequently observed, and which have as great a variety of inclination (the angle between the orbital and equatorial planes) as possible. This variety of inclination is necessary in order to separate the effects of different harmonics whose perturbations have the same periodicity.

The accumulated data are now sufficient to determine the coefficients C_{lm}, S_{lm} up to about l,m = 16,16, with varying degrees of accuracy, dependent not only on the degree l but also on the number of different terms with the same periodicities and on the closeness of the orbit to resonance with the Earth's rotation.

The geodetic application of the Moon is to determine station locations and wobbles of the Earth's rotation by laser ranging to retroreflectors landed by the Apollo project. The Moon's orbit is now known accurately enough that simultaneity of observations is unnecessary.

An auxiliary measurement of geodetic interest is that of the rate of precession of the Earth's rotation axis with respect to the fixed stars (arising from the torque exerted on the Earth's equatorial bulge by the Sun and the Moon), used to determine the dynamical ellipticity $(C-A)/C$.

Important space techniques include (1) very-long-baseline radio interferometry (VLBI), which utilizes differences in phase of radio signals from stellar sources to determine differences in location; (2) radar altimetry, which senses geoidal undulations directly by radar measurements from an orbiting satellite; and (3) satellite-to-satellite Doppler tracking, which yields much more coverage than tracking from the ground.

Numerical values of parameters. For the principal geodetic parameters described above, there are conflicting needs for standardization and accuracy. Hence it is appropriate to list both values adopted as standard by the International Union of Geodesy and Geophysics (IUGG) and the International Astronomical Union (IAU) and values indicated as the most likely by recent data.

The primary quantities GM, a, and γ_e are related through Eq. (2). GM is determined directly from the mean motion and semimajor axis of satellites by Kepler's equation; it is determined most accurately from distant satellites, such that, primarily, the proportionate error of the length measurement is minimized and, secondarily, the clutter of other perturbations from the Earth are reduced. Consequently there are two principal types of determinations: those using the Moon itself and those using lunar probes.

The acceleration of gravity γ_e is probably the second most accurately determined of the three quantities. Recent determinations of absolute gravity indicate that the adopted Potsdam standard is in error by about 0.014 cm/sec². The main limitation on accuracy is the statistical problem of determining the mean value of gravity over the Earth from the nonuniformly distributed data available.

The equatorial radius can be determined independently from astrogeodetic data: a combination of the horizontal control and astronomical positions systems previously described. The principal errors are scale error in the triangulation, from both uncertainty of measured lengths and error propagation through angle measurements, and interpolation error, dependent on astronomic station spacing, which affects the calculation

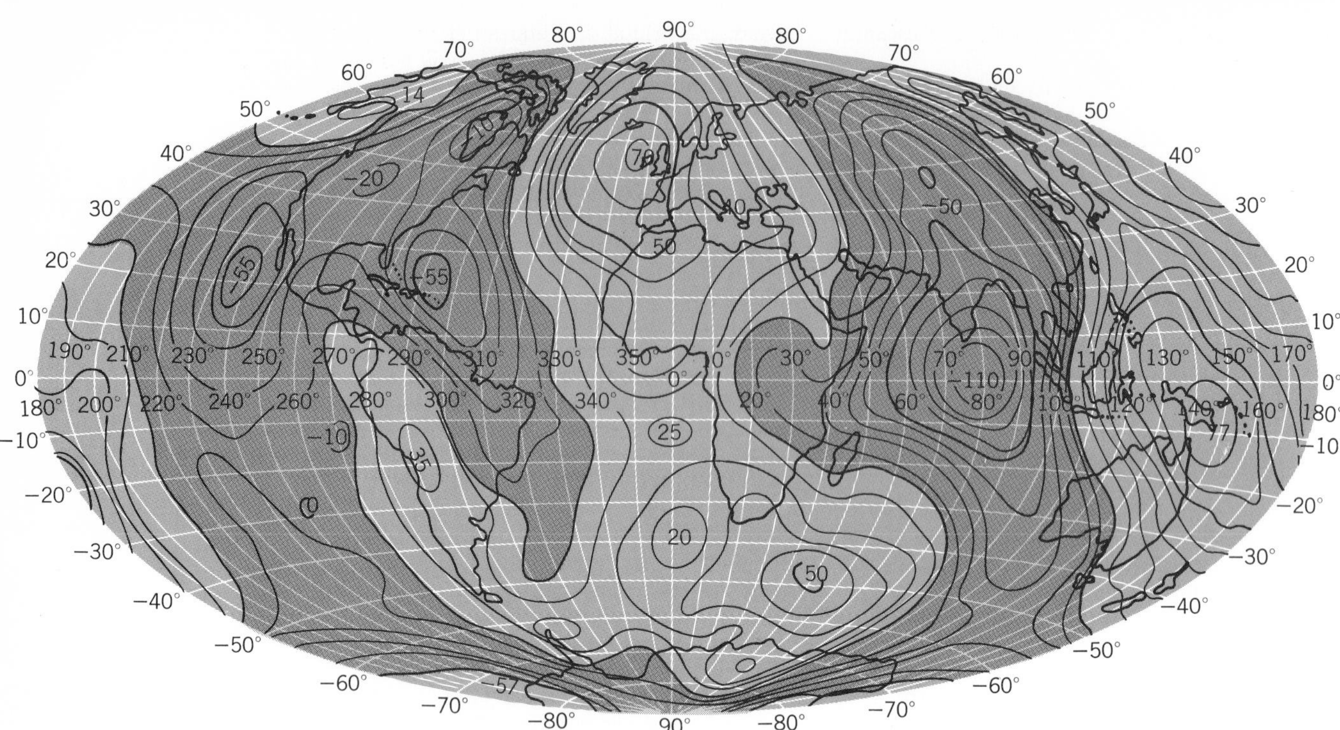

Fig. 2. Geoid heights, in meters, which are referred to an ellipsoid of flattening 1/298.25. (*From F. J. Lerch et al., NASA Goddard Space Flight Center Tech. Rep. X-921-74-145, 1974*)

of ΔN. Probably the most accurate way of determining a, given GM, is from the radial coordinator of tracking stations located by dynamical satellite orbit analysis.

Another way of determining a, given GM, is from the mean radial change in tracking-station position in close-satellite orbit analysis.

Parameters determined with more than sufficient accuracy include the rotation rate ω, the oblateness J_2, and the dynamical ellipticity $(C-A)/C$. Numerical values are given in the table.

The root-mean-square magnitude of the gravity anomalies Δg obtained from observations is about ± 0.035 cm/sec²; of the geoid heights N, ± 33 m; and of the deflections of the vertical ξ,η, $\pm 5.7''$.

In the numerical determination of the coefficients C_{lm}, S_{lm} of the gravitational potential, Eq. (14), there are generally used spherical harmonics normalized so that their mean square value over the unit sphere is one; that is, there is applied a normalization factor N_{lm} which satisfies Eq. (17),

Geodetic parameters

Parameter	Standard values, IUGG/IAU, 1964	Best estimate (standard deviations), 1975
ω	—	$0.7292115085 \times 10^{-4}$ sec^{-1}
J_2	0.0010827	0.00108264 ± 3
γ_e	—	9.780321 ± 10 m sec^{-2}
a	6,378,160m	$6.378144 \pm 4 \times 10^6 m$
GM	3.986030×10^{14} m³/sec^{-2}	$3.986013 \pm 3 \times 10^{14}$ m³ sec^{-2}
m	—	0.0034678
J_4		0.00000162 ± 6
f	1/298.25	1/298.257
β_2	—	0.0053025
β_4	—	−0.0000059
$(C-A)/C$	—	0.0032732
C/MA^2	—	0.33076

$$N_{lm}^2 \int_{-\pi/2}^{\pi/2} P_{lm}^2(\sin\psi)\cos\psi\, d\psi = 2\left(2-\delta_{om}\right) \quad (17)$$

where P_{lm} is defined by Eq. (6) and δ_{om} is the Kroneker delta function. The root-mean-square magnitude by degree l of these normalized coefficients $\overline{C}_{lm}, \overline{S}_{lm}$ from satellite orbits follows very closely a rule expressed in Eq. (18).

$$\sigma\left\{\overline{C}_{lm}, \overline{S}_{lm}\right\} = \pm 10^{-5}/l^2 \quad (18)$$

Figure 2 shows a geoid determined from a combination of satellite and gravimetric data. The major features depend almost entirely on the satellite data and are not expected to change appreciably in future determinations.

Time-varying quantities measurement. These measurements include latitude variation, rotational variation, crustal motion, horizontal crust motion, and Earth tides.

Polar wobble. This involves variations of latitude arising from shifts of the rotation axis with respect to the Earth's crust caused by geophysical effects. The amplitude of this variation is always less than 0.5″, as shown in Fig. 3. The spectrum of the variation has two peaks: one annual, a forced oscillation, and one of about 14½ months, a free oscillation.

Rotational variation. Time observations yield a secular slowing down of the Earth, caused by tidal friction, plus periodic variations with semiannual and annual peaks, caused by meteorological effects or core-mantle coupling. *See* TIME.

Vertical secular motions. Releveling every few years indicates in several areas changes of a few millimeters a year. The most extensive such areas are associated with postglacial crustal rebound in

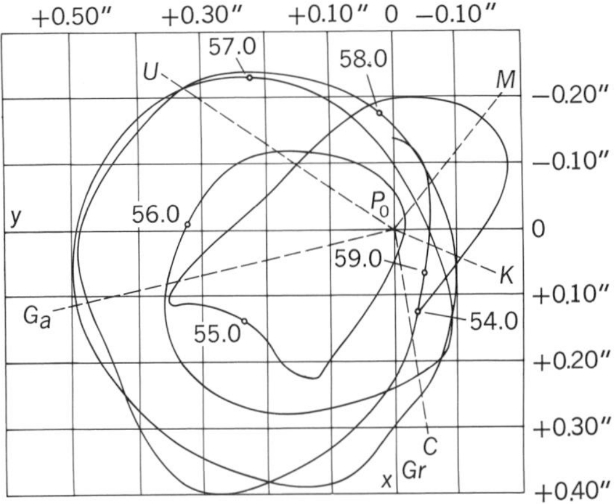

Fig. 3. Motion of the instantaneous pole 1954.0 to 1959.2 referred to the mean pole 1900–1905. (*From W. A. Markowitz, in S. K. Runcorn, ed., Methods and Techniques in Geophysics, vol. 1, 1960*)

Canada and Fennoscandia; others are caused by withdrawal of water or tectonic activity.

Horizontal secular motions. Remeasurements—by triangulation in earlier years, now by laser ranging—every few years across certain faults, such as the San Andreas in California, indicate relative horizontal motion up to some centimeters per year.

Earth tides. Continuous measurements of gravity and of the tilt of the Earth's surface indicate modifications of the tidal attraction of the Sun and Moon in amplitude, dependent mainly on variations in loading by ocean tides, but also on the elastic properties of the Earth, and of the phase lag, dependent on the energy-dissipating properties of the Earth. *See* EARTH TIDES.

[WILLIAM M. KAULA]

Bibliography: G. Bomford, *Geodesy*, 3d ed., 1971; G. D. Garland, *The Earth's Shape and Gravity*, 1965; W. A. Heiskanen and H. Moritz, *Physical Geodesy*, 1967; M. Hotine, *Mathematical Geodesy*, 1965.

Geologic thermometry

The measurement or estimation of temperatures at which geologic processes take place. Methods used can be divided into two groups, nonisotopic and isotopic. The isotopic methods involve the determination of distribution of isotopes of the lighter elements between pairs of compounds in equilibrium at various temperatures and application of these data to problems of the temperature of formation of these compounds (commonly minerals) in nature.

NONISOTOPIC METHODS

Earth temperatures can be measured directly at surface and near-surface features or indirectly from various properties of minerals and fossils.

Direct measurement. Temperatures can be measured directly in hot springs, in fumaroles, in flows of lava, and in artificial openings such as mines, boreholes, and wells.

Hot springs. The temperatures of hot springs range from slightly above the mean annual temperature of the region in which they occur to the boiling point of water at the elevation of the outlets. In other words, in temperate regions and at moderate altitudes the temperatures of hot springs range from about 20 to 100°C.

Fumaroles. At temperatures above its boiling point, water issues from vents as steam called fumaroles; temperatures of these fumaroles have been measured up to 560°C near Vesuvius, and up to 645°C in the Valley of Ten Thousand Smokes. Fumaroles in lavas may reach temperatures of 700–800°C.

Lava. Lava is molten rock coming out on the Earth's surface. Its temperature on extrusion ranges from about 700–900°C for andesitic and dacitic lava to 1200°C for basaltic lava. The viscosity of lava increases with decreasing temperature, and basaltic lava ceases to flow when it cools to 700–800°C. Intrusive magmas of similar compositions are probably intruded at similar temperatures, as indicated by such things as their effects on coal beds into which they are intruded, and the forms and assemblages of the first minerals to crystallize.

The temperature to which rocks around an intrusion (country rocks) are heated depends on many factors: temperature of the magma; temperature, composition, and structure of the country rock; abundance and nature of solutions given off by the intrusion; and size of the intrusion. In general, the temperature of the country rocks at the contact will be much lower than that of the magma. For example, an intrusive sheet of dolerite at 1100°C may heat the contact rocks to 600–700°C.

Mines, boreholes, and wells. Temperatures have been measured in enough artificial openings in the Earth's crust so that the temperature distribution is well known in many areas to depths of several thousand feet. Measurements of gradients range from about 40 to 170 ft/°C in nonvolcanic areas. Highest temperature yet encountered in such an area is 154°C from a well 20,521 ft deep in Sublette County, Wyo.

Gradients in volcanic areas are much higher near the surface (up to 1.3 ft/°C), but at depths of 750–1000 ft a temperature of about 250°C is commonly reached and persists to much greater depths (to at least 5500 ft in Tuscany).

Calculations and extrapolations give greatly different pictures for temperature distribution from the zone of measurements to the center of the Earth, depending on the assumptions made. Estimates of the temperature at the center of the Earth range from 1600 to 76,000°C, but most of the estimates made during 1945–1958 are in the 5000–10,000°C range. An estimate made in 1959 is based on experimental determination of the variation of melting point of iron with pressure up to almost 100,000 atm, and gives 2600°C as the temperature of the center of the Earth and 2340°C as the minimum temperature at the boundary of the core.

Indirect methods. Some indirect nonisotopic methods appear to give estimates with accuracy of the same order of magnitude as direct measurements; others place the temperature within a certain range; still others tell only that a given process took place above or below a certain temperature.

As phase-equilibrium relationships in systems analogous to those in nature become more accurately known, it will be possible to make more accurate estimates of geologic temperatures. The relationships commonly employed are listed below.

Melting points. The melting point of a mineral, corrected for the pressure under which it was formed, gives a maximum temperature of formation for the assemblage in which it grew because other substances, especially water, lower the temperature at which a mineral will crystallize. For example, if realgar, AsS, occurs in a vein with other minerals in such a way that they must have crystallized simultaneously, then the whole assemblage must have formed at a temperature lower than 320°C, the melting point of realgar. Lists of melting points and other phase relationships important in geologic thermometry are discussed in the E. Ingerson (1955) citations listed in the bibliography.

Transformation temperatures (inversions). Many minerals have two or more crystalline modifications which form, or exist, in different temperature ranges. For example, under certain conditions marcasite forms at temperatures below 300°C, but at about 450°C it transforms to pyrite at an appreciable rate. Therefore, a coprecipitated mineral assemblage including marcasite was certainly formed below 450°C and probably below 300°C.

Other pairs of minerals transform in either direction at a definite temperature; for example, low (α) quartz changes to high (β) quartz when it is heated to 573°C, and high quartz changes to low when it is cooled below 573°C. Therefore, phenocrysts of high quartz in lavas were formed above 573°C, crystals of low quartz in veins, below 573°C.

Dissociation and decomposition temperatures. Many minerals break up when they are heated. If one of the products is a gas, the temperature of decomposition changes rapidly with pressure; the pressure at the time of formation must be known or estimated before such a mineral can be used as a geologic thermometer. For example, calcite dissociates into lime and carbon dioxide, CO_2, at 885°C under atmospheric pressure, but under a sufficiently high pressure of CO_2 (1025 atm) it melts at 1339°C without decomposing.

When only solids and liquids are involved, however, pressure is relatively unimportant and can be neglected for processes that take place near the Earth's surface. For example, danburite decomposes to two liquids at about 1000°C, so this mineral and others immediately associated with it in pegmatities must have formed at temperatures below 1000°C.

Solid solutions and exsolution pairs. Many pairs of minerals with similar structures form homogeneous solid solutions at high temperatures, but on cooling separate (exsolve) into lamellae of the two minerals. This process commonly produces a characteristic texture that can be recognized. When two such minerals occur in a rock or ore in an exsolution relationship, this is evidence that the temperature of formation was above their temperature of homogenization.

Some common exsolution pairs and their temperatures of homogenization are magnetite-spinel, 1000°C; ilmenite-hematite, 600–700°C; chalcopyrite-bornite, 500°C; chalcopyrite-cubanite, 450°C; bornite-tetrahedrite, 275°C; and bornite-chalcocite, 225°C. These temperatures depend upon the composition of the host phase, but the ones given here are for compositions commonly encountered in nature.

This method has been used to estimate the temperatures of formation of many ore deposits. Some examples are sulfide replacement at Gilman, Colo., 150–300°C; sulfide mineralization at Pine Vale, Queensland, 475–500°C; scheelite veins, Australia, 350–600°C; and sphalerite-stannite-chalcopyrite mineralization, Tasmania, about 600°C.

In some mineral pairs there is a limited amount of solid solution; the amount depends upon the temperature of formation. For example, in the system FeS-ZnS the amount of FeS in sphalerite is a function of the temperature of formation if there is an excess of FeS (pyrrhotite) present when the sphalerite crystallizes. For instance, sphalerite formed at 200°C under these conditions will contain about 7 mole % FeS; at 500°C, 18 mole %. Other associations that can be used to indicate temperature of formation in this way are scandium in biotite (in a given petrologic province), TiO_2 in magnetite (with coexisting ilmenite), iron and magnesium in coexisting olivines and pyroxenes, albite in potassium feldspar (with coexisting plagioclase), and $MgCO_3$ in calcite (with coexisting dolomite).

By using the FeS-ZnS relationship, the temperature of formation of sphalerite in various settings has been estimated: in graphite schist near a granite contact (Norway), 440°C; replacement deposits, Gilman, Colo, 380–600°C; Broken Hill, Australia, 600°C; and in uranium deposits, Colorado Plateau, $<$138°C. TiO_2 in the magnetite of the northwestern Adirondacks indicates a temperature of formation of 475–600°C. The composition of alkali feldspars in granite indicates a final consolidation temperature of about 600°C. A metamorphic calcite with 10% $MgCO_3$ forms at about 650°C; with 20%, at 830°C.

Eutectics. When two or more minerals crystallize simultaneously at a eutectic, a so-called eutectic texture may be produced. However, it is difficult to be certain that a natural intergrowth of minerals was produced by eutectic crystallization; therefore this method has been little used thus far in geologic thermometry. It can be used in the same way that melting points can, that is, to indicate a temperature that cannot have been exceeded during crystallization of the assemblage. Some eutectic temperatures of common minerals are iron-nickel, 1435°C; anorthite-diopside, 1270°C; albite-nephelite, 1068°C; orthoclase-albite, 1070°C; orthoclase-silica, 990°C; quartz-orthoclase-albite, 937°C; silver-copper, 785°C; chalcocite-galena-argentite, 400°C; and sulfur-selenium, 100°C. Some of these temperatures are affected markedly by volatile components of a magma. They may be lowered hundreds of degrees by 1000 atm or more of water-vapor pressure.

Mineral assemblages. Certain types of mineral assemblages, such as eutectics and exsolution pairs, have been discussed. Other types that do not belong to one of these classes can also give indications of temperatures of formation. These indications may be based on syntheses, including hydrothermal experiments, known stability ranges of the individual minerals of the assemblage, and effect of pressure (where volatiles such as H_2O and CO_2 occur).

For example, hydrothermal experiments have

shown that analcite forms at temperatures of about 100–380°C under moderate water-vapor pressures, but in runs of the same compositions albite forms above 380°C. Therefore, in mineral assemblages that formed near the Earth's surface, that is, in cavities in volcanic rocks or shallow intrusions, the presence of analcite can be taken to indicate formation temperatures below about 400°C, the presence of albite to indicate higher temperatures.

Good crystals of potassium feldspar (variety adularia) have been formed hydrothermally at 245°C and of quartz down to about 200°C. Well-formed crystals of these minerals, even in sedimentary rocks, indicate growth temperatures of at least 100°C and probably 200°C or higher.

Clay minerals can also be important indicators of temperatures of formation. Their stability ranges are affected by such factors as pH of the solutions, water-vapor pressure, and concentrations of constituent cations in the solutions, but some useful generalizations can be made without quantitative evaluation of these variables. Kaolin forms in acid solutions up to about 350°C if aluminum is high and potassium is low, but when the pH is significantly above 7, montmorillonite forms from the same compositions over the same temperature range. Coprecipitated gels of alumina and silica (neutral) produce kaolin up to a little over 300°C, dickite at about 345–360°C, and beidellite at 360–390°C.

Sepiolite and attapulgite are decomposed hydrothermally at temperatures at least as low as 200°C, so the presence of either of these in a mineral assemblage indicates temperature of formation not over 200°C and possibly below 100°C. The lower limit of formation of these minerals is not known.

Similar experiments give analogous results for other minerals and mineral groups. Sericite, for example, forms at about 200–525°C in slightly basic to somewhat more acid solutions if aluminum and potassium are both high. Pyrophyllite forms at about 300–550°C if aluminum and potassium are both low. Serpentine cannot form above about 500°C even under very high pressures of water vapor, and most varieties of chlorite crystallize below 500°C. Muscovite is stable from about 400 to 800°C at pressures likely to be involved in rocks formed at depths small enough so they can be brought to the surface by erosion. Phlogopite forms from about 800 to 1100°C in the same pressure range. Talc does not form above about 825°C.

Where two or more minerals were formed in equilibrium, it may be possible to narrow the temperature range considerably. Although talc is stable at a temperature up to 825°C, in equilibrium with enstatite at moderate pressures the assemblage is stable only from about 670 to 800°C. Likewise, serpentine can form in equilibrium with brucite only below 450°C at moderate pressures or below 400°C at low pressures.

Similar relationships have been established for hydrous aluminum oxides and silicates, carbonates, and evaporites (E. Ingerson, 1955).

Other indirect methods. Other methods commonly used in geologic thermometry are discussed in this section.

Fossil assemblages. By determining the temperatures of the water in which certain types of organisms grow, it is possible to infer the temperature of the water at the time strata containing fossils of the same species, or perhaps closely similar species, were laid down. Thus far, this method has been confined to differentiation between cold- and warm-water assemblages, but as more information is obtained about temperatures at which certain species lived, it should be possible to assign temperature ranges to the water in which the formations containing them were laid down.

Properties dissipated by heating. Properties such as thermoluminescence, radiation colors, and metamictization, which are exhibited by many minerals (and thermoluminescence by some rocks), are dissipated by heating. Most thermoluminescence is dissipated below 250°C, although a few materials have been reported which retain some capacity for thermoluminescence up to about 500°C. Likewise, most radiation colors in minerals are dissipated at temperatures below 300°C, but in a few they persist up to 500°C or higher.

Metamictization is the destruction of the regular internal structure of a mineral produced by emanations from contained radioactive elements. The metamictization of minerals can be dissipated and the original structure restored by heating them to about 450–900°C, depending upon the mineral and rate and time of heating. *See* METAMICT STATE.

Possession of these properties by minerals does not mean that they were formed at temperatures lower than those at which the properties are dissipated, but that they have not been heated to higher temperatures since the properties were acquired. In deducing the thermal history of such materials, therefore, the problem of when the property was acquired becomes important.

Crystallography. The generalization has been made that crystals grown at relatively low temperatures are likely to be simple in habit; those grown at high temperatures, complex. The composition and pH of the solution, presence of impurities, rate of growth, and other factors can affect crystal habit.

Potassium feldspar is a good example of the change of crystal habit with temperature of formation. Phenocrysts of potassium feldspar in porphyries (800°C±) are dominated by base, clinopinacoid, and orthodomes, giving crystals elongate parallel to the *a* axis. Crystals in pegmatites (500°C±) are elongate parallel to the *c* axis and are dominated by (010), (001), and (110). In high-temperature veins such as those of the Alps (350°C±) the (110) form becomes more prominent and the crystals are simpler. In very-low-temperature veins the potassium feldspar is adularia (150°C±) and only the two forms (110) and (101) remain, so the crystals are as simple as possible. Other examples are quartz, calcite, and fluorite (E. Ingerson, 1955).

Liquid inclusions. Minerals crystallizing from aqueous solutions commonly have imperfections that retain samples of the solution as liquid inclusions in the final crystals. When the crystal cools, the solution contracts and a vapor bubble appears in each liquid inclusion. By heating plates of the mineral on a heating stage on a microscope and determining the temperature at which the solution just fills the cavities, it is possible to estimate temperature of formation if the pressure at formation was essentially the same as the vapor pressure of the solution. For significantly higher pressures it is necessary to estimate what the pressure was and apply a correction.

The following necessary assumptions appear to be justified in most, if not all, cases. (1) The inclusion cavities were just filled with fluid under the temperature and pressure prevailing during crystallization. (2) Change of volume of the mineral itself is not significant. (3) Changes in volume and concentration brought about by deposition of material during cooling are such as not to affect the result. (4) Primary and secondary liquid inclusions can be distinguished under the microscope. (5) There has been no leakage from or into the inclusions. (6) The liquid is an aqueous solution containing no carbon dioxide or other gas in large concentration. (7) Pressure-volume-temperature relations are near enough to those of pure water or chloride solutions that have been studied so that no serious errors are introduced by using data that are available.

Temperature ranges that have been estimated by this method for some common vein and pegmatite minerals are as follows: calcite, 40–362°C; sphalerite, 75–275°C; fluorite, 83–350°C; vein quartz, 100–440°C; pegmatite quartz, 200–530°C; and topaz, 275–500°C. [EARL INGERSON]

ISOTOPIC METHODS

In many cases the ratio of the stable isotopes of the lighter elements is not uniform in nature. The nonuniformity in the isotope ratio is a result of many natural processes and the temperatures at which these processes occur. Because temperature is a factor in the determination of isotope distribution, the isotope distribution provides information about the temperature conditions at which certain natural processes occur. For example, the knowledge of the isotopic composition of oxygen of minerals in certain rocks can give information about the temperature at which the rock was formed. Thus, measurements of relative isotope abundance can be useful in geologic thermometry.

Principles. The method of measuring temperature by isotopic analysis is based on the fact that the chemical and physical properties of a chemical compound depend not only upon the elements which form it but also upon the isotopic composition of the elements from which it is constituted. For example, the standard free energy of formation for $C^{12}O^{16}$ is not identical with that of $C^{13}O^{16}$. The differences in properties between the two molecules are sufficiently small so that the natural variations in the isotopic composition of the elements can be ignored (for all practical purposes) when dealing with the usual chemical problems. On the other hand, these small differences can be utilized in studying many problems, among them the determination of temperature at which many natural processes take place.

For a specific illustration, consider a system at some temperature composed of carbon dioxide, CO_2, and liquid water, H_2O. The two components are in chemical and isotopic equilibrium, that is, the isotopes of the element common to both components have distributed themselves between CO_2 and H_2O so that the system is at a minimum free energy. Such a system can be represented by an exchange reaction, Eq. (1).

$$CO_2^{16}(g) + 2H_2O^{18}(l) = CO_2^{18}(g) + 2H_2O^{16}(l) \quad (1)$$

The equilibrium constant K for such a reaction is given by Eq. (2).

$$K = \frac{[CO_2^{18}]/[CO_2^{16}]}{[H_2O^{18}]^2/[H_2O^{16}]^2} \quad (2)$$

It can be readily shown that Eq. (3) holds, where the $[O^{18}]/[O^{16}]$ ratio in the carbon dioxide is given by notation (4).

$$K^{1/2} = \frac{[O^{18}]/[O^{16}] \quad (\text{in the } CO_2)}{[O^{18}]/[O^{16}] \quad (\text{in the } H_2O)} \quad (3)$$

$$\frac{2[CO_2^{18}] + [CO^{16}O^{18}]}{[CO^{16}O^{18}] + 2[CO_2^{16}]} \quad (4)$$

The ratio of ratios given in Eq. (3) is referred to as the fractionation factor α. If the standard free energies of formation of CO_2^{16} and CO_2^{18} were identical and the same were true for H_2O^{16} and H_2O^{18}, the standard free energy change ($\Delta F°$) for Eq. (1) would be zero, and from the well-known thermodynamic equation, Eq. (5), K and α would

$$\Delta F° = -RT \ln K \quad (5)$$

be equal to unity. However, $\Delta F°$ is not equal to zero, and $\alpha = 1.04$ at 25°C.

Equation (5) also shows that α changes with temperature. A knowledge of the relationship between α (or the $[O^{18}]/[O^{16}]$ ratios of the two components) and temperature will then permit the determination of temperatures from the measurements of the isotopic composition of oxygen. Isotope exchange reactions have been considered for many systems involving oxygen compounds as well as a variety of other elements and their isotopes such as H_2, B, C, and N. In principle, for each of these systems an α can be evaluated, and for every α a potential thermometer can exist. In some cases the fractionation factors are calculable by the methods of statistical mechanics, using spectroscopic data. This was done by H. C. Urey and his students as early as 1934.

First application. The first isotope thermometer was developed by a group headed by Urey in 1951. This thermometer was based on the calcium carbonate–water system, in which the change with temperature, given by Eq. (6), is used for the thermometer.

$$\alpha = \frac{[O^{18}]/[O^{16}]_{\text{carbonate}}}{[O^{18}]/[O^{16}]_{H_2O}} \quad (6)$$

In this case the thermometer is used primarily for determining temperatures of localities at which marine animals deposit their calcareous skeletons. Under these conditions of deposition, the amount of calcium carbonate is very small compared to the amount of water with which the carbonate is in equilibrium. Any change in the fractionation factor will, therefore, affect the $[O^{18}]/[O^{16}]$ ratio of the carbonate only. If, for example, α decreases as a result of increase in temperature, the O^{18} in the carbonate will decrease to cause the $[O^{18}]/[O^{16}]$ ratio to decrease. The quantity of O^{18} atoms transferred to the water will be so small compared to the O^{18} present in the water that the $[O^{18}]/[O^{16}]$ ratio of the water will not be significantly changed. It follows therefore that temperature will affect only the $[O^{18}]/[O^{16}]$ ratio of the skeleton material.

Experimental procedure. The techniques involved can be briefly summarized. A Nier-type mass spectrometer was modified to permit the measurements of the $[O^{18}]/[O^{16}]$ ratio to sufficient accuracy.

The $[O^{18}]/[O^{16}]$ ratio is difficult to measure to a sufficiently high accuracy. Because it is of interest to know the change of ratio from sample to sample, it is necessary to measure this difference accurately. This can be done with the necessary precision by the use of the modified mass spectrometer. The isotope data are reported in terms of change in δ per mil of the ratio relative to a standard gas, as shown by Eq. (7), where δ equals 10 means the

$$\delta = \frac{[O^{18}]/[O^{16}]_{sam.} - [O^{18}]/[O^{16}]_{st}}{[O^{18}]/[O^{16}]_{st.}} \times 1000 \qquad (7)$$

$[O^{18}]/[O^{16}]$ ratio of the sample (sam.) is 10 per mil or 1% greater than that of the standard (st.). The term δ equals −10 means that the ratio is smaller than that of the standard by 10 per mil or 1%.

Carbon dioxide from calcium carbonate precipitated under equilibrium conditions at 16.5°C has been used as a standard. Carbon dioxide is the gas used in the mass spectrometer for the isotope measurements. For the present purposes, the ratio of mass 46 ($C^{12}O^{16}O^{18}$) to mass 44 ($C^{12}O^{16}O^{16}$) gives the $[O^{18}]/[O^{16}]$ ratio. Hence δ can be measured to an accuracy of 0.1 per mil. With this accuracy, a change in the fractionation factor caused by 1°C change could be measured.

The chemical techniques required to extract the oxygen (in form of CO_2) from the carbonates involve the acidification of the calcium carbonate with 100% phosphoric acid. If the acidification procedure is the same for all the samples, variation in the $[O^{18}]/[O^{16}]$ ratio of the extracted CO_2 is the same as that in the calcium carbonate.

The oxygen isotopic composition of the water in which the calcium carbonate grows is measured by equilibrating a small quantity of carbon dioxide

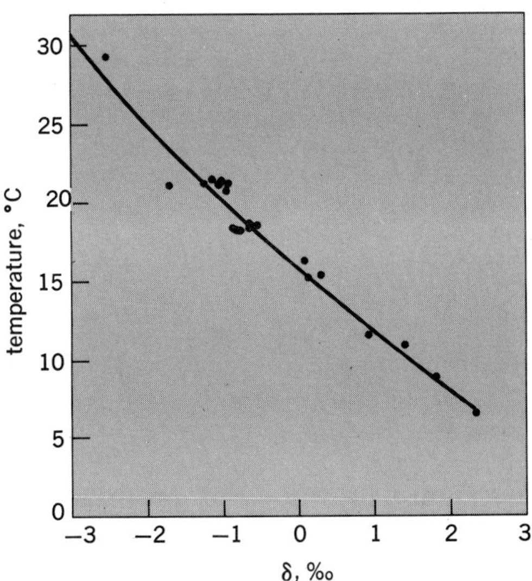

Fig. 1. Isotope temperature scale. (*From S. Epstein et al., Bull. Geol. Soc. Amer., 64(11):1315–1326, 1953*)

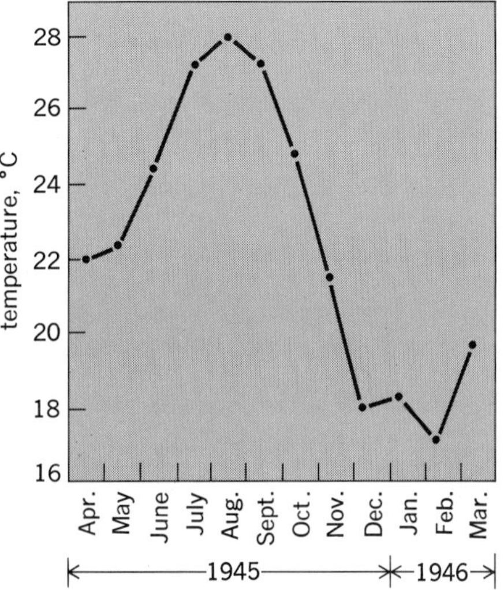

Fig. 2. Sea-surface temperatures in Ferry Reach. (*After H. B. Moore, reported in S. Epstein and H. A. Lowenstam, J. Geol., 61:424–438, 1953*)

with the water and then analyzing the resulting carbon dioxide. Carbon dioxide equilibrated in this way with mean ocean water at 25.3°C gives a δ value of zero.

The relationship between temperature and the value of CO_2 from calcareous skeletons grown in mean ocean water is shown in Fig. 1. The curve is represented by the relationship in Eq. (8).

$$T = 16.5 - 4.3 \times \delta_{carbonate} \qquad (8)$$

The various points on the curve were obtained by analyzing calcareous skeletons, either grown in tanks under controlled conditions or found in natural habitats of known temperature.

A present-day assemblage of marine animals living off the coast of Bermuda was studied for the growth temperature of the skeletons. The temperature records of the localities were well known, with a temperature range from 18 to 29°C (Fig. 2). These temperatures, recorded by the isotopes of the shell material compared to the temperature of the water, provided valuable information in determining the reliability of isotope temperatures recorded by marine calcareous skeletons. Figure 3 shows the temperature recorded by the skeleton of a marine snail (*Strombus gigas*). The data in Fig. 3*a* were obtained by analyzing successive small increments of shell in the sequence laid down by the animal. Each increment represents a short growth period. These animals appear to lay down shells practically the year around. The equivalent fossil shell which grew some hundreds of thousands of years ago shows a similar record (Fig. 3*b*). The pelecypod analyzed, *Chama acrophylla*, retains the shell grown only during a narrow temperature range. Information of this type is relevant to interpretation of fossil material, because accurate temperature records must be interpreted with realization that certain skeletons will not record the complete range of temperatures existing in a habitat. The data on recent shells verify the validity of the isotope temperature method and indicate some possible biological studies associated with marine animal growth.

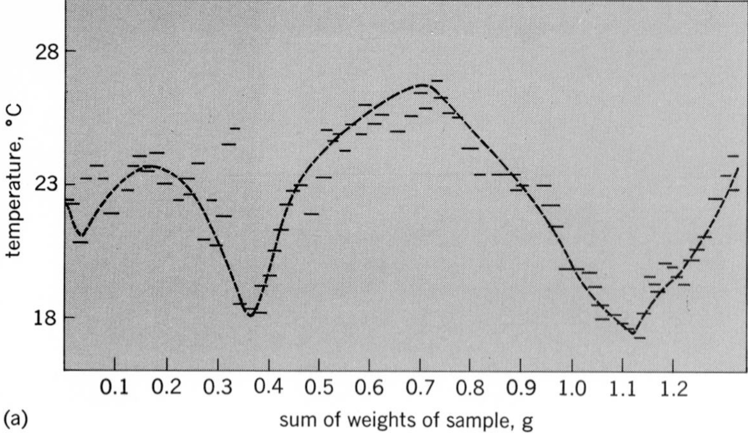

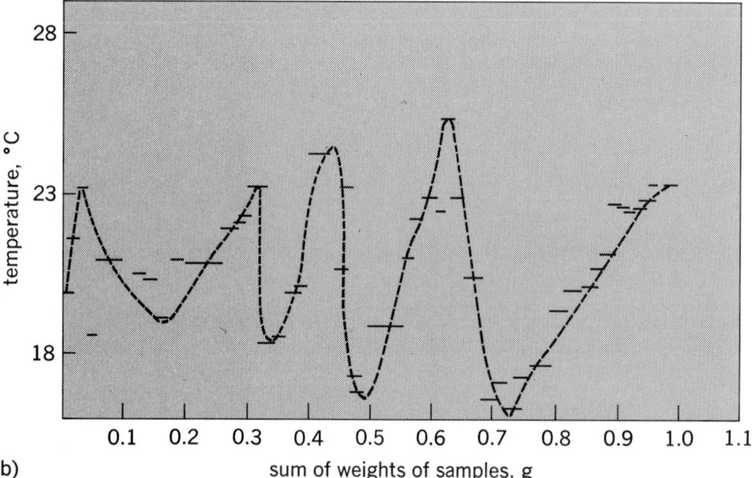

Fig. 3. Isotope temperature recorded by marine calcareous skeletons. (*a*) Seasonal growth temperature of *Strombus gigas*; Recent; North Reef, Bermuda. (*b*) Seasonal shell-growth temperatures of *Strombus gigas*; post-Pleistocene; Kindley Field, Bermuda. (*After S. Epstein and H. A. Lowenstam, J. Geol., 61:424–438, 1953*)

Paleotemperatures in the ocean. The determination of ocean temperatures by using the isotope thermometer of fossil material has several difficulties not inherent in studies of present-day temperatures. The fossil calcium carbonate skeleton must be one that was originally laid down, and laid down in isotopic equilibrium. The isotopic composition of the oxygen of the calcium carbonate depends not only upon the temperature but also upon the $[O^{18}]/[O^{16}]$ ratio of the ocean water. Present-day ocean water can be analyzed, but fossil ocean water is not available. The first problem is the less serious one because it is possible to recognize original material by the structure of the crystals of the calcium carbonate.

Skeletal remains originating as far back as the Paleozoic Era (older than 200,000,000 years) have been found with their original carbonate preserved. Younger fossil remains have a higher probability of preservation, so that a large time span of the Earth's climatic history is available for the application of the isotope thermometer. *See* PALEOECOLOGY, GEOCHEMICAL ASPECTS OF.

The second problem, involving the uniformity of the $[O^{18}]/[O^{16}]$ ratio of the ocean waters, is somewhat more serious. If the $[O^{18}]/[O^{16}]$ variations in the oceans of the past were similar to those of the present oceans, large errors in isotope temperatures could be made. The $[O^{18}]/[O^{16}]$ ratio of the present oceans varies sufficiently to cause errors of temperatures as large as 10°C. The variations in the $[O^{18}]/O^{16}]$ ratio of the present oceans appear to be due primarily to the existence of glaciers, which are a source of water of very low $[O^{18}]/[O^{16}]$ ratio. Melt water is added to the cold currents of the oceans, and water vapor of low $[O^{18}]/[O^{16}]$ ratio is removed from the warmer ocean surfaces, causing heterogeneity in the isotope ratios of the oceans. Because there is little evidence that glaciers existed during the large portion of the Earth's post-Paleozoic history, it can be surmised that the $[O^{18}]/[O^{16}]$ ratio variation of the oceans of the past was markedly less and that errors of only a few degrees are introduced in the isotope thermometer. It can be expected that large errors would occur in habitats near mouths of rivers or areas of very high evaporation, but these would be anomalous areas and easily recognizable from other evidence. The problem of the variations of the $[O^{18}]/[O^{16}]$ ratio of ocean water could be entirely eliminated if another fossil oxygen compound could be used in association with calcium carbonate, so that the thermometer could be based on the fractionation factor between the two available compounds. The $[O^{18}]/[O^{16}]$ ratio of the ocean water of the past would then not enter into the temperature determination. The possibility of using calcium phosphate and calcium carbonate as a pair has been considered, but has not as yet been developed.

The first application of the carbonate temperature scale to fossil material was the determination of temperatures from a Jurassic belemnite skeleton. This skeleton is cigar-shaped and is composed of solid calcite. It was internally grown by the animal, laying down layer upon layer of calcite, so that a cross section of the skeleton has a pattern similar to the ring pattern of a cross section of a tree trunk. Successive layers of a belemnite skeleton were ground off and each increment of powder was analyzed for the $[O^{18}]/[O^{16}]$ ratio. Each increment represented about one-tenth of 1 year's growth.

Figure 4 shows the temperatures recorded by successive increments of skeleton. It can be seen that a seasonal variation in the temperature is recorded and that the animal lived about 4 years. Had the original skeleton been modified while buried, then it would be expected that the variation found here would be obliterated. These results

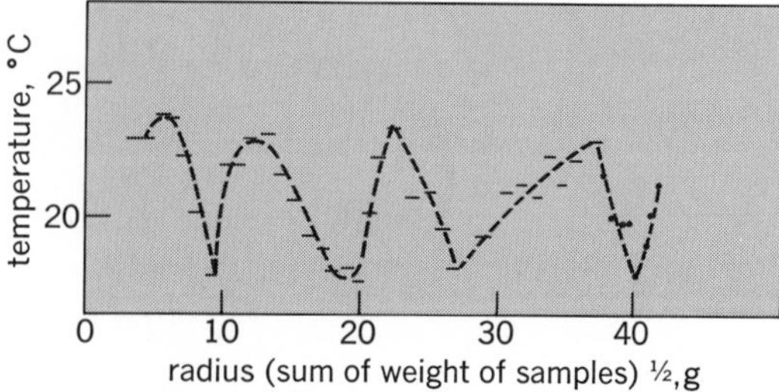

Fig. 4. Record of seasonal variation in temperature of Jurassic belemnite. (*From H. C. Urey et al., Bull. Geol. Soc. Amer., 62(4):399–416, 1951*)

gave excellent evidence for the feasibility of applying the temperature scale, even as far back as Jurassic times.

This work was extended to include the study of the Upper Cretaceous and some of the temperatures determined are shown in Fig. 5.

The data for Fig. 5 were obtained from pieces of belemnite and brachiopod skeletons from the Upper Cretaceous (90,000,000–120,000,000 years ago). Each datum represents the average yearly growth temperature. The samples were collected from a number of localities in Sweden, Denmark, Holland, England, France, and the southeastern United States. The temperature record of the samples indicates that the Upper Cretaceous was, on the whole, warmer than the present, a fact well known by geologists from other lines of evidence. In addition, it appears that the early and later parts of the Upper Cretaceous were cooler than the middle part of the period. The lowering of the temperature as the period approached its end is of interest because it was at the end of the Upper Cretaceous that a number of drastic changes took place in the evolution of life, including the extinction of several important forms of life, such as the dinosaurs. Any theory dealing with these evolutionary aspects must be compatible with the climatic record.

Pleistocene climate. An interesting application of the isotope thermometer was made on relatively young skeletal remains. These are some of the Foraminiferida, small calcareous protozoa, which constitute a large fraction of ocean bottom sediments in some of the oceanic areas. These skeletons, of the species which float near the surface when alive, should preserve a record of surface temperature of open oceans, starting from the present and going back in time. Climatically, the last few hundred thousand years were very interesting. Field evidence indicates that there were a number of great temperature fluctuations, resulting in several blanketings of ice over more than one-half of the North American continent. Results are shown in Fig. 6, and indeed there is an excellent and continuous record of reasonable temperature in the Foraminiferida tests. The maximum southward ice advances, designated by the name ice ages, are accompanied by a record of minimum temperatures. *See* MARINE SEDIMENTS.

The above examples of the application of the carbonate isotope thermometer are but a few of many interesting researches dealing with climatic temperatures that are of geological interest.

Isotope temperatures for rocks. The thermometer based on the $[O^{18}]/[O^{16}]$ distribution between calcium carbonate and water is but one example of oxygen isotope thermometry. In principle the temperature dependence of the $[O^{18}]/[O^{16}]$ distribution (α) between any two chemical compounds or minerals can be used as an oxygen isotope thermometer. The main limitations are the variation of α with temperature as compared to the relative accuracy with which the $[O^{18}]/[O^{16}]$ ratio of the minerals can be measured. In addition calibration of the thermometer requires very careful laboratory work involving high-temperature and high-pressure experiments. Several high-temperature oxygen isotope thermometers using several mineral pairs commonly found in igneous and metamorphic rocks, as well as in hydrothermal deposits, have

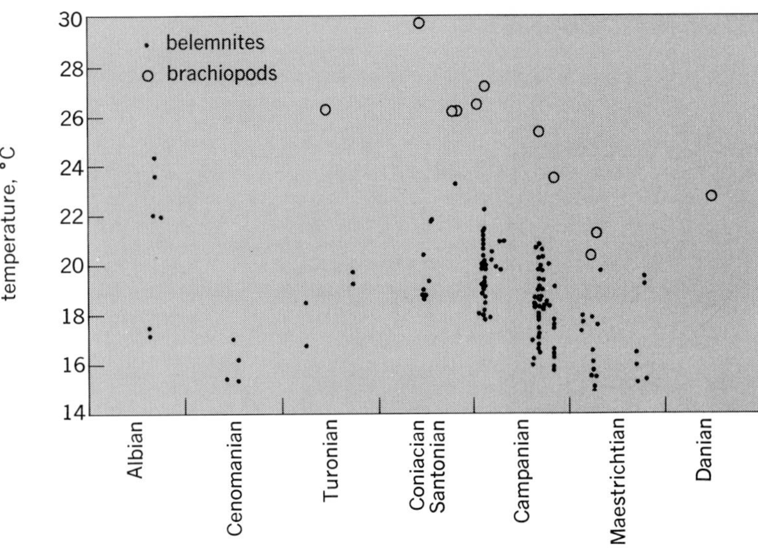

Fig. 5. Mean-temperature distribution of belemnites and brachiopods from the Albian to the Danian of the Upper Cretaceous in western Europe. *(From H. A. Lowenstam and S. Epstein, J. Geol., 62(3):207–248, 1954)*

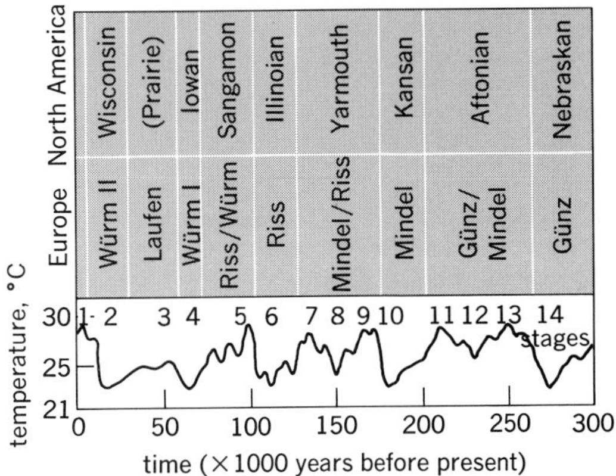

Fig. 6. Generalized temperature curve for tropical surface waters and continental correlations. *(From C. Emiliani, J. Geol., 66(3):264–275, 1958)*

become available. Figure 7 shows some of the relationships between α and temperature of several mineral-water systems. The α between two minerals A and B is simply the ratio of the $\alpha_{(\text{mineral A–water})}$ and $\alpha_{(\text{mineral B–water})}$. Thus the relationships between α and temperature for any two minerals shown in Fig. 7 can be calculated.

The description of the experimental procedures for the equilibration of the minerals and water are too lengthy to be considered here; the reader is referred to the original papers for details. Considerably more two-mineral systems can still be investigated to determine their αs, and the change of α with temperature.

The extraction of oxygen from silicates involves the reaction of the mineral at about 500°C with either F_2 or BrF_5 in a nickel reaction vessel previously purged of oxygen compounds. The silicate reacts with F_2 to form SiF_4, O_2, and metal fluorides. After purification the oxygen is reacted with hot graphite and the resulting CO_2 frozen out in a liquid nitrogen trap. The oxygen of silicate and

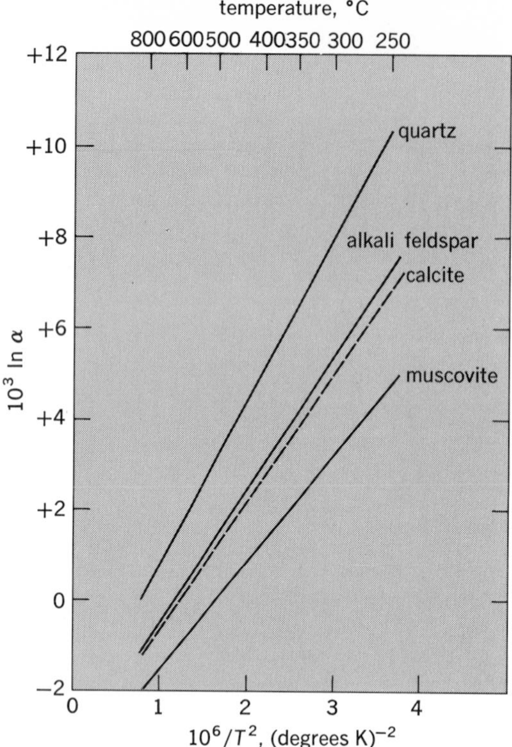

Fig. 7. O^{18} fractionation curves for mineral-water systems. Experimentally determined relationships between fractionation factor (plotted as $10^3 \ln \alpha$) and temperature ($10^6/T^2$) for quartz-water (*R. N. Clayton et al., 1966*), alkali feldspar–water (*J. R. O'Neil and H. P. Taylor, Jr., 1965*), calcite-water (*R. N. Clayton, 1961*), and muscovite-water (*J. R. O'Neil and H. P. Taylor, Jr., 1966*). (*From S. Epstein and H. P. Taylor, Jr., Variations of O^{18}/O^{16} in minerals and rocks, in P. H. Abelson, ed., Researches in Geochemistry, vol. 2, copyright © 1967 by John Wiley & Sons, Inc.; reprinted by permission*)

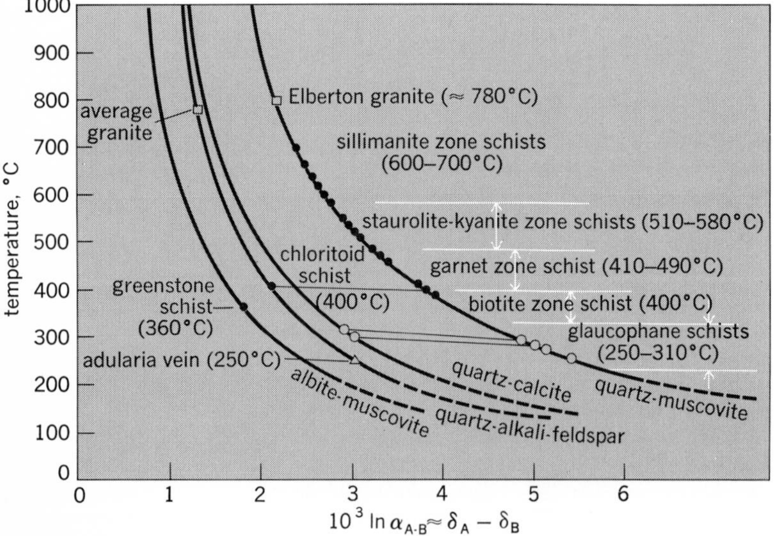

Fig. 8. Oxygen isotope "temperatures of formation" of various metamorphic and igneous rocks, obtained by plotting oxygen isotope Δ values for various mineral pairs on the experimentally determined calibration curves for quartz-alkali feldspar, quartz-calcite, quartz-muscovite, and albite-muscovite (see Fig. 7). Coexisting pairs from the same rock are connected by horizontal lines. Isotope data on natural mineral pairs are from G. D. Garlick and S. Epstein (1966), H. P. Taylor and R. G. Coleman (1965), H. P. Taylor et al. (1963), H. P. Taylor, Jr., and S. Epstein (1962), and H. P. Taylor (unpublished data). (*From S. Epstein and H. P. Taylor, Jr., Variations of O^{18}/O^{16} in minerals and rocks, in P. H. Abelson, ed., Researches in Geochemistry, vol. 2, copyright © 1967 by John Wiley & Sons, Inc.; reprinted by permission*)

oxide minerals can thus be converted to CO_2 for mass spectrometer analyses.

The thermometers in Fig. 7 have been used to measure temperatures recorded by a variety of minerals in a number of rock types; these are summarized in Fig. 8. As is probably true for any geothermometer, the exact meaning of the measured temperature is sometimes difficult to ascertain. For example, is the temperature of the solidification of the minerals being measured? Is this temperature the same as the temperature at which significant oxygen isotope exchange between minerals stops? It is not certain that both events occur at the same time in various rocks. In addition, it is not certain that the minerals measured for the oxygen isotope composition formed at a narrow range of temperatures. Of course if concordant temperatures are obtained by using several coexisting minerals in a rock, this is an important criterion for probable isotope equilibrium between minerals, and probably a meaningful temperature for the rock's temperature of crystallization.

It is therefore interesting to note (Fig. 8) that concordant temperatures of about 400°C are recorded for a chloritoid schist (metamorphic rock) by using the quartz-muscovite and quartz-alkali feldspar isotope geothermometer. In general, the isotope temperatures recorded for the variety of metamorphic rocks are in the range of the temperatures of formation derived by means of ion isotope methods. [SAMUEL EPSTEIN]

Bibliography: P. H. Abelson (ed.), *Researches in Geochemistry*, 1959, 1967; R. N. Clayton, Oxygen isotope fractionation between calcium carbonate and water, *J. Chem. Phys.*, 34:724, 1961; R. N. Clayton and S. Epstein, The relationship between O^{18}/O^{16} ratios in coexisting quartz, carbonate, and iron oxides from various geological deposits, *J. Geol.*, 66(4):352–373, 1958; C. Emiliani, Paleotemperature analysis of core 280 and Pleistocene correlations, *J. Geol.*, 66(3):264–275, 1958; A. E. J. Engel, R. N. Clayton, and S. Epstein, Variations in isotopic composition of oxygen and carbon in Leadville limestone (Mississippian, Colorado) and in its hydrothermal and metamorphic phases, *J. Geol.*, 66(4):374–393, 1958; S. Epstein et al., Revised carbonate-water isotopic temperature scale, *Bull. Geol. Soc. Amer.*, 64(11):1315–1326, 1953; S. Epstein and T. Mayeda, Variation of O^{18} content of waters from natural sources, *Geochim. Cosmochim. Acta*, 4:213–224, 1953; S. Epstein and H.P. Taylor, Jr., Variations of O^{18}/O^{16} in minerals and rocks, in P. H. Abelson (ed.), *Researches in Geochemistry*, vol. 2, 1967; G. D. Garlick and S. Epstein, Oxygen isotope ratios in coexisting minerals of regionally metamorphosed rocks, *Geochim. Cosmochim. Acta*, 31:181–214, 1966; E. Ingerson, Geologic thermometry, *Crust of the Earth*, Geol. Soc. Amer. Spec. Pap. no. 62, 1955; E. Ingerson, Methods and problems of geologic thermometry, *Econ. Geol.*, 50th anniversary vol., pp. 341–410, 1955; H. A. Lowenstam and S. Epstein, Paleotemperatures of the post-Aptian Cretaceous as determined by the oxygen isotope method, *J. Geol.*, 62(3): 207–248, 1954; J. M. McCrea, On the isotopic chemistry of carbonates and a paleotemperature scale, *J. Chem. Phys.*, 18(6):849–857, 1950; J. R. O'Neil and H. P. Taylor, Jr., Oxygen isotope chemistry of feldspars (abstract), *Trans. Amer. Geophys.*

Union, 46:170, 1965; J. R. O'Neil and H. P. Taylor, Jr., Oxygen isotope fractionation between muscovite and water (abstract), *Trans. Amer. Geophys. Union*, 47:212–213, 1966; H. P. Taylor, Jr., A. L. Albee, and S. Epstein, O^{18}/O^{16} ratios of coexisting minerals in three mineral assemblages of kyanite-zone peletic schist, *J. Geol.*, 71:513, 1963; H. P. Taylor, Jr., and S. Epstein, Relationship between O^{18}/O^{16} ratios in coexisting minerals of igneous and metamorphic rocks, *Bull. Geol. Soc. Amer.*, 73(1):461–480, 1962; H. C. Urey et al., Measurement of paleotemperatures and temperatures of the Upper Cretaceous of England, Denmark, and the southeastern United States, *Bull. Geol. Soc. Amer.*, 62(4):399–416, 1951.

Geological time scale

The assignment of time in years to each geological period. The fossil record of life (with the implied irreversibility of evolution) and the deposition and orogenic sequence of earth history events provide the basis for the relative time scale universally accepted by geologists.

Development of absolute time scale. The actual time in years before present (B.P.) is based on the regular rate of decay of radioactive isotopes found in minerals. This decay results in a series of elements, each having fewer atomic numbers. Decay rates are ratios of the amount of daughter isotopes to parent isotopes in a rock sample. The first rate-of-decay sequence was worked out by B. Boltwood in 1907 based on the uranium-lead series, and by 1913 A. Holmes was able to construct a time scale using this method (only a few samples were available). By the 1960s a large number of localities had been sampled using the uranium-lead series along with rubidium-strontium and ^{14}C-^{14}N, resulting in a new time scale by J. L. Kulp in 1961. The time scale of Table 2 is an updated version of Kulp's; only relatively minor adjustments have been made though more data are available.

There are several problems affecting the accuracy of isotopic dating. One is that the isotopes may be altered due to such factors as tectonic activity so that the ratios will not reflect a correct date. Another problem is that the decay process itself is not strictly uniform, so that only average decay rates can be obtained. The best technique is to use several series as cross-checks on each other. The half-life of parent isotopes of elements used in isotopic dating varies from just under 10^9 to nearly 50×10^9 years, with the exception of ^{14}C, which has a half-life of about 5700 years and can yield dates only up to 70,000 years B.P. *See* DATING METHODS.

Source of samples. Isotopic age determinations are made primarily on igneous and metamorphic rocks. Some potassium-argon dates can be made on glauconite, a mineral which is formed in sedimentary rocks but is not abundant. Uranium-lead measurements have been made on uranium-bearing black shales such as the Chattanooga Shale at the Carboniferous-Devonian boundary (see Table 1). The uranium-lead series is used mainly in determining the age of such igneous rocks as granites. Potassium-argon and rubidium-strontium methods are used on micas found in, for example, granites which have intruded fossiliferous strata. In that case a minimum age can be established for the intersected strata. If mica is found in a lava flow above a sedimentary unit or in metamorphic rocks underlying such a unit, a closer or maximum date is possible. Zircon and feldspars are also used in isotopic dating.

Table 2 gives some examples of dates used to assign absolute time to relative time for the Phanerozoic, which includes all rock units containing abundant fossils. The oldest known rocks occur in Minnesota and southwestern Greenland and are about 3.8×10^9 years old. Based on isotopic dates from Moon rocks and on meteorites, the age of the solar system, and hence the Earth, is believed to be about 4.5×10^9 years old. Astronomers theorize that the Galaxy is about 10^{10} years old and that the Sun is 6×10^9 years old. *See* PLANETOLOGY; UNIVERSE.

Table 1. Definitive points in the geological time scale

Locality	Method of determination	Stratigraphic position	Age, 10^6 years
Central City, Colorado	U-Pb (pitchblende veins)	Late Paleocene (post–Fort Union pre-Eocene)	61 ± 5
Caucasus, U.S.S.R.	K-Ar (granite)	Cuts upper Eocene, overlain by lower Oligocene	40 ± 5
Coast Range, California	K-Ar (granites)	Mid-Upper Cretaceous	80 ± 4
Sierra Nevada	K-Ar (granites)	Late Upper Jurassic (cuts Kimmeridgian Stage)	139 ± 4
Kelasury, U.S.S.R.	K-Ar (granite)	Cuts Middle Jurassic, overlain by Upper Jurassic	161 ± 5
Solikamsk, U.S.S.R.	K-Ca (sylvite)	Lower Middle Permian	241 ± 7
Dartmoor, England	K-Ar (granite)	Lowermost Permian (post-Westphalian, pre–Middle Permian)	275 ± 10
Vosges, France; Schwarzwald, Germany	K-Ar, Rb-Sr (granite)	Lower Carboniferous (pre-Visean, post-Dinantian)	330 ± 10
Chattanooga, Tennessee	U-Pb (black shale)	Devonian-Carboniferous boundary	350 ± 10
Shap, England	K-Ar (granite)	Post-Silurian (probably Lower Devonian)	390 ± 10
Maine	K-Ar, Rb-Sr (granite)	Post-Silurian, pre–Upper Devonian	390 ± 15
Vastergotland, Sweden	U-Pb (black shale)	Mid-Upper Cambrian	500 ± 10 (min.)
Wichita Mts., Oklahoma	K-Ar, Rb-Sr, U-Pb (granite)	Pre–Upper Cambrian	520 ± 20

Table 2. Geological time scale

ERA Period Epoch	Age of beginning of period, 10^6 years, new scale	Period length, 10^6 years
CENOZOIC		
Quaternary		
Recent	0.011	
Pleistocene	1.5*	
Tertiary		65
Pliocene	12	
Miocene	20	
Oligocene	35	
Eocene	55	
Paleocene	65	
MESOZOIC		
Cretaceous		65
Upper (Base Santonian)	100	
Middle (Base Albian)	120	
Lower	130	
Jurassic		55
Upper (Base Callovian	155	
Middle (Base Bajocian)	170	
Lower	185	
Triassic		45
Upper	200	
Middle	215	
Lower	230	
PALEOZOIC		
Permian		35
Upper (Base Ochoan)	245	
Middle (Base Guadalupian)	260	
Lower	265	
Pennsylvanian		45
Upper		
Middle		
Lower	310	
Mississippian		45
Upper		
Middle		
Lower	355	
Devonian		58
Upper	365	
Middle	385	
Lower	413	
Silurian		12
Upper		
Middle		
Lower	425	
Ordovician		50
Upper	440	
Middle	460	
Lower	475	
Cambrian		95
Upper (Base Croixian)	500	
Middle	540	
Lower	570	
Eocambrian	680	110
PRECAMBRIAN		
Epiproterozoic	1000	320
Neoproterozoic	1900	900
Upper	1300	
Middle	1600	
Lower	1900	
Mezoproterozoic	2600	700
Paleoproterozoic	3500	900
Archaean	?	?

*This date is according to H. Berggren and V. Couvering, Corrected age of the Pliocene–Pleistocene boundary, *Nature*, 269:483–488, Oct. 6, 1977.

There is no consensus regarding the subdivision of the Precambrian. Knowledge of sequential geologic events is restricted to the scattered regions where Precambrian rocks are found. Few global Precambrian events are known because, for example, there are no meaningful fossils consistently found for use in correlation of rocks. A subdivision of the Precambrian, generally accepted by Canadian geologists, has as the oldest unit the Archaean Eon, ranging from 3.5×10^9 to 2.4×10^9 years B.P.; followed by the Aphebian Era, ranging from 2.4 to 1.8×10^9 years B.P., followed by the Helikian Era, with a range of 1.8 to 0.7×10^9 years B.P.; and finally the Hadrynian Era, with a range of 0.7 to 0.6×10^9 years B.P. In the Great Lakes states, geologists use three subdivisions: the Archeozoic ($3.5-2.4 \times 10^9$ years), the Proterozoic (2.4–0.7), and the Eocambrian (0.7–0.6). Smaller subdivisions within these larger units mark either orogenic events such as the Algomen orogeny (2.4×10^9 years B.P.) or a rock type or terrain such as the Timiskamian system (3×10^9 years B.P.). In 1973 L. J. Salop proposed a classification which is based on a synthesis of rock characteristics and distribution, geochemistry, paleontology, and mineralogy, and is included as the Precambrian section of Table 2. [ROGER L. BATTEN]

Bibliography: R. H. Dott, Jr., and R. L. Batten, *Evolution of the Earth*, 1976; E. I. Hamilton, *Applied Geochronology*, 1965; W. B. Harland, *The Phanerozoic Time Scale*, Geol. Soc. (London) Publ. 5, 1971; A. Holmes, The construction of a geological time scale, *Trans. Geol. Soc. Glasgow*, 21:118–152, 1947; J. L. Kulp, Absolute age determination of sedimentary rocks, *Proc. 5th World Petrol. Congr., N.Y., June 1–5, 1956*, 1960; L. J. Salop, *Precambrian of the Northern Hemisphere*, 1977.

Geology

The study or science of the Earth. Geology is one of several related subjects commonly grouped as geoscience. Geologists are concerned primarily with rocks that make up the outer part of the Earth. An understanding of these materials involves principles of physics and chemistry; geophysics and geochemistry, now important scientific disciplines in their own right, have become essential allies of geology in exploring the visible and deeper parts of the Earth. Study and mapping of surface forms are shared by geology with geodesy. The study of the Earth's waters as related to geologic processes is shared by hydrology and oceanography. Paleontology, the study of records left by animals and plants that lived in past ages, is an essential part of geology, though it involves many fundamental aspects of biologic science.

Geology, to contribute its part to geoscience, has developed a number of branches which may be outlined as follows:

I. Physical geology
   - *A.* Mineralogy, study and classification of minerals
   - *B.* Petrology, study of rocks, their physical and chemical properties, and their modes of origin
   - *C.* Weathering and erosion, study of processes that alter exposed bedrock and shape landforms
   - *D.* Sedimentation, origin and deposition of modern sediments
   - *E.* Structural geology, study of geometry of rock masses, with emphasis on crustal deformation
   - *F.* Economic geology, application of geologic knowledge and principles to solution of practical problems

II. Historical geology
   - *A.* Stratigraphy, systematic study of bedded rocks and their relations in time

B. Paleontology, study of fossils and their locations in a sequence of bedded rocks

III. Geologic mapping, graphic representation of bedrock units defined by physical characteristics and geologic age

Nearly all geologic study seeks to determine an order of events, and a major objective is to work out the full history of the Earth and its inhabitants.

PHYSICAL GEOLOGY

An understanding of the composition of the Earth and the physical changes occurring in it is derived from a study of the rocks, minerals, and sediments; their structures and formations; and their processes of origin and alteration.

Mineralogy and petrology. Minerals are the basic units in the composition of most rocks. Although many hundreds of mineral species are known, comparatively few are important in kinds of rocks that are abundant in the visible part of the Earth crust. Thus a few related minerals known as feldspars, together with quartz, are the essential ingredients of granite and its near relatives. Limestone, widely distributed on all continents, consists largely of the one mineral calcite. Each mineral has its peculiar atomic structure, which under ideal conditions is expressed in crystal form. But when a substance crystallizes in bulk, crowding of grains growing from neighboring centers prevents formation of recognizable crystals. Modern laboratories have effective devices for resolving the mineral content of rock materials; even the ultramicroscopic particles in clays are clearly defined under the electron microscope, which gives images enlarged by a factor of tens of thousands. For further explanation of minerals *see* MINERAL; MINERALOGY.

Rocks of Earth's crust. The term crust, commonly applied to the outer part of the solid Earth, is inherited from speculative concepts that the globe was once a molten mass and still is molten below a shell of solid rock. Though the latter view is no longer tenable, evidence from seismology indicates that an outer part of the Earth, no more than a few tens of kilometers thick, differs in physical properties from deeper zones, and the convenient term crust is retained for the distinctive outer shell. *See* EARTH; SEISMOLOGY.

Known rocks are divisible into three groups: igneous rocks, which have solidified from molten matter (magma); sedimentary rocks, made of fragments derived from preexisting rocks, of chemical precipitates, or of organic products; and metamorphic rocks derived from igneous or sedimentary rocks under conditions that brought about changes in mineral composition, texture, and internal structure. *See* PETROGRAPHY; PETROLOGY; ROCK.

Igneous rocks. Igneous rocks are formed as either extrusive or intrusive masses, that is, solidified at the Earth surface or deep underground. Both kinds range widely in composition; silica, the most abundant ingredient, varies from about 40% to more than 75%. Intrusive bodies (plutons) that formed at various depths are most numerous in mountain zones for two reasons: first, mountain belts have been much deformed, and abundant evidence indicates that crustal disturbance has favored igneous action; second, uplifts in mountain lands have permitted erosion to depths at which plutonic masses were formed. Doubtless some large plutons solidified in reservoirs that formerly supplied volcanoes in action.

Volcanic materials are erupted through two kinds of openings—central vents and long fissures. Central eruptions build up conical mountains; the materials are in large part products of explosion and consist of cinders and ash, with interspersed lava flows. Lavas issuing from fissures have built up vast fields of volcanic rock, chiefly the dark type known as basalt. *See* IGNEOUS ROCKS; MAGMA; PLUTON; VOLCANO.

Sedimentary rocks. Bedrock exposed to air and moisture is broken into pieces, large and small, which are moved by running water and other agents to lower ground, and spread in sheets over river floodplains, lake bottoms, and sea floors. Dissolved matter is carried to seas and other water bodies, and some of it is precipitated chemically and by action of organisms. The material deposited in various ways becomes compacted, and in time much of it is cemented into firm rock. Generally the deposition is not continuous but recurrent, and sheets of sediment representing separate events come to form distinct layers of sedimentary rock. The individual layers are beds or strata, and the rocks are described as stratified.

Large areas in every continent are underlain by sedimentary rocks that represent deposits during many periods of the Earth's history. In part these bedded rocks are nearly horizontal, as they were originally (Fig. 1); but in many places, particularly in mountain belts, they have been deformed.

The principal kinds of sedimentary rock are conglomerate, sandstone, siltstone, shale, limestone, and dolomite. Many other kinds, less important quantitatively but with large practical value, include common salt, gypsum, phosphate, iron oxide, and coal. *See* SEDIMENTARY ROCKS.

Metamorphic rocks. These rocks have been developed from earlier igneous and sedimentary rocks by heat and pressure, most effectively in mountain zones and near large masses of intrusive igneous rock. Thermal metamorphism results from rising temperature, often with addition of new elements by circulating fluids. Common effects are hardening and crystallizing of the affected rock, with changes in mineral composition. Dynamic metamorphism results from shearing stresses in rocks subjected to high pressures; effects are development of cleavage planes and growth of platy (platelike) minerals in parallel arrangement.

The common metamorphic rocks are in the two general classes, foliated (including slate, phyllite, schist, and gneiss) and nonfoliated (including marble and quartzite). *See* METAMORPHIC ROCKS; METAMORPHISM.

Weathering and erosion. Bedrock at and near the Earth's surface is subject to mechanical and chemical changes in a complex process called weathering. Blocks and small chips that become detached are especially vulnerable to chemical attack, which makes radical changes in the mineral content. Some soluble products of alteration are removed by percolating water, and in the less soluble residue the most common constituent is clay, which is the basis of soil. The effectiveness of weathering and the nature of its products are controlled by climate, topography, kinds of bedrock, and other variables. Organic processes play a ma-

Fig. 1. Nearly horizontal sedimentary rocks approximately 4000 ft thick along Colorado River canyon, in Arizona. Some tilting of layers is discernible in the left section of the view. (*Spence Air Photos*)

jor role in chemical weathering, which is most effective under conditions that favor development of bacteria, plants, and ground-dwelling animals. *See* ROCK CONTROL (GEOMORPHOLOGY); WEATHERING PROCESSES.

Weathering prepares the way for removal of rock materials and reshaping of land surfaces by several agents of erosion. The most obvious of these agents is running water, which during a single rainstorm may cut deep gullies into plowed fields and sweep vast quantities of soil, sand, and coarser debris into brooks and eventually into channels of major streams. Abrasion by such moving loads deepens and widens stream channels in hard bedrock. Study of drainage systems brings conviction that even the largest and deepest valleys (Fig. 1) have been fashioned by the action of running water.

Soil on slopes, even those covered with grass and other vegetation, creeps slowly downward. Blocks dislodged from cliffs build steep masses of sliderock which slowly migrate downslope. Frequently in mountain lands great masses, including loose material and bedrock, rush down as landslides. *See* MASS WASTING.

Water moving through underground openings dissolves and carries away great quantities of material. Caverns, large and small, are a conspicuous result. In high latitudes and in some mountain regions glaciers (Fig. 2) are powerful eroding agents. In arid regions quantities of sand and dust are moved by wind with some consequent abrasion of bedrock. Large-scale erosion along the coasts of seas and lakes is performed by waves and currents. *See* DESERT EROSION FEATURES; GLACIATED TERRAIN; GROUNDWATER; NEARSHORE PROCESSES.

Each major agent of erosion fashions characteristic features in landscapes. The net tendency of erosion is to reduce the height of land masses. Various stages in the history of reduction are indicated by forms of valleys and slopes and by relations of land surfaces to the underlying bedrock. Some wide regions have been uplifted and the streams have been rejuvenated after an advanced stage was reached in a cycle of erosion. *See* GEOMORPHOLOGY.

Sedimentation. An understanding of sedimentary rock is all important in geology, and sediments now being deposited provide an essential key. On the basis of depositional environment, sediments are assigned to three categories: terrestrial, those laid down on lands; marine, those deposited on sea floors; and mixed terrestrial-marine, those laid down in transitional zones such as deltas, marine estuaries, and areas between high and low tide. In each major group the sediments are further described as clastic (consisting of rock fragments) and chemical (formed either as inorganic precipitates or partly through organic agencies). Study of modern sediments in the several environments takes account of physical peculiarities and also the included remains of organisms. *See* SEDIMENTARY ROCKS; SEDIMENTARY STRUCTURES; SEDIMENTATION.

Terrestrial sediments, widespread and highly varied, are laid down chiefly through the agencies of mass wasting, running water, glacier ice, and

Fig. 2. Effects of erosion and transport by valley glaciers, southern Alaska. (*Alaskan Aerial Survey, U.S. Navy*)

wind. The deposits are in large part temporary, as the tendency is for them to shift seaward in continued erosion of the lands. Some subsiding basins under semiarid climates and with no throughflowing streams are retaining all sediments they receive.

A classification of marine sediments according to their source follows:

1. Derived from lands and contributed by (*a*) streams, (*b*) wave erosion of coasts, (*c*) winds, and (*d*) floating ice.

2. Formed in the sea by (*a*) shells and skeletons of marine animals and plants, and (*b*) chemical precipitation.

3. Fragmental material erupted from volcanoes.

4. Particles of meteorites from outside the Earth. *See* MARINE SEDIMENTS.

Structural geology. Geometric study of rocks distinguishes primary structures, acquired in the genesis of a rock mass, and secondary structures that result from later deformation. Significant features in sedimentary rocks make them especially valuable for registering later changes in form.

Within historical time abrupt movements have occurred along large breaks (faults), with instantaneous displacements as much as tens of feet (Fig. 3). Such movements are attended by strong earthquakes. Another type of movement, now in progress but at an extremely slow rate, is doming up the surface, as in Scandinavia, on a regional scale. Similar ancient warping is evident in the gentle bending of sedimentary strata, although the land surfaces that were deformed have been destroyed by erosion. In addition to broad warps, the principal kinds of structural features that record deformation are folds, joints, faults, cleavage, and unconformities. *See* STRUCTURAL GEOLOGY.

Evidence of the most pronounced crustal deformation is found in mountain belts, where erosion has exposed very thick sections of sedimentary rocks. These rocks record long histories of slow subsidence and sedimentation, interrupted by large-scale deformation and uplift. *See* DIASTROPHISM; GEOLOGY, EVOLUTION OF; OROGENY.

Major relief features of the Earth reflect differences in density of the underlying rocks. Continental rocks, diverse but including great masses with granitic composition, have appreciably lower average density than the basaltic rocks of ocean floors. Great mountain blocks, such as the Alps and the Himalayan chain, represent thickened parts of continental masses. The condition of approximate balance among diverse parts of the crust is called isostasy (equal standing). *See* ISOSTASY.

Economic geology. A general knowledge of geology has many practical applications, and large numbers of geologists receive special training for service in solving problems met in the mining of metals and nonmetals, in discovering and producing petroleum and natural gas, and in engineering projects of many kinds. *See* ENGINEERING GEOLOGY; PETROLEUM GEOLOGY.

HISTORICAL GEOLOGY

The legible history of the Earth is read chiefly from sedimentary rocks, which record a sequence of events, changing physical environments, de-

Fig. 3. Fault near Great Bear Lake, Northwest Territories, Canada. This fault zone, etched by erosion, is traceable for 80 mi. (*Royal Canadian Air Force*)

Geologic column and scale of time

Periods of time/ systems of rocks	Epochs of time/ series of rocks	Distinctive records of life	Isotopic dates (in years before present)*
		CENOZOIC ERA	
Quaternary	Recent	Modern man†	11,000
	Pleistocene	Early man	1,500,000‡
Tertiary	Pliocene	Large carnivores	10,000,000
	Miocene	Whales, apes, grazing animals	27,000,000
	Oligocene	Large browsing animals	38,000,000
	Eocene	Rise of modern floras	55,000,000
	Paleocene	First placental mammals	70,000,000
		MESOZOIC ERA	
Cretaceous		Last of dinosaurs Last of ammonites Rise of flowering plants	130,000,000
Jurassic		Toothed birds Flying reptiles First primitive mammals	180,000,000
Triassic		Rise of dinosaurs Rise of ammonites Rise of cycads	225,000,000
		PALEOZOIC ERA	
Permian		Primitive reptiles Last of trilobites Glossopteris flora	270,000,000
Carboniferous periods Upper (Pennsylvanian)		Spread of amphibians Great coal forests Climax of spore-bearing plants	315,000,000
Lower (Mississippian)		Abundant sharks Climax of crinoids and blastoids	350,000,000
Devonian		First forests Rise of ferns Earliest known amphibians	405,000,000
Silurian		Appearance of land plants First known scorpions Expansion of brachiopods and corals	440,000,000
Ordovician		Appearance of primitive fishes Climax of trilobites Rise of cephalopods	490,000,000
Cambrian		Abundant trilobites Many kinds of shelled invertebrates	575,000,000
		PRECAMBRIAN TIME§	
No basis for worldwide subdivisions		Marine algae, worm burrows,	1,200,000,000
		other simple forms	2,100,000,000
		Abundant carbon of organic origin	2,980,000,000
		Oldest dated rocks (Congo)	3,200,000,000
		Earliest known record of life	3,490,000,000
		Estimated date, beginning of Earth history	4,500,000,000–5,000,000,000

*Dates are for beginning of time divisions. †Recent and Late Pleistocene.

‡This date is according to H. Berggren and V. Couvering, Corrected age of the Pliocene–Pleistocene boundary, *Nature*, 269:483–488, Oct. 6, 1977.

§Figures for Precambrian time are selected from a large number of available values.

velopments in plant and animal life, and effects of crustal movements. Additional records are supplied by volcanic rocks, which in many areas are interlayered with, and grade into, sedimentary strata; by relations of intrusive igneous bodies to older and younger rocks; and by erosion surfaces, some displayed in present landscapes, others revealed in exposures of unconformities. The long history includes radical changes in physical geography, featuring an endless contest between land and sea; the birth, rise, and wasting away of successive mountain systems; and the evolution of living forms in seas and on lands. A succession of events through eons of time is clearly shown in the stratigraphic record, and increasing numbers of absolute dates are supplied by geochemical studies of critical isotopes.

Stratigraphy is the systematic study of stratified rocks. Paleontology is the study of fossilized plants and animals, with regard to their distribution in time. These two complementary disciplines provide much of the basis for geologic history.

Stratigraphy. Studies of consolidated sedimentary rocks in comparison with the many kinds of modern sediments provide a basis for recognizing conditions under which the older deposits were formed—whether on land, on sea floors, or in transitional zones such as deltas or lagoons. Generally a much closer interpretation is possible; bodies of rock are confidently classified as deposits on floodplains, at margins of glaciers, in large lakes, in shallow seas near shore, or in deep-sea troughs. Each distinctive type of deposit represents a facies. A kind of rock that is essentially uniform over a considerable area constitutes a lithofacies. An assemblage of fossils that is nearly uniform in a large unit of sedimentary deposits, indicating an environment suited to certain forms of life, marks this unit as a biofacies. Deposits formed at the same time may differ greatly both in lithofacies and in biofacies, reflecting differences in topography, climate, and other items of environment.

A fundamental principle in stratigraphic studies, known as the law of superposition, states that in a normal sequence of strata any layer is older than the layer next above it (Fig. 1). This elementary law is of fundamental importance in a study of many mountain zones where thick sections of strata have been overturned, even completely inverted, and can be resolved only through criteria that indicate original tops of beds. Close matching of the many kinds of modern sediments with materials in sedimentary rocks formed over an immense span of time has established the uniformitarian principle, which holds that processes now operating on the Earth have operated in fairly uniform fashion through the ages. *See* STRATIGRAPHY.

Each continent (other than ice-buried Antarctica) has at least one wide lowland or platform that has been occupied repeatedly by seas and is now mantled with little-deformed marine strata. The total deposit on each platform represents a long span of time but ranges from only a few hundred to a few thousand feet thick. Along a margin of each platform the section of strata is much thicker, more varied in character, and strongly deformed. This belt of thick deposits records long-continued subsidence of a geosyncline; the strata then were folded and faulted, and elevation of the belt formed high mountains. Examples are the Appalachian Mountains, the younger Rocky Mountains, and the still younger Alpine-Himalayan chains. *See* GEOSYNCLINE; TECTONIC PATTERNS.

Paleontology and scale of time. At any one locality a sequence of sedimentary beds, from older to younger, can be determined through physical evidence. Persistence of some peculiar units may establish approximate correlations through moderate distances, occasionally hundreds of miles. But the accepted key to relative dating of stratigraphic units and to confident worldwide correlations is supplied by fossils of animals and plants, which record progressive evolution in living forms from ancient to recent times. Many of the oldest known sedimentary rocks are now highly metamorphosed, and any fossils they may have held must have been obliterated. Some thick sections of very old strata that are not appreciably altered have yielded only sparse indications of life, such as patterns of marine algae and burrows made by worms or other lowly forms. Successively younger groups of strata hold abundant fossils of marine invertebrates, marine fishes, land plants that progressed from primitive to more modern kinds, reptiles, birds, small mammals, followed by diverse and generally more advanced kinds of mammals, primitive men, and finally modern man. Some forms evolved rapidly, and short-lived species that were equipped to become widely dispersed are of greatest value for correlation. *See* PALEOBOTANY; PALEONTOLOGY.

A geologic scale of time (see table) is based partly on physical evidence but more largely on the paleontologic record as is suggested by names of the major divisions (Paleozoic, ancient life; Mesozoic, medieval life; Cenozoic, recent life). Absolute dates are determined by analyses of critical isotopes, collected at strategic locations. This dating is known as geochronometry. *See* GEOCHRONOMETRY; GEOLOGICAL TIME SCALE; see also articles on the divisions of geologic time named in the table.

GEOLOGIC MAPPING

A geologic map represents the lithology and, so far as possible, the geologic age of every important geologic unit in a given area. Each distinctive unit that can be shown effectively to the scale of the map is a geologic formation. A good topographic base map is essential for representing relations of bedrock to land surface forms. Cooperation of workers with specialized qualifications—for example, in petrology, paleontology, or structural geology—is required for accurate mapping and description of complex areas. Large organizations, such as Federal geological surveys and some commercial firms, have diversified personnel, laboratories equipped for varied analyses, and special field equipment. Aerial photographic surveys serve as a guide in field work and help in plotting accurate locations. Photogeology is a technique that uses photographs for constructing preliminary maps. These maps are then corrected by geologists on the ground.

A completed geologic map should indicate important structural details, such as inclinations of strata, locations of faults, and axial traces of folds. Usually the map is supplemented by vertical sections on which structural features seen at the surface are projected to limited depths. Maps of small scale may represent sedimentary rocks only according to the systems to which they belong. With larger scale a given system may be represented by several formations, each recording an important episode in the history of the region. In some European countries geologic mapping has been completed to fairly large scales; but in all continents great areas are still unexplored geologically or have been mapped only in reconnaissance fashion.

[CHESTER R. LONGWELL]

Bibliography: T. H. Clark and C. W. Stearn, *The Geological Evolution of North America*, 1960; E. C. Dapples, *Basic Geology for Science and Engineering*, 1959; A. Holmes, *Principles of Physical Geology*, 2d ed., 1965; C. S. Hurlbut, *Dana's Manual of Mineralogy*, 17th ed., 1959; J. A. Jacobs, R. D. Russell, and J. T. Wilson, *Physics and Geology*, 1959; P. H. Kuenen, *Marine Geology*, 1950; L. D. Leet and S. Judson, *Physical Geology*, 3d ed., 1965; F. J. Pettijohn, *Sedimentary Rocks*, 2d ed., 1957; J. S. Shelton, *Geology Illustrated*, 1966; A. O. Woodford, *Historical Geology*, 1966; F. E. Zeuner, *Dating the Past*, 4th ed., 1958.

Geology, evolution of

The evolution of geology as a science has stemmed from the successive development of three contrasting doctrines. Each views the Earth from a fundamentally different analytical perspective.

Catastrophism. This geological doctrine attributes all major features of the Earth's crust, such as folds and faults, as well as the lateral and vertical succession of rocks exposed in a particular region, and the dramatic secular changes in fossil flora and fauna to sudden, somewhat repetitious, short-lived, violent, and essentially global events, that is, catastrophes. In ancient and medieval times, geological thought was dominated by the doctrine of catastrophism. Catastrophic events occurred as a result of forces which were beyond human understanding, and were therefore considered to be "unnatural" and mysterious. As long as catastrophism was the prevailing theory, the development of geology as a science was hindered—the Earth remained a mysterious place which was shaped and continually changed by unknown, seemingly capricious "external" forces.

Uniformitarianism. An alternative view of the Earth was developed early in the 19th century: the doctrine of uniformitarianism. This concept was originally developed by J. Hutton in Scotland in 1788, and was popularized by C. Lyell in his classic work, *Principles of Geology*, in 1830. In essence, uniformitarianism advocates a noncatastrophic view of the Earth and its features. Rocks, changes in the fossil record, and geological features such as landscapes and structures are thought to be produced by the same natural processes and laws which operate at present to modify the Earth. These processes and forces also operate with essentially the same intensity today as they have in the past. Consequently, an understanding of how these processes and laws operate today can be applied to the origin and evolution of the Earth.

Uniformitarianism had replaced catastrophism as the governing doctrine of geology by 1850, and that change led, in turn, to the development of geology as a modern science.

Actualism. Present understanding of the Earth is based on the doctrine of actualism. Actualism accepts the basic premise of uniformitarianism, that physical and chemical laws are constant over geological time, but proposes that the rate and intensity of the resulting processes may vary appreciably. Some important widespread events are indeed catastrophic, such as earthquakes and epochs of glaciation. Also, because the Earth is continually changing in terms of its atmosphere, hydrosphere, and lithosphere, the relative importance of different processes taking place on and within an "evolutionary" Earth has varied significantly in time. *See* GEOLOGY.

[FREDERIC L. SCHWAB]

Bibliography: American Geological Institute, *Glossary of Geology*, 1974; R. H. Dott, Jr., and R. L. Batten, *Evolution of the Earth*, 2d ed., 1976.

Geomagnetism

A term signifying both the magnetism of the Earth and the branch of science that deals with the Earth's magnetism. The term geomagnetism is now given some preference over the older and longer term terrestrial magnetism.

Main geomagnetic field. This is specified at any point O by its vector magnetic intensity F. Its direction is that of the line $P'OP$ from the negative end P' to the positive end P of a magnetized needle $P'P$, perfectly balanced before it is magnetized, and freely pivoted about O, when in equilibrium. The positive pole P is the one that at most places on the Earth takes the more northerly position. Over most of the Northern Hemisphere, P is below O; the needle is said to dip below the horizontal by an angle I, called the magnetic inclination.

Over about half the Earth, however, P will be above O; the inclination I (or magnetic dip) is then reckoned as negative. This is the case over most of the Southern Hemisphere. The value of I thus ranges from 90 to −90°. A point where $I = \pm 90°$ is called a magnetic pole of the Earth.

Magnetic poles and Equator. There are two main magnetic poles; their approximate positions in 1975 were 76.1°N, 100°W and 65.8°S, 139°E. In a few places, near strongly magnetized mineral deposits, there may be local magnetic poles.

The distribution of I over the Earth's surface can be indicated on a globe, or on a map (with any kind of projection), by lines called isoclinic lines or isoclines, along each of which I has the same value. The isocline for which $I = 0$ (where the balanced magnetized needle rests horizontal) is called the magnetic or dip equator. Figure 1 shows isoclines over a large part of the Earth for the epoch 1975.

Magnetic declination. A compass needle is magnetized and pivoted at a slightly noncentral point so as to rest and move in the horizontal plane. The deviation of its direction from geographic (gg) north is called the magnetic declination D. This is reckoned positive to the east and negative to the west; alternatively, declinations can be specified by a positive (eastward) value ranging up to 360°.

Over the greater part of the Earth D is numerically less than 90°, but along any small circuit around the magnetic poles it takes all values. At the magnetic poles themselves the compass needle takes no definite direction. At the gg poles the needle takes a definite direction, but as the northward direction changes through a whole revolution along any small circuit around the pole, D likewise takes all values around the circuit.

The distribution of the declination over the Earth's surface can be indicated by lines along each of which D is constant. These are called isogonic lines, or isogones. For the reasons just stated, the two magnetic poles and the two geographical poles are points toward which isogones converge, for all values of D. Figure 2 shows isogones for a large part of the Earth, for the epoch 1975. The isogones for which $D = 0$ are called agonic lines. Along these lines, the compass points to true north.

The distribution of D over the Earth can also be indicated by magnetic meridians, which at each point P have the direction of the compass needle at P. These lines extend from the south magnetic pole to the north and have no complication (as do the isogones) at the gg poles. They indicate, more clearly than the isogones, the general distribution

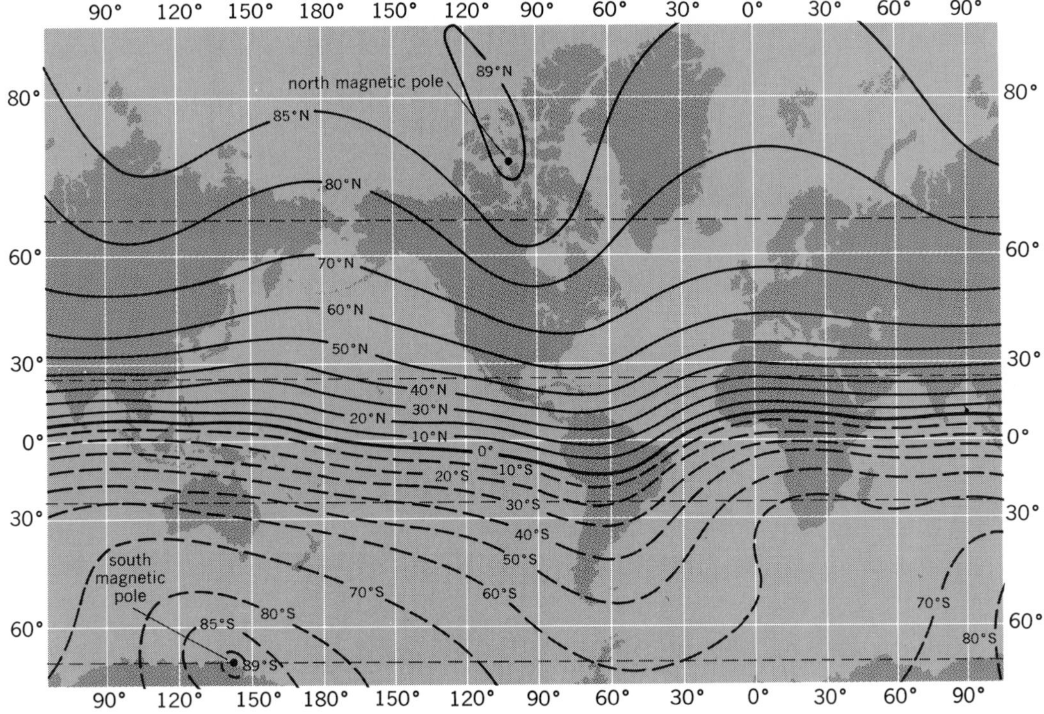

Fig. 1. Lines of equal geomagnetic inclination *I* for 1975. (*U.S. Naval Oceanographic Office*)

of the compass direction over the Earth. But the isogonic map is more convenient in enabling the compass direction at any point to be read or estimated without using an angle measurer. Hence it is the one used by navigators and travelers on sea and land and in the air.

Figure 3, which shows both isogones and magnetic meridians in the region around the north geographic and magnetic poles, illustrates this contrast between them.

Intensity patterns. The strength *F* of the vector intensity **F** is the third quantity which, together with *I* and *D*, completely specifies **F**. It is expressed in a unit which in geomagnetic literature is called the gauss, after the German mathematician, astronomer, and physicist K. F. Gauss (1777–1855), who first showed how *F* could be measured in units of length, mass, and time. This

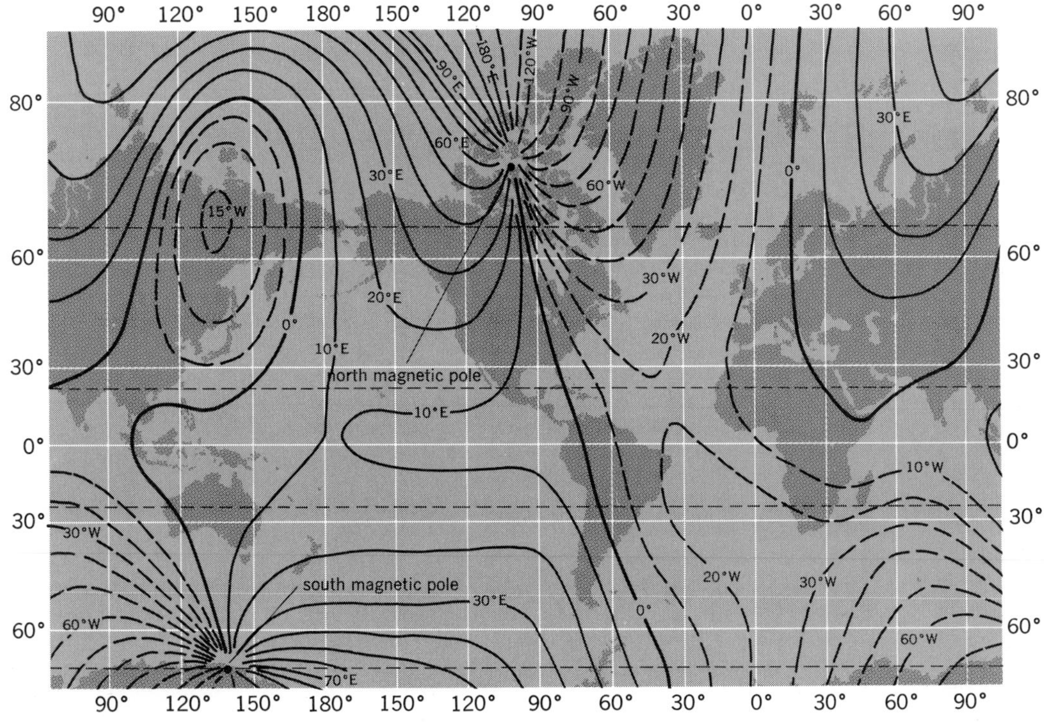

Fig. 2. Lines of equal geomagnetic declination (isogones) for 1975. (*U.S. Naval Oceanographic Office*)

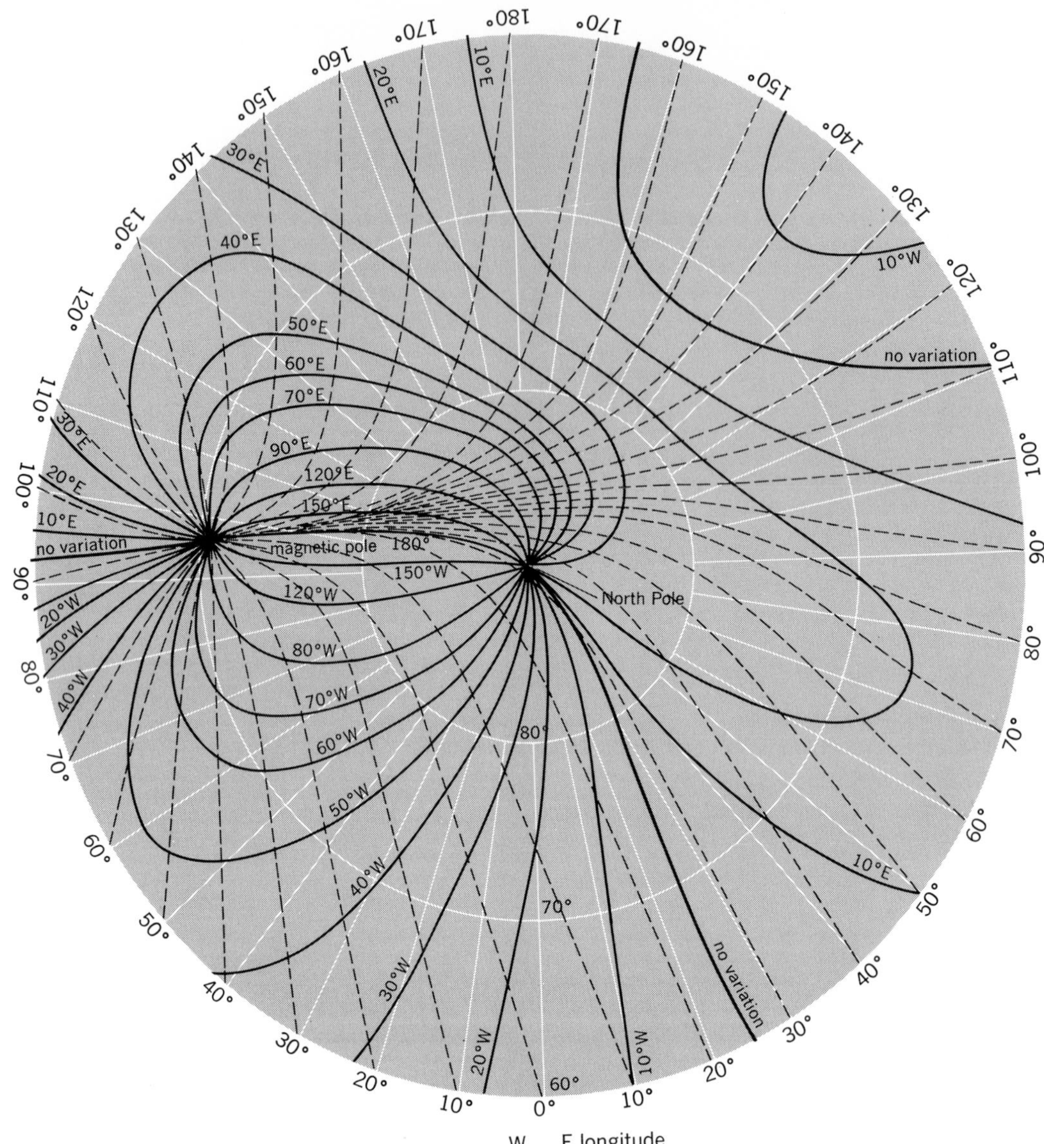

Fig. 3. Lines of equal declination (full lines) and magnetic meridians (broken lines), for 1922, for the Arctic region north of 60° latitude. The agonic lines (isogones for which magnetic declination equals zero) are marked "no variation." *(After H. Spencer Jones, in S. Chapman and J. Bartels, Geomagnetism, 2 vols., Oxford, 1940)*

was the first nonmechanical entity so to be measured; Gauss' innovation was a landmark in the history of physical measurement. Physicists and electrical engineers distinguish the gauss from the oersted, the unit by which they would express the magnetic force. The numerical difference between the two, in geomagnetism, is negligible.

The symbol for the gauss is Γ; much use is made in geomagnetism of a unit called the gamma (symbol γ), which is 10^{-5} Γ. The International System of Units (SI) was adopted in 1960, defining magnetic flux density in terms of a tesla and magnetic field strength as amperes per meter. [One oersted = 7.96 amperes per meter; 1 gauss (Γ) = 10^{-4} tesla; and 1 γ = 1 nanotesla (nT).]

The distribution of the magnetic intensity F over the Earth can be indicated, as for I and D, by lines of constant F. They are called isodynamic lines. They are shown for the epoch 1975 in Fig. 4. The value of F in general increases with increasing latitude north or south; its distribution, however, is not simple. In low latitudes it has a minimum of about 0.23 Γ off the east coast of South America. In the Northern Hemisphere there are two maxima, each about 0.6 Γ, one in northern Canada and the other in Asia. The highest intensity, near 0.7 Γ, occurs near the southern dip pole.

The vector intensity **F** can also be specified by its vertical component Z and its horizontal vector component **H**; the latter can be specified by its direction (given by D) and the horizontal intensity H or, alternatively, by its north and east components X, Y. Clearly $Z = F \sin I$, $H = F \cos I$, $X = H \cos D$, and $Y = H \sin D$; Z, like I, is thus reckoned positive when downward, and Y, like D, is positive when eastward.

Geomagnetic elements. All the magnitudes F, H, X, Y, Z, I, and D are called geomagnetic elements. Lines on a map along which any element has a constant value are called isomagnetic lines; the

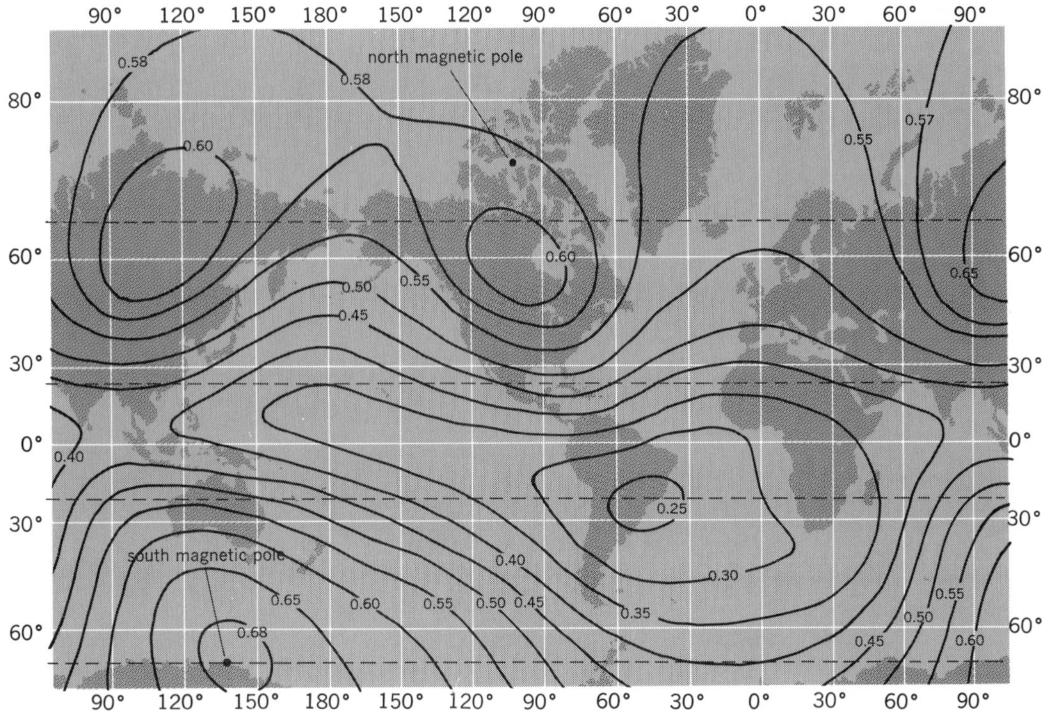

Fig. 4. Lines of equal geomagnetic intensity *F* for 1975. (*U.S. Naval Oceanographic Office*)

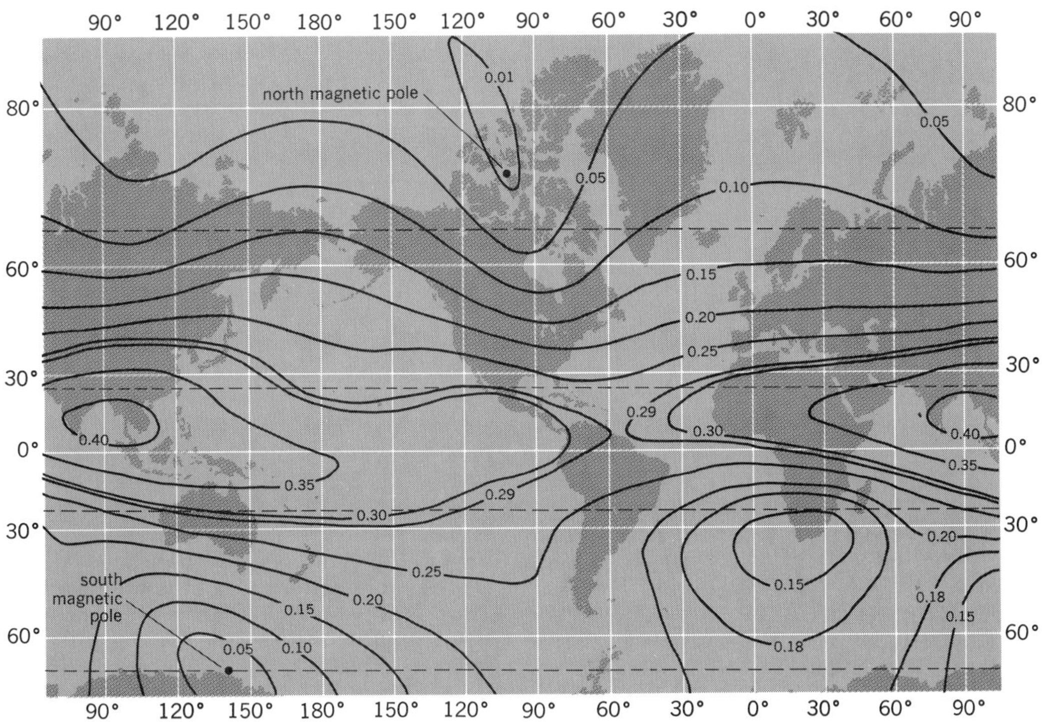

Fig. 5. Lines of equal horizontal geomagnetic force for 1975. (*U.S. Naval Oceanographic Office*)

map is called an isomagnetic chart. Figure 5 shows an isomagnetic chart for *H* for the epoch 1975.

Isomagnetic maps for the whole Earth or any large part of it cannot show the finer details of the field distribution. Maps for smaller areas, such as a European country or a state of the United States, can show such details if the areas have been closely surveyed magnetically. In some parts of such maps the lines may depart considerably from the smooth spacing indicated on world maps, for example, in regions of local magnetic anomaly, where the field is disturbed by magnetic minerals not far below the Earth's surface. The most striking local anomalies occur in two narrow strips in the Kursk region of the Soviet Union. They are 60 km apart and run in parallel from NE to SW. The

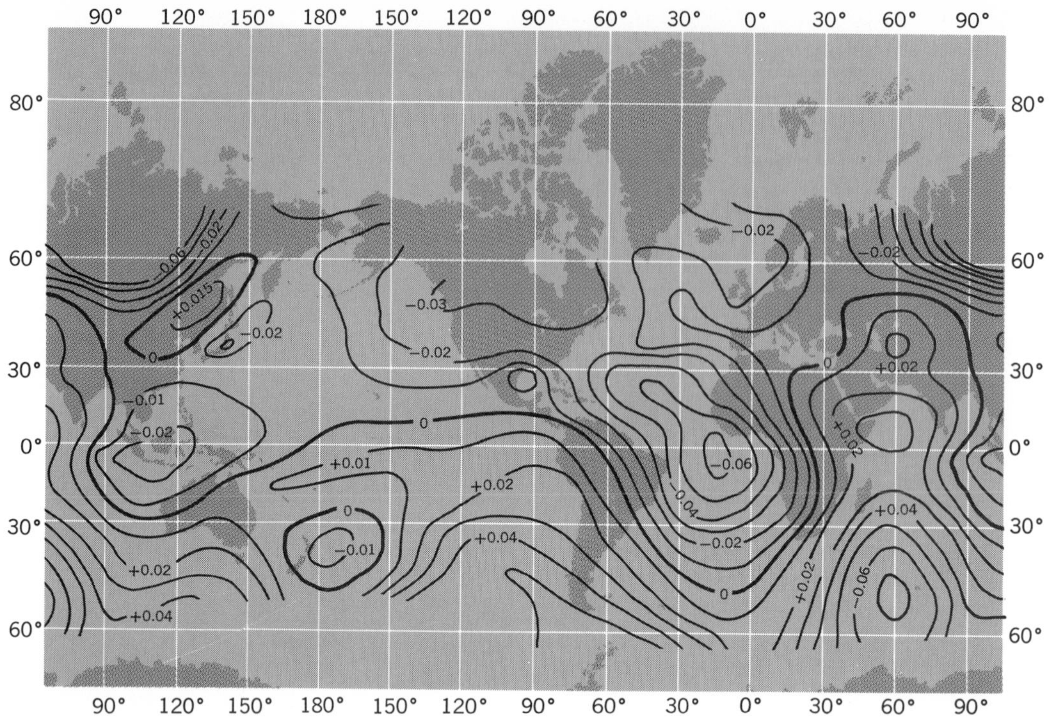

Fig. 6. Lines of equal rate of change (isopors) of the vertical geomagnetic force *Z*, over the interval from 1885 to 1922. The isopors are drawn at 500-γ intervals; increase of downward magnetic force is considered positive. (*After A. G. McNish, in S. Chapman and J. Bartels, Geomagnetism, 2 vols., Oxford, 1940*)

most disturbed part of the major (northerly) strip is only 2 km wide, although the strip is 250 km long; Z is everywhere above normal and ranges up to 1.9 Γ.

Spherical harmonic analysis. The Earth's surface magnetic field can be expressed in mathematical terms by a process called spherical harmonic analysis. The data used are magnetic survey measurements of the magnetic elements at a sufficient number of points on the Earth's surface. The analysis provides separate expressions for the parts of the surface field that originate, respectively, within the Earth and above it. It proves that the main field comes almost entirely from within. The part that is of external origin has not been accurately determined, but is less than 100 γ at the Earth's surface and may average only about 20 γ.

The main term in the spherical harmonic expression corresponds to the field of a uniformly magnetized sphere, or of a point dipole at the Earth's center. Hence this part of the geomagnetic field is called the dipole field. The intensity of magnetization for a sphere the size of the Earth, to give this dipole field, would be about 0.08 Γ. The field could also be produced by a surface distribution of electric current around the Earth, of total amount 1.5×10^9 amp, distributed proportionately to the cosine of the latitude. The moment of the equivalent dipole is 8.1×10^{25} Γ-cm^3. These three possible systems that could produce the dipole field all differ from the actual source of the field.

The Earth's diameter along the direction of the dipole (which is also the direction of magnetization for the alternative model of a uniformly magnetized sphere) is called the dipole (dp) axis. Its ends on the Earth's surface are called the dp or axis poles. The polarity of the northern pole is negative; that of the southern is positive. On maps the northern dp pole is conveniently marked B and the southern, A; these letters may stand for boreal and austral. The gg position of B is 78.5°N, 69°W, and of A, 78.5°S, 111°E. The obliquity of the dp to the gg axis is thus 11.5°. The dp pole B is 1280 km from the N gg pole and 1160 km from the north magnetic pole; the corresponding southern distances are 1280 km and 1350 km. The dp axis has shown no change of direction large enough to be reliably estimated since it was first determined nearly 150 years ago. The position of any point P may be expressed in dp or dipole coordinates relative to the dp axis; the dp colatitude is the angle BOP (O being the Earth's center); the dp longitude is measured eastward from the part of the common gg and dp meridian through B that lies south of B. The dp equator, from which dp latitude is measured north toward B or south toward A, is in the diametral plane perpendicular to the dp axis.

A closer approximation to the Earth's surface field is the field of a point dipole not at the Earth's center. The eccentric dipole must have the same direction and magnetic moment as the centered dipole, but is displaced from O by about 460 km, toward a point in the Pacific Ocean north of New Guinea; it is moving NNW. The eccentric dipole field is only a moderate improvement on the centered dipole field, as an approximation to the actual surface field; there still remain regional deviations from it, amounting in several places to over 0.1 Γ in the horizontal component.

The spherical harmonic expression for the potential of the geomagnetic field enables the field to be calculated at any point above the Earth, except insofar as electric currents in the ionosphere and beyond modify the field. These currents can be explored by rockets and satellites, which confirm that the field does not extend to infinity, varying inversely as the cube of the distance, as formerly supposed. Instead it is confined to a region—the magnetosphere—within a continuing but variable flow of plasma from the Sun. Also, a ring current flows within this region and greatly modifies the field during geomagnetic storms. This current may always be present, though weak when the plasma flow is gentle. The origin of the geomagnetic field is discussed later.

Secular magnetic variation. This slow change of the Earth's field necessitates continual redrawing of the isomagnetic maps. In any magnetic element at a particular place, the variation may be an increase or a decrease; its rate is unpredictable and is not constant in either magnitude or sign. The distribution of the rate for any element can be indicated on maps called isoporic, by lines (isopors) along which the rate is constant.

The pattern of such isopors is more complex than that of the isomagnetic lines for the same element. There may be several regions of maximum or minimum change, producing numerous oval systems of isopors (see Fig. 6, giving Z isopors based on an interval of 38 years). Whereas the main field is a planetary property, the secular variation is regional. Moreover isoporic maps change much more than isomagnetic maps from one decade to the next. Regions of maximum change move, disappear, or newly appear. An example of long-continued changes of the field direction is given in Fig. 7, for London. From 1576 to about 1800 the compass there turned from 11°E to 24°W, while the dip first increased from 71° by 3°, until 1700; since 1700 it has decreased again, to about 67°. Since 1800 the compass has swung eastward through about 16°. Figure 7 shows these changes, with some directions and trends for earlier centuries (indicated by numbers, that is, 15 indicates 15th century), estimated from the magnetization of bricks in kilns whose period can be dated. This is an example of archeomagnetic research, which aims to extend knowledge of the past geomagnetic changes by centuries or even by millennia.

Figure 7 suggests a possible cycle of change of direction at London, with a period of 4–5 centuries. But the changes elsewhere are inconsistent with any regular period of secular geomagnetic change.

Another long-continued secular trend is the movement of the agonic line, which in Fig. 2 crosses the Equator at about 74°W. The course of this agonic line has been followed longer than that of the more complicated looped agonic line that traverses Asia. In 1550 it crossed the Equator at about 20°E, and since then it has moved continuously westward through about 90° in about 4 centuries. This is one token of a noticeable tendency—the westward drift—of isomagnetic and isoporic features. Another large-scale feature of the secular magnetic variation is a decrease of the Earth's magnetic moment by about 5% during the last century.

The study of paleomagnetism, the remanent magnetization of rocks and sediments, extends

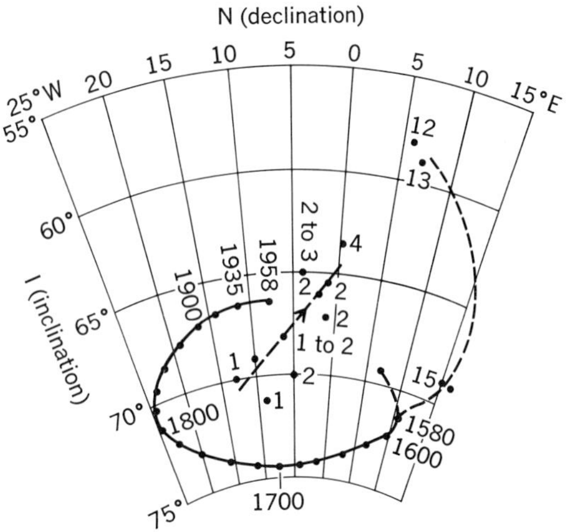

Fig. 7. Geomagnetic direction at London, as observed from 1580 to 1958; the magnetic declination and the dip are shown, mostly at 20-year intervals. In addition, points are shown indicating the direction for earlier centuries, inferred from studies of magnetized bricks in old Roman-British and later kilns. Broken lines are conjectural; the straight broken line shows the estimated changes of direction in Roman times. (*After L. A. Bauer, R. M. Cook, and J. C. Belshe*)

man's knowledge (though with some uncertainties) of the Earth's magnetic history to past geologic ages and indicates that the field has been reversed more than once. *See* ROCK MAGNETISM.

Source of field. The source of the main geomagnetic fild and of its secular variation is the Earth's core, according to accepted theories. Although slow by some standards, the secular variation is very rapid by geologic standards. It cannot reasonably be ascribed to changes in or not far below the Earth's crust. It seems probable that the main field is caused by electric currents in the Earth's liquid core, below about 2900 km (1800 mi). By slow convective movements, electric currents are produced in the core; these maintain the magnetic field, as in a self-exciting dynamo. Rather large-scale eddies in the convective motion produce the regional features of the main field, and their changes produce the secular magnetic variation. These changes have no apparent relation to the broad features of geography and geology.

According to this theory, the geomagnetic field will remain of mainly dipole character nearly down to the surface of the core, with the lines of magnetic force lying nearly in planes through the geomagnetic axis. The magnetic intensity will increase inversely as the cube of the distance from the center, to nearly 5 Γ at the axis poles of the core; but the regional irregularities will become more prominent as the core is approached.

Inside the core the lines of force are probably much twisted around the Earth's axis, and the intensity may exceed 100 Γ. There will be a reaction between the magnetic field and the electrical currents in the core, tending to produce a westward drift of the core (and hence of the secular magnetic variations) relative to the outer solid part of the Earth.

The estimates of the electrical resistivity σ of

the core required by these theories range from 10^{-1} to 10^{-3} or even less, in ohm-cm units. This is far less than that of the upper (dry) layers of the solid Earth, for whose materials σ ranges from 10^4 to 10^{10}. For sea water σ ranges from 15 to 40, depending on the salinity and temperature. At a depth of 600–700 km, σ appears to begin to decrease from about 10^4 to a value of about 0.5 at the core-mantle boundary.

Geomagnetic measurements. To keep the knowledge of the Earth's magnetic field abreast of its secular changes, the world must be magnetically surveyed from time to time. In the past this has generally been done, on land and sea, by measuring D, H, and I. To measure D, it is necessary to determine true north by astronomical observation; for I, the horizontal direction must be determined by a good level. Then a weighted compass needle (declinometer) or a dip circle with balanced needle will indicate D and I; for real accuracy, great care and suitable reversals are required. To measure H, one uses a magnetometer. The original method devised by Gauss was likewise by means of a magnet, whose oscillations and power to deflect another magnet were observed.

Another form of inclinometer to measure I which has almost entirely displaced the dip circle has a rotating coil whose axis of rotation is turned until rotation in the Earth's field produces no induced alternating current in the coil. The axis then lies along the field direction; this enables D and I to be measured. This instrument is called a dip inductor, earth inductor, induction inclinometer, or earth inductor compass. [SYDNEY CHAPMAN]

Geomagnetic field surveys for geophysical exploration or geologic studies are now usually made with optical pumping or nuclear precession magnetometers, which have an accuracy of the order of a nanotesla. Such instruments give no readings of the field direction, but only of its absolute intensity. These devices have been towed by aircraft, near the ocean surface by ships, and even near the ocean floor. The process of reducing these data to anomaly maps representing only the proportion of the field due to the crustal sources is still more of an art than a science. The reason is that the fraction of observed field that comes from the core and the portion that is due to induced or remanent magnetization of the rocks are yet unsettled. *See* GEOPHYSICAL EXPLORATION.

World magnetic survey. A effort to map the global magnetic field was undertaken by the International Association of Geomagnetism and Aeronomy as a project to be carried out during the International Years of the Quiet Sun (1964–1965). The World Magnetic Survey (WMS) Board was established to coordinate this activity, which eventually involved participation by scientists in 41 member countries.

One significant product of the WMS was the agreement on a magnetic reference field for epoch 1965.0. The WMS group met in Washington, DC, in 1968, evaluated the various analyses of the magnetic survey data made by different organizations, and adopted an international geomagnetic reference field (IGRF). This standard is one of the more ambitious attempts at scientific coordination since the field changes in unpredictable ways and must be revised every few years.

Mapping the filed. The purpose of the WMS was to obtain data to be used in compiling charts and models for practical applications and for studies of the sources of the field. The objecive then was to map the smooth "main" field by eliminating from survey data the contribution both from the local anomalies in the Earth's crust and the transient changes caused by the ionosphere and magnetosphere. During and prior to the WMS, significant contributions to the magnetic survey data base were made by such organizations as the U.S. Navy and the Dominion Observatory (Canada) performing aeromagnetic surveys, and by the Soviet Union using the nonmagnetic ship *Zarya*. Many more detailed contributions were made by countries which took magnetic surveys within and near their borders.

However, even with this organized international effort, accurate global mapping was not possible in a time short enough to take a crisp "snapshot" of the field before the secular variation blurred the picture. For this reason, one of the first suggested scientific uses of orbiting spacecraft was the mapping of the geomagnetic field.

Satellite surveys. The pioneering attempt to make a satellite survey was by the Soviet experimentalist Sh. Dolginov, who flew flux-gate (vector) magnetometers on *Sputnik 3* in 1958. The vector-component data were never successfully reduced, so that the final analysis was made combining the three observations into a total field of about 100-nT accuracy. The main reason that such vector data are difficult to analyze is that the orientation, or attitude, of the instrument must be known very accurately. For example, a 1° error in attitude yields a 1000-nT error in components measured in a 57,000-nT field. Improvements in spacecraft attitude determination using star cameras now allow the possibility of making accurate vector measurements, and a new survey was planned by the United States in 1979 or 1980.

However, to date, all low-altitude satellites that have attempted absolute magnetic measurements, except the *Sputnik 3*, have carried only total field magnetometers. Key details of these surveys are given in the table. The Soviet *Cosmos 49* and the American *OGO-2* were the two flights made especially for the United States–Soviet Union bilateral agreements and as contributions to the World Magnetic Survey.

Analysis of magnetic surveys. The techniques of spherical harmonic analysis noted earlier have been significantly improved since 1960 due to the use of high-speed computers. The first application of modern equipment to this problem was made by D. C. Jensen and J. C. Cain in 1962 when they made a direct least-squares analysis of all survey data taken since 1940. They used the classical representation of the field expressed in spherical harmonic coefficients. The magnetic intensity **F** is given by $-\Delta V$, where the scalar potential function V is computed from

$$a \sum_{n=0}^{n*} \left(\frac{a}{r}\right)^{n+1} \sum_{m=0}^{n} (g_n{}^m \cos m\phi + h_n{}^m \sin m\phi) P_n{}^m(\theta)$$

where $a = 6371$ (average radius of Earth in kilometers); r = radial distance from Earth's center; θ = geographic colatitude; ϕ = east longitude; P

Low-altitude satellite geomagnetic measurements 1958–1975

Spacecraft	Orbit: Inclination, degrees	Orbit: Altitude range, km	Interval	Magnetometer	Coverage
Sputnik 3	65	440–600	May–June 1958	Flux gates	Soviet Union
Vanguard 3	33	510–3750	Sept.–Dec. 1959	Proton precession	Near ground stations
Cosmos 26	49	270–403	Mar. 1964	Proton precession	Whole orbit
Cosmos 49	50	261–488	Oct.–Nov. 1964	Proton precession	Whole orbit
1964-83C	90	1040–1089	Dec. 1964–June 1965	Rubidium vapor	Near ground stations
OGO-2	87	413–1510	Oct. 1965–Sept. 1967	Rubidium vapor	Whole orbit
OGO-4	86	412–908	July 1967–Jan. 1969	Rubidium vapor	Whole orbit
OGO-6	82	397–1098	June 1969–June 1971	Rubidium vapor	Whole orbit
Cosmos 321	71	270–403	Jan.–Mar. 1970	Cesium vapor	Whole orbit

represents Schmidt's spherical functions; g and h are the spherical harmonic coefficients determined in the analysis; and n^* is the truncation level of the expansion. The smallest scale size of the field structure represented by this expression is given by $2\kappa/n^*$, or in distance by the circumference of Earth (40,000 km) divided by n^*.

Since the present satellite surveys are only of total field, the analysis depends on having a very accurate set of observations to produce a good vector model. Fortunately this is possible from spacecraft since they operate above the altitudes where local anomalies add a high noise level (about 200–1000 nT) to the observations. One model was developed which matched a 5-year span of Orbiting Geophysical Observatory (OGO) data everywhere to better than 10 nT. Since the model is constrained only by the strength of the field, there is no direct limit to the error in the component perpendicular to the field. However, due to the properties of a source-free field, it is not possible to distort field direction at one point without changing field intensity elsewhere. It has been learned that the transverse error in the models based only on satellite data can be an order of magnitude greater than the longitudinal error. Thus such a model accurate to 50 nT in intensity could have transverse errors of 500 nT, which would correspond to a 1° error in angle near the Equator. A further consideration regarding the use of the results of satellite data is that errors at a few hundred kilometers' altitude magnify by a factor of 2 or more (depending on n^*) when the model is extrapolated to the surface.

In spite of these problems, comparisons with surface data taken near the same epoch indicate that satellite models have generally been at least as accurate as those computed using only surface observations. For example, E. Dawson and P. Serson concluded that satellite-derived models agree with the Canadian and Scandinavian aeromagnetic survey data as well as do local surfaces.

One of the most interesting and unexpected results from satellite surveys is the discovery of long-wavelength crustal features in the magnetic field. This was unexpected because it was previously thought that geologic anomalies could not produce structures more than about 100 km in size. Part of the reasoning lay in the knowledge that the crust cannot sustain magnetization at temperatures above the Curie point. Such temperatures are reached at least by 10 km under the oceans and 50 km under the continents. However, analysis of the field models from the OGO data appear to show that the field due to the harmonics above $n=11$ or 12 (wavelengths shorter than 3500 km) are caused by magnetization of rocks in the lithosphere, whereas the longer structure comes from the Earth's core. In addition to these features, other intense anomalies such as the one previously mentioned at Kursk are also detected at satellite altitude. Although work is continuing in this area, it appears that most of this structure is seen over continents or regions of the ocean which may contain sunken continental material.

Secular variation and the dipole. The use of satellite observations has made it possible to determine the secular variation over the whole Earth for a short interval of time. This has been done with the same least-squares analysis which is used to determine the g and h coefficients; for example, $g=g_0+\dot{g}(t-t_0)$, where t_0 is the epoch of the coefficients, and $\dot{g}$ is the secular change of g in nT/year.

One such analysis of satellite data has determined the global secular variation (SV) to an accuracy of 4 nT/year over a 5-year period. Prior analyses using the year-to-year differences of accurate observations were made at magnetic observatories or special sites called repeat stations. The other studies gave detailed and accurate information concerning the secular variations in the regions of the stations. However, these observing points are located mainly in inhabited land areas, leaving large gaps, such as at the poles and oceans, where data are sparse or nonexistent.

Such studies of global SV revealed that there is a relation between the behavior of the eccentric dipole and the rate of the Earth's rotation. It appears that an irregularity in the length of day causes a disturbance in the flow of material near the core-mantle interface which then propagates in about 7 years to the Earth's surface and changes the rate of westward drift of the eccentric dipole. Such changes may be related to the occurence of major earthquakes. *See* EARTHQUAKE.

The satellite data have also helped to show an irregularity in the SV which occurred over Asia and the Indian Ocean during the interval 1965–1971. Until about 1965 the field intensity was increasing over this whole region. From 1965 to about 1971 the field decreased over Asia and held constant over the Indian Ocean. Later reports from isolated stations appeared to show that the intensity is decreasing rapidly (about 100 nT/year) over the southern Indian Ocean, but is once again increasing over Asia.

Also during 1965–1971, the rate of the main dipole decrease has accelerated from its century average of about 0.05%/year to almost 0.09%/year.

[JOSEPH C. CAIN]

Bibliography: J. C. Cain, Structure and secular change of the geomagnetic field, *Rev. Geophys. Space Phys.*, 13:203–204, 1975; S. Chapman, *Solar Plasma, Geomagnetism and Aurora*, 1964; S. Chapman and J. Bartels, *Geomagnetism*, 2 vols., 1940; S. Matsushita and W. H. Campbell (eds.), *Physics of Geomagnetic Phenomena*, 2 vols., 1968; H. Odishaw, *Research in Geophysics*, vol. 1, 1964; *Rev. Geophys. Space Phys.*, 13:35–51, 174–241, 1975; A. J. Zmuda (ed.), *The World Magnetic Survey*, IAGA Bull. no. 28, Paris, 1971.

Geomorphology

The primary relief elements of the Earth's surface are formed by movements of the crust. Geomorphology is the study of the origin of secondary topographic features which are (1) carved by erosion in the primary elements and (2) built up of the erosional debris. Most of the basic concepts of this branch of geology were formulated by William Morris Davis, of Harvard University, in the early part of this century. Davis held that the morphology of landforms is controlled by the interplay of three factors: structure, process, and stage. *See* DIASTROPHISM; FLUVIAL EROSION CYCLE.

Structure. In its Davisian sense, structure embraces all the characteristics of rocks which influence topographic form. The flat benches and vertical cliffs of the sides of the Grand Canyon are etched by erosion in layers of sandstone, shale, and limestone which are nearly horizontal. The sharply contrasted linear ridges and valleys of the Appalachians correspond in position with the edges of folded strata. Because most erosional processes are in the highest degree selective, differences in permeability, solubility, and other properties of rocks, so subtle as to be difficult to measure in the laboratory, are commonly reflected in surface form. The geomorphologist looks first for structural causes of landform anomalies before considering other explanations. *See* ROCK CONTROL (GEOMORPHOLOGY).

Process. The processes of erosion differ markedly from place to place depending on climate and other circumstances, and each process tends to develop its characteristic landforms. A cirque is a distinctive feature, carved by the head of a valley glacier. A wave-cut cliff is similarly unique in form and uniquely associated with a specific erosional process. In the absence of a vegetative cover, gully washing may cut an area into a plexus of badland channels and knife-edged divides, wholly different from the smoothly contoured landscape with broadly convex hill tops that is modeled in the same material by downhill creep of plant-held soil. A knowledge of the process, preferably obtained by direct observation, contributes a great deal to an understanding of the form, and the reverse is equally true.

The relationship between process and form, once established, becomes the basis for a rich variety of interpretations. Cirques prove the former existence of glaciers where there are none now. A wave-cut cliff high above the modern strand line proves a relative change in level of land and sea. A smoothly modeled hillslope cut by gullies is a compound landform, indicating a change in process. Most present-day landforms are compound, with relict elements formed by Pleistocene processes being modified or destroyed by processes suited to present conditions.

Stage. Stage expresses change in landforms with the passage of time. If the relative level of land and sea remains the same long enough, the Colorado River will eventually reduce its gradient to just that slope which will enable the river to transport to the sea all of the erosional debris shed into it from its valley sides. Continuous loss of material means that the valley sides will be gradually reduced in slope and height until the Colorado Plateau comes to be a peneplain, that is, an erosional lowland of faint relief drained by sluggish rivers meandering on broad valley floors. This is the concept of the geomorphic cycle. The progress of the cycle cannot be seen in any one place because the changes are exceedingly slow, but over the face of the Earth, assemblages of landforms representing every phase of it can be observed. The state of maturity is characterized by an integration of all parts of the area under consideration in a system of slopes graded to the sea. Disadjustments of various types, as lakes and waterfalls that interrupt the profiles of the streams, and analogous breaks in slope in the interstream areas are symptoms of "youth." The term "old" suggests the greatly reduced vigor of the erosional processes as the peneplain is approached.

Validity. There has been debate as to many of the details of the Davis geomorphic cycle, but there is no question as to the validity of its central theme: During a period of crustal stability, landforms develop in an orderly and predictable manner with respect to sea level. Actually, sea level oscillated markedly during Pleistocene time under the influence of glacioeustatism, and the crust has been exceptionally unstable in late Cenozoic time. Far from reducing the value of the cycle concept, these circumstances make it indispensable, because they mean that most assemblages of landforms include elements of two or more cycles that would in most cases not be distinguished without the insight provided by this concept.

The geomorphic cycle, as outlined above, applies primarily to the geomorphic features of the humid temperate zone where the science had its birth. It differs in detail and degree rather than in kind when applied to the sequences of landforms developed under polar and tropical conditions and in arid regions where there is no drainage to the sea. The same basic scheme has proved its worth in analysis of the wholly different kinds of geomorphic features formed by shoreline processes, glaciers, groundwater, and other agents of erosion.

Every erosional landform expresses in its shape the influence of structure, process, and stage. Because each of these factors is complex, it is understandable that erosional landforms are almost infinitely varied. Depositional landforms—deltas, glacial moraines, offshore bars, sand dunes, and other forms—are equally varied in details of morphology. This helps to explain why, until recently, the approach to geomorphic problems has been chiefly qualitative. Since about 1950 the most significant development has been an increase in the application of quantitative and experimental methods in geomorphic research. These methods will do much to clarify and refine the understanding of geomorphic processes and forms. *See* COASTAL LANDFORMS; DESERT EROSION FEATURES; FLUVIAL EROSION LANDFORMS; GLACIATED TERRAIN; KARST TOPOGRAPHY.

[J. HOOVER MACKIN/H. T. U. SMITH]

Bibliography: A. L. Bloom, *The Surface of the Earth*, 1969; R. J. Chorley, A. J. Dunn, and R. P. Beckinsale, *The History of the Study of Landforms*, vol. 1: *Geomorphology before Davis*, 1964; W. M. Davis, *Geographical Essays*, 1909, reprint 1954; C. A. M. King, *Techniques in Geomorphology*, 1966; L. B. Leopold, M. G. Wolman, and J. P. Miller, *Fluvial Processes in Geomorphology*, 1964; W. D. Thornbury, *Principles of Geomorphology*, 2d ed., 1959.

Geophysical exploration

Making, processing, and interpreting measurements of the physical properties of the earth with the objective of practical application of the findings. Most exploration geophysics is conducted to find commercial accumulations of oil, gas, or other minerals, but geophysical investigations are also employed with engineering objectives, in studies aimed at predicting the nature of the earth for the foundations of roads, buildings, dams, tunnels, nuclear power plants, and other structures, and in the search for geothermal areas, water resources, archaeological ruins, and so on.

Geophysical exploration is commonly called applied geophysics or geophysical prospecting. The physical properties and effects of subsurface rocks and minerals that can be measured at a distance include density, electrical conductivity, thermal conductivity, magnetism, radioactivity, and elasticity. Perhaps other properties not used now could also be measured. Occasionally, prospective features can be mapped directly, such as iron deposits by their magnetic effects, but most features are studied indirectly by measuring the properties or the geometry of rocks that are commonly associated with certain mineral deposits.

Exploration geophysics is often divided into subsidiary fields according to the property being measured, such as magnetic, gravity, seismic, electrical, thermal, or radioactive properties. This article will first discuss principles applicable to most of the methods and then in subsequent sections the respective methods, the physical principles involved, instrumentation and field techniques employed, data processing, and geologic interpretation. *See* GEOPHYSICS.

General principles. A number of principles apply to most of the different types of geophysical exploration. Ordinarily, one looks for an anomaly, that is, a departure from the uniform geologic characterics of a portion of the earth (Fig. 1). The primary objective of a survey is usually to determine the location of such departures. If the source of an anomaly is deep in the earth, the anomaly is spread over a wide area, and the anomaly magnitude is small at any given location. Sometimes, areas of anomalous data are obvious, but more often they are elusive because the anomaly magnitude is small compared to the background noise or because of the interference of the effects of different features. A variety of averaging and filtering techniques are used to accentuate the anomalous regions of change. An anomaly usually seems smaller as the distance between the anomalous source and the location of a measurement increases (Fig. 2). Hence, a nearby source usually produces a sharp anomaly detectable only over a limited region, although possibly of large magnitude in this region. The detail of measurement required to locate anomalies must be compatible with the depth of the sources of interesting anomalies. As the depth of the anomaly increases, more sensitive instruments are needed because the effects become much smaller. Hence, the depth of

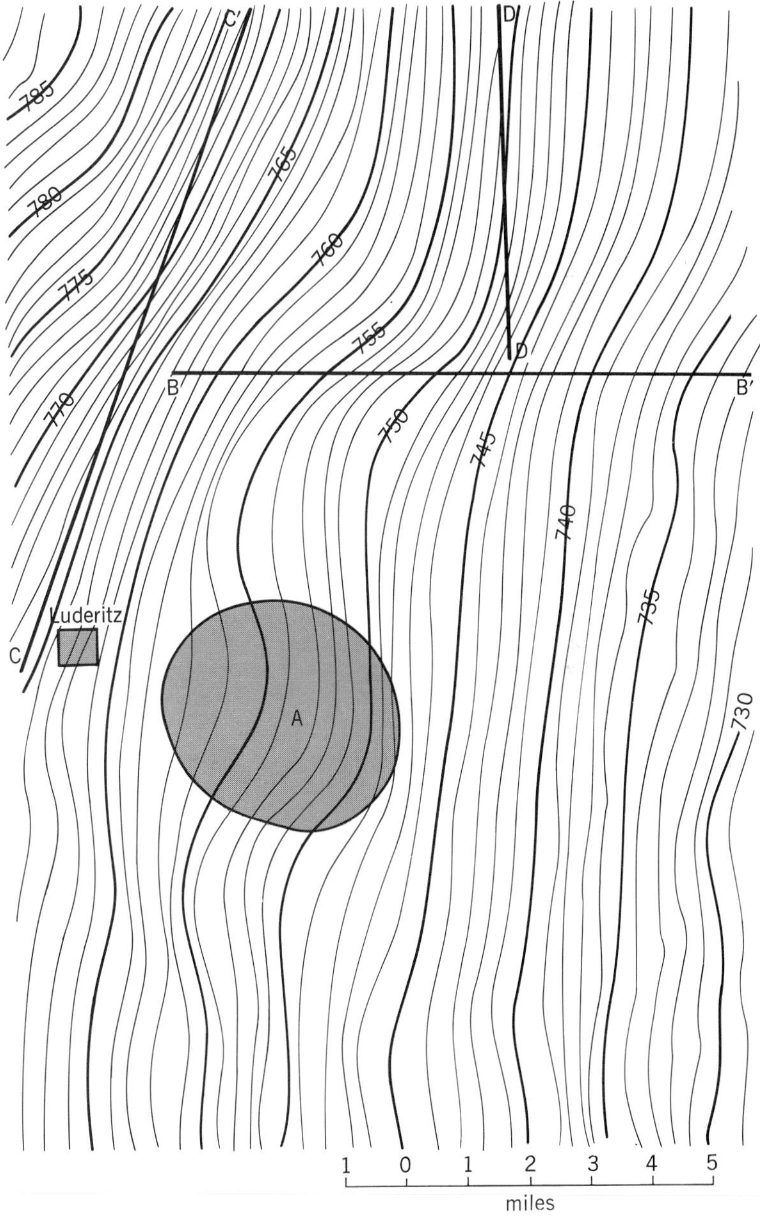

Fig. 1. Bouguer gravity of a portion of the Perth Basin, Western Australia. Contour interval is 1 mGal; datum is arbitrary. Departures from regularity are called anomalies. The bulge A results from an uplift area, the contour offset along BB′ is an east-west trending fault, downthrown to the south. Other faults (CC′ and DD′) are indicated by a tightening of the contours; both are downthrown to the east (*Western Australian Petroleum Pty. Ltd.*)

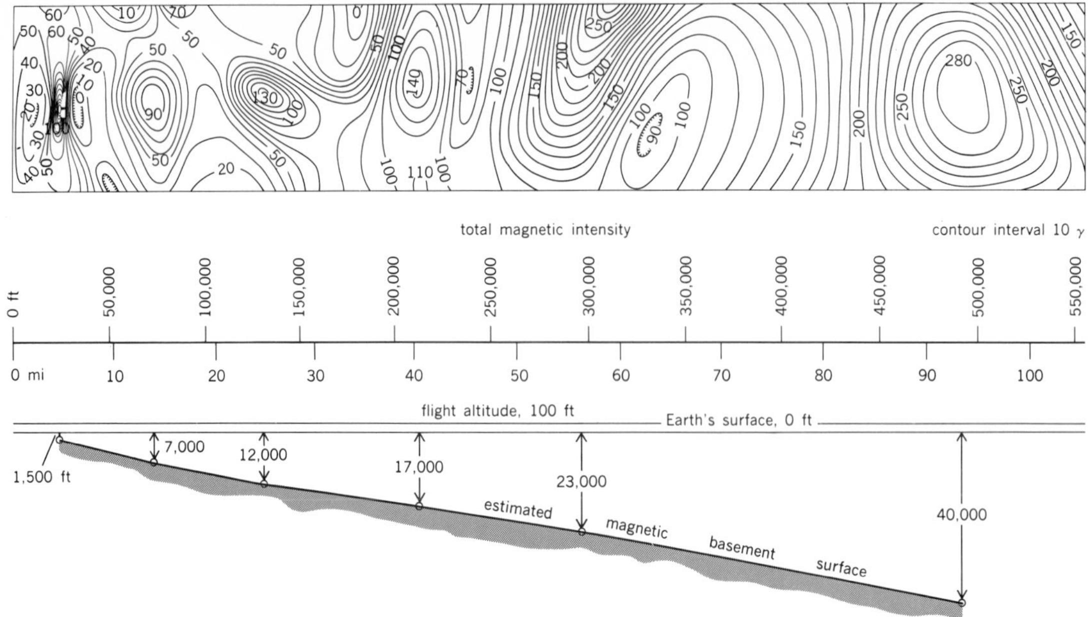

Fig. 2. Portion of a magnetic map (top) and interpretation (below). On the map, sharper features at the left indicate basement rocks are shallow, whereas broad anomalies to the right indicate deep basement. (*After S. L. Hammer, 4th World Petroleum Congress Proceedings, Rome, Sec. 1,1955*)

the feature sought governs both amount of detail and precision required in measurements. Many of the differences in geophysical methods derive from the different depths of interest. Engineering, mineral, and groundwater objectives are usually shallow, whereas petroleum and natural-gas accumulations are usually quite deep – 1 to 6 km.

Geophysical data usually are dominated by effects that are of no interest, and such effects must be either removed or ignored to detect and analyze the anomalous effects being sought. Noise caused by near-surface variations is especially apt to be large. The averaging of readings is the most common way of attenuating such noise.

The interpretation of geophysical data is almost always ambiguous. Since many different configurations of properties in the earth can give rise to the same data, it is necessary to select from among many possible explanations those that are most probable, and to select from among the probable explanations the few that are most optimistic from the point of view of achieving set objectives. Optimistic interpretations are desired in order to prevent overlooking a prospect. Additional measurements can test an interpretation hypothesis so that overoptimism will not be misleading in the final analysis. It is better to be wrong in interpreting a prospect than to miss one.

Geologic features affect the various types of measurements differently, hence more can be learned from several types of measurements than from any one alone. Combinations of methods are particularly useful in mining exploration. In petroleum surveys, magnetic, gravity, and seismic explorations are apt to be performed in that sequence, which is the order of relative cost. First, cheaper methods are used to narrow down the region to be explored by more expensive methods.

Usually the properties which geophysicists are able to measure are not directly related to objectives of interest, hence one must rely on some association between the measured properties and features of interest. Interpretation thus involves much inferential reasoning. For example, a serpentine dike often produces a magnetic anomaly, so that an ore that is often associated with a dike might be found by looking near magnetic anomalies, while realizing that most magnetic anomalies do not have associated ore bodies. Similarly, one might infer that the same factors that produced a particular structural feature also affected sedimentation, so locating such a structural feature may lead to the discovery of a stratigraphic accumulation.

A defect of relying on chains of inference is that one is apt to be wrong, although one is even more apt to be wrong using no inference. Geophysical exploration is thus justified by lessening the risk. The cost of geophysical work rarely exceeds a few percent of exploration costs. Cost effectiveness is a continuing concern. Different geophysical methods usually compete with each other along with nongeophysical methods, such as random drilling, for exploration funds.

Magnetic exploration. Rocks and ores containing magnetic minerals become magnetized by induction in the Earth's magnetic field so that their induced field adds to the Earth's field. Magnetic exploration involves mapping variations in the Earth's field with the objective of determining the location, size, and shape of such bodies. *See* GEOMAGNETISM.

The magnetic susceptibility of sedimentary rock is generally orders of magnitude less than that of igneous or metamorphic rock. Consequently, the major magnetic anomalies observed in surveys of sedimentary basins usually result from the underlying basement rocks. Determining the depths of the tops of magnetic bodies is thus a way of estimating the thickness of the sediments. *See* ROCK MAGNETISM.

Except for magnetite and a very few other minerals, mineral ores are only slightly magnetic. However, they are often associated with bodies such as dikes that have magnetic expression so that magnetic anomalies may be associated with minerals empirically. For example, placer gold is often concentrated in stream channels where magnetite is also concentrated.

Instrumentation. Several types of instruments are used for measuring variations in the Earth's magnetic field. Because the magnetic field is a vector quantity, both its magnitude and direction can be measured or, alternatively, components of the field can be measured in different directions. Often, however, only the magnitude of the total field is measured, and airborne and marine measurements usually are of the total field.

Optically pumped proton and fluxgate magnetometers are used extensively in magnetic exploration. Magnetometers used in exploration typically are accurate to 1 to 10 nanotesla (1 to 10 γ), an accuracy compatible with uncertainties in elevation for airborne measurements or noise background for measurements on land.

Field methods. Most magnetic surveys are made by aircraft, because a large area can be surveyed in a short time, and thus the cost per unit of area is kept very low. Hence, aeromagnetic surveying is especially adapted to reconnaissance, to locate those portions of large, unknown areas that contain the best exploration prospects so that future efforts can be concentrated there.

The spacing of measurements must be finer than the size of the anomaly of interest. Petroleum exploration is usually interested only in large anomalies, hence a survey for such objectives usually involves flying a series of parallel lines spaced 1 to 3 km apart, with tie lines every 10 to 15 km to assure that the data on adjacent lines can be related properly. The flight elevation is usually 300 to 1000 m. In mineral exploration, on the other hand, lines are usually located much closer—sometimes less than 100 m apart—and the flight elevation is as low as safety permits. Helicopters are sometimes used for mineral exploration.

The immediate product of aeromagnetic surveys is a graph of the magnetic field strength along lines of traverse. After adjustment, the data are usually compiled into maps on which magnetism is shown by contours (isogams, which connect points of equal magnetic field strength).

Sometimes, two magnetometers are towed by an aircraft so that they are displaced with respect to each other either horizontally or vertically or both. Thus, they detect magnetic anomalies at different times or in different magnitudes, whereas they simultaneously see variations in the Earth's magnetic field caused by diurnal effects and magnetic storms. This duplication increases confidence that anomalies represent local magnetic concentrations.

In ground mineral exploration, the magnetic field is measured at closely spaced stations. The effects of near-surface magnetic bodies is accentuated over measurements made in the air.

Magnetic surveying is often done in conjunction with other geophysical measurements, because it adds only a small increment to the cost and the added information often helps in resolving interpretational ambiguities. A magnetometer towed behind a marine seismic ship, for example, would distinguish between a volcanic plug and a salt diapir, features which might look nearly alike in the seismic data.

Data reduction and interpretation. The reduction of magnetic data is usually simple. Often, measurement conditions vary so little that the data can be interpreted directly, or else only network adjustments (to minimize differences at line intersections) are necessary because of location uncertainties or instrumental drift. In surveys of large areas, the variations in the Earth's overall magnetic field may be removed (magnetic latitude correction). In exceptional cases, such as in land surveys made over very irregular terrain, as in bottoms of canyons where some of the magnetic sources may be located above the instrument, reduction of the data can become very difficult.

Ths sharpness of a magnetic anomaly depends on the distance to the magnetic body responsible for the anomaly. Inasmuch as the depth of the magnetic body is often the information being sought, the shape of an anomaly is the most important aspect. Modeling is used to determine the magnetic field that would result from bodies of certain shapes and depths. The model anomalies are examined for a parameter of shape that is proportional to the depth (Fig. 3.) The shape parameter is measured on real anomalies and scaled to indicate how deep the body responsible for the anomaly lies. Such estimates are typically accurate to 10–20%, sometimes better.

Iterative modeling techniques are used in more detailed studies. The field indicated by the model is subtracted from the observed field to give an error field. Then the model is changed to obtain a new error field. This process is repeated until the error field is made sufficiently small. Inasmuch as many models can give the same magnetic field, additional constraints must be imposed to make the process meaningful. The significance of the conclusion depends greatly on how realistic the contraints are.

Gravity exploration. Gravity exploration is based on the law of universal gravitation: the gravitational force between two bodies varies in direct proportion to the product of their masses and in inverse proportion to the square of the distance between them.

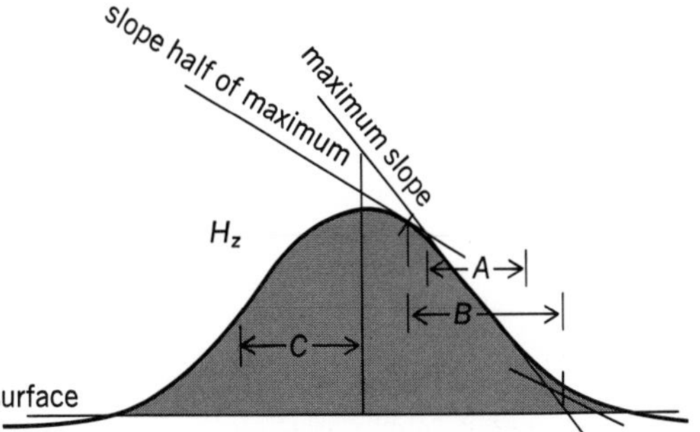

Fig. 3. Variation of measurements across an anomaly. Among the shape factors sometimes measured are *A*, the distance over which the slope is maximum; *B*, the distance between points where the slope is half of the maximum slope; *C*, half the width of the anomaly at half the peak magnitude. Such shape measurements multiplied by index factors give estimates of the depth of the body responsible for the anomaly.

Because the Earth's density varies from one location to another, the force of gravity varies from place to place. Gravity exploration is concerned with measuring these variations to deduce something about rock masses in the immediate vicinity. The variations attributable to such local causes are superimposed on the much larger field determined by the mass, size, and shape of the Earth as a whole. *See* TERRESTRIAL GRAVITATION.

Vertical density changes affect all stations the same and so do not produce easily measured effects. Gravity field variations are produced by lateral changes in density. Absolute density values are not involved, only changes in density. The product of the volume of a body and the difference between the density of the body and of the horizontally adjacent rocks is called the anomalous mass.

Gravity surveys are used more extensively for petroleum exploration than for metallic mineral prospecting. The size of ore bodies is generally small; therefore, the gravity effects are quite small and local despite the fact that there may be large density differences between the ore and its surroundings. Hence, gravity surveys to detect ore bodies have to be very accurate and very detailed. In petroleum prospecting, on the other hand, the greater dimensions of the features more than offset the fact that density differences are usually smaller.

Instrumentation. The most common gravity instrument in use is the gravity meter or gravimeter. The gravimeter basically consists of a mass suspended by springs comprising a balance scale. The gravimeter can be balanced at a given location. When the gravimeter is moved to another location, the minute changes in gravitational force, which require the instrument to be adjusted, can be measured. Hence the gravimeter measures differences in a gravity field from one location to another rather than the gravity field as a whole.

A gravimeter is essentially a very sensitive accelerometer, and extraneous accelerations affect the meter in the same way as the acceleration of gravity affects it. Typically, gravimeters read to an accuracy of 0.01 milligal, which amounts to 1/100,000,000 of the Earth's gravitational field. A mGal is 10^{-5} newton/kg. Anomalies of interest in petroleum exploration are often of the magnitude of 1/2 to 5 mGal.

Early gravity measurements were made with a torsion balance, a device that responds to the gradient of the gravity field. Absolute gravity measurements are made with the pendulum whose period varies with the gravity field. Occasionally two gravimeters are used, or one gravimeter is read successively at different elevations to measure the vertical gradient of gravity, an arrangement called a gradiometer. Other types of instruments for measuring gravity or gravity gradients are sometimes used.

Field surveys. Almost all gravity measurements are relative measurements; differences between locations are measured although the absolute values remain unknown. Ordinarily, the distance between the stations should be smaller than one-half the depth of the structures being studied.

Gravity surveys on land usually involve measurements at discrete station locations. Such stations are spaced as close as a few meters apart in some mining or archeological surveys, about 1/2 km for petroleum exploration, and even 10 to 20 kms for some regional geology studies. While one would like to have gravity values on a uniform grid, often this is not convenient, and so stations are located on traverses around loops. For petroleum exploration, the gravimeter might be read every 1/2 km around loops of about 6 by 10 km.

The gravity field is very sensitive to elevation. An elevation difference of 3 m represents a difference in gravity of about 1 mGal. Hence, elevation has to be known very accurately, and the most critical part of a gravity survey often is determining elevations to sufficient accuracy.

Gravity measurements can be made by ships at sea. Usually the instrument is located on a gyrostabilized platform which holds the meter as nearly level as possible. The limiting factor in shipboard gravity data is usually the uncertain velocity of the meter, especially east to west, since the ship is moving. The velocity of a ship traveling east adds to the velocity because of the rotation of the Earth. Consequently, centrifugal force on the meter increases and the observed gravity value decreases (Eötvos effect).

Gravity measurements are also sometimes made by lowering a gravimeter to the ocean floor and balancing and reading the meter remotely. Gravity measurements have been made by aircraft using techniques like those used at sea, but are not sufficiently accurate to be useful for exploration.

Specialized gravimeters are used to make measurements in boreholes. The main difference between gravity readings at two depths in a borehole is produced by the mass of the slab of earth between the two depths; this mass pulls downward on the meter at the upper level and upward at the lower level. Thus the difference in readings depends on the density of this slab. In sedimentary rocks, the borehole gravimeter is used primarily for measuring porosity.

Data reduction. Gravity measurements have to be corrected for factors other than the distribution of the Earth's mass. Meters drift or change their reading gradually because of various reasons. The Sun and Moon pull on the meter in different directions during the course of a day. The gravitational force varies with the elevation of the gravimeter both because at greater elevations the distance from the Earth's center increases (free-air correction) and because mass exists between the meter and the reference elevation, which is usually mean sea level (Bouguer correction). Gravity varies with latitude because the Earth's equatorial radius exceeds its polar radius and because centrifugal force resulting from the Earth's rotation varies with latitude. Nearby terrain affects a gravimeter; mountains exert an upward pull, valleys cause a deficit of downward pull.

Gravity measurements that have been corrected for all of these effects are called Bouguer anomaly values. They therefore represent the effects of masses within the earth, that is, effects for which corrections have not been made. Most gravity maps display contours (isogals) of Bouguer anomaly values. Sometimes the correction for mass intervening between the meter elevation and the reference elevation is not made; the results are then called free-air anomaly values.

Data interpretation. The most important part of gravity interpretation is locating anomalies that can be attributed to mass concentrations being sought, isolating these from other effects. Separating the main part of the gravitational field, which is not of interest (the "regional"), from the parts attributed to local masses, the "residuals," is called residualizing (Fig. 1).

Many techniques for gravity data analysis are similar to those used in analyzing magnetic data. Shape parameters are used extensively to determine the depth of the mass's center. Another widely used technique is model fitting: a model of an assumed feature is made, its gravity effects are calculated, and the model is compared with measurements of the mass.

Continuation is a process by which one calculates from measurements of a field over one surface what values the field would have over another surface. A field can be continued if there is no anomalous mass between the surfaces. Continuing the field to a lower surface produces sharper anomalies as the anomalous mass is approached. However, if one carries the process too far, instability occurs when the anomalous mass is reached. The technique, however, is very sensitive to measurement uncertainties and often is not practical with real data.

The anomalous mass can also be calculated by integrating the residual anomaly. Anomalous mass calculation is useful in mining exploration in determining how much ore is present when the anomalous mass either is the ore or is related to the ore quantitatively. The accuracy of the calculation depends on the accuracy with which the residual has been defined. This calculation is not usually valuable in petroleum exploration.

Seismic exploration. Seismic exploration is the predominant geophysical activity. Seismic waves are generated by one of several types of energy sources and detected by arrays of sensitive devices called geophones or hydrophones. The most common measurement made is of the travel times of seismic waves, although attention is being directed increasingly to the amplitude of seismic waves or changes in their frequency content or wave shape.

Seismic exploration is divided into two major classes, refraction and reflection. Classification depends on whether the predominant portion of wave travel is horizontal or vertical, respectively.

Principles of seismic waves. A change in mechanical stress produces a strain wave that radiates outward as a seismic wave, because of elastic relationships. The radiating seismic waves are like those that result from earthquakes, though much weaker. Most seismic exploration involves the analysis of compressional waves in which particles move in the direction of wave travel, analogous to sound waves in air. Shear waves are occasionally studied, but most exploration sources do not generate very much shear energy. Surface waves, especially Rayleigh waves, are also generated, but these are mainly a nuisance because they do not penetrate far enough into the earth to carry much useful information. Recording techniques are designed to discriminate against them. *See* SEISMOLOGY.

The amplitude of a seismic wave reflected at an interface depends on the elastic properties, often expressed in terms of seismic velocity and density on either side of the interface. When the direction in which the wave is traveling is perpendicular to the interface, the ratio of the amplitudes of reflected and incident seismic waves is given by the reflection coefficient R, as shown in Eq. (1), where

$$R=\Delta(\rho V)/2(\overline{\rho V}) \qquad (1)$$

$\Delta(\rho V)$ is the change in the product of velocity and density and $(\overline{\rho V})$ is the average of the product of velocity and density on opposite sides of the interface. The ratio of the energy of the reflected to incident waves is R^2.

Seismic waves are bent when they pass through interfaces, and Snell's law holds, shown in Eq. (2),

$$(\sin\sigma_1)/V_1=(\sin\sigma_2)/V_2 \qquad (2)$$

where σ_i is the angle between a wavefront and the interface in the ith medium where the velocity is V_i. Because velocity ordinarily increases with depth, seismic-ray paths become curved with concave-upward curvature.

The resolving power with seismic waves depends inversely on their wavelength λ and is often thought of as of the order of $\lambda/4$. The wavelength is often expressed in terms of the wave's velocity and frequency f: $\lambda=V/f$. Most seismic work involves frequencies from 20 to 50 Hz, and most rocks have velocities from 1500 to 6000 m/s, so that wavelengths range from 30 to 300 m. Usually, the frequency becomes lower and the velocity higher as depth in the earth increases, so that wavelength increases and resolving power decreases. Very shallow, high-resolution work involves frequencies higher than those cited above, and long-distance refraction (and earthquakes) involve lower frequencies.

Refraction exploration. Refraction seismic exploration involves rocks characterized by high seismic velocity. Wavefronts are bent at interfaces (Fig. 4) so that appreciable energy travels in high-velocity members and arrives at detectors some distance from the source before energy that has traveled in overlying lower-velocity members. Differences in arrival time at different distances from the energy source yield information on the velocity and attitude (dip) of the high-velocity member.

A variant of refraction seismic exploration is the search for high-velocity masses in an otherwise low-velocity section, by looking for regions where seismic waves arrive earlier than expected. Such arrivals, called leads, were especially useful in locating salt domes in Louisiana, Texas, Mexico, and Germany in the late 1920s and 1930s.

Refraction seismic techniques are used in engineering geophysics and mining and groundwater studies to map bedrock under unconsolidated overburden, with objectives such as foundation information or locating buried stream channels in which heavy minerals might be concentrated or where water might accumulate. Refraction techniques are also used in petroleum exploration and for crustal studies.

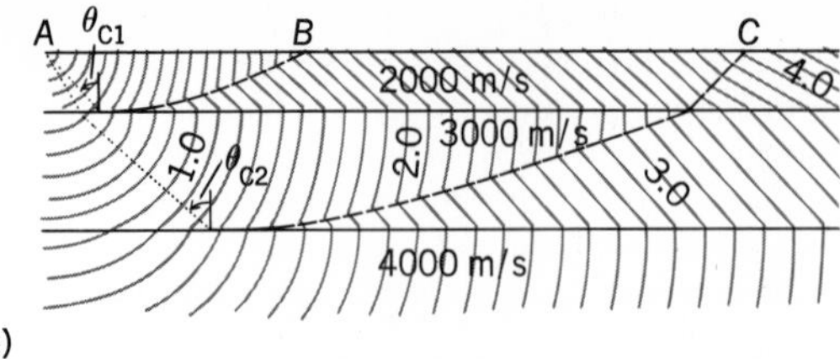

(a)

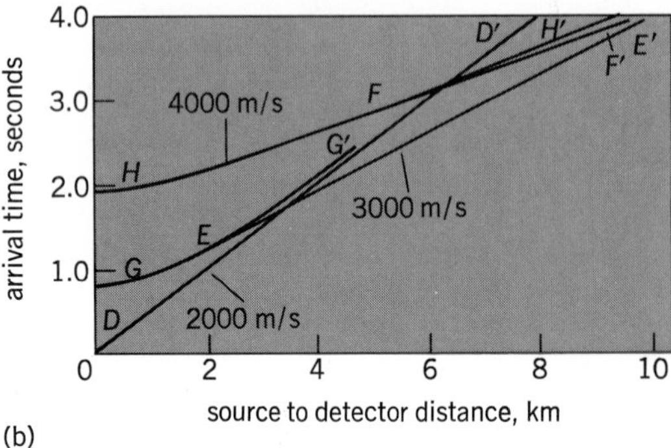

(b)

Fig. 4. Refractive seismic exploration data. (*a*) Section through model of layered earth. Curves (wavefronts) indicate location of seismic energy at successive times after a shot at *A*. Beyond *B*, energy traveling in the second layer arrives first; beyond *C*, that in the third layer. (*b*) Arrival time as a function of source-to-detector distance. *DD′*, direct wave; *EE′*, refracted wave (head wave) in the second layer; *FF′*, refracted wave in the third layer; *GG′*, reflection from interface between first and second layers; *HH′*, reflection from that between second and third layers.

Reflection exploration. Seismic-wave energy partially reflects from interfaces where velocity or density changes. The measurement of the arrival times of reflected waves (Fig. 5) thus permits mapping the interfaces that form the boundaries between different kinds of rock. This, the predominant geophysical exploration method, can be thought of as similar to echo sounding. In Fig. 5*a*, a seismic source *S* generates seismic energy which is received at detectors located at intervals from *A* to *B*. The distance to the reflector *RR′* can be obtained from the arrival time of the reflection if the velocity is known. If the reflector dips as shown, the reflection will arrive sooner at *B* than at *A*; the difference in arrival times is a measure of the amount of dip. The angles between ray paths and perpendiculars to the reflector (*i*, *i*) are equal at any reflecting point. The image point *I* is used as an aid in constructing the diagram.

GEOPHYSICAL EXPLORATION

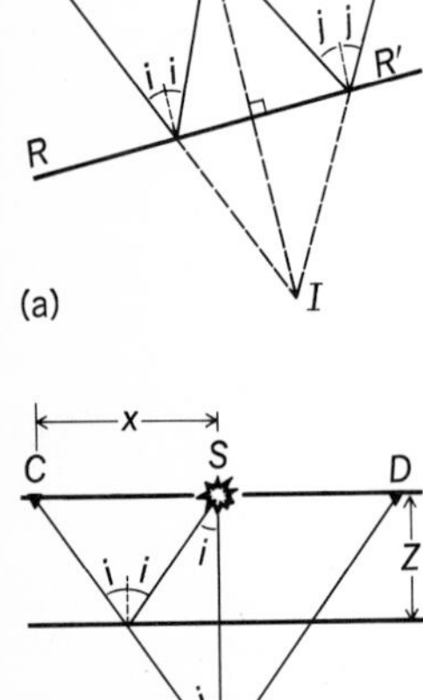

(a) (b)

Fig. 5. Measuring reflected seismic wave energy. (*a*) Dipping reflector. (*b*) Flat reflector.

For a flat reflector (Fig. 5*b*) and constant velocity V, the arrival time at detector C is t_c and the arrival time at a detector at the source S is $t_0=2Z/V$. From the Pythagorean theorem for the triangle *CSI*, Eqs. (3) and (4) are obtained. Similar relationships can be used for nonflat reflectors or nonconstant velocity to yield velocity information.

$$(Vt_c)^2=(Vt_0)^2+X^2 \tag{3}$$

$$V=X(t_c^{\,2}-t_0^{\,2})^{1/2} \tag{4}$$

Usually a number of detector groups are used, and the arrival of reflected waves is characterized by coherency between the outputs. Thus, if all of the detectors in a line move in a systematic way, a seismic wave probably passed. Multiple detectors make it possible to detect coherent waves in the presence of a high noise level and also to measure distinguishing features of the waves.

Instrumentation. Detectors of seismic energy on land (geophones) are predominantly electromechanical devices. A coil moving in a uniform magnetic field generates a voltage proportional to the velocity of the motion. Often the coil is an inertial element that tends to remain at rest as the case and magnetic field move in response to a seismic wave. Usually the coil has only one degree of freedom and is used so it will be sensitive to vertical motion only. Three mutually perpendicular elements in a three-component detector are sometimes used to determine the direction from which waves come or to distinguish the type of waves (compressional, shear, Rayleigh, and such).

Detectors in water are usually piezoelectric, and pressure changes produced as a seismic wave passes distort a ceramic element and induce a voltage between its surfaces. Such detectors are not directionally sensitive.

Detectors are usually arranged in groups (arrays) spread over a distance and connected electrically so that, in effect, the entire group acts as a single large detector. Such an arrangement discriminates against seismic waves traveling in certain directions. Thus a wave traveling horizontally reaches different detectors in the group at different times so that wave peaks and troughs tend to cancel, whereas a wave traveling vertically affects each detector at the same time so that the effects add.

The signal from the detectors is transmitted to recording equipment over a cable or streamer and is amplified and recorded. The output level from a geophone varies tremendously during a recording. Seismic recording systems are linear over ranges of 100 dB or more. Seismic amplifiers employ various schemes to compress the range of seismic signals without losing amplitude information. They also incorporate adjustable filters to permit discriminating on the basis of frequency.

Most recording systems are multichannel. In 1975, 24 or 48 channels were most common in petroleum exploration, but some 1024 channel systems were in use. Most engineering, mining, and groundwater applications employ one to six channels. The signals are usually digitized and recorded on magnetic tape so that computer analysis can be carried out subsequently. The signals are also displayed as a function of arrival time, on recorded paper. Sometimes signals are displayed on small cathode-ray tubes, and the time from source to first-energy arrival may be timed automatically and displayed on a counter. The latter type is used especially in engineering, groundwater, and mineral-refraction work.

Many sources are used to generate seismic energy. The classical energy source is an explosion in a borehole drilled for the purpose, and solid explosives continue to be used extensively for work on land and in marshes. The explosion of a gas mixture in a closed chamber, a dropped weight, a hammer striking a steel plate, and other sources of impulsive energy are used in land work. An air gun, which introduces a pocket of high-pressure air into the water, is the most common energy source in marine work. Other marine energy sources involve the explosion of gases in a closed chamber, a pocket of high-pressure steam introduced into the water, the discharge of an electrical arc, and the sudden mechanical separation of two plates (imploder).

An oscillatory mechanical source is also used, especially on land. Such a source introduces a long wavetrain so that individual reflection events cannot be resolved without subsequent processing (correlation with the input wavetrain), which, in effect, compresses the long wave train and produces essentially the same result as an impulsive source.

Field techniques. Most petroleum exploration geophysics is carried out by survey methods. These surveys are often run parallel to each other at right angles to the geological strike with occasional perpendicular tie lines, often run on a regular grid. Long lines, placed many kilometers apart, are sometimes run for regional information, but lines are often concentrated in regions in which anomalies have been detected by previous geophysical work. Most seismic work has the objective of mapping interfaces continuously along the seismic lines to map the geological structure.

Geophone groups are spaced 50 to 100 m apart with 24 to 48 adjacent groups of 6 to 24 geophones each being used for each recording. The source is sometimes located at the center of the active groups ("split spread"), sometimes at one end ("end-on spread"), and sometimes at a different location.

Following a recording, the layouts and sources are advanced down the line by half the distance over which the geophone groups are disposed (the "spread length") for continuous coverage or by some smaller multiple of the group interval for redundant coverage (Fig. 6). The active geophone groups are advanced usually by electrical switching rather than physical movement of the geophones and cables. Sometimes some geophone groups are laid out perpendicular to the seismic line on a cross-line to measure components of dip perpendicular to the line.

With surface sources such as vibrators, gas exploders, and weight droppers, source trucks stop to deliver energy into the ground for a recording. They then move forward a few meters and repeat, and so on down the line. Several source trucks located a few meters apart are often used simultaneously; they are synchronized by radio from the recording unit. Often several individual recordings are summed together to make one field record. A typical seismic crew can survey from 1 to 10 km per day.

Small marine operations, often called profiling, consist of an energy source and a short streamer containing a number of hydrophones and feeding a single-channel recorder. Larger marine operations (Fig. 7) involve ships 60 m or more in length towing a streamer 2 to 3 km long with 24 or 48 groups of hydrophones spaced along the streamer. The streamer is typically towed at a depth of 10 to 15 m. An energy source is towed near the ship. Recordings are made as the ship is continuously underway at a speed of about 6 knots (3 m/s).

Marine operations require precise knowledge of position. Observations of navigation satellites are made whenever such a satellite passes overhead, yielding fixes every couple of hours with an uncertainty of the order of 50 m. Locations between satellite fixes are obtained by such techniques as Doppler-sonar navigation, various radio-location methods, and inertial navigation. Most marine seismic operations incorporate an integrated navigation system in which a number of sensors feed a computer that uses their data synergetically to determine the most probable location.

Data reduction and processing. Seismic data are corrected for elevation and near-surface variations on the basis of survey data and observations of the travel time of the first energy from the source to reach the detectors, which usually involve either travel in a direct path or in shallow refractors.

If data are not processed by computers, the arrival times of energy are plotted on graphs for use in interpretation. It is essential to make certain that the same point on the same reflection or refraction is always picked, because interpretation depends much more heavily on differences between travel times than on the magnitude of travel times.

If data are processed by computers, the first step is often editing, wherein data are merged with identifying data, rearranged, checked for being either dead or wild (with bad values sometimes replaced with interpolated values), time shifted in accordance with elevation and near-surface corrections that have been determined in the field, scaled, and so on.

Following the editing, different processing sequences may follow, including (1) filtering

GEOPHYSICAL EXPLORATION

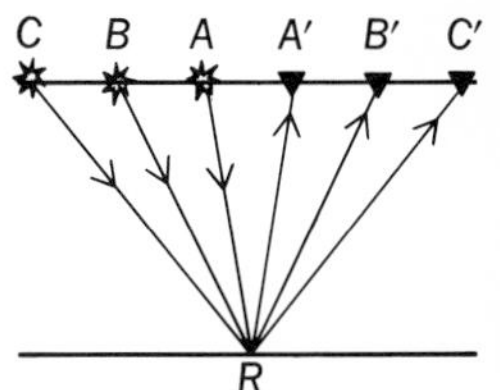

Fig. 6. Reflection of seismic energy obtained from reflection point R (common-depth point) by positioning a source at A and a detector at A'. The same point is involved with a source at B and detector at B', or C and C'. The redundancy in measuring reflections from the same point many times permits sorting out different kinds of waves.

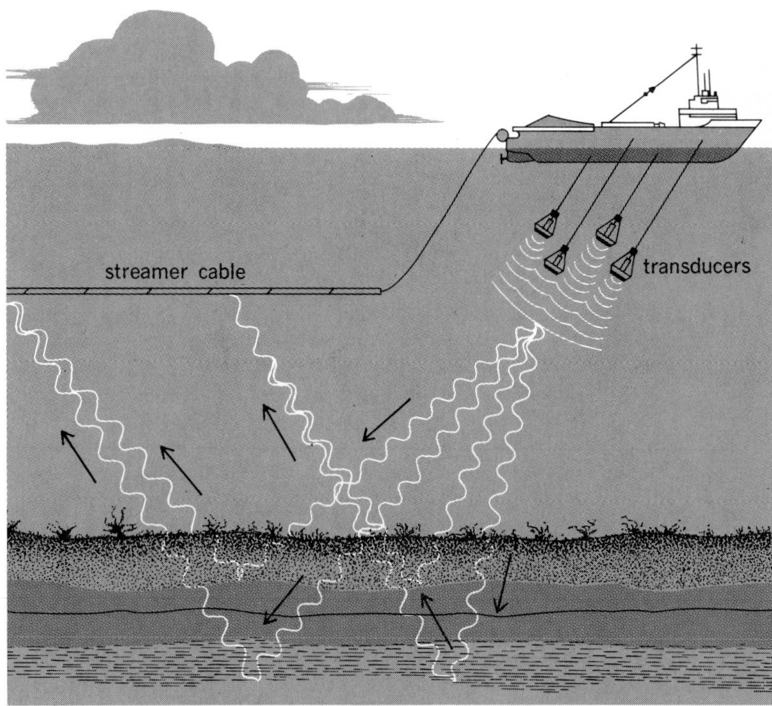

Fig. 7. Seismic surveying at sea is done by towing a long streamer containing detectors that sense seismic energy. Seismic energy is generated by several transducers that are also towed from the ship.

(deconvolution) to remove undesired natural filter effects, trace-to-trace variations, variations in the strength or wave shape of the source, and so on; (2) grouping according to common-depth point (Fig. 6) or some other arrangement; (3) analyzing to see what velocity values will maximize coherency as a function of source-to-detector distance; (4) statistically analyzing to see what trace shifts will maximize coherency; (5) trace-shifting according to the results of steps 3 or 4; (6) stacking by adding together a number of individual traces; (7) automigrating by rearranging and combining data elements in order to position reflection events more nearly under the surface locations where the appropriate reflecting surface is located; (8) another filtering; and (9) displaying of the data.

A number of data outputs and displays are made during the processing sequence. These are analyzed for control and to determine parameters for subsequent processing steps.

Data interpretation. The travel times of seismic reflections are usually measured from record section displays (Fig. 8), which result from processing. Appropriate allowance ("migration") must be made because reflections from dipping reflectors appear at locations downdip from the reflecting points. Allowance must also be made for variations in seismic velocity, both vertically and horizontally. Seismic events other than reflections must be identified and explained.

In petroleum exploration the objective is usually to find traps, places in which porous formations are high relative to their surroundings and in which the overlying formation is impermeable. If oil or gas, which are lighter than water, were present, they would float on top of the water and accumulate in the pores in the rock at the trap. Seismic exploration determines the geometry, hence where traps might be located. However, one cannot usually tell from seismic data whether oil or gas was ever generated, whether the rocks have porosity, whether overlying rocks are impermeable, or whether oil or gas might have escaped or been destroyed, even if they were present at one time.

To reconstruct the geologic history, often several reflections at different depths are mapped, and attempts are made to reconstruct where the deeper reflectors were at the time of deposition of shallower rocks.

After some experience has been developed in an area, features in the seismic data that distinguish certain reflectors or certain types of structure often can be recognized. Seismic velocity measurements are helpful here.

In relatively unconsolidated sediments and in some other circumstances, gas and oil containing considerable gas may lower the seismic velocity or rock density or both sufficiently to produce a distinctive reflection, usually evidenced by strong amplitude (a "bright spot") and other distinguishing features (Fig. 9). This is often called direct detection, although such anomalies do not necessarily correspond to accumulations of commercial importance. Coal and peat beds are also characterized by reflections of strong amplitude.

Electrical and electromagnetic exploration. Variations in the conductivity or capacitance of rocks form the basis of a variety of electrical and electromagnetic exploration methods, which are used primarily in metallic mineral prospecting. Both natural and induced electrical currents are measured. Direct currents and low-frequency alternating currents are measured in ground surveys, and ground and airborne electromagnetic surveys involving the lower radio frequencies are made.

Some mineral deposits give rise to spontaneous earth currents. The attendant "self-potential" field can often be mapped so that the source of the self-potential can be found.

Natural currents in the earth, called telluric currents, affect large areas. They are believed to be related to ion-current circulations in the upper atmosphere. The current density of telluric cur-

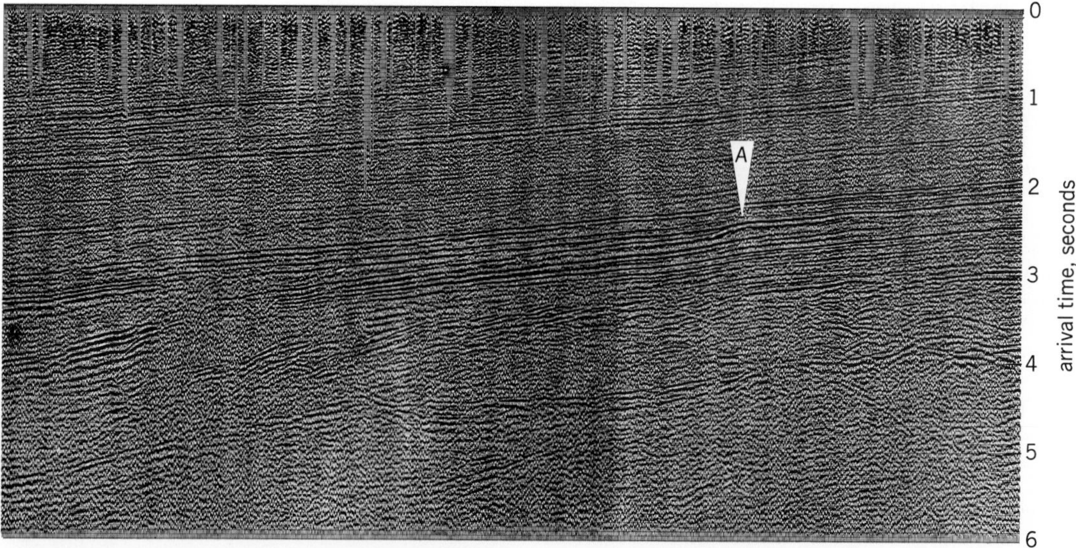

Fig. 8. Seismic record section in East Texas after processing. The reflection event at *A* is attributed to the Edwards Reef, which formed a barrier at the time the adjacent sediments were deposited. Downdip to the left of the reef formations the Woodbine sands can be seen pinching out in the updip direction. These are productive of oil and gas in this area. (*Seiscom Delta Inc.*)

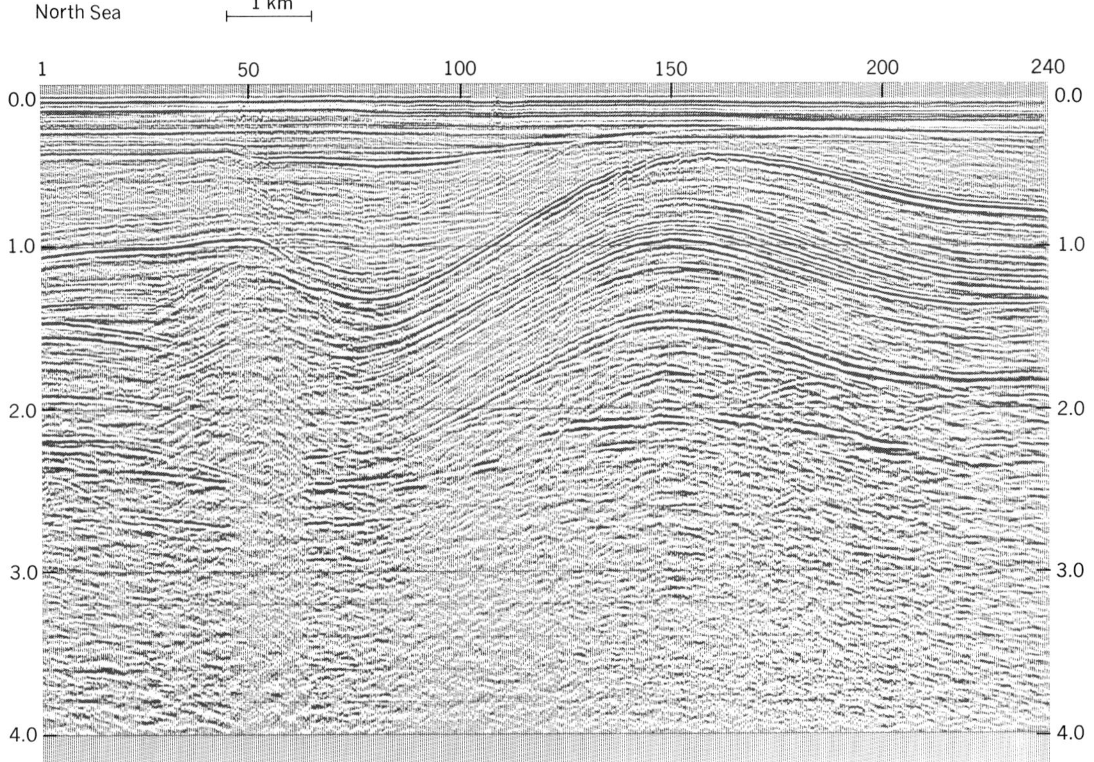

Fig. 9. Seismic record section from the North Sea. The layer from about 2 to 2.3 s at the right edge is composed primarily of salt that has undergone plastic flow as part of the folding of the rock formations. The upfolding (an anticline) is not productive here, but the formations at the left just below 1 s are believed to contain gas. This is a trap bounded at the right by a down-to-the-right fault. (*Seiscom Delta Inc.*)

rents varies with rock conductivity. Comparisons are made between readings at various locations, and the readings are observed simultaneously at a reference location.

Changes in electrical current flows give rise to associated magnetic fields, and the converse is also true, according to Maxwell's equations. Natural currents are somewhat periodic. Magnetotellurics involves the simultaneous measurement of natural electrical and magnetic variations from which the variation of conductivity with depth can be determined.

Certain mineral ores store energy as a result of current flow and, after the current is stopped, transient electrical currents flow. This phenomenon is called induced polarization. The storage mechanism is not only capacitative, but the exact mechanism is not understood. Observations of the rate of decay of these transient currents are studied in time-domain methods.

Alternating currents tend to flow along the surface of conductors rather than in their interior. The thickness which contains most of the current is called the skin depth. The skin depth, in meters, is given by $(2/\sigma\mu\omega)^{1/2}$, where σ is the conductivity in mhos per meter, μ is permeability in henrys per meter and ω is angular frequency in radians per second.

Since the skin depth becomes greater as frequency becomes lower, measurements at different frequencies give information on the variation of conductivity with depth. Methods in which apparent resistivity is determined as a function of frequency are called frequency-domain methods.

Direct-current and low-frequency alternating-current ground surveys are carried out with a pair of current electrodes, by which electrical current is introduced into the ground, and a pair of potential electrodes across which the voltage is measured. The equipment is often simple, consisting essentially of a source of electrical power (battery or generator), electrodes and connecting wires, ammeter, and voltmeter. A key problem here is providing equipment that will generate enough electrical or electromagnetic energy in the ground but is reasonably portable. The voltages are usually measured in a potentiometric arrangement so that current flow in the measuring circuit does not produce distortions. Various arrangements of electrodes are used. The depth of current penetration depends on the geometry of disposition of instruments, on the frequency used, and on the conductivity distribution. There are two basic types of measurement: (1) electrical sounding or electrical drilling, wherein apparent resistivity is measured as dimensions between the electrodes are increased—these measurements depend mainly on the variation of electrical properties with depth; (2) electrical profiling, in which variations are measured as the electrode array is moved from location to location.

Electromagnetic methods generally involve a transmitting coil, which is excited at a suitable frequency, and a receiving coil, which measures one or more elements of the electromagnetic field at a number of observation points. The receiving coil is usually oriented in a way that minimizes its direct coupling to the transmitter, and the residual effects are then caused by the currents that have been induced in the ground. A multitude of configurations of transmitting and receiving antennae are used in electromagnetic methods, both in ground surface and airborne surveys. Among the methods used are long linear antennae, large rectangular loops, and small portable loops, which are varied both in orientation and location. In airborne surveys, the transmitting and receiving coils and all associated gear are carried in an aircraft, which normally flies as close to the ground as is safe. Airborne surveys often include multisensors, which may record simultaneously electromagnetic, magnetic, and radioactivity data along with altitude and photographic data. Sometimes several types of electromagnetic configurations or frequencies are used.

The effective penetration of most of the electromagnetic methods into the earth is not exceptionally great, but they are used extensively in searching for mineral ores within about 100 m of the surface. Electrical methods are effective in exploring for groundwater and in mapping bedrock, as at dam sites. They are used also for detecting the position of buried pipelines and in land-mine detection and other military operations.

Radioactivity exploration. Natural radiation from the earth, especially of gamma rays, is measured both in land surveys and airborne surveys. Natural types of radiation are usually absorbed by a few feet of soil cover, so that the observation is often of diffuse equilibrium radiation. The principal radioactive elements are uranium, thorium, and potassium; radioactive exploration has been used primarily in the search for uranium and other ores, such as columbium, which are often associated with them. Radioactive methods are sometimes used in the search for potash deposits. The Geiger counter and scintillation counter are instruments generally used to detect and measure the radiation.

Remote sensing. Measurements of natural and induced electromagnetic radiation made from high-flying aircraft and earth satellites are referred to collectively as remote sensing. This comprises both the observation of natural radiation in various spectral bands, including both visible and infrared radiation, such as by photography and measurements of the reflectivity of infrared and radar radiation.

Well logging. A variety of types of geophysical measurements are made in boreholes, including self-potential, electrical conductivity, velocity of seismic waves, natural and induced radioactivity, and temperature variations. Borehole logging is used extensively in petroleum exploration to determine the characteristics of the rocks which the borehole has penetrated, and to a lesser extent in mineral exploration.

Measurements in boreholes are sometimes used in combination with surface methods, as by putting some electrodes in the borehole and some on the surface in electrical exploration, or by putting a seismic detector in the borehole and the energy source on the surface.

[R. E. SHERIFF]

Bibliography: M. B. Dobrin, *Introduction to Geophysical Prospecting*, 3d ed., 1976; F. S. Grant and G. F. West, *Interpretation Theory in Applied Geophysics*, 1965; D. H. Griffiths and R. F. King, *Applied Geophysics for Engineers and Geologists*, 1965; L. L. Nettleton, *Gravity and Magnetic Prospecting*, 1975; D. S. Parasnis *Principles of Applied Geophysics*, 2d ed., 1972; R. E. Sheriff, *Encyclopedic Dictionary of Exploration Geophysics*, 1973; W. M. Telford et al., *Applied Geophysics*, 1975.

Geophysics

Those branches of earth science in which the principles and practices of physics are used to study the Earth. Geophysics is considered by some to be a branch of geology, by others to be of equal rank. It is distinguished from the other earth sciences largely by its use of instruments to make direct or indirect measurements of the parts of the Earth being studied, in contrast to the more direct observations which are typical of geology. *See* GEOLOGY.

SUBDIVISIONS OF GEOPHYSICAL SCIENCE

Geophysics consists of several principal fields plus parallel and subsidiary divisions. These are commonly considered to include plutology, with geodesy, geothermometry, seismology, and tectonophysics as subdivisions; hydrospheric studies, mostly hydrology (groundwater studies) and oceanography; atmospheric studies, meteorology and aeronomy; and several fields of geophysics which overlap one another, including geomagnetism and geoelectricity, geochronology, geocosmogony, and geophysical exploration and prospecting. Planetary sciences, the study of the planets and satellites aside from the Earth, are usually considered a branch of geophysics because the techniques used for such studies are, until man can land on each body, instrumental rather than direct observation.

Plutology. This is a general term covering geophysical methods of studying the solid part of the Earth. The field has the following four major subdivisions.

Geodesy. The science of the shape and size of the Earth is geodesy. This includes consideration of the Earth's mass distribution as determined by measurements of the gravitational-centrifugal field. Because of man's interest in space travel, geodesy is concerned also with the distribution of the gravitational field above the Earth's surface out to the limit of the Earth's detectable effect. *See* CONTINENT FORMATION; GEODESY; ROCK MAGNETISM; TERRESTRIAL GRAVITATION.

Geothermometry. This is the science of the Earth's heat. It includes the study of temperature variation and of heat generation, conduction, and loss in the Earth, and their effects on the materials of which the Earth is composed. Volcanology, the science of volcanoes, uses some of the principles and techniques of geothermometry. *See* GEOLOGIC THERMOMETRY; HIGH-PRESSURE PHENOMENA.

Seismology. The science of earthquakes and other ground vibrations is the field of seismology. Seismology has made particularly important contributions to man's knowledge of the Earth's interior. Study of the times of passage of seismic waves through the Earth gives information on the distribution of different types of rock. *See* EARTH, INTERIOR OF; SEISMOGRAPH; SEISMOLOGY.

Tectonophysics. Sometimes called geodynamics, this is the science of the deformation of rocks. It consists of tectonics, the study of the broader structural features of the Earth and their causes, as in mountain building; and rock mechanics, the measurement of the strength and related physical properties of rocks. *See* OROGENY; ROCK MECHANICS; TECTONOPHYSICS.

Hydrospheric studies. Geophysical study of the hydrosphere has two main branches, hydrology and oceanography.

Hydrology. This groundwater science also includes glaciology, the study of groundwater in the form of snow and ice. *See* GLACIOLOGY; GROUNDWATER.

Oceanography. The scientific study of the oceans includes the study of the shape and structure of the ocean basins; the physical and chemical properties of sea water; ocean currents, waves, and tides; thermodynamics of the oceans; and the relation of these to the organisms which live in the sea. *See* MARINE GEOLOGY; OCEANS.

Atmospheric studies. Because of extending interest from the face of the Earth toward outer space, aeronomy has come to be recognized as a distinct science separate from the science of meteorology.

Meteorology. The science of the Earth's atmosphere yields meteorological information not only used in climatology and weather prediction, but also important in understanding the problems of aircraft flight and air pollution.

Aeronomy. This field is concerned with the phenomena of the upper atmosphere above 100 km, where its electrical behavior is important because the air is strongly ionized. There, where the mean free path of an atom or electron is long, the physical behavior of the material is controlled at least as much by its electrical properties as by its density and other mass properties. The outer boundary of the atmosphere is not a distinct surface; there is instead a gradual transition to the relative emptiness of interplanetary space. In this transition zone there is continuous interaction between the matter of the Earth's atmosphere and the radiations and particles arriving from outside. The Sun, the source of most of these, is a major object of geophysical investigation. The Earth moves within the outer fringe of the Sun's atmosphere. Aeronomy is much concerned with the behavior of the Sun, especially as it influences conditions on the Earth. Understanding of solar phenomena is essential for the safe flight of man beyond the lower reaches of the Earth's atmosphere.

Planetary sciences. Astronomical data on the planets and their satellites come from photographs and analysis of the spectra of their reflected light. Radar reflections give additional information. Observing instruments landed on the Moon and nearby planets or flown past their surfaces radio back data of many types. Although these methods may ultimately be replaced by exploration parties and manned observatories, much of the knowledge of the more distant members of the solar system will depend greatly upon such indirect observations for a long time. *See* MOON.

Overlapping fields of geophysics. In addition to the regional subdivisions, there are several other fields of geophysics which overlap and concern all the others.

Geomagnetism, geoelectricity. The science of the Earth's magnetic field and magnetic properties is termed geomagnetism. Geoelectricity is the science of electrical currents in the Earth and of its electrical properties. Geomagnetism and geoelectricity are closely related because the magnetic field is due largely to electrical currents in the solid earth and atmosphere. *See* GEOMAGNETISM.

Geochronology. This field deals with the dating of events in the Earth's history. The principal technique used is based on radioactive disintegration. The proportions of parent to daughter elements in a mineral or rock are a measure of the age of the material. Other methods depend on the red shift of the spectrum of distant stars, the rate of recession of the Moon, and the rates of erosion and sedimentation. *See* DATING METHODS; GEOCHRONOMETRY.

Geocosmogony. This is the study of the origin of the Earth. The many hypotheses proposed fall into two groups, those which postulate that the Earth is primarily an aggregate of once smaller particles, and those which claim that is is largely a fragment of a larger body. Current speculation favors the former theory. Geocosmogony is intimately linked with the origin of the solar system and our galaxy. Many lines of evidence suggest that the formation of the Earth was a typical minor event in the evolution of the Milky Way or of the universe as a whole, occurring $5-8 \times 10^9$ years ago.

Exploration and prospecting. Geophysical techniques are widely used not only to study the general structure of the Earth but also in prospecting for petroleum, mineral deposits, and groundwater and in mapping the sites of highways, dams, and other structures. Seismic methods are the most widely used, but electrical, electromagnetic, gravity, magnetic, and radioactivity surveying methods are also well developed. Many geophysical surveys can be made by lowering measuring apparatus into bore holes. *See* GEOPHYSICAL EXPLORATION.

Technical literature. Some of the principal geophysical publications largely in English are the *Journal of Geophysical Research* of the American Geophysical Union; *Geophysics*, published by the Society of Exploration Geophysicists; the *Bulletin of the Seismological Society of America*; the *Geophysical Journal of the Royal Astronomical Society*; the *Journal of Meteorology* (now the *Journal of Atmospheric Sciences* and the *Journal of Applied Meteorology*); the *Bulletin of the American Meteorological Society;* the *Quarterly Journal of the Royal Meteorological Society;* the *Journal of Atmospheric and Terrestrial Physics*; and the *Bulletin of the Earthquake Research Institute of Tokyo*. Geophysical papers are commonly found also in the *Bulletin of the Geological Society of America*; *Geochemica and Cosmochemica Acta*, the journal of the Geochemical Society; the *Transactions of the American Institute of Mining, Metallurgical, and Petroleum Engineers*; and the *Proceedings of the National Academy of Sciences of the United States of America.*

[BENJAMIN F. HOWELL, JR.]

PROGRAMS OF GEOPHYSICS

A fundamental feature of geophysics is the necessity to record, collect, exchange, collate, analyze, and synthesize large quantities of data from many sites over extended periods of time. This is a complex and difficult task requiring considerable international cooperation and coordination, large staffs of technicians and scientists, and extensive field programs using sophisticated instrumentation as well as laboratory activities supported by the fastest electronic computers.

The various disciplines of geophysics have one further feature in common, namely, the phenomena they embrace are of fundamental importance to mankind. Whether in the domestic or business environment, rarely a day goes by without the need for an administrative decision based on a meteorological or other geophysical event. As a consequence, geophysical studies are the concern of nations as well as of scientists, and the planning of geophysical programs is becoming a major function of governmental and nongovernmental bodies, particularly in the international sphere. On the international governmental side, specialized agencies of the United Nations have become involved in such programs, especially the United Nations Educational, Scientific, and Cultural Organization (UNESCO) in the fields of hydrology and oceanography and the World Meteorological Organization (WMO) in the field of atmospheric science. On the international nongovernmental side, such activities are coordinated by the International Council of Scientific Unions (ICSU), through its various components, primarily the scientific unions such as the International Union of Geodesy and Geophysics (IUGG), as well as the special scientific and interunion committees which have special responsibilities for individual geophysical programs. A noticeable trend in these programs has been the increasing cooperation between governmental and nongovernmental bodies, often stimulated by specific resolutions from the United Nations General Assembly.

Both the number and extent of large international geophysical programs have increased, and this can be ascribed, at least in part, to the tremendous success of the International Geophysical Year (IGY, 1957–1958). For one thing, IGY captured the interest and imagination of the nonscientific public, in part because it was a truly international program in which virtually all nations collaborated and in part because its more spectacular phases represented readily recognizable advances of major importance. To the scientist, IGY was a success since it led to quasi-global and quasi-simultaneous observational programs; moreover, geophysical studies became the purpose and not merely a sideline of expeditions. The vertical and horizontal probing of the Antarctic continent and of its atmosphere was (and, fortunately, still is) an excellent example of geophysical study being the main purpose of expeditions.

While one may include various justifiable and worthwhile projects in an international geophysical program, one should realize that such a program is not merely an attempt to exploit the enhanced availability of financial and personnel support but is rather a deliberate attempt to schedule, on a global basis, the orderly advance of technology in research and related fields. Because of the requirement for simultaneous or at least coordinated studies, it is simply not effective or economical to introduce new or expanded programs haphazardly in space or in time. This is precisely why each international geophysical program appears to lay major emphasis on relatively new types of observational studies, in order that these programs (whose introduction would have been inevitable, in any case) may be instituted in such a manner (global and coordinated) that significant results should be achievable almost from the start. At the same time, it is necessary to examine existing programs and identify their inadequacies (instrumentation, networks, training, and so on), particularly if they are of fundamental importance for the newer studies, so that a concerted drive to remove or minimize these weaknesses may take place also during the chosen period of observation and study. The international geophysical program has the added advantage of permitting all countries to participate; those which are not prepared to enter the newer (and, generally more expensive) fields will almost certainly be able to discover relevant more conventional fields which are undersubscribed.

Solar-oriented programs. The IGY, from July, 1957, to December, 1958, was an internationally accepted period of concentrated and coordinated geophysical exploration, primarily of the solar and terrestrial atmospheres. It was marked by intense solar activity as well as by intense human activity, and enjoyed an unqualified success partly because of both these factors. It was apparent by 1960 that there would be considerable merit in a second international period of scope, concentration, and coordination, at least comparable to that of the IGY, to take place during the time of the following minimum of solar activity. The period called the International Years of the Quiet Sun (IQSY) was therefore carried out from January, 1964, to December, 1965, to permit a comparison between phenomena and parameters at times of maximum and minimum solar activity.

The IQSY program drawn up by the ICSU special committee for the IQSY on the basis of discussions in nations, in unions, and in the IQSY assemblies comprised activities in the following disciplines: meteorology (large-scale physical and dynamic characteristics of atmosphere above 10 miles), geomagnetism, aurora, airglow, ionosphere (vertical incidence, absorption, and drift studies, plus rocket and satellite data), solar activity, cosmic rays, space research, and aeronomy. Seventy-one nations participated in the IQSY, and the results have appeared in hundreds of research papers in scientific journals and are summarized in the seven-volume *Annals* of the IQSY.

Following the IQSY, cooperation in international and national projects relating to solar-terrestrial disciplines came under the aegis of the Inter-Union Commission on Solar-Terrestrial Physics (IUCSTP), a component body of the ICSU. Initial emphasis was on the period of maximum solar activity (1968–1970), which is referred to as the International Years of the Active Sun (IASY). For that period, specific coordinated programs have been planned in the following fields: monitoring of the solar-terrestrial environment, proton flares, disturbances of the interplanetary magnetic field configuration, characteristics of the magnetosphere, conjugate-point experiments, magnetic storms, low-latitude auroras, upper-atmosphere structure and dynamics, ionospheric chemistry, and ionospheric disturbances.

Solid-earth geophysical programs. One highly effective program, oriented toward solid-earth as well as space problems, has been the world magnetic survey, coordinated on behalf of the ICSU by the World Magnetic Survey Board of IUGG. The aim of this program was to define, with a prior agreement as to precision and detail, the magnetic field of the Earth at the epoch 1965.0. Both surface and airborne data were extrapolated to 1965.0 whenever possible, and many special observation programs were undertaken. A definitive framework was established so that any subsequent observations of geomagnetic phenomena either at the surface or within 1000 km of the surface (by rocket or satellite) could be classified readily as either normal or abnormal.

For pure solid-earth disciplines, the requirement for coordinated programs is considerably less than for other branches of geophysics. Nevertheless, from time to time it becomes apparent that a specific field can be advanced rapidly only if certain studies are carried out in many representative areas. Such was the case for the Upper Mantle Program, with peak activity in the mid-1960s, which was coordinated on behalf of the ICSU by the Upper Mantle Committee of the IUGG, in association with six other ICSU unions. The emphasis was on indirect probing of the upper mantle of the Earth, which is found tantalizingly close to the Earth's surface.

International hydrological decade. Despite its tremendous economic significance, hydrology has long been one of the lesser developed geophysical sciences. To rectify this imbalance and permit hydrology to make a meaningful contribution to global water problems, including especially those of the developing nations, UNESCO took the initiative in launching the International Hydrological Decade (IHD, 1965–1974). A coordinating council, under UNESCO, is the central agency for planning and promoting the Decade, with scientific advice from ICSU's Scientific Committee on Water Research (COWAR). The coordinating council established working groups for various facets of the overall program to deal with network planning and design, global water balance, hydrological maps, representative and experimental basins, hydrology of fractured limestone terrains, nuclear techniques to determine water content in saturated and unsaturated zones, influence of man in the hydrological cycle, floods and their computation, and exchange of information, including publication, standardization problems, and education and training.

Oceanographic programs. The primary coordinating body for oceanographic geophysical programs is the Intergovernmental Oceanographic Commission (IOC), which is an autonomous organization within UNESCO. Scientific advice to IOC is provided by ICSU's Scientific Committee on Oceanic Research (SCOR). A number of international oceanographic programs of somewhat limited extent such as the International Indian Ocean Expedition (IIOE, 1961–1966), studies of the tropical Atlantic and the Kuroshio Current, and so on have been undertaken. A patchwork program of this type, however, is rather unsatisfactory, since priorities for various activities are hard to establish. As a result, SCOR has produced a document, entitled General Scientific Framework for World Ocean Study, which puts regional studies into a global perspective. Almost simultaneously, the United Nations General Assembly adopted Resolution 2172 on Resources of the Sea, which has provided new impetus for oceanic research. As a consequence, international organizations are giving serious consideration to an expanded, accelerated, long-term, and sustained program of exploration of the oceans and their resources.

Atmospheric science programs. Vigorous programs in meteorology were carried out in connection with the IGY and IQSY, but subsequent developments, primarily in space technology, have thrust the atmospheric sciences into the international limelight. Two key United Nations General Assembly resolutions (1721 and 1802) have called upon international governmental (WMO) and nongovernmental (ICSU) organizations to develop significant programs in the atmospheric sciences, both operational (World Weather Watch, WWW) and research (Global Atmospheric Research Program, GARP). A history-making precedent was set in October, 1967, when the ICSU and WMO agreed on the characteristics of GARP and set up a joint organizing committee to plan the program.

The WWW is a global data collecting, processing, and dissemination system whose aim is to provide optimum service to all peoples through collective action within WMO. There are many aspects of WWW that would require research either in the laboratory or in the atmosphere, on a local or global scale. Numerical weather prediction and general circulation research suggest that extended-range prediction should be possible in some detail provided that (1) sufficient initial data on the atmosphere and on the air-Earth boundary are available, (2) all significant atmospheric and boundary processes are adequately included in the physical formulation of the problem, and (3) extremely sophisticated computers and highly accurate and stable numerical procedures are utilized. It is the aim of GARP to test this hypothesis. Preliminary subprograms will tackle individual problem areas and individual geographic areas, with the main phase of GARP, tentatively scheduled for the late 1970s, embracing a truly global and complete atmospheric observation program, to produce the raw data for detailed testing of atmospheric models and atmospheric predictability.

[WARREN L. GODSON]

Bibliography: C. M. Minnis (ed.), *Annals of the IQSY*, vols. 1–7, 1968–1970; H. Odishaw (ed.), *Research in Geophysics*, vol. 1: *Sun, Upper Atmosphere, and Space*, vol. 2: *Solid Earth and Interface Phenomena*, 1964; N. V. Pushkov and B. I. Silkin, *The Quiet Sun*, English transl., 1968.

Geosyncline

A part of the crust of the Earth that sank deeply through time; traditionally, a great trough hundreds of miles long and tens of miles wide that subsided as it received thousands of feet of sedi-

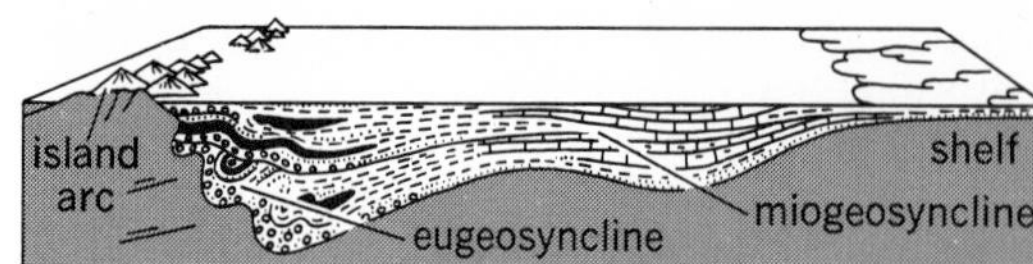

Fig. 1. Diagrammatic section of Cordilleran geosyncline in southeastern Alaska and British Columbia at close of Permian. Volcanic deposits indicated in black. (*After A. J. Eardley, J. Geol., 55:319–342, 1947*)

mentary and volcanic rock through millions of years. A synclinorium, in contrast, is a great downfold that was produced by deformation that occurred later than the deposition of the contained rock. *See* ANTICLINE; SYNCLINE.

Linear geosynclinal belts, or orthogeosynclines, are great subsiding tectonic divisions of the crust which lie between more stable areas, or cratons, the structurally higher continents and the ocean basins of their time. Orthogeosynclinal belts generally have internal volcanic (eugeosynclinal) belts and external nonvolcanic (miogeosynclinal) belts. Although the North American Cordilleran (Fig. 1) and Appalachian eugeosynclinal belts lie beside oceans, the Atlantic Ocean is only half as old as the rocks in the geosynclines. The Appalachian belt seems to have been continuous into the Caledonian belt of the British Isles and Scandinavia prior to the opening of the Atlantic by sea-floor spreading. The eugeosynclines have been deformed in orogenies and intruded by igneous rocks that metamorphosed their sediments and volcanic rocks; they seem analogous to modern volcanic archipelagoes such as the East and West Indies. The deep oceanic troughs associated with these archipelagoes are geosynclines that have not been filled (leptogeosynclines); and some of the rocks in deformed eugeosynclinal belts were originally formed in ancient oceans.

Plate tectonics. The most significant developments during early 1970s have been in relating the types of geosynclines to modern analogs, particularly to the concepts of plate tectonics and global tectonics that have emerged in the period 1965–1975. Until the 1970s the general view was that the ocean basins were floored by dense mafic rocks that had at one time comprised the fundamental crust of the Earth. *See* PLATE TECTONICS.

A succession of observations and discoveries led to a reappraisal of the nature of ocean basins and, ultimately, to a recognition of analogs in the geosynclinal belts. R. Dietz and J. Holden identified the gently seaward-thickening wedges of sediments, such as the Mesozoic-Tertiary of the Atlantic Coastal Plain of the Carolinas, as miogeosynclines. The term miogeosynclinal was applied to nonvolcanic areas, regardless of whether the basement floor was a tilted plane or trough—or an assemblage of troughs and basins, as was the case with H. Stille's original miogeosynclinal belt in the Cordilleran region of North America (Fig. 2). *See* TECTONIC PATTERNS.

Aulacogens. N. S. Schatsky proposed that the major fault-bounded troughs, such as the Dnieper-Donetz Graben, be termed aulacogens. E. S. Belt showed that the conceptual deep downfolds that had been defined as epieugeosynclines were fault-bounded and, hence, also taphrogeosynclines. It was further suggested, by J. M. Bird and J. F. Dewey, that the original opening of ocean basins took place along such rift belts, as in the Red Sea. Aulacogens have been considered to be one part of a three-rayed fault system on the domes above mantle hot spots, the other two rays opening as proto-ocean basins.

Intraplate basins. The intracratonic basins that were termed autogeosynclines seem to relate to the stability of continental plates, for they are present in the North American Paleozoic craton but apparently not in the European Paleozoic plate. The European plate moved over the mantle toward the North American plate in closing the ancient ocean (the Protacadic or Proto-Atlantic Ocean) that was deformed in the early Paleozoic, Taconian, Caledonian, and Acadian orogenies. The main developments have been in the understanding of the eugeosynclinal belts, which were

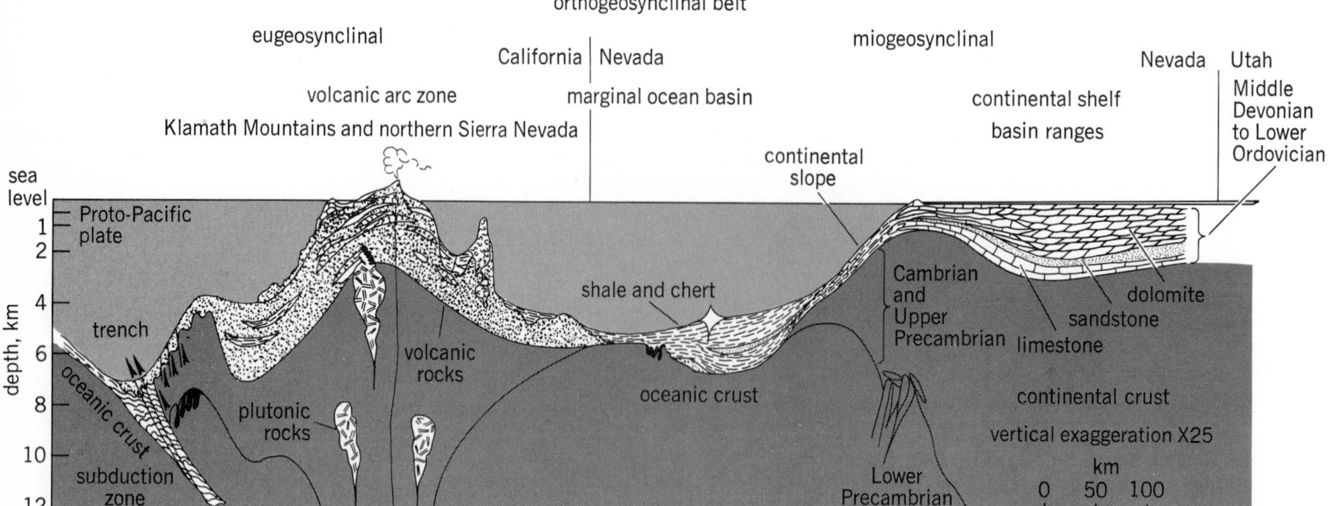

Fig. 2. Middle Paleozoic reconstruction of western margin of North America showing relation of ancient geographic features in a geosynclinal belt. (*From Michael Churkin, Jr., in R. H. Dott, Jr., and R. H. Shaver, eds., Modern and Ancient Geosynclinal Sedimentation, Soc. Econ. Paleont. Mineralog. Spec. Publ. no. 19, 1974*)

characterized by the presence of volcanic and intrusive rocks in lava-floored troughs, volcanic island arcs, and tectonic lands (lands composed of older rocks). According to the concept of plate tectonics, oceans are floored by basaltic lavas that have been extruded along the flanks of a mid-oceanic rift. The rocks found within the eugeosynclinal belts correspond to oceanic rocks. Ocean basins were in themselves eugeosynclinal, but the term was originally applied to rocks in mobile belts where they are associated with volcanic, hypabyssal, and plutonic rocks of types now found in volcanic island arcs. Thus the eugeosynclinal belts contain bands of oceanic rocks that were subducted, that is, descended, beneath island arc crusts. Tectonic lands are now seen as welts raised by processes that took place within the eugeosynclinal belts, associated with intrusions of more silicic rocks and granites, granodiorites, and similar rocks. In some instances, they are attributed to slabs of continental rocks ("microcontinents") isolated during the processes of continental drift and separated from the parent continents by interarc basins of oceanic rocks; the analogies are Japan and the Sea of Japan, although the scale may not have been as great. *See* OCEANIC ISLANDS.

Thus, the orthogeosynclinal belts of Stille and M. Kay have come to be understood as assemblages of crustal features related to continental shelf subsidence, subducted ocean floor, associated island arc or continental margin arc volcanism, and intrusion into island arc assemblages—and perhaps some continental fragments. *See* CONTINENT FORMATION; OROGENY.

[MARSHALL KAY]

Bibliography: R. H. Dott, Jr., and R. H. Shaver (eds.), *Modern and Ancient Geosynclinal Sedimentation*, Soc. Econ. Paleontol. Mineral. Spec. Publ. no. 19, 1974; J. A. Jacobs, R. D. Russell, and J. T. Wilson, *Physics and Geology*, 1973; M. Kay, *North American Geosynclines*, Geol. Soc. Amer. Mem. 48, 1951.

Geyser

A natural spring or fountain which discharges a column of water or steam into the air at more or less regular intervals. It may be regarded as a special type of spring. Perhaps the best-known area of geysers is in Yellowstone Park, Wyo., where there are more than 100 active geysers and more than 3000 noneruptive hot springs. Other outstanding geysers are found in New Zealand and Iceland. The most famous geyser is probably Old Faithful (see illustration) in Yellowstone Park, which erupts about once an hour. Then for about 5 min the water spouts to a height of 100–150 ft. Other geysers are less regular, but some intermittently discharge water and steam to heights of 250 ft or more.

The eruptive action of geysers is believed to result from the existence of very hot rock, the relic of a body of magma, not far below the surface. The neck of the geyser is usually an irregularly shaped tube partly filled with water which has seeped in from the surrounding rock. Far down the pipe the water is at a temperature much above the boiling point at the surface, because of the pressure of the column of water above it. Its temperature is constantly increasing, because of the volcanic heat source below. Eventually the superheated water changes into steam, lifting the column of water out of the hole. The water may overflow gently at first but, as the column of water becomes lighter, a large quantity of hot water may flash into steam, suddenly blowing the rest of the column out of the hole in a violent eruption.

Old Faithful, Yellowstone Park, Wyo. (*National Park Service, U.S. Department of the Interior*)

[ALBERT N. SAYRE/RAY K. LINSLEY]

Gilsonite

An asphaltite that occurs in the Uinta Basin of northeastern Utah. Most deposits lie in Uintah and Duchesne counties. The nearly parallel veins have a northeast-southwest trend, are characterized by nearly vertical, sheer sandstone walls, and may be traced for distances up to 40 mi by surface outcrops. Exposures may range from several inches to 22 ft in width. The gilsonite is essentially free of mineral matter, but veins do occasionally contain floated blocks of wall rock. *See* ASPHALT AND ASPHALTITE.

Gilsonite is used in the manufacture of paints, battery boxes, asphalt floor tiles, brake linings, printing inks, and electrical insulation. It is also used for waterproofing and insulating high-temperature piping.

Gilsonite is black and has high luster, coarsely conchoidal fracture, and a red-brown streak. The material is friable and in mining or crushing breaks to a fine, penetrating dust. It is sold as selects, coming from the interior of a vein, or as seconds, obtained near the contact of the vein with the enclosing rock. The seconds are not as lustrous as the selects, have a higher fusing point and lower solubility in naphtha, and differ in specific gravity, as follows:

	Selects	*Seconds*
Specific gravity at 77°F	1.03–1.05	1.05–1.09
Fusing point	230–240°F	240–280°F
Solubility in naphtha	50–60%	20–50%

The larger deposits of gilsonite have been mined since shortly after their discovery in 1882. Among the better known are the Cowboy, Bonanza, Rainbow, and Black Dragon veins. Some veins have been mined to nearly 1000 ft in depth.

The ultimate analysis of gilsonite is variable: carbon, 85–86%; hydrogen, 8.5–10.0%; sulfur, 0.3–0.5%; nitrogen, 2.0–2.8%; and oxygen, 0–2%. Gilsonite contains a small percentage of hydrocarbons (4–8%) that can be isolated by chromatography. Some of these compounds are naphthenic (60–70%), and others are aromatic (30–40%). Other compounds that have been isolated in very small quantities are unsaturated and have the following approximate empirical formulas: $C_{18}H_{25}O_3N_2$, $C_{10}H_{13}O$, and $C_{63}H_{95}N$. Gilsonite from the Bonanza vein has been found to contain 0.03% porphyrins; the nickel complex of deoxophylloerythroetioporphyrin has been isolated and identified.

Neither the origin of gilsonite nor its peculiar occurrence is completely understood. The material probably originated in sediments of upper Green River age (Eocene) as a lacustrine sapropel derived from both animal and vegetable sources. The sapropel and its subsequent diagenesis to gilsonite were undoubtedly dependent upon the nature of the contributing organisms and the controlling calcareous and dolomitic environment of deposition. Thermal effects do not appear to have been important in the conversion of the sapropel to gilsonite. Gilsonite has been batch-distilled to temperatures of 660°–770°F yielding 50–55% distillate, 15–20% gas, and 30% coke. Analysis of the distillate has shown it to contain acetone, methyl ethyl ketone, diethyl ketone, pyrrole, 2-methylpyrrole, phenol, *o*-cresol, 3,5-dimethylphenol, 1-naphthol, 4-picoline, 2,5-lutidine, 2,6-lutidine, 3,4-lutidine, 3,5-lutidine, 2,3,5-trimethylpyridine, and 3-ethylpyridine. The presence of saturated and unsaturated hydrocarbons and the absence of aromatic hydrocarbons have also been indicated. *See* DIAGENESIS; SAPROPEL.

Early transportation was by horse or mule, and somewhat later the 60-mi narrow-gage Uintah Railroad was constructed. Later the railroad was abandoned and sacked gilsonite has been moved by trucks.

A major development in the large-scale utilization of gilsonite took place in 1957, when the American Gilsonite Co. began operation of a $13,000,000 gilsonite refinery near Grand Junction, Colo.

To supply material for the refinery, gilsonite is mined underground hydraulically with a jet of water and pumped (2000 lb/in.²) to the surface. Rather than reactivate the old Uintah Railroad or truck the product 182 mi, the gilsonite is transported to Grand Junction in a 72-mi pipeline as an aqueous slurry containing 35% gilsonite and 65% water. This pipeline, one of the first to carry solids, has a 6-in. diameter, crosses an 8500-ft mountain pass, and has a capacity of 700 tons per day of 8 mesh gilsonite. On arrival at the refinery the gilsonite is dewatered by filtration, melted, and refined by conventional techniques used in the petroleum industry. The only major products are coke for use in the aluminum industry, gasoline for local consumption, and gas, valuable as a fuel. The rated capacity of the refinery is 1300 bbl of gasoline and 250–275 tons of coke per day. Proven reserves of gilsonite available for processing are equivalent to more than 100,000,000 bbl of crude oil.

[IRVING A. BREGER]

Bibliography: H. Abraham, *Asphalts and Allied Substances*, 6th ed., vol. 1, 1960; R. K. Bond, Designing the gilsonite pipeline, *Chem. Eng.*, 64(10): 249–254, 1957; Gilsonite yields coke plus gasoline, *Chem. Eng. News*, 35(32):28–29, 1957.

Glacial epoch

An informal popular name which, with its synonym Great Ice Age, refers to the prehistoric time characterized by extensive glaciers and related phenomena that constituted a rather recent part of geologic time. It corresponds to the Pleistocene Epoch of stratigraphic literature and an ill-defined portion of pre-Pleistocene time. The primary characteristics of the glacial epoch were the repeated fluctuation of mean temperatures of air and sea water through a range of several degrees Celsius (C) and, in its later part, the repeated invasion of middle-latitude regions by glacier ice. These events were accompanied by conspicuous fluctuations of sea level, regional subsidence of the Earth's crust beneath the extra loads imposed by ice sheets (the largest glaciers), fluctuation of rainfall in wide regions not covered by ice, and extensive migrations of plants and animals. Compared with the general progress of change during geologic time, these events occurred rapidly.

Temperature. The glacial epoch was actually a group of cold glacial episodes alternating with warmer intervals. The record shows evidence of temperatures at times lower than those of today and at other times slightly higher.

Temperatures lower than those of today at the same localities are estimated mainly from four lines of evidence: (1) extent of the former ice sheets, which reached south to latitude 50°N in Europe and to latitude 38°N in North America; (2) difference in altitude of the former climatic snow line on mountains, as approximated by the floors of abandoned cirques, from the snow line of today—measured directly this amounts in places to more than 1000 m; (3) former extent of perennially frozen ground inferred from geologic features, compared with that of today; and (4) occurrence of temperature-sensitive fossil plants and animals, both terrestrial and marine, beyond the present ranges of the same or closely related forms.

Equatorial temperatures 2–5°C lower than those of today are suggested by the criteria indicated in (2) and (4) above. Much greater departures are indicated for continental stations in high latitudes. Temperatures higher than those of today are indicated by (4). At middle-latitude stations, temperatures ~2°C higher than those of today are indicated by (4). Thus the range of temperature throughout the Pleistocene may possibly have been as much as 6°C near the Equator and even greater in higher latitudes.

Precipitation. Little is known about the range of precipitation values during the Pleistocene. Evidence of temporary lakes, some of large size, in areas that are dry today, such as western United States, eastern and northern Africa, and central Asia, indicates increased rainfall and reduced evaporation in those areas, but such evidence does not indicate whether the lakes represent a general increase of precipitation or merely an areal redistribution of an unchanged precipitation total. The belief, once widely held, that increased precipitation values are required to build the extensive Pleistocene glaciers, does not seem necessary because with sufficiently reduced temperatures and with time for snow accumulation, glaciers should have been able to grow with snowfall even below today's values. Meteorologic models have assumed both greater and less general precipitation during the glacial times. In southern Africa, at least, ancient sand dunes, ancient soils, and other evidence suggest former climates slightly drier than that of today, but their chronology has not been established.

Fluctuation of climate. In North America four major cold episodes, with at least three of them marked by subdivisions, have been identified thus far; in Europe the number identified is much greater. The principal evidence consists of successive bodies of glacial drift and loess, separated by mature ancient soils and by nonglacial sediments containing fossils that imply climates as warm as those of today or warmer. The sediments recovered in cores from beneath the tropical Pacific and Atlantic oceans show temperature fluctuations, expressed in various ways, that are compatible with the terrestrial data; these may also indicate still earlier fluctuations not yet recognized on the lands. The fluctuations are commonly referred to as glacial ages and interglacial ages, respectively. *See* MARINE SEDIMENTS.

Distribution of glaciers. The extent of the area formerly covered by glaciers is determined mainly from geologic evidence of erosion and deposition by the ice. In general, the central parts of wide glaciated regions have undergone net erosion, having been denuded of their soil cover. Much bare bedrock is exposed at the surface, polished and striated as a result of the passage of ice over it. In contrast, the peripheral parts have received a general mantle of glacial drift averaging a few meters in thickness. Elongate streamline forms fashioned by the ice from bedrock and drift, positions of rock fragments of known origin in the bedrock, striations, moraines, and eskers afford a basis for determining the directions of movement of the glacier ice. *See* GLACIATED TERRAIN.

Major glaciers, which almost completely covered even the mountainous parts of the terrain on which they lay, included the Antarctic, Greenland, Laurentide (in North America east of the Rockies), and Scandinavian (in northern Europe) ice sheets. Smaller glaciers of various kinds occupied highlands in all latitudes, including the equatorial. The area covered by the recurring Laurentide ice sheet, 13.4×10^6 km², extended from the Arctic Ocean to a line passing through the vicinities of New York City, Cincinnati, St. Louis, Kansas City, and the Dakotas. When at its maximum, that ice sheet was larger than the existing Antarctic ice sheet. The recurring Scandinavian ice sheet, some 6.6×10^6 km², reached southern Britain, northern Germany, central Poland, and southern European Russia. The margins of the successive glaciers are subparallel, suggesting that the same group of controls operated without major changes during each of the recognized glacial ages.

The combined area once covered by glaciers is estimated at 44.4×10^6 km², about 30% of the present land area of the world. Today glaciers cover about 14.9×10^6 km², about one-third as much area (Fig. 1).

The thickness of former glaciers is very difficult to determine; most of the thickness values quoted are rough estimates and some are hardly more than guesses. Indirect evidence suggests that the ice sheet over eastern Scandinavia may have attained a maximum thickness greater than 2500 m. Direct evidence implies that the Laurentide ice sheet had a minimum thickness of 1600 m over central New England, and as deduced from theory, its maximum thickness may have exceeded 3000 m. The world distribution of former glaciers is reasonable in terms of today's atmospheric circulation system, suggesting that no radical departures from that system need be invoked to explain Pleistocene glaciations. *See* GLACIOLOGY.

Effect on Earth's crust. The coasts of Fennoscandia and northern North America (including the Great Lakes) are marked by a succession of shorelines that have been deformed systematically in such a way that when reconstructed they constitute structural domes whose central parts approximate the central parts of the former ice sheets. The shorelines date from the time of the late deglaciation and shortly thereafter. Tide gage records show that the upwarping, with little change in form, is still in progress. The crust is regarded as having been depressed isostatically beneath the excess loads created by the ice sheets, with gradual recovery toward its preglacial position as the ice melted. The shorelines at any place were created during the relatively short intervals between deglaciation and crustal recovery. Altitude relations of the shorelines imply that recovery has been discontinuous, possibly because of variation in the

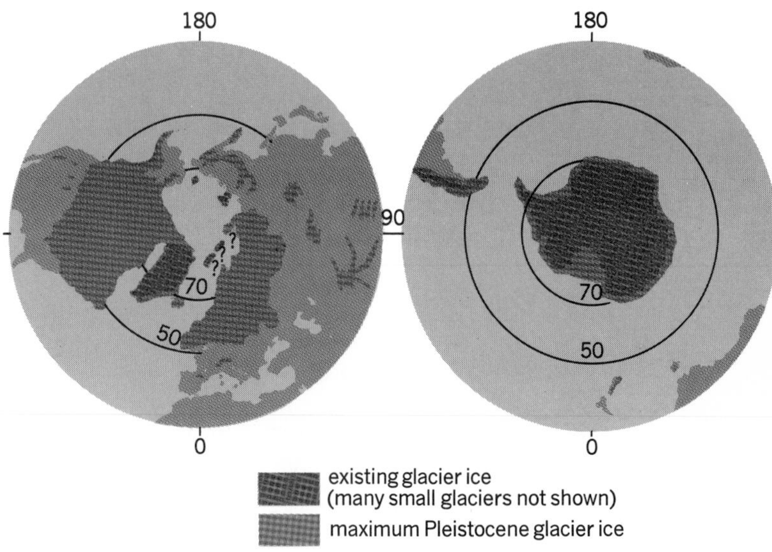

Fig. 1. Glacial maps of North and South Pole hemispheres.

rate of deglaciation and because of temporary expansions of the otherwise shrinking ice sheets. The movement still in progress is occurring at a rate of ~ 1 mm/(100 km)(year) in the Great Lakes region, and is having an effect on the depths of Great Lakes harbors. *See* WARPING OF EARTH'S CRUST.

Sea level changes. Beaches and other marine shorelines of Pleistocene or presumed Pleistocene age occur at various altitudes above sea level, and are detected by echo sounding at depths down to at least 100 m. Submerged stream valleys, deltas, and windblown sand have been identified off various coasts. Tide gage measurements at harbors on the southern Atlantic and Gulf coasts of the United States indicate that sea level is rising relative to the land; between 1930 and 1948 the rate was $\sim$6 mm/year. The rise is believed to be an actual rise of sea sevel, and not a result of sinking of the coast, because of agreement of values at widely separated stations. These data are reasonably explained as an effect of the building and melting of glaciers in response to Pleistocene changes of climate. As temperatures, wind velocities, evaporation, and precipitation vary, equilibrium in the sea water–atmospheric moisture–glacier ice system is established at correspondingly varying positions; as glaciers increase, sea level falls. The current rise of sea level is coincident with worldwide shrinkage of glaciers, established by sampling.

The opinion most widely held is that during glacial maxima, sea level stood at least 100 m below its present position; however, much greater estimates have been offered. On fairly good though not indisputable evidence, it can be said that in interglacial times sea level rose as high as 20 m above its present level. Marine strandlines, probably interglacial, exist at higher positions, but the possibility that they have been deformed casts doubt on the significance of their altitudes. An amplitude of 120 m can therefore be taken as a probable minimum through which the sea level has fluctuated during Pleistocene time. In consequence of this fluctuation continental shelves emerged widely during glacial ages, restricting the habitats of marine organisms, connecting lands now separated, and affording migration routes for land animals and plants. Lands thus formerly connected include England-Ireland, England-France, Alaska-Siberia, Siberia-Japan, Malaya-Sumatra-Java-Borneo, Australia–New Guinea, Australia-Tasmania, and India-Ceylon. All the separating straits are less than 100 m below present sea level.

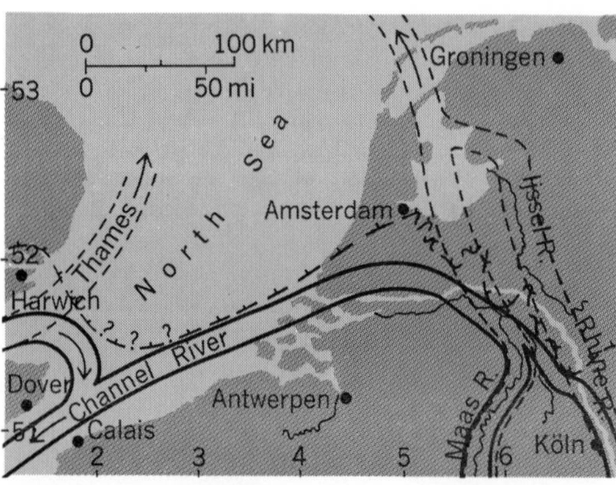

Fig. 2. Former drainage of the Rhine, Maas (Meuse), and Thames rivers. After the Holstein Sea had been lowered far below present sea level, Rhine, Maas, and Thames flowed northward during early Saale time. With the arrival of the Saale ice sheet, the Rhine and Maas rivers took a new, westward course, forming the Channel River. Presumably the Thames River joined them. (*Most data from J. I. S. Zonneveld*)

Changes of drainage. The great ice sheets blotted out, at least temporarily, the streams throughout the regions they covered. In some areas the slopes of ice sheets and adjacent land were so related that new rivers formed along the glacier margins and became so deeply incised that they remained in their new positions after deglaciation. Major examples include the Missouri River through North and South Dakota, the Ohio River from Cincinnati eastward, the Thames River near London, and the lower Rhine–English Channel system, now partly submerged (Fig. 2). Many valleys that existed before glaciation were completely filled with drift by the advancing ice sheets and buried from view. They are being rediscovered by subsurface exploration techniques, and the sand and gravel many of them contain are exploited as sources of water supply.

The Great Lakes basins in the northern United States were developed from major preglacial river valleys, partly by glacial excavation of valley floors and partly by the accumulation of drift to form dams (Fig. 3). During deglaciation water was held between the glacier margins on the north and high ground on the south. The several successive lake levels recorded by former shorelines were controlled by differential upwarping of the crust and by the uncovering of new and lower outlets by deglaciation. The chronology of the lake sequence is known through the carbon-14 dates of driftwood fragments buried in ancient beaches, of wood buried in contemporary glacial drift, and of contemporary peat. *See* DATING METHODS.

Glacial drift. The glaciers eroded the unconsolidated soil material and the underlying bedrock down to various depths. Erosion was deepest, amounting in places to more than 1000 m, in the narrow valleys of highland regions. Over wide areas of central Canada and Finland the average depth of glacial erosion is probably no more than a few meters. The eroded material, broken up and comminuted, is spread irregularly over the glaciated regions, particularly the outer parts near the margins of the former glaciers. Some of it was exported by streams draining away from the ice sheets. The Mississippi River carried away from the ice a huge quantity of sediment, part of which still constitutes an outwash fill of gravel, sand, and silt.

Peripheral to the glaciated region are broad blankets of windblown silt (loess) picked up mainly from the outwash fills of major river valleys while the discharges of meltwater were in operation. In places more than 30 m thick, the loess represents a further sorting and redistribution of glacial debris. *See* LOESS.

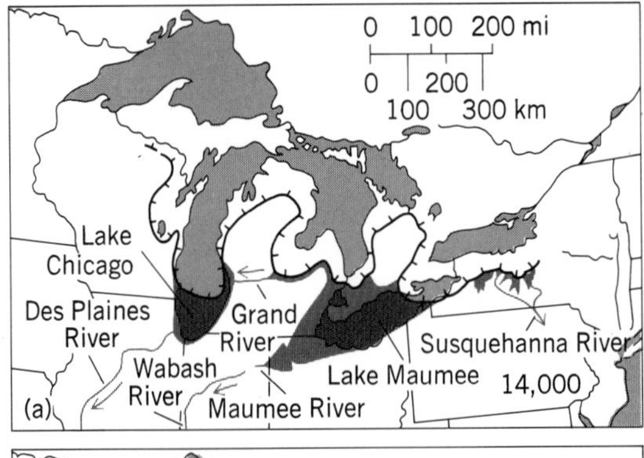

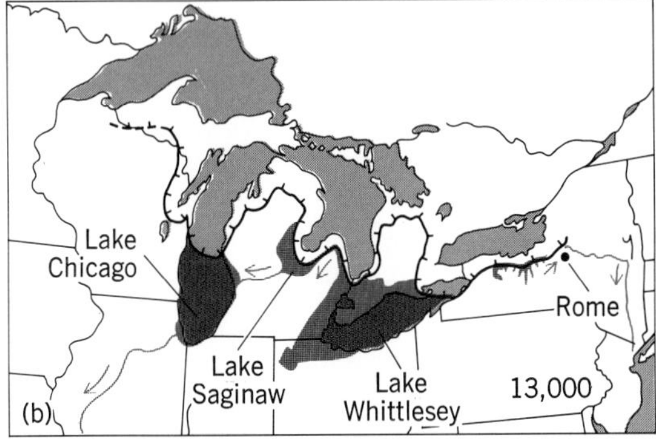

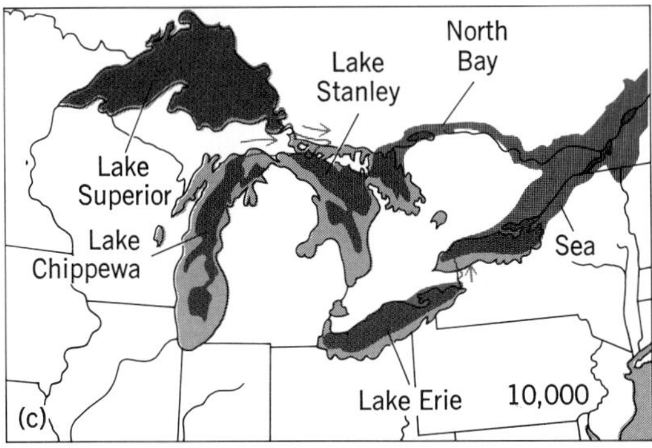

Fig. 3. Evolution of the Great Lakes. Figures indicate the approximate number of years before the present, based on ^{14}C dating techniques. Arrows mark the directions of flow along the outlets from the lakes, (*a*) at the time of Lake Maumee, (*b*) at the time of Port Huron Readvance, and (*c*) at the time of Lake Chippewa.

Ecologic changes. Study of the accumulated collections of Pleistocene fossil plants and animals has made possible rough reconstructions of ecologic zones in Europe and parts of Asia during the last glacial age. The zones lay far south of their present positions, with tundra and park tundra extending from the margin of the Scandinavian ice sheet southward to the Alps and Carpathians and with forests, so extensive today, confined mainly to the Mediterranean and Black Sea coastal regions. Steppe country lay east of the tundra. Tundra, steppe, and forest each had its appropriate groups of animals.

Zone shifts in North America were less drastic, probably because the arrangement of topography and atmospheric circulation permitted the incursion of warm air from the Atlantic and the Gulf of Mexico into the heart of the continent even when the ice sheet was at its maximum. The tundra belt was narrow and perhaps discontinuous, and may have consisted mainly of park tundra; forests of spruce and fir lay south of it, and on the west and southwest was steppe, less dry than that of today.

As deglaciation occurred, plants and animals rapidly occupied the region progressively uncovered by the ice, and ecologic zones shifted toward their present boundaries. Rate of shift can be closely inferred from C^{14} dates of lake and bog sediments containing fossil pollen.

Organism evolution and extinction. Little evidence regarding the Pleistocene evolution of plants has been accumulated, but evolutionary development, although not striking or conspicuous, occurred among land mammals. Subdivision of the Pleistocene strata into three faunal zones has been possible in North America, Europe, and Africa. Large mammals adapted to low temperature, such as the woolly mammoth, woolly rhinoceros, and certain musk-oxen, are popularly associated with the glacial ages, but actually they did not appear until late in the Pleistocene. Many of the mammal assemblages were not unlike those found today in comparable climates, with the difference that the ancient assemblages were richer and more varied than those of the present, owing to intervening extinctions. Such extinctions occurred throughout the epoch, but were most marked at a very late time, 5000–20,000 years ago. Opinion is divided as to whether the extinctions were primarily the direct and indirect result of the hunting activity of man, or whether they should be attributed to climatic change or other environmental causes.

Chronology. During the first 60 years of the 20th century it was generally believed that the Pleistocene Epoch was roughly 1,000,000 years long. However, potassium-argon (K-Ar) dating of volcanic rocks closely related to glacial strata has shown that layers of glacial sediment are far older. Dates as old as 2,500,000 years are thus attributed to strata believed to be early Pleistocene because of their fossils, at places in middle latitudes, whereas in high latitudes glacial strata with K-Ar dates of 10,000,000, and even 20,000,000 years or more, are reported. Although not Pleistocene, such strata indicate that glaciers were present in high latitudes long before the Pleistocene, and they weaken the argument that the Pleistocene should be defined on the basis of its cold climates.

The latest 50,000 years or so of geologic time, including the end of Pleistocene time, is measured by C^{14} dating. The thousands of pertinent C^{14} dates show that the last great glacial invasion of the Great Lakes region began more than 25,000 years ago and reached its maximum extent around 18,000 years ago. From these dates a mean rate of glacier advance through the region of ~160 ft/year is deduced. Thereafter the ice sheet shrank, with at least two conspicuous reexpansions culminating around 12,000–13,000 and 10,000–11,000 years ago, respectively. The later reexpansion reached close to the sites of Milwaukee, Buffalo, and extreme northern New England. By 8000 years ago the ice sheet had cleared out of all the Great Lakes basins, and by 5000 years ago it seems to have disappeared completely, although evidence on the latter event is scanty. One climatic fluctuation, possibly culminating around 5000 years ago, carried mean annual temperatures in middle north

Pleistocene time units

Central North American		Alps region, Europe	
Glacial	*Interglacial*	Glacial	*Interglacial*
Wisconsin		Würm	
	Sangamon		*Riss/Würm*
Illinoian		Riss	
	Yarmouth		*Mindel/Riss*
Kansan		Mindel	
	Aftonian		*Günz/Mindel*
Nebraskan		Günz	
			Donau/Günz
		Donau	

latitudes to as much as 2°C above those prevailing today.

Comparable events in Europe. For the last 12,000 years or so the climatic changes in Europe, as dated by C^{14} and other means, were essentially contemporaneous with those in North America. Although less well established, similar parallelism seems to characterize the entire period of the last 50,000 years as well. C^{14} dates of selected layers in sedimentary cores from ocean floors and from lake sediments in a dry basin in the western United States confirm a cold wet climate changing toward a warmer and drier climate sometimes between 18,000 and 11,000 years ago.

Stratigraphy. Designations of glacial and interglacial time units for central North America and for the Alps region in Europe are shown in the table. Interglacial units are italicized. The Wisconsin and Würm units, about which the most is known by virtue of the excellent state of preservation of the evidence, are further subdivided, but universal agreement on the basis of subdivision is lacking. The time correlation between America and Europe, implied by the chart, is considered probable for the upper units, but for the lower ones is highly speculative. In any case the sequence shown is very incomplete.

Causes of climatic fluctuation. A large number of theories have been advanced to explain the Pleistocene climates, which seem to have fluctuated more widely and rapidly than had those of any time within the preceding 200,000,000–250,000,000 years. Although the fossil record suggests that mean atmospheric temperatures in middle north latitudes fell by some 10°C during the 60,000,000 years preceding the Pleistocene, sharp fluctuation seems to have begun with the Pleistocene, as did the glaciation of middle latitudes. There is no evidence that the fluctuation was periodic, and the evidence of recent millenniums implies that fluctuation was broadly synchronous throughout the world. Present-day temperatures appear to be close to the interglacial highs.

This and other evidence mentioned earlier suggests an extraterrestrial cause or causes. The cause most frequently appealed to is fluctuation in the amount of energy emitted by the Sun; opinions differ concerning the competence of variations in the system of heat transfer within the Earth's atmosphere. Alternative theories, appealing to shifts of the Earth's crust or of individual continents relative to the poles, to geometric variation in elements of the Earth's orbit, to changes in ocean currents, and to variations in the constituents or turbidity of the atmosphere, appear to be less probable from the facts now known. It is clear that most glaciers, including the largest ones, were closely related to highlands, and that prior to the Pleistocene many existing highlands were lower or nonexistent. Hence, whatever the causes of the variations in climate, Pliocene and Pleistocene mountain-making promoted the development of glaciers in an important way. Once formed, glaciers favored their own self-extension by reflecting solar heat and by setting up anticyclonic conditions in the atmosphere above them. The ultimate solution to the problem of causes is likely to come from the field of meteorology.

[RICHARD F. FLINT]

Bibliography: J. K. Charlesworth, *The Quaternary Era*, 1957; R. F. Flint, *Glacial and Pleistocene Geology*, 1957; H. E. Wright and D. G. Frey (eds.), *The Quaternary of the United States*, 1965.

Glaciated terrain

A region distinctive because it once bore great masses of glacial ice, which may have been either mountain or mountain and valley type where high ridges and peaks projected above the ice; or continental, ice cap, or plateau type where the ice spread out over a large area, generally of relatively low relief, and concealed almost all the surface. Since mountain valley glaciers in many places coalesce and spread out over adjacent lowlands in piedmont glaciers, the line of division is not absolutely definite and sharp. The diversely complex regional features of glaciated terrains change with the passage of time and vary from place to place, and these conditions are reflected in considerable diversity of scientific views upon the subject. The first part of this article deals with distinguishing characteristics, including regional evidences of prior glaciers, and the difficulties encountered in regional identification. A second part outlines principal glaciated regions of the world.

Distinguishing characteristics. The regional features characteristic of glaciated terrains and the local patterns associated with prior glaciation are briefly outlined. Modifications of features with the passage of time and other observational difficulties in such regions are then discussed.

Marks of glaciation. That glacial ice extended beyond its present limits or formerly occupied a region now devoid of glaciers is demonstrated by at least a dozen diagnostic features: (1) scratches (striae) on bedrock and loose rocks which are exactly like those observed in and around existing glaciers; (2) rock fragments of all sizes, many of

Fig. 1. A pit cut in bouldery till. Note unstratified and unassorted materials exposed on sides of pit.

them striated, which are unlike adjacent bedrock (hence termed erratics), some of which occur higher than their probable bedrock sources and many of them being too large to have been transported by streams; (3) unstratified and unassorted deposits (till) of clay, sand, gravel, cobblestones, and boulders which were derived in large part by mechanical attrition of bedrock rather than predominantly by chemical weathering; (4) drainage features unlike those due to normal stream work, including valleys out of harmony with the nature of the bedrock and many waterfalls, rapids, and lakes; (5) basins completely rimmed with bedrock which are far larger than those due to weathering and lack the regular boundaries of depressions formed by earth movements; (6) marginal slopes bounding unconsolidated deposits unlike those around the edges of ordinary water-laid deposits; (7) numerous topographic features which must have been built rather than worn like those due to stream work—including streamlined hills (drumlins), knobs, ridges, and depressions, all of which can only be constructional (see Figs. 1 to 5) and can be explained only by glacial and glaciofluvial action; (8) water-laid deposits where such are now impossible because some solid obstruction must have been removed since their formation, there being no indication of erosion to account for this fact; (9) mountain valleys with an irregular longitudinal profile unrelated to the nature of the bedrock; (10) mountain valleys which are wider, straighter, and steeper-sided than is normal for the type of bedrock in which they occur; (11) valley junctions in mountains which are not accordant (hanging) but have falls or rapids; (12) valley heads and mountainside depressions with abnormally steep sides (cirques). *See* CIRQUE; DRUMLIN; ESKER; MORAINE; TILL.

Observational difficulties. The practical application of the above criteria is fraught with many pitfalls. It is rare to find all or even a majority of them in the same area unless it has been glaciated very recently. As the time since glaciation increases, the criteria become more and more difficult to discover. Common errors in application result from (1) confusion of striae with scratches due to faulting (slickensides), particularly on loose rocks (2) failure to find striae, which are quickly obliterated by weathering and require a cover of till for their preservation for any appreciable fraction of geologic time and consequently are rare on exposed rock ledges; (3) difficulty in demonstrating that some loose rocks are erratics; (4) misinterpretation of deposits of landslides and creep which are deceptively like till; (5) resemblance to drumlins of some rounded hills due to weathering and erosion of massive rock; (6) presence of hanging valleys due to other agencies than glacial erosion (7) occurrence of erratics transported by icebergs in standing water (8) removal of glacial deposits by mass movement on hillsides. It is small wonder that opinions concerning the former existence of glaciation in certain localities differ with the personal views of the geologist who observes the evidence.

Principal glaciated areas. Distinctive glaciated terrain are found in every continent beyond the margins of existing glaciers or in regions in which there is no glacial ice at the present time.

Fig. 2. View longitudinally along the summit of an esker, a till plain feature which consists of bouldery gravel. (*Wisconsin Geological Survey*)

North America. Much work on glacial geology has been done in North America. Maps, many of them drawn from air photographs, are available even for some wilderness areas. Continental ice survives in Greenland, and to a lesser extent in some of the Arctic islands. The Canadian Shield of crystalline bedrock is largely scoured rock with thin drift, some eskers, and many lakes and extensive areas of lake sediments due to former obstruction of the drainage. In southern Ontario and the Prairie Provinces, drift is thicker and moraines and drumlins occur. Weathering seems to have destroyed striae on the mountains of Labrador, but it has been thought that this was the point of origin of the continental ice, which spread farther toward the west and southwest than in other directions because of the moisture brought by westerly winds. The western mountains of Canada were largely buried by ice, but the spectacular glacial topography is chiefly the result of later local glaciers. In the United States the glacial drift is thick on sedimentary bedrock, particularly in the region south and southwest of the Great Lakes. The bottoms of all the Great Lakes except Lake Erie ex-

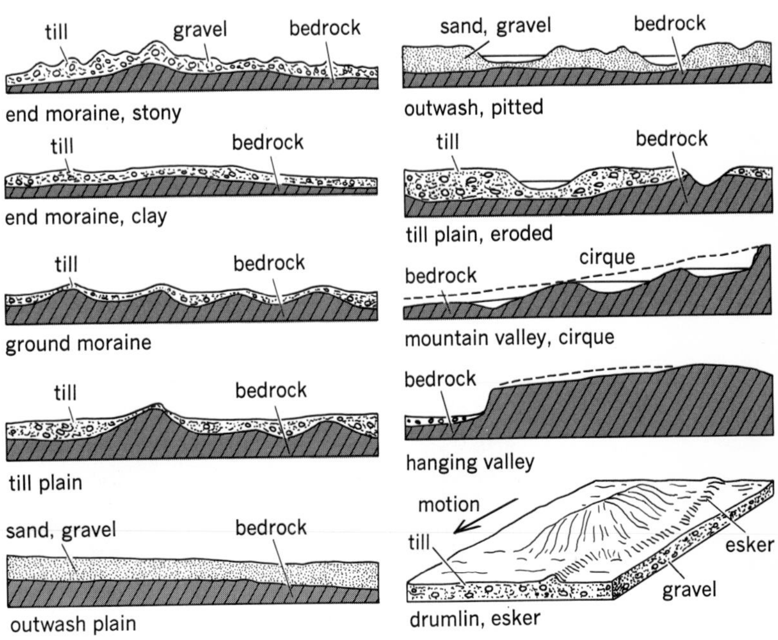

Fig. 3. Cross-section profile and block views of typical features of glaciated terrain distinctly different from those developed by a normal fluvial erosion cycle.

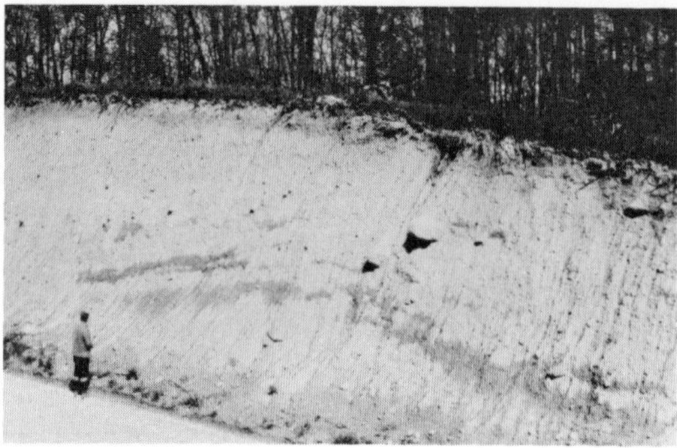

Fig. 4. Cross section along a road cut exposing the till and occasional silt materials (darker) in a drumlin.

tend below sea level, and drilling shows that they lie in rock basins eroded in the weaker sediments and evaporites. Plains of lake sediment are found near the shores of the Great Lakes. Glacial drift of several distinct ages has been identified in the Mississippi Valley. The young drift found in the north is termed Wisconsin. In its type locality and throughout much of its extent, its glacial topography is clearly marked with large rugged moraines, pitted outwash plains, eskers, and vast fields of drumlins. Some of the most striking lake districts are in pitted outwash. The lakes there have more rounded outlines than do those in either the thin drift or the end moraines. In the area of crystalline bedrock in Wisconsin, Michigan, and New England the till is very bouldery. This is due to the fact that such rock breaks into large fragments and does not indicate any special kind of glaciation.

Southwestern Wisconsin contains a considerable area of rugged rock topography in which no erratics have been discovered except those evidently brought from outside by lake or stream waters. This region is called the Driftless Area and it has been generally thought that it never was glaciated. The exact boundary of glaciation around this area is definite only on the east side, where it is an end moraine of Wisconsin drift. On the other sides the drift thins out gradually, and marginal lakes with ice-rafted boulders make tracing of the border most uncertain. In the nearby glaciated terrain much of the region is strewn with drift too thin to smooth out the rock topography. Although drift might have been removed from hillsides by mass movement, it seems impossible to remove all of it, and it is generally recognized that the area escaped glaciation because of its protected position south of the crystalline highland of northern Wisconsin and Michigan.

Fig. 5. A general view over pitted outwash plain. Note lake and lack of well-developed stream and valley system, which will probably develop with the passage of time. (*Wisconsin Geological Survey*)

One of the most spectacular features of glacial origin in Wisconsin is the famous Kettle Interlobate Moraine, which records division of the ice into lobes, one in the Green Bay lowland, the other in the basin of Lake Michigan. The Kettle Moraine is the accumulation, chiefly of gravel, in the reentrant-angle between these two lobes as their margins melted back.

To the south in Illinois the end moraines become less and less prominent, for the till there has a high content of clay. Ground moraine becomes till plain, partly because of the fluidity of the wet till when deposited, and partly because of buried lake clays. Rock hills rise from these plains like islands from a sea. Postglacial erosion is confined to major stream courses. South of this Wisconsin drift, the margin of which is marked by an end moraine, there is more drift plain, more deeply weathered and more extensively eroded, which is designated as the Illinoian drift. West of the Mississippi River in Iowa the amount of erosion increases abruptly and the Kansan drift occurs west of a definite line. With the same relief, material, and climate, the difference can only be due to a longer postglacial time. This conclusion checks with the buried Kansan soil profile. In Iowa there is a soil profile buried below the Kansan drift; this is the top of the Nebraskan drift, the surface exposure of which, if any, is not definitely recognizable. This region of clay till shows little evidence of moraines. *See* GLACIAL EPOCH.

The glacial features of Minnesota and the Dakotas are mainly like those of Wisconsin. The clay content of the till increases to the west, making the moraines less evident and outwash rarer. The continental drift meets drift from the Rocky Mountains not far from the foothills. Scenic features such as glacial cirques, waterfalls, and lakes decrease in the Rockies from north to south. The southernmost glaciation was on the high volcanoes of Mexico.

Because of the greater moisture available, glacial features are especially prominent in the mountains close to the Pacific. Magnificent fiords extend south from Alaska to near Puget Sound. The sound itself shows deep water but the shores are mainly stratified drift. Some enclosed kettles contain fresh-water lakes. The Yosemite Valley of California is famous for its high waterfalls from hanging valleys, but these are not entirely due to glacial erosion. Such spectacular valleys are few; as pointed out by F. E. Matthes, glacial excavation was profoundly influenced by the amount of fracturing of the otherwise massive rock (Fig. 6).

Europe. In spite of the long time during which the glacial geology of Europe has been under investigation, few maps of large areas are available. Ice came from the Scandinavian mountains, the Urals, and minor centers in the British Isles. Ice spread into the lowlands of Germany and Russia.

Asia. Continental ice covered some of northwestern Siberia but its margin is disputed. The dry climate of much of Asia prevented extensive glaciation except in some of the higher mountain ranges.

Fig. 6. View up hanging valley of Bridal Veil Creek, a tributary to the valley carved by the main Yosemite Glacier in the geologically recent past. There is debate whether it is the dimensions of the glacier or the geological control (diagonal joint structure) that predominates in the pronounced form of this hanging valley. *(Photograph by F. E. Matthes, USGS)*

Other areas. Antarctica is covered by the largest continental glacier now in existence. Local glaciers were, or still are, present in New Zealand, Australia, and some of the higher Pacific islands. In South America the high peaks of the northern and central Andes and the lower elevations of the southern Andes display glacial features, including fiords along the coast of Chile and extensive moraines in Patagonia. Glaciation is also reported on high peaks in Africa, as well as in New Zealand. *See* FIORD.

[FREDRIK T. THWAITES/RICHARD F. FLINT]

Bibliography: E. Antevs, Maps of the Pleistocene glaciations, *Bull. Geol. Soc. Amer.*, 40:631–720, 1929; R. F. Flint, *Glacial and Pleistocene Geology*, 1957; R. F. Flint et al. (eds.), *Glacial Map of North America*, Geol. Soc. Amer. Spec. Pap. no. 60, 1945; R. F. Flint et al. (eds.), *Glacial Map of the United States East of the Rocky Mountains*, Geol. Soc. Amer., 1959; H. E. Wright and D. G. Frey (eds.), *The Quaternary of the United States*, 1965.

Glaciology

The study of existing or modern glaciers in their entirety, involving all related scientific disciplines. As glaciology embraces so many interconnecting facets, it is considered a master science. It is largely concerned with present glacial characteristics and processes, as opposed to studies in glacial geology which relate to the nature and effects of former glaciation.

A glacier is a naturally accumulating mass of ice that moves in the process of discharging from head or center to its margins or a terminal dissipation zone. Glaciers are nourished in areas of snow accumulation that lie above the mean climatological or orographical snow line, which on the glacier surface is referred to as the névé line or firn line. The most active glaciers are generally found in regions receiving the heaviest snowfall, such as the maritime flanks of high coastal mountain ranges. Exemplifying this are the great westerly facing glaciers of Mount Saint Elias and the Saint Elias Mountains in south coastal Alaska which are nourished by the heavy precipitation brought in by warm, cyclonic air masses moving across the warm waters of the Gulf of Alaska. Similarly, in New Zealand the vigorous glaciers of the southern Alps lie in the storm tracks of the prevailing westerly wind which brings much greater accumulation to the western than to the eastern slopes. Other such maritime glacial bodies are the Patagonian ice field in the southern Andes; the small ice sheet and ice caps of Iceland; the mountain ice fields of Norway and the Kebnekaise region of Sweden; and the mountain glaciers of eastern Siberia. There are many other glaciers in the high mountains of the middle and equatorial latitudes (as in Peru and East Africa); however, 96% of the world's glacial ice is represented by the vast continental ice sheets of Antarctica and Greenland. Together, these regions contain at least 5,600,000 mi^2 (14,300,000 km^2) of ice cover, or nearly 10% of the total land area of the globe.

The total volume of the world's glaciers, ice fields, and ice sheets can only be estimated, but is at least 24,000,000 km^3. This mass of frozen water, if melted and returned to the sea, would raise the sea level 160–200 ft (±60 m). R. F. Flint in 1957 estimated that, during the Ice Age, this volume was probably 300–400% greater. *See* GLACIAL EPOCH.

Glaciation and deglaciation. In many regions glacier activity has formed a continuous series of events throughout the Pleistocene glacial-pluvial epoch, and in postglacial (post–Ice Age) and modern times. In order to simplify the terminology of past and present glaciation, regions which are presently glaciated are also sometimes referred to as glacierized, glacier-covered, or ice-covered. These terms are used instead of the ambiguous and unqualified term glaciated for reference to all areas formerly or currently ice-covered. In this context, for any one region, the glaciation limit is represented by the lowest elevation mountain summit carrying an existing glacier.

A glacier at any one time presents only partial patterns of its long-term regime. The complex study of glacial regimen, the processes and consequences of growth and decay, is being actively pursued in glacial regions throughout the world today. Other glaciological studies embrace research in many allied disciplines, such as geomorphology, meteorology and climatology, physics of ice deformation (glacier mechanics or continuum mechanics), thermodynamics, survey and mapping, glacier geophysics, lichenometry, palynology, and plant ecology.

Materials. Glaciers are composed of three substances: snow, firn, and ice. The main material of glaciers is bubbly glacier ice of a specific gravity approximating 0.88–0.90 g/cm^3. This glacier ice is composed of myriads of interlocking crystals, hence it is a polycrystalline material containing air pockets and entrapped water bubbles. Because of the proportion of bubbly glacier ice, a mean bulk

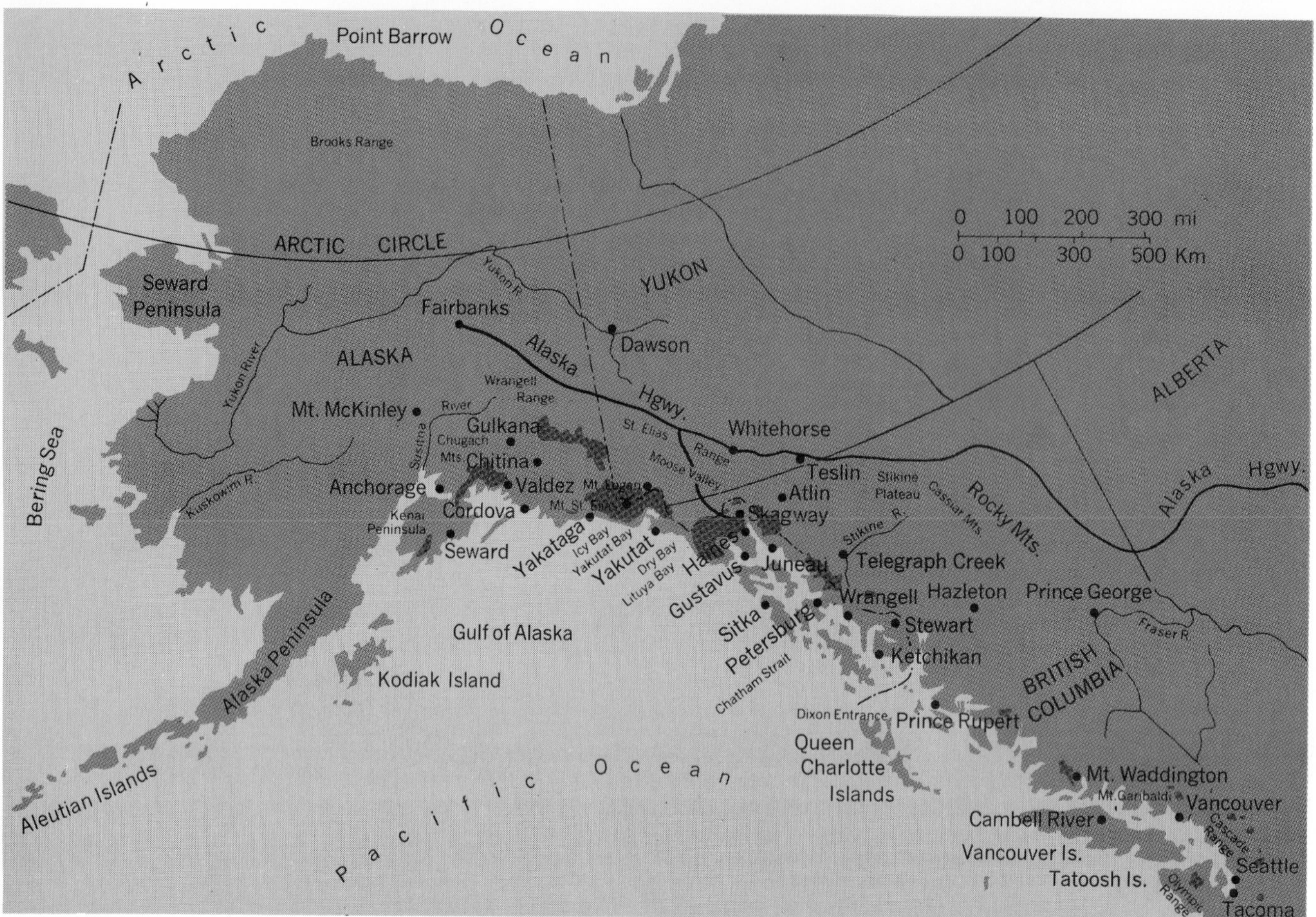

Fig. 1. Map of Alaska showing main centers of existing glaciers. Note that these centers lie along the south and the southeastern coasts, where there is a ready source of moisture and high ranges of mountains, resulting in heavy annual accumulations of snow. In northern Alaska it is too cold and dry to produce the requisite snowfall for glaciation under present climatic conditions. *(Foundation for Glacier and Environmental Research)*

specific gravity may be taken as 0.90 g/cm^3, as opposed to a specific gravity of 0.917 g/cm^3 in dense, solid, and unaerated ice. Below the late-summer glacier snow line (névé line, as defined below), only bubbly glacier ice is exposed. Above the névé line, the other categories of snow and firn (and firn ice) exist to depths of a few to hundreds of feet. It is deeper in polar (colder) firn packs. (Firn is a consolidated granular transition of snow not yet changed to glacier ice. The process of transformation of snow to firn is termed firnification.) The density gradation between firn and ice is usually asymptotic, but with a relatively sharp line of demarcation between a seasonal snowpack and underlying firn. Arbitrary and approximate densities of new snow are 0.1 – 0.3: old snow, 0.3 – 0.45: firn, 0.45 – 0.75: and firn ice, 0.75 – 0.88. Firn ice is not considered to be a separate stage of metamorphosis but rather the result of a mixture of partially altered firn and bubbly glacier ice. The processes by which new snow is transformed into bubbly and dense glacier ice are complex and varied. In middle-latitude regions the refreezing of percolated meltwater, compaction, and flow deformation play substantial roles in the metamorphosis. In polar regions wind packing, mechanical compaction, and flow recrystallization are the prime factors producing glacier ice.

Terminology. The geographical term névé refers to the area covered by perennial snow or firn, that is, the area lying entirely within the zone of accumulation. The word firn refers only to the substance of the material itself. The terms snowpack and firn pack refer to the volume of snow or firn, respectively, at any one point on the névé of a glacier's surface and connote thickness or depth characteristic rather than area. Snow cover and firn cover refer to the blanket of snow or firn over a névé, a glacier, or a bedrock surface, and have areal rather than depth connotations.

The term névé line (firn line) represents the elevation of the periphery of a névé at any point on a glacier's surface. More specifically, the elevation of the névé line's most stable position over a period of several years is referred to as the semipermanent névé line. The term glacier snow line or transient snow line describes the transient outer limit of retained winter snow cover on a glacier. Its elevation gradually rises until the end of the annual ablation season, by which time the old snow has become firn and the glacier snow line becomes the seasonal névé line. In regions such as coastal Alaska, where glaciers descend to sea level, the summer snow line on intervening rock ridges and peaks is often much higher than the snow line on the valley glacier, and in most instances is more

Fig. 2. Oblique air photograph of 4-mi wide tidewater terminus of the Hubbard Glacier, a prototypical valley glacier which reaches tidewater from a source area in the St. Elias Mountains of Alaska and the Yukon. This glacier has advanced several miles since 1894. (*Photograph by H. B. Washburn, August, 1938*)

irregular and indefinite. If it has not disappeared completely from the bedrock surfaces by the end of summer, the lowest limits of retained snow may be connected as an irregular limit and be termed the orographical snow line since it is primarily controlled by local conditions and topography. Variations in the position of the orographical snow line are so great from year to year that only from records over a long period can a meaningful trend be discerned. This snow line is important in the development of nivation hollows and protalus ramparts in deglaciated cirque beds. Of greater present-day significance is the average regional level of the orographical snow line rather than its lower limit. This mean position, based on observations over a number of years, is called the regional snow line or climatological snow line. On adjacent ice sheets and glaciers this coincides with the mean névé line. In regions of extensive present glaciation the regional snow line thus equates to the mean névé line. Specifically, the mean névé line is the statistical average of consecutive annual positions of semipermanent névé lines over a period of at least 10 years.

In glaciers of the polar regions a different term is used to delineate areas of net accumulation from those of net wastage, in consequence of the refreezing of meltwater and drainage on surfaces down-glacier from the transient snow line. The position where this refreezing ceases is referred to as the equilibrium line, a theoretical line separating the area of net gain from the area of net loss. On temperate glaciers this coincides with the névé line. It is an important concept in glacier velocity considerations because it represents the position or zone least subject to seasonal variations in flow resulting from excessive accumulation or ablation.

Morphological categories. Glaciers develop numerous forms. Basically, they are of the mountain (alpine) type and of the plateau (polar) type. Mountain glaciers generally are moderate to small in size and include valley glaciers (main ice streams), icefall glaciers, cirque glaciers, basin glaciers, hanging glaciers, cliff glaciers, and glacierets. These terms are self-descriptive and for the most part relate to strong and varied relief of the kind found along the mountainous southern coast of Alaska (Fig. 1). Valley (ice stream) types are the most common (Fig. 2). They are often in the form of glacier systems fed by cirque-headed tributary valleys and serve as outlets from ice fields, such as those found in the St. Elias and Boundary ranges between Alaska and Canada (Figs. 2 and 3). They are also the main type of glacier in the Alps, the southern Andes, New Zealand, the Caucasus, and the Himalayas. The longest valley glacier in the temperate regions is the Hubbard Glacier (Fig. 2), with a length of about 100 mi in the Alaska-Yukon border area. The Vaughan Lewis Glacier and the Upper Herbert (Camp 16) Glacier on Alaska's Juneau Icefield are typical icefall glaciers derived from high névé basins or plateaus (Figs. 3 and 4). Plateau-type glaciers, dominantly of the ice-sheet and ice-cap form, are characterized usually by vast size with relatively flattened surfaces or low relief. These are typical of the Greenland and Antarctic ice sheets, and are sometimes termed inland or continental ice. Often valley glaciers extend outward as distributary tongues. Intermediate between valley glaciers and ice sheets are piedmont glaciers. These occupy broad lowlands bordering a glacial highland. The best known of this category is the Malaspina Glacier near Yakutat, Alaska, with an area of 1400 mi^2 (Fig. 1).

Geophysical types. Glaciers are classified geophysically into two major groups, polar and temperate, and two transitional groups, subpolar and

Fig. 3. Part of Vaughan Lewis Glacier, Juneau Icefield, Alaska, showing surface bulges and wave ogives in apron area just below icefall. Similar to view looking downvalley in Fig. 4. Series of medial moraines visible upper right. (*Photograph by M. M. Miller, August, 1968*)

subtemperate (Fig. 5). The temperature of a polar glacier is perennially subfreezing, except for a shallow surface zone which may be warmed for a few weeks of each year by seasonal atmospheric variations. The extreme polar condition depicted in Fig. 5 is found in the heart of the Antarctic continent at the South Pole. In temperate glaciers the temperature below a recurring winter chill layer is always at the pressure melting point. This situation is typical of middle-latitude glaciers, such as those in southern Alaska. Because these terms are thermodynamic in meaning but geographical in connotation, it should be pointed out that glaciers of the geophysically polar type can still exist at relatively low latitudes, and that geophysically temperate glaciers are even found at latitudes to the north of the Arctic Circle.

Fig. 4. Oblique air photograph of Upper Herbert (Camp 16) Glacier on the Juneau Icefield, Alaska, showing its plateau névé or accumulation zone at 5000-ft elevation, as well as the ice cascade, seracs, and wave bulges on the apron at the base of the icefall. (*U.S. Forest Service photograph, October, 1962*)

Of the transitional categories, in subpolar glaciers the penetration of seasonal warmth is restricted to a relatively shallow surface layer, but is greater than in polar glaciers. The transitional subtemperate glacier is characterized by a relatively deep zone of annual warming. These subordinate terms are useful because the former may refer to transitional glaciers of dominantly polar character which still have certain temperate characteristics, and the latter to dominantly temperate glaciers having a tendency towards polar characteristics. The significance of such differentiation relates to the close control ice temperatures exert on flow deformation, and so also to glacial fluctuations. This includes a possible relationship to kinematic surges, which are described later. Each geophysical category (polar, subpolar, subtemperate, and temperate ice) can be found in any one glacier system if there is sufficient range in latitude or elevation for the requisite climatological factors. Present-day ice fields and ice sheets which are thermally temperate in some sectors and grade through to thermally polar geophysical conditions in other sectors are referred to as polythermal. Such probably was the geophysical character of the continental ice sheets during the waxing and waning phases of the Pleistocene Epoch. A glacier which is geophysically temperate (0°C) throughout is usually referred to as isothermal.

Thermodynamics. The internal thermodynamic and heat-transfer character of glaciers is more complex than suggested by the foregoing geophysical differentiations. This is because individual glaciers vary much in their structural makeup and related regime histories. Some of the complication is shown by the table of thermal constants for snow, firn, and ice at 0°C.

The problem of thermal changes within glaciers thus requires the approach of the physicist. One supplemental effect, however, must be kept in mind. This relates to the fact that, although normal thermodynamical processes pertain in temperate glaciers, infiltrating meltwater also plays an important role. In polar glaciers the fundamental diffusivity relationship shown in this table dominates, as there is only minimal meltwater effect restricted to the surface zone.

Irregularities in thermal dissipation occasioned by the presence of mobile water and by various physical inhomogeneities in glaciers usually preclude close quantitative agreement with hypothetical temperature curves calculated from assumed bulk diffusivity. Computed values are near to observed conditions only in purely polar glaciers, or in those sections of temperate glaciers in which during the winter months there is no liquid water to prevent changes in internal temperature from

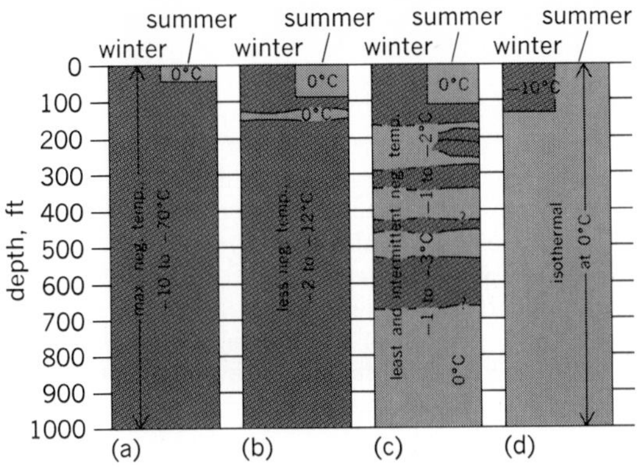

Fig. 5. Thermophysical classification of glaciers or ice sheets. (*a*) Polar. (*b*) Subpolar. (*c*) Subtemperate. (*d*) Temperate. Arbitrary depth-temperature values at surface illustrate seasonal variations. Dark tone, "cold" subfreezing state; light tone, "warm" (0°).

being controlled by conduction. The detailed mathematical analysis of the heat transfer is not considered here, beyond mention of the fundamental thermodynamic properties pertaining to understanding and appreciation of field observations.

In snow and firn, internal temperature changes may be considered as occurring in a semi-infinite, homogeneous, isotropic solid (although individual crystals are structurally anisotropic) influenced by fluctuating external temperatures which are a harmonic function of time. The basic physical factors controlling development of the sinusoidal cold wave are the thermal conductivity k of the medium, its specific heat c, and its density ρ. It can be shown that the transmission of heat actually depends on the diffusivity K, defined by a combination of these quantities in the relationship, $K = k/c\rho$. Here the specific heat is taken as constant at the value for ice. The density and conductivity, however, are variable, depending on the age of the medium and related genetic factors. The resulting diffusivity is therefore a function of these factors.

The relative diffusivity between dense firn and ice is not so large that mass heat transfer and temperature changes in firn are greatly altered by the presence of a few ice strata. On the other hand, a glacier with a substantial covering of snow will display important differences in surface heat transfer compared to one with only solid ice exposed. The general relationships are illustrated in the table. The acute sensitivity of glacier flow to changes in internal temperature, as discussed under structure and movement below, underscores the importance of appreciating the role of these thermal constants in circumstances of pronounced and long-term climatic change.

Regime. Upon the regime, or the annual state of health of a glacier system, depends the glacier's eventual growth or decay, that is, its mass balance. The controlling factors are accumulation (gain) and ablation (loss) over the whole system. On temperate glaciers the critical sector is the névé line, or more properly a névé-line zone which at the end of the annual melt season separates the névé area (accumulator) from the bare-ice area (dissipator). As already noted, it equates to the lower elevation limit of retained winter snowpack as observed at the end of summer. On polar glaciers the equivalent critical sector is, of course, the equilibrium line.

The height of the seasonal névé line, or yearly equilibrium line, shifts greatly according to the regional climatological situation and local conditions. With respect to the Juneau Icefield in southeastern Alaska, this critical line (Fig. 6) varied over a 20-year period between 2400 and 3800 ft (mean 3100 ft). In high-latitude polar regions the equilibrium line is usually at, or close to, sea level. On alpine glacier systems a higher than average névé line means a tendency toward a negative regime (more loss than gain), whereas a lower than average névé line is associated with an increase in the accumulation area and hence a positive regime. This kind of interrelationship is illustrated by the accumulation trend in Fig. 6 and by the typical mass-balance relationship depicted for a healthy glacier situation in Fig. 7. The specific net budget curve in Fig. 7 shows the difference between accumulation and ablation per unit area at every specific elevation. To obtain the actual or total budget (mass balance), this figure is multiplied by the total glacier area at that specific elevation. In alpine glaciers the area ratio of accumulator to

Thermal constants for snow, firn, and ice at 0°C*

Material	Conductivity, cal °C^{-1} cm^{-1} sec^{-1}	Specific heat, cal °C^{-1} g^{-1}	Density, g cm^{-3}	Thermal diffusivity, cm^2 sec^{-1}	Relative diffusivity to ice, approx ratio
New snow	0.0003	0.5	0.20	.0030	0.27
Old snow	0.0006	0.5	0.30	.0040	0.36
Average firn	0.0019	0.5	0.55	.0070	0.64
Firn ice	0.0038	0.5	0.75	.0100	0.91
Ice	0.0050	0.5	0.92	.0110	1
Water, 0°C	0.0014	1.0	1.00	.0014	0.13
Rubber	0.0005	0.40	0.92	.0014	0.13
Steel, mild	0.1100	0.12	7.85	.12	11
Aluminum	0.4800	0.21	2.70	.86	78
Copper	0.9300	0.09	8.94	1.14	104

*The given conductivities for snow and average firn are based on data from U.S. Army Corps of Engineers, Snow, Ice and Permafrost Research Establishment, *Review of the Properties of Snow and Ice*, rep. no. 4, 1951. The other data for snow and ice are from M. M. Miller, *Glaciothermal Studies on the Taku Glacier, Southeastern Alaska*, Tome 4, Publ. no. 39 de l'Association Internationale d'Hydrologie, Union Internationale de Geophysique et Geologie, Assemblée generale de Rome, 1954.

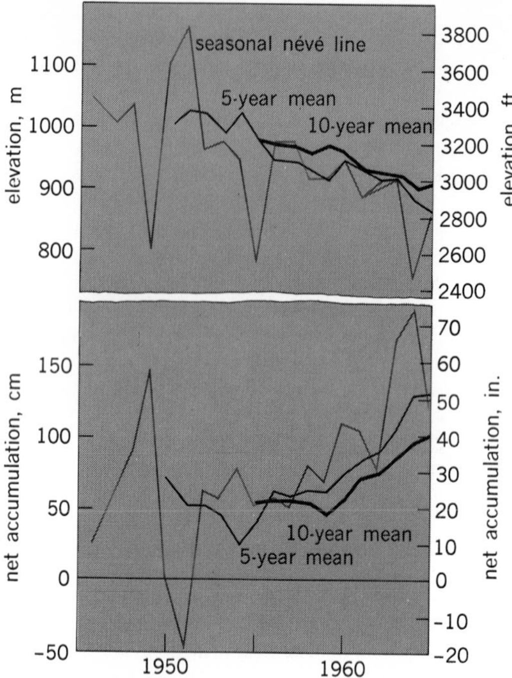

Fig. 6. Comparative névé line and net accumulation trends on the Taku Glacier, Alaska, during 1946–1965. (*Juneau Icefield Research Program*)

dissipator is on the order of 4:1, whereas in the Greenland ice sheet it is in excess of 100:1, and in the Antarctic as much as 1000:1. Also, in the accumulation zone there is an increase of retained accumulation in any one year with an increase in elevation until the height of maximum snowfall is attained. It should be noted that this is not always

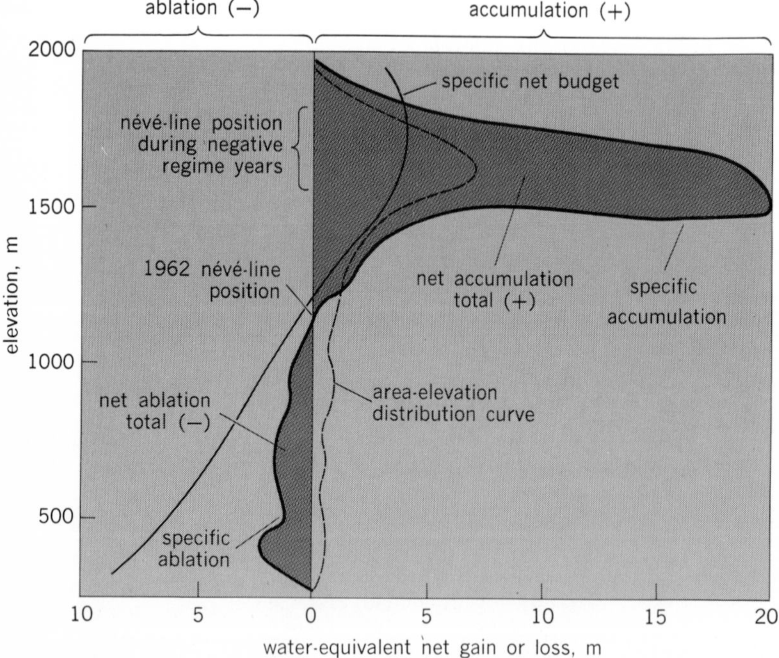

Fig. 7. Mass balance relationship on Nigardsbreen Glacier, Norway, in 1961–1962 budget year. This glacier normally has a negative regime, but it was surplus in this year, calculated to be about 95,000,000 m³ water equivalent. (*After G. Ostrem, Nigardsbreen Hydrologi, 1962, Norsk. Geog. Tidskr., vol. 18, pp. 156–202, 1963*)

at the highest elevation of land, because maximum snowfall is controlled by the mean freezing level, that is, the greatest snowfall occurs at or near 0°C. Thus, in mountain regions or on large ice sheets with a great elevation range at very high elevations, there is commonly less snowfall, the conditions there being generally too cold for much snow.

The rise and fall of the névé line over a period of years also parallels vertical shifts in the zone of maximum accumulation. This vertical shift may be measured by means of soundings, borings, test-pit studies, and observations on crevasse walls. In a healthy glacier the average level of maximum snowfall lies at the elevation of greatest glacier area (Fig. 7). A glacier in a poor state of health has usually experienced a rise in its level of maximum snowfall to a height above the area of the main névé. Also to be considered as an additional increment of positive net accumulation in a glacier system is that portion of summer meltwater in geophysically subpolar or polar (cold) névés recaptured by freezing as it percolates to depth. In contrast, meltwater percolation in temperate glaciers drains away almost entirely in subglacial drainage channels, eventually flowing out at the snout of the glacier as an increment of net loss.

Structure and movement. Because of the many structures in ice comparable to those in sedimentary and metamorphic rocks, a glacier is an ideal field laboratory for the structural geologist. A few such structures are: primary stratification (bedding strata), secondary fracture structures, discontinuous marginal or basal tectonic foliation, (Fig. 8), ablation surfaces overthrust surfaces, faults, folded structures, and a varied group of deformed sedimentary and structural bands, including the wave-ogive bands and surface bulges illustrated in Figs. 3 and 4. There are also subsurface diagenetic structures of both stratiform and transverse ice which result from refreezing of downward percolating meltwater in subfreezing firn. Cross-cutting transverse types are manifest at the surface as ice columns and dikes. Tension and shear fractures are also common as crevasses and bergschrunds (the latter exhibiting an overhanging upper lip), as well as moulins (glacier mills, or deep rounded holes caused by water action on embedded stones), cryoconite holes (thermal pits produced by the inmelting of organic or rock fragments), sastrugi (wind-scoured features), and other surface features. All of these are representations of the combination of processes which affect and control the surface regime and the dynamics of the internal structure and movement of glaciers.

Glacier deformation is a composite of internal and external movement. The internal movement is dominated by a continuous plastic creep (flow deformation), and the external in some cases by a fracture type of discontinuous movement at the bed or near the glacier margins. As a general rule, flow deformation, hence the rate of movement, is greater in the upper portion of a glacier than in its basal section (Fig. 9). In glaciers with a healthy or strongly positive regime in the accumulation zone, a dominant proportion of the total mass transfer may be expected to be via actual sliding of the glacier over its bed. Such glaciers are characterized by rectilinear, pluglike, or Block-Schollen velocity profiles. In plan view the greatest amount of move-

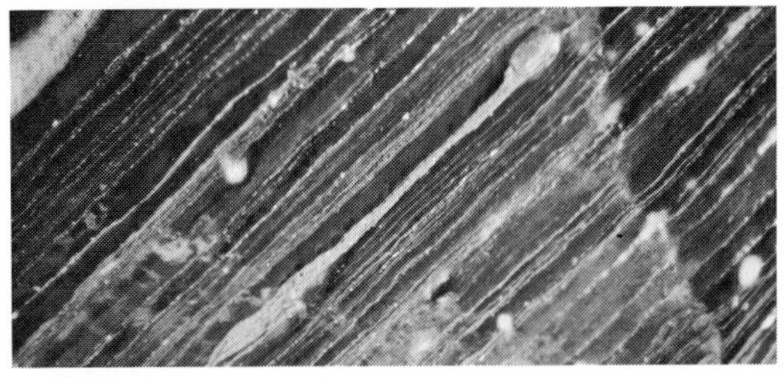

Fig. 8. Basal tectonic foliation (discontinuous laminar shear bands) with entrained fragments of pebble-size rock and fines in sole of Moltke Glacier, a geophysically polar glacier east of Thule. *(Photograph by M. M. Miller)*

ment is expressed in a broad central area. The surface velocity profiles in Fig. 10 were surveyed at stated intervals up-glacier from the terminus and represent the general flow characteristics of this strongly advancing glacier during the 1950s. Sliding glaciers often contain sheared basal ice heavily entrained with rock fragments and serving as effective agents of erosion. In glaciers of equilibrium or negative regime in the névé, the dominant mode of movement is expressed as internal streaming flow. In plan view such velocity profiles have a smooth, arched, parabolic form with associated streamlines. All gradations of movement from continuous laminar (streaming) flow to discontinuous Block-Schollen, or plug flow, may occur in any one glacier system. The relative proportions depend upon the névé regime pattern in recent decades, on internal temperature characteristics of the ice, and on configuration of the bedrock channel.

During the 1950s laboratory and field experiments by M. Perutz, J. Glen, M. Miller, J. Nye, S. Steinemann, and others ushered in a new set of concepts in glaciology, which related rheological and mathematical models of glacier mechanics (deformation and flow) to the phenomena of glacier movement and mass transfer observed in nature. The field measurements were largely based on the deformation of glacier tunnels and elongated metal

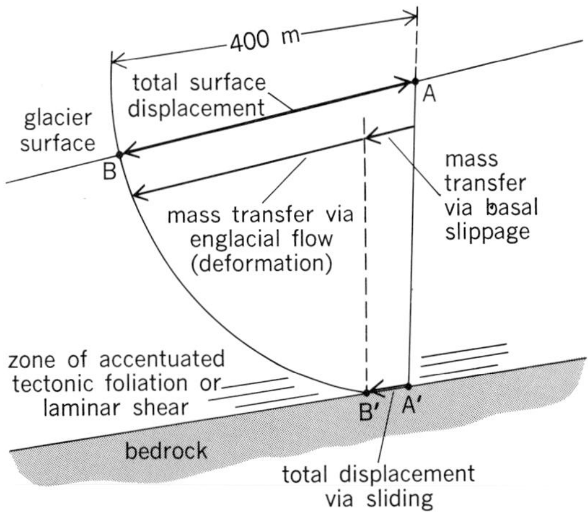

Fig. 9. Sketch of hypothetical case to demonstrate the proportion of total mass transfer across a given profile which can be ascribed to englacial flow deformation plus the proportion resulting from erosive slippage or bed sliding. This case is typical of plug flow or Block-Schollen mass transfer in a healthy advancing glacier.

pipes drilled into the glacier perpendicular to its surface. As a result of this pioneering work, the nature of internal deformation of glaciers is treated as a problem in the physics of shear, with reference to plasticity theory. Thus it is expressed by an exponential relationship between gravitational stress and deformation per unit time (creep velocity) with a simplified power law equation, Eq. (1).

$$\frac{d\gamma}{dt} = k\tau^n \qquad (1)$$

Here γ is the shear strain, τ is the shear stress in bars, k is a constant for any given temperature, and n is an empirical constant, depending in large measure on the physical character of the ice. Also, the exponent n probably depends to some degree on the magnitude of stress, a factor judged as probably significant only in very deep or otherwise highly stressed ice.

Under temperate valley glacier conditions, the factor n in nature is found to be close to a value of 3. In the flow law n represents the slope of the double logarithmic plot of the fundamental shear stress – shear strain relationship. The constant k is obtained from the stress-strain formula at the stress of 1 bar. Since the logarithm of 1 is zero, k is, in fact, the ordinate at the point where the abscissa of any log stress – log strain rate line is zero.

Also $d\gamma/dt$ or $\dot{\gamma}$ is the strain rate per year, usually calculated in radians of angular deformation per year, because it is found to be numerically equivalent to the tangent of the changing tilt angle in slowly deforming englacial pipes from which vertical velocity profiles have been measured. Several examples of field measurements are given in Fig. 11.

It should be noted that the power law used with a universal constant n represents only an average for low-gradient temperate glaciers, and therefore can give no more than a good approximation of the true strain rate in a glacier at depth. Nonetheless, by application of this law, at least for glaciers of simple configuration, a reasonable determination can be made of the movement within a glacier. And by combining the englacial movement data with surface velocities obtained by periodic surveys of across-glacier stakes and with geophysically determined depth records, an assessment can be made of the relative proportion of sliding or slippage on the bed (Fig. 9).

The shear stress at any depth with a glacier is calculated from relation (2) in dynes cm^{-2}, where D is

$$\tau = D\rho g \sin \alpha \qquad (2)$$

the depth in centimeters, ρ the bulk specific gravity of the overlying mass in grams per cubic centimeter, α the surface gradient, and g the acceleration of gravity (980 cm sec^{-2}). In relation (2) the very critical control exercised by slight variations in gradient is well revealed.

Thus in the flow law with all factors determinable in nature, and with n and k calculated for any particular velocity profile, it is possible to extrapolate the differential flow all the way to the base of a glacier. In this way determination is made of the proportion of total mass transfer within a glacier. For this calculation the difference is found between the quasi-plastic flow velocity at any given depth U_D and the surface velocity U_0 at the glacier

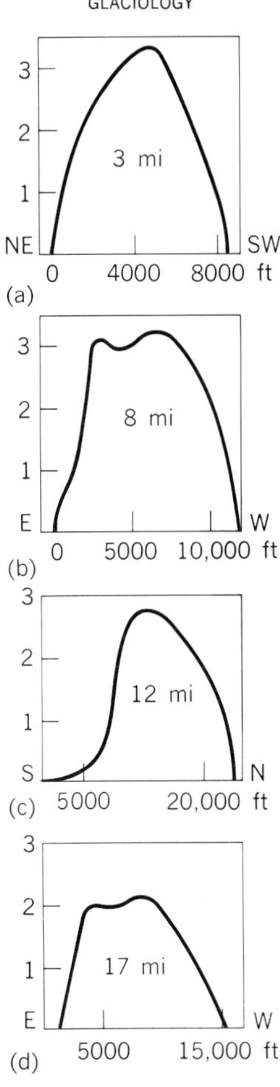

Fig. 10. Horizontal across glacier surface velocity profiles in ft/day from terminus up into névé zone of Taku Glacier, Alaska. *(a, c)* Partial parabolic streaming. *(b, d)* Block-Schollen (plug flow).

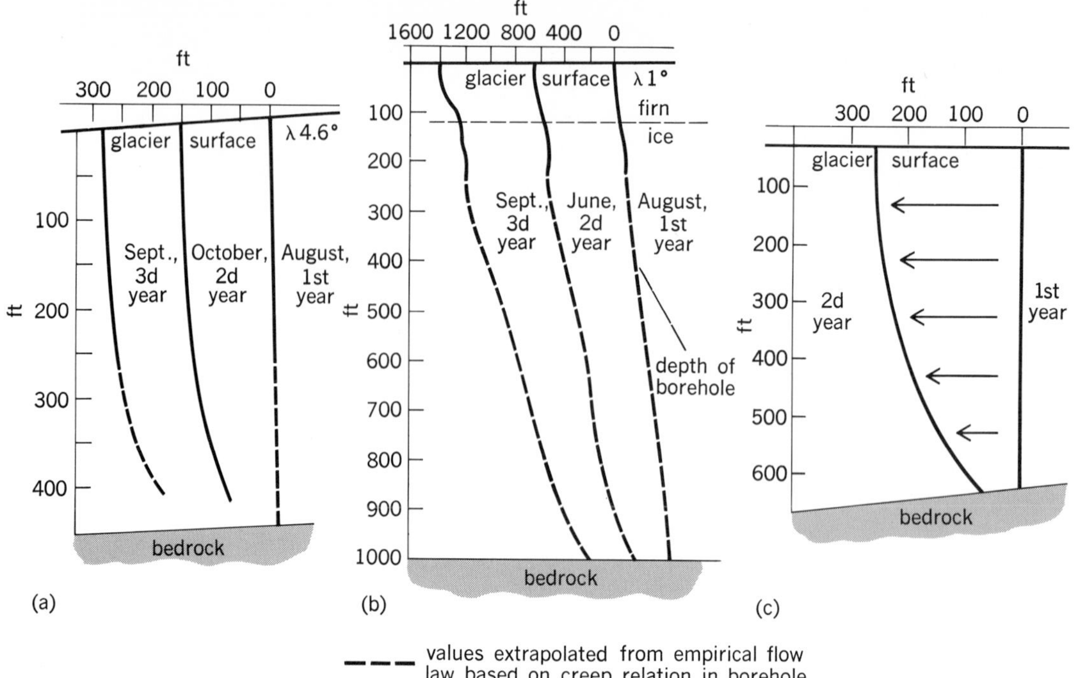

Fig. 11. Vertical velocity profiles, based on measurements made in the field, showing annual changes on the (*a*) Aletsch Glacier, Switzerland (*after M. Perutz*); (*b*) Taku Glacier, Alaska (*after M. Miller*); and (*c*) Saskatchewan Glacier, Canada (*after M. Meier*).

surface. The relation is expressed by combining the empirical constants k and n with the stress at the indicated depth D as in Eq. (3), which was formulated by Nye.

$$U_O - U_D = \frac{k}{n+1} D\tau^n \tag{3}$$

In this manner the approximate value of basal sliding is determined. A hypothetical case is shown in Fig. 9. The actual extent to which bottom sliding takes place depends not only on the surface gradient but on the slope and roughness of the underlying topography, on the width and thickness of the glacier, and on any other blocking factors involved. All these factors vary in their effect on the total stress distribution in different sectors of a glacier, and they are expressed at any particular point in the ice by the creep law, which integrates the gravitational load and englacial temperature. It must be kept in mind that such analyses oversimplify the situation in nature, because ice is not actually a homogeneous plastic material. Instead it should be considered as a quasi-elastoplastic substance, whose behavior under stress is still not fully understood.

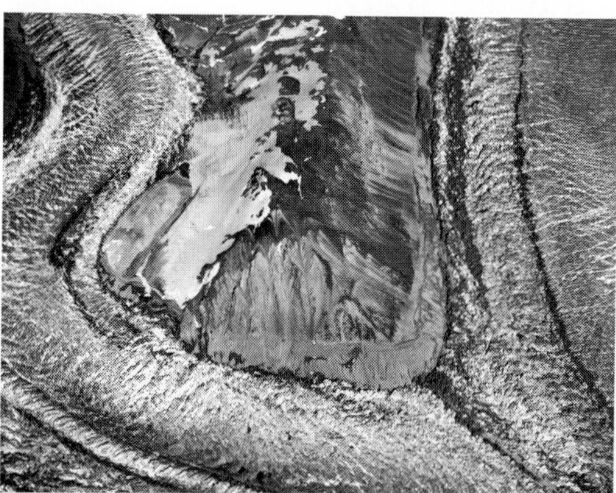

Fig. 12. Massive marginal shearing and lowering of ice along the valley walls of surging Walsh Glacier, in the Yukon Territory of Canada. Note the strongly related development of conjugate shear crevasses. This glacier moves forward tens of feet per day. (*Photograph by H. B. Washburn, August, 1966*)

In glacier ice when the stress conditions are great and substantial bottom sliding takes place, there is also much marginal shearing along the edges of the glacier. This is suggested by the form of transverse surface-velocity curves (Fig. 10) and the field measurements of vertical velocity profiles which have been noted (Fig. 11). The accentuation of laminar shearing or tectonic foliation in the sole of a glacier, observed as ribbon structures along the glacier's margins and bed (Fig. 8), indicates that slippage more usually takes place in a zone of highly sheared basal ice carrying entrained rock fragments and debris, which in turn serve as the key erosive tools of the glacier.

Erosion and transportation. Like rivers, glaciers have distinct regions of erosion and transportation. For example, in the high cirque basins of alpine glaciers, rock fragments fall from cliffs or slide into bergschrunds or marginal moats and become entrained in the ice for transport down the valley. These transported fragments thus become the critical tools of erosion along the ice-bedrock interface (Fig. 8). The larger angular rocks scratch the bedrock and form grooves and striae (Fig. 14), while the finer clastic detritus serves as the abrasive to smooth and polish. In such areas the pres-

Fig. 13. Massive earthquake avalanche of rocks and debris covering approximately 10 mi² of the Schwann Glacier in the Eastern Chugach Range, Alaska, was caused by the 1964 Good Friday earthquake. (*Photograph by M. M. Miller*)

ence of impact structures, striae, grooves, and crescentic features are not only proof of former glaciation but also reveal the former direction of flow.

Other geomorphological characteristics support the role of bottom slippage in erosion, such as deeply entrenched and U-shaped valleys reflecting the parabolic form of englacial creep in an actively advancing glacier, and the vast quantity of rock flour (clay size) carried out from beneath many glaciers by subglacial and proglacial drainage streams. Neither of these conditions could possibly have developed without considerable abrasion from basal sliding of a debris-entrained sole of the glacier.

Plucking is also an important and related agent of erosion. This process involves the penetration of ice or rock wedges into subglacial niches, crevices, and joints in the bedrock. As the glacier moves, it plucks off pieces of jointed rock and incorporates them as supplemental agents of abrasion and further plucking. Down-valley ends of jointed hummocks in the bedrock are produced in this manner and are known as roches moutonnées. Figure 14 shows the glacial grooves and oriented crescentic erosional features on top of a roche moutonnée. A sequence of such bedrock bosses produces a steplike longitudinal profile on a glacial valley floor. The steps often coincide with particularly resistant lithologies or selective low-angled joint surfaces.

Surging glaciers and kinematic waves. Research has suggested that glacier surges, expressed as sudden catastrophic advances or raising and lowering of the ice surface, are relatively common phenomena in some regions. Such abnormal surges are characterized by marked and seemingly anomalous increases in flow velocity and often by a rapid transfer of ice from the névé to the terminus, with much increased pinnacling and crevassing. The upper glacier surface may sink or be substantially lowered with this volume loss, expressed as a comparable thickening in the lower valley sector. Such surges are believed to be kinematic in nature, in that the wave moves through the glacier at a substantially faster rate than the actual discharge of ice. Often much crevassing and geometric folding of in-ice structures and medial moraines occur, as well as strong buckling, shearing, and surface lowering along the valley walls (Fig. 12). Striking examples are the Dusty, Lowell, Walsh, and Steele glaciers in the Yukon, where in 1965–1969 surge velocities of up to 60 ft per day were reported.

Surging glaciers were first reported by R. F. Tarr and L. Martin in Yakutat Bay, Alaska, in 1910–1913, with later reports of vigorous surging in this same area in 1965–1968. Surges of spectacular nature have been reported on many glaciers, including the Black Rapids Glacier, Alaska, in 1937; Bruarjokul Glacier, Iceland, in 1965; Medvezhii Glacier in the Pamirs, Soviet Union, in 1963; and Muldrow Glacier on Mt. McKinley, Alaska, in 1964. Others have been reported in Spitsbergen in 1952, Ellesmere Land in 1964, and even in the Karakoram and Himalayas in 1953 and 1963.

Although this phenomenon is not adequately understood, much research is under way. It is at least known to relate to development of a dynamic instability that usually attenuates within 1 to 5 years, after which the lower glacier area affected by such an advance begins to stagnate. It is not the same phenomenon producing normal discharge wave-bulges and advance, as these may be presumed to relate to annual accumulation variations and to gradual climatic change. One possible origin of the catastrophic surge phenomenon is a very unusual and sudden change in load stress via thickening of the glacier in its upper névés or high-level nourishment basins, with a consequent energy release via a kinematic wave.

Of course such thickening may be associated with abnormal increases of snowfall by way of what appears to be sudden climatic change. In certain unique situations, the suddenness of such change can be accentuated by significant horizontal or vertical shifts in the zone of maximum snowfall across the critical névé level. But there are several other possibilities. One may relate to warming of the englacial ice itself, a factor which probably caused the Brasvalsbreen Glacier in Spitsbergen to slide forward catastrophically in the late 1940s (a 5-mi advance in less than 5 months). This may not have been a true surge but a response to substantial changes in englacial temperature, since it is known that the mean winter temperatures in the northeastern Arctic rose about 20°F during 1920–1960.

The third possibility is the sudden imbalance produced by earthquake-caused rock and ice slides. A number of huge slides were reported on certain Alaskan glaciers following the 1964 Good Friday earthquake. One of these is pictured in Fig. 13, showing apparent flow effects in a deformed medial moraine below a 10-mi^2 debris slide on Schwann Glacier in the Copper River region, Alaska. A fourth, and probably minor, possibility is some actual effect of the earthquake shock in the disarticulation of the glacier structure, which could abet the "sudden-slip" character of a subsequent advance. This would probably be true when the epicenter of the earthquake lies in the vicinity of the glaciers involved. A further possibility is a buildup of excessive pressure melting on a glacier's bed, or in some cases an intensification of geothermal heat at the sole of the glacier, resulting in abnormal quantities of lubricating water (slush) which could accentuate the effects of hasal slip. In some cases the most dynamic surges could be a result of a combination of two or more of these factors. However, glaciologists must remain cautious, because many glaciers surge and even express seemingly less kinematic effects, such as surface buckling and locally increased crevassing, without any evidence of avalanche material being dumped on them. At least one correlation which is of interest is that some of the most spectacular surging glaciers have been reported in regions of the most active tectonic-earthquake activity.

Fig. 14. Glacial pavement on grantic bedrock surface in recently deglaciated mountain region of Alaska, showing orientated abrasion structures resulting from basal sliding of the glacier over its bed. In this view can be seen crescentic gouges (concave in up-glacier direction), lunate furrows (concave in down-glacier direction), grooves (elongate linear furrows), and striae (fine lines of scratches). The direction of the former ice movement is from foreground toward valley in distance. (*Foundation for Glacier and Environmental Research*)

The greatest bed erosion takes place when the regime of a glacier is healthy and its mass transfer substantial. A vigorously advancing glacier exhibiting Block-Schollen or plug flow is the most effective erosive body. This adds credence to the concept that glacial erosion is an indirect consequence of glacioclimatological oscillations affecting the growth and decay, thickening and thinning, and advance and retreat of glaciers.

As to the direct cause of high ratios of bed slippage and the occasional development of anomalous surges, scrutiny of the shear stress equation reveals that flow within a low gradient glacier or ice sheet (for example, at slopes of 1–3°) is so small that in many cases stresses other than those explained by pure gravitational shear must be in effect. Hydrostatic pressure is ruled out on the basis that it is as negligible in ice as in liquids. The most significant supplemental stress may be attributed to a strong down-glacier longitudinal force superimposed on the gravitational stress in quasi-elastoplastic ice whose yield limit is already exceeded. Such supplemental stressing can be the result of any one factor or a combination of the factors which come into play in advancing and surging glaciers. To such stress tensors, the strain effects of which are sometimes accentuated by an increase in temperature within the ice, the phenomenal glacial advances in Alaska (group IV, Fig. 15) may be ascribed.

The repeated oscillation of glaciers in mountain cirques and over the floors of outlet valleys results in effective scouring, transportation, and removal of material. Such continuous sequences of process produce the wide-strath highland glacial basins and deep U-shaped outlet valleys so common in the Alps, the Cascades, and the Rocky Mountains. Turbid glacier rivers and moraines (lateral, medial, and terminal) are living examples of the immense

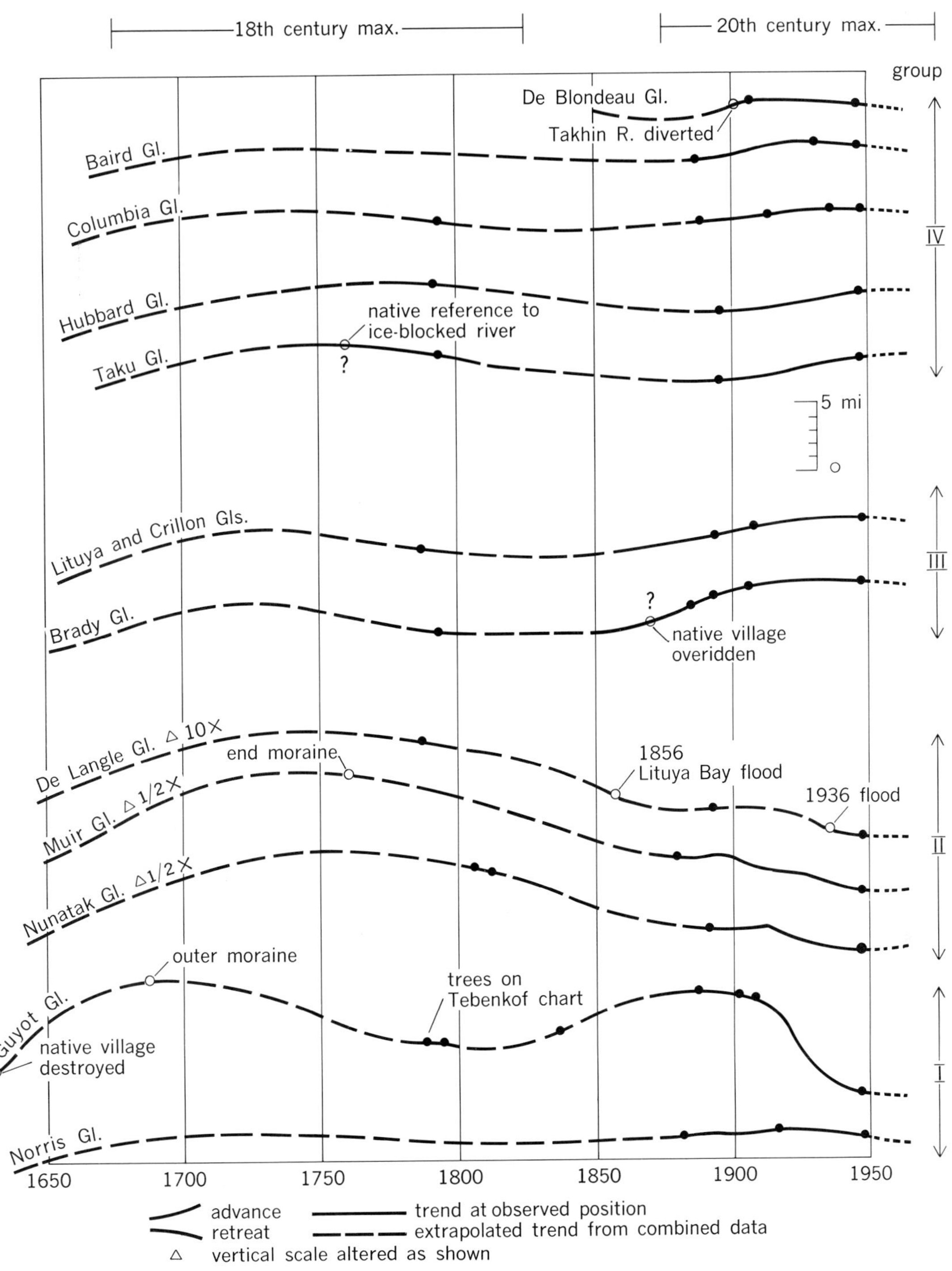

Fig. 15. Two-phased regional fluctuation pattern of the Alaskan Little Ice Age showing how, at any one time, glaciers in a region may reveal patterns of strong different advance simultaneous with significant retreat of others. This out-of-phase pattern reflects the role of differences in size, geographical position, and elevation or orientation of névés, as well as differences in flow lag via differences in size, configuration, and length of the outlet valley glaciers.

eroding and transporting power of glaciers. Festooned arrays of terminal moraines also testify to continuous glacier activity, because they express repeated oscillations most usually related to cyclic changes of accumulation and ablation in the distant névés many years before.

Recent and current fluctuations. Following the post-Wisconsinan maximum about 10,000 years ago, the Ice Age entered its latest waning phase. The result was an almost complete retreat and disappearance of glaciers in the mid-latitudes and tropical regions of the Earth. Coincident with this disappearance was the Thermal Maximum (Hypsithermal Interval or Climatic Optimum) culminating about 5000 years ago. Between 1000 B.C. and A.D. 1 a worldwide recrudescence of glaciers began to take place. This is termed the Neoglaciation and is associated with a return to harsher climatological conditions, which still prevail today. This temperature fluctuation culminated in a colder condition at the beginning of the Christian era. The evidence suggests that in the 5th and 7th centuries the polar seas were freer of ice than they are today, as far north as the pole. Peripheral waters thus remained relatively clear through the 10th century. The records of Norse settlers in Greenland indicate that the climate remained relatively mild until the 14th century. Then, about 5 centuries ago, there began another worldwide expansion of glaciers and a thickening of polar ice, generally referred to as the Little Ice Age. Technically, the Little Ice Age refers to this latest major phase of reglaciation in the latter quarter of the Neoglacial Age.

Little Ice Age fluctuations are double-phased in nature, having produced a worldwide growth of temperate glaciers which reached their culminations in the early to mid-18th century, and again in the late 19th to mid-20th centuries. The Alaska Little Ice Age pattern is illustrated in Fig. 15. This reveals that at any one time the terminal regime of glaciers in adjoining areas may be quite out of phase; that is, one may advance while another simultaneously retreats. The latest advances on a small percentage of high-level trunk glaciers have continued into the 20th century, in spite of a general diminution of ice cover around the periphery of some of the lower Alaskan ice fields. Recent fluctuations in Scandinavia and Patagonia are quite similar to this Little Ice Age pattern and have been shown to have a teleconnectional similarity via upwards of a dozen recessional moraines over the past 200 years. Such evidence supports the global nature of the causal factor, and the acute sensitivity of glaciers as excellent historians of secular climatic change.

In contrast to the behavior of temperate glaciers, the polar glaciers of Antarctica exhibit a fairly stable regime. This suggests that significant volume changes in middle-latitude glaciers, including the sub-Arctic and the sub-Antarctic, and noticeable increases in their internal temperatures have been instrumental factors in accentuating the fluctuation patterns of Neoglacial time. *See* GLACIAL EPOCH; GLACIATED TERRAIN.

[MAYNARD M. MILLER]

Bibliography: L. A. Bayrock, *Catastrophic Advance of the Steele Glacier, Yukon, Canada*, Boreal Inst. Univ. Alberta Occasional Publ. no. 3, 1967; J. A. Gerrard, M. F. Perutz, and A. Roch, Measurement of the velocity distribution along a vertical line through a glacier, *Proc. Roy. Soc. London Ser. A*, 213:546–558, 1952; J. W. Glen, The creep of polycrystalline ice, *Proc. Roy. Soc. London Ser. A*, 228:519–538, 1955; A. E. Harrison, Ice surges on the Muldrow Glacier, Alaska, *J. Glaciol.* 5(39): 365–368, 1964; *J. Glaciol.* 1947 onward; B. Kamb, Glacier geophysics, *Science*, vol. 146, no. 3642, Oct. 16, 1964; M. M. Miller, Alaska's mighty rivers of ice, *Nat. Geogr. Mag.*, 131(2):194–217, February, 1967; M. M. Miller, Phenomena associated with the deformation of a glacier borehole, *Trans. Int. Union Geod. Geophys.*, vol. 4, 1958; M. M. Miller, 1965–1968 Studies of surge activity and moraine patterns on the Dusty Glacier, St. Elias Mountains, Yukon Territory, *Proc. 19th Alaska Sci. Conf.*, AAAS, 1968; J. F. Nye, The flow law of ice from measurements in glacier tunnels: Laboratory experiments and the Jungfraufirn borehole experiment, *Proc. Roy. Soc. London Ser. A*, 219:477–489, 1953.

Glauconite

The term glauconite as currently used has a twofold meaning. It is used as both a mineralogic and morphologic term. The mineral glauconite is defined as an illite type of clay mineral. It is dioctahedral with considerable replacement of aluminum by iron and magnesium. Structural substitutions result in a charge deficiency in both the tetrahedral and octahedral sheets, and interlayer cations seem to balance both of these charges. Calcium and sodium as well as potassium are the interlayer cations. A fundamental characteristic of glauconite is that the unit cell is composed of a single silicate layer rather than the double layer of most other dioctahedral micas. *See* CLAY MINERALS; ILLITE.

Glauconite is known to occur in flakes and as pigmentary materials. When used in the morphological sense, the term glauconite often refers to small, green, spherical, earthy pellets. Some of these pelletal varieties are composed solely of the mineral described above, others are a mixed-layer association of this mineral and other three-layer structures.

The mineral glauconite is formed during marine diagenesis. Because of the frequent association of glauconite with organic residues, it has been generally concluded that the presence of organic material is necessary for the formation of this mineral. Glauconite forms in a marine environment from a variety of raw materials. It is known to form as an alteration product of biotite mica, when the alteration takes place under reducing conditions. The deposition of sediment must be slow during this mineralogic transformation to allow complete alteration before burial. If burial is accomplished before alteration, the biotite mica persists. *See* AUTHIGENIC MINERALS; DIAGENESIS.

Relatively shallow water is another requisite for the formation of glauconite. It has been shown that glauconite forms in the shallow sea in agitated waters which are not highly oxygenated. A reducing environment is also probably necessary for the formation of glauconite.

The magnesium content of glauconite is very uniform, and the ratio of ferric to ferrous iron is rather constant. This suggests that a critical content of magnesium and a particular oxidation-reduction potential might be required for this mineral to form. It has been suggested that certain concentrations of sodium and potassium ions are also necessary for glauconite formation, since the ratio of these cations in the interlayer positions is rather distinctive.

In summary, glauconite forms during marine diagenesis, in relatively shallow water, and at times of slow or negative deposition. In addition, peculiar reducing conditions and particular concentrations of alkalies and magnesium seem to be essential for its formation.

Glauconite readily occurs in pellet form and has been identified in both recent and ancient sediments. It is a major component in some "greensand" deposits and has been used commercially for the extraction of potassium from such sources. *See* MARINE SEDIMENTS.

[FLOYD M. WAHL; RALPH E. GRIM]

Glaucophane

The name given to the monoclinic sodium amphiboles of the general composition $Na_2Mg_3Al_2Si_8O_{22}(OH)_2$. Like most sodium amphiboles, glaucophane is blue to black with marked pleochroism (color change on rotation in plane-polarized light) in thin sections from yellowish green to deep blue. The crystals are prismatic and sometimes fibrous. The mineral is most easily identified by its occurrence. Glaucophane occurs in a restricted group of metamorphic schists that commonly exhibit an unusual mineral association with dense minerals, such as lawsonite, $CaAl_2Si_2O_7(OH)_2$; jadeite; pumpellyite, $Ca_2Al_3Si_3O_{12}(OH)$; garnet; and less dense minerals, such as epidote, mica, chlorite, and albite. The bulk composition of glaucophane schists is not unusual or particularly rich in sodium. Most often the bulk of the sodium of a rock is present in the feldspars, but in the glaucophane schists it is present in glaucophane and jadeite. Because of the higher density of the amphiboles and pyroxenes relative to feldspar, the role of high pressures (at least locally) is thought by many to be important in the formation of glaucophane. It is also thought that glaucophane and jadeite may be volumetrically important minerals in the lower regions of the Earth's crust if sodium is present in the usual amounts. For the glaucophane schist, the mineral associations and the low metamorphic grade of the rocks deny the presence of high temperatures for these rocks. *See* AMPHIBOLE; JADEITE.

[GEORGE W. DE VORE]

Gneiss

An old term in mining and geology that was first applied by the miners of the Erzgebirge, Austria, to designate the country rock adjacent to the ore veins. Gneiss still has a broad definition, and in its widest sense, it is used as a structural rather than compositional term. Thus it applies to a great variety of rocks with a banded or coarsely foliated structure. Precambrian gneiss was once regarded as the material of Earth's first primitive crust. A modern version includes in the definition of gneiss (primary gneiss) that it is of plutonic igneous origin with banding caused by flow movements in a crystallizing heterogeneous magma.

Typical gneisses are highly metamorphic, coarse-grained, irregularly banded or foliated rocks composed mainly of quartz and feldspar with some biotite (= granitic composition). As the grain becomes finer and the amount of mica increases, gneisses pass into pelitic schists. With fine grain and decrease of mica, or garnet substituting for mica, they become less foliated and pass into granulite. *See* GRANULITE; SCHIST.

Primary gneiss is a term used for rocks of metasomatic, migmatitic origin which are related to regional processes of granitization. In the root parts of folded mountains a pore fluid, or ichor, is generated by differential melting (palingenesis or anatexis), and the solid residues and the contiguous sediments are stewed in this magmatic-anatectic liquid which reacts with the sediments and metasomatically transforms them into migmatic gneisses. Lit par lit gneisses and augengneisses belong to this category. *See* METASOMATISM; MIGMATITE.

Paragneiss shows a sedimentary parentage. In the deeper parts of the orogenic belts any sedimentary complex will recrystallize under rather high temperature and pressure (the conditions of the amphibolite or the granulite facies). Aided by emanations or ichors of granitic composition, various types of paragneisses may form. If derived from argillites, the gneisses may be rich in muscovite, biotite, garnet (almandine), cordierite, kyanite, or sillimanite. If calcareous material is present, the gneisses may carry epidote, augite, garnet (grossularite), and wollastonite. Many banded gneisses have a sedimentary parentage.

Orthogneisses are primary igneous rocks in which a gneiss structure has been induced by essentially kinetic metamorphism. Flaser granite and flaser gabbro designate the first stage of the metamorphism; after progressive recrystallization, terms like gabbro gneiss have been used. Ordinarily the term gneiss refers to a granitic composition; anything else should be specially noted. Typical gneisses (of granitic composition) make up large areas in Precambrian regions and in the deeper parts of orogenic belts. *See* OROGENY.

[T. F. W. BARTH]

Goethite

A mineral having composition FeO(OH) and comprising most of the material known as brown iron ore. Goethite crystallizes in the orthorhombic system. Crystals are rare and the mineral is usually in radiating fibrous aggregates with a reniform surface (see illustration). It may also be stalactic or massive. There is one perfect cleavage. The hardness is 5–5.5 on Mohs scale; the specific gravity is 4.37, or lower because of impurities. The luster is adamantine to dull and the color light to dark brown. It is characterized by a yellowish-brown streak.

Limonite commonly appears to be amorphous but is similar in its physical and chemical proper-

Goethite, taken from Negaunee, Mich. (*Specimen from Department of Geology, Bryn Mawr College*)

ties and occurrence to crystalline goethite. Until the study of these minerals by x-rays in the 1920s, goethite was considered to be a relatively rare mineral and limonite common. X-ray analysis showed most of the presumed limonite to be crystalline and therefore goethite.

Goethite is a common mineral formed by the oxidation of iron-bearing minerals. It is the major constituent of the gossan at the surface of metalliferous veins, forms as the residual mantle from the weathering of serpentine, and is deposited as bog iron in marshes and stagnant pools. In some localities it is found in large masses and constitutes a valuable iron ore, for example, the minette ores of Alsace-Lorraine. *See* LIMONITE; ORE AND MINERAL DEPOSITS.

[CORNELIUS S. HURLBUT, JR.]

Graben

A block of the Earth's crust, generally with a length much greater than its width, that has been dropped relative to the blocks on either side (see illustration). The size of a graben may vary; it may

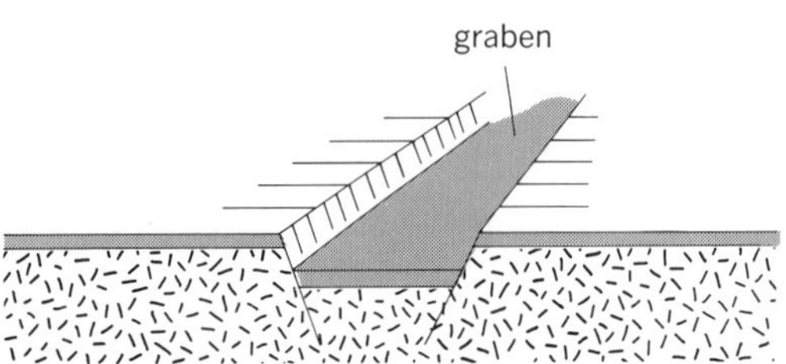

Diagram of simple graben. (*From A. K. Lobeck, Geomorphology, McGraw-Hill, 1939*)

be only a few inches long or it may be hundreds of miles in length. The faults that separate a graben from the adjacent rocks are inclined from 50–70° toward the downthrown block and have displacements ranging from inches to thousands of feet. The direction of slip on these indicates that they are gravity faults. Graben are found in regions where the crust has undergone extension. They may form in the crests of anticlines or domes, or may be related to broad regional warpings. *See* FAULT AND FAULT STRUCTURES; HORST; RIFT VALLEY; WARPING OF EARTH'S CRUST.

[PHILIP H. OSBERG]

Granite

A phaneritic (visibly crystalline) plutonic rock with granular texture, composed largely of quartz and alkali feldspar with subordinate plagioclase and dark-colored (mafic) minerals (biotite and hornblende). More generally the term includes most plutonic rocks rich in quartz and feldspar. In this sense it includes, in addition to true granite, such rocks as granodiorite and quartz monzonite. Commercially the term granite is still more inclusive and is used for any phaneritic rock rich in feldspar and with or without quartz and mafic minerals. Because of its strength, durability, and pleasing colors (gray, pink, and red) granite is an important building and ornamental stone. For a general discussion of textural, structural, and compositional characteristics *see* IGNEOUS ROCKS.

Mineralogy. True granite is distinguished from granodiorite by the preponderance of alkali feldspar over plagioclase. Otherwise the two rocks are similar. Alkali feldspar (microcline, orthoclase, usually perthitic) is generally pink, flesh, or buff, and normally occurs as anhedral (without crystal form) or subhedral (with partial crystal outline) grains. In alkali granites it is more soda-rich than in normal (calc-alkali) granites. The common plagioclase of calc-alkali granites is oligoclase, whereas alkali granites may carry discrete crystals of albite. In granodiorite the plagioclase is commonly as calcic as sodic andesine. It is usually subhedral and may show normal zoning (calcium-rich cores surrounded by successively more sodic shells).

Quartz appears as glassy clear to smoky gray grains or grain aggregates of anhedral form and is generally interstitial. Under the microscope it may show effects attributed to strain. With high magnification it may be seen to contain tiny gas-liquid bubbles or inclusions and hairlike needles of rutile.

Black, flaky biotite mica (microscopically brown) is the principal mafic mineral of calc-alkali granites. It may occur as the only mafic or with more or less amphibole. Biotite of alkali granites is iron-rich. Green hornblende as short subhedral prisms is the usual amphibole, but soda amphibole (hastingsite, arfvedsonite, or riebeckite) is typical of alkali granites. Pyroxene is rare, but diopsidic augite may form cores within hornblende in calc-alkali granites. Alkali granites may carry aegirine-augite or aegirite as interstitial material or as accicular crystals. Diopside and augite are more common and abundant in granodiorite.

Muscovite mica may be a primary or secondary mineral. Primary muscovite does not occur with hornblende or pyroxene but is usually accompanied by biotite. It may be intergrown with biotite or occur as thick hexagonal plates.

The chief accessory minerals of granites are magnetite, ilmenite, apatite, zircon, and sphene as tiny grains or crystals. Dustlike particles of hematite are commonly abundant in the potash feldspar and give that mineral its pink or salmon color. A few granites carry fayalite, an iron-rich olivine, and some contain tourmaline or fluorite which may have formed subsequent to consolidation of the rock.

Texture. Granites display a wide variety of textures and structures. Cuneiform intergrowths of quartz and alkali feldspar forming micrographic or granophyric texture are not uncommon and characterize the special varieties, graphic granite and granophyre, respectively. A fine-grained, sugary-textured rock (aplite) is developed in dikes and locally in some larger granite bodies. Elsewhere the texture becomes extremely coarse (pegmatitic) and the rock passes into granite pegmatite. *See* APLITE; PEGMATITE.

Many granites show porphyritic texture in which large crystals (phenocrysts) of alkali feldspar are embedded in a finer-grained matrix. The phenocrysts are euhedral (with good crystal form) to anhedral and may have formed either early or late in the period of crystallization. In some cases the large crystals formed by metasomatic replacement of solid rock and are more properly called porphyroblasts. Quartz is less common as phenocrysts. Fine-grained granites with abundant phenocrysts are known as granite porphyry. *See* PHENOCRYST; PORPHYRY; RAPAKIVI GRANITES.

Structure. Directional structures are common in many granites and granodiorites (gneissic granites), and may be best developed or confined to margins of the body. These take such forms as banding, schlieren (elongate tabular lenses), and parallel orientation of elongate or platy minerals and inclusions. These features (flow structures) may trend parallel to the margin of the granitic body or may conform closely with similar structures in the surrounding rocks. They may have formed by flowage of the crystallizing granite magma or they may represent relics of bedding or foliation in metasomatic or metamorphic granite. Granites crystallized at shallow depth and, therefore, under relatively low pressure may show miarolitic cavities containing beautifully crystallized minerals. Orbicular structure, although not common, may be well displayed by granites and granodiorites.

Joints or extensive, smooth fractures in parallel arrangement are characteristic structures of most granitic rocks. A special type of joint known as sheeting fracture divides the granite into huge slabs or sheets, resting one above the other. Normally these fractures form parallel to the Earth's surface presumably due to expansion and release of confining pressure as erosion strips away the thick overburden. Closely spaced joints or joints which do not intersect at right angles may make a granite unsuitable for quarrying.

Occurrence. Granitic rocks form bodies of various sizes ranging from small dikes and sills up to huge batholiths covering thousands of square miles and localized in the cores of fold-mountain ranges and the great Precambrian shield areas of the world. Among the smaller- and intermediate-sized bodies are the stocks, sheets, and domes, and a variety of irregularly shaped plutons. *See* PLUTON.

In addition to occurring independently, granite may be associated with great masses of granodiorite and tonalite; with gabbro and intermediate rocks in smaller complex assemblages; with gabbro and related rocks in large, thick, sheetlike bodies; with subordinate amounts of quartz syenite and syenite; and with small masses of quartz syenite and alkali syenite.

Origin. Just how granite forms constitutes a major problem in geology. Three principal types of processes appear to be operative: magmatic, metamorphic, and metasomatic; these may act independently or in various combinations. Magmatic granite forms by slow crystallization of a deeply buried granitic melt (magma). Metamorphic recrystallization (reconstitution by heat, pressure, and volatiles) may transform volcanic or sedimentary rocks into granite. A wide variety of sedimentary or igneous rocks may be changed to granite, in essentially the solid state, by the introduction of certain elements, such as alkalies and silica, and the removal of others, such as iron, magnesium, and calcium. This process of replacement or metasomatism is involved in the phenomenon of granitization. *See* GRANITIZATION; MAGMA; METAMORPHIC ROCKS; METAMORPHISM; PETROGRAPHIC PROVINCE.

[CARLETON A. CHAPMAN]

Experimental petrology. Experimental studies of the phase relationships in systems containing alkali feldspars, plagioclase, and quartz provide a basis for the interpretive petrology of granitic rocks. The presence of water and other volatile components under pressure permits the study of reactions which are too sluggish to be investigated at atmospheric pressure. Phase relationships in the synthetic systems are very similar to those determined by using crushed natural rocks as starting materials. Comparison of experimental phase relationships with the mineralogy of granitic rocks supports the proposals that crystal-liquid reactions were involved in the origin of many rocks. The process could be fractional crystallization of a more basic magma, or partial fusion of rocks containing feldspars and quartz. Experimental determination of the compositions of coexisting feldspars and liquids may eventually provide a calibrant for geothermometers. The pressure-temperature curve for the beginning of melting of granite in the presence of water is generally considered to establish the conditions for migmatite formation during regional metamorphism. Experimental studies of genetically related series of rocks from a batholith (tonalite-granodiorite-granite) are more directly related to the development of the batholiths themselves, and the extension of these experimental studies to pressures corresponding to mantle conditions (greater than 10 kilobars) should elucidate the ultimate source and origin of the calc-alkaline magmas. *See* PETROLOGY.

[PETER J. WYLLIE]

Bibliography: F. J. Turner and J. Verhoogen, *Igneous and Metamorphic Petrology*, 2d ed., 1960; M. Walton, Granite problems, *Science*, 131:635–645, 1960.

Granitization

The process whereby various types of rocks may be converted, essentially in the solid state, to granite or closely related material, such as granodiorite. The process is generally a large-scale change and is considered to have operated in orogenic zones (zones of fold mountains) essentially contemporaneously with metamorphism. Here stratified rocks, undergoing reconstitution and recrystallization, may suffer changes in bulk composition adequate to convert them to rocks of granitic character. Such stratified rocks have thus been granitized, and the granites formed thereby may be indistinguishable from those of magmatic origin. In fact, one type of granite may pass imperceptibly into the other. *See* GRANODIORITE.

Granitization on a small scale, more or less confined to contacts of magmatic bodies at high levels in the Earth's crust, might more appropriately be referred to as contact metasomatism or feldspathization. *See* AUREOLE, CONTACT.

The change in bulk composition during granitization involves the introduction of certain constituents and removal of others. This substitution, in essentially solid rock and with no appreciable overall volume change, is known as metasomatism. Bodily movement of fluids (along fissures and openings) carrying requisite substances in solution may account for the compositional changes in the rocks. Where openings are too small to permit fluid flow, diffusion along intergrain surfaces may occur. For short distances diffusion through the crystal structure may be important. Extensive deformation accompanying metamorphism may maintain adequate channelways for the migrating material. *See* METASOMATISM.

Movement of fluids, perhaps derived from some underlying body of magma or from the deep parts of the Earth, would permit principally a one-way transfer. Elements necessary for granitization, such as alkalies, could be brought up from below; the displaced elements (calcium, Ca; iron, Fe; and magnesium, Mg) could be carried to higher levels and, perhaps, the Earth's surface.

Diffusion along intergranular surfaces or throughout a stationary pore fluid or diffusion through the crystal structure would permit multidirectional transfer. Each chemical component would move more or less independently, prompted by pressure and temperature gradients and by gravity. Consequently, certain components could move upward and others downward; but direction and rate of movement for a specific component could change with changes in physical conditions.

Some geologists believe that, as granitization proceeds, elements driven out may concentrate at the diffusion limit to form a zone or front with distinctive minerals. In some instances a basic front (rich in Ca, Fe, and Mg) appears to have formed in advance of a wave of granitization. According to the concept of two-way migration during granitization, it seems likely that basic material may move downward as alkalies and silica may move upward.

Transformation may not proceed uniformly. It may take place more readily or completely along certain layers or beds so that long fingerlike processes of metamorphic rock are left extending deeply into metasomatic granite. Partially replaced layers may appear as narrow slivers and lenses enclosed by granite. These might resemble xenoliths in magmatic rocks drawn out and oriented by flowage. Evidence for metasomatic origin of granite is perhaps most compelling where the bedding of the metamorphic rocks can be traced, although somewhat vaguely, well into the granite body. *See* XENOLITH.

The gradual transition from metamorphosed sediments to metasomatic granite may be studied by chemical and petrological methods. Mica schist or gneiss, derived from shaly sediments, may gradually gain in plagioclase and potash feldspar largely at the expense of muscovite mica and mafic (iron-magnesium) minerals. There is generally an increase in grain size and commonly a marked development of large feldspar crystals (porphyroblasts). This textural change may be accompanied by partial or complete obliteration of the original schistose or gneissose structure.

Large feldspar crystals in metamorphic inclusions in granite must be of replacement origin. They commonly resemble in detail the large crystals within the granite, which also appear to have formed late. These large feldspars suggest, but do not prove, that the granite also is of metamorphic origin.

Granitization may lead to the formation of migmatites, that is, rocks which appear to be mixtures of igneous and metamorphic material. Selective replacement along certain beds or layers in metamorphic rocks could give rise to veined gneiss or banded varieties of migmatite. Replacement along diversely oriented fractures in a rock would at a certain stage create a migmatite composed of remnant blocks enclosed in granite matrix.

There is a mergence of metamorphic and igneous action where pods and streaks of magmalike fluid, rich in alkalies and silica (granitic), form in place by selective fusion or solution. Upon solidification of this interstitial material, another type of migmatite is developed. If before solidification, however, the rocks become compressed, the melt may be squeezed out to form large independent bodies of magma. It is possible that the whole mass may become mobilized and take on the aspect of a magma. The liquid phase serves to lubricate contacts between the more or less plastic, solid portions. Such mobilized mixtures are sometimes called migmas. Thus, migmatization (the formation of migma and migmatite), granitization, metamorphism, and igneous action may be closely associated phenomena. *See* MAGMA; METAMORPHIC ROCKS; METAMORPHISM; MIGMATITE. [CARLETON A. CHAPMAN]

Bibliography: H. H. Read, *The Granite Controversy*, 1957; H. H. Read and J. Watson, *Introduction to Geology*, 1962; L. E. Spock, *Guide to the Study of Rocks*, 1962.

Granodiorite

A phaneritic (visibly crystalline) plutonic rock composed chiefly of sodic plagioclase (oligoclase or andesine), alkali feldspar (microcline or orthoclase, usually perthitic), quartz, and subordinate dark-colored (mafic) minerals (biotite, amphibole, or pyroxene). Granodiorite is intermediate between granite and quartz diorite (tonalite). Alkali feldspar is dominant over plagioclase in granite but is subordinate to plagioclase in granodiorite. Quartz diorite carries little or no alkali feldspar. For convenience granite and granodiorite are commonly grouped and referred to as granite. *See* GRANITE; IGNEOUS ROCKS.

[CARLETON A. CHAPMAN]

Granulite

In petrology the term granulite has been used with different meanings, denoting simply fine-grained gneisses of more or less granitic composition, and denoting medium- to fine-grained quartzofeldspathic rocks recrystallized in the granulite facies group at temperatures of 700–900°C and in the pressure range 2–14 kb. *See* GNEISS; METAMORPHIC ROCKS.

For historical reasons the first definition is sometimes still applied. For example, a large part of the Highlands of Scotland consists of so-called granulites (Moine granulites) which are simply gneisses derived from arkose, grits, and sandstone.

Typical granulites of the second definition have a distinctive fabric: granulite texture, with feldspar and quartz in particular deformed to thin plates. Diagnostic mineral assemblage is orthopyroxene + clinopyroxene (thus the name "two-pyroxene facies" is synonymous to granulite facies). Other common minerals not mentioned above are cordierite, biotite, sillimanite (kyanite may occur at very high pressure), and characteristically rounded, pale-brown garnets. The potassium feldspar is orthoclase or microcline mesoperthite with

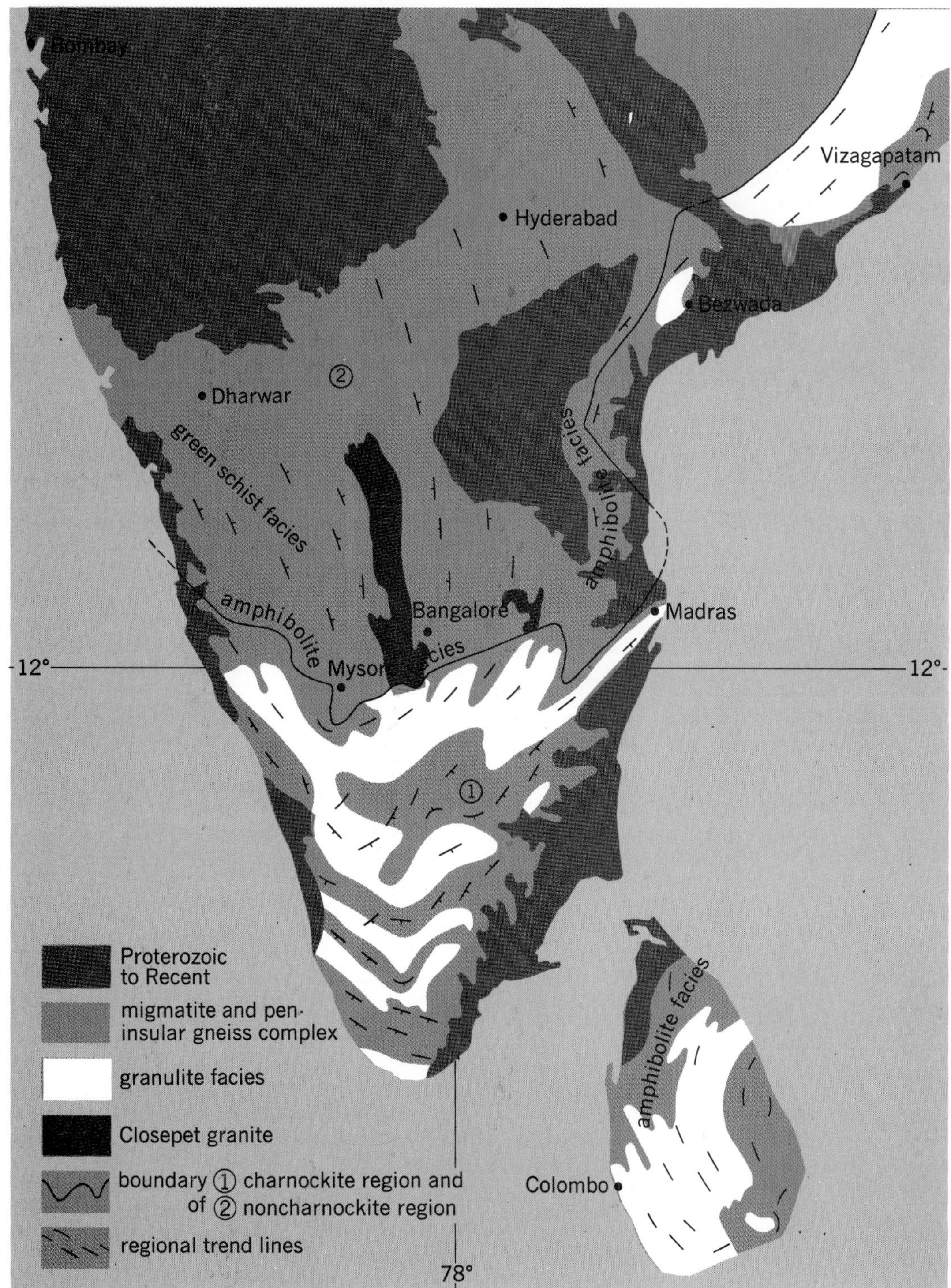

Southern India and Ceylon, showing broad tectonic and metamorphic features. (*After A. P. Subramaniam, Charnockites and granulites of southern India: A review, Dan. Geol. Foren., 17:473–493, 1967*)

distinctive drops of exsolved plagioclase. The plagioclase is often antiperthitic.

Charnockite (and enderbite with plagioclase dominant over alkali feldspar) used to be regarded as a magmatic rock, but this contention cannot be maintained. Therefore, it is no longer possible to make a sharp distinction between granulite and charnockite, both being quartzofeldspathic rocks in the two-pyroxene facies, with charnockite restricted to faintly foliated, nearly massive varieties. The original rock material of granulites and charnockites derives from a geosynclinal association of sediments and volcanics. The temperature of recrystallization was high, actually higher than the minimum melting temperature of granite; therefore, magmatic and metamorphic processes converged and evolved the particular structural and textural features of granulites.

Granulite rocks occur within areas of the two-pyroxene facies of the regional metamorphism;

they are being increasingly recognized in the Precambrian in all parts of the world. The eastern and southern part of peninsular India is the largest and best-studied area (see illustration); it is from this region that charnockites were first described by T. H. Holland in 1893 and 1900. Here and in many other places in the world (for example, Adirondacks, N.Y.; Laramie Mountains, Wyo.; Newfoundland-Labrador south through Quebec, Canada; southern Norway; and Angola, Africa) granulites are associated with anorthosites and congenetic rocks. [T. F. W. BARTH]

Bibliography: R. A. Howie, Charnockites, *Sci. Progr.*, 52:628–44, 1964; *J. Geol. Soc. India*, vol. 8, 1967; K. Parras, On the charnockites in the light of a highly metamorphic rock complex in S.W. Finland, *Bull. Comm. Geol. Finland*, 181:1–137, 1958; P. Quensel, The charnockite series of Varberg district in the S.W. coast of Sweden, *Ark. Mineral. Geol.*, 1:227–332, 1951; A. P. Subramaniam, Charnockites and granulites of southern India: A review, *Dan. Geol. Foren.*, 17:473–493, 1967.

Graphite

A low-pressure polymorph of carbon, the common high-pressure polymorph being diamond. Several other rare polymorphs have been synthesized or discovered in meteorites. The contrast in physical properties between these two polymorphs is remarkable: Graphite is metallic in appearance and very soft, whereas diamond is transparent and one of the hardest substances known.

Graphite is hexagonal, space group $P\ 6_3/m\ mc$, $a = 2.48$ A, $c = 6.80$ A, with 4C in the unit cell. Its atomic arrangement consists of sheets of carbon atoms at the vertices of a planar network of hexagons (Fig. 1). Thus, each carbon atom has three nearest-neighbor carbon atoms. The layer distance between the sheets is $c/2 = 3.40$ A. Diamond, on the other hand, is a three-dimensional framework structure with the carbon atoms in tetrahedral (fourfold) coordination. Crystals of graphite are infrequently encountered since the mineral usually occurs as earthy, foliated, or columnar aggregates often mixed with iron oxide, quartz, and other minerals (Fig. 2). *See* DIAMOND.

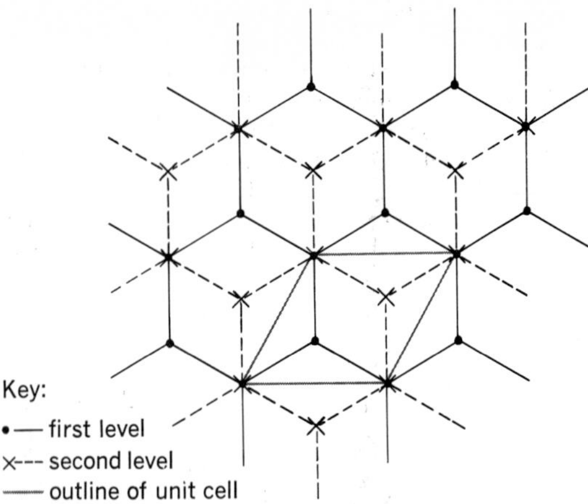

Fig. 1. The graphite atomic arrangement down the **c**-axis. There are two distinct layers of carbon atoms, related by symmetry. The first level consists of carbon atoms (solid circles) at the vertices of a network of hexagons. The second level, also a hexagonal network with carbon atoms at the vertices (crosses), is shifted relative to the first level.

GRAPHITE

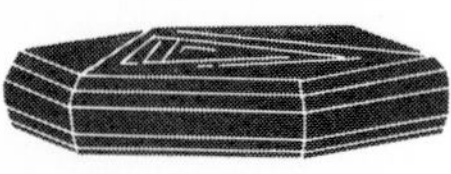

Fig. 2. Graphite. (*a*) Earthy aggregate (*American Museum of Natural History specimen*). (*b*) Hexagonal crystal of graphite with triangular markings on face (*from C. Palache, H. Berman, and C. Frondel, Dana's System of Mineralogy, vol. 1, 7th ed., copyright © 1944 by John Wiley & Sons, Inc.; reprinted by permission*).

Properties. The sheetlike character of the graphite atomic arrangement results in distinctive physical properties. The mineral is very soft, with hardness $1\frac{1}{2}$; it soils the fingers and leaves a black streak on paper, hence its use in pencils. The specific gravity is 2.23, often less because of the presence of pore spaces and impurities. The color is black in earthy material to steel-gray in plates, and thin flakes are deep blue in transmitted light. One perfect cleavage is parallel to the hexagonal sheets, allowing the mineral to be split into thin flexible but nonelastic folia. Graphite is a conductor of electricity, distinguishing it from amorphous carbon (lampblack).

Graphite closely resembles molybdenite, MoS_2, a mineral with similar crystal structure, but the two can be distinguished by the greenish streak and much higher (4.70) specific gravity of the latter mineral. Early confusion with the brittle gray lead sulfide, galena, resulted in the synonymous trade names "plumbago" and "black lead" for graphite. In a similar manner, the misleading term "lead pencil" has persisted and is still in common usage.

Occurrence. Graphite arises from the thermal and regional metamorphism of rocks such as sandstones, shales, coals, and limestones which contained organic products not exposed to an oxidizing environment. It also can form in a strongly reducing environment, such as in serpentinites and limestones where hydrogen gas may reduce carbon dioxide. Platy graphite showing crude crystal surfaces often occurs speckled in coarsely crystallized marbles. The major sources of graphite are in gneisses and schists, where the mineral occurs in foliated masses mixed with quartz, mica, and so on. Noteworthy localities include the Adirondack region of New York, Korea, and Ceylon. In Sonora, Mexico, graphite occurs as a product of metamorphosed coal beds. Graphite is also observed in meteorites, where the mineral was formed under strongly reducing conditions, usually in association with metallic iron. [PAUL B. MOORE]

Synthetic graphite. Graphite has a highly developed crystalline structure, and its softness, high thermal and electrical conductivity, and self-lubricating qualities differentiate it from other forms of carbon.

Carbon in graphitic form has both metallic and nonmetallic properties. Commercially produced synthetic graphite is a mixture of crystalline graphite and cross-linking intercrystalline carbon. Its physical properties are the result of contributions from both sources. Thus, among engineering materials, synthetic graphite is unusual because a wide variation in measurable properties can occur without significant change in chemical composition.

At room temperature the thermal conductivity of synthetic graphite is comparable to that of aluminum or brass. An unusual property of graphite is its increased strength at high temperature. The crushing strength is about 20% higher at 1600°C and the tensile strength is 50–100% higher at 2500°C than at room temperature.

Graphite is resistant to thermal shock because of its high thermal conductivity and low elastic modulus. It is one of the most inert materials with respect to chemical reaction with other elements and compounds. It is subject only to oxidation, reaction with and solution in some metals, and formation of lamellar compounds with certain alkali metals and metal halides.

Uses. Graphite has many uses in the electrical, chemical, metallurgical, nuclear, and rocket fields: electrodes in electric furnaces producing carbon steel, alloy steel, and ferroalloys; anodes for electrolytic production of chlorine, caustic and chlorates, magnesium, and sodium; motor and generator brushes; sleeve-type bearings; seal rings; electronic tube anodes and grids; nuclear reactor moderators, reflectors, and thermal columns (Fig. 3); rocket motor nozzles; missile rudder vanes; metallurgical molds and crucibles; linings for chemical reaction vessels; and in the resin-impregnated impervious form, for heat exchangers, pumps, piping, valves, and other process equipment.

Preparation. Synthetic graphite can be made from almost any organic material that leaves a high carbon residue on heating to 2500–3200°C. In commercial operations, raw materials are carefully selected because not all substances with high carbon content undergo a suitably complete transformation to graphite at these temperatures. Petroleum coke is raw material for the most commonly used production process. After calcining and sizing, the coke is mixed with coal tar pitch, heated to about 165°C, and formed by extrusion or molding to "green" shapes. Baking to 750–1400°C in gas- or oil-fired kilns follows the forming operation. Graphite is produced by heating the baked shapes to 2600–3000°C by passing electricity amounting to 1.6–3.0 kw per pound of graphite through the bed of a furnace made of the shapes laid in granular coke (Fig. 4). The whole bed is covered by an insulating blanket of silicon carbide, coke, and sand. Higher-density synthetic graphite can be obtained by impregnating the baked carbon with pitch prior to graphitization. Graphite with total ash content less than 20 parts per million is needed for a number of nuclear and electrolytic uses, and is obtained by heating the graphite shapes electrically to about 2500°C while bathing them in a purifying gas.

Highly ordered crystalline graphite can be produced up to about 1/4-in. thickness by pyrolyzing organic gases under controlled conditions at 1400–2000°C. Pyrolytic graphite exhibits a high degree of anisotropy (varying properties in different directions). Parallel to the thickness the thermal conductivity is comparable to copper, but perpendicular to the thickness conductivity is about 1/200 the conductivity of copper. Tensile strength and thermal expansion also vary greatly with orientation. The room temperature density of pyrolytic graphite reaches up to 2.22, about 98% of the 2.26 density of the graphite single crystal, whereas the density of graphite electrodes ranges from 1.5 to 1.7. Pyrolytic graphite has been formed into rocket nose cones.

Fig. 3. Graphite reflector for a test reactor.

Graphite can be recrystallized to increase its density and improve other related and significant properties. Starting with an initial level of 1.7 for a conventional grade, bulk densities as high as 2.18 at room temperature have been obtained on a laboratory basis. Recrystallized graphite was introduced in 1960 and is now commerically available at a density range of 1.92–1.97. The permeability of recrystallized graphite is comparable to brass and is 10^5–10^7 less than electrode-grade graphite. Since the degree of anisotropy (always less than pyrolytic graphite) can be controlled, the usefulness of recrystallized graphite is not limited (or restricted by production considerations) to thin sections in the presence of thermal stresses. Initial use of this form of graphite has been in nozzles of rocket motors.

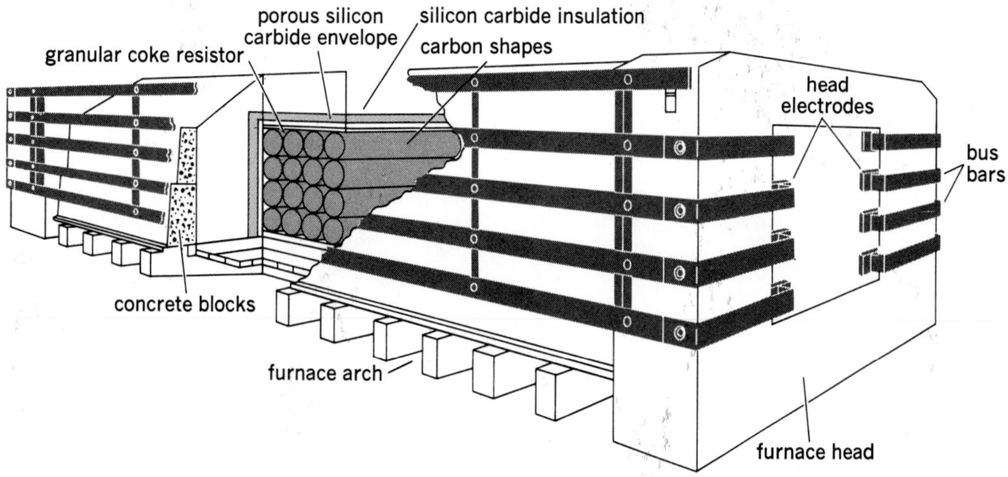

Fig. 4. Graphitizing furnace, for making synthetic graphite.

Table 1. Graphite cloth oxidation in still air

Temperature, °C	Hr to lose 1% weight
266	10,000
360	100
493	1.0
699	0.01

Graphite and carbon fibers. Commercial carbon and graphite fibers are made from any carbonaceous fibrous raw material that pyrolyzes to a char, does not melt, and leaves a high carbon residue. The most commonly used precursor materials are rayon, polyacrylonitrile, and pitch. Working in the laboratory near the triple point of carbon (92 atm, 3650°C), submicron-diameter graphite whiskers of near theoretical strength—100 to 140×10^6 psi modulus—are found inside boules. Vapor-deposited pyrolytic graphite fibers have been made on a small scale.

There are two distinct families of carbonaceous fibers. Rigid classification depends on x-ray diffraction measurements, but fibrous carbon and graphite may be loosely distinguished from each other by the highest temperature reached during their manufacture. Fibers and fabrics heated to less than 2000°C are designated carbon. Graphite fiber products are furnaced above 2500°C.

Fibrous carbon is prepared by pyrolyzing spun, felted, or woven raw materials to 700–1800°C. Prior to the heat treatment, chemical or other treatments are employed in some processes. Graphite articles are made by further processing the carbonized material to 2500–3000°C.

In comparison with natural and synthetic fibers, carbon and graphite filaments have a high modulus of elasticity, over 4,000,000 psi, and low elongation at break. In these respects, they resemble glass filaments and, in general, can be handled on equipment designed for glass yarns.

Graphite cloth and yarn oxidize in air relatively slowly, considering the large surface-area-to-weight ratio, because their relatively high purity excludes most of the oxidation-promoting catalysts found in most synthetic graphite. Table 1 tabulates the results of a typical oxidation experiment.

It can be said that graphite cloth oxidizes in air at about the same rate as common graphite rods and plates. The threshold of oxidation of graphite in steam is about 700°C and in carbon dioxide is 900°C. Depending on its carbon assay, fibrous carbon has less oxidation resistance than graphite.

As with all forms of carbon and graphite, these fibers do not melt at ordinary pressures, and temperatures above 3600°C are required to volatilize them. The strength of general-purpose carbon and graphite cloth and yarn increases with temperature up to about 2000°C, and then declines. High-performance fibers have usable strength at 3000°C.

Cloth. One of the major uses of fibrous carbon products is for ablative composites for rocket motors in the form of rayon cloth converted to carbon and graphite cloth. The most widely used fabrics weigh 7–8 oz/yd² and have filament diameters of 0.0003–0.0004 in. (Table 2).

Both carbon and graphite cloths as reinforcement in resin composites are used in a variety of applications which require short-term high-temperature strengths, good ablation characteristics, and insulation with low density. Since graphite is somewhat more resistant than carbon and physically stable at elevated temperatures, graphite reinforcements are preferred in rocket nozzle throats and ablation chambers. Carbon cloths are used where greater strength and lower thermal conductivity is required, such as in reentry vehicles and rocket nozzle entrance sections and exit cones of spacecraft.

Continuous filament yarns and strands. General-purpose and high-performance continuous filament yarns have yields from 300 yd/lb up to 4100 yd/lb, depending on construction (Table 3). They can be woven, knitted, or braided into fabrics, and are well suited for filament winding or unidirectional tape-winding equipment. Carbon yarns are characterized by higher surface area, generally

Table 2. Typical properties of carbon and graphite fabrics and felts at 20°C*

		Carbon fabric			Graphite fabric		Carbon felt	Graphite felt
Weave:	5 H.S.	8 H.S.	Plain	5 H.S.	8 H.S.	Plain	Needled	Needled
Characteristics								
Width, in.	42–44	33–46	32–42	42.5–45.0	33.0–44	41.5–44.6	44	43.5
Weight, oz/yd²	8	8–9	6–8	8.0	7.2	7.4	15.7	16.6
Gage, in.	0.018	0.020	0.017–0.038	0.018	0.023	0.022	0.2–0.3	0.2
Count, yarns/in. Warp	40	50–52	27–34	40	51	27	–	–
Count, yarns/in. Fill	36	49–51	23–32	36	49	21	–	–
Filaments/yarn, bundle	960	720	1440	960	720	1440	–	–
Filaments, diam., in.	0.0004	0.0005	0.0004	0.0003	0.0004	0.0004	0.0005	0.0005
Tensile strength, lb/in. Warp	38.0	21–59	10–100	38	85	42	2	1
Tensile strength, lb/in. Fill	32.0	11–30	15–90	34	75	25	2	1
Surface area, m²g	6	1	250	3	3	3	20–200	1–2
Density (He), g/cm³	1.47	1.50	1.4–1.8	1.50	1.50	1.42	.08	.08
Carbon assay, w/o	99	86–95	86–99	99.9	99.9	99.9	91–98	99.6
Ash, w/o	0.02	0.9	0.5–10	0.009	0.009	0.009	0.5	0.04
pH	7.0	2.2	7.0–10			7.9	–	–
Resistivity, 70°F, ohm/yd² Warp	0.40	280	–	0.41	0.46	0.38	–	0.35
Resistivity, 70°F, ohm/yd² Fill	0.48	270	0.5–1.6	0.42	0.50	0.50	–	0.55

*H.S. = harness satin; w/o = weight percent.

Table 3. Typical carbon and graphite filament and yarn properties at 20°C

Property	General-purpose carbon	General-purpose graphite	High-modulus graphite*	
Filament				
Tensile strength, lb/in.^{2} (1-in. gage)	120,000	90,000	285,000	360,000
Tenacity, g/denier (1-in. gage)	6.2	5.3	13.7	–
Elongation at break, %	2.0	1.5	0.6	1.0
Elastic recovery, %	100	100	100	100
Modulus of elasticity, lb/in.^{2} (1-in. gage)	6,000,000	6,000,000	50,000,000	30,000,000
Stiffness, g/denier (1-in. gage)	310	350	2300	–
Density, Hg immersion, g/cm^3	1.53	1.32	1.63	1.76
Equivalent diameter, μ	9.5	8.9	6.6	6.9
Carbon assay, w/o†	90.0	98.8	99.9	92.0
Volume electrical resistivity, ohm-cm $\times 10^6$	6000	3500	–	–
Surface area, m^2/g	130	4	1	1
Thermal conductivity, Btu/hr/ft^2/°F/ft	13	22	68	12
Specific heat, 70°F, Btu/lb/°f	0.17	0.17	0.17	0.17
(mean 70 to 2700°F)	0.40	0.40	0.40	0.40
Construction	*2-ply yarn‡*	*2-ply yarn‡*	*2-ply yarn‡*	*Strand*
Yield, yd/lb	3100	4100	6355	2600
Breaking strength, lb (average, 5-in. gage)	12	6	6.4	13.5
Tenacity, g/denier (5-in. gage)	3.8	2.3	4.7	–
Elongation at break, %	2.4	1.3	0.53	1.0
Yarn diameter, in. (approx.)	0.03	0.03	0.015	0.017
pH, ASTM D-1512-60	7.0	6.0	7.5	–

*10,000-, 40,000- and 160,000-filament tows are also commercially available
†w/o = weight percent.
‡Single-, 5-, 10-, and 30-ply yarn constructions have been made commercially.

over 100 m^2/g, lower thermal and electrical conductivity, and lower carbon assay than graphite yarns. They have significant moisture regain and are readily wet with resins to form low-thermal-conductivity composites. Graphite yarns have low surface area, do not absorb moisture, and are more oxidation resistant. Applications for these refractory yarns are in reentry systems and ablative structures requiring greater strength in woven form than that possible in conventional, carbonized broad goods.

Graphite yarns braided into packing for pumps and valves have the leak-free, low-maintenance performance of radial seal rings at stuffing box cost. Carbon- and graphite-fiber-filled materials are being developed for nonlubricated sleeve-bearing applications.

Carbon yarns are many times stronger than metallic wire of the same electrical resistance per unit length. The volume resistivity of carbon yarns at 20°C is about 0.005 ohm-cm, approximately 60 times that of nickel alloy resistance wire, and about 3500 times that of copper. Carbon yarns have a slight negative temperature coefficient of resistance; thus no current surge occurs in heating elements at start-up. An important advantage of carbon yarns is their high electrical resistance, which facilitates design of heating elements connected in parallel in which loss of a single yarn will not affect overall performance significantly.

High-modulus carbon and graphite fibers are available in multifilament yarns, strands, and tows (Table 3). Depending on precursor and processing conditions, a wide range of modulus and tensile properties can be provided. By applying stress to a rayon precursor yarn during pyrolysis, fibers with modulus from 50 to 75×10^6 psi and tensile strength from 285 to 350,000 psi can be produced. Because of the relatively high cost of these products, applications today are primarily in aerospace reentry components. Polyacrylonitrile precursors process through pyrolysis to carbon and graphite fibers with modulus from 30 to 70×10^6 psi. The 30×10^6 psi modulus fiber with tensile strength of 360,000 offers higher elongation and a balance of tension and compression in resin matrix composites. This product has found wide acceptance in aircraft and space structures, sporting goods, and industrial and construction markets. The high specific strength and modulus of the composites provide cost-effective performance in competition with metals and other fibrous composites.

Pitch is a very-low-cost precursor material. Ordinary petroleum or coal tar pitches can now be converted to high-modulus continuous fibers by a process that eliminates hot stretching of the fibers. This technical breakthrough offers the opportunity for very-low-cost high-modulus fibers to exploit volume commercial markets. High-modulus pitch-based fibers are also available in a blown filament product in mat form. They can be chopped and made into high-modulus carbon paper or used as a modulus modifier in polymeric molding compounds. Because they are electrically conductive, they can be used in resistance-heating applications.

Felts. Carbon and graphite felts are used as insulation and heat shields in high-temperature induction and resistance furnaces (Table 2). Graphite-tape electrical heating elements are used in temperatures up to 2500°C in metalworking furnaces. Protective atmospheres or high vacuum are necessary to prevent oxidation. These felts are better insulators than carbon black. Felt density is uniform and from 1/4 to 1/10 that of tamped carbon black. The low density of felt speeds furnace thermal response and reduces energy requirements. The felts do not develop voids characteristic of powders, thus increasing furnace life.

[R. B. FORSYTH; W. M. GAYLORD]

Bibliography: R. Bacon, *J. Appl. Phys.*, 31(2): 283–290, 1960; R. Bacon and W. H. Smith, Tensile behavior of carbonized rayon filaments at elevated temperature, *Society of Chemical Industries Second Conference on Industrial Carbon and Graphite, London, April, 1965*, 1966; E. N. Cameron and P. L. Weis, *Strategic Graphite: A Survey*, Geol. Surv. Bull. no. 1082-E, 1960; L. M. Currie et al., The production and properties of graphite for reactors, *Proc. Int. Conf. Peaceful Uses At. Energy*, 8:451–473, 1956; R. B. Forsyth, *Low cost continuous fiber from a pitch precursor, 20th National SAMPE Symposium*, 1975; J. E. Hove, Graphite as a high temperature material, *Met. Soc. AIME Trans.*, 212(1):7–13, 1958; N. J. Johnson, V. J. Nolan, and J. W. Shea, *Carbon and Graphite in the Metallurgical Industries*, 1961; J. K. Lancaster, *The Effect of Carbon Fibre Reinforcement on the Function and Wear of Polymers*, Roy. Aircraft Estab. Tech. Rep. no. 66378, 1966; National Carbon Co., *The Industrial Graphite Handbook*, 1959.

Gravel

A gravel is a loose, or unconsolidated, deposit of rounded pebbles, cobbles, or boulders, whose size is greater than 2 mm in diameter. Gravels may consist only of pebbles, cobbles, and boulders, with large voids between the particles; these are the openwork gravels. Commonly, however, gravels have sand filling the interstices between the pebbles. The sand filling the pores is referred to as the matrix and commonly makes up about 30% of the rock. The size distribution of gravels has been used as a clue to origin. Beach gravels normally seem to have only one modal size (that size about which most of the particles cluster), but river gravels have two modes, one in the gravel size and one in the sand sizes. The shapes, roundness, and surface textures of pebbles have also been used for guides to origin. Thus the degree of roundness and sphericity reflects the rigors of transportation and may be directly related to distance traveled. The shapes of pebbles may be inherited from fracture, joints, cleavage, and bedding patterns of the parent rock. Faceted pebbles called dreikanters have been produced by wind action. The shape of these pebbles is probably related to the pebble rock type and its bedding and jointing characteristics, both of which affect the original shape of the particle before transportation. *See* BRECCIA; CONGLOMERATE; SEDIMENTARY ROCKS. [RAYMOND SIEVER]

Gravitation

The mutual attraction between all masses and particles of matter in the universe. In a sense this is one of the best-known physical phenomena. During the 18th and 19th centuries gravitational astronomy, based on Newton's laws, attracted many of the leading mathematicians and was brought to such a pitch that it seemed that only extra numerical refinements would be needed in order to account in detail for the motions of all celestial bodies. In the present century, however, Albert Einstein shattered this complacency, and the subject is currently in a healthy state of flux.

Until the 17th century, the sole recognized evidence of this phenomenon was the gravitational attraction at the surface of the Earth. Only vague speculation existed that some force emanating from the Sun kept the planets in their orbits. Such a view was expressed by Johannes Kepler (1571–1630), the author of the laws of planetary motion. But a proper formulation for such a force had to wait until Isaac Newton (1643–1727) founded Newtonian mechanics, with his three laws of motion, and discovered, in calculus, the necessary mathematical tool.

Newton's law of gravitation. Newton's law of universal gravitation states that every two particles of matter in the universe attract each other with a force that acts in the line joining them, the intensity of which varies as the product of their masses and inversely as the square of the distance between them. Put into symbols, the gravitational force F exerted between two particles with masses m_1 and m_2 separated by a distance d is given by Eq. (1), where G is called the constant of gravitation.

$$F = Gm_1m_2/d^2 \tag{1}$$

A force varying with the inverse square power of the distance from the Sun had been already suggested—notably by Robert Hooke (1635–1703) but also by other contemporaries of Newton, such as Edmund Halley (1656–1742) and Christopher Wren (1632–1723)—but this had only been applied to circular planetary motion. The credit for accounting for, and partially correcting, Kepler's laws and for setting gravitational astronomy on a proper mathematical basis is wholly Newton's.

Newton's theory was first published in *The Principia* in 1686. According to Newton, it was formulated in principle in 1666 when the problem of elliptic motion in the inverse-square force field was solved. But publication was delayed in part because of the difficulty of proceeding from the "particles" of the law to extended bodies such as the Earth. This difficulty was overcome when Newton established that, under his law, bodies having spherically symmetrical distribution of mass attract each other as if all their mass were concentrated at their respective centers.

Newton verified that the gravitational force between the Earth and the Moon, necessary to maintain the Moon in its orbit, and the gravitational attraction on the surface of the Earth were related by an inverse square law of force. Let E be the mass of the Earth, assumed to be spherically symmetrical with radius R. Then the force exerted by the Earth on a small mass m near the Earth's surface is given by Eq. (2), and the acceleration of gravity on the Earth's surface, g, by Eq. (3).

$$F = GEm/R^2 \tag{2}$$
$$g = GE/R^2 \tag{3}$$

Let a be the mean distance of the Moon from the Earth, M the Moon's mass, and P the Moon's sidereal period of revolution around the Earth. If the motions in the Earth-Moon system are considered to be unaffected by external forces (principally those caused by the Sun's attraction), Kepler's third law applied to this system is given by Eq. (4).

$$4\pi^2a^3/P^2 = G(E+M) \tag{4}$$

Equations (3) and (4), on elimination of G, give Eq. (5).

$$g = 4\pi^2 \frac{E}{E+M} \frac{a^2}{R^2} \frac{a}{P^2} \quad (5)$$

Now the Moon's mean distance from the Earth is $a = 60.27R = 3.84 \times 10^{10}$ cm and the sidereal period of revolution is $P = 27.32$ days $= 2.361 \times 10^6$ sec. These data give, with $E/M = 81.35$, $g = 977$ cm sec^{-2}, which is close to the observed value. *See* TERRESTRIAL GRAVITATION.

This calculation corresponds in essence to that made by Newton in 1666. At that time the ratio a/R was known to be about 60, but the Moon's distance in miles was not well known because the Earth's radius R was erroneously taken to correspond to 60 mi per degree of latitude instead of 69 mi. As a consequence, the first test was unsatisfactory. But the discordance was removed in 1671 when J. Picard's measurement of an arc of meridian in France provided a reliable value for the Earth's radius.

Gravitational constant. Equation (3) shows that the measurement of the acceleration due to gravity at the surface of the Earth is equivalent to finding the product G and the mass of the Earth. Determining the gravitational constant by a suitable experiment is therefore equivalent to "weighing the Earth."

In 1774 N. Maskelyne determined G by measuring the deflection of the vertical by the attraction of a mountain. This method is much inferior to the laboratory method in which the gravitational force between known masses is measured. In the torsion balance two small spheres, each of mass m, are connected by a light rod, suspended in the middle by a thin wire. The deflection caused by bringing two large spheres each of mass M near the small ones on opposite sides of the rod is measured, and the force is evaluated by observing the period of oscillation of the rod under the influence of the torsion of the wire (see illustration.) This is known as the Cavendish experiment, in honor of H. Cavendish who achieved the first reliable results by this method in 1797–1798. More recent determinations using various refinements yield the results: constant of gravitation $G = 6.67 \times 10^{-11}$ mks units; mass of Earth $= 5.98 \times 10^{27}$g. The uncertainty of these results is probably about one-half unit of the last place given.

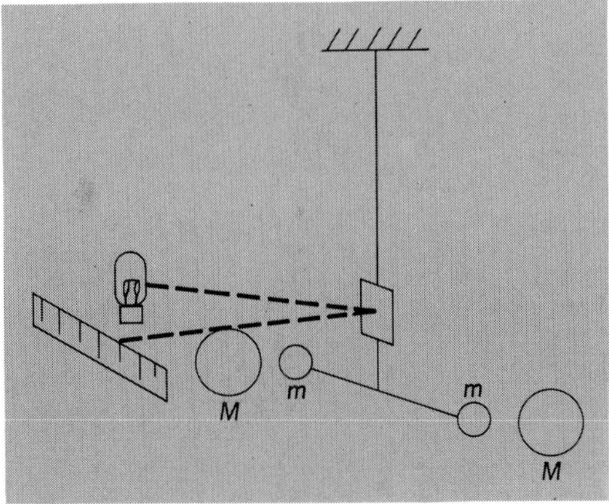

The torsion balance.

In Newtonian gravitation G is an absolute constant, independent of time, place, and the chemical composition of the masses involved. Partial confirmation of this was provided before Newton's time by the experiment attributed to Galileo when different weights released simultaneously from the top of the tower of Pisa reached the ground at the same time. Newton found further confirmation, experimenting with pendulums made out of different materials. Early in this century, R. Eotvos found that different materials fall with the same acceleration to within one part in 10^7, and more recently R. H. Dicke, using aluminum and gold, has extended the accuracy to one part in 10^{11}.

With the discovery of antimatter, there was speculation that matter and antimatter would exert a mutual gravitational repulsion. But experimental results indicate that they attract one another according to the same laws as apply to matter of the same kind.

Mass and weight. A cosmology with changing physical "constants" was first proposed in 1937 by P. A. M. Dirac. Field theories applying this principle have since been proposed by P. Jordan and D. W. Sciama and, in 1961, by C. Brans and R. H. Dicke. In these theories G is diminishing; for instance Brans and Dicke suggest a change of about 2×10^{-11} per year. This would have profound effects on phenomena ranging from the evolution of the universe to the evolution of the Earth. For instance, stars evolve faster if G is greater, so that stellar evolutionary ages computed with constant G at its present value would be too great. The Earth, compressed by gravitation, would expand, having a profound effect on surface features. Planetary orbits would gradually be increasing in size. About 3×10^9 years ago the Sun would have been hotter than it is now, and the Earth and its orbit would have been smaller, so that the temperature on the Earth's surface might have approached the boiling point of water; this would be important for the origin of life on the Earth. Astronomical observations of the planets over the past few hundred years are not accurate enough for the predicted change to be detected; but T. C. Van Flandern has reported that observations of the motion of the Moon agree with the predicted change in G.

In the equations of motion of Newtonian mechanics, the mass of body appears as inertial mass as a factor of the acceleration, and as gravitational mass in the expression of the gravitational force. The equality of these masses is confirmed by the Eotvos experiment. It justifies the assumption that the motion of a particle in a gravitational field does not depend on its physical composition. In Newton's theory the equality can be said to be a coincidence, but not in Einstein's theory, where inertia and gravitation are unified.

While mass in Newtonian mechanics is an intrinsic property of a body, its weight depends on certain forces acting on it. For example, the weight of a body on the Earth depends on the gravitational attraction of the Earth on the body and also on the centrifugal forces due to the Earth's rotation. The body would have lower weight on the Moon, even though its mass would remain the same.

Gravity. This should not be confused with the term gravitation. Gravity is the older term, meaning the quality of having weight, and so came to be applied to the tendency of downward motion on the Earth. Gravity or the force of gravity are today used to describe the intensity of gravitational forces, usually on the surface of the Earth or another celestial body. So gravitation refers to a universal phenomenon, while gravity refers to its local manifestation.

A rotating planet is oblate (or flattened at the poles) to a degree depending on the ratio of the centrifugal to the gravitational forces on its surface and on the distribution of mass in its interior. The variation of gravity on the surface of the Earth depends on these factors and is further complicated by irregular features such as oceans, continents, and mountains. It is investigated by gravity surveys and also through the analysis of the motion of artificial satellites. Because of the irregularities, no mathematical formula has been found that satisfactorily represents the gravitational field of the Earth, even though formulas involving hundreds of terms are used. The problem of representing the gravitational field of the Moon is even harder because the surface irregularities are proportionately much larger.

In describing gravity on the surface of the Earth, a smoothed out theoretical model is used, to which are added gravity anomalies, produced in the main by the surface irregularities.

Gravity waves are waves in the oceans or atmosphere of the Earth whose motion is dynamically governed by the Earth's gravitational field. They should not be confused with gravitational waves, which are discussed below.

Gravitational potential energy. This describes the energy that a body has by virtue of its position in a gravitational field. If two particles with masses m_1 and m_2 are a distance r apart and if this distance is slightly increased to $r+\Delta r$, then the work done against the gravitational attraction is $Gm_1m_2\Delta r/r^2$. If the distance is increased by a finite amount, say from r_1 to r_2, the work done is given by Eq. (6).

$$\begin{aligned} W_{r_1,r_2} &= Gm_1m_2\int_{r_1}^{r_2} dr/r^2 \\ &= Gm_1m_2(1/r_1-1/r_2) \end{aligned} \tag{6}$$

If $r_2 \to \infty$, Eq. (7) holds.

$$W_{r_1,\infty} = Gm_1m_2/r_1 \tag{7}$$

If one particle is kept fixed and the other brought to a distance r from a very great distance (infinity), then the work done is given by Eq. (8).

$$-U=-Gm_1m_2/r \tag{8}$$

This is called the gravitational potential energy; it is (arbitrarily) put to zero for infinite separation between the particles. Similarly, for a system of n particles with masses m_1, m_2, . . . , m_n and mutual distance r_{ij} between m_i and m_j, the gravitational potential energy $-U$ is the work done to assemble the system from infinite separation (or the negative of the work done to bring about an infinite separation), as shown in Eq. (9).

$$-U=-G\sum_{i<j} m_im_j/r_{ij} \tag{9}$$

A closely related quantity is gravitational potential. The gravitational potential of a particle of mass m is given by Eq. (10), where r is distance

$$V=-Gm/r \tag{10}$$

measured from the mass. The gravitational force exerted on another mass M is M times the gradient of V. If the first body is extended or irregular, the formula for V may be extremely complicated, but the latter relation still applies.

A good illustration of gravitational potential energy occurs in the motion of an artificial satellite in a nearly circular orbit around the Earth which is affected by atmospheric drag. Because of the frictional drag the total energy of the satellite in its orbit is reduced, but the satellite actually moves faster. The explanation for this is that it moves closer to the Earth and loses more in gravitational potential energy than it gains in kinetic energy.

Similarly, in its early evolution a star contracts, with the gravitational potential energy being transformed partly into radiation, so that it shines, and partly into kinetic energy of the atoms, so that the star heats up until it is hot enough for thermonuclear reactions to start.

Another related phenomenon is that of speed of escape. A projectile launched from the surface of the Earth with speed less than the speed of escape will return to the surface of the Earth; but it will not return if its initial speed is greater (atmospheric drag is neglected). For a spherical body with mass M and radius R, the speed of escape from its surface is given by Eq. (11). For the Earth V_e is 11.2

$$V_e=(2MG/R)^{1/2} \tag{11}$$

km/sec; for the Moon it is 2.4 km/sec, which explains why the Moon cannot retain an atmosphere such as the Earth's. By analogy, a black hole can be considered a body for which the speed of escape from the surface is greater than the speed of light, so that light cannot escape. P. S. Laplace speculated along these lines; but the analogy is not really exact since Newtonian mechanics is not valid. The question as to whether the universe will continue to expand can be considered in the same way. If the density of matter in the universe is great enough, then expansion will eventually cease and the universe will start to contract. At present the density cannot be found with sufficient accuracy to decide the question.

Application of Newton's law. In modern times Eq. (5), in a modified form with appropriate refinements to allow for the Earth's oblateness and for external forces acting on the Earth-Moon system, has been used to compute the distance to the Moon. The results have only been superseded in accuracy by radar measurements and observations of corner reflectors placed on the lunar surface.

Newton's theory passed a much more stringent test than the one described above when he was able to account for the principal departures from Kepler's laws in the motion of the Moon. Such departures are called perturbations. One of the most notable triumphs of the theory occurred when the observed perturbations in the motion of the planet Uranus enabled J. C. Adams in 1845 and U. J. Leverrier in 1846 independently to predict the existence and calculate the position of a hitherto unobserved planet, later called Neptune. When yet another planet, Pluto, was discovered in 1930, its position and orbit were strikingly similar to predictions based on the method used to discover Neptune. But the discovery of Pluto must be ascribed to the perseverance of the observing astronomers; it is not massive enough to have revealed itself through the perturbations of Uranus and Neptune.

F. W. Bessel observed nonuniform proper motions of Sirius and Procyon and inferred that each was gravitationally deflected by an unseen companion. It was only after his death that these bodies were telescopically observed, and they both later proved to be white dwarfs. More recently P. van de Kamp has accumulated evidence for the existence of some planetary masses around stars. The discovery of black holes (which will never be directly observed) hinges in part on a visible star showing evidence for having a companion of sufficiently high mass (so that its gravitational collapse can never be arrested).

Newton's theory supplies the link between the observed motion of celestial bodies and certain physical properties, such as mass and sometimes shape. Knowledge of stellar masses depends basically on the application of the theory to binary-star systems. Analysis of the motions of artificial satellites placed in orbit around the Earth has revealed refined information about the gravitational field of the Earth and of the Earth's atmosphere. Similarly, satellites placed in orbit around the Moon have yielded information about its gravitational field, and other space vehicles have yielded the best information to date on the mass of Venus.

Newtonian gravitation has been applied without apparent difficulty to the motion in distant star systems. But over very great distance (or over very small distances, when gravitation is swamped by other forces) it has not been confirmed or disproved.

Accuracy of Newtonian gravitation. Newton was the first to doubt the accuracy of his law when he was unable to account fully for the motion of the perigee in the motion of the Moon. In this case he eventually found that the discrepancy was largely removed if the solution of the equations were more accurately developed. Further difficulties to do with the motion of the Moon were noted in the 19th century, but these were eventually resolved when it was found that there were appreciable fluctuations in the rate of rotation of the Earth, so that it was the system of timekeeping and not the gravitational theory that was at fault.

A more serious discrepancy was discovered by Leverrier in the orbit of Mercury. Because of the action of the other planets, the perihelion of Mercury's orbit advances. But allowance for all known gravitational effects still left an observed motion of about 43 seconds of arc per century unaccounted for by Newton's theory. Attempts to account for this by adding an unknown planet or by drag with an interplanetary medium were unsatisfactory, and a very small change was suggested in the exponent of the inverse square of force. This particular discordance was accounted for by Albert Einstein's general theory of relativity in 1916, but the final word on the subject has yet to be said.

Gravitational lens. Light is deflected when it passes through a gravitational field, and an analogy can be made to the refraction of light passing through a lens. It has been suggested that a galaxy situated between an observer and a more distant source might have a focusing effect, and that this might account for some of the observed properties of quasi-stellar objects.

Testing of gravitational theories. One of the greatest difficulties in investigating gravitational theories is the weakness of the gravitational coupling of matter. For instance, the gravitational interaction between a proton and an electron is weaker by a factor of about 5×10^{-40} than the electrostatic interaction. (If gravitation alone bound the hydrogen atom, then the radius of the first Bohr orbit would be 10^{13} light-years, or about 1000 times the radius of the Hubble universe!) The contrast between the accuracy achieved in the laboratory when measuring G and the accuracy required if the possible inconstancy of G is to be investigated shows that gravitation is not a laboratory subject. But the astronomical universe provides a wealth of situations for investigating gravitation. Their main drawback is that they must be taken as they occur: with the exception of some experiments included in the space program, the situations cannot be controlled, modified, or repeated for experimental convenience.

In the solar system there are planetary orbits around the Sun and satellite orbits around the planets. Gravitational effects on electromagnetic radiation as well as those on orbiting bodies can be observed and tested against theory. There are stars that rotate, oscillate, explode, and collapse, including white dwarfs, neutron stars (observed as pulsars), and probably black holes, some with enormous masses. There are binary systems, some with orbital periods as low as 1000 sec. There are star clusters, galaxies (some of them exploding), clusters of galaxies, and finally there is the universe itself.

In the solar system, increased observational accuracy is needed to detect small departures from Newtonian gravitational theory. In other situations, these departures may be very great, but observable effects are attenuated due to great distances. Of greatest interest are systems where there is a high concentration of matter and rapid motion, involving rotation, revolution, or collapse.

Relativistic theories. Before Newton, detailed descriptions were available of the motions of celestial bodies – not just Kepler's laws but also empirical formulas capable of representing with fair accuracy, for their times, the motion of the Moon. Newton replaced description by theory, but in spite of his success and the absence of a reasonable alternative, the theory was heavily criticized, not least with regard to its requirement of "action at a distance" (that is, through a vacuum). Newton himself considered this to be "an absurdity," and he recognized the weaknesses in postulating in his system of mechanics the existence of preferred reference systems (that is, inertial reference systems) and an absolute time. Newton's theory is a superb mathematical one that represents the observed phenomena with remarkable accuracy.

Einstein showed in his special theory of relativity that these postulates were physically unacceptable. In his general theory of relativity he incorporated gravitational phenomena in such a way that there was no longer any preferred reference system. He treated the phenomenon of gravitation as a consequence of the geometric properties of space-time, this geometry being affected by the presence of gravitating matter. The acceleration of a body is determined by the local geometry, and the Newtonian concepts of gravitational force and action at a distance are abandoned. Mathematically the theory is far more complicated than Newton's. Instead of the single potential described above, Einstein worked with 10 quantities that are members of a tensor.

Principle of equivalence. An important step in Einstein's reasoning is his "principle of equivalence": that a uniformly accelerated reference system imitates completely the behavior of a uniform gravitational field. Imagine, for instance, a scientist in a space capsule infinitely far out in empty space so that the gravitational force on the capsule is negligible. Everything would be weightless; bodies would not fall; and a pendulum clock would not work. But now imagine the capsule to be accelerated by some agency at the uniform rate of 981 cm/sec^2. Everything in the capsule would then behave as if the capsule were stationary on its launching pad on the surface of the Earth and therefore subject to the Earth's gravitational field. But after its original launching, when the capsule is in free flight under the action of gravitational forces exerted by the various bodies in the solar system, its contents will behave as if it were in the complete isolation suggested above. Note that this principle requires that all bodies fall in a gravitational field with precisely the same acceleration and that this is confirmed by the Eotvos experiment mentioned earlier. Also, if matter and antimatter were to repel one another, it would be a violation of the principle.

Einstein's theory requires that experiments should have the same results irrespective of the location or time. This has been said to amount to the "strong" principle of equivalence.

Classical tests. The ordinary differential equations of motion of Newtonian gravitation are replaced in general relativity by a nonlinear system of partial differential equations for which general solutions are now known. Apart from a few special cases, knowledge of solutions comes from methods of approximation. For instance, in the solar system, speeds are low so that the quantity v/c (v is the orbital speed and c is the speed of light) will be small (about 10^{-4} for the Earth). The equations and solutions are expanded in powers of this quantity; for instance, the relativistic correction for the motion of the perihelion of Mercury's orbit is adequately found by considering no terms smaller than $(v/c)^2$. This is called the post-Newtonian approximation. (Another approach is the weak-field approximation.)

Einstein's theory has appeared to pass three famous tests. First, it accounted for the full motion of the perihelion of the orbit of Mercury. (Mercury is the most suitable planet, because it is the fastest-moving of the major planets and has a high eccentricity, so that its perihelion is relatively easily studied.) Second, the prediction that light passing a massive body would be deflected has been confirmed with an accuracy of about 5%. Third, Einstein's theory predicted that clocks would run more slowly in strong gravitational fields compared to weak ones; interpreting atoms as clocks, spectral lines would be shifted to the red in a gravitational field. This, again, has been confirmed with moderate accuracy.

I. I. Shapiro has confirmed predictions of the theory in an experiment in which radar waves were bounced off Mercury; the theory predicts a delay of about 2×10^{-4} sec in the arrival time of a radar echo when Mercury is on the far side of the Sun and close to the solar limb. In another test, the precession of a gyroscope in orbit around the Earth is to be studied for evidence of the so-called geodetic precession. The motion of a perihelion is suitable for study since its effects continue to accumulate. Other periodic (noncumulative) orbital effects have until recently been too small to observe. But the current revolution in observational techniques and accuracy has changed the situation; post-Newtonian terms are now routinely included in many calculations of the orbits of planets and space vehicles, and comparison with observations will furnish tests of the theory.

The observation and analysis of gravitation waves, discussed below, will constitute further tests.

Mach's principle. One of the most penetrating critiques of mechanics is due to E. Mach, toward the end of the 19th century. Some of his ideas can be traced back to Bishop G. Berkeley early in the 18th century. Out of Mach's work there has arisen what is known as Mach's principle; this is philosophical in nature and cannot be stated in precise terms. The idea is that the motion of a particle is only meaningful when referred to the rest of the matter in the universe. Geometrical and inertial properties are meaningless for an empty space, and the motion of a particle in such space is devoid of physical significance. Thus the behavior of a test particle should be determined by the total matter distribution in the universe and should not appear as an intrinsic property of an absolute space. If this is so, then the quantitative aspects of physical laws (that is, the various constants involved) should be dependent on position.

Brans-Dicke theory. The field theory developed in 1961 by Brans and Dicke is perhaps the best-known theory in conformity with Mach's principle. For instance, the expansion of the universe causes G to diminish in time. In this theory the gravitational field is described by a tensor and a scalar, the equations of motion being the same as those in general relativity. The addition of a scalar field leads to the appearance of an arbitrary constant, whose value is not known exactly. The Brans-Dicke theory predicts that the relativistic motion of the perihelion of Mercury's orbit is reduced compared with Einstein's value, and also that the light deflection should be less. The experimental determinations of the light deflection are not accurate enough for a distinction to be drawn. With regard to the orbit of Mercury, Dicke pointed out that if the Sun were oblate, this might account for some of the motion of the perihelion. In 1967 he announced that measurements showed a solar oblateness of about 5 parts in 100,000 (or a difference in the polar and equatorial radii of about 34 km). Most of this oblateness is not accounted for by the solar rotation but would require in addition a quadrupole gravitational field; extrapolating the influence of this field out to the orbit of Mercury, he was able to account for a perihelion motion of 3.4 seconds of arc per century, agreeing with the prediction of his theory. However, to account for the quadrupole field of the Sun, it is necessary to have the inner regions much more oblate and more rapidly rotating than the outer regions of the Sun. This may be true if the solar wind acts as a braking force on the outer layers, but interpretations of the situation are still subject to considerable discussion.

There are, of course, many other theories not mentioned here.

Gravitational waves. The existence of gravitational waves, or gravitational "radiation," was predicted by Einstein shortly after he formulated his general theory of relativity. They are now a feature of any relativity theory. Gravitational waves are "ripples in the curvature of space-time." In other words, they are propagating gravitational fields, or propagating patterns of strain, traveling at the speed of light. They carry energy and can exert forces on matter in their path, producing, for instance, very small vibrations in elastic bodies. The gravitational wave is produced by change in the distribution of some matter. It is not produced by a rotating sphere, but would result from a rotating body not having symmetry about its axis of rotation: a pulsar, perhaps. In spite of the relatively weak interaction between gravitational radiation and matter, the measurement of this radiation is now technically possible and has probably already been achieved. This is due to the work of Joseph Weber, whose original and pioneering work has led to a very exciting situation in science. The present situation contains some uncertainties; but gravitational-wave astronomy has been added to other branches of astronomy, and a new window is opening to the universe.

A classical problem, solved by Einstein, concerns the gravitational radiation from a rod spinning about a perpendicular axis through its center. If the rod has moment of inertia about the axis of spin I ($I=Md^2/3$, where M is the mass of the rod in kilograms and $2d$ its length in meters) and angular velocity ω, the power of the radiation in watts (1 W$=10^7$ ergs/sec) is given by Eq. (12), where G

$$P=\frac{32GI^2\omega^6}{5c^5}=1.73\times10^{-52}I^2\omega^6 \qquad (12)$$

is the constant of gravitation in mks units and c is the speed of light in meters per second. A calculation using a steel rod of mass 4.9×10^5 kg (490 metric tons), length 20 m, and angular velocity $\omega=28$ rad/sec, limited by the balance between centrifugal force and tensile strength, gives 2.2×10^{-29} W. So the problem of the generation and detection of gravitational waves in the laboratory is at present somewhat academic.

In electromagnetic theory, electric-dipole radiation is dominant. The gravitational analog of the electric dipole is the mass dipole moment whose time rate of change is the total momentum of the system; since this is constant, there is no gravitational dipole radiation; the principal power is in quadrupole radiation. The radiation has fairly elaborate polarization properties.

Binary systems. Consider a binary star system having period P hr and masses m_1, m_2, where the relative orbit is circular. If $M=m_1+m_2$ and $\mu=m_1m_2/M$, the power output by gravitational radiation is given by Eq. (13), where $M_\odot$ is the mass of

$$P_B=\left(\frac{\mu}{M_\odot}\right)^2\left(\frac{M}{M_\odot}\right)^{4/3}P^{-10/3}\,3.0\times10^{26}\text{ W} \qquad (13)$$

the Sun. For the orbit of the Earth around the Sun, P_B is about 200 W. The gravitation radiation extracts energy from the system. If a binary system has a relative elliptic orbit, then most of the energy is extracted at the closest point of approach and the orbit approaches a circle; then the orbit will gradually shrink, with the bodies colliding after a "spiral time" given by Eq. (14), where a_0 is the

$$\tau_0=\frac{5c^5}{256G^3}\frac{a_0^{\,4}}{\mu M^2} \qquad (14)$$

initial radius of the relative orbit. Under this mechanism the Earth would have fallen toward the Sun less than a centimeter in the lifetime of the solar system!

Clearly one must look outside the solar system for promising sources. Ordinary binaries are not helpful. Sirius and its companion, with a spiral time of 7×10^{21} years, radiate at 10^8 W, the flux received at the Earth being 10^{-31} W. The closer the members of the system are to each other, the more promising they are; some eclipsing binaries can generate power that would be observed on the Earth at about 10^{-20} W, and have spiral times of the order of 10^{10} years. The shortest periods known are for close pairs consisting of a white dwarf and a main sequence star; here spiral times can be as low as 10^9 years, and the predicted flux at the Earth for the most promising candidate, ι Boo, is 18×10^{-18} W. For these binaries, gravitational radiation appears to play an important part in their physical characteristics. Matter flows toward the white dwarf from the companion star, causing flickering and occasional nova outbursts. The stars are very close, and it seems that the contraction of the orbit, caused by gravitational radiation, plays a crucial part in instigating the flow of matter. Closer binaries can at present only be generated by hypothesis. Two neutron stars, having solar masses and with 10^4-km separation, radiate at 3×10^{34} W and have a spiral time of 3 years. With such a system the formulas given above show that the evolution becomes increasingly rapid and the power input increases, so that the final stages of collapse constitute a burst of radiation.

Pulsars. The most rapidly rotating single objects that have been observed are pulsars. These are neutron stars rotating with periods mostly less than 1 sec. From their irregular light curves it is reasonable to suppose that they do not possess symmetry about the axis of rotation. Suppose that they are assumed homogeneous and the equatorial section is an ellipse with axes a and b, and that the ellipticity of the equator is $\epsilon=(a-b)/a$. If the star rotates with angular velocity ω, then the power radiated is given by Eq. (15), where I is the moment

$$P_R=\frac{32G\omega^6I^2\epsilon^2}{5c^5} \qquad (15)$$

of inertia about the axis of rotation. A promising candidate here is the pulsar in the Crab Nebula, remnant of a supernova; the period of rotation is 0.033 sec; the moment of inertia is likely to be of the order of 4×10^{37} kg m², and the power output can be estimated by writing Eq. (15) as Eq. (16),

$$P_R=\left(\frac{I}{4\times 10^{37}\text{ kg m}^2}\right)^2\left(\frac{P}{0.033\text{ sec}}\right)^{-6}\left(\frac{\epsilon}{10^{-3}}\right)^2 10^{31}\text{ W} \quad (16)$$

where P is the period. Clearly it is important to estimate ϵ. The periods of rotation are known to be increasing, and this puts an upper limit on ϵ. It is estimated that the flux received from the Crab pulsar would be less than 3×10^{-20} W. Some pulsars occasionally show sudden changes or glitches in their rotational period. These could be due to starquakes (neutron stars have solid surfaces) and might lead to strong bursts of gravitational radiation.

Explosive events. The gravitational collapse involved in a supernova explosion might produce the strongest radiation that can be observed. The processes involved can only be tentatively estimated, and unfortunately supernova occur about once every 100 years in the Galaxy. But their radiation, probably in short bursts, could be sufficiently powerful for them to be observed from other galaxies. It is possible that stellar collapse takes place without the display of a supernova, so estimates of frequency may be much too low.

Many galaxies show evidence of explosive activity. For quasars, gravitational radiation at 10^{38} W has been suggested, and for explosions in galactic centers, 10^{30} W; but these estimates are not at all definitive.

As matter falls into a black hole, it will release a burst of gravitational radiation; the energy released is proportional to the square of the mass captured and inversely proportional to the mass of the black hole; the time of the outburst is proportional to the mass of the black hole. It has been suggested that there might be a large black hole at the center of the glaxy; if its mass were 10^8 solar masses and it captured a star of 1 solar mass, a burst of energy 10^{37} W might be produced. If the black hole were rotating or the infalling star somehow had a speed greater than that acquired from falling from infinity, then the energy could be greater.

Nature of radiation. The radiation discussed could be continuous or in bursts. The radiation would have a spectrum that might be discrete, as in the case of rotation or orbital revolution, where the fundamental frequency is 2ω (there will also be harmonics), or broadband in the case of explosive events. The longest wavelengths suggested are from the primordial history of the universe, when they could be greater than the size of a galaxy; the shortest, in supernovae and stellar collapse.

Dirac worked out a quantum theory for this radiation; the graviton is a theoretically deduced particle postulated as the quantum of the gravitational field.

Detection of gravitational waves. When a gravitational wave interacts with a system of particles, the particles wiggle slightly; in the case of a solid body, strains are set up in the body; what is actually measured is a sort of tidal effect. Most of the detectors currently under consideration involve strains in solid bodies and they involve the principle of resonance; that is, they react much more strongly to radiation of a given frequency than to other frequencies. When radiation of the correct frequency impinges on the detector it oscillates, as if ringing; radiation at other frequencies is essentially ignored.

Weber's experiment. Weber's detectors principally consist of cylinders suspended in vacuum. They are typically of aluminum, 66 cm in diameter and 153 cm long, weighing 1.5×10^3 kg and resonant at 1661 Hz. They are directional, being most strongly sensitive to radiation traveling perpendicular to the axis of the cylinder. The strains in the cylinder are converted into measurable voltages by quartz strain gages: piezoelectric crystals bonded around the girth of the cylinder. Strains of the order of 1 part in 10^{17} can be detected. There will be continued background thermal "noise," a random effect due to the thermal motion of individual molecules. Since the gravitational signals are of the same order of magnitude as this noise, it is necessary to have two independent receivers, and to look for coincidences in the signals received.

In the principal experiments, Weber analysed signals received 1000 km apart. Since 1969 Weber has reported coincidences at the average rate of about three times a day; typical displacements are around 5×10^{-17} m. He has looked for possible correlations between his coincidences and solar flares, electric storms, surges in the interstate power grid, network television broadcasting, seismic events, and cosmic rays, with no success, and so concludes that the observations are consistent with the detection of gravitational radiation.

Interpretation of Weber's results. In 1970 Weber reported that the coincident pulses were strongly correlated with sidereal time in a manner consistent with the antenna pattern if the gravitational radiation were coming from the center of the Galaxy. This leads to an astrophysically satisfying (in principle) source for the radiation and also a distance for the source: the galactic center. It therefore makes possible the calculation of the energies at the source that would produce the flux observed at Earth. Immediately problems arise. Each burst is what would be expected from a supernova or stellar collapse; if so, these events are more than 1000 times more frequent that had been expected. By using the formula $E=mc^2$, the mass loss that would be the equivalent to this energy can be calculated, and it appears that the Galaxy loses at least 500 solar masses each year; since Weber cannot be observing all events, perhaps this figure should be 10 times as great! But this is not acceptable; the Galaxy would be used up in a time of about one-hundredth of its known age.

The calculation of total energy uses the assumption that the energy is beamed equally in all directions and over a broad band. The band must be at least wide enough to include 1580 Hz, since Weber has observed coincidences at this frequency also. Many suggestions have been made to ease the situation: (1) Gravitational waves have been emitted at this rate for only a small fraction of the age of the Galaxy; this would imply that there is something special about the present age, and that there is something even more special about gravitational waves, since no other observed phenomenon shows the same preference for this era. (2) The source might be much nearer to Earth than the center of the Galaxy. (3) The radiation is strongly focused; into the galactic plane, for instance. (4) There may be some mechanism by which the radiation is magnified between the source and the Earth. (5) There is also the possibility that Weber's events are not caused by gravitational waves. Theories for the origin of the radiation include the possible presence of a very massive black hole at the galactic center, perhaps even disk-shaped, and the possibility of synchrotron-type gravitation radiation.

In another experiment Weber used a disk-shaped antenna. This was designed to search for scalar gravitation radiation. The result was negative, but the disk reacted in a way that was consistent with a source of tensor gravitational waves at the galactic center. So Weber has obtained similar results from two quite different antennas.

Other experiments. A great difficulty in the situation in 1975 was that no other experimenter had reproduced Weber's results. More than a dozen groups in half a dozen countries have been designing experiments. A variety of frequencies are involved, although most of the apparatus is essentially similar to that of Weber. Sensitivity, calibration methods, and methods of data processing have come under discussion, but without satisfactory conclusion. The range of apparatus considered includes different methods of measuring strain; for instance, V. Braginskii of Moscow has "horns" at the ends of his cylinder that nearly meet above the cylinder. The space between their ends serves as the cavity of a parallel plate capacitor which is wired into a resonant L-C circuit. In other cases, the thermal noise will be reduced by cooling the cylinders to millidegree temperatures and suspending them by superconducting magnets. R. Forward and G. Moss at Hughes Laboratories developed a broad-band receiver in which distances between loosely suspended bodies are measured using interferometric techniques.

The Earth has a mass-quadrupole mode with a period of 54 min, and Weber suggested instrumenting the Earth as a gravitation-wave detector using seismographs; unfortunately the Earth is naturally too noisy. But the *Apollo 17* astronauts installed on the Moon a gravimeter designed by Weber and others, and data from these have been transmitted back to the Earth.

With the various improved methods under consideration, it seems safe to predict that doubts will be resolved by 1985. *See* PLANETOLOGY.

[J. M. A. DANBY]

Bibliography: R. H. Dicke, *Gravitation and the Universe*, 1970; G. Gamow, *Gravity*, 1962; J. L. Logan, Gravitational waves: A progress report, *Phys. Today*, 26:44, March 1973. C. W. Misner, K. S. Thorne, and J. A. Wheeler. *Gravitation*, 1973; W. H. Press, and K. S. Thorne, Gravitational wave astronomy, *Annu. Rev. Astron. Astrophys.*, 10:335, 1972; T. J. Sejnowski, Sources of Gravity waves, Phys. *Today*, 27:40, January 1974; D. T. Whiteside, *J. Hist. Astron.*, 1:5, 1970; E. T. Whittaker, *From Euclid to Eddington, a Study of Conception of the External World*, 1949; E. T. Whittaker, *A History of the Theories of the Aether and Electricity*, 1954.

Graywacke

An argillaceous sandstone ("dirty" sandstone) that is characterized by an abundance of unstable mineral and rock fragments and a fine-grained clay matrix binding the larger sand-size detrital fragments. One of the most important characteristics is the clay matrix, which constitutes a minimum of 15% of the rock and may, in extreme cases, make up almost half of the bulk.

Mineral composition. The unstable mineral fragments include feldspar, augite, hornblende, serpentine, biotite, chlorite, and magnetite. Rock fragments include many varieties of low- and high-grade metamorphic rocks (phyllite, slate, quartzite, and granulite), as well as aphanitic (microscopically crystalline) volcanic rocks, typically mafic in composition. The clay matrix is composed of a mixture of micaceous minerals, with biotite and chlorite predominating over muscovite (and illite). Kaolinite, abundant in arkoses, is almost completely absent in the graywackes. Some graywackes have the appearance of water-laid tuffs and may grade into them. These are quartz-poor, and have abundant hornblende and volcanic rock fragments.

Texture. The sorting of graywackes is poor; the presence of a clay matrix inevitably contributes to this. Thus the size distribution curve is skewed toward the fine sizes. Even when the matrix is not considered, the larger detrital grains are poorly sorted. The grains tend to be sharp and angular and may show little or no evidence of abrasion. Many of the grains of metamorphic derivation have low sphericity, either being roughly rod-shaped (quartz, amphiboles, and pyroxenes) or flat (micas). Precipitated mineral cements such as silica or calcite are absent.

Structure. The sedimentary structures of the graywackes differ greatly from those of orthoquartzites and arkoses. Bedding is thin and individual laminae may persist for hundreds of yards. The individual laminae are hard and homogeneous and tend to break with subconchoidal fracture. Graded bedding—that is, repetitive sequences of beds, each bed containing a range of particle sizes grading upward from coarse to fine—is a significant feature. Crossbedding is absent in most beds; if present, it is only a few inches thick. A group of bedding-plane structures, variously called flow casts, groove casts, and lobate rill marks, are commonly found in this rock type. Penecontemporaneous deformation features, small folds and thrusts formed shortly after deposition, are often found in graywackes. Most common is convolute bedding, a plastic deformation of a bed between undeformed beds. Much effort has been devoted to mapping the distinctive repetitive sequences of beds, a few to a few tens of feet thick, that are characteristic of most graywacke sections. Most sequences include a basal massive sandstone with a sharp base, an intermediate finer-grained laminated sandstone, and an upper siltstone or shale.

Graywackes are found in areas of thick sediment accumulation, frequently over 10,000 ft, in association with slates, pillow lavas, cherts, and greenstones. Their occurrence is restricted to geosynclinal areas. Almost all ages are known, from Precambrian to Tertiary. Sediments similar in composition and texture to ancient graywackes are being deposited near the ends of some submarine canyons and in deep basins near tectonically active lands, such as in the Mediterranean and Caribbean seas. See GEOSYNCLINE.

Origin. The genetic significance of graywackes is revealed by the presence of a large amount of unstable mineral and rock fragments, and the distinctive sedimentary structures found in them. Increasingly many geologists are coming to believe that the matrix is largely secondary, having been derived from the mechanical squashing and chemical alteration of rock fragments. The investigation of turbidity (density) currents as agents of transportation has strengthened the hypothesis that the graywackes probably represent a large group of rocks that have been formed by such currents. Further corroboration of this hypothesis is given by the prevalence of graded bedding and other sedimentary structures of the graywackes. Deep-water origin for graywackes has been argued by some geologists, primarily because in deep water normal currents would not be present to sort the material, whereas turbidity currents could be the dominant agent of transport. The unstable mineral and rock fragments imply, as in the arkoses, tectonically active source areas. The penecontemporaneous deformation and association with pillow lavas implies tectonic activity in the basin of deposition as well. See ARKOSE; ORTHOQUARTZITE; SANDSTONE; SEDIMENTARY ROCKS; SUBGRAYWACKE; TURBIDITY CURRENT.

[RAYMOND SIEVER]

Bibliography: A. H. Bouma and A. Brouwer (eds.), *Turbidites*, 1964; S. Dzulynski and E. K. Walton, *Sedimentary Features of Flysch and Greywackes*, 1967.

Greisen

A pneumatolytically altered granite consisting of mainly quartz and a light-green (white) mica; the feldspars and biotites of the original granite have been replaced more or less completely by muscovite, lithium mica, kaolinite, and tourmaline. Other substances usually introduced during alteration are topaz, apatite, fluorite, and ores of iron, and in some places ores of tungsten and tin. See PNEUMATOLYSIS.

Greisen occurs in bands and veins intersecting granite. The bands are rarely more than 2 ft thick with indefinite margins that grade into unaltered granite. They were evidently formed by flow of vapors and gases through fissures.

In principle the formation of greisen is easily understood, being a typical example of pneumatolytic metasomatism. A crystallizing granitic magma gives off gases containing silicon, Si; fluorine, F; boron, B; lithium, Li; and often tin, Sn; or tungsten, W. They cause a complete alteration of the adjacent solid rock, be it the granite itself or its country rock. The usual, completely altered product is a rock consisting only of quartz and mica (that is, greisen). In limestone and dolomite, fluorite often accompanied by humite will develop instead of topaz, and instead of tourmaline, the boron-bearing mineral axinite may develop. Datolite also may occur. See METASOMATISM.

Tin ore deposits (cassiterite, SnO_2) are usually accompanied by greisen and the concomitant impregnation of topaz and tourmaline. The metal content is probably introduced in the form of fluorides by distillation and subsequent hydrolytic decomposition; thus the equation below holds,

$$\underset{\text{Tin(IV) fluoride}}{SnF_4} + \underset{\text{Water}}{2H_2O} \rightarrow \underset{\text{Cassiterite}}{SnO_2} + \underset{\text{Hydrofluoric acid}}{4HF}$$

where the liberation of hydrofluoric acid accounts for the strong alteration of the adjacent rocks.

[T. F. W. BARTH]

Groundwater

The water in the zone in which the rocks and soil are saturated, the top of which is the water table. The zone of saturation is the source of water for wells, which provide about one-fifth of the water supplies of the United States. It is also the source of the water that issues as springs and seeps, and maintains the dry-weather flow of perennial streams. The saturated zone is a great natural reservoir which absorbs and stores precipitation during wet periods and pays it out slowly during dry periods; it is, therefore, a natural regulating mechanism which tempers the severity of floods and droughts. The amount of ground water stored in the rocks in the United States is estimated to be several times as great as that stored in all lakes and reservoirs, including the Great Lakes.

Subterranean water. Water beneath the land surface may be subdivided into two parts, water in the zone of aeration above the water table and water in the zone of saturation below the water table. Water in the zone of aeration, also called vadose water, is divided into soil water or rhizic water, intermediate vadose water or argic water, and water of the capillary fringe or anastatic water.

Water in the capillary fringe is connected with the zone of saturation and is held above it by capillary forces. The lower part of the fringe may be saturated but is not a part of the zone of saturation because the water is under less than atmospheric pressure and will not flow into a well, although the walls of the well are moist. When the well reaches the zone of saturation, water will begin to enter it and will stand at the level of the water table.

Rock formations capable of yielding significant volumes of water are called aquifers. Some wells are artesian; that is, the water rises above the top of the water-bearing bed. This is a special case and will be discussed later. Other wells encounter water above the saturated zone and lose their water if extended through the impermeable bed upon which the water rests. Such bodies of water are said to be perched. This also is a special case.

Subterranean water occurs in a geologic environment, and therefore a knowledge of geology is essential to an understanding of the occurrence of water. For this reason, the study of groundwater is sometimes called hydrogeology or geohydrology. The hydrologist is particularly concerned with the number and size of the openings in rocks and soils and the manner in which they are interconnected. Variations in these openings are almost infinitely diverse. Openings are practically absent in some of the igneous rocks. They are numerous but microscopic in clay. They are large and interconnected in many sands and gravels. There are huge caverns and tubes in many limestones and lavas.

Their distribution and types are as diverse as the geology itself, so that general statements about them applicable to one area may be incorrect for another.

Openings in rocks. These are of two broad types. Primary are those which existed when the rock was formed, and secondary, those which resulted from the action of physical or chemical forces after the rock was formed. Primary openings are found in sedimentary rocks such as sand and clay and certain kinds of limestone composed of triturated shells. Certain openings in lava are formed at the stage when the lava is partly liquid and partly solid and also are considered primary. Most rocks containing primary openings are relatively young geologically. Those which contain primary openings large enough to carry useful amounts of water are represented, for example, by the seaward-dipping strata of the Atlantic and Gulf coastal plains, including the coquina limestone of Florida, the intermontane valleys of the western United States, the glacial deposits of the United States, and the lava rocks of the Pacific Northwest.

Secondary openings are common in older rocks. Sand and gravel that have been cemented by chemical action, limestone indurated by compression or recrystallization, schist, gneiss, slate, granite, rhyolite, basalt and other igneous rocks, and shale generally contain few primary openings; but they all may contain fractures that will carry water. Limestone in particular is subject to solution which, beginning along small cracks, may develop channels ranging from openings a fraction of an inch across to enormous caverns capable of carrying large amounts of water.

Porosity. The property of rocks for containing voids, or interstices, is termed porosity. It is expressed quantitatively as the percentage of the total volume of rock that is occupied by openings. It ranges from as high as 80% in newly deposited silt and clay down to a fraction of 1% in the most compact rocks.

Permeability. This is the characteristic capability of rock or soil to transmit water. The porosity of a rock or soil has no direct relation to the permeability or water-yielding capacity. This capacity is closely related to the size and the degree to which the pores or openings are interconnected. If the pores are small, the rock will transmit water very slowly; if they are large and interconnected, they will transmit water readily. The standard coefficient of permeability (the Meinzer unit) used in the hydrologic work of the U.S. Geological Survey is defined as the rate of flow of water at 60°F, in gallons per day through a cross section 1 ft^2, under a hydraulic gradient of 100% (1 ft of head loss per foot of water travel). Under field conditions, the adjustment to standard temperature is commonly ignored, and permeability is expressed as a field coefficient at the prevailing water temperature.

Transmissibility. Another coefficient that is commonly used, transmissibility, expresses the rate at which water moves through a saturated body of rock. It is expressed as the rate of flow of water at the prevailing temperature, in gallons per day through a vertical strip of aquifer 1 ft wide, extending the full saturated height of the aquifer under a hydraulic gradient of 100%. This coefficient is especially useful for expressing the total yield of an aquifer.

Controlling forces. Water moves through permeable rocks under the influence of gravity from places of higher head to places of lower head; that is, from areas of intake or recharge to areas of discharge, such as wells or springs. Water moving through rocks is acted upon also by friction and by molecular forces. The molecular forces are the attraction of rock surfaces for the molecules of water (adhesion) and the attraction of water molecules for one another (cohesion). When wetted, each rock surface is able to retain a thin film of water despite the effect of gravity. In very fine-grained rocks, such as clay and fine silt, the interstices may be so small that molecular attraction extends from one side of a pore to the opposite side. Molecular force then becomes dominant, and water moves through the rock only very slowly under the gradients typical of natural conditions.

The amount of water that drains from a saturated rock under the influence of gravity, expressed as a percentage of the total volume of the rock, is called the specific yield. The percentage of water retained in the rock is called the specific retention. Specific yield is often called effective porosity because it represents the pore space that will surrender water to wells and so is effective in supplying water for human use. The term porosity is poorly defined and its use should be discontinued. A part of the water stored in the rocks is held by molecular forces and may have only a small share in supplying springs or wells. This latter portion is of special interest to the agriculturalist because it sustains plant life. A soil that is highly permeable permits water to pass through it easily and little is retained for the nourishment of plant life, whereas a soil that is relatively impermeable retains much of its water until it is extracted by plants or by evaporation.

Sources. It has been firmly established that the chief means of replenishment of groundwater is downward percolation of surface water, either direct infiltration of rainwater or snowmelt or infiltration from bodies of surface water which themselves are supplied by rain or snowmelt. Evidence on the replenishment of groundwater is furnished by analysis of data on the downward movement of precipitation through the soil and subsoil, the rise and fall of groundwater levels and spring discharge in response to precipitation and seepage losses from streams, and the slope of the hydraulic gradient from known areas of intake to areas of discharge. Some groundwater may originate by chemical and physical processes that take place deep within the Earth. Such water is called juvenile water or primary water to indicate that it is reaching the Earth's surface for the first time. The available evidence indicates that such water is always highly mineralized. Some water is stored in deep-lying sedimentary rocks and is a relic of the ancient seas in which these rocks were deposited. It is called connate water. The total quantity of water from juvenile and connate sources that enters the hydrologic cycle is insignificant when compared with the quantities of water derived from precipitation (meteoric water). It is balanced to some extent by withdrawal of water from the hydrologic cycle by such processes as deposition

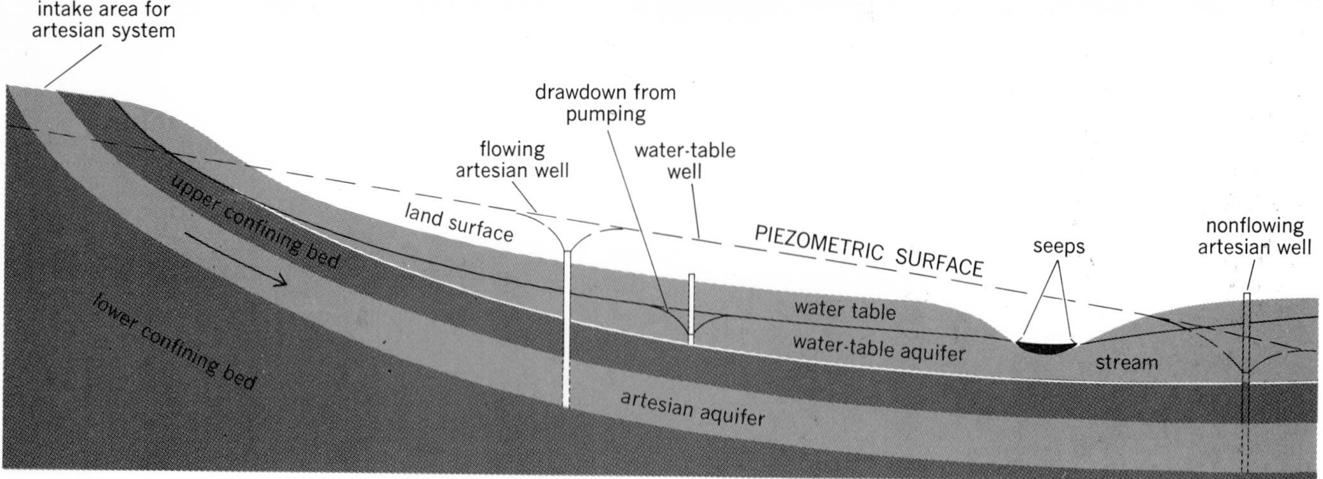

Water-table and artesian conditions.

of minerals that include water in their crystalline structure.

Infiltration. Replenishment of water in the zone of saturation involves three steps: (1) infiltration of water from the surface into the rock or soil that lies directly beneath the surface, (2) downward movement through the zone of aeration of the part of the water not retained by molecular forces, and (3) entrance of this part of the water into the zone of saturation, where it becomes groundwater and moves, chiefly laterally, toward a point of discharge. Infiltration is produced by the joint action of molecular attraction and gravity. The rate of infiltration then becomes chiefly a function of the permeability of the soil. This permeability is almost infinitely diverse. Under conditions of unsaturated flow such as dominate in the zone of aeration, it varies with moisture content as well as with pore size. It varies also with the geology. For example, in the Badlands, S.Dak., where the soil and rocks are of low permeability, the infiltration capacity of the soil is low and is reached quickly after rainfall or snowmelt begins. Hence, there is not much infiltration, and any excess of precipitation or snowmelt over infiltration runs off over the surface and enters the streams. If the excess is large, serious floods and erosion result. On the other hand, the soils of the Sand Hills, Nebr., and the glacial outwash deposits of Long Island, N.Y., are so permeable that they absorb the water of the most violent storms and permit little or no direct runoff.

The permeability of the rock materials beneath the soil zone also is important. Since the soil is commonly formed by weathering of the underlying rock, the permeability of the rocks is generally comparable to that of the soil.

Water-table and artesian conditions. Water that moves downward through the soil and subsoil in excess of capillary requirements continues to move downward until it reaches a zone whose permeability is so low that the rate of further downward movement is less than the rate of replenishment from above. A zone of saturation then forms, its thickness depending on the opportunity for lateral escape of water in relation to replenishment. The top of this zone is the water table (see illustration). Under these circumstances, the water is said to be unconfined, or under water-table conditions. However, since much of the crust of the Earth has a more or less well-defined layered structure in which zones of high and low permeability alternate, situations are common in which groundwater moving laterally in a permeable rock passes between layers of relatively low permeability. Although the permeable layer contains unconfined groundwater in the area where there is no impermeable layer (confining bed) above, the part of the layer, or aquifer, that passes beneath the confining bed contains water that is pressing upward against the confining bed, and if a well is drilled in this area the water in it will rise. It tends to rise to the level of the water in the unconfined area, but fails to reach that level by the amount of pressure head lost by friction as the water moves from the unconfined area to the well. Confined water is also called artesian water, and wells tapping it are called artesian whether or not their head is sufficient for them to flow at the land surface.

Chemical qualities. Water is said to be the universal solvent. When it condenses and falls as rain or snow, it absorbs small amounts of mineral and organic substances from the air. After falling, it continues to dissolve some of the soil and rocks through which it passes. Thus, no groundwater is chemically pure. Its commonest mineral constituents are the bicarbonates, chlorides, and sulfates of calcium, magnesium, sodium, and potassium, in ionized or dissociated form. Silica also is an important constituent. Common also are small, but significant, concentrations of iron, and manganese, fluoride, and nitrate. The concentration of the dissolved minerals varies widely with the kind of soil and rocks through which the water has passed. Ordinarily, water that contains more than 1000 parts per million (ppm) of dissolved solids is considered unfit for human consumption, and water that contains more than 2000 ppm is considered unfit for stock. However, both human beings and animals may become accustomed to greater concentrations.

[A. NELSON SAYRE/RAY K. LINSLEY]

Bibliography: S. N. Davis and R. J. M. DeWiest, *Hydrogeology*, 1966; R. J. Kazmann, *Modern Hydrology*, 1965; D. K. Todd, *Groundwater Hydrology*, 1959; R. C. Ward, *Principles of Hydrology*, 1964.

Gypsum

A mineral with the chemical composition $CaSO_4{\cdot}2H_2O$. Gypsum is the commonest of the sulfate minerals. It occurs in five varieties: (1) rock gypsum; (2) gypsite, an impure earthy form; (3) alabaster, a massive, fine-grained, translucent variety; (4) satin spar, a fibrous silky form; and (5) selenite, transparent crystals. Crystals of gypsum are monoclinic, clear, white to gray, yellowish, and brownish in color, with well-developed cleavages (see illustration). Platy fragments are flexible but not elastic. Luster is subvitreous to pearly. Hardness is 2 on Mohs scale and specific gravity is 2.3. It fuses readily. It is soluble in hydrochloric acid and slightly soluble in water.

Gypsum is calcined in kettles or kilns at temperatures of 190–200°C to remove part of the water of crystallization. The calcined product, known as plaster of paris, sets in 6–8 min after addition of water. The setting time can be extended to 1–2 hr by adding glue, fiber, lime, or other material.

Gypsum and gypsum plaster products are chiefly used in the building industry and as a retarder in portland cement. Ground gypsum and anhydrite are used in the southern United States as soil conditioners to improve permeability. Gypsum and anhydrite, particularly the latter, may be made into ammonium sulfate fertilizer. Calcined gypsum is also used for industrial plasters, such as those used in pottery, molding, dentistry, and statuary. Pure white uncalcined gypsum, known as terra alba, is used as a filler in paper and paints and as a nutrient in growing yeast. Other uncalcined ground gypsum is used as a filler in insecticide. *See* ANHYDRITE.

Gypsum is deposited, in beds, from sea water or brines from salt lakes at temperatures below 42°C. It is also precipitated from aqueous solutions in limestone caves, as in the gypsum flowers of Mammoth Cave, Ky. Large deposits of gypsum are generally associated with limestone, dolomite, shale, sandstone, and salt beds. *See* EVAPORITE, SALINE.

Gypsum. (*a*) A crystal, found in Oxfordshire, England (*American Museum of Natural History specimens*). (*b*) Crystal habits (*from C. S. Hurlbut, Jr., Dana's Manual of Mineralogy, 17th ed., copyright © 1959 by John Wiley & Sons, Inc.; reprinted by permission*).

Gypsum is of world-wide distribution. Thick commercial deposits of calcium sulfate occur in Nova Scotia and in Stassfurt, Germany. In the western United States extensive commercial gypsum deposits occur in California, Nevada, Utah, Texas, Iowa, Oklahoma, Kansas, Montana, Colorado, Arizona, South Dakota, Wyoming, and New Mexico. In the East, large mines are located in Michigan, New York, Ohio, and Indiana.

[EDWARD C. T. CHAO]

Half-life

The time required for one-half of a given material to undergo chemical reactions; also, the average time interval required for one-half of any quantity of identical radioactive atoms to undergo radioactive decay.

Chemical reactions. The concept of the time required for all of the material to react is meaningless, because the reaction goes very slowly when only a small amount of the reacting material is left and theoretically an infinite time would be required. The time for half completion of the reaction is a definite and useful way of describing the rate of a reaction.

The specific rate constant k provides another way of describing the rate of a chemical reaction. This is shown in a first-order reaction, Eq. (1),

$$k=\frac{2.303}{t}\log\frac{c_0}{c} \qquad (1)$$

where c_0 is the initial concentration and c is the concentration at time t. The relation between specific rate constant and period of half-life, $t_{1/2}$, in a first-order reaction is given by Eq. (2). In a

$$t_{1/2}=\frac{2.303}{k}\log\frac{1}{1/2}=\frac{0.693}{k} \qquad (2)$$

first-order reaction, the period of half-life is independent of the initial concentration, but in a second-order reaction it does depend on the initial concentration according to Eq. (3).

$$t_{1/2}=\frac{1}{kc_0} \qquad (3)$$

[FARRINGTON DANIELS]

Radioactive decay. The activity of a source of any single radioactive substance decreases to one-half in 1 half-period, because the activity is always proportional to the number of radioactive atoms present. For example, the half-period of Co^{60} (cobalt-60) is $t_{1/2}=5.3$ years. Then a Co^{60} source whose initial activity was 100 curies will decrease to 50 curies in 5.3 years. The activity of any radioactive source decreases exponentially with time t, in proportion to exp $-0.693\ t/t_{1/2}$. After 1 half-period (when $t=t_{1/2}$) the activity will be reduced by the factor $e^{-0.693}=1/2$. In 1 additional half-period this activity will be further reduced by the factor 1/2. Thus, the fraction of the initial activity which remains is 1/2 after 1 half-period, 1/4 after 2 half-periods, 1/8 after 3 half-periods, 1/16 after 4 half-periods, and so on.

The half-period is sometimes also called the

half-value time or, with less justification, but frequently, the half-life. The half-period is 0.693 times the mean life or average life of a group of identical radioactive atoms. The probability is exactly 1/2 that the actual life-span of one individual radioactive atom will exceed its half-period.

[ROBLEY D. EVANS]

Halite

A mineral with chemical composition NaCl. Sylvite with the composition KCl is similar in structure and occurrence. Their crystal structures consist of large close-packed chloride ions in a face-centered cubic lattice, with Na^+ or K^+ filling the largest interstices, so that each Cl^- is surrounded by six alkali ions and each alkali ion has six Cl^- ions as nearest neighbors (Fig. 1). The bonding is due to electrostatic ionic forces. Consequently, the minerals are poor conductors of electricity and heat, and when pure are transparent to visible light. Rare but scientifically interesting varieties are strikingly colored (for example, dark blue) because of small concentrations of "color centers," which are ionic vacancies or other electronic defects which absorb visible light. Naturally occurring halite and sylvite are commonly colored brown to red by fine inclusions of goethite or hematite, or gray by anhydrite, clay minerals, or carbonaceous matter. The minerals are relatively soft (2.5 and 2.0 for halite and sylvite, respectively, on the Mohs scale) and fracture with perfect cubic (001) cleavage. At elevated temperatures (300–400°C) and pressure they become plastic due to translation gliding [along the dodecahedral or (110) directions]. The densities of both minerals (2.16 for halite and 1.99 for sylvite) are lower than those of most common rock-forming minerals; hence massive salt deposits are less dense than surrounding rocks.

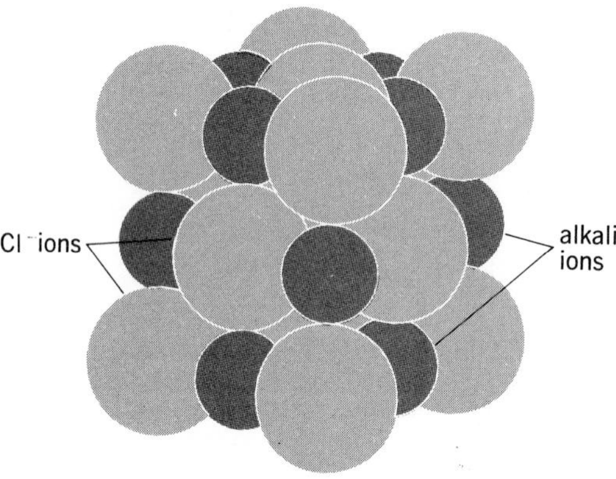

Fig. 1. The crystal structure of halite and sylvite. The close-packed planes of large Cl^- ions lie along the cube-diagonal planes [Miller indices (111)], and alkali ions fill interstices in octahedral six-coordination. *(From R. W. G. Wyckoff, Crystal Structures, vol. 1, 2d ed., copyright © 1963 by John Wiley & Sons, Inc.; reprinted by permission)*

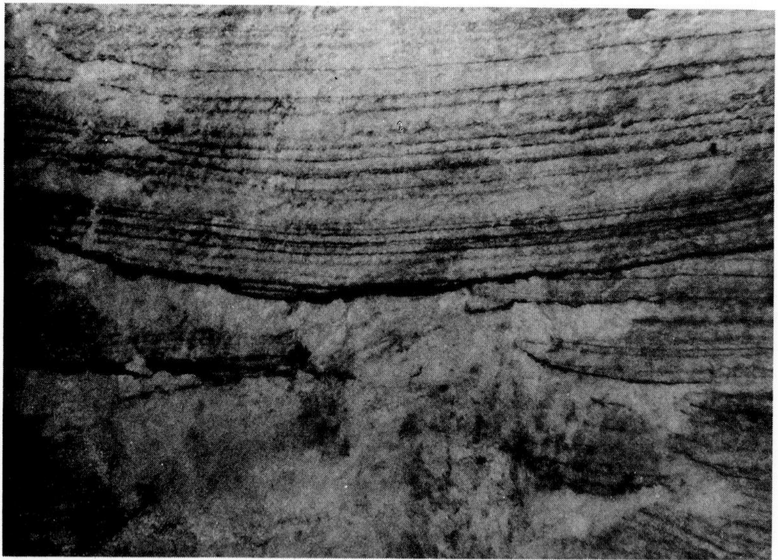

Fig. 2. Beds of salt rock of Silurian age in the Ojibway Mine, Windsor, Ontario. Light layers (1 to 4 cm thick) are composed of nearly pure halite; dark layers are concentrations of anhydrite ($CaSO_4$) and clay minerals. The large mass is halite that has been dissolved by subsurface waters and regrown to crystals several centimeters in diameter. *(From W. T. Holser, Bromide geochemistry of salt rock, in J. L. Rau, ed., Second Symposium on Salt, vol. 1, p. 267, Northern Ohio Geological Society, 1966)*

Formation and occurrence. Both minerals occur primarily as stratified beds in sedimentary rocks of marine origin. Individual crystals in these deposits range in size from about 0.2 to 2 cm as evidenced by well-developed cleavage surfaces on grains (Fig. 2). Their origin is by the evaporation of sea water (or less commonly, by the evaporation of saline lake water). When sea water of present-day composition is evaporated, minerals precipitate in the sequence shown in the table. The geological record indicates that during at least the last billion years the composition of sea water has been substantially constant with this same resulting sequence. The crystallization of halite requires an 11-fold concentration of sea water by evaporation, and this occurs mainly in dry subtropical maritime regions (in some cases on tidal flats or sebkhas), as seen today along the southern coast of the Persian Gulf. In many paleogeographical environments (such as the Jurassic of the Middle East and Gulf of Mexico and the Permian of central Europe and the southwestern United States), bodies of salt rock several kilometers thick and thousands of kilometers in extent were accumulated. This deposition took place, in part, in deep marine basins that received inflows of sea water but were substantially isolated from oceanic circulation.

Salt domes. Where these thick sequences of salt have been buried more than about 5 km by the accumulation of later sediments, geothermal heating raises the temperature into the plastic region, and because of halite's lower viscosity and its low density relative to the overlying rocks, the salt flows upward as domes 1 to 10 km in diameter, piercing the later sedimentary rock sequence. Some columns have risen more than 8 mi (12.9 km). Under common climactic conditions both salt beds and salt domes are "subroded" by fresh groundwater to a few hundred meters below the surface, but in some arid areas (notably southeastern Iran) the domes rise to and above the surface, where they form salt mountains from which the salt glides off in "salt glaciers" over the surrounding countryside. *See* SALT DOME.

Salt resources. Nearly all sedimentary basins have halite somewhere in the sedimentary section below the surface, or intruded as salt domes. In

Crystallization of sea water

Facies	Mineralogy	Evaporation stage			
		Salinity, parts per thousand	Density	Fraction evaporated, wt % H_2O	Concentration × sea water, wt % H_2O
Potash-magnesia	Carnallite, sylvite, kainite, kieserite, halite; anhydrite or polyhalite	380	1.39	98.7	75
Halite	Halite; anhydrite or gypsum	300	1.20	91	11.5
$CaSO_4$	Gypsum or anhydrite; dolomite	150	1.10	72	3.5
Carbonate (normal sea water)	Dolomite, calcite	35	1.02	0	1

many of these basins it is economical to recover halite by underground mining or by brining (underground solution from wells drilled into the salt rock), but in other areas these processes lose out in competition with artificial solar ponds using sea water or saline lake water. Salt domes, which are common in the Gulf Coast of the United States are also economically important as traps for petroleum and natural gas and as a source of native sulfur which is frequently found in material "capping" these domes. All are important sources of salt, not only for domestic use in the preparation and production of food but also for dairy culture, snow and ice removal, and especially as a primary raw material for the chemical industry for the manufacture of sodium, chlorine, soda ash (Na_2CO_2), and hydrochloric acid. Few areas of the world have any shortage of this important natural product.

Sylvite. Sylvite is much less common than halite, representing a much greater evaporative concentration of sea water, and hence is a less frequent geological accident. Under equilibrium conditions sea water precipitates the mineral carnallite, $KMgCl_3 \cdot 6H_2O$, but sylvite is commonly found in association with carnallite, as a result of metastable crystallization, sulfate removal, or alteration of carnallite by subsurface waters.

Sylvite occurs near the top of evaporite sections, but extensive deposits are found in relatively few areas, notably in the Permian of the Soviet Union and East and West Germany and in the Devonian of Saskatchewan, Canada. In the United States significant deposits occur in southeastern New Mexico and eastern Utah. Sylvite is principally recovered by underground mining, but technical problems that had previously precluded brining have recently been overcome. Sylvite, along with carnallite and sometimes other potassium minerals found with them, is the main source of the potash component of fertilizer as well as being a raw material for other chemical manufacturing. These resources are therefore of critical importance. Thus far, crystallization of sylvite or other potash minerals from sea water has not been economical, although they are recovered from some lake brines. [WILLIAM T. HOLSER]

Bibliography: H. Borchert and R. O. Muir, *Salt Deposits, Their Origin and Composition*, 1964; D. W. Kirkland and R. Evans, *Marine Evaporites: Origin, Diagenesis, and Geochemistry*, 1973; S. L. Lafond, *World Salt Resources*, 1970; C. Palache, H. Berman, and C. Frondel, *The System of Mineralogy of James Dwight Dana*, 7th ed., vol. 2, pp. 3–15, 1951.

Halloysite

A group of clay minerals made up of the same structure units as kaolinite. The major use of halloysite clay is in the manufacture of cracking petroleum catalysts. Pure halloysite clays are not abundant; hence their potential economic use has not been explored. *See* CLAY MINERALS; KAOLINITE.

There are two types of halloysite. Both are light-colored; one is porous and friable, while the other is dense, nonporous, and porcelainlike. The porous variety has the same chemical composition as kaolinite, whereas the nonporous hydrated form contains an added amount of water ($2H_2O$). The hydrous variety has a larger *c*-axis spacing than kaolinite and changes upon dehydration to about the same dimension as kaolinite. The transition from the higher hydrated form to the lower hydrated form begins at temperatures as low as 60°C and is not reversible. Temperatures of the order of 400°C are necessary for complete removal of this interlayer water. The nomenclature of the types of halloysite is confusing.

Structure and morphology. The halloysite minerals are made up of the same structure units as kaolinite; that is, layers composed of single silica tetrahedral and alumina octahedral units. They differ in the presence of a single layer of water molecules between each silicate layer in the hydrated form. The basal spacing of the dehydrated form is about 7.2 A, or about the thickness of the kaolinite layer, and the basal spacing of the hydrated form is about 10.1 A. The difference, 2.9 A, is about the thickness of a single-molecule-thick sheet of water molecules.

Halloysite also differs from kaolinite in that there is random displacement of silicate layers in both the *a* and *b* crystallographic directions.

Electron micrographs show that halloysite is composed of elongate particles which appear to be tubes (see illustration). The tubes of the hydrated $4H_2O$ form appear to be made up of overlapping, curled-up sheets of the kaolinite type. The presence of interlayer water molecules and the irregular stacking of units would facilitate this curvature. On dehydration the tubes frequently collapse, split, or unroll.

Electron micrograph of halloysite, British Guiana. (*From R. E. Grim, Clay Mineralogy, McGraw-Hill, 1953*)

Other characteristics. The cation-exchange capacity of the lower hydration form of halloysite is 5–10 milliequivalents per 100 g, whereas that of the hydrated variety is somewhat higher. The major cause of this exchange capacity is broken bonds.

Differential thermal curves for halloysite show a loss of pore water and interlayer water at about 100°C. Above about 200°C the dehydration characteristics are essentially like those for kaolinite. The high-temperature phases formed on firing halloysite are like those for kaolinite, but there are some differences in their development. Thus halloysite does not uniformly vitrify and its fusion point is higher than that of kaolinite.

Halloysite is often found in hydrothermal deposits. It has also been reported in weathering products; however, it is a rare component of such materials. It is generally absent in sedimentary rocks, and it has been suggested that metamorphic processes would change halloysite to kaolinite.

[FLOYD M. WAHL; RALPH E. GRIM]

Halogen minerals

Naturally occurring compounds containing a halogen as the sole or principal anionic constituent. There are over 70 such minerals, but only a few are common and can be grouped according to the following methods of formation.

1. Saline deposition by evaporation of sea water or salt lakes. Halite (rock salt), NaCl, is the most important of this type and is found in beds covering many hundreds of square miles and ranging in thickness from a few feet to over 1000 ft. Of the other minerals associated with halite, sylvite, KCl, and carnallite, $KMgCl_3 \cdot 6H_2O$, are the most important. *See* EVAPORITE, SALINE; HALITE.

2. Hydrothermal deposition. Fluorite, CaF_2, is the chief representative of this type and occurs in veins by itself or associated with metallic ores. Cryolite, Na_3AlF_6, may be of primary deposition or may result from the action of fluorine-bearing solutions on preexisting silicates. *See* CRYOLITE; FLUORITE.

3. Secondary alteration. Chlorides, iodides, or bromides of silver, copper, lead, or mercury may form as surface alterations of ore bodies carrying these metals. The most common are cerargyrite, AgCl, and atacamite, $Cu_2(OH)_3Cl$. *See* CERARGYRITE.

4. Deposition by sublimation. Halides formed as sublimation products about volcanic fumaroles include salammoniac, NH_4Cl; malysite, $FeCl_3$; and cotunnite, $PbCl_2$. At Mount Vesuvius, Italy, is the most noted occurrence of such minerals.

5. Meteorites. Lawrencite, $FeCl_2$, has been found in iron meteorites. *See* METEORITE.

[CORNELIUS S. HURLBUT, JR.]

Bibliography: H. Borchest and R. O. Muir, *Salt Deposits*, 1964; C. W. Correns, *The Geochemistry of the Halogens*, in L. H. Ahrens, K. Rankama, and S. K. Runcorn (eds.), *Physics and Chemistry of the Earth*, vol. 1, 1956; C. Palache, H. Berman, and C. Frondel, *Dana's System of Mineralogy*, vol. 2, 1951.

Heavy minerals

Minerals with a density above 2.9, which is the density of bromoform, the liquid used to separate heavy from light minerals. Heavy minerals are sometimes used in igneous petrology, but their main importance is in the study of sedimentary rocks.

If not modified by weathering or sedimentation processes, the heavy-mineral suites of sedimentary rocks, primarily sandstones, offer clues to the composition of the source rock. For example, amphibole and amphibole-epidote suites suggest an igneous or high-grade metamorphic source, augite suites a volcanic source, kyanite-sataurolite suites a low-grade metamorphic source, and zircon-tourmaline suites a sedimentary source from which all unstable minerals that might originally have been present have been eliminated by one or more cycles of weathering and erosion. *See* SEDIMENTARY ROCKS.

Sediments can be characterized by their heavy-mineral association. The study of the distribution of heavy-mineral associations in a depositional basin permits the tracing of the distribution in space and time of sediment supply from various sources and assists in the unraveling of the erosional history of the source areas. Since frequently several sources supply sediments with different heavy-mineral suites to a single basin, the pattern of heavy-mineral associations can be complex, in particular when mixing occurs. Sophisticated mathematical and statistical techniques have been used to analyze these complex relationships. Thus, heavy minerals are an aid in the study of the depositional history and paleogeography of a basin. In some cases they have been used successfully in correlating sedimentary strata. *See* SEDIMENTATION.

The heavy-mineral suite provided by a source rock can be modified by various processes before and after deposition. Weathering may result in the selective destruction of unstable minerals, for example, augite or olivine. It is effective mainly in hot, humid areas where weathering is intense. During transportation, mechanical destruction of soft minerals may occasionally cccur. Sorting during wind or water transportation tends to concentrate coarse particles in one deposit, fines in another. If certain minerals are restricted to specific size ranges, sorting may modify the mineral composition greatly. In the Rhone Delta, for example, the fine-grained offshore deposits with an epidote-amphibole association are derived from the same source as the coarse-grained augite-bearing sands of the subaerial delta. Selective solution of unstable minerals after burial (intrastratal solution) has been proposed as an important factor in the control of heavy-mineral composition. Such selective solution may be important locally but appears to occur infrequently.

Natural heavy-mineral concentrates, common in beaches and streams, have considerable economic value. The monazite (thorium and rare earths) sands of India and Brazil and the tin sands of the Bangka and Billiton islands, Indonesia, are examples of commercially important heavy-mineral deposits. During the low-sea-level stands of the ice ages, beach and stream deposits were laid down extensively on the continental shelves of the world. Some of these are known to contain valuable mineral concentrates, for example, the offshore diamond sands of South-West Africa, the offshore gold deposits of Alaska, and the marine iron sands of New Zealand; and other widespread occurrences must be expected. Very active leasing, exploration, and mining development programs are underway along the Pacific coast of Alaska, around Australia, and elsewhere, and a spectacular increase in offshore heavy-mineral exploitation is to be expected. [TJEERD H. VAN ANDEL]

Bibliography: Tj. H. van Andel, Reflections on the interpretation of heavy mineral analyses, *J. Sediment. Petrol.*, 29(2):153–163, 1959; G. Baker, *Detrital Heavy Minerals in Natural Accumulates*, Aust. Inst. Min. Met. Monogr. no. 1, 1962; J. Imbrie and Tj. H. van Andel, Vector analysis of heavy mineral data, *Bull. Geol. Soc. Amer.*, 75(11): 1131–1156, 1964; J. L. Mero, *The Mineral Resources of the Sea*, 1965; J. J. Pettijohn, *Sedimentary Rocks*, 1957.

Hematite

The most important ore of iron, with composition α-Fe_2O_3, crystallizing in the rhombohedral system, space group $R\bar{3}c$. Its crystal structure consists of Fe^{3+} cations octahedrally coordinated by hexagonal close-packed oxygen atoms, the arrangement belonging to the corundum (α-Al_2O_3) structure type. A third rare isotype, eskolaite (α-Cr_2O_3), also occurs in nature. Hematite has a dimorph, maghemite (γ-Fe_2O_3), a magnetic defect spinel structure occurring only rarely in nature.

The crystals are thick tabular, usually flattened parallel to the base, and are frequently platy in habit (see illustration). Masses composed of compact platelets are called "specular hematite." Hematite usually occurs as rouge-red earthy masses of finely divided particles. It is the major red-coloring agent in rocks and is a common interstitial cement in sediments. When mixed with quartzite or finely divided quartz, the mixture is called jasper, jaspilite, or taconite. Botryoidal masses are called "kidney ore," and splinters of these masses are "pencil ore." The mineral also occurs in oolitic form.

The color is steel gray, blood red in thin fragments, and streak and powder are rouge red; hardness is 6 on Mohs scale and specific gravity is 5.25. The fracture is conchoidal; though possessing no prominent cleavage direction, basal or pseudocubic parting is common. The mineral is only weakly magnetic.

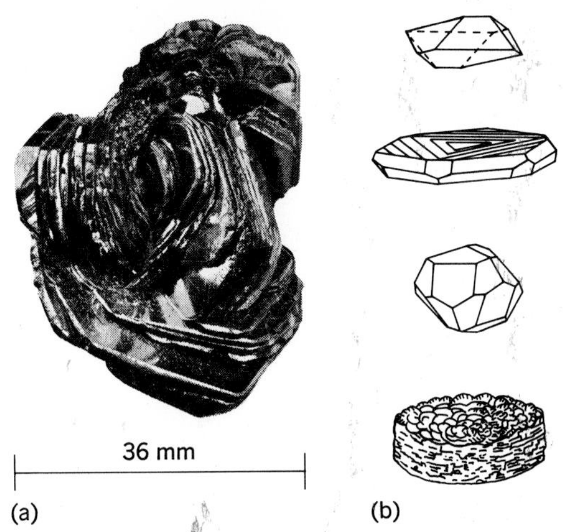

Hematite. (*a*) Crystals, Saint Gothard, Switzerland (*American Museum of Natural History specimens*). (*b*) Crystal habits (*from C. S. Hurlbut, Jr., Dana's Manual of Mineralogy, 17th ed., copyright © 1959 by John Wiley & Sons, Inc.; reprinted by permission*).

Hematite is easily obtained by the oxidation of iron in solution followed by heating the precipitate; by oxidation of magnetite, $Fe^{2+}Fe^{3+}{}_2O_4$; by the dehydration of goethite, α-FeO(OH); and by the oxidation of ferrous chloride sublimates.

Hematite is the most widespread iron mineral. It is not as common in igneous rocks as magnetite and ilmenite, $Fe^{2+}Ti^{4+}O_3$, where it is often formed by the decomposition of iron-bearing minerals, particularly silicates. It occurs associated with magmas poor in FeO, such as granites and syenites. Hematite frequently occurs in fumaroles and with volcanic rocks, derived by the oxidation of volcanic sublimates, and sufficient quantities may constitute ore material.

The most important ores are in low-to-medium-grade metamorphic rocks of sedimentary origin. The original sediments are believed to have obtained a high-iron content through precipitation and accumulation of ferric hydroxides which were deposited in fresh water or in shallow parts of the sea, particularly in Precambrian time. Enormous beds of hematite, jaspilite, taconite, and iron silicates up to 1000 ft in thickness occur in the Great Lakes region of the United States. Hematite also occurs in contact metamorphic and metasomatic deposits, often derived from the oxidation of magnetite and frequently associated with limestones.

Nearly every country in the world mines some hematite ore; the most important occurrences outside of the United States include India, Cuba, China, Chile, North Africa, and the Soviet Union. *See* ILMENITE; REDBEDS.

[PAUL B. MOORE]

Hemimorphite

A mineral sorosilicate having the composition $Zn_4Si_2O_7(OH)_2 \cdot H_2O$; an ore of zinc. It crystallizes in the orthorhombic system, pyramidal class, and thus the prismatic crystals have different forms at top and bottom (see illustration). Also, as a result of the symmetry, the vertical axis is polar and crystals display the properties of pyroelectricity and piezoelectricity. There is perfect prismatic cleav-

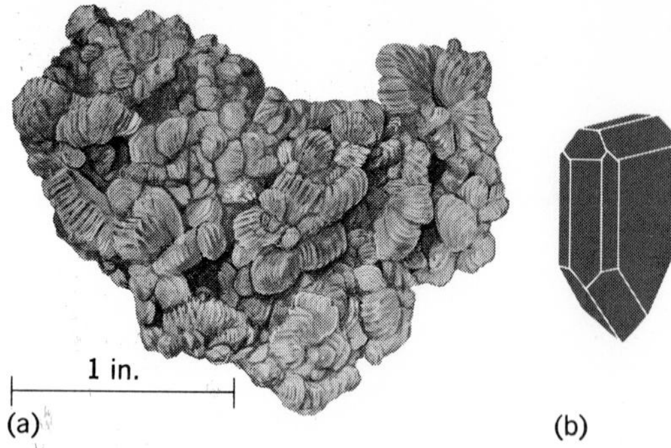

Hemimorphite. (*a*) Crystals forming coxcomb masses, Granby, Mo. (*specimen from Department of Geology, Bryn Mawr College*). (*b*) Crystal habit (*from C. S. Hurlbut, Jr., Dana's Manual of Mineralogy, 17th ed., copyright © 1959 by John Wiley & Sons, Inc.; reprinted by permission*).

age. Botryoidal and staclactic aggregates, showing a crystalline surface and frequently impure, are common. Crystals are usually colorless and the aggregates white, but in some cases there are faint shades of green, yellow, and blue. The mineral has a vitreous luster, a hardness of 4½ to 5 on Mohs scale, and a specific gravity of 3.45. Hemimorphite frequently resembles the zinc carbonate, or smithsonite, but can be distinguished by the reaction with hydrochloric acid. Hemimorphite is slowly soluble but smithsonite effervesces in cold acid. *See* SILICATE MINERALS; SMITHSONITE.

Hemimorphite is a secondary mineral found in the oxidized portion of zinc deposits associated with smithsonite, sphalerite, cerrussite, anglesite, galena, and more rarely the oxidized minerals of copper. It has a wide distribution and has been mined in Belgium, Germany, Romania, England, Algeria, and Mexico. In the United States it is found at Sterling Hill, N.J.; Friedensville, Pa.; and Elkhorn Mountains, Mont.

[CORNELIUS S. HURLBUT, JR.]

Heulandite

A mineral belonging to the zeolite family of silicates and crystallizing in the monoclinic system. It usually occurs in crystals with prominent side pinacoid, often having a diamond shape. There is perfect side pinacoid cleavage on which the luster is pearly; elsewhere the luster is vitreous (see illustration). The crystals often have undulating faces, and are made up of subindividuals in nearly parallel position. In polarized light they show optical anomalies of a sectoral nature. The hardness is 3½ to 4 on Mohs scale; specific gravity is 2.18–2.20. The mineral is usually white or colorless but may be yellow or red. *See* ZEOLITE.

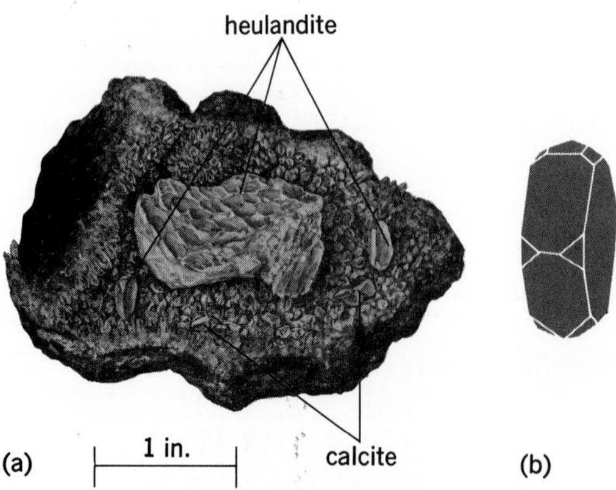

Heulandite. (*a*) Crystals with calcite and quartz, Paterson, N.J. (*specimen from Department of Geology, Bryn Mawr College*). (*b*) Crystal habit (*from C. S. Hurlbut, Jr., Dana's Manual of Mineralogy, 17th ed., copyright © 1959 by John Wiley & Sons, Inc.; reprinted by permission*).

Heulandite is essentially a hydrous calcium aluminum silicate, $Ca(Al_2Si_7O_{18})\cdot 6H_2O$. Small amounts of sodium and potassium usually substitute for calcium. Heulandite is a secondary mineral found in cavities in basalts associated with other zeolites and calcite. Notable localities are in the Faeroe Islands, India, Nova Scotia, and West Paterson, N.J.

[CLIFFORD FRONDEL; CORNELIUS S. HURLBUT, JR.]

High-pressure phenomena

Natural conditions and processes occurring at high pressures and their duplication in the laboratory. Of those variables which have the largest effect on the free energy or chemical activity of any system (defined as a portion of the universe isolated for study) the most important are composition, temperature, and pressure. The vast majority of relevant research in chemistry and physics has been concerned with variations in the first two parameters. There has been a good reason for this: It has been experimentally difficult to change the pressure enough to make an appreciable difference on a system. It can easily be calculated that to cause the equivalent change in an average, or typical, substance it is possible either to cool it by 1°C or to subject it to a pressure of 100 atm. (A bar equals 0.987 atm, and can be considered to be roughly equal to 15 psi.) Every science laboratory equipped with a bunsen burner can heat a substance to about 1000°C; to achieve the same effect with pressure one would need 100,000 atm, or the equivalent of close to 1,500,000 psi. An orientation in the magnitude of high pressures is important. Atmospheric pressures on or above the surface of the Earth range from 0 to 14.7 psi. Certainly 99.99% of the chemical reactions studied have been studied at or just below atmospheric pressure. However, high pressures are not uncommon. They range from the 2–5 atm in automobile tires and pressure cookers, through 100 or so atm in the boilers of ships and power plants, to perhaps 100,000 atm at the point of impact of a high-speed rifle bullet and a hard wall. High pressures up to 200,000 atm can be generated in a very simple apparatus using automotive truck jacks. In the laboratories of chemists and earth scientists the pressures and temperatures attainable for studying chemical reactions have increased from about 10,000 atm in 1900, to 25,000 atm in 1940, and to about 200,000 atm in 1970. In areas of research that are more concerned with physical than chemical phenomena, shock pressures are being utilized in the order of several million atmospheres. This article deals with high-pressure techniques and applications to problems in the earth sciences.

Areas of high-pressure research. The Earth itself is a giant laboratory in which pressures up to about 3,000,000 atm are generated with increasing depth; moreover, in the Earth the temperature and

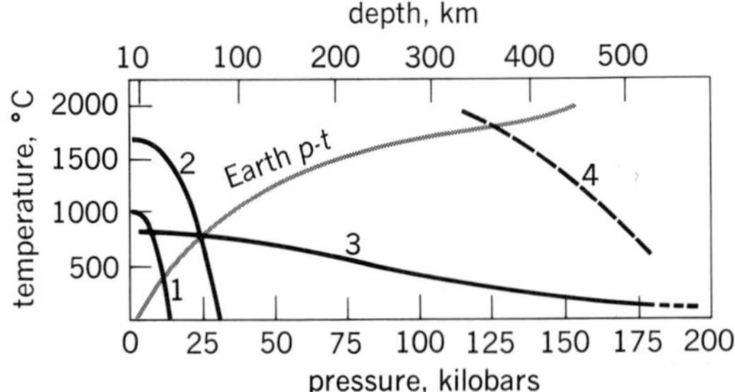

Fig. 1. Experimental limits of various types of apparatus and a generalized pressure-temperature relation of a portion of the Earth. Curve 1—simple, externally heated test-tube or cold-seal vessels. Curve 2—internally heated, hydrostatic-pressure vessels. Curve 3—externally heated, uniaxial pressure devices. Curve 4—internally heated, modified piston and cylinder devices.

pressure increase together (Fig. 1). It is generally agreed, however, that of the rocks actually observed on the surface of the Earth, none has been subjected to pressure greater than 10,000–100,000 atm. Possible exceptions may be some meteorites which could have originated in the interior of a large planet, or a portion of the rocks in and about large craters that were formed by the many meteoritic collisions throughout geologic time.

The interest of the earth scientist, especially the geochemist, in high-pressure inorganic reactions is therefore natural. In attempting to understand natural processes leading to the formation of various rocks, it is essential to duplicate the conditions existing in the Earth. The science of interpretive petrology rests on the accumulation of data obtained at high temperatures, high pressures, or both. Theories concerning the composition and properties of the different layers deep within the Earth, which may explain seismic discontinuities, Earth's magnetism, possible slippage of one layer over another, and other natural phenomena, can be checked only by research of high-temperature and high-pressure processes and reactions.

An additional impetus to research in this field is the ability to produce, under pressure, previously unsynthesized minerals, as well as quite new materials which may be expected to be especially dense and hard. Other phases which contain volatile components, such as water or carbon dioxide, can only be prepared under high pressures of these volatiles. Finally, there is considerable evidence that fluids such as water are excellent catalysts for many inorganic reactions occurring under high pressure.

Apparatus in high-pressure research. Much of the recent significant research in this field has been made possible by new apparatus. This in turn owes its origin in large measure to the new materials, such as hard alloys and carbides, produced by an advancing technology. Below are described three or four major families of apparatus with which the majority of high-pressure chemical research is carried on. In each of these a sample is subjected to fluid or mechanical pressure for a period of time, from minutes up to days or months, at some desired temperature. The reaction is stopped abruptly, or quenched, by rapid cooling and more or less rapid removal of pressure. The resultant products, often in a metastable condition in the laboratory, are examined by using x-ray techniques, the petrographic microscope, differential thermal analysis, infrared absorption spectroscopy, and other more common or more specialized techniques.

The starting materials used in these reactions are often of great importance. The use of amorphous materials such as glasses and gels or other metastable phases is often a decisive factor in the chemical kinetics problem encountered, since time is usually an important limitation with respect to the strength of the equipment.

The types of apparatus commonly used can be divided into four categories, depending on whether or not the pressure is hydrostatic and whether the heating is external or internal. The range of pressure and temperature which each of these types can cover is shown in Fig. 1. Also shown are the estimated pressure and temperature within a portion of the Earth. The complexity and ease of operation are not inversely related to the pressures attainable, both the externally heated devices being quite simple.

In Fig. 2 more detail is given on the first type, the cold-seal test-tube pressure vessel which is the workhorse of much high-pressure research in the

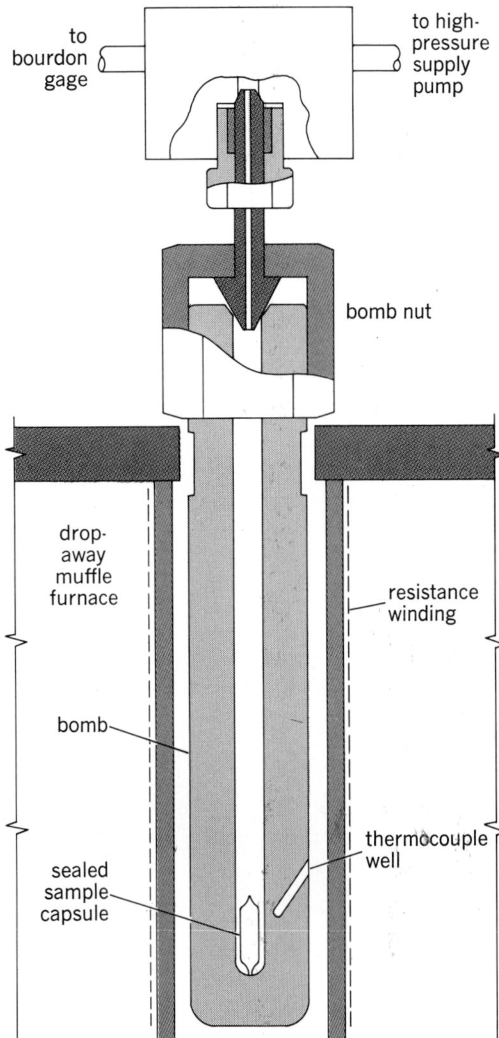

Fig. 2. Cold-seal test-tube pressure vessel. Basic overall size of the test tube is 8 in. length by 1 in. diameter.

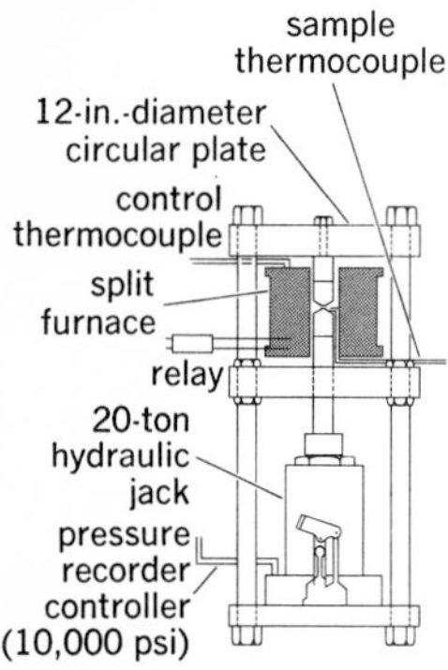

Fig. 3. Layout of uniaxial high-pressure apparatus, based on Bridgman's design.

range up to 5000 atm and 1000°C. Newly developed and commercially available high-strength metals have been used in the construction of such units, which, with modified cold seals, extend the useful range to 10,000 atm at 750°C. This extremely simple device can be used to react a very wide range of materials, whether solid or liquid. The materials, whether simple silicates or concentrated HF or NaOH solutions, are sealed into platinum or gold capsules which transmit the hydrostatic pressure actually supplied by an inert fluid outside the capsule. These vessels are heated by an external furnace as illustrated. A related device, which also uses fluid pressure, has the furnace inside the vessel, which is cooled by flowing water. Such devices are ultimately limited by the fact that virtually all gases are frozen at room temperature when the pressure reaches 25,000–30,000 atm. The pressure-temperature (*p-t*) working range of these types of vessels can be compared by making reference to curves 1 and 2 respectively in Fig. 1.

In general, the working volume of these units can accommodate sample capsules of fairly large size, 1½-in.-diam by 2 in. or more long, which lend themselves to studies of such factors as solubility, viscosity, thermal expansion, and basic solid-liquid-vapor equilibria. With special adaptations all manner of electrical, magnetic, dielectric, and optical studies are possible.

The second and simpler type has been evolved from designs by P. W. Bridgman, a pioneer in high-pressure research. As shown in Fig. 3, this device uses a hydraulic ram to apply a directed or uniaxial pressure on a small wafer of sample surrounded by a nickel ring and platinum foil pressed between appropriate small-area piston faces. Pressures up to 60,000 atm are possible with this setup, and the sample can be heated externally to temperatures approaching 650°C when special tool steels are used for the pistons, or anvils, as they are often called. At lower temperatures the use of "compound" anvils made of combinations of tool steels and tungsten carbides extends the effective range of this useful and versatile apparatus to 180,000–200,000 atm. The working limits of these typical compound externally heated opposed-anvils pressure devices are represented by curve 3 of Fig. 1. Many variants of this apparatus have been found useful: In one, oscillating shearing stresses may be applied continuously on the sample; in another, electrical-resistance heating of a furnace element in direct contact with the sample permits the use of higher temperatures at high pressures; in still another, small diamond anvils make possible x-ray diffraction and optical studies of samples at pressures up to 200,000–300,000 atm. In one design by H. G. Drickamer, which in essence is a modification of an anvil system, pressures of the order of 500,000 atm have been attained by providing support to the most highly stressed (conical) portions of the anvils by means of a material such as sodium chloride under high pressure.

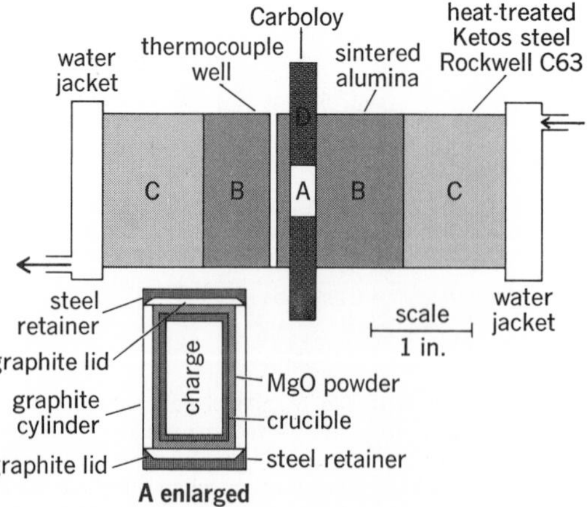

Fig. 4. Coes-type apparatus, which is used in high-temperature high-pressure research.

The third type, the prototype of which was first described fully by L. Coes, is shown in Fig. 4. It is essentially an internally heated chamber of a piston and cylinder arrangement with the cylinder walls buttressed and cooled to support internal pressures of 60,000 atm at sample temperatures up to 2000°C. A number of variants of this type of apparatus have been designed and used to attain even higher pressures and temperatures simultaneously. The type shown in Fig. 5, first described by H. T. Hall and coworkers, may be viewed as a hybrid of the opposed-anvils and the piston-cylinder apparatuses in which ingenious compensation for the weak features of the parents is achieved at the same time that their basic simplicity is retained.

Somewhat more divergent variants use 4, 6, and even as many as 8 anvils arrayed appropriately in three dimensions to apply and confine high pressures on a small chamber hollowed out of carefully shaped blocks of a fine-grained rock known as pyrophyllite. The principle is illustrated in Fig. 6. In these, as in most of the other high-pressure systems, ingenious techniques have been devised for heating the sample and for providing various leads for electrical measurements.

Inevitably, where great diversity of apparatus exists the problem of determining pressure, temperature, and other parameters of the sample under study becomes important; in fact, a considerable portion of the work done with the more complex devices is directed toward calibration. The problem is complicated because no simple or single material is available which can be included in the sample to serve as an internal calibrant and still remain effective over the *p-t* range being utilized. At high pressures and temperatures practically all substances are subject to physical or chemical changes which would make them ineffective as calibrants. In any event, temperatures in the vicinity of 4000°C at moderate pressures have been attained in some applications with this type of apparatus; around 1000–1500°C successful experiments have been made up to 150,000–200,000 atm. (Some reports mention 500,000 atm.)

In the 1960s numerous studies were undertaken with such apparatus. They varied from the synthesis of diamonds and stishovite and solid-liquid-vapor equilibria in petrologic systems to the determination of the electrical, mechanical, or magnetic behavior of materials; with the aid of very clever design, some studies even incorporated x-ray diffraction, optical absorption, and Mössbauer spectra studies of samples under pressure.

Another type of apparatus uses the energy of a shock wave to produce high pressures. Explosive charges or the impacts of projectiles propelled in "guns" by rapidly expanding columns of a gas such as helium or hydrogen are utilized to produce pressures of one-tenth of a million to several million atmospheres; the highest pressures are produced by the explosive charges.

The coincident temperatures and pressures of shock-wave experiments are difficult to measure or calculate. The duration of these conditions is in microseconds, very often much too short to allow rearrangements of atoms or ions in crystalline structures. In addition, in many cases the "residual" temperature of the shocked material may be so high after the pressure pulse has subsided that interesting phase transitions are lost in the ensuing melting or annealing processes. For these reasons shock-wave experiments are of limited interest to the geochemist. For example, although coesite, one of the new dense forms of silica, is made quite readily in hours at 20,000 atm and 400–500°C, shock pressures of 200,000–400,000 atm have failed to accomplish this. However, in one or two large-scale experiments with underground nuclear explosions, traces of coesite have been found in the shocked rocks surrounding the cavity.

Synthesis of minerals and new materials. A good measure of the success of the earth scientist in the high-pressure field is the fact that since the synthesis of diamonds, it can now be claimed that any material which has been observed in nature can be made in the laboratory. To be sure, there are many which have not been tried, but there is no reason to doubt that in every case the laboratory can duplicate the efforts of nature. *See* DIAMOND.

First, one may mention the synthesis of the hard, dense phases which have presumably been thrust up rapidly from great depth. The most spectacular of the recent successes is the synthesis of diamonds in a few laboratories. Perhaps more important to the mineralogist has been the synthesis of the minerals jadeite, lawsonite, sillimanite, kyanite, and garnets such as pyrope, almandine, and andradite. All these phases were first synthesized in a dramatic breakthrough by Coes, using the apparatus shown in Fig. 4. Next is the extensive and systematic work on volatile-containing phases such as micas, clays, and complex carbonates. Under high water pressures all pure or end-member phases have been prepared and their properties studied and defined. Moreover, the extent of the systematic replacement of one ion by another has been studied. Isotopic fractionation effects in minerals have likewise been examined in the laboratory, using deuterium for hydrogen and O^{18} for O^{16}.

Many of the most interesting syntheses achieved have yielded phases which do not occur in nature and have never before been prepared. In some cases a long-expected phase was finally prepared, such as the hard cubic form of BN, analogous to diamond. In others, unexpected phases were found; thus, silica, SiO_2, the most abundant substance on Earth, gives a new form called coesite, some 10% denser than the usual form, quartz. A second new form of silica, called stishovite (61% denser than quartz), also has been found; for its synthesis, conditions of about 100,000 atm and 600°C are required. *See* COESITE; STISHOVITE.

New high-pressure forms of a large number of substances have been prepared. They cover the periodic table: B_2O_3, BN, and BeF_2; Mg_2SiO_4, $Al(OH)_3$, and S; TiO_2, $MnPO_4$, and Fe_2SiO_4; ZnO, Ge, and CdS; PbO_2 and U_3O_8 are only a partial list. It must be remembered that these changes include only those transformations involving such a major rearrangement of strong bonds that, when the temperature is lowered rapidly and the pressure is released, the reverse reaction is prevented

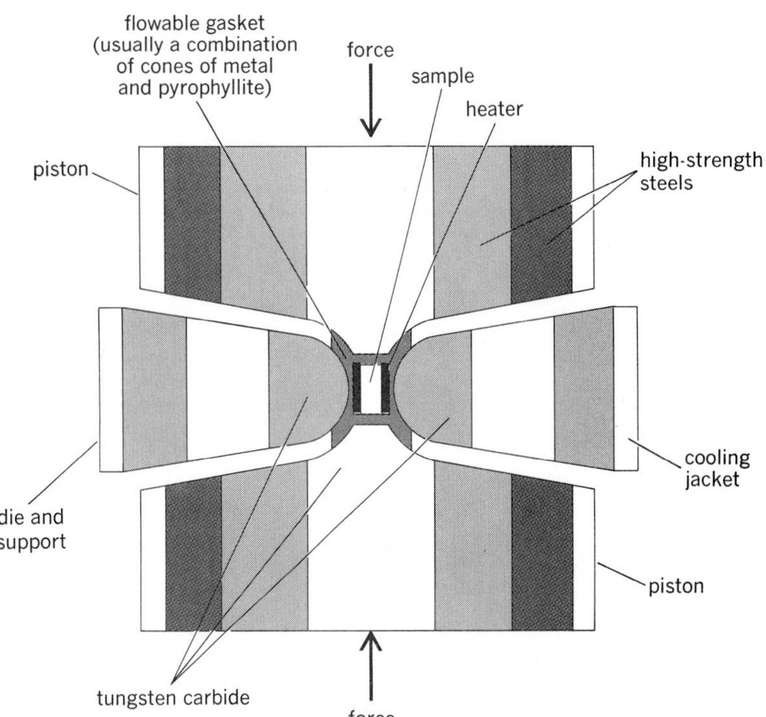

Fig. 5. Section of a "belt" apparatus. Tightly fitted rings of high-strength steels support the pistons and die made of carbides. The most critical support of the working faces of the pistons and die is provided by the highly compressed gasket formed by controlled and limited extrusion.

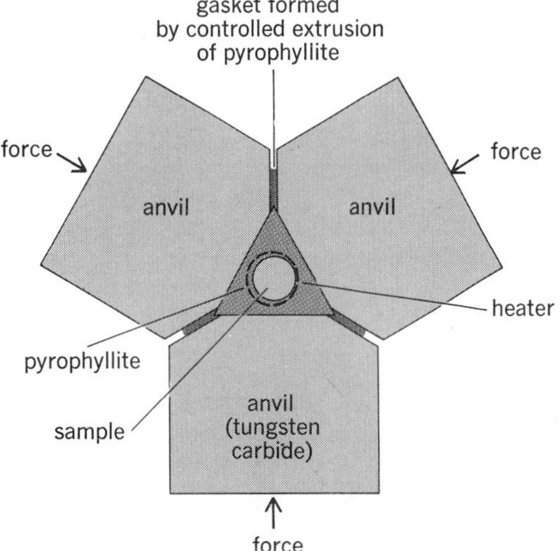

Fig. 6. Principle of the multianvil apparatus. A minimum of four closely fitted anvils are necessary to enclose a volume. In one variant many anvils are nested together to form a sphere which, when enclosed in an impermeable bag, can be squeezed by the fluid of a high-pressure tank.

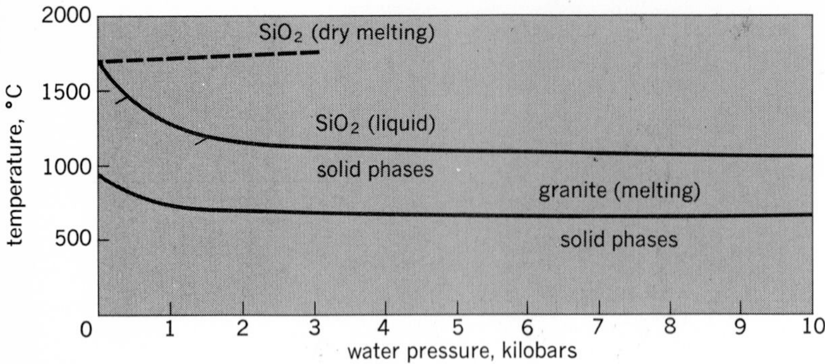

Fig. 7. Effect of water pressure on the melting behavior of silica (SiO_2) and granite with water entering the liquids. The melting behavior of SiO_2 under inert, or dry, pressure is shown for comparison.

and the high-pressure form is preserved metastable to ambient conditions. So far only simple compounds of monovalent cations have failed to yield new phases of this type.

Melting under volatile pressure. It is well known that the addition of one substance to another lowers the freezing point of the latter. This is true of the vast majority of substances; the addition of salt to water is one of the commonest in experience. For many decades the influence of water or other volatile materials on the melting behavior of rocks intrigued geologists. Naturally, to carry out such an experiment the entire assembly of rock and volatile addition would have to be confined under pressure at or near the melting points of the rocks; these temperatures are of the order of 1000–1500°C. In Fig. 7 two examples are shown of the influence of water pressure on common rocks and minerals. The lowering of the melting point of SiO_2 from 1730 to 1250°C under only 1000 atm of water pressure is quite dramatic. The water in this case dissolves in the siliceous liquid. Also shown for comparison is the influence of inert or dry pressure on the melting point of silica. In this case the gas or solid transmitting the pressure does not dissolve in either the liquid or solid silica. The similar dry-pressure curve is also shown in Fig. 9 for the important mineral diopside, to illustrate the fact that in general the melting point is raised slowly with pressure. The distinction between melting in an inert atmosphere and one in which the volatile phase under pressure dissolves in the liquid should be borne in mind. In deeply buried rocks it can be assumed that much of the water is effectively sealed in the system and may enter the liquid phase. The results of the study of melting the assemblage of minerals which form granite are also shown in Fig. 7. The fact that a granite can be melted as low as 660°C at a pressure of 4000 atm (corresponding to a burial of about 10 mi) with only a few percent of water in the liquid is one of the most important clues on the origin of many of the commonest igneous rocks in the Earth.

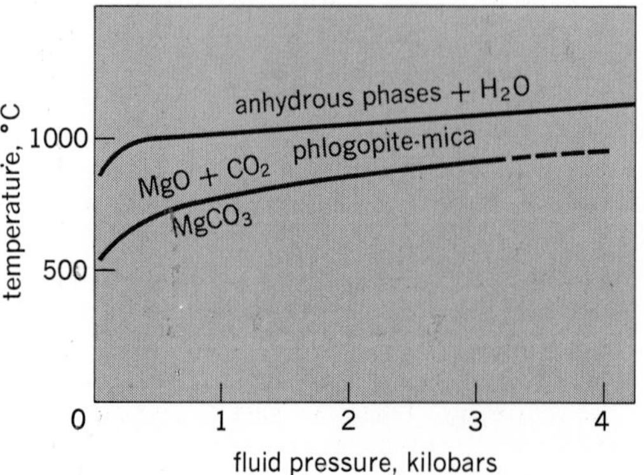

Fig. 8. Typical stability curves for minerals which lose CO_2 or H_2O on heating.

Applications in petrology. Although the synthesis of new or important compounds has a spectacular aspect, much of the research in high-pressure phenomena is concerned only incidentally with synthesis. Instead, the major effort is concentrated on obtaining new data on the pressure, temperature, and composition conditions under which certain mineral assemblages are stable. Nature provides several typical assemblages of minerals as characteristic of certain rock types or families. As the conditions under which each particular assemblage is stable are determined, a partial reconstruction can be provided of the conditions which must have existed at any particular place on the Earth.

Two different types of reactions may be studied in this connection. The first type involves reactions such as decarbonation or dehydration. Two simple reactions of this type are used in the pressure-temperature diagram of Fig. 8 to illustrate the principle of the application of such data. The general form of such curves is seen to be convex toward the high-temperature and low-pressure side, with very steep slopes at pressures above a few thousand atmospheres and an asymptotic approach to the temperature axis at very low pressures. On the low-temperature side of such curves, the hydrates, such as mica, or carbonates, such as magnesite, are stable. Thus, if the pressure from the depth of burial can be estimated by studying the minerals present, it is possible to determine whether or not the rock has been heated to temperatures above the curve. Thus it is possible to explain why the dark micas occur in some high-temperature extrusive rocks (lavas), whereas the white or muscovite micas do not. The dehydration-decomposition curve for the latter, not shown in Fig. 8, lies some 350°C lower than the dark-mica curve shown, while the temperatures of these rocks are intermediate.

In Fig. 9 is shown a compilation on one diagram of several *p-t* curves for assorted yet common compositions. Each curve represents a transformation reaction where the composition of the condensed phases on either side of the *p-t* curve is the same. It may be noted that these curves are quite different from those of Fig. 8. They are all essentially straight lines in the pressure range shown. The dense form is favored by high pressure. The pressure theoretically required to make diamonds is actually quite modest. Transformations that require twice the pressure to give the dense forms shown in Fig. 9 have been studied quantitatively.

In general, Fig. 9 illustrates that solid-liquid transformations such as the melting of ice do not differ basically from solid-solid transitions. These curves may of course be used in a manner similar to that described for Fig. 8 to suggest whether or not a particular rock was exposed to a certain set of *p-t* conditions. In practice, curves of both types are used to refine as far as possible the petrologic predictions which can be made. *See* PETROLOGY.

Geological and geophysical problems. An example of how laboratory results spread out and influence other fields of activity is appropriate here. In 1953 an industrial researcher, L. Coes, synthesized coesite in high-pressure apparatus. A world authority on meteorites, H. H. Nininger, in a book published in 1956, speculated that coesite, as well as other high-pressure phases of minerals, might be found in meteoritic impact craters. In 1960, E. C. T. Chao, E. M. Shoemaker, and B. M. Madsden reported the first natural occurrence of coesite—in the impacted Coconino sandstone of the Barringer or Arizona Meteor Crater. This report stimulated the rediscovery and burgeoning growth of the field of meteoritics, a stimulus practically on a par with that provided by the space program; it has led to the accumulation of so many observations and experimental findings that a whole new field of study has been added to geology—shock metamorphism. Almost anticlimatic was the synthesis of stishovite in 1961 (S. M. Stishov and S. V. Popova) and its discovery in, again, the Meteor Crater in Arizona in 1962 (Chao and others).

Earlier it was mentioned that a trace of coesite was recovered from the site of an underground nuclear explosion. It is one of the sobering findings in the field of meteoritics that the characteristics of an explosion of this kind, in intensity and magnitude, only begin to approach those of the "explosions" caused by the crashing of giant meteorites onto the Earth's surface (or the Moon's or Mars's, for that matter).

From the results of other studies on the changes in phase (from less dense to more dense forms) in those minerals which geologists believe make up most of the mantle of the Earth, it is possible to

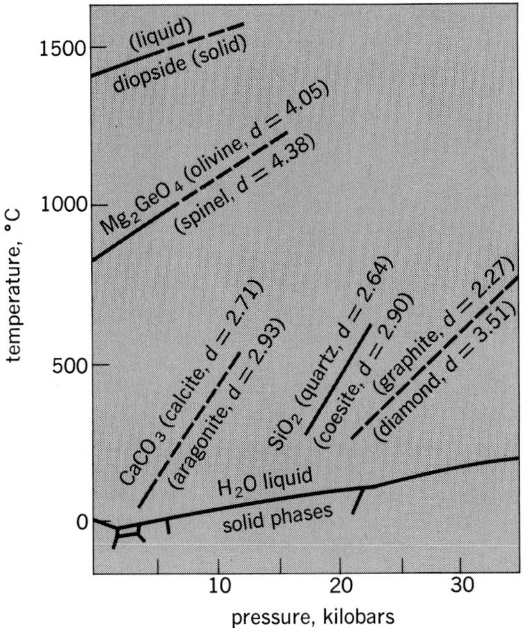

Fig. 9. Curves showing pressure-temperature dependence of typical solid-solid and solid-liquid transformations for single substances; *d* is density in g/cm³.

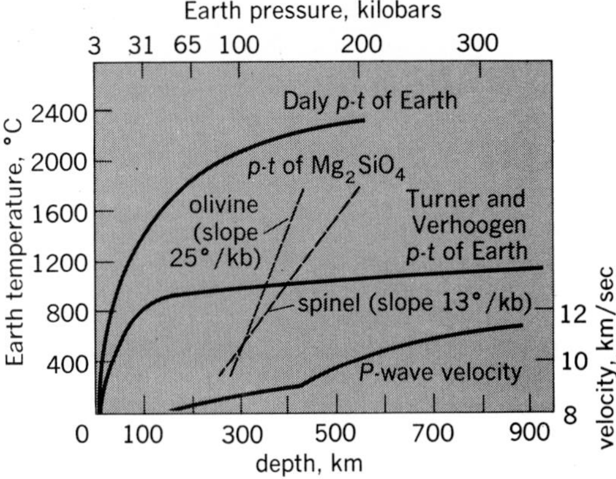

Fig. 10. Pressure, temperature and *P*-wave velocity relations for a portion of the mantle on which are superimposed possible Mg_2SiO_4 pressure-temperature relations deduced from experiments.

explain some of the variations in the velocities of seismic pulses at different depths in the Earth. The present results do, in fact, suggest the possibility that some of the layering in the Earth may not reflect changes in its composition at all. For example, there is good reason to believe that high magnesian olivine (predominantly Mg_2SiO_4) makes up the major portion of the composition of the mantle. From crystal-chemical reasoning the theory was advanced that this silicate should undergo an olivine-spinel transition under sufficient pressure (see Fig. 9 for the analogous transition of Mg_2GeO_4). The results of the first comprehensive experimental study addressed to this problem lead to certain geophysical implications which remain applicable to date. These are summarized in Fig. 10, which shows for a portion of the Earth the variation of the *P*(seismic)-wave velocity, the pressure-temperature relation of F. J. Turner and J. Verhoogen (1951), and the pressure-temperature relation of R. A. Daly (1943). The two *p-t* lines, representing the limits of the uncertainty of the *p-t* dependence of the Mg_2SiO_4 olivine-spinel transition, have slopes of nearly 25 and 13°C per 1000 atm and are plotted passing through 542°C and 100,000 atm (300 km depth).

An inspection of Fig. 10 shows that a transition of olivine to spinel could be the determining factor in the second-order discontinuity just below 400 km depth. If the seismic discontinuity is accurately placed at 413 km depth and if the composition of the mantle is essentially forsterite (Mg_2SiO_4) with up to 10–20 mole % fayalite (Fe_2SiO_4), the main considerations then are the dt/dp of the transition and the *p-t* curve of the Earth.

Clearly, using either of the widely divergent Earth *p-t* curves, the seismic discontinuity can be reasonably reconciled with the transition. It is expected that there will be a direct and accurate determination of the *p-t* relation of the transition for Mg_2SiO_4, thus decreasing the uncertainty of extrapolation. It will then be possible to examine more closely the temperature and compositional variations in the Earth, at least to 500 km depth. *See* EARTH; OLIVINE; SPINEL.

One of the trends observed over the whole field of phase transitions with pressure is that many of the new phases approach, in varying degree, a more metallic state with regard to crystal structure, interatomic bonding, or physical properties. Indeed, measurements of color changes in important minerals such as olivine at pressures over 200,000 atm indicate that such a mineral may become metallic at pressures of less than 1,000,000 atm. Of course, as we have seen above, olivine would undergo at least one and perhaps a few more phase transitions before reaching a metallic state. At sufficiently high pressures it is logical to assume that atomic cores may eventually be stripped of their extranuclear electrons and result in a sort of plasma. Accumulating evidence, however, does not indicate that this process is taking place at the core of the Earth to any noticeable extent.

[FRANK DACHILLE; RUSTUM ROY]

Bibliography: P. W. Bridgman, *Physics of High Pressure*, 1949; F. Dachille and R. Roy, High pressure studies with reference to the olivine-spinel transition, *Amer. J. Sci.*, 258:225–246, 1960; B. French and N. Short (eds.), *Shock Metamorphism of Natural Materials*, 1968; A. A. Giardini and E. C. Lloyd (eds.), *High Pressure Measurement*, 1963; R. Roy and O. F. Tuttle, Investigations under hydrothermal conditions, in L. H. Ahrens, K. Rankama, and S. K. Runcorn (eds.), *Physics and Chemistry of the Earth*, vol. 1, 1956; R. H. Wentorf (ed.), *Modern Very High Pressure Techniques*, 1962.

Hornblende

A general name given to the monoclinic calcium amphiboles that form extensive solid-solution series between the various metals in the generalized formula $(Ca,Na)_2(Mg,Fe,Al)_5(Al,Si)_8O_{22}(OH,F)_2$. Hornblende has a widespread occurrence in metamorphic, igneous (intrusive), and volcanic rocks, ranging from dominantly ferromagnesian rocks to granites. The term common hornblende refers to the middle metamorphic grade hornblendes found in schists. Common hornblende usually forms as long, prismatic needles, dark green to black in color. The hornblendes of the high grades of metamorphism are usually short, stubby prisms, black to brown in color, and high in iron and aluminum, with some sodium (see illustration). See AMPHIBOLE; METAMORPHISM.

HORNBLENDE

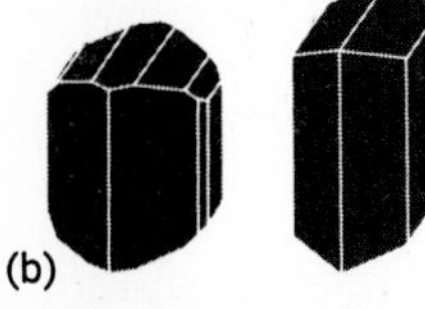

Hornblende. *(a)* Crystals and aggregate with calcite, Franklin, NJ *(specimen from Department of Geology, Bryn Mawr College)*. *(b)* Crystal habits *(from L. G. Berry and B. Mason Mineralogy, copyright © 1959 by W. H. Freeman and Co.)*.

Hornblende exhibits the characteristic 56° amphibole (110) cleavages and, except for tremolite, is green, black, or brown. In thin sections, hornblende is strongly colored (greens and brown) and strongly pleochroic (color change on rotation in plane-polarized light); the sodium hornblendes, when low in iron, are blue or bluish-green. Common hornblende is associated with plagioclase, garnet, epidote, pyroxenes, quartz, chlorite, anthophyllite, and biotite; it also forms monomineralic masses. The aluminum- and iron-rich hornblendes are more often found with the alkali feldspars, quartz and biotite, in granites, gneisses, and pegmatites. The sodium-rich varieties (except glaucophane) are found in the alkali-rich rocks. A textural variety of intertwined needles near tremolite in composition is nephrite a form of jade used for carvings.

The hornblende schists are common metamorphic rocks which formed extensive belts or layers in many areas throughout geologic time. These hornblende-schist layers are interpreted as being original geosynclinal graywackes (paraamphibolites) or altered basaltic lava flows and greenstones (orthoamphibolites) that were recrystallized under metamorphic conditions during mountain-building processes. The strong preferred orientation of the hornblende crystals in the schists in addition to fold structures indicates an extensive regional deformation of the rock materials during their metamorphism. See GLAUCOPHANE; JADE; TREMOLITE.

[GEORGE W. DE VORE]

Hornfels

A common name given to a class of metamorphic rocks produced by contact metamorphism, also known as hornstone. Hornfelses were originally sedimentary rocks. As magma intruded into the sediments, the heat given off induced recrystallization and effected a complete alteration of the primary sedimentary strata into hard, often flinty rocks. Their fine-grained constituent minerals usually can be discerned only with the microscope. Chemical alterations accompanying the formation of normal hornfelses are small, except for the partial removal of certain fugitive constituents of the sediments (such as water and carbon dioxide). Thus, the chemical composition of a hornfels depends only upon the composition of the original sediment, and consequently the mineral composition of hornfelses is predetermined by the nature of the original sediment. See METAMORPHIC ROCKS; METAMORPHISM.

Among the varieties of hornfelses which may develop from various sediments, the continuous series from shale to limestone with some admixture of marl is the most interesting. Chemically this series is made up of the following chief constituents: silicon dioxide, SiO_2; aluminum oxide, Al_2O_3; calcium oxide, CaO; iron(II) oxide, FeO; and magnesium oxide, MgO. Assume first that there is sufficient SiO_2 to form highly silicified minerals and secondly that FeO and MgO can be grouped together because they substitute for each other diadochally. Other than silica this is a system of three variables, alumina, lime, and ferro-

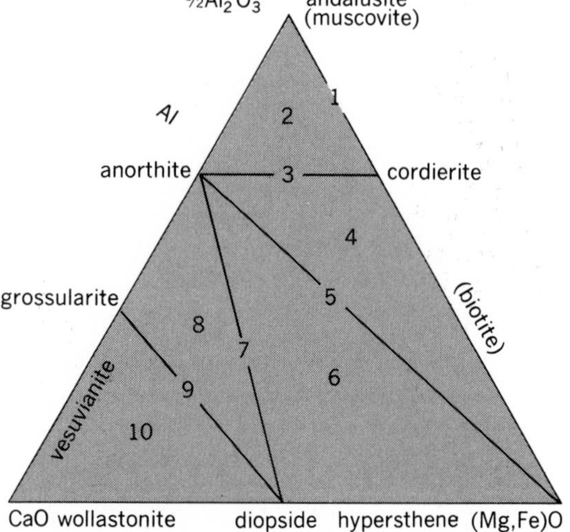

Fig. 1. Diagram of mineral assemblages of the 10 Goldschmidt classes of hornfelses. Other mineral constituents, in addition to those shown in the diagram, are quartz, orthoclase, and accessories such as apatite.

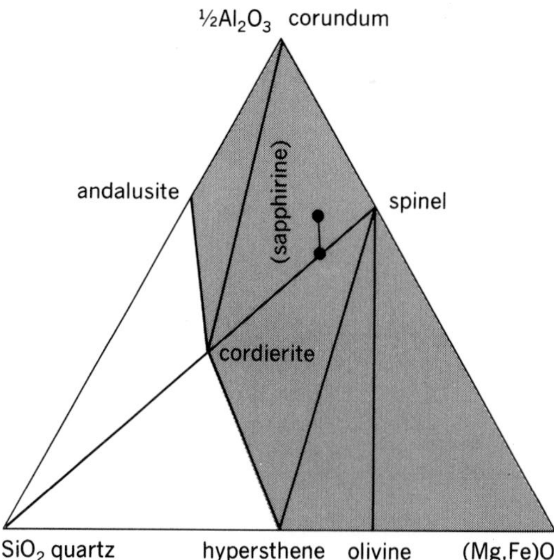

Fig. 2. Diagram of equilibrium assemblages of mineral phases in lime-poor hornfelses. The unshaded area represents the field of normal pelitic sediments. (*After C. E. Tilley, Mineral. Mag, 1923*)

magnesia, as illustrated in Fig. 1. According to the chemical composition of the sediments, there will be 10 classes of silica-rich hornfelses, as outlined below:

1. Pure shale contains alumina and traces of ferromagnesia which form andalusite and cordierite (and an excess of silica which produces quartz).
2. A small addition of lime produces anorthite (in the actual hornfels represented by plagioclase, since some Na_2O is present) in addition to andalusite and cordierite.
3. More lime produces more anorthite, whereas andalusite disappears.
4. Next, hypersthene will form in addition to cordierite and anorthite.
5. With still more lime, the cordierite disappears, with the formation of more anorthite and hypersthene.
6. Lime is now so dominating that it combines with magnesia to form diopside in addition to anorthite and hypersthene.
7. Hypersthene disappears.
8. More lime reacts with anorthite to form grossularite in addition to diopside.
9. Anorthite is completely used up in the grossularite reaction.
10. Pure lime silicate, that is, wollastonite, occurs with diopside and grossularite.

In certain regions, however, limestone is missing as a component in the sediments, and the corresponding hornfelses will be poor in lime, with a composition approaching the ternary system SiO_2-Al_2O_3-(Mg,Fe)O. Figure 2 shows the mineral associations encountered in such hornfelses. Each of the seven triangular areas shown in Fig. 2 corresponds to a possible mineral assemblage. Two of the assemblages, quartz-cordierite-andalusite and quartz-cordierite-hypersthene, are representatives of the usual silica-rich hornfelses. The other mineral assemblages occur in hornfelses of the Harz Mountains, Germany, of the Combrie area, Perthshire, Scotland, and elsewhere.

The natural hornfelses very closely follow the theoretical scheme of mineral associations, showing incontestably that a state of chemical equilibrium was approached in the rocks and that the laws of physical chemistry may be applied to metamorphic processes.

[T. F. W. BARTH]

Horst

A segment of the Earth's crust, generally long as compared to its width, that has been upthrown relative to the adjacent rocks (see illustration).

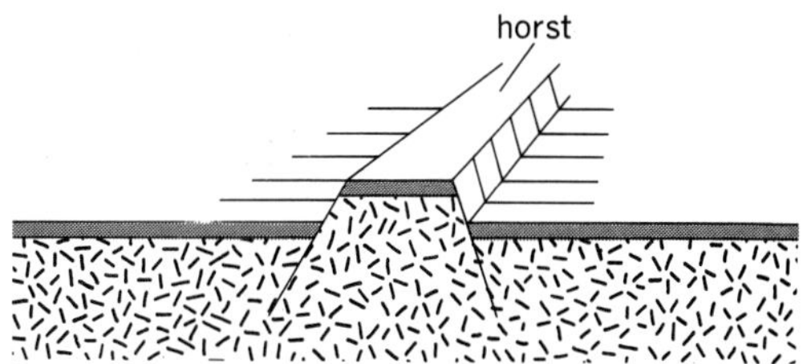

A simple horst with associated faults. (*From A. K. Lobeck, Geomorphology, McGraw-Hill, 1939*)

Horsts range in size from those that have lengths and upward displacement of a few inches to those that are tens of miles long with upward displacements of thousands of feet. The faults bounding a horst on either side commonly have inclinations of 50–70° toward the downthrown blocks, and the direction of movement on these displacements indicates that they are gravity faults. These relationships suggest that horsts develop in regions where the crust has undergone extension. They may form in the crests of anticlines or domes, or may be related to broad regional warpings. *See* FAULT AND FAULT STRUCTURES; GRABEN; WARPING OF EARTH'S CRUST.

[PHILIP H. OSBERG]

Huebnerite

A mineral with the chemical composition $MnWO_4$. Huebnerite is the manganese member of the wolframite solid-solution series. It commonly contains small amounts of iron. It occurs in monoclinic, short, prismatic crystals. Fracture is uneven. Luster varies from adamantine to resinous (see illustration). Hardness is 4 on Mohs scale and specific gravity is 7.2. Huebnerite is transparent and yellowish to reddish-brown in color; streak is brown. It is fusible with difficulty. *See* WOLFRAMITE.

[EDWARD C. T. CHAO]

HUEBNERITE

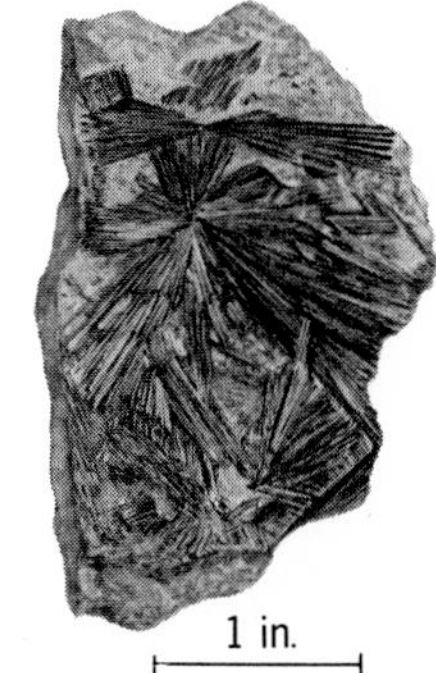

Radiating groups of huebnerite crystals in quartz vein, Silverton, CO. (*Specimen from Department of Geology, Bryn Mawr College*)

Humite

A series of magnesium nesosilicate minerals closely related in crystal structure and chemical composition. The group comprises four minerals as shown in the table.

Norbergite and humite are orthorhombic; chondrodite and clinohumite are monoclinic; all have closely related axial lengths. As shown in the table, the cell dimensions are nearly the same in the *b* and *c* directions, while the *a* dimension is related to the number of magnesium atoms in the formula. *See* SILICATE MINERALS.

The minerals of the humite group have similar physical properties. The luster is resinous and the

Composition and cell dimensions of minerals in humite group

Mineral	Composition	a_0	b_0	c_0
Norbergite	$Mg_3(SiO_4)(F,OH)_2$	8.74	4.71	10.22
Chondrodite	$Mg_5(SiO_4)_2(F,OH)_2$	7.89	4.74	10.29
Humite	$Mg_7(SiO_4)_3(F,OH)_2$	20.90	4.75	10.25
Clinohumite	$Mg_9(SiO_4)_4(F,OH)_2$	13.71	4.75	10.29

color usually light yellow to brown, more rarely white or red. Hardness is 6 to 6½ on Mohs scale; specific gravity is 3.1 to 3.2. Since the several minerals occur under similar conditions, it is difficult to distinguish them by inspection. They are found characteristically in contact zones in limestone. Norbergite, found at Norberg, Sweden, is the rarest of the group. The others have all been found in the ejected material at Mount Vesuvius. Fine crystals of chondrodite have come from the Tilly Foster iron mine at Brewster, N.Y.

[CORNELIUS S. HURLBUT, JR.]

Humus

The amorphous, ordinarily dark-colored, colloidal matter in soil, representing a complex of the fractions of organic matter of plant, animal, and microbial origin that are most resistant to decomposition.

Humus consists of the combined residues of organic materials which have lost their original structure following the rapid decomposition of the simpler ingredients and includes synthesized cell substance as well as by-products of microorganisms. It is not a definite substance and is in a continual state of flux, disappearing by slow decomposition, and being constantly renewed by incorporation of further residual matter. With a balance between these processes, humus, though not static, remains relatively uniform in nature and amount in a given soil. It constitutes a reservoir of stabilizing material which imparts beneficial physical, chemical, and biological properties to soil. Fertile soils are rich in humus.

Humus improves the texture of soils. It exerts a binding effect on sandy soils, and loosens the harder, clayey soils, thus increasing their porosity and permeability. It increases the moisture-holding capacity and improves the granular structure by cementing mineral particles into stable crumbs. This helps soils resist the pulverizing and eroding action of wind, water, and cultivation. As a storehouse of elements important to plants, humus functions as a regulator of soil processes by liberating gradually nutrients that would otherwise drain away. A soil rich in humus provides optimum conditions for the development of beneficial microorganisms and constitutes the best medium for growth of plants.

Humus formation depends upon an adequate supply of raw organic residues and upon suitable conditions for their decomposition. The composition of humus varies with the nature of the surface vegetation which furnishes the bulk of its raw supply. The degree of decomposition of residues, and the nature of the humus formed, depend also on soil and climatic factors, chief of which are temperature, aeration, and moisture. Under optimum conditions bacteria, actinomycetes, and fungi engage in the rapid breakdown of primary organic matter. With excessive moisture and poor aeration the aerobic microorganisms are relatively suppressed and thus decomposition of residues is only partial.

Peat. This is a type of humus that results from the decomposition of plant material under conditions of excessive moisture or in areas submerged in water. It is an organic deposit formed in marshes and swamps by the partial decomposition of countless generations of a variety of plants. The water, by excluding air, prevents rapid oxidation, and decay results from the action of more restricted groups of microbes than in mineral soils. Anaerobic bacteria are prominent, effecting partial mineralization with much liberation of gases. In the course of time large deposits are built up, often in layers as the nature of the marsh vegetation alters. This sedimentary lowmoor peat, slightly acid, may by proper drainage be turned into good agricultural soil. Highmoor peat, formed in circumstances providing for less sedimentation, largely from moss-type plants, is more fibrous, poorer in nitrogen and minerals, and very acid. The microbial content is low and conversion to agricultural use more difficult.

Decomposition and humification. When organic matter such as plant stubble, green manures, farmyard manures, composts, or organic fertilizers is incorporated in soil, microorganisms begin to decompose it. Some ingredients are readily attacked, while others are very resistant. Sugars, starches, organic acids, and alcohols are most rapidly destroyed, followed by fats, cellulose, and hemicelluloses as well as the less-resistant proteins. More resistant to microbial attack are waxes, certain proteins, and lignins, particularly. Lignins constitute increasingly higher proportions of the organic residues as humification proceeds. During decomposition there is considerable synthesis of microbial cell substance as microorganisms increase. As certain groups are superseded by others with the changing nature of the residual matter, large numbers of dead cells are incorporated into the organic complex. This has the effect of increasing the protein content.

During the course of decomposition the carbon-nitrogen ratio, which may vary from 12:1 to 20:1 in green manure crops to as much as 80:1 in straw, gradually narrows with the liberation of carbon dioxide until a level of approximately 10:1 is reached, which, in most mineral soils, is characteristic of humus.

Carbohydrates. These compounds are readily attacked by a wide variety of bacteria, actinomycetes, and fungi in the presence of sufficient oxygen. The disaccharides and starches are first hydrolyzed with formation of monosaccharides which are oxidized, and the carbon is finally liberated as carbon dioxide. With limited air supply anaerobic bacteria are chiefly responsible for decomposition. In this case the rate is slower, the chemical changes less complete, and in addition to carbon dioxide, less completely oxidized products such as methane, organic acids, and alcohols are formed. Under favorable conditions these may in turn be completely oxidized by aerobic microorganisms.

Cellulose. This is a polysaccharide found in the fibrous structure of many plants. Cellulose is decomposed chiefly by bacteria and fungi in a manner similar to that described previously for carbohydrates. However, it is resistant to attack by the majority of soil microorganisms, the types capable of producing cellulose-hydrolyzing enzymes being restricted. Cellulose decomposes much more slowly than sugars or starches but the end products of aerobic and anaerobic decomposition respectively are much the same. During decomposition of nonnitrogenous organic compounds, such as cellulose, the organisms involved demand adequate nitrogen. Consequently the addition of excess amounts of cellulosic material to soil may result in temporary loss of mineral nitrogen to plants through its being assimilated by microbes into cell substance.

Hemicelluloses. These are important constituents of plant material but unlike cellulose, they vary in chemical composition. They consist of polysaccharides combined with sugar acids of the uronic type and are related to the pectins. They are more readily attacked by microorganisms than cellulose, but the rate of decomposition varies with their chemical nature. On hydrolysis they yield monosaccharides in the form of hexose and pentose sugars and organic acids and alcohols. These products are readily decomposed further.

Proteins. These compounds are decomposed by large numbers of soil microorganisms, comprising fungi, actinomycetes, and aerobic and anaerobic bacteria. On hydrolysis by microbial enzymes, proteins are split first into polypeptides and then into simpler amino acids. These are readily attacked further, with the result that ammonia appears as the chief nitrogenous by-product of respiration. For this reason, the process of protein decomposition is referred to as ammonification. This step is a necessary preliminary to nitrification, which results in the oxidation of ammonia to nitrate. Under the aerobic conditions existing in most arable soils the products of oxidation include also carbon dioxide and sulfates. Under anaerobic conditions, where oxidation is incomplete, decomposition products include hydrogen sulfide, amines, mercaptans, and such compounds as indole and skatole, as well as organic acids and alcohols.

Lignin. A polymerized compound of high molecular weight composed of carbon, hydrogen, and oxygen, lignin is of uncertain chemical composition. It represents the plant ingredient most resistant to decomposition. It is an important constituent of straw and wood, forming 5–30% of the tissue, with the percentage rising as the plant matures. Few microorganisms are able to attack lignin and these are chiefly higher fungi. Hydrolysis and oxidation of lignin are slow and the by-products of its degradation are not well known. As plant material decomposes, lignin accumulates to constitute an important ingredient of humus, though in a modified condition. In the process of humification lignin unites with proteins, chiefly those elaborated by microorganisms, to form a lignin-proteinate complex. This renders the protein very resistant to microbial attack. Thus nitrogen in humus is retained and only slowly liberated.

Humus itself undergoes slow decomposition so that in well-aerated soils receiving normal additions of organic material its level remains fairly constant. This slow decomposition is of great practical importance. It constitutes a means whereby nitrogen, as well as simple mineral compounds, can be stored in soil and released gradually for consumption by growing plants.

[ALLAN G. LOCHHEAD]

Humus, geochemistry of

Humus consists principally of the dark-colored constituents in soil, other sediments, and rocks; it also includes light-colored or uncolored organic substances representing or associated with products of organic decomposition.

Related terms are mull, a mixture of organic and inorganic matter in soil, due to mixing action of organisms; mor, partly decomposed forest and meadow litter; highmoor peat, formed of upland moss and heather; lowmoor peat, swamp accumulations; dopplerite, amorphous, organic, water concentrate in or beneath bogs; copropel or gyttja, aquatic, pulpy, coprogenic accumulations; dy, allochthonous aquatic accumulation of humic acids and detritus; pelogloea, marine detrital slime from settled plankton; kerogen, amorphous organic accumulations in lacustrine rocks, particularly of alkaline or saline lakes; thucolite, concentrations of carbonaceous matter in ancient sedimentary rocks; tasmanite, spore-rich organic accumulations in rocks; torbanite, kukersite, coorongite, and n'hangellite, deposits rich in *Botryococcus* and related algae in rocks and sediments; humic acid, that portion of humus extracted with sodium hydroxide solution and precipitated with hydrochloric acid. In this preparation fulvic acid remains in solution; the dried humic acid when treated with alcohol produces a soluble hymatomelanic acid fraction. *See* DOPPLERITE; HUMUS; KEROGEN; TORBANITE.

Methods by which humic components are extracted and analyzed govern to a great extent the properties of the resulting extracts, and humus is commonly classified on the basis of methods of extraction. Physical and chemical properties and the origin of humus are important in agriculture and in geology. Studies on soil humus and aquatic humus have employed to advantage several modern analytical techniques.

Preparation and analysis. The usual methods of extracting humic components have been treatments of humus with sulfuric acid, chromic acid, sodium pyrophosphate, or sodium hydroxide. The more recently developed methods have used electrophoresis, combinations of caustic alkali and calcium-masking reagents, chromatography, infrared analysis, electron paramagnetic resonance, polarography, and various combinations of these analytical procedures.

Electrophoresis. By the use of electrophoresis in the extraction and fractionation process, humus reportedly can be differentiated according to the efficiency of the chemical fractionation, percent distribution of humic acids, ratio of fulvic to humic acids, degree of humification, seasonal variations, microclimates, and abundance of Ca and Mg.

Alkali and calcium-masking reagents. Various combinations of caustic alkali and calcium-masking reagents in humus extraction can be employed. Fulvic acid combined with Fe_2O_3 is easily extracted with $Na_4P_2O_7$ but not with NaOH; the latter is most efficient when fulvic acid is combined with Al_2O_3. One major problem in humus extraction is the formation of insoluble polycondensation products by autoxidation. When added to extraction solution under airtight conditions, $SnCl_2$ resulted in twice as much humus being extracted. Other electron acceptors besides oxygen may also lead to some autoxidation.

Chromatography. Chromatographic analyses of humus have also been carried out. Fulvic and hymatomelanic acids separated with various alcohols yield chromatographic zones. Paper-chromatographic analyses of alkali degradation products of humic acid, podsols, and lignin give similar chromatographic fractions. Similar structural models of humic acid from various sources are suggested by these results.

Infrared analysis. In the conversion of *Phragmites* grass to peat, chemical and infrared studies show that the ratio of *p*-hydroxybenzyl to vanillyl and syringyl groups is higher in the humic acid fraction of *Phragmites* peat than in the original plant lignin. Differential thermal analyses (DTA) of the complete combustion of peats of various types under both oxidizing and inert atmospheres indicate that significant differences exist in the DTA curves of different peats.

Infrared analyses of extract of humus purified by electrodialysis have shown that the various humus compounds are distinguishable by the method. The infrared spectra of fulvic acid are reported to contain carbonyl derivatives (2,4-dinitrophenylhydrazone, semicarbazone) which show presence of C-N bonds in region 1630–1690 cm^{-1}. Pyrolysis of a humic acid showed a decrease in the intensity of OH and CO absorption due to decarboxylation. In an examination of the infrared spectra of humic extracts of Recent sediments, an order of aromaticity has been reported: peat > bay sediments > sedimentary rocks > lake sediments. The absorption peaks at 1640 and 1530 cm^{-1} of humic acid from bay and lake sediments are considered to be due to peptidelike bonds.

Two main fractions of humic substances from natural waters can be obtained by fractionation on Sephadex G-100; one with molecular weight greater than 100,000 and the other with molecular weight less than 10,000. Both contain Fe^{2+} complexes and the lower-molecular weight fraction also has Ca.

Electron paramagnetic resonance and infrared analyses of melanin, tannin, lignin, humic acid, and hydroxyquinones have been cited as showing similar features: All contain stable free-radical moieties; all base-soluble compounds can be reversibly converted to sodium salts accompanied by a 10- to 100-fold increase in unpaired spin content; infrared absorption bands in free acid form and sodium salt form are consistent with hydroxyquinone structures. The existence of trihydrated hydronium ion in the solid state in humic acid has been demonstrated by constant H_2O-CO_2 ratios during hydrolysis and infrared absorption bands at 2900 and 1205 cm^{-1}, with a shoulder at 1750 cm^{-1}.

Table 1. Functional groups found in humus*

Group	Amount	Spectral absorption, μ
OH	2–3 meq/g	3.0
CH_3		3.45
Methoxyl	6.7%, 0.2–0.4 meq/g	6.3
Carboxyl	4–6 meq/g	5.85, 6.22
Carboxyl	2–9 meq/g	
?		7.2
Aromatic	27%	
Clay impurity		9.0
Stable free radicals	10^{18}/g	Electron paramagnetic resonance
Quinone		
Phenolic	3 meq/g	
Catechin, (e.g., *epi*-)		
?		Fluorescence

*From A. D. McLaren and G. H. Peterson (eds.), *Soil Biochemistry*, Dekker, 1967.

Polarography. This method is useful for determination of carboxyl groups in soil humic compounds. The humic compounds are isolated, refluxed with various reagents, and the excess reagent studied by polarography. The results with known aromatic aldehydes and aliphatic and aromatic ketones are said to agree with theoretical values, but values for quinones, except in the case of phenathrenequinone, are low. The humic products agree with the carboxyl values obtained by a reoxidation method. The groups that react were ketonic CO and perhaps quinone groups.

Classification. One example of a geochemical classification of humus is that of G. K. Fraser, as follows:

1. Humin, extracted with acetyl bromide.
2. Fulvic acid, that portion of alkali extract remaining in solution after treatment with acid, for example, pentosans, uronic acids, polyuronides, phenolic glucosides, sugars, and amino acids.
3. Humic acid, precipitated from alkaline solution with acid and composed of a hydrolyzable part—that is, sugar and proteins; and an unhydrolyzable part—that is, phenolic, carboxyl, acetyl, and methoxyl groups (Table 1) and hydroxyquinone.

Physical and chemical properties. Humus has several important physical and physicochemical properties. Depth of color can be used only roughly to indicate humus content because degree of plant decomposition, development of phenolic compounds, and presence of iron sulfide affect color. The original source material, fabric, and diagenetic history affect the structure of humus, with the result that, for example, forest peat and lake peat have very few properties in common. Similarly cohesion of forest and moss peat is much less than that of clay, but cohesion of silty lake peat is equal to, or greater than, clay. Freezing of humic materials results in loss of cohesion. Although the water-holding capacity of fresh humus may be very high, 85–90%, once dried, most humus is not easily wetted again. Shrinkage of lake humus on drying may be 80% or more, but that of moss peat and forest peat is much lower because of their porous fabric. Fresh humus in general also has high permeability, high heat capacity and heat absorption of salts and gases, and high buffering power.

The gross chemical composition of humus ranges widely. In forest soil organic carbon may range from 50% or more in surface litter to 2% or less in the underlying A zone; and, respectively, nitrogen 2 to 0.1%; pH 4.5 to 4; ether-soluble compounds (hydrocarbons, waxes, and so on) 6.5 to 3.0%; hot-water-soluble compounds (organic acids and so on) 7 to 2%; and methanol-soluble compounds (pigments) 4 to 1%.

Antiseptic properties of some types of humus are provided by the presence of phenols, quinones,

Table 2. Major bands and relative absorbances in the infrared spectra of extracted organic matter*

Frequency, cm^{-1}	Relative absorbance: A_0† organic matter	Relative absorbance: B_h† organic matter	Interpretation
3380	Strong	Strong	Hydrogen bonded —OH and bonded —NH groups
2910	Medium	Shoulder	Aliphatic C—H stretching vibrations
2840	Medium	0	Aliphatic C—H stretching vibrations
2600	Shoulder	Shoulder	Carboxyl C—H stretching vibrations
1720	Shoulder	Strong	Carboxylic carbonyl
1620	Strong	Strong	Joint interaction of hydroxyl and carboxyl with carbonyl, also possibly carboxylic groups associated with metals so as to give the carboxylate structure
1450	Strong	0	CH_3 or CH_2 or both in plane-deformation vibrations
1400	0	Medium	Carboxylate
1250	Weak	0	Phenoxy C—O
1200	0	Medium	Carboxyl
1030	Medium	0	Si—O of silica due to presence of clay

*From A. D. McLaren and G. H. Peterson (eds.), *Soil Biochemistry*, Dekker, 1967. †Soil horizon.

phenol carboxylic acids, and phenol glucosides, which retard activity of microorganisms. Mummification of organisms in peat, including humans in Danish peat bogs, is an example of this activity.

The precipitation of mineral matter in solution over pH range 4–8 is impeded by development of alkali humates. Thus phosphorus and other elements may undergo geological transportation and concentration to form ore deposits.

Humus possesses colloidal properties to the extent that humic acids which have particle diameters of about 0.02 μ and are negatively charged can be separated by electrodialysis from a humic solution. The weight of particles of humus colloids is believed to be a direct function of the alkalinity of the solution; the effect is thought to be reversible and to be caused by hydrogen bonding of the particle units. Coacervation of humus particles in an acid solution results in a ζ-potential of −23 mv, while in alkaline solutions $\zeta=0$. Such geological processes as the transportation and deposition of minerals as humates, and the leaching of iron and manganese coatings of sediments are possibly affected in an important way by these phenomena. Humus effects peptization of clays and subsequent humic coatings of the clay minerals, ferric hydroxides, and other minerals. Flocculation of the mineral particles and their precipitation are consequently retarded.

The ultraviolet absorption spectra of humus preparations vary widely with the method employed in preparation and cannot be used for any but the grossest characterizations. Absorption maxima, varying with the pH, occur rather typically at 275–285, 300–305, and 350 mμ, suggesting mono- and dinucleate aromatic and heteroaromatic compounds, furfurals, and so on. Many individual compounds showing unique infrared and ultraviolet-to-visible spectra, however, can be separated from humus by acid-, base-, and petroleum-solvent extraction and preparative chromatography (Table 2). X-ray studies indicate that graphite structure may appear in geologically aged humus as a result of condensation reactions leading first to insoluble humins and then to graphite.

Humus plays a part in the formation of such geologically important materials as underclays, bauxite, and laterites in that they bring about removal of iron, silica, and bases by weathering from associated aluminum-bearing rocks. Other substances that may be concentrated by humic processes are phosphorus, manganese, molybdenum, copper, zinc, silver, arsenic, germanium, sulfur, selenium, fluorine, bromine, iodine, and uranium. The absorption of zinc, for example, on humic acid is dependent on pH, type of exchange site, and nature of the cation; zinc is more strongly bound in humic acid than calcium but less strongly than copper or ferrous iron. *See* BAUXITE; LATERITE; SOIL CHEMISTRY.

Origin. It is generally agreed that microbiological processes are responsible for most humus formation, with the plant carbohydrates furnishing the source of energy for the microorganisms and the latter themselves contributing considerable amounts of humus. Fungi appear to play a major role in humus formation in regions of brown-water lakes and rivers, whereas bacilli are perhaps the most important humus-forming organisms in bogs. Some wholly chemical humus formation may occur: (1) Furfurals may form by action of mineral acids on carbohydrates and may condense to form humus. (2) Amino acids and peptides may condense with carbohydrates to form melanoid humic compounds. (3) Aldol condensation of amino acids with methylglyoxal may produce humus. (4) Phenols, quinones, and other aromatic compounds may become oxidized to form humus, and opening of lignin rings may also lead to humus formation.

In the production of fulvic acids from *Eucalyptus* in acid sandy soil, Fe_2O_3 promotes humus formation, and $CaCO_3$ increases the oxidation of humic acids. Humus formation is said to depend on the linkage of reactive humin structural elements from the substratum to colored humins of a higher molecular weight; the process involves successive passage from fulvic to hymatomelanic to humic acid fractions. In formation in aqueous extracts of *Melandryum silvestre* under conditions excluding microbial action, extracts at 4°C are said to darken and form black deposits much more rapidly in the presence of soil, particularly of clay. The substances in the soil were found to include all forms of humus.

Brown metabolic products resulting from vital activity of *Streptomycetes albidus* in a culture medium yielded, by chromatographic analysis, two humic acid–like fractions one of which was reported to be antagonistic to bacteria and the other to fungi.

Soil humus studies. Chromatographic analyses of humic and fulvic acids in various soils have shown that high fulvic acid content of chernozem was probably due to complex organic-mineral compounds that may include iron and aluminum chelates. With respect to complexing with copper, fulvic acid shows maximum adsorption at about pH 6, while that of humic acid is at about pH 3.

The degree of humification of some organic soils may be judged from their contents of carboxyl and phenolic hydroxyl groups. Treatment of the soil with HCl and HF results in free carboxyl groups that were detected by infrared spectra (1720-cm^{-1} band) and decrease in carboxylate bands at 1625 cm^{-1}. Phenolic hydroxyl groups do not seem to be affected by the treatment.

Only a little of the nitrogen of humic acids seems to be available to cellulose-decomposing microorganisms in soil. That part of the nitrogen in humic acids which occurs as amino acids is noticeably different in different humic acid preparations, but the amino acids of humates seem to be firmly bound to the aromatic nucleus. *Pseudomonas fluorescens* is one of the more effective species in decomposing humic acids. Enzyme preparations, such as peroxidase, seem to degrade humic acid preparations at about the same rate as the more active microorganisms.

Further evidence that the humic acid nucleus is aromatic is believed to lie in the fact that humic acids from certain chernozem soils contain about 20% carboxylic acids, and that the high nitrogen content is indicative of heterocyclic compounds, perhaps pyridine and pyrimidine derivatives.

The humic acid of calcium- and sodium-rich soils has been found, through use of Ca^{45}, Ce^{144}, and Pr^{144} as tracers, to adsorb rare-earth elements strongly.

Aquatic humus studies. The humic acid contents of natural water can be determined, it is reported, by measuring the photometric absorption at 410 μ for low Fe concentration and at 480 μ for high Fe concentration. Ethylenediaminetetraacetic acid is used to mask the effect of Fe, and hydrooxylamine plus hydrochloric acid to prevent oxidation of humic acids by Fe^{3+}. Adjustment to pH 10 is made for colorimetric observation to redissolve the precipitated humic acid brought about by the decomposition of colloidal iron.

The entraining effect of $Ca(HCO_3)_2$ on humic substances during precipitation of the carbonate was observed to be smaller when the solubility of $CaCO_3$ was increased by the presence of other salts.

Mineral waters in the Borzhon area of the Soviet Union are said to contain 43.7–54% of the luminescent organic matter in the form of humus, the remainder being neutral resins (21.4–30.8%) and acid resins (23.0–31.3%). Organic substances in Dnieper River water are reported to attain maxima in spring and minima in winter. Biochemically stable humus of lignin-protein complexes predominate: ratio C:N=16.24; C:P=350; and permanganate O:C=1.3.

Increasing amounts of organic matter in medium-mineralized Estonian lakes of humic character results, it is stated, in greater predominance of fulvic over humic acid, presumably because of complete solubilization; the maximum concentration of organic matter here is in the winter and the minimum in the spring. In several lakes in the Soviet Union (Blado, Bladko, and Mtynek) of high humic content, several of the humic compounds are said to be toxic to *Daphnia pulex*, *Leptodora kindtii*, and *Simocephalus serrulatus*.

Atlantic sea water contains a reported average

Table 3. Organic residues of profile zones of Rossburg Lake and bog deposits, Minnesota*

Zone	pH	Eh, mv	N_2, %	Ether extracts, %	Saturated HC, %	Aromatic HC, %	Asphaltenes, %	Naphthol
Π_o, surface sphagnum and forest peat	4	+400	1	2	.06	.10	.35	none
Π_m, sphagnum moss and forest peat to depth of 2 ft	4.2	+400	1	1.5	.02	.05	.30	present
β_p, moss peat 2–6 ft deep	4–5	+300 to +400	1–3	1–4	.02–.04	.1–.3	.05–2	high
β_{p-c-s}, moss peat with copropel and sapropel layers, 6–12 ft deep	5–6	+200 to +400	1.5–2	.4–.6	0–.015	0–.08	.04–.5	medium
β_{c-s}, copropel and sapropel layers, 12–20 ft deep	6.2–6.8	+150 to +250	2–3	.2–.5	.025–.085	.04–.07	.09–.14	low or absent
M zone, marly clay, and silt, 20–26 ft deep	7–7.2	+80 to +110	1	.2	.01	.02	.1	none
T zone, sand at base of deposit, 26–28 ft deep	7	−10	–	–	–	–	–	none

*From. F. M. Swain, in E. Yochelson and C. Teichert (eds.), *Essays in Paleontology and Stratigraphy*, University of Kansas Press, 1967.

of 1.5 mg/liter of C; the highest content is in the surface waters, due to plankton, and a regular decrease with depth occurs. In the Black Sea the distribution of humus is said to be similar but the average content is twice that of the Atlantic. With reference to the distribution of salinity, oxygen, and phosphates as related to humus formation in bays and estuaries, oxygen is stated to be less and phosphates greater on the outgoing tide than on incoming tide.

The humus of Black Sea deepwater sediments reportedly varies from traces to 4–6%. The humus consists of 39% C, 7.84–10.71% H, and a large amount of ash. Asphaltenes form 29.2% of $CHCl_3$ extracts of this humus; bitumens form 98.1% of alcohol-benzene extracts; the remainder is tar and traces of oils. Humic acid is said to be the main organic component in surface marine sediments in the Pacific Ocean off California, making up 30–60% of total organic matter. Significant amounts of phenols and amino acids occur with humic acids, perhaps by covalent linkage. Lignins from the continents are the most likely source of the humic acid, based on the nature of phenols and carbon isotope studies.

Organic contents of Bering Sea sediments are cited as increasing with decreasing grain size, and are sufficiently active biochemically to support microorganisms.

Residual organic material in lake and bog sediments bears a general relationship to the trophic and climatic history of the water body. Total amino acids from proteins, total hydrolyzable sugars, and chlorinoid pigments show increases related to periods of high organic productivity in the water. Increases of carbohydrates, amino acids, and pigments occur just beneath the surface of many lake and bog sediments; these may be due to microbial synthesis, or downward concentration and sorption of the substances, or other causes. Basic amino acids—lysine, histidine, and arginine—seem to be preferentially preserved in acid bogs, possibly because of hybrid-ion properties of the amino acids. The distribution of individual carbohydrates in some lake sediments showed a natural stability series to be: fairly stable, xylose, glucose, rhamnose, arabinose; moderately stable, ribose, mannose; fairly unstable, galactose; and very unstable, glucuronic acid.

In a moss peat deposit (4 m) overlying lake peat (3 m) in a deglaciated region, saturated and aromatic hydrocarbons and phenolic residues increase from about 2×10^{-4} mg/g at the top of the moss peat to about twice that amount at a depth of 4 ft, below which a decrease occurs (Table 3). Carbonyl groups, including possibly carboxylic acid and enol group residues, have similar distribution according to infrared analysis. There is a 2-naphthol in the moss peat that reaches maximum concentration in the middle of that peat; it may have formed from a plant-growth accelerator (auxin) such as naphthylacetic acid. β-carotene of phytoplanktonic origin is present in the underlying lake peat, but is nearly absent from the moss peat. Pheophytin *a* from chlorophyll, shows more general distribution in the bog, but also is richer in the lake peat. When the moss peat was subjected to low-temperature distillation, the principal product was a phenolic compound having a prominent absorption maximum at 266 mμ. An equivalent product from the lake peat was toluene (λ max 255 mμ) which may have been produced from β-carotene during sample treatment. Protein amino acids show a distribution that is consistent with variations in type of peat and in nitrogen content, and are slightly higher in the moss peat than in the lake peat. Basic amino acids occur throughout the peat, indicating that acid conditions prevailed in the history of the bog. Total carbohydrates are about 100 mg/g in the moss peat but decrease to half that amount in the lake peat. Glucose, arabinose, and xylose are the predominant monosaccharides of

Table 3. Organic residues of profile zones of Rossburg Lake and bog deposits, Minnesota (cont.)

CH absorption	C=O absorption	Total amino acid, $^0/_{000}$	Total carbohydrates, mg/g	Glucose, mg/g	Arabinose, mg/g	Xylose mg/g	Pheophytin mg/g	β-carotene	Phenols	Toluene	Pollen zone
medium	high	75	100	4.5	1.5	2	20	low or absent	–	–	Upper pine zone
high	high	200	75	4.5	1.0	1.5	20	low or absent	–	–	
high	medium to high	200–300	75–100	2.5–3	2.5	1.5–3	20–25	low or absent	present	absent	
medium	medium	200–250	100–150	.5–1	.5–1	.8–1.2	20–40	medium	–	–	
medium	low	200	30–60	.8–2	.8–2	.8–1	20–40	medium to high	absent	present	Oak-herb zone; Pine zone
medium	low	2	10–70	–	–	–	.2–2	none	–	–	Spruce zone
–	–	–	–	–	–	–	2	–	–	–	

the peat. Bitumen and amino acid analyses of individual plant species of the moss peat are not much different from the peat as a whole; carbohydrates of the individual moss species, however, are dominated by galactose, mannose, and glucose, whereas the first two sugars are not plentiful in the peat as a whole. [FREDERICK M. SWAIN]

Bibliography: U. Columbo and G. D. Hobson (eds.), *Advances in Organic Geochemistry*, 1964; E. T. Degens, *Geochemistry of Sediments*, 1965; P. H. Given et al., *Coal Science*, 1966; G. D. Hobson and M. Louis, *Advances in Organic Geochemistry 1964*, 1966; A. D. McLaren and G. H. Peterson (eds.), *Soil Biochemistry*, 1967; F. M. Swain, in E. Yochelson and C. Teichert (eds.), *Essays in Paleontology and Stratigraphy*, 1967; F. M. Swain, in I. A. Breger et al. (eds.), *Organic Geochemistry*, 1963; F. M. Swain, in H. E. Wright and D. G. Frey (eds.), *The Quaternary of the United States*, 1965.

Huntite

A very rare magnesium-calcium carbonate, composition $Mg_3Ca(CO_3)_4$, of low-temperature formation. Huntite has been reported from Nevada, Hungary, and Australia. It occurs in fine-grained masses with a distinct x-ray powder-diffraction pattern. *See* CARBONATE MINERALS.

[ROBERT IAN HARKER]

Hydrosphere, geochemistry of

Those processes that control the introduction, distribution, and removal of dissolved materials in waters of the Earth's atmosphere, oceans, and crust. The unique chemical properties of water make it an effective solvent for many gases, salts, and organic compounds. Circulation of water and the dissolved material it contains is a highly dynamic process driven by energy from the Sun and the interior of the Earth. Each component has its own geochemical cycle or pathway through the hydrosphere, reflecting the component's relative abundance, chemical properties, and utilization by organisms. The introduction of materials by humans has significantly altered the composition and environmental properties of many natural waters.

Rainwater. Rainwater contains small but measurable concentrations of many elements derived from the dissolution of airborne particulate matter and produced by equilibration of rainwater with atmospheric gases.

Total dissolved solids in rainwater range from over 10 parts per million (ppm) in rain formed in marine air masses to less than 1 ppm in rain precipitated over continental interiors. The major dissolved constituents of rainwater are chloride, sodium, potassium, magnesium, and sulfate (Table 1). These salts are derived over oceans and coastal areas from the dissolution of aerosol particles formed during the evaporation of sea spray.

Table 1. Chemical composition of average rainwater, river water, and sea water, in parts per million

Constituent	Average rainwater	Average river water	Sea water
Na	1.98	6.3	10,500
K	0.3	2.3	380
Mg	0.27	4.1	1,300
Ca	0.09	15	400
Cl	3.79	7.8	19,000
SO_4	0.58	11.2	2,650
HCO_3	0.12	58.4	140
SiO_2	–	13.1	6
pH	5.7	–	8.2

A significant portion of the dissolved sodium, potassium, calcium, and sulfate in rain formed over continental areas is introduced by reaction with land-derived dust particles. Additional sulfate comes from the oxidation of sulfur dioxide, produced by the oxidation of hydrogen sulfide and by the burning of fossil fuels and smelting of sulfide ores, shown in reaction (1).

$$2SO_2 + O_2 + 2H_2O \rightarrow 4H^+ + 2SO_4^{2-} \quad (1)$$

A map of the average sodium content of rain in the continental United States (Fig. 1*a*) shows sodium contours subparallel to the coastlines, reflecting mixing of continental air masses with salt-rich marine air. The distribution of sulfate (Fig. 1*b*) is more complex, and reflects significant continental input from dust storms and industrial activity.

Rainwater contains in dissolved state each of the gases present in the lower atmosphere. Carbon dioxide is derived from both biological respiration and the burning of fossil fuels. It reacts with rain to form carbonic acid, as in reaction (2).

$$CO_2 + H_2O \rightarrow H_2CO_3 \quad (2)$$

The presence of free oxygen and carbon dioxide makes rain both a natural oxidizing agent and an acid. Rain equilibrated with normal atmosphere has a pH of 5.7. More highly acidic rains form in areas where the industrial discharge of carbon dioxide or sulfur dioxide is intense.

Rain contains variable trace concentrations of many additional elements and compounds. Some of these, such as heavy metals and radionuclides, are derived from industrial pollution and nuclear testing, respectively. Precipitation of rain is the primary process by which many of these materials are transported from the atmosphere to the continents and oceans.

Soil waters. As rainwater percolates downward and laterally through the soils and surface rocks of the continents, a complex group of reactions occurs.

The release of carbon dioxide and organic acids by bacterial processes increases the chemical reactivity of waters passing through the upper part of the soil zone. Weathering of most carbonate or silicate minerals generally involves acid attack, with carbonic acid (H_2CO_3) dominating [reactions (3) and (4)].

$$\underset{\text{Calcite}}{CaCO_3} + H_2CO_3 \rightarrow Ca^{2+} + 2HCO_3^- \quad (3)$$

$$\underset{\text{Illite}}{2KAl_5Si_7O_{20}(OH)_4} + 2H_2CO_3 + 13H_2O \rightarrow$$

$$\underset{\text{Kaolinite}}{5Al_2Si_2O_5(OH)_4} + 4H_4SiO_4 + 2K^+ + 2HCO_3^- \quad (4)$$

The weathering of carbonates and silicates consumes acid and produces a soil water enriched in cations, bicarbonate, and dissolved silica (H_4SiO_4). Dissolved sulfate is derived from dissolution of sulfate minerals and by reaction between sulfide minerals and dissolved oxygen in soil waters, as in reaction (5). Chloride is derived from the weather-

$$4FeS_2 + 15O_2 + 8H_2O \rightarrow 2Fe_2O_3 + 8SO_4^{2-} + 16H^+ \quad (5)$$

Pyrite → Hematite

ing of fluid inclusions in silicate minerals and dissolution of halite (NaCl). *See* WEATHERING PROCESSES.

The ease with which a particular element can be accommodated in soil waters depends in part on the ionic radius r and charge Z of the cation which it forms (Fig. 2). The ratio of Z to r is known as the ionic potential. Large cations with a small charge, such as K^+ and Ca^{2+}, are usually readily accommodated in aqueous solution. In oxidizing environments, elements which form small, highly charged cations, such as S^{6+}, combine with oxygen to form highly soluble and stable anionic complexes (for example, SO_4^{2-}). Cations of intermediate size and charge, however, including Al^{3+} and Fe^{3+}, are only sparingly soluble. These elements are usually incorporated in the solid products of weathering, for example, aluminum in kaolinite in reaction (4) and iron in hematite in reaction (5). In soil waters depleted in free oxygen, the more highly soluble, reduced form of iron, Fe^{2+}, may go into solution. Other elements may also be solublized in the absence of oxygen or in the presence of suitable complexing agents.

River water. River water represents a variable mix of subsurface waters, which enter the river at the ground water table, and surface runoff. Some of the material in river water is derived from the dissolved sea salts and dust present in rainwater, but most has been introduced through weathering reactions. River waters are higher in bicarbonate and dissolved silica, and the relative abundance of the cations they contain reflects the lithology of the drainage basins from which they are derived (Table 1). Waters draining carbonate terranes are typically enriched in calcium and magnesium [reaction (3)].

Shale terranes, in contrast, will produce waters preferentially enriched in potassium, which is released during weathering of illite [reaction (4)]. The salinity of river waters varies from less than 40 ppm for the Amazon River, which drains a region of exceptionally high rainfall, to over 800 ppm for the Rio Grande, which drains a region of low rainfall and high evaporation. Dissolved organic material is high in tropical streams and rivers, where rates of organic production and decay are high. Organics in many rivers draining the southeastern United States exceed the concentration of dissolved inorganic salts. The composition and salinity of a given river may vary seasonally.

Rivers and streams have been used by humans since earliest history as a source of potable water, a place to discard wastes, and a vehicle for the transportation of goods. The concentrations of many metals and organic compounds deliberately or accidentally introduced as wastes now often exceed natural river levels of these materials. Humans have also introduced compounds such as chlorinated hydrocarbons, which were unknown in the natural environment.

Most river water eventually mixes with marine waters in coastal and estuarine areas. The concentrations of most of the major cations and anions in

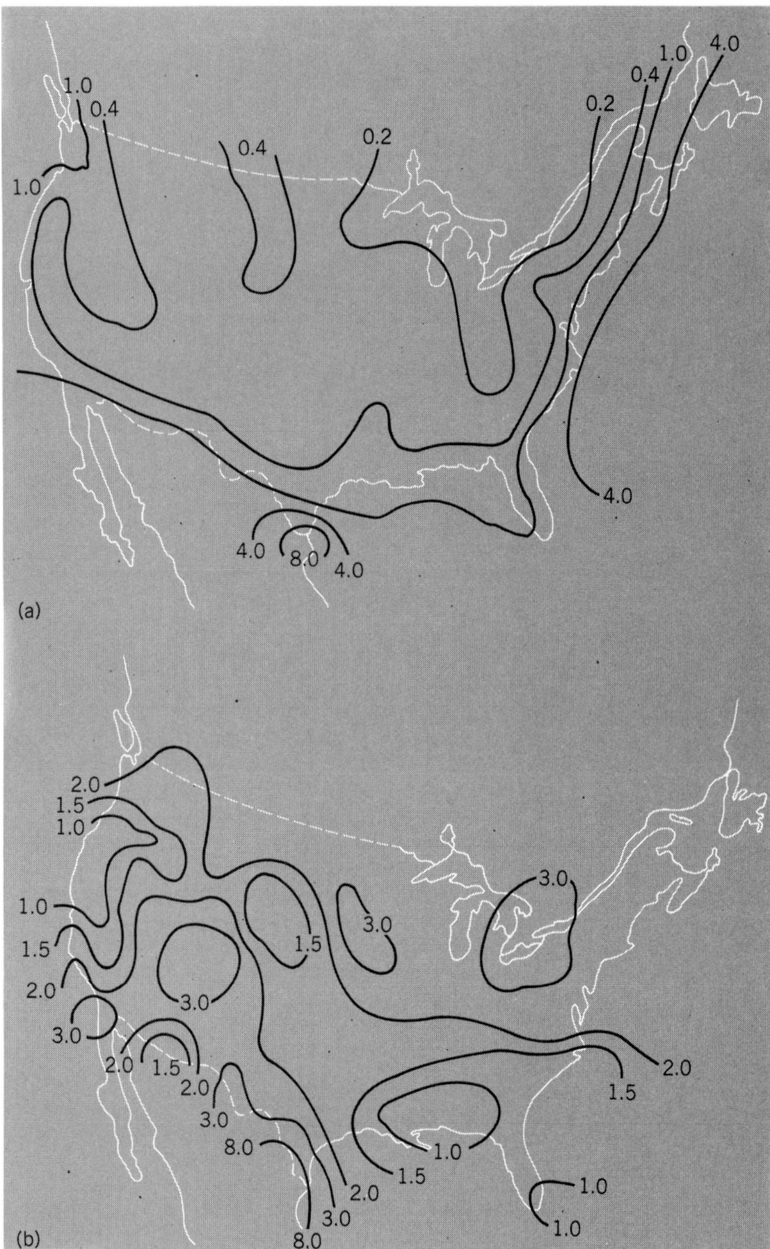

Fig. 1. Dissolved *(a)* sodium and *(b)* sulfate in rain (in parts per million) over the continental United States. *(Adapted from R. M. Garrels and F. T. Mackenzie, Evolution of Sedimentary Rocks, copyright © by W. W. Norton and Co., Inc., 1971)*

these zones of mixing are not affected by processes other than the physical mixing of fresh and marine waters. Such constituents are said to behave conservatively. Many minor and trace constituents, however, behave nonconservatively, and are preferentially introduced into or removed from aqueous solution by chemical or biological processes occurring in the zone of mixing. Significant quantities of barium, for example, are desorbed from river clays when these particles are transported into marine waters. Humic-metal colloids present in river waters are flocculated as they mix with marine waters. The removal of dissolved iron by this process has been extensively documented. Field studies have shown that silica is removed from solution in some river estuaries. However, the question of

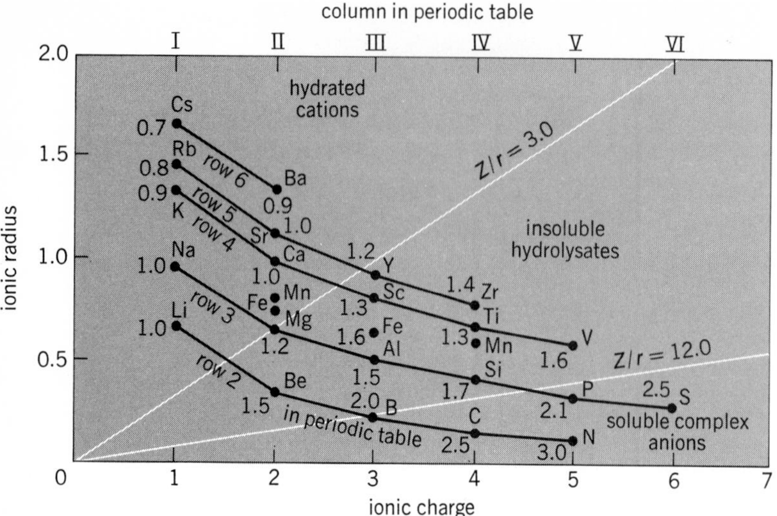

Fig. 2. Accommodation of cations in aqueous solution. (*Adapted from H. Blatt, G. Middleton, and R. Murray, Origin of Sedimentary Rocks, Prentice-Hall, 1972*)

whether this removal is due to biological uptake, reaction with suspended mineral particles, or both, has not been resolved.

Sea water. The dissolved salt content of open ocean water varies between 32,000 and 37,000 ppm. This range reflects dilution of sea water by rain and concentration by evaporation. Chloride, sodium, sulfate, magnesium, calcium, and potassium ions dominate sea salt (Table 1) and, with the exception of calcium, are present in remarkably constant proportions throughout the oceans. Other elements, such as boron, bromine, and fluorine, also show a constant ratio with chloride, but the chloride ratios of many elements vary significantly.

Most variations in the composition of sea water arise from the removal of elements by organisms living in surface sea water and the later release of these elements by the destruction of biologically produced particles which have sunk downward into deeper waters. Exceptions to this general rule are dissolved gases, whose solubility and concentration in surface sea water increase with decreasing temperatures.

Marine plants can live only in surface sea water, where sufficient light is available for photosynthesis. These organisms give off oxygen and extract carbon dioxide and nearly all of the dissolved nitrate and phosphate from sea water to produce organic matter. Some plants, in addition, secrete solid particles of calcium carbonate ($CaCO_3$) or opaline silica ($SiO_2 \cdot nH_2O$). Marine plants are consumed by animals, some of which also extract dissolved calcium, bicarbonate, and silica to make carbonate or opaline shells or tests. During the downward rain of particles produced by plants and animals in surface waters, destruction of organic matter by bacteria and animals releases dissolved nitrogen, phosphorus, and carbon dioxide back into the water column at depth and consumes dissolved oxygen (Fig. 3). Ocean waters are undersaturated with respect to opaline silica, and these particles begin to dissolve, releasing dissolved silica, after the death of their parent organism. Some particles, however, reach the sea floor to accumulate as siliceous oozes. Carbonate particles are stable in surface waters, which are supersaturated, and accumulate readily in shallower areas of the sea floor. Deeper waters are undersaturated because of increased pressure, and below depths of 4000 m the degree of undersaturation is such that carbonate dissolves very rapidly.

Thus, in response to biological processes, nitrogen, phosphorus, and silicon are almost totally depleted in surface waters, and marine plant life can flourish only where upwelling currents renew surface water in these biolimiting elements. Elements which show some lowering in concentration in surface waters are carbon, copper, nickel, and cadmium and the alkaline earths calcium, strontium, barium, and radium. The behavior of strontium, barium, and radium may reflect in part their coprecipitation with calcium in carbonate. Dissolved oxygen is unique in that it is produced at the surface and consumed at depth (Fig. 3). Analytical data for many elements are not precise enough to establish patterns of variation in their concentration.

In closed basins on the sea floor, stagnant bottom waters can become totally depleted in dissolved oxygen. In anoxic waters, anaerobic respiration reduces sulfate and forms hydrogen sulfide. Iron and manganese become more soluble and may increase in concentration, while other metals, such as copper, precipitate out as sulfides.

Mass balance. The uniform relative abundance of minerals in recent and ancient marine evaporites has been interpreted to mean that the concentrations of major constituents in sea water have rarely been more than double or less than half their present concentrations during the past 700×10^6 years. As a first approximation, the oceans have behaved as a steady-state system, with the rate of removal of a particular dissolved element being equal to its combined rate of introduction from the continents, atmosphere, and sea floor. The rate of turnover of a particular element in the oceans is reflected in its residence time, which is the average length of time an atom of the element spends in the sea between the time it is introduced and the time it is removed. Residence time may be calculated by dividing the total mass (grams) of the element in the oceans by its rate of input or, if known, rate of output (grams per year). The longest residence times, for sodium, chlorine, or bromine, are on the order of 10^8 years. The shortest residence times, for aluminum, iron, and titanium, are only 10^2 years. The age of the oceans is at least 3.5×10^9 years, and thus is significantly greater than any of these residence times. Sea water does not store indefinitely the dissolved components entering it, but is simply a temporary way station for materials passing through the hydrosphere.

In constructing a geochemical mass balance for the oceans, a reasonable estimate can be made of the rate at which material enters the oceans from the continents and atmosphere. Many of the processes by which elements are removed from sea water are also well understood. Some dissolved silica and calcium and some trace elements, for example, are permanently removed from sea water by biological precipitation of opaline silica and calcium carbonate. Nitrogen, phosphorus, and copper are directly concentrated by organisms in organic matter and can accumulate in this form in marine sediments. Much of the removal of heavy metals, however, is probably by adsorption on organic particles or mineral oxides. All dissolved components are removed to some degree when sea water becomes trapped in the pores of marine sediments during deposition.

Among the least understood processes which regulate the mass balance of the oceans are those which control the removal of elements with very long residence times, such as the major cations Na, K, and Mg. In the early 1960s, it was proposed that the composition of sea water is controlled by thermodynamic equilibrium between the atmosphere, sea water, calcium carbonate, and a suite of silicate mineral phases. This led to the further suggestion that major cations, dissolved silica, and bicarbonate are continuously removed by processes of "reverse weathering" in which the above components react with aluminum silicate minerals, brought in as suspended load by rivers, to form illite, chlorite, and montmorillonite. Such a reaction would essentially be the reverse of reaction (4). Field and laboratory studies, however, have failed to identify reverse weathering as a significant process, and it is now believed that silicate reactions with sea water at normal marine temperatures (2–30°C) are simply too slow to have a major effect on sea water composition. A more probable site for sea water–silicate interaction is thought to be the mid-ocean rise and ridge system, where high temperatures cause the convective circulation of sea water down and up through newly formed basaltic rocks. Reaction rates are rapid at elevated temperatures, and some elements, such as magnesium and alkalies, are removed from sea water, while others, such as aluminum and calcium, are introduced. These submarine processes may play a profound role in the mass balance of the oceans.

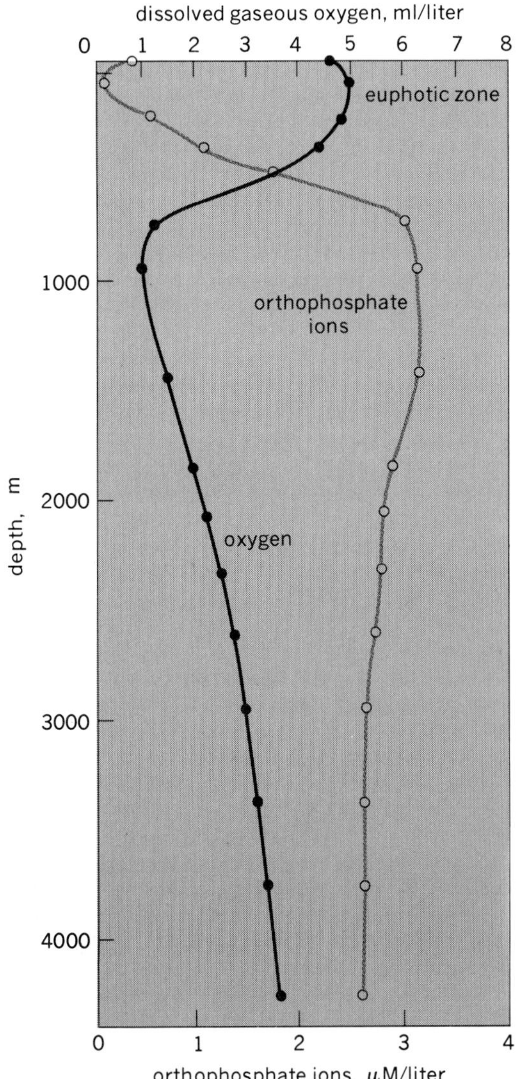

Fig. 3. Distribution of dissolved gaseous oxygen (low values) and nutrient species (high concentrations) of orthophosphate ions at 26°22′.4N and 168°57′.5W in the Pacific Ocean. (*Data from Chinook Expedition of 1956 of the Scripps Institution of Oceanography*)

Mixing studies. Various aspects of sea water geochemistry provide a means of studying mixing processes in the oceans. The distribution of the stable isotope ^{18}O has been found to be useful in evaluating sources and degree of mixing of deep-water masses. The variations in concentration of the natural radioactive isotopes ^{14}C, ^{32}Si, ^{226}Ra, ^{228}Ra, and ^{222}Rn are used to calculate rates of vertical and lateral transport processes. The distribution of artificial radioisotopes, introduced into surface ocean waters as a result of atmospheric nuclear testing, provides a means of determining mixing processes in the upper ocean.

As part of the International Decade of Ocean Exploration in the 1970s, the Geochemical Oceans Sections (GEOSECS) Program was created to obtain high-precision geochemical and hydrographic measurements for the study of circulation and mixing processes in the world oceans.

Surface brines. Evaporation of fresh waters flowing into closed basins on the continents typically produces alkaline brines (Table 2, Soap Lake). Calcium and magnesium precipitate out as insoluble carbonates or hydroxysilicates. Sodium and potassium concentrate continuously, and total carbonate and pH increase. Chemical evolution of marine waters during evaporation follows a different course. Gypsum ($CaSO_4 \cdot 2H_2O$) is the first mineral to precipitate out during continued evaporation, followed by halite (NaCl). Reaction with carbonates to form dolomite [$CaMg(CO_3)_2$] may remove magnesium (Table 2, Dead Sea).

Subsurface waters. Marine waters trapped in the pore spaces of sediments during deposition react with the mineral and organic particles surrounding them and undergo significant changes in composition. Pore waters in organic-rich sediments are quickly depleted in dissolved oxygen, and anaerobic reduction of sulfur destroys dissolved sulfate and produces hydrogen sulfide. Anaerobic reduction of carbon dioxide in the ab-

Table 2. Chemical composition of some saline waters, in parts per million

Constituent	Soap Lake, WA	Dead Sea	Subsurface brine, Louisiana
Na	12,500	39,700	63,900
K	12,500	7,590	869
Mg	23	42,430	1,070
Ca	4	17,180	9,210
Cl	4,680	219,250	124,000
SO_4	6,020	420	153
HCO_3	11,270	220	115
CO_3	5,130	–	–
SiO_2	101	–	16
pH	–		6.3

sence of sulfate produces methane. Changes in the relative proportions of dissolved cations occur as a result of diagenetic reactions with silicate minerals. The vertical variation in concentration of dissolved species in sediment pore waters provides valuable information on the nature and rates of diagenetic reactions in fine-grained marine sediments.

With deeper burial, increases in temperature and compaction of sediment produce additional changes in subsurface water composition. Salinity usually increases with depth, and may reach values in excess of 400,000 ppm. High dissolved salt contents reflect solution of evaporites at depth or infiltration of hypersaline waters formed by evaporation in surface environments. Some increases in salinity may be due to membrane filtration. Compacted shales behave as a semipermeable membrane which allows water molecules to escape upward during burial but which retards the migration of cations and anions. The salinity of the residual pore water is thus increased. Some decreases in salinity with depth have been observed which may be due to dehydration of clay minerals at elevated temperatures.

Each sedimentary basin has its own unique suite of subsurface water compositions. As a general rule, however, subsurface waters are enriched in calcium and strontium and depleted in sodium, potassium, magnesium, and sulfate, relative to sea-water chloride ratios (Table 2). Bicarbonate decreases with depth. Some subsurface waters contain abundant dissolved hydrocarbons, introduced during thermal diagenesis of organic matter. Deep-basin waters are introduced back into surface environments by regional groundwater flow or during crustal deformation.

Some subsurface waters are capable of extracting, transporting, and precipitating significant quantities of metals. In areas of igneous activity these waters may be derived in part from magmas or from the circulation of groundwaters. Metals are leached out of the surrounding rocks or contributed by the magma and are transported in aqueous solution. Other ore-forming fluids closely resemble sedimentary brines in composition, and it seems likely that some metals can be introduced into subsurface waters during normal burial diagenesis.

Ice. Ice is a nearly pure solid, and in contrast to the solvent power of liquid water, few foreign ions can be accommodated in its lattice. Ice does contain particulate matter, however, and the change in the composition of these particles with time, as recorded in the successive layers of ice which have accumulated in polar regions, has provided much information on the progressive input of lead and other materials into the environment by humans.

[JEFFREY S. HANOR]

Bibliography: W. S. Broecker, *Chemical Oceanography*, 1974; Y. Kitano (ed.), *Geochemistry of Water*, 1975; J. D. Riley and G. Skirrow (eds.), *Chemical Oceanography*, 2d ed., 1975.

Ice field

A network of interconnected glaciers or ice streams, with common source area or areas, in contrast to ice sheets and ice caps. The German word *Eisstromnetz*, which translates literally to ice-stream net, is sometimes used for glacial systems of moderate size (such as less than 3000 mi²) and is most applicable to mountainous regions. Being generally associated with terrane of substantial relief, ice-field glaciers are mostly of the broad-basin, cirque, and mountain-valley type. Thus, different sections of an ice field are often separated by linear ranges, bedrock ridges, and nunataks.

Contrast with ice sheet. An ice sheet is a broad, cakelike glacial mass with a relatively flat surface and gentle relief. Ice sheets are not confined or controlled by valley topography and usually cover broad topographic features such as a continental plateau (for example, much of the antarctic ice sheet), or a lowland polar archipelago (such as the Greenland ice sheet). Although ice sheets are generally of very large dimension, in some regions small, rather flat ice bodies have been called ice sheets because they are thinned remnants of once large masses of this form. Small ice sheets and even ice fields are sometimes incorrectly referred to by casual or lay observers as "ice caps," even though their configurations have been well characterized.

Contrast with ice cap. Ice caps are properly defined as domelike glacial masses, usually at high elevation. They may, for example, make up the central nourishment area of an ice field at the crest of a mountain range, or they may exist in isolated positions as separate glacial units in themselves. The latter type is characterized by a distinctly convex summit dome, bordered by contiguous glacial slopes with relatively regular margins not dissected by outlet valleys or abutment ridges.

Similarities and gradations. There are all gradations between ice caps, ice fields, and ice sheets. Over a period of time, a morphogenetic gradational sequence may also develop in any one region. Major ice sheets, for example, probably originate from the thickening and expansion of ice fields and the coalescence of bordering piedmont glaciers. Conversely, ice fields can develop through the thinning and retraction of a large ice sheet overlying mountainous terrane. *See* GLACIATED TERRAIN; GLACIOLOGY.

[MAYNARD M. MILLER]

Iceberg

A large mass of glacial ice broken off and drifted from parent glaciers or ice shelves along polar seas. Icebergs should be distinguished from polar pack ice which is sea ice, or frozen sea water, though rafted or hummocked fragments of the later may resemble small bergs. *See* GLACIOLOGY.

Characteristics and types. The continental or island icecaps of both Arctic and Antarctic regions produce icebergs where the icecaps extend to the sea in the form of glaciers or ice shelves. The "calving" of a large iceberg is one of nature's greatest spectacles, considering that a Greenland berg may weigh over 1,000,000 tons and that Antarctic bergs are many times larger. An iceberg consists of glacial ice which is compressed snow having a variable specific gravity that averages about 0.89. This results in an above-water mass of from one-eighth to one-seventh of the entire mass. However, spires and peaks of an eroded or weathered berg will result in height to depth ratios of between 1–6 and 1–3. Tritium age experiments with melted Greenland berg ice indicate these bergs may be of the order of 50,000 years old. Minute air bubbles imprisoned in glacial ice impart to bergs a snow-white color and cause it to effervesce when immersed.

Icebergs are classified by shape and size. The terms used are arched, blocky, dome, pinnacled, tabular, valley, and weathered for berg discription, and bergy-bit and growler for berg fragments ranging smaller than cottage size above water. The life-span of an iceberg may be indefinite while the berg remains in cold polar waters, eroding only slightly during summer months. But under the influence of ocean currents, an iceberg that drifts into warmer water will disintegrate rapidly, its life being measured in weeks in sea temperatures between 40–50°F and in days in sea temperatures over 50°F. A notable feature of icebergs is their long and distant drift which may carry them into steamship tracks, where they become hazards to navigation. The normal extent of iceberg drift is shown in Fig. 1.

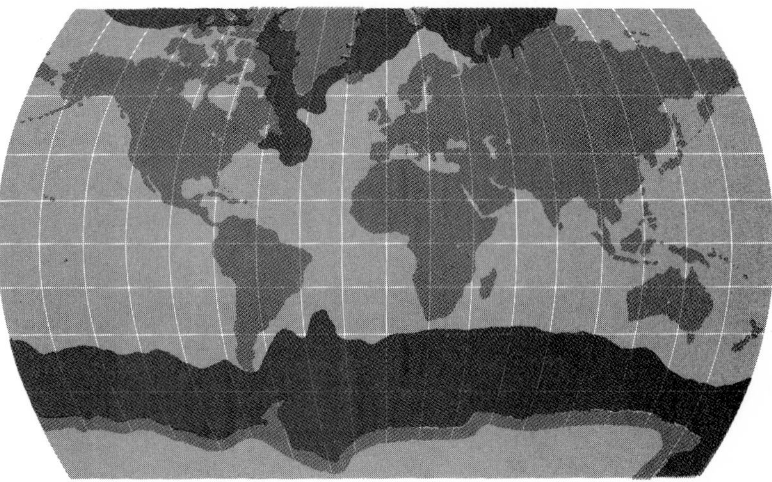

Fig. 1. Normal extent of iceberg drift.

Arctic icebergs. In the Arctic, icebergs originate chiefly from glaciers along Greenland coasts. It is estimated that a total of about 16,000 bergs are calved annually in the Northern Hemisphere, of which over 90% are of Greenland origin; but only about half of these have a size or source location to enable them to achieve any significant drift. The majority of the latter stem from some 20 glaciers along the west coast of Greenland between the 65th and 80th parallels of latitude. The most productive glacier is the Jacobshavn Glacier at latitude 68°N, calving about 1400 bergs yearly, and the largest is the Humboldt Glacier at latitude 79° with a seaward front extending 65 mi. The remainder of the Arctic berg crop comes from East Greenland and the island icecaps of Ellesmere Island, Iceland, Spitzbergen, and Novaya Zemlya, with almost no sizable bergs produced along the Eurasian or Alaskan Arctic coasts. No icebergs are discharged or drift into the North Pacific Ocean or its adjacent seas, except a few small bergs each year that calve from the piedmont glaciers along the Gulf of Alaska. These achieve no significant drift.

Ocean currents of the Arctic and adjacent seas determine the drift and ultimate distribution of icebergs, wind having little effect except on small, sail-shaped fragments. The dominant drift along the East Greenland coast is southward around the tip of Greenland and then northward along the west coast. Here the drifting bergs join the main body of West Greenland bergs and drift in a counterclockwise gyral across Davis Strait and Baffin Bay. The bergs are then swept southward along the coasts of Baffin Island, Labrador, and Newfoundland by the Labrador Current. This drift terminates along the Grand Banks of Newfoundland, where the waters of the Labrador Current mix with the warm Gulf Stream and even the largest of bergs melt within 2–3 weeks. Freak iceberg drifts have been reported where bergs or remaining fragments were sighted off Scotland, Nova Scotia, Bermuda, and even the Azores Islands. Such reports, however, are extremely rare. About 400 bergs each year are carried past Newfoundland as survivors of the estimated 3-year journey from West Greenland. The remainder become stranded along Arctic coasts and shoals and are ultimately destroyed by wave erosion and summer melting.

Icebergs in the Northern Hemisphere rarely reach proportions larger than 2000 ft in breadth or 400 ft in height above the water (Fig. 2). However, true glacial ice islands several miles in extent are occasionally found and have even served as floating bases for scientific studies. The origin of these rare counterparts of the common Antarctic type is uncertain but is thought to be an ice shelf along northern Ellesmere Island.

Fig. 2. Arctic iceberg, eroded to form a valley or dry-dock type; grotesque shapes are common to the glacially produced icebergs of the North. Note the brash and small floes of sea ice surrounding the berg.

Fig. 3. Antarctic iceberg, tabular type. Such bergs develop from great ice shelves along Antarctica and may reach over 100 mi in length. The U.S. Coast Guard icebreaker *Westwind* is in the foreground.

Antarctic icebergs. In the Southern Ocean, bergs originate from the giant ice shelves all along the Antarctic continent. These result in huge, tabular bergs (Fig. 3) or ice islands several hundred feet high and often over a hundred miles in length, which frequent the entire waters of the Antarctic seas. The most active iceberg-producing regions are the Ross and Filchener ice shelves in the Ross and Weddell seas. The large size of these bergs and influence of the Antarctic Circumpolar Current give them an indeterminant life-span. When weathered, Antarctic icebergs attain a deep bluish hue of great beauty rarely seen in the Arctic.

[ROBERTSON P. DINSMORE]

Idocrase

A sorosilicate mineral of complex composition crystallizing in the tetragonal system; also known by the name vesuvianite. Crystals, frequently well formed, are usually prismatic with pyramidal terminations (see illustration). It commonly occurs in

Idocrase. (*a*) Crystal, Christiansand, Norway (*specimen from Department of Geology, Bryn Mawr College*). (*b*) Crystal habits (*from C. S. Hurlbut, Jr., Dana's Manual of Mineralogy, 17th ed., copyright © 1959 by John Wiley & Sons, Inc.; reprinted by permission*).

columnar aggregates but may be granular or massive. The luster is vitreous to resinous; the color is usually green or brown but may be yellow, blue, or red. Hardness is 6½ on Mohs scale; specific gravity is 3.35–3.45. *See* SILICATE MINERALS.

The composition of idocrase is expressed by the formula $Ca_{10}Al_4(Mg,Fe)_2Si_9O_{34}(OH)_4$. Magnesium and ferrous iron are present in varying amounts, and boron or fluorine is found in some varieties. Beryllium has been reported in small amounts.

Idocrase is found characteristically in crystalline limestones resulting from contact metamorphism. It is there associated with other contact minerals such as garnet, diopside, wollastonite, and tourmaline. Noted localities are Zermatt, Switzerland; Christiansand, Norway; River Vilui, Siberia; and Chiapas, Mexico. In the United States it is found at Sanford, Maine; Franklin, N.J.; Amity, N.Y.; and at many contact metamorphic deposits in western states. A compact green variety resembling jade is found in California and is called californite.

[CORNELIUS S. HURLBUT, JR.]

Igneous rocks

Those rocks which have congealed from a molten mass. They may be composed of crystals or glass or both, depending on the conditions of formation. The molten matter from which they come is called magma; where erupted to the surface, it is commonly known as lava. Solidification of the hot rock melt occurs in response to loss of heat. Generated at depth the magma tends to rise. It commonly breaks through the Earth's crust and spills out on the Earth's surface or ocean floor to form volcanic or extrusive rocks. At the surface where cooling is rapid, fine-grained or glassy rocks are formed.

Where unable to reach the surface, magma cools more slowly, insulated by the overlying rocks; and a coarser texture develops. The resulting igneous rocks appear intrusive relative to adjacent rocks. In general, deeply formed (plutonic) rocks display the coarsest texture. Igneous rocks formed at shallow depths (hypabyssal) display features somewhat intermediate between those of volcanic and plutonic types. *See* MAGMA; PLUTON; VOLCANO; VOLCANOLOGY.

Textures. Texture refers to the mutual relation of the rock constituents within a uniform aggregate. It is dependent upon the relative amounts of crystalline and amorphous (glassy) matter as well as the size, shape, and arrangement of the constituents.

Rock textures are highly significant; they shed light on the problem of rock genesis, and tell much about the conditions and environment under which the rock formed.

Crystallinity. This property expresses the proportion of crystalline to amorphous material in an igneous rock. Most igneous rocks, such as granite, are composed entirely of crystalline material and are called holocrystalline. Entirely glassy, or holohyaline, rocks such as obsidian are extremely rare. Many rocks such as rhyolite or vitrophyre

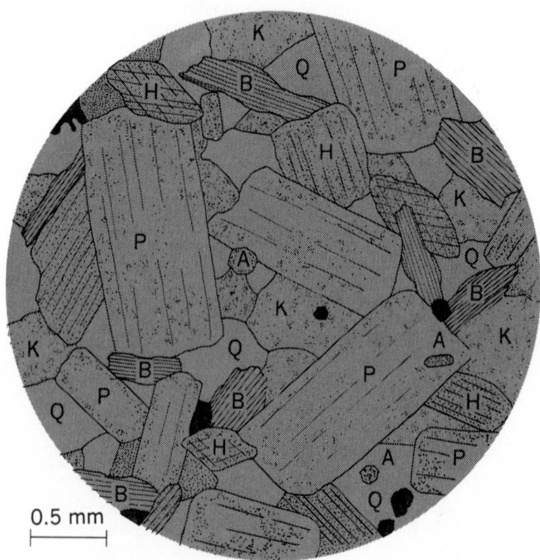

Fig. 1. Hypiodiomorphic granular texture in granodiorite. Euhedral plagioclase (P), subhedral hornblende (H), biotite (B), anhedral quartz (Q), and potash feldspar (K). Accessory minerals include euhedral apatite (A) and subhedral magnetite (black).

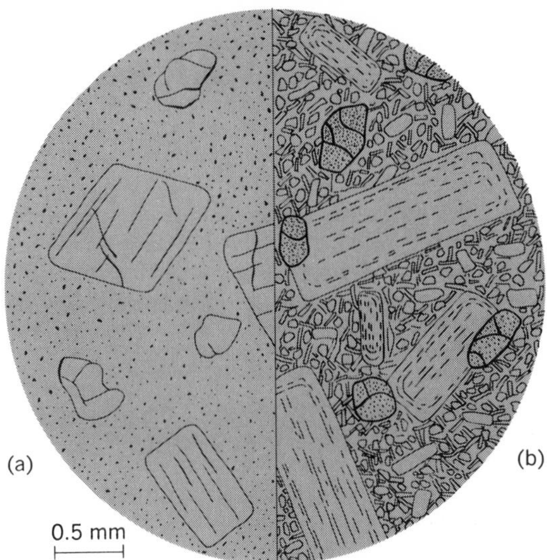

Fig. 2. Porphyritic textures. (*a*) Porphyritic rhyolite showing euhedral phenocrysts of sanidine and resorbed crystals of quartz in a submicroscopically crystalline matrix. (*b*) Porphyritic basalt showing euhedral phenocrysts of plagioclase and subhedral olivine in a matrix of granular pyroxene and feldspar microlites.

contain both glass and crystals and are called hypocrystalline or hypohyaline.

Glass may be considered an amorphous solid with no systematic arrangement of its constituent atoms. Crystals form as the temperature of a magma falls and atoms begin to arrange themselves into orderly, repetitive groups. With rapid cooling there may be no opportunity for crystals to develop, and a magma will congeal as glass.

Granularity or grain size. In igneous rocks grain size ranges widely and depends in large part upon rate of cooling. Rocks are phaneric or phanerocrystalline if their constituent mineral grains can be distinguished as individual entities by the naked eye. All other igneous rocks are aphanitic.

Phaneritic rocks are divided on the basis of their average grain diameter as follows: fine-grained, grains less than 1 mm; medium-grained, grains 1–5 mm; coarse-grained, grains 5–30 mm; very coarse-grained (pegmatitic), grains more than 3 cm.

Aphanitic rocks are microcrystalline if individual constituents can be distinguished only with the microscope. They are cryptocrystalline if constituents are submicroscopically crystalline. Dominantly glassy rocks are considered aphanitic. Aphanitic rocks rich in light-colored (felsic) minerals are termed felsitic. *See* FELSITE.

Grain shape. In igneous rocks grain shape is controlled by many factors. In highly glassy rocks the rate of growth is important. Crystallites are the most rudimentary forms and abound in glassy rocks in which rapid consolidation has arrested further growth. They are too small to polarize light and cannot be identified as to mineral species. These embryonic forms are perhaps most varied and beautifully displayed in the glassy rock, pitchstone. *See* OBSIDIAN.

Microlites are slightly larger, elongate crystals. They polarize light and can usually be identified specifically under the microscope. Many have grown rapidly but imperfectly to form skeletal crystals.

In most rocks, grain shape is controlled largely by sequence of mineral crystallization and the nature and variety of associated minerals. A grain is said to be euhedral if bounded by its characteristic crystal faces and anhedral if crystal faces are absent. Intermediate forms are subhedral. Crystals developed early in a magma tend to be euhedral. Late crystals, however, meet interference from numerous adjacent grains and are forced to assume irregular mutual boundaries.

It is not necessary that all mutually interfering grains develop anhedral forms. Some mineral species possess a greater power of growth (a greater form energy) and are capable of maintaining their characteristic crystal form in competition with adjacent minerals.

Most igneous rocks show a grainy or granular texture in which the majority of crystals are roughly equidimensional. Rarely, grains with euhedral outline dominate and give the rock an idiomorphic granular texture. More commonly, nearly all grains are anhedral and the rock texture is allotriomorphic granular. Most rocks show an intermediate or hypidiomorphic granular texture (Fig. 1).

Porphyritic texture. The grain size of some igneous rocks is extremely uniform (equigranular texture), but that of others may be highly inequigranular. Rocks in which relatively large crystals (phenocrysts) are dispersed in a matrix or groundmass of finer-grained or glassy material are said to be porphyritic or phyric (Fig. 2). Porphyritic glasses with abundant phenocrysts are known specifically as vitrophyres.

Porphyritic rocks may form in a number of ways. (1) Phenocrysts may have grown early and slowly while the magma was deeply buried. The groundmass may have congealed later after the magma was erupted to higher levels where rapid cooling ensued. (2) Phenocrysts in many rocks

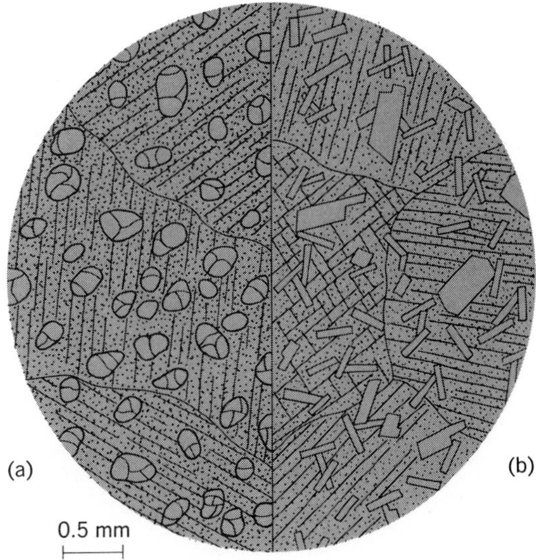

Fig. 3. (*a*) Poikilitic texture with large crystals of hornblende enclosing abundant small grains of olivine. (*b*) Orphitic texture with large crystals of pyroxene enclosing small laths of plagioclase.

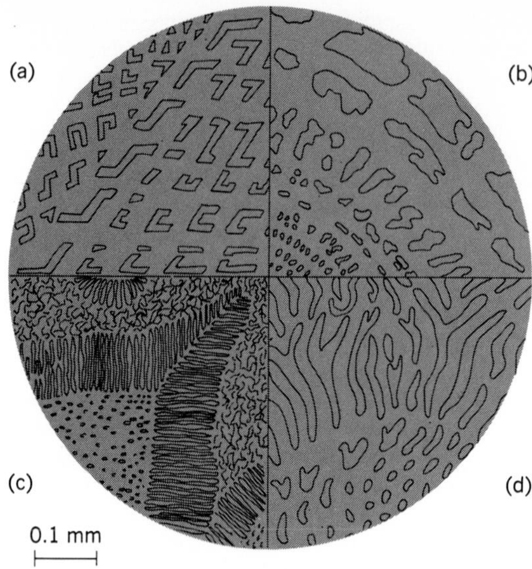

Fig. 4. Implication or intergrown textures. (*a*) Micrographic texture. (*b*) Micropegmatitic texture. (*c*) Granophyric texture. (*d*) Myrmekitic texture.

(some granites) may develop late and still attain large dimensions if their growth rate is sufficiently greater than that of adjacent minerals. (3) The large crystals of some plutonic rocks are probably more properly classed as porphyroblasts. They may have formed essentially in solid rock by recrystallization aided by residual fluids from the solidifying magma. (4) Large crystals in many rocks (certain porphyries and lamprophyres) may not be indigenous. They may have been incorporated during intrusion of the magma. (5) Phenocrysts might develop by inoculation or by disturbance of supersaturated magma. *See* LAMPROPHYRE; PHENOCRYST; PORPHYRY.

Poikilitic texture. This texture involves numerous small grains of one mineral, in random orientation, enclosed by single large crystals of another (Fig. 3*a*). Conditions favoring development of poikilitic texture are not well understood. In some rocks this texture may have developed by direct crystallization of magma. In other rocks this texture may represent recrystallization of magmatic rocks.

Ophitic texture. This is a special type of poikilitic texture and is characteristic of the rock diabase (Fig. 3*b*). The texture involves lath-shaped crystals of plagioclase feldspar enclosed by large anhedral grains or plates of pyroxene (augite or pigeonite). If the length of the feldspar crystals exceeds that of the pyroxene, enclosure is only partial and the texture is called subophitic. *See* DIABASE.

Other textures, more or less related to ophitic, are characteristic of very fine-grained and glassy rocks of basaltic composition. *See* BASALT.

Implication or intergrown textures. These are formed by the mutual penetration of two or more mineral phases. The intergrowth may be so intimate that one phase appears disintegrated into smaller grains which are isolated by the other. Within small domains, however, grains of one phase show optical and crystallographic continuity.

Graphic or micrographic textures may develop between almost any mineral pair where one member, in cuneiform masses resembling runic inscriptions, is enclosed by the other (Fig. 4*a*). Micropegmatitic texture is essentially a micrographic texture involving only quartz and potash feldspar (Fig. 4*b*). If the intergrowth is more varied and involves plumose, fringing, radial, or micropegmatitic patterns, the texture is granophyric (Fig. 4*c*). In myrmekitic texture plagioclase (generally oligoclose) grains enclose vermicular quartz (Fig. 4*d*). Perthitic texture is extremely common in feldspars and takes on a wide variety of forms (Fig. 5). It usually consists of tiny masses of sodic plagioclase enclosed by potash feldspar. Various proportions of the two constituents may exist. Where potash feldspar is more abundant and constitutes the host mineral, the material is known as perthite. Where plagioclase predominates, the material is called antiperthite.

Some implication textures may develop by simultaneous crystallization of two constituents. Others may form by exsolution in the solid state (some perthite). Still other textures may be due to the partial replacement of one mineral phase by another.

Structures. Structure as applied to igneous rocks is easily confused with texture. In general, however, structure refers to a geometrical form or architectural feature in a rock. Structure emphasizes the heterogeneous nature of a rock or mineral aggregate; texture emphasizes homogeneity. Certain large-scale structures, such as faults, folds, and joints, are common to most rock types. They are perhaps more properly classed as geologic structures. Like textures, the structures of igneous rocks may tell much about the history or conditions of formation of the rocks themselves.

Vesicular structures. These structures are common in many volcanic rocks. They form when magma is brought to or near the Earth's surface. Here the low pressure permits partial release and expansion of dissolved water or other volatiles and the formation of steam bubbles which may be preserved as small cavities when the magma congeals. In highly viscous lavas (rhyolitic) much gas may be

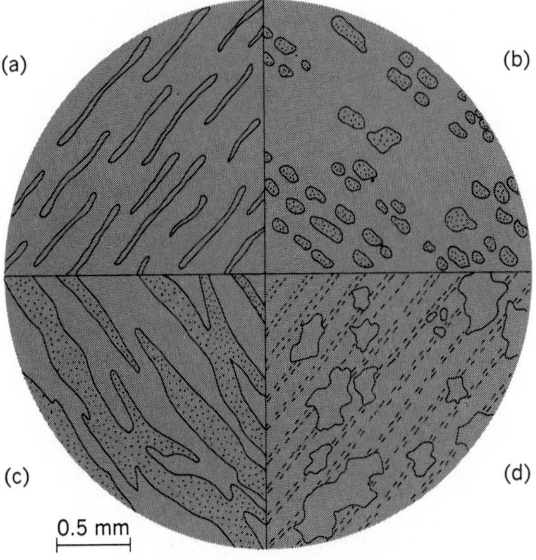

Fig. 5. Perthitic textures. (*a*) String perthite. (*b*) Patch perthite. (*c*) Vein perthite. (*d*) Antiperthite.

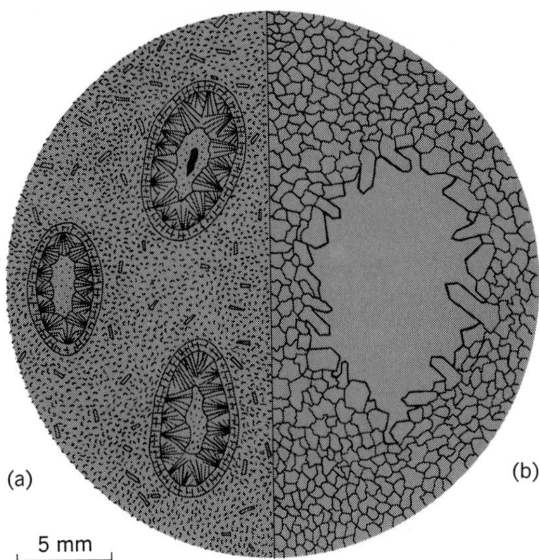

Fig. 6. (*a*) Amygdaloidal structure showing former gas cavities (bubbles) in lava filled with later minerals. (*b*) Miarolitic cavity in fine-grained granite with allotriomorphic granular texture. Euhedral outline is visible only where crystals form cavity boundary.

trapped, but only tiny bubbles may form. Rapid cooling of this frothy liquid produces a pumiceous structure (characteristic of the rock pumice). In less viscous lavas (basaltic) integration of tiny bubbles produces a coarser, spongy, or scoriaceous structure (characteristic of the rock scoria). Vesiculation of some basaltic lavas produces well-formed, ellipsoidal cavities. These may later be filled with minerals (such as quartz, calcite, epidote, zeolite) deposited from fluids which permeated the rock. Such fillings are called amygdules in allusion to their almond shape. The structure is known as amygdaloidal (Fig. 6*a*).

Miarolitic openings. These are the most common cavernous structures found in plutonic rocks. They are irregular, range up to several inches across, and appear crusted with beautifully formed crystals (quartz and feldspar). These crystals are not truly encrusting a cavity wall; their bases constitute an integral part of the rock (Fig. 6*b*). This fact indicates that each cavity (vug) formed as a small interspace in the crystal aggregate and filled with residual magmatic fluid against which bounding grains readily developed their euhedral outline. The presence of muscovite, tourmaline, topaz, and apatite suggests that volatiles (water, fluorine, and boron), which concentrate in residual magmatic fluids, have played an important role in the development of miarolitic structures. These cavities are most likely to form in rocks crystallizing at shallow depth where confining pressure is relatively low.

Zoned crystals. Crystals possessing zonal structures are common and appear to be built up of concentric shells or zones of different composition which follow the general crystal outline (Fig. 7). Though minute, these zonal structures are readily detected in thin sections under the microscope. Individual zones may be thick or thin, and zoned boundaries may be sharp or gradational. Compositional changes from the crystal's center outward may be great or slight; they may be progressive, reversed, interrupted, oscillatory, or repetitive. Zoning is characteristic of minerals belonging to solid solution series (for example, plagioclase, pyroxene, and amphibole).

Numerous theories have been proposed to explain various types of zoning in minerals. These are based on physicochemical principles relating to supersaturation of the magma and changes in composition, pressure, temperature, and volatile content of the magma. Movement of crystals from one part of the magma chamber to another may have been important. The most common type of zoning (progressive) appears because of incomplete reaction between solid and liquid phases during the crystallization of an isomorphous series.

Hourglass stucture. This structure, somewhat related to zoning, is most frequently displayed by crystals of pyroxene. Certain sections through a crystal possessing the structure have the appearance of an hourglass (Fig. 7*c*). This structure probably demonstrates the minute differences in energy involved at different faces of a growing crystal. It may be due to selective adsorption of ions by different faces during crystal growth.

Reaction rims. These rims or zones, in which one mineral envelopes another, are believed to have formed by reaction and are common in some rocks (Fig. 8). They may develop by reaction between early formed crystals and surrounding magma (pyroxene rims on early formed olivine crystals). Reaction between two incompatible minerals, induced by residual fluids in the late stages of magma consolidation, may produce similar effects. Pyroxene or amphibole may form by reaction around olivine crystals where they would otherwise come in contact with plagioclase. Some petrologists refer to rims of primary origin as coronas and those of secondary origin as kelyphitic borders.

IGNEOUS ROCKS

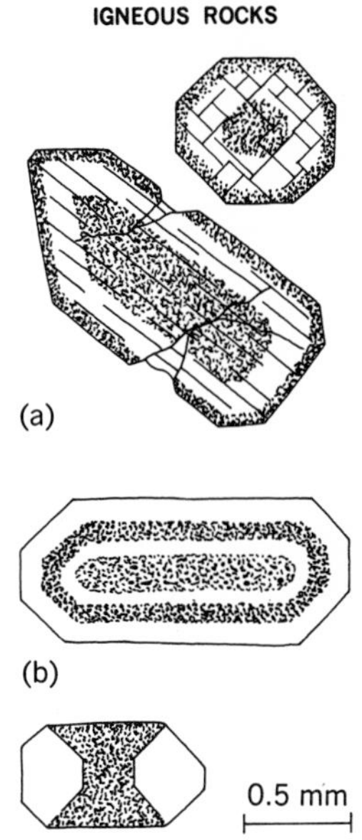

Fig. 7. (*a*) Zoned crystals of pyroxene. (*b*) Zoned plagioclass. (*c*) Hourglass structure as seen in pyroxene.

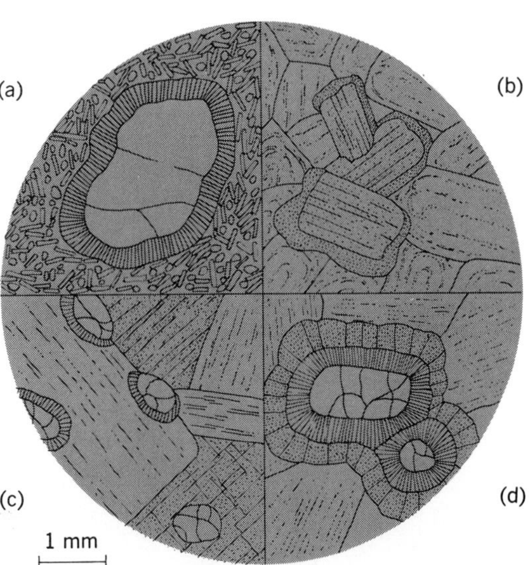

Fig. 8. Reaction rims. (*a*) Pyroxene around quartz grain in basalt. (*b*) Hornblende around three grains of orthopyroxene in norite. Most of rock is composed of plagioclase. (c) Rim of orthopyroxene formed between small olivine grains and plagioclase in diabase. No rim exists between olivine in norite. (*d*) Reaction rims (inner, orthopyroxene; outer, amphibole) around olivine in norite.

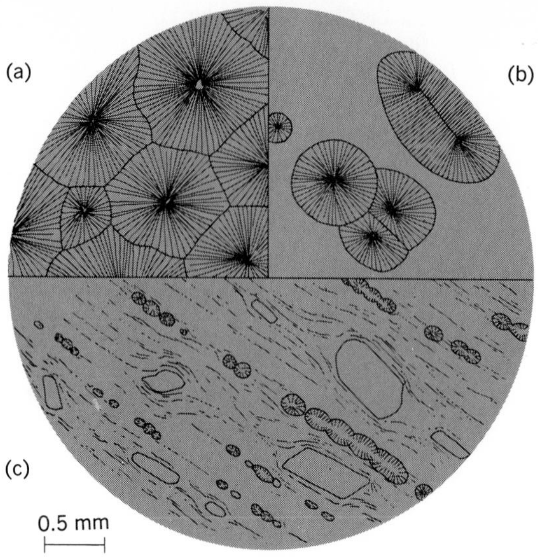

Fig. 9. (*a*) Spherulitic structure. (*b*) Spherulites in volcanic glass. (*c*) Fluidal structure showing trains of spherulites, phenocrysts, microlites, and crystallites.

Spherulites. These are radial aggregates of needlelike crystals. They are roughly spherical and usually less than a centimeter across (Fig. 9). They abound in silica-rich lavas, particularly rhyolitic glass, and are composed principally of quartz, tridymite, and alkali feldspar.

Somewhat similar aggregates in basaltic rocks, called varioles, consist of radial plagioclase crystals with interstitial glass or granules of olivine or pyroxene.

Spherulites consisting of concentric shells with cavernous interspaces are known as lithophysae, or stone bubbles. In many, the tiny annular cavities are lined with delicate crystals of cristobalite, quartz, and feldspar.

Inclusions or enclosures. Inclusions are common in most varieties of igneous rocks. These masses of extraneous-looking material vary widely in size, shape, constitution, and origin. Inclusions demonstrated to be foreign rock fragments enclosed and trapped by congealing magma or lava may be specifically designated as xenoliths. Incorporated foreign crystals are known as xenocrysts. *See* XENOLITH.

If an earlier consolidated portion of a magma is ruptured and fragments of it become enclosed by portions which solidify later, the older rock bodies are known as autoliths. Enclosures formed by selective accretion of minerals, either during or after consolidation of a magma, are termed segregations.

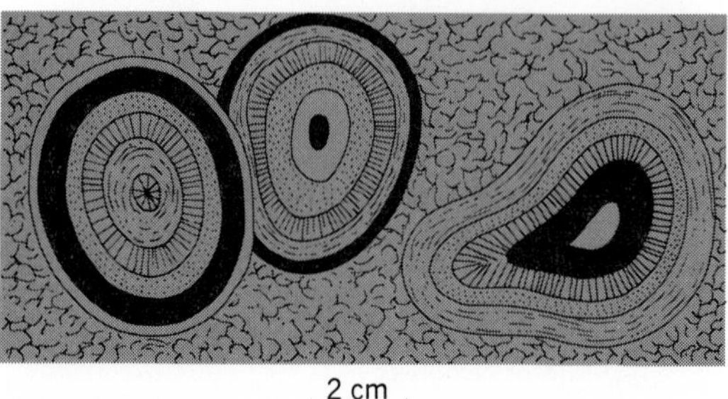

Fig. 10. Orbicular structure.

Orbicular structures. These structures, found in some plutonic rocks (granite, granodiorite, and diorite), are orblike masses generally up to a few inches across. They show concentric shells of different mineral composition and thickness which may envelope xenolithic cores (Fig. 10). Most commonly dark mineral shells (rich in biotite, hornblende, or pyroxene) alternate with light shells (rich in feldspar). Individual shells may be sharply or vaguely defined, and the minerals within may be granular or elongate and in radial or tangential arrangement.

Orbicular structures may develop by reaction (between xenoliths and magma) involving chemical reconstitution of the solid fragments or rhythmic crystallization around xenoliths. Many orbicular structures may represent products of metamorphism and metasomatism of solid rock. *See* METAMORPHIC ROCKS; METAMORPHISM; METASOMATISM.

Pillow or ellipsoidal structure. This type of structure is a peculiar feature of certain lava flows (basalt, spilite). Rocks exhibiting this structure appear to be composed of closely packed, pillow-shaped masses up to several feet across. Individual pillows have a very fine-grained crust or margin which carries abundant vesicles, commonly arranged concentrically with the pillow surface.

The pillows are so perfectly fitted together as to suggest that they were assembled in a plastic state. Relatively little matrix occurs; and it consists commonly of chert, limestone, or shale. The close association of pillow lavas with sedimentary rocks is in agreement with the popular belief that they are of subaqueous origin. Pillow lavas have been observed to form both on dry land and in water, but the precise conditions favoring their formation are still not well understood. *See* PRECAMBRIAN; SPILITE.

Flow structure. This is a nongenetic term for a number of directive features in rocks. The structure may be formed by flowage during crystallization of a magma (primary flow structure). Post-solidification flow (secondary) may develop similar features, but these are classed as metamorphic in origin.

The structure takes the form of parallel streaks or lenses of different minerals or textures; or it may result from parallel arrangement of elongate or platy minerals (mica, hornblende, or feldspar). Some flow structures consist of abundant slabby inclusions or xenoliths in parallel orientation. Flowage may be expressed by flow lines (lineation) or flow layers (some foliation). These may be straight or contorted.

Fluidal structure and fluxion structure. These are genetic terms and specifically imply flowage of lava or magma (Fig. 9*c*).

Schlieren. Schlieren are irregular streaks, patches, or layers having more or less blended outlines and measuring up to many feet in length. They are generally composed of the same minerals as the enclosing rock but in different proportions. Schlieren may represent early segregation drawn out by magma flow. Some may be xenoliths more or less digested and reworked by magma. Others may represent residual magmatic liquors of different composition injected into already crystallized portions. Schlieren formed in solid rocks are more properly metamorphic or metasomatic features. *See* GRANITIZATION.

Banding. Banding is exhibited by rocks composed of alternating layers of different composition, texture, or color. The term is merely descriptive, not genetic. If flowage is to be implied, the term flow banding is used.

Classification. Schemes for classifying igneous rocks are numerous. Prior to the advent of the polarizing microscope (roughly 1870), rock classifications were based on megascopic characteristics, many of them misleading. These systems were gradually improved as chemical analyses were more commonly employed.

Today three principal methods of classification are used. (1) Megascopic schemes are based on the appearance of the rock-in-hand specimen or as seen with a magnifying glass (hand lens). Such schemes are useful in the field study of rocks. (2) Microscopic schemes (largely mineralogical) are employed in laboratory investigations where more detailed information is needed. (3) Chemical schemes are very useful but have more limited application. The mineral content and texture of a rock generally tell much more about the rock's origin than does a bulk chemical analysis. For example, granite, quartz porphyry, rhyolite, and obsidian may all have the same chemical composition; but the geologic conditions under which each forms may be very different. Granite solidifies slowly at depth and under high pressure. The porphyry may crystallize in two stages, one at depth and a later one nearer the surface. The other two rocks are of surficial origin; the obsidian solidifies most rapidly and as glass.

Igneous rocks show great variations chemically, mineralogically, texturally, and structurally with few if any natural boundaries. This accounts in large part for the great disagreement among petrologists as to how igneous rocks should be classified. The following subsections discuss plutonic, volcanic, and hypabyssal types.

Plutonic rocks. Plutonic rocks occur in large intrusive masses (batholiths, stocks, and other large plutons). They form at great depth and, therefore, are often referred to as abyssal rocks. Generated from large bodies of magma which has cooled slowly, they characteristically show a phaneritic, holocrystalline texture.

Under deep-seated conditions where confining pressure is high, volatiles dissolved in the magma are retained until the last stage of crystallization. These act as fluxes and reduce the temperature of crystallization. Consequently plutonic rocks, as compared with volcanic rocks, may carry relatively low-temperature mineral phases.

Volcanic rocks. These are formed as lava flows or as pyroclastic rocks (heterogeneous accumulations of volcanic ash and coarser fragmental matter). They have solidified rapidly to develop an aphanitic texture with more or less glass. Volatiles are readily lost as the lava reaches the Earth's surface. Therefore, crystallization tends to proceed within a relatively high-temperature range, so high-temperature minerals such as sanidine and high-temperature plagioclase are characteristic. Expanding gas bubbles formed by escaping volatiles frequently create highly porous rocks.

Volcanic rocks frequently show evidence of two stages of cooling, an early, deep-seated stage (intratelluric) and a later, effusive stage. Slow cooling in the first stage may produce a few large crystals. These become suspended in the lava and frozen into an aphanitic matrix during the effusive stage. This accounts for the porphyritic texture so commonly encountered in volcanic rocks.

Hypabyssal rocks. These rocks exhibit characteristics more or less intermediate between those of volcanic and plutonic types. They differ from volcanic rocks in that they are intrusive and generally free from glass and vesicular structures. They

Table 1. Classification of igneous rock families*

	Feldspar	Quartz	Quartz or feldspathoid (5%)	Nepheline	Leucite
Alkali feldspar and plagioclase (oligoclase, andesine)	Alkali feldspar 50% of the feldspar	Granite (P) Rhyolite (A)	Syenite (P) Trachyte (A)	Nepheline syenite (P) Phonolite (A)	Leucite phonolite (A)
	Alkali feldspar 5–50% of the feldspar	Granodiorite (P) Quartz latite (A)	Monzonite (P) Latite (A)		
	Alkali feldspar 5% of the feldspar	Quartz diorite (P) (tonalite) Dacite (A)	Diorite (P) Andesite (A)		
			Anorthosite (P)		
Labradorite-anorthite			Gabbro (P) Basalt (A)	Theralite (P) Tephrite (A)	Leucite tephrite (A)
			Diabase (H)	Basanite (A)	Leucite basanite (A)
Plagioclase < 10%			Periodotite (P) Pyroxenite (P) Hornblendite (P)	Nephelinite (A) Nepheline basalt (A)	Leucitite (A) Leucite basalt (A)

*In this table P indicates phaneritic (plutonic) rock; A, aphanitic (volcanic) rock; and H, hypabyssal rock.

Table 2. Average chemical compositions of igneous rocks (totals reduced to 100%)*

Components	Plutonic rocks							
	Granite	Grano-diorite	Quartz diorite	Syenite	Monzo-nite	Diorite	Gabbro	Nepheline syenite
SiO_2	70.18	65.01	61.59	60.19	56.12	56.77	48.24	54.63
TiO_2	0.39	0.57	0.66	0.67	1.10	0.84	0.97	0.86
Al_2O_3	14.47	15.94	16.21	16.28	16.96	16.67	17.88	19.89
Fe_2O_3	1.57	1.74	2.54	2.74	2.93	3.16	3.16	3.37
FeO	1.78	2.65	3.77	3.28	4.01	4.40	5.95	2.20
MnO	0.12	0.07	0.10	0.14	0.16	0.13	0.13	0.35
MgO	0.88	1.91	2.80	2.49	3.27	4.17	7.51	0.87
CaO	1.99	4.42	5.38	4.30	6.50	6.74	10.99	2.51
Na_2O	3.48	3.70	3.37	3.98	3.67	3.39	2.55	8.26
K_2O	4.11	2.75	2.10	4.49	3.76	2.12	0.89	5.46
H_2O	0.84	1.04	1.22	1.16	1.05	1.36	1.45	1.35
P_2O_5	0.19	0.20	0.26	0.28	0.47	0.25	0.28	0.25
Components	**Aphanitic rocks**							
	Rhyolite	Quartz latite	Dacite	Trachyte	Latite	Andesite	Basalt	Phonolite
SiO_2	72.80	62.43	65.68	60.68	57.65	59.59	49.06	57.45
TiO_2	0.33	0.85	0.57	0.38	1.00	0.77	1.36	0.41
Al_2O_3	13.49	16.15	16.25	17.74	16.68	17.31	15.70	20.60
Fe_2O_3	1.45	4.04	2.38	2.64	2.29	3.33	5.38	2.35
FeO	0.88	1.20	1.90	2.62	4.07	3.13	6.37	1.03
MnO	0.08	0.09	0.06	0.06	0.10	0.18	0.31	0.13
MgO	0.38	1.74	1.41	1.12	3.22	2.75	6.17	0.30
CaO	1.20	4.24	3.46	3.09	5.74	5.80	8.95	1.50
Na_2O	3.38	3.34	3.97	4.43	3.59	3.58	3.11	8.84
K_2O	4.46	3.75	2.67	5.74	4.39	2.04	1.52	5.23
H_2O	1.47	1.90	1.50	1.26	0.91	1.26	1.62	2.04
P_2O_5	0.08	0.27	0.15	0.24	0.36	0.26	0.45	0.12

*After Daly.

Table 3. Approximate mineral composition of the common plutonic rocks

Rock	Felsic minerals, %	Total felsic, %	Mafic minerals	%
Granite	Potassium feldspar, 35–45 Sodic plagioclase, 20–30 Quartz, 20–30	80–95	Biotite, hornblende	5–20
Granodiorite	Potassium feldspar, 15–25 Sodic plagioclase, 35–45 Quartz, 15–25	75–90	Biotite, hornblende	10–25
Quartz diorite	Oligoclase, andesine, 55–65 Quartz, 15–25 Potassium feldspar, 0–5	70–85	Hornblende, biotite, pyroxene	15–30
Syenite	Potassium feldspar, 60–70 Sodic plagioclase, 10–20 Quartz or nepheline, 0–5	70–90	Biotite, hornblende, pyroxene	10–30
Monzonite	Potassium feldspar, 20–30 Sodic plagioclase, 45–55 Quartz or nepheline, 0–5	65–85	Biotite, hornblende, pyroxene	15–35
Diorite	Oligoclase, andesine, 60–70 Potassium feldspar, 0–5 Quartz or nepheline, 0–5	60–80	Hornblende, biotite, pyroxene	20–40
Gabbro	Labradorite, bytownite, 45–70 Potassium feldspar, 0–5 Quartz or nepheline, 0–5	45–75	Pyroxene, olivine, hornblende, biotite	25–55

differ from plutonic rocks in that they occur in small bodies (dikes and sills) or in larger bodies formed at shallow depths (laccoliths) and they have textures characteristically resulting from more rapid cooling. Hypabyssal rocks cannot be sharply distinguished from volcanic rocks on the one hand and plutonic rocks on the other. The recognition of such a group, therefore, is perhaps of greater value in field studies, where mode of occurrence is known, than in the laboratory.

The rocks of each subdivision (plutonic, volcanic, and hypabyssal) may be further divided into families (groups with the same or closely allied composition and relatively limited textural variation). Families are sometimes grouped into rock clans. A clan includes all families with the same chemical composition. Thus, the gabbro family (plutonic), diabase family (hypabyssal), and basalt family (volcanic) have the same chemical composition and, therefore, belong to the same clan (gabbro clan). The clan name is derived from the plutonic family member.

The principal families of the phaneritic (plutonic) and aphanitic (volcanic) rocks are shown in Table 1. This scheme expresses something of the bulk chemical composition of the rock and the conditions of formation. For discussion of rock types, see articles under specific names.

More specific rock types may be indicated by prefixing some pertinent mineral, textural, or structural term to the appropriate family name (such as biotite granite, graphic granite, orbicular granite). *See* PEGMATITE.

Chemical composition. By averaging a large number of chemical rock analyses, one may derive a representative composition for each of the rock families. Average analyses for the more common igneous rocks are shown in Table 2. Such average values are very useful standards for comparison in spite of the fact that the variations within families may be greater than those between families.

Mineral composition. Igneous rock-forming minerals may be classed as primary or secondary. The primary minerals are those formed by direct crystallization from the magma. Secondary minerals may form at any subsequent time.

Essential primary minerals. The principal primary minerals are relatively few and may be classed as light-colored (felsic) or dark-colored (mafic) varieties. Felsic is a mnemonic term for feldspathic minerals (feldspar and feldspathoids) and silica (quartz, tridymite, and cristobalite). Mafic is mnemonic for magnesium and iron-rich minerals (biotite, amphibole, pyroxene, and olivine). Felsic minerals are composed largely of silica, alumina, and alkalies. Mafics are rich in iron, magnesium, and calcium.

Table 3 summarizes the essential primary constituents of the more common plutonic rocks. The percentage ranges are highly generalized. Individual rock specimens may depart radically from these values, but the averages are fairly representative and useful for comparison. The mineral composition of the corresponding volcanic rocks is roughly similar to the values in the table. Major departures will be encountered particularly in the glassy rocks.

Accessory minerals. Accessory minerals are those occurring in very small or trace amounts. They consist principally of magnetite, ilmenite, pyrite, hematite, apatite, zircon, rutile, and sphene. Most generally these are widely distributed as tiny grains or crystals.

Secondary minerals. Included in this group are minerals formed by addition of material subsequent to solidification of the rock or by alteration of minerals already present in the rock. The addition of fluorine and boron, which tend to concentrate in the residual magmatic liquids, to already crystallized portions of the rock may form small crystals of fluorite, topaz, or tourmaline. Alteration in which certain minerals become more or less reconstituted is common and widespread. It is generally believed to occur during the last stages of solidification while hot, residual fluids (for example, water and carbon dioxide) permeate the crystal aggregate and convert water-free silicate minerals into hydrous forms. This hydrothermal or deuteric effect may be so intense that virtually all igneous characteristics of the rock are lost. The common alteration products derived from the essential primary minerals are listed as follows.

Primary mineral	*Secondary mineral*
Quartz	Not altered
Potash feldspar	Kaolinite, sericite
Plagioclase	Kaolinite, sericite (paragonite), epidote, zoisite, calcite
Nepheline	Cancrinite, analcite, natrolite
Leucite	Nepheline and potash feldspar
Sodalite	Analcite, cancrinite
Biotite	Chlorite, sphene, epidote, rutile, iron oxide
Hornblende	Actinolite, biotite, chlorite, epidote, calcite
Orthopyroxene	Antigorite, actinolite, talc
Clinopyroxene	Hornblende, actinolite, biotite, chlorite, epidote, antigorite
Olivine	Serpentine, magnetite, talc, magnesite

Density. Density is a significant rock property and is a function largely of mineralogical composition and porosity. Chemical composition alone is not a reliable indication of density because different minerals (with different densities) may form from a single bulk composition.

Table 4 gives the approximate average and common range of density for the more abundant plutonic rocks. Densities of volcanic equivalents are generally slightly lower due to higher porosity and greater amount of glass. Highly porous volcanic rocks (pumice and scoria) may be so vesicular as to float on water. The density of completely glassy rocks is approximately 6% less than that of the corresponding holocrystalline (entirely crystalline) type. [CARLETON A. CHAPMAN]

Table 4. Approximate densities of common plutonic rocks

Rock	Average	Common range
Granite	2.67	2.60–2.73
Granodiorite	2.71	2.65–2.77
Syenite	2.76	2.65–2.85
Quartz diorite	2.82	2.72–2.92
Diorite	2.85	2.75–2.95
Gabbro	2.99	2.85–3.15

Bibliography: B. Bayly, *Introduction to Petrology*, 1968; E. W. Heinrich, *Microscopic Petrography*, 1956; W. W. Moorhaus, *The Study of Rocks in Thin Sections*, 1959; H. H. Read and J. Watson, *Introduction to Geology*, 1962; L. E. Spock, *Guide to the Study of Rocks*, 1962.

Illite

The term illite is not a specific clay-mineral name, but is a general term for the mica-type clay minerals. It is commonly used for any nonexpanding clay mineral with a 10-A *c*-axis spacing. *See* CLAY MINERALS.

Illite clays are used for the manufacture of structural clay products, such as brick and tile.

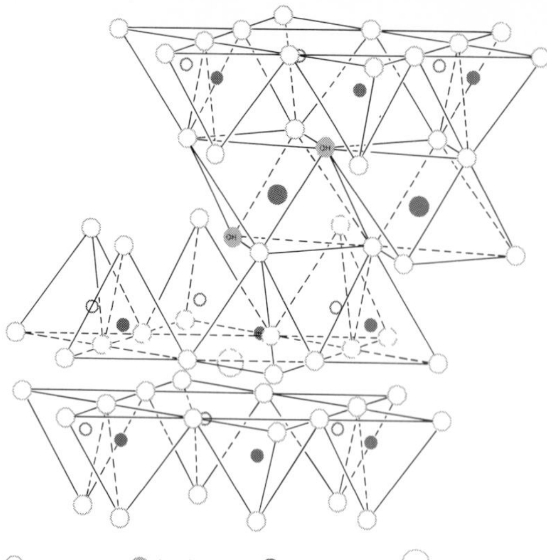

Diagrammatic sketch of the structure of illite. *(From R. E. Grim, Clay Mineralogy, McGraw-Hill, 1953)*

Some degraded high plastic illites are used for bonding molding sands.

Structural characteristics. By definition, all illites have the mica-type structure. The basic structural unit is composed of two silica tetrahedral sheets with a central octahedral sheet. This unit is essentially the same as that of montmorillonite, except that there is always some replacement (15%±) of silicon by aluminum in the tetrahedral sheets. This substitution results in a charge deficiency which is balanced by potassium ions between the illite unit layers. The stacking of adjacent illite layers is such that the potassium just fits and is always surrounded by a total of 12 oxygens, thus balancing the structural charge and binding the layers together without the possibility of expansion (see illustration). Illites may be either dioctahedral or trioctahedral, depending on the population of possible octahedral positions. They differ from well-crystallized micas in that they exhibit less substitution of aluminum for silicon, resulting in a higher silicon-to-aluminum molecular ratio.

Nonstructural characteristics. The size of naturally occurring illite particles is very small, yet they are larger and thicker than montmorillonite particles and have better-defined edges.

The illites have a moderately low cation-exchange capacity (20–30 meq/100 g). It is primarily due to broken bonds, but lattice substitutions may also be a cause in poorly crystallized varieties. The rate of the exchange reaction is likely to be slow, since a small part of the exchange occurs through partial replacement of the firmly held interlayer potassium ions. Dehydration curves for illites show the presence of a small amount of interlayer water which is lost below 100°C. The OH lattice water is lost between 300 and 600°C. Following the loss of structure between 850 and 950°C, spinel and mullite phases form prior to fusion at about 1400°C.

Occurrence. Illite is a common product of weathering if potash is present in the environment of alteration. It is a frequent constituent in many soil types, and may form in soils under certain conditions as a consequence of the addition of potash fertilizers. This is possible because illite has the ability to "fix" potassium. Much illite in soils loses its potassium through leaching and becomes degraded. On the addition of potassium to such material, K^+ goes back into these normal positions and the illite is rebuilt. *See* SOIL CHEMISTRY.

Illite is common in recent sediments and is particularly abundant in deep-sea clays. It probably forms from montmorillonite and other minerals during marine diagenesis. Because of its stability, illite is often found in ancient sediments. It is the dominant clay mineral in many, probably most, shales that have been studied. *See* DIAGENESIS; MARINE SEDIMENTS; SHALE.

[FLOYD M. WAHL; RALPH E. GRIM]

Ilmenite

A rhombohedral mineral, space group $R\overline{3}$, with composition $Fe^{2+}Ti^{4+}O_3$ It is derived from the hematite crystal structure by ordering iron and titanium atoms over the octahedral sites, thus resulting in lower symmetry. The arrangement is such that the iron-centered oxygen octahedrons share edges and faces only with titanium-centered oxygen octahedrons. The hardness is $5\frac{1}{2}$ on Mohs scale, specific gravity 4.72. Color is black, and there is no cleavage. Crystals are thick tabular parallel to {0001} (see illustration) but the mineral usually occurs massive or in thin plates. Mg^{2+} and Mn^{2+} often substitute for Fe^{2+} in ilmenite; two other minerals belonging to the ilmenite structure type are geikielite, $MgTiO_3$, and pyrophanite, $MnTiO_3$.

Ilmenite crystals, from Arendal, Norway. *(Specimen from Department of Geology, Bryn Mawr College)*

Ilmenite is the most abundant titanium mineral in igneous rocks and the most important ore of titanium. It occurs as an early differentiate in anorthosite, norite, and gabbroic rocks, sometimes in large quantities; also in metamorphic rocks such as gneisses and marbles. Ilmenite resists weathering, and alluvial accumulations may occur in sufficient quantity to constitute ore. Important occurrences include the Ilmen Mountains, Soviet Union (whence the name); Kragerø, Norway; and Allard Lake, Quebec. *See* HEMATITE.

[PAUL B. MOORE]

Impsonite

A black, naturally occurring carbonaceous material having specific gravity 1.10–1.25 and fixed carbon 50–85%. Impsonite is infusible and is insoluble in carbon disulfide. A vein of impsonite 10 ft wide occurs in La Flore County, Okla., and there are other deposits in Arkansas, Nevada, Michigan, Peru, Argentina, Brazil, and Australia. The Peruvian impsonite may be derived from grahamite by weathering. The origin of impsonite is not well understood, but it appears to be derived from a fluid bitumen that polymerized after it filled the vein in which it is found. Impsonites are known to contain unusually high percentages of vanadium. *See* ALBERTITE; ASPHALT AND ASPHALTITE; ELATERITE; WURTZILITE.

[IRVING A. BREGER]

Index fossil

The ancient remains and traces of a plant or animal that lived during a particular span of geologic time and that geologically dates the containing rocks.

Index fossils are almost exclusively confined to sedimentary rocks which originated in such diverse environments as open oceans, tropical lagoons, coral reefs, beaches, lakes, and rivers. They represent a variety of organic forms, such as microscopic algae, plant pollen, oyster shells, shark teeth, geometric cavities in the rock that are clearly organic but otherwise unknown, or such large and obvious examples as the teeth of extinct elephants. Index fossils are a very necessary means for comparing the geologic age of sedimentary rock formations and are an everyday tool in the search for petroleum, coal, and metallic ores.

Criteria. The choice of a fossil as an index depends on several criteria. In general, the fossil represents a group that evolved rapidly. Through the mechanisms of evolution, plant and animal forms have appeared in a determinable order, persisted for awhile, and then been replaced by new forms, the old never to return. They thus have formed a general succession of floras and faunas in which individual genera and species serve as detailed time markers. The greater the rate of evolution, the shorter the period of time represented by any given index fossil and the narrower the limits of relative age assigned to the rocks containing the index. Commonly, the span of geologic time during which a fossil lived is referred to as its range, and the thickness of rocks through which a particular index fossil or selected group of fossils occurs is referred to as a faunal zone. Ranges and zones take their names from the fossils whose occurrence they represent, that is, the range of *Fusulinella* or the *Siphonodella duplicata* Zone.

An index fossil also must be present in the rocks in sufficient numbers to be found with reasonable effort, must be relatively easy to collect or identify, and must be geographically extensive so that the zone it defines is widely applicable. These requirements also imply that it must be resistant to burial pressures, to solution by groundwater percolation, or to surface weathering in order to be preserved. Fossils commonly are incompletely preserved or cannot be removed from the rock without considerable effort or possible destruction. Features that permit the fossil to be readily identified under such circumstances are of great advantage. The usefulness of a fossil is increased, however, if it is relatively easy to separate from the rock. Conodonts, spores, and pollen, for example, are resistant to acid and can thus be chemically separated from the rocks. Small size is likewise an advantage, inasmuch as many fossils must be identified in the rock chips of drilled well cuttings. Such fossils, mainly foraminifers, spores, and pollen, are widely used by the petroleum industry.

The presence or absence of the index fossil ideally should not reflect environmental conditions. Many fossil groups that qualify as indices in other respects are unsuitable in application because they were especially adapted to a restricted environment and thus are present only where that particular fossil environment is represented in the rocks. Such "facies" fossils were not under enough environmental pressure to evolve as rapidly as many other groups; their value lies in interpretation of environment rather than in age determination.

Common flora and fauna. The fossil groups most useful as index fossils are generally marine and either floaters or open ocean swimmers, such as cephalopods, or bottom dwellers that had a floating or swimming stage in their life cycles, such as the medusa stage in the brachiopods. Such characteristics are necessary for rapid dispersal of newly evolved forms. On land, such mobile forms as the horses or wind-borne pollen and spores were relatively unrestricted by environmental barriers and became widely dispersed. All of these groups have provided biochronological zones of worldwide extent.

Sequence of evolution. The recognition of evolutionary sequences of fossil species is a prime factor in the determination of precise geologic age. In the 19th century general faunal succession received considerable attention, but present-day workers emphasize the use of phylogenetic lineages wherein species complexes grade from ancestor to descendant, and fossil populations are recognized as possessing the same kinds of inter- and

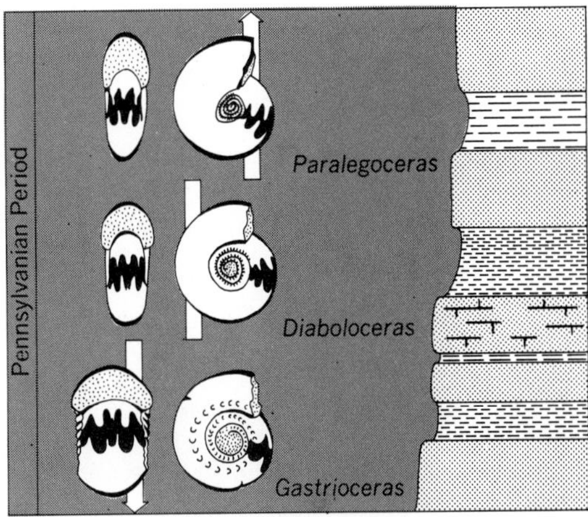

The cephalopod genera *Gastrioceras, Diaboloceras*, and *Paralegoceras* form an evolutional sequence. Their short ranges (indicated by bars) and distinctive suture lines (area between two sutures is shown in black) make them ideal index fossils for the early Pennsylvanian Period. (*Modified from A. K. Miller and W. M. Furnish, Middle Pennsylvanian Schistoceratidae (Ammonoidea), J. Paleontol., 32(2):254, 1958*)

intraspecific variation as living populations. Thus in modern use the individual index fossil is no longer entirely adequate for detailed geologic age determinations, and close examination of multiple phylogenetic sequences is necessary for precise dating and correlation of rock strata. Variations from the normal evolutionary sequence are keys to differences in sedimentation rates, gaps in the geologic record, or recognition of special events. Modern paleontologists, using foraminifera, discoasters, and conodonts in particular, are commonly able to date the occurrence of geologic events to within a million years.

Indices for biochronology. Relatively few groups of fossil organisms embody all of the above characteristics, but several groups have proved to be practical indices and have been widely accepted as a basis for world biochronological systems. Foremost among the forms are spores and pollen, discoasters, foraminifers, graptolites, ammonoid cephalopods (see illustration), and conodonts.

During the Cambrian Period (600,000;000 years before present) the oldest highly developed animals appeared; among them the trilobites provide the first important group of index fossils. They were marine bottom dwellers that evolved rapidly and are keys for both Cambrian and Ordovician time. Small plantlike floating colonial animals called graptolites have proved useful in correlating Ordovician (500,000,000 years B.P.) and Silurian (425,000,000 years B.P.) rocks in areas as widely separated as North America, Sweden, and the Soviet Union. Ammonoids, coiled cephalopods with irregular suture patterns that record the animal's history of growth as well as the species' evolutional relationships, are a classic example of the internationally useful index fossil and are important beginning in the Devonian Period (405,000,000 years B.P.) and extending to the end of the Cretaceous Period (135,000,000 years B.P.). From the Pennsylvanian Period (310,000,000 years B.P.) fusulinids, a family of Foraminiferida, and pollen and spores from the coal forests are important indices. Small phosphatic teethlike fossils known as conodonts, whose zoological affinities are unknown, have been useful for detailed zonation throughout the Paleozoic Era as well as the early part of the Mesozoic. Closer to present time, the bones and teeth of vertebrate animals serve as index fossils for the Tertiary Era while the remains of primitive man himself have been used to date the Recent past. *See* EVOLUTION, ORGANIC; FOSSIL; GEOLOGICAL TIME SCALE; STRATIGRAPHIC NOMENCLATURE; STRATIGRAPHY.

[CHARLES COLLINSON]

Bibliography: J. R. Beerbower, *Search for the Past*, 1968; C. O. Dunbar and J. Rodgers, *Principles of Stratigraphy*, 1958; M. Kay and E. H. Colbert, *Stratigraphy and Life History*, 1965; R. C. Moore, C. G. Lalicker, and A. G. Fischer, *Invertebrate Fossils*, 1952; J. M. Weller, *The Course of Evolution*, 1969.

Index mineral

In metamorphic petrology, a characteristic mineral which by its presence in a rock indicates the mineral facies of the rock. While an index fossil characterizes a sedimentary bed (stratum) in the geological sequence, an index mineral characterizes the mode of origin of a metamorphic rock in regard to temperature, pressure, and composition. A mineral such as quartz is not an index mineral because quartz is found in a great variety of rocks formed from low to high temperature and pressure. However, minerals with stability fields within restricted parts of the geologically important temperature-pressure range (by a given bulk composition of the containing rock) usually may be regarded as index minerals. In a crude way they are also geological thermometers and pressure gages. *See* GEOLOGIC THERMOMETRY; HIGH-PRESSURE PHENOMENA; METAMORPHIC ROCKS.

In going from an area of nonmetamorphic rocks into an area of progressively more highly metamorphic rocks, each zone of progressive metamorphism is defined by an index mineral the first appearance of which marks the outer limit of the zone in question. The sequence of index minerals in progressively metamorphosed pelitic rocks is as follows: chlorite, biotite, almandite, garnet, staurolite, kyanite, and sillimanite. An index mineral for high pressure but relatively low temperature is glaucophane. Pyrope garnet and omphacite pyroxene indicate high values of both pressure and temperature. *See* ECLOGITE. [T. F. W. BARTH]

Isostasy

A theory of hydrostatic equilibrium in the various parts of the Earth's crust. Many scientists judge this theory so well confirmed as to be considered a basic law of geology. B. F. Howell, Jr., in 1959, stated it in such explicit terms as: "All large land masses on the Earth's surface tend to sink or rise so that, given time for readjustment to occur, their masses are hydrostatically supported from below, except where local stresses are acting to upset equilibrium." *See* EARTH; TERRESTRIAL GRAVITATION. [CHARLES V. CRITTENDEN]

Bibliography: W. A. Heiskanen and F. A. Vening Meinesz, *The Earth and Its Gravity Field*, 1958; B. F. Howell, Jr., *Introduction to Geophysics*, 1959; E. N. Lyustikh, *Isostasy and Isostatic Hypotheses*, 1960.

Jade

A name that may be applied correctly to two distinct minerals. The two true jades are jadeite and nephrite. In addition, a variety of other minerals are incorrectly called jade. Idocrase is called California jade, dyed calcite is called Mexican jade, and green grossularite garnet is called Transvaal or South African jade. The most commonly encountered jade substitute is the mineral serpentine. It is often called "new jade" or "Korean jade." The most widely distributed and earliest known true type is nephrite, the less valuable of the two. Jadeite, the most precious of gemstones to the Chinese, is much rarer and more expensive.

Nephrite. Nephrite is one of the amphibole group of rock-forming minerals, and it occurs as a variety of a combination of the minerals tremolite and actinolite. Tremolite is a calcium-magnesium-aluminum silicate, whereas iron replaces the magnesium in actinolite. Although single crystals of the amphiboles are fragile because of two directions of easy cleavage, the minutely fibrous structure of nephrite makes it exceedingly durable. It occurs in a variety of colors, mostly of low intensity, including medium and dark green, yellow, white, black, and blue-gray. Nephrite has a hardness of 6 to 6½ on Mohs scale, a specific gravity near 2.95, and refractive indices of 1.61 to 1.64. On the refractometer, nephrite gem stones show a single index near 1.61. Nephrite occurs throughout the world; important sources include the Soviet Union, New Zealand, Alaska, several provinces of China, and a number of states in the western United States. *See* AMPHIBOLE; GEM; TREMOLITE.

Jadeite. Jadeite is the more cherished of the two jade minerals, because of the more intense colors it displays. It is best known in the lovely intense green color resembling that of emerald (caused by a small amount of chromic oxide). In the quality known as imperial jade, the material is at least semitransparent. White, green and white, light reddish violet, bluish violet, brown, and orange colors are also found. Jadeite also has two directions of easy cleavage, but a comparable random fibrous structure creates an exceptional toughness. Although jadeite has been found in California and Guatemala, the only important source of gem-quality material ever discovered is the Mogaung region of upper Burma. The hardness of jadeite is 6½ to 7, its specific gravity is approximately 3.34, and its refractive indices are 1.66 to 1.68; on a refractometer, only the value 1.66 is usually seen. *See* JADEITE. [RICHARD T. LIDDICOAT, JR.]

Jadeite

The name given to the monoclinic sodium aluminum pyroxene, $NaAl(SiO_3)_2$. Jadeite forms green, fibrous crystals that are colorless in thin sections and exhibit the 87° pyroxene (110) cleavages. Jadeite is a rare metamorphic mineral found in some serpentine masses associated with other dense minerals such as lawsonite, $CaAl_2Si_2O_7(OH)_2$; glaucophane; garnet; and other minerals such as diopside, tremolite serpentine, natrolite (a zeolite), and chlorite. The presence of albite and quartz or albite and nepheline mineral pairs in some jadeite specimens, and the rough similarity of the rock to the eclogites, implies a high-pressure, low-temperature modification of albite and nepheline: albite + nepheline = 2 jadeite. This reaction proceeds to the right at 100°C and 5 kilobars (1 kbar = 10^8 N/m²) pressure, or 500°C and 11 kbar pressure. The reaction albite = jadeite + quartz occurs at 100°C and 10 kbar pressure, or 200°C at 12.5 kbar pressure. The mineral associations in the natural occurrences of jadeite deny the presence of high temperatures. Jadeite could be the sodium mineral in the lower regions of the Earth's crust. Jadeite is rare; the few places where it occurs include Burma, Celebes, Central America, Japan, Sardinia (see illustration), and San Benito County, CA. In plate tectonic theory, jadeite-bearing rocks are interpreted as high-pressure, low-temperature transformations in graywacke and altered basaltic materials in the oceanic plate portion of the subduction zones. The jadeite-bearing rocks may be characteristic of zones of interaction between certain continents and ocean basins. The absence of jadeite in rocks of Precambrian age suggests that crustal temperatures were higher in the past at the depths that jadeite forms in more recent rocks.

Specimen of massive jadeite in its dark-green variety chloromelonite from Susa in Sardinia. (*Specimen from Department of Geology, Bryn Mawr College*)

The albite + nepheline reactions to form jadeite at 1000°C requires 20 kbar pressure or a depth of 60 km. Accordingly, jadeite may be a mineral in the upper mantle of the Earth if albite or nepheline compositions are present. Jadeite is valued as a precious stone for carvings. *See* GLAUCOPHANE; HIGH-PRESSURE PHENOMENA; JADE; PYROXENE. [GEORGE W. DE VORE]

Jasper

An opaque, impure type of massive fine-grained quartz that typically has a tile-red, dark-brownish-red, brown, or brownish-yellow color. The color of the reddish varieties of jasper is caused by admixed, finely divided hematite, and that of the brownish types by finely divided goethite. Jasper has been used since ancient times as an ornamental stone, chiefly of inlay work, and as a semiprecious gem material. Under the microscope, jasper generally has a fine, granular structure, but fairly large amounts of fibrous or spherulitic silica also may be present. *See* GEM; QUARTZ.

Jasper has a smooth conchoidal fracture with a dull luster. The specific gravity and hardness are variable, depending upon particle size and the nature and amount of the impurities present; both values approach those of quartz. The color of jasper often is variegated in banded, spotted, or orbicular types. Heliotrope is a translucent greenish chalcedony containing spots or streaks of opaque red jasper, and jaspagate contains bands of chalcedonic silica alternating with jasper. Jaspilite is a metamorphic rock composed of alternating layers of jasper with black or red bands of hematite. *See* CHALCEDONY.

[CLIFFORD FRONDEL]

Joint

A fracture that traverses a rock and does not show any discernible displacement of one side of the fracture relative to the other. The term joint refers primarily to the actual fracture as represented by a fine line or trace marking the intersection of the fracture and rock surfaces. Commonly, however, the joint is represented superficially by a cleft or fissure resulting from weathering, mechanical separation, or by one face of the fracture on an outcrop (Fig. 1).

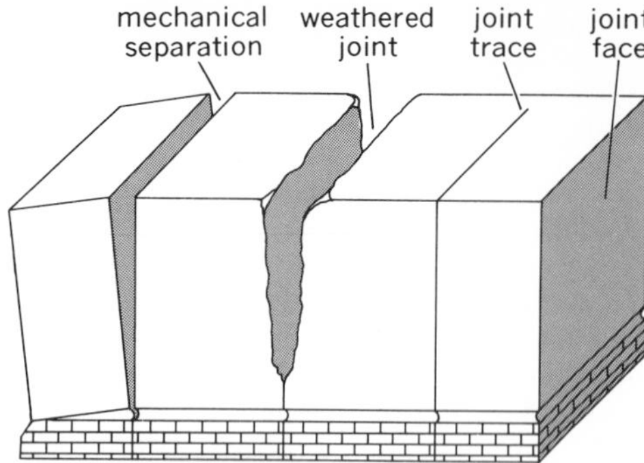

Fig. 1. Aspects of a systematic joint.

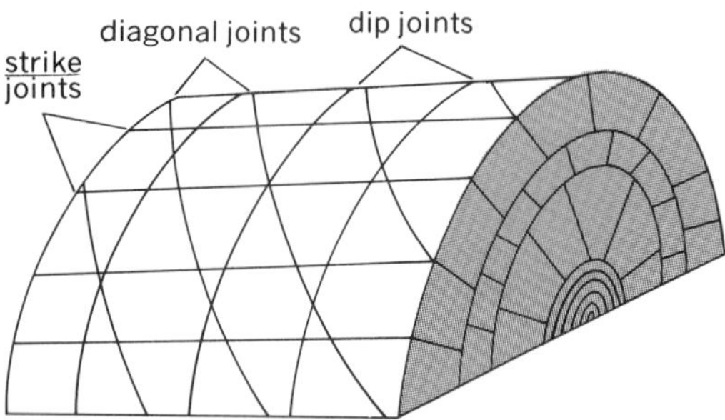

Fig. 2. Diagram showing geometric relations between systematic joints and folded or tilted strata.

Varieties of joints. Joints fall into two major classes based on their spatial relations, geometry, and surface structures: systematic and nonsystematic joints.

Systematic joints. Systematic joints are arranged in groups of regularly spaced parallel or subparallel planar fractures called joint sets. Two or more joint sets comprise a joint system. Two joint sets believed related genetically constitute a conjugate joint set or, if more than two, a conjugate joint system.

Specific terms applied to individual systematic joints are both descriptive and genetic. Descriptive terms are based on the geometric relations between joints and other structures. In sedimentary rocks, joints that parallel the strike or dip of strata are called strike and dip joints, respectively. Joints having an intermediate angular relation are called diagonal or oblique joints (Fig. 2). Where one set of joints appears dominant, the joints may be called primary joints, and those of less prominent sets then are termed secondary. Particularly large or prominent joints of any set are called major or master joints.

In igneous and metamorphic rocks, systematic joints are described on the basis of their geometric relations with linear or planar structures within the rock. Joints lying perpendicular to lineation are called cross or *Q* joints and, where parallel with lineation, longitudinal or *S* joints. Those with intermediate angular relations are diagonal joints.

Genetic terms denote the geometric relations between joints and forces that have deformed the host rock. Where the acute angle of intersection between two sets of joints is bisected by the direction of the maximum principal stress, as inferred from other structural features such as folds or faults, the joints are called shear joints and, where parallel or normal to this stress direction, tension joints. Joints produced directly by bending of strata into folds are termed release joints where parallel to the fold axis, and extension joints where normal to the axis. Joints that originate as a direct result of deformation may be called first-order joints. Joints that result from a readjustment of stresses following first-order jointing, faulting, or folding may be called second-order joints.

Nonsystematic joints. Nonsystematic joints fall into two classes: those related geometrically to systematic joints, and those related to weathering of rock surfaces.

Joints of the first class are highly curved or roughly planar fractures that extend across the interval between systematic joints. They probably result from adjustment of the rock between systematic joints to local tectonic stresses. Joints of the second class commonly are of shallow depth, occur normal to the weathered rock surface, and show random curvilinear patterns (Fig. 3).

Sheeting is a variety of nonsystematic joint that lies roughly parallel to the weathered rock surface. Spacing between these joints is irregular and increases with depth. They are generally believed to result from near-surface rock expansion due to removal of overburden or to chemical weathering. They may possibly result also from stresses related to residual strain energy remaining in the rock from previous deformations. Similar flat-lying joints occur in igneous intrusions and are believed to result from a decrease in magmatic pressure as the magma solidifies, or from removal of overburden.

Size and spacing. Joints of all types have a very great range in size. The surface area of individual fractures ranges from a few square inches to several thousand square feet. Despite this range in size, all classes of joints are entirely similar in their other morphologic aspects. The spacing of joints tends to be more close in thin-layered rocks than in thick-layered ones. Spacing between systematic joints ranges from a few inches to as much as 50 ft.

Columnar or prismatic joints. Columnar joints occur in sheetlike or pluglike intrusive and extrusive igneous rock bodies. They also may be found on occasion in sedimentary rocks that have been heated adjacent to igneous intrusions. Individual columns ordinarily are five- or six-sided but may have from three to eight sides in some examples.

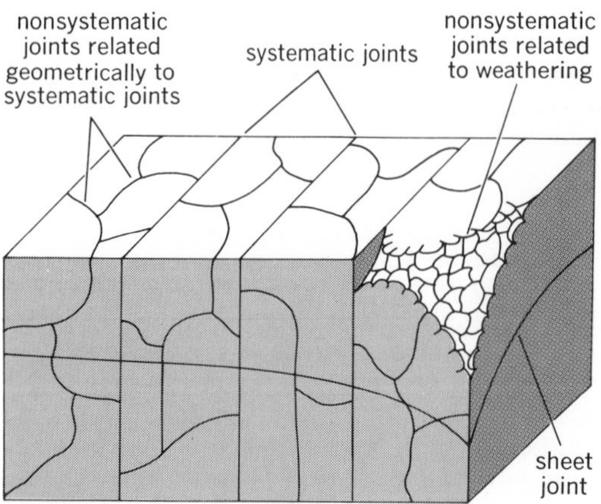

Fig. 3. Major types of joints.

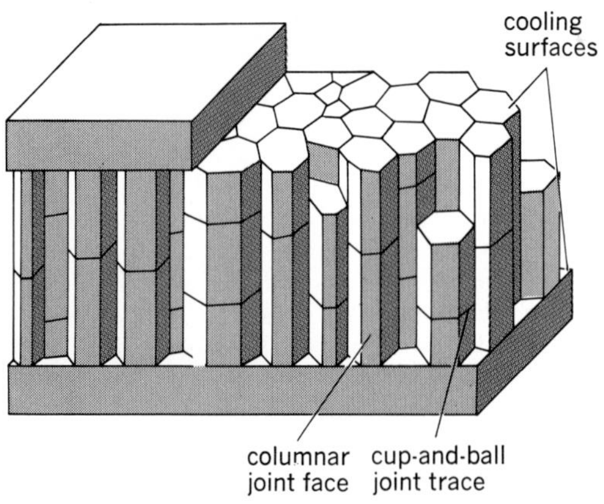

Fig. 4. Columnar jointing.

The columns range from a few inches to several feet in diameter and from a few feet to many tens of feet in length. Dish-shaped cross joints (cup-and-ball joints) divide the columns into segments.

Columnar joints invariably are oriented normal to cooling surfaces and result from the contraction of the igneous rock on cooling (Fig. 4). The columnar structure is what may be expected if the strain energy generated by contraction of the rock is to be dissipated by the least amount of work.

Plumose structure. Joint surfaces commonly show a patterned structure composed of ridges of low relief. These plumose ridges converge at a point or straight axis near the center of the joint face on systematic joints. Curvilinear ridges of proportionally greater relief and disposed concentrically about the central axis or point occur on some systematic joint faces (Fig. 5). These are referred to as conchoidal ridges. Irregular, curvilinear plumes characterize the faces of nonsystematic joints. Plumose structures suggest that joints are tensional brittle fractures that originate at some small structural inhomogeneity within the rock and are propagated outward from that point, until the causal forces are dissipated.

Occurrence. Joints are the most abundant structures in the Earth's crust and are present everywhere. They constitute structural inhomogeneities in rock bodies and as such are influential in determining details of form in topographic relief. Locally, systematic joints determine the azimuth of topographic features such as hills, valleys, stream courses, and cliff faces and underground features such as cave passages. Joints also serve as loci for limonite, quartz, calcite, or other vein and ore minerals deposited from solution.

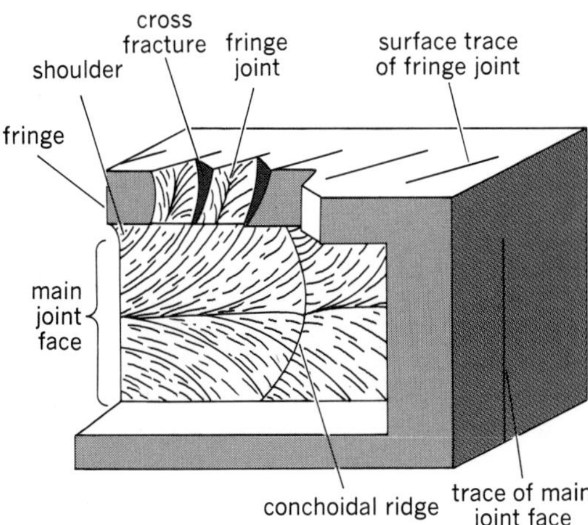

Fig. 5. Schematic block diagram showing primary surface structures of a systematic joint.

Various theories of jointing

Theory	Origin
	Systematic joints
Cleavage	Genetically similar to cleavage in minerals (theory no longer maintained)
Magnetic forces	Genetically related to and controlled by Earth's magnetic field (theory no longer maintained)
Torsion	Result of local and regional warping of Earth's crust
Earthquake	Seismic shock produces and controls direction of jointing
Torsion-earthquake	Rock warped to breaking point; fracturing triggered by earthquakes
Tension	Result of contraction of sediments due to compaction or loss of water or both
Tension	Result of local or regional (or both) vertical forces involved in uplift or folding of Earth's crust
Tension and shear	Result of local or regional (or both) tangential forces involved in uplift or folding of Earth's crust
Tidal	Result of cyclic tidal forces acting tangential to Earth's crust; forces control direction of jointing
Tidal fatigue	Result of rock fatigue engendered by cyclic semidiurnal tidal forces; direction of jointing inherited from preexisting fracture pattern
Residual stresses	Residual rock stresses modified during uplift in such a manner as to produce shear and tension joints
	Nonsystematic joints
Tension	Resulting from contraction of sediments due to compaction or loss of water or both
Tension	Resulting from contraction of igneous rock upon cooling
Compression-tension	Resulting directly or indirectly from removal of overburden or from surface weathering of rock

Origin. At present no one theory for the origin of systematic joints is accepted universally. The majority of theories are based on analyses of systematic joint patterns and their relations to other structures, such as folds and faults. The interpretations of these relations is based primarily on theoretical and experimental work on rock fracture and deformation.

Most geologists believe systematic joints are related genetically to folding, faulting, and uplift and attribute them to tangential compressional or tensional forces acting regionally or locally in the Earth's crust. A minority holds that joints are produced by some universal long-acting force, such as earth tides, and are not directly the result of local folding or faulting. The several major theories of jointing are summarized in the table. *See* FAULT AND FAULT STRUCTURES; FOLD AND FOLD SYSTEMS; STRUCTURAL PETROLOGY; TECTONOPHYSICS. [ROBERT A. HODGSON]

Bibliography: M. P. Billings, *Structural Geology*, 1954; L. U. De Sitter, *Structural Geology*, 2d ed., 1964; E. S. Hills, *Elements of Structural Geology*, 1963; F. H. Lahee, *Field Geology*, 6th ed., 1961; C. R. Longwell and R. F. Flint, *Introduction to Physical Geology*, 2d ed., 1962; N. J. Price, *Fault and Joint Development in Brittle and Semi-Brittle Rock*, 1966.

Jurassic

The system of rocks deposited during the middle part of the Mesozoic Era. The Jurassic System is normally underlain by the Triassic and overlain by the Cretaceous System. It was named after the Jura Mountains in Switzerland and consists mainly of sedimentary rocks. Volcanic rocks are restricted to certain regions, particularly the North American Cordillera. The duration of the Jurassic Period, which according to M. Howarth ended about 135,000,000 years ago, is estimated to be about 55,- to 60,000,000 years.

PRECAMBRIAN	PALEOZOIC							MESOZOIC			CENOZOIC	
					CARBONIFEROUS							
	CAMBRIAN	ORDOVICIAN	SILURIAN	DEVONIAN	Mississippian	Pennsylvanian	PERMIAN	TRIASSIC	JURASSIC	CRETACEOUS	TERTIARY	QUATERNARY

Subdivision. The Jurassic System consists of three main divisions: the Lower, Middle, and the Upper Jurassic. Either Lias, Dogger, and Malm or Black, Brown, and White Jurassic are the terms used in some countries, particularly Germany. Their boundaries coincide largely with those of the Lower, Middle, and Upper Jurassic. Finer subdivisions into stages (chronostratigraphic units) and zones (biostratigraphic units) are based mainly on the northwest European standard section. The Hettangian, Sinemurian, Pliensbachian, and Toarcian stages form the Lower Jurassic; the Bajocian, Bathonian and Callovian form the Middle Jurassic; and the Oxfordian, Kimmeridgian, Portlandian, and Purbeckian form the Upper Jurassic. In some countries, particularly France and Germany, a special stage name, the Aalenian, is used instead of the Lower Bajocian, most commonly applied in the United States, Canada, and Great Britain. Up to the Kimmeridgian these names are in worldwide use, but because of regional differentiation of the faunas in the younger Jurassic, other stage names have been introduced for these beds, for example, Volgian in Russia and the Arctic and Tithonian in Mediterranean regions. The stages are groupings of zones which are based on certain species of ammonites (order Ammonoidea of Cephalopoda) characteristic of certain beds and, in many cases, of worldwide distribution.

The northwest European standard section is now subdivided into about 60 ammonite zones. Local faunal differentiations have resulted in separate zone sequences for certain regions, and some zones may be based on fossils other than ammonites, for example, the pelecypod genus *Buchia* (*Aucella*) in the Upper Jurassic.

Other subdivisions are formations mainly defined by lithology, which form the basis of geological mapping. In North America formation names are extensively used. In Great Britain formation names such as Lias, Inferior Oolite, Cornbrash, and Oxford Clay are used, and as some of these rock units are traceable to other countries, the same names have been applied. The use of formation names for mapping has great advantages; however, worldwide correlations are based on stages and zones.

Paleogeography. Different opinions have been expressed on the configuration and position of continents and oceans during the Jurassic Period. The existence of large continents or land bridges in the region of present-day oceans is now denied by many authorities. W. Arkell considered as possible the existence of an African-Brazilian continent that formed the western half of Gondwanaland. He also supposed that the present-day Indian Ocean was land in the Jurassic. In this interpretation the Australo-Indo-Madagascan continent, or Lemuria or eastern part of Gondwanaland, was separated from the western part of Gondwanaland by a Jurassic sea on the east side of Africa. The distribution of both land and neritic faunas makes land connections between continents—as between Africa and South America—necessary. But as there is no indication of any former continents or land bridges in the Atlantic and Indian oceans, and as most if not all of the ocean floor seems to be younger than Jurassic, the hypothesis of continental drift is now widely accepted. *See* CONTINENT FORMATION.

In recent years continental drift has been seen as a result of movement of lithospheric plates away from oceanic spreading centers. A. Hallam's tenta-

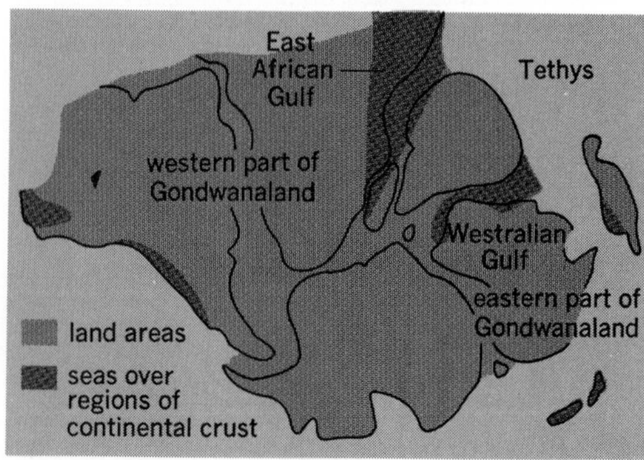

Fig. 1. Tentative reconstruction of Gondwanaland for the Callovian. (*After A. Hallam, The bearing of certain palaeozoogeographic data on continental drift, Palaeogeography, Palaeoclimatology, Palaeoecology, vol. 3, 1967*)

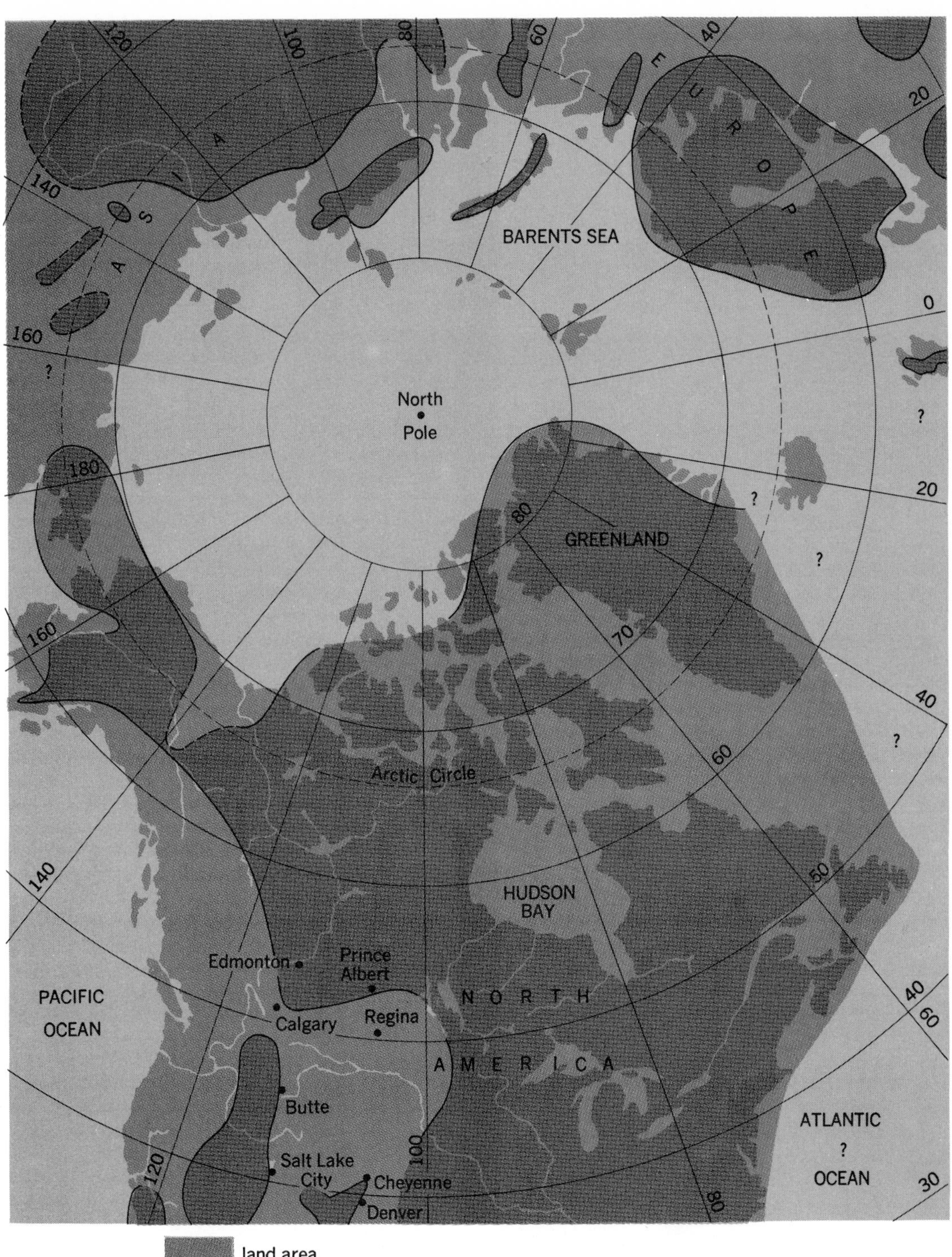

Fig. 2. Paleogeography of the boreal regions for the time of the Oxfordian of the Jurassic Period. (*Modified from H. Frebold, The Jurassic faunas of the Canadian Arctic, Geol. Surv. Can. Bull., no. 74, 1961*)

tive reconstruction of Gondwanaland for the Callovian by reassembly of Gondwanaland illustrates this hypothesis (Fig. 1). Active drifting apart of various parts of Gondwanaland did not take place before the Middle or Late Cretaceous. The distribution of faunas in the Arctic and Scandic suggest that the Arctic area was occupied by an oceanic basin in Jurassic times. The basin extended southward into the North Atlantic, where marine Jurassic sediments are now also known to be present in the offshore areas of eastern Canada. Figure 2 illustrates the paleogeography in these northern areas for the time of the Oxfordian. All known marine Jurassic deposits occur in parts of the continents, where they were deposited in comparatively shallow water. There is hardly a single epeiric (shallow inland) sea which remained in one and the same region throughout all of the estimated 55,- to 60,000,000 years of the Jurassic Period. The duration of the Jurassic marine transgressions is dependent on the character of the part of the continent which was flooded. Stable shelves, such as those in the European part of Russia, generally have fairly thin and incomplete covers of Jurassic sediments, whereas less stable shelves, such as those in northwest Europe, have considerably thicker deposits.

Great thicknesses of sediments were accumulated in the mobile belts—for example, the American Cordillera, which formed part of the circum-Pacific mobile belt. Another mobile belt, the Tethys, extended from the Mediterranean to Persia, the Himalaya, and Indonesia. Such mobile belts included were rapidly and extensively sinking troughs. The erosion of the equally fast-rising adjacent lands made vast quantities of sediments available for filling them. In spite of their rapid sinking, the depths of these geosynclinal seas were not (with some exceptions) considerably greater than those of the more stable areas, as the rapid sinking was compensated by rapid filling. Typical shallow-water deposits were common. *See* GEOSYNCLINE.

The Precambrian shields, which form the oldest and most stable parts of the Earth's crust, were land areas subject to erosion during the Jurassic Period. Thus in North America the Canadian Shield was land bordered by seas both to the north and west, and the Scandinavian Shield was surrounded by seas in Late Jurassic times (Fig. 2). Smaller areas underlain by older rocks, such as the Bohemian Mass, the French Central Plateau, the Belgian Ardennes, and the German Hartz Mountains, were lands, the coastlines of which are indicated in many places by conglomerates and sandstones (Fig 3). *See* PRECAMBRIAN.

The faunas of the Early Jurassic, particularly the ammonites, seem to have been universal, and distinct faunal realms can scarcely be recognized for this time. During the early part of the Middle Jurassic (middle Bajocian) a Pacific province seems to be indicated and, at about the same time and particularly during the Bathonian and Callovian, a boreal realm is developing in the Arctic and parts of Russia. Its ammonite faunas are clearly distinguished from those of the realm of the Tethys. These faunistic differences do not necessarily indicate any differentiation of climate, which apparently was very uniform as indicated by the fossil vegetation. Jurassic glaciations are unknown.

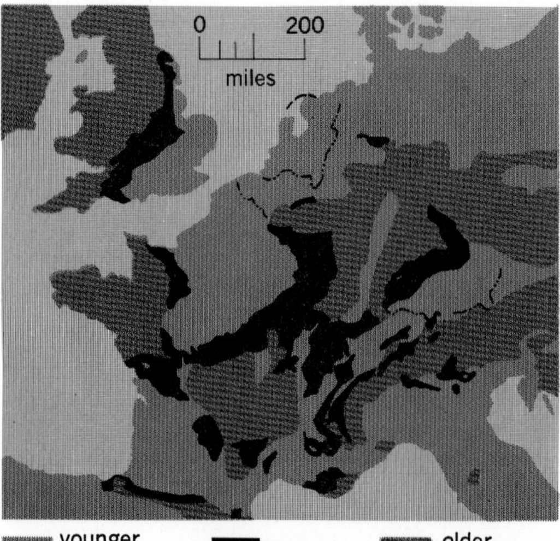

Fig. 3. Distribution of Jurassic outcrops in west central Europe. (*From R. C. Moore, Introduction to Historical Geology, 2d ed., McGraw-Hill, 1958*)

Tectonic history. The Jurassic Period was for most parts of the world a time of comparative tectonic passivity: diastrophism of a major order, connected with the intrusion of batholiths, occurred in the North American Cordillera. Mountain building connected with volcanic activity took place in the Crimea and Caucasus, and minor folding and faulting (the germano-type folding) is known from northwest Germany and other areas. In the North American Cordillera, as the Pacific plate was subducted along the western margin of the American plate, two episodes of volcanism, plutonism, and deformation occurred—the Nassian in the Middle Jurassic and the Nevadan or the Columbian in the Late Jurassic or Early Cretaceous. While the Nassian seems to have been restricted to parts of southern British Columbia, the Nevadan resulted in great uplifts throughout the mobile belt. This new land was consequently subjected to erosion, and the eroded material was deposited in the East, where it contributed to the sediments of the Lower Cretaceous.

Germanotype folding affected many parts of the world within belts of earlier folding which had not yet become entirely stable. Orogenic movements of this type occurred during the Late Triassic and perhaps earliest Jurassic (early Cimmerian phase). The late Cimmerian phase of many regions is about equivalent to the Nevadan orogeny F. *See* BATHOLITH.

Of utmost importance in the history of the Jurassic Period were the epirogenic movements. They were the cause of some transgressions and regressions of the sea, of the openings and closings of seaways, and of the migrations of faunas. They also influenced the rate and character of the sedimentation. *See* DIASTROPHISM; OROGENY; TECTONIC PATTERNS.

Life of the Jurassic. The Jurassic flora is characterized by ferns, which occur abundantly in certain regions, and by numerous gymnosperms. Among these the cycadeoids (Bennettitales), such as *Taeniopteris*, *Nilssonia*, and *Pterophyllum*, extinct palmlike plants related to the modern cycads, were dominant. Ginkgos were widely distributed; pines, cedars, and other conifers were abundant. The presence of angiosperms is indicated by pollen grains. Calcareous algae built reefs in some marine areas. The comparatively great uniformity of the Jurassic land plants does not suggest the existence of climatic zones.

Reptiles were represented by a great number of different forms. On the land the dinosaurs were the dominating forms, represented by the orders Saurischia, with hipbones comparable to those of other reptiles, and the Ornithischia, with more birdlike hipbone structure. Among the Saurischia both small and giant forms were present. *Compsognathus* was about the size of a chicken, while the bipedal carnivore *Allosaurus* attained gigantic size. *Brontosaurus*, an amphibious, four-footed, plant-eating dinosaur, reached a length of more than 60 ft. The plant-eating Ornithischia were represented by bipedal Ornithopoda and the heavily plated quadruped *Stegosaurus*. In the sea *Plesio-*

saurus, the fishlike *Ichthyosaurus*, and crocodiles, such as *Stenosaurus* (up to 18 ft long), were locally abundant; in the air the pterosaurs were represented by *Pterodactylus* and others.

Skeletons of the first bird, *Archaeopteryx*, which still had reptilian jaws and teeth and strong hip and breast girdles like the pterosaurs, were found in the Late Jurassic lithographic stones at Solenhofen, Bavaria. Practically all groups of modern fishes were present in the Jurassic. Of the ganoids, homocercal forms such as *Gyrodus*, *Lepidotus*, and *Dapedius* were abundant; the teleosts were represented by *Leptolepis* and the sharks by *Hybodus* and *Acrodus*. Of the amphibians, frogs and toads were present. In both England and North America fragments, mostly jaws and teeth of small mammals belonging to the Marsupialia, occur.

The most important marine invertebrates of the Jurassic are the ammonites, which developed many new families and genera unknown in the preceding Triassic Period. Their short vertical and wide horizontal range and their appearance in almost every facies have made them the best guide fossils of the Jurassic. Other Jurassic cephalopods are nautiloids and the very abundant belemnoids. Both gastropods and pelecypods are well represented, the latter with a number of genera (such as *Buchia*, *Gryphaea*, *Inoceramus*, and *Trigonia*) of importance for age determinations and correlations. The echinoids are abundant in certain regions, but the crinoids are represented only by a few genera, such as *Pentacrinus*. The brachiopods are restricted to a few groups. Siliceous sponges formed reefs locally, and the lime-secreting hydrozoan colonies were abundant in the Mediterranean region. Corals lived in certain regions at various times during the Jurassic. Land invertebrates are represented by such insects as flies, butterflies, and moths. Among the crustaceans, ostracods are known from both marine and fresh-water beds and are used for age determinations. *See* PALEOBOTANY; PALEONTOLOGY.

Economic products. Iron ores of Jurassic origin are present in western Europe. Those of Lorraine belong to the Middle Jurassic, whereas those in Britain occur in both the Lower and Upper Jurassic and the underlying beds of the Rhaetic in many parts of the world, including Arctic regions. Upper Jurassic coal is known from Scotland, the Lofoten Islands, and northeast Greenland. Not all of the Jurassic coal occurrences are workable under present conditions. Highly bituminous rocks and oil shales occur in western Europe. Lime and cement are made from calcareous beds in England and Germany and bricks from the clays. Lithographic stone comes from the Upper Jurassic of Solenhofen, Bavaria; many good building stones from various beds are quarried in the Jurassic limestones of western Europe. Petroleum is obtained from various Jurassic horizons in the United States, Canada, Mexico, Brazil, France, Germany, Morocco, and Saudi Arabia. [HANS FREBOLD]

Bibliography: W. J. Arkell, *Jurassic Geology of the World*, 1956; R. H. Dott, Jr., and R. L. Batten, *Evolution of the Earth*, 1971; H. Frebold, The Jurassic faunas of the Canadian Arctic, *Geol. Surv. Can. Bull.*, no. 74, 1961; A. Hallam, The bearing of certain palaeozoogeographic data on continental drift, *Palaeogeography, Palaeoclimatology, Palaeoecology*, vol. 3, 1967; H. Hoelder, Jura, in F. Lotze (ed.), *Stratigraphische Geologie*, vol. 4, 1964; M. K. Howarth, The Jurassic period, *Quart. J. Geol. Soc. London*, vol. 122S, 1964; R. W. Imlay, Paleoecology of Jurassic seas in the western interior of the United States, in H. S. Ladd (ed.), Treatise on marine ecology and paleoecology, *Geol. Soc. Amer. Mem.*, no. 67, vol. 2, 1957.

Kame

A moundlike hill of stratified glacial drift, commonly sand and gravel (see illustration). Kames range from a few meters to more than 100 m in height and are built commonly during the process of deglaciation. Some consist of sediment deposited by superglacial streams in crevasses and other openings in nearly stagnant ice, which then melted, leaving the sediment as a mound. Others were deposited as small fans or deltas built outward from ice or inward against ice which, in melting, isolated the body of sediment as an irregular mound. Some kames grade into eskers; many are closely associated with kettles. Large fields of kames are common in western New York and in many parts of New England. *See* GLACIATED TERRAIN.

Kames in southeastern Wisconsin. (*USGS*)

[RICHARD F. FLINT]

Kaolinite

The principal mineral of the kaolinite group of clay minerals. Kaolinite is composed of a single silica tetrahedral sheet and a single alumina octahedral sheet. These two units are combined so that a common layer is formed by the tips of the silica tetrahedrons and one layer of the octahedral sheet (Fig. 1). *See* CLAY MINERALS.

Kaolinite is important in the ceramics industry because of its excellent firing properties and refractoriness. It is also used extensively as a filler in rubber products and for coating and filling paper products.

Structure. In the common structural layer, two-thirds of the atoms become O instead of OH because they are shared by the silicon and aluminum. The aluminum atoms which are present occupy only two-thirds of the possible positions in the octahedral sheet and are hexagonally distributed in a single plane in the center of the sheet. All charges within the structural unit are balanced, and the formula for kaolinite is $(OH)_8Si_4Al_4O_{10}$. Any replacements within the lattice are of very small magnitude.

The sheet units of the kaolinite minerals are

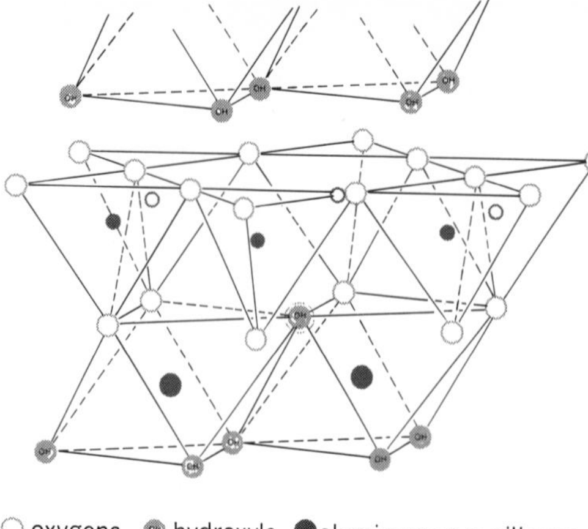

Fig. 1. Diagram of the structure of kaolinite layer. (*From R. E. Grim, Clay Mineralogy, McGraw-Hill, 1968*)

continuous in both the *a* and *b* crystallographic directions. They are stacked one above the other in the *c* direction with superposition of O and OH planes in adjacent units. These units are held firmly together by hydrogen bonding.

All kaolinites do not have the same order of crystallinity. Poorly crystallized kaolinites are disordered along the *b* axis with the unit layers randomly displaced by multiples of $b_0/3$. These poorly crystalline varieties also have a slightly higher first-order spacing than the well-crystallized mineral. Kaolinite is usually referred to as a 7-A clay mineral because its first-order basal spacing is of this approximate magnitude. Dickite and nacrite, although rarely found in clay materials, have structures similar to kaolinite. They are composed of the same structural units and differ only in the stacking of the layers. The unit cell of dickite is made up of two kaolinite layers. The nacrite unit cell is composed of six unit layers.

Morphology. Electron micrographs of well-crystallized kaolinite show well-formed, six-sided flakes, frequently with an elongation in one direction (Fig. 2). Particles with less distinct six-sided flakes have been observed in poorly crystallized

Fig. 2. Electron micrograph of kaolinite, Macon, Ga. (*From R. E. Grim, Clay Mineralogy, McGraw-Hill, 1968*)

kaolinite. The latter have ragged and irregular edges, and a very crude hexagonal outline. In general, poorly crystallized kaolinite occurs in smaller particles than the well-crystallized mineral.

Dickite particles are very well formed and have a distinct hexagonal outline. They differ from kaolinite in that they are much larger and can sometimes be studied with a light microscope.

Nacrite particles are somewhat irregular, rounded, flake-shaped units. Some of them show a crude hexagonal outline.

Properties. The mineral kaolinite has a low cation-exchange capacity (5–15 meq/100 g). Broken bonds around the edges of the silica-alumina units are the major cause of this exchange capacity. Anion-exchange capacity is also low, except when structural hydroxyls are replaced or the anion is adsorbed because of its epitaxial fit. The rate of the exchange reaction is very rapid and the exchange capacity increases as the particle size decreases.

Kaolinite contains no interlayer water between the unit-cell layers. It does, however, have the ability to adsorb water and develop plasticity. This water has a definite configuration and is referred to as nonliquid water.

When kaolinites are heated they begin to lose their OH structural water at about 400°C with the dehydration being complete at 550–600°C. The precise temperature for the loss of this OH water varies from kaolinite to kaolinite, and may be explained by variations in particle size and crystallinity. The loss of OH water in poorly crystallized kaolinite is accompanied by a fairly complete loss of structure, but in well-crystallized kaolinite some structural remnants persist until subjected to a higher temperature.

A sharp exothermic reaction occurs at 950°C in poorly crystallized kaolinite. The explanation for the exothermic reaction has been attributed to the formation of γ-Al_2O_3 by some workers, and to the nucleation of mullite by others. Recent work strongly favors the latter explanation.

Formation. The kaolinite type of mineral forms under acid conditions at low temperatures and pressures. It can form under hydrothermal conditions or through the alteration of other clay minerals. Kaolinite may form from any constituents if the alkalies and alkaline earths are removed as fast as they are liberated from the parent rock or if the environment is acid and the temperature is moderate. It is a principal component of lateritic-type soils. When calcium is present in the environment, the formation of kaolinite is retarded.

The present-day marine environment does not favor the formation of kaolinite. The presence of a kaolinitic marine sediment is evidence of a kaolinitic source area, since this mineral does not form in the sea. It is also indicative of relatively rapid accumulation of material.

[FLOYD M. WAHL; RALPH E. GRIM]

Karst topography

Characteristic associations of minor, third-order, destructively developed land features resulting from subaerial and underground solution of limestone under conditions of humid climate. These pattern features are progressively carved into second-order structural forms, such as plains, plateaus, or even hilly and mountainous uplands, containing limestone layers at or near the surface. Virtually all limestone formations are products of biological or chemical precipitation of calcium carbonate, $CaCO_3$, from aqueous solution. They are therefore susceptible in varying degrees to resolution in surface waters and groundwaters, particularly when such waters are charged with carbon dioxide, CO_2, from vegetative sources or when the waters are under pressure. *See* LIMESTONE.

Favoring combinations of the factors involved produce a landscape unlike those resulting from other agents of land sculpture. Where a karst topography has developed, most surface water disappears quickly by entering sinkholes (see illustration) and other entrance ways into underground tubes, passages, and caverns that are enlarging because of the presence and passage of these waters. Most streams—generally all but a few trunk streams in their deeper valleys—are short and scarce; ill-defined valleys end abruptly in sinks or swallow holes. The lack of a surface stream system is reflected in hills without the linear divides and intervalley character of normally stream-eroded regions. The name karst is taken from the regional designation of a coastal zone of hills and low mountains along the northeastern shore of the Adriatic Sea. In this region of limestone rocks the landscape is nearly everywhere characterized by a predominance of solution-made topography.

Sinkhole types. Two contrasted types of sinkholes, the doline and the ponor, may be present.

Doline type. A doline is a saucer-shaped, solution-made depression, commonly called sinkhole, perhaps hundreds of feet in diameter with soil-covered slopes. It generally has no open entrance to the underground and usually does not discharge into an enterable cave. The bottom may be sufficiently clogged with waste to maintain a pond for years, then may abruptly lose all its water in one abrupt down-gulp.

Ponor type. These generally result from local failure of a solution-chamber roof, whose collapse produces the surface ponor sinkhole. This type tends to be steeply walled, commonly with bare rock, and leads into a cave large enough to have developed instability in roof rock. Coalescence of such collapse openings may eventually expose most of the former cave at the Earth's surface, with remnant portions of the roof in some places as natural bridges or tunnels. A surface valley which, by coincidence, crossed the course of an existing cave may have its stream swallowed down by formation of a ponor and thus become a sinking creek or lost river. Return of this stream water to surface drainage is likely along some major river valley. It there appears as a resurgence which, turbid and contaminated, is unlike a spring of filtered groundwater.

Soil, lapiés, and shore conditions. In a karst topography much limestone is dissolved by rainwater and soil water before runoff reaches sinkholes or streamways. Insoluble constituents like clay, sand grains, and chert nodules may remain on gentler slopes as a continually thickening cover of residual soil, named terra rossa from its red color.

Contact of such soil with the underlying limestone is generally sharp and may be very irregular because of varying structures in the rock, variations in supply of water, and varied solubility of the

Sinkholes in Meade County, Kans. Depth to the underground water is about 50 ft. (*U.S. Geological Survey*)

rock. If erosion strips off this soil mantle, a miniature karst may come to light, a solution-made surface with relief of a few inches to a few feet that is known as lapiés. Even without a soil cover, the attack of rainwater and snow water can produce a kind of lapiés.

Seawater, even when saturated with $CaCO_3$, may produce a ragged lapiés surface on limestone shorelines that are subject to constant wave wash. This may be a result of slight changes in temperature and pressure in breaking waves. Marine algae growing on such wave-swept surfaces may also be a favoring factor.

Related features. A sinkhole-riddled surface in noncalcareous rock may result from extensive groundwater solution in a subjacent limestone or dolomite. Consequent subsidence of the overlying insoluble rock into depressions occurs with solutional removal. This is commonly termed solution subsidence, the filled cavities never having been open caverns. Even the partial removal of buried bodies of gypsum and rock salt may similarly produce a solutional subsidence topography.

Cyclical landform development. The sequence of changes in landforms of karst regions, from initiation of sinkholes through stages of their increase in number, enlargement, and coalescence, with collapse of caverns, and eventual wasting away of intervening hills, leaves broad-bottomed lowlands (uvalas) and scattered surviving elevations. Although this karst cycle is still subject to theoretical debate, the concept seems valid because different stages of landform development are identifiable in various karsts of differing ages.

Noteworthy karst regions are relatively uncommon in spite of extensive limestone terrains on the lands of the Earth. The best showing in Europe are the Karst area of Yugoslavia, the Causse region of France, and areas in Greece and Andalusia. North American examples are found in northern Yucatan, in central Florida, the Appalachian Great Valley in Tennessee and Virginia, in southern Indiana, and in west-central Kentucky. Jamaica, Puerto Rico, and Cuba also have well-developed karsts. *See* CAVE; GEOMORPHOLOGY; GROUNDWATER; WEATHERING PROCESSES. [J. HARLEN BRETZ]

Bibliography: A. H. Doerr and D. R. Hoy, Karst landscapes of Cuba, Puerto Rico, and Jamaica, *Sci. Mon.*, vol. 85, no. 4, 1957; O. D. von Engeln, *Geomorphology*, 1942; B. W. Sparks, *Geomorphology*, 1960; W. D. Thornbury, *Principles of Geomorphology*, 1954.

Kernite

A hydrated borate mineral with chemical composition $Na_2B_4O_6(OH)_2 \cdot 3H_2O$. It is monoclinic with symmetry 2/m. It occurs only very rarely in crystals but is found most commonly in coarse, cleavable masses (see illustration) and aggregates. It has two perfect cleavages, {001} and {100}, at about 71° to each other, which together produce cleavage fragments that are elongated parallel to the *b* crystallographic axis. It is colorless to white; colorless and transparent specimens tend to become chalky white on exposure to air with the formation of tincalconite, $Na_2B_4O_5(OH)_4 \cdot 3H_2O$. The structure of kernite, with space group *P* 2/*a*, contains infinite chains of composition $[B_4O_6(OH)_2]^{2-}$ which run parallel to the *b* crystallographic axis. The chains are linked to each other by bonding to Na^+ and by hydroxyl-oxygen bonds.

Typical cleavage fragment of kernite, from Boron, Kern County, CA. (*Specimen from Department of Geology, Indiana University*)

Kernite was originally discovered at Boron, in the Kramer borate district, in the Mojave Desert of California. It is the second most important ore mineral, after borax, in this borate-producing region. It occurs with borax, $Na_2B_4O_5(OH)_4 \cdot 8H_2O$, and other borates such as ulexite and colemanite in a bedded series of Tertiary clays. Kernite occurs mainly near the bottom of the deposits and is believed to have formed by recrystallization from borax due to increased temperature and pressure. Kernite is also found in borate deposits in Tincalayu, in the province of Salta, Argentina. The three largest producers of crude borates are the United States, Turkey, and Argentina.

Boron compounds are used in the manufacture of glass, especially in glass wool used for insulation purposes. They are also used in soap, in porcelain enamels for coating metal surfaces, and in the preparation of fertilizers and herbicides. *See* BORATE MINERALS.

[CORNELIS KLEIN]

Bibliography: L. F. Aristarain and C. S. Hurlbut, Jr., Boron minerals and deposits, *Mineral. Rec.*, 3:165–172, 213–220, 1972; R. F. Giese, Jr., Crystal structure of kernite, $Na_2B_4O_6(OH)_2 \cdot 3H_2O$, *Science*, 154:1453–1454, 1966; C. S. Hurlbut, Jr., L. F. Aristarain, and R. C. Erd, Kernite from Tincalayu, Salta, Argentina, *Amer. Mineral.*, 58:308–313, 1973; C. S. Hurlbut, Jr., and C. Klein, *Manual of Mineralogy*, 19th ed., 1977; V. Morgan and R. C. Erd, Minerals of the Kramer borate district, California, *Mineral. Inform. Serv. (Calif. Div. Mines)*, 22:143–153, 165–172, 1969.

Kerogen

A name given to the complex organic matter present in carbonaceous shales and oil shales. It is insoluble in all common solvents but on destructive distillation yields oil, gas, and acidic and basic compounds. Kerogen is formed by the biochemical and dynamochemical conversion of plant and animal remains; both may be present in variable proportions. It is the most common form of organic carbon on Earth, and it has been estimated that there is 1000 times as much kerogen as coal. Because of its diversified origin, kerogen varies in composition, consisting of approximately 77–83% carbon, 5–10% hydrogen, 10–15% oxygen, and some nitrogen. *See* OIL SHALE.

[IRVING A. BREGER]

Kutnahorite

A rare carbonate of calcium and manganese, $CaMn(CO_3)_2$. Kutnahorite is never found in nature without some magnesium and iron substituting for manganese. Investigations of mineral synthesis show complete solid solution between calcite and kutnahorite at somewhat elevated temperature.

Kutnahorite has hexagonal (rhombohedral) symmetry and the same structure as dolomite. It is pink and has a specific gravity of 3. It is found in Czechoslovakia, Sweden, and at Franklin, N.J. *See* CARBONATE MINERALS.

[ROBERT I. HARKER]

Kyanite

A nesosilicate mineral, composition Al_2SiO_5, crystallizing in the triclinic system. Crystals are usually long and tabular and commonly occur in bladed aggregates (see illustration). There is one perfect cleavage; the specific gravity is 3.56–3.66. The hardness of kyanite, one of its interesting features, is 5 (Mohs scale) parallel to the length of the crystals but 7 across the length.

The luster is vitreous to pearly, and the color is usually a shade of blue but may be white, gray, or green. *See* SILICATE MINERALS.

Kyanite is one of three polymorphic forms of Al_2SiO_5, the other two being andalusite and sillimanite. All three forms are commonly found in pelitic (aluminous) schists and hornfelses. Experimental work on Al_2SiO_5 has made it possible to use the presence of a specific polymorph as an indicator of the temperature and pressure of formation of the rock in which the polymorph occurs. Kyanite is formed at low temperature and

Kyanite on paragonite schist, St. Gotthard, Switzerland. (*American Museum of Natural History Specimens*)

high pressure relative to andalusite and sillimanite. The transitions from one mineral to another are so sluggish that they may coexist in the same rock. *See* ANDALUSITE; SILLIMANITE.

Kyanite is characteristically found in mica schists in association with garnet, staurolite, quartz, muscovite, and biotite. Fine crystals, some of gem quality, are found at St. Gothard, Switzerland. Kyanite has been mined in India and in the United States in Georgia and North Carolina for the manufacture of highly refractory porcelain.

[CORNELIUS S. HURLBUT, JR.]

Labradorite

A plagioclase feldspar with a composition ranging from $Ab_{50}An_{50}$ to $Ab_{30}An_{70}$, where $Ab = NaAlSi_3O_8$ and $An = CaAl_2Si_2O_8$ (see illustration). In the high-

Labradorite from Labrador, Canada. (*Specimen from Department of Geology, Bryn Mawr College*)

temperature state, labradorite has albite-type structure. In the course of cooling, natural material develops a peculiar structural state which, when investigated by x-rays, shows reflections that indicate the beginning of an exsolution process sometimes accompanied by a beautiful variously colored luster (labradorizing). *See* FELDSPAR; IGNEOUS ROCKS.

[FRITZ H. LAVES]

Laccolith

A geologic name for a body of igneous rock intruded into sedimentary rocks in such a way that the overlying strata have been notably lifted by the force of intrusion. In the most simple example the laccolith is circular or oval in plan, the subjacent sedimentary rocks and the floor of the laccolith are nearly horizontal, and the roof is dome-shaped, with the sedimentary rocks arching over the dome. Type examples in the Henry Mountains, Utah (described by G. K. Gilbert in 1877), and many supposed examples subsequently described in the western United States have been shown by more detailed work to be much more complex than originally described. *See* PLUTON.

[JAMES A. NOBLE]

Lamprophyre

Any of a group of igneous rocks characterized by a porphyritic texture in which abundant large crystals (phenocrysts) of dark-colored (mafic) minerals (biotite, amphibole, pyroxene, or olivine) appear set in an aphanitic (not visibly crystalline) matrix. As a group, these dark rocks are characterized by (1) an abundance of mafic minerals in association with alkali-rich feldspar, (2) the presence of mafics both as phenocrysts and in the matrix, and (3) an abundance of mafic phenocrysts in the absence of feldspar phenocrysts. Lamprophyres are chemically unusual. They have a low silica content and a high iron, magnesium, and alkali content. *See* PHENOCRYST.

Varieties. Many varieties of lamprophyre are known, but only the more common are shown in the table.

Minette, vogesite, kersantite, and spessartite. The four most common types are minette, vogesite, kersantite, and spessartite. The first two are commonly referred to as syenitic lamprophyres, the last two as dioritic lamprophyres. Under the microscope large hexagonal plates of biotite show zonal structure with pale yellow, magnesium-rich centers and red-brown iron-rich borders. In some rocks these crystals are corroded. Slender prisms or needles of green or brown hornblende are common. Pyroxene is generally diopsidic augite, but in some rocks a titanium-rich augite is abundant. Olivine is rare in most types but is common in spessartite.

A large proportion of the rock matrix is composed of feldspar. Plagioclase (albite-andesine) forms irregular grains or poorly developed laths and may show zonal structure with calcium-rich cores and sodium-rich borders. Potash feldspar, usually orthoclase or sanidine, occurs as irregular to rectangular grains and commonly encloses other minerals as abundant tiny grains. Quartz may be a minor constituent and, when present, is interstitial. Accessory minerals include apatite, sphene, and magnetite.

Camptonite. The uncommon lamprophyre camptonite is characterized by the presence of barkevikite and labradorite or andesine feldspar. Pyroxene, biotite, and olivine may or may not be present.

Alnoite and ouachitite. These varieties are rare biotite lamprophyres. They are feldspar-free but carry melilite or feldspathoid. Alnoite is characterized by lepidomelane phenocrysts and by the presence of melilite, perovskite, olivine, and carbonate in the matrix. Ouachitite is devoid of olivine and may carry more or less glass and considerable augite as phenocrysts.

Monchiquite and fourchite. The rare types monchiquite and fourchite lack feldspar but carry more or less barkevikite in addition to augite, biotite, analcite, and glass. Monchiquite is distinguished from fourchite largely by the presence of olivine.

Alteration products. Lamprophyres are highly susceptible to weathering, and many are so completely decomposed that it is impossible to do

Common lamprophyres

Principal or diagnostic mafic	Principal feldspar		Without feldspar	
	Alkali feldspar	Plagioclase	With olivine	Without olivine
Biotite	Minette	Kersantite	Alnoite	Ouachitite
Hornblende	Vogesite	Spessartite		
Barkevikite		Camptonite	Monchiquite	Fourchite

more than approximate their original mineral composition. Common products of alteration include carbonate, chlorite, serpentine, and limonite.

Occurrence. Lamprophyres occur most commonly in small or shallow intrusives (dikes, sills, and plugs) and are frequently associated with large bodies of granite and diorite. Lamprophyre dikes may form parallel swarms or, as at Spanish Peaks, Colo., may form groups which radiate from a common center.

Formation. Lamprophyres form in a variety of ways. Some are products of direct crystallization of lamprophyric magma (rock melt); others represent older rock which has been converted by metamorphic or metasomatic action. Normal basaltic magma may be made lamprophyric by assimilation of foreign material. That such a process has operated is suggested by the presence, in some lamprophyres, of abundant foreign rock and mineral fragments. The mafic phenocrysts commonly show strong resorption, indicating they were not in equilibrium with the adjacent liquid. Early formed mafic crystals may settle out of a deep, slowly crystallizing magma. Clusters of these may be reincorporated in late, alkali-rich fraction of the melt, just before it is erupted, to form lamprophyres at higher levels.

Some normal basaltic dikes appear to have been transformed to lamprophyre after solidification. This metamorphic or metasomatic change could have been accomplished by vapors or fluids which were driven out from the deeper crystallizing portions to permeate and alter the solidified portions above. Similar emanations from deeply buried granitic masses may be channeled along dikes of basalt or diabase in the overlying rocks and convert them to lamprophyres.

Some bodies of lamprophyre which resemble dikes may not actually be intrusive. They may have formed when solution or fluids from depth moved up along fractures and reacted with the adjacent rock and converted it to lamprophyre. *See* DIABASE; IGNEOUS ROCKS; METASOMATISM.

[CARLETON A. CHAPMAN]

Landslide

The perceptible downward sliding or falling of a relatively dry mass of earth, rock, or combination of the two. The term is sometimes extended to cover related flowage movements, including earthflow, mudflow, and debris-avalanche. All belong to the family of mass movement processes which range from slow soil creep to abrupt rockfall.

Landslides have been given greatly increased attention because of a better understanding of the place of mass movement in the shaping of the landscape, and because of increase in magnitude and frequency of landslide problems as ever larger cuts and fills are made for dams, superhighways, foundations, and other engineering projects. Landslides cost the highways and railroads of the United States and Canada over $10,000,000 a year in construction and maintenance and an unknown but large additional amount in indirect losses. In the world at large many lives are lost each year directly or indirectly because of landslides, and major catastrophes sometimes wipe out whole villages.

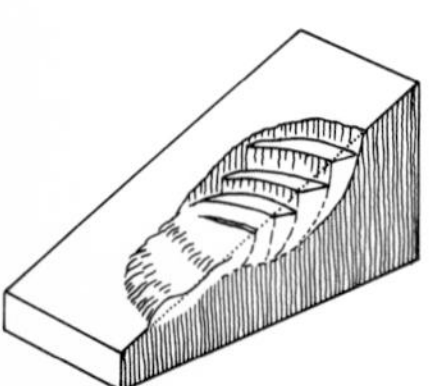

slump with earthflow toe

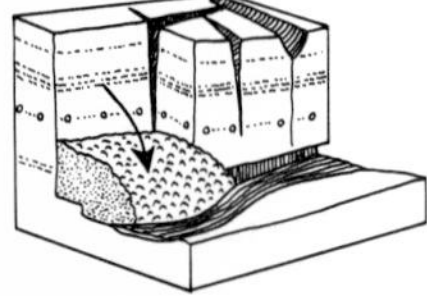

debris-fall

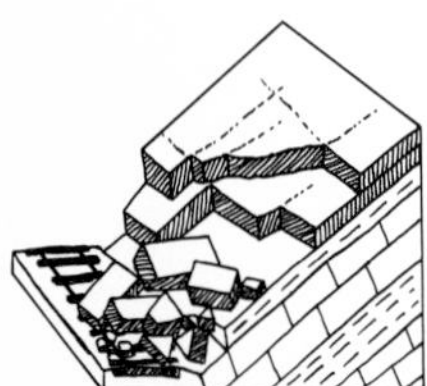

rockslide

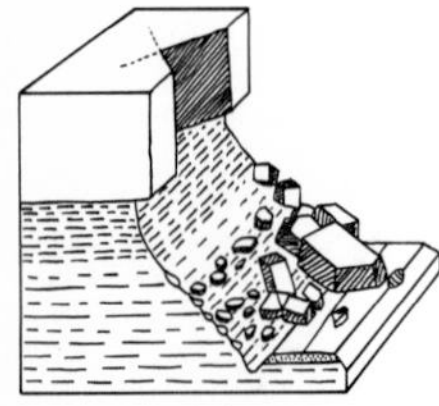

rockfall

Fig. 1. Some principal types of landslides.

Landslides, also called landslips, range from low-angle, rather slow slides to vertical falls. Some flowage may accompany sliding, for example, following initial slippage of a wet soil mass or in the very rapid outward spreading of the tongue of a large rockslide or rockfall which may consist almost entirely of dry rock. *See* EROSION; MASS WASTING.

Types. Based on type of movement, relative rate, and kind of material involved, landslides can be separated into five main types: slump, debris-slide, debris-fall, rockslide, and rockfall (Fig. 1).

Slump. The downward slipping of a mass of rock or unconsolidated debris, moving as a unit or several subsidiary units, characteristically with backward rotation on a horizontal axis parallel to the slope, is known as slump. Movement is usually rather slow and may be intermittent; displacement is small relative to the size of the mass. The major slip surface is typically spoon-shaped and concave toward the slip block in both vertical and horizontal section. The moved area at the head of a slump may be broken into many steplike, irregularly tilted blocks. A marsh, pond, or lake may form between a backward tilted block and the cliff from which it descended. Slumping usually results from removal of support lower on the slope. This may occur by natural or artificial undercutting or by flowage, either subaqueously or as an upward-bulging earthflow. Slumps are common on natural cliffs and banks and on the sides of artificial cuts and fills.

Debris-slide. This type of landslide is a rapid downward sliding and forward rolling of unconsolidated earth and rocky debris, usually with the formation of an irregular hummocky deposit. At the time of movement the material must be fairly dry or the mass would take on the characteristics of an earthflow. Debris-slides are usually rather small and often result from natural or artificial undercutting of a slope.

Debris-fall. A relatively free downward or forward falling of unconsolidated or poorly consolidated earth or rocky debris constitutes this type. Debris-falls are common along undercut banks of rivers, from walls of rapidly eroding gullies, and in steep excavations.

Rockslide. This type applies to any downward and usually rapid movement of newly detached segments of the bedrock, sliding on bedding or any other plane of separation. Rockslides may form wherever dipping strata or jointed rocks are interrupted downslope by any kind of cut. They include some of the greatest of recorded landslides. A rockslide in the valley of the Madison River in Montana accompanying the Hebgen Lake earthquake on Aug. 17, 1959 (Fig. 2), displaced more than 35,000,000 yd^3 of slide material. The slide moved a maximum of about 1300 ft vertically and extended to a distance of 3000 ft, killing 25 persons and damming the Madison River to a depth of over 150 ft.

Rockfall. The relatively free falling of a newly detached segment of bedrock of any size from a cliff or steep slope is called rockfall. Rockfalls are common along headwalls of glacial cirques and on wave-cut cliffs. They are a constant hazard on vertical rock cuts along transportation routes, where the fall of a block weighing even a few pounds may disable a vehicle or kill its occupants.

Fig. 2. Madison Canyon landslide, showing slide scar on south wall and slide debris damming the Madison River, forming a new lake, seen in the foreground. (*J. R. Stacy, USGS*)

Many of the world's largest landslides have been combinations of rockslide and rockfall. Landslides into fiords, rivers, lakes, or reservoirs sometimes produce enormous waves capable of demolishing waterfront villages or doing extensive damage even at a considerable distance. On the night of Oct. 9, 1963, what was probably the largest landslide in Europe in historical time fell from Monte Toc into the reservoir behind the Vaiont Dam, near Belluno in northern Italy. The mass of about one-third billion cubic yards of rock and earth almost filled the reservoir and dashed a 300-ft-high wall of water over the crest of the 858-ft-high concrete dam. Although the dam held, the water rushing from the mountain canyon into the valley of the Piave surged hundreds of feet up the opposite wall, wiping out almost all of the town of Longarone and a number of smaller villages down the valley, with a loss of about 2200 lives.

Prevention and control. This depends primarily on avoidance of unsuitable construction in areas of old slides or recognizable mass movement hazard. Other basic measures for prevention and control include excavation to remove fallen or unstable material; drainage of unstable or potentially unstable material to reduce weight and increase shear resistance, and to prevent additional water from gaining access to dangerously placed masses; placement of restraining structures, such as piling, buttresses, retaining walls, cribbing, and wire fences or netting, to keep fallen rocks off communication routes. Warning devices are sometimes used to close railroad blocks when fallen rocks or debris-slides enter a right of way.

[C. F. STEWART SHARPE]

Bibliography: *Landslides and Engineering Practice*, NAS-NRC Publ. 544, Highway Research Board Spec. Rept. 29, 1958; C. F. S. Sharpe, *Landslides and Related Phenomena*, 1938; reprint, 1960.

Laterite

The name given by F. Buchanan in 1807 to the iron-rich weathering product of basalt in southern India. The term is now used in a compositional sense for weathering products composed principally of the oxides and hydrous oxides of iron, aluminum, titanium, and manganese. Iron-rich or ferruginous laterite is largely hematite, Fe_2O_3, and goethite, $HFeO_2$, and may be an ore of iron and nickel (Cuba, New Caledonia). Aluminous laterite is composed of gibbsite and boehmite, and is the principal ore of aluminum. Clay minerals of the kaolin group are typically associated with, and are genetically related to, laterite. Laterites range from soft, earthy, porous material to hard, dense rock.

Concretionary forms of varying size and shape commonly are developed. The color depends on the content of iron oxides and ranges from white to dark red or brown, commonly variegated. *See* BAUXITE; CLAY MINERALS; KAOLINITE; WEATHERING PROCESSES.

Origin. Laterite is formed by weathering under conditions that lead to the removal of silica, alkalies, and alkaline earths. The resulting concentrations of iron and aluminum oxides sharply differentiate lateritization from temperate-climate weathering in which the end product is largely clay minerals (hydrous aluminum silicates). Early workers from temperate regions considered lateritization as profound leaching (desilication) in a stage beyond ordinary kaolinization. Studies in tropical regions, however, show that weathering of alkaline silicates may yield gibbsite directly without passing through an intermediate clay stage.

Investigations in many parts of the world stress certain genetic factors. A tropical to subtropical climate with high temperature and abundant rainfall, seasonal or at least with periods of marked dryness, is fundamental. Relief sufficient to ensure good drainage is requisite: lateritic soils are permeable and do not erode by sheet wash like clay or shale. Aluminous laterite forms above the water table and may grade to clay in depth. It may be found on hills and on well-drained slopes, but the residual deposits in adjacent valleys usually are kaolin.

Parent materials. The parent material controls or greatly influences the composition of laterites, which may be developed from a variety of igneous, metamorphic, and sedimentary rocks. Iron-rich rocks (peridotite) yield iron ore; aluminous rocks (syenite) produce bauxite; whereas andesite or basalt give intermediate products. Commonly textural and structural features of the parent material are preserved and the more resistant insoluble minerals remain.

Mature lateritic soils lack fertility for most systems of agriculture. Savannas or parklike grasslands are typical on laterite. Clay, not laterite, is found beneath rainforests and jungle vegetation.

[SAMUEL S. GOLDICH]

Bibliography: *See* BAUXITE.

Latite

An aphanitic (not visibly crystalline) rock of volcanic origin, composed chiefly of sodic plagioclase (oligoclase or andesine) and alkali feldspar (sanidine or orthoclase) with subordinate quantities of dark-colored (mafic) minerals (biotite, amphibole, or pyroxene). Latite is intermediate between trachyte and andesite. Plagioclase is dominant over alkali feldspar in latite but is subordinate to alkali feldspar in trachyte. Andesite carries little or no alkali feldspar. *See* ANDESITE; TRACHYTE.

[CARLETON A. CHAPMAN]

Lava

Molten rock material that reaches the Earth's surface through volcanic vents and fissures; also, the igneous rock formed by consolidation of such molten material. Relatively rapid cooling at the Earth's surface may transform fluid lava into a dense-textured volcanic rock composed of tiny crystals or glass or both. Molten rock material below the Earth's surface, however, is usually known as magma and, upon cooling, gives rise to coarse-grained igneous rock, such as granite or gabbro.

Magma and lava are mutual solutions of silicate minerals with more or less dissolved gases. When magma is brought to the surface from regions of high pressure within the Earth, the gases expand and form bubbles in the fluid lava. If this lava quickly congeals, many of these bubbles may be trapped to form a highly porous rock.

The temperature of liquid lava ranges widely but generally does not exceed 1200°C. Basaltic lavas are usually hotter than rhyolitic ones. The viscosity of lava depends largely upon the temperature, composition, and gas content. Lavas poor in silica (basaltic) are the most fluid and may flow down very gentle slopes for many miles. The Hawaiian flows advance commonly at a rate of about 2 mph. When descending the courses of steep valleys, local velocities of up to 40 mph may be attained. Silica-rich lavas (rhyolite) are highly viscous. They move slowly and for relatively short distances. As lava cools, it becomes more viscous and the rate of flow decreases. Rapid cooling, as at the surface of a flow, promotes the formation of glass. Slower cooling, as near the center of a flow, favors the growth of crystals.

During many volcanic eruptions the lava is so rapidly ejected that it is blown to bits by the explosive force of expanding gases. The small masses rapidly congeal and settle to the earth to form thick blankets of volcanic tuff and related pyroclastic rock. Lava flows and volcanic tuffs cover large areas of the Earth's surface and may form more or less alternating layers totaling many thousands of feet in thickness. *See* IGNEOUS ROCKS; MAGMA; PYROCLASTIC ROCKS; TUFF; VOLCANIC GLASS; VOLCANO. [CARLETON A. CHAPMAN]

Lazurite

The chief mineral constituent in the ornamental stone lapis lazuli. It crystallizes in the isometric system, but well-formed crystals, usually dodecahedral, are rare (see illustration). Most commonly, it is granular or in compact masses. There is imperfect dodecahedral cleavage. The hardness is 5–5.5 on Mohs scale, and the specific gravity is 2.4–2.5. There is vitreous luster and the color is a deep azure, more rarely a greenish-blue. Lazurite is a tectosilicate, the composition of which is expressed by the formula $Na_4Al_3Si_3O_{12}S$, but some S may be replaced by SO_4 or Cl. Lazurite is soluble in HCl with the evolution of hydrogen sulfide.

Lazurite is a feldspathoid but, unlike the other members of that group, is not found in igneous rocks. It occurs exclusively in crystalline limestones as a contact metamorphic mineral. Lapis lazuli is a mixture of lazurite with other silicates and calcite and usually contains disseminated pyrite. It has long been valued as an ornamental material. Lazurite was formerly used as blue pigment, ultramarine, in oil painting. Localities of occurrence are in Afghanistan; Lake Baikal, Siberia; Chile; and San Bernardino County, Calif. *See* FELDSPATHOID; SILICATE MINERALS.

[CORNELIUS S. HURLBUT, JR.]

Lazurite from Afghanistan. (*Specimen from Department of Geology, Bryn Mawr College*)

Lead isotopes, geochemistry of

The study of the isotopic composition of lead in minerals and rocks in order to relate it to past associations of the lead with uranium and thorium. Lead has four isotopes, of relative mass 204, 206, 207, and 208. Pb^{206}, Pb^{207}, and Pb^{208} are produced by the radioactive decay of uranium and thorium. The decay relations are shown in Eqs. (1), where α

$$\begin{aligned} &U^{238}(\text{half-life } 4.5\times 10^9 \text{ years})\rightarrow \\ &\qquad\qquad Pb^{206}+8\alpha+6\beta \\ &U^{235}(\text{half-life } 0.71\times 10^9 \text{ years})\rightarrow \\ &\qquad\qquad Pb^{207}+7\alpha+4\beta \\ &Th^{232}(\text{half-life } 13.9\times 10^9 \text{ years})\rightarrow \\ &\qquad\qquad Pb^{208}+6\alpha+4\beta \end{aligned} \tag{1}$$

denotes an alpha particle (doubly charged helium nucleus) and β denotes a beta particle (electron). For a given system such as a mineral or rock, primary lead is defined as the lead present in the system at the time it was formed. Radiogenic lead is lead produced in the system by the decay of uranium and thorium since it was formed. Since Pb^{204} is not produced by the decay of any naturally occurring radioactive species, it can be used as an index to determine the amount of primary lead in a system. Primary lead will include all of the Pb^{204} and variable amounts of Pb^{206}, Pb^{207}, and Pb^{208}, depending on its origin.

Common rocks, for example, granites and basalts, or a system such as the crust of the Earth have sufficiently high ratios of thorium and uranium to lead to cause measurable changes in the isotopic composition of their lead over time intervals as short as some tens of millions of years. Some minerals have practically all radiogenic lead, while others have such low ratios of uranium and thorium to lead that the isotopic composition of their lead will not have changed appreciably over thousands of millions of years. Since the isotopic composition of lead observed in rocks and minerals is related directly to the amounts of uranium and thorium the lead was associated with in the past and the length of time of the association, it is possible to gain information from coordinated studies of the geochemistry of uranium, thorium, and lead which could not be obtained from elements not involved in radioactive-decay chains. Some examples will illustrate the unique value of this type of investigation.

Variation with time. If the uranium, thorium, and lead contents and lead isotopic composition are known for a particular mineral, it is possible to calculate the isotopic composition of the lead at any time in the past if the mineral represents a closed system. A closed system is one in which the uranium-lead and the thorium-lead ratios have not been changed in the past by processes other than radioactive decay; that is, these ratios were never changed by chemical processes. It is possible in principle to treat the outer portion of the Earth as a geologic system and to study the isotopic composition of its lead as a function of time. Meaningful results can be obtained if the uranium-lead and thorium-lead ratios are reasonably uniform throughout the system or if sampling is extensive enough to average out the local variations which may exist. In either case it is possible then to determine the uranium-lead and thorium-lead ratios of the system producing the lead.

This time variation has been studied in two different sources, potassium feldspar and galena. Potassium feldspar ($KAlSi_3O_8 + NaAlSi_3O_8$) is a common mineral in igneous rocks, such as granites and pegmatites. Galena, PbS, commonly occurs in veins of hydrothermal origin. Both minerals have such low uranium-lead and thorium-lead ratios that the isotopic composition of their lead will not have changed appreciably since the time of crystallization, providing there has been no contamination from external sources. Although potassium feldspar contains relatively small amounts of lead (5–100 ppm), it has the great advantage that the time of formation of the mineral can be accurately determined by the rubidium-strontium method or by measuring the age of cogenetic minerals, such as mica, uraninite, or zircon. On the other hand, the veins from which galena is obtained do not as a rule contain minerals which are suitable for age determination, so that the time of formation of a galena must be inferred from general geological relations. Since lead is easily obtained from galena, many more data exist for it than for igneous minerals, although the data given by the latter can be interpreted with greater certainty. *See* ROCK, AGE DETERMINATION OF.

The isotopic composition of lead has been determined in about 50 potassium feldspars of known age, covering a time span of 3×10^8 to 27×10^8 years. Representative data are illustrated in Fig. 1, in which the ratios of Pb^{206}, Pb^{207}, and Pb^{208} to Pb^{204} are plotted as a function of age. The experimental uncertainties in the ratios are about 1% or less; thus the variations are many times the errors. It is seen that the more recently a feldspar was formed, the more Pb^{206}, Pb^{207}, and Pb^{208} it contains with respect to Pb^{204}, indicating an accumulation of radiogenic lead in the source environment.

All points in Fig. 1 represent feldspars except those at 0 and 4.5×10^9 years. The points at the present (modern primary lead) represent the range in isotopic composition of lead from three basalts, a gabbro, red clay from the Pacific Ocean, and 26 manganese nodules from the Atlantic and Pacific oceans. The basalts and the gabbro probably originated from partial fusion of silicate minerals at great depths in the Earth. Defining the crust as the region above the Mohorovičić discontinuity, which has a depth of about 35 km under the continents and 10 km under the oceans, and the mantle as the region between the Mohorovičić discontinuity and the boundary of the core at 2900 km, basalts and gabbros originate from the deep crust or outer mantle. The red clay and manganese nodules, on the other hand, represent a sampling of lead from a large number and variety of sources on the surface of the Earth. Their lead may thus represent a reasonable average for the isotopic composition of lead at the Earth's surface today. Note that the oceanic and basaltic leads have quite similar isotopic compositions, indicating that average surface lead is not greatly different in isotopic composition from lead of deep-seated origin.

The points at 4.5×10^9 years represent lead in troilite, FeS, from two iron meteorites. The troilite contains so little uranium with respect to lead that the isotopic composition of its lead would not have changed measurably in the last 4.5×10^9 years, the presently accepted age of the Earth. The data from igneous rocks are indeed compatible with the assumption that the isotopic composition of terrestrial lead 4.5×10^9 years ago is represented by the troilite lead. That is, starting 4.5×10^9 years ago

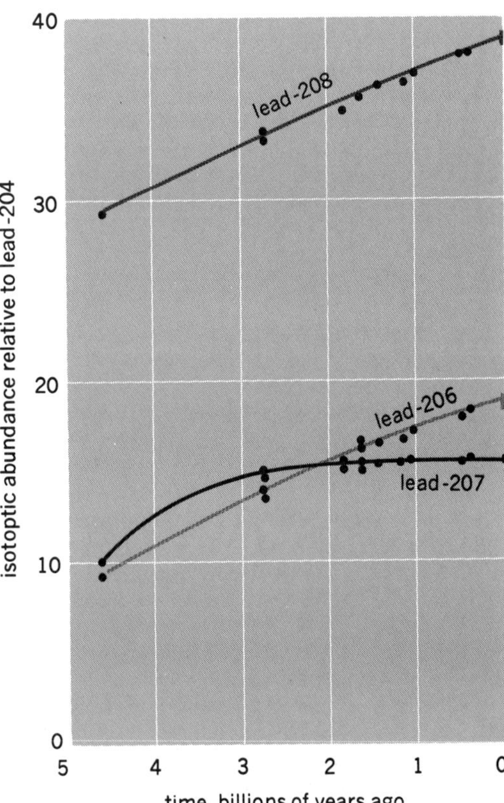

Fig. 1. Isotopic composition of lead in igneous rocks plotted as a function of age.

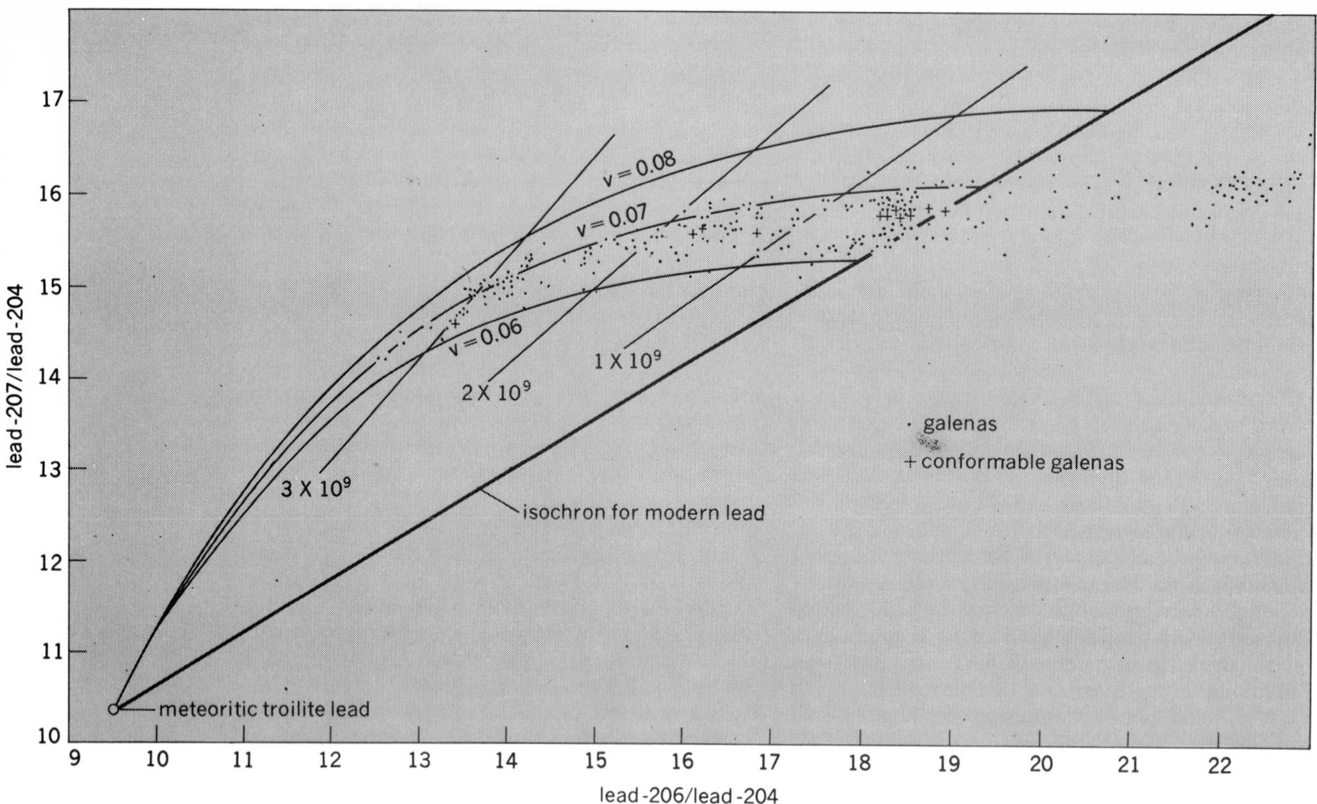

Fig. 2. Pb^{206}/Pb^{204} as a function of Pb^{207}/Pb^{204} for lead from galenas. Conformable galenas are defined in the text. The age of the Earth is taken as 4.55×10^9 years. Diagonal lines are isochrons giving model ages in years; $v = U^{235}/Pb^{204}$ (atom ratio) at the present time in hypothetical sources of ores. The growth curves are calculated for various values of v, assuming closed systems prior to the time of ore formation.

with lead of the same isotopic composition as troilite lead, it is possible to select values for the uranium-lead and thorium-lead ratios which will give growth curves for the lead isotopes which fit the observed data quite closely. The calculated growth curves in Fig. 1 correspond to present-day ratios in the source environment shown in Eqs. (2).

$$\begin{aligned} U^{238}/Pb^{204} &= 9.3 \text{ (atom ratio)} \\ U/Pb &= 0.18 \text{ (weight ratio)} \\ Th^{232}/U^{238} &= 4.0 \text{ (atom ratio)} \end{aligned} \qquad (2)$$

This simple model assumes a closed-system source environment in the sense defined above. *See* EARTH, AGE OF; METEORITE.

It is not entirely clear just what source material the feldspar lead characterizes. The feldspars occur in pegmatites which were associated with processes of mountain building in the past. The lead may have been derived either from lead brought toward the surface from the mantle, from sediments which were buried and melted or partially melted during the process, or from both of these sources. However, it has been shown above that lead in rocks of deep-seated origin, such as basalts, does not differ greatly in isotopic composition from lead in oceanic materials which are of surface origin and have undergone erosion. Thus the curves in Fig. 1 approximate the past variation of isotopic composition of lead in both the outer mantle and average crustal material.

Model age. The isotopic composition of lead in several hundred galenas has been published. Some difficulty arises in determining the time of formation of galena, as already mentioned; accordingly, a different method of plotting these leads is used. If Pb^{206}/Pb^{204} is plotted against Pb^{207}/Pb^{204} from the curves in Fig. 1, a new curve will result on which each point corresponds to a model age for a particular lead. The model age is the time at which the lead appears to have been separated from uranium in the source environment. This treatment may be expanded to allow for varying uranium-lead ratios in the source environment. Assume that a group of galenas originated at a time $t \times 10^6$ years ago from source environments which all had the isotopic composition of meteoritic troilite lead 4.5×10^9 years ago and which were closed systems with differing uranium-lead ratios. Since the ratio of radiogenic Pb^{207} to radiogenic Pb^{206} will be the same in all the leads, regardless of the uranium-lead ratios in the sources, a plot of Pb^{207}/Pb^{204} against Pb^{206}/Pb^{204} for these leads will yield a straight line. Such a line of constant model age and different uranium-lead ratios is called an isochron. Figure 2 shows isochrons for modern lead and for leads which were separated from uranium 1×10^9, 2×10^9, and 3×10^9 years ago. Knowledge of the Pb^{207}/Pb^{204} and Pb^{206}/Pb^{204} ratios for any lead will determine its model age.

The model age is based on a number of assumptions, the most stringent being the requirement of a closed system in the source environment up to the time of separation of the lead and absence of contamination of the lead with lead of a different age after separation. It is not surprising that these conditions are not always met (see following section in this article on anomalous leads). In favorable cases, model ages agree with the ages of supposedly contemporaneous igneous rocks within $\pm 100 \times 10^6$ years, about the limit of accuracy of the method. The method is better suited for distinguishing between two leads of greatly differing age than for determining precise ages.

The Pb^{206}/Pb^{204} and Pb^{207}/Pb^{204} ratios for approximately 200 galena specimens taken from all continents except Antarctica are shown in Fig. 2. To a first approximation the galenas plot on a closed-system growth curve, as do the feldspar leads in Fig. 1. The U^{238}/Pb^{204} ratio of 9.3 used in Fig. 1 corresponds to a value of 0.067 for v in Fig. 2. In general there is a close relationship between the isotopic composition of lead in igneous rocks and in lead ores of the same age. In fact, detailed studies in several localities have shown that the isotopic composition of lead in ores and in neighboring, approximately contemporaneous igneous rocks are nearly identical. It is striking that the galena leads form such a closely defined curve, since they represent random sampling on a worldwide scale.

The conformable galenas, delineated in Fig. 2, show much less scatter about a single closed-system growth curve than do the remaining samples, and deserve special mention. Conformable galenas are recognized from geological criteria and are believed to be derived from volcanic activity in island-arc environments off the margins of continents. The Aleutian Islands of Alaska are an example of such an environment at the present time. If the theory is correct, lead in conformable galenas is the same as that in the igneous rocks formed by the volcanism. Whether the theory is correct or not, the conformable galenas closely fit a single closed-system growth curve. Their model ages agree with the estimated times of emplacement within 100×10^6 years in all cases—substantially better agreement than is obtained from ores in general. It is widely believed that the igneous rocks in island arcs originate in the outer mantle of the Earth, because in these areas it is difficult to generate enough heat within the crust to melt rocks. For this reason the conformable galenas probably record the evolution of the isotopic composition of lead in some portion of the Earth's outer mantle.

Anomalous leads. These leads give model ages which are clearly incorrect. For instance, if the model used to determine ages is correct, no leads should be found to the right of the modern lead isochron in Fig. 2 (negative model age). Figure 2 does not cover the full range of anomalous leads; Pb^{206}/Pb^{204} ratios as high as 93 and Pb^{207}/Pb^{204} ratios as high as 24 have been observed. These extreme cases are rare, however. Anomalous leads are generally characterized by high Pb^{206}/Pb^{204} and sometimes by high Pb^{208}/Pb^{204} ratios as well. Pb^{207}/Pb^{204} may not be high since the abundance of U^{235} is presently so low that only small amounts of Pb^{207} with respect to Pb^{206} have been made over the past 2×10^9 years. Note that the Pb^{207}/Pb^{204} curve in Fig. 1 is nearly flat over that time interval.

A significant feature of anomalous leads is that the isotopic composition varies considerably from one sample to another. The Mississippi Valley leads of Arkansas, Missouri, Illinois, Iowa, and Wisconsin have variable Pb^{206}/Pb^{204} ratios; they range from 18.5 to 23. Several examples are known of variable isotopic composition in different parts of the same mine. Nonanomalous lead, on the other hand, does not display this variation, giving almost the same isotopic composition throughout a mine or even a geologic province. Thus, some sort of localized mixing is indicated to be the cause of anomalous lead.

An occurrence of anomalous lead at Lake Athabaska, Saskatchewan, has been well studied; in this article it will be used to illustrate a solution to the problem. Uraninite in a pegmatite in the district is reliably dated by concordant lead ages at 1.9×10^9 years. Pitchblende of hydrothermal origin occurs widely, and isotopic ages, all discordant, have been determined for 28 of these. A systematic study of the discordant pitchblende ages indicates that they could have been formed from a 1.9×10^9-year-old uraninite by dissolving parts of it 1.1×10^9 to 1.2×10^9 years ago and again about 0.2×10^9 years ago, transporting the uranium hydrothermally to new sites in veins and losing variable fractions of the accumulated radiogenic lead in the process. Three kinds of lead are found in lead ores, an ordinary lead and two kinds of anomalous lead. Examples of their isotopic compositions are given in Table 1.

The ordinary galena may be used as a basis to calculate the ratios of excess or radiogenic Pb^{207} and Pb^{206}, the values of which appear in Table 1. Also, it has been calculated that a uranium mineral formed 1.9×10^9 years ago would contain radiogenic lead with a Pb^{207}/Pb^{206} ratio of 0.17, 1.2×10^9 years ago and 0.12, 0.2×10^9 years ago. Thus, the uraninite, pitchblendes, and lead ores tell a consistent story: Pitchblendes were formed 1.2×10^9 and 0.2×10^9 years ago from 1.9×10^9-year-old uranium minerals, probably uraninite. Varying proportions of highly radiogenic lead separated from uranium during the formation of the pitchblendes, the freed radiogenic lead finding its way into lead ores which now contain anomalous lead. The mixing of radiogenic lead with ordinary lead might be expected to have varied considerably on a local scale, which fits observation.

Other occurrences of pitchblendes together with lead ores having anomalous lead are the Colorado Plateaus and the Blind River district in Ontario, north of Lake Huron. These districts have not been studied as completely as that at Lake Athabaska, but a similar mechanism undoubtedly applies. The source of radiogenic lead for the anomalous Mississippi Valley leads has not yet been identified. No pitchblendes are observed, and the

Table 1. Isotopic composition of lead from the Lake Athabaska district, Saskatchewan

	Atom ratios			
Type	Pb^{206}/Pb^{204}	Pb^{207}/Pb^{204}	Pb^{208}/Pb^{204}	Radiogenic Pb^{207}/Radiogenic Pb^{206}
Ordinary	14.36	14.96	34.49	
Anomalous	40.01	19.36	37.10	0.17
Anomalous	43.5	18.7	35.7	0.12

Pb^{208}/Pb^{204} ratios are high, along with the Pb^{206}/Pb^{204} ratios. The radiogenic lead in the galena is probably derived from the Precambrian (1.4×10^9 years old) granitic rocks in the area. Such a source is plausible according to data given in the following section.

Lead isotopes in a granite. Another application of isotopic-lead data to geochemical problems is illustrated by a study of the distribution of uranium, thorium, and lead isotopes in minerals of a Precambrian granite from the Canadian Shield, collected near Tory Hill, Ontario. (Precambrian rocks are those having ages greater than 6×10^8 years). Zircon contained lead which was entirely radiogenic, enabling the mineral to be dated accurately at 1.05×10^9 years. Perthite, plagioclase, and quartz contained lead which appeared to be entirely primary. Determination of the uranium-lead and thorium-lead ratios in perthite indicated that the ratios were so low that the isotopic composition of the lead would not have changed appreciably in the last 10^9 years if the mineral was a closed system. The other minerals studied, sphene, apatite, and magnetite, all had mixtures of primary and radiogenic lead. The composite rock was analyzed for uranium, thorium, and lead concentration and lead isotopic composition.

From the age of the rock, it is possible to make material balance calculations to study possible migrations of lead, uranium, and thorium within the rock. If each mineral has been a closed system, then the isotopic composition of lead calculated for the rock 1.05×10^9 years ago from the present-day uranium, thorium, and lead data should be the same as that found in the feldspars. This comparison is shown in Table 2. It is obvious that some type of migration has occurred. The leads in the feldspars have model ages of about 10^8 years, which suggests that they are contaminated with radiogenic lead and are anomalous. A large crystal of perthite from a neighboring pegmatite was found to contain ordinary lead with a model age of about 10^9 years. Several galenas with similar lead are also known from the district.

The uranium-lead ratio found for the granite gives calculated Pb^{206}/Pb^{204} and Pb^{207}/Pb^{204} ratios in the rock 1.05×10^9 years ago, which agree rather closely with the isotopic composition of lead in the feldspar crystal in a pegmatite. This could indicate that the rock as a whole closely approximated a closed system since it was formed, with respect to uranium, Pb^{206}, and Pb^{207}. However, a similar test of the Pb^{208}/Pb^{204} ratio shows that the granite cannot have been a closed system for thorium and Pb^{208}. The close balance calculated for uranium and lead may thus be accidental. The fact that lead from both feldspars in the granite has an isotopic composition close to that of an ordinary modern lead is further suggestive of external contamination. At any rate, the lead in the feldspars is anomalous, although it is not clear whether the source of contamination was from within the granite or outside it. Only two granites have been studied in this manner by use of lead isotopes. However, many analogous studies based on the radioactive decay of rubidium—87 to strontium—87 have been made. Many times these studies can show whether or not the rubidium and strontium in granites have been subject to contamination from external sources.

Table 2. Isotopic composition of lead in a granite and an associated pegmatite

	Atom ratios		
Source of lead	Pb^{206}/Pb^{204}	Pb^{207}/Pb^{204}	Pb^{208}/Pb^{204}
Granite, today	20.3	15.7	48.7
Granite, 1.05×10^9 years ago	16.4	15.4	30.0
Granite, perthite	18.6	15.7	39.5
Granite, plagioclase	18.2	15.5	40.0
Pegmatite, perthite crystal	16.8	15.3	36.0

Note that the present-day lead in the granite is anomalous; that is, it has a negative model age when plotted in an isochron diagram such as Fig. 1. All Precambrian granites for which the isotopic composition of lead is known contain anomalous lead. This indicates that granites have higher uranium-lead ratios than that which exists in the source environment which produced the lead in the galenas and potassium feldspars plotted in Figs. 1 and 2. Granites are possible sources of anomalous lead if a mechanism exists for concentrating their lead into ore bodies or pegmatites.

Lead isotopes and magmatic differentiation. Isotopic lead studies have been used to demonstrate a genetic relation between several different rock types in the southern California batholith. Gabbro, tonalite, granodiorite, and granite were analyzed. The Pb^{206}/Pb^{204} ratios varied from 18.72 in the gabbro to 19.44 in the granite. The age of the batholith is accurately known to be 10^8 years. Treating each of the above rocks as closed systems and calculating the isotopic composition of the lead 10^8 years ago from the uranium-lead ratios, it was found that all of the Pb^{206}/Pb^{204} ratios agree within limits of error at 18.6 to 18.7. Thorium was not determined, so that similar calculation for Pb^{208}/Pb^{204} ratios could not be made. The data suggest that this sequence of rocks originated from a common reservoir in which the isotopes of lead were homogeneously distributed. The rubidium-87 – strontium-87 system can also be applied to this kind of problem.

[GEORGE R. TILTON]

Bibliography: H. Brown, The age of the solar system, *Sci. Amer.*, April, 1957; H. Faul, *Ages of Rocks, Planets and Stars*, 1966; R. D. Russell and R. M. Farquhar, *Lead Isotopes in Geology*, 1960; G. R. Tilton et al., Isotopic composition and distribution of lead, uranium and thorium in a Precambrian granite, *Geol. Soc. Amer. Bull.*, 66(9):1131 – 1148, 1955.

Lepidolite

A mineral of variable composition that is also called lithium mica and lithionite, $K_2(Li,AL)_{5-6}$-$(Si_{6-7},Al_{2-1})O_{20-21}(F,OH)_{3-4}$. Rubidium, Rb, and cesium, Cs, may replace potassium, K; small amounts of Mn, Mg, Fe(II), and Fe(III) normally are present; and the OH/F ratio varies considerably. Polithionite is a silicon- and lithium-rich, and thus aluminum-poor, variety of lepidolite.

Lepidolite is uncommon, occurring almost exclusively in structurally complex granitic pegmatites, commonly in replacement units. Common associates are quartz, cleavelandite, alkali beryl,

Group of lepidolite crystals found in Zinnwald, Czechoslovakia. *(Specimen from Department of Geology, Bryn Mawr College)*

and alkali tourmaline. Lepidolite is a commercial source of lithium, commonly used directly in lithium glasses and other ceramic products. Important deposits occur in the Karibib district of South-West Africa and at Bikita, Rhodesia.

The structural modifications show some correlation with lithium content: the six-layer monoclinic form contains 4.0–5.1% Li_2O; the one-layer monoclinic, 5.1–7.26% Li_2O. A three-layer hexagonal form is also found. There is a compositional gradation to muscovite, intermediate types being called lithian muscovite, containing 3–4% Li_2O, and having a modified two-layer monoclinic muscovite structure.

Lepidolite usually forms small scales or fine-grained aggregates (see illustration). Its colors, pink, lilac, and gray, are a function of the Mn/Fe ratio. It is fusible at 2, yielding the crimson (lithium) flame. It has a perfect basal cleavage. Hardness is 2.5–4.0 on Mohs scale; specific gravity is 2.8–3.0. *See* MICA; SILICATE MINERALS.

[E. WILLIAM HEINRICH]

Leucite

A framework structure silicate of the feldspathoidal mineral group. It is commonly found as euhedral phenocrysts in K_2O-rich and SiO_2-poor volcanic rocks, consistent with its chemical composition, $(K,Na)AlSi_2O_6$, where K is more abundant than Na. The crystals are typically white, and the luster varies from dull to vitreous. Mohs hardness lies between 5.5 and 6. Density ranges from 2.45 to 2.5. Crystals, exhibiting the characteristic isometric trapezohedral form (see illustration), as large as 7 cm are common in the lavas of Nyiragongo volcano, Congo. Other well-known areas include: West Kimberley, Australia; Mount Vesuvius, Italy; Highwood Mountains and Bearpaw Mountains, Montana; and the Leucite Hills, Wyoming.

At low temperatures leucite is tetragonal; heating to 625°C results in a gradual change to the isometric symmetry consistent with the external morphology. This is presumably due to structural collapse of the (Al,Si)-O framework around the potassium ions which are not large enough to fill the cavities in the structure under low-temperature conditions.

Leucite found in the natural environment is usually nearly pure $KAlSi_2O_6$. Experimental phase equilibrium studies have demonstrated that large amounts of Na can substitute for K at high temperatures. If cooled rapidly, these crystals can persist metastably and finally react to form pseudoleucite, a mixture of nepheline (nearly $[Na_{.75}K_{.25}]AlSiO_4$) and potassium feldspar ($KAlSi_3O_8$). Pseudoleucite may also form through reaction of magmatic leucite crystals with residual silicate liquid, suggested by the existence of crystals that have cores of leucite and rims of pseudoleucite. Leucite, as pure $KAlSi_2O_6$, is not thermodynamically stable at low temperature and high pressure, a mixture of potassium feldspar and kalsilite ($KAlSiO_4$) being favored. The boundary between the fields of leucite and kalsilite plus potassium feldspar is defined by the equation $P = 3.230T - 1583$, where pressure, P, is in kilobars, and the temperature, T, is in degrees celsius, with leucite existing only on the high-temperature and low-pressure side of the boundary. This observation, based on experimental data, is consistent with the absence of leucite in plutonic rocks that have crystallized at relatively high pressures.

[WILLIAM LUTH]

LEUCITE

(a)

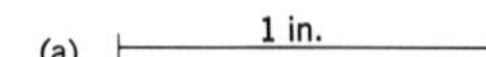

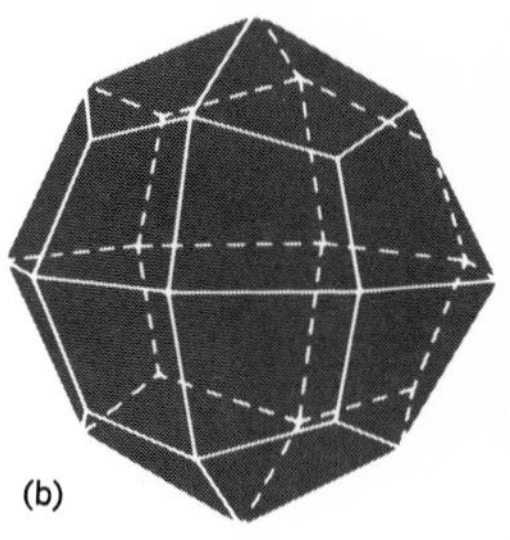

(b)

Leucite. (*a*) Crystals from Wiesenthal, Saxony (*specimen from Department of Geology, Bryn Mawr College*). (*b*) Crystal habit (*from C. S. Hurlbut, Jr., Dana's Manual of Mineralogy, 17th ed., copyright © 1959 by John Wiley & Sons, Inc.; reprinted by permission*).

Leucite rock

Igneous rocks rich in leucite but lacking or poor in alkali feldspar. Those types with essential alkali feldspar are classed as phonolites, feldspathoidal syenite, and feldspathoidal monzonite. The group includes an extremely wide assortment both chemically and mineralogically.

The rocks are generally dark-colored and aphanitic (not visibly crystalline) types of volcanic origin. They consist principally of pyroxene and leucite and may or may not contain calcic plagioclase or olivine. Types with plagioclase in excess of 10% are called leucite basanite (if olivine is present) and leucite tephrite (if olivine is absent). Types with 10% or less plagioclase are called leucitite (if olivine is absent) and olivine leucitite or leucite basalt (if olivine is present).

The texture is usually porphyritic with large crystals (phenocrysts) of augite and leucite in a very fine-grained or partly glassy matrix. If plagioclase occurs as phenocrysts, it is generally labradorite or bytownite and is slightly more calcic than that of the rock matrix. It may be zoned with more calcic cores surrounded by more sodic margins. Leucite appears in two generations. As large phenocrysts it forms slightly rounded to octagonal grains with abundant tiny inclusions of glass or other minerals zonally arranged. Small, round grains of leucite with tiny glass inclusions also occur in the rock matrix. Augite or diopside (sometimes rimmed with aegirine-augite) and aegirine-augite form the mafic phenocrysts. Pyroxene of the matrix is commonly soda-rich. Olivine may occur as well-formed phenocrysts. Other minerals present may include nepheline, sodalite, biotite, hornblende, and melilite. Accessories include sphene, magnetite, ilmenite, apatite, and perovskite.

Leucite rocks are rare. They occur principally as lava flows and small intrusives (dikes and volcanic plugs). Well known are the leucite rocks of the Roman province, in Italy, and the east-central African province. In the Italian area the feldspathoidal lavas are essentially leucite basalts and may have developed by differentiation of basalt magma (rock melt). In the African province the leucitic rocks are associated with ultramafic (peridotite) rocks and may have been derived from peridotitic material. This may have been accomplished by the abstraction of early formed crystals from a peridotitic magma or by the mobilization of peridotitic rocks by emanations from depth. Assimilation of limestone by basaltic magma may help to decrease the silica content and promote the formation of leucite instead of potash feldspar. The crystallization of leucite, however, is in large part a function of temperature and water content of the magma. Conditions of formation, therefore, may strongly influence the formation of leucitic rocks. *See* IGNEOUS ROCKS; NEPHELINE SYENITE; PETROGRAPHIC PROVINCE; PHONOLITE.

[CARLETON A. CHAPMAN]

Life, origin of

The origin of life refers to the means by which living organisms arose on the Earth, which contained no life when it was formed. While there is little difficulty in telling whether a higher organism is alive, there is no agreement as to what characteristics would be required for the most primitive organisms in order to call them living. The most outstanding characteristic of living organisms is their ability to reproduce. Therefore, the definition used here for a living organism will be an entity which is capable of making a reasonably accurate reproduction of itself, with the copy being able to accomplish the same task. Furthermore, this organism must be subject to a low rate of changes (mutations) that are transmitted to its progeny.

Spontaneous generation. Until the 19th century it was almost universally held that life could arise spontaneously, as well as by sexual or nonsexual reproduction. The appearance of various insects and animals from decaying organic materials in the presence of warmth or sunlight was a common observation, but this observation was misinterpreted to mean that the insects and animals arose spontaneously.

The Tuscan physician Francesco Redi performed the first effective experiments (1668) to disprove the hypothesis of spontaneous generation. He showed that the development of maggots in meat did not occur when the flask containing the meat was covered with muslin so that flies could not lay their eggs on the meat. Lazzaro Spallanzani performed similar experiments (1765) showing that microorganisms did not appear in various nutrient broths if the vessels were sealed and boiled. Objections were raised that the heating had destroyed in the broth and air the "vital force" which was postulated as necessary for life to develop. By readmitting air it was possible to show that the broth could still support the growth of microorganisms. But Spallanzani could not demonstrate that the air in the sealed flask had not been altered, and the doctrine of spontaneous generation persisted widely.

The problem was finally solved by Louis Pasteur (1862) who used a flask with broth, but instead of sealing the flask, drew out a long S-shaped tube with its end open to the air. The air was free to pass in and out of the flask, but the particles of dust, bacteria, and molds in the air were caught on the sides of the S-shaped tube. When the broth in the flask was boiled and allowed to cool, no microorganisms developed, but when the S-shaped tube was broken off at the neck of the flask, microorganisms developed. These experiments were extended by Pasteur and by J. Tyndall to answer all objections that were raised, and the doctrine of spontaneous generation was finally disproved.

Theories. Shortly before 1858, Charles Darwin and A. R. Wallace had published, simultaneously and independently, the theory of evolution by natural selection. This theory could account for the evolution from the simplest single-celled organism to the most complex plants and animals, including man. Therefore, the problem of the origin of life was no longer how each species developed, but only how the first living organism arose. To answer this question a number of proposals have been advanced.

It has been proposed that life was created by a supernatural event. This has been a common belief of many people based on a literal interpretation of the first chapter of Genesis, which describes the creation of all living organisms by a direct act of God. This type of proposal is not considered by most scientists since it is not subject to scientific investigation.

In 1903 S. Arrhenius offered a second theory, that life developed on the Earth as a result of a spore or other stable form of life coming to the Earth in a meteorite from outer space or by the pressure of sunlight driving the spores to the Earth. One form of this theory assumes that life had no origin, but like matter, has always existed. The presence of long-lived radioactive elements shows that the elements were formed anywhere from 5×10^9 to 12×10^9 years ago. If the elements have not always existed, it is difficult to understand how life could have always existed. Another form of this theory assumes that life was formed on another planet and traveled to the Earth. This hypothesis does not answer the question of how life arose on the other planet. In addition, most scientists doubt that any known form of life could survive for very long in outer space and fall through the Earth's atmosphere without being destroyed. Therefore, while this theory has not been disproved, it is held to be highly improbable.

A third hypothesis held that the first living organism arose from inorganic matter by a very improbable event. This organism, in order to grow in an inorganic environment, would have to synthesize all of its cellular components from carbon dioxide, water, and other inorganic nutrients. Present knowledge of biochemistry shows that even the simplest bacteria are extremely complex and that the probability of the spontaneous generation of a cell from inorganic compounds in a single event is much too small to have occurred in the approximately 5×10^9 years since the Earth's formation.

A more plausible proposal is that life arose spontaneously in the oceans of the primitive Earth, which contained large quantities of organic compounds similar to those which occur in living organisms. This theory was outlined mainly by A. I. Oparin in 1938 and forms the basis of most of the present ideas on the origin of life. Oparin argued that, if large amounts of organic compounds were in the oceans of the primitive Earth, these organic compounds would react to form structures of greater and greater complexity, until a structure would form which would be called living. In other words, the synthesis of the first living organism would involve many nonbiological steps, and none of these steps would be highly improbable.

Oparin also proposed that organic compounds might have been formed on the primitive Earth if there were a reducing atmosphere of methane, ammonia, water, and hydrogen, instead of the present oxidizing atmosphere of carbon dioxide, nitrogen, oxygen, and water. In 1952 H. C. Urey placed the reducing-atmosphere theory on a firm foundation by showing that methane, ammonia, and water are the stable forms of carbon, nitrogen, and oxygen if an excess of hydrogen is present. Cosmic dust clouds, from which the Earth is believed to have been formed, contain a great excess of hydrogen. The planets Jupiter, Saturn, and Uranus are known to have atmospheres of methane and ammonia. Oxidizing conditions have developed on Mercury, Venus, Earth, and Mars due to the escape of hydrogen followed by the production of oxygen by the photochemical splitting of water. There has not been sufficient time for the hydrogen to have escaped from Jupiter, Saturn, Uranus, Neptune, and Pluto owing to their lower temperatures and higher gravitational attraction.

Origin of organic compounds. Many attempts have been made to synthesize organic compounds under oxidizing conditions using carbon dioxide and water with various sources of energy, such as ultraviolet light, electric discharges, and high-energy radiation (see illustration). All of these experiments failed to give organic compounds except in extremely small yield.

Support for the ideas of Oparin and Urey came from experiments by S. L. Miller in 1953. He showed that a mixture of methane, ammonia, water, and hydrogen, when subjected to an electric discharge, gave significant yields of simple amino acids, hydroxy acids, aliphatic acids, urea, and possibly sugars. Ultraviolet light gives similar results.

On the basis of these experiments, it is thought that various organic compounds (such as amino acids, hydroxy acids, aliphatic acids, and sugars) were present in the oceans of the primitive Earth. It is still necessary to show mechanisms for the formation of polypeptides, purines and pyrimidines (the bases which occur in nucleic acids), nucleotides and polynucleotides, a source of phosphate bond energy (as in adenosinetriphosphate), a synthesis of polypeptides with catalytic activity (enzymes), and the development of polynucleotides capable of self-duplication.

Nature of first organisms. In present living organisms the duplication proceeds by duplicating the genes, followed by synthesis of more enzymes and other cell constituents, and division of this cell into two fragments. The genes, which are located on the chromosomes, are composed of polynucleotides called deoxyribonucleic acid.

Since the essential characteristic of living organisms is the ability to duplicate, it has been proposed that the first living organisms were simply strips of deoxyribonucleic acid (or perhaps strips of ribonucleic acid) which, together with the necessary enzymes, could duplicate. This organism would be similar to a virus except that it would have the necessary enzymes for its reproduction.

The virus consists of several strips of nucleic

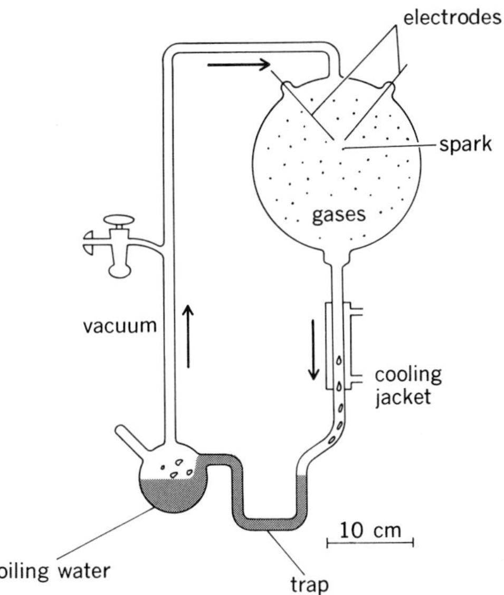

Apparatus used for production of amino acids by electric discharges. Water is boiled in small flask and mixes with gases in large flask where spark is located. Products of the discharge are condensed, pass through U-tube, and accumulate in small flask.

acid surrounded by a coat of protein. The virus is capable of duplication, but only within another living cell where the virus makes use of the cell's enzymes and metabolism. Since the virus can reproduce itself, but only within a living cell, there is considerable dispute as to whether a virus should be called living or not. In any case many virologists believe that the virus was not the first type of living organism, but that it is a regression from a more advanced type of organism.

It has been proposed by Oparin that the first organisms were coacervate particles instead of strips of polynucleotides. A coacervate is a type of colloid which forms two phases, one of the solution and one of the coacervate, instead of a uniform dispersion as with most colloids. It is assumed that there would be some coacervate particles which could absorb proteins and other substances from the environment, grow in size, and then split into two or more fragments which would repeat this process. In time the duplication would become more accurate, and the genetic apparatus of nucleic acids would then develop.

These ideas about the nature of the first organisms are based on the assumption that the first forms of life were similar to present organisms in their basic chemical composition. All present organisms contain proteins, nucleic acids, carbohydrates, and lipids. While the assumption may be incorrect, it is the best working hypothesis until it can be shown to be inadequate. Other possibilities for the development of the first organisms can be enumerated, but so little is known of the composition of the primitive oceans that such speculation is not profitable.

Heterotrophic organisms. While little can be said about the development of the first living organisms, reasonable hypotheses can be made for the development of the simple bacteria, algae, and protozoa from the most primitive organisms. The theory that the primitive oceans contained large quantities of organic compounds implies that the first organisms were heterotrophic. Heterotrophic organisms do not synthesize their basic constituents, such as amino acids, nucleotides, carbohydrates, and vitamins, but obtain them from the environment. Autotrophic organisms synthesize all their cell constituents from CO_2, H_2O, and other inorganic materials. Heterotrophic organisms are simpler than autotrophic organisms in that they contain fewer enzymes and specialized structures to carry out their metabolism. Hence heterotrophic organisms would have been formed first.

A mechanism by which heterotrophic organisms could acquire various synthetic abilities was proposed by N. H. Horowitz in 1945. It has been found that the presence of an enzyme in an organism is often dependent on a single gene. This is known as the one gene–one enzyme hypothesis. Suppose that the synthesis of A involves the steps

$$D \xrightarrow{c} C \xrightarrow{b} B \xrightarrow{a} A$$

where a, b, and c are the enzymes and A, B, and C are compounds which the organism cannot synthesize. If the necessary compound A becomes exhausted from the environment, then the organism must synthesize A in order to survive. But it is extremely unlikely that there would be three simultaneous mutations to give the enzymes a, b, and c. However, a single mutation to give enzyme a would not be unlikely. If compounds D, C, and B were in the environment when A was exhausted, the organism with enzyme a could survive while the others would die out. Similarly, when compound B was exhausted, enzyme b would arise by a single mutation, and organisms without this enzyme would die out. By continuing this process the various steps of a biosynthetic process could be developed, the last enzyme in the sequence being developed first, and the first enzyme last.

Energy and biosynthetic processes. It is necessary for all living organisms to have a source of energy to drive the biochemical reactions that synthesize the various structures of the organism. The quantity that measures the available energy for a chemical reaction at constant temperature and pressure is termed the free energy. Animals obtain their free energy from the oxidation of organic compounds by molecular oxygen. Plants and other photosynthetic organisms obtain their free energy from the energy of light. There are also many microorganisms that obtain their free energy from fermentation reactions. For example, the lactic acid bacteria obtain their free energy from glucose as shown in reaction (1). Yeasts also ferment glu-

$$\underset{\text{Glucose}}{C_6H_{12}O_6} \rightarrow \underset{\text{Lactic acid}}{2CH_3CH(OH)COOH} \qquad (1)$$

cose, but they produce ethyl alcohol and CO_2 instead of lactic acid, and there are bacteria which carry out many other types of fermentation reactions. It is likely that organisms obtained their free energy from fermentation reactions until the supply of fermentable compound was exhausted. At that point the development of photosynthetic organisms which obtain their free energy from light would become necessary. The porphyrin substance chlorophyll seems to be necessary for all types of photosynthesis and would have had to be present in the environment or its biosynthesis developed in some way. The first type of photosynthesis was probably similar to that of the sulfur bacteria and blue-green algae which carry out reactions (2*a*) or (2*b*), where (H_2CO) means carbon on

$$2H_2S + CO_2 \xrightarrow{\text{Light}} 2S + (H_2CO) + H_2O \qquad (2a)$$

$$H_2S + 2CO_2 + 2H_2O \xrightarrow{\text{Light}} H_2SO_4 + 2(H_2CO) \qquad (2b)$$

the oxidation level of formaldehyde (carbohydrate). It is much easier to split H_2S than to split H_2O, so that it would seem likely that organisms would develop photosynthesis with sulfur first. When the H_2S and S were exhausted, it would become necessary to split water and evolve O_2, the hydrogen being used for the reduction of CO_2.

When the methane and ammonia of the primitive atmosphere had been converted to carbon dioxide and nitrogen by photochemical decomposition in the upper atmosphere, water would be decomposed to oxygen and hydrogen. The hydrogen would escape, leaving the oxygen in the atmosphere and thereby resulting in oxidizing conditions on the Earth. It is likely, however, that most of the oxygen in the atmosphere was produced by the photosynthesis of plants rather than by the photochemical splitting of water.

Multicellular organisms. The evolution of multicellular organisms probably occurred after the development of photosynthesis. The evolution of primitive multicellular organisms to more complex types and the development of sexual reproduction can be understood on the basis of the theory of evolution. *See* EVOLUTION, ORGANIC.

[STANLEY L. MILLER]

Bibliography: E. S. Barghoorn, *Origin of Life*, Geol. Soc. Amer. Mem. no. 67, 2:75–86, 1957; J. D. Bernal, *The Physical Basis of Life*, 1951; G. Ehrensvärd, *Life: Origin and Development*, 1962; J. Keosian, *The Origin of Life*, 1964; S. L. Miller, A production of amino acids under possible primitive earth conditions, *Science*, 117:528–529, 1953; S. L. Miller, Production of some organic compounds under possible primitive earth conditions, *J. Amer. Chem. Soc.*, 77:2351–2361, 1955; S. L. Miller and H. C. Urey, Organic compound formation on the primitive earth, *Science*, 130:245–251, 1959; A. I. Oparin, *The Chemical Origin of Life*, 1964; M. Rutten, *Geological Aspects of the Origin of Life on Earth*, 1962; G. Wald, The origin of life, *Sci. Amer.*, 191(2):44–53, 1954; Symposium: Modern ideas on spontaneous generation, *Ann. N.Y. Acad. Sci.*, 69(art. 2):257–376, 1957.

Lignite

Soft and porous carbonaceous material intermediate between peat and subbituminous coal, with a heat value less than 8300 Btu on a mineral-matter-free (mmf) basis. In the United States, lignitic coals are classified as lignite A or lignite B, depending upon whether they have calorific value of more or less than 6300 Btu. Elsewhere than in North America, lignites are called brown coals, two major classes being recognized, namely, hard and soft brown coals. The soft brown coals are described as either earthy, resembling peat, or as fragmentary; the hard brown coals are regarded as dull (matte) or bright (glance). The soft brown coals correspond in a general way with lignites of class

B, and the hard brown coals with lignites of class A, but the two classifications should not be regarded as strictly interchangeable. Brown coals generally have a brown color, compared with the usual black color of bituminous coal. At the present time there is a vigorous effort being made by the International Committee of Coal Petrology that will make possible a satisfactory differentiation on a petrologic basis of bituminous and lignitic coals and provide a satisfactory classification of varieties of lignite on a basis other than calorific.

Reserves. The lignite of North America is mainly of lignite class A. The known reserves of lignite in 15 states have been estimated as 224,000,000,000 short tons, as compared with an estimated total coal reserve of 828,000,000,000 tons. About 98% of the lignite reserves in the United States are in the northern Great Plains area, which includes the western half of North Dakota, eastern Montana, northeastern Wyoming, and northeastern South Dakota. A single lignite mine in California operates for the production of montan wax. The reserves in the northern Great Plains are believed to represent a total of 99% of the total reserves of lignite in the United States. The production of lignite in the United States in 1963 was 2,700,000 tons, excluding the production from two mines in Texas, amounting to less than 1% of the world production. Of these 2,700,000 tons, all but approximately 7,000 tons was recovered by strip mining operations.

Uses. The main use of lignite in the United States is in residential and industrial heating and in the generation of power. Because of the competition with bituminous coal, the market area of lignite in the United States is relatively small. When used at any great distance from its source, lignite requires dehydration and even briquetting.

Some investigation is under way that may lead to the production of pipeline gas from lignite. The U.S. Bureau of Mines regards the total gasification of lignite a potential for a large tonnage outlet for this variety of coal. *See* COAL.

[GILBERT H. CADY]

Bibliography: *Mineral Facts and Problems*, U.S. Bur. Mines Bull. no. 360, 1965.

Limestone

A sedimentary rock composed dominantly of carbonate minerals, principally carbonates of calcium and magnesium. Limestones are the most abundant of the nonclastic rocks. They are overwhelmingly the largest reservoir of the element carbon at or near the surface of the Earth. Much knowledge of invertebrate paleontology and consequently of the evolution of life and earth history comes from the fossils contained in them. Although the word limestone is used in the general sense above, specifically it refers to carbonate rocks dominated by the mineral calcite, $CaCO_3$, as opposed to dolomite, a term for carbonate rocks dominated by the mineral dolomite, $CaMg(CO_3)_2$. Although the mineralogical composition of most limestones is similar, their textures are not because the limestones are formed under a great variety of conditions. *See* DOLOMITE ROCK.

Chemical composition. The chemical composition of limestones is largely calcium oxide, CaO, and carbon dioxide, CO_2. Magnesium oxide, MgO, is a common constituent; if it exceeds 1 or 2%, the rock may be termed magnesian limestone. Small amounts of silica and alumina may also be present as a result of the presence of clastic materials, quartz, and clay. Iron oxide may be present, either as carbonate (siderite, ferroan-dolomite) or in other minerals, such as clays. Strontium may be present as an important trace element, probably derived from original fossil materials in which it was incorporated into aragonite, a form of $CaCO_3$.

Mineralogy. The chief minerals of limestones are calcite and aragonite, and in the dolomitic limestones, dolomite. Calcite and aragonite have the same composition but different crystal structures. Aragonite is unstable with reference to calcite in surface environments and is transformed into the calcite with time. Even though aragonite is unstable, it forms as a precipitate from sea water and some fresh waters inorganically, for example, oolites, and biogenically through the intermediate mechanism of biological secretion by shell-forming organisms. The conversion of aragonite to calcite is accomplished in a relatively short time—just a few years for some fossils. Large crystals may require a much longer amount of time. Ancient rocks always contain calcite rather than aragonite. *See* ARAGONITE; CALCITE.

Although dolomite is stable in surface environments, it is formed as a primary precipitate only under special conditions, such as high salinity or high alkalinity. Petrographic and O^{18}/O^{16} isotopic evidence suggests that most dolomite is formed postdepositionally by the interaction of magnesium ions in interstitial waters with calcite. Studies of modern environments in which dolomitization is taking place imply that much of the alteration of dolomite must have taken place relatively soon after its original deposition.

Siderite, the iron carbonate, is found in some limestones, but iron occurs in carbonate form chiefly as ferroan-dolomite. Silica may be present, either finely disseminated throughout the rock or segregated into nodules of chert. Silica also occurs as small crystals of quartz that have grown in place during diagenesis. Feldspar occurs in the same way but is a little less common than silica. Other minerals found in limestones are glauconite, collophane, and pyrite. A host of other minerals may be found as small amounts of clastic material brought in by currents, including almost always a small amount of fine-grained clay. Organic matter may also be found. *See* DIAGENESIS.

Textures. The textures of limestones reveal much about their origin. The chief textural elements in a great many limestones are fossils, whole or fragmental. The species represented and their state of preservation are a guide to the ecology and environment of the site of deposition. In addition, the nature of the fossil debris, whether complete articulated shells or fragments in all stages of disarticulation, fragmentation, and comminution, is a guide to the presence of bottom currents. Shells may be broken or destroyed by bottom-dwelling organisms that ingest sediment in search of organic matter for food or by parasitic boring algae. Scavenging organisms turn the sediment over and destroy much of its original structure while producing tubes or burrows and leaving behind spherical or ovoid fecal pellets of fine-

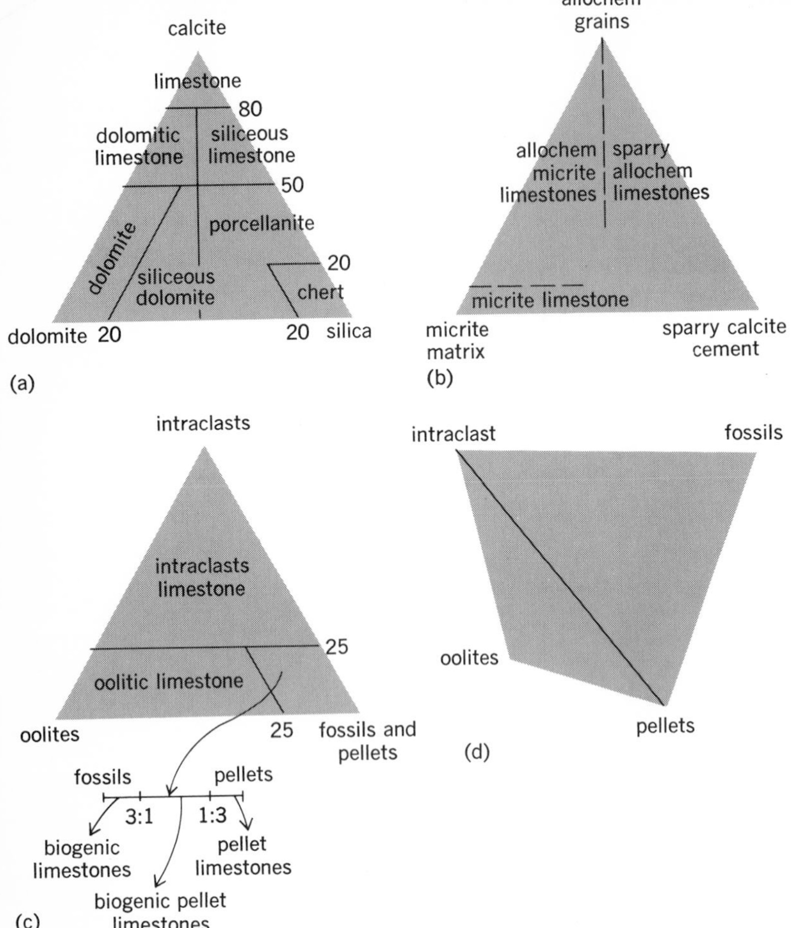

Generalized end-member groupings of carbonate rocks. (*a*) Limestone-dolomite-chert grouping (*adapted from an unpublished chart by F. J. Pettijohn, 1944*). (*b*) Grouping based on coarse and fine components; (*c*) Folk's basic classification of limestones (*b and c from R. L. Folk, 1959*). (*d*) Tetrahedral grouping of Folk's end members. (*From W. C. Krumbein and L. L. Sloss, Stratigraphy and Sedimentation, 2d ed., copyright © 1963 by W. H. Freeman and Co.*)

grained lime mud. All of these textures can be recognized in ancient limestones.

Some mechanically deposited limestones show well sorted particles of calcite cemented by clear intergranular calcite. If the particles are sand-size, the rock is termed calcarenite. In general, the mechanically deposited limestones show the same kinds of sedimentary structures as do clastic rocks. Cross-bedding, stratification, current lineation, and even graded bedding may be displayed. Oolitic and pisolitic textures are most abundant in limestones. *See* OOLITE AND PISOLITE.

A prominent textural constituent of many limestones is a featureless, finely crystalline mosaic of calcite. R. L. Folk termed this material micrite and emphasized its importance as the lithified equivalent of the fine-grained lime mud found in many modern carbonate depositional environments.

Secondary textural features that are found in limestones include stylolites, cross-cutting veinlets filled with calcite, and replacement effects (typically that of rhombohedrons of dolomite replacing calcite). Because the carbonate minerals are relatively soluble in aqueous solutions and because of the transformation of aragonite to calcite and calcite to dolomite, recrystallization and other diagenetic effects are particularly common in limestones. *See* STYLOLITES.

Classification. Many schemes of limestone classification have been proposed and used, but since 1959, when R. L. Folk showed the utility of textural elements as a primary basis for classification, most workers have used systems similar to his (see illustration). Folk used the general concepts developed in 1949 by F. J. Pettijohn, who saw limestones as mixtures of three pure end-member components, calcite, dolomite, and silica. The textural classification recognized limestones as made up of varying proportions of allochem grains (oolites, intraclasts, fossils, and pellets), micrite (fine-grained mosaic), and sparry calcite cement (clear, coarsely crystalline, postdepositional pore fillings). Recognition of these textures in ancient limestones and comparison with their analogs in modern carbonate sediments have lead to vigorous activity in mapping of ancient limestone terrains in order to reconstruct paleogeographies and deduce environments of deposition.

Origin. Limestones may be referred to two types of origin, autochthonous and allochthonous. Autochthonous limestones have been formed in place by biogenic precipitation from the water of the environment, usually sea water; allochthonous limestones have been transported from the site of original precipitation by current action. The total distance of transport of allochthonous particles may not be great; the transportation is mainly a process of moving a chemical precipitate from one part of a sedimentary basin to another. Primary agents in the formation of autochthonous limestones are the lime-secreting organisms. Some of the most important of these are the calcareous coralline algae, such as *Lithothamnion*, which, together with corals, make up the major part of modern reef limestones. Foraminiferans also contribute large quantities of carbonate. Other groups of carbonate-rock-forming invertebrates are mollusks and echinoderms (in particular the crinoids).

The commonest autochthonous limestone is the normal marine limestone. Although whole fossils or parts of them are plentiful, the bulk of the rock may not be recognizable as fossil debris. Much of the now unrecognizable material probably came from calcareous algae or other biogenic carbonate broken up into very small particles by the action of bottom-dwelling scavengers. Inorganic precipitation may account for some of the fine-grained micrite. The rock is light-colored and normally has moderately well-developed bedding. Dolomitization is common, as are chert nodules.

Biohermal limestones are reefs or reeflike mounds of carbonate that accumulated in much the same fashion as the modern reefs and atolls of the Pacific Ocean. The mounds may range from a few feet up to several thousand feet in diameter and hundreds of feet thick. Some of the best described fossil reefs are those of the Silurian system in Illinois, Indiana, and Wisconsin. The central core of the reef is fossiliferous, dolomitized, and massive. It grades radially outward into sparsely fossiliferous, well-bedded reef flank strata that commonly dip away from the core. Farther away from the core, the reef flank beds grade into fine-grained, well-bedded, relatively unfossiliferous, normal carbonate interreef sediment. *See* BIOHERM; REEF.

Biostromal limestones are biogenic carbonate accumulations that are laterally uniform in thickness, in contrast to the moundlike nature of bioherms. The fossils may be of many different kinds or they may be dominated by a single group. Particularly common are crinoidal and algal biostromes. The algal biostromes may show very few recognizable fossils, but stromatolites and algal laminations are common. Many of the biostromes are of mixed autochthonous and allochthonous origin; that is, some of the fossil debris of many biostromes shows evidence of transport. *See* BIOSTROME; STROMATOLITE.

Pelagic limestones are formed from the accumulation of the limy parts of pelagic, or floating, organisms such as foraminiferans. The resulting limestones are fine-grained and contain very few fossils of bottom-dwelling faunas. Since the foraminiferans are chiefly responsible for pelagic limestones and the lime-secreting pelagic foraminiferans did not evolve until the Cretaceous, pelagic limestones are restricted to Cretaceous and later systems.

The allochthonous, or transported, limestones show clastic textures typical of detrital rocks. The clastic particles may be of fossils, as in coquinas or coquinoid limestones; of inorganically precipitated carbonate particles, as in oolites; or of earlier deposited limestones. *See* CALCARENITE; CHALK; COQUINA; MARL; SEDIMENTARY ROCKS; SEDIMENTATION; TRAVERTINE; TUFA.

[RAYMOND SIEVER]

Bibliography: W. E. Ham (ed.), *Classification of Carbonate Rocks*, 1962; W. C. Krumbein and L. L. Sloss, *Stratigraphy and Sedimentation*, 2d ed., 1963; F. J. Pettijohn, *Sedimentary Rocks*, 2d ed., 1957.

Limonite

A natural amorphous-appearing material constituting in part the material known as brown iron ore. Limonite occurs in mammillary, stalactitic, and earthy masses (see illustration). The hardness is 5–5.5 (Mohs scale) and the specific gravity is 3.6–4. The luster is vitreous and the color dark brown to black. Limonite is essentially $FeO(OH) \cdot nH_2O$, but with admixed hematite, clay, and manganese oxides. X-ray analysis has shown that much of the material formerly thought to be limonite is goethite. Limonite forms by the oxidation of preexisting iron minerals and occurs with goethite. It is the pigmenting material in yellow soils and mixed with fine clays is known as yellow ocher. *See* GOETHITE; ORE AND MINERAL DEPOSITS.

[CORNELIUS S. HURLBUT, JR.]

Lithosphere, geochemistry of

The study of the distribution of the elements and of their isotopes in the lithosphere and of the processes, both past and present, which affect this distribution. The lithosphere contains more than 99% of the mass of the Earth; it is therefore by far the most important reservoir of almost all of the chemical elements, and the processes in the lithosphere exert a dominant influence on many surface phenomena.

Knowledge of the distribution of the elements and of their isotopes in the lithosphere is very uneven. Almost all of the samples that have actually been analyzed have come from the present surface of the Earth or from a depth of less than 8 km. Some materials from depths of several kilometers have been brought to the surface in kimberlite pipes and are yielding important clues regarding the nature of the upper mantle. The average radius of the Earth is, however, 6371 km; analyzed samples therefore represent only scratchings from the outermost portions of the globe. The evidence for the chemical composition of the Earth as a whole is largely indirect and is based on the data of geophysics and inferences from the composition of meteorites. Since the mass and volume of the Earth are known, its average density can be calculated. On the basis of travel-time curves for sound waves generated by earthquakes and man-made explosions, convincing models of the changes with depth in the density and in the velocity of compressional and shear waves can be constructed. These data, together with information on the magnetic field of the Earth and the mechanical and thermal properties of the Earth, contribute to the fund of information used in constructing the presently accepted model for the chemical structure of those portions of the Earth which cannot be observed directly. *See* SEISMOLOGY.

Data regarding the chemical composition of meteorites complement the geophysical data. Models based on geophysical data and those based on meteorite data are surprisingly similar, and present concepts of the major features of the chemistry of the Earth's interior are almost certainly correct for the major elements.

Such a degree of certainty cannot be claimed for most of the minor elements. Fewer than 10% of the chemical elements account for over 98% of the mass of the lithosphere. The abundance of the majority of the elements in the lithosphere is therefore well below 0.1%. The advances in trace element analysis since 1920 have opened the study of the distribution even of elements present in rocks in amounts less than 1 part per million (ppm), or 10^{-4}%, so that the geochemistry of these elements in the upper few kilometers of the lithosphere is reasonably well known. However, geophysical techniques are not suitable for such studies in the deeper parts of the Earth, and the use of meteorite data is always beset by uncertainties regarding the assumption that there is necessarily a close correspondence between the abundance of elements in the analyzed meteorites and in the Earth's interior.

Chemical composition of crust. Analyses numbering in the tens of thousands have been made on geologic materials. All or nearly all of the analyzed samples have come from the crust of the Earth, that is, from above the Mohorovičić discontinuity, which lies at a depth of about 30–50 km under the continents and at a depth of about 5–10 km under the oceans. Although many of these chemical analyses are of somewhat questionable accuracy, the large mass of available chemical data is amply sufficient to give an accurate description of the average composition and of the compositional variation of the various types of geologic materials. This information can be combined with data on the relative abundance of the various rock types in the Earth's upper crust to give a fairly complete description of the average chemical composition of directly accessible parts of the lithosphere.

Some data on the chemical composition of vari-

LIMONITE

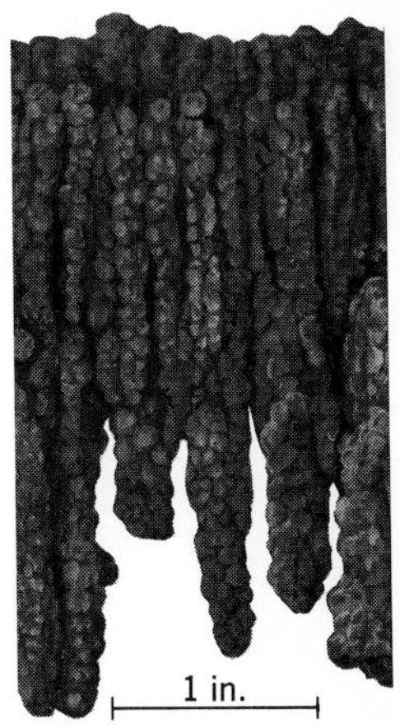

Limonite in stalactitic masses, which are found in the Tintic District of Utah. *(Specimen from Department of Geology, Bryn Mawr College)*

Average chemical composition of sediments and rocks*

Sediments and rocks	Index	SiO_2	TiO_2	Al_2O_3	Fe_2O_3	FeO	MnO	MgO	CaO	Na_2O	K_2O	P_2O_5	CO_2	H_2O§
Sediments and sedimentary rocks														
Calcareous sands, oozes†	1	18.8	0.3	5.1	3.8		0.4	1.4	39.0	0.5	0.7	0.2	29.8	
Red clay†	2	52.8	0.9	16.4	9.0		1.0	2.9	7.7	1.7	2.7	0.3	4.6	
Siliceous oozes†	3	55.5	0.5	15.1	5.8		0.7	2.3	9.7	1.0	2.2	0.3	6.9	
Average pelagic sediment†	4	28.5	0.4	8.1	5.0		0.6	1.8	30.5	0.8	1.2	0.2	22.9	
Composition of suboceanic sediments†	5	43.2	0.7	11.6	4.6	0.6	0.3	2.4	21.1	1.1	1.8	0.1	12.5	
Average sediment†	6	44.5	0.6	10.9	4.0	0.9	0.3	2.6	19.7	1.1	1.9	0.1	13.4	
Average shale‡	7	58.10	0.65	15.40	4.02	2.45		2.44	3.11	1.30	3.24	0.17	2.63	5.00
Average sandstone‡	8	78.33	0.25	4.77	1.07	0.30		1.16	5.50	0.45	1.31	0.08	5.03	1.63
Average limestone‡	9	5.19	0.06	0.81	0.54			7.89	42.57	0.05	0.33	0.04	41.54	0.77
Average sediment‡	10	57.95	0.57	13.39	3.47	2.08		2.65	5.89	1.13	2.86	0.13	5.38	3.23
Igneous rocks														
546 granites† 4	11	70.8	0.4	14.6	1.6	1.8	0.1	0.9	2.0	3.5	4.1	0.2		
137 granodiorites† 6	12	67.2	0.6	15.8	1.3	2.6	0.1	1.6	3.6	3.9	3.1	0.2		
635 intermediate igneous rocks† 13	13	54.9	1.5	16.7	3.3	5.2	0.2	3.8	6.6	4.2	3.2	0.4		
198 basalts† 19	14	49.9	1.4	16.0	5.4	6.5	0.3	6.3	9.1	3.2	1.5	0.4		
182 ultramafic rocks† 25	15	44.0	1.7	6.1	4.5	8.8	0.2	22.7	10.2	0.8	0.7	0.3		
Average igneous rocks† 11	16	60.1	1.1	15.6	3.1	3.9	0.1	3.5	5.2	3.9	3.2	0.3		
Metamorphic rocks														
250 quartzofelds-pathic gneisses† 50	17	70.7	0.5	14.5	1.6	2.0	0.1	1.2	2.2	3.2	3.8	0.2		
103 mica schists† 61	18	64.3	1.0	17.5	2.1	4.6	0.1	2.7	1.9	1.9	3.7	0.2		
61 slates† 65	19	61.8	0.7	19.1	3.3	5.4	0.2	2.9	1.0	1.7	3.8	0.1		
200 amphibolites† 68	20	50.3	1.6	15.7	3.6	7.8	0.2	7.0	9.5	2.9	1.1	0.3		
Crust of the Earth														
Average composition†	21	55.2	1.6	15.3	2.8	5.8	0.2	5.2	8.8	2.9	1.9	0.3		
Average composition‡	22	60.18	1.06	15.61	3.14	3.88		3.56	5.17	3.91	3.19	0.30		

*Based on summaries by A. Poldervaart, *Crust of the Earth*, Geol. Soc. Amer. Spec. Pap. no. 62, 1955, and by B. Mason, *Principles of Geochemistry*, 3d ed., Wiley, 1966.

†From Poldervaart; figure, if given, refers to index number as used in Poldervaart.

‡From Mason.

§Values omitted where samples were calculated water-free.

ous sediment and rock types are given in the table. A comprehensive statement of all the elements in each geochemical phase is given elsewhere. *See* ELEMENTS, GEOCHEMICAL DISTRIBUTION OF.

One of the most striking aspects of these data is the relatively small range of chemical composition in the upper part of the lithosphere. The same dozen or so elements predominate in all of the major sediment and rock types. All of these, except the carbonate rocks, contain roughly between 40 and 75% silica, SiO_2. The carbonate rocks are almost always calcium or calcium-magnesium carbonates, often containing an admixture of siliceous material. The crustal abundances of the major elements, and also of the minor elements, therefore do not depend in a very critical way on the relative proportions assigned to the various sediment and rock types in the Earth's crust. Nevertheless, it should be remembered that the crustal abundances listed in the table and in Figs. 1 and 2 must be regarded as averages subject to revision.

The crustal abundance data have been plotted on a linear scale in Fig. 1 and on a logarithmic scale in Fig. 2. The first plot emphasizes the fact of the preponderance of certain of the elements of low atomic number. No element with atomic number greater than 26 is present in the Earth's crust with an abundance greater than 0.1%. The logarithmic plot of Fig. 2 permits the representation of the abundance of all of the naturally occurring elements and emphasizes the frequency of crustal abundance values between 10 ppm (10^{-3}%) and 0.1 ppm (10^{-5}%), particularly among the elements of higher atomic number. The broad features of these figures are now reasonably well understood. They are the product of the processes of element formation, of the processes operating during the formation of the Earth, and of the processes within the earth, which have operated both in the direction of homogenization and diversification of the chemical composition of the various parts of theEarth.

Chemical composition of mantle. The Earth's mantle consists of material between the Mohorovičić discontinuity, marking the base of the crust, and the discontinuity at a depth of 2898 km, marking the outer limit of the Earth's core. Present

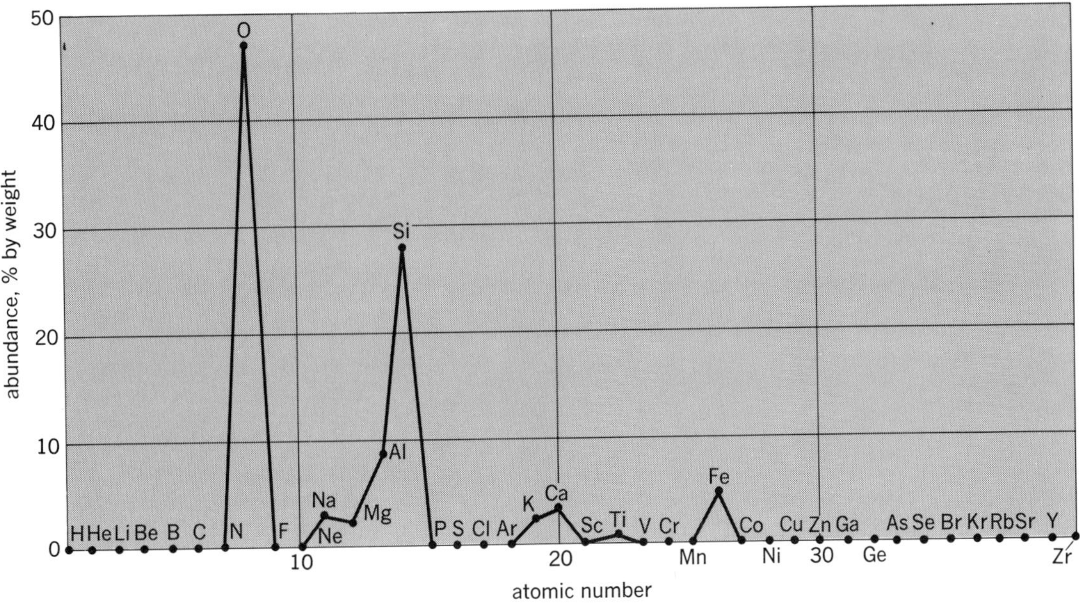

Fig. 1. Crustal abundance of elements of atomic number 1–40.

thinking regarding the chemical composition of this portion of the Earth is based largely on geophysical evidence and on the circumstantial meteorite evidence. The latter line of evidence suggests that the composition of the mantle is not very different from that of ultramafic igneous rocks (see index 15 in the table); the geophysical evidence is in substantial agreement with this hypothesis. However, the evaluation of the density and elastic wave propagation data is at present somewhat tenuous, since observations on the state of silicate minerals at temperatures and pressures characteristic of the middle and lower portions of the mantle are still fragmentary. It is very likely that silicon, oxygen, magnesium, iron, and calcium are the major elements in the Earth's mantle and that its bulk composition corresponds roughly to a 1:3 mixture of basalt and periodotite. It is also likely that the mantle, at least in its upper portion, is more heterogeneous than has usually been assumed, and that its structure may be almost as complicated as that of the crust.

Knowledge of the abundance of the minor elements in the mantle is even more fragmentary than that regarding the major elements. Only in the case of the radioactive elements can something be said independent of the meteorite model. The abundance of these elements must be low enough to prevent melting of the mantle, and this must mean that the abundance of potassium, uranium, and thorium must be very much less in the mantle than in the crust. In theory, heat flow measurements can give more definite indications about maximum concentrations of these elements in the mantle. However, the complexities introduced by radiative and possibly convective heat flow in the mantle introduce serious complications into the interpretation of surface heat flow measurements. *See* EARTH, HEAT FLOW IN.

Chemical composition of core. The core of the Earth consists of the material between a depth of 2898 km and the center of the Earth. Problems similar to those besetting the study of the composition of the mantle also limit study of the core. Again, all data are circumstantial. The meteorite data suggest a core of metallic iron with an admixture of nickel. The geophysical data indicate that this is distinctly probable, but that the average atomic number of core material is less than that of iron. Silicon has been suggested as an important core constituent, but several other alternatives, including sulfur, are at least possible, and the problem remains open. A small amount of potassium may be present and may supply the energy for convection in the core. The concentration of other minor constituents in the Earth's core can be estimated roughly from the chemistry of meteorites. *See* METEORITE.

Important crystalline phases. The major portion of the lithosphere is solid. In the crust the solids are predominantly well crystallized, and this is almost certainly true for the mantle as well. Oxygen and silicon are the most abundant elements in the crust and in the mantle, and the solid phases are almost all either oxides or silicates. Since the processes described below depend to a considerable extent on the properties of these oxide and silicate compounds, the crystal chemistry of the major mineral phases is briefly reviewed at this point. *See* MINERALOGY.

Oxide structures. The structure of most of the rock-forming oxides is simple. In all of them O^{--} is the largest ion and determines the structural framework. In periclase, MgO, the O^{--} ions are cubic close-packed and the Mg^{++} ions occupy the octahedrally coordinated holes between neighboring O^{--} ions. In the spinel group, $R''R'''_2O_4$, the divalent R'' ions and the trivalent R''' ions again occupy either tetrahedrally or octahedrally coordinated holes between cubic close-packed O^{--} ions. Hexagonal close packing of O^{--} ions is found in corundum, Al_2O_3, and in related minerals such as ilmenite, $FeTiO_3$, and geikielite, $MgTiO_3$. In these structures the cations again occupy 4- or 6-coordinated holes in the O^{--} framework.

Silicate structures. The rather improbable for-

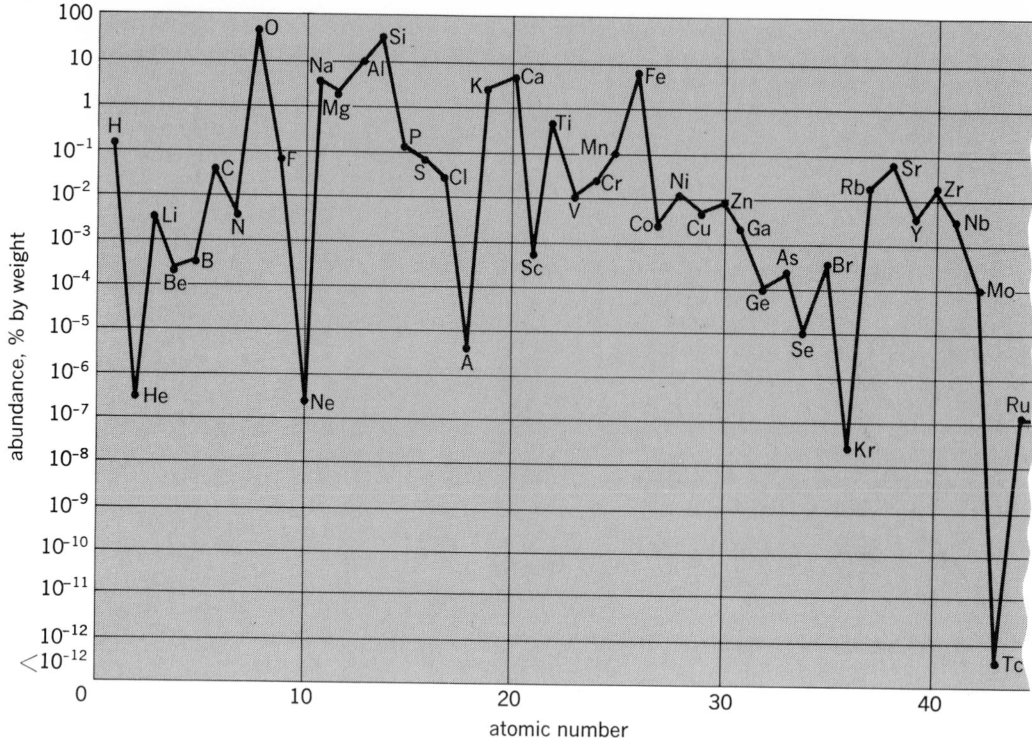

Fig. 2. Crustal abundance of elements of atomic number 1–92.

mulas suggested for many silicate minerals by chemical analyses before the advent of x-ray crystallography were shown to be quite natural results of their often complicated silicon-oxygen framework structures, and the development of the crystal chemistry of the silicates has been one of the most significant contributions of mineralogy to geology since 1920. The structure of silicate minerals is normally more complicated than that of the oxide. The basic unit of structure is the silicon-oxygen tetrahedron, SiO_4^{4-}, consisting of a Si^{4+} ion at the center of a tetrahedron whose corners are occupied by the center of O^{--} ions. These tetrahedrons may be joined by cations, as in the olivines $(Mg, Fe)_2SiO_4$, zircon, $ZrSiO_4$, or the garnets, $R''_3R'''_2(SiO_4)_3$. In these structures the $[SiO_4^{4-}]$ group serves a function similar to that of the $[SO_4^{2-}]$ group in the sulfates. The structure of the olivines is similar to that of the spinels; in fact, olivines invert to a spinel form at high pressures.

In the majority of silicate compounds the $[SiO_4^{4-}]$ groups are polymerized. Such polymerization may lead to groups of two, three, or six tetrahedrons, to effectively infinite chains or double chains, to sheet structures, or to three-dimensional networks of $[SiO_4^{4-}]$ groups. The physical properties of silicate minerals are often related in a simple manner to the type of polymerization of the $[SiO_4^{4-}]$ groups. The cleavage planes of the pyroxenes, the amphiboles, and the feldspars are parallel to the effectively infinite chains or double chains characteristic of these mineral groups. Sheets of $[SiO_4^{4-}]$ tetrahedrons are parallel to the pronounced cleavage of the micas, clays, and chlorite group minerals. The uniform, three-dimensional network of quartz is reflected in the lack of cleavage in this mineral.

At pressures in excess of roughly 10,000 atm the structure of many silicates transforms to higher-density modifications. Quartz collapses to coesite and then to stishovite; it is very likely that the mineralogy of the mantle is strongly influenced by such phase changes. *See* SILICATE MINERALS.

Processes involving lithosphere. The Earth is not at rest. Earthquakes, volcanism, and mountain building, the three major manifestations of disequilibrium at depth, are evidence of widespread physical and chemical change below the surface of the Earth. Erosion and chemical weathering are evidence of the pronounced disequilibrium between the atmosphere and the lithosphere. Chemical changes thus are taking place at various levels in the lithosphere and at the boundary between the lithosphere and the atmosphere, biosphere, and hydrosphere. The processes taking place at the Earth's surface can be observed directly and therefore are more completely understood than the processes taking place within and below the Earth's crust. Yet it is just the processes at depth that are of primary interest as a key to the history of the crustal evolution of the Earth.

For most of the elements, the Earth can be considered a closed or an almost closed system. The two major exceptions to this rule are hydrogen and helium, which escape at significant rates from the upper atmosphere. Hydrogen escape has probably played an important role in the evolution of the atmosphere. A number of elements are entering the atmosphere as constituents of meteorites and as cosmic-ray particles, or as interplanetary material which is swept up by the Earth. It seems unlikely that these additions have affected the chemistry of the Earth materially.

The Earth therefore can be regarded as an es-

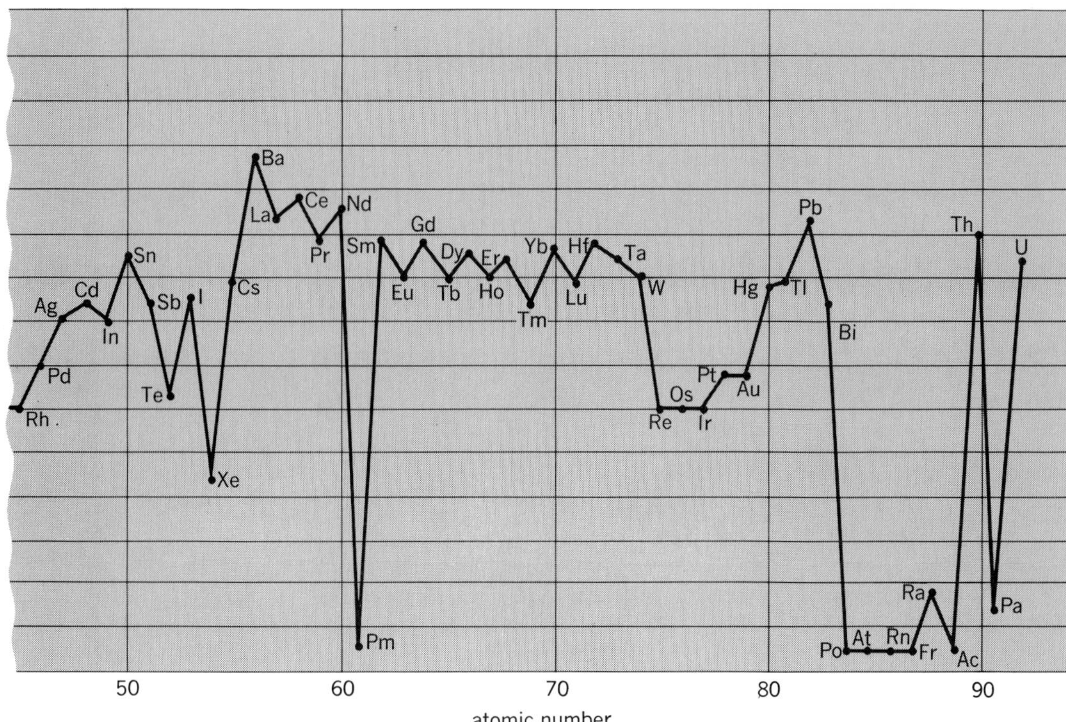

sentially closed system consisting of a number of reservoirs. The exchange between these reservoirs may be appreciable and is often a complex function of time. During the course of history, the mass and composition of various reservoirs have changed radically. The basic problem in the study of the dynamics of geochemistry consists of defining suitable reservoirs and of determining the nature and the rates of the processes that are operating on them today and have done so in the past.

The primary division of the Earth into the lithosphere, hydrosphere, and atmosphere is convenient since the limits of these reservoirs are reasonably well defined and their contents are appreciably different. The biosphere has a somewhat different status since in space it overlaps the other spheres completely. The primary division into four spheres is too coarse for many problems in earth science. The subdivision of the lithosphere into crust, mantle, and core is evidence of this, and the further subdivision of the crust into sediments and sedimentary rocks, metamorphic rocks, and igneous rocks was implied in the discussion of the average chemical composition of materials in these categories. Further subdivisions of these large units into regional reservoirs, or local reservoirs, is demanded by many problems. For instance, in studies of the age of an igneous rock unit the systems of interest are individual crystals of the mineral or minerals used for dating purposes. The problem, in general terms, thus consists of determining the events that have taken place within the system of interest and the extent of exchange that has occurred between the system and its environment. *See* ROCK, AGE DETERMINATION OF.

Changes in systems are of various kinds. Certain systems probably have changed rather little in size and composition during much of earth history. The core of the Earth may well be one such system. Other systems probably have grown steadily throughout earth history. The hydrosphere is almost certainly one of these, and the Earth's crust itself may well have increased appreciably in volume at the expense of the Earth's mantle. Still other systems, the biosphere for one, have probably fluctuated appreciably in volume during geologic time. All these systems are at least partially interdependent, so that the dynamics of the Earth are exceedingly complicated and are generally difficult to reduce to manageable proportions.

The major cycle. Many of these processes participate in what has come to be known as the major geochemical cycle, whose most important aspects are shown in Fig. 3. The weathering of igneous, sedimentary, and metamorphic rocks at the lithosphere-atmosphere boundary produces sediments. These are buried and compacted at depth to produce sedimentary rocks. At more elevated temperatures or pressures or both, sedimentary rocks recrystallize to metamorphic rocks. At still higher temperatures partial or complete melting may produce rock melts, which cool to form igneous rocks. Exposure of these rocks to weathering processes then renews the cycle.

Weathering processes. The chemical disequilibrium between the atmosphere and many of the rock types exposed to it is one of the important facts of geochemistry. The reactions of hydration, solution, oxidation, and carbonation, characteristic of the zone of weathering, are part of the continuing process of the approach to equilibrium at the lithosphere-atmosphere boundary.

The effect of the atmosphere, rain, ice, and groundwater on various minerals depends on their

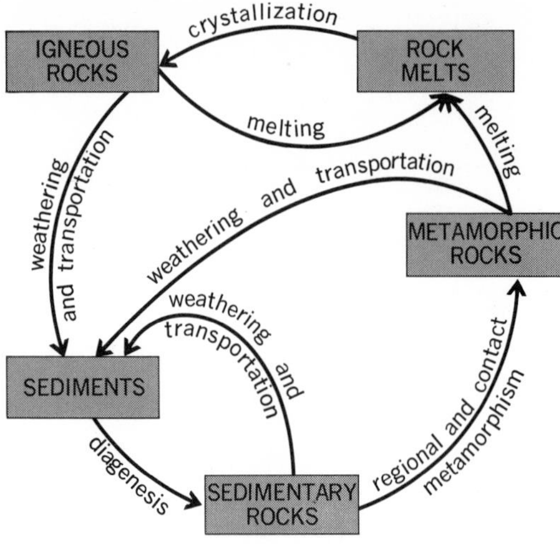

Fig. 3. Important features of major geochemical cycle.

stability in this type of environment and on the rate at which unstable minerals disappear in favor of stable ones. The feldspars, pyroxenes, and olivine are among the most readily destroyed rock-forming minerals. In this type of environment, the removal of the larger cations (Na^+, K^+, Ca^{++}, Mg^{++}) is usually rapid, and the remaining minerals are usually hydrated aluminum silicates containing none of these cations or containing them in appreciably smaller quantities than the original minerals. Some unstable rock-forming minerals react much more slowly with the atmosphere. Magnetite, Fe_3O_4, is one of these. The stable oxide of iron in equilibrium with atmospheric oxygen is hematite, Fe_2O_3, yet magnetite is a common constituent of beach sands, accumulates of the residue of weathering processes. Finally, minerals such as quartz, SiO_2, and calcite, $CaCO_3$, are stable in the zone of weathering. These minerals will only dissolve in rainwater and groundwater until these waters are saturated with respect to SiO_2 and $CaCO_3$. Weathering and transport processes thus alter exposed units of the lithosphere in the direction of equilibrium with the atmosphere but often do not reach this state completely. *See* WEATHERING PROCESSES.

Sedimentation and diagenesis. The processes that take place as sediments enter the oceans are complicated but normally alter their overall chemistry and mineralogy only slightly. However, major ore deposits of iron, potentially major deposits of copper and nickel in manganese nodules, and very large, low-grade uranium deposits have formed in the oceans, and evaporation in closed or partially closed basins has led to the formation of evaporite beds of composition very different from that of the normal detrital sediments. Sequences of sulfates, carbonates, and halides formed in such environments therefore are interspersed with sediments consisting largely of silicates and oxides. The overall composition of sedimentary rocks is appreciably richer in carbon dioxide, water, chlorine, nitrogen, sulfur, and boron than average igneous rock. This is not unexpected since CO_2 and H_2O, as well as geologically important compounds of the other elements, are in a gaseous state at the temperature of rock melts; they are undoubtedly lost preferentially to the Earth's surface and finally incorporated in sediments. *See* EVAPORITE, SALINE.

After deposition most sediments become buried. They therefore are subjected to progressively higher temperatures and pressures. At temperatures up to 200°C and pressures up to 2000 atm cementation, a loss of porosity, and a certain amount of recrystallization take place, and sandstones, limestones, and shales (the major types of sedimentary rocks) are formed. Such changes are perhaps most pronounced in the case of carbonaceous sediments. Peat becomes dehydrated and passes through the various grades of coal, approaching the composition of graphite with increasing depth of burial. Organic compounds in marine sediments can break down to form hydrocarbons; in suitable settings, these accumulate to form gas and oil fields. *See* DIAGENESIS.

Metamorphism. At higher temperatures and pressures important changes take place in the mineralogy and texture of most sedimentary rocks. These changes are termed metamorphism and lead to the formation of metamorphic rocks. There is no sharply defined boundary between diagenetic and metamorphic processes. Metamorphism may be either local in nature, as in the vicinity of a body of intrusive igneous rock, or may be regional, occasioned by a rise in temperature, pressure, or both over a considerable area. Reactions in both contact and regional metamorphism are determined by the prevailing pressure and temperature, by the composition of the rock units, and by the kinetics of the possible reactions. Understanding of these reactions has been increasing, especially since the introduction of water carbon dioxide, and salts as components in systems studied in the laboratory. Many metamorphic reactions are nearly isochemical, but volatile compounds such as H_2O, CO_2, CH_4, and H_2S are frequently lost, and cation-exchange reactions involving the chlorides of Na^+, K^+, Ca^{++}, and Mg^{++} may be important. *See* METAMORPHISM.

Melting and crystallization. At temperatures somewhat above 600°C, melting of rock units begins. The composition of rock melts depends on the composition of the parent material, on the ratio of liquid volume to that of the solid residuum, on the temperature and pressure during melting, and on the degree of equilibrium attained between the liquid and solid phases. The liquid fraction may move into higher levels of the Earth's crust and may break through to the surface as lava or other volcanic products. Crystallization normally takes place at the lower temperatures higher in the crust; the nature of the resulting solids is determined largely by the initial composition of the liquid and by the conditions prevailing during cooling. With the aid of pertinent laboratory studies the relationships between many different igneous rock types from the highly silicic granites through the intermediate gabbros to the ultramafic peridotites and dunites have been satisfactorily established, although the major question of the origin of the bulk of the siliceous rocks of the upper crust is still largely unanswered. *See* MAGMA.

This question leads directly to the problem of the nature and extent of exchange between mantle and crustal material. The validity of the proposal of ocean-floor spreading by H. H. Hess has been supported dramatically by studies of the intensity and direction of magnetization of oceanic rocks. F. J. Vine was the first to demonstrate the marked symmetry of the magnetization of rocks on opposite sides of mid-oceanic ridges and to suggest that this symmetry is related to ocean-floor spreading due to upwelling of mantle material under mid-oceanic ridges. W. J. Morgan has shown that this process is consistent with the present evidence of worldwide faulting and crustal movement.

The rapidity of this ocean-floor spreading implies that exchange of material between the mantle and the crust may be quantitatively important. The data suggest that sediments and volcanics on the ocean floor are returned to the mantle, that the reaction of sea water with oceanic basalt is quantitatively important, and that mantle-crust exchange during geologic time must be considered a two-way process. [HEINRICH D. HOLLAND]

Bibliography: L. H. Ahrens (ed.), *Origin and Distribution of the Elements,* 1969; W. Eitel, *Structural Conversions in Crystalline Systems and Their Importance for Geological Problems,* Geol. Soc. Amer. Spec. Pap. no. 66, 1958; V. M. Goldschmidt and A. Muir (eds.), *Geochemistry,* pt. 1, 1954; K. Krauskopf, *Introduction to Geochemistry,* 1967; B. Mason, *Principles of Geochemistry,* 3d ed., 1966; A. Poldervaart, *Crust of the Earth,* Geol. Soc. Amer. Spec. Pap. no. 62, 1955; A. E. Ringwood, in P. M. Hurley (ed.)., *Advances in Earth Science,* 1966; H. J. Rösler and H. Lange, *Geochemical Tables,* 1972; O. F. Tuttle and N. L. Bowen, *Origin of Granite in the Light of Experimental Studies in the System $NaAlSi_3O_8$-$KAlSi_3O_8$-H_2O,* Geol. Soc. Amer. Mem. no. 74, 1958; K. H. Wedepohl (exec. ed.), *Handbook of Geochemistry,* 1969; H. G. F. Winkler, *Petrogenesis of Metamorphic Rocks,* 2d ed., 1968.

Loess

An essentially unconsolidated, unstratified, calcareous silt. Most commonly it is homogeneous, permeable, buff to gray in color, and contains calcareous concretions and fossils. In natural and artificial excavations the loess maintains notably stable, vertical faces.

Texture and composition. Mechanical analyses of loess give the following composition: fine sand (grains > 0.074 mm), 0–10%; silt (0.074–0.005 mm), 50–85%; and clay (grains < 0.005 mm), 15–45%. Variations in published analyses may result in part from the different grade-size limits used by the investigators. Although dominantly silt and often referred to as well sorted, loess is actually only moderately well sorted. The silt (and sand) grains are usually angular to subangular. Clay occurs as silt-size aggregates, as coatings on silt grains, and as interstitial filling.

The silt and sand fraction of the loess has the following mineral composition: quartz, 50–70%; feldspars, 15–30%; carbonates (mainly calcite), 0–11%; and heavy minerals, 5–15%. X-ray identification of the minerals in the < 0.005-mm fraction show abundant quartz plus varying proportions of montmorillonoids, illites, and chlorites. In texture and mineral composition, loess is comparable to the average mudstone or shale. It is susceptible to authigenic changes such as the transformation of clay minerals, formation of chlorites, and the movement of soluble materials, especially carbonates. *See* AUTHIGENIC MINERALS; CLAY MINERALS; SEDIMENTARY ROCKS.

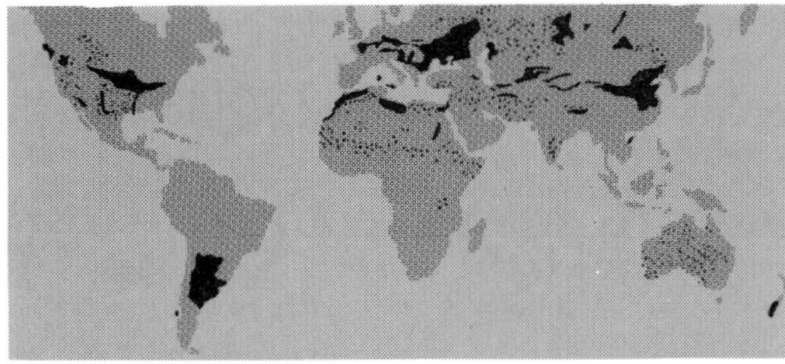

Loess deposits of the world. (*From A. K. Lobeck, Geomorphology, McGraw-Hill, 1939*)

Occurrence and origin. Loess is typically developed in areas peripheral to those covered by the last ice sheets in North America and Europe (see illustration). Greatest thicknesses (more than 150 ft) occur in the uplands bordering the valleys of the major streams. There is a general, but irregular, thinning of the loess in one or more directions away from each valley. Other notable areas are northern China and Argentina. Most of the loess seems to have been deposited during the latest (Wisconsin) glacial stage. Little or no loess is found associated with earlier glacial deposits or covering the area actually occupied by the latest ice sheet. No loess deposits older than the Pleistocene are known.

Many theories have been proposed to explain the deposition of loess. The most widely accepted theory is that the materials were transported and deposited by wind. Source areas were nearby floodplains and till plains. Controversy continues, however, for some investigators present evidence and arguments in favor of the alluvial origin of some loess. Others hold that primary deposits, by wind or water, have been greatly modified by colluviation to produce loess. *See* SEDIMENTATION.

[CHALMER J. ROY]

Bibliography: W. C. Krumbein and L. L. Sloss, *Stratigraphy and Sedimentation,* 2d ed., 1963; A. K. Lobeck, *Geomorphology,* 1939; F. J. Pettijohn, *Sedimentary Rocks,* 2d ed., 1957.

Macroevolution

The larger course of evolution by which the categories of animal and plant classification above the species have been evolved from each other and have differentiated into the forms within each. Only that part of evolution for which concrete evidence exists will be considered.

The organisms observed, alive or as fossils, can be classified into large groups or phyla. Within each of these the body is organized in variations of a single fundamental type. Such groups are the annelids, mollusks, echinoderms, and other phyla. Macroevolution is the process by which evolution has taken place within these phyla. In some cases evidence can be given that two or more such phyla have earlier evolved from a single ancestral group.

The evidence for the reality of macroevolution is derived from many sources. Paleontology provides evidence from fossils of the course of evolution. Fossils preserved at intervals of geological time may be compared and deductions made of their relationships. Thus, phylogenetic trees may be constructed. Studies of comparative morphology and embryology give evidence of many kinds from the similarities and dissimilarities of living animals. Geographical distribution of forms, together with the known history of the lands, gives data which should be in agreement with beliefs concerning the evolution of the organisms. *See* PALEONTOLOGY.

Speciation or microevolution has continued throughout evolution and provides its background. But the question remains whether the course of change in the long progress of macroevolution has been controlled in other ways besides the mechanisms considered in speciation. Another question for discussion is whether macroevolution is merely long-continued speciation. Before the evidence on this question can be considered, some characters of the general nature of macroevolution must be noted.

GENERAL CONSIDERATIONS

Throughout the course of evolution, change has occurred during the life of organisms. These organisms must always be competent to survive in competition with other organisms. The whole biology evolves; evolution is a phenomenon in natural history. Also, change of habit and habitat normally accompanies the evolution of structure. It is probably without meaning to ask which of these precedes the other.

Types of change. In all evolution one may distinguish two types of change, cladogenesis and phyletic evolution. Cladogenesis is the type of evolution associated with altered habit and habitat, usually in populations of the species separated from the rest. Adaptation to the new conditions results in division of the originally homogeneous species into distinguishable parts. Phyletic evolution is the gradual evolution of a population in its environment without division into isolated parts. This consists partly in improved adaptation to the conditions, partly in readaptation when the conditions change, as they frequently do in natural environments. Usually change in environmental conditions is gradual, but sudden and irreversible change sometimes occurs and organisms often spread into new habitats where conditions differ from those to which they are adapted. The organisms must then adapt quickly to the new conditions if they are to survive. In such circumstances evolution will be rapid under the changed selection due to the new conditions. This process is known as quantum evolution and is considered as a special although extreme case of phyletic evolution. It may lead to new groups on any taxonomic level. The distinctions between these types of evolution are illustrated diagrammatically in Fig. 1. The evolutionary changes in the organisms are similar in all the types.

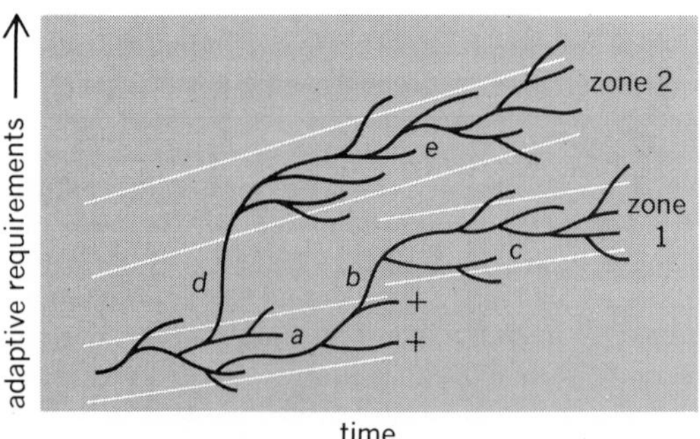

Fig. 1. Diagram of the main types of evolution: *a* represents phyletic evolution in an adaptive zone or niche (zone 1); *b* indicates sudden alteration of conditions in this zone, with the death (+) of most of the lines inhabiting it and the readaptation of one line to the new conditions, that is, quantum evolution; *c* shows phyletic evolution in the new conditions; *d* shows migration to a new niche (zone 2, cladogenesis); *e* indicates phyletic evolution in this zone. Rapid (tachytelic) evolution occurs at *b* and *d* and horotelic evolution at *a*, *c*, and *e*. (*Modified from G. G. Simpson, Tempo and Mode in Evolution, Columbia University Press, 1944*)

Adaptive radiation. The course of evolution differs between successful groups, such as the vertebrates or the insects, which have enlarged their areas of dominance with time and evolved into many smaller groups, and other groups, such as many forms among the lower invertebrates, which show no progressive evolution and have often survived through long ages with little change.

The knowledge of vertebrate evolution is the most detailed. When the evolution of this successful group is examined it is found that a limited number of dominant types (several groups of fishes, amphibians, reptiles, birds, and mammals) have followed each other, each in a new mode of vertebrate life, and that each was evolved from the previous dominant group. Further, each dominant group divided soon after its appearance into a large number of subsidiary types adapted to more restricted modes of life within the range of the larger group. This diversification into different adaptive zones is known as adaptive radiation. The radiation of the primates, the mammalian order including the monkeys, apes, and man, is illustrated in the diagram in Fig. 2. In adaptive radiation the greater part of the changes of form are due to alterations of the relative sizes of the parts of the body by modifications of their growth rates during development and not to the evolution of new organs. In the mammals the radiating lines appear today as the orders of bats, primates, rodents, and so on. All of these can be traced back to near the beginning of the dominance of the mammals. Each line continues to radiate throughout its successful history. Formation of new organs, which are not modifications of organs previously present, occurs in the radiation but accounts for only a small part of the changes. The development of horns in the mammals is an example.

The evolution of a major dominant group from a preceding group, for example, the mammals or birds from the reptiles, or the amphibians from the fishes, demands much more fundamental changes than those of adaptive radiation. The whole form and function of the body is reorganized. It takes a considerable time, for example, about 40,000,000 years in the mammals and birds, and proceeds by successive alterations of one feature after another toward the new type. It occurs in one of the radiating lines of the preceding group, while its other lines, though they continue to radiate, clearly remain members of the preceding group.

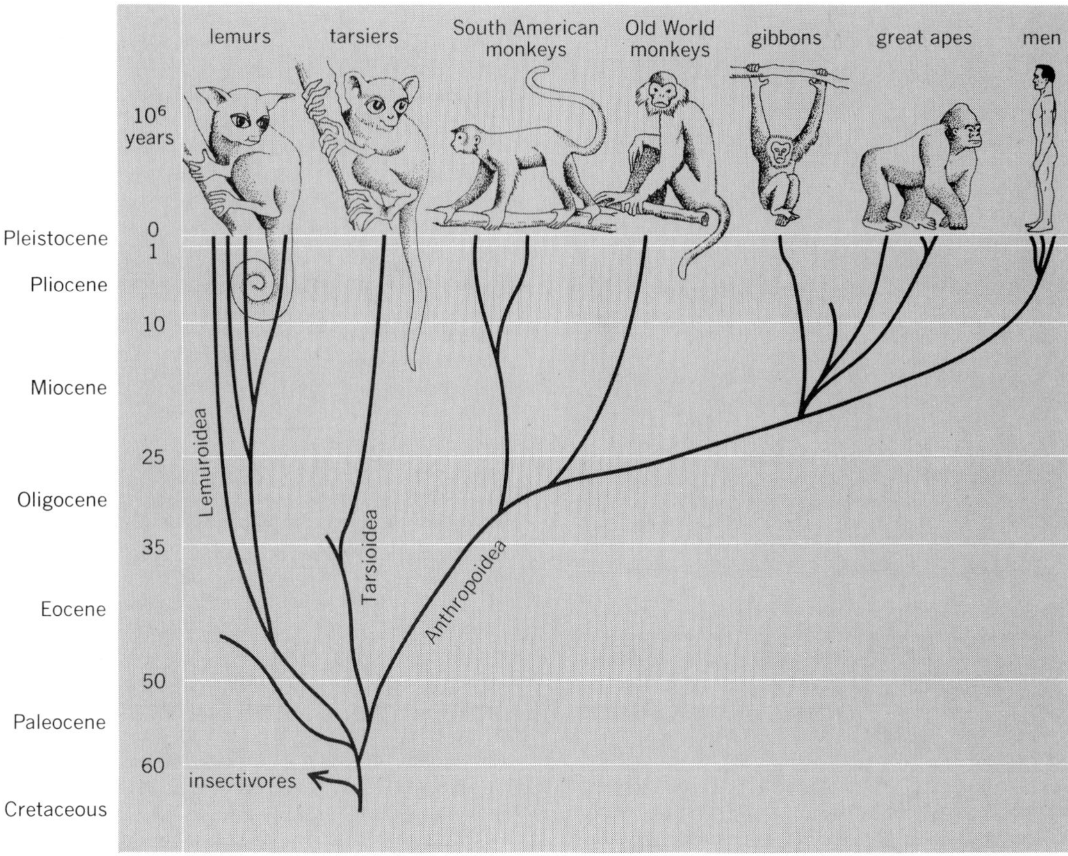

Fig. 2. Evolution of the primates. (*Modified from J. Z. Young and P. P. Grasse*)

SPECIAL PROBLEMS

Some special problems raised by the study of macroevolution are considered next.

Adaptation. Adaptation is the process by which, as the result of selection, the biology of an organism becomes capable of satisfying the needs of its life. It includes besides the general adaptation that all organisms must possess to be viable, many more special forms such as mimicry and protective resemblance. The conditions to which adaptation is necessary are of many kinds. The physical conditions of its surroundings, the biotic conditions of the organism's contacts with other species, the intraspecific conditions of its contacts within its own species, and the conditions within the organism's body needed to maintain its life and health, all require adaptation. If distantly related animals live similar lives under like conditions, their adaptations may be similar, and they are said to be convergent. Thus the body forms of whales and porpoises are convergent with those of many fishes and ichthyosaurs (Fig. 3). When related forms adapt to the same needs, they may evolve similar adaptations by parallel evolution (Fig. 4). Many different birds have evolved webbed feet for swimming, and the parallel adaptations of marsupials and placental mammals to various habits of life are well known. In parallel evolution related organisms evolve in the same direction by similar genetic changes; in convergence, organs originally unlike come to resemblance by adaptation to similar conditions. In both phenomena the resemblances will not be exact, for the genetic changes on which they depend are unlikely to be identical.

Specialization and survival. Organisms are often adapted to narrow ranges of conditions in small environments, known as niches, and may be confined to these conditions. Though natural environments are continually changing and change of conditions requires constant readaptation, this specialization may be maintained indefinitely, provided the niche remains available. This close specialization will be of selective advantage, for the organism so specialized is likely to be more efficient in its niche. Its specialization is therefore likely to become closer with time. But specialized forms are always open to the danger that as conditions change their niches may cease to exist and the organisms may be exterminated. Close specialization is therefore a trap into which organisms may be led by the force of selection. Probably for these reasons, new successful groups have generally been evolved from the less specialized members of earlier groups, and the forms which have survived longest have been those adapted to a wide range of conditions.

Preadaptation. In spite of the dangers of specialization, animals specialized to one mode of life have sometimes in evolution passed to another mode of life and become readapted to it. For instance, horses at first lived in forests browsing on leaves and later took to grazing on grassy plains; the fishes ancestral to the amphibians passed from

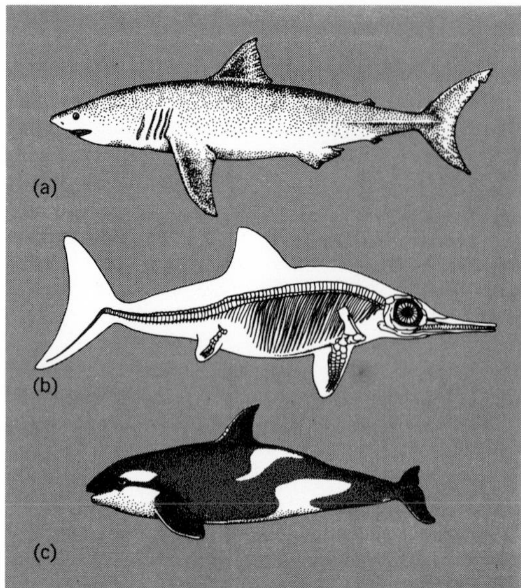

Fig. 3. Convergent evolution of body form in three actively swimming marine vertebrates. (*a*) A shark (*from J. Z. Young, The Life of Vertebrates, Oxford University Press, 1950*). (*b*) An ichthyosaur (Reptilia). (*c*) The killer whale (*Orcinus*, Mammalia) (*from J. Z. Young, The Life of Vertebrates, Oxford University Press, 1950*).

aquatic to amphibious life; many pests have taken to feeding on crops which may be very different from their natural food. Large changes of these kinds probably took place in many steps, but even so they raised the difficulty of explaining how animals specialized to one mode of life have been able to survive in a life in which different adaptations would seem to be required. It is thought that the large changes are made possible when, and perhaps only when, adaptations already made before the change happened to be adaptive after the change, that is, when the animals were preadapted

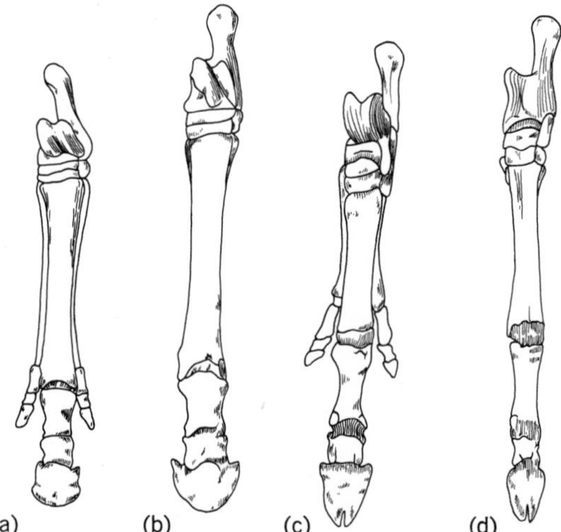

Fig. 4. Parallel evolution as illustrated by the reduction of digits in horses and the Litopterna, a South American group of mammals. (*a*) Three-toed horse (*Protohippus*). (*b*) Single-toed horse (*Equus*). (*c*) Three-toed litopternan (*Diadiaphorus*). (*d*) Single-toed litopternan (*Thoatherium*). (*From G. S. Carter, Animal Evolution, Sidgwick and Jackson, 1951*)

to the change. Since this correlation between conditions before and after the change could only occur by chance, it is unlikely to be frequent; indeed, large changes of this type are rare in evolution.

Correlation. In order that the organism may remain viable and efficient enough to survive under competition, it is necessary that all parts of the body should be so correlated in structure and activities as to be able to work together as a single, efficient whole. This correlation must always be maintained in evolution. Changes which may be advantageous for the functioning of one part may be disadvantageous to the functioning of the whole body unless other parts are modified. Thus, in mammals large horns may be valuable in fighting but they require strengthening of the bones of the skull and of the neck muscles to support them; elongation of the horse's legs required simultaneous lengthening of the neck to enable it to graze. When the change in an organ is small, correlation may be maintained by the required changes in other organs taking place during the individual's life history, in response to the demands made on these organs, for example, the blacksmith's arm becomes stronger with use. This principle is known as organic selection. Such phenotypic changes will not be inherited, and therefore will not be of evolutionary value, but they may allow the organism to survive until mutations giving the required result occur. If the functional change is large and sudden, organic selection will not be able to maintain correlation. This is probably one of the reasons why large and sudden changes are at least rare in evolution.

The need to maintain correlation reduces the speed of evolutionary change, but since variation occurs in all parts of the body, it need not prevent it. It implies that evolution should be looked upon as a process of change which occurs in the whole organization of the body, rather than in its parts separately.

Hypertely. In numerous animals organs are found so fantastically overdeveloped that it is hard to believe that they are not harmful to the activities of the animal. Such organs are called hypertelic. The peacock's tail and the horns of the Irish elk are examples among many. These organs are not useless; that is, they have their uses in sexual activities, fighting, or other activities, and the survival of the animals, now or in the past, shows that they are not so harmful as to render the animals nonviable. The organs are specializations to definite functions in the life of the animals, and as such their development will be favored by selection until they so far impair viability as to lead to extinction. In some cases they may be genetically bound to other and valuable characters. In deer the relative size of horns increases with body size. If large size is valuable to the elk, the horns should increase in size beyond the size that is optimum for them. A balance would be struck between too large horns and the advantage of large size.

Irreversibility. It is often said that evolution is irreversible. This is an overstatement, for often organs previously evolved have been lost; for example, snakes have lost their legs and birds the reptilian teeth. Their evolution has thus been reversed. But it is true that when an organ has been lost it is hardly ever regained in later evolution in the same form. This rule, proposed by L. Dollo, is an immediate inference from the fact that the genetic backgrounds of all the organs are complex, many genes determining the structure and functions of each. If this complex of genes is lost, it is not likely to be exactly regained by new mutation. A new organ evolved to serve the same purpose will differ in its details. Dollo's rule is not true of characters controlled by single genetic factors or even by a few genes. These may be regained by new mutation.

Parallelism. One surprising fact in some examples of macroevolution is that distantly related organisms sometimes evolve very similar organs in parallel, in spite of the complex genetic backgrounds of organs. Birds and mammals have evolved separately since Paleozoic times but have evolved very complex and detailed similarities in the structure of their hearts, ears, and other organs. No full explanation of this aspect of evolution has been given.

Extinction. The fauna and flora of the world change with time either by the extinction of the earlier forms or by their further evolution. Organisms rarely survive indefinitely. Extinction may be due to various causes. Catastrophies may sometimes have caused the extinction of species, especially soon after their evolution. Failure to readapt to changed conditions has probably been a more general cause; the extinction of many groups of reptiles in the Late Cretaceous and of many other groups in the Permian was probably caused by changes of climatic conditions. Competition with forms later evolved has very often led to the extinction of earlier types. In general, evolution has progressed toward better organization and greater efficiency. It is to be expected that the earlier forms should most frequently be eliminated.

In view of the progressive character of evolution it is at first sight surprising that organisms are found alive today on all levels of organization from the simplest unicellular forms upward. Many of the simpler organisms are common and widespread and even dominant in restricted habitats. One might expect them to have been exterminated by organisms later evolved. The explanation seems to be that these simple forms are well adapted to the life they lead and either are able to avoid the competition of more complex forms or, like many coelenterates, have evolved organs of defense against their attacks. In some cases they have been able to hold their own by adaptation to special habits of life such as parasitism or other forms of association with the later forms.

Relicts. Besides the simple organisms that are still widespread, a few representatives are often found of groups that were once dominant but have become otherwise extinct. The lampreys and hagfishes alone represent a very early vertebrate group, the Agnatha; the king crab (*Limulus*) is the only representative of a formerly widespread group of arachnids; *Peripatus* is the only living type of the Onychophora, an early arthropod group. These relicts are forms that have evolved slowly, and have escaped extinction by close adaptation to the conditions of their life. Geographical relicts are also found surviving in restricted regions where they have found refuge. The shrimp *Mysis relicta* is found in lakes of northern Europe and is probably a marine form isolated when the lakes were closed and later adapted to fresh-water life; *Peripatus* occurs today in several forest regions of the warmer countries, these being probably relict parts of a much wider distribution.

Rates of evolution. The rates at which organisms evolve vary from almost zero in a few forms that have survived as genera from the Paleozoic to Recent times (*Lingula*, Brachiopoda, Cambrian to Recent, 450,000,000 years; *Ostrea*, Lamellibranchiata, Carboniferous to Recent, 250,000,000 years) to the rapid evolution of the char (*Salvelinus*), a fish that has evolved to at least subspecific level in European lakes in the 12,000 years since the end of the Ice Age. Even more rapidly, the Faeroe house mouse has evolved a new subspecies in historic times, in the last 500 years.

Between these extremes lie the great majority of evolutionary rates, but these also vary widely. F. E. Zeuner has estimated that the normal rate in the mammals, a progressive group, is for a specific difference to evolve in 1,000,000 years and a generic difference in 4,000,000 or 5,000,000 years. Many groups evolve more slowly. G. G. Simpson in 1953 found that the mean survival time for genera of bivalves, the Lamellibranchiata, is about 50,000,000 years. The rate may vary during the history of a group. Often it is rapid when the group is radiating actively shortly following its appearance; subsequently it settles down to a considerably slower rate.

Both the slow (bradytelic) and rapid (tachytelic) exceptional rates seem to differ fundamentally in their causes from the normal or horotelic rates. It is thought that the differences are not due to different mutation rates. Probably bradytelic rates occur in forms which are almost perfectly adapted to modes of life in abnormally constant environments, so that any change is likely to be harmful. Tachytelic evolution is probably due to rapid change in the environment, leading to changes in the direction of selection. In some cases, such as that of the char in European lakes, it is probable that isolation has led to independent evolution and has encouraged differentiation.

One cannot expect to reach so complete an interpretation of macroevolution as of speciation, for the evidence is less complete. But the phenomena observed in it do not often conflict with the modern theory of the causes of evolution. *See* ANIMAL EVOLUTION; EVOLUTION, ORGANIC.

[GEORGE S. CARTER]

Bibliography: G. S. Carter, *Animal Evolution*, 1951; G. S. Carter, *Structure and Habit in Vertebrate Evolution*, 1967; W. K. Gregory, *Evolution Emerging*, 2 vols., 1951; J. S. Huxley, *Evolution: The Modern Synthesis*, 1943; J. S. Huxley, *Problems of Relative Growth*, 1932; G. L. Jepson, E. Mayr, and G. G. Simpson (eds.), *Genetics, Palaeontology and Evolution*, 1949; G. G. Simpson, *The Major Features of Evolution*, 1953; F. E. Zeuner, *Dating the Past*, 2d ed., 1950.

Magma

The hot material, partly or wholly liquid, from which igneous rocks form. Besides liquids, solids and gas may be present in magma. Most observed magmas are silicate melts with associated crystals and gas, but some inferred magmas are carbonate, phosphate, oxide, sulfide, and sulfur melts.

Strictly, any natural material which contains a finite proportion of melt (hot liquid) is a magma. However, magmas which contain more than about 60% by volume of solids generally have finite strength and fracture like solids.

Hypothetical, wholly liquid magmas which develop by partial melting of previously solid rock and segregation of the liquid into a volume free of suspended solids and gas are called primary magmas. Hypothetical, wholly liquid magmas which develop by crystallization of a primary magma and isolation of rest liquid free of suspended solids are called parental (or secondary) magmas. Although no unquestioned natural examples of either primary or parental magmas are known, the concepts implied by the definitions are useful in discussing the origins of magmas.

Bodies of flowing lava and natural volcanic glass prove the existence of magmas. Such proven magmas include the silicate magmas corresponding to such rocks as basalt, andesite, dacite, and rhyolite as well as rare carbonate-rich magmas and sulfur melts. Oxide-rich and sulfide-rich magmas are inferred from textural and structural evidence of fluidity as well as mineralogical evidence of high temperature, together with the results of experiments on the equilibrium relations of melts and crystals. *See* IGNEOUS ROCKS.

Temperatures. The temperatures observed for silicate magmas flowing on the surface (lavas) range from slightly greater than 1200°C for basalts to about 800°C for rhyolites. The lower temperatures reflect both the nonequilibrium persistence of supercooled liquid and a lower temperature of melting at equilibrium for rhyolite than for basalt. Temperatures inferred for intrusive magmas vary widely and are both higher and lower than those observed for lavas. Both relations are possible, because some very hot magmas may cool before reaching the surface and because pressure enhances the solubility in melts of H_2O and other volatile substances which lower the equilibrium temperature of freezing. The volatile substances may largely escape during crystallization, and the minerals in a rock may interact during slow cooling; therefore, it is generally impossible to infer, with certainty, the temperatures of intrusive magmas.

Density. The mass density of silicate magmas (about 2.7 to 2.4 g/cm^3) depends upon composition and temperature. In general, the density of the melt is less than that of crystals forming from it. Thus, buoyancy forces lead to gravitative settling of crystals. In hot fluid basaltic magmas, millimeter-sized crystals may settle at rates of centimeters per hour. Indeed, some pillows of basalts 20 cm in diameter are known to contain accumulations of olivine crystals in their lower parts. In some iron-rich melts, feldspar crystals may float. Flotation of feldspar may occur on Earth rarely and probably has occurred on the Moon.

Many basaltic magmas are denser than continental crustal material and are necessarily forced to the surface either by hydrostatic head in the mantle, where the encasing rock is denser than the melt, or by gas propulsion. Most other magmas are less dense than most rock types and are buoyant everywhere within the Earth. *See* BASALT.

Viscosity. Silicate magmas are variably viscous, ranging from about 100 poise (1 poise = 0.1 N s/m^2) to more than 100,000 poise. Viscosity of magmas increases with decrease in temperature, with decrease in dissolved H_2O, with increase in crystal fraction, and with increase in fraction of SiO_2. In natural magmas the above variables are not independent; the viscosity of silicate magmas generally increases with fraction of SiO_2 in the melt. Thus, basaltic magmas produce flat-lying lava fields, whereas rhyolitic magmas commonly yield mounds and domes. Magmas with more than a few percent of crystals may have a finite yield strength.

Structure. Silicate melts are ionic solutions and have high electrical conductivity (100 ohm^{-1}cm^{-1}). The ions are difficult to characterize, however, and little is known regarding their molecular weights or sizes. Melts rich in silica are highly polymerized, which fact accounts for their high viscosity. Following the practice used in studies of slags, siliceous melts are commonly termed acidic (low activity of oxygen anion), and melts rich in basic oxides such as MgO and CaO are termed basic (high activity of oxygen anion).

Crystallization. Magmas crystallize as a consequence of cooling or effervescence, or both. Minerals grow in crystallizing magmas according to mass-action principles whereby the most abundant, least soluble constituents crystallize first, followed by less abundant, more soluble substances. Most magmas display progressive enrichment of both crystals and residual melts in soda relative to lime and ferrous iron relative to magnesia with crystallization. In many systems an immiscible sulfide melt develops at an early stage. Minerals (such as apatite and zircon) which contain high concentrations of minor elements generally crystallize late and tend to have comparatively high ratios of surface area to volume. However, minerals which crystallize early in one magma may form late or not at all in other magmas. *See* PETROLOGY.

Magmas which cool and crystallize quickly (in a few decades or less) develop small crystals (a few millimeters in diameter or smaller). Magmas which cool and crystallize slowly (hundreds of thousands of years) develop large crystals (up to a few centimeters, rarely many tens of centimeters).

Occurrence. Magma is presumed to underlie regions of active volcanism and to occupy volumes comparable in size and shape to plutons of eroded igneous rocks. However, it is not certain that individual plutons existed wholly as magma at one time; therefore, actual bodies of magma may be smaller than plutons. Because some catastrophic eruptions have yielded up to about 100 km^3 of material on a void-free basis in a single event of short duration (few weeks or years), it is certain that some large bodies of magma have existed at one time, and presumably exist today. Magma may underlie some regions where no volcanic activity exists, because many plutons appear not to have vented to the surface.

Remote detection of bodies of magma is difficult because the properties of magma are closely approximated by hot fractured rock filled with water or water vapor. Seismic, gravimetric, and magnetic data are consistent with the existence of major bodies of magma under Yellowstone Park: Katmai-Trident, AK; regions in Kamchatka; and Japan.

Origin. Diverse origins are probable for various magmas. Basaltic magmas because of their high temperatures probably originate within the mantle several tens of kilometers beneath the surface of the Earth. Experiments at high pressure have led many scientists to conclude that most basaltic magmas are modified by crystallization and separation of residual liquid from crystals between the site of melting in the mantle and the place of crystallization within the crust. Rhyolitic magmas may originate through crystallization of basaltic magmas or by melting of crustal rock. Intermediate magmas may originate within the mantle or by crystallization of basaltic magmas, by melting of appropriate crustal rock, and also by mixing of magmas or by assimilation of an appropriate rock by an appropriate magma. *See* PETROGRAPHIC PROVINCE; VOLCANO.

[ALFRED T. ANDERSON]

Bibliography: I. S. E. Carmichael, F. J. Turner, and J. Verhoogen, *Igneous Petrology*, 1974; G. A. Macdonald, *Volcanoes*, 1972; L. R. Wager and G. M. Brown, *Layered Igneous Rocks*, 1968.

Magnesite

The mineral form of magnesium carbonate, $MgCO_3$. Iron, manganese, and cobalt may replace some magnesium in magnesite.

Magnesite has hexagonal (rhombohedral) symmetry and exhibits the same structure as calcite. It is usually massive and white, but iron impurities may give it a brownish tint (see illustration). The specific gravity is 3 and the hardness is 4 on Mohs scale. Magnesite is stable up to 740°C at 10,000 psi and up to 850°C at 30,000 psi of carbon dioxide. The equilibrium replacement of magnesium by calcium increases with temperature up to 2% at 900°C.

Magnesite crystal in chlorite schist; the sample is from Tyrol, Austria. (*Specimen from Department of Geology, Bryn Mawr College*)

Magnesite may be associated with peridotites, soapstones, and dolomites, or it may occur as sedimentary deposits. Deposits are in Austria, Manchuria, California, Nevada, and Washington. Magnesite is a source of magnesia. *See* CARBONATE MINERALS.

[ROBERT I. HARKER]

Magnetite

A cubic mineral and member of the spinel structure type, space group Fd3m, with composition $[Fe^{3+}]^{IV}$ $[Fe^{2+}Fe^{3+}]^{VI}$ O_4. It possesses the inverse spinel structure, in which half the ferric iron is tetrahedrally coordinated and the remaining half as well as all ferrous iron are octahedrally coordinated by the cubic close-packed oxygen atoms. The color is opaque iron-black and streak black, the hardness is 6 (Mohs scale), and the specific gravity is 5.20. The habit is octahedral (see illustration), but the mineral usually occurs in granular to massive form, sometimes of enormous dimensions. Though possessing no cleavage, some specimens show an octahedral parting. Limited Al^{3+} may substitute for Fe^{3+}; and Ca^{2+}, Mn^{2+}, and Mg^{2+} for Fe^{2+}.

Magnetite. (*a*) Crystal on block of mica schist with small quartz crystals on left, Binnenthal, Switzerland (*American Museum of Natural History specimens*). (*b*) Crystal habits (*from C. S. Hurlbut, Jr., Dana's Manual of Mineralogy, 17th ed., copyright © 1959 by John Wiley & Sons, Inc.; reprinted by permission*).

Magnetite is a natural ferrimagnet, but heated above 578°C (the Curie temperature) it becomes paramagnetic. It can be prepared by heating hematite (α-Fe_2O_3) in a reducing atmosphere, by heating hematite in air above 1400°C, or by the oxidation of iron at high temperatures. Oxidation of magnetite leads to maghemite (γ-Fe_2O_3), which gradually inverts to hematite.

The major magnetic ore of iron, magnetite may be economically important if it occurs in sufficient quantities. It often occurs as an early magmatic segregation in basic rocks, such as anorthosites and norites where it has accumulated in bands by gravity settling. It is also a contact metamorphic product, occurring with limestones, leptites, and so forth. Moderately resistant to weathering, magnetite sometimes accumulates in detrital sands. The most spectacular ore body occurs at Kiruna in northern Sweden, where the magnetite fluid differentiate was injected into syenitic rocks. Other important occurrences are in Norway, the Soviet Union, and Canada. Magnetite is common as an accessory mineral in igneous rocks throughout the world. *See* EMERY; HEMATITE; ILMENITE; ORE AND MINERAL DEPOSITS; SPINEL.

[PAUL B. MOORE]

Malachite

A basic carbonate of copper with the chemical formula $Cu_2(OH)_2(CO_3)$. Malachite is normally associated with the more important copper ore deposits. Large quantities of malachite have been found in Siberia and used for ornamental stone. It has been mined as an ore of copper near Kolwezi, Republic of the Congo. It is common in much smaller amounts in the western mining districts of the United States.

Malachite is monoclinic but usually occurs in massive forms or in bundles of radiating fibers (see illustration). It is invariably green. It has a specific gravity of 4.05 and a hardness of 3.5–4 on Mohs scale. Malachite can be readily synthesized in a number of different ways.

[ROBERT I. HARKER]

Manganite

A mineral having composition MnO(OH) and crystallizing in the orthorhombic system in prismatic crystals with deep vertical striations (see illustration). Manganite crystals are often arranged in radiating masses. There is one perfect cleavage. The hardness is 4 on Mohs scale, and the specific gravity is 4.3. The luster is metallic and the color iron black. Manganite is associated with other manganese oxides, and pseudomorphs of pyrolusite after manganite are common. Fine crystals have been found in the Harz Mountains; in Cornwall, England; and in the United States at Negaunee, Mich. Manganite is a minor ore of manganese. *See* PYROLUSITE.

[CORNELIUS S. HURLBUT, JR.]

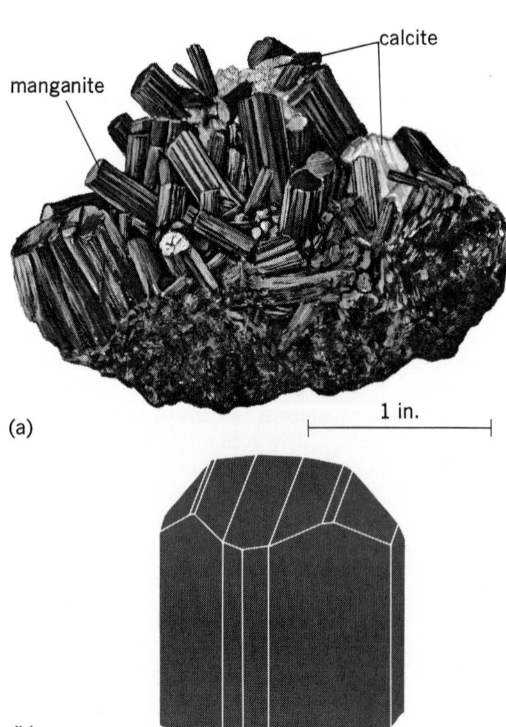

Manganite. (*a*) Crystals with calcite from Ilfeld, Harz, Germany (*specimen from Department of Geology, Bryn Mawr College*). (*b*) Crystal habit (*from L. G. Berry and B. Mason, Mineralogy, copyright © 1959 by W. H. Freeman and Co.*).

Marble

A term applied commercially to any limestone or dolomite taking polish. Marble is extensively used for building and ornamental purposes. *See* LIMESTONE.

In petrography the term marble is applied to metamorphic rocks composed of recrystallized calcite or dolomite. Schistosity, often controlled by the original bedding, is usually weak except in impure micaceous or tremolite-bearing types. Calcite (marble) deforms readily by plastic flow even at low temperatures. Therefore, granulation is rare, and instead of schistosity there develops a flow structure characterized by elongation and bending of the grains concomitant with a strong development of twin lamellae. In calcite the twinning plane is the flat rhombohedron $\{01\bar{1}2\}$; in dolomite twinning is markedly less common and the plane is the steep rhombohedron $\{02\bar{2}1\}$. *See* METAMORPHIC ROCKS; MINERALOGY.

Regional metamorphic marbles are more or less deformed and therefore texturally marked by characteristic deformation patterns—parallel elongation of irregularly bounded lensoid grains of calcite (less typically of dolomite)—and by preferred orientation of mica flakes, if mica is present, producing a tightly woven, close fabric. This makes the rock, particularly the pure calcite (not dolomitic) marble, well suited for building purposes.

Contact metamorphism of limestones and of dolomites produces granoblastic rocks composed mainly of a mosaic of equant grains of calcite. The fabric is loose and the rock not suitable for use as building material.

Pure marbles attaining 99% calcium carbonate, $CaCO_3$, are often formed by simple recrystallization of sedimentary limestone. Dolomite marbles are usually formed by metasomatism. More interesting petrographically are impure marbles in which various reactions between silicates and carbonates have taken place.

If an impure carbonate rock is subjected to slowly increasing temperature at constant pressure, a series of reactions involving progressive elimination of carbon dioxide takes place. Each of these reactions is determined by the temperature, pressure, and bulk composition of the reacting system. In natural rocks, water and traces of other fugitive compounds are usually present. These facilitate the metamorphism and permit certain reactions to take place at low temperatures. The pressure usually has no great influence.

If at the beginning of the reaction there is a dolomitic limestone at low temperature composed of the phases calcite + dolomite + quartz (+ traces of water), certain mineral sequences will develop at given stages in the metamorphism. If the temperature is raised only a moderate amount, talc will form, followed by tremolite, forsterite, diopside, and brucite, until eventually wollastonite forms according to the reaction shown below. This reac-

$$\underset{\text{Calcite}}{CaCO_3} + \underset{\text{Quartz}}{SiO_2} \rightarrow \underset{\text{Wollastonite}}{CaSiO_3} + \underset{\text{Carbon dioxide}}{CO_2}$$

tion has been carefully studied, and it can be shown that it takes place at about 450°C (assuming no great buildup of CO_2 pressure). At higher temperatures rare lime-silicates, such as monticellite, akermanite, tilleyite, spurrite, merwinite, and others, develop.

Examinations of natural occurrences show that at the relatively low temperatures of regional metamorphism the diopside stage is rarely passed. At the somewhat higher temperatures of the contact zones about granites, which come from the coolest of magmas, the wollastonite stage may be attained but not passed. It is only at the hotter contacts of syenitic or granodioritic masses that members higher in the series begin to form, and the highest members are associated only with basic, and usually basaltic, rocks.

[T. F. W. BARTH]

Marcasite

A mineral having composition FeS_2 and crystallizing in the orthorhombic system. Marcasite crystals are common, usually tabular parallel to the basal plane, and characteristically twinned, giving cockscomb groups (see illustration). Marcasite frequently has a radiating structure and may be globular or stalactitic. There is poor prismatic cleavage. The hardness is 6–6.5 on Mohs scale and the specific gravity is 4.89. The luster is metallic and the color pale bronze-yellow to nearly white on a fresh fracture. Marcasite and pyrite are dimorphous; both have the composition FeS_2. Because marcasite is whiter, it is called white iron pyrite. It is much less common and less stable than pyrite. Specimens may in a year or two completely disintegrate, with the formation of ferrous sulfate and sulfuric acid.

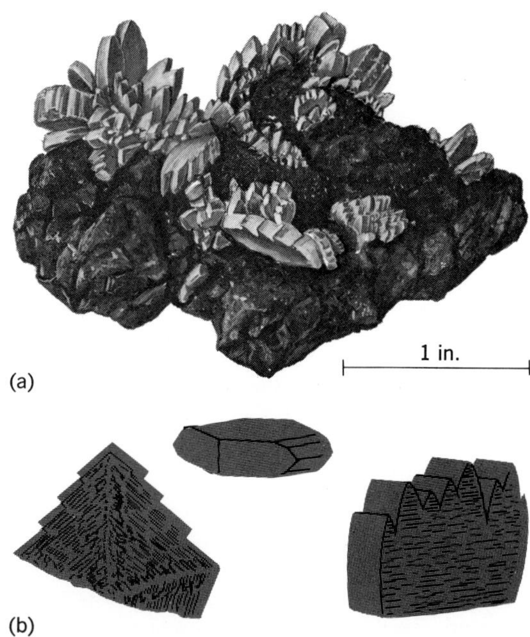

Marcasite. (*a*) Typical cockscomb groups of crystals found in galena-sphalerite ore, Carterville, Mo. (*specimen from Department of Geology, Bryn Mawr College*). (*b*) Crystal habits (*from C. S. Hurlbut, Jr., Dana's Manual of Mineralogy, 17th ed., copyright © 1959 by John Wiley & Sons, Inc.; reprinted by permission*).

Marcasite is usually found in surface or near-surface deposits in which it was formed at low temperatures from acid solutions. Most marcasite is supergene, but in some places it is believed to have been deposited near the surface from ascending vein solutions. Under these conditions it is usually the last mineral to be deposited.

It is found in metalliferous deposits associated with lead and zinc ores, as replacement deposits in limestone, and in concretions in clays and shales. The nodular and lenticular masses in coal known as brasses are in part marcasite and in part pyrite. *See* PYRITE.

[CORNELIUS S. HURLBUT, JR.]

Marine geology

The study of the portion of the Earth beneath the oceans. More than 70% of the Earth's surface is covered by marine waters. Of the oceanic area (361×10^6 km^2) approximately 300×10^6 km^2 is contributed by the deep-sea floor; the remaining 60×10^6 km^2 represents the submerged margins of the continents. The distribution of elevations on the Earth is shown in Fig. 1.

Soundings. Soundings are measurements of ocean depth made from ships. Early soundings were made with a lead attached to a hemp line; about 1875 the hemp line was replaced by piano wire. Since about the middle of the 1920s, virtually all deep-sea soundings have been made by echo sounding. The echo-sounding machine sends out a sound pulse (10–20 Hz) and then times the interval from the sound pulse to the returning echo. The early sounders required manual operation, but since about 1935 automatic recording sounders, which plot a graph of depth versus time or distance, have been used almost exclusively. Since 1953 precision, high-resolution echo sounders have been used in increasing numbers.

Sounding corrections. Wire and hemp line soundings require a correction for wire angle, for stretch of the wire, and for calibration of the me-

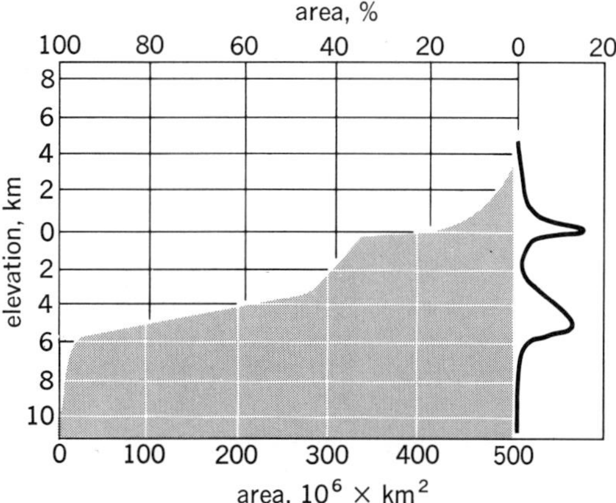

Fig. 1. Hypsographic curve showing area of Earth's solid surface above any given level of elevation or depth. Curve at the right shows frequency distribution of elevations and depths for 2-km intervals.

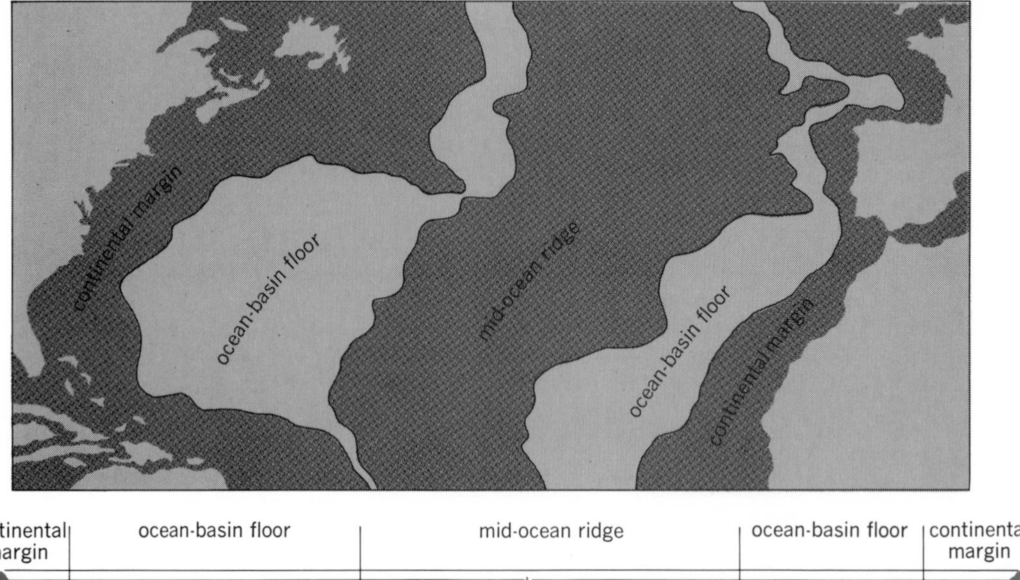

Fig. 2. Major morphologic divisions of North Atlantic Ocean. The profile is from New England to Sahara.

tering counters used. Echo soundings require a correction for sound velocity, since the average vertical velocity is not constant, and for slope of the bottom, as the point from which the first echo returns is not always directly beneath the ship. In addition, corrections for inaccuracies of timing and mechanical imperfections must be made for most sounders. The position of the sounding lines is generally determined by standard astronomical fixes and dead reckoning. Errors of a few miles are the rule in deep-sea sounding surveys.

PHYSIOGRAPHIC PROVINCES

The Earth relief lies at two dominant levels (Fig. 1); one, within a few hundred meters of sea level, represents the normal surface of the continental blocks; the other, between 4000 and 5000 m below sea level, and comprising more than 50% of the Earth's surface, represents the ocean-basin floor. The topographic provinces beneath the sea can be included under three major morphologic divisions: continental margin, ocean-basin floor, and mid-oceanic ridge. These are indicated on a typical transoceanic profile taken from the North Atlantic in Fig. 2. Each of these major divisions can be further divided into categories of provinces and those into individual physiographic provinces (Fig. 3).

Continental margins. The continental margin includes those provinces associated with the transition from continent to ocean floor. The continental margin in the Atlantic and Indian oceans is generally composed of continental shelf, continental slope, and continental rise. A typical profile off the northeastern United States is shown in Fig. 4.

Gradients on the continental shelf average 1:1000, while on the continental slope gradients range from 1:40 to 1:6, and occasionally local slopes approach the vertical. The continental rise lies at the base of the continental slope. Continental rise gradients average 1:300, but individual slope segments may be as low as 1:700 or as steep as 1:50. The continental slopes are cut by many submarine canyons. Some of the larger canyons such as the Hudson extend across the continental rise (Fig. 4). Submarine alluvial fans extend out from the seaward ends of the larger canyons. *See* CONTINENTAL SHELF AND SLOPE; SUBMARINE CANYON.

The continental margin can be divided into three categories of provinces. Category I includes the continental shelf, marginal plateaus, and shallow epicontinental seas, all slightly submerged portions of the continental block. Category II includes the continental slope, marginal escarpments, and the landward slopes of marginal trenches, all expressions of the outer edge of the continental block. Category III includes the continental rise, the ridge-basin complex, and the ridge-trench complex. The continental slope of the northeastern United States can be traced directly into the marginal escarpment (Blake Escarpment) off the southeastern United States (Fig. 5) and the landward slope of the Antilles marginal trench (Puerto Rico). The continental rise off New England can be traced into the Antilles Outer Ridge. Seismic refraction studies show that a trench filled with sediments and sedimentary rocks lies at the base of the continental slope off New England. Thus the main difference in morphology between the trenchless continental margins and continental margins with a marginal trench is that in the former the trench has been filled with sediments.

In the continental margins of the Atlantic, Indian, Arctic, and Antarctic oceans and the Mediterranean Sea, the continental rise generally represents the category III provinces. The Pacific, however, is bounded by an almost continuous line of marginal trenches. The high seismicity, vulcanism, and youthful relief of the Pacific borders suggest a very recent origin. In contrast, the nonseismic, nonvolcanic character, as well as the lower relief, of the Indo-Atlantic margins suggests

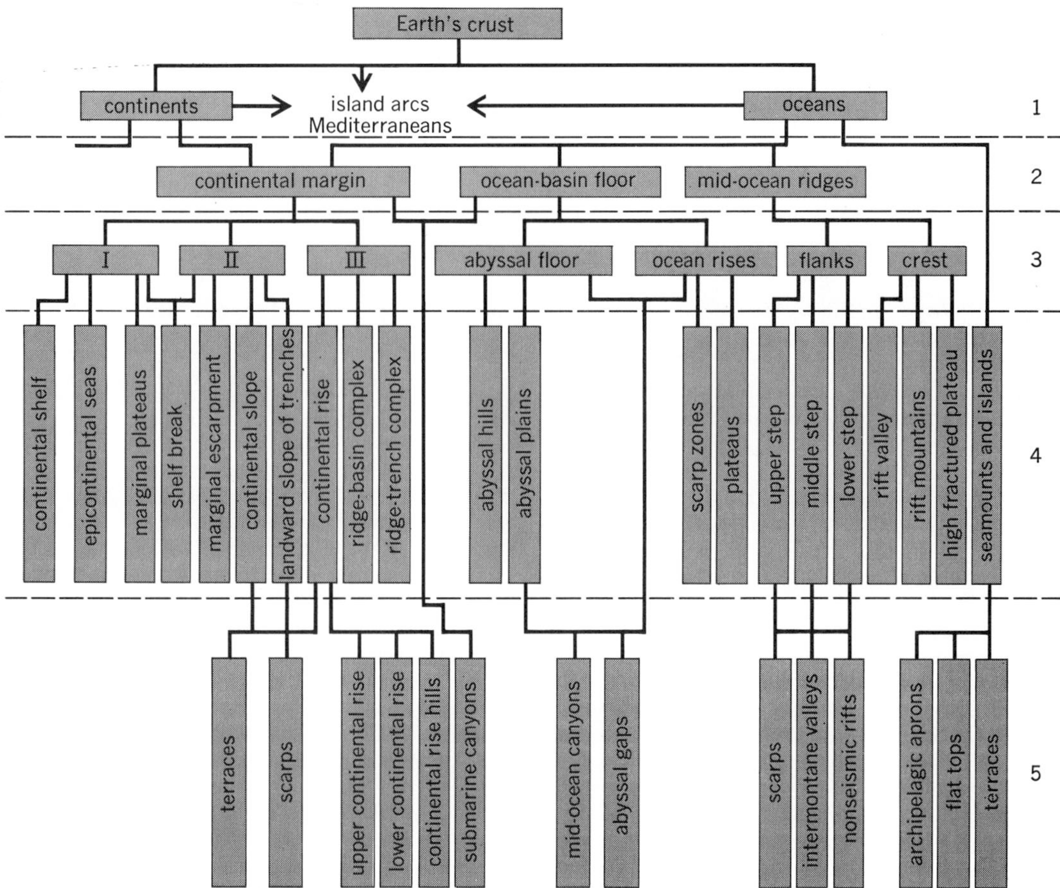

Fig. 3. Outline of submarine topography. Line 1, first-order features of the crust; line 2, major topographic features of the ocean; line 3, categories of provinces and superprovinces; line 4, provinces; line 5, subprovinces and other important features.

a greater age. Thus on the old, stable, continental margins the deposition of sediment derived from the land has filled the marginal trench and produced the continental rise. The local relative relief on the continental margin rarely exceeds 20 fathoms, with the major exception of submarine canyons and occasional seamounts. *See* TECTONOPHYSICS.

Submerged benches. Submerged marine beach terraces have been identified throughout the world. Since the beaches seem to correlate well between areas of vastly different tectonic development, it has been concluded that those which are listed in the accompanying table represent submerged late Pleistocene beaches. *See* COASTAL LANDFORMS.

Structural benches. Structural benches, the topographic expression of outcropping beds, have been identified on the continental slope. Near Cape Hatteras, N.C., the structural benches have been dated by extrapolating data obtained in several test borings near the coastline (Fig. 6). Benches on Georges Bank have been dated from bottom samples obtained by dredging. Through the action

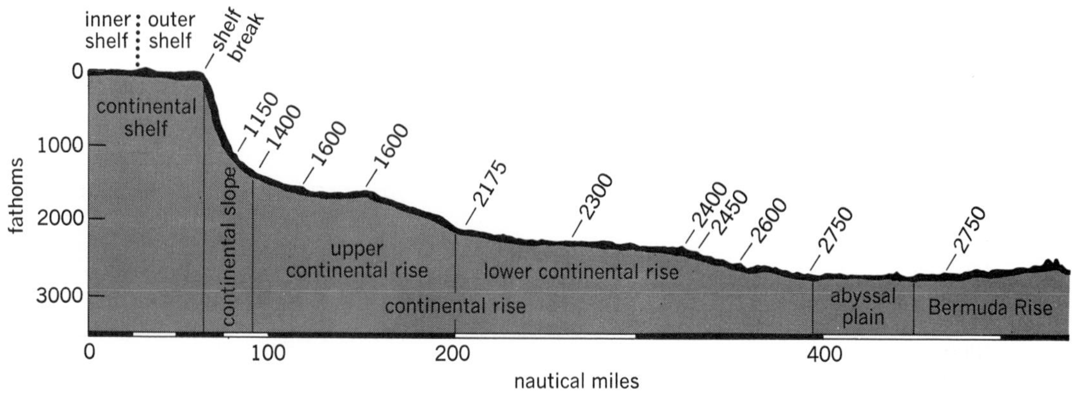

Fig. 4. Continental margin provinces: type profile off northeastern United States.

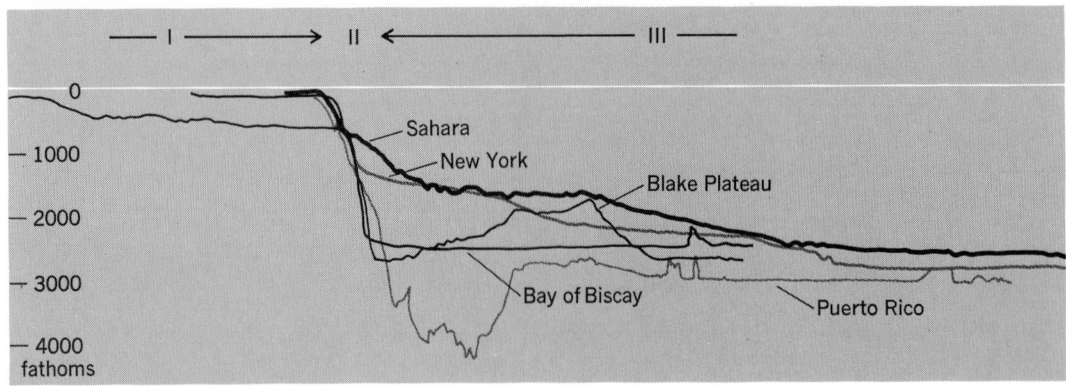

Fig. 5. Three categories of continental-margin provinces.

of slumps, bottom currents, and turbidity currents, sediments are continually removed from the continental slope. Thus, there is no cover of recent sediments to obscure the outcrops of the ancient formations. *See* TURBIDITY CURRENT.

Ocean-basin floor. Excluding the marginal trenches and mid-oceanic ridges, the deepest portions of the ocean are included in this division. Approximately one-third of the Atlantic and three-fourths of the Pacific fall under this heading. The ocean-basin floor can be divided into three categories of provinces; the abyssal floor, oceanic rises, and seamounts and seamount groups.

Abyssal floor. The abyssal floor includes the broad, deep areas of the central portion of the ocean. In the Atlantic, Indian, and northeast Pacific oceans, abyssal plains occupy a large part of the abyssal floor. An abyssal plain is a smooth portion of the deep-sea floor where the gradient of the bottom does not exceed 1:1000. Abyssal plains adjoin all continental rises and can be distinguished from the continental rise by a distinct change in bottom gradient. At their seaward edge, most of the abyssal plains gradually give way to abyssal hills. Individual abyssal hills are 50–200 fathoms high and 2–6 mi wide. In the Atlantic, the abyssal hill provinces only locally exceed 50 mi in width. Abyssal plains in the same area range from 100 to 200 mi in width. Core samples of sediment obtained from the Atlantic abyssal plain invariably contain beds of sand, silt, and gray clay intercalated in the red or gray pelagic clay which is generally characteristic of the deep-oceanic environment. These deep-sea sands were transported by turbidity currents from the continental margin. Some of the currents probably descended along a broad front, while others certainly followed the submarine canyons and spread out fanwise from their submarine alluvial cones. *See* MARINE SEDIMENTS.

The abyssal hills are thought to represent tectonic or volcanic relief of a type identical with that buried beneath the abyssal plains. Abyssal plains are also found in the marginal trenches, marginal basins, and in epicontinental marginal seas. Features of exactly the same morphology and origin are found in some lakes. Of similar origin are archipelagic aprons, which spread out from the base of oceanic islands. *See* OCEANIC ISLANDS.

Oceanic rises. Oceanic rises are areas slightly elevated above the abyssal floor which do not belong to the continental margin or the mid-oceanic ridges. In the North Atlantic, the Bermuda Rise is the best-known example (Fig. 7). In contrast to the mid-oceanic ridges, oceanic rises are nonseismic; their relief is more subdued, and they are asymmetrical in cross section. The western and central Bermuda rise is characterized by gentle, rolling relief. The average depth gradually decreases toward the east. In the eastern third, the rise is cut by a series of scarps, 500–1000 fathoms in height, from which the sea floor drops to the level of the abyssal plain on the east. The series of eastward-facing scarps suggest block faulting. Situated ap-

Depth in fathoms of prominent continental-shelf terraces*

Placentia Bay, Newfoundland	Norfolk, Va.	Charleston, S.C.	Bimini, B.W.I.	St. Vincent, Cape Verde Is.	Dakar, Senegal	San Pedro, Calif.
10		12	10	8	10	10
				15	15	15
20	18	20	20	24	20	20
	30	30	28	28	28	28
35	35	35		32		
40				38	38	38
42		45	42	42	45	45
	50					
55	58			54	55	55
				60		
68		68	65			
80	80	80	85	80	78	80

*Each column based on a single nonprecision echogram.

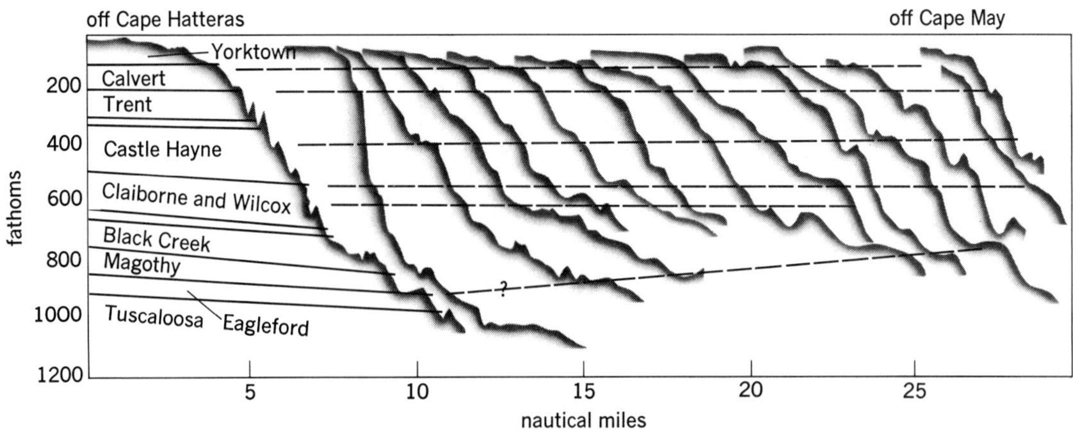

Fig. 6. Correlation of structural benches (outcropping beds) on the continental slope from Cape Hatteras, N.C., to Cape May, N.J. The soundings were taken by the U.S. Coast and Geodetic Survey.

proximately in the center of the Bermuda Rise is the volcanic pedestal of Bermuda. A small archipelagic apron surrounds the pedestal. The turbidity-current origin of the smooth apron is supported by cores containing shallow-water carbonate clastic sediments in depths of 2300 fathoms. In the Pacific, extending for more than 3000 mi west of Cape Mendocino, Calif., is an asymmetrical rise with a southward-facing scarp, which has been named the Mendocino Fracture Zone. The Bermuda Rise is less than one-fourth as long as the Mendocino Rise, but otherwise the relief of both features is quite similar. Although the circum-Pacific seismic belt crosses its trend, the Mendocino Escarpment is nonseismic. Other nonseismic fracture zones, which probably can be classified as oceanic rises, have been reported from the eastern Pacific. The Rio Grande Rise of the South Atlantic and the Mascarene Ridge of the Indian Ocean are similar in form.

Seamounts and seamount groups. A seamount is any submerged peak more than 500 fathoms high. This discussion, however, is limited to the larger, more or less conical peaks more than 1000 fathoms in height. Seamounts are distributed through all the physiographic provinces of the oceans. Seamounts sometimes occur randomly scattered but more often lie in linear rows. It seems safe to conclude that virtually all conical seamounts are extinct or active volcanoes. The Kelvin seamount group, a line of seamounts 800 mi long, stretches out from the vicinity of the Gulf of Maine toward the Mid-Atlantic Ridge. The Atlantis – Great Meteor seamount group extends for 400 mi along a north-south line, south of the Azores. In the southwestern Pacific, many lines of islands and seamounts crisscross the ocean. In the mid-Pacific, southwest of Hawaii, is a large area of seamounts whose flat summits range from 50 to 850 fathoms beneath sea level. These tablemounts have been termed guyots. From the flat summits, shallow-water fossils of Cretaceous age have been dredged. Such sunken islands are not limited to the Pacific. Several of the Kelvin seamounts are flat-topped at 650 fathoms, and the seamounts of the Atlantis – Great Meteor group have flat summits at 150 – 250 fathoms. *See* SEAMOUNT AND GUYOT.

Mid-Oceanic Ridge. The middle third of the Atlantic, Indian, and South Pacific oceans is occupied by a broad, fractured swell known as the Mid-Oceanic Ridge. In the Atlantic, it is known as the Mid-Atlantic Ridge; in the southern Indian Ocean, the Mid-Indian Ridge; in the Arabian Sea, the Carlsberg Ridge and Murray Ridge; and in the South Pacific, the Easter Island Ridge.

The Mid-Atlantic Ridge can be divided into distinctive physiographic provinces which can be identified on most transatlantic profiles (Figs. 2 and 8).

Crest provinces. The rift valley, rift mountains, and high-fractured plateau which constitute this category form a strip 50 – 200 mi wide. The rift valley is bounded by the inward-facing scarps of the rift mountains. The floor of the rift valley lies 500 –

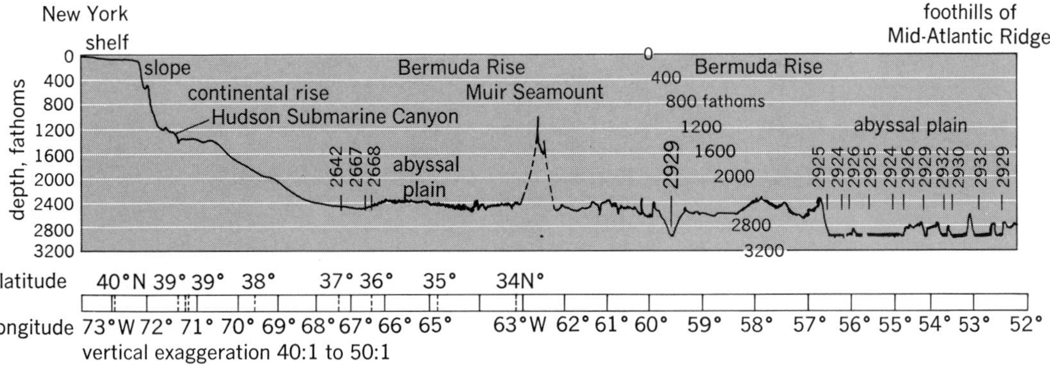

Fig. 7. Precision-depth-recorder profile between Mid-Atlantic Ridge and New York.

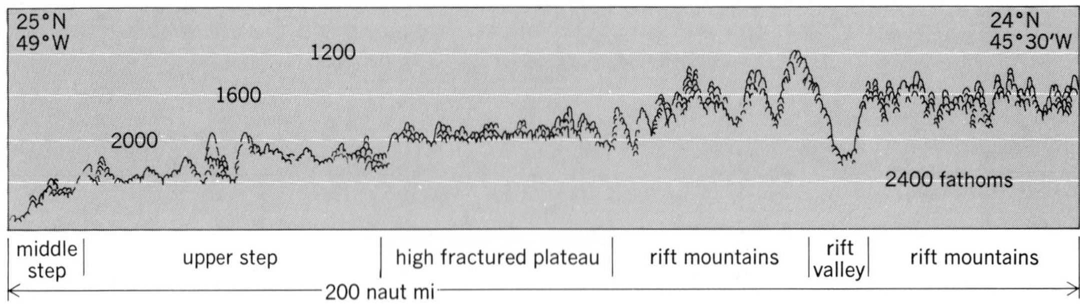

Fig. 8. Tracing of a precision-depth-recorder record showing crest and western flank of Mid-Atlantic Ridge.

1500 fathoms below the adjacent peaks of the rift mountains, which drop abruptly to the high fractured plateau, lying at depths of 1600–1800 fathoms on either side of the rift mountains. The topography of the crest provinces is the most rugged submarine relief. An earthquake belt accurately follows the rift valley through a distance of over 40,000 mi. Heat-flow measurements in the crest provinces give values several times greater than have been obtained in normal ocean or continental areas. A large, positive magnetic anomaly and a moderate (−20 mgal) negative gravity anomaly are associated with the rift valley. Seismic refraction measurements indicate a crust intermediate in composition between the oceanic crust and the mantle. The crest provinces of the Mid-Oceanic Ridge can be traced directly into the rift valleys, rift mountains, and high plateaus of Africa. These features of African geology are clearly the result of extensional forces in the Earth's crust. The Mid-Oceanic Ridge is probably similar in all essential characteristics, including origin, to the African rift valley complex. *See* RIFT VALLEY.

Flank provinces. The flank provinces of the Mid-Oceanic Ridge can be divided into several steps or ramps, each bounded by scarps somewhat larger than those which characterize the entire area.

Parts of the flank provinces, particularly the Upper Step south of the Azores, are characterized by smooth-floored intermontane valleys. Photographs, cores, and dredging indicate that the crest of the Mid-Atlantic Ridge north of the Azores is being denuded of its sediments. As these sediments are eroded from the crest provinces and deposited on either side, they are gradually filling the intermontane basins and smoothing the relief of the flanks. [BRUCE C. HEEZEN]

UNDERLYING STRUCTURE

Because approximately 70% of its surface is covered by the oceans, the typical structure of the Earth is found in the oceanic and not in the land areas. Statistical examination shows that most of the Earth's solid surface is either at the elevation of the ocean floors or at the elevation of the continents. The anomalous areas, those of extreme or of intermediate elevation, are long, narrow features—the mountain ranges, island arcs, deep-sea trenches, and continental margins. With the exception of a few intermediate areas, such as the Red Sea, the crustal structure of the ocean basins is distinctly different from the crustal structure of the continents, and it appears that this has been the case for most of the Earth's history.

To have a rough model of a section through the Earth, one draws a circle about 5 in. in diameter and a concentric one of about half that diameter. Inside the smaller circle is the core, probably of a nickel-iron composition. The part between the circles is the mantle, a crystalline, basic rock with density about 3.3 g/cm^3. The line forming the outer circle, if made with an average pencil point, will include all of the crust of the Earth. The crust has a density of about 2.7 g/cm^3. The oceanic crust is about 6 km thick compared to about 36 km for the continents. The boundary between the crust and the mantle is called the M-discontinuity (Mohorovičić discontinuity) by seismologists. *See* MOHO (MOHOROVIČIĆ DISCONTINUITY).

Instruments and techniques. Unlike the continental areas, the ocean floors cannot be studied directly; hence, most of the information about the structure of the Earth beneath the oceans comes from geophysical measurements, from samples dredged or cored from the ocean floor, and, to a smaller extent, from observations made from deep submersibles. Principally employed geophysical techniques are earthquake seismology, explosion seismology, and measurements of the variations in the Earth's gravitational and magnetic fields. Seismology is the study of the propagation of sound waves or elastic waves in the Earth. The source of these waves can be natural, which is the case in earthquake seismology, or man-made. Commonly used man-made sound sources include explosions, high-energy electrical sparkers, and pneumatic devices. The speed of sound waves traveling in the various layers, considered together with gravity and magnetic measurements, makes it possible to estimate certain physical properties such as density and elastic constants. These in turn suggest the type of rock constituting each layer. The travel time of sound waves reflected or refracted by a particular layer provides a measure of the depth and thickness of the layer. In marine investigations, a relatively new device for measuring the thickness and structure of the sediments is the continuous seismic reflection profiler. A sound source, towed behind the ship, emits periodic pulses of acoustic energy. These pulses travel through the water and are reflected back to the surface from the interfaces between the layers of sediment and rock. The reflected pulses are received by a towed array of hydrophones, amplified, and printed by a scanning recorder. This technique produces a cross-sectional view of the sediments on the ocean floor and usually shows the topography of the basement beneath the sediments as well (Fig. 9). *See* MARINE SEDIMENTS; SEISMOLOGY.

The sea gravimeter is essentially a highly damped, extremely sensitive spring balance. Changes in the displacement of the springs reflect minute variations in gravitational force. Total magnetic field intensity is measured with a magnetometer towed well astern to minimize the magnetic effects of the ship. The gravitational and magnetic anomalies (variations) when combined with seismological data are indications of compositional changes, or structural features such as folds or faults, and have provided much information about the structure of the Earth. *See* TERRESTRIAL GRAVITATION.

Paleomagnetism is the study of the remanent magnetization in rocks. The direction of remanent magnetization in sedimentary rocks is parallel to the Earth's magnetic field at the time that the rock was deposited. The remanent magnetic vector in igneous rocks indicates the direction of the Earth's field at the time the rock cooled through the Curie temperature. Paleomagnetic studies of rocks of various ages and wide geographic distribution appear to indicate that the magnetic poles have shifted with respect to the geographic poles through geologic time or that the continents have shifted with respect to each other. Data have been accumulated that prove there have been many reversals of the Earth's magnetic field. These reversals have been used to explain linear magnetic anomaly patterns in some parts of the oceans and to support the hypothesis of continental drift by spreading of the sea floor away from the axes of the mid-ocean ridges.

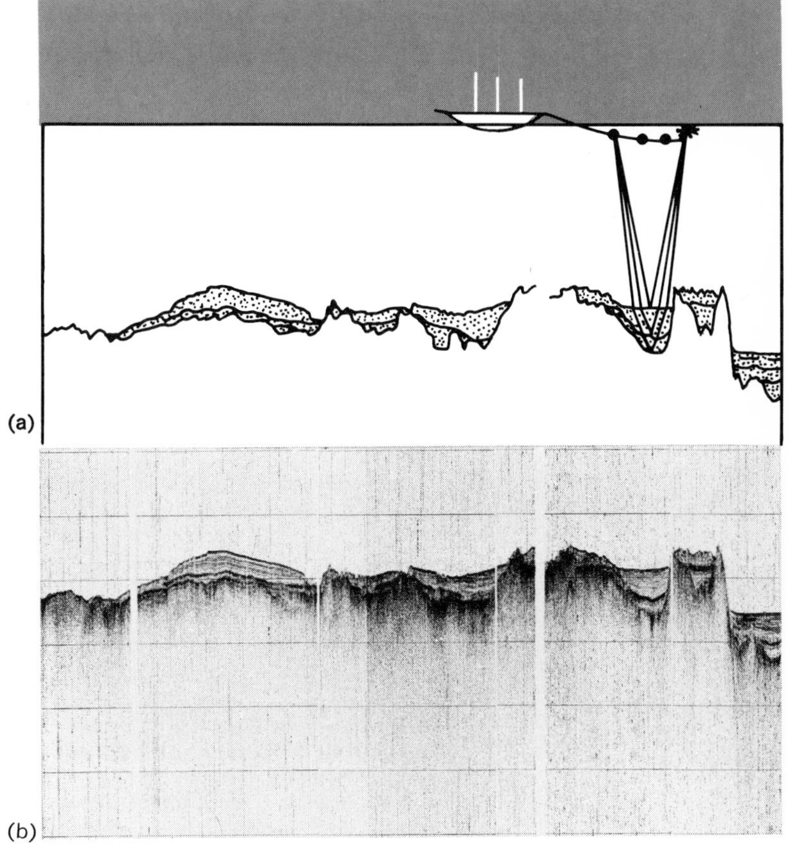

Fig. 9. Continuous seismic profiling: (*a*) technique and (*b*) typical record. Record shows approximately 200 km of traverse. The average thickness of these ocean-bottom sediments is about 500 m. Vertical exaggeration × 25.

Ocean basins and continental margins. Figure 10 is a structure section across an Atlantic type (rifted) margin based on an interpretation of geologic and geophysical data within the framework of plate tectonics and sea-floor spreading. On the continent and continental shelf, the rocks beneath the sedimentary layers are mainly of the acidic type, such as granite, gneiss, or schist. Beneath the rocks is an intermediate layer believed to be gabbroic or basaltic. The mantle is probably composed of ultramafic rocks, such as peridotite, enstatite, or eclogite. This is the most prominent layer in the Earth, extending from near the surface approximately halfway to the center. There are some variations in the upper parts of the mantle between continental and oceanic areas, but the most apparent difference is in the thickness and composition of the crust. The continental crust is six or seven times thicker than the oceanic crust and contains almost all of the acidic rocks, such as granites, whereas the oceanic crust is almost entirely composed of basic rock. *See* EARTH.

The average depth of the ocean basins is about 4.8 km. The topography of the ocean floor is rough in the majority of the explored parts, although there are broad areas, particularly in the Atlantic, where the bottom is almost completely flat. These abyssal plains are thought to have been formed by turbidity current deposition where sediments, set in motion during underwater landslides and thrown into suspension in the water, flow to the deepest parts of the basins.

Away from the abyssal plains and continental rises (see section above on physiographic provinces) the sediment layers are mainly composed of clay-sized particles (less than 2μ) and of the skeletons of plankton that lived and died in the waters above. Pelagic sediments of this type generally form an approximately uniform blanket over preexisting topography. In some areas the sea-floor topography is shaped by nonuniform deposition and erosion by bottom currents, particularly by the flow of the cold, dense water generated in the polar regions. The thickness of the unconsolidated sediments varies from tens of meters in areas which receive no turbidities and where the planktonic population is sparse to thousands of meters in other areas. The average is less than 1 km.

Layer 2, immediately below the layer-1 sediments, has a seismic velocity between 4 and 6 km/s. This range of velocities encompasses those

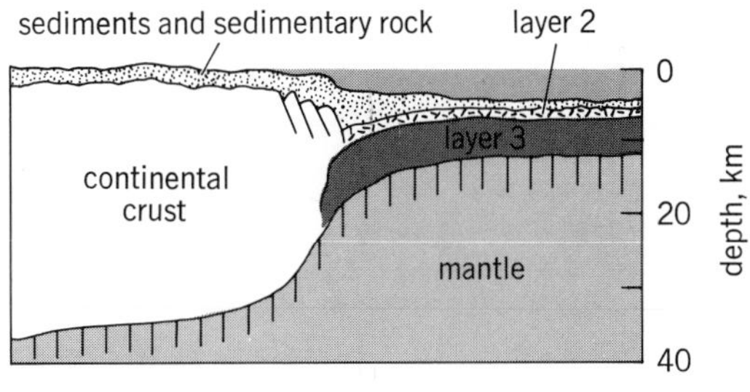

Fig. 10. Structure section across Atlantic-type (rifted) margin.

appropriate to compacted or metamorphosed sediments or volcanic rocks. Shallow penetrations into layer 2 by deep-sea drilling indicate that its upper part consists of a mixture of pillow basalts, dikes, volcanic debris, and sediments. The lower part is probably massive basalt or metabasalt or both.

The principal layer of the oceanic crust, layer 3, shown by check marks in the structure sections (Figs. 10–12), has been found by most of the numerous seismic refraction measurements made in the Atlantic, Pacific, and Indian oceans. The average velocity of sound in this layer is near 6.7 km/s, and the thickness is about 5 km. Variations from these average are often found in the neighborhood of anomalous areas such as seamounts, trenches, or continental margins. Laboratory sound-velocity measurements of various rock types dredged from sea-floor escarpments suggest that layer 3 is composed of gabbro and metagabbro. Whether or not a similar layer exists under the continents is in dispute. There is much evidence that the velocity in the lower part of the continental crust is intermediate between that in the upper part and that in the mantle, and this could indicate the presence of material similar to that in the oceanic crust.

Continental shelves. These are the submerged borders of the continents. The water depth is on the order of a few hundred meters, and the width of the shelf varies from a few kilometers in some places to a few hundred kilometers in others. The thickness of the crust is intermediate between that of the continents and oceans, and its composition is continental. In some places, such as parts of the east coast of North America and South America, the continental shelf and continental slope are broad areas where erosion of the continental masses has resulted in the deposition of many thousands of meters of sediment during the past several million years. In other areas which do not receive much sediment, notably the west coast of the Americas, there is only a narrow continental shelf.

Submarine ridges. There are two types of oceanic ridges, seismic and aseismic. The mid-ocean ridge system is of the former type and is the largest single feature on the Earth. It is more than 80,000 km long and completely encircles the globe. In many places the ridge is offset by large fracture zones, along which seismic activity is generally higher than elsewhere on the ridge system. Over its entire length the ridge is characterized by a narrow zone of shallow-focus earthquake epicenters. Usually associated with this zone are a rift valley and a large positive magnetic anomaly. In addition, parallel linear bands of alternating positive and negative magnetic anomalies are found on the flanks in many areas. The sediment cover is typically thin and in some places in the crest region is entirely absent in a zone 100–500 km wide. On the flanks an appreciably thicker cover is found; in some places it covers the basement rock more or less uniformly, while in others it is collected in pockets separated by exposed basement peaks or ridges.

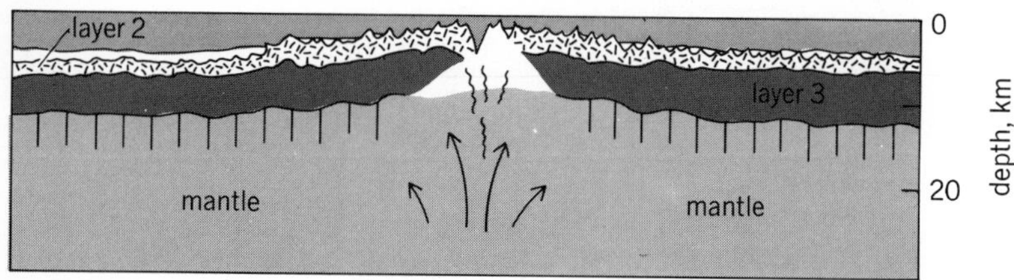

Fig. 11. Structure section across Mid-Atlantic Ridge, indicating crustal accretion and divergence.

The tops and flanks of seismic ridges have been dredged and have yieled basaltic volcanic rocks with some inclusions of gabbroic or ultramafic materials. Figure 11 shows a structure section across the Mid-Atlantic Ridge based on seismic refraction and gravity measurements.

In addition to the mid-ocean ridge system, there are other oceanic ridges which are seismically inactive. Typical examples of these are the Walvis Ridge off southwestern Africa and the Hawaiian Ridge, Emperor Seamount Chain, and Line Island Chain in the Pacific. These ridges are thought to have been formed by extrusion of large volumes of volcanic material as the oceanic crust drifted over "hot spots" in the lower part of the mantle. Another class of aseismic ridge is exemplified by the Lomonosov Ridge in the Arctic Ocean and the Jan Mayen Ridge in the Norwegian Sea. These ridges appear to be thin slivers of continental rocks which were separated from the larger continental masses by rifting associated with sea-floor spreading.

Deep-sea trenches. Deep-sea trenches are important structural features associated with some continental margins, island arcs, earthquake belts, and areas of volcanic activity. Most of them are confined to the margins of the Pacific Ocean, although there are trenches in the Atlantic and Indian oceans. The greatest depths in the oceans are found in these trenches, the deepest in the Pacific being about 10.7 km in the Marianas Trench, and the deepest in the Atlantic about 8.4 km in the Puerto Rico Trench. The trench is formed by the depression of the high-velocity crustal layer and the mantle by several kilometers. In the bottom are layers of sediments which are generally thickest landward of the trench axes. The great depth of the underlying dense layers causes a pronounced deficiency of gravity, a characteristic feature of all deep-sea trenches. Earthquake foci tend to lie on inclined planes dipping landward from the trenches.

Several hypotheses have been advanced to account for the existence of island arcs and the associated trenches and to relate them to a major processes going on in the Earth. The theory receiving much attention currently is that there are convection cells in the mantle which cause upwelling of new crustal material along the mid-ocean ridges and convergence at the trenches. The convergent flow causes the oceanic crust to be thrust underneath the continents or island arcs, creating the negative gravity anomalies, the earthquake foci along the shear zones, and the volcanic and orogenic activity landward of the trench itself. Figure 12 illustrates the concept of trench structure at a convergent margin.

Continental drift and sea-floor spreading. The parallel Atlantic borders of the Americas and Africa first led scientists to speculate that these conti-

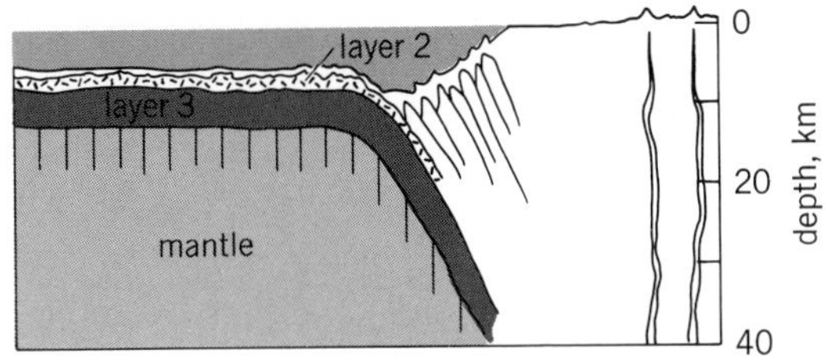

Fig. 12. Structure section across Pacific-type (convergent) margin.

nents were originally a single unit which was split apart by drifting. Alfred Wegener first popularized the idea in 1910. Throughout this century, the theory has been the subject of much debate, and although the early evidence was confined to the continents themselves, an impressive body of geological, climatological, and paleontological observations was brought to bear on the theory, including the existence of truncated mountain ranges and similar fossils and mineral assemblages found on opposite sides of the Atlantic. Paleomagnetic data also have been interpreted as indicating that magnetic poles have moved and that the continents have drifted relative to each other. Opposition to the theory has been due largely, but not entirely, to the difficulty in finding a suitable mechanism for drift that was not in conflict with concurrent theories on the composition of the Earth's interior. *See* CONTINENT FORMATION.

Convection cells. A concept of continental drift by spreading of the sea floor has attracted much attention. An important concept in the sea-floor spreading hypothesis is that heat generated in the Earth causes convective overturn of mantle material in a number of convection cells. The mid-ocean ridges are thought to lie along the juncture of the upwelling limbs of separate cells. At the surface, the upwelling material diverges and flows away from the ridge axes, carrying the continental blocks along with it until they come to rest over or near a downflowing convergence. It is suggested that reversals of the magnetic field produce areas of reversely magnetized crustal rock, symmetrical with respect to the ridge axis. The times of reversals are known for the last 5,000,000 years, and on the basis of this hypothesis values of 1–4 cm/year have been deduced for spreading rates. The descending flow is thought to create the deep-sea trenches and earthquake fault planes mentioned previously, and the resulting compression to be responsible for the mountain building and volcanism associated with the trenches. This mechanism has also been offered as an explanation for the fact that no sediments or rocks older than Jurassic have yet been found in the ocean basins, based on the assumption that all older ones have been swept into the mantle at the trenches. Such a description is greatly simplified, and many complexities are encountered in attempting to fit the entire surface of the Earth into a plausible pattern and in visualizing convection cells with the great horizontal dimensions implied by this theory and the small vertical dimensions implied by mantle stratification. But there are many arguments in favor of it.

Mid-ocean ridge system. Modern investigations in submarine geology and geophysics have supplied many new facts with which to test the theory. Among these was the discovery of the mid-ocean ridge system and the fact that, particularly in the Atlantic Ocean, the ridge lies almost exactly equidistant between the continents on either side. In addition to the median position of the ridge in the Atlantic, several other features about it have been discovered which seem to support the concept of continental drift. The great majority of earthquake epicenters in the oceans are concentrated in the axial zone of the ridge. An axial rift valley characterizes much of its length and appears to be a tensional feature, such as would be produced if the crust were being pulled apart. In many places long fracture zones are found, perpendicular to the ridge axis, which may represent flowlines of the continents' movements. There are distinct concentrations of earthquake activity near the intersections of the fracture zones and the ridge axis. Directions of first motion for these earthquakes appear to be best explained by transform faults associated with crustal spreading away from the ridge axis. Measurements of the thermal gradient in the bottom sediments have shown that the crest of the ridge is, in many places, characterized by abnormally high heat flow, which would be expected if convection cells were bringing new, hot material to the surface at this point. *See* TRANSFORM FAULT.

Rock magnetism. Widely acclaimed evidence cited in support of sea-floor spreading has come from the observed bands of alternating positive and negative magnetic anomalies parallel to and symmetric about the ridge axis. Figure 13 shows an example of this magnetic pattern on the Mid-Atlantic Ridge southwest of Iceland. It has been proposed that the positive anomaly bands (including the central one) correspond to rocks that cooled through the Curie temperature with the Earth's magnetic field in its present polarity. The negative bands correspond to rocks magnetized by a reversed field. According to the sea-floor spreading hypothesis, the banded pattern was formed by upwelling and outward spreading of

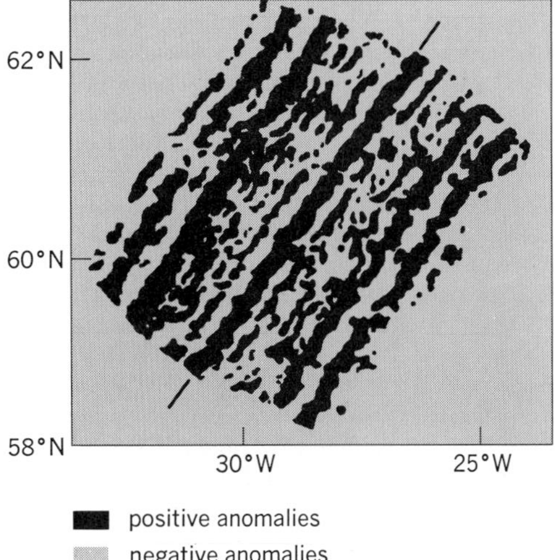

Fig. 13. Pattern of magnetic anomalies on the Reykjanes Ridge southwest of Iceland. (*After Heirtzler, Le Pichon, and Baron, 1966*)

new material at the ridge crest, each newly generated strip of crust acquiring permanent magnetization in the direction of the ambient field. The spatial arrangement of the positive and negative bands near the ridge crest, normalized for spreading rate, is in agreement with that predicted by the history of field reversals. *See* ROCK MAGNETISM.

Sediment distribution. Since the mid-1960s many hundreds of thousands of kilometers of seismic reflection measurements have shown the general pattern of sediment thickness in the oceans. On and near the crests of the mid-ocean ridges the sediment cover is very thin, becoming progressively thicker down the ridge flanks and into the basins. The total amount of sediment on the ridges varies from area to area because sediment productivity varies, but the increase of thickness with increasing distance from the ridge crest is a general characteristic of the pattern. The deep-sea sediment drilling program has now provided the additional evidence that the age of the sediment lying on the igneous basement (that deposited earliest) increases with increasing distance from the ridge crest. Thus, both the patterns of sediment thickness and sediment age are consistent with a young crust and progressively older flanks, concepts inherent in the sea-floor spreading hypothesis.

[MAURICE EWING; JOHN EWING]

Bibliography: C. A. Burk and C. L. Drake (eds.), *The Geology of Continental Margins*, 1974; A. E. Maxwell (ed.), *The Sea*, vol. 4, pts. I and II: *New Concepts of Sea Floor Evolution*, 1970; W. C. Pitman and J. R. Heirtzler, Magnetic anomalies over the Pacific-Antarctic Ridge, *Science*, 154:1164–1171, 1966; L. R. Sykes, Mechanism of earthquakes and nature of faulting on the mid-oceanic ridges, *J. Geophys. Res.*, 72:2131–2153, 1967; Symposium on Continental Drift, *Phil. Trans. Roy. Soc. London, Ser. A*, 258(1088):41–75, 1965; F. J. Vine, Spreading of the ocean floor; New evidence, *Science*, 154:1405–1415, 1966; T. J. Wilson, A new class of faults and their bearing on continental drift, *Nature*, 267:343–347, 1965.

Marine sediments

The accumulation of minerals and organic remains on the sea floor. Marine sediments vary widely in composition and physical characteristics as a function of water depth, distance from land, variations in sediment source, and the physical, chemical, and biological characteristics of their environments. The study of marine sediments is an important phase of oceanographic research and, together with the study of sediments and sedimentation processes on land, constitutes the subdivision of geology known as sedimentology. *See* MARINE GEOLOGY; SEDIMENTATION.

ENVIRONMENTS OF DEPOSITION

Traditionally, marine sediments are subdivided on the basis of their depth of deposition into littoral (0–20 m), neritic (20–200 m), and bathyal (200–2000 m) deposits. This division overemphasizes depth. More meaningful, although less rigorous, is a distinction between sediments mainly composed of materials derived from land, and sediments composed of biological and mineral material originating in the sea. Moreover, there are significant and general differences between deposits formed along the margins of the continents and large islands, which are influenced strongly by the nearness of land and occur mostly in fairly shallow water, and the pelagic sediments of the deep ocean far from land.

Sediments of continental margins. These include the deposits of the coastal zone, the sediments of the continental shelf, conventionally limited by a maximum depth of 100–200 m, and those of the continental slope. Because of large differences in sedimentation processes, a useful distinction can be made between the coastal deposits on one hand (littoral), and the open shelf and slope sediments on the other (neritic and bathyal). Furthermore, significant differences in sediment characteristics and sedimentation patterns exist between areas receiving substantial detrital material from land, and areas where most of the sediment is organic or chemical in origin.

Coastal sediments. These include the deposits of deltas, lagoons, and bays, barrier islands and beaches, and the surf zone. The zone of coastal sediments is limited on the seaward side by the depth to which normal wave action can stir and transport sand, which depends on the exposure of the coast to waves and does not usually exceed 20–30 m; the width of this zone is normally a few miles. The sediments in the coastal zone are usually land-derived. The material supplied by streams is sorted in the surf zone; the sand fraction is transported along the shore in the surf zone, often over long distances, while the silt and clay fractions are carried offshore into deeper water by currents. Consequently, the beaches and barrier islands are constructed by wave action mainly from material from fairly far away, although local erosion may make a contribution, while the lagoons and bays behind them receive their sediment from local rivers. The types and patterns of distribution of the sediments are controlled by three factors and their interaction: (1) the rate of continental runoff and sediment supply; (2) the intensity and direction of marine transporting agents, such as waves, tidal currents, and wind; and (3) the rate and direction of sea level changes. The balance between these three determines the types of sediment to be found. *See* DELTA.

On the Texas Gulf Coast, rainfall and continental runoff decrease gradually in a southwesterly direction. The wind regime favors considerable wave action and a southwesterly drift of the near-shore sand from abundant sources in the east. Since sea level has been stable for several thousand years, the conditions have been favorable for the construction of a thick and nearly closed sand barrier which separates a large number of bays from the open Gulf. This barrier is constructed by marine forces from sediments from distant sources and varies little in characteristics along its length. The bays, on the other hand, receive local water and sediment. In the east, the supply of both is fairly abundant, and since the streams are small, the sediment is dominantly fine; the bays have muddy bottoms and brackish waters. Conditions are fairly stable, and a rich, but quantitatively not large fauna is present, including oyster banks. At the southwestern end, continental runoff and sediment supply are negligible.

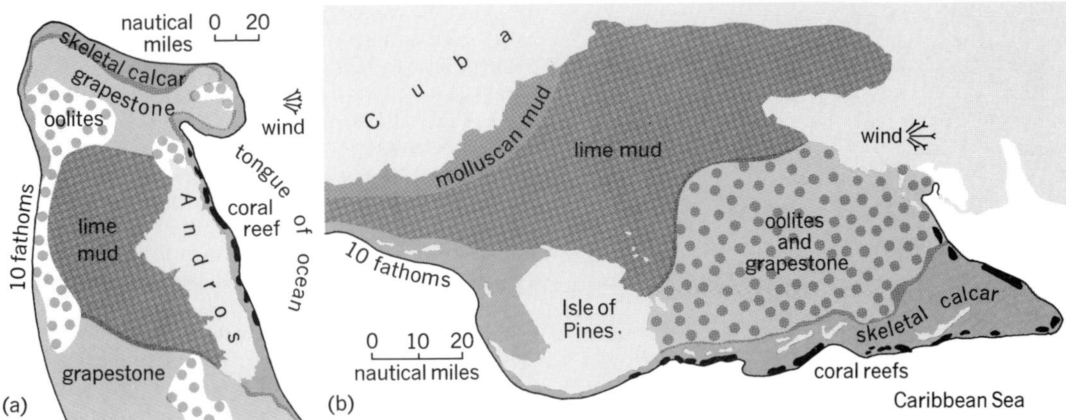

Fig. 1. Shoal water sediments as examples of sedimentation without supply of land-derived sediment. (*a*) The Bahamas. (*b*) The Gulf of Batabano, which is in southwestern Cuba. Both of these shoals border directly on the deep ocean.

The only sediment received by the bays comes from washovers from the barriers and is therefore mainly sandy, and the virtually enclosed bays with no runoff are marine to hypersaline. Locally, this yields chemical precipitates such as gypsum and calcium carbonate, and is also conducive to the development of a restricted but very abundant fauna, which produces significant deposits of calcareous material. The sediments of bays and lagoons are often more stratified than those of the open sea as a result of fluctuating conditions. The textural and compositional characteristics depend on local conditions of topography, shore development, and wave and current patterns. They range from coarse gravel and cobbles on rocky beaches fronting the open sea to very fine clayey silt in the interior of quiet lagoons.

The effects of sea level changes are imperfectly known, but it is easily comprehended that the development of coastal sediments is to a large extent a function of the duration of this environment in a particular place. Thus if sea level rises or falls rapidly, there is no time for extensive development of beach, barrier, and lagoon deposits, and discontinuous blankets of nearshore sands, with muds behind them, are formed. As the rate of change decreases, open barriers, consisting of widely spaced low sand islands, tend to form, which imperfectly isolate open shallow lagoons in which essentially marine conditions prevail. A prolonged stability is required to produce thick, closed barriers and completely isolate the lagoon environment.

Entirely different nearshore deposits are found on shoals where supply of sediment and fresh water from the land is absent, either because land areas are small (Bahamas), the drainage is directed elsewhere (southern Cuba), or there is no rainfall (Arabian Peninsula). Calcareous muds and coarse calcareous sands then make up the lagoon and beach deposits. If, in addition, the shoal borders directly on the deep ocean without transitional shelf, as is the case in Fig. 1, cool water is driven onto the shoal, where it warms up and precipitates calcium carbonate. In the turbulent water of the shoal this precipitation either takes place in the form of oolites or it cements organic debris together in small aggregates (grapestone). At the edge of the shoal, the presence of cool, nutrient-rich ocean water is favorable for the growth of coral and algal reefs which are bordered by a zone of skeletal sand derived from broken calcareous organisms. The inner sheltered portions receive the finest calcareous sediments. The types of sediment and their distribution patterns are mainly controlled by the shape of the shoal, in particular the position of its edge, by the prevailing wind, and by the location of sheltered or somewhat deeper quiet areas.

Shelf and slope sediments. The continental shelf is a gently seaward sloping plain of greatly varying width, ranging from less than a mile along steep rocky coasts to several hundred miles, for example, in the western Gulf of Mexico. A distinct break in slope at 100–500-m depth marks the transition to the continental slope which descends somewhat more steeply (5–10°) to the deep-sea floor. In areas of active tectonism, for example, off the coast of southern California, the shelf is narrow and separated from the continental slope by a wide zone of deep basins alternating with shallow banks and islands. Submarine canyons cut the edge of the continental shelf in many places and sometimes reach back into the nearshore zone. *See* CONTINENTAL SHELF AND SLOPE; SUBMARINE CANYON.

During the Pleistocene, the continental shelf was subjected to repeated transgressions and regressions. During each interglacial, sea level was high and the shoreline was located near its present position; during each glacial period, much water was withdrawn from the ocean and the shoreline occurred near the edge of the shelf. The last low sea level stand occurred approximately 19,000 years ago, and the present shoreline was established as recently as 3000–5000 years ago. On most shelves, equilibrium has not yet been fully established and the sediments reflect to a large extent the recent rise of sea level. Only on narrow shelves with active sedimentation are present environmental conditions alone responsible for the sediment distribution.

Sediments of the continental shelf and slope belong to one or more of the following types: (1) biogenic (derived from organisms and consisting mostly of calcareous material); (2) authigenic (precipitated from sea water or formed by chemical replacement of other particles, for example,

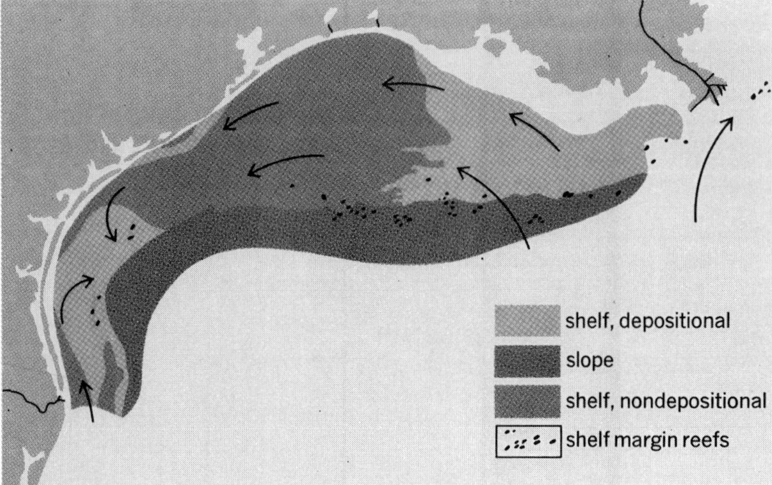

Fig. 2. Sediment distribution of the northwestern Gulf of Mexico as an example of the sediments of a shelf with abundant land-derived sediment. Shelf depositional sediments and slope deposits are silty clays; nondepositional area is covered with relict sediments. Arrows show generalized circulation.

glauconite, salt, and phosphorite); (3) residual (locally weathered from underlying rocks); (4) relict (remnants of earlier environments of deposition, for example, deposits formed during the transgression leading to the present high sea level stand); and (5) detrital (products of the weathering and erosion of the land, supplied by streams and coastal erosion, such as gravels, sand, silt, and clay).

On shelves with abundant land-derived sediment, the coastal zone is composed of deltas, lagoons, bays, and beaches and barriers. Outside the beaches and barriers, a narrow strip of wave-transported sand, usually less than 2 or 3 mi wide, fringes the coast. On the open shelf, the sediment deposited under present conditions is a silty clay, which, near deltas, grades imperceptibly into its bottomset beds. Usually, the silty clay, which results from winnowing near the coast, is carried no more than 20–30 mi offshore by marine currents, so that the zone of active deposition is restricted. If the shelf is narrow, all of it will fall into this zone, but if it is wide, the outer part will be covered by relict sediments resulting from the recent transgression. These relict sediments were deposited near the migrating shoreline and consist of beach sands and thin lagoonal deposits. They have been extensively churned by burrowing animals and wave action, resulting in a mottled structure, and authigenic glauconite has formed in them.

On many shelves, small calcareous reefs (shelf-edge reefs) occur at the outer edge. These reefs apparently depend on the presence of deep water for their growth, although it is not certain that they are growing vigorously at the present time. In the Gulf of Mexico, where they are particularly abundant, they mark the tops of salt domes in the subsurface. Beyond the reefs begins the zone of slope deposition, where in deeper and quiet water silty clays with abundant calcareous remains of open water organisms are being slowly deposited. Thus, there are in principle four parallel zones on each shelf: an inner sandy zone; an intermediate zone of clay deposition; an outer shelf zone of no deposition, where relict sediments occur, terminating in edge reefs; and a slope zone of calcareous clays. This parallel zonation is often strongly modified by special current patterns, which carry fine sediments farther out across the shelf, as in the western Gulf of Mexico; by rapidly advancing deltas which provide a sediment source far out on the shelf, as in the Mississippi delta; or by exposure to unusually vigorous wave action which prevents fine sediments from being deposited, even though a supply is present.

The fine-grained deposits that are being formed tend to be deposited more rapidly near the source than farther away, and as a result contain more biogenous material with increasing distance from the source, so that they become more calcareous. An example of shelf sediments with abundant supply of land-derived material is shown in Fig. 2. On shelves with little or no land-derived material, the only available sources of sediment are biogenous and authigenic. These sources provide far less material than rivers do, and as a result sedimentation rates are much lower. Even on shelves with abundant supply of land-derived material, the areas of nondeposition are extensive, often 40–50% of the total area. On the calcareous shelves, relict sediment may cover up to 75% of the entire area. Near the shelf edge these relict sediments are shallow-water deposits formed during the last low stand of the sea, and consist of small algal reefs and oolites as described for the Bahamas.

Landward of this zone, the deposits, formed when sea level rose rapidly, are thin blankets of calcareous debris, consisting of shell material, bryozoa, coral debris, and so forth (skeletal calcarenites). The only active sedimentation zones occur very near the shore, where calcareous and sometimes land-derived sand, silt, and clay are being deposited at present, and at the outermost shelf margin and continental slope, where a blanket of fine-grained calcareous mud very rich in planktonic Foraminiferida is found (foraminiferid calcilutite and calcarenite). Figure 3 shows two examples of calcareous shelves; similar are the Persian Gulf and many of the shelves of Australia.

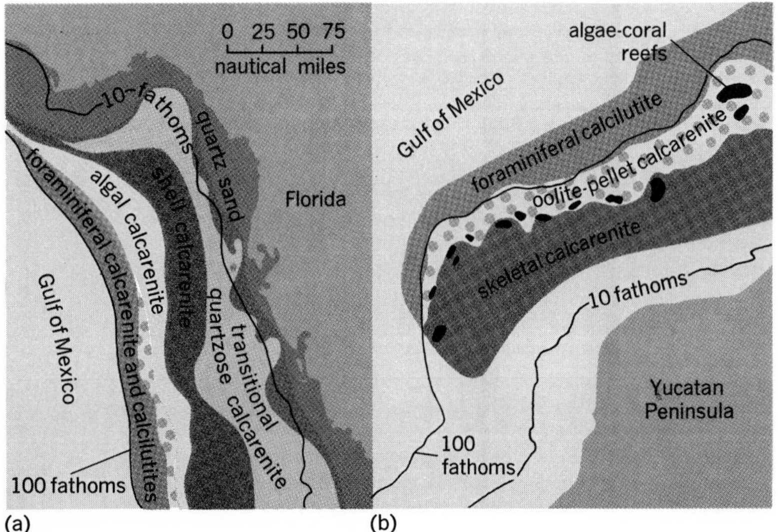

Fig. 3. Sediments on shelves with little land-derived sediment supply. (*a*) Off western coast of Florida. (*b*) Off northwestern Yucatan Peninsula.

The rise of sea level has, in many regions, severely restricted the supply of sediment to the continental shelf. Along the eastern coast of the United States, the valleys of many rivers have been flooded, and during the present stable sea level, barriers have been built across them which restrict the escape of sediment from the estuaries. As a result, sedimentation on this shelf is slow, and relict transgressive sands occur nearly everywhere at the surface.

Much of the fine-grained sediment transported into the sea by rivers is not permanently deposited on the shelf but kept in suspension by waves. This material is slowly carried across the shelf by currents and by gravity flow down its gentle slope, and is finally deposited either on the continental slope or in the deep sea. If submarine canyons occur in the area, they may intercept these clouds, or suspended material, channel them, and transport them far into the deep ocean as turbidity currents. If the canyons intersect the nearshore zone where sand is transported, they can carry this material also out into deep water over great distances.

Complex sediment patterns form in areas of considerable relief, for example, the borderland off southern California, where very coarse relict and residual sediments on shallow banks alternate rapidly with silty clays and calcareous deposits in the deep troughs. Such cases, however, are rare along the continental margins. Unimportant, but striking and geologically interesting, are the calcareous sediments associated with coral reefs and atolls. Usually, they occur on islands in mid-ocean, where clear water with abundant nutrients is available, and land-derived sediments are absent; but fringing and barrier reefs with associated calcareous sediments also occur along coasts with low sediment supply. *See* REEF.

[TJEERD H. VAN ANDEL]

Deep-sea sediments. Sediments covering the floor of the deep sea were first systematically described and classified during the late 19th century by J. Murray and A. F. Renard (1884, 1891) after their observations during the Challenger Expedition (1872–1876). Their classification included two principal sediment types, terrigenous (sediments deposited near to and derived from continental areas) and pelagic (sediments, principally fine grained, accumulated slowly by settling of suspended material in those parts of the ocean farthest from land). Both categories include biogenic and nonbiogenic material, as well as sediment derived from continents, that is, terrigenous, making the classification, at best, difficult to apply. Some of the terms, such as pelagic red clay, however, have remained in general usage but with meanings somewhat modified from the original definitions. Some of the more widely quoted classifications have been proposed subsequently by H. U. Sverdrup and coworkers (1942), R. R. Revelle (1944), P. H. Kuenen (1950), F. P. Shepard (1963), and E. D. Goldberg (1964). A problem with most classifications has been in distinguishing descriptive categories, for example, red clay, from genetic categories, that is, those that include an interpretation of sediment origins, for example, volcanic mud. In addition, classifications are difficult to apply because so many deep-sea sediments are widely ranging mixtures of two or more end-member sediment types. The following sections will briefly describe the most important end members, their manner of origin, and some of the factors that control their distribution.

Biogenic sediments. Biogenic sediments, those formed from the skeletal remains of various kinds of marine organisms, may be distinguished according to the composition of the skeletal material principally either calcium carbonate or opaline silica. The most abundant contributors of calcium carbonate to the deep-sea sediments are the planktonic foraminiferids, coccolithoforids and pteropods. Organisms which extract silica from the sea water and whose hard parts eventually are added to the sediment are radiolaria, diatoms, and to a lesser degree, silicoflagellates and sponges. The degree to which deep-sea sediments in any area are composed of one or more of these biogenic types depends on the organic productivity of the various organisms in the surface water, the degree to which the skeletal remains are redissolved by sea water while settling to the bottom, and the rate of sedimentation of other types of sediment material. Where sediments are composed largely of a single type of biogenic material, it is often referred to as an ooze, after its consistency in place on the ocean floor. Thus, depending on the organism, it may be called a foraminiferal ooze (or globigerina ooze, after one of the most abundant foraminiferid genera, *Globigerina*), coccolith ooze, pteropod ooze, or for the siliceous types, diatomaceous or radiolarian ooze. The distribution map of major sediment types (Fig. 4*a*) does not distinguish the several types of organism, but indicates that the siliceous oozes are most abundant at high latitudes and in the highly productive areas of upwelling such as the Equatorial Pacific. Calcareous oozes are also abundant beneath zones of upwelling and along topographic highs such as the Mid-Atlantic Ridge. Their relative abundance in shallower water is due to an increased tendency for calcium carbonate (calcite) to redissolve at ocean depths greater than about 4000 m because of lower water temperature, greater pressure, and increased dissolved carbon dioxide in the water

Nonbiogenic sediments. The nonbiogenic sediment constituents are principally silicate materials and, locally, certain oxides. These may be broadly divided into materials which originate on the continents and are transported to the deep sea (detrital constituents) and those which originate in place in the deep sea, either precipitating from solution (authigenic minerals) or forming from the alteration of volcanic or other materials. The coarser constituents of detrital sediments include quartz, feldspars, amphiboles, and a wide spectrum of other common rock-forming minerals. The finer-grained components also include some quartz and feldspars, but belong principally to a group of sheet-silicate minerals known as the clay minerals, the most common of which are illite, montmorillonite, kaolinite, and chlorite. The distributions of several of these clay minerals have yielded information about their origins on the continents and, in several cases, clues to their modes of transport to the oceans. For example, the distribution map of kaolinite concentrations in the less than 2-μ fraction (Fig. 4*b*) shows that this mineral occurs in the deep sea principally adjacent to continental areas,

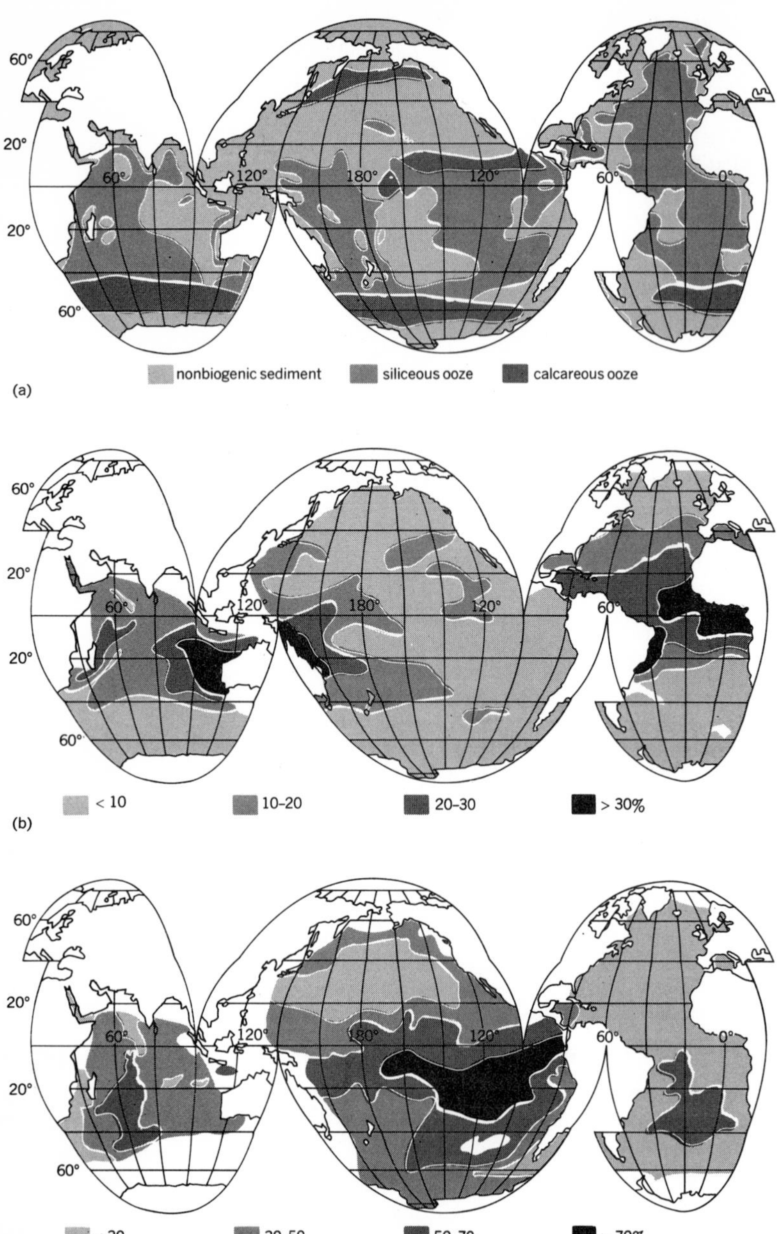

Fig. 4. Distribution of deep-sea sediments. (*a*) Major types generalized after H. U. Sverdrup and coworkers, 1942, and K. K. Turekian, 1964. (*b*) Kaolinite concentration. (*c*) Montmorillonite concentrations. Parts *b* and *c* are generalized after J. J. Griffin and coworkers, 1968, and P. E. Biscaye, 1965. The concentrations are in the <2-μ size fraction on a carbonate-opaline-silica-free basis.

mostly at low latitudes, where it is a common constituent of tropical soils. Its distribution indicates that it occurs only in the deep sea as a detrital phase and, in the western Equatorial Atlantic, appears to be carried to the ocean only by rivers. On the eastern side of the Equatorial Atlantic, however, its more widespread distribution corresponds with a region of frequent dust storms and suggests that its transport there is both by rivers and by the Northeast Trade Winds.

One of the common clay minerals, montmorillonite, has both a detrital and volcanogenic origin in different parts of the deep sea. In the Atlantic Ocean its distribution is controlled by its relative abundance in the soils of adjacent continental areas, whereas in the South Pacific Ocean its great abundance results from its formation, along with authigenic minerals called zeolites, as an alteration product of widespread volcanic glass (Fig. 4*c*). The reason these materials—both the volcanic glass shards and their alteration products—are so abundant in the South Pacific is that they are relatively undiluted by detrital constituents, the South Pacific being so large and rimmed by numerous deep trenches which act as traps for seaward-moving detrital sediments.

An authigenic constituent of deep-sea sediments that has received much recent attention, because of the potential economic value of trace elements that occur with it (principally copper), is manganese oxide. This material occurs in many areas as manganese-iron nodules. These are spheroidal pellets of MnO_4 and Fe_2O_3, ranging in size from microscopic to tens of centimeters in diameter and covering thousands of square miles of the deep-sea floor, principally in areas where both biogenic and nonbiogenic sedimentation rates are very slow.

Both in the case of biogenic and nonbiogenic sedimentation, the old concept of a slow, gentle "snowfall" of sediment particles to the sea floor where they remain forever buried, has been shown to be untrue in many parts of the ocean. Agencies are at work within the deep sea capable of distributing and redistributing sediment from its point of origin or entry in the ocean. Turbidity currents, high-density clouds of entrained sediment, are responsible for rapidly moving downward vast amounts of sediment from the continental slopes and other topographic highs. In addition, the deep oceanic current systems have been shown to be capable of eroding and transporting sediment along the ocean floor. The evidence for this comes from deep-sea photographs which show current ripple marks and scour marks in many parts of the ocean as well as measurements of relatively turbid zones of water traveling near the bottom (the nepheloid layer) in many parts of the ocean; this research suggests a rather continuous process of sediment erosion and redeposition. This process is discussed in a later section on transport and rate of sedimentation.

Ancient sediments. Because intensive studies of deep-sea sediments are a relatively recent phenomenon in the study of the Earth, and because the surface layer of sediment covers nearly two-thirds of the Earth's surface, investigators have only begun to understand the factors which presently control the distribution of sediments. An entire additional dimension, however, is introduced by extending these studies down through the sediment column, that is, back through geologic time. Samples of sediment have been taken in deep-sea cores back through the Pleistocene, Tertiary, and as old as Jurassic, or more than 130,000,000 years old. The few studies made to date (1969) on these older deep-sea sediments show that sedimentary conditions were very different from those that now obtain. Several studies in the Pacific Ocean and scattered results from the Atlantic Ocean suggest that sediments derived from volcanic sources were possibly much more important in late Mesozoic and during most of Tertiary time, and that the importance of detrital sediment sources is a relatively recent phenomenon, beginning as late as Pliocene or Pleistocene time.

[PIERRE E. BISCAYE]

Bibliography: Tj. H. van Andel, Morphology and sediments of the Sahul Shelf, northwestern Australia, *Trans. N.Y. Acad. Sci.*, 28:81–89, 1965; Tj. H. van Andel and J. R. Curray, Regional aspects of modern sedimentation in northern Gulf of Mexico and similar basins, in F. P. Shepard (ed.), *Recent Sediments, Northwestern Gulf of Mexico*, Amer. Ass. Petrol. Geol. Spec. Publ., 1960; P. E. Biscaye, Mineralogy and sedimentation of Recent deep-sea clay in the Atlantic Ocean and adjacent seas and oceans: *Geol. Soc. Amer. Bull.*, 76:803–832, 1965; K. O. Emery, *The Sea of Southern California*, 1960; D. B. Ericson and G. Wollin, *The Deep and the Past*, 1964; E. D. Goldberg, The oceans as a geological system, *Trans. N.Y. Acad. Sci.*, 27:7–19, 1964; J. J. Griffin, H. Windom, and E. D. Goldberg, The distribution of clay minerals in the world ocean, *Deep-Sea Res.*, 15:433–459, 1968; F. P. Shepard, *Submarine Geology*, 2d ed., 1963; H. U. Sverdrup, M. W. Johnson, and R. H. Fleming, *The Oceans*, 1942; K. K. Turekian, The geochemistry of the Atlantic Ocean Basin, *N.Y. Acad. Sci. Trans.*, ser. 2, 26:312–330, 1964.

PHYSICAL PROPERTIES

Physical properties of marine sediments, such as density and the elastic constants, depend on many factors. These include grain shapes, sizes, and compositions; the amount of interstitial fluid and its properties; the nature of grain-to-grain contacts; the degree of compaction and consolidation; and the age. Measurable physical quantities appear to be affected to a much greater extent by the fractional volume of fluid in the sediment (porosity) than by sediment type. This is to be expected since density and elastic constants do not differ much for the principal constituents of sediments (silica, calcium carbonate, clay minerals). Water depth affects physical properties only to a slight extent.

Some measurements of physical properties are made on samples recovered from the ocean bottom by coring devices or in sufficiently shallow water by divers. Such observations are limited to sediments lying within a few tens of feet of the water-sediment interface. Properties of deeper-lying sediments are known from seismic refraction measurements of the velocities of elastic waves, by inference from dispersion of surface waves, or from gravity data. A few typical measurements are given in the table. These are merely illustrative and are not in any sense average or most likely values.

Definitions and some useful interrelationships among measurable quantities are given by Eqs. (1)–(5), where ρ_1 and ρ_2 are fluid and average particle densities, respectively, ϕ is the porosity ($\phi = 1$ at 100% fluid, $\phi = 0$ at 0% fluid), μ is rigidity, and σ is Poisson's ratio.

$$\rho = \rho_1 \phi + \rho_2 (1 - \phi) \quad \text{(bulk density)} \qquad (1)$$

$$\alpha = \sqrt{(k + 4/3\mu)/\rho} \quad \text{(compressional wave velocity)} \qquad (2)$$

$$\beta = \sqrt{\mu/\rho} \quad \text{(shear wave velocity)} \qquad (3)$$

$$(\alpha/\beta)^2 = 2(1 - \sigma)/(1 - 2\sigma) \qquad (4)$$

$$k = 1/C \quad \text{(incompressibility} = 1/\text{compressibility)} \qquad (5)$$

Several general conclusions may be drawn from observations. Density and porosity are nearly linearly related and for most ocean sediments observations lie between the two lines, as in Eq. (6).

$$\rho_{\text{cgs}} = 1.03 + (1.67 \pm 0.05)(1 - \phi) \qquad (6)$$

Compressional wave velocities for $\phi > 0.6$ agree well with the predictions, or equation, of A. B. Wood. This is obtained by inserting into Eq. (2) $\mu = 0$ and $C = 1/k = C_1\phi + C_2(1 - \phi)$, where C_1 and C_2 are compressibilities of fluid and particles, respectively. At smaller porosities compressional velocity rises more steeply, the $\phi = 0$ limit being near 6

Typical measurements of selected properties of marine sediments

Property measured	Fine sand, 17-station average[a]	Clayey fine silt[a]	Gray clay or silt[b]		Cream calcisiltite[c]		Gray clay[c]	Artificially compacted globigerina ooze pressure[d], kg/cm²		
								512	768	1024
Medium grain diameter, mm	0.19	0.02			[0.01][e]	[0.01]				
ρ, g/cm³	1.93	1.60	1.72	1.58	1.60	1.71	1.46	2.14	2.22	2.26
ϕ, %	46.2	65.6	[56]	[65]	65	57	74	[32]	[28]	[26]
α, km/sec	[1.68]	[1.46]			1.59	1.68	1.49	2.68	2.89	3.06
β, km/sec								1.20	1.42	1.57
σ (Poisson's ratio)	0.44[f]	0.50[f]						0.37	0.34	0.32
k, 10^{-11} dyne/cm²	[0.472]	[0.342]						1.13	1.25	1.38
μ, 10^{-11} dyne/cm²	[0.06][g]	[0.00][g]						0.31	0.45	0.56
Thermal conductivity, 10^{-4} cal/(cm)(°C)(sec)			26.8	23.1						
Approximate water depth, fathoms	15	15	1000	1000	2300	1600	2550			

[a]After E. L. Hamilton et al., 1956. [b]After E. Bullard, 1954. [c]After G. H. Sutton et al., 1957. [d]After A. S. Laughton, 1957. [e]Brackets indicate conversion of units from those used in the original publication. [f]Lower limit. [g]Upper limit.

km/sec. Poisson's ratio may be expected to vary from a value near 0.25 at zero porosity to 0.5 at $\mu = 0$. Good agreement of α with the Wood equation for $\phi > 0.6$ indicates that σ reaches its upper limit near $\phi = 0.6$.

Seismic refraction measurements indicate that compressional velocity increases with depth in the sedimentary column, the gradient being from 0.5 to 2.0/sec. Thus ρ, μ, k, β should also increase with depth and ϕ and σ decrease. [JOHN E. NAFE]

Bibliography: E. Bullard, The flow of heat through the floor of the Atlantic Ocean, *Proc. Roy. Soc. London, Ser. A*, 222:408–429, 1954; E. L. Hamilton et al., Acoustic and other physical properties of shallow-water sediments off San Diego, *J. Acoust. Soc. Amer.*, 28:1–15, 1956; A. S. Laughton, Sound propagation in compacted ocean sediments, *Geophysics*, 22:233–260, 1957; J. E. Nafe and C. L. Drake, Variation with depth in shallow and deep water marine sediments of porosity, density and the velocities of compressional and shear waves, *Geophysics*, 22:523–552, 1957; C. B. Officer, Jr., A deep-sea seismic reflection profile, *Geophysics*, 20:270–282, 1955; G. H. Sutton, H. Berckhemer, and J. E. Nafe, Physical analysis of deep sea sediments, *Geophysics*, 22:779–821, 1957; A. B. Wood, *A Textbook of Sound*, 3d ed., 1955.

TRANSPORT AND RATE OF SEDIMENTATION

Rivers, glaciers, wind, and ocean waves and currents carry particles from continents and continental margins to the various environments of deposition in the ocean. The rates of sedimentation can be determined by study of the micropaleontology, radioactivity, or paleomagnetism of the sediments.

Transport of sediments. Sedimentary particles derived from the continents by weathering are brought to the sea by rivers and streams. Most of these particles never reach the deep sea but are (1) trapped in estuaries; (2) deposited in or near deltas; (3) concentrated on beaches or other littoral deposits; or (4) carried to and deposited on the inner continental shelves by waves and longshore currents. A small fraction of the clay-sized particles, especially near the mouths of major rivers, is carried off of the continental shelf by currents. *See* DELTA; NEARSHORE PROCESSES.

Low stands of the sea during glacial epochs exposed only the inner margins of deeper continental shelves to erosion, while their outer portions became zones of littoral deposition. Rivers crossed shallower shelves through valleys cut in the shelf plain and debauched directly down the continental slopes. As glaciation waned and sea level rose, the littoral deposits of the outer shelf were sometimes reworked and their finer clasts carried inward by wave action to contribute, in part, to present coastal sediments.

On most continental shelves the outer sections are receiving no contemporary continental sedimentation; therefore, they retain relict littoral sedimentary deposits which document former low stands of the sea. Some continental shelves in high latitudes are floored with glacial deposits: terminal moraines, recessional moraines, or glacial marine sediments deposited during Pleistocene glacial epochs. These deposits, too, are relict, except for the Antarctic and Greenland continental shelves where present glaciers terminate on the shelf and icebergs raft glacially derived sediments seaward before melting. Tropical shelves which lie between the 18°C sea-water isotherms often have carbonate sediments, in part derived by wave and current action from barrier and fringing reefs at the shelf margin, and in part biochemically precipitated by other invertebrates associated with carbonate reef-bank ecologies. Under optimum conditions reefs grow upward at rates up to 50 m/1000 years. *See* GLACIAL EPOCH; GLACIOLOGY; REEF.

Sediments accumulate on the continental slopes until slumping occurs to carry the deposits down to the abyssal floor. These slumps often occur by rotational faulting, but sometimes the slump transforms by mixing with water to become a mud flow (landslide) or density current. The transport of very fine-grained material by density currents, also known as turbidity currents, has been hypothesized as being a major transportive mechanism from continental slopes to abyssal plains, especially in the Northern Hemisphere. During the low sea level stands of glacial epochs when rivers discharged directly down the continental slopes, density currents were much more prevalent and may have been instrumental in eroding submarine canyons, forming the great submarine alluvial cones which occur at their bottoms and building abyssal valleys which show natural levies hundreds of miles from the continental margins. *See* SUBMARINE CANYON; TURBIDITY CURRENT.

There is increasing evidence that geostrophic ocean currents are responsible for considerable sediment transport, even of sand-sized sedimentary particles, in the deep sea (Fig. 5). Such currents have been demonstrated to cause erosion, not only from submarine ridges and sea mounts, but also across abyssal plains, and to deposit sediments on the protected side of obstructions and in submarine valleys.

Large portions of the ocean floor that are far from land and lie under zones of high surface biological productivity are covered by biological deposits which consist of the calcareous or siliceous tests of plankton. These deposits constitute the organic oozes which cover more than 60% of the world's ocean floors (Fig. 4).

Much smaller contributions to abyssal sediments are made by wind transport of silt, clay, and

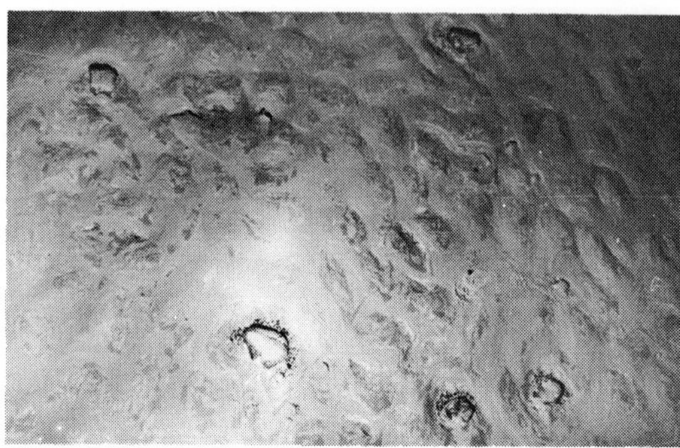

Fig. 5. On southern margin of the Burdwood Bank, in northern Drake Passage. Rhomboidal ripple marks in foraminiferid ooze attest to current velocities in excess of 1 knot. Scour around ice-rafted boulders occurs on down-current side. Lat. 55-59°S; long. 61-43°W; 4060 m.

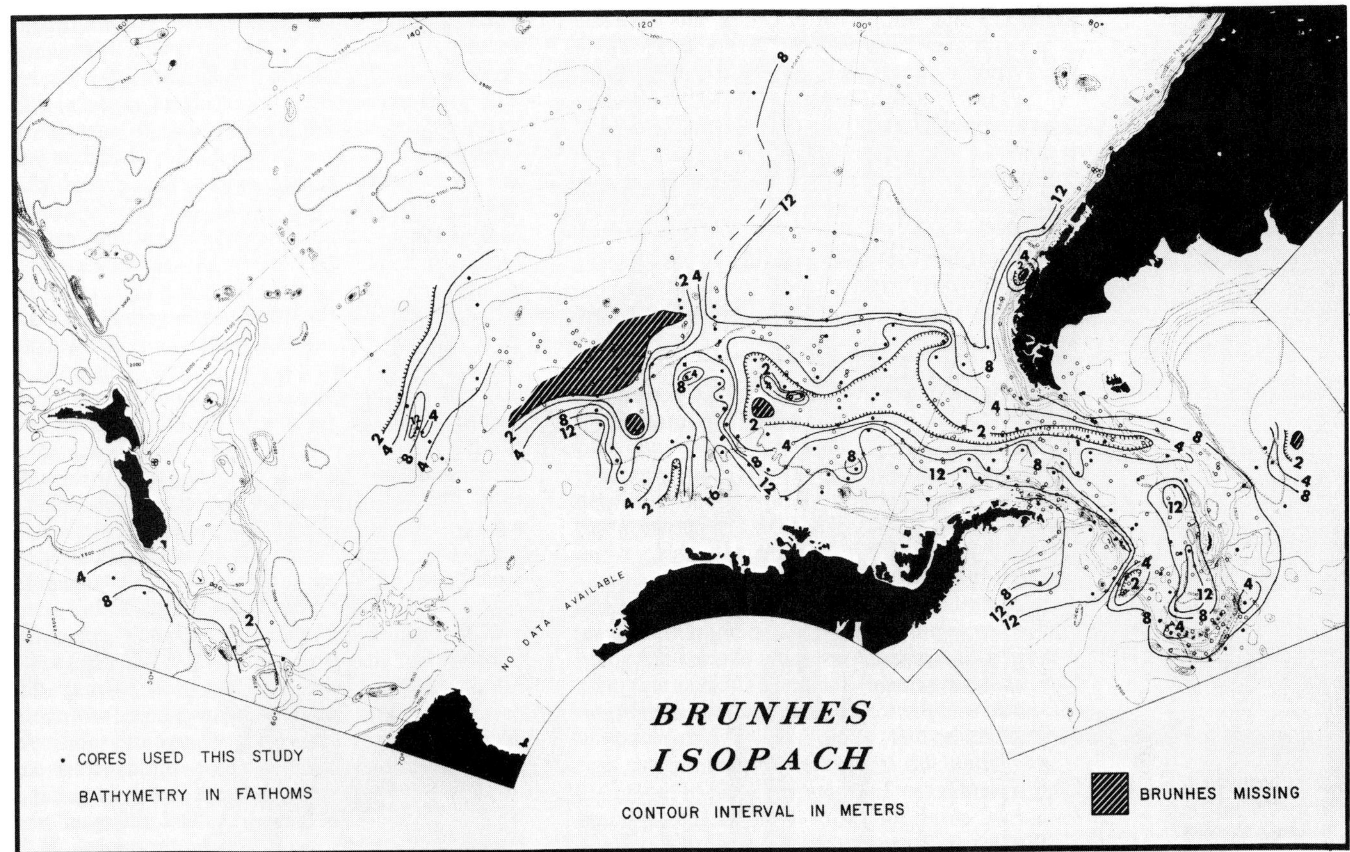

Fig. 6. Thickness map of Brunhes-age sediments. Reference horizons are sediment-water interface and Brunhes-Matuyama polarity reversal boundary 700,000 years ago. The area where the Brunhes-age sediments are missing, as well as the area of the Brunhes-age sediments outlined by 2-m thickness or less, across the Southern Pacific is a zone of currents, documented by bottom photography, as in Fig. 5.

volcanic ash from the continents, cosmic dust and micrometeorites, and precipitation of inorganic constituents directly out of sea water. Of still less importance is biological transport of detritus by mammals, fish, birds, and kelp.

Rates of sedimentation. Early measures of rates of sedimentation on nearshore deposits were estimated from historical filling of bays and estuaries, and progradation and erosion of coastlines. Direct observation in the deep sea was limited to the apparent depth of burial of such objects as deep-sea cables, which had been on the sea floor for several years. Since such objects could have partially sunk into the sediment, rates of burial derived from sedimentation alone were conjectural.

Other early attempts revolved about the recognition in deep-sea cores of the end of the last glacial epoch as documented by temperature-sensitive pelagic Foraminiferida. However, such climatic changes are usually not synchronous worldwide and in tropical regions may not even have occurred. Furthermore, the time of the end of the last glacial epoch was not established until recently, and in the deepest parts of the sea, solution of the calcareous tests erased the record.

Modern methods of calculating rates of sedimentation necessitate the determination of time boundaries in deep-sea cores, using radioactivity or paleomagnetism. Radioactive determinations utilize the rate of change of surface disequilibrium to equilibrium-with-depth in cores in isotope pairs of the U^{238}, U^{235}, or Th^{232} families. These methods depend upon numerous assumptions, including the chemical separation in sea water of uranium from thorium by the precipitation of thorium hydroxide into the sediment, as well as the retention and nonmigration of radiodisintegration daughter products within the sediment column. Theoretically, these methods cover the time span back to more than 300,000 years. Other methods have used C^{14}, Be^{10}, Al^{26}, or Si^{31}, all of which are produced in the outer atmosphere by cosmic ray bombardment. Carbon-14 can be used only in carbonate sediments and has a limited half-life. The remainder occur in very low quantities, requiring sophisticated ion-exchange techniques for concentration or low-level counting for long periods of time. *See* ROCK, AGE DETERMINATION OF.

Sedimentation rates based on paleomagnetic chronology depend upon the determination of the paleomagnetic polarity of deep-sea sediments and upon the correlation of these polarities with the paleomagnetic polarity time scale developed from terrestrial lavas. It has been demonstrated that fine particles settle through a water column to align themselves with the Earth's field, rather than doing so hydrodynamically, and acting as tiny magnets, they represent the Earth's field at the time they are deposited. If only the vertical orientation of a core is known (that is, which end is up), the direction of the vertical vector of the detrital remanent magnetism is sufficient to determine whether or not the sediment is normally magnetized (deposited during a time when the Earth's magnetic field is oriented N–S, as it is today), or if there is reversed magnetism (sediment deposited when the Earth's field was in the opposite sense). Inasmuch as the times when the Earth's magnetic field has reversed are known with some precision back to almost 3,500,000 years, reversals in remanent magnetism in sedimentary cores provide time lines for correlation, against which rates of sedimentation can be measured. Reversals in the Earth's magnetic field are worldwide and geologically synchronous and therefore provide a chronological datum everywhere in fine-grained deep-sea deposits. Where sufficient cores are available, it has been possible to draw maps of the thickness of sediment deposited during any selected paleomagnetic time interval. These maps are also maps of rates of sedimentation (Fig. 6). *See* ROCK MAGNETISM.

Rates of sedimentation around deltas and in estuaries are as high as 50,000 cm/1000 years; on continental slopes and rises, 100 cm/1000 years; and on the abyssal floors, 0.01–2 cm/1000 years. These rates have varied considerably in the past, depending upon climate, tectonism, and organic productivity in the ocean. Rates have been highest when the continents and mountain ranges were high, during glacial stands of low sea level when continental drainage was dumped directly onto abyssal floors, and when the wind and ocean circulation patterns were such as to ensure maximum biological productivity. Rates have been lowest when widespread epeiric seas have inundated the continents, when the continental margins consisted of basins or geosynclines which trapped sediments, and prior to the advent of calcareous Foraminiferida in the late Mesozoic. *See* GEOSYNCLINE. [H. G. GOODELL]

Bibliography: G. Arrhenius, Pelagic sediments, in M. N. Hill (ed.), *The Sea*, vol. 3, 1963; R. A. Bagnold, Beach and nearshore processes, pt. 1: Mechanics of marine sedimentation, in M. N. Hill (ed.), *The Sea*, vol. 3, 1963; A. Cox, Geomagnetic reversals, *Science*, 163:237–245, 1969; K. O. Emery, Organic transportation of marine sediments, in M. N. Hill (ed.), *The Sea*, vol. 3, 1963; H. G. Goodell and N. D. Watkins, The paleomagnetic stratigraphy of the Southern Ocean: 20° west to 160° east longitude, *Deep-Sea Res.*, 15:89–112, 1968; B. C. Heezen, Turbidity currents, in M. N. Hill (ed.), *The Sea*, vol. 3, 1963; D. L. Inman and R. A. Bagnold, Beach and nearshore processes, pt. 2: Littoral processes, in M. N. Hill (ed.), *The Sea*, vol. 3, 1963; F. F. Koczy, Age determination in sediments by natural radioactivity, in M. N. Hill (ed.), *The Sea*, vol. 3, 1963.

CLIMATIC RECORD IN SEDIMENTS

An undisturbed section of fossiliferous pelagic sediments contains important information on the geophysical and environmental conditions prevailing in the Earth's interior, on the ocean floor, in the water column above, at the ocean's surface, in the atmosphere, on adjacent continents, and even in outer space, during the time of sediment deposition. The sediments are sampled, to a depth of about 20 m below the sea floor, by piston coring and, at greater depths, by drilling and wire-line coring inside the drill string. Rates of sedimentation for fossiliferous pelagic sediments of globigerina-ooze facies range from one to a few centimeters per 1000 years. Sediments as old as 10^6 years have been reached by piston coring, and sediments older than 150×10^6 years have been sampled by the drilling vessel *Glomar-Challenger*.

Methods of study. Several parameters characterizing pelagic sediments are directly or indirectly related to climate. Among these are clay mineralogy, carbonate content, ice-rafted detritus, the composition of planktonic and benthic fossil microfaunas, and the oxygen isotopic composition of calcareous skeleta and skeletal elements of foraminifera and coccolithophoridae.

Clay mineralogy. Of the common clay minerals, kaolinite is typical of weathering in the tropics while chlorite is a more common product of weathering in temperate and cold regions. The ratio of these two minerals in pelagic Atlantic sediments is indicative of climate in the surrounding continents.

Calcium carbonate content. Calcium carbonate content depends upon the productivity of shelled carbonate plankton (mostly coccolithophoridae and planktonic foraminifera) and upon the influx of noncarbonate particles (mostly clay minerals and quartz). Both variables are climate-dependent. Productivity generally increases during glacial ages because of the more vigorous vertical oceanic circulation. In the Atlantic, which drains most of the lands that were glaciated during the ice ages, the influx of noncarbonate particles increases during glacial ages even more than that of the carbonate material, so that the percentage of carbonate in bulk sediment is directly related to temperature. In the Pacific, where the influx of noncarbonate material to the open ocean is smaller because of the more limited drainage and also because of sediment trapping in offshore trenches, the carbonate percentage is inversely proportional to temperature.

Ice-rafted detritus. When ice expands on Earth, icebergs are more abundant and reach lower latitudes. The arrival of ice-rafted detritus in an area of the ocean floor previously free of such material is a clear indication of climatic deterioration.

Composition of fossil microfaunas and microfloras. The more abundant fossil components of pelagic sediments are, in order of decreasing global abundance, coccolithophoridae, planktonic foraminifera, diatoms, radiolarians, pteropods, and benthic foraminifera. Only the first two taxa have produced important information on climatic change. In terms of species diversity, the standing crop amounts to about 60 for coccolithophoridae and 30 for planktonic foraminifera. Sampling by means of water samplers and opening-and-closing plankton nets has shown that different species of both coccolithophoridae and planktonic foraminifera live at different latitudes as well as at different depths (largely within the euphotic zone). Because of the latitudinal preferences, a change in the fossil assemblage is indicative of climatic change at the ocean surface.

Oxygen isotopic composition. When a crystal is precipitated in isotopic equilibrium from a solution, the isotopes of the more common elements in

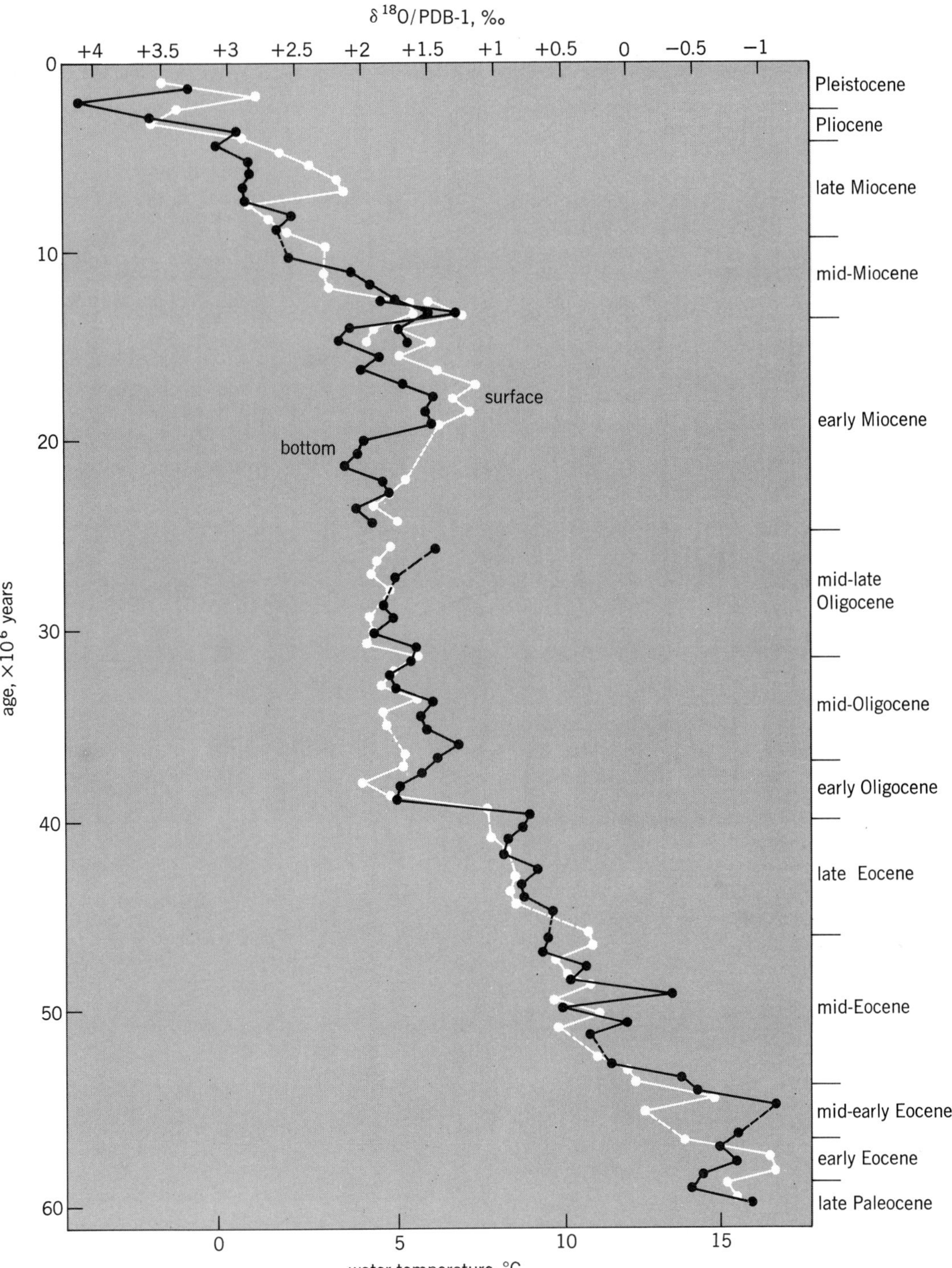

Fig. 7. Oxygen isotopic data (deviation per mil from the Chicago standard PDB-1) and early and middle Cenozoic paleotemperatures at three subantarctic DSDP sites (277, 279, and 281) for planktonic (white curve) and benthic (black curve) foraminifera. (*From J. P. Kennett, Cenozoic evolution of Antarctic glaciation, the Circum-Antarctic Ocean, and their impact on global paleoceanography, J. Geophys. Res., 82:3843–3860; copyright 1977 by American Geophysical Union*)

the system will distribute themselves among the phases present so as to reduce free energy to a minimum. Partition is temperature-dependent, being greater at lower temperatures and less at higher ones. In the H_2O-CO_2-$CaCO_3$ system, the temperature effect on the oxygen isotopic composition of the calcium carbonate being deposited is 0.23 per mil per degree Celsius. This effect is independent of the oxygen isotopic composition of the water. If ice increases on Earth, sea water is slightly enriched in ^{18}O (because evaporation favors the lighter H_2O^{16} molecule) and, in addition, the calcium carbonate of both foraminifera and coccolithophoridae is further enriched in the heavier oxygen isotope because of the lower temperature produced by the expanding ice. Oxygen isotopic

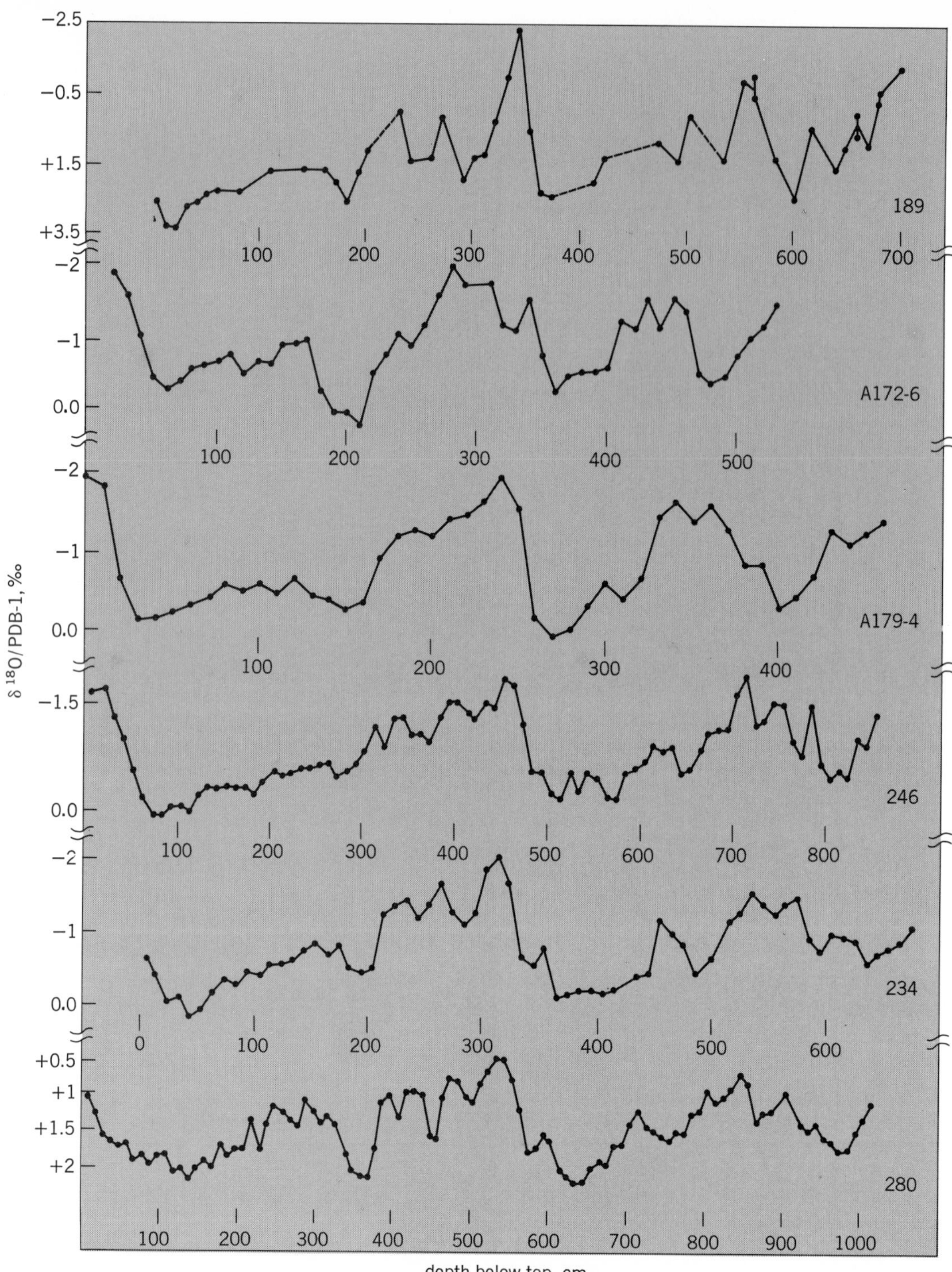

Fig. 8. Isotopic curves of six deep-sea cores: core 280, North Atlantic (34°57′N, 44°16′W, 4256 m); core 234, equatorial Atlantic (5°45′N, 21°43′W, 3577 m); core 246, equatorial Atlantic (0°48′N, 31°28′W, 3210 m); core A179-4, Caribbean (16°36′N, 74°48′W, 2965 m); core A172-6, Caribbean (14°59′N, 68°51′W, 4160 m); core 189, Mediterranean (33°54′N, 28°29′E, 2664 m). (*From C. Emiliani, Paleotemperature analysis of core 280 and Pleistocene correlations, J. Geol., 66:264–275, 1958*)

analysis of foraminiferal and coccolith shells by mass spectrometry yields the total isotopic effect and, if the change in the oxygen isotopic composition of sea water can be estimated, the climatic change can be assessed.

Tertiary climates. Pelagic sediments of Tertiary age (63 to 1.8×10^6 years ago) have been abundantly sampled by the *Glomar-Challenger*. Oxygen isotopic analysis of planktonic foraminifera from subantarctic sediments (south of New Zealand–Tasmania) has shown (Fig. 7*a*) that the surface temperature of the ocean in that area decreased 10°C from 55 to 34×10^6 years ago, abruptly dropped 4°C within 100,000 years about 34×10^6 years ago, remained constant or even slightly increased from 34 to 16×10^6 years ago, and then decreased, reaching 1 or 2°C at the beginning of the Quarternary. A similar temperature change is shown by the benthic foraminifera (Fig. 7*b*). These impressive changes in both surface and bottom temperatures provide information on the glaciation of Antarctica: the abrupt temperature drop about 34 × 10^6 years ago has been interpreted as indicating the development of significant sea-level freezing around Antarctica and more abundant formation of bottom water; and the temperature drop about 16 × 10^6 years ago has been interpreted as due to the rapid formation of a major, permanent ice sheet over East Antarctica. Ice-rafted detritus appeared 25 × 10^6 years ago in sediments deposited at 77°S in the Ross Sea, expanding to 53°S between Antarctica and Australia about 3 × 10^6 years ago. During this period, the northward march of Antarctic ice-rafted detritus averaged 1.3° of latitude per 10^6 years. *See* TERTIARY.

Quaternary climates. The beginning of the Quaternary, which includes the Pleistocene and the Holocene, has been established, by convention, at the time when the benthic foraminiferal species *Hyalinaea baltica* appeared in the section at Le Castella, Calabria, Italy. This event has been found to approximate in time the disappearance of discoasters from the world ocean and has been dated at about 1.8 × 10^6 years ago. Climatically, the Quaternary is dominated by the great, repeated glaciations over the northern continents. Mass spectrometric analysis of deep-sea cores from the Atlantic, the Caribbean, and the Mediterranean (Fig. 8) revealed strong oscillations in the oxygen isotopic composition of the shells of planktonic foraminifera. It has been estimated that at least one-third of the amplitude of the observed oscillations is due to temperature changes related to glaciation and the balance to the waxing and waning of the ice sheets on the northern continents. A time scale has been provided by ^{14}C, $^{231}Pa/^{230}Th$, and paleomagnetic measurements. Strong isotopic oscillations began at least 3 × 10^6 years ago, that is, in late Pliocene time, before the beginning of the Quaternary, and have continued to the present with increasing amplitude (Fig. 9). Carbon-14 dating has shown that the most recent minimum in the isotopic curves (Fig. 8) represents the last major glaciation, the classical Wisconsin of North America and Main Würm of Europe. It follows that earlier minima must represent major, earlier glaciations.

Oxygen isotopic analysis of a late Pliocene core (Fig. 10) has demonstrated that the last time of sustained warm climate was before 3.2 × 10^6 years ago. At that time, the Central American isthmus was formed, blocking the circulation between Atlantic and Pacific and profoundly altering the circulation in the North Atlantic, the Arctic, and the northern Pacific. It is possible that glaciation in Greenland and elsewhere in the northern continents was initiated by this event.

The detailed study of Quaternary cores has shown (Fig. 11) that major glaciations occurred at approximately 100,000-year intervals; that pleniglacials (times of full glaciation) and hypsithermals (times as warm as today) were of relatively short duration in the Quaternary, amounting to no more than 5–10% of Quaternary time; and that important secondary oscillations occurred during the longer periods between hypsithermals and pleniglacials.

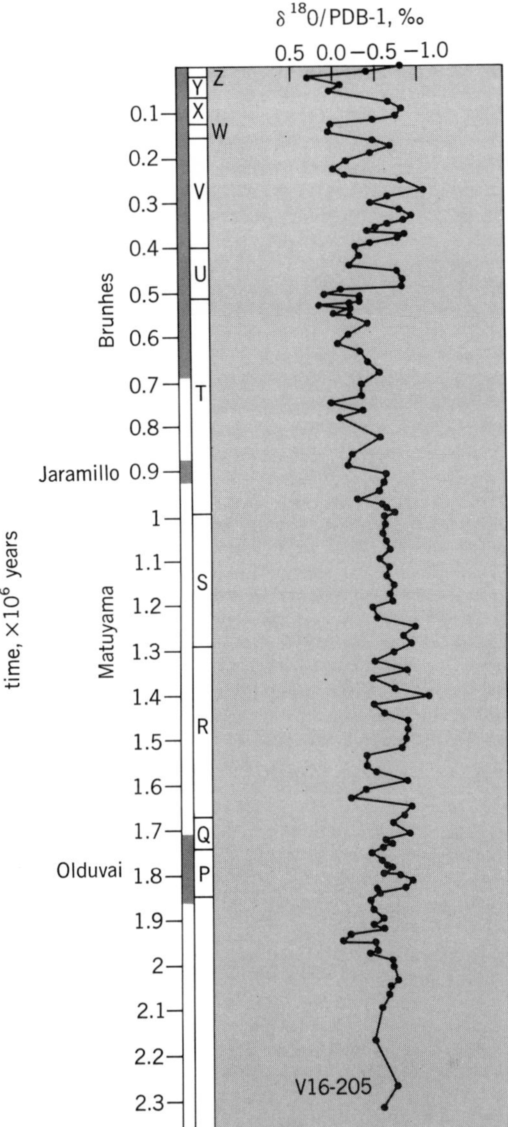

Fig. 9. Oxygen isotopic record (based on *Globigerinoides sacculifera*) from core V16-205 from the equatorial Atlantic (15°14′N, 43°24′W). (*From J. van Donk, O¹⁸ record of the Atlantic Ocean for the entire Pleistocene epoch, Geol. Soc Amer Mem., 145:147–163, 1976*)

The oxygen isotopic oscillations are paralleled by changes in the composition of the planktonic foraminiferal faunas (Fig. 12). These changes have been studied in detail by using modern statistical methods, and the results have been compiled in surface-water isotherm maps showing the seasonal temperatures and average salinity during the last pleniglacial (18,000 years ago; Figs. 13 and 14).

The planktonic foraminiferal species *Globigerinoides sacculifera* from equatorial Atlantic and Caribbean cores has 1.8 per mil less ^{18}O in its shell if formed in hypsithermal times than if formed in pleniglacial times. Benthic foraminifera from the same cores as well as from Pacific cores have 1.0 per mil less ^{18}O (Fig. 15). If the assumption is made that ocean-bottom temperatures did not change between glacial and interglacial ages, the

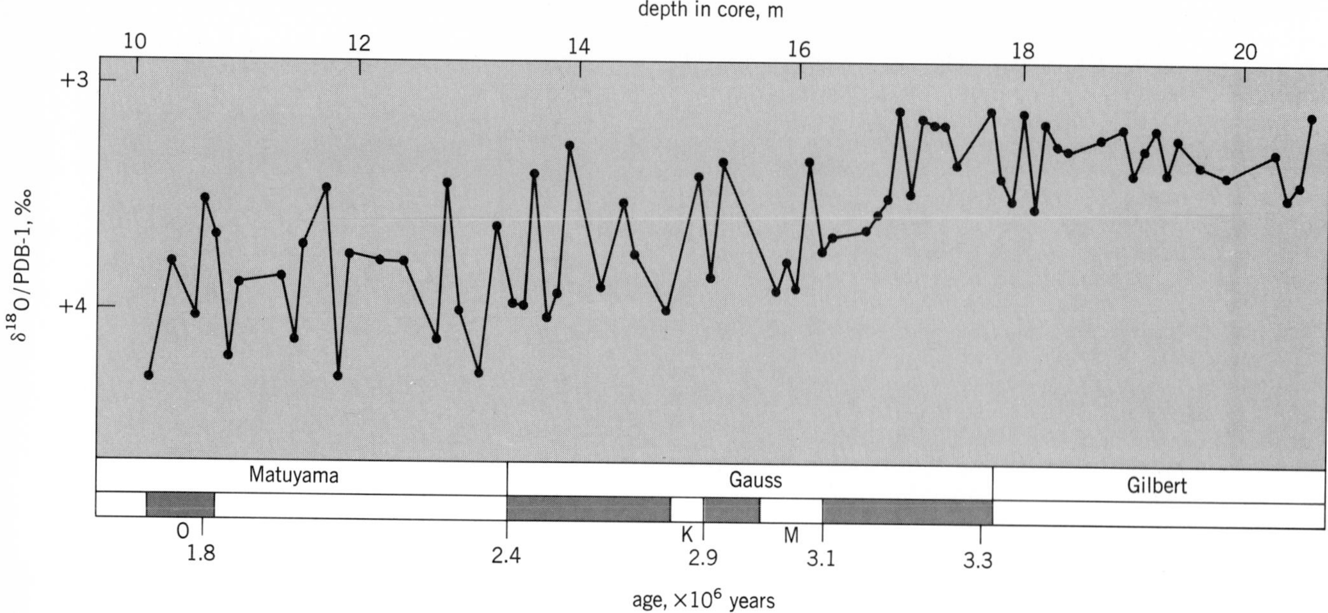

Fig. 10. Oxygen isotopic composition of *Globocassidulina subglobosa* in core V28-179. *G. subglobosa* living on the sea floor at the coring site today would have an ^{18}O content of about $+3.50°/_{oo}$. Magnetic events are indicated by O (Olduvai), K (Kaena), and M (Mammoth). Ages are based on estimates of the chronology for the magnetic record. (*From N. J. Shackleton and N. D. Opdyke, Oxygen isotope and paleomagnetic evidence for early Northern Hemisphere glaciation, Nature, 270:216–219, 1977*)

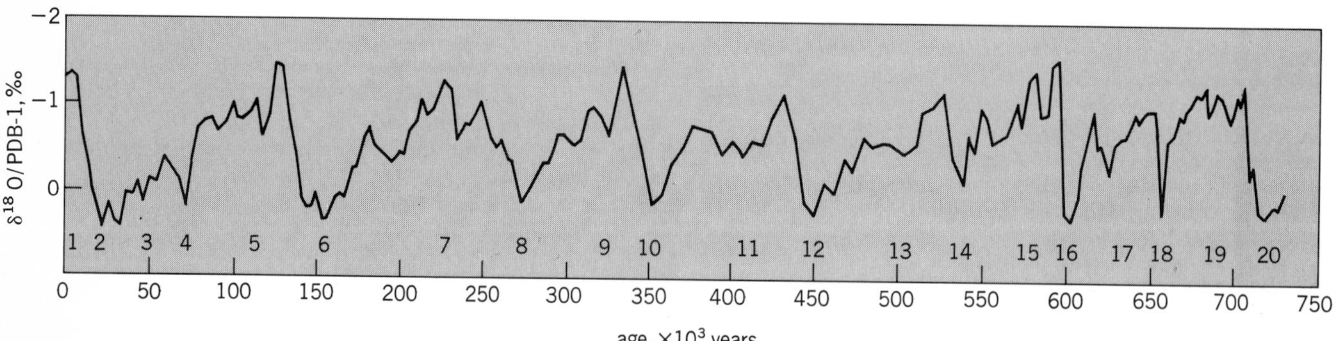

Fig. 11. Composite oxygen isotope curve based on cores P6304-9, 0–72,000 years; P6408-9, 72,000–153,000 years; 280, 153,000–275,000 years; P6408-9, 275,000–730,000 years, showing $\delta^{18}O$ (per mil with respect to Chicago standard PDB-1) of *Globigerinoides saçculifera* (including correction of $2.0°/_{oo}$ for the section from core 280). Absolute time scale based on age 0 for top of core; 125,000 years for peak of stage 5; and 700,000 years for peak of stage 19, representing the Brunhes/Matuyama boundary. Interpolations based on average stage thicknesses from Caribbean (stages 1–6) and on the thicknesses of stages 6–20 in core V28-238. (*From C. Emiliani, The cause of the ice ages, Earth Planet. Sci. Lett., 37:349–352, 1978*)

excess ^{18}O change in the shells of planktonic foraminifera from the Atlantic and the Caribbean may be assigned to the isotopic temperature effect, indicating a glacial/interglacial surface temperature change of 3.5°C for the equatorial Atlantic and the Caribbean. In the equatorial Pacific, on the other hand, isotopic change in *G. sacculifera* is similar to that in the benthic foraminifera (1.0 per mil), indicating that *G. sacculifera* did not occupy in the Pacific the constantly shallow habitat it occupied in the Atlantic and Caribbean throughout the Quaternary, but rather a deeper habitat during interglacial ages (as it does today) and a shallower one during glacial ages. In spite of this difficulty, Pacific cores have provided an isotopic record which, if less precise than that of the equatorial-Atlantic-Caribbean, is stratigraphically longer (Fig. 16). The two records complement each other and lead to the conclusion that approximately 30 glaciations occurred during the past 3×10^6 years, apparently increasing in sever-

---

Opposite page:
Fig. 12. Caribbean core P6304-8. (*a–i*) Ratios of the relative abundance of planktonic foraminiferal species or groups of species diagnostic of warm and cold periods. (*j*) $\delta^{18}O$/PDB-1 ($°/_{oo}$) in pelagic foraminiferal species *Globigerinoides triloba sacculifera*. Vertical scales are proportional to original micropaleontological and isotopic ratios. (*After L. Lidz, Deep sea Pleistocene biostratigraphy, Science, 154:1448–1452; copyright © 1966 by the American Association for the Advancement of Science*)

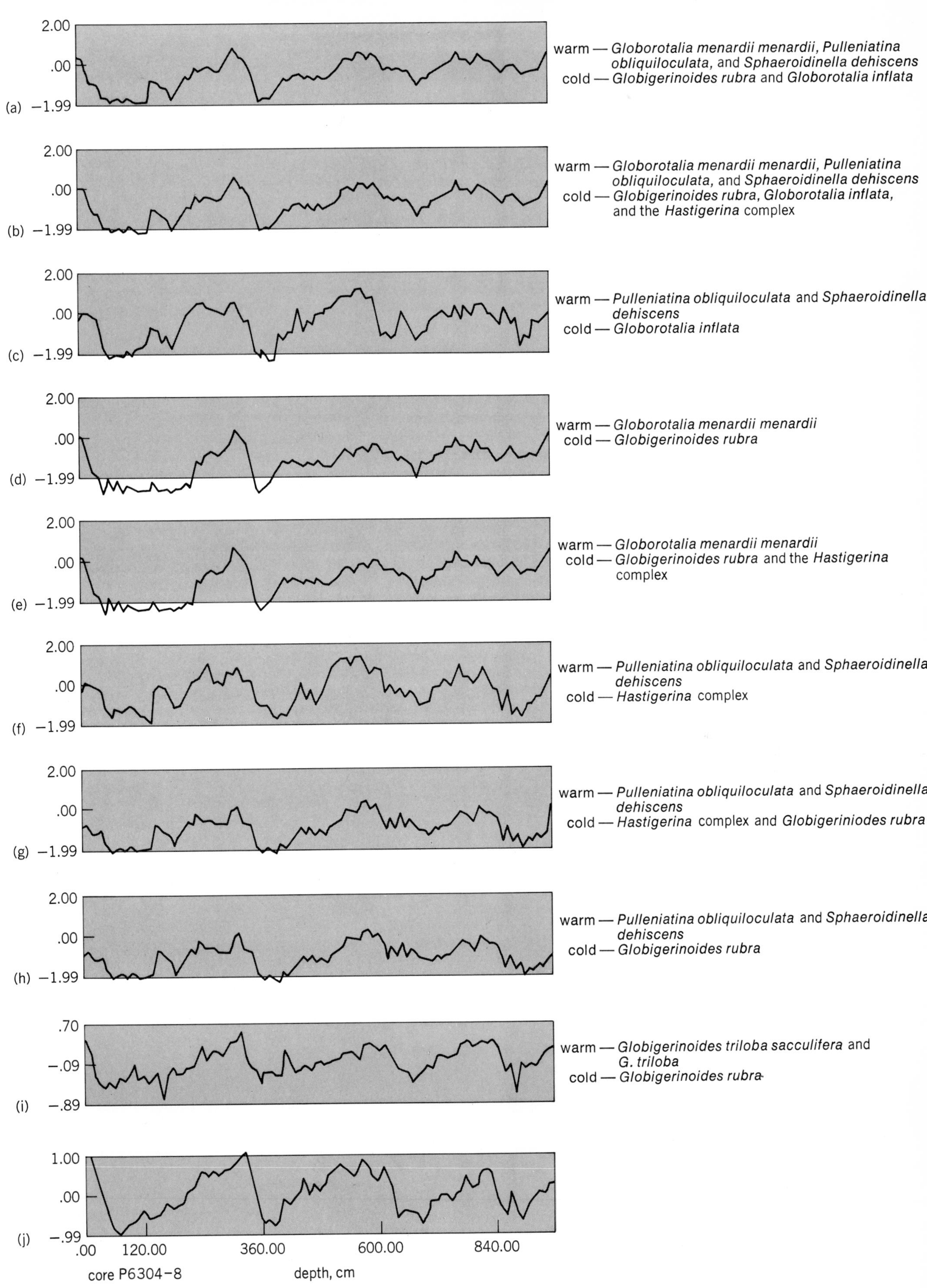
2.00
.00
(a) −1.99
warm — Globorotalia menardii menardii, Pulleniatina obliquiloculata, and Sphaeroidinella dehiscens
cold — Globigerinoides rubra and Globorotalia inflata
2.00
.00
(b) −1.99
warm — Globorotalia menardii menardii, Pulleniatina obliquiloculata, and Sphaeroidinella dehiscens
cold — Globigerinoides rubra, Globorotalia inflata, and the Hastigerina complex
2.00
.00
(c) −1.99
warm — Pulleniatina obliquiloculata and Sphaeroidinella dehiscens
cold — Globorotalia inflata
2.00
.00
(d) −1.99
warm — Globorotalia menardii menardii
cold — Globigerinoides rubra
2.00
.00
(e) −1.99
warm — Globorotalia menardii menardii
cold — Globigerinoides rubra and the Hastigerina complex
2.00
.00
(f) −1.99
warm — Pulleniatina obliquiloculata and Sphaeroidinella dehiscens
cold — Hastigerina complex
2.00
.00
(g) −1.99
warm — Pulleniatina obliquiloculata and Sphaeroidinella dehiscens
cold — Hastigerina complex and Globigeriniodes rubra
2.00
.00
(h) −1.99
warm — Pulleniatina obliquiloculata and Sphaeroidinella dehiscens
cold — Globigerinoides rubra
.70
−.09
(i) −.89
warm — Globigerinoides triloba sacculifera and G. triloba
cold — Globigerinoides rubra
1.00
.00
(j) −.99
.00
120.00
360.00
600.00
840.00
core P6304-8
depth, cm

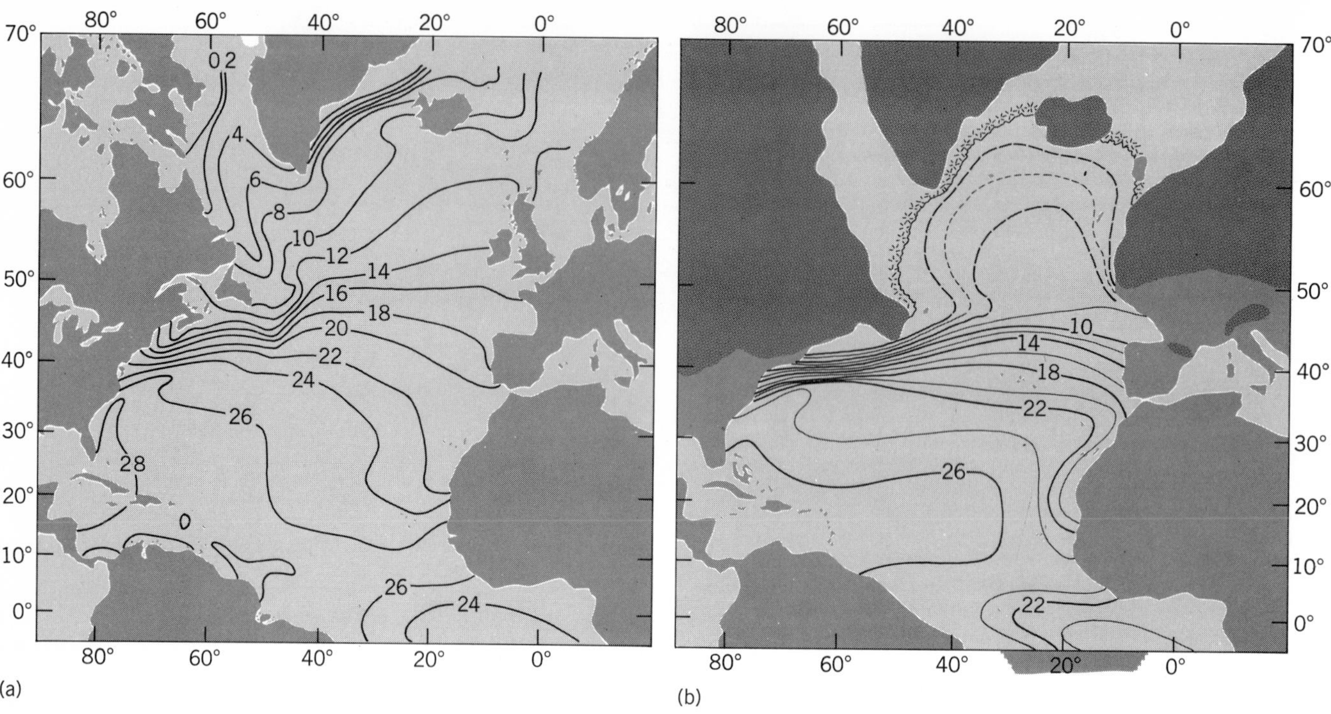

Fig. 13. Surface-water isotherm map for (*a*) August today and (*b*) August 18,000 years ago (*From A. McIntyre et al., Glacial North Atlantic 18,000 years ago: A CLIMAP reconstruction, Geol. Soc. Amer. Mem., 145:43–76, 1976*)

ity during the past 700,000 years (Fig. 10). Barring human interference with climate, it is likely that glaciation will continue for some millions of years in the future. *See* QUATERNARY.

Causes of glaciation. A polar or circumpolar distribution of the continents appears to be a necessary, but not sufficient, prerequisite for glaciation on Earth. A major glaciation occurred in Permo-Carboniferous time, when South Africa was located at the South Pole and was surrounded by the other southern continents and India. Since then, all continents except Antarctica have moved north. North America and Eurasia are now jammed against each other around the North Pole, and another major glaciation is in progress. During the Mesozoic the Earth seems to have remained free of ice, even though Antarctica continued to occupy a polar position. Polar temperatures at that time seem to have been about 15°C, as indicated by oxygen isotopic analysis of high-latitude fossils. Increasing glaciation in Antarctica during the Cenozoic may have been related to orogenesis and epeirogenesis, and the initiation of glaciation in the Northern Hemisphere may have resulted from the closing of the Central American isthmus (as previously mentioned). The oxygen isotopic curve of Fig. 12 clearly shows that Quaternary glaciation has an apparent periodicity of 100,000 years. Time series analysis of these curves indicates that the dominant periodicities

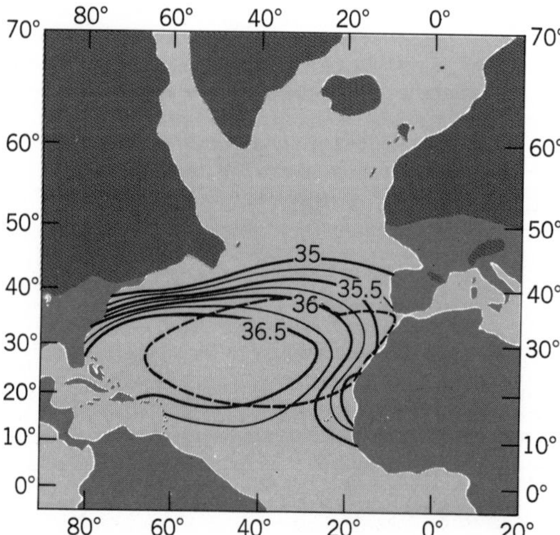

Fig. 14. One-hundred-meter isohaline map for February 18,000 years ago in parts per mil. Contour interval is 0.25°/₀₀. Line shows position of the modern 36.5°/₀₀ isohaline. (*From A. McIntyre et al., Glacial North Atlantic 18,000 years ago: A CLIMAP reconstruction, Geol. Soc. Amer. Mem., 145:43–76, 1976*)

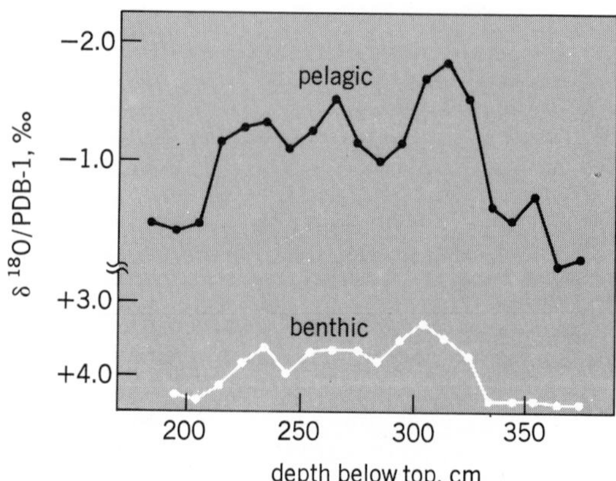

Fig. 15. Oxygen isotopic composition of benthic and pelagic foraminifera in the equatorial Atlantic during the last interglacial.

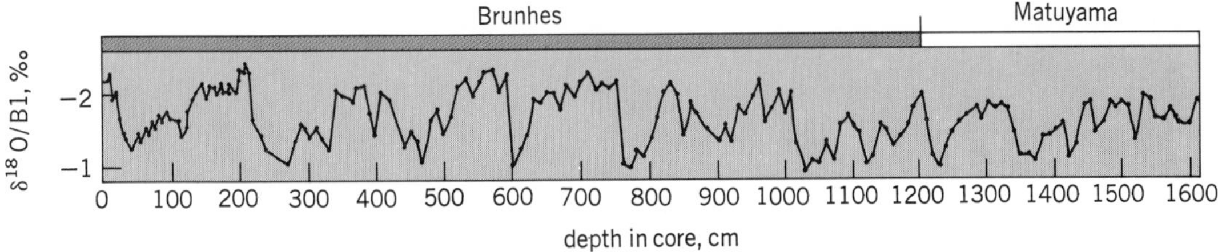

Fig. 16. Oxygen isotopic composition of *Globigerinoides sacculifera* in core V28-238 (Western Equatorial Pacific), expressed as deviation (°/₀₀) from Emiliani B1 standard. *(From N. J. Shackleton and N. D. Opdyke, Oxygen isotope and paleomagnetic stratigraphy of equatorial Pacific core V28-238: Oxygen isotope temperatures and ice volumes on a 10^5 and 10^6 year scale, Quaternary Res., 3:39–55, 1973)*

are 23,000, 42,000, and 100,000 years, which are the periodicities of, respectively, the precession of the equinoxes, the obliquity of the ecliptic, and the eccentricity of the Earth's orbit. These three motions combine to produce periods of warm summers and cold winters alternating with cool summers and warmer winters. Although the climatic effect of the astronomical motions may be opposite in the two hemispheres, it is the Northern Hemisphere that determines global climate because of the large land masses occupying high northern latitudes. A period of cool summers in the Northern Hemisphere is believed to produce a glacial age, although the climatic effect of the astronomical motions remains poorly understood because of the large climatic effect of the ice itself and the complex heat exchange processes between ice and ocean. The development of a full glaciation, in fact, may be a largely self-sustaining process leading to thermal equilibrium between expanded ice and ocean. Deglaciation at the end of each glacial cycle may have resulted from surface freezing of the northern North Atlantic and the northeastern Pacific, together with time-delay effects introduced by plastic flow of ice sheets, heat absorption by ice melting, and downbuckling of the lithosphere under the ice sheets. Oxygen isotopic analysis of late Quaternary cores from the Gulf of Mexico has shown that the end of the last glaciation was very rapid, with a catastrophic flood down the Mississippi Valley about 11,600 years ago. The oxygen isotopic record also indicates that the Earth is now at the end of a hypsithermal, and that a downward trend leading toward cooler climate has been already established. *See* GLACIAL EPOCH; PALEOCLIMATOLOGY.

[CESARE EMILIANI]

Bibliography: W. L. Donn and D. M. Shaw, Model of climate evolution based on continental drift and polar wandering, *Geol. Soc. Amer. Bull.*, 88:390–396, 1977; C. Emiliani, The cause of the ice ages, *Earth Planet. Sci. Lett.*, 37:349–352, 1978; C. Emiliani, Paleotemperature analysis of the Caribbean cores P6304-8 and P6304-9, and a generalized temperature curve for the past 425,000 years, *J. Geol.*, 74:109–126, 1966; C. Emiliani, Quaternary paleotemperatures and the duration of the high-temperature intervals, *Science*, 178: 398–401, 1972; C. Emiliani et al., Paleoclimatological analysis of late Quaternary cores from the northeastern Gulf of Mexico, *Science*, 189: 1083–1088, 1975; C. Emiliani and N. J. Shackleton, The Brunhes epoch: Paleotemperatures and geochronology, *Science*, 183:511–514, 1974; J. D. Hays, J. Imbrie, and N. J. Shackleton, Variations in the Earth's orbit: Pacemaker of the ice ages, *Science*, 194:1121–1132, 1976; J. P. Kennett, Cenozoic evolution of Antarctic glaciation, the Circum-Antarctic Ocean, and their impact on global paleoceanography, *J. Geophys. Res.*, 82:3843–3860, 1977; J. P. Kennett and N. J. Shackleton, Laurentide ice sheet meltwater recorded in Gulf of Mexico deep-sea cores, *Science*, 188:147–150, 1975; N. G. Kipp, New transfer function for estimating past sea-surface conditions from sea-bed distribution of planktonic foraminiferal assemblages in the North Atlantic, *Geol. Soc. Amer. Mem.*, 145:3–41, 1976; L. Lidz, Deep-sea Pleistocene biostratigraphy, *Science*, 154:1448–1452, 1966; A. McIntyre et al., Glacial North Atlantic 18,000 years ago: A CLIMAP reconstruction, *Geol. Soc. Amer. Mem.*, 145:43–76, 1976; N. J. Shackleton and N. D. Opdyke, Oxygen isotope and paleomagnetic evidence for early Northern Hemisphere glaciation, *Nature*, 270:216–219, 1977; N. J. Shackleton and N. D. Opdyke, Oxygen isotope and paleomagnetic stratigraphy of equatorial Pacific core V28–238: Oxygen isotope temperatures and ice volumes on a 10^5 and 10^6 year scale, *Quaternary Res.*, 3:39–55, 1973; N. J. Shackleton and N. D. Opdyke, Oxygen-isotope and paleomagnetic stratigraphy of Pacific core V28-239: Late Pliocene to latest Pleistocene, *Geol. Soc. Amer. Mem.*, 145:449–464, 1976.

SAMPLING AND CORING DEVICES

The bottom-sampling devices used to collect sediments from the ocean floor are of three main types: snappers, dredges, and coring tubes.

Fig. 17. Bottom dredge, showing bottom sample. *(U.S. Navy Hydrographic Office)*

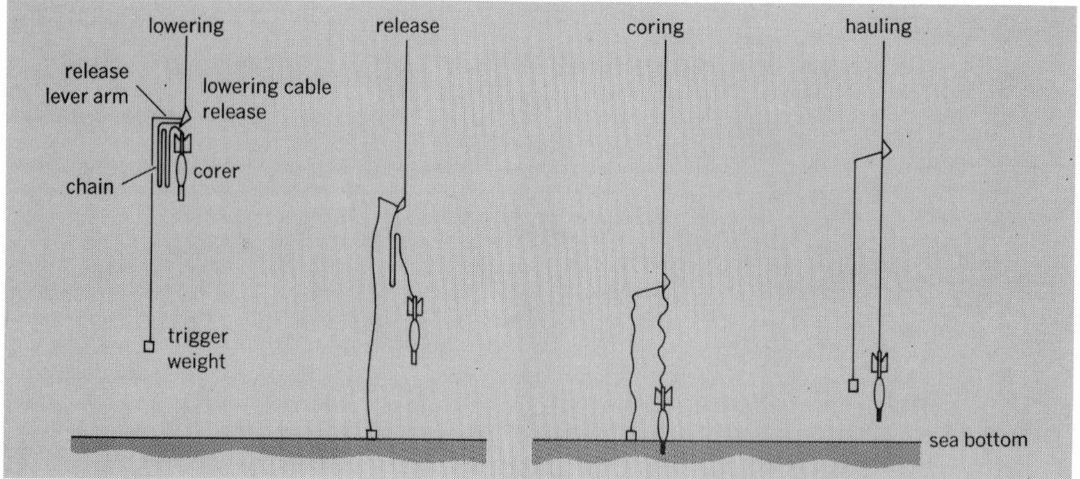

Fig. 18. Principle of operation of free-fall corers. (*From U.S. Navy H.O. Publ. no. 607, 2d ed., 1955*)

Snapper-type samplers, with closing jaws actuated by a tension spring and trigger mechanism, are used to obtain small samples from the sediment surface. Snapper, or grab, samplers are generally used in shallow waters where it is desirable to gather a large number of samples as rapidly as possible. The dredge is essentially a bag made of steel links; its mouth is held open by a rectangular frame provided with a bail to which the ship's trawl wire is shackled. When it is dragged along the bottom, the dredge collects relatively large objects such as manganese oxide and phosphate nodules, sharks' teeth, and ice-rafted rocks lying on or near the sediment-water interface (Fig. 17).

Coring devices are used to obtain samples of bottom material in place. This device consists essentially of a steel tube, a glass or plastic liner that may be removed without disturbing the sample, attached weights, and a core-catcher ring and cutting edge which fit the bottom, or penetration edge, of the tube. The amount of sample collected depends on the length of the corer, its weight, and the penetrability of the bottom. Core samples up to 1.2 m in length may be obtained when the smaller types of corer are used. Modifications in coring devices include the piston corer and free-fall release mechanism. Both are utilized in the heavier coring equipment which is used to obtain longer cores.

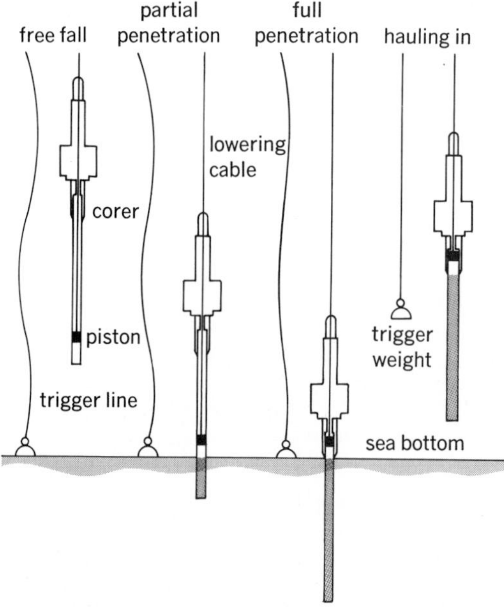

Fig. 19. Principle of operation of piston corers. (*From U.S. Navy H O. Publ. no. 607, 2d ed., 1955*)

Piston corer. The piston corer, a steel tube measuring about 60 mm in inside diameter, is capable of recovering undistorted vertical sections of sediment as much as 20 m in length. The corer is driven into the sediment by a free fall of about 5 m and by 500–1000 kg of lead attached to the upper end. Free fall is effected by a trigger weight attached by a wire line to the end of a trigger arm, which rises and releases the coring tube when the trigger weight comes to rest on the sediment surface (Fig. 18). Connection between the ship's cable and the apparatus is maintained by a length of wire rope which passes from the end of the ship's cable down through the coring tube to a piston which is initially at the bottom of the tube. The length is so adjusted that the wire rope becomes taut just when the bottom of the coring tube reaches the sediment surface (Fig. 19). As the tube sinks into the sediment, the piston, immobilized by the cable to the ship, makes hydrostatic pressure outside the tube effective in overcoming frictional resistance between the entering sediment and the inner wall of the tube.

The apparatus is raised by the wire rope attached to the piston which is retained by shoulders near the top of the tube. The bottom of the tube is armed with a removable steel cutting edge. Just above this is the core catcher, a circular set of leaves of spring bronze which close as the tube is withdrawn from the sediment.

The original piston corer of B. Kullenberg (Sweden) is provided with a metal liner in which the core is removed. A simplified corer designed by M. Ewing is used without a liner, and the core must therefore be pushed out of the tube with a plunger and rod. A pilot core about 30 cm long is usually taken with each long core by a small tube with plastic liner attached to the trigger weight.

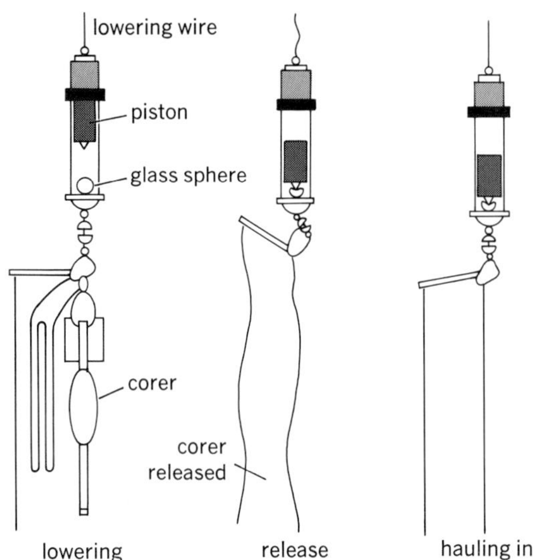

Fig. 20. The ball-breaker principle of operation. (*From U.S. Navy H.O. Publ. no. 607, 2d ed., 1955*)

Bottom contact. In bottom-sampling operations it is necessary to know when bottom contact is made because any excessive wire that is payed out usually kinks. Close attention must be given to the sonic depth, the wire angle, and the length of wire payed out.

In shallow waters contact with the bottom may be detected by a slack in the wire and a jerk in the meter wheel. In deeper waters a spring scale or dynamometer may be used to register variations in wire tension. A bottom-signaling device, or ball-breaker, is sometimes used for this purpose. Upon contact with the bottom a small glass ball is crushed (Fig. 20). The resulting implosion signal may be seen on the pen trace of the echo sounder or heard as an audible signal on suitable monitoring equipment.

[DAVID B. ERICSON]

Bibliography: H. Barnes, *Oceanography and Marine Biology*, 1959; J. B. Hersey, Acoustically monitored bottom coring, *Deep-Sea Res.*, 6:170–172, February, 1960; *Instruction Manual for Oceanographic Observations*, U.S. Navy H.O. Publ. no. 607, 2d ed., 1955; B. Kullenberg, The piston core sampler, *Svenska Hydrograf. Biol. Komm. Skrifter*, [3]1(2):1–46, 1945; D. A. McManus, A large-diameter coring device, *Deep-Sea Res.*, 12:227–232, 1965; D. G. Moore, The free-corer: Sediment sampling without wire and winch, *J. Sediment. Petrol.*, 31:627–630, 1961; A. F. Richards and H. W. Parker, Surface coring for sheer strength measurements, in *Proceedings of the Conference on Civil Engineering in the Oceans*, 1968.

Marl

An argillaceous, nonindurated calcium carbonate deposit that is commonly gray or blue-gray. It is somewhat friable, and in some respects resembles chalk, with which it is interbedded in some localities. It is formed in some fresh-water lakes, partially by the action of some aquatic plants. The plants extract carbon dioxide for photosynthesis from bicarbonate in the water and locally reduce solubility of calcium carbonate next to the leaves, where it precipitates. The precipitate forms a scale, which then falls to the bottom and accumulates. The clay content of marls varies, and all gradations between small amounts of clay (marly limestones) and large amounts (marly clay) are found. Marlstone, or marlite, is an indurated rock of the same composition as marl. The marlstones are not fissile but blocky and massive with subconchoidal fracture. *See* LIMESTONE; SEDIMENTARY ROCKS.

[RAYMOND SIEVER]

Mass wasting

The downslope movement of masses of bedrock or soil and rock debris under the influence of gravity. This process, also known as mass movement, is distinct from the transport of soil and rock dispersed in moving water, air, or ice. Mass wasting is a leading process of degradation in areas untouched by stream erosion, in rugged mountains, and in certain areas of particularly weak rock or clayey soil. There are four main types: slow flowage, rapid flowage, landslide, and subsidence.

Slow flowage, or creep is the most widely distributed but least spectacular form of mass wastage. Soil creep is active in the large interstream areas and proceeds even beneath a grass or forest cover (Fig. 1). Its presence is revealed by such signs as broken retaining walls and tilted posts, poles, and monuments. Rock creep is evident in the downhill bending of layers of bedded or foliated rock and in the slow downslope migration of large blocks of rock away from their parent outcrop, sometimes spoken of as block glide. Frost-

Fig. 1. Hillside creep in the valley of a tributary of the Yukon, Yukon Territory, Canada. (*USGS*)

Fig. 2. Earthflow near Berkeley, Calif. (*USGS*)

Fig. 3. Landslide of slump type from bank of cutoff of Mississippi River. (*U. S. Soil Conservation Service*)

controlled creep is especially important in regions of permanently frozen ground. *See* FROST ACTION.

Rapid flowage, characteristic only of certain regions and typically in individual occurrences rather than of wide areal extent, includes earthflow, mudflow, and debris-avalanche.

Earthflow is particularly common in areas of weathered shale bedrock or clay. In hillside earthflows (Fig. 2), an area of soil and mantle rock to a depth of several feet is unable when saturated to maintain its form on the slope and flows a short distance downhill, often confined beneath a tight sod. Movement slows and stops when the bulging toe cracks and some of the water escapes. Because earthflow removes support from the area directly upslope, a landslide of slump type (Fig. 3) usually results. Earthflow may be moderately rapid or slow and intermittent over a long period.

Mudflow is a phenomenon of semiarid lands, subalpine mountains, or volcanic slopes. Mudflows usually follow a drainageway and often recur along the course of previous mudflows. The material is wetter and the movement faster than in earthflow. Large blocks of rock carried in a thick mudflow can be moved far out from a mountain front.

Debris-avalanche is the equivalent of the mudflow in mountains of humid regions. It usually involves the entire thickness of the soil mantle down to bedrock but does not follow a previous avalanche track. Deposits are heaps, fans, or valley trains of heterogeneous rock debris, usually cluttered with trees and other vegetation.

On Jan. 10, 1962, the breaking away of several million tons of glacial ice from near the top of Peru's 21,834-ft Mount Huascarán started a great ice avalanche. Hurtling down the slope, the ice picked up soil, rock, vegetation—anything in its path—and the mass became more fluid

Fig. 4. The lower part of the debris-avalanche from Mount Huascarán, Peru. Emerging from the narrow valley in the background, the flowing mud spread out to cover an area about 1 mi wide. (*Wide World Photos*)

as the ice was crushed and churned in the flowing debris. This debris-avalanche traveled a 9-mi course, descending almost 2 mi to the Río Santa in about 7 min. On its way it destroyed all or large parts of eight towns and killed 3500 people (Fig. 4).

The term landslide includes sliding and falling movements of relatively dry masses of earth and rocks. It is sometimes used broadly to cover all forms of rapid mass movement. *See* LANDSLIDE.

Subsidence is a downward movement without any forward or outward component, as in the settling of ground over mines.

In recent years many man-made deposits have suddenly flowed disastrously, with considerable loss of life. Twenty persons were reported killed near Liège, Belgium, in February, 1961, when, following heavy rains, a 300-ft-high slag heap from a coal mine collapsed and swept over nearby houses. On Mar. 13, 1961, according to *Pravda*, 145 people were killed and 143 were injured in the outskirts of Kiev, U.S.S.R., when streetcars, buses, public buildings, and 22 small private homes were overwhelmed or washed away by a mudflow. The accident occurred when a ravine was being filled for use as a park. Water collected in the fill, and a heavy wind caused the liquid silt to wash over and breach the retaining earth dam, releasing the wet fill. On June 26, 1965, in Kawasaki, Japan, a rapid flow from a 50-ft-high mound of cinders soaked by heavy rains destroyed 15 houses and killed 24 persons. On Oct. 21, 1966, at Aberfan, Wales, following heavy rains, a rapid flow of about 2,000,000 tons of wet sludge from a high pile of coal-mine waste demolished 17 dwellings and part of a school, with a loss of 116 children and 28 adults.

[C. F. STEWART SHARPE]

Bibliography: C. F. S. Sharpe, *Landslides and Related Phenomena*, 1938, reprint 1960.

Massif

A block of the Earth's crust commonly consisting of crystalline gneisses and schists, the textural appearance of which is generally markedly different from that of the surrounding rocks. Common usage indicates that a massif has limited areal extent and considerable topographic relief. Structurally, a massif may form the core of an anticline or may be a block bounded by faults or even unconformities. In any case, during the final stages of its development a massif acts as a relatively homogeneous tectonic unit which to some extent controls the structures that surround it. Numerous complex internal structures may be present; many of these are not related to its development as a massif but are the mark of previous deformations. *See* TECTONIC PATTERNS.

[PHILIP H. OSBERG]

Mesozoic

In geology, the system of rocks younger than the Paleozoic and older than the Cenozoic; also, the geologic era during which those rocks were formed. The era is now believed to have extended from about 225,000,000 to 64,000,000 years ago. The name Mesozoic was given by early workers in geology to indicate the intermediate nature of its life between the ancient forms in the Paleozoic Era and the more modern life of the Cenozoic Era. The Mesozoic is commonly referred to as the Age of

PRECAMBRIAN	PALEOZOIC							MESOZOIC			CENOZOIC	
					CARBONIFEROUS							
	CAMBRIAN	ORDOVICIAN	SILURIAN	DEVONIAN	Mississippian	Pennsylvanian	PERMIAN	TRIASSIC	JURASSIC	CRETACEOUS	TERTIARY	QUATERNARY

Reptiles inasmuch as these animals dominated the land, sea, and air. The era is divided into three systems or periods which are, from oldest to youngest, Triassic, Jurassic, and Cretaceous. *See* CRETACEOUS; JURASSIC; TRIASSIC.

[WILLIAM A. COBBAN]

Metamict state

The amorphous state of substances that have lost their original crystalline structure because of the radioactivity of uranium or thorium. The process of rendering crystalline substances partly or wholly amorphous is known as metamictization. Minerals whose crystal structure has been disrupted by this process are known as metamict minerals. In 1893 W. C. Broegger recognized that some minerals which show crystal form are nevertheless structurally amorphous but have attained this state in a different manner than glasses or rigid gels. For this third class of amorphous substances he proposed the name metamict (mixed otherwise). Broegger contended that "the reason for the amorphous rearrangement of the molecules might perhaps be sought in the lesser stability which so complicated a crystal molecule . . . must have in the presence of outside influences."

Properties. Many features of metamict minerals were recognized before 1893, but some were not recognized until later. The more significant features are as follows:

1. Anomalous optical behavior. Many metamict substances are heterogeneous, being partly isotropic and partly anisotropic (doubly refractive).

2. Pyrognomic behavior. The minerals readily become incandescent on heating. This varies greatly. In some cases glowing is not observable. Even strongly pyrognomic minerals, such as gadolinite, can be annealed below the temperature of glowing with complete loss of the pyrognomic quality.

3. Glasslike properties. The minerals lack cleavage but exhibit conchoidal fracture, some being particularly brittle.

4. Density. The density is increased by heating. The change, however, may be slight.

5. Reconstitution of crystalline structure by heating. This may be a complex process. Even if a single phase results, it is usually polycrystalline.

6. Resistance to attack by acids. This is increased by heating. It is the opposite of the effect produced by heating most substances.

7. Presence of uranium (U) or thorium (Th). The content may be low, for example, 0.41% ThO_2

in gadolinite from Ytterby, Sweden. The presence of rare earths and related elements has been emphasized by some observers.

8. Existence of some minerals in both the crystalline and metamict states. In these cases little, if any, chemical difference can be found. There is evidence of hydration attending isotropization, but no direct or general correlation has been established.

9. Amorphous behavior when exposed to x-rays. In 1916 L. Vegard reported the absence of x-ray diffraction in thorite. This has since been noted for many other metamict minerals and is now regarded as the crucial test. Some traces of x-ray diffraction may be observed, though the minerals have become optically quite isotropic.

Causes. It is now generally agreed that α-particles and recoil nuclei from the radioactive disintegration of uranium and thorium are the cause of isotropization. In fact, some physicists have used the expression metamict state to refer to radiation damage. Metamictization is doubtless similar to the more rapid processes of radiation damage in technological materials.

The structural damage is done largely by the recoil atom and by the α-particle in the final part of its trajectory. This is believed to cause a dumbbell-shaped pair of damaged regions for each α-particle event, involving the order of 10^3 atomic dislocations. The damage in each region is equivalent to a local source of intense heat which, on cooling, leaves the structure as a glass. The total damage inflicted is therefore proportional to the α-particle dosage, which is the α-particle rate per unit volume times the age of the mineral. Minerals such as zircon therefore undergo damage toward the metamict or final state of disorder over geologic periods of time. The greater the uranium content, the shorter the time.

Superimposed on this effect is a natural annealing, or recrystallization of the glassy domains, the rate of which is strongly dependent on the ambient temperature and the nature of the crystal structure. For example, zircon anneals so slowly at Earth surface temperatures that its degree of damage (determined by its unit cell dimensions) may be used to determine its approximate age if the α-particle activity is also measured. On the other hand, monazite, commonly much richer in radioactive elements, anneals quickly enough so that it seldom appears greatly damaged.

Occurrence of metamict minerals. Scores of different mineral species in the metamict state are known. Many of these minerals are characteristic of pegmatites but are also known in hydrothermal deposits. Some have been concentrated in placers, possibly even in ancient placers. A few metamict minerals, notably brannerite, which is essentially UTi_2O_6, and davidite, are important constituents of uranium ores in Ontario and Australia. These and a number of other complex oxides such as euxenite and fergusonite are known only in a metamict or largely metamict condition. Others, such as thorite, allanite, gadolinite, and pyrochlore, may or may not be metamict.

Degree of metamictness. X-ray methods do not permit a close estimate of the degree of metamictness, since the remnants of structure may be confined to small volumes that may be disoriented. P. Pellas estimated in 1954 that a material would appear x-ray amorphous when 20% or more of its atoms have been displaced. In a few minerals it has been possible to detect a lattice expansion in specimens that are partly metamict. This might be taken as a measure of the degree of metamictness in its early stages. The effect is pronounced in zircons, which vary between about 4.7 and 3.9 in density. *See* ZIRCON.

Reconstitution of metamict minerals. The pyrognomic effect observed on heating some metamict minerals is evidence of the exothermic character of the reconstitution, which in these cases takes place rapidly. Metamict materials can be considered to have stored energy derived from radiation. The stored energy is released when the material recrystallizes. This energy was measured for gadolinite by A. Faessler in 1942.

Most metamict minerals can be recrystallized, usually reconstituted, at easily detectable rates at temperatures between 500 and 900°C. Some metamict minerals can be slowly reconstituted at lower temperatures. Though characteristic differential thermal analysis curves can be obtained from some metamict minerals, the peaks are controlled by heating conditions. Even where the original structure is finally reestablished by heating, usually as a crystalline aggregate, the process may involve several intermediate steps. Thus when metamict thorite is heated, cryptocrystalline ThO_2 is formed before or with the reconstituted $ThSiO_4$.

Because of such side reactions and the fact that the results of annealing may in some cases be modified by the surrounding atmosphere, whether air or an inert gas, annealing may not really reconstitute the original structure or its cryptocrystalline equivalent. There should be no doubt in those cases in which it has been possible to reconstitute single crystals with lattices conforming strictly to the morphology of the initial substances. Such perfect reconstitution may occur in minerals that are only slightly metamict. When material that is fully x-ray amorphous can be annealed to single crystals, it must be presumed that sufficient remnants of the original structure persisted to determine the orientation of the newly crystallized material. *See* GEOLOGIC THERMOMETRY; PLEOCHROIC HALOS.

[PATRICK M. HURLEY]

Bibliography: H. Faul (ed.), *Nuclear Geology*, 1954; E. W. Heinrich, *Mineralogy and Geology of Radioactive Raw Materials*, 1958; J. Orcel, L'état métamicte, *Bull. Soc. Belge Géol.*, 65:165–194, 1956; A. Pabst, The metamict state, *Amer. Mineral.*, 37:137–157, 1952; P. Pellas, Sur la formation de l'état métamicte dans le zircon, *Bull. Soc. Fr. Mineral. Crist.*, 77:447–460, 1954; K. Rankama, *Isotope Geology*, 1954.

Metamorphic rocks

One of the three major groups of rocks that make up the crust of the Earth. The other two groups are igneous rocks and sedimentary rocks. Metamorphic rocks are preexisting rock masses in which new minerals, or textures, or structures are formed at higher temperatures and greater pressures than those normally present at the Earth's surface. *See* IGNEOUS ROCKS; SEDIMENTARY ROCKS.

Two groups of metamorphic rocks may be distinguished: cataclastic rocks, formed by the operation of purely mechanical forces; and recrystallized rocks, or the metamorphic rocks properly so called, formed under the influence of metamorphic pressures and temperatures.

Cataclastic rocks are mechanically sheared and crushed. They represent products of dynamometamorphism, or kinetic metamorphism. Chemical and mineralogical changes generally are negligible. The rocks are characterized by their minute mineral grain size. Each mineral grain is broken up along the edges and is surrounded by a corona of debris or strewn fragments (mortar structure, Fig. 1*a*). During the early stages of this alteration process the metamorphosed product is known as flaser rock (Fig. 1*b*). Eventually the original mineral grains are entirely gone, as in the mylonites. When seen through the microscope, the comminuted particles consist of a mixture of finely powdered quartz, feldspar, and other minerals with an incipient recrystallization of sericite or chlorite. Pseudotachylite is an extreme end product of this process. *See* METAMORPHISM; MYLONITE.

Structural relations. Metamorphic rocks, properly so called, are recrystallized rocks. The laws of recrystallization are not the same as those of simple crystallization from a liquid, because the crystals can develop freely in a liquid, but during recrystallization the new crystals are encumbered in their growth by the old minerals. Consequently, the structures which develop in metamorphic rocks are distinctive and of great importance, because in many ways they reflect the physiochemical environment of recrystallization and thereby the genesis and history of the metamorphic rock.

Crystalloblastic structure. A crystalloblast is a crystal that has grown during the metamorphism of a rock. The majority of the minerals in metamorphic rocks are irregular in outline (xenoblasts), but some minerals are frequently bounded by their own crystal faces (idioblasts). Larger crystals are often packed with small inclusions of other minerals exhibiting the so-called sieve structure (poikilitic or diablastic structure).

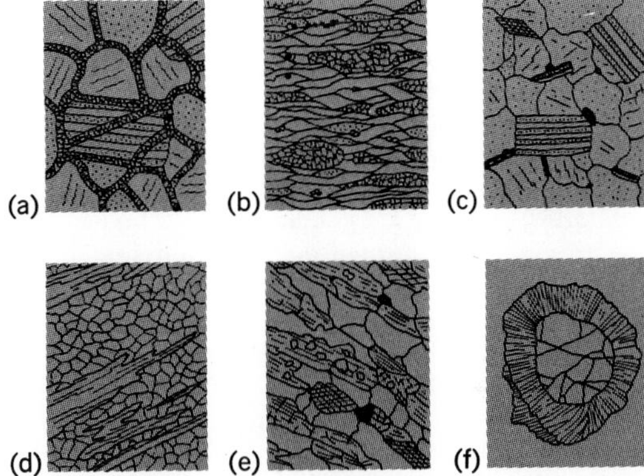

Fig. 1. Fabrics of metamorphic rocks under microscope. (*a*) Mortar fabric; (*b*) flaser or mylonitic fabric; (*c*) granoblastic fabric (*from E. E. Wahlstrom, Petrographic Mineralogy, copyright © 1955 by John Wiley & Sons, Inc.*). (*d*) Lepidoblastic fabric; (*d*) nematoblastic fabric; (*f*) porphyroblast with reaction rim (*from T. F. W. Barth, Theoretical Petrology, copyright © 1952 by John Wiley & Sons, Inc.; reprinted by permission*).

Granoblastic refers to a nondirected rock fabric, with minerals forming grains without any preferred shape or dimensional orientation (Fig. 1*c*). Lepidoblastic (Fig. 1*d*), nematoblastic (Fig. 1*e*), and fibroblastic refer to rocks of scaly, rodlike, and fibrous minerals, respectively.

The metamorphic minerals may be arranged in an idioblastic series (crystalloblastic series) in their order of decreasing force of crystallization as follows: (1) sphene, rutile, garnet, tourmaline, staurolite, kyanite; (2) epidote, zoisite; (3) pyroxene, hornblende; (4) ferrogmagnesite, dolomite, albite; (5) muscovite, biotite, chlorite; (6) calcite; (7) quartz, plagioclase; and (8) orthoclase, microcline. Crystals of any of the listed minerals tend to assume idioblastic outlines at surfaces of contact with simultaneously developed crystals of all minerals of lower position in the series.

Preferred orientation. Certain minerals have a tendency to assume parallel or partially parallel crystallographic orientation. The shape and spatial arrangement of minerals, such as mica, hornblende, or augite, show a definite relation to the foliation in the schist or gneiss; that is, both foliation and fissility of a metamorphic rock are directly related to the preferred position assumed by the so-called schist-forming minerals, such as mica, hornblende, and chlorite. *See* GNEISS; SCHIST.

Students of structural petrology distinguish between preferred orientation of inequidimensional grains according to their external crystal form, and preferred orientation of equidimensional grains according to their internal or atomic structure. *See* PETROFABRIC ANALYSIS; STRUCTURAL PETROLOGY.

A special microscope technique (universal stage technique) is necessary in most cases to demonstrate in detail the preferred orientation of the mineral grains according to their atomic structure. *See* PETROGRAPHY.

Relic structures. Mineral relics often indicate the temperature and pressure that obtained in the preexisting rock. If a mineral, say quartz, is stable in the earlier rock and is also stable in the later rock, it will be preserved (unless stress action sets in) in its original form as a stable relic. However, when a mineral or a definite association of minerals becomes unstable, it may still escape alteration and appear as an unstable relic. These relics are proterogenic, that is, representative of an earlier, premetamorphic rock, or of an earlier stage of the metamorphism. Hysterogenic products are of later date, and are formed in consequence of changed conditions after the formation of the chief metamorphic minerals.

A common phenomenon, fairly illustrative of the tendency toward equilibria, is the formation of armors or reaction rims around minerals (Fig. 1*f*), which have become unstable in their association but have not been brought beyond their fields of existence in general (the armored relics). Thereby the associations of minerals in actual contact with one another become stable. If, however, the constituent minerals of a rock containing armored relics are named without noting this phenomenon, it may be taken as an unstable association. *See* PORPHYROBLAST.

Structure relics are perhaps of still more importance, directly indicating the nature of the preexisting rock and the mechanism of the metamorphic deformation. The interpretation of relics has been compared to the reading of palimpsests, parchments used for the second time after original writing was nearly erased. Every trace of original structure is important in attempting to reconstruct the history of the rock and in analyzing the causes of its metamorphism.

In sedimentary rocks the most important structure is bedding (stratification or layering) which originally was approximately horizontal. In metamorphic rocks deformed by folding, faulting, or other dislocations, the sum of all deformations can be referred to the original horizontal plane, and the deformations can be analyzed.

Fissility and schistosity. One of the earliest secondary structures to develop in sediments of low metamorphic grade is that of slaty cleavage (also referred to as flow cleavage or fissility), which grades into schistosity which is different from fracture cleavage, or strain-slip cleavage. Slaty cleavage is developed normal to the direction of greatest shortening of the rock mass, and cuts the original bedding at various angles. Tectonic forces acting on a book of sediments of heterogeneous layers will throw them into a series of folds, and slaty cleavage develops in response to the stresses imposed on the rock system as a whole, because of the differential resistance of the several layers. Consequently, folding and slaty cleavage have a common parentage, as illustrated in Fig. 2.

In the rock series slate-phyllite-schist the slaty cleavage will grade into schistosity. It is a chemical and recrystallization phenomenon, as well as a mechanical one, and the directions of the schistosity become the main avenues of chemical transport. *See* CLEAVAGE, ROCK; SCHISTOSITY, ROCK; SLATE.

Contact-metamorphic rocks. Igneous magma at high temperature may penetrate into sedimentary rocks, it may reach the surface, or it may solidify in the form of intrusive bodies (plutons). Heat from such bodies spreads into the surrounding sediments, and because the mineral assemblages of the sediments are adjusted to low temperatures, the heating-up will result in a mineralogical and textural reconstruction known as contact metamorphism. *See* PLUTON.

The width of the thermal aureole of contact metamorphism surrounding igneous bodies varies from almost complete absence in the case of small

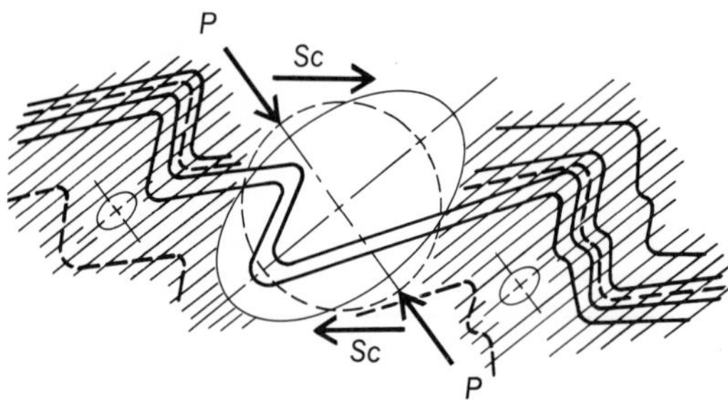

Fig. 2. Diagram showing the general relationship between deformation folding and slaty cleavage caused by pressure *PP* or the couple *ScSc*. Heavy black lines denote the original bedding deformed into folds. Thin lines indicate slaty cleavage which may grow into schistosity (false bedding). (*After G. Wilson, Proceedings of the Geological Association, 1946*)

intrusions (basalt dikes or diabase sills) to several kilometers in the case of large bodies. *See* AUREOLE, CONTACT.

The effects produced do not depend only upon the size of the intrusive. Other factors are amount of cover and the closure of the system, composition and texture of the country rock, and the abundance of gaseous and hydrothermal magmatic emanations. The heat conductivity of rocks is so low that gases and vaporous emanations become chiefly responsible for the transportation and transfer of heat into the country rock. *See* ORE DEPOSITS, GEOCHEMISTRY OF.

Alteration of stratified rocks. Stratified rocks are altered in the contact zone to what is commonly called hornfels or hornstone. They are hardened, often flinty rocks, usable for road material, and so fine-grained that the mineral components can be discerned only with the microscope. Hornfels used to be regarded as "silicified" sediments. However T. Kjerulf, in the latter half of the 19th century, analyzed sedimentary shale and "silicified" shale of the Oslo region and found that, chemically, they were identical (except for water and carbon dioxide content). Then geologists realized that the "hardening" of the shale took place without appreciable change in the chemical composition. Kjerulf summarized his results by saying that the composition of a hornfels depended on the original sediment (the shale) and was independent of the kind of adjacent igneous rock.

Later H. Rosenbusch arrived at the same conclusion and pronounced that no chemical alterations accompany the formation of hornfelses except for the removal of fugitive constituents. The Kjerulf-Rosenbusch rule is useful but needs modification, because chemical changes may ensue from hydrothermal and pneumatolytic action.

The next problem then is to see how the mineral assemblages of the hornfelses depend upon the chemical composition of the original sediments. The chief types of sedimentary rocks are sandstone (sand), shale (clay), and limestone. Among the varieties of hornfelses which may develop from different mixtures of these components, the continuous series from shale to limestone is the most interesting.

Most shales contain some iron- and magnesia-bearing constituents in addition to feldspar and clay minerals. Quartz, SiO_2, is always admixed. Consequently, sufficient SiO_2 is often present in the hornfelses to form highly silicified minerals. Other than SiO_2, the four chief chemical constituents are Al_2O_3, CaO, FeO, and MgO. The last two constituents are grouped together to define a system of three components: alumina, lime, and ferromagnesia.

By applying the mineralogical phase rule, which states that the number of stable minerals in a rock shall not be larger than the number of components, it follows that, except for quartz and some alkali-bearing minerals listed below, no more than three additional minerals should occur in any one (variety) of these hornfelses. Observations have verified this. Thus from alumina, lime, and ferromagnesia, seven minerals will form that are stable under the conditions of contact metamorphism: anadalusite, Al_2SiO_5; cordierite, $Mg_2Al_4Si_5O_{18}$; anorthite, $CaAl_2Si_2O_8$; hypersthene, $(Mg,Fe)SiO_3$; diopside, $Ca(Mg,Fe)Si_2O_6$; grossularite, $Ca_3Al_2Si_3$-

O_{12}; and wollastonite, $CaSiO_3$. Only three (or fewer) of these minerals can occur together. In this way different mineral combinations develop, each combination (plus quartz and an alkali-bearing mineral) representing a natural hornfels. There are 10 such combinations, corresponding to hornfelses of classes 1–10 of V. M. Goldschmidt's terminology. *See* HORNFELS.

Variations from the above scheme are easily explained. Usually enough water and potash are present to produce mica; muscovite may form instead of, or together with, andalusite, and in the hornfelses of classes 4 and 5, biotite is usually present inducing a characteristic chocolate-brown color into the rocks. In hornfelses of class 10 some lime-rich hydrous silicates may develop, for example, vesuvianite (idocrase). The presence of ferric iron may produce andradite, $Ca_3Fe_2Si_3O_{12}$, a yellow to dark-green garnet which will form mixed crystals with grossularite.

Pneumatolysis and metasomatism. Other factors of importance in contact metamorphism are chemical changes that ensue from pneumatolytic and hydrothermal action. These changes are brought about by the magmatic gases and high temperatures that accompany igneous intrusions. The surrounding rocks are deeply penetrated not only by the heat but also by water and other volatile compounds. Because chemical alterations take place in this so-called pneumatolytic or hydrothermal contact zone, the Kjerulf-Rosenbusch rule is not applicable. The width of the affected zone varies from nil to thousands of feet. *See* METASOMATISM; PNEUMATOLYSIS.

The primary magmatic gases are acid and in consequence show high reactivity. If the contact rock is basic, especially limestone, the acid gases will react effectively with it. Limestone acts as a filter, capturing the escaping gases. As a result, a great variety of reaction minerals is formed. The corresponding rocks are known as skarns. If the reaction rocks are limestones composed of lime silicates, the reaction minerals are mainly garnet and pyroxene, often accompanied by phlogopite and fluorite. Sulfides of iron, zinc, lead, or copper may be present, and in some occurrences magnetite is formed. *See* SKARN.

Summary. Contact metamorphism caused by deep-seated magma intrusions is very common, and the products (disregarding the pneumatolytic action) vary regularly in accordance with the chemical compositions of the preexisting contact rock. Another factor of equal importance is the variation in temperature as influenced by the nature of the intruding rock and the distance from the contact. Thus it is possible to distinguish between an inner and an outer contact zone. The zones grade into each other by imperceptible transitions, but the mineral associations in the typical inner contact zone, the only zone considered so far, are markedly different from the associations in the outer contact zones.

These problems involve a consideration of the general relationships between minerals and mineral associations, on the one hand, and temperature and pressure, on the other. They are discussed further in connection with the facies principle and the general process of regional metamorphism. However, it is important to realize that contact metamorphism, although it appears to be well defined and seems to stand out as an isolated natural phenomenon, is complex and variegated and passes by gradual transitions into other kinds of metamorphism. Geologically, contact metamorphism should be considered in connection with, and as a part of, the general system of rock metamorphism and metasomatism.

Regional metamorphic rocks. Crystalline schists, gneisses, and migmatites are typical products of regional metamorphism and mountain building. If sediments accumulate in a slowly subsiding geosynclinal basin, they are subject to down-warping and deep burial, and thus to gradually increasing temperature and pressure. They become sheared and deformed, and a general recrystallization results. However, subsidence into deeper parts of the crust is not the only reason for increasing temperature. It is not known what happens at the deeper levels of a live geosyncline, but obviously heat from the interior of the Earth is introduced regionally and locally, partly associated with magmas, partly in the form of "emanations" following certain main avenues, determined by a variety of factors. From this milieu rose the lofty mountain ranges of the world, with their altered beds of thick sediments intercallated with tuffs, lava, and intrusives, all thrown into enormous series of folds and elevated to thousands of meters. Thus were born the crystalline schists with their variants of gneisses and migmatites. *See* EARTH, HEAT FLOW IN; OROGENY.

A. Michel-Lévy (1888) distinguished three main étages in the formation of the crystalline schists; F. Becke and U. Grubenmann (1910) demonstrated that the same original material may produce radically different metamorphic rocks according to the effective temperature and pressure during the metamorphism. Grubenmann distinguished three successive depth zones, epizone, mesozone, and katazone, corresponding to three consecutive steps of progressive metamorphism. In eroded mountain ranges, rocks of the katazone are, generally speaking, encountered in the central parts; toward the marginal parts are found rocks of the mesozone and epizone.

It is of paramount importance to obtain better information about the temperature-pressure conditions of the recrystallization, and thus to show the relation between the chemical and mineralogical composition of all varieties of rocks. A large-scale attempt in this direction was the development of the facies classification of rocks.

Mineral facies. As defined by P. Eskola (1921), a mineral facies "comprises all the rocks that have originated under temperature and pressure conditions so similar that a definite chemical composition has resulted in the same set of minerals, quite regardless of their mode of crystallization, whether from magma or aqueous solution or gas, and whether by direct crystallization from solution . . . or by gradual change of earlier minerals . . . ." To learn which mineral associations were characteristic of high temperature or of low temperature, and to determine which associations combined with high pressure and with low pressure, Eskola studied the mineral associations in the rocks.

It has long been known that in an area of progressive metamorphism each successive stage, or each new zone of metamorphism, is reflected in the appearance of characteristic rock types (G. Barrow, 1893). Rocks within the same zone may be called isofacial, or isograde as proposed by C. E. Tilley (1924) who, furthermore, proposed the term "isograd" for a line of similar degree of metamorphism.

In going from an area of unmetamorphosed sedimentary rocks into an area of progressively more highly metamorphic rocks, new minerals appear in orderly succession. Thus, in a series of argillaceous rocks subjected to progressive metamorphism, the first index mineral to appear is usually chlorite, followed successively by biotite, garnet (almandite), and sillimanite. A line can be drawn on the map indicating where biotite first appears. This line is the biotite isograd. The less metamorphosed argillites on one side of this line lack biotite, whereas the more metamorphosed rocks on the other side contain biotite. An isograd can be drawn for each mineral. Actually the isograds are surfaces, and the lines drawn on the map are the intersections of these surfaces with the surface of the Earth.

Further work along these lines resulted in the conclusion that it was possible to single out a well-defined series of mineral facies. Sedimentary rocks of the lowest metamorphic grade recrystallized to give rocks of the zeolite facies. At slightly higher temperatures the greenschist facies develops—chlorite, albite, and epidote being characteristic minerals. A higher degree of metamorphism produces the epidote-amphibolite facies, and a still higher degree the true amphibolite facies in which hornblende and plagioclase mainly take the place of chlorite and epidote. Representative of the highest regional metamorphic grade is the granulite facies, in which most of the stable minerals are water-free, such as pyroxenes and garnets. Any sedimentary unit will recrystallize according to the rules of the several mineral facies, the complete sequence of events being a progressive change of the sediment by deformation, recrystallization, and alteration in the successive stages: greenschist facies → epidote-amphibolite facies → amphibolite facies → granulite facies. The mineral associations of these rocks are summarized in the next section.

During regional metamorphism a stationary temperature gradient is supposed to be established in the mountain masses. Usually, the outer parts of a geosynclinal region are less affected, and in the ideal case the marginal parts contain unmetamorphosed sediments, clay, sand, and limestone, which gradually change into metamorphic rocks of successively higher facies as they extend into the central and deeper parts. *See* GEOSYNCLINE.

The table summarizes the metamorphic series of rocks that develop from the several types of common sediments and usually converge toward a granitic composition regardless of the nature of the original material. Basic igneous rocks (gabbros, basalts) show a composition related to that of marl and yield analogous metamorphic products. Not listed are ultrabasites (peridotites, and others) which by metamorphism become serpentine, chlorite or talc schist, soapstone, hornblende schist, pyroxene, or olivine masses. Original acid igneous rocks (granite, diorite, rhyolite) show a composition related to that of arkose and yield analogous products. Leptite is primary fine-grained, usually showing tufaceous or blastopor-

Metamorphic rock series

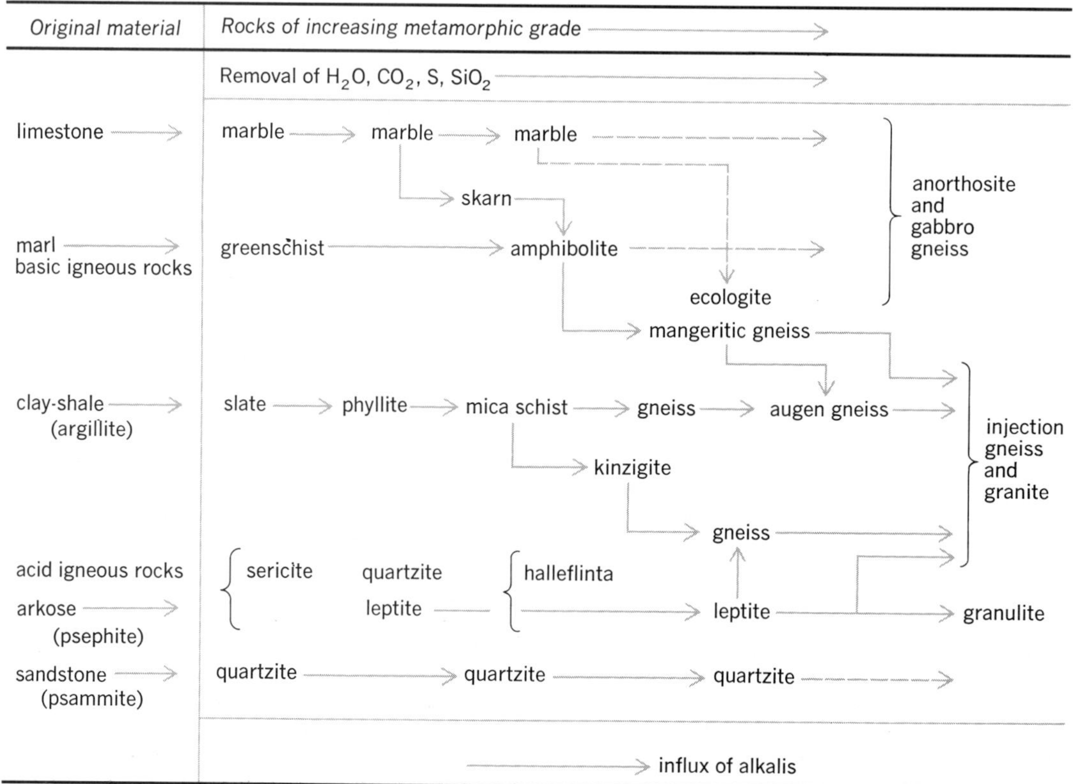

phyric relic structures; or it is derived from argillaceous sediments. Hälleflintas are dense rocks of conchoidal fracture, genetically related to leptites. Kinzigites, characterized by containing aluminum silicates and usually also rich in magnesia, are metasomatic gneisses, but probably argillites also enter into their constitution. Granulite is a gneiss recrystallized in the high-temperature mineral facies group. *See* GRANULITE.

Chemical alterations. The chemical changes in the progressive series are complicated. When a sediment is heated, it obviously loses water and other volatile compounds, carbon dioxide, halogens, and so forth. These vapors act as carriers of several of the nonvolatile elements, for example, Si, Fe, and Mn. As heat emanates from the central parts of a geosynclinal region, it is heralded by a cloud of these vapors migrating centrifugally through the surrounding sediments. Usually, however, no major chemical alterations take place. Petrographers used to believe that the vapors were rich in alkalis and that large quantities, particularly of sodium, gradually would be deposited and fixed in the sediments. The argument was that the sediments, concomitant with a progressive change by increasing metamorphism, seem to exhibit an increase in their sodium content. However, this is not a real increase, but due to an erroneous sampling of the sedimentary material. Normally the lower stages of metamorphism tend to conserve the original composition.

Mineral associations of the facies. Rocks of the greenschist facies recrystallized at low temperatures and often under high shearing stress. These include chlorite schists, epidote-albite schists, and actinolite schists, all of which are green, hence the name. Other common rocks in this group are serpentinites, talc schists, phyllites, and muscovite (sericite) schists. Plagioclase is not stable in green schists but breaks up into epidote and albite. Glaucophane schists are rare and probably they represent greenschists formed under conditions of high stress.

Rocks of the epidote-amphibolite facies recrystallized in a somewhat higher temperature range. Amphibolites with epidote and either albite or oligoclase are typical. Mica schists (garnet), biotite schists, staurolite schists, and kyanite schists are common. Cordierite-antophyllite (gedrite) schists and chloritoid schists also occur. Sodic plagioclase is stable in these rocks.

Rocks of the amphibolite facies recrystallized at about 600°C. Amphibolites of hornblende and plagioclase are typical and often carry quartz or biotite or both, or garnet. Sillimanite-muscovite schists (gneisses) are common. Andalusite and staurolite occur at low pressure; kyanite and staurolite at intermediate and high pressure.

Rocks of the granulite facies have their greatest extension in old Precambrian areas, but are also found in younger deeply eroded mountain chains. Diagnostic association is ortho- and clinopyroxene; hence the alternative name, two-pyroxene facies group. Hypersthene and augite (garnet) plagioclase gneisses, usually with quartz but occasionally with olivine or spinel, are typical. Amphibolites with hypersthene or diopside or both, sillimanite gneisses, and at high pressure kyanite gneisses are found. The temperature range is probably around 800°C. The experiments by H. S. Yoder (1952) in the system $MgO-Al_2O_3-SiO_2-H_2O$ indicated that at approximately 600°C and 1200 atm it is possible to have different mineral assemblages suggestive of every one of the now accepted metamorphic facies in stable equilibrium. These different mineral assemblages (artificial facies) observed by Yoder are the result of differences in the water content, and are not related to variation in temperature and pressure.

As an example, the mineral clinochlore, corresponding to one of the most common rock-making chlorites, shows an upper limit of stability, either alone or in association with talc, of 680°C. This is, indeed, a high temperature for any kind of metamorphism, almost a magmatic temperature. But according to the tenets of the mineral facies, chlorite is strictly limited to the greenschist facies, about 200°C.

Although Yoder has proved that there is no absolute relation between temperature and facies, it appears likely that, to the field geologist and to the laboratory man as well, the facies will still remain the best system of classification of metamorphic rocks; and in a majority of cases the facies will indicate the temperature-pressure conditions under which the several rocks recrystallized. Generally speaking, there is a regular relation between the chemical activity of water and the facies of the metamorphic rock.

Water content and mineral facies. The role of water in metamorphism is determined by at least four variable, geologically related parameters: rock pressure, temperature, water pressure, and the amount of water present. During a normal progressive regional metamorphism, rock pressure and temperature are interdependent. The amount of water and the pressure of water are related to the encasing sediments and to the degree of metamorphism in such a way that, generally speaking, the low-grade metamorphic facies are characterized by the presence of an excess of water, the medium-grade by some deficiency in water, and the high-grade by virtual absence of water.

In the usual diagrammatic illustration of the mineral facies of rocks, temperature and pressure (depth) are taken as coordinates; in regional-metamorphic rocks a third, dependent coordinate may be added, the activity of water running upward approximately along the geothermal gradient.

Facies series and groups. Metamorphic facies may be divided into facies series depending mainly on pressure, and facies groups depending mainly on temperature.

Three facies series have been proposed (A. Miyashiro 1961): low-pressure, intermediate-pressure, and high-pressure series. The low-pressure series dominates throughout the Hercynian and Svecofennian of Europe, the Paleozoic of Australia, and part of the paired belts in Japan and New Zealand. The intermediate-pressure series (the original Barrowian zones) occurs in the European Caledonides, the Appalachians, the Precambrian Belt series of Idaho, the Himalayas, and parts of Africa. The high-pressure facies series is found in the Alps and the circumpacific region—Japan, New Zealand, Celebes, and the United States. Thus each of the intracontinental orogenic belts is characterized by one facies series which reflects

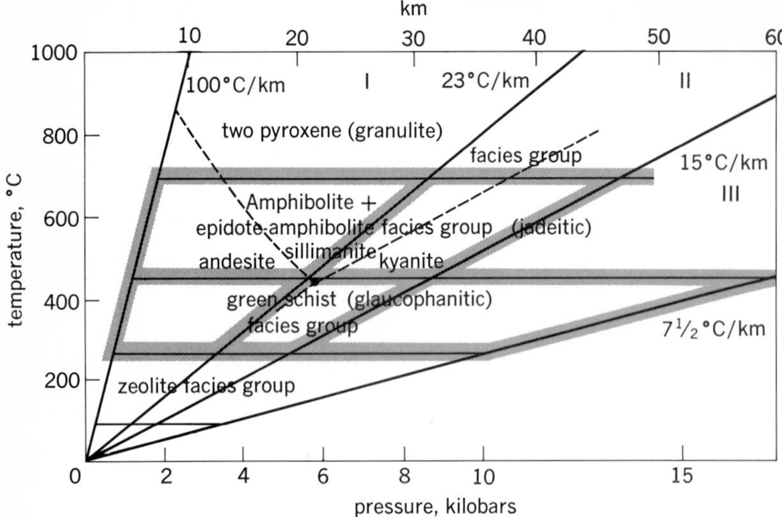

Key to facies series:
I = low pressure (23–100°C/km) II = intermediate pressure (15–23°C/km) III = high pressure (7½–15°C/km)

Fig. 3. The mineral facies groups of regional metamorphic rocks showing the temperature and pressure of metamorphism.

the pressure that prevailed during metamorphism, whereas the circum-Pacific region with its paired metamorphic belts exhibits two facies series.

Four facies groups are recognized from low to high temperature (H. J. Zwart 1967): laumontite and prehnite-pumpellyite; greenschist, including glaucophane schist; amphibolite, including epidoteamphibolite; and the two-pyroxene (granulite) facies group.

This scheme is presented in Fig. 3. The "normal" geothermal gradient is in the range 15–23°C per kilometer depth. The intermediate-pressure facies series is found in areas with this gradient; the high-pressure series (the Alpine series) in areas with lower thermal gradients or with high overpressure (orogenic pressure); and the low-pressure facies series in areas with steep thermal gradients. The phase boundaries of the polymorphic forms of Al_2SiO_5 (andalusite, sillimanite, kyanite) have a central position in this scheme, the triple point being located approximately at 450°C and 6 kilobars. The "minimum" melting of granite under water pressure occurs approximately along the boundary between the amphibolite and the two-pyroxene facies group. A separate eclogite facies is not recognized by this scheme. *See* ECLOGITE.

Figure 3 illustrates the distributions and interrelations of the various metamorphic rocks. It also represents a schematic profile through the continental crust down to 60-km depth, that is, down to the Moho discontinuity. Thus the normal continental crust is entirely made up of metamorphic rocks; where thermal, mechanical, and geochemical equilibrium prevails, there are only metamorphic rocks. Border cases of this normal situation occur in the depths where ultrametamorphism brings about differential melting and local formation of magmas. When equilibrium is restored, these magmas congeal and recrystallize to (metamorphic) rocks. At the surface, weathering processes oxidize and disintegrate the rocks superficially and produce sediments as transient products. Thus the cycle is closed; petrology is without a break. All rocks that are found in the continental crust were once metamorphites. *See* AMPHIBOLITE; EPIDIORITE; EPIDOSITE; GREISEN; MARBLE; MICA SCHIST; MIGMATITE; PHYLLITE; PHYLLONITE; QUARTZITE; SANIDINITE; SCAPOLITE; SERPENTINITE; SOAPSTONE.

[T. F. W. BARTH]

Bibliography: T. F. W. Barth, *Theoretical Petrology*, 1952; T. F. W. Barth, C. W. Correns, and P. Eskola, *Die Entstehung der Gesteine*, 1939; W. H. Bucher, Fossils in metamorphic rocks, *Geol. Soc. Amer. Bull.*, 64:275–300, 1953; W. S. Fyfe, F. J. Turner, and J. Verhoogen, *Metamorphic Reactions and Metamorphic Facies*, Geol. Soc. Amer. Mem. no. 73, 1958; A. Harker, *Metamorphism*, 1932; A. Hietanen, The facies series in various types of metamorphism. *J. Geol.*, 75:187–214, 1967; A. Miyashiro, Evolution of metamorphic belts, *J. Petrol.*, 2:277–311, 1961; H. Ramberg, *Origin of Metamorphic and Metasomatic Rocks*, 1952; B. Sander, *Einführung in die Gefügekunde der Geologischen Körper*, 1948; F. J. Turner, *Mineralogical and Structural Evolution of Metamorphic Rocks*, Geol. Soc. Amer. Mem. no. 30, 1948; F. J. Turner and J. Verhoogen, *Igneous and Metamorphic Petrology*, 1951; H. G. F. Winkler, *Petrogenesis of Metamorphic Rocks*, 1965; H. S. Yoder, The $MgO-Al_2O_3-SiO_2-H_2O$ system and the related metamorphic facies, *Amer. J. Sci.*, Bowen vol., no. 2, pp. 569–627, 1952; H. J. Zwart et al., A scheme of metamorphic facies for cartographic representation, *Geol. Newslett.*, 2:57–74, 1967.

Metamorphism

The alterations and transformations in preexisting rock masses effected by temperature and pressure, but excluding changes produced by weathering and sedimentation. The changes may include the production of new minerals, structures, or textures, or all three. They give a distinctive new character to the rock as a whole, but they do not involve the loss of individuality of a rock mass, such as changes brought about by fusion. Quantitatively, the metamorphic rocks, including gneisses and migmatites, are the most important group of rocks in the crust of the continents.

James Hutton (1726–1797) of Edinburgh was first to advance the opinion that some of these rocks were ordinary sedimentary strata which were subsequently so altered by subterranean heat as to assume a new texture. Sir Charles Lyell, who adopted the Huttonian theory, proposed in the first edition of his *Principles of Geology* (1833) the term metamorphic for the altered strata.

Both igneous rocks and sediments are subject to alteration. The former are fusion products of processes which operated at high temperatures, whereas sediments are deposits which accumulated under conditions of low temperature. Once formed, the rocks of both classes respond to moderately elevated temperatures and changes in pressure, or shearing stresses, by reaction and recrystallization.

Metamorphic processes cannot be directly observed in natural rocks, nor have they been adequately investigated by experiments. For this reason the mechanism and thermodynamic relations of metamorphic processes are not completely understood. The study of transitional rock series ranging from unaltered sediments to completely recrystallized rocks provides the data on which the principles of metamorphic geology are formulated.

Metamorphism versus metasomatism. The alteration of rocks by recrystallization is generally accompanied by a change in the chemical composition, that is, by loss and gain of chemical matter resulting in the so-called allochemical or metasomatic metamorphism, or simply metasomatism. One can often find statements to the effect that metamorphism without chemical change is the normal thing, and supposed cases of such isochemical metamorphism are still proposed as normal metamorphism. In such cases the recrystallized minerals may be of the same kind as the old ones. However, they are usually different, indicating that a relocation of the chemical constituents has taken place within the rock without affecting the total chemical composition. Therefore in isochemical metamorphism the chemical elements have migrated over short distances and in allochemical metamorphism, over large distances. In harmony with this nomenclature it has become customary, and is perhaps expedient in an attempt to survey and classify the various petrographic types, to distinguish between the normal metamorphism, with only small changes in the total chemical composition, and the metasomatic, or allochemical metamorphism, with large changes in the chemical composition. From a genetic point of view, however, this distinction is illogical and obscures the broader relations. Metamorphic rock always has undergone a change in composition, for transportation of chemical matter and transfer of heat are not only concomitant with, but essential parts of, the earth processes constituting metamorphism.

In geologic discussions the fact that sediments at the very early stages of metamorphism lose volatile constituents (water and carbon dioxide) and consequently regularly change their chemical composition has often been neglected. However, these changes should not be neglected, for the water and carbon dioxide in rocks are just as important as the many other rock-making constituents and serve to characterize the rocks in which they occur.

Large-scale chemical transport in and through large volumes of rock is the normal; gneiss-granites, migmatites, and mica schists exhibit a chemical composition different from that of the preexisting rock complex. In a similar manner, the mineralogical composition changes with the changing chemical composition and with variations in temperature and pressure. Thermodynamic equilibrium is approached during the course of the metamorphism.

The mechanisms that produced the changes are not readily explained, but investigations bear out the conclusion that large-scale chemical changes in solid rocks did take place. Studies carried out on fossils in metamorphic rocks demonstrate that in many specimens the most delicate structures of the fossil organisms were preserved, although the original calcareous shell substance might have been altered chemically into silica, iron sulfide, or other compounds. It is a reasonable position to base the discussion of recrystallization and evolution of metamorphic rocks on the postulate that free interchange of matter (atoms or ions) can take place over large volumes of rocks existing as essentially solid masses at a temperature well below that of fusion. *See* MAGMA.

Theory. The simplest kind of transformation in rocks involves changes occurring in individual minerals, that is, the polymorphic transformation of one mineral to another. Thermodynamically, changes occur in a mineral that cause a decrease in the free energy. Thus if a chemical compound such as silicon dioxide, SiO_2, occurs in a number of polymorphic forms such as quartz, tridymite, and cristobalite, then all polymorphic forms tend to transform into the one stable form which at a given temperature and pressure is characterized by the minimum free energy, for example, into tridymite at 900°C and 1 atm. This statement may be clarified by a graph (see illustration) and the useful, elementary thermodynamic equation shown below, where A denotes the (Helmholtz) free en-

$$A = U - TS$$

ergy of the mineral, U is the internal energy, and the quantity TS is the bound energy (T is the absolute temperature and S is the entropy). Thus in the equation free energy equals internal energy less bound energy.

At absolute zero the TS term vanishes, and the free energy becomes equal to the internal energy U of the crystal. Thus, at very low temperatures the internal energy term is dominant and the polymorphic form with the least internal energy tends to be the stable one. With increasing temperature the TS term becomes increasingly important. It may happen, therefore, that because of the possibility of larger entropy in a second form, its TS term so reduces its free energy that, in spite of greater U, the differences between these two terms are greater than for the first polymorphic form. If this occurs, the second form becomes the stable one, and the first form tends to transform into it. The temperature at which the free energies become equal is the transition temperature.

Analogously, the fact that a mineral assemblage changes into another shows that the new assem-

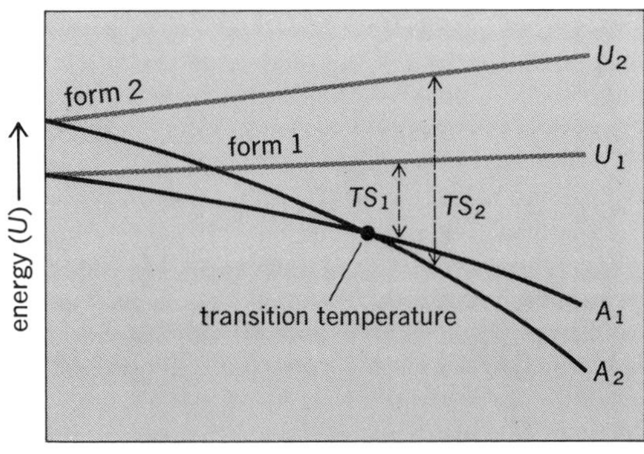

Diagram of the relation $A = U - TS$, showing how a high-temperature form (form 2), having high internal energy U_2 because of its high entropy S_2 at elevated temperatures, corresponds to the state of lowest free energy, A_2. *(After M. J. Buerger, Amer. Mineral., copyright 1948 by the Mineralogical Society of America)*

blage, as such, has a lower total free energy than the old. Under the conditions of the metamorphic environment, a new set of minerals V, W, X, . . . has a lower net free energy than the old set of minerals A, B, C, ... It is important to note that the energy drop represents the physicochemical force which drives the metamorphic process and that its potential comes from a difference in the free energy of the old set of minerals and the free energy of the new set of minerals.

The energy relations, therefore, are the controlling factors in metamorphism, and the geological agents determining these relations are mainly temperature and pressure.

It is worthy of note that, in general, the free energy of polymorphic transitions or of reactions between silicates is small, of the order of a few hundred or a few thousand calories per mole. This fact leads to formation or persistence of metastable assemblages. Crystalline schists from various localities are known in which, for example, both sillimanite and kyanite, two forms of Al_2SiO_5, have been found together.

Types. Different kinds of metamorphism may be defined according to genetic criteria, such as the geologic processes that were assumed to have caused the metamorphism, or the physical and chemical conditions that appear to have been predominant in determining the course of metamorphism. Using these criteria three general kinds of metamorphism are noted.

1. Dislocation, mechanical, or dynamic metamorphism is the result of pressure (or stress) along dislocations in the Earth's crust. The deformed rocks commonly show marked zones of extremely fine-grained rocks, such as mylonites, whose structures are determined by crushing and movement of the grains without important recrystallization of old, or growth of new, minerals. This type of metamorphism is local and restricted in occurrence.

2. Contact or thermal metamorphism occurs in response to increased temperature induced by adjacent intrusions of magma. Chemical reconstitution of the rocks is due to magmatic exhalation; other conditions, such as confining pressure, exert subordinate influence.

3. Regional metamorphism, the most widespread type, is brought about by an increase in both temperature and pressure in orogenic regions, which are vast segments of the crust represented by the folded mountain ranges. Heat and pressure are mainly consequences of downwarping and deep burial. Pressure is also generated by shearing stresses accompanying the orogenic movements. *See* OROGENY.

Other systems of nomenclature employed in classifying metamorphism may depart from the ideal simple categories noted above.

Autometamorphism is a term applied to chemical adjustment in newly congealed igneous rocks, brought about by a decrease in temperature. The residual hydrothermal fluids are then able to react with the igneous minerals.

Diaphtoresis, or retrograde metamorphism, is a general term denoting any metamorphism that takes place as a result of decreasing temperatures.

Static metamorphism (load or geothermal metamorphism) is the term used to describe changes in rocks supposedly brought about by a regular increase in temperature and hydrostatic pressure through deep burial. However, such conditions are hardly sufficient to induce notable metamorphic effects in silicate rocks; but oceanic salt deposits, famous for their great varieties of minerals, have suffered profound changes in this way. Their metamorphism has been carefully studied and represents one of the best-known physicochemical processes in geology. It has been shown that the famous salt deposits at Stassfurt, Germany, recrystallized under a superincumbent layer several kilometers thick and at about 80°C. *See* EVAPORITE, SALINE.

A distinction must be made between metamorphism in the superstructure or *oberbau* of the orogenic mountain chains and the infrastructure or *unterbau* lying below the usual depth range of regional metamorphism. Migmatization takes place as a result of the physicochemical conditions of the infrastructure. Thus migmatites are usually grouped with metamorphic rocks. *See* MIGMATITE.

[T. F. W. BARTH]

Bibliography: *See* METAMORPHIC ROCKS.

Metasomatism

A process by which the chemical composition of rocks is altered. As originally defined by C. F. Naumann, the term was applied to certain kinds of pseudomorphs, namely, those formed by chemical replacement at the expense of some original mineral. Every transformation of minerals by chemical replacement may now be designated as metasomatism, and in this sense most of the newly formed constituents of the metamorphic rocks would be metasomatic minerals. *See* METAMORPHIC ROCKS; MINERAL.

Metasomatism is fundamentally a change, atom for atom, in a rock mineral. It postulates a reaction between a solid mineral and a disperse or fluid phase. It corresponds to what several United States authors have called replacement. It may be well, therefore, to distinguish between mineral metasomatism and rock metasomatism. The mechanism of the mineral metasomatism in a metamorphic rock is the same as that of the rock metasomatism of large bodies, but in mineral metasomatism the fluid phase, which is generated of and by the metamorphic rock, would cause only a relocation of the chemical elements within the rock without changing the chemical composition of the whole rock. In the metasomatic rock, however, the fluid is introduced into the rock and acts metasomatically, thereby changing the initial chemical composition of the rock. Thus the mineral metasomatism taking place in a metamorphic rock may be considered as a metabolism of rock, that is, a relocation of the chemical elements which have to migrate over only short distances; whereas true metasomatism of rocks should be restricted to only those replacements by which the chemical composition of the whole rock, as well as that of the individual minerals, is essentially changed through an immigration of foreign chemical elements and an expulsion of other elements over long distances, up to several kilometers. Metasomatism in this sense would include, for example, the alteration of limestone into dolomite. Distinction may also be made between mutative, additive, and expulsive metasomatism. *See* DOLOMITE ROCK.

Metasomatic processes take place over a wide range of temperature and pressure and in rocks that are to all intents and purposes solid. They may be accompanied by a change in volume, but in most cases the replacements are effected without any appreciable change in volume (the "volume law" of Lindgren), and that is why original structures and textures often are well preserved. In general, metasomatic reactions obey the law of mass action, and in order to effect replacement the introduced fluids must attain a minimum concentration. This explains why gradational passage from one type of rock into another is rare; sharp contacts are usual in metasomatic rocks.

A fine example of simple metasomatism is the formation of dolomite from limestone (calcite) in contact with sea water. Magnesium ions dissolved in the sea are slowly extracted through metasomatic reaction with limestone according to the schematic relation shown below.

$$2CaCO_3 + Mg^{++} \rightarrow CaMg(CO_3)_2 + Ca^{++}$$

Many dolomite rocks appear to have this mode of formation. The dolomitization of many coral reefs can be followed step by step (borings in the atoll Funafuti, Ellice Islands, Pacific). Phosphatization of limestones is another example.

Metasomatism is widespread in silicate rocks. In contact zones of igneous intrusions the replacements may affect not only the intruded rocks but also the crystallized magma from which the fluids are derived. Such magmatic fluids are often gaseous, giving rise to the so-called pneumatolytic metasomatism with formation of skarn rocks. These are lime-silicate felses (such as andradite garnet or hedenbergite rocks) which often contain special minerals rich in boron, fluorine, and sulfur. Valuable ore deposits are often associated with rocks of this character (copper, lead, silver). *See* ORE AND MINERAL DEPOSITS; PNEUMATOLYSIS; SKARN.

During the regional metamorphism, metasomatism is ubiquitous. Regional granitization occurs and is characterized by an increase in the alkali metals and a decrease in magnesium and iron. *See* METAMORPHISM. [T. F. W. BARTH]

Meteorite

Any natural solid object that falls to the Earth from space and retains its primordial characteristics. Such objects are rare, and they differ from all terrestrial rocks in their composition. There are only about 2200 meteorites known over the entire world. Meteorites are named after the town or post office near where they fell or were found.

Meteoroids are attracted to the Earth if they are moving through space in orbits which bring them into the Earth's gravitational field. Any such mass that is moving on a trajectory which will bring it to the Earth must first pass through the outer atmosphere, then through the denser air. Because most of these incoming bodies are relatively small, they are consumed in flight, but very large ones may pass through the atmosphere and strike the Earth. Fortunately, the atmosphere is an excellent shield against such falling objects.

Most meteorites are stony masses and lack the durability to survive a passage through the air. These objects usually have internal fractures, so during their fall they break along preexisting planes of weakness. Their external surfaces, prior to the entry into the atmosphere, are angular and rough, so after entering the atmosphere these bodies are soon ablated; thus recovered meteorites generally have rounded forms.

Anything entering the atmosphere from space is traveling at high speed, but something moving in the same direction as the Earth and overtaking it is traveling at a higher speed than the Earth. But these overtaking bodies fall though the air at slower speeds than those which hit the Earth's atmosphere head on. The reason is that the overtaking bodies' speed of fall through the atmosphere is the difference between their velocity and the velocity of Earth moving on its orbit. But a body that comes directly toward the Earth enters the atmosphere at much higher speeds: to whatever velocity that object has must be added the speed at which the Earth is traveling.

Overtaking objects fall between noon and midnight, while oncoming ones fall between midnight and noon. The accounts of the meteoritic material recovered from observed falls show that the vast majority of meteorites come from the falls that occur between noon and midnight.

Due to the speed at which any meteorite comes through the atmosphere, the air in front of it gets compressed as well as heated because the object is moving faster than the air can be pushed aside. As this hot air jets over the sides of the object, the surfaces are heated and the heat-softened material is carried off. The turbulent vapor trails which extend behind for such long distances contain vast quantities of ablated material. Falling meteorites resemble fireballs, and frequently they break and scatter pieces along their line of flight.

Studies of the distribution of helium in large iron meteorites showed that some of the irons lost about 50% of their bulk during their fall. Helium-3, which is a product of cosmic bombardment in space, progressively decreases inward from the surface. By mapping zones of equal helium-3 content, it is possible to approximately determine where the original surface of the object was located.

Meteorite falls. A falling meteorite, if the sky is clear, may be visible over thousands of square miles, but its sound can be heard over perhaps a hundred square miles. When a fireball is seen but no sound is heard, it is far away, possibly a hundred or more miles. Fireballs are impressive sights, and are better understood than they were in ancient times. Fireballs or meteorite falls once terrified people, and thus these events became the topics for legends. From about 400 B.C. to nearly A.D. 300, in some areas the populace struck off metallic coins commemorating these events.

Today millions of people regard the Black Rock of Mecca, which probably is a meteorite (most newly fallen meteorites are black), as a sacred symbol. All followers of the Muslim faith strive to make a pilgrimage to Mecca to pay respect to this celebrated object.

The Casa Grande meteorite from Chihuahua, Mexico, weighed 1545 kg when dug from the middle of a room in a temple. It had been wrapped in coarse linen cloth, like an Egyptian mummy, indicating the Montezuma Indians regarded it as something special. This specimen, being iron and unlike familiar rocks, was a new kind of substance which was unique to the Indians.

The oldest witnessed meteorite fall, now preserved, fell at Ensisheim, France, on Nov. 18, 1492. But 302 years passed before science accepted the fact that stones could fall from the sky. When a stone fell at Luce, France, on Sept. 13, 1768, Father Barchley reported the event to the Royal Academy of Science in Paris, and the Academy appointed a commission to investigate this fall. A. L. Lavoisier was on that commission. After some study the Academy reported that the Luce stone was a terrestrial rock. When the Luce stone fell, it weighed 3.5 kg, but apparently those academicians did not think it was important enough to save. Today there is less than 200 g of it in existence, and none is in France. Another stone fell in Barbotan, France, on July 24, 1790; another at Siena, Italy, on June 16, 1794; and another at Wold Cottage, England, on Dec. 13, 1795. Finally, a shower consisting of thousands of stones fell at L'Aigle, France, on Apr. 26, 1803. This event changed the thinking of the Royal Academy, and the new commission that was appointed reported that these stones did come from the sky.

Witnessed falls and recovered meteorites. There have been 762 meteorite falls witnessed from within the United States, and portions of 562 of them are preserved in the Smithsonian Institution's meteorite collection. More meteorites fall each year than are recovered, and countless more enter the atmosphere than survive the passage through the air. Statistics about the recovery of meteorites from observed falls are good, but many fireballs are sighted each year from which no meteorites are recovered. The number of meteorites recovered from observed falls for the entire world is given in 5-year intervals, starting with 1800, in Table 1. The upward trend in recoveries between 1800 and 1934 coincides with the improvements in communications and with the dissemination of people into less settled areas. The peak number of falls occurred in the 1930–1934 interval; since then a significant decrease has occurred. This downward trend started about the time science became seriously interested in all phases of meteoritic research, also at the time when worldwide communication was greatly improved. This possibly means that Earth is moving through a section in space where less material is available for capture.

Many pieces fall from some fireballs, but since most of the fireballs approach the Earth at a low angle the pieces are scattered over an oval-shaped area. The larger masses, which have the most momentum, usually are carried to the far end of the strewn field. The smaller pieces tend to drop at the end of the strewn field which lies toward the direction in which the fireball approached the Earth.

To recover newly fallen metorites, the searcher should get reports from witnesses, preferably from those who saw the fireball from different directions. The searcher should evaluate a number of accounts to establish the terminal point of the fireball. Reports from people who saw the fireball from a moving car or who were in open spaces with an unobstructed view of the sky generally are the least reliable accounts; such people rarely can relocate the exact spot where they sighted the fireball, and thus their directions are likely to be inaccurate. When a number of accounts indicate that the end point is confined to a limited area, the searcher can go to that place and check roofs for punctures and ask occupants if they heard sounds indicating something falling on their roofs. Local people and the weather bureau can be asked if there were noticeable winds at the time of the fireball; as relatively small bodies approach the end of their flight, they have lost most of their momentum and so crosswinds can blow them off course. A field search requires numerous traverses across the estimated end point of the fireball. The searcher must look for broken limbs on trees, for small patches of freshly disturbed ground, and for any object that resembles a meteorite.

No one can estimate how many pieces will fall from a fireball or predict how widely they will be scattered. After the Allende meteorite fell in Chi-

Table 1. Witnessed meteorite falls from which a specimen has been preserved

Interval	Number of meteorites	Interval	Number of meteorites
(Jan. 1) (Dec. 31)		(Jan. 1) (Dec. 31)	
1800–1804	7	1885–1889	21
1805–1809	11	1890–1894	20
1810–1814	14	1895–1899	28
1815–1819	9	1900–1904	29
1820–1824	13	1905–1909	25
1825–1829	13	1910–1914	29
1830–1834	9	1915–1919	32
1835–1839	16	1920–1924	34
1840–1844	19	1925–1929	33
1845–1849	13	1930–1934	50
1850–1854	15	1935–1939	33
1855–1859	25	1940–1944	23
1860–1864	20	1945–1949	29
1865–1869	34	1950–1954	31
1870–1874	19	1955–1959	22
1875–1879	31	1960–1964	26
1880–1884	20	1965–1969	21
		1970–1974	17

huahua, Mexico, on Feb. 8, 1968, various-size pieces were collected from an oval-shaped area that was about 50 km long and enclosed an area exceeding 300 km². More than 1000 kg of meteorites was collected from this unusually large strewn field. In future years additional pieces will be found, but these specimens will be weathered and thus have a different color and luster from the original specimens.

Distribution of falls. There is no reason to suspect that certain places are preferred targets for meteorites to fall on. A localized concentration of hexahedrites, an unusual type of iron meteorite, occurs in southeastern United States and in northern Chile. However, investigations have shown that there are minor differences in chemical composition, particularly in the trace elements, so these hexahedrites are no longer considered as having a common origin. But the unsolved question remains: Why have so many meteorites, of the rare type, been found so close together? Two stony meteorites fell almost on the same spot; the Honolulu, Hawaii, stone fell on Sept. 24, 1849, and the Palolo Valley stone fell in the same city on Apr. 24, 1949.

Interesting relationships appear when the places where meteorites fell or where they were found are located on maps. The states of Ohio and Kentucky have about the same areas, but Ohio has 9 meteorites, while Kentucky has 23 (Table 2). This difference could be due to the fact that Ohio was glaciated, while Kentucky was not.

It takes people to observe and find a meteorite, hence some relationship must exist between population density and the number of meteorites reported from any area. Another very important factor is the use that is made of the land. The majority of meteorites are found in areas where the soil is cultivated. The old-fashioned plow was a very efficient instrument in locating meteorites.

Table 3 gives data about meteorites that were either observed to fall or are chance finds from four widely separated places. Although these localities vary in size and population densities and have been settled for different periods of times, interesting relationships show up. (1) Australia and the United States have far more iron meteorites than the other two places. The reason is that the aboriginal peoples in Australia and the Indians of North America in Pre-Columbian times lived in the stone age, hence these people had no use for iron meteorites. But the peoples in Asia and in Europe who had advanced into the iron age centuries earlier could make use of their iron meteorites. (2) The quotients obtained by dividing the number of witnessed falls of stony meteorites by the witnessed falls of irons show how much more frequently stony meteorites fall than iron meteorites. (3) The quotients obtained by dividing the number of witnessed falls for the stony meteorites by the iron meteorites in each of these different places show how rare any type of meteorite is when one compares the areas. (4) Further study of these data, Table 3, shows that there must be quantities of meteorites on the surface of Earth that have not been found. Some may think this indicates the possibility that there are unrecognized meteorites, but all the 791 witnessed meteorites can be fitted into the classification.

About 75% of the Earth's surface is under water, and perhaps such a percentage was covered in former geological times; thus, the largest proportion of the meteorite falls certainly went into the sea. Few of those will ever be recovered, but because so many fell in the sea some must have fallen in shallow water, where they would be quickly covered either by mechanically or chemically deposited materials. Thus, some meteorites possibly were sealed and thereby rather well protected from complete disintegration.

Geological history. Coal beds at one time were extensive surface swamps into which some meteorites probably fell. These meteorites would go deep enough to get into an oxygen-deficient environment, favorable for the preservation of the meteorites. Later, when the coal is mined and processed to remove the foreign material, a meteorite may be recovered. Although no meteorite has been reported from any of the coal-mining operations, this does not mean that no meteorites were there to be found. The most likely explanation is that those examining the rejected material were not familiar with meteorites or with the scientific importance of such a discovery. Iron meteorites have densities between 7.7 and 8, which means they are about twice as heavy as the pebbles in gravel deposits. The 43-kg iron from Aggie Creek, Alaska, was dredged from under 4 m of alluvial material; yet it had no scars indicating it had been dragged over the basement rocks, and it lacked scars showing it hit that gravel when it fell. Iron meteorites are sufficiently heavy that they would work their way to the bottom of the gravel if much movement occurred in the gravel. It is illogical to assume that the Aggie Creek meteorite penetrated 4 m of the gravel.

The two Klondike irons from Canada have similar histories. The first of these irons weighed 483 g and was found in Pleistocene gravel at Gay Gulch. The second meteorite weighed 16 kg and was discovered at another location under 18.5 m of Pleistocene gravel. The engineers operating the dredge concluded these irons were older than the gravel; if true, the terrestrial age of the irons is over a million years.

Perhaps the Sardis, Georgia, iron meteorite is older. This badly altered 880-kg iron meteorite was recovered from below the surface in some Miocene beds on a farm in eastern Georgia and is the largest iron meteorite so far found in the United States east of the Mississippi River. An iron mass of this size would require a long time to become altered to the degree this meteorite was altered. Had the Sardis iron meteorite fallen in recent times on these same beds, most likely some impact scar would have been made. However, a shallow impact scar could easily have been obliterated within the last geological period.

Probably the date of this fall will never be established, but investigations to get its terrestrial age show it has been on Earth longer than existing methods can measure time. Thus if this iron fell in Miocene times, it is the oldest meteorite and has a terrestrial age of about 20,000,000 years before present.

Fusion crusts. Objects entering the atmosphere from space are cold, but as soon as they start through the air their surfaces are heated. As a rela-

Table 2. Distribution of meteorites in the United States

State	Witnessed falls: Stones	Witnessed falls: Irons	Total known meteorites	Number of meteorites in Smithsonian Institution
Alabama	6	0	16	12
Alaska	0	0	4	4
Arizona	1	0	30	17
Arkansas	4	2	12	11
California	1	0	21	12
Colorado	2	0	59	42
Connecticut	3	0	3	2
Delaware	0	0	0	0
Florida	0	0	3	3
Georgia	2	0	22	20
Hawaii	2	0	2	2
Idaho	0	0	3	1
Illinois	3	0	7	4
Indiana	3	0	11	9
Iowa	4	0	5	5
Kansas	5	0	90	73
Kentucky	4	0	23	23
Lousiana	0	0	1	1
Maine	4	0	4	3
Maryland	2	0	4	4
Massachusetts	0	0	0	0
Michigan	2	0	8	7
Minnesota	1	0	4	4
Mississippi	2	0	3	2
Missouri	6	0	20	17
Montana	0	0	3	2
Nebraska	1	1	32	26
Nevada	0	0	4	4
New Hampshire	0	0	0	0
New Jersey	1	0	2	2
Nex Mexico	2	0	54	34
New York	2	0	9	6
North Carolina	9	0	29	25
North Dakota	1	0	6	5
Ohio	2	0	9	7
Oklahoma	5	0	24	18
Oregon	0	0	5	4
Pennyslvania	3	0	8	5
Rhode Island	0	0	0	0
South Carolina	2	0	6	6
South Dakota	2	0	9	6
Tennessee	2	1	22	20
Texas	9	0	137	78
Utah	1	0	9	7
Vermont	0	0	0	0
Virginia	3	0	11	10
Washington	1	0	2	1
West Virginia	0	0	3	3
Wisconsin	3	0	12	8
Wyoming	1	0	11	7
District of Columbia	0	0	0	0
TOTAL	107	4	762	562

Table 3. Statistical data about meteorites from four widely separated areas

Area	Country	Square miles	Comparative size using smallest area as unity	Total of stony mete-orites	Total of iron mete-orites	Total of palla-sites	Total of meso-siderites	Witnessed falls of stony meteorites	Witnessed falls of iron meteorites	Total falls of stones / Total falls of irons
1	Australia	2,968,000	5.3	109	71	4	3	7	3	2.3
2	United States	3,608,672	6.5	359	393	11	4	107	4	26.7
3	Indian subcontinent	1,571,997	2.8	131	6	1	1	115	6	19.1
	France	211,207		58	1	0	0	55	0	
	Spain	194,884		20	3	0	0	20	1	
4	Portugal	35,553	1.0	2	2	0	0	2	1	36.0
	Italy	116,303		28	1	0	0	25	1	

tively small body nears the end of its fall, its speed is so retarded, by the resistance of the air, that it is being cooled rather than heated. The heated zone on these falling bodies is thin, so the heat will quickly be lost. Thus far, no meteorite has been hot when it fell, but perhaps one might have felt warm to the touch. Heat has more difficulty penetrating stony meteorites than the irons because silicates are poor heat conductors. Also, stony meteorites are usually aggregrates of small silicate grains which are often bonded together by troilite, a sulfide of iron that has a low melting point. As the troilite melts, the silicate grains are freed and can be carried off by the airflow. Thus, the heated layer on stony meteorites is thin.

A paper-thin layer of molten material clings to the surface of these falling objects. This layer is thinner on the leading face and on the forward parts of the side faces, but on the aft, trailing surface, a thicker layer of the fusion crust accumulates (Figs. 1 and 2). This molten material cools to form a black glaze which always contains grains of partly melted silicates and gas bubbles. The fusion crust on the forward face is thin because this surface is under considerable air pressure so that the bubbles that form there are small. On the rear face, where more of the slag accumulates, the gas bubbles are much larger, because immediately behind a falling object there is an area of low air pressure, almost a vacuum.

When a metallic body falls through the air, the general conditions are essentially the same as for a stony mass, but heat is more readily conducted into a metallic object. As the metal on the surface is heated, it softens and the airflow moves the heated layer toward the rear face and much of it passes off into the vapor trail. Enough heat penetrates the metal to modify the internal structure. Thus, when an iron meteorite is cut, polished, and etched, the depth of the thermally treated metal can be seen.

The fusion crust on stony meteorites generally is black when they fall because the most common minerals in these meteorites contain considerable iron. However, there are a few stony meteorites in which the minerals are essentially free of iron, so on these meteorites the crust will have just a tinge of brownish color.

As a falling meteorite streaks across the night sky, the observer gets the impression that it is a ball of fire. Actually, the object itself is not hot, but the air surrounding it is very hot. Many meteorites have fallen on buildings without starting fires, or have fallen on dried grass without charring it.

Two interesting situations illustrate how cool these masses can be when they reach the Earth. On July 14, 1860, at Dhurmasala, India, and on July 14, 1917, at Colby, Wisconsin, stones fell, and both were quickly coated with frost. The humidity at those places was high, and the object that fell was cold enough to condense the moisture from the air and form a frosty coat.

Composition. Meteorites contain the same elements that occur in terrestrial rocks, and these elements combine in the same manner to form essentially the same minerals that make up terrestrial rocks. However, meteoritic minerals have slightly different characteristics than the minerals in terrestrial rocks, because they formed in different conditions from those existing when the minerals in terrestrial rocks formed. But there is a group of stony meteorites, achondrites (see type B in the outline below), which are aggregrates of minerals that almost duplicate the fabric and texture of some terrestrial rocks. However, the meteorites often have nickel-iron inclusions, and these are unknown in terrestrial rocks.

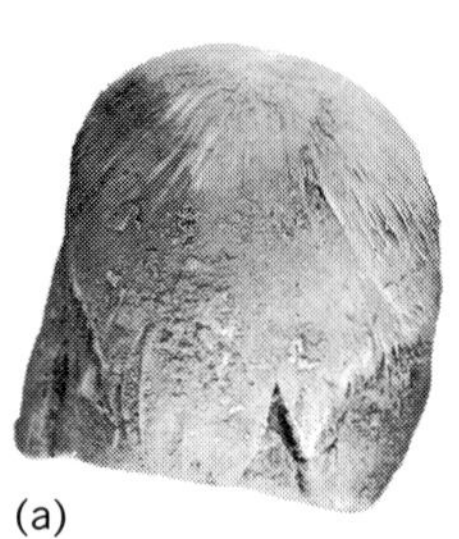

(a)

(b)

Fig. 1. Freda, N. Dak., meteorite showing (*a*) front face with flow lines and (*b*) crust on rear. Pattern of flow lines shows specimen fell in fixed position.

Chemically, some stony meteorites are similar to some basic igneous rocks which are rich in magnesium and calcium but contain almost no quartz. Table 4 gives the average composition of three rock types and three types of stony meteorites; the similarity is apparent. Table 5 shows the average abundance of the common elements in iron meteorites. A characteristic feature of iron meteorites is the presence of nickel, but the nickel percentage in the different types of iron meteorites varies. Gallium and germanium, both minor elements, occur in all iron meteorites, along with traces of other elements. *See* IGNEOUS ROCKS.

The number of minerals found in meteorites has grown since 1885, when Gustav Tschermark listed 16 different meteoric minerals. Today, due to modern instruments, there are over 105 recognized minerals, and 25 of these occur only in meteorites (Table 6).

Craters. Meteorites which fall and form craters are never buried in their craters (Fig. 3). The greater part of that impacting mass, which is not

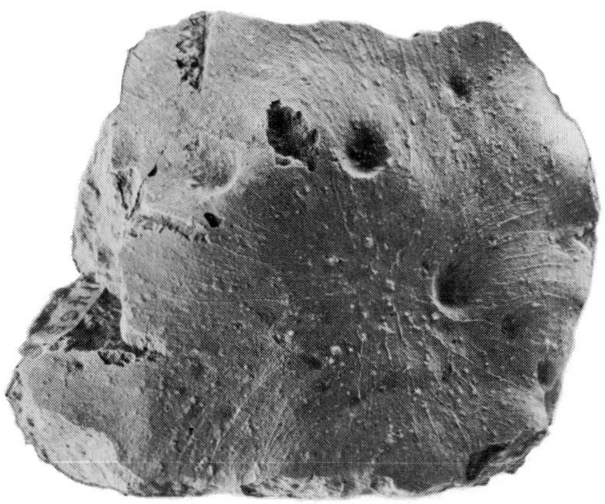

Fig. 2. Amherst (No. 2), Neb., meteorite showing dome-shaped forward face with flight markings which radiate from center. Deflection of flow lines around pits indicates pits complicated airflow over this leading face.

Table 4. Average compositions (wt %) of some terrestrial rocks and meteorites

	Rocks			Meteorites		
	Basalts from mid-Atlantic ridge	Olivine basalts (surface rocks)	Dunites (surface rocks)	Unequilibrated chondrites	Calicum-rich achondrites	Calcium-poor achondrites
Number averaged:	15	28	9	28	20	11
Compound or element						
SiO_2	49.38	47.90	40.16	38.95	48.99	52.27
Al_2O_3	16.43	11.84	0.84	2.40	10.77	1.22
Fe_2O_3	2.02	2.32	1.88	–	1.55	–
FeO	6.98	9.80	11.87	14.33	16.80	10.60
MgO	8.34	14.07	43.16	24.63	10.47	31.11
CaO	11.26	9.29	0.75	1.96	9.21	1.17
NaO	2.74	1.66	0.31	0.91	0.82	0.43
K_2O	0.28	0.54	0.14	0.10	0.42	0.13
MnO	0.15	0.15	0.21	0.31	0.60	0.45
TiO_2	1.32	1.65	0.20	0.12	0.46	–
P_2O_5	0.15	0.19	0.04	0.26	0.08	0.17
H_2O^+	0.63	0.59	–	0.91	{ 0.43	–
H_2O^-	0.45	–	–	0.20		–
Fe	–	–	–	2.02	0.64	3.12
Ni	–	–	–	1.28	–	0.17
Co	–	–	–	0.06	–	–

Table 5. Abundance of common elements in five types of iron meteorites

		Octahedrites*			
Element	Hexahedrites*	Coarse	Medium	Fine	Ataxite†
Fe	93.59	91.22	90.67	86.75	79.72
Ni	5.57	7.39	8.22	11.65	19.14
Co	0.66	0.54	0.59	0.61	0.83
Cu	0.35	0.18	0.03	0.11	–
P	0.29	0.18	0.18	0.24	0.25
S	0.06	0.08	0.08	0.63	–
C	0.09	0.21	0.61	0.01	0.05

*Data from J. D. Buddhue, *Pop. Astron.*, 49(3), 1946.
†Data from M. H. Hey, *Catalogue of Meteorites*, British Museum (Natural History), 1966.

atomized, is tossed out of the crater almost immediately by the rebound in the impacted rock (see Table 7). The rejected fragments generally are hurled outward along straight lines, at right angles to the rim of the crater, and these lines are called rays. Rim rocks surrounding impact craters are upturned, and frequently parts of the rim are flipped over. In time, crater rims may be eroded away, so a geologically old crater may not have a prominent rim. About 45 craters are known which have yielded no meteoritic fragments, yet the surrounding rocks contain shock features typical of the rocks around meteorite craters (see Table 8). The object which hit the Earth does not have to have the size of the crater, because much of the country rock around the impact point is broken to bits and is tossed from the crater.

Types. There are three major types of meteorites: stony, intermediate, and irons.

I. Stony meteorites.
 A. Chondrites, the most abundant stony meteorite, are characterized by small, spherical silicate particles known as chondrules. These escaped recrystallization at the place where they were formed, but by some mechanical process were transported and latter accu-

Table 6. Minerals found in meteorites

Mineral	Formula or symbol
Native elements, metallic compounds	
*Awaruite	$Ni_{13}Fe$
Carbon	C
Chaoite	C
*Lonsdalite	C
Copper	Cu
Diamond	C
Gold	Au
Sulfur	S
*Kamacite	NiFe (less than 6%Ni)
*Taenite	NiFe (more than 6% Ni)
Carbides, phosphides, nitrides, silicides	
Cohenite	$(FeNi)_3C$
*Carlsbergite	CrNi
*Haxonite	$Fe_{23}C$
Moissanite	SiC
*Osbornite	TiN
*Perryite	$(NiFe)_5(SiP)_2$
*Rhabdite	$(NiFe)_3P$
Schreibersite	$(FeNi)_3P$
*Senoite	Si_2N_2O
Sulfides	
Alabandite	(MnFe)S
Bornite	Cu_5FeS_4

Table 6. Minerals found in meteorites (cont.)

Mineral	Formula or symbol
Bravoite	$(FeNi)S_2$
*Brezinaite	Cr_3S_4
Chalcopyrite	CuFeS
Chalcopyrrhoite	$CuFe_4S_5$
Cubanite	$CuFe_2S_3$
*Daubreelite	$FeCr_2S_4$
Djerfisherite	$K_3CuFe_{12}S_{14}$
*Gentnerite	$Cu_8Fe_3Fe_{12}S_{14}$
Heazelwoodite	Ni_3S_2
Mackenawite	FeS
*Niningerite	(MgFe)S
*Oldhamite	CaS
Pentlandite	$(FeNi)_9S_8$
Pyrrhotite	$Fe_{1-x}S$
Sphalerite	ZnS
Troilite	FeS
Vallerite	$CuFeS_2$
Violarite	Ni_2FeS_4
Chlorides	
Lawrencite	$FeCl_2$
Oxides	
Chromite	$FeCr_2O_4$
Christobalite	SiO_2
Hematite	Fe_2O_3
Ilmenite	$FeTiO_3$
Maghemite	Fe_2O_3
Perovskite	$CaTiO_3$
Quartz	SiO_2
Rutile	TiO_2
Spinel	$MgAl_2O_4$
Tridymite	SiO_2
Wustite	FeO
Carbonates, phosphates, sulfates	
Apatite	$Ca(PO_4)_3Cl$
Bloedite	$Na_2Mg(SiO_4)_24H_2O$
Breunnerite	$(MgFe)CO_3$
*Brianite	$Na_2MgCa(PO_4)_2$
Calcite	$CaCO_3$
Cassidyite	$Ca_2(NiMg)(PO_4)_2H_2O$
Collinsite	$Ca_2(MgFe)(PO_4)_2$
Dolomite	$CaMg(CO_3)_2$
Epsomite	$MgSO_47H_2O$
Farringtonite	$Mg_3(PO_4)_2$
Graftonite	$(FeMn)_3(PO_4)_2$
Gypsum	$CaSO_42H_2O$
Lipscomite	$Fe_3(OH)_2(PO_4)_2$
Magnesite	$MgCO_3$
Merrillite	$Ca_3(PO_4)_2$
*Panethite	$(CaNa)_2(MgFe)_2(PO_4)_2$
Siderite	$FeCO_3$
*Stanfieldsite	$Ca_4(MgFe)_5(PO_4)_6$
Whitlockite	$Ca(PO_4)_2$
Organic minerals, oxalates	
Whewellite	$CaC_2O_4H_2O$
Silicates	
Andradite	$Ca_3Al_2Si_3O_2$
Augite	$Ca(MgFeAl)(AlSi)_2O_6$
Clinopyroxene	$(CaMgFe)SiO_3$
Chlorite	$(MgFeAl)_6(AlSi)_4O_{10}(OH)_8$
Cordiorite	$Mg_2Al_4Si_5O_{15}$
Diopside	$(CaMg)(SiO_3)_2$
Enstatite	$MgSiO_3$
Gehlenite	$Ca_2Al_2SiO_7$
Grossularite	$Ca_3Al_2Si_3O_{12}$
Hypersthene	$(MgFe)SiO_3$
*Krinovite	$NaMg_2CrSi_3O_{10}$
*Majorite	$Mg_3(MgSi)Si_3O_{12}$
Melilite	$Ca_2(MgAl)(SiAl)_2O_7$
*Merrihueite	$(KNa)_2(FeMg)_5Si_{12}O_{30}$
Monticellite	$Ca(MgFe)SiO_4$
Nepheline	$(KNa)_3(AlSiO_4)_4$
Olivine	$(MgFe)_2SiO_4$
Plagioclase	$(CaNa)(AlSi)AlSi_2O_8$
Potash feldspar	$(KNa)AlSi_3O_8$
Richterite	$Na_2CaMg_5Si_8O_{22}F$
Rhonite	$CaMg_2TiAl_2SiO_{10}$
*Roedderite	$(NaK)_2(MgFe)_5Si_{12}O_{30}$
*Ringwoodite	$(MgFe)_2SiO_4$
Sodalite	$Na_8Al_6Si_6O_{24}Cl_2$
Wollastonite	$CaSiO_3$
*Ureyite	$NaCrSi_2O_6$
*Yagiite	$(KNa)_2(MgAl)_5(SiAl)_{12}O_{30}$
Hydrous oxides	
Akaganeite	FeOOH
Garnierite	$(NiMg)_3Si_2O_5(OH)_4$
Goethite	FeOOH
Lepedrosite	FeOOH
Limonite	$FeOOHH_2O$
Opal	SiO_2H_2O
Reevesite	$Ni_6Fe_2(OH)_{16}(CO_3)_4H_2O$

*Found only in meteorites.
SOURCE: From B. Mason, Mineralogy of meteorites, *Meteoritics*, 7(3).

Fig. 3. Canyon Diablo, Arizona, meteorite crater. The rocks immediately surrounding this impact crater are turned upward. Meteorites from this crater were found on the plains around the depression.

Table 7. Meteoritic craters (meteorite fragments have been recovered around these craters)*

Name	Country	Latitude	Longitude	Number of craters	Diameter, m
Aourelloul	Mauritania	20°15′N	12°41′W	1	250
Barringer	Arizona	35°02′N	111°01′W	1	1200
Boxhole	N.T., Australia	22°37′S	135°12′E	1	175
Campo del Cielo	Argentina	27°28′S	61°30′W	9	70
Dalgaranga	W. Australia	27°45′S	117°05′E	1	21
Haviland	Kansas	37°37′N	99°05′W	1	11
Henbury	N.T., Australia	24°34′N	113°10′E	14	150
Kaälijarvi	Estonian S.S.R.	58°24′N	22°40′E	7	110
Odessa	Texas	31°48′N	102°30′W	3	168
Sikhote Alin	E. Siberia	46°07′N	134°40′W	22	26.5
Wabar	Saudi Arabia	21°30′N	50°28′E	2	90
Wolf Creek	W. Australia	19°18′S	127°47′E	1	850

*From M. R. Dence, *Canada Earth Physics Branch Contribution*, no. 393.

Table 8. Probable impact craters by continents*

Site	Latitude	Longitude	Diameter, km
North America			
Brent, Ontario, Canada	46°05′N	78°29′W	4
Carswell, Saskatchewan	58°27′N	109°30′W	30
Charlevoix, Quebec	47°32′N	70°18′W	35
Clearwater Lake, E. Quebec	56°05′N	74°07′W	15
Clearwater Lake, W. Quebec	56°13′N	74°30′W	30
Crooked Creek, Montana	37°50′N	91°23′W	5
Decaturville, Montana	37°54′N	92°43′W	6
Deep Bay, Saskatchewan	56°24′N	102°59′W	9
Flynn Creek, Tennessee	36°16′N	85°37′W	3.6
Holleford, Ontario	44°28′N	76°35′W	2
Kentland, Indiana	47°13′N	10°58′E	5
Lac Couture, Quebec	60°08′N	75°18′W	10
Manicounagan, Quebec	51°23′N	68°42′W	65
Manson, Iowa	42°35′N	94°31′W	30
Middlesboro, Kentucky	36°37′N	83°44′W	7
Mistasten, Labrador	55°53′N	63°18′W	20
New Quebec, Quebec	61°17′N	73°40′W	3.2
Nicholaon, N.W.T.	62°40′N	102°41′W	12.5
Pilot Lake, N.W.T.	60°17′N	111°01′W	5
St. Martin, Manitoba	51°47′N	98°33′W	24
Serpent Mound, Ohio	39°02′N	83°25′W	6.4
Sierra Madera, Texas	30°36′N	102°55′W	13
Steen River, Alberta	59°31′N	117°38′W	25
Sudburry, Ontario	46°36′N	81°11′W	100
Wanapitie, Ontario	46°44′N	80°44′W	8.5
Wells Creek, Tennessee	36°23′N	87°40′W	14
West Hawk Lake, Manitoba	49°46′N	95°11′W	2.7
Europe			
Dellen, Sweden	61°50′N	16°45′E	12
Lappajarvi, Finland	63°10′N	23°40′E	10
Mien, Sweden	56°25′N	14°55′E	5
Ries, Germany	48°53′N	10°37′E	24
Rochechouart, France	45°50′N	0°56′E	15
Siijan, Sweden	61°05′N	15°0′E	45
Steinheim, Germany	48°02′N	10°4′E	3
Africa			
Bosumtwi, Ghana	6°32′N	1°23′W	10.5
Tenourmer, Muritana	22°55′N	10°24′W	1.8
Vredefort, South Africa	27°0′S	27°30′E	100
Australia			
Gosses Bluff, N.T.	23°48′S	132°18′E	22
Kofels	47°13′S	10°58′E	5
Liverpool, N.T.	12°24′S	134°3′E	1.6
Strangwaya, N.T.	15°12′S	135°36′E	16
South America			
Monturaqui, Chile	23°56′S	68°17′W	0.48

*From M. R. Dence, *Earth Physics Branch Contribution*, no. 393.

mulated before being bonded together. Thus, chondrites are probably a product of the original process by which the planets were made from the solar nebula, so this type of meteorite is important scientifically. Sometimes metal inclusions occur within a single chondrule, but usually the metallic inclusions, which often are rather abundant, occur immediately around the chondrule. Most chondrites are composed of pyroxene and olivine with minor amounts of feldspar and of interstitial glass of a feldspathic composition (Fig. 4). Texture within a chondrule can vary, but there are two types of chondrites, equilibrated and unequilibrated. Equilibrated chondrules are the most abundant, and in these the silicates within the chondrules and the interstitial matrix are essentially the same; the chondrules are usually small and so well bonded that their outlines are not well defined. The major silicates in the unequilibrated group have variable composition, and the aggregate is so friable that it tends to crumble; the chondrules occur in well-defined, rounded forms, but the matrix is fine-grained. There are five types of chondrites.

1. *Enstatite chondrites*, which are usually nearly pure $MgSiO_3$; metal inclusions may be abundant, and usually are low in nickel. These stones are highly reduced, often containing silicon.
2. *Olivine-bronzite chondrites* are more abundant than the enstatite chondrites and contain about equal amounts of olivine and bronzite.
3. *Olivine-hypersthene chondrites* are the most abundant chondritic meteorites and generally contain more olivine than pyroxene. This hypersthene contains from 12 to 20% Fe, hence these stones have a darker color than the bronzite-chondrites. Their metal grains usually contain between 7 and 12% nickel.
4. *Olivine-pigeonites* differ from the olivine-hypersthene group by having pigeonite rather than hypersthene, with olivine as the predominant mineral. The metal is usually rich in nickel, and troilite, an iron sulfide common in other types of chondrites, may be replaced by pentlandite.
5. *Carbonaceous chondrites* have dark colors, due to the carbon content.

Unlike other meteorites, carbonaceous chondrites contain considerable water. The carbon is present in complex compounds. There are three types.

a. *Type I* is strongly magnetic, contains sulfates, and has a carbon content of about 3.5%.
b. *Type II* is weakly magnetic, and most of its sulfur is present as free sulfur; this type contains about 2.5% of carbon.
c. *Type III* contains the lowest percentage of water, has the highest density, and often contains pigeonite.

B. Achondrites are coarsely crystalline meteorites which are very similar to terrestrial rocks such as basalts and dunites. They usually contain less olivine than the chondrites, and their feldspar is generally more calcic.

1. *Calcium-poor achondrites.*
   a. *Aubrites* usually contain almost pure $MgSiO_2$, with minor quantities of accessory minerals.
   b. *Diogenites* contain hypersthene and modest amounts of nickel iron and other accessory minerals.
   c. *Chassignites*, a type with only one example, consist almost entirely of olivine.
   d. *Ureilites* are a type of achondrites which contains graphite and occasionally small diamonds. The metal is low in nickel.
2. *Calcium-rich achondrites.*
   a. *Nakhlites* contain diopside and olivine.
   b. *Angrites* are augite-bearing achondrites.
   c. *Eucrites* are achondrites composed of plagioclase and pigeonite.
   d. *Howardites* are achondrites composed of hypersthene and plagioclase.
   e. *Whitleynite* has only one example, which is composed of enstatite and dark angular chrondritic inclusions.

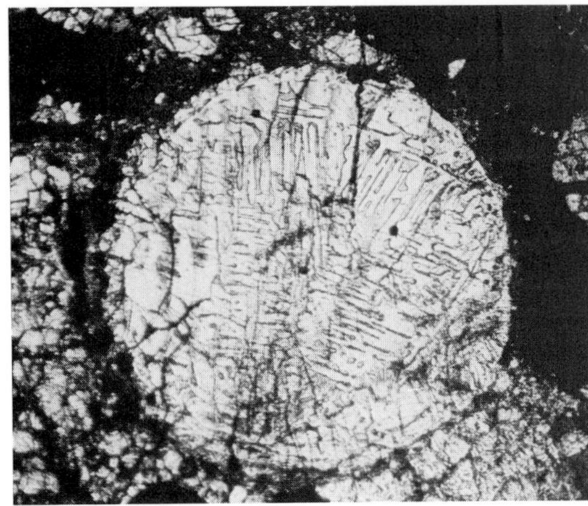

Fig. 4. Ioka, Utah, meteorite: a cross section through an olivine chondrule shown in transmitted light. The lighter bands are olivine and the darker are glass. Since each has different refractive indices and colors, their outlines can be observed. Dark area outside this chondrule is due to troilite or nickel-iron, both of which are opaque.

II. Intermediate types (stony and iron meteorites).

A. *Pallasites* are the most abundant of this group and consist of olivine in a nickel-iron matrix.
B. *Siderophyres* consist of bronzite and tridymite in nickel-iron.
C. *Lodranites* consist of bronzite and olivine in nickel-iron.
D. *Mesosiderites* have about equal amounts of silicates and nickel-iron.
E. *Sorotites* are similar to the pallasites, with troilite replacing olivine.

III. Iron meteorites (see Table 9).

A. *Hexahedrites* are the simplest type of meteorite, consisting of one phase of nickel-iron, kamacite (alpha iron). These irons often contain accessory minerals and usually display Neumann lines (Fig. 5). The nickel content ranges between 5 and 6.5%.

METEORITE

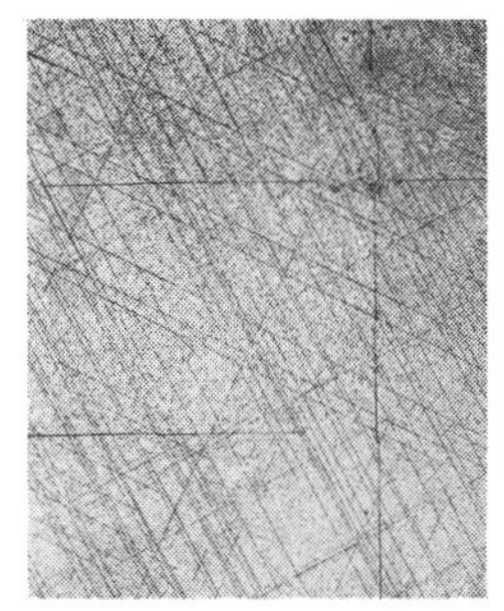

Fig. 5. Etched surface of Edmonton, Canada, meteorite. Parallel Neumann lines of kamacite are found in all hexahedrites. Two prominent lines intersecting at right angles are cleavage directions and parallel a cube face.

Table 9. Classification of iron meteorites

Class	Common symbol	Kamacite band widths, mm	Remarks
Hexahedrites	H	–	Neumann lines abundant, cubic cleavage
Coarse octahedrites	Ogg	> 3.3	Large grains of kamacite, taenite rare, usually no well-developed octahedral pattern
Coarse octahedrites	Og	1.3–3.3	–
Medium octahedrites	Om	0.5–1.3	–
Fine octahedrites	Of	< 0.2–0.5	–
Finest octahedrites	Off	< 0.2	Well-developed octahedral structure, narrow kamacite bands
Ataxites	D	–	Fine structure, spindles of kamacite phosphide inclusions
Anomalous	Anomalous	–	All meteorites which require special descriptions

METEORITE

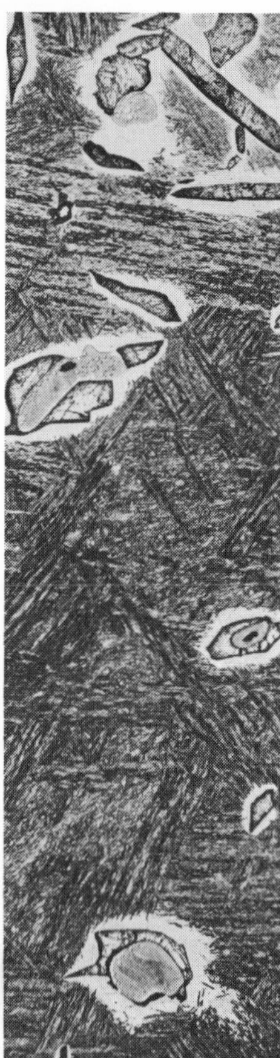

Fig. 6. Freda, N. Dak., meteorite. Etched surface is typical of ataxites. Inclusions are kamacite and schreibersite in acicular groundmass of kamacite and taenite. Light areas around the inclusions are taenite.

B. *Octahedrites* have nickel values from about 6.5 to about 16%. These are often grouped into coarse, medium, and fine octahedrites, depending upon the widths of the kamacite lamellae. No sharp distinction exists between the hexahedrites and octahedrites, or between the other three subgroups. Octahedrites contain two distinct phases of nickel-iron, kamacite and taenite.

C. *Ataxites* have structures which are usually fine-grained, and unlike either of the other groups of iron meteorites. Some of the ataxites have nickel contents similar to hexahedrites; others have nickel contents higher than the octahedrites. Ataxites with high nickel percentage usually contain a considerable amount of schreibersite inclusions (Fig. 6).

The structures used in typing iron meteorites become apparent when a surface through the meteorite is highly polished; however, these structures are best seen after a polished surface is etched with nital, a reagent with attacks the two nickel-iron phases differently. The pattern that develops depends upon the amount of nickel that is present and on the rate at which the metal cooled. When the nickel content is below about 6%, the predominating nickel-iron phase is kamacite, alpha-iron; however, some of the high-nickel phase, taenite, gamma-iron, may form at the edges of the kamacite grains. As the percentage of nickel increases, more taenite forms and the widths of the kamacite lamellae decrease.

Iron meteorites formed as the core of a sizable body or perhaps as large metallic masses in a sizable silicate body. Temperature changes in any large body proceed at a very slow rate, but the cooling rates of all iron meteorites have not been the same.

Identification. Meteorites have features which distinguish them from the rocks of this Earth, but it is impossible to give a simple and conclusive test that will identify all types of meteorites. When something is suspected of being a meteorite, evaluators must make certain that it is different from the other rocks in that area. If it is different, it possibly is a meteorite.

Stony meteorites are heavier than the average rock, but an iron meteorite is almost three times heavier, but their metallic nature makes them easy to recognize. Metallic iron is almost unknown in terrestrial rocks, but discarded pieces of manufactured metal are widespread. Usually, manufactured metal can be recognized by shape, but it can always be identified by its structures on a polished surface. All meteoritic iron contains nickel, but manufactured iron may also have nickel.

The crust on freshly fallen stony meteorites is generally black, but within a few years the black color gives way, due to weathering, and the outside surface takes on a brownish color. Newly fallen stones usually have localized places where the crust is broken off; these places give the finder a chance to see the internal nature of the specimen. Most stony meteorites have small silver-gray-colored inclusions of metal. Such inclusions add weight to the specimen, and there often are enough of these iron inclusions to make the specimen cling to a magnet. The fact that magnetite is a common mineral in many terrestrial rocks makes this useful test inconclusive proof that a specimen is a meteorite.

When a meteorite is found remote from any of its type, the tendency is to give it a name. If later studies show that it is chemically and mineralogically similar to a previously known meteorite from that same general area, the two named meteorites are called paired meteorites. As previously mentioned, meteorites tend to weather once they fall on Earth, and their silicate minerals become stained with secondary brown hydrous iron oxides. Some iron meteorites become so altered that little of their metal remains. When polished surfaces of some highly altered iron meteorites are examined, small specks of achreibersite can be seen. This phosphide is very resistant to weathering. Its presence in a highly altered mass is evidence of the meteoritic nature of that sample since this mineral occurs only in meteorites. Occasionally these highly altered irons will retain vestiges of their original Widmanstatten structure, but such features show up best after a surface has been cut and polished and then, generally, these structures are preserved in limited areas.

[EDWARD P. HENDERSON]

Mica

Any one of a group of hydrous potassium-aluminum silicate minerals, some of which also contain Mg, Fe(II), Mn, Li, Fe(III), and Ti as major constituents. All the micas are characterized by perfect, easy basal cleavage (micaceous cleavage). The main series and species are muscovite-lepidolite, phlogopite-biotite [containing Mg and Fe(II)], manganophyllite (Mn mica), paragonite (Na mica), and zinnwaldite (Li-Fe mica). Less common species are roscoelite (V mica), and taeniolite (Li-Mg mica). *See* BIOTITE; CHLORITOID; LEPIDOLITE; MUSCOVITE; PHLOGOPITE; SILICATE MINERALS.

The micas of economic importance are muscovite and phlogopite. Their technological significance stems from combinations of the following properties: perfect basal cleavage flakes that are flexible, elastic, tough, and translucent to transparent; and low electrical and heat conductivity. Both species have high dielectric strengths, that is, the ability to withstand high voltages without puncture. The power factor of mica, which determines its suitability for use in capacitors, is the ratio of the total power loss in a capacitor in which mica is the dielectric to the total volt-amperes supplied to the capacitor.

Structure. The mica group is structurally complex; muscovite occurs in three distinct structural types, lepidolite in at least four, phlogopite in three, and biotite in at least six, representing stacking variations of successive layers. Most micas are monoclinic pseudohexagonal, but a few are hexagonal or triclinic. Crystals commonly are tabular with prominent basal planes and hexagonal or lozenge-shaped outlines (edges intersecting at nearly 60°, or 120°). The mica structure is based on sheets of linked $(Si,Al)O_4$ tetrahedrons (phyllosilicates), which accounts for the characteristic cleavage into sheets.

Thin cleavage sheets of unaltered micas are both flexible and elastic. Percussion figures may be developed on cleavage plates by striking the surface sharply with a dull-pointed tool to yield a six-rayed star, two lines of which are parallel to the prism edges. The third, most prominent ray is parallel to the trace of the clinopinacoid. Optically, most micas are biaxial and levorotatory, with moderate birefringence.

Upon heating in a closed tube, micas yield water; they fuse with difficulty. Specific gravities are in the range 2.8–3.2; hardnesses are between 2 and 3 on Mohs scale.

Occurrence. The three major species, muscovite, biotite, and phlogopite, are widely distributed rock-forming minerals, occurring as essential constituents in a variety of igneous, metamorphic, and sedimentary rocks and in many minerals deposits. In igneous rocks, biotite is the most common, appearing in some gabbros, diorites, tonalites, granodiorites, syenites, nepheline syenites, and granites. Phlogopite occurs in some peridotites, and muscovite appears in a few granites. Both muscovite and biotite are widespread and abundant constituents of pegmatites, commonly occurring in crystals 1 in. to several feet across, called books. Pegmatites are the chief sources of commercial muscovite mica.

In metamorphic rocks, muscovite and biotite appear chiefly in schists and gneisses in association with such species as chlorite, garnet, kyanite, staurolite, quartz, and feldspar. Both biotite and muscovite occur as detrital minerals in sands and sandstone, but biotite is chemically altered much more readily than muscovite and tends to disappear. Muscovite is an exceedingly stable mineral in the weathering and sedimentary environments. Micas formed within the sedimentary environment (authigenic) are apparently chiefly of the hydromica (illite) type in which the oxonium ion (H_3O^+) is substituted isomorphously for the potassium ion, K^+. Thus, authigenic micas are species generally regarded as belonging to the clay minerals group.

Muscovite, especially as the fine-grained variety commonly called sericite, is a widespread gangue mineral in many hydrothermal mineral deposits, either in the deposits themselves or in the adjacent altered wall rocks. The process of sericitization commonly accompanies the development of metalliferous ore deposits formed within the intermediate temperature-pressure range (mesothermal deposits). Sericite also is formed as a common low-temperature replacement mineral (in many instances pseudomorphous) of many species, for example, sillimanite, andalusite, kyanite, cordierite, staurolite, beryl, potash feldspars, plagioclase, tourmaline, and topaz.

Uses. Commercial mica is of two main types: (1) sheet and punch, and (2) scrap or flake. High-quality sheet muscovite is essential as the dielectric in radio and radar circuit capacitors, in magnetocapacitors and coils, and as insulation in radio tubes and airplane spark plugs. Muscovite of lower quality (electric mica) is extensively employed as insulator elements in hot plates, toasters, irons, and other electrical home equipment. Built-up mica plate (micanite), which is made by bonding thin layers of mica splittings together with shellac or glyptal binders and drying under pressure, is extensively used as insulation between copper segments of generators and motors.

Scrap and flake mica is ground, both wet and dry, for use in paints, decorative inks, rubber filler, and coatings on roofing materials and waterproof fabrics. The chief producers of mica are the United States (North Carolina, New England states), India, and Brazil; the United States is the largest consumer. Synthetic fluorphlogopite is produced commercially as a substitute for natural mica in some applications.

[E. WILLIAM HEINRICH]

Bibliography: W. A. Deer, J. Zussman, and R. A. Howie, *Rock Forming Minerals*, 5 vols., 1962–1963; E. W. Heinrich et al., *Studies in the Natural History of Micas*, Univ. Mich. Eng. Res. Inst. Final Rep., Proj. M978, 1953; E. W. Heinrich, *Microscopic Petrography*, 1956; E. H. Kraus, W. F. Hunt, and L. S. Ramsdell, *Mineralogy*, 5th ed., 1959.

Mica schist

A widely distributed group of rocks of medium to high metamorphic grade, composed essentially of mica and quartz and exhibiting a foliated or schistose structure which is easily revealed by the parallel orientation of the mica flakes. Varieties are biotite schists, biotite-sericite schists, biotite-chlorite schists, and albite-biotite schists. Of lower metamorphic grade and of fine grain are phyllites or chlorite-sericite schists, without biotite but otherwise similar to mica schists, into which they pass by gradual transitions. *See* METAMORPHIC ROCKS; PHYLLITE.

The simple mineral composition of the ordinary mica schist (quartz and micas with or without feldspars) demonstrates the rule of the paucity of mineral phases. Micas are members of a chemically complex mineral group, and during the metamorphism and the crystalloblastic growth of mica, all possible ions try to accommodate themselves in its crystalline structure; those that cannot are simply carried away. However, this process cannot go on indefinitely. Eventually the composition is overstrained and if, for example, the alumina concentration is high, new minerals like andalusite, sillimanite, cordierite, or almandine garnet may develop. Rare types contain corundum, paragonite, and fuchsite. Minor accessories are rutile, zircon, tourmaline, and iron ores. When feldspar minerals are abundant, mica schist grades into gneiss; with an increase of quartz it grades into micaceous quartzite.

Mica schists are most widely distributed in Precambrian areas and in the younger eroded mountain ranges where they usually represent metamorphic equivalents of aluminum-rich sediments (clays and argillites). However, tuffs and certain acid igneous rocks (rhyolites, granites) may form part of the primary constituents. Contrary to earlier belief, metasomatic alterations are insignificant; the chemistry is conservative and reflects rather well the original composition.

[T. F. W. BARTH]

Microcline

The name for a triclinic potassium-rich feldspar. The microcline structure is built up of rather pure $KAlSi_3O_8$. Most of any $NaAlSi_3O_8$ found in the bulk composition of microcline crystals is present as exsolved albite (frequently up to 25 mole %) in the form of microcline-perthite. Microcline is usually twinned with typical crosshatching, indicating that it usually grew (stably or metastably) as sanidine (monoclinic) and subsequently changed during geological time into the triclinic modification by passing through a state called normal orthoclase (see illustration). It can invert into sanidine under prolonged heat treatment at high temperature; however, all attempts to effect the change from sanidine to microcline in the laboratory or to synthesize microcline have been unsuccessful. This is explained by the fact that microcline has an ordered Al/Si distribution in contrast to the disordered one in sanidine, and that the ordering process is a very sluggish one at low temperature. Microcline is typically found in granites and pegmatites. Green microcline is known as amazon stone or amazonite. *See* FELDSPAR; GEM; IGNEOUS ROCKS; PERTHITE.

[FRITZ H. LAVES]

Migmatite

Rocks originally defined as of hybrid character due to intimate mixing of older rocks (schist and gneiss) with granitic magma. Now most plutonic rocks of mixed appearance, regardless of how the granitic phase formed, are called migmatites. Commonly they appear as veined gneisses.

Several modes of origin have been proposed. (1) Granitic magma may be intercalated between thin layers of schist (lit-par-lit injection) to form a banded rock called injection gneiss. (2) The granitic magma may form in place by selective melting of the rock components. (3) The granitic layers may develop by metamorphic differentiation (redistribution of minerals in solid rock by recrystallization). (4) The granitic layers may represent selectively replaced or metasomatized portions of the rock.

Veined gneisses include two genetic types: arterites, in which the vein material was injected, and veinites, in which the vein material was secreted from the rock itself. There are many other types of migmatites. Some consist of blocklike masses of various shapes enclosed in granitic rock and resembling fragments of a breccia.

Migmatites are found in zones marginal to intrusive granite and in deep zones of ultrametamorphism. *See* GRANITIZATION; METAMORPHISM; METASOMATISM. [CARLETON A. CHAPMAN]

Mineral

A naturally occurring substance with a characteristic chemical composition expressed by a chemical formula. Although organic substances such as coal and oil are usually listed under mineral resources, they are not minerals, being complex mixtures without definite chemical formulas. Minerals constitute an extremely important natural resource. Most metals and inorganic chemicals, and many other products essential to civilization, are derived from minerals. Both forests and farms are dependent upon soils, which are composed chiefly of minerals.

Occurrence. Minerals may occur as individual crystals or they may be disseminated in some other mineral or rock. Most minerals at the face of the earth occur in rocks. Minerals may be attached to the surface of some opening. Geodes are cavities in a rock which are lined with a mineral, or in some cases, completely filled. Quartz geodes in limestone are common. They sometimes form well-developed crystals, or they may form fine-grained layers called agate or onyx. The term vug applies to irregular openings containing ore minerals. Veins are fissures of varying size which have been filled with one or more minerals. In some cases veins are banded, indicating a sequence of deposition. The term lode is used for a vein or group of veins in a definite area. Gangue minerals are the worthless minerals associated with a valuable mineral or ore. The term ore is usually restricted to minerals from which a metal is obtained. A placer is a concentration of relatively heavy and durable minerals which have been transported and redeposited in a stream bed where the water velocity is lowered. Beach sands may contain a concentration of certain heavy minerals. *See* ORE AND MINERAL DEPOSITS.

Names. Minerals usually have both a chemical name and a mineral name. Thus lead sulfide occurring in nature is called galena, and sodium chloride is called halite. Some old mineral names are of unknown or uncertain origin, while many come from the Latin, as orpiment (*auri pigmentum*) or from the Greek, as chalcocite (*chalcos*, copper or brass).

Most modern names end in "ite." Some have a chemical connotation, as molybdenite, MoS_2, and zincite, ZnO. A crystallographic derivation is illustrated by tetrahedrite and hemimorphite; geographic by labradorite and vesuvianite. Physical properties are the basis for the names magnetite, graphite (to write), rhodonite (rose color), cryolite (ice stone), and azurite. Many minerals have been named after individuals, as scheelite, smithsonite and goethite. Some minerals have variety names, as amethyst, agate, and jasper for quartz. There are also colloquial names, such as fool's gold for pyrite, heavy spar for barite, and tin stone for cassiterite.

Classification. Minerals are classified first with respect to chemical composition, and then so far as possible by isomorphism, or similarity of crystalline form. In general the cation is of less significance than the anion. Thus the iron minerals pyrrhotite, FeS; pyrite, FeS_2; hematite, Fe_2O_3; magnetite, $FeFe_2O_4$; and siderite, $FeCO_3$, are not grouped together, in spite of the common cation Fe. $FeCO_3$ is classified with the hexagonal carbonate group including calcite, $CaCO_3$, and rhodochrosite, $MnCO_3$. Fe_2O_3 is grouped with the isomorphous corundum, Al_2O_3, and $FeFe_2O_4$ with spinel, $MgAl_2O_4$.

The major groups in mineral classification are:

Native elements	Sulfate type
Sulfide and sulfo minerals	Chromates
Oxides and hydrated oxides	Molybdates
Halogen minerals	Tungstates
Nitrates	Phosphate type
Carbonates	Arsenates
Borates	Vanadates
	Silicates

See BORATE MINERALS; CARBONATE MINERALS;

HALOGEN MINERALS; NATIVE ELEMENTS; NITRATE MINERALS; SILICATE MINERALS.

Formation. Minerals may be formed by four general processes: (1) from a gas by sublimation, (2) from a liquid (aqueous solution), (3) from a liquid (molten rock or magma), and (4) from a solid by metamorphism. In nature these processes may be intimately related.

Sublimation. This process is comparatively rare. In volcanic eruptions minor amounts of certain minerals may form from gases escaping from vents or fumaroles. Examples are salammoniac, NH_4Cl, sulfur, S, and boric acid, H_3BO_3. Scales of hematite may form by the reaction in Eq. (1). The only

$$\underset{\text{Vapor}}{FeCl_3} + \underset{\text{Steam}}{H_2O} \rightarrow Fe_2O_3 + HCl \qquad (1)$$

common example of sublimation is unrelated to volcanoes. It is the formation of snow flakes from water vapor.

Aqueous solution. Aqueous solutions are the source of important minerals. Dissolved material may be precipitated from solution by several processes: (1) evaporation of the solvent, (2) decrease in temperature and pressure, (3) loss of carbon dioxide, CO_2, or (4) action of organisms.

1. The ocean is a vast reservoir of dissolved materials. During various geologic periods, portions of oceans were cut off, and subsequent evaporation caused deposition of enormous quantities of the dissolved material. The most abundant of these evaporites are halite (NaCl) and gypsum ($CaSO_4{\cdot}2H_2O$). In some cases evaporation proceeded further, and the more soluble minerals magnesium, Mg, and potassium, K, were also precipitated. In desert regions where occasional rain produces temporary lakes, playa deposits are formed in which, in addition to halite and gypsum, sodium sulfate and carbonate and various borates occur. *See* EVAPORITE, SALINE.

2. In regions of hot springs and geysers, the hot water under pressure dissolves material from the underlying rocks and brings it to the surface. In Yellowstone Park the geysers bring up siliceous sinter (opal), and the hot springs deposit travertine (calcite). *See* SILICEOUS SINTER; TRAVERTINE.

3. Loss of CO_2. This process is mostly limited to a single mineral, but has occurred on a large scale. Calcium carbonate, $CaCO_3$, is relatively insoluble in water, but is slightly soluble if dissolved CO_2 is present. The reaction, Eq. (2), produces

$$CaCO_3 + H_2O + CO_2 \rightleftharpoons CaH_2(CO_3)_2 \qquad (2)$$

$CaH_2(CO_3)_2$, which is soluble. The many caves in limestones all over the world are formed this way. However, this reaction is reversible. Upon loss of the CO_2, $CaCO_3$ is redeposited in the form of stalactites, stalagmites, and other formations. Calcareous tufa may be deposited around springs and streams by this same process. *See* STALACTITES AND STALAGMITES; TUFA.

4. Ocean water does not normally contain enough dissolved $CaCO_3$ or silicon dioxide, SiO_2, to allow much direct precipitation of these substances. But various living organisms are able to extract them from the water and then secrete them to form the hard parts of their bodies. Corals, crinoids, mollusks, and foraminiferans secrete $CaCO_3$, while diatoms, sponges, and radiolarians secrete $SiO_2{\cdot}xH_2O$. Large deposits of limestone, chalk, and diatomaceous earth have been formed this way.

Crystallization from magma. At moderate depths in the Earth's crust, previously existing rocks may be melted. This molten rock (magma) tends to work its way upward. If it reaches the surface there will be volcanic eruptions or lava flows. This quickly cooled magma forms either glassy or very fine-grained rocks (extrusive) which are of minor mineralogical interest. However, if the magma does not reach the surface, cooling takes place very slowly, allowing time for a complex series of reactions to occur.

In an average magma the major chemical constituents are oxides of silicon, SiO_2; aluminum, Al_2O_3; iron, Fe_2O_3 and FeO; calcium, CaO; magnesium, MgO; sodium, Na_2O; and potassium, K_2O. The most abundant is SiO_2, or silica. In addition, there are substances which would be gases, except for the high pressure, such as water, chlorine, fluorine, carbon dioxide, boron, and sulfur compounds. These increase the fluidity of the magma. Moreover, they do not enter appreciably into the earlier minerals forming from the magma, and are finally concentrated in a phase which is no longer magma, but rather a hot, highly concentrated aqueous solution, called magmatic water. The general order of crystallization from the magma is as follows: (1) basic minerals (low in silica), such as olivine and basic plagioclase, together with nonsilicate minerals, which are oxides and sulfides of metals such as iron, copper, nickel, chromium, platinum, titanium, and diamond; (2) intermediate minerals (medium silica); and (3) acid minerals (high in silica). The igneous rock called granite is the last to crystallize from the magma. *See* GRANITIZATION; MAGMA.

This progressive change in the kind of minerals crystallizing is called magmatic differentiation. Since the magma tends to move upward, these successive stages may be separated in space.

The residual magmatic liquid is still very hot and concentrated, and under high pressure, but is much less viscous than the magma. It forms pegmatite dikes, frequently as offshoots from the main magmatic mass. It may contain large masses of quartz, big crystals of microcline, and sheets of mica. There may be minerals containing relatively rare elements, such as beryl (beryllium), spodumene (lithium), uraninite (uranium), wolframite (tungsten), and columbite (columbium, or niobium).

The final stage of magmatic differentiation is the hydrothermal stage, with high, medium, and low temperature phases. These occur in veins, usually somewhat removed from the main magmatic mass. Many important minerals occur in these veins, including ores of tin, tungsten, molybdenum, gold, copper, zinc, mercury, and antimony, as well as pyrite, marcasite, quartz, calcite, fluorite, barite, and opal.

Contact metamorphism. Adjacent to the ascending magma the country rock may be profoundly changed by the heat and by chemical reaction with the magma. This is known as local or contact metamorphism. Many new minerals are formed, especially when the magma intrudes into impure limestones. *See* AUREOLE, CONTACT; METAMORPHISM.

Regional or dynamic metamorphism. Heat and pressure associated with mountain-forming processes, and aided by any water present in the rocks, may cause recrystallization of existing minerals or formation of new minerals. Examples of recrystallization are the change of limestone to marble, and a sandstone with siliceous cement to quartzite. Schists may be formed, consisting of minerals with a platy or prismatic habit, such as mica, talc, chlorite, and hornblende. *See* METAMORPHIC ROCKS.

Alteration. As soon as minerals are formed they are subject to change as chemical processes act upon them. These changes may take place rapidly or they may proceed very slowly.

Weathering. Weathering may be important in the forming of new minerals, as well as in concentrating them. Deposits of carbonates of copper, zinc, and lead have been formed from primary ores by ground water containing CO_2. Weathering in some tropical regions has produced important deposits of iron, aluminum, and manganese ores. With alternating wet and dry seasons, a special type of leaching occurs. In quite different rocks, practically all constituents may be dissolved except iron and aluminum oxides. The laterite deposits of India are iron ores, while bauxite deposits formed by this process are the chief source of aluminum. Some deposits of manganese oxides have formed this way. Secondary enrichment of certain copper ores has occurred when veins of lean ore have been weathered, with the copper minerals being dissolved and reprecipitated at lower levels. As the surface was gradually lowered by erosion, this process continued until rich deposits were formed. *See* BAUXITE; LATERITE.

Usually weathering is a destructive force. It is divided into physical and chemical weathering. The first involves changes in temperature, with alternate expansion and contraction; the expansion of water freezing in crevices; the abrading action of rock particles carried by wind, water, and ice. These tend to break down rocks and minerals into finer particles. Chemical weathering involves solution, oxidation, reduction, hydration, dehydration, and carbonation. *See* WEATHERING PROCESSES.

Pseudomorphs. A special type of weathering may result in pseudomorphs, or false forms. A crystal may be chemically altered, but retain its original shape. Thus a cube of pyrite, FeS_2, may alter to fine-grained goethite, $HFeO_2$, but retain the original cube shape. This is called a pseudomorph of goethite after pyrite, and is an alteration pseudomorph. There may be a complete change in composition, as quartz after fluorite, giving a subsitution pseudomorph, In polymorphous substances, crystals of one form may alter to another, but retain the original crystal shape, as rutile, TiO_2, after brookite, TiO_2. These are called paramorphs. In a somewhat different category are crystals of one mineral which have become coated with another. These are called encrustation pseudomorphs.

Metasomatism. Metasomatism is a special type of mineral transformation. For example, in hydrothermal veins where galena has been deposited, galena has been found partly replacing the limestone wall rock, with the original rock texture preserved. *See* METASOMATISM.

Deuteric alteration. This occurs chiefly in pegmatites, where earlier-formed minerals may react with later pegmatite fluids to form new minerals. *See* PEGMATITE.

Groups. Any complete classification of minerals must be based on chemical composition. However, limited groups of minerals may be of interest for various special purposes. Such groups may be based on origin, type of occurrence, certain physical properties, or use.

Primary minerals are those formed from the magma, including the pegmatite and hydrothermal phases. Others are secondary. Rock-forming minerals are those which make up the great bulk of the igneous, sedimentary, and metamorphic rocks. Thus quartz, feldspar, and mica are found in granite and are rock-forming minerals. They are also called essential minerals, in contrast to the occasional tiny crystals of pyrite, zircon, or apatite which might occur in granite, and which are called accessory minerals.

Classes of minerals are those which make up a certain chemical group, as carbonates, sulfates, or oxides.

Isomorphous groups are those whose members are strictly isomorphous, as the garnet group.

Families of minerals include certain closely related minerals which have close chemical and physical similarities, but are not necessarily isomorphous, as the feldspars or pyroxenes.

Economic minerals are those which are of economic importance, and include both metallic (ore minerals) and nonmetallic minerals, as cryolite and sulfur, and gem minerals. *See* GEM.

Clay minerals have certain common physical and chemical properties. They are fine-grained, plastic when wet, and become hard when dried or fired. They are chiefly hydrous silicates of aluminum. *See* CLAY MINERALS.

Stable minerals are those resistant to both chemical and mechanical weathering, being both insoluble and hard. Heavy minerals are those which collect in placers and beach sands because of their higher specific gravities, or those which can be separated in the laboratory by gravity methods. Detrital minerals are rock fragments which are essentially unaltered. Authigenic minerals are those generated in the place of formation, and allogenic minerals are those which have been transported. *See* AUTHIGENIC MINERALS.

Mineral associations refer to minerals formed by the same process and hence closely associated. Mineral sequence refers to a series of associated minerals formed at successive stages.

A mineral suite is a general term which may apply to a group of associated minerals in one deposit; a representative group from a certain locality; or a group of specimens showing variations, as in color or form, in a single mineral species. *See* MINERALOGY. [LEWIS S. RAMSDELL]

Mineralogy

The science which concerns the study of natural inorganic substances, whether of terrestrial or extraterrestrial origin, called minerals. Mineralogy has so dramatically changed within the past few decades that the science can no longer be easily

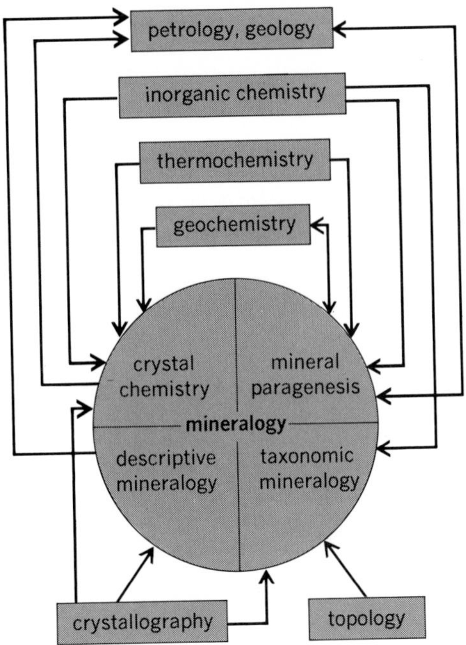

Fig. 1. Diagram showing the transmission of information between mineralogy and some other sciences. Arrows imply the direction of information.

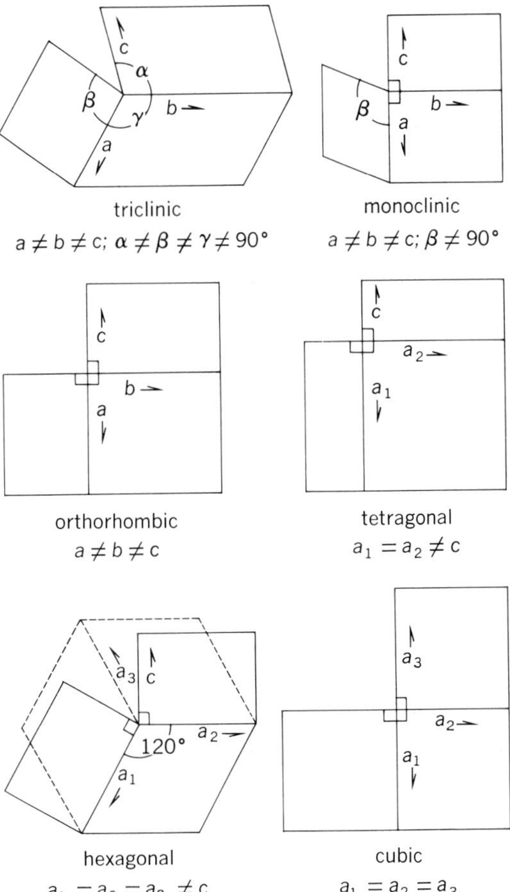

Fig. 2. The cell shapes of the six crystal systems shown as their principal projections.

defined. It is most properly a branch of inorganic chemistry, but the discipline concentrates on the origin, description, and classification of minerals.

Thus, four main categories may be considered: crystal chemistry (composition and atomic arrangement of minerals), paragenetic mineralogy (the study of mineral association and occurrence both in natural and synthetic systems), descriptive mineralogy (the study of the physical properties of minerals and the means for their identification), and taxonomic mineralogy (mineral classification, systematization, and nomenclature). *See* MINERAL.

Crystal chemistry. This is the most vital aspect of mineralogy because it is the basis for the other studies. The fields of mineralogy, crystallography, inorganic chemistry, geochemistry, petrology, and geology are connected in the domain of crystal chemistry (Fig. 1).

A particular mineral, called a mineral species, is defined on the basis of a specified chemical composition and a specified crystal structure (atomic arrangement). These two criteria provide almost sufficient knowledge for characterization of a mineral since in principle all other properties can be derived from them. Since minerals are nearly always crystalline inorganic substances, a mineralogist seeks to know a mineral's crystal chemistry.

Crystallography and crystal structure. This part of mineralogy is closely allied to inorganic chemistry and certain aspects of solid-state physics and draws heavily from these fields. Crystallography once fell within the realm of mineralogy but is now a science more closely related to solid-state physics.

By application of group theory it can be shown that crystalline substances, that is, periodic arrangements of matter in three dimensions, can be divided into six crystal systems: triclinic, monoclinic, orthorhombic, tetragonal, hexagonal (including the trigonal and rhombohedral subdivisions), and cubic. These crystal systems can be considered as structure cells, each of which contains a certain integral number of atoms of a substance. Figure 2 depicts these systems and offers the criteria for distinguishing them. Furthermore, there are 32 crystal classes among these six systems, also derived by strict mathematical arguments. Finally, there are 230 distinct ways of periodically arranging objects in space; these are the space groups. *See* CRYSTALLOGRAPHY.

If the arrangement of atoms which make up a crystalline substance is known, it is said that the crystal structure of that substance is known. A particular crystal structure has a characteristic collection of symmetry elements in three dimensions, hence it must belong to a particular space group. A unique structural arrangement, regardless of the atomic elements present, is called a structure type.

Knowledge of a mineral's crystal structure is obtained by standard crystallographic techniques, in particular, x-ray diffraction analysis. This very important aspect of mineralogy allows for classification of minerals and partial explanation of a mineral's physical properties. Often, it is used because minerals tend to be highly complicated inorganic substances which cannot be understood even if a complete chemical analysis is available.

Chemical composition. Any crystalline substance has a certain integral number of its essential chemical formula units (loosely called molecules) in its structure cell. Thus, each unique atomic position in the structural unit can be occupied by a particular kind (or kinds) of atom(s). The word unique is used because a particular position may occur more than once in a unit cell, the number of occurrences stipulated by the symmetry of the space group. Consider now the ideal cell formula $(A_a)(B_b) \cdot \cdot \cdot (P_p)$. Each of the parentheses specifies a unique atomic position. The capital letters specify the element present and the small subscripts the number of times it occurs in the cell. If the small subscripts have a factor in common, it is factored out and what remains is the formula unit. The ideal formula is further defined in terms of the atomic element which occurs in excess of 50 mole % within each of the parentheses. The structure type, along with the ideal formula unit, defines a mineral species.

This strict definition of a species is required since a particular mineral may have a range of compositions. The range of compositions is called a series. The ideal limiting compositions are called end members and each of the end members has a specific name.

This concept can be applied to the olivine series of minerals. The olivine structure type has the formula $M(1)M(2)TO_4$, where M(1) and M(2) are the two crystallographically unique octahedral centers or sites, T is the unique tetrahedral center, and O is the anionic hexagonal close-packed framework. The voids left after packing the O atoms as closely as possible are octahedral (with six nearest neighbors) and tetrahedral (with four nearest neighbors), and these voids can accommodate cations.

The olivine series has two end members: fayalite ($FeFeSiO_4$ or Fe_2SiO_4) and forsterite (Mg_2SiO_4). The species fayalite is ferrous orthosilicate with the olivine structure type and the species forsterite is magnesium orthosilicate, also with the olivine structure type. The olivine series is simply written $(Fe,Mg)_2SiO_4$ if Fe is in excess of Mg molewise, or $(Mg,Fe)_2SiO_4$ if Mg is in excess of Fe. If Fe and Mg mix in all proportions over the two distinct octahedral positions, the formula is written as the series $(Fe,Mg)_2SiO_4$ or $(Mg,Fe)_2SiO_4$ and the octahedral cations are said to be disordered. If Fe prefers one octahedral site and Mg the other, then at 50 mole % Fe and Mg, the formula is written $FeMgSiO_4$ and the octahedral cations are said to be ordered. At specified temperatures and pressures, the olivine crystal may be disordered, ordered, or even partly ordered (limited mixing). Studies on order-disorder in crystals are very important for a mineralogist's understanding of the substances which make up the Earth. These studies are pursued by x-ray diffraction analysis and mineral synthesis, guided by an understanding of chemical thermodynamics.

For a particular structure type, different kinds of atomic elements may participate in the structure. Some, like Fe and Mg in olivine, form an isomorphous (solid-solution) series; others do not. The latter are called isotypes of each other. Thus, $CaMgSiO_4$ (monticellite) and $LiFePO_4$ (triphylite) are isotypes. Their structure type is that of the ordered olivine.

Studies of the thermochemical properties of a mineral and the crystal structure of a mineral go hand in hand. Mineral synthesis, aided by thermodynamic arguments, is usually considered in terms of a phase diagram. A phase diagram shows the regions in temperature, pressure, and composition of components (and perhaps even oxygen fugacities, water pressure, and so forth) where a mineral is stable and assists in defining its composition more precisely. Regions, or fields, of coexisting minerals are of great value in understanding mineral paragenesis and petrological problems. Knowledge of the genesis of rocks and rock types is greatly aided by this kind of study, particularly in silicate systems.

A particular chemical composition may be represented by more than one structure type, depending on thermochemical criteria. Thus, forsterite at a suitably high pressure becomes the spinel structure type. Both have composition Mg_2SiO_4. Whereas the former is based on hexagonal close-packed oxygen atoms, the latter is based on cubic close-packed oxygen atoms. Minerals with one composition but of different crystal structures are distinct species, called polymorphs of each other.

The chemical composition of a mineral is obtained in many ways. First a qualitative analysis is usually performed to determine the elements present. Then a good approximation to the exact atomic ratios is obtained by a quantitative analysis. Formerly, "wet" chemical analysis, which requires destruction of the sample, was exclusively used. Modern techniques, such as x-ray fluorescence analysis, emission spectrographic analysis, and electron microprobe analysis, are now playing a more significant role. The last technique is the most efficient since it requires only small grains (hence, higher sample purity) and no sample destruction. It is capable of qualitatively and quantitatively analyzing all elements from sodium to uranium. Other highly sophisticated techniques are coming into use. These include Mössbauer resonance (for valence state and site preference determination), nuclear magnetic resonance (for site preference determination), and infrared absorption (for determining the structural state of water molecules) techniques.

Finally, the crystal chemistry of a mineral is obtained by a crystallochemical calculation, combining both chemical and structural data. An example of this computation is given in the table for an olivine mineral which crystallizes in the orthorhombic system, space group *Pbnm*. Measured values for this mineral are: density $(\rho) = 4.07$ g/cm³; $a = 4.79$ A; $b = 10.33$ A; $c = 6.06$ A. This gives volume $V = abc = 299.85$ A³; and molecular weight in cell $= \rho V/1.66 = 735.17$. As the table shows this olivine has four units of $(Fe_{0.68}, Mg_{0.32})_2[SiO_4]$ in the unit cell.

Computation for an olivine mineral

Compound	Wt, %	Molecular weight in cell	Moles in cell		Ideal
SiO_2	32.73	240.45		4.00	4.00
FeO	53.22	390.99	5.44	8.01	8.00
MgO	14.12	103.73	2.57		
	100.07	735.17			

Paragenetic mineralogy. Paragenetic mineralogy is the study of mineral paragenesis, or the association and order of crystallization of minerals. The problem may concern mineral association within a single hand specimen or may embrace a much larger region, such as an entire ore body, in which case many representative specimens are judiciously collected. This study usually accompanies the analysis of the general geological structures within and around the ore body, such as the bedding, folding, and faulting. *See* GEOLOGY; PETROLOGY.

Included among the important aspects of paragenetic mineralogy are ore mineralogy, the mineralogy of a sequence of phases crystallized from a parent magma, the sequence of minerals crystallized in a vein, and so forth. In every instance the order of mineral crystallization is determined and the minerals are usually listed from oldest to youngest. Establishing the correct sequence is not always simple; many minerals after crystallizing break down into other phases (species) and knowledge of the systems involving these phases is important. Thus, mineral paragenesis is sometimes established by the study of synthetic systems.

Textural evidence is very important but often may be misleading: The relative crystal size of two associated minerals does not necessarily mean that the larger crystallized first. Often two compounds crystallizing at the same time may have different rates of growth. But if a euhedral (all faces) crystal of mineral A is wholly embedded in subhedral (some faces) or anhedral (no faces) crystals of B, it can be stated that A crystallized earlier, even if B may have a larger grain size.

Mineral paragenesis is usually considered in relative time and the absolute difference in time between the oldest and youngest minerals is often not known. Absolute age differences can be obtained in some instances, for example, by lead isotope age dating of a sequence of crystallized lead-bearing minerals. Even without this knowledge, mineral paragenesis can greatly assist in understanding the geochemistry of cations and anions over a range of temperatures, pressures, and so forth. Often the parent fluid from which the mineral crystallized can be more fully understood and possibly even synthesized under laboratory conditions to the point of duplicating the sequence of crystallized minerals. For this reason, paragenetic mineralogy is intimately related to crystal chemistry. *See* ROCK, AGE DETERMINATION OF.

Descriptive mineralogy. Mineral recognition directly by the senses is very subjective and requires considerable experience. Gross features of a mineral such as color, form, hardness, and specific gravity are important criteria for identification in the field where a well-equipped laboratory is usually not available. More objective criteria such as the optical properties and x-ray powder diffraction spectra of a mineral require specialized equipment, but the results are usually certain since these data are known for most mineral species and are extensively tabulated. Formerly, such methods as fusibility, flame tests, and blowpipe analysis were extensively used but have been largely abandoned because they are unreliable. A good mineral description gives all the properties which can be used to identify a mineral. Two categories are considered: the crystallographic and the physical properties. Data for the former are obtained directly from a crystallochemical investigation, discussed earlier. For the latter, no single criterion can completely ensure the correct identity of a mineral. Nor can any tabulation suffice since these properties are highly subjective and may vary from individual to individual.

Crystallographic properties. Crystal morphology, or crystal form, is an outward expression of the internal atomic arrangement. A crystal is a convex polyhedron and the planes bounding its surface are called faces. Analysis of the symmetry relations of the faces enables a mineralogist to determine its crystal class. Since each mineral species by definition belongs to only one space group, it must also belong to only one crystal class. Furthermore, the relative size of the faces and the shape of the crystal may vary considerably, even within one crystal class. The crystal's appearance is called its development and can be an important criterion for mineral identification.

A crystal's morphology is usually obtained from a two-circle reflecting goniometer, an instrument which has a fixed light source, a fixed telescope to receive the reflected light signal from a crystal face, and a coordinate system of two rotating axes upon which the crystal is mounted. Coordinates are obtained for each crystal face and can be used to reconstruct the crystal as a model or to display the crystal as a series of projections on paper. By examining several crystals and averaging the data, the crystal class can also be obtained. Since a particular mineral species has a unique structure cell and since crystal faces have rational plane intercepts with that structure cell, different crystals of a particular species must have the same interfacial angles for any given set of faces. These interfacial angles are often used for mineral identification.

Powder x-ray diffraction. The diffraction pattern of any crystalline substance depends on the size and shape of the unit cell, the space group, the atomic coordinates, and the kinds of atoms present. Since each mineral species is unique by definition, each species must also yield a characteristic diffraction pattern. The diffraction pattern can be used to identify a mineral. The powder method is probably the most reliable and efficient method for mineral identification.

A pure fragment of a mineral is ground into a powder and then formed into a rod by gluing it to a thin glass fiber with some amorphous adhesive. The sample is then rotated continuously in the center of a cylindrical camera which has x-ray film secured against the walls of the cylinder. Sometimes a spherical sample of the powder is used, mitigating preferred orientation of the grains. X-rays are then allowed to impinge on the sample and in time the diffraction pattern is recorded on film. The pattern can then be interpreted in terms of interplanar spacings which are based on the structure cell. Each diffraction cone has a particular spacing and a particular relative intensity. Spacing and intensity (darkness of film) are then recorded, usually in the order of decreasing interplanar spacing. The powder diffraction patterns of most minerals are known and are on file with the American Society for Testing and Materials.

Other techniques for obtaining diffraction patterns are in use. An x-ray powder diffractometer gives similar results, but a Geiger counter or scintillation counter is used instead of film and the diffraction spectra are recorded on a chart. The diffractometer has become more popular than the powder camera. Perhaps the most powerful methods are the single-crystal methods, in which diffraction patterns are obtained on film using a single crystal (usually about 0.1 mm in dimension) instead of a powder. From these patterns, the structure cell can be obtained. This method requires special cameras and involves a detailed and often difficult study. For this reason, structure cell data for mineral species are not as extensively known and tabulated as are the powder data.

Optical properties. Optical properties of minerals are obtained by examining small fragments in transmitted plane-polarized light for nonopaque minerals and in reflected plane-polarized light for opaque minerals. These properties are related to the crystal structure, the kinds of atoms present, and their electronic arrangement. Amorphous (noncrystalline) and cubic substances are isotropic; that is, they have only one index of refraction for given monochromatic radiation. Hexagonal and tetragonal minerals are uniaxial and are doubly refracting; orthorhombic, monoclinic, and triclinic minerals are biaxial and are also doubly refracting. The terms isotropic, uniaxial, and biaxial refer to the mathematical description for the range of refractive indices of a crystal, often described in terms of an ellipsoid called the optical indicatrix. Thus, isotropic substances are referred to a spheroid, uniaxial substances to an ellipsoid with one circular section and one axis normal to that section, and biaxial substances to an ellipsoid with two circular sections and two axes normal to these sections. Isotropic substances have one refractive index, usually written n, for any given monochromatic light radiation and this value is the radius of the spheroid. The uniaxial indicatrix can be defined by two indices, ω (omega), which is in the plane of the circular section, and ϵ (epsilon), which is normal to this plane. The biaxial indicatrix has three indices, usually written α (alpha), β (beta), and γ (gamma), which are the three orthogonal axes of the ellipsoid, and $2V$, the angle between the optic axes. The optical properties are related to crystallographic properties since the indicatrix is oriented according to the point symmetry of the crystal.

A mineralogist seeks these values for a mineral by use of optical instruments such as the petrographic microscope, spindle stage, and universal stage. Optical properties of most minerals are known and are extensively tabulated. In certain instances, optical data can be used to assess chemical composition of a mineral series, but standardized curves are needed in advance. *See* PETROGRAPHY.

Physical properties. The physical properties of a mineral are those features which aid in identifying it by sight, touch, heft, fabric, and so forth. These properties result from the crystal structure, the atomic composition, and the kinds of chemical bonds present. Thus, most properties of a mineral can be predicted if knowledge of its internal arrangement is available.

Cleavage is the tendency of a mineral to preferentially break along particular crystallographic planes which, like the crystal faces, are rational. Cleavage is easiest along the plane of weakest chemical bonds in the structure. If all bonds are of the same kind in three dimensions (such as in framework silicates), then the plane with the least bonds per unit area is the preferred cleavage plane. Cleavages are classified according to their degree of perfection: perfect, good, distinct, poor, or no cleavage. Mica, for example, has one perfect cleavage direction. Its atomic arrangement consists of sheets of silicate units. The bonds within the sheets are very strong and those between the sheets are relatively weak, so the perfect cleavage plane is parallel to the plane of the silicate sheets. Fracture, a gross surface feature, is the way a mineral breaks other than cleavage. Typical terms are even, splintery, hackly, conchoidal (shell-like), rough, and smooth.

A mineral's hardness is related to the kinds of chemical bonds present. Substances with strong chemical bonds are harder than those with weaker bonds. Diamond is harder than its polymorph graphite. Though both consist of carbon-carbon bonds, the carbon-carbon bond in diamond is stronger. Bromellite (BeO) and zincite (ZnO) both belong to the same structure type but bromellite is much harder because the Be-O bonds is stronger than the Zn-O bond. Absolute hardness can be measured by a microhardness tester which places an indentation on the surface of a crystal beyond a certain pressure value. This value is in terms of force per unit area. Mineralogists often use a relative scale, or Mohs scale, based on 10 minerals in order of increasing hardness: talc (1), gypsum (2), calcite (3), fluorite (4), apatite (5), potash feldspar (6), quartz (7), topaz (8), corundum (9), and diamond (10). Thus, a mineral of hardness 4.5 will scratch fluorite but not apatite. Alternatively, the mineral will be scratched by apatite but not fluorite.

Color is more difficult to assess. Some minerals may occur in many different colors but usually the pure end member has a particular color. Impurities often color minerals. Thus, a little chromium in the pyroxene mineral diopside (CaMg $[Si_2O_6]$) colors it bright green, whereas pure diopside is colorless. Chromium is said to be a chromophore. Careful description of a mineral's color may be obtained by the use of standard color charts. Streak is the color of finely divided mineral grains and is usually obtained by pressing the mineral across an unglazed porcelain plate.

Luster is a rather subjective quality. It is the appearance of the mineral surface in reflected light and is related to the index of refraction in a direct way. Luster increases with increasing index of refraction. Two important categories are metallic and nonmetallic. Metals like silver (Ag) and many sulfides like galena (PbS), stibnite (Sb_2S_3), and pyrite (FeS_2) are metallic in appearance. Nonmetallic lusters may be described as vitreous (quartz, SiO_2), resinous (sphalerite, ZnS), or adamantine (diamond, C). The physical character of the crystals also influences luster. Fibrous minerals may have a silky luster and platy minerals may be pearly in appearance.

Density is a very important property and plays a significant role in crystallochemical calculations

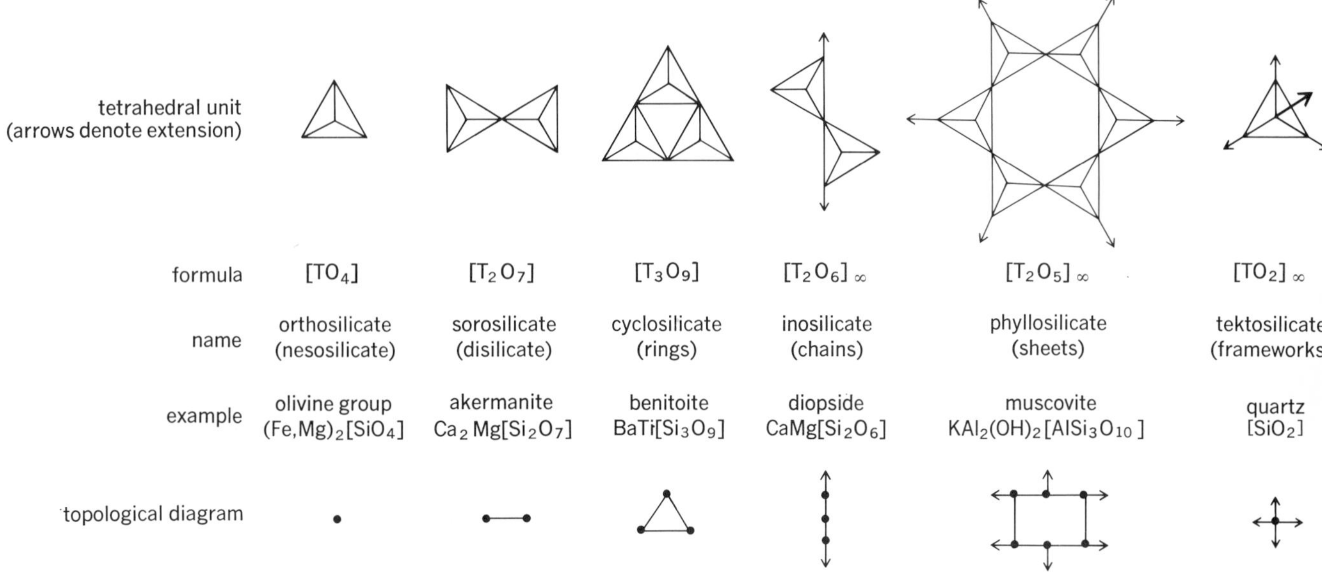

tetrahedral unit
(arrows denote extension)
formula
$[TO_4]$
$[T_2O_7]$
$[T_3O_9]$
$[T_2O_6]_\infty$
$[T_2O_5]_\infty$
$[TO_2]_\infty$
name
orthosilicate
(nesosilicate)
sorosilicate
(disilicate)
cyclosilicate
(rings)
inosilicate
(chains)
phyllosilicate
(sheets)
tektosilicate
(frameworks)
example
olivine group
$(Fe,Mg)_2[SiO_4]$
akermanite
$Ca_2Mg[Si_2O_7]$
benitoite
$BaTi[Si_3O_9]$
diopside
$CaMg[Si_2O_6]$
muscovite
$KAl_2(OH)_2[AlSi_3O_{10}]$
quartz
$[SiO_2]$
topological diagram

Fig. 3. Classification of the silicates on the basis of tetrahedral polymers.

since it is related to the structure cell and the cell formula. It is in units of grams per cubic centimeter. Specific gravity is the ratio of the weight of a substance to the weight of an equal volume of water at 4°C. Many instruments are available for determining specific gravity, including the pycnometer, the Berman microbalance, and the Jolly balance.

Taxonomic mineralogy. There are approximately 3000 distinct mineral species known to science. Over 20,000 names exist in the literature but most of these are varieties or incompletely studied mixtures of well-known mineral species. About 60 new mineral species are discovered each year. For a new species to be properly defined, the chemical analysis, structure cell and space group, crystal morphology, powder pattern, optical data, all physical properties, and paragenesis must be given as completely as possible and this study may involve months of labor. Most names end in "ite"; the name is usually based on the locality, some mineralogist, or possibly some unusual physical property. The basing of mineral names on some acronym of the chemical composition is becoming more frequent.

There are many ways to classify minerals. One way could be a paragenetic classification which may include the minerals crystallized from oldest to youngest. For example, an ore body may consist of primary ore minerals, skarn (reaction) products, replacement minerals, fissure and hydrothermal vein minerals, and finally, surface oxidation products. Displays of specimens from extraordinary localities, such as Franklin, N.J. are often shown this way.

Most mineralogy reference books favor a crystallochemical classification. Thus, a mineral is classified under its anionic group. A typical scheme is: elements; sulfides and sulfosalts; oxides; carbonates; sulfates; phosphates and arsenates; borates; and silicates, in order of increasing complexity.

Another approach is gaining some support, but the mathematical formalism is as yet incomplete. This is the topological classification, which concerns the hierarchy of structure types and is influenced by that branch of mathematics known as topology. An example is furnished in Fig. 3 for the classification of the silicate polymeric units. These units may be even considered more abstractly, for instance, as branches on a topological tree. This kind of classification has the advantage of placing known structure types within the realm of all geometrically possible arrangements, determined beforehand by mathematical arguments.

[PAUL B. MOORE]

Bibliography: L. G. Berry and B. Mason, *Mineralogy*, 1959; W. A. Deer, R. A. Howie, and J. Zussman, *Rock-forming Minerals*, 5 vols., 1962; I. Kostov, *Mineralogy*, 1968; C. Palache, H. Berman, and C. Frondel, *Dana's System of Mineralogy*, 2 vols., 7th ed., 1944–1951; A. F. Wells, Topological approach to structural inorganic chemistry, *Chemistry*, 40:12–18, 22–27, 1967; J. Zussman, *Physical Methods in Determinative Mineralogy*, 1967.

Miocene

The next to the youngest of the five major worldwide divisions (epochs) of the Tertiary Period (Cenozoic Era); the epoch of geologic time extending from the end of the Oligocene to the beginning of the Pliocene; and the rocks (series) formed during this epoch and the fossils therein. *See* CENOZOIC; OLIGOCENE; PLIOCENE; TERTIARY.

Miocene time embraces two widespread reexpansions of the sea, separated by an interval of localized mountain making and shiftings of the seaways. Miocene strata include all the normal marine and terrestrial deposits. Extrusive volcanic rock is common; Miocene plutonic rocks are locally (Philippines) exposed.

The term Miocene was proposed in 1833 by Charles Lyell for the middle subdivision of the Tertiary, with "rather less than" 18% living species. Though it was first conceived with reference to the sequence at Superga Hill near Turin, Italy,

PRECAMBRIAN	PALEOZOIC							MESOZOIC			CENOZOIC	
					CARBONIFEROUS							
	CAMBRIAN	ORDOVICIAN	SILURIAN	DEVONIAN	Mississippian	Pennsylvanian	PERMIAN	TRIASSIC	JURASSIC	CRETACEOUS	TERTIARY	QUATERNARY

TERTIARY					QUATERNARY	
Paleocene	Eocene	Oligocene	Miocene	Pliocene	Pleistocene	Recent

as type, other fossils and stratigraphic relationships in the following places were also originally involved: additional Piedmont localities, Touraine, Bordeaux, Dax, and the Vienna Basin.

Subdivisions. Neither subdivisions nor exact limits of the Miocene were defined by Lyell. In Europe only in Jutland was marine deposition continuous from Oligocene into Miocene; nonmarine beds often delimit top and bottom elsewhere. Since degree of zoogenetical affinity is not in itself indicative of time, and geological changes in environment do not affect all organismal lineages equally, marine invertebrate paleontologists and terrestrial mammal specialists in particular have differed in their placing of the two Miocene boundaries. In Europe a threefold stage subdivision has developed: Burdigalian below, Helvetian and Tortonian above. A Vindobonian Stage (Vienna Basin) embraces both Helvetian and Tortonian ages; the deepwater Langhian (northern Italy) includes Burdigalian and Helvetian equivalents. The Aquitanian Stage (pre-Burdigalian) is often included as Miocene, especially its upper beds; and though even the sub-Aquitanian Oligocene proper has also sometimes locally been included with the Neogene (as, for example, in California), the supra-Aquitanian Miocene proper is invariably included within the Neogene. The supra-Vindobonian Sarmatian Stage is also so included by invertebrate paleontologists, whereas mammal specialists tend to consider these Pliocene (see illustration). In faunal provinces beyond the European, local stage and zone subdivisions are in use.

In Europe Burdigalian occasionally lies on pre-Aquitanian strata with little or no discordance. In California early Miocene lavas, ash, and breccias (San Onofre) are contemporaneous with conspicuous block faulting (Lompocan epeirogeny of Dibblee); a rising Sierra Nevadan fault block interrupted westward drainage and began depositing terrestrial detritals (Barstovian Age and younger) there, permitting biogenetic Monterey siliceous shale deposition to the west. Early Miocene strata generally reflect the following extensive marine transgressions and a widespread rejuvenation of the oceans of the world approaching in extent those of the marine Eocene: the First Mediterranean Stage of E. Suess, from Europe to Turkestan; from Cyrenaica through Egypt (*Globigerina* marl) toward Iran; Gaj and equivalent formations from India to Zanzibar, Mozambique Channel, and Ceylon; the Beboeloe Transgression in the East Indies affecting Australia and New

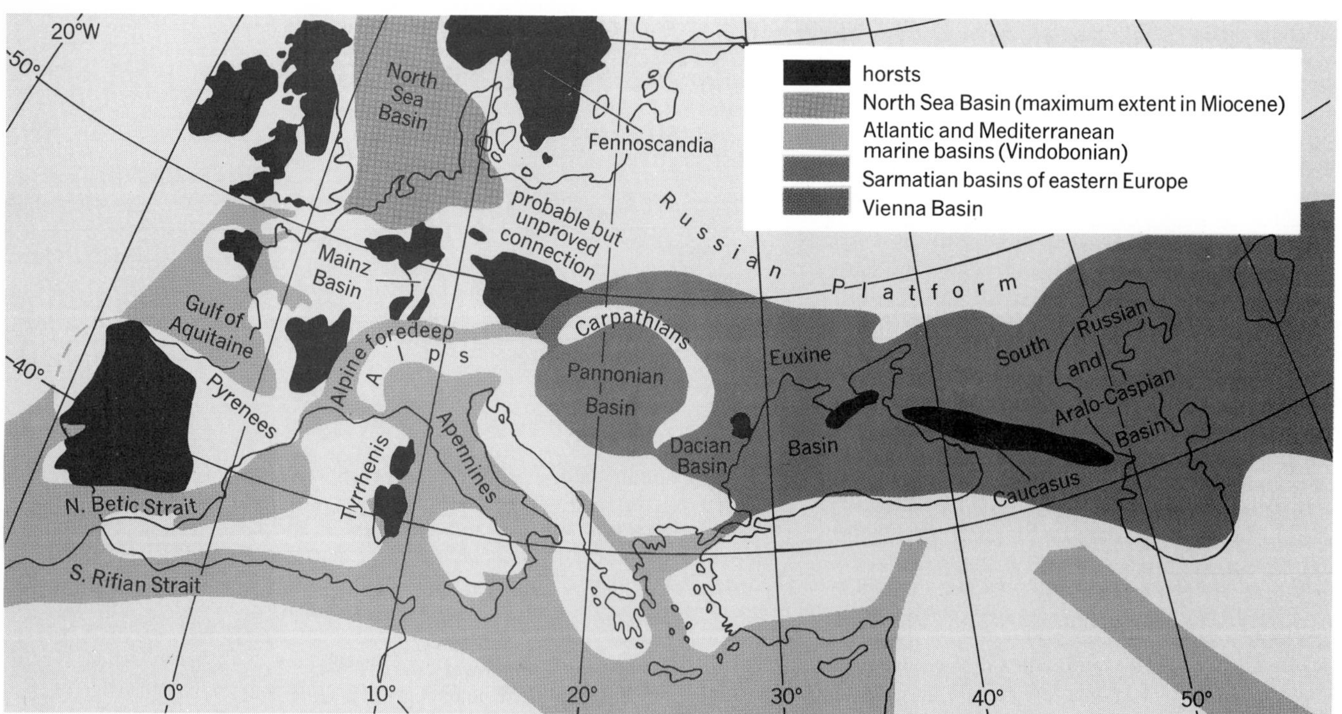

Miocene geography of Europe. The map does not represent any particular age within the Miocene Epoch. *(Modified after M. Gignoux, from A. M. Davies, Tertiary Faunas, vol. 2, Thomas Murby, 1934)*

Zealand; and the Relizian Stage in the California-Oregon-Washington area. Mid-Miocene mountain-making movements followed (Styrian orogenies in Europe, earliest Luisian Stage and closing Zuman orogeny in California, and East Indian Tertiary f2 disturbances), producing such conspicuous geologic phenomena as a Rhone foredeep, Swiss marine molasse, deepwater Schlier (rock typical in upper Austria), and much terrestrial deposition in Middle America; and the Indo-Gangetic plain was raised above sea level to inaugurate, in middle Miocene times, continuous Neogene terrestrial Siwalik deposition (with apelike and manlike fossils) south of the rising Himalayas, while a sinuous mountain chain arose from southeastern Asia around the Banda Arc through the Celebes to the Philippines. A second major marine transgression followed (the Second Mediterranean Stage of Suess in Europe, the Mohnian Stage of California, and f3 of the East Indies), reviving once more in late Miocene time the widespread marine condition of the early Miocene, to be succeeded by large-scale Mio-Pliocene withdrawals of the sea, orogeny (such as Rafaelan and Zacan orogenies of Dibblee in California), increased surface relief, land uplift, and Arctogaeic continental reconnections.

Life. In Miocene seas certain gastropod families (nassid, columbellid, thaiid, and cancelariid) suddenly diversified and multiplied, and other well-established gastropods (*Turritella*) persisted; large scallops of Oligocene inspiration, and clams (*Dosinia* and others) expanded; oysters attained great size; cake urchins and sand dollars diversified and multiplied; diatoms, corals, cephalopods, crustacea, insects, and large-toothed sharks were locally important; offshore small foraminifers (especially rotalids and buliminids) proliferated; and large foraminifer holdovers, *Lepidocyclina* and *Miogypsina*, dominated the reexpanded tropical shallow seas, then suddenly died out at the close of the Miocene. Mammalian marine adaptations (seals, sea lions, walruses, and sea-cowlike extinct desmostylids) appeared, or became modernized (sea cows, dugongs, and whales). Several Oligocene-erected barriers to marine dispersal (isthmian links and marine deeps) persisted; at high latitudes thermal barriers still adversely affected current-borne tropical life, and no cosmopolitan shellfish fauna reappeared. Though elements of Eocene and Oligocene origins are still discernible in them today, Miocene endemic evolution largely determined the major three to six distinct Recent marine shellfish faunas.

On land, except for rodents, all but two or three recognized mammal families already existed. Eurasian faunas included deer, hyenas, the earliest giraffes, and the bovines. A huge bear-dog, *Amphicyon*, was conspicuous, and mastodonts spread into North America for the first time. As grasslands spread, grazing life became dominant over browsing life; horses and camels, especially, evolved and spread rapidly. Australian, South American, and Madagascar mammal faunas remained isolated.

Ancestral redwoods (*Metasequoia*) flourished in China and California. Responding to a progressive cooling of climates through the Cenozoic, floras shifted equatorward along the north-south–trending mountain ranges bordering the Pacific, though species piled up where strandlines interfered (Malaysia and the Caribbean). Where mountains run east-west through the Old World, however, a Miocene Mediterranean flora was trapped and is ancestral to the Recent flora upon which human agricultural, high civilizations first developed. *See* PALEOBOTANY; PALEONTOLOGY.

[ROBERT M. KLEINPELL]

Mississippian

A large division of late Paleozoic geologic time, varyingly considered to rank as an independent period or as a main subdivision (termed epoch) of the Carboniferous Period. It is named Mississippian for a succession of highly fossiliferous marine strata consisting largely of limestones found along the Mississippi River between southeastern Iowa and southern Illinois. The rocks formed during the time defined as Mississippian are classed as a geologic system by most American geologists, and this practice is increasingly accepted outside of North America even though foreign usage favors Lower Carboniferous. Widespread marine transgression of the continents, accompanied especially by accumulation of limestones, characterized Mississippian time. Locally, this was preceded or succeeded by mountain building. *See* CARBONIFEROUS.

PRECAMBRIAN	PALEOZOIC							MESOZOIC			CENOZOIC	
	CAMBRIAN	ORDOVICIAN	SILURIAN	DEVONIAN	CARBONIFEROUS		PERMIAN	TRIASSIC	JURASSIC	CRETACEOUS	TERTIARY	QUATERNARY
					Mississippian	Pennsylvanian						

[RAYMOND C. MOORE]

Moho (Mohorovičić discontinuity)

A seismic discontinuity at the base of the Earth's crust inferred from travel time curves indicating that seismic waves undergo a sudden increase in velocity. Immediately below the discontinuity the velocity of seismic compressional waves increases to a little over 8 km/sec. The discontinuity is named after its discoverer, the Yugoslavian geophysicist A. Mohorovičić, who noticed the sudden change in velocity when examining earthquake records of the Yugoslavian Kulpa Valley earthquake of Oct. 8, 1909. Such a discontinuity seems to exist every place that geophysicists seismologically investigate the Earth's structure. Under the ocean basins its depth is generally 10–12 km; under the continents it is usually at a depth of 33–35 km (Fig. 1). It usually lies deeper under mountain masses and shallower under lowlands and plains, sloping upward near coastal margins. Islands and island arcs are also marked by a downward bend of the Moho under the ocean basins. *See* SEISMOLOGY.

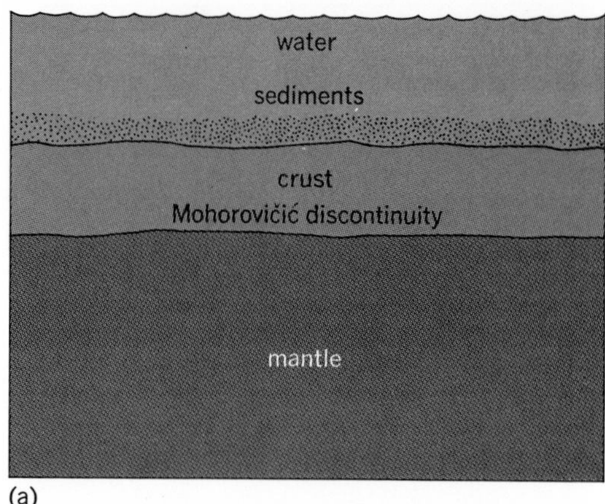

(a)

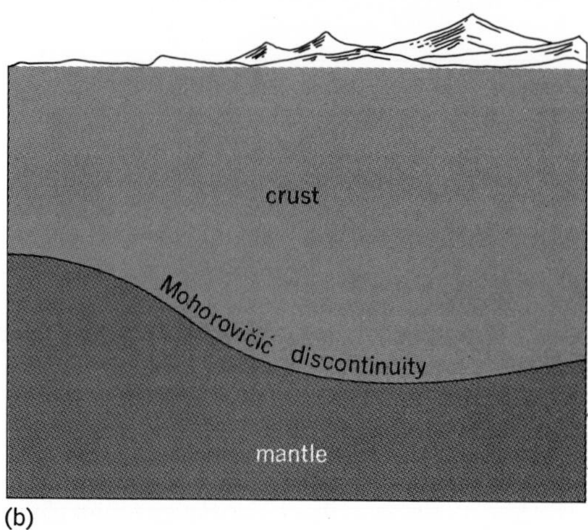

(b)

Fig. 1. Mohorovičić discontinuity. (*a*) At average depths of about 10–12 km beneath ocean basins. (*b*) At average subcontinental depths of 33–35 km.

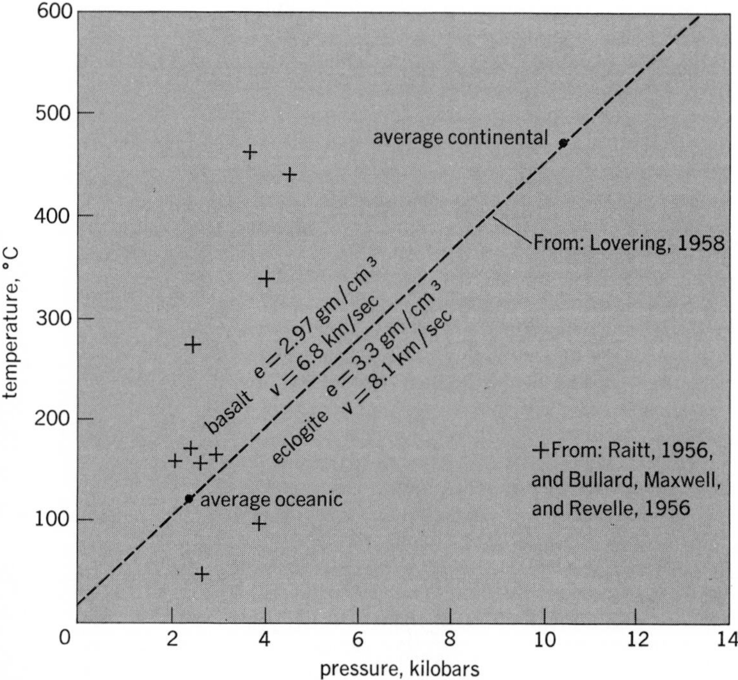

Fig. 2. Temperature-pressure relationship relative to Moho as suggested by T. S. Lovering; Pacific seismic and geothermal data reported by R. W. Raitt and by E. C. Bullard et al.; *e* is density, *v* is compressional velocity of seismic waves.

PROMINENT FEATURES

The following sections give the characteristics of the discontinuity including density, composition, temperature, pressure, and depth.

Rock density. The increase in velocity is because of an increase in the density of the rocks as they change from crustal to mantle material. The density of the crust varies from 2.2 to 2.7 g/cm^3, and the density in the surface of the mantle is 3.3. The crust is composed of light, silicic rocks which surround, or ride upon, the denser, basic mantle.

Composition. Although the Moho is generally accepted as the dividing zone between the crust and mantle, there is still considerable speculation over its composition and configuration. M. Ewing (Lamont Geophysical Laboratory) noted that in some localities a peculiar scattering takes place in the seismic waves which would indicate a rough surface, perhaps the remnants of an original surface of the Earth that existed prior to continental differentiation and subsequent sedimentation. It is known that the surface undulates under the varying load of the Earth's crust. Since the Moho represents, at least partly, the beginning of the mantle, its composition is in question by some authorities and accepted as peridotite, eclogite, dunite, or serpentine by others. Those who accept the composition of the outer mantle base their reasoning upon laboratory seismic velocity measurements, theory, and, in one instance, the outcropping of St. Paul's Rocks in the mid-Atlantic. At St. Paul's Rocks peridotite is found above sea level along a tremendous vertical fault. It was postulated by H. Hess (Princeton) that the crust is broken through and heaved up along one side of the fault to the extent that the mantle is exposed. In spite of these theories, the nature of the Moho and outer mantle will not be determined completely until one or more core holes are drilled through the crust and samples are collected.

In certain areas of the ocean basins the Moho is relatively close to the surface (sea level). In the area just north of San Juan, Puerto Rico, on the rise to the north of the Puerto Rican Trench, the Moho was observed, by J. Nafe, to lie at a depth of 9+ km. At the southern end of the Gulf of California it was observed, by G. Shor, at a depth of 9+ km. It might be possible to reach the Moho by drilling in either of these locations. Contemplation of tentative planning for such drilling brought about the coining of the term Mohole.

Temperature and pressure. The question as to whether presence of the Moho is evidence of chemical changes in the rocks of the crust and mantle or only of a physical phase change is still unanswered. Evidence is cited in favor of the phase change, suggesting that the Moho results from conditions of temperature and pressure which cause the transformation of a basalt crust into an eclogite mantle. Using the average for crustal thicknesses beneath the continents and oceans and assuming that the heat flow is similar in the two locations, a relation has been derived, illustrated by the dashed line in Fig. 2. However, seismic and geothermal results from the Pacific, which are also shown in Fig. 2, tend to controvert this idea.

Table 1. Moho depths*

Location	Seismic velocity, km/sec	Depth to Moho km	mi
East coast, U.S.	8.10	33	20.6
New York – Pennsylvania	8.15	35	21.8
Central Appalachian	8.03	40	25.0
Canada	8.10	36	22.5
Wisconsin	8.10	40	25.0
Southern California valleys	8.10	32	20.0
South Africa	8.20	36	22.5
Deep Atlantic (average)	8.10	10	6.2
Pacific Basin (average)	8.20	11	6.8

*After B. Gutenberg.

They indicate temperatures at the Moho that are far in excess of those which had been previously assumed.

Depth summary. Table 1 presents a summation of some continental and oceanic seismic velocities and Moho depths from refraction measurements.

[GORDON G. LILL]

MOHOLE PROJECT

The Mohole Project was a national scientific project with the broad and important research objective of studying the Earth as a planet, rather than studying only a portion of it – a continent, an ocean, a deep-sea canyon, or a high mountain peak. This was to be accomplished under the direction of the National Science Foundation through exploration and sampling of all layers of the Earth's crust and the unknown mantle beneath by core drilling near Hawaii (Fig. 3) through 17,000 ft (5181.6 m) of rock under 15,000 ft (4572 m) of water to an approximate total depth of 32,000 ft (9753.6 m). All required equipment, machinery, and instrumentation had been designed and some of it had been fabricated at the time of the termination of the project by Congress in 1966.

Phase I. About 85% of the Earth lies in the mantle beneath the crust. By studying and understanding the mantle, the forces at work there, its chemical composition, its age, and possibly its origin, scientists hoped to achieve a firsthand understanding of the planet Earth. These objectives prompted prominent American earth scientists in 1957 to initiate the idea of drilling through the Mohorovičić discontinuity at the upper boundary of the mantle.

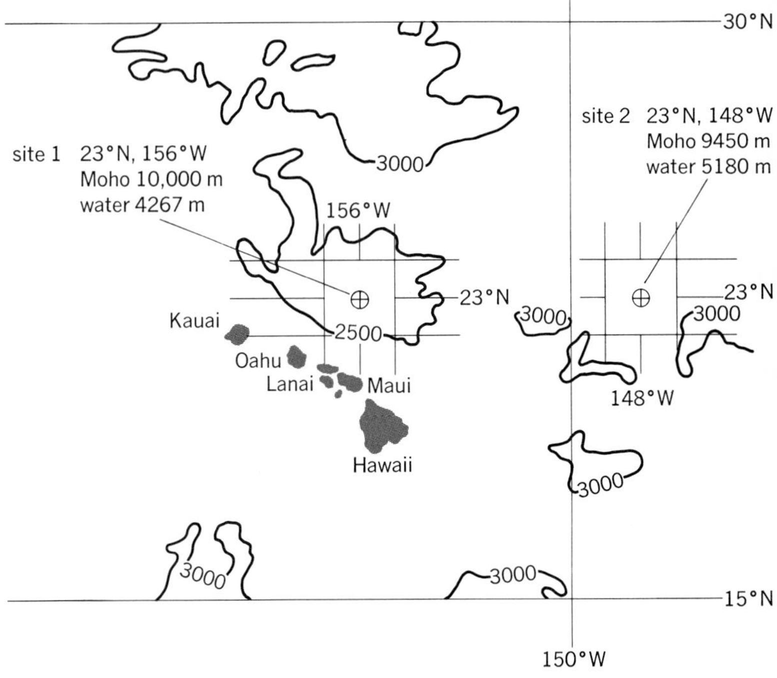

Fig. 3. Planned core drilling sites near Hawaii.

After preliminary investigations sponsored by the National Academy of Sciences and the National Science Foundation, phase I of Project Mohole was conducted in 1961. This phase consisted of drilling in approximately 12,000 ft (3657.6 m) of water and into several hundred feet of the Earth's crust off the coast of California from a floating and dynamically positioned ship. Existing offshore oil-industry equipment and technology were used. Core samples were successfully retrieved and it was demonstrated that the major undertaking of drilling at sea in deep water was feasible.

Phase II. This phase of the Mohole Project began in March, 1962, with the award of a contract to an industrial organization. From 1962 through 1965 extensive research, development, and tests in multiple technologies and disciplines were carried out by some of the most competent firms and individuals in the United States to solve the complex problems of creating a successful system to drill in the deep ocean and into the Earth to the depths required. Investigations of all types of ships and

Table 2. Mohole drilling platform characteristics

Characteristic	Measurement	
Size of upper platform	85.0 × 71.3 m	(279 ft × 234 ft)
Height of upper platform	7.0 m	(23 ft)
Diameter of supporting columns	9.5 m	(31 ft)
Total height (keel to upper deck)	41.1 m	(135 ft)
Derrick height	59.7 m	(196 ft)
Overall height (keel to top of radar mast)	114.3 m	(375 ft)
Diameter of lower hulls	10.7 m	(35 ft)
Length of lower hulls	118.9 m	(390 ft)
Light ship displacement	12,250 tonneaux at 8.5-m draft	(13,500 tons at 28 ft)
Normal drilling displacement	19,500 tonneaux at 18.3-m draft	(21,500 tons at 60 ft)

floating vessels and of the stresses on the long drill string demonstrated that the drilling had to be accomplished from an extremely stable platform. Such a platform, designed and extensively model tested, was a major breakthrough in marine engineering and naval architecture. Model tests demonstrated minimum motion in 25-ft (7.62 m) seas and in over 30 knots of wind, with a resulting safe stress on the drill pipe suspended below the platform. The drilling platform was designed to be self-propelled so that it could move from location to location—ocean to ocean—and so that it could handle great loads with minimum motion. Table 2 gives characteristics for the final platform.

The platform and its associated complex systems were never completed. However, all research, designs, and technology developed during the life of the project have been preserved, and such information can be obtained from the Clearinghouse of the Department of Commerce in Springfield, Va. [DANIEL HUNT, JR.]

Bibliography: *Design of a Deep Ocean Drilling Ship*, Nat. Acad. Sci.–Nat. Res. Coun. Publ. no. 984, 1962; *Experimental Drilling in Deep Water at La Jolla and Guadalupe Sites*, Nat. Acad. Sci.–Nat. Res. Coun. Publ. no. 914, 1961; Floating platform may drill the mohole, *Oil Gas J.*, vol. 61, no. 18, May 6, 1963; B. Gutenberg, Wave velocities in the earth's crust, *Crust of the Earth*, Geol. Soc. Amer. Spec. Pap. no. 62, 1955; H. Jeffreys, *The Earth: Its Origin, History and Physical Constitution*, 4th ed., 1959; D. Lambert, Will the U.S. lose race to "Inner Space"?, *World Oil*, July, 1966; H. E. Landsberg (ed.), *Advances in Geophysics*, vol. 3, 1956; G. G. Lill and A. E. Maxwell, The earth's mantle, *Science*, 129(3360):1407–1410, 1959; A. C. McClure, *Development of the Project Mohole Drilling Platform*, Soc. Nav. Architects Mar. Eng. Publ. no. 1, Nov. 11–12, 1965; Office of Naval Research, The effects of ship motions on the mohole drilling string, *Nav. Res. Rev.*, vol. 16, June, 1963; Office of Naval Research, Experimental deep-ocean drilling, *Nav. Res. Rev.*, vol. 16, June, 1963; Office of Naval Research, Mohole: A review, *Nav. Res. Rev.*, vol. 16, June, 1963; *Project Mohole: A Report Bibliography*, Clearinghouse for Federal Scientific and Technical Information, U.S. Department of Commerce, June, 1965; G. G. Shor, Jr., and D. D. Pollard, Mohole site selection studies north of Maui, *J. Geophys. Res.*, vol. 69, no. 8, Apr. 15, 1964.

Molybdenite

A mineral having composition MoS_2. Molybdenite is the chief ore of molybdenum. It crystallizes in the hexagonal system, but crystals are rare and when found are hexagonal plates. It is commonly in scales or foliated masses. There is one direction of perfect cleavage yielding flexible but nonelastic folia. The mineral is sectile and has a greasy feel. The hardness is 1.5 (Mohs scale) and the specific gravity is 4.7. The luster is metallic and the color lead gray (see illustration). Molybdenite and graphite have long been confused because of their nearly identical physical properties. They can be distinguished by the streak left on glazed paper, black for graphite and green for molybdenite. Molybdenite has been used as a lubricant.

MOLYBDENITE

Molybdenite on biotite mica from Edison, N.J. (*American Museum of Natural History specimens*)

Molybdenite is found as an accessory mineral in certain granites and pegmatites but more commonly is associated with fluorite in vein deposits of tin and tungsten. It is also present in some contact metamorphic deposits. It occurs in various places in Norway, Sweden, Australia, England, China, and Mexico. In the United States, molybdenite is found in small amounts at many localities but the most important occurrence is at Climax, Colo., where it is found in quartz veinlets in a silicified granite with topaz and fluorite. Molybdenite as an ore of molybdenum is mined there on a large scale, making Climax the world's largest producer of molybdenum.

[CORNELIUS S. HURLBUT, JR.]

Monazite

A phosphate (mineral) of the cerium metals, (Ce,La,Y,Th) (PO_4). Ordinarily lanthanum, La, is present in about 1:1 ratio with cerium. Small amounts of the yttrium, Y, earths substitute for Ce and La. Thorium substitutes for Ce and La and generally ranges up to 10% ThO_2. A series of monazite minerals ranging up to 30% ThO_2 probably exists. Thorium-free monazite is rare. Uranium, U, in small amounts has been reported.

Monazite crystallizes in the monoclinic system. Crystals are prismatic and generally minute but occasionally large (see illustration). Colors range from white through shades of yellow, green, and brown.

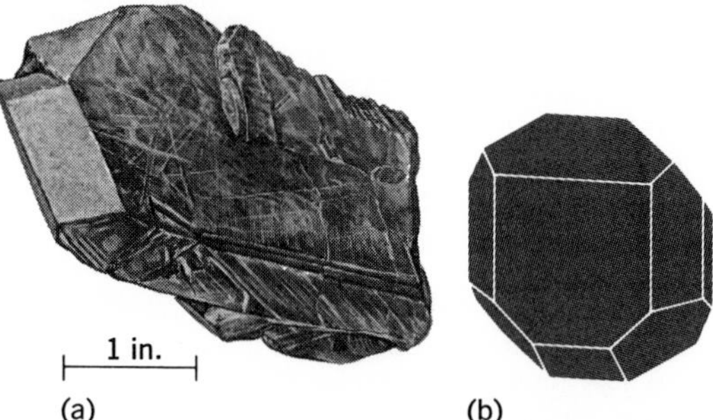

Monazite. (*a*) Crystals from Minas Gerais, Brazil (*specimen from Department of Geology, Bryn Mawr College*). (*b*) Crystal habit (*from L. G. Berry and B. Mason, Mineralogy, copyright © 1959 by W. H. Freeman and Co.*)

Monazite is widely disseminated as accessory grains and crystals in granitic and pegmatitic rocks and in metamorphic gneissic rocks. In regions of such rock types, fluviatile and beach sands may contain commercial quantities of monazite. Monazite occurs in many regions of the world, but the major production comes from placer deposits in Idaho, South Carolina, and Florida in the United States, and in India, Brazil, and the Union of South Africa. Recovered monazite concentrates are sources of thorium, thorium compounds, and cerium metals. *See* RADIOACTIVE MINERALS.

[WAYNE R. LOWELL]

Montmorillonite

A group name for all clay minerals with an expanding structure, except vermiculite, and also a specific mineral name for the high alumina end member of the group. *See* CLAY MINERALS; VERMICULITE.

Montmorillonite clays have wide commercial

use. The high colloidal, plastic, and binding properties make them especially in demand for bonding molding sands and for oil-well drilling muds. They are also widely used to decolorize oils and as a source of petroleum cracking catalysts.

Structure. Because of the extremely small particle size of the montmorillonite minerals, there is still some uncertainty regarding details of their structure. According to the structural concept which is currently accepted, montmorillonite is composed of units made up of two silica tetrahedral sheets with a central alumina octahedral sheet. The atoms in these layers which are common to both sheets become O instead of OH. Montmorillonite is thus referred to as a three-layer clay mineral with tetrahedral-octahedral-tetrahedral layers making up the structural unit (see illustration).

These silica-alumina-silica units are continuous in the *a* and *b* crystallographic directions and are stacked one above the other in the *c* direction. In the stacking of these units, the oxygen layers of neighboring units are adjacent. This causes a very weak bond and an excellent cleavage between the units. Water and other polar molecules can enter between the unit layers and cause an expansion of the structure in the *c* direction. Thus montmorillonite does not have a fixed *c*-axis dimension, but can vary considerably depending on the absence or presence of interlayer molecules. The *c*-axis spacing also varies with the nature of the interlayer cation present between the silicate layers. A montmorillonite in an air-dried condition with sodium as the exchange ion frequently has one molecular water layer and a *c*-axis spacing of about 12.5 A. Under similar conditions there are two molecular water layers with calcium, giving a *c*-axis spacing of about 15.5 A. The expansion properties of montmorillonite are reversible; however, reexpansion may be difficult after complete structural collapse by removal of all interlayer polar molecules.

Atomic substitution. The theoretical formula for montmorillonite without structural substitutions is $(OH)_4Si_8Al_4O_{20} \cdot nH_2O$ (interlayer). However, montmorillonite always differs from the above theoretical formula because of structural substitution. In the tetrahedral sheet, aluminum and possibly phosphorus substitute for silicon, whereas ions such as magnesium, iron, and lithium substitute for aluminum in octahedral coordination. Total replacement of aluminum by magnesium yields the mineral saponite; replacement of aluminum by iron yields nontronite. If all octahedral positions are filled by ions, the mineral is trioctahedral; if only two-thirds are occupied, the mineral is dioctahedral.

The montmorillonite structure is always unbalanced by the substitutions noted above. The resulting positive net charge deficiency is balanced by exchangeable cations adsorbed between the unit layers and around their edges. The cation-exchange capacity of montmorillonite is normally quite high (100± milliequivalents per 100 g) and is not appreciably affected by particle size. Substitutions within the structure cause about 80% of the total exchange capacity, and broken bonds are responsible for the remainder.

Other properties. Montmorillonite particles are extremely small and may further disperse in water

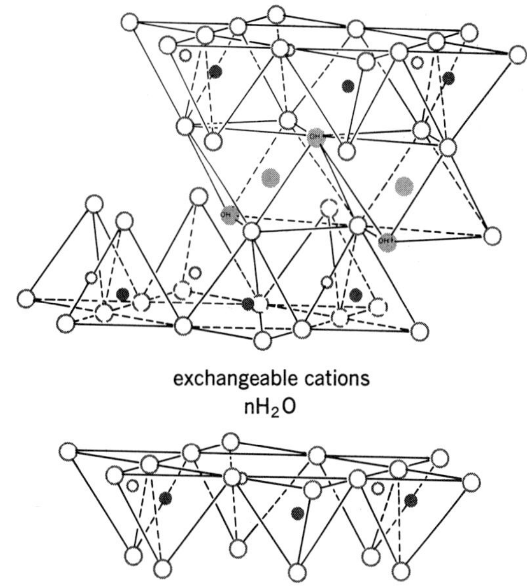

Diagrammatic sketch of structure of montmorillonite. *(From R. E. Grim, Clay Mineralogy, McGraw-Hill, 1953)*

to units approaching single-cell-layer dimensions. Most montmorillonite units are equidimensional flakes. However, nontronite tends to occur in elongate lath-shaped units, and hectorite, the fluorine-bearing magnesium-rich montmorillonite, is found in thin laths.

There is general agreement that the adsorbed interlayer water between the silicate layers has some sort of definite configuration, but the precise nature of this configuration is not agreed upon. The extent and nature of the orientation of the adsorbed water varies with identity of the adsorbed cations.

When montmorillonite is dehydrated, the interlayer water is lost at a relatively low temperature (100–200°C). The loss of structural (OH) water begins gradually at about 450–500°C, ending at 600–750°C. These temperatures vary with the type and amount of structural substitution. The structure of montmorillonite usually persists to temperatures of the order of 800–900°C. On further heating, a variety of phases form, such as mullite, cristobalite, and cordierite, depending on the composition and structure prior to fusion at 1000–1500°C.

Organic ionic compounds enter into cation-exchange reactions with montmorillonite. Polar organic compounds, like glycerol, react by replacing the interlayer water, causing a shift in the *c*-axis spacing of the montmorillonite units. Thus, the identification of montmorillonite by x-ray diffraction is greatly simplified by preliminary treatment with certain organic reagents. The reaction of montmorillonite and organic material is the base of considerable economic use of montmorillonite clays.

Occurrence. Members of the montmorillonite group of clay minerals vary greatly in their modes of formation. Alkaline conditions and the presence of magnesium particularly favor the formation of these minerals. Montmorillonites are stable over a wide temperature range and have formed by low-temperature hydrothermal processes, as well as by weathering processes. Several important modes of occurrence are in soils, in bentonites, in mineral veins, in marine shales, and as alteration products of other minerals. Recent sediments have a fairly high montmorillonite content. *See* BENTONITE; MARINE SEDIMENTS.

[FLOYD M. WAHL; RALPH E. GRIM]

Monzonite

A phaneritic (visibly crystalline) plutonic rock composed chiefly of sodic plagioclase (oligoclase or andesine) and alkali feldspar (microcline, orthoclase, usually perthitic), with subordinate amounts of dark-colored (mafic) minerals (biotite, amphibole, or pyroxene). Monzonite is more or less intermediate between syenite and diorite. Plagioclase is dominant over alkali feldspar in monzonite but is subordinate to alkali feldspar in syenite. Diorite contains little or no alkali feldspar. *See* SYENITE.

[CARLETON A. CHAPMAN]

Moon

The Earth's natural satellite, target of the first attempts by humans to reach another world. *Apollo 11* astronauts landed on the Moon on July 20, 1969. Of all the satellites in the solar system, the Moon is the largest in proportion to its primary. Despite centuries of observation from Earth and elaborate space flight experiments, the Moon is still a mysterious object. Theories of its origin include (1) independent condensation and then capture by the Earth, (2) formation in the same cloud of preplanetary matter with the Earth, and (3) fission from the Earth. Some of the major properties of the Moon and their values are listed in Table 1.

Table 1. Characteristics of the Moon

Characteristics	Values and remarks
Diameter (approximate)	3476 km
Mass	1/81.301 Earth's mass, or 73.49×10^{24} g
Mean density	0.604 Earth's, or 3.34 g/cm^3
Mean surface gravity	0.165 Earth's, or 162 cm/sec^2
Surface escape velocity	0.213 Earth's, or 2.38 km/sec
Atmosphere	Surface pressure 10^{-12} torr (1.3×10^{-10} Pa); hints of some charged dust particles and occasional venting of volatiles
Magnetic field	Dipole field less than $\sim 0.5 \times 10^{-5}$ Earth's; remanent magnetism in rocks shows past field was much stronger
Dielectric properties	Surface material has apparent dielectric constant of 2.8 or less; bulk apparent conductivity is 10^{-5} mho/m or less
Natural radioactivity	Mainly due to solar- and cosmic-ray-induced background (about 1 mr/hr for quiet Sun)
Seismic activity	Much lower than Earth's; deep moonquakes occur more frequently when Moon near perigee; subsurface layer evident
Heat flow	3×10^{-2} W/m^2 (*Apollo 15* site)
Surface composition and properties	Basic silicates, three sites (Table 4); some magnetic material present. Soil grain size is 2–60 μm; 50% less than 10 μm. Soil-bearing strength 1 kg/cm^2 at depth of a few centimeters
Rocks	All sizes up to tens of meters present, concentrated in strewn fields. Rock samples from Mare Tranquillitatis include fine- and medium-grained igneous and breccia
Surface temperature range	At equator 400 K at noon; 80–100 K night minimum; 1 m below surface, 230 K; at poles ~150 K

The apparent motions of the Moon, its waxing and waning, and the visible markings on its face (Fig. 1), are reflected in stories and legends from every early civilization. At the beginning of recorded history on the Earth, it was already known that time could be reckoned by observing the position and phases of the Moon. Attempts to reconcile the repetitive but incommensurate motions of the Moon and Sun led to the construction of calendars in ancient Chinese and Mesopotamian societies and also, a thousand years later, by the Maya. By about 300 B.C., the Babylonian astronomer-priests had accumulated long spans of observational data and so were able to predict lunar eclipses. Major events in the subsequent development of human knowledge of the Moon are summarized in Table 2.

Space flight experiments have now confirmed and vastly extended understanding of the Moon; however, they have also opened many new questions for future lunar explorers.

Motions. The Earth and Moon now make one revolution about their barycenter, or common center of mass (a point about 4670 km from the Earth's center), in $27^d\ 7^h\ 43^m\ 11.6^s$. This sidereal period is slowly lengthening, and the distance (now about 60.27 earth radii) between centers of mass is increasing, because of tidal friction in the oceans of the Earth. The tidal bulges raised by the Moon are dragged eastward by the Earth's daily rotation. The displaced water masses exert a gravitational force on the Moon, with a component along its direction of motion, causing the Moon to spiral slowly outward. The Moon, through this same tidal friction, acts to slow the Earth's rotation, lengthening the day. Tidal effects on the Moon itself have caused its rotation to become synchronous with its orbital period, so that it always turns the same face toward the Earth.

Tracing lunar motions backward in time is very difficult, because small errors in the recent data propagate through the lengthy calculations. Nevertheless, the attempt is being made by using diverse data sources, such as the old Babylonian eclipse records and the growth rings of fossil shellfish. At its present rate of departure, the Moon would have been quite close to the Earth about 4×10^9 years ago, a time which other evidence suggests as the approximate epoch of formation of the Earth. The Moon could not have been a single body inside a distance of about 3 earth radii (the Roche limit), where tidal forces would have been stronger than the mutual gravitational attraction of the Moon's material.

The Moon's present orbit (Fig. 2) is inclined about 5° to the plane of the ecliptic. Table 3 gives the dimensions of the orbit (in conventional coordinates with origin at the center of the Earth, rather than the Earth-Moon barycenter). As a result of differential attraction by the Sun on the Earth-Moon system, the Moon's orbital plane rotates slowly relative to the ecliptic (the line of nodes regresses in an average period of 18.60 years) and the Moon's apogee and perigee rotate slowly in the plane of the orbit (the line of apsides advances in a period of 8.850 years). Looking down on the system from the north, the Moon moves counterclockwise. It travels along its orbit at an average speed of

key:

- • Apollo landing site and mission number
- O passive seismometer operating
- □ heat flow probe
- △ laser ranging retroreflector

Fig. 1. Map of near side of Moon, showing principal features and Apollo landing sites.

nearly 1 km/sec or about 1 lunar diameter per hour; as seen from Earth, its mean motion eastward among the stars is 13°11′ per day.

As a result of the Earth's annual motion around the Sun, the direction of solar illumination changes about 1° per day, so that the lunar phases do not repeat in the sidereal period given above but in the synodic period, which averages $29^d\ 12^h\ 44^m$ and varies some 13 hr because of the eccentricity of the Moon's orbit. *See* EARTH, ROTATION AND ORBITAL MOTION OF.

When the lunar line of nodes (Fig. 2) coincides with the direction to the Sun and the Moon happens to be near a node, eclipses can occur. Because of the 18.6-year regression of the nodes, groups of eclipses recur with this period. When it passes through the Earth's shadow in a lunar eclipse, the Moon remains dimly visible because of the reddish light scattered through the atmosphere around the limbs of the Earth. When the Moon passes between the Earth and Sun, the solar eclipse may be total or annular. As seen from Earth, the angular diameter of the Moon (31′) is almost the same as that of the Sun, but both apparent diameters vary because of the eccentricities of the orbits of Moon and Earth. Eclipses are annular when the Moon is near apogee and the Earth is near perihelion at the time of eclipse. A partial solar eclipse is seen from places on Earth that are not directly along the track of the Moon's shadow.

The Moon's polar axis is inclined slightly to the pole of the lunar orbit (Fig. 2) and rotates with the same 18.6-year period about the ecliptic pole. The rotation of the Moon about its polar axis is nearly

Table 2. Growth of human understanding of the Moon

Prehistory	Markings and phases observed, legends created connecting Moon with silver, dark markings with rabbit (shape of maria) or with mud.
~300 B.C.	Apparent lunar motions recorded and forecast by Babylonians and Chaldeans.
~150 B.C.	Phases and eclipses correctly explained, distance to Moon and Sun measured by Hipparchus.
~A.D. 150	Ancient observations compiled and extended by Ptolemy.
~700	Ephemeris refined by Arabs.
~1600	Empirical laws of planetary motion derived by Kepler.
1609	Lunar craters observed with telescopes by Harriot and Galileo.
1650	Moon mapped by Hevelius and Riccioli; features in system named by them still in use.
1667	Experiments by Hooke simulating cratering through impact and vulcanism.
1687	Moon's motion ascribed to gravity by Newton.
1692	Empirical laws of lunar motion stated by Cassini.
1700–1800	Lunar librations measured, lunar ephemeris computed using perturbation theory by Tobias Mayer. Secular changes computed by Lagrange and Laplace. Theory of planetary evolution propounded by Kant and Laplace. Many lunar surface features described by Schroeter and other observers.
1800–1920	Lunar motion theory and observations further refined, leading to understanding of tidal interaction and irregularities in Earth's rotation rate. Photography, photometry, and bolometry applied to description of lunar surface and environment. Lunar atmosphere proved absent. New disciplines of geology and evolution applied to Moon, providing impetus to theories of its origin.
1924	Polarization measured by B. F. Lyot, showing surface to be composed of small particles.
1927–1930	Lunar day, night, and eclipse temperatures measured by E. Pettit and S. B. Nicholson.
1946	First radar return from Moon.
1950–1957	Renewed interest in theory of lunar origin, new methods (for example, isotope dating) applied to meteorites, concepts extended to planetology of Moon. Low subsurface temperatures confirmed by Earth-based microwave radiometry.
1959	Absence of lunar magnetic field (on sunlit side) shown by *Luna 2*.
1960	Eastern far side photographed by *Luna 3*. Slower cooling of Tycho detected during eclipse.
1961	United States commitment to manned lunar flight.
1962	Earth-Moon mass ratio measured by *Mariner 2*.
1964	High-resolution pictures sent by *Ranger 7*. Surface temperatures during eclipse measured by Earth-based infrared scan.
1965	Western far side photographed by *Zond 3*.
1966	Surface pictures produced by *Luna 9* and *Surveyor 1*. Radiation dose at surface measured by *Luna 9*. Gamma radioactivity measured by *Luna 10*. High-resolution, broad-area photographs taken by *Lunar Orbiter 1*. Surface strength and density measurements made by *Luna 13*.
1967	Mare soil properties and chemistry measured by *Surveyor 3, 5*, and *6*. Whole front face mapped by *Lunar Orbiter 4*, sites of special scientific interest examined by *Lunar Orbiter 5*. Particle-and-field environment in lunar orbit measured by *Explorer 35*.
1968	Highland soil and rock properties and chemistry measured by *Surveyor 7*. Mass concentrations at circular maria discovered.
1968	Astronauts orbit Moon, return with photographs.
1969	Astronauts land and emplace instruments on Moon, return with lunar samples and photographs.
1969–1972	Lunar seismic and laser retroreflector networks established. Heat flow measured at two sites. Remanent magnetism discovered in lunar rocks. Geologic traverses accomplished. Orbital surveys of natural gamma radioactivity, x-ray fluorescence, gravity, magnetic field, surface elevation, and subsurface electromagnetic properties made at low latitudes. Metric mapping photos obtained. Samples returned by both manned (United States) and automated (Soviet) missions; sample analyses confirmed early heating and chemical differentiation of Moon, with surface rocks enriched in refractory elements and depleted in volatiles. Age dating of lunar rocks and soils showed that most of the Moon's activity (meteoritic, tectonic, volcanic) occurred more than 3×10^9 years ago.
1975–to date	The multiplicity of theories of lunar origin, though much more closely limited by Apollo and Luna data, continues, with no model being accepted as unique.

uniform, but its orbital motion is not, owing to the finite eccentricity and Kepler's law of equal areas, so that the face of the Moon appears to swing east and west about 8° from its central position every month. This is the apparent libration in longitude. The Moon does rock to and fro in a very small oscillation about its mean rotation rate; this is called the physical libration. There is also an apparent libration in latitude because of the inclination of the Moon's polar axis. The librations make it possible to see about 59% of the Moon's surface from the Earth, and they also permit stereoscopic determination of elevations on the Moon, though the precision of the latter is low.

The lunar ephemeris, derived from precise astronomical observations and refined through lengthy computations of the effects perturbing the movements of the Moon, has now reached a high degree of accuracy in forecasting lunar motions and events such as eclipses. Laser ranging to retroreflectors landed on the Moon, aided by radio ranging to spacecraft, provides measurements of Earth-Moon distances to a precision of the order of meters.

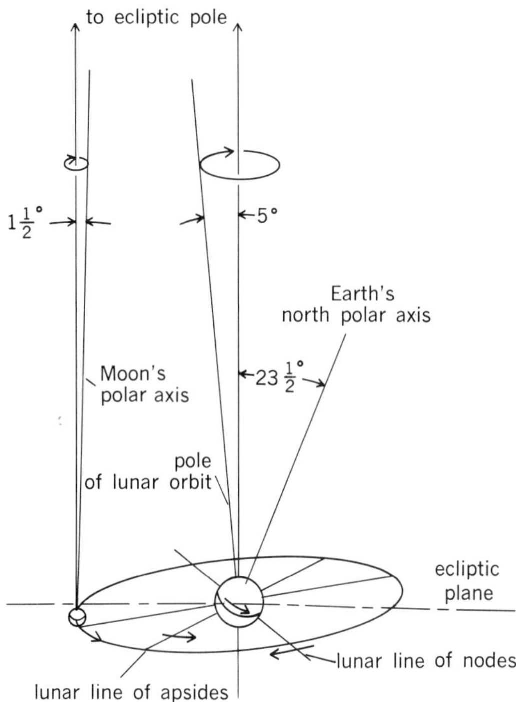

Fig. 2. Sketch of Moon's orbit.

Selenodesy. The problem of determining the Moon's true size and shape and its gravitational and inertial properties has been under attack by various methods for centuries (Tables 1 and 2). However, results from space flights have invalidated some of the premises on which the earlier methods were based, and have revealed discrepancies in the older data. The relation between the Moon's shape and its mass distribution is very important to theories of lunar origin and the history of the Earth-Moon system. Radio-tracking data from Lunar Orbiters indicate that the Moon's gravitational field is ellipsoidal, with the short axis being the polar one (as expected for any rotating body), and with the equatorial section being an ellipse possibly slightly elongated in the Earth-Moon direction. But the Earth-based radar measurements and tracking data from Rangers and Surveyors showed that the Moon's actual surface at the points of landing is about 2 km farther from the

Table 3. Dimensions of Moon's orbit

Characteristics	Values
Sidereal period (true period of rotation and revolution)	(27.32166140 + 0.00000016T) ephemeris days, where T is in centuries from 1900
Synodic period (new Moon to new Moon)	(29.5305882 + 0.00000016T) ephemeris days
Apogee	406,700 km (largest); 405,508 km (mean)
Perigee	356,400 km (smallest); 363,300 km (mean)
Period of rotation of perigee	8.8503 years direct ("direct" meaning that the motion of perigee is in the direction of Moon's motion about the Earth)
Period of regression of nodes	18.5995 years
Eccentricity of orbit	0.054900489 (mean)
Inclination of orbit to ecliptic	5°8′43″ (oscillating ±9′ with period of 173 days)
Inclination of orbit to Earth's Equator	Maximum 28°35′, minimum 18°21′
Inclination of lunar equator	
to ecliptic	1°32′40″
to orbit	6°41′

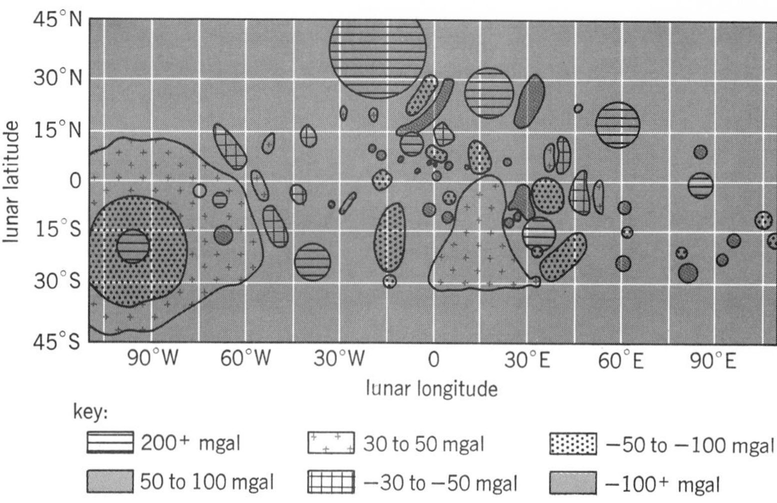

Fig. 3. Gravity anomalies on the lunar near side and limb regions. 1 mgal = gravitational acceleration of 10^{-5} m/s^2. Circular areas correspond to mass concentrations in circular maria. (*W. L. Sjogren, Jet Propulsion Laboratory, NASA*)

Earth than expected. This difference is much too large to be either an ephemeris bias or a tracking error and must represent a real difference in the distance from the landing sites to the Moon's center of mass. Further evidence of an anomalous relationship between mass and shape for the Moon is provided by the mass concentrations in circular maria, discovered through analysis of short-term variations in the Lunar Orbiter tracking data and then mapped in detail by Apollo tracking (see Fig. 3). By radio altimetry, Apollo confirmed that the Moon's surface on the far side is higher on the average than the near side; that is, the center of mass is offset from the center of figure. The offset is about 2 km toward the Earth.

Body properties. The Moon's small size and low mean density (Table 1) result in surface gravity too low to hold a permanent atmosphere, and therefore it was to be expected that lunar surface char-

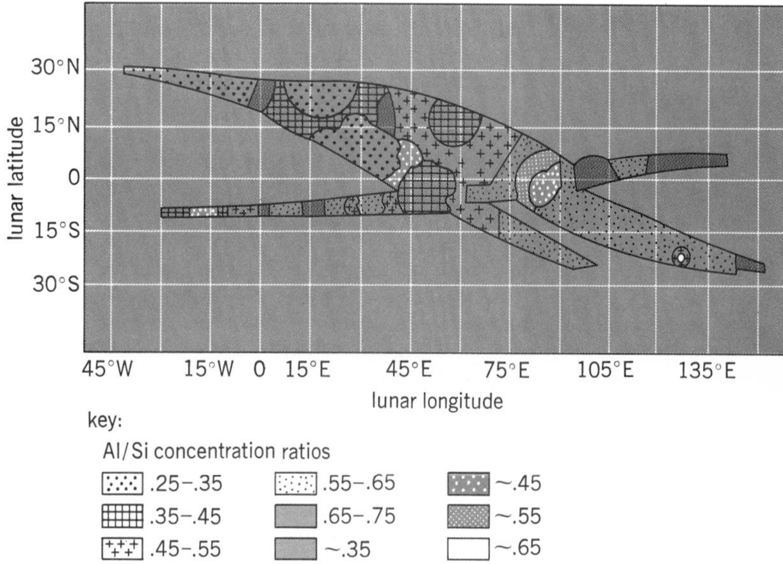

Fig. 4. Aluminum-silicon concentration ratios as detected by x-ray experiments on *Apollo 15* and *16*. (*I. Adler, University of Maryland*)

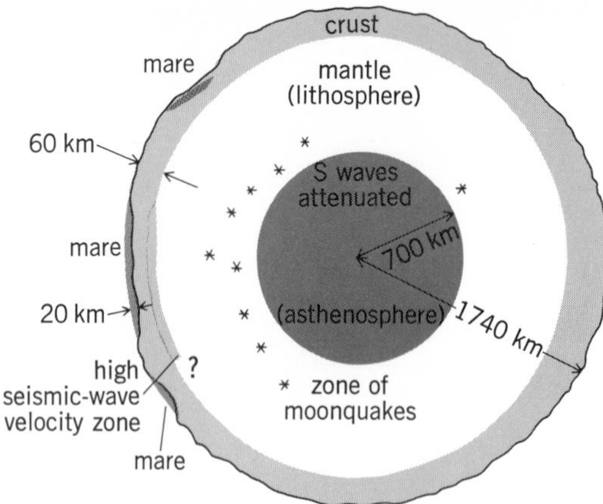

Fig. 5. Schematic diagram of lunar structure. The near side of the Moon is to the left of the figure. *(From S. R. Taylor, Lunar Science: A Post-Apollo View, © 1975 by Pergamon Press; reprinted with permission)*

acteristics would be very different from those of Earth. However, the bulk properties of the Moon are also quite different—the density alone is evidence of that—and the unraveling of the Moon's internal history and constitution is a great challenge to planetologists. *See* EARTH.

The Earth, with its dense metallic fluid core, convective mantle, strong and variable magnetic field with trapped radiation belts, widespread seismic tremors, volcanoes, and folded mountain ranges, moving lithospheric plates, and highly differentiated radioactive rocks, is plainly a planet seething with inner activity. Is the Moon also an active, evolving world or is it something very different? The answer lies in a group of related experiments now starting to be done on the Moon: seismic investigations, heat-flow measurements, surface magnetic and gravity profiles, determination of abundances and ages of the radioactive isotopes in lunar material, and comparison of the latter with those found in the Earth and meteorites. Present theories and preliminary data yield the following clues to the problem.

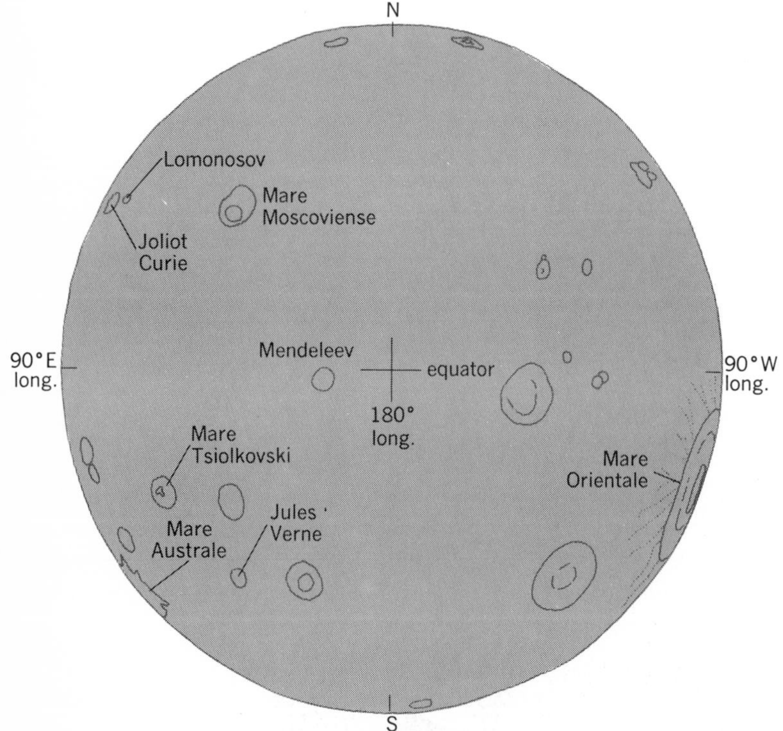

Fig. 6. Map of far side of Moon.

1. The Moon is too small to have compressed its silicates into a metallic phase by gravity; therefore, if it has a dense core at all, the core should be of nickel-iron. But the low mass of the whole Moon does not permit a large core unless the outer layers are of very light material; available data suggest that the Moon's iron core may have a diameter of at most a few hundred kilometers.

2. The Moon has no radiation belts, only a slight magnetic field, and behaves as a nonconductor in the presence of the interplanetary field.

3. The Moon's natural radioactivity from long-lived isotopes of potassium, thorium, and uranium, expected to provide internal heat sufficient for partial melting, was roughly measured from orbit by *Luna 10*, and the component above the cosmic-ray-induced background radiation was found to be at most that of basic or ultrabasic earthly rock, rather than that of more highly radioactive, differentiated rocks such as granites. *Apollo 11* and *12* rock samples confirmed this result; *Apollo 15*, *16*, and *17* mapped lunar composition and radioactivity from orbit (Fig. 3). X-ray experiments showed higher aluminum-silicon concentration ratios over highland areas and lower values over maria (Fig. 4), while magnesium-silicon ratios showed a converse relationship—higher values over maria and lower values over highlands.

When all of the Apollo observations are taken together, it is evident that the Moon was melted to an unknown depth and chemically differentiated about 4.5×10^9 years ago, leaving the highlands relatively rich in aluminum and an underlying mantle relatively rich in iron and magnesium, with all known lunar materials depleted in volatiles. The subsequent history of impacts and lava flooding is subject to varying interpretations, but there is general agreement on the final result: a thick, rigid crust with only minor evidence of recent basaltic extrusions. The temperature profile and physical properties of the Moon's deep interior are, despite the Apollo seismic and heat-flow data, under active debate. Figure 5 shows a rough sketch of the Moon as revealed by the data.

Large-scale surface features. As can be seen from the Earth with the unaided eye, the Moon has two major types of surface: the dark, smooth maria and the lighter, rougher highlands (Figs. 1 and 6). Photography by spacecraft shows that, for some unknown reason, the Moon's far side consists mainly of highlands (Fig. 7). Both maria and highlands are covered with craters of all sizes. Craters are more numerous in the highlands than in the maria, except on the steeper slopes, where downhill movement of material apparently tends to obliterate them. Numerous different types of craters can be recognized. Some of them appear very similar to the craters made by explosions on the Earth; they have raised rims, sometimes have central peaks, and are surrounded by fields of hummocky, blocky ejecta. Others are rimless and tend to occur

Fig. 7. Mare Tsiolkovski on far side of Moon. Crater, partly flooded by dark mare material, is about 200 km across. *(Langley Research Center, NASA)*

in lines along cracks in the lunar surface. Some of the rimless craters, particularly those with dark halos, may be gas vents; others may be just the result of surface material funneling down into subsurface voids. Most prominent at full moon are the bright ray craters (Fig. 1) whose grayish ejecta appear to have traveled for hundreds of kilometers across the lunar surface. Observers have long recognized that some erosive process has been and may still be active on the Moon. For example, when craters overlap so that their relative ages are evident, the younger ones are seen to have sharper outlines than the older ones. Bombardment of the airless Moon by meteoritic matter and solar particles, and extreme temperature cycling, are now considered the most likely erosive agents, but local internal activity is also a possibility. Rocks returned by the Apollo astronauts are covered with tiny glass-lined pits, confirming erosion by small high-speed particles.

The lunar mountains, though very high (8000 m or more), are not extremely steep, and lunar explorers see rolling rather than jagged scenery (Fig. 8). There are steep slopes (30–40°) on the inside walls and central peaks of recent craters, where the lunar material appears to be resting at its maximum angle of repose, and rocks can be seen to have rolled down to the crater bottoms.

Though a widespread network of fault traces is visible, there is no evidence on the Moon of the great mountain-building processes seen on the Earth. There are some low domes suggestive of volcanic activity, but the higher mountains are all part of the gently rolling highlands or the vast circular structures surrounding major basins. Figure 9 shows one of these, the Mare Orientale, as revealed by *Lunar Orbiter 4*. This large concentric structure is almost invisible from Earth because it lies just past the Moon's western limb; at favorable librations parts of its basin and mountain ramparts can be seen. The great region of radial sculpture surrounding the Orientale basin strongly suggests a catastrophic origin, with huge masses of matter thrown outward from the center. Note, however, the gentle appearance of the flooding by the dark mare material, which seems to lie only in the lowest parts of the concentric rings. Other basins, namely, Imbrium, Serenitatis, and Crisium, appear more fully flooded (Fig. 1). Were these maria created by giant impacts, followed by subsidence of the ejecta and upwelling of lava from inside the Moon, or is there another explanation? Examination of small variations in Lunar Orbiter motions has revealed that each of the great circular maria is the site of a positive gravity anomaly (excess mass), shown in Fig. 3. The old argument about impact versus vulcanism as the primary agent in forming the lunar relief, reflected in lunar literature over the past 100 years, appears to be entering a new, more complicated phase with the confirmation of extensive flooding of impact craters by lava.

In some of the Moon's mountainous regions bordering on the maria are found sinuous rilles (Fig. 10). These meandering valleys, some of them known since the 18th century, were shown in Lunar Orbiter pictures to have an exquisite fineness of detail. Some of them originate in small circular pits and then wriggle delicately across the Moon's gentle slopes for hundreds of kilometers, detouring around even slight obstacles, before vanishing on the plains. Though their resemblance to earthly rivers is strong, the sinuous rilles have no tributaries or deltas. What are they? No explanation offered (for example, dust flows, lava channels, or subsurface ducts made by water eroding ice) has proved entirely convincing.

Fig. 8. The crater Copernicus, showing the central peaks, slump terraces, patterned crater walls, and (background) slopes of the Carpathian Mountains. *(Langley Research Center, NASA)*

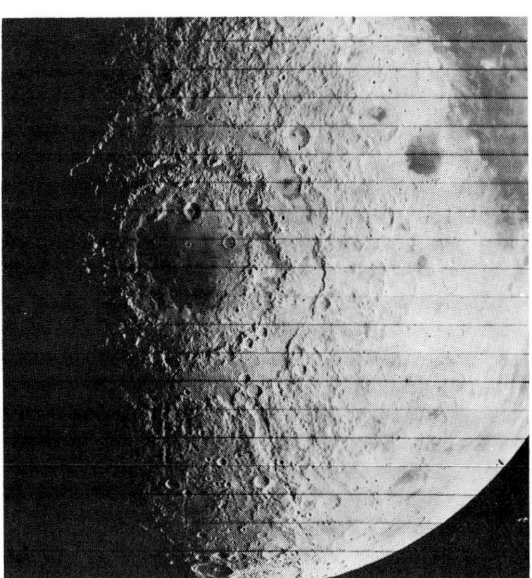

Fig. 9. Mare Orientale. *(Langley Research Center, NASA)*

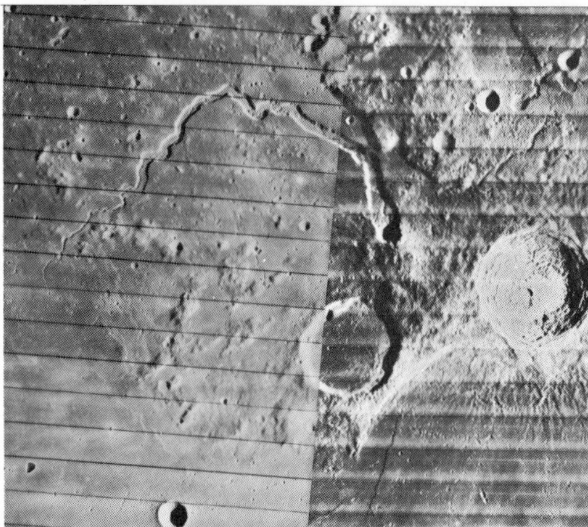

Fig. 10. Schroeter's Valley, a large sinuous rille, and craters Herodotus and Aristarchus. Aristarchus, brightest of the ray craters, is about 40 km across and 3600 m deep. *(Langley Research Center, NASA)*

Fig. 12. Lunar patterned ground, a common feature on moderate slopes. *(Langley Research Center, NASA)*

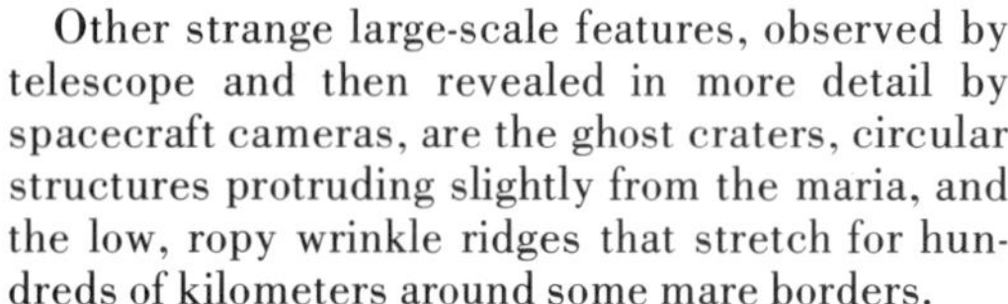

Other strange large-scale features, observed by telescope and then revealed in more detail by spacecraft cameras, are the ghost craters, circular structures protruding slightly from the maria, and the low, ropy wrinkle ridges that stretch for hundreds of kilometers around some mare borders.

Small-scale surface features. Careful observations, some of them made decades before the beginning of space flight, revealed much about the fine-scale nature of the lunar surface. Since the smallest lunar feature telescopically observable from the Earth is some hundreds of meters in extent, methods other than direct visual observation had to be used. Photometry, polarimetry, and later radiometry and radar probing gave the early fine-scale data. Some results of these investigations suggested bizarre characteristics for the Moon. Nevertheless, many of their findings have now been confirmed by spacecraft. The Moon seems to be totally covered, to an unknown depth, by a layer of rubble and soil with very peculiar optical and thermal properties, but otherwise not too different from certain soils on the Earth. The observed optical and radio properties are as follows.

1. The Moon reflects only a small portion of the light incident on it (the average albedo of the maria is only 7%, darker than any familiar object except things like soot or black velvet).

2. The full moon is more than 10 times as bright as the half moon.

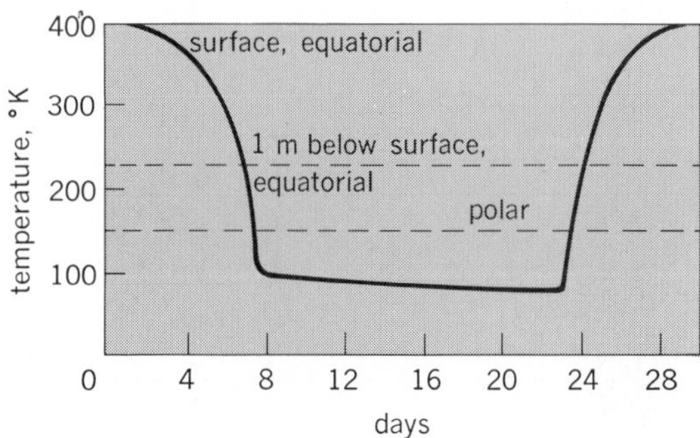

Fig. 11. Lunar surface and subsurface temperatures.

3. At full moon, the disk is almost equally bright all the way to the edge; that is, there is no "limb darkening" such as is observed for ordinary spheres, whether they be specular or diffuse reflectors.

4. Color variations are slight; the Moon is a uniform dark gray with a small yellowish cast. Some of the maria are a little redder, some a little bluer, and these differences do correlate with large-scale surface morphology, but the visible color differences are so slight that they are detectable only with special filters. (Infrared spectral differences are more pronounced and have provided a method for mapping variations in the Moon's surface composition.)

5. The Moon's polarization properties are those of a surface covered completely by small, opaque grains in the size range of a few microns.

6. The material at the lunar surface is an extremely good thermal insulator, better than the

Fig. 13. Lunar soil and rocks, and the trenches made by *Surveyor 7*. *(Jet Propulsion Laboratory, NASA)*

Fig. 14. Crater showing appearance of upwelling on its floor. (*Langley Research Center, NASA*)

Fig. 15. Region near crater Tycho, including *Surveyor 7* landing site. (*Langley Research Center, NASA*)

most porous terrestrial rocks. The cooling rate as measured by infrared observations during a lunar eclipse is strongly variable; the bright ray craters cool more slowly than their surroundings.

7. The Moon emits thermal radiation in the radio wavelength range; interpretations of this and the infrared data yield estimates of surface and shallow subsurface temperatures as shown in Fig. 11.

8. At wavelengths in the meter range, the Moon appears smooth to radar, with a dielectric constant lower than that of most dry terrestrial rocks. To centimeter waves, the Moon appears rather rough, and at visible light wavelengths (a conclusion from observations 1–5 above), it is extremely rough.

These observations all point to a highly porous or underdense structure for at least the top few millimeters of the lunar surface material. The so-called backscatter peak in the photometric function, which describes the sudden brightening near full Moon, is characteristic of surfaces with deep holes or other roughness elements which are shadowed when the lighting is oblique.

The Ranger, Luna, Surveyor, and Lunar Orbiter missions made it clear that these strange electromagnetic properties are generic characteristics of the dark-gray, fine soil that appears to mantle the entire Moon, softening most surface contours and covering everything except occasional fields of rocks (Figs. 12 and 13). This soil, with a slightly cohesive character like that of damp sand and a chemical composition similar to that of some basic silicates on the Earth, is a product of the radiation, meteoroid, and thermal environment at the lunar surface. Figure 12 shows a surface texture, called patterned ground, that is common on the moderate slopes of the Moon. This widespread phenomenon is unexplained, though there are some similar surfaces developed on the Earth when unconsolidated rock, lava, or glacial ice moves downhill beneath an overburden. At many places on the Moon, there is unmistakable evidence of downward sliding or slumping of material and rolling rocks. There are also a few instances of apparent upwelling (Fig. 14), as well as numerous "lakes" where material has collected in depressions. Figures 15, 16, and 17 show, at different scales, the landing site of *Surveyor 7* near the great ray crater Tycho (Fig. 1). *Surveyor 7* found the soil of the highlands near Tycho to be rockier than, and slightly different in chemical composition to, the mare materials sampled by earlier Surveyors. Figure 13 shows trenches made in the lunar soil during soil mechanics experiments on *Surveyor 7*. Magnets on the Surveyors collected magnetic particles from the soil, demonstrating the presence of either meteoritic or native iron minerals at the sites examined. Meteoroid experiments on the Lunar Orbiters showed about the same flux of small particles as is observed at the Earth, so that the lunar soil would be expected to contain a representative sample of meteoritic and possibly also cometary matter. Apollo results confirmed and extended the Surveyor data and also indicated that glassy particles are abundant in and on the soil. Evidence of

Fig. 16. *Orbiter* photograph of *Surveyor 7* site. (*Langley Research Center, NASA*)

Fig. 17. Surface view at *Surveyor 7* site. (*Jet Propulsion Laboratory*)

micrometeoroid bombardment is seen in the many glass-lined microcraters found on lunar rocks (Fig. 18).

Chemical, mineral, and isotopic analyses of minerals from the *Apollo 11* site showed that mare rocks there are indeed of the basic igneous class and are very ancient ($3-4 \times 10^9$ years). The *Apollo 12* samples are significantly younger, suggesting that Mare Tranquillitatis and Oceanus Procellarum were formed during a long and complex lunar history. The *Apollo 12* astronauts visited *Surveyor 3* (Fig. 19) and brought back parts of that spacecraft to permit analysis of the effects of its $2\frac{1}{2}$-year exposure on the surface of the Moon. The lunar rock and soil samples returned by the Apollo and Luna missions have yielded much new information on the composition and history of the Moon. Among the dominant characteristics of these rocks are enrichment in refractories, depletion in volatiles, much evidence of repeated breaking up and rewelding into breccias, and ages since solidification extending back from the mare flows of $3-4 \times 10^9$ years ago into the period of highland formation more than 4×10^9 years ago, but not as yet including the time of the Moon's original accretion. Some characteristics of the lunar samples are summarized in Table 4.

As the Apollo missions progressed, each new landing site was selected with the aim of elucidating more of the Moon's history. A main objective was to sample each of the geologic units mapped by remote observation, either by landing on it or by collecting materials naturally transported from it to the landing site. Although this process did result in collection of both mare and highland materials with a wide range of ages and chemical compositions, it did not result in a complete unraveling of the history of the Moon. Apparently, the great impacts of $3-4 \times 10^9$ years ago erased much of the previous record, resetting radioactive clocks and scrambling minerals of diverse origins into the complicated soils and breccias found today. *See* METEORITE.

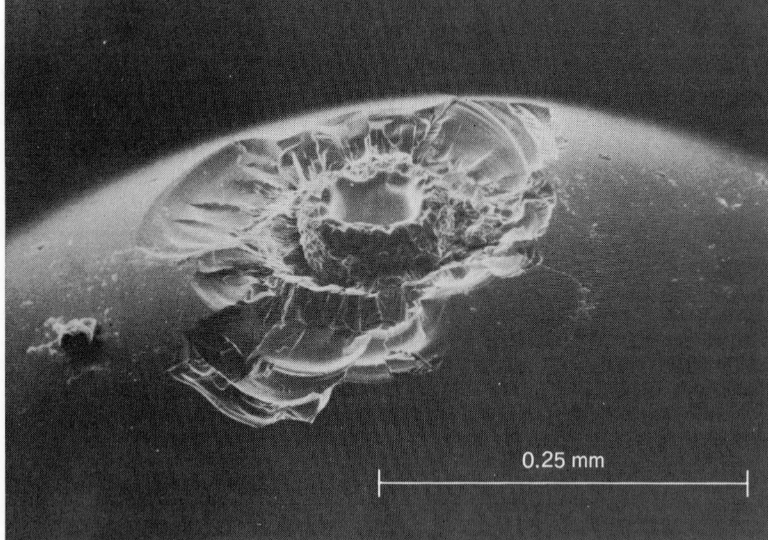

Fig. 18. Example of microcratering, caused by hypervelocity impact of tiny particles, on a dark-brown glass sphere. The diameter of the sphere is approximately 0.75 mm, and the diameter of the inner crater, inside the raised rim, is about 50 μ. This photograph was taken through a scanning electron microscope.

Atmosphere. Though the Moon may at one time have contained appreciable quantities of the volatile elements and compounds (for example, hydrogen, helium, argon, water, sulfur, and carbon compounds) found in the meteorites and on the Earth, its high daytime surface temperature and low gravity would cause rapid escape of the lighter elements. Solar ultraviolet and x-ray irradiation would tend to break down volatile compounds at the surface, and solar charged-particle bombardment would ionize and sweep away even the heavier gas species. Observations from the Earth, looking for a twilight glow of the lunar atmosphere just past the terminator on the Moon, and watching radio-star occultations have all been negative, setting an upper limit of 10^{-12} times the Earth's sea-level atmospheric density for any lunar gas envelope. Therefore, either the lunar volatile compounds have vanished into space or they are trapped beneath the surface. The samples returned by Apollo are enriched in refractory elements, depleted in volatiles, and impregnated with rare gases from the solar wind. No water appears to have ever been present at any Apollo site, and carbonaceous materials were present, if at all, only in very small amounts.

Occasional luminescent events reported by reliable observers suggest that some volcanic gases are vented from time to time on the Moon, particularly in the regions of the craters Aristarchus and Alphonsus. A slight, transient atmosphere does exist on the night side of the Moon as a result of the trapping and release of gas molecules at the very low temperatures prevailing there; also frozen liquids or gases could exist in permanently shadowed crater bottoms near the lunar poles. No experiment to detect such accumulations of volatiles has been made. *Lunokhod 2*, a Soviet roving spacecraft, measured a slight glow attributed to a very thin cloud of small particles near the surface, which could explain the Surveyor observations of a slight horizon glow after sunset. Also, the ALSEP

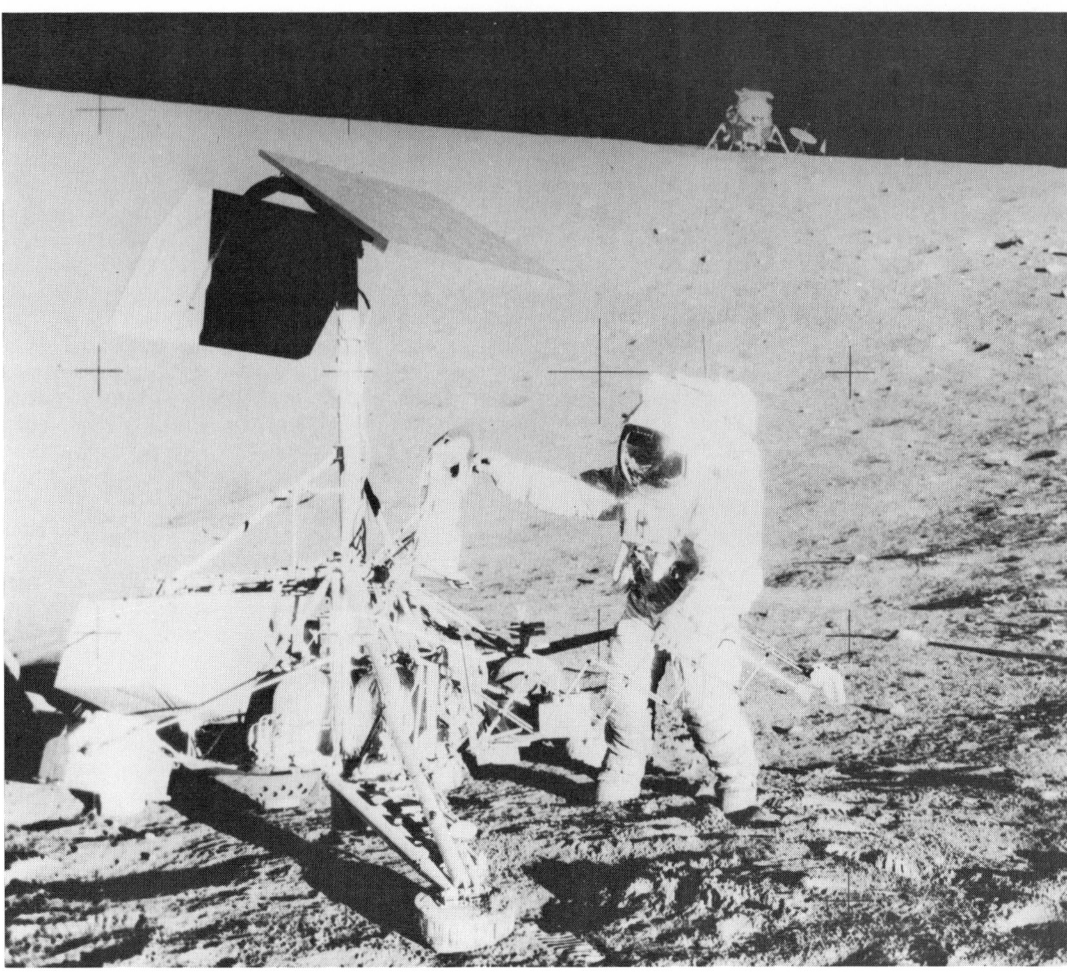

Fig. 19. *Apollo 12* astronaut examining *Surveyor 3*. Lunar module in background. (*NASA*)

Table 4. Some selected data from Apollo and Luna missions

Mission	Main sample properties	Other data
Apollo 11	Mare basalts, differentiated from melt at depth 3.7×10^9 years ago. Some crystalline highland fragments in soils. Unexpected abundance of glass. Much evidence of impact shock and microcratering. No water or organic materials.	Study of seismic properties showed low background, much scattering, and low attenuation.
Apollo 12	Basalts 3.2×10^9 years old. One sample 4.0×10^9 years old includes granitic component. Some samples with high potassium, rare-earth elements, and phosphorus (KREEP) may be Copernicus crater ejecta.	Surveyor parts returned showed effects of solar and cosmic bombardment.
Apollo 13	Spacecraft failure—no samples.	Despite emergency, some lunar photos returned.
Luna 16	Basalt 3.4×10^9 years old, relatively high Al content.	
Apollo 14	Shocked highland basalts, probably Imbrium ejecta, 3.95×10^9 years old, higher Al and lower Fe than mare materials.	Deep moonquakes.
Apollo 15	Highland anorthosites including one sample 4.1×10^9 years old, mare basalts similar to *Apollo 11* samples.	Orbital remote sensing began mapping of surface compositions.
Luna 20	Possibly Crisium ejecta, 3.9×10^9 years old.	
Apollo 16	Highland anorthosite breccias $3.9–4 \times 10^9$ years old, also possibly Imbrium ejecta.	Seismic network began recording locations of impacts and deep moonquakes; orbital compositional mapping extended.
Apollo 17	Variety of basalts and anorthosites $3.7–4 \times 10^9$ years old, possibly volcanic glass, few dunite fragments 4.48×10^9 years old, possibly surviving from before the great highland bombardment.	Orbital mapping, and study of seismic, particle-and-field, and subsurface electrical properties yielded comprehensive (but still unexplained) picture of Moon.

(Apollo lunar surface experiments packages) experiments landed by Apollo have occasionally detected small gas emanations, including water. It is clear, however, that any lunar atmosphere is exceedingly tenuous. [JAMES D. BURKE]

Bibliography: *Apollo 15 Preliminary Science Report*, NASA SP-289, 1972; *Apollo 16 Preliminary Science Report*, NASA SP-315, 1972; *Apollo 17 Preliminary Science Report*, NASA SP-330, 1973; Apollo Preliminary Examination Team, Preliminary examination of lunar samples from *Apollo 11*, *Science*, 165(3899):1211, 1969; R. B. Baldwin, *The Face of the Moon*, 1949; R. B. Baldwin, *The Measure of the Moon*, 1963; V. de Callatay, in A. de Visscher (ed.), *Atlas of the Moon*, transl. by R. G. Lascelles, 1964; A. Dollfus (ed.), *Moon and Planets I: 7th International Space Science Symposium, Cospar, IAU, Vienna*, May 10–18, 1966; A. Dollfus (ed.), *Moon and Planets II: A Session of the 10th Plenary Meeting of Cospar, London*, July 26–27, 1967; F. El-Baz, The Moon after Apollo, *Icarus*, 25:495–537, 1975; G. Fielder, *Lunar Geology*, 1965; W. Hess (ed.), *The Nature of the Lunar Surface: IAU-NASA Symposium*, 1965; Z. Kopal and Z. K. Mikhailov (eds.), *The Moon*, 1962; G. P. Kuiper (ed.), *Photographic Lunar Atlas*, suppl. 2: *Rectified Lunar Atlas*, 1963; R. A. Lyttleton, *Mysteries of the Solar System*, 1968; A. Pannekoek, *A History of Astronomy*, 1961; S. F. Singer (ed.), *Physics of the Moon: Proceedings of AAS/AAAS Symposium, Washington, Dec. 29, 1966*, vol. 13, 1967; *Surveyor 7: A Preliminary Report*, NASA SP-173, May, 1968; S. R. Taylor, *Lunar Science: A Post-Apollo View*, 1975; USSR Academy of Science, *Atlas of the Far Side of the Moon*, pts. 1 and 2, 1960, 1967; USSR Academy of Science, *First Panoramas of the Surface of the Moon*, 1966; F. L. Whipple, *Earth, Moon and Planets*, 3d ed., 1968.

Moraine

An accumulation of glacial drift deposited chiefly by direct glacial action, and possessing initial constructional form independent of the floor beneath it. Moraine is divided into two chief kinds: ground moraine and end moraine. Ground moraine is thin and possesses only small relief, devoid of linear elements. Its surface is an undulating plain marked by swells and basins, all with gentle slopes. It consists chiefly of till.

End moraines. (*a*) Wisconsin (*Wisconsin Geological Survey*). (*b*) Pennsylvania (*U.S. Geological Survey*).

End moraine is a ridgelike accumulation of drift built at the margin of an active glacier (see illustration). Its form is primarily constructional. It is subdivided into three types: terminal moraine (built at the terminal margin of a glacier lobe), lateral moraine (built at the lateral margin of a lobe), and interlobate moraine (built along the line of junction of two adjacent lobes). The three types grade into one·another. End moraines can consist of any combination of till and stratified drift. They range in height to as much as 300 m, and in topographic detail consist of complexes of knolls and basins with steep slopes. Their scientific value lies principally in the fact that they define significant positions of the margins of glaciers, either the limits of glacial advances or stable positions during periods of deglaciation. *See* GLACIATED TERRAIN. [RICHARD F. FLINT]

Mountain systems

Defined by some authorities as cordilleras, cordilleran belts, fold belts, or mountain chains. A major part of each continent is made up of mountain systems, although some mountains are geologically old and much subdued by erosion. The youthful and high mountain systems of the world are in two great chains: the east-west Mediterranean system, which stretches from Spain and Morocco to Malaya and the East Indies, and the circum-Pacific belt, which embraces the cordilleras of South and North America, the Alaskan ranges, the eastern Asiatic island archipelagos, and the New Zealand–New Guinea mountains. *See* CORDILLERAN BELT.

Other authorities take a more restricted view of a mountain system and recognize it as being composed of several mountain ranges tied together by common geological features. Several systems compose a cordillera in this concept, and each system stands apart from the others by its special geological and topographic features. The systems of the North American cordillera number about 20, depending on various classifications, and are most complex in the western United States. There the systems from California eastward are the Coast Ranges and the Great Valley, the Sierra Nevada, the Basin and Range system (Great Basin), the Columbia Plateaus, the Colorado Plateaus, and the Rocky Mountains of Wyoming and Colorado (see illustration).

Some systems are young and have incessant earthquake activity, such as the Coast Ranges. Some have been built in two or more stages, such as the Sierra Nevada. The first stage consisted of the intrusion of vast volumes of molten rock into the upper crust, to cool and solidify into a granitelike rock; the second stage consisted of the elevation and tilting of a block of the Earth's crust consisting largely of the granite (tilted fault block).

Another type of mountain system may consist of huge piles of volcanic rocks that have been built around volcanic vents. An example is the Cascade Range of Oregon and Washington.

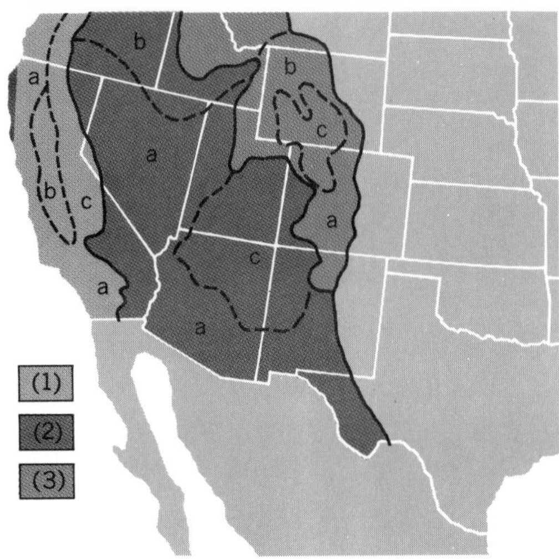

Mountain systems of southwestern United States. (1) Pacific Mountain system: a = Coast Ranges, b = Great Valley (basin), c = Sierra Nevada; (2) Intermontane Plateau systems: a = Basin and Range (Great Basin), b = Columbia Plateaus, c = Colorado Plateaus; (3) Rocky Mountain systems: a = Southern Rocky Mountains, b = Middle Rocky Mountains, c = Wyoming Basin.

Still another type of mountain system occurs where large basins have been filled with thick series of sediments in near-horizontal layers which subsequently have been deformed so that the strata are intensely folded. Such is the case in the Appalachians, the Alps, and the Himalayas. *See* TECTONIC PATTERNS.

[ARMAND J. EARDLEY]

Mud volcano

A conical landform composed of clay and silt with variable admixtures of sand and rock fragments, the whole resulting from eruption of wet mud and impelled upward by fluid or gas pressure. The form of such features depends upon the ratio of mineral matter to the fluids and gases in the mobile material. More liquid mud produces oval or round cakelike forms known as mud fladens, a few square feet to an acre or more in area and with heights of only 5–10 ft above their surroundings. More viscous mud constructs mud cones with slopes occasionally as steep as 40° and diameters ranging upward to several hundred yards. With repeated eruptions such cones may be built up to heights of many hundred feet (Figs. 1 and 2).

Fig. 1. Mud volcanoes near Douglaston, N.Y. Photograph taken at low tide, Sept. 12, 1903. (*USGS*)

Fig. 2. A group of mud volcanoes, which are now submerged beneath the Salton Sea, 6 mi northwest of Imperial Junction, Calif. (*USGS*)

Associated features include mud pits, developed where the mobile material contains excessive amounts of fluid and gas, and pitch lakes, formed where the bitumens of petroleum residues are at least as voluminous as the clay particles in the viscous mud. Mud pits range in size from mere pinholes to depressions covering several acres. Minor amounts of petroleum residues are often found in mud volcanoes and associated features, but pitch lakes are comparatively rare phenomena. Mud volcanoes arising out of the sea do on occasion form short-lived islands, as in Trinidad, B.W.I. *See* ASPHALT AND ASPHALTITE; PETROLEUM GEOLOGY.

Occurrence. Mud volcanoes are most commonly observed in regions of geologically recent and tectonically intense orogeny, as in Trinidad, B.W.I.; the Caucasus; Java; Burma; Romania; and Colombia; or in areas of unusually rapid sedimentation with compaction lag, as along the Gulf Coast of North America or in intermontane basins at many places throughout the world.

Dynamics. The eruptive energy responsible for the formation of mud volcanoes and associated features is that of the pressure of fluids and gases accumulated in sedimentary deposits. No igneous (pyrogenetic) activity is involved. Layers or lenses of incompetent clay or silt with at least average puddling properties are ordinarily essential. Eruptions occur when the accumulated pressure in the reservoir rocks exceeds the overburden or hydrostatic pressure. This may result from orogenetic or compactional stresses, or may be due to the removal of critical amounts of overburden by erosion. Compaction may be abruptly expedited by earthquake vibrations; fluid and gas pressures may be increased or decreased by various geochemical or geophysical processes.

The rate, intensity, frequency, and duration of eruptions depend on the accumulated stress potential, on the permeability and saturation of the source material, and on the physical properties of the clay or silt under stress. Activity is commonly intermittent over a period of years, but this is by no means always the case. Some occurrences, such as those on the island of Trinidad, are known to have been active since late Pleistocene time.

[HANS H. SUTER]

Bibliography: G. E. Higgins and J. B. Saunders, Report on 1964 Chatham Mud Island, Erin Bay, Trinidad, West Indies, *Amer. Ass. Petrol. Geol.*, 51: 55–64, 1967; A. I. Levorsen, *Geology of Petroleum*, 1954.

Muscovite

One of the mica group of minerals with the composition $K_2Al_4(Si_6Al_2)O_{20}(OH)_4$. It is also called white, or potash, mica.

Muscovite occurs in some granites and is especially abundant in pegmatites, the chief commercial deposits. It is a widespread constituent of slates, phyllites, schists, and gneisses. Secondary fine-grained muscovite (sericite), usually of hydrothermal origin, replaces many silicates, particularly feldspars. Detrital muscovite persists in sandstones.

Minor amounts of sodium (Na), barium (Ba), and rubidium (Rb) may substitute for potassium (K); some magnesium (Mg), ferrous iron [Fe(II)], ferric iron [Fe(III)], manganese (Mn), and chromium (Cr, in fuchsite) for aluminum (Al); and minor amounts of fluorine (F) for hydroxyl (OH).

Most muscovites have the two-layer monoclinic structure; a few have the one-layer monoclinic type or the three-layer hexagonal type. Crystals are tabular sheets with prominent base and hexagonal or rhomboid outline (see illustration). Aggregates of these sheets are large to small books in radial, parallel, or random orientation; plumose to vermicular intergrowths with quartz; globular to hemispherical clusters of concentric plates; and fine-grained masses of irregular or pseudomorphous form. Physical properties include easy perfect basal cleavage to sheets that are flexible and elastic; specific gravity of 2.7–3.1; and hardness of 2–2.5 on Mohs scale. Colors in sheets are in shades of brown, green, and yellow; viewed microscopically, muscovite is colorless. *See* MICA; SILICATE MINERALS.

Muscovite in aggregate of tabular sheets, taken from Washington Heights, New York City. (*American Museum of Natural History specimens*)

[E. WILLIAM HEINRICH]

Mylonite

A rock that has suffered extreme mechanical deformation and granulation but has remained chemically unaltered. In the early stages of this process, cataclasites or flaser rocks are produced. These rocks are characterized by the fact that the nature of the parent rock is easily recognized. The mylonites, however, are altered beyond recognition. In spite of their pulverized condition they are hard, coherent, and often black and glassy-looking rocks, because the crushing took place under confining pressures so high that the comminuted particles have been welded tightly together. Pseudotachylite is the ultimate product of mechanical grinding. It is black, glassy, and partly isotropic when viewed under the microscope. X-ray analyses show that it consists of submicroscopically comminuted dust of various minerals, mostly quartz and feldspar. *See* METAMORPHIC ROCKS.

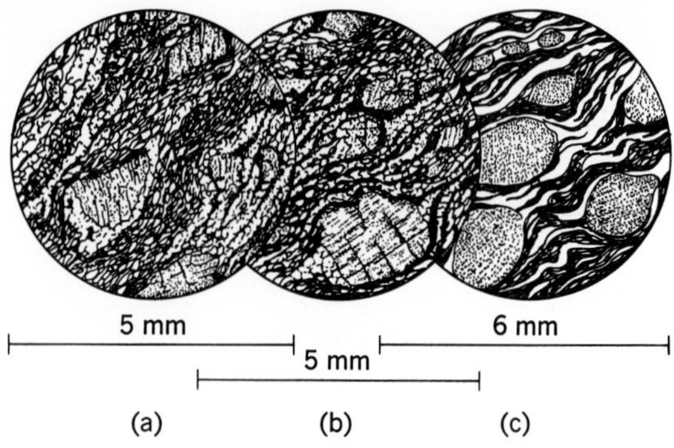

Mylonites. (*a*) Strained and broken coarse crystals (porphyroclasts) of feldspar and a train of garnet granules set in fine-grained schistose matrix of quartz and feldspar veined with granoblastic quartz; San Gabriel Mountains, Calif. (*b*) Granite mylonite; coarse, strained, partially granulated crystals of plagioclase, microcline, and quartz; granulated matrix composed of quartz, feldspar, and biotite; San Gabriel Mountains, Calif. (*c*) Mylonitic augen gneiss; ovid relic crystals of plagioclase and of potash feldspar in a matrix of muscovite, chlorite, and quartz and traversed by swarms of stringers of later undeformed quartz; Deadman Lake, British Columbia. (*From H. Williams, F. J. Turner, and C. M. Gilbert, Petrography: An Introduction to the Study of Rocks in Thin Sections, copyright © 1954 by W. H. Freeman and Co.*)

True mylonites still contain relics of uncrushed parent rock in lensoid patches parallel to the direction of movement. But microscopically the minerals of these patches of relic material exhibit obvious strain effects, such as marginal bending and granulation, cracking, and undulatory extinction. Quartz in particular exhibits undulatory extinction, often in combination with the so-called Boehm lamellae. The fine lamellae probably are produced by intracrystalline gliding and are often emphasized by dusty inclusions which are subparallel to the basal plane [0001]. Such quartz grains, as well as feldspars, often appear as insets or eyes in the granulated matrix. This has been denoted as porphyroclastic texture (see illustration).

Quartz and feldspar (alkali feldspar and acid plagioclase) are the chief minerals in the most common types of mylonite, because they are brittle and chemically stable over a wide range of temperatures. Thus belts of mylonite in granite, gneiss, and quartzites are common. Mylonites are found in practically all kinds of rock; the dunite mylonites are found in many parts of the world.

Laminated fabric is frequently found in mylonites; it can be correlated with flow movements (or creep movements) in the rock during its deformation. A breccia structure also may be present; that is, the mylonite mass is made up of angular fragments. The fragments vary in their development according to whether the recrystallization occurred during or after the period of granulation. Rocks which have recrystallized after the granulation are called blastomylonites.

[T. F. W. BARTH]

Native elements

Those elements which occur in nature uncombined with other elements. Aside from the free gases of the atmosphere there are about 20 elements that are found as minerals in the native state. These are divided into metals, semimetals, and nonmetals. Gold, silver, copper, and platinum are the most important metals and each of these has been found abundantly enough at certain localities to be mined as an ore. Native gold and platinum are the major ore minerals of these metals. Rarer native metals are others of the platinum group, lead, mercury, tantalum, tin, and zinc. Native iron is found sparingly both as terrestrial iron and meteoric iron. *See* ORE AND MINERAL DEPOSITS.

The native semimetals can be divided into (1) the arsenic group, including arsenic, antimony, and bismuth; and (2) the tellurium group, including tellurium and selenium. The members of the arsenic group crystallize in the hexagonal system, scalenohedral class; those of the tellurium group in the hexagonal system, trigonal trapezohedral class. Only rarely do the semimetals occur abundantly enough to be mined as ores of their respective elements. *See* MINERALOGY.

The native nonmetals are sulfur, and carbon in the forms of graphite and diamond. Native sulfur is the chief industrial source of that element.

[CORNELIUS S. HURLBUT, JR.]

Native metals

Elements belonging to the metals group which occur as minerals in the native state in the rocks of the Earth's crust. The most common native metals, with rather simple internal (atomic) structures and isometric symmetry, constitute three groups: the platinum group (space group *Fm3m*), comprising platinum, palladium, iridium, and osmium, all of which are isostructural; the gold group (space group *Fm3m*), comprising gold, silver, copper, and lead, all of which are isostructural; and the iron group, comprising iron and nickel-iron, of which pure iron, as well as one variety of nickel-iron (kamacite), has space group *Im3m* and the more nickel-rich variety of nickel-iron (taenite) has space group *Fm3m*.

Platinum is the most abundant element of the platinum group, although its average abundance in the Earth's crust is only 0.01 part per million (ppm). Very rare members of the platinum group are palladium, platiniridium, and iridosmine. Platiniridium and iridosmine are alloys of iridium and platinum and iridium and osmium, respectively, which differ from other platinum group minerals by having hexagonally close-packed structures and space group $P6_3/mmc$.

The elements of the gold group are of the same group in the periodic table and therefore have somewhat similar chemical properties. All occur in their elemental state in nature and exhibit identical structures in which atoms are held together by rather weak metallic bonds. These elements are rather soft, malleable, ductile, and sectile. They are excellent conductors of heat and electricity, display metallic luster and hackly fracture, and have relatively low melting points. All exhibit high, isometric symmetry (4/m $\bar{3}$ 2/m) and have high densities.

Members of the iron group include pure iron (αFe) and two species of nickel-iron, kamacite and taenite. Elemental iron occurs only sporadically on the Earth's surface, and the two nickel-iron species are found only in meteorites. All three species are isometric, with pure iron and kamacite exhibiting space group *Im3m*, and taenite, space group *Fm3m*. The composition of kamacite ranges from pure iron to iron with about 5.5 wt % nickel, and the composition of taenite ranges from iron with 27 wt % nickel to iron with 65 wt % nickel in solid solution.

Platinum. This mineral is composed essentially of elemental platinum and exhibits isometric symmetry (4/m $\bar{3}$ 2/m). Platinum is not generally found in good crystal forms but rather in small grains and scales, and also in irregular masses and nuggets. Hardness is 4–4½, and specific gravity ranges from 21.45 when pure to 14–19 in the native state with other elements in solid solution. It is malleable and ductile and has a steel gray color and bright luster. Most native platinum contains small amounts of Fe, Ir, Os, Rh, Cu, Au, and Ni in solid solution. The structure of platinum, with space group *Fm3m*, is based on cubic closest packing of platinum atoms.

Most platinum occurs in the native state in ultrabasic igneous rocks such as peridotites, pyroxenites, and dunites. It also occurs in placer deposits which are usually close to the platinum-bearing igneous parent rock. The three largest platinum producers are the Republic of South Africa, the Soviet Union, and Canada. Platinum in South Africa is produced from the Merensky Reef in the ultrabasic rocks of the Bushveld igneous complex and as a by-product of gold mining in the Witwa-

Fig. 2. Native silver from Chañarcillo, Atacama, Chile. (*From C. S. Hurlbut, Jr., and C. Klein, Manual of Mineralogy, 19th ed., copyright © 1977 by John Wiley & Sons, Inc.; reprinted by permission*)

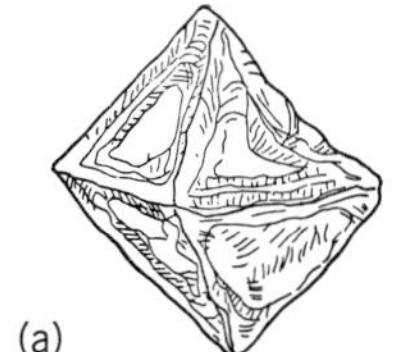

(a)

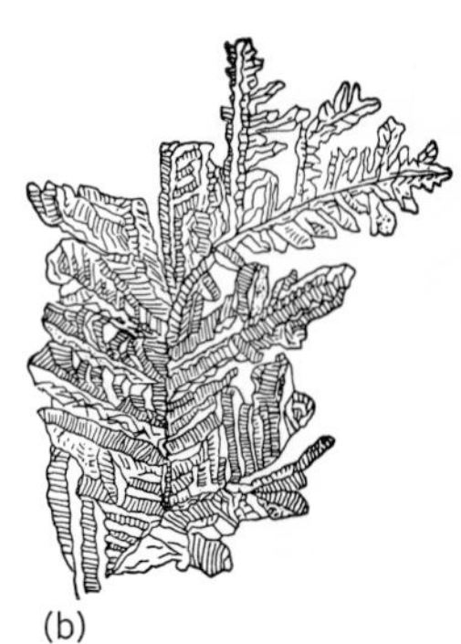

(b)

Fig. 1. Crystal forms of gold: (*a*) malformed octahedron and (*b*) dendritic. (*From C. S. Hurlbut, Jr., and C. Klein, Manual of Mineralogy, 19th ed., copyright © 1977 by John Wiley & Sons, Inc.; reprinted by permission*)

Fig. 3. Native copper crystals in arborescent groupings, Keweenaw Peninsula, MI. *(From C. S. Hurlbut, Jr., and C. Klein, Manual of Mineralogy, 19th ed., copyright © 1977 by John Wiley & Sons, Inc.; reprinted by permission)*

tersrand conglomerate. The Canadian platinum comes mainly from sperrylite ($PtAS_2$), which occurs in the copper-nickel ore in Sudbury, Ontario. *See* DUNITE; PERIDOTITE; PYROXENITE.

The uses of platinum are directly related to its high melting point (1755°C), resistance to chemical attack, and superior hardness. It is used in the chemical, automotive, and electrical industries, in dentistry, and for jewelry and surgical equipment.

Gold. This chemical element occurs as a mineral, with isometric symmetry (4/m $\bar{3}$ 2/m). When exhibiting crystal form, gold is commonly octahedral (Fig. 1*a*); it may occur in arborescent crystal groups (Fig. 1*b*), but is usually found in irregular plates, scales, or masses. Hardness is 2½–3, and specific gravity is 19.3 when pure. The presence of other metals (such as silver) in solid solution in the structure may lower specific gravity to 15. Gold is very malleable and ductile. Color is various shades of yellow, depending on the amount of silver in solid solution. The structure of gold consists of gold atoms in cubic closest packing with space group *Fm*3*m*. A complete solid solution series exists between gold and silver because of the identical atomic radii of these two metals (both 1.44 A or 0.144 nm). Most gold contains small amounts of silver; California gold contains between 10 and 15%. Small amounts of copper may also be present in solid solution. The purity of gold (termed its fineness) is expressed in parts per thousand. Much gold contains about 10% of other metals and thus has a fineness of 900.

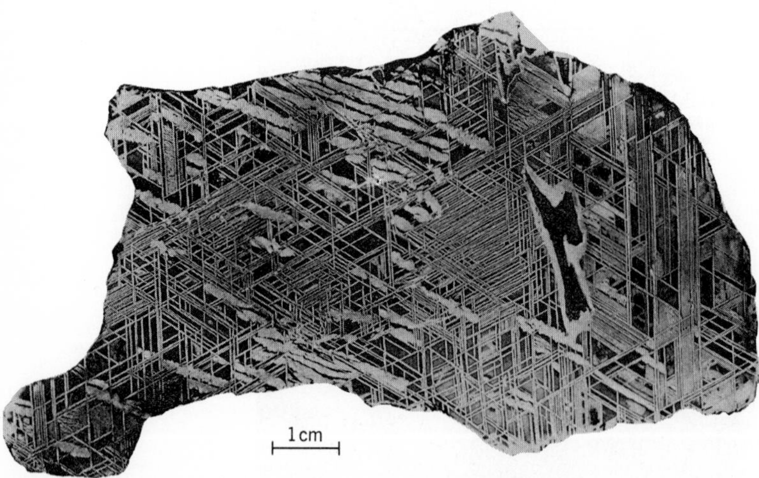

Fig. 4. The Edmonton (Kentucky) iron meteorite showing Widmanstätten pattern outlined by coarse blades of kamacite and very thin taenite lamellae. *(From C. S. Hurlbut, Jr., and C. Klein, Manual of Mineralogy, 19th ed., copyright © 1977 by John Wiley & Sons, Inc.; reprinted by permission)*

Gold has an average abundance of 0.004 ppm in the Earth's crust, and is therefore a rare element. It is widely distributed in nature in very small amounts but is found in minable concentrations mainly in veins and eluvial (placer) deposits, which are the weathering products of vein-type occurrences. The vein-type deposits consist mainly of quartz, with pyrite and other sulfides and variable concentrations of elemental gold. The placer deposits consist of sands and gravels between which fine particles or larger nuggets of gold are concentrated. Such placer deposits can be of recent origin, in which case the deposit is surficial and consists of loose fluvial materials, or they can be found as much older conglomeratic rock units. The largest producer of gold is the Precambrian Witwatersrand conglomerate, "the Rand" (a fossil placer), in the Transvaal of South Africa. Similar gold-carrying conglomerates occur in the Orange Free State of South Africa. The Republic of South Africa produces approximately two-thirds of the world's supply. Although no accurate figures are available, the Soviet Union is believed to be the second-ranking country in gold production. In the United States the most important gold-producing states are South Dakota, Nevada, Utah, Arizona, Colorado, and Montana.

Gold is used mainly as a monetary standard. Other uses include jewelry, small gold bars for investment purposes, scientific instruments, gold leaf, and dental appliances.

Silver. This chemical element occurs as a mineral with isometric symmetry (4/m $\bar{3}$ 2/m). Crystals are not commonly observed; when present they may be malformed or in branching or reticulated groups. Silver most frequently is found as irregular masses and plates, or in coarse and fine wire (Fig. 2). Hardness is 2½–3, and specific gravity ranges from 10.5 when pure to 10–12 when impure (other metals in solid solution may be Au, Hg, and Cu). Malleable and ductile, silver displays a metallic luster and silver-white color which frequently tarnishes to brown or gray-black. Native silver may contain considerable amounts of Au or Hg and, less commonly, small amounts of As, Sb, Pt, and Cu in solid solution. The structure of silver, with space group *Fm*3*m*, is based on cubic closest packing of silver atoms (isostructural with gold).

Native silver is widely distributed in small amounts, mainly in the oxidized zones of ore deposits. However, the larger deposits are the result of deposition from hydrothermal solutions. Such deposits can be grouped into three types: (1) Those associated with various silver minerals, sulfides, zeolites, calcite, barite, fluorite, and quartz. An example of such a deposit is located in Kongsberg, Norway. (2) Those associated with arsenides and sulfides of nickel and cobalt and other silver minerals, as well as with calcite or barite. An example of this type was the silver mines at Cobalt,

Ontario. (3) Those associated with uraninite and nickel-cobalt minerals such as are found in the still-active mines of Joachimsthal in Bohemia and at Great Bear Lake, Northwest Territories, Canada. *See* ARGENTITE; ARGYRODITE; CERARGYRITE; PYRARGYRITE.

In the United States small amounts of native silver are associated with native copper on the Keweenaw Peninsula, MI. Most of the world's supply comes from silver sulfides and sulfosalts. Large producers of silver are Canada, Peru, Mexico, and the United States.

Silver is used in photographic film emulsions, sterling and electroplated ware, electrical and electronic components, and alloys and solder. Other metals, such as nickel and copper, are often substituted for silver because of its high price.

Copper. This chemical element occurs in an uncombined state as a mineral and exhibits isometric symmetry (4/m $\bar{3}$ 2/m). Copper is usually found in a malformed state in branching and arborescent groups (Fig. 3), mostly as irregular masses and plates and in twisted and wirelike forms. Hardness is 2½–3, and specific gravity is 8.9. Malleable and ductile, with a metallic luster and copper-red color on the fresh surface, it may tarnish to a darker color and dull luster. The native material generally contains small amounts of Ag, As, Bi, Sb, and Hg in solid solution. The structure of copper, with space group *Fm3m*, is based on cubic closest packing of copper atoms (isostructural with gold).

Native copper is most commonly found in basaltic lava flows, but also occurs in the oxidized zones of copper sulfide deposits. The largest known deposit is in the Lake Superior district on the Keweenaw Peninsula. The rocks of this area consist of interbedded basic lavas, sandstones, and conglomerates in which the copper occurs either as fillings in cavities of the lava flows or as interstitial fillings to or replacing pebbles in the sedimentary rocks. *See* AZURITE; BORNITE; BROCHANTITE; CHALCOCITE; CHALCOPYRITE; CHRYSOCOLLA; COVELLITE; ENARGITE; MALACHITE.

Native copper is only a secondary source of copper; copper sulfides are the primary source of the metal. Copper is used extensively for electrical purposes and also in brass and bronze alloys.

Iron. This mineral is composed of elemental iron (αFe). Native iron is very rare on Earth, but kamacite (with about 5.5 wt % nickel in solid solution) is common in meteorites, as is taenite, which contains 27 to about 65 wt % nickel. Iron exhibits isometric symmetry (4/m $\bar{3}$ 2/m). Its crystals are rare; it occurs most commonly in blebs and masses. Hardness is 4, and specific gravity ranges from 7.3 to 7.9 for αFe. It has a hackly fracture, is malleable, has a metallic luster and steel-gray to black color, and is strongly magnetic. Naturally occurring αFe always contains some Ni and commonly small amounts of Co, Cu, and Mn. The structure of αFe, as well as kamacite, is based on cubic packing of iron atoms with space group *Im3m*. The structure of taenite is based on a different lattice type with space group *Fm3m*.

Native iron is very rare terrestrially, but kamacite and taenite are common in meteorites. The most important terrestrial source is Disko Island, Greenland, where masses of native iron are embedded in basalts. Iron meteorites are composed primarily of a regular intergrowth, known as Widmanstätten pattern, of kamacite and taenite (Fig. 4). Stony-iron meteorites consist mainly of mixtures of silicates, kamacite, taenite, and troilite (FeS). *See* METEORITE.

The massive world production of iron metal is from iron oxides and hydroxides, such as hematite (Fe_2O_3), magnetite (Fe_3O_4), and goethite (FeO · OH). *See* GOETHITE; HEMATITE; LATERITE; MAGNETITE; PYRRHOTITE; TACONITE.

[CORNELIS KLEIN]

Bibliography: C. R. Anhaeusser, Archean metallogeny in southern Africa, *Econ. Geol.*, 71: 16–44, 1976; C. A. Cousins and C. F. Vermaak, The contribution of southern African ore deposits to the geochemistry of the platinum group metals, *Econ. Geol.*, 71:287–305, 1976; C. S. Hurlbut, Jr., and C. Klein, *Manual of Mineralogy*, 19th ed., 1977; C. Palache, H. Berman, and C. Frondel, *The System of Mineralogy*, vol. 1, 1944; P. Ramdohr, *The Ore Minerals and Their Intergrowths*, 1969.

Natrolite

A fibrous or needlelike mineral belonging to the zeolite family of silicates. It crystallizes in the monoclinic system in pseudo-orthorhombic prismatic crystals which are often acicular (see illustration). Most commonly it is found in radiating

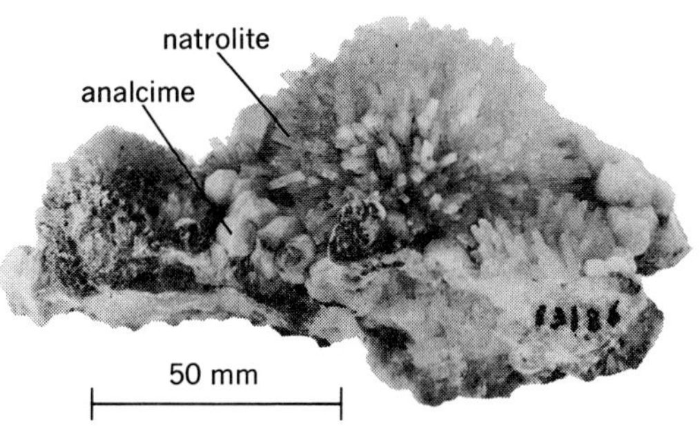

Natrolite crystals on analcime, Cape d'Or, Nova Scotia. (*American Museum of Natural History specimen*)

fibrous aggregates. There is perfect primatic cleavage, the hardness is 5–5½ on Mohs scale, and the specific gravity is 2.25. The mineral is white or colorless with a vitreous luster that inclines to pearly in fibrous varieties. The chemical composition is $Na_2(Al_2Si_3O_{10})\cdot 2H_2O$ but some potassium is usually present substituting for sodium.

Natrolite is a secondary mineral (low-temperature hydrothermal mineral) found lining cavities in basaltic rocks, where it is associated with other zeolites, calcite, apophyllite, and prehnite. Its outstanding locality in the United States is at Bergen Hill, N.J. *See* ZEOLITE. [CLIFFORD FRONDEL; CORNELIUS S. HURLBUT, JR.]

Natural gas

A combustible gas that occurs in porous rock of the Earth's crust and is found with or near accumulations of crude oil. Being in gaseous form, it may occur alone in separate reservoirs. More commonly it forms a gas cap, or mass of gas, entrapped between liquid petroleum and impervious capping rock layer in a petroleum reservoir. Under conditions of greater pressure it is intimately mixed with, or dissolved in, crude oil.

Composition. Typical natural gas consists of hydrocarbons having a very low boiling point. Methane, CH_4, the first member of the paraffin series, and with a boiling point of −254°F, makes up approximately 85% of the typical gas. Ethane, C_2H_6, with a boiling point of −128°F, may be present in amounts up to 10%; and propane, C_3H_8, with a boiling point of −44°F, up to 3%. Butane, C_4H_{10}; pentane, C_5H_{12}; hexane; heptane; and octane may also be present. Structural formulas of four of these compounds are given here.

```
        H              H  H
        |              |  |
     H—C—H         H—C—C—H
        |              |  |
        H              H  H
     Methane        Ethane

   H  H  H  H  H         H  H  H  H
   |  |  |  |  |         |  |  |  |
H—C—C—C—C—C—H     H—C—C—C—C—H
   |  |  |  |  |         |  |  |  |
   H  H  H  H  H         H  H  |  H
                                |
                             H—C—H
                                |
                                H
 Normal pentane          iso-Pentane
```

Whereas normal hydrocarbons having 5–10 carbon atoms are liquids at ordinary temperatures, they have a definite vapor pressure and therefore may be present in the vapor form in natural gas. Carbon dioxide, nitrogen, helium, and hydrogen sulfide may also be present.

Types of natural gas vary according to composition and can be dry or lean (mostly methane) gas, wet gas (considerable amounts of so-called higher hydrocarbons), sour gas (much hydrogen sulfide), sweet gas (little hydrogen sulfide), residue gas (higher paraffins having been extracted), and casinghead gas (derived from an oil well by extraction at the surface). Natural gas has no distinct odor. Its main use is for fuel, but it is also used to make carbon black, natural gasoline, certain chemicals, and liquified petroleum gas. Propane and butane are obtained in processing natural gas.

Distribution and reserves. Gas occurs on every continent (see table), and estimates of world resources indicate that they may be 10 times proved reserves. Wherever oil has been found, a certain amount of natural gas is also present. In production and known reserves the United States stands first among nations. During 1976, 31 states produced natural gas, and of these, Texas and Louisiana account for more than 70% of the total gas produced.

World gas reserves, Dec. 31, 1975

Region	Volume, 10^{12} ft^3
United States	228
Canada	57
Netherlands	65
Other Market Economy countries	1220
Soviet Union	710
Other Central Economy countries	50
World total	2330

The estimated known reserves in the United States at the end of 1975 were 228×10^{12} ft^3. Consumption in the United States in 1976 was 19×10^{12} ft^3. In the 1960s natural gas was discovered in western Canada at a rapid rate; at the beginning of 1976 the proven reserves were 57×10^{12} ft^3. During the period 1972–1975, 99% of United States gas imports were from Canada.

Fluctuations in supply and consumption are expected to be small through 1980. Long before supplies of natural gas run out or become expensively scarce, it is expected that some process of coal gasification will produce a gas which is completely interchangeable with natural gas and at a competitive price. This is important because coal makes up a majority of the world's known fossil fuel reserves. But since energy consumers have indicated in the marketplace their preference for fluid and gaseous fuels over the solid forms, coal gasification research will be given additional impetus.

In estimating gas reserves, the volumetric method is preferred. The volume of the reservoir is determined by means of the thickness, porosity, and permeability of the producing zones. A study of many depleted fields suggests that about 85% of all gas in dry-gas reservoirs is recovered. Some engineers use the production versus pressure-decline method. They calculate future production by plotting past production against the decline in reservoir pressure.

In California 60% of the gas is associated with oil, but in western Texas the percentage is about 40%. The percentage figure for the United States as a whole is only 23%. This means that a large proportion of the reserves is stored in such dry-gas fields as the Hugoton-Panhandle area of Kansas, Oklahoma, and Texas; Monroe, Louisiana; Carthage, northeastern Texas; Bethany-Waskom, Texas; Big Sandy, Kentucky; Sligo, Louisiana; Blanco Mesaverde, New Mexico; Red Oak–Norris, Oklahoma; Long Lake, Texas; Katy, Texas; San Salvador, Texas; Joaquin-Logansport, Texas; Lake Arthur, Louisiana; Big Piney, Wyoming; Chocolate Bayou, Texas; and Kenai and Cook Inlet, Alaska; there are also about 20 other so-called giant fields in the United States. In addition there are at least 12 giant (10^{12} ft^3 or more) gas fields already discovered in offshore Louisiana in the Gulf of Mexico. Furthermore, prospects of offshore activity in other parts of the world, especially Africa, Asia, the North Sea, and South America, are tremendous for the development of future giant gas fields. This is made possible by the almost incredible technological advances in deep-water drilling, which has enabled explorers to drill in waters in excess of 10,000 ft deep as compared with some 200 ft in depth in the late 1950s. In western Canada some of the large gas fields are the Pincher Creek, the Waterton, and the Jumping Pound. The largest dissolved-gas area in the United States lies along the Gulf Coast of Texas and Louisiana and contains about 50% of the total known reserves of associated gas. Offshore drilling in the waters of the Gulf will add considerably to these reserves. In an average year slightly over 91% of the gas produced is marketed, while 7.7% is used for repressuring, and 1.4% is vented or wasted. In earlier years a much larger percentage was piped away from oil fields and burned.

Geological associations. Natural gas is present in every system of rocks down to the Cambrian. The first gas deposits found in the United States were those in the Eastern states. In New York and Pennsylvania 85% of the gas came from Devonian rocks. In West Virginia, Kentucky, and eastern Ohio, Devonian and Mississippian rocks rank nearly equally, but Silurian rocks are also important. In Indiana and Illinois, Pennsylvanian rocks outrank the Mississippian. The Hugoton-Panhandle field is one of the largest in the world, and the Permian dolomites produce gas from five different levels. The fact that oil is found lower down in Pennsylvanian and older rocks proves the superior migratory capacity of gas. This field has a high percentage of nitrogen (almost 15%).

Oklahoma and the western part of Texas have gas in many stratigraphic zones, from the Permian down to the Cambrian. Most of it is associated with crude oil, either in solution or in the form of gas-cap accumulations. The Carthage field is in northeastern Texas, where 10 different layers in the Trinity division of the Cretaceous have been found productive. Cretaceous rocks are the principal reservoir rocks in northern Louisiana and in Mississippi. Throughout the Rocky Mountain states various layers in the Cretaceous system account for most of the gas. There are many dry-gas pools, of which the outstanding are the Blanco, northwestern New Mexico; the Baxter Basin, southwestern Wyoming; and the Cedar Creek, southwestern Montana. In California gas production is derived from various layers in the Tertiary system. Although about 60% of the gas is associated with oil reservoirs, there are a number of dry-gas fields. *See* PETROLEUM GEOLOGY.

Of much interest is the trend toward low-temperature transportation and storage of liquid petroleum gas and methane. The ability to store frozen gas (as a liquid) underground will enable pipeline companies to more efficiently meet the cyclical demands of the seasons. Storage of gas close to markets will do away with the necessity of having the large pipeline capacity, which is needed to take care of peak seasons but which lies more or less idle during periods of lessened demand. The ability to condense 600 ft^3 of gas into 1 ft^3 of liquid opens great possibilities for the movement of gas across oceans in tankships. Through such transportation remote areas of the world can become consumer areas, fuel-short consumer areas will have access to needed supplies; and producing areas will benefit from new revenues.

Helium, which has many industrial uses, is a by-product of natural gas and is present in some fields. The Rattlesnake field in New Mexico contains 7.5%, the highest percentage of helium to total gas content found up to 1968.

[MICHEL T. HALBOUTY]

Bibliography: American Gas Association, Inc., American Petroleum Institute, and Canadian Petroleum Association, *Reserves of Crude Oil, Natural Gas Liquids, and Natural Gas in the United States and Canada*, vol. 22, 1968; American Petroleum Institute, *Petroleum Facts and Figures*, 1967; J. A. Clark, *The Chronological History of Petroleum and Natural Gas*, 1963; R. Epp and A. G. Fowler, *J. Hydraul. Div. ASCE*, vol. 96, no. HYl, 1970; A. M. Leeston, J. A. Chrichton, and J. C. Jacobs, *The Dynamic Natural Gas Industry*, 1963; E. J. Neuner, *The Natural Gas Industry*, 1960; H. H. Rachford, *Oil Gas J.*, pp. 93–96, July 16, 1973; H. H. Rachford and T. Dupont, *J. Petrol. Eng. AIME*, vol. 14, no. 2, 1974; M. A. Stoner, *J. Petrol. Eng. AIME*, vol. 12, no. 1, 1972; M. A. Stoner and M. A. Karnitz, *Oil Gas J.*, pp. 97–100, Dec. 10, 1973; M. A. Stoner and M. A. Karnitz, *Transp. Eng. J. ASCE*, vol. 100, no. TE3, 1974; E. B. Wylie, V. L. Streeter, and M. A. Stoner, *J. Petrol. Eng. AIME*, vol. 14, no. 1, 1974.

Nearshore processes

The processes that shape the shore features of coastlines and begin the mixing, sorting, and transportation of sediments and runoff from land; particularly those interactions among waves, winds, tides, currents, and land that relate to the waters, sediments, and organisms of the continental shelf and nearshore areas.

The energy for nearshore processes comes from the sea and is produced by the force of winds blowing over the ocean, by the gravitational attraction of Moon and Sun acting on the mass of the ocean, and by various impulsive disturbances at the atmospheric and terrestrial boundaries of the ocean. These forces produce waves and currents that transport energy toward the coast. The configuration of the land mass and adjacent shelves modifies and focuses the flow of energy and determines the intensity of wave and current action in coastal waters. Rivers and winds transport erosion products from the land to the coast, where they are sorted and dispersed by waves and currents.

The dispersive mechanisms operative in the nearshore waters of oceans, bays, and lakes are all quite similar, differing only in intensity and scale, variables that are determined primarily by the nature of the wave action and the dimensions of the surf zone. The most important mechanisms are the orbital motion of the waves, which is the basic mechanism by which wave energy is expended on the shallow sea bottom, and the currents of the nearshore circulation system that produce a continuous interchange of water between the surf zone and offshore. The dispersion of water and sediments near the coast and the formation and erosion of sandy beaches are some of the more common manifestations of nearshore processes.

Erosional and depositional nearshore processes play an important role in determining the configuration of coastlines. Erosion is usually dominant off headlands and along coastal sections backed by alluvium and other unconsolidated material, whereas deposition is most common along indentations between headlands. The overall effect of such processes is usually a straightening and smoothing of the coastline. However, this is not always the case; differential wave erosion may cause a rapid erosion of material between headlands and thus cause irregularities in the coastline. *See* COASTAL LANDFORMS.

Whether deposition or erosion will be predominant in any particular place depends upon a number of interrelated factors: the amount of available beach sand and the location of its source; the configuration of the coastline and of the adjoining ocean floor; and the effects of wave, current, wind, and tidal action. The establishment and persistence of natural sand beaches are often the result of a delicate balance among a number of these factors, and any changes, natural or man-made, tend to upset this equilibrium.

Waves. Waves and the currents that they generate are the most important factor in the transportation and deposition of nearshore sediments. Waves are effective in moving material along the bottom and in placing it in suspension for weaker currents to transport. In the absence of beaches, the direct force of the breaking waves erodes cliffs and sea walls.

Wave action along most coasts is seasonal in nature in response to the changing wind systems over the waters where the waves are generated. The height and period of the waves depend on the speed and duration of the winds generating them and the fetch, or length, over which the wind blows. Consequently, the nature and intensity of wave attack against coastlines varies considerably with the size of the water body, as well as with latitude and exposure. Waves generated by winter storms in the Southern Hemisphere of the Pacific Ocean may travel more than 5000 mi before breaking on the shores of California, where they are common summer waves for the Northern Hemisphere. Using sensitive instruments, W. H. Munk and others have recorded waves off San Clemente Island, Calif., that have traveled more than 10,000 nautical miles from their generation by storms in the South Indian Ocean.

The profiles of ocean waves in deep water are long and low, approaching a sinusoidal form. As the waves enter shallow water the wave velocity and length decrease, the wave steepens, and the wave height increases, until the wave train consists of peaked crests separated by flat troughs. Near the breaker zone the process of steepening is accelerated, so that the breaking waves may attain a height several times greater than the deep-water wave. This transformation is particularly pronounced for long-period waves from a distant storm. The profiles of local storm waves and the waves generated over small water bodies such as lakes show considerable steepness even in deep water, so that the shallow-water steepening is not as pronounced as in the case of ocean swell.

The shallow-water transformation of waves commences at the depth where the waves "feel bottom." This depth is equal to one-half the deep-water wave length, where the wave length is the horizontal distance from wave crest to crest. The deep-water wave length is given by the relationship $L = gT^2/2\pi$, where g is the acceleration of gravity and T is the wave period in seconds. Upon entering shallow water, waves are also subjected to refraction, a process in which the wave crests tend to parallel the depth contours. For straight coasts with parallel contours, this decreases the angle between the approaching wave and the coast and causes a spreading of the energy along the crests. The wave height is decreased by this process, but the effect is uniform along the coast (Fig. 1). A submarine canyon or depression causes waves to be refracted, or bent, in such a manner that waves over the canyon will diverge and decrease in height and the line of wave crests will be convex toward the shore. Waves will converge on either side of the canyon over a ridge, causing the wave height to increase and the line of wave crests to be concave toward the shore. The amount of wave refraction and the consequent change in wave height and direction at any point along the

Fig. 1. Longshore currents, generated when waves approach the beach at an angle. In this photograph at Oceanside, Calif., the longshore current is flowing toward the observer. (*Department of Engineering, University of California, Berkeley*)

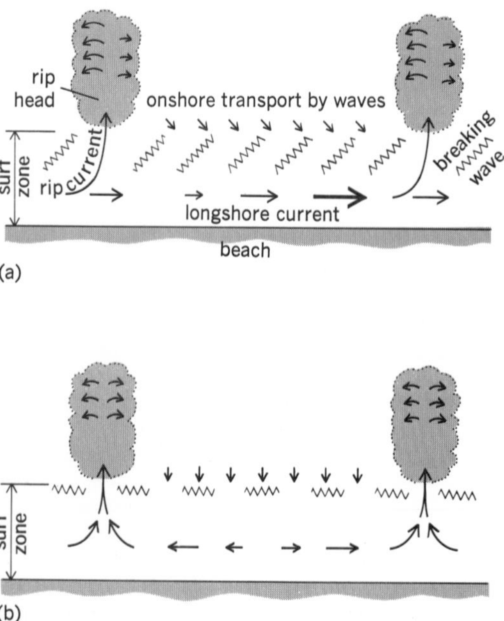

Fig. 2. Definition sketches for the nearshore circulation cells. (*a*) Asymmetrical cell for breakers oblique to the shore. (*b*) Symmetrical cell for breakers parallel to the shore.

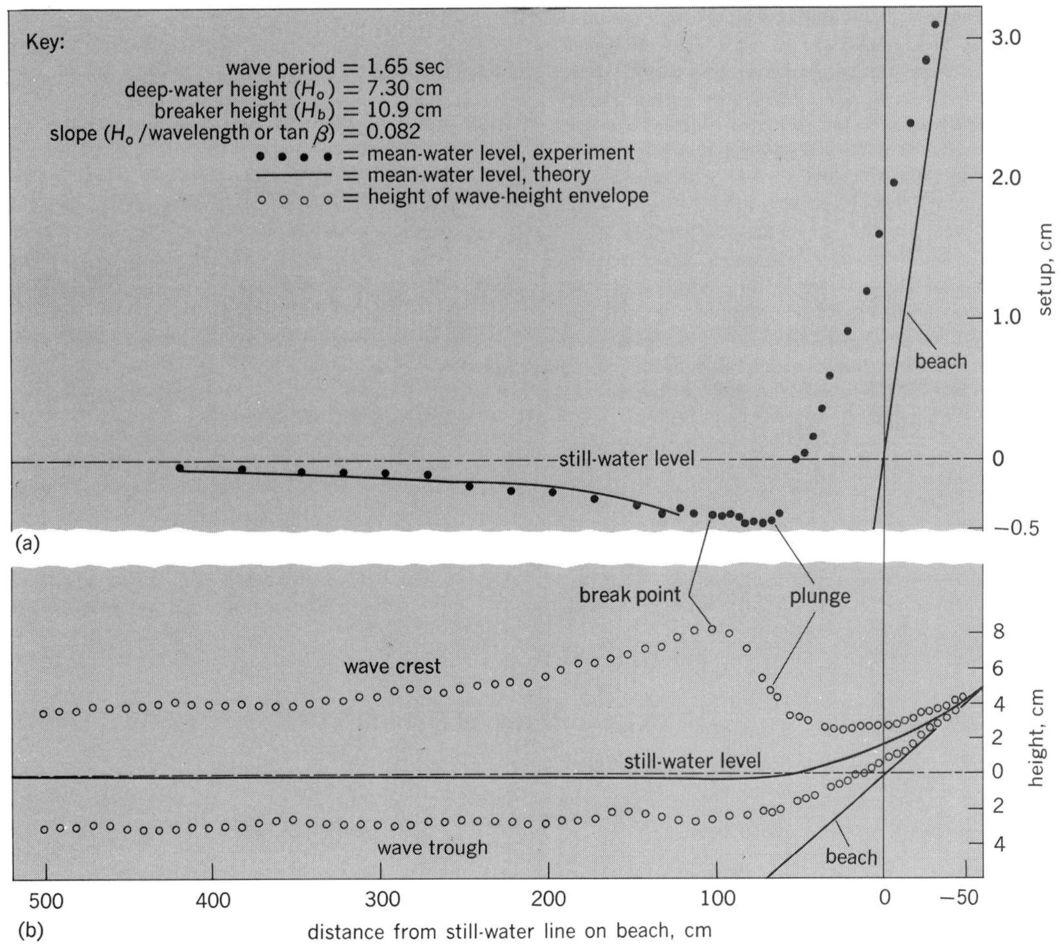

Fig. 3. Measurements of (*a*) wave setup and (*b*) envelope of wave height for waves breaking on a plain beach in the laboratory. The measurements are relative to still-water level and show the decrease in mean-water level near the break point of the waves (setdown) and increase in level over the beach face (setup). The vertical exaggeration is ×80 for *a* and ×10 for *b*. (*Measurements from A. J. Bowen, D. L. Inman, and V. P. Simmons, J. Geophys. Res., 73(8), 1968; theoretical curve from M. S. Longuet-Higgins and R. W. Stewart, Deep-Sea Research, vol. 11, copyright © 1964 by Pergamon Press; reprinted with permission*)

coast is a function of wave period, direction of approach, and the configuration of the bottom topography.

Waves from distant storms may have periods as great as 20 sec or more when they reach ocean coasts. Since refraction commences when waves reach a depth equal to one-half the wave length, these long waves will be refracted by topographic features on the ocean floor in depths as great as 300 m. Thus, the formation of beaches and the effect of waves on a coastline may be influenced by the topography of the bottom many miles from the coast. When submarine ridges cause a concentration of wave energy at certain points along a coast, severe erosion and damage to coastal structures often result. This occurs periodically along the California coast when the wave period, deep-water direction of approach, and height are such as to focus energy on coastal structures.

Currents in the surf zone. When waves break so that there is an angle between the crest of the breaking wave and the beach, the momentum of the breaking wave has a component along the beach in the direction of wave propagation. This results in the generation of longshore currents that flow parallel to the beach inside of the breaker zone (Fig. 2*a*). After flowing parallel to the beach as longshore currents, the water is returned seaward along relatively narrow zones by rip currents. The net onshore transport of water by wave action in the breaker zone, the lateral transport inside of the breaker zone by longshore currents, the seaward return of the flow through the surf zone by rip currents, and the longshore movement in the expanding head of the rip current all constitute the nearshore circulation system. The pattern that results from this circulation commonly takes the form of an eddy or cell with a vertical axis. The dimensions of the cell are related to the width of the surf zone; the spacing between rip currents is usually two to eight times the width of the surf zone.

When waves break with their crests parallel to a straight beach, the flow pattern of the nearshore circulation cell becomes symmetrical (Fig. 2*b*). Longshore currents occur within each cell, but there is no longshore exchange of water from cell to cell.

The nearshore circulation system produces a continuous interchange between the waters of the surf and offshore zones, acting as a distributing mechanism for nutrients and as a dispersing mechanism for land runoff. Offshore water is transported into the surf zone by breaking waves, and particulate matter is filtered out on the sands of the beach face. Runoff from land and pollutants introduced into the surf zone is carried along the shore and mixed with the offshore waters by the seaward-flowing rip currents.

There is no evidence of undertow in the surf zone, other than the instantaneous motion occurring as the backwash flows under a wave breaking on the face of a steep beach. The principal danger to swimmers is from rip currents that may carry them seaward unexpectedly. Longshore currents may attain velocities in excess of 5 knots, while rip current velocities in excess of 3 knots have been measured.

Periodicity or fluctuation of current velocity and direction is a characteristic of flow in the nearshore system. This variability is primarily due to the grouping of high waves followed by low waves, a phenomenon called surf beat that gives rise to a pulsation of water level in the surf zone.

Formation of circulation cells. Nearshore circulation cells result from differences in mean water level in the surf zone associated with changes in breaker height along the beach. Waves transmit momentum in the direction of their travel, and their passage through water produces second-order pressure fields that significantly change the mean water level near the shore. Near the surf zone the presence of the pressure field produces a decrease in mean water level termed wave setdown. The setdown is proportional to the square of the wave height and for waves near the surf zone has a maximum value that is about one-sixteenth that of the breaker height. Shoreward of the break point the onshore flux of momentum against the beach produces a rise in mean water level over the beach face referred to as setup. A profile of the envelope of wave height and the experimental and theoretical values of wave setdown and setup are shown in Fig. 3.

If the wave height varies along a beach, the setup will also vary, causing a longshore gradient in mean water level within the surf zone. Longshore currents flow from regions of high water to regions of low water and thus flow away from zones of high waves. The longshore currents flow seaward as rip currents where the breakers are lower. Pronounced changes in breaker height along beaches usually result from wave refraction over irregular offshore topography. However, on a smaller scale, uniformly spaced zones of high and low breakers occur along straight beaches with parallel offshore contours. It has been shown that these alternate zones of high and low waves are due to the interaction of the incident waves traveling toward the beach from deep water with one of the many possible modes of oscillation of the nearshore zone known as edge waves. Edge waves are trapped modes of oscillation that travel along the shore (Fig. 4). Circulation in the nearshore cell is enhanced by edge waves having the period of the incident waves, or that of their surf beat, because these interactions produce alternate zones of high and low breakers whose positions are stationary along the beach. It appears that the edge waves can be either standing or progressive. In either case, the spacing between zones of high waves (and hence between rip currents) is equal to the wave length of the edge wave.

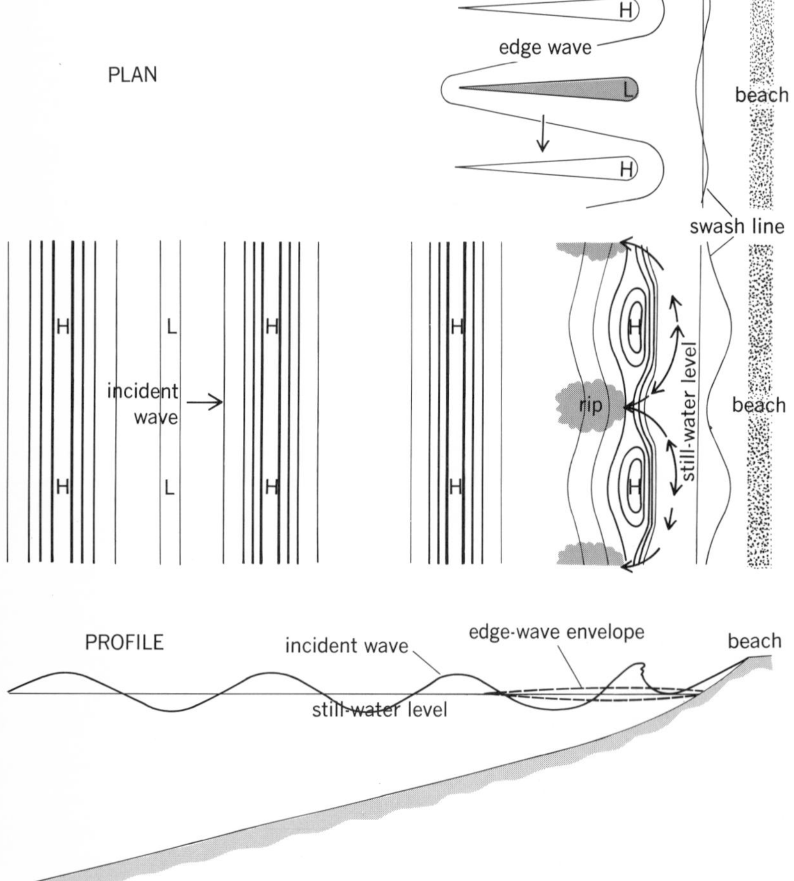

Fig. 4. Diagram showing the formation of rip currents. The interaction of incident waves from deep water with edge waves traveling along the beach produces alternate zones of high (H) and low (L) breakers along beach. Longshore currents flow away from zones of high waves where setup is maximum and converge on points of low waves, causing rip currents to flow seaward.

The position and spacing of rip currents can be predicted under certain controlled conditions in the laboratory if the height and period of the incident wave and the slope and length of the beach are known. However, because of variability in incident waves and in the beach configuration, prediction of rip current spacing on natural beaches is uncertain. Where beaches are irregular, or where offshore topography produces irregularities in the wave refraction pattern, the location of cells is also dependent upon the regional gradation of breaker height along the beach.

Points, breakwaters, and piers all influence the circulation pattern and alter the direction of the currents flowing along the shore. In general, these obstructions determine the position of one side of the circulation cell. In places where a relatively straight beach is terminated on the down-current side by points or other obstructions, a pronounced rip current extends seaward. During periods of large waves having diagonal approach, these rip currents can be traced seaward for 1 mi or more.

Beach types. Beaches consist of transient clastic material (unconsolidated fragments) that reposes near the interface between the land and the sea and is subject to wave action. The material is in dynamic repose rather than a stable deposit, and thus the width and thickness of beaches is subject to rapid fluctuations, depending upon the amount and rigor of erosion and transportation of beach material. Beaches are essentially long rivers of sand that are moved by waves and currents and are derived from the material eroded from the coast and brought to the sea by streams.

The geometry of beaches is also dependent on coastal history, and there is a close relationship between beach characteristics and type of coast. Long straight beaches are typical of low sandy coasts; shorter crescent-shaped beaches and small pocket beaches are more common along mountainous coastlines. The coast may be cliffed as shown in Fig. 5, or it may contain a ridge of windblown sand dunes and be backed by marshes and water. Along many low sandy coasts, such as the East and Gulf coasts of the United States, the beach is separated from the mainland by water or by a natural coastal canal. Such beaches are called barrier beaches. A beach that extends from land and terminates in open water is referred to as a spit, while a beach that connects an island or rock to the mainland or another island is a tombolo.

While differing in detail, beaches the world over have certain characteristic features which allow application of a general terminology to their profile (Fig. 5). The beach or shore extends landward from the lowest water line to the effective limit of attack by storm waves. The region seaward is termed the offshore; that landward, the coast. The beach includes a backshore and foreshore. The backshore is the highest portion and is only acted upon by waves during storms. The foreshore extends from the crest of the berm to the low-water mark and is the active portion of the beach traversed by the uprush and backwash of the waves. The foreshore consists of a steep seaward dipping face related to the size of the beach material and the rigor of the uprush and of a more gentle seaward terrace, sometimes referred to as the low-tide terrace or step, over which the waves break and surge. In some localities the foreshore face and terrace merge into one continuous curve; in others, there is a pronounced discontinuity at the toe of the beach face. The former condition is characteristic of fine sand beaches and of coasts where the wave height is equal to or greater than the tidal range; the latter is typical of coastlines where the tidal range is large compared with the wave height, as along the Patagonian coast of South America and portions of the Gulf of California.

The offshore zone frequently contains one or more bars and troughs that parallel the beach; these are referred to as longshore bars and longshore troughs. Longshore bars commonly form on the bottom at the plunge point of the wave, and their position is thus influenced by the breaker height and the nature of the tidal fluctuation.

Beach cycles. Waves are effective in causing sand to be transported laterally along the beach by longshore currents and in causing movements of sand from the beach foreshore to deeper water and back again to the foreshore. These two types of transport, although interrelated, are more conveniently discussed in separate sections. The offshore

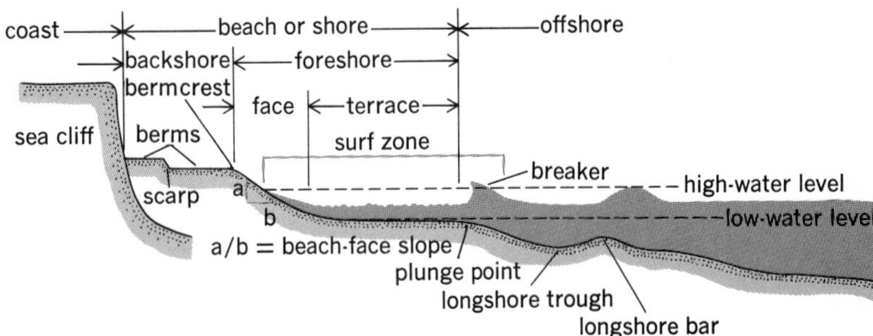

Fig. 5. Beach profile, showing characteristic features.

and onshore transport of sand is closely related to the beach profile and to the cycles in beach width and will be discussed here and under the mechanics of beach formation.

Along most coasts there is a seasonal migration of sand between the beaches and deeper water, in response to the changes in the character and direction of approach of the waves. In general, the beaches build seaward during the small waves of summer and are cut back by high winter storm waves. There are also shorter cycles of cut and fill associated with spring and neap tides and with nonseasonal waves and storms. Bottom surveys indicate that most offshore-onshore interchange of sand occurs in depths less than 10 m but that some effects may extend to depths of 30 m or more.

Figure 5 shows the profile of a typical summer beach which has been built seaward by low waves. During stormy seasons the beach foreshore is eroded, frequently forming a beach scarp. Subsequent low waves build the beach foreshore seaward again. The beach face is a depositional feature and its highest point, the berm crest, represents the maximum height of the run-up of water on the face of the beach. The height of wave run-up usually varies between one and three times the height of the breaking wave. Since the height of the berm depends on wave height, the higher berm, if it is present, is sometimes referred to as the winter or storm berm, and the lower berm as the summer berm.

The entire beach may be cut back to the country rock during severe storms. Under such conditions the waves erode the coast and form the well-known sea cliffs and wave-cut terraces, which are frequently preserved in the geologic record and serve as markers to the past relations between the levels of the sea and the land.

Beach cusps. A series of regularly spaced scallops, called beach cusps, sometimes forms on the beach foreshore (Fig. 6). Beach cusps consist of short transverse valleys formed in the beach face and separated by ridges with cuspate points. The spacing between cuspate points ranges from several to several hundred meters, while the depths of the valleys range from a few centimeters on fine sand beaches to several meters on pebble and cobble beaches. The formation of cusps appears to be associated with a wave-wave interaction similar to that operative in the formation of the nearshore circulation cell. In the case of cusps the run-up on the beach face and the longshore flow modify the beach face where the longshore flow converges, causing a depression or valley. When low waves

Fig. 6. The effect of headlands on the accretion of beach sand, shown in photograph of Point Mugu,.Calif. The point forms a natural obstruction which interrupts the longshore transport of sand, causing accretion and a wide beach to form (foreground). The regularly spaced scallops are beach cusps. (*Department of Engineering, University of California, Berkeley*)

break directly on narrow, steep beaches, the cusps assume the dimensions of the nearshore circulation cell from which they are derived (Fig. 6). However, where the surf zone is wide, cusps that form on the beach face are small compared to the spacing between rip currents. In this case it seems likely that the formation of cusps is in response to an interaction of the bore from the breaking wave with an edge wave disturbance traveling within the surf zone. Cusps occur with greatest frequency during neap tides when the water level remains nearly constant.

Mechanics of beach formation. Wherever there are waves and an adequate supply of sand or coarser material, beaches form. Even man-made fills and structures are effectively eroded and reformed by the waves. The initial and most characteristic event in the formation of a new beach from a heterogenous sediment is the sorting out of the material, with coarse material remaining on the beach and fine material being carried away. Concurrent with the sorting action, the material is rearranged, some being piled high above the water level by the run-up of the waves to form the beach berm, some carried back down the face to form the foreshore terrace. In a relatively short time, the beach assumes a profile which is in equilibrium with the forces generating it.

The back-and-forth motion of waves in shallow water produces stresses on the bottom that place sand in motion. The interaction of the wave stresses with the bottom also induces a net boundary current flowing in the direction of wave travel. The most rapidly moving layer of water is near the bed, and for waves traveling over a horizontal bed the interaction of wave stresses and the boundary current produces a net transport of sand in the direction of wave travel. Thus waves traveling toward the shore tend to contain sand against the shore.

The action of waves on an inclined bed of sand eventually produces a profile that is in equilibrium with the energy dissipation associated with the oscillatory motion of the waves over the sand bottom. If an artificial slope exceeds the natural equilibrium slope, an offshore transport of sand will result from the gravity component of the sand load, until the slope reaches equilibrium. Conversely, if an artificial slope is less than the natural equilibrium slope, a shoreward transport of sand will result and the beach slope will steepen. Thus, the equilibrium slope is attained when the up-slope and down-slope transports are equal.

Since the equilibrium slope steepens with increasing energy dissipation and with increasing speed of the boundary current, it is usually gentle in deeper water where wave dissipation is slight and currents are low, and steeper near the surf zone where both dissipation of energy and the currents are greatest. This dependence of slope on wave stress and boundary currents causes the slight steepening of slope found just outside of the surf zone in many beach profiles (Fig. 5).

The slope of the beach face is related to the dissipation of energy by the swash and backwash over the beach face. Percolation of the swash into a permeable beach reduces the amount of flow in the backwash and is thus conducive to deposition of the sand transported by the swash. If in addition the beach is dry, this action is accentuated. Coarse sands are more permeable and consequently more conducive to deposition and the formation of steep beach faces. Large waves elevate the water table in the beach. When the beach is saturated, the backwash has a higher velocity, a condition that is conducive to erosion. From the foregoing it follows that the slope of the beach face, sometimes referred to as the foreshore slope, increases both with increasing sediment size and with decreasing wave height. Some typical average values for the face slopes of ocean beaches are given in the table.

Oscillatory ripples, which form on the sandy bottom, are an important factor in the mechanism of transportation and sorting of beach sands. Be-

Relationship between size of beach sediment and slope of beach face

Beach sediment		Slope of beach face, degrees
Type	Size, mm	
Very fine sand	1/16–1/8	1
Fine sand	1/8–1/4	3
Medium sand	1/4–1/2	5
Coarse sand	1/2–1	7
Very coarse sand	1–2	9
Granules	2–4	11
Pebbles	4–64	17
Cobbles	64–256	24

cause turbulence and lift force are most intense at the crest of the ripple, only the coarsest material is deposited there. The fine material is placed in suspension and removed from the area, while the coarse material moves with the ripple toward the beach. As a consequence of the sorting action by ripples and by the uprush and backwash on the beach face, beach sands are the best sorted sedimentary deposits.

The beach face is frequently characterized by laminations (closely spaced layers) that show slight differences in the size, shape, or density of the sand grains. The laminations parallel the beach face and represent "shear sorting" within the granular load as it is transported by the swash and backwash of the waves over the beach face (Fig. 7). Detailed examination shows that each lamina of fine-grained minerals delineates the plane of shear between moving and residual sand. The mechanism concentrating the heavy fine grains at the shear plane is partly the effect of gravity acting during shearing, causing the small grains to work their way down through the interstices between larger grains, and partly the dependence of the normal dispersive pressure upon grain size. When grains are sheared, the normal dispersive pressure between grains, which varies as the square of the grain diameter, causes large grains to drift toward the zone of least shear strain, that is, the free surface, and the smaller grains to drift toward the zone of greatest shear strain, that is, the shear plane.

Longshore movement of sand. The movement of sand along the shore occurs in the form of bed load (material rolled and dragged along the bottom) and suspended load (material stirred up and carried with the current). Suspended load transportation occurs primarily in the surf zone where the turbulence and vertical mixing of water are most effective in placing sand in suspension and where the longshore currents which transport the sediment-laden waters have the highest velocity. Concentrations of suspended sand in the surf zone, as high as 30 g/liter, have been measured along the coast of the Black Sea, suggesting that suspension is one of the most important processes in littoral drift. On the other hand, it is probable that suspension becomes less important for very coarse sand and cobble beaches.

The volume of littoral transport along oceanic coasts is usually estimated from the observed rates of erosion or accretion, most often in the vicinity of coastal engineering structures such as groins or jetties. In general, beaches build seaward upcurrent from obstructions and are eroded on the current lee, where the supply of sand is diminished. Such observations indicate that the transport rate varies from almost nothing to several million cubic meters per year, with average values of 150,000–600,000 m^3/year. Along the shores of smaller bodies of water, such as the Great Lakes of the United States and the Mediterranean Sea, the littoral transport rate can be expected to range about 7000–150,000 m^3/year. In general these are conservative estimates, since the volume of material moved commonly exceeds that indicated either by deposition or erosion.

The large quantity of sand moved along the shore and the pattern of accretion and erosion that occurs when the flow is interrupted pose serious problems for the coastal engineer. The problem is particularly acute when jetties are constructed to stabilize and maintain deep navigation channels through sandy beaches. A common remedial procedure is to dredge sand periodically from the accretion on the up-current side of the obstruction and deposit it on the eroding beaches in the current lee. Another method is the installation of permanent sand bypassing plants which continually remove the accreting sand and transport it by hydraulic tube to the beaches in the lee of the obstruction.

Mechanics of longshore transport of sand. The longshore transport of sediment is caused by the action of waves and currents. Waves and surf dissipate power on a granular bed by the stress of the motion of the water over the bed. The immersed weight of sediment placed in motion is proportional to the power dissipated by the waves. The presence of any net current, such as the longshore current, will produce a net transport of sediment in the direction of the current. In the absence of a current, the wave power is simply expended in a back-and-forth motion of sediment, and there is no net transport of sediment. Thus the transport rate should be proportional to the product of the power dissipated by the waves and the speed of the current. The power dissipated by waves can be approximated by assuming that all of the wave energy is expended in the surf zone. Unfortunately, reliable relations for predicting the longshore current have yet to be developed.

The longshore transport rate of sand is directly proportional to the longshore component of wave power estimated at the breaker point. Thus, the longshore transport rate of sand can be estimated from a knowledge of the total budget of wave energy incident upon the beach. The longshore transport rate is given by the relation $I = KP$, where I is the immersed weight transport rate in dynes per second, K is a constant of proportionality having a dimensionless value of about 0.77, and P is the longshore component of wave power per unit of beach length computed at the breaker point of waves in ergs per second per centimeter of beach length.

Fig. 7. Laminations in the beach face at La Jolla, Calif. Black laminae are heavy minerals and the light ones are quartz. Scale is 15 cm long. *(From D. L. Inman, Beach Erosion Board, U.S. Eng. Corps Tech. Mem. no. 39, 1953)*

Source of beach sediments. The principal sources of beach and nearshore sediments are the rivers, which bring large quantities of sand directly to the ocean; the sea cliffs of unconsolidated material, which are eroded by waves; and material of biogenous origin (shell and coral fragments and skeletons of small marine animals). Occasionally sediment may be supplied by erosion of unconsolidated deposits in shallow water. Beach sediments on the coasts of the Netherlands are derived in part from the shallow waters of the North Sea. Wind-blown sand may be a source of beach sediment, although winds are usually more effective in removing sand from beaches than in supplying it. In tropical latitudes many beaches are composed entirely of grains of calcium carbonate of biogenous origin. Generally the material consists of fragments of shells, corals, and calcareous algae growing on or near fringing reefs. The material is carried to the beach by wave action over the reef. Some beaches are composed mainly of the tests of foraminifera that live on sandy bottom offshore from the reefs.

Streams and rivers are by far the most important source of sand for beaches in temperate latitudes. Cliff erosion probably does not account for more than about 5% of the material on most beaches. Wave erosion of rocky coasts is usually slow, even where the rocks are relatively soft shales. On the other hand, retreats greater than 1 m a year are not uncommon in the unconsolidated sea cliffs. C. Lyell in 1873 showed that the Ganges and Brahmaputra rivers carry a volume of sediment into the Bay of Bengal each year that is 780 times greater than the material eroded by wave action from the 36-mi stretch of cliffs in the vicinity of Holderness, England. The Holderness cliffs, which are on the exposed North Sea coast, are noted for their rapid rate of erosion. According to J. A. Steers, they are 40 ft high and recede at the rate of about 7 ft/year.

Surprisingly, the contribution of sand by streams in arid countries is quite high (Fig. 8). This is because arid weathering produces sand-size material and results in a minimum cover of vegetation, so that occasional flash floods may transport large volumes of sand. The maximum sediment yield occurs from drainage basins where the mean annual precipitation is about 30 cm/year, as has been shown by W. B. Langbien and S. A. Schumm.

Fig. 8. Sand delta at Rio de la Concepción on the arid coast of the Gulf of California. Such deltas are important sources of sand for beaches. (*Courtesy of D. L. Inman*)

Following initial deposition at the mouths of streams entering the ocean, much of the sand-size fraction of terrestrial sediments is carried along the coast by longshore currents. The sand carried by these littoral currents may be deposited in continental embayments, or it may be diverted to deeper water by submarine canyons which traverse the continental shelf and effectively tap the supply of sand. Recent observations by H. W. Menard suggest that most of the deep sediments on the abyssal plains along a 250-mi section of the California coastline are derived from two submarine canyons, Delgada Submarine Canyon in northern California and Monterey Submarine Canyon in central California. *See* DELTA; MARINE SEDIMENTS; SEDIMENTATION.

Biological effects. The rigor of wave action and the continually shifting substrate make the sand beach a unique biological environment. Because few large plants can survive, the beach is occupied largely by animals and microscopic plants. Much of the food supply for the animals consists of particulate matter that is brought to the beach by the nearshore circulation system and trapped in the sand. The beach acts as a giant sand filter that strains out particulate matter from the water that percolates through the beach face.

Since the beach-forming processes and the trapping of material by currents and sand are much the same everywhere, the animals found on sand beaches throughout the world are similar in aspects and habits, although, according to E. Dahl, different species are present in different localities. In addition, since the slopes and other physical properties of beaches are closely related to elevation, the sea animals also exhibit a marked horizontal zonation. Organisms on the active portion of the beach face tend to be of two general types, insofar as the procurement of food is concerned: those that burrow into the sand, using it for refuge while they filter particulate matter from the water through siphons or other appendages that protrude above the sandy bottom, and those that remove organic material from the surface of the sand grains by ingesting them or by "licking." There are usually few species, but those which are present may be very abundant; for example, *Thoracophelia nucronata*, a beach worm which ingests sand grains, was estimated to have a population of 100,000/m³ in the fine sand beach at La Jolla, Calif., and the bean clam *Donax gouldii* has peak populations of more than 25,000/m² on the same beach.

A black layer is frequently found at depths of 5–75 cm below the surface of the beach foreshore. Chemically this is a reducing layer which has pH values greater than 8.0. J. R. Bruce has shown that the discoloration is caused by presence of ferrous sulfide which oxidizes to a reddish yellow on exposure to air. The formation of the layer is apparently related to the activity of bacteria on the organic material in the beach. This reducing layer is conducive to the deposition of calcium carbonate and may play an important role in cementing the beach sand and forming beach rock and nodules.

In tropical seas the entire shore may be composed of the cemented and interlocking skeletons of reef-building corals and calcareous algae. When this occurs, the nearshore current system is controlled by the configuration of the reefs that the organisms form. Where there are fringing reefs, breaking waves carry water over the edge of the reef, generating currents that flow along the shore inside of the reef and then flow back to sea through deep channels between reefs. Under such conditions beaches are usually restricted to a berm and foreshore face bordering the shoreward edge of the reef. See REEF. [DOUGLAS L. INMAN]

Bibliography: R. A. Bagnold, *An Approach to the Sediment Transport Problem from General Physics*, USGS Prof. Pap. no. 422–1, 1966; A. J. Bowen, D. L. Inman, and V. P. Simmons, Wave set-down and set-up, *J. Geophys. Res.*, 73(8), 1968; D. L. Inman and R. A. Bagnold, Littoral processes, in M. N. Hill (ed.), *The Sea: Ideas and Observations*, vol. 3, 1963; D. L. Inman and A. J. Bowen, Flume experiments on sand transport by waves and currents, in J. W. Johnson (ed.), *Proceedings of the Eighth Conference on Coastal Engineering*, 1963; D. L. Inman, G. C. Ewing, and J. B. Corliss, Coastal sand dunes of Guerrero Negro, Baja California, Mexico, *Geol. Soc. Amer. Bull.*, 77(8), 1966; D. L. Inman and J. D. Frautschy, Littoral processes and the development of shorelines, *Coastal Engineering*, Santa Barbara Conference of the American Society of Civil Engineers, 1965; M. S. Longuet-Higgins and R. W. Stewart, Radiation stress in water waves: A physical discussion with application, *Deep-Sea Research*, vol. 11, 1964; F. P. Shepard, *Submarine Geology*, 2d ed., 1963; U.S. Army, *Shore Protection, Planning, and Design*, Coastal Eng. Res. Center Tech. Rep. no. 4, 1966; R. L. Wiegel, *Oceanographical Engineering*, 1964.

Nepheline syenite

A phaneritic (visibly crystalline) plutonic rock with granular texture, composed largely of alkali feldspar (orthoclase, or microcline, usually perthitic), nepheline, and dark-colored (mafic) minerals (biotite, soda-amphibole, and soda-pyroxene). If sodic plagioclase exceeds the quantity of alkali feldspar, the rock may be called nepheline monzonite. Nepheline syenites have many features in common with syenites into which they grade, but chemically, mineralogically, and texturally they are much more variable. *See* SYENITE.

Composition. The alkali feldspar is soda-rich and usually exhibits perthitic texture (intergrown potash and soda feldspars). Barium-rich feldspar cores may be surrounded by barium-poor shells to give a zonal structure. If plagioclase occurs as discrete grains, it is usually albite or sodic oligoclase. More calcic types are found in nepheline monzonite. In some rock types the perthite is rimmed by albite.

Nepheline, a gray mineral with a greasy appearance, is also a major constituent. It may be highly altered and is commonly converted to bright-yellow cancrinite. Additional feldspathoids (including sodalite, analcite, and leucite) may occur in minor quantities. *See* FELDSPATHOID.

Biotite mica rich in iron and titanium frequently shows a zonal structure with light-colored cores fringed by darker borders. The amphiboles are usually soda-rich (arfvedsonite, hastingsite, and riebeckite), but zoned brown hornblende occurs in some varieties. Pyroxenes are also soda-rich and show zoning with more diopsidic cores and aegirine-augite or aegirite borders. Some crystals show hourglass structure.

The most common accessory minerals are sphene, zircon, apatite, ilmenite, and magnetite. Rarely fluorite, garnet, corundum, and a variety of unusual minerals may be present.

Texture and structure. Equigranular texture (uniform grain size) is most common, and locally very coarse (pegmatitic) phases occur. Porphyritic texture (large crystals in finer-grained matrix) is almost confined to the nepheline syenite porphyries. The phenocrysts, where present, however, are usually sanidine and mafics. Feldspars may be anhedral (without crystal outline) to give allotriomorphic texture, or they may be nearly euhedral (with crystal outline) and associated with interstitial nepheline or mafics. Poikilitic texture is found where large alkali feldspar grains enclose nearly euhedral crystals of nepheline. Various combinations of graphic (cuneiform) intergrowths occur between alkali feldspar, feldspathoids, and mafics. Tabular feldspar crystals and streaks of mafic minerals may show subparallel arrangement producing a flow structure in the rock.

Occurrence. Nepheline syenite and related rocks are rare and generally occur in small bodies (dikes, sills, laccoliths, stocks, and small irregular plutons). Only a few large bodies are known. Three of the largest bodies are in southern Greenland, Pilaansberg in South Africa, and Kola Peninsula, Soviet Union. Nepheline syenites may be associated with alkali syenites, with other feldspathoidal rocks, or with some alkali granites.

Formation. The origin of nepheline rocks is still a much debated problem. Many of these rocks may have formed from magma (molten rock material) of nepheline syenitic composition. This magma may have been derived from a basaltic one by fractional cystallization in which certain early-formed crystals were removed from the melt. By assimilation of abundant limestone, certain magmas may be desilicated to yield nepheline syenites. Some nepheline syenites and related rocks are believed to be of metamorphic and metasomatic origin. They may have formed from other rock types by introduction of certain elements and removal of others. *See* IGNEOUS ROCKS; MAGMA; METASOMATISM. [CARLETON A. CHAPMAN]

Nephelinite

A dark-colored, aphanitic (very finely crystalline) rock of volcanic origin, composed essentially of nepheline (a feldspathoid) and pyroxene.

The texture is usually porphyritic with large crystals (phenocrysts) of augite and nepheline in a very-fine-grained matrix. Augite phenocrysts may be diopsidic or titanium-rich and may be rimmed with soda-rich pyroxene (aegirine-augite). Microscopically the matrix is seen to be composed of tiny crystals or grains of nepheline, augite, aegirite, and sodalite with occasional soda-rich amphibole, biotite, and brown glass. If leucite becomes the dominant feldspathoid, the rock becomes a leucitite. If calcic plagioclase exceeds 10%, the rock passes into tephrite and basanite. If olivine is present, the rock is an olivine nephelinite (nepheline basalt). Accessories usually include magnetite, ilmenite, apatite, sphene, and perovskite.

Nephelinite and related rocks are very rare. They occur as lava flows and small, shallow intrusives. A great variety of these feldspathoidal rocks is displayed in Kenya. *See* FELDSPATHOID; IGNEOUS ROCKS; LEUCITE; LEUCITE ROCK.

[CARLETON A. CHAPMAN]

Niccolite

A minor ore of nickel. Niccolite is a mineral having composition NiAs and crystallizing in the hexagonal system. Crystals are rare, and niccolite usually occurs in massive aggregates with metallic luster and pale copper-red color. Because of the color, not the composition, it is called copper nickel. The hardness is 5.5 on Mohs scale and the specific gravity is 7.78. Niccolite is frequently associated with other nickel arsenides and sulfides in massive pyrrhotite. It is also found in vein deposits with cobalt and silver minerals, as in the silver mines of Saxony, Germany, and Cobalt, Ontario, Canada. *See* PYRRHOTITE.

[CORNELIUS S. HURLBUT, JR.]

Niter

A potassium nitrate mineral with chemical composition KNO_3. Niter crystallizes in the orthorhombic system, generally in thin crusts and delicate acicular crystals; it also occurs in massive, granular, or earthy forms. It has good cleavage in three directions; fracture is subconchoidal to uneven; it is brittle; hardness is 2 on Mohs scale; specific gravity is 2.109. The luster is vitreous, and the color and streak are colorless to white. *See* NITRATE MINERALS.

Niter is commonly found, usually in small amounts, as a surface efflorescence in arid regions and in caves and other sheltered places. It is usually associated with soda-niter, epsomite, nitrocalcite, and gypsum. The mineral may occur as an efflorescence on soils rich in organic matter from the action of certain bacteria on nitrogenous or animal matter. Niter occurs associated with soda-niter in the desert regions of northern Chile, and in similar occurrences in Italy, Egypt, the Soviet Union, the western United States, and elsewhere. It was formerly found in some abundance in limestone caves in Tennessee, Kentucky, Alabama, and Ohio, and was used for the manufacture of gunpowder during the War of 1812 and the Civil War.

[GEORGE SWITZER]

Nitrate minerals

These minerals are few in number and with the exception of soda niter are of rare occurrence. Normal anhydrous and hydrated nitrates occurring as minerals are soda niter, $NaNO_3$; niter, KNO_3; ammonia niter, NH_4NO_3; nitrobarite, $Ba(NO_3)_2$; nitrocalcite, $Ca(NO_3)_2 \cdot 4H_2O$; and nitromagnesite, $Mg(NO_3)_2 \cdot 6H_2O$. In addition there are three known naturally occurring nitrates containing hydroxyl or halogen, or compound nitrates. They are gerhardtite, $Cu_2(NO_3)(OH)_3$; buttgenbachite, $Cu_{19}(NO_3)_2Cl_4(OH)_{32} \cdot 3H_2O$; and darapskite, $Na_3(NO_3)(SO_4) \cdot H_2O$. *See* NITER; SODA NITER.

The natural nitrates are for the most part readily soluble in water. For this reason they occur most abundantly in arid regions, particularly in South America along the Chilean coast.

[GEORGE SWITZER]

Obsidian

A volcanic glass, usually of rhyolitic composition, formed by rapid cooling of viscous lava. The color is jet-black because of abundant microscopic, embryonic crystal growths (crystallites) which make the glass opaque except on thin edges. Iron oxide dust may produce red or brown obsidian.

Obsidian usually forms the upper parts of lava flows. Well-known occurrences are Obsidian Cliffs in Yellowstone Park, Wyo.; Mount Hekla, Iceland; and the Lipari Islands off the coast of Italy. Less commonly, obsidian forms selvages of dikes and sills. *See* IGNEOUS ROCKS; VOLCANIC GLASS.

[CARLETON A. CHAPMAN]

Oceanic islands

Those islands which rise from the deep-sea floor rather than from shallow continental shelves. Most islands in gulfs and seas that fringe the great ocean basins are geologically similar to the nearby continents. On the other hand, almost all islands that rise from the ocean basins are volcanoes with or without coral reef and, geologically, bear little relation to the continents. Volcanic islands are only the tops of much larger undersea volcanoes, most of which are associated with great submarine structures, such as submarine ridges and fractures in the Earth's crust (Fig. 1).

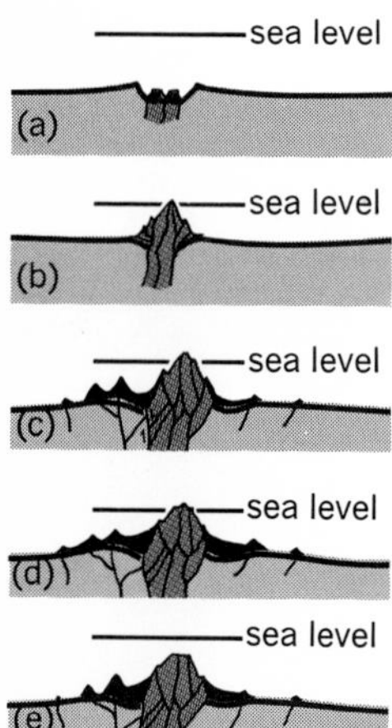

Fig. 1. Development of submarine volcano or ridge. Center areas are initial volcanic extrusion; black areas are later deposits of volcanic material and erosional debris. Sequence (*a*) to (*e*) is explained in text.

Submarine volcanoes. On the deep-sea floor, volcanoes begin as lava flows from fissures in the Earth's crust under 2 or 3 mi of water. Gradually they build upward through the water, but about nine-tenths of the submarine volcanoes become inactive and stop their growth before they reach the sea surface. The others burst from the deep water into a new realm, where wave and subaerial erosion combat their upward growth. At first the volcanoes tend to produce ash and cinders, which are easily eroded. Falcon Island, an active volcano in the Tonga group, has several times been reduced to a submarine bank within a few years after an eruption built up an island. Gradually as the pile becomes broader, volcanoes rise above the waves and more resistant fluid lava flows build a solid island. Where several nearby volcanoes merge together, as in Hawaii, a great island may form. *See* VOLCANO; VOLCANOLOGY.

Volcanoes are active for no more than a few million years, however, and inactive volcanoes are inevitably worn down to shallow submarine banks by erosion which never stops. In addition to these worn-down volcanoes, drowned former islands called guyots or tablemounts have been discovered in all the ocean basins, mostly at depths of 1000–7000 ft. Reef coral as old as the Cretaceous and volcanic erosional debris have been dredged on some guyots. Also, drilling on atolls shows that the coral is a capping several thousand feet thick on former volcanic islands. *See* ATOLL; SEAMOUNT AND GUYOT.

Associated submarine structures. Most submarine volcanoes are associated with great submarine structures: long, straight, narrow features,

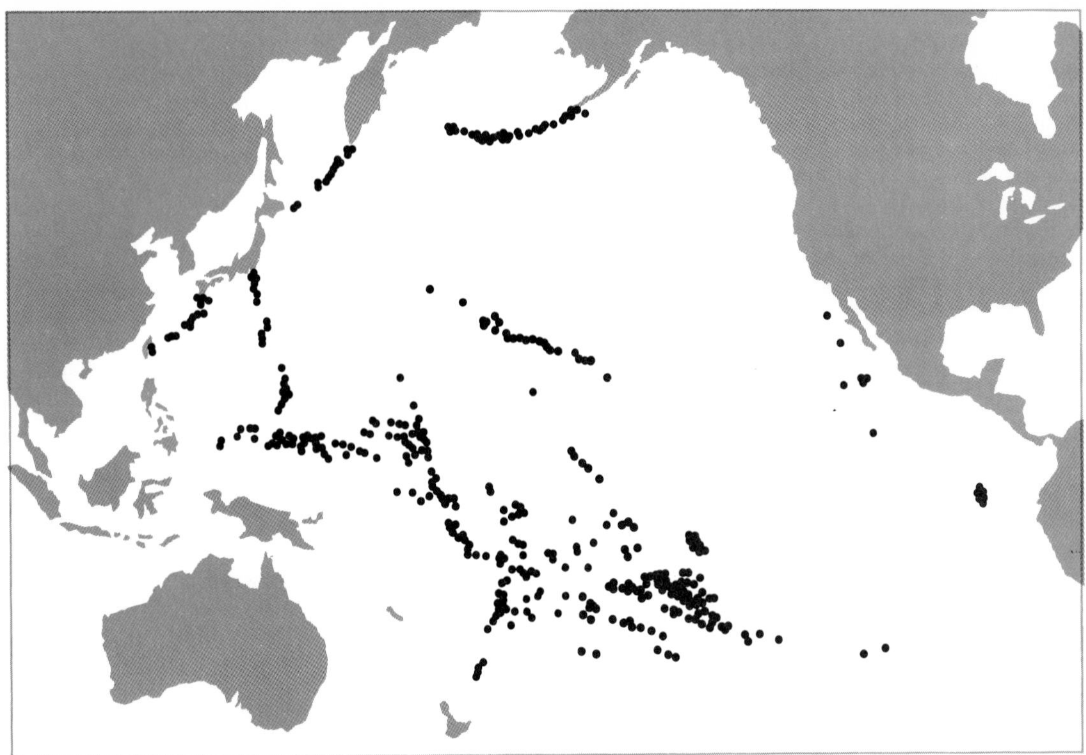

Fig. 2. Distribution of volcanic islands, banks, and atolls in the Pacific Basin.

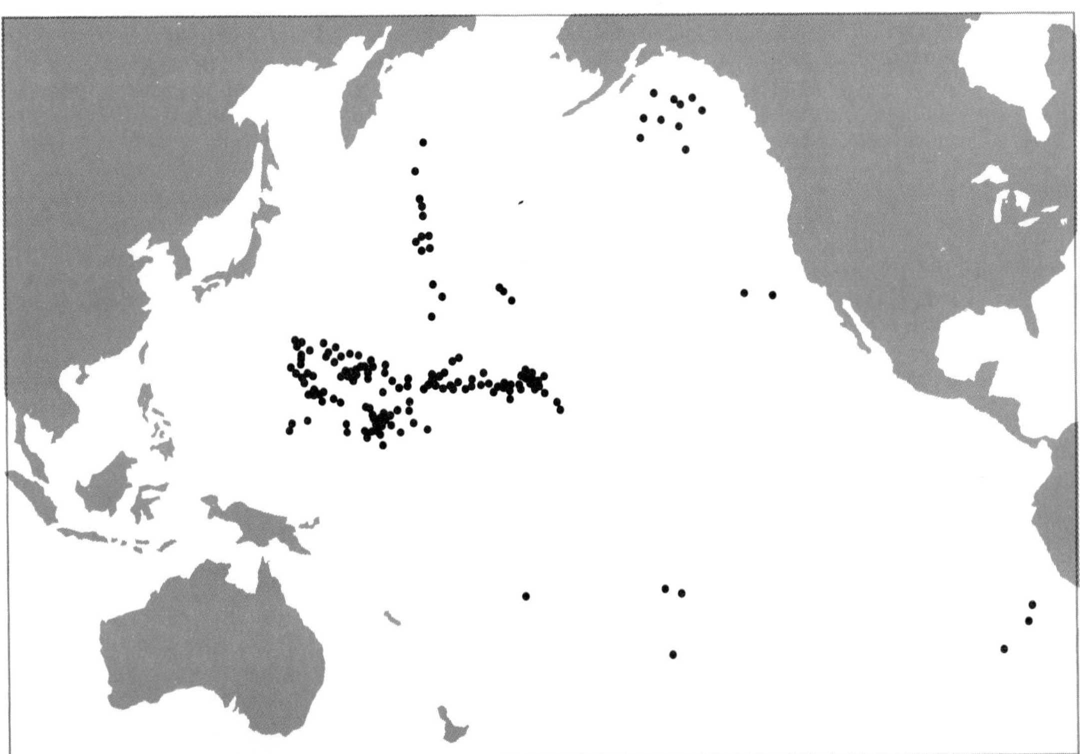

Fig. 3. Distribution of guyots, or former islands, in the Pacific Basin.

such as the Hawaiian Ridge and Murray Fracture Zone; broad oceanic rises, such as the Mid-Atlantic Ridge; and island arcs and trenches, such as the Aleutian Arc. *See* MARINE GEOLOGY.

Long straight structures. Lines of volcanoes occur in the Atlantic and Indian oceans but are relatively rare. A linear group of guyots extends southeast from Cape Cod, and other groups may be undiscovered in the less well-surveyed parts of these oceans.

It is the Pacific, however, that is the type area for linear archipelagoes, and there they are extremely common. Existing linear groups are largely confined to the southwestern and central Pacific (Fig. 2), but, in the past, islands were present in the northern part of the basin (Fig. 3). Some very large archipelagoes, like the Tuamotu Islands, and former archipelagoes, such as the Mid-Pacific Mountains, consist of individual volcanoes only a few thousand feet high rising as peaks above great steep-sided ridges. Other archipelagoes, including the Hawaiian, Samoan, and Marquesas islands, have large volcanoes rising from lower ridges. The occurrence of volcanoes in a straight line suggests an underlying linear fracture in the Earth's crust. The association of volcanoes with a fracture in the crust (Fig. 1*a*) is clearly demonstrated along the Clarion Fracture Zone, a submarine feature in the east-central Pacific. A long straight trough forms the western part of the fracture zone. Toward the east it is interrupted by seamounts until the trough disappears and is replaced by a line of shallow banks, and even farther to the east it is replaced by the volcanic Revillagigedo Islands.

The Earth's crust is able at first to support the load of a new volcano or volcanic ridge, but as the structure becomes larger, a point is reached when crustal strength is insufficient and elastic downbowing begins (Fig. 1*b*). The topographic expression is that of a central ridge surrounded by a depression or moat outside of which an arc may occur. A single volcano may be encircled by a small individual moat. As downbending continues, tension fractures permit volcanism on the arches (Fig. 1*c*). Volcanism and erosion debris commence filling the marginal depressions and faulting may occur. These depressions may become filled with volcanic material and sediment to form smooth archipelagic aprons sloping away from the island, seamount, or ridge, as around the Marquesas Islands (Fig. 1*d*). If eroded to a flat bank and relatively sunk below the surface, the seamount becomes a guyot (Fig. 1*e*); if coral was present at the surface and kept pace with the sinking, an atoll forms.

Broad oceanic rises. Submarine ridges or rises are the locus of solitary volcanic islands and seamounts, such as Ascension, Reunion, and Easter islands. Typically volcanoes are located near, but not on, the crests of the broad submarine rises. In addition, volcanic islands occur in clusters. Examples are the Azores and Galapagos islands. The latter group is surrounded by a thin archipelagic apron, but moats have not been found around this type of island cluster.

Island arcs. Groups of islands which follow more or less curved lines are called island arcs. They are associated with deep trenches, large gravity anomalies, and earthquakes; hence they are in a region of great crustal instability. Typically there is an inner arc of active volcanoes and an outer arc of nonvolcanic islands which may contain sediments of the types now found on the deep-sea floor, thus indicating uplift of several miles. Uplift is also shown by raised sea cliffs and deposits of coral. Drowned former islands do not occur along island arcs. *See* TECTONIC PATTERNS.

[EDWIN L. HAMILTON; HENRY W. MENARD, JR.]

Bibliography: R. S. Dietz, Marine geology of Northwestern Pacific, *Bull. Geol. Soc. Amer.*, 65: 1199, 1954; E. L. Hamilton, *Sunken Islands of the Mid-Pacific Mountains*, Geol. Soc. Amer. Mem. no. 64, 1956; P. H. Kuenen, *Marine Geology*, 1950; H. W. Menard, Archipelagic aprons, *Bull. Amer. Assoc. Petrol. Geol.*, 40:2195, 1956; H. W. Menard, *Marine Geology of the Pacific*, 1964; F. P. Shepard, *Submarine Geology*, 1963.

Oceans

The vast salt-water bodies surrounding the continents and filling the great basins of the Earth's crust. Some 70% of the Earth's surface is covered by the oceans, with a total volume of approximately 1.35×10^9 km^3. The world ocean can be more aptly described as a hemispheric ocean; that is, the Southern Hemisphere has a greater aerial distribution of sea water than the Northern Hemisphere. In fact, more than two-thirds of all land lies north of the Equator.

The oceanic regions of the Earth comprise not only the bodies of water which fill the large deep-ocean basins, but also the shallower seas that cover the slightly submerged edges and interiors of continents. The distribution of elevations of the Earth's surface relative to sea level can be plotted on a graph called a hypsographic curve. This curve (Fig. 1) indicates that two dominant elevations characterize the Earth's surface: one approximately 100 m above sea level and the other 5000 m below sea level. These two average elevations are separated by a sharp transition zone. The distribution indicates that the oceans form a geographic province entirely distinct from that of the continents. The great oceanic depth is a reflection of the vastly different compositional makeup of the Earth's crust beneath the sea compared with that which underlies the continents.

An understanding of the oceans includes the history of formation, composition, and origin of the ocean basins; the distribution and origin of sediment and mineral deposits; the composition, distribution, and circulation of the water masses; and the study and distribution of life within the oceans.

Features of ocean floor. Three major morphologic provinces characterize the ocean floor: the continental margins, the deep-ocean basins, and the mid-oceanic ridges (Fig. 2).

Around the Atlantic Ocean, and along other similar ocean margins, the continental margins consist of shelf areas which are submerged edges of the continents and the continental slopes and rises which are transition zones between the continents and deep-ocean basins. Continental shelves comprise about one-sixth of the Earth's surface. Their width varies from several kilometers (southern California) to several hundred kilometers (Gulf Coast, Grand Banks), averaging about 65 km. The shelves dip seaward at a very gentle gradient of about 0.07° and attain depths of about 130 m at their outer margin, called the shelf break. The shelf edge may be a depositional or erosional feature. It is subject to modification by the sculpting and scouring force of strong currents and erosion during low sea-level stands. Active faulting occurs along unstable margins, such as that off the coast of southern California and southern Alaska. Seaward of the shelf break is the more steeply dipping (4°) continental slope, an approximately 20-km-wide zone characterized by either seaward-dipping sediments or slumping of sediments downslope. The continental rise, at the base of the continental slope, has a seaward dip of less than 0.5° and varies in width from 100 to 1000 km. Rises consist of thick wedges of sediment which accumulate by gravitational gliding of semifluidized material down the steeper continental slope or by the deposition of land-derived clastics in the form of great aprons of sediment on the deep-ocean floor, termed submarine fan deposits. *See* CONTINENTAL SHELF AND SLOPE.

Some continental margins are so unlike the conventional Atlantic-type margins (with continental shelves, rises, and slopes) that they constitute an additional oceanic morphological realm, the western Pacific type and Andean types of continental margin. These may include offshore volcanic island arcs like those fringing the western Pacific (Japanese Islands, Philippine Islands, and so on) and eastern Caribbean. These island arcs are paralleled offshore by deep narrow trenches which represent the greatest depths in the oceans. The deepest trench is the Challenger Deep, part of the Marianas Trench in the Pacific Ocean, where a depth of 11,022 m below sea level has been recorded. In general, trenches are hundreds of kilometers wide and 3–4 km deeper than the surrounding ocean floor. They are continuous for thousands of kilometers and are generally V-shaped, with narrow flat floors due to sediment infilling. Small ocean basins which commonly occur behind the trench-bounded chains of volcanic islands (Sea of Japan, Philippine Sea) are called marginal seas. Other small ocean basins such as the Mediterranean Sea, Gulf of Mexico, and Caribbean, Bering, and Black seas, which are partially separated from the main ocean bodies, are often also termed marginal seas. *See* MARINE GEOLOGY.

The deeper-realm ocean floor (excluding the continental margins) covers 56% of the Earth's surface. Much of the deeper-realm ocean floor consists of the abyssal plains (30% of the Earth's surface area), which are broad, flat, featureless plains with an average depth below sea level of 4–6 km. Surface slopes do not exceed 0.05 degrees. The subdued relief is produced by the absence of tectonic activity and by the blanketing effect of sedimentary cover.

The topography of the deeper-realm ocean floor is dominated by an extensive mid-oceanic ridge system which is continuous for more than 55,000 km through the Atlantic, Pacific, Arctic, and Indian oceans. The ridges of this system are characterized by a central, 25–50-km-wide rift valley whose average depth is 2.7 km. The rift is bordered by steep mountains standing 1–3 km above the sea floor, peaks of which emerge above sea level as islands (Iceland and Bouvet and St. Paul islands). Numerous fracture zones which trend perpendicular to the ridge and offset the ridge crest (transform faults), and the many volcanoes and relict volcanoes associated with the ridge system produce irregular and rugged morphologic surface in some portions of the usually monotonous abyssal plain areas. Some volcanic islands (Hawaii, Line Islands) and seamounts (New England and Emperor seamounts) form linear chains of emergent or submergent volcanoes. Submerged seamounts

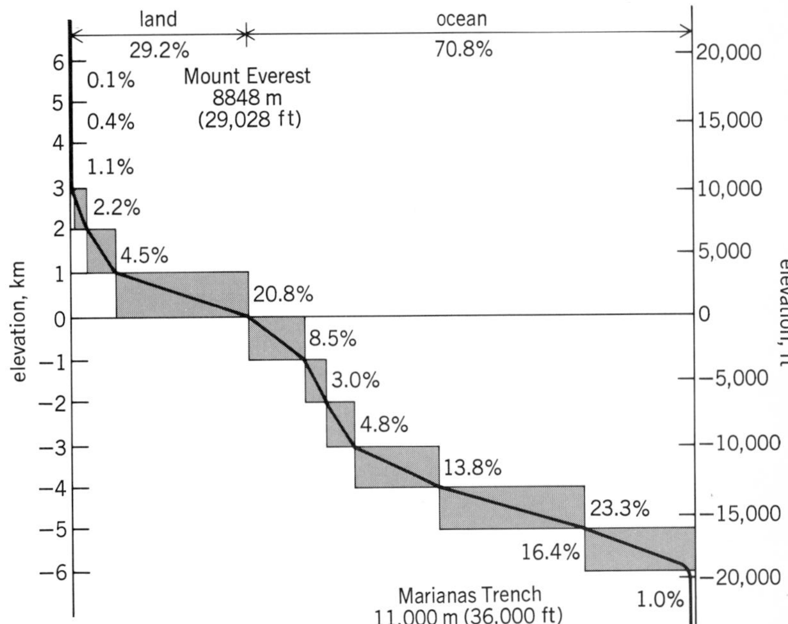

Fig. 1. Hypsographic curve showing the relative proportions of the Earth's surface at elevations above and below sea level. (*From A. N. Strahler, Physical Geography, 2d ed., copyright © 1960 by John Wiley & Sons, Inc.; reprinted by permission*)

with flat tops are called guyots. These beveled peaks were eroded flat when the seamounts were near sea level. Coral atolls sometimes fringe submerged seamounts or guyots in the equatorial regions. Rises and swells are broad regions of higher elevation on the ocean floors, such as the Bermuda Rise in the Atlantic and the Shatsky Rise in the Pacific. These higher regions presumably reflect an extensive period of past volcanic outpouring over a wide region of the ocean floor. *See* OCEANIC ISLANDS; SEAMOUNT AND GUYOT.

Evolution of ocean floor. The difference in elevation of the continents and ocean basins attests to the differences in composition between continental and oceanic crust. Continental crust is thick (35–60 km) and composed primarily of granitic-type rocks with density = 2.7 (Fig. 3). Oceanic crust is denser (density = 3.0) and only 5–10 km thick, consisting of a carpet of sediment (layer 1) overlying thick piles of extrusive volcanic rocks (layer 2). These volcanic rocks of the ocean crust are basaltic in composition and are characterized by tubular and bulbous "pillow lavas" and ropy-textured plains flooded by "pahoehoe" lavas. The extrusive flows cap a third component of the volcanic crust (layer 3), intrusive-type volcanic rocks composed of densely spaced vertical dikes fed from subjacent plutonic complexes and the dense, underlying, coarsely crystalline rocks of the mantle.

The oceanic crust is also much younger than the continental crust. The bulk of the continental crust is at least 10^9 years old, much is $2-3 \times 10^9$ years old, and continental rocks as old as 3.9×10^9 years have been described. However, the oldest in-place oceanic crust is about 200×10^6 years old. No oceanic crust older than this exists within the ocean basins, although older slices of ultramafic (mantle) rocks, sheeted dike complexes (densely spaced vertical dikes), and basaltic rocks capped

Greenland
Jan Mayen Fracture Zone
Jan Mayen Ridge
Iceland
Faeroe Islands
Scandinavia
Hudson Bay
Mid-ocean Canyon
Reykjanes Ridge
Lousy Bank
Rockall
North Sea
Baltic Sea
British Isles
Canadian Shield
St. Lawrence River
Newfoundland
Flemish Cap
Europe
North America
Great Lakes
Milne Seamount
Laurentian Cone
Rift Valley
Biscay Abyssal Plain
Alps
Adriatic Sea
Mississippi River
Appalachian Mountains
Kelvin Seamount
Sohm Abyssal Plain
Oceanographer Fracture Zone
Azores
Iberian Peninsula
Balearic Islands
Aegean Sea
coastal plain
continental shelf
Hatteras Abyssal Plain
Bermuda Rise
Corner Seamounts
Ampere Seamount
Atlas Mountains
Mediterranean Ridge
Blake Plateau
Madeira Islands
Mississippi Cone
Gulf of Mexico
Bahama Islands
Mid-Atlantic Ridge
Canary Islands
Nares Abyssal Plain
Sigsbee Knolls
Puerto Rico Trench
Greater Antilles
Cape Verde Islands
Central America
Caribbean Sea
Atlantic Ocean's deepest point
Krylov Seamount
Lesser Antilles
Africa
Demerara Abyssal Plain
Gambia Abyssal Plain
Sierra Leone Rise
Amazon Cone
continental shelf
Ceara Abyssal Plain
St. Peter and St. Paul Rocks
Fernando Po
Cocos Ridge
Guiana Highlands
Galapagos Islands
Equator
São Tomé
Romanche Fracture Zone
Congo Basin
Amazon Basin
Congo Canyon
Ascension
Great Rift Valley
South America
Pernambuco Abyssal Plain
Angola Abyssal Plain
Andes
Brazilian Highlands
Stocks Seamount
St. Helena
Peru-Chile Trench
Nazca Ridge
Columbia Seamount
Walvis Ridge
Namib Desert
Kalahari Desert
Rift Valley
Juan Fernandez Islands
Aconcagua South America's highest point
Rio Grande Rise
Wust Seamount
Vema Seamount
South Pacific Ocean
continental shelf
continental slope
continental rise
Tristan da Cunha Group
Agulhas Plateau
Argentine Abyssal Plain
Mid-Atlantic Ridge
Herdman Seamount
Mornington Abyssal Plain
Merz Seamount
Meteor Seamount
Falkland Islands
Falkland Plateau
South Georgia
Scotia Ridge
South Sandwich Trench

by thin, deep marine sediments are found on land. These have been tectonically thrust onto the continents and incorporated into highly deformed mountain belts. Known as "ophiolites," these sequences are interpreted to represent fragments of older oceanic terranes that no longer exist.

The contrasts between the lithology and age of continental and oceanic crust is presently explained largely by the theories of sea-floor spreading and plate tectonics. These theories also provide an interpretative basis for differences in the geological characteristics of continental margin types, the abyssal plains, and the mid-ocean ridge system.

The sea-floor spreading theory postulates that new oceanic crust is continually being produced by the upwelling of magmatic material along the mid-ocean ridge system. As new magma rises and solidifies, older ocean crust is pushed aside and the ocean basins grow. This spreading of the sea floor also causes continental drift because the granitic continental blocks become further and further apart as the ocean floor grows. The high topography of the mid-ocean ridge system is directly related to the presence of anomalously hot rocks which expand in volume. As newly formed oceanic crust cools with progressively more time (and further from the ridge crests), oceanic crust volumetrically contracts, producing the topographically lower, subdued abyssal plain areas. Formation of new crust at ridges occurs at the rate of 1 – 10 cm per year on each side of a ridge.

A variety of data support this model for sea-floor spreading. Magnetometer surveys indicate that the ocean floor exhibits a pattern of linear stripes parallel to the mid-ocean ridges. These stripes are actually belts of magnetic anomalies, that is, alternating strips of crust which exert greater or less magnetism than the average. These magnetic anomaly patterns characterize many major ridge systems. In general, the strips or belts parallel the ridge axes and are symmetric in width across the ridge system. For years they remained a mystery, until in 1963 F. J. Vine and D. H. Matthews proposed that they represented evidence for the continuous generation of new ocean crust at the ridges by the extrusion of liquid basaltic lava. They reasoned that as the lava cooled, it became magnetized in the Earth's magnetic field. Because the Earth's magnetic field apparently alternates through geologic time between normal (as it is now) and reverse polarization, there is a reinforcing or subtractive effect exerted by the basaltic oceanic crust on the present-day magnetic field. Magma extruded during times of normal polarity operates in coincidence with the present-day magnetic field to produce a stronger than normal, or positive, anomaly. Magma extruded during times of reversed polarity produces a small component of magnetic strength opposite to that at present, and results in a negative (less than normal strength) magnetic anomaly.

Opposite page:

Fig. 2. Physiography of the floor of the North Atlantic Ocean, showing major features. (*From M. G. Gross, Oceanography: A View of the Earth, 2d ed., copyright © 1977 by Prentice-Hall, Inc.; reprinted by permission*)

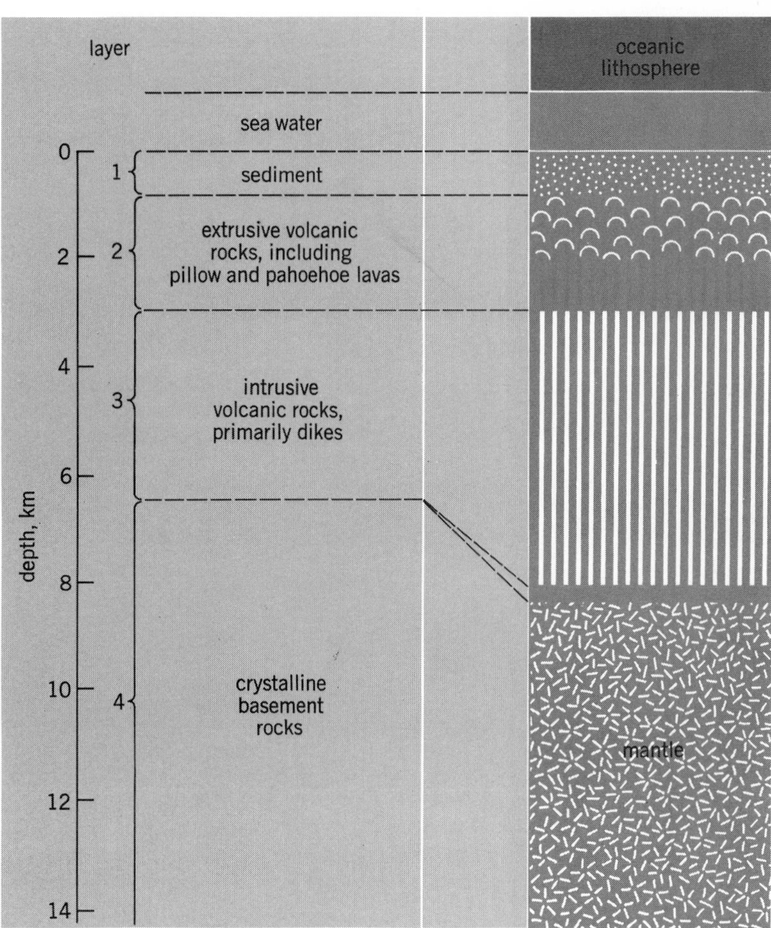

Fig. 3. Structure of oceanic crust. (*From P. A. Rona, Plate tectonics and mineral resources, Sci. Amer., 229(1):86–95; copyright © 1973 by Scientific American, Inc.; all rights reserved*)

A paleomagnetic time scale had been developed by dating continental basalts that displayed magnetic properties that were interpreted as the results of polar reversals through time. These were then correlated with the stripelike magnetic anomalies on parts of the ocean floor. By estimating spreading rates from these data, it was possible to extrapolate the paleomagnetic time scale further into the past and predict the ages of the ocean floors. The oldest oceanic crust is believed to lie in the northwest Pacific and is about 200×10^6 years old. Spreading rates of about 2 – 20 cm per year are adequate to produce all the ocean basins within this time span. *See* CONTINENT FORMATION.

The exact nature of the forces behind sea-floor spreading and continental drift is not clear. It is now known that long, linear mid-oceanic ridges form along belts of higher-than-average heat flow. The accretion of the oceanic crust is evidently linked to partial melting of the deep primitive mantle of the aesthenospheric layer of the Earth along linear zones which coincide with the mid-ocean ridge system. This melting liberates basaltic composition magma, which rises to the surface and either chills rapidly to form glass in contact with sea water or moves slowly in dike swarms or through crystal settling in deep cumulate chambers. Local "hot spots" also occur where plumes of hot magma rise from the mantle to produce lin-

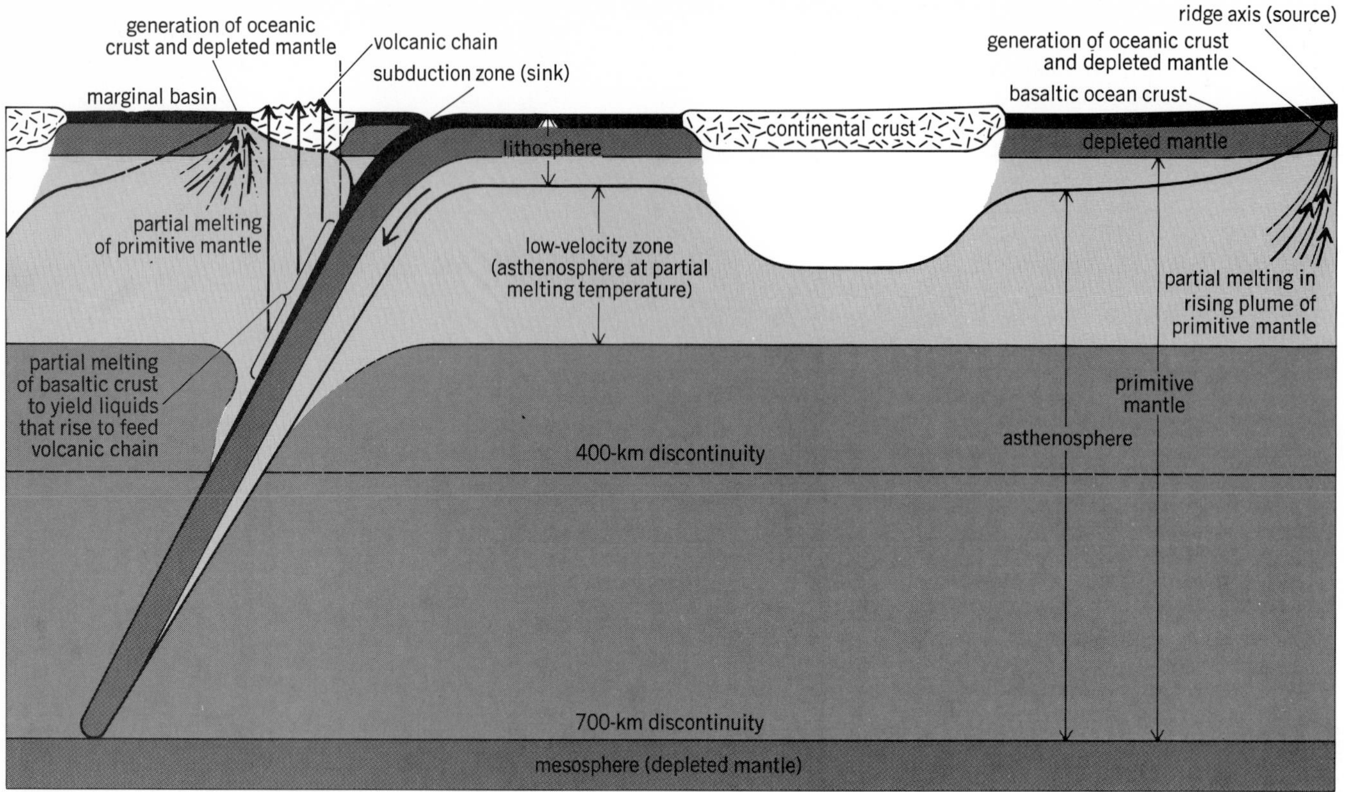

Fig. 4. Cross section of the oceanic crust, showing the mode of formation at mid-oceanic ridges and the subsequent destruction along deep-sea trenches or subduction zones. (*From J. F. Dewey, Plate tectonics, Sci. Amer., 226(5):62; copyright © 1972 by Scientific American, Inc.; all rights reserved*)

ear chains of volcanoes as they are overridden by a moving rigid plate of ocean crust (Hawaiian-Emperor Seamount chain).

Because the Earth is believed not to be expanding, the formation of new ocean crust must be accompanied by simultaneous destruction of ocean crust elsewhere (Fig. 4). This allows the origin of trenches, island arcs, and inland seas to be explained in the context of sea-floor spreading. The deep-sea trenches are believed to be areas where older, cooler oceanic crust is consumed as it slips beneath adjacent oceanic or continental crustal blocks. Regions of descending lithospheric slabs are the locus of destructive earthquakes, caused as the downgoing rock internally deforms. These earthquake regions, called Benioff zones, define the configuration of the downgoing lithospheric plate. Partial melting of descending rocks, as well as melting of rocks overlying the descending plate, causes magma rich in sodium and potassium to rise to the surface, forming the chains of volcanic islands which parallel the overlying deep-sea trenches. The frictional drag exerted on subjacent, continentally capped lithosphere probably produces marginal ocean basins. Compressive forces along these zones of convergence also cause deformation of the Earth's crust and the formation of mountain belts, such as the Andes.

Deep-sea trenches and mid-oceanic ridges are the main (convergent and divergent, respectively) boundaries of the rigid plates of the Earth's crust which move with respect to each other (Fig. 5). A third type of boundary is a transform fault, where the plates on either side are neither created nor destroyed but slide past one another, such as along the San Andreas Fault. Ridge-ridge transform faults of the mid-oceanic ridge system join different portions of a spreading ridge axis. The concept that the Earth's crust consists of plates that are created in one area, move, and are destroyed in another is termed plate tectonic theory. *See* FAULT AND FAULT STRUCTURES; PLATE TECTONICS.

Sediments, mineral deposits, and energy resources. The ocean floor is a repository (Fig. 6) for a continuous rain of material manufactured in the sea itself (shells, skeletons, teeth, scales, and animal and plant tissue), transported from the surrounding land (gravel, sand, silt, and mud derived from rivers and coastal beaches; windblown dust from deserts; ash and pumice from volcanic eruptions; boulders carried by melting icebergs; and spores and pollen of terrestrial plants), impacted from space (meteorites and tektites), products chemically precipitated from sea water or subsea thermal springs (halite, gypsum, metalliferous shales, polymetallic nodules, and crusts), or derived from alteration or abrasion of the ocean bedrock and previous generations of sediment (volcanic sands, zeolites, some iron/aluminum silicate clay minerals and phosphates). These various kinds of bottom sediments can be grouped into a number of genetically distinct types, of which three are most important: the terrigenous sediments, biogenous sediments, and red clay. Deep-sea sediment without significant amounts of terrigenous (land-derived) detritus are referred to as

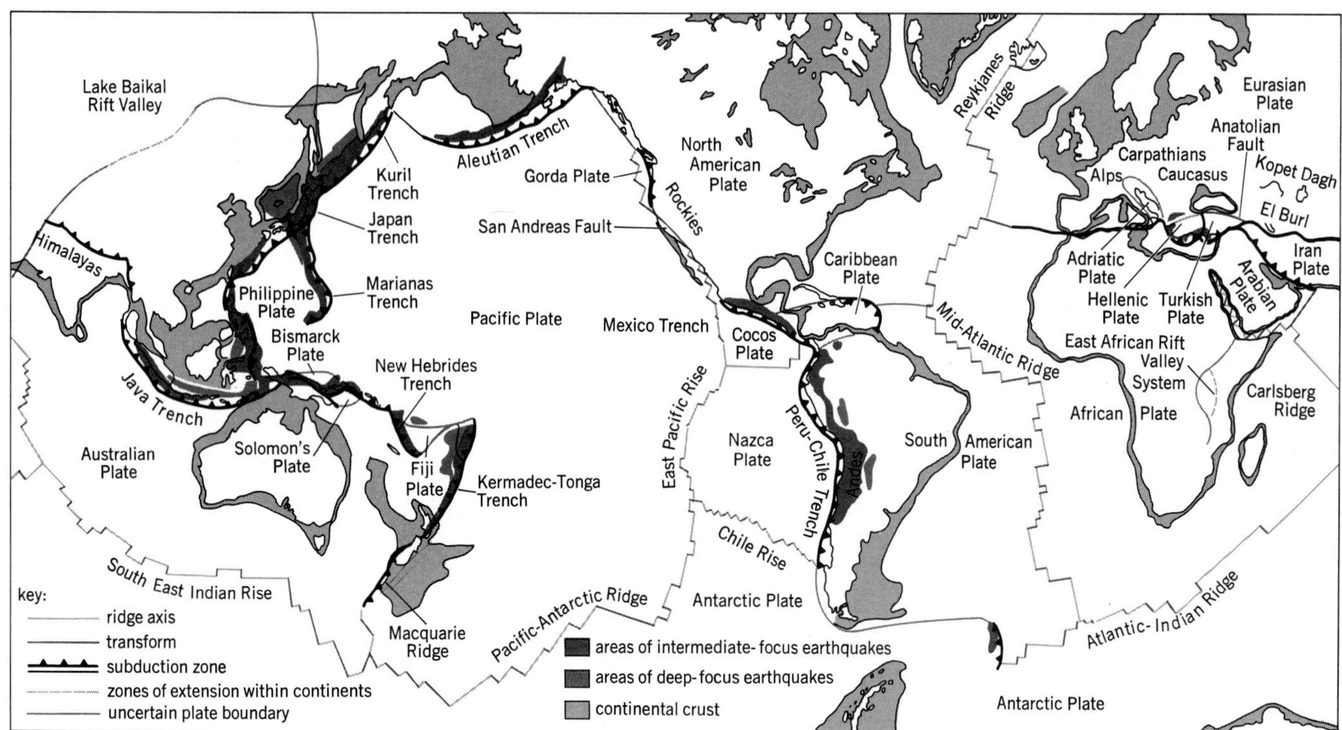

Fig. 5. Configuration of plates which segment the Earth's crust, showing locations of ocean ridges, trenches or subduction zones, and transform faults. Triangles along subduction zones are on upper, overriding plate. *(From J. F. Dewey, Plate tectonics, Sci. Amer., 226(5): 56–68; copyright © 1972 by Scientific American, Inc.; all rights reserved)*

pelagic deposits.

Terrigenous sediment. Land-derived material is sometimes carried to the ocean basins in episodic, swiftly flowing sediment suspensions known as turbidity currents which are generated by underwater avalanches, often initiated in the heads of submarine canyons carved into the continental slope. Deposits of turbidity currents, or turbidites, consist of size-graded sand, silt, and clay layers, with the coarsest components at the base of the layer, and are commonplace on continental rises and the adjacent abyssal plains. Fine-grained materials are often carried long distances in currents before settling to the seabed. Clouds of diffuse sediment in suspension in currents hugging the ocean floor are known as nepheloid layers. Violent explosions from land volcanoes in island arcs blast ash and glass shards to heights reaching 30 km in the upper atmosphere. Winds can carry this sand-sized volcanic tephra seaward for distances exceeding 500 km and the finer dust up to several thousand kilometers. Flowing continental glaciers (such as those found today in Greenland and Antarctica) and mountain glaciers (such as those in Alaska and Scandinavia) rip up bedrock and soil at their basal shear planes and incorporate this debris into the ice flow. Upon reaching the land's edge, the glaciers collapse into icebergs which, prior to melting, raft this material, known as erratics, for hundreds of kilometers out to sea. During the last ice age, glacial boulders were rafted in the Northern Hemisphere as far south as the coast of southern Portugal and almost to the vicinity of Bermuda.

Biogenous sediment. Microscopic skeletons of marine plankton make up layers of biogenic ooze which can be either calcareous (composed of pteropods, foraminiferans, coccolithophores, and discoasters) or siliceous (consisting of radiolarians, diatoms, and silicoflagelletes). Siliceous oozes are dominant in regions of high latitude and are abundant in equatorial and coastal regions of strong upwelling. In areas where the ocean is very deep, the calcareous material dissolves in corrosive bottom waters. The level of dissolution, known as the carbonate compensation surface (CCS), is shallowest near the margins of the ocean basins and deepest in the equatorial belt of high ocean-surface fertility. Although the mean depth of the CCS is about 5 km today, it has varied significantly in the past, being several hundred meters deeper in the Pacific during the ice ages and several kilometers shallower prior to about 40×10^6 years ago.

Red clay. Bright reddish-brown to chocolate brown, fine-grained (particle diameters less than 4 μm) pelagic material containing less than 10% calcium carbonate and amorphous silica is referred to as red clay. Red clay is polygenic in origin, consisting of windblown terrigenous and volcanogenic material, cosmic material such as tektites, and fine organic dust derived from the abrasion of biogenous material such as shark teeth.

The areal distribution of the various kinds of bottom sediment on the ocean floor is largely a function of oceanic depth, water chemistry and temperature, organic productivity, and the distance from terrigenous source areas. Terrigenous sediments form the dominant sediment cover in

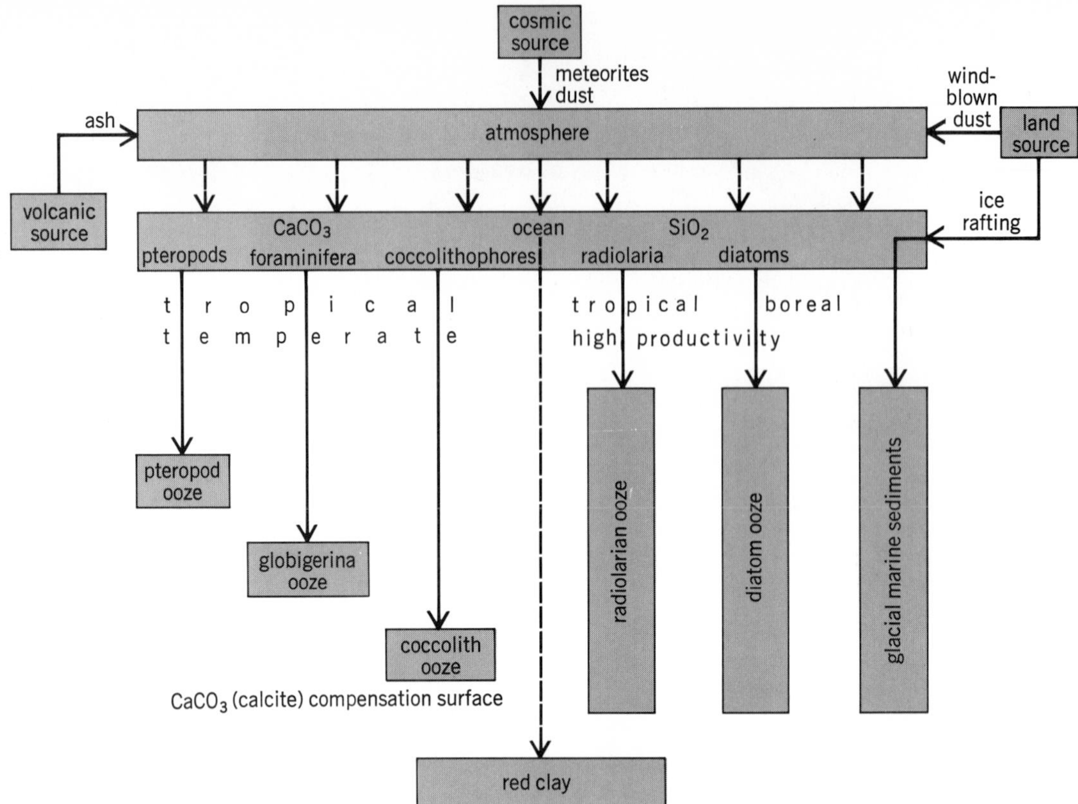

Fig. 6. Schematic representation of pelagic sedimentation in the ocean. (*From W. W. Hay, ed., Introduction, Studies in Paleo-Oceanography, Soc. Econ. Paleontol. Mineral. Spec. Publ. no. 20, pp. 1–5, 1974*)

belts bordering all the major continental masses and along the flanks of some portions of the mid-ocean ridge system. Away from terrigenous sources, biogenous sediments predominate when organic activity is sufficiently high, as is typical in tropical latitudes. Where the water depth in such areas exceeds the critical carbonate compensation depth, the dominant sediment is siliceous ooze. Where the water depth is shallower, calcareous oozes predominate. Abyssal plains which are both remote from terrigenous sources and in areas of low organic productivity are covered with a mantle of red clay.

Minerals. In areas characterized by either very slow deposition or erosion from bottom currents, slabs or nodules of iron and manganese oxides are accreted in concentric shells to eventually construct a nearly continuous indurated pavement (hydrogenous sediments). Some nodules reach 10 cm or more in diameter and contain economically interesting abundances of copper, nickel, and cobalt. Important nodule fields extend over vast tracts of the equatorial Pacific between the Clipperton and Clarion fracture zones, but also occur in the Cape Basin of the South Atlantic and in the Mozambique Basin of the Indian Ocean.

High heat flow in the crestal area of the mid-oceanic ridge induces hydrothermal circulations in permeable strata of layer 2 of the oceanic crust. Leaching of the basaltic rocks in the absence of oxygen scavenges iron and manganese ions into solutions that are ejected in hot thermal springs within the rift valley floor. Under such circumstances, metalliferous shales are precipitated onto the top of the volcanic pile and metallic oxides build hollow cylinders centered over the thermal vents. Unusual animal communities have adapted themselves to the springs by feeding upon sulfate-reducing bacteria in the emissions. Where thick layers of salt and associated evaporites are present such as in the Red Sea, pools of anoxic brine as hot as 60°C become trapped in rift valley depressions.

Persistent winds, such as the trades, blow surface water offshore, allowing cold subsurface water to upwell and replace it. Belts of upwelling (for example, the Benguela Current off southwestern Africa, Canary Current off northwestern Africa, and Peru-Chile Current off western South America) are recognized by their anomalously high nutrient content, great abundances of life, and generally reduced level of dissolved oxygen in underlying intermediate waters. Phosphates are an important economic resource of upwelling areas, and organic rich muds (sapropels) are present only where the sediment carpet of the continental slope intersects water masses which have been depleted in oxygen by the consumptive processes related to high productivity.

Carbonate platforms and barrier reefs (the Bahamas and northeastern Australia) occur at the continent's edge in warm surface waters and in isolation from coastal deltas. The seaward escarpments of these limestone terraces are among the steepest continuous slopes on the Earth's surface.

Gas and oil. Upon deep burial, the organic compounds in marine strata thermally maturate and

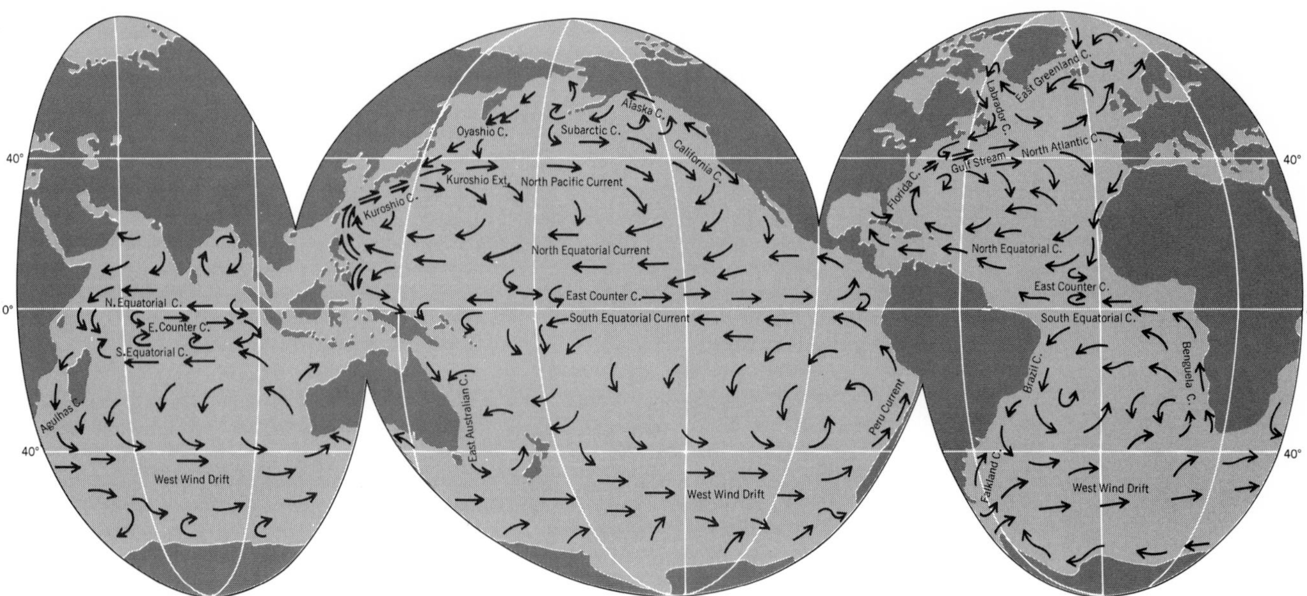

Fig. 7. Map showing major surface currents of the world's oceans. Winter conditions are shown for the Indian Ocean. (*From D. A. Ross, Introduction to Oceanography, copyright © 1970 by Prentice-Hall, Inc.; reprinted by permission*)

generate natural gas, liquid petroleum, or both. Under favorable conditions, gas and oil migrate along permeable beds to become stored in geologic reservoirs which can include ancient strandline (beach) sands or reef tracts whose matrices have been enlarged by solution-related processes. Continental shelves and slopes are most likely to provide fossil energy resources.

General ocean circulation. The atmosphere and hydrosphere interact to circulate the ocean's various water masses. Principal energy inputs are wind shear and heat. For the most part, wind shear drives the surface-ocean gyres, and heat, the deep-ocean boundary currents. Complications are introduced by uneven solar insolation as the result of both latitudinal differences and the Earth's orbital tilt and by deflections due to the Earth's rotation (Coriolis forces) which cause moving masses to turn to the right (clockwise) in the Northern Hemisphere and to the left (counterclockwise) in the Southern Hemisphere.

Equatorial regions are dominated by the westerly flow of the North and South Equatorial currents driven primarily by the tradewinds and bisected by a narrow easterly flowing countercurrent (Fig. 7). The equatorial currents form gyres elongated east-west and enclosing the subtropical regions centered on 30°N and 30°S latitude. Prominent easterly directed currents include the West Wind Drift in the Southern Hemisphere and the North Pacific Current and North Atlantic Current (Gulf Stream) in the Northern Hemisphere. Seasonally variable winds in the Indian Ocean dominated by the heating and cooling of surrounding land areas are responsible for the monsoon currents. Small counterclockwise surface gyres are present in the north subpolar and polar regions.

The deep thermohaline circulation (Fig. 8) is a global convective process in which dense cold waters in high latitudes sink and spread toward the Equator. In the Northern Hemisphere most of the deep water is created in the Norwegian Sea and, flowing southward as North Atlantic Deep Water, it hugs the western boundary of the ocean basin, reaching velocities of 10–20 cm/s. These bottom-hugging, contour-following currents are capable of sculpting the ocean bottom and retransporting and depositing fine-grained sediment in thin laminated deposits termed contourites. Cold, dense water produced around Antarctica forms both the Antarctic Bottom Water and the Antarctic Intermediate Water. The Antarctic Bottom Water flows through the Argentine and Brazil basins in the Atlantic and through the Samoa Passage in the Pacific, and crosses the Equator to spread out onto the Bermuda Rise and reach as far north as the Aleutian Abyssal Plain. Bottom-water circulation is strongly influenced by sea-floor topography.

Chemistry and origin of sea water. Sea water is a weak solution of various substances dissolved in water (see table). The term "salinity" is used to define the total weight of dissolved substances. Salinity is customarily expressed in grams per kilogram, parts per thousand (ppt), or parts per million (ppm). The average salinity of sea water is 35,000

Major elemental abundances in sea water

Element	Abundance, mg/liter	Principal species
Oxygen (O)	857,000	H_2O
Hydrogen (H)	108,000	H_2O; O_2; SO_4^{--}
Chlorine (Cl)	19,000	Cl^-
Sodium (Na)	10,500	Na^+
Magnesium (Mg)	1,350	Mg^{++}; $MgSO_4$
Sulfur (S)	885	SO_4^{--}
Calcium (Ca)	400	Ca^{++}; $CaSO_4$
Potassium (K)	380	K^+
Bromine (Br)	65	Br^-
Carbon (C)	28	HCO_3^-; H_2CO_3; CO_3^{--} and organic compounds

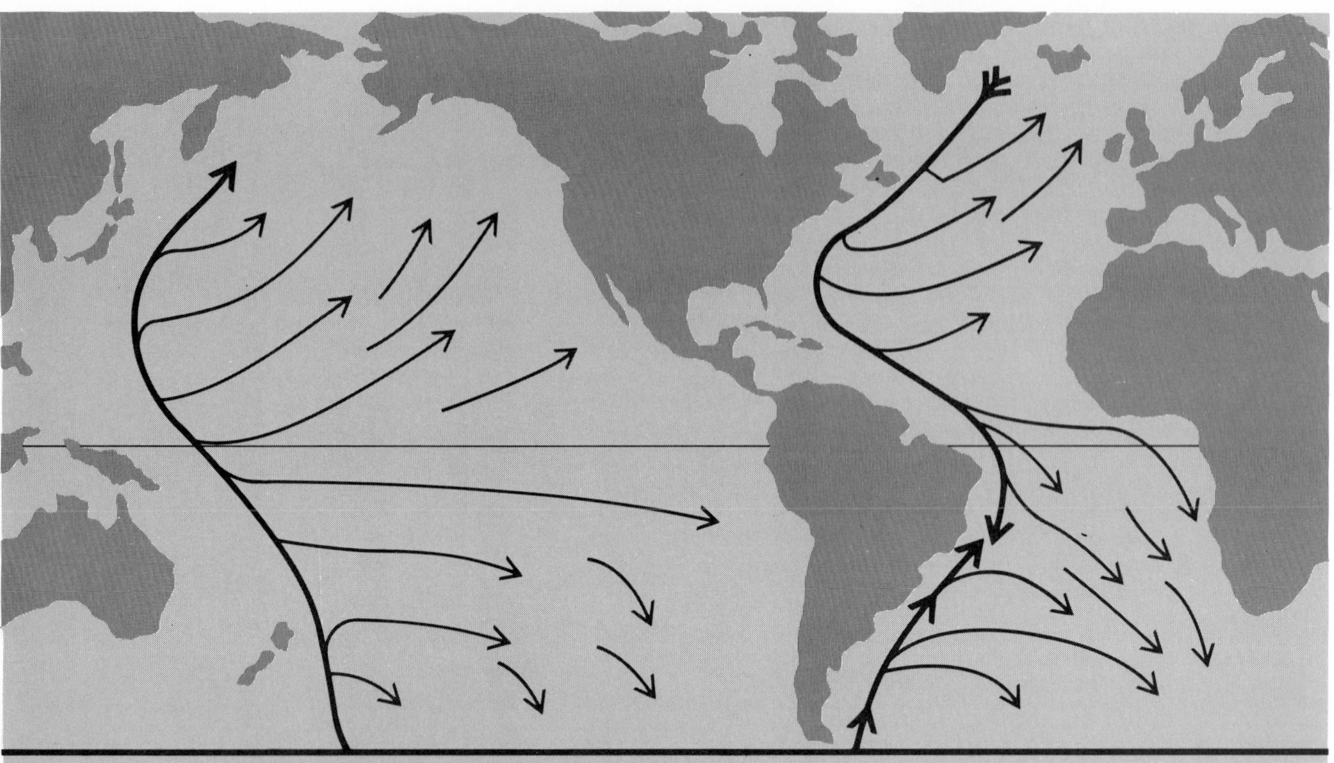

Fig. 8 Approximate direction of deep-water circulation. Details of deep circulation are mostly unknown. (*From R. W. Stewart, The atmosphere and the ocean, Sci. Amer., 221(3):76–86; copyright © 1969 by Scientific American, Inc.; all rights reserved*)

ppm. The bulk of this salinity, which gives sea water its "saltiness," is dissolved sodium and chlorine.

Speculation on the origin of sea water and the source of sea water salinity is based largely on comparisons between the chemistry of sea water, river runoff, volcanic gases, and geysers. The water itself is thought to represent the accumulation of condensed steam and juvenile gases "exhaled" from the Earth during the early stages of Earth history. This degassing is also thought to have produced the bulk of the constituents in the Earth's atmosphere. Several constituents dissolved in sea water must have had a similar origin, because they are not contributed to the sea by river runoff (examples are chlorine, boron, and bromine).

Many of the other dissolved constituents in sea water (carbonate, sulfate, magnesium, calcium, potassium, and sodium) evidently represent the cumulative buildup of components provided to the sea by the dilute solutions of river runoff. These substances are put into solution by the chemical decomposition which occurs during the weathering of preexisting rocks and minerals. The discrepancy in the proportion of some of these constituents in sea water versus runoff is largely due to likelihood of their extraction once in sea water. Some constituents have a very brief "residence time" in sea water. For example, calcium, carbonate, and silica are readily extracted from sea water by organisms and eventually incorporated as shell fragments in siliceous and calcareous oozes. Still other constituents such as iron and potassium are quickly extracted from solution and absorbed on clay minerals or become involved in diagenetic chemical reactions. Conversely, other constituents, particularly sodium and chlorine, are not organically extracted, nor do the proper conditions for their chemical precipitation occur frequently. Consequently, they tend to accumulate as the most important dissolved constituents in sea water.

The distribution of life in the oceans is controlled by the biolimiting nutrients (oxygen, silica, nitrogen, and phosphorus) dissolved in sea water. Dissolved oxygen is abundant in the surface ocean, depleted at intermediate levels, and enriched in bottom waters advected from the cold polar surface ocean. The Black Sea is the only major ocean basin whose center is totally deficient in oxygen (that is, euxinic), although oxygen levels are extremely low in silled basins on the continental margin of California and in certain coastal fiords. Silica, nitrogen, and phosphorus are removed from the surface ocean by biological factories and are enriched in deep-water masses by the downward settling and decay of dead marine organisms. Enhanced fertility is therefore brought about by upwelling phenomena. Sunlight is attenuated in sea water, and for all purposes is negligible at depths in excess of 250 m.

Life in the ocean. The surface ocean is the most prolific in terms of biomass alone. The sunlit realm, known as the photic region, supports a microscopic plant kingdom, predominantly algae, consumed by floating protozoa (plankton), swimming fish (nekton), and large marine mammals (whales and porpoises). Life on the ocean floor is much less diverse. Of the many major groups of marine organisms, only three dozen have abyssal representatives. Animals that live within the seabed are benthic infauna, and those that crawl on, swim closely above, or are attached to the substratum are benthic epifauna. With increasing depth, the number of animals deriving their nourishment from the sediment carpet, or filtering it from sea water, increases as the number of scavengers and carnivores decreases. Life on the ocean floor is sparse beneath temperate ocean surface gyres, but abundant beneath polar and coastal waters. Whereas 5000 g of living organisms can be found per square meter of inshore sea floor and 200 g might be recovered from a square meter of the continental shelf, in the mid-ocean abyss the seascape is stark and barren. Species diversity in the deep sea is nevertheless remarkably great, possibly because biological disturbances such as predation keep the population of potential competitors at low enough densities to preclude competitive exclusion.

Rates of biological processes in the deep sea are shown to be extremely slow, as supported by studies of growth rates, metabolic activity, and recolonization. *See* OCEANS AND SEAS.

[ELIZABETH L. MILLER; WILLIAM B. F. RYAN]

Bibliography: C. A. Burk and C. L. Drake (eds.), *The Geology of Continental Margins*, 1974; D. B. Ericson and G. Wollin, *The Ever-Changing Sea*, 1970; M. G. Gross, *Oceanography: A View of the Earth*, 2d ed., 1977; B. C. Heezen and C. D. Hollister, *The Face of the Deep*, 1971; B. Mason, *Principles of Geochemistry*, 1958; F. P. Shepard, *Submarine Geology*, 3d ed., 1973.

Oceans and seas

The interconnecting body of salt water that covers 70.8% of the surface of the Earth is called the world ocean, or simply the ocean. Its major subdivisions, corresponding to the continents, are oceans. Subdivisions of oceans in turn are called seas; these range all the way from vague regions with no fixed limits (such as Sargasso Sea) to almost completely landlocked bodies (Black Sea). The terms bight, strait, gulf, and bay are often used interchangeably with sea (Great Australian Bight, Denmark Strait, Gulf of Mexico, Bay of Bengal). Salt lakes lacking outlets to the ocean are also usually called seas (Salton Sea, Dead Sea, Caspian Sea).

The ocean has a total area of 362×10^6 km^2 and an average depth of 3729 m, with a total volume of 1.35×10^{18} m^3. It has a mean temperature of 3.90°C and a mean specific gravity of 1.045, giving a total mass of 141×10^{16} metric tons. Since the mean salinity is 34.75‰ (3.475% by weight), there are 136×10^{16} tons of water and 4.93×10^{16} tons of salt in the ocean.

The world ocean is commonly divided into four oceans (see table). The Arctic Ocean is separated from the Pacific at Bering Strait and from the Atlantic by Fury and Hecla straits, Davis Strait, and a line from Greenland to Svalbard, Bear Island, and the north cape of Norway. It includes the Arctic Mediterranean, consisting of Hudson Bay, Baffin Bay, and the Northwestern Passages or Canadian Straits. The meridian of Cape Agulhas (20°E) divides the Atlantic and Indian oceans. The Indian and Pacific oceans are separated by the Andaman Islands; Indonesia; a line from Timor to Cape Talbot, Australia; the western end of Bass Strait; and the meridian of Southeast Cape, Tasmania (146°52′E). All of Magellan Strait is part of the Pacific; to the south the boundary between Atlantic and Pacific is a line from Cape Horn to the South Shetland Islands and thence to the Antarctic Peninsula.

Characteristics of the oceans and adjacent seas

Oceans and adjacent seas	Area, 10^6 km^2	Volume, 10^6 km^3	Mean depth, m
Pacific	166.241	696.189	4188
Asiatic Mediterranean	9.082	11.366	1252
Bering Sea	2.261	3.373	1492
Sea of Okhotsk	1.392	1.354	973
Yellow and East China seas	1.202	0.327	272
Sea of Japan	1.013	1.690	1667
Gulf of California	0.153	0.111	724
Pacific and adjacent seas, total	181.344	714.410	3940
Atlantic	86.557	323.369	3736
American Mediterranean	4.357	9.427	2164
Mediterranean	2.510	3.771	1502
Black Sea	0.508	0.605	1191
Baltic Sea	0.382	0.038	101
Atlantic and adjacent seas, total	94.314	337.210	3575
Indian	73.427	284.340	3872
Red Sea	0.453	0.244	538
Persian Gulf	0.238	0.024	100
Indian and adjacent seas, total	74.118	284.608	3840
Arctic	9.485	12.615	1330
Arctic Mediterranean	2.772	1.087	392
Arctic and adjacent seas, total	12.257	13.702	1117
Totals and mean depths	362.033	1349.929	3729

The Gulf of Mexico and Caribbean Sea together make up the American Mediterranean, and the seas between the Andaman Islands, East Indies, New Guinea, the Phillipines, and Formosa collectively make up the Asiatic Mediterranean. Using these limits, H. W. Menard and R. Smith (1966) computed the characteristics of the oceans and their principal seas listed in the table.

[JOHN LYMAN]

Oil field waters

Waters of varying mineral content which are found associated with petroleum and natural gas or have been encountered in the search for oil and gas. They are also called oil field brines, or brines. They include a variety of underground waters, usually deeply buried, and have a relatively high content of dissolved mineral matter. These waters may be (1) present in the pore space of the reservoir rock with the oil or gas, (2) separated by gravity from the oil or gas and thus lying below it, (3) at the edge of the oil or gas accumulation, or (4) in rock formations which are barren of oil and gas. Brines are commonly defined as water containing high concentrations of dissolved salts. Potable or fresh waters usually are not considered oil field waters but may be encountered, generally at shallow depths, in areas where oil and gas are produced.

Oil field waters or oil field brines differ widely in composition and concentration. They may differ from one geologic province to another, from one formation to another within a given geologic province, or from one part of a specific geologic horizon to another. They range from slightly salty water with 1000–3000 parts of dissolved substances in 1,000,000 parts of solution to very nearly saturated brines with dissolved mineral content of more than 270,000 parts per million (ppm).

The most common and abundant mineral found in oil field waters is sodium chloride, or common table salt. Calcium chloride is next in order of abundance. Carbonates, bicarbonates, sulfates, and the chlorides of magnesium and potassium are present in lesser quantities. In addition to the above mentioned salts, salts of bromine and iodine are also found. Traces of strontium, boron, copper, manganese, silver, tin, vanadium, and iron have been reported. Barium has been reported in many of the Paleozoic brines of the Appalachian region. The commercial value of a brine depends upon the concentration of salts, purity of the products to be recovered, and value and practicability of by-product recovery. Concentrations less than 200,000 ppm are seldom of commercial interest.

Slightly salty waters, while not suitable for human consumption, may be used in some industrial processes or may be amenable to beneficiation in areas lacking fresh waters.

Classified genetically, oil field waters are generally considered connate; that is, they are sea waters which (presumably) originally filled the pore spaces of the rock in which they are now confined. However, few analyses of these waters correspond to present-day sea water, thus indicating some mixing and modification since confinement. Dilute solutions suggest that rainwater has percolated into the rocks along bedding planes, fractures, faults, and other permeable zones. Presence of carbonates, bicarbonates, and sulfates in an oil field water further suggests that at least some of the water had its origin at the surface. Concentrations of dissolved solids greater than that of modern sea water suggest partial evaporation of the water or addition of soluble salts from the adjacent or enclosing rocks.

Waters in most sedimentary rocks increase in mineral concentration with depth. This increase may be due to the fact that, since salt water is heavier than fresh water, the more dense solution will eventually find a position as low as possible in the aquifer. An additional factor would be the longer exposure of the deeper waters to the mineral-bearing rocks. Exceptions have been noted and probably are due to the presence of larger quantities of soluble salts in some geological formations than in others.

Probably the most important geological use of oil field water analyses is their application to the quantitative interpretation of electrical and neutron well logs, particularly micrologs. In order to compute the connate water saturation of a formation in a quantitative manner from electrical data, it is necessary to know with accuracy the connate water resistivity.

Naturally mineralized waters are frequently the only waters available for water-flooding operations. Water analyses are useful in predicting the effect of the water on minerals in the reservoir rock and on the mechanical equipment employed on the project. Waters which exert a corrosive action on the lines and pumps or which tend to plug up the pay zone are not suitable for water flooding.

Oil field water composition may be an important factor in the determination of the source of water in oil wells which have leaky casings or improper completions with resulting communication between wells, and in identifying and correlating reservoirs in multipay oil pools, particularly in those containing lenticular sand bodies.

Industrial wastes, including mineralized water produced with oil, may be disposed of in underground reservoirs. Between the zone of potable water and the horizon of commercial brines, there commonly are rock formations, the waters of which contain chemicals in amounts sufficient to make the waters unsuitable for domestic, municipal, industrial, and livestock consumption, but not in sufficient quantity to be considered as a source for recovery of chemicals. Provided there is sufficient porosity and permeability, these rock formations could receive industrial wastes which would otherwise contaminate surface streams and shallow, fresh groundwater horizons into which they might be discharged. *See* GEOPHYSICAL EXPLORATION; PETROLEUM GEOLOGY.

[PRESTON MC GRAIN]

Oil sand

A loose sand or a semiconsolidated sandstone impregnated with a heavy asphaltic crude oil too viscous to be processed by conventional methods; also known as tar sand or bituminous sand. *See* ASPHALT AND ASPHALTITE.

Distribution. Oil sands are distributed throughout the world but occur primarily in Canada, Venezuela, the United States, Madagascar, Albania, Trinidad, Rumania, and the Soviet Union. The known world reserves of heavy hydrocarbons approximate 1.7×10^9 bbl (210×10^9 m^3). Estimates of the amount that will ultimately prove recoverable—generally of the order of 10 to 33%—are highly speculative and depend on the development of successful technologies at competitive costs.

The Alberta Oil Sands, comprising four enormous deposits—including the famous Athabasca Tar Sands, location of the only existing commercial oil sand mining development—and a number of lesser deposits, contain the largest and best-known of the oil sand reservoirs. Established reserves exceed 900×10^9 bbl (143×10^9 m^3).

Ranking closely behind the Alberta Oil Sands in total reserves volume are the tar sands and heavy oil deposits of the Orinoco Petroleum Belt of Venezuela. Here an estimated 700×10^9 bbl (111×10^9 m^3) of 8–15° API oil occur in reserves stretching along the northern bank of the Orinoco River for a distance of 375 mi (600 km). In 1975 about 80,000 bbl (12,700 m^3) per day were being produced by injecting light oil into the formation to dilute the heavy crude. Numerous on-site recovery methods have been applied in mainly unsuccessful attempts to boost the sagging production.

In the United States, oil sands are found in Utah,

California, Texas, Kentucky, Missouri, and Kansas. Over 90% of the total reserves, or some 28×10^9 bbl (1.3×10^9 m^3) are situated in 24 Utah tar sand deposits. The majority of these are located in the Uinta Basin of northeast Utah. The thickness of the reservoirs varies between 0 and 300 ft (100 m), and the depth ranges from 0 to 2000 ft (670 m).

One interesting characteristic distinguishes much of the heavy oil in Utah. The sulfur content of less than 0.5% is about one-tenth that of the Canadian and Venezuelan tars.

Far back in second place is California, which contains an estimated 200×10^6 bbl (31.8×10^6 m^3) of oil sand reserves. The Edna deposit, located midway between Los Angeles and San Francisco, is the largest. It is a rarity among oil sand deposits in that it is considered a marine facies. Virtually all of the other major oil sand deposits in the world occur in fresh-water fluviatile and deltaic environments. The Edna formation is fossiliferous and consists largely of diatomaceous sandstone.

Though there are claims of a large deposit in the Olenek anticline in northeastern Siberia, the only other substantial oil sands reservoir about which much is known is the Bemolanga deposit in western Madagascar. It covers approximately 150 mi^2 (388 km^2) and contains reserves estimated at 1.75×10^9 bbl (278×10^6 m^3).

Alberta deposits. While the energy crisis is focusing attention on all oil sand deposits, much of the experimental and developmental activity currently is directed toward the Alberta Oil Sands. The deposits range across the northern part of the province (see illustration).

The Athabasca deposit is the largest known oil field in the world. The aerial extent of this deposit is about 13,000 mi^2 (33,670 km^2). The McMurray formation, in which the oil-impregnated sands reside, belongs to the Lower Cretaceous age. The formation outcrops along the Athabasca River, north of Fort McMurray. Elsewhere, the deposit is buried under a variable layer of overburden which reaches up to 1700 ft (almost 600 m) in thickness. As a fortunate circumstance, some of the richest parts of the deposit are covered by the thinnest overburden.

It is in this area that Great Canadian Oil Sands (GCOS), a subsidiary of Sun Oil Company, was producing 50,000 bbl (7950 m^3) per day of synthetic crude from mined tar sands in 1975. On an adjacent lease, the immense Syncrude Canada Ltd. mining project was scheduled to start production in 1978. This 125,000 barrels-per-day (19,860 m^3) project was severely affected by an unprecedented worldwide escalation in process-industry construction costs, and the original capital cost estimates of $1,000,000,000 doubled by 1975.

Surface mining. Both the GCOS and Syncrude projects employ open-pit mining methods to remove the overburden and underlying tar sands, hot water extraction units to recover 90% or more of the tarry bitumen, and upgrading units to convert it to high-quality synthetic crude. Differences in the two projects, aside from the size—the Syncrude project with a peak mining rate of 336,000 tons (302,400 metric tons) of tar sand per day will be the largest mine in the world—lie mainly in the mining and upgrading methods employed. GCOS uses large bucket-wheel excavators to mine the abrasive tar sand, whereas Syncrude has purchased large draglines for this purpose; GCOS converts the bitumen to lighter products in a delayed coking unit, whereas Syncrude is building two large fluid coking units. In each case, the high-sulfur-content coker distillates are upgraded by high-pressure hydrogenation to "sweet" 34–38° API gravity synthetic crudes. The economically recoverable, or "proved," minable reserves in the Athabasca deposit have been calculated at 38×10^9 bbl (6×10^9 m^3) of bitumen, corresponding to 26.5×10^9 bbl (4.2×10^9 m^3) of synthetic crude.

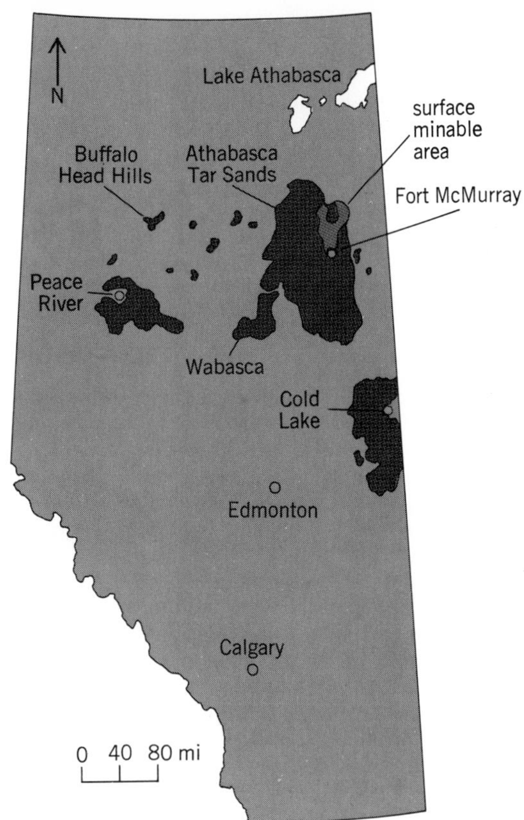

Tar sand deposits in Alberta.

Recovery of deep deposits. Many experimental programs have been undertaken in an attempt to devise a commercially feasible method of recovering the 90% of the bitumen which is buried too deeply for surface mining. Amoco Canada refers to its process as COFCAW, for "combination of forward combustion and waterflood." Most of the other companies have utilized some form of steam stimulation. Imperial Oil was producing 4000 bbl (637 m^3) per day at its Cold Lake pilot project in 1975. The heavy oil at Cold Lake is 2–3° lighter than the 8° API Athabasca oil, and this has a significantly beneficial effect on the ease of recovery. To date, however, recovery efficiencies of all on-site schemes have been too low to justify a commercial project. [G. RONALD GRAY]

Bibliography: M. A. Carrigy (ed.), *Guide to the Athabasca Oil Sands Area*, Alberta Research, 1973; G. R. Gray, *Conversion of Athabasca Bitumen*, American Society of Chemical Engineers Conference, 1971; L. V. Hills (ed.), *Oil Sands: Fuel of the Future*, Canadian Society of Petroleum Geologists, Calgary, Alberta, 1974; *Properties of Utah Tar Sands*, U.S. Bur. Mines Rep. no. 7923, 1974; *Reserves of Crude Oil, Gas, Natural Gas Liquids and Sulfur*, Province of Alberta, ERCB Rep. 75–18, 1974.

Oil shale

A sedimentary rock containing solid, combustible organic matter in a mineral matrix. The organic matter, often called kerogen, is largely insoluble in petroleum solvents, but decomposes to yield oil when heated. Although "oil shale" is used as a lithologic term, it is actually an economic term referring to the rock's ability to yield oil. No real minimum oil yield or content of organic matter can be established to distinguish oil shale from other sedimentary rocks. Additional names given to oil shales include black shale, bituminous shale, carbonaceous shale, coaly shale, cannel shale, cannel coal, lignitic shale, torbanite, tasmanite, gas shale, organic shale, kerosine shale, coorongite, maharahu, kukersite, kerogen shale, and algal shale.

Origin and mineral composition. Oil shale is lithified from lacustrine or marine sediments relatively rich in organic matter. Most sedimentary rocks contain small amounts of organic matter, but oil shales usually contain substantially larger amounts. Specific geochemical conditions are required to accumulate and preserve organic matter, and these were present in the lakes and oceans whose sediments became oil shale. R. M. Garrels and C. L. Christ define these conditions in terms of oxidation-reduction potential (Eh) and acid-base condition (pH) of the water in and around the sediment. Organic matter accumulates under the strongly reducing conditions and neutral or basic pH present in euxinic marine environments and organic-rich saline waters. The organic-rich sediments which became oil shale accumulated slowly in water isolated from the atmosphere, a condition relatively rare in natural waters. This isolation was achieved by stagnation or stratification of the water body and the accompanying protection of its sediments.

Quartz, illite, and pyrite (sometimes with marcasite and pyrrhotite) occur in virtually every oil shale. Feldspars and other clays, particularly montmorillonite, are found in many oil shales. Most oil shale deposits contain small amounts of carbonate minerals, but some, notably the Green River Formation in Colorado, Utah, and Wyoming, contain large amounts of dolomite and calcite. The oil shale minerals were probably formed in the sediment by chemical processes related, at least in part, to the presence of organic matter.

Some oil shales, particularly those called black shales because of the coallike color of their organic matter, have tended to become enriched in trace metals. The reducing conditions necessary to preserve organic matter were conducive to precipitating available trace metals, frequently as sulfides. The Kupferschiefer of Mansfield, Germany, contains an unusually high content of copper, and the Swedish Alum shale has been exploited for its uranium content. The Devonian Chattanooga Shale of Tennessee and neighboring states contains an average of close to 0.006 wt % uranium and has been extensively studied as a potential low-grade source of this element. Vanadium in potentially commercial amounts occurs in the Permian Phosphoria Formation of Wyoming and Idaho. Enrichment of As, Sb, Mo, Ni, Cd, Ag, Au, Se, and Zn has also been noted in black oil shales.

Physical properties. Oil shales are fine-grained rocks generally with low porosity and permeability. Many are thinly laminated and fissile. On outcrop, some oil shales weather to form stacks of thin organic-rich layers called paper shale. The colors of oil shales range from black to light tan and are produced or altered by organic matter.

The physical properties of oil shale are strongly influenced by the proportion of organic matter in the rock. Its decrease in density with increasing organic content illustrates this most graphically. The mineral components have densities of about 2.6–2.8 g/cm^3 for silicates and carbonates and 5 g/cm^3 for pyrite, but the density of organic matter is near 1 g/cm^3. Larger fractions of organic matter produce rocks with appreciably lower density. The equation below quantifies this relationship. Here,

$$D_T = \frac{D_A D_B}{A(D_B - D_A) + D_A}$$

A = weight fraction of organic matter, B = weight fraction of mineral matter, and $A + B = 1$. The organic fraction has an average density D_A, and the mineral fraction an average density D_B. This relationship is applicable to any relatively uniform oil shale deposit when appropriate values for D_A and D_B are known. For Green River Formation oil shales, D_A is 1.06–1.07 g/cm^3 and D_B is about 2.7 g/cm^3. By incorporating a factor for conversion of organic matter to oil, the equation yields a relationship between oil yield and oil shale density which is particularly useful in calculating resources and reserves in an oil shale deposit.

The volume of organic matter in an oil shale rock affects its physical strength properties and its crushability. The rapid increase in volume of organic matter in the shale with increasing organic content weight fraction can be demonstrated by calculations based on the equation above. For example, in the Green River Formation, in oil shale containing 4 wt % organic matter (a very lean shale yielding 2.6 wt % oil), the organic matter makes up 10 vol % of the rock. In richer oil shales, the organic matter is the largest volume component of the rock, and the physical properties of the organic matter predominate. In richer Green River Formation oil shales the organic matter makes the rock tough and resilient; under load, the shales will deform plastically rather than break.

Organic composition and oil production. The organic matter in oil shales and other sedimentary rocks has been extensively studied by organic geochemists, but a specific description of it has not been produced. Although some oil shales contain recognizable organic fragments like spores or algae, most do not, because the basic reducing conditions associated with oil shale development digested and homogenized the organic debris. The resulting organic matter (kerogen) is best described as a high-molecular-weight organic mineraloid of indefinite composition. This composition varies from deposit to deposit and is influenced by the depositional conditions and the nature of the organic debris. Variations in the hydrogen content

Table 1. Relationship between organic carbon-organic hydrogen ratio and conversion of oil shale organic matter to oil by heating

Deposit sampled	Carbon-hydrogen value	Organic carbon recovered, wt %
Pictou County, Nova Scotia, Canada	12.8	13
Top Seam, Glen Davis, Australia	11.5	26
New Albany Shale, Kentucky, United States	11.1	33
Ermelo, Transvaal, South Africa	9.8	53
Cannel Seam, Glen Davis, Australia	8.4	60
Garfield County, Colorado, United States	7.8	69

of this organic matter are significant, however, because the fraction of organic matter converted to oil on heating increases as the amount of hydrogen available in the organic matter increases. To illustrate this relationship, Table 1 compares the proportion of organic carbon recovered as oil during Fischer assay with the weight ratio of organic carbon to organic hydrogen in several oil shales. For petroleum, the carbon-hydrogen values range from 6.2 to about 7.5; for coal, they range upward from 13. The carbon-hydrogen values for organic matter in oil shales range from near petroleum to near coal.

Analytical determination of the elemental composition of the organic matter has been difficult because of the heterogeneous nature of oil shales. Carbon, hydrogen, sulfur, oxygen, and nitrogen are the major elements of the organic matter; but (except for nitrogen) they also occur in the mineral material for oil shales. The organic matter and the mineral matter in oil shales are difficult to separate either physically or chemically. Analytical techniques designed to distinguish between organic and mineral forms of elements, specialized organic matter enrichment techniques, and specialized evaluation techniques have been and are still being developed to aid in the study of oil shales.

The Fischer assay is the best known of the specialized analytical procedures. It was developed by the U.S. Bureau of Mines for oil shale evaluation. The method, employing a modified Fischer retort, determines the quantities of liquid oil and other products recoverable from an oil shale sample heated under prescribed conditions. Although the procedure does not measure the total amount of organic matter in the sample, it approximates the oil available by commercial operations. This simple procedure has proved to be suitable for most oil shale evaluation purposes. Resource information for United States oil shales is based on Fischer assay oil-yield data accumulated by the Laramie Energy Research Center of the U.S. Energy Research and Development Administration.

World oil shale resources. The world's organic-rich shale deposits represent a vast store of fossil energy. They occur on every continent in sediments ranging in age from Cambrian to Tertiary. D. C. Duncan and V. E. Swanson estimated the shale oil represented by the world's oil shale deposits. Their evaluations are summarized in Table 2. The values given for known resources refer only to evaluated resources. To these values Duncan and Swanson added possible extensions of known resources and geologically based estimates of undiscovered and unappraised resources to obtain their estimate of order-of-magnitude values for the total in-place oil resource in the world's oil shale deposits.

A barrel of oil, the 42-gal volume unit used in Western petroleum commerce, has no direct equivalent in metric countries. A barrel of oil represents 0.159 kiloliter. Specification of oil density, a variable, is necessary to convert the barrels into the tonne, the metric ton. A conversion factor agreed on at the World Energy Conference in 1974 defines a barrel of oil as 0.145 tonne, ignoring the density conversion. The Western shale grade unit, gallons per ton, is also a volume unit equivalent to 4.172 liters per metric ton (tonne). The relationship 1 gal/ton × 0.29 = 1 kg/tonne is an approximation agreed on at the 1975 World Energy Conference.

The size of the potential shale-oil resource is staggering. The richest part of the world's evaluated oil shale resource [9×10^{11} bbl (1.3×10^{11} tonne); see Table 2] alone is equivalent to the world's crude oil reserves in 1975 (7×10^{11} bbl, or 1×10^{11} tonne). These petroleum reserves represent only 4% of the projected total resource of rich oil shale (25–100 gpt, or gallons per ton). Since the 1965 estimates shown in Table 2 were made, some deposits have been moving from unknown to known resource classification.

The resource estimates in Table 2 are separated into three grades according to oil yield, recognizing that the richest deposits are more amenable

Table 2. Shale-oil resources of the world

Continents	Known resources*			Order of magnitude of total resources*		
Range in grade (oil yield in gal/ton):	25–100	10–25	5–10	25–100	10–25	5–10
Africa	100	Small	Small	4,000	80,000	450,000
Asia	90	14	†	5,500	110,000	590,000
Australia and New Zealand	Small	1	†	1,000	20,000	100,000
Europe	70	6	†	1,400	26,000	140,000
North America	600	1,600	2,200	3,000	50,000	260,000
South America	50	750	†	2,000	40,000	210,000
Totals	910	2,400	2,200	17,000	325,000	1,750,000

*In 10^9 bbl. †Not estimated.

to economic development. Since many factors besides richness affect the economics of development, the grade designations in Table 2 have limited significance.

World oil shale developments. Although the oil potential of the world's oil shales is great, commercial production of this oil has generally been considered uneconomic. Oil shales are lean ores, producing only limited amounts of oil which historically has been low in price. Mining and heating 1 ton of 25 gal/ton (104 liters/tonne) oil shale produces only 0.6/bbl (0.087 tonne) of oil.

In special situations when other fuels were in short or uncertain supply, or when energy transportation was difficult, energy development from oil shales has been carried out commercially. The 1694 English patent granted for a process "to distill oyle from a kind of stone" is the earliest such record, although medicinal oils were apparently produced from oil shales earlier. A French operation initiated in 1838 is probably the earliest energy development recorded; and Scotland, Canada, and Australia produced shale oil commercially before 1870. From 1875 to 1960, energy equivalent to about 250,000,000 bbl (31×10^6 tonne) of oil was produced from Europe's oil shale deposits. Most of this production was derived from deposits in Estonia, S.S.R., and Scotland. Low-priced oil from the Near East and improving oil transport systems stopped most oil shale developments.

World War II caused sharp increases in petroleum demand and disrupted both petroleum production and petroleum distribution, reactivating interest in oil shale development. Oil shale production operations during and since World War II have been conducted in Germany, France, Spain, Manchuria (China), Estonia and other areas of the Soviet Union, Sweden, Scotland, South Africa, Australia, and Brazil.

Two modern developments, the Manchurian and the Estonian, are relatively large. The Manchurian shale development is near the city of Fushun. The Oligocene oil-shale deposit averages about 450 ft (150 m) of shale, yielding approximately 15 gpt (63 liters/tonne). The deposit overlies a thick coal seam. Removal of the oil shale deposit to enable the coal to be mined by open-pit methods has resulted in the development of the world's largest oil shale industry. Production information has been difficult to obtain, but a daily output of 40,000 bbl of oil (5800 tonne) has been reported. Successful development at Fushun appears to have generated other oil shale developments in the area.

Broad Soviet areas in Estonia and the adjacent Leningrad region are underlain by Ordovician oil shale (kukersite) beds at shallow depths. These shales, reaching 10 aggregate feet (3.3 m) of 50 gpt (210 liters/tonne) shale, are being used to generate electricity and large quantities of low-heating-value gas for domestic and industrial purposes in Leningrad and Tallin. Production exceeded 30,000,000 tons (27×10^6 tonne) of shale in 1973. Most of this was burned directly to generate electricity.

Several smaller-scale oil shale operations have been conducted during and since World War II. Australia operated an oil shale plant at Glen Davis, New South Wales, during World War II. Problems associated with mining thin seams caused this plant to close about 1950. Brazil, always short of domestic petroleum, has intensively investigated two major deposits for shale-oil production. Petrobras, a corporation partly sponsored by the Brazilian government, collaborated with a United States firm to develop and apply the Petrosix retort to the Permian Irati shale. France pioneered destructive distillation of oil shale at Autun, and operated plants on three other deposits after World War II. All had ceased to operate by 1957. Germany operated oil shale plants on the Jurassic deposits in Württemberg during World War II, but these developments did not survive postwar economics. Scotland, an oil shale pioneer, continued to produce shale oil by mining and processing the Carboniferous Lothians oil shale deposits until 1963. In South Africa, the South African Torbanite Mining and Refining Company, Ltd., began operations on a deposit near Ermelo, Transvaal, in 1935. This grew into a large-scale operation which exhausted the 20,000,000–30,000,000-ton (tonne) oil shale deposit. An integrated company operating at Puertollano in Spain's Ciudad Real Province has produced gasoline, diesel and fuel oils, lubricants, and other by-products on a small scale since about 1922. An enlarged company created in 1942 by the National Institute of Spain built a new installation which incorporated a low-temperature hydrogenation plant to upgrade shale oil. The Swedish government built a large plant at Kvarntorp in Närke Province during World War II to produce oil from the Alum black shale. This plant was in full production by 1947 but closed in the 1950s. In conjunction with this plant, Sweden tested underground gasification of oil shale by electrical heating. In this procedure, known as the Ljungstrom method, hydroelectric power available during times of low demand was used. Its proponents claimed that the calorific value of the oil and gas vaporized was about three times that of the energy used to produce them. Developments being investigated in other areas have been continually reported. Like the more extensive efforts, these came to an end because of the economic pressures caused by the abundant and inexpensive petroleum supplies from the Near East.

Two major laboratories working primarily on oil shales exist, both government-sponsored. The Oil Shale Institute at Kohtla-Javre, Estonia, S.S.R., was founded in 1950 to investigate Estonian and other oil shales. Two additional Soviet laboratories devote part of their effort to other organic shales of the country. The Laramie Energy Research Center in Laramie, WY, began work on oil shale in 1944 under the U.S. Department of Interior's Bureau of Mines. The center's primary interests are the Green River Formation and other United States oil shales. In 1975 the center became part of the U.S. Energy Research and Development Administration.

United States oil-shale resources. Organic-rich sedimentary deposits underlie about 20% of the United States land area. They range in age from Cambrian to late Tertiary, and most have not been evaluated or have shown only limited shale-oil potential. The Cretaceous Niobrara Formation in Wyoming, Colorado, Nebraska, and South Dakota;

the Tertiary Humboldt Formation in Nevada; and several Alaskan occurrences are examples of unevaluated, organic-rich deposits.

In the United States, the largest deposit in terms of area is the Devonian-Mississippian black shale composite, which extends from Texas to New York and from Alabama to the Canadian border. This vast area, estimated at 250,000 mi^2 (65,000,000 hectares), is underlain by a time-transgressive continuum of black shale marine sediments occurring in formations locally referred to by names such as Chattanooga, New Albany Shale, Antrim, Ohio, Sunbury, Marcellus, Middlesex, Rhinestreet, Genessee, and Woodford. *U.S. Geological Survey Bulletin no. 523* indicates that the combined deposits offer about 1×10^{12} bbl (145×10^9 tonne) of oil from shale yielding 10 gpt (42 liters/tonne). Approximately 20% of this total resource is classed as known.

Organic matter in these Devonian-Mississippian black shales tends to be low in hydrogen, yielding only a small fraction (from zero to about one-third) of its weight as oil on heating (see Table 2). The heating value of the organic matter ranges from 8500 to 8900 cal/g. Although the black shales underlie a large area and the organic matter disseminated in them represents a huge amount of fossil energy, the deposits are low-grade and the resource in any area is relatively small. In Kentucky, where the deposit seems richest, the resource reaches 50×10^6 bbl of oil per square mile (28,000 tonne oil/hectare) in a 100-ft-thick (30 m) section, with an average oil yield of 10 gal/ton (42 liters/tonne). Organic matter in the New Albany Shale in this area is 10 wt %, representing 22×10^6 tons of organic matter per square mile (32,000 tonne/hectare).

The world's largest oil shale resource is the Eocene Green River Formation in Colorado, Utah, and Wyoming (Fig. 1). The oil potential of oil shales in this 16,500-mi^2 (1 mi^2 = 259 hectares) deposit exceeds 2×10^{12} bbl (290×10^9 tonne). Of this, at least 600×10^9 bbl (87×10^9 tonne) occur in deposits yielding 25 gal or more of oil per ton of shale (104 liters/tonne) in continuous sections of oil shale at least 10 ft (3 m) thick. In the Piceance Creek Basin of Colorado, the oil shale beds reach a thickness of 2100 ft (630 m) and contain about 500×10^6 tons of organic matter per square mile (700,000 tonne/hectare). In Utah and Wyoming, the shale beds are not as thick and are sometimes separated by lean or barren rock.

Minerals ubiquitous in the Green River Formation oil shales are dolomite, quartz, sodium and potassium feldspars, illite, and pyrite. In some locations, the Green River Formation also contains large amounts of sodium carbonate minerals, including trona ($Na_2CO_3 \cdot NaHCO_3 \cdot 2H_2O$), nahcolite ($NaHCO_3$), dawsonite [$NaAl(OH)_2CO_3$], and shortite ($Na_2CO_3 \cdot 2CaCO_3$). More than 60% of the soda ash supply in the United States was produced from Green River Formation trona in 1975. Nahcolite and dawsonite occur in some rich Colorado oil shales. Nahcolite can yield soda ash, and dawsonite can yield alumina; consequently, production of soda ash and alumina together with shale oil is being investigated. The mineral shortite, unique to the Green River Formation, has no known commercial value but occurs in huge amounts in Wyoming and Utah oil shales.

The entire Green River Formation is characterized by a remarkable lateral homogeneity, showing only very gradual geographic changes in its organic and mineral composition. Vertically, however, the formation is extremely variable, most notably in its organic content. Although the oil shale may be vertically continuous, its oil yields range from a few gallons per ton to nearly 100 gpt (417 liters/tonne).

The Mahogany Zone, a particularly organic-rich bed in the Green River Formation of Colorado and Utah, has been investigated intensively as a source of fossil energy. Organic matter in the Mahogany Zone has the following average elemental composition, in wt %: C, 80.5; H, 10.3; N, 2.4; S, 1.0; and O, 5.8, with a gross heating value of 9500–9600 cal/g. The high hydrogen content of this organic matter correlates with the large fraction of oil recovered when the shale is heated (Table 1). Hydrogen-rich organic matter is characteristic of all the Green River Formation oil shales.

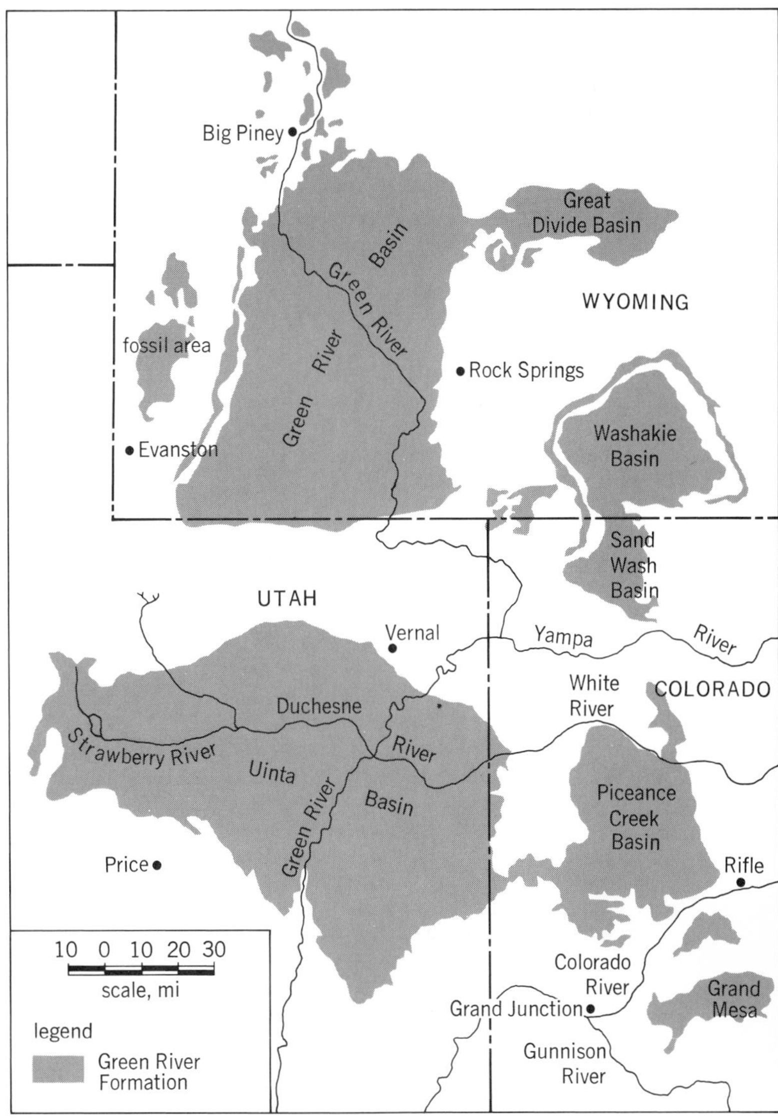

Fig. 1. Location of Green River Formation oil shale deposits.

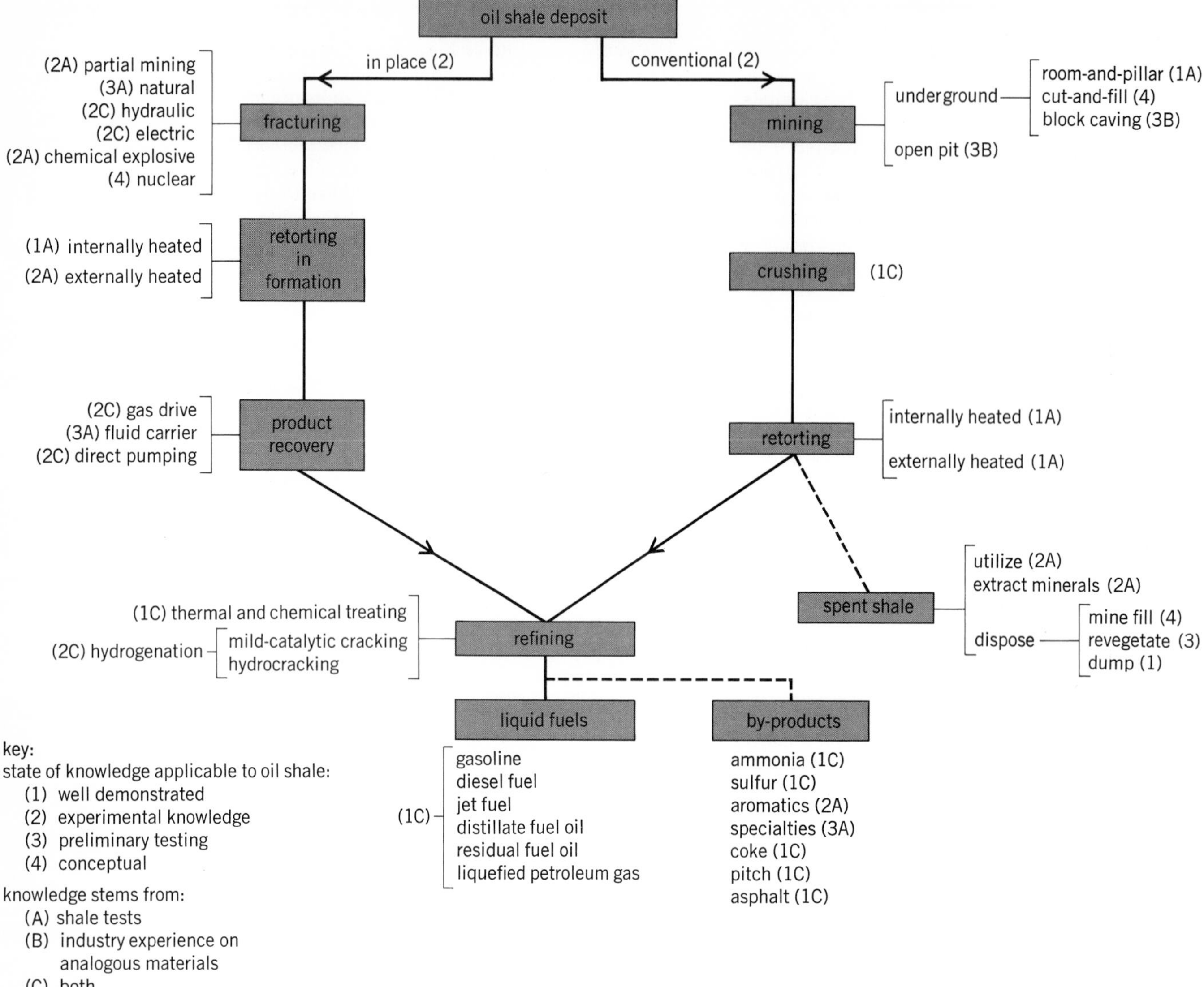

Fig. 2. Oil shale technology routes and states of knowledge.

Shale oil. Shale oil is produced from the organic matter in oil shale when the shale rock is heated in the absence of oxygen (destructive distillation). This heating process is called retorting, and the equipment that is used to do the heating is called a retort.

The rate at which the oil is produced depends upon the temperature at which the shale is retorted. For example, if Mahogany Zone oil shale is heated rapidly to 500°C and held at that temperature, it will take about 10 min for the reaction to reach completion. However, if the shale is heated rapidly to only 340°C, it will take more than 100 hr; when the shale is heated rapidly to a temperature of 660°C, the reaction takes only seconds. Most references report retorting temperatures as being about 500°C.

Retorting temperature affects the nature of the shale oil produced. Low retorting temperatures produce oils in which the paraffin content is greater than the olefin contents; intermediate temperatures produce oils that are more olefinic; and high temperatures produce oils that are nearly completely aromatic, with little olefin or saturate content.

In those retorting systems not capable of rapidly heating the oil shale to a constant temperature, the nature of the oil is determined by the rate at which the oil shale is heated. Thus at heating rates of about 1°C/min, the reaction is essentially completed by the time the temperature reaches 425°C, and the oil is principally paraffinic. At heating rates of about 100°C/min, the reaction is not complete until the temperature reaches more than 600°C, and the oil is more olefinic. As heating rates are increased above 100°C/min, both the paraffin and olefin contents of the oil decrease, and its aromatic content increases.

In general, shale oils can be refined to marketable products in modern petroleum refineries. There is no really typical shale oil produced from Green River oil shale, but the oils do have many properties in common. They usually have high pour points, 20–32°C; high nitrogen contents, 1.6–2.2 wt %; and moderate sulfur contents, about 0.5 wt %. High pour points make necessary some processing before the oils are amenable to pipeline transportation. The high nitrogen contents make hydrogenation necessary to reduce the nitrogen contents so that the oils can be processed into fuels. Hydrogenation also reduces the sulfur content to an acceptable level.

United States technology. The two general approaches to recovering shale oil from Green River Formation oil shales are (1) mining, crushing, and aboveground retorting, called conventional processing; and (2) in-place processing. The basic problems facing conventional processing are handling and heating huge amounts of low-grade ore and disposing of huge volumes of spent shale, the residue remaining after oil production. The in-place approach largely avoids the problems of handling and disposal, with its attendant environmental questions, but faces a different basic problem—the impermeability of the oil shale beds. Progress toward solving the basic problems of both approaches has been made. Figure 2 summarizes the state of knowledge of each step required with the two development approaches.

With the conventional approach, oil shale mining by the room-and-pillar technique developed by the U.S. Bureau of Mines appears capable of producing the huge amounts of ore necessary to operate a large production plant. The procedure has also been tested by industry in Mahogany Zone shales. Outputs on the order of 2500 tons of oil shale per manshift have been reported from the highly mechanized operation. Crushing technology is well demonstrated. Retorting must be done continuously in order to reach the throughput necessary for economic production of shale oil. Two general systems for heating a continuous stream of oil shale are outlined in Fig. 3. In the internally heated system the oil shale furnishes its own heat because part of its organic matter is burned inside the retort. The U.S. Bureau of Mines gas combustion retort and one form of the Paraho retort of Development Engineering, Inc., are examples of this system. In the externally heated systems, heat generated outside the retort is carried inside. The Oil Shale Corporation's TOSCO II retort, in which preheated ceramic balls heat the oil shale stream, is an example of an externally heated retort. Several retorting systems, including both internally and externally heated designs, have been tested on pilot or semiworks scales, but a full-scale retort capable of operating in a commercial oil shale development had not been built by 1975.

Spent shale disposal has been studied intensively. More than 80% of the mined oil shale remains as residue after oil production. An oil shale plant producing 50,000 bbl of oil (7250 tonne) per day from Mahogany Zone shale might mine 75,000 tons (68,000 tonne) of rock and dispose of 60,000 tons (54,000 tonne) of spent shale daily. In the vast and largely unpopulated areas of the Green River Formation, dumping these volumes of spent shale is not as large an environmental problem as it appears to be. The spent shale is virtually insoluble, and contoured dumping to control water flow will minimize leaching, already low in an arid region. Native vegetation will establish itself on spent shale dumps, and revegetation procedures can accelerate this process. Returning spent shale to the mine to furnish roof support may permit recovery of additional ore, but this approach has not been tested.

Research and development efforts toward in-place processing have concentrated on creating permeability in the impermeable oil shale (Fig. 2). In-place processing may be accomplished by two means: (1) a borehole technique in which oil shale is first fractured undergound and heat is applied, and (2) a process in which some rock is first removed by mining, then the remaining oil shale is fragmented into the voids created by mining, and finally heat is applied. These two methods are referred to as in-place (no mining) and modified in-place (some mining) processing. Several investigators have tested in-place methods, and Occidental Oil Shale Corporation has tested the vertical-burn type of modified in-place production. Both of the heating methods outlined for retorts (Fig. 3) have also been applied to in-place processing.

Less than one-third of the total Green River Formation oil shale resource is considered minable for conventional processing. In-place processing may make the remainder of the resource available to production. Conventional processing offers process control advantages, including ready adaptation of many existing industrial procedures, but it is capital-intensive, requiring huge investments before production begins. In-place processing is less easily controlled and evaluated but requires less capital outlay before production begins.

Prospects for United States oil-shale development. By 1975, there was still no commercial production of shale oil. Oil shale is a major fossil fuel resource. The balance between energy supply and energy demand controls the position of oil shale in the United States energy market. Shale oil can compete in the petroleum market when the price of oil exceeds the cost of producing shale oil. The Middle East oil embargo in the winter of 1973–1974, together with subsequent petroleum price increases by OPEC members, intensified interest in oil shale as a domestic supply of fossil energy. Consumption of domestic petroleum reserves and the ever-increasing domestic demand for petroleum products are additional forces acting toward bringing oil shale into the energy market.

Federal land ownership is one of several factors

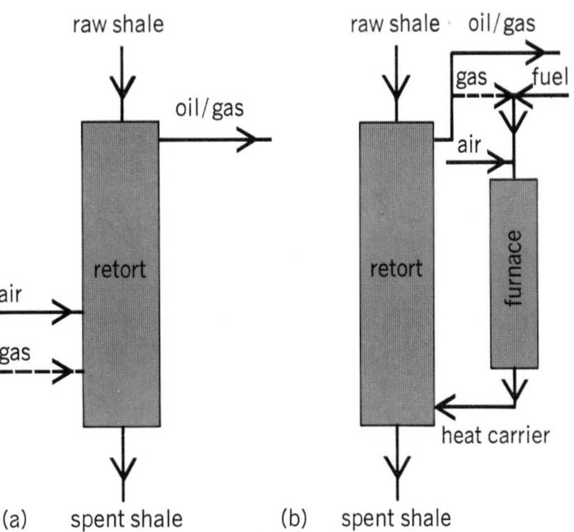

Fig. 3. Oil shale retorting systems. (*a*) Internally heated. (*b*) Externally heated.

affecting oil shale development. More than 80% of the Green River Formation oil shale reserves, including the richest and thickest deposits, are federally owned. President Herbert Hoover "temporarily" withdrew oil shale from leasing pending evaluation of the resource. The National Petroleum Council believes that oil shale development would be unlikely without leasing of Federal lands. Four tracts for commercial shale-oil production were leased in 1974 by bonus bidding, and leasing of two more tracts restricted to in-place techniques is planned. The current Federal attitude is that no more oil shale land will be leased until these prototype tracts are in production. Of the privately held land, two-thirds is held by five major oil companies. Although the total resource is large enough to support any reasonable level of shale-oil production, its legal availability may be a major deterrent to development.

Environmental and socioeconomic impacts of oil shale operations also complicate oil shale development. It has been estimated that 7 years is required between the time a lease is offered and the time mature production is reached. Most of this time must be spent preparing detailed environmental-impact statements, obtaining Federal, state, and local licenses and permits, and preparing to meet housing and service requirements.

Water availability has been frequently cited as a problem in oil shale development. The Federal Energy Administration Project Independence oil shale report concludes that water is available for normal development, and that the water requirement of a sharply accelerated development could be met by additional water-storage facilities.

Many forms of government support for developing synthetic fuel industries, including an oil shale industry, have been proposed to augment the energy supply available in the United States. Implementation of these programs is being evaluated. However, the continually increasing demand for petroleum is the largest spur toward development of oil shale.

[JOHN WARD SMITH; HOWARD B. JENSEN]

Bibliography: D. C. Duncan and V. E. Swanson, *U.S. Geological Survey Circular no. 523*, 1965; R. M. Garrels and C. L. Christ, *Solutions, Minerals, and Equilibria*, 1965; T. A. Hendrickson, *Synthetic Fuels Data Handbook*, 1975; D. K. Murray, *Energy Resources of the Piceance Creek Basin*, Rocky Mountain Association of Geology, 1974; *Oil Shale: Prospects and Constraints*, Federal Energy Administration Project Independence Blueprint, Govt. Print. Office stock no. 4118-00016, 1974; M. P. Rogers, *Bibliography of Oil Shale and Shale Oil*, O.S.R.D. 59, Laramie Energy Research Center, 1974.

Oligocene

The third of the five major worldwide divisions (epochs) of the Tertiary Period (Cenozoic Era); the epoch of geologic time extending from the end of the Eocene to the beginning of the Miocene. Oligocene also denotes the series of rocks formed during the epoch and the fossils therein. *See* CENOZOIC; EOCENE; MIOCENE; TERTIARY.

Strata. In 1833 Charles Lyell subdivided the Tertiary into Eocene (oldest), Miocene, and Pliocene. Subsequently, it was realized that certain contemporaneous deposits were being classified as upper Eocene by some geologists and as lower Miocene by others. Accordingly, as a result of studies of these particular strata in Germany and Belgium, Ernst von Beyrich in 1854 proposed the name Oligocene and indicated that it was intermediate between the older Eocene and the younger Miocene.

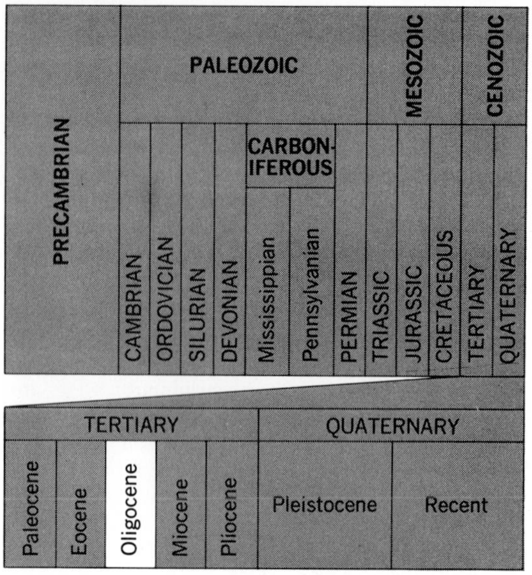

No Lyellian percentages of living molluscan species were designated by von Beyrich as diagnostic for strata of Oligocene age, though the superpositional criteria he used were clearly employed in direct relationship to older and younger strata diagnosed as to age by the Lyellian principle. Subsequent usages have designated as Oligocene strata having from as low as 1.8% living species, as in Venzuela where beds bearing 1% or less have been designated Eocene and those bearing 4.3% lower Miocene, to as high as 15% living species in other areas. Such usages usually stem from series correlations based first on criteria other than Lyellian, as in the case of the Philippine Vigo fauna for which more careful biostratigraphic studies have shown the principle of Lyellian age correlation to be valid for series-magnitude interregional age correlations rather than invalid, as for many years has been erroneously supposed.

The Oligocene of von Beyrich is typically well developed along the east side of the Rheinische Schiefergebirge from the area surrounding the basaltic Vogelsberg, where it consists of partly marine, partly fresh-water deposits (the latter containing lignites), through the Hessian area of Homberg, Fritzlar, Melsingen, Kassel, and Hofgeismar, and up to the Weser River in the vicinity of Karlshafen. More northerly extensions of smaller scattered deposits occur at Alfeld, Hildesheim, farther westward at Lemgo, Bünde, and Osnabrück; and connected deposits are distributed in the eastern part of the so-called Mainz Basin and the Salza Valley (especially Eckardroth and Romsthal) at the southeastern edge of the Vogelsberg on the south side of the Taunus Mountains. The Count of Münster, in 1835, considered some of these strata to be of Pliocene age, as did Philippi in 1843 in reference to the fossiliferous Kassel beds. Other early correlations linked them with Tertiary strata in France (Bordeaux, the Touraine, the Paris Basin), England (the Barton Clay), Belgium (the Bolder Mountains, the Clay at Boom, the Limburg Beds—Tongrian and Rupelian—of Dumont), the Septarien Clay at Hermsdorf, the subsurface Sternberg "Beds" around Mecklenburg, the fossiliferous clay between Landwehrhagen and Lutterberg on the way from Kassel to Münden, at Diekholzen, and in much of the Magdeburg district, and clay that is widespread (though less fossiliferous) in the Mark of Brandenburg—all variously ranging in age from late Eocene to Miocene and even Pliocene. Many additional localities were also noted by von Beyrich in 1854. When a sharply marked change in fossil faunas across the Oligocene-Miocene boundary was noted by Austrian paleontologists farther to the southeast, a two-fold subdivision of Cenozoic strata was proposed: Paleogene (including the Oligocene) for the older beds and Neogene for the younger. But the conspicuous differences in these stratigraphically adjacent fossil faunas has proved to be more of facies than of time significance. This, together with the earlier confusion of Oligocene with late Eocene and Miocene beds at different localities in Europe (and subsequently elsewhere also), has led to the Oligocene being included as the highest Paleogene in some areas, and as the lowest Neogene in others.

Beyrich's Oligocene represented a single cycle of marine deposition on the northwest European plain. Lowermost are the fossiliferous marine glauconitic sandstones overlying lignites at Latdorf (south of Magdeburg) and extending eastward to the amber-bearing echinoid-rich beds of Samland (East Prussia) and southwestward to the petroliferous marls and sandstones of Pechelbronn (Alsace). At Latdorf the sandstones bear much the same fauna as the marine lower beds at Tongres (Belgium) and correlate in age with the supragypsiferous marls of the Paris Basin (Fig. 1). Above the sandstones the second layer is a marine concretionary claystone (Septarien-ton) with volcanic ash, widespread from Brandenburg across Germany to the North Sea, to Boom on the Rupel River in Belgium (known there as the Rupel-ton or Argile de Boom) and to Alsace. In the Mainz basin the claystone occurs with a lower sandstone facies (Meeressand) and with lignite in the Wetterau region, all correlating in age with the Fontainebleau Sandstone (sable superieur) of the Paris Basin.

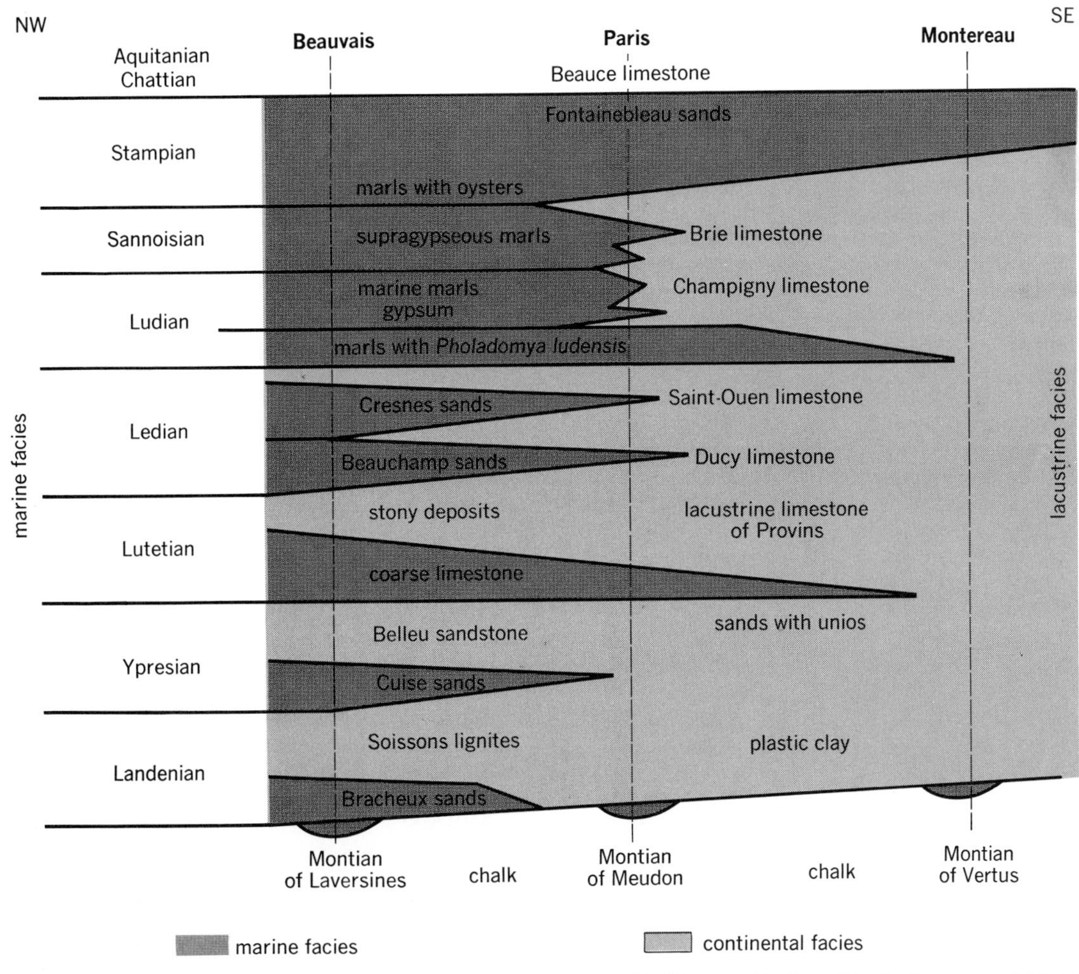

Fig. 1. Diagram showing C. Lyell's original type Eocene sequence in Paris Basin, including (top) local age equivalents of E. von Beyrich's more northerly Oligocene. *(From M. Gignoux, Stratigraphic Geology, copyright © 1955 by W. H. Freeman and Co.)*

The highest layer in Beyrich's Oligocene cycle of deposition is the mollusk- and echinoid-bearing marine yellow Kassel sands of Hessia. These sands pass laterally into the estuarine *Cyrena*-marls and fresh-water limestones that bear the same land snails as the lower and middle members of the uppermost formation (Calcaire de Beauce) in Lyell's original Paris Basin type Eocene sequence. *See* EOCENE.

A threefold subdivision of the Oligocene has subsequently developed: (1) Lattorfian, the oldest stage, from Latdorf (corresponding to the Tongrian of Belgium and the Sannoisian of the Paris Basin); (2) Rupelian, from Dumont's earlier designation in Belgium (corresponding to the Stampian Stage of the Paris Basin); and (3) Chattian, the youngest stage (named for the ancient inhabitants of Hesse-Kassel). Both lower and upper limits of the Oligocene are controversial even in Europe. It is thought by some that the Ludian Stage (uppermost restricted Eocene) overlaps in age the Lattorfian and Tongrian in part and that the Chattian is equivalent in age to the Aquitanian (Miocene?) Stage or at least to the lower portion of the Aquitanian Stage.

Marine fossils in the north European Oligocene Series indicate a progressive cooling from bottom to top (with a few cool-water bivalves already present in the oldest fauna), while reef corals were invading the Mediterranean (Castel Gomberto north to Styria, Carinthia, Salzburg, the Tyrol-Bavarian border). These marine corals were closely related to Caribbean corals although provincially distinct from those of India (Nari), in keeping with a breaking up of eastwest Tethyan connections during Oligocene time.

The lateral differences in rock type and differences in environmental optima of contemporaneous fossils so conspicuous in Europe are characteristic of the Oligocene the world around. Marked and sudden lateral changes in thickness of strata are common as are interfingering marine and terrestrial (often red-bed) deposits. In many areas of Tertiary deposition Oligocene strata are missing entirely. Large-scale mountain-making movements are widely reflected (Ynezan orogeny of Dibblee in the New World, beginnings of the "Late Alpine" orogenies in the Old), with greatly increased surface relief, land emergence, widespread erosion, and equatorward shrinking of the tropics. Consequently, local time subdivisions are used in other faunal regions: Tertiary c, d, e1 to e3 (East Indies); Duntroonian, Waitakian, Otaian stages (New Zealand); Magellanian (Patagonia); upper Refugian, Zemorrian, lower Saucesian stages (California northward); Chadronian, Orellan, Whitneyan, lower Arikareean (mammalian ages, North America); "Vicksburg" (Gulf Coast); and "Culebra" (Caribbean).

Fig. 2. Early Oligocene brontothere (*Brontotherium platyceras*), an archaic browsing mammal. (*From R. A. Stirton, Time, Life and Man, copyright © 1959 by John Wiley & Sons, Inc.; reprinted by permission*)

Life. Marine faunal provinces were fragmented in the Oligocene. Conversely, some isolated Northern Hemisphere landmasses became reconnected, permitting interchange of some mammals (opossums, anthracotheres, chalicotheres, tapirs, rhinocerids). Cool-water marine shellfish moved equatorward, certain pelecypod and gastropod stocks arriving for the first time in habitats they still occupy today. Scallops increased greatly in size; uvigerine and siphogenerine small foraminiferids diversified; nummulites persisted into mid-Oligocene, their habitats later being taken over by lepidocyclines, spiroclypeids, and miogypsinids.

The marine planktonic foraminiferal faunas of the preceding Eocene have been shown to have suffered an entire eclipse, having become restricted to a single lineage, much as among the ammonites at a comparably critical horizon within the Mesozoic. With the rejuvenation in the extent of the world's ocean surfaces that followed in Miocene time, the surviving lineage proceeded to diversify iteratively into a great number of Miocene species, paralleling in this radiation many of the earlier morphotypes, much as happened previously during the Paleocene and into Eocene time. The cause of these phenomena, especially since they did not recur during the comparably orogenic Pleistocene, is not clear; some significant change in the composition of the oceanic waters during Paleocene and Oligocene times has been postulated, though the nature of such a hypothetical change is not understood.

With the rise of grasses the earlier huge browsers, such as the brontotheres (Fig. 2) and other archaic mammals, made their last stand. In the Egyptian Fayum other great grotesque extinct mammals such as the horned *Arsinoitherium* flourished with the earliest mastodonts. Hyracoids appeared and have left a lone relic descendant coney or "rock rabbit" which lives today in the same general region. Primitive primates (lemurs) and insectivores carried on in isolated Madagascar. Ten living genera of birds already existed. The first New and Old World monkeys and Great Apes appeared. Carnivora modernized (dogs, cats, saber-toothed cats) alongside their disappearing creodont ancestors. Beavers, pangolins, three-toed horses, and pig, peccary, and tapir families appeared and other hoofed mammals diversified, among them the 18-ft high *Baluchitherium*, largest land mammal known. *See* PALEOBOTANY; PALEONTOLOGY. [ROBERT M. KLEINPELL]

Oligoclase

A plagioclase feldspar with a composition ranging from $Ab_{90}An_{10}$ to $Ab_{70}An_{30}$ where $Ab = NaAlSi_3O_8$ and $An = CaAl_2Si_2O_8$ (see illustration). Natural material in the range from Ab_{98} to Ab_{83} is usually submicroscopically unmixed into domains of An_2 and $An_{25\text{-}30}$ composition. Oligoclases and albites that exhibit a blue luster as a consequence of this unmixing are called peristerite. If Fe_2O_3 is present as thin flakes that are oriented parallel to certain

Oligoclase from North Carolina. (*Specimen from Department of Geology, Bryn Mawr College*)

structurally defined planes, such oligoclase is called aventurine or sunstone. *See* FELDSPAR; IGNEOUS ROCKS. [FRITZ H. LAVES]

Olivine

A name given to a group of magnesium-iron silicate minerals crystallizing in the orthorhombic system. Crystals are usually of simple habit, a combination of the dipyramid with prisms and pinacoids. The luster is vitreous and the color olive-green, giving rise to the name olivine. Hardness is 6½–7 on Mohs scale; specific gravity is 3.27–3.37, increasing with increase in iron content. *See* SILICATE MINERALS.

Olivine is a nesosilicate with composition $(Mg, Fe)_2SiO_4$. It comprises a complete solid solution series from the pure iron member fayalite, Fe_2SiO_4, to the pure magnesium member forsterite, Mg_2SiO_4. Minerals of intermediate composition have been given their own names but are usually designated simply as olivine. The magnesium-rich varieties are more common than those rich in iron. The minerals tephroite, Mn_2SiO_4, monticellite, $CaMgSiO_4$, and larsenite, $PbZnSiO_4$, although not in this chemical series, are of the olivine structure type.

Occurrence. Olivine is found in some crystalline limestones but occurs chiefly as a rock-forming mineral in igneous rocks. It varies greatly in amount from an accessory to the main rock-forming constituent. Although it may be present in granites and other light-colored rocks, it is found chiefly in the dark rocks such as gabbro, basalt, and peridotite. The rock dunite is composed almost completely of olivine.

Olivine is one of the first minerals to form upon crystallization of a magma. It is believed that this early-formed olivine accumulated through the process of magmatic differentiation to form the large dunite masses. The type locality is at Dun Mountain, New Zealand; the rock is also found with corundum deposits in North Carolina. *See* MAGMA; PERIDOTITE.

Olivine alters readily to serpentine, a hydrous magnesium silicate. The alteration may take place on a large scale to form great masses of the rock serpentine, or on a small scale to form pseudomorphs of serpentine after single crystals of olivine. *See* SERPENTINE; SERPENTINITE.

At a few localities, notably on St. John's Island in the Red Sea and in Burma, olivine is found in transparent crystals. These are cut into gem stones which go under the name of peridot. Olivine is a major constituent of many stony meteorites. *See* METEORITE. [CORNELIUS S. HURLBUT, JR.]

Spinel transition. When subjected to very high pressures, minerals with olivine structure are transformed to denser polymorphs with spinel structure..For the composition Fe_2SiO_4, the univariant curve for the olivine-spinel transformation has been determined experimentally. The divariant transition interval for compositions in most of the series $Fe_2SiO_4 - Mg_2SiO_4$ has also been determined at pressures up to about 100 kilobars. Olivine is believed to constitute a high proportion of the upper mantle, estimates ranging between 60 and 90%. The composition of the olivine is generally assumed to be about 90% forsterite (Mg_2SiO_4) and 10% fayalite (Fe_2SiO_4). The experimental results, combined with estimates of the temperature distribution in the mantle, indicate that olivine of this composition would begin to transform to a more Fe-rich spinel phase at a depth of about 370 km, with transformation being completed at a depth of about 435 km. Gradients of seismic-wave velocities in the transition zone of the upper mantle are abnormally high in two layers 50–100 km thick, one beginning at 350-km depth, and the other at about 630-km. The olivine-spinel transformation appears to correlate well with the 350-km layer. The spinel present below this layer may be replaced in turn by even more dense phases at the 630-km depth. *See* LITHOSPHERE, GEOCHEMISTRY OF; SPINEL. [PETER J. WYLLIE]

Onyx

The name onyx is applied correctly to banded chalcedonic quartz, in which the bands are straight and parallel, rather than curved, as in agate. Unfortunately, in the colored-stone trade, gray chalcedony dyed in various solid colors such as black, blue, and green is called onyx, with the color used as a prefix. Because the color is permanent, the fact that it is the result of dyeing is seldom mentioned.

The natural colors of true onyx are usually red or brown with white, although black is occasionally encountered as one of the colors. When the colors are red-brown with white or black, the material is known as sardonyx; this is the only kind commonly used as a gemstone. Its most familiar gem use is in cameos and intaglios, in which the figure is carved from one colored layer and the background in another. *See* CAMEO; CHALCEDONY; GEM; QUARTZ.

[RICHARD T. LIDDICOAT, JR.]

Oolite and pisolite

Oolites are small, more or less spherical particles commonly found in limestones and dolomites. Most oolites are 0.5–1.0 mm in diameter, but their size range is much greater. Pisolites are similar particles that are greater than 2.0 mm in diameter. Ooolites show varying degrees of departure from sphericity; some may be ellipsoidal, others may be appreciably flattened or distorted. The term oolite has been used to denote both the small, spherical bodies and the rock composed of an aggregate of these bodies. Some geologists prefer to call the particles oolite and the rock by its common lithologic name, prefixed by the word oolitic, for exam-

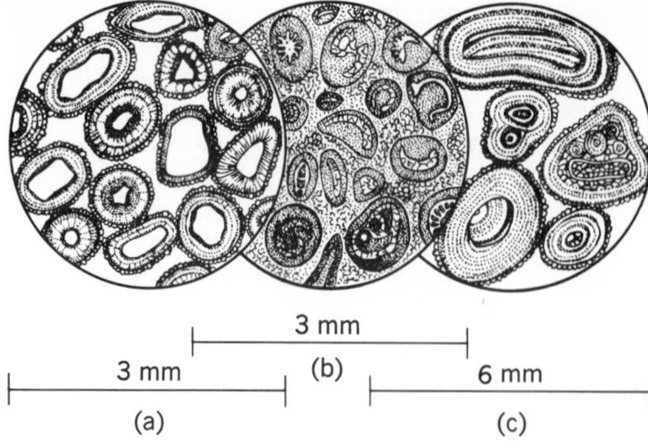

Oolitic limestones. (*a*) Pleistocene oolites, Great Salt Lake, Utah. Oolites consist of subangular detrital quartz grains enclosed by carbonate having both concentric and radial fibrous structure. Radial fibrous carbonate is calcite; at least some of the concentric carbonate (right center and top) is aragonite. An incipient cement composed of finely granular calcite rims the oolites, but rock is very porous. (*b*) Oolitic limestone, Völksen, Deister Mountains, Germany. Oolites consisting of shell fragments encased by microcrystalline calcite (dark stippling) are firmly cemented by a matrix of fine-grained calcite having somewhat variable grain size. (*c*) Composite oolites (Pleistocene), Pyramid Lake, Nev. Large calcareous oolites consisting of cryptocrystalline (stippled) and radial fibrous (clear) concentric layers. The fibrous layers are calcite; the cryptocrystalline layers are at least partly aragonite. The nuclei are fragments of broken oolites, clusters of tiny oolites (right and center), and bits of granular carbonate (lower right). Incipient cementation is as in *a*. (*From H. Williams, F. J. Turner, and C. M. Gilbert, Petrography, An Introduction to the Study of Rocks in Thin Sections, Freeman, 1954*)

ple, oolitic limestone (see illustration).

Sectioned oolites show either radial or concentric structures or both. They commonly have cores that are of material other than the bulk of the oolite; frequently they are pieces of shell or detrital quartz grains. The appearance of oolites suggests that they have grown outward from the core by successive precipitations of calcium carbonate in thin concentric shells. Although oolites may be composed of many materials, mainly calcite, aragonite, silica, hematite, and dolomite, by far the most common in the geologic column are the calcareous ones. Siliceous and dolomitic oolites are formed by the replacement of an original calcareous oolite. Phosphatic and hematitic oolites seem to have formed as primary oolites. The explanation for the origin of oolites generally given is that they represent inorganic precipitation in turbulent waters, where the small grains roll with the current as they gradually pick up more and more layers of precipitate. *See* CALCARENITE; CHERT; DOLOMITE; LIMESTONE; SEDIMENTARY ROCKS.

[RAYMOND SIEVER]

Opal

A natural hydrated form of silica. There are many different varieties of opal, but the best known are those which are highly prized as gemstones. Precious opal displays the property of opalescence, a fine play of spectral colors resulting from the interference of light rays within the stone. Fire opal shows intense orange-to-red reflections against a yellow-to-orange body color. Black opal has a black background against which the colors are displayed. Common opal is milk white, yellow, green, or red, but without opalescence. The variety hyalite is clear and colorless with a globular surface. Fine precious opals are found in Hungary, Mexico, Honduras, and New South Wales, Australia. In the United States opal has been found in Nevada and Idaho.

Opal is amorphous and usually occurs in botryoidal or stalactitic masses. It has a conchoidal fracture and a hardness of 5–6 on Mohs scale. The specific gravity varies from 1.9 to 2.2, depending upon the water content. Opal is found in cavities in igneous and sedimentary rocks and in fossil wood in which it is the petrifying substance. As geyserite or siliceous sinter it is deposited from geysers in Yellowstone National Park. Its largest deposits are in sedimentary beds as diatomite which result from the accumulation of tiny opalean tests of diatoms. Such a deposit at Lompoc, Calif., is 4000 ft thick. *See* DIATOMACEOUS EARTH; GEM; SILICATE MINERALS; SILICEOUS SINTER.

[CORNELIUS S. HURLBUT, JR.]

Ordovician

The second period of the Paleozoic Era, and the system of rocks deposited during this time—the succession of rocks overlying the Cambrian System and underlying the Silurian System. The Ordovician Period had a duration of some 60,000,000 to 80,000,000 years.

PRECAMBRIAN	PALEOZOIC							MESOZOIC			CENOZOIC	
					CARBONIFEROUS							
	CAMBRIAN	ORDOVICIAN	SILURIAN	DEVONIAN	Mississippian	Pennsylvanian	PERMIAN	TRIASSIC	JURASSIC	CRETACEOUS	TERTIARY	QUATERNARY

The system of rocks was named by C. Lapworth, an English geologist, in 1879, for the Ordovices, an aboriginal tribe that occupied parts of Wales before the coming of the Romans. The system included parts of the original Cambrian System of A. Sedgwick and of the Silurian System of R. Murchison, specifically those strata which succeeded the Tremadoc and underlay the Landovery beds, and were characterized by distinctive fossil graptolites. This nomenclature is followed in most of the world, but in some countries it is the practice to continue the Silurian in its original extended sense, assigning the Lower Silurian or Untersilur to the Ordovician, and the Upper Silurian to the Gotlandian System, named from a Swedish island in the Baltic Sea.

Rocks. The typical Ordovician rocks of Wales are extremely variable but generally consist of thousands of feet of graywacke and argillite with associated lava and volcanic fragmental rocks. The sequence has been divided into several series; in ascending order they are the Arenigian, Llanvirnian, Llandeilan, Caradocian, and Ashgillian. Each is composed of one or more fossil zones, characterized by assemblages of graptolites. The sequence of graptolite zones established in Britain has been found to be generally adaptable to rocks of similar shaly facies throughout the world, as in Bohemia, Australia, and North America. A greater variety of fossils has been found in the sandy and calcareous rocks, the shelly facies.

In North America, rocks having the characteristic Ordovician graptolites are found in many localities from northeast Newfoundland to eastern Tennessee and southeastern Oklahoma along the Atlantic and Gulf coasts, and from southeastern Alaska and eastern British Columbia to central Idaho and central Nevada along the Pacific side of the continent. They are in sequences of graywackes, argillites, and volcanic rocks as in Wales. On the other hand, graptolites are few or lacking in carbonate rocks and associated shales and sandstones that are classified as Ordovician in the interior of the continent; however, these are placed in the Ordovician System because graptolites are occasionally found in association with the other fossils and some of the carbonate rocks can be traced stratigraphically into graptolite-bearing shales.

The North American classification is based on the rocks of shelly facies, in which brachiopods and cephalopods are generally the most definitive fossils. The rocks have been divided into series such as the following in ascending order: Canadian, Chazyan, Bolarian, Trentonian, and Cincinnatian (Fig. 1), though other terms have been applied. Mohawkian is used frequently for the Bolarian and Trentonian; Champlainian has been used for the whole system and for the three middle series as listed above. The base of the Caradocian of Britain is about the base of the Bolarian, and the Ashgillian is about equivalent to the Cincinnatian.

Life. The Ordovician contrasts with the older and underlying Cambrian in having a greater abundance and variety of fossils. Trilobites and brachiopods are common in many Cambrian rocks, but other organisms are rarely found. In the Ordovician rocks of shaly facies, graptolites are generally abundant only on occasional beds. But in the calcareous rocks or shelly facies, fossils abound in many places. Brachiopods are perhaps the most generally present; they are of great variety and differ from stage to stage. Though less frequent, cephalopods and trilobites are quite useful in recognition of ages of beds, particularly in the Canadian and Chazyan series. Corals became plentiful enough to form small patch reefs in the Chazyan; and bryozoans first appeared in some abundance in that series. Well-developed crinoids and cystids are common in a few limited zones. Gastropods and pelecypods are sometimes abundant, particularly in argillaceous rocks, and ostracods are first known in profusion and great variety in Chazyan sediments. The Ordovician rocks yield the first great variety of conodonts, forms that seem quite useful in classifying some rocks that

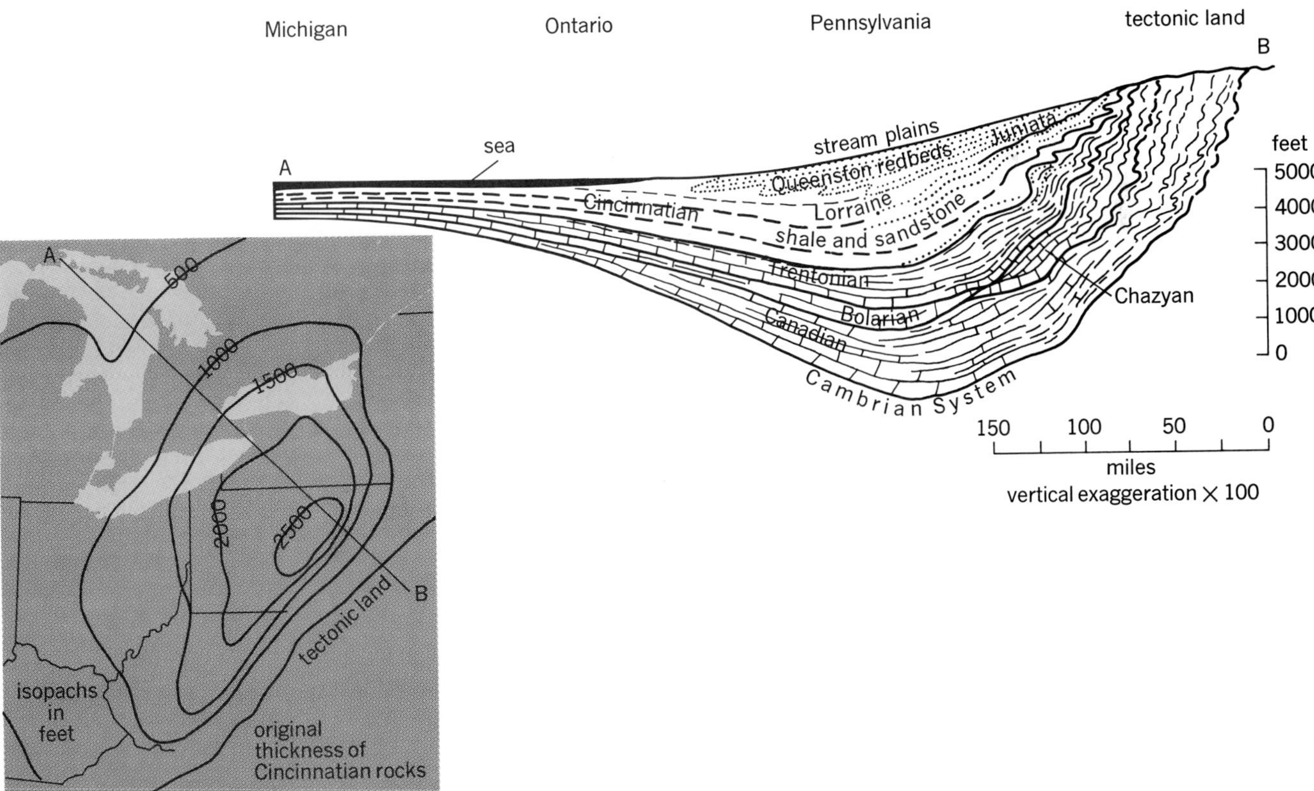

Fig. 1. Section through the rocks of the Ordovician System from Pennsylvania northwestward at the close of the Ordovician Period. The line of section is shown on the map with the lines of equal thickness (isopachs) of the Cincinnatian Series. The deposits accumulated in a subsiding basin on the margin of the more stable continental interior. Late in the period, the early stages of the Taconian orogeny raised lands in the geosynclinal belt on the southeast, and the eroded sediments spread northwestward into the subsiding basin.

are otherwise sparsely fossiliferous. Sponge fossils are occasionally common and distinctive; and there are representatives of other fossil invertebrate classes and orders. The advent of the first vertebrate may have preceded the Ordovician, but at least there are scales of primitive fishes in some abundance in rocks of about Bolarian age, particularly in the Harding sandstone of Colorado. Calcareous algae are the only plant fossils of consequence. Of all the Ordovician animal life, graptolites are probably the most distinctive, for though they lived from Cambrian to Carboniferous, they are almost entirely limited to the Ordovician and Silurian systems.

Tectonic provinces. The Ordovician rocks of North America fall into several tectonic provinces (Fig. 2). The central part of the continent or hedreocraton accumulated a few hundred feet to a thousand feet or so of sedimentary rocks, principally limestone and dolomite; it was a relatively stable area. Along the eastern and western borders of the craton, separated by zones of crustal flexure, were belts having much thicker sections (a mile or more) of sedimentary rocks, again principally carbonate rocks, but with terrigenous sediments locally prevalent; these are the miogeosynclines, belts of greater subsidence than the hedreocraton. Beyond them to the edge of the continent, Ordovician rocks are almost entirely terrigenous and volcanic, consisting of graywacke and argillite with associated lava flows and fragmental volcanic rocks; these belts of subsidence had associated islands that rose and were eroded, as well as volcanic centers, and are eugeosynclinal belts. The rocks in these belts were greatly deformed by later mountain making, and in many cases were so invaded by plutonic igneous rocks and were so metamorphosed as to be difficult to identify and date, or to arrange in stratigraphic order. The typical Ordovician of Great Britain is eugeosynclinal, as is that of coastal Scandinavia, whereas that of Sweden, Estonia, and Poland is similar to that of interior North America. *See* GEOSYNCLINE.

Deformation in North America. The most striking structural changes in the continent were along the eastern margin, from Newfoundland to the southern Atlantic Coast. Initially, in the Canadian Epoch, graptolite-bearing rocks of the eugeosynclinal belt were laid in submerged troughs adjoining islands along the coast, while carbonates, which were thick in the miogeosynclinal belt, thinned toward the cratonal interior area. The interrelations of the two facies are obscured by later folding and thrust faulting along the zone of change; the carbonate rocks were laid in shallow water and probably graded into the argillites laid in deeper marginal troughs. In later epochs, at different times along the 1000-mi length of the belt in eastern United States, lands rose in the eugeosynclinal belt and shed sediment into troughs extending into the margin of the hedreocraton. The effect of the rising lands became pronounced in the Cincinnatian Epoch, when sands and muds spread inland as far as Ohio and Michigan in a great delta that filled an elongate basin centered in Pennsylvania to a depth of several thousand feet. The later Ordovician rocks were folded, and Silurian rocks lie unconformably on them at localities from Newfoundland to Pennsylvania. Moreover, there are great thrust faults attributed to this, the Taconian orogeny, and intrusions of granitic rocks in maritime Canada and New England that were unroofed (exposed by erosion) by Silurian time; intrusions in the Carolinas have also been dated as Ordovician by geochemical methods. *See* GEOCHRONOMETRY; UNCONFORMITY.

In western North America there is similar contrast between volcanic-bearing argillaceous and graywacke sequences, with graptolites as the most frequent fossils from southeastern Alaska and Yukon to central Nevada and the southern Sierra Nevada of California, and carbonate rocks of a mile or so thickness extending eastward to the cratonal margin. Though the zone of contact between the facies is again obscure, in some areas the two facies are known to grade into each other, as though the carbonate rocks were laid in shallow water passing over a flexure into deeper sinking troughs that received terrigenous sediment from lands raised in the eugeosynclinal belt to the west. Orogeny is not recognized in the West during this period. A third geosynclinal belt trends through the Arctic islands of northern Canada.

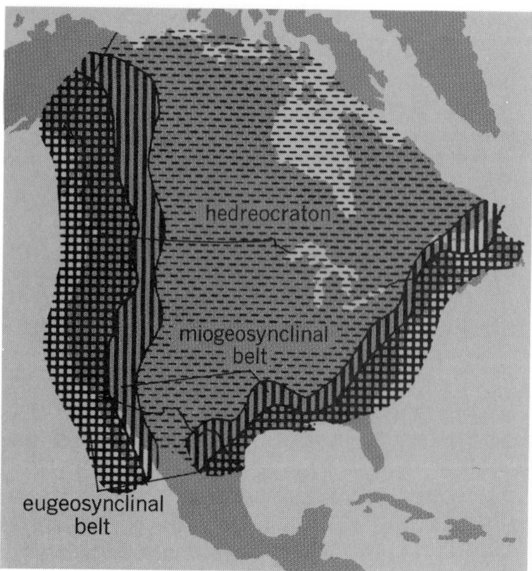

Fig. 2. North American Ordovician tectonic provinces.

In the continental interior, carbonate rocks are prevalent, and in the Late Ordovician (Late Trentonian and the Cincinnatian) seas covered all but a very small part of the continental interior from the Gulf of Mexico to the Arctic. During the Late Ordovician there were islands and volcanoes along the present borders of North America, but the interior was largely beneath the sea. The interior, however, was not fully stable, for the thicknesses of rocks in the stages of the several series thicken into basins of greater subsidence and thin to disappearance along the margins of other areas that subsided little or not at all. At one stage in the Chazyan Epoch, sands from dunes in the northern interior drifted into shallow seas in the Mississippi Valley region, forming the few hundred feet of remarkably pure St. Peter sandstone, the source of silica for glass manufacture and other chemical industries.

Other continents. Ordovician rocks are known from each of the continents, but knowledge of them comes principally from Europe and North America. In South America, Ordovician fossils have been described from areas scattered from Venezuela and Colombia to northwestern Argentina along the west slope of the Andes. In Asia, the Ordovician of Manchuria has been a source of faunas, but little is known of the paleogeography of the continent. In Australia, the best-known Ordovician is that in Victoria, having an excellent graptolite succession, and similar faunas are known in New Zealand. The principal Paleozoic sections in Africa are in the countries bordering the western Mediterranean and adjacent Atlantic; Ordovician fossils have been found in the Cape Group of South Africa.

[MARSHALL KAY]

Bibliography: W. H. Twenhofel et al., Correlation of the Ordovician formations of North America, *Bull. Geol. Soc. Amer.*, 65(3):247–298, 1954.

Ore and mineral deposits

Ore deposits are naturally occurring geologic bodies that may be worked for one or more metals. The metals may be present as native elements, or, more commonly, as oxides, sulfides, sulfates, silicates, or other compounds. The term ore is often used loosely to include such nonmetallic minerals as fluorite and gypsum. The broader term, mineral deposits, includes, in addition to metalliferous minerals, any other useful minerals or rocks. Minerals of little or no value which occur with ore minerals are called gangue. Some gangue minerals may not be worthless in that they are used as by-products; for instance, limestone for fertilizer or flux, pyrite for making sulfuric acid, and rock for road material.

Mineral deposits that are essentially as originally formed are called primary or hypogene. The term hypogene also indicates formation by upward movement of material. Deposits that have been altered by weathering or other superficial processes are secondary or supergene deposits. Mineral deposits that formed at the same time as the enclosing rock are called syngenetic, and those that were introduced into preexisting rocks are called epigenetic.

The distinction between metallic and nonmetallic deposits is at times an arbitrary one since some substances classified as nonmetals, such as lepidolite, spodumene, beryl, and rhodochrosite, are the source of metals. The principal reasons for distinguishing nonmetallic from metallic deposits are practical ones, and include such economic factors as recovery methods and uses.

Concentration. The Earth's crust consists of igneous, sedimentary, and metamorphic rocks. Table 1 gives the essential composition of the crust and shows that 10 elements make up more than 99% of the total. Of these, aluminum, iron, and magnesium are industrial metals. The other metals are present in small quantities, mostly in igneous rocks (Table 2).

Most mineral deposits are natural enrichments and concentrations of original material produced by different geologic processes. To be of commercial grade, the metals must be present in much higher concentrations than the averages shown in Table 2. For example, the following metals must be concentrated in the amounts indicated to be considered ores: aluminum, about 30%; copper, 0.7–10%; lead, 2–4%; zinc, 3–8%; and gold, silver, and uranium, only a small fraction of a percent of metal. Therefore, natural processes of concentration have increased the aluminum content of aluminum ore three or four times, and even a low-grade gold ore may represent a concentration of 20,000 times. Economic considerations, such as the amount and concentration of metal, the cost of mining and refining, and the market value of the metal, determine whether the ore is of commercial grade.

Table 1. Elemental composition of Earth's crust based on igneous and sedimentary rocks*

Element	Weight, %	Atom, %	Volume, %
Oxygen	46.71	60.5	94.24
Silicon	27.69	20.5	0.51
Titanium	0.62	0.3	0.03
Aluminum	8.07	6.2	0.44
Iron	5.05	1.9	0.37
Magnesium	2.08	1.8	0.28
Calcium	3.65	1.9	1.04
Sodium	2.75	2.5	1.21
Potassium	2.58	1.4	1.88
Hydrogen	0.14	3.0	

*From T. F. W. Barth, *Theoretical Petrology*, John Wiley & Sons, Inc., 1952 (recalculated from F. W. Clarke and H. S. Washington, 1924).

Forms of deposits. Mineral deposits occur in many forms depending upon their origin, later deformation, and changes caused by weathering. Syngenetic deposits are generally sheetlike, tabular, or lenticular, but may on occasion be irregular or roughly spherical.

Epigenetic deposits exhibit a variety of forms. Veins or lodes are tabular or sheetlike bodies that originate by filling fissures or replacing the country rock along a fissure (Fig. 1). Replacement bodies in limestone may be very irregular. Veins are usually inclined steeply and may either cut across or conform with the bedding or foliation of the enclosing rocks. The inclination is called the dip, and is the angle between the vein and the horizontal. The horizontal trend of the vein is its strike, and the vertical angle between a horizontal plane and the line of maximum elongation of the vein is the plunge. The veins of a mining district commonly occur as systems which have a general strike, and one or more systems may be present at some angle to the main series. In places the mineralization is a

ORE AND MINERAL DEPOSITS

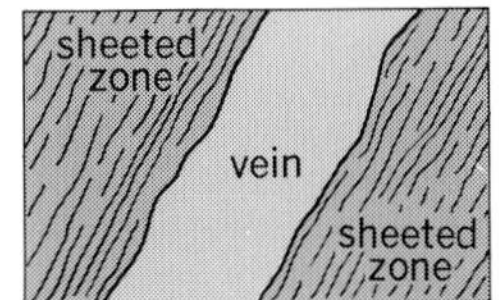

Fig. 1. Vein developed in fissured or sheeted zone.

Table 2. Abundance of metals in igneous rocks

Element	%	Element	%
Aluminum	8.13	Cobalt	0.0023
Iron	5.00	Lead	0.0016
Magnesium	2.09	Arsenic	0.0005
Titanium	0.44	Uranium	0.0004
Manganese	0.10	Molybdenum	0.00025
Chromium	0.02	Tungsten	0.00015
Vanadium	0.015	Antimony	0.0001
Zinc	0.011	Mercury	0.00005
Nickel	0.008	Silver	0.00001
Copper	0.005	Gold	0.0000005
Tin	0.004	Platinum	0.0000005

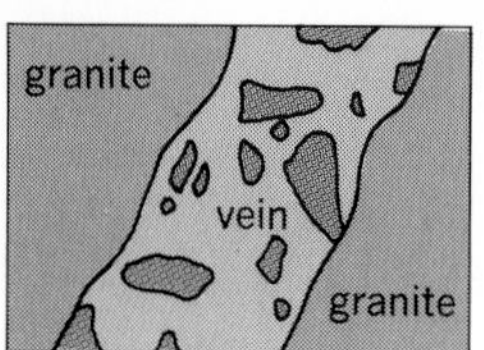

Fig. 2. Brecciated vein in granite.

network of small, irregular, discontinuous veins called a stockwork.

Mineral deposits are seldom equally rich throughout. The pay ore may occur in streaks, spots, bunches, or bands separated by low-grade material or by gangue. These concentrations of valuable ore are called ore shoots; if roughly horizontal they are called ore horizons, and if steeply inclined they are called chimneys. After their formation mineral deposits may be deformed by folding, faulting, or brecciation (Fig. 2).

Metasomatism, or replacement. Metasomatism, or replacement, is the process of essentially simultaneous removal and deposition of chemical matter. A striking manifestation of this process in mineral deposits is the replacement of one mineral by another mineral or mineral aggregate of partly or wholly different composition. A large volume of rock may be transformed in this manner, and the resulting deposit is generally of equal volume. Commonly the original structure and texture of the replaced rock is preserved by the replacing material.

Replacement, evidence for which is found in many mineral deposits, operates at all depths under a wide range of temperature. The evidence indicates that the new minerals formed in response to conditions that were unstable for the preexisting ones.

Usually the replacing material moves to the site of metasomatism along relatively large openings such as faults, fractures, bedding planes, and shear zones. It then penetrates the rock along smaller cracks and finally enters individual mineral grains along cleavage planes, minute fractures, and grain boundaries where substitution may take place on an atomic scale until the entire mass has been transformed (Fig. 3). After gaining access to individual grains, the replacement may proceed by diffusion of ions through the solid, controlled in large part by imperfections in the crystal structure. In many deposits repeated movement has opened and reopened channelways, which would otherwise have become clogged, to permit continued and widespread replacement. The process may take place through the action of gases or solutions or by reactions in the solid state.

Classification. Mineral deposits are generally classified on the basis of the geologic processes responsible for their formation as magmatic, contact metasomatic, pegmatitic, hydrothermal, sedimentary, residual, and regional metamorphic deposits.

Magmatic deposits. Some mineral deposits originated by cooling and crystallization of magma, and the concentrated minerals form part of the body of the igneous rock. If the magma solidified by simple crystallization, the economically valuable mineral is distributed through the resulting rock; diamond deposits found in peridotite are believed by some geologists to be of this type. However, if the magma has differentiated during crystallization, early-formed minerals may settle to the bottom of the magma chamber and form segregations such as the chromite deposits of the Bushveld in South Africa. Late-formed minerals may crystallize in the interstices of older minerals and form segregations like the Bushveld platinum deposits. Occasionally, the residual magma becomes enriched in constituents such as iron, and this enriched liquid may form deposits, such as the Taberg titaniferous iron ores of Sweden. It is also possible that during differentiation some of the crystals or liquid may be injected and form sills or dikes. The iron ores of Kiruna, Sweden, have been described as early injections, and certain pegmatites are classed as late magmatic injections. Magmatic deposits are relatively simple in mineral composition and few in number.

Contact metasomatic deposits. During the crystallization of certain magmas a considerable amount of fluid escapes. This fluid may produce widespread changes near the contacts of magma with the surrounding rocks (Fig. 4). Where such changes are caused by heat effects, without addition of material from the magma, the resulting deposits are called contact metamorphic. If appreciable material is contributed by the magma, the deposits are termed contact metasomatic. The term skarn is applied to the lime-bearing silicates formed by the introduction of Si, Al, Fe, and Mg into a carbonate rock; some skarns contain ore bodies. The magmas that produce these effects are largely silicic in composition and resulting mineral deposits are often irregular in form.

A complicating case exists where little fluid escaped from the magma but the heat of the intrusion was great enough to cause dissolution and movement of certain metals from the surrounding rocks. It is believed by some investigators that solutions formed in this manner may become concentrated in metal content and subsequently deposit these metals near the contact of the intrusion and the surrounding rocks. In this case, the ore minerals were deposited by replacing preexisting rocks but the source of the ore is the surrounding rocks, not the magma. To further complicate matters, the ore in some deposits apppears to consist of material derived from both the intrusion and the surrounding rocks. In such deposits the source of the ore is generally controversial, and the size, amount, and composition of the mineralization would depend upon the relative contributions from the intrusion and the associated rocks.

Under contact metasomatic conditions, the (ore-forming) fluids extensively replace the country rock to produce a variety of complex minerals. Contact metasomatic deposits include a number of important deposits, whereas contact metamorphic deposits are rarely of economic value. Many gar-

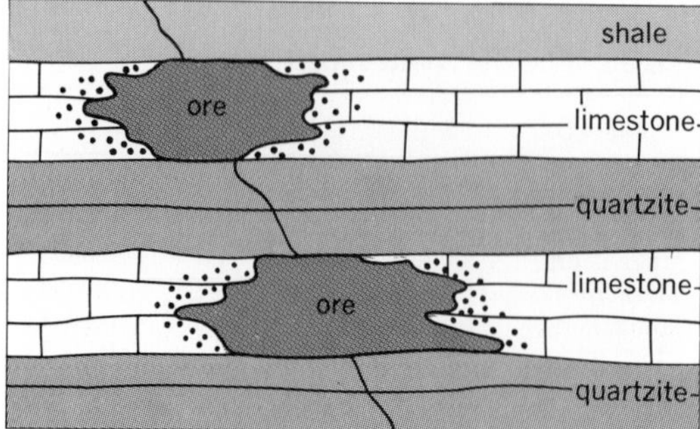

Fig. 3. The replacement of limestone by ore along a fissure. Disseminated ore, indicated by the dots, is forming in advance of the main body.

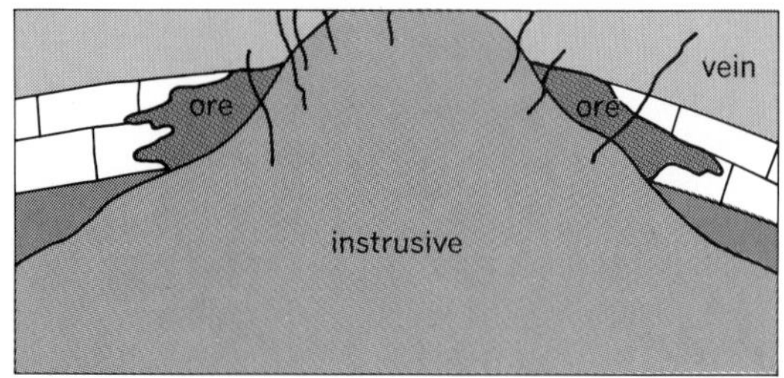

Fig. 4. Association of contact metasomatic and vein deposits with intrusive magmas.

net, emery, and graphite deposits are classed as contact metasomatic, as are such metalliferous deposits as the iron ores of Cornwall, Pa., Iron Springs, Utah, and Banat, Hungary; many copper ores of Utah, Arizona, New Mexico, and Mexico; the zinc ores of Hanover, N.M.; and various tungsten ores of California and Nevada.

Pegmatite deposits. Pegmatites are relatively coarse-grained rocks found in igneous and metamorphic regions. The great majority of them consist of feldspar and quartz, often accompanied by mica, but complex pegmatites contain unusual minerals and rare elements. Many pegmatites are regular tabular bodies; others are highly irregular and grade into the surrounding rocks. In size, pegmatites range from a few inches in length to bodies over 1000 ft long and scores of feet across. Some pegmatites are zoned, commonly with a core of quartz surrounded by zones in which one or two minerals predominate.

Pegmatites may originate by various igneous and metamorphic processes. Fractional crystallization of a magma results in residual solutions that are generally rich in alkalies, alumina, water, and other volatiles. The volatiles lower the temperature of this liquid and make it unusually fluid; the low viscosity promotes the formation of coarse-grained minerals. The rare elements that were unable by substitution to enter into the crystal structure of earlier-formed minerals, principally because of differences in size of their atomic radii, are concentrated in the residual pegmatite solutions. Late hydrothermal fluids may alter some of the previously formed pegmatite minerals.

Some pegmatites develop by replacement of the country rock and commonly these are isolated bodies with no feeders or channels in depth. They occur in metamorphic regions usually devoid of igneous rocks and contain essentially the same minerals as those in the country rocks. In some regions small pegmatites have grown by forcing apart the surrounding metamorphic rock, and others have formed by filling a fissure or crack from the walls inward. In both cases growth is believed to have taken place by diffusion and consolidation of material in the solid state.

Hydrothermal deposits. Most vein and replacement deposits are believed to be the result of precipitation of mineral matter from dilute, hot ascending fluids. As the temperature and pressure decrease, deposition of dissolved material takes place. It is not altogether certain how important the gaseous state is in the transport of ore material. It may be that at relatively shallow depth and high temperature gaseous solutions transport significant amounts of ore-forming material.

W. Lindgren, who developed the hydrothermal theory, divided these deposits into three groups on the basis of temperature and pressure conditions supposed to exist at the time of formation. Deposits thought to form at temperatures of 50–200°C at slight depth beneath the surface are called epithermal. Many ores of mercury, antimony, gold, and silver are of this type. Deposits formed at 200–300°C at moderate depths are known as mesothermal and include ores of gold-quartz, silver-lead, copper, and numerous other types. Hypothermal deposits are those formed at 300–500°C at high pressures; certain tin, tungsten, and gold-quartz ores belong to this type.

The nature of hydrothermal fluids is inferred by analogy with laboratory experiments, and by investigation of deposits forming around volcanoes and hot springs at the present time. Studies of liquid inclusions in minerals, of mineral textures, and of inversion temperatures of minerals indicate that mineralization takes place at elevated temperatures. Layers of minerals on the walls of open fissures with crystal faces developed toward the openings suggest deposition from solution. In some of these cavities later crystals were deposited on earlier ones in a manner that suggests growth in moving solutions. Certain secondary replacement phenomena, such as weathering and oxidation of mineral deposits, also indicate deposition from liquid solutions. Studies of wall rock alteration where hydrothermal solutions have attacked and replaced rock minerals indicate that these solutions change in character from place to place. Sulfur in such solutions may react with or leach metals from the surrounding rocks or partly solidified magma to form certain kinds of mineral deposits. On the basis of geochemical data it has been estimated that most hydrothermal ore-forming solutions had a temperature in the range 50–600°C, formed under pressures ranging from atmospheric to several thousand atmospheres, commonly contained high concentrations of NaCl and were saturated with silica but were not highly concentrated in ore metals, were neutral or within about 2pH units of neutrality; and that the metals probably were transported as complexes.

The principal objections to the hydrothermal theory are the low solubility of sulfides in water and the enormous quantities of water required. W. Lindgren realized this and, for some deposits, favored colloidal solutions as carriers of metals. Laboratory synthesis of sulfide minerals by G. Kullerud shows that some ore-bearing solutions must have been considerably more concentrated than is generally believed. *See* ORE DEPOSITS, GEOCHEMISTRY OF.

Two common features of hydrothermal deposits are the zonal arrangement of minerals and alteration of wall rock.

1. Zoning of mineralization. Many ore deposits change in composition with depth, lateral distance, or both, resulting in a zonal arrangement of minerals or elements. This arrangement is generally interpreted as being due to deposition from solution with decreasing temperature and pressure, the solution precipitating minerals in reverse order of their solubilities. Other factors are also involved such as concentration, relative abundance, de-

crease in electrode potentials, and reactions within the solutions and with the wall rocks as precipitation progresses.

Zonal distribution of minerals was first noted in mineral deposits associated in space with large igneous bodies, and has since been extended to include zoning related to sedimentary and metamorphic processes in places where no igneous bodies are in evidence. Although many geologists interpret zoning as a result of precipitation from a single ascending solution, others believe deposition is achieved from solutions of different ages and of different compositions.

The distribution of mineral zones is clearly shown at Cornwall, England, and at Butte, Mont. At Cornwall, tin veins in depth pass upward and outward into copper veins, followed by veins of lead-silver, then antimony, and finally iron and manganese carbonates. Such zoning is by no means a universal phenomenon, and, in addition to mines and districts where it is lacking, there are places where reversals of zones occur. Some of these reversals have been explained more or less satisfactorily by telescoping of minerals near the surface, by the effects of structural control or of composition of the host rock in precipitating certain minerals, and by the effects of supergene enrichment on the original zoning, but many discrepancies are not adequately explained.

2. Wall rock alteration. The wall rocks of hydrothermal deposits are generally altered, the most common change being a bleaching and softening. Where alteration has been intense, as in many mesothermal deposits, primary textures may be obliterated by the alteration products. Chemical and mineralogical changes occur as a result of the introduction of some elements and the removal of others; rarely a rearrangement of minerals takes place with no replacement.

Common alteration products of epithermal and mesothermal deposits are quartz, sericite, clay minerals, chlorite, carbonates, and pyrite. Under high-temperature hypogene conditions pyroxene, amphibole, biotite, garnet, topaz, and tourmaline form. In many mines sericite has been developed nearest the vein and gives way outward to clay minerals or chlorite. The nature and intensity of alteration vary with size of the vein, character of the wall rock, and temperature and pressure of hydrothermal fluids. In the large, low-grade porphyry copper and molybdenum deposits associated with stocklike intrusives, alteration is intense and widespread, and two or more stages of alteration may be superimposed.

Under low-intensity conditions, the nature of the wall rock to a large extent determines the alteration product. High-intensity conditions, however, may result in similar alteration products regardless of the nature of the original rock. Exceptions to this are monomineralic rocks such as sandstones and limestones. Wall rock alteration may develop during more than one period by fluids of differing compositions, or it may form during one period of mineralization as the result of the action of hydrothermal fluids that did not change markedly in composition. Alteration zones have been used as guides to ore and tend to be most useful where they are neither too extensive nor too narrow. Mapping of these zones outlines the mineralized area and may indicate favorable places for exploration.

Sedimentary and residual deposits. At the Earth's surface, action of the atmosphere and hydrosphere alters minerals and forms new ones that are more stable under the existing conditions. Sedimentary deposits are bedded deposits derived from preexisting material by weathering, erosion, transportation, deposition, and consolidation. Different source materials and variations in the processes of formation yield different deposits. Changes that take place in a sediment after it has formed and before the succeeding material is laid down are termed diagenetic. They include compaction, solution, recrystallization, and replacement. In general, the sediment is consolidated by compaction and by precipitation of material as a cement between mineral grains. Sedimentation as a process may itself involve the concentration of materials into mineral deposits. *See* DIAGENESIS.

The mineral deposits that form as a result of sedimentary and weathering processes are commonly grouped as follows: (1) sedimentary deposits, not including products of evaporation; (2) sedimentary-exhalative deposits; (3) chemical evaporities; (4) placer deposits; (5) residual deposits; and (6) organic deposits.

1. Sedimentary deposits. Included in this group are the extensive coal beds of the world, the great petroleum resources, clay deposits, limestone and dolomite beds, sulfur deposits such as those near Kuibyshev, Soviet Union, and the deposits of the Gulf Coast region, and the phosphate of North Africa and Florida. Metalliferous deposits such as the minette iron ores of Lorraine and Luxembourg, and Clinton iron ores of the United States, and the manganese of Tchiaturi, Georgia, and Nikopol in the Ukraine also belong here. There are other deposits of metals in sedimentary rocks whose origin remains an enigma, such as the uranium of the Colorado Plateau, the Witwatersrand in South Africa, and Blind River in Ontario; and the copper deposits of Mansfeld, Germany, and of the Copperbelt of Zambia and the Democratic Republic of the Congo. These deposits have characteristics of both syngenetic and epigenetic types. A controversy centers around the genesis of these and similar deposits of the world.

2. Sedimentary-exhalative deposits. Many large stratiform deposits are found in marine sedimentary rocks associated with volcanic rocks. It is well known that volancoes and fumaroles carry in their gases a number of metals. On land these gases escape into the atmosphere. Under water the gases, if they carry matter which is insoluble under the existing conditions, will precipitate their metals as oxides, sulfides, or carbonates in the vicinity of the gas emission. If the gases contain matter that is soluble, the metal content of the sea water will increase, and upon reaching saturation level will precipitate an extensive disseminated ore deposit. Where submarine emissions take place in a large ocean basin, they may be deposited over the floor of the basin as part of the sedimentation process.

Deposits exemplified by lead-zinc-barite-fluorite mineralization, most commonly found in carbonate rocks, occur in the Mississippi Valley region of North America and also on other continents. These ores are included with the sedimentary-exhalative type, but could also be discussed under several other classes of deposits since they are very difficult to categorize. They have been considered by various geologists to be true sediments, diagenetic deposits, lateral secretion deposits, deposits formed by downward leaching of overlying lean ores, deposits formed by solutions that descended and subsequently acended, deposits resulting from magmatic-hydrothermal processes, and sea-floor deposits from thermal springs. Most geologists favor either a syngenetic-sedimentary hypothesis or an epigenetic-hypogene one. Some studies hypothesize a source of metal-bearing waters similar to those in deep brines which rise and move through fissures in overlying rocks or are poured out on the sea floor and are added to accumulating sediments. A single generally acceptable hypothesis of origin, if such eventually emerges, must await the accumulation and interpretation of additional geological and geochemical data.

3. Chemical evaporites. These consist of soluble salts formed by evaporation in closed or partly closed shallow basins. Deposits of salt or gypsum that are several hundred feet thick are difficult to explain satisfactorily. Oschsenius suggested that they formed in basins which were separated from the ocean by submerged bars except for a narrow channel (inlet); such barriers are common along coastal areas. Intermittently, sea water flowed over the barrier and was concentrated into saline deposits by evaporation. Modifications of this theory have been proposed to account for the omissions of certain minerals and the interruptions in the succession.

Deposits of gypsum and common salt (halite) are found in many countries, whereas the larger concentrations of potash salts, borates, and nitrates are much more restricted in occurrence. *See* EVAPORITE, SALINE.

4. Placer deposits. Placers are the result of mechanical concentration whereby heavy, chemically resistant, tough minerals are separated by gravity from light, friable minerals. Separation and concentration may be accomplished by streams, waves and currents, and air, or by soil and hill creep. The most important economic placer deposits are those formed by stream action (Fig. 5).

Stream and beach placers are widespread in occurence and include the famous gold placers of the world, as well as deposits of magnetite, ilmenite chromite, wolframite, scheelite, cassiterite, rutile, zircon, monazite, and garnet. Placer deposits of diamond, platinum, and gemstones are less common.

5. Residual deposits. Complete weathering results in distribution of the rock as a unit and the segregation of its mineral constituents. This is accomplished by oxidation, hydration, and solution, and may be accelerated by the presence of sulfuric acid. Some iron and manganese deposits form by accumulation without change, but certain clay and bauxite deposits are created during the weathering of aluminous rocks. Residual concentrations form where relief is not great and where the crust is stable; this permits the accumulation of material in place without erosion. *See* WEATHERING PROCESSES.

Large residual deposits of clay, bauxite, phosphate, iron, and manganese have been worked in many parts of the world, as have smaller deposits of nickel, ocher, and other minerals.

6. Organic deposits. Plants and animals collect and use various inorganic substances in their life processes, and concentration of certain of these substances upon the death of the organisms may result in the formation of a mineral deposit. Coal and peat form from terrestrial plant remains and represent concentration by plants of carbon from the carbon dioxide of the atmosphere. Petroleum originates by the accumulation of plant and animal remains. Many limestone, phosphate, and silica deposits also form by plant and animal activity. Hydrated ferric oxide and manganese dioxide are precipitated by microorganisms; anaerobic bacteria can reduce sulfates to sulfur and hydrogen sulfide. There is considerable controversy, however, as to whether microorganisms are responsible for the formation of certain iron, manganese, and sulfide deposits. Some uranium, vanadium, copper, and other metalliferous deposits are considered to have formed, in part at least, by the activity of organisms.

Deposits formed by regional metamorphism. Regional metamorphism includes the reconstruction that takes place in rocks within orogenic or mountain belts as a result of changes in temperature, pressure, and chemical environment. In these orogenic belts, rocks are intensely folded, faulted, and subjected to increases in temperature. The changes that occur in this environment affect the chemical and physical stability of minerals, and new minerals, textures, and structures are produced, generally accompanied by the introduction of considerable material and the removal of other material.

Some geologists believe that the water and metals released during regional metamorphism can give rise to hydrothermal mineral deposits. Along faults and shear zones movement of fluids could take place by mechanical flow, though elsewhere movement might be by diffusion. The elements released from the minerals would migrate to low-pressure zones such as brecciated or fissured areas and concentrate into mineral deposits. It has been suggested that the subtraction of certain elements during metamorphism also can result in a relative enrichment in the remaining elements; if this process is sufficiently effective, a mineral deposit may result. Certain minerals also may be

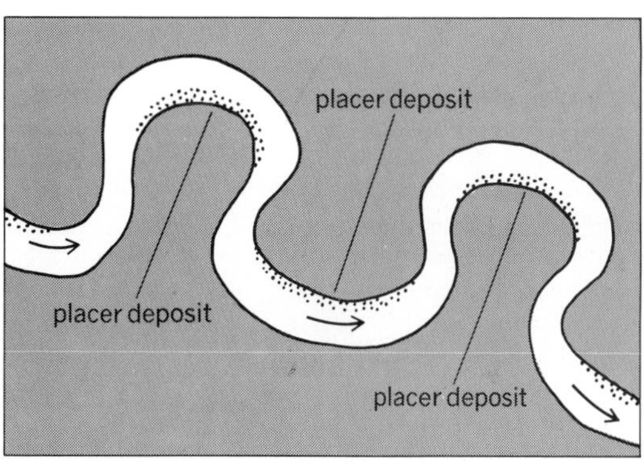

Fig. 5. Diagram, showing deposition of placer by stream action on the inside of meander bends.

concentrated during deformation by flow of material to areas of low pressure such as along the crests of folds.

Deposits of magnetite, titaniferous iron, and various sulfides may form in metamorphic rocks, as well as deposits of nonmetallic minerals such as kyanite, corundum, talc, graphite, and garnet.

Opponents of the concept of mineral formation by regional metamorphism believe that a dispersal of minerals, rather than a concentration, would result from the processes operative. However, if movement of material were confined to specific channelways, this objection would not necessarily hold.

Oxidation and supergene enrichment. Many sulfide minerals form at depth under conditions differing markedly from those existing at the surface. When such minerals are exposed by erosion or deformation to surface or near-surface conditions, they become unstable and break down to form new minerals. Essentially all minerals are affected.

The oxidation of mineral deposits is a complex process. Some minerals are dissolved completely or in part, whereas elements of others recombine and form new minerals. The principal chemical processes that take place are oxidation, hydration, and carbonation. The oxidation of pyrite and other sulfides produces sulfuric acid, a strong solvent. Much of the iron in the sulfides is dissolved and reprecipitated as hydroxide to form iron-stained outcrops called gossans. Metal and sulfate ions are leached from sulfides and carried downward to be precipitated by the oxidizing waters as concentrations of oxidized ores above the water table. Oxides and carbonates of copper, lead, and zinc form, as do native copper, silver, and gold. The nature of the ore depends upon the composition of the primary minerals and the extent of oxidation. If the sulfates are carried below the water table, where oxygen is excluded, upon contact with sulfides or other reducing agents they are precipitated as secondary sulfides. The oxidized zone may thus pass downward into the supergene sulfide zone. Where this process has operated extensively, a thick secondary or supergene-enriched sulfide zone is formed. Enrichment may take place by removal of valueless material or by solution of valuable metals which are then transported and reprecipitated.

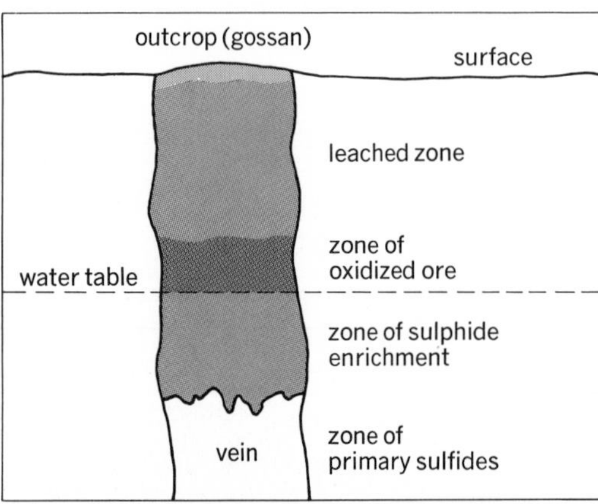

Fig. 6. Vein deposit of sulfide ore, showing changes due to oxidation and supergene enrichment.

This enrichment process has converted many low-grade ore bodies into workable deposits. Supergene enrichment is characteristic of copper deposits but may also take place in deposits of other metals. Beneath the enriched zone is the primary sulfide ore (Fig. 6).

The textures of the gossan minerals may give a clue to the identity of the minerals that existed before oxidation and enrichment took place. These have been used as guides in prospecting for ore.

Sequence of deposition. Studies of the relationships of minerals in time and space have shown that a fairly constant sequence of deposition, or paragenesis, is characteristic of many mineral deposits. This sequence has been established largely by microscopic observations of the boundary relationships of the minerals in scores of deposits. Subsequent experimental studies of mineral phases have contributed to the knowledge of paragenesis. In magmatic and contact metasomatic ores, silicates form first, followed by oxides and then sulfides. W. Lindgren presented the paragenesis for hypogene mineral associations, and others have discussed the problems involved. The sequence of common minerals starts with quartz, followed by iron sulfide or arsenide, chalcopyrite, sphalerite, bornite, tetrahedrite, galena, and complex lead and silver sulfo salts. It indicates the existence of some fundamental control but attempts to explain the variations in it have been largely unsuccessful, or are applicable to only part of the series or to specific mineralized areas. Local variations are to be expected since many factors such as replacement, unmixing, superimposed periods of mineralization, structural and stratigraphic factors, and telescoping of minerals may complicate the order of deposition.

Paragenesis is generally thought to be the result of decreasing solubility of minerals with decreasing temperature and pressure. It has also been explained in terms of relative solubilities, pH of the solutions, metal volatilities, decreasing order of potentials of elements, free energies, and changing crystal structure of the minerals as they are deposited. R. L. Stanton has reevaluated paragenetic criteria as applied to certain stratiform sulfide ores in sedimentary and metamorphic rocks. He proposes that the textures of such ores do not represent sequences of deposition but are the result of surface energy requirements during grain growth, or annealing of deformed minerals. To explain mineral paragenesis more satisfactorily, many additional experiments must be made to determine phase relations at different temperatures and pressures. *See* MINERAL.

Mineralogenetic provinces and epochs. Mineral deposits are not uniformly distributed in the Earth's crust nor did they all form at the same time. In certain regions conditions were favorable for the concentration of useful minerals. These regions are termed mineralogenetic provinces and they contain broadly similar types of deposits, or deposits with different mineral assemblages that appear to be genetically related. The time during which these deposits formed constitutes a mineralogenetic epoch; such epochs differ in duration, but in general they cover a long time interval that is not sharply defined. Certain provinces contain mineral deposits of more than one epoch.

During diastrophic periods in the Earth's history mountain formation was accompanied by plutonic and volcanic activity and by mineralization of magmatic, pegmatitic, hydrothermal and metamorphic types. During the quieter periods, and in regions where diastrophism was milder, deposits formed by processes of sedimentation, weathering, evaporation, supergene enrichment, and mechanical action.

During the 1960s numerous studies were made of the regional distribution of mineral deposits associated with long subsiding belts of sediments, or geosynclines, and with platform areas of relatively thin sediments adjoining the thick geosynclinal wedge. Geosynclinal areas commonly suffer folding and later uplift and become the sites of complex mountain ranges. It has been proposed that the outer troughs and bordering deep-seated faults contain ore deposits of subcrustal origin, the inner uplifts contain deposits of crustal origin, and the platforms contain ores derived from subcrustal and nonmagmatic platform mantle rocks. V. I. Smirnov and others have summarized available information on types of mineral deposits characteristic of the processes most active during the evolutionary stages of geosynclinal and platform regions. In the early prefolding stage of subsidence, subcrustal juvenile basaltic sources of ore fluids prevail, and the characteristic metals are Cr, titanomagnetite, Pt metals, skarn Fe and Cu, and deposits of pyritic Cu and Fe and Mn. In the folding episode, rocks of the geosyncline are melted to produce magma from which ore components are extracted or leached by postmagmatic fluids. The most typical ores of this stage are Sn, W, Be, Ni, Ta, and various polymetallic deposits. The late stage is characterized by ore deposits associated with igneous rocks and other deposits with no apparent relationship to igneous rocks. Smirnov believes these ores originated by the combined effect of subcrustal, crustal, and nonmagmatic sources of ore material. Typical metals of this stage include Pb, Zn, Cu, Mo, Sn, Bi, Au, Ag, Sb, and Hg. In the tectonically activated platform areas, deposits of Cu-Ni sulfides, diamonds, various magmatic and pegmatitic deposits, and hydrothermal ores of nonferrous, precious, and rare metals are found. In addition, there are nonmagmatic deposits of Pb and Zn. Some ore material is believed to be both subcrustal and nonmagmatic in origin. Relative proportions of types of mineralization differ from one region to another.

The relationship between mineral deposition and large scale crustal movements permits a grouping of mineralogenetic provinces by major tectonic features of the continents such as mountain belts, stable regions, and Precambrian shields. The Precambrian shield areas of the world contain the Lake Superior, Kiruna, and Venezuelan iron provinces, the gold provinces of Kirkland Lake and Porcupine in Canada, the gold-uranium ores of South Africa, the gold deposits of western Australia, and the base metals of central Australia. In the more stable regions are the metalliferous lead-zinc province of the Mississippi Valley and provinces of salt and gypsum, iron, coal, and petroleum in different parts of the world. The mountain belts are the location of many diverse kinds of mineral provinces such as the gold-quartz provinces of the Coast Range and the Sierra Nevadas, various silver-lead-zinc provinces of the western United States, the Andes, and elsewhere, and numerous base-metal provinces in the Americas, Africa, Australia, and Europe. *See* TECTONIC PATTERNS.

Localization of mineral deposits. The foregoing discussion has shown that mineral deposits are localized by geologic features in various regions and at different times. Major mineralized districts within the shield areas and mountain belts are often localized in the upper parts of elongate plutonic bodies. Specific ores tend to occur in particular kinds of rocks. Thus tin, tungsten, and molybdenum are found in granitic rocks, and nickel, chromite, and platinum occur in basic igneous rocks. In certain regions mineral deposits are concentrated around plateau margins. Tropical climates favor the formation of residual manganese and bauxite deposits, whereas arid and semiarid climates favor the development of thick zones of supergene copper ores. Major mineralized districts are also localized by structural features such as faults, folds, contacts, and intersections of superimposed orogenic belts. The location of individual deposits is commonly controlled by unconformities, structural features, the physical or chemical characteristics of the host rock (Fig. 7), topographic features, basins of deposition, groundwater action, or by restriction to certain favorable beds (Fig. 8). *See* PLUTON.

Source and nature of ore fluids. Widely divergent views have been expressed as to the original source and mode of transport of mineral deposits. Each view has certain advantages when applied to specific types of deposits. However, the complex nature of some mineralizations and the highly diverse physicochemical environments in which mineral deposits form make it impossible to select one theory to account for the source of all ore-forming materials.

According to one view, the source of the ore material was a juvenile subcrustal basaltic magma from which mineral deposits crystallized by simple crystallization, or in some cases were concentrated by differentiation. Most of the ore deposits associated with such magmas show a close spatial relationship to the enclosing igneous rocks and are similar in composition from one province to another. The exceptions to this are certain pyritic copper and skarn ores that apparently were introduced into sedimentary rocks and are now re-

ORE AND MINERAL DEPOSITS

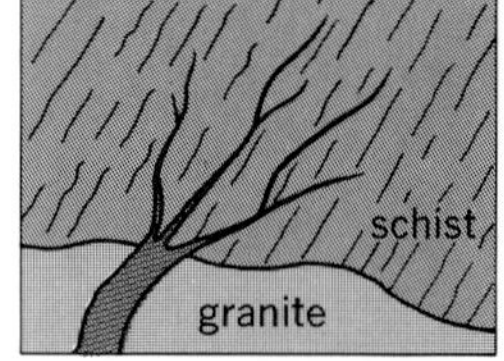

Fig. 7. Strong vein in granite dividing into stringers upon entering schist.

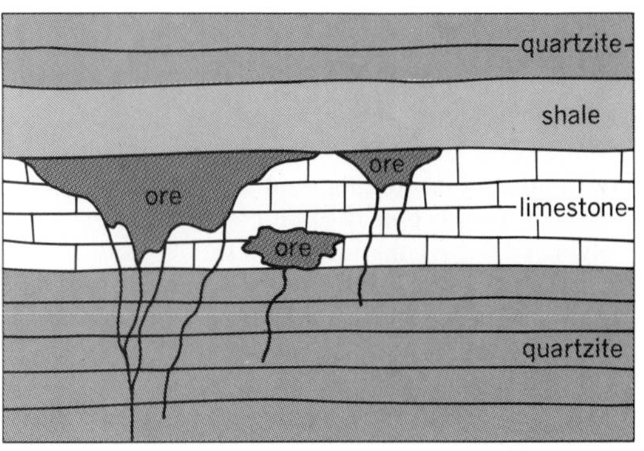

Fig. 8. Ore in limestone beneath impervious shale.

moved from the postulated source magma.

Another hypothesis holds that many ore deposits associated with granitic rocks were derived from magmas generated by remelting of deep-seated sedimentary rocks, followed by movement of the magma into higher levels of the Earth's crust. As an end product of crystallization and differentiation, an ore fluid was produced containing concentrations of metals originally present in the magma. Commonly, such deposits are confined to the apical portions of granitic plutons that have been altered by postmagmatic fluids. An increasing number of geologists ascribe to the view that the ore material was removed from the solidified magma by these late-stage fluids. Such ore deposits are complex, and their composition, dependent in part on the composition of the remelted rocks, is diverse and variable from one region to another. For certain ores associated with major deep-seated faults or with intersections of extensive fault or fissure systems, an ore source in deeper, subcrustal, regions has been advocated.

Circulation of surface waters may have removed metals from the host rocks and deposited them in available openings; this is the lateral secretion theory. The metals were carried either by cool surface waters or by such waters that moved downward, became heated by contact with hot rocks at depth, and then rose and deposited their dissolved material.

As sediments are compacted and lithified, huge volumes of water may be expelled. It has been suggested that the ore-forming fluid in some sedimentary deposits contains metals that were in solution before the sediment was buried, plus metals acquired during diagenesis of the sediments. For certain ores with colloform textures, it is believed that movement took place as a finely divided suspension, and that the ore minerals initially precipitated as gels. The metals could have been held as adsorbates on clays and other colloids and then released for concentration during later crystallization of the colloids. Crystallization would exclude the metals as finely divided material which could combine with released water and move to places favorable for precipitation.

A considerable amount of experimental work has been done on the geochemistry of ore fluids in an attempt to determine the source, nature, solubility, transport, and deposition of these fluids. Some of the results of these investigations and the problems that exist are discussed in the references by H. L. Barnes and K. B. Krauskopf listed in the bibliography. Studies of metal-bearing saline waters and of thermal waters and brines in igneous, sedimentary, and metamorphic rocks have also contributed to the knowledge of this complex subject. D. E. White has analyzed and summarized these studies, and he stresses that four mutually interdependent factors must be considered: a source of the ore constituents, the dissolving of these constituents in the hydrous phase, migration of the ore-bearing fluid, and the selective precipitation of the ore constituents in favorable environments. The ore-bearing fluids are Na-Ca-Cl brines that may form by magmatic or connate processes, solution of evaporates by dilute water, or membrane concentration of dilute meteoric water. *See* ORE DEPOSITS, GEOCHEMISTRY OF.

During regional metamorphism large quantities of hydrothermal fluids may be released from rocks in deep orogenic zones. These fluids remove metals and other minerals from the country rock and redeposit them at higher levels along favorable structures. Elements may also move by diffusion along chemical, thermal, and pressure gradients.

A number of the famous mineralized districts of the world that have characteristics of both epigenetic and syngenetic deposits have been modified by later metamorphism, thereby further obscuring their origin. In some of these districts the fissure and joint systems in the rocks reflect the pattern in deeper-seated rocks. H. Schneiderhohn (Stuttgart) has suggested that repeated rejuvenation of these systems by tectonic movements, accompanied by the dissolving action of thermal waters on old ore deposits in depth, would result in upward movement and reprecipitation of metals in higher formations; Schneiderhohn calls these deposits secondary hydrothermal ores. Elsewhere old folded rocks and ore deposits have been greatly deformed, and the ores taken into solution and transported to higher and younger strata; such deposits Schneiderhohn terms regenerated ores. Controversy centers around suitable criteria for epigenetic and syngenetic deposits, the problems of solubility of metals in thermal waters, their transport over long distances, and whether such rejuvenated and regenerated ores would be dispersed or concentrated by the processes envisaged by Schneiderhohn. [A. F. HAGNER]

Bibliography: H. L. Barnes (ed.), *Geochemistry of Hydrothermal Ore Deposits*, 1967; A. M. Bateman, *Economic Mineral Deposits*, 2d ed., 1950; A. M. Bateman (ed.), *Economic Geology: Fiftieth Anniversary Volume, 1905–1955*, 1955; R. L. Bates, *Geology of the Industrial Rocks and Minerals*, 1960; J. S. Brown (ed.), *Genesis of Stratiform Lead-Zinc-Barite-Fluorite Deposits*, Econ. Geol. Monogr. no. 3, 1967; A. B. Edwards, *Textures of the Ore Minerals*, 2d ed., 1954; K. B. Krauskopf, *Introduction to Geochemistry*, 1967; J. Kutina (ed.), *Symposium: Problems of Postmagmatic Ore Deposition*, 1963; C. A. Lamey, *Metallic and Industrial Mineral Deposits*, 1966; W. Lindgren, *Mineral Deposits*, 4th ed., 1933; H. E. McKinstry, *Mining Geology*, 1948; C. J. Park, Jr., and R. A. MacDiarmid, *Ore Deposits*, 1964; J. D. Ridge (ed.), *Ore Deposits in the United States, 1933–1967*, 1968.

Ore deposits, geochemistry of

Geochemistry in general deals with the amounts and distribution of the elements and isotopes of the Earth, and the nature of the processes affecting them. Although ore bodies are formed by many different processes, only those deposits of the heavy metals believed to be formed by precipitation from heated water-rich fluids (hydrothermal) are discussed in detail here. Most of the world's supply of base metals, silver, and gold originates in such deposits. *See* ORE AND MINERAL DEPOSITS.

Mineral and chemical composition. The minerals in ore deposits frequently are divided into two groups, ore and gangue; the former constitute those for which the deposit is mined, the latter are the waste minerals associated with the ore. The same mineral may be an ore in some deposits and a gangue in others. The common minerals of hy-

Table 1. Some common primary minerals of hydrothermal ore deposits

Element	Common minerals	Idealized formulas	Significant minor elements occurring in these minerals, underlined where economically important
Iron	Hematite	Fe_2O_3	
	Magnetite	Fe_3O_4	Mn
	Pyrite	FeS_2	Au,* Co, Ni
	Pyrrhotite	$Fe_{1-x}S$	Ni, Co
	Siderite	$FeCO_3$	Mn, Ca, Mg
	Arsenopyrite	FeAsS	Sb, Co, Ni
Copper	Chalcopyrite	$CuFeS_2$	Ag, Mn, Se
	Bornite	Cu_5FeS_4	
	Chalcocite	Cu_2S	Ag
	Enargite	Cu_3AsS_4	Ag, Sb
	Tetrahedrite	$Cu_{12}Sb_4S_{13}$	Ag, Fe, Zn, Hg, As
Zinc	Sphalerite	ZnS	Fe, Mn, Cd, Cu, Ga, Ge, Sn, In
Lead	Galena	PbS	Ag, Bi, As, Tl, Sn, Se, Sb
Bismuth	Native bismuth	Bi	
	Bismuthinite	Bi_2S_3	
Silver	Native silver	Ag	Au
	Argentite	Ag_2S	
	Various sulfo salts		
Gold	Native gold	Au	Ag, Cu
	Various tellurides of gold and silver		
Mercury	Cinnabar	HgS	
Tin	Cassiterite	SnO_2	
Uranium	Uraninite	UO_2	Ra, Th, Pb
Cobalt	Cobaltite	CoAsS	
	Smaltite	$CoAs_2$	
Nickel	Pentlandite	$(Fe,Ni)_9S_8$	
Tungsten	Scheelite	$CaWO_4$	Mo
	Wolframite	$(Fe,Mn)WO_4$	Mo
Molybdenum	Molybdenite	MoS_2	Re
Manganese	Rhodochrosite	$MnCO_3$	Fe, Mg, Ca
	Rhodonite	$MnSiO_3$	Ca
Others	Calcite	$CaCO_3$	Mn
	Dolomite (and ankerite)	$CaCO_3 \cdot MgCO_3$	Fe, Mn
	Barite	$BaSO_4$	
	Fluorite	CaF_2	
	Quartz	SiO_2	
	Sericite, chlorite, feldspars, clays, and various other silicates		

*Intimately associated as minute particles of metallic gold, but not in the crystal structure of the pyrite.

drothermal deposits (Table 1) are sulfides, sulfo salts, oxides, carbonates, silicates, and native elements, although sulfates, a fluoride, tungstates, arsenides, tellurides, selenides, and others are by no means rare. Many minor elements which seldom occur in sufficient abundance to form discrete minerals of their own may substitute for the major elements of the ore minerals and thus be recovered as by-products. For example (as shown in Table 1), the ore mineral of cadmium, indium, and gallium is sphalerite; the major ore mineral of silver and thallium is galena; and pyrite is sometimes an ore of cobalt. *See* ELEMENTS, GEOCHEMICAL DISTRIBUTION OF.

Ore deposits consist, in essence, of exceptional concentrations of given elements over that commonly occurring in rocks. The degree of concentration needed to constitute ore varies widely, as shown in Table 2, and is a complex function of many economic and sometimes political variables. The quantity of these elements in the total known or reasonably expected ore bodies in the world is infinitesimal when compared with the total amounts in the crust of the Earth. Thus, each and every cubic mile of ordinary rocks in the crust of the Earth contains enough of each ore element to make large deposits (Table 2). Although there is a large number of geologic situations that are apparently favorable, only a very few of them contain significant amounts of ore. Thus it is evident that the processes leading to concentration must be the exception and not the rule, and obviously any understanding or knowledge of these processes should aid in the discovery of further deposits.

It is apparent from the above that each step in the process of ore formation must be examined carefully if this sporadic occurrence of ore is to be placed on a rational basis. In order for ores to form, there must be a source for the metal, a medium in which it may be transported, a driving force

Table 2. Approximate concentration of ore elements in Earth's crust and in ores

Element	Approximate concentration in average igneous rocks, %	Tons per cubic mile of rock	Approximate concentration in ores, %	Concentration factor to make ore
Iron	5.0	560,000,000	50	10
Copper	0.007	790,000	0.5–5	70–700
Zinc	0.013	1,500,000	1.3–13	100–1000
Lead	0.0016	180,000	1.6–16	1000–10,000
Tin	0.004	450,000	0.01*–1	2.5–250
Silver	0.00001	1,100	0.05	5000
Gold	0.0000005	56	0.0000015*–0.01	3–2000
Uranium	0.0002	22,000	0.2	1000
Tungsten	0.003	340,000	0.5	170
Molybdenum	0.001	110,000	0.6	600

*Placer deposits.

to move this medium, a "plumbing system" through which it may move, and a cause of precipitation of the ore elements as an ore body. These interrelated requirements are discussed below in terms of the origin of the hydrothermal fluid, its chemical properties, and the mechanisms by which it may carry and deposit ore elements.

Source of metals. It is not easy to determine the source for the metals in hydrothermal ore deposits because, as shown above, they exist everywhere in such quantities that even highly inefficient processes could be adequate to extract enough material to form large deposits.

Fluids associated with igneous intrusion. In many deposits there is evidence that ore formation was related to the intrusion of igneous rocks nearby, but in many other deposits intensive search has failed to reveal any such association. Because the crystal structures of the bulk of the minerals (mostly silicates) crystallizing in igneous rocks are such that the common ore elements, such as copper, lead, and zinc, do not fit readily, these elements are concentrated into the residual liquids, along with H_2O, CO_2, H_2S, and other substances. These hot, water-rich fluids, remaining after the bulk of the magma has crystallized, are the hydrothermal fluids which move outward and upward to areas of lower pressure in the surrounding rocks, where part or all of their contained metals are precipitated as ores. A more detailed discussion of the composition of these fluids follows.

Fluids obtained from diagenetic and metamorphic processes. Fluids of composition similar to the above also could be obtained from diagenetic and metamorphic processes. When porous, water-saturated sediments containing the usual amounts of hydrous and carbonate minerals are transformed into essentially nonhydrous, nonporous metamorphic rocks, great quantities of water and carbon dioxide must be driven off. Thus, each cubic mile of average shale must lose about 3×10^9 tons of water and may lose large amounts of carbon dioxide on metamorphism to gneiss. The great bulk of the water presumably comes off as connate water (entrapped at time of rock deposition) under conditions of fairly low temperature. In many respects this water has the same sea-water composition as it had to start with. However, as metamorphism proceeds, accompanied by slow thermal buildup from heat flow from the Earth's interior and from radioactivity, the last fluids are given off at higher temperatures and are richer in CO_2 and other substances. These fluids would have considerably greater solvent power and can be expected to be similar to those coming from cooling igneous rocks.

Role of surface and other circulating waters. It is very likely that the existence of a mass of hot rock under the surface would result in heating and circulation of meteoric water (from rain and snow) and connate water. The possible role of these moving waters in dissolving ore elements from the porous sedimentary country rocks through which they may pass laterally and in later depositing them as ore bodies has been much discussed. The waters may actually contribute ore or gangue minerals in some deposits. The test of this theory of lateral secretion on the basis of precise analyses of the average country rocks around an ore body would involve an exceedingly difficult sampling job. It also would require analytical precision far better than is now feasible for most elements, as each part per million uncertainty in the concentration of an element in a cubic mile of rock represents about 10,000 tons of the element or 1,000,000 tons of 1% ore.

Movement of ore-forming fluids. In addition to the high vapor pressures of volatile-rich fluids acting as a driving force to push them out into the surrounding country rocks and to the surface, there may well be additional pressures from orogenic or mountain-building forces. When a silicate magma has an appreciable percentage of liquid and is subjected to orogenic forces, it moves en masse to areas of lower pressure (it is intruded into other rocks). But if the magma has crystallized 99% or more of its bulk as solid crystals and has only a very small amount of water-rich fluid present as thin films between the grains, and then is squeezed, this fluid may be the only part sufficiently mobile to move toward regions of lower pressure. (If the residual fluid, containing the ore elements, stays in the rock, it reacts with the early formed, largely anhydrous minerals of the rock to form new hydrated ones, such as sericite, epidote, amphibole, and chlorite, and its ore elements precipitate as minute disseminated specks and films along the silicate grain boundaries.)

The ore-bearing fluid leaves the source through a system comprised of joints, faults, porous volcanic plugs, or other avenues. As the fluid leaves the source, it moves some appreciable but generally unknown distance laterally, vertically, or both, and finally reaches the site of deposition. This system of channels is of utmost importance in the process of ore formation.

Localization of mineral deposits. It is stated frequently that ore deposits are geologic accidents; yet there are reasons, however abstruse, for the localization of a mineral deposit in a particular spot. One reason for localization is mere proximity to the source of the ore-forming fluids, as in rocks adjacent to an area of igneous activity or near a major fracture system which may provide plumbing for solutions ascending from unknown depths. Zones of shattering are favored locales for mineralization since these provide plumbing and offer the best possibility for the ore solution to react with wall rock, mix with other waters, and expand and cool, all of which may promote precipitation. Some types of rock, particularly limestone and dolomite, are especially susceptible to replacement and thus often are mineralized preferentially. The chemical or physical properties which cause a rock to be favored by the replacing solutions often are extremely subtle and certainly not understood fully.

Zoning and paragenesis. Mineral deposits frequently show evidence of systematic spatial and temporal changes in metal content and mineralogy that are sufficiently consistent from deposit to deposit to warrant special mention under the terms zoning and paragenesis. Zoning may be on any scale, though the range is commonly on the order of a few hundred to a few thousand feet, and may have either lateral or vertical development. In mining districts, such as Butte, Mont., or Cornwall, England, where zoning is unusually well developed, there is a peripheral zone of manganese minerals grading inward through successive, overlapping silver-lead, zinc, and copper zones (and in the case of Cornwall, tungsten, and finally tin). The same sequence of zones appears in many deposits localized about intrusive rocks, suggesting strongly that the tin and tungsten are deposited first from the outward-moving hydrothermal solutions and that the copper, zinc, lead, and silver were deposited successively as the solutions expanded and cooled. In other districts the occurrences of mercury and antimony deposits suggest that their zonal position may be peripheral to that of silver or manganese. The paragenesis, or the sequence of deposition of minerals at a single place, as interpreted from the textural relations of the minerals, follows the same general pattern as the zoning, with the tin and tungsten early and the lead and silver late. With both zoning and paragenesis there are sometimes reversals in the relative position of adjacent zones, and these are usually explained as successive generations of mineralization. Some metals, such as iron, arsenic, and gold, tend to be distributed through all of the zones, whereas others, such as antimony, tend to be restricted to a single position.

The sequence of sulfide minerals observed in zoning and paragenesis matches in detail the relative abilities of the heavy metals to form complex ions in solution. This observation strongly supports the hypothesis developed later that most ore transport occurs through the mechanism of complex ions, since no other geologically feasible property of the ore metals or minerals can explain the zoning relations.

Environment of ore deposition. Important aspects of the environment of ore deposition include the temperature, pressure, nature, and composition of the fluid from which ores were precipitated.

Temperatures. Although there is no geological thermometer that is completely unambiguous as to the temperatures of deposition of ores, there is a surprising number of different methods for estimating the temperatures that prevailed during events long since past that have been applied to ores with reasonably consistent results. Those ore deposits which had long been considered to have formed at high temperatures give evidence of formation in the range of 500–600°C, or possibly even higher. Those that were thought to be low-temperature deposits show temperatures of formation in the vicinity of 100°C or even less, and the bulk of the deposits lie between these extremes. *See* GEOLOGIC THERMOMETRY.

Pressures. It would be useful to know the total hydrostatic pressure of the fluids during ore formation. Most of the phenomena used for determination of the temperatures of ore deposition are also pressure-dependent, and so either an estimate of the correction for pressure must be made, or two independent methods must be used to solve for the two variables.

Pressures vary widely from nearly atmospheric in hot springs to several thousand atmospheres in deposits formed at great depth. Maximum reasonable pressures are considered to be on the order of that provided by the overlying rock; conversely, the minimum reasonable pressures are considered to be about equal to that of a column of fluid open to the surface. Pressures therefore range from approximately 500 to 1500 psi per 1000 ft of depth at the time of mineralization. *See* HIGH-PRESSURE PHENOMENA.

Evidence of composition. Geologists generally concede that most ore-forming fluids are essentially hot water or dense supercritical steam in which are dissolved various substances including the ore elements. There are three lines of evidence bearing on the composition of this fluid. These are fluid inclusions in minerals, thermal springs and fumaroles, and the mineral assemblage of the deposit and its associated alteration halos.

1. Fluid inclusions in minerals. Very small amounts of fluid are trapped in minute fluid-filled inclusions during the growth of many ore and gangue minerals in veins, and these inclusions have been studied intensively for evidence of temperature and composition (F. G. Smith, 1953). Although the relative amounts may vary widely, these fluids will have 5–25 or even more weight percent soluble salts, such as chlorides of Na, K, and Ca, plus highly variable amounts of carbonate, sulfate, and other anions. Some show liquid CO_2 or hydrocarbons as separate phases in addition to the aqueous solution. A few show detectable amounts of H_2S and minor amounts of many other substances. After losing some CO_2 and H_2S through release of pressure and oxidation when the inclusions are opened, the solutions are within 2 or 3 pH units of neutral. There is little evidence of sizable quantities (> 1 g/liter) of the ore metals in these

solutions, and the evidence indicates that the concentrations of the ore elements must generally be very low (< 0.1 g/liter). Even if the concentrations were in the range of 0.1 g/liter, there should be analytical evidence in the fluid inclusion studies, but this is lacking. In addition, if fluids of such composition were trapped in fluid inclusions in transparent minerals and on cooling precipitated even a fraction of their metal content as opaque sulfides, these should be visible (under the microscope) within the inclusions, but none are seen. If the concentrations of ore elements are much less than 0.001 g/liter, the volume of fluids that must be moved through a vein to form an ore body becomes geologically improbable.

2. Thermal springs and fumaroles. These provide the closest approach to a direct look at the processes of ore deposition as some ore and gangue minerals form within the range of direct observation. The solutions from these springs give diluted and possibly contaminated, partly oxidized and partly devolatilized samples of the sort of fluid that presumably forms ore bodies at greater depths. Isotopic studies show that the solutions have been diluted by local meteoric water until less than 5% (if any) of the fluid emitted at the surface is of deep-seated origin. The compositions of these thermal springs, after correction for such dilution, are in good agreement with the data from fluid inclusions.

3. Mineral assemblage. The assemblage of minerals that occurs within a deposit provides a great deal of information about the chemical nature of the fluid from which the ores were precipitated. There are a great number of stable inorganic compounds of the heavy metals known, yet unaltered ore deposits contain only a relatively small number of minerals. For example, lead fluoride, lead chloride, lead carbonate, lead sulfate, lead oxide, lead sulfide, and many others are known stable compounds of lead, yet of these, primary ore deposits contain only the sulfide (galena). Some elements, such as calcium, which occur in combination with several types of anions, for example, the carbonate, fluoride, sulfate, and numerous silicates, are found with the ore minerals. A quantitative approach to the compositional problem may be made by considering such reactions as shown in Eq. (1).

$$CaCO_3 + 2F^- = CaF_2 + CO_3^{--} \quad (1)$$

The equilibrium constant for this reaction is $(CO_3^{--})/(F^-)^2 = 10^{1.4}$ at 25°C. Thus when calcite and fluorite are in equilibrium, the requirements for the constant are met, and the $(CO_3^{--})/(F^-)^2$ ratio is known. A large number of such equations can be evaluated and from comparison with the mineral assemblage known to occur in ores, limits on the possible variation of the composition of the ore-forming fluid may be estimated. Unfortunately, calculations of this sort involving ionic equilibria are limited to fairly low temperatures (less than 100–200°C) since there are few reliable thermodynamic data on ionic species at high temperature. At any temperature, reactions such as shown in Eq. (2), can be used to evaluate or place limits on

$$2Ag + \tfrac{1}{2}S_2 = Ag_2S \quad (2)$$

the possible variation of the chemical potential of some components in the ore-forming fluid.

The composition of the ore fluid tends to become adjusted chemically by interaction with the rocks with which it comes in contact, and these changes may well contribute to the precipitation of the ore minerals. Thus, the K^+/H^+ ratio may be controlled by such reactions as Eq. (3), where the equilibrium

$$\underset{\text{Muscovite}}{4KAl_2(AlSi_3)O_{10}(OH)_2} + 6H_2O + 4H^+ = \underset{\text{Kaolin}}{6Al_2(Si_2O_5)(OH)_4} + 4K^+ \quad (3)$$

constant has the form shown in Eq. (4). Likewise,

$$K = \frac{(K^+)^4}{(H_2O)^6 \cdot (H^+)^4} \quad (4)$$

the quantitatively small but nevertheless important partial pressures of sulfur and oxygen may be governed by such reactions as Eq. (5). Such

$$Fe_3O_4 + S_2 = Fe_2O_3 + FeS_2 + \tfrac{1}{2}O_2 \quad (5)$$

changes in the wall rock come under the general heading of wall-rock alteration and may be of many types, only a few of the more common of which are mentioned below.

High-temperature alteration of limestones usually results in the formation of water-poor calcium silicates, such as garnet, pyroxenes, idocrase, and tremolite, and the resulting rock is termed skarn. At lower temperatures in the same types of rock, dolomitization and silicification are the predominant forms of alteration, because the partial pressure of CO_2 is too high to permit calcium silicate to form.

At high temperatures in igneous and metamorphic rocks near granite in composition, the solutions are approximately in equilibrium with the primary rock-forming minerals, and thus there is little alteration except development of sericite and occasionally topaz and tourmaline. At lower temperatures, the characteristic sequence of alteration from fresh rock toward the vein is first an argillic zone, then a sericitic zone, and finally a silicified zone bordering the vein.

Summary. Summarizing the environment of ore deposition, there are various lines of evidence to show that most hydrothermal ore deposits were formed at temperatures of 100–600°C and at pressures ranging from nearly atmospheric to several thousand atmospheres. The solutions were dominantly aqueous and were fairly concentrated in sodium chloride and potassium chloride; however, they were relatively dilute in terms of the ore metals.

Mechanisms of ore transport and deposition. The ore minerals, principally the sulfides, are extremely insoluble in pure water at high temperatures as well as low; the solubility products are so low, in fact, that literally oceans of water would be required to transport the metal for even a small ore body. Thus, it is not easy to explain the mechanism whereby the minerals are solubilized to the extent necessary for ore transport.

Crystals of ore and gangue minerals frequently exhibit evidence of repeated partial re-solution (or leaching) and regrowth. This demonstrates that the process of ore formation may, in at least in some instances, be reversible. In such cases studies of artificial systems at equilibrium are applicable.

This re-solution of ore minerals is important in another connection. Some geologists have advocated colloidal solutions, or sols, as an alternative to true solutions for ore transport. This was based on the belief, now known to be generally false, that colloform textures in ore minerals are a result of original deposition as a colloidal gel. Colloidal solutions were attractive also because they permitted ore metal concentrations – even in the presence of sulfide – many orders of magnitude higher than true solutions. The re-solution of ore minerals precludes the process of colloidal ore transport, as colloidal solutions are supersaturated and therefore cannot redissolve a crystal of the dispersed phase.

In addition to the fact that the absolute solubilities, calculated from the solubility products, are extremely low, the relative solubilities of the sulfides are radically different. For example, according to the solubility products, FeS is many, many times more soluble than PbS (about 10^{10} times at 25°C), yet the two minerals occur together in ore deposits and behave as if galena were slightly more soluble than pyrrhotite. From this and other lines of evidence, it appears necessary to conclude that the solubilities of the various contemporaneous minerals in a given deposit could not have differed among themselves by more than a few orders of magnitude.

The only geologically and chemically feasible mechanism by which these solubilities may be equalized approximately is the formation of complex ions of the heavy metals. Such complexes can increase the solubilities of heavy metals tremendously. As an example, the activity (thermodynamic concentration) of Hg^{++} in a solution saturated with HgS (cinnabar) and H_2S at 25°C, 1 atm pressure, and pH 8, is only about 10^{-47} mole/liter, representing a concentration much less than 1 atom of mercury in a volume of water equal to the entire volume of the oceans of the world. However, in the same solution is formed a very stable sulfide complex of mercury, HgS_2^{--}, which increases the total concentration of mercury in solution by the impressive factor of about 10^{42}, giving a concentration on the order of 0.001 g/liter. Not only does complex formation provide a means to achieve adequate solubility for ore transport, but the relative tendency for metals to form certain types of complexes matches in detail the commonly observed zoning and paragenetic sequences mentioned previously. The metals whose sulfides are the least soluble tend to form the most stable complexes, and metals whose minerals are comparatively soluble form weaker complexes.

There are many kinds of complexing ions or molecules (ligands) of possible geologic importance; a few of the more significant are sulfide (S^{--}), hydrosulfide (HS^-), chloride (Cl^-), polysulfides (S_x^{--}), thiosulfate ($S_2O_3^{--}$), sulfate (SO_4^{--}), and carbonate (CO_3^{--}), with the first three being most frequently considered. One of the major unsolved problems concerns the behavior of sulfur: What is its oxidation state and concentration relative to metals? If solutions were rich in reduced sulfur species, then the sulfide or hydrosulfide complexes would be dominant. On the other hand, solutions poor in reduced sulfur may transport the metals as chloride complexes.

The precipitation of minerals from complexed solutions takes place either by shifts in equilibrium caused by changing (usually cooling) temperature or by a decrease in the concentration of the ligand, thereby reducing the ability of the solution to carry the metals. This latter alternative can take place in several ways, as by reaction with wall rock, by mixing with other solutions, or by formation of a gas phase through loss of pressure.

Oxidation and secondary enrichment. When ore deposits are exposed at the surface, they are placed in an environment quite different from that in which they were formed, and the character of the deposit is changed through the processes of oxidation and weathering. The sulfides give way to oxides, sulfates, carbonates, and other compounds which are more or less soluble and tend to be leached away, leaving a barren gossan of insoluble siliceous iron and manganese oxides. Some minerals, such as cassiterite and native gold, may leach away at a less rapid rate than does the surrounding material; thus they are concentrated as a surficial residuum.

Where the country rock is relatively inert to the acid solutions generated by the oxidizing sulfides, as in the case of quartzites and some hydrothermally altered rocks, copper and especially zinc are leached away readily; lead and silver may be retained temporarily in the oxidized zone as the carbonate or sulfate, and the chloride or native metal, respectively; but eventually these too are dissolved away. The various metallic ions are carried downward until they reach unoxidized sulfides in the vicinity of the water table, where the solutions interact with these sulfides to form a new series of supergene sulfide minerals. Copper sulfide is the least soluble sulfide of the base and ferrous metals in the solution, and hence the zone of supergene sulfide enrichment is predominantly a copper sulfide zone with occasional rich concentrations of silver. Zinc nearly always remains in solution and is lost in the groundwater.

In reactive wall rocks, such as limestones, reaction with the wall rock prevents the solutions from becoming acid enough for large amounts of metal to be removed in solution; the base metals are retained almost in place as carbonates, sulfates, oxides, and halides, and there is no appreciable sulfide enrichment.

The behavior of some elements is governed by the availability of other materials. Thus, for example, uranium is readily leached from the oxidized zone in many deposits; however, when the oxidizing solutions contain even very small amounts of potassium vanadate, the extremely insoluble mineral carnotite precipitates and uranium is immobilized. Highly soluble materials, such as uranium in the absence of chemicals that precipitate it, may be temporarily fixed in the oxidized zone by adsorption on colloidal materials such as freshly precipitated ferric oxides.

Trends in investigation. There has been a great increase in the degree to which the experimental methods and principles of physical chemistry have been applied to aid in understanding the processes by which ores have formed, and this approach can be expected to be even more fruitful in the future. Several avenues appear promising and are under active investigation in numerous laboratories.

Among these are the following studies:

1. Phase equilibrium studies of both natural and synthetic ore and gangue minerals.
2. Distribution coefficients for trace elements between coexisting phases, and between various forms on the same crystal.
3. Experimental solubility studies in dominantly aqueous solutions.
4. Studies of the composition and origin of thermal spring waters and fluid inclusions in minerals.
5. Thermodynamic properties of minerals.
6. Isotopic fractionation during transportation and deposition processes.
7. Rate studies on crystal growth, habit, diffusion, reaction, and transformation, as well as studies of sluggish homogeneous reactions, such as the reduction of sulfate.
8. Crystal structure determinations and crystal chemical studies of ore and gangue minerals.
9. Distribution of elements in the Earth's crust and in various rock types.
10. Detailed field studies of the relations between minerals in ore deposits.

For a discussion of sensitive chemical analytical techniques used in the search for ore deposits *see* GEOCHEMICAL PROSPECTING. For further discussion of chemical principles involved in ore deposition *see* GEOLOGIC THERMOMETRY; LEAD ISOTOPES, GEOCHEMISTRY OF; LITHOSPHERE, GEOCHEMISTRY OF.

[PAUL B. BARTON, JR.; EDWIN ROEDDER]

Bibliography: H. L. Barnes (ed.), *Geochemistry of Hydrothermal Ore Deposits*, 1967; P. B. Barton, Jr., P. M. Bethke, and P. Toulmin, III, *Equilibrium in Ore Deposits*, Mineral. Soc. Amer. Spec. Pap. no. 1, pp. 171–185, 1963; R. M. Garrels and C. L. Christ, *Solutions, Minerals, and Equilibria*, 1965; H. C. Helgeson, *Complexing and Hydrothermal Ore Deposition*, 1964; H. D. Holland, Some applications of thermochemical data to problems of ore deposits, *Econ. Geol.*, 54:184–233, 1959, and 60:1101–1166, 1965; E. Roedder, Ancient fluids in crystals, *Sci. Amer.*, 207:38–47, 1962.

Orogeny

The process by which mountainous tracts such as the Alpine-Himalayan, Appalachian, and Cordilleran orogenic belts are formed, also known as orogenesis. Characteristically, orogenic belts are long and linear or arcuate, with distinctive zones of sedimentary, deformational, and thermal patterns that are, in general, parallel but asymmetric to the belts. Orogenic belts have complex internal geometry, involving extensive mass transport of very dissimilar rock sequences of dominantly marine sediments and continental crust (sial) and oceanic crust and mantle (ophiolite suite). Intensive deformation and metamorphism of a particular orogenic belt is relatively short-lived when compared to the time during which much of the sedimentary rock of the belt was deposited. Typically, orogenic belts are sites of abnormally thick accumulations of sedimentary and volcanic rock and severe deformation and thermal alterations.

Mountain belts are objects of intense study, argument, and mystery for geologists. Their observable features, entirely land-based, are exceedingly complex, primarily because of severe deformation; their internal features for the most part have been interpreted from surface data, although over the past several decades seismological studies have provided much new information about the interior of orogenic belts. However, it is now apparent that results of ocean-based study of the Earth provide fundamental keys to understanding orogeny.

Seismological and oceanographic studies during the past several decades have revealed a globe-encircling, seismically active belt of mountains, called oceanic ridges, and trenches with associated volcanic island arcs; this oceanic ridge and trench system is fundamentally different from orogenic belts as perceived by Gilbert. Models propose that orogeny is a consequence of the evolution of oceanic ridges and trenches and continental drift. Although oceanic ridges and trench–island arc systems are mountainous, they are not orogenic belts in the classical sense which restricts orogenic belts to continental masses. Orogeny results from interactions of this global, continuously evolving system of oceanic ridges and trenches, according to concepts of sea-floor spreading or, currently, lithosphere plate tectonics.

Lithosphere plate tectonics. Lithosphere plates are spherical segments of upper mantle and crust, varying in thickness from approximately 5 km at ridges to 150 km under central areas of continents, that are generated by growth of crust and mantle at oceanic ridges (accreting plate margins) and consumed in trenches (consuming plate margins or subduction zones). Plates are generated at accreting plate margins such as the Mid-Atlantic Ridge between the American and African plates (Fig. 1*a*), and consumed in subduction zones such as the Peru-Chile Trench, just west of the Andes, and the Japan Trench (Fig. 1*b*). Island arcs (unpatterned in Fig. 1) develop above subduction zones within entirely oceanic portions of plates. Various combinations of subduction zones, island arcs, and microcontinents, such as the Lord Howe Rise, are shown in Fig. 1*c*–*f*. The collision of the Indian continent with the Tibetan continental mass, with the resulting termination of subduction and development of the Himalayan collisional mountain belt, is shown in Fig. 1*g*.

Plates move symmetrically away from ridges, over the low-velocity channel of the mantle, with rates ranging from about 1 to 10 cm/year. They behave as more or less rigid masses; their motions are described by poles of rotation with respect to one another. Continental masses such as Africa and Australia are merely passive passengers on evolving plates; continental drift is prescribed by plate evolution. Vectors of motion (magnitude and direction) of plates are determined from F. J. Vine and D. H. Matthews's interpretation of oceanic magnetic anomaly patterns about oceanic ridges. Ocean evolution, such as Atlantic opening and Mediterranean (Tethyan) closing of the past approximately 200,000,000 years, results from plate evolution.

Fundamental concepts of orogeny based on plate tectonics are that continental margins of the Atlantic type, which develop during ocean opening, are converted to island arc–cordilleran-type (Andean) orogenic belts as lithosphere is consumed in subduction zones, below trenches, along the continental margin during closing of an ocean. A collision-type (Himalayan) orogenic belt is superimposed on a cordilleran belt upon final ocean

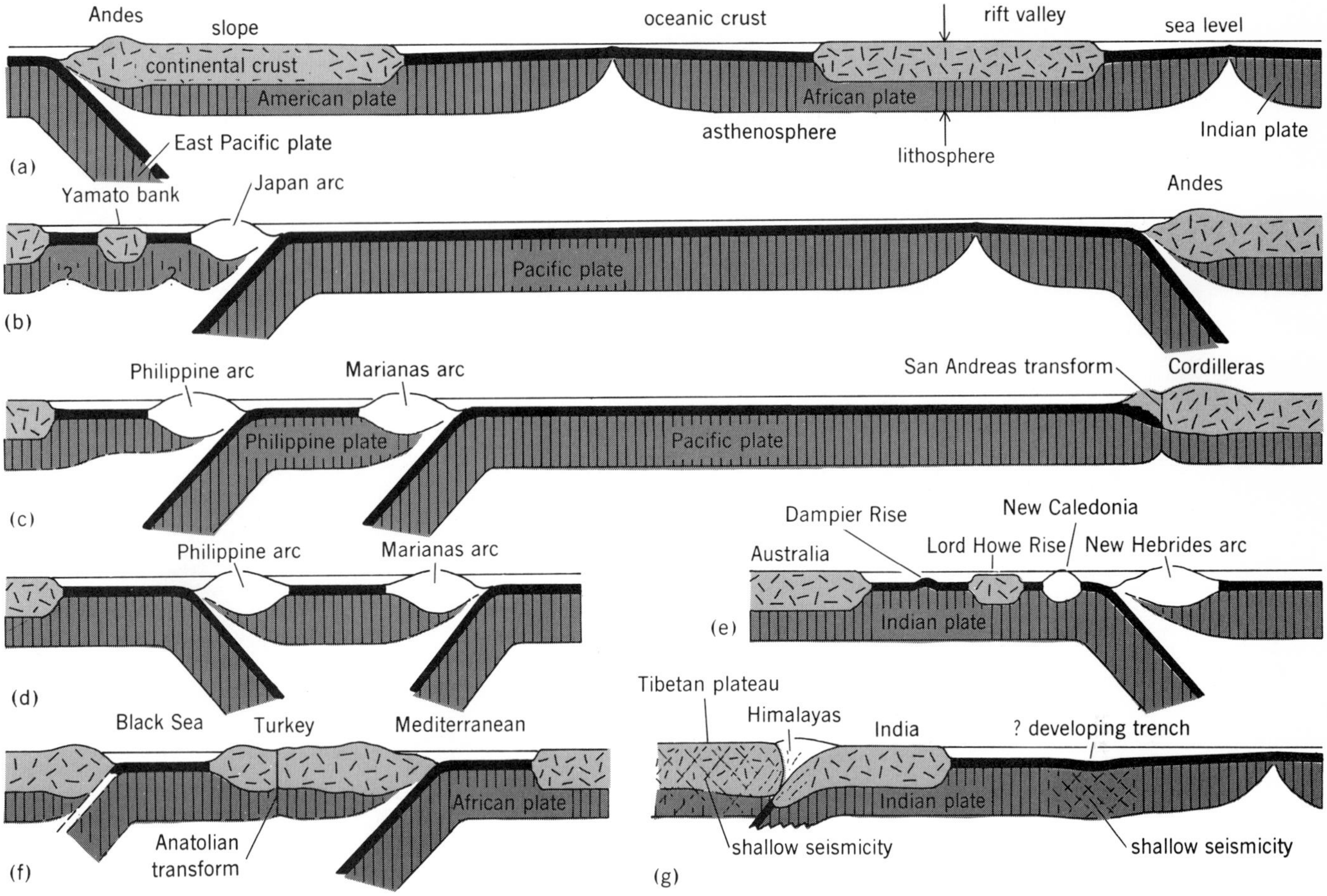

Fig. 1. (*a–g*) Schematic sections showing world lithosphere plate, ocean, continent, and island arc relationships. (*From J. F. Dewey and J. M. Bird, Mountain belts and the new global tectonics, J. Geophys. Res., 75:2625–2647, copyright 1970 by American Geophysical Union*)

closing, as opposing continental margins suture. *See* PLATE TECTONICS.

Ocean sediment accumulation. Fundamental to this mechanism of orogeny is the evolution of the whole assemblage of continental margin–deep oceanic sediments and new oceanic crust of an opening ocean. The Atlantic is taken as typical and is shown schematically in Fig. 2. Here lithosphere is being generated at the oceanic ridge and moves symmetrically away from the ridge axis. Upon opening of an ocean, overlying continental crust separates and, as it is carried away from the ridge, its trailing edges become continental margins of the continuously opening ocean and the sites of the bulk of sedimentation in the ocean.

Essentially, the sediment accumulation in the ocean is synchronous with plate accretion at the ridge. After initial rupture and distension of conti-

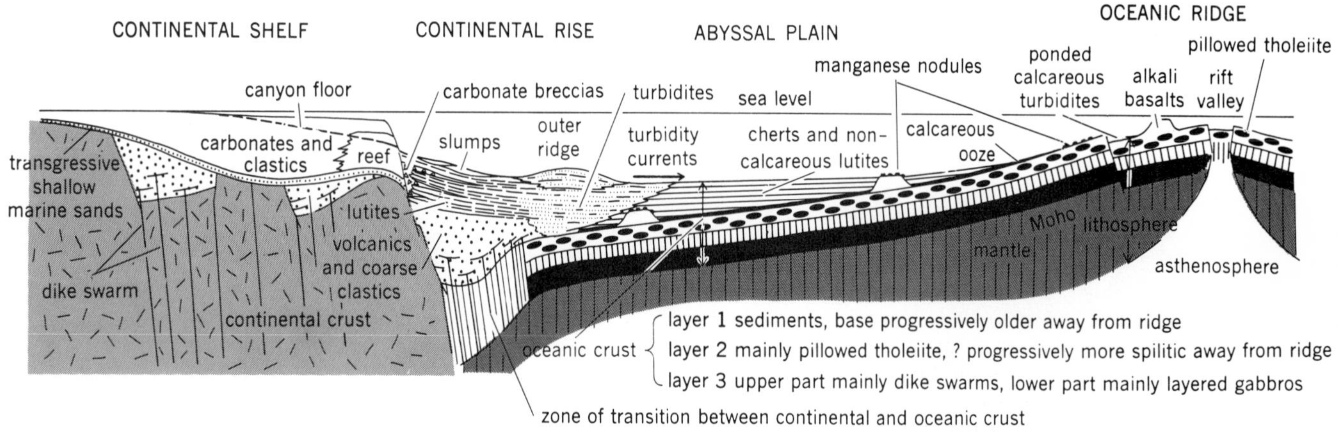

Fig. 2. Schematic section of Atlantic-type half-ocean, relationships of continental and oceanic crust, and sediments. (*From J. F. Dewey and J. M. Bird, Mountain Belts and the new global tectonics, J. Geophys. Res., 75:2625–2647, copyright 1970 by American Geophysical Union*)

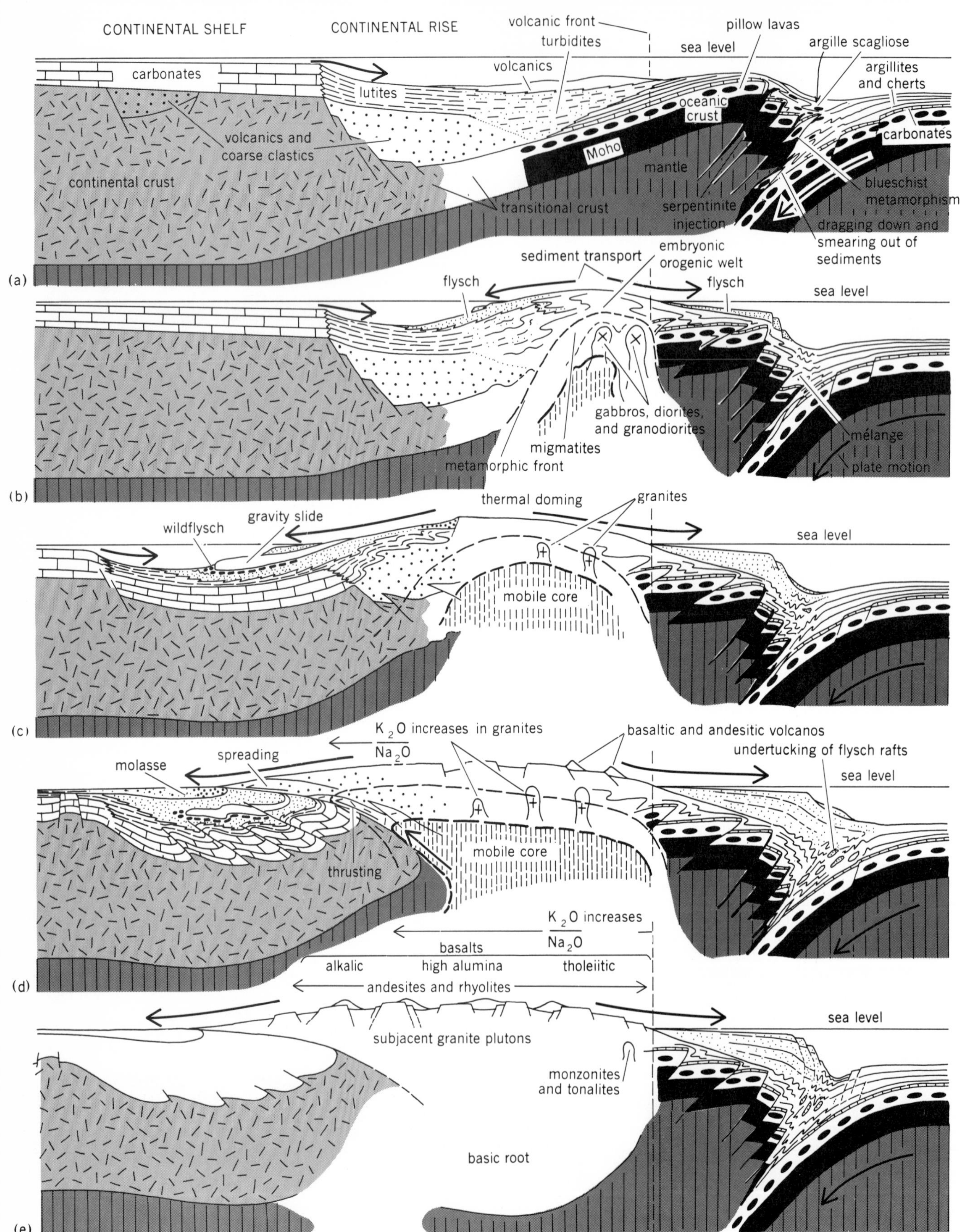

Fig. 3. (*a–e*) Schematic sections showing conversion of continental margin to cordilleran-type mountain belt by the progressive subduction of lithosphere. (*From J. F. Dewey and J. M. Bird, Mountain belts and the new global tectonics, J. Geophys. Res., 75:2625–2647, copyright 1970 by American Geophysical Union*)

nental crust over an evolving accreting plate margin (the Red Sea is an example), the continental margins are stabilized as the ocean widens. Vast amounts of sediment may accumulate along the continent-ocean interface as the ocean becomes large. If a consuming plate margin (subduction zone) develops along the continental prism of sediment (as has happened along the west coast of the Americas), lithosphere plate is consumed, the ocean commences to close, vulcanism develops above the subduction zone, and the continental margin is converted into a cordilleran-type orogenic belt. This is shown schematically in Fig. 3; lithosphere is consumed in a subduction zone as the ocean closes. If the subduction zone develops near the continental margin, as shown in Fig. 3*a*, the margin is progressively converted to an orogenic belt, dominantly by thermally driven processes originating in the subduction zone. Schematic sections of this progressive conversion that is synchronous with lithosphere consumption follow in Fig. 3*b* through *e*. The trench, above the subduction zone, is the site of sedimentation of debris from the orogenic belt; high pressure–low temperature deformation occurs in the upper few kilometers of the subduction zone. Sedimentation and synchronous deformation occur on the continent-ward side of the orogenic belt.

Marginal basins such as the Philippine Sea may develop behind the trench by several mechanisms of diffuse, inter-arc sea-floor spreading which complicate time-space relationships of sedimentation, orogeny, and island arc vulcanism. Furthermore, depending on the direction of dip of the subduction zone, island arcs may collide with Atlantic-type margins. Several schematic relationships are shown in Fig. 4. If ocean closing commences with the lithosphere plate consumption direction as shown in Fig. 4*a*, in contrast to that shown in Fig. 3*a*, the resulting island arc and associated trench sediments and the continent, with its associated continental margin assemblage of sediments, approach one another by the consumption of the intervening ocean. Sediments fill the small ocean remaining just before collision and then are incorporated into the deformed belt resulting from final suture (Fig. 4*c*). However, because the main ocean continues to close, a "flip" in plate descent direction occurs (Fig. 4*d*), and continued cordilleran-type orogeny is superimposed on the preexisting belt. As lithosphere is consumed in a closing ocean, the opposing continental margin will arrive at the island arc–cordilleran belt, resulting in a collision (Fig. 5). This results in a Himalayan-type orogeny, and because continental crust is less dense than mantle and therefore not significantly consumed in the subduction zone, plate consumption ceases (Fig. 1*g*). If plate consumption occurs on only one side of a closing ocean (Fig. 3), the trench-bearing margin may be associated with an existing or developing cordilleran-type orogen or a margin resulting from an island arc–continent collision (Fig. 4). The opposite continental margin will arrive at this margin as the ocean finally closes. The resulting collision, primarily mechanically driven, superimposes an added orogenic "event" on the margin. Ultimately the ocean disappears, the margins are sutured, and due to buoyancy restraints of the continental crust, plate consumption ceases.

Fundamental mechanism. The plate tectonics model indicates that mountain building results from two fundamental mechanisms. Island arc–cordilleran orogeny, for the most part thermally driven, occurs along leading lithosphere plate edges, above subduction zones. Continent–island arc or continent-continent collision orogeny is for the most part mechanically driven and occurs subsequently to island arc development or during final closing of an ocean or both. Actual orogenic belts are usually the result of complex combinations of these basic mechanisms. Widely differing rates of consumption, vulcanism, and deformation along consuming plate margins result because of varying distances from the poles of rotations of the involved plates. Further complexities result from expanding and contracting transform offsets of consuming plate margins. Diachronous events, more or less normal to the plate margin, may migrate along the orogenic belt as the result of oblique intersection of triple junctions (for example, the southward migration of the East Pacific Rise intersection with the Americas Trench off California).

The foregoing concerns active orogenic belts such as the Alpine system. Older, extinct orogenic belts such as the Appalachians and Urals, by analysis of the time-space relations of their rock assemblages, also resulted from ocean evolution. Although they are presently within continental masses, they mark the site of preexisting oceans that resulted from plate evolution.

As yet, no satisfactory mechanism has been found to account for the driving forces of plates. However, the existence of the world's present system of lithosphere plates cannot be denied. The Joint Oceanographic Institutions for Deep Earth Sampling (JOIDES) results indicate that practically all of the oceanic crust of the Earth underlying

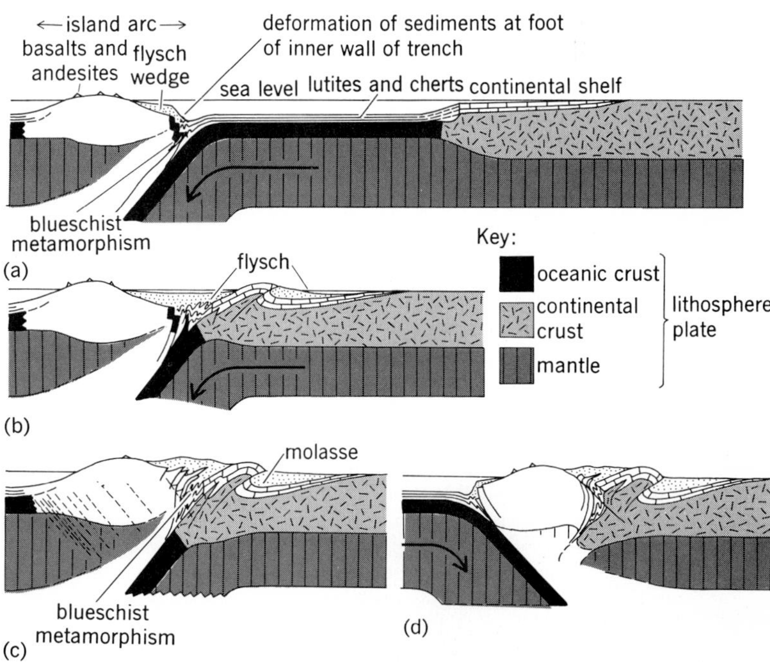

Fig. 4. (*a–d*) Schematic sections showing collision of Atlantic-type continental margin and island arc, followed by change in direction of lithosphere plate consumption. (*From J. F. Dewey and J. M. Bird, Mountain belts and the new global tectonics, J. Geophys. Res., 75:2625–2647, copyright 1970 by American Geophysical Union*)

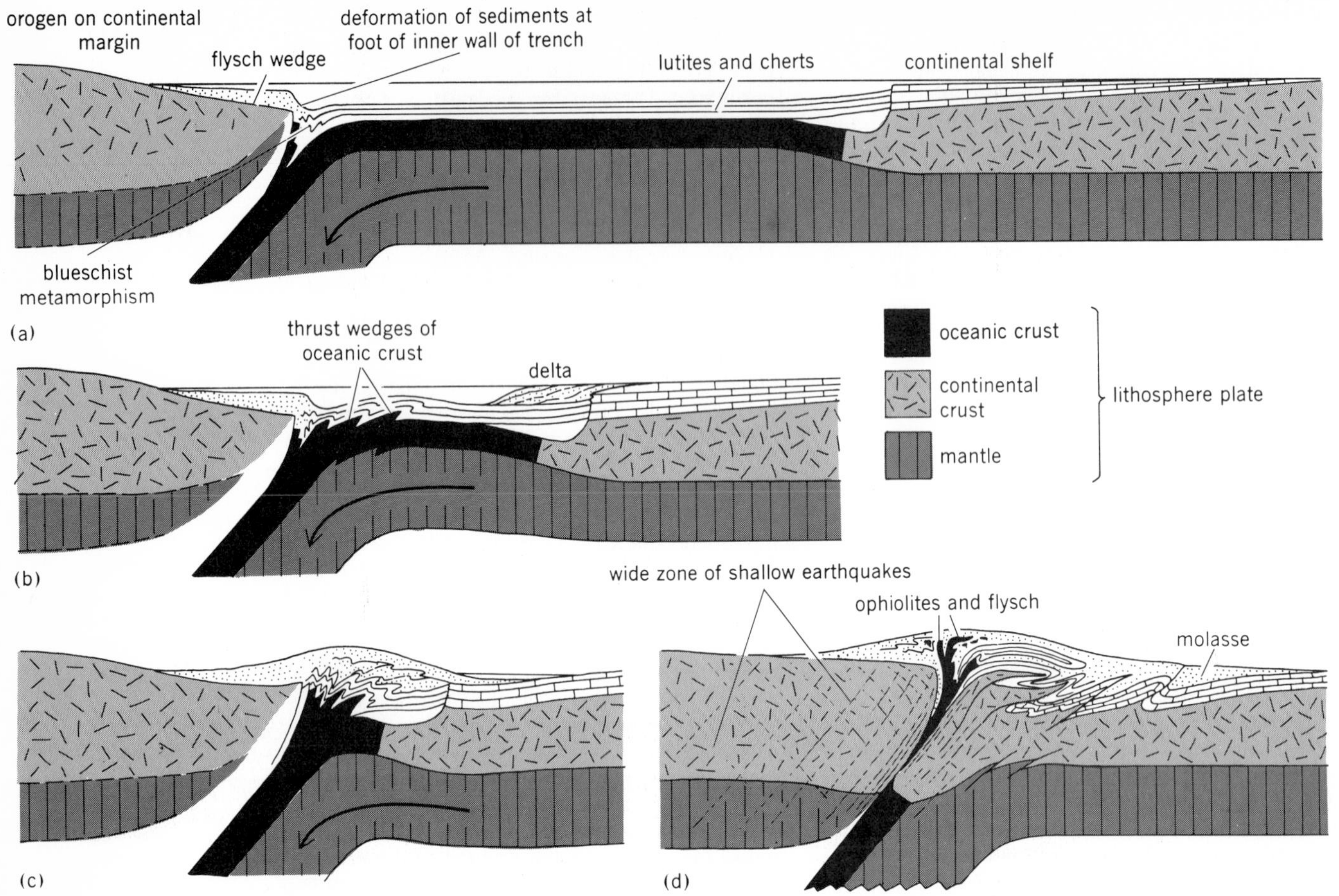

Fig. 5. (*a–d*) Schematic sections showing collision of two continental margins. (*From J. F. Dewey and J. M. Bird, Mountain belts and the new global tectonics, J. Geophys. Res., 75:2625–2647, copyright 1970 by American Geophysical Union*)

the oceans has been generated over the past 200,000,000 years (about 6% of geologic time) at the globe-encircling oceanic ridge system. All of the world's active orogenic belts and island arcs are along consuming plate margins, where oceanic crust is being consumed in subduction zones at rates commensurate with rates of accretion.

Therefore, either plates now in existence evolved from about Permian times concurrently with about a 1.8/1.0 global expansion and no significant consumption, or plates existed in Paleozoic and even during some Precambrian time and that plate evolution has been the mechanism by which even ancient (Precambrian) linear orogenic belts were formed. Analysis of pre-Permian orogenic belts such as the Appalachians and Urals strongly indicate they were the consequence of plate evolution. Plate tectonics models have not yet been applied to Precambrian (pre-600,000,000 years before present) orogenic belts. However, initial study suggests that plates have existed on the Earth at least to about 2,000,000,000 years before present and that their evolution is independent of any factor of global expansion or contraction. *See* CONTINENT FORMATION; CORDILLERAN BELT; OCEANIC ISLANDS. [JOHN M. BIRD; JOHN F. DEWEY]

Bibliography: J. M. Bird and J. F. Dewey, *Bull. Geol. Soc. Amer.*, 81:1031–1060, 1970; J. F. Dewey and J. M. Bird, *J. Geophys. Res.*, 75:2625–2647, 1970; J. F. Dewey and J. M. Bird, *J. Geophys. Res.*, 76:3179–3206, 1971; W. R. Dickinson, *Rev. Geophys. Space Phys.*, 8:813–851, 1970; W. Hamilton, *Bull. Geol. Soc. Amer.*, 81:2553–2576, 1970; B. J. Isacks, J. Oliver, and L. R. Sykes, *J. Geophys. Res.*, 73:5855–5900, 1968.

Orpiment

A mineral having composition As_2S_3 and crystallizing in the monoclinic system. Crystals are small, tabular, and rarely distinct; the mineral occurs

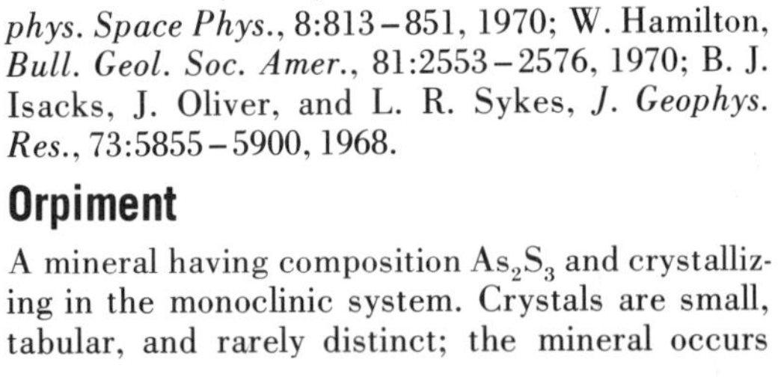

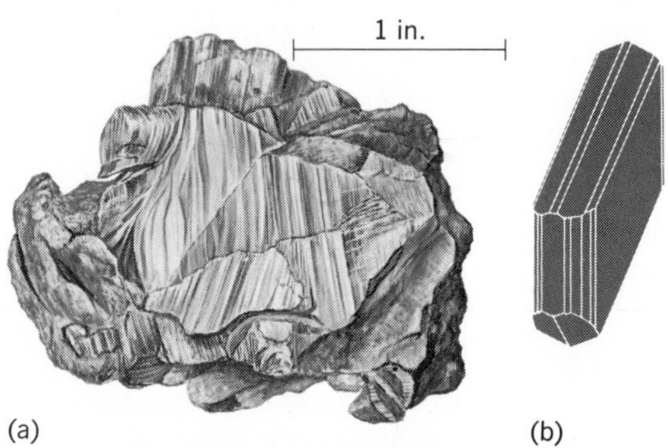

Orpiment. (*a*) Foliated masses found at Mercer, Utah (*specimen from Department of Geology, Bryn Mawr College*). (*b*) Crystal habit (*From C. S. Hurlbut, Jr., Dana's Manual of Mineralogy, 17th ed., copyright © 1959 by John Wiley & Sons, Inc.; reprinted by permission*).

more commonly in foliated or columnar masses (see illustration). There is one perfect cleavage yielding flexible folia which distinguishes it from other minerals similar in appearance. The hardness is 1.5–2 (Mohs scale) and the specific gravity is 3.49. The luster is resinous and pearly on the cleavage surface; the color is lemon yellow. Orpiment is associated with realgar and stibnite in veins of lead, silver, and gold ores. It is found in Romania, Peru, Japan, and the Soviet Union. In the United States it occurs at Mercer, Utah; Manhattan, Nev.; and in deposits from geyser waters in Yellowstone National Park. *See* REALGAR; STIBNITE.

[CORNELIUS S. HURLBUT, JR.]

Orthoclase

A name for potassium feldspar, $KAlSi_3O_8$, that usually contains some sodium feldspar (up to about 50 mole % $NaAlSi_3O_8$) either in solid solution or exsolved as relatively pure $NaAlSi_3O_8$. The latter can be present as albite or analbite or in structural states intermediate between albite and analbite. If exsolution is detectable, such material is called cryptoperthite, microperthite, or perthite according to increasing size of the exsolved areas.

Orthoclase. (*a*) Twinned crystal from Gunnison County, Colo. (*American Museum of Natural History specimens*). (*b*) Crystal habits (*from C. S. Hurlbut, Jr., Dana's Manual of Mineralogy, 17th ed., copyright © 1959 by John Wiley & Sons, Inc.; reprinted by permission*).

The symmetry of orthoclase may be truly monoclinic (sanidine) or can only appear to be monoclinic (normal orthoclase) according to the Al-Si distribution within the $AlSi_3O_8$ framework. The illustration shows a twinned crystal and examples of crystal habits. *See* ALBITE; FELDSPAR; PERTHITE.

[FRITZ H. LAVES]

Orthoquartzite

A rock (also known as quartzose sandstone) composed almost entirely of detrital quartz grains; it is generally considered that over 95% of the detrital grains must be quartz for it to fall in this group. Orthoquartzite is a sedimentary rock distinct from metamorphic quartzite (metaquartzite, a metamorphic rock formed at high temperature or pressure).

Mineral composition. Any clay that is present in orthoquartzites is insignificant in amount. Feldspar is either absent or present in very small amounts. Other unstable mineral grains are not common. The minor accessory minerals are only the most stable ones, zircon and tourmaline. The texture of orthoquartzites, aside from the lack of clay matrix, tends to be distinctive. Typically, the grains are extremely well rounded and tend to have high sphericity as well. The sorting of the various sizes is very good. Some of the best-sorted sands that have been described belong in this category. Any rock fragments are almost invariably chert.

The orthoquartzites are dominantly cross-bedded, either of windblown or subaqueous type, and ripple-marked. Many orthoquartzite beds are massive, that is, without many bedding planes. The local variation in mineral composition and texture laterally (along beds) and vertically (from bed to bed) is small. Regionally, orthoquartzite formations tend to be homogeneous, showing only small changes in heavy mineral suites that may be due to contributions from different source areas.

Mineral cements. The particles of orthoquartzites are bound together by precipitated mineral cements. The most abundant mineral cement is quartz. Normally quartz cement occurs as secondary overgrowths on detrital grains, the overgrowth being deposited as a single crystal, optically continuous with the detrital grain crystal. The overgrowths may show euhedral crystal faces if they have grown into open pore spaces. Those sandstones that are so completely cemented by secondary quartz that practically no pores remain show boundaries between adjacent overgrowths that are irregular as the result of interference during crystal growth. In some orthoquartzites the silica cement is chert or opal. Not all orthoquartzites are well cemented; many are friable and almost unindurated.

Carbonate minerals are important as cementing agents in orthoquartzites. Calcite and dolomite are most abundant but siderite, iron-rich dolomite, and aragonite may also be present. Carbonate cement frequently has replaced some detrital quartz; that is, some of the original detrital quartz has been dissolved and carbonate precipitated in its place. In some orthoquartzites the carbonate cement is so abundant that quartz grains do not touch each other and appear to float in the mineral cement. This may happen as a result of extensive replacement of detrital quartz or as a result of original sedimentation conditions under which a mixture of quartz grains and carbonate grains settled together, the carbonate grains later having lost their identity by recrystallization.

Occurrence. Typically, orthoquartzites are associated with fossiliferous limestone and calcareous shale beds. Orthoquartzites are most abundant in thin sedimentary sections, primarily in the interior of continents, but may also be found as the early deposits in some geosynclines. They range in age from Precambrian to Tertiary but appear to be more abundant in the Precambrian and Paleozoic than in the Mesozoic and Tertiary.

Origin. The chief inference from the mineralogical composition (all quartz) and texture (well-sorted, well-rounded) is that the orthoquartzites represent material that has been subjected to a great deal of chemical weathering at the source and to abrasion and sorting during transportation and deposition. One way by which material of this composition and texture can be produced is by the slow erosion of a low-lying, tectonically stable source area. Under these conditions chemical weathering has had sufficient time to eliminate all unstable minerals, and extensive abrasion and sorting by shoreline processes would give a well-sorted, well-rounded sandstone. A second way in which orthoquartzites can be produced is by repeated reworking of older sediments. As the sandstones go through several cycles of erosion, transportation, and deposition, they become progressively better sorted, and most unstable minerals are lost. Second-cycle quartz grains, grains that show abraded secondary quartz overgrowths, are evidence of derivation from preexisting sediments. *See* ARKOSE; GRAYWACKE; SANDSTONE; SEDIMENTARY ROCKS; SUBGRAYWACKE.

[RAYMOND SIEVER]

Orthorhombic pyroxene

A group of minerals having the general chemical formula $XYSi_2O_6$, in which the Y site contains Fe or Mg and the X site contains Fe, Mg, Mn, or a small amount of Ca (up to about 3%). They are characterized by orthorhombic crystal symmetry, which makes it impossible for significant amounts of larger cations such as Ca or Na to enter the X position. The most important compositional variation in the orthopyroxene series is variation in the Fe-Mg ratio, leading to a solid solution series bounded by the end members enstatite ($Mg_2Si_2O_6$) and ferrosilite ($Fe_2Si_2O_6$). Names used for intermediate members of the series are enstatite ($Fs_{0\text{-}10}$; Fs = ferrosilite molecule), bronzite ($Fs_{10\text{-}30}$), hypersthene ($Fs_{30\text{-}90}$), and orthoferrosilite ($Fs_{90\text{-}100}$). In addition to the major elements noted above, minor components in orthopyroxene may include Al, Ti, Cr, Ni, and Fe^{3+}, although only Al occurs in substantial amounts. Conditions of formation may be important in determining orthopyroxene compositions: for example, Al content is greater at higher temperatures, and the most Fe-rich compositions are stable only at very high pressures. Further, there is a limited solid solution of orthopyroxene toward calcic clinopyroxene, noted above, which increases with increasing temperature.

Crystal structure. Orthopyroxenes crystallize in a primitive orthorhombic cell, space group *Pbca* (a high-temperature Mg-rich form called protoenstatite has slightly different symmetry and is in space group *Pbcn*). The *b* and *c* unit-cell dimensions are about the same as for monoclinic pyroxene (such as diopside), but the *a* dimension is doubled. This suggests that the absence of large cations in orthopyroxene causes a slight structural shift so that the smallest repeated unit contains twice the number of atoms as in a monoclinic pyroxene unit cell. The crystal chemistry is otherwise similar to that of diopside, with single chains of (SiO_4) tetrahedra alternating in the structure with strips of octahedral sites that accommodate larger cations. As in monoclinic pyroxene, Al is small enough to enter both octahedral and tetrahedral sites. *See* DIOPSIDE.

Properties. Many of the physical and optical properties of orthopyroxene are strongly dependent upon composition, and especially upon the Fe-Mg ratio. Enstatite is commonly pale brown in color; the color becomes much darker for more Fe-rich members of the series. The specific gravity and refractive index both increase markedly with increasing Fe content. The common variety bronzite is recognized by its characteristic bronze iridescence, which is due either to very fine exsolved lamellae of high-Ca monoclinic pyroxene which diffract light rays (an effect similar to the iridescence of labradorite feldspar) or to very thin oriented plates of ilmenite, an iron-titanium oxide. In hand specimens, orthopyroxene can be distinguished from amphibole by its characteristic 88° cleavage angles, and from augite by color—augite is typically green to black, while orthopyroxene is more commonly brown, especially on slightly weathered surfaces. In rock thin sections, orthopyroxene is usually recognized by its parallel extinction and blue-green to pink pleochroism. *See* PETROGRAPHY.

Occurrence. Orthopyroxene is a widespread mineral in metamorphic rocks. It is characteristic of granulite facies metamorphism, in both mafic and feldspathic gneisses. In feldspathic rocks such as charnockite it occurs with garnet, biotite, augite, hornblende, alkali feldspars, and quartz, while in mafic gneisses it usually coexists with augite, hornblende, garnet, and calcic plagioclase. In very-high-grade metamorphic rocks, especially in contact metamorphism, orthopyroxene plus alkali feldspar results from the dehydration of biotite. Orthopyroxene is typical of metamorphosed ultramafic rocks, in which it coexists with olivine, spinel, Mg-rich chlorite, anthophyllite, and hornblende. In this occurrence it may represent dehydration of talc or amphibole with increasing temperature. Orthopyroxene is common in peridotites of the Earth's upper mantle (found as nodules in basalts and kimberlites), where it is accompanied by olivine, augite, and spinel or garnet.

Orthopyroxene occurs in many basalts and gabbros, particularly those of tholeiitic composition, and in many meteorites, but is notably absent from most alkaline igneous rocks. It is relatively common in intermediate volcanic and plutonic rocks such as andesite and diorite, and rarer in silicic rocks such as rhyolite and granite. Orthopyroxene is an essential constituent of the type of gabbro called norite which consists of only orthopyroxene and plagioclase. The greatest abundance of orthopyroxene is in ultramafic rocks, especially those in large layered intrusions such as the Bushveld Complex of southern Africa or the Stillwater Complex of Montana. In these areas, along with olivine and augite, it represents the earliest crystallized minerals from the magma, which were denser than the magma and settled to the bottom of the intrusion to form "cumulate" rocks. The typical orthopyroxene of this occurrence is bronzite. *See* PYROXENE.

[ROBERT J. TRACY]

Bibliography: N. L. Bowen, *The Evolution of the Igneous Rocks*, 1928; W. A. Deer, R. A. Howie, and J. Zussman, *Rock-forming Minerals*, vol. 2: *Chain Silicates*, 1963.

Paleobiochemistry

The study of chemical processes used by organisms that lived in the geological past. Most information on the nature of life in the geological past comes from the study of fossils; a record of biochemical processes that occurred can be found in the organic molecules of sedimentary rocks and fossils. The organic matter in fossil fuel deposits (coal, petroleum, and oil shale) and finely dispersed in shales and limestones represents the debris of cells which have been chemically altered to a more stable form. The relatively reactive organic constituents of cells are subject to bacterial and chemical transformation on the death of the organism. Most organic matter of living organisms is consumed in the metabolism of other organisms and returned to the seas and to the atmosphere. A small amount is deposited in the sediments, where it is preserved or converted to more stable entities. A comparison of the molecular structure of these preserved organic compounds with that of components of living cells enables the researcher to identify similarities and dissimilarities between past and present biochemistry. *See* PETROLEUM, ORIGIN OF.

Alkanes. Saturated hydrocarbons (alkanes) are among the most stable organic compounds and presumably represent transformation products of the lipid fraction of cells. Two major classes of alkanes in sedimentary rocks are the normal hydrocarbons with linear arrays of carbon-to-carbon bonds, and isoprenoid hydrocarbons with combinations of five-carbon branched-chain units (Fig. 1).

Normal and isoprenoid carbon chains are used preferentially in present-day lipids. There are many thousands of possible spatial arrangements (isomers) of alkanes ($C_{20}H_{42}$); yet these two structural types are the predominant ones both in living cells and in sedimentary rocks. Thus the evidence indicates that this mode of biochemical synthesis has been in use for over 1,500,000,000 years.

Tetrapyrrole compounds. Molecules with tetrapyrrole ring structures, such as chlorophyll (Fig. 2) and hemin, are used by all living organisms as either photosynthetic or respiratory pigments. In a sedimentary rock, such compounds are transformed into more stable metalloporphyrins and have been found in rocks as old as 1,100,000,000 years. The use of tetrapyrrole-containing molecules in energy-transformation processes obviously evolved very early in the history of terrestrial life.

Asymmetrical molecules. The ability to synthesize compounds with molecular asymmetry is a characteristic property of living organisms and is due to the use of enzymes as catalysts in biochemical processes. Such asymmetrical compounds display optical activity; that is, their solutions can rotate the plane of polarization of light. Optically active hydrocarbons have been detected in petroleums as old as 400,000,000 years. It is believed that the optically active preserved hydrocarbons are the steranes, which have been derived from steroids of cells (Fig. 3). If correct, this indicates that the use of enzymes and steroids is an ancient feature of terrestrial life.

Amino acids. The structural units of proteins are amino acids, of which there are about 25. While the proteins and most of the amino acids are not very stable in the geological environment, seven of the amino acids are stable, can persist for great lengths of time, and have been found in fossils several hundred million years old. These amino acids are alanine, glycine, glutamic acid, leucine, isoleucine, proline, and valine. If any of the other common amino acids, such as serine, threonine, and arginine, which are not very stable, were to be detected in a very ancient fossil, contamination by recent organic matter would be suspected. Such contamination is a serious problem in paleobiochemical studies.

Paleobiochemical studies have shown that a number of the common chemical processes used by living organisms today have been in use for a very great length of time. Paleobiochemical tech-

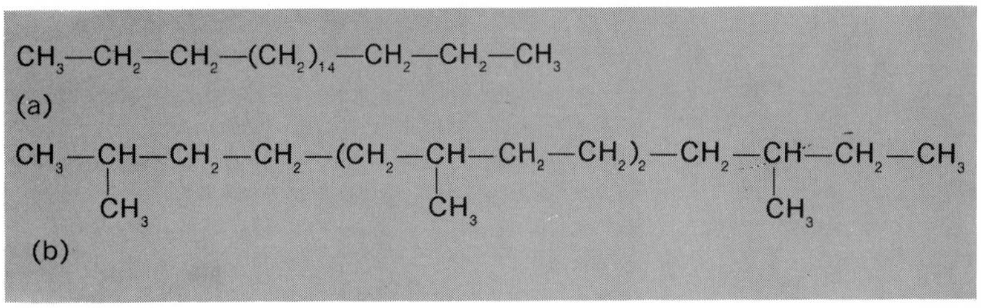

$$CH_3{-}CH_2{-}CH_2{-}(CH_2)_{14}{-}CH_2{-}CH_2{-}CH_3$$

(a)

$$CH_3{-}\underset{\displaystyle CH_3}{\underset{|}{CH}}{-}CH_2{-}CH_2{-}(CH_2{-}\underset{\displaystyle CH_3}{\underset{|}{CH}}{-}CH_2{-}CH_2)_2{-}CH_2{-}\underset{\displaystyle CH_3}{\underset{|}{CH}}{-}CH_2{-}CH_3$$

(b)

Fig. 1. Alkane isomers in sedimentary rock. (*a*) Normal alkane (*n*-eicosane). (*b*) Isoprenoid alkane (phytane).

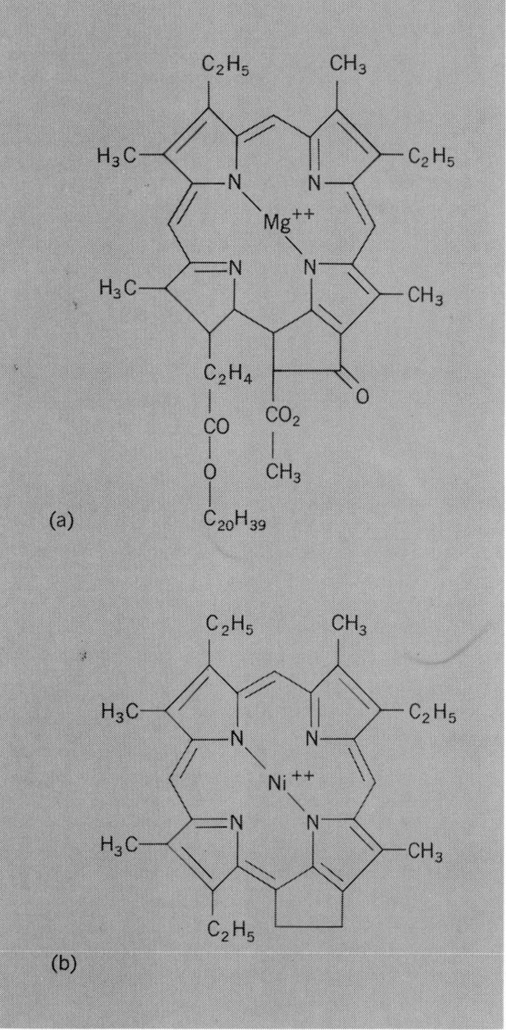

Fig. 2. Structural formulas of two pigment molecules. (*a*) Chlorophyll *a*. (*b*) Metalloporphyrin.

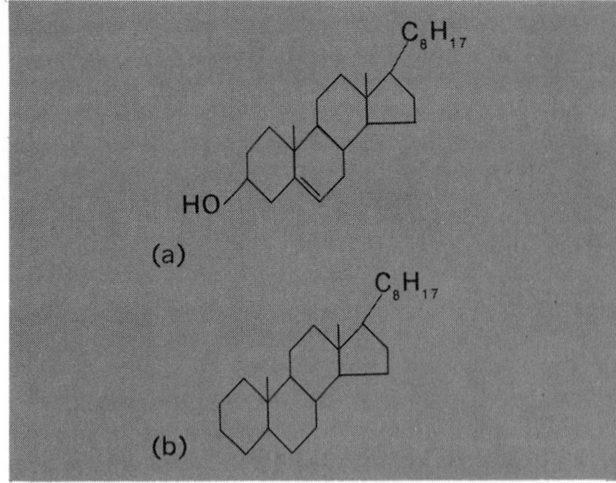

Fig. 3. Structural formulas for a steroid compound and for its possible sterane derivative. (*a*) Cholesterol (steroid). (*b*) Cholestane (sterane).

niques will be used on materials returned from extraterrestrial sources to determine whether life exists outside of Earth, and if so, to see how it may differ from life on Earth. *See* FOSSIL; PALEOECOLOGY, GEOCHEMICAL ASPECTS OF; PALEONTOLOGY.

[THOMAS C. HOERING]

Bibliography: P. H. Abelson, Paleobiochemistry, *Sci. Amer.*, 195(1):83–92, 1956; P. H. Abelson (ed.), *Researches in Geochemistry*, 2 vols., 1959, 1967.

Paleobotany

The study of fossil plants and vegetation of the geologic past. As a branch of paleontology, it combines a knowledge of both geology and botany. Its materials are the fossilized remains of prehistoric plants preserved in the rocks, including fossil leaves, seeds, pollen, spores, and wood, and occasionally fruits and flowers. From such materials the paleobotanist attempts to reconstruct whole plants as they actually grew, as well as the ancient forests of which they were a living part. The fossil plant record shows the succession of plants which have inhabited the Earth, from the relatively simple aquatic forms of the older geologic eras to forms progressively more complex and better adapted to land habitats in the younger geologic eras. Paleobotany involves not only collecting, describing, and naming fossil plants, but also aims at their interpretation in terms of the evolutionary history of plant groups, the relationships between extinct and living forms, the reconstruction of the environmental conditions under which the ancient forests lived, and the relative geologic ages of the rocks in which the fossils were found.

PLANT FOSSILIZATION

In living forests accumulations of plant debris normally disappear as the result of the natural processes of disintegration and decay. However, when forests grow in or adjacent to areas of sedimentary deposition, such as swamps, lakes, streams, and estuaries, plant debris may be buried along with the sediments. In the course of time and under the proper geologic conditions such sediments become sedimentary rocks, and the buried plant remains become fossilized. The plant fragments most likely to be preserved as fossils are those possessing hard tissues that resist decay. The most resistant of these are the wax-coated spores and pollen grains; they are followed in relative order by nuts and seeds, wood, leaves, and finally flowers. *See* FOSSIL SEEDS AND FRUITS.

Not all environments of deposition are equally conducive to the preservation of fossil plants. Deposits laid down in upland lakes, streams, or swamps, for example, are easily eroded away in the course of time, whereas lowland deposits are much less apt to be removed; the majority of fossil plants occur in deposits of the latter type. The deposits of the ocean bottoms, on the other hand, rarely contain the remains of land plants. This is mainly because of the destructive action of waves and currents along coastlines where plant remains might otherwise enter the sedimentary record. The remains of seaweeds and related aquatics are therefore usually the only kinds of plant fossils found in marine deposits.

Sources. Most fossil plants are found in shales and fine-grained siltstones, which are the lithified muds of ancient deposits. The best collections are obtained from rocks originally laid down as deposits in floodplains, lakes, swamps or bogs, and coastal lagoons or estuaries. Volcanic ash, the lithified form of which is known as tuff, is also an excellent medium for the rapid burial and subsequent preservation of plant remains. Only rarely do associated volcanic lavas preserve the casts of tree trunks or the impressions of leaves, due to their destructive high temperature at the time of eruption.

Coal and associated beds. Since coal is known to be altered plant materials originally accumulated in swamps, it is natural that much knowledge of ancient plants is derived from a study of coal itself and of the fossil plants that occur both in coal and in the roof shales and underclays associated with coal beds. *See* COAL; COAL PALEOBOTANY.

Fossilization processes. After burial by sediments, plant remains may undergo any one of several different processes leading to ultimate fossilization: (1) They may remain essentially unaltered or only partially altered, forming compressions; (2) they may be completely removed, leaving only impressions or molds, and sometimes casts; or (3) they may be partially altered and subsequently permeated by mineral-bearing solutions, producing petrifactions.

Compressions. Most plant fossils occur as compressions formed when leaves, seeds, flowers, or other plant remains are compressed between layers of sediments or their lithified equivalents. Only rarely do plant materials remain essentially unaltered (mummified): Examples include specimens of fresh plant debris from frozen muds of polar regions and the asphalt-impregnated leaves, cones, seeds, and wood fragments found in the Rancho La Brea tar pits of California. Also essentially unaltered are spores and pollen grains of microscopic size, whose waxy outer coats enable them to be preserved in countless numbers in many coals and nonmarine shale beds. Slightly more altered are compressions in which only the resistant carbon is left. Such carbonized leaves, woody trunks, and branches are common in the impervious clays of New Jersey despite their age of more than 80,000,000 years. In the still older paper coals of the

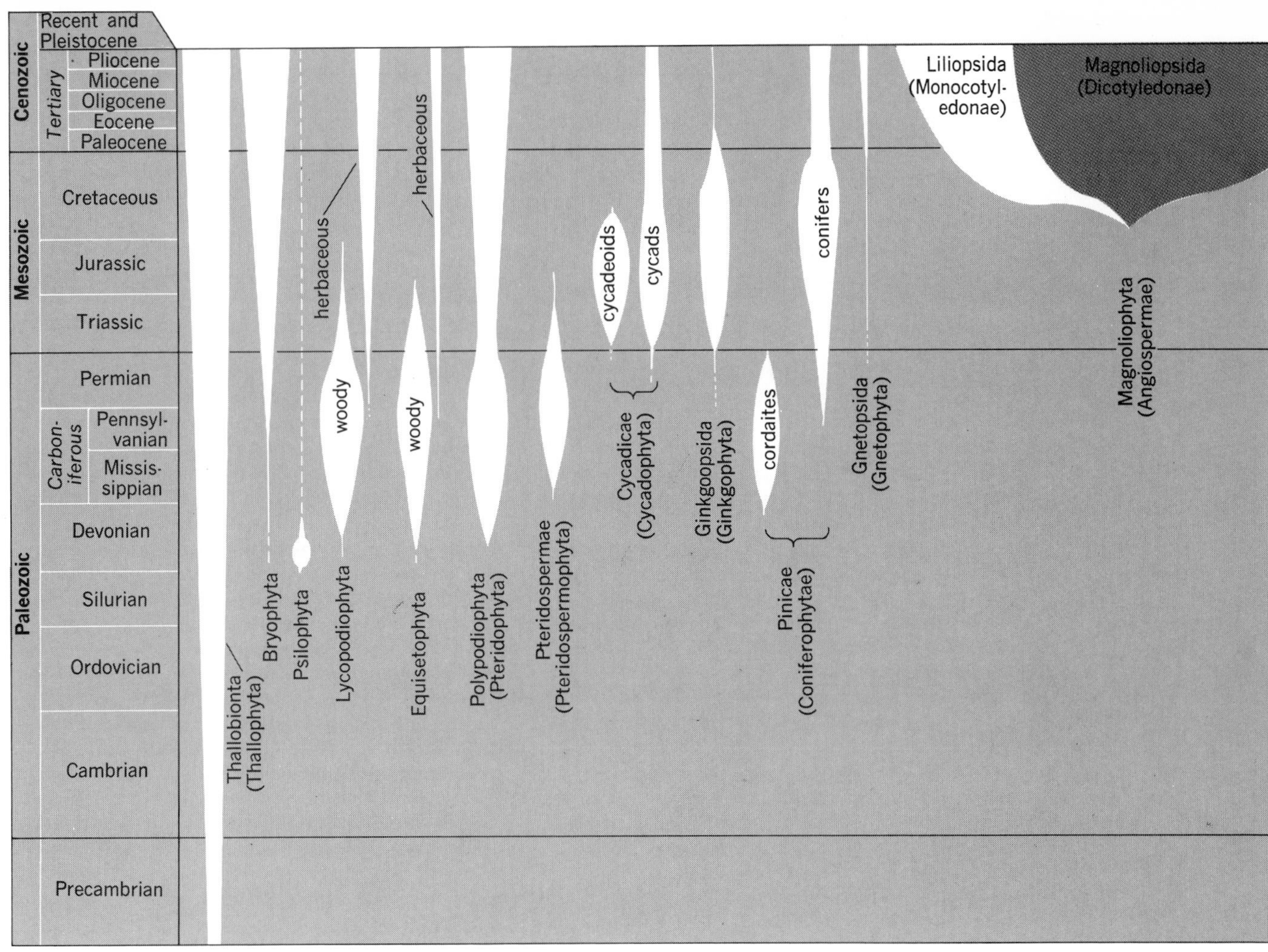

Fig. 1. Geologic distribution of the important plant groups.

Soviet Union, in deposits 240,000,000 years old, carbonized leaves retain their outside cuticle and may be peeled from their clay medium and mounted either in plastic or between two pieces of glass.

Molds, casts, and impressions. More or less complete decay and removal of plant remains from their enclosing rock may leave only an empty cavity or mold. If such molds subsequently fill in with sediments or mineral deposits, casts may be formed producing replicas of the original plant material. Examples of these are common in the vicinity of hot springs where standing trees are killed and entombed by the deposition of travertine. Molds of plant fragments including delicate flowers are also found in amber, which is hardened resin of ancient gum-bearing trees. *See* AMBER; TRAVERTINE.

In some cases leaves, needles, seeds, or flowers may be completely dissipated and leave only imprints in their enclosing medium. These are known as impressions, which are common in the coarser, pervious rocks in which active groundwater movement has facilitated decay and removal of the buried plant materials.

Petrifactions. In some instances groundwater may carry high concentrations of mineral matter in solution. These mineral-bearing solutions may circulate through plant materials buried in sediments or rocks and gradually produce petrifactions. Although it was formerly believed that petrifactions are fossils in which the original plant material is replaced, molecule by molecule, by mineral matter, this is actually not the case. The process of petrifaction is now believed to involve the slow infiltration of solutions through the buried plants and the gradual filling in of the minute cell cavities and intercellular spaces with mineral matter. The original woody components of the cell walls are not replaced but are rather surrounded by mineral matter; they remain so nearly intact that they can be recovered by the chemical removal of the petrifying minerals. In the case of silicified wood, removal of the silica by immersion in hydrofluoric acid leaves a soft, spongy mass of woody tissue which can be embedded and sectioned for study by the same techniques botanists use to study modern wood.

The more common minerals producing petrifactions are quartz (silica SiO_2), calcite (calcium carbonate, $CaCO_3$), pyrite and marcasite (iron sulfides), and various iron oxides.

Significance of fossil plants. To the geologist, the record of fossil plants added to that of fossil animals supplies the knowledge of the changing panorama of life in Earth history. The study of the many diverse groups of plants (floras) which successively inhabited the Earth's lands and seas can also be used by the geologist to determine the geologic ages of the rocks in which the fossils are found. Where the geologic time range of a plant group has been determined in rock sequences whose relative ages have been established, occurrences of the same plant group elsewhere will indicate equivalence in age. The ranges of the important plant groups during the eras, periods, and epochs of the geologic past are shown in Fig. 1. In the study of historical geology the fossil plant record helps to complete the total picture of the ancient geography (paleogeography) of each of the successive chapters of the past.

Fossil plants are also reliable indicators of past environmental conditions and are often called the thermometers of the past. It is a well-known fact that in modern floras certain plants are restricted to rather specific climatic and topographic conditions. If fossilized plants can be shown to be closely related to such living plants of restricted range, it may be assumed that the fossil plants lived under similar restricted conditions. The finding of fossil palms, laurels, magnolias, peppers, and cycads in the early Tertiary rocks of southeastern Alaska, for example, is evidence of subropical to warm temperate, lowland conditions there at that time.

Major subdivisions of the plant kingdom

Scientific names	Common names
Thallobionta (Thallophyta)	Thallophytes
Algae	Seaweeds and allies
Fungi	Fungi
Bryophyta	Bryophytes
Marchantiopsida (Hepaticae)	Liverworts
Bryopsida (Musci)	Mosses
Psilophyta	
Psilotophyta (Psilophytales)*	Psilophytes
Lycopodiophyta	
Lepidodendrales*	Scale trees
Pleuromeiales*	
Lycopodiales	Lycopods
Equisetophyta	
Hyeniales*	
Sphenophyllales*	Sphenophylls
Calamitales*	Calamites
Equisetales	Horsetail rushes
Polypodiophyta (Pteridophyta)	Pteridophytes
Coenopteridales*	Ancient ferns
Polypodiales (Filicales)	Modern ferns
Pteridospermae (Pteridospermophyta)*	Seed ferns
Cycadicae (Cycadophyta)	Cycadophytes
Cycadeoidales*	Cycadeoids
Cycadales	Cycads
Ginkgoopsida (Ginkgophyta)	Ginkgophytes
Pinopsida (Coniferophyta)	
Cordaitales	Cordaites
Pinales (Coniferales)	Conifers
Gnetopsida (Gnetophyta)	
Magnoliophyta (Angiospermophyta)	Angiosperms (flowering plants)
Liliopsida (Monocotyledonae)	Monocots
Magnoliopsida (Dicotyledonae)	Dicots

*Known only in the fossil state.

To the botanist, fossil plants supply a knowledge of the numerous past developments within the vegetable kingdom which have led to present conditions. The fossil plant record has furnished many missing chapters to the story of the slow, unending evolution of plants from the simpler forms of earlier geologic ages to the more complex plants of the present day. The recognition that living vegetation is merely the end result of major changes in distribution, as well as of evolution in the past, has also led to a better understanding of modern plant geography. The present restricted distribution of such forms as the redwoods (*Sequoia*), the dawn redwoods (*Metasequoia*), and the Oriental *Ginkgo* can be properly appreciated only in terms of their widespread distribution in the past. Still other plants, including several large groups such as the seed ferns (Pteridospermae or Pteridospermophyta) and the cycadeoids (Cycadeoidales), are shown by their fossil record to have flourished for many millions of years and then declined to complete extinction, a phenomenon of special interest to plant geneticists.

The fossil plant record has also been responsible for a better understanding of plant classification (taxonomy). The most widely accepted scheme of classification at present has been based almost as much on a knowledge of fossil plants as of living plants.

CLASSIFICATION OF FOSSIL PLANTS

Insofar as possible, fossil plants are classified according to the natural classification used by both botanists and paleobotanists. The basis for this classification consists essentially of three morphological characters: (1) the nature of relationship of leaf and stem; (2) the anatomy of the stem; (3) the arrangement and position of the spore-bearing organs and fruits. The classification presented in the table reflects the influence of a knowledge of fossil plants on concepts of classification. The fossil record shows, for example, that the production of seeds, previously considered to be an evidence of affinity, has been developed independently in such diverse groups as seed ferns, conifers, cycads, and flowering plants.

The assignment of fossil plants to their proper groups within the plant kingdom often presents difficulties not encountered with modern plants. For example, the identification of species, genera, and families among the living flowering plants depends in large part on characteristics of the flowers, which occur only very rarely as fossils. In this group, therefore, the identification of fossil forms must rely on those portions which are actually found fossilized, such as leaves, wood, fruits, and pollen.

Another difference encountered in the classification of fossil plants and modern plants is a result on the fragmented condition of most fossil plants. Such detached parts, particularly those belonging to extinct plant groups, often cannot be certainly associated or reconstructed into a complete plant. This makes it necessary and convenient to institute so-called form genera for different portions of what may later prove to be a single plant. For example, in Carboniferous rocks segmented, ribbed stems may be referred to the form genus *Calamites*, the foliage to *Annularia*, and the spore-bearing cones to *Calamostachys.*

Form genera of another type may refer to fossil plants whose morphological characters show only a family relationship to modern forms; *Magnoliophyllum*, for example, implies a general resemblance to the magnolia family without a commitment of closer relationship to any particular genus within the family.

GEOLOGIC DISTRIBUTION OF PLANT GROUPS

The fossil record of the plant kingdom shows that the major plant groups have varied considerably in time of origin, geologic range, and period of maximum development and decline (Fig. 1). Clearly demonstrating the theory of evolution, the earliest plants belong to the simpler groups and the later ones become progressively more complex. Modern land vegetation is thus a composite made up chiefly of the most highly complex plants, the flowering plants, living in association with a lesser number of survivors of simpler types of plants.

Thallobionta. This group, also known as the Thallophyta, is composed in part of the seaweeds (algae) and their allies and in part of the numerous bacteria and fungi. The great majority of them are rare in the fossil record because of their soft, perishable nature. It has, however, been demonstrated quite recently that very simple Thallophytes are the oldest known evidences of life on the Earth, going back to roughly 3×10^9 years ago. More commonly found, especially in marine limestones, are the lime-secreting algae, whose record also extends far back into the Precambrian eras (Fig. 1). Best known, perhaps, are the rounded, concentrically laminated forms which make up the *Cryptozoon* reefs in the Upper Cambrian limestones near Saratoga Springs, N.Y.

Bryophyta. Belonging to the simplest of the living land plants, the bryophytes are as soft and perishable as seaweeds and so are equally scarce as fossils. As seen in Fig. 1, the earliest problematical moss (*Sporongonites*) occurs in the Lower Devonian beds. Less doubtful are forms which occur in rocks of Pennsylvanian age: *Hepaticites*, a liverwort, from the Coal Measures of England, and *Muscites*, a true moss, from the rocks of the Carboniferous Period of France.

Psilotophyta. This is a little-known group of plants, also known as Psilopsida, represented among living plants by the tropical species of *Psilotum.* These are believed to be remotely related to the Psilophytales of Late Silurian and Devonian age. Both living and fossil groups are characterized by their small size, their absence of roots, their leafless stems, and their simple spore-bearing organs borne on short lateral shoots.

Lycopodiophyta. The living forms of this group are the inconspicuous, herbaceous species of *Lycopodium* and *Selaginella.* The Lycopodiophyta, also called Lycopsida, include some of the largest, most abundant, and most characteristic trees of the late Paleozoic swamp forests. Beginning in the Devonian, the group reached its climax in the Pennsylvanian Period, after which its woody forms declined rapidly and became extinct during the Jurassic Period. Among the best-known forms are the tall scale trees whose flattened trunks are especially common in the roof shales of coal mines. Their surfaces are marked by closely spaced scalelike leaf cushions and leaf scars, which are diamond-shaped and spirally arranged in the genus *Lepidodendron* and rounded or hexagonal in shape and vertically arranged in *Sigillaria* (Fig. 2).

Equisetophyta. Plants of this group are characterized by jointed, usually ribbed stems and short, linear leaves arranged in whorls. The only living forms are the so-called horsetail rushes (*Equisetum*) which are essentially small, herbaceous replicas of the tall, woody *Calamites* of the Pennsylvanian coal swamps. None of the arborescent forms survived beyond the Jurassic Period.

Polypodiophyta. The true ferns have undergone remarkably little change during their long geologic history which extends back to the Devonian Period. The modern ferns, Polypodiales (Filicales), so common as underbrush in living forests, can trace their ancestry back to the coal swamps of the late Paleozoic Era. The ancient ferns (Coenopteridales), differing mainly in stem structure and arrangement of spore-bearing organs, were dominant in the Pennsylvanian forests. Some of the living Ophioglassaceae and Marattiaceae may be descendants of the ancient ferns, most of which became extinct by the end of the Permian Period.

Pteridospermae. These ancient plants, also known as Pteridospermophyta and commonly called seed ferns, differ from true ferns in the possession of well-defined seeds usually borne at the ends of branches of the fernlike fronds. They were common during the late Paleozoic Era, after which they declined steadily and became extinct before the end of the Jurassic Period.

Cycadeoidales. The cycadeoids were essentially restricted to the Mesozoic Era. They resembled most living cycads in both foliage and internal stem structure but differed greatly from them in the nature of their flowerlike inflorescence. These developed laterally on the trunks among the leaf bases, often as many as 200 "flowers" on a plant.

Cycadales. The modern cycads are widespread in the present tropics where they are often called sago palms. In contrast to the cycadeoids their reproductive structures resemble more closely the cones of living conifers. The group has a long geologic history extending as far back as the Permian Period; they were among the dominants of the Mesozoic Era but have dwindled in importance since the middle of the Cretaceous Period.

Ginkgoopsida. This group originated in the Late Paleozoic Era and reached its peak in the Early Cretaceous Period. Since then it has been reduced to near extinction, being represented by only a single living species, *Ginkgo biloba*, known in the native state only in Chekiang Province, eastern China.

Cordaitales. Members of this division are believed to be ancestral to the later Pinales, from which they differ chiefly in their straplike leaves and their more open, conelike, seed-bearing

PALEOBOTANY

(a)

1 cm

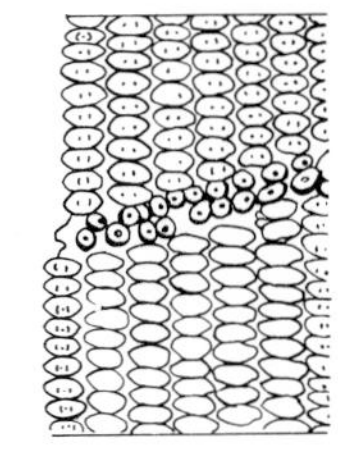

(b)

Fig. 2. Scalelike markings on portions of fossilized trunks. (*a*) *Lepidodendron.* (*b*) *Sigillaria.*

PALEOBOTANY

Fig. 3. Impression of an angiosperm leaf, a fan palm, from the Eocene of Wyoming.

organs. The cordaites were common among the forest trees of the late Paleozoic Era; they are not known to have survived into the Mesozoic Era.

Pinales. The conifers, also called Coniferales, are chiefly evergreens with needle-shaped or awl-shaped leaves and with seeds borne in true cones. They are common and widespread in the temperate forests of both the Northern and Southern hemispheres. The earliest conifers, belonging to extinct genera, occurred only rarely in the late Paleozoic Era. Increasing in abundance and becoming widespread in the Triassic Period, the group attained its greatest development in the Jurassic and Cretaceous periods (Fig. 1). Since then the conifers have declined somewhat, presumably as the result of competition from the flowering plants.

Gnetopsida. This little-known group is represented in the modern flora by only 70 species, belonging to three genera, two of which are native only in the Southern Hemisphere. The fossil record is mainly of the genus *Ephedra*, whose pollen is doubtfully known in the Permian and becomes common in beds of Cenozoic age (Fig. 1).

Magnoliophyta. The flowering plants (Angiospermae), in addition to having flowers, are characterized by enclosed seeds and by broad, flat leaves like those of palms, oaks, maples, and elms (Fig. 3). They are not reliably known before the Cretaceous Period. They increased and spread slowly during the Early Cretaceous, more rapidly during the Late Cretaceous, and reached almost their present dominance before the beginning of the Cenozoic Era.

GEOLOGIC ERAS OF THE FLORAS

Fossil records for land and marine flora date back to bacteria identified from the Archeozoic Era of the Precambrian Period. The ancestry and evolution of modern plants can be traced from this point in time.

Precambrian floras. The fossil remains of both plants and animals are extremely rare in rocks of Precambrian age, which are usually referred to an older Archeozoic Era and a younger Proterozoic Era. There is no reliable evidence of the existence of land plants during this portion of the Earth's history; even marine plants are scarce and are restricted chiefly to the later phases of the Proterozoic Era. However, two distinct kinds of microorganisms have been reported from Archeozoic rocks of South Africa believed to be about 3×10^9 years old. One of these, *Archaeosphaeroides*, is a spheroidal, unicellular, algalike organism possibly distantly related to modern coccoid blue-green algae. The other, *Eobacterium*, is described as a minute, bacteriumlike, rod-shaped organism. These two forms are at present regarded as the oldest known remains of life on the Earth. Other types now known from the Archeozoic Era include calcareous algae from Rhodesia (2.6×10^9 years old), probable blue-green algae and rod-shaped and coccoid bacteria from Ontario (1.9×10^9 years old), and both blue-green and green algae from Australia (1×10^9 years old).

PALEOBOTANY

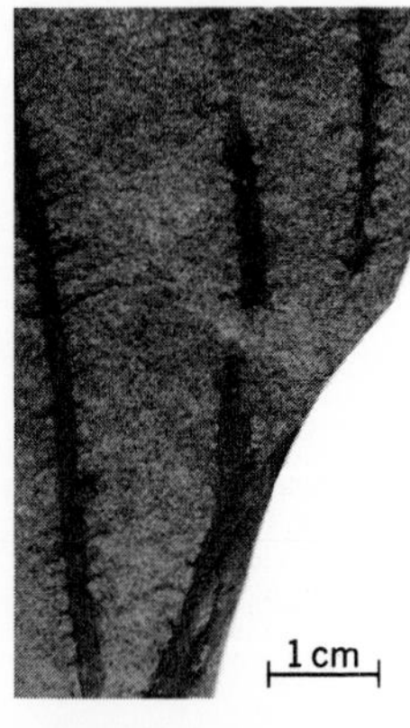

Fig. 4. Spiny stems of *Psilophyton* from the lower Devonian of Gaspe, Canada.

In the younger Proterozoic limestones of Montana and adjacent Canada numerous spherical, dome-shaped, and columnar masses of concentric laminations are believed to be the remains of ancient algal reefs. Similar forms, referred to the genus *Collenia*, are widely known from rocks of the same age in Alaska, Ontario, Greenland, and the Grand Canyon in Arizona. Specimens of blue-green algae and bacteria have been described from the Lake Superior iron ores. In the Adirondack region, beds of Precambrian graphite are interpreted as metamorphosed coal beds which were originally layers of algal debris.

Paleozoic floras. The floras of this era are discussed in chronological sequence from early to late Paleozoic time.

Cambrian and Ordovician. Plant remains continue to be rare in rocks belonging to the Cambrian and Ordovician periods. Most common are the marine calcareous algae such as *Cryptozoon* from the Cambrian of New York, *Solenopora* from the Ordovician at numerous localities, *Primicorallina* from the Ordovician and Silurian, and the widespread *Girvanella*, which ranged from the Cambrian to the Permian.

Silurian and Devonian. Rocks of latest Silurian age in Bohemia have yielded the oldest reliable record of land plants, such as *Cooksonia* and *Taeniocrada*. These consisted of small, leafless stems terminated by simple spore-bearing organs. All other pre-Devonian records of true land plants are considered doubtful.

In the Early Devonian small, primitive land plants, belonging mainly to the extinct psilophytes, increased in abundance and in geographic distribution (Fig. 1). Most characteristic was the genus *Psilophyton*, whose small spiny stems (Fig. 4) with laterally arranged spore cases are known from localities in eastern Canada, Maine, Wyoming, China, Spitsbergen, and numerous localities in western Europe. Other characteristic taxa include *Zosterophyllum*, *Bucheria*, *Drepanophycus*, *Dawsonites*, and the lycopodium-like *Baragwanathia* of Australia.

By Middle Devonian time more advanced types of plants made their appearance. The lycopods were represented by *Asteroxylon*, whose stems were clothed in short, leaflike scales. The equisetophytes are known from specimens of thin, jointed stems bearing whorled leaves. These belong to the genera *Hyenia* and *Calamophyton* of the extinct order Hyeniales. Fernlike plants, *Protopteridium* and *Aneurophyton*, were simple branched axes with rudimentary frondlike leaves.

By the beginning of Late Devonian time one of the major changes in the history of plant life had taken place. The earlier shrubby vegetation, dominated by the primitive psilophytes, had evolved into low forests dominated by woody lycopodiophytes and equisetophytes with underbrush of ancient ferns and sphenophylls. Most abundant in Upper Devonian rocks are the ferns, of which the most widespread and characteristic are species of *Archeopteris*, *Aneurophyton*, and *Sphenopteris*. In none of these were the fronds and pinnules as well developed as they were in late Paleozoic forms. The lycopodiophytes were represented by the scale trees, *Archeosigillaria* and *Cyclostigma*, characterized by small scalelike leaves on dichotomously branched axes. Large, jointed woody equisetophytes included mainly species of *Archeocalamites* and *Pseudobornia*, the latter bearing whorls of finely divided, featherlike leaves. Sphenophylls are of rare occurrence; more common are the petrified stems of the most primitive gymnosperm, *Callixylon.*

Mississippian and Pennsylvanian. The development of widespread, dense, lowland forests was attained by gradual floristic evolution in the late Paleozoic Coal Age, referred to in Europe as the Carboniferous Period. Prominent in the forests were the stately lycopodiophytes, the small-leaved scale trees *Lepidodendron* and *Sigillaria,* with their relatively unbranched trunks reaching heights of over 100 ft. Beside them towered species of *Cordaites* with their long, straplike leaves and lax cones. Somewhat lower in stature were many types of *Calamites* with thick, jointed, ribbed trunks and stems and whorls of linear, pointed leaves (*Annularia* and *Asterophyllites*) (Fig. 5). The first true conifers developed by Pennsylvanian time; *Walchia* and *Lebachia* are known from numerous impressions of shoots bearing short, closely spaced needles. Throughout the Carboniferous the shrubby undergrowth was apparently made up chiefly of ferns and seed ferns (Fig. 6), whose spreading fronds of delicately sculptured pinnules are among the most common fossils found in the roof shales of many of the numerous coal mines which exploit the extensive coal beds of this age. Among the most widespread are species of *Neuropteris, Alethopteris, Sphenopteris,* and *Pecopteris,* which are characteristic of the Pennsylvanian of North America and the Upper Carboniferous of Europe. Other shrubby plants were the climbers and trailing species of *Sphenophyllum* with thin, jointed stems and whorls of six to nine small, wedge-shaped leaves (Fig. 7).

Permian. During the Permian Period many of the great Carboniferous groups of plants began to decline, on their way to ultimate extinction before the end of the Paleozoic Era or early in the Mesozoic (Fig. 1). Greatly reduced in the Permian, the tall, woody lepidodendrons and sigillarias as well as the stately cordaites did not survive beyond this period. Also disappearing before the end of the Permian were the ancient ferns (Coenopteridales), the delicate sphenophylls, and the jointed, woody calamites. The seed ferns (Pteridospermae) were severely affected but managed to continue as a minor group into the Mesozoic Era. The conifers, cycads, and ginkgos, which had barely begun their development at the beginning of the Permian Period, increased in importance during this period of decline in other groups. It is generally believed that the great changes which occurred at this time in both the plant and animal kingdoms were the result of worldwide physical and environmental changes, which also produced a long episode of continental glaciation, mainly confined to the Southern Hemisphere.

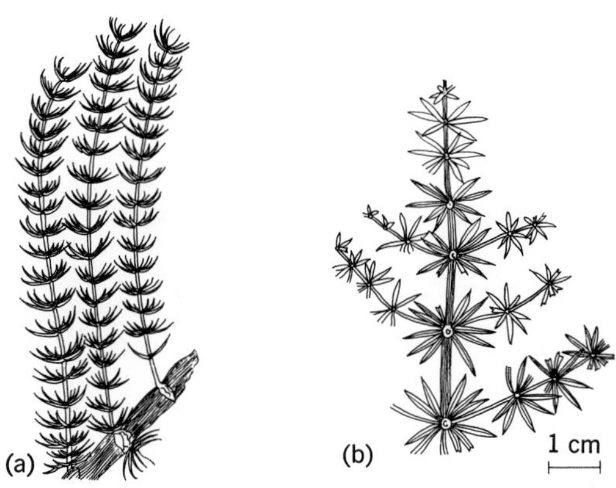

Fig. 5. Examples of the leafy foliage of the calamites. (*a*) *Asterophyllites.* (*b*) *Annularia.*

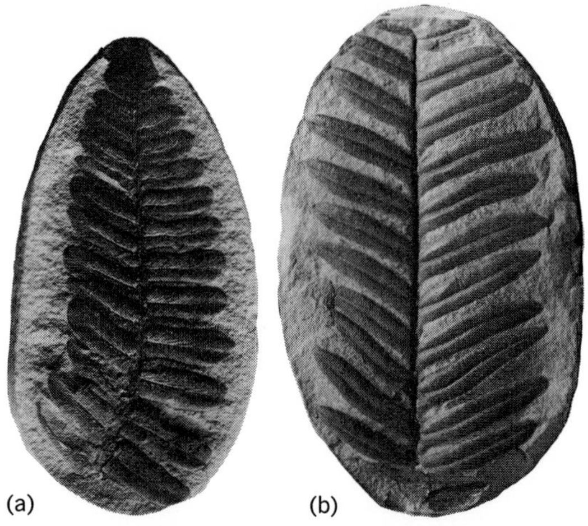

Fig. 6. Pennsylvanian ferns from Illinois, showing the typical features of leaflets and fronds. (*a*) *Neuropteris.* (*b*) *Alethopteris.* (*Field Museum of Natural History, Chicago*)

Mesozoic floras. Changes in floras are described for the three major subdivisions of Mesozoic time.

Triassic. During the Triassic Period the great transformation in the Earth's vegetation, which had begun in the Permian, resulted in the further restriction of the ancient Paleozoic groups of plants and the introduction of many new groups. Ferns continued to be dominant, though most of them belonged to modern families. The larger forest trees were mainly conifers, cycads, and cycadeoids along with lesser numbers of ginkgos. Most of the conifers belonged to extinct genera; a few, including *Araucarioxylon* from the Petrified Forest region of Arizona, were related to groups still living.

The cycadeoids, represented by such well-known forms as *Wielandiella* and *Williamsonia,* are completely extinct at the present time. The Triassic ginkgos were characterized especially by the deeply dissected leaves of *Sphenobaiera* and *Baiera* (Fig. 8). The few remaining seed ferns were mainly species of *Glossopteris, Thinnfeldia,* and *Neuropteridium.*

Jurassic. The vegetation of the Jurassic Period was quite similar to that of the Triassic. Most characteristic are the cycads and cycadeoids, whose broad fronds closely resembled those of the living cycads. In the cycadeoids the numerous petrified stems differ from the stems of the cycads chiefly in their possession of "flowers" borne on short lateral branches and largely enclosed by persistent leaf bases and scales. Both the conifers and the ginkgos continued to increase in abundance. Ancestors of both the living pines and redwoods are believed to occur as far back as the Late Jurassic. Mild climates are indicated by the occurrence of typical

PALEOBOTANY

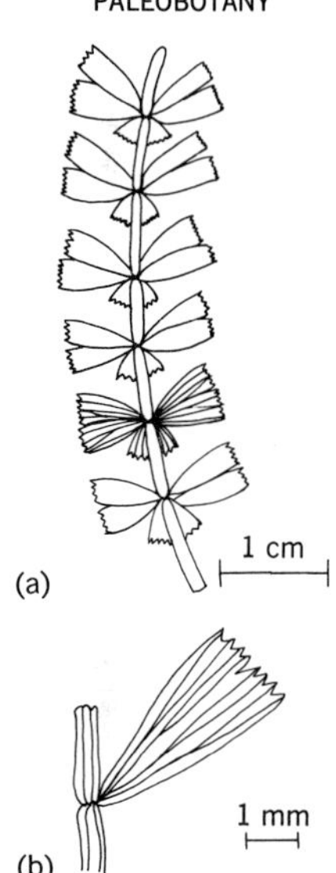

Fig. 7. *Sphenophyllum.* (*a*) Branch and leaves. (*b*) Leaf venation.

PALEOBOTANY

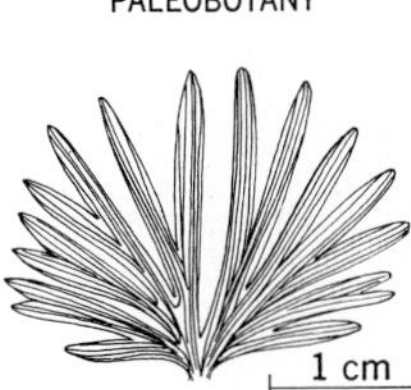

Fig. 8. Leaf of *Baiera,* which was characteristic of the Triassic and Jurassic periods.

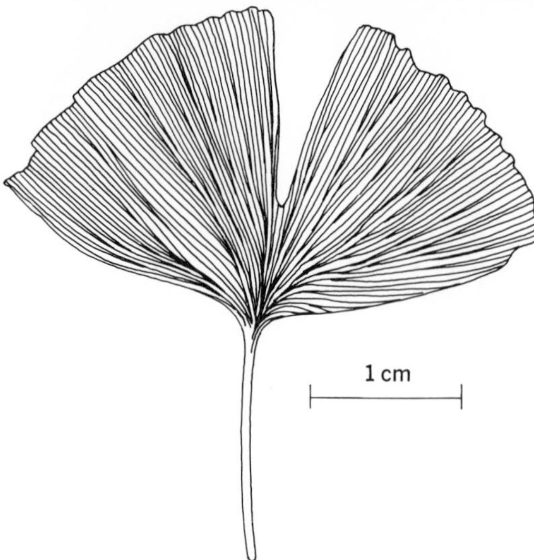

Fig. 9. Leaf of *Ginkgo*, characteristic of Cretaceous and Cenozoic, the only modern survivor of Ginkgoales.

Jurassic floras as far north as northern Alaska and Siberia and as far south as Graham Land in Antarctica.

Cretaceous. In general aspect the floras of the Early Cretaceous were similar to those of the Jurassic. Conifers, ferns, and ginkgos continued to flourish; cycads and cycadeoids declined, the latter group becoming extinct by the beginning of Late Cretaceous time. Of about 35 genera of conifers known from Lower Cretaceous rocks, 15 are believed to belong to modern groups including the pines, firs, redwoods, and cedars. The numerous ferns were almost exlusively modern in aspect. The ginkgos reached their climax with species belonging to 11 genera, including the sole living survivor, *Ginkgo* (Fig. 9). The cycadeoids are represented by both the typical cycadean foliage and by numerous petrified trunks, especially from the Lower Cretaceous of the Black Hills in South Dakota and the Chesapeake Bay region.

Of greatest significance during the Cretaceous Period was the apparent origin and rise to worldwide dominance of the flowering plants. Magno-

5 cm

(a)

(b)

(c)

Fig. 10. Angiosperm (Magnoliophyta) leaves and fruits. (*a*) Sycamore leaf. (*b*) Sycamore fruits. (*c*) Katsura leaf.

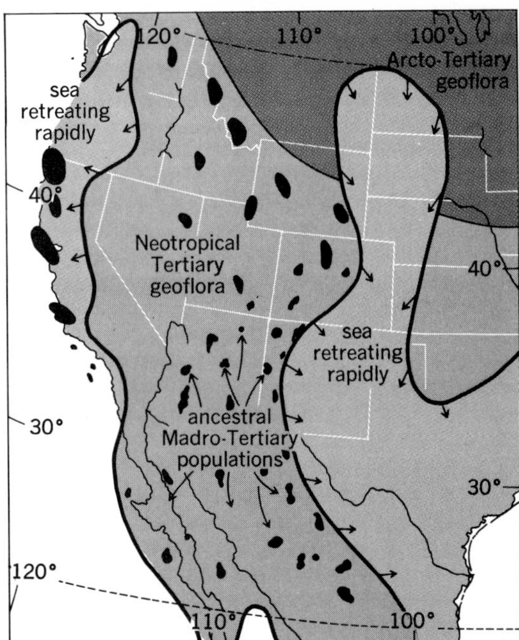

Fig. 11. Inferred distribution and migration of ancestral Madro-Tertiary plants in pre-Eocene time. (*From D. I. Axelrod, Evolution of the Madro-Tertiary geoflora, Bot. Rev., 24(7):433–509, 1958*)

liophyta (Angiospermae). Although pre-Cretaceous angiosperms have often been reported, their authenticity still remains in doubt. Among the earliest possible angiosperms are several forms from the Lower Cretaceous of Maryland, Alaska, China, and Siberia. In Maryland, for example, broad net-veined leaves believed to be primitive angiosperms make up about 6% of the flora of the Patuxent formation. The overlying units (Arundel and Patapsco formations) contain 8 and 40%, respectively, of angiosperm leaves (Fig. 10). By Late Cretaceous time the angiosperms, including both the monocots (Liliopsida) and the dicots (Magnoliopsida), normally make up 70 to 90% of the known fossil floras.

[ERLING DORF]

Cenozoic floras. The dominant plants during the past 60×10^6 years have been angiosperms and conifers, with ferns, cycads, and ginkgos reduced in number. No major groups have appeared or become extinct during this time. However there have been noteworthy changes in distribution, with an equatorward shifting of forest units (geofloras). Cenozoic plants more often occur as impressions than as petrifactions. As a result, attention has been directed toward the distribution of vegetation in time and space, rather than to a study of plant anatomy and phylogeny. The discussion will center in North America, where north-south mountain ranges, in contrast with the predominantly east-west ranges of Eurasia, have provided a terrain favorable for migration with survival.

Most Cenozoic plants are assignable to existing families and genera on the basis of their leaves, with confirming evidence of fruit, stems, and pollen. Two floristic units are designated: The Arcto-Tertiary Geoflora, dominated by deciduous trees whose nearest living relatives are temperate in occurrence; and the Neotropical-Tertiary Geoflora, whose principal members are broad-leaved evergreens like those now living at low latitudes. The Madro-Tertiary Geoflora, characterized by small-leaved shrubs and trees, may have been derived from these in response to progressive aridity in western North America (Fig. 11).

Evidence of wide intercontinental connections is afforded by the uniform composition of the Arcto-Tertiary Geoflora at high northern latitudes during the Eocene. Alder, chestnut, elm, maple, and katsura were forest dominants, with deciduous conifers such as Chinese redwood (*Metasequoia*) more numerous than firs and pines. By Miocene time when this geoflora occupied middle latitudes, new trees were appearing, notably black oaks; several genera which have survived in Asia were becoming rare or extinct in North America. Later Cenozoic forests show further elimination of summer-wet trees from western America, as emergence and orogeny brought continental climate. During the Pleistocene, conifers whose present southern limits are in the Great Lakes region occupied the Gulf states, and the coast redwood ranged into southern California. This represents the climax of the Cenozoic trend toward colder and dryer climate.

The Neotropical-Tertiary Geoflora, best known in the lower Mississippi Basin and from California to Washington during the Eocene, shows a corresponding southward migration. At the middle of the era, its palms, laurels, and legumes were still surviving in coastal California. Reduced temperature and rainfall are probable causes for the subsequent restriction of this subtropical forest to its present low latitude occurrence.

The boundary between the principal geofloras during the Cenozoic bends southward across the northern continents from west to east and turns northward over the oceans, as do existing isotherms. Throughout the era these primary relief features appear to have controlled the pattern of temperature and vegetation as they do today, a relationship which provides much evidence to refute the theory of continental drift during later geologic time.

Thus it is seen that migrations of Cenozoic forests toward the Equator have established a trend from trees with large, evergreen leaves to those with small, deciduous leaves. This sequence provides a basis for dating plant-bearing rocks with the same degree of accuracy as recognition of first appearances and extinctions in older floras, or as evolutionary development in land mammals. Reconstruction of climate and topography is made possible by comparisons between living trees and their immediate ancestors in Cenozoic rocks.

[RALPH W. CHANEY]

Bibliography: H. N. Andrews, Jr., *Ancient Plants and the World They Lived In*, 1947; C. A. Arnold, *Introduction to Paleobotany*, 1947; C. A. Arnold, *Geological History and Evolution of Plants*, 1967; D. I. Axelrod, Evolution of the Madro-Tertiary geoflora, *Bot. Rev.*, 24(7):433–509, 1958; W. C. Darrah, *Principles of Paleobotany*, 1960; A. Poldervaart (ed.), *Crust of the Earth*, Geol. Soc. Amer. Spec. Pap. no. 62, pp. 575–592, 1955; A. C. Seward, *Plant Life Through the Ages*, 2d ed., 1933; D. H. Scott, *Studies in Fossil Botany*, 3d ed., 1962.

Paleocene

The oldest of five major worldwide divisions (epochs) of the Tertiary Period (Cenozoic Era); the epoch of geologic time extending from the end of the Cretaceous Period (Mesozoic Era) to the Eocene Epoch.

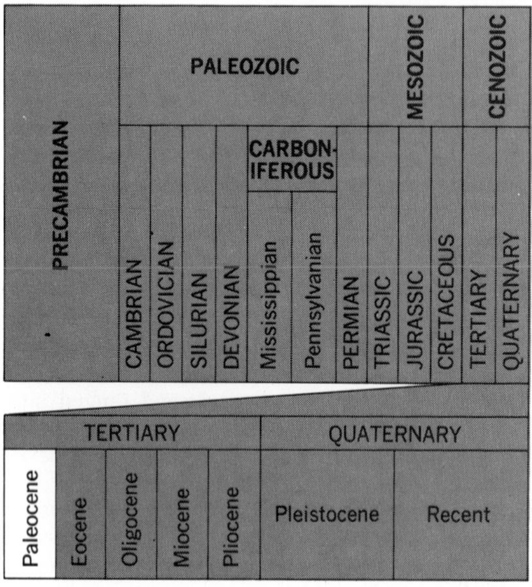

Paleocene strata are similar to the marls, limestones, mudstones, sandstones, and conglomerates of the underlying Cretaceous, but the formerly widespread chalk deposition had ceased; greensands (with glauconite) formed widely, brown coal locally, and phosphates in Tunis and Algeria; Paleocene deposits generally are geographically patchy. Igneous rocks seem not to be extensive, but Paleocene tuff is not lacking in Europe.

Stratigraphy. The term Paleocene was introduced by the paleobotanist W. P. Schimper in 1874. Schimper's fossil plants (from the Bracheux sandstones, travertines of Sezanne, and lignites and sandstones of the Soissons district east and northeast of Paris) included birch, elm, cottonwood, and willow, which are characteristic of the Northern Hemisphere in contrast to the Cretaceous floras which are predominant in the Southern Hemisphere. The distinction was local and geographical; yet these plant-bearing strata bear or pass laterally into beds bearing terrestrial mammals and marine invertebrates. The mammalian data led to intercontinental use of the term Paleocene about 1920, the marine mollusks permitting incorporation within the Lyellian scheme of Tertiary classification. *See* EOCENE; TERTIARY.

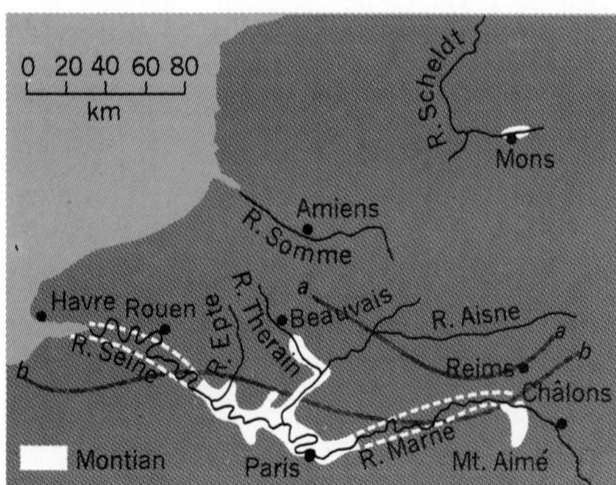

Fig. 1. Franco-Belgian Paleocene. Successive margins of Landenian sea shown by *aa* (zone of *Pholadomya konincki*) and *bb* (zone of *Cyprina scutellaria*). (*From A. M. Davies, Tertiary Faunas, vol. 2, George Allen & Unwin Ltd., 1934*)

Usages differ as to both lower and upper boundaries of the Paleocene. Slightly older beds in the Lowlands and France making up the Montian Stage (from Mons, Belgium) have generally been included as Paleocene (Fig. 1); since about 1950 the Danian Stage—marls in and around Denmark previously considered Cretaceous, although lacking in ammonites diagnostic of that period—is also included in the Paleocene Series. Schimper's lower beds correlate with a Thanetian Stage (Isle of Thanet mouth of the Thames); his upper (Soissonais) beds are incorporated in a Sparnacian Stage which bears distinctively younger (Eocene in mammalian usage) mammals in France, though relics of the older fauna lingered on in England. In the Lowlands Paleocene a Landenian Stage nonsequentially succeeds the Montian (Fig. 1) and encompasses strata of both Thanetian and Sparnacian age. Comparable facies problems, inherent in the Paleocene Series as a whole, have led to local terms for its subdivisions elsewhere: "Midway" (American Gulf states); Cheneyian, Ynezian, Bulitian, and "Martinez" (California); Rocanean (Argentina); Wangaloan (New Zealand); Tertiary al (East Indies); and Ranicot (Indus Valley). Terrestrial sequences based on mammalian evolution are also given local names (Puercan-Dragonian-Torrejonian-Tiffanian-Clarkfordian sequence, interior North America). *See* CENOZOIC.

Seas and mountain building. Paleocene time encompassed two intervals of regional uplift and marine regression, separated by an interval of marked changes in the distribution of temporarily expanding seas. The preceding Cretaceous was brought to a close by a shrinking of the world's previous seaways, giving rise to local fresh-water (England, Haute-Garonne, and Japan) or estuarine (southern France, Dalmatia, and lower Indus) deposition, to extensive terrestrial deposits, and to widespread erosion through regional uplifts. True mountain-making movements (orogenies) were few though locally intense, such as the Laramide Disturbance (Rocky Mountains). Areas of known continuous marine deposition from Cretaceous to Paleocene are scattered and restricted: Denmark and eastward, the river valleys of northern France and the Lowlands, parts of Egypt, the American Gulf states, California, New Zealand, Tenimber, and south Celebes. In some areas (Tibet and North American interior) retreating Paleocene seas deposited the last local marine strata.

Life. Paleocene marine life was characterized by the spread of pelecypods, gastropods, echinoids, and foraminifers already present but rare in earlier seas. Great limestone-forming foraminifers (nummulites and discocyclines) dominated tropical shallow clear marine waters, and the cocklelike "*Venericardia planicosta*" stock, so-called "finger-

Fig. 2. Paleocene condylarth, *Tetraclaenodon*. This five-toed ungulate was near the ancestry of the first horses. (*From R. A. Stirton, Time, Life, and Man, copyright © 1959 by John Wiley & Sons, Inc.; reprinted by permission*)

post of the Eocene," had already appeared and spread widely. At a rather sharply defined biostratigraphic horizon (end of the Cretaceous) termed the time of "the Great Dying," previously dominant cephalopod mollusks (true belemnites and ammonites) and marine reptiles (ichthyosaurs, plesiosaurs, and mososaurs) became suddenly extinct, as did the huge dinosaur reptiles on land. While marine cephalopod squid, octopus, and pearly nautilus persisted, reptilian turtles, snakes, lizards, alligators, and crocodiles carried on along with the many trees and shrubs that also came from Cretaceous into Tertiary times. The outstanding feature of Paleocene land life, however, was the appearance, proliferation, and explosive evolution (Fig. 2) of small archaic placental mammals (insectivores, creodont carnivores, condylarths, ancestral rodents, and primates), while an older marsupial mammal fauna became isolated in Australia.

Marine faunas fragmented. Old World (Tethyan) warm-water faunas were rich in nummulites; New World tropical faunas (with the west African) lacked them but had some affinities with a north European cooler-water fauna, which from Denmark eventually spread westward (Landenian and Thanetian) as far as eastern Greenland, extended eastward to the Don-Volga area with possible warm-water faunal connections in Bavaria, western Austria, and the Crimea. The New Zealand fauna (Wangaloan) had affinities with other provincially distinct faunas in Patagonia (Rocanian, also related to the Landana fauna of the Congo) and California (which had also both east Tethyan and Gulf state connections).

In the closing phases of the Paleocene, mammalian faunas indicate reestablishment of some intercontinental land connections. *See* PALEOBOTANY; PALEONTOLOGY.

[ROBERT M. KLEINPELL]

Paleoclimatology

That aspect of the study of climates which deals with ancient, long-term patterns of weather, usually treated on a global or regional scale. The study of present-day weather is obtained by meteorologists through the use of measuring instruments or procedures: thermometers, barometers, satellite-based cloud photography, and so on; and from these data the climatic patterns are established on a seasonal or annual basis, aided by cartography, statistics, and computer-based techniques. No such measuring devices were available for recording ancient weather conditions or the climatic patterns of the past, that is, prior to the last few centuries. Paleoclimatology requires the use of "climatic indicators" that are identified and analyzed by geologists; this is one aspect of paleogeography, the study of past environments. The science of paleoclimatology thus requires knowledge of both geology and atmospheric physics. *See* GEOLOGY; PALEOGEOGRAPHY.

Indicators. Examples of paleoclimatic indicators abound in the records of geology. Many were first discovered during the 19th century and constituted some of the many marvels of the past that helped to bring an understanding that the present planet Earth has had a changing and evolving history. Fossil palm leaves discovered in Spitsbergen and Antarctica suggested tropical conditions there at one time. The distinctive grooving of rock surfaces by glacier ice can be recognized in India and in the Sahara Desert. Bones of mammoths are found in New York State. Continental drift and plate tectonics explain some of the anomalies, but not all. From what humans know about it, the universe is also an evolving, non-steady-state system, partly cyclic and predictable (like the motion of the Earth around the Sun), and partly random and unpredictable. *See* CONTINENT FORMATION; PLATE TECTONICS.

Environmental systems. The geologic history of the Earth, from its beginning about 4.6×10^9 years ago to the present, has been one of slow evolution, interrupted by threshold stages that led to revolutionary changes in certain aspects of the ongoing environmental systems. These systems involve both the living and physical worlds, which constantly interact. The climates of the past were in part directly related to these planetary conditions, but also in part to extraplanetary controls.

Both modern and ancient climates are dependent upon these two variables. The extraterrestrial irregularities are due, on the one hand, to the slight changes in the relationship of the Earth to the Sun (such as the shape of Earth's orbit and the tilt of its spin axis), and, on the other, to extremely subtle variations of the Sun's emissions (for example, the Sun's luminosity has doubled during Earth history, and there are 11- and 22-year and longer cycles in solar activity). Variables within the planet itself depend on such things as the amount of dust thrown into the atmosphere by volcanic eruptions, changes in the height of mountains, and changes in the oceanographic circulation. In this way, a major eruption can lower the mean global temperature by 1°C for a year or two. Mountain building is irregular in geological history, and high mountains lead to climatic extremes. As a third example, the closing of the Panama and Suez isthmuses several million years ago blocked the circulation of warm equatorial currents and played a major role in starting the contemporary ice age. In some ways the Earth's atmosphere and oceans are like a heat engine: If the circulation of a car radiator is allowed to become blocked with rust, the system gives trouble. In the car, the motor boils; in the case of the Earth, which operates within the freezing realm of outer space, the result is an ice age.

Paleoclimatic cycles. The Earth's environments have thus passed through both repetitive events and evolutionary, noncyclic stages of development. The principal cyclic control is probably the rotation of the Galaxy, which takes place about once every 200–250 × 10^6 years. Its effects are not yet well understood, but it almost certainly affects the Earth's gravitational field and many dependent factors, not the least of which is climate. A correlation has not been proven, but major ice ages and stages of mountain building recur at rather similar intervals. Large-scale glaciations coincide with less than 10% of recorded geological history and must therefore be regarded as something like "climatic accidents" in long-term history. Each ice age lasts approximately 10–20 × 10^6 years and is itself subdivided into "glacial" and "interglacial" intervals following a roughly 90,000-year cycle (the planetary precession, first identified as a climatic control by the Yugoslav mathematician M. Milankovitch). Progressively shorter cycles can be postulated, down to the fundamental 12-month cycle of the terrestrial year. The 11-year sunspot cycle seems to be a solar tidal effect loosely tied to the Jovian (Jupiter) year.

Noncyclic evolution of terrestrial environments. Climatic history is bound up with evidences of the Earth's atmosphere and its progressive changes. The climatic indicators are partly found in the minerals or sediments, and partly in the evidence of animals and plants.

First stage. Very little is known about the Earth's earliest beginnings (4.6 to 3.5 × 10^9 years ago) except by deduction and evidence from astronomy, but it is likely that the Earth had a reducing atmosphere, rich in ammonia and methane, because certain minerals are preserved that are unstable in the present oxygen-rich environment.

Second stage. This stage covers the period of time from 3.5 to 2.9 × 10^9 years ago. First life probably appeared around 3.5 × 10^9 as primitive bacteria or algae, which began to photosynthesize atmospheric carbon dioxide to generate oxygen, although at first the oxygen was all taken up again by the oxidation of metals such as iron. The climates were warm enough for biologic metabolism and for running water to exist, thus temperatures were probably in a mean range of 10–25°C.

Third stage. This immense time (2.9 to 0.57 × 10^9 years ago) was occupied by the very slow initial evolution of primitive life, and by the gradual growth and thickening of continental crust on the evolving globe. The first oxygen was in very small amounts, but it gradually increased to a level where primitive animals could evolve and were able to use it for their metabolism; increasing quantities led to the worldwide appearance of oxidized iron ores. This and other geochemical reactions still required mean global temperatures in the range 10–25°C. In sedimentary deposits there are widespread limestones (indicators of an annual mean temperature in the shallow ocean of over 15°C), but in places there are also conglomerates (boulder accumulations) that suggest the moraines of former glaciers; periodic glaciation is certainly proved in certain places (such as Norway and Africa) by the preservation of the distinctive and unique rock grooving created by moving ice. *See* GLACIOLOGY.

Fourth stage. This is the stage from 570 to 250 × 10^6 years ago. A critical threshold was reached in global geochemistry about 570 × 10^6 years ago, when all over the world marine organisms suddenly began to secrete calcium carbonate shells which must have played a protective role in a Darwinian selection, whereby they were defended from predators. Such calcification could have been favored by warming climates and rising pH (alkalinity) of sea water that may have been due to the increasing amount of biological weathering on land surfaces and therefore to increasing salts in the sea.

Land plants were at first only unicellular, but later there were club mosses and swamp plants, so that by 400 × 10^6 years ago, larger plants such as palms and tree ferns were evolving. Today similar vegetation requires mean temperatures in the range 25 ± 10°C and humidity around 50% or more. The first coal swamps started to evolve around 340 × 10^6 years ago, and a vast withdrawal of carbon started that must have changed the atmospheric balance.

Because of oscillatory ("eustatic") rise and fall of sea level (mainly due to accelerations or slowing down of sea-floor spreading), the withdrawal of $CaCO_3$ from the oceans was irregular and paralleled a long history of climatic change marked by alternating global oceanic conditions and continental (low sea level) conditions. Glaciation was restricted to the stages of continentality, that is, about 450 and 250 × 10^6 years ago). At other times (with high "oceanicity") there were worldwide mild, warm, and wet conditions.

Fifth stage. This is the stage from 250 to 100 × 10^6 years ago. Another ice age preceded the great Mesozoic age of mild conditions. This was the "age of reptiles," and it is possible that the atmospheric carbon dioxide level was somewhat higher than today. Only after its termination did mammals become dominant. Phases of occasionally accelerated volcanic activity must have raised the CO_2 of the atmosphere-ocean-sediment reserve, whereas progressive burial of carbon (coal) and $CaCO_3$ (limestone) during sedimentation tended to keep a balance.

Extreme changes in the global carbon dioxide budget were almost certainly buffered by the bicarbonate reaction in sea water and the presence of limestone ($CaCO_3$) which would protect the atmosphere from catastrophic lowering of oceanic pH.

Sixth stage. This is the last major cycle in Earth history, from 100 × 10^6 years ago till today, marked by steady deterioration of climates and increasing continentality and planetary extremes. The present ice age began 2 × 10^6 years ago and may well continue (with ups and downs) for a long time.

Summary. It is known that although there have been very large climatic changes in the past, they could not have surpassed certain limits because the fossil record of extinct life shows that favorable metabolic environments (about 25 ± 10°C) persisted in spite of all revolutionary events. There is a "law of biologic continuity" that has never once been violated throughout geologic time. There has never been a time when the entire global biota was wiped out by some climatic, physical, or geochemical catastrophe. Life has continued to evolve, albeit unevenly. There seems little doubt that climatic changes, aided and abetted by changes in the Earth's atmosphere, physical geography, and geophysical fields (notably magnetic), have been a major force in organic evolution. The geologic record shows that major exinctions and new speciation tend to coincide in time with major events in paleogeography and paleomagnetic and climatic history. The specific mechanisms of all these changes constitute one of the major challenges of science. [RHODES W. FAIRBRIDGE]

Bibliography: C. E. P. Brooks, *Climate through the Ages*, 1949; R. W. Fairbridge (ed.), *Encyclopedia of Atmospheric Sciences and Astrogeology*, 1967; K. A. Kvenvolden (ed.), *Geochemistry and the Origin of Life*, 1975; H. H. Lamb, *The Changing Climate*, 1966; A. E. M. Nairn (ed.), *Problems in Paleoclimatology*, 1968; M. Schwarzbach, *Climates of the Past*, 1963; G. E. Williams, *Geol. Mag.*, 112(5):441–544, September 1975; A. E. Wright and F. Moseley (eds.), *Ice Ages: Ancient and Modern*, 1975.

Paleoecology

The ecology of the geologic past, a study of the relations of fossil organisms to each other and to the environments in which they lived. Paleoecology is a study based on inferences and interpretations and on the basic assumption that the animals and plants of the past lived under essentially the same environmental conditions as do living relatives.

Scope and aims. As in studies dealing with living organisms, all types of environments are considered in paleoecology: marine, brackish and fresh waters, and land areas. Marine environments have received a major share of attention because most of the fossils that make up the paleontological record are contained in sediments that were deposited in the sea.

The aim of paleoecology is to infer, in terms of present-day conditions, the physical environments in which fossil organisms lived and their relations to each other in the changing environments of the past. In studying the record as preserved in the rocks, the paleoecologist attempts, as does the ecologist, to make a complete list of the animals and plants that inhabited a particular area and to obtain an estimate of their relative abundance. This is a difficult task even for the ecologist. For example, to compile such a census along a given stretch of coast today is not easy, because the most intensely studied shores continually yield new occurrences during the course of investigations. The chief cause for this is the rarity of some species, but there are other difficulties: Some species burrow far below the surface and others appear only at night or at certain seasons of the year. The comparable task of a paleoecologist is still more difficult. He has access only to a limited amount of sea bottom, and in the beds that were laid down at any particular time the remains of only a small fraction of the population then living are preserved. Furthermore, this meager record may contain a mixture of forms, including some that lived at the spot of burial and some that were brought in from other and perhaps quite different environments.

The first written observations on the subject now recognized as paleoecology were made by the Greeks about 500 B.C., but these were few and widely scattered. What might be called modern ecology did not begin to take form until the early part of the 18th century and paleoecology began to develop at about the same time.

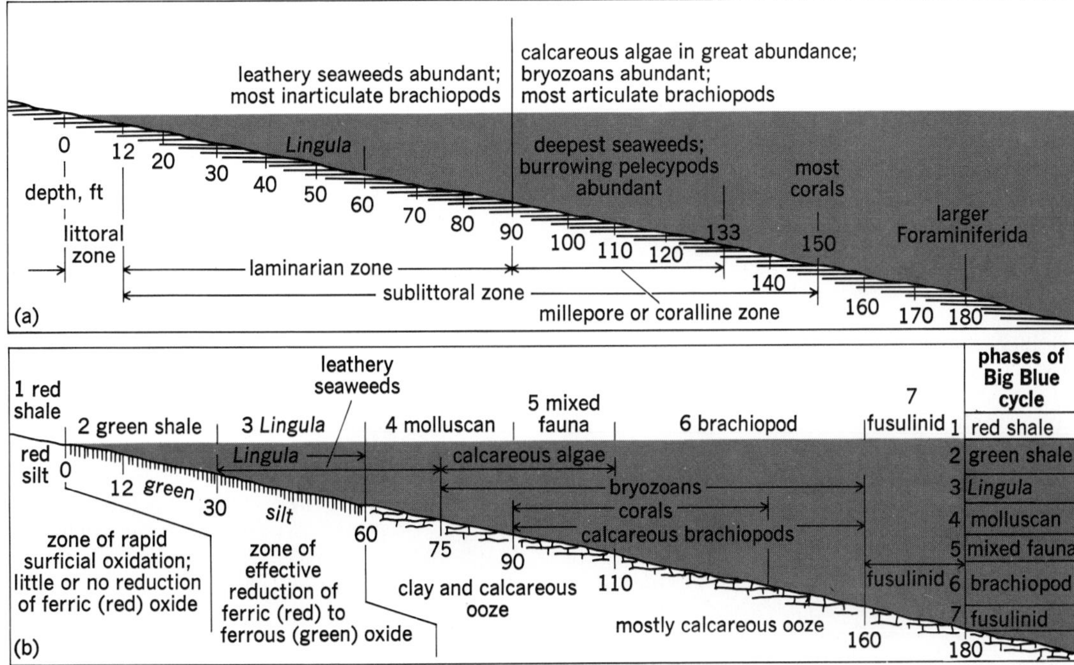

Fig. 1. Sea-bottom zones in existing and ancient seas. (*a*) Ideal distribution of benthonic organisms in the shallow waters of existing seas. (*b*) Restoration of the ancient sea-bottom zones in Big Blue (Permian) time in Kansas. (*After M. K. Elias, Depth of deposition of the Big Blue, Late Paleozoic, sediments in Kansas, Bull. Geol. Soc. Amer., 48(3):403–432, 1937*)

Terms. Studies of the distribution of living plant and animal life, both on the land and in the sea, have led to the recognition of communities or assemblages, each of which is adjusted to the physical and biological conditions that make up its environment. In the sea, for example, the assemblages of animals and plants that live at intertidal levels differ from those found at greater depths. These fairly obvious distinctions were recognized before 1860 and names were given to major bathymetric and biogeographic zones. The major subdivisions of the sea are divided into smaller units. All such units are referred to as facies. Thus, a deep-water assemblage is referred to as a bathyal facies, and an assemblage from the inner shelf area is known as inner neritic facies. Paleoecologists attempt to recognize comparable facies in the fossiliferous rocks that make up a large part of the Earth's crust (Fig. 1).

A group of organisms that live together as a community is termed a biocenose by ecologists. This useful term cannot be carried over directly into paleoecology for two reasons: (1) Many of the organisms that made up an important part of a given community do not lend themselves to fossilization and are rarely preserved; and (2) before an assemblage is buried and thus fossilized, the organic composition may be radically altered by the addition of shells or other skeletal parts of animals and plants that lived elsewhere under different conditions. A fossil assemblage is properly referred to as a death assemblage (thanatocenose).

An area occupied by a recognizable community of organisms is called a biotope. Such areas may be easily recognized and delimited in the sea or on the land today, but are more difficult to determine in ancient sediments because the extraneous elements mentioned must be recognized and eliminated.

Nature of evidence. Some paleoecological interpretations are based upon data obtained from the fossils themselves, others on features of the rocks that contain the fossils; in many cases data from both sources are used.

Data from fossils. The shape of a cephalopod shell or other swimming or crawling form may suggest a mode of locomotion; the thin and fragile shells of certain foraminifers or gastropods may point clearly to a pelagic existence. Corals and sedentary or burrowing pelecypods and brachiopods may be found in position of growth, indicating that they probably were buried in the place where they lived. When the members of a fossil assemblage are compared with living relatives whose life habits are known (Fig. 2), it may become apparent that the fossil assemblage is a mixed one with representatives from more than one environment. Such an interpretation may be confirmed if some members of the group exhibit signs of breakage or wear or if only the lighter unattached valves of sedentary forms are present. Studies of the isotopic composition of the shells may indicate the temperature or some other aspect of the past environment. *See* PALEOECOLOGY, GEOCHEMICAL ASPECTS OF.

Many of the older rocks contain fossils that have no close living relatives. Among such extinct groups are the conodonts, the archaeocyathid "corals," and the graptolites. Speculation about the conditions under which such organisms lived may, of necessity, be based largely on the supposed living habits of associated organisms and partly on the lithology and structures of the sediments containing the fossils.

Knowledge of the living habits of existing organisms is not always a sure clue to the habits of ancestral types. Today, for example, the stalked crinoids are solitary forms found at great depths but their numerous ancestors in Paleozoic times were gregarious and inhabited shallow waters (Fig. 3).

Data from rocks. The rock containing the fossils may indeed be the major source of interpretative data. The texture of the enclosing sediments may reveal much about the site of deposition. In water-

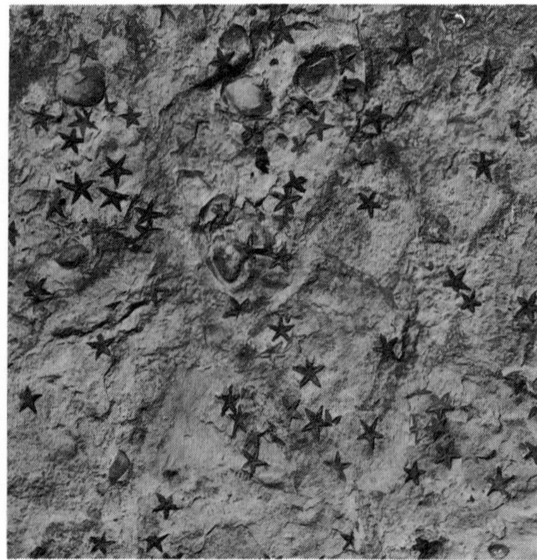

Fig. 2. Devonian sandstone, showing starfishes and clams. It has been suggested that the starfishes were feeding on the clams when they were buried. The width of the area shown is about 18 in. (*Photograph from D. W. Fisher, New York State Museum*)

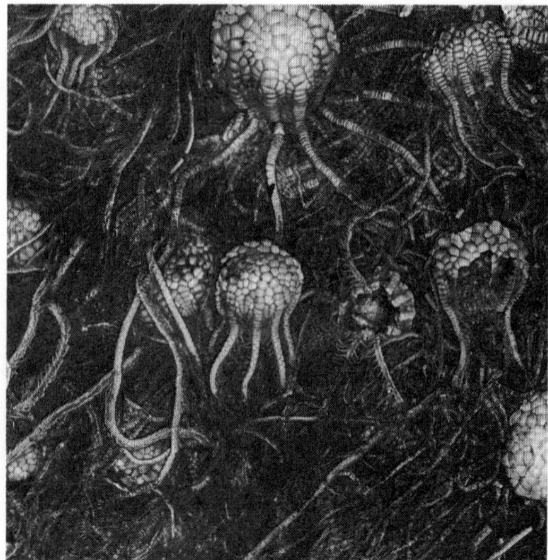

Fig. 3. Gregarious, free-swimming crinoids (*Uintacrinus socialis*) from the Cretaceous rocks of Kansas. Rounded cups are $1\frac{1}{2}-2\frac{1}{2}$ in. in diameter. (*Photograph from H. S. Ladd, ed., Geol. Soc. Amer. Memo no. 67, vol. 2, 1957, courtesy of Smithsonian Institution*)

Fig. 4. Manganese nodules. (*a*) At a depth of 3000 fathoms in the North Atlantic. Largest nodules are 5 in. in diameter (*photograph by D. M. Owen, Woods Hole Oceanographic Institution, from H. S. Ladd, ed., Geol. Soc. Amer. Memo. no. 67, vol 2, 1957*). (*b*) In red deep-sea clay of Cretaceous age on Island of Timor, Indonesia (*photograph by H. G. Jonker, courtesy of J. G. Ubaghs, Mineralogisch Geologisch Museum, Delft, from H. S. Ladd, ed., Geol. Soc. Amer. Memo. no. 67, vol. 2, 1957*).

laid sediments the occurrence of graded beds, that is, beds in which the texture grades upward from coarse to fine, may point to landslides or some form of turbidity current that threw the sediment into suspension before permitting final settlement. The presence of certain types of ripple marks, mud cracks, rain prints, or other sedimentary structures may aid in determining the nature of the original environment. Other interpretations may be based on a high content of organic matter or lime carbonate, or on the presence of oolites, glauconite, or nodules of phosphorite or manganese oxides (Fig. 4). The occurrence of such materials is significant because limnologists and oceanographers have studied the areas in which they are being formed today. *See* SEDIMENTARY ROCKS.

Data from fossils and rocks. Combinations of the above-mentioned types of data may give a suggestion or a clear indication, in the case of water-laid sediments, about such features as type of bottom, nearness to land, depth, agitation, and turbidity. Specific examples to illustrate this type of interpretation are given below.

Common sediments such as mud, silt, sand, and gravel are deposited under a variety of conditions. When these deposits are elevated at a later geologic time as shale, siltstone, sandstone, or conglomerate, the exact type of the original sediment may mean very little except to indicate that the lake or sea bottom at the time of deposition was muddy, sandy, or covered with boulders. However, when the beds in a given elevated section show variable texture, they preserve a record of changing times, possibly a record of unusual events. Such records are preserved in sediments deposited during Tertiary and Pleistocene times in basins along the coast of California. With the aid of subsidence, many thousands of feet of beds were deposited. Large parts of the section are fine-grained shales and siltstones whose contained Foraminifera (when compared with living forms collected from known depth zones) indicate depths at which

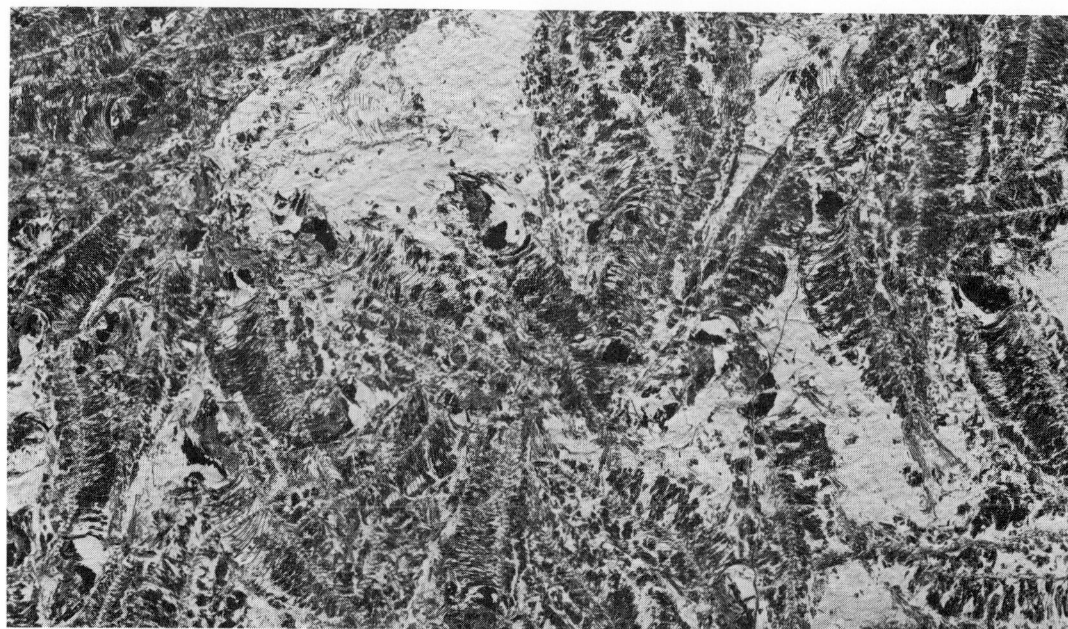

Fig. 5. Catastrophic death in the Miocene. Skeletons of herring preserved on a bedding plane of diatomaceous earth in the Monterey shale of Lompoc, Calif. The skeletons are 6–8 in. long. (*Photograph from A. B. Cumings, Johns-Manville Co., in H. S. Ladd, ed., Geol. Soc. Amer. Memo. no. 67, vol. 2, 1957*)

deposition occurred as great as 4000 ft. Sandstones alternate with the fine-grained sediments and many of these are graded, indicating that they probably were transported into the basin by turbidity currents. Beds of gravel and massive conglomerate also occur and these are believed to have been emplaced by landslides. During parts of the Pleistocene the basins were filled with marine sediments, and terrestrial sediments containing the remains of land vertebrates were laid down. *See* SEDIMENTATION.

Other examples involve specialized types of sediments such as black shales and limestones. Ecologists have described many areas in which black muds are being deposited today, in which there is little or no circulation, in which the supply of oxygen is low, and consequently, in which there is practically no benthonic (bottom-dwelling) life. Many of the black shales of the geologic column record deposition under such conditions in past times. R. Ruedemann has cited specifically the dark shales that are widely developed in eastern New York and concluded that planktonic (floating and weakly swimming) organisms were brought in freely by currents, but that at deeper levels circulation was poor and toxic conditions impoverished or prevented the existence of benthonic life. *See* BLACK SHALE.

An environment comparable to that of the black shales described by Ruedemann is recorded in the La Luna limestone and its equivalents, which constitute a unique lithologic unit 500–3000 ft thick widely developed over much of northern South America and some nearby regions in Cretaceous time. H. Hedberg has showed that the La Luna, a dark carbonaceous limestone, is composed almost entirely of the tests of pelagic Foraminiferida and that large fossils are rare. He expressed the opinion that at the time of deposition the sediment was a *Globigerina* ooze and speculated that life in the La Luna seas was almost exclusively planktonic. Thus, according to this theory, the bottom waters developed a toxicity (because of lack of circulation) that rendered them uninhabitable by marine

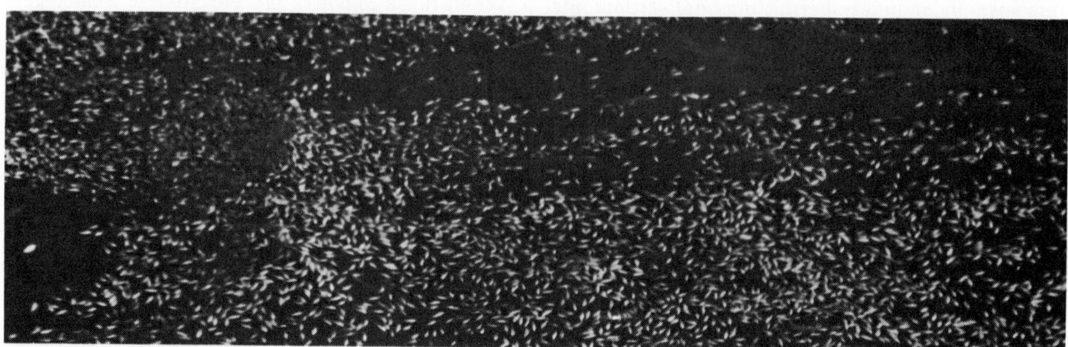

Fig. 6. Fishes killed on a catastrophic scale by a red tide in the Gulf of Mexico, 5 mi south of Sanibel Island, November, 1953. (*Photograph by Kenneth Marvin, U.S. Fish and Wildlife Service, from H. S. Ladd, ed., Geol. Soc. Amer. Memo. no. 67, vol. 2, 1957*)

benthonic animals. Such conditions would permit the accumulation of dead plankton on a sea floor undisturbed by bottom scavengers. The toxicity and the lack of oxygen were thought to have prevented rapid bacterial decay and permitted the accumulation of soft organic matter. Much of this material is believed to be still present in the form of carbonaceous-bituminous matter with some free petroleum. The rock appears in many ways to be an ideal petroleum source bed. *See* LIMESTONE.

Relations to other sciences. The interpretations made by the paleoecologist have direct applications to other fields of earth science, especially paleontology and stratigraphy. In paleontology, for example, the investigator is faced with the problem of determining the significance of concentrations of fossils on individual bedding planes in many parts of the geologic column. At Lompoc, Calif., the Miocene shales contain a bed the surface of which is covered with skeletons of a species of herring (Fig. 5). This bed once formed the bottom of a bay comparable to those along the coast today. In the bay, over an area of 4 mi^2, more than 1,000,000,000 herring died almost simultaneously. Ecological studies show that catastrophic death on a comparable scale occurs in the sea today, brought on, in many instances, by the appearance of red water or the red tide (Fig. 6). The upwelling of cold waters along a coast brings a rich supply of nutrients to the surface and this may lead to the development of noxious blooms of microscopic flagellates and dinoflagellates fatal to fishes. Ecological studies show that such red water is only one of several causes that may bring about mass mortality in the sea today and may have been responsible for the preservation of crowded layers of fossils in the past. Other causes include volcanic eruptions, tidal waves, and rapid changes in temperature and salinity.

Paleoecology is closely allied to stratigraphy, the branch of geology dealing with the sedimentary rocks and their order of superposition (chronological sequence). The paleoecologist, familiar with the distribution and living habits of life in existing environments, may be able to show, for example, that dissimilar faunas in two separated geologic sections may be contemporaneous or that two assemblages that resemble each other may merely reflect a similar environment, one being appreciably older than the other. *See* STRATIGRAPHY.

Paleoecological interpretations assist the geologist in reconstructing ancient landscapes, shorelines, and sedimentary basins. For this reason they are of value to those engaged in the search for oil and gas. The same is true in investigations of mineral deposits, especially those that have a stratigraphic control, such as phosphate and associated trace elements. *See* PALEOGEOGRAPHY; PALEONTOLOGY. [HARRY S. LADD]

Bibliography: D. B. Ager, *Principles of Paleoecology*, 1963; C. O. Dunbar and J. Rodgers, *Principles of Stratigraphy*, 1957; R. F. Hecker, *Introduction to Paleoecology*, 1965; H. D. Hedberg, Cretaceous limestone as petroleum source rock in northwestern Venezuela, *Bull. Amer. Ass. Petrol. Geol.*, 15(3):229–246, 1931; J. W. Hedgpeth (ed.), *Treatise on Marine Ecology and Paleoecology*, Geol. Soc. Amer. Memo. no. 67, vol. 1, 1957; G. E. Hutchinson, *A Treatise on Limnology*, vol. 1, 1957; J. Imbrie and N. Newell (eds.), *Approaches to Paleoecology*, 1964; H. S. Ladd (ed.), *Treatise on Marine Ecology and Paleoecology*, Geol. Soc. Amer. Memo. no. 67, vol. 2, 1957; H. Ladd, Ecology, paleontology, and stratigraphy, *Science*, 129(3341):69–78, 1959; R. Ruedemann, Ecology of black mud shales of eastern New York, *J. Paleontol.*, 9(1):79–91, 1935.

Paleoecology, geochemical aspects of

The study of the chemical and mineral composition of fossil organisms as these relate to the ecological and geochemical features of ancient environments. Chemical studies of fossils, particularly those that are abundant and widespread and that have related modern forms restricted either to marine or to fresh-water environments, yield valuable evidence concerning the paleoecologic and evolutionary aspects of skeletal building materials. For further treatment of the diagnostic use of fossils *see* FOSSIL; PALEOBIOCHEMISTRY; PALEOECOLOGY.

Chemical studies on the carbonate skeletons of Recent marine organisms have shown that the temperature and chemistry of sea water in which the organisms grow may influence the contents of strontium, Sr, and magnesium, Mg; the ratios of oxygen-18 to oxygen-16 (O^{18}/O^{16} ratios); and the aragonite-calcite ratios of the skeletons. Specific information about the relationships between the two ecologic factors and the characteristics of the skeletons is given in examples on studies done on present-day organisms and is followed by certain paleoecological applications of the data.

Skeletal mineralogy. In recent species of certain bryozoa, serpulid worms, pelecypods, and gastropods, the skeletons are commonly composed of aragonite and calcite. The two mineral modifications always form separate microarchitectural units of the skeletons. The aragonite-calcite ratios for the skeletons as a whole among individuals of a given species in a given environment vary with the age of the individual. The variations in the aragonite-calcite ratios within a species may be correlated with temperature; the higher the temperatures, the higher is the aragonite-calcite ratio. The dependence of the aragonite-calcite ratios upon temperature is most clearly defined in individual skeletons in which skeletal growth is entirely peripheral, such as those of the calcareous tubes of serpulid worms. Figure 1 shows the aragonite-calcite ratios for consecutive growth increments of

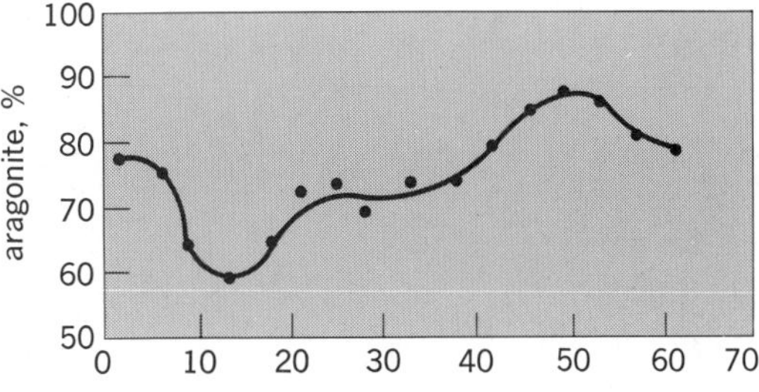

Fig. 1. Cyclic changes in aragonite-calcite ratios for consecutive growth increments of a worm tube from Bermuda inshore waters, with yearly temperature variations.

a worm tube collected from the Bermuda inshore waters where the yearly temperature variations are large, ranging from 16 to 30C°. The specimen was taken alive in early summer prior to the seasonal miximum of water temperatures. The aragonite-calcite ratios are plotted against the length of successive growth increments, with the last increment shown at the left of the diagram. The effect of temperature is shown by the cyclic changes in the aragonite-calcite ratios. Summer growth zones are 90% aragonite, whereas the minimum value for winter growth is 60% aragonite.

The physiology of the organisms also appears to influence the aragonite-calcite ratios. Individuals of different species grown in the same temperature may show differences in aragonite-calcite ratios. A more limited effect of temperature on the skeletal mineralogy is observed in certain gastropods and pelecypods which range from tropical or subtropical to temperate waters. The shells of these mollusks consist of 100% aragonite in all but those that live in the coldest waters. Traces of calcite, ranging up to 10%, are found at the coldest temperatures of their range. Green, brown, and certain red algae illustrate dependence of carbonate precipitation on temperature. These aragonite-precipitating species are essentially limited to the tropical belt bounded by the 16°C isotherms for the coldest month of the year. Evidently the aragonite is a passive warm-water precipitate.

The aragonite-calcite ratios may also depend upon the water chemistry. This is indicated in species which range from mean ocean to brackish waters. The few aragonite-calcite ratios determined for individuals from waters of low salinity appear to be noticeably higher than the ratios for individuals in the same temperature regime from normal marine waters.

Magnesium content. The Mg concentrations of calcareous skeletons in Recent species depend on several factors: the mineral form, the phylogenetic level of the species, the water temperature, and the water chemistry. The calcitic structure can accommodate a considerably larger amount of Mg in solid solution than can the aragonitic structure. Biological aragonite precipitates therefore rarely have over 1 mole % $MgCO_3$, whereas the calcitic ones may contain as much as 19 mole %.

In calcitic skeletons of individuals within the same phylogenetic class the $MgCO_3$ content shows an increase with higher environmental temperatures. The relative concentrations of Mg and the slopes of the magnesium-temperature curves differ for different classes, as shown in Fig. 2. The slopes of the magnesium-temperature curves tend to become lower as the phylogenetic rank of the class becomes higher. However, data on an echinoid species indicate that the magnesium-temperature curve for a single species may deviate from the curve for the class as a whole.

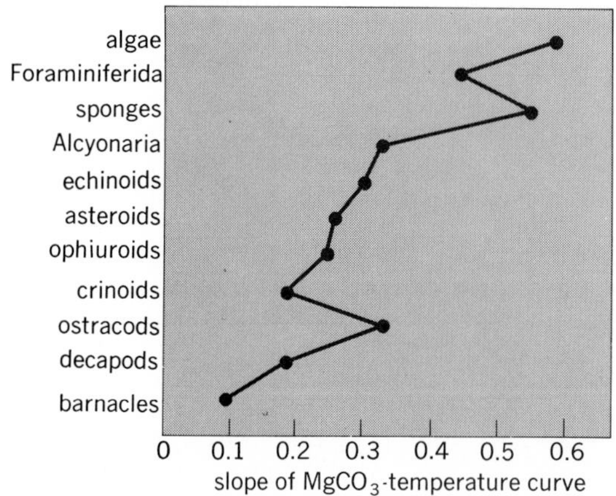

Fig. 2. Relation between slope of temperature-magnesium curve and organic complexity.

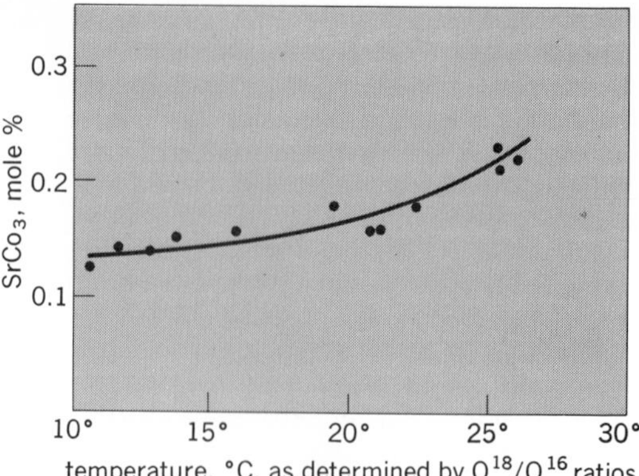

Fig. 3. Relationship between strontium content and temperature in articulate brachiopods.

The magnesium content of carbonate skeletons is also affected by changes in the chemistry of the sea water. In the same temperature regime echinoids from hyposaline waters have lower magnesium contents, and articulate brachiopods from hypersaline waters have higher magnesium contents than conspecific forms or related species from mean ocean waters.

Strontium content. The Sr content of the calcareous skeletons in present-day species is affected by the same factors which influence the Mg content. The effect of crystal form, however, differs from that of Mg in that Sr is more readily accommodated in the aragonitic than in the calcitic structure. The $SrCO_3$ content of calcitic skeletons rarely exceeds 0.4 mole %, whereas in aragonitic skeletons its content may be as high as 1.3 mole %.

An effect of temperature upon the Sr content in calcareous skeletons has been demonstrated so far only in two groups of organisms, the articulate brachiopods and the echinoids. The crystal form in both groups is calcite. Figure 3 shows the relationship between Sr content and temperature in articulate brachiopods. The temperature values are based on determinations of the O^{18}/O^{16} ratios of total shells, corrected for the O^{18}/O^{16} ratios of the waters from which the specimens were derived. *See* GEOLOGIC THERMOMETRY.

The $SrCO_3$ content is shown to increase with higher environmental temperatures. The curve is based on data from samples of species from several superfamilies. The samples include, however, pairs of conspecific and congeneric forms from different temperature regimes. In the echinoid *Dendraster* the relationship between Sr content and temperature differs from that in the articulate brachiopods. The Sr content in the echinoids studied decreases with elevation in environmental temperatures.

Organisms grown under either controlled or natural conditions show that the Sr content of calcareous skeletons is also dependent upon the chemistry of the waters. Fresh-water gastropods (*Physa*) grown in waters of increasing Sr content show a corresponding increase of strontium in their shells. The shells of articulate brachiopods taken from the hypersaline waters of the Mediterranean have a higher Sr content than species which grew in the same temperature regime but in mean ocean waters.

Paleoecologic applications. The foregoing demonstrates that certain chemical and mineralogical properties of the skeletons of marine organisms are influenced by the temperature and chemistry of sea water. It should therefore be possible, by studying the skeletal carbonates of marine fossils, to determine the temperature and chemistry of the oceans of the past. Because both environmental factors influence organisms simultaneously, neither should be studied independently. It is essential to determine that the chemistry of the skeletons is unaltered by diagenetic change. *See* DIAGENESIS.

These difficulties may be largely overcome by considering, in single fossil specimens, all the properties that are known to be influenced by the two environmental factors. Fossils chosen from a single class have been studied in this manner. The O^{18}/O^{16} ratios, the Mg concentrations, and the Sr concentrations in fossil articulate brachiopods ranging in age from the Pliocene to the Late Mississippian have been determined. The temperature values based on these determinations agree within 3°C in each specimen. The results indicate that the shells are chemically unaltered and that the organisms grew in waters chemically similar to mean ocean water today. If the organisms had lived in insolated water, the temperature values indicated by the O^{18}/O^{16} ratios would be noticeably lower than those indicated by the trace elements. A Permian brachiopod which has been studied shows this effect. In samples that had been diagenetically altered by fresh water, the temperature values determined by oxygen isotopes would be higher than those indicated by the trace elements. Several late Paleozoic brachiopod samples appear to have been altered in this way. The close agreement of the temperature determination in the bulk of the samples studied indicates that the chemistry of mean ocean water has been essentially the same for the last 2.5×10^8 years.

Several of the unaltered fossil brachiopods used in the above study are from assemblages containing belemnites which were used for paleotemperature determinations by means of the O^{18}/O^{16} method alone. The close agreement between temperature values determined for the brachiopods and the belemnites adds to one's confidence in paleotemperature determinations, based largely on belemnite rostra, for the Middle and Upper Cretaceous.

Serpulid worm tubes from the Upper Cretaceous of the Coon Creek formation in southern Tennessee illustrate the application of mineralogical studies to paleoecology. The worm tubes consist of an outer calcitic and an inner aragonitic layer. Longitudinal cross sections of the tubes show rhythmic variations of the diameters of the calcitic and aragonitic layers similar to those found in Recent species in response to environmental temperatures. [HEINZ A. LOWENSTAM]

Bibliography D. B. Ager, *Principles of Paleoecology*, 1962; K. E. Chave, Aspects of the biogeochemistry of magnesium, 1: Calcareous marine organisms, *J. Geol.*, 62(3):266–283, 1954; R. F. Hecker, *Introduction to Paleoecology*, 1965; H. A. Lowenstam, Systematics, paleoecologic and evolutionary aspects of skeletal building materials, in status of invertebrate paleontology, 1953, *Bull. Mus. Comp. Zool.*, 112:287–317, 1954; H. A. Lowenstam and S. Epstein, Paleotemperatures of the post-Aptian Cretaceous as determined by the oxygen-isotope method, *J. Geol.*, 62(3):207–248, 1954; H. T. Odum, Biogeochemical deposition of strontium, *Inst. Mar. Sci.*, 4(2):38–114, 1957; O. H. Pilkey, Effects of water temperature and salinity on skeletal magnesium and strontium uptake by *Dendraster* (abstract), *Geol. Soc. Amer., Rocky Mt. Sect.*, pp. 18–19, 1959.

Paleogeography

The geography of the geologic past; although the term is commonly associated with maps, it concerns all physical aspects of an area that can be determined from the study of the rocks.

Geography deals with the face of the Earth at a particular time, the present or some interval in the historic past. Paleogeography in the strictest sense concerns the geography of moments or very short time spans in the past. It considers the distributions of lands and seas, their elevations, depths, and forms. A number of analogous matters commonly are treated as a part of paleogeography, though really stratigraphic, involving spans of time during which a thickness of sediments was laid. *See* STRATIGRAPHY.

Time and correlations. Paleogeography generally involves the correlation of the rock record and events at the time being considered. In a local area or province, the criteria may be lithic, inasmuch as conditions may have been such that a single kind of sediment was synchronously deposited throughout a considerable area of land or sea. The identification of time is substantiated by study of organisms, for fossils are the principal means of carrying time correlations among distant places. Isotope ratios among uranium-lead, thorium-lead, potassium-argon, strontium-rubidium, and carbon isotopes are used increasingly in varying degrees of accuracy for differing spans of time. The problems of correlation are in the domain of stratigraphy, but they are basic to paleogeography. *See* ROCK, AGE DETERMINATION OF.

Map projections. The preparation of paleogeographic maps involves not only the same problems of distortion of shapes, forms, and areas that are encountered in all map projections, but many additional difficulties. Most available geologic maps fail to present the proper relative positions of the rocks at the time that the paleogeographic map portrays. When rocks are deformed by folding or faulting after their deposition, the positions and directions between points on opposite sides of each fold or fault have been changed. If a continent has moved relative to another, or a great mountain range has risen through compression, significant deviations develop in the subsequent and present geography from that prior to the deformation. Maps that reconstruct the original relative positions of rocks are known as palinspastic maps. Generally the limits of error in the paleogeography are so great that the present geographic base maps are reasonably satisfactory.

Paleogeographic maps. The simplest forms of paleogeographic maps show the distribution of lands and seas. The study of sedimentary petrology permits determination of the nature of the source lands of the time. Orientations of depositional structures such as ripple marks, cross stratification, flow casts, and elongations of particles and organisms reveal directions of streams, currents, and winds, aspects of paleoclimatology. The elevations of lands can be shown approximately by hypsometric contours of elevation, with colors or shades as used for present geography; more often, lands are shown by standard geomorphic landform symbols that give better impressions of the character of lands and less emphasis on elevations. Paleogeologic maps showing the pattern of rocks on the surface at a past time aid in the interpretation of landforms. Paleolithologic maps showing bottom sediment patterns suggest whether rocks were laid in depths of strong wave action or in quieter water of deeps or broad shoals; lines of equal sediment property, isoliths, can be drawn for many parameters. The organisms in the strata give strong indication of the environments of deposition, the paleoecology. From the study of these and other data, judgments can be reached of water depths and current flows that suggest bathymetry such as is expressed in bathymetric maps. Comparisons between the kinds of rocks and the interpretations of elevations of sources and depths of sites of deposition lead to judgments on regional stability and tectonism. *See* PALEOCLIMATOLOGY; PALEOGEOLOGY.

Stratigraphic maps. In addition to studies that concern but instants of time, paleogeography in its broader sense may include the regional distribution of stratigraphic data, such as thicknesses of rocks, representing a considerable span of time. The plotting of thicknesses of sedimentary rocks lying between two surfaces of deposition is by isopachs, or lines of equal thickness, on isopach maps. If the surfaces of deposition at the top and at the base of the sequence mapped were horizontal planes, the thickness would represent the amount of warping or deformation of the lower plane prior to the formation of the upper, assuming a stable sea level. As depositional surfaces deviate from horizontality, isopach maps only approach being measures of structural change. The sequence within a time span or stratigraphic interval in an area includes rocks that may differ appreciably both laterally and vertically. Maps showing the ratios of rocks of different kinds, or their constituents, within a stratigraphic interval of some thickness representing a considerable span of time are lithofacies maps, in comparison with paleolithologic maps representing a single surface of deposition. Lithofacies maps can show such ratios as those of sand to silt, or calcium carbonate to calcium magnesium carbonate, or terrigenous sediment to indigenous or precipitated sediment. As long as the average represents conditions that prevailed through the time, lithofacies give evidence of land sources, depths, and other geographic factors. The difficulties of determining precise planes of synchroneity are such that lithofacies maps are the most commonly used expressions of sediment distributions. *See* FACIES.

Tectonic interpretation. A succession of paleogeographic maps showing changes in distributions of the many aspects derived from the study and interpretation of the rocks provides a basis for tectonic interpretations. Seas can spread or retreat because of rise or fall of sea level, that is, as an effect of eustatic movements. A spread over great areas may involve rise of only a few scores of feet but entail a tremendous volume of marine water. Such changes have been attributed to the addition of water through melting of waning glaciers; some of them may, however, be due to structural changes in ocean basins. On the other hand, the sea can spread because the crust of the Earth subsides along the coasts of lands or can retreat because the lands rise; such changes in relative elevation of the land through warping movements are called epeirogenic. Eustatic movements cause universal advances and retreats, but epeirogenic movements can be provincial or local. The distribution of sea and land depends on the balance between these two sorts of movements. These are the most general concern of paleogeography. *See* WARPING OF EARTH'S CRUST.

Knowledge of the geography of the past has not accumulated to a degree that there are world atlases of paleogeography; in fact, there are very few maps showing world paleogeography, and these are rather simple and crude. Series of maps have been prepared for such limited areas as the British Isles, in some detail for single systems in the United States, and in more general form for the paleogeography of North America. Paleogeography represents an end toward which the stratigraphic geologist directs his investigations. [MARSHALL KAY]

Bibliography: P. E. Potter and F. J. Pettijohn, *Paleocurrents and Basin Analysis*, 1963.

Paleogeology

The geology of the past, but a term applied particularly to the interpretation of the rocks at a surface of unconformity, that is, an old erosion surface concealed by the deposition of overlying sedimentary rocks. A paleogeologic map showing the distribution pattern of rocks beneath an unconformity can be interpreted like a geologic map of the present surface, permitting recognition of such structural features as anticlines and synclines having older and younger rocks on their axes. Such maps suggest the possible sites of petroleum reservoirs beneath concealing sediments, and indicate likely channels of fluid migration. *See* UNCONFORMITY.
[MARSHALL KAY]

Paleomagnetics

The study of the direction and intensity of the Earth's magnetic field throughout geological time. Paleomagnetic studies have been important in investigations designed to study the past position of the continents and have helped to reopen the discussion concerning the reality of continental drift. Paleomagnetism has also made a major contribution in demonstrating that the magnetic field of the Earth has two stable polarities, one in which the north-seeking end of the compass points toward the present north magnetic pole, which is regarded as the "normal" direction of the field, and the second in which the polarity of the field is reversed so that the north-seeking end of a compass would point toward the south. Paleomagnetic-intensity studies seem to indicate that the intensity of the Earth's magnetic field has not varied by much more than a factor of 2 throughout geological time.

Magnetization of rocks. Paleomagnetic studies are possible because certain rocks become permanently magnetized at the time of their formation and retain an accurate memory of this magnetic direction for long periods of time, in some cases for billions of years. Igneous rocks, such as lava flows, and intrusive rocks, such as granite and gabbro, become magnetized as the rock cools from the molten state. In cooling, the magnetic minerals in the rocks, such as magnetite (Fe_3O_4), pass through the Curie point. This is the temperature at which the mineral becomes permanently magnetized in the direction of the Earth's field. The Curie point of magnetite, for example, is 575°C. This type of magnetization is called thermal remanent magnetization (TRM). Its intensity of magnetization is proportional to the intensity of the Earth's magnetic field, a fact basic to paleomagnetic-intensity studies.

A second type of permanent magnetization is chemical remanent magnetization (CRM). This type of magnetization is acquired when a magnetic mineral such as hematite is grown at low temperature through the oxidation of some other iron mineral, such as magnetite or goethite. As the mineral grows, it becomes magnetized in the direction of any field which is present. Under natural conditions this is the magnetic field of the Earth.

Detrital remanent magnetization (DRM) occurs when a magnetic mineral which has previously acquired a TRM or CRM is disaggregated by erosion from its original lost rock and redeposited within a sedimentary rock. As the magnetic particle settles in still water, it tends to align itself in the direction of the Earth's field producing a net magnetization in the sediment in the direction of this field.

The types of magnetization mentioned above form the principal stable component of magnetization found in rocks. However, the original direction of magnetization in a rock is often masked by secondary or unstable components of magnetization which grow in the rock after its formation. The most common secondary component is viscous remanent magnetization (VRM). In most rocks there are portions of the magnetic minerals which are very easily realigned so that, if the direction of the Earth's magnetic field changes, these components realign themselves with the changing field. This process leads to the masking of the original direction of magnetization in the rock. A second very important way in which the original direction of magnetization is altered is when the outcrop from which the specimen is obtained is struck by lightning. In this case the direction of magnetization in the rock is remagnetized by the large magnetic fields produced by the lightning stroke. This secondary magnetization is called isothermal remanent magnetization (IRM).

Field sampling and measurement. Samples are collected from an outcrop of the rock unit to be investigated. The samples in modern investigations usually consist of drill cores. These cores are left standing attached to the parent rock and are oriented usually either with a magnetic compass or a sun compass before the core is marked and removed from its position. The number of samples obtained from a single site depends on the type and magnetic stability of the rock being sampled. In most cases a single site is regarded as representing the field for a single point in time and 4–10 samples are usually taken. For purposes of calculating a palaeomagnetic pole position it is necessary to sample extensively enough to average out the scatter due to secular variations. Most investigators believe that the secular variation averages out in periods of time of the order of 10^4 or 10^5 years; therefore most investigators try to spread the sites they sample so that the entire length of time represented in the rock unit is covered.

The samples are brought back to the laboratory and cut into cylinders. The direction of magnetization of these cylinders is then measured on a magnetometer, usually of the rock generator or astatic varieties. Very sensitive modern magnetometers can measure intensities down to the 1×10^{-8} electromagnetic unit per cubic centimeter (emu/cm^3) level. For a basis of comparison it can be noted that strongly magnetized lava flows have intensities of magnetization of 1×10^{-3} emu/cm^3.

Each direction is represented by the north-seeking vector which has a horizontal component or declination D and inclination I. The declination is measured as an angle from 0 to 360°. The inclination is the angle below the horizon (+) or above the horizon (−) from 0 to 90°. Using the values obtained from the magnetometer, it is possible to calculate algebraically the mean direction from within sites and formations. Adequate statistics for describing the scatter and distribution of these directions are available.

In any paleomagnetic study it is of utmost importance to try to determine whether the natural remanent magnetization (NRM) of any rock unit is stable for long periods of time. One way is through the fold test (Fig. 1). In this test, sites are collected from both limbs of an anticline or a syncline and the mean direction of magnetization is determined. The fold is then restored to the horizontal. If the magnetization has preceded the folding, the directions coincide after the beds are restored to the horizontal. If the magnetization has occurred after folding, the directions diverge on restoration to the horizontal. A second field test for stability is the conglomerate test. In this test a conglomerate

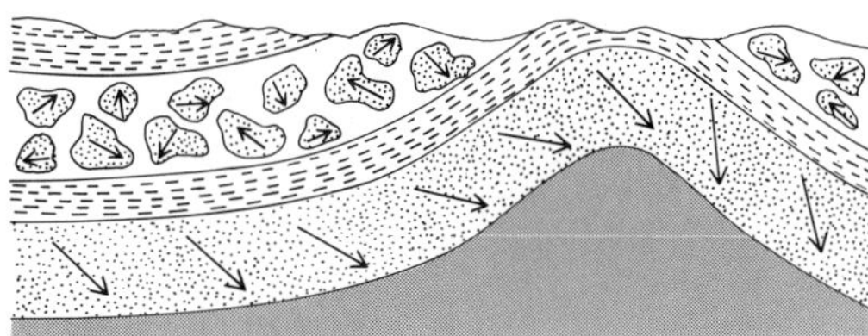

Fig. 1. Field relationships indicating stable magnetizations in an anticline using the conglomerate (upper) and fold tests. *(After Cox and Doell)*

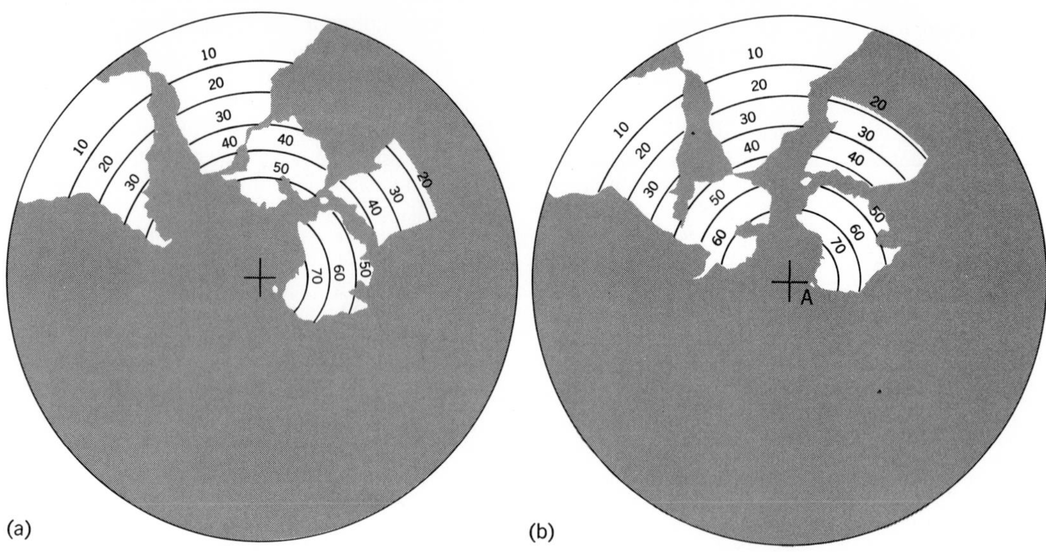

Fig. 2. Two possible reconstructions of the Southern Hemisphere for middle Mesozoic times using paleomagnetic data alone. (*a*) The paleomagnetic poles from all the southern continents are brought into coincidence. (*b*) The paleomagnetic pole from Antarctica is slightly displaced. (*After M. W. McElhinny*)

which contains boulders from the rock unit to be tested is sampled; if the directions of magnetization of the boulders is scattered, one can assume that the directions in the parent formation are stable over periods of time as long as the elapsed time since the formation of the conglomerate.

In paleomagnetic studies techniques are often employed to remove unwanted secondary components (VRM, IRM) of magnetization. The most widely used technique is alternating-field partial demagnetization (af demagnetization). In this technique the specimen is measured and then placed inside a solenoid in a specimen holder which rotates the specimen about two or three axes. The field of the solenoid is raised to a preselected level, held there for several seconds, then reduced smoothly to zero. The Earth's magnetic field is often canceled in the region of the specimen. The specimen is then removed and the direction is again measured in the magnetometer. Any change is noted and the specimen is replaced in the solenoid and treated in still higher fields. This technique is very effective in removing unwanted VRM or IRM for specimens, and aids in isolating the components with very high stability. In many studies this technique succeeds in removing much of the variation in direction both within and between sites.

A second laboratory technique often used is partial thermal demagnetization. In this technique specimens are placed in an oven and the Earth's field in the oven is canceled. The specimen is heated to a preselected level and cooled in field free space, and the direction is then measured. The process is repeated at increasingly high temperatures until the Curie temperature of the rock is exceeded. This technique serves to isolate the stable components of magnetization and is very effective in the treatment of redbeds where af demagnetization is often ineffective because of the high stability of the secondary components.

Very early in the history of the study of the Earth's magnetic field it was realized that the external field of the Earth could be described as a dipole field such as would arise from a bar magnet roughly aligned along the axis of rotation. The present dipole field of the Earth is not aligned precisely along the axis of rotation of the Earth. However, there are good theoretical reasons for believing that, on the average, the dipole field of the Earth aligns itself along the axis of rotation. Using the dipole formula and knowing the D and I for a site or a formation, it is possible to calculate the paleomagnetic pole position which gave rise to the measured declination and inclination. This is of great value when a comparison of directions from sites with a wide geographic distribution is made.

In the 1950s, when modern paleomagnetic work had just begun, it soon became clear that, although pole positions from Pleistocene and Recent rocks clustered around the present rotational pole of the Earth, poles from older and older rocks diverged more and more from the present axis of rotation so that, by lower Paleozoic time, poles from European rocks were falling near the Equator in the Pacific.

It also became clear that the pole positions from a stable continental area for selected periods of time gave results and pole positions which were in agreement with one another. However, pole positions from different continents for rocks of the same age did not always coincide. One possible way to reconcile these results from the different continents was the resurrection of the hypothesis of continental drift (Fig. 2). Geophysical observations from the ocean also lend support to this theory. *See* ROCK MAGNETISM.

[NEIL D. OPDYKE]

Bibliography: C. S. Gromme, R. T. Merrill, and J. Verhoogen, Paleomagnetism of Jurrassic and Cretaceous plutonic rocks in the Sierra Nevada, California, and its significance for polar wandering and continental drift, *J. Geophys. Res.*, 72:5661–5684, 1967; E. Irving, *Paleomagnetism and its Application to Geological Problems*, 1964; M. W. McElhinny et al., Geological and geophysical implications of paleomagnetic results from Africa, *Rev. Geophy.*, 6:160–175, 1968.

Paleontology

The study of animal history as recorded by fossil remains.

The fossil record includes a very diverse class of objects ranging from molds of microscopic bacteria in rocks more than 3×10^9 years old to unaltered bones of fossil men in ice-age gravel beds formed only a few thousand years ago. Quality of preservation ranges from the occasional occurrence of soft parts (skin and feathers, for example) to barely decipherable impressions made by shells in soft mud that hardened to rock. *See* FOSSIL.

The most common fossils are hard parts of various animal groups. Thus the fossil record is not an accurate account of the complete spectrum of ancient life but is biased in overrepresenting those forms with shells or skeletons. Fossilized worms are extremely rare, but it is not valid to make the supposition that worms were any less common in the geologic past than they are now.

The data of paleontology consist not only of the parts of organisms but also of records of their activities: tracks, trails, and burrows. Dinosaur footprints, for example, are very common in the Connecticut Valley. Even chemical compounds formed only by organisms can, if extracted from ancient rocks, be considered as part of the fossil record. Artifacts made by men, however, are not termed fossils, for these constitute the data of the related science of archeology, the study of human civilizations. *See* PALEOBIOCHEMISTRY.

Paleontology lies on the boundary between two disciplines, biology and geology. Various scientists have tried to place it more firmly in one camp than in the other, but such designations emphasize only a small area of the science's domain and do not acknowledge its entire range. If all types of paleontological research are regarded as equal in importance, then the geological and biological aspects must be granted equal weight. *See* GEOLOGY.

GEOLOGICAL ASPECTS

A major task of any historical science, such as geology, is to arrange events in a time sequence and to describe them as fully as possible. Geology, the study of Earth history, did not become a modern science until the 19th century, when a worldwide time scale based on fossils was established; Earth history could not be deciphered until events that occurred in different places were related to one another by their position in a standard time sequence (Fig. 1). The data provided by fossils are used to accomplish these tasks in the following ways.

Chronology. Many geologists of the 18th century tried to use rock type as a criterion for judging the relative age of rocks, saying that heavy rocks such as granites were older than lighter sandstones and limestones. But granites formed throughout Earth history and Cenozoic granites cannot be distinguished from Cambrian granites by mineral content. A species of organisms, however, is unique; it lives for only a short time before becoming extinct or evolving into something else, and once gone, it never reappears. Thus fossils can be used to correlate rocks; for if species A is found in both a Nevada limestone and a Colorado sandstone, it may be concluded that the rocks are approximately the same age (Fig. 2). Fossils of certain species, or index fossils, are ideally suited for correlation because they were abundantly distributed over a large geographic area, yet persisted for a very short time before becoming extinct. *See* INDEX FOSSILS; STRATIGRAPHY.

Of course, fossils only tell that a rock is older or younger than another; they do not give absolute age. The decay of radioactive minerals may provide an age in years, but this method is expensive and time-consuming, and cannot always be applied since most rocks lack suitable radioactive minerals. Correlation by fossils remains the standard

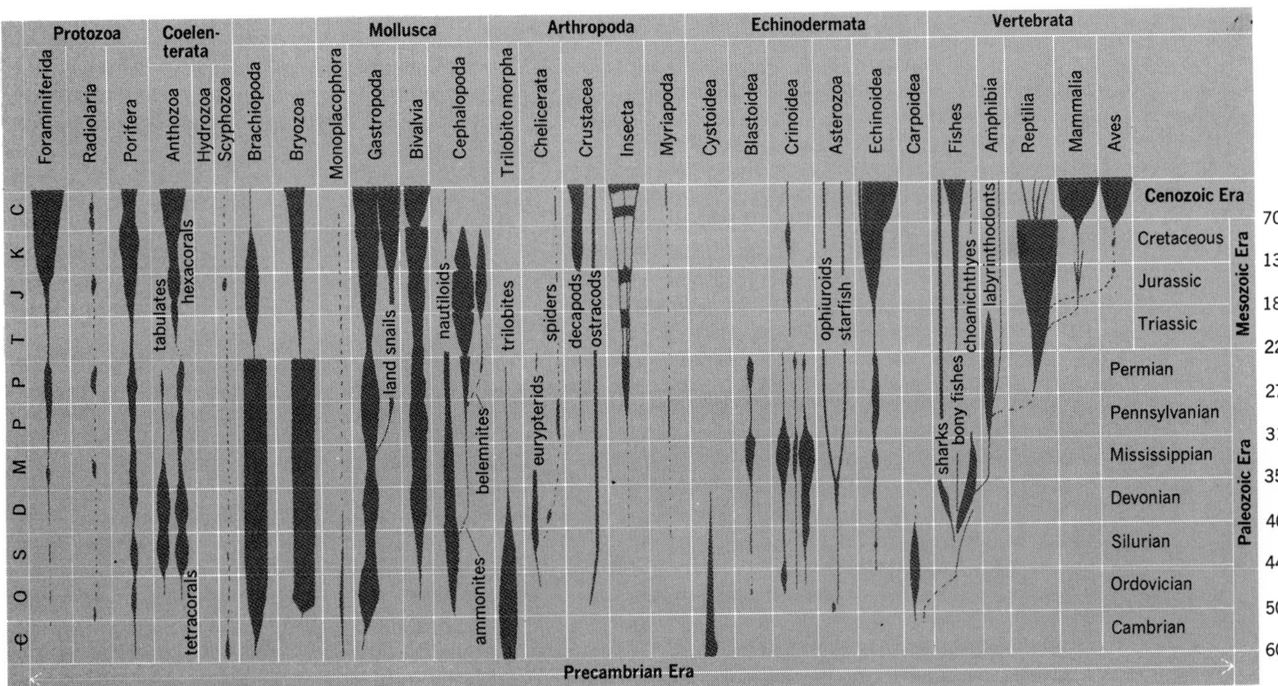

Fig. 1. Geologic distribution of animal life since the beginning of the Cambrian Period.

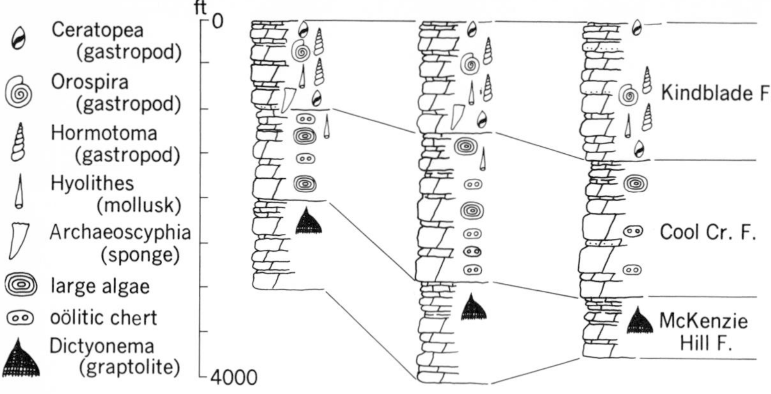

Fig. 2. Correlation of rocks by means of fossils. Although these three rock columns of Ordovician rocks in southern Oklahoma are all very similar in lithology, they can be divided and correlated on the basis of fossils.

method for comparing ages of events in different areas.

Determining ancient Earth's appearance. The physical appearance and climate of the Earth during a given period of the geologic past can be described from compilation and analysis of the data which is obtained through studies of the habitats of extant fauna, the geographic distribution of fossils, and the climatic preferences of ancient forms of life.

Local conditions. Life habits of fossils can be inferred by comparing them with modern organisms to which they are closely related. For example, all modern sea urchins live in marine environments. Since the Cretaceous limestones of southern England contain numerous fossil sea urchins, the conclusion is that this area was covered by ocean waters when these rocks were deposited. The study of fossils in relation to their physical and biological environment is called paleoecology. *See* PALEOECOLOGY.

Paleogeography. Fossils are indispensable guides for determining the positions of continents and seas in former times. The formation of the Isthmus of Panama can be dated, for example, by studying the distributions of marine and terrestrial fossils. Before North and South America were connected by this land bridge, Atlantic and Pacific marine faunas were very similar but the mammals of the two continents were completely different. However, in South American rocks deposited after the Isthmus was formed, there are fossils of North American mammals which had migrated over the newly formed land. Likewise, Atlantic and Pacific marine faunas, isolated from each other by the rise of the Isthmus, began to evolve in different directions; this increasing difference can be traced in the fossils of successively younger rocks. Fossils also help the scientist to decide whether continental drift has occurred. If South America and Africa were once united, their faunas should be similar during that time. As they drifted apart, there should be stronger and stronger faunal differences.

Even more striking is the difference between Northern and Southern Hemisphere faunas and floras. Many comparisons, particularly those of unbalanced biotas of Australia and the Pacific oceanic islands, can be reasonably explained by continental drift, when Gondwanaland (uniting South America, Africa, Australia, the oceanic islands, and Antarctica) began to drift apart. For example, one explanation for the almost exclusive marsupial fauna in Australia is that the marsupials may have originated in North America; by Cretaceous-Paleocene time they migrated to and diversified in South America, and eventually migrated to Australia via Antarctica before continental breakup and before the placentals could reach Antarctica. *See* PALEOGEOGRAPHY.

Paleoclimatology. The natural occurrence of polar bears always implies a cold climate. In the same way, it is possible to learn about ancient temperatures if the climatic preferences of fossilized organisms are known. There is, for example, a one-celled animal which tends to coil its shell one way when the water temperature is cold and the other way when it is warmer.

During the last million years, the oceans have been successively cooled and warmed as giant glaciers of the ice ages grew and melted. The most recent warming of the oceans (melting of the last continental ice sheet) can be dated by finding the age of the sediments in which the change of the fossil's coiling direction is noted. Cold- and warm-water planktonic faunas and floras recently identified in deep-oceanic sediments also record temperature changes. Extensive coring of oceanic sediments has helped reconstruct paleoclimates as far back as the Cretaceous when modern groups first appeared and, further, has helped date sea-floor spreading. *See* GEOLOGIC THERMOMETRY; PLATE TECTONICS.

BIOLOGICAL ASPECTS

Paleontology as a branch of the life sciences may be called the biology of fossils. Any technique applied by biologists to the study of living organisms is theoretically relevant to fossils, although many cannot be applied in practice because of the limitations of paleontological evidence. The rarity of preserved soft parts, for example, severely limits the possibilities for a physiology of fossil animals.

The most fundamental fact of paleontology is that organisms have changed throughout earth history and that each geological period has had its characteristic forms of life. Before Charles Darwin won acceptance for his theory of evolution, various explanations were offered for the differences between ancient and modern life. Some claimed that fossils were not organic remains, but manifestations of a plastic force in rocks; others attributed fossils to the devastation of Noah's flood; still others envisaged a whole series of catastrophes that destroyed life and led to the creation of an entirely new set of organisms. When Darwin published his theory of evolution in 1859, paleontological evidence for evolution was very meager but it began to accumulate quickly. The "feathered reptile" *Archaeopteryx*, an almost perfect link between reptiles and birds, was discovered early in the 1860s and the first fossil men were found a few years later. In the light of this evidence and the purview of a more scientific culture, evolution was accepted as the only reasonable explanation for change in the history of life. To a paleontologist, life's outstanding attribute is its evolution.

Evolutionary process and life history. An evolutionist has two major interests: first, to know how the process of evolution works; this is accomplished by studying the genetics and population structure of modern organisms; second, to reconstruct the events produced by this process, that is, to trace the history of life. This is the paleontologist's exclusive domain. Any modern animal group is merely a stage, frozen at one moment in time, of a dynamic, evolving lineage. Fossils give the only direct evidence of previous stages in these lineages. Horses and rhinoceroses, for example, are very different animals today, but the fossil history of both groups is traced to a single ancestral species that lived early in the Cenozoic Era. From such evidence, a tree of life can be constructed whereby the relationships among organisms can be understood.

Life properties. Evolution is responsible for life's outstanding properties: its diversity and its adaptation to environment.

Diversity. A paleontologist studies diversity by classifying organisms into basic units known as species. The number and geographic distribution of species in a more inclusive taxonomic group, such as vertebrates, can then be tabulated for different periods of geologic history. When this is done for all major groups, very definite patterns emerge. At the close of the Permian and Cretaceous periods, for example, extensive extinction of species occurred in many major groups at about the same time. Although the cause for these mass extinctions is unknown (theories range from large-scale shifts of sea level to bursts of cosmic radiation), they are major events in the history of life.

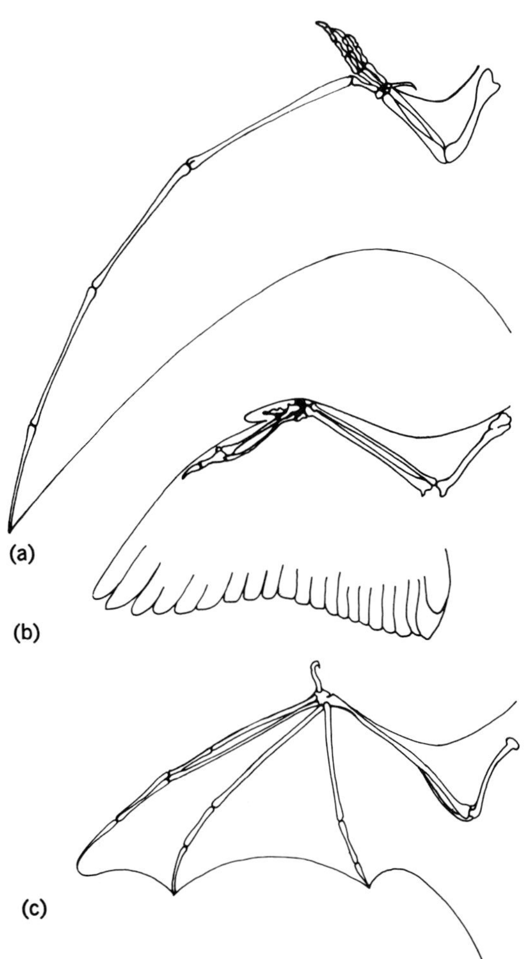

Fig. 3. Adaptation. Three groups of vertebrates independently developed wings for flight. (*a*) Pterodactyl (extinct reptilian pterosaur). (*b*) Bird. (*c*) Bat.

Adaptation. The term adaptation refers to the way organic form is fitted to function (Fig. 3). Given the raw materials of life, an engineer could scarcely design a better flying machine than a bird, yet the bird evolved naturally from reptilian ancestors which did not fly. Paleontologists study the development of new adaptations by a variety of techniques. For example, consider the increase in brain size between man's apelike ancestors and modern man. To study this a paleontologist employs the following: (1) studying the rocks in which fossil bones are found to look for signs of any environmental change which might have made increased brain size advantageous; (2) determining the advantages of a larger brain if possible; for example, it may, among other things, have conferred upon ancient man the ability to learn to control fire for warmth and cooking (charred bones of edible animals in caves inhabited by ancient men could be used as evidence for such an assertion); and (3) applying mathematical techniques to obtain a precise measurement of the rate of evolution.

Current research is directed toward the reconstruction of past environments by analyzing organisms according to their function and adaptation to particular environments. For example, clams, such as the mussel and oyster, are cemented to hard substrates in shallow water and are associated with other restricted organisms. Counterparts of this assemblage can be found throughout the fossil record, implying a similar habitat. *See* EVOLUTION, ORGANIC; EXTINCTION (BIOLOGY).

CAPSULE HISTORY OF LIFE

Although life was not present on the Earth when it first formed about 4.5×10^9 years ago, the raw materials necessary for its natural development were available. Electrical discharges and ultraviolet radiation induced the components of the original atmosphere to form complex organic compounds which later combined to form the prototype of a living cell. The oldest recognizable organisms so far discovered are bacterialike cells from South African cherts dated at 3.1×10^9 years old. By the time the Gunflint cherts of Canada were deposited 2×10^9 years ago, complex algae had evolved. With the exception of a few forms found in rocks just slightly older than 570×10^6 years, invertebrate animals are not found until the base of the Cambrian Period, 570×10^6 years ago. It is not known why so many groups of invertebrates made their first appearance as fossils in rocks of the same age; perhaps the base of the Cambrian marks a time during which animals first developed hard parts. The first vertebrate fossils, primitive fishes, occur in Lower Ordovician rocks. Vertebrates invaded the land with the evolution of amphibians in the Late Devonian, about 350×10^6 years ago. Reptiles evolved soon afterward in Mississippian times; the first true mammals are Jurassic in age. Man is a newcomer, a product of the last few million years. He is merely a single species, a natural product of an unplanned evolution; yet from a geological perspective, he has altered the Earth far more than it had changed in any comparable time since life evolved.

In addition to the first appearance of animal groups, another observation can be made of Fig. 1. The Cambrian is dominated by trilobites in terms of diversity of contained groups, that is, numbers of species, and actual abundance. These are not necessarily correlated because many animals are more prolific than others, with invertebrates tending to be more abundant than vertebrates. For example, Cretaceous reptiles appear to be as diverse as some invertebrate groups, yet in terms of actual abundance the Mollusca and Foraminifera are dominant.

In the Ordovician, trilobites are slightly less diverse than in the Cambrian but are numerically less abundant than the brachiopods and bryozoans. The most noticeable changes in the Paleozoic involve the dominance of one brachiopod order over another: the orthid brachiopods are important in the lower Paleozoic; the spiriferids in the middle Paleozoic; and the productids in the upper Paleozoic. Some animal groups form distinct global assemblages. A worldwide Devonian fauna is recognized by spiriferid brachiopods, trepostome bryozoans, tabulate sponges, and solitary and colonial corals. This fauna can be further subdivided into provinces which appear to be climate-dependent. The upper Paleozoic is characterized by the increased diversity of crinoids, blastoid echinoderms, cryptostome bryozoans, and productid brachiopods.

Mollusks first diversified in the Ordovician when nautiloids became abundant. By Devonian time the allied ammonoids were established, and became increasingly common during the upper Paleozoic. They gave rise to the ammonites, which were the most important invertebrate group of the Mesozoic. By the end of the Mesozoic the invertebrate fauna took on a modern appearance with the rapid expansion of the gastropods and clams. Dinosaurs and ammonites became extinct, while the mammals were poised for a striking diversification in the Cenozoic.

The first known plants, in the form of aqueous algae, are found in rocks dated about 10^9 years before present. Up until the lower Paleozoic, plants were marine, quite simply constructed, and relatively rare. By the middle Paleozoic, vascular plants had emerged and land plants were established. The flora quickly diversified and broad climatic zones were formed by late Paleozoic time. However, it was not until the late Mesozoic that the predominant Cenozoic and Recent flowering plants (angiosperms) appeared and rapidly expanded. Along with flowering plants, insects also experienced a high degree of diversification to become the largest group of all living forms. *See* ANIMAL EVOLUTION; PALEOBOTANY.

[ROGER L. BATTEN]

Bibliography: J. R. Beerbower, *Search for the Past*, 2d ed., 1968.

Paleosol

A soil that formed on a landscape of the geologic past, that is, an ancient soil. There are three kinds of paleosols: relict, buried, and exhumed.

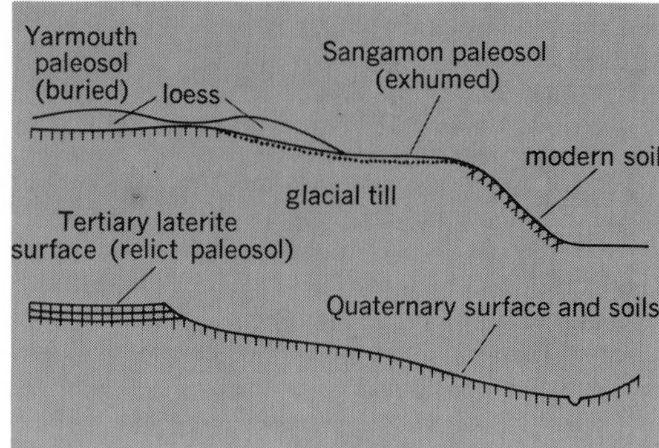

Fig. 1. Diagrammatic landscape profiles illustrating relict, buried, and exhumed paleosols.

Relict soils formed on preexisting landscapes but subsequently were not buried under younger sediments. For example, the Western Australia soils containing laterite generally are considered to be paleosols formed on landscapes during the Tertiary Period. Subsequent dissection of the region has caused the relict laterite landscapes to stand above younger land surfaces on which other kinds of soils have formed (Fig. 1). *See* SOIL.

Buried soils are soils formed on preexisting landscapes and subsequently buried by younger sediment or rock. These soils crop out in excavations, either natural, such as stream banks, or man-made, such as road cuts. Buried soils are common in many regions. For example, they occur beneath eolian sediments (loess) and glacial deposits (till) in the midcontinent region of North America, beneath lithified dunes in Bermuda, beneath alluvial-fan gravel in the deserts of southwestern United States (Fig. 2), and between basaltic lava flows in Hawaii.

Exhumed paleosols are soils that were buried but have been reexposed on the present land surface by erosion of the covering mantle. These soils occupy appreciable parts of the present land surface and are juxtaposed geographically to other

Fig. 2. Buried soil in the Las Cruces area in New Mexico. Horizons of buried soil (designated Ab, Bb, Cb) are readily discernible. The paleosol is buried by the younger alluvial-fan gravel (Cca).

soils more in harmony with the present environment.

Most paleosols are similar in their morphologies and properties to soils on the present land surface but are not necessarily analogs of the soils of the same area or region. Analyses of paleosols and comparison with present surface soils and their environments aid in reconstructing possible environments of landscapes of the geologic past. [ROBERT V. RUHE]

Paleozoic

One of the major divisions of Earth history, with a time span of approximately 340,000,000 years according to radioactive age measurements. The Paleozoic Era (signifying ancient life era) comprises the earlier two-thirds of the so-called Phanerozoic Eon (evident life), which is characterized by relatively abundant records of the plant and animal life of the Earth's past preserved as fossils in successive layers of sedimentary rocks. The deposits of Paleozoic age contrast very strongly with older rocks, collectively termed Cryptozoic (nonevident life), not only in richness of organic content but in their prevailing lack of metamorphic alteration and generally much more simple structure. Paleozoic formations are well represented by

PRECAMBRIAN	PALEOZOIC							MESOZOIC			CENOZOIC	
					CARBONIFEROUS							
	CAMBRIAN	ORDOVICIAN	SILURIAN	DEVONIAN	Mississippian	Pennsylvanian	PERMIAN	TRIASSIC	JURASSIC	CRETACEOUS	TERTIARY	QUATERNARY

wide distribution and great aggregate thickness on all of the continents, but classification of their main divisions, designated as systems, is based on early geologic studies in Europe. *See* STRATIGRAPHIC NOMENCLATURE.

Throughout most of the world, the Paleozoic Era is considered to embrace six geologic periods that correspond to, and are designated by, the names of geologic systems. In order from oldest to youngest, with estimates of their respective duration in millions of years given in parentheses, these are Cambrian (80), Ordovician (70), Silurian (20), Devonian (50), Carboniferous (75), and Permian (45). Each of these is treated by a separate article. *See* GEOLOGY. [RAYMOND C. MOORE]

Peat

A dark-brown or black residuum produced by the partial decomposition and disintegration of mosses, sedges, trees, and other plants that grow in marshes and other wet places. Forest-type peat, when buried and subjected to geological influences of pressure and heat, is the natural forerunner of most coal.

Peat may accumulate in depressions such as the coastal and tidal swamps in the Atlantic and Gulf Coast states, in abandoned oxbow lakes where sediments transported from a distance are deposited, and in depressions of glacial origin. Moor peat is formed in relatively elevated, poorly drained moss-covered areas, as in parts of Northern Europe. *See* COAL.

In the United States, where the principal use of peat is for soil improvement, the estimated reserve on an air-dried basis is 13,827,000 short tons. In Ireland and Sweden peat is used for domestic and even industrial fuel. In Germany peat is the source of low-grade montan wax. [GILBERT H. CADY]

Pectolite

A mineral inosilicate with composition $Ca_2NaSi_3O_8(OH)$, crystallizing in the triclinic system. Crystals are usually acicular in radiating aggregates (see illustration). There is perfect cleavage

Globular masses of needlelike pectolite crystals with calcite at left, found in Paterson, N. J. (*American Museum of Natural History specimen*)

parallel to the front and basal pinacoids yielding splintery fragments elongated on the *b* crystal axis. The hardness is 5 on Mohs scale, and the specific gravity is 2.75. The mineral is colorless, white, or gray with a vitreous to silky luster. Pectolite, a secondary mineral occurring in cavities in basalt and associated with zeolites, prehnite, apophyllite, and calcite, is found in the United States at Paterson, Bergen Hill, and Great Notch, N.J. *See* SILICATE MINERALS. [CORNELIUS S. HURLBUT, JR.]

Pegmatite

Exceptionally coarse-grained and relatively light-colored crystalline rock composed chiefly of minerals found in ordinary igneous rocks. Extreme variations in grain size also are characteristic (Fig. 1), and close associations with dominantly fine-grained aplites are common. Pegmatites are widespread and very abundant where they occur, especially in host rocks of Precambrian age, but their aggregate volume in the Earth's crust is small. Individual bodies, representing wide ranges in shape, size, and bulk composition, typically occur in much larger bodies of intrusive igneous rocks or in terranes of metamorphic rocks, from most of which they are readily distinguished by their unusual textures (Fig. 2) and often by concentrations of relatively rare minerals. Many pegmatites have been

Fig. 1. Contrasting textures in pegmatite of quartz monzonite composition, New Hampshire. Upper part is relatively fine grained and granitoid, and remainder comprises large, roughly faced crystals of alkali feldspar (light) set in slightly smoky quartz (dark).

economically valuable as sources of clays, feldspars, gem materials, industrial crystals, micas, silica, and special fluxes, as well as beryllium, bismuth, lithium, molybdenum, rare-earth, tantalum-niobium, thorium, tin, tungsten, and uranium minerals. *See* APLITE.

Compositional types. Most abundant by far are pegmatite bodies of granitic composition, corresponding in their major minerals to granite, quartz monzonite, granodiorite, or quartz diorite. In some regions or districts they contain notable quantities of the less common or rare elements such as As, B, Be, Bi, Ce, Cs, La, Li, Mo, Nb, Rb, Sb, Sn, Ta, Th, U, W, Y, and Zr. Diorite and gabbro pegmatites occur in many bodies of basic igneous rocks, but their total volume is very small. Syenite and nepheline syenite pegmatites are abundant in a few regions characterized by alkaline igneous rocks. Some of them are noted for their contents of As, Ce, La, Nb, Sb, Th, U, Zr, and other rare elements.

Shape and size. Individual bodies of pegmatite range in maximum dimension from a few inches to more than a mile, and in shape from simple sheets, lenses, pods, and pipes to highly compex masses with many bulges, branches, and other irregularities. Tabular to podlike masses of relatively small size commonly occur as segregations or local injections within bodies of cogenetic igneous rocks, where they represent almost the full spectrum of known pegmatite compositions. Granitic and syenitic pegmatites also occur in metamorphic rocks, some of them distributed satellitically about igneous plutons. Their contacts with the enclosing rocks may be sharp or gradational. Many bodies formed in strongly foliated or layered rocks are conformable with the host-rock structure (Fig. 2); others are markedly discordant, as if their emplacement were controlled by cross-cutting fractures or faults.

Mineral composition. Essential minerals (1) in granitic pegmatities are quartz, potash feldspar, and sodic plagioclase; (2) in syenitic pegmatites, alkali feldspars with or without feldspathoids; and (3) in diorite and gabbro pegmatites, soda-lime or

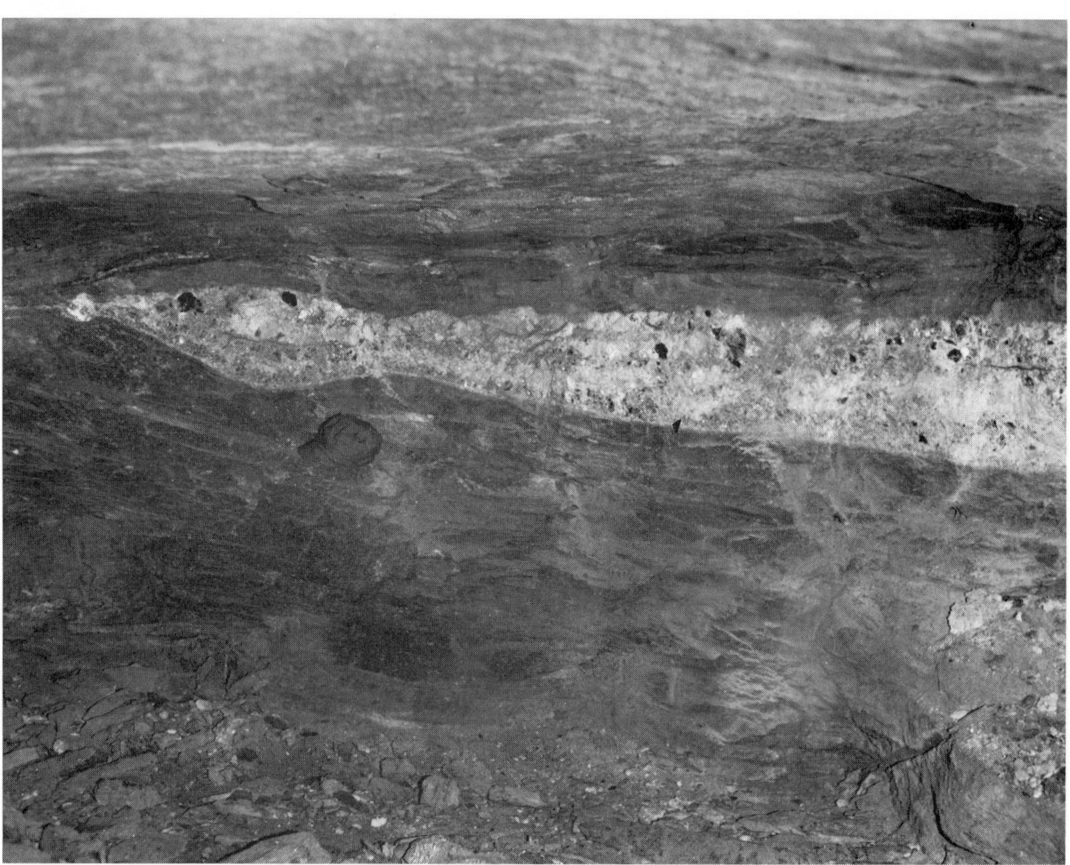

Fig. 2. Sill of faintly layered quartz diorite pegmatite, 6 to 12 in. (15 to 30 cm) thick, in fine-grained granitic gneiss, North Carolina. Sodic feldspar (light), quartz (darker), and muscovite (very dark) are the principal pegmatite minerals. (*From R. H. Jahns, The study of pegmatites, Econ. Geol., 50th Anniv. Vol., pp. 1025–1030, 1955*)

Fig. 3. Four contrasting zones in upper part of thick, gently dipping dike of very coarse-grained granite pegmatite overlain by dark-colored metamorphic rocks, New Mexico. From top down, the zones consist of: quartz with alkali feldspars, apatite, beryl, and muscovite (light band); quartz (dark band); quartz with giant lathlike crystals of spodumene (thick band with comblike appearance); and quartz, alkali feldspars, spodumene, lepidolite, and microlite (mined-out recesses). *(From E. N. Cameron et al., Internal structure of granitic pegmatites, Econ. Geol., Monogr. no. 2, 1949)*

lime-soda plagioclase. Varietal minerals such as micas, amphiboles, pyroxenes, black tourmaline, fluorite, and calcite further characterize the pegmatites of specific districts. Accessory minerals include allanite, apatite, beryl, garnet, magnetite, monazite, tantalite-columbite, lithium tourmaline, zircon, and a host of rarer species.

Most pegmatites are mineralogically simple, with only one or two varietal constituents and a few sparse accessories. More complex pegmatites typically contain much albite (commonly the variety cleavelandite) and groups of minerals that reflect unusual concentrations of rare elements such as beryllium (beryl, chrysoberyl, gadolinite, phenakite), boron (axinite, tourmaline), and lithium (amblygonite, lepidolite, petalite, spodumene, triphylite-lithiophilite, zinnwaldite). Cavities and pockets in some pegmatite bodies contain beautifully formed crystals, and such occurrences have yielded transparent gemstones of beryl, garnet, quartz, spodumene, topaz, tourmaline, and other minerals.

Texture. The most distinctive features of pegmatites are coarseness of grain and great variations in grain size over short distances (Figs. 1 and 3). The average grain size for all occurrences is on the order of 4 in. (10 cm), but mica and quartz crystals 10 ft (1 ft = 0.3 m) across, beryl and tourmaline crystals 10 to 20 ft long, feldspar crystals 30 ft long, and spodumene crystals nearly 50 ft long have been found, as have individual crystals of allanite, columbite, monazite, and topaz weighing hundreds of pounds.

On much smaller scales, sodic plagioclase is intergrown with host potash feldspar to form perthite, and quartz with feldspars in cuneiform fashion to form graphic granite. More or less regular intergrowths of hematite and magnetite in muscovite, and of muscovite, garnet, or tourmaline in quartz also are common. Evidence of corrosion and replacement of some minerals by others typically is present over a wide range of scales. *See* IGNEOUS ROCKS.

Internal structure. Many pegmatite bodies are grossly homogeneous, whereas others, including nearly all of those with economic mineral concentrations, comprise internal units of contrasting composition or texture. Such internal zoning is systematically expressed as successive layers in nearly flat-lying bodies with tabular form (Figs 2 and 3), and elsewhere as layers or shells that also reflect the general form of the respective bodies (Fig. 4). In horizontal section the zones tend to be arranged concentrically about a single or segmented core, but in vertical section they are typically asymmetric and less continuous (Fig. 4). Some pegmatite bodies are distinguished by relatively fine-grained outer zones that are more regular and continuous in three dimensions. Associations of major minerals within successive zones consistently follow a definite sequence of worldwide application.

Fracture-filling masses of tabular form transect parts of some pegmatite bodies. They range in thickness from less than an inch to nearly 10 ft, and most are composed of quartz or of quartz, feldspars, muscovite, and other common minerals in various combinations. Less regular masses of lenticular, branching, or network form, and a few inches to several hundred feet in maximum dimension, appear to have developed wholly or in part at the expense of earlier pegmatite. Typical examples are quartz, potash feldspar, and muscovite replaced by aggregates of cleavelandite or sugary

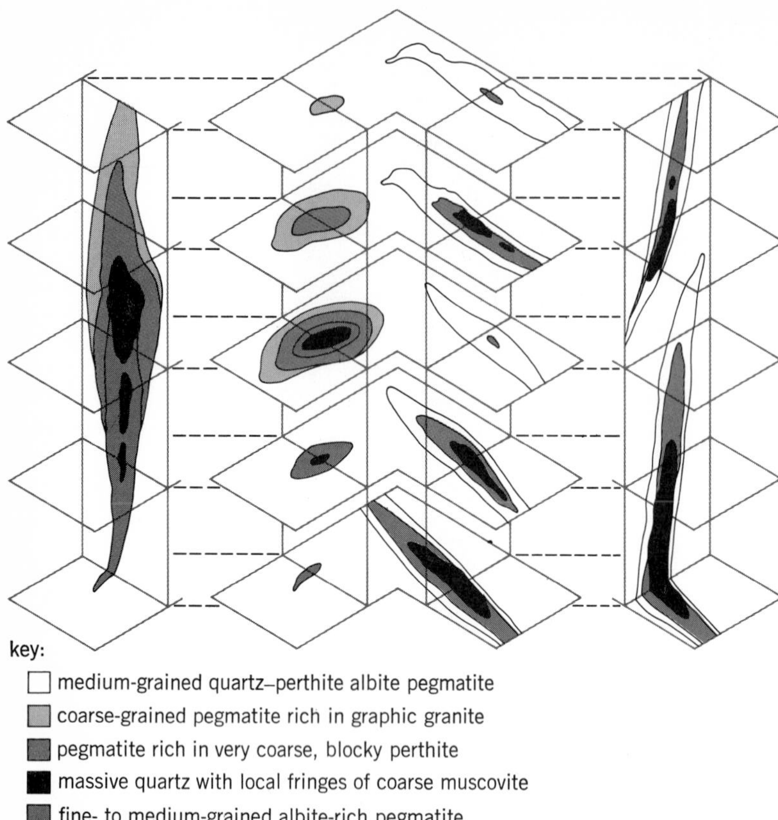

Fig. 4. Isometric diagram comprising vertical sections of a cucumber-shaped pegmatite body (left) and parts of two pegmatite dikes (right), together with successive horizontal sections of all three bodies (center). Internal zoning is shown in idealized form without a specific scale, and its configuration within relatively horizontal bodies can be visualized through a figurative vertical compression of the bodies shown here.

albite, alkali feldspar and quartz by muscovite, and potash feldspar and spodumene by lepidolite. Some replacement masses also contain notable concentrations of rare minerals.

Origin. Many pegmatites must have formed from the residual fractions of crystallizing rock melts (magmas) or from the initial fractions of deeply buried crustal materials undergoing partial melting. Such fractions would be silicate liquids relatively rich in water, halogens, and other volatile constituents, and also in many of the less common elements. They could remain within the crystallized or partly melted parent rocks to form bodies of segregation pegmatite with irregular and gradational boundaries, or they could be forced away to form homogeneous or zoned bodies of pegmatite elsewhere in the parent rocks or in rocks beyond. *See* MAGMA.

Other pegmatites, generally regarded as metamorphic in origin, consist of materials probably derived from the enclosing rocks via aqueous pore fluids and concentrated at favorable sites in the form of secretionary or replacement bodies. Such transfer of materials could account for numerous veins, pods, and irregular bodies of pegmatite with relatively simple composition and internal structure, and for the development of pegmatite on a grand scale in Precambrian terranes of metamorphism and granitization in regions such as Fennoscandia, India, central Africa, and eastern Canada. *See* GRANITIZATION; METAMORPHIC ROCKS; METAMORPHISM.

Long-time studies of pegmatites and experimental investigations of pertinent silicate systems indicate that the unifying factor in pegmatite genesis is the presence of a high-temperature aqueous fluid. When such a fluid exsolves from a crystallizing water-rich magma, it serves as a low-viscosity medium into which other constituents can be preferentially transferred from the magma, and through which they can diffuse rapidly but selectively in response to temperature-induced concentration gradients. Thus it is an effective vehicle for the nourishment of large crystals and development of coarse textures, and for the gross segregation of constituents to form zonal structure in many pegmatite bodies. Upward gravitational rise of the low-density aqueous fluid also can account for the observed vertical asymmetry of zonal structure (Fig. 4).

The aqueous fluid promotes exsolution of alkali feldspars, mobilization of the released albitic constituents, and extensive reactions among earlier-formed crystals both before and after all magma has been used up. Thus pegmatites of igneous origin reflect metasomatic activities as well. Moreover, the transfer of materials and the mineral replacements implied in the origin of metamorphic pegmatites can be expected to occur if an interstitial aqueous fluid is available in the hot host rocks and a temperature or pressure gradient is present, whether or not a magma is at hand. *See* METASOMATISM. [RICHARD H. JAHNS]

Bibliography: E. N. Cameron et al., Internal structure of granitic pegmatites, *Econ. Geol.*, Monogr. no. 2, 1949; R. H. Jahns, The study of pegmatites: *Econ. Geol.*, 50th Anniv. Vol., pp. 1025–1130, 1955; R. H. Jahns and C. W. Burnham, Experimental studies of pegmatite genesis: I. A model for the derivation and crystallization of granitic pegmatites, *Econ. Geol.*, 64:843–864, 1969; K. K. Landes, Origin and classification of pegmatites, *American Mineralogist*, vols. 18 and 20, 1933, 1935; H. Ramberg, *The Origin of Metamorphic and Metasomatic Rocks*, 1952.

Pennsylvanian

A major division of late Paleozoic time, the Pennsylvanian is considered to be either an independent period or the younger subperiod of the Carboniferous. In North America it is widely recognized as a geologic period and derives its name from coal-bearing strata in Pennsylvania. Radiometric dates place the beginning of the Pennsylvanian at approximately 320,000,000 years ago and the end at about 280,000,000 years ago. Outside of

PRECAMBRIAN	PALEOZOIC							MESOZOIC			CENOZOIC	
					CARBONIFEROUS							
	CAMBRIAN	ORDOVICIAN	SILURIAN	DEVONIAN	Mississippian	Pennsylvanian	PERMIAN	TRIASSIC	JURASSIC	CRETACEOUS	TERTIARY	QUATERNARY

North America, rocks of equivalent age are commonly designated Upper Carboniferous or Coal Measures. The Pennsylvanian was a time of extensive coal deposition in coastal and lowland swampy regions in central and eastern North America, Europe, and many parts of Asia. Coal of this age has great economic importance, and the geographical distribution of these coal-bearing strata strongly influenced the location and development of the industrial revolution in the mid-1800s. As a source of readily available combustible fuel, these deposits remain of great importance. Marine strata of Pennsylvanian age, particularly beach sandstones and limestone reefs and mounds, are important traps for petroleum.

During Pennsylvanian time North America and northern Europe were joined as one large low-lying continent. Africa, South America, India, Australia, and Antarctica were also joined into an even larger low-lying continent called Gondwana. During the Pennsylvanian the northwestern part of Africa and the northern part of South America were in the process of colliding against southern (Ouachita orogenic belt) and eastern (Appalachian orogenic belt) North America. Before the end of the period these large continental masses joined to form an extensive land mass called Pangaea. Several additional island continents lay to the east and north of Pangaea and subsequently collided to form much of Asia.

The absence of fossil tree rings suggests that the paleoequator for the Pennsylvanian Period extended across the southwestern United States, southeastern Canada, north-central Europe, Ukraine, and parts of China. The presence of fossil tree rings in the central and northeastern parts of Siberia and in much of southern Africa, Australia, Antarctica, southern South America, and India suggests that these areas were located in temperate to polar latitudes. These shifts in the position of the ancient supercontinental landmasses are also suggested by paleomagnetic data. Glacial deposits of Pennsylvanian age are widely distributed in the Gondwanan continents, presumably as a consequence of their peripolar position. Some of the late Paleozoic glaciation is older (Mississippian) and some is younger (early Permian). This late Paleozoic glaciation was multiple, and glaciers advanced and receded many times during the period, resulting in repeated worldwide fluctuations in ocean levels. These changes in ocean levels are recorded as thin, repetitious, transgressive and regressive deposits of marine and nonmarine sedimentary rocks (cyclothems) produced by the shift of shorelines as the sea encroached onto or retreated from the edges of the low-lying continents. *See* CARBONIFEROUS; COAL; CONTINENT FORMATION; CYCLOTHEM; PLATE TECTONICS.

[CHARLES A. ROSS]

Pentlandite

A mineral having composition $(Fe,Ni)_9S_8$. Pentlandite is the major ore of nickel. It crystallizes in the isometric system, but crystals are rare. It is usually massive, showing a well-defined octahedral parting. The hardness is 3.5–4 (Mohs scale) and the specific gravity varies from 4.6 to 5.0, depending on the ratio of iron to nickel; greater amounts of iron cause an increase in the specific gravity. The luster is metallic and the color yellowish bronze. Pentlandite is usually associated with pyrrhotite, which it closely resembles in appearance, but the two can be distinguished by the octahedral parting and lack of magnetism of pentlandite. It is found at many localities in small amounts, but its chief occurrence is at Sudbury, Ontario, where it is mined on a large scale as a nickel ore.

[CORNELIUS S. HURLBUT, JR.]

Peridotite

A rock consisting of more than 90% of millimeter- to centimeter-sized crystals of olivine, pyroxene, and hornblende, with olivine predominant. Other minerals are mainly plagioclase, chromite, and garnet. Most of the volume of the Earth's mantle probably is peridotite.

Chemical composition. Most peridotites are rich in magnesium and have high Mg/Fe ratios similar to many meteorites. Peridotites generally are rich in nickel and chromium and are important ores of these elements. Rare peridotites and associated rocks contain platinum or diamond.

Mineralogy. Peridotites have various names depending upon the subordinate minerals: lherzolite has Ca-poor (orthorhombic) pyroxene and Ca-rich (monoclinic) pyroxene, harzburgite has Ca-poor pyroxene, wehrlite has Ca-rich pyroxene. Generally the aluminous mineral is mentioned as a modifier, as in feldspathic lherzolite. Other minerals, if significant, are similarly used as modifiers—hornblende peridotite, for example. *See* PYROXENE.

Structure and texture. Some peridotites are layered or are themselves layers; others are massive. Many layered peridotites occur near the base of bodies of stratified gabbroic complexes. Other layered peridotites occur isolated, but possibly

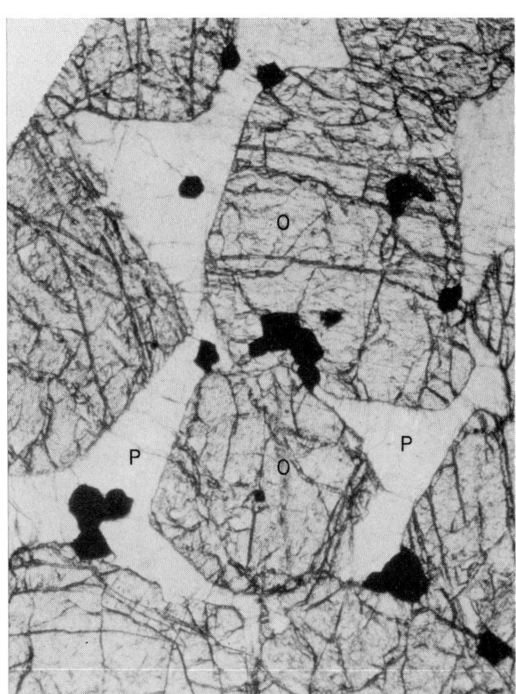

Millimeter-sized crystals of olivine (O) with good crystal form are outlined by plagioclase (P) which fills the space between the accumulated crystals of olivine. Small crystals of black chromite occur in the olivine and plagioclase.

once composed part of major gabbroic complexes. *See* GABBRO.

Both layered and massive peridotites can have any of three principal textures: (1) rather well-formed crystals of olivine separated by other minerals (see illustration); (2) equigranular crystals with straight grain boundaries intersecting at about 120°; and (3) long crystals with ragged curvilinear boundaries. The first texture probably reflects the original deposition of an olivine sediment (or cumulate) from magma. The second texture may result from slow cooling whereby recrystallization leads to a minimization of surface energy. The third texture probably results from internal deformation.

Physical properties. Peridotites are comparatively dense rocks (3.3–3.5 g/cm^3) with rather high velocities ($V_p \approx 8$ km/s) for body-wave propagation. The electrical conductivity is sensitive to oxidation state, lattice defects, and temperature. Remanent magnetization is variable in peridotites, being particularly high in altered varieties rich in small grains of magnetite.

Occurrence. Peridotites have three principal modes of occurrence corresponding approximately to their textures: (1) Peridotites with well-formed olivine crystals occur mainly as layers in gabbroic complexes; so-called Alpine peridotites generally have irregular crystals and occur as more or less serpentinized lenses bounded by faults in belts of folded mountains such as the Alps, the Pacific coast ranges, and the Appalachian piedmont. (2) Peridotite nodules in alkaline basalts and diamond pipes generally have equigranular textures, but some have irregular grains. (3) Peridotite also occurs on the walls of rifts in the deep sea floor and as hills on the sea floor, some of which reach the surface. Many Alpine peridotites occur in the ophiolite association: peridotite, gabbro, diabase sill-and-dike complex, pillow basalt, and red chert. Some peridotites rich in amphibole have a concentric layered structure and form parts of plutons called Alaskan-type zoned ultramafic complexes. Small pieces of peridotite have been found in lunar breccias.

Origin. Layered peridotites are igneous sediments and form by mechanical accumulation of dense olivine crystals (see illustration). Peridotite nodules are pieces of mantle rock more or less modified by partial melting. Alpine peridotites of the ophiolite association probably formed in the oceanic crust and uppermost mantle by transfer of partial melt from the mantle to the crust, followed by tectonic emplacement of both crust and mantle along thrust faults in mountain belts. Peridotites associated with Alaskan-type ultramafic complexes probably formed in the root zones of volcanoes. *See* PLATE TECTONICS.

Use. Peridotite is an important rock economically. Where granites have intruded peridotite, asbestos and talc are common. Pure olivine rock (dunite) is quarried for use as refractory foundry sand and refractory bricks used in steelmaking. Serpentinized peridotite is locally quarried for ornamental stone. Tropical soils developed on peridotite are locally ores of nickel. The sulfides associated with peridotites are common ores of nickel and platinoid metals. The chromite bands commonly associated with peridotites are the world's major ores of chromium. *See* IGNEOUS ROCKS; PETROGRAPHIC PROVINCE.

[ALFRED T. ANDERSON, JR.]

Bibliography: I. S. E. Carmichael, F. J. Turner, and J. Verhoogen, *Igneous Petrology*, 1974; L. R. Wager and G. M. Brown, *Layered Igneous Rocks*, 1968; P. J. Wyllie (ed.), *Ultramafic and Related Rocks*, 1976.

Perlite

A natural glass with abundant spherical or convolute cracks that cause it to break into small pearl-like masses or "pebbles," usually less than a centimeter across. It is commonly gray or green with a pearly luster due to reflections from the thin air films formed along the perlitic fractures. Perlitic cracks are not necessarily confined to perlite but appear sporadically in most natural glasses. Glass is formed by rapid cooling of molten rock material (lava), and the cracks are generally believed to develop by contraction during cooling. The water content of perlite is commonly 3–4% by weight. Most of this moisture is believed to have been absorbed by the glass from its surroundings. Some studies suggest that perlitic cracks may form in response to this hydration.

Under heat treatment (about 1500–2000°F) the contained moisture forms tiny steam bubbles in the softened glass, and the perlite is "popped" or exploded to roughly 15 or 20 times its original volume. Thus, the material is excellent for lightweight aggregate, insulation, fillers, and filters. Notable deposits are worked in California and New Mexico. Perlite may constitute major portions of lava flows or occur in small intrusions (dikes). *See* IGNEOUS ROCKS; LAVA; VOLCANIC GLASS.

[CARLETON A. CHAPMAN]

Permafrost

Perennially frozen ground, occurring wherever the temperature remains below 0°C for several years, whether the ground is actually consolidated by ice or not and regardless of the nature of the rock and soil particles of which the earth is composed. Perhaps 25% of the total land area of the Earth contains permafrost; it is continuous in the polar regions and becomes discontinuous and sporadic toward the Equator. During glacial times permafrost extended hundreds of miles south of its present limits in the Northern Hemisphere.

Permafrost is thickest in that part of the continuous zone that has not been glaciated. The maximum reported thickness, about 1600 m, is in northern Yakutskaya, U.S.S.R. Average maximum thicknesses are 300–500 m in northern Alaska and Canada and 400–600 m in northern Siberia. In Alaska and Canada the general range of thickness in the discontinuous zone is 50–150 m and in the sporadic zone less than 30 m. Discontinuous permafrost in Siberia is generally 200–300 m. *See* GLACIAL EPOCH.

Temperature of permafrost at the depth of no annual change, about 10–30 m, crudely approximates mean annual air temperature. It is below −5°C in the continuous zone, between −1 and −5°C in the discontinuous zone, and above −1°C in the sporadic zone. Temperature gradients vary horizontally and vertically from place to place and from time to time. Deep temperature profiles record past climatic changes and geologic events from several thousand years ago.

Ice is one of the most important components of permafrost, being especially important where it exceeds pore space. Physical properties of permafrost vary widely from those of ice to those of normal rock types and soil. The cold reserve, that is, the number of calories required to bring the material to the melting point and melt the contained ice, is determined largely by moisture content. Ice occurs as individual crystals ranging in size from less than 0.1 mm to at least 70 cm in diameter. Aggregates of ice crystals are common in dikes, layers, irregular masses, and ice wedges. These forms are derived in many ways in part when permafrost forms. Ice wedges grow later and characterize fine-grained sediments in continuous permafrost, joining to outline polygons. Microscopic examination of thin sections of ice wedges reveals complex structures that change with the seasons.

Permafrost develops today where the net heat balance of the surface of the Earth is negative for several years. Much permafrost was formed thousands of years ago but remains in equilibrium with present climates. Permafrost eliminates most groundwater movement, preserves organic remains, restricts or inhibits plant growth, and aids frost action. It is one of the most important factors in engineering and transportation in the polar regions. *See* FROST ACTION; MASS WASTING.

[ROBERT F. BLACK]

Bibliography: J. B. Bird, *The Physiography of Arctic Canada*, 1967; R. F. Black, Permafrost: A review, *Bull. Geol. Soc. Amer.*, 65:839–856, 1954; R. J. E. Brown, Comparison of permafrost conditions in Canada and the USSR, *Polar Rec.*, 13: 741–751, 1967; *Proceedings of the Permafrost International Conference*, Nat. Acad. Sci.–Nat. Res. Counc. Publ. no. 1287, 1966.

Permian

The name applied to the last period of geologic time in the Paleozoic Era and to the corresponding system of rock formations. *See* PALEOZOIC.

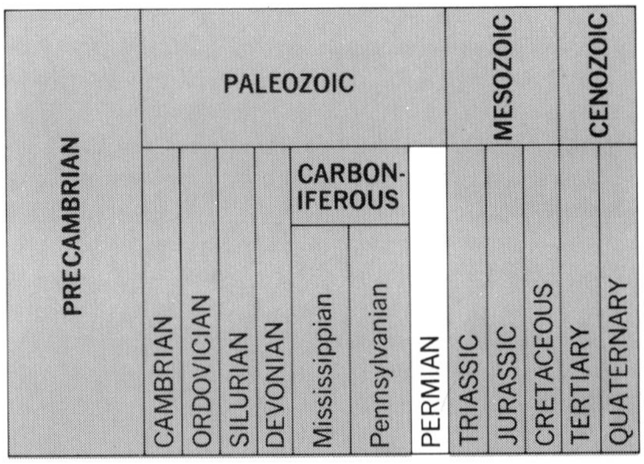

The type Permian. Permian strata occupy a vast area west of the Urals, a region structurally divisible into the Uralian Geosyncline and the Russian Platform (Fig. 1). In the geosyncline the deposits are very thick and are almost all detrital; but as they spread onto the platform they thin rapidly and change to a calcareous facies that is widely exposed in the Ufa Plateau and again, much farther west, in the region of the Samara Bend of the Volga River near Kuibyshev.

When R. Murchison proposed the Permian System in 1841, he drew its lower boundary at the base of the Kungurian Stage. At that time the Artinskan detritals had yielded no fossils, and on a purely lithologic basis Murchison mistakenly assigned these beds to the Lower Carboniferous. At the same time he correlated the limestones of the Ufa Plateau with the Upper Carboniferous. Soviet geologists have since discovered that the Artinskan detritals and the limestones of the Ufa Plateau are contemporaneous (Fig. 2).

By 1889 A. P. Karpinsky had discovered abundant ammonites in the Artinskan detritals and showed that they were related to Permian faunas in other parts of the world. This judgment soon gained wide acceptance, but the limestones of the Ufa Plateau continued to be classified as Upper Carboniferous until after 1930. Then during 1930–1940 Soviet geologists discovered numerous faunal zones in the Ufa Plateau that could be identified in the Artinskan detritals. Among these Rauser-Chernousova recognized the zone of *Pseudoschwagerina* (then called *Schwagerina*) in the lower part of each facies, and Ruzencev recognized prolific ammonites which, like the fusulines, were present in both facies. On this basis Ruzencev, in 1936, proposed to subdivide the original Artinskan, restricting this name to the upper part and introducing the name Sakmarian Stage for the lower part. Thus the type Permian came to embrace five stages (Fig. 1).

The Sakmarian and Artinskan stages include normal marine faunas with abundant fusulines and ammonites, both of which are extremely useful in interregional correlation, but the higher stages have more limited faunas.

During the Kungurian time the sea shrank into the geosyncline where under intensely arid conditions vast deposits of salt and anhydrite were precipitated. In the vicinity of Solikamsk one of the three greatest known deposits of potash salts were formed. Under these conditions the Kungurian deposits are virtually unfossiliferous.

During Kazanian time the sea again spread widely over the platform while nonmarine redbeds were deposited over the geosynclinal basin. Deep intertonguing of the marine and nonmarine facies is now well exposed in the bluffs of the Kama River. The Kazanian faunas are largely formed of bryozoans and brachiopods but, unfortunately, have yielded neither ammonites or fusulines.

The Tatarian Stage is entirely nonmarine and consists largely of redbeds. These have yielded both vertebrate and plant fossils, and of five chief faunal zones the lower two are Permian and the upper three are Triassic. Therefore the upper boundary of the Permian System lies within the original Tatarian Stage. *See* REDBEDS.

Since 1958 Soviet geologists have recognized two major subdivisions of the Permian System—the Lower Permian Series, embracing the Sakmarian, Artinskan, and Kungurian stages; and the Upper Permian, embracing the Kazanian and the lower part of the Tatarian stage. In 1958 some Soviet geologists were inclined to recognize a sixth

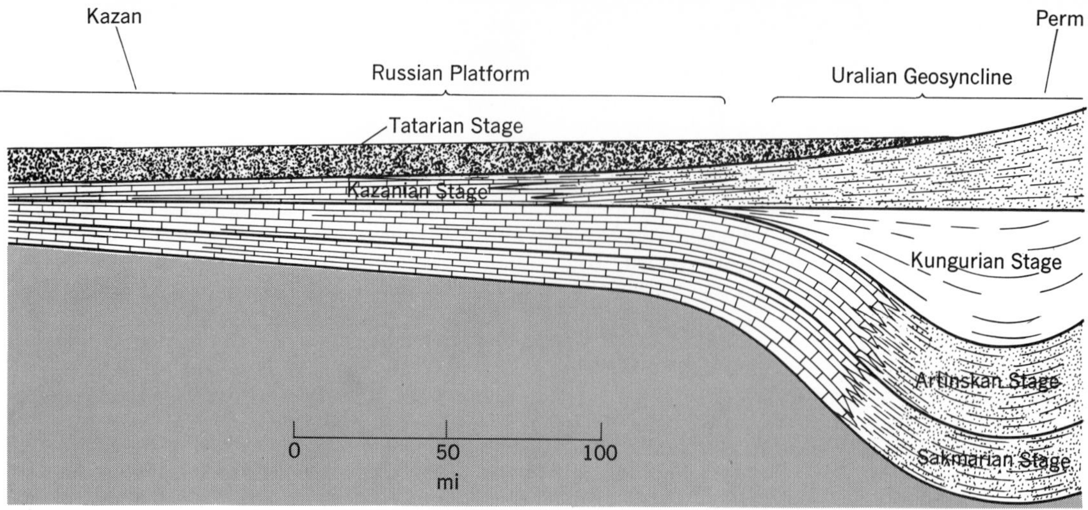

Fig. 1. East-west section of the Permian System in its type region.

stage, the Ufimian, between the Kungurian and the Kazanian stages, but others considered the Ufimian beds as only a local facies of one or both stages which is confined to the eastern margin of the geosynclinal basin.

The American Permian. Probably the finest Permian section known is that in the basin of Trans Pecos Texas and southeastern New Mexico, where the Permian sequence reaches a thickness of more than 12,000 ft. The rocks of the lower 7500 ft are of marine origin and are highly fossiliferous. Although a few fossils from this region were described in 1858, this occurrence remained virtually unknown until 1909, when G. H. Girty published his monograph *The Guadalupian Faunas.* Intensive study followed the discovery of oil in the region about 1920, and it has become the standard section for North America and, to a considerable degree, for the world. It is divided into four stages as follows in descending order: Ochoan Stage about 4500 ft; Guadalupian Stage about 3000 ft; Leonardian Stage about 3000 ft; and Wolfcampian Stage 1500+ ft.

The Wolfcampian correlates closely with the Sakmarian Stage of the type section in the Soviet Union and, like the latter, is the zone of *Pseudoschwagerina*; the Leonardian correlates with the Artinskan but has a larger and more varied fauna; the Guadalupian cannot be correlated in detail with the upper part of the Soviet section, because the latter is largely unfossiliferous. The Ochoan is entirely unfossiliferous except for a thin zone near the top (in the Rustler Dolostone) which contains productid brachiopods and a few other types of Paleozoic invertebrates.

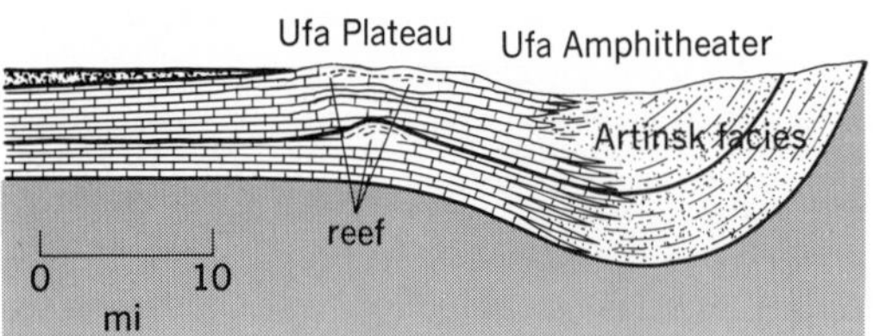

Fig. 2. Idealized section, showing relations of the Artinskan detritals to the limestones of the Ufa Plateau.

Leonardian time. Facies changes in this region are spectacular. During Early Permian time (Wolfcampian Stage), the region was a broad, shallow marine basin in which a rich and varied fauna thrived. By Leonardian time three distinct basins (Fig. 3) were subsiding more rapidly than the surrounding area, which became a broad shelf occuppied by wide, shallow lagoons. Light-colored, fossiliferous limestones (Victorio Peak Limestone) then accumulated on the shelves, while black limestone and black shale accumulated in the basins. Evidently the threshold to the basins, which was somewhere in Mexico, was then so shallow that water in the basins was density-stratified and the bottom was stagnant and foul, so that almost no benthonic organisms could survive. The black Bone Spring Limestone is generally barren of fossils.

Guadalupian time. During Guadalupian time the basins continued to deepen, and as the climate became strongly arid, surface water flowed radially out of the basins onto the platform to replace the water lost by evaporation in the shallow lagoons. As a result, a narrow limy bank grew up along the margins of the basin to form the great Capitan Reef. With this development, three strongly contrasted facies accumulated simultaneously: (1) basin or pontic deposits under normal marine conditions, (2) reef and reef talus, and (3) back-reef or lagoonal deposits. Figure 4 shows the complex relations within the Leonardian and Guadalupian deposits along the face of the Guadalupe Mountains. Massive deposits of reef talus were derived from the growing front of the reef. These dip steeply into the basin, become finer down dip, and grade out into thin tongues of calcarenite. The back-reef deposits are calcareous for distances varying from 1 to 5 mi. The deposits then grade rapidly into gypsiferous shales and anhydrite. Gray sands intertongue from the landward margins of the lagoon. Farther back the sands pass into redbeds.

Ochoan time. Finally, during Ochoan time the marine water withdrew entirely into the basins and, under intensely arid conditions, became a dead sea in which enormous deposits of anhydrite and, later, of halite were precipitated. Interbedded with the salt in the center of the Delaware Basin

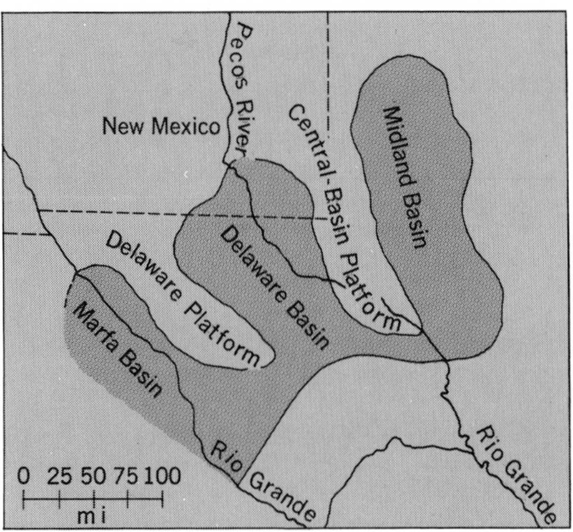

Fig. 3. Map of the Permian basins which occur in West Texas. (*After P. B. King*)

are several lenses of potash salts—sylvite, carnallite, and polyhalite.

Probably no other great reef complex is so well exposed or has been so intensively studied as that of Capitan Reef. Cenozoic faulting uplifted the Guadalupe Mountains, but immediately to the west it depressed the Salt Flat graben. Thus, a great natural section across the reef complex was exposed. The Capitan Reef was essentially a slender limy bank along the margin of the platform, and much of the carbonate may have been precipitated biochemically. Much of it is sparingly fossiliferous; in places, however, it is filled with fossils which locally accumulated in a typical reef environment.

From study of deep well cores, the West Texas formations are known to extend eastward under the Staked Plains to central Texas where they reappear at the surface. Thence their outcrops stretched northeastward across Oklahoma and Kansas. The Wolfcamp equivalents are largely of marine origin in central Texas and Kansas. Marine conditions persisted well into Leonardian time in central Texas, but in Kansas a very large salt deposit (in the Wellington Shale) is followed by a thick redbed sequence. By Guadalupian time most of the deposits in central Texas were nonmarine redbeds with several thin marine dolostones intertonguing from the west. In Oklahoma all but the lower part of the Wolfcampian is in the redbed facies because of the local influence of the Oklahoma mountains.

In 1939 a committee of Permian specialists (J. E. Adams and coworkers) proposed to subdivide the American Permian section into four series, and this usage was widely accepted until 1960 when, after a major conference of Permian specialists, the U.S. Geological Survey issued a directive to its staff:

"The Chief Geologist of the U.S. Geological Survey has approved the recommendation of the Names Committee that a two-fold subdivision of the Permian System and Period be adopted for use by the U.S. Geological Survey. The divisions (Lower and Upper Series and Early and Late Epochs) are to coincide as nearly as possible with those recognized in the type Permian and are to be drawn according to existing concepts of biotal correlation with the type section. The reference section for the United States shall be the Permian outcrops of northwestern Trans Pecos Texas (Delaware Mountains, Guadalupe Mountains, and Sierra Diablo Mountains) where the approximate faunal boundary is taken as that between the Cherry Canyon and Bell Canyon formations" (Fig. 4). With this change the four subdivisions of the system become stages instead of series.

Other regions. Permian deposits are thick and widespread in many other parts of the world, particularly in South Africa, Australia, India, China, Indochina, Japan, Europe, and South America.

South Africa. The Permian of South Africa is largely nonmarine and is notable for its extensive glacial deposits as well as its fossil reptiles. Here the section begins with the widespread Dwyka Tillite, followed by the Ecca Series and the lower part of the Beaufort Series. These three constitute the lower part of the Karoo System. The Dwyka Tillite is discussed below under Permian glaciation.

Both the Ecca and Beaufort series are very thick in the old geosyncline which paralleled the southern coast of Cape Colony, but each thins northward and spreads widely over the southern part of Africa. The Ecca Series consists of dark shales and sandstones and includes the Coal Measures of Africa. The Beaufort Series was spread over the

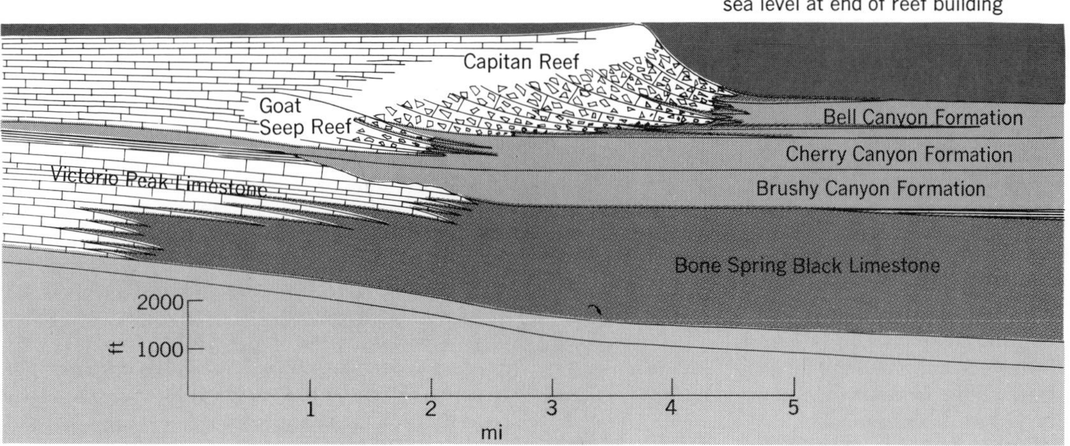

Fig. 4. Section across the Capitan Reef Complex. (*After P. B. King*)

Ecca Series after an interval of erosion. It differs from the Ecca Series in having interbedded yellowish sandstones, some members of red shale, and districtive reptilian faunas.

Australia. The Australian continent was largely emergent during Permian time except for a series of marginal basins along both the western and eastern coasts. These basins subsided independently; although the Permian deposits are thick, it is impossible to treat them adequately within the limits of this article. The most significant fact is that the Permian here, as in South Africa, begins with widespread glacial deposits. In the Gascoyne Basin of northwest Australia these include tillite which represents a ground moraine resting on an old erosion surface. In most of Western Australia, however, the glacial deposits appear as boulder beds intercalated in marine sandstone or shale, a clear indication that the boulders were dropped from floating ice. In the western basins, interbedded marine zones in the Lyons Sandstone bear the ammonite *Metalegoceras jacksoni* which dates the deposits as Sakmarian or early Artinskan. After deposition of thick glaciomarine deposits of this sort, each of the western basins was occupied by a warmer marine incursion in which widespread and fossiliferous, but rather thin, calcareous formations accumulated (Fossil Cliff Limestone of Irwin River Basin, Callytharra Limestone of Gascoyne and Minilya basins, and Nura Nura Limestone of the Kimberley District). Their faunas are of Artinskan age. In both the Irwin River and Gascoyne basins, glacial erratics recur for some distance above this limestone horizon, which proves that glaciation persisted here at least into early Artinskan time.

In southeastern Australia where Permian rocks are very thick they have been subdivided into four major units, a lower and an upper marine unit and Lower and Upper Coal Measures. Glacial erratics occur in both of the marine units.

Europe. The Permian strata in Germany comprise two major units. The lower is the Rothliegendes Sandstone, consisting of nonmarine redbeds; the upper is the Zechstein Formation, consisting of marine limestones and shales which grade upward into salt and gypsum. The potash deposits at Stassfurt lie within the upper part of the Zechstein. The Magnesian Limestone of England is a thin western tongue of the Zechstein Limestone.

Glaciation. Early in Permian time, large areas in the southern continents were covered by glacial ice. The widespread Dwyka Tillite indicates that most of Africa south of lat 23°S was ice-covered. It includes abundant striated boulders and in places rests on a spectacularly grooved and striated floor. In its northern outcrops (northern Karoo, Natal, and Zululand), it is a typical ground moraine resting on an undulating pre-Permian surface; but farther south it thickens greatly and grades into a glaciomarine deposit formed of material dropped from floating icebergs. Orientation of glacial striae and distribution of boulders from known source areas indicate that the ice moved westward into the province of South Africa and generally south from the Transvaal. N. Boutakoff (1940) has reported that tillite (presumably Dwyka) is widespread in the Congo Basin, even within 4° of the Equator.

Glacial deposits are also widespread in western and southern Australia and in Tasmania. In Western Australia such deposits originally covered an area of about 200,000 mi^2. Locally, in the Canning Basin the glacial deposits are typical ground moraine resting on a striated floor, but for the most part they are glaciofluvial in Western Australia and occur within a thick marine sequence (the Lyons Formation of the Irwin and Canning basins, and the Kungangie and Grant Range formations of the Kimberley District).

Permian glacial deposits are also widespread in South America (in Uruguay, in the Precordillera of northern Argentina and Bolivia, and in southern Brazil). An extensive ice sheet in India is recorded by the Talchir tillite of the Salt Range and central India (Rewah Province). The ice sheet is believed to have stretched for some 600 mi from east to west and 1000 mi from south to north and to have moved northward into the Salt Range region.

Curiously, no glaciation is known to have occurred elsewhere in the Northern Hemisphere.

Date of ice age. In each of the regions mentioned above, the glacial deposits are at the base of the Permian section and the immediately overlying fossiliferous formations are of Artinskan age; hence it appears that the glaciation occurred early in the period, during either Sakmarian or early Artinskan time. In India the Talchir Tillite is overlain by the Lower Productus Limestone which bears a prolific marine fauna including fusulines of Artinskan age. In South Africa the tillite is succeeded by the Ecca Formation which bears vertebrate fossils of mid-Permian age, and in Western Australia the glaciomarine deposits include the ammonite, *Metalegoceras jacksoni*, which is correlated with the lower Permian or early Artinskan faunas of Timor.

The reptiles in the Ecca Formation of South Africa indicate a mild climate, as do the marine faunas of the Lower Productus Limestone of India. It appears therefore that the glacial episode occupied but a relatively short part of Permian time. The glacial deposits of South America are not well dated by fossils and have been classified by most South American geologists as Upper Carboniferous; but because their occurrence so closely resembles that of Australian glacial deposits, it seems likely that they are of early Permian age.

The Glossopteris flora. Throughout the glaciated regions of the Southern Hemisphere the nonmarine Permian formations are commonly characterized by the tongue ferns, *Glossopteris* and *Gangamopteris*. In South Africa *Glossopteris* has been found between the glaciated floor and the Dwyka Tillite, and in numerous places in India, South Africa, and Australia its distinctive spores have been found in the tillite. It is therefore believed to have been adapted to a cold climate. Unfortunately, the biologic relations of these peculiar plants are still uncertain.

It was the distribution of Permian glacial deposits in the southern landmasses (and India) that led to the theory of continental drift. When it was advanced by Alfred Wegener in 1912, he argued that in Permian time these landmasses were united in a major continent, which he called Pangea, and which was centered near the South Pole. He postulated that it later broke into the present units which slowly drifted apart to their present positions. The idea was opposed by many geologists and for the next half century it developed into one of the great controversies in geology. But spectacular discoveries in geophysics during the 1960s, especially in paleomagnetism and in the study of the ocean floors, have become so convincing that continental drift is now a major working hypothesis of most geologists throughout the world.

Deserts. Desert conditions probably were more widespread in the Permian than at any other time before the late Cenozoic. The vast deposits of salt and anhydrite in the Permian Basin of West Texas and New Mexico, the salt of Kansas, and the extensive deposits of dune sand in the eastern part of the Colorado Plateau indicate a great arid basin in the west-central part of the United States. Ralph King has estimated that, if it required 300,000 years to precipitate the salines of the Ochoan series, the rate of evaporation over the entire basin must have averaged 9.5 ft per year, which is only about 2 ft less than it is in the modern Death Valley of California.

Similar conditions must have obtained during deposition of the Kungurian salts in the Permian Basin of the Soviet Union. Western Europe was also the scene of great aridity, as indicated by the thick salt deposits at Stassfurt, Germany, and by widespread dune sands in Germany and England. The three greatest known deposits of potash salts lie within the Permian System, one in West Texas, one about Solikamsk in the Permian Basin of the Soviet Union, and one about Stassfurt, Germany. Although these basins were of regional extent, they were local in the world scene, and in some regions (notably South Africa, Australia, and south China) the Permian System includes much coal and abundant evidence of humid conditions. *See* EVAPORITE, SALINE.

Orogeny. The Permian was a time of continental uplift and of widespread orogeny. At this time the Appalachian Mountain system was formed and the Uralian geosyncline of the Soviet Union was folded and uplifted into a great mountain chain. In Europe the Variscan Alps stretched from southern England across central Germany and from Normandy into the central Massif of France and thence northeastward through the area of the modern Schwartzwald and northeastward through the Erzgebirge to beyond Vienna. The folding of the Variscan Mountains occurred in three stages, the first in the Lower Carboniferous, the second in the Upper Carboniferous, and the final movement in Permian time. There was Permian orogeny also in the area of the modern Kuen-Lun Range in the northern flank of the Himalayas. In none of these regions is the uplift within the Permian closely dated, but before the end of the period the continents were almost completely emergent; the youngest known Permian deposits are generally nonmarine. This may, in part at least, account for the profound change in so many groups of animals and plants at the close of the Paleozoic Era. *See* OROGENY.

Life. The invertebrate faunas of the Permian developed from those of the Pennsylvanian with gradual change and marked specialization. Brachiopods, bryozoans, and fusulines and goniatites continued to dominate the faunas, but at the close of the period each of these groups suffered a great decline. Among the brachiopods the Productidae were especially varied and gave rise to highly specialized offshoots such as the leptodids, the richtofenids, and the scacchinellids, all of which were associated with reef facies. By the end of the period all the productids were extinct. The bryozoans were prolific and highly varied, but by the close of the period two of the chief Paleozoic orders, the Trepostomata and the Cryptostomata, were extinct. The fusulines and neoschwagerines were extraordinarily abundant until near the end of Permian time and reached their maximum size, but none survived beyond this period. Goniatites expanded rapidly into several families, in some of which complex sutures reached the typical ammonitic stage; but late in the period they suffered near extinction; only a few genera of two families survived to start a spectacular new evolution in the succeeding Triassic Period. *See* TRIASSIC.

On land the insects showed a great advance over those of the Coal Measures, and several of the modern orders emerged, among them the Mecoptera, Odonata, Hemiptera, Copegnathia, Hymenoptera, and Coleoptera. Extensive insect faunas have been found in the Lower Permian rocks of Kansas and Oklahoma, the Permian Basin of the Soviet Union, and the Upper Permian of Australia.

Land plants displayed at first a gradual, and eventually a profound, change as the dominant lowland plants—the lepidodendrons, sigillarias, and cordaites—of the moist coal swamps declined, and the conifers advanced to a dominant position.

Of the vertebrates, the labyrinthodont amphibians were common and varied, but the reptiles showed the most significant advances. Reptiles have been found in abundance in the lower half of the system in Texas, throughout most of the system in the Permian Basin of the Soviet Union, and in the Karoo Series of South Africa. Nearly all of the Permian reptiles were short-legged sprawlers. Of several orders, the most significant were the Theriodonta or mammallike reptiles that foreshadowed the advent of the mammals. These reptiles carried their bodies off the ground and walked or ran like mammals instead of sprawling. Their teeth became specialized—incisors, canines, and jaw teeth as in the mammals—and all the elements of the lower jaw except the mandibles showed progressive reduction. Most of the known theriodonts are from South Africa and the Soviet Union, but a few typical genera have been found in Permian beds of the Cordilleran region in North America.

[CARL O. DUNBAR]

Bibliography: C. O. Dunbar, Historical Geology, 2d ed., 1960; A. L. DuToit, *Geology of South Africa*, 3d ed., 1954; *Geologicheskoe Stroenie, USSR*, 1:345–372, 1958; E. Irving, *Paleomagnetism*, 1964; P. B. King, *Geology of the Southern Guadalupe Mountains, Texas*, USGS Prof. Pap. no. 215, 1948; N. D. Newell et al., *The Permian Reef Complex of the Guadalupe Mountains Region, Texas and New Mexico*, 1953; Symposium on Continental Drift, *Phil. Trans. Roy. Soc. London*, no. 1088, 1965; K. Teichert, Upper Paleozoic of western Australia: Correlation and paleogeography, *Bull. Amer. Ass. Petrol. Geol.*, 25:371–415, 1941; D. Van Hilten, Presentation of paleomagnetic data, polar wandering, and continental drift, *Amer. J. Sci.*, 260:401–426, 1962.

Perovskite

A natural mineral and a structure type which includes no less than 150 synthetic compounds. The mineral perovskite is ideally $Ca[TiO_3]$ but extensive substitutions occur, such as $Ce^{3+} \rightarrow Ca^{2+}$ in the variety knopite and Na^+, $Ce^{3+} \rightarrow Ca^{2+}$, and $Nb^{5+} \rightarrow Ti^{4+}$ in the varieties dysanalyte and loparite. Other members of the perovskite group include lueshite, $Na[NbO_3]$, and latrappite, $(Ca,Na)[(Nb,Ti)O_3]$.

The perovskite structure type is of great technical interest, since slight distortions away from cubic symmetry result in noncentrosymmetric (polar) arrangements which may have ferroelectric and antiferroelectric properties. One of the most intensely studied of all compounds is $Ba[TiO_3]$, which undergoes at least four temperature-dependent structural transitions: Above 120°C the compound is cubic and nonferroelectric; at 120°C it is tetragonal; at −5°C, orthorhombic; and at −90°C, rhombohedral, the last three being polar and ferroelectric substances.

The perovskite crystal structure is ideally cubic, but distortion leads to lower crystal symmetries. The essential features of the crystal structure include a framework of corner-sharing titanium-centered oxygen octahedrons in which calcium atoms are embedded, coordinated by 12 nearest-neighbor oxygen atoms. Thus, the general formula for the structure type is $A[BO_3]$, where the A are the holes in the octahedral framework and the framework is denoted by $[BO_3]$.

The mineral occurs as rounded cubes modified by the octahedral and dodecahedral forms (see illustration). Hardness is 5½ (Mohs scale), specific gravity 4.0, luster subadamantine to submetallic. The color ranges from yellow, brownish yellow, reddish, and dark brown to black, the darker shades appearing with increasing Nb^{5+} content.

Perovskite crystals, Magnet Cove, Ark. (*Specimens from Department of Geology, Bryn Mawr College*)

Perovskite occurs as an accessory mineral in basic alkalic rocks, such as carbonatites, ijolites, kimberlites, and melilitites; it is also usually associated with apatite, magnetite, melanite garnet, and melilite. Varieties rich in Ce^{3+} and other rare earths and in Nb^{5+} may be mined for these metal oxides. The perovskite structure type may occur as the arrangement for some high-pressure phases crystallizing in Earth's mantle. *See* ILMENITE; SPHENE.

[PAUL B. MOORE]

Perthite

A parallel-to-subparallel intergrowth of potassium and sodium feldspar. Usually the intergrowth occurs in a lamellar fashion. In most cases the lamellae are approximately parallel, but other orientations have also been found. With increasing size of the intergrowths, cryptoperthites, microperthites, and macroperthites may be distinguished. In microperthites and macroperthites the potassium feldspar is usually present as normal orthoclase to microcline, and the sodium feldspar exhibits its most ordered form as albite. In cryptoperthites the potassium feldspar is usually present as sanidine to normal orthoclase, and the sodium feldspar is present as small domains approximating analbite or albite. Connected with the small domain size in cryptoperthites a beautiful blue-to-whitish luster may be developed (moonstone).

It is not yet clear why the lamellar texture of the perthites is so different between specimens, but it is generally agreed that most—if not all—perthites are formed by unmixing of originally monoclinic $(K,Na)AlSi_3O_8$ into potassium- and sodium-rich feldspar domains. The unmixing starts with the formation of cryptoperthite (probably as a spinodal process), at any temperature below approximately 600°C. Depending on the cooling rate, different unmixing textures can be expected. With decreasing temperature or with time, or both, the early formed domains grow continuously until they may become visible with a microscope or even to the naked eye. During this process they change their composition and Al/Si distribution in the direction of microcline and low albite, the triclinic states stable at room temperature. Whether these states are reached or not is predominantly a function of temperature and available time. It is apparent that the process is extremely slow. Water pressure and water content probably have an important influence on degree of equilibration.

It should be kept in mind that the unmixing process has been deduced from several types of observation, and there is no hope to prove it completely by laboratory experiments because the lifetime of humans is very short compared to the time involved in the process. Electron microscopy is an important method for studying the incipient unmixing processes. *See* ALBITE; FELDSPAR; GEM; MICROCLINE; ORTHOCLASE. [FRITZ LAVES]

Bibliography: P. E. Champness and G. W. Lorimer, in H. R. Wenk and G. Thomas (eds.), *Exsolution in Silicates, Applications of Electron Microscopy in Mineralogy*, 1975; I. B. Ramberg, Braid perthite in nepheline syenite pegmatite, Langesundsfjorden, Oslo Region (Norway), *Lithos*, 5:281–306, 1972; J. V. Smith, *Feldspar Minerals in Three Volumes*, 1974.

Petrofabric analysis

The systematic study of the fabrics of rocks, generally involving statistical study of the orientations and distribution of large numbers of fabric elements. The term fabric denotes collectively all the structural or spatial characteristics of a rock mass. The fabric elements are classified into two groups: (1) megascopic features, including bedding, schistosity, foliation, cleavage, faults, joints, folds, and mineral lineations; and (2) microscopic features,

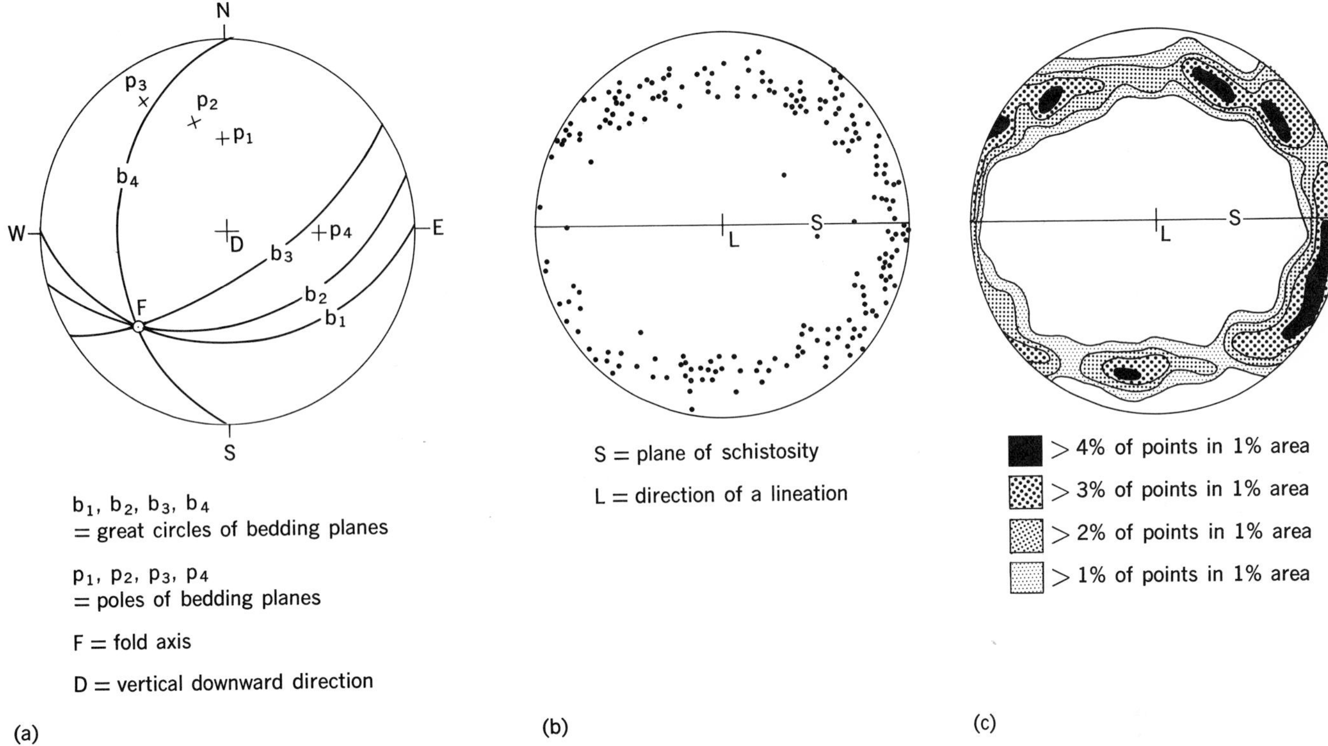

Fig. 1. Representation of the orientations of structures in spherical projections. (*a*) Stereographic projection of the orientations of four bedding planes in a cylindrical fold. The poles of the planes are distributed on a great circle perpendicular to the fold axis. (*b*) Equal-area projection of 200 *c* axes of quartz crystals in a metamorphic quartzite. The crystal axes show a strong preferred orientation. (*c*) Same data as *b*, contoured to show the distribution of the crystal axes.

including the shapes, orientations, and mutual arrangement of the constituent mineral crystals (texture) and of internal structures (twin lamellae, deformation bands, and so on) inside the crystals. The aim of fabric analysis is to obtain as complete and accurate a description as possible of the structural makeup of the rock mass with a view to elucidating its kinematic history. The fabric of a sedimentary rock, for example, may retain evidence of the mode of transport, deposition, and compaction of the sediment in the size, shape, and disposition of the particles; similarly, that of an igneous rock may reflect the nature of the flow or of gravitational segregation of crystals and melt during crystallization. The fabrics of deformed metamorphic rocks (tectonites) have been most extensively studied by petrofabric techniques with the objective of determining the details of the history of deformation and recrystallization. *See* FAULT AND FAULT STRUCTURES; FOLD AND FOLD SYSTEMS; JOINT; STRUCTURAL GEOLOGY; STRUCTURAL PETROLOGY.

Techniques. The methods of investigation of fabrics include mapping the distribution and recording the orientations of megascopic structures in the field, and collecting selected specimens of recorded orientation for laboratory study. Measurements of the orientations of microscopic structures are made on cut, polished, or etched surfaces of oriented specimens with a binocular microscope or in thin sections of known orientation with a polarizing microscope and universal stage. The orientations of planar and linear structures are analyzed by means of spherical (stereographic and equal-area) projections; each plane or line is considered to pass through the center of the sphere, so that the orientations of planes or lines are represented by a great circle or a point, respectively, in the projections (Fig. 1*a*). If the equal-area projection is employed, the distribution of large numbers of points on the projection (representing directions such as poles of bedding, mineral lineations, or crystallographic axes of mineral grains) may be contoured to give a statistical representation of the preferred orientation of the fabric elements (Fig. 1*b* and *c*). The spatial distribution of groups of elements may be analyzed on maps or thin sections by statistical techniques. The fields of homogeneity of subfabrics of different elements must be defined. *See* PETROGRAPHY.

Applications. Significant information may be derived directly from a synthesis of the fabric data. The geometry of a rock mass is described in quantitative terms. In areas that have suffered repeated deformation it is generally possible by association to identify groups of structures assignable to separate phases of deformation. It is also possible in some instances to establish the sequence of the deformations from simple geometric considerations.

Although an exhaustive study of tectonite fabrics on all scales is seldom made, analysis of some elements of the microfabric is commonly done to supplement the information obtained from megascopic structures. Tectonites usually display textures that clearly originated by recrystallization,

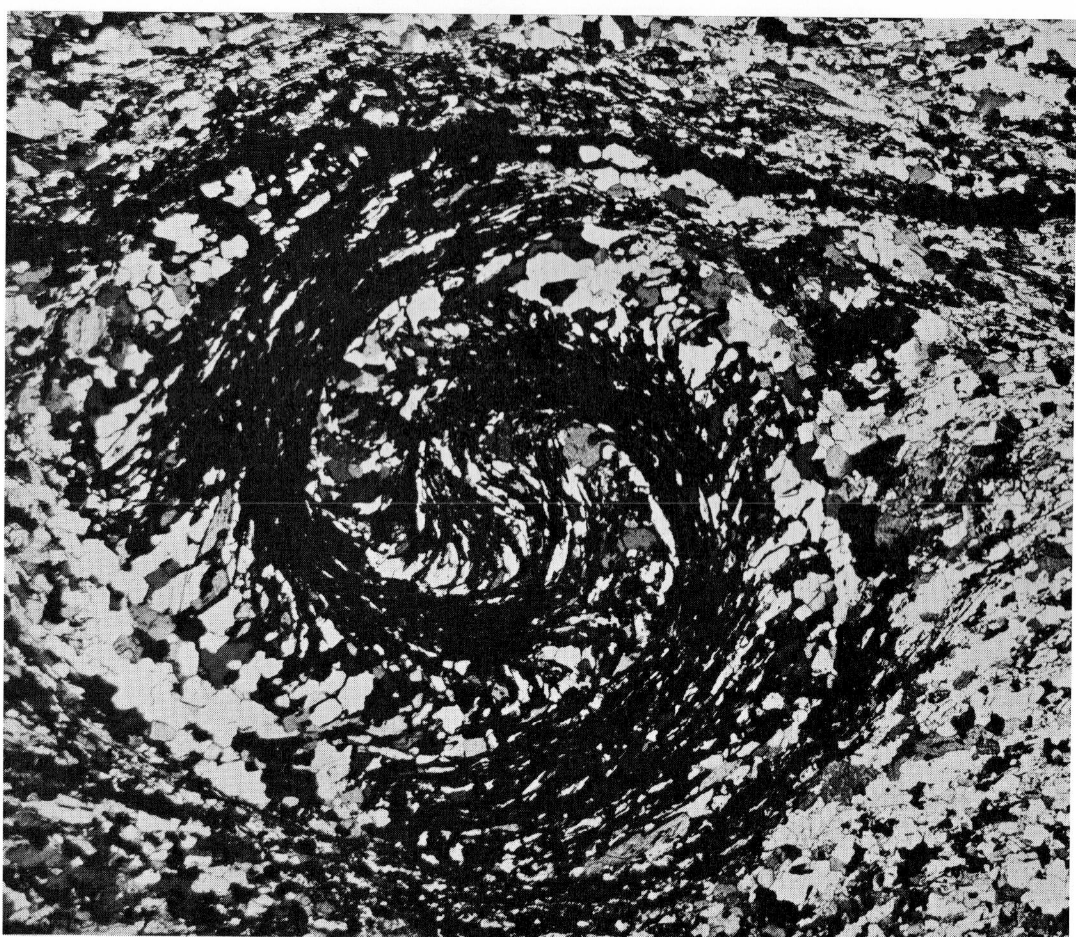

Fig. 2. Photomicrograph of a thin section of schist from Proctorsville, Vt., showing a "snowball" garnet (crossed polarizers). The garnet (circular black area) contains many inclusions of quartz, feldspar, and mica arranged in a helical distribution as a result of rotation of the garnet during its growth. Garnet is 6 mm in diameter. *(Courtesy of J. L. Rosenfeld)*

and many common minerals (such as quartz, micas, calcite, and dolomite) show preferred orientations of their crystal directions (Fig. 1*b* and *c*) analogous to those in deformed metals. Microscopic textures are especially informative in determining the sequence of recrystallization and deformation in a rock, as recrystallization obliterates preexisting textures and internal structures in the mineral grains. Thus microstructures induced after metamorphism are preserved, whereas those produced before or during metamorphism are removed in the process of recrystallization and growth of new mineral phases. Fortunately, a record of the deformation during metamorphism is also retained in the textures and preferred orientations of crystals, though the exact interpretation of these records is still not fully understood. One spectacular example of such a record is seen in the trains of mineral inclusions in "rolled," or "snowball," garnets, which were rotated because of strain of the enclosing rock during their growth (Fig. 2).

Traditionally the strain history has been inferred from fabrics, largely by the use of symmetry arguments of the type introduced by B. Sander in his classic studies of rock fabrics: The overall symmetry of a fabric is considered to be the same as that of the deformation that produced the fabric. The general validity of such arguments is borne out by experiments, if the influence of the predeformational fabric and the sequence of incremental contributions to the strain (that is, the history of strain) are taken into account.

This approach does not yield a unique picture of the strain history, however, and more precise interpretation of fabrics is based on results of experimental deformation of rocks and mineral crystals at high temperatures and pressures. Certain types of folding, microscopic textures, and preferred orientations of minerals have been reproduced in the laboratory under known conditions closely analogous to those existing at depth in the Earth's crust. The results of these experiments provide a reliable basis for specifying the mode of origin of many features of rock fabrics. Theoretical analysis and experiments with dimensionally matched models also contribute appreciably to the understanding of rock structures and development of rock fabrics. *See* HIGH-PRESSURE PHENOMENA; PETROLOGY; SCHISTOSITY, ROCK. [JOHN M. CHRISTIE]

Bibliography: D. Griggs and J. Handin (eds.), *Rock Deformation*, Geol. Soc. Amer. Mem. no. 79, 1960; J. G. Ramsay, *Folding and Fracturing of Rocks*, 1967; B. Sander, *Einführung in die Gefügekunde der geologischen Körper*, vol. 1, 1948, vol. 2, 1950; F. J. Turner and L. E. Weiss, *Structural Analysis of Metamorphic Tectonites*, 1963.

Petrographic province

A region in which the igneous rocks, formed during a limited period, show a certain community of character (chemical, mineralogical, or petrological) that distinguishes them from other rocks in the area. Rocks of a petrographic province are consanguineous in that they are, theoretically at least, derived from a common parental magma. A classic province is represented by the highly potassic rocks around Rome and Naples. Other provinces are characterized by either high or low content in other elements (such as sodium and titanium).

If the chemical analyses of rocks representing a genetically related series are plotted on a variation diagram, many characteristics of the series may be brought out. The illustration shows the percentages of oxides in each rock analysis as plotted against the silica content of that analysis.

All points lie on or close to smooth curves which represent the variation in composition with silica content (or roughly with time, because younger rocks tend to be more silicic). Noteworthy are the positive slopes for potassium oxide (K_2O) and sodium oxide (Na_2O), the arched curve for aluminum oxide (Al_2O_3), and the negative slopes for magnesium oxide (MgO), calcium oxide (CaO), ferrous oxide (FeO), and ferric oxide (Fe_2O_3). The combined alkali ($Na_2O + K_2O$) curve is seen to cross the CaO curve at 56.5% silica. This silica value is known as the lime-alkali index for the rock series. On the basis of lime-alkali indices, the numerous rock series are arbitrarily divided into groups as follows: alkalic < 51, alkali-calcic $> 51 < 56$, calc-alkalic $> 56 < 61$, and calcic > 61.

Certain rock series appear related to certain types of geological environments. In general, the more alkalic types occur in regions subjected to tension and vertical movement (faulting and subsidence), whereas more calcic series are found in compressional regions (fold-mountain belts).

Olivine basalt-trachyte association. Rocks belonging to this association are extremely widespread. The association is well represented in the central Pacific islands (Hawaii, Tahiti, Samoa), the islands along the mid-Atlantic ridge (Ascension, Saint Helena), and islands of the Indian Ocean (Kerguelen). Both mineralogically and petrographically the association is relatively simple. Primary olivine basalt magma has given rise to the sequence

Olivine basalt → basalt → andesite → trachyte

largely through crystallization and sinking of heavy minerals. The last two types are not abundant. Locally the series may be carried beyond trachyte to quartz trachyte (Samoa) or soda rhyolite (Ascension). Elsewhere it may pass through tephrite to phonolite. The extrusion of phonolite may be due to a strongly undersaturated (silica-deficient) parent magma. The quartz-bearing end products may be due in some areas to a saturated parent magma and in others (Ascension) to slight assimilation of older granitic rocks.

The characteristic development of oceanite, ankaramite, and limburgite may be explained by gravitational accumulation of olivine and pyroxene at lower levels in the volcanic reservoir. Density stratification of this type would permit nearly simultaneous eruption of highly contrasting lavas from neighboring vents.

The olivine basalt-trachyte association is also well represented on the continents (Otago, New Zealand; Oslo, Norway; East African rift zone; Midland Valley of Scotland). Here, however, the later members of the series (trachyte, soda rhyolite, and phonolite) are relatively more abundant, and leucite (rare in oceanic areas) may be locally important.

In general, the association occurs in regions of faulting and marked vertical movement. *See* MAGMA.

Flood basalts. These basalts, also known as plateau basalts, form thick accumulations of nearly horizontal flows over vast areas (hundreds of thousands of square miles). Examples include the Columbia-Snake River basalts of Oregon and Washington, the Deccan plateau lavas of western India, and the Keweenawan lavas of Lake Superior. These flows were probably erupted through fissures from great deep-seated supplies of basaltic magma.

Rocks of this association are overwhelmingly basalts, but rare amounts of rhyolite, trachyte, and andesite occur. Some geologists distinguish two types of flood basalts, olivine basalt and tholeiitic basalt. The former is slightly lower in silica but higher in soda, potash, and magnesia than is the latter. Both types may occur in the same area. They may be derived from different earth shells (at different levels), or the tholeiitic type may form from olivine basalt by crystal fractionation. Other geologists consider the two types merely variations of a single primary basalt magma. *See* BASALT.

Intrusive masses (sills and dikes) of basaltic material (diabase) form swarms over tens of thousands of square miles in many parts of the world. Particularly notable are the nearly flat sheets of diabase in the sediments of the Karroo system in South Africa and the flat Palisade sill of New Jersey. The thick, tabular bodies of magma differentiated somewhat as they cooled; and olivine accumulated in a layer near the sill floors. Pyroxene is less abundant and more iron-rich toward the top, and plagioclase increases in abundance and soda content in the same direction. These relations demonstrate the control of crystal

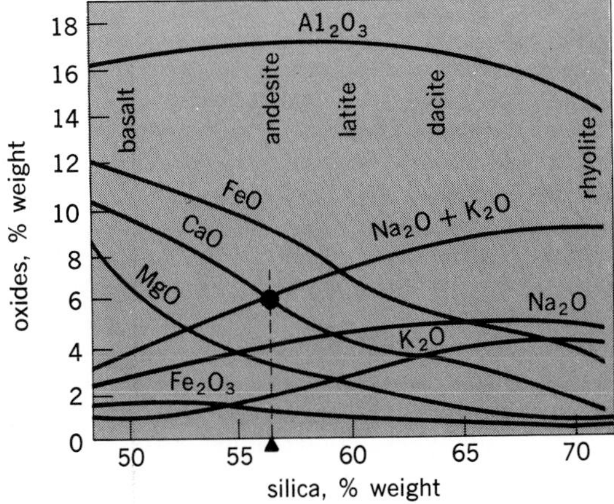

Igneous rock variation diagram of calc-alkali series of volcanic rocks from San Francisco Mountains, Ariz.

fractionation and settling in the process of differentiation. *See* DIABASE.

Basalt-andesite-rhyolite association. The andesite-rhyolite kindred appears on continents and is typical of, but not restricted to, regions of orogeny (fold-mountain belts). Most striking is the circum-Pacific belt of volcanoes (many still active) extensively developed along the western margin of North and South America.

The rocks consist chiefly of andesite and rhyolite with smaller amounts of basalt, dacite, latite, and quartz latite. Both lavas and tuffs are represented. The sequence of eruption of rock types is extremely complicated and varies with location as well as with time. Basaltic magma may have been the parental fluid in most areas. Differentiation of this basic melt has operated by fractional crystallization, but superimposed upon its effects are those of assimilation of crustal rocks and mixing of partly crystallized magmas (consanguineous) at various stages of differentiation. Possibly much rhyolitic and andesitic magma was generated by local melting of the crustal rocks.

Basic-ultrabasic association. This association includes rocks of basic or ultrabasic composition and is found in large sheetlike, saucer-shaped, or conical bodies commonly injected at shallow depths. Examples include the Duluth lopolith of Minnesota, the Stillwater complex of Montana, the Skaergaard body of east Greenland, and the Bushveld complex of South Africa.

The rock types are distributed in flat or somewhat basined layers to form sequences thousands of feet thick. In general the rocks become heavier toward the bottom, but in detail, thin layers of contrasting types may alternate to give a strongly banded appearance. Peridotite and pyroxenite are most abundant near the base. Upward these give way to norite and gabbro with some anorthosite. Uppermost rocks may be dioritic to granitic. Olivine and pyroxene become richer in iron, and plagioclase becomes more sodic from bottom to top in these bodies.

Such distributions suggest that the originally injected basaltic or basic magma solidified in general from the floor upward. Crystals may have initially formed in upper regions and may have settled to build up the floor. Convection currents in the melt may have helped to redistribute and sort the crystals of different size and density. The abundant granitic material commonly found near the roof may represent, in varying amounts, a late differentiate of the original basaltic magma, a product of assimilation of siliceous rocks (sediments and felsites) by the magma, or a crystallized secondary melt generated by the hot basic intrusion.

Granodiorite-granite association. This extensively developed, coarse-grained plutonic assemblage is restricted to continents and may be subdivided into two categories. The first includes the smaller masses (stocks, ring dikes, and so on) commonly widely scattered and formed at relatively shallow depths. These masses usually transgress the structure of surrounding rocks, but some are highly concordant and may have spread or domed the adjacent and overlying rocks. They are usually surrounded by a metamorphic halo or zone of recrystallization which is generally most conspicuous where the rocks have not been previously metamorphosed.

Granite and granodiorite predominate, with minor amounts of diorite, gabbro, and syenite present. Most bodies have probably formed by crystallization of granite or granodiorite magma. Where minor basaltic melts have been involved, crystal fractionation and assimilation of adjacent rock material may have been operative. *See* AUREOLE, CONTACT.

The second category includes immense, deep-seated bodies (batholiths) surrounded by metamorphic rocks and restricted to orogenic zones. The Sierra Nevada batholith of California and the Coast Range batholith of British Columbia are typical examples. The predominant rock type is granite in those masses of Precambrian age and granodiorite in the younger bodies. Quartz diorite is somewhat less abundant, whereas diorite, gabbro, and syenite occur only locally. Batholiths are generally elongate parallel to the orogenic belt, but locally they are highly crosscutting. The granitic rocks may be massive or foliated (showing parallel streaking or layering of minerals) much like the adjacent metamorphic rocks into which they may appear to grade. Many are characteristically associated with pegmatites and migmatites.

Some of these large bodies may form from granitic magmas derived by downbuckling and melting of the Earth's sialic (granitic) layer. Others may represent more or less reconstituted (metamorphosed) sediments in geosynclines. Still other bodies may be products of granitization. *See* GRANITIZATION; METAMORPHIC ROCKS; METAMORPHISM; MIGMATITE; PEGMATITE.

Leucite basalt–potash trachybasalt. This association includes a wide variety of silica-poor, potash-rich rocks of volcanic and near surface origin. The association is confined to the continents. Basic (low in Si, high in Ca, Fe, and Mg) and ultrabasic lavas with leucite are dominant in many areas. Occurrences are restricted but widespread (Rome-Naples, Italy; Uganda, east Africa; West Kimberley, Western Australia; and Leucite Hills, Wyo.). Rock types include leucite basalt, leucite basanite, potash trachybasalt, and melilite basalt. The association appears in regions of faulting and marked vertical movement. *See* LEUCITE ROCK.

Spilite-keratophyre association. This association includes volcanic flows and tuffs with minor intrusives, intimately associated with sediments in geosynclinal regions. The rocks are soda-rich and potash-poor; they are chiefly basaltic (spilites) with some soda trachyte (keratophyre). Many appear altered (metamorphosed), and some are associated with rocks of the basalt-andesite-rhyolite suite. *See* SPILITE.

Ultrabasic rock. The predominantly peridotite and serpentine rock in abundant intrusive bodies is closely associated with the spilitic suite. Together the two rock associations constitute the so-called ophiolites, generally considered to represent the earliest magmatic phenomenon in orogenic regions. *See* PERIDOTITE.

Nepheline syenite association. Nepheline syenite and associated alkali-rich rocks are widespread but rare. They are continental rocks and commonly appear in areas of subsidence and faulting. *See* NEPHELINE SYENITE.

Anorthosite. This rock forms gigantic masses in Precambrian terranes and appears associated with hypersthene granite and norite. It is composed of andesine or labradorite and, therefore, differs from the calcium-rich anorthosite associated with gabbro in large stratiform sheets. *See* ANORTHOSITE.

[CARLETON A. CHAPMAN]

Bibliography: T. F. W. Barth, *Theoretical Petrology*, 2d ed., 1962; F. J. Turner and J. Verhoogen, *Igneous and Metamorphic Petrology*, 2d ed., 1960.

Petrography

A branch of petrology, the study of rocks, that emphasizes the description and systematic classification of rocks, especially by the study of thin sections under the petrographic microscope, by analysis of disaggregated samples, or by analysis of individual mineral components. *See* PETROLOGY.

The megascopic classification of rocks, based on characteristics observed in field exposures or hand specimens, suffices for some purposes; but refined description and classification require determination of the kinds, sizes, shapes, and space interrelationships in the aggregates of original mineral components and the changes and alterations that have affected rocks or their separate components subsequent to their initial formation.

The petrographic investigation of a rock usually entails both an analysis of the rock after disaggregation into mineral or particle size fractions and examination of the rock in thin sections under the petrographic microscope.

Study of disaggregated samples. Techniques for study of disaggregated samples differ depending upon whether the rock is composed of an accumulation of particles and fragments, as in sandstone (clastic rock), or consists of interlocking mineral grains, as in granite (crystalline rock).

Poorly to moderately indurated, or hardened, clastic rocks are disaggregated by crushing or pulverization, by dissolving cementing materials, or by using the disrupting effect of salts precipitating from saturated solutions with which the rock is impregnated. After disaggregation the size distribution of the clastic particles is ascertained by screen analysis or by liquid or air elutriation; and the shapes, roundness, and surface characteristics of the particles are noted under a stereoscopic binocular microscope.

Hard clastic rocks and crystalline rocks, including igneous, metamorphic, and recrystallized or firmly cemented sedimentary rocks, are disaggregated by crushing into fragment sizes that will permit separation into mineral fractions. A variety of methods for separation of disaggregated rocks into mineral fractions are available.

Separation of mineral fractions. Commonly used methods include hand sorting under the microscope, separation in heavy liquids, magnetic separation, and electrostatic separation.

Heavy-liquid separation. Heavy liquids consisting of suspensions of ground metals in liquids are available with densities as high as 7.5 but are infrequently used because of the opacity. Widely used transparent liquids are bromoform, which has a density near 2.9, and acetylene tetrabromide, with a density of 2.96. These are employed to separate heavy minerals from light minerals. Powdered rocks, sometimes sized by screening and washed to remove the very fine fractions, are suspended in a heavy liquid, in a funnel or evaporating dish, and vigorously agitated to cause the light minerals to float and the heavy minerals to sink. After the heavy minerals have been collected and washed, successive crops of the light minerals may be obtained by careful progressive dilution of the heavy liquid with its appropriate solvent.

Magnetic separation. Electromagnetic separators provide mineral fractions differing in magnetic susceptibilities. Satisfactory electromagnetic separators are constructed to allow variations in field intensity at the point or area of separation, variation of the tilt of the poles of the magnet, and variation of the rate of feed. Best results are obtained with clean, free-flowing aggregates of uniformly sized grains. Electromagnetic separation is not satisfactory when minerals contain nonuniformly distributed impurities or alteration products.

Electrostatic separation. Minerals of different electrical conductivities can be segregated by electrostatic separation. Mineral grains of approximately equal size placed on a grounded metal plate are differentially attracted to a charged surface, and separation of the different mineral fractions is accomplished by varying the intensity of the charge or the spacing between the charged surface and the plate supporting the mineral grains. As in magnetic separation, irregularly included impurities and alteration products prevent clean separations.

Study of mineral fractions. After separation, each mineral component is identified by one of the several techniques of determinative mineralogy. Preliminary identification by physical properties and by simple wet and dry tests using the procedures of systematic blowpipe analysis may precede more elaborate tests. Spectrographic analysis, microchemical tests, or partial or complete chemical analyses are made to identify gross or trace elements. Particularly useful in identification are x-ray diffraction patterns of powdered minerals as recorded on film strips or as traced on charts on recording diffractometers. *See* MINERALOGY.

Minerals such as the feldspars, feldspathoids, carbonates, and clay minerals may be treated chemically so that they assume distinctive colors resulting from absorption of dyes or from precipitation of colored chemical compounds. Minerals

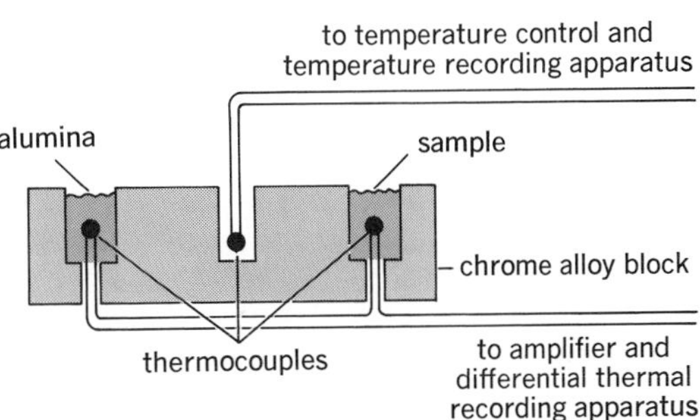

Fig. 1. Idealized diagram of components of sample holder for differential thermal analysis. (*From E. E. Wahlstrom, Petrographic Mineralogy, John Wiley & Sons, Inc., 1955*)

Fig. 2. Four-axis universal stage. (*E. Leitz, Inc.*)

that undergo rapid chemical changes or crystallographic inversions during heating are studied successfully with differential thermal apparatus (Fig. 1) or by determination of change of weight during heating in the case where decomposition with loss of a volatile substance takes place. Differential thermal analysis measures the magnitudes and temperatures of endothermal and exothermal reactions in minerals as they are heated in a furnace from room temperature to 1000°C or more.

Rocks consisting in whole or part of clay minerals or clay-sized particles pose special problems. Mineral fractions or unseparated aggregates of clay minerals are studied by stain tests, x-ray diffraction patterns, differential thermal analysis, base exchange properties, infrared adsorption, and the electron microscope.

Microscopic petrography. For rocks in which the mineral components are large enough to be identified under the microscope, examination under a petrographic microscope of thin sections ground to a thickness of 0.03 mm is standard procedure. The petrographic microscope differs from

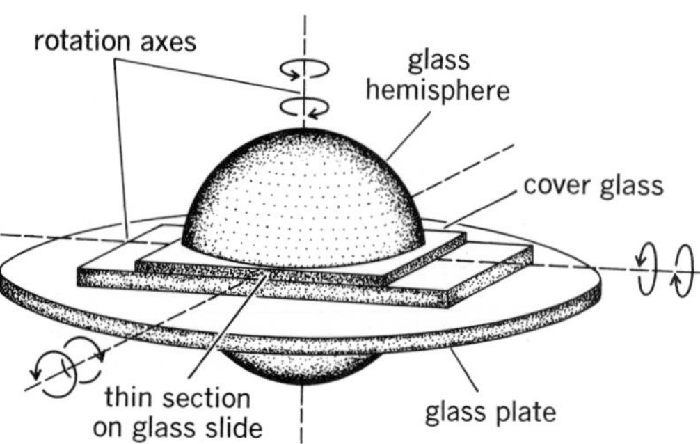

Fig. 3. Thin section mounted between glass hemispheres on universal stage. (*Adapted from E. E. Wahlstrom, Petrographic Mineralogy, John Wiley & Sons, Inc., 1955*)

an ordinary microscope in that it has a rotating stage and polarizing filters or prisms which produce optical phenomena not observable in ordinary light. Minerals are identified by their optical and other physical properties by using tables and charts which correlate physical properties with chemical composition.

Minerals previously separated from disaggregated rocks are further analyzed under the microscope by placing fragments in liquid immersion mediums of known refractive indices and determining the refractive index or indices of the mineral by a matching process. A useful aid in optical studies is the universal stage (Fig. 2), a multiaxis device which is attached to the stage of the microscope and permits rotation of thin sections or mineral fragments mounted in immersion mediums into any desired position for the measurement of optical properties (Fig. 3). Thin sections from which the cover glass has been removed may be etched and stained to aid in the identification of some components by the same reagents used for staining crushed fragments.

A particular advantage of the thin-section technique is that it allows measurement of the dimensions of the mineral components and reveals the nature of the contacts and the manner of intergrowth of the mineral components. Subtle differences in composition within various portions of single crystals, twinning, strain, alterations, and other features that might not be noted in nonoptical techniques become very apparent under the microscope.

Mode. The mode of a rock expresses the mineral composition in weight or volume percentages. The mode of crushed or otherwise disaggregated rocks is obtained by weighing the separated mineral fractions. The volume mode of rocks obtained from thin sections may be converted to a weight mode by calculations using assumed or measured specific gravities for each mineral component. Weight modes are used to calculate bulk chemical compositions of rocks after the composition of each mineral has been determined by optical or other means. The volume mode is determined by micrometric analysis in thin sections. With a mechanical stage several linear traverses are made across the thin section and the mode is calculated on the assumption that the volume of each mineral is proportional to the total of the linear intercepts for the mineral measured by the stage. A more rapid method employs a point counter, a mechanical stage which moves the thin section to a succession of equally spaced points in a linear traverse. The mineral at the intersection of the cross hairs of the microscope is noted at each point, and the volume of each mineral in the rock is assumed to be proportional to the number of points counted for each mineral.

Space arrangement. The space arrangement of the mineral components is analyzed quantitatively by means of petrofabric techniques. The angular relationships to the plane of the thin section, and ultimately to the site of collection of the sample in the field, of crystallographic directions, as determined by optical measurements or by observation of crystal shapes, are measured with a universal stage. Measured angles are used to plot points on an equal-area projection, a special type of projection derived from the spherical projection. The density of points on the projection is indicated by contours, so that the finished plot is a statistical indication of the preferred orientation in one respect or another of one or several mineral components. *See* PETROFABRIC ANALYSIS.

A specialized branch of microscopic petrography deals with opaque minerals, especially ore minerals, as studied under a reflecting microscope. The highly polished surface of a mineral or mineral aggregate is examined by reflected light which is directed onto the surface of the mineral by a vertical illuminator, a device inserted into the microscope between the object lens and the barrel of the microscope. Polarizing plates or prisms permit observations in plane-polarized light. Minerals are identified by their isotropism or anisotropism, color, reflectivity, hardness, and response to a standard set of etch reagents. Small portions of minerals scratched from the polished surface are subjected to systematic microchemical analysis, x-ray analysis, or spectrographic analysis.

Rocks containing both opaque and nonopaque minerals may be studied by making polished thin sections. Part of a section is left uncovered so that the opaque components can be studied in reflected light while the other components are examined by transmitted light by ordinary methods.

Chemical petrography. Crushed samples of rocks and separated fractions of mineral components may be analyzed chemically by routine methods of wet and dry analysis. However, mineral analysis gives only a bulk composition and does not enable determination of variations of composition within single crystals owing to zoning or exsolution. Also, intergrowth of two or more minerals may prevent separation of pure fractions.

These difficulties are overcome by using an electron microprobe, which is especially useful in ascertaining the chemical composition of very small crystals and in the detailed study of variations of composition within single crystals.

A very narrow beam of electrons incident on the smooth, polished surface of a thin slice of a mineral or mineral aggregate interacts not only with the electron shells but also with the nuclei of atoms or ions; it produces characteristic radiations that can be resolved and analyzed by use of spectrometer crystals, a sensitive pickup, and a chart recorder. [ERNEST E. WAHLSTROM]

Bibliography: E. Cameron, *Ore Microscopy*, 1961; A. V. Carozzi, *Microscopic Sedimentary Petrography*, 1960; E. W. Heinrich, *Microscopic Petrography*, 1956; L. W. LeRoy, *Subsurface Geologic Methods*, 2d ed., 1950; W. W. Moorhouse, *The Study of Rocks in Thin Section*, 1959; F. J. Turner and L. E. Weiss, *Structural Analysis of Metamorphic Tectonites*, 1963; E. E. Wahlstrom, *Petrographic Mineralogy*, 1955; H. Williams, F. J. Turner, and C. M. Gilbert, *Petrography*, 1954.

Petroleum, origin of

A complex mixture of hydrocarbons, containing small amounts of oxygen, nitrogen, and sulfur compounds and traces of metal salts. Accumulation of petroleum is believed to involve three steps: generation of oil, primary migration (the movement of oil from source to reservoir rock), and secondary migration (the redistribution of oil within the reservoir rock to form a pool). On the basis of the best available data, these steps can be described as follows.

Generation of oil. Oil is generated in sedimentary basins. These basins are shallow depressions on the continents that have intermittently been covered with sea water, or offshore basins on continental shelves. They are hundreds of square kilometers in area and contain sediments of three types: (1) rock particles varying from sands to clay muds, which were eroded from hills and mountains and were carried to the basins by streams; (2) biochemical and chemical precipitates such as limestone, gypsum, anhydrite, and chert; and (3) organic matter from the plants and animals that lived in the sea or were carried in by rivers. Some of these types of sediments are being laid down today in basins such as the Persian Gulf and the Gulf of Mexico.

The third type of sediment, the organic matter, is the source of petroleum. Evidence for this is the fact that petroleum contains small amounts of several substances that could have come only from living things. Examples of these are porphyrins related to hemin and chlorophyll; optically active compounds (compounds that will rotate the plane of a ray of polarized light); and structures related to phytol, cholesterol, carotene, terpenoids, and several other plant products.

It is believed that oil is generated from organic matter in two ways. A small amount, probably less than 10%, comes directly from the hydrocarbons that marine organisms form as part of their living cells. Evidence for this process is that some of the complex hydrocarbons in petroleum are comparable to those found in modern plankton, kelp, and higher organisms, including fish. When such organisms die in the waters of a sedimentary basin, their remains drop into the material accumulating in the basin. Their hydrocarbon content is low (0.005–0.1% by weight), and some of this is destroyed by bacterial oxidation or recycling through the marine food chain. The total amount of hydrocarbon produced in this manner is so great, however (probably more than 1,000,000 barrels a year), that enough is preserved to account for these hydrocarbons in the world's oil accumulations.

The second process, by which about 90% of the oil is formed, involves the formation of hydrocarbons from the decay and alteration of buried organic matter. Nearly all of the hydrocarbons containing up to 10 carbon atoms and probably over half of those of higher molecular weight are formed in this manner.

In the first few meters of sedimentary burial, the organic matter is converted into more petroleum-like hydrocarbons through biochemical action. At increasing depths simple low-temperature cracking results in the removal of oxygen, sulfur, and nitrogen, eventually to form hydrocarbons. By the time the sediments are buried to depths of 500–700 m, enough hydrocarbons have been generated to enable a commercial oil field to form under favorable accumulation conditions. The formation of hydrocarbons continues to increase with the increasing temperatures occurring at greater depths. Beyond 2000 m there is a tendency to form large quantities of light hydrocarbons containing up to 10 carbon atoms. The increasing depth causes the source rock to be more compacted, so there is a decrease in the ease of the hydrocarbon migration from source to reservoir rock. The result is that vast quantities of hydrocarbons never get out of the source rock. It is estimated there are 6000×10^9 tons of petroleum in the reservoir rocks of the continents and continental shelves of the world. At least 100 times this amount of petroleum still remains in the source beds. As the petroleum and organic matter are buried to still greater depths, beyond 10,000 m, everything is gradually converted to the two end products of all organic matter—methane and graphite.

Some hydrocarbons are formed in nearly every type of sediment, but the greatest quantity are associated with the most reducing environments. The best source beds of petroleum are fine-grained clay or carbonate muds deposited in basins under reducing conditions. Coarse sediments, such as clean sandstones, reefs, and oolites, are not source rocks of petroleum because they are deposited in shallow-water aerated areas, where water movement winnows out the organic matter and the oxygen tends to destroy hydrocarbons. Some present-day basins, such as the Gulf of Mexico, the basins off California, the Orinoco Delta, the Persian Gulf, and the Caspian Sea, provide conditions that are representative of those under which oil was formed in the past.

Primary migration. Primary migration of petroleum from source to reservoir, the second step in accumulation, is caused by the movement of water, which carries oil or its precursor in concentrations of less than 100 ppm out of compacting sediments.

When the source muds are deposited, they contain 70–80% water. The remainder is solids, mostly clay minerals or carbonate particles; as they build up to great thicknesses in a sedimentary basin, water is squeezed out by the weight of the overlying sediments. After compaction, the source muds contain only about 25% water at a depth of 700 m and 10% at 2000 m.

The water and oil move in the direction of least resistance (lowest hydrostatic pressure) at rates of 2–10 cm per year. Early in compaction, the direction of fluid movement is straight up. As compaction progresses, there is lateral as well as vertical movement of the fluids. The lateral movement results primarily from the tendency of the flat clay mineral particles to lie horizontally as they are compressed. This reduces the vertical permeability of the compacting muds. In addition, the long, continuous sands on the edges of basins orient fluid movement laterally as burial progresses.

The mechanism by which the water carries the oil is uncertain. The oil cannot travel as droplets or colloidal particles because neither are capable of penetrating the fine pore openings of a clay mud. The oil probably travels in solution as hydrocarbons, which are soluble in the low-molecular-weight range, and as precursors (nonhydrocarbons), which are soluble in the high-molecular-weight range. The latter could be converted to hydrocarbons, through mild cracking reactions, after entering the reservoir rock. The soluble hydrocarbons or precursors are believed to coagulate in the reservoir rock because of the change in the physical-chemical environment. When such a rock is completely enclosed with compacting muds, the oil particles will be held in by the capillary pressures in the muds and the water will pass through. If there is a porous avenue of escape that bypasses the muds, hydraulic currents created by the moving water will sweep the oil droplets out of the reservoir. As compaction progresses and most of the water is expelled from the muds, they will develop into tight shales or dense carbonates. These form the covering seal for the accumulated oil particles in the reservoir rock. Although hydrocarbons are being formed continuously from organic matter in sediments, the migration process takes 1,000,000 or more years to form a commercially valuable oil accumulation. Consequently, man is removing oil from the earth at a much faster rate than it is accumulating.

Secondary migration. In secondary migration, the last stage in the origin of an oil accumulation, the oil droplets are moved about within the reservoir to form the pool. Secondary migration includes in some instances a second step, during which crustal movements of the earth shift the position of the pool within the reservoir rock.

The position of the accumulating pool is affected by several, sometimes conflicting, factors. Buoyancy causes oil to seek the highest permeable part of the reservoir; capillary forces direct the oil into the coarsest-grained portion first, and into successively finer-grained portions as the reservoir is filled. Any permeability barriers in the reservoir channel the oil into a somewhat random distribution. Oil accumulations in carbonate rocks are often erratic because part of the original void spaces have been plugged by minerals introduced from water solutions after the rock is formed. In large sand bodies, barriers formed by thin layers of dense shale may hold the oil at various levels. When crustal movements of the earth occur, oil pools are sometimes shifted away from the place in which they originally accumulated. Faults sometimes cut through reservoirs, destroying part of the pools or shifting them to different depths. Uplift and erosion bring the pools near the surface, where the lighter hydrocarbons evaporate. Fracturing of the cover rock allows oil to migrate vertically to a much shallower depth. Wherever differential pressures exist and permeable openings such as fractures provide a path, petroleum will move.

The composition and character of the oil ultimately formed is controlled by all phases of its origin and migration. The organic structures from which the oil originated, the increasing formation of light hydrocarbons with depth, the selectivity of the migration phase in carrying only part of the hydrocarbons from the source rock, and the alteration of the oil within the reservoir by subsurface water and ground movements, are believed to play an important part in determining the composition of petroleum. *See* PETROLEUM GEOLOGY; SEDIMENTATION. [JOHN M. HUNT]

Bibliography: H. D. Hedberg, Geological aspects of origin of petroleum, *Bull. Amer. Ass. Petrol. Geol.*, 48:1755–1803, 1964; G. D. Hobson, *Some Fundamentals of Petroleum Geology*, 1954; J. M. Hunt, The origin of petroleum in carbonate rocks, in G. V. Chilingar, H. J. Bissell, and R. W. Fairbridge (eds.), *Carbonate Rocks*, 1967; L. G. Weeks (ed.), *Habitat of Oil*, 1958.

Petroleum geology

The application of geological principles in the discovery and development of oil and gas pools. The geology of petroleum includes the origin, migration, and accumulation of petroleum; the structural and stratigraphic relations of oil and gas pools; and the lithologic and paleontologic characteristics of geological formations and producing horizons. Petroleum geology is strongly influenced by the economic aspects of petroleum exploration.

Geological aspects treated here include the occurrence of petroleum, the character of reservoir rocks, typical reservoir traps, and the general nature of reservoir fluids. For further treatment of the origin and migration of petroleum, *see* PETROLEUM, ORIGIN OF.

OCCURRENCE OF PETROLEUM

Petroleum deposits may be classified as surface occurrences and subsurface occurrences.

Surface occurrences. These occurrences may be thought of as currently active or "live" occurrences, such as seepages, springs, exudates, mud volcanoes, and mud flows. Others may be termed fossil or "dead" occurrences, such as bitumen-impregnated sediments, inspissated deposits, and dike and vein fillings of solid bitumens.

Seepages, springs, and exudates. Petroleum that exudes in any of these forms may reach the surface along fractures, joints, fault planes, unconformities, or bedding planes or through the connected porous openings of the rocks. Most seepages (or springs) are formed by the slow escape of petroleum from accumulations that are close to the surface or have been tapped by faults and fractures. Many pools and producing regions have been discovered by drilling near seepages.

Exudates of asphaltic oils issuing at the surface are likely to be changed to asphalt, partly by the escape of volatile fractions, but mainly by chemical changes such as combination with oxygen or sulfur. The asphalt is black and varies in consistency from a sticky liquid to a substance hard enough to walk on. Outcrops of asphaltic oils are sometimes marked by small pools, some of which contain the bones of animals caught in the sticky material. Asphalt particles may be transported by water, and some deposits consist largely of transported material.

Mud volcanoes and mud flows. These are high-pressure gas seepages that carry with them water, mud, sand, fragments of rock, and occasionally oil. Mud volcanoes are usually confined to regions underlain by incompetent softer shales, boulder and submarine landslide deposits, clays, sands, and unconsolidated sediments. The surface of a mud volcano is often a conical mound or hill, with an opening or crater at the top through which issue mud and water which is usually salty. Only the type which emits gas, with or without oil, in addition to the mud and water, should be considered a surface indication of oil or gas. Mud volcanoes occur chiefly in areas of Cenozoic rocks that have been strongly deformed. *See* MUD VOLCANO.

Solid and semisolid deposits. Tar, asphalt, wax, and hard brittle bitumen (any of the flammable, viscid, liquid or solid hydrocarbon mixtures soluble in carbon disulfide) are popularly regarded as solid although, strictly speaking, some of them are highly viscous liquids. Outcrops of solid petroleum are found in the form of disseminated deposits and as veins or dikelike deposits filling cracks and fissures.

Disseminated deposits are sediments containing petroleum in the form of asphalts, bitumen, pitch, or thick heavy oil, disseminated through the pore spaces of rock either as a matrix or as the bonding material. They are commonly called bituminous sands or bituminous limestones, depending on the nature of the host rock. Two different types of disseminated occurrences are found, inspissated deposits and primary mixtures of rock and bitumen.

Inspissated or dried-up deposits occur in place and were probably once a pool in liquid and gaseous form. They now consist of only the more resistant and heavier residues, the lighter fractions having been lost. An inspissated deposit may be thought of as a fossil oil field. As erosion gradually removed the overburden, bringing the surface closer to the pool, the decreased pressure permitted gases and lighter oil fractions to come out of solution and expand, leaving the heavier hydrocarbon fractions behind. As the pool approached the zone of weathering, the opening of incipient fractures allowed the gases to escape more readily. Oxidizing agents aided in solidifying the heavier oils that remained behind.

Primary mixtures are those in which the sediments were mixed with the oil, asphalt, or tar during their deposition, the whole deposit having later been buried by younger sediments and then exposed by erosion. The Athabasca oil sands in Alberta, Canada, are thought by many to be such a deposit. One theory is that oil seeped up from the underlying and then outcropping Devonian organic limestone and was redeposited during Cretaceous time together with Cretaceous sands in lagoons and barred basins along the shore. These oil sands may be considered a primary disseminated deposit in which the sand was deposited in or with the oil. *See* OIL SAND.

Dike and vein fillings may be regarded as fossil or dead seepages from which the gaseous and liquid fractions have been removed, leaving only the solid residues behind. In inspissated deposits the separation of the lighter constituents occurred in place in the rock. In the primary deposits the separation of the gas from the liquid took place before the contemporaneous deposition of the oil and asphalt with the enclosing sediments. In the solid vein and dike fillings the loss of the gaseous and liquid fractions probably occurred while the petroleum was filling the opening.

Oil shale is rock which yields oil upon being heated. Deposits of oil shale which are suitable for commercial exploitation yield 5–15 gal or more of oil per ton. The organic material, termed kerogen, in oil shale is in solid form prior to being heated. Some of the kerogen decomposes into gaseous and liquid petroleum hydrocarbons when heated to 350°C or more. So many different meanings have been given to the word kerogen that it is not in good scientific standing today. *See* OIL SHALE.

Asphalt is a black, plastic to fairly hard substance, easily fusible and soluble in carbon disulfide. It occurs in nature, but it is also obtained as the residue from the refining of certain petroleums; then it is known as artificial asphalt. Asphalt melts between 150 and 200°F. It may occur as seepages, surface accumulations, and impregnations, and also in large lakes such as the Rancho La Brea deposit in Los Angeles.

Asphaltites are harder solid hydrocarbons which differ from asphalt in being strictly of an intrusive nature. They are found in veins or dikes cutting across the sediments. Asphaltites are fusible, but melt at somewhat higher temperatures and are harder and heavier than the asphalts. *See* ASPHALT AND ASPHALTITE.

Naturally occurring mineral waxes are solid hydrocarbons believed to result from the drying out of a paraffin-base oil. One example is ozokerite, a plastic waxlike paraffin vein material which is found in Utah and near Boryslaw, Poland.

Subsurface occurrences. Underground occurrences of petroleum may be classified as pools, fields, and provinces.

Pools. Underground accumulations of petroleum characterized by a single and separate natural reservoir (usually a porous sandstone or limestone) and a single natural pressure system are called pools. The production of petroleum from one part of a pool affects the reservoir pressure throughout its extent. A pool is bounded by geological barriers in all directions, such as rock structure, impermeable strata, and water in the formations, so that the pool is effectively separated from any other pools that may be present in the same district or on the same geological structure.

Fields. An oil field may be a single pool, or it may consist of two or more pools which are commonly but not necessarily related to the same geological structure. Where more than one pool is present in the same field, the different pools are separated from one another. The different pools may occur at several stratigraphic horizons separated by impermeable strata, and they may partially or completely coincide in their horizontal distribution. Geological features that influence the accumulation of petroleum include salt domes. anticlinally folded strata, and combinations of faulting, folding, and stratigraphic variations.

Provinces. A petroleum province is a region in which a number of oil and gas pools and fields occur in a similar or related geological environment. The term is used to indicate the larger producing regions, such as the mid-continent region of the United States.

RESERVOIR ROCKS

Reservoir rocks are rocks with sufficient porosity and permeability to allow oil and gas to accumulate and be produced in commercial quantities. There are three requisites for a reservoir rock: (1) It must be porous, that is, have enough pore space to contain oil or gas; (2) it must be permeable to allow fluids, including oil and gas, to move through it; and (3) there must be a trap which prevents escape of the oil and gas. Any rock with these characteristics may become a reservoir for oil and gas provided that hydrocarbons are available to migrate into the rock.

The reservoir character of a rock may be an original feature (intergranular porosity of sandstones) or a secondary feature resulting from chemical changes (solution porosity of limestones), or it may be the result of physical changes (fracturing of brittle rock).

Types of reservoir rocks. Reservoir rocks may be classified as fragmental or clastic (broken), chemical and biochemical, or miscellaneous. They may also be classified as marine and nonmarine reservoir rocks.

Fragmental type. Some reservoir rocks are aggregates of particles, that is, fragments of older rocks. They are also called clastic or detrital rocks because they consist of mineral and rock particles derived from eroded areas. The constituent particles of fragmental rocks may range in size from colloidal particles up to pebbles and cobbles. The most common of the fragmental reservoir rocks are sandstones, conglomerates, arkoses, graywackes, and siltstones. Many, however, are carbonate rocks, such as oolitic rocks, calcarenites, and coquinas, which are made up of oolites and skeletal fragments that have been cemented and in some cases recrystallized.

Some sandstone reservoir rocks consist either entirely or in part of loose, uncemented sand grains. The grains tend to be brought to the surface in large quantities along with oil during production. The sand grains in most sandstones, however, are held together by various kinds of cementing material, mostly carbonates, silica, or clays. Some of the cementing materials are primary, having been deposited along with the sand grains. Other cementing materials are secondary, having been precipitated from solutions that entered the rock after it was deposited.

Clastic limestones and dolomites consist of particles of calcite and dolomite that have been transported and deposited in much the same manner as quartz and other mineral grains. Rocks thus formed are commonly recemented with calcite and, if fine grained, may resemble chemically deposited limestones or dolomites. Carbonate rocks thus may form good oil reservoirs because of high porosity and permeability, particularly if the original pore spaces are not completely filled with cement.

Chemical types. These rocks are made up chiefly of chemical or biochemical precipitates. They are composed of mineral matter that was precipitated at the place where the rocks were formed (in contrast to the transported grains in clastic carbonates). The most important chemical reservoir rocks are limestones and dolomites. Some chemically precipitated rocks consist entirely or almost entirely of silica in the form of chert or novaculite, but such rocks provide few reservoirs. The porosity of carbonate rocks is largely the result of solution leaching by percolating groundwaters.

Miscellaneous types. Other reservoir rocks include igneous and metamorphic rocks and mixtures of both. Any porous and permeable igneous rock in close association with sedimentary rock may become a reservoir rock when saturated by oil derived from the sediments. Igneous and metamorphic rocks are only a minor source of oil and gas because, generally, they are not permeable enough, and when they are they are not often associated with suitable source rocks.

Marine and nonmarine types. A distinction may be made between reservoir rocks which were deposited in ancient seas and those deposited in fresh water. Most petroleum occurs in rocks deposited under marine conditions but some occurs in sediments of nonmarine origin. The occurrence of oil in nonmarine sediments is sometimes explained as the result of migration of oil along faults, fractures, or bedding planes from adjacent marine sediments, but some petroleum probably has formed in nonmarine rocks.

Properties of reservoir rocks. The porosity and permeability of reservoir rocks, as well as the nature of the traps, are all factors which regulate the accumulation of petroleum. Porosity is the total space in the rock (pores, voids, interstices) not occupied by solid material. It is expressed as a percentage. Factors which influence porosity are the size of the rock particles, arrangement, sorting, shape, and cementing material. Most oil-producing rocks have porosities above 10% and thicknesses greater than 10 ft.

Total pore space is not the sole determinant of a petroleum reservoir. A reservoir must also have permeability; that is, it must allow fluids to flow through it with relative ease. Pumice, for instance, has a large amount of pore space, but the pores are not connected and it has very low permeability. A rock that is permeable necessarily has interconnected pores or fractures that are greater than capillary in size.

RESERVOIR TRAPS

Reservoir traps contain the accumulated oil or gas so that it cannot escape. The upper boundary of a reservoir trap is called the roof or cap rock; the lower boundary is the oil-water or gas-water contact.

Roof or cap rock is an impermeable layer of rock forming the roof of an oil trap. The connecting pores in the reservoir rock, which are individually minute, are as a rule saturated with water. Since oil and gas are lighter than water, the petroleum rises through the water until it is stopped by the roof rock. If the roof rock is concave (domed, arched, folded, peaked, or roof-shaped), it acts as a trap, keeping the oil and gas from escaping upward or laterally.

The oil-water contact or gas-oil contact generally forms the lower boundary of the accumulation. The water is the water that normally fills the pores of the reservoir rock. The water supports the pool of oil and gas, and the forces of buoyancy impel the petroleum upward against the bounding surfaces of the trap, holding it in place, as shown in the illustration.

Geological structure. The anticlinal theory is a successful theory of petroleum geology. Most of the oil in the major oil fields of the world occurs in anticlines. The fact that oil and gas commonly occur on anticlinal axes was first noted by W. E. Logan in 1842. He observed that oil seeps occurred in the vicinity of anticlinal axes near the mouth of the St. Lawrence River. Although the term anticlinal theory has fallen into disuse, the fundamental principle on which it is based remains generally valid—oil and gas tend to accumulate in the highest places within the reservoir. It is recognized today, however, that other factors affect the accumulation of oil in many pools, and that the anticlinal theory by no means provides an explanation for all oil accumulations.

Classification of traps. Three basic types of traps generally are recognized: structural traps, stratigraphic traps, and combination traps.

Structural traps. A trap whose upper boundary has been made concave by folding or faulting, or both, of the reservoir rock is known as a structural trap. The edges of a pool occurring in a structural trap are determined by the intersection of the underlying water table with the enclosing roof or cap rock. Structural traps include closed anticlines or domes, faulted anticlines with closure, closure against faults, anticlines on downdip sides of faults, and oil and gas accumulations in fractures produced by structural deformation.

Stratigraphic traps. Also known as varying permeability traps, stratigraphic traps are those in which the chief trap-making element is some variation in the stratigraphy or lithology, or both, of the reservoir rock. These include facies change, variable local porosity and permeability, and any up-structure termination of the reservoir rock. Stratigraphic traps include sandstone lenses, channels, bars, and reefs and porosity lenses. Some of the most common stratigraphic traps are strandline pools, shoestring sand traps, biostromes, and bioherms. *See* FACIES.

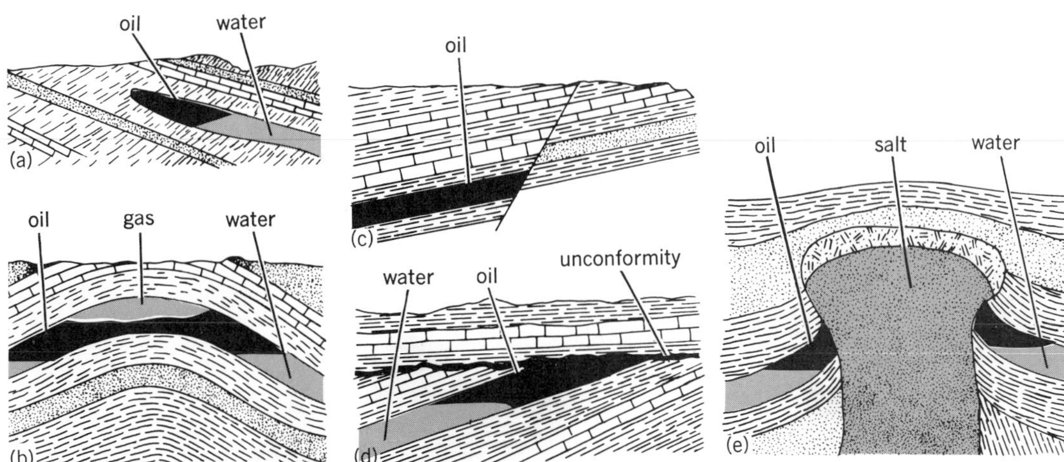

Typical petroleum traps. (*a*) Lens. (*b*) Anticline. (*c*) Fault. (*d*) Unconformity. (*e*) Salt dome.

Shoestring sand traps are long, narrow, discontinuous sandstone deposits which are commonly a few miles wide or less and which may be many miles in extent. Except at their terminal ends they are generally surrounded by impermeable shales and clays. Some sand traps of this nature are believed to be channel fillings and others offshore sandbars.

Two general classes of primary stratigraphic traps occur in rocks of chemical origin, almost all of them carbonate rocks. These are biostromes and bioherms. Biostromes include porous lenses, enclosed by impermeable shales, limestones, or dolomites. Bioherms, or organic reefs, are porous, domelike, moundlike, or otherwise circumscribed masses, built largely by lime-secreting organisms such as corals, algae, brachiopods, mollusks, or crinoids and enclosed in strata of different lithologic character. *See* BIOHERM; BIOSTROME.

Combination traps. These traps result from both structural and stratigraphic conditions. An example of such a trap is the salt dome. Salt domes are cylindrical or steeply conical masses of salt which has flowed plastically under pressure. These masses, called plugs or domes, originate at depths of 20,000 ft or more and pierce the overlying sedimentary strata. Three kinds of traps are associated with salt plugs: cap rock, flanking sands, and supercap sands. Cap rock consists of limestone, gypsum, and anhydrite and occurs as a capping over the tops of the salt plugs. Flanking sands are strata abutting upon and cut off by the salt plug. Supercap sands are sandy strata that arch over the tops of the plugs in the form of structural domes. In many salt domes recurrent movement of the dome during deposition of overlying strata caused variations in the thickness and lithology of the strata, creating traps that are in part stratigraphic traps. *See* SALT DOME.

RESERVOIR FLUIDS

Fluids fill the voids or pore spaces in all reservoir rocks. The fluid may be water, water and oil, water and gas, or a mixture of water, oil, and gas. The fluid content of a gas pool consists of water and gas; that of an oil pool consists of gas, oil, and water. Gas is almost invariably present in solution in oil and, in addition, free gas may be present.

The distribution of gas, oil, and water in the reservoir depends upon relative buoyancy, relative saturation of pore space with each fluid, and capillary and displacement pressures, as well as the porosity, permeability, and composition of the reservoir rock. In traps that contain oil, water, and free gas, the fluids occur in distinct zones. Gas, being the lightest, occurs at the top of the trap. Below the gas, oil occurs, and below that, water. Where there is gas but no oil, the gas is immediately underlain by water and the contact is the gas-water table. Interstitial water (adsorbed water or wetting water which lines the pore walls or occurs on the surfaces of mineral grains) is generally present throughout the reservoir, occupying 10–30% of the pore space.

Oil field waters are waters associated with oil and gas pools. They may be classified as meteoric waters, connate water, and mixed water. Most oil field waters are saline. *See* OIL FIELD WATERS.

Oil saturation is the amount of oil contained in a petroleum reservoir. It is measured as a percentage of the effective pore space.

Gas volume, or natural-gas content, of a petroleum reservoir may range from small quantities dissolved in oil up to 100% of petroleum content. The natural gas in a reservoir may occur as free gas, as gas dissolved in oil, and as gas dissolved in water. *See* NATURAL GAS.

[JOHN W. HARBAUGH]

Bibliography: G. D. Hobson, *Some Fundamentals of Petroleum Geology*, 1954; K. K. Landes, *Petroleum Geology*, 2d ed., 1959; A. I. Levorsen, *Geology of Petroleum*, 2d ed., 1967; W. I. Russell, *Principles of Petroleum Geology*, 2d ed., 1960; E. N. Tiratsoo, *Petroleum Geology*, 1952.

Petrology

The study of rocks, their occurrence, composition, and origin. Petrography is concerned primarily with the detailed description and classification of rocks, whereas petrology deals primarily with rock formation, or petrogenesis. Experimental petrology reproduces in the laboratory the conditions of high pressure and temperature which existed at various depths in the Earth where minerals and rocks were formed. A petrological description includes definition of the unit in which the rock occurs, its attitude and structure, its mineralogy and chemical composition, and conclusions regarding its origin. In a restricted sense, however, petrology has come to emphasize the study of rocks in the field and in hand specimens, without recourse to the microscope. For a discussion of mineral identification, petrographic analysis, and the

Table 1. Types of volcanic structure

Name	Characteristics
Shield	Low height, broad area; formed by successive fluid flows accumulating around a single, central vent
Cinder cone	Cone of moderate size with apex truncated; circular in plan, gently sloping slides; composed of pyroclastic particles, usually poorly consolidated
Spatter cone	Small steep-sided cone with well-defined crater composed of pyroclastic particles, well consolidated (agglomerate)
Stratocone	Composed of interlayered flows and pyroclastics; flows from sides (flank flows) common, as are radial dike swarms; slightly concave in profile, with central crater
Caldera	Basins of great size but relatively shallow; formed by explosive decapitation of stratocones, collapse into underlying magma chamber, or both
Plug dome	Domal piles of viscous (usually rhyolitic) lava, growing by subsurface accretion and accompanied by outer fragmentation
Cryptovolcanic structures	Circular areas of highly fractured rocks in regions generally free of other structural disturbances; believed to have formed either by subsurface explosions or sinking of cylindrical rock masses over magma chambers

Table 2. Characteristics of intrusive igneous rock masses

Name	Shape	Structural relations to wall rocks	Size and other features
Dikes	Tabular, lensoid	Discordant	Few feet to hundreds of miles long
Sills	Tabular, lensoid	Concordant	Up to several hundred feet thick
Laccoliths	Plano-convex or doubly convex lenses	Generally concordant	1–4 mi in diameter; several thousand feet thick
Volcanic necks	Pipelike	Discordant	Few hundred feet to a mile in diameter; cores of eroded volcanoes
Stocks	Irregular, with steep walls	Crosscutting	A small batholith or its upward projection; outcrop area less than 40 mi^2
Batholiths	Irregular, contacts dip steeply or outward; no bottoms known	(1) Discordant (2) Concordant in general, may be crosscutting in detail	Some cover 16,000 mi^2; some are composite intrusives of varied petrology
Plutons	Irregular	Usually crosscutting	Usually large; used as general name for intrusive masses that do not fit other definitions

classification of rocks *see* MINERALOGY; PETROGRAPHY; ROCK.

Igneous rocks. Extrusive (effusive) igneous rocks reach the surface either through fissures of considerable linear extent (fissure eruptions) or through pipelike channelways around which volcanoes are built. Extrusive material may flow out relatively quietly as lava or it may be exploded as pyroclastic material. Fissure eruptions are generally quiet and repeated over long periods of time to build up thick platforms of considerable extent consisting chiefly of basalt. In the northwestern United States the Columbia River Plateau, built in this way, embraces 200,000 mi^2 in Idaho, Oregon, and Washington, and in local areas has an aggregate thickness of 5000 ft. *See* VOLCANO.

Volcanic structures are of a variety of types (Table 1). Lava flows may be characterized by a smooth or ropy surface with prominent flow structure (pahoehoe) or by a jumbled blocky surface (aa). Flows commonly show columnar jointing which has been produced by contraction upon crystallization. A lava tongue solidifies first along its upper surface against the air and along its bottom contact with cooler rock, leaving a central stream which is still liquid, flowing in a tunnel of its own construction. With sufficient slope the streams drain away, leaving cavernous passageways.

Volcanic activity varies greatly in intensity, duration, periods between eruptions, and quantities of gases, liquid rock, and solidified fragments expelled. The important factors influencing these differences are (1) chemical composition of the magma; (2) amount of gas dissolved in it; (3) extent of crystallization or cooling before eruption; and (4) configuration of the conduit and depth to the magma chamber. *See* MAGMA.

Intrusive igneous rocks occur in many different types of units or intrusive masses, which are classified chiefly by their shape and structural relations to their wall rocks (Table 2). Bodies that crystallize at great depths (such as batholiths) are referred to as plutonic; those consolidated under shallow cover are termed hypabyssal. *See* PLUTON.

The crystallization of the larger intrusives may result in profound alterations in the adjacent wall rocks (exomorphism). Where stocks and batholiths have invaded sedimentary rocks an aureole of contact metamorphism is developed. This results from recrystallization under increased temperature and may be accompanied by chemical transformations (pyrometasomatism) produced by hydrothermal solutions generated during the latter stages of

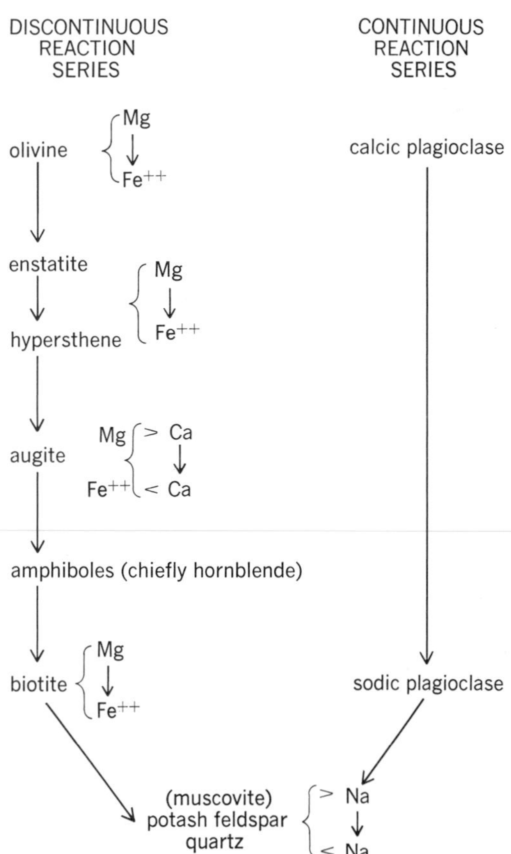

Fig. 1. Reaction series of Bowen (modified).

magmatic differentiation. Where batholiths have been intruded into rocks which are already regionally metamorphosed, the contact rocks formed are injection gneisses or migmatites. *See* AUREOLE, CONTACT.

Igneous rocks make room for themselves by forceful injection (dilatance), by engulfing wall rock blocks (magmatic stoping), or by subsidence of overlying rocks. The hypothesis of granitization maintains that granites result from the wholesale transformation of sedimentary or metamorphic rock layers by solutions operating through mineral replacement or by ionic emanations acting through solid diffusion. *See* GRANITIZATION.

Blocks of wall rock included in an intrusive mass are xenoliths; their partial destruction by reaction may produce irregular clumps of mafic minerals called schlieren. In some instances such endomorphic effects are sufficiently intensive to result in modification of the composition of the magma (syntexis). *See* XENOLITH.

Crystallizing under equilibrium conditions, early magmatic minerals react with remaining fluid to yield new species (Fig. 1). Interruption of the sequence will yield liquid fractions richer in silicon dioxide, alkalies, iron, and water than the original magma and crystalline fractions richer in calcium and magnesium than the parent magma (magmatic differentiation).

Igneous rocks occur in clans or associations which possess characteristic trace elements and appear in specific structural provinces (Table 3). The origins of various igneous rocks are summarized in Table 4. *See* IGNEOUS ROCKS; PETROGRAPHIC PROVINCE.

Sedimentary rocks. With the exception of material deposited by glaciers (till or the consolidated form tillite), sedimentary rocks show bedding or stratification. This separation into generally parallel layers (beds, strata) results from sorting according to grain size during deposition, from differences in composition or texture, or from variations in the rate of deposition. The development of most sedimentary rocks proceeds in the following stages: (1) There is a source rock, any older rock or, for organic sediments, a supply of organically originated material. (2) By weathering, the older rock is mechanically comminuted, chemically altered, or both, to form unconsolidated surficial rock debris called mantle. (3) Particles are transported by streams, ocean and lake currents, wind, glaciers, or by the direct action of gravity which causes particles to slide and roll down slopes. (4) Material moved by rolling, suspension, or solution is deposited. (5) Deposits usually are consolidated by the processes of cementation (sandstones), compaction (shales), and recrystallization (limestones).

Chemical changes accompanying consolidation are termed diagenetic. Weathered material not transported may become a residual sedimentary rock (bauxite). Sedimentary rocks are deposited either on land areas (continental) or in ocean waters (marine). Most marine sedimentation takes place on the submarine extensions of the continents called continental shelves. Examples of types of sedimentary deposits are listed in Table 5. Features characteristically found in sedimentary rocks, in addition to stratification, are cross-bedding, concretions, ripple marks, mud cracks, and fossils. *See* DIAGENESIS; SEDIMENTATION; WEATHERING PROCESSES.

A formation, which is the basic unit of stratigraphy, is a series of rocks deposited during a specific unit of geologic time and consisting either of a particular rock type or of several types deposited in a sedimentary cycle. Such a cycle is the changing sequence of deposits reflecting, for example, advance or retreat of marine waters in a particular area.

However, while sandstone may be deposited at one time in one place in the sedimentary basin, limestone may be formed simultaneously elsewhere. Such lateral variation in a formation is referred to as facies. *See* CYCLOTHEM; FACIES; STRATIGRAPHY.

Table 3. Igneous rock clans or associations

Name	Main rock types	Environment
Oceanic Olivine basaltic	Olivine basalt, minor trachyte, peridotite	Volcanic islands of deep oceanic basins
Alkaline volcanic	(*a*) Olivine basalt, trachyte, phonolite (*b*) Leucite basalt, trachybasalt, trachyte (*c*) Spilite, ketatophyre	(*a, b*) Nonorogenic continental regions (*c*) Orogenic continental regions; former geosynclines
Tholeiitic basaltic	Basalt (generally olivine-free), quartz diabase	Continental plateau areas
Calc-alkalic volcanic	Andesite, rhyolite, basalt	Continental orogenic areas
Lopolithic	Norite, gabbro, anorthosite, peridotite	Lopoliths, thick differentiated sheets
Alpine-type peridotites	Peridotites, serpentinites	Orogenic zones
Precambrian anorthositic	Andesine or labradorite anorthosite, norite, syenite, monzonite	Domed pluton of massifs in Precambrian terrains
Granite batholithic	(*a*) Simple: granite, granodiorite (*b*) Complex: gabbro, tonalite, granodiorite, minor granite	Precambrian shields; cores of mountain ranges
Minor granitic intrusive	Granite (some alkalic), quartz syenite, syenite, diorite	Hypabyssal, in mountain ranges and as their outliers
Nepheline syenitic	Feldspathoidal rocks, carbonatites	(*a*) Simple plutons (*b*) Ring complexes
Lamprophyric	Minette, kersantite, camptonite	Dike swarms

Table 4. Synopsis of magmatic evolution*

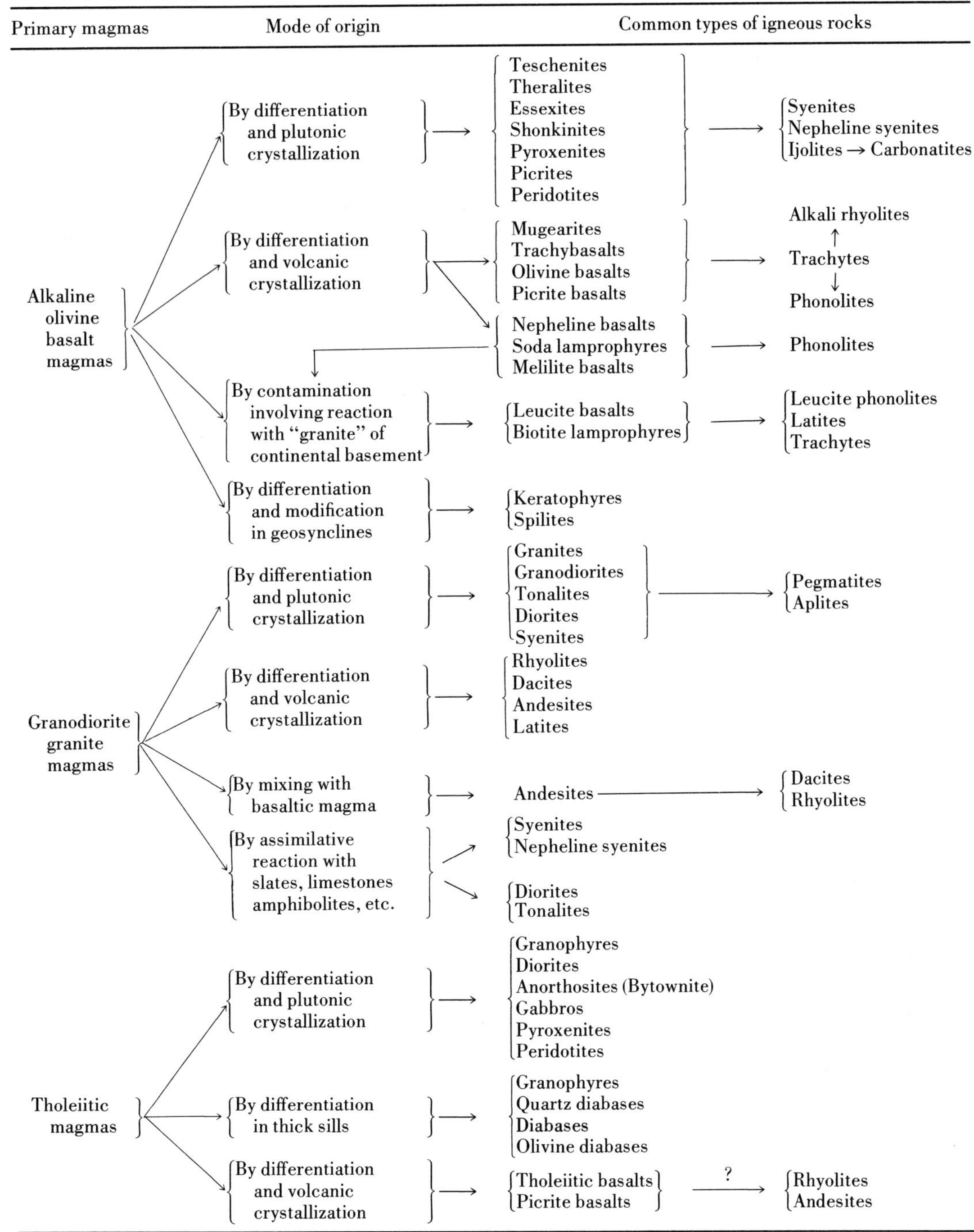

Primary magmas	Mode of origin	Common types of igneous rocks	
Alkaline olivine basalt magmas	By differentiation and plutonic crystallization →	Teschenites Theralites Essexites Shonkinites Pyroxenites Picrites Peridotites	→ Syenites Nepheline syenites Ijolites → Carbonatites
	By differentiation and volcanic crystallization →	Mugearites Trachybasalts Olivine basalts Picrite basalts	→ Alkali rhyolites ↑ Trachytes ↓ Phonolites
	↘	Nepheline basalts Soda lamprophyres Melilite basalts	→ Phonolites
	By contamination involving reaction with "granite" of continental basement →	Leucite basalts Biotite lamprophyres	→ Leucite phonolites Latites Trachytes
	By differentiation and modification in geosynclines →	Keratophyres Spilites	
Granodiorite granite magmas	By differentiation and plutonic crystallization →	Granites Granodiorites Tonalites Diorites Syenites	→ Pegmatites Aplites
	By differentiation and volcanic crystallization →	Rhyolites Dacites Andesites Latites	
	By mixing with basaltic magma →	Andesites	→ Dacites Rhyolites
	By assimilative reaction with slates, limestones amphibolites, etc. ↗	Syenites Nepheline syenites	
	↘	Diorites Tonalites	
Tholeiitic magmas	By differentiation and plutonic crystallization →	Granophyres Diorites Anorthosites (Bytownite) Gabbros Pyroxenites Peridotites	
	By differentiation in thick sills →	Granophyres Quartz diabases Diabases Olivine diabases	
	By differentiation and volcanic crystallization →	Tholeiitic basalts Picrite basalts	?→ Rhyolites Andesites

*From F. J. Turner and J. Verhoogen, *Igneous and Metamorphic Petrology*, 2d ed., McGraw-Hill, 1960.

By means of detailed studies of the fossils of a formation and its lithology, composition, structure, and distribution, the paleoecology of the area may be reconstructed. Correlation of formations is attempted chiefly on the basis of fossils, with supplementary data drom the lithology, stratigraphic position, insoluble residues (in acid-soluble rocks), heavy detrital minerals (in clastic rocks), and in drill holes by electrical conductivity, radioactivity, and seismic-wave velocities. *See* SEDIMENTARY ROCKS.

Metamorphic rocks. Metamorphism transforms rocks through combinations of the factors of heat, hydrostatic pressure (load), stress (directed pressure), and solutions. Most of the changes are in texture or mineral composition; major changes in chemical composition are called metasomatism. The major types of metamorphism are presented in Table 6. Rocks that can serve as parent material for metamorphic derivatives include both igneous and sedimentary types, and, less commonly, older metamorphic rocks as well. The complexity of the possible metamorphic mineral assemblages stems not only from the variety of possible parent rocks and from the imposition of the several kinds of metamorphism but also from variation in the inten-

Table 5. Selected examples of sedimentary deposits under various environments

Agent	Deposit	Resulting rock
Continental		
Streams	Valley fill	Sandstone
	Alluvial fan	Conglomerate
	Delta	Siltstone
Lakes	Varved clay	Shale
Springs		Travertine
		Siliceous sinter
Swamps	Peat	Coal
Wind	Dune	Sandstone
	Dust	Loess
	Volcanic ash	Tuff
Glaciers	Moraine	Tillite
Groundwater	Stalactite	Dripstone
Gravity	Talus	Breccia
	Avalanche	Conglomerate
	Landslide	
Marine		
Breakers and alongshore currents	Beach	Sandstone Conglomerate
Longshore currents		Sandstone Shale
Marine organisms	Reefs and other shell deposits	Shell limestone Coquina Diatomite
Marine water	Evaporites	Rock salt Rock anhydrite
Marine water	Colloidal precipitates	Phosphorite Manganese oxide concretions Chert

sity of particular types of metamorphism (grade), and from the difficulty of readily achieving chemical equilibrium through solid-state reactions. Various features characteristic of metamorphic rocks include foliation (slaty cleavage, schistosity, and gneissic structure), lineation, banding, and relict structures. *See* METAMORPHISM; METASOMATISM.

The facies principle is employed in attempting to reconstruct the environment under which a metamorphic rock was developed. A metamorphic facies consists of all rocks, without respect to chemical composition, that have been recrystallized under equilibrium within a particular environment of stress, temperature, load, and solutions. The first two factors are considered critical. The facies are named after metamorphic rocks

Table 6. Types of metamorphism and their factors

Type	Factors	Changes in rock
Cataclastic	Stress, low hydrostatic pressure	Fragmentation, granulation
Contact (thermal)	Heat, low to moderate hydrostatic pressure	Recrystallization to new minerals or coarser grains; rarely melting
Pyrometa-somatism	Heat, additive hydrothermal solutions, low to moderate hydrostatic pressure	Reconstitution to new minerals; change in rock composition
Regional (dynamic)	Heat, weak to strong stress, moderate to high hydrostatic pressure, ± nonadditive solutions	Recrystallization to new minerals or coarser grains; parallel orientation of minerals to produce foliation

deemed diagnostic of such restricted conditions. In practice a group of related rocks of different compositions is assigned to a particular facies upon the presence of such a key assemblage. Facies and their type descriptions are given in the list below.

A. Facies of contact metamorphism. Load pressure low, generally 100–3000 bars. Water pressure highly variable, in some cases possibly exceeding load pressure, in a few cases very low. Facies listed in order of increasing temperature for given range of pressure conditions.
   1. Albite-epidote hornfels (formerly albite-epidote amphibolite facies, actinolite-epidote hornfels subfacies)
   2. Hornblende hornfels (formerly amphibolite facies, cordierite-anthophyllite subfacies)
   3. Pyroxene hornfels
   4. Sanidinite—corresponds to minimum pressures (load, P_{H_2O}, P_{CO_2}) and maximum temperatures—pyrometamorphism

B. Facies of regional metamorphism. Load and water pressures generally equal and high (3000–12,000 bars). Facies listed in order of increasing temperature and pressure.
   1. Zeolitic (hitherto not recognized)
   2. Greenschist
      a. Quartz-albite-muscovite-chlorite (formerly muscovite-chlorite)
      b. Quartz-albite-biotite (formerly biotite-muscovite)
      c. Quartz-albite-almandine (formerly albite-epidote amphibolite facies, chloritoid almandine subfacies)
   3. Glaucophane schist (hitherto of uncertain status; previously equated with the greenschist facies; the glaucophane schists and their associates seem to represent a divergent line of metamorphism conditioned by development of unusually high pressures at low temperatures)
   4. Almandine amphibolite
      a. Staurolite-quartz (formerly staurolite-kyanite)
      b. Kyanite-muscovite-quartz (formerly staurolite-kyanite)
      c. Sillimanite-almandine
   5. Granulite
      a. Hornblende granulite
      b. Pyroxene granulite
   6. Eclogite

Regional variations in grade may be mapped by means of isograds, lines formed by the intersection of planes of isometamorphic intensity with the Earth's surface. These are defined on the appearance of a specific mineral known to reflect a major increase in the intensity of metamorphism.

The primary cause of stresses acting during regional metamorphism is diastrophism of the mountain-building type. The higher temperatures may result from deep burial, owing to the geothermal gradient of the Earth, in part to concentrations of radiogenic heat, or in part to heat supplied by cooling masses of magma. In contact metamorphism this last is the sole heat source. *See* GEOLOGIC THERMOMETRY.

Once formed, metamorphic rocks are subject to further changes through folding and crumpling of

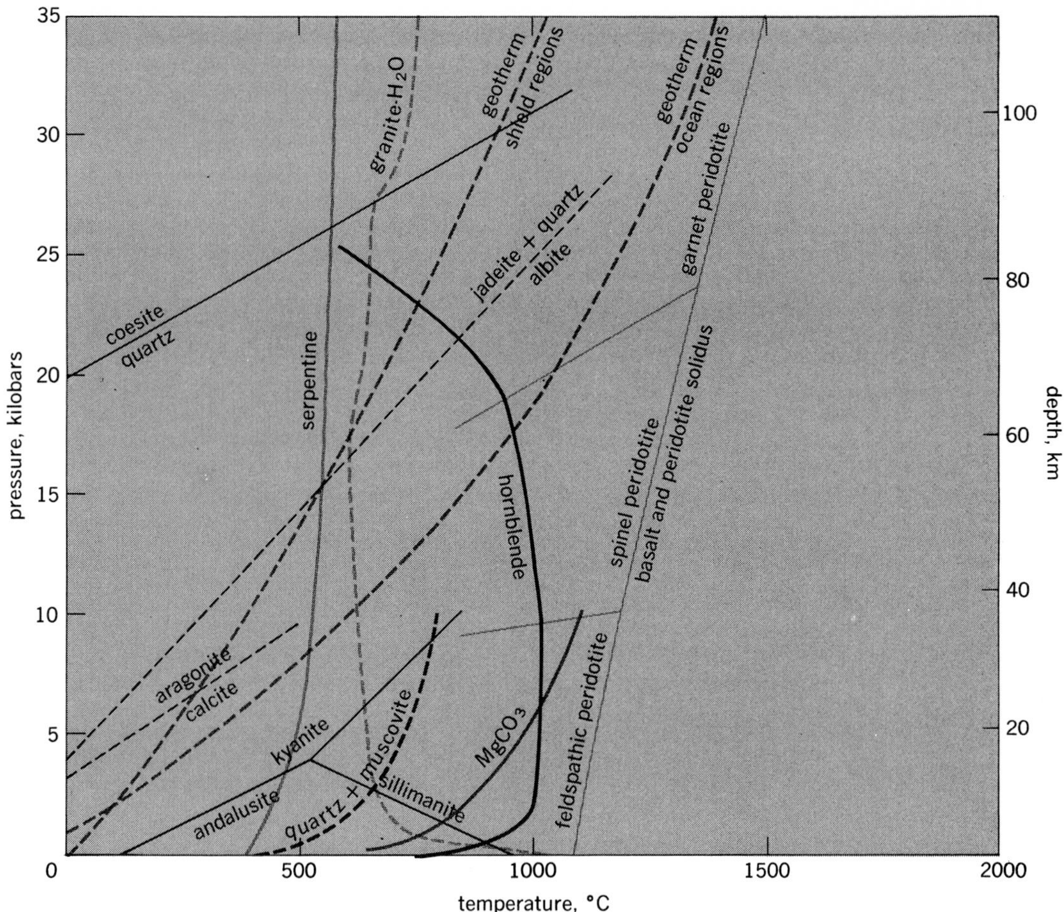

Fig. 2. Selected results in experimental mineralogy and petrology, averaged from many sources. The two geotherms show the temperature distribution with depth in the Earth. The solidus for dry basalt and peridotite provides the upper temperature limit for magma generation to occur in the mantle. The mineral facies of a peridotite upper mantle are shown below the solidus. The lower limit for magma generation is given by the curve granite-H_2O. Solid-solid mineral transitions and reactions plotted are quartz-coesite, the breakdown of albite, calcite-aragonite, and the polymorphic transitions among kyanite-andalusite-sillimanite. A decarbonation reaction is shown by the curve for $MgCO_3$ (magnesite), and dehydration reactions are shown for serpentine, for the assemblage muscovite + quartz, and for hornblende in mafic or ultramafic rocks. Note the effect of pressure on the hornblende reaction in the garnet-peridotite facies.

the foliation and through extensive injection of igneous material to form migmatites. *See* METAMORPHIC ROCKS. [E. WILLIAM HEINRICH]

Experimental mineralogy and petrology. One aim of mineralogy and petrology is to decipher the history of igneous and metamorphic rocks. Detailed study of the field geology, the structures, the petrography, the mineralogy, and the geochemistry of the rocks is used as a basis for hypotheses of origin. The conditions at depth within the Earth's crust and mantle, the processes occurring at depth, and the whole history of rocks once deeply buried are deduced from the study of rocks now exposed at the Earth's surface. One of the approaches used to test hypotheses so developed is termed experimental petrology; the term experimental mineralogy refers to similar studies involving minerals rather than rocks (mineral aggregates).

The experimental petrologist reproduces in the laboratory the conditions of high pressure and high temperature encountered at various depths within the Earth's crust and mantle where the minerals and rocks were formed. By suitable selection of materials he studies the chemical reactions that actually occur under these conditions and attempts to relate these to the processes involved in petrogenesis.

The experiments may deal with the stability range of minerals and rocks in terms of pressure and temperature; with the conditions of melting of minerals and rocks; or with the physical properties or physical chemistry of minerals, rocks, and rock melts (magmas), or of the vapors, gases, and solutions coexisting with the solid or molten materials. These experiments may thus be related to major geological processes involved in the evolution of the Earth: the conditions of formation of magmas in the mantle and crust and their subsequent crystallization either as intrusions or lava flows; the evolvement of gases by the magmas during their crystallization and the precipitation of ore deposits from some of them; the processes leading to the development of volcanic arcs and mountain ranges; and the metamorphism and deformation of the rocks in the mountain chains, and thus the origin and development of the continents. Representative experimental results are shown in Fig. 2.

Experimental probes into the Earth. The pioneer work of Sir James Hall of Edinburgh on rocks and minerals sealed within gun barrels at high pressures and temperatures earned him the title "father of experimental petrology." His experiments, published early in the 19th century, heralded an era of laboratory synthesis in mineralogy and petrology. Experimental petrology gained tremendous impetus in 1904, when the Geophysical Laboratory was founded in Washington, D.C., and new techniques were developed for the study of silicate melts at carefully controlled temperatures. The results obtained by N. L. Bowen, J. F. Schairer, and others have persuaded most petrologists that physicochemical principles can be successfully applied to processes as complex as those occurring within the Earth, and they have formed the basis for much of the work performed at high temperatures and high pressures. The design of equipment capable of maintaining high temperatures simultaneously with high pressures has been successfully achieved only since about 1950, with a few notable exceptions. The extent of the experimental probe into the Earth varies with the type of apparatus used. The Tuttle cold-seal pressure vessel, or high-pressure test tube, reproduces conditions at the base of the average continental crust. Internally heated, hydrostatic pressure vessels reproduce conditions similar to those just within the upper mantle. Opposed-anvil devices, or simple squeezers, readily provide pressures of 200 kilobars or more, but temperatures attainable are considerably lower than those existing at depths in the Earth corresponding to these pressures.

The deepest experimental probe is provided by a variety of internally heated opposed-anvil devices which can produce pressures of 200 kilobars with simultaneous temperatures of 2000°C or more, although the pressure and temperature measurements become less accurate with the more extreme conditions. This deepest probe produces conditions equivalent to depths of 300–400 km in the upper mantle, which are not very deep when compared with the 2900-km depth at the core, but are well within the outer 500 km of the Earth where many petrological processes have their origins. *See* HIGH-PRESSURE PHENOMENA.

Studies related to Earth structure. Geophysical studies and inferences about the compositions of deep-seated rocks based on petrological studies of rocks now exposed at the surface, together with laboratory measurements of physical properties of selected materials, provide the basis for theories concerning the composition of the Earth and those parts of the Earth which may be treated as units. *See* EARTH, INTERIOR OF.

At high pressures, most silicate minerals undergo phase transitions to denser polymorphs, and it is believed that polymorphic transitions within the mantle contribute to the high gradients of seismic-wave velocities occurring at certain depths. Experimental studies of the olivine-spinel transition have been successfully correlated with the beginning of the transition zone of the upper mantle at a depth of about 350 km. It has been widely held that the Mohorovičić discontinuity marks a change in chemical composition from basalt of the lower crust to peridotite of the upper mantle, but recent experimental confirmation that basalt undergoes a phase transition to eclogite, its dense chemical equivalent, suggests that the M discontinuity could be a phase transition. This now appears to be unlikely for several reasons, but more detailed experimental studies are required because there remains a strong possibility that this phase transition is involved in tectonically active regions of high heat flow where the M discontinuity is poorly defined. *See* MOHO (MOHOROVIČIĆ DISCONTINUITY).

Igneous petrology. The Earth's mantle transmits shear waves and is therefore considered to be crystalline rather than liquid; but the eruption of volcanoes at the Earth's surface confirms that melting temperatures are reached at depth in the Earth from time to time and from place to place. One group of experiments is therefore concerned with the determination of fusion curves for minerals believed to exist in the Earth's mantle and of melting intervals for mineral aggregates, or rocks. The position of the solidus curve provides an upper temperature limit for the normal geotherm, and the experiments dealing with melting intervals at various pressures have direct bearing on the generation of basaltic magmas at various depths within the mantle. They also provide insight into the processes of crystallization and fractionation of these magmas during their ascent to the surface. This is one of the major problems of igneous petrology. *See* IGNEOUS ROCKS; MAGMA.

Many reactions among silicate minerals are very sluggish, and the presence of water vapor under pressure facilitates the experiments. Water is the most abundant volatile component of the Earth, with carbon dioxide a poor second, and the water in the experiments often plays a role as a component as well as a catalyst. In the presence of water vapor under pressure, melting temperatures of silicate minerals and rocks are markedly depressed, and curves for the beginning of the melting of rocks in the presence of water provide lower limits for the generation of magmas. It is widely believed that many of the granitic rocks constituting batholiths, the cores of mountain ranges, were derived by processes of partial fusion of crustal rocks in the presence of water. The hypothesis appears to be consistent with experimental results on feldspars and quartz in the presence of water at pressures up to 10 kilobars and with similar experiments on plutonic igneous rock series ranging in composition from gabbro to granite. However, evidence suggests that some magmas forming batholiths, and andesite lavas of equivalent composition in tectonically active regions, originated directly from mantle material. Experimental petrologists are therefore extending their studies of the minerals and rocks to granitic and andesitic compositions to pressures greater than 10 kilobars, corresponding to upper mantle conditions, with and without water present.

The path of crystallization of a magma and the mineral reaction series thus produced are apparently quite sensitive to many variables, including pressure, pressure or fugacity of water, and oxygen fugacity. Techniques have been developed for studying the effect of oxygen fugacity on crystallization paths and also for controlling oxygen fugacity at very low values while the water pressure is simultaneously maintained at very high values.

Metamorphic petrology. When rocks are metamorphosed, they recrystallize in response to changes in pressure, temperature, stress, and the passage of solutions through the rock. The facies classification is an attempt to group together rocks that have been subjected to similar pressures and temperatures on the basis of their mineral parageneses. The metamorphic facies are arranged in relative positions with respect to pressure (depth) and temperature scales, and the experimental petrologist attempts to calibrate these scales by delineating the stability fields of minerals and mineral assemblages under known conditions in the laboratory. Potentially the most useful reactions for this purpose are solid-solid reactions involving no gaseous phase, such as the reactions among the polymorphs of Al_2SiO_5, kyanite, sillimanite, and andalusite.

Progressive metamorphism of sedimentary rocks produces a series of dehydration and decarbonation reactions with progressive elimination of water and carbon dioxide from the original clay minerals and carbonates. Experimental determination of these reactions, along with the solid-solid reactions, provides a petrogenetic pressure-temperature grid, in which mineral assemblages occupy pigeonholes bounded by specific mineral reactions. If mineral assemblages in metamorphic rocks are matched with the grid which has been calibrated in the laboratory, this provides estimates of the depth and temperature of the rock during metamorphism. Unfortunately, the temperatures of dissociation reactions are very sensitive to partial pressures of the volatile component involved in the reaction, and therefore there are complications introduced in application of the grid to metamorphic conditions. However, continued experimental studies of the stability of minerals and mineral stabilities under a wide range of laboratory conditions (in the presence of H_2O and CO_2 gas mixtures, for example, and with oxygen fugacity varied as well) may eventually provide a guide to the composition of the pore fluid that was present during metamorphism. If this is known, then the experimental data can be applied with greater confidence. *See* METAMORPHIC ROCKS.

[PETER J. WYLLIE]

Bibliography: T. F. W. Barth, *Theoretical Petrology*, 2d ed., 1962; B. Bayly, *Introduction to Petrology*, 1968; G. V. Chilinger, H. J. Bissel, and R. W. Fairbridge (eds.), *Carbonate Rocks*, 2 vols., 1967; E. Wm. Heinrich, *Microscopic Petrography*, 1956; H. H. Hess and A. Poldervaart (eds.), *Basalts*, 2 vols., 1967; R. C. Newton, The status and future of high static-pressure geophysical research, in R. S. Bradley (ed.), *Advances in High-Pressure Research*, vol. 1, 1966; R. Roy and O. F. Tuttle, Investigations under hydrothermal conditions, in L. H. Ahrens, K. Rankama, and S. K. Runcorn (eds.), *Physics and Chemistry of the Earth*, vol. 1, 1956; F. J. Turner, *Metamorphic Petrology*, 1968; F. J. Turner and J. Verhoogen, *Igneous and Metamorphic Petrology*, 2d ed., 1960; O. F. Tuttle and N. L. Bowen, *Origin of Granite in the Light of Experimental Studies in the System $NaAlSi_3O_8$-$KAlSi_3O_8$-SiO_2-H_2O*, Geol. Soc. Amer. Mem. no. 74, 1958; H. G. F. Winkler, *Petrogenesis of Metamorphic Rocks*, 2d ed., 1967; P. J. Wyllie, Applications of high-pressure studies to the earth sciences, in R. S. Bradley (ed.), *High-Pressure Physics and Chemistry*, vol. 2, 1963; P. J. Wyllie, High-pressure techniques, in R. S. Bradley (ed.), *Methods and Techniques in Geophysics*, vol. 2, 1966.

Phenocryst

A relatively large crystal embedded in a finer-grained or glassy igneous rock. The presence of phenocrysts gives the rock a porphyritic texture (see illustration). Phenocrysts are represented most commonly by feldspar, quartz, biotite, hornblende, pyroxene, and olivine. Strictly speaking, phenocrysts crystallize from molten rock material (lava or magma). They commonly represent an earlier and slower stage of crystallization than does the matrix in which they are embedded. Phenocrysts are to be distinguished from certain relatively large crystals (porphyroblasts) which develop late in solid rock as the result of metamorphism or metasomatism. If the origin of a large crystal is in question, the nongenetic term megacryst should be used. *See* AUREOLE, CONTACT; IGNEOUS ROCKS; PORPHYROBLAST; PORPHYRY; RAPAKIVI GRANITES. [CARLETON A. CHAPMAN]

Granite (quartz monzonite) from the Sierra Nevada of California showing numerous phenocrysts of microcline feldspar in parallel orientation with banded structure of the rock. Hammer is 10 in. long. (*United States Geological Survey photograph by W. B. Hamilton*)

Phlogopite

A mineral of the mica group, also called bronze mica. Its composition is $K_2[Mg,Fe(II)]_6(Si_6,Al_2)$-$O_{20}(OH)_4$, including minor amounts of sodium, Na, that substitute for potassium, K, and containing small amounts of Mn, Fe(III), and Ti. With an increase in Fe(II), it grades into biotite from which there is no sharp distinction. It is the Mg-rich half of the biotite-phlogopite series known to the mica industry as amber mica.

Phlogopite is stable at higher temperatures and has a higher power factor than muscovite, with about the same voltage breakdown. It is widely used as an electrical insulator.

It occurs in disseminated flakes, foliated masses, or large crystals (see illustration). The basal cleavage is easy and perfect; specific gravity is 2.8–3.0; hardness is 2.5–3.0 on Mohs scale. Thin sheets are transparent in shades of light brown and green. Reddish-brown reflections are characteris-

Sheets of phlogopite from Otty Lake, Canada. (*Specimen from Department of Geology, Bryn Mawr College*)

tic of cleavage surfaces. It may be colorless to weakly pleochroic.

The structures are monoclinic (one-layer, two-layer, and three-layer pseudorhombohedral). Many phlogopites display asterism in transmitted light because of oriented exsolved rutile needles.

Phlogopite occurs chiefly in certain peridotites (kimberlites), in carbonatites, in serpentinized peridotites, in marbles derived from impure dolomitic limestones, and as very large crystals of commercial importance in coarse-grained plagioclase-apatite-calcite-pyroxene rocks of pegmatitic affinity (Ontario and Quebec). It alters to vermiculite. *See* MICA; SILICATE MINERALS.

[E. WILLIAM HEINRICH]

Phonolite

A light-colored, aphanitic (not visibly crystalline) rock of volcanic origin, composed largely of alkali feldspar, feldspathoids (nepheline, leucite, sodalite), and smaller amounts of dark-colored (mafic) minerals (biotite, soda amphibole, and soda pyroxene). Phonolite is chemically the effusive equivalent of nepheline syenite and similar rocks. Rocks in which plagioclase (oligoclase or andesine) exceeds alkali feldspar are rare and may be called feldspathoidal latite. *See* FELDSPATHOID.

Rapid cooling at the surface causes lavas to solidify with very fine-grained textures. Most phonolitic lavas, however, carry abundant large crystals (phenocrysts) when they are erupted, and these are soon frozen into the dense matrix to give a porphyritic texture. Generally very little material congeals as glass. The phenocrysts, many visible to the naked eye, include alkali feldspar, feldspathoids, and mafics. These may be well-formed (euhedral) or moderately well-formed (subhedral).

Most other features of phonolites can be seen only microscopically. The alkali feldspar is principally soda-rich sanidine and orthoclase. It generally occurs in the rock matrix, but if abundant it may also form as phenocrysts. Plagioclase is not abundant except in nepheline latites where it may form abundant phenocrysts.

Nepheline may occur as euhedral crystals (square or hexagonal), some of which may be phenocrysts. Otherwise it is irregular (anhedral) and interstitial. Nosean, hauyne, and sodalite, as euhedral or partly corroded crystals, may occur as phenocrysts and matrix grains. These twelve-sided (dodecahedral) crystals generally show hexagonal outlines in thin sections of the rock. Eight-sided euhedral crystals of pseudoleucite may occur as phenocrysts in potash-rich rocks. More rounded grains of leucite may form part of the matrix. Leucite is commonly altered to pseudoleucite, but the euhedral outline is retained. Analcite occurs principally as matrix material but in some rocks it is abundant and as large euhedral phenocrysts.

Biotite is not common but may form large strongly resorbed phenocrysts. Amphiboles are usually soda-rich (riebeckite, hastingsite, and arfvedsonite). They may occur as phenocrysts or as interstitial clusters. They may show resorption or may be replaced by pyroxene. The most important mafic is soda pyroxene. As phenocrysts it is commonly zoned with cores of diopside surrounded by progressively more sodic shells of aegirine-augite and aegirite. Aegirite is the common pyroxene of the rock matrix.

Accessory minerals are varied and include sphene, magnetite, zircon, and apatite.

The structures and textures of phonolite are similar to those of the more common rock trachyte. Fluidal structure, formed by flowage of solidifying lava and expressed by lines or trails of phenocrysts, may be seen without magnification. Under the microscope, flowage is shown by subparallel arrangement of elongate feldspar crystals. *See* TRACHYTE.

Phonolites are rare and highly variable rocks. They occur as volcanic flows and tuffs and as small intrusive bodies (dikes and sills). They are associated with trachytes and a wide variety of feldspathoidal rocks.

The origin of phonolites and related rocks constitutes an interesting problem. There is still a considerable difference of opinion as to how the phonolitic magma (molten material) originates. One theory assumes an origin from basaltic magma by differentiation. Certain early formed crystals are removed (perhaps by settling), causing the residual magma to approach the composition of phonolite. Another theory supposes these peculiar magmas to form when a more normal rock melt assimilates large quantities of limestone fragments. Volatiles, notably carbon dioxide, are considered by many to play an important role in transferring and concentrating certain constituents (like potassium) in the magma. The great variety of rock types and modes of association strongly suggests that several different mechanisms may operate to form these feldspathoidal rocks. *See* IGNEOUS ROCKS; MAGMA.

[CARLETON A. CHAPMAN]

Phosphate minerals

Any naturally occurring inorganic salts of phosphoric acid, $H_3[PO_4]$. All known phosphate minerals are orthophosphates since their anionic group is the insular tetrahedral unit $[PO_4]^{3-}$. Mineral salts of arsenic acid, $H_3[AsO_4]$, have a similar crystal chemistry. There are over 150 species of phosphate minerals, and their crystal chemistry is often very complicated. Phosphate mineral paragenesis can be divided into three categories: primary phosphates (crystallized directly from a melt or fluid), secondary phosphates (derived from the primary phosphates by hydrothermal activity), and rock phosphates (derived from the action of water upon buried bone material, skeletons of small organisms, and so forth). See MINERAL.

Primary phosphates. These phosphate minerals have crystallized from fluids, usually aqueous, during the late stage of fluid and minor element segregation and concentration. Certain rocks, particularly granite pegmatites, frequently contain phosphate minerals which sometimes occur as enormous crystals. Giant crystals up to 10 ft across are known for apatite, $Ca_5(F,Cl,OH)[PO_4]_3$; triphylite-lithiophilite, $Li(Fe,Mn)[PO_4]$; amblygonite, $(Li,Na)Al(F,OH)[PO_4]$; and graftonite, $(Fe,Mn,Ca)_3[PO_4]_2$. Rare-earth phosphates, such as monazite, $(La,Ce)[PO_4]$, and xenotime, $Y[PO_4]$, are known from some pegmatites and are mined for rare-earth oxides. The $[PO_4]^{3-}$ anionic group is often considered a "mineralizer"; that is, it contributes to decreased viscosity of the fluid and promotes the growth of large crystals. Primary phosphates are segregated since they usually do not combine with silicate-rich phases to form phosphosilicates. See AMBLYGONITE; APATITE; PEGMATITE.

Primary phosphates also occur with other rock types, notably ultrabasic rocks such as nepheline syenites, jacupirangites, and carbonatites. Enormous quantities of apatite occur with nepheline syenite in the Kola Peninsula, Soviet Union. Carbonatites often contain rare-earth phosphates such as britholite, $(Na,Ce,Ca)_5(OH)[(P,Si)O_4]_3$. Primary phosphates also occur in metamorphosed limestones, usually as apatite. Apatite also occurs with magnetite, $FeFe_2O_4$, segregated from basic rocks such as norites and anorthosites. Rarely, phosphates occur in meteorites, including apatite, whitlockite, β-$Ca_3[PO_4]_2$, and sarcopside, $(Mn,Fe,Ca)_3[PO_4]_2$. See CARBONATITES; NEPHELINE SYENITE.

Secondary phosphates. A large spectrum of secondary phosphates is known, particularly because they have formed at low temperatures and over a range of pH and pO_2 conditions. Over 50 species are known to have been derived from the action of water on triphylite-lithiophilite. Their crystal chemistry is very complicated; most contain Fe^{2+}, Fe^{3+}, Mn^{2+}, and Mn^{3+} octahedrally coordinated by $(OH)^-$, (H_2O), and $[PO_4]^{3-}$ ligand groups and can be considered coordination complexes. The most common species are strengite, $Fe[PO_4](H_2O)_2$; ludlamite, $Fe_3[PO_4]_2(H_2O)_4$; and vivianite, $Fe_3[PO_4]_2(H_2O)_8$. Because they contain transition-metal cations, they are often beautifully colored and are highly prized by mineral fanciers.

Members of another group, known as the fibrous ferric phosphates, arise by reaction of phosphatic water with goethite, α-FeO(OH), and occur in certain limonite beds, sometimes in generous quantities. These include rockbridgeite, $Fe^{2+}Fe_4^{3+}(OH)_5[PO_4]_3$, and dufrenite, $Fe^{2+}Fe_5^{3+}(OH)_5[PO_4]_4(H_2O)_2$. There are several other species, but this group of interesting minerals is poorly understood. See LIMONITE.

Rock phosphates. These are phosphates that are derived at very low temperatures by the action of water on buried organic material rich in phosphorus, such as bones, shells, and diatoms. Also included are phosphates obtained by the reaction of phosphatic waters with carbonate materials, such as corals. These waters were enriched in the phosphate anion by previous action upon organic material.

Large beds of phosphatic oolites are known. The mineralogy is poorly understood because of the small grain size. The major mineral is carbonate-apatite, but minor amounts of monetite, $CaH[PO_4]$, and brushite, $CaH[PO_4](H_2O)_2$, are also encountered, as well as many gel-like substances which are not strict mineral entities.

Crystals of several hydrated magnesium phosphates have been recovered from bat guanos which were suitably preserved. Common minerals include newberyite, $MgH[PO_4](H_2O)_3$, and struvite, $NH_4Mg[PO_4]\cdot 6H_2O$. Guanos are often marketed for fertilizer because of their high phosphorus content.

Phosphate minerals of this category are also known to occur as urinary calculi. Most frequent is apatite. Other species are struvite, newberyite, and whitlockite. Struvite crystals have also been found in canned sardines. [PAUL B. MOORE]

Phyllite

A large group of regional metamorphic rocks derived from argillaceous sediments and recrystallized in the greenschist facies (low degree of metamorphism), essentially composed of white mica and quartz. Phyllites are fine-grained, strongly schistose rocks, and the schistosity surfaces exhibit a glittery sheen given off by mica. They are widely distributed and easily recognized. In the past, geologists called them the lustrous schists of the mountain ranges (the *schistes lustrés* of the French and Swiss Alps). With very low metamorphism the phyllites pass into slates, and with increasing metamorphism they pass into mica schists. See METAMORPHIC ROCKS; SCHIST; SLATE.

The simple mineral composition of the ordinary phyllite (quartz and muscovite) is explained by the rule of the paucity of mineral phases. With increasing content of iron-magnesia, new minerals like chlorite, almandine-spessartite garnet, or chloritoid may develop. Among the garnetiferous phyllites may be mentioned the whetstones of the Ardennes. Chloritoid minerals are present in the ottrelite schists of the Alps. Paragonite, the sodium-mica, is commonly present in phyllites.

Phyllite has a wide distribution in the crystalline mountain ranges of the world: the Highlands of Scotland, mountains of Norway, Harz Mountains and mountains of Saxony, Alps, Appalachians, and Great Lakes district of the United States. The schistosity of phyllites is sometimes flat but usually is crumpled and so imperfect as to render the rock unsuitable for roofing material.

[T. F. W. BARTH]

Phyllonite

A metamorphic rock—the name is a combination of phyllite and mylonite; the phyllonites occupy a position between the rock types representing the component parts of the name. There are two distinct stages of their development. In the first stage the original rock is granulated by extreme deformation and pulverized to a mylonite. In the second stage, but frequently overlapping the first stage in time, new minerals recrystallize and grow (this is called crystalloblastesis). B. Sander originally introduced the name for phyllitelike rocks which had suffered a deformation after recrystallization, regardless of whether they were derived from argillites, like the phyllites, or from orthocataclasites. *See* METAMORPHIC ROCKS; MYLONITE; PHYLLITE.

[T. F. W. BARTH]

Physiographic provinces

Portions of the Earth's surface that display a similar array of landforms and relief features. Relief refers to those forms of surface roughness which determine and define the highest and lowest elevations within the province. The origins of the individual landforms within a given physiographic province may be determined through the geologic subscience of geomorphology. In general, adjacent regions assigned to different physiographic provinces differ not only in landscape, but in the nature of their underlying geological structure, in the variety of rock types exposed, and perhaps even in climate. However, physiographic provinces are recognized and separated only on the basis of appearance, not origin. *See* GEOMORPHOLOGY.

Subtle variations in relief features within physio-

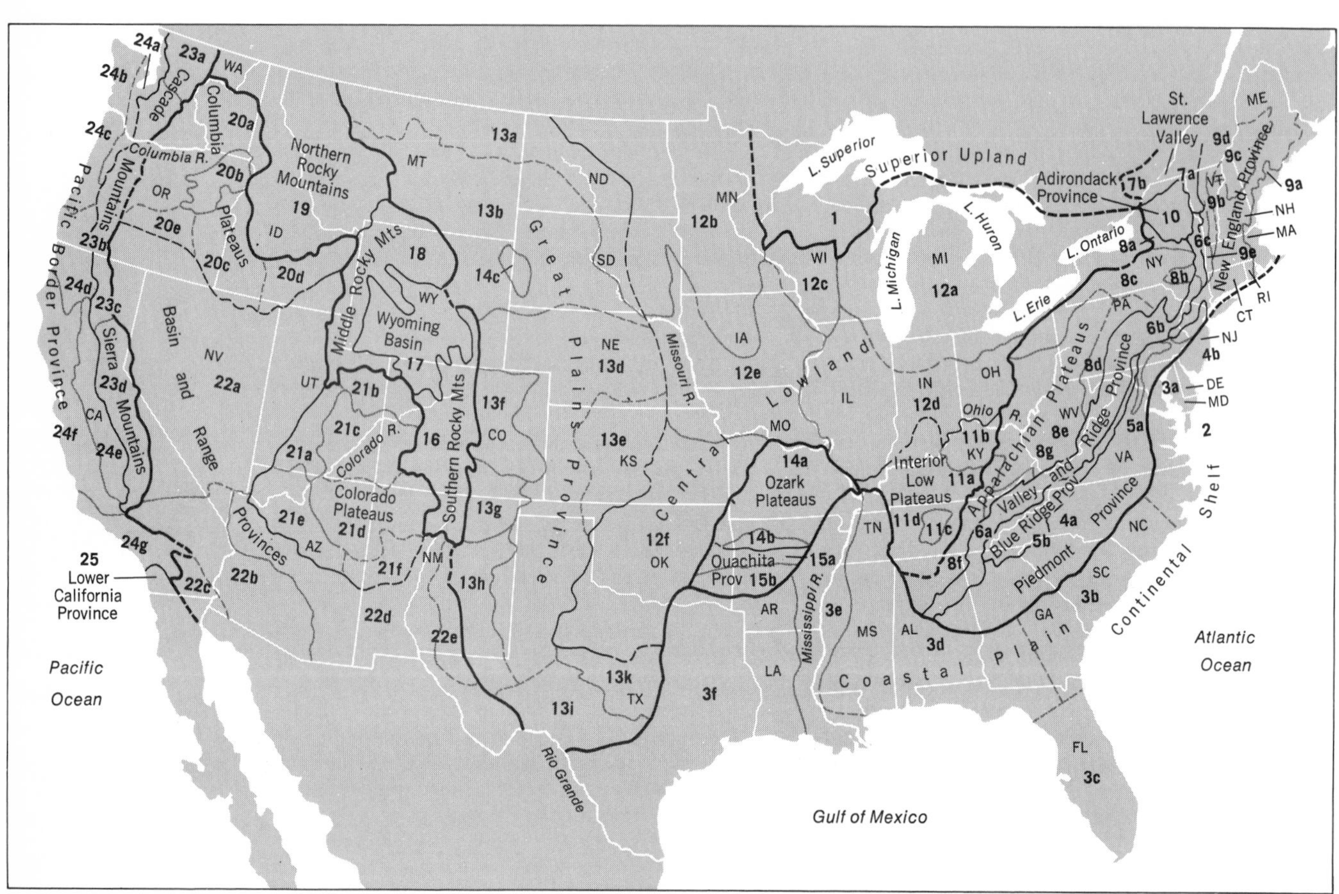

Fig. 1. United States physiographic divisions and provinces. (*U.S. Geological Survey, Dept. of the Interior, 1962*)

Key to map of physiographic provinces in Fig. 1

Major division	Province	Section
Laurentian Upland	1. Superior Upland	
Atlantic Plain	2. Continental shelf	
	3. Coastal plain	a. Embayed Section
		b. Sea Island Section
		c. Floridian Section
		d. East Gulf Coastal Plain
		e. Mississippi Alluvial Plain
		f. West Gulf Coastal Plain
Appalachian Highlands	4. Piedmont Province	a. Piedmont Upland
		b. Piedmont Lowlands
	5. Blue Ridge Province	a. Northern Section
		b. Southern Section
	6. Valley and Ridge Province	a. Tennessee Section
		b. Middle Section
		c. Hudson Valley

Key to map of physiographic provinces in Fig. 1 (cont.)

Major division	Province	Section
Appalachian Highlands (cont.)	7. St. Lawrence Valley	a. Champlain Section
		b. Northern Section
	8. Appalachian Plateaus	a. Mohawk Section
		b. Catskill Section
		c. Southern New York Section
		d. Allegheny Mountain Section
		e. Kanawha Section
		f. Cumberland Plateau Section
		g. Cumberland Mountain Section
	9. New England Province	a. Seaboard Lowland Section
		b. New England Upland Section
		c. White Mountain Section
		d. Green Mountain Section
		e. Taconic Section
	10. Adirondack Province	
Interior Plains	11. Interior Low Plateaus	a. Highland Rim Section
		b. Lexington Plain
		c. Nashville Basin
		d. Possible western section (not delimited)
	12. Central Lowland	a. Eastern Lake Section
		b. Western Lake Section
		c. Wisconsin Driftless Section
		d. Till Plains
		e. Dissected Till Plains
		f. Osage Plains
	13. Great Plains Province	a. Missouri Plateau, glaciated
		b. Missouri Plateau, unglaciated
		d. High Plains
		e. Plains Border
		f. Colorado Piedmont
		g. Raton Section
		h. Pecos Valley
		i. Edwards Plateau
		k. Central Texas Section
Interior Highlands	Black Hills	
	14. Ozark Plateaus	a. Springfield Salem Plateaus
		b. Boston Mountains
	15. Ouachita Province	a. Arkansas Valley
		b. Ouachita Mountains
Rocky Mountain System	16. Southern Rocky Mountains	
	17. Wyoming Basin	
	18. Middle Rocky Mountains	
	19. Northern Rocky Mountains	
Intermontane Plateaus	20. Columbia Plateaus	a. Walla Walla Plateau
		b. Blue Mountain Section
		c. Payette Section
		d. Snake River Plain
		e. Harney Section
	21. Colorado Plateaus	a. High Plateaus of Utah
		b. Uinta Basin
		c. Canyonlands
		d. Navajo Section
		e. Grand Canyon Section
		f. Datil Section
	22. Basin and Range Provinces	a. Great Basin
		b. Sonoran Desert
		c. Salton Trough
		d. Mexican Highland
		e. Sacramento Section
Pacific Mountain System	23. Cascade Mountains	a. Northern Cascade Mountains
		b. Middle Cascade Mountains
		c. Southern Cascade Mountains
	Sierra Mountains	Sierra Nevada
	24. Pacific Border Province	a. Puget Trough
		b. Olympic Mountains
		c. Oregon Coast Range
		d. Klamath Mountains
		e. California Trough
		f. California Coast Ranges
		g. Los Angeles Ranges
	25. Lower Californian Province	

Table 1. United States physiographic units*

Major physiographic division	Province
Laurentian Upland	Superior Upland
Atlantic Plain	Continental shelf
	Coastal plain
Appalachian Highlands	Piedmont Province
	Blue Ridge Province
	Valley and Ridge Province
	St. Lawrence Valley
	Appalachian Plateaus
	New England Province
	Adirondack Province
Interior Plains	Interior Low Plateaus
	Central Lowland
	Great Plains Province
Interior Highlands	Ozark Plateaus
	Ouachita Province
Rocky Mountain System	Southern Rocky Mountains
	Wyoming Basin
	Middle Rocky Mountains
	Northern Rocky Mountains
Intermontane Plateaus	Columbia Plateaus
	Colorado Plateaus
	Basin and Range Province
Pacific Mountain System	Cascade-Sierra Mountains
	Pacific Border Province
	Lower California Province

*After N. M. Fenneman.

graphic provinces are often used to delineate minor physiographic differences; for example, the Ozark Province in the central United States includes the Springfield and Salem plateaus, the Boston Mountains, and the St. Francis Mountains. A widely accepted physiographic classification of North America in use at mid-century recognized six regions comprising some 30 provinces and more than 150 major province subdivisions. Clearly, the number of landscape distinctions that can be made is great and depends upon the amount of detail desired. At some point, however, continued refinement of a classification brings one to basic units which are individual landforms. Distinctions between such landforms must be based on factors that are not always outwardly apparent and that relate to origin. At this point geomorphologic analysis must replace physiographic description.

BASIS FOR CLASSIFICATION

Physiographic provinces are usually natural areas defined on the basis of a certain obvious landscape characteristic or set of characteristics. Sometimes the basis for classification is entirely topographic and pertains to the "lay of the land," as in the Appalachian Highlands Province of A. K. Lobeck. But Lobeck also relies on geologic characters, for example, in his Canadian Shield group, defined by the large expanse of exposed older crystalline rocks in northeastern North America. Alternatively, geographic factors have been used—as in the Pacific Borderlands Province of N. M. Fenneman and W. W. Atwood—which indicate the location of the province but not its characteristics. In general, however, the most well-defined provinces are differentiated by outward physical characteristics, and are termed plains, mountains, uplands, lowlands, and plateaus.

Although the 1948 physiographic classification of Lobeck is one of the most detailed for North America, the slightly earlier attempt of Fenneman (1946) is perhaps the most widely cited for the United States (Fig. 1). Fenneman's classification is outlined here (Table 1) because its simplified number of provinces (25) is adequate for most purposes, and one can avoid the misconception that increased classification detail necessarily gives one a better understanding of landscape genesis. In most such classifications, some province subdivisions are not topographically distinct from one another but are merely geographically separated. In other cases, provinces differ from one another less externally than internally.

None of the physiographic provinces listed in Table 1 ends precisely at national borders, and each continent has its own set of physiographic provinces. In North America, for example, the Great Plains, Rocky Mountains, and Central Lowlands of the United States all extend northward into Canada. Also, the Lower California Province, Sonora Desert, Colorado Plateau, Coastal Plain, and Southern Rocky Mountains have counterparts or extensions, respectively, in the Mexican Baja Peninsula, Mexican Plateau, east and west Sierra Madre, and West Gulf Coastal Plain. These overlappings emphasize the natural (nonpolitical) character of the provinces, although political boundaries have, on occasion, been drawn to coincide with the edges of physiographic units.

One of the provinces in early physiographic classifications anticipates physiographic refinements made in the past quarter-century—the continental shelf. Study of the ocean floors has disclosed a series of submarine physiographic provinces of which the continental shelf is the most nearly emergent (Table 2).

Table 2. Submarine physiographic provinces

Type	Example
Continental slope	North American Atlantic Slope
Abyssal plains	Ceylon Abyssal Plain
Mid-oceanic ridges	Mid-Atlantic Ridge
Oceanic trenches	Tonga Trench
Submarine plateaus	Madagascar Plateau
Volcanic piles	Hawaiian Islands
Submarine basins	Natal Basin
Fracture zones	Mendocino Fracture Zone

DESCRIPTIONS

Descriptions of physiographic provinces tend to be colored by words referring to theoretical concepts of landscape genesis prevalent at the time in which they were written. This is particularly true of descriptions developed near the turn of the century when it was supposed that certain landforms indicated certain ages or stages of development and certain types of origin. If pure physiography rather than geomorphology is the aim of such descriptions, it is best to avoid terms with theoretical overtones.

The following is a description in modern terms of the major United States physiographic provinces following the Fenneman format with slight modifications. A summary of submarine provinces is also included. Degrees of relief are characterized as low (at most a few tens of meters), moderate (a few score to several hundred meters), and

Fig. 2. View of a portion of the Laurentian Upland near the village of Nipigon, Ontario, just north of Lake Superior. *(From H. F. Garner, The Origin of Landscapes, Oxford University Press, 1974)*

high (several hundred to a few thousand meters). The terms "dissection" or "incision" are used to describe linear erosion by running water that forms gully-, ravine-, valley-, canyon-, or gorgelike features with intervening ridges or more or less broad drainage divides. Land dissection may be local or general and slight, moderate, or deep in relation to corresponding degrees of relief. Much land-surface relief is a result of stream dissection to a degree essentially limited by overall land elevation (stream erosion ceases not far below sea level). Some forms of land relief are the result of crustal movements.

Laurentian Upland Division. This is a broad area of slightly elevated older crystalline rocks situated mainly in northeastern Canada south of Hudson Bay. The topography is generally rolling, and elevations rarely exceed 300 m (Fig. 2). The region has been intensely glaciated, and bedrock is widely exposed at the surface. One portion of this division extends slightly southward into the United States.

The Superior Upland is rolling, glaciated terrain developed on crystalline rocks which have considerable resistance to erosion and only moderate stream dissection. Former drainage patterns have been obscured or altered by glacier scour and deposition.

Atlantic Plain Division. This division is a low, shelving, sediment-covered plains region extending from southern New England south and west past Florida and around the Gulf of Mexico to its western margins in Mexico. All drainages crossing this plain enter the Atlantic Ocean, and its margins are under water (continental shelf). Its inner edges abut a variety of interior uplands exposing older, consolidated bedrock.

Continental shelf. This is a gently sloping plain which has been occupied by shelf seas during the past 20,000 years. Areas of sediment accumulation alternate with those of marine current scour and sediment transport; the latter areas are prevalent on the heads of submarine canyons that notch the shelf margins at intervals, often opposite the mouths of rivers. *See* CONTINENTAL SHELF AND SLOPE.

Coastal plain. The coastal plain is a complex, slightly dissected plain composed of deposits produced by seas retreating from the continental interior during its last emergence; river valley alluvium and delta sediment accumulations; beach deposits made by longshore marine waves and currents; local deposits of windblown dust and sand (in dunes); and slightly elevated beaches and terraces and (in Florida) local effects of limestone solutioning (that is, sinks). The transition with the continental shelf occurs through beach, marsh, swamp, and river mouth configurations. Inland margins of the plain are generally marked by more elevated areas of eroded older rock. *See* COASTAL PLAIN.

Appalachian Division. This is a broad, linear region of subdued mountains and related depressions extending from northern Georgia and Alabama northeastward into the maritime provinces of Canada. Most uplands are rounded in profile, although only northern provinces experienced glaciation. Watercourses south of the glacial borders tend to be situated in V-shaped valleys devoid of alluvium, and secondary water divides are often sharp ridges.

Piedmont Province. This is a moderate- to low-relief, slightly elevated area of deformed rock with general dissection by streams flowing on rocks of varying resistance to erosion. More elevated areas developed on tougher rocks, but most rock types are covered with thick, residual soils under heavy forest cover. This forest cover extends into other provinces of the southern Appalachian highlands.

Blue Ridge Province. This province is composed of deeply dissected mountains of moderate elevation on deformed, altered crystalline rocks with strongly developed ridge-ravine topography under dense forest cover.

Valley and Ridge Province. This province's elongate ridges and valleys developed on strongly folded, layered sedimentary rocks. Rock layers vary in resistance to erosion and ridge-forming potential; an example is commonly massive sandstone. Ridges are more numerous than valleys in mid-portions of the region, whereas valleys are more prevalent to the south. Major valleys (for example, the Great Valley) are nearly devoid of soil cover over wide areas. Ridges in the northern portion have been rounded off by glacier erosion, and valleys contain glacial deposits.

St. Lawrence Valley. Characterized by rolling, glaciated lowland developed on recent marine sediments in an embayment opening toward the northeast, this area is mantled with unsorted glacial debris (till) and local glacial lake accumulations of clay and sand plus an assortment of gravel and sand deposited by glacier meltwater. Several knobs of crystalline rock of volcanic origin rise above the lowland at Montreal and to the east (for example, Mount St. Hillaire).

Appalachian Plateaus. This complex of dissected plateau areas, including the Catskill, Allegheny, and Cumberland regions, developed on stratified sedimentary rocks that are locally warped into broad open folds. Mountainous relief is locally developed by erosion of less resistant rock types along watercourses. Many areas are deeply and finely dissected by stream erosion, producing moderate relief, and much close-set ridge-ravine topography exists. The northern quarter of this province was glaciated until about 10,000 years ago and is mantled with glacial debris. Uplands

Fig. 3. View of an elevated plateau in the New England Province. Arrow points to Mount Monadnock, in New Hampshire, an isolated erosion remnant that rises above the plateau. *(Photograph by G. Theokritoff, in H. F. Garner, The Origin of Landscapes, Oxford University Press, 1974)*

have been rounded by glacier scour and abrasion. Ridge forms to the south are sharper and more angular. Extensive broad flattened areas are comparatively rare.

New England Province. This is a good example of a "classic" physiographic province that includes several types of topography. Totally glaciated until about 10,000 years ago, this region includes dissected planar areas near the coast generally less than 150 m above sea level; more elevated and deeply dissected planar areas situated somewhat inland that are surmounted by isolated bedrock knobs (Fig. 3); larger and essentially isolated mountain masses of crystalline rock (all elevations affected by glacier erosion); narrow zones of linearly folded and glaciated rock of mountainous relief; and deeply dissected and glaciated mountains with somewhat equivalent summit elevations. Rock types vary from resistant crystalline varieties to several types of deformed and altered, layered sedimentary rocks. Evidence suggests that this generally mountainous region experienced at least one episode of mountain building in addition to that occurring in the southern part of the Appalachian Division. This province probably includes topographic elements that are divided into several separate physiographic provinces farther southwest.

Adirondack Province. This province has subdued, moderately dissected mountains with rounded crests which developed on complexly deformed crystalline rocks more than 1,000,000,000 years old. Rocks are very resistant to erosion but have been intensely glaciated by overriding continental ice sheets.

Interior Plains Division. This division includes a group of provinces of low to intermediate elevation that are broadly planar in appearance although locally dissected. Some of the more elevated portions are plateaulike, although extensively dissected by rivers for the most part. Lower areas are generally less dissected, in part because they were formed by recent glacier deposits. *See* PLAINS.

Interior Low Plateaus. These slightly to moderately dissected, somewhat elevated plains developed on compartively nonresistant sedimentary rocks of nearly flat-lying aspect. Northern margins show influences of glacier deposition and ice-margin drainage development from several glacial episodes. Many of the valley bottoms related to Ohio River drainage are flat because of silt or alluvial fills. Portions of this province resemble the Ozark Plateaus to the west.

Central Lowland. This is another example of a "classic" physiographic province that includes several distinct types of topography: (1) A central east-west-trending belt of till plain extends from eastern Nebraska through Iowa, Illinois, and Indiana into Ohio. These same glacial deposits define the limits of the Corn Belt, and parts of the till plain extend into Kansas, Missouri, the Dakotas, Minnesota, Wisconsin, and Michigan. Relief varies from none to low, and stream dissection is mostly local and slight. Exceptions occur in marginal areas on older glacial deposits. Generally, the youngest till plains and those in drier regions are least dissected. The oldest till plains and those in the most humid (eastern) areas are most deeply dissected. (2) A region generally lacking glacial deposits in Wisconsin (the "driftless area") includes portions of a moderately dissected plateau and adjacent lowland. (3) Low-relief, lake-studded areas of the Dakotas, Wisconsin, and Michigan are developed on rolling and knobby but planar topography created by melting glaciers under stagnant ice. The flattest areas are till plains and lake deposits. Dissection is greatest in humid eastern areas and least in dry western regions, and varies from slight and local to slight but general. (4) The Osage Plains of the southwestern part of the Central Lowland are low-relief surfaces that bevel gently inclined sedimentary rock strata and display low, curvilinear escarpments. Local drainage lines are weakly incised; through-flowing lines from more elevated areas to the west are the most deeply entrenched. This region lies to the south of glaciated areas and exhibits localized strings of sand dunes to the northeast of the main drainage lines (for example, the Cimarron River) and local accumulations of windblown dust (loess) in more eastern margins. Vegetal cover varies from deciduous forest in the east to grassland in the west.

Great Plains Province. This is a complex plains region extending from Texas into Canada and inclined eastward from the Rocky Mountains to the Interior Lowland. Slope on the plains in the central United States averages about 1 m/km (one-sixth of a degree). Northern portions in the vicinity of the Missouri River (Missouri Plateau) were glaciated and include local, mountainous elevations. Drainage lines crossing the province flow eastward and are only moderately incised, though often complexly terraced and usually alluviated. Exposures of poorly consolidated, fine-grained sedimentary rock are locally dissected into badlands, but regionally, dissection by streams over wide areas is slight and surfaces are gently rolling. The High Plains portion consists of remnants of a formerly extensive alluvial covermass preserved on flattened divide areas between present east-west-trending drainage lines. The alluvium was apparently derived from the Rocky Mountains by streams that lost sediment transport capacity as they flowed eastward, perhaps in a desert. The High Plains alluvium deposits become thinner to the east. The extent of present stream dissection from local precipitation increases eastward from local to general, as does the regional humidity and vegetation density. The more deeply entrenched streams (such as the Arkansas River) flow from mountainous headwaters where there is snow meltwater in the spring. The lateral continuity of plains and plateaulike areas increases to the west and southwest along with aridity. Higher-level plains remnants in the Raton section are capped by lava flows and volcanic cones.

Interior Highlands Division. This division includes relatively isolated highland areas that rise several hundred meters above surrounding lowland plains and include a variety of landforms and bedrock exposures. The Black Hills Province has been included by some physiographers in the High Plains Division but more closely resembles the other provinces discussed here.

Black Hills Uplift. This highland lies in the north-central portion of the Great Plains and is the northernmost of at least three elevated areas that can be called interior highlands. The Black Hills area is a domelike uplift of mountainous proportions surrounded by lower expanses of the High Plains. It is deeply dissected, and within the encircling cuestas and escarpments exposes erosionally resistant older crystalline rocks. Notable badlands occur in less-elevated marginal plains areas that expose soft sedimentary rocks, particularly to the east of the uplift.

Ozark Plateaus. These moderately elevated plateaus developed on flat-lying sedimentary rock strata at several distinct elevations in stair-stepped fashion across a broad area of crustal uplift with local exposures of older crystalline rocks (St. Francis Mountains). The most continuous planar areas (Springfield and Salem plateaus) show slight to moderate dissection that is most pronounced along the plateau margins. These plateaus are separated from each other by erosional escarpments several tens of meters high in some areas. The Boston Mountains are actually erosion remnants of a plateau standing an average of about 700 m above sea level and commonly incised by stream valleys as deep as 300 m (Fig. 4). Carbonate rocks are widespread at the surface, especially in northern areas, where they commonly display topographic effects due to solution. Most of the plateaus are capped by water-deposited gravel and thin layers of windblown dust. Valleys are generally V-shaped and show stepped sides reflecting rock intervals of varying resistance to erosion. Most valleys have small amounts of alluvium in their bottoms.

Fig. 4. View of a portion of the Boston Mountain Plateau in the vicinity of Heber Springs, AK. *(From H. F. Garner, The Origin of Landscapes, Oxford University Press, 1974)*

Ouachita Province. This province is divisible into (1) a northern, east-west-trending depression (Arkansas Valley) developed on moderately folded sedimentary rocks dominated by easily eroded shale with alternating sandstone intervals that stand up as low sinuous ridges; and (2) the east-west-trending Ouachita Mountains to the south which display a "valley and ridge" topography and strongly folded and otherwise deformed sedimentary rocks that include alternating "tough" and "weak" rock types. Rivers in this province are more or less extensively alluviated and display a series of alluvial terraces.

Rocky Mountain Division. This structurally and topographically complex region extends westward from the Great Plains Province for several hundred kilometers and trends north-south for several thousand kilometers. Elevations commonly exceed 4000 m, and local relief of more than 2000 m is common. Typically the region consists of localized mountainous uplifts exposing crystalline rocks and intervening lower depressions, basins, plateaus, and major valleys. Northern portions show extensive glacial effects, whereas southern areas do not. Surface accumulations of glacial debris occupy northern hillslopes and valley bottoms. Southern plains, basins, and valley bottoms display alluvial deposits of several types, and there

Fig. 5. A view of a U-shaped valley located in the northern portion of the Southern Rocky Mountains, Dolores County, CO, showing glacial effects. *(Photograph by W. Cross, U.S. Geological Survey, in H. F. Garner, The Origin of Landscapes, Oxford University Press, 1974)*

are extensive lake deposits in some areas. *See* MOUNTAIN SYSTEMS.

Southern Rocky Mountains. These mountains consist of uplifted and eroded crustal blocks including both crystalline rocks and surrounding sedimentary rocks exposed along margins of intervening intermontane basins and high plateaus. The basins and plateaus are extensively alluviated and are locally fringed with extensive alluvial fans and pediments. Higher elevations to the north show glacial effects (Fig. 5).

Wyoming Basin. This region is composed of elevated plains showing variable degrees of alluviation and subsequent stream dissection surmounted by isolated low mountain masses of moderate relief.

Middle Rocky Mountains. These are isolated mountainous uplifts of high relief and locally deep dissection, with folded and fault-bounded mountain types. Mountain crests are intensely glaciated. Alluviated intermontane basins separate uplifts.

Northern Rocky Mountains. These deeply dissected mountainous uplifts are separated by intermontane basins and plateaus that exhibit a variety of alpine glacial features and lake deposits. These mountains extend to the north into Canada (Fig. 6).

Intermontane Plateaus Division. This division includes a variety of intermontane plateaus which developed on several types of rock of different ages and origins and were subsequently modified by a number of different processes which are expressed in minor topography on the plateau surfaces. *See* PLATEAU.

Columbia Plateaus. These are actually erosional remnants of a single lava plateau, for the most part, with an overall area of about 200,000 mi² (5.18×10^{11} m²). Northern portions around Walla Walla, WA, were strongly modified by extensive flooding following the rupture of an ice dam that impounded Glacial Lake Missoula to the east. The resulting channels, waterfall scars, scour pits, and fluvial deposits developed on flat-lying basaltic lavas of the plateau surface. The plateaus are deeply incised by the Snake and Columbia rivers and their major tributaries, and are locally capped by thick accumulations of windblown dust which display a rolling topography. There are some features of recent volcanic origin in southern portions.

Colorado Plateaus. The high plateaus of western Colorado, Utah, Arizona, and New Mexico have an average elevation of nearly 3500 m. Utah areas are partly lava-capped and show some terracing. The Uinta Basin region is deeply dissected and shows lhgh relief, whereas the Navajo area in the central portion shows little dissection. Other parts are locally and very deeply incised by the Colorado River and major tributaries such as the San Juan River to produce the Canyonlands section and adjacent areas of the Kaibab Plateau (Fig. 7).

Basin and Range Province. This is a generally elevated region of isolated fault-block mountains

Fig. 6. View of an extension of the Northern Rocky Mountains into Canada (Mount Assiniboine in foreground). *(Photograph by Austin Post, University of Washington, in H. F. Garner, The Origin of Landscapes, Oxford University Press, 1974)*

in various states of dissection with intervening alluviated intermontane basins occupied by alluvial fans, pediments, and lake deposits. Relief is primarily due to the interplay between vertical crustal uplifts of mountainous blocks and local sediment filling of adjacent basins, many of which are undrained (Fig. 8). Stream dissection away from the Colorado River is minor and local.

Pacific Mountain Division. Four areas of mountainous character compose this diverse physiographic region. Each differs in types and ages of bedrock and in times of development.

Cascade Mountains. The Cascades are isolated, high volcanic mountains of the Pacific Northwest, generally having broadly accordant summits; they are not deeply dissected for the most part, and some are still active for example, Mounts Baker and Rainier).

Sierra Nevada. This tilted block mountain range developed on strongly deformed crystalline rocks and subsequently was deeply and locally dissected and glaciated. Higher peaks are to the east, as well as the strongest glacial effects, particularly in the north.

Pacific Border Province. In this province elongate depressions and troughs separate parallel ranges of complex mountains. More deeply subsided fault-bounded troughs are submerged (Puget Trough) or deeply filled with sediment (Los Angeles Basin), and interior basins (Great Valley of California) are arid, alluvial plains.

Lower California Province. This dissected, westward-sloping upland developed on crystalline rocks extending southward into the Lower California (Baja) Peninsula.

Submarine Physiographic Division. The areas of the Earth occupied by salt water are occupied primarily by shelf seas along the continental margins and are underlain by the continental shelf, the land-locked (epiric) seas on the continents (such as

Fig. 7. View of the curves of the San Juan River Valley where it entrenches the Colorado Plateau in the Canyonlands area. *(Photograph by W. N. Gilliland, in H. F. Garner, The Origin of Landscapes, Oxford University Press, 1974)*

Hudson Bay), and the deep oceans. The physiographic divisions of the deep oceans belong in this division and include several areas of distinctive relief and topography beginning with the outer edge of the continental shelf and extending to the abyssal depths. The undersea-scapes that exist there are primarily the result of movements of the Earth's crust, volcanism, and the slow accumulation of sediment. *See* MARINE GEOLOGY.

Fig. 8. View of Amargosa Mountains and adjacent depression from the west, north of Las Vegas, NV, in the Basin and Range Physiographic Province. *(From H. F. Garner, The Origin of Landscapes, Oxford University Press, 1974)*

Continental slope. The continental slope inclines rather steeply from the outer edge of the continental shelf to the abyssal depths several thousand meters below sea level. There the slope variously merges with the more gentle slopes that characterize most submarine plains, basins, and plateaus. In a few instances, trenches, volcanic piles, or fracture zones cut across or otherwise modify the slope, and the smooth inclination developed by accumulating sediment may give way to steeper notched or stepped ramplike surfaces. Unlike the other physiographic provinces of the deeper-ocean basins, the names of continental slopes generally consist of only the continent and ocean of which they are a part. One can therefore speak of the North American Atlantic Continental Slope or the African Indian Continental Slope. Although they are broadly smooth surfaces of sediment accumulation, continental slopes are locally notched by submarine canyons that extend from the outer shelf to the abyssal depths. Fanlike deposits of sediment occur at the lower ends of some canyons. *See* SUBMARINE CANYON.

Abyssal plains. These are large expanses of nearly level ocean floor at depths of 3 to 5 km occupied by relatively unconsolidated sediment. Locally, flat-topped submarine mountains (guyots) rise above the abyssal plains and appear to mark the positions of former volcanoes. Abyssal plains are the deepest extensive planar areas of the deep oceans and are usually rather distant (several hundred kilometers) from ocean ridges. In general, their depths tend to gradually increase away from such ridges. *See* SEAMOUNT AND GUYOT.

Mid-oceanic ridges. These make up a complex system of submarine mountains situated more or less midway between adjacent continents in each ocean basin, although some terminate against continental margins. Many of the ridge mountains are volcanic and most are intermittently active. The ridge exhibits little or no sediment cover and generally exposes black volcanic rock (basalt), much of which was erupted beneath the sea. A continuous length of mid-oceanic ridge extends beneath the seven oceans for more than 65,000 km. Over much of this length, the ridge consists of a paired range of mountains with a rather flat-floored valley between. Volcanic eruptions occur frequently on the median valley floor, and the ridge itself is cut transversely and locally offset by major fractures in the ocean crust. The transverse fractures appear as linear depressions in the ocean floor at right angles to the trend of the ridge and are sites of crustal movements and earthquakes in the axial zone of the ridge.

Oceanic trenches. These are elongate, often somewhat curved deep depressions in the ocean floor. The deepest parts of the oceans occur in trenches (more than 10 km below sea level). Trenches nearest to continents may contain considerable sediment, but most are asymmetrically V-shaped in cross section. The steepest margin of the trench is usually surmounted by volcanoes that extend to the ocean surface, where the volcanoes form an arcuate row of islands. At least some of the volcanoes are active in most "island arcs." Fracture systems have been noted on the volcanic side of oceanic trenches, and the zones occupied by trenches are frequent sites of earthquakes.

Submarine plateaus. These are rather isolated, somewhat elevated, flattened areas of the ocean floor, usually capped by a thickness of unconsolidated sediment. Submarine plateaus often separate abyssal plains from the more confined basins but may extend out beneath the ocean surface from the continental margin (such as the Blake Plateau east of North America).

Submarine basins. Submarine basins are localized depressions in the ocean floor more or less isolated from abyssal plains or one another by fracture zones, ridges, plateaus, or volcanic piles. Most basins have rather flat floors and contain some sediment.

Volcanic piles. These are rather isolated accumulations of volcanic debris on the ocean floors that extend above the ocean surface. The Hawaiian Islands are portions of such a volcanic pile in the Pacific Ocean. Iceland is a larger example in the North Atlantic Ocean.

Fracture zones. These are elongate breaks in the Earth's crust, and those on the ocean floors are commonly marked by deep linear depressions. Many fracture zones are at right angles to and offset the mid-oceanic ridge, although some seem to originate where ridges terminate. Fracture zones are most pronounced near ridges and generally disappear under sediment cover away from mid-oceanic ridges.

STATUS OF THE PHYSIOGRAPHIC PROVINCE CONCEPT

It has been suggested that purely descriptive physiography is essentially obsolete, insofar as it has been replaced by the more analytic and scientific geomorphology. If this is true, what is the value of learning about physiographic provinces? It seems that the "creations" of physiography have outlived the discipline. This is not surprising, for the Earth still has surface regions that are characterized by suites of landforms which are similar in appearance (if not in origin). Together, these landforms make up landscapes of a particular aspect, and knowledge of these landforms is valuable even if one lacks the details of landform genesis. For example, a farmer is more apt to want to cultivate a plain than a mountainside, although it may later prove to be of value to know about variations in that plain. Conversely, a farmer in southern France may wish to grow grapes on a well-drained south-facing steep slope. Thus, those concerned with any of the myriad aspects of land use can begin to learn the lay of the land by examining physiographic provinces.

[H. F. GARNER]

Bibliography: W. W. Atwood, *The Physiographic Provinces of North America*, 1940; N. M. Fenneman, *Ass. Amer. Geogr. Ann.*, 18:261–353, 1928; H. F. Garner, *The Origin of Landscapes*, 1974; A. K. Lobeck. *The Physiographic Division of North America*, 1948.

Pigeonite

The name given to the monoclinic pyroxenes of the general formula (Mg,Fe)SiO_3 with some augite in solid solution. Pigeonite bears the same relation to the orthorhombic pyroxenes as augite does to the diopside-hedenbergite series. Pigeonite is the orthorhombic pyroxene equivalent in the volcanic rocks. Most high-temperature metamorphic and igneous orthorhombic pyroxenes were probably originally pigeonite. The small optic angle (2V) distinguishes the mineral from augite and the inclined extinction distinguishes it from the orthorhombic pyroxenes. Pigeonite forms black, brown, or dark-green short stubby crystals with the 87° pyroxene (110) cleavages. The slower cooling rates of the igneous and metamorphic rocks usually permit the augitic materials in solution to exsolve and the remaining monoclinic pyroxene to invert to the orthorhombic form. The original augitic material is evident by the oriented exsolution lamellae in the host orthorhombic pyroxene. The faster cooling rates of the volcanic rocks quenches in the augitic material and thereby preserves the metastable pigeonite at surface temperatures. *See* AUGITE; DIOPSIDE; ENSTATITE; ORTHORHOMBIC PYROXENE; PYROXENE. [GEORGE W. DE VORE]

Pitchstone

A natural glass with dull or pitchy luster and generally brown, green, or gray color. It is extremely rich in microscopic, embryonic crystal growths (crystallites) which may cause its dull appearance. The water content of pitchstone is high and generally ranges from 4 to 10% by weight. Only a small proportion of this is primary; most is believed to have been absorbed from the surrounding regions after the glass developed. Pitchstone is formed by rapid cooling of molten rock material (lava or magma) and occurs most commonly as small dikes or as marginal portions of larger dikes. *See* IGNEOUS ROCKS; VOLCANIC GLASS.

[CARLETON A. CHAPMAN]

Plains

The relatively smooth sections of the continental surfaces, occupied largely by gentle rather than steep slopes and exhibiting only small local differences in elevation. Because of their smoothness, plains lands, if other conditions are favorable, are especially amenable to many human activities. Thus it is not surprising that the majority of the world's principal agricultural regions, close-meshed transportation networks, and concentrations of population are found on plains. Large parts of the Earth's plains, however, are hindered for human use by dryness, shortness of frost-free season, infertile soils, or poor drainage. Because of the absence of major differences in elevation or exposure or of obstacles to the free movement of air masses, extensive plains usually exhibit broad uniformity or gradual transition of climatic characteristics.

Distribution and varieties. Somewhat more than one-third of the Earth's land area is occupied by plains. With the exception of ice-sheathed Antarctica, each continent contains at least one major expanse of smooth land in addition to numerous smaller areas. The largest plains of North America, South America, and Eurasia lie in the continental interiors, with broad extensions reaching to the Atlantic (and Arctic) Coast. The most extensive plains of Africa occupy much of the Sahara and reach south into the Congo and Kalahari basins. Much of Australia is smooth, with only the eastern margin lacking extensive plains.

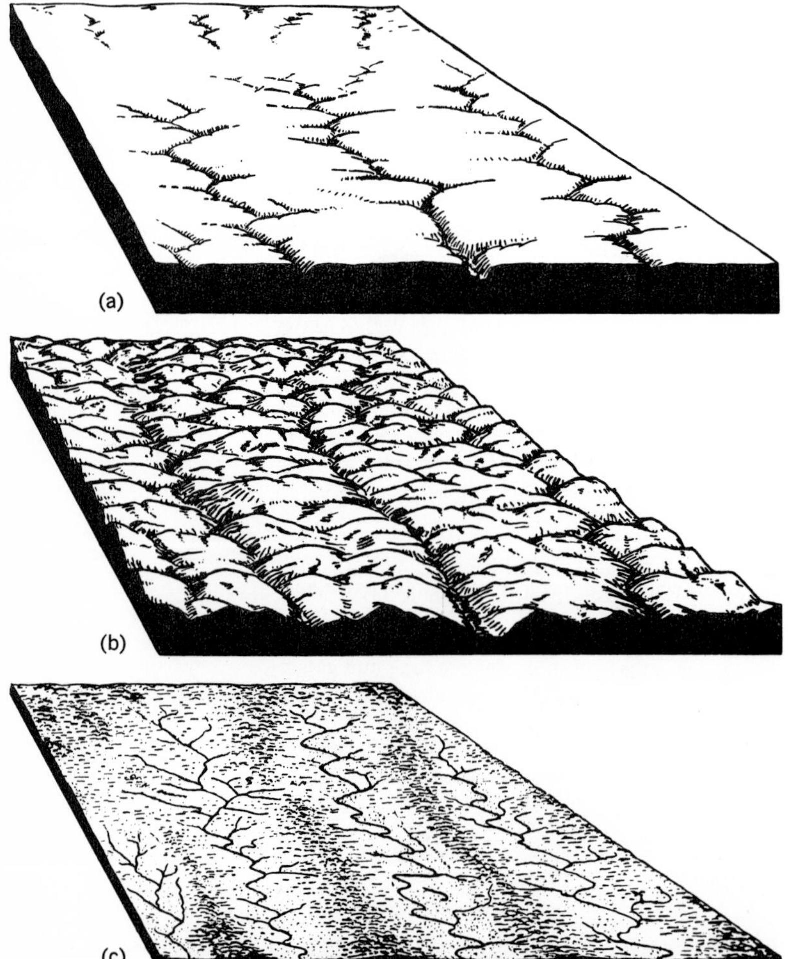

Fig. 1. Ideal stages in the progressive development of a stream-eroded landscape. (a) Youth. (b) Maturity. (c) Old age. (*From G. T. Trewartha, A. H. Robinson, and E. H. Hammond, Elements of Geography, 5th ed., McGraw-Hill, 1967*)

Surfaces that approach true flatness, while not rare, constitute a minor portion of the world's plains. Most commonly they occur along low-lying coastal margins or the lower sections of major river systems. Some occupy the floors of inland basins where extensive stream deposition has occurred. The majority of plains, however, are distinctly irregular in surface form, as a result of valley-cutting by streams or of irregular erosion and deposition by continental glaciers.

Plains are sometimes designated by the situations in which they occur. In common speech a coastal plain is any strip of smooth land adjacent to the shoreline, though in geology the term is often restricted to such a plain that was formerly a part of the shallow sea bottom. An example that fits both definitions is the South Atlantic and Gulf margin of the United States. Intermontane plains lie between mountain ranges, and basin plains are surrounded by higher and rougher land. Upland plains (sometimes loosely termed plateaus) lie at high elevations, or at least well above neighboring surfaces, while lowland plains are those lying near sea level, or distinctly below adjacent lands.

Types of plains are also sometimes designated according to the processes that have produced their distinctive surface features. These differences are discussed below.

Origin. The existence of plains terrain generally indicates for that area a dominance of the erosional and depositional processes over the forces that deform the crust itself. Most of the truly extensive plains, such as those of interior North America or that of northwestern Eurasia, represent areas which have experienced nothing more severe than slow, broad warping of the crust over a long period of geological history. Throughout that time the gradational processes have been able to maintain a relatively subdued surface. Certain other areas, including the upland plains of central and south-central Africa and eastern Brazil, have suffered moderate general uplift in late geologic time and have not yet been subjected to deep valley cutting.

Many plains of lesser extent, however, have been formed in areas where crustal deformation has been intense. Most of these represent depressed sections of the crust which have been partially filled by smooth-surfaced deposits of debris carried in by streams from the surrounding mountains. Examples are the Central Valley of California, the Po Plain of northern Italy, the plain of Hungary, the Mesopotamian plain, the Tarim Basin of central Asia, and the Indo-Gangetic plain of northern India and Pakistan.

SURFACE CHARACTERISTICS

The detailed surface features of plains result mainly from local erosional and depositional activity in relatively recent geologic time. Each of the major gradational agents – running water, glacial ice, and the wind – produces its own characteristic set of features, and any given section of plains terrain is characterized predominantly by features typical of one particular agent.

Features associated with stream erosion. Plains sculptured largely by stream erosion, which are far more widespread than any other class, are normally irregular rather than flat. Integrated valleys and the divides between them are the hallmarks of stream sculpture, and the differences among stream-eroded plains are generally expressible in terms of the size, shape, spacing, and pattern of these features.

The shallow depth of valleys on plains usually indicates that the surface has not been uplifted, in late geologic times, far above the level to which the local streams can erode. Wide valleys commonly indicate weak valley-wall materials, active erosion on the valley sides by surface runoff and soil creep and, in many instances, a long period of development. Resistant materials and permeable upland surfaces, on the other hand, favor narrower valleys and steeper side slopes.

Differences in valley spacing or in the degree to which the plain has been dissected by the development of valley systems are especially striking. Ideally, tributary growth is considered to proceed progressively headward from the major streams into the intervening uplands. For this reason, plains on which tributary valleys are few, major valleys are widely spaced, and broad areas of uncut upland remain (Fig. 1) are often termed youthful. If tributary valleys have extended themselves into all parts of the surface so that the area is almost wholly occupied by valley-side slopes, the surface is called mature. If valleys have so widened and coalesced that only a few narrow, low divide remnants are left, the surface is said to have reached old age. However, not all stream-eroded surfaces follow the ideal sequence, and many factors other than time affect their development.

Some plains, for example, remain persistently youthful because gentle or inadequate rainfall, excessive flatness, dense vegetation cover, or unusual permeability or resistance of the surface materials either fail to provide the surface runoff that could carve tributary valleys or else inhibit the erosive capacity of those smaller streams that do form. An example which combines several of these causal factors is provided by the High Plains, which stretch from southwestern Nebraska into the Panhandle of Texas. By way of contrast, some other plains appear to have reached maturity very quickly as a result of the generation of copious surface runoff by heavy rains and an initially rolling surface of low permeability, combined with readily erodible materials. The Dissected Till Plains of northern Missouri and southern Iowa are an example.

Although valleys of great breadth are not uncommon, extensive plains of ideal old age are rare. Valley widening is at best a slow process, and the reduction of low, gentle-sloped, well-vegetated divides requires immense lengths of time without interruption by uplift or environmental changes that initiate a renewal of valley deepening. Some modern scholars suggest that certain wide-valleyed, long-sloped, or gently undulating plains with an unbroken vegetation cover may in fact represent plains on which erosional development has practically ceased. *See* FLUVIAL EROSION CYCLE.

The pattern of valleys on plains depends chiefly upon the pattern of outcrop of rock materials of contrasting resistance. In the absence of strong contrasts the pattern is usually branching and tree-like as in Fig. 1. Where there is great differential resistance to erosion, the unusually resistant rocks form drainage divides or uplands, whereas weak rock belts are soon excavated into broad valleys

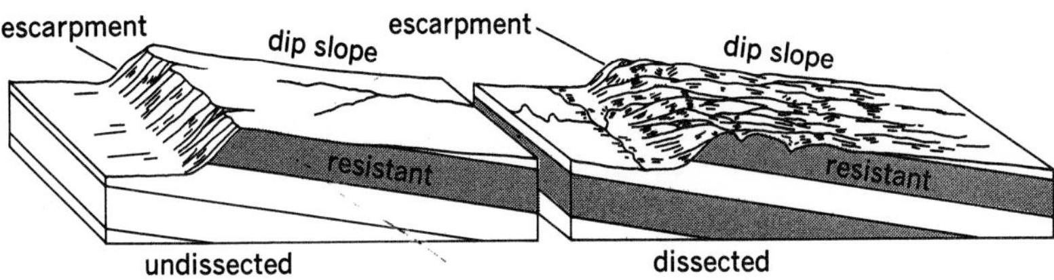

Fig. 2. Surface form and structure of cuestas. Dissected form at right is more typical, especially in humid regions. (*From G. T. Trewartha, A. H. Robinson, and E. H. Hammond, Elements of Geography, 5th ed., McGraw-Hill, 1967*)

or lowlands. Where erosional plains bevel across gently warped rock strata of varying resistance, the belts of outcrop of the more resistant strata form strips of higher, rougher country, with an abrupt escarpment on one margin and a more gradual dip-slope in the direction toward which the strata are inclined (Fig. 2). These features, called cuestas, are common in the Middle West and Gulf Coastal Plain and in western Europe. The various wolds and downs of England and the côtes of northeastern France are cuesta ridges. *See* ROCK CONTROL (GEOMORPHOLOGY).

Most plains that develop in dry climates are characterized predominantly by stream-produced landforms, in spite of infrequent rain. The development of valley systems and erosional features follows the same general rules as it does in humid regions. However, some differences in relative rates and relative significance of certain of the developmental processes produce distinctive landscape characteristics in arid lands. First, rock decomposition is very slow, so that the surface accumulation of weathered material is normally thin and coarse textured. Second, the sparse vegetation affords to the naked surface little protection against the battering and washing of the occasional torrential rains. As a result, the upper slopes become strongly gullied and often partially stripped of their covering material, leaving much bedrock exposed. Because of the short duration and local nature of the rains and hence the intermittent character of stream flow, however, most of the debris load is dropped in the neighboring basins and valley floors, "drowning" broad areas beneath plains of silt, sand, or gravel. Hence denuded and gullied upper slopes and broad depositional flats in the lowlands are characteristic features of desert plains. (Figs. 5 and 10). *See* DESERT EROSION FEATURES.

Features produced by solution. Features resulting from underground solution characterize several rather extensive areas of plains. The principal features of this class are depressions, or sinks, produced by collapse of caverns underneath. Solution is also an active process in erosion by surface streams, but its effects upon valley form have been little studied.

Significant groundwater solution is largely confined to areas underlain by thick limestones. As subsurface cavities are progressively enlarged by solution, more and more drainage is diverted to subterranean channels. Surface streams become fewer, often disappearing into the ground after a short surface run. Eventually solution cavities near the surface collapse, forming surficial depressions of various sizes. Some are shallow and inconspicuous; others are great steep-walled pits or elongated enclosed valleys. Some of the small depressions contain lakes, because their outlets are plugged with clay. The most extensive areas of solution-featured plains in the United States are in central and northern Florida and in the Panhandle of Texas. In both of these areas shallow sink holes, some of them lake-filled, are numerous. Some of the areas of most active solution work have developed surfaces far too rough to be called plains. This is true of the Mammoth cave area of west-central Kentucky and especially so for the mountainous area of great sinks and solution valleys in the Dalmatian Karst of Yugoslavia. *See* KARST TOPOGRAPHY.

Features associated with stream deposition. As a group, alluvial plains (so called from the term alluvium, which refers to any stream-deposited material) are among the smoothest and flattest land surfaces known. Stream-deposited plains fall into three classes: (1) floodplains, which are laid down along the floors of valleys; (2) deltas, formed by deposition at the stream mouths; and (3) alluvial fans, deposited at the foot of mountains or hills.

Floodplains. The flat bottomlands so common to valley floors develop wherever a given segment of a stream system is fed more sediment by its tributaries or upstream reaches than it is able to carry. Most of the excess sediment is deposited in the stream bed, usually in the form of bars. In flood time, some may be strewn across the whole width of the valley floor. Because of continued deposition and choking of the channel, repeated flooding, and the ease with which banks composed of loose alluvium collapse, streams continually shift their channels on floodplains. On silty plains the channel is usually highly sinuous or meandering, while on a coarse sandy or gravelly floodplain the channel is braided, that is, broad, shallow, and intricately subdivided by innumerable sandbars (Fig. 3). In either case, many loops or strands of abandoned channels, now mostly dry and partially filled with sediment, scar the floodplain surface. On broad silty floodplains slightly elevated strips, called natural levees, are formed by active deposition immediately adjacent to the channel when the velocity of flow is abruptly checked as the water leaves the swift current of the channel to spread thinly over the plain during floods.

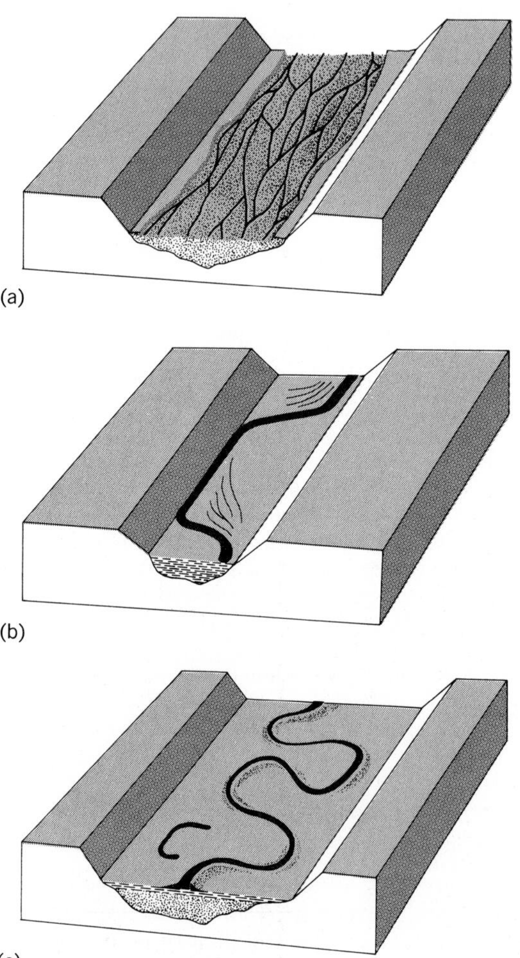

Fig. 3. Types of floodplains. (*a*) With braided channel. (*b*) With sinuous channel but too narrow for free meandering. (*c*) With freely meandering channel. (*From G. T. Trewartha, A. H. Robinson, and E. H. Hammond, Elements of Geography, 5th ed., McGraw-Hill, 1967*)

On major floodplains the groundwater table is everywhere close to the surface, and swampy land is common in the abandoned channels and shallow swales. For this reason, any slightly higher features, such as natural levees, are especially sought after for cultivation, town sites, and transportation routes. Though harassed by a high water table, recurrent floods, and shifting channels, floodplains, especially silty ones, are often prized agricultural lands because of the level, easily tilled surface. In some instances the alluvium is also more fertile than the soils of the surrounding uplands.

Here and there along the sides of valleys, benches and terraces of alluvium stand above the level of the present floodplain. These are remnants of earlier floodplains that have been largely removed because of a renewal of downcutting by the streams. These alluvial terraces may be valuable agricultural lands and serve well for town sites and transportation routes because they stand above flood levels. *See* FLOODPLAINS.

Deltas. The surface features of deltas are essentially the same as those just described. As a rule, however, the delta surface is even less well drained than the floodplain surface and at its outer margin may merge with the sea through a broad belt of marshy land. As the delta grows, continual clogging of the stream mouth produces repeated diversion and bifurcation of the channel, so that the stream commonly discharges through a spreading network of distributaries (Fig. 4).

Many deltas, like those of the Mississippi, the Nile, the Danube, or the Volga, are immense fan-shaped features that have produced broad coastal bulges by their growth. Others, like those of the Colorado, the Po, or the Tigris-Euphrates, though no less extensive, are less apparent on the map because they have been built in large coastal embayments. Some great rivers have no true deltas because their sediment load has been dropped in some interior settling basin. For example, the Great Lakes remove most of the sediment from the St. Lawrence system and the Congo deposits most of its load in its broad upland basin.

Like floodplains, deltas are sometimes highly valued as agricultural lands, though they have the

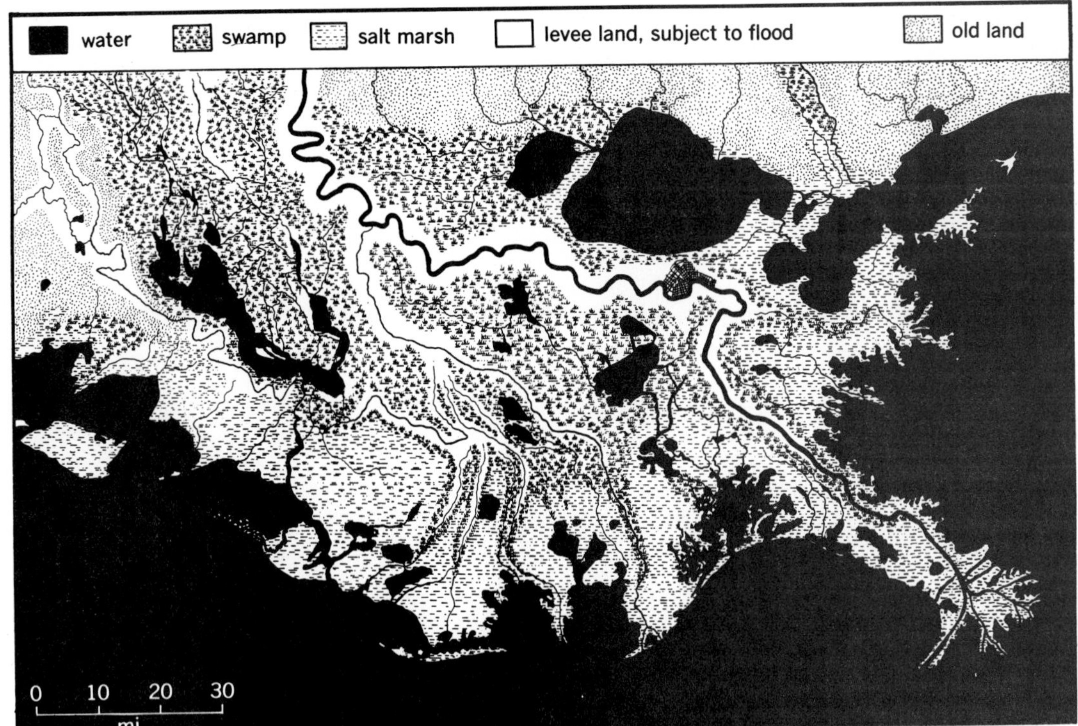

Fig. 4. Salt marsh, swamp, and natural levee lands in the Mississippi River delta. (*After V. C. Finch, adapted from G. T. Trewartha, A. H. Robinson, and E. H. Hammond, Elements of Geography, 5th ed., McGraw-Hill, 1967*)

same problems of poor drainage and frequent flooding. The Nile delta and the huge, silty delta plain of the Hwang Ho, in north China, are famous centers of cultivation. Many deltas, of which that of the Mississippi is a good example, are too swampy to permit tillage except along the natural levees. The Netherlands, occupying the combined deltas of the Rhine and Maas, stands as an example of what can be done toward reclamation when population pressure is great. *See* DELTA.

Alluvial fans. If a stream emerges upon a gentle plain from a steeply plunging mountain canyon, its

Fig. 5. Alluvial fans in the Mojave Desert, southeastern California. (*J. L. Balsley, U.S. Geological Survey*)

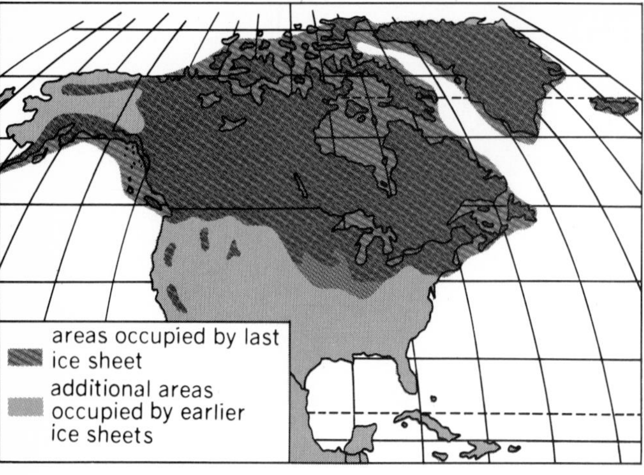

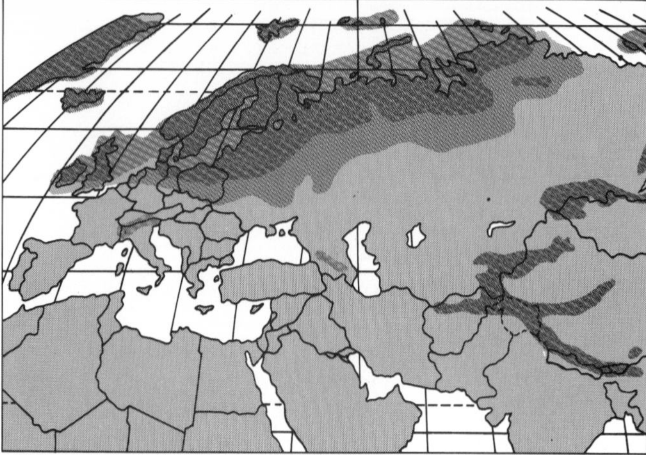

Fig. 6. Extent of former continental glaciers in North America and Eurasia. (*After R. F. Flint, from G. T. Trewartha, A. H. Robinson, and E. H. Hammond, Elements of Geography, 5th ed., McGraw-Hill, 1967*)

velocity is abruptly checked, and it deposits most of its load at the mouth of the canyon. Because of the tendency toward repeated choking and diversion of the channel, the deposit assumes the form of a broad, spreading alluvial fan, essentially similar to a delta, even to the diverging distributary channels (Fig. 5). Usually, however, the gradients developed are steeper than those on a delta because sediments are coarser.

Small individual alluvial fans are common features in mountainous country, especially where the climate is dry except for occasional torrential showers. Particularly significant, however, are the rows of alluvial fans that have coalesced to form an extensive, gently sloping piedmont alluvial plain along the foot of long, precipitous mountain fronts. The city of Los Angeles is built on such a plain. Still larger ones occupy much of the southern part of the Central Valley of California and stretch eastward from the Andes in northwestern Argentina, Paraguay, and eastern Bolivia.

Because of their smoothness and ease of tillage, alluvial fans, like other alluvial surfaces, are often especially amenable to cultivation. They are particularly significant in drier areas, partly because of the ease with which water may be conducted by gravity from the mountain canyon to any part of the fan, and partly because the thick, porous alluvium itself serves as a reservoir in which ground water is naturally stored.

Lake plains and coastal plains. Closely allied to stream-deposited plains are surfaces that represent former lake bottoms or recently exposed sections of the former shallow sea floor. These nearly featureless plains have been formed by the deposition of sediments carried into the body of water by streams or wave erosion and further distributed and smoothed by the action of waves and currents. Former beach lines and other shore features often provide the only significant breaks in the monotonous flatness. In some places, shallow valleys have been cut by streams since the surface became exposed. The lower parts of lake plains and the outer margins of coastal plains are often poorly drained.

The flat surfaces upon which Detroit, Toledo, Chicago, and Winnipeg stand are all lake plains, as are the famed Bonneville Salt Flats of western Utah. The south Atlantic and Gulf margins of the United States and much of the Arctic fringe of Alaska and Siberia are examples of newly emerged coastal plains. Some of these plains represent valuable agricultural land; others are excessively swampy or sandy. *See* COASTAL LANDFORMS; COASTAL PLAIN.

Features due to continental glaciation. On several occasions during the last 1,000,000 to 2,000,000 years (Pleistocene Epoch), immense ice caps, comparable to the one which now covers Antarctica, developed in Canada and Scandinavia and spread over most of northern North America and northwestern Eurasia (Fig. 6). The last such ice sheet (called in North America the Wisconsinan) reached its maximum extent about 18,000 years ago and did not disappear until 5000–6000 years ago. *See* GLACIAL EPOCH.

These great glaciers significantly modified the land surfaces over which they moved. Except near their melting margins, they were able to remove and transport not only the soil but also much weathered, fractured, or weak bedrock. This material was subsequently deposited beneath the ice or at its edge, wherever melting released it. The marks of glacial sculpture are unsystematically distributed swells and depressions; numerous lakes, swamps, and aimless streams (Fig. 7); and an irregular mantle of debris (glacial drift), some of which is clearly not derived from the local bedrock upon which it now lies. Such surfaces often contrast strongly with the systematic valley and divide patterns characteristic of stream erosion.

Most of the area that was covered by ice during the Pleistocene is now more or less thickly covered with drift, the thickest cover generally occurring in regions of weak rocks. Characteristic glacial landforms, however, are largely confined to those areas that were occupied by the ice during middle and late Wisconsinan times. Earlier glacial features have been modified beyond recognition by surface erosion and soil creep. Hence several large areas, including parts of the midwestern United States and southern Soviet Union, though drift-mantled, display the surface features characteristic of stream erosion.

Drift-covered plains offer a wide range of potentialities for human use, the critical characteristics being chiefly the adequacy of drainage and the texture of the soils. Excessive roughness is rarely a problem. *See* GLACIATED TERRAIN.

Till. Much of the drift represents mixed rock and soil deposited directly by melting beneath the ice sheet or at its edge. This material, called till, is as a rule most thickly deposited in the valleys and thinly over the ridge tops, thus reducing terrain irregularity. The surface of the till sheet itself is usually gently rolling, with poor and unsystematically patterned drainage. Hummocky, often stony ridges, called marginal moraines, mark lines along which the fluctuating edge of the ice remained stationary for a considerable time. In several localities, notably in eastern Wisconsin and western New York, are swarms of smooth, low drift hills called drumlins, all elongated in the direction of ice movement.

The surfaces of stony till plains are usually more irregular than those on clay till. Northeastern Illinois has a remarkably smooth surface developed on clay till, much of it eroded from the Lake Michigan basin. Eastern Wisconsin, northern Michigan, western New York, and southern New England, on the other hand, have more rolling surfaces underlain by till having a high content of stone and sand. In a few areas, especially in southern New England and in the marginal moraines elsewhere, the till is so stony as to impede cultivation.

Outwash. Some of the debris transported by the ice is carried out beyond the glacial margin by streams of meltwater. This material, called outwash, may be deposited as a floodplain (here called a valley train) along a preexisting valley bottom, or it may be spread broadcast over a preexisting plain. In either case the surface will usually be smooth, with features typical of alluvial plains. Unlike the heterogeneous, unsorted, and unstratified till, outwash material is usually layered and sorted in size. The fine material of silt and clay size is carried out downstream, leaving the coarser sands and gravels to form the outwash deposits. Most of the gravel and sand pits that abound in glaciated areas are developed in outwash plains.

Where outwash was laid down over already deposited till surfaces after the ice had melted back from its maximum extent, the surface of the outwash plain is sometimes pitted and lake-strewn, the depressions having formed as a result of the melting of relict ice masses that were buried by outwash deposition.

Patches and ribbons of outwash are common in glaciated areas, and in a few places, into which unusual quantities of meltwater were funneled, there are very extensive sandy plains. Noteworthy in this respect are the southern Michigan and northern Indiana area and Europe immediately south of the Baltic.

Lacustrine plains. Also present in and around the glaciated areas are numerous plains marking the beds of former lakes that resulted from the blocking of rivers by the glacial ice itself. During the melting of the last ice sheet, while the St. Lawrence and other northward-flowing rivers were still ice-dammed, the Great Lakes basins were much fuller than now and overflowed to the southward. When lake levels were eventually lowered, lacustrine plains were exposed, notably about Chicago, at the western end of Lake Erie, and about Saginaw Bay in Michigan. One of the most featureless plains of North America occupies the former bed of an immense lake (Lake Agassiz) that was

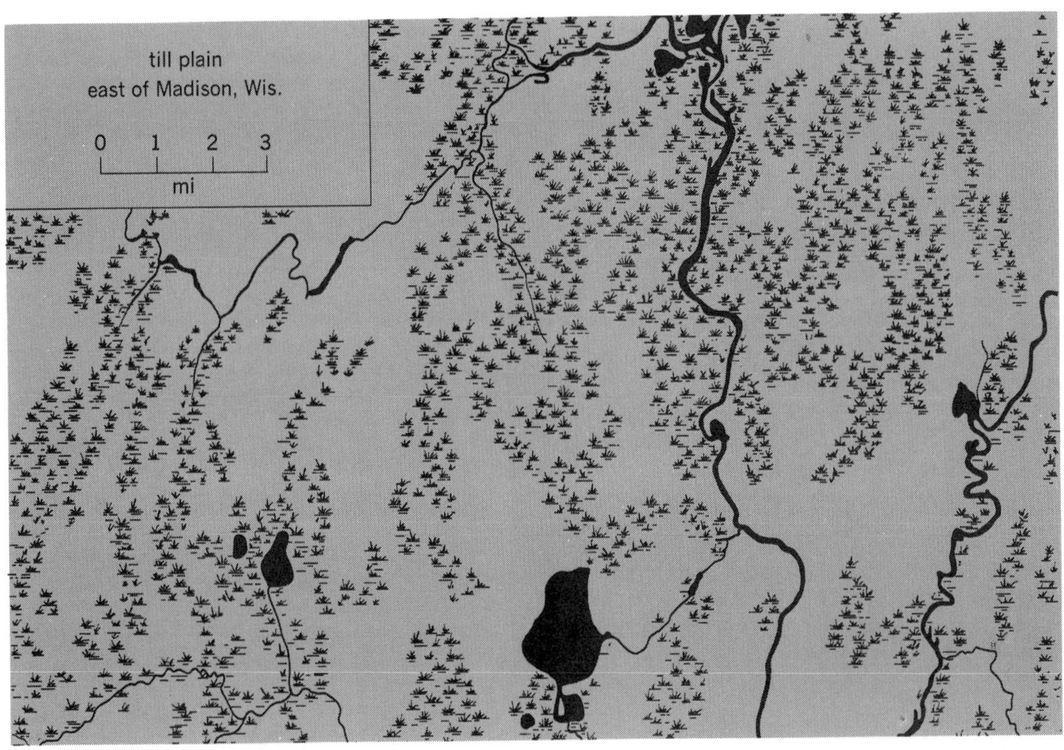

Fig. 7. Diagram of the drainage pattern of a glaciated plain in eastern Wisconsin. (*From G. T. Trewartha, A. H. Robinson, and E. H. Hammond, Elements of Geography, 5th ed., McGraw-Hill, 1967*)

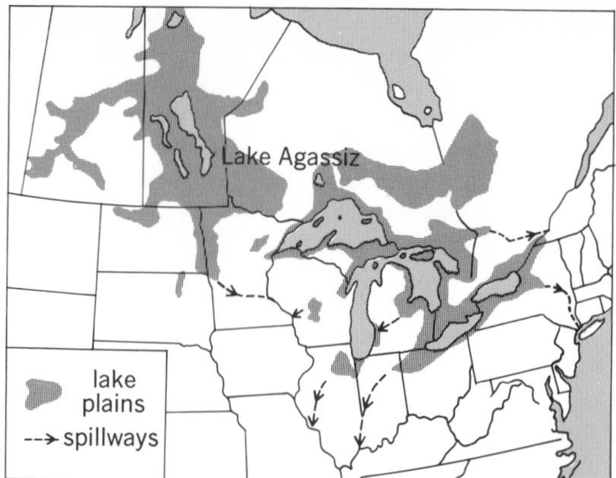

Fig. 8. Plains of former glacial lakes in North America. (*From G. T. Trewartha, A. H. Robinson, and E. H. Hammond, Elements of Geography, 5th ed., McGraw-Hill, 1967*)

present in late glacial time in southern Manitoba, northwestern Minnesota, and eastern North Dakota (Fig. 8).

Thin-drift areas. Considerable areas in central and eastern Canada and northern Fennoscandia have only a thin and discontinuous drift cover. These hard-rock regions yielded little debris and, especially in the rougher sections, the small drift supply was thinly strewn in the valleys and depressions. Lakes and swamps occupy drift-blocked valleys and shallow erosional depressions, rapids and waterfalls are common, and broad expanses of scoured bedrock are exposed. Such areas offer little opportunity for agricultural use and usually support only an open, patchy, and often stunted forest growth.

Features reflecting wind action. Features formed by the wind are less widespread and, as a rule, less obtrusive than the forms produced by streams and glaciers. Since the wind can attack only where the surface is almost free of vegetation, its work is strongly evident only in arid regions and in the vicinity of beaches and occasionally exposed river beds in more humid lands. Today plowed fields are also important prey for wind erosion.

The most striking and significant wind-produced features are sand dunes. Where sand is exposed to strong winds, it is moved about for short distances and accumulates in heaps in the general vicinity of its place of origin. Sand dunes assume many forms, from irregular mounds and elongated ridges to crescent-shaped hills and various arrangements of sand waves, depending apparently upon the supply of sand, the nature of the underlying surface, and the strength and directional persistence of the wind.

Nearly all the truly extensive areas of sand dunes are found in the Eastern Hemisphere, especially in the Sahara and in Arabia, central Asia, and the interior of Australia. Most of the dunes have been whipped up from alluvium that has been deposited in desert basins and lowlands.

In several regions of the world, notably in north-central Nebraska and in the central and western Sudan south of the Sahara, extensive areas of dunes have become covered by vegetation and fixed in position since they were formed, suggesting the possibility of climatic change. *See* DUNE.

Other wind-formed features are polished and etched outcrops of bedrock that show the effects of natural sand-blasting; gravel "pavements" resulting from the winnowing out of finer material from mixed alluvium; and shallow blowouts, which are depressions formed by local wind erosion.

The finer silty material moved by the wind is spread as a mantle over broad expanses of country downwind from the place of origin. Though usually thin, this mantle in place reaches a thickness of a few tens of feet, thus somewhat modifying the form of the surface. The extensive deposits of unstratified, buff-colored, lime-rich silty material known as loess are believed to have originated from such wind-laid deposits. Loess is abundant in the central United States, eastern Europe and southern Soviet Union, and in interior northern China. It yields productive soils but is subject to gullying erosion and has the facility of maintaining remarkably steep slopes, so that erosional terrain developed on deep loess is often unusually rough and angular. *See* LOESS.

INTERRUPTED PLAINS

Plains interrupted by some features of considerable relief occur widely and merit independent treatment. They may be divided into two contrasting groups: (1) tablelands, which are upland plains deeply cut at intervals by steep-sided valleys or broken by escarpments, and (2) plains with spaced hills or mountains, in which the excessive relief is afforded by steep-sided eminences that rise above the plain. Both types of surface, with their combination of plain and rough land, suggest histories of development that combine extensive gradation and strong tectonic activity.

Tablelands. These are essentially youthful plains that have been cut to unusual depth by valleys (Fig. 9). This requires that the plain shall have been brought to a level hundreds or even a few thousands of feet above the level to which streams can erode. In most cases, this elevation has been accomplished by the broad uplifting of an erosion-

Fig. 9. View of an ideal tableland: Canyon de Chelly National Monument, northeastern Arizona. (*Spence Air Photos*)

al or alluvial plain, but in a few instances the plain has been built to high level by the deposition of many thick sheets of lava.

It is also necessary that while a few major streams have cut deep canyons, large areas of the upland plain shall have remained largely uncut by tributary development, a condition requiring special circumstances. Such inhibited tributary growth is usually the result of either (1) slight local surface runoff of water because of aridity, extreme flatness of the upland, or highly porous material; or (2) the presence, at the upland level, of very resistant rock strata that have permitted only the most powerful streams to cut through. Hence tablelands are predominantly a dry-land terrain type. Those that do occur in rainier climates usually indicate the presence of an exceptionally resistant cap-rock layer.

The cliffs and escarpments that are common features of tableland regions are sometimes fault scarps, produced by the breaking and vertical dislocation of the crust during uplift. More often, however, they are simply steep valley sides that have been worn back long distances from their original positions under the attack of weathering and erosion. Sometimes a once extensive tableland will have been so encroached upon by the retreat of its bordering escarpments that nothing remains but a small, flat-topped mesa or butte.

Tablelands are the least widespread of the major terrain types, presumably because of the restrictive circumstances under which they can develop. The most extensive examples occur in the American continents. Especially noteworthy are the Colorado Plateaus, largely in northern Arizona and southern Utah, which represent a complex erosional plain that has been greatly uplifted and then deeply carved by the Colorado River and its major tributaries. Preservation of large sections of the upland plain has been favored by dryness and by the presence of nearly horizontal resistant rock strata. The few major streams that have cut deep canyons are all fed from moister mountainous areas round about.

The Columbia Plateau in eastern Washington is a porous lava plain, cut by the Columbia River and a few tributaries. The northern Great Plains, lying east of the Rocky Mountains from Nebraska north-

Fig. 10. Death Valley, Calif., a basin of interior drainage in the Basin-and-Range province. (*John S. Shelton*)

ward into Alberta, are an old erosional plain on weak rocks, now crossed by valleys of moderate depth that have been cut by streams issuing from the Rockies. The Cumberland Plateau in Tennessee is maintained by a resistant sandstone caprock. The Patagonian Plateau of southern South America is a somewhat similar surface, locally reinforced by extensive lava flows. Parts of the upland of interior Brazil, though in a moist environment, retain a tableland form because of the resistance of thick sandstone beds and, locally, lavas at the upland level.

Though there are significant exceptions, tablelands as a group suffer, in their economic development, from the difficulty of passage through their narrow gorges and across their numerous escarpments, and in most cases also from the dryness of their upland surfaces.

Plains with spaced hills or mountains. These are much more widespread than tablelands and are extensively represented in each continent. Surfaces included under the general heading vary from plains studded with scattered small hills and hill groups to mountain-and-plain country in which high, rugged ranges occupy almost as much space as the plains between them (Fig. 5).

Surfaces of the general type can be produced by two quite different lines of development. (1) They may be mountain, hill, or tablelands that have been brought to the erosional stage of early old age, in which case the isolated hills represent the only remnants of a once extensive highland. (2) They may represent areas in which separated hills or mountains have been constructed by volcanic eruption or by folding or buckling of the crust, the land between them having remained smooth from the outset, or having been smoothed by processes of erosion and deposition since the mountains were formed.

Examples of the first course of development are limited in the United States to small areas in southern New England and in the Appalachian Piedmont just east of the Blue Ridge. Many patches occur in the glacially scoured sections of northern and eastern Canada and similarly in northern Sweden and Finland. Extensive areas are found in the southern regions of Venezuela and Guiana, in the upland of eastern Brazil, and especially on the uplands of western, central, and southern Africa.

In such areas the plains are typical late-stage erosional-depositional surfaces, usually gently undulating. The remnant hills (monadnocks) rise abruptly from the plain, like islands from the sea. Although monadnocks commonly represent outcrops of unusually resistant rocks that have withstood erosion most effectively, many owe their exsistence solely to their position at the headwaters of the major stream systems, where they are the last portions of the highland to be reduced.

Surfaces on which spaced mountains have been constructed by crustal deformation or volcanic activity occur extensively in the great cordilleran belts of the continents, and rarely outside of those belts. The largest of all such regions is the Basin-and-Range section of western North America, which extends without interruption from southeastern Oregon through the southwestern United States and northern Mexico to Mexico City. The majority of the rugged mountain ranges of this section are believed to represent blocks of the crust that have been uplifted or uptilted and then strongly eroded. The plains between them are combination erosional-depositional surfaces, some of which have clearly expanded at the expense of the adjacent mountains. As the mountains have been reduced by erosion, there have evolved at their bases smooth, gently sloping plains that are in part erosional pediments, closely akin to old-age plains, and in part piedmont alluvial plains. Many of the basins have interior drainage, and in these the floor is likely to be especially thickly alluviated and to contain a shallow saline lake or alkali deposit in its lowest part (Fig. 10).

Other areas of somewhat similar terrain are found in the central Andes, in Turkey and the Middle East, and in Tibet and central Asia. The Tibetan and Andean sections are noteworthy for the extreme elevation (12,000–15,000 ft) of their basin floors. In general the ranges and basins in the Asiatic areas are developed on a grander scale than those in North America.

It is a curious circumstance that practically all these regions are dry and exhibit the intensely eroded mountain slopes and alluvially drowned basin floors characteristic of that climatic realm. For this reason they are only locally useful to man, in spite of the large amount of smooth land that they afford. There are, however, many important oases, usually near the bases of the mountains or along the courses of the few streams that have wandered in from moister adjacent regions.

[EDWIN H. HAMMOND]

Bibliography: W. D. Thornbury, *Principles of Geomorphology*, 2d ed., 1969; W. D. Thornbury, *Regional Geomorphology of the United States*, 1965; G. T. Trewartha, A. H. Robinson, and E. H. Hammond, *Fundamentals of Physical Geography*, 2d ed., 1968.

Planetology

The science of planetology studies all condensed matter in the solar system except the Earth (the other eight planets and their natural satellites, asteroids, and meteorites). The objective of planetology is to understand the nature, origin, and history of extraterrestrial bodies.

Many of the principles and techniques of geology, geochemistry, and geophysics are used in planetology. Of particular importance is a broad variety of remote-sensing methods (such as infrared imagery), chemical and mineralogical analyses of objects that occur on Earth but are probably of extraterrestrial origin (meteorites and tektites), and a multitude of data obtained from crewed and uncrewed space missions.

The growth of planetology as a science has been beneficial to both geologists and astronomers. For geologists, planetology provides a different perspective from which to view the Earth: not as an isolated body, but in relation to the rest of the solar system. Contrasts between the physical and chemical characteristics of the Earth and extraterrestrial condensed matter provide essential clues to the Earth's origin and early evolution, as well as to the nature of its interior. For astronomers, all theories which seek to explain the origin of stars, this solar system, and other solar systems, have in common the theoretical constraints imposed by the differences in density, chemistry, and orbital radius of the various condensed bodies of this solar system.

GENERAL CHARACTERISTICS OF THE SOLAR SYSTEM

The solar system consists of a single star (the Sun), its system of nine planets, their satellites (or moons), and numerous asteroids (small celestial minor planets, or "planetoids," in orbit around the Sun). The solar system is only one of billions of star-centered systems in the universe. All stars, star groups, and nebulae (clouds of interstellar dust and gas) occur in clustered groups called galaxies. The solar system is situated near the outer edge of the spirallike disk-shaped Milky Way Galaxy.

Planets and their satellites. The nine planets are distributed about the Sun at orbital distances which vary in an orderly fashion as a simple geometrical progression. This relationship is known as Bode's law or the Titius-Bode Law. If the asteroids in the belt between Mars and Jupiter are grouped together and considered as the remnants of a disrupted tenth planet (which seems likely), they occur precisely where Bode's law predicts they should be. In fact, it was Bode's law which, in part, led to their discovery.

More than 99% of the total mass of the solar system is concentrated in the sun, but almost 98% of the angular momentum is concentrated in the orbital motion of the large outer planets (Jupiter, Saturn, Uranus, and Neptune). All planets except Uranus rotate about their axes in a direction parallel to the axial rotation of the Sun; Uranus rotates in the reverse direction. Mercury and Venus actually rotate little, if at all, probably as a result of the drag exerted on them by solar tidal friction. All planets except Pluto and most of the planetary satellites revolve about the Sun or their principal in nearly circular, almost coplanar orbits which also mimic the manner of the Sun's rotation. The anomalous orbital behavior of Pluto and its eccentric small size and high density are puzzling. Pluto's bizarre character and that of some of the planetary satellites (notably the outer satellites of Jupiter and Saturn and all those of Neptune, which have either irregularly shaped orbits or retrograde motion) probably denote that they were captured from other principals after the initial formation of the solar system.

On the basis of their size, density, and orbital radius, all planets can be placed in one of two groups, the "inner" and "outer" planets (see the table). The inner planets (Mercury, Venus, Earth, and Mars) are Earth-like (terrestrial) in character. They are small in comparison with the outer planets, are dense, and are composed mainly of nonvolatile materials, especially silicon, magnesium, and iron (all in varying states of oxidation). They either almost totally lack an atmosphere or possess atmospheres which consist largely of heavier gases such as carbon dioxide, nitrogen, and oxygen. The outer planets (Jupiter, Saturn, Uranus, and Neptune) are much larger, but are less dense than the four terrestrial planets. They are commonly referred to as the Jovian (Jupiter-like) planets because Jupiter typifies their physical and chemical characteristics. The Jovian planets have densities so low that they are thought to consist mainly of light volatile materials, especially hydrogen. Their atmospheres are comparatively dense mixtures of hydrogen and helium.

Pluto does not fit neatly into either category. Like the outer Jovian planets, Pluto possesses a large orbital radius; it is the most remote planet in the system. However, Pluto is small and dense, like the four inner terrestrial planets. Pluto may actually at one time have been a satellite of Neptune, which was later torn from its original orbit and placed into a solar orbit. In any case, despite its remoteness from the Sun, Pluto is most commonly grouped with the four inner planets.

The planetary satellites also vary in number, size, and density. Some planets have no satellites, the Earth has its single Moon, and several planets have a number of satellites. In addition to its 10 moons, Saturn possesses a system of rings. The Moon is by far the largest satellite in relation to its primary, and the Earth-Moon system can be considered as a double planet rather than a planet and satellite. Generally the relationship between satellite size, density, and orbital radius imitates that of the planets themselves. Satellites with small orbital radii about their primary tend to be small and dense; more remote satellites have larger diameters and are less dense.

Planetary data

Body	Mean distance from Sun, 10^6 km	Orbital radius relative to Earth's	Equatorial radius, km	Mass (Earth = 1)	Mean density, g/cm^3	Composition of atmosphere	Number of satellites
Sun			696,000	332,958	1.41		
Mercury	58	0.382	2,420	0.054	5.41	Little, if any	0
Venus	108	0.723	6,150	0.815	4.99	CO_2(90%), H_2O, N_2	0
Earth	150	1	6,378	1	5.52	N_2(78%), O_2(21%), CO_2, H_2O	1
Moon	0.384 (from Earth)		1,738	0.012	3.34		
Mars	228	1.524	3,395	0.107	3.94	CO_2(80%), H_2O, O_2	
Asteroids	404	2.7			3.9		
Jupiter	778	5.203	71,400	317.89	1.33	H_2, He, CH_4, NH_3	12
Saturn	1427	9.539	59,650	95.14	0.71	H_2, He, CH_4, NH_3	10 plus rings
Uranus	2869	19.18	23,550	14.52	1.7	H_2, He, CH_4, NH_3	5
Neptune	4497	30.06	22,400	17.46	2.26	H_2, He, CH_4, NH_3	2
Pluto	5896	40? (eccentric)	2,950	0.1	5.5?	?	0

Asteroids, meteorites, and tektites. Theories about asteroids and the solar system in general are based largely on the study of meteorites (the science of meteoritics). Meteorites are small iron and stone objects which are found on the Earth but which originated elsewhere in the solar system. Based on their composition, their estimated arrival velocity, and the implications of Bode's law, meteorites are believed to be almost exclusively derived from the asteroid belt which lies between Mars and Jupiter and consists of innumerable small (diameters less than 800 km) subspherical bodies orbiting the Sun.

Two principal varieties of meteorites exist: iron (10% of all observed falls) and stony (90% of all observed falls). Iron meteorites (siderites) consist mainly of nickeliferous iron, a solid solution of iron and up to 40% nickel. Iron meteorites are thought to be similar in composition to the Earth's core. Stony meteorites (aerolites) resemble ultramafic rocks such as peridotite in composition. They consist largely or entirely of silicate minerals, mainly olivine, pyroxene, and plagioclase. Stony meteorites are thought to be similar in composition to the Earth's mantle. Stony meteorites can be further subdivided into chondrites and achondrites, depending on whether or not they contain chondrules (spheroidal, radially crystallized granules about 1 mm in diameter, thought to have originated as molten silicate droplets).

Similarities in composition and density between meteorites and the inner terrestrial planets suggests that asteroids should be collectively grouped with the terrestrial planets. Asteroids and meteorites are considered to be the remains of a disrupted planet or planets which formerly orbited the Sun between Mars and Jupiter. The existence of two main types of meteorites suggests that the parent body or bodies underwent an early history of density segregation analogous to that inferred for the Earth. A nickel-iron phase presumably segregated toward the center of the body or bodies to form a core; ultramafic silicates segregated outward and upward to form a less dense mantle. However, despite the general similarities between stony meteorites and the Earth's mantle and iron meteorites and the Earth's core, an exact equivalency probably does not exist. Variations in the relative proportion and exact composition of nickel-iron and silicate phases undoubtedly occur from planet to planet and between individual planets and average meteorites.

Tektites constitute an additional category of body which may be a type of meteorite. Tektites are small (usually 1–4 cm in diameter), variously shaped (for example, teardrop- or dumbbell-shaped) pieces of nonvolcanic, silica-rich glass found on Earth but believed to be extraterrestrial in origin. Tektites have not been explained to the complete satisfaction of all planetologists, but are believed to be fused splash droplets produced when meteorites impacted with the Moon. Some may represent splash droplets produced by meteorites impacting with the Earth. *See* METEORITE.

AGE OF THE SOLAR SYSTEM AND UNIVERSE

Most astronomers believe that the planets, their satellites, and the asteroids of the solar system formed at approximately the same time as the Sun, about 4.5×10^9 years ago. Several lines of evidence suggest such a date. The age of the Sun can be independently estimated by comparing its luminosity (brightness) and surface temperature (estimated on the basis of the Sun's surface color) with that of other stars, because astronomers generally believe that all stars pass through a series of evolutionary stages whose development is time-dependent. The size, luminosity, and surface temperature of the Sun suggest that its age (and that of the rest of the solar system) is between 4.5 and 5×10^9 years.

Conventional absolute age dating of both terrestrial rocks and meteorites supports this estimate. The oldest known rocks on Earth range in age between 3.4 and 3.5×10^9 years. However, these are crustal rocks, and if models inferred for the early Earth are correct, early-formed crust would have been continually remelted and recrystallized due to the initially large amounts of heat produced by large concentrations of radioactive isotopes. The age of the Earth's mantle should more closely coincide with the age of the Earth because theoretical considerations imply that segregation and stabilization of a core and mantle took place almost immediately after the formation of the planet. The age of the mantle can be indirectly estimated by analyzing the ratio of uranium isotopes and their decay products of lead (^{206}Pb and ^{207}Pb) in recent (post-Mesozoic), mantle-derived lead mineral deposits. The major uncertainty in this method is in predicting the amount of primeval (original) lead in such deposits. However, it is possible to estimate the abundance of primeval lead in either of two ways: by assuming that it is similar in abundance to the amount in iron meteorites or by studying local variations in ^{204}Pb, ^{206}Pb, and ^{207}Pb in various mantle-derived rocks. Both methods yield ages for the mantle of 4550×10^6 years. Independent radioactive age dating of meteorites, whose age is assumed to correspond to that of the Earth and other planets, also averages 4550×10^6 years. *See* EARTH, AGE OF; ROCK, AGE DETERMINATION OF.

The universe is thought to be considerably older than the solar system, having an age estimated at between 10 and 18×10^9 years. However, these age estimates are extremely tenuous. They are based on the assumption that the universe is expanding, an assumption which in turn requires acceptance of only one plausible explanation for the red shift phenomenon which is observed when analyzing light emitted from other stars. Red shift refers to the observation that spectral lines of light reaching the Earth after emission from stars in other galaxies shift systematically toward the red (longer wavelength) end of the spectrum. This red shift can be used to postulate a universe which is systematically expanding if recession of light-emitting sources from one another produces an apparent slowing in the velocity of light (visible as a lengthening of wavelength). The effect is comparable to the Doppler effect in sound waves. Acceptance of such an explanation allows for further extrapolation: Variation in the degree of red shift with distance suggests that the velocity at which other galaxies are receding from the Milky Way is directly proportional to distance. This situation is analogous to the separation of points stenciled on the surface of an inflating balloon. Assuming that the universe began with a cosmic explosion concentrated at a point source (the "big bang"), one simply extrapolates back in time, using present expansion rates, the amount of time required to return all galaxies to the single point source. Such a procedure yields age estimates for the universe ranging from 10 to 18×10^9 years. *See* UNIVERSE.

ORIGIN AND EVOLUTION OF THE SOLAR SYSTEM

The conspicuous regularities displayed by the solar system impose rigorous constraints within which theories to explain its origin and early evolution must be framed. These regularities include the identical manner of rotation and revolution exhibited by most planets and their satellites in imitation of the Sun's rotation, the coplanarity and circularity of planetary and satellite orbits, and the systematic differences among the planets and their satellites in size, orbital radius, mass, density, composition, and atmosphere, all of which have been previously discussed.

A number of theories have been proposed to explain these regularities and to document the origin of the solar system, but all fall into two major groupings: catastrophic theories and evolutionary theories. All catastrophic theories regard the solar system as a rare, almost unique phenomenon in the universe, invoking for its origin a single, catastrophic event: the close passage or collision of the Sun with another celestial body. In contrast, all evolutionary theories regard the solar system as one of many in the universe, explaining its origin as the logical consequence of the simple, sequential process believed to characterize the birth and growth of all stars: condensation of contracting masses of interstellar dust and gas (nebulae).

Catastrophic theories. Several different catastrophic theories have been proposed. Leading original proponents include G. L. L. Buffon (1749), T. C. Chamberlain (1901), F. R. Moulton (1905), J. A. Jeans (1917), H. Jeffreys (1918, 1929), and R. A. Lyttleton (1936). All versions specify an episode during which the Sun either collided with or passed very close to another body, probably another star. The resulting interaction purportedly pulled or drew off a filament, cloud, or tidal wave of molten, gaseous "planetesimal" or "protoplanet" material from the Sun, suspending it in a haphazard orbit about the Sun. With time the material cooled, condensed, and solidified. Denser portions coalesced to form planet nuclei, and individual nuclei subsequently grew by gravitationally accreting adjacent planetesimal material to their surface. The regularity in shape of planetary orbits is explained as a response to the resistivity exerted against planet revolution by the remaining, uncondensed portion of the filament.

Despite the large number of variations offered to improve catastrophic theories, a number of problems remain unresolved. For example, most astrophysicists believe that any filament of material drawn out of the Sun by its collision or close contact with another body would be quickly vaporized or dissipated. Also, none of these theories satisfactorily explains the systematic differences between the inner and outer planets. Most also fail to satisfactorily account for the concentration of the bulk of the angular momentum of the solar system in the orbital revolution of the outer planets. With so many objections, the various catastrophic theories find little current acceptance.

Evolutionary (nebular) theories. Versions of the evolutionary (or contracting nebula) theories were originally proposed by I. Kant in 1755 and modified by P. S. Laplace in 1796. Further refinements and modifications were proposed in the 20th century by C. von Weizsacker, F. Whipple, and G. Kuiper, among others. All versions assume the early existence of a large cloud of interstellar gas and dust which may have been initially produced by a massive cosmic explosion (the "big bang") more than 10^{10} years ago. Individual stars were produced periodically over the course of time as portions of the interstellar gas and dust coalesced into rotating globules dense enough to be gravitationally unstable. The gradual collapse of these individual globules released enormous amounts of heat, enough to cause each newborn star to radiate and be luminous. Eventually, internal temperature would be raised high enough to initiate self-perpetuating nuclear fusion reactions. The development of planetary systems around many of these stars would be a logical consequence of the process. Disks of material thrown off the rotating central globule, or secondary smaller eddies developed within the flattening mass itself, would eventually condense to form planets.

These theories are appealing in their simplicity and attractive in form because they describe processes which astronomers believe can be observed occurring elsewhere in the universe presently. However, they are somewhat disappointing in their vagueness and general descriptiveness. Nevertheless, many of the conspicuous regularities visible within the solar system can be explained by postulating such an origin. For example, the circularity and coplanarity of planetary orbits, and the coincidence between planetary orbital motion and axial rotation with that of the Sun's rotational motion are predictable, inheritable traits of a system which could have originated as a rotating, disk-shaped body of gas and dust which periodically threw off minor wisps of material. Theoretical models concerning the size of vortices developed in such a rotating, turbulent nebula indicate that the size of such vortices should increase outward from the center in a regular, orderly fashion, that is, in a pattern identical with the regularity specified by the Titius-Bode relationship. A major obstacle to the nebular hypothesis was how to account for the transfer of angular momentum originally concentrated in the Sun outward to the Jovian planets. This problem has been theoretically resolved by invoking the effects of viscous drag on the internal portion of the rotating ball or the possible interaction between the ionized part of the nebula and the solar magnetic field.

The nebular theories also allow for considerable chemical differentiation within the condensing solar nebula, a process obviously required by the contrasts in size, composition, and density of the inner and outer planets. Less volatile elements such as magnesium, iron, and silica would largely have been concentrated in the inner portion of the nebula, eventually condensing to form the inner terrestrial planets. More volatile constituents such as hydrogen and helium would have been selectively moved outward to be concentrated in the external portions of the nebula, gradually coalescing to form the low-density Jovian planets. Contrasts in atmospheric composition are probably explainable in terms of differences in planetary gravitational fields. The smaller, denser inner planets would have lost most of the lighter volatile constituents present in the original nebular material, retaining only heavier gases such as carbon dioxide, nitrogen, and oxygen (the oxygen in the Earth's atmosphere may have been largely produced organically from carbon dioxide). The larger, less dense outer planets retained an atmosphere rich in the gases which are abundant in the original solar nebula, hydrogen and helium. Finally, heat supplied by radioactive decay and the conversion of gravitational energy to thermal energy in the dense inner planets would have resulted in the rapid separation of an inner iron-nickel-rich core from a less dense silicate mantle, a condition found in the Earth's interior and suggested for terrestrial planets in general by contrasts in stony and iron meteorites.

[FREDERIC L. SCHWAB]

Bibliography: American Geological Institute, *Glossary of Geology*, 1972; M. H. Bott, *The Interior of the Earth*, 1971; P. Harris, The composition of the Earth, in I. C. Gass, P. J. Smith, and R. C. L. Wilson (eds.), *Understanding the Earth*, 1972; H. S. Jones, The origin of the solar system, in L. H. Ahrens, K. Rankama, and S. K. Runcorn (eds.), *Physics and Chemistry of the Earth*, vol. 1, pp. 1–16, 1956; K. Keil, Meteorite compositions, in K. H. Wedepohl (ed.), *Handbook of Geochemistry*, vol. 1, pp. 78–115, 1969; C. K. Seyfert and L. A. Sirkin, *Earth History and Plate Tectonics*, 1973; F. D. Stacey, *Physics of the Earth*, 1969; H. C. Urey, Boundary conditions for theories of the origin of the solar system, in L. H. Ahrens et al. (eds.), *Physics and Chemistry of the Earth*, vol. 2, pp. 46–76, 1957.

Plate tectonics

Plate tectonics theory provides an explanation for the present-day tectonic behavior of the Earth, particularly the global distribution of mountain building, earthquake activity, and volcanism in a series of linear belts. Numerous other geological phenomena such as lateral variations in surface heat flow, the physiography and geology of ocean basins, and various associations of igneous, metamorphic, and sedimentary rocks can also be logically related by plate tectonics theory.

The theory is based on a simple model of the Earth in which a rigid outer shell 50–150 km thick, the lithosphere, consisting of both oceanic and continental crust as well as the upper mantle, is considered to lie above a hotter, weaker semiplastic asthenosphere. The asthenosphere, or low-velocity zone, extends from the base of the lithosphere to a depth of about 700 km. The brittle lithosphere is broken into a mosaic of internally rigid plates which move horizontally across the Earth's surface relative to one another. Only a small number of major lithospheric plates exist, which grind and scrape against each other as they move independently like rafts of ice on water. Most dynamic activity such as seismicity, deformation, and the generation of magma occur only along plate boundaries, and it is on the basis of the global distribution of such tectonic phenomena that plates are delineated.

The plate tectonics model for the Earth is consistent with the occurrence of sea-floor spreading and continental drift. Convincing evidence exists that both these processes have been occurring for at least the last 600 million years (m.y.). This evidence includes the magnetic anomaly patterns of the sea floor, the paucity and youthful age of marine sediment in the ocean basins, the topographic features of the sea floor, and the indications of shifts in the position of continental blocks which can be inferred from paleomagnetic data on paleopole positions, paleontological and paleoclimatological observations, the match-up of continental margins and geological provinces across present-day oceans, and the structural style and rock types found in ancient mountain belts.

Plate motion and plate boundaries. Geological observations, geophysical data, and theoretical considerations support the existence of three fundamentally distinct types of plate boundaries, named and classified on the basis of whether immediately adjacent plates move apart from one another (divergent plate margins), toward one another (convergent plate margins), or slip past one another in a direction parallel to their common boundary (transform plate margins). Figure 1 shows the major plates of the lithosphere, the major plate margins, and the type of motion between plates. Plate margins are easily recognized because they coincide with zones of seismic and volcanic activity; little or no tectonic activity occurs away from plate margins. The boundaries of plates can, but need not, coincide with the contact between continental and oceanic crust. The nature of the crustal material capping a plate at its boundary may control the specific processes occurring there, particularly along convergent plate margins, but in general plate tectonics theory considers the continental crustal blocks as passive passengers riding on the upper surface of fragmenting, diverging, and colliding plates.

The velocity at which plates move varies from plate to plate and within portions of the same plate, ranging between 2 and 20 cm per year. This rate is inferred from estimates for variations in the age of the sea floor as a function of distance from mid-oceanic ridge crests. Ocean-floor ages can be directly measured by using paleontological data or radiometric age-dating methods from borehole material or can be inferred by identifying and correlating the magnetic anomaly belt with the paleomagnetic time scale.

Divergent plate margins. As the plates move apart from the axis of the mid-oceanic ridge system, the new volcanic material welling up into the void forms a ribbon of new material which gradually splits down its center as the boundary of plate separation continues to develop. Each of the separating plates thus accretes one-half a ribbon of new lithosphere, and in this way new lithosphere and hence new surface area are added. The process is considered to be continuous, and the boundary at which separation is taking place always maintains itself in the center of the new material.

The accretion at any spreading boundary is usually bilaterally symmetric. The morphology of the ridges is also quite symmetric and systematic. The new material that wells up at the ridge axis is hot and therefore expanded and less dense than the surrounding older material. Consequently, the new material is topographically highest. As new material divides and moves away from the ridge axis, it cools, contracts, becomes more dense, and subsides. The densification is caused by the combined effect of pure thermal contraction and thermally driven phase changes. Subsidence is fastest for newly generated oceanic crust and gradually decreases exponentionally with time. This observation explains the fact that in cross section the shape of the slope of the ridges is steepest at the ridge axis and gradually decreases down the flanks beneath the abyssal sediments and to the bounding continents. Since all known oceanic lithosphere has been generated by the spreading at a ridge axis, all oceanic lithosphere is part of the mid-oceanic ridge system. Because of the systematic way in which the morphology of the ridges is formed, almost all oceanic crust follows the same time-dependent subsidence curve within an error range of about 100 m. This means that the same age versus depth curve fits nearly all parts of the mid-oceanic ridge system. The ridge axis is found at a depth of 2.75 km ± 100 m, and oceanic lithosphere that is 30 m.y. old is found at a depth of 4.37 km ± 100 m. There are exceptional areas such as Iceland where the ridge axis actually is above sea level.

Magnetic lineations and age of oceans. Much of the evidence which leads to the development of the concept of sea-floor spreading, and in turn to an understanding of divergent plate margins and plate tectonics theory, came from analyses of the magnetic properties of the sea floor. Magnetometer surveys across the sea floor near mid-oceanic ridges reveal a pattern of alternating positive and negative magnetic anomalies (Fig. 2). The characteristics of these magnetic anomaly patterns (parallel to and in symmetrically matching widths across the ridge crest) and present-day general knowledge of the Earth's magnetism logically support the conclusion that sea-floor spreading occurs.

The Earth's magnetic field as measured at the surface may be approximately represented by an axial geocentric dipole, in other words, as if a simple two-pole magnet existed coincident with the Earth's rotational axis. The Earth's magnetic field reverses polarity episodically. Over an interval of time the strength of the dipole field gradually decreases to zero and then gradually increases in the opposite direction. The total transit time from full strength in one direction (polarity) to full strength in the opposite direction may be less than 6000 years. The residence time in any one particular polarity ranges from 10,000 years to tens of millions of years. Thus the polarity behavior of the Earth's magnetic field is approximately analogous to a randomly triggered flip-flop circuit.

The top 200 to 500 m of the lithosphere is the oceanic basalt, which is accreted at separating boundaries and which contains magnetizable iron minerals. As the newly injected material cools through its Curie point (~570°C), it becomes permanently magnetized in the direction of the Earth's magnetic field at that time and place. Because the oceanic crust is continuously being formed and transported in a bilaterally symmetric pattern away from the line of rifting, it is similar to a continuous magnetic tape record of the Earth's magnetic polarity events. At the mid-oceanic ridge axis the present polarity of the field is recorded. Down the ridge flanks progressively older oceanic crust is traversed, hence the crust contains in its remanent magnetism a continuously older record of the polarity of the field.

As a result of sea-floor spreading and polar reversals, the oceanic crust appears magnetically as a set of alternately normally or reversely magnetized strips of basalt arranged in a bilaterally symmetrical pattern about the mid-oceanic ridge axes. Because the rate of separation at most segments of the mid-oceanic ridge axes has varied only slowly with time, the spatial distribution of the stripes is proportional to the temporal history of the polarity of the field. Thus the polarity of the oceanic crust may be inferred indirectly from the pattern of magnetic anomalies.

Because the occurrence of polarity events is random, they form a unique sequence which is reflected directly in the magnetic anomaly patterns. A unique pattern of magnetic anomalies has been correlated throughout most of the world's oceans. From this pattern a magnetic reversal time scale from the present to 160 m.y. ago has been developed. This geomagnetic time scale has been calibrated by drilling on key anomalies in the oceans and paleontologically determining the age of the sediment-basalt interface (the JOIDES *Glomar-Challenger* Project). Thus a magnetic polarity time scale has been developed, and with it identifiable magnetic anomalies that can be correlated with part of the time scale may be assigned an absolute age. In this way, the spatial distribution of the magnetic anomalies is used to calculate the rates of separation of the plates in the past.

Convergent (destructive) plate margins. Because the Earth is neither expanding nor contracting, the increase in lithosphere created along divergent boundaries must be compensated for by the destruction of lithosphere elsewhere. The rates of global lithosphere construction and destruction must be equal, or the radius of the Earth would change. Compensatory destruction or removal of lithosphere occurs along convergent plate margins (subduction zones) and is accomplished by two mechanisms, plate subduction and continental collision.

Along subduction zones, one plate plunges beneath another (Fig. 3). The downgoing slab is usually oceanic because the relatively buoyant continental lithosphere cannot be subducted beneath the relatively more dense oceanic lithosphere. However, the upper or overriding plate may be continental lithosphere, as in the case of South America which is overriding the Nazca Plate (Fig. 1), or it may be oceanic lithosphere, as in the case of the Bering Sea which is overriding the Pacific Plate (Fig. 1).

The dip of the downgoing, underthrust slabs varies, but averages 45°. Earthquake foci along individual subduction zones define this zone of plate underthrusting, often referred to as the Benioff plane or Benioff zone. This plane dips away from oceanic trenches toward adjacent island arcs and continents, and marks the surface of slippage be-

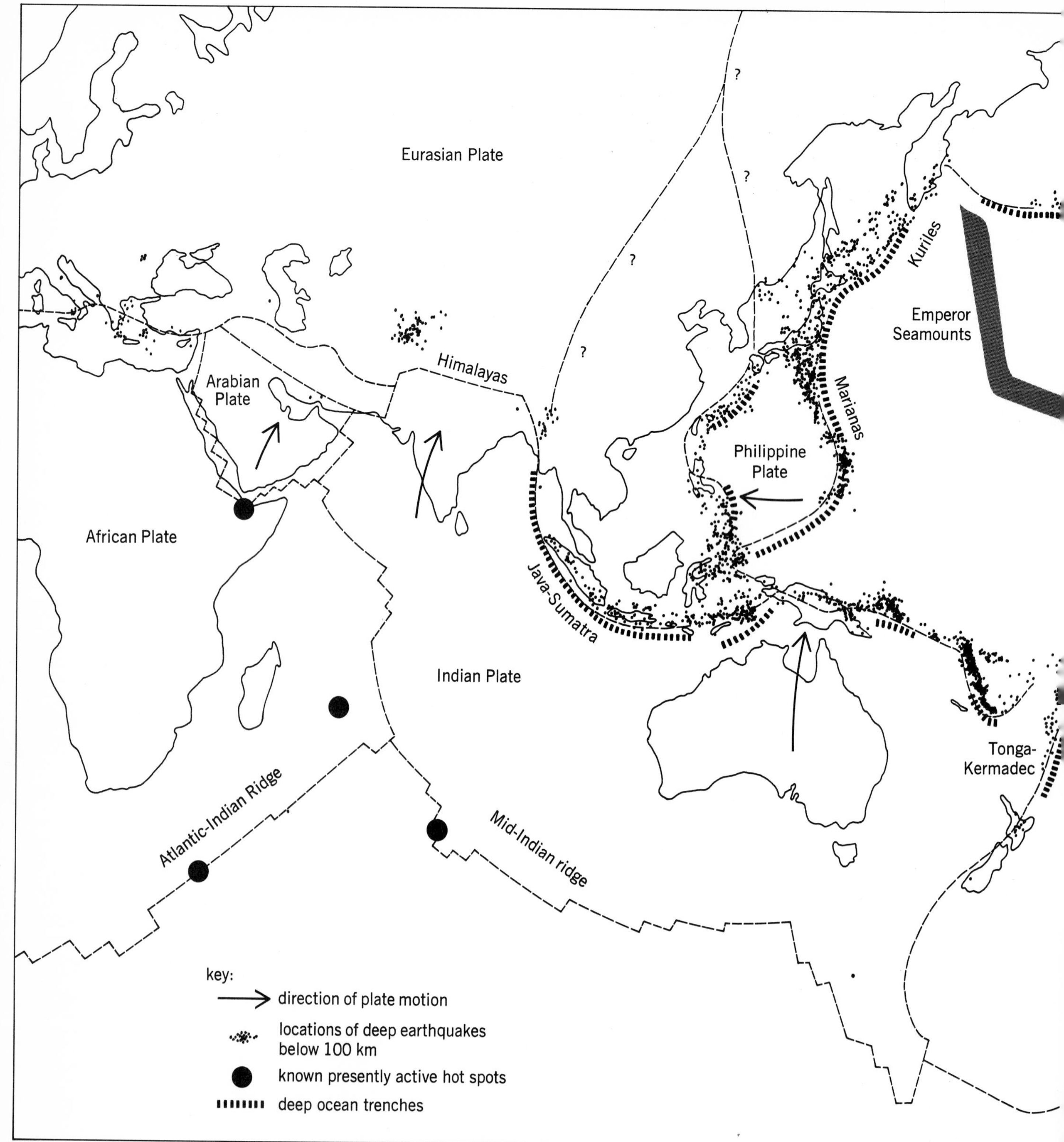

Fig. 1. Tectonic map of Earth. (*From M. Nafi Toksoz, The subduction of the lithosphere, Sci. Amer., 233(5):88–98, November 1975; copyright © 1975 by Scientific American, Inc.; all rights reserved*)

tween the overriding and descending lithospheric plates. Successive belts of shallow (less than 70 km), intermediate (70–300 km), and deep-focus (300–700 km) earthquakes define all subduction zones. Subduction zones are also characterized by active volcanism, and the development of deep-ocean trenches. These features occur today along the western border of South America and along most margins of the Western Pacific.

A line of andesitic volcanoes usually occurs on the upper plate, forming a chain that is parallel to the trench. The volcanism occurs at that point

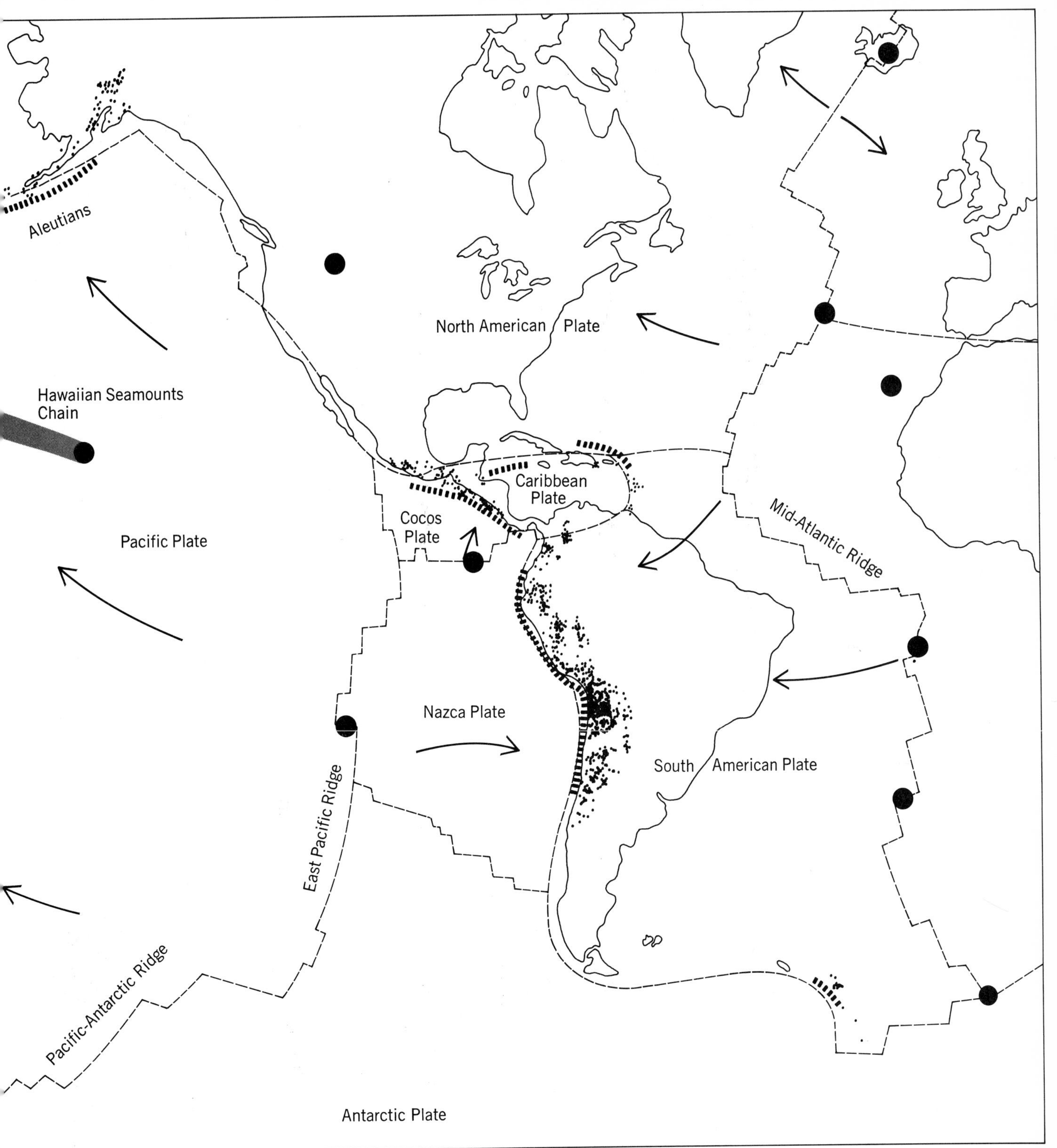

above the subduction zone where the upper surface of the downgoing lithosphere has reached a vertical depth of approximately 120 km. The volcanic materials arise from partial melting of both the downgoing and the upper slabs. The volcanoes of the Andes chain of South America and island arcs in the Pacific such as Japan and the Philippines have largely been formed in this manner.

Convergent plate margins which develop island arc systems adjacent to trench systems are geologically complex. The region between the volcanic island arc and the trench, the arc-trench gap, consists of masses of complexly folded and thrusted slabs of metamorphosed oceanic lithosphere and sediments. The slivers of oceanic lithosphere are probably derived from the downgoing slab, with

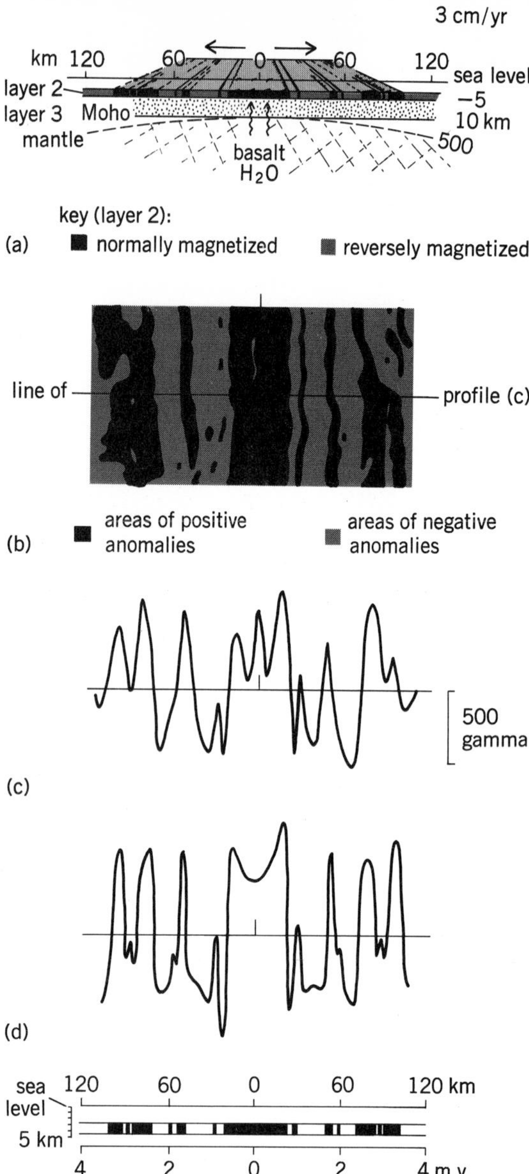

Fig. 2. The tape-recorder conveyor-belt scheme for sea-floor spreading. (*a*) Schematic representation of the crustal model, applied to the Juan de Fuca Ridge, southwest of Vancouver Island. (*b*) Part of a summary map of magnetic anomalies recorded over the Juan de Fuca Ridge. (*c*) Total-field magnetic anomaly profile along the line indicated in *b*. (*d*) Computed profile assuming the model and reversal time scale. Intensity and dip of the Earth's magnetic field taken as 54,000 gamma (0.54 oersted) and +66°; magnetic bearing of profile 087°; 10 km horizontally is equivalent to 100 gamma (10^{-3} oersted) vertically. Normal or reverse magnetization is with respect to an axial dipole vector, and the assumed effective susceptibility is ±0.01 except for the central block and ridge crest (+0.02). (*From F. J. Vine, Magnetic anomalies associated with mid-ocean ridges, in R. A. Phinney, ed., The History of the Earth's Crust, p. 75, copyright © 1968 by Princeton University Press; reprinted by permission*)

sediments either scraped off the downgoing slab or incorporated from the trench axis. Slumping and turbidity transport may cause some of the sediment to be transported back into the trench, but the general pattern of accumulation, thrusting, compression, and lift that occurs within the arc-trench gap often causes continental accretion.

Because of density consideration, for subduction to occur, at least one of the two converging plates must be oceanic. If both converging plates consist of continental lithosphere, neither plate is subducted and continental collision occurs instead. However, even continental collision provides the required compensatory reduction in lithosphere by folding and compressing lithosphere into narrower, linear mobile belts. In such collisions, sediments deposited along the continental margins and within the closing ocean basins are compressed into a series of tight folds and thrusts. Fragments of oceanic crust may be thrust up onto adjacent continental rocks (obduction) as ophiolite successions. A classic example of a continental collision belt is the Himalayan belt, produced during the Cenozoic Era by the convergence of the Indian continent with Eurasia.

Like plate divergence, plate convergence produces a distinctive suite of igneous rock types. Subduction zones are marked by the belts of predominantly andesitic volcanoes located either in island arcs located landward of the trench system (Japan and the Philippines) or along the rim of overriding continental blocks (the Andes belt). These andesitic volcanic terranes are commonly intimately associated with plutonic igneous rocks, mainly granodiorites. The origin of andesitic magmas, and the predominance of granodiorite plutons within continental blocks, to the exclusion of most other igneous rock varieties, was perplexing until the development of plate tectonics theory. Now both seem to be directly related to the generation of parent magmas by the frictional melting of ocean-floor basalt and overlying sediment cover along subduction zones. Partial melting of the lower crust and upper mantle probably also occurs.

Plate subduction and continental collision can also explain the origin of two other puzzling rock sequences commonly found within mountain belts: mélange and blueschist terranes. Mélange, a heterogeneous assemblage of intensely sheared, poorly sorted, angular blocks set in a fine-grained matrix, is probably generated at shallow depths along subduction zones as the oceanic crust and overlying sediment cover of the descending plate are scraped and crushed against the overriding plate. Blueschist terranes—so called because of the dark blue color imparted by the presence of various low temperature–high pressure metamorphic minerals such as glaucophane, lawsonite, and jadeite—occur in belts within mountain chains, parallel with but external to (toward the ocean) the more conventional paired metamorphic facies of the greenschist terrane. The peculiar physical conditions required by the blueschist facies, great burial depth (in excess of 20 km) but moderate temperature (200–450°C), should be generated along subduction zones when the descending lithospheric plate is underthrust at a greater rate than the local geothermal gradient can heat it.

Transform plate margins. Transform faults are always strike-slip faults. They occur where the relative motion between the two plates is parallel to the boundary that separates the plates. They may join a ridge to a ridge, a ridge to a trench, or a

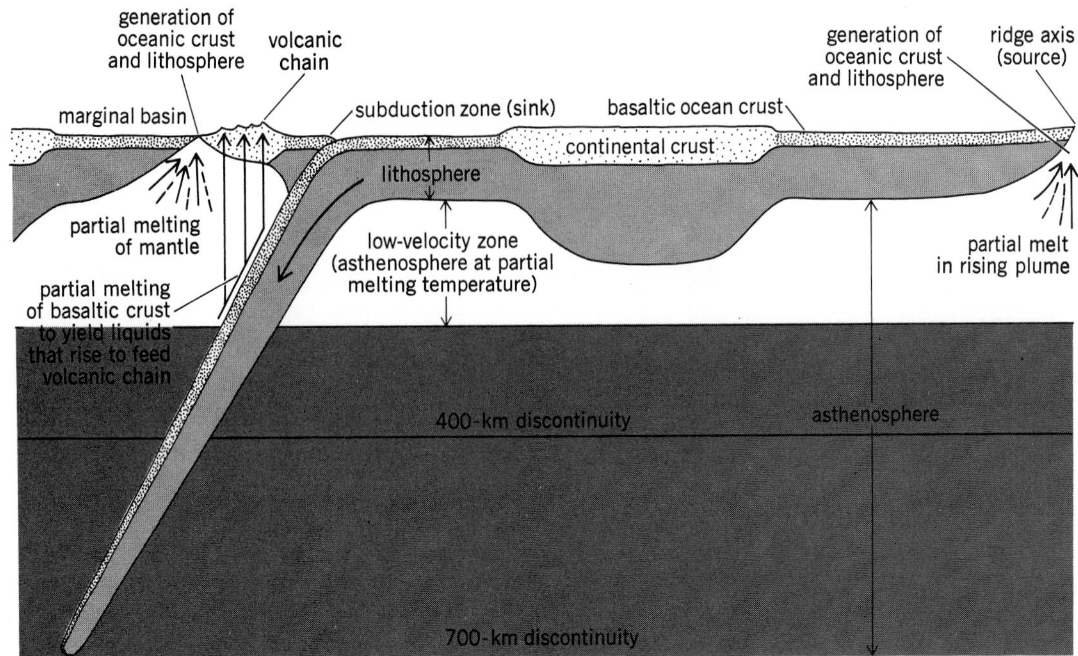

Fig. 3. Cross section of the upper mantle and crust showing a lithospheric plate riding on the asthenosphere. The continent is embedded in the plate and moves with it. Divergent plate margin is shown on the right; convergent plate margin with subduction is shown on the left, with Benioff zone defined by plate of shallow-, intermediate-, and deep-focus earthquakes produced where one plate underthrusts another. (*From J. F. Dewey, Plate tectonics, Sci. Amer., 226:56–68; copyright © 1972 by Scientific American, Inc.; all rights reserved*)

trench to a trench. Ridge-trench transforms will always change length with time. A trench-trench transform may either lengthen, shrink, or remain constant, depending on which of the plates that form the subduction system is the downgoing plate. A transform that joins two ridge axes will not change in length with time.

First consider ridge-ridge transforms (Fig. 4). Earthquake epicenter data show that earthquakes occur only along the ridge axes and the connecting transform. Studies of earthquakes that occur at a ridge-ridge transform fault show that the first motion was strike-slip parallel to the direction of the transform and opposite to the sense of offset of the ridge axis. If the ridge axes are offset left laterally, the relative motion across the transform that joins the ridges will be right-lateral as the plates separate. The motion of the fault takes place spasmodically. Deformation occurs on either side of the fault until the elastic limit is reached and the rupture occurs, causing an earthquake.

Fracture zones are morphologic scars that are fossilized transform faults. As such, they may be used to determine the past direction of relative motion of the bounding plates. In Figs. 5 and 6 the transform fault A-B offsets two ridge axes, and there is right-lateral strike-slip motion across the transform. All the material to the left of the ridge-transform-ridge boundary A′ABB′ belongs to the torsionally rigid plate I, and all the material to the right of this boundary, is part of plate II. Thus, beyond the ends of the transform, to the left of A′ and to the right of B, there is no relative horizontal motion.

Because of the way in which fracture zones are formed, they also preserve a record of the past direction of relative motion of the plates. The transform fault between plates I and II in Figs. 5 and 6 is in theory a vertical plane between these plates along which horizontal strike-slip motion occurs. The face of plate I that abuts against plate II in effect defined this transform plane. But as plate I moves to the left this transform plane or face passes by the end of ridge A′A at point A, and new, younger material is intruded against it. Together the transform plane of plate I and the new, younger material move off to the left. This process is continuous. The new, younger material is more elevated and forms an escarpment whose horizontal trace records the direction of relative movement. Because the rate of subsidence decreases with time, the difference in elevation across a fracture zone decreases with age.

A very narrow and deep rift valley is also associated with the transform. However, because of time-dependent cooling, the oceanic lithosphere contracts horizontally. This horizontal contraction causes cracking of the lithospheric plate and also causes widening and splaying of the rift associated with the fracture zone. Young fracture zones are characterized by an escarpment and a deep fissure along their length. The magnitude of the vertical offset decreases with age, and the fissure tends to widen. Down the lower flanks of the ridges, turbidites from the abyssal plains often finger into the fracture zone. The fracture zones form long, curvilinear features throughout the oceans and are mappable on the basis of morphology alone.

The numerous strike-slip faults which offset segments of the mid-oceanic ridge system are classical examples of transform plate boundaries. The San Andreas system of California, which offsets portions of the East Pacific Rise, is probably the best-known example. Transform faults show ap-

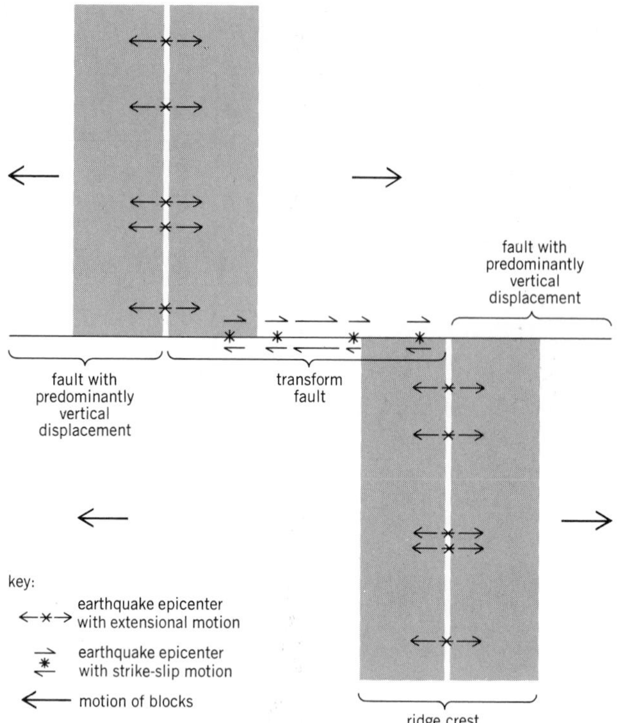

Fig. 4. Idealized map view of a transform fault. The ridge appears to be offset in a left-lateral sense, but the motion on the transform fault between the ridge crests is right-lateral. *(From C. K. Seyfert and L. A. Sirkin, Earth History and Plate Tectonics, copyright © 1973 by Carl K. Seyfert and Leslie A. Sirkin; used by permission of Harper & Row, Publishers, Inc.)*

parent lateral displacements of many tens or even hundreds of kilometers. Occasionally extension, and consequent igneous activity, may occur at a transform fault.

Transform faults are parallel to the direction of relative motion between the bounding plates. According to a theorem of spherical geometry, if plate 1 is moving with respect to plate 2 on a sphere, the instantaneous relative motion may be represented by the rotation of one plate with respect to the

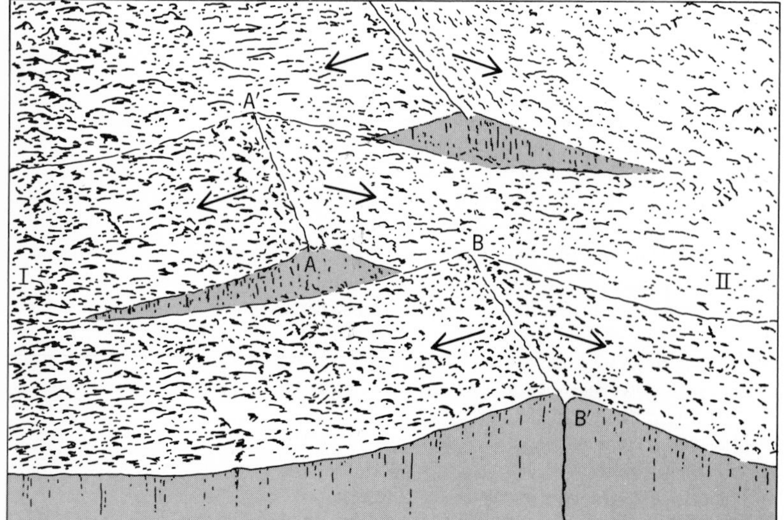

Fig. 5. Ridge-ridge transform fault appears between two segments of ridge that are displaced from each other. *(From H. W. Menard, The deep-ocean floor, in J. Tuzo Wilson, ed., Continents Adrift and Continents Aground, W. H. Freeman and Co., 1976)*

other about a single stationary pole. Circles drawn on the sphere concentric to the pole of relative motion will be parallel to the direction of instantaneous relative motion. Great circles drawn perpendicular to these concentric circles will intersect at the pole of instantaneous relative motion. The transform faults that separate block 1 from block 2 are parallel to the direction of present-day relative motion and therefore must lie on circles that are concentric to a pole of relative motion of 1 with respect to 2. Great circles drawn perpendicular to the transform faults will intersect at the pole of relative motion. In this way the pole of relative motion may be found for any two plates whose common boundary consists in part of transform faults.

The instantaneous rate of relative motion is an angular rate of rotation of 1 with respect to 2 about an axis through the pole of relative motion and the center of the Earth. The linear rate of relative motion R (in centimeters per year) will then be proportional to the sine of the colatitude with respect to the instantaneous pole, as shown in the equation below, where a = the radius of the Earth, 6×10^8 cm; θ = the colatitude of the point at which the rate is to be determined with respect to the instantaneous pole; and $d\phi/dt$ = the angular rate of motion of 1 with respect to 2.

$$R = a \sin \theta \frac{d\phi}{dt}$$

In this way, spreading-rate data calculated from the distribution of magnetic lineations near the axis of a spreading ridge-transform system may be used to compute a pole of relative motion. Note that in this case the pole is not strictly instantaneous because magnetic lineations covering a finite time period must be used. It has been found that the poles and rates of relative motion tend to remain nearly constant, perhaps changing quite slowly over long time intervals. Thus, poles and rates computed in the above fashion may often be used to describe the relative motion for several tens of millions of years.

The fracture zones may often be treated as fossil transform faults. To obtain paleopoles of relative motion, segments of fracture zones spanning the same age and from the same plate must be used. Perpendicular great circles drawn to these segments will give a pole of relative motion for that time interval in a reference frame fixed with respect to that plate. The magnetic anomaly lineations that abut the fracture zone may be used to determine the rate of motion. By calculating a time sequence of poles and rates using time sequential segments of fracture zones, the spreading history of an entire ocean basin may be determined.

Triple junctions. The point where the boundaries between three pairs of plates join is called a triple junction. Because the Earth's lithosphere shell is segmented into a mosaic, the boundary between any two plates must end in a triple junction. Quadruple junctions are theoretically possible, but except under very unusual circumstances they degenerate immediately into two triple junctions. Examination of Fig. 1 shows that there are a number of triple junctions which vary as to the types of boundaries that meet and as to the geometry.

When three bodies are moving with respect to

each other across a spherical surface, if the relative motion of each of two pairs of the plates is known, the relative motion of the third pair may be determined. In the case of a spherical surface, the relative motion between two plates may be represented by a vector through the pole of instantaneous relative motion, with the length of the vector proportional to the rate of rotation. The relative motion of three plates or bodies on a sphere may be represented by three such vectors. Each vector gives the instantaneous relative motion of two of the three plates. Thus three such vectors describe completely the relative motion of the three plates and must add to zero.

The orientation of the boundary between these plates will determine the type of interaction at the boundary (that is, spreading, subduction, or transform). There are a number of possible triple junctions. The Cocos, Pacific, and Nazca plates join at a ridge-ridge-ridge triple junction (Fig. 1). This is an example of a stable triple junction, that is, the triple junction moves in a constant direction and at a constant speed with respect to each of the plates. Other triple junctions are unstable or transitional, such as a ridge-ridge-transform triple junction which will change immediately to a transform-ridge-transform triple junction. The most geologically important triple junctions are those that migrate along a boundary. As a consequence, the boundary along which the triple junction moves will experience a change in the direction of relative motion. If the change in relative motion is radical, there may be an accompanying change in tectonic style. This type of plate interaction occurred at the western boundary of the North American Plate during the Cenozoic, and has resulted in a time transgressive change, from south to north, from subduction to strike slip. This is reflected in the geologic record as a change from compressive tectonics with thrusting, folding, and Andean or island arc volcanism to strike slip, as occurs along the San Andreas Fault.

Paleogeography. Because the accretion of new lithosphere is bilaterally symmetric, the associated magnetic lineations must also be bilaterally symmetric across a ridge axis. Because, once formed, oceanic crust is rarely disturbed by horizontal shear, each magnetic lineation preserves the shape of the mid-oceanic ridge axis and offsetting transforms at the time of its formation. The fact that lineations of the same age from opposite sides of a ridge axis may be fitted together is considered to be evidence of the undisturbed nature of the oceanic crust.

The North and South Atlantic oceans were formed by the rifting and drifting of the continents that now surround the Atlantic. These continents have been rafted as part of the separating lithospheric plates since the Triassic as the Atlantic Ocean slowly opened. The magnetic lineation pattern in the North Atlantic records this history of separation of Africa, Europe, and North America. When magnetic lineations of the same age from opposite sides of the ridge axis are fitted together, the relative positions of the continents at the time the lineations formed are obtained. This is equivalent to reversing the process of sea-floor spreading. In effect, the material younger than the two lineations is removed, and the older portions of the plates with the continents move toward each other until the two lineations fit together.

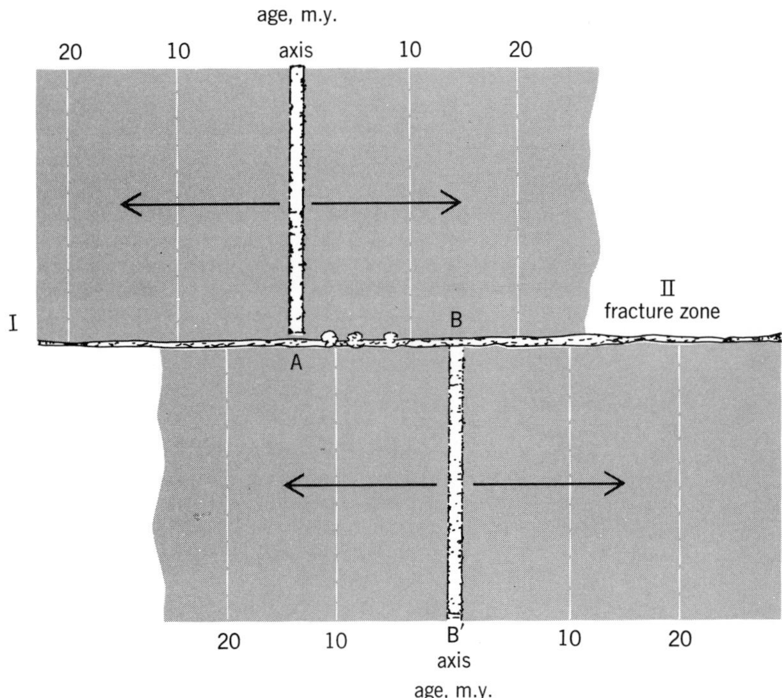

Fig. 6. Molten rock wells up from the deep Earth along a spreading axis, solidifies, and is moved out (shown by arrows). The axis is offset by a transform fault. Between two offset axes, material on each side of the transform fault moves in opposite directions, causing shallow earthquakes. *(From J. R. Heirtzler, Sea-floor spreading, Sci. Amer., vol. 219, no. 6, December 1968, copyright © 1968 by Scientific American, Inc.; all rights reserved)*

This technique is applicable to any ocean where separation at ridge axes has led to the passive rafting of the surrounding continents, such as the southeast and central Indian Ocean, the South and North Atlantic, the Labrador and Norwegian seas, and the Arctic Ocean.

The above technique gives the position of the continental land masses relative to each other, not their position relative to the rotational axis or paleomagnetic pole. This must be determined from paleomagnetic data. It is assumed that the Earth's magnetic field, now approximated by an axial geocentric dipole, has been similarly oriented throughout geologic time. Subaerial volcanic rocks acquire a remanent magnetism in the same way as oceanic crust, by cooling below the Curie point. Subaerial sedimentary rocks may also be remanently magnetized by the process of detrital remanent magnetization. As iron-bearing particles settle from moving currents, their orientation is influenced by the prevailing magnetic field. By measuring the direction of the remanent magnetism of both igneous and sedimentary rocks whose ages can be determined, the position of the paleomagnetic pole for that age can be found. From these data, the paleolatitude may be found for the set of continental land masses whose position relative to each other is known (Fig. 7).

Sea-level changes. Many important geologic phenomena may in part be related to plate tectonics processes, for example, changes in sea level and climate and changing patterns of evolution. The geologic record contains evidence of almost continuous eustatic sea-level changes throughout

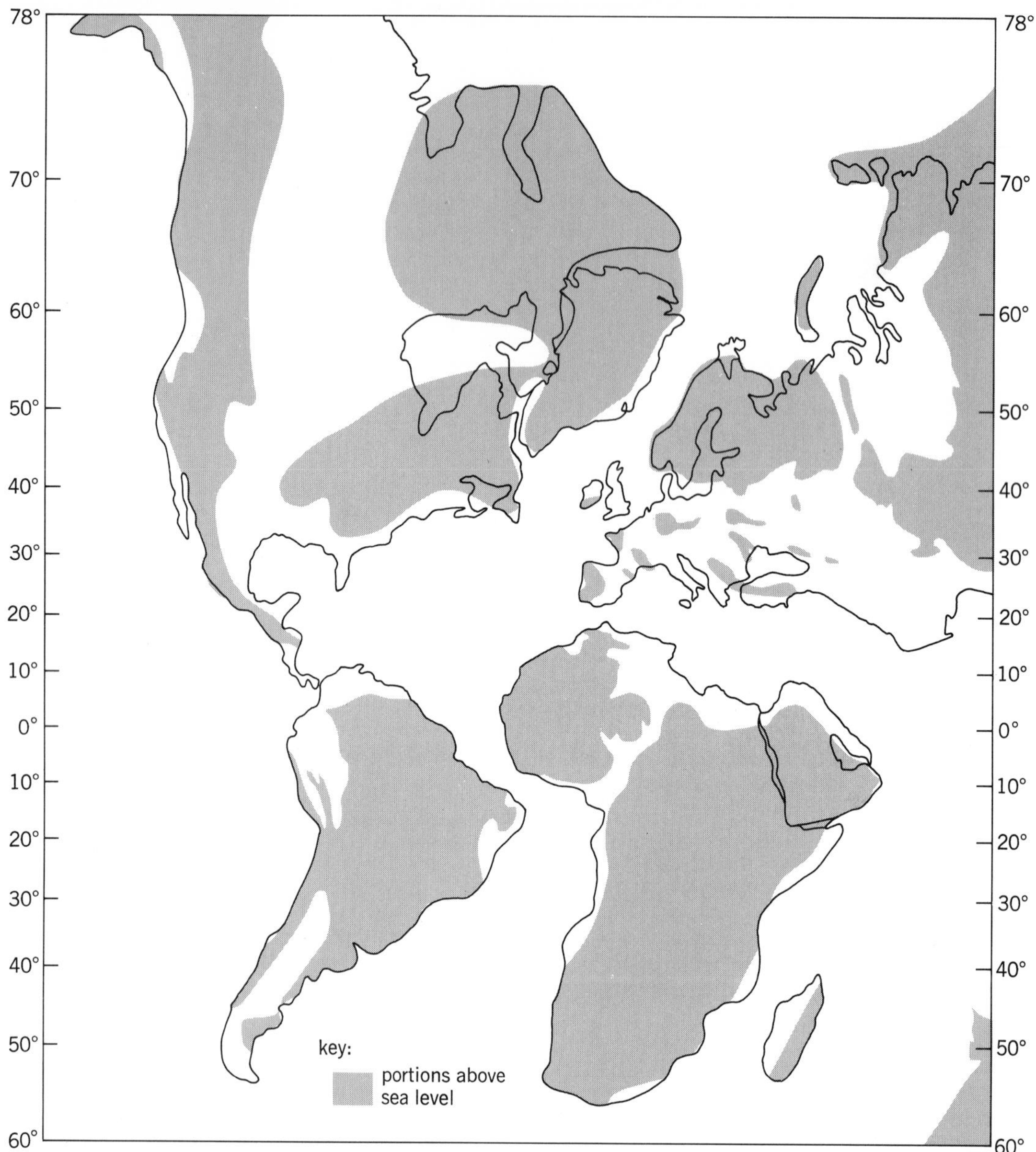

Fig. 7. Map illustrating the paleogeographic arrangement of the Atlantic continents during Coniacian time. Sea level may have been over 300 m above present level at that time because of the volume of the mid-oceanic ridge system.

the Phanerozoic. Several large and rapid sea-level changes can be attributed to fluctuations in the volume of massive continental glaciers. However, it is apparent that large sea-level changes (greater than 100 m) also occurred during periods when continental glaciers were small or nonexistent.

Several factors affect the volume of the ocean basins or the ocean waters and can thus alter eustatic sea level. One of the most important of these in terms of magnitude and rate of sea-level change is variation in the volume of the mid-oceanic ridge system. As explained above, once formed, the sea floor subsides systematically with time, with all oceanic crust following the same exponential subsidence curve. The volume of the mid-oceanic ridge system is quite large. If sea-floor spreading were to cease today, 70 m.y. from now the ridges would have subsided sufficiently to decrease the depth of the water over the abyssal area by about 450 m. The continental freeboard would increase by 320 m.

The volume of the mid-oceanic ridge system may be altered in several ways, one of which is to change the spreading rate. Because all ridges follow the same subsidence curve (which is a function of time only), age versus depth relationships are the same for all ridges. This means that if two ridges have been spreading at a constant but different rate for 70 m.y., the ratio of their volumes per unit length will equal the ratio of their spreading rates. If the spreading rate of the faster ridge is reduced to that of the slower ridge, volume/unit length of the larger ridge will gradually be reduced to that of the slower ridge. If spreading rates decrease, ridge volume also decreases, increasing the freeboard of the continents. The converse is also true: by increasing spreading rates, ridge volume will increase and the freeboard of the

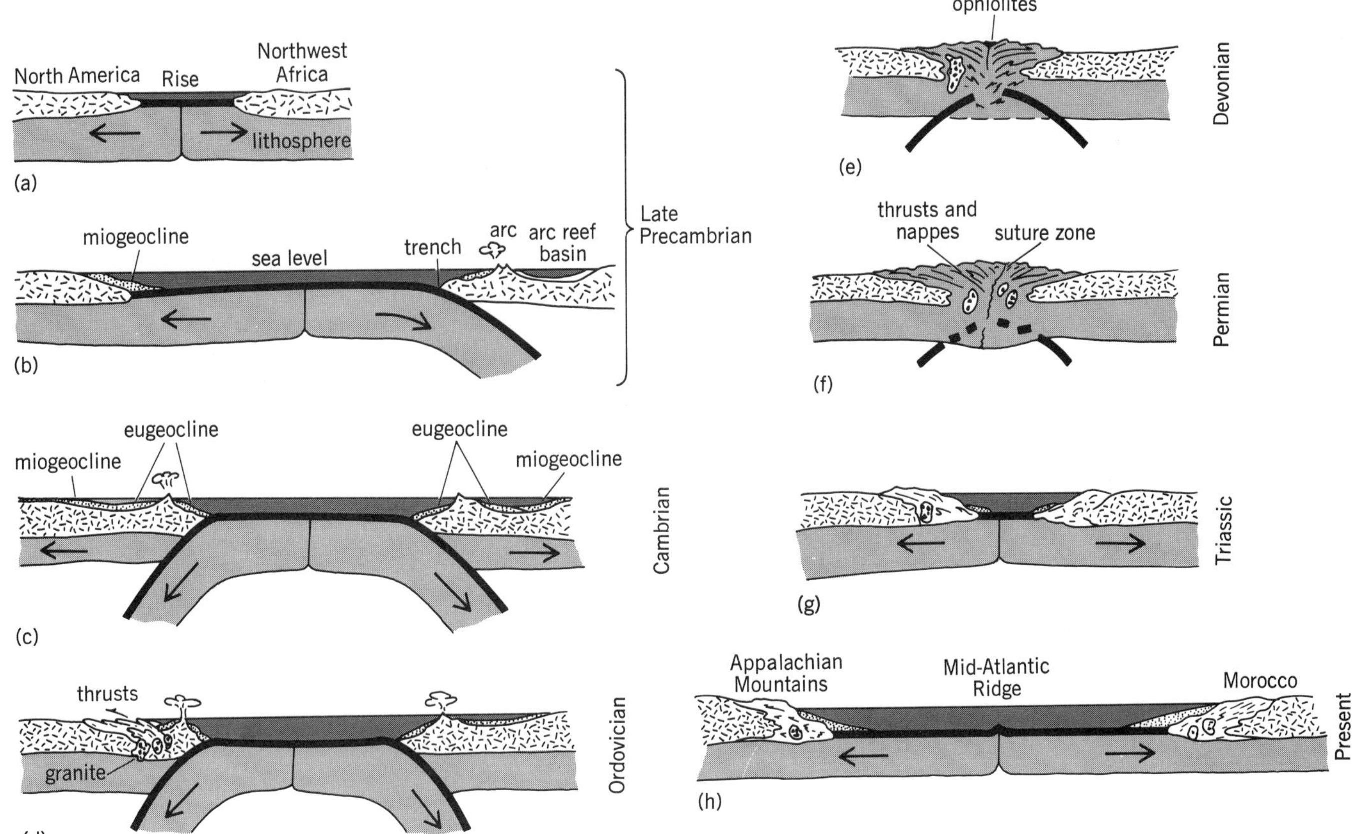

Fig. 8. Schematic representation (*a–h*) of the plate tectonics history of the Appalachian orogenic belt. (*From K. C. Condie, Plate Tectonics and Crustal Evolution, Pergamon, 1976*)

continents will be reduced. Several other ways of changing the volume of the mid-oceanic ridge system exist: ridges may be destroyed; segments of the ridge system may be subducted; and new ridges may be created by continental rifting and new rifting within ocean basins.

Both the length of the mid-oceanic ridge system and the spreading rate at its various segments have varied considerably during Phanerozoic time. These changes have caused large variations in sea level. For example, by using magnetic anomaly data to calculate spreading rates and ridge lengths back to the Upper Cretaceous (Coniacian), it has been estimated that sea level then may have been as much as 300 m above present (Fig. 8).

Hot spots and mantle plumes. The existence of convective plumes originating in the deep mantle from below the level of the asthenosphere and rising to the bottom of the lithosphere has been proposed. About 14 major convective plumes are believed to exist today (see Fig. 1). These plumes are believed to be nearly stationary with respect to each other, and hence they may be used as reference points, with respect to which all the plates are moving. Mantle plumes, rather than convection currents, may also be the driving mechanism for plate motion. The latter two hypotheses are not necessary conditions for the validity of the first. In other words, there may be mantle plumes that rise to the bottom of the lithosphere but they may not be stationary with respect to each other and may not furnish the plate driving mechanism. The major clue to the existence of hot spots is the lines of intraplate volcanoes that are left as a trace of the passage.

As a plate passes over a hot spot, the hot spot burns its way through the plate. This releases volatile and eruptive magma on the plate surface. A classic example of the "volcanic" consequences of passage of a plate over a hot spot is the Emperor Seamount—a Hawaiian Island chain. Hawaii and the islands immediate to it are volcanically active at present. The rest of the islands and seamounts of the Hawaiian and Emperor chains are inactive, but all are of volcanic origin. The islands and seamounts are found to be sequentially older west-northwestward along the Hawaiian chains and are progressively older north-northwestward along the Emperor Seamount chain. As an explanation for this, it has been hypothesized that the Pacific Plate moved first north-northwesterly and then west-northwesterly over a single hot spot. Other seamount chains in the Pacific such as the Line Islands, the Tuamotos, and the Austral Seamount chain are presumed to be of similar origin.

In the Atlantic, features that are regarded as major hot spot traces are Iceland and the Iceland Faroes Ridge, the New England Seamount chain, the Columbia Seamount chain, the Rio Grande Rise, and Walvis Ridge; and in the Indian Ocean, the Ninety-East Ridge is an analogous feature. The age progression along each of these features has been estimated; the theory that these hot spots have remained fixed in position relative to each other has been tested. The results indicate that some relative motion may occur, but it appears to be less by an order of magnitude than the rate of interplate motion that is generally observed.

The hot spot hypothesis is a significant complement to plate tectonics. There seems to be little doubt that hot spots or plumes of some type do occur. Intraplate volcanism often seems to be a manifestation of their existence. If, in addition, the hot spots prove to be essentially stable with respect to the rotational axis, they may be useful as a reference frame. They could serve as a latitudinal constraint in addition to paleomagnetic data. They would also be very useful in finding the relative motion between plates that have been separated by subduction zones or transform faults (that is, Pacific – North America relative motion).

Plate tectonics and Earth history. Not only does plate tectonics theory explain the present-day distribution of seismic and volcanic activity around the globe and physiographic features of the ocean basins such as trenches and mid-oceanic rises, but most Mesozoic and Cenozoic mountain belts appear to be related to the convergence of lithospheric plates. Two different varieties of modern mobile belts have been recognized, cordilleran type and collision type. The Cordilleran range, which forms the western rim of North and South America (the Rocky Mountains, Pacific Coast ranges, and the Andes) have for the most part been created by the underthrusting of an ocean lithospheric plate beneath a continental plate. Underthrusting along the Pacific margin of South America is causing the continued formation of the Andes. The Alpine-Himalayan belt, formed where the collision of continental blocks buckled intervening volcanic belts and sedimentary strata into tight folds and faults, is an analog of the present tectonic situation in the Mediterranean, where the collision of Africa and Europe has begun.

Abundant evidence suggests that sea-floor spreading, continental drift, and plate tectonics have occurred for at least the past 600 m.y., that is, during Phanerozoic time. Furthermore, it is probable that plate tectonics phenomena have dominated geologic processes for at least 2.5 billion years (b.y.). Present-day ocean basins are young (post-Paleozoic) features, and their origin is adequately explained by sea-floor spreading. The origin and evolution of the continental blocks, substantial portions of which are Precambrian in age, may probably be explained by accretion at subduction zones. Continental blocks, though largely composed of granite and granodiorite, appear to be mosaics of ancient mobile belts which progressively accreted laterally, vertically, or both, through time. Many late Precambrian and Paleozoic mobile belts, such as the Appalachian-Caledonian belt of the North Atlantic region, are interpretable in terms of plate tectonics mechanisms (Fig. 8). Initial rifting breaks a preexisting supercontinental landmass into two or more fragments which drift apart at the flanking margins of the growing ocean basin. At the margin of the growing basin, continental shelves, continental rises, and abyssal plains accumulate sediments. Cessation of sea-floor spreading and the development of new convergent margins along one or both sides of the ocean basin lead first to cordilleran-type orogeny and subsequently to continental collision. These orogenies suture or stitch together the original continental fragments and laterally buckle the intervening sediments and volcanic belts into folds and faults, initiate metamorphism, and generate volcanic and plutonic magmas. Belts may later be bisected by the development of new divergent plate margins which produce ocean basins whose axes may cut across the older mountain chains. This seems to be the case with the Appalachian-Caledonian belt, where the Mesozoic-Cenozoic development of the Atlantic Ocean has separated the chain into the Appalachians in North America and the Caledonides in Greenland, Great Britain, and Scandinavia. Most Paleozoic mobile belts, and many Precambrian belts at least 2.5 b.y. old, can be interpreted in an analogous fashion, for they contain the various petrogenetic associations of ocean opening and suture: ophiolites, blueschist terranes, mélanges, and a suite of sedimentary rocks whose composition, texture, and distribution suggest deposition in trenches and inland seas, on the abyssal plains of ocean basins, and along continental shelves and rises. Because the Earth itself has evolved, chemical and petrologic aspects of these more ancient mobile belts often differ in detail from more modern analogs. Their organization leaves little doubt that they were formed by plate tectonics processes.

Plate tectonics is considered to have been operative as far back as 2.5 b.y. Prior to that interval, evidence suggests that plate tectonics may have occurred, although in a markedly different manner, with higher rates of global heat flow producing smaller convective cells or more densely distributed mantle plumes which fragmented the Earth's surface into numerous small, rapidly moving plates. Repetitious collision of plates may have accreted continental blocks by welding primitive "greenstone island arc terranes" to small granitic subcontinental masses.

[WALTER C. PITMAN, III]

Bibliography: K. C. Condie, *Plate Tectonics and Crustal Evolution*, 1976; C. K. Seyfert and L. A. Sirkin, *Earth History and Plate Tectonics*, 1973; B. F. Windley, *The Evolving Continents*, 1977; P. J. Wyllie, *The Dynamic Earth: Textbook in Geosciences*, 1971.

Plateau

Any elevated area of relatively smooth land. Usually the term is used more specifically to denote an upland of subdued relief that on at least one side drops off abruptly to adjacent lower lands. In most instances the upland is cut by deep but widely separated valleys or canyons. Small plateaus that stand above their surroundings on all sides are often called tables, tablelands, or mesas. The abrupt edge of a plateau is an escarpment or, especially in the western United States, a rim. In the study of landform development the word plateau is commonly used to refer to any elevated area, especially one underlain by nearly horizontal rock strata, that once had a smooth surface at high level, even though that surface may since have been largely destroyed by valley cutting. An example is the now-hilly Appalachian Plateau of western Pennsylvania, West Virginia, and eastern Kentucky. *See* ESCARPMENT; PLAINS.

Among the extensive plateau lands of the world are the Colorado and Columbia plateaus of the western United States, the plateau of southeastern Brazil, the Patagonian Plateau of southern South America, the Central Siberian Plateau, and the Deccan Plateau of peninsular India.

[EDWIN H. HAMMOND]

Playa

The low, essentially flat part of a basin or other undrained area in an arid region. In heavy rains the playa may be temporarily covered with a shallow sheet of water and is then a playa lake.

Five principal types of playa have been recognized by Richard O. Stone. The dry playa develops where the water table is below capillary reach of surface; its surface is hard, flat, and smooth, composed of silt and clay. The moist playa or salina occurs where the water table is within capillary reach of surface; this type may be further subdivided into (1) salt-encrusted playa, where the water table is at the surface or so near it that salt water evaporating on the surface leaves a salt crust (see illustration); and (2) clay-encrusted playa, where the water table is near the limit of capillary action and salt brought to the surface is mixed with silt and clay, forming a puffy surface (self-rising ground). The crystal-body playa is essentially a massive body of crystalline salt at or very close to the surface. The compound playa results from a water table at different levels in different parts; these exhibit characteristics of dry playa in one part and moist playa in another. The lime-pan playa is formed in basins receiving drainage from limestone terrain and has a floor of hard travertine. *See* DESERT EROSION FEATURES; EVAPORITE, SALINE. [THOMAS CLEMENTS]

Light-colored saline deposits encrust lowest parts of the basin or bolson in Death Valley, Calif.

Bibliography: T. Clements et al., *A Study of Desert Surface Conditions*, U.S. Army Tech. Rep. EP-53, 1957; A. S. Leopold and Life (eds.), *The Desert*, 1961.

Pleistocene

A term referring to a sequence of geologic deposits (the Pleistocene Series) and also the time (the Pleistocene Epoch of the Quaternary Period) during which those deposits were made. Introduced by the British geologist Sir Charles Lyell in 1839, it has been defined on various bases, such as its content of fossil mollusks, its fossil mammals, and its evidence of glacial climate. No single definition is universal. In most European countries and by some authorities in the United States, the Pleistocene is considered to begin with the first appearance of the horse, cattle, and elephant in the specific sense. According to the U.S. Geological Survey, the Pleistocene includes the Great Ice Age, or Glacial Epoch, and possibly some preglacial time. In sedimentary cores raised from the deep-sea floor, the Pleistocene Series is identified by some on a basis of inferred water temperatures. Others prefer to attempt identification on a basis of marine microfossils. However, general agreement on a dividing line between Pleistocene and Pliocene has not yet been reached. Pleistocene time has been regarded as including the present, but more commonly as ending with the end of the Ice Age (arbitrarily taken as 11,000 or 10,000 years ago, after which Recent, or Holocene, time began. *See* GLACIAL EPOCH; MARINE SEDIMENTS; PLIOCENE; QUATERNARY; RECENT.

PRECAMBRIAN	PALEOZOIC							MESOZOIC			CENOZOIC	
					CARBONIFEROUS							
	CAMBRIAN	ORDOVICIAN	SILURIAN	DEVONIAN	Mississippian	Pennsylvanian	PERMIAN	TRIASSIC	JURASSIC	CRETACEOUS	TERTIARY	QUATERNARY

TERTIARY					QUATERNARY	
Paleocene	Eocene	Oligocene	Miocene	Pliocene	Pleistocene	Recent

The Pleistocene deposits include a very large variety of sediments. In middle and high latitudes glacial deposits are prominent among them. In most places these are unconsolidated or semiconsolidated, and blanket the underlying bedrock over wide areas. The length of Pleistocene time has not yet been measured accurately. It was long thought to be about 1,000,000 years, but on a basis of K-Ar dating it is now believed to be considerably more.

[RICHARD F. FLINT]

Pleochroic halos

Small halos of color or color differences that are sometimes observed around inclusions in minerals. They were noted by Harry Rosenbusch (1873) around cordierite and were later reported by many observers in many minerals. If halos occur in doubly refracting substances, they may show pleochroic discoloration, and the term pleochroic halos is loosely applied to such small colored halos generally. *See* CORDIERITE.

Distribution and description. The halos are found only around minute inclusions of certain minerals, especially zircon, allanite, monazite, and others known to contain minor amounts of uranium or thorium. Halos have been reported in many rock-making minerals, including amphiboles, pyroxenes, and micas. *See* RADIOACTIVE MINERALS.

The halos are usually spheroidal (circular, as seen in microscopic section), sometimes consisting of several concentric rings, and are fairly sharply bounded. The outermost ring in biotite does not attain a diameter of more than 0.04 mm.

Origin and interpretation. J. Joly in 1907 first ascribed pleochroic halos to the effect of irradiation with α-particles. He and others later were able to explain the details of the ring structure by the ranges (length of tracks) of α-particles from the several sources in the uranium or thorium series in the mineral affected. Changes in refringence or birefringence (either increase or decrease) may be associated with the coloring, and it has been suggested that the formation of pleochroic halos is comparable to the radiation damage in minerals called metamictization. The darkening that causes the ringlike appearance, rather than equal darkening over the entire range, is probably due to the annealed structural damage caused by knock-on atoms near the end of the α-particle trajectory, where the particle has slowed to a point at which it interacts strongly with the nuclei of surrounding atoms. Pleochroic halos are most widespread in minerals such as biotite, which is never known in the metamict state. The phenomenon is doubtlessly closely related to other types of coloration induced by radiation. Pleochroic halos have been produced artificially in various materials by E. Rutherford and others. *See* METAMICT STATE.

It has been suggested that the halos might offer a means for estimating the ages of minerals. This seems unlikely, since there is a limit to the coloration that can be produced in any material; moreover, there is some indication that reversal can occur on prolonged exposure and that the color can be dissipated by heat.

[PATRICK M. HURLEY]

Pliocene

The youngest of the five major worldwide divisions (epochs) of the Tertiary Period (Cenozoic Era), extending from the end of the Miocene to the beginning of the Pleistocene (Quarternary). Pliocene also denotes the series of rocks and their fossils formed during this epoch. Pliocene rocks include most normal sedimentary types. Those of a mechanical (clastic) rather than chemical or biogenetic origin are the more common. High surface relief was characteristic of Pliocene times. Marine deposits are patchy though often thick, totaling 15,000 ft in one California sequence where oil-bearing sandstone lenses are widely productive, as are the Pliocene sandstones in the East Indies (Tertiary h1-h2) and elsewhere. *See* CENOZOIC; MIOCENE; PLEISTOCENE; TERTIARY.

PRECAMBRIAN	PALEOZOIC							MESOZOIC			CENOZOIC	
					CARBONIFEROUS							
	CAMBRIAN	ORDOVICIAN	SILURIAN	DEVONIAN	Mississippian	Pennsylvanian	PERMIAN	TRIASSIC	JURASSIC	CRETACEOUS	TERTIARY	QUATERNARY

TERTIARY					QUATERNARY	
Paleocene	Eocene	Oligocene	Miocene	Pliocene	Pleistocene	Recent

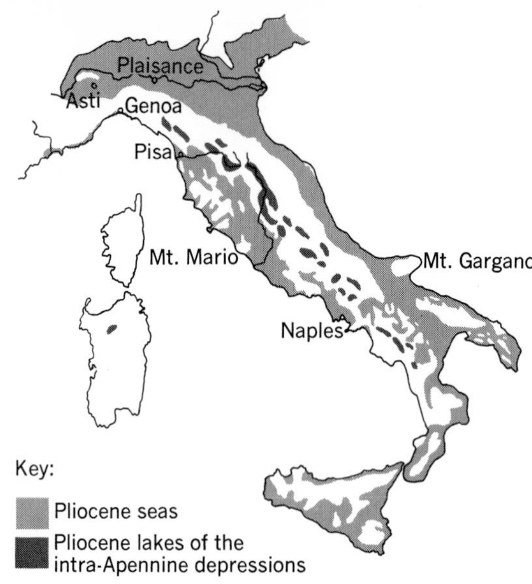

Fig. 1. Paleogeographic map of Italy in the Pliocene after F. Sacco, 1919. After this map was made, some marine Pliocene was discovered in Sardinia. *(From M. Gignoux, Stratigraphic Geology, copyright © 1955 by W. H. Freeman and Co.)*

Sir Charles Lyell in 1883 introduced the terms Older Pliocene (for strata bearing "upwards of a third to somewhat more than half" living species), and Newer Pliocene (with 90–96% living species) which he later changed to Pleistocene, retaining Pliocene for his Older Pliocene designation. Typical Pliocene were the strata of the Subappenine hills of northern Italy (and also Pliocene were the formations of Tuscany and the English Crag). Current usage of the term, however, often includes the older part of Lyell's Newer Pliocene, originally from the isle of Ischia, near Naples, to the Val di Noto, Sicily (Fig. 1).

Subdivisions. Subdivisions subsequently recognized have been a lower Plaisancian Stage (blue marls of Plaisance, or Piacenza) and an upper Astian (yellow sands of Asti). To these have been added a higher Calabrian and still higher Sicilian Stage (their cooler-water faunas variously considered Pliocene or lowermost Pleistocene) and lower, in eastern Sicily, one, two, or three stages (often termed Mio-Pliocene) corresponding in age to nonmarine strata transitional from typical Miocene to typical Pliocene. These lower stages, are: (1) Zanclean, from marine white marls beneath Plaisancian strata; under this stage is a (2) gypsiferous sulfur-bearing Messinia (with similar strata in Calabria, Pisa, Siena, and Genoa); and (3) a diatomaceous Sahelian named from Sahel d'Oran, Algeria, where abyssal and diatomaceous strata embrace more than the entire transitional interval. The more restricted Sahelian strata in Sicily have equivalents in Crete and the Piraeus. From the Vienna basin eastward a Sarmatian Stage with brackish-water fossils overlies Vindobonian Miocene and is succeeded by a nonmarine Pontian Stage which is further subdivided into Meotian below, Cimmerian above. Local marine Mio-Pliocene occurs also in northern Italy (Tabianian), Brittany (Redonian), Cornwall (?), Belgium (Anversian), northern Germany (glimmerton), the East Indies (Tertiary g), Australia (Mitchellian), New Zealand (Tongaporotuan, Kapitean), California (Delmontian, Repettian), and the Atlantic Seaboard (Choctawhatchee ?). These local marine beds are all essentially equivalent to extensive terrestrial beds such as Clarendonian in North America, with lower Cerrotejonian and upper Montediablan subdivisions.

Both lithofacies and biofacies problems, created by the approach of the mid-Pleistocene Ice Age peak of the Late Alpine orogeny, have rendered interregional correlations within the Mio-Pliocene and Pliocene difficult. Extensive studies of planktonic foraminifers, diatoms, and nannoplankton hold promise of a more refined late Tertiary timescale than is at present available. But planktonic as well as other modes of life reveal inherent biofacies phenomena, and planktonic age determinations have leaned heavily on the index-fossil and purely theoretical evolutionary bioseries approach to age correlation that tends to ignore extensions in the vertical range of planktonic species through stratigraphical sequences. Thus interpretive classifications of time based on the distribution of these organisms vary greatly, and await further discipline, especially in the fields of facies, stratigraphic ranges of species, and the details of parallel evolution. *See* OLIGOCENE.

Geography and climate. In its broadest usage, Pliocene time encompassed two intervals of widespread orogenic movements separated by an interval of general marine expansion, the Third Mediterranean Stage of E. Suess. During Pliocene time the ocean withdrew from eastern Europe and Turkestan, never to return. It lingered longest in Volhynia, with probable connections across Poland to the North Sea. Nearby, marine waters turned brackish, then fresh, then saline, then fluviatile, though not always at the same time in the several local basins. Remnants were left such as the Aral, Caspian, and Black Seas. A later Pliocene North Sea retreated progressively northward after deposition of the Lenham-Diestian-Sylt beds and the Coralline Crag. Through the Pliocene worldwide temperatures dropped steadily. By mid-Pliocene time world climates had attained temperatures essentially those of today though generally more arid. Since then climates have become colder (peaks of Ice Age glaciation), then again warmer than those prevalent during historic time.

Life. Pliocene marine shellfish look modern. Sand dollars proliferated. Equatorward expansion of bivalves, gastropods, and foraminiferids (especially cassidulinids) having cold-water environmental optima was general. A chill was even upon the tropics; the former limestone-forming large foraminiferids were mostly gone (endemic relics remained in the Indo-Pacific), largely replaced by inshore small foraminiferids (*Elphidium*, *Rotalia*)

Fig. 2. Pliocene in the San Joaquin Valley, Calif., showing the camel (*Parocamelus*), the horse (*Pliohippus*), and the pronghorn (*Sphenophalos*). (*From R. A. Stirton, Time, Life, and Man, copyright © 1959 by John Wiley & Sons, Inc.; reprinted by permission*)

and medium-sized tropical holdovers.

Pliocene mammals were larger than those in preceding epochs. Grasslands were widespread (Fig. 2). Mio-Pliocene land emergences permitted *Hipparion*, a late Miocene genus of North American horse, to spread intercontinentally. Interchanges of other mammals also occurred before endemic evolution on land once more set in. Afro-Asian Pliocene seaways (Zanzibar, Red Sea, Persia, Mekran coast) semi-isolated much of the earlier Pliocene fauna in Africa. Late Pliocene and early Pleistocene land reconnections (Old World Villafranchian Stage, New World Blancan-Irvingtonian) permitted further intercontinental mammal interchanges; northern forms went south, some such as elephants, zebras and hippopotamuses to Africa; others migrated to the newly reconnected South America from which armadillos, capibaras, porcupines, and ground sloths went north, while many distinctive South American mammals such as notoungulates, horselike litopterns, and others died out. Much of the climacteric Plio-Pleistocene mammal fauna remained in the modern Oriental Region, but north and northwest of the Himalayas selective scouring during the subsequent Ice Age was rigorous, and this fauna was modified both in Old World and New. *See* PALEOBOTANY; PALEONTOLOGY.

[ROBERT M. KLEINPELL]

Pluton

A rock body formed by the consolidation of magma (molten rock) below the surface of the Earth, but now exposed due to erosion.

Methods of observation. The subsurface shape of larger plutons is usually difficult to determine. Generally the shape is deduced from observations at the surface of the Earth, but projecting such data to any depth is hazardous. The shape of at least a portion of small plutons may be observed in exposures a few hundred feet across. The three-dimensional shape of medium-sized plutons may be partially observed in mountainous regions, but the three-dimensional shape of large plutons, thousands of feet or miles in diameter, must be deduced from indirect evidence. Contacts may be observed in mountains or deduced from the relations to topography. Bore holes and mine workings may give valuable information, but few plutons have been adequately investigated by such means. In recent years geophysical data, especially those

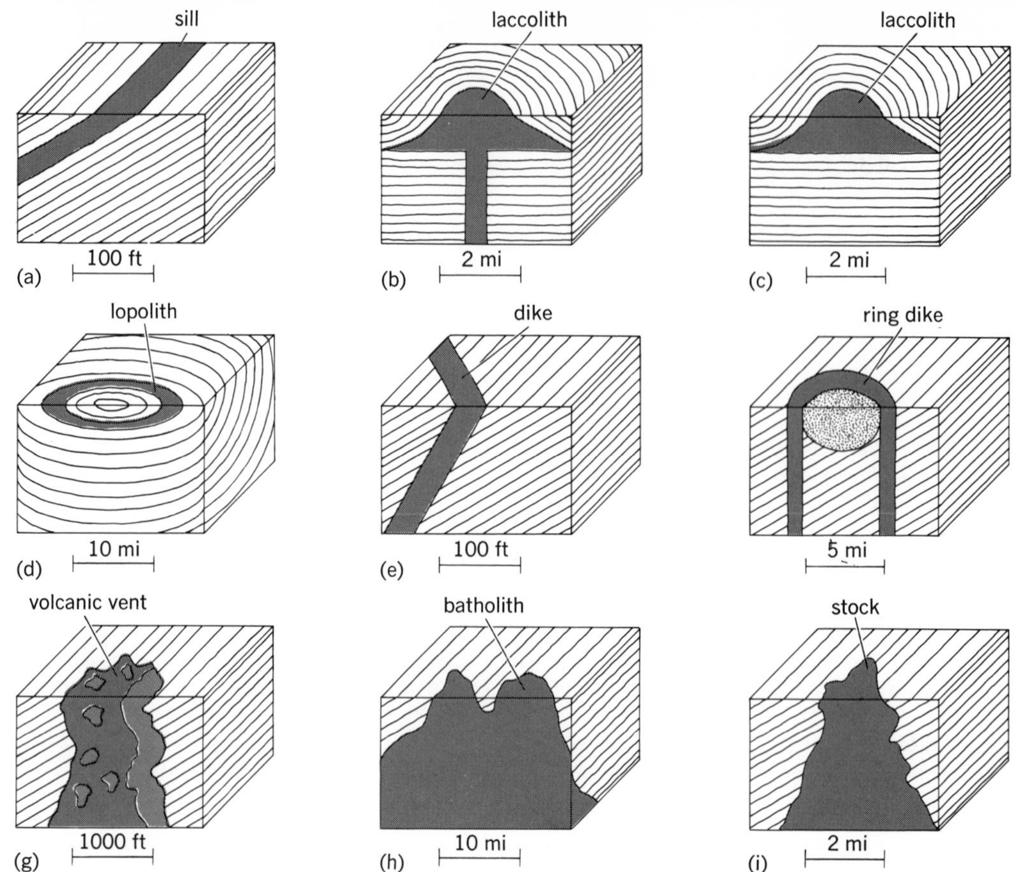

Plutons. (*a-i*) Scales are shown merely to give some idea of the magnitude of the various bodies.

obtained by gravitational and magnetic methods, have been used. Measurements of terrestrial heat flow have also been utilized to a very limited extent. But even under the most favorable conditions the geophysical methods give only qualitative results. *See* GEOPHYSICAL EXPLORATION; IGNEOUS ROCKS; PETROLOGY.

Types. A rather elaborate terminology based on the supposed three-dimensional shape of plutons has evolved. Some of the more common terms are defined below. But because of the difficulty in determining the real shape, the modern tendency is to use the general term pluton, except for relatively small bodies.

Considerable emphasis has been placed on whether the body is concordant or discordant. In concordant bodies the contacts are parallel to the planar features, generally bedding, of the adjacent rocks. In discordant bodies the contacts cut across the planar structures. Locally, of course, the contacts of a body that is essentially concordant may be discordant, and conversely, the contacts of a discordant body may be locally concordant.

Sill. A sill is a concordant sheet of igneous rock that is younger than the adjacent rocks, having been injected along the bedding (illustration *a*). Sills may range in thickness from a few inches to thousands of feet.

Laccolith. A laccolith is a concordant, mushroom-shaped body, younger than the adjacent rocks. It is thousands of feet in diameter and hundreds of feet thick. The overlying rocks are domed up as the viscous magma is injected parallel to the bedding. Although some laccoliths may have been fed through a small cylindrical conduit below them (illustration *b*), many appear to have been fed from the side (illustration *c*).

Lopolith. A lopolith is a large concordant sheet of igneous rock, younger than the adjacent rocks, in the form of a great saucer. Lopoliths are generally many miles in diameter and thousands of feet thick. They are injected along the bedding and are apparently fed from beneath the center of the body; subsidence accompanied injection.

Dike. A dike is a discordant sheet of igneous rock cutting the adjacent rocks. Dikes form when magma injects a fracture and displaces the walls. Although most dikes range in thickness from a few inches to tens of feet, some are hundreds or even thousands of feet thick. Some dikes have been traced on the surface of the Earth for miles, even tens of miles.

Ring dike. Ring dikes are arcuate in plan and the contacts dip steeply. The radii of the arcs are 1–10 mi, and the widths of the ring dikes are 500–15,000 ft. They form when magma rises into a cylindrical fracture.

Volcanic vent. Volcanic vents are subcircular bodies composed of lava and fragmented volcanic rocks. Contacts are generally steep and the vents range in diameter from a few tens to thousands of feet. The vents fed volcanic cones that have since been eroded away.

Batholith. The term batholith has been defined in many ways. There is general agreement that batholiths are large, covering (by definition) an area of 40 mi² or more, and are composed of coarse-grained igneous rocks. They are emplaced in various ways, such as stoping, a process whereby magma eats its way upward by breaking off small blocks that sink back into the magma to be assimilated at depth; underground cauldron subsidence, whereby magma fills a potential cavity developing above a large subsiding cylindrical block miles in diameter; forceful injection, in which magma pushes the older rocks aside; and perhaps metasomatic replacement, that is, atom-by-atom replacement of the older rock by moving solutions.

Stock. Stocks are similar to batholiths, but by definition cover less than 40 mi².

[MARLAND P. BILLINGS]

Bibliography: M. P. Billings, *Structural Geology*, 2d ed., 1954; C. B. Hunt, P. Averitt, and R. W. Miller, *Geology and Geography of the Henry Mountains Region, Utah*, USGS Prof. Pap. no. 228, 1953.

Pneumatolysis

The alteration of rocks by the action of magmatic gases. The gases accompany magmatic intrusions and permeate the intruded rocks along fissures and other lines of least resistance. *See* METAMORPHISM; METASOMATISM.

Primary magmatic gases are acid and therefore react readily with limestone to form skarn rocks. Appreciable amounts of heavy metals (usually present as chlorides, fluorides, or sulfides) are associated with the magmatic gases. These are captured by the limestone and retained as deposits of skarn rocks or ores. In other places the heavy metals are deposited in granite or schist, causing the formation of greisen. *See* GREISEN; SKARN.

The following list indicates common pneumatolytic minerals concentrated in limestone of the Oslo region.

Metals	*Minerals*
Fe	Andradite, hedenbergite, oxidic and sulfidic iron ore
Zn, Cu, Pb	Sphalerite, ZnS; chalcopyrite, $CuFeS_2$; galena, PbS
Mn	Andradite; hedenbergite; rhodonite, $(Mn,Fe,Ca)SiO_3$
Bi, Ag	Bismuthinite, Bi_2S_3; galena; sphalerite
Mo, W	Molybdenite, MoS_2; scheelite, $CaWO_4$
Co, As, Sb	Cobaltite, CoAsS; arsenopyrite, FeAsS; bismuthinite
Be, Ce	Helvite, $(Mn,Fe,Zn)_4Be_3Si_3O_{12}\cdot S$; vesuvianite; allanite, $(Ca,Fe)_2(CeAl,Fe)_3Si_3O_{12}OH$
Metalloids	
Si	Silicates in skarn, quartz
F, Cl, S	Fluorite, CaF_2; scapolite; sulfidic ore
B, P, Ti	Axinite, apatite, sphene

Additional pneumatolytic minerals are those of the humite group and, by pneumatolysis of shale, tourmaline, topaz, muscovite, phlogopite, and lithium micas. The mineral luxullianite is a tourmalinized granite.

In conclusion it may be emphasized that limestone is especially susceptible to pneumatolytic contact metamorphism, resulting in formation of skarn rocks (see illustration) and often useful ore deposits (Iron Springs, Utah; Macay, Idaho; Yerrington, Nev.; Concepción del Oro, Mexico). The minerals often develop as well-formed crystals, and the deposits belong to the best-known mineral occurrences in the world (Franklin Furnace, N.J.; Clifton-Morenci, Ariz.; Auerbach, Germany; Berggieshübel, Switzerland; Banat, Hungary; Concepción del Oro, Mexico). Silicate rocks are not usually as intensively altered, but by introduction of fluorine and lithium, and by the formation of greisen, feldspars may change into topaz, zinnwaldite, or other micas. Tin is sometimes introduced, forming cassiterite, SnO_2. Other introduced minerals in argillites are, for instance, molybdenite, apatite, or beryl.

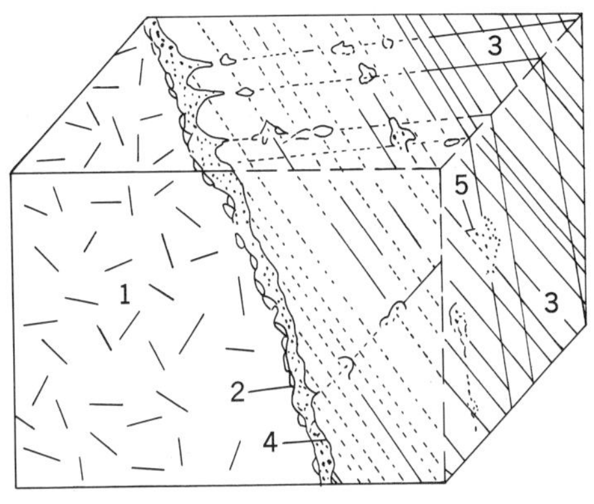

Skarn rocks and pneumatolytic ore at Aranzazu, Mexico: 1, granodiorite; 2, garnetized border of granodiorite (exaggerated); 3, limestone; 4, garnet rock at immediate contact carrying some ore; 5, bodies of andradite-wollastonite-copper ore localized along intersections of fissures and bedding planes. (*After A. Knopf, in W. H. Newhouse, Ore Deposits as Related to Structural Features, Princeton University Press, 1942*)

Sometimes intense boron pneumatolysis may produce datolite, axinite, and rare minerals like kotoite, $Mg_3(BO_3)_2$; fluoborite, $Mg_3(BO_3)(F,OH)_3$; ludwigite, $(Mg,Fe'')_2Fe'''BO_5$; and others.

[TOM F. W. BARTH]

Porphyroblast

A relatively large crystal formed in a metamorphic rock. The presence of abundant porphyroblasts gives the rock a porphyroblastic texture. Minerals found commonly as porphyroblasts include biotite, garnet, chloritoid, staurolite, kyanite, sillimanite, andalusite, cordierite, and feldspar. Porphyroblasts are generally a few millimeters or centimeters across (see illustration), but some attain a diameter of over 1 ft. They may be bounded by well-defined crystal faces, or their outlines may be highly irregular or ragged. Very commonly they are crowded with tiny grains of other minerals that occur in the rock.

Some porphyroblasts appear to have shoved aside the rock layers (foliation) in an attempt to provide room for growth. Others clearly transect the foliation and appear to have replaced the rock. The presence of ghostlike traces of foliation, in the form of stringers and trails of mineral grains, passing uninterruptedly through a porphyroblast is further evidence of replacement.

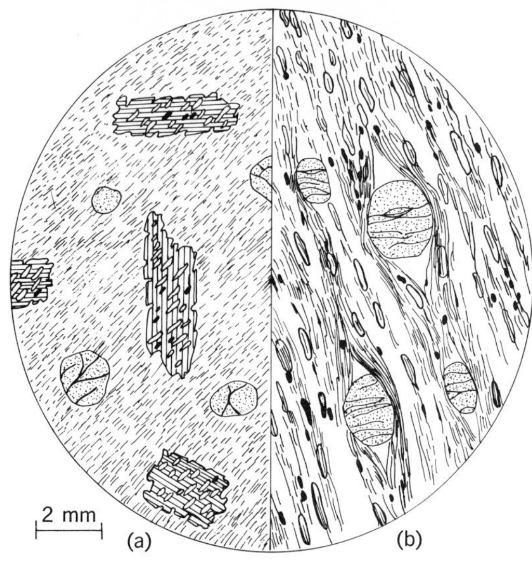

Porphyroblasts. (*a*) Porphyroblastic mica schist. Large crystals (porphyroblasts) of biotite mica formed late and replaced the rock except for the enclosed grains (quartz) aligned parallel to foliation. Smaller porphyroblasts of garnet replaced the rock completely. (*b*) Porphyroblastic quartz-mica schist. Large porphyroblasts of garnet have grown by spreading apart mica-rich layers of schist. Elongate grains (gray) are flakes of biotite mica. Finer flakes of muscovite mica are widespread and give a pronounced schistosity. They are closely packed where crowded by garnet crystals. Elongate grains of iron and titanium oxide (black) are also aligned.

Porphyroblasts have many features in common with phenocrysts but are to be distinguished from the latter by the fact that they have developed in solid rock in response to metamorphism. Most commonly they develop in schist and gneiss during the late stages of recrystallization. As the rock becomes reconstituted, certain components migrate to favored sites and combine there to develop the large crystals. *See* GNEISS; METAMORPHIC ROCKS; PHENOCRYST; SCHIST.

[CARLETON A. CHAPMAN]

Porphyry

An igneous rock characterized by porphyritic texture, in which large crystals (phenocrysts) are enclosed in a matrix of very fine-grained to aphanitic (not visibly crystalline) material.

Porphyries are generally distinguished from other porphyritic rocks by their abundance of phenocrysts and by their occurrence in small intrusive bodies (dikes and sills) formed at shallow depth within the earth. In this sense porphyries are hypabyssal rocks. *See* PHENOCRYST.

Compositionally, porphyries range widely, but varieties may be distinguished by prefixing to the term the common rock name which the porphyry most closely resembles (such as granite porphyry, rhyolite porphyry, syenite porphyry, and trachyte porphyry). *See* IGNEOUS ROCKS.

Porphyries are gradational to plutonic rocks on the one hand and to volcanic rocks on the other. In the granite clan, for example, six porphyritic types may be recognized: (1) porphyritic granite, (2) granite porphyry, (3) rhyolite porphyry, (4) porphyritic rhyolite, (5) vitophyre, and (6) porphyritic obsidian or porphyritic pitchstone.

A rock of granitic composition with abundant large phenocrysts of quartz and alkali feldspar in a very fine-grained matrix of similar composition is a porphyry, or more specifically a granite porphyry. Granite porphyry passes into porphyritic granite as the grain size of the matrix increases and abundance of phenocrysts decreases. Thus the rock becomes a granite, or more specifically a porphyritic granite. Granite porphyry passes into rhyolite porphyry as the grain size of the matrix and abundance of phenocrysts decrease. The principal distinction between rhyolite porphyry and porphyritic rhyolite is the mode of occurrence. Rhyolite porphyry is intrusive; porphyritic rhyolite is extrusive. Porphyritic rocks with a glass matrix are known as vitrophyres. With decrease in number of phenocrysts, vitrophyre passes into porphyritic obsidian or porphyritic pitchstone.

The phenocrysts of porphyries consist largely of quartz and feldspar. Quartz occurs as well-formed (euhedral) hexagonal bipyramids, which in thin section under the microscope exhibit a diamond, square, or hexagonal outline. Individual phenocrysts may show more or less rounding or resorption with deep embayments. Alkali feldspar is usually euhedral sanidine, orthoclase, or microperthite. Plagioclase occurs more in association with phenocrysts of hornblende or other dark-colored (mafic) minerals. Porphyritic rocks with predominantly mafic phenocrysts (olivine, pyroxene, amphibole, and biotite) are commonly classed as lamprophyres. *See* LAMPROPHYRE.

Outside the United States, it is common to further restrict the term porphyry to those rocks in which the feldspar phenocrysts are principally alkali feldspar. Rocks with dominantly plagioclase phenocrysts are called porphyrites.

Porphyries occur as marginal phases of mediumsized, igneous bodies (stocks, laccoliths) or as apophyses (offshoots) projecting from such bodies into the surrounding rocks. They are also abundant as dikes cutting compositionally equivalent plutonic rock or as dikes, sills, and laccoliths injected into the adjacent older rocks.

[CARLETON A. CHAPMAN]

Precambrian

The term used to designate rocks and time older than the Cambrian, which is the oldest geologic period from which abundant fossils have been recovered. Age determinations by means of radioactive elements place the upper boundary of the Precambrian at about 600,000,000 years ago. The oldest dated rocks appear to be 3,800,000,000–

PRECAMBRIAN				PALEOZOIC							MESOZOIC			CENOZOIC	
ARCHEAN (ARCHEOZOIC)		PROTEROZOIC (ALGONKIAN)						CARBONIFEROUS							
1	2	3	4												
EARLY	MIDDLE		LATE	CAMBRIAN	ORDOVICIAN	SILURIAN	DEVONIAN	Mississippian	Pennsylvanian	PERMIAN	TRIASSIC	JURASSIC	CRETACEOUS	TERTIARY	QUATERNARY

4,000,000,000 years old. All evidence indicates that these rocks are not remnants of the original Earth crust, and it follows that the Precambrian interval of Earth history lasted well over 3,500,000,000 years. *See* CAMBRIAN; RADIOACTIVE MINERALS; ROCK, AGE DETERMINATION OF.

Subdivision. Formerly, the Precambrian was divided into the Archean (or Archeozoic) and the Proterozoic (or Algonkian). Other names were also used, especially in Europe and elsewhere, but the names Archean and Proterozoic have been most widely used. Students of the Precambrian, however, are discarding this terminology. Some geologists still retain the terms, but generally restrict the meaning of the Archean and expand the Proterozoic into Lower, Middle, and Upper. The majority of Precambrian investigators prefer to discard the former terms entirely and to make use of Early, Middle, and Late Precambrian, or of some numerical subdivision, until a more precise and better-documented time scale can be worked out. The present state of knowledge is such that a twofold subdivision appears highly inadequate and construction of a general time scale to represent the sequence of Precambrian events for the entire world is yet impossible. Lack of continuous outcrops, lack of fossils, and destruction of primary characteristics of rocks by subsequent metamorphism make correlation of Precambrian rocks and events much more difficult than correlation of later strata and crustal disturbances. Data from radioactive determinations may solve the problem, but they are yet too few. Data are accumulating in increasing numbers, however, and have begun to show patterns of possible major events in Precambrian history. The world picture is still developing and is not yet clearly delineated. Continental subdivisions are emerging but cannot yet be considered more than tentative, and worldwide correlation is further in the future. The fact that certain dates, or comparatively narrow ranges of dates, are being obtained much more frequently than others, appearing to reflect more widespread and significant events, has led the U. S. Geological Survey to adopt an interim scheme for subdividing Precambrian time. The scheme is based strictly on age dating (without regard to formational boundaries, tectonic episodes, erosional intervals), designed solely for the United States, and meant to serve as a convenient temporary vehicle for classification until a satisfactory subdivision is developed. The Precambrian is divided into four units, somewhat as other geologists have suggested, but here called Precambrian W (oldest), Precambrian X, Precambrian Y, and Precambrian Z (youngest), with dividing lines set at 2,500,000,000, 1,600,000,000, and 800,000,000 years. If older subdivisions are desired as information is acquired, they may be added as Precambrian V, Precambrian U, and so forth, for progressively older intervals.

The variety of rocks found in Precambrian terranes is great, but one continent differs little from another in this regard. The Precambrian rocks of Australia, Canada, Scandinavia, and other parts of the world show the same range in variety and abundance. They are represented by all kinds of sedimentary rocks (conglomerates, sandstones, shales, and limestones) and extensive lava flows, some still relatively undeformed and unmetamorphosed, others strongly deformed, highly metamorphosed, and invaded by great masses of granite or granitelike rocks. In general, the older rocks exhibit higher degrees of metamorphism, but this is not invariably true. The degree of rock deformation and metamorphism depends on nearness to a center of mountain-making activity rather than on time. Degree of metamorphism should never be used as a criterion of age.

Several episodes of sediment accumulation, lava eruption, deformation, and metamorphism, with emplacement of great batholiths and subsequent erosion, are recorded in the rocks. In places, the site of one episode overlaps part of that of a succeeding episode, and mutual relations may be observed.

Occurrence. Precambrian rocks underlie all other rocks of continents. They occur beneath a great blanket of flat or deformed sedimentary and other rocks, are near the surface covered by a thin veneer of strata or unconsolidated material, or are exposed. The exposed areas occupy 8–9% of the Earth's land surface. They occur in two principal types of geologic setting: shields and mountain cores.

Shields are large, low-lying, relatively stable parts of continental masses, which have remained low and relatively undisturbed since Precambrian time. They form the stable foundations or nuclei of the continents. Gneisses, schists, and cross-cutting granites and granodiorites are common. Gabbros and related rocks form small isolated masses, though some are relatively large. Sedimentary and volcanic rocks may be preserved in downwarped or downfolded segments. By far the greatest mass of exposed Precambrian rocks is found in shield areas, of which each continent has at least one (Fig. 1).

Precambrian rocks are also found in numerous but smaller exposed masses in the cores of many

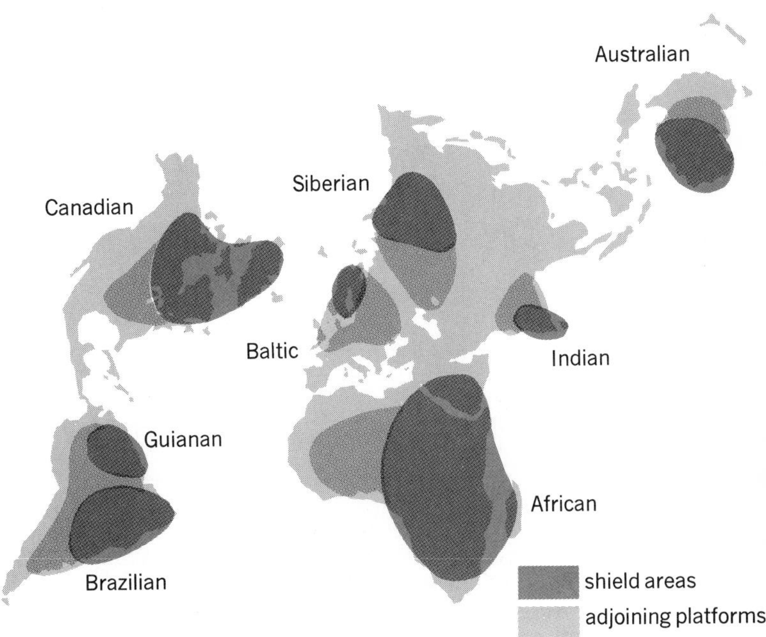

Fig. 1. Shield areas and adjoining platforms composed of Precambrian rocks. (*From R. C. Moore, Introduction to Historical Geology, 2d ed., McGraw-Hill, 1958*)

mountain ranges—zones of the Earth's crust that have been warped upward, or sharply folded or faulted. Erosion has been intense here because of the elevation. In places, major streams have cut down deeply to expose Precambrian rocks in steep-walled gorges. Elsewhere, with more extensive erosion, the higher parts of the range have been removed, the flanking sedimentary strata have been stripped away, and the entire mountain core of Precambrian rocks has been exposed.

North America. The main shield area, called the Canadian shield, occupies most of eastern Canada east of the Great Plains, and extends into the United States in Minnesota, Wisconsin, northern Michigan, and New York. (Radioactive age determinations have shown the oldest rocks yet identified to be the Amitsoq gneiss on the west coast of Greenland, and corresponding rocks on the Labrador coast, with different analytical techniques yielding values from 3,620,000,000 to 3,980,000,000 years; and the Montivideo gneiss in southwestern Minnesota, determined to be 3,800,000,000 years old.)

The Southern and Central Rockies and many of the fault-block mountains of southern New Mexico, Arizona, Utah, Nevada, and California are cored by crystalline Precambrian rocks. The Northern Rockies contain a great thickness of Precambrian rocks, many of them relatively undeformed and unmetamorphosed, thrust to the surface along gently inclined fault planes. Glacier Park displays majestic vistas of such Precambrian rocks. Highly metamorphosed Precambrian rocks (gneisses, schists, and granites) overlain unconformably by nearly flat-lying unmetamorphosed Precambrian rocks are exposed in the inner gorge of the Grand Canyon (Fig. 2). The Blue Ridge and the New England areas of the eastern United States represent worn mountain ranges with exposed Precambrian crystalline rocks. *See* UNCONFORMITY.

Europe. The principal shield area is the Scandinavian or Baltic shield, including eastern Norway, Sweden, Finland, Karelia, and the Kola Peninsula. Precambrian rocks are also found in the Highlands of Scotland, in the Ukraine, and in several mountain ranges of central and western Europe. The Scottish Highlands represent overthrust blocks of gneisses and intrusive rocks overlain unconformably by conglomerate and sandstone. Gneisses and granite of a partially covered shield area are exposed in the Ukraine.

Asia. The Siberian shield area includes the Angaran or Aldan (Lake Baikal region) and the Anabar (northern Siberia). Peninsular India is also a shield area. Precambrian rocks are found in abundance in the cores of many mountain ranges of Central Asia.

Australia. The western half of Australia is all shield area. Folded mountain ranges occur in eastern Australia, but no Precambrian cores are exposed.

South America. The eastern half of South America is mainly shield area, most of the bulge of Brazil constituting the Brazilian shield, and the area of French Guiana, Guyana, Surinam, and much of Venezuela making up the Guianan shield. These are separated by only the narrow zone of the Amazon basin. The Platian shield in central Argentina is covered mainly by Cenozoic strata, but small isolated outcrops of Precambrian rocks are found.

Antarctica. From the growing store of informa-

Fig. 2. Uniformly dipping Late Precambrian (Algonkian) sedimentary rocks in the lower part of the Grand Canyon Sequence. The rocks dip eastward at about 30° as a result of post-Precambrian block faulting. They rest with profound unconformity on highly deformed and metamorphosed Early Precambrian (Archean) rocks. (*N. W. Carkhuff, U.S. Geological Survey*)

tion about this continent it is known that most of the eastern part (from 0° eastward to 180° longitude) is chiefly a shield area of gneiss, schist, and granite. The smaller western part is a zone of folded Paleozoic and Mesozoic strata.

Africa. This continent contains the best display of Precambrian rocks in the world. More than half the area of the continent has Precambrian rocks at the surface. Except for the northern and southern extremities, the continent has been stable and relatively undisturbed since Precambrian time. Granites, gneisses, and schists are abundant, but layers of sandstone, shale, conglomerate, and volcanic rock, only slightly or moderately deformed and metamorphosed, are thick and extensive. The main shield area has been called the Ethiopian. South Africa displays a basement complex of crystalline rocks about 3,000,000,000 years old, overlain by six to eight systems of sedimentary and volcanic rocks. The famous Bushveld complex of granite and gabbroic rock is intrusive into the lower systems (about 2,000,000,000 years old).

Economic geology. Precambrian rocks are important sources of metalliferous ores, particularly iron, nickel, gold, uranium, and copper. Tremendous stores of iron ore have been mined and are still being mined from the Canadian shield, the Baltic shield, the Ukraine, and the Brazilian and Guianan shields. The ores are mostly sedimentary, but large masses of magnetite of probably igneous origin are also productive.

Almost 75% of the world's production of nickel comes from a large intrusive mass at Sudbury, Ontario, in the Canadian shield. Other large deposits have also been found in the Canadian shield, and similar deposits are being exploited in the Baltic shield.

More gold has been produced from Precambrian rocks than from all others combined. Half the world's present production comes from Precambrian conglomerates of the Witwatersrand of South Africa. Other production is obtained from the Canadian shield, Australia, and India.

Rich deposits of uranium ore are being mined in the Canadian shield and in South Africa. Copper in great quantities has been mined from Precambrian rocks of northern Michigan. Other valuable resources are platinum (Soviet Union and Canada), silver (Canada), lead and zinc (Canada, United States, Australia), chromium (South Africa and Soviet Union), cobalt (Canada), manganese (India, Africa, and Brazil), graphite (eastern Asia), mica (Soviet Union and India), and talc (eastern U.S.).

Life. Microorganisms, possibly of algal type but probably more primitive, have been found in the Gunflint Formation of the Lake Superior region. This formation has been dated by radioactive means as 2,000,000,000 years old. Calcareous algae of later Precambrian time were extensive and abundant. They have been found in the Belt series of Montana, the Animikian rocks of Michigan, the iron formation of Labrador, and elsewhere. Life was probably abundant in the seas of late Precambrian time, since Early Cambrian formations contain abundant fossils, but forms with shells capable of being preserved were still few.

Climate. The atmosphere undoubtedly underwent great change in Precambrian time, but evidence from the rocks indicates that climatic variation from dry to wet and from hot to cold took place then as later. Glaciation was widespread. Some conglomeratic material, long considered to be glacial, may prove to be landslide material, but some, especially where it rests on a striated rock surface, is undoubtedly glacial (Fig. 3).

Fig. 3. Striated and faceted pebble from Precambrian conglomerate in the Kekeko Hills, western Quebec, Canada, interpreted as glacial till but believed by some to be landslide material. Despite variations in interpretation of some deposits, evidence strongly favors the view that widespread glaciation did take place in Precambrian time. *(M. E. Wilson, Geological Survey, Canada)*

[J. PAUL FITZSIMMONS]

Bibliography: A. L. DuToit, *The Geology of South Africa*, 3d ed., 1954; S. S. Goldich et al., *The Precambrian Geology and Geochronology of Minnesota*, 1961; K. Rankama (ed.), *The Precambrian*, 4 vols., 1963–1968; M. E. Wilson, Precambrian classification and correlation in the Canadian Shield, *Bull. Geol. Soc. Amer.*, 69:757–774, 1958; O. A. Woodford, *Historical Geology*, 1965.

Precious stones

The materials found in nature that are used frequently as gemstones, including amber, beryl (emerald and aquamarine), chrysoberyl (cat's-eye and alexandrite), coral, corundum (ruby and sapphire), diamond, feldspar (moonstone and amazonite), garnet (almandite, demantoid and pyrope), jade (jadeite and nephrite), jet, lapis lazuli, malachite, opal, pearl, peridot, quartz (amethyst, citrine, and agate), spinel, spodumene (kunzite), topaz, tourmaline, turquois and zircon. *See* GEM.

The terms precious and semiprecious have been used to differentiate between gemstones on a basis of relative value. Because there is a continuous gradation of values from materials sold by the pound to those valued at many thousands of dollars per carat, and because the same mineral may furnish both, a division is essentially meaningless. All but a few of the gems listed cost many dollars per carat in fine quality. Gold, the symbol of concentrated value and the measure of preciousness, brings less than $0.30 per carat. Thus, almost all gemstones merit the use of the term precious.

[RICHARD T. LIDDICOAT, JR.]

Prehnite

A mineral sorosilicate, composition $Ca_2Al_2Si_3O_{10}$-$(OH)_2$, crystallizing in the orthorhombic system. Distinct crystals are rare and it occurs usually in reniform and stalactitic aggregates with crystalline surface (see illustration). Hardness is 6–

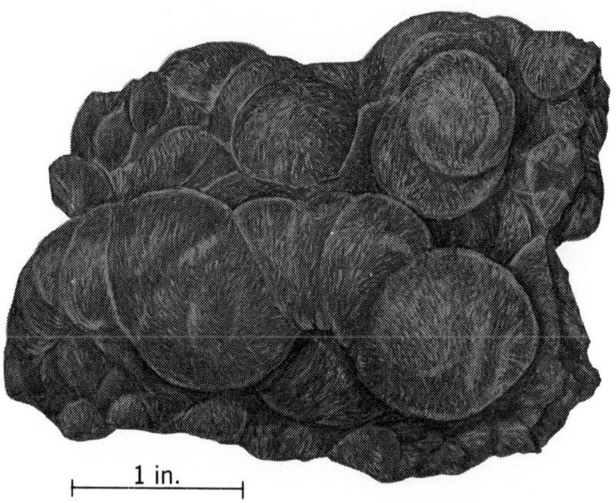

Prehnite crystals in reniform aggregates, Paterson, NJ. *(Specimen from Department of Geology, Bryn Mawr College)*

6½ on Mohs scale; specific gravity is 2.8–2.9. It has a vitreous luster with a light-green to white color. Prehnite is characteristically found lining the cavities in basaltic rocks. It is associated with datolite, zeolites, calcite, and pectolite. Noted localities in the United States are at Paterson and Bergen Hill, N.J.; Westfield, Mass.; and Farmington, Conn. *See* SILICATE MINERALS.

[CORNELIUS S. HURLBUT, JR.]

Proterozoic

A term, more or less synonymous with Algonkian, widely used for later Precambrian time in the twofold subdivision of the Precambrian. Many students of the Precambrian find this division unsuitable and no longer use it. The Geological Society of Canada has divided it into Lower, Middle, and Upper, and has endeavored to fit later Precambrian rocks and history into this framework. *See* PRECAMBRIAN.

[J. PAUL FITZSIMMONS]

Bibliography: J. E. Gill (ed.), *The Proterozoic in Canada*, 1957.

Proustite

A mineral having composition Ag_3AsS_3 and crystallizing in the hexagonal system. It occurs in prismatic crystals terminated by steep ditrigonal pyramids (see illustration), but is more commonly massive or in disseminated grains. There is good rhombohedral cleavage. Hardness is 2–2.5 (Mohs scale) and specific gravity is 5.55. The luster is adamantine and the color ruby red. It is called light ruby silver in contrast to pyrargyrite, dark ruby silver. Proustite is less common then pyrargyrite but the two minerals are found together in silver veins. Noted localities are at Chañarcillo, Chile; Freiberg, Germany; Guanajuato, Mexico; and Cobalt, Ontario, Canada. *See* PYRARGYRITE.

[CORNELIUS S. HURLBUT, JR.]

PROUSTITE

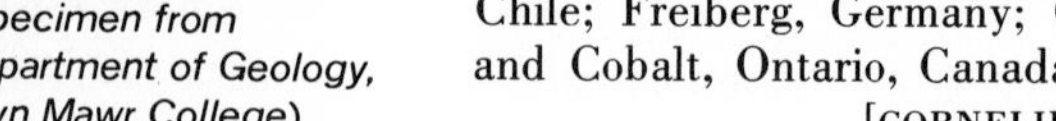

Proustite crystals, Marienberg, Germany. *(Specimen from Department of Geology, Bryn Mawr College)*

Psilomelane

A basic oxide of barium and manganese with the idealized composition $BaMnMn_8O_{16}(OH)_4$. X-ray structural studies have shown that psilomelane is orthorhombic in crystallization. The mineral itself is not well crystallized, and typically occurs

Psilomelane from Thuringia, Germany. *(American Museum of Natural History specimen)*

as fine-grained masses and crusts with a botryoidal or reniform structure (see illustration). The color is iron black to dark steel gray. The hardness is about 5½ on Mohs scale, and the specific gravity is 4.71.

Psilomelane frequently occurs admixed with other manganese oxides, chiefly pyrolusite, and with clay and hydrated iron oxides. The recognition of psilomelane and the proper use of the name has been attended by much confusion. The name formerly was used in part in a generic sense for ill-defined, hard, fine-grained manganese oxides, which often contained little or no barium, and the true status of many psilomelane-like minerals described in the literature or preserved in museum collections still remains uncertain. Psilomelane is a secondary mineral formed under surface or near-surface conditions of temperature and pressure. *See* PYROLUSITE.

[CLIFFORD FRONDEL]

Pumice

A rock froth, formed by the extreme puffing up (vesiculation) of liquid lava by expanding gases liberated from solution in the lava prior to and during solidification. Some varieties will float in water for many weeks before becoming waterlogged. Typical pumice is siliceous (rhyolite or dacite) in composition, but the lightest and most vesicular pumice (known also as reticulite and thread-lace scoria) is of basaltic composition. *See* LAVA; VOLCANIC GLASS.

[GORDON A. MACDONALD]

Pyrargyrite

A mineral having composition Ag_3SbS_3 and crystallizing in the hexagonal system. Pyrargyrite crystals are prismatic with hemimorphic development and are usually distorted (see illustration). The mineral also occurs in massive form and in disseminated grains. There is good rhombohedral cleavage; the hardness is 2.5 on Mohs scale and specific gravity is 5.85. The luster is adamantine and the color a deep ruby red to black, giving it the name dark ruby silver. Pyrargyrite is an important silver ore when it is found in veins associated with proustite and other silver minerals. It has been mined as silver ore at Chañarcillo, Chile; Freiberg, Germany; Guanajuato, Mexico; and Cobalt, Ontario, Canada. *See* PROUSTITE.

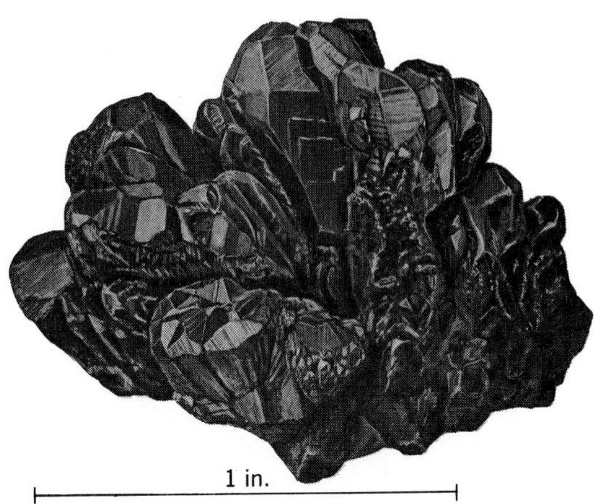

Pyrargyrite crystals, Harz Mountains, Germany. (*Specimen from Department of Geology, Bryn Mawr College*)

[CORNELIUS S. HURLBUT, JR.]

Pyrite

A mineral having composition FeS_2 and crystallizing in the isometric system. Pyrite, or iron pyrites, is more commonly well crystallized than any other sulfide mineral; the cube is the dominant form (see illustration). Striae are usually present on the cube faces running at right angles to the edges. The pyritohedron and octahedron are frequently present. A penetration twin of two pyritohedrons is known as the iron cross. Pyrite is also massive, granular, and stalactitic.

Pyrite has a hardness of 6–6.5 on Mohs scale and a specific gravity of 5.02. The luster is metallic and the color brass yellow; in very-fine-grained, compact aggregates the color may be greenish. Pyrite is the most common "fool's gold" but is hard and brittle, whereas gold is soft and sectile. Its high hardness also distinguishes it from softer chalcopyrite. In the pure mineral, iron makes up 46.6% and sulfur 53.4%. When ignited, the high percentage of sulfur permits pyrite to support its own combustion. Small amounts of nickel and cobalt may be present, and some analyses show considerable nickel, indicating the possibility of a complete solid-solution series between pyrite and bravoite, $(Ni,Fe)S_2$. Gold, copper, nickel, and arsenic reported in some analyses are probably the result of mechanical mixtures of other minerals, and the elements present are not substituting for iron or sulfur in the structure. FeS_2 is dimorphous and crystallizes in an orthorhombic modification as marcasite. Marcasite is distinguished from pyrite by its lighter color, lower specific gravity, and crystal form. *See* MARCASITE.

Pyrite is the most common as well as the most widespread of the sulfide minerals. It occurs under almost all conditions of mineral deposition, from the high temperatures of an igneous magma to the temperature of the ocean bottom, near 0°C. It is present as an accessory mineral in many igneous rocks and in some has formed large deposits as a magmatic segregation. Large masses are found in contact metamorphic ore deposits and are mined for the associated copper minerals, chiefly chalcopyrite and chalcocite. Pyrite is a common mineral in most sulfide veins, whether formed at great depths and high temperature or at shallow depths and low temperature. In veins it is associated with many minerals but most frequently with chalcopyrite, sphalerite, and galena. Pyrite of both primary and secondary origin is common in sedimentary rocks and forms nodular and banded masses in coal known as brasses. It is also found in metamorphosed sedimentary rocks, in places in well-formed crystals. *See* CHALCOPYRITE; GALENA; SPHALERITE.

Pyrite is so universal in its occurrence that only a few major localities will be noted here. Fine crys-

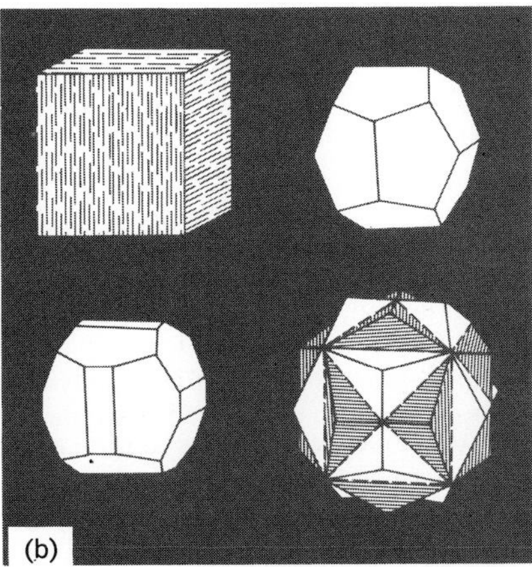

Pyrite. (*a*) Crystal from Rio Marina, Elba, Italy (*American Museum of Natural History specimens*). (*b*) Crystal habits (*from C. S. Hurlbut, Jr., Dana's Manual of Mineralogy, 17th ed., copyright © 1959 by John Wiley & Sons, Inc.; reprinted by permission*)

tals have been found at Waldenstein, Austria; St. Gotthard, Switzerland; on the island of Elba; at Saint-Pierre-de-Mésage, France; and in Cornwall, England. Large deposits occur at Rio Tinto and elsewhere in Spain where the pyrite and associated chalcopyrite are mined on a large scale. In the United States fine crystals have been reported from many localities, notably Rossie, N.Y.; Leadville, Colo.; Clifton, Ariz.; and Bingham Canyon, Utah. Well-formed cubes are found in a chlorite schist at Chester, Vt. Massive deposits are found at Louisa, Va.; Ducktown, Tenn.; and Rico, Colo.

Under oxidizing conditions pyrite readily alters to various iron sulfates and eventually to goethite and limonite. These iron oxides form the chief minerals of gossan, which is the surface expression of pyrite-rich mineral deposits. *See* ORE AND MINERAL DEPOSITS.

Because of the high percentage of sulfur (53.4%) in pyrite, this mineral is one of the sources of sulfur used in the manufacture of sulfuric acid. In places it is mined for sulfur alone; elsewhere sulfur is recovered as a by-product in smelting of ores rich in pyrite. In the United States, native sulfur supplies most of the demand for the element. Elsewhere, however, pyrite assumes a more important role as a source of sulfur. *See* GOETHITE; LIMONITE. [CORNELIUS S. HURLBUT, JR.]

Pyroclastic rocks

Rocks of extrusive (volcanic) origin, composed of rock fragments produced directly by explosive eruptions. Pyroclastic fragments may represent shattered and comminuted older rocks (volcanic, plutonic, sedimentary, or metamorphic) or solidified lava droplets formed by violent explosion (see table). *See* TUFF; VOLCANO.

[CARLETON A. CHAPMAN]

Classification of pyroclastic materials

Grain size, mm	Unconsolidated rock	Consolidated rock
< 4	Ash	Tuff
> 4 < 32	Lapilli	Lapilli tuff
> 32	Blocks (angular)	Breccia
	Bombs (rounded)*	Agglomerate†

*Masses erupted as large blobs of liquid lava.
†This is essentially a breccia with abundant bombs and huge blocks, found within or near the volcanic vent.

Pyrolusite

A mineral having composition MnO_2. Pyrolusite is the most important ore of manganese. It crystallizes in the tetragonal system but well-developed crystals (polianite) are rare. It is usually in radiating fibers or reniform coatings. The hardness is 1–2 on the Mohs scale (often soiling the fingers) and the specific gravity is 4.75. Crystals of polianite show a perfect prismatic cleavage and have a hardness of 6 and a specific gravity of 5.1. The luster is metallic and the color iron-black. It frequently forms pseudomorphs after other manganese minerals, notably manganite.

Pyrolusite is a secondary mineral formed by the alteration of other manganese minerals such as manganite, psilomelane, rhodochrosite, and rhodonite. Manganese dissolved from rocks by surface solutions may be redeposited as pyrolusite as dendritic coatings on the walls of fractures, as nodules on the sea bottom, and as beds in residual clays. Pyrolusite is extensively mined as a manganese ore in many countries, chiefly in the Soviet Union, Ghana, India, the Republic of South Africa, Morocco, Brazil, and Cuba. The chief use of manganese is in making spiegeleisen and ferromanganese, employed in steel manufacture. It is also used as an oxidizer in the production of chlorine, bromine, and oxygen; in electric cells and batteries; and as a decolorizer in glass. *See* MANGANITE; PSILOMELANE; RHODOCHROSITE; RHODONITE.

[CORNELIUS S. HURLBUT, JR.]

Pyroxene

A family of diverse and important rock-forming minerals crystallizing over a wide range of temperature, pressure, and composition and occurring as constituents of a host of rock types. The crystal chemistry of the pyroxene family is very complicated: Many chemical and crystallographic problems occur for these compounds, and their study often requires meticulous and detailed investigation. Since the pyroxenes are so ubiquitous and since their compositions range extensively, research on these compounds reveals important information about the origin and thermal history of certain rocks. A brief discussion of their crystallography, chemistry, and nomenclature follows.

Crystallography. All pyroxenes have infinite $(Si_2O_6)_\infty$ inosilicate chains as their principal motif. In crystallographic discussion, it is often convenient to think of the oxygen atoms as situated at the corners of polyhedrons whose centers are the positively charged cations. A pyroxene chain is constructed by joining silicate tetrahedrons at corners such that two out of four tetrahedral corners are shared with other tetrahedrons (Fig. 1). For pyroxenes, the chain repeats itself every 5.2 A. The related pyroxenoids, also chain silicates, are distinct in having longer chain repeats. These chains in turn link to sheets of two kinds of edge-sharing polyhedrons, whose centers are called M(1) and M(2) (Fig. 1). In this manner, a three-dimensional edifice is constructed. M(1) is the smaller polyhedron approximating an octahedron, and the M(1) site is coordinated to six vertices (oxygen atoms). M(2) is not only larger but also more complicated: It may possess 6-, 7-, or 8-coordination by oxygen, depending on which pyroxene structure is depicted.

There are several ways of stacking the chains and sheets in three dimensions while preserving the general pyroxene formula; each arrangement corresponds to a distinct crystal structure. For the pyroxenes, at least four arrangements are known; of these four arrangements, two have orthorhombic symmetry and two monoclinic symmetry.

Chemistry. Following H. Hess (1949), the general pyroxene formula can be written $(W)_{1-p}(X,Y)_{1+p}(Z_2O_6)$, where $W = Ca^{2+}$, Na^+; $X = Mg^{2+}$, Fe^{2+}, Mn^{2+}, Ni^{2+}, Li^+; $Y = Al^{3+}$, Fe^{3+}, Cr^{3+}, Ti^{4+}; and $Z = Si^{4+}$, Al^{3+}. Extensive solid solution often exists, such as the series $CaMg(Si_2O_6)$ (diopside)–$CaFe(Si_2O_6)$ (hedenbergite).

The pyroxenes can be subdivided on the basis of

their crystal structures and chemical compositions. Each end member and structure type has an associated specific name (see table).

The phase relations among the pyroxenes are complicated and only incompletely understood. Briefly, the composition $Mg_2(Si_2O_6)$ possesses several polymorphs. The low-temperature polymorph is orthoenstatite. At about 1000°C, the mineral inverts to protoenstatite. Quenching of the protoenstatite leads to metastable low clinoenstatite. Since the monoclinic structure cell of low clinoenstatite can be geometrically related to the orthogonal cell of protoenstatite, the inverted low clinoenstatite shows polysynthetic twinning. Adding iron to the system leads to the high-temperature inversion orthoenstatite (orthoferrosilite) → high clinoenstatite (high clinoferrosilite), which also shows polysynthetic twinning. Rapid quenching leads to low clinoenstatite.

A diagram of the pyroxene nomenclature, modified from A. Poldervaart and H. H. Hess (1951), appears in Fig. 2. It is seen that the addition of calcium to the series enstatite-ferrosilite produces the pigeonite structures with space group $P2_1/c$. Further addition of calcium leads to pyroxenes having space group $C2/c$. The influence of the large calcium atom upon the crystal structure is attributed to changes in the environment about the M(2) site. Thus, different arrangement of the chains can better accommodate the large cation, and the orthorhombic pyroxenes proceed to the monoclinic pyroxenes when calcium is added.

Cooling of pyroxenes under certain conditions can lead to complex exsolution textures. For example, orthorhombic pyroxenes close to the $Mg_2(Si_2O_6)-Fe_2(Si_2O_6)$ join may exsolve polysynthetically twinned augite. Pigeonites may also exsolve augite. In certain instances, further phase inversion of the host or the exsolved lamellae may lead to very complex crystallographic interrelationships. For example, exsolved lamellae of a pigeonite structure may further invert to an orthorhombic pyroxene structure, with the augite host retaining its original structure. Exsolution is often attributed to the effect of strain on the crystal upon cooling, often caused by the presence of a large cation such as calcium whose solubility in orthopyroxene and pigeonite structures increases with increasing temperature. The study of exsolution textures and twinning requires careful single-crystal x-ray diffraction experiments.

Since the M(1) and M(2) sites in pyroxenes are crystallographically distinct, preferential distribution of cations is possible. For example, in the hypersthenes $(Fe,Mg)_2(Si_2O_6)$ iron prefers the M(2) site. Furthermore, its fractional occupancy varies with temperature. Such information, established by Mössbauer resonance experiments, provides means of assessing cooling histories of pyroxenes in natural systems and, consequently, the cooling histories of their host rocks. *See* AUGITE; DIOPSIDE; ENSTATITE; JADEITE; PIGEONITE.

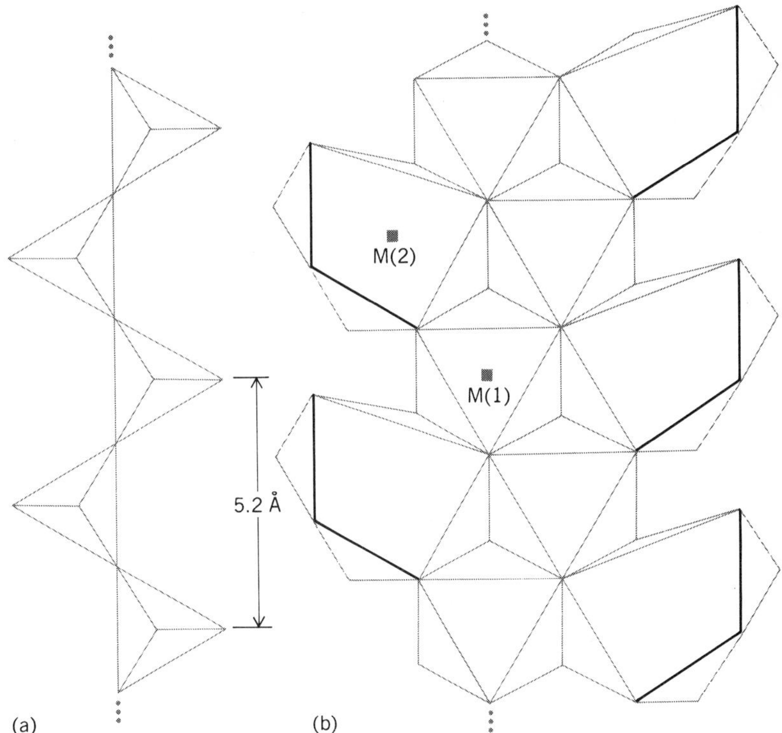

Fig. 1. Principal structural features of the C-centered clinopyroxenes. (*a*) The inosilicate chain, made up of corner-linked silicate tetrahedrons. (*b*) Portion of the large cation sheet, with octahedral M(1) and the less regular 8-coordinated M(2) sites. The sheet is constructed by joining dashed edges of M(2) polyhedrons with other M(2) polyhedrons. Inosilicate chains join to sheets by sharing the edges indicated by heavy lines.

The pyroxene family of minerals

Name	Formula	Space group
Orthorhombic pyroxenes		
Orthoenstatite-orthoferrosilite	$(Mg,Fe)_2(Si_2O_6)$	*Pbca*
Protoenstatite	$Mg_2(Si_2O_6)$	*Pbcn*
Monoclinic pyroxenes		
Diopside-hedenbergite	$Ca(Mg,Fe)(Si_2O_6)$	*C2/c*
Johannsenite	$CaMn(Si_2O_6)$	*C2/c*
Schefferite	$Ca(Mg,Mn)(Si_2O_6)$	*C2/c*
Augite	$(Ca,Mg,Fe,Al)_2((Si,Al)_2O_6)$	*C2/c*
Aegirine	$NaFe(Si_2O_6)$	*C2/c*
Jadeite	$NaAl(Si_2O_6)$	*C2/c*
Spodumene	$LiAl(Si_2O_6)$	*C2/c*
High clinoenstatite	$Mg_2(Si_2O_6)$	*C2/c*
Clinoferrosilite	$Fe_2(Si_2O_6)$	$P2_1/c$
Low clinoenstatite	$Mg_2(Si_2O_6)$	$P2_1/c$
Pigeonite	$(Mg,Fe,Ca)(Mg,Fe)(Si_2O_6)$	$P2_1/c$

Physical properties. Common pyroxenes range in color from white (pure diopside and enstatite) through shades of yellow and green to deep brown and greenish-black (ferrosilite, hedenbergite, and aegirine). Spodumene of a lilac-pink color and of gem quality is called kunzite; sea-green varieties are called hiddenite. Specific gravity ranges from 3.2 (enstatite) to 4.0 (ferrosilite). The hardness on Mohs scale ranges from $5\frac{1}{2}$ to 6. The orientation of the chains dictates the cleavage directions, since good cleavage planes should not pass through the silicate chains. In the orthopyroxenes, this direction is (210); for the clinopyroxenes, it is (110). The angle between these planes is nearly 90°, distinguishing the pyroxenes from the related amphibole double-chain structures, which possess an angle of about 55°. *See* AMPHIBOLE.

Occurrence. Pyroxenes occur in most rock types of medium to high grade, particularly in basalts, pyroxenites, charnockites, and ultrabasic rocks derived from gravity settling of ferromagnesian minerals. As constituents in ultrabasic rocks, they occur often with olivine and spinel. Diopside

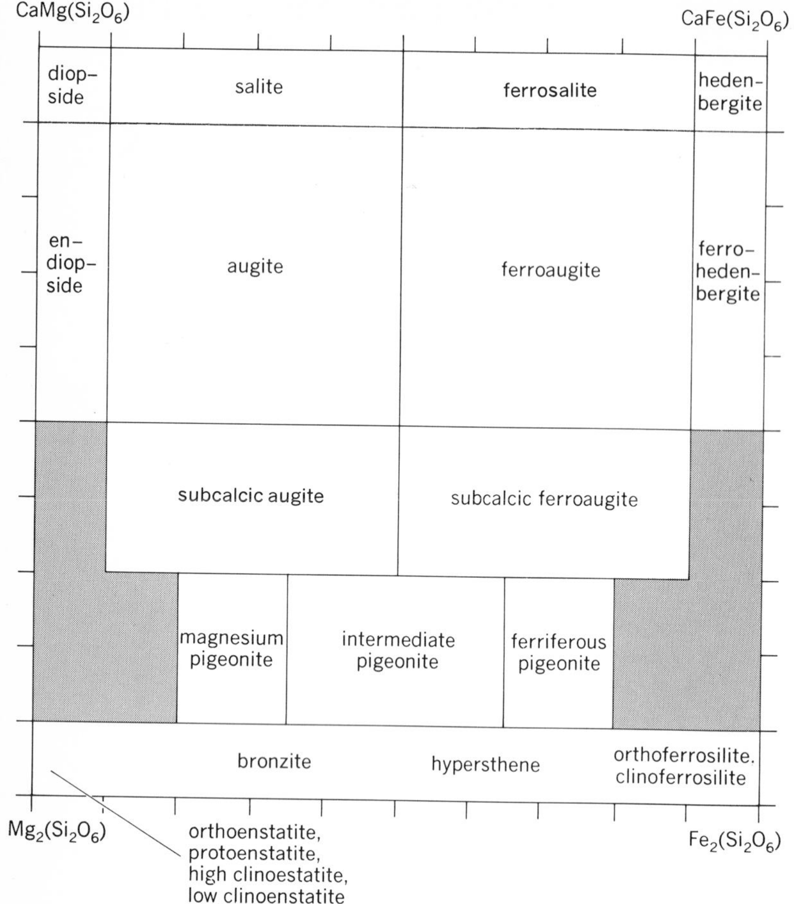

Fig. 2. Pyroxene nomenclature. Shaded regions have no natural counterparts. *(After A. Poldervaart and H. H. Hess, 1951)*

and hedenbergite are typical metamorphic and skarn products. A common reaction is given below. Augites occur as constituents of basalts and

$$CaMg(CO_3)_2\ \text{(dolomite)} + 2SiO_2\ \text{(quartz)} \rightleftarrows CaMg(Si_2O_6)\ \text{(diopside)} + 2CO_2$$

charnockites, and as common metamorphic products, often associated with hornblende amphibole. Clinoenstatite is very rare in terrestrial environments, but it frequently occurs in stony meteorites.

Aegirine is a typical pyroxene in nepheline syenites, where it occurs as nearly black elongated prismatic crystals, often called acmite. Jadeite is a high-pressure mineral and is one of the varieties of the gem-quality jade. Related is omphacite, which occurs with pyrope garnet in the high-pressure rock called eclogite. Johannsenite is a rare pyroxene occurring as a skarn product, associated with schefferite, rhodonite, and high-grade manganese oxide ores. Spodumene is practically restricted to coarse-grained lithium-bearing granite pegmatites. Crystals of this mineral, often of large size, are mined as an ore of lithium. *See* SPODUMENE.

Pyroxenes are frequently found more or less altered to hydrated minerals. A typical process of pyroxene alteration involves retrograde metamorphism where water has been introduced. Typical products are amphiboles, chlorites, and serpentines. Spodumene is known to alter to eucryptite and albite; further alteration leads to albite and micas, the assemblage being called cymatolite.

[PAUL B. MOORE]

Bibliography: W. A. Deer, R. A. Howie, and J. Zussman, *Rock-Forming Minerals*, vol. 2: *Chain Silicates*, 1963; H. H. Hess, Chemical composition and optical properties of common clinopyroxenes, *Amer. Mineral.*, 34:621–666, 1949; A. Poldervaart and H. H. Hess, Pyroxenes in the crystallization of basaltic magma, *J. Geol.*, 59:472–489, 1951.

Pyroxenite

A heavy, dark-colored, phaneritic (visibly crystalline) igneous rock composed largely of pyroxene with smaller amounts of olivine or hornblende. Pyroxenite composed largely of orthopyroxene occurs with anorthosite and peridotite in large, banded gabbro bodies. These rocks are formed by crystallization of gabbroic magma (rock melt). Some of these pyroxenite masses are rich sources of chromium. Certain pyroxenites composed largely of clinopyroxene are also of magmatic origin, but many probably represent products of reaction between magma and limestone. Other pyroxene-rich rocks have formed through the processes of metamorphism and metasomatism. *See* GABBRO; IGNEOUS ROCKS; PERIDOTITE; PYROXENE.

[CARLETON A. CHAPMAN]

Pyrrhotite

A mineral with composition $Fe_{1-x}S$ ($x = 0$ to 0.2), possessing an atomic arrangement based on the NiAs (niccolite) structure type in which the sulfur atoms reside in approximate hexagonal close-packing. Eskebornite, $Fe_{1-x}Se$, is the selenium analog. All pyrrhotites have vacancies over the iron positions and the formula should be written $\square_x Fe_{1-x}S$, where $\square$ are holes. These vacancies are ordered; that is, they occur periodically in the pyrrhotite crystal structure. Writing $A (= 3.4\ A)$, $B (= \sqrt{3}\ A)$, and $C (= 5.7\ A)$ as the niccolite cell parameters, the following pyrrhotite structure cells have been found on the basis of extensive single crystal studies: B, $2C$ (hexagonal); $2A$, $4C$ (hexagonal or lower, typical pyrrhotite); $2B$, $2A$, $4C$, $\beta = 90°40'$ (monoclinic); $2A$, $5C$ ("hexagonal"); and $2A$, $3C$ (hexagonal). More than one of these arrangements may occur on the same specimen. The iron-deficient pyrrhotites are ferrimagnetic at room temperature, but at some higher temperature they become paramagnetic, presumably because of vacancy disorder.

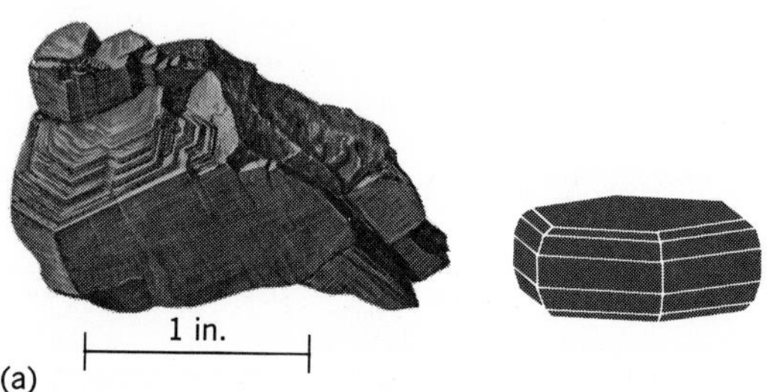

Pyrrhotite. (*a*) Crystals from Chihuahua, Mexico (*specimen from Department of Geology, Bryn Mawr College*). (*b*) Crystal habit (*from C. S. Hurlbut, Jr., Dana's Manual of Mineralogy, 17th ed., copyright © 1959 by John Wiley & Sons, Inc.; reprinted by permission*)

The mineral occurs as rounded grains to large masses, more rarely as tabular pseudohexagonal crystals (see illustration) and rosettes. Color is brownish bronze-yellow with dark grayish-black streaks. Hardness is 4 on Mohs scale and specific gravity 4.6 (for the composition Fe_7S_8). It can be synthesized by heating iron and sulfur or by heating pyrite, FeS_2, in a stream of H_2S vapor at a temperature of 550°C.

Pyrrhotite occurs in basic igneous rocks as a late-stage fractional differentiate, particularly in norities and gabbros, and sufficient quantities may constitute an ore of iron, such as in Sweden, Finland, Norway, and Canada. At Sudbury, Ontario, it occurs with chalcopyrite, $CuFeS_2$, and pentlandite, $(Fe,Ni)_9S_8$, in gabbroic rocks, the latter mineral constituting an important ore of nickel. Pyrrhotite also occurs with magnetite and chondrodite in contact metamorphic marbles, and in low-temperature veins with calcite and other sulfides and sulfosalts. Troilite, FeS, has been found in some meteorites but it is rare as a terrestrial mineral. Various other iron sulfides include pyrite and marcasite, FeS_2; smythite and greigite, Fe_3S_4; and mackinawite, FeS. *See* PENTLANDITE; PYRITE.

[PAUL B. MOORE]

Quartz

The most abundant and widespread of all minerals. Quartz is an important rock-forming mineral that occurs as a subordinate constituent of many igneous, metamorphic, and sedimentary rocks; it is the principal constituent of sandstone and quartzite and of unconsolidated sands and gravels. It does not vary much in its chemical composition (SiO_2, silicon dioxide). Trace amounts of aluminum and of alkalies, chiefly sodium and lithium, are commonly present in solid solution. Ferric iron and titanium are characteristically present in traces in amethyst and rose quartz, respectively. Quartz crystallizes in the trigonal trapezohedral class of the rhombohedral subsystem; the lattice type is hexagonal. It often occurs as well-formed crystals, which typically are short prismatic in shape and are bounded by the hexagonal prism $m\{10\bar{1}0\}$ with terminal faces of the rhombohedrons $r\{10\bar{1}1\}$ and $z\{0111\}$. Small modifying faces of the trigonal pyramid $s\{11\bar{2}1\}$ and of the trigonal trapezohedron $x\{51\bar{6}1\}$ are often present (see illustration). The hardness is arbitrarily designated as 7 on Mohs scale, and the specific gravity is 2.650. The fracture of crystals is subconchoidal, but more or less distinct cleavages are sometimes observed on the rhombohedrons $(10\bar{1}1)$ and $(01\bar{1}1)$ and on the prism $(10\bar{1}0)$. Ordinarily, quartz is colorless and transparent, and the luster is vitreous. Amethyst is a purple or bluish-violet gem variety, and citrine is an orange-brown gem variety; most citrine sold commercially is produced by heat treatment of amethyst. Rose quartz is a massive type found in pegmatites. Quartz also may have a smoky yellow to dark smoky brown color, varying to brownish-black and almost opaque.

So-called blue quartz is found as irregular grains in some igneous and metamorphic rocks. The color is due to the scattering of light (Tyndall effect) from microscopic inclusions of rutile. Translucent to nearly opaque white or grayish-white types, usually found massive in veins or pegmatites, are known as milky quartz. The term rock crystal is applied to colorless and flawless crystals of quartz used for carving ornamental objects. The term crystal as applied to fine glassware is derived from this use. Small, brilliant quartz crystals resemble diamonds and are familiarly known as Herkimer diamonds, Cape May diamonds, and so on from the locality where they are found. *See* AMETHYST.

Quartz. (*a*) Smoky quartz crystal with microcline, found in Florissant, Colo. (*American Museum of Natural History specimen*). (*b*) Left-handed crystal form. (*c*) Right-handed crystal form.

Properties. Quartz is enantiomorphous; crystals are either right-handed or left-handed, and correspondingly rotate the plane of polarization of transmitted light. It also is piezoelectric. Twinning is common in quartz. The chief twin laws are the Dauphiné law, in which the twinned parts are related by a rotation of 180° around the *c* axis, and the Brazil law (or optical twinning), in which the twinned parts are enantiomorphs of different hand. In twins on the Japan law, the *c* axes are inclined at 84°33′, and a pair of prism faces are coplanar. Quartz, also known as α-quartz or low quartz, is one of seven known polymorphs of silica. It is stable below about 573°C; heated above this temperature, it inverts reversibly to high quartz.

Large flawless crystals of quartz are employed for the manufacture of quartz oscillator plates, as prisms in optical spectrographs, and as other optical devices. Quartz also is employed as an abrasive and in the manufacture of firebrick and other refractories and of glass. Natural quartz crystals of commercial grade are principally obtained from Brazil, but in recent years crystals of adequate quality have been synthesized. Synthetic crystals up to several pounds in weight are produced commercially by crystallization from alkaline solutions contained in steel pressure vessels at temperatures from 200 to 500°C and pressures up to 25,000 psi.

Varieties. The varieties of quartz may be classed into two broad categories: coarse-crystalline, including euhedral crystals and massive granular types, in which the individual grains are visible to the unaided eye; and fine-crystalline, comprising massive types in which the individual grains or fibers are distinctly seen only under the microscope. Coarse-crystalline quartz includes varieties based on color, structure, and inclusions. The fine-crystalline types usually have a waxy to dull luster and are weakly translucent to opaque. They are characterized by a slightly diminished specific gravity, due to porosity, and by greater ease of attack by chemical agents. The fine-crystalline types of quartz are loosely divided into two main groups on the basis of particle shape. These comprise the fibrous kinds, including chalcedony, carnelian, agate, and their color variants; and fine-granular kinds, chiefly flint, together with jasper and its variants. These materials have a very extensive varietal nomenclature that has arisen for the most part from their use since ancient times as gem or ornamental materials. *See* AGATE; CHALCEDONY; GEM; JASPER.

[CLIFFORD FRONDEL]

Quartzite

A metamorphic rock consisting largely or entirely of quartz. Most quartzites are formed by metamorphism of sandstone; but some have developed by metasomatic introduction of quartz, SiO_2, often accompanied by other chemical elements, for example, metals and sulfur (ore quartzites). The geological relations and the shape of quartzite bodies serve to distinguish between them (see illustration). The metasomatic quartzites are often found as contact products of intrusive bodies. *See* METAMORPHIC ROCKS; METASOMATISM; SANDSTONE.

The transition from sandstone to quartzite is gradational. All stages of relic clastic structures are encountered. Some sandstones are soon completely metamorphosed. Others are very resistant, and in many highly metamorphic quartzites of the Precambrian, there are relic structures still to be observed.

Pure sandstones yield pure quartzites. Impure sandstones yield a variety of quartzite types. The cement of the original sandstone is in quartzite recrystallized into characteristic silicate minerals, whose composition often reflects the mode of development. Even the Precambrian quartzites correspond to types that are parallel to present-day deposits.

Carbonate cement reacts with silica to produce silicates during high-temperature metamorphism. However, ferric silicate has never been observed. The Fe_2O_3 pigment in deposits of desert sand resists any degree of metamorphism. Therefore, old Precambrian quartzites exhibit the same red color as the present-day sand of Sahara.

Under the conditions of regional metamorphism, cement composed of clay gives rise to sillimanite or kyanite, potash-rich cement yields potash feldspar or mica, lime and alumina yield plagioclase or epidote, dolomitic cement yields diopside or tremolite, and siderite cement yields gruenerite. Wollastonite will crystallize from pure calcite cement. Such quartzites occur as folded layers alternating with layers of other sedimentary rocks.

In some feldspathic quartzites, thin dark micaceous layers parallel to the foliation superficially suggest relic bedding. Many of the Moine schists of Scotland illustrate this. True bedding is marked, however, by thin strings of zircon, iron ore, or other inherited concentrations of the heavy minerals.

Under the condition of contact metamorphism, the cement of the original sandstone will recrystallize and minerals of the hornfels facies will develop. Quartz itself is usually stable. However, in very hot contacts, against basic intrusions, quartz may invert to tridymite, which again has reverted to quartz; or it may even melt. Such vitrified sandstones resulting from partial fusion of inclusions in volcanic rocks are called buchites. They often contain fritted feldspar fragments, corroded grains of quartz, and a matrix of slightly colored glass corresponding to the fused part of the sandstone. *See* HORNFELS; QUARTZ; SILICATE MINERALS.

[T. F. W. BARTH]

Waterloo quartzite, with schistosity developed by shearing on bedding planes, Dodge County, WI. (*USGS*)

Quaternary

The latest major division of Cenozoic time (Cenozoic Era); the name refers collectively to all the geologic deposits (Quaternary System) that overlie or are younger than the Tertiary deposits, and also to the time (Quaternary Period) during which these deposits accumulated. The name was proposed by J. Desnoyers in 1829 as an addition to the standard names of rock groups (Primary, Secondary, and Tertiary) that had been in use since 1760.

Although Primary and Secondary have since been generally abandoned, Tertiary and Quater-

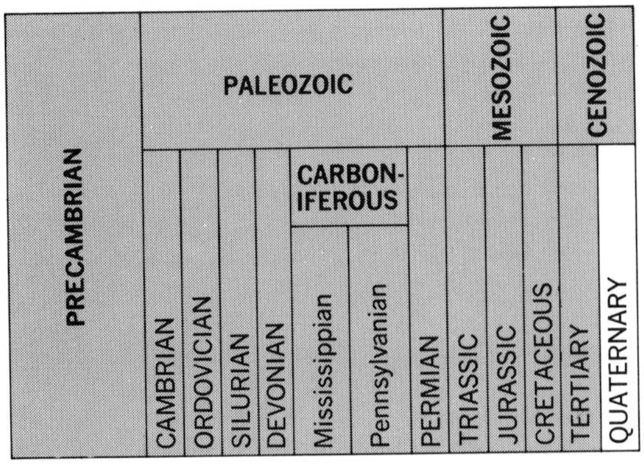

PRECAMBRIAN	PALEOZOIC							MESOZOIC			CENOZOIC	
					CARBONIFEROUS							
	CAMBRIAN	ORDOVICIAN	SILURIAN	DEVONIAN	Mississippian	Pennsylvanian	PERMIAN	TRIASSIC	JURASSIC	CRETACEOUS	TERTIARY	QUATERNARY

nary still remain in rather wide use. The Quaternary System is commonly subdivided into a Pleistocene Series and, above it, a Recent Series. *See* CENOZOIC; PLEISTOCENE; RECENT.

[RICHARD F. FLINT]

Radioactive minerals

The so-called radioactive minerals loosely comprise species that contain uranium or thorium as an essential part of their chemical composition, together with minerals in which these elements are sometimes present in solid solution, usually in small and variable amounts. About 150 minerals fall into the first category, including many that are rare or that are imperfectly known. The principal uranium minerals from the point of view of their economic importance are the oxide uraninite and its variety pitchblende, the vanadates carnotite and tyuyamunite, the silicate coffinite, the phosphates autunite and torbernite, and the complex oxides brannerite and davidite. The chief thorium minerals of economic interest are the silicates thorite and thorogummite and the oxide thorianite.

The minerals that contain uranium or thorium in small amounts as vicarious constituents are relatively numerous. Some of them are of present or potential economic interest as sources of uranium or thorium, particularly when these elements can be obtained as by-products of the recovery of associated elements or minerals. The chief source of thorium in the past has been the rare-earth phosphate mineral monazite, which commonly contains thorium as a vicarious constituent in the range from 3 to 10% ThO_2. Uranium often is an important accessory constituent in the niobate-tantalates, and is present sometimes in significant amounts in many other minerals including allanite, zircon, and the apatite of some phosphate-rock deposits. Thorium occurs as an accessory constituent chiefly in minerals containing zirconium, cerium, calcium, or uranium.

The radioactive decay of the uranium and thorium present in minerals is accompanied by the emission of α-particles, electrons, and γ-radiation (x-rays), and a variety of methods of study of such minerals and of prospecting for them are based on the detection and measurement of this radiation. The crystalline structure of minerals containing uranium and thorium may be broken down in part or entirety by the internal absorption of radiation, chiefly α-particles, released within the crystal. In the final stages of radiation damage the crystal may become transformed into an amorphous, glassy body that is optically isotropic and does not diffract x-rays. Such minerals are said to be metamict. The structural change usually is accompanied by chemical alteration. Lead accumulates in all radioactive minerals, since it is the end product of the radioactive decay of the uranium and thorium, and measurement of the ratio of the lead to uranium and thorium, or of lead isotope ratios, has been widely used as a method of measuring geologic time.

From the crystallochemical viewpoint, minerals that contain uranium fall into two broad categories: those that contain uranium in its quadrivalent state, and those that contain it in its hexavalent state as the so-called uranyl ion. The minerals that contain quadrivalent uranium generally are black in color, do not fluoresce in ultraviolet radiation, and occur as primary or hypogene deposits. They include uraninite and coffinite among the uranium minerals proper, together with virtually all the minerals that contain uranium as a vicarious constituent. Most of the known uranium minerals contain the hexavalent uranyl ion instead of quadrivalent uranium. These uranyl compounds are characterized by a relatively bright lemon-yellow to green or orange color, and commonly fluoresce a bright lemon-yellow color in ultraviolet radiation; the colors may be modified by the nature of other cations present in the mineral in addition to uranium. Virtually all of the uranyl minerals are secondary in origin, and form from solution at relatively low temperatures and pressures. *See* ALLANITE; CARNOTITE; LEAD ISOTOPES, GEOCHEMISTRY OF; MONAZITE; ROCK, AGE DETERMINATION OF; URANINITE; ZIRCON.

[CLIFFORD FRONDEL]

Bibliography: H. Craig (ed.), *Isotopic and Cosmic Chemistry*, 1964; H. Faul, *Ages of Rocks, Planets and Stars*, 1966; C. Frondel, Systematic mineralogy of uranium and thorium, *U.S. Geol. Surv. Bull.*, no. 1024, 1958; J. W. Frondel, M. Fleischer, and R. S. Jones, Glossary of uranium- and thorium-bearing minerals, *U.S. Geol. Surv. Bull.*, no. 1250, 1967; E. W. Heinrich, *Mineralogy and Geology of Radioactive Raw Materials*, 1958; F. Kirchheimer, *Das Uran und seine Geschichte*, 1963.

Radiocarbon dating

A method of estimating the age of carbon-bearing materials which have formed within the last 70,000 years and which have utilized carbon dioxide from the atmosphere or some other portion of the Earth's dynamic carbon reservoir. The method, discovered by W. F. Libby and coworkers in 1947, is based on the decay of the naturally occurring radioactive isotope of carbon which has a mass of 14 atomic mass units (C^{14}). This method has been applied to wood and other plant remains, charcoal, marine and freshwater shells and other carbonate deposits, and carbon dissolved in groundwater and the ocean. Radiocarbon dating is invaluable in geology and anthropology, helping date climatic change and giving information concerning sequences and rates of geological events and of animal and cultural evolution. It also aids in understanding ocean circulation and groundwater movement.

More recently, radiocarbon, both naturally

occurring and produced through nuclear weapons testing, has been used as an isotopic tracer in oceanography, hydrology, and meteorology. Natural variations in atmospheric radiocarbon concentration have geophysical implications.

RADIOACTIVE DECAY OF CARBON-14

Carbon-14 is one of the three naturally occurring isotopes of the element carbon. Unlike C^{12} and C^{13}, which are stable isotopes, C^{14} undergoes a nuclear transformation to nitrogen-14. The average C^{14} atom exists 8000 years before it ejects a beta particle (converting one of its neutrons into a proton), giving the atom the chemical characteristics of nitrogen. As with all radioactive nuclides, C^{14} may be characterized by a half-life (5730 ± 40 years), the time required for half of a given number of atoms to undergo radioactive decay. Since the rate of transformation cannot be altered by any physical conditions on the surface of the Earth, the rate of disappearance of C^{14} from a sample bears an absolute relationship to time.

Since the Earth is about 4.5×10^9 years old, any C^{14} produced initially with the rest of the nuclides making up the solar system would have long since disappeared. C^{14} is observed on the surface of the Earth only because it is continuously being produced. Primary cosmic-ray particles upon entering the atmosphere produce neutrons which react with the abundant N^{14} atoms of the air. A neutron enters the N^{14} nucleus and knocks out a proton, converting the atom to C^{14}, which rapidly is oxidized to a $C^{14}O_2$ molecule. If this production has proceeded at a constant rate for many thousands of years, then the amount of C^{14} present on the surface of the Earth should reach a constant value. This can be most easily understood by considering the flow of water through a funnel. If water is poured into a funnel at a uniform rate, the level or amount of water in the funnel will increase to some constant amount and remain unchanged as long as the rate of pouring is not altered. Likewise, since the rate of decay of C^{14} atoms is a function only of the number of atoms present, the total number of C^{14} atoms in the reservoir (atmosphere, biosphere, and hydrosphere) must be constant if the production rate does not change. *See* EARTH, AGE OF.

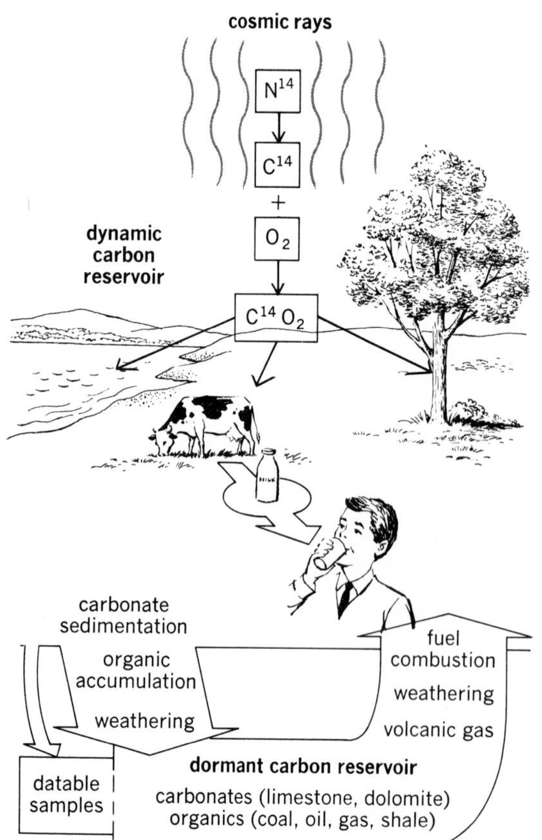

Fig. 1. The Earth's carbon.

The C^{14} produced in the atmosphere gradually mixes with all the other carbon in the dynamic reservoir (Fig. 1). Since the time required for mixing is less than 5730 years, the concentration of C^{14} in the carbon in the CO_2 of the Earth's atmosphere, in the Earth's hydrosphere (largely as bicarbonate ion), and in the Earth's biosphere as organic carbon is relatively uniform. When a material which received its carbon from this dynamic reservoir is isolated, its C^{14} concentration begins to decrease at a rate of half every 5730 years. Thus, if the C^{14} concentration in the carbon from a plant of unknown age were found to be half that in a plant living today, its age would be estimated as 5730 years.

MEASUREMENT OF RADIOCARBON

The concentration of radiocarbon in contemporary materials is so small that direct detection is impossible. Only one carbon atom in 10^{12} in living wood, for example, is C^{14}. This is far below the sensitivity limit of an isotope measurement device such as the mass spectrometer. Fortunately, the presence of C^{14} can be demonstrated through its radioactivity. The beta particle which is emitted during the nuclear transformation process can be detected with a counter. Since the number of beta particles emitted by a carbon sample in a given interval of time is directly proportional to the number of radiocarbon atoms present, an age determination can be made by comparing the radioactivity of carbon from a sample of unknown age with that from a contemporary sample. The relationship between the age t and the radioactivities is given by the equation below, where A_0 and A

$$t = 8268 \ln (A_0/A) \quad \text{years}$$

are, respectively, the radioactivity of the carbon prepared from the contemporary and from the unknown sample.

In order to obtain precise measurements of the radioactivity of carbon, very elaborate chemical and radiochemical procedures had to be developed. This is because the number of beta particles to be detected is extremely small (13 disintegrations per minute per gram contemporary carbon) and because they lack penetrating power (maximum energy, 0.15 Mev). To record these radiations, the sample has to be placed within an extremely sensitive radiation counter which has been adequately shielded from extraneous radioactivity (Fig. 2).

The original measurements were made by converting the carbon in the sample to carbon black. The sample was mounted on the inside of a steel cylinder which could be inserted into a sensitive

Geiger counter. Subsequently it was found that more precise measurements could be made if the sample was converted into a gas such as carbon dioxide, acetylene, or methane. The sample in this case is used as the filling gas for the counter.

The reduction of background radiation resulting from cosmic rays and natural radioactivity in the materials which make up the laboratory and its surroundings is accomplished as follows. The counter itself is constructed of materials as free as possible of radioactivity. The bulk of the external gamma radiation is eliminated by placing the counter within an iron or lead shield from 8 in. to as much as 30 in. thick. Additional mercury or lead shielding is often placed immediately around the counter to eliminate gamma radiation originating in the iron shield itself. H. DeVries showed that the addition of neutron shielding, such as paraffin plus boric acid, produces an additional reduction in background. Background from highly penetrating cosmic-ray mesons which cannot be physically shielded is eliminated by surrounding the sample counter with a ring of permanent Geiger counters. These counters record the mesons entering the shield, allowing electronic cancellation of any pulses produced by mesons in the sample counter (Fig. 2). A further reduction in the background is possible in some cases through the use of an internal anticoincidence system.

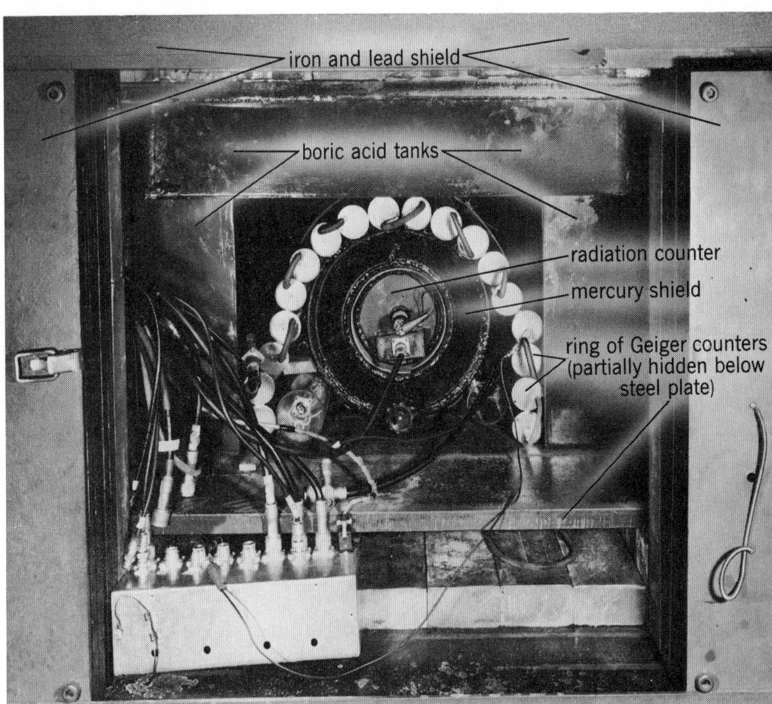

Fig. 2. Large-volume proportional counter, which is used for carbon-14 measurements. The outer shield is closed by rolling doors. The sample is introduced into the radiation counter in the form of carbon dioxide gas. (*Geochemical Laboratory, Lamont-Doherty Geological Observatory, Columbia University*)

A highly efficient technique for converting the gas carbon dioxide, CO_2, into the liquid benzene has made the use of an entirely different type of counter practical for radiocarbon dating. This technique of liquid scintillation counting includes the addition to the liquid sample of an organic compound which emits light when struck by a charged particle. This flash of light is "seen" by a light-sensitive phototube which records the count. Because the liquid sample occupies much less volume than does the gas, many fewer cosmic rays pass through the sample, and the background from this source is much reduced. The characteristics of this counter are compared to those of a typical high-sensitivity gas counter and those of the original Libby black-carbon counter in Table 1.

Such factors as sample size, precision on contemporary materials, and practical range for age estimation dating are also summarized. Whereas the limitation of counting sensitivity allows samples up to only about 40,000 years in age to be dated directly, preenrichment of the C^{14} in the sample gas by thermal diffusion raises the experimental sensitivity by as much as a factor of 16, thereby extending the potential age range to 70,000 years.

LIMITATIONS OF RADIOCARBON METHOD

The radiocarbon method should be regarded as a means of estimating the age rather than a method of age determination. Naturally one would like this estimation to be as close as possible to the real age, although the necessity to be assured of this depends in part on what use is to be made of the estimate.

The limitations of the method are of three types. One is the predictable uncertainty arising from statistical considerations, which is usually referred to as the precision of the method. The second arises from violations of the assumptions used in the dating method, and this problem affects the accuracy of the age, or how closely the estimation approximates the true age.

It should be emphasized that the true uncertainty in the age stems not from the predictable uncertainties in the measurement of radiocarbon but in how close the assumptions used in calculating that age are to reality. This is usually not reflected in the uncertainty limits placed on the age.

The third type of uncertainty or error is associated with the interpretation or use of the radiocarbon date.

Uncertainty in measurement. The measurement uncertainty results largely from the statistical error in the counting. Just as 10 flips of a coin will not always produce 5 heads and 5 tails, the same number of radioactive transformations will not always occur in a given sample in one 5-min period as in the next. As in the case of the coin,

Table 1. Characteristics of some typical radiocarbon counters

Characteristic	Solid-carbon counter*	Gas proportional counter	Liquid scintillation counter
Sample size (grams C)	6	2	3.5
Contemporary sample (counts/min)	5.5	15.2	30.0
Shielded background (counts/min)	4.0	0.8	8.0
Counting precision for contemporary samples (years)†	200	100	100
Maximum age limit (years)†	23,000	38,000	33,000
Minimum age limit (years)†	500	370	360

*This type of counter is no longer used
†Counting precision and age limits are given for one counting period.

the deviations are random and occur with a probability which can be mathematically predicted. Also in both cases the deviations will decrease as the number of events (hence radioactive transformations) observed increases. Thus most radiocarbon activity measurements are carried out for periods of from 1–4 days. Reduction of the error by counting for longer periods is impractical and improvements in the technology to reduce counting uncertainties are extremely difficult.

The measurement uncertainty is customarily converted into an age uncertainty and expressed, for instance, as 11,000±200 years before the present. (The "present" by custom is A.D. 1950.) This means, in effect, that the estimated age is most probably (two chances out of three or a 67% probability) between 10,800 and 11,200 years old. There is still a 25% chance that it is older or younger than these limits by another 200 years and a much reduced probability of being even younger than 10,600 years or older than 11,400 years.

The statistical uncertainty also in practice determines the age limits of the method. When the radioactivity of the sample is so low that its count rate is within the uncertainty limit of the normal background of the counter, the age is then given as a minimum. (For example, ≥37,000 years reads as "greater than or equal to" 37,000 years.) The statistical uncertainties as a function of age for a typical gas counter are shown in Fig. 3.

Table 1 compares characteristics of several types of counters using typical examples. The solid-carbon counter has clear disadvantages, requiring larger samples but yielding a limited range of dating. The gas counter appears to be better than the liquid scintillation type, especially for dating older samples, but in practice, each type has some advantages over the other.

Customarily, laboratories count several aliquots of a sample, and the average result provides a more reliable date than that for a single sample count. Also, counting arrays vary widely in their characteristics, some being better than others.

Systematic or random sources of laboratory error are to be expected. These include variations in counter efficiency and background, contamination by extraneous radioactivity, and uncertainties in calibration of a counter.

Radiocarbon dates are calculated by comparison with modern material. Man-made and natural fluctuations make the choice of representative modern material difficult. All radiocarbon laboratories use a standard distributed by the U.S. National Bureau of Standards, which itself is tied to the analysis of 1890 tree rings. This standard represents a hypothetical 1950 atmosphere with all cultural effects eliminated. Thus all laboratories should present comparable results.

Many interlaboratory comparisons such as those shown in Table 2 assure that different laboratories report consistent and comparable results.

Errors in assumptions. Deviations from the basic assumptions of the method limit the absolute accuracy which can be assigned to an age. These assumptions, which the method shares with all other methods of age estimation using natural radioactivity, are (1) that the half-life is known and constant; (2) that the initial radiocarbon concentration of the datable material is known; and (3) that no postdepositional alteration has occurred which would alter the radiocarbon concentration.

Increased study of radiocarbon for dating and other applications has shown that these assumptions are frequently violated. This has been accompanied by a widespread feeling that radiocarbon dates are less reliable than originally assumed. A sense of frustration is particularly evident among archeologists whose time scale begins with historical records and who are disturbed by the uncertainties implicit in the radiocarbon date.

Quite to the contrary, radiocarbon age estimates are now more reliable than ever because their limitations are better understood and because corrections are becoming available to improve the accuracy of the ages.

Considerations of the half-life. One should note that radiocarbon dates are reported in radiocarbon years, which do not necessarily have a well-defined relationship to solar years. The ages are customarily calculated using a half-life of 5565 years (the Libby half-life giving a mean life of 8033 years) instead of the more recently determined "best" value of 5730 years. The use of an "incorrect" half-life keeps all ages on a standard basis regardless of when they were analyzed, and serves to emphasize the difference between radiocarbon years and solar years. Corrections may be applied to dates as needed and as necessary information becomes available.

There is a very high probability that the half-

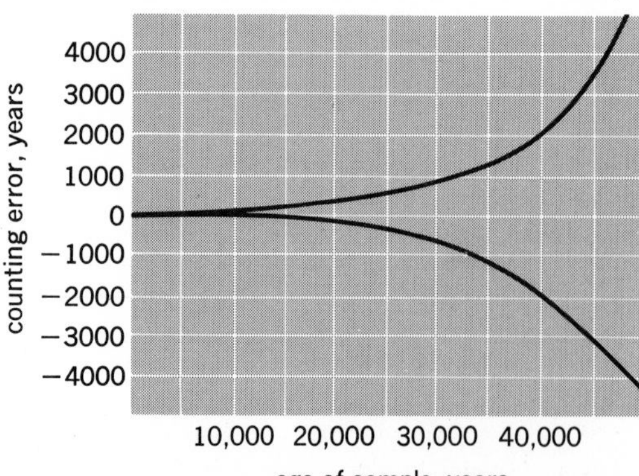

Fig. 3. Measurement error for sensitive gas counting system plotted as function of sample age.

Table 2. Interlaboratory comparisons

Laboratory	Sample number	Age*
Groningen	GRO 1172	2885±60
Groningen	GRO 1512	2770±50
Columbia University	L 427	2770±90
University of Michigan	M 290	5130±150
Columbia University	L 214	5090±300
Copenhagen	K 101	10,890±240
U.S. Geological Survey	W 82	10,260±200
U.S. Geological Survey	W 84	10,510±180
Heidelberg	H 105–87	11,500±300
Columbia University	L 289M	11,700±200
U.S. Geological Survey	W 442	12,050±400
Columbia University	L 185B	28,200±1000
U.S. Geological Survey	W 85	27,500±1200

*Years before present.

life is constant and cannot be affected by conditions of temperature, pressure, or chemical association prevailing on the surface of the Earth. The question of whether half-life is constant is nevertheless raised from time to time.

Initial radiocarbon concentration. The initial concentration of radiocarbon in a sample may be difficult to know either because different kinds of carbon-bearing material draw their carbon from different reservoirs or because the production ratio of C^{14} may have varied with time. These deviations are caused by three different processes: (1) isotope fractionation between various carbon compounds which make up the reservoir; (2) depletion of radiocarbon by decay in parts of the reservoir more rapidly than the newly formed radiocarbon from the atmosphere can become mixed with it; and (3) dilution of local reservoirs with "old carbon" released from the dormant carbon reservoir (mainly limestone).

Of these, isotope fractionation is the least serious since its magnitude can be accurately estimated by determining the ratio of C^{13} to C^{12} in the sample. The C^{14} fractionation will be twice the C^{13} fractionation. This has been verified by numerous measurements on different materials receiving their carbon from the same reservoir (Table 3).

Fresh-water deposits (Table 4) are greatly affected by all three processes mentioned above. The dissolved carbon of groundwater commonly contains contributions from the atmosphere, from decaying organic material which could be several hundred years old, and from limestone containing no C^{14} at all. Most importantly, groundwater may be equilibrated with the atmosphere, or nearly so initially, but after traveling underground, it "ages" because no new radiocarbon can enter the water from the atmosphere. This latter effect makes radiocarbon useful in studying rates of groundwater movement. *See* GROUNDWATER.

When the water returns to the surface again, atmospheric CO_2 may mix with the water, but the rate is sometimes too slow to equilibrate the C^{14} concentration with the atmosphere. As shown by the examples in Table 4, rivers and lakes are often quite low in radiocarbon, imparting an apparent age of several hundred years or more to materials growing in them. It is obvious that the ages of fresh-water materials must be considered very carefully.

A similar but more predictable situation exists in the ocean. Most of the deep ocean is in slight contact with the surface and mixes slowly relative to the decay rate of C^{14}. Parts of the deep Pacific Ocean are depleted in radiocarbon by as much as 25%. The surface of the ocean which is in contact with the atmosphere contains much more radiocarbon and is about 5% below the atmosphere in radiocarbon concentration. Most datable material found in the ocean is produced in this surface water.

Constancy of radiocarbon in atmosphere. Beyond the uncertainties introduced by variations in the assay of contemporary material, the possibility of time variations must be considered.

The importance of this problem can be understood if one imagines that, 5730 years ago (one radiocarbon half-life), the atmosphere contained twice as much radiocarbon as it does now. A tree which grew then would now have as much radiocarbon as a modern tree and would not appear to have aged at all.

Variations in atmospheric C^{14} content can be studied in many ways. Tree rings are the most suitable for such studies. A tree-ring chronology is obtained through dendrochronological studies where each consecutive annual growth layer is assigned to the calendar year in which it was formed. For a living tree, the outermost ring has a precisely known date, and successive annual growth layers are assigned to sequentially earlier years by counting inward from the bark layer. Cross dating among different specimens is obtained by matching identical patterns of wide and narrow rings between specimens. Through this technique, fossil trees that partly overlap each other timewise can be used for extending the chronology. In addition, information is obtained on missing and "double" rings. The oldest living bristlecone pine is about 4500 years old but, through cross dating of fossil specimens, C. W. Fergusson at the University of Arizona has been able to establish an 8200-year bristle cone chronology.

Table 3. Examples of C^{14} fractionation during sample formation and use of C^{13} measurements to correct for these deviations

Materials receiving carbon	δC^{14}*	δC^{13}†	ΔC^{14}‡
From Pyramid Lake			
Tufa	−3.6	+0.63	−4.9
Algae	−8.8	−2.27	−4.3
Plants	−11.0	−2.63	−5.7
Fish	−5.9	−0.59	−4.7
Dissolved HCO_3^-	−9.3	−2.23	−4.8
From atmosphere			
Wood	+0.8	−2.50	+5.8
Atmospheric CO_2	+3.6	−0.90	+5.4

*Difference (%) between C^{14} concentration in sample and standard.

†Difference (%) between C^{13} concentration in sample and standard.

‡Results normalized to same C^{13} concentration (eliminates differences introduced by isotope fractionation).

Table 4. C^{14}/C^{12} ratios for contemporary samples from fresh-water systems

University laboratory	Material	Locality	ΔC^{14}*
Yale	Several types of materials	Queechy Lake, N.Y.	~ −20†
	Several types of materials	Lake Zoar, Conn.	~ −10†
Columbia	Fish	Walker Lake, Nev.	−0.3
	Shell	Truckee River, Truckee, Calif.	−0.9
	Plants	Bear River, Woodruff, Utah	−15.0
	Dissolved HCO_3^-	Mono Lake, Calif.	−15.4

*Difference (%) from C^{14} concentration in atmospheric CO_2 over water body. All results have been normalized to the same C^{13} concentration to eliminate differences introduced by fractionation.

†Average of a number of results.

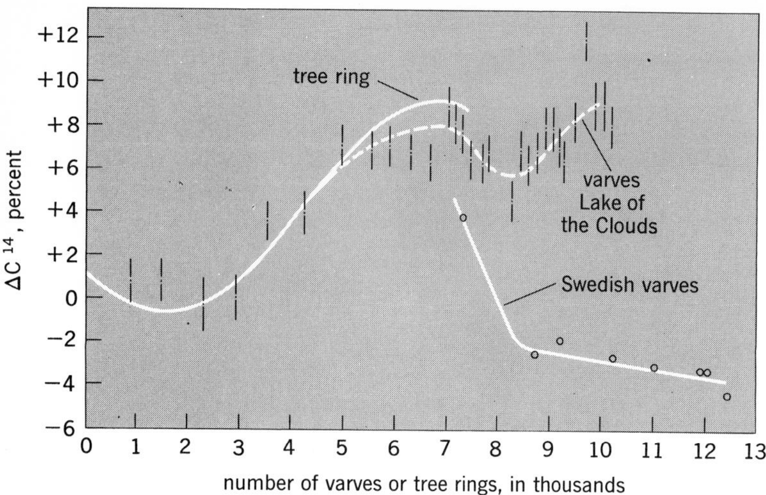

Fig. 4. Percent deviation of atmospheric C^{14} content plotted against varve years and tree rings. Lines are approximate. (*From M. Stuiver, Tree ring, varve and carbon-14 chronologies, Nature, 228:454–455, 1970*)

By measuring the present C^{14} concentration in wood samples whose ages are dendrochronologically determined, it is possible to calculate the initial C^{14} concentration in the sample at the time of formation. The atmospheric C^{14} activity is calculated from the initial wood C^{14} activity by applying a correction for the isotope fractionation between wood and atmospheric CO_2. Basic to this method is that each ring is formed during one year only and afterward ceases to accumulate or exchange carbon.

A large number of tree-ring samples has been analyzed by several laboratories and both long- and short-term variations in the C^{14} concentration of wood samples have been proven. Figure 4 gives a generalized graph of radiocarbon variations in the atmosphere as deduced from tree-ring data from several sources. Figure 5 is a more detailed example showing a large number of data points. There are approximately three times as many data available from the laboratories of P. Damon and E. Ralph as well as H. Suess, showing the great importance attached to the understanding of radiocarbon variations. Short-term variations involve changes for single consecutive years, as well as oscillations lasting several hundred years. Within one solar cycle, large changes in the C^{14} production rate occur in the upper atmosphere through modulation of the cosmic-ray flux by the magnetic field associated with the solar wind. This would, theoretically, result in only small C^{14} changes in the large reservoirs (atmosphere, oceans, and biosphere) on Earth. However, it has recently been suggested that the C^{14} production of several years is stored in the stratosphere and released to the troposphere at a certain stage of the 11-year solar cycle. This would explain year-to-year changes of a few percent in atmospheric C^{14} content experienced in the 20th century.

Many data are available for the short-term oscillations of a few hundred years. These oscillations appear to correlate significantly with solar cycle modulation of the cosmic-ray flux. Sunspot maxima often form patterns; for instance, a series of several 11-year cycles may have low sunspot maxi-

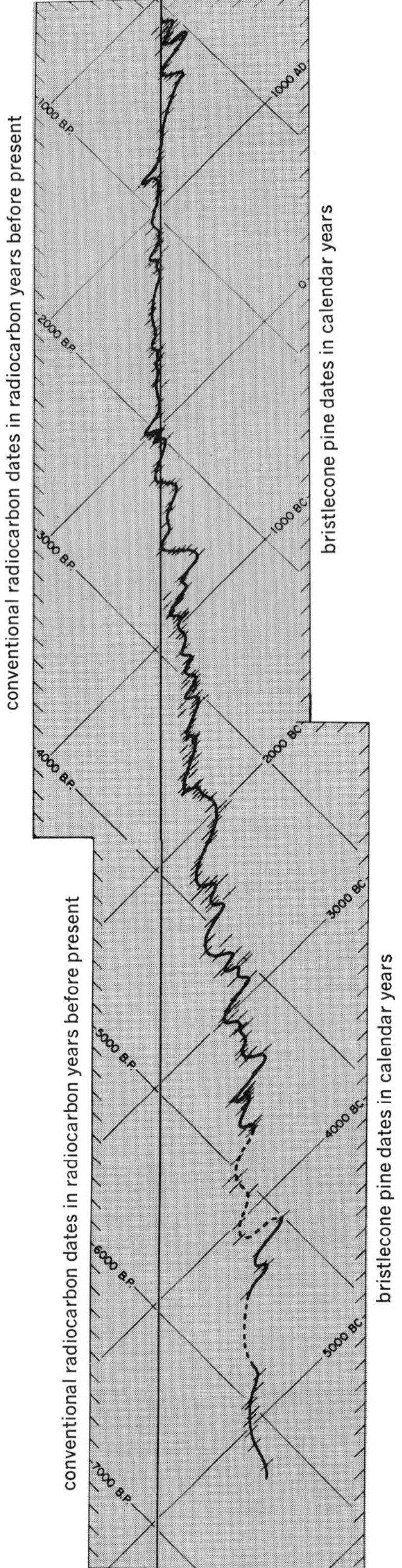

Fig. 5. Conventional radiocarbon ages of tree-ring-dated wood versus tree-ring dates for the last 7400 years. Conventional radiocarbon ages are calculated with a 5568-year half-life; zero age is 1950 A.D. The coordinate is tilted 45° in order to save space. (*From H. E. Suess, Bristlecone pine calibration of the radiocarbon time scale 5200 B.C. to the present, in I. U. Olsson, ed., Radiocarbon Variation and Absolute Chronology: Nobel Symposium 12, John Wiley & Sons, Inc., 1970*)

ma and be followed by a series with high maxima. This results in a few percent variation in atmospheric C^{14} content over intervals of the order of 100 years. An additional 400-year cycle in atmospheric C^{14} content, probably also associated with solar modulation of the cosmic-ray flux, has been suggested.

A long-term change in atmospheric C^{14} concentration is also evident. For the past 2500 years the average C^{14} concentration has been close to the baseline (standard) activity, but it was nearly 10% higher for the interval 7500–6000 before present (B.P.). Between 6000 and 2500 years B.P., C^{14} content was gradually reduced from this high level.

Geomagnetic effect. Earth magnetic field changes are the most likely cause of a major portion of the C^{14} change in the atmosphere. Measurements of paleomagnetic field intensities show changes over the last 8000 years amounting to 50% of present-day values. The less energetic component of the cosmic radiation is subject to greater deflection than the more energetic, and when the Earth magnetic field is stronger, a smaller portion reaches the upper atmosphere where C^{14} production takes place.

Climatic effect. In addition to this shielding effect, there is the possibility that climatic factors are responsible for part of the long-term change in C^{14} concentration. Sea-level changes, pH changes of ocean water associated with temperature changes, and changes in exchange rates between the surface layer and deep ocean can all be associated with climatic changes and can influence the C^{14} concentration in the atmosphere. At present, it is not entirely clear what portion of the long-term trend should be assigned to geomagnetic or climatic effects.

The relationship between calendar years and radiocarbon years is given in Fig. 5, which provides corrections for variations in radiocarbon and half-life. The vertical line gives the correction for half-life only. To use this diagram, find the reported radiocarbon date on the left-hand scale and follow the value diagonally downward to the line defined by the data. The line running diagonally downward from the right which also intersects the data line at this point will give the calendar age.

Superimposed on this major trend are the short-term variations of a few percent which correspond to age errors of 100–200 years. Thus even with this calibration curve, the basic inaccuracy of a date has to be reckoned in centuries.

A portion of the calibration curve obtained from tree-ring studies has been confirmed by analysis of historically dated samples. C^{14} analysis of materials from the various dynasties of Egypt going back to about 5000 years B.P. results in the proper historical age when corrected for the known C^{14} variations (Table 5).

A study of the natural radiocarbon variations becomes more difficult as 10,000 years are approached because of the difficulty in extending the tree-ring record accurately. Other, less satisfactory methods will have to be employed to continue the record to earlier times.

Varves also are useful for the study of long-term C^{14} variations. Varves are produced in sedimentary environments by annual deposition of layers of different composition or texture. The most extensively studied varve series consists of silt and clay laminae deposited as couplets in proglacial lakes. Prominent in these investigations is the Swedish varve chronology of G. DeGeer. Although the Swedish varves contain insufficient organic material for C^{14} measurements, it is possible to correlate C^{14}-dated climatic episodes with varve-dated rates of recession and geomorphological features in southern Sweden. In addition, pollen-zone boundaries that are varve-dated can be correlated with C^{14} pollen-zone boundaries in nearby peat bogs. The bottom curve in Fig. 4 gives the calculated C^{14} deviations resulting from the differences in C^{14} age and the Swedish varve age.

Another varve series involves fresh-water sediments in Lake of the Clouds in northern Minnesota. These varves contain sufficient organic material for a direct C^{14} analysis, permitting a comparison of the C^{14} ages with the number of varves deposited (Fig. 4). Agreement between the tree-ring and Lake of the Clouds varve series is excellent, but is lacking between the two varve series. This discrepancy may have been caused by errors in either the Swedish or the Lake of the Clouds varve chronology, or both. However, both varve sequences show that deviations in atmospheric C^{14} content are maximally 10%. Thus conventional radiocarbon ages are expected to deviate less than 800 years from the true age over the 7500–12,000 year B.P. interval for which tree-ring data are lacking.

At present, comparisons of radiocarbon ages with those obtained by other methods indicates that the variation of C^{14} in the atmosphere within the last 35,000 years was not much greater than the variation already observed within the last 10,000 years.

Although it is often necessary to make corrections for the half-life and variations of radiocarbon in the atmosphere, note that these two problems affect all samples equally and, except for a few minor ambiguities, do not affect the capacity to rank samples in order of age or to correlate materials of similar age. Thus, for many purposes, corrections are not really necessary.

Table 5. Comparison of radiocarbon ages of some Egyptian samples with historical ages*

Material	Historical age	Radiocarbon age (relative to 1950 A.D.)	Corrected radiocarbon age (calendar years)
Wood from coffin of Tutankamun	1340 + 10 B.C. (3290 B.P.)	2960 + 50 B.P.	1340 + 100 B.C.
Wood from funerary monument of Sesostris II	1870 + 10 B.C. (3820 B.P.)	3500 + 70 B.P.	1900 + 150 B.C.
Wood from coffin of Aha-nakht	2110 + 30 B.C. (4060 B.P.)	3700 + 100 B.P.	2150 + 100 B.C.
Wood from tomb of Neferirkare	2450 + 30 B.C. (4400 B.P.)	3950 + 150 B.P.	2450 + 300 B.C.
Wood from tomb of King Djoser	2590 + 40 B.C. (4540 B.P.)	4100 + 30 B.P.	2800 + 300 B.C.

*Data compiled by Säve-Söderberbergh and Olsson in I. U. Olsson (ed.), *Radiocarbon Variations in the Atmosphere*, 1970.

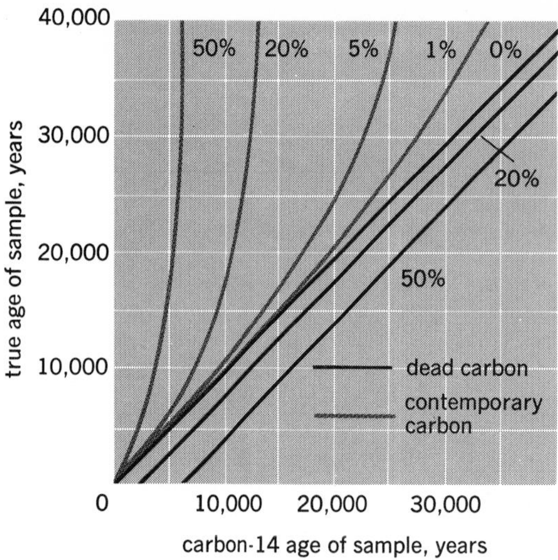

Fig. 6. Effect of contamination of samples of various ages with varying amounts of contemporary and ancient carbon causing difficulty in dating.

The study of the variations in the contemporary distribution of C^{14} has been complicated by two man-made processes. H. E. Suess has shown that the CO_2 released by the burning of fossil fuels has diluted the C^{14} in the atmospheric reservoir. This Suess effect has been recently confirmed by M. Baxter and A. Walton, who find from analysis of distilled spirits that plants growing in 1890 had about 2% more radiocarbon than those living in 1950. The additional CO_2 may eventually present a serious pollution problem, with implications concerning climatic variations and biological activity.

The explosion of thermonuclear devices created sufficient C^{14} to raise the atmospheric concentration by 100% between 1954 and 1965. This concentration is now dropping as the atmospheric CO_2 mixes with that in the ocean or the biosphere.

Postdepositional alteration. There are three possibilities for postdepositional alteration of the C^{14}/C^{12} ratio in the sample: (1) preferential removal of one carbon isotope with respect to another during physical or chemical decomposition; (2) fossilization of material with incorporation of recent carbon; and (3) physical or chemical intrusion of extraneous carbon. Of these, only the third is serious in most cases. Fractionation of isotopes during diagenesis or even during sample processing is possible, but a correction can be applied by making C^{13}/C^{12} stable isotope ratio determinations. Chemical alteration of organic material to incorporate more recent carbon seems to be rare and unlikely. Recrystallization of carbonate materials is common, however.

Contamination with recent carbon presents a serious problem in dating very old samples. As can be seen in Fig. 6, the addition of 1% recent carbon to a sample 40,000 years old would give it an apparent age of about 33,000 years. Such contamination can occur in organic materials by physical intrusion (rootlets) or by precipitation of humic materials leached from overlying soils. In the case of carbonates, precipitation of groundwater carbonate, exchange of surface CO_2 atoms with those in the atmosphere, and physical mixing of old with young material are all serious problems. In general, finite ages of carbonates 25,000 years old or more must be considered minimum ages because of the likelihood of contamination.

Bones contain both organic material, which may be destroyed rapidly, and carbonate, which invariably becomes contaminated by exchange. Bones can be dated directly only by carefully extracted organic material or indirectly by association with datable materials.

In the absence of intercalibrations with other reliable absolute methods, internal crosschecks must be made. Ages on various materials from the same horizon or on various chemical fractions of the same material provide the necessary information. The following types of checks have been made: (1) The ages obtained on carbonate and organic materials in the same horizon have been compared; (2) the ages from the surface fraction of carbonate samples have been compared with those from the interior of the same sample; (3) the ages obtained for the base soluble fraction (which should contain any precipitated contaminants derived from organic materials) have been compared with those obtained on the insoluble fraction; and (4) the ages for the cellulose fraction of organic materials have been compared with those on the lignin fraction.

The results of the checks are summarized in Table 6. In general agreement is excellent, suggesting that the contamination levels are quite low and only rarely do they lead to significant errors for samples with ages less than 40,000 years. Extension of the method to samples greater than 40,000 years in age, however, requires careful sample selection and extensive chemical pretreatment. Contamination will undoubtedly place an upper limit on the extension of the method. Reliable C^{14} ages are not expected for samples greater than 70,000 years, regardless of experimental advances.

While some problems such as isotopic fractionation and uncertainty in the initial concentration of the radiocarbon may have serious implications for the calculation of ages of young materials, they become less serious for older samples relative to the inherent decrease in precision. Contamination becomes a more important concern for older samples, and in some cases will determine the reliable upper limit of the dating method.

Errors in interpreting dates. It should be noted that frequently the age of a piece of wood or clam shell or other material has little intrinsic importance but is valuable for the associations or events that it dates. The stratigraphic association between the dated sample and the climatic, geological, or cultural event must be carefully demonstrated. There are sometimes rather subtle factors involved. A tree always predates the object it constructed, frequently by an unknown length of time. In dating the tomb of an Assyrian prince by a piece of wood from an entombed table, one must wonder if the owner had been a collector of antique furniture.

One should not ignore the limitation placed on the age by its limits of precision. From data in Table 5, one can deduce that radiocarbon dating cannot distinguish the coffin of King Djoser from that of his great-grandson. Placing a random object precisely into a historical framework is even more difficult.

Historians, archeologists, paleontologists, den-

Table 6. Internal checks of C^{14} ages

Organic-carbonate comparisons

Laboratory	Material	Locality	Age*
U.S. Geological Survey	Plant remains	Utah soil	8330 ± 300
U.S. Geological Survey	Shell	Utah soil	7720 ± 300
Columbia University	Wood	Vancouver delta sands	11,850 ± 250
Columbia University	Shells	Vancouver delta sands	12,000 ± 250
Copenhagen	Wood	Denmark lake deposits	10,890 ± 240
Copenhagen	Marl	Denmark lake deposits	10,930 ± 300
Columbia University	Dispersed organic	Great Salt Lake sediments	26,300 ± 1100
Columbia University	Dispersed $CaCO_3$	Great Salt Lake sediments	25,300 ± 1000
U.S. Geological Survey	Wood	Searles Lake core	26,700 ± 2000
U.S. Geological Survey	Na_2CO_3	Searles Lake core	23,000 ± 1400
U.S. Geological Survey	Dispersed organic	Searles Lake core	29,500 ± 200

Surface-interior $CaCO_3$ comparison

Laboratory	Material	Age surface	Age interior
Columbia University	Tufa	9550 ± 250	9450 ± 250
Columbia University	Tufa	12,200 ± 300	13,000 ± 400
Columbia University	Tufa†	8800 ± 200	10,700 ± 400

Base soluble – base insoluble comparison

Laboratory	Material	Age soluble fraction	Age insoluble fraction
Columbia University	Peat	4700 ± 150	4650 ± 150
Columbia University	Peat	8350 ± 200	7350 ± 650
Columbia University	Lignitized wood	39,000 ± 2000	39,000 ± 2600

Lignin-cellulose comparisons

Laboratory	Material	Age lignin	Age cellulose
Columbia University	Wood	25,850 ± 500	25,900 ± 300
Columbia University	Peat	25,050 ± 300	23,450 ± 300

*Years before present. †Very high surface area.

drochronologists, and varve counters can frequently rank events in order more reliably and construct more precise time scales than can radiometric dating. Frequently these scales are "floating" in time. Just as a person in the woods, possessor of a fine watch which has stopped, may have to set the watch by a crude estimate from the position of the Sun, so radiocarbon dates are useful in placing events or sequences of events into their approximate positions in time.

APPLICATIONS

The radiocarbon method has been applied to numerous problems. Perhaps the most important is that of establishing the chronology of the climatic changes which characterize the Quaternary Period. Carbon-14 ages of samples from trees knocked over by advancing ice masses record the chronology of advancing continental glaciers. Peat samples from bogs and driftwood from the shorelines of proglacial lakes make it possible to establish a time table for the retreat of glaciers. Carbonate samples from the shorelines of the great pluvial lakes (which once covered many of the desert areas in the Great Basin of the western United States) allow the absolute history of the climate fluctuations which produced these changes to be established. Radiocarbon dates on the shells of planktonic animals found in deep-sea sediments define the chronology of fluctuations in oceanic conditions and related climatic conditions. The results for these various systems yield a consistent picture of worldwide climate changes which have occurred over the past 40,000 years. As shown in Fig. 7, in each case the data suggest a warm climate during the past 10,000 years preceded by a cold period extending back to about 27,000 years. Prior to 27,000 years conditions were intermediate between those characterizing the interglacial climate of the present and the glacial climate for the period preceding 10,000 years ago.

Radiocarbon dates on charcoal from the hearths of ancient man have been a great aid in working out man's history and relating it to the cli-

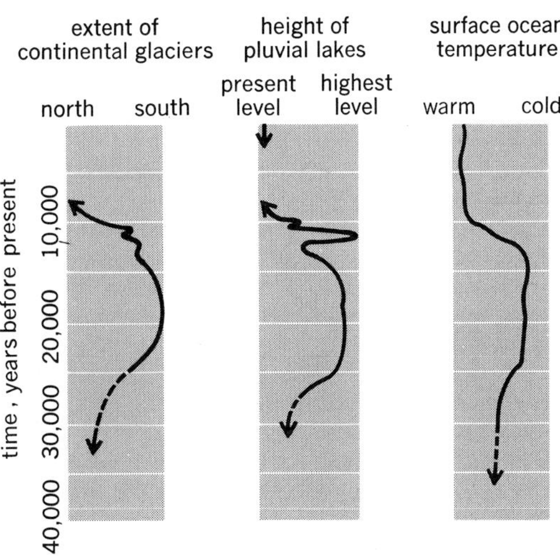

Fig. 7. Climate chronologies obtained from C^{14} measurements on samples from continental glacier deposits, pluvial lake deposits, and ocean sediments. Right side of each graph represents cool, moist conditions of periods of glaciation.

matic fluctuations mentioned above. Volcanic eruptions have been dated by radiocarbon measurements on materials covered by the lava or ash; age determinations on hydrocarbons extracted from soils yield valuable information concerning the origin of petroleum; dates on the remains of extinct animals such as the mastodon and giant sloth allow estimates to be made of the time and cause of extinction; and dates on charcoal and artifacts left by modern man in remote areas allow much of their history to be related to that of cultures which have provided historic records.

Radiocarbon ages are reported in the journal *Radiocarbon*, which is especially dedicated to this cataloging. [DAVID L. THURBER]

Bibliography: W. S. Broecker and J. Van Donk, Insolation changes ice volumes and the O^{18} record in deep-sea sediments, *Rev. Geophy.*, 8:169–198, 1970; R. Chatters (ed.), *Proceedings of the International Carbon-14 and Tritium Dating Conference*, 1965; W. F. Libby, *Radiocarbon Dating*, 2d ed., 1955; I. U. Olsson (ed.), *Radiocarbon Variations in the Atmosphere*, 1970; A. Rafter and T. Grant-Taylor, (eds), *Proceedings of the 8th International Conference on Radiocarbon Dating*, Royal Society of New Zealand, 1972.

Radiolarian earth

A porous, earthy, unconsolidated sediment formed from the opaline silica skeletal remains of Radiolaria. It is formed from radiolarian oozes that accumulate on the deep ocean floor. The indurated equivalent with pores filled by silica is radiolarite. The radiolarian earths and radiolarites are white and cream-colored. Minor components in radiolarian earths are manganese nodules, shark teeth, and other resistant remains of vertebrates. Radiolarites are known from Devonian to Tertiary age. *See* CHERT; SEDIMENTARY ROCKS.

[RAYMOND SIEVER]

Rapakivi granites

A term originally applied to those granites with abundant, large, ovoid crystals of potash feldspar (orthoclase or microcline), commonly mantled by sodic plagioclase (oligoclase or albite) and embedded in a matrix of quartz, potash feldspar, plagioclase, biotite, and hornblende. The ovoid cores may contain grains of the matrix minerals arranged in concentric bands. The term rapakivi is now commonly applied to any granite with relatively large mantled crystals of potash feldspar (rapakivi structure). More specifically, rapakivi structure embraces an ovoid core mantled by small prisms of oligoclase in more or less radical or tangential arrangement. Best known are the immense bodies of Precambrian rapakivi granite in Finland and Sweden.

Many theories have been advanced to explain these puzzling rocks. Very likely, mantled crystals may form by various means. Rapakivi structure may be attributed to (1) direct crystallization of granite magma in which a shift in equilibrium conditions has resulted from changes in pressure, temperature, composition, or volatile content; (2) late-magmatic or postmagmatic reconstitution of the rock; or (3) metasomatic replacement of older rocks, in some cases in combination with granitization. *See* GRANITE; GRANITIZATION; MAGMA; METASOMATISM. [CARLETON A. CHAPMAN]

Realgar

A mineral having composition AsS and crystallizing in the monoclinic system. Realgar is found in short, vertically striated crystals (see illustration),

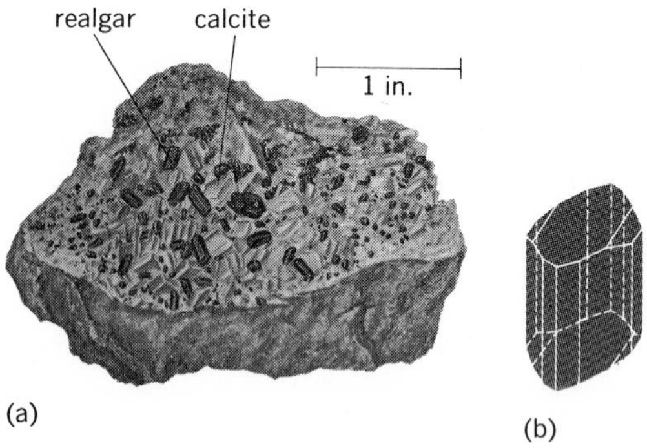

Realgar. (*a*) Crystals with calcite, Felsöbanya, Rumania (*specimen from Department of Geology, Bryn Mawr College*). (*b*) Crystal habit (*from C. S. Hurlbut, Jr., Dana's Manual of Mineralogy, 17th ed., copyright © 1959 by John Wiley & Sons, Inc.; reprinted by permission*)

but more frequently is granular and in crusts. There is one pinacoidal cleavage; the hardness is 1.5–2 (Mohs scale) and the specific gravity is 3.48. The luster is resinous and the color red to orange. Realgar is found in ores of lead, silver, and gold associated with orpiment and stibnite. It occurs with the silver and lead ores in Hungary, Czechoslovakia, and Germany. Good crystals have come from Binnenthal, Switzerland, and Allchar, Macedonia. In the United States it is found at Manhattan, Nev.; Mercer, Utah; and as deposits from geyser waters in Yellowstone National Park. *See* ORPIMENT; STIBNITE.

[CORNELIUS S. HURLBUT, JR.]

Recent

A term roughly defining a sequence of geologic strata (the Recent Series) and also the time (late Quaternary) during which the strata were deposited. Introduced by the British geologist Sir Charles

PRECAMBRIAN	PALEOZOIC							MESOZOIC			CENOZOIC	
	CAMBRIAN	ORDOVICIAN	SILURIAN	DEVONIAN	CARBONIFEROUS		PERMIAN	TRIASSIC	JURASSIC	CRETACEOUS	TERTIARY	QUATERNARY
					Mississippian	Pennsylvanian						

TERTIARY					QUATERNARY	
Paleocene	Eocene	Oligocene	Miocene	Pliocene	Pleistocene	Recent

Lyell in 1833, the term is now widely considered to represent the sediments postdating the Ice Age. It is recognized, however, that the last deglaciation of middle and high latitudes has been progressive, and various events have been suggested as an arbitrary basis of separation. By others "recent" has been employed in an informal sense without specific definition. In several European countries the term Holocene is preferred to Recent, but is used with essentially the same meaning as Recent. *See* QUATERNARY.

Recent strata represent virtually every environment of deposition, as they include all the sediments that are being deposited at present. Although Recent time (in the formal sense) falls well within the range of C^{14} dating, it lacks an accepted value for its duration because of uncertainty in fixing its inception. It is held by some to represent a lapse of the order of 10,000 years, by others 11,000 years. The term Postglacial is applied by pollen stratigraphers in northern Europe to the sediments represented by pollen zones IV–IX, approximately the last 10,000 years as determined by C^{14} dating. *See* RADIOCARBON DATING.

[RICHARD F. FLINT]

Redbeds

Detrital sedimentary rocks pigmented by ferric oxide that coats grains, fills pores as cement, or is dispersed in a muddy matrix. These conspicuously colored rocks commonly constitute thick sequences of nonmarine to shallow marine deposits. Detrital redbeds accumulated in many parts of the globe during the past 1,000,000,000 years of Earth history. They were among the first sedimentary deposits to be considered climatic indicators, because their color was assumed to reflect a unique, identifiable condition of deposition. On the other hand, detailed development of red pigment in most redbeds was complex and is difficult to decipher. Ferric oxides also pigment marine chert, limestone, and iron formations, but these chemical deposits are not usually included among redbeds.

Pigmentation. Redbeds common in late Precambrian and Phanerozoic nonmarine to paralic (marginal marine) facies range between two rather distinctive end members. One includes extensive formations of uniformly pigmented quartz-rich sandstone and siltstone, many of which are associated with tongues of marine limestone, and with eolian sandstone and evaporites as evidence of aridity. The other end member consists of thick sequences of immature sandstone and mudstone with laterally equivalent border conglomerate. Many of these deposits are arranged in fining-upward cycles with lower sandstone and upper mottled and reddish-brown mudstone. Redbeds of this sort commonly contain fossils of warm, moist floras and vegetarian vertebrates. Although fining-upward cycles are most common in alluvial deposits, similar successions also develop on prograding tidal flats. *See* EVAPORITE, SALINE; GYPSUM; ROCK SALT.

Some redbeds contain abundant grains of sedimentary and low-grade metamorphic rocks and relatively few grains of iron-bearing minerals. Most of them, however, contain feldspar and relatively abundant grains of opaque black oxides derived from igneous and high-grade metamorphic source rocks. In older redbeds the black grains are predominantly specular hematite replacing magnetite; in younger ones magnetite is more common. Dark grains of iron-bearing silicate minerals are rather rare in red sandstones; but they are more common in early-cemented concretions and in modern analogs, pointing to their postdepositional destruction which released pigmenting ferric oxide. Clay minerals in redbeds, as in most other detrital deposits, are predominantly illite and chlorite, thus providing no specific clue to the climate in the source area or at the place of deposition. *See* ARKOSE.

In many of the younger redbeds the pigmenting ferric oxide mineral cannot be identified easily because of its poor crystallinity. In most of the older ones, however, hematite is the pigment. As seen under the scanning electron microscope, the hematite is in the form of hexagonal crystals scattered over the surface of grains and clay mineral platelets. In red mudstones most of the pigment is associated with the clay fraction.

Redbeds do not contain significantly more total iron or ferric oxide than nonred sedimentary rocks. Normally, both increase with decreasing grain size of redbeds. Moreover, the amount of iron in the pigments is very small compared with that in opaque oxides, dark silicates, and clay minerals. These facts demonstrate that chemical and mineralogical data cannot differentiate redbeds formed in moist climate from those formed in deserts.

Source areas. Actively eroded source areas of most redbeds supplied relatively abundant grains of magnetite and iron-bearing silicate minerals, whether in humid or dry climates. Accordingly, the common grains of specular hematite in redbeds must have been produced by an oxidizing burial regime. Similarly, the sparse crop of dark silicates commonly rimmed with red pigment resulted from postdepositional alteration, emphasizing its role in pigmentation and as a source of some of the clay mineral matrix as well.

Regardless of climate or color of their soils, source areas generally deliver brown sediments. Accordingly, the inherited brown muddy matrix that contained amorphous to very poorly crystalline hydrated ferric oxide must have transformed slowly to red hematite pigment after deposition. This process is favored by persisting oxygenating conditions of burial in nonmarine to paralic environments. Postdepositional conversion of brown hydrated ferric oxide to red pigments has also operated in shallow- to deeper-water marine deposits. In all of these red deposits the final color is no clue to the climate in the source area.

Most redbeds were made from rather ordinary alluvium that accumulated in nonmarine to shallow-marine environments, under a pervasive oxygenating condition of burial within a common range of water chemistry. None was red at the time of deposition, and for each the environment during and after accumulation was crucial. Criteria such as associated fauna, flora, eolian sands, evaporites, or coal measures provide the most reliable evidence about the climate.

Tectonics and continental drift. A genetic association of major redbeds with tectonic activity has long been recognized. As one tectonic end member, extensive uniform redbed-evaporite sequences accumulated on stable cratons, whereas some thick wedges of desert redbeds with border conglomerates accumulated in rift valleys. Overall, the tectonic background most commonly associated with redbeds is a late to postorogenic framework, as noted in the late Hercynian deposits of Europe and northwestern Africa, the late Alpine deposits of southern Europe, and the late Andean deposits of South America. Repeated episodes of deformation along the Appalachian belt produced redbeds at the close of three Paleozoic orogenies. *See* TECTONIC PATTERNS.

On a global scale, paleomagnetic evidence of the distribution of redbeds relative to their pole position corroborates paleogeographic data, suggesting that most redbeds, evaporites, and eolian sandstones accumulated less than 30° north and south of a paleoequator where hot, dry climate generally prevailed. Moreover, continental drift reconstructions reveal that the most widespread redbeds in the geologic record developed near the Equator in late Paleozoic and early Mesozoic time when the continents were assembled in a great landmass, Pangea. *See* CONTINENT FORMATION; PALEOMAGNETICS.

Modes of origin. Several quite different facies of desert to savanna redbeds were produced by different combinations of factors. Some accumulated in basins where adjacent uplands had essentially the same climate; others had relatively remote source areas whose climate may have been quite different from that of the basin.

Sediments deposited in a hot, dry climate were derived either from arid uplands nearby or from more distant sources that may have been moister. Desert redbeds composed of sandy alluvium, dune sand, and fanglomerate, and relatively little mudstones, were pigmented by postdepositional destruction of iron-bearing grains. Examples include late Paleozoic redbeds of Colorado and eastern Canada, Permian upper Rotliegendes of central Europe, and Penrith eolian sandstone of Great Britain.

Extensive, well-sorted fine-grained redbeds that were derived from distant sources and that accumulated on evaporitic tidal flats and in saline lagoons were also pigmented largely by postdepositional destruction of iron-bearing minerals. These deposits include the Permian redbeds of Texas, Oklahoma, New Mexico, and Arizona; Triassic redbeds of Wyoming and South Dakota; and Jurassic redbeds on the Colorado Plateau. Thick and extensive red claystone in some of these sequences suggests derivation from a distant soil-mantled upland with a moister climate, and implies a reddening of the mud by postdepositional dehydration of inherited brown pigment.

Sediments that accumulated in savannas were derived either from adjacent uplands with seasonally humid climate or from more distant sources that may have been even moister. Savanna redbeds were supplied with detritus that included both fresh grains of iron-bearing minerals and a soil-derived clay fraction pigmented with brown hydrated ferric oxide. Some of these deposits bear reliable evidence of a warm, moist climate, as in the late Paleozoic coal measures redbeds of Great Britain and eastern Canada, the lower Rotliegendes of central Europe, and the early Cenozoic deposits of the Rocky Mountain region. Other deposits are suggestive of somewhat drier climatic conditions, for example, the Devonian Catskill redbeds of New York and Pennsylvania, the early Carboniferous Mauch Chunk Formation of Pennsylvania, and the Permian New Red Sandstone of Scotland.

Some of the red mudstone and siltstones that formed before the development of advanced land plants in mid-Paleozoic time accumulated in offshore marine environments, as in the Silurian marine redbeds of Great Britain and the United States, either on quite deep shelves or in basins of rapid sedimentation and burial. Moreover, some deepwater marine redbeds are resedimented nearshore deposits that were swept into deep basins by turbidity currents, as in the early Paleozoic of eastern Canada. *See* SEDIMENTARY ROCKS.

[FRANKLYN B. VAN HOUTEN]

Bibliography: E. F. McBride, Significance of color in red, green, purple, olive, brown and gray beds of Difunta group, northeastern Mexico, *J. Sed. Petrol.*, 44:760–773, 1974; F. B. Van Houten, Origin of red beds: A review—1961–1972, *Annu. Rev. Earth Planet. Sci.*, 1:36–61, 1973; D. L. Woodrow, F. W. Fletcher, and W. F. Ahrnsbrah, Paleogeography and paleoclimate at the deposition sites of the Devonian Catskill and Old Red facies, *Geol. Soc. Amer. Bull.*, 84:3051–3064, 1973; A. M. Ziegler and W. S. McKerrow, Silurian marine red beds, *Amer. J. Sci.*, 275:31–56, 1975.

Reef

A mass or ridge of rock or rock-forming organisms in a water body, a rock trend on land or in a mine, or a rocky trend in soil. Usually the term reef means a rocky menace to navigation, within 6 fathoms of the water surface. Various kinds of calcium carbonate–secreting animals and plants create biogenic, or organic, reefs throughout the warmer seas (Figs. 1 and 2). Naturally cemented sand ridges make reefs along the coast of Brazil and elsewhere. Rocky shores of seas, lakes, and navigable rivers commonly exhibit reefs of rock types similar to those of the adjacent land, for example, the *Felsenriffe* of the Lorelei legend.

Biogenic reefs. Reefs designated as biogenic, or organic, consist of the hard parts of organisms, or of a biogenically constructed frame enclosing detrital particles; the hard parts of free-living organisms; and precipitated calcium carbonate. Most biogenic reefs are made of corals and associated organisms, but some entire reefs and important parts of others consist mainly of lime-secreting algae (Fig. 3), hydrozoans, annelids, oysters, or sponges. In nautical language a rocklike organic mass must be a menace to navigation before it can be classed as a reef. However, the term may also be accurately applied to any sizable biogenic eminence or buildup that grows, or once grew, upward from the floor of a water body, ordinarily the sea.

Coral reef. The most widespread and, volumetrically, the most important kind of biogenic reef is the coral reef, consisting of corals and associated calcium carbonate–secreting organisms. Coral reefs first attracted wide scientific attention through the accounts of Charles Darwin, who divided them into three principal types: fringing reef, barrier reef, and atoll. Darwin considered that these developed in the order named as a result of persistent and profound subsidence. Modern

Fig. 1. A biogenic reef, the Ine Anchorage Reef, Arno Atoll, Marshall Islands.

Fig. 2. Coral reef at Ine Anchorage, Arno Atoll, Marshall Islands.

Fig. 3. Algal buttresses and surge channels of eastern peripheral reef at Onotoa Atoll, in the Gilbert Islands.

reef theory is more complicated but still retains subsidence as an important feature. In modern seas, coral reefs are important as hazards to navigation, as natural breakwaters surrounding boat passages and harbors, and as sites of complex life associations and high biological productivity. Fossil biogenic reefs are common reservoir rocks for oil.

Corals are exclusively marine, and the typical reef-building types are restricted to shallow warm water because of their symbiotic association with microscopic algae, known as zooxanthellae (Fig. 4). Although fossil representatives are known from Middle Ordovician time, Paleozoic coral reefs were made by organisms quite different from those responsible for the Mesozoic and Cenozoic reefs.

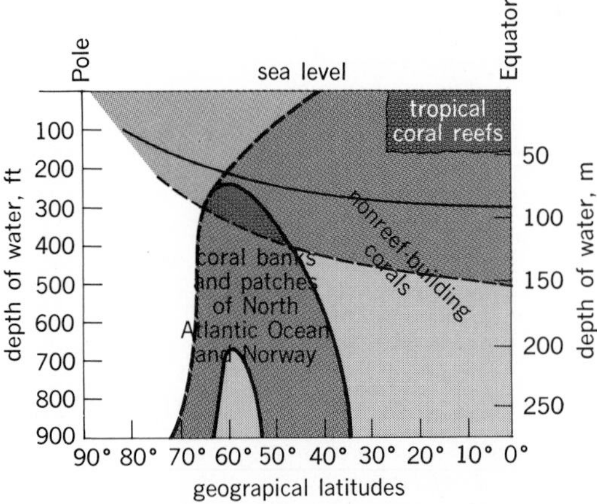

Fig. 4. The generalized depth ranges of the existing corals and calcareous algae. (*After C. Teichert, Cold- and deep-water coral banks, Amer. Ass. Petrol. Geol. Bull., 42(5):1064–1082, 1958*)

Kinds and origins. A fringing reef growing against the shore may, with subsidence and continued upward growth, become separated from the beach by a lagoon to become a barrier reef (Fig. 5). Continued subsidence and upgrowth can produce an atoll after all preexisting land has disappeared beneath a central lagoon, which is surrounded and defined by the peripheral atoll reef. Filling of an atoll lagoon, independent upward growth, or emergence and planation may produce a table reef, which is less common than the other major types. These are all open-sea reef types, characteristically rising from oceanic depths. Small reefs associated with these major categories or forming complex reef communities in quiet shallow waters that lack the larger reef types fall into the general categories of reef patches, pinnacles, and knolls.

Because the typical coral-reef association (Fig. 6) rarely grows with vigor below a depth of 20 m and largely dies out below 100 m, a reef frame thicker than this implies a subsidence of the sea floor or a rise of sea level. It was demonstrated by drilling that Eniwetok Atoll has subsided about 1200 m since first reef growth occurred beneath it. This finding confirms the importance of subsidence in the evolution of central Pacific atolls. Change of sea level with growth and melting of the

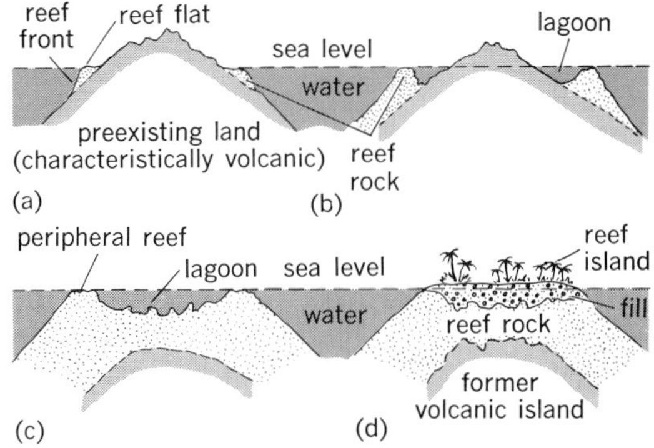

Fig. 5. Diagram showing reef sequence according to Darwin (table reef added). (*a*) Fringing reef. (*b*) Barrier reef. (*c*) Atoll. (*d*) Table reef.

Pleistocene ice sheets is equally important in the explanation of existing reef features. As an added elaboration, biogenic reefs may also grow upward from, or form a veneer over, existing platforms, even inheriting their general configuration from preceding topographic forms. *See* OCEANIC ISLANDS.

Geographic distribution. Existing coral reefs occur widely at tropical latitudes. They are most common in the Pacific and Indian oceans but are also found in the Atlantic Ocean and the Red Sea. They are uncommon on the western sides of continents because of the upwelling of cold water from the depths along these coasts. Coral reefs extend farther north than south and locally range well beyond the tropics because of displacement of the controlling 22°C isotherm. Some calcium carbonate buildups made by corals and brachiopods are also found in cold, relatively deep water in Scandinavian fiords and the northern Mediterranean. The corals that build these reeflike masses are of types not dependent upon the microscopic algal symbionts (zooxanthellae) of typical reef corals.

Geologic range. Coral reefs are known to range from Middle Ordovician to modern times. They became abundant during the Middle Silurian, when their distribution included regions now within the Arctic Circle—a fact that has been explained in a variety of ways. Devonian reefs ranged across central Europe and much of central and western North America and Asia. Biogenic reefs, but not coral reefs, are prominent Permian features. A great change in the coral faunas and their associates took place near the end of Paleozoic time. No Early Triassic coral reefs or corals are known, but coral-reef associations of modern types are known from the Middle Triassic onward and locally make up much of the rock sequence.

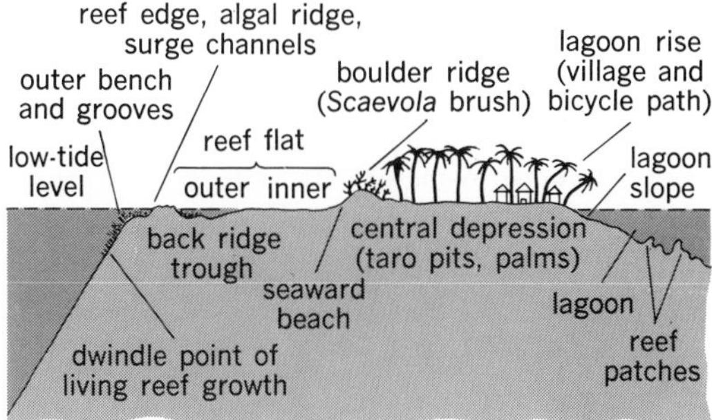

Fig. 6. Diagram of zonation of Pacific windward reef.

Economic and human significance. Tens of thousands of people live on Pacific and Indian Ocean islands that are built of reef formations or have reef foundations. In the reef waters and on the islands live animals and plants that provide food, shelter, and tools to maintain life, and copra for the markets of the world. The air bases and harbors at many reef islands provide the only stopovers in broad expanses of water. Rich phosphate deposits occur on some elevated reef islands, such as Ocean, Nauru, and Angaur. The porous nature and organically rich environment of reef rock make it a potential source of petroleum.

The fringing reef and the barrier reef are discussed below. For a discussion of the third main form in Darwinian reef evolution, the atoll, *see* ATOLL.

Fringing reef. Fringing reef refers to a coral or other biogenic reef that fringes the edge of the land. A fringing reef is ordinarily divided into a steeply descending seaward front and a flat, broad or narrow pavementlike surface that is awash at low tide. Although surfaces and fronts of such reefs may show vigorous growth of algae, corals, or other lime-secreting organisms, such growth is commonly only a veneer over an erosional sea-level bench. Charles Darwin's idealized sequence of reef development begins with the fringing reef, but studies of fossil and recent reef development show that it need not precede, or be followed by, more complex reef types.

Barrier reef. Barrier reef refers to a reef, ordinarily of corals or other organisms, that parallels the shore at the seaward side of a natural lagoon. The surface may be regularly awash at low tide or may break water only at times of strongest swell. Ordinarily, the lagoonward slope is gentle, the seaward slope abrupt. Subsidiary reef patches are common in barrier-reef lagoons and locally beyond the ends or front of the continuous reef. Barrier reefs may develop by continuous upward growth or veneer ridges at the seaward edges of submerged erosional benches, but sinking bottom or rising water is probably essential at some stage of their development.

[PRESTON CLOUD]

Bibliography: L. Barnett, The coral reef, *Life*, The World We Live In, pt. 8, pp. 74–94, Feb. 8, 1954; L. Barnett, The mystery of coral isles, *Life*, Darwin's World of Nature, pt. 7, pp. 54–68, July 20, 1959; P. E. Cloud, Jr., Facies relationships of organic reefs, *Bull. Amer. Ass. Petrol. Geol.*, 36: 2125–2149, 1952; P. E. Cloud, Jr., Nature and origin of atolls, *Proc. 8th Pac. Sci. Congr. Pac. Sci. Ass.*, 3-A:1009–1024, 1958; R. A. Daly, Coral reefs: A review, *Amer. J. Sci.*, 246:193–270, 1948; R. W. Fairbridge et al., *Selected Bibliography on the Geology of Organic Reefs*, Int. Comm. Reef Terminol., Pac. Sci. Board, Nat. Acad. Sci.–Nat. Res. Counc. Circ. no. 3, 1958; H. S. Ladd et al., Drilling on Eniwetok Atoll, Marshall Islands, *Bull. Amer. Ass. Petrol. Geol.*, 37:2257–2280, 1953; F. S. MacNeil, The shapes of atolls: An inheritance from subaerial erosion forms, *Amer. J. Sci.*, 252: 402–427, 1954; W. E. Pugh, *Bibliography of Or-*

ganic Reefs, Bioherms, and Biostromes, Seismograph Service Corp., 1950; C. Teichert, Cold and deep-water coral banks, *Bull. Amer. Ass. Petrol. Geol.*, 42:1064–1082, 1958; J. I. Tracey, Jr., *Natural History of Ifaluk Atoll: Physical Environment*, Bernice P. Bishop Mus. Bull. no. 222, 1961.

Regolith

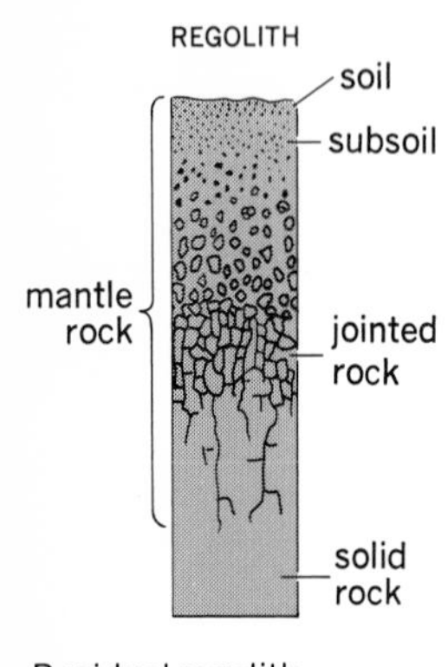

Residual regolith.

The mantle rock or blanket of unconsolidated rocky debris of any thickness that overlies bedrock. Undisturbed regolith may grade from agricultural soil at the surface, through fresher and coarser products of rock weathering, to solid bedrock tens or even hundreds of feet beneath (see illustration). Elsewhere, bedrock may be covered by transported soil and rock debris deposited in such forms as floodplains and deltas, sand dunes, beaches and bars, moraines, and gravity accumulations at the foot of steep slopes and cliffs. Such transported regolith may bear no relation to the bedrock on which it rests, and the contact may be abrupt rather than transitional. *See* BASEMENT ROCK; SOIL; WEATHERING PROCESSES.

[C. F. STEWART SHARPE]

Rhodochrosite

The mineral form of manganese carbonate. Calcium, iron, magnesium, and zinc have all been reported to replace some of the manganese. The equilibrium replacement of manganese by calcium has been determined and found to increase with the temperature of crystallization.

Rhodochrosite is sometimes found in low-temperature veins near deposits of copper, lead, zinc, and silver, or it may occur with other manganiferous minerals of higher temperature origin. It has also been found in sediments and in pegmatites. Well-known occurrences of rhodochrosite are in Europe, Asia, and South America. In the United States large quantities occur at Butte, Mont. As a source of manganese, rhodochrosite is also important at Chamberlain, S.Dak., and in Aroostook County, Maine.

Rhodochrosite crystals, Alicante, Colo. (*Specimen from Department of Geology, Bryn Mawr College*)

Rhodochrosite has hexagonal (rhombohedral) symmetry and the calcite-type crystal structure (see illustration). It occurs more often in massive or columnar form than in distinct crystals. The color ranges from pale pink to brownish pink. Hardness is 3.5–4 on Mohs scale, and specific gravity is 3.70. Rhodochrosite is stable up to 690°C at 10,000 psi and 790°C at 29,000 psi of carbon dioxide. It can be synthesized in its stability field. *See* CARBONATE MINERALS.

[ROBERT I. HARKER]

Bibliography: J. R. Goldsmith and D. L. Graf, The system $CaO\text{-}MnO\text{-}CO_2$: Solid solution and decomposition relations, *Geochim. Cosmochim. Acta*, 11:310–334, 1957; I. M. Varentsov, *Sedimentary Manganese Ores*, 1964.

Rhodonite

A mineral inosilicate with composition $MnSiO_3$. Rhodonite crystallizes in the triclinic system in crystals that are commonly tabular parallel to the base (see illustration). More often it is in cleavable to compact masses or in embedded grains. Crystallographically, rhodonite is closely related to the pyroxenes and thus has two cleavage directions at about 88 and 92°. Hardness is 5.5–6 on Mohs scale, and specific gravity is 3.4–3.7. The luster is vitreous and the color is rose red, pink, or brown. Rhodonite is similar in color to rhodochrosite, manganese carbonate, but it may be distinguished by its greater hardness and insolubility in hydrochloric acid. It has been found at Langban, Sweden; near Sverdlovsk in the Ural Mountains; and at Broken Hill, Australia. Fine crystals of a zinc-bearing variety, fowlerite, are found at Franklin, N.J. *See* SILICATE MINERALS.

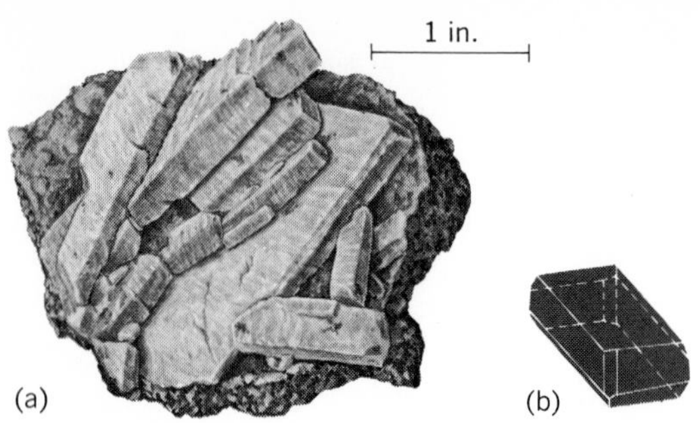

Rhodonite. (*a*) Crystals of variety fowlerite with limestone, Franklin, N.J. (*specimen from Department of Geology, Bryn Mawr College*). (*b*) Crystal habit (*from C. S. Hurlbut, Dana's Manual of Mineralogy, 17th ed., copyright © 1959 by John Wiley & Sons, Inc.; reprinted by permission*)

[CORNELIUS S. HURLBUT, JR.]

Rhyolite

A light-colored, aphanitic (not visibly crystalline) rock of volcanic origin, composed largely of alkali feldspar and free silica (quartz, tridymite, or cristobalite) with minor amounts of dark-colored (mafic) minerals (biotite, hornblende, or pyroxene). If sodic plagioclase exceeds the amount of alkali feldspar the rock is a quartz latite. Rhyolite is chemically equivalent to granite. Quartz latite is chemically equivalent to granodiorite.

Texture. The high silica content gives rhyolitic lava a high viscosity which hinders crystallization and promotes the formation of glass. Rhyolites composed almost entirely of glass are called obsidian, pitchstone, perlite, or pumice. If the glass carries abundant scattered crystals, the rock is a vitrophyre. Typical rhyolites are mixtures of microcrystalline aggregates and glassy material, but many are entirely crystalline, and some grade into microgranite. Porphyritic varieties are those in which numerous large crystals (phenocrysts) are disseminated throughout an aphanitic matrix. The phenocrysts are usually quartz and sanidine. Plagioclase, if present, also occurs as phenocrysts and is usually zoned oligoclase. In quartz latite, plagioclase (oligoclase or andesine) is the most common phenocryst. Quartz phenocrysts are well-formed (euhedral), bipyramidal crystals, but they commonly show corrosive effects including deep embayment and rounded crystal corners. As the abundance of phenocrysts increases the rock passes into porphyry. *See* PORPHYRY; VOLCANIC GLASS.

The finer features of rhyolite must be studied microscopically. The crystalline portion of the groundmass consists of quartz and tridymite more or less intergrown with alkali feldspar. Cristobalite is less abundant. Two types of rhyolite are recognized, normal (potassic) and alkali (sodic). In the potassic type the principal feldspars are sanidine and orthoclase; in the sodic type the principal feldspars are soda-rich (soda sanidine and albite). Normal rhyolite most commonly carries flakes of brown biotite mica which may be so strongly resorbed by the lava that nothing remains but patches of dusty iron oxide which outline the earlier crystals. Green hornblende, much as resorbed crystals, is less abundant; augite, if present, appears typically as phenocrysts. In alkali rhyolite the mafics are soda-rich and generally confined to the matrix as patches of riebeckite, arfvedsonite, hastingsite, or aegirite. Diopside phenocrysts with rims of aegirite or aegirineaugite may also occur. Accessories almost universally present are zircon, sphene, magnetite, and apatite. Hematite as dusty material in the feldspar may be sufficiently abundant to color the rock red.

Structure. Many rhyolites are streaked or banded owing to flowage of solidifying lava. The fluidal structures consist of curving or wavy trails of tiny crystals or of alternating layers of glass and microcrystalline material. Spherulitic structures are characteristic of rhyolite. These spheroidal bodies consist of radial aggregates of alkali feldspar needles, sometimes with intergrown quartz, tridymite, or cristobalite. Spherulites may coalesce or may be arranged in trains parallel to fluidal structure. Tiny cavities within spherulites (lithophysae) may be common. The high volatile content of many rhyolitic lavas promotes the formation of abundant gas cavities in the solidified rock. These are usually tiny, irregular vugs more or less filled or coated with minerals, as in the case of lithophysae. *See* IGNEOUS ROCKS; OBSIDIAN.

Devitrification, the conversion of glass to a crystalline aggregate, has occurred in many rhyolites. In this process crystallization begins commonly along tiny cracks or around phenocrysts and gradually spreads until the whole mass is converted to a minutely crystalline aggregate of quartz or tridymite and alkali feldspar.

Occurrence and origin. Next to basalt and andesite, rhyolite is considered the most common volcanic rock. It commonly forms lava flows and tuffs. Some small dikes and sills or the chilled margins of larger bodies may be composed of rhyolite. The rock is widespread and occurs in large quantities associated with andesite, basalt, and small amounts of dacite and quartz latite. The extremely explosive nature of rhyolitic eruptions reduces greatly the amount of rhyolite formed as lava flows and correspondingly increases the amount formed as fragmental or tuffaceous material. Great quantities of rhyolite may have been erupted as volcanic ash and dust and later incorporated in the thick sedimentary accumulations of geosynclines. *See* TUFF.

Rhyolitic magma may form by differentiation from basaltic magma perhaps in combination with more or less assimilation of siliceous material (sedimentary and granitic rocks). It may also be generated by melting portions of the earth's sialic layer. *See* MAGMA; PETROGRAPHIC PROVINCE.

[CARLETON A. CHAPMAN]

Rift valley

An elongated, relatively narrow depression caused by the subsidence of a crustal block between two or more faults. The boundary faults are steeply inclined down toward the downthrown block, and where the direction of displacement has been ascertained, it indicates that the dislocations are gravity faults. Thus, rift valleys are the surface expression of large graben (see illustration). The term rift is used by some geologists as a synonym for graben. Others define a rift as a strike-slip fault that parallels the trend of the regional structure. *See* FAULT AND FAULT STRUCTURES; GRABEN.

Rift valleys commonly have lengths measured in hundreds of miles with relief at their margins of hundreds or thousands of feet. Rift valleys cut across broadly arched regions and are produced by the lateral extension of the rocks of these areas. The association of basaltic lavas with many rift valleys suggests that the boundary faults are major breaks in the crust and pass downward into the subcrustal region of the Earth. Rift valleys, therefore, have been interpreted to be a part of the major tectonic pattern of the Earth and to be the result of deep-seated deforming pressures.

Rift valleys are associated with many mid-oceanic ridges, where they are interpreted to be graben in an uplifted segment of oceanic crust over a rising column of mantle material. These rift valleys occupy positions where the oceanic crust has split apart and new crust is forming. Basaltic volcanic activity accompanied by shallow-seated seismicity are concentrated along these rift valleys. Smaller, less extensive rift valleys occur in regions underlain by continental crust.

In terms of global tectonics, oceanic crust represents mantle-derived crust younger than continental crust, where, at some previous times in geologic history, continental edges at the margins of an ocean must have been in juxtaposition. In the initial stages of continental breakup and the formation of an ocean basin, a series of mantle diapirs produced regional uplifts along the line of crustal extension. The stretching of the crust above these diapirs produced rift valleys arranged roughly at 120°. Two of these rift valleys propagated to join similarly arranged rift valleys in neighboring uplifted regions, and the connecting of these rift valleys produced a zigzag zone of rifting which later became the continental margins bordering the newly formed and expanding ocean basin. The third arm of the rift valley systems, because it did not link up with other rift valleys, has restricted

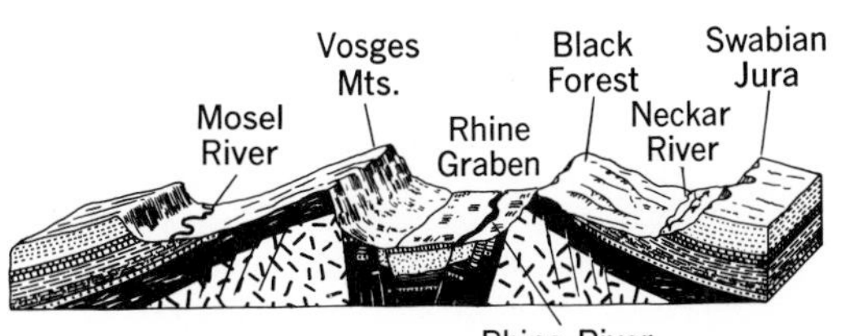

A generalized cross section of the rift valley of the Rhine and the adjoining block mountains of the Vosges and the Black Forest. (*From P. E. James, An Outline of Geography, 2d ed., copyright 1943 by Ginn and Co.*)

length. These became the rift valleys of the continental regions. *See* TECTONIC PATTERNS.

[PHILIP H. OSBERG]

River

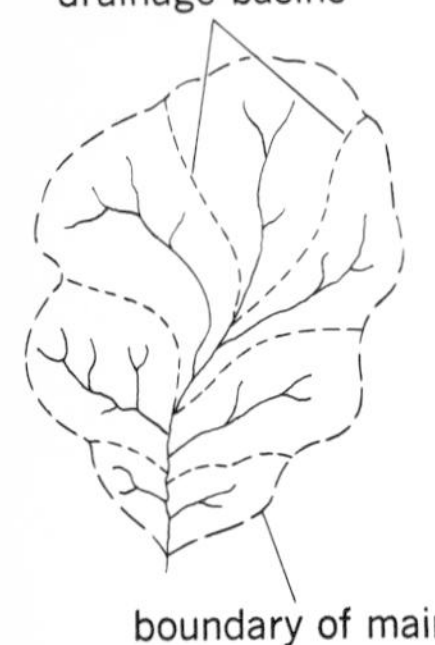

Fig. 1. Maplike diagram of a drainage basin. Note that such basins are composed of a system of secondary basins.

A water stream of natural origin which flows across the surface of a continent or island. A river is part of a river system which drains a topographically related section of land surface known as a river basin (Fig. 1). The system begins in the precipitation which falls on a rock-, soil-, or vegetation-covered surface and immediately becomes surface runoff, or eventually appears as snow and ice meltwater or underground drainage. Such a system may be divided into headwater streams, tributary streams, and the main stem. The headwaters are in springs, marshes, lakes, or small upper streams, generally in the highest relative elevation in a basin. A river ends in a mouth, where it may discharge into a major lake, a dry basin of interior drainage (playa), an inland sea, or the ocean.

Terminology. Like many words which have long been in general use, the term river is somewhat elastic in meaning. In English usage the main stem of a stream system is nearly always designated as a river, but so are all important tributaries and even many secondary tributaries. A tributary may also be known as a fork, branch, or creek, and may have the same volume of flow as other streams called rivers. Smaller headwater streams are usually creeks or brooks.

Rivers flow in channels or watercourses and develop many distinctive valley features by erosion and deposition. For details of form and character of these valley patterns *see* FLOODPLAINS; FLUVIAL EROSION LANDFORMS.

Rivers may be described by the pattern of the system of which they are part and by their length, velocity, volume of discharge, and the nature of water flowing within them. Most rivers are part of a dendritic drainage pattern (Fig. 2), but some, responding to the underlying geologic structure, are in radial, annular, rectangular, or trellised (lattice-like) pattern. In some limestone regions a karst (enclosed depression) drainage may be found, with associated underground rivers. A few rivers, such as the Nile in its lower reaches, are exotic and flow for considerable distances without receiving drainage of any consequence from tributaries. Such river reaches always occur in arid regions.

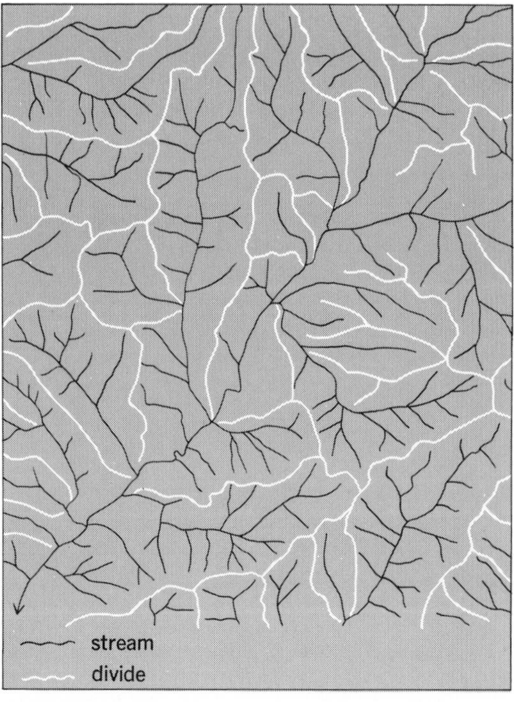

Fig. 2 Cartographic diagram illustrating stream, divide, and basin patterns in a dendritic drainage system.

Regime and flow patterns. The regime is directly dependent on the climate of the region or regions involved. It also is influenced by the size of the drainage basin funneling upon the stream; the direction of flow; the conditions of vegetative cover; and the nature of the surface geology (Fig. 3), topography, and soil conditions in the basin. Few if any streams have completely stable conditions of flow; the rule is variation from day to day, season to season, and year to year. Study of these variations and their causes is an important part of the science of hydrology.

In arid regions, intermittent streams are common. The flow of an intermittent stream may fluctuate markedly from nothing to flood stage within a matter of minutes if a storm of sufficient extent and intensity covers part or all of its drainage area. Normally dry channels of these streams are called arroyos or wadis. *See* DESERT EROSION FEATURES.

Under more humid climatic conditions, the channels of streams of sufficient volume to be called rivers are only occasionally dry. Fluctuations of flow nonetheless are found everywhere. For example, natural flow near the mouth of the Tennessee has varied between 4500 and 500,000 ft^3/sec. In middle latitudes the season of low flow is generally summer, when evaporation and transpiration within the basin are greatest. High water may come during autumn, winter, or spring, depending on temperature conditions and the time of heaviest precipitation. Storage of large volumes of water, such as snow over frozen ground, characteristically causes early spring floods in the Great Plains region of the United States when a rapid thaw takes place.

Within high latitude areas of the Northern Hemisphere, high water inevitably occurs in spring on the northward flowing rivers because melting progresses from headwater to mouth, and the flow of water released upstream is barred by ice dams remaining downstream. The rivers of Siberia are notable examples of this condition, with broad flooding over lowland plains.

In low latitude areas, on the other hand, high water is directly related to seasonal maxima of rainfall, but high altitude conditions may complicate the regime in most zones. Where there is a pronounced dry season, as on the Indian peninsula, high water occurs soon after the onset of the rainy season, when the moisture requirements of hitherto dormant vegetation are still low. In all parts of the world, altitudinal conditions may influence the regime of a stream in another manner. Where headwaters are in extensive high mountain areas with heavy winter accumulations of snow and ice, high water occurs at the season of heaviest melt, early summer or midsummer. The

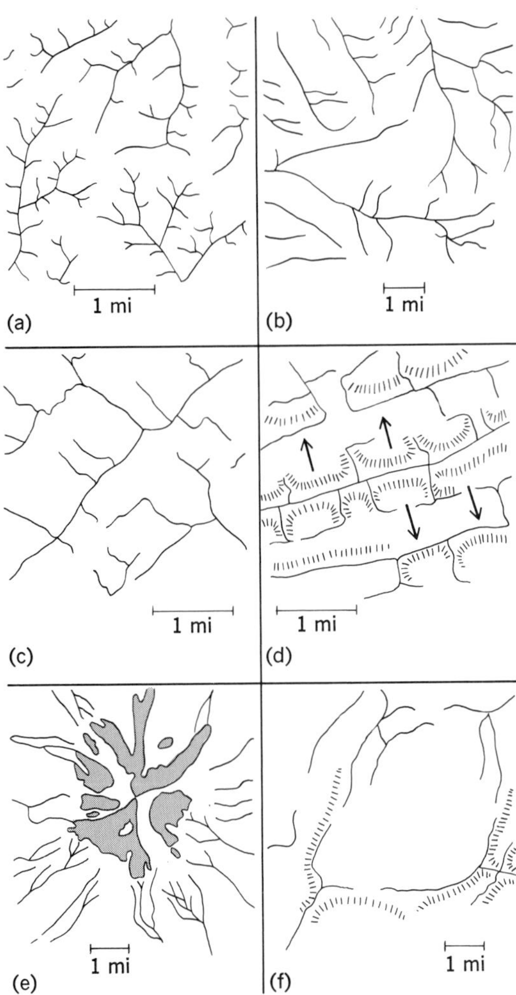

Fig. 3. Stream patterns. (*a*) Dendritic drainage in horizontal rocks, West Virginia. (*b*) Dendritic drainage in crystalline rocks, Rocky Mountains. (c) Rectangular drainage in jointed crystalline rocks, Adirondacks. (*d*) Trellis drainage in folded rocks, Pennsylvania. (*e*) Radial drainage on a volcano, Mount Hood, Ore. (*f*) Annular drainage on dome, Turkey Mountain, N.Mex. (*After A. K. Lobeck, Geomorphology, McGraw-Hill, 1939*)

Columbia, Ganges, Indus, and Rhine rivers show this influence in their regimes.

Surface materials and the nature of vegetative cover also influence the regime. The more continuous the forest or grass cover, in general the more stable the volume of discharge. Soil conditions which favor easy infiltration also promote more equitable flow, as on the sands of the Atlantic and Gulf Coastal Plain of the United States.

Water qualities. Every river is an agent of erosion, as well as an agent of drainage. Many mineral materials other than water consequently are constantly in motion where a river flows. These materials are transported by water in solution, in suspension, and as bed load. For discussion of stream erosion, transport, deposition, and associated landforms *see* STREAM TRANSPORT AND DEPOSITION.

The high capacity of water as a solvent imparts many different qualities to river water as a solution. The great majority of rivers are fresh water, but a few are saline (relatively high salt content). All rivers, however, contain perceptible amounts of mineral material in water solution. In most cases this is calcium, the most common cause of "hard" water, but any of the elements soluble in water may be found, such as magnesium, potassium, sodium, silicon, nitrogen, and the elements which combine with them to form salts. The content of salts in solution is highest in the rivers of regions under desert or semiarid climates, but calcareous materials derived from limestone may yield hard water in humid regions.

Like most other bodies of water on the Earth's surface, a river also is a medium for the support of life, from bacteria and simple forms of plant life to fish, and amphibian, mammal, and bird wildlife. This is related not only to the capacity of water to carry nutrient minerals in solution but also dissolved gases, particularly oxygen.

Management. The characteristics of rivers have made them important to human society. No other natural feature, except the soil, has been more closely tied to the past progress of civilization for the majority of human beings. Means of counteracting the vagaries of flow have been an important part of civil engineering for centuries. This has been true in part because of the attractiveness of floodplains to agricultural occupance, and the consequent need to avoid natural flooding. It has also followed from man's need for water storage in order to live through drought seasons. In modern times the problem of river management or river control has become much more difficult because of the rapid increase of population, its concentration in dense settlements, the extent to which manufacturing and other economic functions have encroached on floodplains, the vastly increased disposal of wastes in rivers, and the larger number of purposes that rivers must serve simultaneously. The general objects of river management are the conservation of natural flow for release at the times needed by man, the confinement of flood flow to the channel and planned areas of floodwater storage, and the maintenance of water quality at a level which will yield optimum benefit through multiple use. The techniques of river management are well understood; their practice is still very incomplete, in part because the economics of river development is not well known. Domestic river development is an important responsibility of the U. S. Army Corps of Engineers. It also is the central responsibility of the Tennessee Valley Authority and is an important objective of the Bureau of Reclamation of the Department of the Interior.

Of the major rivers in the world (see table) none is yet controlled or managed in the manner which modern engineering, administrative, and biological techniques would permit. The closest approach to such management is made on some medium-sized streams, the Tennessee, the Rhine, and the Rhône, for example. Some other rivers, such as the San Joaquin in California, have been fully developed for a single purpose, irrigation. Commencing in the 1930s, the greatest river-regulation works of all history were undertaken. The United States, the Soviet Union, and (since 1946) France have been foremost in supporting work of this kind. Among the notable achievements have been the series of great dams on the Columbia, Missouri, and Colorado and the regulation of the Tennessee in the

Discharge, basin area, and length of some of the world's major rivers

River	Average discharge, ft³/sec	Basin area, mi²	Length, mi
Amazon	4,000,000	2,772,000	3900
La Plata-Paraná	2,800,000	1,198,000	2450
Congo	1,400,000	1,425,000	2900
Yangtze	770,000	750,000	3100
Brahmaputra	700,000	361,000	1680
Ganges	660,000	450,000	1640
Mississippi-Missouri	620,000	1,243,000	3892
Yenisei	615,000	1,000,000	3550
Orinoco	600,000	570,000	1600
Lena	547,000	1,169,000	2860
St. Lawrence	500,000	565,000	2150
Ob	441,000	1,000,000	2800
Mekong	390,000	350,000	2600
Volga	350,000	592,000	2325
Amur	338,000	787,000	2900
Mackenzie	280,000	682,000	2525
Columbia	256,000	258,200	1214
Zambesi	250,000	513,000	2200
Danube	218,000	347,000	1725
Niger	215,000	584,000	2600
Indus	196,000	372,000	1700
Yukon	180,000	330,000	2100
Huang	116,000	400,000	2700
Nile	100,000	1,293,000	4053
Sâo Francisco	100,000	252,000	1811
Euphrates	30,000	430,000	1700

United States; the Volga-Don Canal, the lower Volga dams, and the very large dams on the Angara and Yenisei in the Soviet Union; and the Rhône regulation in France.

The greatest and potentially most productive works remain for the future. These include plans for important work on the three largest streams of all the Amazon, the La Plata-Paraná, and the Congo. These basins contain storage and power-generation sites of several times the capacity of the largest hitherto developed. Of the eight rivers having basins of 1,000,000 mi² or more in extent, only the Mississippi and Nile have more than minor control works.

Still other great streams which offer major possibilities for physical development are the Yenisei, Yangtze, Huang, Amur, Mekong, Chao Phraya, Tigris-Euphrates, Niger, Zambesi, Orinoco, Sâo Francisco, Danube, Mackenzie, and Yukon. The extent and timing of such development will depend upon economic need, availability of investment funds, and political cooperation. The need is patent for development of the Yangtze, Huang, Nile, Niger, Tigris-Euphrates, Danube, Sâo Francisco, and lesser streams in densely settled, underdeveloped areas. It is therefore probable that the latter half of the 20th century will be a period of extending control of these streams, as political conditions permit.

[EDWARD A. ACKERMAN; DONALD J. PATTON]

Bibliography: S. Leliavsky, *An Introduction to Fluvial Hydraulics*, 1966; L. B. Leopold, M. G. Wolman, and J. P. Miller, *Fluvial Processes in Geomorphology*, 1964; M. Morisawa, *Streams: Their Dynamics and Morphology*, 1968; R. J. Russell, *River Plains and Sea Coasts*, 1967; United States President's Water Resources Policy Commission, *Ten Rivers in America's Future*, vol. 2, 1950.

Rock

A relatively common aggregate of mineral grains. Some rocks consist essentially of but one mineral species (monomineralic, such as quartzite, composed of quartz); others consist of two or more minerals (polymineralic, such as granite, composed of quartz, feldspar, and biotite). Rock names are not given for those rare combinations of minerals that constitute ore deposits, such as quartz, pyrite, and gold. In the popular sense rock is considered also to denote a compact substance, one with some coherence; but geologically, friable volcanic ash also is a rock. A genetic classification of rocks is shown in the following list.

- Igneous
  - Intrusive
    - Plutonic (deep)
    - Hypabyssal (shallow)
  - Extrusive
    - Flow
    - Pyroclastic (explosive)
- Sedimentary
  - Clastic (mechanical or detrital)
  - Chemical (crystalline or precipitated)
  - Organic (biogenic)
- Metamorphic
  - Cataclastic
  - Contact metamorphic and pyrometasomatic
  - Regional metamorphic (dynamothermal)
- Hybrid
  - Metasomatic
  - Migmatitic

Exceptions to the requirement that rocks consist of minerals are obsidian, a volcanic rock consisting of glass; and coal, a sedimentary rock which is a mixture of organic compounds. *See* COAL; VOLCANIC GLASS.

Table 1. Simplified classification of major igneous rocks on the basis of composition and texture

	SiO_2-rich (acidic) ⟵			⟶ SiO_2-poor (basic)	
	Light colored ⟵		Gray ——— Dark colored		⟶ Black
Mineral composition	Quartz, potash feldspar, biotite	Potash feldspar, biotite, or amphibole	Sodic plagioclase, hornblende, or augite	Augite, olivine, hypersthene, calcic plagioclase	Olivine, enstatite, augite
Intrusive					
Medium-grained	Granite*	Syenite	Diorite	Gabbro	Peridotite
Extrusive					
Fine-grained to aphanitic	Rhyolite	Trachyte	Andesite	Basalt	
	⟵ Felsite ⟶				
Porphyritic	Rhyolite porphyry	Trachyte porphyry	Andesite porphyry	Basalt porphyry	
Glassy	Obsidian				
Vesicular	Pumice			Scoria	
Fragmental	Tuff and agglomerate of each type				

*Exceptionally coarse-grained rock of general granitic composition is pegmatite.

Table 2. Size classification of clastic sedimentary particles and aggregates

Size, mm	Particle	Aggregate
Greater than 256	Boulder	Gravel, conglomerate (psephite, rudite)
256–64	Cobble	Breccia (angular)
64–4	Pebble	
4–2	Coarse sand	Sandstone (psammite, arenite)
2–1/16	Sand	
1/16–1/256	Silt	Siltstone (pelite, lutite)
Less than 1/256	Clay	Clay Shale

Igneous rocks. Igneous rocks are those that have solidified from a molten condition. The parent material is magma—a natural, hot, mutual solution of silicates with minor amounts of water and other volatiles. Igneous rocks are divided into those which crystallized before magma reached the Earth's surface (intrusive rocks) and those that solidified at the surface, some as layers of lava (the extrusive flow rocks) and others as pyroclastic debris in explosive eruption (Table 1). *See* MAGMA; PYROCLASTIC ROCKS.

Chief elements in igneous rocks are oxygen, O; silicon, Si; aluminum, Al; iron, Fe; magnesium, Mg; calcium, Ca; sodium, Na; and potassium, K. Compositions range from about 40% SiO_2 (peridotites) to as much as 70% SiO_2 (granites). Silica-poor rocks contain relatively large amounts of Ca, Mg, and Fe^2 (basic rocks), whereas silica-rich types contain larger amounts of Na and K (acidic rocks). *See* IGNEOUS ROCKS.

Sedimentary rocks. Clastic sedimentary rocks (consisting of mechanically transported particles) are subdivided on the basis of particle size (Table 2). Those having intermediate and fine-grain sizes are further subdivided on the basis of composition (see illustration). Other significant clastic rocks are those consisting of detrital calcite, the calcarenites and calcilutites. *See* CALCARENITE.

Textures of clastic rocks derive from grain size, sorting, form, and arrangement. Form includes sphericity (shape), the degree to which a particle approximates a sphere; and roundness, the measurement of the sharpness of edges and corners. Only glacial sedimentary rocks (tillite) do not show layering or stratification.

Chemical sedimentary rocks are those precipitated from ocean, lake, and groundwater. The most important ones are shown in the following list.

Rock	*Chief mineral*
Chert	Chalcedony, SiO_2
Limestone	Calcite, $CaCO_3$
Travertine (spring deposit)	Calcite, $CaCO_3$
Dolomite	Dolomite, $CaMg(CO_3)_2$
Phosphorite	Apatite, $Ca_{10}(PO_4)_5(CO_3)F_3$
Salines (evaporites)	
Rock salt	Halite, NaCl
Rock anhydrite	Anhydrite, $CaSO_4$
Rock gypsum	Gypsum, $CaSO_4 \cdot 2H_2O$

Organic sedimentary rocks include (1) siliceous types made up of opaline tests of diatoms, diatomite, or radiolaria, radiolarite; (2) calcareous types,

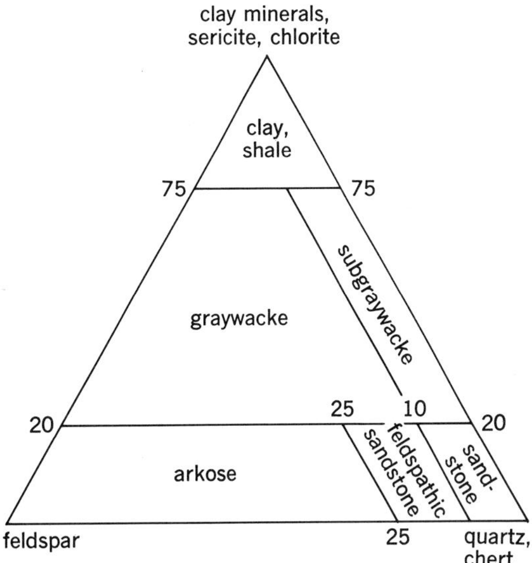

Classification of psammitic and pelitic rocks based on proportions of three most common clastic minerals: quartz, feldspars, and clays—excluding calcilutites and calcarenites. (*From E. W. Heinrich, Microscopic Petrography, McGraw-Hill, 1956*)

Table 3. Simplified classification of metamorphic rocks, with selected examples

Parent rocks	Contact metamorphism	Regional metamorphism		
		Low grade	Intermediate grade	High grade
Sandstone, arkose	Quartzite (quartz)	Quartzite and quartz-feldspar gneiss		
Shale	Hornfels (andalusite, cordierite)	Slate, phyllite, (chlorite, muscovite)	Mica schist (biotite, garnet, kyanite, staurolite)	Sillimanite gneiss (sillimanite, almandite)
Limestone	Marble (calcite)	Marble and calc-silicate gneiss		
		(Calcite, tremolite)	(Calcite, wollastonite)	(Calcite, diopside, anorthite)
Basalt	Hornfels (plagioclase, hypersthene)	Greenschist (chlorite, albite, epidote)	Amphibole schist, (plagioclase, actinolite)	Amphibolite (andesine, hornblende, garnet)

consisting of calcite shell, namely, shell limestone and coquina; and (3) carbonaceous types, consisting of coal and other accumulations of altered plant debris. *See* SEDIMENTARY ROCKS; SEDIMENTATION.

Metamorphic rocks. Metamorphic rocks owe their complexity of composition and texture not only to the existence of several types of metamorphism but also to the application of these types under different intensities to a variety of parent rocks; thus, both sedimentary and igneous rocks may be metamorphosed (Table 3). Regional metamorphic rocks are distinguished by foliation, a parallel orientation of platy or prismatic minerals. *See* METAMORPHIC ROCKS; METAMORPHISM.

Physical properties and behavior. When stresses are applied to rocks, either natural (those active in mountain building) or man-made, resulting from loading with structures (such as dams), deformation (strain) may result. Rocks subjected to stress normally undergo three deformation stages: elastic—the rock returns to its original size or shape upon withdrawal of stress; plastic—beyond a limiting stress (elastic limit) there is only partial restoration upon stress removal; and fracture—breakage with further increase in stress. With increases in confining pressure (load of overlying rocks) and temperature, the interval between the elastic limit and fracture increases. Thus rocks that behave as brittle substances near the Earth's surface, failing by fracturing, will at depth be deformed plastically by solid flow. The effects of stress depend on physical properties (Table 4). *See* ENGINEERING GEOLOGY; HIGH-PRESSURE PHENOMENA; ROCK MECHANICS; STRUCTURAL GEOLOGY.

Rock cycle. Igneous rocks exposed at the Earth's surface are subject to weathering, which alters them chemically and physically. Such material when transported, deposited, and consolidated becomes sedimentary rock, which, through heat and pressure, may be converted to metamorphic rock. Both sedimentary and metamorphic rocks also may be weathered and transformed into younger sediments. Deeply buried metamorphic rocks may be remelted to yield new igneous material. *See* LITHOSPHERE, GEOCHEMISTRY OF; PETROGRAPHY.

[E. WILLIAM HEINRICH]

Table 4. Physical properties of some common rocks

Rock	Specific gravity	Porosity, %	Compressive strength, psi	Tensile strength, psi
Igneous				
Granite	2.67	1	30,000–50,000	500–1000
Basalt	2.75	1	25,000–30,000	
Sedimentary				
Sandstone	2.1–2.5	5–30	5,000–15,000	100–200
Shale	1.9–2.4	7–25	5,000–10,000	
Limestone	2.2–2.5	2–20	2,000–20,000	400–850
Metamorphic				
Marble	2.5–2.8	0.5–2	10,000–30,000	700–1000
Quartzite	2.5–2.6	1–2	15,000–40,000	
Slate	2.6–2.8	0.5–5	15,000–30,000	

Rock, age determination of

The determination of the time of formation of crustal rocks. Absolute rock age determination became possible, in principle, with the discovery of radioactivity. Prior to this discovery, the methods of stratigraphy and paleontology provided the only means of determining the time of rock formation. These methods were and still are useful in establishing the sequence of events and in defining the successions of strata. They aid in piecing together the geologic time scale, but provide only a relative geochronology. *See* DATING METHODS; FISSION TRACK DATING; GEOLOGICAL TIME SCALE.

Theory. All modern isotopic methods of age determination depend on the following principles. The average rate of disintegration of a large number of radioactive atoms of a given kind is constant. The mathematical relationship is $-dN/dt = \lambda N$, where N is the number of atoms of this kind present and λ is the disintegration constant. This rate can be accurately measured in the laboratory by determining the number of atoms decaying per unit time for a given total number of atoms. The rate is not altered by any physical or chemical conditions to which a crustal rock could have been subjected since its origin. In the simplest case a mineral containing radioactive parent atoms A, but not stable daughter atoms B, crystallizes from a silicate melt during the formation of an igneous rock. The longer the time interval between mineral formation and measurement, the larger the ratio of B/A. The age relationship is given precisely by Eq. (1), where T is the age (generally given in

$$T = \frac{1}{\lambda} \ln \frac{B+1}{A} \quad (1)$$

millions of years). The rate of radioactive disintegration is also given by the half-life of the isotope, that is, the time required for half of the existing number of atoms to disintegrate. The half-life ($t_{1/2}$) is related to the decay constant by $t_{1/2} = 0.693/\lambda$. For practical purposes the half-life of a radioactive isotope must be the same order of

Table 1. Methods of quantitative geochronometry

Method	$T_{1/2}$, years	Effective range in years	Applicable minerals or rocks
Rb^{87}-Sr^{87}	5.0×10^{10}	$T_0{}^* - 10^8$	Muscovite, biotite, K-feldspar, lepidolite, glauconite
K^{40}-Ar^{40}	1.3×10^9†	$T_0 - 10^4$	Muscovite, biotite, glauconite
U^{238}-Pb^{206}	4.5×10^9		
U^{235}-Pb^{207}	0.71×10^9	$T_0 - 10^7$	Uraninite, monazite, zircon, black shale
Th^{232}-Pb^{208}	1.39×10^{10}		
C^{14}	5.6×10^3	50,000 – present	Carbon-bearing materials once in contact with the atmosphere-biosphere system
U, Th-He		$T_0 - 10^6$(?)	Possibly sulfides and magnetite
K^{40}-Ca^{40}	1.3×10^9†	$T_0 - 10^8$(?)	Sylvite
Pb^{207}-Pb^{206}		$T_0 - 10^8$	Galena, lead in pyrite
Radiation damage		Uncertain	Zircon, samarskite
Cl^{36}	2×10^5	Uncertain	Chlorine-rich rocks exposed to cosmic-ray neutrons
H^3	12	50 – present	Groundwater
Ionium	8×10^4	4×10^5 – present	Deep-sea sediments
Re^{187}-Os^{187}	6×10^{10}(?)	T_0 – ?	Old molybdenites
Lu^{176}-Hf^{176}	2×10^{10}	T_0 – ?	Old rare-earth minerals

*T_0 is the age of the Earth (about 4.6×10^9 years). †Total half-life.

magnitude as the time span to be measured. Thus isotopes of interest for geochronology must have half-lives ranging from thousands of years (for the study of recent geological and archeological events) to hundreds of millions of years (for ancient rocks and the age of the Earth). Providing the parent (A) and daughter (B) atoms are quantitatively separated at the time of mineral formation and that the mineral has remained a closed chemical system during its history, an absolute age can be obtained, the accuracy of which is only limited by the uncertainties in the chemical and isotopic analyses of A and B and in the half-life. Under the most favorable conditions the age determinations can be made to within a few percent even for a rock thousands of millions of years old.

The early phase of quantitative age determination based on radioactivity occurred between 1900 and 1938. During this period work was restricted to the measurement of lead-uranium ratios in uranium minerals and in helium-uranium ratios in a variety of mineral and rock types. This pioneer work established the order of magnitude of the geologic time scale and provided the first age measurements for many geologic provinces. It suffered from the relatively crude analytical methods that were available, inadequate knowledge of the nuclear phenomena involved, and absence of reliable criteria for identifying chemical alteration or the incorporation of daughter product of the isotopic clocks at the time of mineral formation.

Modern isotopic geochronometry began with the first comprehensive and precise measurement of the lead isotopes in uranium and lead minerals by A. O. Nier and his coworkers in 1939. The development of the instrumentation and analytical procedures for microassay by the isotope dilution method and the precise measurement of the isotopic composition made possible the discovery and application of a number of isotopic chronometers apart from the uranium-lead system. The possibility of obtaining independent age estimates from different mineral phases provided the necessary criteria for demonstrating a closed chemical system and for detecting primary contaminants. In addition, valuable geochemical information could be obtained from partially open mineral systems.

The most important age determination methods are summarized in Table 1. The U-Pb (uranium-lead), Rb-Sr (rubidium-strontium), K-Ar (potassium-argon), and C^{14} (carbon-14) methods have been shown to yield reliable results on suitable samples and have permitted the construction of absolute geologic history in many areas of the world. These methods can span all of Earth history, although the analytical errors in the K-Ar method for ages below 10^7 years become rather large. All other methods require further research before they can be established as useful geochronometers.

The U-Pb, Rb-Sr, and K-Ar methods are discussed in some detail later. Table 2 gives the isotopic ages obtained by the three primary chronometers on unaltered rocks of various ages. The results agree within the uncertainties of analysis and decay constants. For the C^{14} method *see* RADIOCARBON DATING.

Rubidium-strontium method. The radioactive transformation of Rb^{87} to Sr^{87} by β-decay provides one of the most reliable and useful isotopic geochronometers, particularly for older rocks. The method can be applied to Cenozoic rocks, providing a sufficiently high Rb/Sr ratio is present, as in lepidolite. Ordinary biotite in granites as young as Mesozoic can be dated by the Rb-Sr transformation. The soft energy of the Rb^{87} β-particles produces negligible radiation damage, so that mineral

Table 2. Comparison of isotopic ages, 10^6 years, obtained by different methods*

Locality	K-Ar	Rb-Sr	U^{238}-Pb^{206}	U^{235}-Pb^{207}
Beartooth Mts., Mont.	2500 M	2740 M	2600 U	2640 U
Keystone, S.D.	1570 M	1690 M	1610 U	1620 U
Wilberforce, Ontario	960 B	1020 B	1020 U	1020 U
Goodhouse, South Africa	920 B		930 Mo	915 Mo
Wichita Mts., Okla.	460 B	510 B	517 Z	525 Z
Redstone, N.H.	168 B	190 B	187 Z	184 Z

*Decay constants same as in Table 1: B = biotite, M = muscovite, Mo = monazite, U = uraninite, Z = zircon.

phases which can be dated by this method are intact even in the oldest rocks. The Rb-Sr method has been applied successfully to muscovite and biotite in a great variety of igneous and metamorphic rocks and to all potash-micas, pollucite, rhodizite, amazonite, and perthite in pegmatites. There are many other potassium minerals in igneous and metamorphic rocks that may be used, including orthoclase, sanidine, leucite, and phlogopite, provided the ratio of rubidium to common strontium is large enough for the age involved. Glauconite appears promising for dating sedimentary rocks. The ratio of rubidium to common strontium in some stony meteorites is high enough for age determinations to be made on the whole rock.

Assuming a half-life of 5.0×10^{10} years, which is probably accurate to at least ±6%, the age formula for the Rb-Sr method is given in Eq. (2), where T is

$$T = 7.2 \times 10^4 \ln \left[\frac{Sr^{87*}}{Rb^{87}} + 1\right] \qquad (2)$$

the isotopic age in millions of years, Rb^{87} and Sr^{87*} are the number of radioactive rubidium and radiogenic strontium atoms, respectively. The isotopic age is the true age if no alternation has occurred during the history of the mineral. The total rubidium and strontium contents of the sample are obtained with accuracies of 1–3% by routine, but sophisticated, isotope dilution techniques. The isotopic composition of the strontium in the sample must be analyzed separately. The Rb^{87} can then be readily obtained from the known isotopic composition of natural rubidium; that is, $Rb^{85} = 72.15\%$ and $Rb^{87} = 27.85\%$. The calculation of the radiogenic Sr^{87*} content is more complex because all rubidium-bearing minerals also contain measurable common strontium which was incorporated at the time of mineral formation. The isotopic composition of strontium in modern ocean water is $Sr^{88} = 82.5\%$, $Sr^{87} = 7.02\%$, and $Sr^{86} = 9.85\%$. In the rock-forming environment the isotopic composition of the common strontium has generally not changed significantly with time, so that the Sr^{86} abundance in the sample whose age is to be determined can be used to determine the fraction of the total Sr^{87} that was incorporated at the time of mineral formation. The difference is the radiogenic Sr^{87*}. Should there be a question about the isotopic composition of the common strontium incorporated into the rubidium-rich mineral at the time of formation, a rubidium-poor but strontium-rich phase, such as plagioclase or apatite, may be analyzed isotopically for strontium.

It is evident that as the ratio of rubidium to common strontium in a rock increases, the error in the determination of the radiogenic Sr^{87*} will increase. Thus in order to obtain an isotopic age that is analytically accurate to ±5%, assuming that the isotopic composition of the strontium can be measured to ±0.5%, the ratio of rubidium to common strontium would have to be at least 10 for a mineral 1,800,000,000 years old, but at least 100 for a mineral 180,000,000 years old.

The Sr^{87*}/Rb^{87} ratio appears to remain unaffected in feldspar unless recrystallization occurs. In biotite, however, there is some evidence that alteration may take place at lower temperatures, possibly involving base exchange phenomena. If Rb-Sr ages on cogenetic mica and feldspar agree, it is strong evidence of a real date of the last metamorphic or igneous event.

Potassium-argon method. The radioactive isotope of potassium, K^{40}, decays by β-emission to Ca^{40} and by K-electron capture to Ar^{40}. The decay to Ca^{40} has only restricted value in geochronology, because common calcium is largely Ca^{40} and most potassium minerals contain significant amounts of calcium. The argon part of the decay is particularly attractive because potassium minerals appear to form without incorporating any primary argon from their environment and the analytical methods for argon determination are extremely sensitive. The widespread occurrence of potassium minerals suggests that the method can be applied to virtually all rock complexes. By using the decay constants for the separate branches of the disintegration of K^{40}, $\lambda_e = 0.584 \times 10^{-4}/10^6$ years and $\lambda_\beta = 4.72 \times 10^{-4}/10^6$ years, the age is given by Eq. (3).

$$T(\text{in } 10^6 \text{ years}) = 1885 \ln(1 + 9.10\, Ar^{40}/K^{40}) \qquad (3)$$

The K^{40} isotopic abundance is constant (0.0119%) in natural potassium so that this isotope may be determined by standard wet-chemical methods. The Ar^{40} is determined by the isotope dilution method after it is released quantitatively from the mineral by fusion. Correction is made for contamination by Ar^{40} from air by monitoring the Ar^{36} in the sample gas. Isotopic ages may be determined by this method over most of geologic time with an accuracy of a few percent. For young minerals the atmospheric correction becomes a limiting factor.

At low temperatures mica appears to retain essentially all of its radiogenic argon (Table 2). At elevated temperatures diffusion may cause partial loss of radiogenic argon. This has been observed in field tests along the border of a younger metamorphic belt that is superimposed on a preexisting basement.

Other minerals such as feldspar, lepidolite, and glauconite do not appear to hold all of the radiogenic argon. It may be, however, that criteria can be developed for estimating the degree of retention for various structural types. Measurements on these minerals and whole rocks (particularly basalts and siliceous effusives) yield minimum ages only, but these may be of great value for certain geological problems.

Uranium-lead method. The method based on the decay of the thorium and uranium to lead is the oldest and most elegant; it involves two or three radioactive isotopes, U^{238}, U^{235}, and Th^{232}, to three distinct isotopes of lead. In pitchblende the thorium content is negligible, but in most other uranium or thorium minerals both elements are present in sufficient quantities for analysis. Each of the above isotopes decays through a series of 8–12 isotopes until a stable lead isotope is produced, Pb^{206}, Pb^{207}, and Pb^{208}, respectively. The intermediate isotopes all have much shorter half-lives than the parent isotopes, so that except for alteration effects the chronometers may be considered as simple parent-daughter decay systems of U^{238} to Pb^{206}.

The presence of two uranium isotopes with different half-lives and different chemical intermediates with distinctive nuclear properties provides a mutual check on the reliability of the ages obtained. Any chemical alteration in the system will affect the apparent uranium-to-lead isotopic ratios in such a way that the calculated ages will differ. Such a discordance can be used to evaluate the nature, time, and extent of alteration if sufficient data are available. Concordance among these ratios is strong evidence that a true age has been obtained.

The age determination consists of analyzing the sample for total uranium, thorium, and lead by the isotope dilution method and for its lead isotopic composition. There is only one significant isotope of thorium (Th^{232}) and two of uranium (U^{238} and U^{235}) in constant proportion so that these quantities are readily calculated from the chemical analysis. The lead presents a more complex problem because nearly all uranium and thorium minerals incorporate at least small quantities of common rock lead at the time of formation. This rock lead contains Pb^{204}, Pb^{206}, Pb^{207}, and Pb^{208}. All of these isotopes were present in the primeval lead of the Earth but additional Pb^{206}, Pb^{207}, and Pb^{208} have been added throughout Earth history as a result of the radioactive decay of U^{238}, U^{235}, and Th^{232} in the crust and mantle. The ratios of Pb^{206}/Pb^{204}, Pb^{207}/Pb^{204}, and Pb^{208}/Pb^{204} are therefore a function of time. They have increased only slowly, and therefore it is generally sufficient to use the average crustal value at the approximate time of mineral formation in order to make the correction for the incorporated lead. If necessary, a more precise correction can be made by analyzing the lead isotopic composition in a uranium-free mineral that is cogenetic with the uranium mineral. Thus if the lead isotopic ratios in the sample and in the contaminating common lead are known, the quantity of the radiogenic isotopes Pb^{206}, Pb^{207}, and Pb^{208} may be derived. The ages are then calculated by using equations analogous to Eq. (2) for the simple decay of Rb^{87} to Sr^{87}. *See* LEAD ISOTOPES, GEOCHEMISTRY OF.

In some cases such as those in Table 2, the isotopic ages obtained from the uranium isotopes agree within the experimental error. This appears generally true for large fresh uraninite crystals from pegmatites in areas that have not been subjected to any later thermal effects. In many cases, however, significant discordance occurs. Most often this is due to selective lead loss, but loss of intermediate decay products may also play a significant role. By examining different minerals and different samples of the same mineral, it is possible to reconstruct the geochemical history, as well as the original age, of the mineral.

Although the original use of the U-Pb method was to date pegmatite crystals or pitchblende veins in metallic ores, the most important application appears to be to accessory zircon crystals in igneous and metamorphic rocks. Some attempts have been made to apply the method to date uranium-rich black shales, but the high common-lead content and ease of alteration make a precise determination difficult.

The variation of the average lead isotopic composition in the crust has been useful in obtaining the approximate ages of ore deposits. The U-Pb method has also been found applicable to the age of meteorites. *See* EARTH, AGE OF; METEORITE; RADIOACTIVE MINERALS. [J. LAURENCE KULP]

Bibliography: L. T. Aldrich and G. W. Wetherill, Geochronology by radioactive decay, *Annu. Rev. Nucl. Sci.*, 8:257–298, 1958; E. I. Hamilton, *Applied Geochronology*, 1965; J. L. Kulp, *Quantitative Geochronometry*, 1962; Laboratorio di Geologia Nucleare, *Summer Course on Nuclear Geology, Varenna 1960*, pt. 5: *Geochronology*, 1961.

Rock, electrical properties of

The effect of changes in pressure and temperature on electrical properties of rocks. There has been increasing interest in the electrical properties of rocks at depth within the Earth and the Moon. The reason for this interest has been consideration of the use of electrical properties in studying the interior of the Earth and its satellite, particularly to depths of tens or hundreds of kilometers. At such depths pressures and temperatures are very great, and laboratory studies in which these pressures and temperatures are duplicated have been used to predict what the electrical properties at depth actually are. More direct measurements of the electrical properties deep within the Earth have been made by using surface-based electrical surveys of various sorts. An important side aspect of the study of electrical properties has been the observation that, when pressures near the crushing strength are applied to a rock, marked changes in electrical properties occur, probably caused by the development of incipient fractures. Such changes in resistivity might be used in predicting earthquakes, if they can be measured in the ground. *See* GEOPHYSICAL EXPLORATION.

Electrical zones in the Earth. Attempts to measure the electrical properties of rocks to depths of tens or hundreds of kilometers in the Earth indicate that the Earth's crust is zoned electrically. The surface zone, with which scientists are most familiar, consists of a sequence of sedimentary rocks, along with fractured crystalline and metamorphic rocks, all of which are moderately good conductors of electricity because they contain relatively large amounts of water in pore spaces and other voids. This zone, which may range in thickness from a kilometer to several tens of kilometers, has conductivities varying from about 1/2 ohm-m in recent sediments to 1000 ohm-m or more in weathered crystalline rock.

The basement rocks beneath this surface zone are crystalline, igneous, or metamorphic rocks which are much more dense, having little pore space in which water may collect. Since most rock-forming minerals are good insulators at normal temperatures, conduction of electricity in such rocks is determined almost entirely by the water in them. As a result, this part of the Earth's crust is electrically resistant. There are numerous experimental difficulties involved in trying to measure the electrical properties of this part of the Earth's crust, but it appears that the resistivity lies in the range 10,000–1,000,000 ohm-m.

At rather moderate depths beneath the surface of the second zone, resistivity begins to decrease with depth. This decrease is considered to be the result of higher temperatures, which almost certainly are present at great depths. High temperatures lead to partial ionization of the molecular structure of minerals composing a rock, and the ions render even the insulating minerals conductive. As temperature increases with depth, this effect becomes more pronounced and resistivity decreases markedly. The maximum resistivity in the Earth's crust probably occurs at depths of only 5–10 km below the surface of the crystalline basement. At depths of 30–40 km, corresponding to the top of the mantle, resistivity is only about a few hundred ohm-meters. Beyond this depth, resistivity seems to decrease slightly with increasing depth, but again drops sharply at depths around 700 km. At greater depths the resistivity is less than 1 ohm-m. A profile of resistivity as it varies downward into the Earth is shown in Fig. 1.

At relatively shallow depths information is gained by lowering equipment into bore holes to measure the electrical properties of rocks directly. Another approach is the use of low-frequency currents introduced into the Earth at the surface. These methods are restricted to use in studying resistivities at depths less than 10 km. Information from greater depths is obtained by studying slow variations in the Earth's magnetic field and the currents induced in the Earth by these variations. Fluctuations in the magnetic field with periods measured in months penetrate to distances up to 1000 km, and they may be used in estimating conductivity to these depths.

Temperature effects. Laboratory investigations have also been pursued to find how the electrical properties of rocks depend on the pressure and temperature deep in the Earth. Inasmuch as the temperature at depths of 40–60 km is probably in the range 800–1200°C and the pressure is in the range 10,000–15,000 kg/cm², these conditions are difficult to duplicate in the laboratory. Both electrical conductivity and dielectric constant have been measured by several researchers at pressures to 50,000 kg/cm² and temperatures to 1200°C, using dry rock samples. The most obvious conclusion from such studies is that for this range of conditions the effect of temperature on electrical properties is much more profound than that of pressure. An increase in temperature from 20 to 1000°C will reduce the resistivity of a dry rock by a factor of 1,000,000. On the other hand, an increase in pressure from 0 to 50,000 kg/cm² changes the resistivity of a dry rock by less than 100%. Typical behavior for the resistivity of dry rocks as temperature is raised is shown by curves in Fig. 2. At temperatures up to about half the melting point, conduction increases relatively slowly with increasing temperature, and it is thought that conduction is

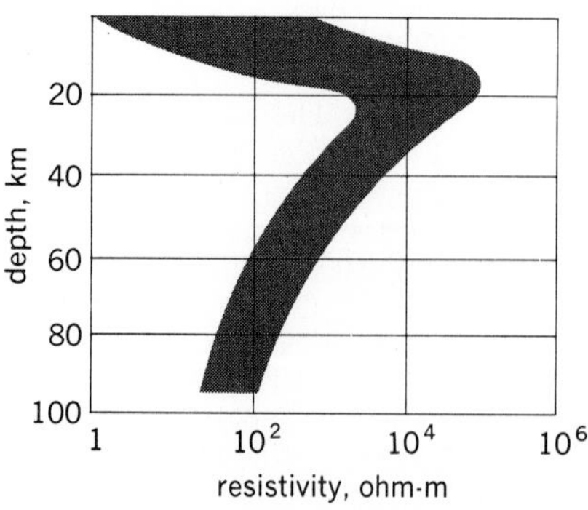

Fig. 1. Probable range of electrical resistivity through the crust and into the upper mantle.

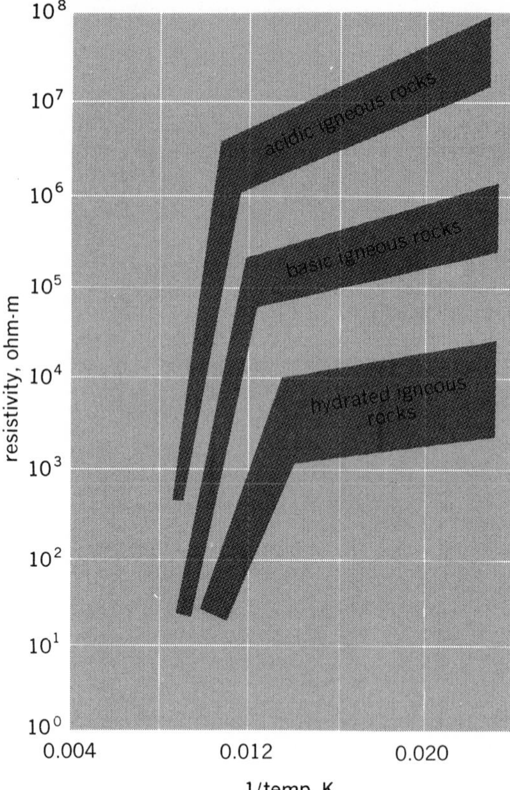

Fig. 2. Variation of resistivity with temperature in dry igneous rocks. Hydrated samples contained only water of hydration, not free water. At the melting point, conduction increases abruptly.

due to impurities, which contribute ions under weak thermal excitation. At temperatures above half the melting point, conduction increases rapidly with increasing temperature, with conduction being caused by ions torn from normal lattice positions in the crystals by violent thermal agitation. At the melting point, conduction increases abruptly as a rock melts and becomes highly ionized.

It has been observed that silicic rocks, such as granite, granodiorite, and rhyolite, are poorer conductors of electricity at all temperatures than other rock types. Rocks, such as gabbro, basalt, and dunite, which are rich in ferromagnesian minerals, are more conductive than the silicic rocks by about an order of magnitude. This is believed to result from the fact that the principal ions which contribute to conduction are the small ferromagnesian ions, which are more abundant in the dark rocks.

Pressure effects. Typical curves for the behavior of resistivity and of dielectric constant with the application of pressure are shown in Figs. 3 and 4. The dielectric constant is found to increase by 10–20% as the pressure on a rock sample is increased to 100–200 kg/cm² and then to increase more slowly as the pressure is elevated further. The increase in dielectric constant is due in part to an increase in density of the rocks under pressure.

Two forms of behavior have been recognized in studies of the behavior of resistivity at pressures up to 50,000 kg/cm². With some rocks resistivity decreases uniformly with increasing pressure; in

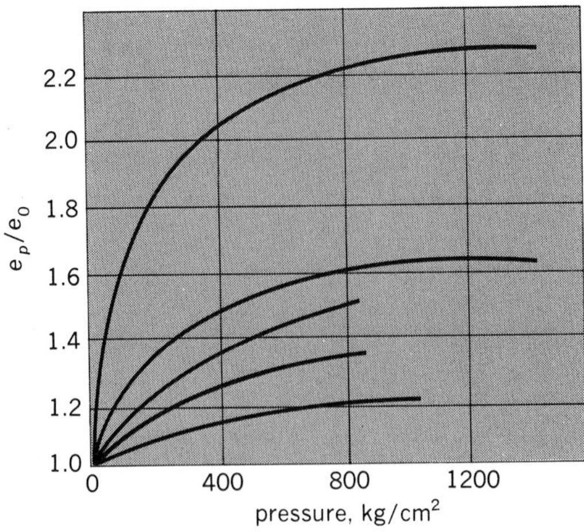

Fig. 3. Effect of pressure on the dielectric constants of five samples of granite. The ratio e_p/e_0 is the ratio of the dielectric constant measured at elevated pressure to dielectric constant measured at zero pressure.

other rocks resistivity decreases at low applied pressures but increases at higher pressures. A uniform decrease has been observed in rocks such as basalt, diabase, and amphibolite, while an increase in resistivity over part of the pressure range has been observed in microcline-rich rocks. The increase in resistivity may be associated with a decrease in the mobility of charge carriers as the crystal lattice is compressed.

The effect of pressure on the resistivity of rocks in which conduction is determined by water content rather than by the crystal matrix is quite different. Several investigators have studied the effect of having greater pressure on the crystal framework than on the pore fluid. As an overpressure is applied to the rock skeleton, the framework starts to close in on the void spaces, reducing their volume and increasing the resistivity. It has also been observed that, if the overpressure is made great enough, resistivity begins to increase again, a phenomenon which is attributed to the opening of small cracks prior to rock failure, with the movement of water into these cracks to contribute to conduction. The behavior of resistivity for a number of rock specimens as a function of frame overpressure is shown by the curves in Fig. 5. According to W. F. Brace and A. S. Orange, observations of such changes in resistivity might be used to detect minute changes in the structure of rock under

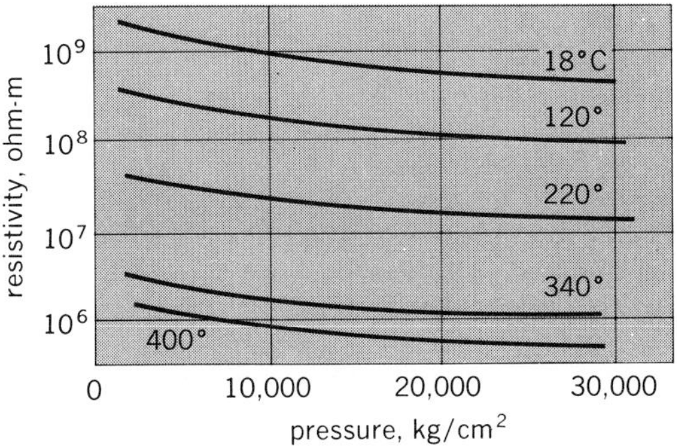

Fig. 4. Resistivity of a basalt sample measured as a function of pressure at various temperatures.

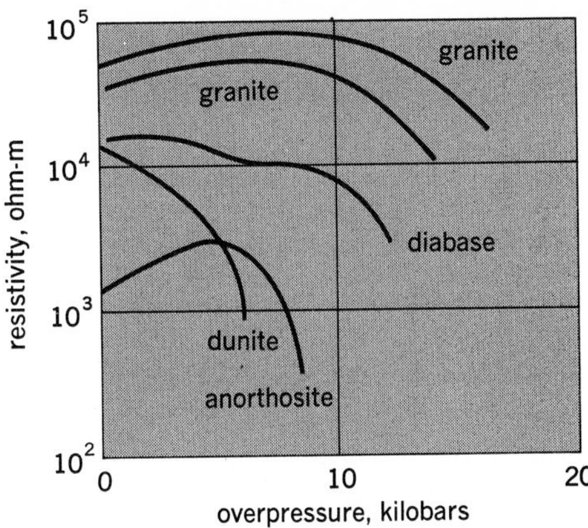

Fig. 5. Effect of greater pressure on framework of a rock than on fluid in pores in water-bearing rocks.

stress. They observe that the buildup of strain in the Earth preceding earthquakes ought to be accompanied by changes in resistivity, and that these changes might warn of an impending earthquake. *See* HIGH-PRESSURE PHENOMENA; PETROLOGY.

Electrical zones in the Moon. In planning the exploration of the Moon, some thought has been given to the use of measurements of electrical properties as a means for studying the Moon's interior. Although there is little information on the electrical properties of the Moon, the knowledge gained about the Earth can be used to make an intelligent guess. The major difference is the lack of an atmosphere and moisture, which means that the surface rocks on the Moon must be quite different from those on Earth. Measurements with radar and infrared sensors have indicated that the surface of the Moon must be covered with a rock froth or soil which has a very low density—the rocks may be only about 10% solid material. Inasmuch as there is probably no moisture in this rock, it seems very likely that the electrical resistivity must be very large, perhaps 10^{10} ohm-m.

It is believed that temperature increases with depth inside the Moon, though perhaps not as rapidly as in the Earth. The temperature at the center of the Moon may be 2000 K and, as a result, the rocks in the interior can be expected to be quite conductive, just as they are on the Earth. The resistivity at the center of the Moon may be of the order of 1–100 ohm-m.

The change in resistivity between the highly resistant surface and the conductive interior depends very much on the composition of the Moon. Some scientists believe that the interior contains as much water as the rocks within the Earth, at depths beyond which evaporization into space has been able to take place. Others believe that the Moon had a different origin from the Earth, and thus never contained very much water. If water is present, one would expect the resistivity to drop very rapidly over the first few kilometers in depth, as water contributes to electrical conduction. If the Moon is devoid of water, the resistivity should change more gradually with depth and become low only at depths of many hundreds of kilometers, where the temperatures are high. The presence of water at moderate depths would be very important to future lunar exploration—it might be used as fuel for trips deeper into space, or even to provide a tenuous atmosphere to protect man's installations on the Moon. Thus it is likely that electrical surveys for water on the Moon will have a high priority in the early stages of lunar exploration. *See* MOON. [GEORGE V. KELLER]

Bibliography: W. F. Brace and A. S. Orange, Electrical resistivity changes in saturated rock under stress, *Science*, 153:1525–1526, 1966; S. P. Clark, Jr., *Handbook of Physical Constants*, Geol. Soc. Amer. Mem. no. 97, 1966; A. W. England, G. Simmons and D. Strangway, Electrical conductivity of the Moon, *J. Geophys. Res.*, 73:3219–3226, 1968; G. V. Keller, Electrical prospecting for oil, *Colo. Sch. Mines Quart.*, vol. 63, no. 2, 1968; G. V. Keller, L. A. Anderson, and J. I. Pritchard, Geological survey investigations of the electrical properties of the crust and upper mantle, *Geophysics*, 31: 1078–1087, 1966; E. I. Parkhomenko, *Electrical Properties of Rocks*, 1967; S. H. Ward, G. R. Jiracek, and W. I. Linlor, Electromagnetic reflection from a plane layered lunar model, *J. Geophys. Res.*, 73:1355–1372, 1968.

Rock control (geomorphology)

The influence of differences in earth materials on development of landforms. The materials out of which landforms are carved are not homogeneous; even in a small area they differ from one another in composition, physicochemical properties, mechanical behavior, and various other ways. These differences and variations are visibly reflected in landforms. Although the concept of cyclic evolution of landforms proposed by W. M. Davis has become less popular, that of rock control, the first of his trilogy (structure, process, and stage), is still regarded as valid. As examples of the problems of rock control, many books on geomorphology quote such instances as cuestas, structural benches, knickpoints, dike ridges, mesas, hogbacks, structural plains, and karst and inversion of topography. *See* GEOMORPHOLOGY.

A. Strahler (1952) laid stress on the consideration of the properties of materials and on the types of strain and failure as the dynamic bases of geomorphology. J. Hack (1960) denied the concept of the evolutionary development through time altogether, and accentuated the importance of rock properties and processes in the formation of landforms. According to Hack, the differences in form between one area and another, including relief, form of stream profiles, valley cross sections, width of floodplains, shape of hilltops, and other forms, are explicable solely in terms of differences in the bedrock and the manner in which it breaks up into different component parts when subjected to exogenetic (external) forces such as weathering. Moreover, even with exogenetic forces certain processes are closely related to the nature of rock itself. According to E. Yatsu (1967), mass wasting, such as flowslides and earthflows, occur exclusively in areas containing expandable clay minerals, while landslides of the slope-rupture variety take place under other suitable conditions.

Aerial photographs are being used in the study of geological structure. Interpretation of such photographs and deduction of geological structure entirely depend on the fact that the topography is controlled by the distribution and arrangement of rocks with different properties. Therefore, an intensive study of these properties is indispensable. It is urgent that the most rigorous scientific research methods be adopted for studying rock control problems, rather than the somewhat haphazard methods of the past. *See* FLUVIAL EROSION CYCLE; MASS WASTING; ROCK MECHANICS; STRUCTURAL GEOLOGY. [EIJU YATSU]

Bibliography: J. T. Hack, Interpretation of erosional topography in humid temperate regions, *Amer. J. Sci.*, 258-A:80–97, 1960; A. N. Strahler, Dynamic basis of geomorphology, *Geol. Soc. Amer. Bull.*, 63:923–938, 1952; E. Yatsu, *Geogr. Ann.*, ser. A, 2(4):396–401, 1967; E. Yatsu, *Rock Control in Geomorphology*, 1966.

Rock magnetism

The natural remanent magnetization (NRM) of igneous, metamorphic, and sedimentary rocks has enabled inferences to be made concerning changes in the direction and intensity of the geomagnetic field in past geological times and the relative movements of the poles and the continents.

The magnetic properties of rocks result from the iron oxide minerals present to a few percent in many rocks. Two groups are especially important: the magnetite (Fe_3O_4)–ulvospinel (Fe_2TiO_4) solid solution series and the hematite (Fe_2O_3)–ilmenite ($FeTiO_3$) solid solution series. Intergrowths between magnetic minerals are of frequent occurrence.

Magnetization induced by the present geomagnetic field is also important in the interpretation of air or ground geomagnetic surveys. Laboratory experiments, applying magnetic fields to rocks at different temperatures, are important not only in the interpretation of NRM but also as a petrological and mineralogical tool. *See* PALEOMAGNETICS.

Primary magnetization. Three types of remanent magnetization may be acquired during the process of formation of an igneous or sedimentary rock.

Thermoremanent magnetization (TRM). The permanent or natural remanent magnetization of igneous rocks, such as lavas, sills, and dykes, is due to the magnetization of these minerals during cooling in the geomagnetic field from the temperatures at which they solidify, which are usually above the Curie points of their magnetic minerals (Fig. 1).

Depositional remanent magnetization (DRM). In sediments the magnetization may arise through the orientation of the detrital iron oxide grains by the geomagnetic field during deposition in water.

Chemical remanent magnetization (CRM). This form of magnetization may arise during chemical change, as in the growth of grains of hematite in the red sandstones, during or perhaps slightly later than deposition.

Secondary magnetization. On these original magnetizations is often superposed a secondary magnetization, acquired in more recent times. This is due in part to magnetically unstable iron oxide grains, which may be in the original rock or may result from weathering. According to the theory of L. Néel, these grains are either too small or too finely divided by intergrowths of different chemical composition to retain a permanent magnetization indefinitely, and if placed in zero magnetic field will gradually lose their magnetization logarithmically with time. Alternatively, such grains will, according to the same law, acquire a magnetization parallel to the field in which they lie. This is known as viscous magnetization. Rocks are found with varying amounts of this secondary magnetization, which sometimes changes appreciably during laboratory storage but more often possesses a relatively stable component along the present geomagnetic field.

Secondary magnetization of this type may sometimes be proved to be absent. This is best done by showing that specimens of the rock tilted, folded, or dispersed in a conglomerate since the time of their original magnetization have consistent directions of magnetization when allowance is made for these subsequent movements.

After a correction is made for the geological tilt, supposing that the blocks have not been rotated around the vertical, the paleomagnetic directions should become more closely grouped. If they become more scattered, the presence of a component of magnetization imposed by the ambient geomagnetic field subsequent to the tectonic movements is demonstrated. The magnetization of pebbles from one geological formation should be randomly directed; if not, remagnetization has occurred since the laying down of the conglomerate bed. Secondary magnetization can sometimes be removed by the process of ac demagnetization or by heating to a few hundred degrees and cooling in zero magnetic field.

Measurement of the direction of the original component of magnetization, making allowance if necessary for geological tilting, enables changes in the geomagnetic field to be traced throughout the geological record in different places.

It is impossible to make any straightforward interpretation of directions of magnetization when components, acquired through different processes at different stages in the history of the rock from fields of varying direction, are present. Little headway has therefore been made in the study of the magnetization of metamorphic rocks, although there are cases, for example, in granites from California, where a good grouping of directions of NRM suggests that magnetization has been acquired mainly at one time.

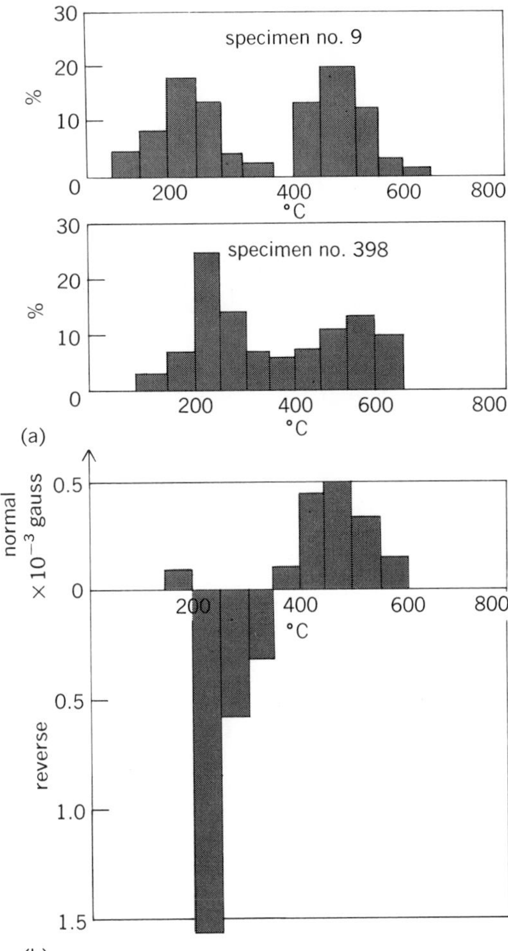

Fig. 1. Graphs showing partial thermoremanent magnetization. (*a*) Icelandic specimens of Tertiary plateau basalt with reverse natural permanent magnetization. Intensities are expressed in percent of the sum: for no. 9 it is 13.2×10^{-3} gauss and for no. 398, 5.6×10^{-3} gauss. The partial thermoremanent magnetization is in the direction of the field for all temperature intervals. (*b*) A specimen of dacite pumice from Mount Haruna, Japan. "Normal" denotes that the magnetization is in the direction of the external field, "reverse" that it is in the opposite direction.

Recent field directions. In historic times the secular variation has been studied with archeological specimens, dated varve clays, and lavas from dated eruptions. The first two investigations show that the geomagnetic field has had a secular variation similar to that observed in the last few hundred years. The third method, applied to the Mount Etna lavas, has shown that the directions of magnetization of the lava flows agree with the observations of the field of the same date. *See* GEOMAGNETISM.

Mean geomagnetic field. Spherical harmonic analysis of the global field at epochs since 1830 shows it to be predominantly a geocentric dipole. The higher harmonics change rapidly by comparison with the geological time scale, containing periods of a few hundred years. Because the core is loosely coupled with the mantle and rotates with varying velocity relative to it, theory requires that the components of the field not symmetrical about the rotation axis average out over some thousands of years. Paleomagnetic observations in the Quaternary and upper Tertiary from every continent, during which time movements of the continents relative to the pole have not been large enough to be a complication, support this prediction by giving mean declination $D=0$, and in addition show that the mean field is an axial dipole for which the angle of magnetic dip or inclination I is given by $\tan I = 2 \tan \lambda$, where λ is the latitude of the site. Since this property of the magnetic dynamo in the Earth's core results from the dominance of the Coriolis force on fluid motions within it—a dominance not likely to be appreciably altered for most of the Earth's history—and is not affected by frequent reversals of polarity, the axial dipole hypothesis is used to interpret the paleomagnetism of all other geological periods.

The oldest known NRM is that of the Stillwater complex, which suggests that the Earth had a magnetic field 2,600,000,000 years ago.

Geomagnetic reversals. For most of Cenozoic time (Tertiary and Quaternary periods), worldwide paleomagnetic observations show that al-

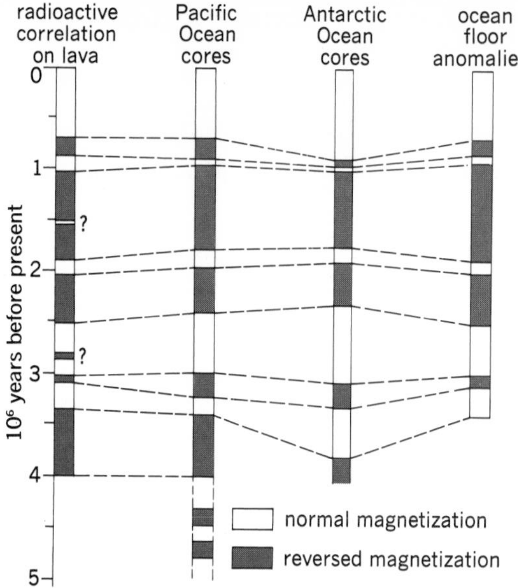

Fig. 2. The sequence of reversals of the geomagnetic field in late Cenozoic time.

though the mean geomagnetic field (in the sampling of a series of rocks the averaging is done usually over 10^3-10^6 years) is that of a geocentric dipole along the present geographical axis, successive parts of the stratigraphical column have magnetizations of opposed directions. A few such reversed magnetizations may result from anomalous properties of certain of the iron oxide minerals, which are ferrimagnetic, or through chemical changes subsequent to original magnetization. Accumulated evidence suggests that they are mostly explained by the geomagnetic dipole reversing its polarity at irregular intervals of the order of 10^6 years (Fig. 2).

Between many of the zones of opposed magnetization are found horizons of scattered, and often weaker, magnetizations, suggestive of the interval

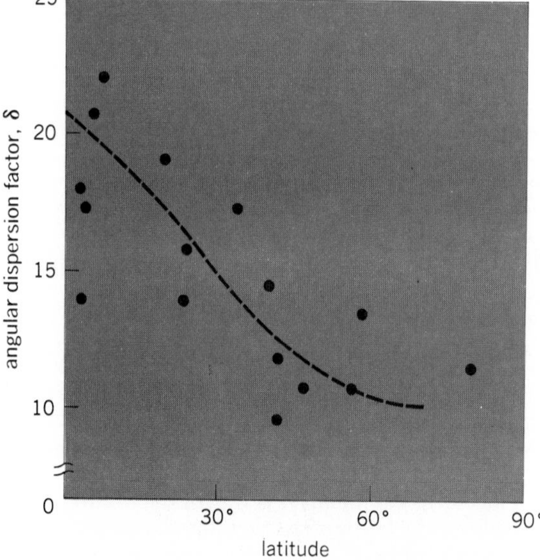

Fig. 3. Graph of secular variation based on Phanerozoic igneous rocks, plotted against paleomagnetic latitude. (*After K. M. Creer*)

in which the field is changing polarity by rotation, by disappearing and growing again, or by the relative strengthening of nondipole field components.

Where the interval between successive reversals is long, paleontological correlation enables a test to be made of the worldwide character. For the whole of Permian time the field in at least three continents is directed southward.

Reversed magnetizations occur in sedimentary and igneous rocks in spite of their different magnetization mechanisms. Many examples exist of dykes and lavas reheating, and of adjacent sediments or lavas which always become magnetized in the same sense as the baking rock. These objections are hard to explain by self-reversal mechanisms in the iron oxide minerals, but they fit the field-reversal hypothesis.

In the last few million years field reversals have been proved to occur, because accurate potassium argon dates on lavas show that reversed and normal fields correlate in different areas. The same sequence of reversals has also been found in the continuous time record found in ocean-bottom sedimentary cores and in the magnetic anomaly patterns produced by the ocean floor spreading away from the ocean ridges, which appears to occur at a uniform rate. The time intervals between reversals show up proportionally to vertical distances in the cores and to horizontal distances from the ridges. *See* MARINE GEOLOGY.

In addition to reversals which last for about 1,000,000 years, there appear geomagnetic events, in which the polarity reverses for only 100,000 years before returning.

Over the whole of geological time, the Earth does not appear to have had any preferred polarity. The comparatively short interval between reversals in the late Cenozoic is not unparalleled in other parts of the geological record. Time intervals between reversals of direction of magnetization appear to be random.

Geomagnetic secular variation. The scatter of a set of paleomagnetic observations is often in part due to experimental errors or inherent poor recording of the original field directions. Where steps can be taken to average these errors, the remaining dispersion arises from the secular variation. From igneous rocks back to the Permian, K. M. Creer showed that the dispersion decreases with increasing latitude, as simple theory predicts and as is found for the present field (Fig. 3).

Polar wandering and continental drift. As well as providing a better record of the properties of the geomagnetic field and the dynamo in the Earth's core, results of paleomagnetic observations have led to a new appreciation of movements in the crust. The paleomagnetic results from early Cenozoic, Mesozoic, and Paleozoic and late Precambrian rocks show magnetizations which can be interpreted as arising from a dipole field not aligned along the present geographical axis. Very similar pole positions for any one geological period are obtained from rocks within one continental mass but differ from those of other geological periods and from other continents. Thus a plot of the path of the pole through geological time can be drawn for each continent and is one convenient way of summarizing the paleomagnetic results obtained there (Fig. 4). The striking differences between these paths are interpreted in terms of continental

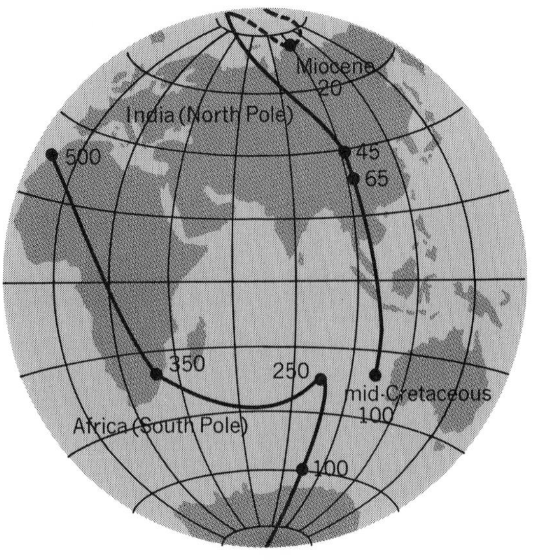

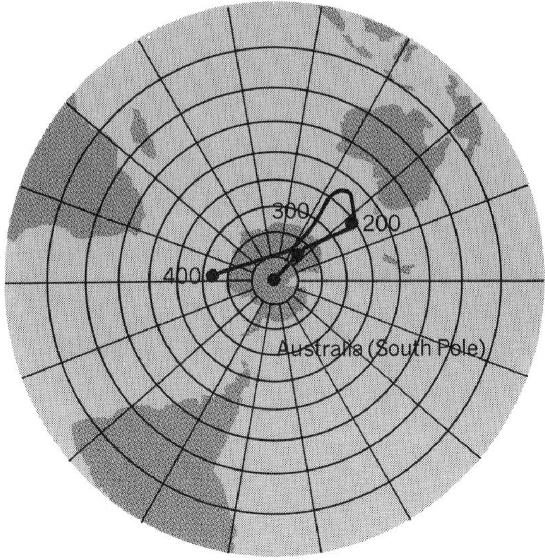

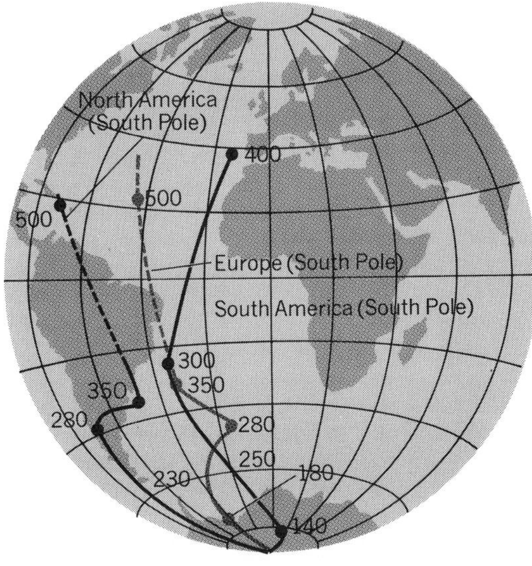

Fig. 4. Polar wandering curves for different continents obtained from the directions of remanent magnetization of their rocks. Numbers denote millions of years before present. (*After K. M. Creer et al., Geophysical interpretation of palaeomagnetic directions from Great Britain, Phil. Trans. Roy. Soc. London, vol. 250, Cambridge University Press, 1957*)

Geological age of rocks yielding evidence of positions of Australia

Position (Fig. 5)	Geological age
PPR	Pliocene, Pleistocene, and Recent
E	Lower Tertiary, probably Eocene
J	Mesozoic, probably Jurassic
Tr	Triassic, probably Lower Triassic
P_2	Permian, Upper Marine Series
P_1	Permian, Lower Marine Series
C	Upper Carboniferous
D	Devonian, probably Lower Devonian
S	Upper Silurian
ϵ_2	Middle Cambrian
ϵ_1	Lower Cambrian
Pre-ϵ_3	Top of Upper Proterozoic
Pre-ϵ_2	Upper Proterozoic
Pre-ϵ_1	Lower part of Upper Proterozoic

drift, occurring in relatively late geological times. If, on the basis of the results for Cenozoic times and from the theory of the geomagnetic field, the mean field is assumed to have an axis coincident with the axis of rotation, the latitudes calculated can be compared with those expected on paleoclimatic evidence. Within any one continent there is good agreement. Polar wandering is therefore also inferred from the slow change of the direction of magnetization through the geological column.

Another representation of the paleomagnetic data is useful. From the relations $D=0$ and $\tan I = 2 \tan \lambda$, it follows that the mean paleomagnetic field lay in the then geographic meridian, and the pole lay at an angular distance $(90° - \lambda)$ from the site. Thus the motion of a continent relative to a latitude grid fixed to the axis of rotation may be inferred. Its motion in longitude is not directly determinable but may be guessed.

Paleomagnetic surveys of the continents indicate that North America and Europe, and South America and Africa have drifted apart, and that India and Australia (see table) have moved north since early Mesozoic times (Fig. 5). A. Wegener's reconstruction of the continents and the formation

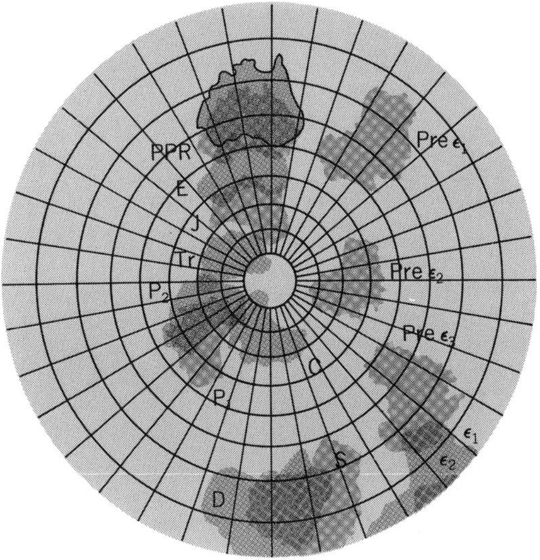

Fig. 5. Movement of Australia relative to South Pole of rotation. Heavy outline denotes present position; see table for geological age of rocks yielding evidence.

of the Atlantic and Indian ocean basins has thus received quantitative proof.

It is possible to argue that the discrepancy between the polar wandering paths from the different continents does not prove that continental drift occurred, but that the mean geomagnetic field in earlier geological times was not dipolar. Analysis of the observations does not support this explanation.

The original magnetization of rocks may be controlled to an important degree by causes other than the ambient field, such as flow in igneous rocks or the orienting action of water currents in sediments or stress patterns. Field evidence can test this by relating magnetization directions and texture. Many rocks have an anisotropic susceptibility, show magnetostriction, and have very high coercivities.

Past intensity of geomagnetic field. Igneous rocks usually have intensities of magnetization between 10^{-2} and 10^{-4} gauss and sedimentary rocks between 10^{-4} and 10^{-7} gauss. The intensity of the geomagnetic field at the time of magnetization is only one of many factors influencing remanent magnetization; yet it appears that the geomagnetic dipole moment could not have changed by orders of magnitude during geological time (Fig. 6).

Attempts have been made to determine the strength of the past field by comparing the NRM of igneous rocks and baked clays to the TRM acquired by cooling from above to Curie point in laboratory fields, and rejecting samples in which changes in the iron oxide minerals occur during the experiment. In relatively young material where natural decay of intensity is negligible, reliable values of the ancient field strength are obtainable. In sediments a statistical study of the ratio of the NRM to the susceptibility has yielded a curve of intensity with geological time.

[STANLEY K. RUNCORN]

Bibliography: P. M. S. Blackett, *Lectures on Rock Magnetism*, 1956; D. W. Collinson et al., Palaeomagnetic investigations in Great Britain, *Phil. Trans. Roy. Soc. London*, vol. 250, 1957; A. Cox and R. R. Doell, *Geol. Soc. Amer. Bull.*, 71:645–768, 1960; E. Irving, *Palaeomagnetism*, 1964; T. Nagata, *Rock Magnetism*, 1953; L. Néel, Some theoretical aspects of rock magnetism, *Advan. Phys.*, 4:191–243, 1955; G. D. Nicholls, The mineralogy of rock magnetism, *Advan. Phys.*, 4:113–190, 1955; S. K. Runcorn, Magnetization of rocks, in S. Fluegge (ed.), *Handbuch der Physik*, vol. 47, 1956; S. K. Runcorn, Rock magnetism: Geophysical aspects, *Advan. Phys.*, 4:245–291, 1955; R. L. Wilson, *Earth Sci. Rev.*, 1:175–212, 1966.

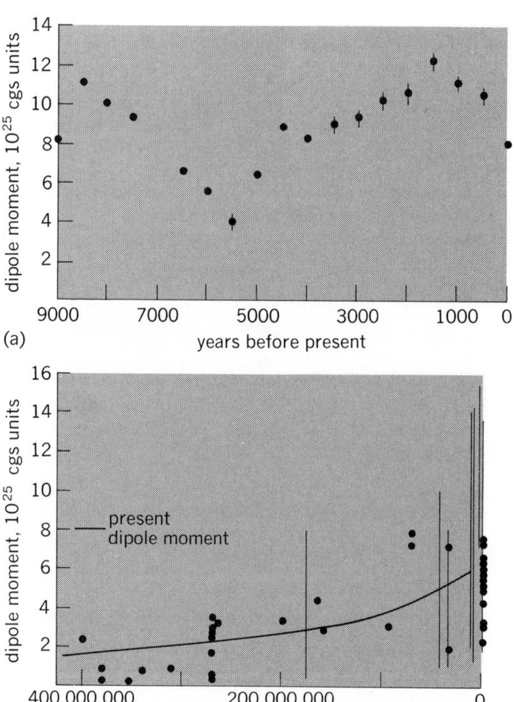

Fig. 6. Dipole moment of the Earth. (*a*) Determined by archeomagnetism. (*b*) Based upon igneous rocks. (*After P. J. Smith*)

Rock mechanics

Application of the principles of mechanics and geology to quantify the response of rock when it is acted upon by environmental forces, particularly when human-induced factors alter the original ambient forces. Rock mechanics is an interdisciplinary engineering science that requires interaction between physics, mathematics, and geology, and civil, petroleum, and mining engineering (Table 1). The present state of knowledge permits only limited correlations between theoretical predictions and empirical results; therefore, the most useful rock mechanics principles are based upon data obtained from laboratory and in-place measurements and from prototype behavior (behavior of the full-size engineering structure). *See* ENGINEERING GEOLOGY; TECTONOPHYSICS.

Some of the empiric principles now used in rock mechanics long have been employed to solve problems encountered with the rock in the Earth's crust. For example, Stone Age humans used them to select and work the stone that produced the most durable spear points, as did the Romans to obtain and work the stone for constructing roads, buildings, and aqueducts, and to dig rock tunnels to transport water and to use as catacombs. One of the first recognitions of rock mechanics as a coherent modern science in the United States was the first professional conference on the subject in 1956: "Symposium on Rock Mechanics" at the Colorado School of Mines. At Salzburg, Austria, in 1963 the International Society for Rock Mechanics was organized and now is issuing test standards. Attention focused on the value of rock mechanics studies because of catastrophic failures of the Malpasset Dam, in France, in 1959, apparently because of weakening of abutment rock, and of the Vaiont Dam, in Italy, in 1963 when reservoir water apparently triggered massive slides that pushed most of the water over the top of the dam. The exact causes for these catastrophes are debatable; however, there is little argument that the intrinsic properties of the rock system played a significant role.

Definitions. Because of the comparative youth of rock mechanics as a science and because of its interdisciplinary aspects, there still is no standardization of terminology; but a listing of "standard"

Table 1. Scope of rock mechanics*

General	Examples
Fundamentals	Theories of failure, creep, fracture, and so on
Measurements	Basis for predicting behavior of a rock system under a static or a dynamic load
Laboratory (static or dynamic)	
Field, in place (static or dynamic)	
Applications	
Surface foundations	Support for dams, multistory buildings, powerhouses, or microwave, radio, or electrical transmission-line towers
Surface excavations Man-made or natural, temporary or permanent	Open-pit mining; transportation routes; construction of buildings, dams, or reservoirs; canals; protective construction to house weapon systems (for example, missile silos) or civil defense facilities
Underground openings	Housing for power plants or military facilities; storage of valuable records or petroleum products at low or high pressures; for extraction of ore; for highway, railroad, or parking tunnels; and for passage of water at gravity heads or under pressure
Boring (drilling)	Design of equipment associated with extraction of ore or petroleum; excavation of shafts or tunnels by rotary drilling machines; and obtaining samples (core) of the rock system
Comminution	Intentional breaking or crushing of rock to produce useful sizes or to remove it from excavations
Construction material	Facing stone for buildings; riprap (protective layer) for earth dams; rockfill for dams; concrete aggregate; stone for breakwaters and other protective structures that minimize destructive action of water bodies on harbor or flood-control facilities
Geological processes	Surface subsidence caused by underground extraction of ore or fluids; theories to explain the structure of the Earth; earthquake prediction or control

*Adapted from W. R. Judd (ed.), *Rock Mechanics Research*, National Academy of Sciences – National Research Council, Committee on Rock Mechanics, 1966.

symbols and units is available from the International Society for Rock Mechanics, Lisbon. To minimize confusion, the following terms and definitions are used in this article.

Environmental factors. These are the natural factors and human influences that require consideration in engineering problems in rock mechanics. The major natural factors are geology, ambient stresses, and hydrology. The human influences derive from the application of chemical, electrical, mechanical, or thermal energy during construction (or destruction) processes.

Ambient stress field. This is the distribution and numerical value of the stresses in the environment prior to its disturbance by humans. These stresses also are called residual, primary, or in-place stresses; they may result from ancient orogenies.

Rock system. This includes the complete environment that can influence the behavior of that portion of the Earth's crust that will become part of an engineering structure. Generally, all natural environmental factors are included.

Rock element. This is the coherent, intact piece of rock that is the basic constituent of the rock system and which has physical, mechanical, and petrographic properties that can be described or measured by laboratory tests on each such element. There is no agreed-upon definition of rock that is usable for engineering purposes. Analysis of over 70 definitions indicates that to the engineer, a "rock" is a natural part of the Earth's crust that is constituted of one or more minerals, is a coherent piece of matter, and has some degree of hardness. In some research, a coherent rock specimen may in itself be regarded as a rock system that has as elements the minerals and their relationships, dissolved or pore water, microdefects such as fractures, and the ambient stresses locked in the minerals. The concepts of rock system and rock element enable the concomitant engineering design to be optimized according to the principles of system engineering.

"Rock failure" occurs when a rock system or element no longer can perform its intended engineering function. Failure may be evidenced by fractures, distortion of shape, or reduction in strength.

"Failure mechanism" includes the causes for and the manner of rock failure. Most of the present failure theories are based upon modifications of the Griffith theory.

Analysis of rock behavior. Optimal analysis of a rock mechanics problem requires an understanding of all mechanisms that influence the behavior of the rock system. The first clues are in the atomistic or molecular relationships in a rock element; that is, the strength and certain other properties of a rock element can in part depend upon the kind and arrangement of the chemical bonds between the atoms. For example, the great differences in strength between the carbon minerals, graphite and diamond, can be attributed to the fact that each carbon element in diamond is bonded covalently to its four nearest neighbors at tetrahedral angles, thus forming a space-filling network of strong bonds that does not have any slip planes; but graphite has only three covalent bonds around each carbon element at 120° angles, resulting in two-dimensional sheets not bonded together except for very weak Van der Waals forces. Subsequent analytical steps progress from studies of the influence of microfractures in the rock element to macro-sized weaknesses in the rock system, such as visible joints, fractures, and layers. The influence of such fissures on the stress-strain relationships for a rock is depicted in Fig. 1.

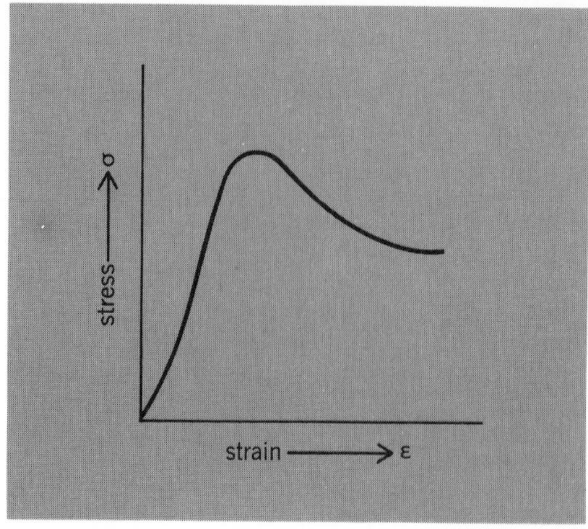

Fig. 1. Behavior of a fissured rock system. (*After J. Bernaix, Properties of rock and rock masses, in Proceedings of the 3d International Congress on Rock Mechanics, vol. IA, 1974; reproduced with permission of the National Academy of Sciences*)

Thus the behavior of a fractured rock can range from brittle failure to pseudo-plastic sliding. Also significant are hydrologic factors, as they may cause buoyancy, or hydrostatic or flow pressure in the joints, or may reduce the internal coefficient of friction in a rock element or along the contact between two rock elements; and ambient stresses. *See* PETROGRAPHY.

Humans' disturbance of the rock system can (1) modify the ambient stress field because of the removal (excavation) of a portion of the rock system or the imposition of explosion shock waves on an underground opening; (2) introduce chemical effects that alter the rock elements or produce new rock elements; and (3) produce thermal effects. Changes in the ambient stress field can produce unusually high stress concentrations around an underground opening that will result in its collapse, severe deformation of its supports, or rock bursts. Examples of undesirable chemical effects are: water with a high gypsum content depositing gypsum in fractures and subsequent drying and rewetting, producing expansion forces that create new fractures; or the air- or water-slaking of a rock such as shale, during the excavation of a tunnel or structure foundation, with resulting decomposition of the rock. Another important property of a rock system is its tendency toward dilatancy when certain load conditions occur. The result can be an arching action that may improve the stability of an underground opening.

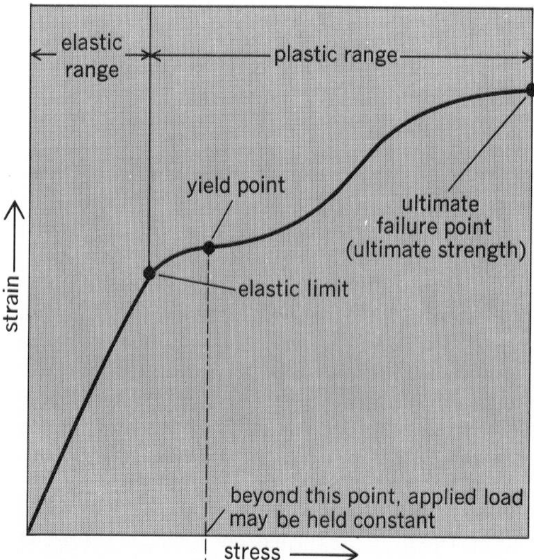

Fig. 2. Rheological response of rock element under unconfined compressive load.

Rock has very low coefficients of heat conductivity and transmissibility. Thermal properties of rock are considered in the development of geothermal energy; in earthquake prediction; in studies of effects of nuclear explosions; in the use of a rock system as a heat sink for underground reservoirs needed to cool underground power plants; in the design of ventilation systems for deep mines; and in the fracturing of rock sometimes produced by the underground storage of cryogenic substances. *See* EARTH, HEAT FLOW IN.

The little available knowledge of the electromagnetic properties of rock systems resulted from the use of the rock system as a ground for electrical power plants; from research on use of the rock system as a carrier for electromagnetic or seismic signals in communication systems; and from the severe damage to electrical systems associated with nuclear explosions (the "electromagnetic pulse"). Research on the acoustical properties of rock has been termed geoacoustics and can provide information on stress conditions and on mechanical properties of the rock system. For information on the seismological properties of rock *see* SEISMOLOGY; for information on electrical resistivity and magnetic characteristics *see* GEOPHYSICAL EXPLORATION; ROCK, ELECTRICAL PROPERTIES OF.

Quantitative definition of the properties of a rock element or system requires direct measurements in the laboratory or in place. The significant considerations in performing such measurements are: (1) The low reliability in correlating laboratory results with prototype behavior may require high safety factors and, consequently, costly designs. (2) Rock is heterogeneous in composition and seldom is isotropic or elastic in response to load (Figs. 1 and 2). (3) Lack of standardized test procedures can increase frequency of test errors.

Test errors. Random errors can be minimized by performing numerous tests on each rock type. Systematic errors are caused by low voltage, loss of calibration, and other problems in the test equipment; eccentricity of load application because the ends of the specimen are not at right angles to the longitudinal axis of the applied load (recommended angle is $90 \pm 1/4°$); unknown or unintentional changes in loading rates during one or several tests on the same rock type; lack of smoothness in the ends of the test specimen (recommended practice is that the maximum roughness not exceed 0.001 in.; 0.00254 cm); lack of a smooth surface where gages are cemented to the rock surface or poor bonding of the gages; a low coefficient of friction between the ends of the specimen and the load plates; and the loaded specimen having nonuniform stress distribution when the computation of results requires uniformity.

The resonant-frequency and the pulse-velocity tests, commonly used to determine Young's modulus of elasticity and Poisson's ratio, may incorporate errors caused by lateral expansion of the rock when it is vibrated; by improper placement of the transducers and failure to correct for their mass; by irregularities in the specimen shape that cause errors in measuring its volume or mass; and by undetected fractures and anisotropy. Environmental errors can result from changes in the ambient humidity or temperature during the test or from unintentional vibrations in the test equipment.

Conventional statistical analysis methods can be used to minimize errors and to maximize reproducibility of results. Test results indicate that the minimum standard deviation among samples of the same rock type range from 3.5 to 10 and that a ±30% variability in measured value of specimens is acceptable.

Another consideration is the increasing evidence that some rock properties are altered by changes in size of test specimens; for example, compressive strength has been found to decrease with increase in volume of the test specimen. This phenomenon is termed the scale effect or factor.

Preparation of laboratory and field models and tests and their correlation with prototype behavior requires use of dimensional analysis. These principles permit the data collection to be systematized, reduce the number of variables to be tested, and are particularly significant in rock mechanics testing because of the numerous different measurement units that generally are used. For example, in the resonant frequency test, the specimen may be measured in centimeters, its weight in kilograms, and the frequency in hertz, but the modulus is expressed in kilograms or kilonewtons per square centimeter.

Laboratory tests. The physicomechanical properties of rock commonly determined by laboratory tests are specific gravity, porosity, compressive strength as determined by uniaxial or triaxial loads, Young's modulus of elasticity, and Poisson's ratio. The triaxial compressive-strength test (also referred to as the triaxial shear test) used for rock does not produce true triaxial loading because only two of the applied stresses (σ_A and σ_R) can be varied during the test; that is, it generally is assumed that $\sigma_A = \sigma_1$ and $\sigma_R = \sigma_2 = \sigma_3$, where σ_A is the axial stress and σ_R is the radial stress. Figure 3 is a sketch of a commonly used triaxial-test procedure; σ_A is produced by a hydraulic ram on the end of the specimen, and σ_R by hydraulic pressure applied to the fluid (oil or water) that surrounds all but the ends of the specimen. The specimen and the gages are encased in a rubber, neoprene, or plastic sleeve that prevents the surrounding fluid from permeating the specimen or gages. Tests for the other physical properties noted above generally are similar to those used for concrete. A difficulty with the conventional test for unconfined compressive strength is that the steel loading system stores stress as the ram loads the specimen. And if this system has a stiffness less than the rock, the rock may fail explosively, and thus complete evaluation of the total rock strength is not obtained. This can be overcome by use of a stiff testing machine. The machine is designed so the specimen failure can be controlled beyond its actual rupture point, as

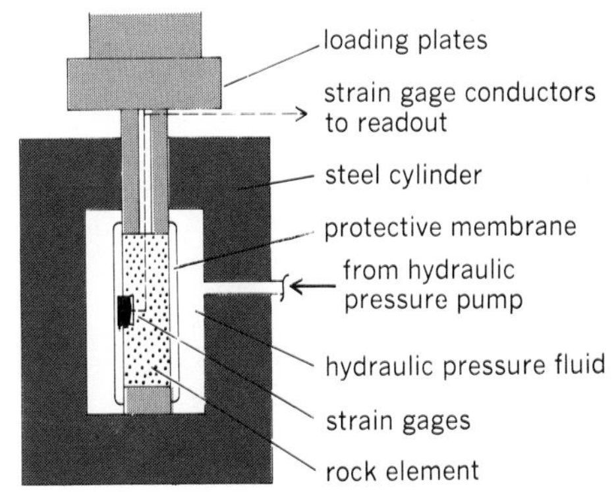

Fig. 3. Triaxial compression test.

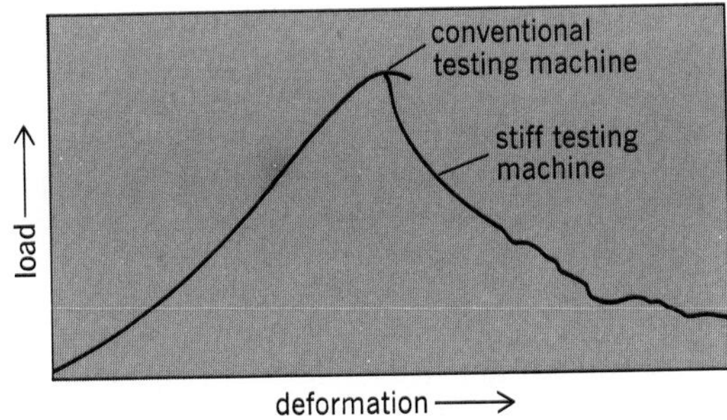

Fig. 4. Comparison of failure curves from conventional and stiff testing machines.

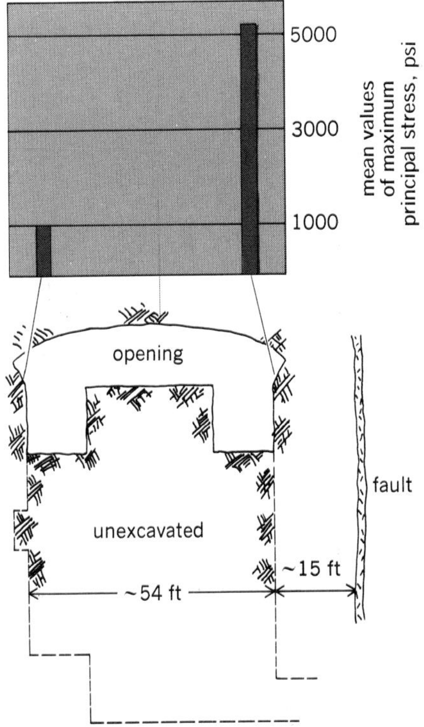

Fig. 5. Influence of geology on stress distribution around an underground opening. (*Adapted from J. L. Serafim, Internal stresses in galleries, in 7th Congress on Large Dams, sect. R-1, Q-25, 1961*)

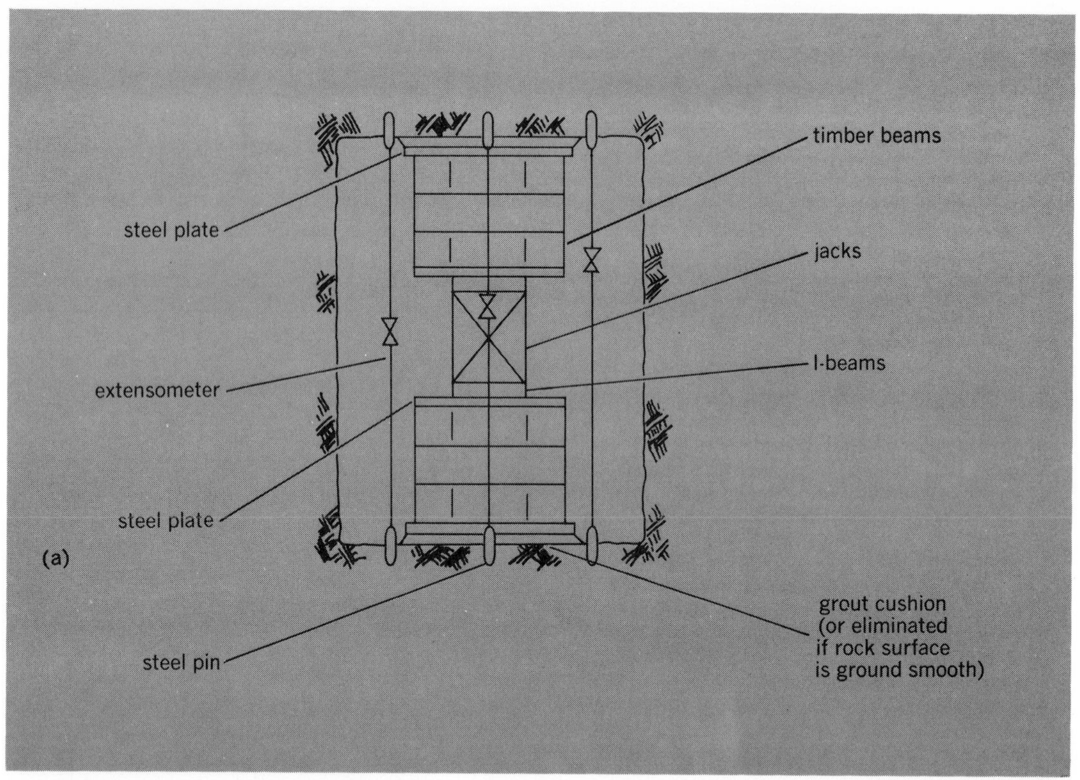

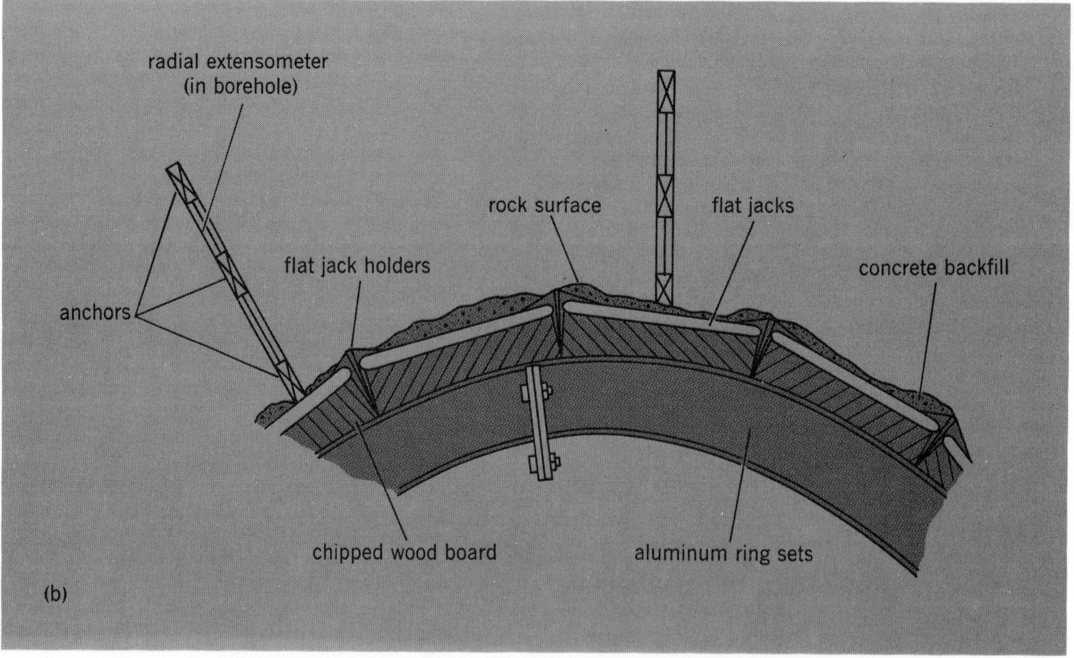

Fig. 6. In-place moduli tests. (*a*) Tunnel jacking test. (*b*) Radial jacking test. (*After G. B. Wallace et al., Radial jacking test for arch dams, in the 10th Symposium on Rock Mechanics, Society of Mining Engineers, © 1972 AIME*)

indicated in Fig. 4. Statistical analysis of thousands of test results has indicated a danger in generalizing a value for any specific physical property. For example, the average for 1801 values of Poisson's ratio under various loading conditions was 0.15, but the range was from less than 0 to 0.94. (These latter values are impossible according to the definition of Poisson's ratio, but they were reported in laboratory tests.) Another example is that the mean unconfined compressive strength for tests on 1151 granites was 16,200 psi, but the range was from 100 to 99,900 psi (1 psi = 6895 N/m²). (Values for some 11,000 sets of mechanical and thermophysical properties of rocks in a computerized format are available from the Underground Exploration and Rock Properties Information Center at Purdue University.)

Field measurements. The major objective in field

measurements is to determine (1) the deformation modulus and shear strength of the rock system, which is important in the design of arch dams; and (2) the ambient stress field (and shear strength), which is important in design of underground openings (Fig. 5). The deformation modulus requires a test of as large a portion of the rock system as is economically possible; the objective is close simulation of the prototype load on the system. Frequently used tests for deformation modulus employ either a system of hydraulic jacks (Fig. 6*a*), a pressurized chamber, or a system of radial flat jacks (Fig. 6*b*). The test is performed in a tunnel in that part of the rock system believed to have the same characteristics as the rock system that will be loaded by the prototype structure. In the radial jacking test, a continuous series of flat jacks are installed around the periphery of the test tunnel; when pressurized, they react against a circular steel or aluminum rib support. The deformations are measured by extensometers in boreholes radiating out from the test section. One type of borehole extensometer has a series of anchors at specified intervals in the borehole. A wire connects each anchor to a cantilevered spring in the measurement head at the rock surface. As each anchor is displaced by changes in strain along the borehole, the connecting wires deflect the corresponding cantilever; the amount of this deflection is calculated from the resulting strain on an electrical strain gage mounted on the cantilever. Figure 7 depicts the maximum amount of such displacements that occurred along three extensometers at a tunnel section in Wyoming.

The pressure-chamber test usually is conducted in a tunnel that has a length-to-width ratio of 2 or 3 to 1. The tunnel can be lined with a water-impervious, low-modulus, elastic material such as rubber or sprayed with an epoxy compound; strain gages are affixed to one or more grids of extensometer bars placed perpendicular to the long axis of the tunnel or the chamber walls, or both. A watertight bulkhead is constructed at the open end of the tunnel, and the chamber then is pressurized with water. The circumferential deformation of the tunnel is measured by electrical readout instruments connected to strain gages. A major advantage of the pressure-chamber and the radial jack tests is that sufficient readings can be obtained on the circumference of the tunnel to construct a polar plot of deformation that will disclose anisotropy in the rock system. (Fig. 8). The in-place Young's modulus of elasticity also can be obtained by inserting a jacking device into boreholes as small as 4 in. (10.16 cm) diameter. This device has tiny plates or steel points that are forced against the borehole walls by a measurable hydraulic-mechanical force in the interior of the jack. Strain gages in the device measure the resulting deformation of the rock in contact with the plates or points.

There is no satisfactory device for the direct measurement of the original ambient stresses, but numerous methods have been developed to measure differential ambient strains. These methods require insertion of the measuring device into a drilled or sawed hole, but which, by its presence, alters ambient stress conditions. In major surface or underground excavations these devices may remain until the construction is complete so that changes in strain caused by the excavation process can be monitored. These data aid in prediction and prevention of excavation failures, in the design of temporary and permanent supports and lining, and in the formulation of efficient designs for future excavation projects. The operational principle of these devices is restoration of the initial ambient

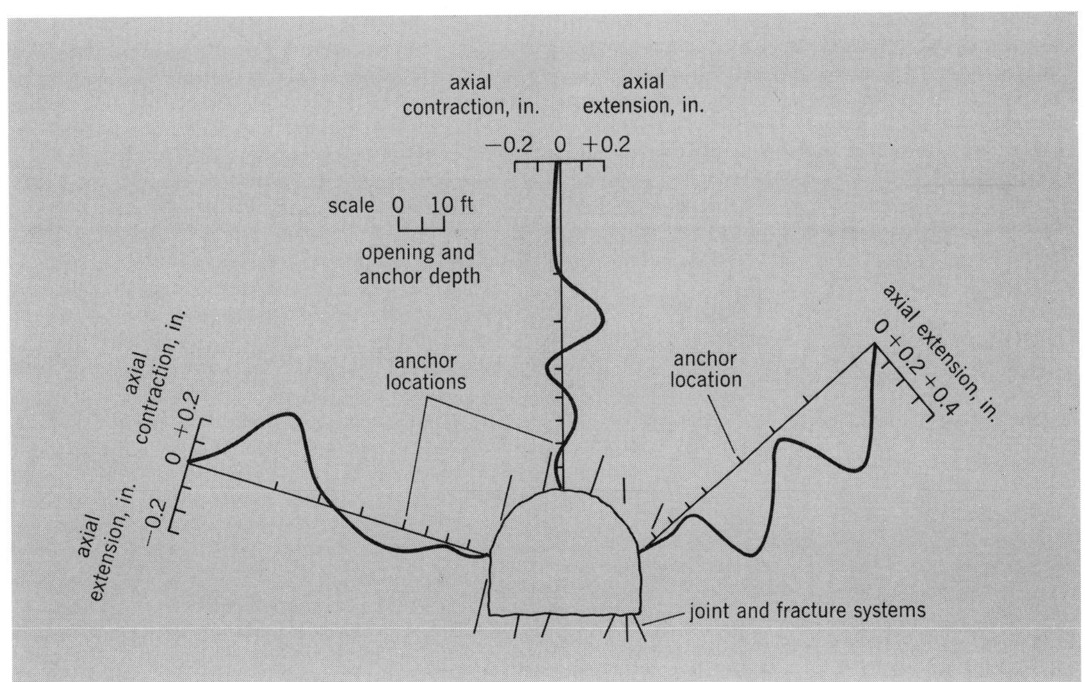

Fig. 7. Maximum displacements measured on multiple-position borehole extensometers (Green River North Tunnel, WY). (*From W. R. Judd and W. H. Perloff, Comparison between predicted and measured displacements, Defense Documentation Center Rep. AD 723532, 1971*)

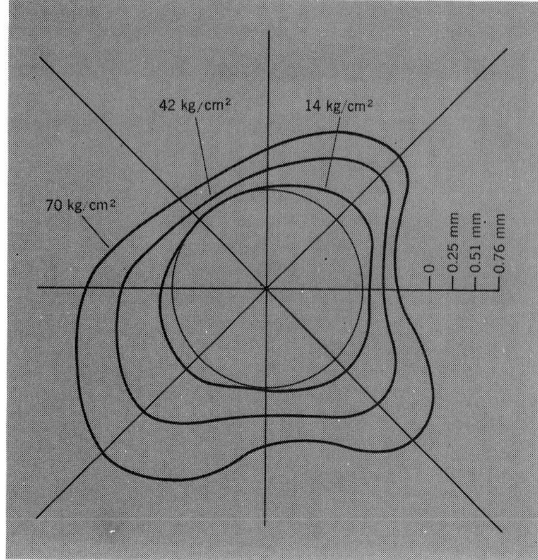

Fig. 8. Radial jacking test results (amount of deformation around a circular tunnel). (*After G. B. Wallace et al., Radial jacking test for arch dams, 10th Symposium on Rock Mechanics, Society of Mining Engineers, © 1972 AIME*)

strain by jacking in a hole or relief of the initial ambient strain by drilling or sawing a hole (sometimes termed the strain-relief method). The first type of test uses a flat jack (Fig. 9), which may be two square or triangular steel plates from about 30.5 cm to 91.5 cm on a side. The plates are edge-welded to leave a small interior opening for the injection of oil under pressure. Strain gages or extensometers are fixed to the surface immediately adjacent to the periphery of a slot to be cut into the rock, or such gages or hydraulic pressure cells may be inserted into boreholes above and below the slot; a diamond drill or saw, or a jackhammer then cuts a slot having dimensions slightly greater than the flat jack; the flat jack is inserted into the slot, and high-modulus cement grout or epoxy is pumped around the flat jack to assure a complete contact between the flat jack and the rock. The strain-measuring devices on the rock surface are returned to their null or zero point (their value prior to cutting of the slot) by pumping hydraulic fluid into the jack. Thus a measurement is obtained of the amount of ambient strain that theoretically has been relieved by the cutting of the slot, and Young's modulus of elasticity of the rock system immediately adjacent to the slot can be computed because the hydraulic load on the jack is measurable when the gages are nulled. A variation of this technique uses a diamond saw to cut a slot large enough to permit insertion of a semicircular flat jack; this results in only a minimum of (or no) grout being required to achieve contact between the jack and the rock on the sides of the slot.

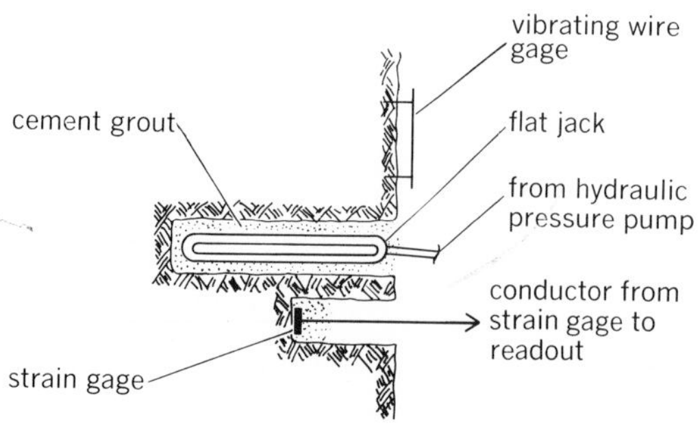

Fig. 9. Flat jack test. The strain induced by the jack is measured either by a vibrating wire gage or by an alternate method developed by the U.S. Bureau of Mines, wherein strain gages are placed in boreholes above and below the flat jack and at the center of the jack.

In the strain-relief method, a rosette of strain gages is cemented to a small area on the rock surface. The ambient strain then is relieved by drilling out the core of rock that has the strain gages cemented to its end or around the outside diameter of the core hole. (The former is termed the overcoring strain-relief method.) Where the gages are overcored, the electrical leads from the gages are inserted through the center of the core barrel and the drill rods up through a special water swivel on the drilling machine; these conductors are then connected to a conventional strain-readout bridge. A variation of this method uses a borehole gage inserted into a drilled hole. On the sides of this gage are movable pistons that maintain contact with the walls of the borehole by means of a spring or hydraulic system in the interior of the gage; inside the borehole gage, strain gages are attached to each piston. The ambient strain is relieved by drilling a larger-diameter hole over the core hole containing the gage. The disadvantage in the strain-relief methods is that ambient stresses cannot be computed until Young's modulus is determined by laboratory tests on a rock element from the test site. Such a determined modulus may be considerably higher than that of the rock system, because the tests are performed on rock elements having a minimum of defects (unlike the rock system). Another variation uses a disk of material that is either sensitive to polarized light (and the strain is computed from the fringe pattern) or contains a rosette of strain gages; this disk is cemented to the bottom of a cored hole and is overcored. This variation has the advantage of providing strain measurements in the direction of two principal stresses.

A major disadvantage of the preceding methods is that they can be used only at relatively shallow depths—the flat jack near the rock surface and the overcoring method perhaps down to 50 m. A variation of the hydraulic-fracturing method (used in petroleum recovery) has been used successfully at a depth of 1915 m. A portion of the borehole is sealed by rubber packers, and the sealed-off portion then is pressurized by water until the annulus of rock fractures. After release of the pressure, a special ("impression") packer is inserted and pressurized in the newly fractured part of the borehole; this packer is retrieved and the impression of the fractures remains on its surface, so their azimuth and inclination can be measured to permit extrapolation of the directions of the principal stresses. Equations incorporating the fracture pressure, Poisson's ratio, and a porous elastic parameter which is determined in the laboratory permit calculation of the amounts of the principal stresses.

Geophysical seismic methods can be used to approximate Young's modulus for a large segment of a rock system. These methods are based upon the equation below, where γ is the unit weight

$$E=\gamma V^2\left[\frac{(1-2\mu)(1+\mu)}{g(1-\mu)}\right]$$

of the rock element, V is the velocity of propagation of the longitudinal wave produced by a mechanical impactor or a dynamite explosion and recorded by geophones connected to an oscillograph, g is the acceleration of gravity, and μ is the laboratory-determined Poisson's ratio of the rock element. The accuracy of the result is dependent upon how closely the rock system simulates an isotropic, homogeneous, linearly elastic material. The use of nondestructive test methods to measure ambient strain or stress has not been too successful. However, measurable changes in velocities in such methods, while not accurate for calculating stresses, occasionally can provide an approximation of the thickness of rock loosened by the excavation process (the "destressed" zone). An analogous test is to use geoacoustics to measure the frequency of "noises" in the rock system; increases in rock noise may indicate potential failure of a tunnel roof or may locate a failure plane in an unstable rock slope.

Time factor. The development of theories and experiments that accurately measure and predict rock behavior requires full consideration of the influence of time on such behavior. The rheological models shown in Fig. 10 are a possible simulation of rock behavior, as they incorporate a time rate-of-strain factor. When the loading rate reaches the

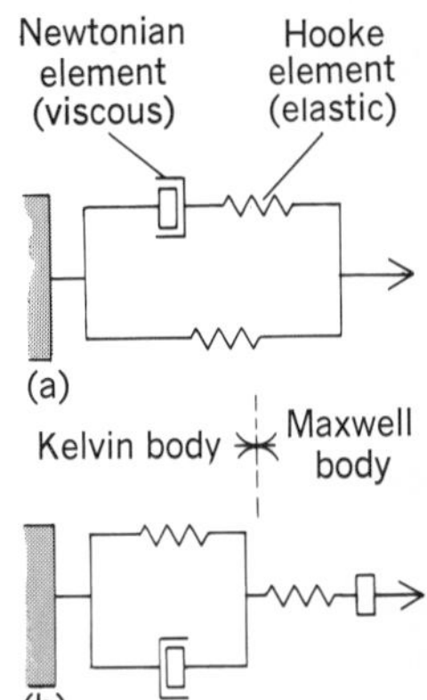

Fig. 10. Rheological models of rock behavior. (a) General linear substance. (b) Burgher model.

Table 2. Loading rates for tests on rock*

Type of load	Loading rate, psi/sec	General formula for rock response	Examples
Creep or constant	$10^{-1} \rightarrow 10^{-4}$	$\epsilon=\epsilon_e+\epsilon\,(t)+At+\epsilon_T(t)$	Dams during and after construction; underground openings after completion
Static or conventional compressive	$10 \rightarrow 10^2$	$E=\frac{\sigma}{\epsilon}$	Laboratory or field uniaxial unconfined compression test
Dynamic, explosive but nonnuclear	$10^4 \rightarrow 10^{10}$	$E=\rho V^2(\frac{1-\mu-2\mu^2}{1-\mu})$	Laboratory resonant frequency test; field seismic determination of modulus
Nuclear	$10^4 \rightarrow > 10^{12}$	$e-e_0=\frac{1}{2}p(v_0-v)$	Explosion of nuclear weapons (critical parameters are displacement, acceleration, and particle velocity in the the rock system).

Explanation of notations:
ϵ = unit strain
ϵ_e = amount of strain within the elastic limits
$\epsilon\,(t)$ = decelerating creep of the primary stage (numerous relations dependent upon temperature and amount of pressure and time)
ρ = density
A = constant
t = time
At = strain linear with time after transient creep stage
$\epsilon_T(t)$ = Accelerating creep of the tertiary stage of loading
E = Young's modulus of elasticity
σ = unit stress
μ = Poissons ratio
e_0 = amount of original energy available
e = final energy available
v_0 = initial volume of material
v = final volume of material

*Adapted from L. Obert and W. I. Duvall, *Rock Mechanics and the Design of Structures in Rock*, John Wiley & Sons, Inc., 1967; E. C. Robertson, Viscoelasticity of rocks, in W. R. Judd (ed.), *State of Stress in the Earth's Crust*, American Elsevier, 1964.

Table 3. Rock classification by physical properties*†

Strength		Response to load§		Geologic structure	
Descriptor	Criteria‡	Descriptor	Criteria	Descriptor	Criteria
Very strong	> 25,000 psi	Elastic	75% of total strain is recoverable	Massive	6 ft between layers or joints or other evident discontinuities
Strong	10,000 → < 25,000				
Weak	5000 → < 10,000	Elastoplastic			
Very weak	< 5000	Plastic	75% of total strain is irrecoverable	Layered	Bonding between evident layers appears less than within a layer
				Blocky	> 1 ft < 6 ft between layers, joints, or other evident discontinuties
				Broken	> 3 in. < 1 ft between evident discontinuities
				Very broken	< 3 in. between evident discontinuities

*Adapted from D. F. Coates, *Rock Mechanics Principles*, Dep. Energ. Mines Resour. (Can.), Mines Br. Monogr. no. 874, 1970.

†The table is used by selecting the appropriate descriptor from each of the three major columns and using these three words to describe the material. If available, the petrographic name of the material can be used in combination with these descriptors.

‡Estimated or measured unconfined compressive strength. §Estimated or determined by tests.

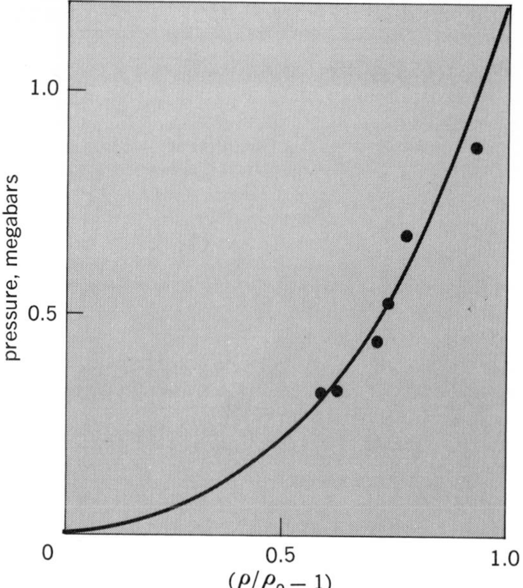

Fig. 11. Hugoniot curve for granite. Here ρ is density and ρ_0 is density at pressure = 0. *(From T. R. Butkovich, Calculation of the shock wave from an underground explosion in granite, in Engineering with Nuclear Explosives, U.S. At. Energy Comm., Div. Tech. Inform. Rep. no. TID-7695, 1964)*

Table 4. Relationship of RQD index to rock quality*

RQD index	Rock quality
90–100	Excellent
75–90	Good
50–75	Fair
25–50	Poor
<25	Very poor

*From C. Jaeger, *Engineering Geology and Rock Mechanics*, 1972.

magnitudes produced by high explosives or nuclear devices, the rock system behavior adjacent to the explosion can best be approximated by a hydrodynamic model, with gradual change to a viscoelastic model, and then to an elastic model as the distance from the explosion source increases. In the hydrodynamic zone, behavior is approximated by Hugoniot curves, which are shown in Fig. 11 for granite.

The rheological models generally apply to a rock element but may not provide an accurate analog of a rock system that contains discontinuities and anisotropy. Therefore, to maximize the accuracy of laboratory and in-place tests, the loading rate selected should closely simulate the type of loading that will be induced by the prototype (Table 2). For example, in determining the elastic modulus of a rock element from a proposed dam site, the best practice is to impose several cycles of loading on the element; this simulates the cyclic loading induced in the rock system by fluctuations in reservoir level and in temperature in a concrete arch dam.

Rock slopes. One of the problems in rock mechanics is to accurately model the potential stability of excavated rock slopes; the problem exists because the strong influence of geologic factors is difficult to quantify and ambient stresses are difficult to measure. One model uses the soil-mechanics approach, which includes the circular (Fellenious or Bishop) and the noncircular (Morgenstern-Price or Janbu) sliding-surface methods; these are based on the limiting equilibrium principle and the Coulomb-Navier failure criterion, with the factor of safety defined as the ratio of driving over resisting forces or moments. (It can also be defined by ratios of cohesion, friction, or height.) Other approaches include the discontinuum methods, where the statics and kinematics of individual rock blocks or wedges are examined in stereographic projections; and the continuum methods,

Table 5. Rock classification by physical properties and ability to sustain excavations*†

Competent				Incompetent
Rock can sustain underground opening without aid of artificial (man-made) supports				Artificial supports required to sustain opening; considered generally inelastic so classification is not subdivided
Structure		Response		
Descriptor	Criteria	Descriptor	Criteria	
Massive	Linearly elastic, isotropic, homogeneous (spacing between discontinuities is large compared with critical dimensions of the opening or the bond strength across discontinuities is > strength of rock)	Elastic	Time-independent response	
		Inelastic	Time-dependent response	
Laminated (bedded)	Separation between approximately parallel planes of weakness is small compared with critical dimensions of the opening			
Jointed	A variation of massive where the discontinuities form evident polyhedral blocks that are relatively large			

*Compiled from L. Obert and W. I. Duvall, *Rock Mechanics and the Design of Structures in Rock*, John Wiley & Sons, Inc., 1967.

†The table is used by selecting one descriptor from each of the two columns under "Competent" and combining these terms with the word competent. The term incompetent is used as a sole descriptor.

where the stresses and strains in the rock slope for homogeneous, anisotropic, inelastic, and layered material in an ambient stress field are compared to the existing strength patterns of the slope as affected by excavation, groundwater fluctuations, earthquakes, and so forth. The comparisons are made by use of photoelasticity, finite differences or finite-element methods in two or three dimensions.

Rock classification. Because of the wide scope of rock mechanics, as yet there is no system of rock classification that is acceptable to all users of rock mechanics. Presently used systems range from classical petrographic description based upon the genesis of the rock and its fabric, to those based upon one or more physical properties of the rock element or rock system such as presented in Table 3. The Rock Quality Designation (RQD) is another system based upon the total length of core pieces exceeding 4 in. (10.16 cm) that are recovered in one core run. Thus, in a core run of 60 in., if there is a total of 25 in. of core pieces with lengths greater than 4 in., the RQD would be $(25/60) \times 100$ or 41%. These latter values, then, are related to rock quality as presented in Table 4.

Another system used for underground excavations is given in Table 5. This system presumes that the competency can change with depth because of the greater loads to be supported; and the size and depth of an opening also can cause the rock class to change from massive to laminated and from elastic to inelastic. The differences in these systems illustrate the importance of specifying the classification system being used for a particular engineering design. *See* PETROLOGY; ROCK.

[WILLIAM R. JUDD]

Bibliography: A. B. Cummins and I. A. Given (eds.), *SME Mining Engineering Handbook*, chaps. 6, 7, 9, 13, 1973; D. U. Deere and R. D. Miller, *Engineering Classification and Index Properties for Intact Rock*, AFWL Tech. Rep. TR-65-116, 1966; *Determination of the In Situ Modulus of Deformation of Rock*, ASTM Spec. Tech. Publ. 477, 1970; I. W. Farmer, *Engineering Properties of Rocks*, 1968; J. Handin and D. Griggs, *Rock Deformation*, Geol. Soc. Amer. Mem. no. 79, 1967; J. A. Hudson et al., Soft, stiff and servocontrolled testing machines, *Eng. Geol.*, vol. 6, no. 3, 1972; W. Hustrulid and F. Robinson, A simple stiff machine for testing rock in compression, in *14th Symposium on Rock Mechanics*, 1973; C. Jaeger, *Engineering Geology and Rock Mechanics*, 1972; W. R. Judd, *Statistical Relationships for Certain Rock Properties*, Purdue Research Foundation Report to the U.S. Corps of Engineers, 1971; *Proceedings of the 1st, 2d, and 3d International Congresses on Rock Mechanics*, 1966 (Lisbon), 1970 (Belgrade), 1974 (Denver); *Proceedings of the Symposium on Rock Mechanics*, annually; J. G. Ramsay, *Folding and Fracturing of Rocks*, 1967; K. Szechy, *The Art of Tunnelling*, 1966; *Testing Techniques for Rock Mechanics*, ASTM Spec. Tech. Publ. no. 402, 1966.

Rubellite

The red to red-violet variety of the gem mineral tourmaline. Perhaps the most sought-for of the many colors in which tourmaline occurs, it was named for its resemblance to ruby. The color is thought to be caused by the presence of lithium. Fine gem-quality material is found in Brazil, Madagascar, Maine, southern California, the Ural Mountains, and elsewhere. Gem material is almost exclusively a product of pegmatite dikes. Although rubellite is relatively inexpensive, it is regarded by many as one of the loveliest of gemstones. It has a hardness of 7–7.5 on Mohs scale, a specific gravity near 3.04, and refractive indices of 1.624 and 1.644. *See* GEM; TOURMALINE.

[RICHARD T. LIDDICOAT, JR.]

Ruby

The red variety of the mineral corundum, in its finest quality the most valuable of gemstones. Only medium to dark tones of red to slightly violet-red or very slightly orange-red are called ruby; light reds, purples, and other colors are properly called sapphires. In its pure form the mineral corundum, with composition Al_2O_3, is colorless. The rich red of fine-quality ruby is the result of the presence of a minute amount of chromic oxide, usually well under 1%. The chromium presence permits rubies to be used for lasers producing red light. *See* CORUNDUM; SAPPHIRE.

The mineral corundum is commonly a constituent of basic igneous rocks, but it rarely occurs in a transparent form suitable for gem use. It is also known as a constituent of the type of marble formed in the contact zone of an igneous intrusion into an impure limestone. It is in this type of deposit that the finest rubies (those of Mogok, Burma) were formed. Today, most of those mined in the Mogok region are taken from the famous gem gravels of that area. There are only two other sources of significance, Ceylon and Thailand. In each of these countries, the finest quality obtained is far less valuable than fine Burma material, for the Ceylon ruby is too light in color and the so-called Siam ruby is a darker red. The finest ruby is the transparent type with a medium tone and a high intensity of slightly violet-red, which has been likened to the color of pigeon's blood. Star rubies do not command comparable prices, but they, too, are in great demand. *See* GEM.

The ruby was among the first of the gemstones to be duplicated synthetically and the first to be used extensively in jewelry. A French chemist, A. Verneuil, announced successful reproduction in 1902 by a flame-fusion process. (However, it is certain that flame-fusion synthetic rubies were available much earlier.) Many years later, C. Chatham produced synthetic ruby by flux fusion. More recently, several others have been successful. Bell Laboratories has made synthetic rubies by a hydrothermal process. Some synthetics are now being marketed.

[RICHARD T. LIDDICOAT, JR.]

Rutile

The most frequent of the three polymorphs of titania, TiO_2, crystallizing in the tetragonal system, space group *P4/m nm*. The two other polymorphs are brookite (orthorhombic) and anatase (tetragonal). The rutile structure type consists of alternate chains of edge-sharing oxygen octahedrons. The titanium atoms reside in the centers of these distorted octahedrons, and the oxygen atoms occur in distorted hexagonal close packing. A

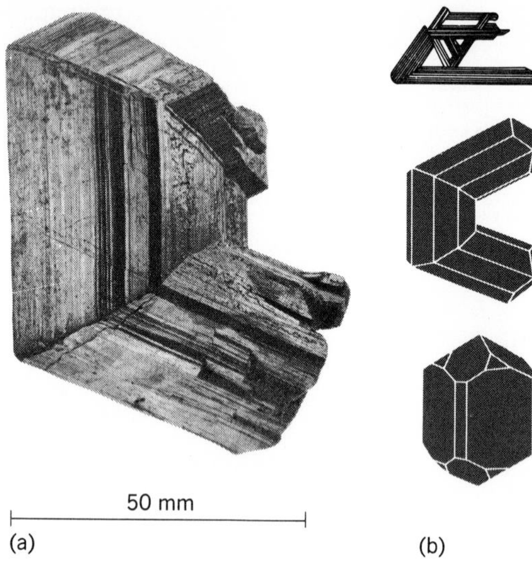

Rutile. (*a*) Crystal specimen taken from Parksburg, Pa. (*American Museum of Natural History specimens*). (*b*) Crystal habits (*from C. S. Hurlbut, Jr., Dana's Manual of Mineralogy, 17th ed., copyright © 1959 by John Wiley & Sons, Inc.; reprinted by permission*)

number of quadrivalent (4+) metal oxides crystallize in the rutile structure type, including pyrolusite, MnO_2; cassiterite, SnO_2; and plattnerite. PbO_2. Silica at sufficiently high pressures crystallizes in this structure type as stishovite, SiO_2, an observation similarly noted for germania, GeO_2. Rutile is also the basis of other structures: Ordering of the iron and niobium atoms in columbite, $FeNb_2O_6$, results in the "trirutile" structure type.

The mineral occurs as striated tetragonal prisms and needles, commonly repeatedly twinned (see illustration). The color is deep blood red, reddish brown, to black, rarely violet or yellow. Specific gravity is 4.2, and hardness 6.5 on Mohs scale. Cleavage is {110} distinct. Melting point is 1825°C. It is synthesized from $TiCl_4$ at red heat, and colorless crystals are grown by the flame-fusion method, which because of their adamantine luster are used as gem material. Rutile can also be formed by the breakdown of other titanium minerals such as ilmenite, $FeTiO_3$, and sphene, $Ca(TiO)[SiO_4]$. Natural rutile can contain Fe^{2+}, Fe^{3+}, Nb^{5+}, and Ta^{5+} ions.

Rutile occurs as an accessory in many rock types, ranging from plutonic to metamorphic rocks, and even as detrital material in sediments and placers because of its resistance to weathering. Large crystals have been found in some granite pegmatites; in Brazil it often occurs as inclusions in clear quartz crystals (rutilated quartz). Other rocks include contact metamorphic limestones, alpine gneisses and schists, and pyrophyllite-lazulite schists. Rutile is commonly associated with apatite in high-temperature veins. In sufficient quantities, it is marketed as an ore of titanium. *See* CASSITERITE; ILMENITE; PYROLUSITE; STISHOVITE.

[PAUL B. MOORE]

Salt dome

An intrusive body of rock salt which has penetrated large thicknesses of overlying sedimentary rock. Salt domes are distinguished from other geological deformations involving salt in being roughly circular or elliptical in cross section and in having horizontal dimensions of the same order of magnitude or less than their vertical dimensions.

Salt domes are best known along the Gulf Coast of the United States, where they are important economically because of their association with oil and sulfur deposits. Most Gulf Coast domes have horizontal dimensions of 1–6 mi and vertical dimensions of 3–6 mi, depending on the depth of burial of the source salt bed. Depths to the top vary from surface exposure to many thousands of feet (see illustration).

Formation of domes. The geological process by which domes are formed is not completely understood, but the general mechanism seems fairly clear. The domes occur in areas underlain by evaporite deposits containing large thicknesses of salt (primarily sodium chloride), which flow into the dome from the immediately surrounding area. The salt, being relatively plastic, is highly deformed by the flowage. The flowage apparently results from the fact that the salt is of lower density than the overlying sediments. This density difference is demonstrated by the fact that salt domes produce definite negative gravity anomalies. A model to illustrate this process can be made by filling a glass-sided box nearly full of a high-density viscous liquid, such as boiled-down corn syrup, and then filling the small remainder (10% or less) with a low-density viscous liquid, such as soft asphalt, to represent the salt. When such a model is inverted so that the low-density fluid is on the bottom, the asphalt will flow up through the syrup and take a form which is quite similar to the general form known (by drilling) to be that of many salt domes. Dimensional analysis indicates that the ratios of the physical properties of the fluids in the model and those of the prototype in nature are approximately correct for dynamical similitude.

The model explains in a general way the mecha-

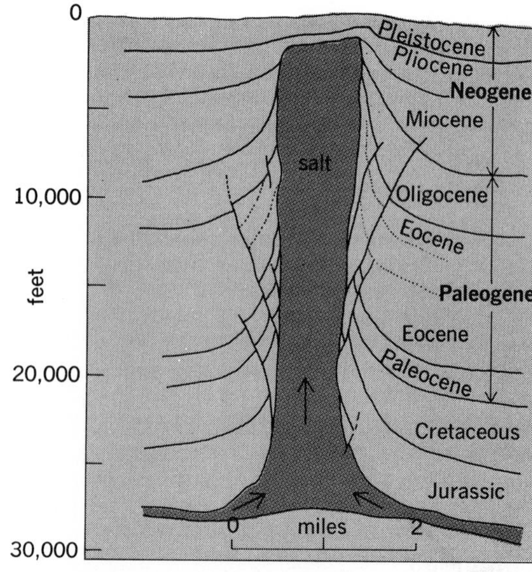

Section of a salt dome in the Gulf region. The drawing is a composite of data obtained from several domes, for no wells have been drilled deeply enough to penetrate all the rock divisions shown. The base of the salt plug is hypothetical. (*From R. C. Moore, Introduction to Historical Geology, 2d ed., McGraw-Hill, 1958*)

nism by which salt flows into the domes because of the difference in density, but in many cases the movement of salt is significantly influenced by the geological conditions and tectonic forces of the area. It appears that the salt must be covered by a minimum of some 10,000 ft of overburden before flowage into domes takes place. This suggests that the salt may become more plastic under the pressure of a thick overburden, but a clear demonstration of such a property of salt has not been made. The increase of plasticity must be caused, in part, by the natural increase in rock temperature with depth since experimental investigation has shown a marked decrease of long-time strength of salt with increase of temperature.

Distribution of domes. The Gulf Coast salt-dome belt of Texas and Louisiana, including those domes known to exist as a result of oil exploration offshore, contains approximately 150 known domes. The northeast Texas basin has 20 domes, the interior basin of northern Louisiana and southern Mississippi has about 75 domes, and the Rio Grande basin of southern Texas has 6 domes. Extension of geophysical exploration and drilling has found a large number of domes under the continental shelf off Texas and Louisiana, and important oil production is localized by these domes. Well over 100 such domes were known as of 1969, and the number is increasing rapidly. The existence of salt domes that were indicated by experimental seismic testing in the Sigsbee Deep in the central part of the Gulf of Mexico, where the water depth is nearly 2 mi, was confirmed in 1968 by drilling of the ship *Glomar Challenger* of JOIDES (Joint Oceanographic Institutions Deep Earth Sampling) program. The Gulf Coast basins are the only true salt dome basins in the United States. The Paradox Basin of western Colorado and eastern Utah contains salt structures, but these are elongated anticlines rather than domes.

Salt domes are known in many other parts of the world. They are found in northern Germany, Denmark, Rumania, northeastern Spain, southwestern France, in the Persian Gulf and along its northeasterly coast in northern Iran, in the Emba basin of the Soviet Union just north of the Caspian Sea, in several smaller basins in northern Pakistan, along the western coast of Africa near the Equator and off Nigeria, and on the Isthmus of Tehauntepec in Mexico. Extensive geophysical exploration in the North Sea has revealed a very large salt basin with domes and salt "pillows" (relatively broad salt structures). Numerous domelike structures are known in the Arctic islands of northern Canada; these domes are thought to have gypsum rather than salt cores.

Cap rock. In many of the domes in the southeastern United States and to a certain extent in other parts of the world, particularly in Germany, a cap rock of limestone, $CaCO_3$, anhydrite, $CaSO_4$, or both, occurs on top of the salt. The process by which the cap rock is formed is not completely understood. The most acceptable theory supposes that anhydrite, which is usually a disseminated constituent of the salt, is concentrated as salt is dissolved away by circulating meteoric waters.

Economic importance. Where the salt has intruded petroliferous sediments, the resulting deformation often forms traps in which oil accumulation occurs. Current production from the salt-dome belt of Texas and Louisiana (including offshore) is about 800,000,000 bbl per year; a large fraction of this production, probably about two-thirds, is from salt-dome fields. Salt-dome oil fields are also important in Germany, Rumania, Mexico, and the Emba basin of the Soviet Union. *See* PETROLEUM GEOLOGY.

A relatively small number of Gulf Coast domes are a very important source of sulfur, which occurs in the pores of the limestone cap rock. The sulfur is recovered by the Frasch process, in which superheated water is circulated to heat the body of cap rock above the melting point of sulfur, which is pumped out as a liquid. Of the large number of salt domes in the Gulf Coast region of the United States, only about 18 have had sufficient sulfur for economical production. The only other cap rock sulfur production is from the Isthmus of Tehauntepec in Mexico. The Frasch process of sulfur production from salt domes recovers about 7,000,000 long tons per year or approximately 40% of the world's sulfur supply.

In a few domes salt is mined either by opening shafts into the salt or by drilling wells through which water is circulated to dissolve the salt. Salt production from domes is about 5,000,000 tons per year. *See* DIAPIRIC STRUCTURES.

[LEWIS L. NETTLETON]

Bibliography: D. C. Barton and G. Sawtelle (eds.), *Gulf Coast Oil Fields*, 1936; H. Borchert and R. O. Muir, *Salt Deposits*, 1964; M. T. Halbouty, *Salt Domes*, 1967; R. C. Moore (ed.), *Geology of Salt Dome Oil Fields*, 1926; New Orleans Geological Society, *Salt Domes of South Louisiana*, vol. 1 (rev.), 1963, and vol. 2, 1962.

Sand

A loose material consisting of small mineral particles, or rock and mineral particles, distinguishable by the naked eye. Most sands are formed by natural agencies. Many deposits of such sands contain clay and silt in varying amounts; some deposits contain pebbles. The mineral composition of sands varies as does the size of the grains composing them. Sands are widely distributed and have many industrial uses. In 1977, 898,000,000 tons of sand having a value of $1,900,000,000 were sold or used by producers in the United States. All states reported 1976 production (see table).

The term sand also is applied, especially commercially, to small mineral or rock particles produced by crushing larger materials; for example, limestone sand made by crushing limestone, slag sand from slag, or sand made by crushing quartzite. Various granular materials, not necessarily of inorganic composition, likewise may be called sand because they consist of sand-size particles.

Natural sands result primarily from the disintegration of rocks by weathering or erosion. Streams and the waves and currents of lakes and oceans are major agencies eroding rock into sand. The grinding action of glaciers is another important sand-producing agency.

Some sand deposits are formed in place by the weathering of rocks, such as sandstone. Others are the result of the sorting out and concentration of sand from particulate material (composed of particles of various sizes) by the running water of streams or by the waves and currents of lakes and seas. Wind also concentrates sand from certain

Sand sold or used by producers in the United States in 1976*

Class of operation and use	Quantity, 10^3 short tons	Value, 10^3 dollars
Commercial		
Construction		
Processed sand	418,495	654,389
Unprocessed sand	†	–
Industrial		
Unground		
Glass	11,467	66,551
Molding	6,896	37,264
Grinding and polishing	76	304
Blast sand	1,498	9,946
Fire or furnace	301	1,164
Engine	752	2,315
Filtration	183	1,060
Oil hydrofac	660	4,759
Metallurgical	2,146	7,365
Other	2,251	14,430
Total	26,230	144,159
Ground	3,440	24,968
Total industrial	29,670	169,127

*From *Sand and Gravel in 1976*, U.S. Bureau of Mines, Mineral Industry Surveys, 1978.

†Processed and unprocessed sand are no longer separated.

materials. Deposits of sand accumulate as bars in rivers and streams, in river deltas, as beaches and bars of lakes or seas, and as dunes built up by wind. They may be designated by the name of the place where they are found as river sand, lake sand, or dune sand. More technical terms may be used, as fluvial, lacustrine, and eolian sand, for the same three occurrences. *See* DUNE.

The mineral quartz is probably the most common major constituent of sands, but locally other materials may predominate as in coral sands, gypsum sands, or black sands (composed of fragments of volcanic rocks). The sands in which quartz is the major component usually contain various amounts of other mineral grains. Coarse-grained sands also may contain small rock fragments. The nonquartz particles are of diverse character and in many cases reflect the mineral composition of the original source materials. *See* QUARTZ.

Sand grains vary from almost spherical to angular. Industrially, angular sands are sometimes referred to as sharp sands. The degree of rounding of sand grains is in large measure an indication of the amount of wear to which they have been subjected. There are no generally accepted size limits for sands. Sands used in the construction industry are commonly finer than 1/4 in., or will pass a 4-mesh sieve but will be retained on a 200-mesh sieve. Geologists use maximum and minimum limits of 2 mm and 1/16 mm.

Sand has many commercial uses. It is extensively utilized in construction as fine aggregate for concrete, mortars, plasters and for many other purposes. Black sands, such as those in Florida, contain ilmenite ($FeTiO_3$) and rutile (TiO_2) in such quantities that they can be recovered for commercial use. Green sands contain the mineral glauconite and have been employed as fertilizers because of their potash content.

The term silica sand is applied to sands composed almost exclusively of grains of the mineral quartz (SiO_2). There are no exact limits for the silica content of silica sands but they commonly contain more that 95% SiO_2 and some of them more than 99%. Silica sand is used in glassmaking, as molding sand, refractory sand, filter sand, grinding and polishing sand, and for many other industrial purposes. Silica sands are often referred to as industrial sands.

Natural-bonded molding sand contains sufficient clay and other bonding material so that it can be used for making molds in which metal is cast. Synthetic molding sand consists of silica sand to which is added a controlled amount of fireclay, bentonite, or other bond. *See* SEDIMENTARY ROCKS; SEDIMENTATION. [JOHN E. LAMAR]

Bibliography: P. G. Cotter, *Mineral Facts and Problems*, U.S. Bur. Mines Bull. no. 630, 1965; J. W. Gillson (ed.), *Industrial Minerals and Rocks*, 3d ed., 1960; R. B. Ladoo and W. M. Myers, *Nonmetallic Minerals*, 2d ed., 1951; F. J. Pettijohn, *Sedimentary Rocks*, 2d ed., 1957.

Sandstone

The most informative variety of sedimentary rocks in terms of the inferences which can be drawn about origin, depositional environment, source area, paleogeographic setting, and so on; thus classification schemes for sandstones involve a long history of debate, analysis, and revision.

Classification. There are three major groups of sandstone: terrigenous, carbonate, and pyroclastic. Terrigenous sandstones consist of clasts derived from the erosion of preexisting rocks located outside the depositional basin. They are transported by moving currents of wind and water and are deposited in a wide variety of environments. Carbonate sandstones consist mainly of fragments of calcium carbonate in the form of skeletal debris, oolites or other layered carbonate pellets, and locally derived (intraformational) carbonate rock debris (intraclasts). Most carbonate sands are deposited in marine environments and are conventionally classified and analyzed as limestones. Pyroclastic sandstones are composed of rock fragments which were produced directly by explosive volcanic activity. They are volumetrically the least important sandstone variety and are therefore seldom studied in great detail.

The classification schemes and nomenclature applied to carbonate and pyroclastic sandstones are rather straightforward. Both are based on the relative abundance of the major types of framework fragments. For example, carbonate sandstones can easily be separated into bioclastic, oolitic, pelletoid, intraclastic, and intermediate varieties. Likewise, pyroclastic sandstones can be subdivided into a number of categories separated by the relative proportion of volcanic glass (shards), mineral crystals, and volcanic rock fragments of which they are composed.

Conversely, the nomenclature and classification schemes developed for terrigenous sandstones are rather complex. These sandstones not only are the most abundant group, but also provide the most information about depositional environment, source area, and paleogeographic setting. Thus, most work done on ancient sandstones involves the terrigenous varieties, and it is therefore appropriate to outline schemes of nomenclature and classification.

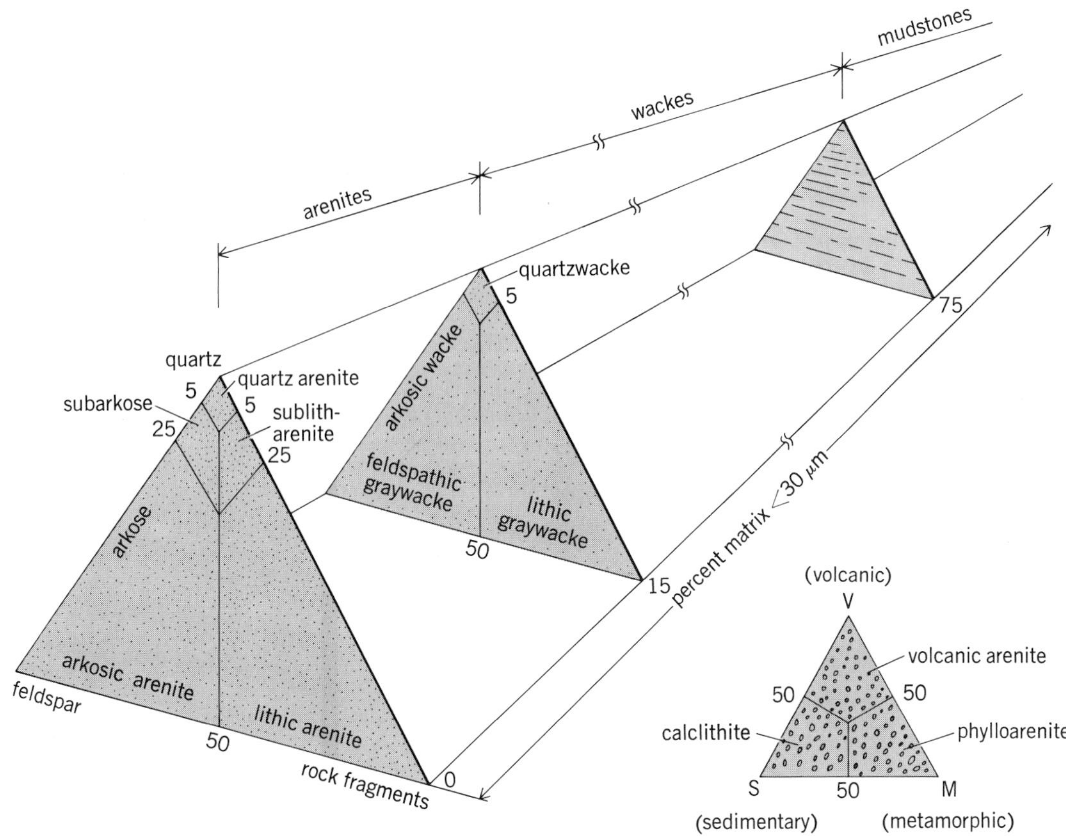

Modern scheme for the classification of terrigenous sandstones, based on texture and mineral composition. *(From F. J. Pettijohn, P. E. Potter, and R. Siever, Sand and Sandstone, Springer-Verlag, p. 158, 1973)*

Modern schemes for classifying terrigenous sandstones derive from the work of P. D. Krynine and F. J. Pettijohn. Most modern classification schemes are based only on texture and mineral composition (see illustration).

Texture. Sandstones consist of two major components: the framework fraction composed of all sand-sized clasts, regardless of composition, and the interstitial areas between the grains. In both modern sands and ancient sandstones, these interstices may be empty voids. However, in most ancient sandstones the interstices are commonly filled with either of two essentially mutually exclusive materials: crystalline chemical cement (usually silica or calcium carbonate) or fine-grained (less than 0.03-mm-diameter particles) matrix. This dichotomy produces the two principal sandstone groups: arenites and wackes. Arenites, the so-called clean sandstones, have less than 15% matrix. They are produced by the lithification of sands which originally had empty interstitial areas. Wackes, the so-called dirty sandstones, have more than 15% matrix. Wackes are thought to have been deposited by turbidity currents, rather than by conventional currents of wind or water which transport arenites. However, modern turbidite sands have less than 15% matrix, suggesting that much of the matrix in ancient sandstones is diagenetically produced.

Mineral composition. The principal mineral components of the framework fraction are most often sand-sized particles of quartz, feldspar, and rock fragments. It is therefore easy to subdivide each of the two major sandstone groups (arenites and wackes) into three sandstone families or classes based only on the relative abundance of the three principal framework components. This approach generates the sandstone names most frequently used: quartz arenite (or orthoquartzite), feldspathic arenite (or arkose and subarkose), lithic arenite (or subgraywacke and protoquartzite), quartz wacke, feldspathic wacke, and lithic wacke. Wackes are commonly undifferentiated by composition and are often referred to simply as graywackes. *See* GRAYWACKE.

Occurrence. Sandstones represent between 10 and 20% of the total volume of sedimentary rocks in the Earth's crust. Most sandstone is found within the continental blocks rather than in ocean basins, and most sandstone in continental blocks occurs within the mobile mountain belts rather than in the stable cratons. Various estimates have been made of the relative abundance of the major sandstone families. Pettijohn's estimates are representative: quartz arenite (34%), feldspathic arenite (15%), lithic arenite (26%), wackes (20%), and other sandstones (5%).

Each of the major sandstone families is also systematically distributed unevenly in time and space, tends to be selectively produced and to be deposited in specific environments, and also tends to be associated only with certain other sedimentary rock types. For example, quartz arenites are most commonly associated with shallow marine carbonates and shales. They almost invariably are beach, offshore bar, tidal, or dune deposits. Also, most quartz arenites are either late Precambrian or Early Paleozoic in age. Conversely, wackes are associated mainly with deep-water abyssal plain shales. Most wackes appear to be submarine fan deposits or their derivatives. The wacke sandstone-shale-chert "flysch" association constitutes an important portion of the sedimentary assemblage of mobile belts of almost every geologic age. Lithic arenites make up a large part of the clastic wedge deposits produced in continental areas during and after the uplift of adjacent mountain belts. They also are represented by modern river sands as well as by the Mesozoic and Cenozoic deposits of the continental coastal plains. Feldspathic arenites are deposited mainly in deep rift-valley areas produced by tensional faulting within continental blocks. *See* SEDIMENTARY ROCKS.

[FREDERIC L. SCHWAB]

Bibliography: R. H. Dott, Jr., *J. Sediment. Petrol.*, 34:625–632, 1964; F. J. Pettijohn, *Chemical Classification of Sandstones*, U.S. Geol. Surv. Prof. Pap. 440S, 1963; F. J. Pettijohn, *Sedimentary Rocks*, 3d ed., 1976; F. J. Pettijohn, P. E. Potter, and R. Siever, *Sand and Sandstone*, 1973.

Sanidinite

A group of rocks formed by high-grade contact metamorphism (pyrometamorphism) and pneumatolysis of certain sediments, mostly argillites, which have been either trapped by the lavas in a volcanic vent or thrown out of the vent as ejecta. The most famous sanidinites are those of the Laacher See (Lake) area, West Germany, which have given name to the sanidinite facies of P. Eskola, a facies characterized by extreme high temperature and low pressure. The original material was mostly an alumina-rich schist with quartz, feldspar, kyanite, staurolite, and garnet. This rock has been altered by simple remelting to a glass in some places; in other places it has been completely recrystallized to form hypersthene, cordierite, corundum, and other minerals. The pneumatolytic introduction of soda and other gases has produced large crystals of sodium-rich sanidine and various minerals containing chlorides, sulfates, and carbonates, such as cancrinite, noselite, haüyne, scapolite, apatite, and calcite. Similar mineral assemblages are known from ejecta of many other volcanoes, for example, those of Vesuvius (Monte Somma). *See* METAMORPHIC ROCKS; METAMORPHISM; PNEUMATOLYSIS. [T. F. W. BARTH]

Sapphire

The name given to all gem varieties of the mineral corundum, except those that have medium to dark tones of red that characterize ruby. Although the name sapphire is most commonly associated with the blue variety, there are many other colors of gem corundum to which sapphire is applied correctly; these include yellow, brown, green, pink, orange, purple, colorless, and black. The lovely orange variety of gem corundum is known by the exotic name of padparadsha and as orange sapphire. In addition to transparent varieties of sapphire, an equal number of translucent types are fashioned in high-domed cabochons (curved cuttings) to bring out the six-rayed stars for which sapphire is famous. Asterism, the star effect, is the result of reflections from tiny, lustrous, needlelike inclusions of the mineral rutile, plus the domed form of cutting. The minute rutile crystals are oriented in three sets parallel to the base of the corundum crystal, with one set parallel to each of the three pairs of parallel prism faces. *See* CORUNDUM; RUTILE.

The most famous source of blue sapphires is Kashmir, the northernmost state in the Indian Peninsula; however, this deposit appears to have been exhausted. The most important source today is the Mogok area of Burma, which is also the most important source of ruby. Mogok produces a number of other colors, as well as blue and red. Thailand and Ceylon are fairly important sources, the former particularly of inky, dark-blue stones, and the latter of stones too light in tone to achieve maximum value. Australia is the source of very dark-blue transparent sapphires, black-star material, and transparent golden sapphires. Light- to medium-blue sapphires are mined from a basic igneous dike in Montana. A new source of sapphires of all light colors is Tanzania. Most of the gem-sapphire sources occur either in alluvial deposits or in the type of marble which results from the intrusion of an igneous mass into an impure limestone.

Blue sapphire is most valuable when it is a medium to medium-dark tone of a slightly violet-blue; this is often referred to as a cornflower blue. The Kashmir grade has a slightly "sleepy" appearance, caused by inclusions that reduce transparency somewhat. Blue sapphire is much more valuable than any other color. The best of the transparent stones are slightly more expensive than the finest stars. Sapphire has a hardness of 9, a specific gravity near 4.00, and refractive indices of 1.76–1.77. Sapphire in blue and other colors is synthesized by the Verneuil flame-fusion process, but no announcement has yet been made of an effort to synthesize it either by flux fusion or hydrothermally as a gem substitute, as has been the case with ruby. *See* GEM.

[RICHARD T. LIDDICOAT, JR.]

Sapropel

A mud, slime, or ooze deposited in more or less open water. Sapropel may vary widely in composition depending upon relative contributions from decomposing substances derived from plants and animals. Hydrogen sulfide, produced during the initial biochemical degradation of these substances, promotes preservation of the more resistant parts of the organisms. Most marine sapropelic deposits contain no appreciable contribution from humic substances of terrestrial origin, but certain carbonaceous marine shales appear to contain some humic matter that was part of the original sapropelic deposit. Metamorphism of a sapropelic deposit leads to such products as torbanite, oil shales, and asphaltites. *See* ASPHALT AND ASPHALTITE; OIL SHALE; TORBANITE.

[IRVING A. BREGER]

Scapolite

An aluminosilicate of Ca and Na, containing the anions Cl, CO_3, and SO_4. It commonly occurs as light-colored translucent prisms with a glassy luster which are often terminated with simple tetragonal pyramid faces (Fig. 1). There is a good 90° cleavage parallel to the *c*-axis, and the hardness is 5–6 on Mohs scale. The color is usually white-buff, but many other colors are known and a dark-blue variety resembling sodalite is not uncommon. *See* SILICATE MINERALS.

The composition approximates a solid solution of marialite ($Na_4Al_3Si_9O_{24}Cl$) and (Me) meionite ($Ca_4Al_6Si_6O_{24}CO_3$), but the end members have not been found and natural scapolites fall in the range Ca/(Ca + Na) = 0.2 to 0.9. The series is divided into meionite, $Me_{80}-Me_{100}$; mizzonite, $Me_{50}-Me_{80}$; dipyre, $Me_{20}-Me_{50}$; marialite, Me_0-Me_{20}. With substitution of SO_4 for the anions and K for the cations, the series becomes rather complex.

Although the first structural studies showed a body-centered structure I4/*m*, refinements now favor the primitive space group $P4_2/n$. The structure (Fig. 2) consists of a framework of Si-O tetrahedra, with Si substituted in part by Al, containing oval cavities in which the cations Ca and Na are located and larger cavities which hold the anions Cl and CO_3. The cell volume and density increase with Ca content.

Scapolite is a common mineral in metamorphic terranes rich in carbonates, that is, in schists, gneisses, and contact rocks (skarns); however, it must form much less than 0.1% of the Earth's crust. Feldspar, carbonate, pyroxene, sphene, amphibole, and many other Ca-Na minerals are common associates. Yellow gem varieties are known. It also occurs in igneous rocks in inclusions derived from deep within the Earth.

Experiments have shown that meionite is stable up to 1500°C at 20 kbar (1 kbar = 10^8 N/m²) pressure and could form as an igneous mineral by deep-seated crystallization from basic magma. In the presence of NaCl and $CaCO_3$, various assemblages of scapolite, plagioclase, calcite, and halite have been synthesized at 750°C and 4 kbar. However, many scapolitic rocks were probably formed at still lower temperatures under conditions not yet experimentally reproduced. [DENIS M. SHAW]

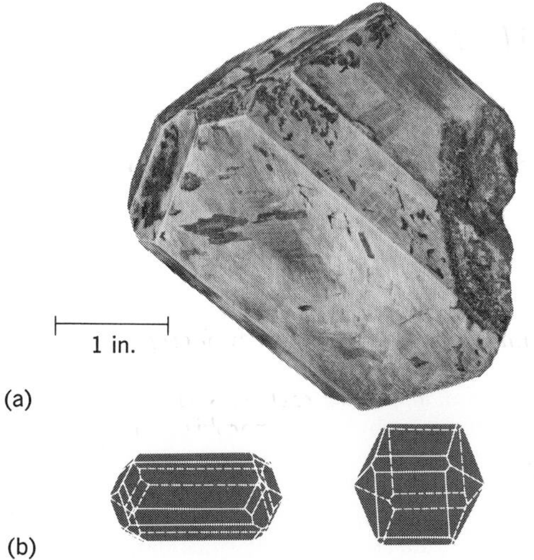

Fig. 1. Scapolite. (*a*) Crystal taken from Pierrepont, NY. (*specimen from Department of Geology, Bryn Mawr College*). (*b*) Crystal habits (*from C. S. Hurlbut, Jr., Dana's Manual of Mineralogy, 17th ed., copyright © 1959 by John Wiley & Sons, Inc.; reprinted by permission*).

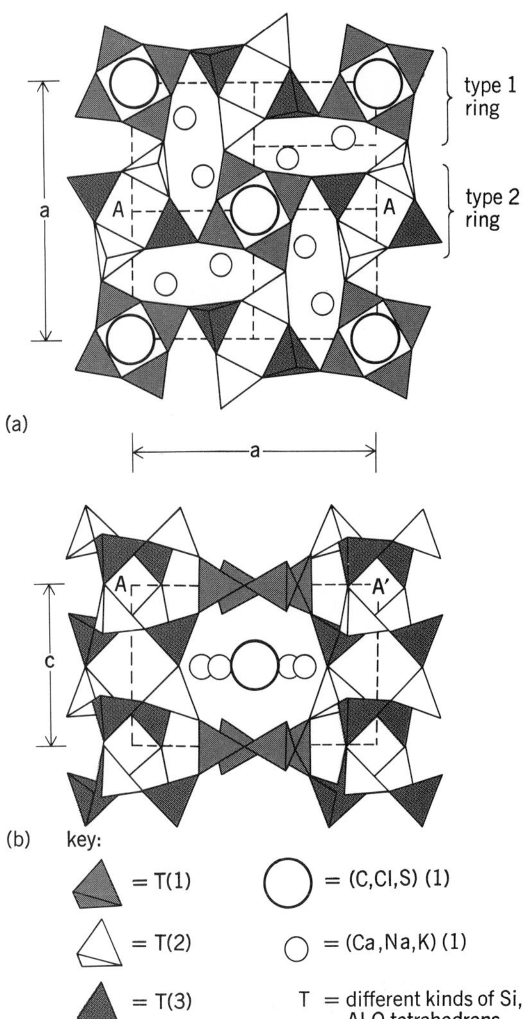

Fig. 2. The crystal structure of scapolite, (*a*) perpendicular to and (*b*) parallel to the tetragonal c-axis. The dimensions designated a and c are lengths of the unit cell edges. (*After S. B. Lin and B. J. Burley, The crystal structure of an intermediate scapolite-wernerite, Tschermaks Min. Petr. Mitt., 21:196–215, 1974*)

Bibliography: W. A. Deer, R. A. Howie, and J. Zussman, *An Introduction to the Rock-forming Minerals*, 1966; S. B. Lin and B. J. Burley, The crystal structure of an intermediate scapolite-wernerite, *Tschermaks Min. Petr. Mitt.*, 21:196–215, 1974; P. M. Orville, Stability of scapolite in the system Ab-An-NaCl-$CaCO_3$ at 4 kb and 750°C, *Geochim. Cosmochim. Acta.* 39:1091–1106, 1975.

Scheelite

A mineral consisting of calcium tungstate, $CaWO_4$. Scheelite occurs in colorless to white, tetragonal crystals (see illustration); it may also be massive and granular. Its fracture is uneven, and its luster is vitreous to adamantine. Scheelite has a hardness of 4.5–5 on Mohs scale and a specific gravity of 6.1. Its streak is white. The mineral is transparent and fluoresces bright bluish-white under ultraviolet light.

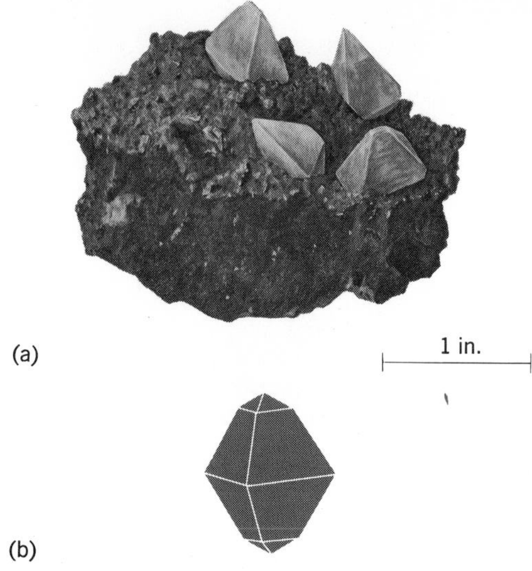

Scheelite. (*a*) Crystals with chalcopyrite ore from Piedmont, Italy (*specimen from Department of Geology, Bryn Mawr College*). (*b*) Crystal habit (*from C. S. Hurlbut, Jr., Dana's Manual of Mineralogy, 17th ed., copyright © 1959 by John Wiley & Sons, Inc.; reprinted by permission*).

Scheelite may contain small amounts of molybdenum. It is an important tungsten mineral and occurs principally in contact metamorphosed deposits (known as tactite) associated with garnet, diopside, tremolite, epidote, wollastonite, sphene, molybdenite, and fluorite, with minor amounts of pyrite and chalcopyrite. It also occurs in small amounts in vein deposits. The most important scheelite deposit in the United States is near Mill City, Nev.

[EDWARD C. T. CHAO]

Schist

A large group of rocks which by deformation during regional metamorphism have acquired a schistosity; that is, they show a more or less perfect cleavage along which they easily split up in flaggy slabs. There are orthoschists derived from igneous rocks and paraschists derived from sedimentary rocks. Crystalline schist is a common designation for all rocks which have recrystallized during regional metamorphism. *See* METAMORPHISM.

Composition. The chemical composition of a schist is similar to that of most silicate rocks, but mineralogically all schists have one feature in common, namely, that one or more of the major mineral constituents are flaky or fibrous (having the crystal structure of phyllosilicates or inosilicates). These minerals are arranged as flakes parallel to the schistosity planes or as subparallel fibers in the schistosity planes, thus emphasizing the schistosity or inducing a linear element in addition to the schistosity. With increase of feldspar and quartz and decrease of schist-forming minerals, the schists pass into the less schistose and more irregular foliated gneisses. *See* GNEISS; SILICATE MINERALS.

Development of schistosity. The metamorphism of the rocks has a direct relation to the orogenic movements in the mountain ranges. At great depths differential movements have taken place by which the constituent minerals of the rocks have acquired a certain spatial arrangement. If this arrangement results in the alignment of flaky, prismatic, or fibrous crystals, a schistose structure is produced. It may be platy or linear. The preferred mineral orientation is either according to the external crystal form (*Formregelung*) or according to the crystal structure (*Gitterregelung*); for an explanation *see* METAMORPHIC ROCKS.

Rocks that show any kind of preferred mineral orientation are called tectonites by B. Sander. However, not all tectonites are truly schistose. The following differential movements will cause preferred mineral orientation.

Laminar gliding. Laminar gliding is similar to the relative displacement of the leaves of a paperbound book which is bent or folded. Laminar gliding is caused by shearing stress. The flaky minerals align on the shear planes, but even the atomic structure of crystals may be shorn and will then slip; that is, a whole layer of ions parallel to so-called glide planes or glide lines in the structure will be displaced relative to the rest of the crystal. Atoms, therefore, move in one preferred direction, resulting in a transport along the greatest stress component. The mechanism is a type of directed self-diffusion and results in creep in this direction. Slip and creep are therefore factors in gliding.

Homogeneous irrotational strain. When flattened by homogeneous irrotational strain (analogous to the flattening of a ball of dough to a cake), the fibrous and flaky minerals are arranged perpendicularly to the pressure. Flattening is probably of great importance in deformations related to vertical or steeply inclined schistosity planes, whereas laminar gliding is more important in thrusts and other shearing movements in the nappes of the mountain ranges. Usually neither type of movement occurs alone, but always in various, and often complicated, combinations.

In mechanically sheared rocks the individual crystals are deformed; this is called fracture cleavage which may pass into flow cleavage. However, recrystallization and blastesis of preferred minerals during the shearing results in undeformed crystals, and the rock then shows crystallization cleavage. The deformation may be paracrystalline, postcrystalline, or precrystalline. *See* AMPHIBOLITE; CLEAVAGE, ROCK; MICA SCHIST; PETROFABRIC ANALYSIS; PHYLLITE; SLATE.

[T. F. W. BARTH]

Schistosity, rock

A type of cleavage characteristic of metamorphic rocks, notably schists and phyllites. The rocks tend to split along parallel planes defined by the distribution and parallel arrangement (preferred orientation) of platy mineral crystals such as muscovite, biotite, chlorite, talc, and graphite or, less commonly, rodlike crystals such as tremolite, actinolite, and hornblende. The preferred orientation of these minerals is produced during deformation and recrystallization (metamorphism) of the rocks. The schistosity may have a consistent attitude over large areas but local folding is also common.

The detailed mechanism of development of schistosity is still uncertain, but it has been demonstrated experimentally that parallel arrangements of micaceous minerals are produced by recrystallization during deformation. According to the several hypotheses that have been proposed, the schistosity may originate (1) perpendicular to the axis of greatest compressive stress, (2) perpendicular to the axis of greatest shortening, (3) parallel to one of the planes of maximum shear stress, or (4) parallel to one of the planes of maximum shear strain. The limited experimental evidence obtained to date favors hypothesis 1 or 2, but 3 and 4 cannot be eliminated and may account for some examples. *See* PETROFABRIC ANALYSIS.

[JOHN M. CHRISTIE]

Seamount and guyot

A seamount is an isolated submarine mountain rising 3000 ft or more above the ocean floor. In the Pacific Basin there are at least 10,000 such mountains, which occur as volcanic peaks on ridges, or rises, or as individual peaks.

Flat-topped seamounts are called guyots, or tablemounts (see illustration). They are present on all ocean floors but are most common in the Pacific. Bottom samples dredged from several guyots include reef corals and rounded volcanic cobbles. Both the coral and volcanic erosion debris indicate that the flattops were once at sea level though they are now 1000–7000 ft below the ocean surface. Thus guyots are ancient islands which were truncated to sea level by erosion. *See* OCEANIC ISLANDS.

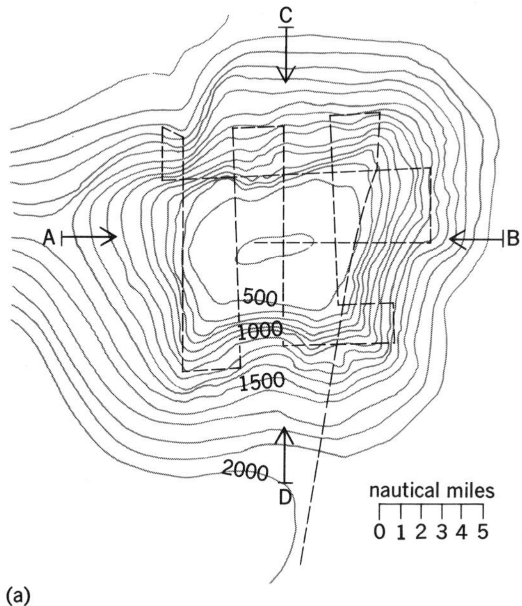

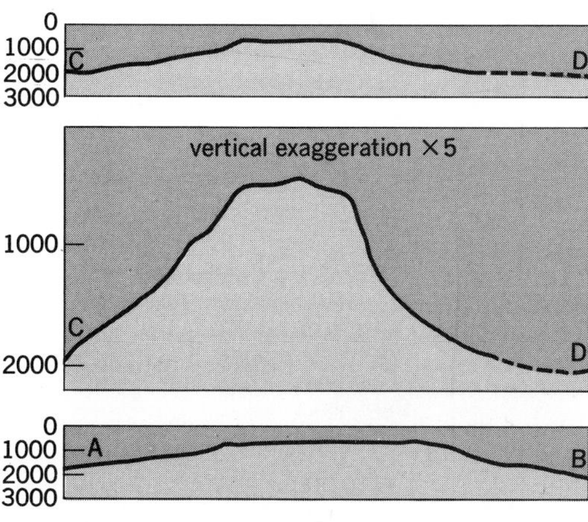

Pratt Seamount, an isolated flat-topped seamount (guyot) in the Gulf of Alaska, 142°30′W 56°20′N. (*a*) Plan. Contour interval = 100 fathoms. (*b*) Profiles.

Although the relative subsidence of guyots, atolls, and seamounts is established, the causes of subsidence are still subject to speculation. Among the possible causes are local, regional, or general sinking of the sea floor, fluctuations of sea level, and an unusual increase in the volume of the oceans since formation of the features. Although combinations of causes are likely, the most important is probably the elevation and subsidence of broad regions of the sea floor during the evolution of oceanic rises. The guyots of the central Pacific and the Gulf of Alaska may have formed in this manner. Some guyots and seamounts lie in or near regions of tectonic instability and apparently subsided as the result of large-scale faulting. *See* ATOLL; MARINE GEOLOGY; REEF.

[EDWIN L. HAMILTON; HENRY W. MENARD]

Bibliography: H. W. Menard, *Marine Geology of the Pacific*, 1964; F. P. Shepard, *Submarine Geology*, 1963.

Sedimentary rocks

One of the three major groups of rocks that make up the crust of the Earth, the other two being igneous and metamorphic. Most sedimentary rocks are layered, and, as is implied by the name, have originated by the sedimentation, or settling, of particles. Thus layering, or stratification, is the most important single characteristic of sediments and sedimentary rocks, even though there are some igneous and metamorphic rocks that show some kind of stratification or pseudostratification. *See* IGNEOUS ROCKS; METAMORPHIC ROCKS.

The distinction between sedimentary and other rocks is understood best by considering their origin. Sediments are formed at or very near the surface of the Earth as a result of processes operating at the surface, at normal surface temperatures and pressures. Most igneous and metamorphic rocks, on the other hand, are formed as a result of conditions deep in the crust of the Earth, where temperatures and pressures may be very high. Some overlapping between the three rock families exists; for example, it is difficult to classify rocks that originate as volcanic ash falls (igneous) but are then transported and become interlayered with normal sediments. It may also be difficult to distinguish between a hard, compacted sedimentary rock and a weakly metamorphosed rock.

Though sedimentary rocks account for only 5% of the Earth's outer crust (a shell 10 mi thick), they make up 75% of the exposed rocks at the surface. From this relationship alone it becomes apparent that, in general aspect, sediments are distributed as a rather thin layer at the surface. The thickness of this thin layer may vary greatly from place to place; the thickness of the total sedimentary volume may be only a few tens of feet at the edges of some old igneous mountain masses such as the

Ozark or Adirondack mountains, but may be well over 30,000 ft in some places where the crust is subsiding rapidly, such as the delta of the Mississippi River. Though sediments are quantitatively relatively unimportant as crustal constituents, they have been the chief means of elucidating the history of the Earth and, with their contained fossils, the development and evolution of life forms. Sedimentary rocks are also important as the source of many of our major mineral resources, notably coal, oil and gas, iron ores, and limestone. *See* DELTA; GEOSYNCLINE.

Origin. Sedimentary rocks originate primarily as the result of the fragmentation and destruction of preexisting rocks. As rocks are weathered by the action of water, wind, frost, and organic decay, large masses become mechanically broken into finer sizes and some of the constituents dissolve in rain or soil water. The solid fragments, ions in solution, and colloids in suspension, are transported, primarily by running water and secondarily by wind and groundwater, from the site of weathering, the source area, to this site of deposition. Transportation of detritus may be temporarily interrupted by sedimentation in streams or lakes, resulting in river bars, alluvial fans, or lake deltas. Eventually, however, most of the material reaches the site of the lowest gravitational potential energy on the Earth's surface, the bottom of the sea. After final deposition the soft, water-saturated muds, silts, and sands become buried under successive layers of later sediment, the water is squeezed out, the sediments become compacted, and chemical changes result in cementing the original unconsolidated material to a rock. *See* DIAGENESIS; SEDIMENTATION.

Sedimentary petrologists commonly divide the sedimentary rocks into two large groups, the detrital and the chemical. This division is based on the differing origins of the two groups. The detrital (sometimes called clastic) rocks are formed by the sedimentation of mineral or rock fragments that were derived from the mechanical disintegration of preexisting rocks in the source area and have been transported, as solids, to the site of deposition. The chemical rocks originate as chemical precipitates at the site of deposition; they may be inorganic precipitates formed from supersaturated solutions or they may be formed by the biochemical action of organisms, as are the calcium carbonate shells of mollusks. A great many sediments are mixtures of detrital and chemical components; many chemically precipitated limestones contain some fine-sized grains of quartz and clay minerals, most of which probably originated as wind-blown or animal-carried material. Predominantly detrital rocks, such as sandstones and shales, commonly contain some amount of chemical precipitate, calcium carbonate and silica being the two most abundant; the chemical components may have been introduced at the time of deposition or during postdepositional changes (diagenesis). *See* PETROLOGY; ROCK.

The three most abundant kinds of sedimentary rocks, shale, sandstone, and limestone, together account for over 95% of all sediments. Of these, limestone, the chemical rock, composes only about 20% of the total volume of sedimentary rocks in the crust. Estimates of the relative proportions of the two detrital rocks, shale and sandstone, vary, but it appears that shales are between two and three times as abundant as sandstones.

Textural characteristics. Because the majority of sediments are dominantly mechanical mixtures of detrital mineral and rock fragments, floccules of colloidal materials such as clay, and chemically precipitated particles, it is important to determine the geometrical properties of the individual particles and their relationships to each other. Textural analysis has led to a fuller understanding of the genetic factors involved in the formation of sediment by settling of particles through a fluid (water) or gaseous (air) medium, for the textures are directly related to the hydrodynamics of the medium. A number of different textural properties have been defined.

Size. Perhaps the most important textural property is the size of individual particles and the size distribution of all the particles in the sediment. Size of particles is the basis for the division of the detrital rocks into lutite, or shale (fine); arenite, or sandstone (medium); and rudite, or conglomerate (coarse). Because it is manifestly impossible to measure the sizes of all of the particles in a sediment, statistical analysis has been used extensively to characterize a distribution of sizes. By statistical analysis the sedimentary petrographer can characterize four properties of size distributions based on counts of frequency of grains in several size grades. The first property is the calculation of some kind of average size, either the arithmetical or geometrical mean, or a median (50 percentile) size. The most common measure used by geologists is median size. A second property of the size distribution is the spread or dispersion of size values about the mean or median size. Standard deviation is one measure of this spread. One of the more common measures is quartile deviation (evaluating spread between the 75 and 25 percentiles) or sorting index (Fig. 1). A third property of size distributions is the skewness or asymmetry of the distribution about the median value. This measure indicates whether a sediment has much more material finer than the median size as compared with the fraction coarser than the median size, or vice versa. A fourth property of the size distribution is kurtosis, which measures the number of grains that cluster in size around the median as compared with the number of grains that are

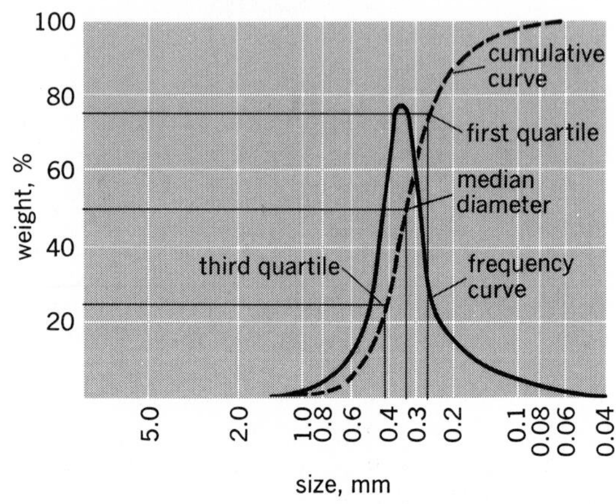

Fig. 1. Size distribution curves.

much finer or coarser than the median. The median size of a sediment has been used to estimate the competence of a current to transport sediment: The coarser the size, the stronger the current. A variant of this approach has been the use of coarsest grain size fraction as an indication of current strength. Sorting has been used by itself and together with other statistical measures to distinguish between beach, river, and dune sands, but there is some uncertainty in the validity of this application of statistical measures. With the development of computers, more sophisticated statistical analysis of size data has become possible and many more data can be analyzed. Multiple correlations of all size parameters have been used not only to distinguish between depositional environments but to make deductions on the hydraulic characteristics of the depositing current.

Shape and roundness. In addition to the size distribution, textural analysis includes the study of the shape and roundness of the particles. Shape is defined as the degree of sphericity, or approach to a sphere. Roundness is defined as the degree of sharpness of corners or edges of a particle. A particle may have many sharp, small projections and have a low roundness value and yet be very close to a sphere in shape. On the other hand, a particle may be long and rodlike, very far in shape from a sphere (low sphericity), and yet be very smooth and rounded (high roundness). Shape and roundness are related to mechanical abrasion during transportation of the detritus prior to deposition; the greater the abrasion, the higher the roundness, and, in general, the higher the sphericity (Fig. 2).

Surface configuration. Another element of texture is the surface configuration of the grains. Some sand grains show frosting and pitting, which has been interpreted as the result of being windblown and having gone through a great many collisions with other grains. Collisions under water are softened by the fluid medium and do not result in this texture.

Philip Kuenen concluded that surface frosting of many eolian sands was the result of the chemical action of dew. Such grains became smooth and polished when subjected to a different kind of chemical process in rivers or beaches. A striking example is the frosted sand of the Kalahari Desert of Southern Rhodesia, which, after being blown into the Zambesi River, loses its frosting in a distance of only 60 km.

Packing and fabric analysis. In the late 1940s and 1950s two new kinds of textural analysis were introduced, packing analysis and grain-shape fabric analysis. Packing analysis relates to the way in which the particles are arranged in the rock to give more or less dense aggregates. The density of packing is related to the weight of overlying sediments and perhaps also to lateral compressive forces operating during mountain-building episodes. Grain-shape fabric analysis is the study of the degree of preferred orientation of the long axis of elongate particles and the direction of the orientation. A rock with a high degree of preferred orientation would show all of the long particles lined up in the same direction. This direction of preferred orientation is related to the average direction of current flow of the medium from which the sediment settled. See PETROFABRIC ANALYSIS.

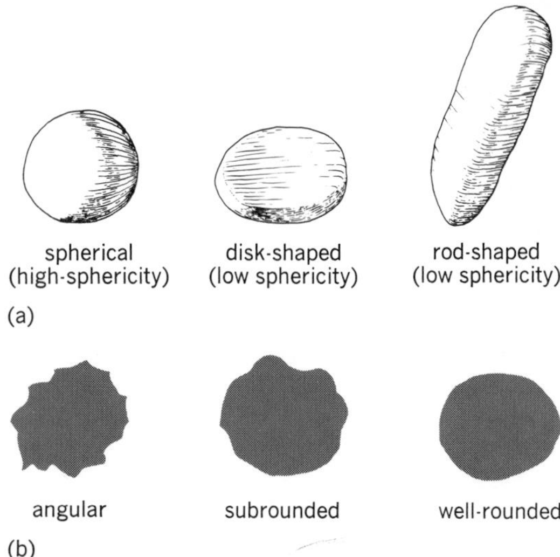

Fig. 2. (*a*) Degree of sphericity. (*b*) Roundness or degree of sharpness of corners or edges.

Chemical composition. Chemical composition of rocks is expressed in terms of the oxides of the elements. Determinations are made normally by wet chemical analysis and the assumption is made that the elements are present as oxides, the oxygen not being determined directly. Although the chemical composition of sediments varies widely with lithology and grain size, the average sedimentary rock contains about 58% silicon dioxide (SiO_2), 13% aluminum oxide (Al_2O_3), 6% calcium oxide (CaO), 5% ferrous oxide (FeO) and ferric oxide (Fe_2O_3), 5% carbon dioxide (CO_2), and smaller amounts of magnesium oxide (MgO), potassium oxide (K_2O), sodium oxide (Na_2O), and many trace elements. This average sediment may be thought of as a combination of sandstone (mainly SiO_2), shale (mainly Al_2O_3), and limestone (mainly CaO and CO_2). The average sedimentary rock differs from the average igneous rock in having much more CO_2, much lower Na_2O, and much more ferric than ferrous iron. The CO_2 is added primarily from the atmosphere, in the process of weathering. Much sodium is lost, during weathering, as soluble salts, which partly accumulate in the oceans, and much ferrous iron is changed to ferric under the oxidizing conditions prevalent over most of the Earth's surface. See WEATHERING PROCESSES.

Mineralogical composition. The composition of sediments is best expressed in terms of the relative abundances of minerals present. This is so because it is not always possible to calculate mineralogy from gross chemical composition without knowing what minerals are present; however, the chemical composition can easily be calculated once the mineral assemblage is known. Also, because sediments are often a mechanical mixture of minerals of many different chemical origins, the interpretation of mineralogy proceeds in a much more direct fashion than that of gross chemical composition. Although a great number of mineral species have been found in sedimentary rocks, usually in very small amounts, only some 20 minerals account for over 99% of the bulk of the sedimentary rocks. These, in general, are the minerals that are fairly stable chemically in Earth surface environments. The minerals of sediments, like the rocks themselves, are divided into two groups, the detrital and the chemical. The detrital minerals are mechanically transported and deposited, and, normally, are considered to have originated by the mechanical disintegration of the parent rock in a weathering source area of sediments. The chemical minerals are considered to have formed by precipitation, either inorganic or biogenic by the action of organisms, at the site of deposition.

Many minerals may be either detrital or chemical or both in origin. Quartz, for example, is dominantly detrital and yet it is frequently found as a chemically precipitated mineral that formed in rocks after they were deposited. Although calcite (calcium carbonate) is normally considered to be a chemical precipitate, it is recognized in many limestones as an essentially detrital mineral, having been transported some distance from its original place of formation and deposited mechanically in the same way as a quartz sand grain. The most abundant detrital minerals in sediments are quartz and clay minerals. Less abundant, but still quantitatively important in many sediments, are feldspar, rock fragments, and coarse-grained micas. A host of other detrital minerals are found in many sedimentary rocks, the sum of them rarely making up more than 1% of the rock. Most of these are grouped under the name "heavy minerals" because they have specific gravities greater than 2.85, the specific gravity of bromoform, the liquid commonly used to separate heavy from light minerals (quartz, feldspar) by a sink or float method. Some of the most common detrital heavy minerals are zircon, tourmaline, garnet, hornblende, epidote, rutile, staurolite, and magnetite. *See* AUTHIGENIC MINERALS; MINERALOGY; PETROGRAPHY.

The major chemically precipitated minerals are the carbonates—calcite, aragonite, dolomite, and siderite. Less important are chert, gypsum and anhydrite, other saline residues, such as common rock salt (halite), and a number of phosphates, such as collophane. A number of heavy minerals may be chemical in part; zircon and tourmaline often show chemically precipitated secondary additions to the original detrital grain, and anatase seems always to be formed as a chemical precipitate subsequent to deposition.

Significance of detrital minerals. The detrital minerals are guides to the composition of the parent rocks of the sediment and indexes of the degree of weathering of those parent rocks. If a source area terrain is subjected to rapid mechanical erosion and little chemical action, most of the major minerals and rock fragments will be transported as such to the depositional area. In such cases the sediment may show a high percentage of minerals that are unstable at the Earth's surface. Such mineral assemblages in the sediment may differ, depending on the kinds of rocks exposed to erosion in the source area. Also, low-grade metamorphic may be distinguished from high-grade metamorphic or igneous terrains. Heavy minerals are particularly useful for this purpose. If mechanical erosion is at a minimum and chemical weathering is intense in the source area, most unstable minerals will tend to dissolve or alter, leaving behind a residue rich in quartz, the common constituent of most rocks that is stable in sedimentary environments. Thus a sediment that contains only quartz as a detrital mineral may be interpreted as the product of intense chemical weathering in the source area. The ratio of quartz to feldspar in true surface sediments is a rough index of source area weathering, as quartz is stable and feldspar unstable. The higher the quartz-feldspar ratio, the more intense the chemical weathering of the parent rocks. One cannot always interpret detrital mineralogy strictly in terms of the above ideas, however, for there may be changes in mineralogy after deposition. It appears that certain of the heavy minerals, in particular, olivine, augite, and other ferromagnesians, tend to dissolve if the rock has been buried a long time, and feldspars are found as postdepositional chemical precipitates in sediments.

Significance of chemical minerals. The chemical precipitates are reflectors of the chemical environment of deposition. The chemical controls that determine if and what kinds of minerals are precipitated are the composition of the water solutions at the depositional site, that is, sea water, brackish water, or fresh water; the oxidation-reduction potential (Eh) and the acidity-alkalinity (pH) of the solutions; and to a lesser extent pressure and temperature. In addition to the inorganic chemical controls, there are biological factors controlling mineral precipitates. Many invertebrates and plants, in particular mollusks and algae, are prime agents for the precipitation of carbonates. Indeed, some authorities believe that originally all sedimentary carbonates were biogenic, and that lack of fossil structures is chiefly due to postdepositional recrystallization or solution and reprecipitation. Carbonate minerals can be interpreted mainly as the result of biological activity and the pH of the solutions. Iron and sulfide minerals may reflect pH, Eh, and biological activity. Chemically precipitated minerals that form postdepositionally in a rock are often called authigenic; some of the most common abundant authigenic minerals are quartz, calcite, and dolomite.

Clay minerals. The clay minerals occupy a position between the detrital and the chemical. Basically the clays are detrital, and their fundamental crystallographic structures are normally preserved through mild weathering, transport, deposition, and diagenesis. But the clay minerals are very susceptible to exchange of alkali and alkaline earth cations with their environment, and the clay that is finally produced in a sediment may be of a composition and overall structure quite different from that originally supplied from the parent rocks in the source area. *See* CLAY; CLAY MINERALS.

Sedimentary structures. These are the larger textural features of sediments, such as bedding, ripple marks, and concretions, that were formed during or shortly after deposition, as distinct from the still larger elements of structure, folds and faults, which were produced much later than deposition. Sedimentary structures include the mechanical, made by the currents that transport sediment; the chemical, produced by inhomogeneous precipitation; and the organic, produced by organisms living in the environment. *See* STRUCTURAL GEOLOGY.

Mechanical structures. These structures include bedding of various kinds, such as cross-bedding,

Fig. 3. Structures in sedimentary rocks. (*a*) Current ripple marks in copper ridge dolomite, south of Bluefield, Va.; (*b*) Mud cracks in Mississippian limestone, northeast of Bluefield, Va.; (*c*) concretions in Pennsylvanian shale, near Montgomery, W.Va. (*Virginia Division of Mineral Resources*). (*d*) Oscillation ripples in a Cretaceous sandstone; (*e*) dune bedding in the Navajo sandstone near entrance to Zion Canyon; (*f*) ball and pillow structure in the Chemung sandstone at Chemung Nose, south of Elmira, N.Y. (*Carl O. Dunbar*).

inclined bedding, graded bedding, and ripple marks, and slump structures caused by small landslides more or less contemporaneous with sedimentation (Fig. 3). Sand beds that appear to have foundered into soft underlying muds show ball and pillow structure. Fossil mud cracks, caused by temporary desiccation of a mud bottom, raindrop impressions, and frost crystal casts, are included in this category. Results of work done in the late 1940s and 1950s on different kinds of sedimentary structures indicate that such structures are related to the environment of deposition, and that many of the structures, cross-bedding in particular, are a clue to the direction of sediment transport. Many

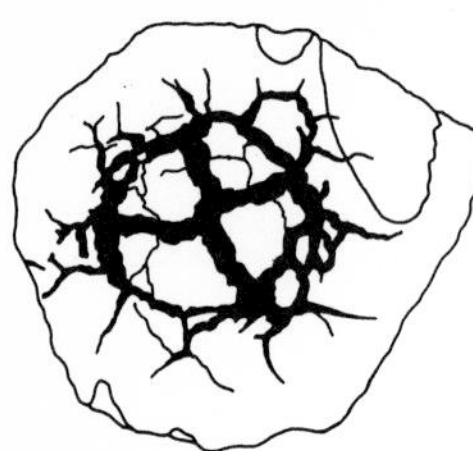

Fig. 4. Septarian structure. Length of specimen about 5 in. *(From F. J. Pettijohn, Sedimentary Rocks, 3d ed., copyright © 1949, 1957 by Harper & Row, Publishers, Inc., copyright © 1975 by Francis J. Pettijohn)*

different kinds of linear structures on bedding plane surfaces have been described, including rill marks and swash marks, made by the retreat of waves on a beach; groove casts and load casts, made by movement of a pebble or cobble on a muddy bottom; and more problematical structures of unknown origin. Most of these have been ascribed to current origin and the lineation parallels the current.

Chemical structures. These structures form as segregations of originally dispersed chemical substances. They include oolites and pisolites, concretions, geodes, nodules, and septaria, as well as stylolites, a solution feature. They take many forms, from highly irregular, perhaps even branched forms, to regular spheres or ellipsoids, and range in size from dumbbell-like objects 3 ft in diameter to tiny spherulites only a few millimeters in diameter. Some of the structures apparently form at the same time as the sediment, others may form very soon after sedimentation, or before compaction, and others form after compaction and consolidation of the sediment, perhaps quite late in its history.

Nodules are concretionary structures, with a great diversity of irregular shapes, composed of material different from that of the rock in which they occur. Most frequently nodules are flattened or elongated in a direction parallel to that of the bedding. Sometimes they tend to coalesce to form almost continuous layers. The most common nodules are of chert (silica); others are of iron oxides, phosphates, iron carbonates, and clay ironstones. They occur most typically in limestones but may also be found in shales and sandstones.

Septaria are rather large nodules, normally greater than 3–4 in., that display a system of polygonal cracks at the center and which tend to die out toward the edges of the nodule (Fig. 4). In almost all of these nodules the cracks are filled with a crystalline mineral, normally calcite. The septaria was originally a gel concretion that hardened or dehydrated on the outside first; shrinkage caused by dehydration of the gel was responsible for the cracking. Later, mineral solutions filled the cracks. *See* CONCRETION; GEODE; OOLITE AND PISOLITE; STYLOLITES.

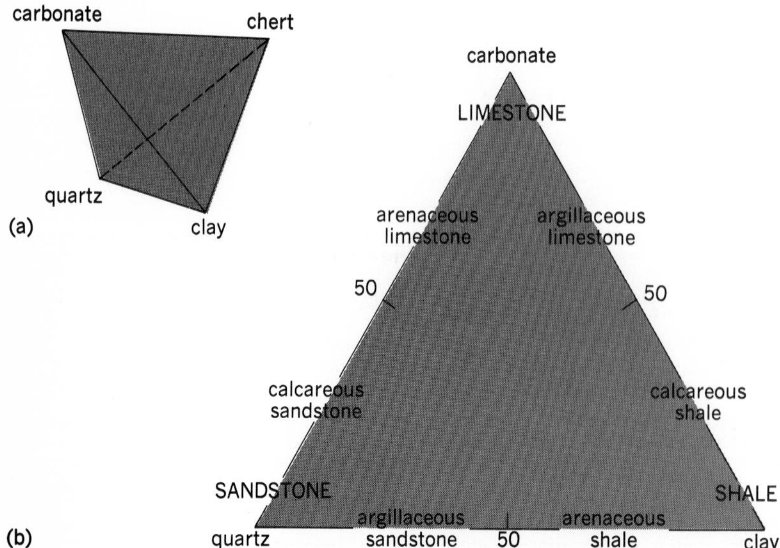

Fig. 5. Classification of sedimentary rocks. (*a*) A simplified tetrahedron for classifying sedimentary rocks. The chert and carbonate apices represent the main chemical fractions and quartz and clay the main detrital fractions. (*b*) The carbonate-quartz-clay face is shown in detail. *(From W. C. Krumbein and L. L. Sloss, Stratigraphy and Sedimentation, 2d ed., copyright © 1963 by W. H. Freeman and Co.)*

Organic structures. Preserved in sedimentary rocks are textural elements that are either the remains of organisms that lived during the time the sediment was being laid down or that have resulted from the activities of those organisms. Of greatest importance are the former, the fossils, which, in the main, are the preserved hard parts of the plants or animals. The evidences of organism activity are less frequent but consist of worm borings and tubes, fecal pellets of many different kinds, and larger structures of excretory origin, the coprolites.

Fossils may consist of calcareous or phosphatic shells, of siliceous shells, of chitinous materials, or of carbonaceous films or impressions. Typically the most abundant fossils are found in limestones and dolomites. They may, in fact, constitute the bulk of the rock as in the variety of limestone called coquina. All traces of original shell materials may be gone but the shape and even fine details of the fossil may be preserved in a mold of the fossil in the surrounding rock. Algal structures, produced by the action of various kinds of lime-secreting algae are often found in limestones. Some of these structures, laminated in a variety of ways, are called stromatolites. *See* FOSSIL; STROMATOLITE.

Classification and nomenclature. Sedimentary rocks have been classified on two bases, the purely descriptive, and the genetic. Thus, in one group might be found all rocks that are colored red or all rocks that contain a certain proportion of any mineral, or, on the other hand, all rocks that were formed as river floodplain deposits. The disadvantage of the purely descriptive approach is that rocks of widely divergent origin might be lumped together. The disadvantage of the genetic approach is that the origin of the rock must be known before it can be classified; this requirement is often one of the most difficult problems of sedimentary geology.

The more fruitful approach has been to combine the two into a classification based on objective descriptive characteristics that have some genetic significance. Using this as a basis, geologists subdivide the sedimentary rocks into two broad groups, the clastic or detrital, and the nonclastic or chemically precipitated rocks. The clastic rocks are characterized by individual grains, often of heterogeneous composition, derived from the erosion of preexisting rocks; the grains may be more or less rounded by abrasion during transport from the erosional area to the site of deposition. The chemical precipitates are normally fairly homogeneous in composition and are characterized by an interlocking crystalline texture, where the crystal sizes may vary greatly.

The chemical precipitates form at the site of deposition and are not derived directly from preexisting rocks (Fig. 5).

Clastic rocks. Rocks of detrital origin are further subdivided on the basis of particle size into three classes, the rudites (conglomerates), the arenites (sandstones), and the lutites (shales). Siltstones are intermediate in grain size between arenites and lutites and may be considered as very fine-grained arenites or grouped with the lutites. Rudites are composed of particles larger than 2 mm in diameter; arenites, $2-\frac{1}{16}$ mm; siltstones, $\frac{1}{16}-\frac{1}{256}$ mm; and lutites finer than $\frac{1}{256}$ mm. In practice, since all of the particles of any rock are rarely in the same size group, the dominant particle size is selected, usually the median size.

Chemical rocks. The nonclastics are subdivided by chemical composition into the carbonates, the siliceous rocks, the evaporites, the phosphorites, ferruginous and manganiferous rocks, carbonaceous rocks, and miscellaneous rare chemically precipitated rocks. The carbonate rocks, the limestones and dolomites, are much the most important in terms of total quantity in the crust and areal extent. Some of the chemically precipitated rocks are formed directly by the agency of organisms, the most prominent being the biogenic limestones. Others are formed by inorganic precipitation, such as the iron formations. *See* ARENACEOUS ROCKS; ARGILLACEOUS ROCKS; CHERT; CONGLOMERATE; LIMESTONE; SANDSTONE; SHALE.

[RAYMOND SIEVER]

Bibliography: C. O. Dunbar and J. Rodgers, *Principles of Stratigraphy*, 1957; W. C. Krumbein and L. L. Sloss, *Stratigraphy and Sedimentation*, 2d ed., 1963; F. J. Pettijohn, *Sedimentary Rocks*, 2d ed., 1957; F. J. Pettijohn and P. E. Potter, *Atlas and Glossary of Primary Sedimentary Structures*, 1964; H. Williams, F. J. Turner, and C. M. Gilbert, *Petrography*, 1954.

Sedimentary structures

Internal megascopic features exhibited by sediments and sedimentary rocks. In contrast to the larger structural features produced in rocks by later deformational stress (folds and faults), most sedimentary structures form during or shortly after deposition. Based on mode of origin, three principal classes of sedimentary structures have been recognized: primary or mechanical; secondary or chemical; and biogenic or organic.

Primary sedimentary structures are, for the most part, produced hydrodynamically by the currents which transport and deposit the sediment in which the sedimentary structure is found. Common varieties include bedding (stratification), cross-bedding, graded bedding, and ripple marks. Secondary sedimentary structures are produced essentially by chemical processes of precipitation, solution, segregation, and recrystallization, which occur after deposition and produce such features as styolite seams and concretions. Biogenic sedimentary structures are produced by the activity of organisms, and include actual preserved hard parts (fossils) as well as organically produced tracks, trails, burrows, and so on (ichnofossils).

Analysis of sedimentary structures commonly leads to a better understanding of sedimentary rock age and genesis. For example, sedimentary structures provide a useful guide to stratigraphic order because several structures clearly indicate bedding tops and bottoms (for example, rain prints, graded bedding, and cross-bedding). Also, particular types of depositional environments commonly are associated with sediments which exhibit a unique and diagnostic suite of sedimentary structures. Finally, many sedimentary structures have an orientation which is selectively controlled by the currents producing them, and measurement of their orientation therefore serves as a guide to the current system which prevailed during deposition (paleocurrent analysis).

Primary structures. The most common variety of primary sedimentary structure is bedding or stratification—the distribution of sedimentary material in systematic layers or bands which are readily visible because of obvious changes in texture and composition. Stratification directly reflects secular changes in the nature of the material deposited as a result of changes in current velocity, depositional depth, source area composition, and so on. Most other primary sedimentary structures express themselves as aspects of bedding: some are largely features which developed on the top of bedding planes (rain prints and ripple marks), some are largely confined to the base of bedding planes (sole markings), and still others occur within the individual beds (cross-bedding, parting lineation, and graded bedding). *See* BEDDING; STRATIGRAPHY.

Many types of primary sedimentary structures other than stratification are produced by the shearing stress exerted by moving currents of ice, water, wind, or sediment-charged density flows upon the surface across which they move. In response to that shear, the underlying surface may deform from a planar bedding surface into a duned or rippled surface. Subsequent slippage of unconsolidated material down the lee side of ripples or dunes may produce cross-bedding or cross-lamination. Conversely, the moving currents may scour erosional channels into the top of the underlying bed, producing simple channels (or a miniaturized version, termed scour-and-fill structures). Sediment-charged density (turbidity) currents often cut a variety of depressions into the underlying surface. These depressions may be elongate (grooves), spoon-shaped (flutes), or nondescript (tool marks). They commonly serve as molds within which sediment is almost immediately deposited—producing a series of variously shaped protuberances on the base of the overlying bed. These bottom or sole marks include groove and flute casts. Those types of current-produced sedimentary structures which have a preferred orientation systematically related to the direction of current flow are collectively referred to as directional sedimentary structures. They are conventionally utilized for reconstructing ancient current systems (paleocurrent analysis), from which it is often possible to infer depositional environment, location of sedimentary source areas, and overall paleogeographic setting.

A variety of other mechanisms also produce primary sedimentary structures. Droplets of rain falling on unconsolidated mud may produce rain prints. Desiccation of muds may produce polygonally shaped mud cracks. Still other primary sedimentary structures are produced by hydroplastic deformation; that is, dense, water-saturated sand may intrude overlying sediments (sandstone dikes) or sink into underlying, unstable mud or silt (load casts and ball-and-pillow structures). Semiconsolidated sediments deposited on slopes may undergo folding (convolute bedding) or slip en masse downslope (slump structures).

Secondary structures. Secondary structures are produced by the precipitation or segregation of minerals within an existing sedimentary rock. While some chemical structures may be produced penecontemporaneously with the sedimentary rock in which they are found, most are related to diagenetic changes in the distribution of the original material within a rock. The most common varieties are nodules and concretions, both of which are small-scale (centimeters) segregations of minor constituents within a sedimentary rock of different composition, such as chert nodules in limestone or calcareous concretions in shale. Nodules are irregular in shape, have a knobby surface, and are internally structureless. Concretions are normally rounded on the surface and are subspherical in shape. Specialized varieties include septaria (large oblate nodules traversed by a series of radiating cracks), cone-in-cone structures, sand crystals, barite rosettes, and geodes.

Biogenic structures. A wide variety of structures preserved in sedimentary rocks are produced by organic activity. These include fossils, which are the hard parts of organisms, often preserved due to selective replacement of organic material by more resistant siliceous or calcareous material. In addition to fossils, many sedimentary rocks exhibit trace fossils (ichnofossils). These are organically produced textural elements such as worm burrows, plant root tubes, fecal pellets, vertebrate footprints (tracks), and invertebrate markings (trails). Fossils and trace fossils often serve as useful guides to sedimentary rock age and environment of deposition. *See* FOSSIL; ROCK, AGE DETERMINATION OF; SEDIMENTARY ROCKS; SEDIMENTATION. [FREDERIC L. SCHWAB]

Bibliography: F. J. Pettijohn, *Sedimentary Rocks*, 3d ed., 1975; F. J. Pettijohn and P. E. Potter, *Atlas and Glossary of Primary Sedimentary Structures*, 1964; R. C. Selley, *An Introduction to Sedimentology*, 1976.

Sedimentation

The processes that operate at or near the Earth's surface to deposit rock-forming material, or sediment. Sedimentation includes the weathering processes that act mechanically and chemically to break up preexisting rocks, the processes of transportation by which the material is carried from its source to the depositional site, the processes of deposition in the sedimentary environment, and the postdepositional processes or diagenesis by which the sediment is compacted and hardened.

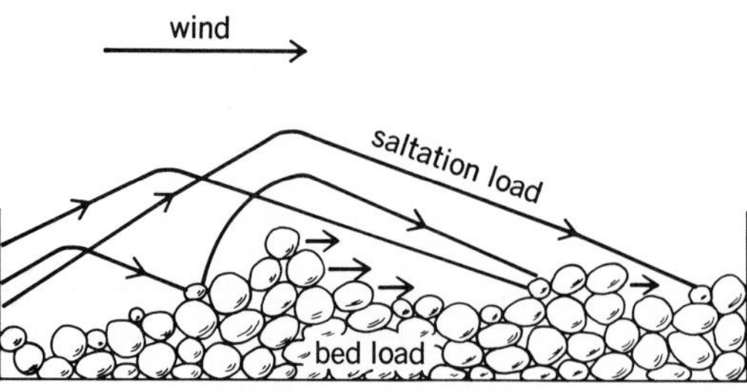

Fig. 1. Detail of saltation and bed load for wind. The range of grain sizes shown is from 0.2 to 2 mm. With increase of wind velocity the smaller particles of the rolling load become part of the saltation load. (*From R. M. Garrels, A Textbook of Geology, Harper & Row, Publishers, Inc., 1951*)

Source of material. The raw materials of sedimentation are the products of weathering of igneous, metamorphic, and sedimentary rocks. Weathering may be primarily chemical, mechanical, or both. In chemical weathering minerals are dissolved or altered by solution and the more soluble salts are carried away by running water, leaving behind a residue of insolubles. In mechanical weathering the rock is physically disintegrated by the action of water, wind, freezing and thawing, and temperature extremes. As the whole spectrum of rock types may be involved in weathering, the detritus supplied from the source may be extremely heterogeneous in mineralogical and chemical composition. *See* WEATHERING PROCESSES.

Transportation and deposition. Material supplied by weathering in the source area is carried by water, wind, and mass movements, such as landslides, to the site of deposition. Transportation and deposition may be intermittent, and may alternate with each other until the detritus reaches its final resting place. A sand grain may be carried for a short distance and dropped temporarily many times in the course of traveling down a river to the sea. In the course of transportation the composition and texture of the sediment may be slightly or drastically altered by the transporting medium. More soluble minerals may dissolve, softer minerals may wear out by abrasion, and material may become sorted by size.

Transportation by wind. Turbulent motion of air close to the ground is responsible for lifting small particles and transporting them. Dust is carried in suspension but sand may be carried both in suspension and along the surface. Movement of sand grains may take place by saltation, a process of moving in discrete jumps, or by surface creep, in which the grains are rolled or pushed forward by the force of the wind and impact of landing grains. In saltation the grains are thrown into the air by collisions of rolling grains. Once thrown into the air the grain follows a parabolic course and comes back to the ground a short distance away. As it hits, it may bounce back into the air or it may knock another grain upward (Fig. 1). Wind transportation operates over all land areas but is most effective in arid regions. Much evidence accumulated in the 1960s showed that wind transportation extending over the world's oceans is an important factor in distributing the fine-grained detritus that becomes a large part of deep-sea sediment. Although dust may be carried to great heights by the wind, sand grains usually remain only a few feet above the surface. Wind normally transports only fine-grained sand, and the sand is well sorted. When the velocity of the wind decreases to the point where it can no longer carry particles, material is dropped, and a windblown sediment is formed. The most common windblown deposits are sand dunes, which have a variety of shapes and sizes. They are common in desert areas and along coastlines and some river valleys, where there is a ready source of sand. Loess is a windblown dust and fine silt deposit. *See* DUNE; LOESS.

Transportation and deposition by ice. Sedimentary particles may be trapped in glaciers and icebergs and transported long distances before the ice melts and the particles are deposited. The material is not sorted in size, and deposits from glaciers are commonly unstratified. Deposits from melting icebergs may be mixed with marine sediment; the resulting glaciomarine sediment may be stratified and slightly sorted by size but contains large boulders and cobbles, called erratics. The most common deposit of glaciers is till, a very heterogeneous mixture of finely ground-up rock flour, clay, sand, and pebbles and boulders. Glaciers pick up avalanche material and rock material by engulfing particles pried loose by frost action and by plucking or quarrying large blocks from the bottom over which the glacier rides. Glacial erosion and transportation do not in general round sharp corners and edges of particles, and thus till particles tend to be sharp and angular. *See* GLACIOLOGY; TILL.

Transportation and deposition by water. Moving waters account for most of the transportation of rock materials on the Earth's surface. Streams and rivers carry tremendous tonnages of materials daily from the erosion areas to the sea. The currents in the sea shift material from place to place on the sea floor. Clastic sediments commonly retain characteristics produced by the transporting currents. Thus the sorting by size in response to the current velocity and density of the water and the rounding of particles by abrasion are both properties of clastic sediments produced during transportation.

Movement of water. Water may move either by laminar or turbulent flow. When the water moves in straight lines parallel to the confining channel surfaces, without mixing of adjacent layers of water, the flow is characterized as laminar. When the water moves in irregular lines in eddies and swirls, and different layers mix, the flow is turbulent. Water in streams is dominantly turbulent, and it is this kind of flow that causes erosion, abrasion, and transportation of all but the finest particles.

Methods of transportation. Transportation by running water is accomplished in three ways. Ions and compounds are carried in solution and end up in lakes or the sea, where they may accumulate or take part in reactions that precipitate solids. Insoluble particles move either by suspension in the water or by being moved along the bottom. These are called, respectively, suspended load and traction load. The suspended load includes colloidal sizes, which need very low velocities to keep them in suspension, and larger particles, which may be suspended only temporarily by the action of high-velocity turbulent currents. The traction or bottom load moves by rolling and sliding of grains. Saltation is an important process in stream transportation, just as in wind movement (Fig. 2).

Causes of deposition. As soon as water is unable to continue transport of material, deposition starts. Also, ions carried in solution may participate in chemical reactions that produce precipitates that settle to the bottom. *See* FLUVIAL EROSION LANDFORMS; STREAM TRANSPORT AND DEPOSITION.

Flocculation and deflocculation. Colloidally suspended materials, dominantly clay minerals, are of small size and therefore need only low velocities and slight turbulence to keep them in suspension. Only when water movement practically ceases can these materials settle by the action of gravity alone. But the colloids are susceptible to flocculation, a process of coagulation or aggregation of the individual particles into larger sized clumps, or floccules. The floccules, being larger

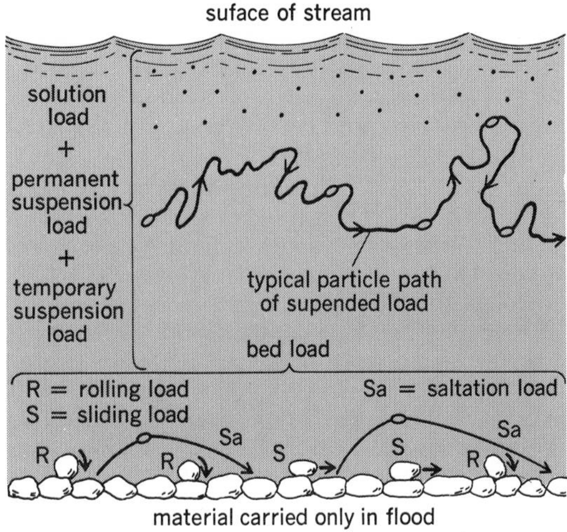

Fig. 2. Diagram of types of load carried by a stream. (*From R. M. Garrels, A Textbook of Geology, Harper & Row, Publishers, Inc., 1951*)

and heavier, will settle to the bottom by gravity even though some current continues. Flocculation may occur when a fresh-water suspension mixes with a salt solution, as when a river enters the sea. Deflocculation, or peptization, is the process of dispersing and breaking up the floccules into smaller colloidal-sized particles.

Settling velocities. When a current is no longer able to keep a particle in suspension, the particle settles to the bottom with a velocity proportional to its size, shape, density, and the viscosity of the suspended medium (in this case, water). Stokes' law states this dependence quantitatively for spherical particles at constant temperature as $v = kr^2$, where v is velocity, r is the radius of the sphere, and k is a constant. The constant $k = \frac{2}{9}(\rho_1 - \rho_2)g/\eta$, where ρ_1 and ρ_2 are the densities of the particle and fluid respectively, g is the acceleration due to gravity, and η is the viscosity of the fluid. This law holds for particles smaller than 0.08 mm in diameter, and appreciable deviations from the law are found with particles over 0.2 mm in diameter. For the larger particles, viscosity forces become unimportant and the impact law becomes applicable. The impact law can be stated as $v = k_2\sqrt{r}$, where the constant depends on the same factors as in Stokes' law. The operation of the laws of settling velocities accounts for the size sorting that is the result of transportation and deposition by running water. Sorting by shape is also important because spherical grains will offer less frictional resistance to settling and will fall to the bottom faster than irregular ones. The fluid through which the grains settle will have its effect on settling velocities. As the viscosity and density of the fluid medium increase, settling velocities decrease and the sorting action of the current decreases. Turbidity or density currents, in which suspended fine material significantly increases the density and viscosity, are poor in ability to sort material by size; they deposit a mixture of a wide range of sizes. Glaciers may be considered as a high-viscosity transporting medium that deposits extremely poorly sorted material. *See* TURBIDITY CURRENT.

Transportation and deposition in the sea. Transportation of suspended and dissolved material in the sea is accomplished by currents caused by wind, tides, and density differences. Terrigenous sediments, derived directly from matter eroded from land surfaces and transported to the sea by rivers, are first deposited in deltas or are moved along the shore by littoral currents, currents that operate near the shoreline at shallow depths. Other more slowly moving currents may redistribute material on the continental shelves, the shallow aprons surrounding the continents. Turbidity currents distribute some terrigenous sediment to abyssal depths. Pelagic sediments are derived from extremely fine windblown mineral particles and organisms that dwell in the sunlit zone of the seas down to about 600 ft; these sediments are spread in thin layers over all of the ocean bottoms. A most important factor in marine sedimentation is the role of organisms that secrete calcium carbonate, silica, or phosphatic shells (tests). These biogenic precipitates may settle to the bottom or may be caught up in currents and transported, eventually to be deposited in some other spot. *See* DELTA; MARINE SEDIMENTS.

Environments of deposition. The sedimentary environment is usually defined in terms of a complex of physical and chemical conditions associated with a particular geomorphologic unit. For example, a lacustrine environment includes all of the physical and chemical forces at work in a lake, from the beach to the deep water center. Some environments, such as the swamps, may be fairly homogeneous; others, such as the littoral (nearshore) marine, may be very variable and heterogeneous. Other concepts of environment are used, such as tectonic, chemical, or hydrodynamic; here each controlling factor is segregated and the total environment is a combination of all of the individual components. Whereas source materials primarily influence detrital mineralogy of a sediment, the environments of deposition primarily control the textures of clastic sediments and the composition of chemical sediments.

An extensive classification of environments by W. H. Twenhofel is as follows:

- Continental environments
  - Terrestrial
    - Desert
    - Glacial
  - Aqueous
    - Fluvial
      - Piedmont
      - Valley flat
    - Paludal
      - Lake swamps
      - River swamps
      - Flatland swamps
      - Paralic swamps
    - Lacustrine
      - Fresh
      - Salt
    - Spelean-cave
- Mixed continental and marine
  - Littoral
  - Delta
  - Marginal lagoon
  - Estuary
- Marine environments
  - Neritic
  - Bathyal
  - Abyssal

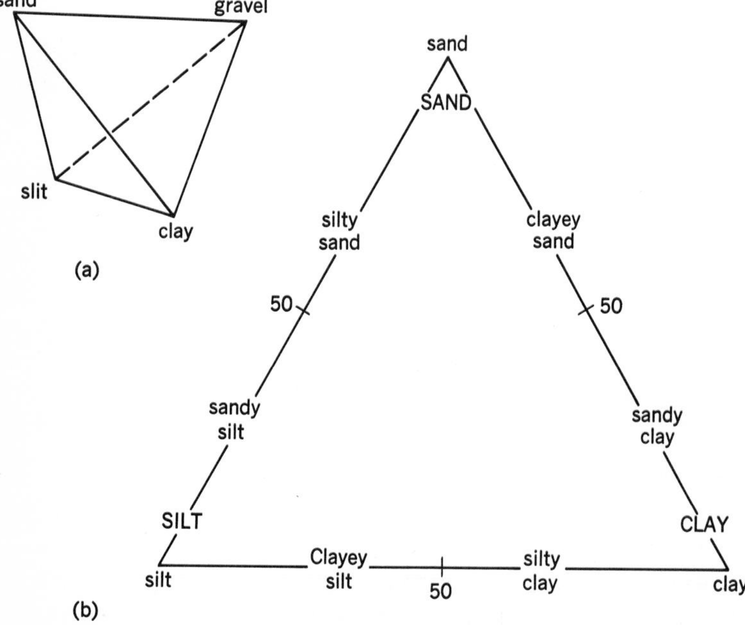

Fig. 3. The use of size to classify the physically deposited sediments. (*a*) A tetrahedron with gravel, sand, silt, and clay end members. The boundary of particle size between sand and gravel is 2 mm; between silt and sand, 1/16 mm; and between clay and silt, 1/256 mm. (*b*) The sand-silt-clay face is shown in detail. (*From W. C. Krumbein and L. L. Sloss, Stratigraphy and Sedimentation, 2d ed., copyright © 1963 by W. H. Freeman and Co.*)

Sediments of physical deposition. Clastic, or detrital, sediments are those that have been deposited by mechanical action, by currents of one kind or another. It is most convenient to subdivide them on the basis of particle size, for the size and sorting are dependent on a genetic factor, the current type and strength (Fig. 3). The current regime is reflected in the sediment not only by size and sorting but by its bedding characteristics and sedimentary structures, such as cross-bedding, ripple marks, and current lineation.

Coarse-grained clastics. Large particles, such as pebbles, cobbles, and boulders, can be carried only by high-velocity or high-density currents. The coarse clastics—gravels, and their indurated equivalents, conglomerates—are those whose particle size is greater than 2 mm in diameter. Rivers transport such coarse materials in times of flood; on land they may also be transported by landslides and mudflows. Submarine currents are in general unable to carry such materials, but they may be found in some sediments deposited by turbidity currents, and to a minor extent, by being rafted by ice or caught in the roots of marine plants, they may be carried far out to sea. Glaciomarine sediments commonly carry coarse detritus. *See* CONGLOMERATE; GRAVEL.

Medium-grained clastics. Sands and their consolidated equivalents, sandstones, are medium-grained clastics, and range in diameter from $\frac{1}{16}$ to 2 mm. Most streams and rivers, as well as nearshore littoral ocean currents, have velocities competent to carry this size particle. Windblown sands are common but tend to be fine-grained and very fine-grained sand, $\frac{1}{4}-\frac{1}{8}$ mm and $\frac{1}{8}-\frac{1}{16}$ mm in diameter, respectively. Sands accumulate as river bars, dunes, alluvial fans, beach deposits, barrier islands, and offshore marine bars, and are a major component of deltas. *See* NEARSHORE PROCESSES; SAND; SANDSTONE.

Fine-grained clastics. The fine-grained clastics are the silts and muds and their indurated equivalents, shales, siltstones, and mudstones. The particle size of silt is $\frac{1}{16}-\frac{1}{256}$ mm in diameter. Clay-size particles are finer than $\frac{1}{256}$ mm. The fine-grained clastics are by far the most abundant sediment type and are found either alone or mixed with other sediment types in almost every environment of deposition. Because of their fine size, down to colloidal dimensions, they may be carried in suspension by very slow-moving currents and so may be distributed far and wide over the ocean bottoms. The clay minerals, hydrous aluminosilicates, all tend to have colloidal-sized particles and make up, on the average, about one-third of the fine clastic sediments. *See* ARGILLACEOUS ROCKS; CLAY MINERALS; SHALE.

Sediments of chemical deposition. The chemical sediments are formed by reactions between the dissolved components in the water of the environment to form precipitates. The precipitation may be inorganic or it may take place by the action of organisms, such as mollusks, which secrete calcium carbonate shells (Fig. 4). By far the most important chemical sediments quantitatively are the carbonates, dominantly calcium carbonate.

Carbonate sediments. A great many invertebrates are able to extract calcium carbonate from sea water for their shells. After death the shells accumulate to form sediment. In some areas very fine-grained calcium carbonate muds accumulate on the bottom; the mud, when lithified, becomes a dense, finely crystalline limestone. There is some dispute over the origin of the mud, some workers claiming it to be an inorganic precipitate and others finding evidence for biogenic origin in the ground-up fragments of small shells and the presence of needles of aragonite, one of the crystalline forms of calcium carbonate, that are secreted by blue-green algae. A variety of calcium carbonate deposits is associated with coral or algal reefs. Globigerina and pteropod oozes are deep-sea carbonate deposits formed from the shells of the animals for whom the deposits are named. These pelagic organisms live in surface layers of the sea and, at death, sink to the bottom, where they accumulate. Chalk is a lithified deposit of this type. Carbonate sediments may contain an admixture of clay, in which case they are called marl. Dolomite is a carbonate sediment composed largely of $CaMg(Co_3)_2$, the mineral dolomite. *See* CHALK; DOLOMITE; LIMESTONE; MARL.

Siliceous sediments. The sediments composed largely or completely of chemically precipitated silica are formed mainly in environments in which the supply of clastic material is small or nonexis-

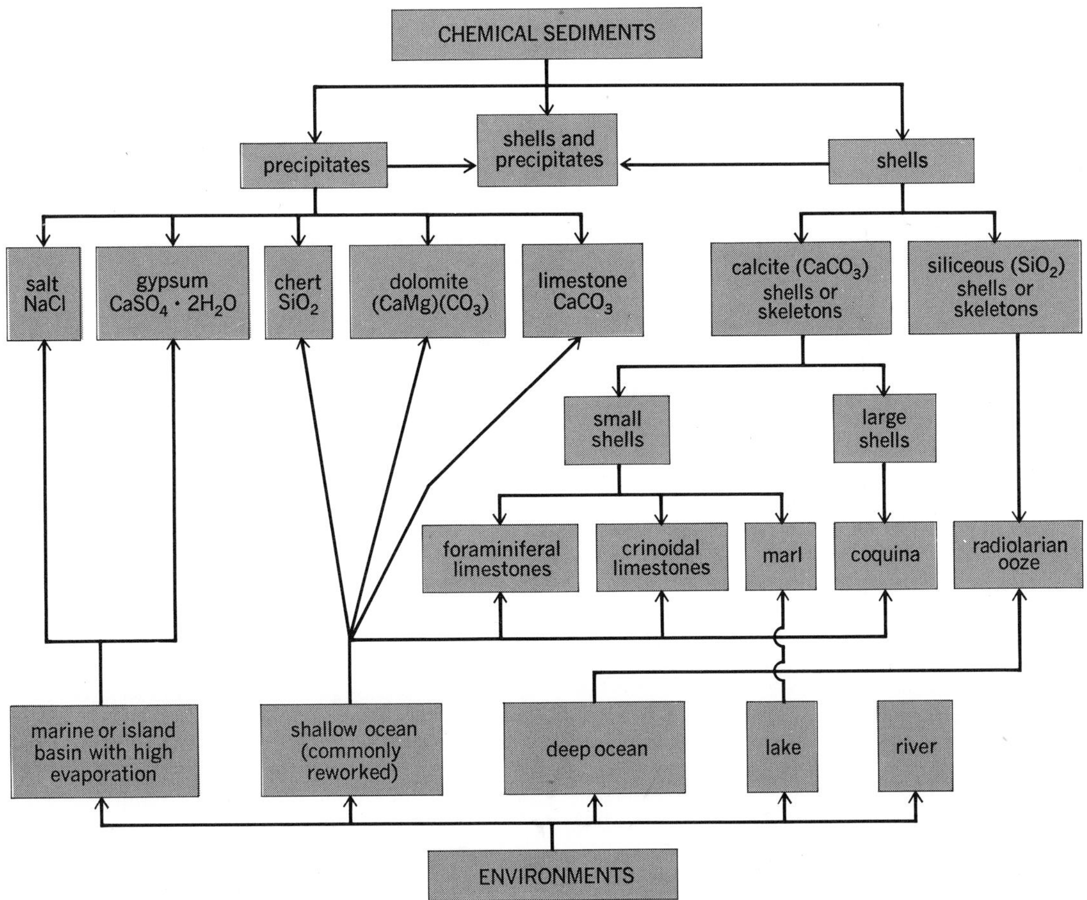

Fig. 4. Interrrelation of composition and environment indicated for some major types of chemical and organo-chemical sedimentary rocks. (*From R. M. Garrels, A Textbook of Geology, Harper & Row, Publishers, Inc., 1951*)

tent. Siliceous sediments being formed today are practically limited to accumulations of tests, or shells, of silica-secreting pelagic organisms, such as the diatoms and radiolaria. Their lithified equivalents are radiolarites, radiolarian earths, diatomites, and diatomaceous earths. Inorganic precipitation of silica around hot springs forms the siliceous sinters. Chert is a siliceous sediment that occurs chiefly as nodules in carbonate rocks, probably of postdepositional origin by replacement of calcium carbonate. Some chert is bedded and may represent primary precipitation from sea water, either inorganically or by a biochemical mechanism, such as the formation of diatom oozes. Sediments that contain large amounts of volcanic ash commonly are siliceous, the silica being contributed from the alteration of volcanic glass. *See* CHERT; DIATOMACEOUS EARTH; RADIOLARIAN EARTH; SILICEOUS SINTER; VOLCANIC GLASS.

Ferruginous sediments. The iron-rich sediments are often formed, as the siliceous sediments are, in the absence of large amounts of clastic material. Many ferruginous sediments are formed in reducing environments where ferrous compounds predominate. Pure iron mineral sediments are rare in the geologic record, most being limited to the Precambrian. Mixtures of ferruginous minerals and clastic components give ferruginous shales and sandstones. Ferruginous limestone may be formed by the replacement of calcium carbonate by siderite, the iron carbonate. Replacement of limestone by hematite formed the Clinton iron ores, which contain oolites and fossils of hematite. Bog iron ores are formed in fresh-water lakes; they are yellow and brown concretionary or irregular masses containing sand and clay.

Phosphatic sediments. Shales and limestones may contain a mixture of different calcium phosphate minerals, some of them precipitated from sea water and others formed in the sediment after deposition by replacement. Phosphate nodules are common in some limestones. Bedded phosphates occur interbedded with other marine sediments. Guano, a deposit of bird excreta, is found on some desert islands. Other phosphate accumulations are residues from the solution of a phosphatic limestone. *See* APATITE.

Manganese sediments. Manganese carbonate concretions and thin layers occur in some marine sediments, many times in association with ferruginous sediments. Manganese oxide is precipitated in the sea as finely divided pigment in some deep-sea sediments, coatings, and nodules and concretions.

Sedimentary sulfur. Native sulfur is found as small particles, lenses, and irregular bodies in some sediments, commonly the argillaceous ones. It is precipitated by the partial reduction of sulfate or partial oxidation of sulfides in solution. Sulfides are precipitated in reducing environments, such as at the bottom of the Black Sea. The sulfate-reducing bacteria play an important role in this precipitation.

Carbonaceous sediments. The carbonaceous sediments are formed by the accumulation of vegetable and animal organic debris, either pure or mixed with some clastic material. The most important deposits are peats and coals. The preservation of the organic material depends on rapid burial or lack of oxygen in the environment. The original remains are partially decomposed and altered in part by bacterial action. If mixed with clastic material, the deposit becomes a carbonaceous shale or sandstone.

Decomposition. Vegetation is transformed into peat and then coal by a combination of bacterial and chemical decay and alteration. The end result of this process, called incoalation, is the reduction of proportion of oxygen and hydrogen and the enrichment of carbon. *See* DOPPLERITE.

Environments of carbonaceous deposition. Peat deposits are formed in fresh-water swamps, bogs, or moors, where vegetation is lush and growth is rapid, and where there is little bacterial activity decomposing the organic matter. Other deposits rich in organic matter are formed at the bottoms of lakes, lagoons, and some inland seas, where the combination of organic productivity and reducing environments is such as to prevent oxidation and decomposition of the organic matter. Bottom deposits of the Black Sea, whose bottom waters are reducing, are notably rich in organic material.

Classification of carbonaceous sediments. The carbonaceous sediments may be divided into the humic (coaly) and sapropelic (bituminous), with all gradations between. The humic deposits are low in hydrogen and high in oxygen, whereas the sapropelic are high in hydrogen and low in oxygen. The best-known representative of the humic class is coal; true cannels exemplify the sapropelic class.

Peat deposits. Peat deposits are composed of two different types of altered plant materials, anthraxylous and attrital. The anthraxylous is composed largely of anthraxylon, the visible ligneous parts of plants, such as wood, leaves, stems, and bark, impregnated with colloidial humic decomposition products. The attrital is composed mainly of attritus, the finely divided decay-resistant plant materials, particularly the waxy, resinous, and fatty materials. Peats of one type are gradational with the other and both may grade into sapropelic deposits, such as cannel. *See* BLACK SHALE; COAL; SAPROPEL.

Diagenetic processes and lithification. After deposition the unconsolidated water-saturated sediment undergoes compaction, squeezing out of water, alteration of minerals and precipitation of new ones, and solution; the end result of the combination of these processes is a lithified sediment, a sedimentary rock. Collectively the processes are known as diagenesis. Extensive changes and modifications of mineralogy and texture of the original sediment may take place. *See* AUTHIGENIC MINERALS; DIAGENESIS.

Provenance and dispersal. The provenance of sediments includes the evaluation of the composition and location of the source rocks whose erosion produces the raw material of sedimentation. Dispersal is the process by which the sedimentary materials are carried by various transporting agents from source area to depositional area and distributed in the environments of sedimentation. The kind of source rock is of primary importance in determining the nature of the clastic material, but major modifications result from the effects of climate and topography. Only if mechanical erosion dominates will the source composition be unaltered. If, owing to climate or topography, chemical erosion is significant, the source materials may be drastically altered before they become transported to the sedimentation area. Changes in texture (size distribution, roundness, sphericity) and mineral composition may result from dispersal conditions.

Nature of the source area. If chemical weathering of source rocks is at a minimum, then reconstruction of source rock types from the sediment is relatively simple, subject mainly to modifications by dispersal. The major problems arise in reconstruction of source areas in which there has been chemical weathering. One way of assessing the amount of chemical weathering is by chemical composition. The residues of weathering tend to be enriched in silica, alumina, and iron and impoverished in alkalies and alkaline earths. Another way of assessment is by consideration of mineral stabilities under weathering conditions. The most stable minerals, quartz, muscovite, and the clay minerals, dominate the residue, whereas the most unstable ferromagnesian minerals such as olivine, pyroxenes, and amphiboles are lost. The degree of weathering may be determined by a mineral stability series, in which the most unstable species are lost first and the most stable are preserved in the residue. Olivine is one of the most unstable, followed by augite and hornblende. Kyanite, staurolite, and garnet are intermediate in stability. The most stable, besides quartz, are zircon, tourmaline, and rutile. Caution is required in using mineral stabilities for source rock evaluation, for there is evidence that over long periods of time some of the unstable minerals may disappear by solution during diagenesis, a process called intrastratal solution.

Mineralogical maturity. The maturity of a sediment is the degree to which the sedimentary material has had its more unstable minerals removed before deposition. A sediment with many unstable minerals is an immature one; one with only stable minerals is mature. Thus maturity is a measure of weathering in the source area, which is in turn a function of climate and topography.

Determining transport distance and direction. The size distribution of the particles in a clastic sediment is strongly affected by transportation agents. The maximum size and median size tend to decrease with travel downstream; thus this property can be used in a general way to deduce distance of transport. Size reduction may be the result not only of abrasion and splitting of grains but also of selective sorting, so that only finer material is carried on while coarser material is dropped. Rate of size reduction is affected by mineral composition, for the softer minerals will wear down more quickly than the hard ones. Roundness and sphericity increase in downstream direction. Since the late 1930s roundness has been successfully used as an indication of abrasion during transportation (Fig. 5*a*). Relation of compositional change to transport distance can also be used (Fig. 5*b*). Feldspar, with a hardness of 6 and good cleavage, apparently wears out somewhat faster than quartz, with a hardness of 7 and no cleavage. Thus changes in quartz-feldspar ratio may indicate changes in transportation. The effect, however, seems to be small and has not as yet been fully evaluated.

Dilution by other sources. Material coming from one source area may become diluted by material from other sources as it travels downstream to the sea. Areal mapping of mineralogical and textural composition is necessary to sort out the contributions of different source areas.

Sedimentary structures and source area. A development of the 1950s has been the use of cross-

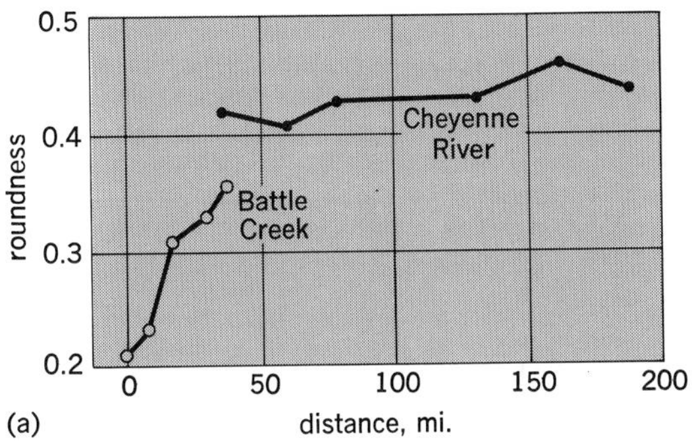

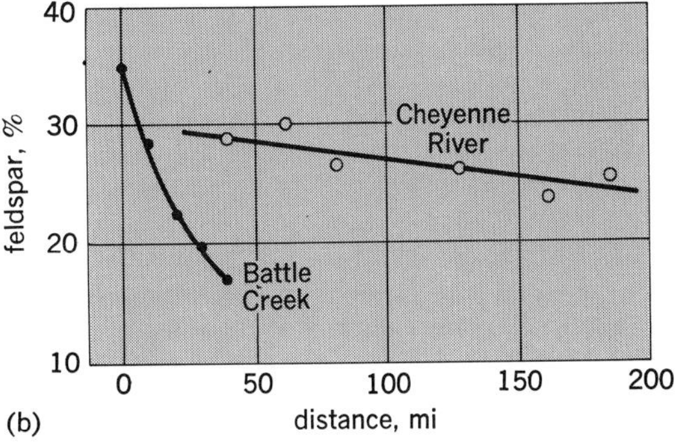

Fig. 5. Relation of textural and compositional change to distance of transport in South Dakota streams. (*a*) Increase in roundness of quartz sand in the 1- to 1.414-mm class; perfect roundness = 1.0. (*b*) Percentage of feldspar in the 1- to 1.414-mm class. (*From W. J. Plumley, Black Hills terrace gravels: A study of sediment transport, J. Geol., 56(6):526–577, 1948*)

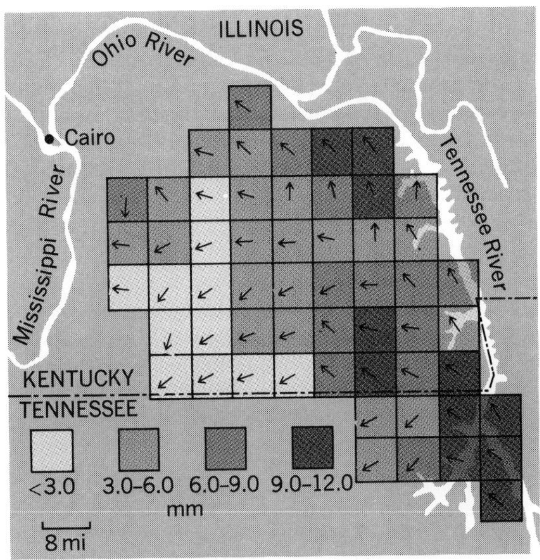

Fig. 6. Average of cross-bedding direction per 64-mi² and median size in Lafayette gravel. (*After P. E. Potter, Petrology and origin of the Lafayette gravel: 1, Mineralogy and petrology, J. Geol., 63(1):1–38, 1955*)

bedding and other directional sedimentary structures to determine the directions of source areas (Fig. 6). Such structures are mapped to get the direction of transport at each place and thus to ascertain regional trends. The direction of the source is the upstream direction of transport. Studies of marine sedimentary structures have been used to determine directions of marine transport.

Sedimentary petrologic provinces. A sedimentary petrologic province is an area in which the sediments have a more or less uniform composition and come from the same source (Fig. 7). If two or more provinces overlap in time, the rock succession represents provincial alteration, or provincial succession. The latter may represent continuing changes in source area contributions.

Sedimentary facies. The areal variation of lithology in a single bed or stratigraphic unit leads to the concept of facies. Facies involve particular lithologic characteristics of a bed at one place that grade into different characteristics in the same bed at other places. Thus a bed may vary in grain size from a shale facies to a sand facies, even though all of the material may come from the same source. Facies maps show the areal distribution of clastics versus chemical sediments, sand-shale ratios, or other lithologic variations in sediments of the same age. *See* FACIES.

Diastrophic movement and sedimentation. Diastrophism, earth deformation, has direct and indirect effects on sedimentation. The tectonic formation of sedimentary basins provides a place for the deposition of material. The depth of water influences sedimentation. The depth is related to downwarping of the basin and sea level, which are influenced by tectonics. The relief and climate of the source area are fundamentally controlled by earth movements. High mountains are the scene of vigorous mechanical erosion. Long-continued earth stability leads to lowlands with chemical erosion predominating. Besides these general considerations, specific sedimentary associations are connected with mountain-building episodes.

Flysch and molasse. The terms flysch and molasse originated as stratigraphic names in the northern Alps, where the orogenic sediments (those involved in overthrusting and recumbent folds) were the flysch, and the postorogenic sediments (later than folding and thrusting) were the molasse. Since their original use, an extension of their lithologies and orogenic and postorogenic significance has resulted in the use of flysch as synonymous with graywacke, and molasse with subgraywacke. Typically the flysch, which is interbedded shale and graywacke, increases in coarseness upward (a reflection of beginning orogeny). The molasse, mainly shale and subgraywacke sandstone, decreases in size upward (a reflection of dying-out orogeny). The molasse is a product of paralic sedimentation.

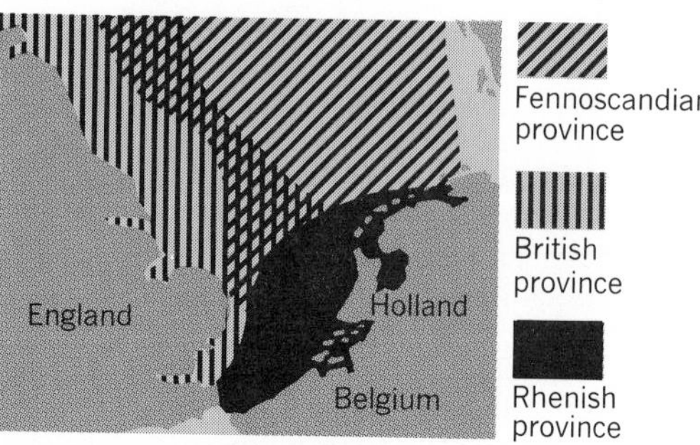

Fig. 7. Sedimentary petrologic provinces of southern North Sea. (*After J. A. Baak, in P. D. Trask, ed., Recent Marine Sediments, reprint, 1955*)

Paralic sedimentation. This sedimentation takes place in areas within or peripheral to continents, where a great amount of terrigenous alluvial deposits is laid down. Deposits may be partially marine or brackish water but are dominantly terrestrial.

Geosynclinal cycle. Many geologists have recognized that geosynclines not only may have distinctive suites of rock types, but that the total prism of sediments that was deposited in the geosyncline represents an evolution of rock types more or less characteristic of all geosynclines, and that such a succession of rock types is linked to the tectonic development of the geosynclines. F. J. Pettijohn characterizes the major cycle as beginning with orthoquartzitic and carbonate sediments deposited on the flooded craton (the stable core of the continent) and the cratonic border of the geosyncline. Following this comes mild uplift of part of the geosyncline and deposition of cherts, black shale, and phosphorite facies. This is succeeded by strong upwarp in adjacent areas and the deposition of flysch, increasing in coarseness upward and including submarine extrusives and tuffs. The next stage is that of postorogenic molasse sedimentation and paralic sedimentation, dominated by nonmarine deposits. Finally the geosyncline is deformed and uplifted into mountain belts. There are many variations in geosynclinal development and therefore in the geosynclinal cycle; thus, no two are alike. The general concept of the cycle, however, seems to have some validity. *See* GEOSYNCLINE; OROGENY; PALEOGEOGRAPHY; PALEOGEOLOGY; SEDIMENTARY ROCKS; TECTONIC PATTERNS.

[RAYMOND SIEVER]

Bibliography: E. T. Degens, *Geochemistry of Sediments*, 1965; C. O. Dunbar and J. Rodgers, *Principles of Stratigraphy*, 1957; W. C. Krumbein and L. L. Sloss, *Stratigraphy and Sedimentation*, 2d ed., 1963; F. J. Pettijohn, *Sedimentary Rocks*, 2d ed., 1957; P. E. Potter and F. J. Pettijohn, *Paleocurrents and Basin Analysis*, 1963; F. P. Shepard, *Submarine Geology*, 2d ed., 1963; W. H. Twenhofel, *Principles of Sedimentation*, 2d ed., 1950.

Sediments, organic geochemistry of

A study of the factors that control the molecular nature and distribution of organic compounds in sedimentary materials throughout geologic time represents the prime subject matter of organic geochemistry. A wide range of problems have been studied by organic geochemists. For example, investigations on calcification phenomena in biological systems have shown that the shell organic matrix is responsible for the deposition of the mineral phase, and furthermore that it provides a clue to environment and molluscan phylogeny. Much activity has also been displayed in the fields of petroleum generation and accumulation, coal formation, microbial activities in sediments and soils, diagenetic stabilities of organic molecules, nature and origin of extraterrestrial organic matter, condensation phenomena, and interactions of organic and inorganic materials. This article discusses the factors governing the fate of organic matter once it has been incorporated into the surrounding sediments.

Geochemical balance of organic matter. Sediments have been the principal depositories of posthumous organic debris throughout the Earth's history. The quantity carried in natural waters in the molecular-dispersed or particulate state is comparatively small. Trace amounts of organic chemicals may also be found in crystalline rocks; even meteorites contain a wide spectrum of "biochemical" constituents.

By grouping sediments into three principal classes, one can say that the average organic matter content is 2.1% in shales, 0.29% in carbonates, and 0.05% in sandstones. It has been calculated that the organic matter entrapped in sediments totals above 3.8×10^{15} metric tons; substantially all, about 3.6×10^{15} metric tons, is present in the shales. This means that most of the organic matter that has come to rest occurs in a finely disseminated state and is associated with fine-grained sediments. For comparison, the coal deposits of the world have been estimated at about 6×10^{12} metric tons. This is 1/500 of the disseminated organic matter. Estimates of the ultimate primary petroleum reserves run to 0.2×10^{12} metric tons. This is only 1/16,000 of the total organic matter entrapped in sediments.

To demonstrate more vividly the ratio of inorganic to organic matter present in the stratigraphic column, the following generalized picture is offered. From an evaluation of the total thickness of sedimentary materials that have been formed and deposited over the last $3-4 \times 10^9$ years, it can be assumed that a sediment layer of approximately 1000 m around the Earth's surface has been laid down. Roughly 2% of this layer, namely, ~20 m, is organic, and the rest is inorganic. Of this 20-m organic part, coal makes up ~5 cm and crude oil only a little more than 1 mm. The bulk of it represents the finely disseminated organic matter in shales, limestones, and sandstones.

The total amount of organic matter that can be expected in various rocks and natural waters is thus reasonably well known. More interesting from a geochemical point of view, however, would be a knowledge of the precise chemical nature and distribution pattern of individual organic species, and a general knowledge of processes and mechanisms causing the preservation, alteration, or destruction of organic matter in geological history.

Principally, the organic materials embedded in sediments can be grouped into compounds that are survivors of diagenesis and products of diagenesis. Diagenesis includes all postdepositional transformation processes, both physical and chemical, that alter sedimentary matter under low-temperature and low-pressure conditions. The first group includes all those organic molecules which are chemically similar or identical to living matter. In contrast, compounds that arose during diagenesis from the organic debris but are generally not part of plant and animal matter belong to the second group.

Survivors of diagenesis. The organic pigments from plants and animals show a surprising stability in a wide range of diagenetic habitats. There is ample evidence for the long-term preservation and antiquity of fossil porphyrins as old as the Precambrian. Similarly, intact peptides have been isolated from molluscan shells of the late Paleozoic. Amino acids and other biologically interesting monomers, such as sugars or the bases of the purines and pyrimidines, are common biochemical fossils in ancient sediments. This attests to the perfect preservation of many an organic compound throughout Earth history.

Products of diagenesis. Although most carbon in the Earth's crust has cycled through organisms, the bulk of the organic matter has lost its biological identity in the course of diagenesis. Polymers gradually break up into their monomeric building blocks, and these in turn can be modified severely by elimination of functional groups: for example, $—COOH$, $—OCH_3$, $—OH$, $—C{=}O$, $—NH_2$, hydrogenations and isomerizations, cleavage reactions, and, in general, processes which destroy the ordered building pattern of biochemical compounds.

These breakdown products can reorganize and polymerize into geochemically stable configurations; typical examples are the humic acids, coals, and kerogens. Humic acids are hereby defined as the materials that can be extracted from a sediment with 0.3 *N* NaOH and that subsequently can be precipitated upon acidification. Kerogen, on the other hand, is the so-called insoluble organic matter that is left in a sediment after organic solvent extraction and acid and base hydrolysis. Thus, both terms refer to a heterogeneous organic complex rather than to a well-defined organic constituent. Other constituents are stable on their own; hydrocarbons belong to this group of compounds. Altogether, more than 800 different organic consitituents have been isolated so far from sedimentary rocks. *See* DIAGENESIS.

Geochemical cycle. In the early states of diagenesis, the bulk of the biochemical macromolecules—for instance, proteins, polysaccharides, fats, or the nucleic acids—will be rapidly eliminated. This is largely a result of microbial activity. Breakdown products, such as amino acids, sugars, bases of the purines and pyrimidines, fatty acids, and phenols, can interact and give rise to constituents commonly termed humic acids, fulvic acids, humins, and ulmins. Inasmuch as most of these complex humic materials are no longer of nutritional value, they may accumulate as organic residues even in the zone of microbial activity. Interactions with the surrounding mineral matter may stabilize the organic complex or may catalyze certain reactions.

Although plants represent the main precursor for the organic matter in sediments, microorganisms have extensively modified the organic debris in the early stages of diagenesis. It is thus a matter of opinion whether plants or microbial metabolic and decay products should be considered the chief source of the organic matter in sediments. After termination of bacterial action, largely as a result of the depletion in available food materials, the dia-

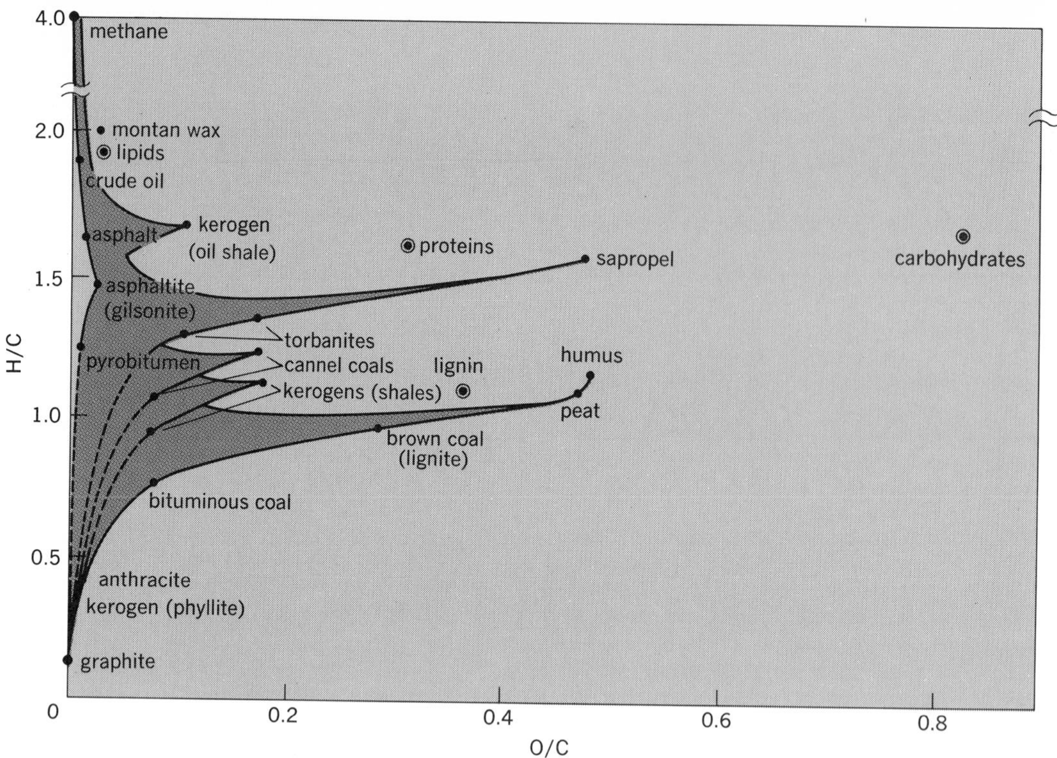

Metamorphism of organic sediments.

genetic history of the organic matter is that of slow inorganic maturation and redistribution (migration of hydrocarbons). Loss in functional groups may result in the formation of hydrocarbons, phenols, amines, and so on, and may cause a reduction in aliphatic side chains in former humic acids. The resulting coal or kerogen-type material will become more and more aromatic in nature. The number-one factor causing the diagenetic alteration is thermal degradation.

It is inferred that practically all former biochemical matter, independent of chemical nature, origin, or environment of deposition, is diagenetically reduced to either coal, kerogen, or the various crude oil components. If given sufficient time, the organic residues will gradually acquire the structural characteristics of graphite, whereas the petroleum will become more and more paraffinic. In its final stage all organic matter will largely end up as graphite, methane, carbon dioxide, ammonia, and water. These are essentially the same products from which the organic precursor materials for life were synthesized in the primitive Earth atmosphere under favorable environmental circumstances in terms of energy supply and reducing conditions. An illustration of the main path of diagenesis of the principal organic constituents is presented in the figure using O/C (oxygen-carbon) and H/C (hydrogen-carbon) ratios as a convenient plot scheme.

Some organic building blocks of living organisms can be recovered from ancient rocks as old as 3×10^9 years. This suggests that diagenesis of organic matter has not reached its final stage or, in other words, that equilibrium has not been attained. *See* BIOSPHERE, GEOCHEMISTRY OF; PETROLEUM, ORIGIN OF; SEDIMENTATION.

[EGON T. DEGENS]

Bibliography: I. A. Breger (ed.), *Organic Geochemistry*, 1963; U. Colombo and G. D. Hobson (eds.), *Advances in Organic Geochemistry*, 1964; E. T. Degens, *Geochemistry of Sediments*, 1965; G. Eglinton and M. T. J. Murphy (eds.), *Organic Geochemistry: Methods and Results*, 1969; J. M. Hunt, Organic sediments, in R. W. Fairbridge (ed.), *Encyclopedia of Earth Sciences*, 1966; G. Larsen and G. V. Chilingar (eds.), *Diagenesis in Sediments*, 1966.

Seismograph

A device for detecting, amplifying, and recording motion of the ground. Ground motion varies greatly, depending on the manner of excitation. Seismic waves from explosions and earthquakes occur in the frequency range from about 100 to 1/3000 hertz

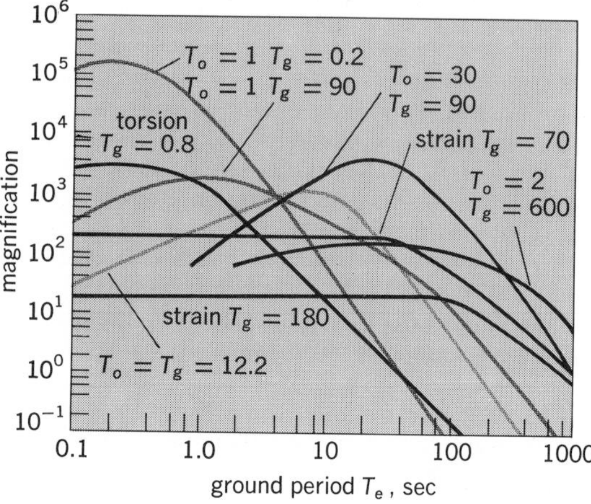

Fig. 1. Response curves for some seismograph systems. T_o is pendulum period; T_g, galvanometer period.

(Hz = cycle per second). Three components of ground motion are involved in the propagation of seismic waves. Defined in terms of direction of wave advance, they include push-pull (forward and backward), left-right (at right angles), and up-down (vertical) motions. Ground motion can vary by a factor of about 10^9 from the smallest to the largest recorded earthquakes. For these reasons many different types of seismographs have been designed. Two major categories exist: pendulum seismographs and strain seismographs.

Pendulum seismographs. This type measures the relative motion between the ground and a loosely coupled inertial mass. In some instruments optical magnification is used, whereas other devices exploit the advantages of electromagnetic transducers, photocells, galvanometers, and electronic amplifiers to achieve higher magnification.

Most widely used in seismograph stations is the electromagnetic pendulum seismograph with galvanometric registration. A coil is attached to the pendulum, and a sensitive galvanometer is connected to the coil. Motion of the coil in a magnetic field induces an electromotive force which activates the galvanometer movement. A mirror on the galvanometer deflects a light beam so as to produce a record on photographic paper. An alternate form uses a variable reluctance transducer in which movement of the pendulum varies the reluctance of a magnetic circuit. The resulting magnetic flux variations induce an electromotive force in a coil surrounding an armature in the magnetic circuit.

The equations for a pendulum-galvanometer seismograph system are given in Eqs. (1) and (2),

$$\left\langle \frac{d^2x}{dt^2} + 2\epsilon_o \frac{dx}{dt} + \omega_o^2 x - \sigma_o \frac{d\theta}{dt} = -\frac{d^2z}{dt^2} \right\rangle \quad (1)$$

$$\left\langle \frac{d^2\theta}{dt^2} + 2\epsilon_g \frac{d\theta}{dt} + \omega_g^2 \theta - \sigma_g \frac{dx}{dt} = 0 \right\rangle \quad (2)$$

after S. K. Chakrabarty. Here, x is the pendulum displacement, z is the ground displacement, and θ is the galvanometer deflection; ϵ and ω are, respectively, damping constant and natural (circular) frequency. The subscript o refers to pendulum and g to galvanometer; σ_o is the coupling factor of galvanometer to pendulum, σ_g that of pendulum to galvanometer, and t is time. For any given ground displacement Z, θ (hence the trace motion on the seismogram) can be computed. Also, Z can be found from θ and its derivatives and integrals. By selecting appropriate values for the constants ϵ, ω, and σ, a seismograph system can be designed for specified sensitivity, frequency response, and phase distortion. For example, the frequency response R of a seismograph system is found by solving for θ in Eqs. (1) and (2) after setting $Z = C \sin \omega_e t$, where C is the maximum amplitude and ω_e is the circular frequency of an assumed steady sinusoidal ground displacement. The frequency response is given by Eq. (3).

$$R = \{[(\omega_o^2 - \omega_e^2)(\omega_g^2 - \omega_e^2) - 4\omega_e^2\epsilon_o\epsilon_g]^2 + [2\omega_e\epsilon_o(\omega_g^2 - \omega_e^2) + 2\omega_e\epsilon_g(\omega_o^2 - \omega_e^2)]^2\}^{-1/2} \quad (3)$$

Need for diverse types. In order to achieve adequate dynamic range and the ability to record over a wide frequency range, several different types of seismograph systems are used simultaneously. Strong-motion instruments of low magnification record ground accelerations for severe shaking which would disable more sensitive apparatus. The background noise (microseisms) in the Earth varies with frequency. Thus, for frequencies near 1/6 Hz, the maximum usable magnification of a seismograph located almost anywhere in the world is about 2000. For frequencies near 5 Hz, magnifications of 1×10^6 are usable at certain quiet locations. For this reason broadband instruments with uniform response over the entire seismic frequency range are not used. Rather, seismographs covering limited parts of the spectrum are designed. This not only improves the signal-to-noise ratio but also makes possible the discrimination between seismic waves of different frequencies which arrive at the same time but are controlled by different mechanisms of propagation. Response curves for several seismograph systems in current use are shown in Fig. 1. The Benioff vertical seismometer (pendulum period, 1 sec) is shown in Fig. 2, and the Press-Ewing vertical seismometer (pendulum period, 30 sec) is shown in Fig. 3.

Strain seismographs. The principle of the Benioff linear strain seismograph is shown in Fig. 4. Two piers, P_1 and P_2, are tied to bedrock and separated by distances of about 100 ft. L is a rigid fused-quartz tube attached to P_1 and suspended so as to have a single degree of freedom in the longitudinal direction. Strains in the ground produce proportional variations in the distance between P_1 and P_2 which may be detected by a sensitive transducer in the gap between the free end of L and P_2. The strain seismograph detects secular strains related to tectonic processes and tidal yielding of the solid Earth. Strains associated with propagating seismic waves are also recorded. The response Y of a strain seismograph to longitudinal motions is given by Eq. (4). Here V represents the response

$$\left\langle Y = -V \frac{L}{c} \cos^2\alpha \frac{d\xi}{dt} \right\rangle \quad (4)$$

of the transducer, L is the length of the rod, c is the apparent surface velocity of the seismic waves, α is the angle between the direction of the rod and the propagation direction, and ξ is the ground particle displacement. Under favorable conditions of stable sites, strains as small as 10^{-9} to 10^{-10} are recorded. A typical installation is shown in Fig. 5.

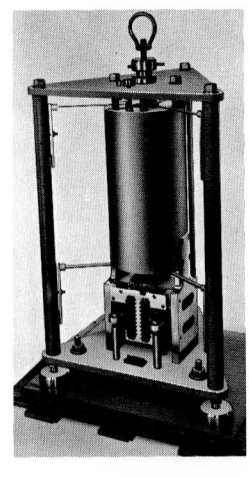

SEISMOGRAPH

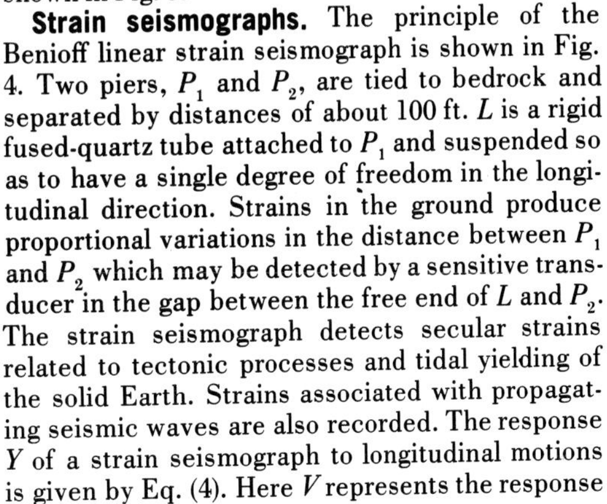

Fig. 2. The Benioff vertical seismometer. Pendulum period is 1 sec.

Fig. 3. The Press-Ewing vertical seismometer. The pendulum period is 30 sec.

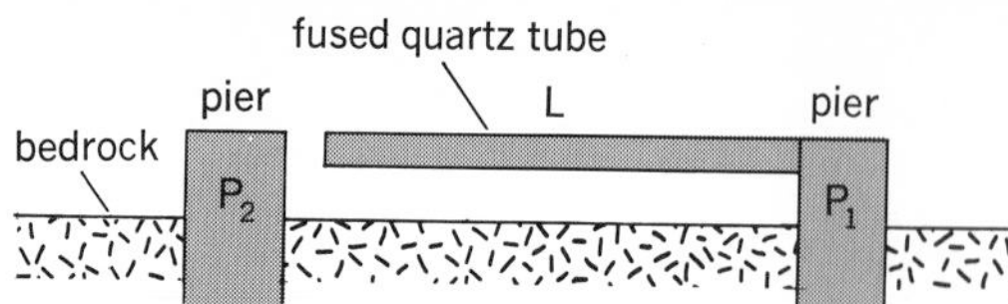

Fig. 4. Principle of Benioff linear strain seismograph.

SEISMOGRAPH

Fig. 5. Benioff linear strain seismograph.

Improved accuracy and longer base lines have proven feasible in more recent installations by using optical interferometry with laser illumination. For these installations (see Fig. 4), the tube L is replaced by an optical path, and P_1 and P_2 support the laser interferometer elements.

Seismic prospecting apparatus. In seismic exploration for petroleum, artificial mechanical excitation such as explosions or dropped weights provide the source of seismic waves. The frequency range is 2–20 Hz for refraction shooting and 20–300 Hz for reflection shooting, depending on local conditions. The principal field problem is to obtain maximum signal-to-noise ratio and to survey large areas rapidly. This requires portable or truck-mounted apparatus and the use of arrays of detectors, filtering techniques, and specially designed recording methods.

The detectors (geophones), often weighing less than 1 lb, contain a mass supported by a spring and constrained to move in the vertical direction. An electromagnetic transducer transforms the mechanical motion into voltages. Special amplifiers are designed for small size, large gain, and freedom from noise; these incorporate band-pass filters with adjustable low- and high-frequency cutoff. Automatic gain control provides usable signals from the arrival of the initial large-amplitude waves until signal strength drops below the level of background noise.

Photogalvanometric recording was almost universally used during the first two decades of seismic exploration. By 1955 this technique was being displaced by magnetic tape recording which, in turn, is now frequently replaced by direct digital recording. The magnetic-tape and digital-recording systems have the basic advantage of recording the information in a format suitable for subsequent application of sophisticated signal-processing techniques at a central data-analysis facility. In prospecting, seismometers are rarely used alone; various combinations of fixed and phasable arrays are employed in order to enhance signal recovery from the raw data. *See* GEOPHYSICAL EXPLORATION.

Developments. A number of new instruments for earthquake and explosion seismology are in various stages of development. These include rotation seismographs, tiltmeters, high-gain (magnification of 10^5 at carefully selected locations) wide-frequency-band long-period seismographs, lunar and planetary seismographs, ocean-bottom seismographs, direct digital-recording seismographs, and the utilization of arrays of seismographs in a manner analogous to usage in exploration seismology. *See* EARTHQUAKE; SEISMOLOGY.

[KER C. THOMSON]

Bibliography: M. Båth, *Introduction to Seismology*, 1973.

Seismology

The science of strain-wave propagation in the planets and their naturally occurring satellites. Strain waves propagate in the interior from a source and are recorded by seismographs distributed over the surface. From wave analysis it is possible to infer the subsurface structure and sometimes the mechanism of the source.

Instrumentation. As a science, seismology began about 1880 with the development of the seismograph for quantitatively measuring earthquake phenomena. Several developments in the deployment of seismographs have made possible very rapid advances in seismology. One is the World-Wide Standard Seismograph Network (WWSSN); the signals from an earthquake or nuclear explosion can be recorded all over the world at 125 stations on instruments of exactly the same characteristics so that accurate comparison of changes in seismic waveform with distance can be carried out (Fig. 1). Another is the development of large-aperture seismic arrays such as LASA in Montana, NORSAR in Norway, and ALPA in Alaska. Still another is the development and deployment of very-long-period, very-high-gain seismographs at a few specially selected low-noise sites throughout the world. *See* SEISMOGRAPH.

Earthquake. An earthquake is the abrupt release of strain waves through the action of natural forces within the Earth. The depth of occurrence of the earthquake focus (that is, the region of the Earth at which the earthquake originates) provides a basis for classification. Shallow-focus earthquakes occur at depths down to 60 km, intermediate focus at 60–300 km, and deep focus at 300–700 km. Most earthquakes have shallow foci.

A map of the world's shallow-focus earthquakes is shown in Fig. 2. Note that most occur in a narrow rim surrounding the Pacific Ocean. A secondary wedge of somewhat lesser activity has the broad base of the wedge in China while the tip lies off the Portuguese coast; the body of the wedge passes through southern Asia, Asia Minor, and the Mediterranean region. The mid-Atlantic ridge is a source of earthquakes of tertiary significance.

The circum-Pacific belt is found to have a well-defined structure in which patterns are observed where a volcanic chain, an island arc, a deep-sea trench, a seismic belt, and a gravity anomaly are found in association, as summarized in Fig. 3. Focal mechanism studies of earthquakes from this region indicate that the entire rim of the Pacific is under compression, and geodetic measurements suggest that the Pacific basin is slowly rotating counterclockwise with respect to the continents.

The new global tectonics provides an explanation for the worldwide distribution of earthquakes. The Earth's outer shell is regarded as being divided into a series of rigid plates that are undergoing slow motion relative to each other. The major active processes of geology such as volcanism and earthquakes are concentrated at or near these plate boundaries. Stresses build up in the vicinity of plate boundaries where the plates are opposing each other's motion. When the stresses exceed the lithospheric breaking strength, an earthquake occurs.

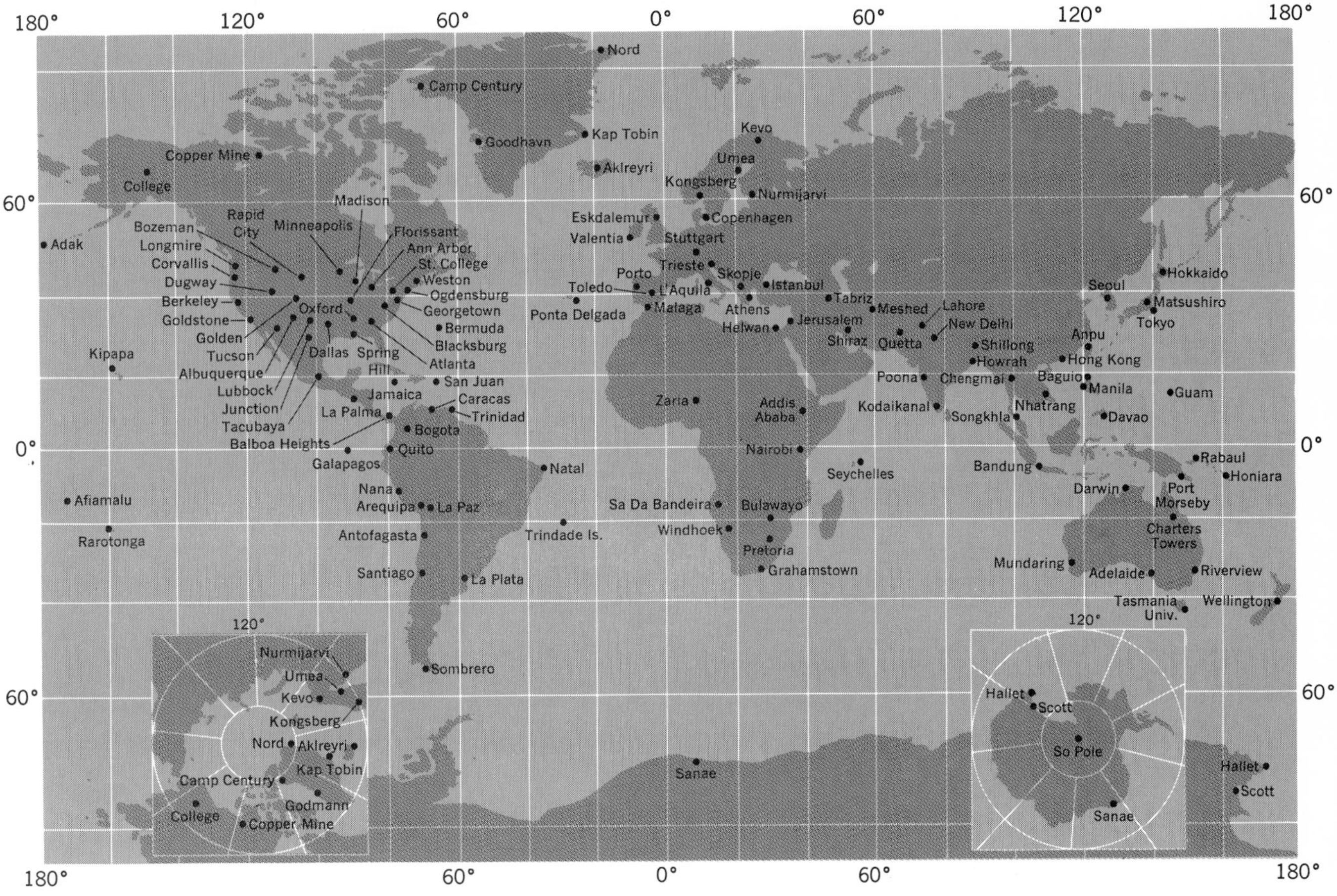

Fig. 1. World-Wide Standard Seismograph Network.

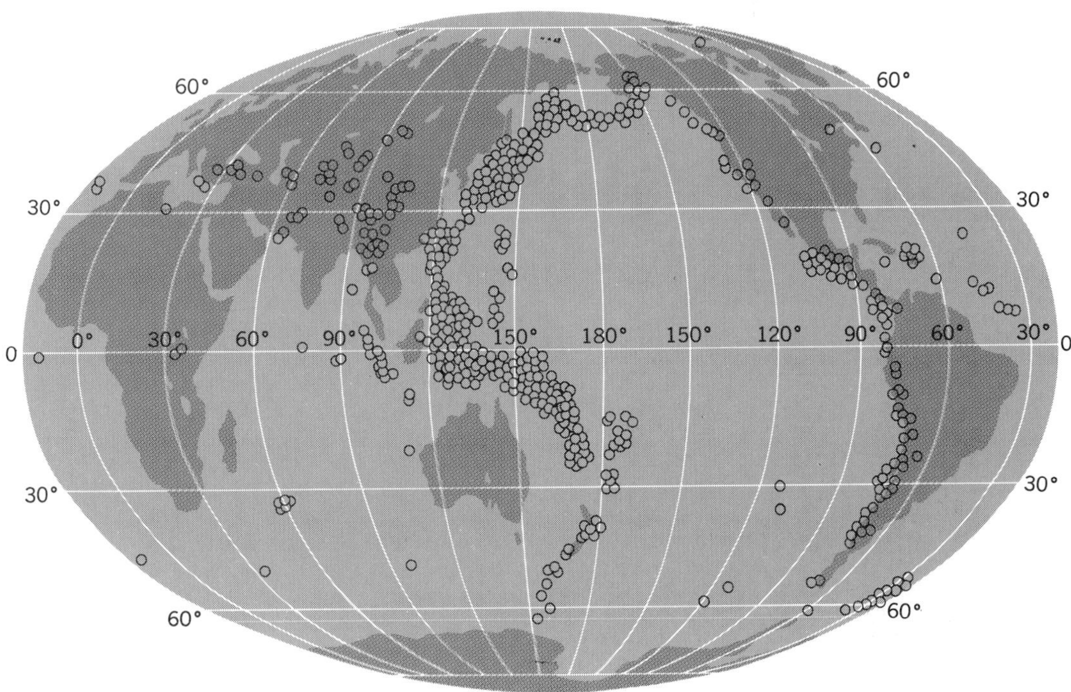

Fig. 2. Earthquake belts as shown by epicenters of shallow-focus earthquakes in 1918–1952. Most shallow-focus earthquakes occur in a narrow rim surrounding the Pacific Ocean. (*After B. Gutenberg and C. F. Richter*)

The "size" of an earthquake can be viewed in two ways, either subjectively by considering the intensity of shaking at a given location or objectively by considering the magnitude of the earthquake source. A semi-quantitative approach to local seismic intensity is provided by the MSK scale of 1964, which characterized observed shaking phenomena in a scale from I (not felt) to XII (major landscape changes). The magnitude scale is instrumental and characterizes the earthquake magnitude quantitatively, independently of individual sensations, and in a manner roughly related to the total energy output from the earthquake focus. If body waves are used to determine magnitude, the symbol m (or m_b) is used. The largest earthquake recorded had an m value of 8.9. Very small earthquakes with negative magnitudes have been studied. *See* EARTHQUAKE; FAULT AND FAULT STRUCTURES; OROGENY; PLATE TECTONICS; ROCK MECHANICS.

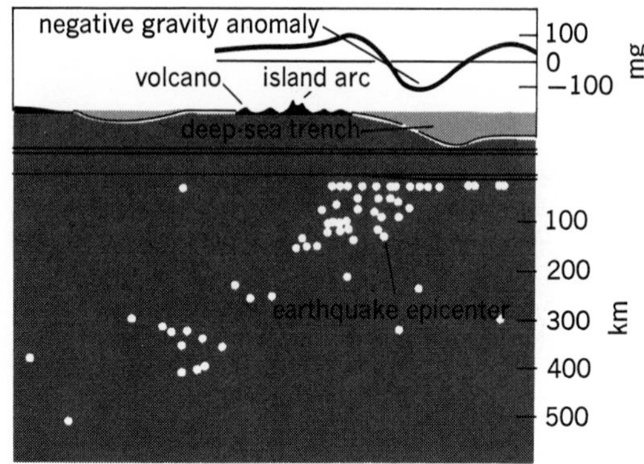

Fig. 3. Arcuate tectonic-seismic structures showing pattern of deep-sea trench, seismic belt, gravity anomaly, and volcanism. (*From B. Gutenberg and C. F. Richter, Seismicity of the Earth and Associated Phenomena, 2d ed., Princeton University Press, 1954*)

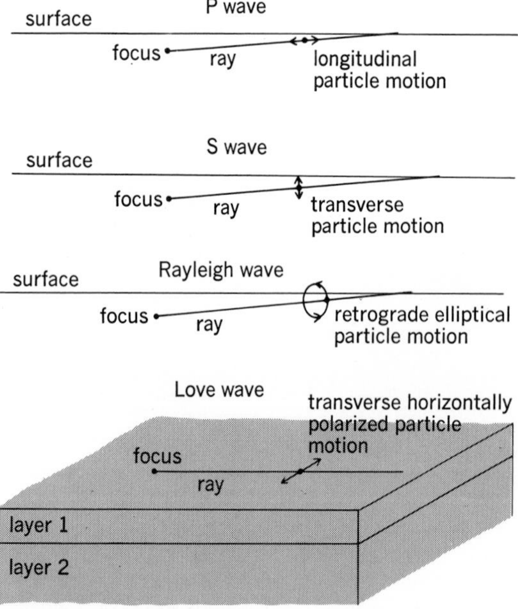

Fig. 4. Fundamental seismic waves.

Propagation theory. Understanding of seismic phenomena in the Earth relies heavily on the theory of strain propagation of solids. Observational experience indicates that the stress-strain relationship of the Earth for infinitesimal strains and seismic frequencies is highly elastic (with the possible exception of a surficial weathered layer less than 1000 ft thick) so that Hooke's law is applicable for seismic propagation theory to a high degree of approximation.

Consider a large solid body made up of regions which are individually homogeneous and isotropic. Applying Hooke's law and Newton's second law to an infinitesimal element of the solid body, one is led to the vector equation of motion, Eq. (1). Here

$$\rho\frac{\partial^2\bar{s}}{\partial t^2}=\left(k+\frac{1}{3}\mu\right)\operatorname{grad}\theta+\mu\nabla^2\bar{s} \qquad (1)$$

$\bar{s}$ is the particle displacement vector, $\theta=\operatorname{div}\bar{s}$ is the dilation, ρ is the density, μ is the rigidity, and k is the bulk modulus. This equation can be separated into two wave equations, Eqs. (2) and (3), which

$$\nabla^2\theta=\frac{\rho}{k+\frac{4}{3}\mu}\cdot\frac{\partial^2\theta}{\partial t^2} \qquad (2)$$

$$\nabla^2\bar{\psi}=\frac{\rho}{\mu}\cdot\frac{\partial^2\bar{\psi}}{\partial t^2} \qquad (3)$$

are generally more tractable than Eq. (1). Here $\bar{\psi}=\operatorname{curl}\bar{s}$.

Equation (2) establishes the existence of irrotational waves (sometimes called dilational or compressional waves) traveling with a velocity $\sqrt{(k+4/3\,\mu/\rho}$, whereas Eq. (3) indicates the existence of equivoluminal waves (sometimes called shear or distortional waves) traveling at the velocity $\sqrt{\mu/\rho}$. Since the first velocity is always greater than the second, irrotational waves arrive first and are designated by the symbol P (for primus), while equivoluminal waves coming in second are designated S (secundus). These are the basic body-wave types. Particle displacement in P is orthogonal to the wavefront whereas in S it is parallel (Fig. 4).

S waves may show polarization phenomena. If the particle motion is parallel to prominent planes in the medium, it is designated SH; if the particle motion is vertical it is referred to as SV.

Considerations so far have been limited to the interior of an elastic solid. Lord Rayleigh was the first to show that special wave types can propagate in the vicinity of the boundary. For seismic purposes the boundary of greatest interest is the free surface of the Earth. Rayleigh waves are characterized by particle trajectories which are retrograde ellipses; that is, when the particle is at the top of its ellipse, it is moving in a direction opposite to that of the propagation of the wave (Fig. 4). There is no component of particle motion which is both transverse to the ray path and parallel to the free surface of the solid supporting the Rayleigh-wave propagation. The speed of these waves is usually about 0.9 of the shear-wave velocity in the medium.

If the boundary value problem under consideration is further complicated by the introduction of a

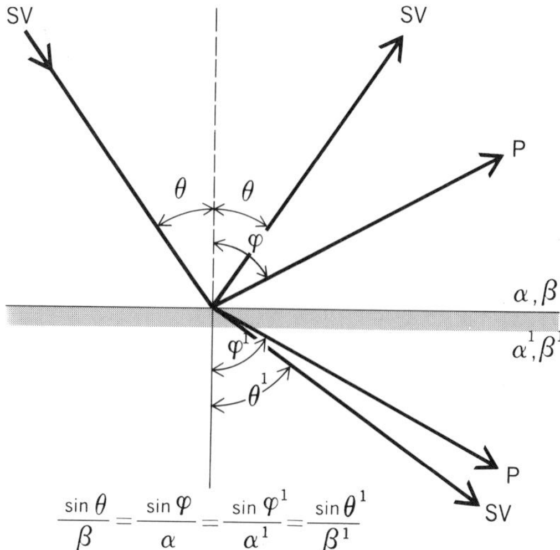

Fig. 5. Reflection and refraction of elastic waves according to Snell's law. α and β are compressional and shear velocities, respectively.

layer over the half space, an additional wave type will be found to propagate within the upper layer but with its particle motion horizontally polarized with respect to the bedding planes (Fig. 4). The Love wave is dispersive with phase velocities lying between the shear velocities for the two media involved.

Elastic waves incident upon an interface between two media having differing elastic properties undergo reflection and refraction analogously to electromagnetic and acoustic waves. In the elastic case the phenomena are more complicated because of the occurrence of wave conversions from *P* to *S* at the interface. In Fig. 5 the extension of Snell's law to cover these conversions is shown for an incident *SV* wave. Note that there are two reflected and two refracted waves produced in general.

Stonely waves are associated with an interface in the interior of a solid. By inserting the condition that amplitudes decrease with distance from the interface into the solution for reflected and refracted waves, the existence of Stoneley waves may be established.

As the number of layers is increased, the wave types observed at the surface (other than *P* and *S*) all become dispersive and group (*V*) and phase velocity (*c*) concepts become important in the analysis of these waves. Theoretical Love-wave dispersion curves for a single layer overlying a half space are shown in Fig. 6.

With the application of digital computers to seismic problems it has become feasible to study wave propagation in multiple horizontally layered structures with a large number of layers. With the increase in the number of layers comes an increase in the number of possible normal modes of excitation. For each type of dispersive wave and for each mode there will be separate group- and phase-velocity curves. Rayleigh-wave dispersion curves for the first two modes of the case of a liquid layer over a solid half space are shown in Fig. 7.

Using available computational programs, it is now possible to predict the seismogram which would result from an Earth model consisting of multiple horizontal layers or from an unlayered sphere, either liquid or solid.

Earth's interior. The primary application of seismology has been to determine the internal structure of the Earth. In seismic prospecting the Earth is excited by explosives, dropping weights, eccentric wheels, and so forth, and the reflections and refractions of the seismic energy produced are observed and analyzed in terms of the subsurface

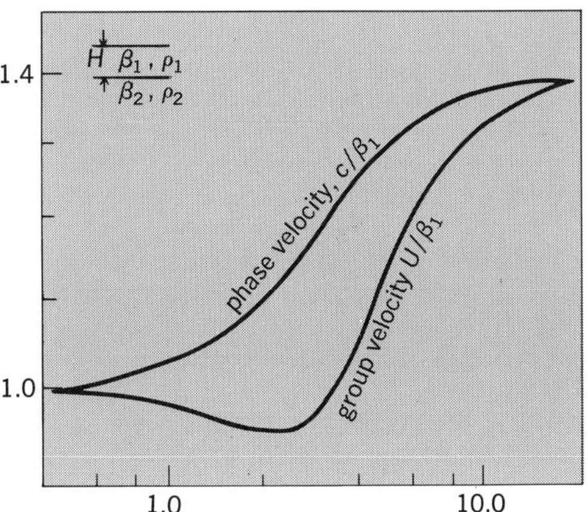

Fig. 6. Theoretical single-layer Love-wave dispersion curves. In the figure, β = shear velocity; ρ = density; $\beta_1/\beta_2 = 1.40$; and $\rho_2/\rho_1 = 1.20$.

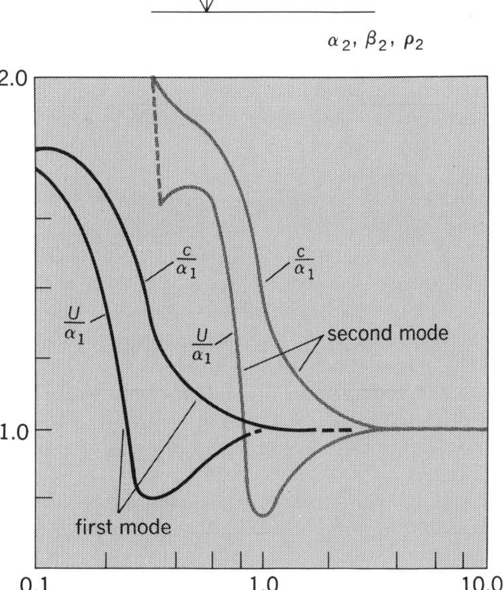

Fig. 7. First two modes of Rayleigh-wave dispersion for an ocean over a granite half space. In the figure, α = compressional velocity; β = shear velocity; ρ = density; $\rho_2/\rho_1 = 2.5$; $\alpha_2/\beta_2 = \sqrt{3}$; $\beta_2/\alpha_1 = 2$. (*From W. M. Ewing, W. S. Jardetzky, and F. Press, Elastic Waves in Layered Media, McGraw-Hill, 1957*)

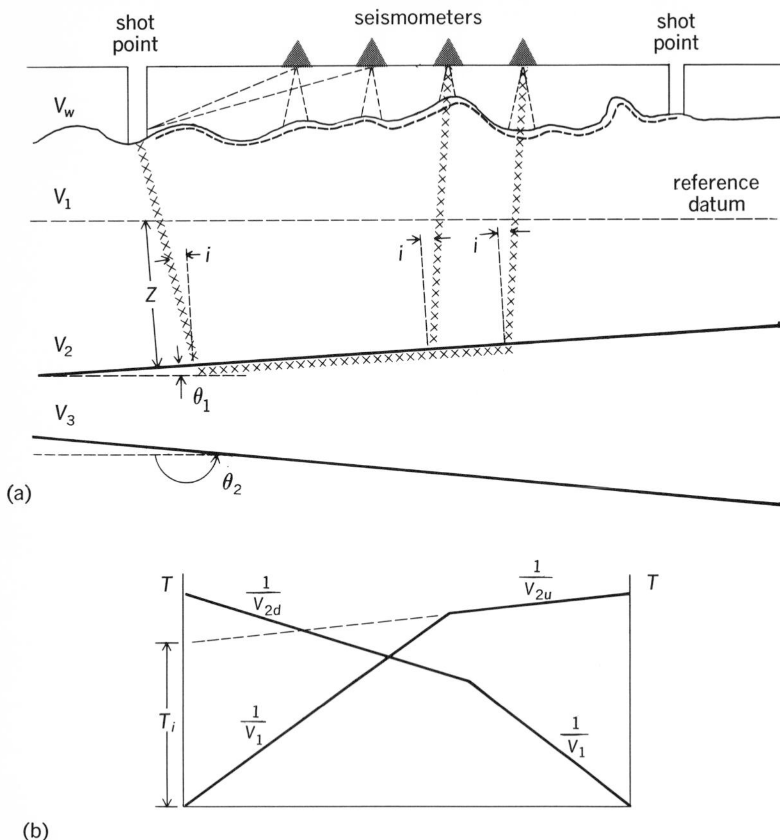

Fig. 8. Refraction shooting. (*a*) Profile showing boreholes, seismometers, and layers. (*b*) Corrected time-distance curve.

structure. The motivation is usually the determination of sedimentary formations favorable for the occurrence of natural deposits of commercially valuable materials (usually petroleum). Two techniques are used: refraction and reflection.

Refraction techniques. The principles of refraction shooting are illustrated in Fig. 8. The near surface is characterized by a thin low-velocity layer of very irregular velocity (V_w) and thickness underlain by a more regular structure which can frequently be idealized as a set of plane dipping interfaces. The depths and orientations of these interfaces are sought. Approximate estimates of the velocity (which are usually accurate enough) can be obtained from shooting in boreholes and by making time-distance plots for events traveling the dotted ray paths. By using head waves shot in two directions (dashed ray paths) at a number of positions in an area, the value of the velocity V_1 can be obtained and also the vertical travel time in the weathered layer under each seismometer. From this information the observed travel times for rays which have penetrated deeper into the sedimentary section (such as the ray path marked with the letter X) can be obtained. The result is the corrected time-distance graph (Fig. 8*b*). From the slopes and intercepts of the straight-line segments of the time-distance curves, the orientation and depth of the V_1-V_2 interface can be obtained from Eqs. (4), (5), and (6). In the equations i, θ, Z, and

$$\sin(i+\theta_1)=\frac{V_1}{V_{2d}} \tag{4}$$

$$\sin(i-\theta_1)=\frac{V_1}{V_{2u}} \tag{5}$$

$$Z=\frac{V_0T_i}{2\cos i} \tag{6}$$

T_i are as shown in Fig. 8, and d and u refer to rays down and up in the V_2 layer. By correcting the travel time of rays which have penetrated below the V_1-V_2 interface to this interface, the procedure can be repeated to obtain depth and orientation of the V_2-V_3 interface, and so forth. Refraction techniques are susceptible to errors produced by unknown lateral variations in velocity at depth (due to the large offset between seismometer and shot point required) and are affected by the existence of low-velocity layers between layers of high-velocity material.

Reflection techniques. Reflection prospecting is not nearly as subject to inhomogeneity and low-velocity-layer limitations as refraction techniques. It is essentially an echo-sounding procedure, since the source and seismometer spread is small relative to the depths usually measured. By accurately timing the arrival of successive reflections from the firing of the shot, the depth to the reflecting interfaces can be found if the velocity is known. The most accurate velocity data are obtained by direct measurement of velocities at (or in) a well which penetrates the geological section under study. Information from refraction shooting in the area is a second choice for velocity information; however, even if neither of these choices is available, another possibility is to make use of the reflection information itself. For any one reflector, a plot of the square of the seismometer offset distance against the square of the reflection time will yield a straight line, the slope of which is the average velocity to the reflector. Reflection depths are generally more precise to the same interface than refraction depths; however, reflections are more sensitive to irregularities of the interface, since an irregular surface sometimes fails to produce a reflection but gives a refracted return. *See* GEOPHYSICAL EXPLORATION.

Structural shells in the Earth. Initially from the study of P-wave arrivals from earthquakes and more recently by the use of refraction shooting techniques, it has been possible to show that the Earth has an outer crust which is separated from the underlying material, referred to as the mantle, by a surface called the Mohorovičić discontinuity (Moho). The crustal velocities increase with depth but are of the order of 6 km/sec, whereas the upper-mantle velocity is about 8 km/sec. The depth to the Moho is found to be about 35 km under continental shields, 55 km under high mountains, and 10 km in the deep ocean basins. A simplified idealization of the crustal structure is shown in Fig. 9. *See* MOHO (MOHOROVIČIĆ DISCONTINUITY).

From the viewpoint of historical seismology the Earth may be approximated as a spherically symmetric body in which the elastic velocity is a function of distance from the center only. There are continuous and abrupt changes in velocity, with the result that many reflections and wave conversions occur and many different phases can consequently be observed. Study of these phases indi-

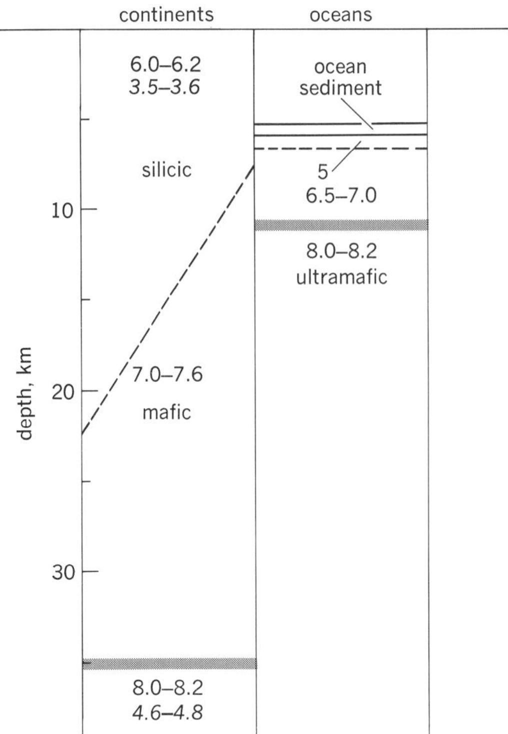

Fig. 9. Seismic velocities (compressional waves) and crustal structure for continents and oceans. Italic numbers represent shear velocity.

cates four major structural shells within the Earth; in descending sequence these are the crust, mantle, core, and inner core. The principal features of the Earth's structure and important seismogram phases which have been used to identify it are shown in Fig. 10. The angle Δ, used to measure distances in Fig. 10, is the angle subtended at the Earth's center by the source receiver arc. *P* is the first arriving phase from an earthquake, *PP* is the once-reflected compressional wave from the Earth's surface; similarly *S* and *SS* are the direct and once-reflected shear waves. From a distance of about 103–142° there is a shadow zone where *P* is not recorded, whereas at distances of about 26–103° the reflected wave sequence *PcP, ScS,* and *PcS* is observed from the same discontinuity that produces the shadow zone (that is, the core). It is found that shear waves will not traverse the core, from which it is inferred that the core is liquid. Compressional waves (*K*) do, however, pass through it, and phases *PKP, PKS, SKS,* and *SKP* have been observed. Reflections such as *PKiKP* have been identified from a discontinuity within the core which, consequently, is called the inner core. Transmission through this inner core is given the symbol *I* as in *PKIKP*. Typical seismograms at various ranges are shown with their phases identified in Fig. 11.

Velocity-depth distribution. Empirical tables and curves giving the observed travel times of the principal seismic phases as a function of distance have been prepared (Fig. 12). These tables were originally prepared from earthquake data where the origin time was not accurately known. Nuclear explosions have, however, confirmed *P* travel times to within an accuracy of 2 sec.

Utilizing the observed travel times and the first derivative of the travel-time curve, G. Herglotz developed analysis techniques by which it has been possible to calculate a smoothed average-velocity-depth distribution in the Earth (Fig. 13).

Numerous types of surface and guided waves have been observed propagating within the Earth. The dispersion of these provides further information on the Earth's internal structure. Love and Rayleigh waves are associated with the crust. *G* waves are long-period Love waves controlled by the velocity structure of the mantle, as is the case for mantle Rayleigh waves. Higher-mode Love and Rayleigh waves have been identified and are found to be particularly sensitive to intracrustal velocity variations. L_g is a short-period wave guided by the gradual velocity increase with depth near the top of the continental crust. In utilizing surface waves for structural information, the basic procedure is to prepare experimental dispersion-curve data such as that shown in Fig. 14 and compare it with theoretical dispersion curves such as Figs. 6 and 7. By using digital computers, a multiplicity of the theoretical curves is prepared, under constraints provided from other sources of information, until a match with the observational data is obtained.

Once the *P*-, *S*-, and surface-wave velocity-depth distribution has been obtained, theoretical techniques are available to permit calculation of the density and pressure distribution within the Earth (Fig. 15).

Typically seismic body waves from distant earthquakes and underground nuclear shots have predominant periods in the range 1–15 sec, with *P* and *S* respectively appearing at the shorter- and longer-period ends of the intervals. Surface waves have been studied in the range 5–500 sec, with the period range 5–50 sec being used for crustal structure and 50–500 sec for studies of the mantle. At the 95% confidence level, free oscillations of the Earth as a whole have been observed in the range 5–120 min. *See* EARTH; EARTH, INTERIOR OF.

Microseisms. The disturbances which are the background unrest of the Earth's surface are called microseisms. They can be classified into three groups on the basis of frequency: Long-period microseisms with periods in excess of 10 sec; storm microseisms with periods from 2–10 sec;

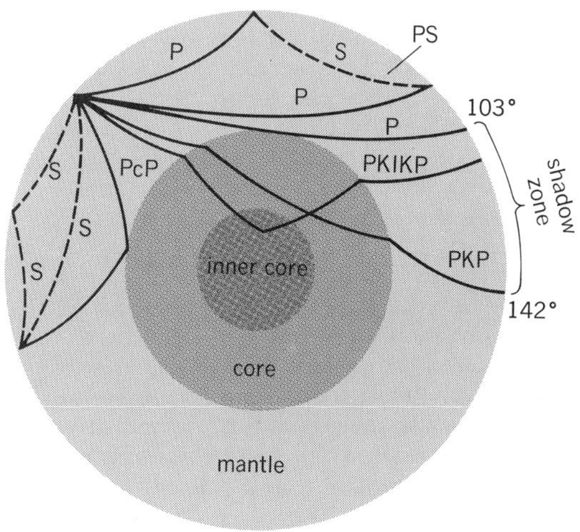

Fig. 10. Paths of seismic waves in the Earth.

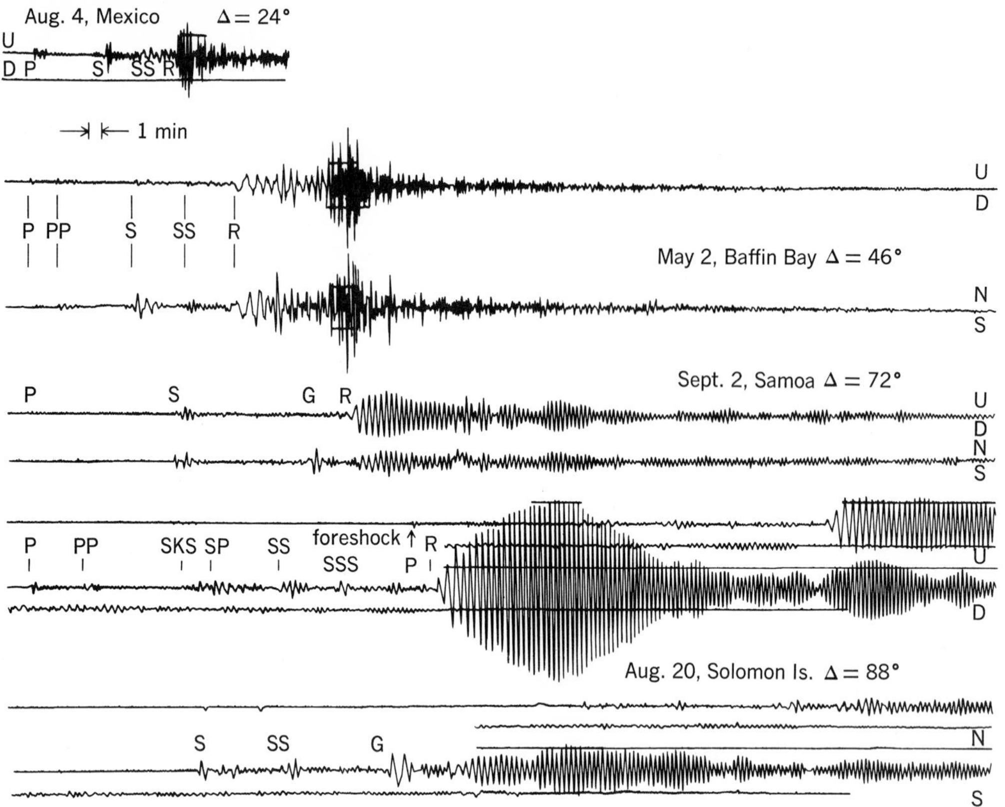

Fig. 11. Typical seismograms at various ranges, shown with seismic phases. (*From C. F. Richter, Elementary Seismology, copyright © 1958 by W. H. Freeman and Company*)

and short-period microseisms with periods less than 2 sec.

Storm microseisms showing a spectral peak of 6–10 sec are found everywhere and continuously on the Earth's surface. The amplitude varies greatly, 1 micron (μ) being representative and 100 μ representing unusually "noisy" conditions. Evidence has been produced linking the production of these waves to storms on the open ocean and to surf breaking on seacoasts. Efforts to use these microseisms to track storms at sea have not been successful. Microseisms of 6–10-sec periods can be correlated across continents, but microseisms with periods less than this are usually incoherent.

Short-period microseisms are usually local phenomena associated with industry, vehicle traffic, and wind effects on trees, structures, and even grass. The longer-period microseisms are generally found to be multimodal Rayleigh waves with an increasing proportion of body-wave content at the shorter periods. The nature and origin of microseisms is poorly understood.

Monitoring nuclear explosions. The prime research goal of seismology since 1960 has been the development of adequate techniques for distinguishing underground nuclear explosions from earthquakes. The problem of remote monitoring of underground nuclear explosions by the detection and analysis of the seismic signals they produce has been reduced essentially to the questions of how weak a signal can be detected with confidence and whether the discrimination criteria for separating explosions from earthquakes, which have been established for moderate and strong events, are applicable to those yielding low levels of radiated energy. Seismic monitoring consists of three steps: detection of the signal from an event, location of the source, and identification of the source as an earthquake or an explosion.

Detection. Signal detection depends on recording waves generated by the event that can be distinguished from ambient, ever-present seismic noise. Advances in detection capability have resulted from improved methods of processing the signals from short-period (about 1 sec) arrays and from the development of stable instruments with long-period (30–40 sec and longer) sensitivity two orders of magnitude greater than that of conventional seismographs. The significance of these new instruments for seismic monitoring is that they make possible the detection of the weak long-period signals from low-magnitude events, by virtue of their high sensitivity and the fact that they take advantage of a widely observed, and perhaps worldwide, minimum in Earth noise at about 30-sec periods.

The detection capabilities of the global network of seismograph stations for both short- and long-period waves have been determined by system analyses. The general conclusion of these analyses is that, under reasonable assumptions about the station distribution and with present instrumentation, the *P*-wave of body-wave magnitude m_b 4.0 or greater (a shallow seismic event) occurring within the coterminous United States or Mexico will be detected, with at least a 90% probability, by four or more stations. Global contours of the detection threshold are shown in Fig. 16.

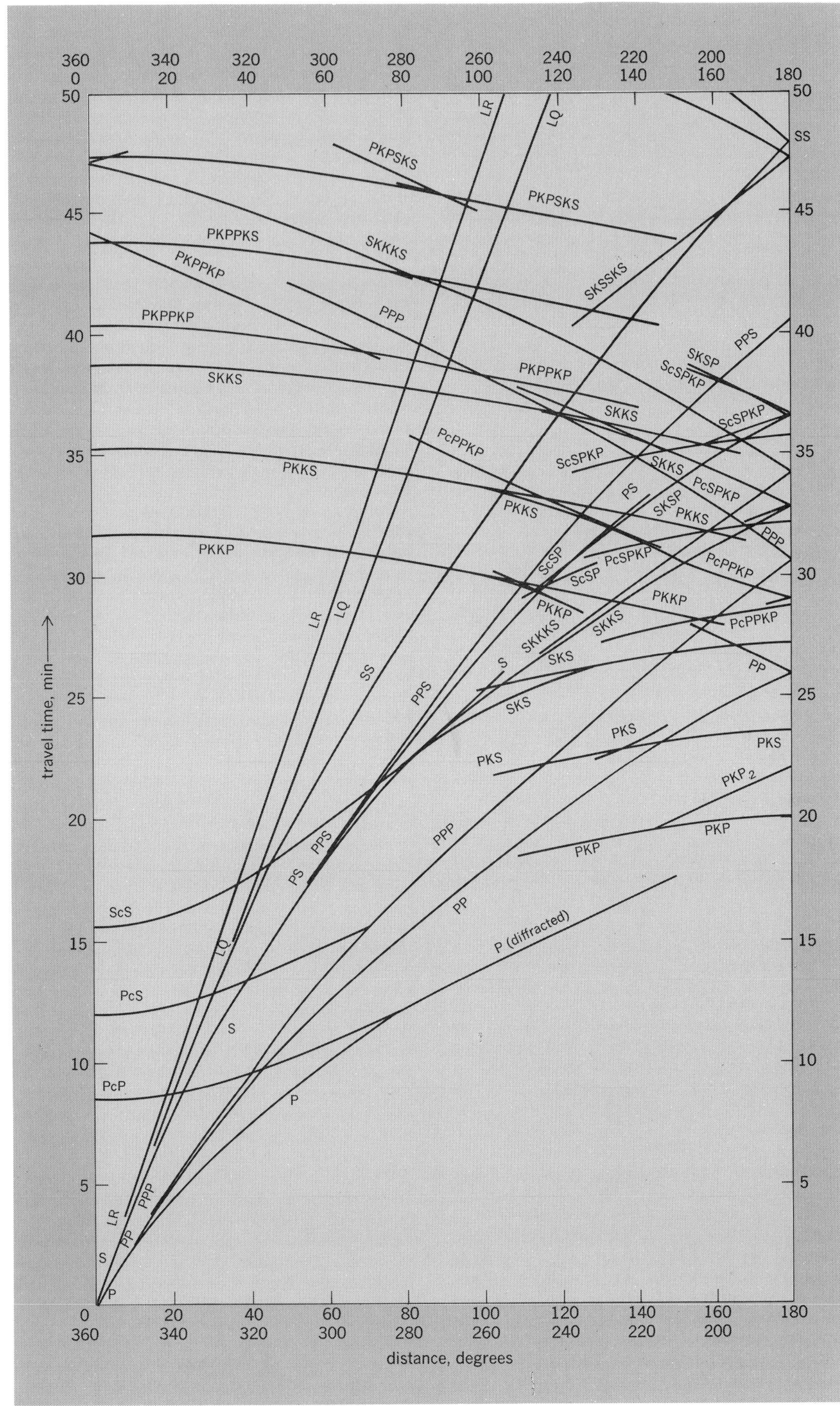

Fig. 12. Seismic travel-time curves. (*After H. Jeffreys and K. E. Bullen*)

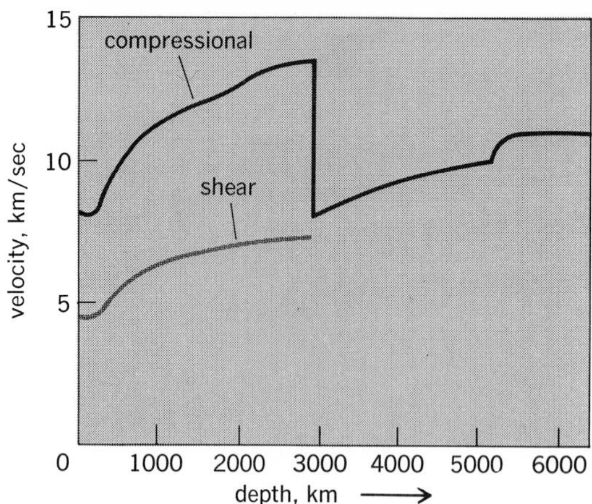

Fig. 13. Seismic velocities in the mantle and core of the Earth. (*After B. Gutenberg*)

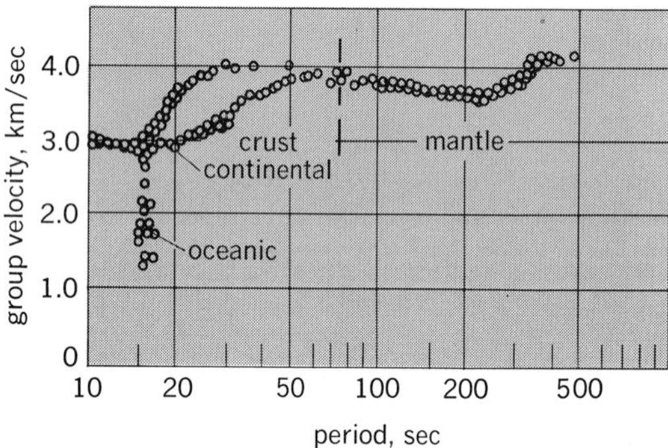

Fig. 14. Crustal and mantle Rayleigh-wave group-velocity curves. (*After W. M. Ewing and F. Press*)

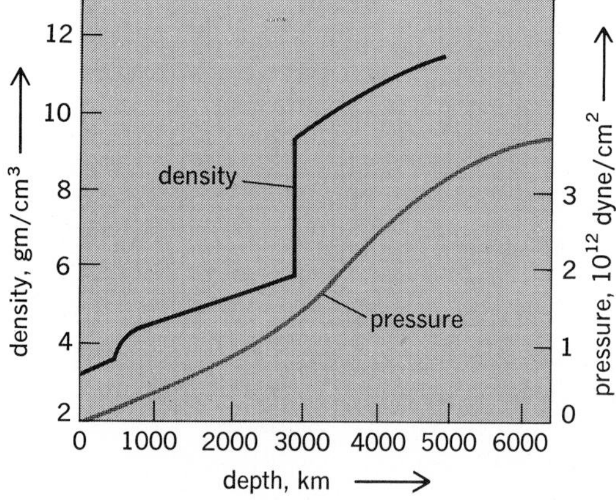

Fig. 15. Density and pressure variation in the Earth. (*After K. E. Bullen*)

*Discriminants and location.*The discrimination of explosions from earthquakes requires the detection of not only the short-period body waves but also of the longer-period surface waves, especially of the Rayleigh waves. The magnitudes of shallow earthquakes for which the Rayleigh waves will be detected by a representative distribution of long-period seismographs have been found to be somewhat higher than for body-wave detection. As shown in Fig. 17, the magnitudes are about m_b 4.5 for earthquakes in central North America and up to m_b 5.2 for the most poorly covered parts of the Northern Hemisphere. A shallow earthquake with the indicated body-wave magnitude will have a 90% probability of Rayleigh-wave detection by 4 or more of the 51 stations. Because Rayleigh-wave amplitudes generated by explosions are much smaller than those from earthquakes with the same body-wave magnitude, these would not be detected as reliably, if at all, by the same network of observatories.

Location of an event recorded at four or more stations is achieved by determining the geographic position, depth within the Earth, and origin time that correspond to the best fit of the observed arrival times of the various seismic waves at all the stations to standard tables of the travel times of these waves through the earth. In many cases, an accurate location alone is a sufficient indicator that an event was an earthquake, either because the place is one at which a nuclear test would be unlikely (for example, near a populated area) or because the depth is so great as to rule out an artificial source.

Unfortunately this latter, potentially powerful discriminant, depth of source, has not proven to be very useful because of the inherent difficulties in determining the depth of events in the Earth's crust sufficiently accurately, within a kilometer or two, from teleseismic data.

The discrimination between earthquakes and explosions on the basis of seismic data must depend on characteristic differences in the two phenomena as wave generators. A number of such differences are expected in view of the known properties of the two source types, and numerous diagnostic features have been proposed and evaluated in the past. Of these, the one that has proven most effective is the difference in the spectral characteristics of the two.

The relative excitation of low- and high-frequency energy may be measured in several ways. One way, which was proven very effective in discriminating between explosions and earthquakes down to at least m_b 4.5 is the ratio of the surface wave magnitude M_s to the body-wave magnitude m_b. The body-wave magnitude, as conventionally calculated, depends on waves with frequencies near 1–2 Hz. The surface-wave magnitude, on the other hand, is derived from the long-period motion that arrives much later. M_s was originally calculated from the 20-sec period surface waves, but relationships giving equivalent values for other periods in the surface-wave train have been developed. As may be seen in Fig. 18, for the same m_b, M_s is greater for earthquakes than explosions.

Two main difficulties remain in the application of this method of explosion identification. The first is that occasional crustal earthquakes are recorded that are explosionlike with regard to surface-wave generation, and if discrimination is based on identifying all earthquakes as such, some events may be incorrectly retained as explosions, leading to a possible false alarm.

The second problem is that by the nature of the M_s/m_b criterion, very small explosions will generate surface waves too small to be detected at all, so that an M_s value cannot be determined. This in turn leads to the need to appeal to "negative evidence." The absence of surface waves accom-

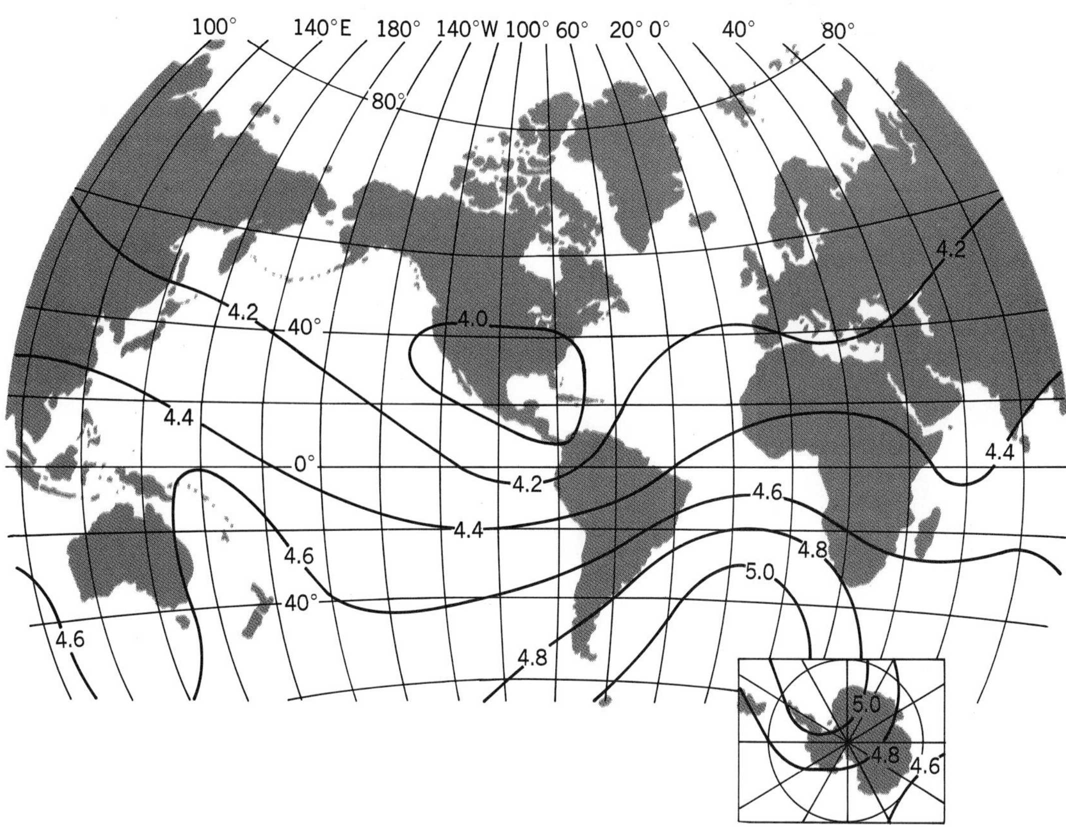

Fig. 16. Global contours of the *P*-wave detection threshold for a network of short-period vertical seismographs. (*From P. W. Basham and K. Whitham, Seismological Detection and Identification of Underground Nuclear Explosions, Publ. Earth Phys. Branch, vol. 41, no. 9, Department of Energy, Mines, and Resources, Canada, 1971*)

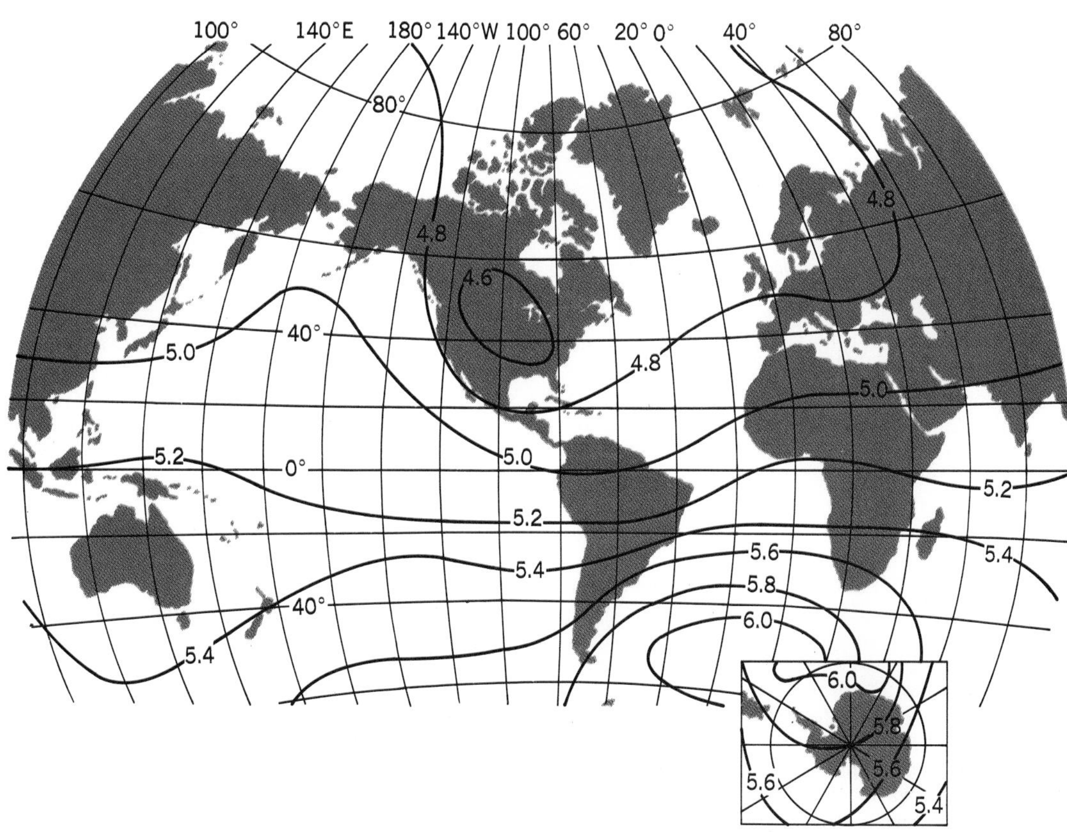

Fig. 17. Global contours of the earthquake Rayleigh-wave detection threshold for a network of long-period seismographs. (*From P. W. Basham and K. Whitham, Seismological Detection and Identification of Underground Nuclear Explosions, Publ. Earth Phys. Branch, vol. 41, no. 9, Dept. of Energy, Mines, and Resources, Canada, 1971*)

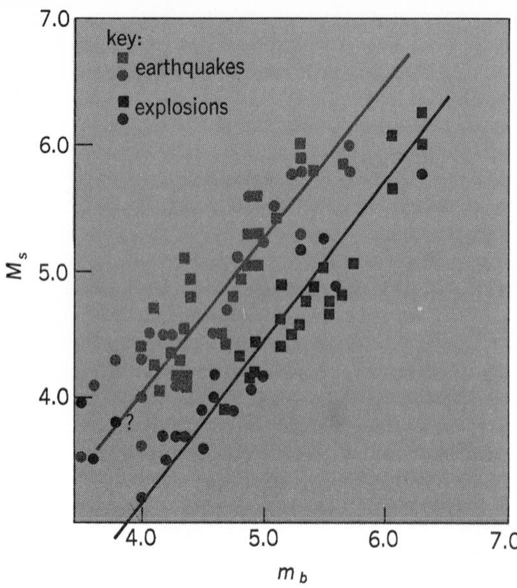

Fig. 18. Surface-wave magnitude versus body-wave magnitude for earthquakes and underground nuclear explosions in the United States. *(From J. F. Evernden et al., Discrimination between small-magnitude earthquakes and explosions, J. Geophys. Res., 76(32):8042–8055, © 1971 by American Geophysical Union)*

panying a detected body-wave signal of sufficient amplitude would in itself be indicative that the source might have been an explosion. Such negative evidence is valid, but not as satisfactory as the results of analysis of a recorded signal. The very-high-gain long-period instruments, recently installed at several locations in the world, have decreased the minimum surface-wave magnitude for which the signal is observable and thus lowered the threshold for which a small M_s/m_b ratio can be determined rather than taken as zero.

Evasion techniques. Several conceivable evasion techniques have been studied, including detonation of the device in a large, mined-out cavity that would decouple the energy from the surrounding rock, thereby reducing the seismic signal; detonation of several properly deployed devices simultaneously, to create a signal that is earthquakelike; and detonation during a large earthquake, when the seismograms would be saturated with the earthquake signal. The first method requires the undetected creation of the cavity, probably in a salt dome for engineering reasons, at great expense. The second requires the existence of alluvial deposits sufficiently thick that the depth of burial can be great enough to prevent the formation of a subsidence sink, which is easily detectable by satellite photography. At the same time the device must be far enough above the water table for the bedrock to achieve the decoupling effect.

With regard to the third method of evasion, it should be possible to combine a number of detonations to simulate a selected feature of an earthquake signal at some distances and at some azimuths. However, neither simulation studies nor field experiments indicate that all features of an earthquake signal can be produced at all distances and azimuths by a combination of explosions.

Although an explosion signal might be very difficult to separate from that of a simultaneous strong earthquake, the practical use of this method of evasion would require that a test series be readied and put in a standby condition, to be activated almost instantaneously when a suitable earthquake occurs. The time frame within which the detonation would have to be carried out is very limited, because once the high-frequency body waves from the earthquake are past, the task of separating the explosion signal from that of the earthquake is somewhat easier, even though very large ground motion of long duration will be in progress. The task of picking out an explosion from the large number of after shocks that usually follow a large earthquake would be challenging, but could be accomplished by painstaking application of present signal-recognition techniques.

Research. In the preceding sections those aspects of seismology on which most seismologists are in agreement have been outlined. It should be emphasized, however, that research in seismology is very active, and undoubtedly the fruits of this work will be a changed and improved understanding of the origin, propagation, and nature of earthquakes, seismic waves, and the internal structure and constitution of not only this planet but also the other satellite members of the solar system. Some account of the trends in this research is given here.

Most, if not all, of the focal region of earthquakes is buried deep within the Earth and is inaccessible for direct observation. Any theory then on the earthquake focal mechanism must rely on indirect observations—specifically, on the movement of surface rocks in the earthquake epicentral region; the high-temperature, high-pressure properties of rocks determined in laboratory experiments; and the observed radiation of seismic waves from earthquakes.

There are two main theories considered as possible explanations for the earthquake focal mechanism by geophysicists. These are the phase-change theory and the elastic-rebound theory. The phase-change concept is that accumulating heat, pressure, or strain within the Earth act to trigger a phase change which is already metastable, so that rapid propagation of the change occurs and energy is released as seismic waves. This is a minority viewpoint among geophysicists. H. F. Reid was led to the elastic-rebound theory after inspecting the large surface displacements (faults) which accompanied the great San Francisco earthquake of 1906. According to the elastic-rebound concept, stresses are being built up in the Earth which, at discrete intervals, are suddenly relieved by the earthquake movement along the fault.

Laboratory studies of rocks at high temperatures and pressures are showing the importance of fluids contained in pores in the rocks and of stick-slip phenomena when rock surfaces move over each other, but much more work of this type is required. Improvements in propagation theory are enabling seismologists to use distant observations to estimate close-in source parameters, such as the velocity of propagation of the discontinuity surface, its orientation, the time function of the displacement, and the equivalent force system at the focus.

From a scientific viewpoint any hope of predicting earthquakes must rest on the observation of some physical phenomenon or groups of phenomena usually found to precede the main shock. Observations do indicate that the Earth's surface in the vicinity of a developing earthquake undergoes anomalous deformation in the day and hours preceding the quake, with the rate of change of these effects accelerating just before the quake itself. Laboratory breaking studies of rock under pressure indicate that a preliminary set of minor breaks (foreshocks?) occur before the main break, and observable changes in other mechanical and electrical properties are also found to occur just before failure. Current studies of the "Chandler wobble," a perturbation of the position of the Earth's poles with a period of 14 months, indicate small but abrupt discontinuities occurring in the instantaneous position of the Earth's poles within the few days preceding a major earthquake. Improved theoretical computations and confirming observational data in larger quantities and at higher precision levels are being obtained with the hope of developing a technique for indicating impending earthquakes. *See* EARTH, DEFORMATIONS AND VIBRATIONS IN; EARTH TIDES.

Current consideration is being given to instrumenting much of the state of California with strain-sensitive devices. These would be individual instruments of various types recording automatically for long periods of time and operating over distances from 1 m to 1 km. It is not expected that this instrumentation will in itself predict earthquakes, but it is expected that it would lead to a much clearer understanding of the nature of earthquakes, of the possibility of predicting them, and of the nature of the instrumentation which an adequate earthquake-prediction system might require.

Project VELA-UNIFORM is concerned with the problem of distinguishing underground nuclear bomb blasts from earthquakes. It is hoped ultimately that a treaty can be negotiated to limit or end underground nuclear testing. With seismic capabilities improved beyond current levels, it is hoped that practical methods of policing such a treaty will be developed. Some of the criteria which are being studied for distinguishing underground nuclear explosions from earthquakes are source depth; radiation patterns of *P*, *S*, and surface waves; and the relative excitation of wave types.

Wave-propagation theory so far applied to seismic problems has assumed Earth models consisting of either plane horizontal layers or concentric spheres. Mathematical techniques are being developed to cover propagation in structures in which the planarity and horizontality are perturbed.

An understanding of the Earth's interior is undoubtedly biased by the fact that almost all seismic observations are made on land whereas most of the Earth's surface is ocean, now known definitely to have a different subsurface structure from the continents. Submarine seismic stations, which can record automatically and unattended on the ocean floor, will soon open up the planet's watery surface to seismic observation.

The use of LASA is greatly improving knowledge of the fine structure of the Earth. As outlined earlier, the ray parameter p (the derivative of the time-distance curve) is usually determined by numerical differentiation of the travel-time curve of Fig. 12. The value of p obtained therefrom is grossly smoothed, and consequently the resulting velocity distribution given by the Herglotz formula is also a broad average. Using LASA, it is possible to measure $p(\Delta)$ directly, so that both vertical and azimuthal velocity information can be obtained with greatly improved precision. The fine structures of both mantle and core are being outlined by this method, and estimates of lateral inhomogeneity in the Earth are being obtained. These results are being incorporated with new results from other geophysical disciplines to produce improved Earth models. Improved computational techniques are permitting the fitting of the resulting Earth model to the many constraints required by the several geophysical disciplines.

A second LASA is under construction in Norway. With radio communication between these two stations it should be possible to set the second LASA's instruments to respond optimally to a signal picked up by the first one. The result will be great improvement in the ability to detect or identify and utilize seismic signals. The theory of time-series analysis as applied to seismology is still actively under development, and further improvements in detection capability can be expected from this source.

Instrument development for passive and active seismic experiments on the Moon are continuing. The passive experiment will telemeter back to Earth the effects of meteor bombardment or other natural seismic sources, as observed by a rocket-placed seismograph. In the active seismic experiment astronauts will use explosive sources to obtain reflections and refraction seismic data.

[KER C. THOMSON]

Bibliography: M. Båth, *Introduction to Seismology*, 1973; W. M. Ewing, W. S. Jardetsky, and F. Press, *Elastic Waves in Layered Media*, 1957; Institute of Electrical and Electronic Engineers, Special issue on nuclear test detection, *Proc. IEEE*, December 1965; F. Press, Earthquake prediction, *Science*, 152:1575–1584, 1966; F. Press, Earthquake prediction, *Sci. Amer.*, 232(5):14–23, 1975; C. F. Richter, *Elementary Seismology*, 1958; E. A. Robinson, *Statistical Communication and Detection*, 1967.

Sepiolite

A complex hydrated magnesium silicate mineral named for its resemblance to cuttlefish bone, alternately named meerschaum (sea foam). The ideal composition, $Mg_8(H_2O)_4(OH)_4Si_{12}O_{30}$, is modified by some additional water of hydration, but is otherwise quite representative. The ideal density, 2.26 g/cm^3, is never realized in aggregates. Interlaced disoriented fibers aggregate into a massive stone so porous that it floats on water. These stones are easily carved, take a high polish with wax, and harden when warmed. They have been valued from antiquity in the eastern Mediterranean region for ornaments and make attractive smoking pipes. *See* CLAY MINERALS; SILICATE MINERALS.

Two related types of fibers exist; either of these may occur as the massive interlacings, as paper-like mattings, as bedded clayey deposits, or as more pronounced asbestoslike fibers. They make up a group related in crystal structure to amphibole. Palygorskite and attapulgite are used as names when the amphibole chain relationship is in double units, and sepiolite when it is in triple units. *See* AMPHIBOLE.

Extensive bedded fibrous deposits of either type have been found useful as dispersoids in oil-well drilling fluids for operations that penetrate high-salt formation waters. They are also active as absorptive agents for decolorizing oils, as carriers for insecticides, and as fillers. The bedded palygorskite (attapulgite) variety is produced in quantity in Florida and Georgia in the United States, and in the Ukraine and the Caucasus in the Soviet Union; much of the sepiolite variety is produced in Spain, North Africa, and Arabia.

A third fiber, loughlinite, contains up to four sodium ions in place of two magnesium of the sepiolite formula but is changed to asbestoslike fibers of sepiolite by soaking in magnesium chloride.

[WILLIAM F. BRADLEY]

Serpentine

The name applied to the hydrous magnesium silicate mineral assemblage of the rock serpentinite. Serpentinization alters host rocks to compositions approximating $3MgO{\cdot}2SiO_2{\cdot}2H_2O$, and the names serpentine, chrysotile, antigorite, and others are applied with regard to degree of crystallinity achieved locally in given serpentinized groundmasses. *See* SERPENTINITE.

The state of the solid matter in massive serpentine has certain aspects in common with synthetic polymers. It is permeated with crystalline nucleations insufficiently articulated to meet the criteria of true crystals. Associated with the massive assemblage, substantially the same compositions, possibly under the influence of some mechanical stress, become organized into varying degrees of improved alignments and have received the various names of the serpentine minerals group.

The most important serpentine mineral is the fibrous variety, chrysotile (Fig. 1), which accounts for as much as 95% of the asbestos of commerce. Although chrysotile is not a crystal in the sense of having plane-bounding faces, it nevertheless has a crystal structure consisting of regularly arrayed sheets of component atoms wrapped about a less well-organized or possibly even unoccupied core. The high tensile strength of these tubular fibers permits the shredding, spinning, and fabrication of useful asbestos products (Fig. 2). *See* ASBESTOS.

Fig. 1. Veins of chrysotile asbestos in serpentine. (*Canadian Johns-Manville Co., Ltd.*)

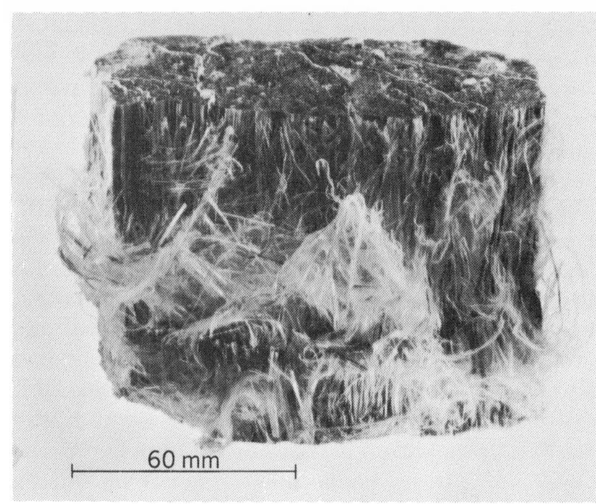

Fig. 2. Serpentine, variety chrysotile, taken from the Thetford Mines, Quebec, Canada. (*American Museum of Natural History specimen*)

Other named varieties, such as antigorite, bastite, marmolite, and picrolite, consist of comparable regularly arrayed sheets, more or less corrugated and stacked colinearly to present a more platy, or at least lathlike, aspect. Some of these platy varieties may involve a pseudomorphism after host crystals which imposed some control onto the product which they altered. The platy varieties are not currently known to have any value in themselves.

Serpentine is easily cut and polished for ornamental stone. It may be blackish-green through leek green to nearly white, or brownish or yellowish, and exhibits attractive textures because of the disseminated filamentous crystalline nucleations. Serpentinized carbonate rocks sometimes show an attractive clouded green color and are then called verde antique, or serpentine marble.

[WILLIAM F. BRADLEY]

Serpentinite

An abundantly occurring, fine-grained massive rock, generally considered to have been derived by pneumatolytic or hydrothermal processes from preexisting basic and ultrabasic igneous intrusions. The inferred processes are called serpentinization and the product serpentine. The serpentinization of metamorphic rocks, and of limestones and dolomites, is common. The composition of serpentine approximates $3MgO{\cdot}2SiO_2{\cdot}2H_2O$. The groundmass itself and the varietal associated minerals of the same composition are called the serpentine minerals. *See* METAMORPHIC ROCKS; PNEUMATOLYSIS.

Serpentine differs from the more conventional rocks – those composed of separable grains of various minerals – for serpentine has no idiomorphic crystal grains of its own. When grains are apparent, they are relicts from the preexisting crystalline rock which was altered to serpentine while retaining the old grain boundaries (called pseudomorphs after the older grain). Pseudomorphs are most frequent after chrysolite (olivine), common after amphibole and pyroxenes, and have been noted after other minerals.

Massive serpentinites often form layers, dikes, or necks (composed more or less of pure serpentine) in various positions in crystalline schists. Many serpentinites develop by the low-grade metamorphism, or hydrothermal alteration, of olivine-rich rocks, especially peridotites and dunites in mountain ranges. Experiments have shown that at low pressure, magnesian olivine is stable only above 400°C when it is in contact with water vapor; below that temperature it changes into serpentine and brucite. Iron-rich olivine is stable in the presence of water at much lower temperatures, and does not readily change into serpentine.

In many rocks where serpentine has formed pseudomorphs after olivine, traces of the original mineral may still be present as mesh structure. Other rocks are composed wholly of fibrous chrysotile serpentines and still others are made up of flaky antigorite which developed under stress conditions. Sometimes the flakes and blades of antigorite are arranged at random in sheaflike bundles. *See* SERPENTINE. [T. F. W. BARTH]

Shale

A fine-grained laminated or fissile sedimentary rock made up of silt- and clay-sized particles. The average shale consists of about one-third quartz, one-third clay minerals, and one-third miscellaneous minerals, including carbonates, iron oxides, feldspars, and organic matter. The complete mineral composition of shales was poorly known until the 1960s because ordinary microscopy is unsatisfactory with such fine-grained materials, and routine x-ray diffraction methods do not give analyses of all of the material.

Chemical composition. The chemical composition of shales, because of the difficulty of mineral identification, has been studied more than their mineral composition. The average shale, as calculated by F. W. Clarke, consists of about 58% silicon dioxide, SiO_2; 15% aluminum oxide, Al_2O_3; 6% iron oxides, FeO and Fe_2O_3; 2% magnesium oxide, MgO; 3% calcium oxide, CaO; 3% potassium oxide, K_2O; 1% sodium oxide, Na_2O; 5% water, H_2O; and lesser amounts of other metal oxides and anions. But the average shale reflects only a lumping of analyses of widely varying shale types and compositions. Chemical composition varies with grain size, the coarser fractions having more silica and the finer having more alumina, iron, potash, and water. Shales normally contain a large amount of quartz silt, up to 60% in some analyses. Some shales have abnormally high silica content that cannot be ascribed to silt content. Most of the excess silica is in the form of very finely crystalline quartz, chalcedony, or opal. Some of this is probably the result of large numbers of diatoms or volcanic ash in the sedimentary environment. Some silica is secondary and may come from chemical alteration of primary silicate minerals. Iron-rich shales may contain much pyrite or siderite, or iron silicates, all of which imply at least mildly reducing conditions in the original environment and possibly a deficiency of alumina. Calcareous shales may contain several carbonate minerals and may be fossiliferous, grading laterally into limestones in some cases. High lime content may also be associated with the presence of gypsum and thus indicate hypersaline conditions. Potash is almost always more abundant than soda, possibly as a result of fixation in illitic clay minerals. Some shales that are extraordinarily rich in alkalies contain large amounts of authigenic feldspar. Shales rich in organic matter signify high organic productivity and lack of oxygen in the sedimentary environment. Organic matter may be carbonaceous, in which oxygen is chemically bound to carbon and hydrogen, or petroliferous, in which oxygen is essentially absent and the compounds are hydrocarbons of various kinds. The color of shales is not in general due to distinctive chemical composition but to pigmentation by small quantities of colored material. Red color in redbeds is associated with pigmentation by ferric iron, black color with carbonaceous material, green or dark gray with ferrous iron.

Fissility. The tendency for shales to split along bedding planes is called fissility; it is a product of the subparallel orientation of the individual grains of micaceous minerals along bedding planes. Part of this orientation may be introduced during sedimentation by slow settling of particles, but most of the effect is due to postdepositional compaction and reorientation of the platy minerals under pressure.

Lamination in shales, which ranges from 0.05 to 1.0 mm, is due to grain size variations, mineralogical variations, or color variations resulting from minor amounts of mineral or organic matter pigment. The laminations may be due to seasonal climatic factors that affect the source of supply and to sedimentation currents, factors similar to those that influence the formation of varves in varved clay. Lack of lamination is chiefly the result of reworking of soft sediment by bottom-dwelling scavenging organisms. Such reworking can also result in mottled mixtures of clay and silt. *See* CLAY; CLAY MINERALS; VARVE.

Classification. No general agreement on a classification scheme for shales has been reached. On the basis of the composition of the silt fraction, W. C. Krumbein and L. L. Sloss have divided shales into groups roughly paralleling sandstone types: quartzose shale (orthoquartzite), feldspathic shale (arkose), chloritic shale (graywacke), and micaceous shale (subgraywacke). Their interpretation of the origin of these groups roughly is the same as for corresponding sandstone groups, stipulating that the shales were deposited in quieter waters than were the sandstones. Shales may also be subdivided on the basis of origin, as was done by F. J. Pettijohn, into residual (from reworked soils), transported, and hybrid. The hybrid shales are mixtures of clastic and nonclastic materials (carbonate, organic matter, iron oxides). The hybrid shales represent slow rates of clastic sedimentation that give the opportunity for chemical and biochemical processes to dominate the sedimentary environment. In this group belong the black shales, which may contain up to 25% organic matter and may even grade laterally into purely organic beds such as coal. Black shales are commonly phosphatic; any fossils present are thin carbonized impressions or pyritized. The faunas represented may be dwarfed or depauperate. Any calcareous shells are thin, normally of the *Lingula* type. Siliceous, alumina-rich, and iron-rich shales represent special conditions under which the appropriate elements are incorporated into the clay minerals or precipitated as oxides.

Origin. The general concept of origin includes the ideas of quiet water, perhaps deeper than 50 ft. Depths of water and distance from shorelines would be greater than for associated sandstones. Though transportation by water is still largely favored, wind transport of large quantities of fine-grained material has now been demonstrated to be a prime agent in the formation of muddy sediments over the world's oceans. Distinctive chemical conditions are probably ultimately related to the restrictedness of the environment, which may make for hypersaline marine or alkaline-evaporite lake conditions or oxygen-deficient waters. Secondarily, the presence of larger than normal amounts of some elements is related to the composition of the source detritus. *See* ARGILLACEOUS ROCKS; ARGILLITE; BENTONITE; BLACK SHALE; LOESS; RED-BEDS; SEDIMENTARY ROCKS.

[RAYMOND SIEVER]

Siderite

SIDERITE

Rhombohedral siderite crystals with quartz from Cornwall, England. *(Specimen from Department of Geology, Bryn Mawr College)*

The mineral form of ferrous carbonate, often containing appreciable amounts of magnesium and manganese substituting for iron.

Siderite may occur in sedimentary deposits or in hydrothermal veins. It may be formed by the action of iron-bearing solutions on limestones. The equilibrium replacement of calcium for iron has been determined and found to increase as a function of temperature. Siderite is found in England, Greenland, Spain, and North Africa. In the United States siderite occurs in Connecticut, Pennsylvania, and in mining districts in the Middle and Far West.

Siderite has hexagonal (rhombohedral) symmetry and the same structure as calcite. Individual crystals are often rhombohedral in shape, sometimes with curved faces (see illustration). Massive varieties also occur. Siderite is often brownish and sometimes gray or greenish. The specific gravity is 3.9 and the hardness is 4 on Mohs scale. The stability of siderite is dependent on the partial pressure of oxygen. *See* CARBONATE MINERALS.

[ROBERT I. HARKER]

Silica minerals

Silica (SiO_2) occurs naturally in at least nine different varieties (polymorphs), which include tridymite (high-, middle-, and low-temperature forms), cristobalite (high- and low-temperature forms), coesite, and stishovite, in addition to high (β) and low (α) quartz. These forms are characterized by distinctive crystallography, optical characteristics, physical properties, pressure-temperature stability ranges, and occurrences (see table).

The transformations between the various forms are of two types. Displacive transformations, such as inversions between high-temperature (β) and low-temperature (α) forms, result in a displacement or change in bond direction but involve no breakage of existing bonds between silicon and oxygen atoms. These transformations take place rapidly over a small temperature interval and are reversible. Reconstructive transformations, in contrast, involve disruption of existing bonds and subsequent formation of new ones. These changes are sluggish, thereby permitting a species to exist metastably outside its defined pressure-temperature stability field. Two examples of reconstructive transformations are tridymite $\rightleftharpoons$ quartz and quartz $\rightleftharpoons$ stishovite.

Crystal structure. The crystal structures of all silica polymorphs except stishovite contain silicon atoms surrounded by four oxygens, thus producing tetrahedral coordination polyhedra. Each oxygen is bonded to two silicons, creating an electrically neutral framework. Stishovite differs from the other silica minerals in having silicon atoms surrounded by six oxygens (octahedral coordination. *See* CRYSTAL STRUCTURE.

Ideal high tridymite is composed of sheets of SiO_4 tetrahedra oriented perpendicular to the *c* crystallographic axis (Fig. 1) with adjacent tetrahedra in these sheets pointing in opposite directions. The apical oxygens of the tetrahedra are bonded to silicons in neighboring sheets, thus generating a continuous framework with hexagonal ($P6_3/mmc$) symmetry. Naturally occurring meteoritic high tridymite deviates somewhat from the ideal structure because adjacent sheets are slightly offset, producing orthorhombic ($C222_1$) symmetry (Fig. 2). Low tridymite has a similar structure, but displacive transformations alter the geometry of the sheets, producing a lower symmetry. Terrestrial low tridymite contains three types of sheets in a complex stacking sequence resulting in triclinic (pseudoorthorhombic) symmetry. Meteor-

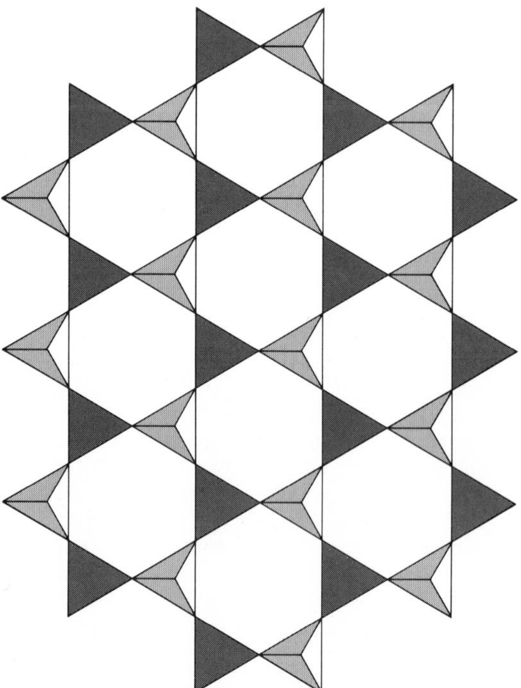

Fig. 1. Portion of an idealized sheet of tetrahedrally coordinated silicon atoms similar to that found in tridymite and cristobalite. Sharing of apical oxygens (which point in alternate directions) between silicons in adjacent sheets generates a continuous framework. *(From J. J. Papike and M. Cameron, Crystal chemistry of silicate minerals of geophysical interest, Rev. Geophys. Space Phys., 14:37–80; © 1976 by American Geophysical Union)*

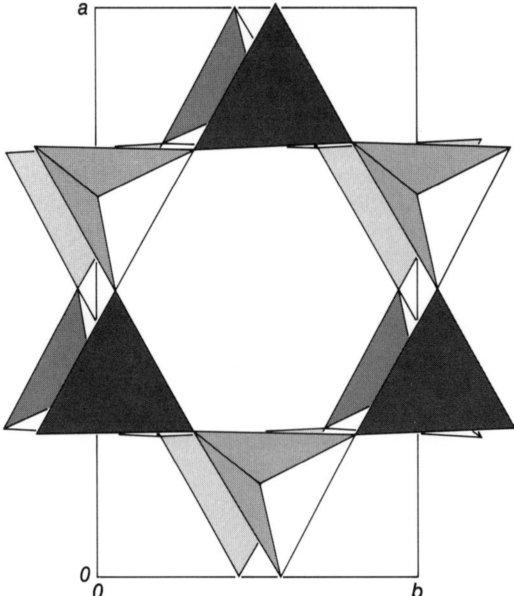

Fig. 2. Portion of the orthorhombic high tridymite structure viewed down the c axis showing the slight displacement between adjacent sheets. In ideal hexagonal high tridymite, neighboring sheets would be superimposed. *(From J. J. Papike and M. Cameron, Crystal chemistry of silicate minerals of geophysical interest, Rev. Geophys. Space Phys., 14:37–80; © 1976 by American Geophysical Union)*

itic low tridymite has a monoclinic (*Cc*) cell caused by a distortion of the hexagonal rings shown in Fig. 1. Little is known about the structure of middle tridymite, but it is believed to be hexagonal.

High cristobalite, like tridymite, is composed of parallel sheets of SiO_4 tetrahedra with neighboring tetrahedra pointing in opposite directions. However, the hexagonal rings are distorted and adjacent sheets are rotated 60° with respect to one another, resulting in the geometry shown in Fig. 3. The structure is cubic (*Fd3m*), with the layers of tetrahedra parallel to (111). Further distortion of these sheets at low temperature causes an inversion to tetragonal ($P4_12_12$ or $P4_32_12$) low cristobalite.

Coesite also contains silicon atoms tetrahedrally coordinated by oxygen. These polyhedra share corners to form chains composed of four-membered rings. Bonding between chains creates a continuous framework of monoclinic symmetry (*C*2/*c*) in which each oxygen is shared between two silicons. The structure, part of which is shown in Fig. 4, is significantly more dense than tridymite and cristobalite.

Silicon in stishovite is octahedrally coordinated by oxygen (Fig. 5). These coordination polyhedra share edges and corners to form chains of octahedra parallel to the *c* crystallographic axis. The resultant structure is tetragonal ($P4_2/mnm$) and is similar to the structure of rutile (TiO_2).

Properties. The physical, optical, and chemical characteristics of the silica polymorphs vary in relation to their crystal structures. Stishovite, which has the most dense packing of constituent atoms, has the highest specific gravity ($G = 4.35$), whereas tridymite and cristobalite, which have relatively open structures, are the silica polymorphs with the lowest specific gravities ($G_{Tr} = 2.26$, $G_{Cr} = 2.32$). In all instances the β forms of the polymorphs have less densely packed structures, hence lower specific gravities, than α counterparts. The refractive indices of the polymorphs are also related to crystal structure, and the mean refractive index is proportional to specific gravity, ranging from 1.47 for tridymite to 1.81 for stishovite. Specific optical and physical properties for the various silica minerals are given in the table. Twinning is common in most of these minerals and, as in the case of quartz, frequently results from the inversion of a high-temperature, high-symmetry form to a low-temperature, low-symmetry form.

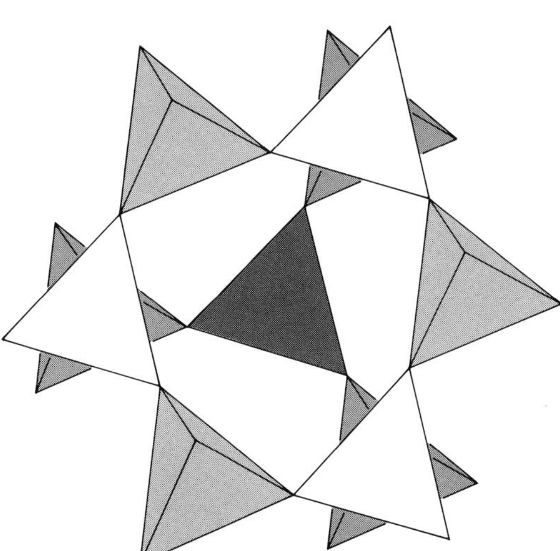

Fig. 3. Portion of cubic high cristobalite illustrating the distortion of tetrahedral sheets, which are oriented parallel to (111), and the 60° rotation of adjacent sheets. *(From J. J. Papike and M. Cameron, Crystal chemistry of silicate minerals of geophysical interest, Rev. Geophys. Space Phys., 14:37–80; © 1976 by American Geophysical Union)*

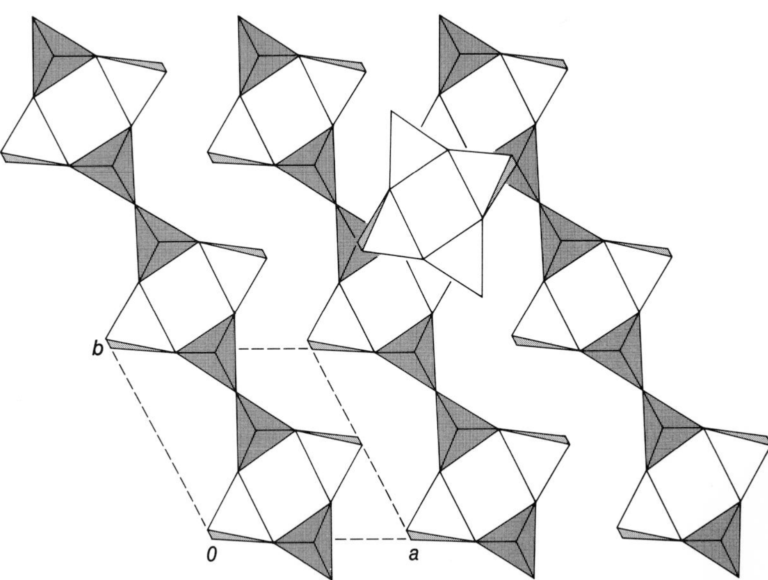

Fig. 4. Four-membered tetrahedral rings in coesite as seen in the plane perpendicular to the twofold axis of symmetry. Bonding between rings in this plane and adjacent planes (as indicated by the single ring at an elevated level) produces the dense coesite structure. *(From J. J. Papike and M. Cameron, Crystal chemistry of silicate minerals of geophysical interest, Rev. Geophys. Space Phys., 14:37–80; © 1976 by American Geophysical Union)*

Optical and physical properties of silica minerals

Mineral	Specific gravity	Optic sign	Mean index	Nα or Nω‡	Nβ‡	Nγ or Nε‡	Cleavage	Color	Hardness (Moh's scale)	Twinning
Stishovite	4.35	Uniaxial (+)	1.81	1.799		1.826	N.R.*	Pale gray	N.R.	N.R.
Coesite	3.01	Biaxial (+)	1.59	1.594	1.596	1.597	None	Colorless	8	{021}
Low (α) quartz	2.65	Uniaxial (+)	1.55	1.544		1.553	{1011}, {0111}, {1010}†	Colorless, white to variable	7	c axis {1120}, {1122}, and others
Low (α) cristobalite	2.32	Uniaxial (−)	1.48	1.487		1.484	None	Colorless, white, or yellowish	6½	{111}
Low (α) tridymite	2.26	Biaxial (+)	1.47	1.471–1.479	1.472–1.48	1.474–1.483	None to poor prismatic	Colorless, white	7	{110}

*N.R. = not reported.
†Cleavages not usually observed so that quartz is frequently described as having no cleavage.
‡Nα, Nβ, and Nγ are the three principal refractive indices for biaxial crystals: Nω and Nε are principal refractive indices for uniaxial crystals.

Chemically, all silica polymorphs are ideally 100% SiO_2. However, unlike quartz which commonly contains few impurities, the compositions of tridymite and cristobalite generally deviate significantly from pure silica. This usually occurs because of a coupled substitution in which a trivalent ion such as Al^{3+} or Fe^{3+} substitutes for Si^{4+}, with electrical neutrality being maintained by monovalent or divalent cations occupying interstices in the relatively open structures of these two minerals. Such substitutions may decrease the SiO_2 content to as little as 95 wt %.

Phase relationships and occurrences. The pressure-temperature stability fields for the silica minerals are shown in Fig. 6. Low quartz is the stable form at low temperatures and pressures. At 1 atm (101,325 N/m^2) high quartz is stable from 573 to 870°C, at which temperature tridymite becomes stable. The stability field for cristobalite ranges from 1470 to 1720°C, at which temperature SiO_2 melts. Coesite and stishovite are stable only at exceedingly high pressures. However, because of the sluggishness with which reconstructive changes take place, various polymorphs can exist and occasionally crystallize metastably outside their defined limits of stability. Thus high tridymite (β_2) can persist to 163°C, at which temperature it transforms displacively to middle tridymite (β_1), which exists metastably to 117°C, at which low (α) tridymite forms. Impurities or disordering of the stacking sequence can cause these transformation temperatures to vary and possibly merge into a single high/low transformation. High cristobalite can exist metastably below 1450°C to 268°C (or lower), at which temperature it transforms displacively to low cristobalite. Variation in the temperature of inversion is caused by compositional variability and stacking disorder.

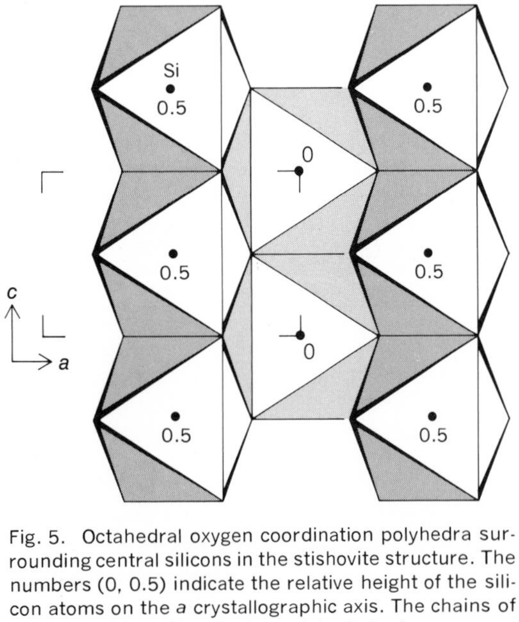

Fig. 5. Octahedral oxygen coordination polyhedra surrounding central silicons in the stishovite structure. The numbers (0, 0.5) indicate the relative height of the silicon atoms on the *a* crystallographic axis. The chains of octahedra formed by sharing polyhedral edges run parallel to the c axis. Chains at different levels are interconnected by silicon-oxygen bonds. *(From J. J. Papike and M. Cameron, Crystal chemistry of silicate minerals of geophysical interest, Rev. Geophys. Space Phys., 14:37–80; © 1976 by American Geophysical Union)*

Tridymite and cristobalite crystallize as primary minerals in siliceous volcanic rocks such as rhyolites, trachytes, and andesites in which both minerals may occur in vesicles (cavities) or in the groundmass. Less frequently they are found in basaltic igneous rocks. They have been identified in siliceous sedimentary rocks, such as sandstones and arkoses, subjected to high temperatures during contact thermal metamorphism. Both minerals

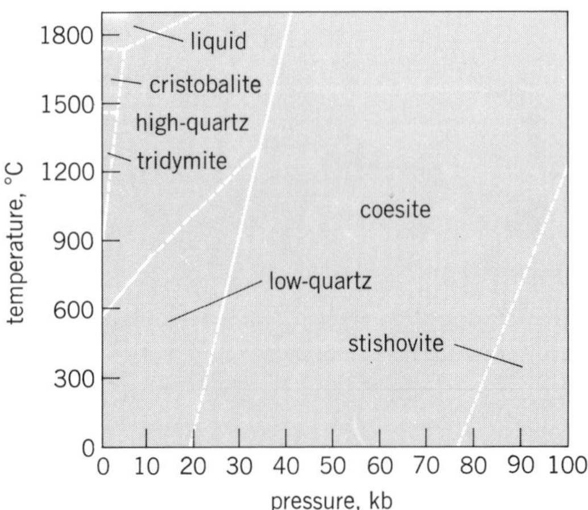

Fig. 6. Stability field of the silica polymorphs. *(From C. S. Hurlbut, Jr., and C. Klein, Manual of Mineralogy, 19th ed.; copyright © 1977 by John Wiley & Sons, Inc.; reprinted by permission)*

also occur in the silicate portions of meteorites and in some lunar igneous rocks. Cristobalite, and to a much lesser extent tridymite, is a common intermediate product in the transformation of amorphous biogenic silica (opal A) in marine sediments to quartz during the process of diagenesis (the lithification and alteration of unconsolidated sediments at low temperatures and pressures). The diagenetic sequence is as follows: opal A → opal CT (= porcellanite = cristobalite containing nonessential water in which the stacking sequence of silica sheets is disordered) → chalcedony (microcrystalline quartz) → quartz.

Naturally occurring coesite and stishovite were both first discovered in the Coconino sandstone adjacent to Meteor Crater, AZ, by E. T. C. Chao and coworkers. These minerals are considered mineralogical indicators of the high pressures associated with the shock wave that resulted from the crater-forming meteorite impact. Since their discovery at Meteor Crater, they have both been found in the highly shocked rocks surrounding Ries Crater in Germany. Coesite has also been identified in rocks associated with several other impact or suspected impact craters, in tektites, in material ejected from human-caused explosion craters, and as inclusions in diamonds from kimberlite originating at great depth (hence at great pressure) within the Earth. Stishovite and coesite, therefore, are of great geological significance as indicators of very high pressure due either to depth of formation within the Earth or to shock processes such as those occurring during meteorite impact. *See* COESITE; QUARTZ; STISHOVITE.

[JOHN C. DRAKE]

Bibliography: W. A. Deer, R. A. Howie, and J. Zussman, *Rock Forming Minerals*, vol. 4: *Framework Silicates*, 1963; C. Frondel, *Dana's The System of Mineralogy*, 7th ed., vol. 3: *Silica Minerals*, 1962; J. J. Papike and M. Cameron, Crystal chemistry of silicate minerals of geophysical interest, *Rev. Geophys. Space Phys.*, 14:37–80, 1976; D. Stöffler, Coesite and stishovite in shocked crystalline rocks, *J. Geophys. Res.*, 76:5474–5488, 1971.

Silicate minerals

All silicates are built of a fundamental structural unit, the so-called SiO_4 tetrahedron. The crystal structure may be based on isolated SiO_4 groups or, since each of the four oxygen ions can bond to either one or two silicon (Si) ions, on SiO_4 groups shared in such a way as to form complex isolated groups or indefinitely extending chains, sheets, or three-dimensional networks. Mixed structures in which more than one type of shared tetrahedra are present also are known.

Classification. Silicates are classified according to the nature of the sharing mechanism, as revealed by x-ray diffraction study. An abbreviated form of such a classification is given in the table. The sharing mechanism gives rise to a characteristic ratio of Si to O, but it is possible for oxygen ions that are not bonded to Si to be present in the structure, and sometimes some or all of any aluminum present must be counted as equivalent to Si. The constitution and classification of the silicates were controversial before the advent of x-ray structure analysis methods in 1912. The silicates were then usually considered as salts of silicic acids, many of them hypothetical, and a chemical classification, such as orthosilicates, metasilicates, and the like, was applied.

Structure. A dominant feature of the crystal chemistry of the silicates that in large part determines the chemical complexity of these species is the dual role played by aluminum in the crystal structure. The radius ratio of this ion with oxygen is near the critical value between four-coordination and six-coordination, and the aluminum (Al) ion can occur in one or the other or both roles simultaneously. When in four-coordination, the trivalent Al ion substitutes for the quadrivalent Si ion, introducing a valence deficiency of one unit and requiring a concomitant substitution of another cation elsewhere in the crystal structure to provide valence compensation. This mechanism usually involves the coupled substitution of a divalent for a monovalent cation, as of calcium (Ca) for sodium (Na), less frequently of a trivalent for a divalent ion. Other less common mechanisms involve coupled omissions or substitutions among the anionic units of structure. In some silicates, such as those with a silicon-oxygen framework based on a polymorph of silica, the serial substitution of Al for Si is compensated by the entrance of a cation, such as Na, into vacant interstices of the crystal.

The detailed crystallographic and physical properties of the various silicates are broadly related to the type of silicate framework that they possess (see illustration). Thus, the phyllosilicates as a group typically have a platy crystal habit, with a cleavage parallel to the plane of layering of the structure, and are optically negative with rather high birefringence. The inosilicates, based on an extended one-dimensional rather than two-dimensional linkage of the SiO_4 tetrahedra, generally form crystals of prismatic habit; if cleavage is present, it will be parallel to the direction of elongation. The tectosilicates commonly are equant in

Silicate structures and their characteristic Si/O ratios

Type	Nature of Si-O linkage	Si/O ratio	Examples
Nesosilicates	Isolated SiO_4 groups	1:4	Olivine, garnet
Sorosilicates	Isolated compound groups, Si_2O_7, Si_6O_{18}, and so on	2:7, 6:18, and so on	Thortveitite, beryl
Inosilicates	One-dimensional extended chains and bands	1:3, 4:11, and so on	Amphiboles, pyroxenes
Phyllosilicates	Two-dimensional extended sheets	2:5	Mica, clays, talc, chlorite
Tectosilicates	Three-dimensional networks	1:2	Feldspars, feldspathoids, zeolites

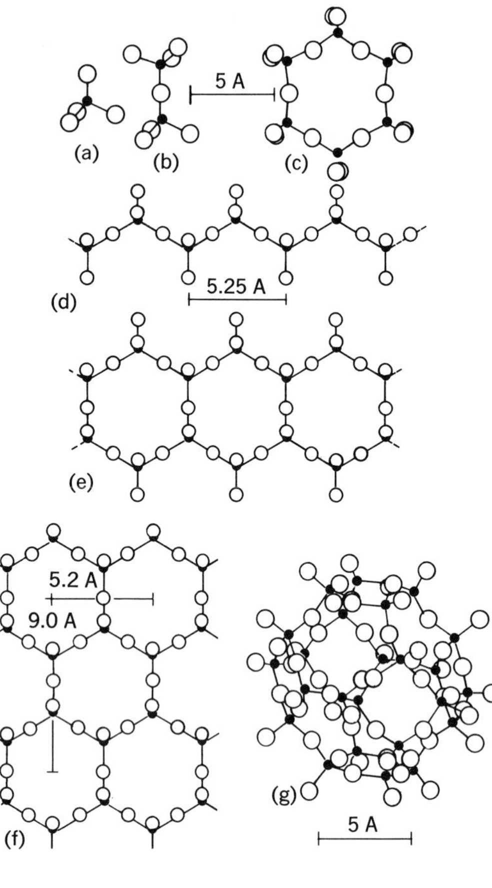

Forms of silicon-oxygen linkage. (*a*) Nesosilicate $(SiO_4)^{4-}$. (*b*) Sorosilicate $(Si_2O_7)^{6-}$. (*c*) Sorosilicate, or cyclosilicate $(Si_6O_{18})^{12-}$. (*d*) Inosilicate $(SiO_3)^{2-}$, showing chain structure of pyroxenes. (*e*) Inosilicate $(Si_4O_{11})^{6-}$, showing band structure of amphiboles. (*f*) Phyllosilicate $(Si_2O_5)^{2-}$, showing extended sheets. (*g*) Tectosilicate, showing three-dimensional structure of lazurite. (*From W. L. Bragg, Atomic Structure of Minerals, Cornell, 1937*)

habit, without marked preference for cleavage direction, and tend to have a relatively low birefringence.

Important minerals. Silicate minerals make up the bulk of the outer crust of the Earth and form in a wide range of geologic environments. Many silicates are of economic importance. Among the clays, feldspars, and refractory minerals, andalusite and wollastonite are used in the ceramic industries, mica as an electrical insulating agent, asbestos and exfoliated vermiculite as thermal insulating agents, and garnet as an abrasive. Talc is a constituent of facial powder. Other silicates are important as ore minerals, beryllium being obtained from beryl, zirconium and hafnium from zircon, and thorium from thorite. Some silicates such as jadeite and nephrite are prized as ornamental materials, and peridot, garnet, tourmaline, and aquamarine are well-known gemstones. *See* CLAY MINERALS.

For discussions of certain silicate mineral groups *see* AMPHIBOLE; ANDALUSITE; CHLORITE; CHLORITOID; EPIDOTE; FELDSPAR; FELDSPATHOID; GARNET; HUMITE; MICA; OLIVINE; PYROXENE; SCAPOLITE; SERPENTINE; ZEOLITE.

[CLIFFORD FRONDEL]

Siliceous sinter

A porous silica deposit formed around hot springs. It is white to light gray and sometimes friable. Geyserite is a variety of siliceous sinter formed around geysers. The siliceous sinters are deposited as the hot subterranean waters cool after issuing at the surface and become supersaturated with silica that was picked up at depth. The sinters are frequently deposited on algae that live in the pools around the hot springs. *See* GEYSER; THERMAL SPRING.

[RAYMOND SIEVER]

Sillimanite

SILLIMANITE

1 in.

Sillimanite crystals from Chester, CT. (*Specimen from Department of Geology, Bryn Mawr College*)

A nesosilicate mineral, composition Al_2SiO_5, crystallizing in the orthorhombic system. It commonly occurs in slender crystals or parallel groups and is frequently fibrous, hence the synonym fibrolite (see illustration). There is one direction of perfect cleavage; the luster is vitreous and the color brown, pale green, or white. The hardness is 6–7 on Mohs scale and the specific gravity is 3.23. *See* SILICATE MINERALS.

Sillimanite, kyanite, and andalusite are all polymorphic forms of Al_2SiO_5. They are metamorphic minerals, found in highly aluminous gneisses and schists. Each mineral is stable under different conditions but the transitions from one to another are so sluggish that they may coexist in the same rock. Sillimanite is less common than the others and is found in the highest-grade metamorphic rocks associated with quartz, corundum, garnet, and muscovite. It is rarely found as a contact metamorphic mineral. *See* ANDALUSITE; KYANITE.

[CORNELIUS S. HURLBUT, JR.]

Silurian

A geologic period of time during which a system of rocks, the Silurian System, was deposited. The strata are identified in terms of the principal forms of life extant at that time as determined by means of the fossils.

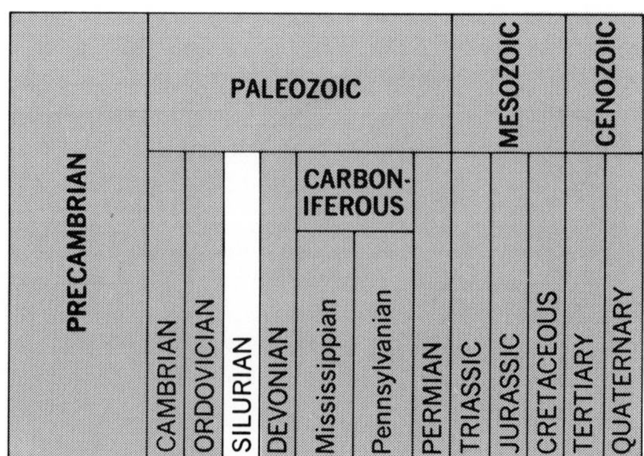

This system was named by R. I. Murchison for exposures in Wales, the name being derived from an ancient Celtic tribe (Silures) that occupied this region during the Roman conquest. Murchison described the stratigraphic character and fossil content of these rocks in his monograph *The Silurian System* in 1839. In this and later publications the Silurian system was defined to include some older strata which were later removed to the Ordovician system by C. Lapworth in 1879. *See* ORDOVICIAN.

Silurian series. Silurian rocks are present in many parts of the world (Figs. 1 and 2), but three of the best-known sections are in Great Britain, eastern United States, and central Bohemia. In the type region of the British Isles the Silurian is divided into four series, in ascending order: Llandoverian, Wenlockian, Ludlovian, and Downtonian (Skala). The Silurian rocks of the eastern United States are divided into three series, the Lower or Medinan (sometimes Alexandrian) series, the Middle Silurian or Niagaran series, and the Upper Silurian or Cayugan series. The Silurian strata in western Czechoslovakia (Bohemia) were first studied by Joachim Barrande, who published a series of paleontological monographs on the fossils of this region (*Système Silurien du centre de la Bohême*, 1865–1907). These rocks are now divided into three series, Liteň, Kopanina, and Přídolí (see table).

In addition to the areas mentioned above, the island of Gotland in the Baltic Sea has a well-known Silurian section; several of the fossils from this island were described by the famous Swedish naturalist Karl von Linné (Linnaeus). Other regions that have been studied in some detail include Poland, southern Norway, the Urals in the Soviet Union, China, and Australia.

Silurian facies. Silurian deposits are represented by both terrestrial and marine strata, but the latter are predominant. The marine strata are composed largely of carbonate facies, either limestone or dolomite, but there are sandstones and shales, including dark graptolite shales. The marine strata commonly contain a large invertebrate fossil fauna. This is especially true of the limestones and dolomites. Terrestrial Silurian deposits are much less common. Strata of this type are present in southeastern New York and eastern Pennsylvania, mostly as conglomeratic sandstones bearing a sparse eurypterid fauna. Westward these strata grade into a typical marine facies as shown in Fig. 3. Terrestrial Silurian strata have been reported on the southern tip of Africa, although the age of these rocks is in question.

In Late Silurian times the seaways of eastern North America began to dry up, leading to the formation of salt beds. Thick beds of Cayugan salt underlie parts of New York, Pennsylvania, and Michigan, and in places are mined on a large scale. Lower and Middle Silurian strata bear extensive iron deposits, mostly in the form of hematite. Rich iron deposits of this type are mined around Birmingham, Ala. Some oil and gas are obtained from beds of Silurian age in Ohio, western New York, and other areas. *See* EVAPORITE, SALINE.

Climate. The climate during the Middle Silurian must have been mild. The invertebrate faunas have a cosmopolitan aspect. Similar and even identical species have a wide geographic distribution from equatorial regions into latitudes north of the Arctic Circle. Moreover, extensive Niagaran carbonate deposits containing numerous organic reefs are known from strata at least as far north as 50°N lat. In the latter part of the Silurian Period more severe conditions appear to have prevailed. The extensive salt deposits of the Upper Silurian indicate increasing aridity and probably decreasing temperature, most likely associated with the Caledonian mountain building which affected parts of arctic North America, Scandinavia, France, Germany, northern Africa, and Siberia.

Series of the Silurian System

United States	Great Britain	Czechoslovakia (Bohemia)
Cayugan	Downtonian (Skala)	Přídolí
Niagaran	Ludlovian	Kopanina
	Wenlockian	Liteň
Medinan	Llandoverian	

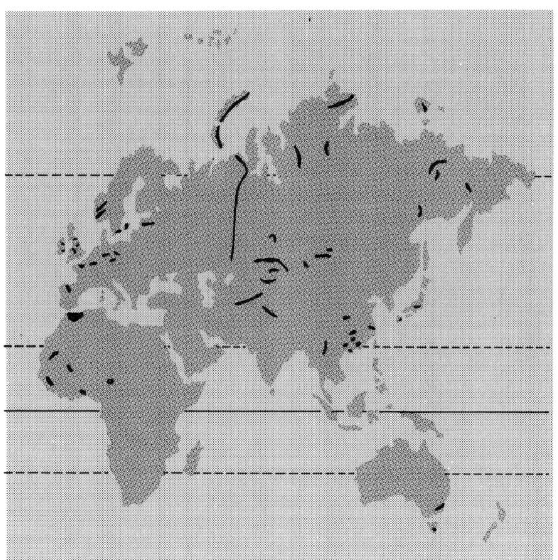

Fig. 1. Map showing generalized distribution of Silurian outcrops in Eastern Hemisphere.

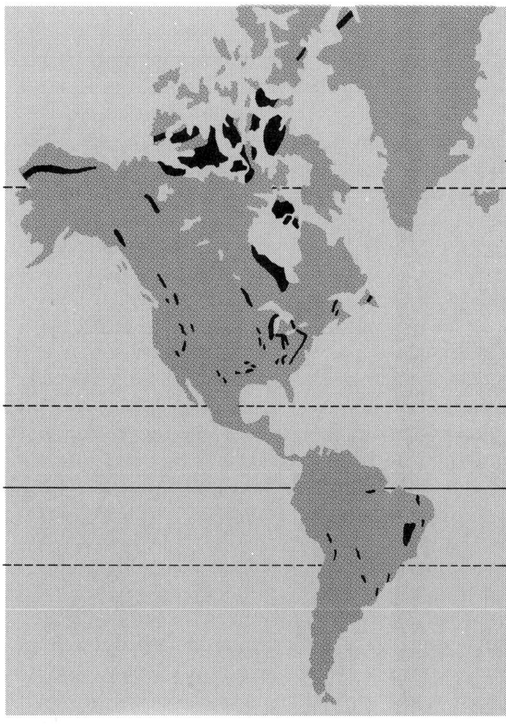

Fig. 2. Map showing generalized distribution of Silurian outcrops in Western Hemisphere.

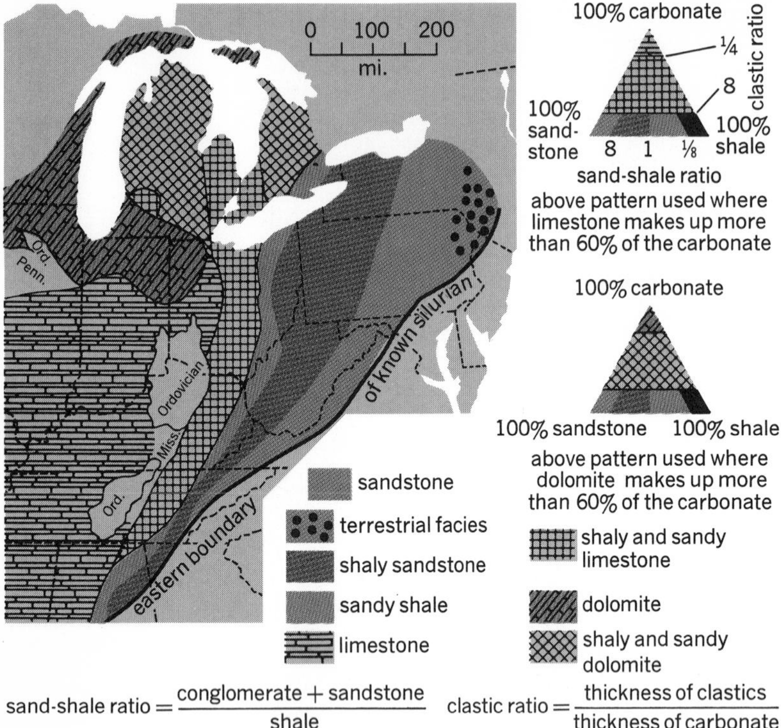

Fig. 3. Lithofacies map of Lower Silurian (Medinan) series in the eastern United States. Terrestrial deposits are present in this area. Pure shale facies are not present. Triangular diagrams show how lithofacies map was constructed. (*After T. W. Amsden, Bull. Amer. Ass. Petrol. Geol., 39(1):60–74, 1955*)

Silurian life. Silurian rocks contain a prolific invertebrate fossil assemblage. Among the best-represented groups were the brachiopods belonging to the class Articulata (Pygocaulia). Pentameroid brachiopods such as *Conchidium* and *Pentamerus* and dalmanelloids such as *Rhipidomelloidea* were especially abundant. Another common group, the spire-bearing brachiopods, became numerous for the first time. The coral faunas also were prolific and included both solitary and colonial tetracorals. The tabulate corals were even more numerous. Such genera as *Favosites* and *Halysites* are among the more common of Silurian fossils. Locally the corals were associated with the stromatoporoids. Bryozoa, and other sedentary reef-building organisms. The straight-shelled nautiloid cephalopods were abundant, but other mollusk groups were well developed only in local areas.

Crinoids (echinoderms) became common for the first time. Complete articulated crowns are seldom found, but the stems and isolated plates are abundant. The stems and plates are so concentrated in some beds that they become the dominant rock constituent (encrinite). Trilobites and graptolites are present, although fewer species are found than in Ordovician rocks. The eurypterids are an extinct group of Arachnoidea that flourished during the Silurian and Devonian periods; a few Silurian species attained a length of almost 9 ft, making them the largest of all known Arthropoda. Scorpions with a body form similar to that of living species have been found in Middle Silurian strata. If these were air breathers they represent the oldest known terrestrial animals; evidence concerning their mode of respiration is inconclusive.

Primitive fish are the only known Silurian vertebrates. One group, the Agnatha (ostracoderms), had no jaws and were early relatives of the modern cyclostome fish. A second group, the Placodermi, had a primitive jaw apparatus and made their appearance late in the Silurian, although they did not become common until the Devonian Period. The head and front part of the body of most of these early fish were protected by an armor of bony plates.

Calcareous algal structures are present in many Silurian beds. These primitive aquatic plants are present in ancient Precambrian strata, and are locally common in many Paleozoic rocks. Of greater significance is the presence of vascular plants belonging to the Pteridophyta (Tracheophyta). These plants, which are representatives of the Psilopsida and Lycopsida, undoubtedly lived on land and are the oldest known terrestrial plants.

[THOMAS W. AMSDEN]

Bibliography: C. O. Dunbar, *Historical Geology*, 2d ed., 1960; A. Martinsson, The series of the redefined Silurian System, *Lethaia*, 2:153–161, 1969; R. C. Moore, *Introduction to Historical Geology*, 2d ed., 1958; E. Neaverson, *Stratigraphical Palaeontology*, 2d ed., 1955; C. Schuchert, *Stratigraphy of the Eastern and Central United States*, Historical Geology of North America, vol. 2, 1943; H. Termier and G. Termier, *Histoire géologique de la biosphère*, 1952.

Skarn

A rock term generally reserved for rocks composed entirely, or almost so, of lime-bearing silicates, and derived from nearly pure limestones and dolomites into which large amounts of silicon, Si; aluminum, Al; iron, Fe; and magnesium, Mg, have been introduced. *See* METAMORPHIC ROCKS; PNEUMATOLYSIS.

Skarn is an old term in mining, and was originally used by Swedish miners to designate dark silicate minerals occurring in seams or masses adjacent to an ore-bearing vein. Now it is applied to a coarse-grained rock or association of minerals formed by reaction between hot silica-rich solutions or acid gases and a limestone or dolomite. The reaction rock is made up of coarse masses of minerals of various kinds, depending upon the temperature and the composition of the reacting gases or solutions. The most important skarn minerals are andradite-garnet, hedenbergite-diopside, iron-rich hornblende, and actinolite-tremolite. Various ore minerals, oxides, and sulfides, as well as fluorite, often occur in connection with skarn. The common occurrence of fluorite, CaF_2, supports the idea that silica and metal halogenides have reacted in the limestone. Thus the formation of andradite may be explained by Eq. (1). Not only

$$2FeF_3 + 3SiO_2 + 6CaCO_3 \rightarrow \underset{\text{Andradite}}{Ca_3Fe_2Si_3O_{12}} + \underset{\text{Fluorite}}{3CaF_2} + 6CO_2 \quad (1)$$

are fluorides active, but also chlorides may be introduced, in which case $CaCl_2$ is formed and reacts with formation of scapolite. Large masses and dikes of scapolite are common in many skarn rocks.

Oxidic ore is very common in skarns and forms in many places useful deposits of skarn ore. Ore formation will take place especially if silica is not introduced, as in Eq. (2). Thus were formed, for

$$2FeF_3 + 3CaCO_3 \rightarrow Fe_2O_3 + 3CaF_2 + 3CO_2 \quad (2)$$

example, the famous hematite deposits in Elba containing large masses of andradite plus the calc-iron silicate lievrite at the contact between granite and limestone. Among other classical contact deposits with skarn formation are those at Campiglia Maritima, Tuscany; Banat, Rumania; Franklin Furnace, N.J.; and Clifton-Morenci, Ariz. *See* ORE AND MINERAL DEPOSITS. [T. F. W. BARTH]

Slate

A group name for various very fine-grained rocks derived from mudstones, siltstones, and other clayey sediments as a result of low-degree regional metamorphism. Highly characteristic of slates is the perfect fissility or slaty cleavage which is a regular and perfect planar schistosity, the slates themselves thus grading into phyllites. *See* CLEAVAGE, ROCK; METAMORPHIC ROCKS; PHYLLITE; SCHISTOSITY, ROCK.

Development of slaty cleavage. The manner in which slaty cleavage develops is best illustrated by ordinary argillaceous sediments and by some fine volcanic tuffs. Other types of sediments, calcareous or quartzitic, may occasionally show a similar cleavage but of a less perfect type. Original semispherical bodies (for example, round fossils) which have been sheared during the deformation of the rock afford by their distortion some measure of the deformation. In a usual roofing slate the ratio of the semiaxis $a{:}b{:}c$ is frequently about 1.5:1:0.4. There has been, therefore, a great compression of the rock in the direction perpendicular to the cleavage planes and a certain elongation along the cleavage planes in the direction of the dip

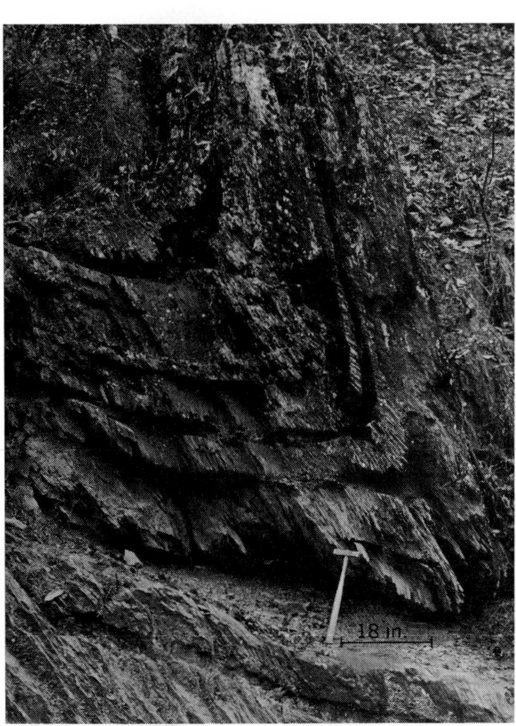

Fig. 1. Cleavage in banded slate. Folded rock structure near Walland, Blount County, Tenn. (*USGS*)

Fig. 2. Detail of cleavage in folded structure beds showing small horizontal bed of quartzite in westwardly overturned plications and quartz vein in slaty cleavage dipping east; from Rutland County, Vt. (*USGS*)

(Figs. 1 and 2). The mineral constituents are fine-grained and cannot be discerned by the naked eye. With the microscope it can be seen that a large part of the rock is made up of thin flaky plates of muscovite or chlorite set in subparallel position. These minerals have recrystallized; that is, they have grown during the low-grade metamorphism suffered by the slates. It is of general interest in the study of the fabric of rocks that crystals grow with the axis of best thermal conductivity perpendicular to the isotherms and, more importantly, with the axis of largest compressibility along the pressure gradient. Therefore, the newly formed crystalloblasts of muscovite, chlorite, stilpnomelane (all being phyllosilicates exhibiting the characteristic sheetlike nature in their crystal structure), all orient themselves, as they grow, in flaky crystals sitting on the planes of slaty cleavage, thus highly accentuating the cleavage. One may distinguish between, on the one hand, paracrystalline and precrystalline deformation commonly identified by undeformed aligned flakes of mica and chlorite which crystallized during or later than the slip movements and, on the other hand, postcrystalline movements partially aligning favorably situated flakes of mica and chlorite which become bent and twisted in the process. *See* SILICATE MINERALS.

Mineral constituents. The chief minerals of slates are muscovite (as sericite), chlorite, and quartz. Common accessories are tourmaline, rutile, epidote, sphene, hematite, and ilmenite. Stilpnomelane has been shown to be a major constituent in many slatelike rocks in southern New Zealand. Stilpnomelane has a position between chlorite and the clay minerals, and may have a wider general distribution in low-metamorphic rocks than has usually been recognized.

Uses. Slates are widely used for roofing purposes. They are easily prepared and fixed, are weatherproof and durable, and in many areas are cheaper and better than any other thatching materials.

The active slate-producing districts in the United States are the Monson (Maine) district; the New York–Vermont district (including Washington and Rutland counties); the Lehigh district, Peach Bottom district, and Berks County, all in Pennsylvania; Harford County, Md.; and Buckingham and Albemarle counties, Va. There are also important quarries in England (Devon, the Lake district), North Wales, Scotland (Ballachulish), Ireland (Kilkenny), France (the Ardennes, and Bohemia, Germany (near Coblenz).

The making of slates is still performed by hand using chisel and mallet. Big slabs are split into separate slates, the thickness of which varies with the size required and the quality of the rock. An average roofing tile of the best kind of rock is about 5 mm thick. The slates are afterward trimmed to size either by hand or by machine-driven rotating knives.

[T. F. W. BARTH]

Smithsonite

The natural form of zinc carbonate. Smithsonite may contain small amounts of iron, calcium, cobalt, copper, manganese, cadmium, and magnesium. It occurs chiefly as a secondary mineral formed by the oxidation of primary zinc ores, such as sphalerite.

Smithsonite has hexagonal (rhombohedral) symmetry and has the calcite structure type. It is rarely found in well-formed crystals and is more often in compact or porous masses (see illustration). Its color is often dirty white, but a variety of tints, including greens, blues, and browns, may be found, depending on the impurities present. For pure smithsonite the hardness is about 4 or $4\frac{1}{2}$ on Mohs scale and the specific gravity is 4.43.

Very pure smithsonite can be synthesized by heating zinc oxide to moderate temperatures at elevated pressures of carbon dioxide. The thermal stability of smithsonite at various pressures of carbon dioxide has been found by experiment to range from 350 to 450°C between 15,000 and 50,000 psi.

Botryoidal aggregate of smithsonite in limestone, Larium, Greece. (*Specimen from Department of Geology, Bryn Mawr College*)

Smithsonite has been referred to as calamine, but the term is confusing as it has also been used as a synonym for hemimorphite and hydrozincite. Smithsonite occurs at many localities in Europe and Africa and throughout the United States. Large deposits occur at Monarch and Leadville, Colo. *See* CARBONATE MINERALS.

[ROBERT I. HARKER]

Soapstone

A soft talc-rich rock. Soapstones are rocks composed of serpentine, talc, and carbonates (magnesite, dolomite, or calcite). They represent original peridotites which were altered at low temperatures by hydrothermal solutions containing silicon dioxide, SiO_2; carbon dioxide, CO_2; and other dissolved materials (products of low-grade metasomatism). Among the rock products thus formed are antigorite schists, actinolite-talc schists, and talc-carbonate rocks. To the last belongs the true soapstone, but the whole group of rocks may loosely be referred to as soapstones because of their soft, soapy consistency. Such rocks were selected by prehistoric men for making primitive vessels and pots, and for making rough carvings for ornamental purposes. *See* TALC. [T. F. W. BARTH]

Soda niter

A nitrate mineral having chemical composition $NaNO_3$ (sodium nitrate). Soda niter is by far the most abundant of the nitrate minerals. It crystallizes in the rhombohedral division of the hexagonal system. It sometimes occurs as simple rhombohedral crystals but is usually massive granular (see illustration). The mineral has a perfect rhombohedral cleavage, conchoidal fracture, and is rather sectile. Its hardness is $1\frac{1}{2}$ to 2 on Mohs scale, and its specific gravity is 2.266. It has a vitreous luster and is transparent. It is colorless to white, but when tinted by impurities, it is reddish brown, gray, or lemon yellow. *See* NITRATE MINERALS.

Soda niter is a water-soluble salt found principally as a surface efflorescence in arid regions, or in sheltered places in wetter climates. It is usually associated with niter, nitrocalcite, gypsum, epsomite, mirabilite, and halite.

The only large-scale commercial deposits of soda niter in the world occur in a belt roughly 450 mi long and 10–50 mi wide along the eastern slope of the coast ranges in the Atacama, Tarapaca, and Antofagasta Deserts of northern Chile. The deposits consist of a thin bed of nitrates and associated minerals, varying from a few inches to a few feet in thickness, overlain by a shallow overburden of sand and gravel. The crude soda niter, known as caliche, is about one-fourth sodium nitrate, admixed with other salts, notably bloedite, anhydrite, gypsum, polyhalite, halite, glauberite, and darapskite, together with minor amounts of various iodates, chromates, and borates. Because of the presence of the iodate minerals lautarite and dietzeite, these deposits yield most of the world's supply of iodine.

The origin of these deposits is controversial. It is generally agreed that the nitrates were transported

Soda niter from La Noria, Chile. (*Specimen from Department of Geology, Bryn Mawr College*)

by groundwater and deposited by evaporation. The source of the nitrate has been attributed to (1) guano; (2) nitrogen fixation by electrical storms; (3) the bacterial fixation of nitrogen from vegetable matter; and (4) a volcanic source in nearby Triassic and Cretaceous rocks. The last source seems the most probable.

Chilean nitrate had a monopoly of the world's fertilizer market for many years, but now occupies a subordinate position owing to the development of synthetic processes for nitrogen fixation which permit the production of nitrogen from the air. This has led to the commercial production of artificial nitrates and has reduced the need to import Chilean nitrates for use in the manufacture of fertilizers and explosives. *See* CALICHE.

Small deposits of soda niter similar to those in Chile are found in Bolivia, Peru, North Africa, Egypt, the Soviet Union, India, and the western United States.

[GEORGE SWITZER]

Sodalite

A mineral tectosilicate of the feldspathoid group, crystallizing in the isometric system, with chemical composition $Na_4Al_3Si_3O_{12}Cl$. Crystals, which are rare, are usually dodecahedrons. Sodalite is most commonly massive or granular. There is poor dodecahedral cleavage. The hardness is 5–6 on Mohs scale and the specific gravity 2.2–2.4. The luster is vitreous and the color, usually blue, may also be white, gray, or green. Sodalite has been cut and polished for use as an ornamental stone. *See* FELDSPATHOID; SILICATE MINERALS.

Sodalite is a relatively rare mineral found in nepheline syenites and leucite-bearing rocks. It is thus associated with nepheline, leucite, cancrinite, and feldspar. It is chiefly a primary mineral, although some has formed by the alteration of nepheline. Transparent crystals have been found in the lavas of Vesuvius, and the massive blue variety occurs in Ontario, Quebec, and British Columbia, Canada; and Litchfield, Maine.

[CORNELIUS S. HURLBUT, JR.]

Soil

Freely divided rock-derived material containing an admixture of organic matter and capable of supporting vegetation. Soils are independent natural bodies, each with a unique morphology resulting from a particular combination of climate, living plants and animals, parent rock materials, relief, the groundwaters, and age. Soils support plants, occupy large portions of the Earth's surface, and have shape, area, breadth, width, and depth. Soil, as used here, differs in meaning from the term as used by engineers, where the meaning is unconsolidated rock material. *See* SOIL CHEMISTRY.

This article is divided into four parts: origin and classification of soils, physical properties of soil, soil management, and soil erosion.

ORIGIN AND CLASSIFICATION OF SOILS

Soil covers most of the land surface as a continuum. Each soil grades into the rock material below and into other soils at its margins, where changes occur in relief, groundwater, vegetation, kinds of rock, or other factors which influence the development of soils. Soils have horizons, or layers, more or less parallel to the surface and differing from those above and below in one or more properties, such as color, texture, structure, consistency, porosity, and reaction (Fig. 1). The horizons may be thick or thin. They may be prominent, or so weak that they can be detected only in the laboratory. The succession of horizons is called the soil profile. In general, the boundary of soils with the underlying rock or rock material occurs at depths ranging from 1 to 6 ft, though the extremes lie outside of this range.

Origin of soils. Soil formation proceeds in stages, but these stages may grade indistinctly from one into another. The first stage is the accu-

Fig. 1. Photograph of a soil profile showing horizons. The dark crescent-shaped spots at the soil surface are the result of plowing. The dark horizon lying 9–18 in. below the surface is the principal horizon of accumulation of organic matter that has been washed down from the surface. The thin wavy lines were formed in the same manner.

mulation of unconsolidated rock fragments, the parent material. Parent material may be accumulated by deposition of rock fragments moved by glaciers, wind, gravity, or water, or it may accumulate more or less in place from physical and chemical weathering of hard rocks. *See* WEATHERING PROCESSES.

The second stage is the formation of horizons. This stage may follow or go on simultaneously with the accumulation of parent material. Soil horizons are a result of dominance of one or more processes over others, producing a layer which differs from the layers above and below.

Major processes. The major processes in soils which promote horizon differentiation are gains, losses, transfers, and transformations of organic matter, soluble salts, carbonates, silicate clay minerals, sesquioxides, and silica. Gains consist normally of additions of organic matter, and of oxygen and water through oxidation and hydration, but in some sites slow continuous additions of new mineral materials take place at the surface or soluble materials are deposited from groundwater. Losses are chiefly of materials dissolved or suspended in water percolating through the profile or running off the surface. Transfers of both mineral and organic materials are common in soils. Water moving through the soil picks up materials in solution or suspension. These materials may be deposited in another horizon if the water is withdrawn by plant roots or evaporation, or if the materials are precipitated as a result of differences in pH (degree of acidity), salt concentration, or other conditions in deeper horizons.

Other processes tend to offset those that promote horizon differentiation. Mixing of the soil occurs as the result of burrowing by rodents and earthworms, overturning of trees, churning of the soil by frost, or shrinking and swelling. On steep slopes the soil may creep or slide downhill with attendant mixing. Plants may withdraw calcium or other ions from deep horizons and return them to the surface in the leaf litter.

Saturation of a horizon with water for long periods makes the iron oxides soluble by reduction from ferric to ferrous forms. The soluble iron can move by diffusion to form hard concretions or splotches of red or brown in a gray matrix. Or if the iron remains, the soil will have shades of blue or green. This process is called gleying, and can be superimposed on any of the others.

The kinds of horizons present and the degree of their differentiation, both in composition and structure, depend on the relative strengths of the processes. In turn, these relative strengths are determined by the way man uses the soil as well as by the natural factors of climate, plants and animals, relief and groundwater, and the period of time during which the processes have been operating.

Composition. In the drier climates where precipitation is appreciably less than the potential for evaporation and transpiration, horizons of soluble salts, including calcium carbonate and gypsum, are normally found at the average depth of water penetration.

In humid climates some materials normally considered insoluble may be gradually removed from the soil or at least from the surface horizons. A part of the removal may be in suspension. The movement of silicate clay minerals would be an example. The movement of iron oxides is accelerated by the formation of chelates with the soil organic matter. Silica is removed in appreciable amounts in solution or suspension, though quartz sand is relatively unaffected. In warm humid climates free iron and aluminum oxides and silicate clays accumulate in soils, apparently because of low solubility relative to other minerals.

In cool humid climates solution losses are evident in such minerals as feldspars. Free sesquioxides tend to be removed from the surface horizons and to accumulate in a lower horizon, but mixing by animals and falling trees may counterbalance the downward movement.

Structure. Concurrently with the other processes, distinctive structures are formed in the different horizons. In the surface horizons, where there is a maximum of biotic activity, small animals, roots, and frost action keep mixing the soil material. Aggregates of varying sizes are formed and bound by organic matter, microorganisms, and colloidal material. The aggregates in the immediate surface tend to be loosely packed with many large pores among them. Below this horizon of high biotic activity, the structure is formed chiefly by volume changes due to wetting, drying, freezing, thawing, or shaking of the soil by roots of trees swaying with the wind. Consequently, the sides of any one aggregate, or ped, conform in shape to the sides of adjacent peds.

Water moving through the soil usually follows root channels, wormholes, and ped surfaces. Accordingly, materials that are deposited in a horizon commonly coat the peds. In the horizons that have received clay from an overlying horizon, the peds usually have a coating or varnish of clay making the exterior unlike the interior in appearance. Peds formed by moisture or temperature changes normally have the shapes of plates, prisms, or blocks.

Horizons. Pedologists have developed sets of symbols to identify the various kinds of horizons commonly found in soils. The nomenclature originated in Russia, where the letters A, B, and C were applied to the main horizons of the black soils of the steppes. A designated the dark surface horizon of maximum organic matter accumulation, C the unaltered parent material, and B the intermediate horizon. The usage of the letters A, B, and C spread to western Europe, where the intermediate or B horizon was a horizon of accumulation of free sesquioxides or silicate clays or both. Thus the idea developed that a B horizon is a horizon of accumulation. Some, however, define a B horizon by position between A and C. Subdivisions of the major horizons have been shown by either numbers or letters, for example, Bt or B2. No internationally accepted set of horizon symbols has been developed. In the United States the designations shown in Fig. 2 have been widely used since about 1935, with minor modifications made in 1962. Lowercase letters were added to numbers in B horizons to indicate the nature of the material that had accumulated. Generally, "h" is used to indicate translocated humus, "t" for translocated clay, and "ir" for translocated iron oxides. Thus, B2t indicates the main horizon of clay accumulation.

Classification. Systems of soil classification are influenced by concepts prevalent at the time a sys-

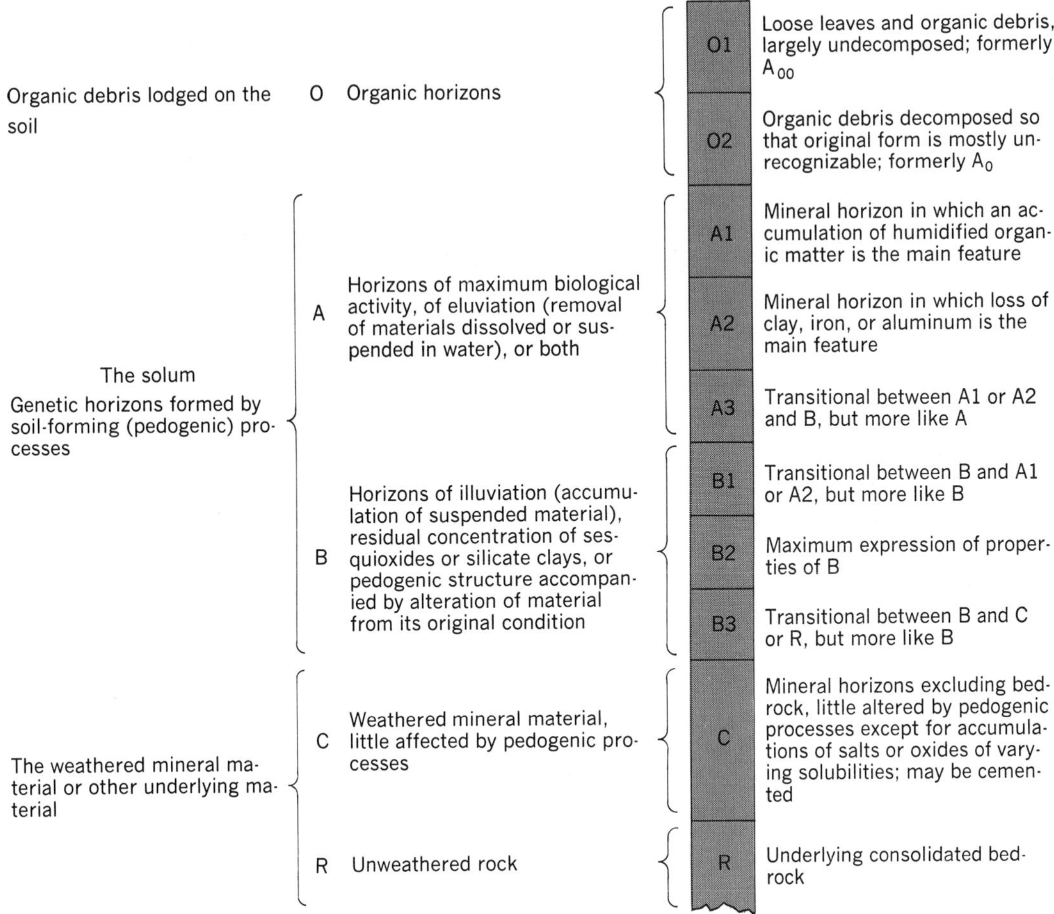
Organic debris lodged on the soil
O Organic horizons
O1
Loose leaves and organic debris, largely undecomposed; formerly A_{00}
O2
Organic debris decomposed so that original form is mostly unrecognizable; formerly A_0
The solum
Genetic horizons formed by soil-forming (pedogenic) processes
A Horizons of maximum biological activity, of eluviation (removal of materials dissolved or suspended in water), or both
A1
Mineral horizon in which an accumulation of humidified organic matter is the main feature
A2
Mineral horizon in which loss of clay, iron, or aluminum is the main feature
A3
Transitional between A1 or A2 and B, but more like A
B Horizons of illuviation (accumulation of suspended material), residual concentration of sesquioxides or silicate clays, or pedogenic structure accompanied by alteration of material from its original condition
B1
Transitional between B and A1 or A2, but more like B
B2
Maximum expression of properties of B
B3
Transitional between B and C or R, but more like B
The weathered mineral material or other underlying material
C Weathered mineral material, little affected by pedogenic processes
C
Mineral horizons excluding bedrock, little altered by pedogenic processes except for accumulations of salts or oxides of varying solubilities; may be cemented
R Unweathered rock
R
Underlying consolidated bedrock

Fig. 2. A hypothetical soil profile having all principal horizons. Other symbols are used to indicate features subordinate to those indicated by capital letters and numbers. The more important of these are as follows: ca, as in Cca, accumulations of carbonates; cs, accumulations of calcium sulfate; cn, concretions; g, strong gleying (reduction of iron in presence of groundwater); h, illuvial humus; ir, illuvial iron; m, strong cementation; p, plowing; sa, accumulations of very soluble salts; si, cementation by silica; t, illuvial clay; x, fragipan (a compact zone which is impenetrable by roots).

tem is developed. Since ancient times, soil has been considered as the natural medium for plant growth. Under this concept, the earliest classifications were based on relative suitability for different crops, such as rice soils, wheat soils, and vineyard soils.

Early American agriculturists thought of soil chiefly as disintegrated rock, and the first comprehensive American classification was based primarily on the nature of the underlying rock.

In the latter part of the 19th century, some Russian students noted relations between the steppe and black soils and the forest and gray soils. They developed the concept of soils as independent natural bodies formed by the influence of environmental factors operating on parent materials over time. The early Russian classifications grouped soils at the highest level, according to the degree to which they reflected the climate and vegetation. They had classes of Normal, Abnormal, and Transitional soils, which later became known as Zonal, Intrazonal, and Azonal. Within the Normal or Zonal soils, the Russians distinguished climatic and vegetative zones in which the soils had distinctive colors and other properties in common. These formed classes that were called soil types. Because some soils with similar colors had very different properties that were associated with differences in the vegetation, the nature of the vegetation was sometimes considered in addition to the color to form the soil type name, for example, Gray Forest soil and Gray Desert soil. The Russian concepts of soil types were accepted in other countries as quickly as they became known. In the United States, however, the name soil type had been used for some decades to indicate differences in soil texture, chiefly texture of the surface horizons; so the Russian soil type was called a Great Soil Group.

Many systems of classification have been attempted but none has been found markedly superior; most systems have been modifications of those used in Russia. Two bases for classification have been tried. One basis has been the presumed genesis of the soil; climate and native vegetation were given major emphasis. The other basis has been the observable or measurable properties of the soil. To a considerable extent, of course, these are used in the genetic system to define the great soil groups. The morphologic systems, however, have not used soil genesis as such, but have attempted to use properties that are acquired through soil development.

The principal problem in the morphologic systems has been the selection of the properties to be used. Grouping by color, tried in the earliest systems, produces soil groups of unlike genesis.

The Soil Survey staff of the U.S. Department of Agriculture and the land-grant colleges adopted a new classification scheme in 1965. Although the new system has been widely tested, only time can tell how much more useful it will be than earlier systems. As knowledge of soil genesis increases, modifications of classification systems will continue to be necessary.

The system differs from earlier systems in that it may be applied to either cultivated or virgin soils. Previous systems have been based on virgin profiles, and cultivated soils were classified on the presumed characteristics or genesis of the virgin soils. The new system has six categories, based on both physical and chemical properties. These categories are the order, suborder, great group, subgroup, family, and series, in decreasing rank.

The nomenclature. The names of the taxa or classes in each category are derived from the classic languages in such a manner that the name itself indicates the place of the taxa in the system and usually indicates something of the differentiating properties. The names of the highest category, the order, end in the suffix "sol," preceded by formative elements that suggest the nature of the order. Thus, Aridisol is the name of an order of soils that is characterized by being dry (Latin *aridus*, dry, plus *sol*, soil). A formative element is taken from each order name as the final syllable in the names of all taxa of suborders, great groups, and subgroups in the order. This is the syllable beginning with the vowel that precedes the connecting vowel with "sol." Thus, for Aridisols, the names of the taxa of lower classes end with the syllable "id," as in Argid and Orthid.

Suborder names have two syllables, the first suggesting something of the nature of the suborder and the last identifying the order. The formative element "arg" in Argid (Latin argillus, clay) suggests the horizon of accumulation of clay that defines the suborder.

Great group names have one or more syllables to suggest the nature of the horizons and have the suborder name as an ending. Thus great group names have three or more syllables but can be distinguished from order names because they do not end in "sol." Among the Argids, great groups are Natrargids (Latin *natrium*, sodium) for soils that have high contents of sodium, and Durargids (Latin *durus*, hard) for Argids with a hardpan cemented by silica and called a duripan.

Subgroup names are binomial. The great group name is preceded by an adjective such as "typic," which suggests the type or central concept of the great group, or the name of another great group, suborder, or order converted to an adjective to suggest that the soils are transitional between the two taxa.

Family names consist of several adjectives that describe the texture (sandy, silty, clayey, and so on), the mineralogy (siliceous, carbonatic, and so on), the temperature regime of the soil (thermic, mesic, frigid, and so on), and occasional other properties that are relevant to the use of the soil.

Series names are abstract names, taken from towns or places near where the soil was first identified. Cecil, Tama, and Walla Walla are names of soil series.

Order. In the highest category 10 orders are recognized. These are distinguished chiefly by differences in kinds and amounts of organic matter in the surface horizons, kinds of B horizons resulting from the dominance of various specific processes, evidences of churning through shrinking and swelling, base saturation, and lengths of periods during which the soil is without available moisture. The properties selected to distinguish the orders are reflections of the degree of horizon development and the kinds of horizons present.

The orders, the formative elements in the names, and the general nature of the included soils are given in Table 1.

Suborder. This category narrows the ranges in soil moisture and temperature regimes, kinds of horizons, and composition, according to which of these is most important. Moisture or temperature or soil properties associated with them are used to define suborders of Alfisols, Mollisols, Oxisols, Ultisols, and Vertisols. Kinds of horizons are used for Aridisols, compositions for Histosols and Spodosols, and combinations for Entisols and Inceptisols.

Great group. The taxa (classes) in this category group soils that have the same kinds of horizons in the same sequence and have similar moisture and temperature regimes. Exceptions to horizon sequences are made for horizons so near the surface that they are apt to be mixed by plowing or lost rapidly by erosion if plowed.

Table 1. Soil orders

Order	Formative element in name	General nature
Alfisols	alf	Soils with gray to brown surface horizons, medium to high base supply, with horizons of clay accumulation: usually moist, but may be dry during summer
Aridisols	id	Soils with pedogenic horizons, low in organic matter, and usually dry
Entisols	ent	Soils without pedogenic horizons
Histosols	ist	Organic soils (peats and mucks)
Inceptisols	ept	Soils that are usually moist, with pedogenic horizons of alteration of parent materials but not of illuviation
Mollisols	oll	Soils with nearly black, organic-rich surface horizons and high base supply
Oxisols	ox	Soils with residual accumulations of inactive clays, free oxides, kaolin, and quartz; mostly tropical
Spodosols	od	Soils with accumulations of amorphous materials in subsurface horizons
Ultisols	ult	Soils that are usually moist, with horizons of clay accumulation and a low supply of bases
Vertisols	ert	Soils with high content of swelling clays and wide deep cracks during some season

Subgroup. The great groups are subdivided into subgroups that show the central properties of the great group, intergrade subgroups that show properties of more than one great group, and other subgroups for soils with atypical properties that are not characteristic of any great group.

Family. The families are defined largely on the basis of physical and mineralogic properties of importance to plant growth.

Series. The soil series is a group of soils having horizons similar in differentiating characteristics and arrangement in the soil profile, except for texture of the surface portion, and developed in a particular type of parent material.

Type. This category of earlier systems of classification has been dropped but is mentioned here because it was used for almost 70 years and many references to it are found in the literature about soils. The soil types within a series differed primarily in the texture of the plow layer or equivalent horizons in unplowed soils. Cecil clay and Cecil fine sandy loam were types within the Cecil series. The texture of the plow layer is still indicated in published soil surveys if it is relevant to the use of the soil, but it is now considered as one kind of soil phase. Soil surveys are discussed in the next section of this article.

Classifications of soils have been developed in several countries based on other differentia. The principal classifications have been those of the Soviet Union, Germany, France, Canada, Australia, and New Zealand, and the United States. Other countries have modified one or the other of these to fit their own conditions. Soil classifications have usually been developed to fit the needs of a government that is concerned with the use of its soils. In this respect soil classification has differed from classifications of other natural objects, such as plants and animals, and there is no international agreement on the subject.

Many practical classifications have been developed on the basis of interpretations of the usefulness of soils for specific purposes. An example is the capability classification, which groups soils according to the number of safe alternative uses, risks of damage, and kinds of problems that are encountered under use.

Surveys. Soil surveys include those researches necessary (1) to determine the important characteristics of soils, (2) to classify them into defined series and other units, (3) to establish and map the boundaries between kinds of soil, and (4) to correlate and predict adaptability of soils to various crops, grasses, and trees; behavior and productivity of soils under different management systems; and yields of adapted crops on soils under defined sets of management practices. Although the primary purpose of soil surveys has been to aid in agricultural interpretations, many other purposes have become important, ranging from suburban planning, rural zoning, and highway location, to tax assessment and location of pipelines and radio transmitters. This has happened because the soil properties important to the growth of plants are also important to its engineering uses.

Soil surveys were first used in the United States in 1898. Over the years the scale of soil maps has been increased from 1/2 or 1 in. to the mile, to 3 or 4 in. to the mile for mapping humid farming regions, and up to 8 in. to the mile for maps in irrigated areas. After the advent of aerial photography, planimetric maps were largely discontinued in favor of aerial photographic mosaics. The United States system has been used, with modifications, in many other countries.

Two kinds of soil maps are made. The common map is a detailed soil map, on which soil boundaries are plotted from direct observations throughout the surveyed area. Reconnaissance soil maps are made by plotting soil boundaries from observations made at intervals. The maps show soil and other differences that are of significance for present or foreseeable uses.

The units shown on soil maps usually are phases of soil series. The phase is not a category of the classification system. It may be a subdivision of any class of the system according to some feature that is of significance for use and management of the soil, but not in relation to the natural landscape. The presence of loose boulders on the surface of the soil makes little difference in the growth of a forest, but is highly significant if the soil is to be plowed. Phases are most commonly based on slope, erosion, presence of stone or rock, or differences in the rock material below the soil itself. If a legend identifies a phase of a soil series, the soils so designated on a soil map are presumed to lie within the defined range of that phase in the major part of the area involved. Thus, the inclusion of lesser areas of soils having other characteristics is tolerated in the mapping if their presence does not appreciably affect the use of the soil. If there are other soils that do affect the use, inclusions up to 15% of the area are tolerated without being indicated in the name of the soil.

If the pattern of occurrence of two or more series is so intricate that it is impossible to show them separately, a soil complex is mapped, and the legend includes the word "complex," or the names of the series are connected by a hyphen and followed by a textural class name. Thus the phrase Fayette-Dubuque silt loam indicates that the two series occur in one area and that each represents more than 15% of the total area.

In places the significance of the difference between series is so slight that the expense of separating them is unwarranted. In such a case the names of the series are connected by a conjunction, for example, Fayette and Downs silt loam. In this kind of mapping unit, the soils may or may not be associated geographically.

It is possible to make accurate soil maps only because the nature of the soil changes with alterations in climatic and biotic factors, in relief, and in groundwaters, all acting on parent materials over long periods of time. Boundaries between kinds of soil are made where such changes become apparent. On a given farm the kinds of soil usually form a repeating pattern related to the relief (Fig. 3).

Because concepts of soil have changed over the years, maps made 30–50 years ago may use the same soil type names as maps made in recent years, but with different meanings. The older maps must therefore be interpreted with caution.

[GUY D. SMITH]

Nutrient element losses. Losses of most elements are normal to soil formation. Losses for a pair of soil orders are described below and the magnitudes indicated for others.

Ultisols are formed in strongly weathered rego-

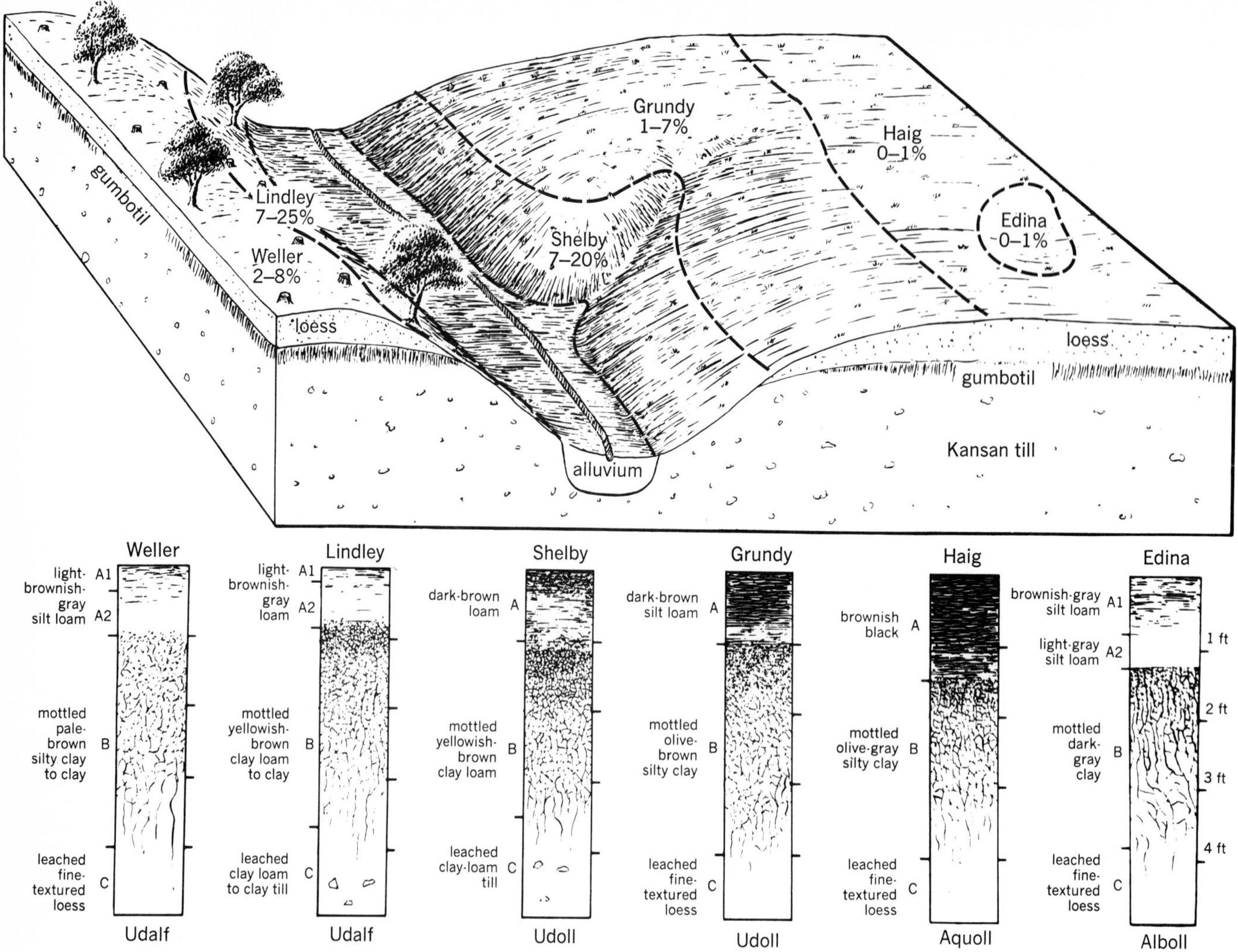

Fig. 3. Sketch showing the relation of the soil pattern to relief, parent material, and native vegetation on a farm in south-central Iowa. The soil slope gradient is expressed as a percentage. (*Modified from R. W. Simonson, F. F. Riecken, and G. D. Smith, Understanding Iowa Soils, Brown, 1952*)

liths from a variety of rocks, chiefly in warm, humid regions. The soils occupy old land surfaces. Major areas are in southeastern Asia and the United States.

Chemical and mineralogical composition of specimen Ultisols and their source rocks suggests that as much as 90% of the calcium, magnesium, and potassium disappears at the weathering front, where the rock decomposes. Losses continue as the soils form. Quantities are eventually reduced to very low levels. Because of the low levels, people have thought that Ultisols were worn out by long cropping, whereas they were really "worn out" while being formed.

Table 2. Amounts of four nutrient elements carried by four rivers in one year*

River (state)	Ca	Mg	K	K+Na	N
Neuse (NC)	20	7	–	32	1.0
Hiwassee (NC)	33	11	9	–	–
Cedar (IA)	86	32	–	22	1.4
Iowa (IA)	78	29	–	22	1.0

*Expressed as pounds per acre of watershed.

The approximate amounts of four nutrient elements carried by a pair of streams draining Ultisols in North Carolina are given in Table 2. Amounts of Ca and Mg are very low, whereas those of K, Na, and N are moderate to low.

Mollisols are formed in slightly weathered regoliths, chiefly in cool-temperate grasslands. The soils occupy young land surfaces. Major areas are in the north-central United States and adjacent Canada, the Ukraine and adjacent parts of the Soviet Union, and the pampas of Argentina.

Roughly a third of the Ca and K, a larger share of the Mg, and a very small part of the P in the source rocks disappear during formation of Mollisols. Hence, the soils have relatively high levels of these elements. Moreover, they have high levels of exchangeable Ca, Mg, and K, which are readily available to plants. Expressed as milliequivalents per 100 g of soil, average figures are 15 of Ca, 6 of Mg, and 0.8 of K. Amounts of phosphorus are also high, with much in plant-available form.

High levels of nutrient elements in Mollisols are reflected in data for a pair of rivers in Iowa, given in Table 2. Quantities of Ca and Mg are about three times as large as in the streams in North Carolina. Quantities of K and Na are slightly lower, whereas that of N is about the same. A share of the amounts in the streams comes from the underlying rock, especially for Ultisols.

Ultisols and Mollisols are opposite extremes in losses of nutrient elements during soil formation. If the average chemical composition of the soils to a depth of 5 ft is compared, the ratios between Mollisols and Ultisols are 10:1 for Ca and Mg, 3:1 for K, 2:1 for P, and 5:1 for N. The ratios for elements in exchangeable form are even larger. As the ratios suggest, Mollisols are much more naturally fertile than Ultisols. Mollisols are at the top of the list, Ultisols near the bottom.

Losses of nutrient elements during formation of soils is well bracketed by the Mollisols and Ultisols. Similar to Mollisols in losses during their formation are Aridisols, Inceptisols of cold or dry regions, and Vertisols. If anything, losses are smaller for these soils than for Mollisols. Collectively, these groups of soils occupy about 40% of the Earth's land surface. Also similar to Ultisols in losses are the Oxisols and Inceptisols of the tropics and subtropics. Collectively, these soils plus Ultisols occupy about 25% of the land surface.

The remaining groups of soils, that is, Alfisols, Entisols, Spodosols, and some Inceptisols, fall between the two extremes. Mountain regions with their great variety of soils belong to this middle class. These soils are all intermediate in losses and also in present fertility levels. Collectively, these soils occupy about 35% of the land surface.

Losses of nutrient elements generally occur during formation of soils. The losses are small enough to be negligible for a few kinds of soils, very large for others, and intermediate for still others. The magnitude of past losses is reflected in the present fertility levels of all soils. The magnitude also directly affects soil usefulness for food and fiber production as well as the contributions of dissolved substances to lakes and streams.

[ROY W. SIMONSON]

PHYSICAL PROPERTIES OF SOIL

The physical properties of soil are important in agriculture because of their influence on plant growth and on the management requirements of the land. They influence plant growth from seeding to maturity by regulating the supply of air, water, and heat. The absorption of essential nutrients by plant roots is dependent upon an available supply of oxygen, water, and heat. Thus, physical properties indirectly regulate the nutrition of plants and their response to liming and fertilization. The more favorable the supply of air, water, and heat in each soil layer or horizon, along with the absence of mechanical impedance to root growth, the greater is the potential rooting system zone for plants.

Physical properties of the soil also determine the kind, amount, and ease of tillage, the runoff and erosion potential, and the type of plants which can or should be grown on a given soil.

Many people use the word tilth in referring to the physical condition of the soil. Tilth has been defined as the physical condition of the soil in its relation to plant growth. The physical condition of the soil is controlled by, or is the result of, whatever set of physical properties the soil has at any given time.

Soil physics is that branch of soil science which is concerned with the study of the physical properties of the soil. These physical properties include texture, particle density, structure, bulk density, porosity, water, air, temperature, consistency, compactibility, and color. Just as important as the amount of water, air, and heat in the soil at any one time is the soil's conductivity for these constituents. All of these properties are interrelated.

The four major components of the soil are inorganic particles, organic matter, water, and air. The proportions of these components vary greatly from place to place in a field, from one layer or horizon to another, and in different parts of the world. The amount of air, water, and heat in the soil changes from day to day and from season to season.

Soil texture. About one-half of the total volume of mineral soils consists of solid matter, of which 80–99% is inorganic and 1–20% is organic material. The inorganic fraction consists of rock and mineral particles of many sizes and shapes. They are classified into five major size groups called separates. The two largest separates are stone and gravel. Stone particles are greater than 76 mm (3 in.) and gravel particles are 2–76 mm along their greatest diameter. Sand particles are 0.05–2.00 mm in diameter. Sand particles may be graded by size as very fine, fine, medium, coarse, and very coarse. Silt has particles 0.002–0.005 mm in diameter. Clay, the smallest of the soil particles, has a diameter of less than 0.002 mm.

After separating the coarser separates by sieving, the amount of silt and clay is determined by methods that depend upon the rate of settling or sedimentation (based on Stokes' law) of these two separates from a water suspension in which they have been well dispersed with the aid of a dispersing agent. The stone, gravel, and sand separates of

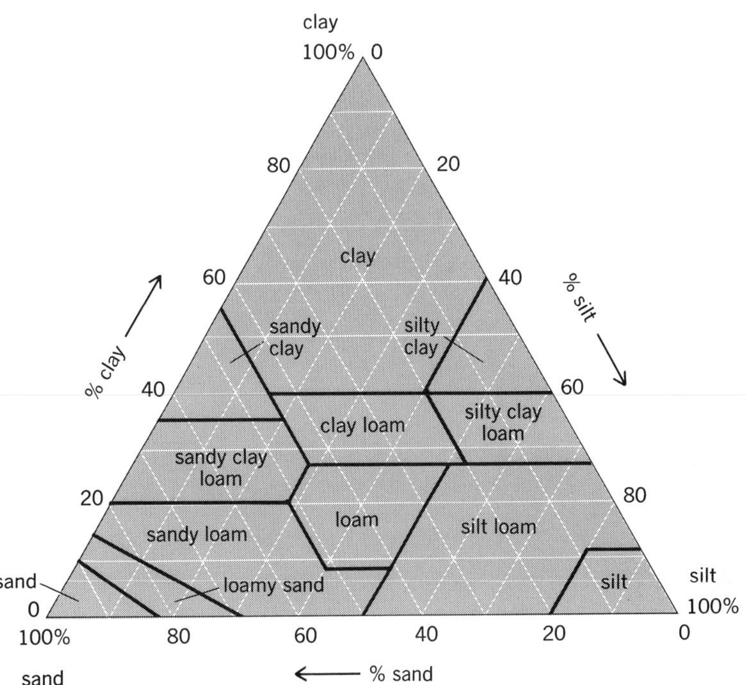

Fig. 4. Triangle showing percentage of sand, silt, and clay in each textural class.

a soil can be seen with the naked eye. Clay can be examined only with an electron microscope.

Determination of the particle-size distribution in a soil is called a particle-size or mechanical analysis. The texture of a soil is determined by its content of sand, silt, and clay. The percentages of sand, silt, and clay in the 12 textural classes are shown in Fig. 4. With this texture triangle one can determine the textural class of a soil from its percentages of sand, silt, and clay. The textural class is combined with the series name of a soil to give the soil type, such as Sassafras sandy loam, Miami silt loam, or Houston clay loam.

The stone, gravel, and larger sand particles usually act as separate particles. They may be rounded, angular, or platelike in shape. They are composed of rock fragments and of primary minerals such as quartz. Soils with large amounts of stone, gravel, and coarse sand have low plant-nutrient and water-holding capacities, and permit rapid air, water, and heat movement through them. A high content of fine sand may increase water-holding capacity, but it also often increases a soil's susceptibility to wind and water erosion. Sand imparts a grittiness to the feel of a soil.

The clay fraction controls most of the important properties of a soil. In soils of the cold and temperate regions, clay is composed chiefly of secondary crystalline alumina silicates. These consist of the kaolinite, illite, and montmorillonite groups of clay minerals. Hydrated oxides of iron and aluminum are the main components of the clay in the more highly weathered soils typical of many parts of the tropics.

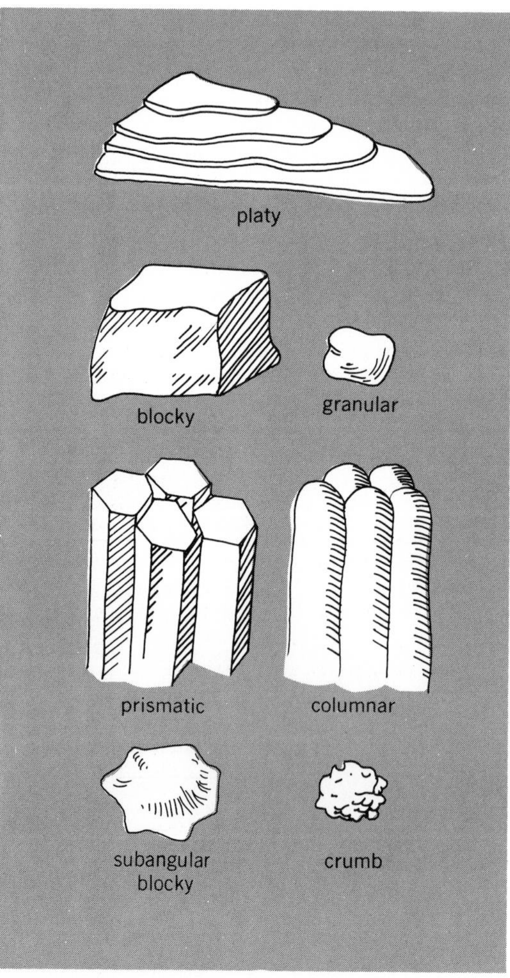

Fig. 5. Types of soil structure.

Because of their extremely small size, clay particles have a very large specific surface which is responsible for the great adsorptive capacity of clay soils for water, gases, ions, and organic molecules. Clays are well known for their plasticity and stickiness when wet. They also expand or swell with wetting and contract or shrink upon drying. Movement of air and water through clay soils is often very slow because of the small size of the pores between the clay particles. Clay particles are platelike in shape.

Silt particles exhibit some of the properties of sand and clay. They are usually angular in shape with quartz being the dominant mineral. The available water-holding capacity of soils often is proportional to their silt content. Many silt particles have a coating of clay particles. Without this clay coating silt has a floury or a talcum-powder feel when dry and loose. Soils with large amounts of silt and clay have very poor air and moisture relations and are very difficult to manage. They are often very erodible. The loam soils generally have the most desirable texture for crop growth and ease of management.

It is seldom feasible to try to change the texture of a soil in the field. However, sand often is mixed with clay soils to change their texture to a sandy loam for special uses, as in greenhouses. The texture of surface soils may change as a result of removal of the smaller particles by wind and water erosion or by eluviation (that is, movement within the soil).

Organic matter. The organic matter in the soil is made up of the partially decomposed remains of plant and animal tissues as well as the bodies of living soil microorganisms and plant roots. Humus is the more or less stable fraction of the organic matter or its decomposition products remaining in the soil. Many good and some bad effects accompany the decomposition of organic matter. During decomposition of organic matter by the soil microorganisms, gluelike soil-aggregate bonding substances are produced. With knowledge of the great importance of these natural soil-conditioning materials, the chemical industry has produced a number of synthetic soil conditioners.

Much of the soil organic matter has colloidal properties. It has two to three times the absorptive capacity for water, gases, ions, and other colloids as the same amount of clay. Its superior water- and nutrient-holding capacity makes it an ideal substitute for clay in improving droughty, infertile sandy soils, and its good tilth-promoting qualities make it the universally recognized ameliorator of tight, sticky, or hard and lumpy clay soils. *See* HUMUS, GEOCHEMISTRY OF.

Density of particles. The inorganic soil particles may consist of many kinds of minerals with a wide range in particle density. The average particle density for most mineral soils varies between 2.60 and 2.75 g/cm^3. The average density of humus particles ranges between 1.2 and 1.4 g/cm^3. For general calculations the average particle density of soil is taken to be 2.65 g/cm^3. The pycnometer is used to determine soil particle density. The plowed layer weighs about 2,000,000 lb/acre.

Soil structure. Soil structure refers to the arrangement of soil particles into aggregates or peds of different sizes and shapes. Pure sands have single grain structure. Because of the adhesive and cohesive properties of clay and organic matter, the inorganic and organic particles have been combined to form the following types of structure as found in the A and B horizons of most soils: platy, prismatic, columnar, blocky, subangular blocky, granular, and crumb (Fig. 5).

These types of structure have been developed from the bonding together of the individual particles (accretion) or the breakdown of large massive mixtures of gravel, sand, silt, clay, and organic matter (disintegration). The formation or genesis of a given type of structure and the stability of the aggregate produced seem to be associated with (1) the contraction and expansion resulting from hydration and desiccation of the clay-organic matter upon wetting and drying, as well as freezing and thawing; (2) the physical activity of roots and soil animals; (3) the influence of humus and decomposing organic matter and of the slimes and mycelia of the microorganisms that provide bonding substances with which aggregates are held together; and (4) the effect of absorbed cations which bring about flocculation or dispersion of the colloidal matter.

The prism, columnar, block, and sometimes the platelike types of structure are found mostly in subsoils. Granules and crumbs are found in largest numbers in surface soils (Fig. 6). Compacted layers in the soil often have a platy structure.

The size, shape, arrangement, and particularly the amount of overlap of soil aggregates and individual sand, gravel, and stone particles are extremely important because they largely determine the size, shape, arrangement, and continuity of pores in the soil.

There are a number of ways of attempting to characterize the structure of a soil. The first and most direct is by visual examination of an undisturbed section of soil. Much can be learned about the size, shape, and arrangement of the soil particles, and the pore space, by close inspection of each horizon with the naked eye or with a magnifying glass. Micromorphology is the microscopic observation and photography of soil structure.

A second method is to measure how much of the soil has been aggregated into granules or crumbs with diameters above a given dimension, 0.25 mm being the most common. In well-granulated soils 70–80% of the total mass may be aggregated into granules or crumbs greater than 0.25 mm, as determined by wet-sieving a sample of soil through a 60-mesh screen. Aggregation values of 40–50% are more commonly found in soils under ordinary management. Sandy soils or clay soils having poor structure may have only 10–20% aggregation. Except in very sandy soils, such a low amount of aggregation usually forecasts a physical condition very unsatisfactory for plant growth.

Measurement of the permeability of the soil to water and air provides another means of evaluating its structure.

Bulk density. Bulk density is the mass (weight) of a unit volume of dry soil usually expressed in grams per cubic centimeter. It is determined by the density of the particles and by their arrangement.

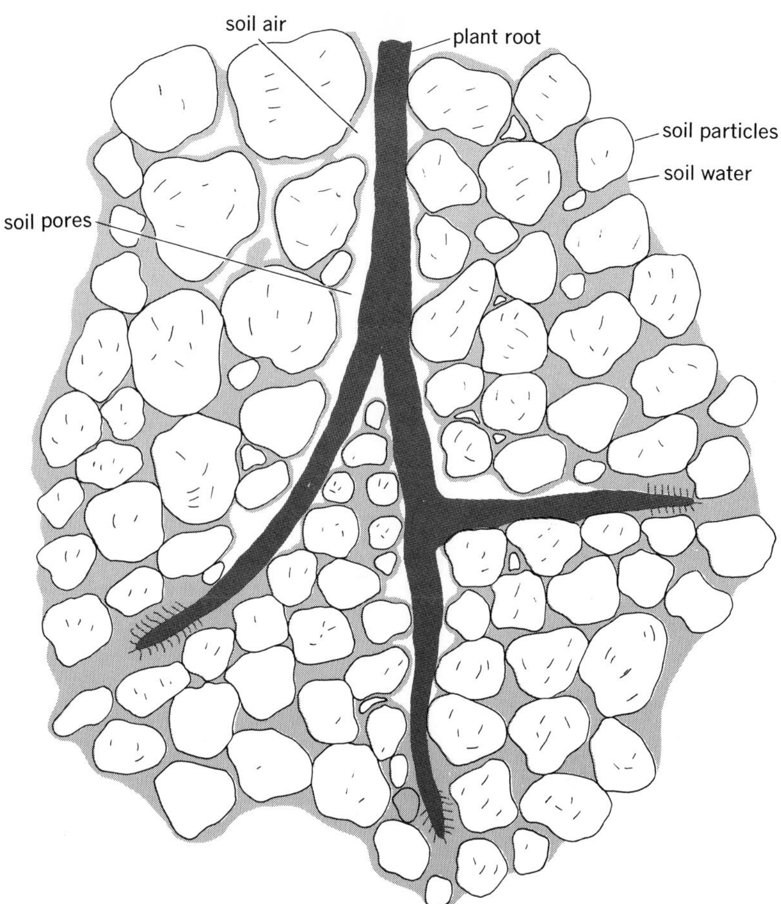

Fig. 6. Portion of surface soil with granular structure.

The soil structure is the major factor in accounting for changes in the bulk density of a soil from time to time or from layer to layer in the profile. Soils with many particles closely packed together have high bulk densities and correspondingly low total pore space. Bulk density is a measure of the amount of compaction in soils. Traffic by farm machinery in intensively cultivated soils, trampling of cattle in heavily grazed pastures, and foot traffic on lawns and recreational areas result in severe compaction as reflected in bulk densities of 1.7–2.0 g/cm³. Bulk densities of uncompacted, porous soils are about 1.2–1.3 g/cm³. In undisturbed forest or grassland soils densities may be 0.9–1.0 g/cm³. High amounts of organic matter will lower the bulk density.

Pore space. The voids or openings between the particles of the soil are spoken of collectively as the pore space. It makes up roughly one-half the volume of the soil. In very loose, fluffy soils with low bulk density it may occupy 60–65% of the total volume. In very compact soil layers it may be reduced to 35–40%. Pore space is calculated by Eq. (1).

$$\text{\% total pore space} = 100 - \left(\frac{\text{bulk density}}{\text{particle density}} \times 100\right) \qquad (1)$$

Pore space in a soil is occupied by air and water in reciprocally varying amounts. Very dry soils have most of their pore spaces filled with air. The opposite is true for very wet soils.

There is considerable variation in the size,

shape, and arrangement of pores in the soil. The effective size of a pore can be estimated by the amount of force required to withdraw water from the pore. These suction values, expressed in centimeters of water, can be translated into equivalent pore diameters using the capillary rise equation, Eq. (2), where r = radius of pore in centimeters,

$$r=\frac{2T}{hdg} \tag{2}$$

T = surface tension of water, d = density of water, g = acceleration of gravity, and h = suction force in centimeters of water.

The ideal soil should have the proper assortment of large, medium, and small pores. A sufficient number of large or macro pores (with diameters greater than 0.06 mm), connected with each other, are needed for the rapid intake and distribution of water in the soil and for the disposal of excess water by drainage into the substratum or into artificial drains. When without water they serve as air ducts. Cracks, old root channels, and animal burrows may serve as large pores. Soils with insufficient functional macro porosity lose a great deal of rainfall and irrigation water as runoff. They drain slowly and often remain poorly aerated after wetting. One of the first effects of compaction is the reduction of the size and number of the larger pore spaces in the soil.

The primary purpose of the small pores (less than 0.01 mm in diameter) is to hold water. It is through medium-sized pores (0.06–0.01 mm in diameter) that much of the capillary movement of water takes place. Loose, droughty, coarse, sandy soils have too few small pores. Many tight clay soils could well afford to have a greater number of larger pores.

Soil water. The movement and retention of water in the soil is related to the size, shape, continuity, and arrangement of the pores, their moisture content, and the amount of surface area of the soil particles. Movement and retention of water may be characterized by the energy relationships or forces which control these two phenomena.

Water retention. Some water is held in the soil pores by the force of adhesion (the attraction of solid surfaces for water molecules) and by the force of cohesion (the attraction of water molecules for each other). Water held by these two forces keeps the smaller pores full of water and also maintains relatively thick films on the walls of many larger pores. Not until the pores in one layer of soil are filled with all the water they can hold does water move into the layer below.

Water is found in the soil in both the liquid and vapor state. The soil air in all the pores, except those in the surface inch or so of very hot, dry soils is saturated with water vapor.

The liquid water may be characterized by the suction force or tension with which it is held in the soil by adhesion or cohesion. These suction or tension values may be expressed as (1) height in centimeters of a unit water column whose weight just equals the force under consideration; (2) pF, the logarithm of the centimeter height of this column; (3) atmospheres or bars; or (4) pounds per square inch (psi). For example, 1000 cm water tension = pF_3 = 1 atm = 14.7 psi. The moisture content of the soil is determined by drying the soil at 105°C until it reaches constant weight and then dividing the weight of water lost by the weight of oven-dry soil. This value times 100 equals the percentage of water in the soil on a dry-weight basis. The percentage of moisture on the volume basis for a given depth of soil is calculated by Eq. (3). Tensiometers,

$$\begin{aligned}&\%\text{ soil moisture (dry weight basis)} \div 100\\ &\quad\times \text{bulk density} \times \text{depth soil (in.)}\\ &\quad\quad= \text{in. of water per in. soil depth}\end{aligned} \tag{3}$$

electrical resistance blocks, and neutron gages are used to measure the moisture content of the soil in place. Thus changes in moisture content in the soil can be followed within the effective range of each instrument.

There are several soil-moisture "equilibrium" points. Water remaining at oven dryness is held at tensions above 10,000 atm. The hygroscopic coefficient is a rough measurement of the water held by air-dry soil at a tension of about 31 atm. The wilting point, or wilting percentage, represents that moisture content or moisture tension (15 atm) at which plant roots cannot absorb water rapidly enough to offset losses by transpiration, causing the plant to wilt, first temporarily and then permanently. Certain plants of desert and dry farming regions are able to stay alive and even grow on water held at tensions up to 25–30 atm by the soil. The field capacity represents the water remaining in a soil layer 2 or 3 days after having been saturated by rain or irrigation when the rate of downward movement of water held at low suction forces (0–1 atm) has decreased to a progressively slower rate of water removal. A definite tension value cannot be assigned to this equilibrium point, although the water held at a tension of 0.33 atm often is used to estimate the upper limit of a soil's available water-holding capacity. The maximum retentive capacity is the moisture content of a soil when all of its pores are filled or saturated with water, and under zero tension.

The moisture in a soil which is available for plant use is usually assumed to be that held between field capacity and the permanent wilting percentage. This is called the available soil moisture. Sandy loams hold 1–1½, loams 1½–2, and

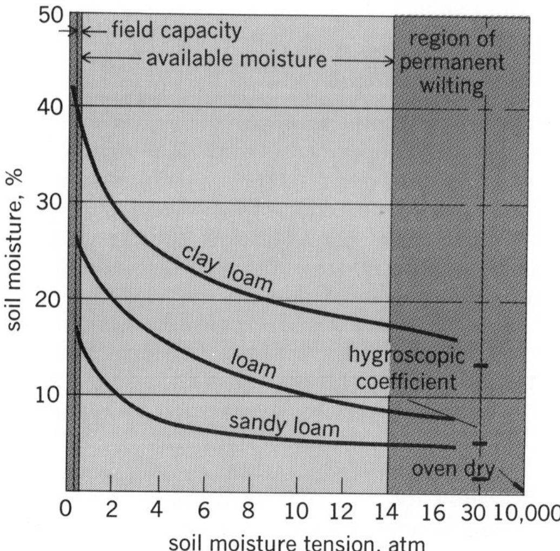

Fig. 7. Soil moisture tension curves.

clay loams $1\frac{3}{4}-2\frac{1}{2}$ in. of available water per foot of soil. The retentivity of soils of different textures for moisture at different tensions is shown in Fig. 7. These are called soil moisture tension curves. Much of the water in sandy soils is held at low tensions. The opposite is true for clay soils.

Water held at tensions less than 1–2 atm is the most easily available for root absorption and plant growth. An adequate supply of this water should be maintained in the root zone by rainfall or irrigation, especially during periods of critical water need by plants.

Water movement. Water moves in the soil as a gas and as a liquid. Vapor transfer takes place by diffusion in response to a vapor pressure gradient. Vapor movement is through air-filled pores from a moist to a dry layer and from a warm to a cool layer. Drying of a wet soil surface on a hot, dry, windy day and the condensation of water droplets on the undersurface of a plastic mulch on a cool, summer morning are the result of vapor transfer.

Liquid movement may be expressed by the equation $V = Ki$. V is the volume of water crossing the unit area perpendicular to the flow in the unit time. The proportionality factor K is the hydraulic conductivity or the permeability of the soil to water. It is controlled by the size, shape, arrangement, and moisture content of the soil pores. The value i is the water-moving force. It has two force components, the force of gravity and a suction or tension gradient force. The force or pull of gravity is of constant magnitude and always acts in a downward direction. The drainage or removal of most of the water from large pores is by this force, and the drainage is sometimes referred to as gravitational water. The suction or tension gradient force may vary both in magnitude and direction. The flow or capillary conduction of water in unsaturated soil is due to the gradient or difference in suction between two points in the soil. The flow is in the direction of the increase in suction. This accounts for the movement of water toward roots which have depleted the supply of water held at low tension in the soil at the soil-root interface, for the upward capillary conduction of water from an underlying saturated layer (water table), and for the slow downward movement of water after a rain or irrigation. Rate and amount of water movement by capillary conduction to the root system is not usually sufficient to meet the demands of the plant for water, except when a sufficient amount of water held at suctions less than 1 atm is present around the root. Since capillary conductivity decreases rapidly as the soil becomes drier, the water needs of plants are satisfied also by an extension of their root systems into fresh supplies of water held at low tension in hitherto untapped or recently refilled soil pores. It is very important, therefore, that soil structure be such as to permit the rapid extension of the root system through the whole soil mass.

Soil air and aeration. Soil air differs from the atmosphere above the soil in that it usually contains 5–100 times as much carbon dioxide (0.15–0.65%) and slightly less oxygen, and is saturated with water vapor. In deep, poorly drained soil layers or in heavily manured soils the CO_2 content may reach 10% and the O_2 content decrease to 1%. In water-logged soils, anaerobic conditions may result in methane and hydrogen sulfide production.

Aeration refers to the movement of gases in and out of each soil layer or horizon. The movement of gas within the soil as well as to and from the atmosphere is by diffusion in a direction determined by its own partial pressure. The rate of diffusion of each gas in and out of the soil depends on differences in concentration of each gas in the soil and in the atmosphere, and on the ability of the soil to transmit the gases. Diffusion or aeration is proportional to the volume of air-filled pores in any soil layer.

Temperature. The temperature of field soils shows rather definite changes at different depths, at different times during the day and night, and at different seasons of the year. These changes are determined by the amount of radiant energy that reaches the soil surface and by the thermal properties of the soil. Only that part of the heat energy which is absorbed causes changes in soil temperature. Heat produced by intense microbial decomposition of fresh organic matter mixed with the soil will also increase soil temperature. Dark-colored soils capture a much higher proportion of the radiant energy than do light-colored soils. The insulating effects of vegetative cover and surface mulches keeps the soil cooler than bare, fallow soil. Wet soils warm more slowly than dry soils because the heat capacity of water is five times that of the mineral soil particles. The energy absorbed by the soil surface is disposed of in one or more of the following ways: radiation to the atmosphere, heating of the air above the soil by convection, increasing the temperature of the surface soil, or conduction to the deeper soil layers.

Consistency and compactibility. As the moisture content of a soil changes from air dryness to saturation, its consistency varies from a state of hardness or brittleness, to loose, soft, or friable, to tough or plastic, to sticky or viscous. The reaction of the soil to physical manipulation, as in tillage, is primarily an expression of the properties of cohesion, adhesion, and plasticity. These properties are largely determined by the structure, organic matter content, kind and amount of clay, nature of adsorbed bases, and moisture content, which regulates the thickness of the water films around the soil particles. Tillage should be done only after the soil attains a soft, friable condition—when it breaks apart or can be worked into granules 1–5 mm in diameter. This is a very desirable range of particle size for good seed and root bed conditions.

Each soil has a critical moisture range, often near field capacity, at which pressure by foot or machinery traffic results in maximum compaction. Bulk density, permeability, porosity, and penetrometer measurements are used to indicate the degree of compaction as found in traffic pans in the surface soil, the plow sole, or natural hardpans. Soil compaction is a very serious problem because it reduces the permeability of the soil to air and water, and increases the resistance of the soil to root penetration. Hard, dry surface crusts may also prevent seedling emergence.

Color. Soil color may be influenced by, and indicates the kind of, parent material, chemical composition, organic matter content, drainage, aeration, or oxidation. A blotched or mottled yellow, gray, and blue subsoil indicates poor drainage, aeration, or oxidation. A clear red, yellow, or brown color indicates good drainage. The color of some soils is inherited from the parent material. Organic matter gives a brown to black color to that horizon where it is concentrated. Color of soils is determined by comparison with standard colors of known hue, value, and chroma in the Munsell soil color charts.

[RUSSELL B. ALDERFER]

SOIL MANAGEMENT

Soil management may be defined as the preparation, manipulation, and treatment of soils for the production of crops, grasses, and trees. Good soil management involves practices which will maintain a high level of production on a sustained basis. Ideally, these practices should provide the crop with an adequate supply of air, water, and nutrients; maintain or improve the fertility of the soil for subsequent crops; and prevent the development of conditions which might be injurious to plants.

Several systems of land-use classification have been developed which help a farmer to know the kinds of soils he has on his farm and their suitability for various types of farming. One of these systems, developed by the U.S. Soil Conservation Service, involves land-use capability ratings.

Land capability survey maps. Land capability survey maps are worked out in conjunction with the soil survey and serve as a guide to the suitability of land for cultivation, grazing, forestry, wildlife, watersheds, or recreation, with primary consideration given to erosion control. There are eight capability classes which describe the characteristics of the land and the difficulty or risk involved in using it for one kind of crop production or another. These eight classes are sometimes distinguished on land capability survey maps by roman numerals as well as by standard colors. Four of the eight classes include land that is suited for regular cultivation with varying degrees of erosion control measures and management practices required; three classes of land are not suited for cultivation but require permanent vegetation and impose severe limitations on land use; and one class includes lands suited only for wildlife, recreation, or watershed purposes.

Cropping system. A cropping system refers to the kind and sequence of crops grown on a given area of soil over a period of time. It may be a regular rotation of different crops, in which the sequence of crops follows a definite order, or it may consist of a single crop grown year after year in the same location. Other cropping systems include different crops but have no definite or planned sequence.

Cropping systems that involve the systematic rotation of different crops generally include hay and pasture crops, small grains, and cultivated row crops. Legumes, such as alfalfa, clover, and vetch, are usually grown alone or mixed with grasses in the hay and pasture sequence in the rotation because they supply nitrogen and contribute to good soil tilth. The beneficial effect of legumes and grasses on tilth may be attributed to the fact that (1) the soil is not tilled while these crops are being grown, and (2) the organic matter returned to the soil by the extensive root systems and in the plowed-under top growth is particularly suited to the development of a stable, porous soil structure.

Small grains function somewhat like legumes and grasses in giving protection against soil erosion, but they add no nitrogen and remove moderate quantities of plant food from the soil. Since small grains do not provide maximum economic return from the high nitrogen residues left in the soil by legumes, and are likely to lodge owing to the stimulation of growth from these residues, they are not planted in the rotation following legumes. Small grains are generally planted either at the end of a rotation following row crops or as a companion crop for legumes.

Row crops, such as corn, potatoes, cotton, and sugarbeets, are an excellent choice to follow legumes because they utilize the nitrogen supplied by the legumes and bring good cash returns. Since row crops in early stages of growth provide little protection against erosion and require considerable cultivation which breaks down soil structure, it is not considered desirable to plant them continuously.

A cropping system that involves growing the same crop year after year generally depletes the soil and results in lower crop yields. This is particularly true if the crop is cultivated frequently and returns little crop residue to the soil. Weeds, diseases, and insects also become more of a management problem when the cropping system does not involve rotation. Thus the farmer who intends to grow one crop year after year becomes completely dependent on disease-resistant varieties of plants, chemical insecticides and fungicides, soil fumigation, and other methods of controlling diseases, insects, and pests. Through the appropriate use of improved varieties, pesticides, and adequate amounts of fertilizers, farmers have succeeded in maintaining a high level of production on land repeatedly planted with the same crop. While such results are causing farmers to take another look at cropping systems that do not involve rotation, they are well aware that more intensive practices and costly supplements are required to maintain production.

Organic matter and tilth. The value of adding organic matter to the soil in the form of animal manures, green manures, and crop residues for producing favorable soil tilth has been known since ancient times. Research has provided information that helps to explain the mechanisms for this effect.

Experiments reveal that during the decomposition of organic matter in the soil, microorganisms synthesize a variety of gumlike substances, at least partly polysaccharide in nature which, when dried with the soil, bind the soil particles together into a porous, water-stable structure. While these binding substances are produced in relatively large quantities during stages of rapid organic-matter decomposition, they may in turn be decomposed by other organisms. Thus, to maintain a continuous supply of binding substances, organic matter must be added to the soil frequently.

In addition to the beneficial action of microbially synthesized binding substances, roots and fungal mycelia also contribute to the development of favorable soil tilth by molding the smaller gum-cemented granules into still larger aggregates. The aggregates adhering together form large pores that permit the rapid movement of air and water and form small pores that store water. Both conditions are essential features of good tilth.

Unfortunately not all the effects of organic matter on tilth are desirable. The growth of organisms in a fine-textured soil may interfere with the downward movement of water whenever the soil pores become clogged with microbial bodies. This condition is of particular significance where water is ponded to recharge underground water supplies or to leach excessive amounts of soluble salts from the soil.

Even the characteristic property of organic matter of promoting aggregate formation is not always desirable. Some surface mulches during decomposition induce the formation of a layer of small surface aggregates which are more susceptible to wind erosion than the fine soil particles initially present. In such cases, aggregation intensifies the hazard of severe wind erosion.

In spite of these negative effects of organic matter on the physical properties of soil, the incorporation of organic matter with soil is the most suitable and practical way of developing and maintaining good tilth.

Conditioners and stabilizers. Soil conditioners and stabilizers include a wide variety of natural and synthetic compounds that, upon incorporation with the soil, improve its physical properties. The term soil amendment also is applied to these compounds but is a more general term since it includes any material, exclusive of fertilizers, that is worked into the soil to make it more productive, regardless of whether it benefits the physical, chemical, or microbiological properties of the soil.

Soluble salts of calcium, such as calcium chloride and gypsum, or acid and acid formers, including sulfur, sulfuric acid, iron sulfate, and aluminum sulfate, have been used as conditioners to improve the physical properties of soils that were made unfavorable by excessive quantities of sodium ions adsorbed on the soil colloids.

The tilth of dense clay soils, which are slow to take water and have a marked tendency to become cloddy, may be improved by the addition of gypsum and by-product lime from sugarbeet processing factories. Limestone has improved the physical condition of acid soils, apparently by stimulating the activity of microorganisms to synthesize substances that bind soil particles into aggregates.

The discovery that soil microorganisms synthesize substances that improve soil structure stimulated the search for synthetic compounds that would be more effective than the natural products. While a wide variety of compounds have improved soil structure temporarily, three water-soluble, polymeric electrolytes of high molecular weight which are very resistant to microbial decomposition have been developed commercially for use in ameliorating poor soil structure. These are modified hydrolyzed polyacrilonitrile (HPAN), modified vinyl acetate maleic acid (VAMA), and a copolymer of isobutylene and maleic acid (IBMA). High cost relative to yield increase has limited these materials to experimental use.

Mixed with the soil in amounts ranging from 0.02 to 0.2% of soil by weight, these compounds are readily adsorbed by moist soils and tend to stabilize or fix the existing structure. They are therefore synthetic binding agents and should be added only to soils that have previously been worked into a desirable physical condition. These materials are not equally effective on all soils, and if improperly used can stabilize a poor physical condition.

Fertility. Soil fertility may be defined as that quality of a soil which enables it to provide nutrient elements and compounds in adequate amounts and in proper balance for the growth of specified plants, when other growth factors such as light, moisture, temperature, and the physical condition of the soil are favorable.

Testing. Even though relatively fertile and of good physical condition, a soil may be lacking in one or more of 16 elements presently known to be essential to plant growth, or it may be strongly acidic, alkaline, or salty, and thus unsuitable for plant growth. Fortunately, soil tests are available that indicate the existence of possible deficiencies or excesses in the soil. In most instances, these tests involve the use of various reagents for extracting from the soil the total or proportionate amount of the nutrient or compound in question. The amount of material extracted is then compared with values that have been correlated previously with crop response on the same or similar soil. No single test is reliable for all crops on all soils.

Control of pH. The availability of soil nutrients for plants is influenced greatly by the reaction of the soil. Soils may be classified as acid, neutral, or alkaline in reaction. The method commonly used in measuring and expressing degrees of acidity or alkalinity is in terms of pH. The pH value of soil may range from less than 4 to more than 8; the lower the value, the more acid the soil. Under most conditions lime is applied to acid soils to maintain their pH between 6.5 and 7.0. Under special conditions it may be desirable to maintain pH values either higher or lower than these. In any case the desirability of applying lime should be determined by the pH of the soil and the requirements of the plants to be grown.

It is occasionally necessary to make soils more acid. Materials commonly used to decrease pH are sulfur, sulfuric acid, iron sulfate, and aluminum sulfate.

Control of salinity. Restricted drainage caused by either slow permeability or a high water table is the principal factor in the formation of saline soils. Such soils may be improved by establishing artificial drainage, if a high water table exists, and by subsequent leaching with irrigation water to remove excess soluble salts.

Soils can be leached by applying water to the surface and allowing it to pass downward through the root zone. Leaching is most efficient when it is possible to pond water over the entire surface.

The amount of water required to leach saline soils depends on the initial salinity level of the soil and the final salinity level desired. When water is ponded over the soil about 50% of the salt in the root zone can be removed by leaching with 6 in. of water for each foot of root zone; about 80% can be removed with 1 ft of water per foot of soil to be leached; and 90% can be removed with 2 ft of water per foot of soil to be leached.

Because all irrigation waters contain dissolved salts, nonsaline soils may become saline unless water is applied in addition to that required to replenish losses by plant transpiration and evaporation, to leach out the salt that has accumulated during previous irrigations and through the addition of fertilizer.

Regulating nutrient supplies. The nutrients supplied to crops can be regulated by modifying the availability of nutrients already present in the soil. This can be accomplished by changing soil reaction, turning under green manure crops, including legumes which add nitrogen, and adding fertilizers.

By changing soil reaction through the addition of lime, acidulating agents such as sulfur, or residually acid fertilizers such as ammonium sulfate, solubility and availability of compounds of phosphorus, iron, manganese, copper, zinc, boron, and molybdenum can be increased or decreased. Phosphorus compounds are generally more available in the slightly acid to neutral pH range, whereas compounds of iron, manganese, zinc, and copper become more available as the acidity of the soil increases. The activity of microorganisms responsible for the transformation of nitrogen, sulfur, and phosphorus compounds into forms available for plants also is influenced by soil reaction. A reaction which is too acid or too alkaline retards the activities of these organisms.

The decay of turned-under green manures and plant residues produces carbon dioxide, rendering soluble the nutrients from soil particles, and the nutrients which were absorbed from the soil during the growth of these crops are also made available. Although the turning under of a green manure affects the availability of nutrients, it does not add to the total nutrient supply unless the green manure is a legume which fixes atmospheric nitrogen.

The system of farming determines to a considerable extent the manner in which fertilizers are used to regulate nutrient supplies. Each system of farming depends upon the crop, soil, climate, kinds and rates of fertilizers applied, and available equipment; and for each system of farming there are many ways of applying fertilizers.

Common methods of applying fertilizers include broadcasting, banding, deep placement, and foliar applications. Broadcasting fertilizer on the soil is usually less desirable than localized placement of the fertilizer in relation to the seed or plant. Banding fertilizers to the side of the rows in furrow bottoms or beds and drilling fertilizer with the seed give the best response from limited quantities of fertilizer. Deep placement of fertilizers is effective in arid regions where soils dry out to a considerable depth or where deep-rooted crops are grown. Foliar applications of fertilizers, particularly those containing micronutrients, circumvent soil interactions. Such interactions within the soil may render the applied fertilizer unavailable to the crop. Foliar applications also make it possible for the farmer to supply his crops directly with a number of essential plant nutrients at critical stages of growth.

[DANIEL G. ALDRICH JR.]

Soils of the tropics. The tropics may be defined as that part of the Earth's surface between the Tropic of Cancer and the Tropic of Capricorn, about 23-1/2° north and south of the Equator. In this region three broad ecological zones may be recognized: evergreen and deciduous forests which cover about one-fourth of the land area; savanna and grasslands, about one-half; and semidesert and desert areas, about one-fourth. Because of the wide range of climate, vegetation, parent materials, and other factors affecting soil formation, there are as many if not more different kinds of soils in the tropics as in temperate regions. They range from fertile soils of alluvial valleys, through deep, highly weathered infertile acid soils of uplands through shallow stony soils of steep mountain slopes, to high lime soils of deserts. Though many soils of the tropics have formed from recently deposited alluvium and volcanic ash, most have developed from older weathered parent materials, and consequently are generally much less fertile than soils in temperate regions which are mostly formed from relatively recent glacial and loessial deposits.

Resources. Of the 12×10^9 acres (5×10^9 hectares) of land area in the tropics, about 2×10^9 acres are estimated to be potentially arable but as yet uncultivated. This is more than is now cultivated and represents about half of the world's uncultivated land which is potentially arable. Most of the potential for bringing new land under cultivation is in Africa and South America, where only about 22 and 11%, respectively, of the potentially arable soils are cultivated. Probably most potentially arable soils in the tropics not under cultivation are relatively infertile and require addition of nutrients and good management for economic crop production. The UN Food and Agriculture Organization and UNESCO have published soil maps of South America and Africa at a scale of 1:5,000,000 with accompanying texts. These are two volumes in a ten-volume series, *Soil Map of the World.* More detailed soil resource appraisals and expanded research efforts are required to provide necessary information for development of new lands and for improvement of soil management practices for intensification of production on presently cultivated soils. Substantial progress has been made in many tropical countries in making inventories of their soil resources.

Misconceptions. There are many misconceptions about soils of the tropics. One relates to the localized occurrence of laterite (irreversibly hardened ironstone) associated with the extensive red soils common in the tropics. These red soils have been called lateritic soils, and the mistaken notion arose that these soils when cleared of vegetation and cultivated would turn to hardened laterite and become worthless in a short time. Actually, extensive areas of red (so-called lateritic) soils in the tropics have been farmed for centuries without laterite forming. Probably only about 5% of soils in the tropics have laterite, and most are relics of previous geologic eras and not related to present use. Laterite forms under a particular set of conditions in localized areas and is of minor importance in the tropics.

A second prevalent misconception is that soils of the tropics have much less organic matter than those of temperate regions. This mistaken idea arose because of (1) warmer temperature in most of the tropics which accelerates organic matter decomposition and (2) widespread occurrence of reddish-colored soils in contrast to the generally dark-colored soils of the temperate regions. Analyses of soils from many different areas in the tropics have shown that, generally, the organic matter content is as high if not higher than that of soils in temperate regions. The long dry season in much of the tropics inhibits biological activity, just as winter temperatures do in temperate climates. In tropical areas with abundant rainfall there is much greater vegetative growth, which balances the increased biological activity in organic matter decomposition because of the warmer temperatures. Hence the soil organic matter content is maintained.

A third common misconception is that luxuriant forest growth in the low humid tropics is an indicator of fertile soils. More often than not, soils in tropical forests are relatively infertile. The nutrients are contained mostly in the vegetation and recycled as the decomposing organic matter releases them. The extensive roots of the forest vegetation prevent leaching of nutrients, and with abundant rainfall throughout the year luxuriant vegetation can be maintained on relatively infertile soils.

Agricultural systems. Though large areas of soils in the tropics are intensively cultivated for commercial export crops, most of the food for much of the population is produced under traditional, largely subsistence, shifting cultivation (slash-and-burn) systems. These are extensive systems supporting low-density populations, as the soil is usually cultivated with mixed crops for only a year or two followed by natural fallow for 5 to 15 years or more. With increased population pressure in many areas, the introduction of modern technology for more intensive use of the soils becomes necessary. Though substantial progress in this direction has been made, the intensification of traditional agriculture in the tropics within the limited resources of the farmers poses a major challenge to agricultural scientists. [MATTHEW DROSDOFF]

Determining irrigation needs. Ideally, farmers should irrigate periodically before drought cuts crop growth and yield and then apply only the water required to refill the soil occupied by roots to its capacity to retain water against drainage. When necessary, more water (about 10%) is applied to leach excess salts to maintain low salinity in the root zone.

Three general approaches have been followed to assist farmers in the irrigation decision-making process: (1) measure soil water content or suction, (2) use plants as indicators, and (3) maintain a soil water budget.

Soil water content or suction. Various methods have been described for measuring soil water content or suction: the tensiometer and sorption block, the electrical resistance unit, the electrothermal unit, and the neutron moderation technique. The neutron meter measures soil water on a volume basis (cm^3/cm^3), which means that readings can be expressed in surface centimeters or inches of water per acre, a distinct advantage when considering water budgets for scheduling irrigations. However, the meter embodies a radiation hazard from fast neutrons and requires a competent technician to use it effectively.

Tensiometers equipped with vacuum gages measure soil water suction up to 0.8 bar (where 1 bar is equivalent to 10^{-5} Pa) and can be read quickly. Irrigation water is applied when the vacuum gage registers a prescribed limit of soil water suction at a specified depth for a given crop. For example, tensiometers used to irrigate avocados are placed in the active root zone and near the bottom of the root zone. Irrigation applied when the soil water suction at the 30-cm depth approaches 50 centibars does not penetrate to the 60-cm depth until the application time is nearly doubled (July 5, August 26, and September 16). Between irrigations, soil water suction at the lower depth gradually increases, indicating that loss of water below root zone is being controlled.

Plants as indicators. Farmers are using change of leaf color (from light to bluish green) as a practical guide for scheduling irrigations on field beans, cotton, and peanuts.

Measurement of the internal plant water condition by sophisticated techniques as a criterion for scheduling irrigation is impractical for two reasons: plant water status is difficult to measure, and variation in plant water stress with time of day is difficult to interpret.

Soil water budget. Use of the soil water budget approach to determine need for irrigation requires a knowledge of: (1) short-term evapotranspiration (ET) rates at various stages of plant development. (2) soil water retention characteristics. (3) permissible soil water deficits in relation to evaporative demand, and (4) the effective rooting depth of the crop grown.

Evaporation pans are being used to develop soil water budget schemes for scheduling irrigations for sugarcane in Hawaii and for cotton and orchard and vegetable crops in Israel. Evaporation from pans to schedule irrigations must be calibrated for a specific crop and geographic area.

Another advance in techniques to schedule irrigations has been made by using the modified Penman equation to estimate potential ET. Four basic meteorological parameters are required: solar radiation, mean temperature, wind speed, and vapor pressure (dew point). Crop characteristic curves are required.

Sophisticated solid-set and traveling-type sprinkler systems, dead-level automated surface irrigation systems, and graded furrow systems with tail-water reuse facilities provide a high degree of water control and are readily adapted to the irrigation-scheduling techniques described. Farmer acceptance of the computerized meteorological approach to irrigation scheduling has been far greater than any of the other methods discussed. Professional scheduling services were provided for a fee on about 250,000 acres of irrigated lands in Arizona, California, Idaho, Nevada, Washington, Nebraska, Kansas, and Colorado in 1974.

[HOWARD R. HAISE]

Soil conservation. Numerous studies continue to indicate that most herbicides and insecticides, when applied to soils at recommended dosage rates, exert little effect on most soil microorganisms or on soil properties. However, certain fumigants and fungicides may temporarily kill or reduce the numbers of nonparasitic soil organisms and may temporarily influence soil chemical properties. Generally the side effects of most pesticides are not detrimental or they may actually be beneficial, but occasionally they may retard or inhibit plant growth for a few weeks or a few months. The magnitude of these effects depends on dosage, soil type, temperature, moisture, and other factors.

Influence on soil organisms. When applied to soils or crops at normal field application rates, most herbicides or insecticides have little effect on the soil microbes, or they may slightly stimulate growth of some species. However, herbicides may kill soil algae, and insecticides may kill nonparasitic soil insects. Volatile soil fumigants, such as D-D (a mixture of dichloropropane and dichloropropene) and methyl bromide and many soil fungicides, on the other hand, kill numerous nonparasitic soil organisms, including bacteria, fungi, actinomycetes, yeasts, algae, protozoa, earthworms, and insects. After the initial kill, numbers of soil bacteria soon reach much higher numbers than were present in the original soil. In acid soil, fungi may quickly return in greater numbers than were originally present, whereas in alkaline soils they may return quickly or their numbers may be reduced for a year or longer. Although the total number of species may be greatly reduced, with time additional species recolonize the soil and total numbers slowly decline to normal. Fumigants and insecticides exert a similar effect on soil insects.

Factors responsible for the increased numbers of certain microbes following a soil treatment which initially kills large numbers of soil organisms are: (1) The pesticide chemical may be utilized as a food source by one or more organisms; (2) The bodies of the organisms killed by the treatment are utilized as a food source by surviving forms or by species that first recolonize the soil. (3) The organisms which survive or first become reestablished can reach very high numbers because the environment is less competitive.

Influence on soil chemistry. Pesticides which reach the soil or which are applied directly to the soil may influence soil chemical properties in the following ways: (1) The pesticide chemical or partial decomposition products represent a change in the chemical composition of the soil. (2) Upon decomposition of the pesticide chemical, the constituent elements are released as simple inorganic compounds such as ammonia, phosphate, hydrogen sulfide, sulfate, chloride, and bromide. Carbon is released as carbon dioxide. Some of the elements, especially carbon and nitrogen, are utilized for synthesis of cells and organic products of the soil population. Eventually about 10–20% or more of the added carbon is incorporated into the relatively resistant soil humus. The new humus, however, decomposes faster than older, stabilized humus. (3) Pesticidal chemicals which kill an appreciable percentage of the soil organisms often increase the water-soluble salt content of soils. Soluble calcium is especially increased; usually more soluble magnesium, potassium, and phosphorus are noted. (4) Fumigants and fungicides may increase soluble or extractable micronutrient elements, including manganese, copper, and zinc. (5) Most fumigants and fungicides and certain other pesticides kill or reduce numbers of the relatively sensitive nitrifying bacteria in the soil. These organisms oxidize ammonia to nitrites and nitrates. Until these bacteria become reestablished, relatively more of the available soil nitrogen will be in the form of ammonia. Reestablishment of nitrifying bacteria occurs generally within a few weeks to a few months. (6) Pesticides containing the benzene ring may be detoxified and altered with respect to side chains or groups, and may then undergo polymerization reactions with plant and microbial phenolic substances. In this way, parts of pesticide molecules may serve as constituent units in the formation of the beneficial soil humus. *See* SOIL CHEMISTRY.

Increased growth response. The microbiological and chemical effects (side effects) of pesticides in soils generally exert little influence on plant growth. Sometimes, however, growth may be temporarily enhanced or retarded. Increased growth may be related to fertilizer value of nitrogen or phosphorus released during decomposition of specific pesticides and cells of organisms killed by the pesticide treatments; increased availability of soil manganese, phosphorus, and other plant nutrient elements; a plant auxin-type action of some pesticide chemicals or of their partial decomposition products; and recolonization of the soil by microbial species that exert a strong antagonism against plant root parasites.

Reduced plant growth. Occasionally something seems to go wrong following application of a pesticide to soils, and plants may be injured or growth may be retarded for short periods of time. These unexpected results may be caused by various factors:

1. Many pesticide chemicals are toxic to some plant species, and a few, such as the fumigants, are toxic to most plants. If time is not allowed for these chemicals to decompose in the soil or volatilize from the soil, the residual chemical may injure or even kill crop plants. Continuous or frequent use of pesticides which decompose very slowly in the soil may increase soil levels of the pesticide to a point that growth of sensitive crops will be retarded.

2. Simple inorganic substances released during decomposition of certain pesticides may injure sensitive plants. Avocado plants, for example, are highly sensitive to soil chloride. Treatment of the soil with D-D, chloropicrin, or other chloride-containing chemicals may cause or increase chloride injury to this plant. Similarly, several plant species, including onions, carnations, and citrus, are sensitive to bromide. Bromide residues from certain pesticides may temporarily retard growth of these species. Other possibly toxic inorganic decomposition residues of pesticides include arsenic, iodine, copper, and mercury.

3. In greenhouse or ornamental operations, in which fertilization rates are high, the killing of bacteria which oxidize ammonia to nitrites and nitrates may result in the accumulation of toxic concentrations of ammonia from decomposing organic nitrogenous fertilizers.

4. In soils high in manganese, fumigation may increase the soluble manganese to toxic levels for a short period of time. Although extractable soil manganese is generally increased, soil fumigation may sometimes cause manganese deficiencies of cauliflower, brussel sprouts, and broccoli which may be related to increases in available potassium

or other elements which reduce manganese uptake.

5. Sometimes treatment of the soil with any pesticide which kills large numbers of microbes will cause a temporary plant growth inhibition manifested by reduced absorption of phosphorus, zinc, and sometimes copper. The growth inhibition is quite spotty, and healthy plants may grow next to severely injured ones. Young citrus, peach, and certain other tree seedlings are especially sensitive to this phenomenon.

Studies have shown that a primary factor in this type of plant growth inhibition is the killing of endotrophic mycorrhizal fungi which aid the plant roots in absorbing certain plant nutrient elements, especially phosphorus. The condition can be corrected, or partially corrected, by proper fertilization with phosphorus, zinc, and sometimes copper, by inoculation of seed or seedlings with an effective mycorrhiza strain, or by delaying the planting of a sensitive crop until the mycorrhizal fungi have become reestablished in the soil.

[JAMES P. MARTIN]

Fertilizer in semiarid grasslands. In the semiarid grasslands of temperate regions, lack of available nitrogen often limits production as much as does lack of available water. Consequently, use of nitrogen fertilizer is increasing on millions of acres of grasslands, particularly in the northern Great Plains of the United States and Canada. This practice immediately raises concern about the ecological impact of extensive nitrogen fertilization on pollution of surface and ground waters with nitrate. Results of recent research on the fate of fertilizer nitrogen applied to semiarid grasslands have greatly reduced the uncertainty that has surrounded this subject.

In addition to the nitrogen absorbed and translocated into plant tops, a semiarid grassland ecosystem can immobilize a fairly definite quantity of fertilizer nitrogen in the roots, mulch, residues, and soil organic matter. The quantity of nitrogen immobilized in these pools varies with soil type and texture, water availability, and possibly temperature, but is not influenced greatly by either grass species or most management schemes. Fertilizer nitrogen immobilized in these organic forms may later be mineralized by soil microorganisms and recirculated through the ecosystem. A relatively small quantity (10 to 40 lb per acre; 1 lb per acre equals 1.12 kg per hectare) of fertilizer nitrogen seems to be absorbed directly into the cells of the soil microbes and in a few weeks or months is mineralized and recirculated as successive generations of microbes are produced and die. Much more fertilizer nitrogen (up to about 200 pounds per acre) is immobilized in grass roots and mulches and seems to be recycled in 3 to 5 years. Typically, up to 350 pounds of fertilizer nitrogen per acre can be immobilized in various organic pools in the soil-plant system.

Nitrogen cycle. The nitrogen cycle in semiarid grassland ecosystems is essentially a closed system; that is, losses of nitrogen from the soil-plant system are relatively low. Ordinarily, no fertilizer nitrogen is leached below the root zone in semiarid grasslands, so leaching losses are generally inconsequential. Losses of fertilizer nitrogen in gaseous form (by ammonia volatilization or denitrification) also seem usually to be relatively small, except perhaps where urea-containing fertilizers are applied to semiarid grasslands at rates exceeding 80 to 100 pounds of nitrogen per acre. In such instances, available data suggest that volatilization losses may be relatively high.

Typical data on the fate of fertilizer nitrogen applied to semiarid grasslands emerged in an experiment in which 80 pounds of nitrogen per acre (as ammonium nitrate) was applied annually for 11 years to mixed prairie (primarily *Agropyron smithii, Stipa viridula,* and *Bouteloua gracilis*) grazed by yearling steers. After 11 years, approximately 35% of the fertilizer nitrogen applied was found in the roots (19%) and vegetative mulch (16%) on the soil surface. A slightly larger quantity remained in the soil as inorganic (ammonium 2%, nitrate 39%) nitrogen, indicating that the fertilizer applied exceeded the nitrogen required by the ecosystem. Less than 3% of the fertilizer nitrogen was physically removed from the pasture in the form of beef. In total, about 82% (including standing tops 2%, crown 1%) of the fertilizer nitrogen applied was accounted for. The 18% not accounted for was immobilized in soil organic matter or lost to the atmosphere as gaseous nitrogen. Other research suggests that the gaseous loss was about 5% of that applied. Therefore, losses from the nitrogen cycle are relatively low, and a major part of the nitrogen applied to grasslands remains in forms that can be recycled and used for plant growth in later years. All of the nitrogen in the inorganic pool plus a major part of that in roots, residues, and mulches, and some in the soil organic nitrogen pool (included in the unaccounted-for fraction), may be recycled.

Plant debris. Research using the ^{15}N isotope shows that within hours after application, the isotope is found primarily in the senescent or dead plant material – mulches and decaying root materials. This suggests that fertilizer nitrogen is absorbed into the cells of the microorganisms as they rapidly multiply after addition of nitrogen fertilizer. The increased microbial population then decomposes the senescent and dead plant materials, mobilizing the nitrogen they contain. Thus, nitrogen immobilized in plant material is recirculated through the nitrogen cycle, illustrating the importance of plant debris both above and below the soil surface as a pool of potential plant-available nitrogen in semiarid grasslands.

[J. F. POWER]

SOIL EROSION

Soil erosion is that physical process by which soil material is weathered away and carried downgrade by water or moved about by wind. Two categories of erosion are recognized. The first, called geologic erosion, is a natural process that takes place independent of man's activities. This kind of erosion is always active, wearing away the surface features of the Earth. The second kind, referred to as accelerated erosion, occurs when man disturbs the surface of the Earth or quickens the pace of erosion in any way. It produces conditions that are abnormal and poses a problem for the future food supply of the world. To combat erosion successfully, it is important that man recognize the erosion processes and have a knowledge of the factors which affect erosion.

Fig. 8. Sheet erosion showing how soil has been brought from entire cultivated hillside. A large soil deposit has collected on flat area at bottom of slope. (*USDA*)

Fig. 9. Rill erosion showing how water has followed the old corn rows. (*USDA*)

Fig. 10. Large gully could be repaired by plowing in and seeding to grass. (*USDA*)

Types. Erosion by running water is usually recognized in one of three forms: sheet erosion, rill erosion, and gully erosion.

Sheet erosion. The removal of a thin layer of soil, more or less uniformly, from the entire surface of an area is known as sheet erosion. It usually occurs on plowed fields that have been recently prepared for seeding, but may also take place after the crop is seeded. Generally only the finer soil particles are removed. Although the depth of soil lost is not great, the loss of relatively rich topsoil from an entire field may be serious (Fig. 8). If continued for a period of years, the entire surface layer of soil may be removed. In many parts of the world only the surface layer is suited for cultivation.

Rill erosion. During heavy rains runoff water is concentrated in small streamlets or rivulets. As the volume or velocity of the water increases, it cuts narrow trenches called rills. Erosion of this type can remove large quantities of soil and reduce the soil fertility rapidly (Fig. 9). This type of erosion is particularly detrimental because all traces of the rills are removed after the land is tilled. The losses which occurred are often forgotten and adequate conservation measures are not taken to prevent further loss of soil.

Gully erosion. This type of erosion occurs where the concentrated runoff is sufficiently large to cut deep trenches, or where continued cutting in the same groove deepens the incision. Gullying often develops where there is a water overfall. The stream bed is cut back at the overfall and the gully lengthens headward or upslope. Once started, gullying may proceed rapidly, particularly in soils that do not possess much binding material. Gully erosion requires intensive control measures (Fig. 10), such as terracing or the use of diversion ditches, check dams, sod-strip checks, and shrub checks.

Affective factors. The rate and extent of soil erosion depend upon such interrelated factors as type of soil, steepness of slope, climatic characteristics, and land use.

Type of soil. Soil types vary greatly in physical and chemical composition. The amounts of sand, silt, and clay constituents, colloidal material, and organic matter all have a bearing upon the ease with which particles or aggregates can be detached from the body of the soil. Such detachment is caused chiefly by the beating action of raindrops. The particles are then transported downgrade by moving water. Sandy or gravelly soils often have little colloidal material to bind particles together, and hence these materials are easily detached. However, because of their size, sand particles are more difficult to move than fine particles. For this reason sand particles are moved chiefly by rapidly flowing water on steep slopes or by streams at flood stage. Finer particles, such as silt, clay, and organic matter, can be carried by water moving at a slower rate. On gently sloping fields there is a tendency for more of the fine particles to be carried away, leaving the heavier sand particles behind. However, if rainfall is intense and the volume of runoff great, sand may be moved even on gentle slopes.

Slope. The relation of slope to the amount of erosion on different classes of soil is illustrated in

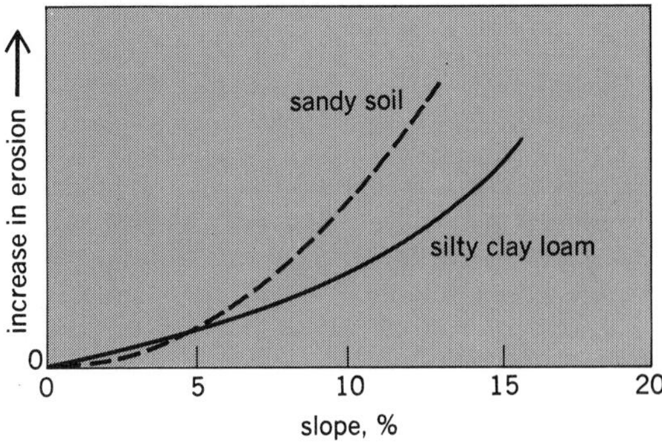

Fig. 11. Generalized diagram illustrating greater loss of fine-grained soil (silty clay loam) on gentle slopes (0–5%) and greater loss of sandy soil on steeper slopes.

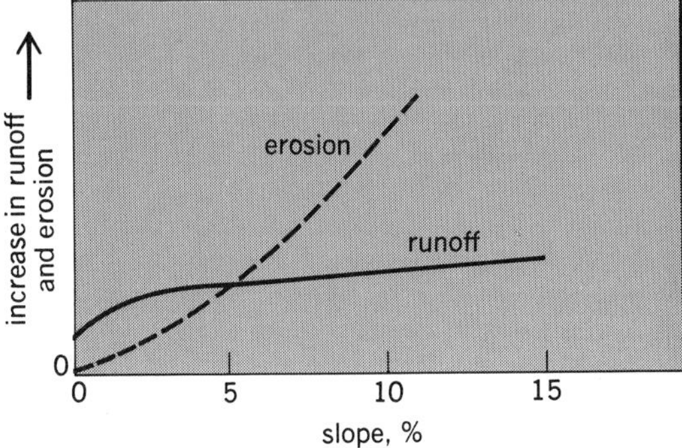

Fig. 12. Effect of slope on total amount of runoff and on rate of runoff and soil erosion.

Fig. 11. The amount of total runoff from rainfall increases only slightly with increase in the slope of the land above 1–2%, but the speed of the flowing water, or rate of runoff, may increase greatly. Since the capacity of moving water to transport soil particles increases in geometric ratio to the rate of flow, the amount of erosion increases greatly with increase in the slope of the land (Fig. 12).

Climate. In cold climates the frozen soil is not subject to erosion for several months of the year. However, if such areas receive heavy snow, serious erosion may take place when the snow melts. This is particularly true if the snow melts as the ground gradually thaws. As the water moves over the thin unfrozen layer of soil, it transports much of this soil material downgrade.

In warm climates soils are susceptible to erosion any time there are heavy rains. Such soils are particularly vulnerable to erosion if rains fall in winter and there is little vegetative cover.

The amount of rainfall is an important factor in determining the erosion that occurs in a given region. However, the character of the rainfall is usually a much more important factor than the total amount in determining the seriousness of erosion. A rain falling at the rate of 2 in./hr may cause three to five times as much erosion as a rainfall of 1 in./hr. Regions where most of the precipitation comes in the form of mist or gentle rain may undergo little erosion, even though the total rainfall may be high and other conditions conducive to erosion.

In some areas of dry climates strong winds cause soil movement and serious loss of soil. Wind erosion is more common on sandy soils, but it is by no means confined to them. Heavier soils, which have a fluffy physical condition produced by freezing and thawing or drying, may be moved in great quantities by the wind.

Land use. The type of crops and the system of management influence the amount and type of erosion. Bare soils, clean uncultivated soils, or land in intertilled row crops permit the greatest amount of erosion. Crops that give complete ground cover throughout the year, such as grass or forests, are most effective in controlling erosion. Small-grain crops, or those that provide a fairly dense cover for only part of the year, are intermediate in their effect on erosion. Table 3 gives results of some of the earliest experiments in the United States on differences in land use and the effect on runoff and erosion. These results show that cultivated land, especially without a crop or protective cover of vegetation, is particularly vulnerable to erosion. In addition, excessive erosion usually occurs where cultivated crops like corn, cotton, and tobacco are grown on hilly or sloping land that is subjected to increased rates of runoff. In some areas where row crops have been grown continually the soil has been removed to the depth of the plow layer within a lifetime.

Pastures in humid areas usually have a tough continuous sod that prevents or greatly reduces sheet or surface erosion. Natural range cover, if in good condition, is usually effective in controlling erosion, but in areas of limited rainfall, where bunch grasses form most of the cover on range land, occasional heavy rain may cause severe erosion of the bare soil exposed between the bunches of grass. Forest lands, with their overhead canopies of trees and surface layers of decaying organic matter, have much greater water intake and much less surface runoff and erosion.

Wind erosion. In the western half of the United States and in many other parts of the world, great quantities of soil are moved by the wind. This is particularly so in arid and semiarid areas. Sandy soils are more subject to wind erosion than silt loam or clay loam soils. The latter, however, are easily eroded when climatic conditions cause the soil to break into small aggregates, ranging from 0.4 to 0.8 mm in diameter. The coarse particles usually are moved relatively short distances, but the fine dust particles may be carried by strong winds for hundreds or even thousands of miles.

In some areas the coarse, or sand, particles are moved by the wind and deposited over extensive

Table 3. Relative runoff and erosion from soil under different land uses, with mean rainfall of 35.87 in.*

Land use and treatment	Runoff, %	Tons soil per acre eroded annually	Years required to lose surface 7 in. of soil
Plowed 8 in. deep, no crop; fallowed to keep weeds down	28.4	35.7	28
Plowed 8 in., corn annually	27.4	17.8	56
Plowed 8 in., wheat annually	25.2	6.7	150
Rotation: corn, wheat, red clover	14.1	2.3	437
Bluegrass sod	11.6	0.3	3547

**Missouri Res. Bull*, no. 63, 1923.

areas as dunes. The dunes move forward in the same direction as the prevailing winds, the particles being moved from the windward side of the dune to the lee side. If dunes become covered with grass or other vegetation, they cease to move. The sandhill region of Nebraska is a good example of such an area.

Control. The following are a few fundamental principles which will help control erosion and greatly reduce the damage done by soil erosion.

1. Keep land covered with a growing crop or grass as much time as possible. Cover increases intake of water and reduces runoff. The extent of erosion control will be roughly in proportion to the effective cover.

2. When there is no growing crop, retain a cover of stubble or crop residue on the land between crops and until the next crop is well started. This can be done by using a system known as stubble-mulch farming. It ultilizes the idea of preparing a seedbed for a new crop without burying the residue from the previous crop. Tillage tools that work beneath the surface and pulverize the soil without necessarily inverting it or burying the residue are used instead of moldboard plows. This system is best adapted to regions of low rainfall or warm climates.

3. Avoid letting water concentrate and run directly downhill. By doing this the soil is protected against water at its maximum cutting power. Construct terraces with gentle grades to carry the runoff water around the hill at slow speeds. These diversions should be designed to empty onto grassed waterways or on meadowland to prevent creation of gullies.

4. Plant crops and till the soil along the contours.

5. Control wind erosion by keeping land covered with sod or planted crops as much of the time as possible. Maintain crop residue on the land between crops and while the next crop is getting started.

6. If wind erosion begins on a bare field or where a crop is just getting started, the soil drifting may be stopped temporarily by cultivation. An implement with shovels that will throw up clods or chunks of soil to give a rough surface is usually effective. Often, only strips through the field need be so treated to stop erosion on the whole area.

7. Moving dunes may require artificial cover or mechanical obstructions on the windward side, followed by vegetative plantings, depending on climatic conditions. Along shorelines, beach grasses followed by woody plants and forests may be required.

[FRANK L. DULEY]

Bibliography: L. D. Baver and W. H. Gardner, *Soil Physics*, 4th ed., 1972; P. W. Birkeland, *Pedology, Weathering, and Geomorphological Research*, 1974; C. E. Black (ed.), *Methods of Soil Analysis*, Amer. Soc. Agron. Monogr. no. 9, 1964; F. E. Broadbent and D. S. Mikkelsen, *Agron. J.*, 60: 674–677, 1968; E. T. Cleaves, A. E. Godfrey, and J. K. Coulter, Soil management systems, in *Soils of the Humid Tropics*, 1972; Food and Agriculture Organization–UNESCO, *Soil Map of the World*, 1971–1976; H. D. Foth and L. M. Turk, *Fundamentals of Soil Science*, 5th ed., 1972; R. M. Hagan, H. R. Haise, and T. W. Edminster (eds.), *Irrigation of Agricultural Lands*, 1967; H. R. Haise and R. M. Hagan, in R. M. Hagan, H. R. Haise, and T. W. Edminster (eds.), *Irrigation of Agricultural Lands*, 1967; *International Rice Research Institute Annual Report*, 1970; D. D. Kaufman, in W. D. Guenzi (ed.), *Pesticides in Soil and Water*, pp. 135–202, 1974; C. E. Kellogg and A. C. Orvedal, Potentially arable soils of the world and critical measures for their use, *Advan. Agron.*, 21:109–170, 1969; G. D. Kleinschmidt and J. W. Gerdeman, *Phytopathology*, 62:1447–1453, 1972; H. Kohnke, *Soil Physics*, 1968; J. P. Martin, in C. A. I. Goring (ed.), *Organic Chemicals in the Soil Environment*, vol. 2, pp. 733–792, 1972; National Academy of Sciences, *Soils of the Humid Tropics*, 1972; C. H. Pair, in C. H. Pair et al. (eds.), *Sprinkler Irrigation*, 4th ed., 1975; J. F. Parr, in W. D. Guenzi (ed.), *Pesticides in Soil and Water*, pp. 315–340, 1974; W. H. Patrick and D. S. Mikkelsen, in *Fertilizer Technology and Use*, 1971; F. N. Ponnamperuma, in *Advances in Agronomy*, 1972; President's Science Advisory Committee, *The World Food Problem*, vol. 2, 1967; R. W. Simonson, Loss of nutrient elements in soil formation, in O. P. Englestad (ed.), *Nutrient Mobility in Soils: Accumulation and Losses*, Soil Sci. Soc. Amer. Spec. Publ. no. 4, 1970; Soil Survey Staff, *Soil Taxonomy: A Basic System of Soil Classification for Making and Interpreting Soil Surveys*, USDA Handb. 436, 1975; A. R. Thompson and C. A. Edwards, in W. D. Guenzi (ed.), *Pesticides in Soil and Water*, pp. 341–386, 1974; L. M. Thompson and F. R. Troeh, *Soils and Soil Fertility*, 3d ed., 1973; U. S. Department of Agriculture, *Soil: The Yearbook of Agriculture*, 1957.

Soil chemistry

The study of the composition and chemical properties of soil. Soil chemistry involves the detailed investigation of the nature of the solid matter from which soil is constituted and of the chemical processes that occur as a result of the action of hydrological, geological, and biological agents on the solid matter. Because of the broad diversity among soil components and the complexity of soil chemical processes, the application of a wide variety of concepts and methods employed in the chemistry of aqueous solutions, of amorphous and crystalline solids, and of solid surfaces is required. For a general discussion of the origin and classification of soils *see* SOIL.

Elemental composition. The elemental composition of soil varies over a wide range, permitting only a few general statements to be made. Those soils that contain less than 12–20% organic carbon are termed mineral. (The exact percentage to consider in a specific case depends on drainage

Table 1. Average percentages of total carbon, total nitrogen, and organic phosphorus in selected soils

Soil	% C	% N	% P
Sand	2.5	.23	.04
Fine sandy loam	3.3	.23	.06
Medium loam	2.3	.22	.05
Clay loam, well drained	4.6	.36	.10
Clay loam, poorly drained	8.0	.43	.05
Peat	46.1	1.32	.03

Table 2. Average percentages of the major and some micro elements in subsurface soil clays and crustal rocks

Soil order:	Alfisol	Inceptisol	Mollisol	Oxisol	Spodosol	Ultisol	Crustal rocks
Si	19.20	24.69	23.01	12.43	5.79	16.02	27.72
Al	12.38	19.61	10.29	19.33	15.86	17.49	8.13
Fe	8.04	3.81	6.83	10.83	3.29	11.96	5.00
Ca	.69	.00	3.59	.10	.29	.15	3.63
Mg	1.26	.40	1.62	.46	.15	.08	2.09
Na	.18	2.52	.04	.00	.27	.06	2.83
K	3.63	n.d.	1.20	.07	.40	.22	2.59
Ti	.40	.28	.44	1.32	.16	.50	.44
Mn	.06	n.d.	.06	.08	.06	.05	.10
P	.14	n.d.	.14	.27	.17	.12	.11

characteristics and clay content of the soil.) All other soils are termed organic. Carbon, oxygen, hydrogen, nitrogen, phosphorus, and sulfur are the most important constituents of organic soils and of soil organic matter in general. Carbon, oxygen, and hydrogen are most abundant; the content of nitrogen is often about one-tenth that of carbon, while the content of phosphorus or sulfur is usually less than one-fifth that of nitrogen (Table 1). The number of organic compounds into which these elements are incorporated in soil is very large, and the elucidation of the chemistry of soil organic matter remains a challenging problem. *See* HUMUS, GEOCHEMISTRY OF.

Besides oxygen, the most abundant elements found in mineral soils are silicon, aluminum, and iron (Table 2). The distribution of chemical elements will vary considerably from soil to soil and, in general, will be different in a specific soil from the distribution of elements in the crustal rocks of the Earth. Often this difference may be understood in terms of pedogenic weathering processes and the chemical reactions that accompany them. Some examples are the illuvial accumulation of aluminum and iron oxides in the B horizon of a Spodosol and of $CaCO_3$ in the calcic horizon of a Mollisol. The most important micro or trace elements in soil are boron, copper, manganese, molybdenum, and zinc, since these elements are essential in the nutrition of green plants. Also important are cobalt, which is essential in animal nutrition, and selenium, cadmium, and nickel, which may accumulate to toxic levels in soil. The average distribution of trace elements in soil is not greatly different from that in crustal rocks (Table 3). This indicates that the total content of a trace element in soil usually reflects the content of that element in the soil parent material and, generally, that the trace element content of soil often is not affected substantially by pedochemical processes.

The elemental composition of soil varies with depth below the surface because of pedochemical weathering. The principal processes of this type that result in the removal of chemical elements from a given soil horizon are: (1) soluviation (ordinary dissolution in water), (2) cheluviation (complexation by organic or inorganic ligands), (3) reduction, and (4) suspension. Soluviation, cheluviation, and reduction include leaching by water into lower horizons; suspension involves removal by erosion or by translocation downward along soil pores. The principal effect of these four processes is the appearance of illuvial horizons in which compounds such as aluminum and iron oxides, aluminosilicates, or calcium carbonate have been precipitated from solution or deposited from suspension. *See* WEATHERING PROCESSES.

Minerals. The minerals in soils are the products of physical, geochemical, and pedochemical weathering. Soil minerals may be either amorphous or crystalline. They may be classified further, approximately, as primary or secondary minerals, depending on whether they are inherited from parent rock or are produced by chemical weathering, respectively.

Primary minerals in soil. The bulk of the primary minerals that occur in soil are found in the silicate minerals, such as the olivines, garnets, pyroxenes, amphiboles, micas, feldspars, and quartz. The feldspars, micas, amphiboles, and pyroxenes commonly are hosts for trace elements that may be released slowly into the soil solution as weathering of these minerals continues. Chemical weathering of the silicate minerals is responsible for producing the most important secondary minerals in soil. The general scheme of the weathering sequence is shown in Fig. 1. *See* SILICATE MINERALS.

Secondary minerals in soil. The important secondary minerals that occur in soil are found in the clay fraction. These include aluminum and iron hydrous oxides (usually in the form of coatings on other minerals), carbonates, and aluminosilicates. The term "allophane" is applied to the x-ray amorphous, hydrous aluminosilicates that are

Table 3. Average amounts of trace elements commonly found in soils and crustal rocks

Trace element	Soil, ppm*	Crustal rocks, ppm
As	6	1.8
B	10	10
Cd	.06	.2
Co	8	25
Cr	100	100
Cu	20	55
Mo	2	1.5
Ni	40	75
Pb	10	13
Se	.2	.05
V	100	135
Zn	50	70

*ppm = parts per million.

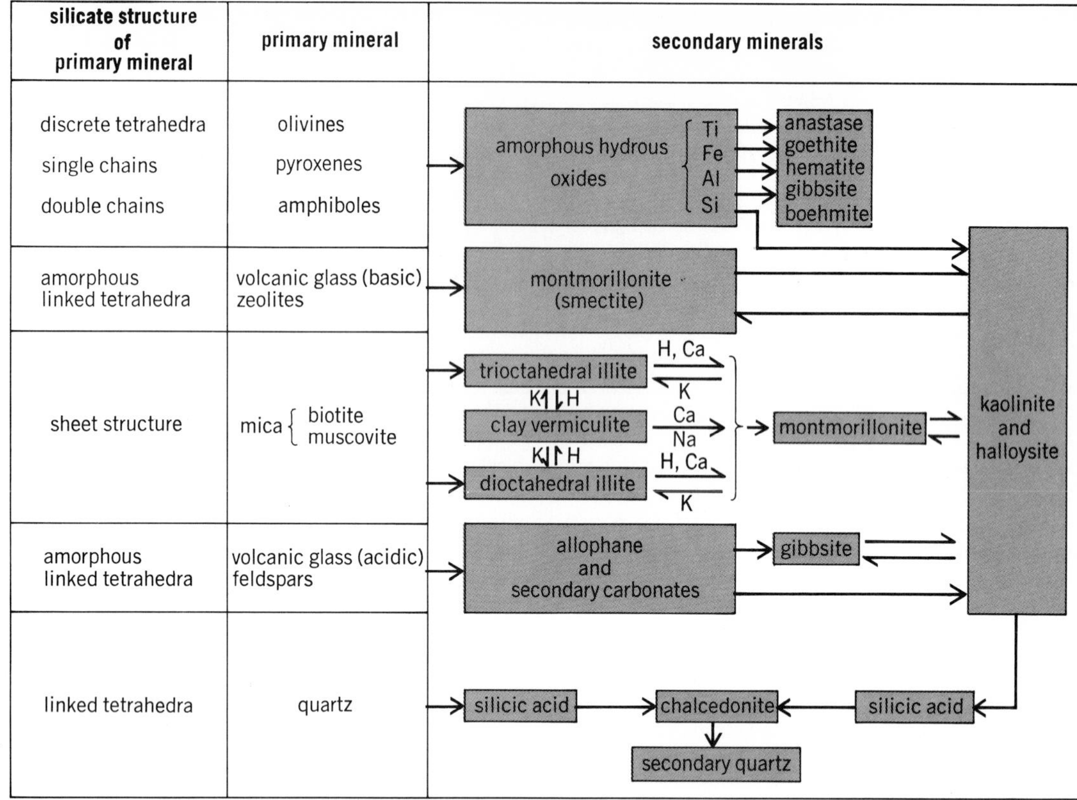

Fig. 1. The scheme of chemical weathering to form secondary minerals from primary silicate minerals. (*Modified from M. Fieldes and L. D. Swindale, Chemical weathering of silicates in soil formation, J. Sci. Tech. New Zealand, 56:140–154, 1954*)

characterized by variable composition, the presence of Si-O-Al bonds, and a differential thermal analysis curve displaying only a low-temperature endotherm and a high-temperature exotherm. The significant crystalline aluminosilicates possess a layer structure; they are chlorite, halloysite, kaolinite, montmorillonite (smectite), and vermiculite. These clay minerals are identified in soil by means of the characteristic x-ray diffraction patterns they produce after certain pretreatments, although their positive identification may be difficult if two or more of the minerals are present at once. *See* CLAY MINERALS.

The distribution of secondary minerals varies among different soils and changes with depth below the surface of a given soil. However, under a leaching, well-oxidized environment, soil minerals do possess a differential susceptibility to decomposition, transformation, and disappearance from a soil profile. This has made possible the arrangement of the clay-sized soil minerals in the order of increasing resistance to chemical weathering. Those minerals ranked near the top of the following list are present, therefore, in the clay fractions of slightly-weathered soils; those minerals near the bottom of the list occur in extensively weathered soils.

Weathering index	*Clay-sized minerals*
1	Gypsum, halite
2	Calcite, apatite
3	Olivine, pyroxene
4	Biotite, mafic chlorite
5	Albite, microcline
6	Quartz
7	Muscovite, illite, sericite
8	Vermiculite
9	Montmorillonite, Al-chlorite
10	Kaolinite, allophane
11	Gibbsite, boehmite
12	Hematite, goethite
13	Anatase, rutile, zircon

In zonal soils of humid-cool to subhumid-temperate regions, illite is the predominant clay mineral. Mixtures of kaolinite, vermiculite, and interstratified clay minerals are found in humid-temperate regions. In humid-warm regions, kaolinite, halloysite, allophane, gibbsite, and amorphous analogs of goethite are found. The mineralogical composition of the highly weathered and leached soils of the humid tropics is a subject of active investigation, in part because these soils (the Oxisols and Ultisols) constitute approximately one-third of the world's potentially arable land. The soil minerals are dominated by iron and aluminum hydrous oxides, kaolinite, halloysite, and quartz. Amorphous weathering residues of kaolinite and halloysite also are found in thin coatings on clay particle surfaces, and Al-chlorite (montmorillonite with interlayer Al-hydroxy polymers) is frequently encountered.

The chemical conditions favoring the genesis of kaolinite are the removal of Na^+, K^+, Ca^{2+}, Mg^{2+}, and Fe^{2+} by leaching, the addition of H^+ in fresh water, and a high Al-Si molar ratio. Smectite (montmorillonite) is favored by the retention of the basic cations (arid conditions or poor drainage) and of silica.

Cation exchange. A portion of the chemical elements in soil is in the form of cations that are not components of inorganic salts but that can be replaced reversibly by the cations of leaching salt solutions or acids. These cations are said to be exchangeable, and their total quantity, usually expressed in units of milliequivalents (meq) per 100 g of dry soil, is termed the cation exchange capacity (CEC) of the soil. The CEC ordinarily is measured by leaching a known amount of soil with a salt solution, such as 1 *N* ammonium acetate at pH 7.0, followed by an additional leaching with isopropyl alcohol to remove the residual salt, then determining the quantity of replacing cation (such as NH_4^+) in the soil. However, this is an arbitrary procedure, since the quantity of cation remaining after such treatment does not have a unique value characteristic of the soil alone, but depends as well on the concentration, the ionic composition, and the pH of the leaching solution. The CEC of a soil generally will vary directly with the amounts of clay and organic matter present and with the distribution of clay minerals (Table 4).

Table 4. CEC values in meq/100 g for some soil textural classes and clay mineral compounds

Soil texture	CEC	Soil mineral	CEC
Sand	1–5	Allophane	25–70
Fine sandy loam	5–10	Chlorite	10–40
Loam or silt loam	5–25	Halloysite	5–50
Clay loam	15–30+	Illite	10–40
Clay (mineral soil)	≳25	Kaolinite	3–15
Clay (14% organic)	23	Smectite	80–150
Clay (39% organic)	76	Vermiculite	100–150+
Clay (100% organic)	150–600	Al, Fe hydrous oxides	≃4

Soils which are less weathered because of recent origin, low precipitation, or temperate to cold climate have as exchangeable cations largely calcium and magnesium. Some soils of dry areas contain significant amounts of exchangeable sodium. Extensively weathered soils, unless formed from basic parent material, have 20–95% of their exchangeable cations as aluminum. Prolonged leaching with fresh water supplies H^+ ions that eventually penetrate and disrupt the structures of soil aluminosilicates, thereby releasing aluminum cations, some of which remain in exchangeable form. The distributions of exchangeable cations for representative soils are shown in Fig. 2.

The chemical equilibrium between exchangeable cations and cations in a leaching solution may be expressed by Eq. (1), where ν is a stoichiometric

$$\nu_A A(ex) + \nu_B B \leftrightarrows \nu'_A A + \nu'_B B(ex) \qquad (1)$$

coefficient, and A or B refers to a cation species, such as Na^+ or Ca^{2+}. Generally speaking, the equilibrium will shift to the right if cation B has a greater charge or a smaller hydrated radius than cation A. The relative affinity of a soil for cation species B may be described formally by the law of mass action as applied to exchange equilibrium equation (2), where K is an equilibrium constant and a is a

$$K = \frac{a_{B(ex)}^{\nu'_B}\, a_A^{\nu'_A}}{a_{A(ex)}^{\nu_A}\, a_B^{\nu_B}} \qquad (2)$$

thermodynamic activity. The applicability of this equation to soils depends on whether it is possible (1) to divide the soil-solution system into "solution" and "exchanger" phases and (2) to define unambiguously the meaning of the activity ratio $a_{B(ex)}^{\nu'_B}/a_{A(ex)}^{\nu'_A}$. Different forms of the expression for K have been developed on the basis of different interpretations of requirements 1 and 2. For example, if the activities of the exchangeable cations are set equal to mole fractions and the activity ratio $a_A^{\nu'_A}/a_B^{\nu_B}$ is set equal to its value in the leaching solution, the expression for K is known as Vanselow's equation. The parameter K then is termed a selectivity coefficient and, in principle, may vary with the solution activity ratio. If the exchangeable cation activities are replaced by equivalents per unit mass of dry soil and the activities in solution are replaced by molar concentrations, the expression for K (again a selectivity coefficient) becomes Gapon's equation. Published data on these kinds of special cases demonstrate that the selectivity coefficients take on constant values only over a limited range of solution concentrations. The development of practicable cation exchange equations with broad applicability remains a challenging problem in soil chemistry.

Isomorphic substitution in clay minerals. One of the most important sources of cation exchange capacity in soils is the negative charge that occurs on clay mineral surfaces because of isomorphic substitution (see Fig. 3). The replacement of Si^{4+} in the tetrahedral sheet by Al^{3+} or of Al^{3+} in the octahedral sheet by Mg^{2+} and Fe^{2+} results in a permanent, negative surface charge on the clay mineral structure. This charge can be balanced by the creation of structural OH groups out of O ions or by the filling of more than two-thirds of the cation positions in the octahedral sheet; however, more commonly charge neutrality is achieved by the adsorption of cations between structural layers on the mineral cleavage surfaces. In kaolinite and illite, isomorphous substitution occurs primarily in the tetrahedral sheet. Because the extent of this substitution is very small in kaolinite and because, in general, cations bonded through a charge deficit in the tetrahedral sheet are held very strongly, the

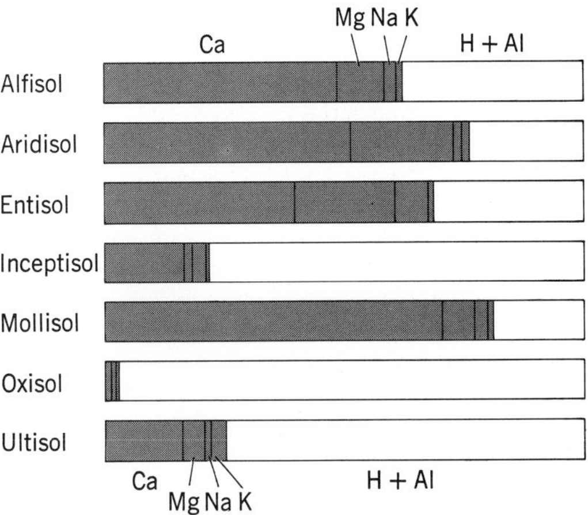

Fig. 2. The typical distributions of exchangeable cations in some soil orders. The shaded regions refer to exchangeable bases.

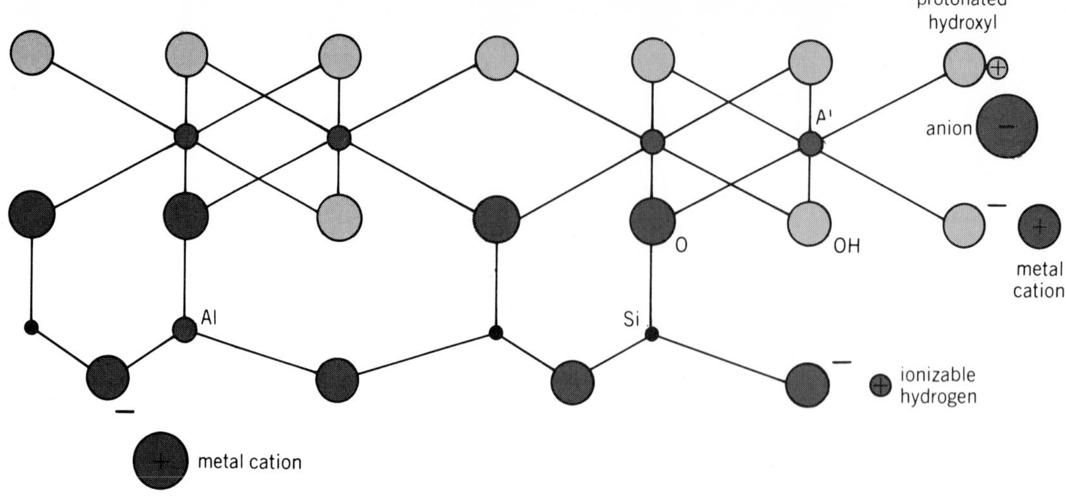

Fig. 3. An idealized diagram of the kaolinite structure showing exchange sites at constant-charge absorbing surfaces (isomorphous-substitution) and constant-potential absorbing surfaces (pH-dependent).

CEC of kaolinite and illite is not large. On the other hand, the CEC of the smectites and vermiculite is relatively large. Extensive isomorphous substitution occurs in the octahedral sheets of these clay minerals, and the resultant electrostatic force bonding the cations adsorbed on the cleavage surface is weak, since it must act through a relatively greater distance than in the case of substitution in a tetrahedral sheet. Interlayer potassium ions in soil micas, illite, and vermiculite, however, usually are not readily exchangeable, both because of the source of the negative charge binding them and because K^+ does not hydrate easily and fits almost perfectly into the hexagonal cavities in the cleavage surfaces of these minerals.

Ionizable hydroxyl groups in clay minerals. At the edges of the structural layers in the crystalline clay minerals and on the surfaces of the amorphous clay minerals, OH groups may be found bonded to exposed Si^{4+} or Al^{3+} cations (Fig. 3). These hydroxyl groups can act as weak Brönsted acids, ionizing in aqueous solution when the pH is sufficiently high. Thus a pH-dependent, negative surface charge can develop that will contribute to the CEC. The mineral surfaces for which this pH-dependence occurs are termed constant-potential adsorbing surfaces, as opposed to the constant-charge adsorbing surfaces generated by isomorphous substitution. The pH value at which the surface charge on a constant-potential adsorbent vanishes is called the zero point of charge (ZPC). ZPC for any soil mineral depends on its composition and defect structure, as well as on the nature of the surrounding electrolyte solution; a value of ZPC between 4 and 6 is common. The maximum negative charge generally is observed at about pH 10. The magnitude of the negative charge developed at a given pH value depends on the extent to which the clay mineral particles possess surfaces bearing ionizable OH groups. For montmorillonite the pH-dependent CEC may be 20% of the total CEC; for illite it may be 50%, while for kaolinite and allophane it may be essentially 100%.

Ionizable functional groups in organic matter. The organic matter in soil contains two types of functional groups that contribute importantly to the CEC: aromatic and aliphatic carboxyls and phenolic hydroxyls. When the pH is greater than 6, these acidic groups ionize in aqueous solution and provide a source of negative charge for metal cation adsorption. The cations of metals in groups IA and IIA (for example, Na^+ and Ca^{2+}) of the periodic table are readily exchangeable after adsorption by soil organic matter, whereas those of the transition metals and of the metals in groups IB and IIB (for example, Fe^{3+} and Cu^{2+}) often are not. The contribution to the CEC from organic functional groups can be as large as 600 meq per 100 g organic matter at high pH values.

Anion exchange. The stoichiometric exchange of the anions in soil for those in a leaching salt solution is a phenomenon of relatively small importance in the general scheme of anion reactions with soils. Under acid conditions ($pH < 5$) the exposed hydroxyl groups at the edges of the structural sheets or on the surfaces of clay-sized particles become protonated and thereby acquire a positive charge. The degree of protonation is a sensitive function of pH, the ionic strength of the leaching solution, and the nature of the clay-sized particle. The magnitude of the anion exchange capacity (AEC) usually varies from near 0 at pH 7 for any soil colloid to as much as 50 meq per 100 g of allophanic clay at about pH 4. Smectite and other clay minerals with high, pH-invariant CEC values do not adsorb exchangeable anions to any degree unless the ionic strength of the leaching solution is very large. The AEC may be measured conveniently by shaking a 5 gm sample of soil for 1 hr in 20 ml 0.17 N $AlCl_3$, filtering, and determining the amount of Cl in the filtrate. AEC is then the difference, per 100 g of soil, between the initial amount of Cl and that in the filtrate.

Negative anion adsorption. Soils whose CEC is approximately independent of pH often display a significant negative adsorption of anions: The concentration of anions in a solution separated from a soil suspension by a membrane permeable to electrolyte is larger than that of the anions in the liquid phase of the suspension. This phenomenon may be understood simply on the basis of the presence of a permanent negative charge on the surface of the solid-phase particles in the suspension. This surface charge attracts cations and repels anions. The principal effects of this repulsion are to reduce the AEC and to increase the ease with which anions may be leached from a soil. If the solution concentration of anions becomes very high, the ionic strength will be high and the concentrations of anion in solution and in the suspension will tend to become equal, if the activity of the anion in solution decreases more than that of the anion in the suspension.

Specific anion adsorption and reprecipitation. Anion exchange in the classic sense applies primarily to halide and nitrate ions. For other ions, in particular, borate, molybdate, sulfate, and orthophosphate, the reaction with the solid matter in soil involves an irreversible specific adsorption or decomposition-reprecipitation. The *o*-phosphate ion, for example, can react with the accessible aluminum hydroxy ions of clay minerals and with hydrous aluminum and iron oxides in soil to form x-ray amorphous analogs of the known crystalline aluminum and iron phosphate minerals. The nature and extent of this reaction are strongly dependent on pH, the metal cations in solution, the acidity of the added phsophate compound, and the structure of the solid phase with which the phosphate ion reacts. Under conditions of a relatively high pH, low acidity of the added phosphate, or high degree of crystallinity of the solid phase reactant, PO_4^{3-} will tend to be "specifically adsorbed" by Al or Fe ions at the surface of the solid. If these conditions are reversed, the combination of PO_4^{3-} with the solid phase may result in a nearly complete destruction of the reactant solid and the formation of an amorphous Al- or Fe-phosphate. In this case, the reaction is more properly termed a "reprecipitation" of the Al or Fe. With either case, the fundamental chemical process is the same. This is an area of active research in soil chemistry.

Soil solution. The solution in the pore space of soil acquires its chemical properties through time-varying inputs and outputs of matter and energy that are mediated by the several parts of the hydrologic cycle and by processes originating in the biosphere (Fig. 4). The soil solution thus is a dynamic and open natural water system whose composition reflects the many reactions that can occur simultaneously between an aqueous solution and an assembly of mineral and organic solid phases that varies with both time and space. This type of complexity is not matched normally in any chemical laboratory experiment, but nonetheless must be amenable to analysis in terms of chemical principles. An understanding of the soil solution in terms of chemical properties has proven to be essential to progress in the maintenance of soil fertility and the quality of runoff and drainage waters.

Chemical speciation of macrosolutes. The macrosolute composition of a soil solution will vary depending on pH, pε (negative common logarithm of the electron activity), organic matter content, input of chemical elements from the biosphere (including humans), and effectiveness of leaching. Under conditions of near-neutral pH, high pε, low organic matter content, no solute input from agriculture, and good but not excessive drainage, the expected macrosolutes are Ca, K, Mg, Na, Cl, HCO_3, $Si(OH)_4$, and SO_4. If the pH is low, H and Al should be added to this list; if it is high, CO_3 should be added. If the soil has been fertilized, NO_3 and H_2PO_4 become important. If the drainage is excessive, Al may be abundant and one or more of the solutes in the original list may be insignificant. If the drainage is poor and, therefore, the pε is low, SO_4 will be replaced by S and CO_3 should be added. If the organic matter content is high, organic solutes become important. Combinations of these different environmental conditions will change the original list of solutes in still other ways (for example low pε and nitrogen fertilizer addition would add NH_4 to the list).

The chemical speciation of the macrosolutes (that is, their distribution among the free-ionic, complexed, precipitated, and adsorbed forms) depends on the nature of the solid matter in the soil, the composition of metals and ligands in solution and their concentration, and the pε value. Clearly the macrosolutes themselves are interdependent in determining their speciation, and even the three factors just mentioned cannot be considered in complete isolation from one another. Nevertheless, it is possible to make some very broad statements about the macrochemical species to be expected by employing the principle of hard and soft acids and bases (HSAB).

The macrosolute metal cations are hard acids. This means that they generally tend to form chemical bonds of a simple electrostatic type and, there-

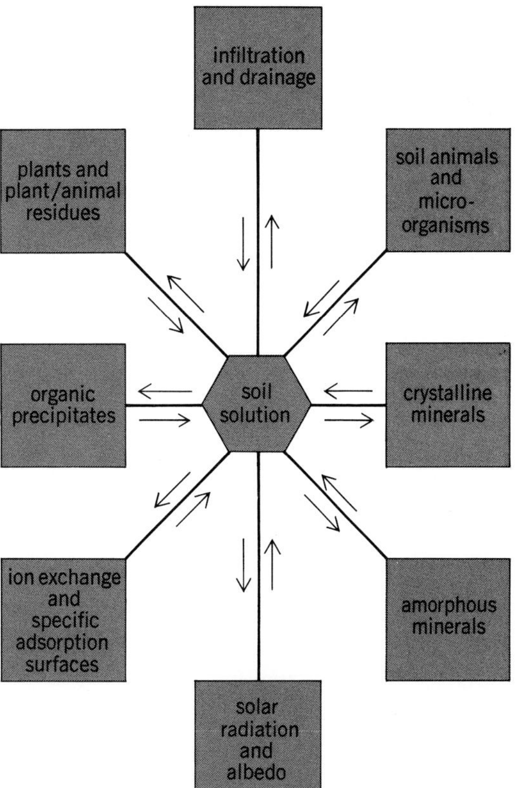

Fig. 4. The factors influencing the chemistry of the soil solution. (*Modified from J. F. Hodgson, Chemistry of the micronutrients in soils, Adv. Agron., 15:141, 1963*)

Table 5. Chemical speciation in percent of the macrosolutes in the soil solution of a well-drained, sandy soil as computed from data on total concentrations and pH using complex stability constants

		Complexed with		
Metal	Free	Cl	HCO_3	SO_4
Ca	68.9	0	1.0	30.1
K	94.1	0	0	5.9
Mg	73.5	0	1.0	25.5
Na	96.8	0	0.2	3.1

*Input data: $pCa_T = 1.47$, $pK_T = 4.10$, $pMg_T = 1.74$; $pNa_T = 1.94$, $pH = 6.8$, $pCl_T = 1.92$; $pCO_{3T} = 2.51$, $pSO_{4T} = 1.35$.

fore, that their reactivities with any ligand (including a mineral surface) should be predictable on the basis of ionic charge and radius along with their common property of low polarizability. In a soil solution, the macrosolute metal cations will tend to: (1) form complexes and sparingly soluble precipitates primarily with oxygen-containing ligands (hard bases), such as H_2O, OH^-, HCO_3^{2-}, CO_3^{2-}, $H_2PO_4^-$, PO_4^{3-}, SO_4^{2-}, and constant-charge; nonhydroxylated mineral surfaces, the stabilities of these complexes will tend to increase with the ratio of ionic charge to ionic radius (or hydration radius) of the metal; usually no reaction will occur with soft or nearly soft bases such as S^{2-} and Cl^-; (2) form complexes with carboxyl groups, but not with organic ligands containing only nitrogen and sulfur electron donors.

These generalizations make it possible to enumerate the probable complexes and precipitates of the macrosolute metals in a soil solution of known composition (including adsorbing surfaces). The exact speciation of the metals then can be calculated if stability constants are available for the expected chemical reactions. A simple example of this type of calculation is shown in Table 5. Usually a large set of nonlinear equations must be solved on a digital computer and ionic strength corrections must be performed.

Chemical speciation of microsolutes. The important microsolutes in the soil include the trace metals, such as Fe, Cu, and Zn, and the trace element oxyanions, such as those formed by As, B, Mo, and Se. The common ranges and median values of trace element concentrations in soil solutions are listed in Table 6. The tableau of microsolutes in a given soil solution is more dependent on inputs from the lithosphere and the biosphere (particularly humans) and less on proton or electron activity and hydrologic factors than is the composition of macrosolutes. The trace metals present, for example, usually are derived from the chemical weathering of specific parent rocks, from the application of fertilizers, pesticides, and urban wastes, and from air pollution.

Table 6. Trace element concentration ranges and median values in the saturation extracts of 30 California soil series

Element	Range, ppm*	Median value, ppm
B	<.1–26.0	<.1
Co	<.01–.14	<.01
Cr	<.01–.017	<.01
Cu	<.01–.2	.03
Fe	<.01–.8	.03
Hg	.0002–.011	.001
Mn	<.01–.95	<.01
Mo	<.01–22.0	<.01
Ni	<.01–.09	<.01
Pb	<.01–.30	<.01
Ti	<.1	<.1
Zn	.01–.40	.04

*ppm = parts per million.

The most general features of the chemical speciation of the microsolutes also may be predicted on the basis of the HSAB principle. The trace metal cations are soft or nearly soft acids, and the trace element anions are hard or nearly hard bases. The exception to this statement is Fe^{3+}, which is a hard acid. This means that the trace metal cations, except for Fe^{3+}, will tend to form chemical bonds of a covalent type whose strength will depend much more on detailed electronic structural considerations than on cationic size or charge. For the anions, the implication is that they will combine strongly with hard-acid metal cations, just as do other oxygen-containing ligands. The trace metal speciation thus presents a more complicated problem than does that of the macrosolute metals. Generally, the trace metal cations in a soil solution will: (1) form complexes and insoluble precipitates more readily with the inorganic ligands Cl^- and S^{2-} and with sites on mineral surfaces that can bind covalently, than with oxyanions; stronger complexes also will form more readily with organic functional groups containing S, P, and N donors than with carboxyl groups; (2) tend to follow the Irving-Williams order in regard to the stabilities of strong complexes: $Mn^{2+} < Fe^{2+} < Co^{2+} < Ni^{2+} < Cu^{2+} > Zn^{2+} > Cd^{2+}$.

These broad predictions imply that the trace metal speciation in soil solutions will depend sensitively on the content and type of organic matter, the percentage of kaolinitic and amorphous hydrous oxide minerals, the pH, the pε, and the ionic strength. For example, a low solubility of the micronutrients Cu and Zn should occur for soils high in immobile organic matter (but not too low in pH) or for soil solutions with low pε values. Moreover, the solubilities of the trace metals should increase significantly with an increase in chloride concentration or in organic solutes. Since the complexes formed in these cases would reduce the ionic charge on the soluble trace metal species, a decrease in the affinity of a constant-charge adsorbing surface (for example, that of montmorillonite) should also occur (Fig. 5). It is evident that the detailed speciation of the trace metals presents a formidable problem. Much information still is needed on the important reactions and corresponding stability constants before definitive predictions can be made.

Clay-organic complexes. The clay minerals in soils often are observed to be intimately associated with carbonaceous materials. These materials may be residues from plant or animal decomposition, herbicides or other pesticides, organic polymers and polyelectrolytes, surfactant compounds, or microbial metabolites. The complexes which they form with clay minerals bear importantly on soil fertility, soil structure, soil moisture and aeration

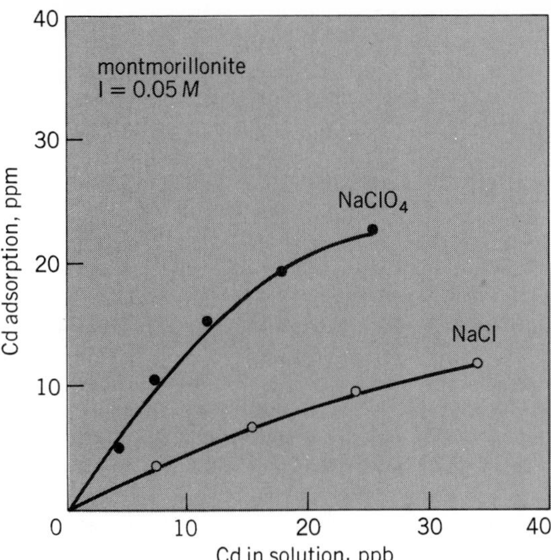

Fig. 5. The adsorption of Cd by montmorillonite at fixed pH and ionic strength as influenced by chloride complexing. In the NaCl system, the amount of adsorption is reduced because of the formation of $CdCl^+$, $CdCl_2^0$, and $CdCl_3^-$, which have smaller positive charges than Cd^{2+}, the only Cd species present in the $NaClO_4$ system.

characteristics, the biological activity of organic compounds applied or disposed on soil, and the degradation of solid and liquid organic wastes in the soil environment. Generally, the organic component of a naturally occurring clay-organic complex will be of a very complicated nature that defies a conclusive structural determination. Therefore, in order to obtain fundamental information about the mechanisms of bonding between clay minerals and organic matter in soil, a major line of research has involved the study of the reactions of known organic compounds with single types of clay mineral. On the basis of these studies, some important bonding mechanisms have been identified. They are expected to apply to the associations between clay minerals and organic matter in nature whenever the appropriate mineral species and organic functional groups are present.

Bonding mechanisms. The principal mechanisms through which organic compounds may bind to clay minerals have been elucidated largely by spectroscopic and x-ray diffraction studies. They may be classified as follows:

1. Organic cation adsorption can occur, through protonated amine or carbonyl groups, onto any constant-charge clay mineral surface. The protonation of the functional groups may be either a pH effect or an acceptance of a proton that was formerly occupying an exchange site, was associated with a water molecule hydrating a metal cation, or was bound on another adsorbed organic cation. The affinity of an organic cation for a constant-charge surface depends on the molecular weight, the nature of the functional groups present, and the molecular configuration. Steric effects can be particularly significant because of the localized character of exchange sites and the quasi-rigid hydration envelope built up on a clay mineral surface. The stability of the water structure on a smectite (montmorillonite) surface is, in fact, great enough to require interstratified layers of either adsorbed metal cations or adsorbed organic cations when the clay mineral surfaces are only partially saturated with the organic compound. A mixture of the two types of cation in a single interlayer region disrupts the water structure too much to be stable.

2. Polar organic functional groups can bind to adsorbed cations through simple ion-dipole forces or complex formation involving covalent bonds. The ion-dipole mechanism is to be expected, of course, for hard-acid metal cations, such as Ca^{2+} and Al^{3+}, while the formation of covalent bonds is to be expected for soft-acid metal cations, such as Cu^{2+}. As the organic functional groups often would have soft-base character, the strength of binding by this mechanism should be greatest for exchangeable transition metal cations. A sharp exception to this rule could occur with "complexable" Al^{3+} (or Fe^{3+}) in amorphous aluminosilicates that bind organic matter containing large numbers of carboxyl groups.

3. Large organic molecules can bind effectively to a clay mineral surface through hydrogen bonding. This bonding can involve a water bridge from a hydrated exchangeable cation to an oxygen-containing functional group, a hydrogen bond from a more acidic functional group adsorbed directly on the clay mineral surface to a less acidic free one containing oxygen, or a direct hydrogen bond to a surface oxygen or hydroxyl plane in the clay mineral. If the exchangeable metal cation is a hard acid, the first type of hydrogen bond will by far dominate the third type in importance. Direct hydrogen bonds to a plane of surface atoms would be accompanied by weaker dipole-dipole (that is, van der Waals) interactions, in general. This type of binding should be most important when very large organic molecules associate with a clay mineral surface containing relatively few exchange sites.

Catalysis reactions. Constant-charge surfaces of clay minerals have been shown often to catalyze reactions involving organic compounds. This catalytic function appears to be connected intimately with the presence of exchangeable metal cations and may be separated into two distinct types. The first type relates to the fact that the water molecules hydrating the exchangeable cations tend to dissociate very readily and, therefore, to endow the clay mineral surface with a pronounced acidity that increases markedly with desiccation. The enhanced proton-donating capability of the clay mineral, which will be greater the harder an acid the exchangeable cation is, serves a catalytic function in, for example, the formation of unsaturated and saturated hydrocarbons during transalkylation of alkylammonium cations and the surface protonation of amines and amino acids.

A second type of catalytic function derives from the formation of organic complexes with the exchangeable cations. This mechanism, which should be more significant the softer an acid the adsorbed metal cation is (again excepting Al^{3+} and Fe^{3+}), appears to play a basic role in, for example, the decarboxylation reaction of fatty acids, carboxyl activation prior to the polymerization of amino acids, and the stabilization of humic compounds against degradation. These and other reactions catalyzed by clay minerals may prove to be very important in understanding how soil organic matter forms and how molecules of biological significance can be synthesized abiotically.

[GARRISON SPOSITO]

Nutrients. Plants tolerate larger amounts of molybdenum than do animals and, except for legumes, do not need cobalt for growth. Ruminant animals require a minimum of about 0.07 ppm (parts per million) of cobalt in feed and, under grazing conditions, cannot tolerate much more molybdenum than 10–20 ppm. Areas where common forage plants have too little cobalt for grazing animals occur principally in the eastern United States, and areas of plants with too much molybdenum are in the West (Fig. 6). Soil and geologic materials determine the distribution of cobalt deficiency and molybdenum toxicity in grazing animals.

Molybdenum. Molybdenum toxicity is a problem among ruminant animals principally in parts of California, Nevada, Oregon, and Montana. The problem areas are largely wet, narrow floodplains and alluvial fans of small streams. The extent of the problem areas tends to be exaggerated because the problem areas are interspersed locally with broad areas of productive soils.

Size of streams and the rock areas they drain determine how much molybdenum is present in alluvium. Most areas of molybdenum toxicity in the western United States are on granitic alluvium that is not mixed with materials from other streams. Small areas also occur on alluvium derived from some shales in northwestern Oregon. Because all alluvium from granite and shales does not give rise to molybdenum-toxic areas, there must be a source of the molybdenum in the higher-lying areas that the streams drain. Broad floodplains of large rivers have materials from many streams, and the large amounts of molybdenum that any stream may contribute are diluted.

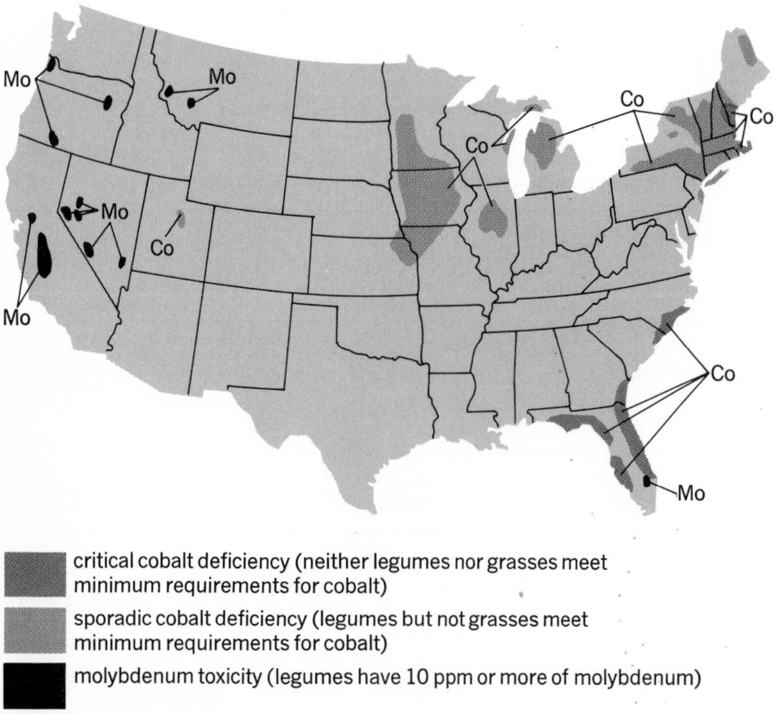

critical cobalt deficiency (neither legumes nor grasses meet minimum requirements for cobalt)

sporadic cobalt deficiency (legumes but not grasses meet minimum requirements for cobalt)

molybdenum toxicity (legumes have 10 ppm or more of molybdenum)

Fig. 6. Areas of cobalt deficiency and molybdenum toxicity.

Thus, molybdenum toxicity is not a problem on broad floodplains.

The molybdenum in alluvium is readily released to plants if soils are wet. If soils are well drained, the release of molybdenum to plants is slow; and plants do not accumulate large amounts of molybdenum, even though the soil may contain large amounts. Molybdenum is also more available to plants in alkaline than in acid soils. Thus, the incidence of molybdenum toxicity is greatest in wet, neutral-to-alkaline soil areas; but molybdenum toxicity can also occur in wet, acid soil areas if the soils have enough molybdenum. The release of molybdenum from these acid soils may not be as rapid as from alkaline soils, but plants may have from 10 to 20 ppm or more of molybdenum.

Cobalt. Cobalt deficiency occurs in the eastern United States. Unlike molybdenum-toxic areas, cobalt-deficient areas cover broad glacial-drift plains in New England and the lower coastal plain from North Carolina to Florida.

Geology also plays an important role in the distribution of areas where deficiencies of cobalt occur. In New England the glacier that overrode the White Mountains left drift deposits on broad plains to the southeast that contribute very little cobalt to soils. The small amounts of cobalt contributed by granites of the White Mountains are today traceable through the low cobalt content in soils on floodplains of the Saco and Merrimac rivers, both of which originate in the White Mountains. Cobalt deficiency in this part of New England was a problem for the earliest settlers who tried to raise cattle.

Only very small amounts of cobalt were contributed by the sandy coastal-plain deposits of the southeastern United States. Here the cobalt deficiency probably resulted from weathering and loss of cobalt as the sandy materials which were weathered in the Piedmont and mountains were carried to the sea to form coastal-plain deposits.

The development of spodosols (podzols) in granitic deposits of New England and humaquods (humus groundwater podzols) in sandy coastal-plain deposits has caused further loss of whatever cobalt these sandy materials had. Cobalt is leached from these soils much as are iron and organic matter. The humaquods especially have very small amounts of reactive forms of cobalt left because of intense leaching. The cobalt leached from humaquods contrasts strikingly with an apparent biopedogenic recycling of soil cobalt from the subsoil to the ground surface in ultisols (red-yellow podzolic soils) of the southeastern United States. Cobalt deficiency is not a problem in ruminant animals on farms with ultisols, because plants have adequate amounts of cobalt to meet their nutritional requirements.

[JOE KUBOTA]

Bibliography: F. E. Bear (ed.), *Chemistry of the Soil*, 2d ed., 1964; J. J. Fripiat and M. I. Cruz-Cumplido, Clays as catalysts for natural processes, *Annu. Rev. Earth Planet. Sci.*, 2:239–256, 1974; J. E. Gieseking (ed.), *Soil Components*, vol. 1: *Organic Components*, vol. 2: *Inorganic Components*, 1975; M. M. Mortland, Clay-organic complexes and interactions, *Adv. Agron.*, 22:75–117, 1970; J. J. Mortvedt, P. M. Giordano, and W. L. Lindsay (eds.), *Micronutrients in Agriculture*,

1972; M. Schnitzer and S. U. Khan, *Humic Substances in the Environment*, 1972; W. Stumm and J. J. Morgan, *Aquatic Chemistry*, 1970; K. Wada and M. E. Harward, Amorphous clay constituents of soils, *Adv. Agron.*, 26:211–260, 1974.

Sphalerite

A mineral, β-ZnS, also called blende. It is the low-temperature form and more common polymorph of ZnS. Pure β-ZnS on heating inverts to wurtzite, α-ZnS, at 1020°C. Sphalerite crystallizes in the hextetrahedral class of the isometric system with a structure similar to that of diamond. Zinc atoms occupy the positions of half the carbon atoms of diamond, and sulfur atoms occupy the other half. Each zinc atom is bonded to four sulfur atoms, and each sulfur atom is bonded to four zinc atoms. The common crystal forms of sphalerite are the tetrahedron, dodecahedron, and cube, but crystals are frequently complex and twinned (see illustration).

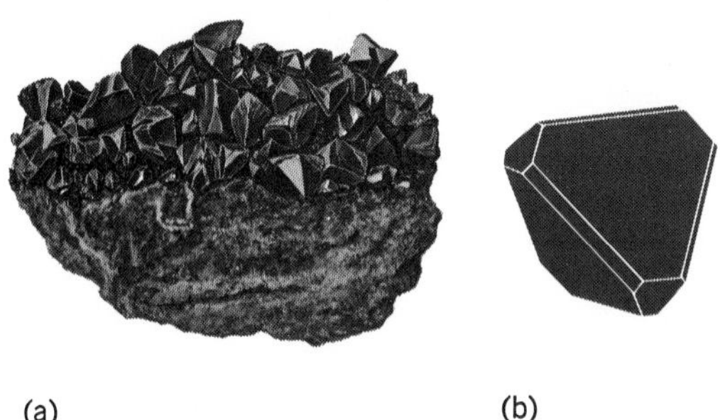

Sphalerite. (*a*) Crystals in limestone from Joplin, Mo. (*specimen from Department of Geology, Bryn Mawr College*). (*b*) Crystal habit (*from C. S. Hurlbut, Jr., Dana's Manual of Mineralogy, 17th ed., copyright © 1959 by John Wiley & Sons, Inc.; reprinted by permission*).

The mineral is most commonly in coarse to fine, granular, cleavable masses. The luster is resinous to submetallic; the color is white when pure, but is commonly yellow, brown, or black, darkening with increased percentage of iron. There is perfect dodecahedral cleavage; the hardness is 3½ on Mohs scale; specific gravity is 4.1 for pure sphalerite but decreases with increasing iron content.

Pure sphalerite contains 67% zinc and 33% sulfur, but iron is usually present, substituting in the structure for zinc and may amount to 36%. The amount of iron in sphalerite varies directly with temperature at the time of crystallization. Cadmium and manganese may be present in small amounts.

Sphalerite is a common and widely distributed mineral associated with galena, pyrite, marcasite, chalcopyrite, calcite, and dolomite. It occurs both in veins and in replacement deposits in limestones. As the chief ore mineral of zinc, sphalerite is mined on every continent. The United States is the largest producer, followed by Canada, Mexico, Soviet Union, Australia, Peru, the Congo, and Poland. *See* WÜRTZITE.

[CORNELIUS S. HURLBUT, JR.]

Sphene

A nesosilicate mineral, composition $CaTiSiO_5$, also known as titanite. It is monoclinic and usually occurs as well-formed crystals with characteristic wedge shape (see illustration). There is prismatic cleavage and frequently a well-developed parting. Hardness is 5–5½ on Mohs scale; specific gravity is 3.4–3.5. The luster is resinous to adamantine and the color brown, green, yellow, gray, or black. Sphene is a common accessory mineral in igneous rocks, particularly syenites and nepheline syenites; it is also present in gneisses, schists, and crystalline limestones. Fine crystals are found at Binnental and St. Gothard, Switzerland; Arendal, Norway; and in Ontario and Quebec, Canada. Sphene, associated with nepheline and apatite, occurs in huge masses on the Kola Peninsula, U.S.S.R., where it is mined as a source of titanium. *See* SILICATE MINERALS.

Sphene. (*a*) Crystal from Eganville, Ontario (*American Museum of Natural History specimen*). (*b*) Crystal habits (*from C. S. Hurlbut, Jr., Dana's Manual of Mineralogy, 17th ed., copyright © 1959 by John Wiley & Sons, Inc.; reprinted by permission*).

[CORNELIUS S. HURLBUT, JR.]

Spilite

An aphanitic (microscopically crystalline) to very fine-grained igneous rock, with more or less altered appearance, somewhat resembling basalt but composed of albite or oligoclase, chlorite, epidote, calcite, and actinolite.

In spite of the highly sodic plagioclase, spilites are generally classed with basalts because of the low silica content (about 50%). They also retain many textural and structural features characteristic of basalt.

Under the microscope, laths of albite or oligoclase usually appear clouded or closely associated with abundant epidote, calcite, and chlorite. Actinolite and additional chlorite and epidote appear to have formed from augite, and small relic grains of augite may survive. Olivine is uncommon and has usually been changed to serpentine. Certain textures of basaltic rocks may still survive; others may be completely destroyed.

Spilites form small intrusive masses (dikes and sills) as well as lava flows. Intrusive spilites appear to grade into diabase; flows grade into basalts. Spilites are most commonly and abundantly associated with stratified rocks of geosynclines and may have had a submarine origin. Pillow structure, in which the rock appears composed of closely packed, pillow-shaped masses up to a few feet across, is typical and more perfectly developed than in other rock types. Vesicules, commonly filled with various minerals, may give the rock an amygdaloidal structure. *See* AMYGDULE; GEOSYNCLINE.

Spilites are generally believed to represent rocks of basaltic composition in which calcic plagioclase (labradorite) has been converted largely to albite. Albitization may have been accomplished during the late stages of crystallization of the basaltic lava or shortly thereafter. Sodium from the sea water may have gradually replaced calcium in the plagioclase to form albite, and some of the displaced calcium may have gone to form epidote and calcite. Another source for requisite sodium may have been emanation from deeper masses of molten rock.

A less popular theory supposes spilite to form by direct crystallization of spilitic magma (rock melt), but the origin of such a magma poses something of a problem. Many so-called spilites may be of metamorphic and metasomatic origin; they may have formed by reconstitution and partial replacement of normal basaltic rocks. *See* BASALT; IGNEOUS ROCKS; METAMORPHISM; METASOMATISM; PETROGRAPHIC PROVINCE. [CARLETON A. CHAPMAN]

Spinel

The mineral of ideal composition $Al^{VI}_2[Mg^{IV}O_4]$, crystallizing with the space group $Fd3m$, $a = 8.08$ A. The spinel structure type is one of the most important and most comprehensively studied of all structure types. Its general formula can be written $M^{VI}_2[T^{IV}O_4]$, where the Roman numerals indicate the coordination numbers: the M atoms are octahedrally coordinated and the T atoms are tetrahedrally coordinated. The oxygen atoms are in cubic closest-packed arrangement, and the structure is quite dense for this reason. Most spinels involve divalent (2+) and trivalent (3+) cations. In normal spinels, the divalent cations occupy the tetrahedral positions and the trivalent cations the octahedral positions. In inverse spinels, the trivalent cations occupy the tetrahedral positions, and equally apportioned divalent and trivalent cations the octahedral positions. Typical normal spinels are Al_2MgO_4, Al_2ZnO_4, Al_2FeO_4, and Al_2MnO_4; inverse spinels include $MgFeFeO_4$, $FeTiFeO_4$, and $FeFeFeO_4$, the last member being the important mineral magnetite. Thus, the magnetite formula should be written $Fe^{2+}Fe^{3+}[Fe^{3+}O_4]$.

No less than 200 different spinels have been synthesized, usually by fusion of the component oxides. Synthetic spinels have many uses. Drawn boules of single crystals of Al_2MgO_4 doped with appropriate impurity ions are manufactured as gem material. The ferrimagnetic properties of some spinels are of great importance in the solid-state industry. Typical divalent cations include Pb, Co, Cd, Ca, Ni, Mn, Mg, Fe, and Zn; some trivalent cations are V, Cr, Ti, Mo, Al, Fe, Sn, Ga, In, Co, Mn, and Ge. Other cations include Li^{1+}, Sb^{5+}, and Ti^{4+}. Thiospinels, involving sulfur instead of oxygen, are also known. Defect spinels include maghemite, γ-Fe_2O_3, which can also be written

$$\square_{1/3}Fe_{2\text{-}2/3}O_4$$

where the $\square$ are holes or vacancies in the structure. There are many examples of structures which are based on distortion of the spinel structure type. Hausmannite, Mn_2MnO_4, is an example, and it is crystallographically tetragonal. The spinel structure type may also occur as components or blocks in some very complex structures, such as the barium ferrites.

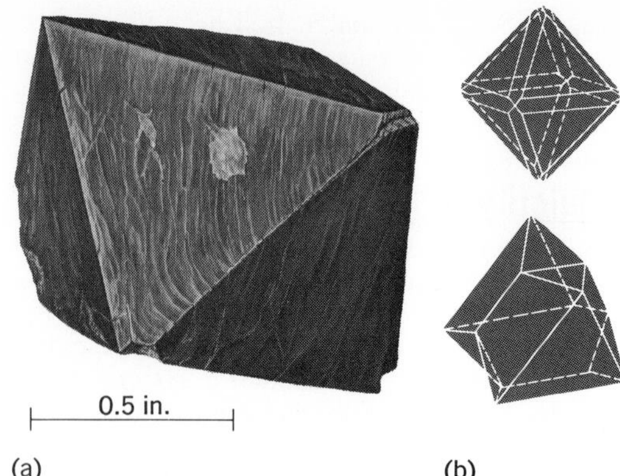

Spinel. (*a*) Crystal from Franklin, N.J. (*specimen from Department of Geology, Bryn Mawr College*). (*b*) Crystal habits (*from C. S. Hurlbut, Jr., Dana's Manual of Mineralogy, 17th ed., copyright © 1959 by John Wiley & Sons, Inc.; reprinted by permission*).

Mineralogically, the spinels are divided into three series: the spinel (Al^{3+}) series, the magnetite (Fe^{3+}) series, and the chromite (Cr^{3+}) series. Some members of these series, such as magnetite and chromite, are valuable ores of their principal metals. The Al_2MgO_4 spinel occurs as simple octahedral crystals, more rarely as dodecahedrons, cubes, or complex modifications thereof (see illustration). It frequently occurs as rounded grains. The hardness is 8 on Mohs scale and the specific gravity 3.5, but extensive substitutions can significantly alter these values. Thus spinel may be blue, gray, green, yellow, red, brown, black, or colorless.

Spinels occur in high-temperature rocks, such as contact metamorphic marbles, where large crystals are sometimes found; in high-grade aluminous silica-deficient rocks along with sapphirine and cordierite; and as early differentiates in basic to ultrabasic rocks such as norites, gabbros, and serpentinites.

It has been discovered that the important rock-forming mineral olivine, $(Mg,Fe)_2[SiO_4]$, inverts at high pressure to the spinel structure type. Olivine is based on a hexagonal closest-packed array of oxygen atoms which, like spinel, geometrically conserves space most efficiently. However, at high pressures the cation-cation repulsion effects favor the spinel arrangement. This olivine-spinel transition is believed to represent an important structural inversion occurring in the upper portion of the Earth's mantle. *See* MAGNETITE; OLIVINE.

[PAUL B. MOORE]

Spodumene

SPODUMENE

Spodumene crystal with pegmatite, Doshen, Mass. (*Specimen from Department of Geology, Bryn Mawr College*)

The name given to the monoclinic lithium pyroxene $LiAl(SiO_3)_2$. Spodumene commonly occurs as white to yellowish prismatic crystals, often with a "woody" appearance, exhibiting the 87° pyroxene (110) cleavages (see illustration). It is easily identified during heating in a flame by the red color given off, accompanied by a marked swelling of the fragment. Spodumene usually contains an appreciable quantity of hydrogen substituting for lithium. At 720° C, spodumene inverts to a tetragonal form, β-spodumene, which is accompanied by a 30% increase in volume. Spodumene is capable of forming immense crystals in nature. A single crystal 47 ft in length and 5 ft in diameter, as well as others almost as large, has been found at the Etta mine in South Dakota. This implies the remarkable ability of a single crystal to replace a large variety of preexisting minerals and yet maintain the integrity of a single crystal, a crystal growth that is unequaled elsewhere in nature. *See* PYROXENE.

Spodumene is usually found as a constituent in certain granitic pegmatites in association with quartz, alkali feldspars, mica, beryl, phosphates, and a large variety of rare minerals. It is also known to occur as disseminated grains in some granite gneisses. Spodumene often alters to a fibrous mass composed of eucryptite $LiAlSiO_4$ and albite, or eucryptite and muscovite. The emerald-green variety, hiddenite, and a lilac variety, kunzite, are used as precious stones. Spodumene from pegmatites is used as an ore for lithium.

[GEORGE W. DE VORE]

Stalactites and stalagmites

Stalactites, stalagmites, dripstone, and flowstone are travertine deposits in limestone caverns (see illustration), formed by the evaporation of waters bearing calcium carbonate. Stalactites grow down from the roofs of caves and tend to be long and thin, with hollow cores. The water moves down the core and precipitates at the bottom, slowly extending the length while keeping the core open for more water to move down. Stalactites are banded concentrically to the center.

Stalagmites grow from the floor up and are commonly found beneath stalactites; they are formed from the evaporation of the same drip of water that forms the stalactite. Stalagmites are thicker and shorter than stalactites and have no central hollow core. The banding of stalagmites is parallel to the surface at the time of deposition of each band.

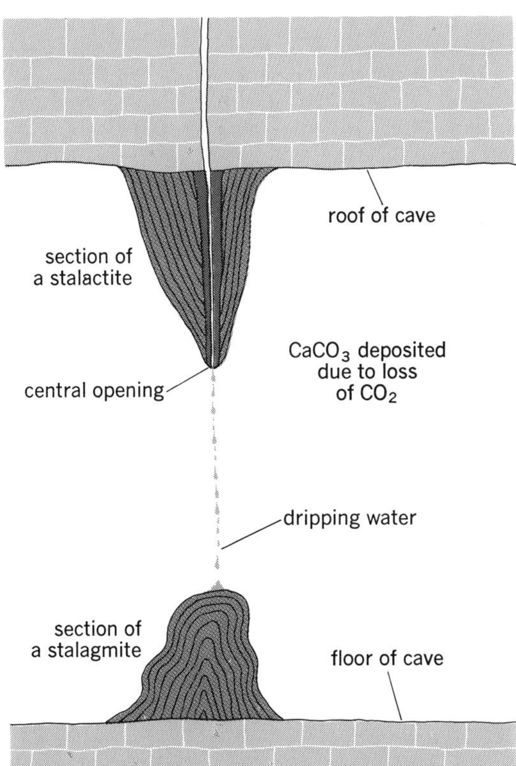

Stalactite and stalagmite deposits. Along the central opening are deposits made by water as it dripped from the crack in the limestone. Outside layers were deposited by water flowing over the surface. Water may drip to the floor and create a stalagmite. (*After E. B. Branson et al., Introduction to Geology, McGraw-Hill, 3d ed., 1952*)

Dripstone and flowstone are travertine accumulations of much the same origin as stalactites and stalagmites (evaporation from cave waters) but are irregular. Their irregularity is due to deposition on irregular sloping walls or surfaces in the cave rather than drips from the roof. *See* CALCITE; CAVE; LIMESTONE; TRAVERTINE. [RAYMOND SIEVER]

Bibliography: V. C. Allison, The growth of stalagmites and stalactites, *J. Geol.*, 31(2):106–125, 1923.

Staurolite

A nesosilicate mineral, $FeAl_4(SiO_4)_2(OH)_2$, that crystallizes in the orthorhombic system. It is frequently in crystals, usually a combination of the vertical prism with the basal and side pinacoid. Equally common are two types of cruciform penetration twins (see illustration). In one type the two individuals cross at approximately 90°; in the other type they cross at about 60°. The hardness is 7–7½ on Mohs scale; specific gravity is 3.7. The luster is resinous to vitreous but may be dull when the mineral is impure or altered; the color is reddish brown to black. Staurolite is a metamorphic mineral found in schists associated with garnet, kyanite, and tourmaline. At St. Gothard, Switzerland, it is found on kyanite in parallel orientation. In the United States it is found in schists in many states, notably New Hampshire, Massachusetts, North Carolina, Georgia, Virginia, and New Mexico. *See* SILICATE MINERALS.

[CORNELIUS S. HURLBUT, JR.]

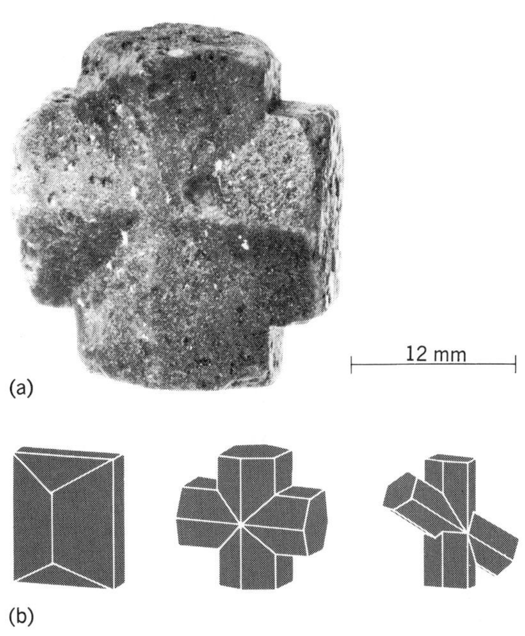

Staurolite. (*a*) Crystal from Taos County, N.Mex. (*American Museum of Natural History specimen*). (*b*) Common crystal habits (*from C. S. Hurlbut, Jr., Dana's Manual of Mineralogy, 17th ed., copyright © 1959 by John Wiley & Sons, Inc.; reprinted by permission*).

Stibnite

A mineral with composition Sb_2S_3 (antimony trisulfide), the chief ore of antimony. It crystallizes in the orthorhombic system, in slender, prismatic, vertically striated crystals which may be curved or bent (see illustration). It is often in bladed, granular, or massive aggregates. There is one direction of perfect cleavage showing cross striations. The hardness is 2 on Mohs scale and the specific gravity 4.5–4.6. The luster is metallic and the color lead-gray to black. It is one of few minerals that fuses easily in the match flame (525°C).

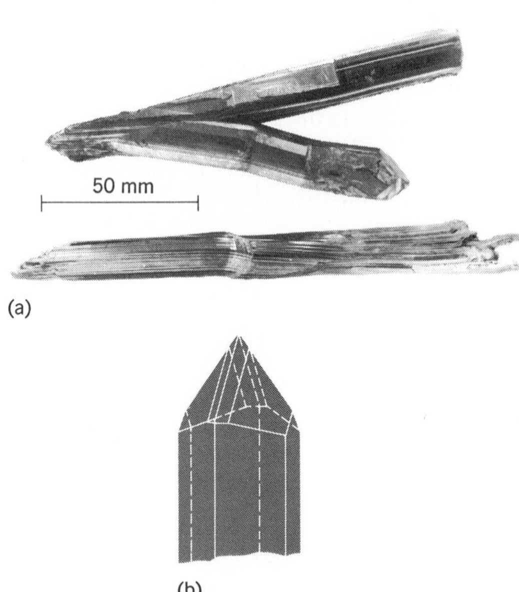

Stibnite. (*a*) Crystals, Ichinokawa, Japan (*American Museum of Natural History specimens*). (*b*) Crystal habit (*from C. S. Hurlbut, Jr., Dana's Manual of Mineralogy, 17th ed., copyright © 1959 by John Wiley & Sons, Inc.; reprinted by permission*).

Stibnite is found in quartz veins in granite and gneiss with few other minerals present. Elsewhere it may be associated with cinnabar, realgar, orpiment, gold, galena, and sphalerite. Replacement deposits in limestone are probably the result of deposition by hot springs. It has been found in various mining districts in Germany, Rumania, France, Bolivia, Peru, and Mexico. Before World War II the most important commercial deposits were in the province of Hunan, China. In the United States the Yellow Pine mine at Stibnite, Idaho, is the largest producer. Other deposits are in Nevada and California. The finest crystals have come from the island of Shikoku, Japan.

[CORNELIUS S. HURLBUT, JR.]

Stilbite

A mineral belonging to the zeolite family of silicates. It crystallizes in the monoclinic system in crystals that are tabular parallel to the side pinacoid. Most characteristic are sheaflike aggregates of thin tabular crystals (see illustration). There is perfect cleavage parallel to the side pinacoid, and here the mineral has a pearly luster; elsewhere the luster is vitreous. The color is usually white but may be brown, red, or yellow. Hardness is $3\frac{1}{2}$–4 on Mohs scale; specific gravity is 2.1–2.2. *See* ZEOLITE.

Sheaflike aggregates of thin crystals of stilbite, found at Two Islands, Nova Scotia. (*American Museum of Natural History specimens*)

Stilbite is a calcium-sodium aluminum silicate, $Ca(Al_2Si_7O_{18}) \cdot 7H_2O$, but usually contains some sodium replacing calcium.

Stilbite is a secondary mineral usually found in cavities in basalts and related rocks. Much less commonly it is found in granites, gneisses, and metal-bearing veins. It is associated with other zeolites, datolite, prehnite, and calcite. Some notable localities are in Iceland, India, Scotland, Nova Scotia, and in the United States at Bergen Hill, N.J., and the Lake Superior copper district in Michigan. [CORNELIUS S. HURLBUT, JR.]

Stishovite

Naturally occurring stishovite, SiO_2, is a mineral formed under very high pressure and the only mineral in which the silicon atom is in sixfold, or octahedral, coordination instead of the usual fourfold, or tetrahedral, coordination. It was first discovered by E. C. T. Chao and others in 1962 in the same type of Coconino sandstone from the Meteor Crater of Arizona in which coesite was discovered (Chao and others, 1960). Stishovite has since been found in rocks from the Ries Crater of Germany on the basis of x-ray studies (Chao and J. Littler, 1962), providing further evidence of the impact origin of this giant German crater. The presence of stishovite indicates formation pressures in excess of 75 kilobars (about 1,000,000 psi), and it is likely that natural occurrence aside from meteorite craters would be at great depths in the Earth, some 400–500 km below the surface. The possibility of the existence of stishovite at great depths stongly influences the interpretations of geophysicists and solid-state physicists regarding the phase transitions of mineral matter, as well as the interpretation of seismic data in the study of such regions of the interior of the Earth. The finding of stishovite has stimulated further research in the phenomenon of shock. *See* COESITE.

Synthesis. In 1961, a few months before the discovery of the natural occurrence, the six-coordinated form of silica was synthesized in the high-pressure laboratory by S. M. Stishov and S. V. Popova of the Soviet Union. They reported that a high-density polymorph of silica was obtained at 1200–1400°C and at a corrected pressure of 125

X-ray powder data for stishovite*

		Natural†		Synthetic‡	
hkl	d_{hkl}, $A_{calc.}$	d_{hkl}, $A_{meas.}$	Intensity	d_{hkl}, $A_{meas.}$	Intensity
110	2.955	2.959	100	2.95	100
101	2.247	2.246	18	2.24	30
200	2.089			2.09	1
111	1.979	1.981	35	1.98	42
210	1.869	1.870	13	1.87	21
211	1.530	1.530	50	1.53	72
220	1.478	1.478	18	1.476	30
002	1.332	1.333	9		
310	1.322	1.322	4	1.326§	15
221	1.292	1.291	1	1.293§	2
301	1.234	1.235	25	1.234	42
112	1.215	1.215	9	1.215	11
311	1.184	1.185	2	1.184	4
320	1.159	1.159	0.7		
202	1.123	1.123	1		
212	1.085	1.084	1	1.086	3
321	1.063	1.062	2	1.063	4
400	1.045	1.045	2	1.044	5
410	1.014	1.013	1	1.014	5
222	0.9895	0.9900	6		
330	0.9850	0.9850	2	0.987§	15
411	0.9473	$0.9475\alpha_1$ $0.9470\alpha_2$	4 2	0.947	11
312	0.9383	$0.9382\alpha_1$ $0.9375\alpha_2$	3 2	0.940	7
420	0.9345				
331	0.9239				
421	0.8818	$0.8824\alpha_1$ $0.8813\alpha_2$	0.7 0.7	0.881	3
103	0.8689	$0.8691\alpha_1$ $0.8683\alpha_2$	2 1	0.876	3
113	0.8507	$0.8506\alpha_1$ $0.8501\alpha_2$	1 0.7	0.851	2
430	0.8358				
402	0.8221	$0.8220\alpha_1$ $0.8215\alpha_2$	2 1	0.823	11
510	0.8196	$0.8194\alpha_1$ $0.8193\alpha_2$	2 1	$0.819\alpha_1$ $0.819\alpha_2$	15 7
412	0.8067	$0.8065\alpha_1$ $0.8065\alpha_2$	1 0.7	$0.8065\alpha_1$ $0.8065\alpha_2$	9 4
213	0.8023	$0.8023\alpha_1$ $0.8022\alpha_2$	6 2	$0.8023\alpha_1$ $0.8023\alpha_2$	21 11
501, 431	0.7975	$0.7975\alpha_1$ $0.7973\alpha_2$	4 1	$0.7972\alpha_1$ $0.7971\alpha_2$	21 11
332	0.7921	$0.7920\alpha_1$ $0.7919\alpha_2$	4 2	$0.7919\alpha_1$ $0.7919\alpha_2$	21 11

*Tetragonal, $a = 4.1790$ A, $c = 2.6649$ (both ±0.0004 A), $c/a = 0.6377$.

†Stishovite from Meteor Crater, Ariz., USGS film no. 17035, camera diameter 114.59 mm Cu $K\alpha$, radiation ($\lambda = 1.5418$ A), Ni filter, lower limit 2θ measurable, approximately 6.3° or 14.0 A. Weak lines of zircon and rutile were observed but are not listed.

‡Synthetic stishovite (Stishov and Popova, 1961).

§Broad line.

kilobars (originally reported to be at a pressure of 160 kilobars).

This new compound was subsequently successfully synthesized by three different high-pressure laboratories in the United States, which essentially confirmed the results of Stishov and Popova. According to the common practice in mineralogy, the natural mineral was named by Chao and others in honor of the senior author Stishov, who, with his coauthor Popova, first succeeded not only in making it but also in describing its physical properties. *See* HIGH-PRESSURE PHENOMENA.

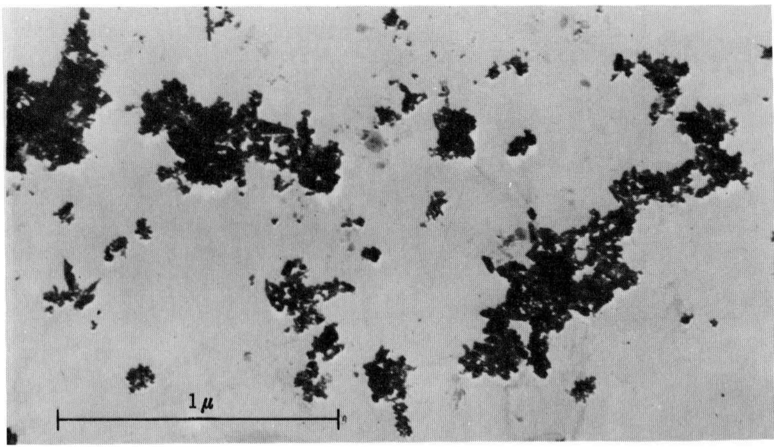

Fig. 1. Electron micrograph of stishovite crystals from Meteor Crater, Ariz.

Natural occurrence. Stishovite occurs in submicron size in very small amounts (less than 1% of the rock) in samples of Coconino sandstones from the Meteor Crater of Arizona, which contains up to 10% of coesite, the other high-pressure polymorph of silica. Besides stishovite and coesite, these rocks consist mainly of quartz, with small amounts of silica glass (lechatelierite) and traces of zircon and, in some cases, rutile. Because of its extremely fine grain size and because of the sparsity of this mineral in the rock, stishovite cannot be readily identified in thin sections under the petrographic microscope. Positive identification of the mineral is possible only by the x-ray diffraction method after chemical concentration. *See* METEORITE.

Properties. Stishovite is much more resistant to attack by concentrated hydrofluoric acid than coesite. It can therefore be extracted or concentrated by prolonged treatment of the stishovite-bearing sandstone in hydrofluoric acid (Chao and others, 1962). Concentrates of stishovite, all of submicron size, are pale gray and have a roughly prismatic habit (Fig. 1). Because of its fine-grained habit, only the mean index of refraction of the natural material can be measured. It is 1.80. The index of refraction reported by Stishov and Popova on the synthetic material is $n_g = 1.826 \pm 0.002$ and $n_p = 1.799 \pm 0.002$.

Crystallographically, stishovite is tetragonal, with $a = 4.1790$ and $c = 2.6649$ A (with a standard error of ±0.0004 A). The entire x-ray diffraction pattern is shown in the table, where the observed values are in excellent agreement with the calculated d spacings. All the reflections have been successfully indexed on the assumption that stishovite has a rutile structure. The data are consistent with the space group $P4/mnm$. The axial ratio c/a is 0.6377, and the volume of the tetragonal unit cell is 46.541 A^3. The specific gravity of stishovite, calculated from the x-ray data, is 4.28, as reported by Chao and others, compared with the value of 4.35 for the synthetic material reported by Stishov and Popova. It is 46% denser than coesite and much denser than other modifications of silica (Fig. 2).

Studies. Since the publication of Stishov and Popova and the work of Chao and his co-workers,

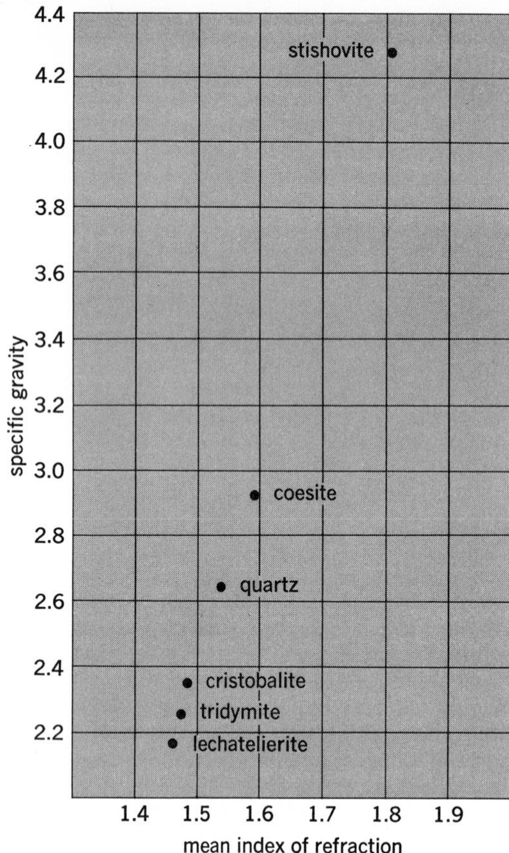

Fig. 2. Comparison of the mean index of refraction and specific gravity of stishovite and other polymorphic modifications of silica. (*Modified after Stishov and Popova*)

many studies on stishovite have appeared. For example, in 1962 the structure of stishovite was described by S. M. Stishov and N. V. Belov in *Geokhimiya* and by Anton Preisinger in *Die Naturwissenshaften.* Infrared studies by R. J. P. Lyon confirm that silicon is in sixfold coordination in the stishovite structure.

In somewhat the same way that diamond, the high-pressure polymorph of ordinary carbon, is of great value to man, stishovite, the very-high-pressure polymorph of ordinary quartz, is proving to be of great worth in scientific studies. *See* MINERALOGY.

[EDWARD C. T. CHAO]

Bibliography: E. C. T. Chao et al., Stishovite, SiO_2, a very high pressure new mineral from Meteor Crater, Arizona, *J. Geophys. Res.*, 67(1):419–421, 1962; E. C. T. Chao and J. Littler, *Additional Evidence for the Impact Origin of the Ries Basin, Bavaria, Germany*, Geol. Soc. Amer. Spec. Pap. no. 73, March, 1963; F. Dachille et al., Coesite and stishovite: Stepwise reversal transformation, *Science*, 140:991–993, 1963; R. J. P. Lyon, Infra-red confirmation of 6-fold coordination of silicon in stishovite (SiO_2), *Nature*, 196(4851):266, 1962; R. G. McQueen et al., On the equation of state of stishovite, *J. Geophys. Res.*, 68:2319–2320, 1963; A. Preisinger, Struktur des Stishovits, Höchstdruck-SiO_2, *Naturwissenschaften*, 49(15):345, 1962; C. B. Sclar et al., Synthesis and optical crystallography of stishovite, *J. Geophys. Res.*, 67(10):4049–4054, 1962; C. B. Sclar, L. C. Carrison, and G. G. Cocks, Stishovite: Thermal dependence of the crystal habit, *Science*, 144:833–835, 1964; B. J. Skinner and J. J. Fahey, Observations on the transition stishovite-silica glass, *J. Geophys. Res.*, 68:5595–5604, 1963; S. M. Stishov, Equilibrium line between coesite and the rutile like modification of silica (in Russian), *Dokl. Nauk. SSSR*, 148:1186–1188, 1963; S. M. Stishov and N. V. Belov, O kristallicheskoe strukture novoi plotnoi modifikatzii kremnezema SiO_2, *Geokhimiya*, 143(4):951–954, 1962; S. M. Stishov and S. V. Popova, New dense polymorphic modification of silica, *Geokhimiya*, 10:837–839, 1961; R. H. Wentorf, Jr., Stishovite synthesis, *J. Geophys. Res.*, 67(9):3648, 1962.

Strand line

A line at the margin or shore of a sea or lake. In geology, strand lines of ancient seas can be identified by recognizing sedimentary structures and textures and the organisms that characterize shores. It is difficult to recognize shores of identical time. Seas spread and retreat in time, causing shore facies to migrate laterally through rock sequences. Organisms are more sensitive to environments than to small time differences, so planes of synchroneity are more difficult to determine than sites of deposition. Changes in sea level (eustatic movements) are universal, so their effects on strand lines have been applied in making worldwide correlations in time. *See* FACIES; WARPING OF EARTH'S CRUST.

[MARSHALL KAY]

Bibliography: Society of Economic Paleontologists and Mineralogists, *Finding Ancient Shorelines*, Spec. Publ. no. 3, 1955.

Stratigraphic nomenclature

A system of naming used by geologists in classifying the rock record of the geologic history of the Earth. Sedimentary rocks, laid down layer by layer in seas, lakes, river floodplains, and elsewhere, are the principal record of that history. The record is complex, for rocks vary greatly not only vertically from layer to layer but also, though over greater distances, horizontally along the layers.

From the first it was realized that any succession of such layers represents a succession in time, the lower layers being the older (law of superposition, N. Steno, 1669). Accordingly, stratigraphic nomenclature has always attempted to express the time relations of the various layers as well as their intrinsic physical character. Extension of such time relations laterally from one succession to another is called correlation. Within a local area the rocks can be correlated by tracing individual layers or

Table 1. Stratigraphic units proposed by International Geological Congress, 1900

Units of time	Units of rock
Era	——*
Period	System
Epoch	Series
Age	Stage
Phase	Zone

*"Group" in an older version; "erathem" later proposed.

groups of layers, but for great distances, especially from continent to continent, correlation depends on the proved generalization that the fossils in the layers of a given age differ from those in layers of all other ages (law of faunal succession, W. Smith, 1799; explained by theory of organic evolution, Darwin, 1859). *See* STRATIGRAPHY.

Categories of stratigraphic subdivisions. Schemes of stratigraphic nomenclature grew up rather haphazardly through the 19th century, but attempts to standardize them began late in that century. At present, two main schemes are prevalent—a North American and a European—though considerable divergences remain in the usage on each continent. In North America a code of stratigraphic nomenclature was published in 1933, and an American Commission on Stratigraphic Nomenclature was established in 1946. The Commission published a revised code in 1961 and continues to work to bring about agreement on principles and uniformity in practice. Similar commissions have now been formed in many countries, such as Australia, the Soviet Union, Venezuela, France, and South Africa; and some of them have also issued codes. Since 1952 the International Commission on Statigraphy has been preparing lexicons of stratigraphic terms used throughout the world and has been formulating grounds for international agreement on principles and categories of units.

Continental European usage emphasizes time (as recorded by fossil content) as the chief basis for stratigraphic nomenclature (Table 1). The stratigraphic subdivisions recognized are intended to embrace all layers laid down during a given time interval; in practice they include all layers containing the fossils of one fauna in the succession of fossil faunas. Stratigraphic units based directly on the physical character of the rocks without reference to time are recognized in European usage, but they are considered informal or preliminary in nature.

North American usage, on the other hand, considers that units based directly on the characters of the rocks—rock-stratigraphic units—should have a separate formal status equal to that of the time-stratigraphic units (Table 2). The American Commission has sponsored a trend in North American usage to distinguish between time-stratigraphic units, based on time in the abstract, and biostratigraphic units, based only on fossil content; zone and related terms are placed in the latter category and stage, series, and system in the former. Dissenters argue that, beyond local basins of deposition, fossils are the only satisfactory clue

Table 2. Stratigraphic units proposed by H. G. Schenck and S. W. Muller, 1941

Geologic-time units	Time-stratigraphic units	Rock-stratigraphic units
Era	—	Group
Period	System	Formation
Epoch	Series	Member
Age	Stage	Bed
—	Zone*	

*American Commission on Stratigraphic Nomenclature recognizes "zone" as a biostratigraphic unit distinct from the other three kinds.

to the time relations of rock successions, and that stage, series, and system are as much based on fossil content as zone.

Usage in Australia and in most major oil-producing areas tends to agree with North American usage. There also are advocates of this usage in Europe, especially in Great Britain and Scandinavia. Soviet usage, codified in 1956, is based on the European tradition, though with a few changes in terms. It recognizes the need for local units in addition to the units of the main time-stratigraphic scale. In theory, it assigns them only an auxiliary place, and the largest local unit used in any area is subordinated to the smallest unit of the main scale there used.

In practice, however, usage seems to approximate North American usage, although the words used are different (series, suite, and bundle, instead of group, formation, and bed).

Names of stratigraphic subdivisions. The stratigraphic nomenclatures discussed above were worked out in and for the generally fossiliferous rocks deposited during the last 600,000,000 years (the Phanerozoic Eon, the time since the appearance of the olenellid-archeocyathid fauna at the beginning of the Cambrian Period). The rocks deposited in earlier times (Cryptozoic Eon or Precambrian time) contain few if any fossils that can be used in correlation. Attempts to classify those rocks into time-stratigraphic units, though numerous, have been mutually contradictory. Methods have been developed since the 1920s for determining rock ages in years by measuring radioactive minerals for parent and daughter elements. The scarcity of suitable minerals in unaltered sedimentary rocks means, however, that ages so determined generally indicate the dates of igneous intrusions, of widespread metamorphism, or simply of regional uplift rather than of original sedimentary deposition. *See* RADIOACTIVE MINERALS; ROCK, AGE DETERMINATION OF.

Table 3 gives the subdivision of the Phanerozoic Eon into eras and periods as now accepted almost throughout the world (the principal divergences are noted).

The subdivision of the periods or systems into epochs or series is less uniform. Table 3 also gives a common subdivision of the Cenozoic Era into epochs, but Paleocene and especially Holocene have met with much opposition. Some of the older periods (Jurassic, Triassic, Devonian, Cambrian) are almost universally divided into three subdivisions, but for others (Cretaceous, Carboniferous) a division into two parts is preferred, or no agreement has been reached. Some of the epochs have their own special names; others are simply designated Late, Middle, and Early (the series are Upper, Middle, and Lower).

Stages and zones are in general much more provincial, though in a few systems, notably the Jurassic and Cretaceous, the stages established in western Europe have been recognized practically around the world. Stages mostly bear the names of places, commonly with the ending "-ian" (Moscovian, Oxfordian, Delmontian stages); each zone bears the name of a characteristic fossil (the zones of *Pseudoschwagerina*, of *Cardioceras cordatum*, of *Hyracotherium*).

The rock-stratigraphic units of the North Ameri-

Table 3. Internationally accepted subdivisions of geologic time (youngest at top)

Eras	Periods		Epochs
Cenozoic	Quaternary (era*)		Holocene (Recent)
			Pleistocene
	Tertiary (era*)	Neogene*	Pliocene
			Miocene
		Paleogene*	Oligocene
			Eocene
			Paleocene
Mesozoic	Cretaceous		
	Jurassic		
	Triassic		
Paleozoic	Permian		
	Carboniferous	Pennsylvanian†	
		Mississippian†	
	Devonian		
	Silurian	Silurian‡	Gothlandian‡
	Ordovician		Ordovician‡
	Cambrian		

*Common European usage.
†Current North American usage.
‡Older European usage, rejected in 1960 by the International Geological Congress.

can scheme are defined as lithogenetic units, that is, units formed under essentially uniform (or uniformly alternating) conditions. Since their principal purpose is to serve as units for detailed geologic mapping or local description (as in studies of cuttings or cores recovered from oil wells), objectivity is of prime importance in their designation. They are named for geographic localities near which they are typically exposed; such type sections play much the same role that type specimens play for the units of biological nomenclature. No two units in the same country are supposed to bear the same geographic name, and priority is generally accepted as a principle of nomenclature, though more exceptions are permitted than in biology. The term for the principal rock type present may also be part of the name (Knox Dolomite Group, Austin Chalk, Monterey Formation). For groups and formations, formal naming according to this scheme is now considered obligatory, but for members it is optional, and the names of beds are considered outside formal stratigraphic nomenclature.

The general principles of rock-stratigraphic nomenclature are also extended to igneous and metamorphic rocks. The units are in all ways comparable to rock-stratigraphic units in sedimentary rocks. Nonsedimentary rocks are also assigned to time-stratigraphic units (for metamorphic rocks, based on the time of their original formation, not their metamorphism). Except for the volcanic igneous rocks, however, it is rarely possible to assign them with any precision. [JOHN RODGERS]

Bibliography: American Commission on Stratigraphic Nomenclature, Code of stratigraphic nomenclature, *Bull. Amer. Ass. Petrol. Geol.*, 45: 645–665, 1961; International Subcommission on Stratigraphic Terminology, *Stratigraphic Classification and Terminology*, in Report of 21st session of the International Geologic Congress, pt. 25, 1961.

Stratigraphy

The branch of the science of geology that studies layered or stratified rocks. Chiefly it concerns sedimentary rocks, but its principles may also be applied to layered igneous rocks, such as lavas and tuffs, and to metamorphic rocks that were formed from sedimentary or volcanic rocks. It deals with the observed interrelations of the layers of such rocks and with the historical conclusions that can be inferred from those interrelations. Other aspects of sedimentary geology are sedimentary petrography (the study of the materials composing sedimentary rocks) and sedimentation (the study of the processes by which sediments are formed at present). No sharp line can be drawn between these subjects, however, and each depends in part on the conclusions of the others. *See* SEDIMENTARY ROCKS; SEDIMENTATION.

Objectives. The first task of stratigraphy is the description of local sequences of strata; from these descriptions local geologic history can be inferred by using the law of superposition, which states that in a local sequence of rock layers, the lower ones are the older. First deduced by N. Steno (1669), the law is amply established by studies of sedimentation.

The second task of stratigraphy is the correlation of these local sequences, that is, the determination of their mutual time relations and the integration of the local histories into a regional or worldwide chronologic framework. Correlation can be accomplished in several different ways, but historically the most important method has been by fossils, using the law of faunal succession, which states that rocks with the same fossil fauna or flora in different parts of the world are of roughly the same age (the converse is only partially true). First empirically worked out by W. Smith (1799), this law has been verified in all parts of the world and is now explained by the theory of organic evolution as the expression of the gradual development of organic life on Earth (Darwin, 1859). The age determination of rocks and minerals by measuring the ratio of parent and daughter elements in a radioactive decay chain has supplemented the use of fossils, especially for the youngest and oldest strata. *See* ROCK, AGE DETERMINATION OF.

The third task of stratigraphy is interpretation of the geologic history of the Earth from the scattered data of local sequences and criteria of correlation. The conclusions of sedimentation, sedimentary petrology, and plant and animal ecology provide clues for this interpretation in accordance with the principle of uniformity. Other branches of geology providing pertinent data include structural geology, geomorphology, and igneous and metamorphic petrology. This aspect of the subject merges with (and is often called) historical geology.

Applications. Stratigraphy is of great practical value. Many of its basic principles were first discovered by copper miners in central Germany and coal miners in Great Britain, and they are obviously applicable to bedded mineral deposits of all sorts (for example, coal, iron ore, and phosphate). Stratigraphy's greatest extension has come about, however, through its application to petroleum exploration. Oil and natural gas occur almost exclusively in stratified rocks; the oil and gas are entrapped mainly where permeable strata are overlain and laterally surrounded by impermeable strata, the favorable configuration of strata being produced by original deposition differences, by later changes in porosity or permeability, by later warping or breaking of the strata, or by various combinations of these. Wells drilled to tap such trapped accumulations yield much information on the strata penetrated, and new methods for obtaining more information from them are constantly being devised. As a result, the stratigraphy of oil-bearing regions is no longer dependent on observations at the Earth's surface but can be studied in three dimensions in considerable detail. *See* PETROLEUM GEOLOGY.

Theoretical conclusions. Several important general theoretical conclusions about the history of the Earth can be drawn from stratigraphy; most of these are expressions of the principle of uniformity, of which stratigraphic succession is one of the best exemplifications. In general, the same conditions of sedimentation—marine and nonmarine—have prevailed throughout the decipherable stratigraphic record (probably roughly 3,000,000,000 years). At one time the oldest rocks were thought to display unique characters, but this view has now been fairly well discredited. As a corollary, liquid water has always been a major agent of erosion and sedimentation, and hence the average temperature of the Earth's surface must have been within the range of liquid water over the same time span. Earth's climatic pattern in the past could not have been very different from today's although it has become accepted that the Earth's atmosphere has changed with time.

Stratigraphy is also responsible for the major generalization that great mountain ranges coincide with former belts of exceptionally thick sedimentary rocks, deposited in long but relatively narrow linear troughs called geosynclines. The nature of the genetic relation between geosynclines and the subsequent mountain ranges is still controversial and is the subject of continuing investigation. Also, for many years the record was believed to show that virtually all marine deposits now found on the continents, including those in geosynclines, were formed in shallow seas spreading over the continental blocks rather than in deep ocean basins or troughs, and hence that continents and ocean basins are permanent features of the Earth's crust, geosynclines being accidents within the continental masses. Studies during the 1950s showed, however, that many sedimentary rocks involved in mountain ranges were deposited in deep water, in deep linear troughs associated with island arcs off the edges of continents, like those of the present East and West Indies (but not in open ocean basins far from the continents). The further conclusion has been drawn by many that the continents have accreted spasmodically through geologic time by the incorporation of such island arc trough systems into the continent in the process of mountain building, followed by the development of new systems along the new continental borders. These ideas are now being incorporated in the new tectonic theories whose point of departure has been the demonstration of sea-floor spreading from mid-ocean ridges. *See* GEOSYNCLINE; MARINE GEOLOGY; OROGENY; TECTONIC PATTERNS.

Finally, stratigraphy provides a record, though incomplete, of the ever-increasing complexity and development of life during the last 600,000,000 years of the Earth's history, including some of the most impressive evidence for the theory of organic evolution. *See* EVOLUTION, ORGANIC.

[JOHN RODGERS]

Bibliography: C. O. Dunbar and J. Rodgers, *Principles of Stratigraphy*, 1957; W. C. Krumbein and L. L. Sloss, *Stratigraphy and Sedimentation*, 2d ed., 1961; J. M. Weller, *Stratigraphic Principles and Practice*, 1960.

Stream transport and deposition

The sediment debris load of streams is a natural corollary to the degradation of the landscape by weathering and erosion. Eroded material reaches stream channels through rills and minor tributaries, being carried by the transporting power of running water and by mass movement, that is, by slippage, slides, or creep. The size represented may vary from clay to boulders. At any place in the stream system the material furnished from places upstream either is carried away or, if there is insufficient transporting ability, is accumulated as a depositional feature. The accumulation of deposited debris tends toward increased ease of movement, and this tends eventually to bring into balance the transporting ability of the stream and the debris load to be transported.

Stream loads. Because streams form and adjust their own channels, the debris load to be carried and the ability to carry load tend to reach and maintain a quasi-equilibrium. A reach of stream (part of the course) which attains this equilibrium is considered graded.

Much has been written concerning the concept of the graded stream. At one time absence of waterfalls or other discontinuities of longitudinal profile was considered necessary and, in fact, evidence for the condition of grade. Because much remains to be learned about the mechanics of debris transportation, the criteria for the graded condition may be expected to be extended and revised. In the present state of knowledge, however, it appears acceptable to think of reaches or segments of channel being graded, even when separated by reaches not so adjusted. A graded stream is one in which, over a period of years, slope and channel characteristics are delicately adjusted to provide, with available discharge, the shear forces required for the transportation of the load supplied from the drainage basin.

Two terms which have been useful to geologists and engineers dealing with rivers are competence and capacity. Competence was used by G. K. Gilbert to mean the ability to move debris, and its measure is the maximum size of material which can barely be moved. Capacity of a stream is the total load which it can carry under given conditions and is measured as weight of debris moved per unit of time. The usefulness of these terms has lessened with demand for increasingly quantitative description of stream action. Sampling equipment now in general use measures only the suspended portion of the debris and not that moving close to the streambed. Thus, except in special situations, the carrying capacity of a stream cannot be precisely measured, and available theory allows only an approximation of total load by computation.

The maximum size of debris which can be carried varies, depending on subtle variations of several factors. Thus competence, a highly useful concept, cannot be determined with satisfaction either in the field or by computation. The concepts implied by these terms will gain even greater value and importance as both theory and field measurement techniques improve. The following review of the present status of theory of debris transport will perhaps indicate how the usefulness of these concepts depends greatly on ability to determine quantitative values for them.

Debris transport theory. Debris transport is inextricably associated with the hydromechanics of flow in open channels. It is now known that the introduction of sediment grains into a fluid alters in an important manner many of the hydraulic relationships which applied to a fixed bed. For example, in a movable-bed channel, boundary roughness is not merely the rugosity of the nonmoving bed and banks. Once the particles begin to move, the shear-resisting flow is altered. Particles can assume many different configurations, among which are dunes or ripples or a plane, and these bed forms depend on the transportation process. Thus the resisting shear at the buundary depends on the debris transport itself.

When shear applied by water to a grain bed of uniform-size particles becomes sufficient to move a layer of grains, successive layers do not progressively peel off indefinitely. After some layers are put in motion, an equilibrium is reached. Transport then continues without further degradation. R. A. Bagnold showed by theory and experiment that the grains in transport add a new force normal to the bed which holds the particles exposed at the bed against the stress of the overlying fluid-grain mixture. This force, the dispersive stress between sheared grains, makes a fundamental difference in the stress structure between fixed and movable-bed channels.

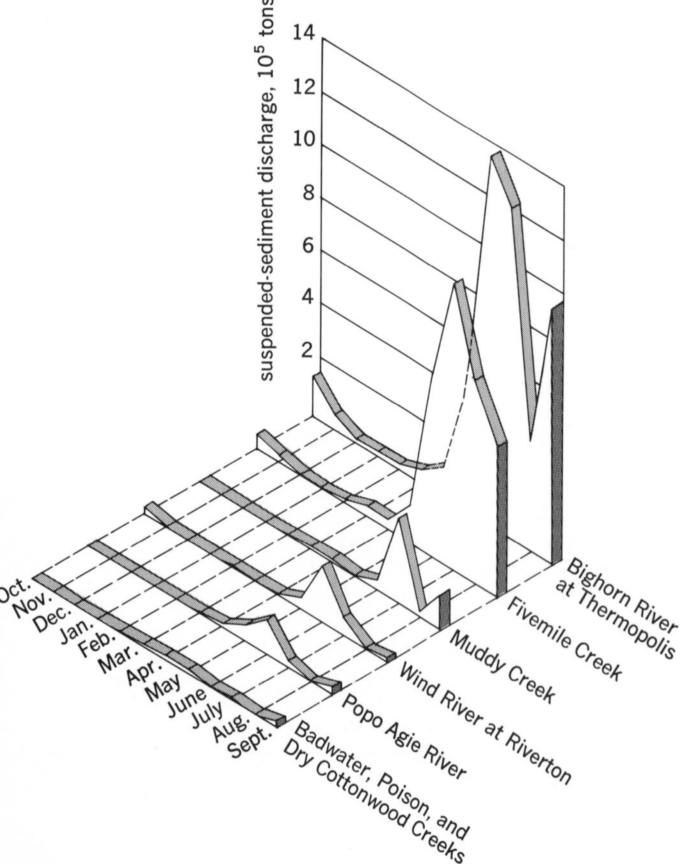

Comparison of sediment loads for several streams in the Wind River drainage basin, in Wyoming, 1950 water year. (*From B. R. Colby, C. H. Hembree, and F. H. Rainwater, Sedimentation and Chemical Quality of Surface Waters in the Wind River Basin, Wyoming, USGS Water-Supply Pap. no. 1373, 1956*)

Rates of suspended sediment production

Region	Sediment yield, tons/mi²-year
Streams in western U.S.*	1200–1400
Cheyenne Basin, Wyo.†	3900
Streams in midwestern U.S.‡	240–850

*From C. B. Brown, *Rates of Sediment Production in Southwestern United States*, U.S. Soil Conserv. Serv. Tech. Pap. no. 58, 1945.

†From R. F. Hadley and S. A. Schumm, *Hydrology of the Upper Cheyenne River Basin*, USGS Water-Supply Pap. no. 1531-B, 1961.

‡From G. M. Brune, *Rates of Sediment Production in Midwestern United States*, U.S. Soil Conserv. Serv. Tech. Pap. no. 65, 1948.

Of all the theories put forward to elucidate the physics of sediment transport, the most objectively derived from first principles rather than from empirical data is that of Bagnold. This theory is based on the general idea that work involved in sediment movement comes from the energy expended by the flowing water. The equations are predicated on the idea that only a portion of the total power spent by the water is used for debris transport, that is, that the available power times an efficiency factor equals that part utilized for carrying clastic load.

The formulation most widely used for computing sediment transport is one derived by H. A. Einstein which involves both theory and several empirically derived coefficients. His computational procedure has been simplified by B. R. Colby and C. H. Hembree for application to field problems.

Field measurement. Nearly without exception, the field measurement of sediment in transport is capable of sampling only the suspended part of the load. That which is of size larger than sand and that transported in the 2 in. nearest the bed are not measured. At a few experimental sites it has been possible to measure both the suspended load at a normal river cross section and the total load. In streams carrying material in the sand-size range and smaller, but no gravel, the total load may be measured with the usual suspended-load sampler by creating turbulence sufficient to throw all debris in transit into a condition of suspension for a brief time. Such a procedure was used on the Niobrara River near Cody, Neb., by Colby and Hembree, who reported that for that location the suspended load concentration averaged 51% of the concentration represented by the total sediment load.

In western rivers flowing through alluvial valleys, suspended-load concentrations tend to occur in the range of 100 to 5000 ppm (parts per million). The concentration increases geometrically, not linearly, with increased water discharge. Such concentrations can result in large amounts of sediment in a day, a month, or a year. An example is shown in the figure. A river of moderate size, Bighorn River at Thermopolis, Wyo., averages 3.6 in. of runoff from the drainage basin of 8080 mi.^{2} This amounted to 1,360,000 acre-feet annually during a 41-year period. In the figure it can be seen that in 1950 about 1,400,000 tons of sediment passed Thermopolis in the month of June alone.

In the humid East, sediment concentrations are considerably smaller, but runoff is larger than in the semiarid West. Some average values of suspended sediment contributed from different areas are shown in the table. [LUNA B. LEOPOLD]

Bibliography: R. A. Bagnold, *An Approach to the Sediment Transport Problem from General Physics*, USGS Prof. Pap. no. 422-I, 1966; C. B. Brown, *Rates of Sediment Production in Southwestern United States*, U.S. Soil Conserv. Serv. Tech. Pap. no. 58, 1945; G. M. Brune, *Rates of Sediment Production in Midwestern United States*, U.S. Soil Conserv. Serv. Tech. Pap. no. 65, 1948; B. R. Colby and C. H. Hembree, *Computations of Total Sediment Discharge Niobrara River Near Cody, Nebraska*, USGS Water-Supply Pap. no. 1357, 1955; B. R. Colby, C. H. Hembree, and F. H. Rainwater, *Sedimentation and Chemical Quality of Surface Waters in the Wind River Basin, Wyoming*, USGS Water-Supply Pap. no. 1373, 1956; H. A. Einstein, *The Bed-Load Function for Sediment Transportation in Open Channel Flows*, USDA Tech. Bull. no. 1026, 1950; G. K. Gilbert, *Transportation of Debris by Running Water*, USGS Prof. Pap. no. 86, 1914; R. F. Hadley and S. A. Schumm, *Hydrology of the Upper Cheyenne River Basin*, USGS Water-Supply Pap. no. 1531-B, 1961.

Stromatolite

A structure in calcareous rocks consisting of concentrically laminated masses of calcium carbonate and calcium-magnesium carbonate which are believed to be of calcareous algal origin. These structures are irregular to columnar and hemispheroidal in shape and range from 1 mm to many meters in thickness (see illustration). They may be as small as buttons or biscuits or have areal extents as great as 1 km^2 or more. Stromatolites are found in rocks ranging from Precambrian to Recent age.

Stromatolite structures, developed in growth position, have convex surfaces upward. These structures are of organic origin, but are not themselves fossil remnants. Stromatolites have been recognized as forming biostromes, bioherms, and organic reefs in close association with, and as a direct component of, the marine ecologic community of the shallow seas on sedimentary shelves and platforms. *See* BIOHERM; BIOSTROME; REEF.

[SHERMAN A. WENGERD]

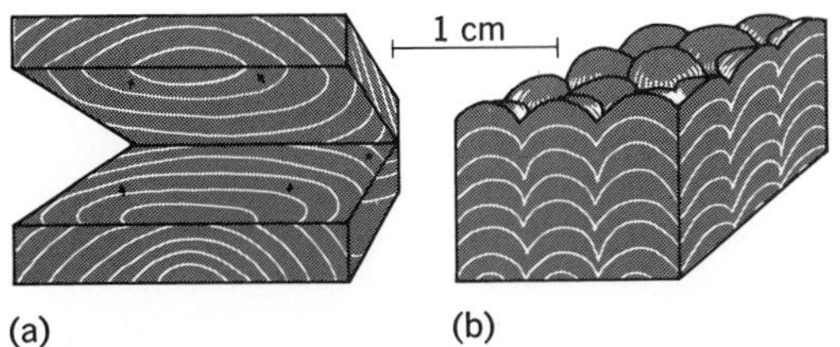

Sectioned stromatolites. (*a*) A large, strongly laminated stromatolite. Laminae dip away from the center of the hemispheroidal mass if viewed from above, and toward the center if viewed from beneath. (*b*) A small stromatolite consisting of columns with highly arched laminae. Lower Ordovician of western Wisconsin. (*From R. R. Shrock, Sequence in Layered Rocks, McGraw-Hill, 1948*)

Strontianite

The mineral form of strontium carbonate, usually with some calcium replacing strontium. It characteristically occurs in veins with barite or celestite or as masses in certain sedimentary rocks. Strontianite has orthorhombic symmetry and the same structure as aragonite. It is normally prismatic with the development of pseudohexagonal form (see illustration), but it may also be massive. It may be colorless or gray with yellow, green, or brownish tints. The hardness is 3½ on Mohs scale, and the specific gravity is 3.76. It occurs at Strontian, Scotland, and in Germany, Austria, Mexico, and India and, in the United States, in the Strontium Hills of California. *See* CARBONATE MINERALS. [ROBERT I. HARKER]

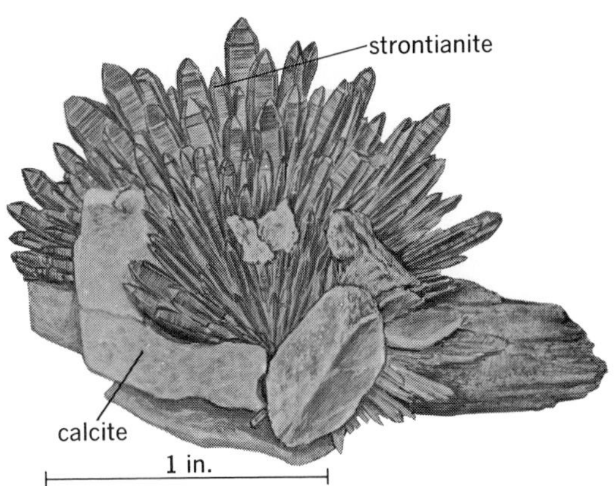

Strontianite crystals with calcite. (*Specimen from Department of Geology, Bryn Mawr College*)

Structural geology

The branch of geology which, in the broadest sense, deals with the description and analysis of the forms and interrelations of rock bodies. Structural features that came into being during the formation of the rocks are termed primary; those that were imposed upon the rocks after their formation, secondary. Although it also considers primary structures such as bedding and igneous features, the major role of structural geology is the description and analysis of secondary structures. Tectonics is synonymous with structural geology. Investigations of tectonic features involve their descriptions and interrelations, the analysis of the mechanics by which they were formed, and finally the synthesis of all structural data. *See* FAULT AND FAULT STRUCTURES; FOLD AND FOLD SYSTEMS.

A large body of information describing the general forms of various secondary structures has accumulated. However, most of these investigations have been qualitative rather than quantitative. For example, few studies have been designed to measure systematically the forms of folds, the directions of minute displacements on fractures, or the internal distortion in rocks. Considerably more progress has been made in understanding the interrelationships of various structures, and undoubtedly this is the direction in which structural geology has made its greatest advance. A vast amount of information has been compiled on the relationships of minor to major structures, and certain of these relations have been sufficiently generalized to allow their use in inferring the presence of major tectonic features. *See* TECTONIC PATTERNS.

Structures have time as well as space relations, and the study of the sequential development of tectonic features generally proceeds hand in hand with the study of their spatial relations. Such studies attempt to ascertain the relative time of formation of various structures as well as to integrate the sequence of tectonic events into a unified geological history of the Earth. *See* GEOLOGY.

The study of the mechanics by which various secondary structures are formed has taken three general directions. One line of investigation has sought to determine the physical characteristics of rocks under various environmental conditions by appropriately designed experiments. A second approach has attempted to mimic, either by actual experiment or by theoretical analysis, the form of structures as they occur in nature. And finally, attempts have been made to discover how the crystalline fabric of a rock deforms so as to accommodate itself to the general structural form. *See* HIGH-PRESSURE PHENOMENA; ROCK MECHANICS; STRUCTURAL PETROLOGY.

Experimental tests. Of fundamental importance in understanding the deformation of rocks is the design of experiments to obtain information concerning the physical properties of rocks under various conditions. These data provide the basis for the mathematical theory of elasticity and plasticity in rocks. Toward this end compression tests, extension tests, and experiments involving the punching of disks have been performed under conditions of various pressures and temperatures on both minerals and rocks. These experiments have furnished considerable information on the strengths of minerals and rocks and on the conditions under which they behave elastically or plastically. However, other factors are also important. Experiments in which the test block is immersed in a solvent yield different results from those obtained from tests in which the specimen is dry. The rate of application of the deforming pressure is also critical in controlling the nature of the deformation. Pressures and temperatures which are insufficient to cause plastic deformation in experiments in which the deforming pressure is applied rapidly are sufficient for plastic deformation when the same pressure is applied slowly. The information gained from these experiments is extremely important but as yet incomplete.

Experimental models. In mimetic investigations the correspondence of detail of form between the natural structure and the experimental or theoretical form serves to identify the mechanism of deformation. In the simplest experiments the gross features of the natural structure are duplicated by a model in which the character of the material and the magnitude and distribution of pressures are controlled. Most of these experiments have limited significance because of the difficulty in scaling the factors involved in the natural structure down to the size of the model. An alternative approach utilizes certain assumptions concerning the stress distribution and the physical characteristics of rocks in the equations of elasticity and plasticity to compare the resulting theoretical form with that of the natural structure. This sort of device overcomes many of the problems of the experimental approach. However, in practice, considerable difficulties are encountered. Much of the basic data concerning the physical characteristics of the rocks and the environmental conditions are inadequately known, and the equations of elasticity and plasticity are not sufficiently general to be applied to the complex structures in nature.

Fabric analysis. Studies aimed at the elucidation of how the crystalline fabric accommodates itself to various structural forms have been based upon detailed statistical descriptions of the fabrics of natural rocks. Such investigations have demonstrated that minerals in deformed rocks may have shape orientation, internal structural orientation, or both. Additional information has been obtained from studies of fabrics imposed on rocks in the laboratory. These experiments have shown that intracrystalline deformation involving twinning and gliding has contributed to the overall distortion of the rock, although without doubt intercrystalline deformation is also an important factor. In addition, observations of the fabrics of natural rocks suggest that the process of solution and recrystallization may be an important factor. Few experiments have been designed to test the effectiveness of this process, but attempts to approach the orientation of the internal structure of minerals from the thermodynamic standpoint have had limited success in certain simple cases, and this theoretical approach implies a process of solution and recrystallization. *See* PETROFABRIC ANALYSIS; PETROLOGY.

The ultimate goal of structural geology is the synthesis of all structural knowledge into a unified whole. In the light of present knowledge, such a synthesis is somewhat unsatisfactory because it is necessarily based on incomplete and faulty data. The geology of large tracts of the Earth's surface is incompletely known, geophysical techniques provide but scanty information concerning the depths of the Earth, and the mechanical basis for the formation of structures is only approximate. Nonetheless, attempts to compile the available data have value in that they help to bring the whole field of structural geology into focus and to stimulate investigations of critical problems. *See* ENGINEERING GEOLOGY; TECTONOPHYSICS.

[PHILIP H. OSBERG]

Bibliography: M. P. Billings, *Structural Geology*, 2d ed., 1954; L. U. DeSitter, *Structural Geology*, 2d ed., 1964; A. J. Eardley, *Structural Geology of North America*, 2d ed., 1962; E. S. Hills, *Elements of Structural Geology*, 1963; C. M. Nevin, *Principles of Structural Geology*, 4th ed., 1949; G. J. Ramsay, *Folding and Fracturing of Rocks*, 1967.

Structural petrology

The study of the structural aspects of rocks, as distinct from the purely chemical and mineralogical studies that are generally emphasized in other branches of petrology. The term was originally used synonymously with petrofabric analysis, but is sometimes restricted to denote the analysis of only microscopic structural and textural features. *See* PETROFABRIC ANALYSIS; PETROGRAPHY.

[JOHN M. CHRISTIE]

Stylolites

Stylolites are irregular surfaces, mostly parallel to bedding planes, in which small toothlike projections on one side of the surface fit into cavities of like shape on the other side (see illustration). A cross section of a typical stylolite would be similar

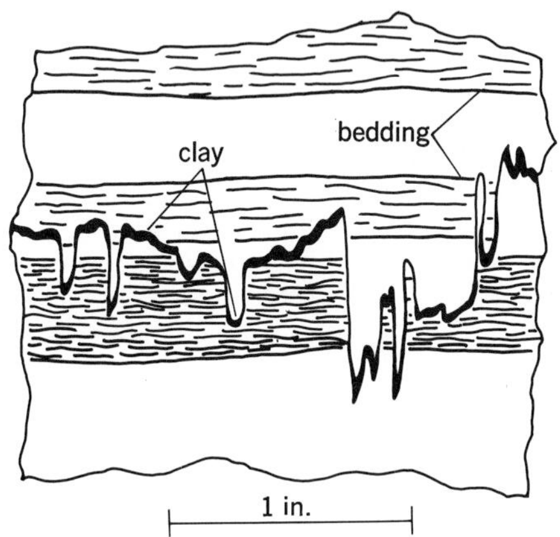

Stylolite in limestone.

to a profile of a series of high ridges and low valleys with many of the peaks and valleys being about the same amplitude. Stylolites are most common in limestones and dolomites but are also present in many other kinds of rock, including sandstones, gypsum beds, and cherts. Along almost all stylolite seams there is a thin layer of clay, quartz silt, or iron oxides, highly insoluble materials as compared with the rock proper. The amplitudes of the peaks and valleys range from a few millimeters to many centimeters.

Stylolites cut across many structures in the rocks they traverse, including fossils and oolites. Commonly these structures show truncation or partial solution along the stylolite seam. The most probable origin for stylolites is pressure solution in an already lithified rock. Large quantities of rock seem to have been dissolved along some stylolites, and the thin insoluble coatings along the seams are apparently the insoluble residues. Insoluble residues are thicker at peaks and valleys than in between, which also suggests direct relationship to solution residues. *See* DOLOMITE; LIMESTONE; SEDIMENTARY ROCKS. [RAYMOND SIEVER]

Bibliography: P. B. Stockdale, Stylolites: Their nature and origin, *Ind. Univ. Stud.*, 11(55):1–97, 1922.

Subgraywacke

An argillaceous sandstone with a composition intermediate between graywacke and orthoquartzite (low-rank graywacke of P. D. Krynine, lithic sandstones of F. J. Pettijohn). A clay matrix is usually present but in amounts less than 15%. Unstable mineral and rock fragments are less than 25%. Precise definitions of the boundaries of this group vary among sedimentary petrologists, but there is general agreement that these rocks contain moderate to large amounts of rock fragments, some clay matrix, and at least a small amount of feldspar.

The rock fragments in subgraywackes may be dominated by chert and other sedimentary species rather than metamorphic or igneous rocks. The pore spaces are filled with a combination of clay matrix and mineral cement, usually quartz and carbonate. The clay matrix is mainly muscovite (illite) with smaller amounts of kaolinite and, in some few cases, biotite and chlorite.

Subgraywackes are better sorted than the graywackes, partly because of the smaller amount of clay matrix and partly because the detrital sand-size fraction is well sorted. Detrital grains vary in roundness but tend to be rounded, in contrast to the angular grains of graywackes and the well-rounded grains of orthoquartzites. Sedimentary structures are similar to those of the orthoquartzites, but sometimes primary current lineation and groove and flute casts are found.

Subgraywackes are probably the most abundant sandstone type and are found in deposits of all ages. They occur in moderately thick stratigraphic sections and are of wide lateral extent. They may be found both in geosynclinal and platform areas. The subgraywackes are found in association with micaceous and carbonaceous shales, thin biogenic limestones, and coal beds. They seem to be characteristic of coastal plain and deltaic sedimentation and may be of mixed marine and nonmarine origin. The source material, as is evidenced by the presence of some unstable minerals and sedimentary rock fragments, is a mixture of older sediments and perhaps some low-grade metamorphics. This implies moderate tectonic activity in the source area but insufficient uplift to have allowed rigorous mechanical erosion to expose large areas of igneous and metamorphic rocks. *See* ARKOSE; GRAYWACKE; SANDSTONE; SEDIMENTARY ROCKS.

[RAYMOND SIEVER]

Submarine canyon

A relatively narrow, deep depression in the sea floor with steep slopes, the bottom of which grades continuously downward. The sea floor has many puzzling features, but none has aroused so much controversy as the great submarine canyons, cut into the continental slopes off most coasts of the world. Many of these have rocky walls thousands of feet high. They have narrow inner gorges, winding courses, numerous tributaries and are, in fact, quite comparable to the great canyons of the land (see illustration). Some of the canyons are direct continuations of land canyons, and others occur off large rivers which flow through broad flat-floored valleys on land. The submarine canyons extend outward down the slope virtually to the deep ocean floor. The outer portions of these sea valleys are of modest dimensions and extend across broad, gently sloping fans, comparable to the piedmont fans along the fronts of mountain ranges in arid regions. *See* CONTINENTAL SHELF AND SLOPE.

The deep floors of the canyons contain sediments alternating between sands, which resemble shallow water deposits, and normal deep-sea mud deposits. It seems probable that landslides, occurring at the canyon heads, stir sediment into the water and produce a heavy suspension which sets up a current, called a turbidity current. This

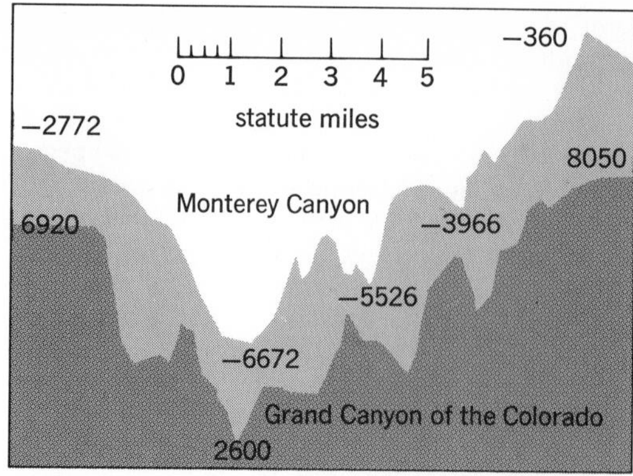

Transverse profile of the submarine canyon in Monterey Bay compared to a profile of the Grand Canyon of the Colorado River in Arizona. Elevations are in feet and the vertical is magnified five times.

moves along the canyon floors, leaving behind the sand deposits when the current loses velocity. These slides occur at rather frequent intervals, changing the depths and often breaking cables laid across the canyons. Thus cable companies avoid laying cables across submarine canyons where possible.

The cause of submarine canyons is much disputed. Their close resemblance to river-cut canyons on land has convinced some geologists that they are due to river cutting followed by submergence of the valleys. The widespread distribution of submarine canyons has caused other geologists to object to this idea. Turbidity currents have been cited as an alternative cause. An unknown factor is the speed of turbidity currents, which may at times be very great although the evidence is not clear. If the canyons are caused by turbidity currents, it is difficult to understand why they should so closely resemble river-cut canyons. Furthermore, the existence of submarine canyons with hard rock walls, such as granite, has caused some dissatisfaction with the turbidity current hypothesis. Most geologists now agree that, however formed, submarine sliding of material and turbidity currents at least prevent the filling of the canyons of the sea floor. *See* MARINE GEOLOGY; TURBIDITY CURRENT.

[FRANCIS P. SHEPARD]

Sulfur

A mineral composed of elemental sulfur. This native element belongs to the group of naturally occurring native nonmetals of which the two forms of carbon, diamond and graphite, are also members. Sulfur most commonly occurs in orthorhombic crystals with symmetry $2/m\ 2/m\ 2/m$ and space group *Fddd*. There are also two monoclinic polymorphs that are commonly produced synthetically but are very rare in nature. Sulfur crystals generally show a pyramidal habit (illustration *a*), often in combination with other forms (illustration *b*). Sulfur occurs most generally in irregular masses, which may have stalactic forms, and also as encrustations. It has a hardness of $1^1/_2$–$2^1/_2$ and a specific gravity of 2.05–2.09. Its yellow color ranges, on account of impurities, through yellowish shades of green, red, and gray. Native sulfur may contain small amounts of selenium in solid solution with sulfur. The structure of the most common polymorph of sulfur is orthorhombic with 18 S_8 rings per unit cell.

SULFUR

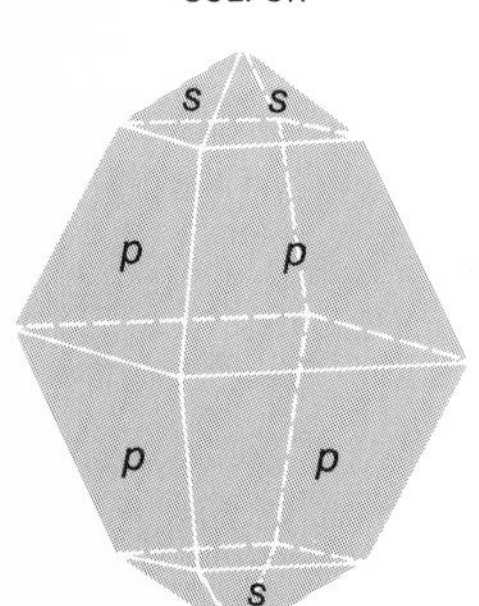

(a)

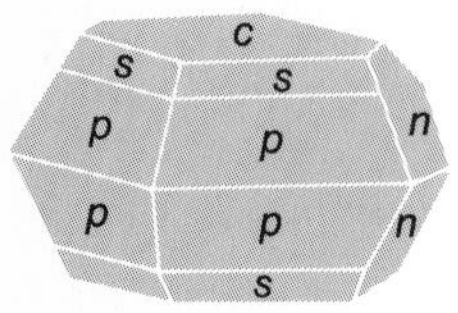

(b)

Sulfur crystals.
(*a*) Pyramidal habit.
(*b*) Combined form. (*From C. S. Hurlbut, Jr., and C. Klein, Manual of Mineralogy, 19th ed., copyright © 1977 by John Wiley & Sons, Inc.; reprinted by permission*).

Sulfur commonly occurs as a sublimation product near the crater edges of active and extinct volcanoes, where it has been deposited from gases in fumaroles. Examples of such occurrences are found in Yellowstone Park, WY. Sulfur of volcanic origin is of economic importance in Japan, Mexico, and Argentina. It is also formed by the reduction from sulfates, especially gypsum, by the action of sulfur-forming bacteria. Extensive sulfur concentrations which are believed to have resulted from bacterial action occur in Tertiary sedimentary sequences associated with anhydrite, gypsum, and limestone. In the United States very productive sulfur deposits, associated with limestone and anhydrite formations as cap rock over salt domes, are located in Texas and Louisiana. The sulfur from such deposits is recovered by pumping superheated steam down to the sulfur horizon, where it melts the sulfur, and by subsequently forcing the sulfur up with compressed air.

Sulfur is used primarily for the manufacture of sulfur compounds, such as sulfuric acid (H_2SO_4) and hydrogen sulfide (H_2S). Large quantities of elemental sulfur are used in insecticides, artificial fertilizers, and the vulcanization of rubber. Sulfur compounds are used in the manufacture of soaps, textiles, leather, paper, paints, dyes, and other products. [CORNELIS KLEIN]

Bibliography: W. E. Colony and B. E. Nordlie, Liquid sulfur at Volcan Azufre, Galapagos Islands, *Econ. Geol.*, 68:371–380, 1973; C. S. Hurlbut, Jr., and C. Klein, *Manual of Mineralogy*, 19th ed., 1977; K. H. Wedepohl (ed.), Sulfur, in *Handbook of Geochemistry*, vol. II/3, pp. 16-A-1–16-A-19, and vol. II/4, pp. 16-L-1–16-L-19, 1974.

Syenite

A phaneritic (visibly crystalline) plutonic rock with granular texture composed largely of alkali feldspar (orthoclase, microcline, usually perthitic) with subordinate plagioclase (oligoclase) and dark-colored (mafic) minerals (biotite, amphibole, and pyroxene). If sodic plagioclase (oligoclase or andesine) exceeds the quantity of alkali feldspar, the rock is called monzonite. Monzonites are generally light to medium gray, but syenites are found in a wide variety of colors (gray, green, pink, red), some of which make the material ideal for use as ornamental stone.

Composition. Syenites may be classed as normal (calc-alkali) syenites or alkali syenites. In the latter the alkali feldspar and mafics are soda-rich. Intergrowths of potash and soda feldspar (perthite) are of various types and are strikingly developed. Feldspar grains usually show fair crystal outlines (subhedral) or may be irregular (anhedral). Many are highly interlocking. Sanidine occurs in some of the finer-grained varieties and in syenite porphyry. Normal syenites generally carry crystals of soda plagioclase (oligoclase) which are usually subhedral and may be zoned (with calcic cores and sodic rims). Some alkali syenites contain discrete grains of albite. Plagioclase of monzonites may be as calcic as sodic andesine.

Black flakes (microscopically brown) of biotite mica and irregular to stubby prisms of green hornblende are characteristic of normal syenite. Diopsidic augite is the most common pyroxene and frequently forms cores within hornblende crystals. In alkali syenite the mafic minerals show wide variation. Biotite is deeply colored and iron-rich. Amphiboles are soda-rich (arfvedsonite, hastingsite, or riebechite) and are commonly zoned. Diopsidic and titanium-rich augite crystals are commonly encased by shells of aegerine-augite and aegerite.

Minor constituents may include quartz which is usually interstitial. When present in amounts between 5 and 10%, the rock is called quartz syenite; in excess of this amount, the rock becomes a granite. Small quantities of feldspathoid (nepheline, sodalite, or leucite) may be present; but if in excess of 10%, the rock becomes a feldspathoidal syenite (nepheline syenite).

Accessory minerals include zircon, sphene, apatite, magnetite, and ilmenite. Accessories in special varieties of syenite include iron-rich olivine, corundum, fluorite, spinel, and garnet.

Texture. The texture of syenite is most commonly even-grained. Very coarse or pegmatitic textures are local. In some syenites numerous, relatively large crystals (phenocrysts) of alkali feldspar give the rock a porphyritic texture. These may be of early or late generation and may range from euhedral (well-formed crystals) to anhedral. They are particularly abundant in the finer-grained varieties and in syenite porphyries.

Structure. A variety of directive structures may be present. Banding and parallel wavy streaks (schlieren) of different minerals are seen in some syenites; flow structures due to clustering and parallel orientation of elongate minerals may be present. Euhedral tabular feldspar crystals in parallel arrangement give the rock a distinctive appearance. In some cases these directive features represent effects of magma flow; in others they represent vestigial bedding or foliation in metasomatic or metamorphic rocks.

Occurrence and origin. Syenite is an uncommon plutonic rock and usually occurs in relatively small bodies (dikes, sills, stocks, and small irregular plutons). Normal syenite may be associated with monzonite, quartz syenite, and granite, whereas alkali syenites are associated with alkali granites or feldspathoidal rocks.

Many syenites have crystallized directly from syenitic magma (rock melt); others may have formed by reaction between magma of nonsyenitic composition and abundant contaminating rock fragments. Still others may have formed metasomatically as alkali-rich emanations, perhaps escaping from deeply buried magmas, have permeated rocks of special composition, and have replaced them with abundant alkali feldspar. *See* IGNEOUS ROCKS; MAGMA; METAMORPHISM; METASOMATISM.

[CARLETON A. CHAPMAN]

Syncline

A fold in which the beds are inclined down and toward the axis. Synclines may be symmetrical, asymmetrical, overturned, or recumbent. Most have elongate trends, with axes that plunge from the extremities toward interior points along the axes. Others, called basins, have no distinct trend. In general, the stratigraphically younger beds are found toward the center of curvature, but in complexly deformed regions such simple concepts may not apply.

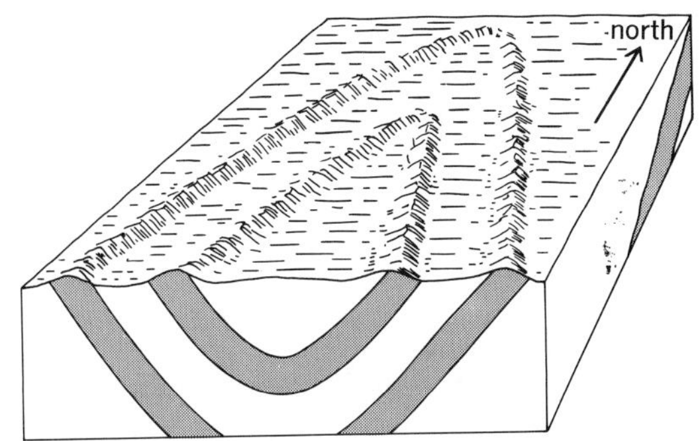

Block diagram showing relation between structural syncline and topography. The topography indicates that the syncline plunges south. (*From M. P. Billings, Structural Geology, 2d ed., © 1954 by Prentice-Hall, Inc.; reprinted by permission*)

Stratigraphic synclines are those folds which, regardless of their observed forms, are inferred from stratigraphic data to have been synclines originally. Structural synclines (see illustration) are those which have synclinal form regardless of their stratigraphic relations. *See* ANTICLINE; FOLD AND FOLD SYSTEMS.

[PHILIP H. OSBERG]

Taconite

The name given to the siliceous iron formation from which the high-grade iron ores of the Lake Superior district have been derived. It consists chiefly of fine-grained silica mixed with magnetite and hematite. As the richer iron ores approach exhaustion in the United States, taconite becomes more important as a source of iron. To recover the ore mineral in a usable form for the production of iron, taconite must be finely ground, and the magnetite or hematite concentrated by a magnetic or other process. Finally, the concentrate must be agglomerated into chunks of size and strength suitable for the blast furnace. *See* ORE AND MINERAL DEPOSITS.

[CORNELIUS S. HURLBUT, JR.]

Talc

A hydrated magnesium silicate approximating in composition to $Mg_3Si_4O_{10}(OH)_2$. Chemical analyses frequently show percentages of calcium oxide, CaO, aluminum oxide, Al_2O_3, and other oxides, but these arise mainly from impurities. Small amounts of Al_2O_3 may enter the talc structure. The structure is closely related to that of the micas, and consists of electrically neutral magnesium silicate layers bonded together by weak, secondary valences. The mineral is therefore extremely soft (hardness 1 on Mohs scale) and possesses a perfect basal cleavage (see illustration). The atoms within the layers are strongly bonded so that the mineral is highly stable, both to acids and to thermal treatment. *See* SILICATE MINERALS.

The word talc or soapstone is applied also to massive materials which may contain many other mineral constituents than talc itself. The more common associated minerals are other magnesian silicates such as serpentine, chlorite, tremolite,

Talc from the Ural Mountains, Soviet Union. (*American Museum of Natural History specimens*)

and the carbonate minerals, calcite, dolomite, and magnesite. Steatite is a term applied to relatively pure compact and massive materials containing mainly the mineral talc; steatites should contain not more than about 1.5% CaO, 1.5% combined FeO, Fe_2O_3, and 4% Al_2O_3. *See* SOAPSTONE.

Talc occurs as a secondary mineral resulting from the hydration of magnesium-bearing rocks and the alteration of minerals such as pyroxenes, amphiboles, and olivine. Fibrous talc is often closely associated with, and probably derived from, tremolite, and the fibrous character is inherited from the parent mineral. The major talc-producing states are New York, California, and North Carolina. The major talc-producing areas of the world are the United States, France, Italy, Austria, and Japan (precise data for the Soviet Union and China are not available).

Talc is a widely used raw material. Figures for talc sold or used by the producers in the United States in 1968 show that 30% was used in ceramic applications, 22% in paints, 11% in roofing materials, 5% in paper, 5% in insecticides, 4% in toilet materials, and 3% in rubber; the remaining 22% was distributed over a wide variety of applications. In the ceramic industry, talc is used in many whiteware bodies, tableware, electrical porcelain, high-frequency insulation, and glazed wall tiles. Talc apparently imparts greater resistance to mechanical stresses arising from temperature differences, and may also prevent crazing. In paints, talc is used as an extender and as a pigment. It is used as a filler for paper, rubber, and asphalt. In the cosmetics industry, it is used in toilet powders, soaps, and creams. Massive talc is cut into slabs and used for laboratory tables, sinks, sanitary appliances, acid tanks, electrical switchboards, mantels, and hearthstones.

[GEORGE W. BRINDLEY]

Talus

A heap or sloping sheet of loose rock waste at the base of a cliff or steep slope (see illustration). Talus and its English equivalent, scree, are terms properly applied to the entire form and not to the fragmental material itself, which is rock waste or sliderock. Where the supply of rock waste is funneled downward through a notch or channel, a conelike talus usually develops.

An extensive talus, or slope of rock waste, on the north side of Mud Creek Canyon, above the timberline, in Siskiyou County, Calif. (*USGS*)

Chemical weathering of susceptible layers or zones in a cliff weakens the support of overlying rock. Recurrent temperature changes and the freezing of water in narrow cracks tend to force joint blocks outward from their original positions. Loosened by these processes, blocks topple from the cliff and fall to the talus, where they slide or bound to a position of rest.

The upper part of a talus is characteristically steep, consists of coarse, angular rock waste, and is easily set in motion by falling blocks or by climbers. In cold climates its interstices may contain snow or ice for much of the year. The lower part is usually less steep and may be partly filled with finer rock debris and soil on which vegetation can gain a foothold. *See* LANDSLIDE.

[C. F. STEWART SHARPE]

Tantalite

A mineral with composition $(Fe,Mn)Ta_2O_6$, an oxide of iron, magnesium, and tantalum. Columbium (niobium) substitutes for tantalum in all proportions; a complete series extends to columbite $(Fe,Mn)Cb_2O_6$. Pure tantalite is rare. Iron and manganese vary considerably in their relative proportions. Tantalite crystallizes in the orthorhombic system and is common in short prismatic crystals (see illustration). There is perfect side pinacoid cleavage. The hardness is 6 on Mohs scale, and the specific gravity 7.95 (pure tantalite). The luster is submetallic and the color iron black. Tantalite is the principal ore of tantalum. It is found chiefly

Tantalite. (*a*) Crystal in pegmatite from Raade, Norway (*specimen from Department of Geology, Bryn Mawr College*). (*b*) Crystal habits (*from C. S. Hurlbut, Jr., Dana's Manual of Mineralogy, 17th ed., copyright © 1959 by John Wiley & Sons, Inc.; reprinted by permission*).

in granite pegmatites and as a detrital mineral, in some places in important amounts, having weathered from such rocks. The chief producing areas are the Congo and Nigeria.

[CORNELIUS S. HURLBUT, JR.]

Tectonic patterns

The large-scale structural features of the Earth—continents, ocean basins, mountain chains, island arcs, basins, high plateaus, oceanic trenches, and oceanic ridges—are thought to be produced by forces within the Earth. These tectonic features are arranged systematically on the Earth's surface.

The first-order features of the Earth—continents and ocean basins—can be subdivided into structurally stable and unstable regions. The continental blocks consist of stable nuclei called cratons which have been subject only to broad, slow, essentially vertical motion since the end of the Precambrian (600,000,000 years ago), and mobile belts comprising linear mountainous terrains subject to horizontal compression, folding, and faulting during Phanerozoic time. Similarly, the ocean basins consist of comparatively stable abyssal plain areas, seamounts and seamount chains, and linear mobile areas such as the trench-island arc systems and the mid-ocean ridge system.

LITHOSPHERIC PLATES

The lithosphere, the outer 100 km of the Earth, is divided into seven major plates and several smaller ones (Fig. 1). A plate may include continental and oceanic material, as well as some of the upper mantle, or it may contain only oceanic and upper mantle material (Fig. 2).

Plates are constantly being created and destroyed. Their boundaries consist of oceanic ridges, where the crust is being pulled apart and oceanic crust is being formed (spreading centers); oceanic trenches (subduction zones), where oceanic crust is reabsorbed into the mantle; modern mountain belts, where two crustal plates collide; and transform faults, where different spreading rates are compensated for along plate boundaries or within plates by slippage parallel with the direction of spreading (Fig. 2).

The materials making up plates are of different densities. The heaviest is mantle material (density 3.3), which occupies the lowest portions. Oceanic crust is less dense (density 3), while continental crust is lightest (density 2.65) and occupies the topographically highest portions of plates. Continental crust is not reabsorbed into the mantle because of its low density and tendency to float. This density difference explains the relative difference in ages of oldest oceanic and continental crust. The oldest continental crust is about 40,000,000,000 years old, while the oldest oceanic crust is only 200,000,000 years old. Therefore the continents have to be used to study the older tectonic patterns and features of the Earth.

Most of the Earth's major tectonic patterns, including mountain chains, oceanic ridges, trenches, and even ocean basins, may be directly linked to plate tectonic processes and interactions at plate boundaries. Interactions along active boundaries are indicated through earthquakes, volcanic activity, and more subtle and less catastrophic phenomena as uplift of parts of continents and deepening of portions of the sea floor. Conversely, ancient plate boundaries may be recognized by the distributions of roots of ancient mountains, presence of

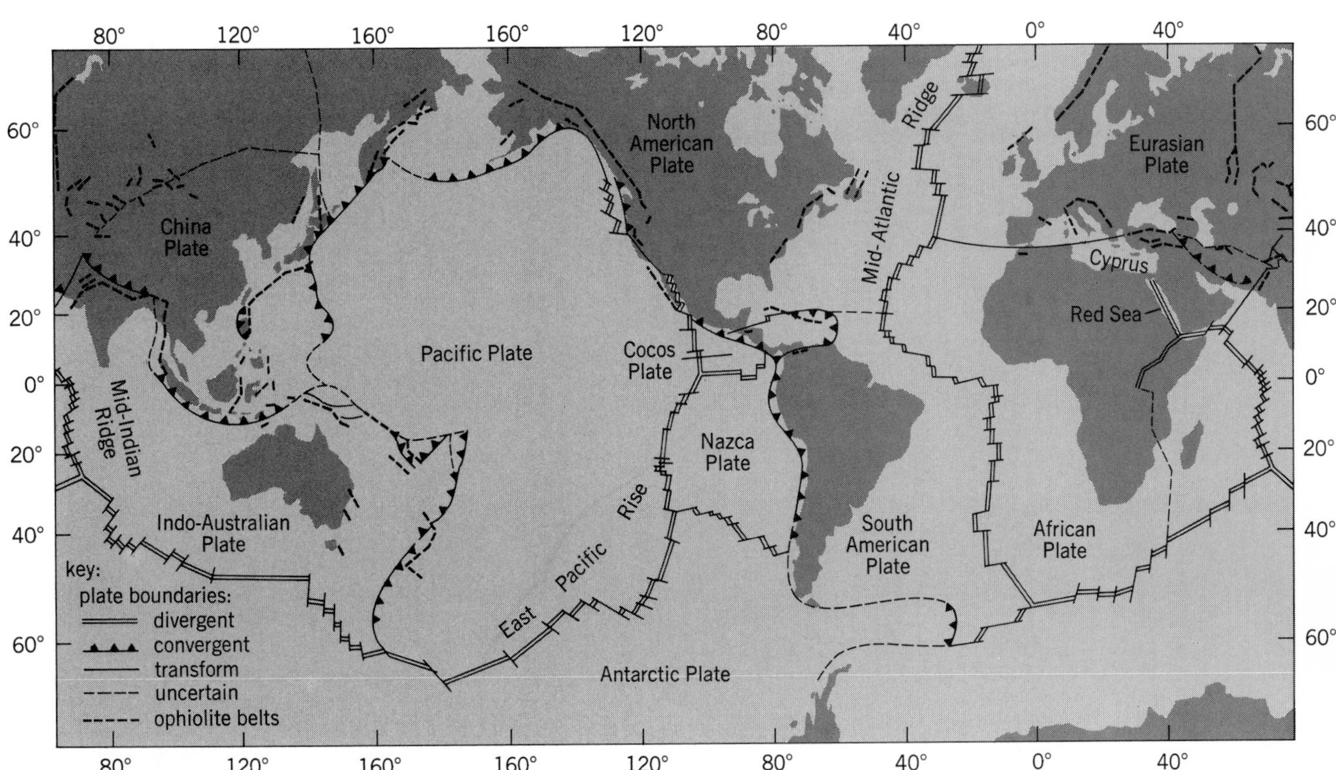

Fig. 1. Major plates and plate boundaries of the Earth. *(From P. A. Rona, Plate tectonics, energy and mineral resources: Basic research leading to payoff, Trans. Amer. Geophys. Union, 58(8):629–639; © 1977 by American Geophysical Union)*

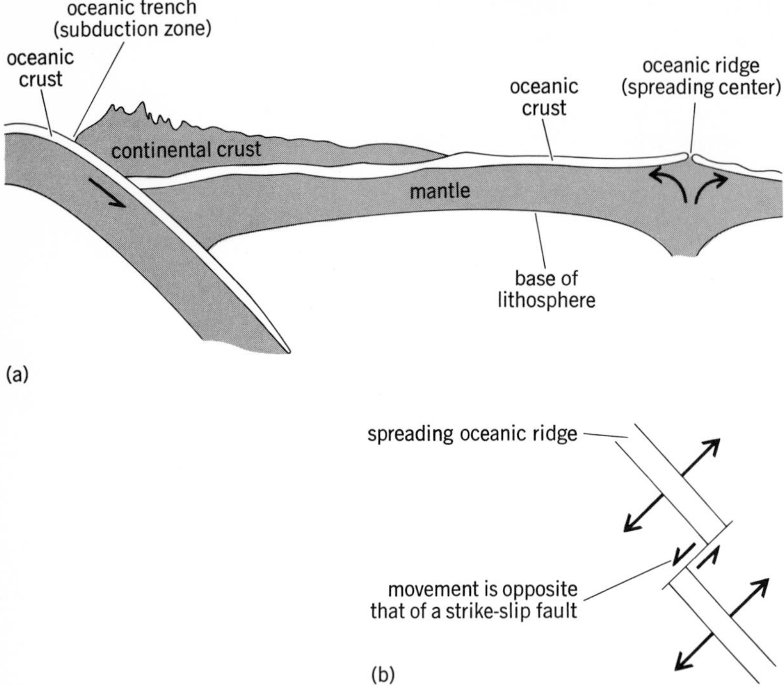

Fig. 2. Lithospheric plates. (*a*) Internal structure. (*b*) Nature of transform faults in *a*.

suspected oceanic crust or mantle rocks in the continents, and the locations of very large faults, some of which may delineate ancient transforms or sutures.

The present position of continents and ocean basins is the result of plate motion over the past 200,000,000 years. The present Mid-Atlantic Ridge is a spreading center which has generated the crust of the present Atlantic Ocean and part of the interconnected worldwide oceanic ridge system. Other ridges in the Pacific, Indian, Arctic, and Southern oceans have generated the crust which forms the floors of those oceans (Fig. 1). Small- to moderate-sized earthquakes produced from extensional forces and volcanic activity along the ridges demonstrate that spreading is still taking place. *See* EARTHQUAKE; PLATE TECTONICS; VOLCANOLOGY.

Oceanic crust is being consumed today in a number of active trenches in the Pacific, the Caribbean, and the East Indies (Fig. 1). The Aleutians, Peru-Chile, Puerto Rico, and the Japan and Philippines trenches are subduction zones where the largest earthquakes occur. *See* MARINE GEOLOGY.

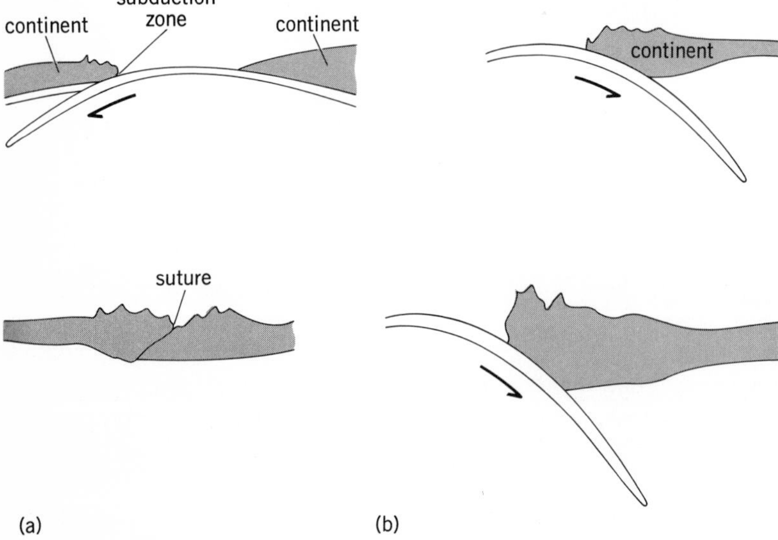

Fig. 3. Formation of (*a*) collisional and (*b*) cordilleran mountains.

MOUNTAIN CHAINS

During the course of the movement of plates, ocean basins may be closed as continents collide, and the leading edges of continents may become crumpled as oceanic crust is pushed beneath the edge of a continent in a subduction zone. In either case, mountain chains are created. The former process creates "collisional" Alpine-Himalayan type mountains and the latter, cordilleran type mountains (Fig. 3). Modern examples of both types exist. The Himalayas are of the collisional type, because India collided with southern Asia. The Andes are cordilleran mountains produced by compressional forces which developed as the leading edge of the westward-moving American Plate was forced against the eastward-directed segment of oceanic crust (Nazca Plate) from the East Pacific Ridge. *See* CORDILLERAN BELT.

Some mountain chains, such as the Appalachians, experienced a composite history of early cordilleran activity and subsequent collisional tectonics. The mountain-building cycle is terminated by collision, since continental crust will not descend into the mantle. The two continents are thus joined or "sutured," and the plate system must reform in another configuration.

The linear, arcuate character of mountain chains is obvious from any physiographic map of the world. It is also interesting that all the youngest mountains occur near the edges of continents. This pattern is real, for mountains, such as the Urals, Appalachians, and Altai Mountains of Asia, which are found inland are also arcuate, linear features. Still older mountain chains are found in the central, now stable, portions of the continents. In these areas there is no longer any topographic expression of ancient mountain chains, but their roots are still recognizable. This pattern of older mountain chains toward the centers and younger chains on the edges of continents is related to continental growth and evolution: when the first continents formed early in the Earth's history, mountain chains formed around their edges and were added to continents, thereby increasing their size (Fig. 4). The cycle has repeated itself for at least 2,000,000,000 years, building new mountains from the eroded debris of older chains and the opening and closing of oceans. The pattern for North America (Fig. 4) suggests that such continental accretion has been a straightforward, simple lateral addition of progressively younger mobile belts to a continental nucleus. However, the other continents show a more complex pattern, with old belts near their exteriors and young belts crisscrossing older nuclei. Continental accretion may in large part have been vertical (thickening of continental crust) rather than lateral (increasing the real extent of continental crust). *See* CONTINENT FORMATION.

Mountain chains have an internal arrangement

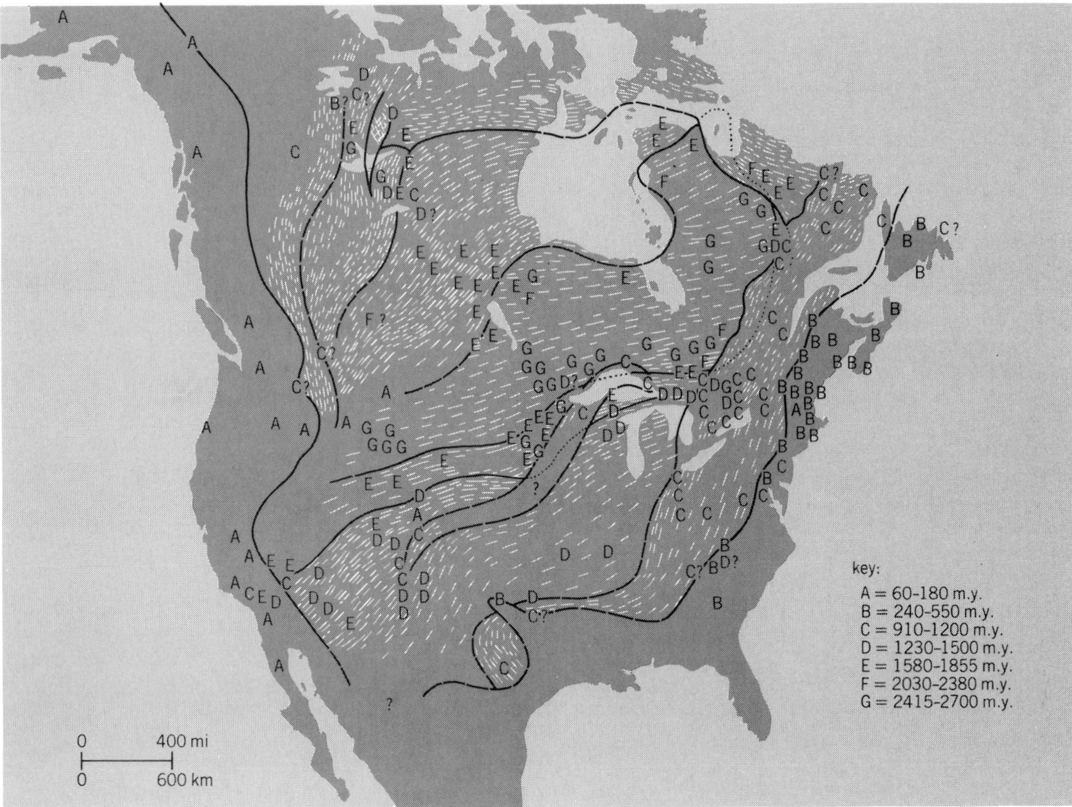

Fig. 4. Map of North America showing the distributions of ages of rocks. Each belt represents a different cycle of orogenic activity. (*From G. Gastil, The distribution of mineral dates in time and space, Amer. J. Sci., 258:1–35, 1960*)

that produces many of their linear arcuate features. Many mountain chains have a belt of low-angle thrusts (marginal thrust belt) and folds directed toward the continent and located closest to the continental interior (Fig. 5). Toward the center of the chain, large thrust faults and nappes (sheetlike, structurally displaced rock units) give way to great plastic folds of the metamorphic core zone. Next is a zone of igneous plutons which gives way to and overlaps with a volcanic belt. The volcanic belts of many mountain chains have proved to be older volcanic island arcs like those of the modern western Pacific (such as Japan, the Philippines, and the Marianas) which have been driven into the continent as two plates collided and the small ocean separating the arc from the continent collapsed. *See* MOUNTAIN SYSTEMS; OROGENY.

Three classic mountain chains. Portions of three mountain chains will be described: the Western Alps, southern Appalachians, and southern Canadian Cordillera. Each has been reasonably well studied, and all contain most of the characteristics of the idealized mountain chain (Fig. 5). However, there are important differences: The Alps and Appalachians are composite cordilleran-collisional mountains, while the Cordillera is a cordilleran type. Cross sections through all three may appear similar, but there is a significant difference in relative size. The Alps are only 100 km across in Switzerland, but the southern Appalachians are about 250 km across, and the southern Canadian Cordillera is about 1000 km across. Moreover, the Alps are a Tertiary chain, the Cordillera is primarily Mesozoic–early Tertiary, and the Appalachians are Paleozoic.

Western Alps. The Western Alps are divisible into several zones, most of which correlate to the idealized mountain chain described above (Fig. 6).

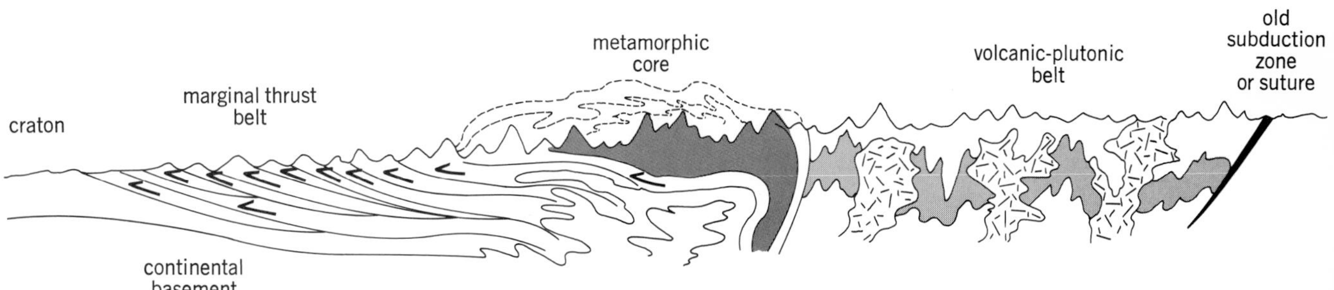

Fig. 5. Cross section through the idealized mountain chain.

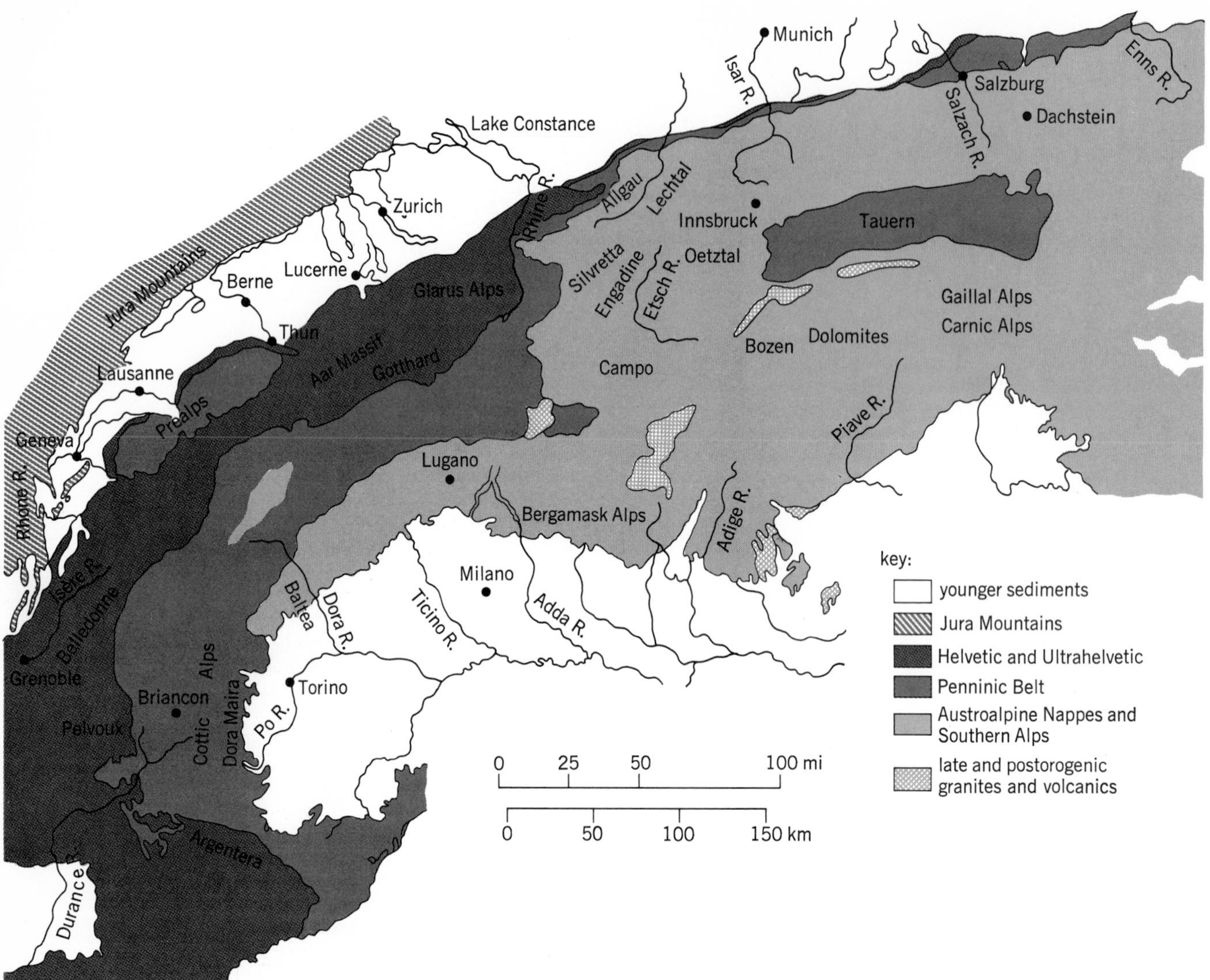

Fig. 6. Map indicating the subdivisions of the Western and Central Alps. *(From R. Trümpy, Paleotectonic evolution of the central and western Alps, Geol. Soc. Amer. Bull., Amer. Bull., 83(9):2735–2760, 1972)*

The outer zone of the Western Alps is the Jura Mountains. Mesozoic platform limestones, shales, and sandstones have been deformed into broad synclines and narrow anticlines, many of which are broken by thrust faults (Fig. 7). The intensity of deformation increases southeastward from flat-lying sediments on the northwest side of the Jura to highly folded and faulted rocks to the southeast where the undeformed Tertiary Molasse deposits of the Swiss Plain cover the older deformed rocks of the Jura.

The Swiss Plain consists of a broad syncline containing a series of sandstones and shales which coarsen to the southeast and are derived from erosion of the Alps. This sedimentary series, known as molasse, is largely undeformed except on the east side of the basin, where it is highly folded and has been thrust over by the Helvetic nappes.

The Pre-Alps is a zone of unmetamorphosed but highly folded and faulted Mesozoic and Tertiary sedimentary rocks which do not relate, either structurally or stratigraphically, to the Helvetic nappes to the east. Although rocks within the Pre-Alps are of the same age as those in the Helvetics, their character is so different that they had to have been deposited in quite different environments. Several nappes have been recognized within the Pre-Alps, and the whole mass is thought to be derived by thrusting from far to the east of the Helvetic nappes, possibly in the Pennine Zone.

The Helvetic Alps, also known as the High Calcareous Alps and Limestone Alps, consist of a series of complexly folded and thrusted rocks, predominantly carbonates, which originated by gravitationally sliding off the basement massifs to the east as they were being uplifted. A series of nappes which have overridden each other has been recognized here.

The Jura, Pre-Alps, and Helvetic nappes correspond generally to the marginal thrust zone of the idealized mountain chain. However, the placement of the Molasse Basin between the Jura and other zones to the southeast has created several controversies over the role of compression and the involvement of basement in Jura tectonics.

Separating the Helvetic nappes from the Pennine Zone in much of Switzerland are two crystalline basement massifs. The northwestern massif is the Aar Massif, and that to the southeast is the Gothard Massif. Crystalline rocks of the massifs

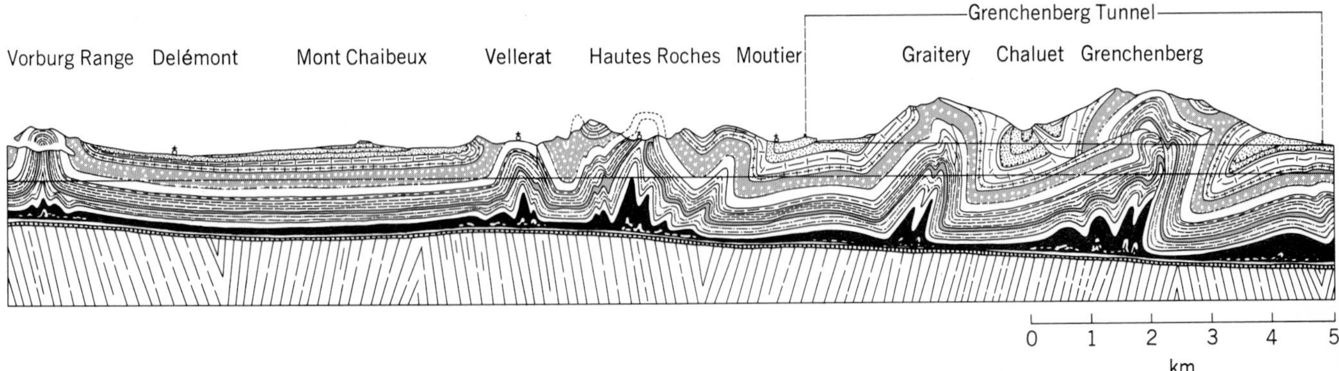

Fig. 7. Cross section through part of the Jura Mountains of Switzerland. Mesozoic platform limestones, shales, and sandstones have been deformed into broad synclines and narrow anticlines. *(From L. W. Collet, Structure of the Alps, Edward Arnold Publishers, 1927)*

were formed in a Paleozoic tectonic cycle, long before the Alpine cycle. Separating the two massifs is a long, narrow belt of near-vertical Paleozoic and Mesozoic sedimentary rocks called the Urseren Zone. These rocks are highly deformed and metamorphosed, yet they are similar in many ways to rocks of the Helvetic Alps. Several geologists believe that some of the Helvetic nappes were squeezed out of and off the Aar and Gothard massifs as the two crystalline masses were shoved together and upward. The Urseren Zone is a thin, squeezed remnant of rocks which were once spread over a wider area between the two massifs.

The Pennine Zone is the metamorphic core of the Alps. It consists of a series of nappes of metasedimentary rocks cored by Paleozoic basement rocks that were remobilized during the Tertiary-Alpine orogenic cycle. Rocks of this zone were in a plastic condition when they were moved to their present site. As in the other zones to the northwest, the nappes here have moved toward the craton to the northwest. However, the thick, nearly horizontal nappes in the northwest part of the Pennine Zone become nearly vertical and very thin in the vicinity of Locarno, where these nappes are "rooted." The Insubric line of highly thinned, flattened, and deformed rocks separates the Pennine Zone from the Dinarides to the south. The Dinarides contain rocks that are comparably little deformed.

The Pennine Alps are overridden by a great thrust sheet in eastern Switzerland and Austria. The thrust sheet contains basement and a series of overlying sedimentary rocks. Pennine nappes are visible in several windows in the thrust sheet, the largest being the Tauern and Engadine windows.

Southern Canadian Cordillera. The southern Canadian Cordillera of Alberta and British Columbia is a composite orogen having most of the subdivisions of the ideal mountain chain, but several subdivisions are duplicated (Fig. 8). This chain was formed during the Mesozoic and early Tertiary, but tectonic activity is still occurring along its western edge.

The marginal thrust belt of the southern Canadian Cordillera comprises the foothills and Rocky Mountains. Here Precambrian, Paleozoic, Mesozoic, and some Tertiary platform sedimentary rocks have been thrust to the northeast. Numerous folds and some normal faults were also formed in this belt. It has been estimated that the total amount of horizontal displacement on these thrusts is as much as 200 km, although it may be much less.

West of the Rocky Mountains lies the Omineca Crystalline Belt. Rocks are highly deformed and range from slightly metamorphosed in the eastern and western portions to partially remelted in portions of the central zone. The central core is known as the Shuswaps Metamorphic Complex. Deformation here was by plastic folding with development of structures similar to those of the Pennine Zone of the Alps and eastern Blue Ridge and Inner Piedmont of the southern Appalachians. Large nappes are present, but have not been traced over large areas, as in the Alps and Appalachians.

The Intermontane Belt lies to the west of the Omineca Crystalline Belt. It contains some Paleozoic and Mesozoic sedimentary rocks, but volcanic and plutonic rocks of Mesozoic age dominate. The

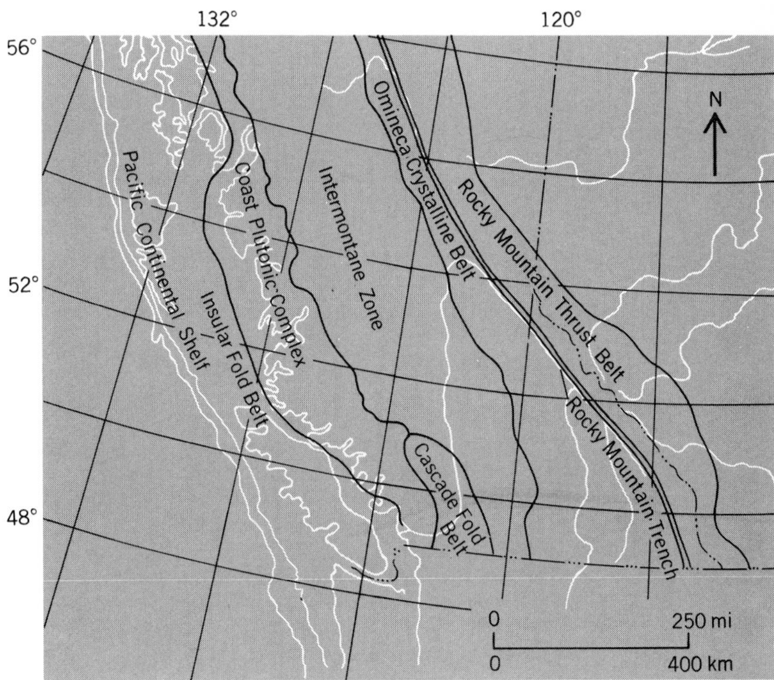

Fig. 8. Subdivisions of the southern Canadian Cordillera. *(From R. J. W. Douglas, ed., Geology and Mineral Resources of Canada, Geol. Surv. Can., 1968)*

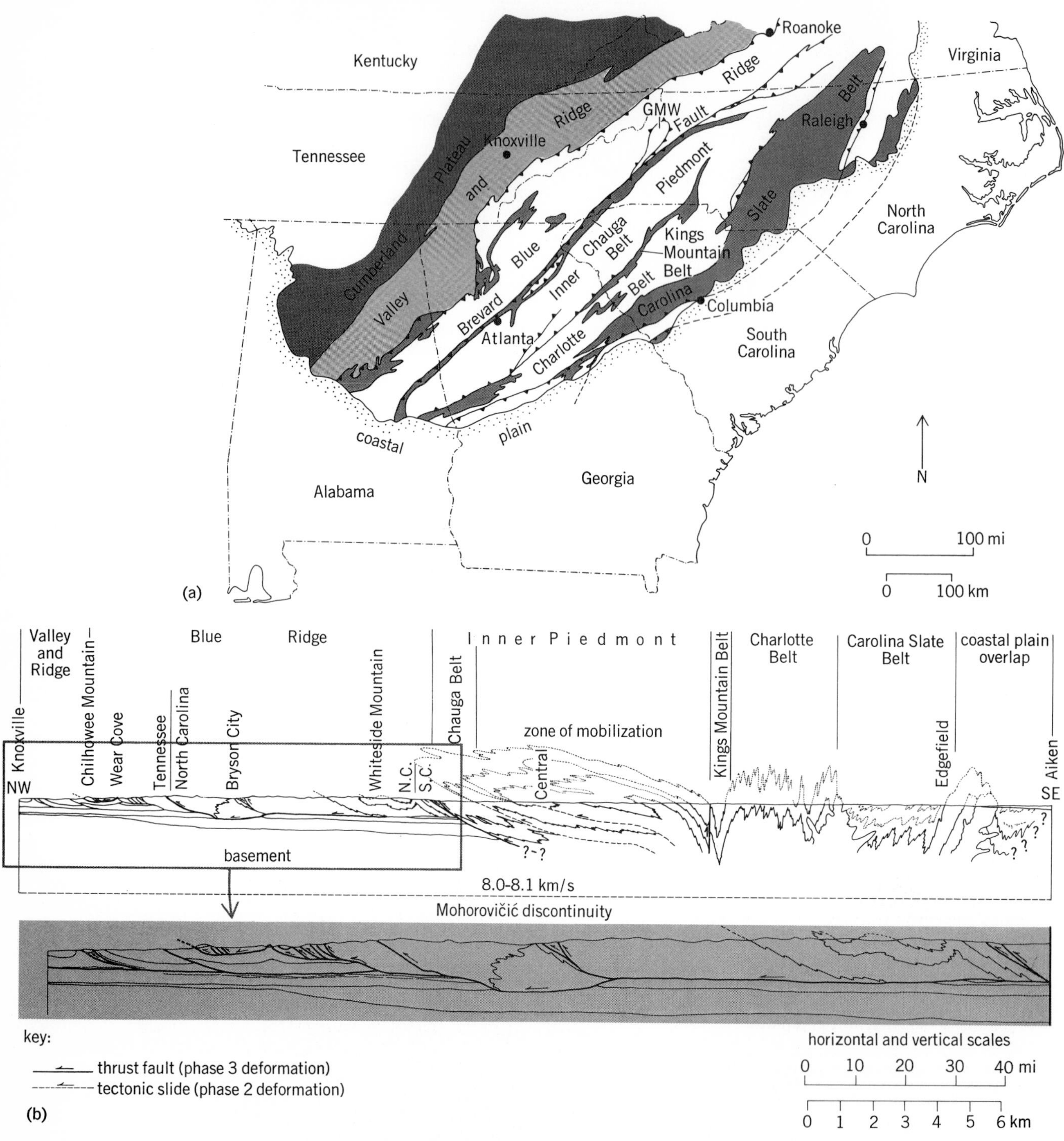

Fig. 9. Southern Appalachians, showing (*a*) subdivisions and (*b*) cross section. *(From R. D. Hatcher, Jr., Developmental model for the southern Appalachians, Geol. Soc. Amer, Bull., 83(9):2735–2760, 1972)*

Intermontane Belt corresponds to the volcanic-plutonic belt of the idealized mountain chain. The rocks here are gently folded and faulted and only slightly metamorphosed.

Another Mesozoic-Tertiary fold belt is present west of the Intermontane Zone. The Cascade Fold Belt of southern British Columbia and Washington contains folds and thrusts, some of which have moved toward the east while others moved west. This belt also has a metamorphic core zone.

West of the Cascades is the Coast plutonic complex of plutons and volcanic rocks, most of which were tightly folded. Several large faults occur in this complex. This belt may represent a deeper crustal level of the Cascades Fold Belt, or it may have a completely different origin.

Southern Appalachians. The southern Appalachians are divisible into several zones which correspond to zones in the idealized mountain chain (Fig. 9). However, its easternmost extremities are buried beneath its erosional debris in the Atlantic and Gulf Coastal Plain, so many questions about its probable earlier connection to Africa remain unanswered.

The Cumberland Plateau–Valley and Ridge comprise the marginal thrust belt of this mountain chain. Early to late Paleozoic sedimentary rocks have been folded and thrust northwestward. Deformation dies out westward in the rocks of the Cumberland Plateau. Most of the older sediments derived from the continent were deposited on the ancient margin of North America. From Middle Ordovician time to the end of the Paleozoic, sediments were derived from the rising mountain chain to the east.

The western Blue Ridge is part of the marginal thrust belt, but here late Precambrian and earliest Paleozoic sedimentary rocks are involved in thrusting, along with some basement. The central and eastern Blue Ridge consists of metamorphosed sedimentary rocks folded in a plastic condition, early Paleozoic granites and Precambrian basement rocks. Several large faults also occur here. Large Pennine-type nappes have been identified in the eastern Blue Ridge.

The entire Blue Ridge from north of Roanoke, VA, to Alabama has moved westward on a great thrust sheet, possibly as much as 125 km. The principal line of evidence supporting this is the Grandfather Mountain window in the eastern Blue Ridge of North Carolina. The Blue Ridge is a crystalline thrust sheet like that of the Eastern Alps of Austria, but the Blue Ridge is on the west side of the orogen, and windows expose rocks that are less metamorphosed than those in the thrust sheet. Windows in the Eastern Alps expose the intensely metamorphosed rocks of the Pennine Zone.

The Brevard Zone separates the Blue Ridge

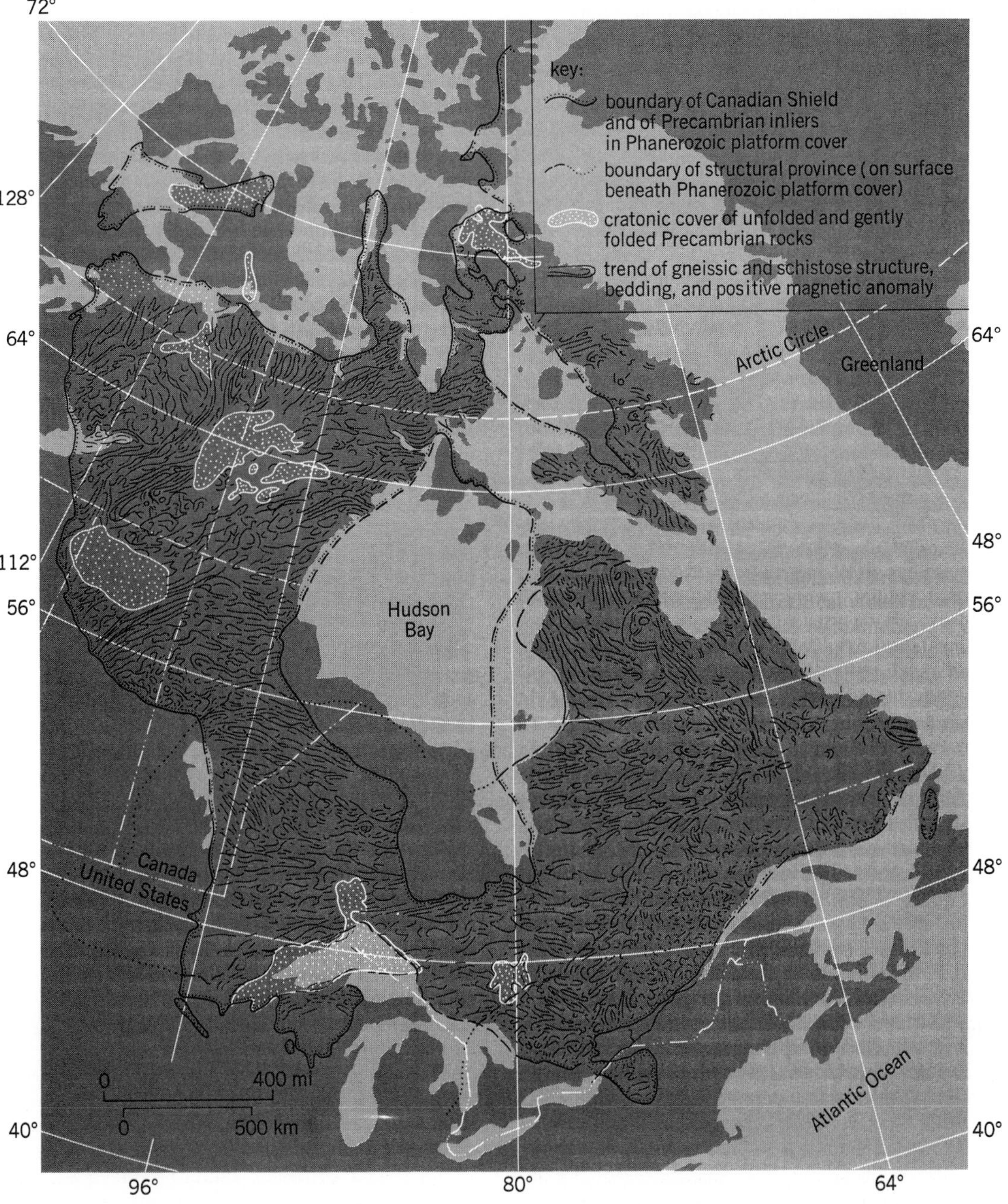

Fig. 10. Map showing structural trends in the Canadian Shield of North America. (*From R. J. W. Douglas, ed., Geology and Mineral Resources of Canada, Geological Survey of Canada, 1968*)

from the Inner Piedmont. The Brevard Zone is a fault with a long and complex movement history. It contains rocks that have been highly deformed but not as intensely metamorphosed as those on either side. It also contains masses of rocks that appear unmetamorphosed, probably derived from the rocks beneath the Blue Ridge thrust sheet. It is probably the root zone for many of the faults farther west.

Southeast of the Brevard Zone is another zone of high-grade metamorphism, the Inner Piedmont. Large plastic nappes have been identified here. The Inner Piedmont is flanked on both sides by lower-grade metasedimentary and metavolcanic rocks in the Chauga Belt (northwest) and the Kings Mountain Belt (southeast). The nappes of the Inner Piedmont are probably rooted just west of the Kings Mountain Belt.

The metamorphic core of the southern Appalachians is very broad, extending from the western Blue Ridge to east of the Inner Piedmont. It contains at least three high-grade zones, suggesting a composite nature to the orogen.

East of the Kings Mountain Belt is the plutonic-volcanic belt of the southern Appalachians, the Charlotte Belt–Carolina Slate Belt. Late Precambrian to early Paleozoic volcanic and sedimentary rocks occur here and were metamorphosed, folded, faulted, and intruded by middle to late Paleozoic plutons. This zone may have been part of an island arc system which was joined to North America during mountain building.

Sutures and cryptic (hidden) sutures. Where two plates are joined together, the junction is called a suture. Sutures are generally located within mountain chains. Some have been recognized, for example, the Indus suture in the Himalayas and the Insubric-Pusteria line of the Alps, but many are not easily discerned, and are termed cryptic sutures.

Mountain chains containing most or all of the belts outlined above (Fig. 5) would contain at least one cryptic suture joining the metamorphic core zone to the plutonic-volcanic belt. This would not be a suture of two major plates but one joining a former island arc to the continent. The possibility thus exists that several cryptic sutures are present within one segment of a mountain chain.

Sutures, either distinct or cryptic, are commonly large faults. However, not all large faults are sutures, hence the difficulties in recognizing sutures in many areas. Large faults which juxtapose distinctly contrasting rock types and structural orientations over great distances may have formed by suturing of two plates. The active San Andreas Fault in California is now a boundary between the American and Pacific plates. However, its early history is that of a major strike-slip fault, and it is still regarded as a transform. The Brevard Fault of the southern Appalachians is thought by some to be a suture, but many of its characteristics are those of a large fault within a plate or a continental mass. Therefore, whether the Brevard Fault is a suture is under dispute. *See* FAULT AND FAULT STRUCTURES.

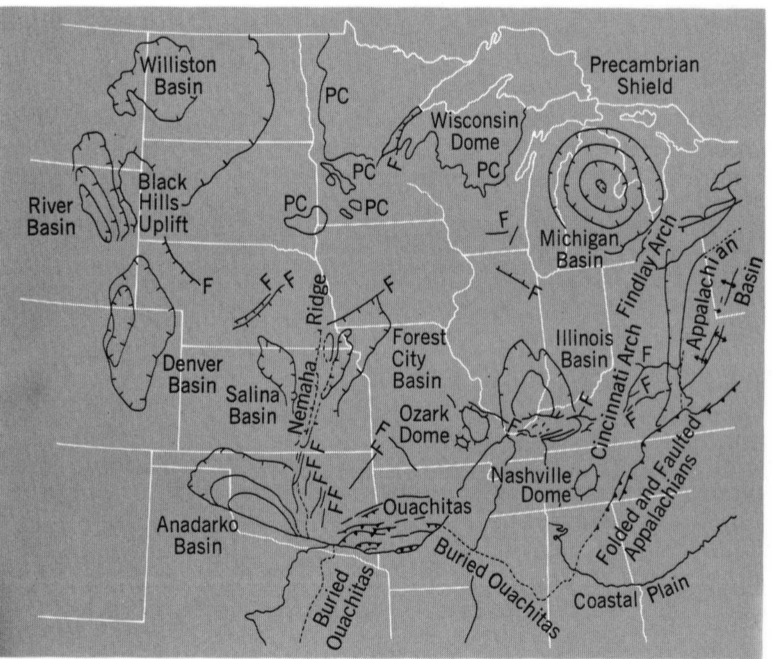

Fig. 11. Tectonic features of the continental interior of the United States. F = fault; PC = rocks of Precambrian age. *(From E. W. Spencer, Introduction to the Structure of the Earth, 2d ed., McGraw-Hill, 1977)*

CRATONS

Cratons, as opposed to mobile belts, have been subject to little internal deformation in the course of the last 600,000,000 years, that is, since the end of Precambrian time. These stable interior portions of the continents can be subdivided into two components, shields and platforms. Shield areas such as the Canadian Shield of North America or the Baltic Shield of northern Europe are broad, rigid areas of Precambrian rocks which represent the most stable nuclei of continents (Fig. 10). Shield areas are surrounded on most margins by platform areas, regions where flat-lying or gently tilted post-Precambrian sediments rest on a buried basement of Precambrian crystalline rocks. Obviously cratons, particularly platform areas, have not been totally inactive structurally for the past 600,000,000 years. The platforms have been subjected to variable but long-term subsidence sufficient to produce the thin cover of sedimentary rocks by which they are differentiated from shield areas. Most of the instability in cratonic regions is epeirogenic, consisting of broad, reversible, essentially vertical movements. The result of vertical movement in the cratons is expressed as domes or arches and basins most readily visible in the platform areas (Fig. 11). The domes and arches are recognized by sedimentary facies changes, by a deletion of part of the sedimentary record, or by thinning of units. Basins are characterized by abnormally thick accumulations of sediments.

The domes and basins of the cratons are mostly subtle features with beds that are gently dipping. A regional study of the thicknesses of rock units is necessary to determine their presence. However, these domes and basins are very important, being major reservoirs of petroleum and natural gas.

Large faults are uncommon in the craton; nevertheless, they are present. However, normal faults are more common and their displacement is generally small. Faults in mountain chains are principally thrust and strike-slip faults produced by compression, although normal faults produced by extension may be important in some mountain chains.

PLATEAUS

Plateaus of different elevations exist in different parts of the world, but not all fit comfortably into a classification scheme which divides continents into cratons and mobile belts. Low plateaus are common in the craton and are due to the presence of some resistant rock type which produces a topographically higher region. These plateaus range in elevation up to a few hundred meters above the surrounding area. Other plateaus up to 1000–2500 m above their surroundings occur in many areas in the craton, adjacent to or within mountain chains. The Roraima Plateau of Venezuela and Guyana is such an area, but its existence is due to the presence of a resistant sandstone.

There are very few high plateaus (>3000 m) in the world, the most notable of which are the Tibetan and Pamir plateaus of Asia and the Andean Altiplano of South America. These are adjacent to or within mountain chains to which they may be genetically related. For example, the very high Tibetan Plateau (3000–5000 m) may be caused by subduction of continental crust (India) beneath continental Asia, that is, by essentially stacking two slices of continental crust on one another. The resulting doubly thick continental crust is lighter than the adjacent oceanic crust and mantle material into which it is moving, and it attempts to float upward, thereby raising everything above it.

The Colorado Plateau of North America has an average elevation of 1800 m, but maximum elevations on the plateau are near 3000 m. The Colorado Plateau resides just east of the compressionally generated Cordillera and west of the southern Rocky Mountains, products of vertical uplift. The origin of the Colorado Plateau is uncertain. *See* PLATEAU. [ROBERT D. HATCHER, JR.]

Bibliography: M. Gary, R. McAfee, Jr., and C. L. Wolf (eds.), *Glossary of Geology*, 1972; B. E. Hobbs, W. D. Means, and P. F. Williams, *An Outline of Structural Geology*, 1976; Tectonics and mountain ranges, in J. H. Ostrom and P. M. Orville (eds.), *Amer. J. Sci.*, vol. 275A (Rodgers vol.), 1975; J. T. Wilson (ed.), *Continents Adrift and Continents Aground*, 1976.

Tectonophysics

The science of the physical processes involved in forming geological structures. It is part of an older branch of geology called tectonics. The application of physics to geological problems has enabled tectonophysics to provide a deeper understanding of the Earth in three ways: Description is now possible of the divisions of the Earth's crust and the substrata of the upper mantle as well as of the land surface; (2) some theories of the nature and rates of processes within the Earth have been proposed; and (3) the location of tectonic forces has been discovered to lie at depths of tens or hundreds of kilometers rather than at the surface. For a discussion of these topics *see* CONTINENT FORMATION; FAULT AND FAULT STRUCTURES; FOLD AND FOLD SYSTEMS; MARINE GEOLOGY; SEISMOLOGY.

[DAVID E. FOGARTY]

Tektite

A collective term applied to certain objects of natural glass of debatable origin that are widely strewn on land and in sediments under the oceans. Tektites have long been exposed to geologic weathering processes, and most are found as broken pieces of glass. Experiments have demonstrated that the shapes and surface sculpture of substantially whole tektites can be reproduced in the aerodynamic laboratory by a two-step process: (1) by aerodynamically ablating tektite glass at hypervelocities of such magnitude as correspond to entry into the Earth's atmosphere from space; and (2) by following this with a small amount of chemical decomposition (to simulate decomposition by ground solutions). Tektites are thus believed to represent a shower of glass objects that long ago plunged into the Earth's atmosphere from space. The overall shapes of most tektites resemble splash forms: spheroids, elongates, dumbbells, teardrops, disks, and irregular lumps. All tektites are glassy and brittle; many exhibit conchoidal fracture. Land tektites range from the size of a pinhead to that of a man's head. They appear dark in reflected light, but thin edges or thin sections transmit light and reveal various colors ranging from yellow to green and from olive brown to a nearly opaque dark brown. Microtektites have been found thus far only in ocean sediments. These are micro forms of glass of the same age and composition as land tektites, with essentially the same splash-form shapes as land tektites, but ranging in size from a pinhead down to about 40 microns (μ).

Groups, ages, and occurrence. Tektites are grouped mainly by age as determined by concordant values from fission-track and K-Ar methods of age dating. Current age values demonstrate that there are three distinct groups widely separated in time: (1) North American, 34,000,000 years age, found in Texas (bediasites), Georgia (georgiaites), and one specimen reported found at Martha's Vineyard, Mass.; (2) Czechoslovakian, 15,000,000 years age, found in Bohemia and Moravia (moldavites); (3) Australasian, 700,000 years age, found as normal-size tektites on land in Tasmania, Australia, Indonesia, Southeast Asia, the Philippines, and as microtektites in sediments from under the Indian and Pacific oceans from near Madagascar to south of Japan.

In addition there is another small group of land tektites found in the Ivory Coast of Africa and corresponding microtektites in the Atlantic just off the Ivory Coast. According to current age values these appear to be a separate group slightly older (1,000,000 years age) than the Australasian group. Thus far several million specimens have been recovered from the Australasian group, the most abundant and widely spread of the known groups. The estimated mass in the Australasian shower, as computed from the abundance of microtektites, is on the order of 10^7–10^8 tons.

Composition. The composition of tektites varies widely. Refractive indexes range from 1.48 to 1.62. Specific gravity varies from 2.3 to 2.6 for land tektites and up to higher values for some microtektites. The overall range in composition presently known for tektites and microtektites, as expressed in weight percent for the major oxides, is as follows: SiO_2, 48 to 85 wt %; Al_2O_3, 8 to 18 wt %; FeO, 1.4 to 11 wt %; MgO, 0.4 to 28 wt %; CaO, 0.3 to 10 wt %; Na_2O, 0.3 to 3.9 wt %; K_2O, 1.3 to 3.8 wt %; TiO_2, 0.3 to 1.1 wt %. If either the natural glass found at Mount Darwin, Tasmania, or that found in the Libyan Desert of Africa were included as tektites, as some have suggested, then the overall compositional range would extend to higher SiO_2 and to correspondingly lower values for the other oxides.

Inclusions. Bubble cavities sized from microns to millimeters are common, but a few cavities range up to several centimeters. Glassy inclusions of lechatelierite (nearly pure silica) are essentially ubiquitous in tektites. These can be bubbly, frothy, or solid and are often twisted and contorted. Small grains of coesite, a high-pressure polymorph of silica, have been reported in a few tektites from Thailand. Small grains of baddeleyite have also been found in a few tektites. All of these crystalline inclusions are only a very minute fraction of the mass of tektites. Thus tektites are virtually 100% glass. Small metallic inclusions of meteoritic iron nickel have been found in some tektites from the Philippines and Vietnam. These inclusions of meteoritic metal, together with those of coesite, are evidence that tektites were formed by the impact of a meteorite. It is generally believed that tektites were formed as the splash from a large impact crater. *See* COESITE; METEORITE.

Place of origin. Although it is now widely accepted that the mode of origin of tektites is from impact somewhere on the Earth or on the Moon, there is no general acceptance as to which of these two celestial objects is the true source. Planets far from the Earth or Moon are excluded mainly by evidence that Al^{26}, which would be produced by long exposure to cosmic rays in space, is absent in tektites. The aerodynamic evidence for the Australasian tektites, as evaluated from the amount of ablation on Australites, indicates that their velocity of atmosphere entry corresponds to that of trajectories which originated from the Moon. The Earth geographic distribution pattern of the Australasian shower is identified with the particular pattern that material would spread over Earth if ejected from one of the prominent rays of the lunar crater Tycho. But age evidence pertaining to the Ivory Coast and Czechoslovakian tektites suggests to some an Earth origin inasmuch as the Ivory Coast tektites appear to be approximately the same age as Bosumtwi crater in Ghana, and the Czechoslovakian tektites approximately the same age as the Reis crater in Germany. Yet no rocks with tektite composition have been found at either of these two craters. There is no known rock province on Earth for which the rock chemistry matches the compositional range of any of the various tektite groups; and no analysis of samples returned from the Moon matches a tektite composition (as of October, 1969). The precise place of origin of tektites, therefore, is a subject of scientific debate and contention. Material returned by later Apollo missions may shed critical light upon this scientific question. *See* MOON.

[DEAN R. CHAPMAN]

Bibliography: D. R. Chapman and H. K. Larson, On the lunar origin of tektites, *J. Geophys. Res.*, 68:4305–4358, 1963; B. Glass, Microtektites in deep-sea sediments, *Nature*, 214:372–374, 1967; J. O'Keefe (ed.), *Tektites*, 1963; Papers on tektites, *J. Geophys. Res.*, vol. 74, no. 27, 1969; Third International Tektite Symposium, *Geochim. Cosmochim. Acta*, vol. 33, no. 9, 1969.

Terrestrial gravitation

A term signifying the effect of gravitational attraction of the Earth. Since the small centrifugal force due to the Earth's rotation is inseparably superimposed on the attraction, usually the combined effect of these two forces is considered.

CONCEPTS

The gravitational potential V at a point P is the potential energy, due to the Earth's gravitational attraction, of a unit mass situated at P. In other words, V is equal to the work done if a unit mass is brought from infinity to the point P under the influence of the Earth's gravitational field. A mathematical expression is obtained by integrating over the Earth the formula for the potential of a point mass, but this expression is almost useless for practical application. For the potential outside the Earth, an expansion into an infinite series (of spherical harmonics, discussed later in this article) is useful; the principal term of this series is GM/r. This term, in which G is the Newtonian gravitational constant (6.67×10^{-8} cm^{3}g^{-1}sec^{-2}), M is the total mass of the Earth, and r is the distance of P from the Earth's center of mass, represents formally the attraction of a spherically symmetric Earth; the consideration of this term only is not, in general, sufficient. The gravitational force $\mathbf{F}$ on a unit mass is the gradient vector of V, $\mathbf{F} = \text{grad } V$; that is, the components of $\mathbf{F}$ are $(\partial V/\partial x, \partial V/\partial y, \partial V/\partial z)$.

Gravitation and gravity. The resultant of gravitation (pure attraction) and centrifugal force is called gravity. Gravity is the force that acts on a body at rest with respect to the Earth since the effects of attraction and of centrifugal force cannot be separated because of the equivalence of gravitational and inertial mass; thus gravity determines the weight of a body. The gravity potential W is the sum of the gravitational potential V and the potential of centrifugal force, which is given by a simple analytical expression and may be considered as known.

A body moving with respect to the Earth is also affected by the Coriolis force. Like centrifugal force, the Coriolis force is an inertial force due to the Earth's rotation, but unlike centrifugal force, it does not possess a potential and hence cannot be easily incorporated into the gravity field. Therefore Coriolis force is not considered in the context of terrestrial gravitation. This is perfectly adequate since this force is zero for bodies at rest with respect to the Earth, and almost all measuring systems are at rest. *See* GRAVITATION.

Gravity vector. The gravity vector $\mathbf{g}$ represents the force of gravity on a unit mass. It is the gradient vector of the gravity potential, $\mathbf{g} = \text{grad } W$. The magnitude of the gravity vector is the intensity of gravity, or briefly, gravity g. The dimension of g is force per unit mass, or acceleration. The unit is the gal (1 gal = 1 cm/sec^2), named after Galileo. Often the milligal (1 mgal = 10^{-3} gal) is used. Gravity g on the Earth's surface varies from about 978 gals at the Equator to about 983 gals at the poles. The direction of the gravity vector defines the vertical, or plumb line.

Geoid. The surfaces of constant gravity potential, W = const, are called equipotential surfaces or level surfaces. The surface of a quiet lake is part of a level surface. So is the surface of the oceans, after some obvious idealization; the whole level surface so defined is called the geoid. After C. F. Gauss, the geoid is considered as the mathematical surface of the Earth, as opposed to the visible topographical surface. The plumb lines intersect the level surfaces orthogonally (Fig. 1); they are not quite straight but very slightly curved.

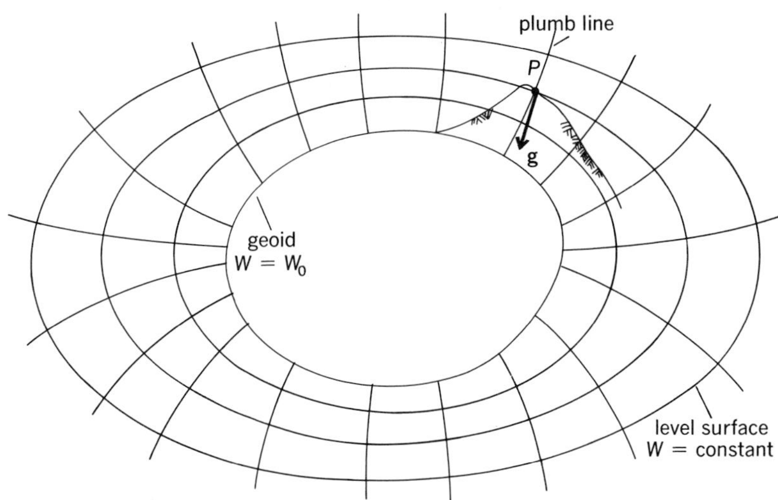

Fig. 1. Level surfaces and plumb lines. The arrow gives the direction of the gravity vector **g** at point *P*. *W* is the gravity potential.

MEASUREMENT

The quantity that is measured most commonly is the gravity *g*. The determination of *g* as such is called an absolute gravity measurement. Usually only relative gravity measurements are performed, determining the difference between, or the quotient of, the gravity values at two different points. The direction of the gravity vector, which gives the plumb line in space, is measured by astronomical methods. Differences in the gravity potential *W* are obtained by geodetic leveling. Finally, certain derivatives of *g* and similar quantities are measured by instruments such as the torsion balance.

Absolute gravity measurement. Pendulums and falling-body experiments are used in making absolute gravity measurements.

Pendulums. For a simple pendulum, consisting of a point mass suspended by a weightless rod or cord of invariable length, the period *T* of oscillation for small amplitudes is given by the well-known formula Eq. (1), which gives Eq. (2), where *l* is the

$$T = 2\pi \sqrt{\frac{l}{g}} \qquad (1)$$

$$g = 4\pi^2 \frac{l}{T^2} \qquad (2)$$

cord length and *g* is gravity. Such a simple pendulum is a mathematical fiction approximately realized by suspending a heavy sphere by a long thin wire or rod. Such a pendulum was used by F. W. Bessel in 1827. The application of a simple pendulum of enormous cord length, suspended in a shaft, to precise absolute gravity measurement is again being investigated.

The above expressions for *T* and *g* hold not only for the simple pendulum but formally also for a pendulum of arbitrary shape; but then, in general, *l* has no longer a geometrical meaning and cannot be determined by a measurement of length. It is possible, however, to construct even a solid pendulum in such a way that the pendulum constant *l* is geometrically defined and measurable. This is the reversible pendulum (Fig. 2). Let the point of suspension be O. Then there exists a second point O' such that the pendulum's center of mass C lies on the straight line OO', $OC \neq O'C$, and the period of oscillation T is the same whether the pendulum is suspended at O or at O'. Then $l = OO'$, so that l can be obtained by measuring the distance between the two knife-edges by which the points O and O' are realized. Until the middle of the 20th century reversible pendulums were the standard instruments for absolute gravity measurements; however, falling-body experiments are coming into prominence.

Even though pendulum measurements are simple in principle, it is extremely difficult to reach the required precision of 1 mgal or better. In addition to accurate length measurement, numerous corrections (for shape of knife-edge, friction, air drag, temperature, and so forth) must be carefully considered. To get a precise value of the period *T*, the duration of a great number of successive oscillations (of the order of several hours) is measured.

Falling-body experiments. If a body, originally at rest, is dropped in vacuum, the vertical distance *s* covered at time *t* is given by Eq. (3), which gives

$$s = \frac{1}{2} g t^2 \qquad (3)$$

Eq. (4). By measuring *s* and *t* it is thus possible to

$$g = \frac{2s}{t^2} \qquad (4)$$

obtain *g*. Various modifications of this principle, which dates back to Galileo, are being used. The required accuracy of at least 1 mgal has become possible only with the development of methods for precise measurement of short time intervals (to

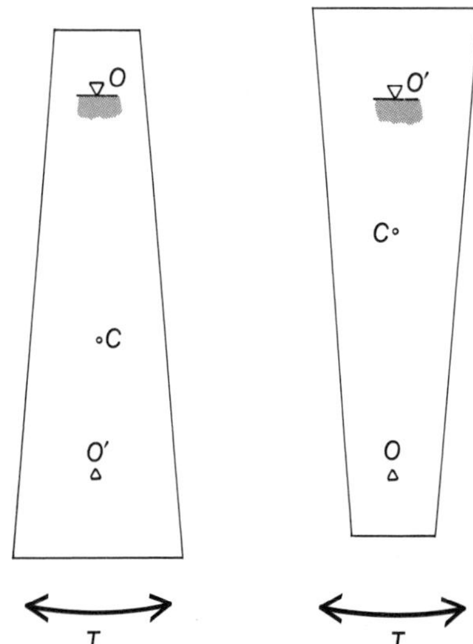

Fig. 2. Principle of the reversible pendulum. The period of oscillation *T* is the same whether the pendulum is suspended at *O* (left) or at *O′* (right).

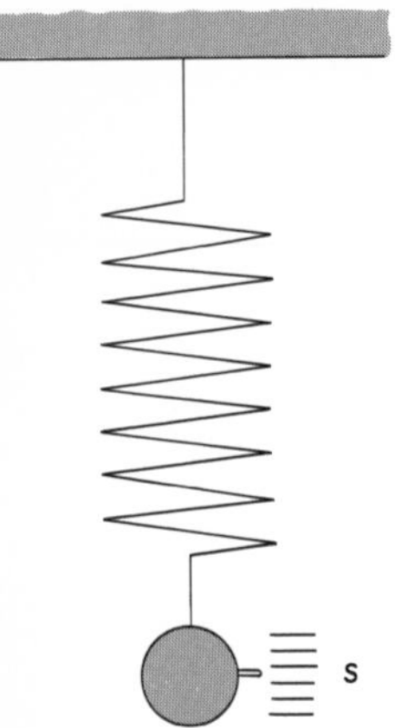

Fig. 3. Principle of a spring gravimeter. The variable length of the spring is s.

fractions of a microsecond) using quartz clocks and electronic counters. Such measurements are still extremely difficult and laborious.

Potsdam system. By international agreement, all gravity observations are referred to an absolute gravity measurement performed at the Geodetic Institute at Potsdam, Germany, about 1900, by means of a reversible pendulum. The reference value is $g_{\text{Potsdam}} = 981.274$ gals. Absolute measurements carried out by various institutions employing different methods indicate that this reference value is too great by about 12 or 13 mgals.

Since relative gravity measurements are much simpler, absolute determinations of gravity are performed only at a few points on the Earth's surface. At other points, gravity is determined relatively to these absolute stations by methods described in the following section.

Relative gravity measurement. Relative measurements of gravity are made by pendulums or gravimeters.

Pendulums. Consider two points P_1 and P_2 at which gravity has the values g_1 and g_2. Assume g_1 to be known; then g_2 is obtained from the pendulum formula, Eq. (2), which gives Eq. (5). Thus the

$$g_2 = g_1 \frac{T_1^2}{T_2^2} \tag{5}$$

pendulum constant l has dropped out. Since l, the determination of which is so difficult, need not be known here, an ordinary solid pendulum may be used, provided the pendulum constant will not change. Thus relative pendulum observations are much simpler than absolute measurements. Accuracies of a few tenths of a milligal are achieved with suitably constructed pendulum apparatuses, which may contain as many as four pendulums. To minimize temperature effects, materials such as quartz or invar are used because of their small thermal expansion coefficient. Pendulum measurements are time-consuming; several hours are needed for one station.

Gravimeters. In principle, a gravimeter may be visualized as a spring balance (Fig. 3). The elongation s of the spring may be considered proportional to g, $s = (1/k)g$, so that Eq. (6) is the basic formula for relative measurements by a gravimeter.

$$g_2 - g_1 = k(s_2 - s_1) \tag{6}$$

Ingenious modifications of this simple principle are required for high precision. Modern gravimeters give an accuracy up to 0.01 mgal for field instruments and 0.001 mgal for stationary instruments. To visualize what this means, consider the simple "gravimeter" of Fig. 3, and let the length of the spring $s = 10$ cm. Then a change in g of 0.001 mgal corresponds to a change in the length s of 10^{-8} cm, which is the diameter of the hydrogen atom.

Therefore sensibility must be increased by extreme optical or electrical magnification, by using complicated systems of springs and lever arms (Fig. 4). Often the principle of astatization is used; that is, with any small deviation from a central position, which is a position of unstable equilibrium, an additional torque tending to increase this deviation becomes effective. Temperature effects are minimized by placing the instrument into a thermostat or by using quartz springs. Gravimeters are also constructed with torsion springs and torsion fibers so that changes in gravity can be read by the associated rotation.

Since gravimeters are static instruments, an observation takes only a few minutes. Gravimeters are therefore the principal tools for routine gravity measurement. Relative pendulum observations are used to obtain a worldwide system of gravity base stations, which serve as starting points of gravity surveys and form calibration lines needed for determining the gravimeter constant k.

Pendulum apparatuses have been installed in submarines to measure gravity at sea. Ship-borne and airborne gravimeters are also used. Moving systems, especially airborne ones, are affected by inertial disturbances (Coriolis forces and the like), the separation of which from true gravitational effects is difficult.

Basically, an instrument that measures gravity (a gravity sensor) is nothing but a special form of accelerometer. The principle of the spring balance is also utilized for accelerometers, and inversely vibrating-string accelerometers have been proposed as gravity sensors since the frequency of a loaded vibrating string is a function of g. Accelerometers have been especially proposed for use in gradiometers, as discussed in the section below on gravity gradients.

Direction of gravity vector. This direction, the plumb line, is defined by two angles: φ, geographical latitude, and λ, geographical longitude (Fig. 5). These angles are determined by astronomical methods to an accuracy of 0.1 second of arc or better. If g, φ, and λ are known, then the gravity vector $\mathbf{g}$ is completely determined.

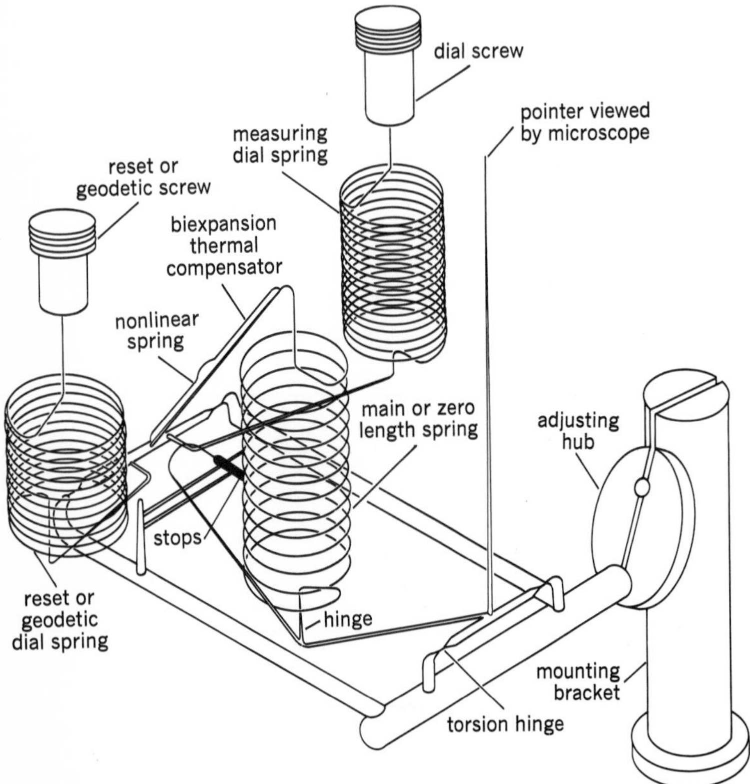

Fig. 4. The complicated spring system of this gravimeter aids in increasing sensibility. Enlarged about five times. (*Houston Technical Laboratories*)

Potential differences. Differences in the potential W can also be measured. If dn is the distance between two neighboring level surfaces of potential difference dW, then $dW=-g\,dn$, so that Eq. (7)

$$W_B - W_A = -\int_A^B g\,dn \qquad (7)$$

holds. The geodetic operation of leveling (spirit leveling or differential leveling) determines dn. The leveling increments dn are multiplied by gravity g and summed along the leveling line that connects points A and B. In this way the difference of potential W between A and B is determined.

Gravity gradients. The gradient of gravity along a certain direction s, $\partial g/\partial s$, is the component of the vector grad g along this direction. Instruments measuring gravity gradients are called gradiometers.

Horizontal gradients (gradients along any horizontal direction) are obtained by means of the torsion balance invented by R. Eötvös about 1900. This instrument consists essentially of a vertically suspended torsion wire carrying a light horizontal beam at which two equal masses are suspended at different heights. The torsion balance gives horizontal gradients of g and, in addition, certain characteristics of the curvature of equipotential surfaces.

The practical application of the torsion balance has suffered from the competition of modern gravimeters, but the interest in gradiometers is being revived because they can be used for gravitational measurements aboard an artificial satellite (a gravimeter would in this case constantly indicate $g=0$!).

A general gradiometer may be viewed as a combination of two gravity sensors separated by a small distance; the difference of their readings, divided by the distance, gives the gravity gradient. The use of vibrating-string sensors, mentioned above, has been proposed for this purpose.

GRAVITATIONAL FIELD

The gravitational field of a reference ellipsoid of revolution which closely approximates the real Earth is called the normal field. The difference between the real field and the normal field is called the anomalous field.

Normal field. Since the Earth is very nearly an ellipsoid, the gravity field of a suitable ellipsoid of revolution is a good approximation to the gravity field of the Earth and can thus be considered as a normal terrestrial gravity field. The normal gravity potential is denoted by U. An ellipsoid of revolution—the reference ellipsoid—that is close to the geoid is defined to be an equipotential surface of this normal field, where $U=U_0=$ constant.

Normal gravity. Gravity on such a "level ellipsoid" is called normal gravity γ; it is given by the gravity formula, Eq. (8), where γ_e is normal gravity

$$\gamma = \gamma_e(1+\beta \sin^2\varphi - \epsilon \sin^2 2\varphi) \qquad (8)$$

at the equator, φ is geographical latitude, and ϵ is a very small constant (for numerical values see below).

The other constant, β, is related to the geometric flattening of the terrestrial ellipsoid, $f=(a-b)/a$ (where $a=$ semimajor axis and $b=$ semiminor axis) by A. C. Clairaut's formula, Eq. (9), where $m=$ centrifugal force at Equator/grav-

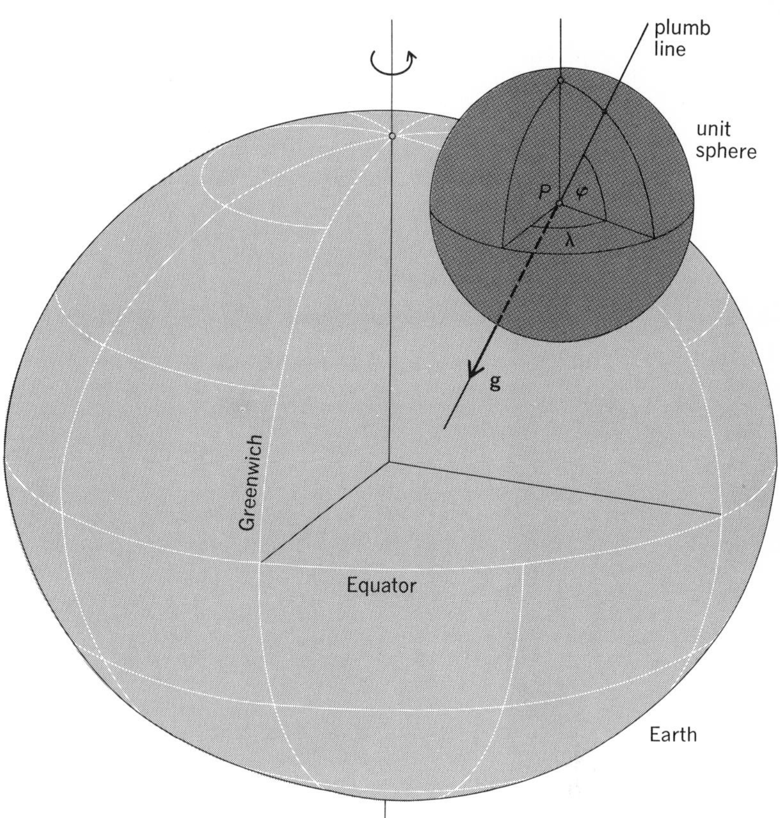

Fig. 5. Direction of plumb line is defined by the geographical coordinates φ and λ. The unit sphere is frequently used for representing spatial directions.

$$f+\beta=\frac{5}{2}m \qquad (9)$$

ity at Equator, in which terms of the second order in f are neglected. This remarkable relation permits the determination of the Earth's flattening f from gravity measurements.

The level ellipsoid and its gravity field are defined by four constants, for example, the set (a, GM, J_2, ω). Here a and GM have been introduced before; J_2 is the "dynamical form factor" or "dynamical ellipticity" of the Earth, defined as $J_2=(C-A)/Ma^2$, where C and A are the Earth's principal moments of inertia; and ω is the Earth's angular spin velocity.

The constants f, β, and γ_e are related to these constants, to the first order in f, by the formulas (10), (11), and (12).

$$f=\frac{3}{2}J_2+\frac{1}{2}m \qquad (10)$$

$$\beta=2m-\frac{3}{2}J_2 \qquad (11)$$

$$\gamma_e=\frac{GM}{a^2}\left(1+\frac{3}{2}J_2-m\right) \qquad (12)$$

The set of values given below is one adopted by the International Astronomical Union in 1964 and by the International Union of Geodesy and Geophysics in 1967.

$$a = 6{,}378{,}160 \text{ m}$$
$$GM = 398{,}603\times10^9 \text{ m}^3\text{ sec}^{-2}$$
$$J_2 = 0.0010827$$
$$\omega = 0.000072921 \text{ sec}^{-1}$$

The corresponding flattening is $f=1/298.247$, and the gravity formula becomes Eq. (13).

$$\gamma = 978.0318\,(1+0.0053024\sin^2\varphi - 0.0000058\sin^2 2\varphi)\text{ gals} \quad (13)$$

This system of numerical values is based on a combination of geodetic measurements (for a), and results from satellite orbit analysis (for GM and J_2) and gravity measurements (for γ_e).

External and internal field. Gravity above the ellipsoid decreases according to formula (14),

$$\gamma_h = \gamma - (0.30877 - 0.00044\sin^2\varphi)h + 0.000072h^2 \quad (14)$$

where γ_h is gravity at elevation h and latitude φ, and γ is the gravity at the same latitude on the surface of the ellipsoid; γ_h and γ are measured in gals, and h is measured in kilometers.

As for the internal field, the internal level surfaces are approximately ellipsoids that become more and more spherical with increasing depth. The theory of the internal field is governed by a differential equation published by Clairaut in 1743. This equation relates the flattening of the level surfaces to the density under the assumption of hydrostatic equilibrium. The solution of this equation offers a possibility of computing the flattening of the Earth from the constant of astronomical precession, which is quite accurately known. Values for f obtained in this way and values obtained more directly from the analysis of satellite orbits, from J_2, see Eq. (10), show discrepancies which have yet to be completely explained.

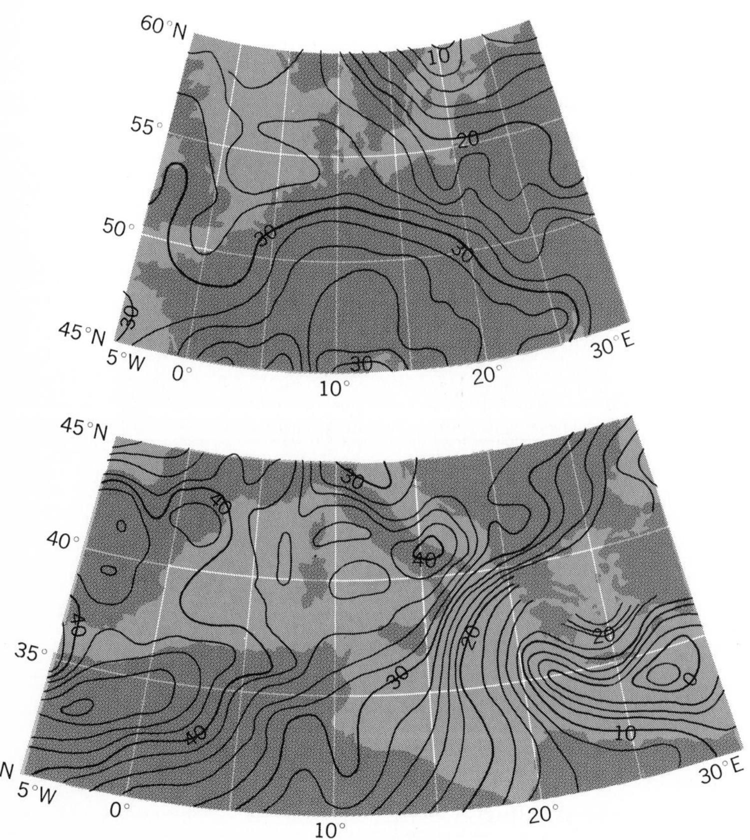

Fig. 6. Gravimetric geoid of Europe, 1957. Geoidal heights in meters, referred to an ellipsoid of flattening 1/297. (*After W. A. Heiskanen*)

Anomalies of the gravitational field. The normal ellipsoidal field incorporates the main part of the Earth's gravitational field. The difference between the real field and the normal field is called the anomalous field. The anomalous potential T is the difference between the actual gravity potential W and the normal gravity potential U. (The circumstance that the symbol T has been previously used to denote the period of oscillation of a pendulum will cause no confusion.)

Other quantities of the anomalous field are the gravity anomaly and the deflection of the plumb line. Both are more readily accessible to observation than the anomalous potential T, and they may be used to determine T.

Gravity anomalies. The gravity anomaly Δg is defined as the difference between gravity at sea level, g, and normal gravity at the reference ellipsoid, γ: $\Delta g = g - \gamma$. Since measurements are not generally made at sea level, the measured gravity value must be reduced from the Earth's surface to sea level, that is, to the geoid. Depending on the way in which this is achieved, different types of gravity anomalies are obtained.

This reduction may be done without considering the masses below the observation station; that is, g is reduced as if these masses did not exist and the station were "in free air." This is the free-air reduction. The amount of this reduction, to be added to the measured gravity, is by formula (14) approximately $0.309h$ mgals, where h is the elevation in meters.

In the process of gravity reduction, the masses above the geoid may also be removed computationally. This is the Bouguer reduction. The amount to be added to measured gravity is now $0.197h$, the units being the same as before; this figure is based on a rock density of 2.67 g/cm³.

Instead of completely removing the masses above the geoid, they may be shifted computationally into its interior in such a way as to get a level homogeneous crust according to some theory of isostasy (see later section of this article). This is an isostatic reduction. There are also a number of other gravity reductions.

Deflections of plumb line. The deflection of the plumb line, or deflection of the vertical, is the deviation of the actual plumb line from the normal to the reference ellipsoid. It is characterized by two components, the component ξ in a northern direction and the component η in an eastern direction. If φ_n and λ_n are the normal (ellipsoidal) coordinates, as opposed to their actual counterparts φ and λ, then ξ and η are given by Eqs. (15 and (16).

$$\xi = \varphi - \varphi_n \quad (15)$$

$$\eta = (\lambda - \lambda_n)\cos\varphi \quad (16)$$

Astronomical observations give φ and λ, whereas φ_n and λ_n may be obtained from geodetic work such as triangulation. The ξ and η so determined are called astrogeodetic deflections of the vertical.

Anomalous potential and the geoid. The height of the geoid above the reference ellipsoid, the geoidal height N, is intimately related to the anomalous potential T by Bruns's formula, $T=\gamma N$, where γ is normal gravity. Therefore, the determination of T is equivalent to the determination of the geoid.

A relative determination can be made by the astrogeodetic method. Differences $T_B - T_A$ or

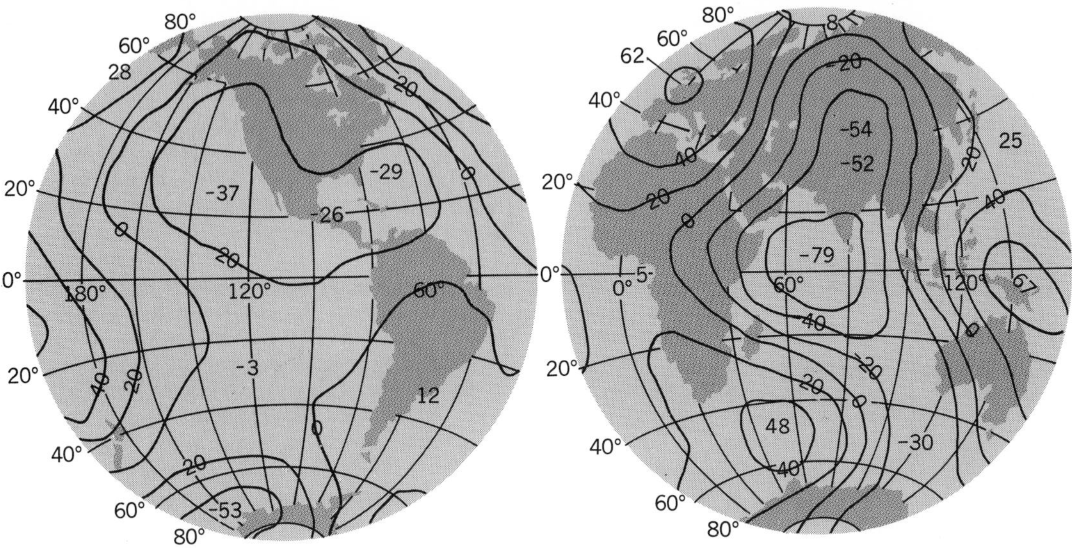

Fig. 7. Generalized geoid from satellite observations, 1966. Geoidal heights in meters, referred to an ellipsoid of flattening 1/298.25. Satellite observation provides a good qualitative picture of large features of the geoid. (*After W. M. Kaula*)

$N_B - N_A$ are obtained by integrating astrogeodetic deflections of the vertical along a line connecting the points A and B. By this method it is possible to determine the geoid in a limited region from data (plumb-line deflections) in this region only. The astrogeodetic method furnishes very detailed and accurate geoidal maps (or maps of the potential T) in areas where good astrogeodetic data are available. However, since this method is only relative, the position of the geoid so obtained is determined only apart from an unknown shift with respect to the Earth's center of mass. The use of the astrogeodetic method is restricted to land areas because the necessary data cannot yet be obtained at sea.

An absolute determination of the anomalous potential or the geoidal height can be achieved from gravity anomalies Δg. This is done by an integral formula due to Stokes (1849). Stokes' formula determines the geoid absolutely; that is, the geoidal heights so obtained refer to an ellipsoid whose center coincides with the Earth's center of mass. It requires the gravity anomalies to be given all over the Earth; this is practicable because gravity can be measured at sea as well as on land areas. Large regions, especially the oceans, have not been satisfactorily surveyed gravimetrically. This brings about certain restrictions on the practical applicability of the gravimetric method; nevertheless it has given important results. Figure 6 shows part of the "Columbus Geoid," computed at Ohio State University.

Harmonic analysis. A periodic function can be expanded into a Fourier series; this is called harmonic analysis. Harmonic analysis may be generalized to functions of more than one variable, such as T or Δg. Instead of Fourier series there is, in this particular case, series of so-called spherical harmonics; such series are applied also in other geosciences, such as geomagnetism.

A series of spherical harmonics is infinite. In practice, there is usually a restriction to series truncated after terms of lower degree. Higher degree terms cannot be determined properly and must be neglected. In this way the general broad features of the expanded function are retained, although fine details are lost.

The spherical-harmonic expansion of T is readily obtained if such an expansion of Δg is available. The practical use of this method is again impaired by the lack of gravity data in parts of the Earth's surface.

The motion of an artificial satellite in the Earth's gravitational field is most strongly affected by the harmonics of lower degree. Therefore satellite observation gives values of the lower degree spherical harmonics of the potential, which provide a good qualitative picture of the general behavior of the terrestrial gravity field and, in particular, of the large features of the geoid (Fig. 7).

Each of the three methods—the astrogeodetic, the gravimetric, and the satellite method—has its own specific merits and shortcomings. In many respects they complement each other. Therefore several attempts are being made to combine them.

Temporal variations. For most purposes, the Earth's gravitational field may be considered invariable in time. However, it is subject to extremely small periodic variations due to tidal effects. These are caused by the attraction of the Sun and Moon. The attraction acts directly by superimposing itself onto the Earth's gravitational attraction; and it acts also indirectly by slightly deforming the Earth and shifting the waters of the oceans, so that the attracting terrestrial masses themselves are modified. The lunar effect on gravity attains a maximum of 0.20 mgal, and the solar effect, a maximum of 0.09 mgal; both are well within the measuring accuracy of modern gravimeters. The results of stationary gravimeters recording variations of gravity may be used to draw conclusions as to the elastic behavior of the Earth under the influence of tidal stresses. *See* EARTH TIDES.

RELATIONS TO OTHER RESEARCH

The determination of Earth's gravitational field depends on other studies, such as the astronomical determination of the plumb line, and has applications in the search for mineral deposits and so forth.

Geophysics and geology. The anomalies of the terrestrial gravitational field are caused by mass irregularities. These may be the visible irregularities of topography such as mountains; or they may be invisible subsurface density anomalies. This is the reason why it is possible to use gravity measurements for investigating the underground structure of the Earth's crust. Thus analysis of gravity is applied by geophysicists and geologists for studying general features of the crust, and by exploration geophysicists for searching for shallow density irregularities that might indicate mineral deposits.

Isostasy. If the mountains were simply superposed on an essentially homogeneous crust and if the ocean depressions were simply hollowed out from such a crust, the irregularities of the gravity field would be almost 10 times as large as they actually are. This indicates that the visible mass anomalies, such as mountains or ocean depressions, are to a large extent compensated: The crustal density underneath the mountains is smaller than normal, and underneath the oceans, greater than normal.

Theories of isostasy have been developed by J. H. Pratt and by G. B. Airy, both about 1850. According to Pratt, the mountains have risen from the underground somewhat like a fermenting dough. According to Airy, the mountains are floating on a fluid lava of higher density, somewhat like an iceberg on water, so that the higher the mountain the deeper it sinks (Fig. 8).

Pratt became aware of isostatic compensation through investigating astrogeodetic deflections of the vertical in the Himalayas. At one station he computed from the visible masses a deflection of 28 seconds of arc, whereas the observed value was only 5 seconds of arc.

Isostasy is also the reason for a conspicuous behavior of the Bouguer anomalies in mountain areas. Here they are systematically negative and increase in magnitude on the average by about 100 mgals/1000 m of mean elevation.

If isostatic compensation were complete, then the isostatic gravity anomalies as defined above would be zero. In general, they are small and fluctuate smoothly around zero; this property makes them particularly well suited for interpolation and extrapolation to poorly observed areas in geodetic applications. Systematically nonzero isostatic anomalies indicate deviation from isostatic equilibrium. Therefore isostatic gravity anomalies, and also Bouguer anomalies, are used to investigate the degree and the mode of isostatic compensation. It appears that the Earth's crust is compensated isostatically to about 90%; the compensation seems in general to be of Airy type rather than Pratt type. *See* EARTH.

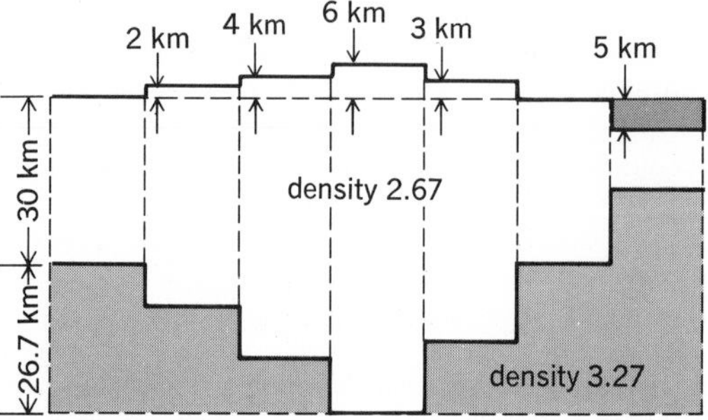

Fig. 8. Airy isostasy model. (*After W. A. Heiskanen*)

Exploration geophysics. Local mass anomalies at shallow depth, which might be related to mineral deposits, are discovered by a local very dense and accurate gravity survey by means of gravimeters. Bouguer anomalies are computed to eliminate as far as possible the effect of visible masses, and various filtering techniques are employed to isolate and localize disturbances. *See* GEOPHYSICAL EXPLORATION.

Geodesy. Geodetic instruments employ spirit levels and other devices to orient them with respect to the horizontal or, what amounts to the same thing, to the plumb line. Since the plumb line is defined by the gravitational field, it can be understood why this field enters essentially into almost all geodetic measurements, even into apparently purely geometric ones. In return, geodetic techniques are among the most efficient means for determining the gravitational field. The "mathematical figure of the Earth" for the purpose of geodesists, the geoid, is defined as a surface of constant potential W. "Heights above the sea level" are heights above the geoid; their determination is therefore a physical as well as a geometric problem. (Geodetic theories have been developed which employ only quantities referred to the Earth's topographical surface; but here the gravitational field enters in an even more complicated way.) Thus geodesy is essentially concerned with the Earth's gravitational field and its determination; the theory of the figure of the Earth is to a large extent equivalent to the theory of terrestrial gravitation. *See* GEODESY.

Astronomy and satellite dynamics. Astronomical observations are indispensable for determining the plumb line, which defines the direction of the gravity vector. Parameters of the Earth's gravitational field such as equatorial gravity γ_e are used to express in kilometers the fundamental astronomical unit, the mean distance of Sun and Earth. The Earth's equatorial radius a, which gives the scale of the geoid, is also used as a basic astronomical constant. *See* EARTH, ROTATION AND ORBITAL MOTION OF.

Terrestrial gravitation affects the motion of the Moon and of the artificial satellites. By observing these satellites, it is therefore possible to determine essential parameters of the gravitational field, such as the geocentric gravitational constant GM, which is the product of the Newtonian gravitational constant and the Earth's mass. The dynamical ellipticity J_2 defined above, and other lower degree coefficients of the spherical-harmonic expansion of the gravitational potential, may also be determined in this way. GM and J_2 are the coefficients of degrees zero and two of this expansion. [HELMUT MORITZ]

Bibliography: W. A. Heiskanen and H. Moritz, *Physical Geodesy*, 1967; W. A. Heiskanen and F. A. Vening Meinesz, *The Earth and Its Gravity Field*, 1958; H. Jeffreys, *The Earth*, 4th ed., 1962; C. A. Lundquist and G. Veis (eds.), *1966 Smithsonian Standard Earth*, Smithson. Astrophys. Observ. Spec. Rep. no. 200, 3 vols., 1966; I. Todhunter, *A History of the Mathematical Theories of Attraction and of the Figure of the Earth*, reprint, 1962.

Terrestrial magnetism

The natural magnetism of the Earth. The designation geomagnetism is now given some preference over the older term, terrestrial magnetism. *See* GEOMAGNETISM.

[CHARLES V. CRITTENDEN]

Tertiary

The older major subdivision (period) of the Cenozoic Era, extending from the end of the Cretaceous (youngest of three Mesozoic periods) to the beginning of the Quaternary (the younger Cenozoic period). The term Tertiary is also applied to all rocks formed during this period and to all the fossils they contain. Tertiary sedimentary rocks include wide-

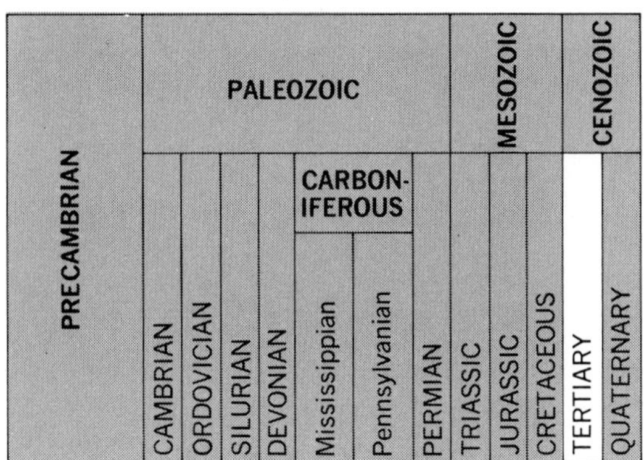

spread limestones, sandstones, marls, mudstones, and conglomerates; igneous rocks include extrusive and intrusive volcanics and locally some rock of more deep-seated (plutonic) origin. Tertiary life was characterized particularly by (1) a diversification and multiplication of pelecypods (clams, cockles, scallops, ark shells) and gastropods (conchs, cones, periwinkles), of sea urchins and sand dollars, of microscopic foraminiferans, and of some other marine shellfish; (2) a great development of shrubs, grasses, and other flowering plants; and (3) the rapid diversification and expansion on both land and sea of birds and mammals, the latter having provided for the Tertiary the more informal designation, Age of Mammals (Fig. 1).

Geography. Configuration of the modern continental land masses developed during Tertiary and Quaternary times. Many major modern seaways are shrunken remnants of the widespread seas of the Late Cretaceous, when the ocean surfaces of the world were at their greatest extent. Great mountain-making movements (orogenies) and occasional uplifts of large segments of the Earth's crust alternated with fluctuating transgressions and reexpansions of the seas over land areas. Orogenies and uplifts nevertheless grew progressively greater at the expense of the earlier seas as Tertiary time wore on and culminated in the great Late Alpine orogenic uplifts which occurred in Pleistocene time. *See* CENOZOIC; CRETACEOUS; PLEISTOCENE.

Rocks. Tertiary sedimentary rocks occur widely as a relatively thin veneer of marine deposits, such as those left north and west of a shrinking Gulf of Mexico and on the nearby Atlantic seaboard of North America. Much greater thicknesses were

Fig. 1. *Baluchitherium*, the largest land mammal known, from the Oligocene of Asia. It was 18 ft (5.4 m) high at the shoulders. *(From R. A. Stirton, Time, Life and Man, copyright © 1959 by John Wiley & Sons, Inc.; reprinted by permission)*

attained in the borderlands around the Pacific Ocean and on the flanks of the rising Alpine-Himalayan chain. In the Santa Barbara–Ventura region of south-central California, Tertiary strata are more than 50,000 ft thick. Nonmarine sedimentary strata are generally thinner and are more widespread in the interiors of the modern continents. Outpourings of Tertiary volcanic rock have been extensive, especially around the rim of the Pacific Ocean, and in the Mediterranean area and Iceland, and also in the intervening submarine region.

Life. Tertiary life is distinguished by fossil animals which, both specifically and in the aggregate, resemble much more nearly those living today and those living during the intervening Quaternary (10% extinct species at most) than they do those of the preceding Cretaceous. However, plant life began taking on a comparably modern aspect earlier, within the Cretaceous Period. Thus the difference between Late Cretaceous and Tertiary life is more conspicuous among fossil animals than among plants.

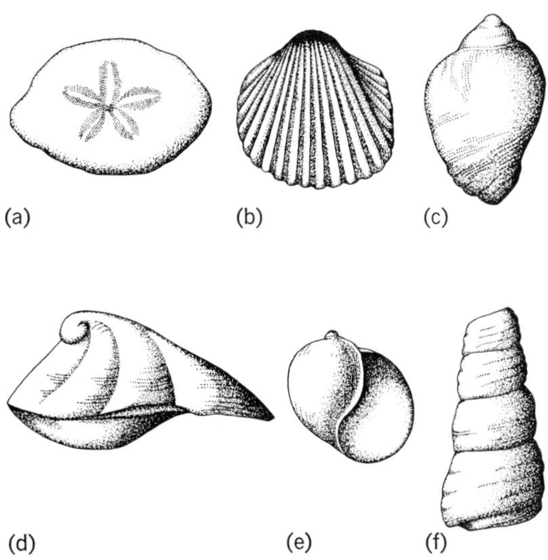

Fig. 2. Marine invertebrates of Tertiary Period. (*a*) *Eoscutella coosensis*, echinoid. (*b*) *Loxocardium brenerii*, pelecypod. (*c*) *Pseudoliva dilleri*, gastropod. (*d*) *Velates schmidelianus*, gastropod. (*e*) *Eocernina hannibali*, gastropod. (*f*) *Turritella uvasana*, gastropod. *(From R. A. Stirton, Time, Life and Man, copyright © 1959 by John Wiley & Sons, Inc.; reprinted by permission)*

Like the stocks which dominate them, the marine shellfish faunas of today first appeared in rudimentary ancestral form in earliest Tertiary times (Fig. 2). Like the bird-mammal faunas on land, which have been more conspicuously modified from time to time, all these faunas attained their present composition and extent through the natural selective interplay of hereditary and environmental forces operating through the fluctuating geographic space relationships of Tertiary and Quaternary time.

Stratigraphy. Hellenic Greeks recognized shells far inland from the Aegean Sea as fossil marine organisms, as did Leonardo da Vinci subsequently in Italy, but the term Tertiary stems from Giovanni Arduino's 18th-century reference to the geologically young strata of northern Italy. In 1810 Georges Cuvier and Alexandre Brongniart applied the term to a formational sequence in the Paris Basin. Local names were applied to the many rock units which comprise the Tertiary System, so that students of the Tertiary are confronted by a bewildering array of formation names. Other formally designated subdivisions which, like the Tertiary System itself, now have a definite time dimension based on life phenomena, are hardly less numerous, yet only five major time-rock subdivisions, or series, are generally recognized. By 1830 Gerard P. Deshayes and other European conchologists had observed that, in ascending the Tertiary stratal column, the percentage of living species in the fossil faunas increased from bottom to top. In 1833 Sir Charles Lyell accordingly recognized three subdivisions: Eocene below, Miocene, and Pliocene above (originally an "Older" and a "Newer" Pliocene; the latter he subsequently changed to Pleistocene, which is post-Tertiary, along with the Holocene added by Paul Gervais in 1860). As provided for by Lyell, in strata previously classed as Miocene by some and Eocene by others in the northern German Plain and in the Lowlands, Heinrich Ernst von Beyrich defined an intermediate Oligocene; and in 1874 the paleobotanist W. P. Schimper set aside the lowermost beds of the original east Paris Basin Eocene as Paleocene (see separate articles with these titles). These Tertiary series and their epoch equivalents can be recognized by Lyellian criteria around the world. Subsequent further subdivisions—the stages and even smaller zones—are, however, limited in geographical extent to faunal provinces. *See* PALEOBOTANY; PALEONTOLOGY.

[ROBERT M. KLEINPELL]

Tetrahedrite

A mineral with composition $(Cu,Fe,Zn,Ag)_{12}Sb_4S_{13}$ (essentially copper, iron, zinc, silver, antimony, and arsenic sulfide), crystallizing in the isometric system. Crystals are commonly in tetrahedrons (see illustration), but the tristetrahedron, dodecahedron, and cube may be present. It is massive or granular. Its hardness is $3\frac{1}{2}$–4 and the specific gravity varies from 4.6 to 5.1, depending on the composition. The luster is metallic and the color grayish-black; thus, in some mining localities, this mineral is called gray copper.

Although analyses of tetrahedrite show varying amounts of copper, iron, zinc, and silver, copper always predominates. The variety rich in silver is called freibergite. Arsenic substitutes for antimony in all proportions, and a complete series extends to the mineral tennantite, $(Cu,Fe,Zn,Ag)_{12}As_4S_{13}$.

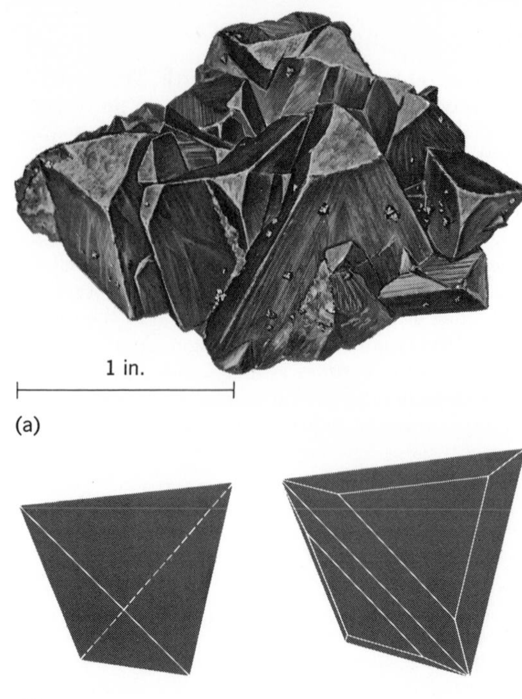

Tetrahedrite. (*a*) Crystals from Bobes, Romania (*specimen from Department of Geology, Bryn Mawr College*). (*b*) Crystal habits (*from C. S. Hurlbut, Jr., Dana's Manual of Mineralogy, 17th ed., copyright © 1959 by John Wiley & Sons, Inc.; reprinted by permission*).

Tetrahedrite is a widely distributed mineral, usually found in silver and copper veins formed at low to moderate temperatures; more rarely it is found in higher-temperature veins and in contact metamorphic deposits. It may be associated with various other copper, lead, and silver minerals, as well as pyrite and sphalerite. In some places it has sufficient silver to be a valuable ore, as at Freiberg, Germany, and silver mines of Peru, Bolivia, and Mexico. It is found in silver and copper mines in the western United States.

[CORNELIUS S. HURLBUT, JR.]

Thermal spring

A spring with water temperature substantially above the average temperature of springs in the region in which it occurs. The average temperature of springs is ordinarily within a few degrees of the mean annual temperature of the atmosphere. Thus waters of thermal springs range in temperature from as low as 60°F, in an area where normal groundwater has a temperature of 40–50°F, to well above the boiling point.

The two main considerations in the origin of thermal springs are the source of the water and the source of the heat. The water may be ordinary groundwater that percolates slowly downward, is heated by the Earth's normal thermal gradient (the temperature of the Earth normally increases about 1°F for each 50–100 ft of depth), and then returns to the surface without losing all the added heat.

The water of thermal springs may be in part juvenile, a product of the crystallization or recrystallization of rock at depth. Since juvenile water is virtually certain to become mixed with connate or meteoric water on its way to the surface, there are no thermal springs whose water can be demonstrated to be wholly juvenile.

Investigations of Warm Springs, Ga., and of other thermal waters in the eastern United States indicate that the water entered the aquifer by normal recharge from precipitation, percolated deep into the Earth by reason of the geologic structure, and there received its heat before returning to the surface. On the other hand, the springs in Yellowstone Park, Wyo., Steamboat Springs, Nev., and many other localities in the western United States may derive part of their water and much of their heat from bodies of superheated rocks, perhaps in the last stages of cooling from the molten state. Many of the springs in the western United States discharge water that is near the boiling point.

Where spring water is above the boiling point it has been tapped to provide steam for power production. Such power installations are found in New Zealand, Italy, and California. Hot springs have also been used to heat homes and swimming pools. *See* GEYSER; GROUNDWATER.

[ALBERT N. SAYRE/RAY K. LINSLEY]

Till

The unstratified portion of glacial drift. The unsorted materials of the till are deposited by the advancing ice, or as a result of melting or evaporation of the ice during the waning stage of glaciation. The term boulder clay refers to a common variety of till containing embedded particles ranging in size from fine grains to boulders.

The texture of till is characterized by extreme variation in grain size (see illustration). The matrix consists of the finer clastic materials, clay, silt, and sand. Randomly embedded in this are larger fragments including boulders of many cubic yards in volume. The coarser fragments, cobbles and boulders, commonly display faceting and striations caused by abrasion during transport by the ice. Careful study may reveal a preferred orientation of the larger fragments. This is usually the only indication of stratification. Lenses of stratified sand, gravel, or silt which occur locally within the till represent the local action of melt water.

Exposure of glacial till at the Black Rocks near Llandudno, Wales. Heterogeneous debris, ranging in size from large boulders to fine powder, displays no assortment or stratification. (*Photograph by K. F. Mather*)

Till consists of physically broken and disintegrated, but essentially undecomposed, rock and mineral fragments. Commonly all types of rocks are represented, but igneous and metamorphic ones predominate. These materials, under favorable climatic conditions, are readily converted into excellent soils. *See* GLACIATED TERRAIN.

[CHALMER J. ROY]

Tonalite

A phaneritic (visibly crystalline) plutonic rock composed chiefly of plagioclase (oligoclase or andesine) and quartz with subordinate dark-colored (mafic) minerals (biotite, amphibole, or pyroxene). The term tonalite is roughly equivalent to quartz diorite. Minor amounts of alkali feldspar may be present, but if this mineral exceeds 5% of the total feldspar, the rock is a granodiorite. As the quartz content decreases, quartz diorite passes into diorite. Tonalite, or quartz diorite, is roughly intermediate between granodiorite and diorite. *See* DIORITE.

[CARLETON A. CHAPMAN]

Topaz

A mineral best known for its use as a gemstone. Crystals are usually colorless but may be red, yellow, green, blue, or brown. The wine-yellow variety is the one usually cut and most highly prized as a

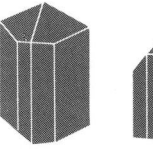

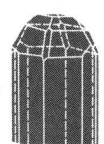

Topaz. (*a*) Crystal from Ramona, Calif. (*American Museum of Natural History specimen*). (*b*) Crystal habits (*from C. S. Hurlbut, Jr., Dana's Manual of Mineralogy, 17th ed., copyright © 1959 by John Wiley & Sons, Inc.; reprinted by permission*).

gem. Corundum of similar color sometimes goes under the name of Oriental topaz. Citrine, a yellow variety of quartz, is the most common substitute and may be sold as quartz topaz. *See* GEM.

Topaz is a nesosilicate with chemical composition $Al_2SiO_4(F,OH)$. The mineral crystallizes in the orthorhombic system and is commonly found in well-developed prismatic crystals with pyramidal terminations (see illustration). It has a perfect basal cleavage which enables it to be distinguished from minerals otherwise similar in appearance. Hardness is 8 on Mohs scale; specific gravity is 3.4–3.6. *See* SILICATE MINERALS.

Topaz is found in pegmatite dikes, particularly those carrying tin. It is also formed during the late stages of the solidification of rhyolite lavas. The minerals characteristically associated are tourmaline, cassiterite, fluorite, beryl, and apatite. It is also found as rolled pebbles in stream gravels. Fine yellow and blue crystals have come from Siberia and much of the wine-yellow gem material from Minas Gerais, Brazil. In the United States topaz has been found near Florissant, Colo.; Thomas Range, Utah; San Diego County, Calif.; and Topsham, Maine. [CORNELIUS S. HURLBUT, JR.]

Torbanite

A variety of coal that resembles a carbonaceous shale in outward appearance. It is fine-grained, brown to black, and tough, and breaks with a conchoidal or subconchoidal fracture. The name torbanite is derived from the initial discovery site of the material in 1850 at Torbane Hill, Linlithgowshire, Scotland. Torbanite is synonymous with boghead coal and is related to cannel coal. It is derived from colonial algae identified with the modern species of *Botryococcus braunii* Kütz and antecedent forms.

Major deposits of torbanite occur in Australia, Tasmania, New Zealand, Scotland, and South Africa. The South African deposit, which is in the Ermelo district of the Transvaal, yields 20–100 gal of oil per ton on retorting. High-assay torbanite yields paraffinic oil, whereas low-assay material yields asphaltic oil. *See* COAL; SAPROPEL.

[IRVING A. BREGER]

Tourmaline

A mineral cyclosilicate with complex chemical composition, long known for its use as a gemstone. *See* GEM; SILICATE MINERALS.

Tourmaline crystallizes in the ditrigonal-pyramidal class of the hexagonal system in prismatic crystals with the trigonal prism dominant. A combination of this prism with a hexagonal prism causes vertical striations and a tendency for the faces to round into each other, giving the crystals a cross section resembling that of a spherical triangle (see illustration). The vertical axis is polar; thus different forms are found at the opposite ends. Because of this polarity, tourmaline is piezoelectric; that is, if pressure is exerted at the ends of the polar axis, one end becomes positively charged and the other end negatively charged. It is also pyroelectric, with the electrical charges developed at the ends of the polar axis on a change in temperature.

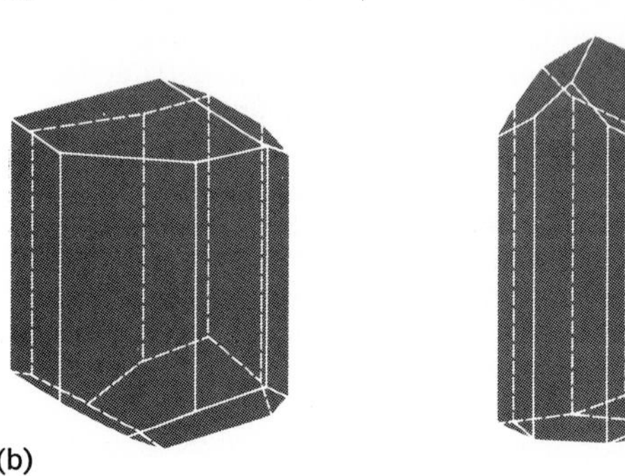

Tourmaline. (*a*) Prismatic crystal from Portland, Conn. (*American Museum of Natural History specimen*). (*b*) Crystal habits (*from C. S. Hurlbut, Jr., Dana's Manual of Mineralogy, 17th ed, copyright © 1959 by John Wiley & Sons, Inc.; reprinted by permission*).

Because of its piezoelectric property, tourmaline is manufactured into gages to measure transient pressures. Plates, cut at right angles to the principal axis, are coated with electrodes from which wires lead to a recording device. The voltage recorded is proportional to the pressure exerted on the plate. Such gages are used to measure the pressures of atomic explosions.

The hardness of tourmaline is $7\frac{1}{2}$ on Mohs scale and the specific gravity 3.0–3.25. The luster is vitreous to resinous and the color, depending on the chemical composition, is variable. Iron-rich tourmaline (shorlite), the most common variety, is black. The magnesium variety is brown. Lithium renders the mineral lighter colored in various shades of red (rubellite), yellow, green, blue (indicolite), and rarely colorless (achroite). If transparent and flawless, these varieties are used as gemstones. Several colors may be present in the same crystal, arranged in layers across the length or in concentric envelopes around the crystal. Some dark varieties are strongly dichoric. The chemical composition is represented by the general formula $XY_3Al_6(BO_3)_3Si_6O_{18}(OH)_4$, where X represents Na or Ca and Y represents Al, Fe^{3+}, Li, or Mg.

Tourmaline is found as an accessory mineral in

igneous and metamorphic rocks, but its most characteristic occurrence is in granite pegmatites. Here the black variety is most common but the light-colored varieties may be present, firmly embedded in the other pegmatite minerals or in cavities known as pockets. Most gem material occurs in this latter form. Noted localities for gem tourmaline are in Minas Gerais, Brazil; Ural Mountains; Madagascar; and, in the United States, Paris and Auburn, Maine; Haddam Neck, Conn.; Mesa Grande and Pala, Calif.

[CORNELIUS S. HURLBUT, JR.]

Trachyte

A light-colored, aphanitic (very finely crystalline) rock of volcanic origin, composed largely of alkali feldspar with minor amounts of dark-colored (mafic) minerals (biotite, hornblende, or pyroxene). If sodic plagioclase (oligoclase or andesine) exceeds the quantity of alkali feldspar, the rock is called latite. Trachyte and latite are chemically equivalent to syenite and monzonite, respectively.

Texture. The extremely fine-grained texture and more or less glassy material are due to rapid cooling and solidification of the lava. Large crystals (phenocrysts) are commonly sprinkled liberally through the dense rock, giving it a porphyritic texture. These may be well formed and 1–2 in. wide. They appear as glassy crystals of sanidine, and in addition small mafic phenocrysts may be present. In latite the phenocrysts are largely plagioclase. As the quantity of glass increases, these porphyric rocks pass into vitrophyre; and as the abundance of phenocrysts increases, these rocks pass into trachyte porphyry. *See* PORPHYRY; VOLCANIC GLASS.

The detailed features of trachyte are best studied microscopically. Sanidine and orthoclase are dominant over oligoclase in normal (potash) trachyte. In alkali (soda) trachyte, both alkali feldspar and mafics are soda-rich.

Composition. Brown biotite mica is the common mafic. It occurs as flakes which may be more or less resorbed by the liquid in the late stages of solidification so that only patches of dusty iron oxide remain. Normal trachyte commonly carries somewhat corroded and resorbed hornblende or diopside. Alkali trachyte usually contains soda-rich amphibole (riebeckite, arfvedsonite, and barkevikite) or pyroxenes (aegirine-augite or aegirite). Zoned crystals with diopsidic cores and progressively more soda-rich margins are common.

Either free silica (quartz, tridymite, or cristobalite) or feldspathoids (leucite, nepheline, or sodalite) may be present in small amounts. With increase in free silica, the rock passes into rhyolite; and with increase in feldspathoids, it passes into phonolite. Accessory minerals as tiny grains and crystals are magmetite, ilmenite, apatite, zircon, and sphene.

Structure. Streaked, banded, and fluidal structures due to flowage of the solidifying lava are commonly visible in many trachytes and may be detected by a parallel arrangement of tabular feldspar phenocrysts. A distinctive microscopic feature is trachytic texture in which the tiny, lath-shaped sanidine crystals of the rock matrix are in parallel arrangement and closely packed. This rather uniform pattern is locally interrupted where the laths more or less deviate or wrap around the phenocrysts. Orthophyric texture is common where tiny feldspar crystals show a stumpy or square outline.

Occurrence and origin. Trachyte is not an abundant rock, but it is widespread. It occurs as flows, tuffs, or small intrusives (dikes and sills). It may be associated with alkali rhyolite, latite, or phonolite.

Trachyte is commonly considered to have been derived from a basaltic magma by differentiation, a process involving removal in large quantities of early formed crystals rich in iron, magnesium, and calcium. A factor of importance in the formation of some trachyte is contamination of the original magma by incorporation of foreign rock material. The chemical transformation of andesite to trachyte may have occurred (in the solid state) where calcium was removed and sodium added metasomatically. This may explain the origin of some keratophyres (a variety of soda-rich trachyte). *See* IGNEOUS ROCKS; MAGMA; PETROGRAPHIC PROVINCE; SPILITE.

[CARLETON A. CHAPMAN]

Transform fault

A fracture along an offset of a spreading mid-ocean-ridge crest. Even over millions of years, the Earth's crust behaves like a simple elastic solid: Fractures in it result from the stress exceeding the breaking stress. On the continental crust, normal and reverse faults were early recognized in which relative vertical movement between the crust on either side had occurred. In strike-slip faults or transcurrent faults the two sides are displaced relatively in a horizontal direction—along the strike of the fault. These vertical and horizontal displacements were determined by correlating the geological strata on either side of the fault or matching geomorphological or topographical features. The problem of correlation when the displacement is large held back the recognition that horizontal movements were often many tens of kilometers. The Great Glen Fault in Scotland was shown by W. Q. Kennedy to have displacement of 100 km by the matching of syenite blocks on either side. M. L. Hill and T. W. Dibblee provided evidence for a continued movement since the mid-Mesozoic of 1 cm per year. Small horizontal displacements could be accommodated without difficulty through vertical or other movements of the rigid crust. The much larger displacements presented considerable difficulty; in each case these faults ended under the sea and mechanics of the ends of the faults remained obscure.

The study of the topography of the ocean floor

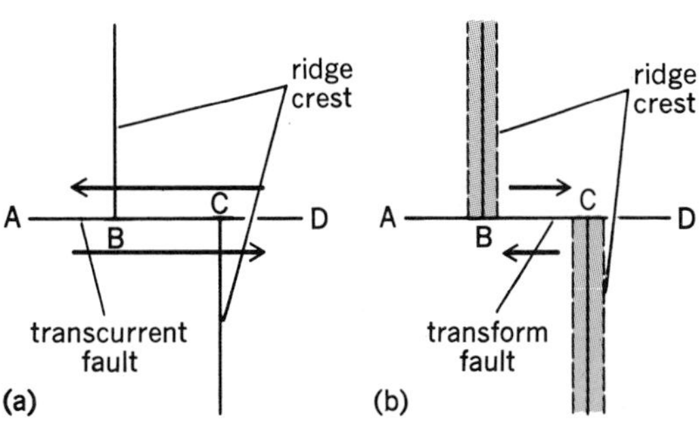

Diagrams of faults. (*a*) Left-lateral transcurrent fault. (*b*) Right-lateral transform fault.

soon showed the existence of escarpments running east-west in the Pacific off North America, and these were identified as faults in the ocean floor. It was found that the pattern of magnetic anomalies over the oceans consisted of north-south elongated anomalies of varying characteristics and, as a result of the matching of these across the faults, very large horizontal displacements in the ocean floor were inferred.

Similar east-west faults were recognized across the ocean ridges, especially the Mid-Atlantic Ridge, and marked east-west displacements were inferred from these topographical misalignments. In an attempt to face the difficulty referred to above, J. T. Wilson recognized that these faults were of an entirely different character. The acceptance of the theory of continental drift required an explanation of the growth of the Atlantic ocean floor as the Americas moved westward relative to Europe and Africa. This was supplied by the complementary theory of sea floor spreading: New sea floor was continually created by lava rising from the mantle into the rift—a continuous feature of the Mid-Atlantic Ridge crest. The recognition that the sea floor could grow only where the ocean crust was broken—along the rift—meant that east-west displacements along the fault occurred only between the offset rifts (BC in the illustration) and not along the entire length of the fault (AD).

Wilson termed this a transform fault. It differs clearly from a transcurrent fault in crucial respects. First, earthquake activity is restricted to the active portion of the fault (BC) between the rifts and not beyond, where the fault is presently inactive (AB, CD). Second, the sense of displacement is opposite in the transform fault to that of the transcurrent fault. Third, the distance BC between the ridge-crest rifts remains constant for the transform fault and increases with time for the transcurrent fault.

These predictions concerning transform faults have been verified by L. Sykes. Since then it has been shown that the central lines of the mid-oceanic ridges are locations of shallow-focus earthquakes. Improved means of locating earthquakes have also shown that the earthquake foci lie along the rift. At the offsets the earthquakes lie along the fault between B and C. From first motion studies, the sense of the displacement along the fault is shown to be in accord with the transform fault interpretation. Strips of normally and reversely magnetized crust have been found on both sides of mid-ocean-ridge rifts. These strips have been interpreted as increments of ocean floor added at the rift during periodic reversals of the Earth's field, thus giving quantitative support for the theory of sea floor spreading (F. J. Vine, 1966). *See* CONTINENT FORMATION; FAULT AND FAULT STRUCTURES; MARINE GEOLOGY; SEISMOLOGY.

[STANLEY K. RUNCORN]

Bibliography: M. L. Hill and T. W. Dibblee, Jr., San Andreas, Garlock Big Pine Fault, California, *Bull. Geol. Soc. Amer.*, 64:433–458, 1953; L. R. Sykes, Mechanism of earthquakes and nature of faulting on the mid-oceanic ridges, *Geophys. Res.*, vol. 72, 1967; F. J. Vine, Spreading of the ocean floor: New evidence, *Science*, 154:1405–1415, 1966; J. T. Wilson, A new class of faults and their bearing on continental drift, *Nature*, 207:343–347, 1965.

Travertine

A rather dense, banded limestone (see illustration), sometimes moderately porous, that is formed either by evaporation about springs, as is tufa, or in caves, as stalactites, stalagmites, or dripstone.

Travertine, Suisun, Calif. Nicols crossed. *(From E. W. Heinrich, Microscopic Petrography, McGraw-Hill, 1956)*

Where travertine or tufa (calcareous sinter) is deposited by hot springs, it may be the result of the loss of carbon dioxide from the waters as pressure is released upon emerging at the surface; the release of carbon dioxide lowers the solubility of calcium carbonate and it precipitates. High rates of evaporation in hot-spring pools also lead to supersaturation. Travertine formed in caves is simply the result of complete evaporation of waters containing mainly calcium carbonate. *See* LIMESTONE; STALACTITES AND STALAGMITES; THERMAL SPRING; TUFA.

[RAYMOND SIEVER]

Tremolite

The name given to magnesium-rich monoclinic calcium amphibole $Ca_2Mg_5Si_8O_{22}(OH)_2$. The mineral is white to gray, but colorless in thin section, and optically negative. It usually exhibits long prismatic crystals with prominent (110) amphibole cleavage. Unlike other end-member compositions of the calcium amphibole group, very pure tremolite is found in nature. Substitution of Fe for Mg is common, but pure ferrotremolite, $Ca_2Fe_5Si_8O_{22}(OH)_2$, is rare. Intermediate compositions between tremolite and ferrotremolite are referred to as actinolites, and are green in color and encompass a large number of naturally occurring calcium amphiboles. The substitution of Na, Al, and Fe^{3+} ions into the amphibole structure is common in actinolites. The nature of these substitutions is complex and leads, under some conditions, to miscibility gaps between actinolites and aluminous calcium amphiboles called hornblendes. For the most part, however, the physical and chemical variations between actinolite and hornblende are continuous so that an arbitrary actinolite-hornblende division is necessary. Most amphibole classifications divide actinolite from the hornblende series on the basis of Al substitution for Si with actinolite containing less than 0.5 atom of Si replaced by Al per formula unit. *See* AMPHIBOLE; HORNBLENDE.

The basic building block for the tremolite-actinolite crystal structure is the silicon tetrahedron. In all amphiboles, double chains of SiO_4 tetrahedrons are formed through joining two or three tetrahedrons at their corners by consecutively sharing two or three oxygens in alternating fashion along the entire chain length. These double chains form two anionic layers: a nearly coplanar layer of oxygens at the base of the tetrahedrons and a second layer of oxygens and associated OH and F along the apices of the tetrahedrons. Silicon, and to a lesser extent aluminum, atoms lie between these basal and apical anionic layers in fourfold coordination. The double chains are arranged so that along the *b* crystallographic direction the basal oxygen layer of one chain is approximately coplanar with the apical anionic layer of the adjacent chains. Along the *a* crystallographic direction, the apical oxygen layers of adjacent chains face each other, as do the basal oxygen layers. The *c* axis parallels the long axis of the double chains. The double-chain structure is held together by bonding apical oxygens of facing double chains to a cation strip of 5Mg and 2Ca which have six- to eightfold coordination. Tremolite is usually described in terms of the (001) face-centered cell (space group C2/m). Typical lattice parameters for tremolite are $a = 9.83$; $b = 18.05$; and $c = 5.27$ (all in angstroms; 10 A = 1 nm); and β angle = 104.5°. Actinolite, due to its higher Fe^{2+} content, has a larger cation strip and hence a proportionately larger *b* dimension.

Tremolite in pure form is a product of thermal and regional metamorphism of siliceous dolomites and marbles as shown by Eq. (1). In similar rocks at

$$\underset{\text{Dolomite}}{5CaMg(CO_3)_2} + \underset{\text{Quartz}}{8SiO_2} + H_2O \rightarrow \underset{\text{Tremolite}}{Ca_2Mg_5Si_8O_{22}(OH)_2} + \underset{\text{Calcite}}{3CaCO_3} + 7CO \quad (1)$$

higher grades of metamorphism, in the presence of both calcite and quartz, tremolite breaks down to form diopside, as shown by Eq. (2). Actinolites,

$$\underset{\text{Tremolite}}{Ca_2Mg_5Si_8O_{22}(OH)_2} + \underset{\text{Calcite}}{3CaCO_3} + \underset{\text{Quartz}}{2SiO_2} \rightarrow \underset{\text{Diopside}}{5CaMgSi_2O_6} + 3CO_2 + H_2O \quad (2)$$

owing to a more variable chemistry, are more ubiquitous than tremolite in occurrence. They are most commonly found in regionally metamorphosed mafic igneous rocks such as basalts and are known to occur in a wide range of pressure conditions. Both tremolite and actinolite can form through the breakdown of olivine and pyroxenes in regionally metamorphosed ultrabasic rocks; associated minerals are talc, chlorite, and carbonates. Actinolitic amphiboles formed as a breakdown product of pyroxenes are referred to as uralite. Intergrown fibrous crystals of tremolite are known as nephrite, a form of jade widely used for centuries in making of artifacts and jewelry. Highly fibrous tremolite is used in commercial asbestos. *See* ASBESTOS; JADE; METAMORPHISM.

[BARRY L. DOOLAN]

Triassic

A geological term designating the lowest rock system of the Mesozoic Era, proposed by F. von Alberti in 1834 for a sequence of strata in central Germany lying above marine Permian (Zechstein) and below marine Jurassic (Liassic). The name is derived from the threefold facies division of these strata into a lower nonmarine redbed facies (Bunter), a middle marine limestone, sandstone, and shale facies (Muschelkalk), and an upper nonmarine continental facies similar to the lower division (Keuper). This sequence and character of facies typify Triassic strata of northern Europe,

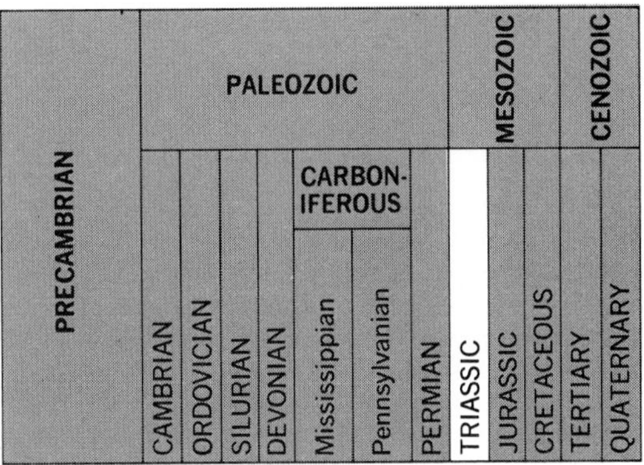

France, Spain, and northern Africa; it is commonly known as the Germanic facies. In England the marine Muschelkalk facies is absent. In contrast to the predominantly continental Germanic facies, there is developed in the Alps a completely marine fossiliferous sequence which includes all of Triassic time. This is commonly called the Alpine facies (Fig. 1); and the sequence of ammonoid zones in these formations, with the addition from southern Asia of some zones in the lower beds, forms the primary standard sequence of stages in the Triassic system. The standard divisions and correlation of the Germanic and Alpine units, plus the ammonite faunal zones, are shown in the table.

It should be noted from the table that French geologists define the term Keuper in a more restricted fashion than do the German geologists; that is, they apply the term to the middle part of the German sequence (Gypskeuper), and refer the Rhaetian (Rhät) to the Jurassic and the Lettenkohle to the Middle Triassic (Muschelkalk). In the French literature the German terms have been translated, in ascending order, as *Grès bigarré* (Bunter), *Calcaire conchylien* (Muschelkalk), and *Marnes irisées* (Keuper, restricted). Additional but minor variations are the use of Virglorian for Anisian, and Werfenian for Scythian.

Paleogeography. The widespread orogenies which began during the Carboniferous Period and continued intermittently until the end of the Paleozoic Era brought to an end the pattern of widespread geosynclines that had characterized the Paleozoic. Thus, during the later phases of the Paleozoic, much of Europe north of the Alps, the Ural

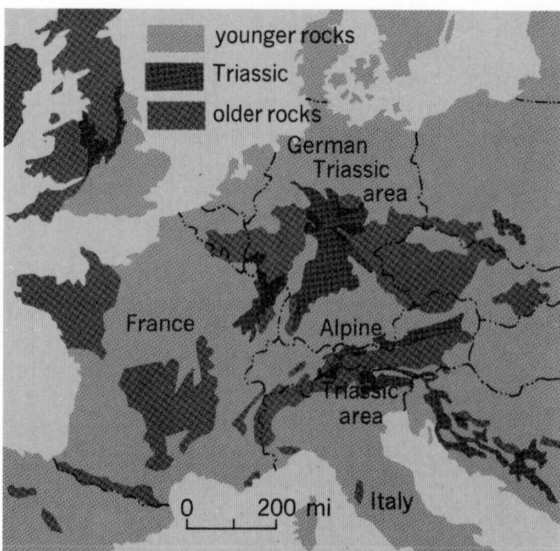

Fig. 1. Distribution of Triassic outcrops in west-central Europe. *(From R. C. Moore, Introduction to Historical Geology, 2d ed., McGraw-Hill, 1958)*

mountain region, eastern North America, and much of Asia north of the Himalayas, which had been active geosynclinal regions, were transformed into rigid continental blocks that became part of the stable portion of the respective continents. With the close of the Permian, epicontinental and geosynclinal seas had retreated from most of the continents and the geography of the lands and seas must have been very similar to what it is today. *See* GEOSYNCLINE; OROGENY.

It is upon this pattern that Triassic history began. Triassic seas were marginal to the continents, except for the Tethyan geosyncline extending from the Alps through the Middle East and the Himalayas to Indonesia. Thus, marine Triassic rocks are found in all the countries on the margins of the Pacific and marginal to the Arctic Ocean. In many of these areas the marine strata grade landward into continental facies. South America, except for its western margin, which is now occupied by the Andes, was land area with local basins of continental deposition. These deposits, consisting mainly of sandstone, are found in southern Brazil, northern and western Argentina, Uruguay, and Paraguay. Africa was emergent throughout the Triassic except for a narrow coastal belt from Libya through Egypt. In South Africa, great thicknesses of continental sandstones and shales accumulated which are well known for their fossil reptile faunas. Madagascar has mainly continental Triassic formations similar to those in South Africa except for a few thin marine tongues in the north. Peninsular India was likewise the site of extensive continental deposition which comprises part of the Gondwana series. These deposits consist mainly of sandstones and shales that accumulated in linear, local, grabenlike basins. Australia was also an emergent continental area, much as today, with local basins wherein stream and lake deposits accumulated. Narrow, restricted marine embayments were present along the western coast of the continent and on the eastern coast in Queensland. The main central area of Eurasia was a land area with only localized sedimentary basins.

It was not until the Jurassic that shallow shelf seas began to spread beyond the peripheral geosynclinal belts and inundated parts of Eurasia and North America.

Life. The great restrictions of epicontinental and geosynclinal seas in the late Paleozoic were ac-

Standard divisions and correlation of Germanic and Alpine Triassic units

Germanic facies			Alpine facies: Series	Stage	Ammonite faunal zones
Keuper	Rhät	Rhétien	Upper Triassic	Rhaetian	*Choristoceras marshi*
	Gypskeuper	Keuper of French geologists *(Marnes irisées)*		Norian	*Sirenites argonautae* *Pinacoceras metternichi* *Cyrtopleurites bicrenatus* *Cladiscites ruber* *Sagenites giebeli* *Discophyllites patens*
	Letten Kohle			Karnian	*Tropites subbullatus* *Carnites floridus* *Trachyceras aonoides* *Trachyceras aon*
Muschelkalk *(Calcaire conchylien)*			Middle Triassic	Ladinian	*Protrachyceras archelaus* *Protrachyceras reitzi*
				Anisian (Virglorian)	*Paraceratites trinodosus* *Paraceratites binodosus* *Nicomedites osmani* *Neopopanoceras haugi*
Bunter or Bundsandstein *(Grès bigarré)*			Lower Triassic	Scythian (Werfenian)	*Prohungarites similis* *Columbites parisianus* *Meekoceras gracilitatis* Unnamed zone *Gyronites frequens* *Otoceras woodwardi*

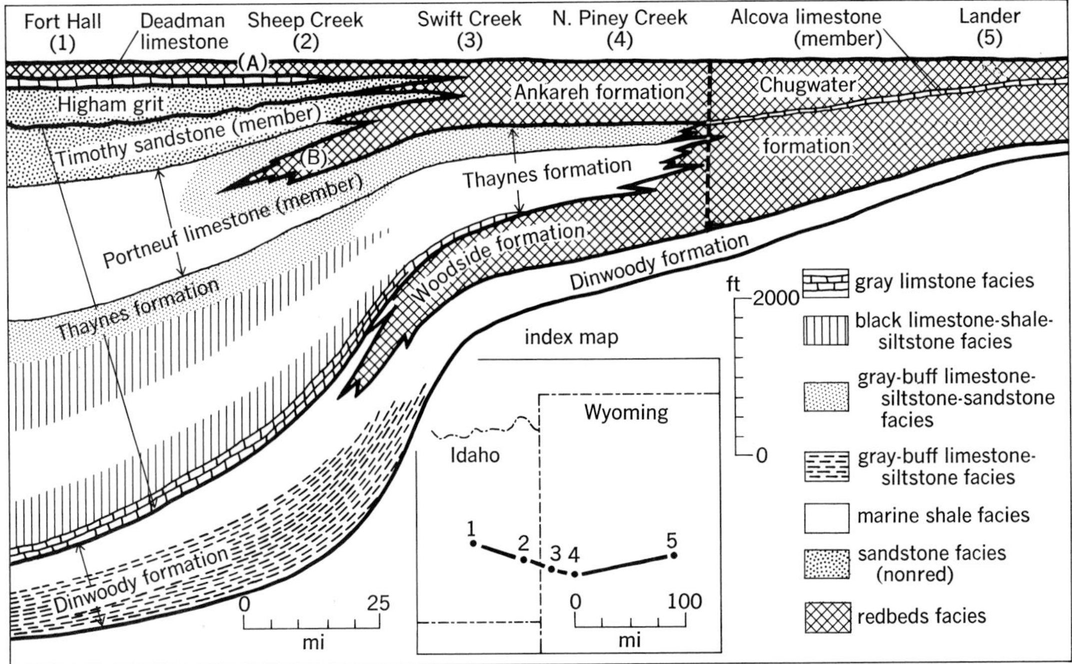

Fig. 2. Triassic deposits of Wyoming and Idaho showing thick marine deposits in the western geosynclinal region and continental redbed deposits in the eastern region of equivalent age. (*Data from B. Kummel, as used in R. C. Moore, Introduction to Historical Geology, 2d ed., McGraw-Hill, 1958*)

companied by the extinction of a large part of the marine faunas characteristic of that era. Major taxa of the corals, echinoderms, arthropods, mollusks, bryozoans, and brachiopods became extinct. This great crisis in the evolutionary history of marine animals is reflected in the composition of Triassic marine faunas. In the Lower Triassic, for instance, corals, foraminifera, and bryozoa are absent or extremely rare. Brachiopods, crinoids, sponges, asteroids, ophuiroids, echinoids, and ostracods are represented by few genera. The dominant marine animals in the Early Triassic seas are mollusca, especially the ammonites. Pelecypods, nautiloids, and gastropods are fairly common but not nearly to the extent of the ammonoids. The most characteristic feature of Early Triassic faunas is their great homogeneity wherever they occur.

Chronology and correlation of Triassic marine rocks are based mainly on the sequence and distribution of ammonites. There are roughly 400 genera of Triassic ammonoids, many of worldwide distribution; in the Lower Triassic alone there are 130 genera. This reflects a phenomenal evolutionary radiation of the group from a single surviving Paleozoic stock. The ammonoids underwent another period of crisis at the end of the Triassic and only a single stock survived into the Jurassic. This again reflects another phase of extensive regression of the seas during the Rhaetian.

The great crisis of mass extinction that affected marine invertebrate life at the end of the Paleozoic and led to such a different character of Triassic marine faunas did not affect the animals living on land. There are no sharp breaks or interruptions in the evolutionary history of the amphibians or reptiles, and none in the evolutionary development of plants.

In the early part of the Triassic both terrestrial and aquatic vertebrates were comparable to those of the late Paleozoic. During the period, however, there were marked changes among both fishes and reptiles, and by the end of the Triassic almost every one of the striking vertebrate groups which were to dominate the Jurassic and Cretaceous had appeared.

The fossil record of early Mesozoic floras is very poor, but the data available indicate that the flora was greatly impoverished, probably because of unfavorable climates, and consisted mainly of survivors from the Paleozoic. It was not until late in the Triassic that the land plants began to reflect a distinctly Mesozoic character. From the Late Triassic through the Early Cretaceous, land floras were surprisingly uniform throughout the world. This was the period of the great evolutionary radiation of the gymnosperms. *See* PALEOBOTANY; PALEOECOLOGY; PALEONTOLOGY.

North America. Geosynclinal and shelf seas were confined to the western part of the continent, covering western Mexico, the Rocky Mountain and Pacific Coast states, western Canada, and Alaska. The Arctic Islands of Canada and northern and eastern Greenland were also inundated by Triassic seas. Continental deposits of this age are known in the eastern United States from Florida (subsurface data) to Nova Scotia and in the western United States in Wyoming, Utah, Colorado, New Mexico, and Texas.

In the geosynclinal area of western North America a highly complex and varied sequence of Triassic facies, many richly fossiliferous, is present. The western portion of this vast geosynclinal tract is characterized by thick volcanic accumulations, along with a varied suite of sedimentary rocks. The eastern half of this geosynclinal region has no volcanic materials and the sedimentary facies consist of shallow-water limestones, sandstones, and shales that grade eastward into nonmarine redbed formations (Fig. 2). This general pattern of facies has persisted since the early Paleozoic. At the end of the Early Triassic the seas retreated from the eastern part of the geosynclinal region in the United States, and through Middle and Late Triassic time the seas were confined to a region west of a north-south line through central Nevada. The area east of this line, which formerly had been covered by geosynclinal seas, became the site of active continental deposition during the Late Triassic and following Jurassic. To illustrate the lithologic character of the strata in the western, volcanic part of the geosyncline, the region of southwestern Nevada is classic. The Early Triassic is represented by 3000 ft of marine shales and limestones with tuffaceous sandstones (Candelaria formation) unconformably overlying Permian and older rocks. The Middle Triassic is represented by a marine limestone and shale formation (Grantsville) and a volcanic sequence with some lenses of shales and limestones (Excelsior formation). This volcanic formation may be in part equivalent to the Grantsville formation. Resting unconformably upon these formations, the Upper Triassic consists of a varied sequence of marine shales and limestones nearly 2 mi thick which grade without change into the overlying Lower Jurassic.

From California north through southern Alaska no Lower or Middle Triassic is present, only Upper Triassic strata consisting of sedimentary and volcanic rocks. From northern Alaska eastward through the Arctic Islands of Canada to Pearyland there are many areas of Triassic rocks, as yet poorly known except that in one place or another a nearly complete sequence of marine faunas has been observed.

Eastern Greenland has various areas with marine Lower Triassic rocks of a nearshore facies that have yielded a very abundant ammonite fauna, nearly identical with that from the classic Himalayan sections.

Continental Triassic deposits are exposed along the Atlantic Coast and in the Rocky Mountain–Colorado Plateau region. The exposures in eastern North America are confined to a series of isolated troughs extending from Nova Scotia south to North Carolina. This is the area of the former Appalachian geosyncline, which underwent its final orogenesis late in the Paleozoic. After this orogenic phase a series of downwarped and faulted troughs developed within this new mountain system. Sediments from the adjoining highlands poured into the troughs, forming very thick deposits of sandstones, shales, and conglomerates representing fan, stream, and lake deposits. Associated with the sedimentary deposits are igneous rocks in the form of flows, sills, and dikes. These Triassic deposits are known as the Newark Group and are all Upper Triassic in age (Fig. 3). No marine fossils are known from the Newark Group, but land plants, fresh-water fish, and dinosaur tracks are fairly abundant.

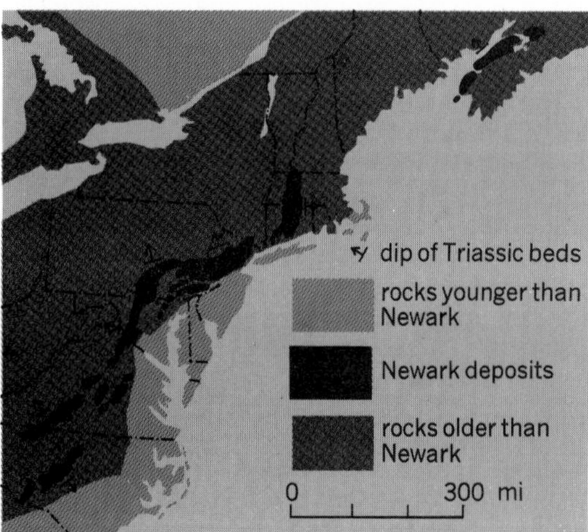

Fig. 3. Distribution of Triassic outcrops in eastern North America. The rocks occupy structurally depressed belts that are distributed from North Carolina to Nova Scotia. *(From R. C. Moore, Introduction to Historical Geology, 2d ed., McGraw-Hill, 1958)*

A sequence of terrestrial or marginal clastic rocks tentatively classified as Triassic were penetrated in 12 scattered oil test wells in southeastern Alabama, southwestern Georgia, and north-central Florida. Lithologically, the subsurface sedimentary rocks are closely similar to rocks of the Triassic Newark Group, which crop out at various localities along the Atlantic seaboard from Massachusetts to North Carolina; also, as in the Newark Group, intrusions and flows of diabase and basalt cut the Triassic (?) strata in several wells.

In the Colorado Plateau area Lower Triassic deposits are termed the Moenkopi formation; to the west and north these grade into marine formations. The Moenkopi consists of dark-red sandstone, siltstone, interbedded gypsum, and some marine limestones in its more western exposures. The vertebrate fauna contained in these beds resembles that of the Early Triassic of Spitzbergen and the Bunter of Germany. Unconformably overlying the Moenkopi and overlapping it extensively to the east is the Chinle formation, with a basal sandstone or conglomerate often separately recognized as the Shinarump. In eastern New Mexico and western Texas, equivalents of the Chinle are termed the Dockum Group. From the Wind River Mountains of Wyoming east into South Dakota, extensive Triassic redbeds are designated as the Chugwater and Spearfish formations. All of these formations are Upper Triassic in age and contain only fair fossil faunas of reptiles, amphibians, and fish. Nonmarine formations of Rhaetian age (Wingate sandstone, Moenave formation, and possibly the Kayenta formation) are widespread in the southwestern United States. [BERNHARD KUMMEL]

Bibliography: B. Kummel, in W. J. Arkell et al., Mollusca, Cephalopoda, Ammonoidea (Triassic ammonoids), *Treatise on Invertebrate Paleontology*, pt. L, 1957; B. Kummel, Paleoecology of Lower Triassic formations of southeastern Idaho and adjacent areas, in H. S. Ladd (ed.), *Treatise on Marine Ecology and Paleoecology*, Geol. Soc. Amer. Mem. no. 67, vol. 2, 1957; E. D. McKee et al., *Paleotectonic maps of the Triassic system*, USGS Misc. Geol. Inv. Map no. 1-300, 1960; F. H. McLearn, Correlation of the Triassic formations of Canada, *Bull. Geol. Soc. Amer.*, 64:1205–1228, 1953; J. B. Reeside, Jr., et al., Correlation of the Triassic formations of North America exclusive of Canada, *Bull. Geol. Soc. Amer.*, 68:1451–1514, 1957.

Tufa

A spongy, porous limestone formed by precipitation from evaporating spring and river waters; also known as calcareous sinter. Calcium carbonate commonly precipitates from supersaturated waters on the leaves and stems of plants growing around the springs and pools and preserves some of their plant structures (see illustration). Tufa tends to be fragile and friable. Tufa deposits are limited in extent and are found mainly in the youngest rocks, Pleistocene or Recent. *See* LIMESTONE; THERMAL SPRING; TRAVERTINE.

[RAYMOND SIEVER]

TUFA

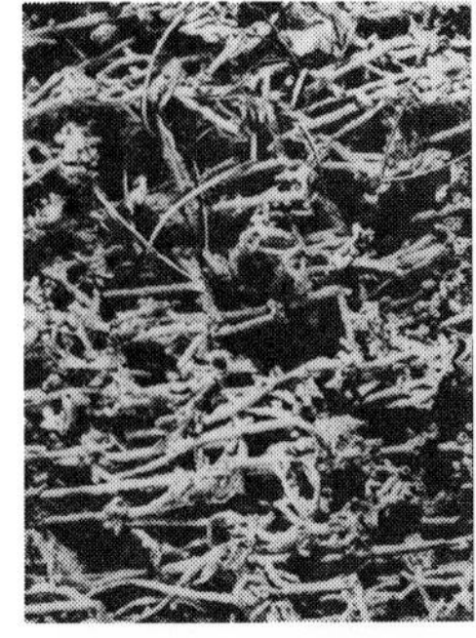

Calcareous tufa deposited on plant stems. (*From F. J. Pettijohn, Sedimentary Rocks, 3d ed., copyright © 1949, 1957 by Harper & Row, Publishers, Inc., copyright © 1975 by Francis J. Pettijohn*)

Tuff

Consolidated volcanic ash, composed largely of fragments (less than 4 mm) produced directly by volcanic eruption. Much of the fragmental material represents finely comminuted crystals and rocks. Much, however, was ejected as blobs of liquid lava which rapidly cooled to form volcanic glass. Lava brought from high pressure, within the Earth, to low pressure, at the volcanic vent, may be explosively disrupted into small masses and droplets because of sudden expansion of abundant dissolved gases. Expansion and solidification to glass may be more or less contemporaneous so that highly vesicular bodies are rapidly shattered to produce vast quantities of tiny particles (ash). Upon consolidation volcanic ash yields tuff. With increase in size of fragments, tuff passes into lapilli tuff, volcanic breccia, and agglomerate. *See* PYROCLASTIC ROCKS; VOLCANIC GLASS.

In general, the quantity of dissolved gas in lava determines the explosiveness of the eruption and, therefore, the size of fragmental particles. The coarser and heavier fragments settle near the volcanic vent, whereas the finer ash may be windborne for hundreds or thousands of miles. Much ash may fall into the ocean and become more or less mixed with ordinary sediments (shales, sandstone, and limestones) which are accumulating there. Ash that falls on dry land is readily washed away by rain and streams and deposited in the sea to form well-stratified rocks. Evidence of such transportation lies in part in the rounded character of the larger fragments. Rocks thus formed are commonly termed hybrid tuffs. *See* MARINE SEDIMENTS; SEDIMENTARY ROCKS.

Tuff may be referred to as rhyolitic tuff, trachytic tuff, or andesitic tuff, depending on the approximate bulk composition. These types may be classed as vitric, crystal, or lithic according to whether the principal constituent is glass, crystals, or rock fragments, respectively (see illustration). These types are gradational.

Vitric tuffs are products of explosive eruption of liquid lava. Comminution of the highly vesicular glass yields irregular fragments (shards) more or less bounded by concave surfaces which represent the walls of ruptured gas bubbles. Numerous vesicles may remain intact in the larger fragments (pumice). Unaltered glass shards may be clear and colorless or clouded with black magnetite dust. Basaltic glass, which is relatively uncommon, is yellow to brown. *See* OBSIDIAN.

Vitric tuffs are generally more characteristic of highly explosive eruptions, and they commonly occur at greater distances from the volcanic source. They are usually rhyolitic, less commonly dacitic, andesitic, and trachytic, and rarely basaltic.

Occasionally, highly gas-charged, viscous lavas erupt with such violence that hot, dense, incandescent clouds of finely divided particles are formed. These *nuées ardentes* emerge from the flanks of the volcanic cone or spill over the crater rim and descend the slope as an avalanche. As the interstitial steam and hot gases escape and the vitric material settles and compacts, individual glass shards are flattened and fused together to form a welded tuff or ignimbrite. *See* VOLCANO.

Crystal tuffs are chiefly products of explosive eruption of lava in which abundant crystals have already developed. Though originally well formed, many of these crystals may be broken by ejection. Many may retain thin films of glass which represent viscous lava adhered to crystal surfaces during eruption. Highly fluid lava may be wiped completely free of the crystals by friction in the atmosphere.

Less abundant constituents of crystal tuffs are crystals and crystal fragments torn from solidified rock.

Lithic tuffs are composed chiefly of relatively angular fragments produced by extensive shattering of solid rock during volcanic eruption. The source rocks are chiefly slightly older lavas and volcanic deposits which are disintegrated during subsequent eruptions. Other rock types (plutonic, sedimentary, or metamorphic) may be present in small quantities. These materials are probably torn loose from the walls of the conduit along which the lava breaks through to the surface.

Volcanic tuff and ash are highly susceptible to alteration. The glassy constituents commonly devitrify or crystallize to extremely fine-grained aggregates of silica and feldspar. Rhyolitic glass is commonly converted to clay minerals (largely montmorillonite) to form the rock bentonite. Brown basaltic glass is converted to yellow palagonite as it combines with water. The breakdown of crystal constituents in tuffs is similar to that of the constituents in most igneous rocks. *See* IGNEOUS ROCKS.

[CARLETON A. CHAPMAN]

TUFF

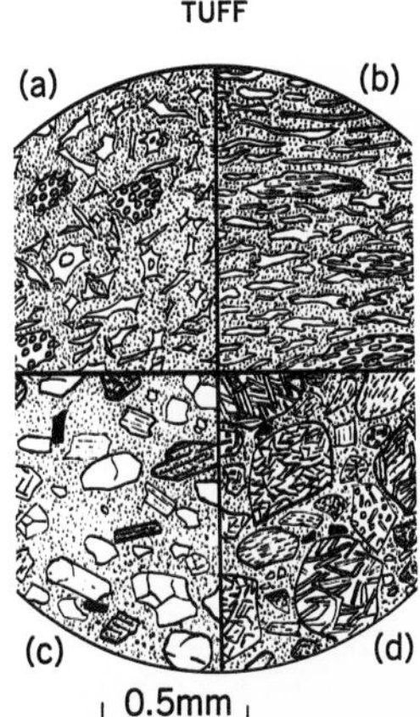

Volcanic tuff. (*a*) Rhyolitic vitric tuff with shards of glass in a matrix of dustlike particles. (*b*) Welded tuff showing deformed and flattened glass shards. (c) Crystal tuff. (*d*) Lithic tuff.

Turbidity current

A submarine flow of sediment or sediment-laden water which occurs when an unstable mass of sediment at the top of a relatively steep slope is jarred loose and slides downslope. As the slide or slump travels downslope, it becomes more fluid because of the loss of internal cohesive strength and because of inmixing of the superadjacent water.

Turbidity currents occur at the edge of the continental slope, in the vicinity of river mouths, and off prominent capes. They are triggered by earthquakes, hurricanes, floods, or simply by the bedload transport of rivers debouching at the edge of continental slopes.

Turbidity currents were first proposed as a hypothesis to account for the erosion of submarine and sublacustrine canyons. Supporting evidence was found in submarine telegraph cable breakage following the Grand Banks earthquake of 1929 and the Orleansville, Algeria, earthquake of 1954. In

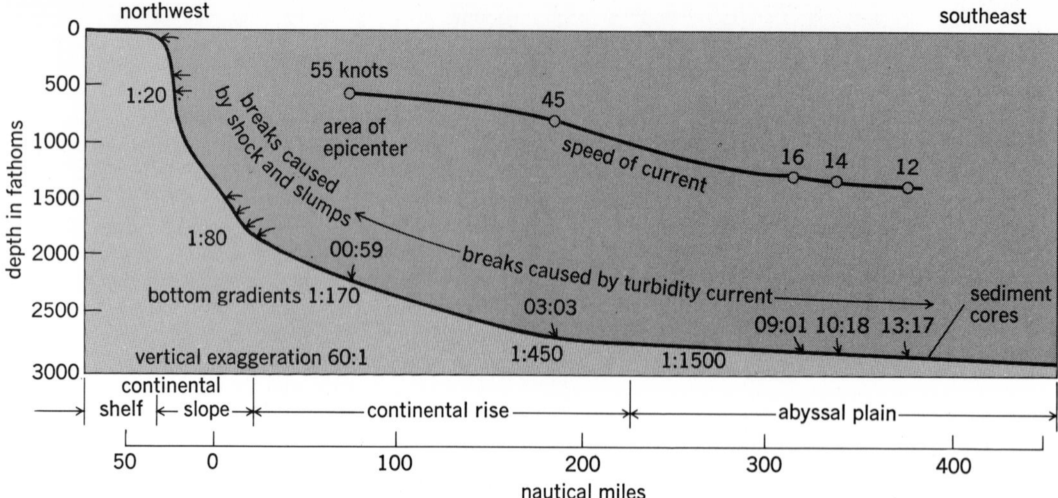

Fig. 1. Profile south of the Grand Banks showing progress of the turbidity current. Positions of 12 submarine cables are indicated by arrows. For each of the last five, the figure above the arrow indicates the time interval between the earthquake and the arrival of the current that broke the cable. Superimposed graph indicates calculated speed of the current as it passed.

both cases submarine cables were broken consecutively in the order of increasing distance downslope. All of the cables were broken in at least two widely separated places. The sections between breaks were swept away or buried beneath sediment deposited by the current (Figs. 1 and 2).

Sediment cores taken in the suspected area of deposition revealed a recently deposited, graded bed of silt and sand on the abyssal plain. Turbidity currents deposit the heaviest grains first and the finest last, and thus their deposits are graded in size from coarse at the base to fine at the top. Graded sands containing shallow-water fossils are found in the beds of submarine canyons, at their mouths, and over the vast abyssal plains of the ocean floor. The turbidity current is now generally considered the agent responsible for the erosion of submarine canyons, the building of mid-ocean canyons, and the smoothing of abyssal plains.

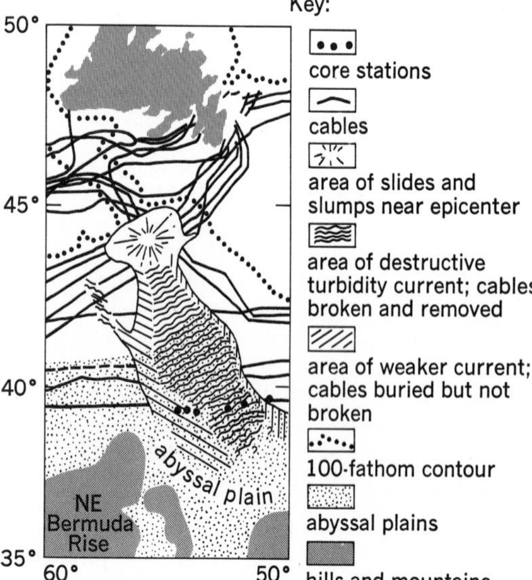

Fig. 2. Area of Grand Banks earthquake of Nov. 18, 1929. Sketch shows areas of cable breakage and cable burial, and relative position of sediment cores (from *Atlantis* cruise A180) and turbidity current.

The telegraph cable breaks which occurred after the Grand Banks and Orleansville earthquakes indicated by the time sequence of the breaks that the current attained velocities of at least 50 knots on a slope of 1:50, but slowed to 12 knots on slopes of 1:1500. Turbidity current breakage of submarine cables off the mouths of the Congo River in Africa and Magdalena River in South America also indicates that turbidity currents occur off major rivers, generally during times of flood. This implies that the high bed-load transport during flood stage can generate turbidity currents.

The lowered sea level of the glacial stages must have forced most rivers of the world to empty at or near the top of the continental slope. Under those conditions turbidity currents must have been much more important than at present, and they undoubtedly transported much of the sediment deposited in ancient geosynclines. *See* CONTINENTAL SHELF AND SLOPE; MARINE GEOLOGY; MARINE SEDIMENTS; SUBMARINE CANYON.

[BRUCE C. HEEZEN]

Bibliography: H. Charnock, Turbidity currents, *Nature*, 183(4662) :657–659, 1959; R. A. Daly, Origin of submarine canyons, *Amer. J. Sci.*, ser. 5, 31: 401–420, 1936; B. C. Heezen, The origin of submarine canyons, *Sci. Amer.*, 195(2) :36–41, 1956; B. C. Heezen and M. Ewing, Orleansville earthquake and turbidity currents, *Bull. Amer. Ass. Petrol. Geol.*, 39(12):2505–2514, 1955; B. C. Heezen and M. Ewing, Turbidity currents and submarine slumps and the 1929 Grand Banks earthquake, *Amer. J. Sci.*, 250:849–873, 1952; P. H. Kuenen, Estimated size of the Grand Banks turbidity current, *Amer. J. Sci.*, 250:874–884, 1952; P. H. Kuenen and C. I. Migliorini, Turbidity currents as a cause of graded bedding, *J. Geol.*, 58:91–126, 1950; F. P. Shepard, *Submarine Geology*, 2d ed., 1963.

Turquois

A mineral consisting of hydrated phosphate of aluminum and copper, having the composition $CuAl_6(PO_4)_4(OH)_8 \cdot 4H_2O$, and prized as a semipre-

Turquois in massive form; sample taken from Santa Fe County, N.Mex. *(Specimen from Department of Geology, Bryn Mawr College)*

cious stone. Fe^{2+} may substitute for some Cu. The name turquois means Turkish, and may have been applied because the mineral was first brought to Europe from Persia by way of Turkey. Bone turquois or odontolite, similar to turquois, is formed in fossil bones or teeth and consists of microcrystalline apatite colored by a hydrated phosphate of iron (vivianite). Bone turquois dissolved in hydrochloric acid will not give a blue color when treated with ammonia as will true turquois. The organic origin of bone turquois is readily apparent when it is observed under a microscope. *See* GEM.

True turquois crystallizes in the triclinic system as short, prismatic crystals, but these are rare. Generally turquois occurs in veinlets or as crusts of massive, dense, finely granular, concretionary, and stalactitic shapes (see illustration). The color of massive turquois ranges through sky blue, bluish green, apple green, and greenish gray. The robin's-egg blue variety is the most valued. The blue color in turquois is due to the presence of a small amount of copper; the presence of Fe^{2+} results in greenish hues. Some turquois loses color upon exposure to a dry atmosphere. The color may also be altered by treatment with ammonia or acids.

Turquois is a secondary mineral, generally formed in arid regions by the action of surface waters during the alteration of high-aluminous igneous and sedimentary rocks. Phosphoric acid may be derived from the alteration of accessory apatite, and copper may be obtained from disseminated copper sulfides to produce turquois.

Turquois of gem quality is found at one or more localities in Arizona, California, Colorado, New Mexico, and Nevada. Large deposits located in the Los Cerillos Mountains, about 20 mi southwest of Santa Fe, N.Mex., were mined very early by Indians and Mexicans and later extensively exploited by Americans. Fine-quality turquois has been mined for at least 800 years from a deposit located on the south slopes of the Ali-Mirsa-Kuh Mountains near Nishapur, Iran. Siberia, Turkistan, Asia Minor, the Sinai Peninsula, Silesia and Saxony in Germany, and France possess turquois deposits.

[WAYNE R. LOWELL]

Ulmin

A term used to designate alkali-soluble organic substances derived from decaying vegetable matter. The term has been replaced by humus and humic substances, with which ulmin is synonymous. *See* COAL; HUMUS; PEAT.

[IRVING A. BREGER]

Unconformity

The relation between adjacent rock strata whose time of deposition was separated by a period of nondeposition or of erosion. The break in the stratigraphic sequence represents a significant change in regimen (in contrast to diastem). Thus marine strata separated by a surface of subaerial erosion are unconformable, but those separated by a surface of mere nondeposition or of erosion by shifting currents which do not represent an overall change of conditions are not. Similarly, strata deposited by a river but separated by an erosion surface representing scour by that river during its normal shifting on its floodplain are not in general unconformable. The word unconformity is sometimes also used to denote the surface between the unconformable strata. *See* DIASTEM.

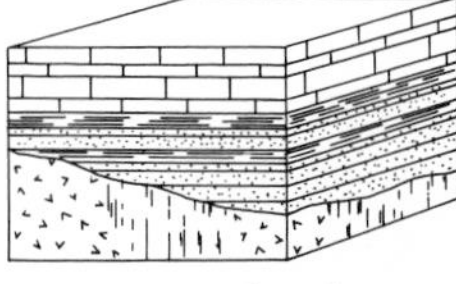

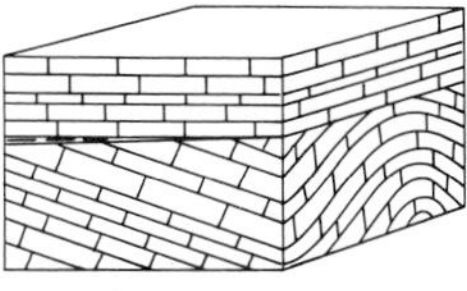

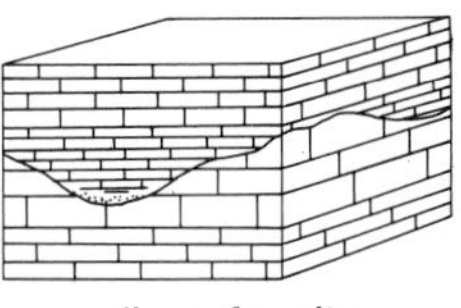

paraconformity

Four types of unconformity. *(From C. O. Dunbar and J. Rodgers, Principles of Stratigraphy, copyright © 1957 by John Wiley & Sons, Inc.; reprinted by permission)*

Types of unconformities. The term unconformity was originally used only where there was an obvious structural discordance between the rocks above and below the break, as where strata below had been tilted and then beveled by erosion before those above were deposited. This restricted usage is still advocated by some. In general, there are four main kinds of unconformable relations (see illustration), although the names of each are not universally accepted:

1. Nonconformity—rocks below the break are not stratified, such as massive crystalline rocks.
2. Angular unconformity—rocks below the break are stratified but lie at an angle to those above.
3. Disconformity—strata below are parallel to those above but are separated by an evident surface of erosion.
4. Pseudoconformity or paraconformity—strata are parallel but surface is hardly distinguishable from a simple bedding plane.

Importance in geologic record. Unconformities are commonly the chief or only record in a given stratigraphic sequence of the time elapsed between the deposition of the rocks above and below. This time is called the hiatus. Unconformities are also important in deciphering the events that took place during that time, such as erosion, deformation, metamorphism, or igneous intrusion. Nonconformable strata often point to major crustal disturbances preceding the erosion represented by the unconformity. The geologic date of the erosion interval and of preceding deformation may be somewhat narrowly defined if there is little difference in the age of the oldest rocks above and youngest rocks below the unconformable surface. *See* STRATIGRAPHY.

[JOHN RODGERS]

Universe

The sum of all the matter and energy that exist. The true scale of the universe is even now only dimly perceived by astronomers, despite tremendous advances in both observing techniques and equipment. The planet Earth seems large to humans, yet it is small compared with the Sun, which is only one of a hundred billion stars in the Milky Way Galaxy. And as large as this "island universe" is, it too is small compared with the clusters of thousands of such galaxies which populate the universe out to and (undoubtedly) beyond the range of astronomical vision.

Distance and time. The immense size of the universe and the great distances between objects introduce some fundamental limitations upon observations of remote galaxies. The basic restriction is caused by the finite speed of light in space. Light travels at 300,000 km/s; it therefore requires some 8.3 min to travel from the Sun to the Earth. Note that light travel time can therefore be used as a measure of distance (the Sun is 8.3 "light-minutes" from the Earth). A more important restriction is the following: the Sun is seen as it was when light left its surface 8.3 min ago. Therefore the past is always being seen, not the present; this "looking into the past" becomes more significant with increasing distance. The nearest large galaxy similar to the Milky Way is almost 2×10^6 light-years (1.89×10^{19} km) distant. A photograph made today of this galaxy records light which began its journey across space at a time when *Homo sapiens* was first beginning to appear on the Earth.

Most stars change very little in a hundred, million, or even a billion years, so the "time lag" caused by their distance is not a problem. Even the Andromeda Galaxy, 2×10^6 (1.89×10^{19} km) light-years away, has not changed significantly during the light travel time. Some stars have died, some new stars have formed, and the galaxy as a whole has rotated perhaps 1/100 of a turn. The overall structure and content of the galaxy have not changed, however, because light travel time is a small fraction of the evolutionary time scale for Andromeda. This is not true for very distant galaxies, however, whose light has been traveling across space for a large fraction of the age of the universe. A galaxy 10^{10} light-years (9.46×10^{22} km) away is seen as it was when the universe was half its present age; such a galaxy may have changed significantly during this interval. The universe and its galaxies can therefore be studied at earlier and earlier epochs merely by observing to greater and greater distances.

It is clear that the calculation of distances to remote galaxies is an important part of modern astronomy, because calculation of distance also fixes the relative age of the object being studied. The observation of spectral lines from distant galaxies shows them to be shifted from their normal laboratory wavelengths toward the red end of the spectrum. This so-called Doppler shift is due to the motion of these galaxies away from the Milky Way; the shift is proportional to the velocity of the galaxy. Most galaxies are moving away from the Milky Way, and fainter ones are moving more rapidly. E. P. Hubble showed that this is due to general expansion of the universe, and demonstrated that there is a relation between velocity v of a galaxy and its distance d, given by $v = H_0 d$, where H_0, the constant of proportionality, is called the Hubble constant, and is a measure of the rate of expansion. It is possible to use this relation to determine the distance to a galaxy through observation of its velocity as shown in the shift of its spectral lines.

Cosmological principle. To be able to understand what is seen at very great distances from the Earth, the following assumption, called the cosmological principle, is required: the same laws of physics operate everywhere throughout the universe. Stated another way, that sector of the universe of which Earth is a part is assumed to be representative of all other sectors (on the scale of galaxies and clusters of galaxies). The force of gravity and the ways in which x-rays are produced are assumed to be the same throughout the universe. There is as yet no experimental basis for this assumption.

Organization. On the scale of atoms, the universe is over 90% hydrogen. Most of these atoms are collected into gaseous stars which, like the Sun, are transforming hydrogen into helium by means of nuclear fusion reactions in their extremely hot interiors. This hydrogen burning is therefore slowly changing the composition of the universe. Stars are collected into vast galaxies; the Milky Way Galaxy alone contains over 10^{11} stars. Galaxies in turn are grouped into clusters of galaxies; each cluster may contain thousands of galaxies, each having billions of stars. Galaxies, or perhaps clusters of galaxies, may be considered the structural blocks of the universe, but stars are the basic unit of these blocks.

Stars and planets. The Sun is a typical middle-aged star, with a mass of 2×10^{33} (that is, some 333,000 times the mass of the Earth). Larger and smaller stars are known. The smallest stars able to sustain hydrogen burning are roughly one-tenth the size of the Sun. The largest normal stars are about 10 times the size of the Sun, although immense, swollen, evolved red giants do exist which

Model of the solar system*

Body	Diameter, km	Representation in model	Diameter in model	Distance from Sun (Earth = 1)	Distance from model "Sun"
Mercury	4,864	Apple seed	0.125 in. (3.2 mm)	0.387	124 ft (37.8 m)
Venus	12,100	Pea	0.311 in. (7.9 mm)	0.723	232 ft (70.8 m)
Earth	12,756	Pea	0.328 in. (8.3 mm)	1	321 ft (97.9 m)
Mars	6,788	Small ball bearing	0.175 in. (4.5 mm)	1.52	489 ft (149.1 m)
Asteroids	<1,000	Large grain of sand	0.026 in. (0.7 mm)	2.8	900 ft (274.5 m)
Jupiter	137,400	Tennis ball	3.533 in. (89.7 mm)	5.2	1,671 ft (509.7 m)
Saturn	115,100	Racquet ball	2.96 in. (75.2 mm)	9.5	3,054 ft (931.5 m)
Uranus	50,000	Cherry	1.29 in. (32.8 mm)	19.2	6,171 ft (1,882.2 m)
Neptune	49,400	Cherry	1.27 in. (32.3 mm)	30.1	9,675 ft (2,950.9 m)
Pluto	?	?	?	39.4	12,664 ft (3,862.5 m)
Comets	<100	Tiny grain of sand	0.03 in. (0.08 mm)	50,000	3,030 mi (4,878 km)
Nearest star		Beachball	3 ft (92 cm)	4.3 light-years	16,324 mi (26,282 km)

*In this model, the Sun is represented by a beachball 3 ft (92 cm) in diameter, and the scale is 3 ft = 1.4×10^6 km.

may become so large as to swallow most of their inner-planet system. Tiny white dwarfs the size of the Earth are also known; these cinders have been slowly cooling off after their nuclear fire died, and represent the end state for the Sun some 6 to 7×10^9 years from now. The lifetime of a star depends on its mass. The Sun will continue to burn hydrogen without significant change in its structure for about 5×10^9 years, thereby doubling its present lifetime. Heavier stars burn out faster and more violently, whereas the smaller, lighter stars will live 100 times longer than the Sun.

Stars are, in general, not isolated in the Galaxy. Most have companions of some kind, and most companions are other stars. Some companions are too small to sustain nuclear reactions; these are called planets. In addition to the nine planets which move about the Sun, there are 33 known satellites of the planets, more than 2000 well-observed asteroids (of 50,000 believed to exist), and an unknown number of comets and meteoroids. So tiny are planets, their satellites, and the comets compared with the Sun that these small companions are nearly undetectable from a distance of 10 light-years (9.46×10^{13} km).

The table shows the solar system in terms of a model in which the Sun is represented by a beachball 3 ft (92 cm) across. On this scale the Earth is the size of a pea, 321 ft (97.9 m) from the Sun (roughly the distance across a football field)! The Moon is only a small grain of sand some 10 in. (25.4 cm) from the Earth. The largest planet, Jupiter, is represented as a tennis ball 3.5 in. (89.7 mm) across and almost 0.3 mi (509.7 m) from the Sun. Comets, which in reality are 10-km-diameter balls of frozen methane, water, and other gases that occasionally come into the inner solar system (but generally lie 50,000 times as far from the Sun as does the Earth), are represented by pinpoints more than 3000 mi (4878 km) from the Sun. This cloud of comets represents the edge of the solar system.

In terms of travel, the solar system seems large: signals from a spacecraft located at the distance of Pluto would take nearly 6 hr to reach the Earth. The radius of the cloud of comets is about 7.5×10^{12} km, or about 290 light-days; this effectively measures the size of the solar system. By comparison, the nearest star (another beachball in the model of the table) is 4.3 light-years from the Sun (about 16,300 mi or 26,282 km in the model). The entire solar system is only about two-tenths the distance to the nearest star.

Galaxies. The number of planetary systems in the Galaxy is not known, but estimates range into the hundreds of millions. But each of these systems is small compared with the galaxy's diameter of 100,000 light-years (9.46×10^{17} km). The 10^{11} stars of the Galaxy are spread out in a thin disk with a central bulge; only a small percentage of the stars lies above or below this plane. For the Galaxy as a whole, the average separation between stars is a few light-years. The Sun lies about 30,000 light-years (2.84×10^{17} km) from the dense galactic center, on the inner edge of one of the spiral arms (which contain hot young stars, gas, and dust) that characterize this type of galaxy. So great is this distance from the center (more than 36,000 times the radius of the solar system) that the Sun requires some 2.5×10^7 years to completely circle the Galaxy. That is, the solar system has orbited only 18 times since the Sun was born (compared with nearly 4.5×10^9 trips of the Earth around the Sun).

Fig. 1. The great spiral galaxy in Andromeda (M31, NGC 224) and its two small elliptical companions (NGC 205 and 221), photographed with the 48-in. Schmidt telescope. (*Hale Observatories*)

While spiral galaxies like the Milky Way Galaxy are composed mostly of stars, perhaps 15% of the total mass (of 10^{44} g) may be in the form of dust and gas (Fig. 1). This is the material from which new stars are formed, a process which continues today in spiral galaxies. No such star formation is observed in elliptical galaxies, however, which are spherical or oval systems of dim reddish stars with almost no gas or dust. Elliptical galaxies outnumber spirals; especially common are dwarf ellipticals of a few million stars spread over 6000 light-years (5.67×10^{16} km) or so. These small galaxies are seen often as satellites to larger systems such as the Andromeda Galaxy. Much rarer are the giant ellipticals with 10^{13} stars (or 100 times the number in the Milky Way). These giants are often found at the center of large clusters of galaxies, and may be three or four times larger than the large spirals.

One simple way to characterize these galaxies is by their mass-to-luminosity ratio (M/L). Both the mass and the brightness of a galaxy depend on the stars in that galaxy; if most of the stars are of the dim, red type, M/L will be a high number. Elliptical galaxies are of this type, and have M/L about 4–10 (M/L for the Sun is 1). Giant ellipticals, with their greater mass, may have $M/L = 30$. A spiral galaxy such as the Milky Way has a large number of hot and bright blue stars, so it is brighter compared with its size than an elliptical, and has $M/L < 1$. The M/L effectively measures the stellar content of a galaxy.

Clusters of galaxies. Essentially all the stars in the universe are found in galaxies (there are effectively none between the galaxies). Most galaxies themselves are not solitary, but are arranged into clusters, which may be composed of a number of subclusterings. The Milky Way Galaxy is part of a collection of about 24 galaxies called the Local Group. This cluster is some 3×10^6 light-years (2.838×10^{19} km) across and is dominated by two large spiral galaxies, the Milky Way and the Andromeda galaxies. A third, somewhat smaller spiral, four irregular, and four regular but small elliptical galaxies, are the major members of the Local Group. There are also an unknown number of dwarf ellipticals present (nine have been detected in a limited search). Observations suggest the presence of two or three wanderers that have penetrated the Local Group; one of them may be a giant elliptical, although observations of it are difficult due to obscuration by the dust of the Milky Way.

Small groups such as the Local Group are probably common, but much larger clusters also exist. The nearest of these, the Virgo Cluster, contains several thousand galaxies of all types within a diameter of about 6×10^6 to 7×10^6 light-years (5.67×10^{19} to 6.62×10^{19} km). The Virgo Cluster, which lies some 5×10^7 light-years (4.73×10^{20} km) from the Local Group, is a typical loose or irregular cluster of galaxies, in which spirals and normal ellipticals are roughly equal in number. In richer, denser regular clusters, spirals are absent, and the number of elliptical galaxies may be 10,000. For such a cluster, the density of galaxies near the center may be a thousand or more times the density of galaxies in the Local Group, making such concentrations the most tightly packed groupings in the universe, relative to the size of their members, as discussed below. The nearest rich regular cluster is called Coma; it lies some 3×10^8 light-years (2.84×10^{21} km) away and is at least 10^7 light-years (9.46×10^{19} km) across.

Thousands of clusters of galaxies have been observed, and some astronomers suggest that these are arranged into clusters of clusters, or superclusters. The Local Supercluster would then have some 100 clusters as members, including the large Virgo Cluster and the Local Group, and it would be some 10^8 to 1.5×10^8 light-years (9.46×10^{20} to 1.419×10^{21} km) across.

One way to compare these objects and how they are grouped is through the computation of a "packing factor," which may be defined as the separation of two objects divided by their mean diameter. This factor shows how closely spaced are objects relative to their size. For example, within the Milky Way Galaxy the average separation between two stars is about 2 light-years, or 1.9×10^{13} km. The diameter of a star such as the Sun is 1.4×10^6 km. Therefore stars are on the average 1.4×10^7 times farther apart than they are large. Compare this with the packing factor for galaxies in the Local Group: The Andromeda spiral galaxy is about 100,000 light-years (9.46×10^{17} km) across and is about 2×10^6 light-years (1.89×10^{19} km) distant. Therefore, within the Local Group spirals are about 20 times as distant as they are large, or 700,000 times more closely packed than the stars in the Milky Way! Inside dense clusters, such as the Coma Cluster, galaxies are spaced a thousand times more closely than this. The same type of computation for packing of clusters of galaxies suggests that clusters are 10 or less times more distant than they are large, at least within the Local Supercluster.

Properties. The universe is dynamic. The early observation that all galaxies are receding led to the now well-accepted conclusion that the universe is expanding. This expansion implies that the universe is evolving. Observations of quasars support this view. These immensely powerful sources of radiation all seem to lie at great distances from Earth. The lack of proved nearby quasars suggests that quasars were a part of an earlier epoch of the universe. Also, very remote clusters of galaxies seem to be denser than nearby clusters, as would be expected if the universe had been smaller in the past, as implied by the expansion.

There are several ways to determine a rough value for the age of the universe. The universe must be older than the Milky Way Galaxy, whose age is estimated at something like 1.4×10^{10} years, based on studies of the oldest known stars. This represents a lower limit to the age of both the Galaxy and the universe. (The Galaxy may, in fact, be somewhat older than the oldest observed stars.) The expansion of the universe itself provides an upper limit to age. Extrapolating the expansion backwards (assuming that the observed rate has been constant since the expansion started), all of the observed galaxies would have started from the same point about 2×10^{10} years ago. The expansion rate was probably greater in the early history of the universe; the age of the universe is therefore somewhat less than the 2×10^{10} years given by the Hubble expansion time.

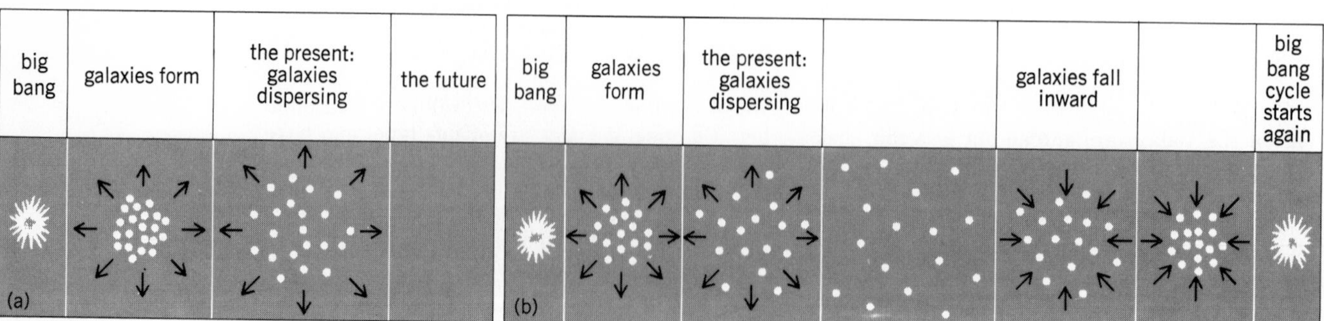

Fig. 2. Comparison of the open and oscillating universes. (*a*) In the open universe, matter continually thins out, and the galaxies continually move farther apart. (*b*) In an oscillating universe, expansion is followed by contraction until another "bang" causes renewed expansion. (*From R. Jastrow and M. H. Thompson, Astronomy: Fundamentals and Frontiers, pp. 269 and 277, 3d ed., copyright © 1977 by John Wiley & Sons; reprinted by permission*)

The universe is at least as large (and as old) as the most distant galaxy or cluster of galaxies observed thus far. But since more and more distant galaxies are still being found, it can only be said that the universe is larger than about 8×10^9 light-years (7.56×10^{22} km) across.

Evolution of the universe. One explanation for the observed expansion of the universe is the "big bang" theory, in which all of the matter and radiant energy of the universe were initially compressed into a single object. At the assumed temperatures of some trillions of degrees, matter could not have existed in its present form, but consisted of subatomic or even subnuclear particles. For reasons unknown, this "primordial egg" (which may have been as small as a planetary system) exploded, hurling matter and radiation outward in a slowly decreasing expansion that led to rapidly dropping temperatures. Ten seconds after the explosion, normal protons, neutrons and electrons could exist in the billion-degree environment. After 100 s of expansion, protons and neutrons began combining at temperatures of 100,000,000 K to form deuterium, a heavy isotope of hydrogen. Reaction with additional protons in the dense conditions produced helium. It has been estimated that some 20% of the original protons were converted into helium in this way within the first several minutes of the universe's existence. After this the expanding matter cooled below 10,000,000 K, making further nucleosynthesis (manufacturing of new elements) impossible. The other 90 natural elements which are part of the universe owe their existence to the nuclear fusion reactions which would occur much later inside stars.

Steadily dropping temperatures accompanied the continuing expansion for the next several million years. The density of the universe remained high enough to trap the radiant energy of the preexplosion time until temperatures fell to about 5000 K. At this point electrons combined with protons to produce neutral hydrogen atoms; as a result radiation was decoupled from matter and was released to flood the universe.

A small remnant of this fireball radiation is in fact observed to be uniformly filling the universe. As with any radiation from a very distant source, the original high-temperature energy has been degraded or red-shifted to much longer wavelengths. The radio observations of this 3 K black-body radiation (as though the universe were at a temperature of 3 K) constitutes the most conclusive proof that some type of "big bang" did indeed occur. The remnant background of the original fireball cannot be explained by alternate theories of the universe which assume a steady state or nonevolutionary history.

Only after the universe had expanded to where the temperature dropped to about 300 K could galaxies begin to form out of gravitational condensations of the expanding gas. Within these large condensations, smaller centers of contraction became the stars which today constitute the major fraction of a galaxy's mass.

Future of the universe. The current expansion of the universe is a product of the "big bang" explosion, and the present rate of expansion should be less than the original rate because of the gravitational slowdown produced by the matter of the universe when it was still small and dense. How long the universe will continue to expand will depend on whether or not the explosion was stronger than the force of gravity due to all the matter of the universe. If the explosion was more powerful than the gravity of the universe, then the universe has "escape velocity" and will expand forever. Such an "open" universe is infinite in both time and space. Galaxies will continue to separate, and the average density of the universe will continue to decrease. The other extreme is given by the reverse situation: the total density of matter in the universe is so great that its gravity is greater than the explosion-induced expansion. In this case the universe will eventually come to rest, then reverse and fall back on itself. Galaxies will approach one another, crowding together until they lose their separate identities. Matter will revert to basic atomic particles and then to more fundamental particles. Radiation will become trapped within the shrinking universe, temperatures will rise, and conditions will approach those of the previous "big bang." It seems reasonable that the universe will then "bounce" or explode in another "big bang" and start over. Such an "oscillating" universe is finite in extent but infinite in time; there is no way of knowing which oscillation is going on at present, as records of the previous oscillations are destroyed in each new "bang" (Figs. 2 and 3). The third but unlikely possibility is that the expansion exactly balances the force of gravity. The universe will then expand and stop, remaining motionless for all time.

To decide which of these three possible futures is most likely, it is necessary to determine the density of matter in the universe. Direct estimates from observed galaxies fall short of the mass needed to close the universe (halt the expansion) by a factor of 1000. Attempts to account for "invisible" mass in the form of nonluminous stars, gas, and black holes (which can be estimated from the motions of galaxies within a cluster) indicate that some clusters of galaxies contain 30 or more times as much matter as would be seen in the galaxies as stars. This still leaves the universe short by more than 10 times. Indirect clues are available from the observed amount of "big bang" deuterium, which depends on the density of matter at the time of formation. Again there seems to have been at least 10 times too little matter to halt the expansion. Finally there is the change in the expansion of the universe itself: the greater the gravity, the greater the slowdown of the expansion rate with time. Observations of the very distant clusters of galaxies show that the expansion was only slightly faster

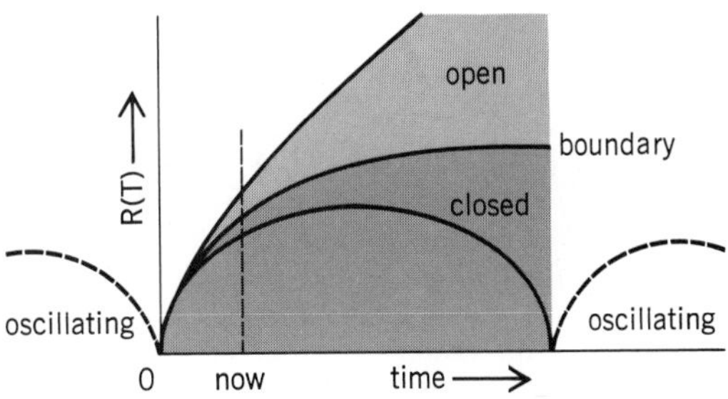

Fig. 3. Comparison of the open and closed universes. $R(T)$ effectively measures the radius of the universe, or the distance between two galaxies. *(From J. M. Pasachoff, Contemporary Astronomy, p. 555, W. B. Saunders, 1977)*

in the past, which implies that the total gravity of the universe is too small to halt the motion of the galaxies. The universe thus appears to be open, and the best evidence suggests that the expansion will continue. Refutation of this conclusion would necessitate the discovery of previously unknown large amounts of mass in the universe.

[HERBERT FREY]

Bibliography: R. Jastrow and M. H. Thompson, *Astronomy: Fundamentals and Frontiers*, 3d ed., 1977; I. R. King, *The Universe Unfolding*, 1976; J. M. Pasachoff, *Contemporary Astronomy*, 1977.

Uraninite

URANINITE

Uraninite crystals in pegmatite, Chester, PA. *(Specimen from Department of Geology, Bryn Mawr College)*

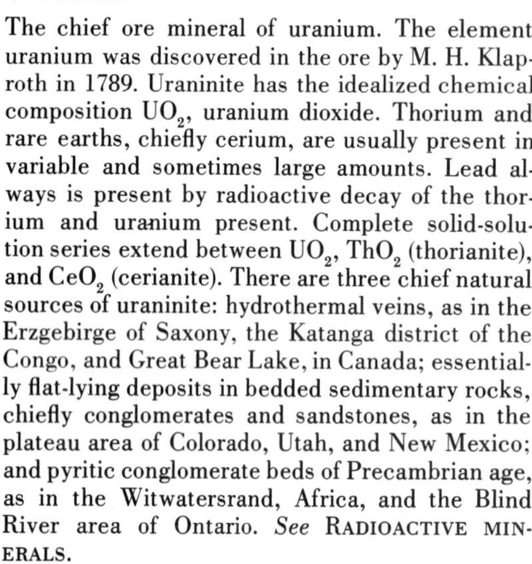

The chief ore mineral of uranium. The element uranium was discovered in the ore by M. H. Klaproth in 1789. Uraninite has the idealized chemical composition UO_2, uranium dioxide. Thorium and rare earths, chiefly cerium, are usually present in variable and sometimes large amounts. Lead always is present by radioactive decay of the thorium and uranium present. Complete solid-solution series extend between UO_2, ThO_2 (thorianite), and CeO_2 (cerianite). There are three chief natural sources of uraninite: hydrothermal veins, as in the Erzgebirge of Saxony, the Katanga district of the Congo, and Great Bear Lake, in Canada; essentially flat-lying deposits in bedded sedimentary rocks, chiefly conglomerates and sandstones, as in the plateau area of Colorado, Utah, and New Mexico; and pyritic conglomerate beds of Precambrian age, as in the Witwatersrand, Africa, and the Blind River area of Ontario. *See* RADIOACTIVE MINERALS.

Uraninite is isometric in crystallization. The crystal structure is of the calcium fluoride type. Natural uraninite always contains U^{6+} in variable amounts, in addition to U^{4+}, and the formula may be written $(U_{1-x}{}^{4+}U_x{}^{6+})O_{2+x}$. The extra oxygen needed for valence compensation is housed interstitially in the vacant eight-coordinated positions of the crystal structure. The mineral sometimes is found as cubic or octahedral crystals, especially in its occurrences in pegmatites (see illustration). It more often occurs as fine-grained masses or disseminations, or as crusts with a botryoidal surface and a layered, radial, fibrous to dense structure. The latter variety, also known under the name pitchblende, occurs chiefly in hydrothermal veins. The color of uraninite is black, grading to brownish black and dark brown in the more highly oxidized material. The luster of fresh material is steel-gray. The hardness is 5½–6 on the Mohs scale. The specific gravity of pure UO_2 is 10.9, but that of most natural material is 9.7–7.5. *See* LEAD ISOTOPES, GEOCHEMISTRY OF; ORE AND MINERAL DEPOSITS.

[CLIFFORD FRONDEL]

Varve

A distinctive, thin annual sedimentary layer, the lower part consisting of coarser, lighter-colored clay and silt that was deposited in summer, and the upper of a finer-grained, darker clay deposited in winter. Numerous successive varves, generally less than 1 in. (2.5 cm) thick, accumulated in temporary lakes near melting glaciers. Thicker and thinner varves at different places can be matched like tree rings. In Sweden, geologic history has been traced back about 18,000 years in this way. Similar layers occur in ancient rocks, but it is difficult to determine whether the layers are annual; many of them represent longer or less regular sedimentary cycles. *See* SEDIMENTATION.

[J. MARVIN WELLER]

Vermiculite

A clay mineral constituent of clay materials. Vermiculites are similar to the montmorillonites in having an expanding structure. They differ, however, in that the expansion can take place only to a limited degree. When rapidly heated, vermiculite produces a lightweight expanded product that is widely used for thermal insulation. *See* CLAY MINERALS.

Structural properties. The structure of vermiculite consists of trioctahedral mica sheets separated by double water layers and is unbalanced by substitution of aluminum for silicon in the tetrahedral layer. The resulting charge deficiency is satisfied by exchangeable cations, usually magnesium, which occur chiefly between the mica layers.

The expansion of the vermiculite structure is restricted to two water layers. The mineral quickly expands by adsorbing two layers of water after heating to temperatures as high as 500°C. This ability to rehydrate disappears gradually above 550°C, and is completely lost at about 700°C.

The adsorbed water layers have a definite structure, but its exact nature has not been determined. As the mineral dehydrates, the structure of the water and its coordination with the oxygens of the mica structure change in stepwise fashion.

Nonstructural properties. Vermiculites show a considerable range in chemical composition. Their composition may be like that for some montmorillonites, in which case the only difference would be in the larger particle size of the vermiculites. Like the montmorillonites, the vermiculites have a high cation-exchange capacity (150 milliequivalents per 100 g). They also adsorb certain organic molecules between the mica layers, but the adsorbed organic layer is thinner and less variable than in montmorillonite.

Vermiculite is frequently listed as an alteration product of biotite mica, and it is often present along with chloritic mica as a minor constituent in ancient sediments.

[RALPH E. GRIM; FLOYD M. WAHL]

Volcanic glass

A natural glass formed by rapid cooling of lava. The material is opaque except on thin edges and occurs in a variety of colors. Shades of red, brown, black, gray, or green may be displayed in a uniform, banded, or variegated fashion. Because of its conchoidal fracture (see illustration), volcanic glass was highly prized among many primitive peoples as a material for tools and weapons.

Types of natural glass. Most natural glasses are chemically equivalent to rhyolite, into which they may grade. Types corresponding to trachyte, dacite, andesite, and latite are much less common. Basaltic glass generally is given the name tachylite. The index of refraction and specific gravity of natural glass (see table) may aid greatly in its identification.

Obsidian is generally a black variety with brilliant luster. Pitchstone, however, has a dull or

Specimen of volcanic glass showing smooth curved, or conchoidal, fracture surfaces. High-silica lavas frequently solidify without crystallization because of the high viscosity and rapid chilling. (*Ward's Natural Science Establishment, Inc.*)

pitchy luster and is frequently brown, green, or gray. Pumice is an amazingly light rock and consists of frothlike glass in which the tiny gas bubbles commonly may not exceed 1 mm in diameter. Perlite is a gray-to-green glass with abundant spherical cracks which cause it to disintegrate into tiny pelletlike masses. *See* OBSIDIAN; PERLITE; PITCHSTONE; PUMICE.

Texture and structure. In addition to crystallites, most volcanic glasses carry microlites (microscopically tiny crystals). Many of these may show skeletal form. Somewhat larger and usually well-formed crystals are called phenocrysts. Their presence gives the glass a porphyritic texture. They consist chiefly of quartz, alkali feldspar, and plagioclase and less commonly of mafics (biotite, hornblende, or pyroxene). As their number increases, these porphyritic glasses pass into vitrophyre. *See* PHENOCRYST; PORPHYRY.

Spherulites and lithophysae are developed strikingly in many glassy rocks. A common feature is fluidal structure resulting from flowage of the viscous lava. It consists of wavy streaks and bands of spherulites, phenocrysts, microlites, and crystallites. In some glasses conspicuous banding is due to alternating layers of different-colored glass or of glass and pumice.

Water content. The water content of natural glass is highly variable. It is usually less than 1% in obsidian, between 3 and 4% in perlite, and between 4 and 10% in pitchstone. The water in obsidian may represent only part of that contained in the original melt. Much water may have been lost as vapor when the lava was erupted. Some of the water in perlite and much of that in pitchstone is believed to have been derived by absorption from the sea or from wet sediments into which the viscous lava came. Since glasses commonly are associated intimately with crystalline material, some of the water driven out of the crystallized portions may have been absorbed by the glassy portions.

Properties of natural glass

Type of glass	Average index of refraction	Average specific gravity
Rhyolitic	1.495	2.37
Trachytic	1.505	2.45
Dacitic and andesitic	1.515	2.50
Basaltic (tachylite)	1.575	2.77

Formation and composition. The formation of glass from magma (rock melt) is favored by rapid cooling and high viscosity. Rapid cooling is most readily attained at or near the Earth's surface; consequently, glassy rocks are found as lava flows and shallow intrusives. High viscosity is characteristic of lavas rich in silica and potassium (rhyolitic); therefore, most glassy rocks are rhyolitic. Chemical composition may control the formation of glass in another respect. The composition of most natural glass is close to the quartz-alkali feldspar cotectic. This permits such lavas to cool to relatively low temperatures at which viscosity is high without crystallizing. If solidification can be forced upon the melt at this stage, crystallization will be impeded and glass will form. Such ideal conditions may be brought about by sudden eruption to the surface, where chilling and loss of volatiles (fluxes), will occur. *See* FELDSPAR; LAVA; MAGMA.

Glass tends to convert (devitrify) spontaneously to a crystalline aggregate. Devitrification commonly starts at the surfaces of phenocrysts and spherulites or along cracks, and it spreads outward. In time the entire mass of glass may be transformed to an aggregate of microscopically fine crystals of quartz, tridymite, and alkali feldspar. This explains why geologically ancient glasses are so rare and why glasses are most common in Tertiary and younger rocks. *See* IGNEOUS ROCKS; RHYOLITE.

[CARLETON A. CHAPMAN]

Volcano

An opening in the Earth's crust through which magma (molten rock) or gases of magmatic origin, or both, issue. Only the products of volcanic activity are described here. For a discussion of processes of volcanism *see* VOLCANOLOGY. *See also* MAGMA; PETROLOGY.

Types of volcanic vents. Fissures in the Earth's crust, produced by orogenic (mountain-making) or other diastrophic forces, commonly constitute the channels through which magma rises toward the surface. Magma consolidating in the fissure forms a dike, but that reaching the surface produces a fissure eruption (Fig. 1). The fissures generally range in width from less than a foot to 10 ft, though many wider ones are known. During fissure eruptions of fluid lava, rows of fountains of liquid lava may extend for thousands of feet, or even several miles, along the fissure. The fissures may be independent of volcanic mountains or may occur on the flanks of large volcanic cones (Fig. 2), especially on shield volcanoes. *See* DIASTROPHISM.

Other volcanic conduits are nearly cylindrical. Consolidation of material in these produces volcanic necks, or bosses. Typically, eruption of material from a cylindrical conduit builds a conical mountain terminating in a summit depression, or crater. The eruption may take place from fissure vents on the flanks of the cone (lateral eruption),

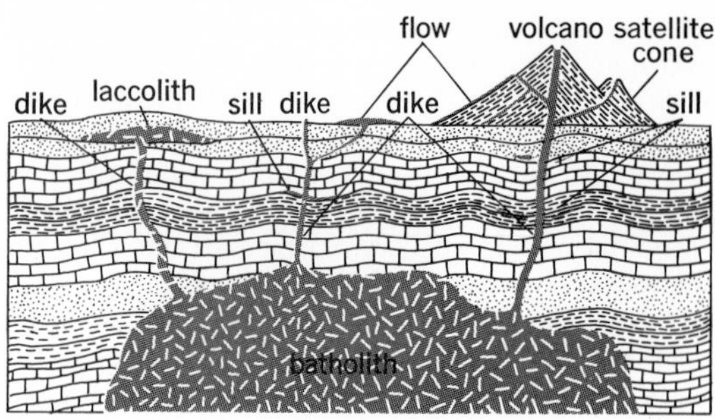

Fig. 1. Cross-sectional diagram illustrating plutons and igneous activity in relation to the geological structures and the surface forms.

Fig. 2. Lava flow issuing from vents, near base of Faralon de Pajaros, an active volcano in the Marianas Islands. (*Official U.S. Navy photo*)

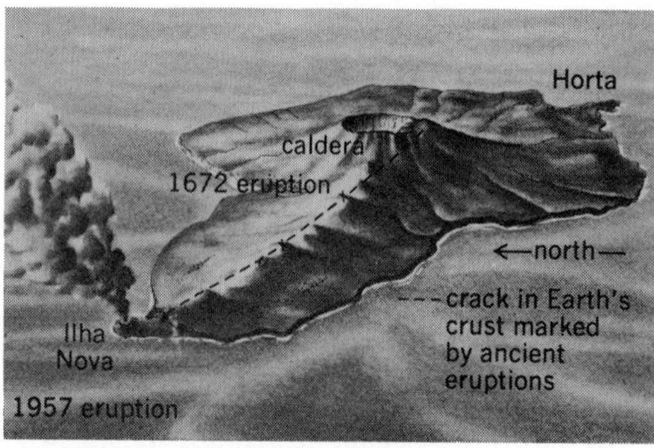

Fig. 3. Volcanic eruptions, localized along a fissure in the Earth's crust, island of Fayal in the Azores. (*National Geographic Magazine*)

but more commonly it occurs at the main, or central, vent (central eruption). Cones built by central eruptions often are arranged in lines, suggesting that they have been localized by fissures in the underlying rocks (Fig. 3).

Volcanic products. Molten rock extruded onto the Earth's surface is called lava. It congeals into either volcanic glass, by quick chilling, or solid rock, by chilling and crystallization. The rock is also termed lava and sometimes lava rock. Depending on its viscosity during extrusion, molten rock may pile up steeply over the vent or pour out as streams known as lava flows. Lava torn into pieces and thrown into the air by explosions is called pyroclastic (fire-broken) material. *See* LAVA; VOLCANIC GLASS.

The character of volcanic eruption depends largely on the viscosity of the liquid lava and the amount and condition of the gas in it. In general, mafic (basaltic and similar) lavas are less viscous and contain less gas than more silicic (dacitic and similar) lavas. Eruptions of mafic lavas commonly are quiet, with little explosion, and produce mostly lava flows; on the other hand, eruptions of silicic lavas commonly are explosive and yield predominantly pyroclastic material. Some eruptions produce only pyroclastic material.

Lava flows. Several types of lava flows are recognized. Flows with smooth or hummocky, gently undulating surfaces and with crust locally wrinkled into ropelike forms are called pahoehoe. Ovoid mounds, a few feet high and a few tens of feet long, formed by buckling up of the crust, are known as tumuli. Flows with very rough irregular surfaces covered with jagged spinose fragments resembling furnace clinker are called aa. Flows in which the fragments that constitute the upper part of the flow are fairly smooth-sided polygons are called block lava. Fluid mafic lavas characteristically form pahoehoe or aa flows, whereas less fluid lavas such as andesites more commonly form block lava flows. All three types commonly are vesicular, and may contain amygdules. *See* AMYGDULE.

Where pahoehoe flows pour into water they may form heaps of irregular ellipsoids, in cross section somewhat resembling so many sacks of grain or pillows. These masses are known as pillow lava.

Emergence at the surface of highly gas-charged silicic lava may be attended by a sudden frothing up as the contained gas comes rapidly out of solution. The rapid expansion of the gas may tear the froth into an almost infinite number of tiny shreds, each of which quickly chills to volcanic glass. Continued ebullition of gas results in a mass of small solid or semisolid fragments, each surrounded by an envelope of still-expanding gas which pushes against all adjacent envelopes. Thus the solid fragments are isolated from contact with each other, and the entire mass is given an expansive quality which is augmented by the expansion of heated air entrapped in the moving mass. The result is a very mobile suspension of incandescent solid fragments in gas which may flow with great speed down steep slopes and spread out to great distances, forming voluminous deposits with nearly horizontal surfaces. When these ash flows come to rest, they often are still so hot that the fragments of glass stick together or even merge in the center of the deposit to form a layer of solid black obsidian. The resulting deposits are known as welded tuffs or ignimbrites. *See* TUFF.

Similar expansive gas-solid suspensions may be formed by explosions, or explosion-induced collapses of active domes at the summit of volcanic cones, producing exceedingly mobile avalanches of hot gas-charged fragments that rush down the mountain side at speeds as great as 100 mph, capped by a spectacular rolling, convoluting cloud of dust. Because of their incandescent nature, these have been called glowing avalanches, or *nuées ardentes*.

Pyroclastic materials. At depth, magma contains gas in solution, but as it rises into regions of lesser pressure near the surface of the Earth, the gas starts to come out of solution and tends to escape from the liquid. From fluid lavas the gas escapes easily and quietly, with little explosion; but in more viscous liquids the gas may accumulate considerable pressure before it escapes and bursts forth in strong explosions. Gradual crystallization of a magma beneath the surface may concentrate the gas in the residual fluid portion and, depending on the strength of the roof of the containing chamber, gas pressures may become very high and lead to violent explosions. The explosions may throw into the air shreds of molten lava (magmatic, or essential, ejecta), fragments of lava of previous eruptions of the same volcano (accessory ejecta), or solid fragments of still older rocks (accidental ejecta).

Pyroclastic materials are classified also by their size, shape, and consistency. Gobs or drops of material still liquid enough to assume rounded or drawn-out forms during flight are known as lapilli if they are 4–32 mm in average diameter and bombs if they are greater than 32 mm. If they are still sufficiently plastic to flatten out when they strike the ground, they are called Hawaiian, or cow-dung, bombs. If they are sufficiently consolidated to retain their shape on impact and have acquired during flight a shape resembling a spindle or a summer squash, they are called spindle, or fusiform, bombs. And if they are shaped like a string or ribbon, they are called ribbon bombs.

Irregular fragments of frothy lava of bomb or lapilli size are called scoria or cinder; if the fragments are sufficiently plastic to flatten or splash as they hit, they are called spatter. The still-molten surfaces of fragments of the spatter type often adhere to each other to form welded spatter, or agglutinate. Angular fragments larger than 32 mm either solid or too viscous to assume rounded forms during flight are known as blocks. They may be essential, accessory, or accidental. Accumulations of blocks are called breccia. Ejecta smaller than 4 mm are called ash, and those smaller than 1/4 mm are called dust. Indurated (hardened) volcanic ash or dust is called tuff. *See* PYROCLASTIC ROCKS.

Volcanic dust. Violent explosions may throw dust high into the stratosphere, where it may drift for thousands of miles. Ash from eruptions of Iceland has fallen in the streets of Moscow. Studies have shown that the particles in the eruption cloud are mostly angular bits of lava, many of them glassy; but some glass spheroids also are present as well as liquid droplets of hydrous solutions of sulfuric acid and sulfates and chlorides. The common volcanic gases H_2S and SO_2 are oxidized to sulfate in the air. Most of the solid fragments in the cloud settle out within a few days, and nearly all within a few weeks, but some finely divided material may remain suspended in the stratosphere for more than a year. Clouds of this sort from the eruption of Agung volcano, in Bali, in 1963, and the still greater explosion of Krakatoa, between Java and Sumatra, in 1883, produced spectacular sunsets all over the Earth for many months. The slow rate of settling of the particles in these clouds and the constancy of the height of the layer that produced the optical effects suggest to M. P. Meinel and A. B. Meinel that the suspension that persists for long periods is largely a sulfate aerosol. At times, volcanic dust in the atmosphere, together with the sulfate aerosol layer, may screen out solar radiation sufficiently to appreciably lower the Earth's surface temperature and affect the weather, possibly even enough to be an important factor in causing world glaciation.

Volcanic mudflows. Mudflows are common on volcanic mountains where pyroclastic material is abundant. They may form by eruptions ejecting the water of a crater lake, or by breakdown of the confining wall of the lake, the water sweeping down the mountainside and forming a slurry of the loose material; or they may form by hot or cold avalanches descending into streams or onto snow or ice. Probably by far the commonest cause, however, is simply heavy rain saturating a thick cover of loose unstable pyroclastic material on the steep slope of the cone, transforming the material into a mobile "mud" which rushes down the mountainside sometimes with a speed as great as 60 mph, sweeping up everything loose in its path, and sometimes taking a heavy toll in property and human lives. Hot volcanic mudflows are sometimes called lahars.

Somewhat related to mudflows are the great floods of water, known in Iceland as "jokulhlaup," that result from rapid melting of ice by volcanic eruption beneath a glacier.

Volcanic landforms. Landscape features of volcanic origin may be either positive forms, the result of accumulation of volcanic materials, or negative forms, the result of lack of accumulation or of collapse.

Positive forms. In certain areas voluminous extrusions of very fluid basaltic lava from dispersed fissure vents have built broad, nearly flat-topped accumulations covering areas as great as 100,000–200,000 mi², with volumes of several tens of thousands of cubic miles. These are known as flood basalts, or sometimes as plateau basalts. Two examples in the United States are the Columbia and Snake river plains. Other examples are known in other parts of the world.

Where, instead of being dispersed, fissure eruptions of fluid lava have occurred repeatedly along the same zone of fissures, there results a broadly rounded dome-shaped hill or mountain known as a shield volcano. Some have volumes of several thousand cubic miles. The fissures extend from the summit down the flanks of the shield and are marked by long walls and rows of conelets of spatter. Shield volcanoes consist almost wholly of innumerable superimposed thin lava flows (Fig. 4).

Fig. 4. Mauna Lua, Hawaii, a typical shield volcano. Lighter and darker streaks are recent lava flows on mountain flank. (*Photo by G. A. Macdonald*)

The eruption of less fluid lava, or lava with a higher gas content, produces more pyroclastic material. Cinder, bombs, and lapilli may accumulate around a central vent to form a conical hill or small mountain known as a cinder cone. Alternations of lava flows and beds of pyroclastic material produce a composite volcano. Most of the well-known volcanic mountains are composite volcanoes; and some of them, such as Mayon in the Philippines, Fuji in Japan, and Shishaldin in the Aleutian Islands, are remarkably symmetrical cones of great beauty. Hills or small mountains composed predominantly or wholly of fine pyroclastic material are termed ash cones, or tuff cones.

Lava too viscous to flow readily may pile up around and over the vent to form a steep-sided heap known as a volcanic dome (Fig. 5). Slender spires thrust through apertures in such a dome are termed spines. The famous spine formed at Mount Pelée, in Martinique, during the eruption of 1902, reached a height of nearly 1000 ft, but like most such spines was very short-lived.

Negative forms. Small bowl-shaped depressions formed by explosion, or by failure of pyroclastic ejecta to accumulate directly above a vent, are known as craters. Most of them are found at the summit or on the flanks of volcanic cones, but some occur away from any cones.

Larger depressions at the summit of volcanic cones are formed by collapse of the summit as the support beneath it is removed by the rapid withdrawal of magma, usually by surface eruption, but probably sometimes by migration of magma within the Earth. A depression of this sort is called a caldera. Probably the best-known example is that containing Crater Lake, in Oregon. Still larger, less regular depressions of similar origin are known as volcano-tectonic depressions. Like many calderas, their formation commonly, if not always, is associated with the eruption of great volumes of welded tuff.

Submarine volcanism. In shallow water, volcanic eruptions are very similar to those on land, though on the average probably somewhat more explosive owing to contact of hot lava with water and resultant violent generation of steam. The glassy ash of cones formed in this way commonly is altered, probably usually by ordinary weathering processes, to brownish palagonite tuff. Such cones, like Diamond Head in Honolulu, usually have broader, flatter profiles than those characteristic of cinder cones.

Fig. 5. Didicas Volcano, north of Luzon, Philippine Islands, in eruption on June 17, 1952. Solid portions of dome protrude near top; lower slopes are mantled with talus formed by crumbling of the surface as the dome grew. (*Official U.S. Navy photo*)

At great depths in the ocean, the pressure of the overlying water may prevent the explosive escape of gas from erupting lava and greatly reduce the vesiculation of the lava itself. At depths greater than 6000 ft, lavas of Kilauea Volcano in Hawaii are almost devoid of vesicles. Much of the Pacific Ocean appears to be floored by basaltic lava which, judging from its apparent density, is much less vesicular than the lavas of the basaltic cones that rise above it to form most of the Pacific Islands. This greater density may be the result of eruption of lava flows under the high pressures existing at the ocean bottom. *See* OCEANIC ISLANDS.

Although pyroclastic materials in the ordinary sense probably seldom, if ever, form in deep water, the lava flows may granulate on contact with the water, forming masses of glassy sand-size fragments resembling ash. Such material, called hyaloclastite, may form in water of any depth, and is often associated with pillow lavas. Great volumes of hyaloclastite were formed in melt water by eruptions beneath glaciers in Iceland. Like ordinary basaltic ash, hyaloclastite commonly alters to palagonite.

Fumaroles and hot springs. Vents at which volcanic gases issue without lava are known as fumaroles. They are found on active volcanoes during and between eruptions and on dormant volcanoes. They may persist long after the volcano has become extinct. The gases include water vapor, sulfur gases, hydrochloric and hydrofluoric acids, carbon dioxide and monoxide, and others in less abundance. They may transport and deposit at the surface small amounts of many of the metals. Temperatures of the escaping gases may reach 500–600°C. The halogen gases and metals generally are found in the high-temperature fumaroles. Lower-temperature fumaroles, in which the sulfur gases predominate along with steam, are called solfataras; and still cooler ones liberating predominantly carbon gases are called mofettes.

Fumaroles grade into hot springs and geysers (intermittently spouting hot springs). The water of most, if not all, hot springs is predominantly of meteoric origin—rainwater which has sunk into the rocks and moves through them, rather than water liberated from magma. Some hot springs appear to result simply from water circulating to warm regions at great depths in the Earth's crust, but in many the heat is of volcanic origin and the water may contain volcanic gases. Indeed, the heat may be derived wholly from rising hot volcanic gases. The thermal springs and geysers of the Yellowstone region are of the latter sort. *See* GEYSER; THERMAL SPRING.

Distribution of volcanoes. Most volcanoes occur in one of three types of geologic setting. It is generally believed that the Earth's solid surface is divided into a dozen or so more or less rigid plates, 35–70 mi thick, which move laterally relative to each other over a zone of low rigidity in the upper part of the Earth's mantle. *See* DIASTROPHISM.

These lithosphere plates have three types of boundaries: those where adjacent plates are moving away from each other (spreading boundaries); those where one plate is sliding past another; and those where the plates are moving toward each other and one is being destroyed. The spreading boundaries lie mostly along the great series of ridges (Mid-Atlantic Ridge, East Pacific Rise, Southeast Indian Rise, and so on) which girdle the Earth, largely on the ocean floors. Beneath these spreading ridges basalt magma is formed by partial melting of rocks in the upper mantle and rises toward the surface in tensional fissures, solidifying there to form dikes or issuing at the surface to form lava flows. In this way new lithosphere is continually formed to heal the spreading rupture. This volcanism is the most voluminous on Earth, but most of it takes place in moderately deep water, and hence is seldom apparent at the ocean surface. Exceptions are Iceland, which is an emergent part of the Mid-Atlantic Ridge, and the less active spreading zones of the East African rift valleys.

The distance between given points on the two sides of the spreading boundaries is constantly increasing, in some instances as fast as 6 in. a year. However, the Earth's circumference is increasing, if at all, much more slowly than would be implied by this rate of formation of new lithosphere, and therefore lithosphere must be destroyed at approximately the same rate it is created. The destruction takes place at the distant edges of the plates, where one of two converging plates turns downward and is subducted into the Earth's mantle. Active subduction zones of this sort surround a large part of the Pacific Ocean. Thus, the Pacific plate is being formed along the East Pacific Rise, moving northwestward at an average rate of about 2.5 in. a year, and disappearing downward beneath the edge of the Pacific plate, while the Nazca plate is moving eastward from the rise and plunging under the edge of the South American plate. Where the plate moves downward, the solid surface of the Earth is depressed, forming deep troughs such as the Japan and Mariana trenches (the latter the deepest on Earth). Downward movement takes place along a shear zone (the Benioff zone) which slopes beneath the edge of the opposing plate at an angle of 45–70°. As the sinking lithosphere enters the hot underlying mantle, it is partially melted, yielding magma that rises through the edge of the overlying plate to produce volcanic activity at the surface. The moderately to highly explosive volcanoes of the "Pacific Rim of Fire," with their predominantly andesitic magmas, have been generated in this way.

Still other volcanoes lie far from the edges of the plates. Typically, these form more or less straight "linear chains." The island chains of the central Pacific, such as the Hawaiian, Society, and Austral archipelagos, are of this sort. The mid-Pacific chains are approximately parallel to the direction of movement of the Pacific plate, as indicated by the belts of differing magnetic attraction on successively formed parts of the ocean floor outward from the East Pacific Rise. For more than a century it has been recognized, on the basis of the degree of weathering and erosion, that the Hawaiian volcanoes decrease progressively in age from the northwest to the southeast. In 1963 J. T. Wilson suggested that this is the result of the northwestward movement of the Pacific plate across a hot, magma-generating spot in the mantle, magma rising through the plate to form a volcano. The latter was then gradually carried northwestward away from the hot spot, its feeding channel ruptured, and another volcano was built behind it over the hot spot, and so on to form the entire chain. Radioactive (K-Ar) dating of the lavas of Hawaiian volcanoes has confirmed the general southeastward decrease in age, though much more detail is needed. Similarly, the youngest volcanoes appear to lie at the southeastern ends of the Society and Austral chains, although present inadequate data indicate that there is no corresponding age progression in the Line Island chain. Furthermore, the rate of northwestward age increases in the Hawaiian chain is about twice the indicated rate of movement of the Pacific plate, leading I. McDougall and others to postulate an approximately equal rate of southeastward movement of the underlying mantle and hot spot, consistant with the necessity for a return flow in the mantle to compensate for the northwestward movement of the plate. Many more data are needed to resolve the uncertainties and the details of the process.

[GORDON A. MACDONALD]

Bibliography: F. M. Bullard, *Volcanoes, in History, in Theory, in Eruption*, 1962; C. A. Cotton, *Volcanoes as Landscape Forms*, rev. ed., 1953; G. A. Macdonald, Forms and structures of extrusive basaltic rocks, in H. H. Hess and A. Poldervaart (eds.), *Basalts: The Poldervaart Treatise on Rocks of Basaltic Composition*, vol. 1, 1968; G. A. Macdonald, *Volcanoes*, 1972; G. A. Macdonald and D. H. Hubbard, *Volcanoes of the National Parks in Hawaii*, 5th ed., 1970; I. McDougall, *Nat. Phys. Sci.*, 231:141–144, 1971; A. Rittmann, *Volcanoes and Their Activity*, 1962; C. K. Wentworth and G. A. Macdonald, *Structures and Forms of Basaltic Rocks in Hawaii*, USGS Bull, no. 994, 1953; H. Williams, Calderas and their origin. *Univ. Calif. Publ. Geol. Sci.*, 25:239–346, 1941; H. Williams, *Crater Lake: The Story of Its Origin*, 1941; H. Williams, *The Geology of Crater Lake National Park, Oregon*, Carnegie Inst. Wash. Publ. no. 540, 1942.

Volcanology

The scientific study of volcanic phenomena. Strictly speaking, it refers only to the surface eruption of magmas and related gases, and the structures, deposits, and other effects produced thereby. Commonly, however, it includes much of plutonic igneous geology, since effects at the Earth's surface are the result of events at depth. This article considers the surface activity of erupting volcanoes and the properties of erupting lavas. For a discussion of the surface structures and deposits which result from the activity of volcanos *see* VOLCANO.

Composition of magma. As formed at depth, magma consists of liquid rock containing dissolved gases. Rising toward the surface, it enters zones of lower temperature and pressure. Decreasing temperature tends to bring about crystallization, which produces solid crystals suspended in the liquid. Other solid fragments are incorporated from the walls and roof of the conduit through which the magma is rising. As crystallization progresses, volatiles and the more soluble silicate components are concentrated in the remaining liquid. *See* MAGMA; PHENOCRYST; XENOLITH.

At some point in the rise of the magma, decreasing confining pressure and increasing concentration of volatiles in the residual liquid initiate the separation of gas from the liquid. From that point on to its final complete consolidation, the magma consists of all three phases: liquid, solid, and gas.

The volcanic gases generally, if not always, are predominantly water. Other gases include sulfur dioxide and trioxide, elemental sulfur, carbon dioxide and monoxide, hydrogen, and hydrochloric and hydrofluoric acids. All collections of volcanic gases contain also nitrogen, argon, and other inert gases, unquestionably derived at least in part from atmospheric contamination at or near the surface, but perhaps also in part of deep-seated origin.

In the laboratory, molten granite at a temperature of 900°C and a pressure of 1000 atm of water vapor will hold in solution nearly 6% (by weight) of water, and basaltic magma under similar conditions may contain more than 4%. However, estimates of the proportion of gas to lava liberated in actual eruptions commonly are much less than that. Thus at the Hawaiian volcanoes, at Hekla in Iceland, at Nyamlagira in central Africa, and at Vesuvius, volatiles constituted less than 1% of the weight of the lava erupted during the same interval. It is unlikely that such magmas are saturated with gas at the higher pressures existing even at depths of only a few kilometers in the Earth. In initially unsaturated magmas, high gas pressures may be developed by the supersaturation in volatiles resulting from their concentration in the residual liquid phase during crystallization.

Part of the gas liberated at volcanoes probably comes from the same deep-seated source as the silicate portion of the magma, but some is of shallower origin. Part of the steam is the result of surficial oxidation of deep-seated hydrogen, and some of the oxidation of the sulfur gases must have taken place close to the surface. At some volcanoes, such as Vesuvius, the carbon gases come in part from reaction of the magma with limestone at shallow depth. Ammonia and hydrocarbon gases present in some collections probably are derived from organic constituents of sedimentary rocks near the surface.

In some eruptions, such as the 1924 eruption of Kilauea, Hawaii, temperatures are low and the gas is wholly or very largely steam. Magmatic material may be entirely absent. In these phreatic eruptions the steam is simply heated groundwater from the rocks adjacent to the volcanic conduit. In other eruptions, such as that of Parícutin, in Mexico, the large volume of steam given off simultaneously with lava and smaller amounts of magmatic gas far exceeds the theoretical saturation limit of the magma, and there is little doubt that it represents volatilized groundwater.

Physical properties of magmas. The temperature of erupting magma has been measured in lava flows, and directly in volcanic vents, by means of thermocouples and optical pyrometers. Reasonably good and consistent measurements have been obtained for basic magmas. Thus, at Hawaiian volcanoes, the temperature of erupting lava in the vents is usually between 1200 and a little below 1100°C, and repeated measurements in cooling lava lakes indicate that the basalt lava there becomes completely solid at about 980°C, although some lava flows that still retain a larger proportion of dissolved volatiles may remain fluid to a lower temperature. At Nyamlagira also, in vents and in flows close to vents, temperatures ranged from about 1040 to 1095°C. Temperatures during the 1950–1951 eruption of Oshima, Japan, were in the same range. Locally, temperatures as high as 1400°C may result from burning of volcanic gases in vents, and some secondary heating of lava flows may result from burning of hydrocarbon gases derived from vegetation buried by the flow.

Temperature measurements on more silicic lavas are much less accurate because of the greater violence of the eruptions and the necessity of working at considerable distances. In general, however, they suggest lower temperatures of eruption.

The viscosity of flowing basic lavas has been measured by means of penetrometers (instruments that measure the rate of penetration into liquid of a slender rod under a given strength of thrust) and by the shearing resistance to the turning of a vane immersed in the liquid, and calculated from observed rates of flow in channels of known dimensions and slope. The best direct measurements are shear measurements in a lava lake at Kilauea, which yielded values of 6500–7500 poises at temperatures of 1130–1135°C. Calculations based on rate of flow at both Kilauea and Mauna Loa, Hawaii, gave viscosities of 3000–4000 poises for lava close to the vents, increasing at greater distances from the vents as the lava cools and stiffens to immobility. At Hekla, Iceland, somewhat more silicic lava in the vent had a viscosity of 10,000 poises; and at Oshima, Japan, the lowest viscosities in two streams near the vent during the 1951 eruption were 5600 and 18,000 poises, respectively. In general, the behavior of more silicic lavas indicates that they are more viscous, though actual measurements of their viscosity have not been made.

Types of volcanic eruptions. The character of a volcanic eruption is determined largely by the viscosity of the liquid phase of the erupting magma and the abundance and condition of the gas it contains. Viscosity is in turn affected by such factors as the chemical composition and temperature of the liquid, the load of solid crystals and xenoliths it carries, the abundance of gas, and whether the gas is dissolved or separated as bubbles. In very fluid lavas small gas bubbles form gradually, and generally are able to rise through the liquid, coalescing to some extent to form larger bubbles, and escape at the surface with only minor disturbance. In more viscous lavas the escape of gas is less easy, and produces minor explosions as the bubbles burst their way out of the liquid. In still more viscous lavas there appears at times to be a tendency for the formation of large numbers of small bubbles more or less simultaneously throughout a large volume of liquid, and the violent expansion of these may tear the frothy liquid to pieces and throw it into the air as a shower of volcanic ash and dust, accompanied by some larger blocks; or it may produce an outpouring of an emulsion of gas and semisolid bits of glass to form an ash flow. Also, gas may accumulate beneath a solid or highly viscous seal in the vent until it acquires enough pressure to burst its way forth in an explosion that hurls out fragments of the disrupted plug.

Types of eruptions customarily are designated

by the name of a volcano or volcanic area that is characterized by that sort of activity (though probably all volcanoes show somewhat different sorts of activity at different times).

Hawaiian eruptions. Eruptions of the most fluid lava, in which relatively small amounts of gas escape freely with little explosion, are given this designation. Most of the lava is extruded as thin flows that spread to distances of several miles from their vents. Clots of lava thrown into the air remain fluid enough to flatten out on striking the ground, and commonly to weld themselves together (see illustration). An occasional feature of Hawaiian activity is the lava lake—a pool of liquid lava with convectional circulation that occupies a shallow depression on the top of more viscous lava in the crater.

Strombolian eruptions. These somewhat more explosive eruptions of lava, with greater viscosity, produce a larger proportion of pyroclastic material. Many of the bombs and lapilli assume rounded or drawn-out forms during flight, but are sufficiently solid to retain these shapes on impact.

Vulcanian eruptions. Generally still more explosive are the vulcanian type. Angular blocks of viscous or solid lava are hurled out, commonly accompanied by voluminous clouds of ash but with little or no lava flow.

Peléean eruptions. These eruptions are characterized by the heaping up of viscous lava over and around the vent to form a steep-sided hill or volcanic dome. Explosions, or collapses of portions of the dome, may result in glowing avalanches (*nuées ardentes*).

Plinian eruptions. These paroxysmal eruptions of great violence are characterized by voluminous explosive ejections of pumice and by ash flows. The copious extrusion of siliceous magma commonly is accompanied by collapse of the top of the volcanic cone, forming a caldera, or by collapse of a broader region, forming a volcano-tectonic depression. The Roman naturalist Pliny was killed A.D. 79 while observing such an eruption of Vesuvius.

Ultravulcanian explosions. In contrast to the foregoing magmatic eruptions, some low-temperature explosions of this type throw out fragments of old volcanic or nonvolcanic rocks, accompanied by little or no magmatic material. Certain explosion pipes and pits, known as diatremes and maars, have been produced by ultravulcanian explosions.

Phreatac explosions. Volatilization of groundwater when it comes in contact with hot lava or hot rocks in the walls of the volcanic conduit causes these occasionally violent disturbances.

Surface deformations from eruptions. Measurements at certain volcanoes have demonstrated that the entire volcanic mountain swells and shrinks, apparently in response to changes in conditions beneath the volcano. The measurements are of two sorts: precise leveling surveys in which the change of altitude of a series of bench marks is related to some point outside the disturbed area, and measurement of the change in inclination of a given point on the slope of the cone by means of instruments known as tiltmeters. Some tiltmeters are capable of detecting changes in inclination measured in fractions of a second of arc, but the total changes of level resulting from tumescence and detumescence of the volcano may amount to several feet. Thus, between 1912 and 1921 a bench mark at the edge of Kilauea caldera, Hawaii, rose about 3 ft, and between 1921 and 1926 the same bench mark sank $3\frac{1}{2}$ ft. During the latter interval a bench mark at the edge of the inner crater, Halemaumau, sank 6 ft.

Liquid lava eruptions during 1955 eruption of Kilauea Volcano, Hawaii. (*a*) Small spatter cone, about 10 ft high, being built. (*b*) Fountain of liquid lava, 200 ft high, building a cone of cinder and spatter. (*G. A. Macdonald, USGS*)

Even more extreme changes have occurred elsewhere. During the 1944–1945 eruption of Usu, Japan, the ground surface near the foot of the mountain was upraised in a broad dome 50 m high before explosions finally burst through its top and a protrusion of viscous lava took place. In this instance it is clear that the deformation resulted from an intrusion of lava beneath the uplifted area. At the Hawaiian volcanoes the mechanism is less clear, and is generally referred to simply as a change in volcanic pressure.

At least in some instances the swelling and shrinking of Hawaiian volcanoes show direct close relationship to eruptions, the mountain swelling before eruption and shrinking during or immediately after it. This suggests swelling as a result of injection of magma into some sort of reservoir in or beneath the cone, and shrinking as a result of drainage of the reservoir.

Similar subsidences have accompanied eruptions in other regions. Thus, during the 1914 eruption of Sakurajima, Japan, an area 47 km across, centered at the volcano, sank more than 15 cm, and the sinking in the central part of the area exceeded 80 cm. For a discussion of the great collapses that form calderas and volcano-tectonic depressions *see* VOLCANO.

Volcanic Earth vibrations. Many eruptions are preceded by numerous earthquakes, and these, together with tumescence of the volcano, constitute the principal evidences of coming eruption. For the most part they closely resemble tectonic earthquakes.

Other volcanic Earth vibrations are termed harmonic tremors, which may go on continuously with little or no change for many hours, days, or weeks; and spasmodic tremor, which is simply a closely spaced succession of minute earthquakes. Volcanic explosions produce shock waves in the atmosphere—sometimes seen as flashing arcs rendered visible by refraction of light in the denser wavefront. Explosions or collapses taking place underwater may cause water waves, sometimes of great size. These fall within the class of tsunamis (seismic sea waves), though most tsunamis have other causes. Waves of this sort accompanied the collapse of Krakatoa Volcano, in 1883, and destroyed many villages on the shores of neighboring Java and Sumatra. *See* EARTHQUAKE; SEISMOLOGY.

[GORDON A. MACDONALD]

Bibliography: L. Civetta et al. (eds.), *Physical Volcanology*, 1974; R. A. Daly, *Igneous Rocks and the Depths of the Earth*, 2d ed., 1933; J. P. Eaton and K. J. Murata, How volcanoes grow, *Science*, 132:925–938, 1960; J. Gilluly, A. C. Waters, and A. O. Woodford, *Principles of Geology*, 2d ed., 1959; T. A. Jaggar, *Origin and Development of Craters*, Geol. Soc. Amer. Memo. no. 21, 1947; G. A. Macdonald, Activity of Hawaiian volcanoes during the years 1940–1950, *Bull. Volcanol.*, 15:119–179, 1954; G. A. Macdonald, *Hawaiian Volcanoes During 1952*, U.S. Geol. Surv. Bull. no. 1021-B, 1955; G. A. Macdonald, Volcanology, *Science*, 133:673–679, 1961; G. A. Macdonald and J. P. Eaton, *Hawaiian Volcanoes During 1955*, USGS Bull. no. 1171, 1964; K. J. Murata et al., *The 1959–60 Eruption of Kilauea Volcano, Hawaii*, USGS Prof. Pap. no. 537 A-D, 1966; A. Rittmann, *Volcanoes and Their Activity*, 1962.

Warping of Earth's crust

Gentle bending of the crust of the Earth without pronounced folds or dislocations. The Earth's crust is being warped at measurable rates believed to be associated with several causes. Measurements have been made by tide gages, tiltmeters, repeated leveling and geodetic surveys, and geological observations.

Melting of former ice sheets has led to uplift, as in northern Finland, estimated to be 476 m in the past 9950 years. Uplift is still proceeding there at a maximum rate of 1.3 cm/year, but sea level is stationary in Denmark. Hudson Bay is likewise rising, and the Great Lakes are tilting southward at 0.1 cm/(km) (year).

As a result of loading in deltas and sedimentation many coasts are sinking elsewhere; for example, the coast of Holland, and the Atlantic and Gulf coasts of North America are sinking at rates of about 0.5 cm/year, which is accentuated by the general rise of sea level of 0.1 cm/year. *See* DELTA.

Mountain uplift and faulting are causing other areas, such as Japan and the north Pacific Coast of North America, to rise at similar average rates with periodic jumps of up to several meters during earthquakes. Relative horizontal movements of the order of 5 cm/year are observed, for example, in California and New Zealand.

Larger tectonic warpings have occurred in the past at unknown rates, as marine sediments high on many mountains and deep in oil wells prove. Drowned beaches and moats which surround the Hawaiian volcanoes demonstrate warping due to loading.

Many authorities have postulated still larger warping movements due to polar wandering, continental drift, or convection currents in the Earth's interior, but the existence, causes, and rates of these phenomena are still being debated.

Over large areas and long periods, the Earth's crust evidently seeks isostatic equilibrium, but it may be warped by disturbances. *See* DIASTROPHISM; OROGENY; STRAND LINE; TECTONOPHYSICS.

[JOHN T. WILSON]

Wavellite

A hydrated phosphate of aluminum mineral with composition $Al_3(OH)_3(PO_4)_2 \cdot 5H_2O$, in which small amounts of fluorine and iron may substitute for the hydroxyl group (OH) and aluminum (Al), respectively. Wavellite crystallizes in the orthorhombic system. The crystals are stout to long prismatic, but are rare. Wavellite commonly occurs as globular aggregates of fibrous structure (see illustration)

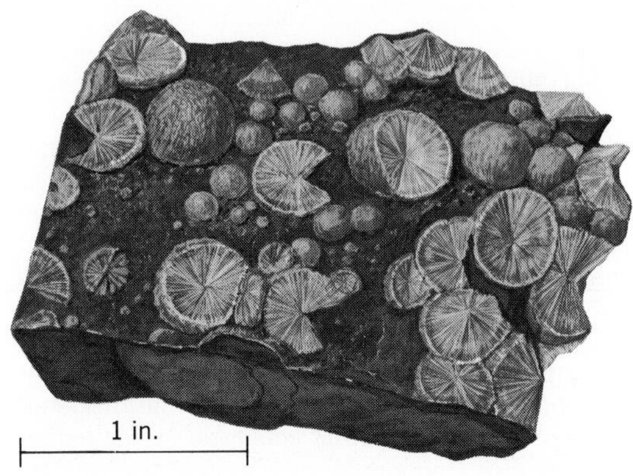

Wavellite in globular aggregates, found in Devonshire, England. (*Specimen from Department of Geology, Bryn Mawr College*)

and as encrusting and stalactitic masses. Wavellite crystals range in color from colorless and white to different shades of blue, green, yellow, brown, and black.

Wavellite is a widespread secondary mineral occurring in small amounts in crevices of low-grade metamorphic aluminous rocks, in limonite, and in phosphate rock deposits. Found at many places in Europe and North America, it is also abundant in tin veins at Llallagua, Bolivia. See PHOSPHATE MINERALS.

[WAYNE R. LOWELL]

Weathering processes

Weathering may be defined as the change of geologic materials (minerals and rocks) at or near the Earth's surface from relatively massive to dispersed states. Climate, plants, bacteria, animals, and such agents as water, wind, ice, gravity, temperature change, and the gases oxygen and carbon dioxide play an important role in weathering.

Types of weathering. There are two general types of weathering: mechanical, in which rocks are disintegrated (broken into smaller fragments by physical forces), and chemical, in which rock materials are decomposed (changed in composition by chemical reactions). In cold and in dry climates the mechanical agents are most active and produce angular-shaped particles and landforms. In warm humid climates chemical reactions and biochemical changes proceed at much faster rates, particularly in the tropics, where weathered zones sometimes extend to great depths.

Products of weathering. Products of mechanical weathering include rock fragments and materials formed by (1) the expansion and contraction in rocks, (2) the action of freezing water, and (3) the shattering effects of downslope movement. See EXFOLIATION, ROCK; FROST ACTION; MASS WASTING.

Products of chemical weathering include such economically and technologically important products as the soil; clays used in making structural products, ceramic whitewares, refractories, paper coating, portland cement, absorbents, catalysts, and fillers; and ores of iron, manganese, uranium, and vanadium. See ORE AND MINERAL DEPOSITS.

Weathered products are commonly designated as relatively insoluble (both those that remain in place and those carried away in suspension) and as soluble (those removed in solution by water). Colloidal products of weathering, however, obscure a sharp division between soluble and insoluble weathering products. The colloidal products are important both quantitatively and qualitatively.

Chemical weathering processes. During chemical weathering, chemical elements of geologic materials assume higher states of oxidation. Silicate rocks weather chiefly by dissolution and hydrolysis, but also simultaneously by oxidation and carbonation, forming hydrates, carbonates, and oxides. For example, iron in silicate minerals combines with oxygen to form Fe_2O_3, thereby removing Fe from the silicate structure and so disrupting that network. The carbonation of calcium and magnesium in silicates also aids in their breakup. The general hydrolysis reaction of silicates may be written as the equation below, where M

$$\mathrm{MAlSiAlO}_n + \mathrm{HOH} \rightarrow \mathrm{M^+OH^-} + \mathrm{HSiO}_n + \mathrm{H(M)AlSiAlO}_n + \mathrm{Al(OH)_3}$$

refers to metal cations (K, Na, Ca, Mg), subscript n to an unspecified ratio of atoms, and the Al following Si substitutes for Si. Thus there are formed in hydrolysis soluble alkali-metal hydroxides, soluble silica (the ionic distribution depends upon pH), and probably a relatively insoluble clay mineral (zeolite) or, less commonly, hydrated alumina. If the hydrolysis takes place at pH 9.5 or higher, both silica and alumina will be relatively soluble and mobile. They may then be separated and form bauxite ($Al_2O_3 \cdot nH_2O$). Under more acid condi-

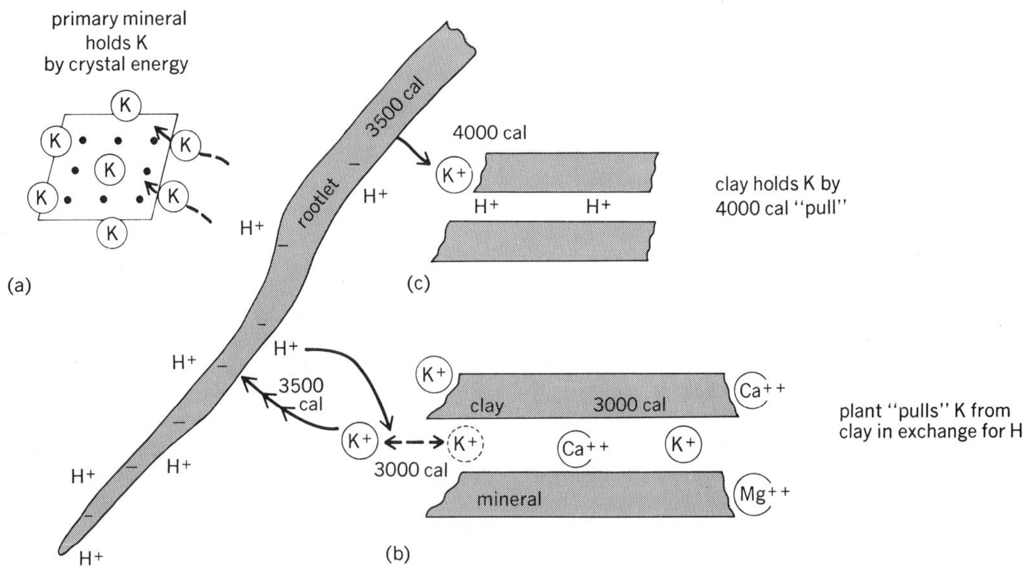

Fig. 1. Exchange-energy relationships between a rootlet and three minerals. (a) A potassium-bearing, primary silicate mineral. (b) A clay mineral well stocked with exchangeable metal cations. (c) A clay mineral scantily stocked with metal cations. The exchange bonding energy (calories per gram-equivalent weight) of K for H in the rootlet exceeds that in only the well-stocked clay mineral which is thus the only one of the three minerals from which nutrient ions can be taken. (*From W. D. Keller, Mineral and chemical alluviation in a unique pedological example, J. Sediment. Petrol., 31:80–86, 1961*)

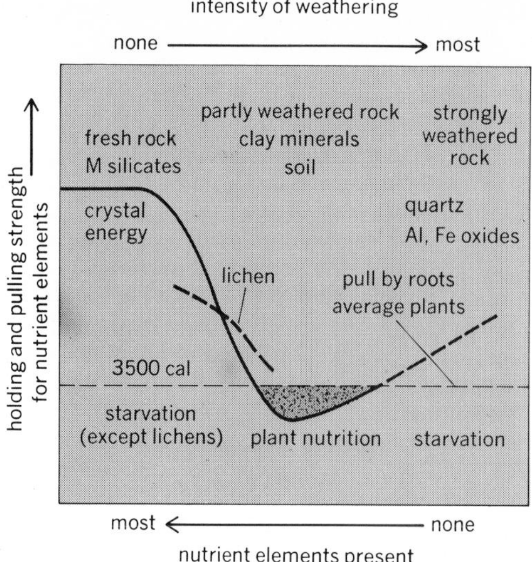

Fig. 2. Comparison of binding energy (calories per gram-equivalent weight) on nutrient metal cation (indicated by M) by rocks and soil minerals with exchange binding energy on cation by plant roots. The relation of plant nutrition to weathering and abundance of nutrients is shown. (*From W. D. Keller, Mineral and chemical alluviation in a unique pedological example, J. Sediment. Petrol., 31:80–86, 1961*)

tions, clay minerals are formed.

Adding hydrogen ions to the hydrolyzing system increases the rate of reaction. Carbonic acid, formed when the carbon dioxide of the air and soil dissolves in water, is a source of hydrogen ions which accelerate the reaction. Calcium bicarbonate, $Ca(HCO_3)_2$, magnesium bicarbonate, $Mg(HCO_3)_2$, and carbonate-complexed compounds of various cations are removed in solution. Organic (humic) and other acids participate in the hydrolysis. Strongly complexing organic acids may mobilize in solution Al more effectively than Si from Al-silicate minerals. Solubilization in and precipitation from organic solutions are sensitive to both Eh and pH. Another major source of hydrogen ions is their production in the ionic atmosphere about the rootlets of growing plants. During plant growth and metabolism, hydrogen ions are evolved; these are exchanged by the roots for nutrient cations (K^+, Ca^{++}, Mg^{++}) present in nearby clay colloids and rocks. Thus the process of nutrition for plants is simultaneously a process of weathering for rocks. Plants that are primitive in development apparently possess higher energies of cation exchange than do those more advanced; lichens derive nutrient cations from fresh rock without intermediary soil.

Plant rootlets may sorb nutrient ions from adjacent soil when the mean free bonding energy of the rootlet exceeds the crystal bonding energy or mean free bonding energies of clay minerals or organic substances for individual nutrient ions in polyionic systems in the soil. Hence plant nutrition and the domain of agriculture occupy an intermediate position in the weathering sequence between fresh rock-forming minerals and intensely weathered "final" products of weathering (Figs. 1 and 2). Chelating organic substances extract cations from rocks, implementing rock breakdown. Partial weathering makes the rock constituents more available to plants, but extended weathering removes the nutrient materials entirely.

The ultimate destination of the soluble products of weathering is the ocean. There the dissolved mineral matter becomes concentrated in solution and in deposits. Potassium, although as soluble as sodium, is more readily absorbed by clay minerals as exchangeable cations and may be incorporated in the crystals of hydrous mica. Potassium is therefore less concentrated than sodium in sea water. Magnesium may be incorporated in chloritic varieties of clay minerals. *See* CLAY MINERALS.

The most abundant weathering products of silicate rocks are the clay minerals. Weathering (hydrolysis) resulting in high concentrations of calcium, magnesium, and iron (ferrous) ions tends to form the montmorillonite group of clays. Such high concentration of ions occurs where evaporation exceeds precipitation, groundwater drainage is poor, or hydrolysis is rapid (as in weathering of volcanic dust). The kaolin group of clay minerals is developed where rainfall exceeds evaporation and leaching is intense. Oxidation of iron is then ordinarily high. Under conditions of very drastic leaching and continual wetting of the rocks, as in a tropical rainforest, silica and most cations dissolve, leaving hydrated oxides of alumina and ferric iron (bauxite and laterite). Rising groundwater solutions may carry Al and Fe upward and leave, because of evaporation or oxidation, deposits of both in the tropical subsoil. A high K^+/H^+ ratio in the aqueous weathering system of Al-silicates yields the illite clay mineral (disordered K-mica). Weathering processes apparently reach a state of near-equilibrium with respect to kaolinite or montmorillonite in environments where thick, valuable deposits of those clays are formed. In contrast, surficial weathering of boulders and outcrops yields highly varied and changing products, quasi-mineral compounds, and rock wreckage.

Limestones and dolomites weather chemically by dissolution of the calcium and magnesium carbonates, leaving behind the less soluble impurities: quartz (sand), chert, iron oxides, and clay minerals. Limestones are vigorously corroded by turbulent water, including rain, containing dissolved CO_2. Building and monument stone of carbonate composition are destructively attacked in this way. Clay minerals are subject also to breakdown during weathering by the removal of (1) exchangeable cations, (2) the more tightly fixed potassium of illite (hydrous mica) and possibly cations (other than Al) in the octahedral layer, and (3) silica. The clay minerals whose crystal structure is partly destroyed are said to be degraded. Entirely desilicated clays become bauxite or laterite. *See* BAUXITE; LATERITE. [WALTER D. KELLER]

Bibliography: D. Carroll, *Rock Weathering*, 1970; W. H. Huang and W. D. Keller, Organic acids as agents of chemical weathering of silicate minerals, *Nature, Phys. Sci.*, 239(96):149–151, 1972; W. D. Keller, *Principles of Chemical Weathering*, rev. ed., 1959; F. C. Loughnan, *Chemical Weathering of the Silicate Minerals*, 1969; P. Reiche, *Survey of Weathering Processes and Products*, Univ. N. Mex. Publ. Geol. Ser. no. 1, 1945; M. M. Sweeting, *Karst Landforms*, 1973; E. M. Winkler, *Stone: Properties, Durability in Man's Environment*, 1973.

Willemite

A nesosilicate mineral, composition Zn_2SiO_4, crystallizing in the hexagonal system. It is usually massive or granular with a vitreous luster; crystals are rare. The mineral has a basal cleavage and may be variously colored, most commonly green, red, or brown. Hardness is 5½ on Mohs scale; specific gravity is 3.9–4.2. Willemite forms a valuable ore of zinc at Franklin, N.J. At this famous zinc deposit willemite fluoresces a yellow-green and is found in crystalline limestone associated with franklinite, zincite, and many other rarer minerals (see illustration). With the exception of the occurrence in New Jersey, willemite is a rare mineral. *See* SILICATE MINERALS.

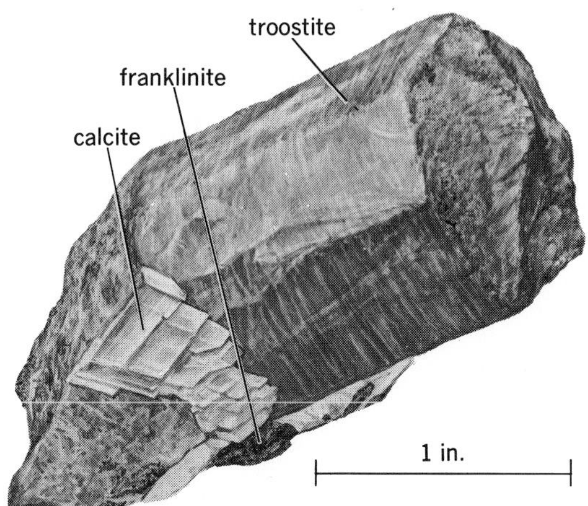

Willemite from Sussex County, N.J. This crystal is the variety troostite, in which manganese replaces a considerable part of the zinc. (*Specimen from Department of Geology, Bryn Mawr College*)

[CORNELIUS S. HURLBUT, JR.]

Witherite

The mineral form of barium carbonate. Witherite has orthorhombic symmetry and the aragonite structure type. Crystals, often twinned, may ap- be white or gray in color, with yellow, brown, or green tints. Its hardness is 3¼ and its specific gravity is 4.3.

Witherite may be found in veins with barite and galena. It is found in many places in Europe, and large crystals occur at Rosiclare, Ill. *See* CARBONATE MINERALS.

[ROBERT I. HARKER]

WITHERITE

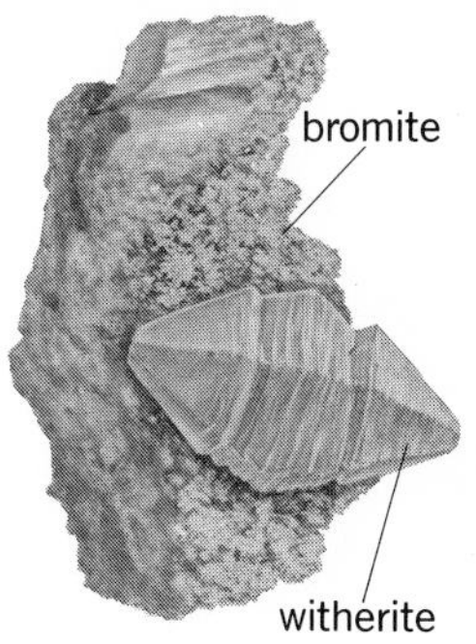

Large crystal with bromite from Fallowfield, Northumberland, England (*specimen from Department of Geology, Bryn Mawr College*).

Wolframite

A mineral with chemical composition (Fe,Mn) WO_4, intermediate between ferberite, the iron tungstate, and huebnerite, the manganese tungstate, which form a complete solid solution series. *See* FERBERITE; HUEBNERITE.

Wolframite occurs commonly in short, brownish-black, monoclinic, prismatic, bladed crystals. It is difficult to dissolve in acids, but wolframite high in iron fuses readily to a magnetic globule.

Wolframite is probably the most important tungsten mineral. A quick and easy test for tungsten is to fuse the mineral powder with charcoal and sodium carbonate and boil the residue in hydrochloric acid with a few grains of granulated metallic tin. The presence of tungsten gives the solution a prussian-blue color. The major use of tungsten is in making ferrous (steel) and nonferrous alloys, tungsten carbide, metallic tungsten, and tungsten chemicals.

Wolframite is found associated with quartz in veins in the peripheral areas of granitic bodies. It is also found in veins associated with sulfide minerals such as pyrite, chalcopyrite, arsenopyrite, and bismuthinite, together with cassiterite, molybdenite, hematite, magnetite, tourmaline, and apatite. It also occurs in placers.

China is the major producer of wolframite. Tungsten minerals of the wolframite series occur in many areas of the western United States; the major producing district is Boulder and northern Gilpin counties in Colorado.

[EDWARD C. T. CHAO]

Wollastonite

A mineral inosilicate with composition $CaSiO_3$. It crystallizes in the triclinic system in tabular crystals (see illustration). More commonly it is massive, or in cleavable to fibrous aggregates. There are two good cleavages parallel to the front and basal pinacoids yielding elongated cleavage fragments. Hardness is 5–5½ on Mohs scale; specific gravity is 2.85. On the cleavages the luster is pearly or silky; the color is white to gray. Wollastonite is the most common of three polymorphic forms of $CaSiO_3$, the other two being pseudowollastonite and parawollastonite. Pseudowollastonite, a high-temperature triclinic form, is very uncommon in rocks but may be a constituent of synthetic $CaO\text{-}SiO_2$ systems and of slags and glasses. Parawollastonite, a monoclinic form, is only rarely found in Ca-rich rocks. Wollastonite, by far the most common polymorph, occurs abundantly in impure limestones that have undergone contact metamorphism. Resulting assemblages may consist of calcite-diopside-wollastonite with variable amounts of tremolite, clinozoisite, and grossularite. Wollastonite occurs sporadically in regionally metamorphosed calcareous sediments as well. It is found in large masses in the Black Forest of Germany; Brittany, France; Chiapas, Mexico; and Willsboro, N.Y., where it is mined as a ceramic material. *See* SILICATE MINERALS.

[CORNELIUS S. HURLBUT, JR.]

WOLLASTONITE

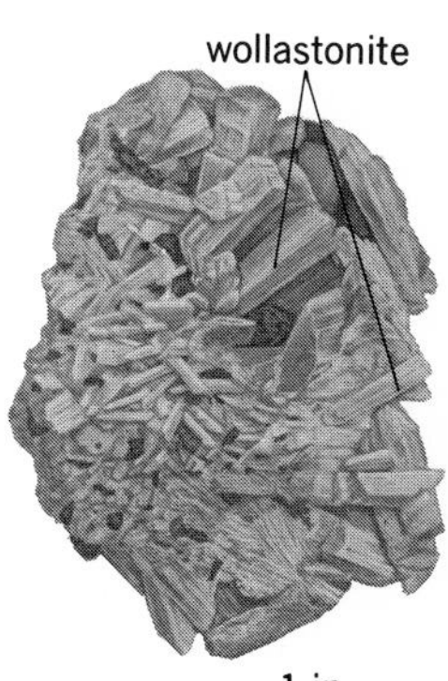

Triclinic tabular crystals embedded in limestone, Santa Fe, Chiapas, Mexico. (*Specimen from Department of Geology, Bryn Mawr College*)

Wulfenite

A mineral consisting of lead molybdate, $PbMoO_4$. Wulfenite occurs commonly in yellow, orange, red, and grayish-white crystals. They may be tetragonal, tabular (see illustration), or pyramidal, with a luster from adamantine to resinous. Wulfenite may also be massive or granular. Its fracture is uneven. Its hardness is 2.7–3 and its specific gravity 6.5–7. Its streak is white. It is easily fusible and is decomposed by hydrochloric or nitric acid with the separation of molybdic oxide.

Wulfenite occurs as a secondary mineral in the oxidized zone (of veins) of lead deposits associated with pyromorphite, cerussite, vanadinite, and other oxide zone minerals such as goethite and calcite.

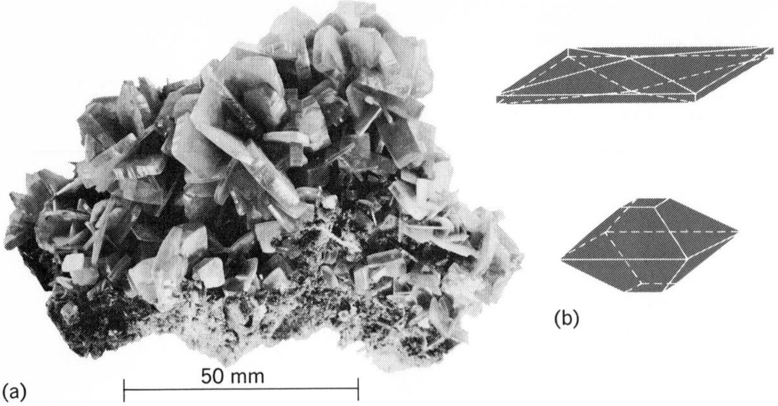

Wulfenite. (*a*) Thin tabular crystals on limonite, Organ Mountains, New Mexico (*American Museum of Natural History specimen*). (*b*) Crystal habits (*from C. S. Hurlbut, Jr., Dana's Manual of Minerology, 17th ed., copyright © 1959 by John Wiley & Sons, Inc.; reprinted by permission*).

Fig. 1. Xenoliths in granodiorite, Sierra Nevada, California, appear soaked and drawn out parallel to the hammer handle. Hammer is 10 in. long. (*USGS photograph by W. B. Hamilton*)

Fig. 2. Xenolith in granite, Sierra Nevada, California. Note sharper contact on one side and gradational contact against contaminated granite on other. (*USGS photograph by W. B. Hamilton*)

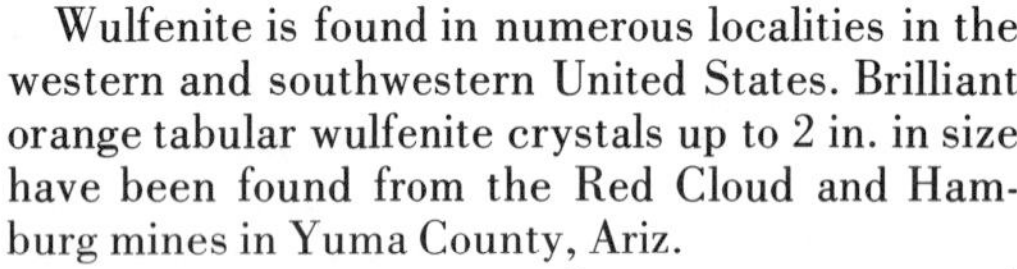

Wulfenite is found in numerous localities in the western and southwestern United States. Brilliant orange tabular wulfenite crystals up to 2 in. in size have been found from the Red Cloud and Hamburg mines in Yuma County, Ariz.

[EDWARD C. T. CHAO]

Wurtzilite

A black, infusible carbonaceous substance occurring in Uinta County, Utah. It is insoluble in carbon disulfide, has a density of about 1.05, and consists of 79–80% carbon, 10.5–12.5% hydrogen, 4–6% sulfur, 1.8–2.2% nitrogen, and some oxygen.

Wurtzilite is derived from shale beds deposited near the close of Eocene (Green River) time. The material was introduced into the calcareous shale beds as a fluid after which it polymerized to form nodules or veins. *See* ALBERTITE; ASPHALT AND ASPHALTITE; ELATERITE; IMPSONITE.

[IRVING A. BREGER]

Wurtzite

A mineral with composition ZnS (zinc sulfide), crystallizing in the hexagonal system (dihexagonal pyramidal class). Crystals are pyramidal with the vertical axis polar. It exists most commonly in fibrous or columnar aggregates and banded crusts with resinous luster and brownish-black color. There is good prismatic cleavage; the hardness is $3\frac{1}{2}$ and its specific gravity 4.0.

Zinc sulfide is dimorphous, crystallizing as both the high-temperature form wurtzite, α-ZnS, and the low-temperature form sphalerite, β-ZnS. Wurtzite is the rarer and at room temperature the less stable form and, unlike sphalerite, is rarely mined as an ore of zinc. Wurtzite usually contains iron and cadmium, and a complete solid solution series extends to greenockite, CdS. *See* SPHALERITE.

[CORNELIUS S. HURLBUT, JR.]

Xenolith

A discrete rock fragment enclosed by magma, lava, or igneous rock. The relation between xenolith and host magma may be close or remote. For example, the consolidated marginal portions of a body of magma may be torn off and incorporated into the unconsolidated central portions of the magma. Xenoliths which are inferred to form in this manner are called cognate xenoliths. Cognate xenoliths in volcanic rocks, like phenocrysts, help scientists find out about the consolidation of bodies of magma underneath volcanoes. On the other hand, many xenoliths are simply pieces of sedimentary and other unrelated igneous rocks. Such xenoliths tend to react with the enclosing magma, so that their constituent minerals become like those in equilibrium with the melt. Reaction is rarely complete, however. Even completely equilibrated xenoliths may be conspicuous because the equilibration process does not require either the texture or the proportions of the minerals in the xenolith to match those in the enclosing rock.

Xenoliths may be angular to round, millimeters to meters in diameter, aligned or haphazard, and sharply or gradationally bounded (see Figs. 1 and 2). Xenoliths are present in most bodies of igneous rock, although they are rare in many igneous rocks. Many rocks are inferred to be igneous because they contain xenoliths, the incorporation of which is taken as evidence of fluidity of the host. Ambiguity arises in the case of some sedimentary rocks such as conglomerate and tillite which may become so thoroughly recrystallized that the sedimentary rock fragments appear to be xenoliths surrounded by crystalline magma.

Some remarkable xenoliths include the metallic

iron-rich rock fragments found in some basaltic rocks which have intruded carbon-rich sedimentary rocks and the peridotite xenoliths found in diamond pipes and in some basaltic rocks. Scientists infer some of the peridotite xenoliths to be actually pieces of the Earth's mantle. *See* IGENEOUS ROCKS; MAGMA; PERIODOTITE.

[ALFRED T. ANDERSON, JR.]

Zeolite

Any mineral belonging to the zeolite family of minerals and synthetic compounds characterized by an aluminosilicate tetrahedral framework, ion-exchangeable large cations, and loosely held water molecules permitting reversible dehydration. The general formula can be written $X^{1+,2+}{}_{y}Al^{3+}{}_{x}Si^{4+}{}_{1-x}O_2 \cdot nH_2O$. Since the oxygen atoms in the framework are each shared by two tetrahedrons, the (Si,Al):O ratio is exactly 1:2. The amount of large cations (X) present is conditioned by the Al:Si ratio and the formal charge of these large cations. Typical large cations are the alkalies and alkaline earths such as Na^+, K^+, Ca^{2+}, Sr^{2+}, and Ba^{2+}. The large cations, which are coordinated by framework oxygens and water molecules, reside in large cavities in the crystal structure; these cavities and channels may even permit the selective passage of organic molecules. Thus, zeolites are extensively studied from theoretical and technical standpoints because of their potential and actual use as "molecular sieves," catalysts, and water softeners. Dehydrated zeolites can absorb other liquids, such as ammonia, alcohol, and hydrogen sulfide, instead of water. Water softening involves the base exchange of Na^+ in the zeolite for Ca^{2+} in the water. *See* SILICATE MINERALS.

Occurrence. Zeolites are low-temperature and low-pressure minerals and commonly occur as late minerals in amygdaloidal basalts, as devitrification products, as authigenic minerals in sandstones and other sediments, and as alteration products of feldspars and nepheline. Phillipsite and laumontite occur extensively in sediments on the ocean floor. Stilbite, heulandite, analcime, chabazite, and scolecite are common as large crystals in vesicles and cavities in the basalts of the Minas Basin Region, Nova Scotia; West Paterson, N.J.; the Columbia River Plateau; Berufjord, Iceland; Poona, India; and many other localities.

Zeolites are usually white, but often may be colored pink, brown, red, yellow, or green by inclusions; the hardness is moderate (3–5) and the specific gravity low (2.0–2.5) because of their rather open framework structures. Their habits are highly variable and depend on the atomic arrangement; some zeolites are fibrous and woolly while others are platy and micaceous and a few are equant in development and lacking in cleavage.

Crystal structure. As a result of detailed crystal structure analysis, the atomic arrangements of most zeolites are known. Important features are loops of 4-, 5-, 6-, 8-, and 12-membered tetrahedral rings which further link to form channels and cages. On the basis of the kinds of loops and chan-

The zeolite family of minerals

Group	Name	Formula	Crystal system	Volume/framework oxygen
Analcime	Analcime*	$Na(AlSi_2O_6) \cdot H_2O$	Cubic, pseudocubic	26.8
	Wairakite	$Ca(AlSi_2O_6) \cdot H_2O$	Monoclinic	
	Pollucite	$Cs(AlSi_2O_6) \cdot xH_2O$	Tetragonal	
Sodalite	Sodalite*	$Na_4(Al_3Si_3O_{12})Cl$	Cubic	30.0
	Linde A	$Na_{12}(Al_{12}Si_{12}O_{48}) \cdot 27H_2O$	Cubic	35.1
	ZK-5	$Na_{24}(Al_{24}Si_{72}O_{192}) \cdot 90H_2O$	Cubic	34.0
	Faujasite	$(Na_2,Ca)_{30}((Al,Si)_{192}O_{384}) \cdot 260H_2O$	Cubic	39.2
Chabazite	Chabazite*	$Ca_2(Al_4Si_8O_{24}) \cdot 13H_2O$	Rhombohedral, pseudorhombohedral	34.2
	Gmelinite	$Na_2(Al_2Si_4O_{12}) \cdot 6H_2O$	Hexagonal	33.8
	Erionite, offretite	$Ca_{4.5}(Al_9Si_{27}O_{72}) \cdot 27H_2O$	Hexagonal	31.9
	Levynite	$Ca(Al_2Si_4O_{12}) \cdot 6H_2O$	Rhombohedral	32.5
Natrolite	Natrolite*	$Na_2(Al_2Si_3O_{10}) \cdot 2H_2O$	Orthorhombic	28.3
	Scolecite	$Ca(Al_2Si_3O_{10}) \cdot 3H_2O$	Monoclinic	28.6
	Mesolite	$Na_2Ca_2(Al_2Si_3O_{10})_3 \cdot 8H_2O$	Monoclinic	
	Edingtonite	$Ba(Al_2Si_3O_{10}) \cdot 3H_2O$	Orthorhombic	30.1
	Thomsonite	$NaCa_2(Al_5Si_5O_{20}) \cdot 6H_2O$	Orthorhombic	28.3
	Gonnardite	$(Ca,Na)_{6-8}((Si,Al)_{20}O_{40}) \cdot 12H_2O$	Orthorhombic	
Phillipsite	Phillipsite	$(K,Na)_5(Al_5Si_{11}O_{32}) \cdot 10H_2O$	Orthorhombic	31.3
	Harmotome	$Ba_2(Al_4Si_{12}O_{32}) \cdot 12H_2O$	Monoclinic	30.9
	Gismondine	$Ca(Al_2Si_2O_8) \cdot 4H_2O$	Monoclinic	32.5
	Garronite	$NaCa_{2.5}(Al_6Si_{10}O_{32}) \cdot 13H_2O$	Tetragonal?	
Mordenite	Mordenite	$Na(AlSi_5O_{12}) \cdot 3H_2O$	Orthorhombic	29.2
	D'achiardite	$(Na_2,Ca)_2(Al_4Si_{20}O_{48}) \cdot 12H_2O$	Monoclinic	28.2
Other	Heulandite*	$Ca(Al_2Si_7O_{18}) \cdot 6H_2O$	Monoclinic	29.3
	Brewsterite	$Sr(Al_2Si_6O_{16}) \cdot 5H_2O$	Monoclinic	28.2
	Epistilbite	$Ca(Al_2Si_6O_{16}) \cdot 5H_2O$	Monoclinic	27.8
	Stilbite*	$Na_2Ca_4(Al_{10}Si_{26}O_{72}) \cdot 28H_2O$	Monoclinic	30.1
	Yugawaralite	$Ca_4(Al_7Si_{20}O_{54}) \cdot 14H_2O$	Monoclinic	29.2
	Laumontite	$Ca(Al_2Si_4O_{12}) \cdot 4H_2O$	Monoclinic	28.7
	Ferrierite	$Na_4Mg_2(OH)_2(Al_6Si_{30}O_{72}) \cdot 18H_2O$	Orthorhombic	28.2
	Paulingite	$(K.Ca)_{120}((Al,Si)_{580}O_{1160}) \cdot 690H_2O$	Cubic	37.5

*See separate article with this title.

nels present, the zeolite minerals can be classified according to groups, as listed in the table. The volume-per-framework oxygen atom is a measure of the packing efficiency of the structure, and zeolites have very high values, usually greater than 25 A^3 per oxygen atom. This can be compared with a typical dense-packed silicate structure, such as olivine, which has a value of about 18 A^3.

Knowledge of the crystal structure also explains the sieving properties of zeolites. The minimum width of a channel is an approximate measure of the maximum diameter of a molecule which can pass through. Thus, chabazite, with a channel of 3.9-A minimum width, allows the passage of straight-chain hydrocarbons but not those with branched chains. Faujasite, with a 12-membered ring and channel diameter of 9 A, even permits the passage of benzene rings. This substance, once a mineralogical curiosity, is now of key importance in the catalytic cracking of petroleum. Large quantities of faujasite and other zeolites are made synthetically, and their desirable qualities can be enhanced by tailoring their compositions.

Zeolites are usually synthesized hydrothermally, starting with gels of appropriate composition in an alkaline environment. The temperature of synthesis may range from 450 to 100°C with the more open framework structures occurring at the lower temperatures. The appearance of zeolites in a metamorphic rock indicates a rock of lowest grade; such assemblages are referred to as the zeolite facies. [PAUL B. MOORE]

Bibliography: K. F. Fischer and W. M. Meier, Kristallchemie der Zeolithe, *Fortschr. Miner.*, vol. 42, 1965; J. V. Smith, *Structural Classification of Zeolites*, Miner. Soc. Amer. Spec. Pap. no. 1, 1963.

Zincite

A mineral with composition ZnO (zinc oxide). It crystallizes in the hexagonal system with a wurtzite-type structure. Thus its principal axis is polar and different forms appear at top and bottom of crystals (see illustration). Such crystals are rare and the mineral is usually massive. Cleavage is prismatic. Its hardness is 4 and its specific gravity 5.6. The mineral has a subadamantine luster and a deep-red to orange-yellow color. The color of the mineral is believed to result partly from the presence of manganese and partly from defects in the crystal structure. Zincite is rare except at the zinc deposits at Franklin and Sterling Hill, N.J. There, associated with franklinite and willemite, it is mined as a valuable ore of zinc.

[CORNELIUS S. HURLBUT, JR.]

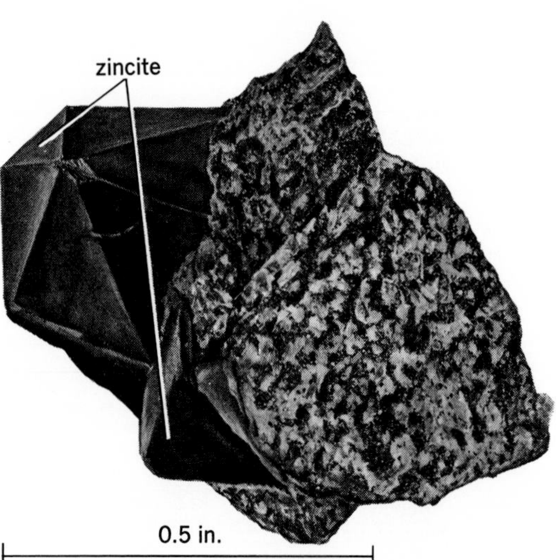

Zincite crystals from Franklin, N.J. (*Specimen from Department of Geology, Bryn Mawr College*)

Zircon

A mineral with the idealized composition $ZrSiO_4$, one of the chief sources of the element zirconium. Trace amounts of uranium and of thorium are often present and the mineral may then be partly or entirely metamict. The name cyrtolite is applied to an altered type of zircon. Structurally, zircon is a nesosilicate, with isolated SiO_4 groups. It is isostructural with the thorium silicate thorite and the yttrium phosphate xenotime. *See* RADIOACTIVE MINERALS; SILICATE MINERALS.

Zircon. (*a*) Crystal from Eganville, Ontario, Canada (*American Museum of Natural History specimen*). (*b*) Crystal habits (*from C. S. Hurlbut, Jr., Dana's Manual of Mineralogy, 17th ed., copyright © 1959 by John Wiley & Sons, Inc.; reprinted by permission*).

Zircon is tetragonal in crystallization. It often occurs as well-formed crystals, which commonly are square prisms terminated by a low pyramid (see illustration). The color is variable, usually brown to reddish brown, but also colorless, pale yellowish, green, or blue. The transparent colorless or tinted varieties are popular gemstones. Hardness is 7½ on Mohs scale; specific gravity is 4.7, decreasing in metamict types. *See* METAMICT STATE.

Because of its chemical and physical stability, zircon resists weathering and accumulates in residual deposits and in beach and river sands, from which it has been obtained commercially in Florida and in India, Brazil, and other countries. It is widespread in small amounts as an accessory mineral in granitic and syenitic igneous rocks, and occurs more abundantly in pegmatite deposits associated with such rocks. Zircon also is a minor constituent in many types of gneissic and schistose metamorphic rocks. *See* HEAVY MINERALS.

[CLIFFORD FRONDEL]

Contributors

List of Contributors

A

Ackerman, Dr. Edward A. *Carnegie Institution of Washington.* RIVER—coauthored.
Alderfer, Dr. Russell B. *Professor of Soils and Crops Department, College of Agriculture and Environmental Science, Rutgers University.* SOIL—in part.
Aldrich, Dr. Daniel G., Jr. *Chancellor, University of California, Irvine.* SOIL—in part.
Amsden, Dr. Thomas W. *Geologist, Oklahoma Geological Survey, Norman.* SILURIAN.
Anderson, Dr. Alfred T., Jr. *Department of the Geophysical Sciences, University of Chicago.* ANDESITE; ANORTHOSITE; articles on other minerals. MAGMA; XENOLITH.

B

Barth, Prof. T. F. W. *Deceased; formerly, Geologic Museum, Oslo.* AMPHIBOLITE; INDEX MINERAL; METAMORPHIC ROCKS; METAMORPHISM; articles on metamorphic petrology.
Barton, Dr. Paul B., Jr. *Geological Survey, U.S. Department of the Interior.* ORE DEPOSITS, GEOCHEMISTRY OF—coauthored.
Batten, Roger L. *Department of Fossil and Living Invertebrates, American Museum of Natural History, New York.* GEOLOGICAL TIME SCALE; PALEONTOLOGY.
Billings, Dr. Marland P. *Geological Museum, Harvard University.* PLUTON.
Bird, Dr. John M. *Department of Geology, State University of New York, Albany.* OROGENY—coauthored.
Biscaye, Dr. Pierre E. *Lamont-Doherty Geological Observatory, Palisades, NY.* MARINE SEDIMENTS—in part.
Black, Dr. Robert F. *Department of Geology, University of Connecticut.* FROST ACTION; PERMAFROST.
Boyd, Dr. F. R. *Geophysical Laboratory, Washington, DC.* DIOPSIDE.
Boyle, Dr. R. W. *Geological Survey of Canada, Department of Energy, Mines and Resources, Ottawa.* GEOCHEMICAL PROSPECTING.
Bradley, Dr. William F. *Professor of Chemical Engineering, University of Texas.* SEPIOLITE; SERPENTINE.
Breger, Dr. Irving A. *U.S. Geological Survey.* ASPHALT AND ASPHALTITE; BITUMEN; related articles.
Bretz, Dr. J. Harlen. *Department of Geophysical Sciences, University of Chicago.* CAVE; KARST TOPOGRAPHY.
Brindley, Dr. George W. *Materials Research Laboratory and Department of Geochemistry and Mineralogy, Pennsylvania State University.* APOPHYLLITE; TALC.
Bullen, Prof. Keith E. *Deceased; formerly, Professor of Applied Mathematics, University of Sydney.* EARTH, INTERIOR OF.
Burke, James D. *Jet Propulsion Laboratory, California Institute of Technology.* MOON.

C

Cady, Dr. Gilbert H. *Deceased; formerly, Consulting Coal Geologist, Urbana, IL.* COAL; LIGNITE; PEAT.
Cain, Dr. Joseph C. *U.S. Geological Survey, Denver.* GEOMAGNETISM—in part.
Cain, Stanley A. *Director, Institute for Environmental Quality, and Charles Lathrop Pack Professor, Department of Resource Planning and Conservation, University of Michigan.* CONSERVATION OF RESOURCES.
Cannon, Helen L. *Branch of Regional Geochemistry, U.S. Geological Survey, Denver.* GEOBOTANICAL INDICATORS.
Carter, Dr. George S. *Fellow of Corpus Christi College, Lecturer in Zoology (retired), Cambridge University.* MACROEVOLUTION.
Chaney, Dr. Ralph W. *Deceased; formerly, Professor Emeritus of Paleontology, University of California.* PALEOBOTANY—in part.
Chao, Dr. Edward C. T. *Geologist, Geological Survey, U.S. Department of the Interior.* ANGLESITE; COESITE; GYPSUM; STISHOVITE; articles on other minerals.
Chapman, Dr. Carleton A. *Department of Geology, University of Illinois.* IGNEOUS ROCKS; LAVA; related articles.
Chapman, Dr. Dean R. *Ames Research Center, Moffett Field, CA.* TEKTITE.
Chapman, Dr. Sydney, *Deceased; formerly, High Altitude Observatory, University of Colorado.* GEOMAGNETISM—in part.
Christ, Dr. Charles L. *Physicist, U.S. Geological Survey.* BORACITE; BORATE MINERALS.
Christie, Dr. John M. *Department of Geology, University of California, Los Angeles.* PETROFABRIC ANALYSIS; SCHISTOSITY, ROCK; STRUCTURAL PETROLOGY.
Clark, Prof. Sydney P., Jr. *Professor of Geology, Yale University.* EARTH, HEAT FLOW IN—in part.
Clements, Dr. Thomas. *Consulting Geologist, Los Angeles.* DESERT EROSION FEATURES; PLAYA.
Clifton, Robert A. *Bureau of Mines, U.S. Department of the Interior.* ASBESTOS.
Cloud, Dr. Preston E. *Department of Geology, University of California, Santa Barbara.* REEF.
Cobban, Dr. William A. *Geologist, U.S. Geological Survey.* CRETACEOUS; MESOZOIC.
Collinson, Dr. Charles W. *Professor of Geology, University of Illinois.* INDEX FOSSIL.
Crittenden, Dr. Charles V. *Geographer, Economic Development Administration, U.S. Department of Commerce.* ISOSTASY; TERRESTRIAL MAGNETISM.

D

Dachille, Dr. Frank. *Materials Research Laboratory and Department of Geochemistry and Mineralogy, Pennsylvania State University.* HIGH-PRESSURE PHENOMENA—coauthored.
Danby, Dr. J. M. A. *Department of Mathematics, North Carolina State University, Raleigh.* GRAVITATION.
Daniels, Dr. Farrington. *Deceased; formerly, Professor Emeritus, Solar Energy Laboratory, University of Wisconsin.* HALF-LIFE—in part.
Dapples, Dr. Edward C. *Department of Geological Sciences, Northwestern University.* DIAGENESIS.
Davis, Prof. Gregory A. *Department of Geological Sciences, University of Southern California.* CORDILLERAN BELT.
Degens, Dr. Egon T. *Senior Scientist, Woods Hole Oceanographic Institution.* SEDIMENTS, ORGANIC GEOCHEMISTRY OF.
Dekeyser, Prof. Willy C. *Laboratory for Crystallography, Ghent, Belgium.* CRYSTAL STRUCTURE; CRYSTALLOGRAPHY.

Delson, Dr. Eric. *Lehman College, New York.* FOSSIL MAN.

De Vore, Dr. George W. *Professor and Chairman, Department of Geology, Florida State University.* AMPHIBOLE; ENSTATITE; GLAUCOPHANE; HORNBLENDE; articles on other minerals.

Dewey, Dr. John F. *Department of Geology, State University of New York, Albany.* OROGENY—coauthored.

Dietz, Dr. Robert S. *Atlantic Oceanographic and Meteorological Laboratories, ESSA, U.S. Department of Commerce, Miami.* CONTINENTAL SHELF AND SLOPE.

Dinsmore, Robertson P. *International Ice Patrol, Woods Hole Oceanographic Institution.* ICEBERG.

Doolan, Barry L. *Department of Geology, Memorial University, St. John's, Newfoundland.* TREMOLITE.

Dorf, Prof. Erling. *Department of Geology, Princeton University.* PALEOBOTANY—in part.

Drake, John C. *Department of Geology, University of Vermont.* SILICA MINERALS.

Duley, Dr. Frank L. *Principal, Agricultural College, University of Peshawar.* SOIL—in part.

Dunbar, Dr. Carl O. *Professor Emeritus of Paleontology and Stratigraphy (retired), Yale University.* FOSSIL; PERMIAN.

E

Eardley, Prof. Armand J. *Deceased; formerly, Professor of Geology, College of Mines and Mineral Industries, University of Utah.* MOUNTAIN SYSTEMS.

Easterbrook, Don J. *Department of Geology, Western Washington University, Bellingham, WA.* AEOLIAN (DESERT) LANDFORMS.

Emiliani, Dr. Cesare. *Rosenstiel School of Marine and Atmospheric Sciences.* MARINE SEDIMENTS—in part.

Engel, Prof. A. E. J. *Professor of Geology, University of California, La Jolla.* CONTINENT FORMATION.

Epstein, Samuel. *Technical Advisor (retired), Bethlehem Steel Company.* GEOLOGIC THERMOMETRY—in part.

Ericson, David B. *Lamont-Doherty Geological Observatory, Palisades, NY.* MARINE SEDIMENTS—in part.

Evans, Prof. Robley D. *Department of Physics, Massachusetts Institute of Technology.* HALF-LIFE—in part.

Ewing, John. *Senior Research Associate, Lamont-Doherty Geological Observatory, Palisades, NY.* MARINE GEOLOGY—in part.

Ewing, Dr. Maurice. *Deceased; formerly, Lamont-Doherty Geological Observatory, Palisades, NY.* MARINE GEOLOGY—in part.

F

Fairbridge, Prof. Rhodes. *Department of Geology, Columbia University.* PALEOCLIMATOLOGY.

Ferm, Dr. John C. *Department of Geology, University of South Carolina.* CYCLOTHEM.

Fitzsimmons, Dr. J. Paul. *Department of Geology, University of New Mexico.* ARCHEOZOIC; BASEMENT ROCK; PRECAMBRIAN; PROTEROZOIC.

Fleischer, Dr. R. L. *Physical Science Branch, General Physics Laboratory, General Electric Company, Schenectady.* FISSION TRACK DATING—coauthored.

Flint, Dr. Richard F. *Department of Geology and Geophysics, Yale University.* CIRQUE; FIORD; GLACIAL EPOCH; PLEISTOCENE; other articles on geology.

Fogarty, David E. *Formerly, Staff Editor, "McGraw-Hill Encyclopedia of Science and Technology," McGraw-Hill Book Company, New York.* DATING METHODS; TECTONOPHYSICS.

Forsyth, R. B. *Carbon Products Division, Union Carbide Corporation, New York.* GRAPHITE—coauthored.

Frebold, Dr. Hans. *Principal Research Scientist (retired), Geological Survey of Canada.* JURASSIC.

Frey, Herbert. *Geophysics Branch—Applications Directorate Code 922-Nasa, Goddard Space Flight Center, Greenbelt, MD.* UNIVERSE.

Frondel, Prof. Clifford. *Department of Geological Sciences, Harvard University.* ALLANITE; DIAMOND—in part; RADIOACTIVE MINERALS; SILICATE MINERALS; related articles.

G

Gaines, Alan M. *Department of Geology, University of Pennsylvania.* ANKERITE; CARBONATE MINERALS; DOLOMITE.

Garner, H. F. *Department of Geology, Newark College of Arts and Sciences, Rutgers University.* PHYSIOGRAPHIC PROVINCES.

Gaylord, W. M. *Project Manager, Union Carbide Corporation, New York.* GRAPHITE—coauthored.

George, Prof. T. Neville. *Department of Geology, University of Glasgow.* EVOLUTION, ORGANIC.

Godson, Dr. Warren L. *Meteorological Service of Canada, Toronto.* GEOPHYSICS—in part.

Goldich, Dr. Samuel S. *Department of Geology, Northern Illinois University.* BAUXITE; LATERITE.

Goodell, Prof. H. G. *Department of Geology and Oceanography, Florida State University.* MARINE SEDIMENTS—in part.

Gray, G. Ronald. *Director, Syncrude Canada Ltd., Edmonton, Alberta.* OIL SAND.

Grim, Dr. Ralph E. *Department of Geology, University of Illinois.* BENTONITE; FULLER'S EARTH; GLAUCONITE; ILLITE; articles on related minerals—all in part.

H

Hagner, Dr. A. F. *Professor of Geology, University of Illinois.* ORE AND MINERAL DEPOSITS.

Haise, Dr. Howard R. *Northern Plains Branch Headquarters, Soil and Water Conservation Research Division, Agricultural Research Service, U.S. Department of Agriculture, Fort Collins, CO.* SOIL—in part.

Halbouty, Dr. Michael T. *Consulting Geologist and Petroleum Engineer, Houston.* NATURAL GAS.

Hamilton, Dr. Edwin L. *U.S. Navy Electronics Laboratory, San Diego.* OCEANIC ISLANDS; SEAMOUNT AND GUYOT—both coauthored.

Hammond, Dr. Edwin H. *Professor and Head, Department of Geography, University of Tennessee.* BASIN; PLAINS; PLATEAU.

Hanor, Jeffrey S. *Department of Geology, Louisiana State University.* HYDROSPHERE, GEOCHEMISTRY OF.

Harbaugh, Dr. John W. *Department of Geology, Stanford University.* PETROLEUM GEOLOGY.

Harker, Dr. Robert I. *University of Pennsylvania.* ARAGONITE; AZURITE; CERUSSITE; MAGNESITE; related articles.

Hatcher, Robert D., Jr. *Department of Chemistry and Geology, Clemson University.* TECTONIC PATTERNS.

Heezen, Dr. Bruce C. *Lamont-Doherty Geological Observatory, Palisades, NY.* MARINE GEOLOGY—in part; TURBIDITY CURRENT.

Heinrich, Dr. E. William. *Professor of Mineralogy, Department of Geology and Mineralogy, University of Michigan.* BIOTITE; MICA; MUSCOVITE; ROCK; related articles.

Henderson, Dr. Edward P. *Curator Emeritus, Division of Mineralogy and Petrology, Smithsonian Institution.* METEORITE.

Hobbs, Dr. Herman H. *Professor of Physics, George Washington University.* CRYSTAL.

Hodgson, Dr. Robert A. *Research Associate, Gulf Research and Development Company, Pittsburgh.* JOINT.

Hoering, Dr. Thomas C. *Geophysical Laboratory, Carnegie Institution of Washington.* PALEOBIOCHEMISTRY.

Holland, Dr. Heinrich D. *Department of Geological and Geophysical Sciences, Harvard University.* ATMOSPHERE, EVOLUTION OF; LITHOSPHERE, GEOCHEMISTRY OF.

Holser, William T. *Department of Geology, University of Oregon.* HALITE.

House, M. R. *Department of Geology, University of Hull, England.* DEVONIAN.

Howell, Dr. Benjamin F., Jr. *Professor of Geophysics, Department of Geology and Geophysics, and Assistant Dean, Graduate School, Pennsylvania State University.* EARTH; GEOPHYSICS — in part.

Hunt, Daniel, Jr. *Head, National Centers and Facilities Operations, National Science Foundation.* MOHO (MOHOROVIČIĆ DISCONTINUITY) — in part.

Hunt, Dr. John M. *Department of Chemistry, Woods Hole Oceanographic Institution.* PETROLEUM, ORIGIN OF.

Hurlbut, Prof. Cornelius S., Jr. *Professor of Mineralogy, Department of Geological Sciences, Harvard University.* ANALCIME; HALOGEN MINERALS; NATIVE ELEMENTS; PYRITE; WOLLASTONITE; series of articles on individual minerals from mineral groups.

Hurley, Prof. Patrick M. *Department of Earth and Planetary Sciences, Massachusetts Institute of Technology.* METAMICT STATE; PLEOCHROIC HALOS.

I

Ingerson, Prof F. Earl. *Department of Geological Sciences, University of Texas.* GEOLOGIC THERMOMETRY — in part.

Inman, Dr. Douglas L. *Professor of Oceanography, Scripps Institution of Oceanography, La Jolla, CA.* NEARSHORE PROCESSES.

J

Jahns, Dr. Richard H. *Department of Earth Sciences, Stanford University.* PEGMATITE.

Jensen, Dr. Howard B. *Research Supervisor, Laramie Energy Research Center, Energy Research and Development Administration.* OIL SHALE — coauthored.

Judd, Prof. William R. *Professor of Rock Mechanics, School of Civil Engineering, Purdue University, and formerly, Chairman, NAS-NAE U.S. National Committee on Rock Mechanics.* ENGINEERING GEOLOGY; ROCK MECHANICS.

Judson, Prof. Sheldon. *Chairman, Department of Geological and Geophysical Sciences, Princeton University.* FALL LINE; FLUVIAL EROSION LANDFORMS.

K

Kanamori, Hiroo. *Department of Geophysics, California Institute of Technology.* BENIOFF ZONE.

Kaula, William M. *Institute of Geophysics and Planetary Physics, University of California, Los Angeles.* GEODESY.

Kay, Marshall. *Deceased; formerly, Newberry Professor of Geology, Columbia University.* GEOSYNCLINE; ORDOVICIAN; PALEOGEOGRAPHY; PALEOGEOLOGY; STRAND LINE.

Keller, Dr. George V. *Colorado School of Mines, Golden.* ROCK, ELECTRICAL PROPERTIES OF.

Keller, Prof. Walter D. *Department of Geology, University of South Florida.* EROSION; WEATHERING PROCESSES.

Kennedy, Dr. George C. *Institute of Geophysics and Planetary Physics, University of California, Los Angeles.* DIAMOND — coauthored.

Klein, Cornelis. *Department of Geology, Indiana University.* BORAX; KERNITE; NATIVE METALS; SULFUR.

Kleinpell, Dr. Robert M. *Department of Paleontology, University of California, Berkeley.* CENOZOIC; EOCENE; OLIGOCENE; series of articles on geologic epochs.

Krumbein, Dr. William C. *Department of Geology, Northwestern University.* FACIES.

Kubota, Dr. Joe. *Research Soil Scientist, Soil Survey Investigations, Plant, Soil and Nutrition Laboratory, U.S. Department of Agriculture, Ithaca, NY.* SOIL CHEMISTRY — in part.

Kulp, Laurence J. *President, Teledyne Isotopes, Westwood, NJ.* GEOCHRONOMETRY; ROCK, AGE DETERMINATION OF.

Kummel, Dr. Bernhard. *Museum of Comparative Zoology, Harvard University.* TRIASSIC.

L

Ladd, Dr. Harry S. *National Museum.* ATOLL; PALEOECOLOGY.

Lamar, John E. *Illinois State Geological Survey, Urbana.* SAND.

Laves, Dr. Fritz H. *Eidg. Technische Hoschschule, Institut für Kristallographie und Petrographie, Zurich.* ALBITE; ANDESINE; ANORTHITE; PERTHITE; articles on related minerals.

Lee, Dr. W. H. K. *Earthquake Research, U.S. Geological Survey, Menlo Park, CA.* EARTH, HEAT FLOW IN — in part.

Leopold, Dr. Luna B. *U.S. Geological Survey.* FLUVIAL EROSION CYCLE; STREAM TRANSPORT AND DEPOSITION.

Liddicoat, Richard T., Jr. *Gemological Institute of America, Los Angeles.* AMBER; AMETHYST; series of articles on various gems. CARAT; PRECIOUS STONES.

Lill, Gordon G. *Coast and Geodetic Survey, Environmental Science Services Administration, U.S. Department of Agriculture.* MOHO (MOHOROVIČIĆ DISCONTINUITY) — in part.

Linsley, Prof. Ray K. *Professor of Civil Engineering, Stanford University.* GEYSER — coauthored. Validator of GROUNDWATER; THERMAL SPRING.

Lochhead, Dr. Allan G. *Canada Department of Agriculture, Microbiology Research Institute, Ottawa.* HUMUS.

Longwell, Dr. Chester R. *Department of Geology, Stanford University.* DIASTROPHISM; GEOLOGY.

Lowell, Prof. Wayne R. *Professor of Geology, Indiana University.* AMBLYGONITE; CARNOTITE; MONAZITE; TURQUOIS; WAVELLITE.

Lowenstam, Dr. Heinz A. *Geological Sciences, California Institute of Technology.* PALEOECOLOGY, GEOCHEMICAL ASPECTS OF.

Luth, Dr. William. *Department of Earth Sciences, Stanford University.* LEUCITE.

Lyman, Dr. John. *Professor of Oceanography, University of North Carolina, Chapel Hill.* OCEANS AND SEAS.

M

Macdonald, Dr. Gordon A. *Institute of Geophysics, University of Hawaii.* AMYGDULE; PUMICE; VOLCANO; VOLCANOLOGY.

McGrain, Preston. *Assistant State Geologist, Kentucky Geological Survey, University of Kentucky.* OIL FIELD WATERS.

Mackin, Prof. J. Hoover. *Deceased; formerly, Department of Geology, University of Texas.* GEOMORPHOLOGY.

Markowitz, Dr. William. *Adjunct Professor of Physics, Nova University, Fort Lauderdale, and Editor, "Geophysical Surveys."* EARTH, ROTATION AND ORBITAL MOTION OF — in part.

Martin, Prof. James P. *Professor of Soil Science, Department of Soils and Plant Nutrition, College of Biological and Agricultural Sciences, University of California, Riverside.* SOIL — in part.

Menard, Dr. Henry W., Jr. *Scripps Institution of Oceanography, La Jolla, CA.* OCEANIC ISLANDS; SEAMOUNT AND GUYOT—both coauthored.

Middleton, Gerard V. *Department of Geology, McMaster University.* BEDDING.

Miles, Thomas K. *Shell Development Company.* ASPHALT AND ASPHALTITE—in part.

Miller, Elizabeth L. *Lamount-Doherty Geological Observatory, Palisades, NY.* OCEANS—coauthored.

Miller, Prof. Maynard M. *Professor of Geology, Michigan State University, and Director, Foundation for Glacial and Environmental Research, Seattle.* GLACIOLOGY; ICE FIELD.

Miller, Stanley L. *Department of Chemistry, University of California, San Diego.* LIFE, ORIGIN OF.

Moore, Dr. Paul B. *Department of the Geophysical Sciences, University of Chicago.* APATITE; MINERALOGY; PHOSPHATE MINERALS; SPINEL; articles on mineralogy.

Moore, Dr. Raymond C. *Deceased; formerly, Department of Geology, University of Kansas.* MISSISSIPPIAN; PALEOZOIC.

Moritz, Prof. Helmut. *Department of Geodesy, Technical University of Berlin.* TERRESTRIAL GRAVITATION.

Mueller, Prof. George. *Institute of Molecular Evolution, University of Miami.* BIOSPHERE, GEOCHEMISTRY OF.

Myer, H. George. *Department of Geology, Temple University.* EPIDOTE.

N

Nafe, Dr. John E. *Lamont-Doherty Geological Observatory, Palisades, NY.* MARINE SEDIMENTS—in part.

Nettleton, Dr. Lewis L. *Consultant, GAI/GMX Division, E.G.&G., Houston.* SALT DOME.

Newell, Dr. Norman D. *Department of Fossil and Living Invertebrates, American Museum of Natural History, New York.* EXTINCTION (BIOLOGY).

Noble, Dr. James A. *Geological Sciences, California Institute of Technology.* BATHOLITH; LACCOLITH.

O

Odom, I. Edgar. *Department of Geology, Northern Illinois University.* AUTHIGENIC MINERALS.

Opdyke, Dr. Neil. *Lamont-Doherty Geological Observatory, Palisades, NY.* PALEOMAGNETICS.

Osberg, Prof. Philip H. *Department of Geological Sciences, University of Maine.* ANTICLINE; FAULT AND FAULT STRUCTURES; FOLD AND FOLD SYSTEMS; RIFT VALLEY; articles on geology.

P

Palmer, Dr. Allison R. *Department of Earth and Space Sciences, State University of New York, Stony Brook.* CAMBRIAN.

Patterson, Dr. Clair C. *Division of Geological Sciences, California Institute of Technology.* EARTH, AGE OF.

Patton, Dr. Donald J. *Carnegie Institution of Washington.* RIVER—coauthored.

Peacor, Dr. Donald R. *Department of Geology, University of Michigan.* CORUNDUM; FELDSPATHOID.

Pitman, Walter C., III. *Lamont-Doherty Geological Observatory, Palisades, NY.* PLATE TECTONICS.

Pollack, Henry N. *Department of Geological Sciences, University of Durham, England.* ASTHENOSPHERE.

Power, Dr. James F. *Northern Great Plains Research Center, Agricultural Research Service, U.S. Department of Agriculture, Mandan, ND.* SOIL—in part.

Price, Dr. P. B. *Metal and Ceramic Physics Branch, General Physics Laboratory, General Electric Company, Schenectady.* FISSION TRACK DATING—coauthored.

Prinn, Prof. Ronald G. *Department of Meteorology, Massachusetts Institute of Technology.* ATMOSPHERIC CHEMISTRY.

Pryor, Prof. Wayne A. *Department of Geology, University of Cincinnati.* CARBONIFEROUS.

R

Ragland, Paul C. *Department of Geology, University of North Carolina, Chapel Hill.* ELEMENTS, GEOCHEMICAL DISTRIBUTION OF.

Ramberg, Prof. Hans. *Professor of Mineralogy, Institute of Geology and Mineralogy, University of Uppsala.* DUNITE.

Ramsdell, Prof. Lewis S. *Professor Emeritus, Department of Mineralogy, University of Michigan.* MINERAL.

Rodgers, Prof. John. *Department of Geology, Yale University.* DIASTEM; STRATIGRAPHIC NOMENCLATURE; STRATIGRAPHY; UNCONFORMITY.

Roedder, Dr. Edwin. *Geologist, U.S. Geological Survey.* ORE DEPOSITS, GEOCHEMISTRY OF—coauthored.

Romer, Prof. Alfred S. *Deceased; formerly, Museum of Comparative Zoology, Harvard University.* ANIMAL EVOLUTION.

Roper, Paul J. *Department of Geology, University of Southwestern Louisiana.* CLEAVAGE, ROCK.

Ross, Dr. Charles A. *Department of Geology, Western Washington State College.* PENNSYLVANIAN.

Roy, Dr. Chalmer J. *Professor of Geology, and Dean, College of Sciences and Humanities, Iowa State University.* LOESS; TILL.

Roy, Dr. Rustum. *Director, Materials Research Laboratory, Pennsylvania State University.* HIGH-PRESSURE PHENOMENA—coauthored.

Ruhe, Dr. Robert V. *Professor of Geology, Indiana University.* PALEOSOL.

Runcorn, Prof. Stanley K. *Department of Geophysics and Planetary Physics, School of Physics, The University, Newcastle upon Tyne.* ROCK MAGNETISM; TRANSFORM FAULT.

Russell, Prof. Richard J. *Deceased; formerly, Director, Coastal Studies Institute, Louisiana State University.* COASTAL LANDFORMS; ESCARPMENT; FLOODPLAINS.

Ryan, William B. F. *Lamont-Doherty Geological Observatory, Palisades, NY.* OCEANS—coauthored.

S

Sayre, Dr. Albert N. *Deceased; formerly, Consulting Groundwater Geologist, Behre Dolbear and Company.* GEYSER; GROUNDWATER; THERMAL SPRING—all coauthored.

Scholz, Christopher H. *Lamont-Doherty Geological Observatory, Palisades, NY.* EARTHQUAKE.

Schwab, Frederic. *Department of Geology, Washington and Lee University.* BLACK SHALE; GEOLOGY, EVOLUTION OF; PLANETOLOGY; SANDSTONE; SEDIMENTARY STRUCTURES.

Scott, Dr. Richard A. *U.S. Geological Survey, Department of the Interior, Denver.* FOSSIL SEEDS AND FRUITS.

Sharpe, Dr. C. F. Stewart. *Falls Church, VA.* AVALANCHE; EXFOLIATION, ROCK; LANDSLIDE; MASS WASTING; articles on geology.

Shaw, Dr. Denis M. *Department of Geology, McMaster University.* SCAPOLITE.

Shepard, Prof. Francis P. *Scripps Institution of Oceanography, La Jolla, CA.* SUBMARINE CANYON.

Sheriff, R. E. *Seiscom Delta, Inc., Houston.* GEOPHYSICAL EXPLORATION.

Shrock, Prof. Robert R. *Department of Earth and Planetary Sciences, Massachusetts Institute of Technology.* EARTH SCIENCES.

Siever, Dr. Raymond. *Department of Geological Sciences, Harvard University.* ARENACEOUS ROCKS; ARKOSE;

CHERT; SEDIMENTARY ROCKS; SEDIMENTATION; series of articles on sedimentary rocks.

Simonson, Dr. Roy W. *Director (retired), Soil Classification and Correlation, U.S. Department of Agriculture, Hyattsville, MD.* SOIL—in part.

Slichter, Prof. Louis B. *Institute of Geophysics, University of California, Los Angeles.* EARTH, DEFORMATIONS AND VIBRATIONS IN; EARTH TIDES.

Sloss, Dr. L. L. *Department of Geological Sciences, Northwestern University.* EVAPORITE, SALINE.

Smith, Dr. Guy D. *Soil Conservation Service, U.S. Department of Agriculture.* SOIL—in part.

Smith, Dr. H. T. U. *Department of Geology, University of Massachusetts.* Validator of GEOMORPHOLOGY.

Smith, Dr. John Ward. *Research Supervisor, Laramie Energy Research Center, Energy Research and Development Administration.* OIL SHALE—coauthored.

Spackman, Dr. William. *Department of Geology, Pennsylvania State University.* COAL PALEOBOTANY.

Sposito, Dr. Garrison. *Department of Soil Science and Agricultural Engineering, University of California, Riverside.* SOIL CHEMISTRY—in part.

Suppe, John. *Department of Geological Sciences, Princeton University.* BLUESCHIST METAMORPHISM.

Suter, Dr. Hans H. *Consultant, Calgary, Canada.* MUD VOLCANO.

Swain, Prof. Frederick M. *Department of Geology and Geophysics, University of Minnesota.* HUMUS, GEOCHEMISTRY OF.

Switzer, Dr. George. *Curator, Department of Mineral Sciences, National Museum of Natural History, Smithsonian Institution.* CALICHE; NITER; NITRATE MINERALS; SODA NITER.

T

Thomson, Dr. Ker C. *Director, Terrestrial Sciences Laboratory, Air Force Cambridge Research Laboratories, Hanscom Air Force Base, Bedford, MA.* SEISMOGRAPH; SEISMOLOGY.

Thurber, Dr. David L. *Lamont-Doherty Geological Observatory, Palisades, NY, and Queens College, City University of New York.* RADIOCARBON DATING.

Thwaites, Fredrik T. *Deceased; formerly, Assistant Professor Emeritus, Department of Geology, University of Wisconsin.* GLACIATED TERRAIN.

Tilton, Dr. George R. *Department of Geology, University of California, Santa Barbara.* LEAD ISOTOPES, GEOCHEMISTRY OF.

Tracy, Robert J. *Department of Geological Sciences, Harvard University.* AUGITE; ORTHORHOMBIC PYROXENE.

Trowbridge, Arthur C. *Deceased; formerly, Professor Emeritus of Geology, University of Iowa.* COASTAL PLAIN.

Turnbull, Prof. David. *Pierce Hall, Harvard University.* CRYSTAL GROWTH.

V

Van Andel, Dr. Tjeerd H. *Department of Oceanography, Oregon State University.* DELTA; HEAVY MINERALS. MARINE SEDIMENTS—in part.

Van Houten, Prof. Franklyn B. *Department of Geology, Princeton University.* REDBEDS.

Vaucouleurs, Prof. Gerard de. *Department of Astronomy, University of Texas.* EARTH, ROTATION AND ORBITAL MOTION OF—in part.

W

Wahl, Prof. Floyd M. *Professor and Chairman, Department of Geology, University of Florida.* BENTONITE; FULLER'S EARTH; GLAUCONITE; ILLITE; articles on silicate minerals—all coauthored.

Wahlstrom, Prof. Ernest E. *Department of Geological Sciences, University of Colorado.* PETROGRAPHY.

Walker, Dr. R. M. *Department of Physics, Washington University, St. Louis.* FISSION TRACK DATING—coauthored.

Weller, Dr. J. Marvin. *Professor Emeritus, Department of Geophysical Sciences, University of Chicago.* VARVE.

Wengerd, Dr. Sherman A. *Professor of Geology, University of New Mexico.* BIOHERM; BIOSTROME; STROMATOLITE.

White, William B. *Department of Geosciences, Pennsylvania State University.* GEOCHEMISTRY.

Wilson, Dr. John T. *Erindale College, University of Toronto.* WARPING OF EARTH'S CRUST.

Wuensch, Dr. Bernhardt J. *Department of Metallurgy and Materials Science, Massachusetts Institute of Technology.* BORNITE.

Wyllie, Dr. Peter J. *Department of Geophysical Sciences, University of Chicago.* ANDALUSITE; CALCITE; GRANITE—all in part. OLIVINE; PETROLOGY.

Y

Yatsu, Prof. Eiju. *Department of Geography, University of Guelph.* ROCK CONTROL (GEOMORPHOLOGY).

Appendix

Table of accepted mineral species

Mineral	Chemical formula	Crystallography class*	Hardness on Moh's scale	Specific gravity
Acanthite	Ag_2S	orth	2–2½	7.2–7.3
Acmite	$NaFe'''(Si_2O_6)$	mon	6–6½	3.5
Actinolite	$Ca_2(Mg,Fe)_5(Si_8O_{22})(OH)_2$	mon	5–6	3.0–3.2
Adamite	$Zn_2(OH)(AsO_4)$	orth	3½	4.32–4.48
Adelite	$CaMg(AsO_4)(OH,F)$	orth	5	3.73
Adularia	$KAlSi_3O_8$	tric	6	2.56
Aegirine	$NaFe(Si_2O_6)$	mon	6–6½	3.40–3.55
Aegirine-augite	$(Na,Ca)(Fe,Mg,Al,Fe)(Si_2O_6)$	mon	6	3.40–3.55
Aenigmatite	$(Na,Ca)_4(Fe'',Fe''',Mn,Ti,Al)_{13}(Si_2O_7)_6$	tric	5½	3.75
Afwillite	$(Ca,Pb)_{10}(OH,Cl)_2(Si_2O_7)_3$	mon	4	2.63
Agate	see quartz			
Aguilarite	Ag_4SeS	iso	2½	7.59
Ahlfeldite	$Ni(SeO_4) \cdot 6H_2O$			
Aikinite	$PbCuBiS_3$	orth	2–2½	7.07
Akermanite	$Ca_2Mg(Si_2O_7)$	tet	5	2.94
Akrochordite	$MgMn_4(AsO_4)_2(OH)_4 \cdot 4H_2O$	mon	3½	3.194
Alabandite	MnS	iso	3½–4	4.0
Alamosite	$PbSiO_3$	mon	4½	6.49
Alaskaite	$Pb(Ag,Cu)_2Bi_4S_8$		2	6.83
Albite	$Na(AlSi_3O_8)-Ab_{90}An_{10}$†	tric	6	2.62
Algodonite	Cu_6As	hex	4	8.38
Allactite	$Mn_7(AsO_4)_2(OH)_8$	mon	4½	3.83
Allanite	$X_2Y_3O(SiO_4)(Si_2O_7)(OH)$‡	mon	5½–6	3.5–4.2
Alleghanyite	$Mn(OH,F)_2 \cdot 2Mn_2(SiO_4)$	mon	5½	4.02
Allemontite	$AsSb$	hex	3–4	5.8–6.2
Allophane	$Al_4(Si_4O_{10})(OH)_8$	amor	2–3	1.85–1.89
Alluaudite	$(Fe^{3+},Mn^{2+})(PO_4)$		5–5½	3.4–3.5
Almandine	$Fe_3Al_2(SiO_4)_3$	iso	6–7½	4.32
Almandite	$Fe_3Al_2(SiO_4)_3$	iso	7	4.25
Alstonite	$CaBa(CO_3)_2$	orth	4–4½	3.71
Altaite	$PbTe$	iso	3	8.16
Aluminite	$Al_2(SO_4)(OH)_4 \cdot 7H_2O$	mon or orth	1–2	1.66–1.82
Alumohydrocalcite	$CaAl_2(CO_3)_2(OH)_4 \cdot 2H_2O$	mon	2½	2.231
Alunite	$KAl_3(SO_4)_2(OH)_6$	hex	4	2.6–2.9
Alunogen	$Al_2(SO_4)_3 \cdot 18H_2O$	tric	1½–2	1.77
Amarantite	$Fe(SO_4)(OH) \cdot 3H_2O$	tric	2½	2.19
Amarillite	$NaFe(SO_4)_2 \cdot 6H_2O$	mon	2½–3	2.19
Amazonite	$KAlSi_3O_8$	tric	6–6½	2.56–2.63
Amblygonite	$(Li,Na)Al(PO_4)(F,OH)$	tric	5½–6	3.11
Amesite	$(Mg_4Al_2)(Si_2Al_2)O_{10}(OH)_8$	hex	2–3	2.75–2.85
Amethyst	SiO_2	hex	7	2.65
Aminoffite	$Ca_8Be_3Al(OH)_3Si_8O_{28} \cdot 4H_2O$			
Ammonia alum	$(NH_4)Al(SO_4)_2 \cdot 12H_2O$	iso	1½	1.65
Ammonioborite	$(NH_4)_2B_{10}O_{16} \cdot 5H_2O$			
Ammoniojarosite	$(NH_4)Fe_3(SO_4)_2(OH)_6$	hex	2½–3½	3.11
Ampangabeite	$A_2B_7O_{18}-AB_6O_{14}$†	orth	4	3.36–4.64
Amphibole	$(Na,Ca)_{2-3}(Mg,Fe^{2+},Fe^{3+},Al)_5(Si,Al)_8O_{22}$-$(OH,O,F)_2$ and $(Mg,Fe^{2+},Fe^{3+},Al)_7$-$(Si,Al)_8O_{22}(OH,O,F)_2$	orth, mon	5	3.0–3.57
Analcime	$Na(AlSi_2O_6) \cdot H_2O$	iso	5–5½	2.27
Anapaite	$Ca_2Fe(PO_4)_2 \cdot 4H_2O$	tric	3½	2.81
Anatase	TiO_2	tet	5½–6	3.90
Anauxite	$(Al,H_3)_4(Si_4O_{10})(OH)_8$	tric	2–2½	2.52
Ancylite	$(Ce,La)_4(Sr,Ca)_3(CO_3)_7(OH)_4 \cdot 3H_2O$	orth	4–4½	3.95
Andalusite	Al_2SiO_5	orth	7½	3.16–3.20
Andersonite	$Na_2Ca(UO_2)(CO_3)_3 \cdot 6H_2O$	hex		2.8
Andesine	$Ab_{70}An_{30}-Ab_{50}An_{50}$†	tric	6	2.69
Andorite	$PbAgSb_3S_6$	orth	3–3½	5.35
Andradite	$Ca_3Fe_2(SiO_4)_3$	iso	7	3.75
Andrewsite	$(Cu,Fe^{2})_3Fe^{3}_6(PO_4)_4(OH)_{12}$	orth	4	3.48
Anglesite	$PbSO_4$	orth	2½–3	6.38
Anhydrite	$CaSO_4$	orth	3½	2.98
Ankerite	$Ca(Fe,Mg)(CO_3)_2$	hex	3½–4	2.85
Annabergite	$(Ni,Co)_3(AsO_4)_2 \cdot 8H_2O$	mon	1½–2½	3.07
Anorthite	$Ab_{10}An_{90}$†	tric	6	2.76
Anorthoclase	$Or_{40}Ab_{40}-Or_{10}Ab_{90}\pm$ up to 20 mole % An†	tric	6	2.58

*For footnotes, see page 889.

Table of accepted mineral species (cont.)

Mineral	Chemical formula	Crystallography class*	Hardness on Moh's scale	Specific gravity
Anthoinite	$Al(WO_4)(OH) \cdot H_2O$		1	4.6
Anthophyllite	$(Mg,Fe)_7(Si_8O_{22})(OH,F)_2$	orth	5½–6	2.85–3.2
Antigorite	$Mg_3(Si_2O_5)(OH)_4$	mon	2½–3½	2.65
Antimony	Sb	hex	3–3½	6.61–6.72
Antlerite	$Cu_3(SO_4)(OH)_4$	orth	3½	3.88
Apatite	$Ca_5(F,Cl,OH)(PO_4)_3$	hex	5	3.15–3.20
Aphthitalite	$(K,Na)_3Na(SO_4)_2$	hex	3	2.656
Apjohnite	$Mn^{2+}Al_2(SO_4)_4 \cdot 22H_2O$	mon	1½	1.78
Apophyllite	$Ca_4K(Si_4O_{10})_2F \cdot 8H_2O$	tet	4½–5	2.3–2.4
Aragonite	$CaCO_3$	orth	3½–4	2.95
Aramayoite	$Ag(Sb,Bi)S_2$	tric	2½	5.60
Arcanite	K_2SO_4	orth		2.66
Ardealite	$Ca_2H(PO_4)(SO_4) \cdot 4H_2O$	mon		2.3
Ardennite	$Mn_5Al_5(V,As)(OH)_2Si_5O_{24} \cdot 2H_2O$	orth	6	3.6–3.65
Arfvedsonite	$Na_3Fe^{2+},Fe^{3+}Si_8O_{22}(OH)_2$	mon	6	3.45
Argentite	Ag_2S	iso	2–2½	7.2–7.4
Argentojarosite	$AgFe_3(SO_4)_2(OH)_6$	hex		3.66
Argyrodite	Ag_8GeS_6	iso	2½	6.1–6.3
Arizonite	$Fe_2Ti_3O_9$	mon	5½	4.25
Armangite	$Mn_3(AsO_3)_2$	hex	4	4.43
Armenite	$BaCa_2Si_8Al_6O_{28}$ $2H_2O$			
Arrojadite	$Na_2(Fe^{2+},Mn^{2+})_5(PO_4)_4$	mon	5	3.55
Arsenic	As	hex	3½	5.63–5.78
Arseniopleite	$(Mn^2,Ca,Pb,Mg)_9(Mn^2,Fe)_2(AsO_4)_6(OH)_6$	hex	3–4	
Arseniosiderite	$Ca_3Fe_4(AsO_4)_4(OH)_4 \cdot 4H_2O$	hex or tet	1½–4½	3.60
Arsenoclasite	$Mn_5(AsO_4)_2(OH)_4$	orth	5–6	4.16
Arsenolamprite	As		2	5.3–5.5
Arsenolite	As_2O_3	iso	1½	3.87
Arsenopyrite	FeAsS	mon	5½–6	6.07
Artinite	$Mg_2(CO_3)(OH)_2 \cdot 3H_2O$	mon	2½	2.02
Arzunite	Pb,Cu sulfate-chloride			
Ashcroftine	$KNa(Ca,Mg,Mn)(Al_4Si_5O_{18}) \cdot 8H_2O$	tet	3	2.61
Astrophyllite	$(K_2,Na_2,Ca)(Fe,Mn)_4(Ti,Zr)(OH,F)_2Si_4O_{14}$	tric	3	3.3–3.4
Atacamite	$Cu_2(OH)_3Cl$	orth	3–3½	3.76
Atelestite	$Bi_3(AsO_4)O_2(OH)_2$	mon	4½–5	6.82
Attapulgite	hydrous Mg,Fe silicate			
Augelite	$Al_2(PO_4)(OH)_3$	mon	4½–5	2.7
Augite	$(Ca,Na)(Mg,Fe,Al)(Si,Al)_2O_6$	mon	5–6	3.2–3.4
Aurichalcite	$(Zn,Cu)_5(OH)_6(CO_3)_2$	orth	1–2	3.64
Aurosmiridium	Ir,Os,Au	iso	>7	20
Austinite	$CaZn(AsO_4)(OH)$	orth	4–4½	4.13
Autunite	$Ca(UO_2)_2(PO_4)_2 \cdot 10{-}12H_2O$	tet	2–2½	3.1–3.2
Avogadrite	$(K,Cs)BF_4$	orth		2.51
Axinite	$Ca_2(Fe,Mn)Al_2(BO_3)(Si_4O_{12})(OH)$	tric	6½–7	3.27–3.35
Azurite	$Cu_3(OH)_2(CO_3)_2$	mon	3½–4	3.8
Babingtonite	hydrous Ca,Fe silicate	tric	5½–6	3.36
Baddeleyite	ZrO_2			
Badenite	$(Co,Ni)(As,Bi)_4$	mon	6½	6.02
Bakerite	$Ca_4B_4(BO_4)(SiO_4)_3(OH)_3 \cdot H_2O$		4½	2.88
Banalsite	$BaNa_2(Al_2Si_2O_8)_2$	orth	6	3.06
Bandylite	$CuB_2O_4 \cdot CuCl_2 \cdot 4H_2O$	tet	2½	2.810
Baraite	$(NH_4)_2SiF_6$	hex	2½	2.152
Barbertonite	$Mg_6Cr_2(OH)_{16} \cdot CO_3 \cdot 4H_2O$	hex	1½–2	2.05–2.15
Barkevikite	$Ca_2(Na,K)(Fe,Mg,Mn)_5Si_{6.5}Al_{1.5}O_{22}(OH)_2$			
Barite	$BaSO_4$	orth	3–3½	4.50
Barroisite	Na,K,Ca,Mg,Fe,Ti aluminosilicate			
Barylite	$BaBeSi_2O_7$	orth	7	4
Barysilite	$Pb_3Si_2O_7$	orth	3	4
Barytocalcite	$CaBa(CO_3)_2$	mon	4	3.66
Basaltic hornblende	$Ca_2(Na,K)_{0.5-1.0}(Mg,Fe)_{3-2}(Fe,Al)_{2-3}$-$Si_6Al_2O_{22}(O,OH,F)_2$	mon	5	3.19–3.30
Basaluminite	$Al_4(SO_4)(OH)_{10} \cdot 5H_2O$	hex		2.12
Bassanite	$2CaSO_4 \cdot H_2O$			
Bassetite	ferrous uranyl phosphate	mon	2–2½	3.10
Bastnaesite	$(Ce,La)(CO_3)F$	hex	4–4½	4.9–5.2
Batavite	hydrous Mg,Al silicate			
Baumhauerite	$Pb_4As_6S_{13}$	mon	3	5.33

Table of accepted mineral species (cont.)

Mineral	Chemical formula	Crystallography class*	Hardness on Moh's scale	Specific gravity
Bauxite	hydrous Al oxides		2–4	2–3.5
Bavenite	$Ca_4(OH)_4Si_9Al_2BeO_{24}$	orth	5½	2.74
Bayldonite	$Cu_3(AsO_4)_2(OH)_2$	mon	4½	5.5
Bayleyite	$Mg_2(UO_2)(CO_3)_3 \cdot 18H_2O$	mon		2.05
Beaverite	$Pb(Cu,Fe,Al)_3(SO_4)_2(OH)_6$	hex		4.36
Beckelite	$(Ca,Ce,La,Nd)_4O_3Si_3O_{12}$	hex	5	3.70–3.76
Becquerelite	$CaU_6O_{19} \cdot 11H_2O$	orth	2–3	5.2
Beegerite	$Pb_6Bi_2S_9$	iso		7.27
Beidellite	$Al_8(Si_4O_{10})_3(OH)_{12} \cdot 12H_2O$	orth?	1½	2.6
Bellingerite	$3Cu(IO_3)_2 \cdot 2H_2O$	tric	4	4.89
Benitoite	$BaTiSi_3O_9$	hex	6½	3.6
Benjaminite	$Pb_2(Cu,Ag)_2Bi_4S_9$		3½	6.34
Beraunite	$Fe^{2+}Fe^{3+}(PO_4)_3(OH)_5 \cdot 3H_2O$	mon	3½–4	2.8–2.99
Beresovite	$Pb_6(CrO_4)_3(CO_3)O_2$			
Berlinite	$AlPO_4$	hex	6½	2.64
Bermanite	$Mn^{2+}Mn_2^{3+}(PO_4)_2(OH)_2 \cdot 4H_2O$	orth	3½	2.84
Berthierite	$FeSb_2S_4$	orth	2–3	4.64
Berthonite	$Pb_2Cu_7Sb_5S_{13}$		4–5	5.49
Bertrandite	$Be_4(OH)_2Si_2O_7$	orth	6	2.6
Beryl	$Be_3Al_2(Si_6O_{18})$	hex	7½–8	2.75–2.8
Beryllonite	$NaBe(PO_4)$	mon	5½–6	2.81
Berzelianite	Cu_2Se	iso	2	6.71
Berzeliite	$(Mg,Mn)_2(Ca,Na)_3(AsO_4)_3$	iso	4½–5	4.08
Betafite	$(U,Ca)(Cb,Ta,Ti)_3O_9 \cdot nH_2$	iso	4–5½	3.7–5
Beta sulfur		mon	2–3	1.96–1.98
Beudantite	$PbFe_3(AsO_4)(SO_4)(OH)_6$	hex	3½–4½	4–4.3
Beyerite	$(Ca,Pb)Bi_2(CO_3)_2O_2$	tet	2–3	6.56
Bianchite	$(Fe,Zn)SO_4 \cdot 6H_2O$	mon	2½	2.07
Bieberite	$CoSO_4 \cdot 7H_2O$	mon	2	1.96
Bilinite	$Fe^{2+}Fe^{3+}(SO_4)_4 \cdot 22H_2O$		2	1.88
Bindheimite	$Pb_2SB_2O_6(O,OH)$	iso	4–4½	4.6–5.6
Biotite	$K_2[Fe(II),Mg]_{6-4}[Fe(III),Al,Ti]_{0-2}\cdot(Si_{6-5},Al_{2-3})O_{20-22}(OH,F)_{4-2}$	mon	2½–3	2.8–3.2
Bischofite	$MgCl_2 \cdot 6H_2O$	mon	1–2	1.6
Bismite	Bi_2O_3	mon	4½	8.6–10.4
Bismoclite	$BiOCl$	tet	2–2½	7.72
Bismuth	Bi	hex	2–2½	9.7–9.8
Bismuthinite	Bi_2S_3	orth	2	6.78
Bismutite	$(BiO)_2(CO_3)$	tet	2½–3½	6.7–7.4
Bismutotantalite	$Bi(Ta,Nb)O_4$	orth	5	8.26
Bityite	$Ca_4(Li,Be)_4Al_8(OH)_{20}[(Si,Al)_4O_{10}]_3$	mon	5½	3.0
Bixbyite	$(Mn,Fe)_2O_3$	iso	6–6½	4.95
Blakeite	ferric tellurite		2½	3.1
Bloedite	$Na_2Mg(SO_4) \cdot 4H_2O$	mon	2½–3	2.25
Bloodstone		hex	6½–7	2.57–2.65
Blue quartz		hex	7	2.65
Bobierrite	$Mg_3(PO_4)_2 \cdot 8H_2O$	mon	2–2½	2.195
Boehmite	$AlO(OH)$	orth		3.01–3.06
Boleite	$Pb_9Cu_8Ag_3Cl_{21}(OH)_{16} \cdot 2H_2O$	tet	3–3½	5.05
Boothite	$CuSO_4 \cdot 7H_2O$	mon	2–2½	2.1
Boracite	$Mg_3B_7O_{13}Cl$	orth	7–7½	2.91–2.97
Borax	$Na_2B_4O_7 \cdot 10H_2O$	mon	2–2½	1.72
Borickite	$CaFe_5(PO_4)_2(OH)_{11} \cdot 3H_2O$		3½	2.69–2.71
Bornite	Cu_5FeS_4	iso	3	5.06–5.08
Bort	C	iso	10	3.5
Botallackite	$Cu_2(OH)_3Cl \cdot 3H_2O$	orth		3.6
Botryogen	$MgFe(SO_4)_2(OH) \cdot 7H_2O$	mon	2–2½	2.14
Boulangerite	$Pb_5Sb_4S_{11}$	mon	2½–3	6.23
Bournonite	$PbCuSbS_3$	orth	2½–3	5.83
Boussingaultite	$(NH_4)_2Mg(SO_4)_2 \cdot 6H_2O$	mon	2	1.722
Brackebuschite	$Pb_4MnFe(VO_4)_4 \cdot 2H_2O$	mon		6.05
Bradleyite	$Na_3Mg(CO_3)(PO_4)$			
Braggite	$(Pt,Pd,Ni)S$	tet		10
Brandtite	$Ca_2Mn(AsO_4)_2 \cdot 2H_2O$	mon	3½	3.67
Brannerite	$(U,Ca,Fe,Y,Th)_3Ti_5O_{16}$‡		4½	4.5–5.43
Braunite	$(Mn,Si)_2O_3$	tet	6–6½	4.72–4.83
Bravoite	$(Ni,Fe)S$	iso	5½–6	4.66

Table of accepted mineral species (cont.)

Mineral	Chemical formula	Crystallography class*	Hardness on Moh's scale	Specific gravity
Brazilianite	$NaAl_3(PO_4)_2(OH)_4$	mon		2.98
Bredigite	Ca_2SiO_4			
Breithauptite	NiSb	hex	5½	8.23
Brewsterite	$(Sr,Ba,Ca)(Al_2Si_6O_{16}) \cdot 5H_2O$	mon	3½–4	2.45
Brittle mica	hydrous Na,Ca,Mg,Al silicates	mon	3½–7	2.6–3.2
Brochantite	$Cu_4(SO_4)(OH)_6$	mon	3½–4	3.97
Bromargyrite	AgBr	iso	2½	6.50
Bromellite	BeO	hex	9	3.02
Bromyrite	AgBr	iso	2½	5.56
Bronzite	$(Mg,Fe)_2(Si_2O_6)$	orth	5½	3.3±
Brookite	TiO_2	orth	5½–6	3.9–4.1
Brucite	$Mg(OH)_2$	hex	2½	2.39
Brugnatellite	$Mg_6Fe(OH)_{13}CO_3 \cdot H_2O$	hex	2	2.14
Brunsvigite	hydrous Fe,Mg,Al silicate	mon	2–3	2.8–3.1
Brushite	$CaHPO_4 \cdot 2H_2O$	mon	2½	2.33
Buetschliite	$K_6Ca_2(CO_3)_5 \cdot 6H_2O$			
Bultfonteinite	$H_2Ca_2F_2SiO_4$	tric	4½	2.73
Bunsenite	NiO	iso	5½	6.89
Burkeite	$Na_6(CO_3)(SO_4)_2$	orth	3½	2.57
Bustamite	$(Mn,Ca,Fe)(SiO_3)$	tric	5½–6½	3.32–3.43
Butlerite	$Fe(SO_4)(OH) \cdot 2H_2O$	mon	2½	2.55
Buttgenbachite	$Cu_{19}(NO_3)_2Cl_4(OH)_{32} \cdot 3H_2O$	hex	3	3.36
Byssolite	hydrous Ca,Mg,Fe aluminosilicate	mon	5–6	3.02–3.44
Bytownite	$Ab_{30}An_{70}-Ab_{10}An_{90}$†	tric	6–6½	2.71–2.75
Cacoxenite	$Fe_4(PO_4)_3(OH)_3 \cdot 12H_2O$	hex	3–4	2.2–2.4
Cadmium oxide	CdO	iso	3	8.1–8.2
Cadwaladerite	$Al(OH)_2Cl \cdot 4H_2O$			1.66
Cahnite	$Ca_2B(OH)_4(AsO_4)$	tet	3	3.16
Calaverite	$AuTe_2$	mon	2½–3	9.24
Calcioferrite	$Ca_3Fe_3(PO_4)_4(OH)_3 \cdot 8H_2O$	hex	2½	2.53
Calciovolborthite	$CuCa(VO_4)(OH)$	orth	3½	
Calcite	$CaCO_3$	hex	3	2.96–3.21
Calclacite	$CaCl(C_2H_3O_2) \cdot 5H_2O$			
Calderite	$Mn_3Fe_2(SiO_4)_3$			
Caledonite	$Cu_2Pb_5(SO_4)_3(CO_3)(OH)_6$	orth	2½–3	5.76
Calomel	HgCl	tet	1½	7.15
Cancrinite	$Na_6Ca[CO_3][(AlSiO_4)_6] \cdot H_2O$	hex	5–6	2.42–2.51
Canfieldite	Ag_8SnS_6	iso	2½	6.28
Caracolite	Na,Pb chloride-sulfate		4½	5.0–5.2
Carbonate-apatite	$CaF(PO_4,CO_3,OH)_3$	hex	5	2.9–3.1
Cardenite	hydrous Mg,Ca,Fe,Al silicate			
Carminite	$PbFe_2(AsO_4)_2(OH)_2$	orth	3½	4.10
Carnallite	$KMgCl_3 \cdot 6H_2O$	orth	2½	1.602
Carnelian		hex	6½	2.57–2.64
Carnotite	$K_2(UO_2)_2(VO_4)_2 \cdot nH_2O$	orth	soft	4.1
Carpholite	$MnAl_2(OH)_4Si_2O_6$	orth	5–5½	2.94
Carphosiderite	$(H_2O)Fe_3(SO_4)_2[(OH)_5H_2O]$	hex	4–4½	2.5–2.9
Carrollite	Co_2CuS_4	iso	4½–5½	4.83
Caryinite	$(Ca,Pb,Na)_5(Mn,Mg)_4(AsO_4)_5$	orth	4	4.29
Cassiterite	SnO_2	tet	6–7	6.99
Catapleite	$Na_2ZrSi_3O_9 \cdot 2H_2O$	hex	6	2.75
Catoptrite	Mn,Al antimonate-silicate	mon	5½	4.5
Celadonite	$(K,Ca,Na)(Al,Fe,Mg)_2(Al_{0.11}Si_{3.89}O_{10})(OH)_2$	mon	1–2	2.6–2.9
Celestite	$SrSO_4$	orth	3–3½	3.97
Celsian	$BaAl_2Si_2O_8$	mon	6	3.37
Cerargyrite	AgCl	iso	2½	5.56
Cerite	$(Ca,Fe)Ce_3Si_3O_{12} \cdot H_2O$	mon	5½	4.65–4.91
Cerussite	$PbCO_3$	orth	3–3½	6.55
Cervantite	Sb_2O_4	orth	4–5	6.64
Cesarolite	$PbMn_3O_7 \cdot H_2O$		4½	5.29
Chabazite	$(Ca,Na_2)[Al_2Si_4O_{12}] \cdot 6H_2O$	rho	4–5	2.05–2.15
Chalcanthite	$CuSO_4 \cdot 5H_2O$	tric	2½	2.29
Chalcedony	SiO_2	hex	6½	2.57–2.64
Chalcoaluminite	$CuAl_4(SO_4)(OH)_{12} \cdot 3H_2O$	tric	2½	2.29
Chalcocite	Cu_2S	orth	2½–3	5.5–5.8
Chalcocyanite	$CuSO_4$	orth	3½	3.65

Table of accepted mineral species (cont.)

Mineral	Chemical formula	Crystallography class*	Hardness on Moh's scale	Specific gravity
Chalcomenite	$CuSeO_3 \cdot 2H_2O$	orth	2–2½	3.35
Chalcophanite	$(Zn,Mn,Fe)Mn_2I_5 \cdot 2H_2O$	hex	2½	4
Chalcophyllite	$Cu_{18}Al_2(AsO_4)_3(SO_4)_3(OH)_{27} \cdot 33H_2O$	hex	2	2.67
Chalcopyrite	$Cu_2Fe_2S_4$	tet	3½–4	4.1–4.3
Chalcosiderite	$CuFe_6(PO_4)_4(OH)_8 \cdot 4H_2O$	tric	4½	3.22
Chalcostibite	$CuSbS$	orth	3–4	4.95
Chamosite	hydrous Fe,Mg,Al silicate	mon	2–3	2.9–3.1
Chenevixite	$Cu_2Fe_2(AsO_4)_2(OH)_4 \cdot H_2O$		3½–4½	3.93
Chert	SiO_2		7	2.65
Childrenite	$(Fe,Mn)Al(PO_4)(OH)_2 \cdot H_2O$	orth	5	3.22–3.28
Chiolite	$Na_5Al_3F_{14}$	tet	3½–4	2.99
Chiviatite	$Pb_3Bi_8S_{15}$		2–3	6.92
Chloanthite	$AlCl_3 \cdot 6H_2O$			
Chloraluminite	$(Ni,Co)As_{3-x}(x=0.5-1)$	iso	5½–6	6.5
Chlorapatite	$Ca_5(PO_4)_3Cl$	hex	5	3.1–3.2
Chlorargyrite	$AgCl$	iso	2½	5.56
Chlorite	$Mg_3(Si_4O_{10})(OH)_2 \cdot Mg_3(OH)_6$	mon	2–2½	2.6–2.9
Chloritoid	$[Fe(II),Mg][Al,Fe(III)](Si,Al)O_5(OH)_2$	mon	6–7	3.5
Chlormanganokalite	K_4MnCl_6	hex	2½	2.31
Chlorocalcite	$KCaCl_3$	orth	2½–3	
Chloromagnesite	$MgCl_2$			
Chlorophoenicite	$(Mn,Zn)_5(AsO_4)(OH)_7$	mon	3½	3.46
Chlorothionite	$K_2Cu(SO_4)Cl_2$	orth	2½	2.67
Chloroxiphite	$Pb_3CuO_2(OH)_2Cl_2$	mon	2½	6.76
Chondrodite	$Mg_5(SiO_4)_2(F,OH)_2$	mon	6–6½	3.1–3.2
Chromite	$FeCr_2O_4$	iso	5½	4.5–4.8
Chrysoberyl	$BeAl_2O_4$	orth	8½	3.75
Chrysocolla	$CuSiO_3 \cdot 2H_2O$	?	2–4	2.0–2.4
Chrysolite	$(Mg,Fe)_2(SiO_4)$	orth	6½–7	3.3–3.6
Chrysoprase		hex	6½	2.57–2.64
Chrysotile	ca. $Mg_3(Si_2O_5)(OH)_4$	mon	2½	2.2
Churchite	$(Ce,Ca)(PO_4) \cdot 2H_2O$	mon	3	3.14
Cinnabar	HgS	hex	2–2½	8.09
Citrine	SiO_2	hex	7	2.65
Clarkeite	$(Na,Ca,Pb)_2U_2(O,OH)_7$		4–4½	6.39
Claudetite	As_2O_3	mon	2½	4.15
Clausthalite	$PbSe$	iso	2½–3	7.8
Clay	chiefly hydrous silicates of Al or Mg	mon	1–3	1.85–3.0
Cliachite	$Al(OH)_3$	amor	1–3	2.5±
Clinoclase	$Cu_3AsO_4(OH)_3$	mon	2½–3	4.38
Clinoenstatite	$Mg_2(Si_2O_6)$	mon	6	3.19
Clinoferrosilite	$FeSiO_3$	mon	6	3.5–3.7
Clinohedrite	$Ca_2Zn_2(OH)_2Si_2O_7 \cdot H_2O$	mon	5½	3.33
Clinohumite	$Mg_9(SiO_4)_4(F,OH)_2$	mon	6	3.1–3.2
Clinohypersthene	$(Mg,Fe)_2(Si_2O_6)$	mon	5–6	3.4–3.5
Clinosklodowskite	$Mg(H_3O)_2UO_2Si_2O_8 \cdot 3H_2O$			
Clinoungemachite	$Na_9K_3Fe(SO_4)_6(OH)_3 \cdot 9H_2O$	mon	2½	2.29
Clinozoisite	$Ca_2Al_3(OH)Si_3O_{12}$	mon	6½	3.21–3.38
Clintonite	$Ca(Mg,Al)_{3-2}Al_2Si_2O_{10}(OH)_2$	mon	3½–6	3–3.1
Cobalt bloom	$CO_3(AsO_4)_2 \cdot 8H_2O$			
Cobaltite	$(Co,Fe)AsS$	iso	5½	6.33
Cobaltocalcite	$CoCO_3$	hex	4	4.13
Cobaltomentite	$CoSeO_3 \cdot 2H_2O$			
Cocinerite	Cu_4AgS		2½	6.14
Coeruleolactite	$Al_3(PO_4)_2(OH)_3 \cdot 4H_2O$		5	2.57
Coesite	SiO_2	mon	7	2.93
Coffinite	$U(Si,H_4)O_4$			
Cohenite	$(Fe,Ni)_3C$	orth	5½–6	7.20–7.65
Colemanite	$Ca_2B_6O_{11} \cdot 5H_2O$	mon	4½	2.42
Collinsite	$Ca_2(Mg,Fe)(PO_4)_2 \cdot 2H_2O$	tric	3½	2.99
Coloradoite	$HgTe$	iso	2½	8.04
Columbite	$(Fe,Mn)(Cb,Ta)_2O_6$	orth	6	5.20
Colusite	$Cu_3(As,Sn,V,Fe,Te)S_4$	iso	3–4	4.50
Common hornblende	$(Ca,Na,K)_{2-3}(Mg,Fe,Al)_5Si_6(Si,Al)_2O_{22}(OH,F)_2$	mon	5–6	3.02–3.45
Conichalcite	$CaCu(AsO_4)(OH)$	orth	4½	4.33
Connellite	$Cu_{19}(SO_4)Cl_4(OH)_{32} \cdot 3H_2O$	hex	3	3.36

Table of accepted mineral species (cont.)

Mineral	Chemical formula	Crystallography class*	Hardness on Moh's scale	Specific gravity
Cookeite	Li-bearing chlorite	mon	2–3	2.69
Cooperite	PtS	tet	4–5	9.5
Copiapite	$(Fe,Mg)Fe^{3}_{4}(SO_4)_6(OH)_2 \cdot 20H_2O$	tric	2½–3	2.08–2.17
Copper	Cu	iso	2½–3	8.95
Coquimbite	$Fe_2(SO_4)_3 \cdot 9H_2O$	hex	2½	2.11
Cordierite	$Mg_2[Al_4Si_5O_{18}]$	orth	7–7½	2.60–2.66
Cordylite	$(Ce,La)_2Ba(CO_3)_3F_2$	hex	4½	4.31
Corkite	$PbFe_3(PO_4)(SO_4)(OH)_6$	hex	3½–4½	4.29
Cornetite	$Cu_3(PO_4)(OH)_3$	orth	4½	4.10
Cornwallite	$Cu_5(AsO_4)_2(OH)_4 \cdot H_2O$		4½	4.17
Coronadite	$MnPbMn_6O_{14}$	tet	4½–5	5.44
Corundophilite	hydrous Mg,Fe,Al silicate	mon	2–3	2.7–2.9
Corundum	Al_2O_3	hex	9	4.0–4.1
Corvusite	$V_2V_{12}O_{34} \cdot nH_2O$		2½–3	2.82
Cosalite	$Pb_2Bi_2S_5$	orth	2½–3	6.76
Cotunnite	$PbCl_2$	orth	2½	5.80
Covellite	CuS	hex	1½–2	4.6–4.76
Crandallite	$CaAl_3(PO_4)_2(OH)_5H_2O$	hex	5	2.78
Crednerite	$CuMn_2O_4$	mon	4	5.01
Creedite	$Ca_3Al_2F_4(OH,F)_6(SO_4) \cdot 2H_2O$	mon	4	2.71
Cristobalite	SiO_2	tet?	7	2.30
Crocidolite	$Na_2Fe^{2+}_3Fe^{3+}_2(Si_8O_{22})(OH)_2$	mon	4	3.2–3.3
Crocoite	$Pb(CrO_4)$	mon	2½–3	5.99
Crookesite	$(Cu,Tl,Ag)_2Se$		2½–3	6.90
Cryolite	Na_3AlF_6	mon	2½	2.96–2.98
Cryolithionite	$Na_3Li_3Al_2F_{12}$	iso	2½–3	2.77
Cryptohalite	$(NH_4)_2SiF_6$	iso	2½	2.004
Cubanite	$CuFe_2S_3$	orth	3½	4.03–4.18
Cumengite	$Pb_4Cu_4Cl_8(OH)_8 \cdot H_2O$	tet	2½	4.67
Cummingtonite	$Fe_7Si_8O_{22}(OH)_2 - Mg_5Fe_2Si_8O_{22}(OH)_2$	mon	6	2.85–3.2
Cuprite	Cu_2O	iso	3½–4	6.14
Cuprocopiapite	$CuFe^{3+}_4(SO_4)_6(OH)_2 \cdot 20H_2O$	tric	2½–3	2.08–2.17
Cuprosklodowskite	$CuU_2O_3Si_2O_8 \cdot 6H_2O$			
Cuprotungstite	$Cu_2(WO_4)(OH)_2$			
Curite	$2PbO \cdot 5UO_3 \cdot 4H_2O$	orth	4–5	7.26
Cuspidine	$Ca_4(F,OH)_2Si_2O_7$	mon	5–6	2.95
Cyanochroite	$K_2Cu(SO_4)_2 \cdot 6H_2O$	mon		2.22
Cyanotrichite	$Cu_4Al_2(SO_4)(OH)_{12} \cdot 2H_2O$	orth		2.95
Cylindrite	$Pb_3Sn_4Sb_2S_{14}$		2½	5.46
Dachiardite	$(Ca,K_2,Na_2)_3Al_4Si_{18}O_{45} \cdot 14H_2O$	mon	3½–4	2.16
Dalyite	$K_2Zr(Si_6O_{15})$			
Danalite	$Fe_4Be_3Si_3O_{12}S$	iso	6	3.28–3.44
Danburite	$Ca(B_2Si_2O_8)$	orth	7	2.97–3.02
Daphnite	hydrous Fe,Al silicate	mon	2–3	2.75–3.1
Darapskite	$Na_3(NO_3)(SO_4) \cdot H_2O$	mon	2½	2.20
Datolite	$CaB(SiO_4)(OH)$	mon	5–5½	2.8–3.0
Daubréeite	$BiO(OH,Cl)$	tet	2–2½	7.72
Daubréelite	Cr_2FeS_4	iso		5.01
Daviesite	Pb oxychloride			
Davisonite	$Ca_3Al(PO_4)_2(OH)_3 \cdot H_2O$		4½	2.85
Davyne	$(Na,K)_6Ca_2(AlSiO_4)_6(SO_4)_2$	hex	5–6	2.42–2.5
Dawsonite	$NaAl(CO_3)(OH)_2$	orth	3	2.44
Dehrnite	$(Ca,Na,K)_5(PO_4)_3(OH)$	hex	5	3.04–3.09
Delafossite	$CuFeO_2$	hex	5½	5.41
Delessite	hydrous Mg,Fe,Al silicate	mon	2–3	2.7–2.9
Delorenzite	$(Y,U,Fe)(Ti,Sn)_3O_8$†	orth	5½–6	4.7
Deltaite	$Ca(Al_2,Ca)(PO_4)_2(OH)_4H_2O$	hex	5	2.95
Delvauxite	$Fe_2(PO_4)(OH)_3 \cdot xH_2O$		2½–4	1.99–2.83
Derbylite	$Fe^{2+}_6Ti_6Sb_2O_{23}$	orth	5	4.53
Descloizite	$(Zn,Cu)Pb(VO_4)(OH)$	orth	3–3½	6.2
Devilline	$Cu_4Ca(SO_4)_2(OH)_6 \cdot 3H_2O$	mon	2½	3.13
Devillite	$Cu_4Ca(SO_4)_2(OH)_6 \cdot 3H_2O$	mon	2½	3.13
Dewindtite	$Pb_3(UO_2)_5(PO_4)_4(OH)_4 \cdot 10H_2O$	orth		5.03
Diabantite	hydrous Mg,Al silicate	mon	2–3	2.7–2.9
Diaboleite	$Pb_2CuCl_2(OH)_4$	tet	2½	5.42
Diadochite	$Fe_2(PO_4)(SO_4)(OH) \cdot 5H_2O$	tric	3–4	2.0–2.4

Table of accepted mineral species (cont.)

Mineral	Chemical formula	Crystallography class*	Hardness on Moh's scale	Specific gravity
Diallage	$Ca_{14}Fe_2Mg_{13}FeAl_5Si_{29}O_{96}$	mon	5½–6½	3.2–3.6
Diamond	C	iso	10	3.50–3.53
Diaphorite	$Pb_2Ag_3Sb_3S_8$	orth	2½–3	6.04
Diaspore	$HAlO_2$	orth	6½–7	3.3–3.5
Dickinsonite	$H_2Na_6(Mn,Fe,Ca,Mg)_{14}(PO_4)_{12} \cdot H_2O$	mon	3½–4	3.38–3.41
Dickite	$Al_2Si_2O_5(OH)_4$	mon	2–2½	2.6
Dietrichite	$(Zn,Fe,Mn)Al_2(SO_4)_4 \cdot 22H_2O$	mon	2	
Dietzeite	$Ca_2(IO_3)_2(CrO_4)$	mon	3½	3.69
Digenite	Cu_9S_5	iso	2½–3	5.5–5.7
Dimorphite	As_4S_3	orth	1½	2.58
Diopside	$CaMg(Si_2O_6)$	mon	5–6	3.2–3.3
Dioptase	$Cu_6(Si_6O_{18}) \cdot 6H_2O$	rho	5	3.3
Dipyre	$meionite_{20-50}$	tet	5–6	2.57–2.69
Djalmaite	$(U,Ca,Pb,Bi,Fe)(Ta,Cb,Ti,Zr)_3O_9 \cdot nH_2O$	iso	5½	5.75–5.88
Dolerophanite	$Cu_2(SO_4)O$	mon	3	4.17
Dolomite	$CaMg(CO_3)_2$	hex	3½–4	2.85
Domeykite	Cu_3As	iso	3–3½	7.2–7.9
Donbassite	hydrous Na,Ca,Mg,Al silicate	tric	2½	2.63
Douglasite	$K_2FeCl_4 \cdot 2H_2O$			
Dravite	$NaMg_3Al_6(OH)_4(BO_3)_3Si_6O_{18}$	hex	7	3.03–3.15
Dufrenite	$Fe^{2+}Fe_4^{3+}(PO_4)_3(OH)_5 \cdot 2H_2O$	mon	3½–4½	3.1–3.34
Dufrenoysite	$Pb_2As_2S_5$	mon	3	5.53
Duftite	$PbCu(AsO_4)(OH)$	orth	3	6.98
Dumontite	$Pb_2(UO_2)_3(PO_4)_2(OH)_4 \cdot 3H_2O$			
Dumortierite	$Al_7O_3(BO_3)(SiO_4)$	orth	7	3.26–3.36
Dundasite	$PbAl_2(CO_3)_2(OH)_4 \cdot 2H_2O$		2	3.25
Durangite	$NaAlF(AsO_4)$	mon	5	3.94–4.07
Dussertite	$BaFe_3(AsO_4)_2(OH)_5H_2O$	hex	3½	3.75
Dyscrasite	Ag_3Sb	orth	3½–4	9.74
Earlandite	$Ca_3(C_6H_5O_7)_2 \cdot 4H_2O$			
Ecdemite	$Pb_6As_2O_7Cl_4$	tet	2½–3	7.14
Edenite	$Ca_2NaMg_5(AlSi_7O_{22})(OH,F)_2$	mon	6	3.0
Edingtonite	$Ba(Al_2Si_3O_{10}) \cdot 3H_2O$	tet	5–5½	2.7–2.8
Eglestonite	Hg_4OCl_2	iso	2½	8.33
Egueite	$CaFe_{14}(PO_4)_{10}(OH)_{14} \cdot 21H_2O$			
Ekermannite	$Na_3Mg_4(Fe^{3+},Al)Si_8O_{22}(OH)_2$	mon	5–6	3.00
Ekmannite	$(Fe,Mg,Mn)_3(OH)_2(Si,Al)Si_3O_{10}$	hex	5–7	5.26
Elbaite	$Na(Li,Al)_3Al_6(OH,F)_4(BO_3)_3Si_6O_{18}$	hex	7	3.03–3.25
Ellestadite	$Ca_5(Si,S,P)_3(Cl,F,OH)$	hex	4½–5½	3.07
Elpasolite	K_2NaAlF_6	iso	2½	2.99
Elpidite	$NaZrSi_6O_{15} \cdot 3H_2O$	orth	7	2.58
Embolite	$Ag(Cl,Br)$	iso	1–1½	5.3–5.4
Emmonsite	$Fe_2(TeO_3)_3 \cdot 2H_2O$	mon	5	4.52
Emplectite	$CuBiS_2$	orth	2	6.38
Empressite	$AgTe$		3–3½	7.5
Enargite	Cu_3AsS_4	orth	3	4.43–4.45
Englishite	$K_2Ca_4Al_8(PO_4)_8(OH)_{10} \cdot 9H_2O$	mon	3	2.65
Enigmatite	$(Ca,Na_2)_2Fe(Al,Fe,Ti)_4(Si_2O_7)_2$	tric	5½	3.74–3.85
Enstatite	$Mg(SiO_3)$	orth	5½	3.2–3.5
Eosphorite	$(Mn^{2+},Fe^{2+})Al(PO_4)(OH)_2 \cdot H_2O$	orth	5	3.25
Ephesite	$(Na,Ca)Al_2[Al(Al,Si)Si_2O_{10}](OH)_2$	mon	5–7	3.0
Epididymite	$NaBe(OH)Si_3O_7$	orth	5½	2.55
Epidote	$Ca_2Fe^{3+}Al_2O \cdot H[Si_2O_7]$	mon	6–7	3.35–3.45
Epigenite	$(Cu,Fe)_5AsS_6$	orth	3½	4.5
Epistilbite	$Ca(Al_2Si_6O_{16}) \cdot 5H_2O$	mon	4	2.1–2.3
Epsomite	$MgSO_4 \cdot 7H_2O$	orth	2–2½	1.75
Erinite	$Cu_5(OH)_4(AsO_4)_2$	orth	4½	4.04
Eriochalcite	$CuCl_2 \cdot 2H_2O$	orth	2½	2.47
Erionite	$Ca_{4.5}(Al_9Si_{27}O_{72}) \cdot 27H_2O$	hex		
Erythrite	$(Co,Ni)_3(AsO_4)_2 \cdot 8H_2O$	mon	1½–2½	3.06
Erythrosiderite	$K_2FeCl_5 \cdot H_2O$	orth		2.37
Eschynite	$(Ce,Ca,Fe,Th)(Ti,Cb)_2O_6$	orth	5–6	5.19
Ettringite	$Ca_6Al_2(SO_4)_3(OH)_{12} \cdot 26H_2O$	hex	2–2½	1.77
Eucairite	$CuAgSe$		2½	7.6–7.8
Euchroite	$Cu_2(AsO_4)(OH) \cdot 3H_2O$	orth	3½–4	3.44
Euclase	$B_2Al_2(SiO_4)_2(OH)_2$	mon	7½	3.1

Table of accepted mineral species (cont.)

Mineral	Chemical formula	Crystallography class*	Hardness on Moh's scale	Specific gravity
Eucryptite	$Li(Al,Si)_2O_4$	hex		2.67
Eudialyte	$(Na,Ca,Fe)_6ZrSi_6O_{18}(OH,Cl)$	tet	5–6	2.8–3.1
Eudidymite	$NaBe(OH)Si_3O_7$	mon	6	2.55
Eulite	$(Fe,Mg)_2(Si_2O_6)$	orth	5–6	3.73–3.86
Eulytine	BiS_3O_{12}	iso	4½	6.6
Euxenite	$(Y,Ca,Ce,U,Th)(Cb,Ta,Tu)_2O_6$‡	orth	5½–6½	5±
Evansite	$Al_3(PO_4)(OH)_6 \cdot 6H_2O$		3–4	1.8–2.2
Fairchildite	$K_2Ca(CO_3)_2$			
Fairfieldite	$Ca_2(Mn,Fe)(PO_4)_2 \cdot 2H_2O$	tric	3½	3.08
Famatinite	Cu_3SbS_4		3½	4.52
Fassaite	$Ca(Mg,Fe,Al,Ti)(Si,Al)_2O_6$	mon	6	2.96–3.34
Faujasite	$(Na_2,Ca)_{30}[(Al,Si)_{192}O_{384}] \cdot 260H_2O$	iso	3–4	1.91–1.93
Fayalite	Fe_2SiO_4	orth	6½	4.14
Feldspar	$Or_xAb_yAn_z$†		6–6½	2.55–3.39
Felsobanyaite	$Al_4(SO_4)(OH)_{10} \cdot 5H_2O$	orth	1½	2.33
Ferberite	$FeWO_4$	mon	4–4½	7.51
Ferghanite	$U_3(VO_4)_2 \cdot 6H_2O$	orth	2½	3.31
Fergusonite	$(Y,Er,Ce,Fe)(Cb,Ta,Ti)_4$‡	tet	5½–6½	5.6–5.8
Fermorite	$(Ca,Sr)_5(P,AsO_4)_3(F,OH)$	hex	5	3.52
Ferrierite	$(Na,K)_2Mg(Al_3Si_{15}O_{36})(OH) \cdot 9H_2O$	orth	3½–4	2.15
Ferrimolybdite	$Fe_2(MoO_4)_3 \cdot 8H_2O$			
Ferrinatrite	$Na_3Fe(SO_4)_3 \cdot 3H_2O$	hex	2½	2.55–2.61
Ferrisepiolite	$(Mg,Fe)_4(H_2O)_3Si_6O_{15}(OH)_2$			
Ferrisicklerite	$(Li,Fe,Mn)(PO_4)$	orth	3½–4½	2.98–3.02
Ferroactinolite	$Ca_2Fe_5Si_8O_{22}(OH,F)_2$	mon	5–6	3.02–3.44
Ferrocarpholite	$FeAl_2(Si_2O_6)(OH)_4$			
Ferroedenite	$NaCa_2Fe_5AlSi_7O_{22}(OH,F)_2$	mon	5–6	3.02–3.45
Ferrohastingsite	$NaCa_2Fe_4(Al,Fe)Si_6Al_2O_{22}(OH,F)_2$	mon	5–6	3.50
Ferrohortonolite	$(Fe,Mg)_2(SiO_4)$	orth	6½–7	4.0–4.3
Ferrohypersthene	$(Fe,Mg)(SiO_3)$	orth	5–6	3.59–3.75
Ferrosalite	$Ca(Fe,Mg)Si_2O_6$	mon	5½–6½	3.2–3.6
Ferruccite	$NaBF_4$	orth	3	2.5
Fersmannite	$Ca_4Na_2Ti_4Si_3O_{18}F_2$	mon	5½	3.44
Fervanite	$Fe_4V_4O_6 \cdot 5H_2O$			
Fibroferrite	$Fe(SO_4)(OH) \cdot 5H_2O$	orth	2½	1.84–2.1
Fiedlerite	$Pb_3(OH)_2Cl_4$	mon	3½	5.88
Fillowite	$H_2Na_6(Mn,Fe,Ca)_{14}(PO_4)_{12} \cdot H_2O$	mon	4½	3.43
Finnemanite	$Pb_5(AsO_3)_3Cl$	hex	2½	7.27
Fizelyite	$Pb_5Ag_2Sb_8S_{18}$		2	
Flajolotite	$FeSbO_4 \cdot 75H_2O$			
Flinkite	$Mn_2^{2+}Mn^{3+}(AsO_4)(OH)_4$	orth	4½	3.87
Flint	SiO_2		7	2.65
Florencite	$CeAl_3(PO_4)_2(OH)_6$	hex	5–6	3.58–3.71
Fluellite	$AlF_3 \cdot H_2O$	orth	3	2.17
Fluoborite	$Mg_3(BO_3)_3(F,OH)_3$	hex	3½	2.98
Fluocerite	$(Ce,La,Nd)F_3$	hex	4–5	6.14
Fluorapatite	$Ca_5(PO_4)_3F$	hex	5	3.1–3.2
Fluorite	CaF_2	iso	4	3.18
Forbesite	$H(Ni,Co)(AsO_4) \cdot 3.5H_2O$		2½	3.13
Formanite	$(U,Zr,Th,Ca)(Ta,Cb,Ti)O_4$	tet	5½–6½	5.6–5.8
Forsterite	Mg_2SiO_4	orth	6½	3.2
Foshallasite	$Ca_3Si_2O_7 \cdot 3H_2O$		2½–3	2.5
Fourmarierite	$PbO \cdot 4UO_3 \cdot 5H_2O$	orth	3–4	6.05
Franckeite	$Pb_5Sn_3Sb_2S_{14}$	orth	2½–3	5.90
Franklinite	$(Zn,Fe,Mn)(Fe,Mn)_2O_4$	iso	5½–6½	5.07–5.22
Freieslebenite	$Pb_3Ag_5Sb_5S_{12}$	mon	2–2½	6.04–6.23
Freirinite	$Na_3Cu_3(AsO_4)_2(OH)_3 \cdot H_2O$			
Friedelite	$(Mn,Fe)_{14}(Si_{14}O_{35})(OH,Cl)_{14}$	hex	4–5	3.06–3.19
Frondelite	$(Mn,Fe^{2+})Fe_4^{3+}(PO_4)_3(OH)_5$	orth	4½	3.3–3.49
Fuloppite	$Pb_3Sb_8S_{15}$	mon	2½	5.23
Gadolinite	$Y_2Fe''Be_2(SiO_4)_2O_2$‡	mon	6½–7	4.0–4.5
Gahnite	$ZnAl_2O_4$	iso	7½–8	4.62
Galaxite	$MnAl_2O_4$	iso	7½–8	4.03
Galena	PbS	iso	2½–2¾	7.4–7.6
Galenobismutite	$PbBi_2S_4$	orth	2½–3½	7.04
Gamma sulfur		mon	1	2.05

Table of accepted mineral species (cont.)

Mineral	Chemical formula	Crystallography class*	Hardness on Moh's scale	Specific gravity
Ganomalite	$(Ca,Pb)_{10}(OH,Cl)_2(Si_2O_7)_3$	hex	3	5.74
Ganophyllite	$(Na,K)(Mn,Al,Mg,Ca)_3(OH)_4(Si,Al)Si_3O_{10}$	mon	4	2.84
Garnet	$A_3B_2(SiO_4)_3$	iso	6–7½	3.58–4.32
Garnierite	$(Ni,Mg)SiO_3 \cdot nH_2O$	amor	2–3	2.2–2.8
Garrelsite	$(Ba,Ca)_2B_3SiO_7(OH)_3$			
Garronite	$NaCa_{2.5}(Al_6Si_{10}O_{32}) \cdot 13H_2O$	mon		
Gaylussite	$Na_2Ca(CO_3)_2 \cdot 5H_2O$	mon	2½–3	1.99
Gearksutite	$CaAl(OH)F_4 \cdot H_2O$	mon	2	2.77
Gedrite	$(Mg,Fe^{2+})_5Al_2(Si_6Al_2)O_{22}(OH)_2$	orth	5½–6	2.85–3.57
Geikielite	$MgTiO_3$	hex	5–6	4.05
Geocronite	$Pb_5(Sb,As)_2S_8$	orth	2½	6.4
Georgiadesite	$Pb_3(AsO_4)Cl_3$	mon	3½	7.1
Gerhardtite	$Cu_2(NO_3)(OH)_3$	orth	2	3.40–3.43
Germanite	$(Cu,Ge)(S,As)$		4	4.46–4.59
Gersdorffite	$(Ni,Fe,Co)AsS$	iso	5½	5.9
Gibbsite	$Al(OH)_3$	mon	2½–3½	2.3–2.4
Gillespite	$FeBaSi_4O_{10}$	tet	3	3.4
Ginorite	$Ca_2B_{14}O_{23} \cdot 8H_2O$	mon	3½	2.09
Gismondine	$Ca(Al_2Si_2O_8) \cdot 4H_2O$	orth	3–4	2.1–2.3
Gladite	$PbCuBi_5S_9$	orth	2–3	6.96
Glauberite	$Na_2Ca(SO_4)_2$	mon	2½–3	2.75–2.85
Glaucocerinite	$Zn_{13}Al_8Cu_7(SO_4)_2(OH)_{60} \cdot 4H_2O$		1	2.75
Glaucochroite	$CaMnSiO_4$	orth	6	3.48
Glaucodot	$(Co,Fe)AsS$	orth	5	6.04
Glauconite	$(K,Na,Ca)_{0.5-1}(Fe^{3+},Al,Fe^{2+},Mg)_2$-$(Si,Al)_4O_{10}(OH_2) \cdot nH_2O$	mon	2	2.3±
Glaucophane	$Na_2Mg_3Al_2(Si_8O_{22})(OH)_2$	mon	6–6½	3.0–3.2
Glockerite	$Fe_4(SO_4)(OH)_{10} \cdot nH_2O$			
Gmelinite	$(Na_2,Ca)(Al_2Si_4O_{12}) \cdot 6H_2O$	hex	4½	2.1
Goethite	$HFeO_2$	orth	5–5½	3.3–4.3
Gold	Au	iso	2½–3	15.0–19.3
Gold amalgam	Au_2Hg_3	iso	3–3½	15.47
Goldfieldite	$Cu_{12}Te_3Sb_4S_{16}$		3–3½	
Gonnardite	$Na_2Ca[(Al,Si)_5O_{10}]_2 \cdot 6H_2O$	orth	5	2.2–2.4
Gonyerite	low-Al chlorite	mon	2–3	3.01
Goongarrite	$Pb_4Bi_2S_7$	mon	3	7.29
Gorceixite	$BaAl_3(PO_4)_2(OH)_5H_2O$	hex	6	3.04–3.19
Gordonite	$MgAl_2(PO_4)_2(OH)_2 \cdot 8H_2O$	tric	3½	2.23
Goslarite	$ZnSO_4 \cdot 7H_2O$	orth	2–2½	1.98
Goyazite	$SrAl_3(PO_4)_2(OH)_5H_2O$	hex	4½–5	3.26
Graftonite	$(Fe,Mn,Ca)_3(PO_4)_2$	mon	5	3.67–3.76
Grandidierite	$H_2Na_2(Mg,Fe)_7(Al,Fe,B)_{15}Si_7Al_7O_{56}$	orth	7–7½	3.3
Graphite	C	hex	1–2	2.09–2.23
Gratonite	$Pb_9As_4S_{15}$	hex	2½	6.22
Greenalite	$(Fe,Mg)_3Si_2O_5(OH)_4$	mon	2–3	2.5–3.5
Greenockite	CdS	hex	3–3½	4.9
Griphite	$(Na,Ca,Fe,Al)_3Mn_2(PO_4)_{2.5}(OH,F)_2$		5½	3.40
Grossular	$Ca_3Al_2(AlO_4)_3$	iso	6–7½	3.59
Grossularite	$Ca_3Al_2(SiO_4)_3$	iso	6½	3.53
Grovesite	$(Mn,Mg,Al)_6(Si,Al)_4O_{10}(OH)_8$			
Gruenlingite	Bi_4TeS_3		2	8.08
Grunerite	$(Mg,Fe)_7Si_8O_{22}(OH)_2$	mon	5–6	3.10–3.60
Guanajuatite	Bi_2Se_3	orth	2½–3½	6.25–6.98
Guarinite	$Ca_2NaZrFSi_2O_8$	tric	5½	3.2–3.5
Gudmundite	FeSbS	mon	6	6.72
Guembelite	$(K,H_2O)(Al_{1.5}Mg_{0.5})(AlSi_3O_{10})(OH,H_2O)_2$	mon	1–2	2.6–2.9
Guildite	$(Cu,Fe)_3(Fe,Al)_4(SO_4)_7(OH)_4 \cdot 15H_2O$	mon	2½	2.72
Guitermanite	$Pb_{10}As_6S_{19}$		3	5.94
Gummite	$UO_3 \cdot nH_2O$		2½–5	3.9–6.4
Gypsum	$CaSO_4 \cdot 2H_2O$	mon	2	2.32
Gyrolite	$Ca_4(OH)_2Si_6O_{15} \cdot 3H_2O$	hex	3–4	2.34–2.45
Haidingerite	$CaH(AsO_4) \cdot H_2O$	orth	2–2½	2.84
Halite	NaCl	iso	2	2.16
Halloysite	$Al_4(Si_4O_{10})(OH)_8$	amor	1–2	2.0–2.2
Halotrichite	$FeAl_2(SO_4)_4 \cdot 22H_2O$	mon	1½	1.89
Hambergite	$Be_2(OH)(BO_3)$	orth	7½	2.36

Table of accepted mineral species (cont.)

Mineral	Chemical formula	Crystallography class*	Hardness on Moh's scale	Specific gravity
Hammarite	$Pb_2Cu_2Bi_4S_9$	mon	3–4	
Hanksite	$Na_{22}K(SO_4)_9(CO_3)_2Cl$	hex	3–3½	2.56
Hannayite	$Mg_3(NH_4)_2H_4(PO_4)_4 \cdot 8H_2O$	tric		1.89
Hardystonite	$Ca_2ZnSi_2O_7$	tet	5–6	3.40
Harkerite	$Ca(Mg,Al)(Si,BH)O_4 \cdot CaCO_3$			
Harmotome	$Ba(Al_2Si_6O_{16}) \cdot 6H_2O$	mon	4½	2.45
Hastingsite	$Ca_2NaMg_4Al_3Si_6O_{22}(OH,F)_2$	mon	6	3.2
Hauerite	MnS_2	iso	4	3.46
Hausmannite	$MnMn_2O_4$	tet	5½	4.84
Haüynite	$(Na,Ca)_{6-8}Al_6Si_6O_{24} \cdot (SO_4)_{1-2}$	iso	5½–6	2.4–2.5
Hectorite	$(Mg,Li)_6Si_8O_{20}(OH)_4$	mon	1–1½	2.5
Hedenbergite	$CaFe(Si_2O_6)$	mon	5–6	3.55
Hedyphane	$(Ca,Pb)_5(AsO_4)_3Cl$	hex	4½	5.82
Heliophyllite	$Pb_6As_2O_7Cl_4$	orth	2	6.89
Heliotrope		hex	6½–7	2.57–2.65
Helvite	$(Mn,Fe,Zn)_4(BeSiO_4)_3S$	iso	6	3.20–3.44
Hemafibrite	$Mn_3(AsO_4)(OH)_3 \cdot H_2O$	orth	3	3.65
Hematite	Fe_2O_3	hex	5–6	5.26
Hematolite	$(Mn,Mg)_4Al(AsO_4)(OH)_8$	hex	3½	3.49
Hematophanite	$Pb(Cl,OH)_2 \cdot 4PbO \cdot 2Fe_2O_3$	tet	2–3	7.70
Hemimorphite	$Zn_4(Si_2O_7)(OH)_2 \cdot H_2O$	orth	4½–5	3.4–3.5
Hercynite	$FeAl_2O_4$	iso	7½–8	4.39
Herderite	$CaBe(PO_4)(F,OH)$	mon	5–5½	2.95
Herzenbergite	SnS			
Hessite	Ag_2Te	mon	2–3	8.4
Hetaerolite	$ZnMn_2O_4$	tet	6	5.18
Heteromorphite	$Pb_7Sb_8S_{19}$	mon	2½–3	5.73
Heterosite	$(Fe,Mn)PO_4$	orth	4–4½	3.2–3.4
Heulandite	$Ca(Al_2Si_7O_{18}) \cdot 6H_2O$	mon	3½–4	2.18–2.20
Hewettite	$CaV_6O_{16} \cdot 9H_2O$			
Hexahydrite	$MgSO_4 \cdot 6H_2O$	mon		1.76
Hieratite	K_2SiF_6	iso	2½	2.67
Hilgardite	$Ca_8(B_6O_{11})_3Cl_4 \cdot 4H_2O$	mon	5	2.71
Hillebrandite	$Ca_2SiO_4 \cdot H_2O$	mon	5½	2.69
Hinsdalite	$(Pb,Sr)Al_3(PO_4)(SO_4)(OH)_6$	hex	4½	3.65
Hodgkinsonite	$MnZn_2(OH)_2SiO_4$	mon	4–4½	3.91
Hoeqbomite	$Mg(Al,Fe,Ti)_4O_7$	hex	6½	3.81
Hoernesite	$Mg_3(AsO_4)_2 \cdot 8H_2O$	mon	1	2.73
Hohmannite	$Fe_2(SO_4)_2(OH)_2 \cdot 7H_2O$	tric	3	2.2
Holdenite	$(Mn,Ca)_4(Zn,Mg,Fe)_2(AsO_4)(OH)_5O_2$	orth	4	4.11
Hollandite	$MnBaMn_6O_{14}$	tet	6	4.95
Holmquistite	$Li_2(Mg,Fe)_3(Al,Fe^{3+})_2-Si_8O_{22}(OH)_2$	orth	5–6	3.06–3.13
Holmilite	$Ca_2FeB_2(SiO_5)_2$	mon	5	3.36
Hopeite	$Zn_3(PO_4)_2 \cdot 4H_2O$	orth	3¼	3.05
Hornblende	$(Ca,Na)_{2-3}(Mg,Fe,Al)_5Si_6(SiAl)_2O_{22}(OH)_2$	mon	5–6	3.2
Horsfordite	$CuSSb$			
Hortonolite	$(Fe,Mg)_2(SiO_4)$	orth	6½–7	3.7–4.1
Howlite	$Ca_2SiB_5O_9(OH)_5$	mon	3½	2.53–2.59
Huebnerite	$MnWO_4$	mon	4–4½	7.12
Hühnerkobelite	$(Na,Ca)(Fe,Mn)_2(PO_4)_2$	orth	5	3.5–3.6
Hulsite	$(Fe^{2+},Ca,Mg)_4(Fe^{3+},Sn^{4+})_2B_2O_{10}$	orth	3	4.28
Humboldtine	$Fe(C_2O_4) \cdot 2H_2O$	orth	1½–2	2.28
Humite	$Mg_7(SiO_4)_3(F,OH)_2$	orth	6	3.1–3.2
Hureaulite	$Mn_5H_2(PO_4)_4 \cdot 4H_2O$	mon	3½	3.19
Hutchinsonite	$(Pb,Ti)_2(Cu,Ag)As_5S_{10}$	orth	1½–2	4.6
Huttonite	$Th(SiO_4)$			
Hyalophane	$(K,Ba)(Al,Si)_2Si_2O_8$	mon	6	2.8
Hyalosiderite	$(Mg,Fe)_2(SiO_4)$	orth	6½–7	3.5–3.75
Hyalotekite	$Ca_3Ba_3Pb_3B_2Si_{12}O_{36}$	orth	5–5½	3.8
Hydrobasaluminite	$Al_4(SO_4)(OH)_{10} \cdot 36H_2O$			
Hydrobiotite	$(K,H_2O)(Mg,Fe,Mn)_3(H_2O,OH)_2AlSi_3O_{10}$	mon	1–2	2.6–3.3
Hydroboracite	$CaMgB_6O \cdot 6H_2O$	mon	2–3	2.17
Hydrocalumite	$Ca_4Al_2(OH)_{14} \cdot 6H_2O$	mon	3	2.15
Hydrocerussite	$Pb_3(CO_3)_2(OH)_2$	hex	3½	6.80
Hydrogrossular	$Ca_3Al_2(SiO_4)_3(OH)_4$	orth	6–7½	3.13–3.59
Hydrogrossularite	$Ca_3Al_2Si_2O_8(SiO_4)_{1-m}(OH)_{4m}$	iso	6–7	3.13–3.59

Table of accepted mineral species (cont.)

Mineral	Chemical formula	Crystallography class*	Hardness on Moh's scale	Specific gravity
Hydrohetaerolite	$Zn_2Mn_4O_8 \cdot H_2O$	tet	5–6	4.6
Hydromagnesite	$Mg_4(OH)_2(CO_3)_3 \cdot 3H_2O$	mon	3½	2.24
Hydromuscovite	$(K,H_2O)Al_2(H_2O,OH)_2AlSi_3O_{10}$	mon	1–2	2.6–2.9
Hydroparagonite	$(Na,H_2O)Al_2(H_2O,OH)_2AlSi_3O_{10}$	mon	1½–2½	2.6–2.85
Hydrophilite	$CaCl_2$			
Hydrophlogopite	$(K,H_2O)Mg_3(H_2O,OH)_2AlSi_3O_{10}$	mon	1–2	2.6–2.9
Hydrotalcite	$Mg_6Al_2(OH)_{16} \cdot CO_3 \cdot 4H_2O$	hex	2	2.06
Hydroxylapatite	$Ca_5(PO_4)_3(OH)$	hex	5	2.9–3.1
Hydroxylherderite	$CaBe(PO_4)(OH,F)$	mon	5–5½	3.01
Hydrozincite	$Zn_5(OH)_6(CO_3)_2$	mon	2–2½	3.5–3.8
Hypersthene	$(Mg,Fe)_2(Si_2O_6)$	orth	5–6	3.4–3.5
Ianthinite	$2UO_2 \cdot 7H_2O$	orth	2–3	
Iddingsite	$H_8Mg_9Fe_2Si_3O_{14}$	orth	3	3.5–3.8
Idocrase	$Ca_{10}(Mg,Fe)_2Al_4(SiO_4)_5(Si_2O_7)_2(OH)_4$	tet	6½	3.35–3.45
Ilesite	$(Mn,Zn,Fe)SO_4 \cdot 4H_2O$			
Illite	$KAl_4Si_7AlO_{20}(OH)_4$	mon	1–2	2.64–2.69
Ilmenite	$FeTiO_3$	hex	5–6	4.7
Ilsemannite				
Ilvaite	$CaAl_2(Si_2O_7)(OH)_2 \cdot H_2O$	orth	5½–6	4.0
Inderborite	$CaMgB_6O_{11} \cdot 11H_2O$	mon	3½	2
Inderite	$Mg_2B_6O_{11} \cdot 15H_2O$	tric	3	1.86
Indialite	$(Mg,Fe)_2Al_4Si_5O_{18} \cdot nH_2O$	hex	7	2.6
Inesite	$Ca_2Mn_7(OH)_2Si_{10}O_{28} \cdot 5H_2O$	tric	6	3.03
Inyoite	$Ca_2B_6O_{11} \cdot 13H_2O$	mon	2	1.88
Iodargyrite	AgI	hex	1½	5.69
Iodobromite	Ag(Cl,Br,I)	iso	1–1½	5.71
Iodyrite	AgI	hex	1½	5.69
Iridium	Ir	iso	6–7	22.7
Iridosmine	Ir,Os	hex	6–7	19.3–21.1
Iron	Fe	iso	4	7.3–7.87
Ishikawaite	$(U,Fe,Y)(Cb,TaO_4)$‡	orth	5–6	6.2–6.4
Isoclasite	$Ca_2(PO_4)(OH) \cdot 2H_2O$	mon	1½	2.92
Jacobsite	$MnFe_2O_4$	iso	5½–6½	4.76
Jadeite	$NaAl(Si_2O_6)$	mon	6½–7	3.3–3.5
Jamesonite	$Pb_4FeSb_6S_{14}$	mon	2½	5.63
Jarlite	$NaSr_3Al_3F_{16}$	mon	4–4½	3.93
Jarosite	$KFe_3(SO_4)_2(OH)_6$	rho	3	3.2
Jasper		hex	6½–7	2.57–2.65
Jeremejevite	$AlBO_3$	hex	6½	3.28
Jezekite	$Na_4CaAl_2(PO_4)_2(OH)_2F_2O$	mon	4–4½	2.94
Joaquinite	$NaBa(Ti,Fe)_3Si_4O_{15}$	orth	5½	3.89
Johannite	$Cu(UO_2)_2(SO_4)_2(OH)_2 \cdot 6H_2O$	tric	2–2½	3.32
Johannsenite	$CaMnSi_2O_6$	mon	6	3.44–3.55
Johnstrupite	$(Ca,Y,Na,Ce)_3(Al,Zr,Ti)(F,OH)Si_2O_8$‡	mon	4	3.3
Jordanite	$Pb_{14}As_7S_{24}$	mon	3	6.38
Joseite	Bi_3TeS		2	8.18
Julienite	$Na_2Co(SCN)_4 \cdot 8H_2O$	tet		1.65
Kaemmererite	Cr,Mg chlorite	mon	2–3	
Kaersutite	Ca,Na,K,Mg,Fe,Ti,Al silicate	mon	5–6	3.2–3.3
Kainite	$KMg(SO_4)Cl \cdot 3H_2O$	mon	2½–3	2.15
Kainosite	$Ca_2(Ce,Y)_2CO_3Si_4O_{12} \cdot 1-2H_2O$‡	orth	5–5½	3.34–3.61
Kaliborite	$KMg_2B_{11}O_{19} \cdot 9H_2O$	mon	4–4½	2.13
Kalicinite	$KHCO_3$	mon	soft	2.17
Kalinite	$KAl(SO_4)_2 \cdot 11H_2O$	mon	2–2½	1.75
Kaliophilite	$K(AlSiO_4)$	hex	6	2.61
Kalkowskite	$Fe_2Ti_3O_9$		3½	4.01
Kalsilite	$K(AlSiO_4)$	hex	6	2.61
Kamarezite	$Cu_3(SO_4)(OH)_4 \cdot 6H_2O$	orth	3	3.98
Kaolinite	$Al_4(Si_4O_{10})(OH)_8$	mon	2–2½	2.6–2.65
Karinthine	$(Na,K)Ca_{2-3}Mg_8Fe_{1-2}(Al,Fe,Ti)_2$-$(Al_{3-4}Si_{13-12}O_{44})(OH)_4$			
Karpinskiite	$(Na,K,Zn,Mg)_2(OH,H_2O)_{1-2}(Al,Be)_2Si_4O_{12}$			
Kasolite	$Pb_2U_2O_4Si_2O_8 \cdot H_2O$	mon	4–5	5.96–6.46
Katophorite	$Na_2CaFe_4(Fe,Al)Si_7AlO_{22}(OH,F)_2$	mon	5	3.20–3.50
Keatite	SiO_2	tet	7	2.50
Kempite	$Mn_2(OH)_3Cl$	orth	3½	2.94

Table of accepted mineral species (cont.)

Mineral	Chemical formula	Crystallography class*	Hardness on Moh's scale	Specific gravity
Kentrolite	$Pb_3Mn_4O_3(SiO_4)_3$	orth	5	6.19
Kermesite	Sb_2S_2O	mon	1–1½	4.68
Kernite	$Na_2B_4O_7 \cdot 4H_2O$	mon	2½	1.95
Kerstenite	$PbSeO_3 \cdot 2H_2O$		3–4	
Kieserite	$MgSO_4 \cdot H_2O$	mon	3½	2.57
Kirovite	$(Fe,Mg)SO_4 \cdot 7H_2O$	mon	2	1.89
Klaprothite	$Cu_6Bi_4S_9$	orth	2½	6.01
Kleinite	Hg,NH_4,Cl,SO_4	hex	3½–4	8.0
Klockmannite	CuSe	hex	3	>5
Knebelite	$(Mn,Fe)_2(SiO_4)$	orth	6½	3.96–4.25
Kobellite	$Pb_2(Bi,Sb)_2S_5$		2½–3	6.33
Koechlinite	$(BiO)_2(MoO_4)$			
Koenenite	$Mg_5Al_2(OH)_{12}Cl_4$	hex	1½	1.98
Koettigite	$Zn_3(AsO_4)_2 \cdot 8H_2O$	mon	2½–3	3.33
Koktaite	$(NH_4)_2Ca(SO_4)_2 \cdot H_2O$			
Kolbeckite	Ca,Be,Al,Fe silicate-phosphate	mon	3½–4	2.39
Kornelite	$Fe_2(SO_4)_3 \cdot 7H_2O$			
Kornerupine	$(Mg,Fe,Al)_4(Al,B)_6(O,OH)_{5-8}(SiO_4)_4$	orth	6½	3.27–3.34
Kotoite	$Mg_3(BO_3)_2$	orth	6½	3.10
Krausite	$KFe(SO_4)_2 \cdot H_2O$	mon	2½	2.84
Kremersite	$(NH_4,K)_2FeCl_5 \cdot H_2O$			
Krennerite	$AuTe_2$	orth	2–3	8.62
Kribergite	$Al_4(PO_4)_2(SO_4)_2(OH)_2 \cdot 8H_2O$			
Kroehnkite	$Na_2Cu(SO_4)_2 \cdot 2H_2O$	mon	2½–3	2.90
Kupletskite	$(K_2,Na_2,Ca)(Fe,Mn)_4(Ti,Zr)(OH)_2Si_4O_{14}$			
Kurnakovite	$Mg_2B_6O_{11} \cdot 7H_2O$	mon	3	1.86
Kutnahorite	$CaMn(CO_3)_2$	rho	3½–4	3.12
Kyanite	Al_2SiO_5	tric	5–7	3.56–3.66
Labradorite	$Ab_{50}An_{50}-Ab_{30}An_{70}$†	tric	6–6½	2.68–2.72
Labuntsovite	$(K,Ba,Na)Ti(Si,Al)_2(O,OH)_7 \cdot H_2O$			
Lacroixite	Na,Ca,Al fluophosphate	mon	4½	3.13
Lamprophyllite	$CaNa_3Ti_3Si_3O_{14}(OH,F)$	mon?	4	3.45
Lanarkite	$Pb_2(SO_4)O$	mon	2–2½	6.92
Landesite	$Fe_6Mn_2O(PO_4)_{16} \cdot 27H_2O$		3–3½	3.03
Langbeinite	$K_2Mg_2(SO_4)_3$	iso	3½–4	2.83
Langite	$Cu_4(SO_4)(OH)_6 \cdot H_2O$	orth	2½–3	3.48–3.50
Lansfordite	$MgCO_3 \cdot 5H_2O$	mon	2½	1.69
Lanthanite	$(La,Ce)_2(CO_3)_3 \cdot 8H_2O$	orth	2½–3	2.69–2.74
Larderellite	$(NH_4)_2B_{10}O_{16} \cdot 5H_2O$			
Larnite	Ca_2SiO_4			
Larsenite	$PbZnSiO_4$	orth	3	5.9
Latiumite	$Ca_6(K,Na)_2Al_4(O,CO_3,SO_4)(SiO_4)_6$			
Laubmannite	$Fe_9(PO_4)_4(OH)_{12}$	orth	3½–4	3.33
Laumontite	$Ca(Al_2Si_4O_{12}) \cdot 4H_2O$	mon	4	2.28
Laurionite	Pb(OH)Cl	orth	3½	6.24
Laurite	RuS_2	iso	7½	6–6.99
Lautarite	$Ca(IO_3)_2$	mon	3½–4	4.59
Lautite	CuAsS	orth	3–3½	4.9±0.1
Lavenite		mon	6	3.5
Lawrencite	$FeCl_2$	hex	soft	3.16
Lawsonite	$CaAl_2(Si_2O_7)(OH)_2 \cdot H_2O$	orth	8	3.09
Lazulite	$(Mg,Fe)Al_2(PO_4)_2(OH)_2$	mon	5½–6	3.08
Lazurite	$(Na,Ca)_8(AlSiO_4)_6(SO_4,S,Cl)_2$	iso	5–5½	2.4–2.45
Lead	Pb	iso	1½	11.37
Leadhillite	$Pb_4(SO_4)(CO_3)_2(OH)_2$	mon	2½–3	6.55
Lechatelierite	SiO_2	amor	6–7	2.2
Lecontite	$Na(NH_4,K)(SO_4) \cdot 2H_2O$	orth	2–2½	
Legrandite	$Zn_{14}(OH)(AsO_4)_9 \cdot 12H_2O$	mon	5	4.01
Lehiite	$(Na,K)_2Ca_5Al_8(PO_4)_8(OH)_{12} \cdot 6H_2O$		5½	2.89
Leifite	$Na_2FSi_5AlO_{12}$	hex	6	2.57
Leightonite	$K_2Ca_2Cu(SO_4)_4 \cdot 2H_2O$	tric	3	2.95
Lengenbachite	$Pb_6(Ag,Cu)_2As_4S_{13}$	tric		5.80–5.85
Leonite	$K_2Mg(SO_4)_2 \cdot 4H_2O$	mon	2½–3	2.20
Lepidocrocite	FeO(OH)	orth	5	4.09
Lepidolite	$K_2(Li,Al)_{5-6}(Si_{6-7}Al_{2-1})O_{20-21}(F,OH)_{3-4}$	mon	2½–4	2.8–3.0
Letovicite	$(NH_4)_3H(SO_4)_2$	mon		1.83

Table of accepted mineral species (cont.)

Mineral	Chemical formula	Crystallography class*	Hardness on Moh's scale	Specific gravity
Leucite	$K(AlSi_2O_6)$		5½–6	2.45–2.50
Leucochalcite	$Cu_2(AsO_4)(OH) \cdot H_2O$			
Leucophane	$(Ca,Na)_2BeSi_2(O,OH,F)_7$	orth	4	2.96
Leucophenicite	$Mn_7(OH)_2Si_3O_{12}$	mon	5½–6	3.85
Leucophosphite	$K_2(Fe,Al)_7(PO_4)(OH)_{11} \cdot 6H_2O$			
Leucoxene	TiO_2			
Levynite	$Ca(Al_2Si_4O_{12}) \cdot 6H_2O$	orth	4½	2.0–2.2
Lewistonite	$(Ca,K,Na)_5(PO_4)_3(OH)$	hex	5	3.08
Libethenite	$Cu_2(PO_4)(OH)$	orth	4	3.97
Liebigite	$Ca_2U(CO_3)_4 \cdot 10H_2O$	orth	2½–3	2.41
Lillianite	$Pb_3Bi_2S_6$	orth	2–3	7.0–7.2
Lime	CaO	iso	3½	3.3
Limonite	$2Fe_2O_3 \cdot 3H_2O$		4–5½	3.6–4.0
Linarite	$PbCu(SO_4)(OH)_2$	mon	2½	5.35
Lindackerite	$Cu_6Ni_3(AsO_4)_4(SO_4)(OH)_4 \cdot 5H_2O$	mon	2–2½	2.0–2.5
Lindgrenite	$Cu_3(MoO_4)_2(OH)_2$	mon	4½	4.26
Lindstromite	$PbCuBi_3S_6$	orth	3–3½	7.01
Linnaeite	Co_3S_4	iso	4½–5½	4.5–4.8
Liroconite	$Cu_2Al(AsO_4)(OH)_4 \cdot 4H_2O$	mon	2–2½	2.9–3.0
Litharge	PbO	tet	2	9.14
Lithidionite	$(Na,K)_2(Si_3O_7)$ (?)			
Lithiophilite	$Li(Mn,Fe)(PO_4)$	orth	4–5	3.34–3.50
Liveingite	$Pb_5As_8S_{17}$	mon	3	5.3
Livingstonite	$HgSb_4S_7$	mon	2	5
Lizardite	ca. $Mg_3(Si_2O_5)(OH)_4$	mon	2½	2.5–2.6
Loellingite	$FeAs_2$	orth	5–5½	7.40
Loeweite	$Na_4Mg_2(SO_4)_4 \cdot 5H_2O$		2½–3	2.37
Lomonossovite	$(Na,Ca)Ti(O,OH)(S,P)O_4$			
Lonsdaleite	C	hex	10	3.3+
Lopezite	$K_2(Cr_2O_7)$		2½	2.69
Lorandite	$TlAsS_2$	mon	2–2½	5.53
Loranskite	$(Y,Ce,Ca,Zr)(Ta,Zr)O_4$‡		5	4.6
Lorettoite	$Pb_7O_6Cl_2$	tet	2½–3	7.39
Loseyite	$(Mn,Zn)_7(CO_3)_2(OH)_{10}$	mon	3	3.27
Ludlamite	$(Fe,Mg,Mn)_3(PO_4)_2 \cdot 4H_2O$	mon	3½	3.12–3.19
Ludwigite	$(Mg,Fe^{2+})_2Fe^{3+}BO_5$	orth	5	4.7
Lueneburgite	$Mg_3B_2(OH)_6(PO_4)_2 \cdot 6H_2O$	mon	2	2.05
Luzonite	Cu_3AsS_4	tet	3–4	4.4
Mackayite	$Fe_2(TeO_3)_3 \cdot nH_2O$	tet	4½	4.86
Maghemite	Fe_2O_3	iso	5	
Magnesiochromite	$MgCr_2O_4$	iso	5½	4.2
Magnesiocopiapite	$(Mg,Fe)Fe^{3+}_4(SO_4)_6(OH)_2 \cdot 20H_2O$	tric	2½–3	2.08–2.17
Magnesioferrite	$MgFe_2O_4$	iso	5½–6½	4.5–4.6
Magnesiokatophorite	$Na_2CaMg_4(Fe,Al)Si_7AlO_{22}(OH,F)_2$	mon	5	3.20–3.50
Magnesite	$MgCO_3$	hex	3¾–4¼	3.0–3.2
Magnesium chlorophoenicite	$Mg_5(AsO_4)(OH)_7$			
Magnesium orthite	$CaCeMgAl_2(OH,F)Si_3O_{12}$	mon	5–6½	3.90
Magnetite	$FeFe_2O_4$	iso	5½–6½	5.18
Magnetoplumbite	$(Pb,Mn^{2+},Mn^{3+})(Fe^{3+},Mn^{3+},Ti)_6O_{10}$	hex	6	5.517
Malachite	$Cu_2(OH)_2(CO_3)$	mon	3½–4	3.9–4.03
Maldonite	Au_2Bi	iso	1½–2	15.46
Malladrite	Na_2SiF_6	hex		2.71
Mallardite	$MnSO_4 \cdot 7H_2O$	mon	2	1.85
Manandonite	hydrous Li,B,Al silicate	mon	2–3	2.69
Manasseite	$Mg_6Al_2(OH)_{16} \cdot CO_3 \cdot 4H_2O$	hex	2	2.05
Manganalluaudite	$(Na,Mn,Fe)(PO_4)$		5–5½	3.4–3.5
Manganberzelite	$(Mn,Mg)_2(Ca,Na)_3(AsO_4)_3$	iso	4½–5	4.08
Manganite	$MnO(OH)$	mon	4	4.3
Manganolangbeinite	$K_2Mn_2(SO_4)_3$			
Manganosite	MnO	iso	5½	5.36
Manganotantalite	$MnO(Ta,Cb)_2O_5$	orth	4½	6.6±
Mansfieldite	$Al(AsO_4) \cdot 2H_2O$	orth	3½–4	3.28
Marcasite	FeS_2	orth	6–6½	4.89
Margarite	$CaAl_2(Al_2Si_2O_{10})(OH)_2$	mon	3½–5	3.0–3.1
Marialite	$(Na,Ca)_4Al_3(Al,Si)_3Si_6O_{24}(Cl,CO_3,SO_4)$	tet	5½–6	2.60±

Table of accepted mineral species (cont.)

Mineral	Chemical formula	Crystallography class*	Hardness on Moh's scale	Specific gravity
Marshite	CuI	iso	2½	5.68
Mascagnite	$(NH_4)_2SO_4$	orth	2–2½	1.79
Massicot	PbO	orth	2	9.56
Matidite	$AgBiS_2$	orth	2½	6.9
Matlockite	PbFCl	tet	2½–3	7.12
Maucherite	$Ni_{11}As_8$	tet	5	8
Meionite	$(Ca,Na)_4Al_3(Al,Si)_3Si_6O_{24}(Cl,CO_3,SO_4)$	tet	5½–6	2.69
Melanite	$Ca_3Fe_3(SiO_4)_3$	iso	7	3.7
Melanostibian	$(Mn,Fe)_6(SbO_3)_2O_3$		4	
Melanotekite	$Pb_3Fe_4O_3(SiO_4)_3$	orth	6½	5.4–6.02
Melanovanadite	$Ca_2V_4^{4+}V_6^{5+}O_{25}$	mon	2½	3.48
Melanterite	$FeSO_4 \cdot 7H_2O$	mon	2	1.9
Mellite	$Al_2C_{12}O_{12} \cdot 18H_2O$	tet	2–2½	1.64
Melonite	$NiTe_2$	hex	1–1½	7.35
Mendipite	$Pb_3O_2Cl_2$	orth	2½	7.24
Mendozite	$NaAl(SO_4)_2 \cdot 11H_2O$	mon	3	1.73
Meneghinite	$CuPb_{13}Sb_7S_{24}$	orth	2½	6.36
Mercallite	$KHSO_4$			
Mercury	Hg		0	13.6
Merwinite	$Ca_3MgSi_2O_8$	mon	6	3.15
Mesolite	$Na_2Ca_2(Al_2Si_3O_{10})_3 \cdot 8H_2O$	mon	5	2.25–2.27
Mesomicrocline	$K(Al,Si)_2Si_2O_8$			
Metacinnabar	HgS	iso	3	7.65
Metahewettite	$CaV_6O_{16} \cdot 9H_2O$			
Metasideronatrite	$Na_4Fe_2(SO_4)_4(OH)_2 \cdot 3H_2O$	orth	2½	2.46
Metastrengite	$FePO_4 \cdot 2H_2O$	mon	3½–4	2.76
Metatorbernite	$Cu(UO_2)_2(PO_4)_2 \cdot 8H_2O$	tet	2½	3.5–3.7
Metauranopilite	$(UO_2)_6SO_4(OH)_{10} \cdot 5H_2O$			
Metavariscite	$Al(PO_4) \cdot 2H_2O$	mon	3½	2.54
Metavauxite	$FeAl_2(PO_4)_2(OH)_2 \cdot 8H_2O$	mon	3	2.35
Metavoltine	$(K,Na,Fe^2)_5Fe_3^{3+}(SO_4)_6(OH)_2 \cdot 9H_2O$	hex	2½	2.5
Metazeunerite	$Ci(UO_2)_2(AsO_4)_2 \cdot 8H_2O$	tet	2–2½	3.64
Meyerhofferite	$Ca_2B_6O_{11} \cdot 7H_2O$	tric	2	2.12
Miargyrite	$AgSbS_2$	mon	2½	5.25
Mica	complex hydrous aluminosilicates	mon	2–4	2.4–3.4
Microcline	$K(AlSi_3O_8)$	tric	6	2.54–2.57
Microlite	$(Na,Ca)_2(Ta,Nb)_2O_6(O,OH,F)$	iso	5–5½	5.48–5.56
Miersite	(Cu,Ag)I	iso	2½	5.64
Milarite	$KCa_2Si_{12}Be_2AlO_{30} \cdot 5H_2O$	hex	5½–6	2.57
Millerite	NiS	hex	3–3½	5.5±0.2
Millisite	$(Na,K)CaAl_6(PO_4)_4(OH)_9 \cdot 3H_2O$	tet	5½	2.83
Mimetite	$Pb_5(AsO_4,PO_4)_3Cl$	hex	3½–4	7.0–7.2
Minasragrite	$(VO)_2H_2(SO_4)_3 \cdot 15H_2O$			
Minium	Pb_3O_4		2½	8.9–9.2
Minnesotaite	$Fe_3Si_4O_{10}(OH)_2$	mon	2½	3.0–3.1
Minyulite	$KAl_2(PO_4)_2(OH) \cdot 3½H_2O$	orth	3½	2.45
Mirabilite	$Na_2SO_4 \cdot 10H_2O$	mon	1½–2	1.49
Misenite	$6KHSO_4 \cdot K_2SO_4$	mon		2.32
Mitridatite	$CaFe_2(PO_4)_2(OH)_2 \cdot H_2O$		2½	
Mitscherlichite	$K_2CuCl_4 \cdot 2H_2O$	tet	2½	2.42
Mixite	$Cu_{11}Bi(AsO_4)_5(OH)_{10} \cdot 6H_2O$	hex	3–4	3.79
Moissanite	SiC	hex	9½	3.1
Molybdenite	MoS_2	hex	1–1½	4.62–4.73
Molybdophyllite	$Pb_2Mg_2(OH)_2Si_2O_7$	hex	3–4	4.72
Molysite	$FeCl_3$	hex	soft	2.90
Monalbite	$NaAlSi_3O_8$	mon		
Monazite	$(Ce,La,Y,Th)(PO_4)$‡	mon	5–5½	4.6–5.4
Monetite	$CaH(PO_4)$	tric	3½	2.93
Monimolite	$(Pb,Ca)_3Sb_2O_8$	iso	4½–6	5.9–7.3
Montanite	$(BiO)_2(TeO_4) \cdot 2H_2O$			
Montebrasite	$(Li,Na)Al(PO_4)(OH,F)$	tric	5½–6	3.11
Montgomeryite	$Ca_4Al_5(PO_4)_6(OH)_5 \cdot 11H_2O$	mon	4	2.53
Monticellite	$CaMgSiO_4$	orth	5	3.2
Montmorillonite	$(OH)_4Si_8Al_4O_{20} \cdot nH_2O$	mon	1–1½	2.5
Montroydite	HgO	orth	2½	11.23
Mooreite	$(Mg,Zn,Mn)_8(SO_4)(OH)_{14} \cdot 4H_2O$	mon	3	2.47

Table of accepted mineral species (cont.)

Mineral	Chemical formula	Crystallography class*	Hardness on Moh's scale	Specific gravity
Mordenite	$(Ca,K_2,Na_2)(AlSi_5O_{12})_2 \cdot 7H_2O$	orth	3–4	2.12–2.15
Morenosite	$NiSO_4 \cdot 7H_2O$	orth	2–2½	1.95
Morinite	Na,Ca,Al fluophosphate	mon	3½–4½	8.0
Mosandrite	$(Ca,Na)_{12}Ce_3(Zr,Ti,Mg)_4F_5Si_{10}O_{40}$	mon	4	2.93–3.03
Moschellandsbergite	Ag_2Hg_3	iso	3½	13.48–13.71
Mosesite	$Hg_6(NH_3)_2Cl_2(SO_4)(OH)_4$	iso	3½	
Mossite	$Fe(Cb,Ta)_2O_6$	tet	6–6½	7.90
Mottramite	$(Cu,Zn)Pb(VO_4)(OH)$	orth	3–3½	5.9
Mullite	$Al_6Si_2O_{13}$	orth	6–7	3.23
Murmanite	$Na_2Ti_2(OH)_4Si_2O_7$	mon	2–3	2.84
Muscovite	$K_4Al_4(Si_6Al_2)O_{20}(OH)_4$	mon	2–2½	2.76–3.1
Muthmannite	(Ag,Au)Te		2½	5.59
Nacrite	$Al_4(Si_4O_{10})(OH)_8$	mon	2–2½	2.6
Nadorite	$PbSbO_2Cl$	orth	3½–4	7.02
Nagatelite	$(Ca,Ce)_2(Al,Fe)_3(OH)(Si,P)_3O_{12}$	mon	5½	3.91
Nagyagite	$Pb_5Au(Te,Sb)_4S_{5-8}$	mon	1–1½	7.41
Nahcolite	$NaHCO_3$	mon	2½	2.21
Nantokite	CuCl	iso	2½	4.14
Narsarsukite	$Na_2(Ti,Fe)(O,OH,F)Si_4O_{10}$	tet	5½–7½	2.75
Nasonite	$(Ca,Pb)_{10}Cl_2(Si_2O_7)_3$	hex	4	5.43
Natroalunite	$NaAl_3(SO_4)_2(OH)_6$	hex	3½–4	2.6–2.9
Natrochalcite	$NaCu_2(SO_4)_2(OH) \cdot H_2O$	mon	4½	3.49
Natrojarosite	$NaFe_3(SO_4)_2(OH)_6$	hex	3	3.18
Natrolite	$Na_2(Al_2Si_3O_{10}) \cdot 2H_2O$	mon	5–5½	2.25
Natromontebrasite	$(Na,Li)Al(PO_4)(OH,F)$	tric	5½–6	3.11
Natron	$Na_2CO_3 \cdot 10H_2O$	mon	1–1½	1.48
Natrophilite	$NaMn(PO_4)$	orth	4½–5	3.41
Naujakasite	$Na_4FeAl_4H_4Si_8O_{27}$			
Naumannite	Ag_2Se	iso	2½	7
Nekoite	$CaH_2(Si_2O_6) \cdot H_2O$			
Nepheline	$(NaK)(AlSiO_4)$	hex	5½–6	2.55–2.65
Neptunite	$(NaK)(Fe'',Mn,Ti)Si_2O_6$	mon	5–6	3.23
Nesquehonite	$MgCO_3 \cdot H_2O$	orth	2½	1.85
Newberyite	$MgH(PO_4) \cdot 3H_2O$	orth	3–3½	2.10
Niccolite	NiAs	hex	5–5½	7.78
Nickel iron	NiFe	iso	5	7.8–8.2
Nickel skutterudite	$(Ni,Co,Fe)As_3$	iso	5½–6	6.5±0.4
Niggliite	$PtTe_3$ or PtSn		3	4
Niter	KNO_3	orth	2	2.09–2.14
Nitrobarite	$Ba(NO_3)_2$			
Nitrocalcite	$Ca(NO_3)_2 \cdot 4H_2O$			
Nitromagnesite	$Mg(NO_3)_2 \cdot 6H_2O$			
Nocerite	$Ca_3Mg_3F_8O_2$	hex		2.96
Nontronite	$Fe_4(AlSi_8)O_{20}(OH)_4$	mon	1–1½	2.5
Norbergite	$Mg_3(SiO_4)(F,OH)_2$	orth	6	3.1–3.2
Nordenskiöldine	$CaSn(BO_3)_2$	hex	5½–6	4.20
Northupite	$Na_3MgCl(CO_3)_2$	iso	3½–4	2.38
Nosean	$Na_8(AlSiO_4)_6(SO_4)$	iso	5½	2.30–2.40
Nosilite	$Na_8(AlSiO_4)_6SO_4$	iso	6	2.3
Okenite	$Ca_2Si_4O_{10} \cdot 4H_2O$	tric	5	2.28–2.33
Oldhamite	CaS	iso	4	2.58
Oligoclase	$Ab_{90}An_{10}-Ab_{70}An_{30}$†		6–6½	2.62–2.67
Oliveiraite	$Zr_3Ti_2O_{10} \cdot 2H_2O$			
Olivenite	$Cu_2(AsO_4)(OH)$	orth	3	4.46
Olivine	$(Mg,Fe)_2SiO_4$	orth	6½–7	3.27–4.37
Omphacite	$(Ca,Na)(Mg,Fe,Fe,Ti,Al)(Si_2O_6)$	mon	5–6	3.29–3.37
Opal	$SiO_2 \cdot nH_2O$	amor	5–6	1.9–2.2
Orientite	$Ca_4Mn_4(SiO_4)_5 \cdot 4H_2O$	orth	4½–5	3.05
Orpiment	As_2S_3	mon	1½–2	3.49
Orthite	$(Ca,Ce,La)_2(Al,Fe,Be,Mg,Mn)_3(OH)Si_3O_{12}$	mon	5–6½	2.8–4.2
Orthoclase	$K(AlSi_3O_8)$	mon	6	2.57
Orthoferrosilite	$Fe_2(Si_2O_6)$	orth	6	3.9
Osbornite	TiN			
Osumilite	$(K,Na,Ca)(Mg,Fe)_2(Al,Fe)_3(Si,Al)_{12}O_{30} \cdot H_2O$			
Otavite	$CdCO_3$			
Ottrelite	$(Fe,Mn)(Al,Fe)(Al_2Si_2O_{10})(OH)_2$	mon	6–7	3.5

Table of accepted mineral species (cont.)

Mineral	Chemical formula	Crystallography class*	Hardness on Moh's scale	Specific gravity
Overite	$Ca_3Al_8(PO_4)_8(OH)_6 \cdot 15H_2O$	orth	3½–4	2.53
Owyheeite	$Pb_5Ag_2Sb_6S_{15}$		2½	6.03
Oxammite	$(NH_4)_2C_2O_4 \cdot H_2O$	orth	2½	1.5
Pachnolite	$NaCaAlF_6 \cdot H_2O$	mon	3	2.98
Paigeite	$(Fe,Mg)_2Fe^3BO_5$	orth	5	4.7
Painite	$Al_2O_3 \cdot Ca_2(Si,BH)O_4$			
Palladium	Pd	iso	4½–5	11.9
Palmierite	$(K,Na)_2Pb(SO_4)_2$	hex		4.5
Palygorskite	hydrous Mg silicate			
Parabutlerite	$Fe(SO_4)(OH) \cdot 2H_2O$	orth	2½	2.55
Paracelsian	$Ba(Al_2Si_2O_8)$	mon	6–6½	3.31–3.32
Paracoquimbite	$Fe_2(SO_4)_3 \cdot 9H_2O$	hex	2½	2.11
Paragasite	$NaCa_2Fe_4(Al,Fe)Al_2Si_6O_{22}(OH)_2$	mon	5½	3–3.5
Paragonite	$NaAl_2(AlSi_3O_{10})(OH)_2$	mon	2	2.85
Parahilgardite	$Ca_8(B_6O_{11})_3Cl_4 \cdot 4H_2O$	tric	5	2.71
Parahopeite	$Zn_3(PO_4)_2 \cdot 4H_2O$	tric	3¾	3.31
Paralaurionite	$Pb(OH)Cl$	mon	soft	6.15
Paramelaconite	$(Cu^2_{1-2x}Cu^1_{2x})O_{1-x}$	tet	4½	6.04
Paratacamite	$Cu_2(OH)_3Cl$	hex	3	3.74
Paravauxite	$FeAl_2(PO_4)_2(OH)_2 \cdot 8H_2O$	tric	3	2.36
Parawollastonite	$CaSiO_3$	orth	5	7.12
Pargasite	$Ca_4Na_2Mg_9Al_4Si_{13}O_{44}(OH,F)_4$	mon	5½	3–3.5
Parisite	$(Ce,La)_2Ca(CO_3)_3F_2$	hex	4½	4.36
Parkerite	NiS_2	mon	2–2½	
Parsettensite	$(K,H_2O)(Fe,Mg,Al,Mn)_3Si_4O_{10}(OH)_2$	mon	3–4	2.59–2.85
Parsonite	$Pb_2(UO_2)(PO_4)_2 \cdot 2H_2O$	mon	2½–3	5.37
Pasarammelsbergite	$NiAs_2$	orth	5	7.12
Pascoite	$Ca_2V_6O_{17} \cdot 11H_2O$	tric	2½	1.87
Paternoite	$MgB_8O_{13} \cdot 4H_2O$			
Paulingite	$(K,Ca)_{120}[(Al,Si)_{580}O_{1160}] \cdot 690H_2O$	cub		
Pearceite	$Ag_{16}As_2S_{11}$	mon	3	6.15
Pectolite	$Ca_2NaH(SiO_3)_3$	tric	5	2.7–2.8
Penfieldite	$Pb_2(OH)Cl_3$	hex		6.61
Pennantite	low-Mg chlorite	mon	2–3	3.06
Penninite	hydrous Mg,Fe,Al silicate	mon	2–3	2.6–3.1
Penroseite	$(Ni,Co,Cu)Se_2$	iso	2½–3	6.9±0.2
Pentahydrite	$MgSO_4 \cdot 5H_2O$			
Pentlandite	$(Fe,Ni)_9S_8$	iso	3½–4	4.6–5.0
Percylite	$PbCuCl_2(OH)_2$	iso	2½	
Periclase	MgO	iso	5½	3.56
Perovskite	$CaTiO_3$	pseudo	5½	4.03
Perrierite	$Ce_2Ti_2Si_2O_{11}$	mon	5½	4.3–4.7
Perthite	$KAlSi_3O_8 + NaAlSi_3O_8$	tric	6–6½	2.56–2.65
Petalite	$Li(AlSi_4O_{10})$	mon	6½	2.41–2.42
Petzite	$(Ag,Au)_2Te$	iso	2½–3	8.7–9.0
Pharmacolite	$CaH(AsO_4) \cdot 2H_2O$	mon	2–2½	2.53–2.73
Pharmacosiderite	$Fe_3(AsO_4)_2(OH)_3 \cdot 5H_2O$	iso	2½	2.79
Phenacite	$Be_2(SiO_4)$	rho	7½–8	2.97–3.00
Phillipsite	$KCa(Al_3Si_5O_{16}) \cdot 6H_2O$	mon	4½–5	2.2
Phlogopite	$KMg_3(AlSi_3O_{10})(OH)_2$	mon	2½–3	2.86
Phosgenite	$Pb_2(CO_3)Cl_2$	tet	2–3	6.0–6.3
Phosphoferrite	$(Fe,Mn)_3(PO_4)_2 \cdot 3H_2O$	orth	3–3½	3.0–3.2
Phosphophyllite	$Zn_2(Fe,Mn)(PO_4)_2 \cdot 4H_2O$	mon	3–3½	3.08
Phosphorroesslerite	$MgH(PO_4) \cdot 7H_2O$	mon	2½	1.73
Phosphuranylite	Ca uranyl phosphate	tet	2–3	
Pickeringite	$MgAl_2(SO_4)_4 \cdot 22H_2O$	mon	1½	1.73–1.79
Picromerite	$K_2Mg(SO_4)_2 \cdot 6H_2O$	mon	2½	2.03
Picropharmacolite	$(Ca,Mg)_3(AsO_4)_2 \cdot 6H_2O$			
Piemontite	$Ca_2(Al,Fe)Al_2(OH)Si_3O_{12}$	mon	6	3.45–3.52
Pigeonite	$(Mg,Fe,Ca)_2(Si_2O_6)$	mon	6	3.30–3.46
Pilbarite	$2PbO \cdot 3ThO_2 \cdot 4UO_2 \cdot 8SiO_2 \cdot 21H_2O$		3	4.6
Pinakiolite	$Mg_3Mn^2Mn^3_2B_2O_{10}$	orth	6	3.88
Pinnoite	$Mg(BO_2)_2 \cdot 3H_2O$	tet	3½	2.27
Pintadoite	$Ca_2V_2O_7 \cdot 9H_2O$			
Pirssonite	$Na_2Ca(CO_3)_2 \cdot 2H_2O$	orth	3–3½	2.35
Pisanite	$(Fe,Cu)SO_4 \cdot 7H_2O$	mon	2	1.89

Table of accepted mineral species (cont.)

Mineral	Chemical formula	Crystallography class*	Hardness on Moh's scale	Specific gravity
Pistacite	$Ca_2(Al,Fe,Mn)_2Al(OH)Si_3O_{12}$	mon	6	3.3–3.49
Pitticite	$Fe_2(AsO_4)(SO_4)(OH) \cdot H_2O$		2–3	2.2–2.5
Plagionite	$Pb_5Sb_8S_{17}$	mon	2½	5.56
Plancheite	$Cu_8(Si_4O_{11})_2(OH)_2 \cdot H_2O$	iso	5½	3.3
Plasma		hex	6½–7	2.57–2.65
Platiniridium	Ir,Pt	iso	6–7	22.65–22.84
Platinum	Pt	iso	4–4½	14–19
Plattnerite	PbO_2	tet	5½	9.42
Platynite	$PbBi_2(Se,S)_3$	hex	2–3	7.98
Plombierite	$Ca_5H_2(Si_3O_9)_2 \cdot 6H_2O$			
Plumboferrite	$PbFe_4O_7$	hex	5	6.07
Plumbogummite	$PbAl_3(PO_4)_2(OH)_5H_2O$	hex	4½–5	4.01
Plumbojarosite	$PbFe_6(SO_4)_4(OH)_{12}$	hex	soft	3.67
Polianite	MnO_2	tet	6–6½	5.0
Pollucite	$Cs_4Al_4Si_9O_{26} \cdot H_2O$	iso	6½	2.9
Polyargyrite	$Ag_{24}Sb_2S_{15}$	iso	2½	6.974
Polybasite	$(Ag,Cu)_{16}Sb_2S_{11}$	mon	2–3	6.0–6.2
Polycrase	$(Y,Ca,Ce,U,Th)(Ti,Cb,Ta)_2O_6$‡	orth	5½–6½	5.00 ± 0.1
Polydymite	Ni_3S_4	iso	4½–5½	4.5–4.8
Polyhalite	$K_2Ca_2Mg(SO_4)_4 \cdot 2H_2O$	tric	3½	2.78
Polymignyte	$(Ca,Fe^2,Y,Zr,Th)(Cb,Ti,Ta)O_4$‡	orth	6½	4.77–4.85
Portlandite	$Ca(OH)_2$	hex	2	2.230
Potarite	Pd_3Hg_2	iso	3½	13.48–16.11
Potash alum	$KAl(SO_4)_2 \cdot 12H_2O$	iso	2–2½	1.76
Potash feldspar	$KAlSi_3O_8$			
Powellite	$Ca(Mo,WO_4)$	tet	3½–4	4.23
Prase		hex	6½–7	2.57–2.65
Prehnite	$Ca_2Al_2(Si_3O_{10})(OH)_2$	orth	6–6½	2.8–2.95
Priceite	$Ca_4B_{10}O_{19} \cdot 7H_2O$	tric	3–3½	2.42
Priorite	$(Y,Er,Ca,Fe^{2+},Th)(Ti,Cb)_2O_6$‡	orth	5–6	4.95 ± 0.1
Probertite	$NaCaB_5O_9 \cdot 5H_2O$	mon	3½	2.14
Prosopite	$CaAl_2(F,OH)_8$	mon	4½	2.88–2.89
Proustite	Ag_3AsS_3	hex	2–2½	5.57
Pseudoboleite	$Pb_5Cu_4Cl_{10}(OH)_8 \cdot 2H_2O$	tet	2½	4.85
Pseudobrookite	Fe_2TiO_5	orth	6	4.33–4.39
Pseudocotunnite	K_2PbCl_4			
Pseudomalachite	$Cu_5(PO_4)_2(OH)_4 \cdot H_2O$	mon	4½–5	4.35
Pseudothuringite	hydrous Fe,Mg,Al silicate			
Pseudowollastonite	$CaSiO_3$	tric	5	2.91
Psilomelane	$BaMn^{2+}Mn_8^{4+}O_{16}(OH)_4$	orth	5–6	3.7–4.7
Pucherite	$BiVO_4$	orth	4	6.57
Pumpellyite	$Ca_4(Mg,Fe^{2+},Mn)(Al,Fe^{3+},Ti)_5O(OH)_3$-$[SiO_4]_2[SiO_7]_2 \cdot 2H_2O$	mon	6	3.18–3.23
Purpurite	$(Mn,Fe)(PO_4)$	orth	4–4½	3.2–3.4
Pyrargyrite	Ag_3SbS_3	hex	2½	5.85
Pyrite	FeS_2	iso	6–6½	5.02
Pyroaurite	$Mg_6Fe_2(OH)_{16} \cdot CO_3 \cdot 4H_2O$	hex	2½	2.12
Pyrobelonite	$MnPb(VO_4)(OH)$	orth	3½	5.38
Pyrochlore	$NaCaCb_2O_6F$	iso	5–5½	4.2–6.4
Pyrochroite	$Mn(OH)_2$	hex	2½	3.25
Pyrolusite	MnO_2	tet	6–6½	4.75
Pyromorphite	$Pb_5(PO_4,AsO_4)_3Cl$	hex	3½–4	6.5–7.1
Pyrope	$Mg_3Al_2(SiO_4)_3$	iso	7	3.51
Pyrophanite	$MnTiO_3$	hex	5–6	4.54
Pyrophyllite	$Al_2(Si_4O_{10})(OH)_2$	mon	1–2	2.8–2.9
Pyrostilpnite	Ag_3SbS_3	mon	2	5.94
Pyroxene	Mg,Fe,Ca,Na,Ti,Al silicates		5–7	2.96–3.96
Pyroxferroite	$Ca_{0.15}Fe_{0.85}SiO_3$	tric		3.7
Pyroxmangite	$(Mn,Fe)(SiO_3)$	tric	5½–6	3.61–3.80
Pyrrhotite	$Fe_{1-x}S$	hex	3½–4½	4.58–4.65
Quartz	SiO_2	rho	7	2.65
Quenselite	$PbMnO_2(OH)$	mon	2½	6.84
Quenstedtite	$Fe_2(SO_4)_3 \cdot 10H_2O$	tric	2½	2.15
Radiophyllite	$Ca_4(Si_4O_{10})(OH)_4 \cdot 2H_2O$		2–3	2.53
Ralstonite	$Na(MgAl_5)_6F_{12}(OH)_6 \cdot 3H_2O$	iso	4½	2.56–2.62
Ramdohrite	$Pb_3Ag_2Sb_6S_{13}$		2	5.43

Table of accepted mineral species (cont.)

Mineral	Chemical formula	Crystallography class*	Hardness on Moh's scale	Specific gravity
Rammelsbergite	$NiAs_2$	orth	5½–6	7.1
Ramsayite	$Na_2Ti_2Si_2O_9$	orth	6	3.43
Rankinite	$Ca_3Si_2O_7$			
Ransomite	$Cu(Fe,Al)_2(SO_4)_4 \cdot 7H_2O$	orth	2½	2.63
Raspite	$PbWO_4$	mon	2½–3	8.46
Rathite	$Pb_{13}As_{18}S_{40}$	mon	3	5.37
Rauvite	$CaU_2V_{12}O_{36} \cdot 20H_2O$			
Realgar	AsS	mon	1½–2	3.56
Reddingite	$(Mn,Fe)_3(PO_4)_2 \cdot 3H_2O$	orth	3–3½	3.0–3.2
Reedmergnerite	$NaBSi_3O_8$			
Renardite	$Pb(UO_2)_4(PO_4)_2(OH)_4 \cdot 7H_2O$	orth	4	
Retgersite	$NiSO_4 \cdot 6H_2O$	tet	2½	2.04
Retzian	$Mn_2Y(AsO_4)(OH)_4$‡	orth	4	4.15
Rezbanyite	$Pb_3Cu_2Bi_{10}S_{19}$		2½	6.24
Rhabdophane	$(Ce,Y,La,Di)(PO_4) \cdot H_2O$‡	tet or hex	3½	3.94–4.01
Rhipidolite	hydrous Fe,Mg,Al silicate	mon	2–3	2.88–3.08
Rhodizite	$NaKLi_4Al_4Be_3B_{10}O_{27}$	iso	8	3.38
Rhodochrosite	$MnCO_3$	hex	3½–4	3.70
Rhodolite	$3(Mg,Fe)O \cdot Al_2O_3 \cdot 3SiO_2$	iso	7	3.84
Rhodonite	$Mn(SiO_3)$	tric	5½–6	3.58–3.70
Rhomboclase	$HFe(SO_4)_2 \cdot 4H_2O$	orth	2	2.23
Richellite	$Ca_3Fe_{10}(PO_4)_8(OH)_{12} \cdot H_2O$		2–3	2
Richterite	$Na_2Ca(Mg,Fe,Mn,Al,Fe)_5Si_8O_{22}(OH,F)_2$	mon	5–6	2.97–3.45
Rickardite	Cu_4Te_3		3½	7.54
Riebeckite	$Na_3Fe_3^{2+}Fe_2^{3+}(Si_8O_{22})(OH)_2$	mon	4	3.44
Ringwoodite	$(Fe,Mg)_2SiO_4$	iso		3.90
Rinkite	$Na(Ca,Ce)_2(Ti,Ce)FSi_2O_8$	mon	5	3.46
Rinneite	NaK_3FeCl_6	hex	3	2.35
Riversideite	$2CaSiO_3 \cdot 3H_2O$		3	2.6
Rockbridgeite	$(Fe^{2+},Mn)Fe_4^{3+}(PO_4)_3(OH)_5$	orth	4½	3.3–3.49
Roeblingite	$2PbSO_4 \cdot H_{10}Ca_7Si_6O_{24}$	orth	3	3.43
Roemerite	$Fe^{2+},Fe_2^{3+},(SO_4)_4 \cdot 14H_2O$	tric	3–4½	2.17
Roesslerite	$MgH(AsO_4) \cdot 7H_2O$	mon	2–3	1.94
Romeite	$(Ca,Fe,Mn,Na)_2(Sb,Ti)_2O_6(O,OH,F)$	iso	5½–6½	4.7–5.4
Rooseveltite	$Bi(AsO_4)$	mon	4–4½	6.86
Rosasite	$(Cu,Zn)_2(CO_3)(OH)_2$	mon	4–5	4.0–4.2
Roscherite	$(Ca,Mn,Fe)_2Al(PO_4)_2(OH) \cdot 2H_2O$	mon	4½	2.92
Roscoelite	$K_2V_4Al_2Si_6O_{20}(OH)_4$	mon	2½	2.97
Roselite	$(Ca,Co)_2(Co,Mg)(AsO_4)_2 \cdot 2H_2O$	mon	3½	3.50–3.74
Rose quartz		hex	7	2.65
Rosenbuschite	$(Na,Ca,Mn)_3(Fe,Ti,Zr)FSi_2O_8$	tric	5–6	3.3
Rossite	$CaV_2O_6 \cdot 4H_2O$	tric	2–3	2.45
Roweite	$(Mn,Mg,Zn)Ca(BO_2)_2(OH)_2$	orth	5	2.92
Ruby spinel	Mg,Al_2O_4			
Russellite	Bi_2WO_6	tet	3½	7.35
Rutherfordine	$(UO_2)(CO_3)$		5½	4.82
Rutile	TiO_2	tet	6–6½	4.18–4.25
Safflorite	$(Co,Fe)As_2$	orth	4½–5	7.2
Sahlinite	$Pb_{14}(AsO_4)_2O_9Cl$	mon	2–3	7.95
Salammoniac	NH_4Cl	iso		1.53
Saléeite	$Mg(UO_2)_2(PO_4)_2 \cdot 10H_2O$	tet	2–3	3.27
Salesite	$Cu(IO_3)(OH)$	orth	3	4.77
Salite	$Ca(Fe,Mg)Si_2O_6$	mon	5½–6½	3.2–3.6
Salmonsite	$Mn_9Fe_2(PO_4)_8 \cdot 14H_2O$	orth	4	2.88
Samarskite	$(Y,Er,Ce,U,Ca,Fe,Pb,Th)(Cb,Ta,Ti,Sn)_2O_6$‡	orth	5–6	5.69
Sampleite	$NaCaCu_5(PO_4)_4Cl \cdot 5H_2O$	orth	4	3.20
Samsonite	$Ag_4MnSb_2S_6$	mon	2½	5.51
Sanbornite	$Ba_2Si_4O_{10}$	orth	5	4.19
Sanidine	$KAlSi_3O_8$	mon	6	2.56–2.62
Sanmartinite	$(Zn,Fe,Ca)WO_4$	mon		6.70
Saponite	$(Mg,Al)_6(Si,Al)_8O_{20}(OH)_4$	mon	1–1½	2.5
Sapphirine	$(Mg,Fe)_{15}(Al,Fe)_{34}Si_7O_{80}$	mon	7½	3.4–3.5
Sarcolite	$(Ca,Na_2)_3Al_2Si_3O_{12}$	tet	6	2.92
Sarcopside	$(Fe,Mn,Ca)_7(PO_4)_4F_2$	mon	4	3.73
Sard		hex	6½	2.57–2.64
Sarkinite	$Mn_2(AsO_4)(OH)$	mon	4–5	4.08–4.18

Table of accepted mineral species (cont.)

Mineral	Chemical formula	Crystallography class*	Hardness on Moh's scale	Specific gravity
Sarmientite	$Fe_2(AsO_4)(SO_4)(OH) \cdot 5H_2O$	mon		2.58
Sartorite	$PbAs_2S_4$	mon	3	5.10
Sassolite	$B(OH)_3$	tric	1	1.48
Sauconite montmorillonite clay		mon	1–2	2.0–3.0
Scacchite	$MnCl_2$	hex	soft	2.98
Scapolite	$(Na,Ca,K)_4Al_3(Al,Si)_3Si_6O_{24}$-$(Cl,F,OH,CO_3,SO_4)$	tet	5–6	2.50–2.75
Scawtite	$2CaCO_3 \cdot Ca_2Si_3O_8$	mon	4½–5	2.77
Schafarzikite	$Fe_5Sb_4O_{11}$	tet	3½	4.3
Schairerite	$Na_3(SO_4)(F,Cl)$	hex	3½	2.61
Schallerite	$(Mn,Fe)_8(OH,Cl)_{10}(Si,As)_6O_{15}$	hex	4½–5	3.37
Scheelite	$CaWO_4$	tet	4½–5	5.9–6.1
Scheteligite	$(Ca,Y,Sb,Mn)_2(Ti,Ta,Cb)_2(O,OH)_7$‡	orth	5½	4.74
Schirmerite	$PbAg_4Bi_4S_9$		2	6.737
Schizolite	$(Ca,Mn)_2NaH(SiO_3)_3$	tric	5–5½	2.97–3.13
Schoepite	$4UO_3 \cdot 9H_2O$	orth	2–3	4.8
Schorl	$Na(Fe,Mn)_3Al_6(OH,F)_4(BO_3)_3Si_6O_{18}$	hex	7	3.10–3.25
Schreibersite	$(Fe,Ni)_3P$	tet	6½–7	7.0–7.3
Schroeckingerite	$NaCa_3(UO_2)(CO_3)_3(O,OH)_7$	hex	2½	2.51
Schultenite	$PbH(AsO_4)$	mon	2½	5.94
Schwartzembergite	$Pb_5(IO_3)Cl_3O_3$	tet	2–2½	7.39
Scolecite	$Ca(Al_2Si_3O_{10}) \cdot 3H_2O$	mon	5–5½	
Scorodite	$Fe^{3+}(AsO_4) \cdot 2H_2O$	orth	3½–4	3.28
Scorzalite	$(Fe,Mg)Al_2(PO_4)_2(OH)_2$	mon	5½–6	3.38
Seamanite	$Mn_3(PO_4)(BO_3) \cdot 3H_2O$	orth	4	3.08
Searlesite	$NaBSi_2O_6 \cdot H_2O$	mon	3½	2.45
Selenium	Se	hex	2	4.80
Selentellurium		hex	2–2½	4.8–6.3
Seligmannite	$PbCuAsS_3$	orth	3	5.44
Sellaite	MgF_2	tet	5	3.15
Semseyite	$Pb_9Sb_8S_{21}$	mon	2½	6.08
Senaite	$(Fe,Mn,Pb)TiO_3$		6	5.301
Senarmontite	Sb_2O_3	pseudo	2–2½	5.50
Sengierite	$Cu(UO_2)(VO_4)(OH) \cdot 4{-}5H_2O$	orth	2½	4
Sepiolite	$Mg_4(Si_6O_{15})(OH)_2 \cdot 6H_2O$	mon?	2–2½	2.0
Serandite	$Mn_2NaH(SiO_3)_3$			
Serendibite	$Ca_2(Mg,Fe)_4Al_6B_2O_{10}Si_4O_{16}$	tric	6½–7	3.42
Sericite	$KAl_2(AlSi_3O_{10})(OH,F)_2$	mon	2½–3	2.77–2.80
Serpentine	$Mg_3(Si_2O_5)(OH)_4$	mon, orth	3–5	2.3–2.6
Serpierite	$(Cu,Zn,Ca)_5(SO_4)_2(OH)_6 \cdot 3H_2O$	orth		2.52
Sharpite	$(UO_2)_6(CO_3)_5(OH)_2 \cdot 6H_2O$	orth	2–3	3.3
Shattuckite	$Cu_5(SiO_3)_4(OH)_2$	orth		3.8
Sheridanite	hydrous Mg,Fe,Al silicate	mon	2–3	2.65–2.80
Shortite	$Na_2Ca_2(CO_3)_3$	orth	3	2.60
Sicklerite	$(Li,Mn,Fe)(PO_4)$	orth	4	3.2–3.4
Siderazot	Fe_5N_2			
Siderite	$FeCO_3$	hex	3¾–4¼	3.83–3.88
Sideronatrite	$Na_2Fe(SO_4)_2(OH) \cdot 3H_2O$	orth	1½–2½	2.15–2.35
Siderotil	$FeSO_4 \cdot 5H_2O$			
Siegenite	$(Co,Ni)_3S_4$	iso	4½–5½	4.5–4.8
Sillenite	Bi_2O_3	iso	soft	8.80
Sillimanite	Al_2SiO_5	orth	6–7	3.23
Silver	Ag	iso	2½–3	10
Simpsonite	$Al_2Ta_2O_8$	hex		5.92–6.27
Sincosite	$CaV_2O_2(PO_4)_2 \cdot 5H_2O$	tet	low	2.84
Siserskite	OS,Ir	hex	6–7	19–21
Sjogrenite	$Mg_6Fe_2(OH)_{16} \cdot CO_3 \cdot 4H_2O$	hex	2½	2.11
Sklodowskite	$MgU_2O_3Si_2O_8 \cdot 6H_2O$			
Skutterudite	$(Co,Ni)As_3$	iso	5½–6	6.5±0.4
Slavikite	$(Na,K)_2Fe_{10}(OH)_6(SO_4)_{13} \cdot 63H_2O$	hex		1.91–1.99
Smaltite	$(Co,Ni)As_{3-x}$	iso	5½–6	6.5
Smithite	$AgAsS_2$	mon	1½–2	4.88
Smithsonite	$ZnCO_3$	hex	4–4½	4.35–4.40
Smoky quartz		hex	7	2.65
Soda alum	$NaAl(SO_4)_2 \cdot 12H_2O$	iso	2½–3½	1.67

Table of accepted mineral species (cont.)

Mineral	Chemical formula	Crystallography class*	Hardness on Moh's scale	Specific gravity
Sodalite	$Na_8(AlSiO_4)_6Cl_2$	iso	5½–6	2.15–2.3
Soda niter	$NaNO_3$	rho	1–2	2.29
Soddyite	$U_5Si_2O_{19} \cdot 6H_2O$	orth	3–4	4.63
Souzalite	$(Mg,Fe^{2+})_3(Al,Fe^{3+})_4(PO_4)_4(OH)_6 \cdot 2H_2O$	mon	5½–6	3.09
Spangolite	$Cu_6Al(SO_4)(OH)_{12}Cl \cdot 3H_2O$	hex	2–3	3.14
Spencerite	$Zn_4(PO_4)_2(OH)_2 \cdot 3H_2O$	mon	3	3.14
Sperrylite	$PtAs_2$	iso	6–7	10.5
Spessartite	$Mn_3Al_2(SiO_4)_3$	iso	7	4.18
Sphalerite	ZnS	iso	3½–4	3.9–4.1
Sphene	$CaTiO(SiO_4)$	mon	5–5½	3.40–3.55
Spherocobaltite	$CoCO_3$	hex	4	4.13
Spinel	$MgAl_2O_4$	iso	7½–8	3.55
Spodumene	$LiAl(Si_2O_6)$	mon	6½–7	3.15–3.20
Spurrite	$CaCO_3 \cdot 2Ca_2SiO_4$	mon	5	3
Stainierite	$CoO(OH)$		4–5	4.13–4.47
Stannite	Cu_2FeSnS_4	tet	4	4.3–4.5
Staurolite	$Fe_2Al_9O_6(SiO_4)_4(O,OH)_2$	orth	7–7½	3.65–3.75
Steenstrupine		hex	4	3.4–3.47
Steigerite	$Al_2(VO_4)_2 \cdot 6.5H_2O$			
Stephanite	Ag_5SbS_4	orth	2–2½	6.2–6.3
Stercorite	$Na(NH_4)H(PO_4) \cdot 4H_2O$	tric	2	1.62
Sternbergite	$AgFe_2S_3$	orth	1–1½	4.10–4.22
Sterrettite	$Al_6(PO_4)_4(OH)_6 \cdot 5H_2O$	orth	5	2.44
Stevensite	$Mg_{5.76}Mn_{0.04}Fe_{0.04}Si_8O_{20}(OH)_4$	mon	1–2	2–3
Stewartite	Mn hydrous phosphate			
Stibiconite	$Sb_3O_6(OH)$	iso	4–5½	5.58
Stibiocolumbite	$Sb(Nb,Ta,Cb)O_4$	orth	5½	5.68
Stibiopalladinite	Pd_3Sb	iso	4–5	9.5
Stibiotantalite	$SbTaO_4$	orth	5½	7.53
Stibnite	Sb_2S_3	orth	2	4.52–4.62
Stichtite	$Mg_6Cr_2(OH)_{16} \cdot CO_3 \cdot 4H_2O$	hex	1½–2	2.16
Stilbite	$Ca(Al_2Si_7O_{18}) \cdot 7H_2O$	mon	3½–4	2.1–2.2
Stillwellite	$(Ce, \ldots)BSiO_5$			
Stilpnochlorane	$(Ca,K,H_2O)(Al,Fe,Mg)_3(OH,O)_2(Si,P)_4O_{10}$		3–4	2.5–3.0
Stishovite	SiO_2	tet	7	4.35
Stokesite	$CaZnSi_3O_9 \cdot 2H_2O$	orth	6	3.19
Stolzite	$PbWO_4$	tet	2½–3	7.9–8.3
Strengite	$Fe(PO_4) \cdot 2H_2O$	orth	4	2.87
Stromeyerite	$CuAgS$	orth	2½–3	6.2–6.3
Strontianite	$StCO_3$	orth	3½	3.7
Struvite	$Mg(NH_4)(PO_4) \cdot 6H_2O$	orth	2	1.71
Stylotypite	$(Cu,Ag,Fe)_3SbS_3$	mon	3	4.79
Suanite	$Mg_2B_2O_5$			
Sulfoborite	$Mg_6H_4(BO_3)_4(SO_4)_2 \cdot 7H_2O$	orth	4–4½	2.38–2.45
Sulfohalite	$Na_6ClF(SO_4)_2$	iso	3½	2.5
Sulfur	S	orth	1½–2½	2.07
Sulvanite	Cu_3VS_4	iso	3½	4
Susannite	$Pb_4(CO_3)_2(OH)_2(SO_4)$			
Sussexite	$(Mn,Mg)(BO_2)(OH)$	orth	3–3½	3.30
Svabite	$(Ca,Pb)_5(AsO_4)(PO_4)_3(F,Cl,OH)$	hex	4–5	3.5–3.8
Svanbergite	$SrAl_3(PO_4)(SO_4)(OH)_6$	hex	5	3.22
Swedenborgite	$NaBe_4SbO_7$	hex	8	4.29
Sylvanite	$(Ag,Au)Te_2$	mon	1½–2	8.161
Sylvite	KCl	iso	2	1.99
Symplesite	$Fe_3(AsO_4)_2 \cdot 8H_2O$	tric	2½	3.01
Synadelphite	$(Mn,Mg,Ca,Pb)_4(AsO_4)(OH)_5$	tric	4½	3.57
Synchisite	$(Ce,La)Ca(CO_3)_2F$	hex	4½	3.90
Syngenite	$K_2Ca(SO_4)_2 \cdot H_2O$	mon	2½	2.60
Szaibelyite	$(Mg,Mn)(BO_2)(OH)$	orth	3–3½	2.62
Szomolnokite	$FeSO_4 \cdot H_2O$	mon	2½	3.03–3.07
Tachyhydrite	$CaMg_2Cl_6 \cdot 12H_2O$	hex	2	1.67
Tagilite	$Cu_2(PO_4)(OH) \cdot H_2O$	orth	3	3.4–3.6
Talc	$Mg_3(Si_4O_{10})(OH)_2$	mon	1	2.7–2.8
Tamarugite	$NaAl(SO_4)_2 \cdot 6H_2O$	mon	3	2.07
Tantalite	$(Fe,Mn)Ta_2O_6$	orth	6–6½	6.5
Tantalum	Ta	iso	6–7	11.2
Tapiolite	$FeTa_2O_6$	tet	6–6½	7.90

Table of accepted mineral species (cont.)

Mineral	Chemical formula	Crystallography class*	Hardness on Moh's scale	Specific gravity
Taramellite	$(Mn,Ba,Ca,Mg)MnSi_2O_6$	orth	3	3.4–3.6
Taramite	$Na_2Ca_2Mg_5Fe_3(Fe,Ti)_2(Al,Fe)_2(OH,F)_4Si_{14}O_{44}$			
Taranakite	$K_2Al_6(PO_4)_6(OH)_2 \cdot 18H_2O$	orth	5½	3.92
Tarapacaite	K_2CrO_4	orth		2.7
Tarbuttite	$Zn_2(PO_4)(OH)$	tric	3¾	4.12
Tauriscite	$FeSO_4 \cdot 7H_2O$			
Taylorite	$(K,NH_4)_2SO_4$		2	
Teallite	$PbSnS_2$	orth	1½	6.36
Teepleite	$Na_2BO_2Cl \cdot H_2O$	tet	3–3½	2.08
Teineite	$CuTeO_3 \cdot 2H_2O$	orth	2½	3.80
Tellurite	TeO_2	orth	2	5.90
Tellurium	Te	hex	2–2½	6.1–6.3
Tellurobismuthite	Bi_2Te_3	hex	1½–2	7.815
Tennantite	$(Cu,Fe)_{12}As_4S_{13}$	iso	3–4½	4.62
Tenorite	CuO	mon	3½	5.8–6.4
Tephroite	$Mn_2(SiO_4)$	orth	6	3.78–4.1
Terlinguaite	Hg_2OCl	mon	2½	8.73
Teschemacherite	$(NH_4)HCO_3$	orth	1½	1.5
Tetradymite	Bi_2Te_2S	hex	1½–2	7.3
Tetrahedrite	$(Cu,Fe)_{12}Sb_4S_{13}$	iso	3–4½	4.97
Thalenite	$Y_2Si_2O_7$‡	mon	6½	4.23–4.45
Thaumasite	$FeSO_4 \cdot 7H_2O$	hex	3½	1.91
Thenardite	Na_2SO_4	orth	2½–3	2.66
Thermonatrite	Na_2CO_3,H_2O	orth	1–1½	2.2
Thomsenolite	$NaCaAlF_6 \cdot H_2O$	mon	2	2.98
Thomsonite	$NaCa_2[(Al,Si)_5O_{10}]_2 \cdot 6H_2O$	orth	5	2.3
Thoreaulite	$SnTa_2O_7$	mon	6	7.6–7.9
Thorianite	ThO_2	iso	6½	9.7
Thorite	$Th(SiO_4)$	tet	5	5.3
Thorogummite		tet	4–4½	4.5
Thorotungstite	$AlFe(Th,Ca,Ce,Zr)WO_3$			
Thortveitite		mon	6–7	3.58
Thulite	$(Ca,Mn)_2Al_3(OH)(SiO_4)_3$	orth	6	3.15–3.37
Thuringite		mon	2–3	2.8–3.3
Tiemannite	HgSe	iso	2½	8.19–8.47
Tilasite	$CaMg(AsO_4)F$	mon	5	3.77
Tilleyite	$Ca_5(CO_3)_2Si_2O_7$			
Tin	Sn	tet	2	7.3
Tincalconite	$Na_2B_4O_7 \cdot 5H_2O$	hex		1.88
Tinticite	$Fe_3(PO_4)_2(OH)_3 \cdot 3.5H_2O$	orth	2–3	2.7–2.9
Titanaugite	$Ca_{14}NaMg_{10}Fe_3Ti_2Fe_2Al_5Si_{27}O_{96}$	mon	5½–6	3.3–3.6
Tobermorite	$Ca_5H_2(Si_3O_9)_2 \cdot 4H_2O$			
Topaz	$Al_2SiO_4(F,OH)_2$	orth	8	3.4–3.6
Torbernite	$Cu(UO_2)_2(PO_4)_2 \cdot 8-12H_2O$	tet	2–2½	3.22
Torendrikite	hydrous Na,Ca,Mg,Fe aluminosilicate	mon	5–6	3.21
Torreyite	$(Mg,Mn,Zn)_7(SO_4)(OH)_{12} \cdot 4H_2O$		3	2.67
Tourmaline	$(Na,Ca)(Li,Mg,Al)(Al,Fe,Mn)_6$-$(BO_3)_3(Si_6O_{18})(OH)_4$	tet	2–2½	3.22
Trechmannite	§	hex	1½–2	
Tremolite	$Ca_2Mg_5(Si_8O_{22})(OH)_2$	mon	5–6	3.0–3.3
Trevorite	$NiFe_2O_4$	iso	5½–6½	5.164
Trichalcite	$Cu_5Ca(AsO_4)_2(CO_3)(OH)_4 \cdot 6H_2O$	orth	2½	
Tridymite	SiO_2	orth	7	2.26
Trigonite	$MnPb_3H(AsO_3)_3$	mon	2–3	6.1–7.1
Trihydrocalcite	$CaCO_3 \cdot 3H_2O$			
Trimerite		mon	6–7	6.99
Triphylite	$Li(Mn,Fe)(PO_4)$	orth	4–5	3.58
Triplite	$(Mn,Fe,Mg,Ca)_2(PO_4)(F,OH)$	mon	5–5½	3.5–3.9
Triploidite	$(Mn,Fe)_2(PO_4)(OH)$	mon	4½–5	3.66
Trippkeite	$CuAs_2O_4$	tet	soft	4.8
Tripuhyite	$FeSb_2O_7$		4½–5	5.82
Tritomite	$Ca_3(La,Ce)_3Zr_3F_6B_3Si_6O_{27}$	hex	5½	4.2
Troegerite	$(UO_2)_3(AsO_4)_2 \cdot 12H_2O$	tet	2–3	
Troilite	FeS	hex	4	4.7
Trona	$Na_3H(CO_3)_2 \cdot 2H_2O$	mon	2½–3	2.14
Trudellite	$Al_{10}(SO_4)_3Cl_{12}(OH)_{12} \cdot 30H_2O$	hex	2½	1.93
Truscottite	$(Ca,Mg)_4(OH)_2Si_6O_{15} \cdot 3H_2O$	hex	3–4	2.34–2.45

Table of accepted mineral species (cont.)

Mineral	Chemical formula	Crystallography class*	Hardness on Moh's scale	Specific gravity
Tschermakite	$Ca_2Mg_3Al_2(Si_7Al_2)OH_2$	mon	5–6	3.02–3.45
Tsumebite	$Pb_2Cu(PO_4)(SO_4)(OH)$	mon	3½	6.13
Tuhualite	$(Na_2,K_2,Mn)(Al,Fe,Mg,Ti)_2Si_{10}O_{24}$			
Tungstenit	WS_2	hex	2½	7.4
Tungstite	$WO_3 \cdot H_2O$	orth	2½	
Turanite	$Cu_2(VO_4)(OH)$	orth	5	
Turgite	$2Fe_2O_3 \cdot H_2O$		6½	4.2–4.6
Turquois	$CuAl_6(PO_4)_4(OH)_8 \cdot 4H_2O$	tric	5–6	2.6–2.8
Tychite	$Na_6Mg_2(SO_4)(CO_3)_4$	iso	3½–4	2.46
Tyrolite	$Cu_5Ca(AsO_4)_2(CO_3)(OH)_4 \cdot 6H_2O$	orth	1½–2½	3.0–3.25
Tyuyamunite	$Ca(UO_2)_2(VO_4)_2 \cdot nH_2O$	orth	2	3.7–4.3
Uhligite	$Ca_3(Ti,Al,Zr)_9O_{20}$	pseudo	5½	4.15
Ulexite	$NaCaB_5O_9 \cdot 8H_2O$	tric	2½	1.96
Ullmannite	NiSbS	iso	5–5½	6.65
Ulvospinel	Fe_2TiO_4	tric	7½–8	4.78
Umangite	Cu_3Se_2		3	5.62
Ungemachite	$K_3Na_9Fe(SO_4)_6(OH)_3 \cdot 9H_2O$	hex	2½	2.29
Uralite		mon	5–6	3.02–3.45
Uraninite	UO_2	iso	5½	7.5–9.7
Uranocercite	$Ba(UO_2)_2(PO_4)_2 \cdot 8H_2O$	tet	2–2½	3.53
Uranopilite	$(UO_2)_6(SO_4)(OH)_{10} \cdot 12H_2O$	mon		3.7–4.0
Uranosphaerite	$Bi_4(UO_2)(AsO_4)O_4 \cdot 3H_2O$	orth	2–3	6.36
Uranospinite	$Ca(UO_2)_2(AsO_4)_2 \cdot 8H_2O$	tet	2–3	3.45
Uranothorite	$(Th,U)(SiO_4)$			
Uvarovite	$Ca_3Cr_2Si_3O_{12}$	iso	7½	3.90
Uvite	$CaMg_4Al_5B_3Si_6O_{29} \cdot 2H_2O$	hex	7	3.05
Valentinite	Sb_2O_3	orth	2½–3	5.76
Valleriite	$Cu_2Fe_4S_7$	orth	soft	3.14
Vanadinite	$Pb_5(VO_4)_3Cl$	hex	2¾–3	6.88
Vandenbrandite	$CuO \cdot UO_3 \cdot 2H_2O$	tric	4	5.03
Vanoxite	$V_4V_2O_{13} \cdot 8H_2O$			
Vanthoffite	$Na_6Mg(SO_4)_4$		3½	2.69
Variscite	$AlPO_4 \cdot 2H_2O$	orth	4½	2.57
Varulite	$(Na,Ca)(Mn,Fe)_2(PO_4)_2$	orth	5	3.5–3.6
Vashegyite	$Al_4(PO_4)_3(OH)_3 \cdot nH_2O$		2–3	1.87
Vaterite	$CaCO_3$			
Vauquelinite	$Pb_2Cu(CrO_4)PO_4(OH)$	mon	2½–3	6.02
Vauxite	$FeAl_2(PO_4)_2(OH)_2 \cdot 7H_2O$	tric	3½	2.39
Veatchite	$Sr_3B_{16}O_{27} \cdot 5H_2O$	mon	2	2.69
Vermiculite	$(Mg,Ca)_{0.3}(Mg,Fe,Al)_{3.0}(Al,Si)_4O_{10}(OH)_4 \cdot 8H_2O$	mon	1–2	2.2–2.4
Vernadskite	$Cu_4(SO_4)_3(OH)_2 \cdot 4H_2O$		3½	3.3
Vesuvianite	$Ca_{10}Al_4(Mg,Fe)_2Si_9O_{34}(OH,F)_4$	tet	6–7	3.33–3.43
Veszelyite	$(Cu,Zn)_3(As,PO_4)(OH)_3 \cdot 2H_2O$	mon	3½–4	3.3–3.5
Villiaumite	NaF	iso	2–2½	2.79
Violarite	Ni_2FeS_4	iso	4½–5½	4.5–4.8
Vishnewite	$(Na,Ca,K)_{6.7}Al_6Si_6O_{24}(SO_4,CO_3,Cl)_{1-5.5} \cdot 1-5H_2O$	hex	5–6	2.32–2.42
Vivianite	$Fe_3(PO_4)_2 \cdot 8H_2O$	mon	1½–2	2.58–2.68
Voglite	$Ca_2CuU(CO_3)_5 \cdot 6H_2O$			
Volborthite	$Cu_3(VO_4)_2 \cdot 3H_2O$	mon	3½	3.5–3.8
Volchonskoite	$(Ca,Mg,Cr,Fe,Al)_2O_3 \cdot 3SiO_2 \cdot nH_2O$	mon	2½	2.2–2.7
Voltaite	$(K,Fe^{2+})_3Fe^{3+}(SO_4)_3 \cdot 4H_2O$	iso	3	2.7
Voltzite	Zn_5S_4O		4–4½	3.7–3.8
Vrbaite	$Tl_4Hg_3Sb_2As_8S_{20}$	orth	3½	5.30
Wad			2–6½	2.8–4.4
Wadeite	$K_2ZrSi_3O_9$			
Wagnerite	$Mg_2(PO_4)F$	mon	5–5½	3.15
Wairakite	$Ca(AlSi_2O_6)_2 \cdot 2H_2O$	mon	5½–6	2.57
Walpurgite	$Bi_4(UO_2)(AsO_4)O_4 \cdot 3H_2O$	tric	3½	6.69
Waltherite			4	5.32
Wardite	$Na_4CaAl_{12}(PO_4)_8(OH)_9 \cdot 3H_2O$	tet	5	2.81–2.87
Warwickite	$(Mg,Fe)_3Ti(BO_4)_2$	orth	3½–4	3.35
Water		hex	1½	.92
Wattevilleite	$Na_2Ca(SO_4)_2 \cdot 4H_2O$			
Wavellite	$Al_3(OH)_3(PO_4)_2 \cdot 5H_2O$	orth	3¼–4	2.36

Table of accepted mineral species (cont.)

Mineral	Chemical formula	Crystallography class*	Hardness on Moh's scale	Specific gravity
Weberite	Na_2MgAlF_7	orth	3½	2.96
Weddellite	$CaC_2O_4 \cdot 2H_2O$	tet	4	1.94
Wehrlite	Bi_3Te_2		1½–2½	8.41
Weibullite	$Pb_4Bi_6S_9Se_4$		2–3	6.97
Weinschenkite	$(Y,Er)(PO_4) \cdot 2H_2O$‡	mon		3.05
Weissite	Cu_5Te_3		3	6
Whewellite	$Ca(C_2O_4) \cdot H_2O$	mon	2½–3	2.23
Whitlockite	$Ca_3(PO_4)_2$	hex	5	3.12
Wilkeite	$Ca_5(P,S,Si,CO_4)_3(OH)$	hex	5	3.1
Willemite	Zn_2SiO_4	rho	5½	3.9–4.2
Witherite	$BaCO_3$	orth	3–3½	4.3
Wittichenite	Cu_3BiS_3	orth	2–3	4.3–4.5
Wittite	$Pb_5Bi_6(S,Se)_{14}$	orth or mon	2–2½	7.12
Woehlerite	$NaCa_2(Zr,Cb)FSi_2O_8$	mon	5½–6	3.42
Wolfachite	$Ni(As,Sb)S$	orth	4½–5	6.37
Wolfeite	$(Fe,Mn)_2(PO_4)(OH)$	mon	4½–5	3.83
Wolframite	$(Fe,Mn)WO_4$	mon	4–4½	7.0–7.5
Wollastonite	$Ca(SiO_3)$	tric	5–5½	2.8–2.9
Woodhouseite	$CaAl_3(PO_4)(SO_4)(OH)_6$	hex	4½	3.01
Wulfenite	$PbMoO_4$	tet	2¾–3	6.5–7.0
Wurtzite	ZnS	hex	3½–4	3.98
Xanthoconite	Ag_3AsS_3	mon	2–3	5.54
Xanthophyllite	$Ca_2(Mg,Al)_{3-2}(Al_2Si_2O_{10})(OH)_2$	mon	3½–6	3–3.1
Xanthoxenite	$Ca_2Fe(PO_4)_2(OH) \cdot 1\frac{1}{2}H_2O$	mon or tric	2½	2.8–2.9
Xenotime	$Y(PO_4)$‡	tet	4–5	4.4–5.1
Xonotlite	$(5CaSiO_3 \cdot H_2O)$	mon	6½	2.71
Yeatmanite	$(Mn,Zn)_{16}Sb_2Si_4O_{29}$	tric	4	5.0
Yttrailite	$(Y,Th)_2(Si_2O_7)$‡	mon	5½–6½	4.3–4.6
Yttrocrasite	$(Y,Th,U,Ca)_2Ti_4O_{11}$‡	orth	5½–6	4.80
Yttrotantalite	$(Y,U,Fe)(Ta,Nb)O_4$‡	orth	5–5½	5.7
Yugawaralite	$Ca_4(Al_7Si_{20}O_{12}) \cdot 14H_2O$	mon		
Zaratite	$Ni_3(CO_3)(OH)_4 \cdot 4H_2O$	iso	3½	2.57–2.69
Zeolite	$(Na_2,K_2,Ca,Ba,Sr)[(Al,Si)O_2]_n \cdot xH_2O$		3–5½	1.9–2.45
Zeophyllite	$Ca_8(OH,F)_{10}Si_6O_{15}$	hex	3	2.76
Zeunerite	$Cu(UO_2)_2(AsO_4)_2 \cdot 10–16H_2O$	tet	2–2½	3.2
Zinc	Zn	hex	2	6.9–7.2
Zincaluminite	$Zn_3Al_3(SO_4)(OH)_{13} \cdot 2.5H_2O$	hex	2½–3	2.26
Zincite	ZnO	hex	4	5.68
Zinkenite	$Pb_6Sb_{14}S_{27}$	hex	3–3½	5.3
Zinwaldite	$K(Li,Fe,Al)(AlSi_3O_{10})(F,OH)_2$	mon	2½–4	2.90–3.02
Zippeite	$(UO_2)_2SO_4(OH)_2 \cdot 4H_2O$	orth	3	
Zircon	$ZrSiO_4$	tet	7½	4.68
Zirkelite	Al,Fe basic chloride	iso	5½	4.74
Zoisite	$Ca_2Al_3(SiO_4)_3(OH)$	orth	6	3.35
Zunyite	$Al_{12}AlO_4(OH,F)_{18}ClSi_5O_{16}$	iso	7	2.88

*amor = amorphous; hex = hexagonal; iso = isometric; mon = monoclinic; orth = orthogonal; pseudo = pseudoisometric; rho = rhombohedral; tet = tetragonal; tric = triclinic.

†Ab = $NaAlSi_3O_8$; An = $CaAl_2Si_2O_8$; Or = $KAlSi_3O_8$.

‡X = Na or Ca; Y = Al, Fe^{3+}, Li, or Mg.

§Qualitative tests indicate presence of Ag, As, and S.

Index

Index

Asterisks indicate page references to article titles.

A

B

C

D

E

F

G

H

N

O

P

S

T

U

V

Mechanical Engineering Series

Frank Kreith and Roop Mahajan - Series Editors

Published Titles

Computer Techniques in Vibration
Clarence W. de Silva
Distributed Generation: The Power Paradigm for the New Millennium
Anne-Marie Borbely & Jan F. Kreider
Elastoplasticity Theory
Vlado A. Lubarda
Energy Audit of Building Systems: An Engineering Approach
Moncef Krarti
Engineering Experimentation
Euan Somerscales
Entropy Generation Minimization
Adrian Bejan
Finite Element Method Using MATLAB, 2nd Edition
Young W. Kwon & Hyochoong Bang
Fluid Power Circuits and Controls: Fundamentals and Applications
John S. Cundiff
Fundamentals of Environmental Discharge Modeling
Lorin R. Davis
Heat Transfer in Single and Multiphase Systems
Greg F. Naterer
Introductory Finite Element Method
Chandrakant S. Desai & Tribikram Kundu
Intelligent Transportation Systems: New Principles and Architectures
Sumit Ghosh & Tony Lee
Mathematical & Physical Modeling of Materials Processing Operations
Olusegun Johnson Ilegbusi, Manabu Iguchi & Walter E. Wahnsiedler
Mechanics of Composite Materials
Autar K. Kaw
Mechanics of Fatigue
Vladimir V. Bolotin
Mechanics of Solids and Shells: Theories and Approximations
Gerald Wempner & Demosthenes Talaslidis
Mechanism Design: Enumeration of Kinematic Structures According to Function
Lung-Wen Tsai
The MEMS Handbook, Second Edition
MEMS: Introduction and Fundamentals
MEMS: Design and Fabrication
MEMS: Applications
Mohamed Gad-el-Hak
Nanotechnology: Understanding Small Systems
Ben Rogers, Jesse Adams & Sumita Pennathur
Nonlinear Analysis of Structures
M. Sathyamoorthy
Practical Inverse Analysis in Engineering
David M. Trujillo & Henry R. Busby
Pressure Vessels: Design and Practice
Somnath Chattopadhyay
Principles of Solid Mechanics
Rowland Richards, Jr.
Thermodynamics for Engineers
Kau-Fui Wong
Vibration Damping, Control, and Design
Clarence W. de Silva
Vibration and Shock Handbook
Clarence W. de Silva
Viscoelastic Solids
Roderic S. Lakes

Vibration Damping, Control, and Design

Vibration Damping, Control, and Design

Edited by

Clarence W. de Silva

The University of British Columbia

Vancouver, Canada

CRC Press is an imprint of the
Taylor & Francis Group, an informa business

This material was previously published in *Vibration and Shock Handbook* © 2005 by CRC Press, LLC.

CRC Press
Taylor & Francis Group
6000 Broken Sound Parkway NW, Suite 300
Boca Raton, FL 33487-2742

Printed in the United States of America on acid-free paper
10 9 8 7 6 5 4 3 2 1

International Standard Book Number-10: 1-4200-5321-3 (Hardcover)
International Standard Book Number-13: 978-1-4200-5321-0 (Hardcover)

Library of Congress Cataloging-in-Publication Data

Vibration damping, control, and design / editor, Clarence W. de Silva.
p. cm. -- (Mechanical engineering series)
Includes bibliographical references and index.
ISBN-13: 978-1-4200-5321-0 (alk. paper)
ISBN-10: 1-4200-5321-3 (alk. paper)
1. Vibration. 2. Damping (Mechanics) I. De Silva, Clarence W. II. Title.

TA355.V5222 2007
620.3'7--dc22 2006100169

Visit the Taylor & Francis Web site at
http://www.taylorandfrancis.com

and the CRC Press Web site at
http://www.crcpress.com

Preface

In individual chapters authored by distinguished leaders and experienced professionals in their respective topics, this book provides for engineers, technicians, designers, researchers, educators, and students, a convenient, thorough, up-to-date, and authoritative reference source on techniques, tools, and data for analysis, design, monitoring and control of vibration, noise, and acoustics. Vibration suppression, damping, and control; design for and control of vibration; system design, application, and control implementation; and acoustics and noise suppression are treated in the book. Important information and results are summarized as windows, tables, graphs, and lists throughout the chapters, for easy reference and information tracking. References are given at the end of each chapter, for further information and study. Cross-referencing is used throughout to indicate other places in the book where further information on a particular topic is provided.

In the book, equal emphasis is given to theory and practical application. Analytical formulations, design approaches, and control techniques are presented and illustrated. Examples and case studies are given throughout the book to illustrate the use and application of the included information. The material is presented in a format that is convenient for easy reference and recollection.

Mechanical vibration is a manifestation of the oscillatory behavior in mechanical systems as a result of either the repetitive interchange of kinetic and potential energies among components in the system, or a forcing excitation that is oscillatory. Such oscillatory responses are not limited to purely mechanical systems, and are found in electrical and fluid systems as well. In purely thermal systems, however, free natural oscillations are not possible, and an oscillatory excitation is needed to obtain an oscillatory response. Sound, noise, and acoustics are manifestations of pressure waves, sources of which are often vibratory dynamic systems.

Low levels of vibration mean reduced noise and improved work environment. Vibration modification and control can be crucial in maintaining high performance and production efficiency, and prolonging the useful life in industrial machinery. Consequently, a considerable effort is devoted today to studying and controlling the vibration generated by machinery components, machine tools, transit vehicles, impact processes, civil engineering structures, fluid flow systems, and aircraft. Noise and acoustic problems can originate from undesirable vibrations and fluid–structure interactions, as found, for example in automobile engines. Noises from engine, environment, and high-speed and high-temperature exhaust gases in a vehicle will not only cause passenger discomfort and public annoyance, but also will result in damaging effects to the vehicle itself. Noise suppression methods and devices, and sound absorption material and structures are crucial under such situations. Before designing or controlling a system for good vibratory or acoustic performance, it is important to understand, analyze, and represent the dynamic characteristics of the system. This may be accomplished through purely analytical means, computer analysis of analytical models, testing and analysis of test data, or by a combination of these approaches.

In recent years, educators, researchers, and practitioners have devoted considerable effort towards studying and controlling vibration and noise in a range of applications in various branches of engineering, particularly, civil, mechanical, aeronautical and aerospace, and production and manufacturing. Specific applications are found in machine tools, transit vehicles, impact processes, civil engineering structures, construction machinery, industrial processes, product qualification and quality control, fluid flow systems, ships, and aircraft. This book is a contribution towards these efforts. In view of these analytical methods, practical considerations, design issues, and experimental techniques are presented throughout the book, and in view of the simplified and snap-shot style presentation of formulas, data, and advanced theory, the book serves as a useful reference tool and an extensive information source for engineers and technicians in industry and laboratories, researchers, instructors, and students in the areas of vibration, shock, noise, and acoustics.

Clarence W. de Silva
Editor-in-Chief
Vancouver, Canada

Acknowledgments

I wish to express my gratitude to the authors of the chapters for their valuable and highly professional contributions. I am very grateful to Michael Slaughter, Acquisitions Editor-Engineering, CRC Press, for his enthusiasm and support throughout the project. Editorial and production staff at CRC Press have done an excellent job in getting this volume out in print. Finally, I wish to lovingly acknowledge the patience and understanding of my family.

Editor-in-Chief

Dr. Clarence W. de Silva, P.Eng., Fellow ASME, Fellow IEEE, Fellow Canadian Academy of Engineering, is Professor of Mechanical Engineering at the University of British Columbia, Vancouver, Canada, and has occupied the NSERC-BC Packers Research Chair in Industrial Automation since 1988. He has earned Ph.D. degrees from the Massachusetts Institute of Technology and the University of Cambridge, England. De Silva has also occupied the Mobil Endowed Chair Professorship in the Department of Electrical and Computer Engineering at the National University of Singapore. He has served as a consultant to several companies including IBM and Westinghouse in the U.S., and has led the development of eight industrial machines and devices. He is recipient of the Henry M. Paynter Outstanding Investigator Award from the Dynamic Systems and Control Division of the American Society of Mechanical Engineers (ASME), Killam Research Prize, Lifetime Achievement Award from the World Automation Congress, Outstanding Engineering Educator Award of IEEE Canada, Yasurdo Takahashi Education Award of the Dynamic Systems and Control Division of ASME, IEEE Third Millennium Medal, Meritorious Achievement Award of the Association of Professional Engineers of BC, and the Outstanding Contribution Award of the Systems, Man, and Cybernetics Society of the Institute of Electrical and Electronics Engineers (IEEE).

He has authored 16 technical books including *Sensors and Actuators: Control System Instrumentation* (Taylor & Francis, CRC Press, 2007); *Mechatronics—An Integrated Approach* (Taylor & Francis, CRC Press, Boca Raton, FL, 2005); *Soft Computing and Intelligent Systems Design—Theory, Tools, and Applications* (with F. Karry, Addison Wesley, New York, NY, 2004); *Vibration: Fundamentals and Practice* (Taylor & Francis, CRC Press, 2nd edition, 2006); *Intelligent Control: Fuzzy Logic Applications* (Taylor & Francis, CRC Press, 1995); *Control Sensors and Actuators* (Prentice Hall, 1989); 14 edited volumes, over 170 journal papers, 200 conference papers, and 12 book chapters. He has served on the editorial boards of 14 international journals, in particular as the Editor-in-Chief of the *International Journal of Control and Intelligent Systems*, Editor-in-Chief of the *International Journal of Knowledge-Based Intelligent Engineering Systems*, Senior Technical Editor of *Measurements and Control*, and Regional Editor, North America, of *Engineering Applications of Artificial Intelligence – the International Journal of Intelligent Real-Time Automation*. He is a Lilly Fellow at Carnegie Mellon University, NASA-ASEE Fellow, Senior Fulbright Fellow at Cambridge University, ASI Fellow, and a Killam Fellow. Research and development activities of Professor de Silva are primarily centered in the areas of process automation, robotics, mechatronics, intelligent control, and sensors and actuators as principal investigator, with cash funding of about $6 million.

Contributors

S. Akishita
Ritsumeikan University
Kusatsu, Japan

Su Huan Chen
Jilin University
Changchun, People's Republic of China

Kourosh Danai
University of Massachusetts
Amherst, Massachusetts

Clarence W. de Silva
The University of British Columbia
Vancouver, British Columbia, Canada

Ebrahim Esmailzadeh
University of Ontario
Oshawa, Ontario, Canada

Seon M. Han
Texas Tech University
Lubbock, Texas

Nader Jalili
Clemson University
Clemson, South Carolina

Takayuki Koizumi
Doshisha University
Kyoto-Hu, Japan

Robert G. Landers
University of Missouri at Rolla
Rolla, Missouri

L.Y. Lu
National Kaohsiung First University of Science and Technology
Kaohsiung, Taiwan

Kiyoshi Nagakura
Railway Technical Research Institute
Tokyo-To, Japan

Teruo Obata
Teikyo University
Totigi-Ken, Japan

Kiyoshi Okura
Mitsuboshi Belting Ltd.
Hyogo-Ken, Japan

Randall D. Peters
Mercer University
Macon, Georgia

H. Sam Samarasekera
Sulzer Pumps (Canada), Inc.
Burnaby, British Columbia, Canada

Y.B. Yang
National Taiwan University
Taipei, Taiwan

J.D. Yau
Tamkang University
Taipei, Taiwan

Contents

1

Vibration Damping

Clarence W. de Silva
The University of British Columbia

Summary

Damping in vibrating systems occurs through the dissipation of mechanical energy. This chapter presents modeling, analysis, and measurement of mechanical damping. The types of damping covered include material internal damping (including viscoelastic damping and hysteretic damping), structural damping, fluid damping, interface damping, viscous damping, Coulomb friction, and Stribeck damping. Representations of various types of damping using equivalent viscous damping models are analyzed. Damping in rotating devices is also studied.

1.1 Introduction

Damping is the phenomenon by which mechanical energy is dissipated (usually by conversion into internal thermal energy) in dynamic systems. Knowledge of the level of damping in a dynamic system is important in the utilization, analysis, and testing of the system. For example, a device with natural frequencies within the seismic range (that is, less than 33 Hz) and which has relatively low damping, could produce damaging motions under resonance conditions when subjected to a seismic disturbance. This effect could be further magnified by low-frequency support structures and panels with low damping. This example shows that knowledge of damping in constituent devices, components, and support structures is important in the design and operation of complex mechanical systems. The nature and the level of component damping should be known in order to develop a dynamic model of the system and its peripherals. Knowledge of damping in a system is also important in imposing dynamic environmental limitations on the system (that is, the maximum dynamic excitation the system can withstand) under in-service conditions. Furthermore, knowledge of a system's damping can be useful in order to make design modifications in a system that has failed the acceptance test.

However, the significance of knowledge of damping levels in a test object for the development of test excitation (input) is often overemphasized. Specifically, if the response spectrum method is used to represent the required excitation in a vibration test, then there is no need for the damping value used in the development of the required response spectrum specification to be equal to the actual damping in the test object. The only requirement is that the damping used in the specified response spectrum be equal to that used in the test response spectrum. The degree of dynamic interaction between the test object and the shaker table, however, will depend on the actual level of damping in these systems. Furthermore, when testing near the resonant frequency of a test object, it is desirable to know about the damping in the test object.

In characterizing damping in a dynamic system it is important, first, to understand the major mechanisms associated with mechanical energy dissipation in the system. Then a suitable damping model should be chosen to represent the associated energy dissipation. Finally, damping values (model parameters) should be determined, for example, by testing the system or a representative physical model, by monitoring system response under transient conditions during normal operation or by employing already available data.

1.2 Types of Damping

There is some form of mechanical energy dissipation in any dynamic system. In the modeling of systems, damping can be neglected if the mechanical energy that is dissipated during the time duration of interest is small in comparison to the initial total mechanical energy of excitation in the system. Even for highly damped systems, it is useful to perform an analysis with the damping terms neglected, in order to study several crucial dynamic characteristics, e.g., modal characteristics (undamped natural frequencies and mode shapes).

Several types of damping are inherently present in a mechanical system. If the level of damping that is available in this manner is not adequate for proper functioning of the system then external damping devices may be added either during the original design or during subsequent design modifications of the system. Three primary mechanisms of damping are important in the study of mechanical systems. They are:

1. Internal damping (of material)
2. Structural damping (at joints and interfaces)
3. Fluid damping (through fluid–structure interactions)

Internal (material) damping results from mechanical energy dissipation within the material due to various microscopic and macroscopic processes. Structural damping is caused by mechanical energy dissipation resulting from relative motions between components in a mechanical structure that has common points of contact, joints or supports. Fluid damping arises from the mechanical energy dissipation resulting from drag forces and associated dynamic interactions when a mechanical system or its components move in a fluid.

Two general types of external dampers may be added to a mechanical system in order to improve its energy dissipation characteristics. They are:

1. Passive dampers
2. Active dampers

Passive dampers are devices that dissipate energy through some kind of motion, without needing an external power source or actuators. Active dampers have actuators that need external sources of power. They operate by actively controlling the motion of the system that needs damping. Dampers may be considered as vibration controllers. In the present chapter, the emphasis will be on damping that is inherently present in a mechanical system.

1.2.1 Material (Internal) Damping

Internal damping of materials originates from the energy dissipation associated with microstructure defects, such as grain boundaries and impurities; thermoelastic effects caused by local temperature gradients resulting from nonuniform stresses, as in vibrating beams; eddy current effects in ferromagnetic materials; dislocation motion in metals; and chain motion in polymers. Several models have been employed to represent energy dissipation caused by internal damping. This variety of models is primarily a result of the vast range of engineering materials; no single model can satisfactorily represent the internal damping characteristics of all materials. Nevertheless, two general types of internal damping can be identified: viscoelastic damping and hysteretic damping. The latter term is actually a misnomer, because all types of internal damping are associated with hysteresis loop effects. The stress (σ) and strain (ε) relations at a point in a vibrating continuum possess a hysteresis loop, such as the one shown in Figure 1.1. The area of the hysteresis loop gives the energy dissipation per unit volume of the material, per stress cycle. This is termed the per-unit-volume damping capacity, and is denoted by d. It is clear that d is given by the cyclic integral

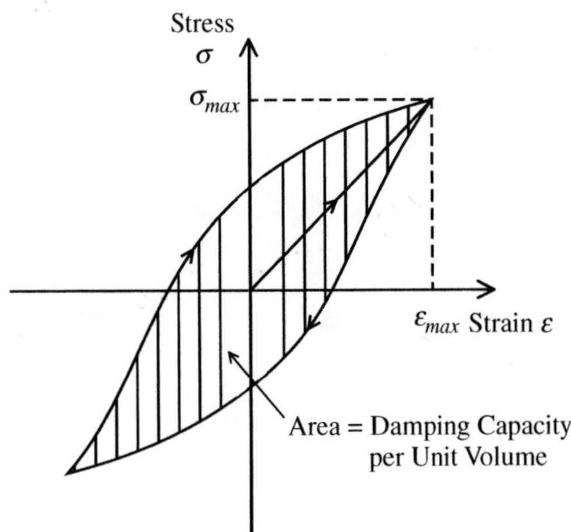

FIGURE 1.1 A typical hysteresis loop for mechanical damping.

$$d = \oint \sigma \, \mathrm{d}\varepsilon \tag{1.1}$$

In fact, for any damped device, there is a corresponding hysteresis loop in the displacement–force plane as well. In this case, the cyclic integral of force with respect to the displacement, which is the area of the hysteresis loop, is equal to the work done against the damping force. It follows that this integral (loop area) is the energy dissipated per cycle of motion. This is the *damping capacity* which, when divided by the material volume, gives the per-unit-volume damping capacity as before.

It should be clear that, unlike a pure elastic force (e.g., a spring force), a damping force cannot be a function of displacement (q) alone. The reason is straightforward. Consider a force $f(q)$ which depends on q alone. Then, for a particular displacement point, q, of the component the force will be the same regardless of the direction of motion (i.e., the sign of $\dot{q}$). It follows that, in a loading and unloading cycle, the same path will be followed in both directions of motion. Hence, a hysteresis loop will not be formed. In other words, the net work done in a complete cycle of motion will be zero. Next consider a force $f(q, \dot{q})$ which depends on both q and $\dot{q}$. Then, at a given displacement point, q, the force will depend on $\dot{q}$ as well. Hence, force in one direction of motion will be different from that in the opposite direction. As a result, a hysteresis loop will be formed, which corresponds to work done against the damping force (i.e., energy dissipation). We can conclude then that the damping force has to depend on a relative velocity, $\dot{q}$, in some manner. In particular, Coulomb friction, which does not depend on the magnitude of $\dot{q}$, does depend on the sign (direction) of $\dot{q}$.

1.2.1.1 Viscoelastic Damping

For a linear viscoelastic material, the stress–strain relationship is given by a linear differential equation with respect to time, having constant coefficients. A commonly employed relationship is

$$\sigma = E\varepsilon + E^* \frac{\mathrm{d}\varepsilon}{\mathrm{d}t} \tag{1.2}$$

which is known as the Kelvin–Voigt model. In Equation 1.2, E is Young's modulus and E^* is a viscoelastic parameter that is assumed to be time independent. The elastic term $E\varepsilon$ does not contribute to damping, and, as noted before, mathematically, its cyclic integral vanishes. Consequently, for the Kelvin–Voigt model, damping capacity per unit volume is

$$d_v = E^* \oint \frac{\mathrm{d}\varepsilon}{\mathrm{d}t} \mathrm{d}\varepsilon\pi \tag{1.3}$$

For a material that is subjected to a harmonic (sinusoidal) excitation, at steady state, we have

$$\varepsilon = \varepsilon_{max} \cos \omega t \tag{1.4}$$

When Equation 1.4 is substituted in Equation 1.3, we obtain

$$d_v = \pi \omega E^* \varepsilon_{max}^2 \tag{1.5}$$

Now, $\varepsilon = \varepsilon_{max}$ when $t = 0$ in Equation 1.4, or when $d\varepsilon/dt = 0$. The corresponding stress, according to Equation 1.2, is $\sigma_{max} = E\varepsilon_{max}$. It follows that

$$d_v = \frac{\pi \omega E^* \sigma_{max}^2}{E^2} \tag{1.6}$$

These expressions for d_v depend on the frequency of excitation, ω.

Apart from the Kelvin–Voigt model, two other models of viscoelastic damping are also commonly used. They are, the Maxwell model given by

$$\sigma + c_s \frac{d\sigma}{dt} = E^* \frac{d\varepsilon}{dt} \tag{1.7}$$

and the standard linear solid model given by

$$\sigma + c_s \frac{d\sigma}{dt} = E\varepsilon + E^* \frac{d\varepsilon}{dt} \tag{1.8}$$

It is clear that the standard linear solid model represents a combination of the Kelvin–Voigt model and the Maxwell model, and is the most accurate of the three. But, for most practical purposes, the Kelvin–Voigt model is adequate.

1.2.1.2 Hysteretic Damping

It was noted above that the stress, and hence the internal damping force, of a viscoelastic damping material depends on the frequency of variation of the strain (and consequently the frequency of motion). For some types of material, it has been observed that the damping force does not significantly depend on the frequency of oscillation of strain (or frequency of harmonic motion). This type of internal damping is known as hysteretic damping.

Damping capacity per unit volume (d_h) for hysteretic damping is also independent of the frequency of motion and can be represented by

$$d_h = J\sigma_{max}^n \tag{1.9}$$

A simple model that satisfies Equation 1.9, for the case of $n = 2$, is given by

$$\sigma = E\varepsilon + \frac{\tilde{E}}{\omega} \frac{d\varepsilon}{dt} \tag{1.10}$$

which is equivalent to using a viscoelastic parameter, E^*, that depends on the frequency of motion in Equation 1.2 according to $E^* = \tilde{E}/\omega$.

Consider the case of harmonic motion at frequency ω, with the material strain given by

$$\varepsilon = \varepsilon_0 \cos \omega t \tag{1.11}$$

Then, Equation 1.10 becomes

$$\sigma = E\varepsilon_0 \cos \omega t - \tilde{E}\varepsilon_0 \sin \omega t = E\varepsilon \cos \omega t + \tilde{E}\varepsilon_0 \cos\left(\omega t + \frac{\pi}{2}\right) \tag{1.12}$$

Note that the material stress has two components, as given by the right-hand side of Equation 1.12. The first component corresponds to the linear elastic behavior of a material and is in phase with the strain. The second component of stress, which corresponds to hysteretic damping, is 90° out of phase. (This stress component leads the strain by 90°.) A convenient mathematical representation is possible, by using the usual complex form of the response according to

$$\varepsilon = \varepsilon_0 \, e^{j\omega t} \tag{1.13}$$

Then, Equation 1.10 becomes

$$\sigma = (E + \mathrm{j}\tilde{E})\varepsilon \tag{1.14}$$

It follows that this form of simplified hysteretic damping may be represented by using a complex modulus of elasticity, consisting of a real part which corresponds to the usual linear elastic (energy storage) modulus (or Young's modulus) and an imaginary part which corresponds to the hysteretic loss (energy dissipation) modulus.

By combining Equation 1.2 and Equation 1.10, a simple model for combined viscoelastic and hysteretic damping may be given by

$$\sigma = E\varepsilon + \left(E^* + \frac{\tilde{E}}{\omega}\right)\frac{\mathrm{d}\varepsilon}{\mathrm{d}t} \tag{1.15}$$

The equation of motion for a system whose damping is represented by Equation 1.15 can be deduced from the pure elastic equation of motion by simply substituting E by the operator

$$E + \left(E^* + \frac{\tilde{E}}{\omega}\right)\frac{\partial}{\partial t}$$

in the time domain.

Example 1.1

Determine the equation of flexural motion of a nonuniform slender beam whose material has both viscoelastic and hysteretic damping.

Solution

The Bernoulli–Euler equation of bending motion on an undamped beam subjected to a dynamic load of $f(x,t)$ per unit length, is given by

$$\frac{\partial^2}{\partial x^2}EI\frac{\partial^2 q}{\partial x^2} + \rho A\frac{\partial^2 q}{\partial t^2} = f(x,t) \tag{1.16}$$

Here, q is the transverse motion at a distance, x, along the beam. Then, for a beam with material damping (both viscoelastic and hysteretic) we can write

$$\frac{\partial^2}{\partial x^2}EI\frac{\partial^2 q}{\partial x^2} + \frac{\partial^2}{\partial x^2}\left(E^* + \frac{\tilde{E}}{\omega}\right)I\frac{\partial^3 q}{\partial t\,\partial x^2} + \rho A\frac{\partial^2 q}{\partial t^2} = f(x,t) \tag{1.17}$$

in which ω is the frequency of the external excitation $f(x,t)$ in the case of steady forced vibrations. In the case of free vibration, however, ω represents the frequency of free vibration decay. Consequently, when analyzing the modal decay of free vibrations, ω in Equation 1.17 should be replaced by the appropriate frequency (ω_i) of modal vibration in each modal equation. Hence, the resulting damped vibratory system possesses the same normal mode shapes as the undamped system. The analysis of the damped case is very similar to that for the undamped system.

1.2.2 Structural Damping

Structural damping is a result of mechanical energy dissipation caused by friction due to the relative motion between components and by impacting or intermittent contact at the joints in a mechanical system or structure. Energy dissipation behavior depends on the details of the particular mechanical system. Consequently, it is extremely difficult to develop a generalized analytical model that would satisfactorily describe structural damping. Energy dissipation caused by rubbing is usually represented by a Coulomb friction model. Energy dissipation caused by impacting, however, should be determined from the coefficient of restitution of the two members that are in contact.

The most common method of estimating structural damping is by measurement. The measured values, however, represent the overall damping in the mechanical system. The structural damping component is obtained by subtracting the values corresponding to other types of damping, such as material damping, present in the system (estimated by environment-controlled experiments, previous data, and so forth) from the overall damping value.

Usually, internal damping is negligible compared to structural damping. A large proportion of mechanical energy dissipation in tall buildings, bridges, vehicle guideways, and many other civil engineering structures and in machinery, such as robots and vehicles, takes place through the structural damping mechanism. A major form of structural damping is the slip damping that results from energy dissipation by interface shear at a structural joint. The degree of slip damping that is directly caused by Coulomb (dry) friction depends on such factors as joint forces (for example, bolt tensions), surface properties and the nature of the materials of the mating surfaces. This is associated with wear, corrosion, and general deterioration of the structural joint. In this sense, slip damping is time-dependent. It is a common practice to place damping layers at joints to reduce undesirable deterioration of the joints. Sliding causes shear distortions in the damping layers, causing energy dissipation by material damping and also through Coulomb friction. In this way, a high level of equivalent structural damping can be maintained without causing excessive joint deterioration. These damping layers should have a high stiffness (as well as a high specific-damping capacity) in order to take the structural loads at the joint.

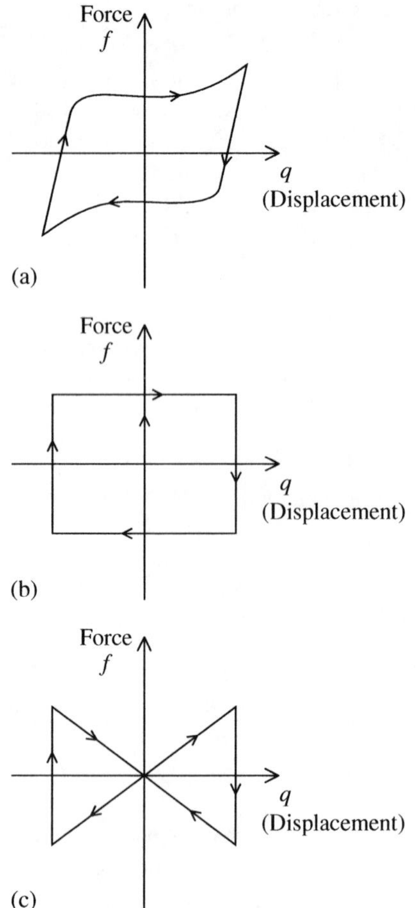

FIGURE 1.2 Some representative hysteresis loops: (a) typical structural damping; (b) Coulomb friction model; and (c) simplified structural damping model.

For structural damping at a joint, the damping force varies as slip occurs at the joint. This is primarily caused by local deformations at the joint, which occur with slipping. A typical hysteresis loop for this case is shown in Figure 1.2(a). The arrows on the hysteresis loop indicate the direction of relative velocity. For idealized Coulomb friction, the frictional force (F) remains constant in each direction of relative motion. An idealized hysteresis loop for structural Coulomb damping is shown in Figure 1.2(b). The corresponding constitutive relation is

$$f = c\,\mathrm{sgn}(\dot{q}) \tag{1.18}$$

in which f is the damping force, q is the relative displacement at the joint and c is a friction parameter. A simplified model for structural damping caused by local deformation may be given by

$$f = c|q|\mathrm{sgn}(\dot{q}) \tag{1.19}$$

The corresponding hysteresis loop is shown in Figure 1.2(c). Note that the *signum function* is defined by

$$\mathrm{sgn}(v) = \begin{cases} 1 & \text{for } v \geq 0 \\ -1 & \text{for } v < 0 \end{cases} \tag{1.20}$$

1.2.3 Fluid Damping

Consider a mechanical component moving in a fluid medium. The direction of relative motion is shown parallel to the y-axis in Figure 1.3. Local displacement of the element relative to the surrounding fluid is denoted by $q(x, y, t)$.

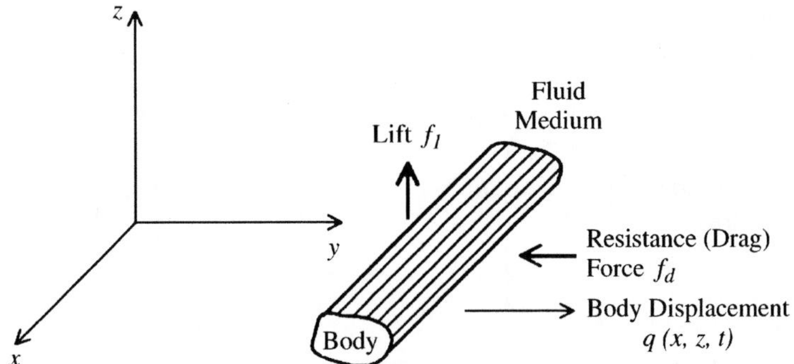

FIGURE 1.3 A body moving in a fluid medium.

The resulting drag force per unit area of projection on the x–z plane is denoted by f_d. This resistance is the cause of mechanical energy dissipation in fluid damping. It is usually expressed as

$$f_\mathrm{d} = \frac{1}{2} c_\mathrm{d} \rho \dot{q}^2 \operatorname{sgn}(\dot{q}) \tag{1.21}$$

in which $\dot{q} = \partial q(x, z, t)/\partial t$ is the relative velocity. The drag coefficient, c_d, is a function of the Reynold's number and the geometry of the structural cross section. A net damping effect is generated by viscous drag produced by the boundary layer effects at the fluid–structure interface, and by pressure drag produced by the turbulent effects resulting from flow separation at the wake. The two effects are illustrated in Figure 1.4. Fluid density is ρ. For fluid damping, the damping capacity per unit volume associated with the configuration shown in Figure 1.3 is given by

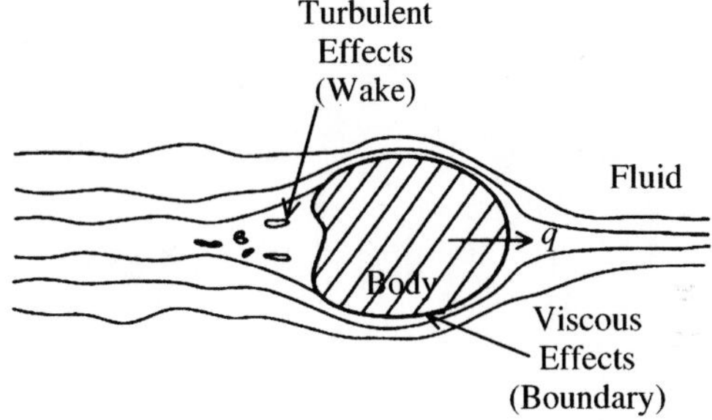

FIGURE 1.4 Mechanics of fluid damping.

$$d_\mathrm{f} = \frac{\oint \int_0^{L_x} \int_0^{L_z} f_\mathrm{d} \, \mathrm{d}z \, \mathrm{d}x \, \mathrm{d}q(x, z, t)}{L_x L_z q_0} \tag{1.22}$$

in which, L_x and L_z are cross-sectional dimensions of the element in the x and y-directions, respectively, and q_0 is a normalizing amplitude parameter for relative displacement.

Example 1.2

Consider a beam of length L and uniform rectangular cross section that is undergoing transverse vibration in a stationary fluid. Determine an expression for the damping capacity per unit volume for this fluid–structure interaction.

Solution

Suppose that the beam axis is along the x-direction and the transverse motion is in the z-direction. There is no variation in the y-direction, and hence, the length parameters in this direction cancel out.

$$d_\mathrm{f} = \frac{\oint \int_0^{L} f_\mathrm{d} \, \mathrm{d}x \, \mathrm{d}q(x, t)}{L q_0}$$

or

$$d_\mathrm{f} = \frac{\int_0^{T} \int_0^{L} f_\mathrm{d} \dot{q}(x, t) \mathrm{d}x \, \mathrm{d}t}{L q_0} \tag{1.23}$$

in which T is the period of the oscillations. Assuming constant c_d, we substitute Equation 1.21 into Equation 1.23:

$$d_f = \frac{1}{2}\frac{c_d \rho}{L q_0}\int_0^L \int_0^T |\dot{q}|^3 dt\, dx \tag{1.24}$$

For steady-excited harmonic vibration at frequency ω and shape function $Q(x)$ (or for free-modal vibration at natural frequency ω and mode shape $Q(x)$) we have

$$q(x,t) = q_{max} Q(x) \sin \omega t \tag{1.25}$$

In this case, with the change of variable $\theta = \omega t$, Equation 1.24 becomes

$$d_f = 2 c_d \rho \frac{q_{max}^3}{L q_0}\int_0^L |Q(x)|^3 dx\, \omega^2 \int_0^{\pi/2} \cos^3\theta\, d\theta$$

or

$$d_f = \frac{4}{3} c_d \rho q_{max}^3 \omega^2 \frac{\int_0^L |Q(x)|^3 dx}{L q_0}$$

Note. The integration interval of $t = 0$ to T becomes $\theta = 0$ to 2π or four times that from $\theta = 0$ to $\pi/2$.

If the normalizing parameter is defined as

$$q_0 = \frac{1}{L} q_{max} \int_0^L |Q(x)|^3\, dx$$

then, we get

$$d_f = \frac{4}{3} c_d \rho q_{max}^2 \omega^2 \tag{1.26}$$

A useful classification of damping is given in Box 1.1.

Box 1.1

Damping Classification

Type of Damping	Origin	Typical Constitutive Relation
Internal damping	Material properties	**Viscoelastic** $\sigma = E\varepsilon + E^* \frac{d\varepsilon}{dt}$ **Hysteretic** $\sigma = E\varepsilon + \frac{\tilde{E}}{\omega}\frac{d\varepsilon}{dt}$
Structural damping	Structural joints and interfaces	**Structural deformation** $f = c\lvert q\rvert\, \mathrm{sgn}(\dot{q})$ **Coulomb** $f = c\, \mathrm{sgn}(\dot{q})$ **General interface** $f = \begin{cases} f_s & \text{for } v = 0 \\ f_{sb}(v)\,\mathrm{sgn}(v) & \text{for } v \neq 0 \end{cases}$
Fluid damping	Fluid–structure interactions	$f_d = \frac{1}{2} c_d \rho \dot{q}^2\, \mathrm{sgn}(\dot{q})$

1.3 Representation of Damping in Vibration Analysis

It is not practical to incorporate detailed microscopic representations of damping in the dynamic analysis of systems. Instead, simplified models of damping that are representative of various types of energy dissipation are typically used. Consider a general n-degree-of-freedom mechanical system. Its motion can be represented by the vector $\mathbf{x}$ of n generalized coordinates, x_i, representing the independent motions of the inertia elements. For small displacements, linear spring elements can be assumed. The corresponding equations of motion may be expressed in the vector matrix form

$$\mathbf{M}\ddot{\mathbf{x}} + \mathbf{d} + \mathbf{K}\mathbf{x} = \mathbf{f}(\mathbf{t}) \tag{1.27}$$

in which $\mathbf{M}$ is the mass (inertia) matrix and $\mathbf{K}$ is the stiffness matrix. The forcing-function vector is $\mathbf{f}(t)$. The damping force vector $\mathbf{d}(\mathbf{x}, \dot{\mathbf{x}})$ is generally a nonlinear function of $\mathbf{x}$ and $\dot{\mathbf{x}}$. The type of damping used in the system model may be represented by the nature of $\mathbf{d}$ that is employed in the system equations. The various damping models that may be used, as discussed in the previous section, are listed in Table 1.1. Only the linear viscous damping term given in Table 1.1 is amenable to simplified mathematical analysis. In simplified dynamic models, other types of damping terms are usually replaced by an equivalent viscous damping term. Equivalent viscous damping is chosen so that its energy dissipation per cycle of oscillation is equal to that for the original damping. The resulting equations of motion are expressed by

$$\mathbf{M}\ddot{\mathbf{x}} + \mathbf{C}\dot{\mathbf{x}} + \mathbf{K}\mathbf{x} = \mathbf{f}(\mathbf{t}) \tag{1.28}$$

In modal analysis of vibratory systems, the most commonly used model is proportional damping, where the damping matrix satisfies

$$\mathbf{C} = c_m\mathbf{M} + c_k\mathbf{K} \tag{1.29}$$

The first term on the right-hand side of Equation 1.29 is known as the inertial damping matrix. The corresponding damping force on each concentrated mass is proportional to its momentum. It represents the energy loss associated with a change in momentum (for example, during an impact). The second term is known as the stiffness damping matrix. The corresponding damping force is proportional to the rate of change of the local deformation forces at joints near the concentrated mass elements. Consequently, it represents a simplified form of linear structural damping. If damping is of the proportional type, it follows that the damped motion can be uncoupled into individual modes. This means that, if the damping model is of the proportional type, the damped system (as well as the undamped system) will possess real modes.

TABLE 1.1 Some Common Damping Models Used in Dynamic System Equations

Damping Type	Simplified Model d_i
Viscous	$\sum_j c_{ij}\dot{x}_j$
Hysteretic	$\sum_j \frac{1}{\omega} c_{ij}\dot{x}_j$
Structural	$\sum_j c_{ij}\lvert x_j\rvert \operatorname{sgn}(\dot{x}_j)$
Structural Coulomb	$\sum_j c_{ij} \operatorname{sgn}(\dot{x}_j)$
Fluid	$\sum_j c_{ij}\lvert\dot{x}_j\rvert\dot{x}_j$

1.3.1 Equivalent Viscous Damping

Consider a linear, single-DoF system with viscous damping, subjected to an external excitation. The equation of motion, for a unit mass, is given by

$$\ddot{x} + 2\zeta\omega_n\dot{x} + \omega_n^2 x = \omega_n^2 u(t) \tag{1.30}$$

If the excitation force is harmonic, with frequency ω, we have

$$u(t) = u_0 \cos \omega t \tag{1.31}$$

Then, the response of the system at steady state is given by

$$x = x_0 \cos(\omega t + \phi) \tag{1.32}$$

in which the response amplitude is

$$x_0 = u_0 \frac{\omega_n^2}{\left[(\omega_n^2 - \omega^2) + 4\zeta^2\omega_n^2\omega^2\right]^{1/2}} \tag{1.33}$$

and the response phase lead is

$$\phi = -\tan^{-1}\frac{2\zeta\omega_n\omega}{(\omega_n^2 - \omega^2)} \tag{1.34}$$

The energy dissipation (i.e., damping capacity), ΔU, per unit mass in one cycle is given by the net work done by the damping force, f_d; thus,

$$\Delta U = \oint f_d \, dx = \int_{-\phi/\omega}^{(2\pi-\phi)\omega} f_d \dot{x} \, dt \tag{1.35}$$

Since the viscous damping force, normalized with respect to mass (see Equation 1.30), is given by

$$f_d = 2\zeta\omega_n\dot{x} \tag{1.36}$$

the damping capacity, ΔU_v, for viscous damping, can be obtained as

$$\Delta U_v = 2\zeta\omega_n \int_0^{2\pi/\omega} \dot{x}^2 \, dt \tag{1.37}$$

Finally, using Equation 1.32 in Equation 1.37 we get

$$\Delta U_v = 2\pi x_0^2 \omega_n \omega \zeta \tag{1.38}$$

For any general type of damping (see Table 1.1), the equation of motion becomes

$$\ddot{x} + \mathbf{d}(x, \dot{x}) + \omega_n^2 x = \omega_n^2 u(t) \tag{1.39}$$

The energy dissipation in one cycle (Equation 1.35) is given by

$$\Delta U = \int_{-\phi/\omega}^{(2\pi-\phi)/\omega} d(x, \dot{x})\dot{x} \, dt \tag{1.40}$$

Various damping force expressions, $d(x, \dot{x})$, normalized with respect to mass, are given in Table 1.2. For fluid damping, for example, the damping capacity is

$$\Delta U_f = \int_{-\phi/\omega}^{(2\pi-\phi)/\omega} c|\dot{x}|\dot{x}^2 \, dt \tag{1.41}$$

By substituting Equation 1.32 in Equation 1.41 for steady, harmonic motion we obtain

$$\Delta U_f = \tfrac{8}{3} c x_0^3 \omega^2 \tag{1.42}$$

TABLE 1.2 Equivalent Damping Ratio Expressions for Some Common Types of Damping

Damping Type	Damping Force, $d(x, \dot{x})$, per Unit Mass	Equivalent Damping Ratio, ζ_{eq}
Viscous	$2\zeta\omega_n\dot{x}$	ζ
Hysteretic	$\frac{c}{\omega}\dot{x}$	$\frac{c}{2\omega_n\omega}$
Structural	$c\lvert x\rvert\,\mathrm{sgn}(\dot{x})$	$\frac{c}{\pi\omega_n\omega}$
Structural Coulomb	$c\,\mathrm{sgn}(\dot{x})$	$\frac{2c}{\pi x_0\omega_n\omega}$
Fluid	$c\lvert\dot{x}\rvert\dot{x}$	$\frac{4}{3\pi}(\frac{\omega}{\omega_n})x_0c$

By comparing Equation 1.42 with Equation 1.38, the equivalent damping ratio for fluid damping is obtained as

$$\zeta_f = \frac{4}{3\pi}\left(\frac{\omega}{\omega_n}\right)x_0c \tag{1.43}$$

in which x_0 is the amplitude of steady-state vibrations, as given by Equation 1.33. For the other types of damping listed in Table 1.1, expressions for the equivalent damping ratio can be obtained in a similar manner. The corresponding equivalent damping ratio expressions are given in Table 1.2. It should be noted that, for nonviscous damping types, ζ is generally a function of the frequency of oscillation, ω, and the amplitude of excitation, u_0. It should be noted that the expressions given in Table 1.2 are derived assuming harmonic excitation. Engineering judgment should be exercised when employing these expressions for nonharmonic excitations.

For multi-DoF systems that incorporate proportional damping, the equations of motion can be transformed into a set of one-DoF equations (modal equations) of the type given in Equation 1.30. In this case, the damping ratio and natural frequency correspond to the respective modal values and, in particular, $\omega = \omega_n$.

1.3.2 Complex Stiffness

Consider a linear spring of stiffness k connected in parallel with a linear viscous damper of damping constant c, as shown in Figure 1.5(a). Suppose that a force, f, is applied to the system, moving it through distance x from the relaxed position of the spring. Then we have

$$f = kx + c\dot{x} \tag{1.44}$$

Suppose that the motion is harmonic, as given by

$$x = x_0 \cos \omega t \tag{1.45}$$

It is clear that the spring force, kx, is in phase with the displacement, but the damping force, $c\dot{x}$, has a 90° phase lead with respect to the displacement. This is because the velocity, $\dot{x} = -x_0\omega \sin \omega t = x_0\omega \cos(\omega t + \pi/2)$, has a 90° phase lead with respect to x. Specifically, we have

$$f = kx_0 \cos \omega t + cx_0\omega \cos\left(\omega t + \frac{\pi}{2}\right) \tag{1.46}$$

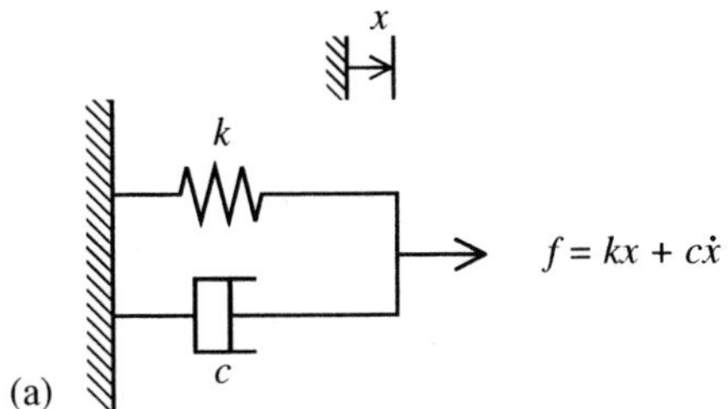

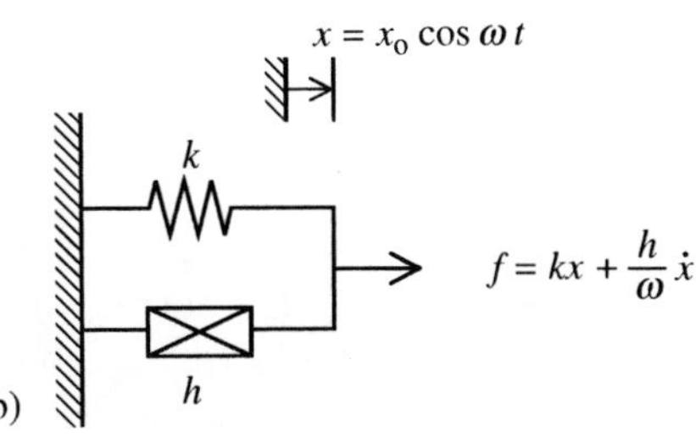

FIGURE 1.5 Spring element in parallel with (a) a viscous damper and (b) a hysteretic damper.

This same fact may be represented by using complex numbers, where the in-phase component is considered as the real part and the 90° phase lead component is considered as the imaginary part with each component oscillating at the same frequency ω. Then, we can write Equation 1.46 in the equivalent form

$$f = kx + j\omega cx \tag{1.47}$$

This is exactly what we get by starting with the complex representation of the displacement

$$x = x_0\, e^{j\omega t} \tag{1.48}$$

and substituting it in Equation 1.44. We note that Equation 1.47 may be written as

$$f = k^* x \tag{1.49}$$

where k^* is a "complex" stiffness, given by

$$k^* = k + j\omega c \tag{1.50}$$

Clearly, the system itself and its two components (spring and damper) are real. Their individual forces are also real. The complex stiffness is simply a mathematical representation of the two force components (spring force and damping force), which are 90° out of phase, when subjected to harmonic motion. It follows that the linear damper may be "mathematically" represented by an "imaginary" stiffness. In the case of viscous damping this imaginary stiffness (and hence, the damping force magnitude) increases linearly with the frequency, ω, of the harmonic motion. The concept of complex stiffness when dealing with discrete dampers is analogous to the use of complex elastic modulus in material damping, as discussed earlier in this chapter.

We have noted that, for hysteretic damping, the damping force (or damping stress) is independent of the frequency in harmonic motion. It follows that a hysteretic damper may be represented by an equivalent damping constant of

$$c = \frac{h}{\omega} \tag{1.51}$$

which is valid for a harmonic motion (e.g., modal motion or forced motion) of frequency ω. This situation is shown in Figure 1.5(b). It can be seen that the corresponding complex stiffness is

$$k^* = k + jh \tag{1.52}$$

Example 1.3

A flexible system consists of a mass, m, attached to the hysteretic damper and spring combination shown in Figure 1.5(b). What is the frequency response function of the system relating an excitation force, f, applied to the mass and the resulting displacement response, x? Obtain the resonant frequency of the system. Compare the results with the case for viscous damping.

Solution

For a harmonic motion of frequency ω, the equation of motion of the system is

$$m\ddot{x} + \frac{h}{\omega}\dot{x} + kx = f \tag{1.53}$$

With a forcing excitation of $f = f_0\, e^{j\omega t}$ and the resulting steady-state response, $x = x_0\, e^{j\omega t}$, where x_0 has a phase difference (i.e., it is a complex function) with respect to f_0. Then, in the frequency domain, substituting the harmonic response $x = x_0\, e^{j\omega t}$ into Equation 1.53 we get

$$\left[-\omega^2 m + \frac{h}{\omega} j\omega + k\right] x = f$$

resulting in the frequency transfer function

$$\frac{x}{f} = \frac{1}{[k - \omega^2 m + jh]} \tag{1.54}$$

Note that, as usual, this result is obtained simply by substituting $j\omega$ for d/dt. The magnitude of transfer function is at a maximum at resonance. This corresponds to a minimum value of

$$p(\omega) = (k - \omega^2 m)^2 + h^2$$

If we set $dp/d\omega = 0$, we get,

$$2(k - \omega^2 m)(-2\omega) = 0$$

Hence, the resonant frequency corresponds to the root of

$$k - \omega^2 m = 0$$

This gives the resonant frequency

$$\omega_r = \sqrt{\frac{k}{m}} \tag{1.55}$$

Note that, in the case of hysteretic damping, the resonant frequency is equal to the undamped natural frequency, ω_n, and, unlike in the case of viscous damping, does not depend on the level of damping itself. For convenience consider the system response as the spring force

$$f_s = kx \tag{1.56}$$

Then, a normalized transfer function is obtained, as given by

$$\frac{f_s}{f} = G(j\omega) = \frac{1}{\left[1 - \omega^2 \frac{m}{k} + j\frac{h}{k}\right]} \tag{1.57}$$

or,

$$\frac{f_s}{f} = \frac{1}{[1 - r^2 + j\alpha]} \tag{1.58}$$

where

$$r = \frac{\omega}{\omega_n} \quad \text{and} \quad \alpha = \frac{h}{k} \tag{1.59}$$

which are the normalized frequency and the normalized hysteretic damping, respectively. The magnitude of the transfer function is

$$\left|\frac{f_s}{f}\right| = \frac{1}{\sqrt{(1 - r^2)^2 + \alpha^2}} \tag{1.60}$$

and the phase angle (phase lead) is

$$\angle f_s/f = -\tan^{-1}\frac{\alpha}{(1 - r^2)} \tag{1.61}$$

These results are sketched in Figure 1.6.

1.3.3 Loss Factor

We define the *damping capacity* of a device (damper) as the energy dissipated in a complete cycle of motion; specifically

$$\Delta U = \oint f_d \, dx \tag{1.62}$$

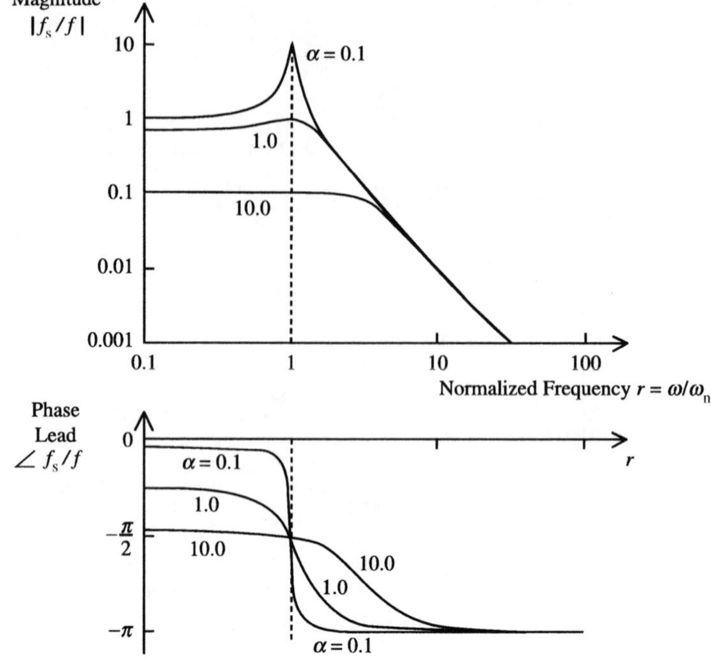

FIGURE 1.6 Frequency transfer function of a simple oscillator with hysteretic damping.

This is given by the area of the hysteresis loop in the displacement force plane. If the initial (total) energy of the system is denoted by U_{max}, then the *specific damping capacity, D*, is given by the ratio

$$D = \frac{\Delta U}{U_{max}} \tag{1.63}$$

The *loss factor,* η, is the specific damping capacity per radian of the damping cycle. Hence,

$$\eta = \frac{\Delta U}{2\pi U_{max}} \tag{1.64}$$

Note that U_{max} is approximately equal to the maximum kinetic energy and also to the maximum potential energy of the device when the damping is low.

Equation 1.38 gives the damping capacity per unit mass of a device with viscous damping as

$$\Delta U = 2\pi x_0^2 \omega_n \omega \zeta \tag{1.65}$$

Here, x_0 is the amplitude and ω is the frequency of harmonic motion of the device, ω_n is the undamped natural frequency and ζ is the damping ratio. The maximum potential energy per unit mass of the system is

$$U_{max} = \frac{1}{2}\frac{k}{m}x_0^2 = \frac{1}{2}\omega_n^2 x_0^2 \tag{1.66}$$

Hence, from Equation 1.64, the loss factor for a viscous damped simple oscillator is given by

$$\eta = \frac{2\pi x_0^2 \omega_n \omega \zeta}{2\pi \times \frac{1}{2}\omega_n^2 x_0^2} = \frac{2\omega\zeta}{\omega_n} \tag{1.67}$$

For free decay of the system, we have $\omega = \omega_d \cong \omega_n$, where the latter approximation holds for low damping. For forced oscillation, the worst response conditions occur when $\omega = \omega_d \cong \omega_n$, which is what one must consider with regard to energy dissipation. In either case, the loss factor is approximately given by

$$\eta = 2\zeta \tag{1.68}$$

For other types of damping, Equation 1.68 will still hold when the equivalent damping ratio, ζ_{eq}, (see Table 1.2) is used in place of ζ.

The loss factors of some common materials are given in Table 1.3. Definitions of useful damping parameters, as defined here, are summarized in Table 1.4. Expressions of loss factors for some useful damping models are given in Table 1.5.

TABLE 1.3 Loss Factors of Some Useful Materials

Material	Loss Factor $\eta \cong 2\zeta$
Aluminum	2×10^{-5} to 2×10^{-3}
Concrete	0.02 to 0.06
Glass	0.001 to 0.002
Rubber	0.1 to 1.0
Steel	0.002 to 0.01
Wood	0.005 to 0.01

TABLE 1.4 Definitions of Damping Parameters

Parameter	Definition	Mathematical Formula
Damping capacity (ΔU)	Energy dissipated per cycle of motion (area of displacement–force hysteresis loop)	$\oint f_d \, dx$
Damping capacity per volume (d)	Energy dissipated per cycle per unit material volume (area of strain–stress hysteresis loop)	$\oint \sigma \, d\varepsilon$
Specific damping capacity (D)	Ratio of energy dissipated per cycle (ΔU) to the initial maximum energy (U_{max}) *Note*: for low damping, U_{max} = maximum potential energy = maximum kinetic energy	$\dfrac{\Delta U}{U_{max}}$
Loss factor (η)	Specific damping capacity per unit angle of cycle. *Note*: for low damping, $\eta = 2 \times$ damping ratio	$\dfrac{\Delta U}{2\pi U_{max}}$

TABLE 1.5 Loss Factors for Several Material Damping Models

Material Damping Model	Stress–Strain Constitute Relation	Loss Factor (η)
Viscoelastic Kelvin–Voigt	$\sigma = E\varepsilon + E^* \dfrac{d\varepsilon}{dt}$	$\dfrac{\omega E^*}{E}$
Hysteretic Kelvin–Voigt	$\sigma = E\varepsilon + \dfrac{\tilde{E}}{\omega}\dfrac{d\varepsilon}{dt}$	$\dfrac{\tilde{E}}{E}$
Viscoelastic standard linear solid	$\sigma + c_s \dfrac{d\sigma}{dt} = E\varepsilon + E^* \dfrac{d\varepsilon}{dt}$	$\dfrac{\omega E^*}{E}\dfrac{(1 - c_s E/E^*)}{(1 + \omega^2 c_s)}$
Hysteretic standard linear solid	$\sigma + c_s \dfrac{d\sigma}{dt} = E\varepsilon + \dfrac{\tilde{E}}{\omega}\dfrac{d\varepsilon}{dt}$	$\dfrac{\tilde{E}}{E}\dfrac{(1 - \omega c_s E/\tilde{E})}{(1 + \omega^2 c_s)}$

1.4 Measurement of Damping

Damping may be represented by various parameters (such as specific damping capacity, loss factor, Q-factor, and damping ratio) and models (such as viscous, hysteretic, structural, and fluid). Before attempting to measure damping in a system, we need to decide on a representation (model) that will adequately characterize the nature of mechanical energy dissipation in the system. Next, we should decide on the parameter or parameters of the model that need to be measured.

It is extremely difficult to develop a realistic yet tractable model for damping in a complex piece of equipment operating under various conditions of mechanical interaction. Even if a satisfactory damping modal is developed, experimental determination of its parameters could be tedious. A major difficulty arises because it usually is not possible to isolate various types of damping (for example, material, structural, and fluid) from an overall measurement. Furthermore, damping measurements must be conducted under actual operating conditions for them to be realistic.

If one type of damping (say, fluid damping) is eliminated during the actual measurement then it would not represent the true operating conditions. This would also eliminate possible interacting effects of the eliminated damping type with the other types. In particular, overall damping in a system is not generally equal to the sum of the individual damping values when they are acting independently. Another limitation of computing equivalent damping values using experimental data arises because it is assumed for analytical simplicity that the dynamic system behavior is linear. If the system is highly nonlinear, a significant error could be introduced into the damping estimate. Nevertheless, it is customary to assume linear viscous behavior when estimating damping parameters using experimental data.

There are two general ways by which damping measurements can be made: using a time–response record and using a frequency–response function of the system to estimate damping.

1.4.1 Logarithmic Decrement Method

This is perhaps the most popular time–response method that is used to measure damping. When a single-DoF oscillatory system with viscous damping (see Equation 1.30) is excited by an impulse input (or an initial condition excitation), its response takes the form of a time decay (see Figure 1.7), given by

$$y(t) = y_0 \exp(-\zeta\omega_{\mathrm{n}} t) \sin \omega_{\mathrm{d}} t \tag{1.69}$$

in which the damped natural frequency is given by

$$\omega_{\mathrm{d}} = \sqrt{1 - \zeta^2}\,\omega_{\mathrm{n}} \tag{1.70}$$

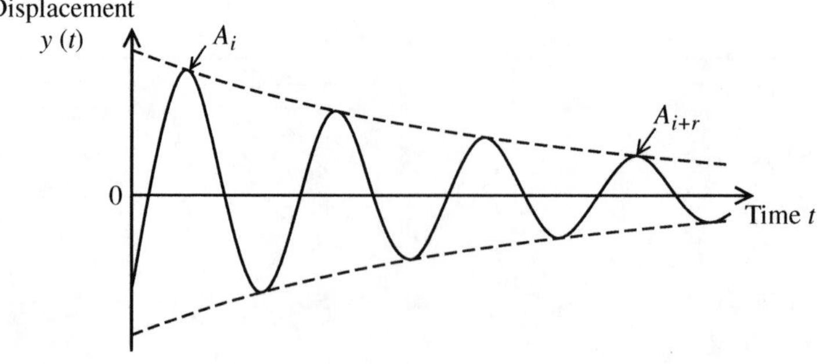

FIGURE 1.7 Impulse response of a simple oscillator.

If the response at $t = t_i$ is denoted by y_i, and the response at $t = t_i + 2\pi r/\omega_d$ is denoted by y_{i+r}, then, from Equation 1.69, we have

$$\frac{y_{i+r}}{y_i} = \exp\left(-\zeta\frac{\omega_n}{\omega_d}2\pi r\right), \qquad i = 1, 2, \ldots, n$$

In particular, suppose that y_i corresponds to a peak point in the time decay function, having magnitude A_i, and that y_{i+r} corresponds to the peak point r cycles later in the time history, and its magnitude is denoted by A_{i+r} (see Figure 1.7). Even though the above equation holds for any pair of points that are r periods apart in the time history, the peak points seem to be the appropriate choice for measurement in the present procedure, as these values would be more prominent than any arbitrary points in a response–time history. Then,

$$\frac{A_{i+r}}{A_i} = \exp\left(-\zeta\frac{\omega_n}{\omega_d}2\pi r\right) = \exp\left[-\frac{\zeta}{\sqrt{1-\zeta^2}}2\pi r\right]$$

where Equation 1.70 has been used. Then, the logarithmic decrement δ is given by (per unit cycle)

$$\delta = \frac{1}{r}\ln\left(\frac{A_i}{A_{i+r}}\right) = \frac{2\pi\zeta}{\sqrt{1-\zeta^2}} \tag{1.71}$$

or the damping ratio may be expressed as

$$\zeta = \frac{1}{\sqrt{1 + (2\pi/\delta)^2}} \tag{1.72}$$

For low damping (typically, $\zeta < 0.1$), $\omega_d \cong \omega_n$ and Equation 1.71 become

$$\frac{A_{i+r}}{A_i} \cong \exp(-\zeta 2\pi r) \tag{1.73}$$

or

$$\zeta = \frac{1}{2\pi r}\ln\left(\frac{A_i}{A_{i+r}}\right) = \frac{\delta}{2\pi} \qquad \text{for } \zeta < 0.1 \tag{1.74}$$

This is in fact the "per-radian" logarithmic decrement.

The damping ratio can be estimated from a free-decay record, using Equation 1.74. Specifically, the ratio of the extreme amplitudes in prominent r cycles of decay is determined and substituted into Equation 1.74 to get the equivalent damping ratio.

Alternatively, if n cycles of damped oscillation are needed for the amplitude to decay by a factor of two, for example, then, from Equation 1.74, we get

$$\zeta = \frac{1}{2\pi n}\ln(2) = \frac{0.11}{n} \qquad \text{for } \zeta < 0.1 \tag{1.75}$$

For slow decays (low damping), we have

$$\ln\left(\frac{A_i}{A_{i+1}}\right) \cong \frac{2(A_i - A_{i+1})}{(A_i + A_{i+1})} \tag{1.76}$$

Then, from Equation 1.74, we get

$$\zeta = \frac{A_i - A_{i+1}}{\pi(A_i + A_{i+1})} \qquad \text{for } \zeta < 0.1 \tag{1.77}$$

Any one of Equation 1.72, Equation 1.74, Equation 1.75, and Equation 1.77 could be employed in computing ζ from test data. It should be noted that the results assume single-DoF system behavior. For multi-DoF systems, the modal damping ratio for each mode can be determined using this method if the initial excitation is such that the decay takes place primarily in one mode of vibration.

In other words, substantial modal separation and the presence of "real" modes (not "complex" modes with nonproportional damping) are assumed.

1.4.2 Step–Response Method

This is also a time–response method. If a unit-step excitation is applied to the single-DoF oscillatory system given by Equation 1.30, its time–response is given by

$$y(t) = 1 - \frac{1}{\sqrt{1-\zeta^2}} \exp(-\zeta\omega_n t) \sin(\omega_d t + \phi) \tag{1.78}$$

in which $\phi = \cos \zeta$. A typical step–response curve is shown in Figure 1.8. The time at the first peak (peak time), T_p, is given by

$$T_p = \frac{\pi}{\omega_d} = \frac{\pi}{\sqrt{1-\zeta^2}\omega_n} \tag{1.79}$$

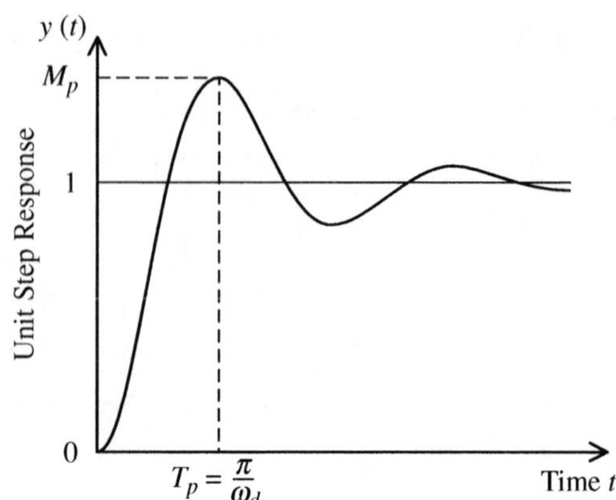

FIGURE 1.8 A typical step–response of a simple oscillator.

The response at peak time (peak value), M_p, is given by

$$M_p = 1 + \exp(-\zeta\omega_n T_p) = 1 + \exp\left(\frac{-\pi\zeta}{\sqrt{1-\zeta^2}}\right) \tag{1.80}$$

The percentage overshoot, PO, is given by

$$\text{PO} = (M_p - 1) \times 100\% = 100 \exp\left(\frac{-\pi\zeta}{\sqrt{1-\zeta^2}}\right) \tag{1.81}$$

It follows that, if any one parameter of T_p, M_p or PO is known from a step–response record, the corresponding damping ratio, ζ, can be computed by using the appropriate relationship from the following:

$$\zeta = \sqrt{1 - \left(\frac{\pi}{T_p\omega_n}\right)^2} \tag{1.82}$$

$$\zeta = \frac{1}{\sqrt{1 + \dfrac{1}{\left[\dfrac{\ln(M_p - 1)}{\pi}\right]^2}}} \tag{1.83}$$

$$\zeta = \frac{1}{\sqrt{1 + \dfrac{1}{\left[\dfrac{\ln(\text{PO}/100)}{\pi}\right]^2}}} \tag{1.84}$$

It should be noted that when determining M_p the response curve should be normalized to unit steady-state value. Furthermore, the results are valid only for single-DoF systems and modal excitations in multi-DoF systems.

1.4.3 Hysteresis Loop Method

For a damped system, the force versus displacement cycle produces a hysteresis loop. Depending on the inertial and elastic characteristics and other conservative loading conditions (e.g., gravity) in the system, the shape of the hysteresis loop will change. But the work done by conservative forces (e.g., inertial, elastic, and gravitational) in a complete cycle of motion will be zero. Consequently, the net work done will be equal to the energy dissipated due to damping only. Accordingly, the area of the displacement–force hysteresis loop will give the damping capacity, ΔU (see Equation 1.62). The maximum energy in the system can also be determined from the displacement–force curve. Then, the loss factor, η, can be computed using Equation 1.64, and the damping ratio from Equation 1.68. This method of damping measurement may also be considered basically as a time domain method.

Note that Equation 1.65 is the work done against (i.e., energy dissipation in) a single loading–unloading cycle per unit mass. It should be recalled that $2\zeta\omega_n = c/m$, where $c =$ viscous damping constant and $m =$ mass. Accordingly, from Equation 1.65, the energy dissipation per unit mass and per hystereris loop is $\Delta U = \pi x_0^2 \omega c/m$. Hence, without normalizing with respect to mass, the energy dissipation per hysteresis loop of viscous damping is

$$\Delta U_v = \pi x_0^2 \omega c \tag{1.85}$$

Equation 1.85 can be derived by performing the cyclic integration indicated in Equation 1.62 with the damping force $f_d = c\dot{x}$, harmonic motion $x = x_0\, e^{j\omega t}$ and the integration interval $t = 0$ to $2\pi/\omega$.

Similarly, in view of Equation 1.51, the energy dissipation per hysteresis loop of hysteretic damping is

$$\Delta U_h = \pi x_0^2 h \tag{1.86}$$

Now, since the initial maximum energy may be represented by the initial maximum potential energy, we have

$$U_{max} = \frac{1}{2} k x_0^2 \tag{1.87}$$

Note that the stiffness, k, may be measured as the average slope of the displacement–force hysteresis loop. Hence, in view of Equation 1.64, the loss factor for hysteretic damping is given by

$$\eta = \frac{h}{k} \tag{1.88}$$

Then, from Equation 1.68, the equivalent damping ratio for hysteretic damping is

$$\zeta = \frac{h}{2k} \tag{1.89}$$

Example 1.4

A damping material was tested by applying a loading cycle of -900 to 900 N and back to -900 N to a thin bar made of the material and measuring the corresponding deflection. The smoothed load vs. deflection curve obtained in this experiment is shown in Figure 1.9. Assuming that the damping is predominantly of the hysteretic type, estimate

1. The hysteretic damping constant
2. The equivalent damping ratio

Solution

Approximating the top and the bottom segments of the hysteresis loop by triangles, we estimate the area of the loop as

$$\Delta U_h = 2 \times \frac{1}{2} \times 2.5 \times 900\ \text{N.mm}$$

Alternatively, we may obtain this result by counting the squares within the hysteresis loop. The deflection amplitude is

$$x_0 = 8.5\ \text{mm}$$

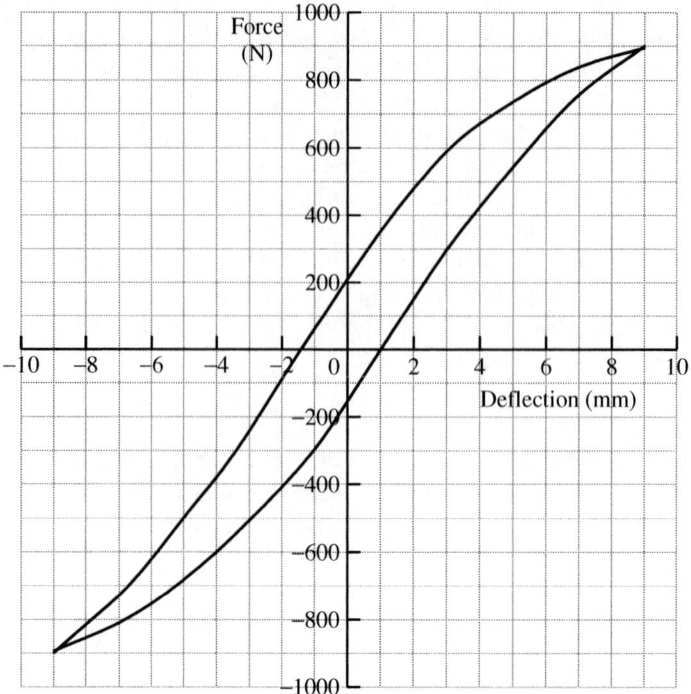

FIGURE 1.9 An experimental hysteresis loop of a damping material.

Hence, from Equation 1.86 we have

$$h = \frac{2 \times \frac{1}{2} \times 2.5 \times 900}{\pi \times 8.5^2} \text{N/mm} = 9.9 \text{ N/mm}$$

The stiffness of the damping element is estimated as the average slope of the hysteresis loop; thus

$$k = \frac{600}{4.5} \text{ N/mm} = 133.3 \text{ N/mm}$$

Hence, from Equation 1.89, the equivalent damping ratio is

$$\zeta = \frac{9.9}{2 \times 133.3} \approx 0.04$$

1.4.4 Magnification Factor Method

This is a frequency–response method. Consider a single-DoF oscillatory system with viscous damping. The magnitude of its frequency–response function is

$$|H(\omega)| = \frac{\omega_n^2}{\left[(\omega_n^2 - \omega^2)^2 + 4\zeta^2 \omega_n^2 \omega^2\right]^{1/2}} \tag{1.90}$$

A plot of this expression with respect to ω, the frequency of excitation, is given in Figure 1.10. The peak value of magnitude occurs when the denominator of the expression is at its minimum. This corresponds to

$$\frac{d}{d\omega}\left[(\omega_n^2 - \omega^2)^2 + 4\zeta^2 \omega_n^2 \omega^2\right] = 0 \tag{1.91}$$

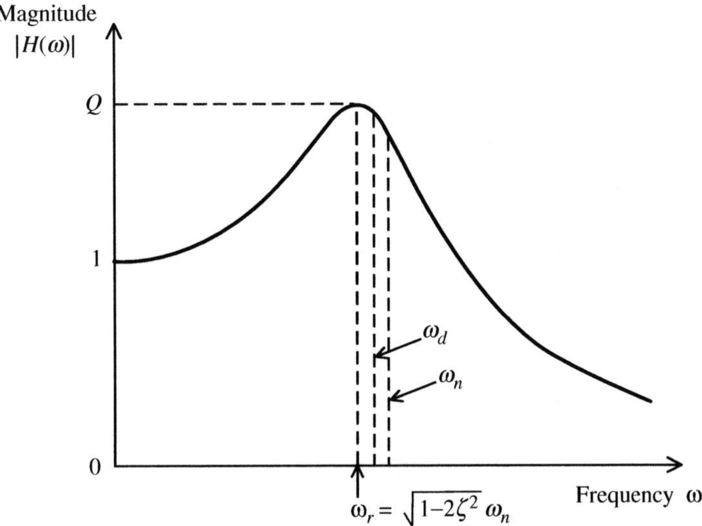

FIGURE 1.10 The magnification factor method of damping measurement applied to a single-DoF system.

The resulting solution for ω is termed the resonant frequency, ω_r

$$\omega_r = \sqrt{1 - 2\zeta^2}\omega_n \tag{1.92}$$

It is noted that $\omega_r < \omega_d$ (see Equation 1.70), but for low damping ($\zeta < 0.1$), the values of ω_n, ω_d, and ω_r are nearly equal. The amplification factor, Q, which is the magnitude of the frequency–response function at resonant frequency, is obtained by substituting Equation 1.92 in Equation 1.90:

$$Q = \frac{1}{2\zeta\sqrt{1 - \zeta^2}} \tag{1.93}$$

For low damping ($\zeta < 0.1$), we have

$$Q = \frac{1}{2\zeta} \tag{1.94}$$

In fact, Equation 1.94 corresponds to the magnitude of the frequency–response function at $\omega = \omega_n$.

It follows that, if the magnitude curve of the frequency–response function (or a Bode plot) is available, then the system damping ratio, ζ, can be estimated using Equation 1.94. When using this method, the frequency–response curve must be normalized so that its magnitude at zero frequency (termed *static gain*) is unity.

For a multi-DoF system modal damping values may be estimated from the magnitude of the Bode plot of its frequency–response function, provided that the modal frequencies are not too closely spaced and the system is lightly damped. Consider the logarithmic (to the base ten) magnitude plot shown in Figure 1.11. The magnitude is expressed in decibels (dB), which is calculated by multiplying the $\log_{10}$(magnitude) by a factor of 20. At the ith resonant frequency, ω_i, the amplification factor, q_i (in dB), is obtained by drawing an asymptote to the preceding segment of the curve and measuring the peak value from the asymptote. Then,

$$Q_i = (10)^{q_i/20} \tag{1.95}$$

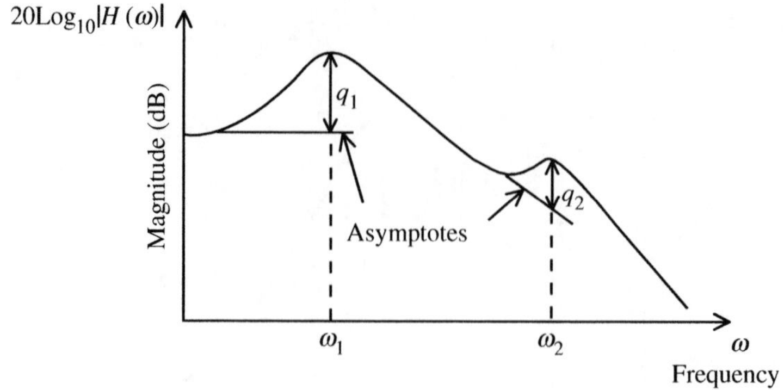

FIGURE 1.11 Magnification factor method applied to a multi-DoF system.

and the modal damping ratio

$$\zeta = \frac{1}{2Q_i}, \qquad i = 1, 2, \ldots, n \tag{1.96}$$

If the significant resonances are closely spaced, curve-fitting to a suitable function may be necessary in order to determine the corresponding modal damping values. The Nyquist plot may also be used in computing damping using frequency domain data.

1.4.5 Bandwidth Method

The bandwidth method of damping measurement is also based on frequency–response. Consider the frequency–response function magnitude given by Equation 1.90 for a single-DoF, oscillatory system with viscous damping. The peak magnitude is given by Equation 1.94 for low damping. Bandwidth (half-power) is defined as the width of the frequency–response magnitude curve when the magnitude is $(1/\sqrt{2})$ times the peak value. This is denoted by $\Delta\omega$ (see Figure 1.12). An expression

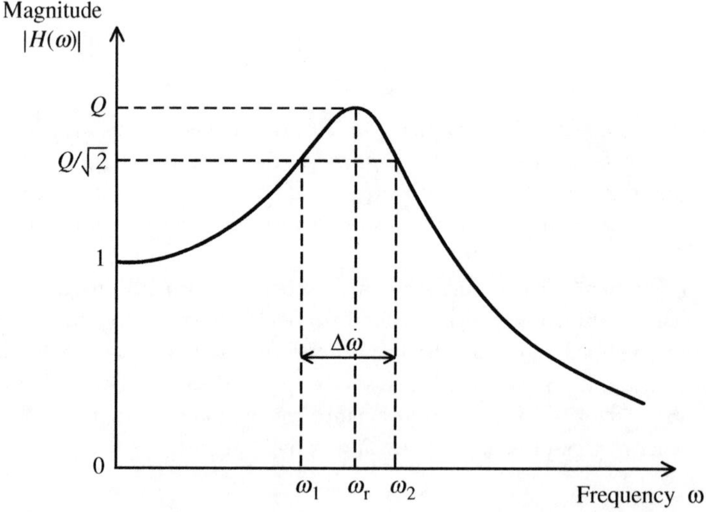

FIGURE 1.12 Bandwidth method of damping measurement in a single-DoF system.

for $\Delta\omega = \omega_2 - \omega_1$ is obtained below using Equation 1.90. By definition, ω_1 and ω_2 are the roots of the equation

$$\frac{\omega_n^2}{\left[(\omega_n^2 - \omega^2)^2 + 4\zeta^2\omega_n^2\omega^2\right]^{1/2}} = \frac{1}{\sqrt{2} \times 2\zeta} \tag{1.97}$$

for ω. Equation 1.97 can be expressed in the form

$$\omega^4 - 2(1 - 2\zeta^2)\omega_n^2\omega^2 + (1 - 8\zeta^2)\omega_n^4 = 0 \tag{1.98}$$

This is a quadratic equation in ω_2, having roots ω_1^2 and ω_2^2, which satisfy

$$(\omega^2 - \omega_1^2)(\omega^2 - \omega_2^2) = \omega^4 - (\omega_1^2 + \omega_2^2)\omega^2 + \omega_1^2\omega_2^2 = 0$$

Consequently,

$$\omega_1^2 + \omega_2^2 = 2(1 - 2\zeta^2)\omega_n^2 \tag{1.99}$$

and

$$\omega_1^2\omega_2^2 = (1 - 8\zeta^2)\omega_n^4 \tag{1.100}$$

It follows that

$$(\omega_2 - \omega_1)^2 = \omega_1^2 + \omega_2^2 - 2\omega_1\omega_2 = 2(1 - 2\zeta^2)\omega_n^2 - 2\sqrt{1 - 8\zeta^2}\,\omega_n^2$$

For small ζ (in comparison to 1), we have

$$\sqrt{1 - 8\zeta^2} \cong 1 - 4\zeta^2$$

Hence,

$$(\omega_2 - \omega_1)^2 \cong 4\zeta^2\omega_n^2$$

or, for low damping

$$\Delta\omega = 2\zeta\omega_n = 2\zeta\omega_r \tag{1.101}$$

From Equation 1.101 it follows that the damping ratio can be estimated from the bandwidth using the relation

$$\zeta = \frac{1}{2}\frac{\Delta\omega}{\omega_r} \tag{1.102}$$

For a multi-DoF system with widely spaced resonances, the foregoing method can be extended to estimate modal damping. Consider the frequency–response magnitude plot (in dB) shown in Figure 1.13.

Since a factor of $\sqrt{2}$ corresponds to 3 dB, the bandwidth corresponding to a resonance is given by the width of the magnitude plot at 3 dB below that resonant peak. For the ith mode, the damping ratio is given by

$$\zeta_i = \frac{1}{2}\frac{\Delta\omega_i}{\omega_i} \tag{1.103}$$

The bandwidth method of damping measurement indicates that the bandwidth at a resonance is a measure of the energy dissipation in the system in the neighborhood of that resonance. The simplified

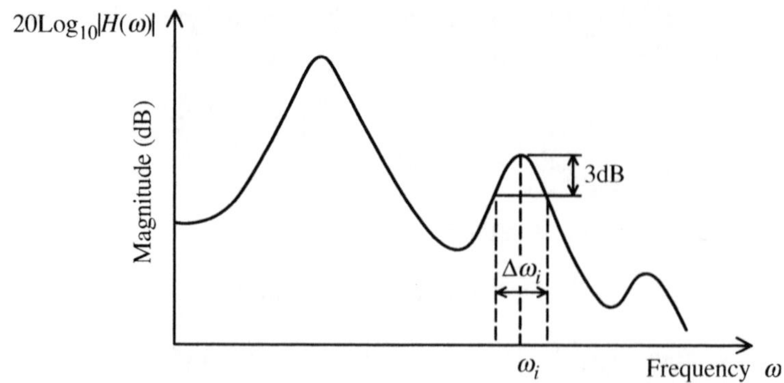

FIGURE 1.13 Bandwidth method of damping measurement in a multi-DoF system.

relationship given by Equation 1.103 is valid for low damping, however, and is based on linear system analysis. Several methods of damping measurement are summarized in Box 1.2.

1.4.6 General Remarks

There are limitations to the use of damping values that are experimentally determined. For example, consider time–response methods for determining the modal damping of a device for higher modes. The customary procedure is to first excite the system at the desired resonant frequency, using a harmonic exciter, and then to release the excitation mechanism. In the resulting transient vibration, however, there invariably will be modal interactions, except in the case of proportional damping. In this type of test, it is tacitly assumed that the device can be excited in the particular mode. In essence, proportional damping is assumed in modal damping measurements. This introduces a certain amount of error into the measured damping values.

Expressions used in computing damping parameters from test measurements are usually based on linear system theory. However, all practical devices exhibit some nonlinear behavior. If the degree of nonlinearity is high, the measured damping values will not be representative of the actual behavior of the system. Furthermore, testing to determine damping is usually performed at low amplitudes of vibration. The corresponding responses could be an order of magnitude lower than, for instance, the amplitudes exhibited under extreme operating conditions. Damping in practical devices increases with the amplitude of motion, except for relatively low amplitudes (see Figure 1.14 illustrating nonlinear behavior). Consequently, the damping values determined from experiments should be extrapolated when they are used to study the behavior of the system under various operating conditions. Alternatively, damping could be associated with a stress level in the device. Different components in a device are subjected to varying levels of stress, however, and it might be difficult to obtain a representative stress value for the entire device. One of the methods recommended for estimating damping in structures under seismic disturbances, for example, is by analyzing earthquake response records for structures that are similar to the one being considered. Some typical damping ratios that are applicable under operating basis earthquake (OBE) and safe-shutdown earthquake (SSE) conditions for a range of items are given in Table 1.6.

When damping values are estimated using frequency–response magnitude curves, accuracy becomes poor at very low damping ratios ($< 1\%$). The main reason for this is the difficulty in obtaining a sufficient number of points in the magnitude curve near a poorly damped resonance when the frequency–response function is determined experimentally. As a result, the magnitude curve is poorly defined in the neighborhood of a weakly damped resonance. For low damping ($< 2\%$), time–response methods are particularly useful. At high damping values, the rate of decay can be so fast that the measurements contain large errors. Modal interference in closely spaced modes can also affect measured damping results.

Box 1.2
Damping Measurement Methods

Method	Measurements	Formulas
Logarithmic decrement method	A_i = first significant amplitude; A_{i+r} = amplitude after r cycles	Logarithmic decrement $\delta = \frac{1}{r}\ln\frac{A_i}{A_{i+r}}$ (per cycle) $\frac{\delta}{2\pi} = \frac{\zeta}{\sqrt{1-\zeta^2}}$ (per radian) or, $\zeta = \frac{1}{\sqrt{1+(2\pi/\delta)^2}}$ For low damping $\zeta = \frac{\delta}{2\pi}$ $\zeta = \frac{A_i - A_{i+1}}{\pi(A_i + A_{i+1})}$
Step response method	M_p = first peak value normalized r.t. steady-state value; PO = percentage overshoot (over steady-state value)	$M_p = 1 + \exp\left[\frac{-\pi\zeta}{\sqrt{1-\zeta^2}}\right]$ $\mathrm{PO} = 100\exp\left[\frac{-\pi\zeta}{\sqrt{1-\zeta^2}}\right]$
Hysteresis loop method	ΔU = area of displacement–force hysteresis loop; x_0 = maximum displacement of the hysteresis loop; k = average slope of the hysteresis loop	Hysteretic damping constant $h = \frac{\Delta U}{\pi x_0^2}$ Loss factor $\eta = \frac{h}{k}$ Equivalent damping ratio $\zeta = \frac{h}{2k}$
Magnification factor method	Q = amplification at resonance, w.r.t. zero-frequency value	$Q = \frac{1}{2\zeta\sqrt{1-\zeta^2}}$ For low damping $\zeta = \frac{1}{2Q}$
Bandwidth method	$\Delta\omega$ = bandwidth at $1/\sqrt{2}$ of resonant peak (i.e., half-power bandwidth); ω_r = resonant frequency	$\zeta = \frac{\Delta\omega}{2\omega_r}$

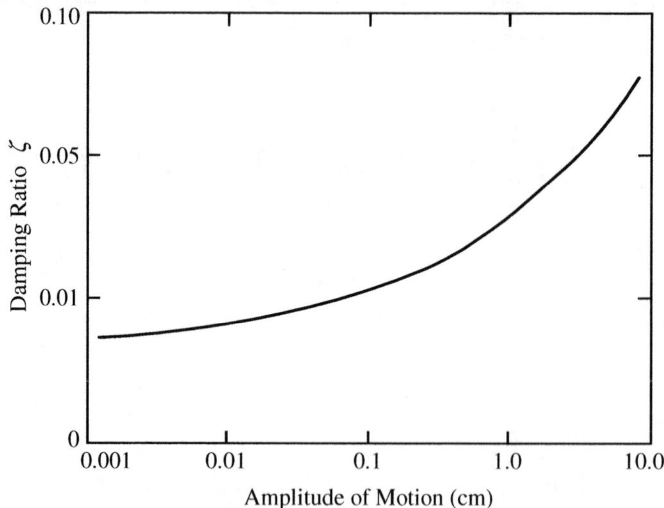

FIGURE 1.14 Effect of vibration amplitude on damping in structures.

TABLE 1.6 Typical Damping Values Suggested by ASME for Seismic Applications

System	Damping Ratio (ζ%)	
	OBE	SSE
Equipment and large diameter piping systems (> 12 in. diameter)	2	3
Small diameter piping systems (≤ 12 in. diameter)	1	2
Welded steel structures	2	4
Bolted steel structures	4	7
Prestressed concrete structures	2	5
Reinforced concrete structures	4	7

1.5 Interface Damping

In many practical applications damping is generated at the interface of two sliding surfaces. This is the case, for example, in bearings, gears, screws, and guideways. Even though this type of damping is commonly treated under structural damping, due to its significance we will consider it again here in more detail, as a category of its own.

Interface damping was formally considered by DaVinci in the early 1500s and again by Coulomb in the 1700s. The simplified model used by them is the well-known Coulomb friction model as given by

$$f = \mu R \operatorname{sgn}(v) \tag{1.104}$$

where

f = the frictional force that opposes the motion
R = the normal reaction force between the sliding surfaces
v = the relative velocity between the sliding surfaces
μ = the coefficient of friction

Note that the signum function "sgn" is used to emphasize that f is in the opposite direction of v. This simple model is not expected to provide accurate results in all cases of interface damping. It is known that, apart from the loading conditions, interface damping depends on a variety of factors such as material properties, surface characteristics, nature of lubrication, geometry of the moving parts, and the magnitude of the relative velocity.

A somewhat more complete model for interface damping, incorporating the following characteristics, is shown in Figure 1.15:

1. Static and dynamic friction, with stiction and stick–slip behavior.
2. Conventional Coulomb friction (Region 1).
3. A drop in dynamic friction, with a negative slope, before increasing again. This is known as the "Stribeck effect" (Region 2).
4. Conventional viscous damping (Region 3).

These characteristics cover the behavior of interface damping that is commonly observed in practice. In particular, suppose that a force is exerted to generate a relative motion between two surfaces. For small values of the force, there will not be a relative motion, in view of friction. The minimum force, f_s, that is needed for the motion to start is the static frictional force. The force that is needed to maintain the motion will drop instantaneously to f_d, as the motion begins. It is as though initially the two surfaces were "stuck" and f_s is the necessary breakaway force. Hence, this characteristic is known as stiction. The minimum force f_d that is needed to maintain the relative motion between the two surfaces is called dynamic friction. In fact, under dynamic conditions, it is possible for "stick–slip" to occur where repeated sticking and breaking away cycles of intermittent motion take place. Clearly, such "chattering" motion corresponds to instability (for example, in machine tools). It is an undesirable effect and should be avoided.

After the relative motion begins, conventional Coulomb type damping behavior may dominate for small relative velocities, as represented in Region 1. For lubricated surfaces, at low relative velocities, there will be some solid-to-solid contact that generates a Coulomb-type damping force. As the relative speed increases, the degree of this solid-to-solid contact will decrease and the damping force will drop, as in Region 2 of Figure 1.15. This characteristic is known as the Stribeck effect. Since the slope of the friction curve is negative in Regions 1 and 2, this corresponds to the unstable region. As the relative velocity is further increased, in fully lubricated surfaces, viscous-type damping will dominate, as shown in Region 3 of Figure 1.15. This is the stable region. It follows that a combined model of interface damping may be expressed as

$$f = \begin{cases} f_s & \text{for } v = 0 \\ f_{sb}(v)\text{sgn}(v) + bv & \text{for } v \neq 0 \end{cases} \tag{1.105}$$

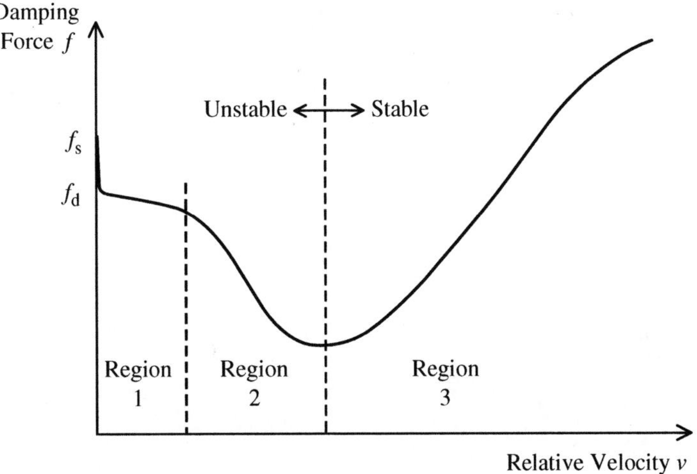

FIGURE 1.15 Main characteristics of interface damping.

Note that $f_{sb}(v)$ is a nonlinear function of velocity that will represent both dynamic friction (for $v > 0$) and the Stribeck effect. Models that have been used to represent this effect include the following

$$f_{sb} = \frac{f_d}{1 + (v/v_c)^2} \tag{1.106}$$

$$f_{sb} = f_d\, e^{-(v/v_c)^2} \tag{1.107}$$

and

$$f_{sb} = (f_d + \alpha|v|^{1/2})\mathrm{sgn}(v) \tag{1.108}$$

Note that f_d represents dynamic Coulomb friction and v_c and α are modal parameters.

Example 1.5

An object of mass m rests on a horizontal surface and is attached to a spring of stiffness k, as shown in Figure 1.16. The mass is pulled so that the extension of the spring is x_0, and is moved from rest from that position. Determine the subsequent sliding motion of the object. The coefficient of friction between the object at the horizontal surface is μ.

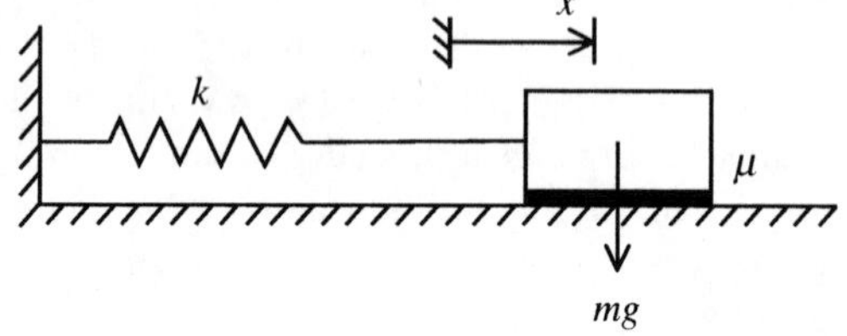

FIGURE 1.16 An object sliding against Coulomb friction.

Solution

Note that, when the object moves to the left, the frictional force, μmg acts to the right, and *vice versa.* Now, consider the first cycle of motion, stating from rest with $x = x_0$ moving to the left, coming to rest with the spring compressed and then moving to the right.

First Half Cycle (Moving to Left)

The equation of motion is

$$m\ddot{x} = -kx + \mu mg \tag{i}$$

or

$$\ddot{x} + \omega_n^2 x = \mu g \tag{ii}$$

where $\omega_n = \sqrt{k/m}$ is the undamped material frequency. Equation ii has a homogeneous solution of

$$x_h = A_1 \sin(\omega_n t) + A_2 \cos(\omega_n t) \tag{iii}$$

and a particular solution of

$$x_p = \frac{\mu g}{\omega_n^2} \tag{iv}$$

Hence, the total solution is

$$x = A_1 \sin(\omega_n t) + A_2 \cos(\omega_n t) + \frac{\mu g}{\omega_n^2} \tag{v}$$

Using the initial conditions $x = x_0$ and $\dot{x} = 0$ at $t = 0$, we get $A_1 = 0$ and $A_2 = x_0 - (\mu g/\omega_n^2)$ Hence, Equation v becomes

$$x = \left(x_0 - \frac{\mu g}{\omega_n^2}\right)\cos(\omega_n t) + \frac{\mu g}{\omega_n^2} \tag{vi}$$

At the end of this half cycle we have $\dot{x} = 0$ or $\sin \omega_n t = 0$. Hence the corresponding time is $t = \pi/\omega_n$. Substituting this in Equation vi, the corresponding position of the object is (note: $cos\ \pi = -1$)

$$x_{l1} = -\left(x_0 - \frac{2\mu g}{\omega_n^2}\right) \tag{vii}$$

Second Half Cycle (Moving to Right)
The equation of motion is

$$m\ddot{x} = -kx - \mu mg \tag{viii}$$

or

$$\ddot{x} + \omega_n^2 x = -\mu g \tag{ix}$$

The corresponding response is given by

$$x = B_1 \sin(\omega_n t) + B_2 \cos(\omega_n t) - \frac{\mu g}{\omega_n^2} \tag{x}$$

Using the initial conditions $x = -(x_0 - (2\mu g/\omega_n^2))$ and $\dot{x} = 0$ at $t = \pi/\omega_n$, we get $B_1 = 0$ and $B_2 = x_0 - (3\mu g/\omega_n^2)$. Hence, Equation x becomes

$$x = \left(x_0 - \frac{3\mu g}{\omega_n^2}\right)\cos(\omega_n t) - \frac{\mu g}{\omega_n^2} \tag{xi}$$

The object will come to rest ($\dot{x} = 0$) next at $t = 2\pi/\omega_n$, hence the position of the object at the end of the present half cycle would be

$$x_1 = x_0 - \frac{4\mu g}{\omega_n^2} \tag{xii}$$

The response for the next cycle is determined by substituting x_1 as given by Equation xii with Equation vi for the left motion and with Equation xi for the right motion. Then, we can express the general response as

$$\text{left motion in cycle } i\text{: } x = [x_0 - (4i - 3)\Delta]\cos \omega_n t + \Delta \tag{xiii}$$

$$\text{right motion in cycle } i\text{: } x = [x_0 - (4i - 1)\Delta]\cos \omega_n t - \Delta \tag{xiv}$$

where

$$\Delta = \frac{\mu g}{\omega_n^2} \tag{xv}$$

Note that the amplitude of the harmonic part of the response should be positive for that half cycle of motion to be possible. Hence, we must have

$x_0 > (4i - 3)\Delta$ for left motion in cycle i
$x_0 > (4i - 1)\Delta$ for right motion in cycle i

Also note from Equation xiii and Equation xiv that the equilibrium position for the left motion is $+\Delta$ and for the right motion is $-\Delta$. A typical response curve is sketched in Figure 1.17.

1.5.1 Friction in Rotational Interfaces

Friction in gear transmissions, rotary bearings, and other rotary joints has a somewhat similar behavior. Of course, the friction characteristics will depend on the nature of the devices and also the loading conditions, but, experiments have shown that the frictional behavior of these devices may be represented by the interface damping model given here. Typically, experimental results are presented as

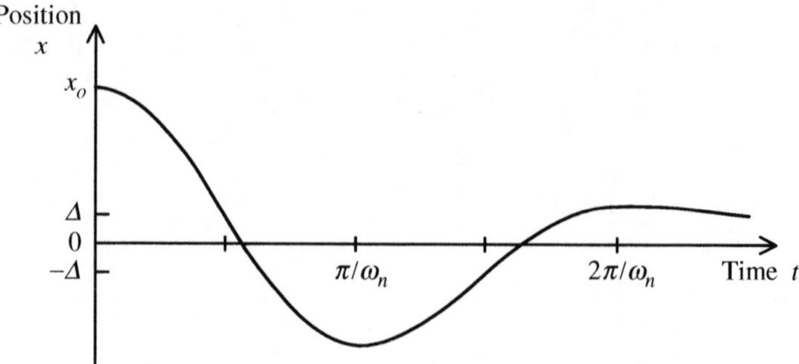

FIGURE 1.17 A typical cyclic response under Coulomb friction.

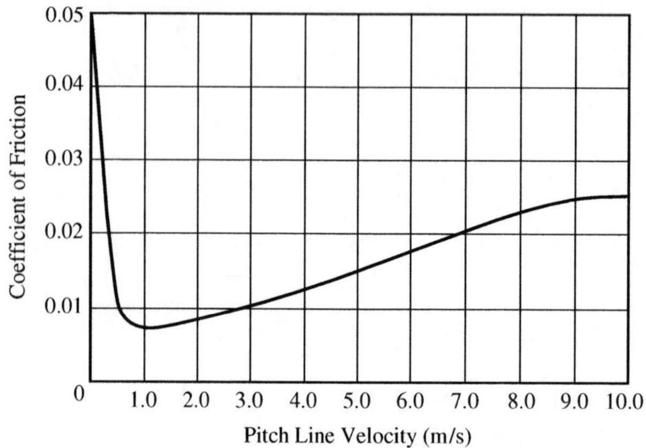

FIGURE 1.18 Frictional characteristics of a pair of spur gears.

curves of coefficient of friction (frictional force/normal force) vs. relative velocity of the two sliding surfaces. In the case of rotary bearings, the rotational speed of the shaft is used as the relative velocity, while for gears; the pitch line velocity is used. Experimental results for a pair of spur gears are shown in Figure 1.18.

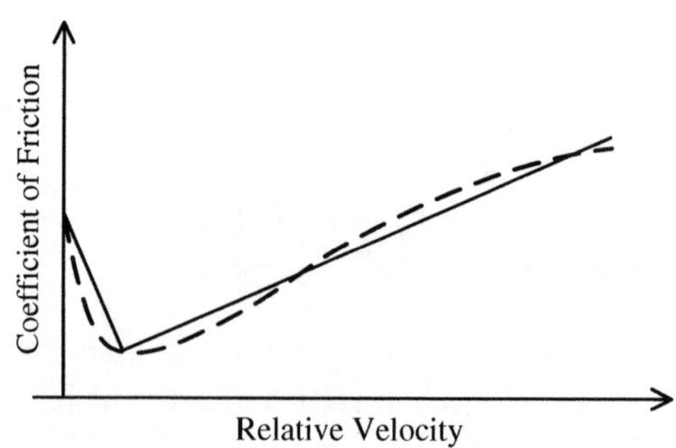

FIGURE 1.19 A friction model for rotatory devices.

What is interesting to notice from the result is the fact that, for this type of rotational device, the damping behavior may be approximated by two straight line segments in the velocity–friction plane; the first segment having a sharp negative slope and the second segment having a moderate positive slope that represents the equivalent viscous damping constant, as shown in Figure 1.19.

1.5.2 Instability

Unstable behavior or self-excited vibrations, such as stick–slip and chatter, that are exhibited by interacting devices such as metal removing tools (e.g., lathes, drills, and milling machines) may be easily explained using the interface damping model. In particular, it is noted that the model has a region of negative slope (or negative damping constant), which corresponds to low relative velocities, and a region of positive slope, which corresponds to high relative velocities. Consider the single-DoF model:

$$m\ddot{x} + b\dot{x} + kx = 0 \tag{1.109}$$

without an external excitation force. Initially the velocity is $\dot{x} = 0$. But, in this region, the damping constant, b, will be negative and hence the system will be unstable. Hence, a slight disturbance will result in a steadily increasing response. Subsequently, $\dot{x}$ will increase above the critical velocity where b will be positive and the system will be stable. As a result, the response will steadily decrease. This growing and decaying cycle will be repeated at a frequency that primarily depends on the inertia and stiffness parameters (m and k) of the system. Chatter is caused in this manner in interfaced devices.

Bibliography

Blevins, R.D. 1977. *Flow-Induced Vibration*, Van Nostrand Reinhold, New York.

den Hartog, J.P. 1956. *Mechanical Vibrations*, McGraw-Hill, New York.

de Silva, C.W., Dynamic beam model with internal damping, rotatory inertia and shear deformation, *AIAA J.*, 14, 5, 676–680, 1976.

de Silva, C.W., Optimal estimation of the response of internally damped beams to random loads in the presence of measurement noise, *J. Sound Vib.*, 47, 4, 485–493, 1976.

de Silva, C.W., An algorithm for the optimal design of passive vibration controllers for flexible systems, *J. Sound Vib.*, 74, 4, 495–502, 1982.

de Silva, C.W. 2005. *MECHATRONICS — An Integrated Approach*, Taylor & Francis, CRC Press, Boca Raton, FL.

de Silva, C.W. 2006. *VIBRATION — Fundamentals and Practice*, 2nd ed., Taylor & Francis, CRC Press, Boca Raton, FL.

Ewins, D.J. 1984. *Modal Testing: Theory and Practice*. Research Studies Press Ltd, Letchworth, England.

Inman, D.J. 1996. *Engineering Vibration*, Prentice Hall, Englewood Cliffs, NJ.

Irwin, J.D. and Graf, E.R. 1979. *Industrial Noise and Vibration Control*, Prentice Hall, Englewood Cliffs, NJ.

Van de Vegte, J. and de Silva, C.W., Design of passive vibration controls for internally damped beams by modal control techniques, *J. Sound Vib.*, 45, 3, 417–425, 1976.

2

Damping Theory

Randall D. Peters
Mercer University

Summary

This introductory chapter synthesizes the many, though largely disjointed attributes of friction as they relate to damping (also see Chapter 1). Among other means, events selected from the history of physics are used to show that damping models have suffered from the inability of physicists to describe friction from first principles. To support fundamental arguments on which the chapter is based, evidence is provided for a claim that important nonlinear properties have been mostly missing from classical damping models. The chapter illustrates how the mechanisms of internal friction responsible for hysteretic damping in solids can lead to serious errors of interpretation. Such is the case even though hysteretic damping often masquerades as a linear phenomenon. One attempt to correct common model deficiencies is the author's work toward a "universal damping model," that is described in Section 2.17. Section 2.17 is developed in a "canonical" damping form. It shows the value of a direct, as opposed to an indirect, involvement of energy in model development. To keep a better perspective on how the treatment of damping is likely to evolve in the future, the last section of the chapter addresses some of the remarkable complexities of damping that are only beginning to be discovered. The manner in which technology has played a role in some of these discoveries is addressed in Chapter 3.

2.1 Preface

The sheer volume of published material on the subject is a testament to the difficulty of selecting topics for inclusion in a chapter on damping. Viscoelasticity alone is the basis for several voluminous engineering handbooks. The present chapter is purposely different from similarly titled chapters of other reference books. There is little repetition of well-known and proven classical methods, for which the reader is referred to excellent other sources, such as de Silva (2006) and Chapter 1 of the present handbook. They provide solution techniques for many of the routine problems of engineering. The goal of the present chapter is to provide assistance with problems that are not routine, problems that are being encountered more frequently as technology advances. It is thought that this goal is best served by revisiting fundamental issues of the physics responsible for damping.

Once a multibody system has come to steady state, its damping treatment can be far less formidable than its description during approach to steady state. When dealing with limit cycles involving aeroelasticity and joints in helicopters, nonlinearity has a profound influence on the transient behavior. Attempts to model it have been largely unsuccessful, forcing the empirical selection of elastomers to reduce the vibration. (In the old days hydraulic dampers were used; Hodges, 2003). At a much lower level of sophistication, our understanding is quite limited on some common phenomena, such as the negative damping character of sound generated by a violin or a clarinet. Historically, when technology "hit the wall" because of too much theoretical handwaving, it became apparent that fundamental assumptions needed to be examined. In physics, a complete alteration of conventional wisdom was sometimes necessary, one of the best examples being the events that gave birth to quantum mechanics. Hopefully, from the multitude of seemingly disparate (but assumed by the author to be connected) observations which follow, the purpose for the architecture of this chapter can be partially realized. The enormous complexity of damping in general makes it unrealistic to hope for complete success.

Physics played a prominent role in developing the classical foundations of damping, starting in the 19th century. Subsequently, engineers uncovered many features of the subject that physicists never even thought about. In recent years, however, physics has been circumstantially forced to reconsider damping fundamentals. With the advent of personal computing, and an increased awareness of the importance of nonlinearity, new discoveries point to serious limitations of the classical foundation. The field of mechanics was severely limited until it began tackling problems of nonlinearity (not of damping type), and became concerned with previously ignored features giving unique system properties. Just as these unique properties could only be solved by techniques more sophisticated than the equations of linear type, there is mounting evidence that nonlinear damping may be the key to understanding some bewildering engineering cases.

It is important to try to identify the major mechanisms responsible for energy dissipation. This is easier said than done, since a host of different friction processes are usually at work. Moreover, the description of friction from first principles remains a daunting task. Thus we are forced to work with phenomenological models. There are also conflicts of nomenclature, with a given word meaning two different things from one profession to another. Thus, much of this chapter will attempt to define carefully terms while focusing on the physics, the treatment of which follows naturally along the lines of historical developments.

Engineers tend to be interested in higher frequencies and higher amplitudes of vibration than are scientists. A perfect damping model would be unconcerned with such differences of application; however, such a model is far from being realized. Because small-amplitude, long-period (low and slow) oscillations provide a valuable means for studying many processes of damping in general, much of this chapter focuses in that direction.

From the multitude of choices available to writers on the subject of damping, this author has selected a single (hopefully) unifying theme — nonlinear damping, especially as found in low and slow oscillations. Because it is a field still in its infancy, many of the ideas that follow are more speculative than one would prefer; however, they deserve discussion because of their perceived importance. To this author's knowledge, damping has not been previously treated in the manner of this chapter. Concerning the earliest relevant paper (Peters, 2001a, 2001b), the following was indicated by oft-cited Prof. A.V. Granato: "I don't know of anyone thinking about internal friction along the lines you have mentioned."

There are two important elements to the unifying theme of nonlinear dissipation: (i) the influence of nonlinear damping on multibody systems in their approach to steady state, and (ii) the close connection between damping and mechanical noise. When vibration decay is not exponential because of nonlinearity, there are significant ramifications and they are only beginning to be appreciated.

The novel features of this chapter are possible because of dramatic improvements in both sensing and data collection/analysis in the last decade. Demonstrating that a decay is not purely exponential requires both (i) a good linear sensor and (ii) the means to study readily long-time records when the damping is small (high Q). The first prerequisite has been met through the use of this author's patented fully differential capacitive sensor. The second has been realized with the availability of good, inexpensive analog-to-digital (A/D) converters having user-friendly, yet powerful Windows-based software. In addition to the "preview" software that comes with Dataq's A/D converter, a proven means for identifying nonexponential decay has been the analysis of records imported to Microsoft Excel. Details of these novel methods will be provided in the various sections that follow.

There are many examples in the engineering literature of nonlinear damping; even Coulomb damping is nonlinear because the friction force involves the algebraic sign of the velocity rather than the velocity itself, as in linear viscous damping. What has been realized for the first time in the course of writing this chapter is the following. As will be shown in the subsequent material, a decay process is not usually a pure exponential. Whatever the reason for a pure exponential, whether fundamentally linear (viscous) or nonlinear (hysteretic present model), the quality factor Q for such a pure exponential decay is constant. When there is a second mechanism, such as amplitude-dependent damping (even if it is the only mechanism), the Q now becomes time dependent. This is significant to mode coupling for the following reason. When a pair of modes couple because of elastic nonlinearity (a process that is impossible assuming linear dynamics), the strength of the coupling is proportional to the product of the individual amplitudes of the pair.

Consequently, variability in Q can influence the evolution to steady state. It is a factor in determining which modes ultimately survive and/or dominate. Moreover, the distribution of the modes which remain depends on initial conditions, including the intensity of excitation.

Long ago, musicians learned to deal with nonlinearity, due in part to properties of the ear that are responsible for aural harmonics. A pair of purely harmonic signals can beat in the ear to produce a "sound" that does not exist when sensed with a linear detector. For example, consider a strong and undistorted 500 Hz signal sounded simultaneously with a pure 1003-Hz sound. The ear will hear a 3-Hz beat due to the superposition of the ear's aural second harmonic of the first with the fundamental of the second. However, there's more to this story. Conductors call for *fortissimo* and *pianissimo* sounds, not only because of the ear's nonlinearity, but also because of nonlinearities inherent to musical instruments. For example, it is easy with a good microphone and LabView (see de Silva, 2006) to demonstrate that the timbre of stringed instruments is intensity dependent. Not only is the mix of harmonics, as displayed in a fast Fourier transform (FFT) power spectrum, different according to volume, but their distribution also changes with time.

Noise is not typically treated in an engineering discussion of damping; however, mechanical noise is an important part of the technical material included in this chapter. Believing that there is a great deal of connectivity among vibration, damping, and noise, evidence will be provided in support of a premise — that the most important and least understood form of internal mechanical damping (material = hysteretic = "universal") is closely allied with the most important and least understood form of noise ($1/f$ = flicker = pink). If this premise is true, then the foundations of damping physics need reconstruction on several counts. Evidence in support of the premise will be provided through tidbits of experimental discoveries from a host of independent investigations. It is hoped that the unusual and lengthy introduction that follows will be beneficial in this regard. Historical elements serve to synthesize the many parts and are offered without apology. Following the introduction, some practical and novel equations of damping will eventually be provided. Even if readers find little identification with the philosophies that birthed them, it is hoped they will at least carefully examine the equations that are presented here in Section 2.17 for the first time.

2.2 Introduction

2.2.1 General Considerations of Damping

The etymology of the word "damping" is difficult to determine. It is obviously allied with the word damper, commonly defined as a "device that decreases the amplitude of electronic, mechanical, aerodynamic or acoustical oscillations," used for centuries, for example, to describe the sound attenuator pedal on the piano. Perhaps the German word *dampfen* (to choke) has had an influence in the evolution of the word. One can only wonder if water, as a moistening agent, played any role. Certainly, liquid water is important to some cases of energy dissipation in oscillators. Moreover, friction determined by the viscosity of a fluid (gas or liquid) is an important type of damping. A curious piece of history, in the celebrated work of Stokes, is why his expression "index of friction" did not take precedence over our modern word, viscosity. Peculiar terminology is also encountered to describe damping, such as the engineering device known as a dashpot, which is a mechanical damper. The vibrating part is attached to a piston that moves in a liquid-filled chamber.

We will see that the number of adjectives used to describe various types of damping is extensive. This multiplicity of terms to describe the loss of oscillatory energy to heat is no doubt an indicator of the complexity of damping phenomena in general. We will attempt (i) to identify similarities and differences among various types of damping, while (ii) explaining some of the physics responsible for the characteristics observed. Conceptual ideas and techniques of both theory and experiment will be provided, targeting the lowest level of sophistication for which semimeaningful results can be obtained. The reader should be aware that a "grand-unified" theory of damping does not exist, nor is it likely that one will ever be created.

Damping causes a portion of the energy of an oscillator, otherwise periodically exchanged between potential and kinetic forms, to be irreversibly converted to heat, sometimes by way of acoustical noise. Whether by suitable choice of materials during design of passive equipment, or by using feedback in active control of a sophisticated system, control of damping is important since mechanical vibrations can be detrimental or even catastrophic. An oft-quoted example of catastrophe is the Tacoma Narrows bridge, which collapsed in high winds on November 7, 1940. Like the vibration of a clarinet reed, this disaster is probably best described by the term negative damping, which can drive parametric oscillations.

The optimal amount of damping for a given system might fall anywhere in a wide range from great to extremely small, depending on system needs. The engineering world frequently wants oscillations to be as close to critically damped as possible. Physics experiments, such as those searching for the elusive gravitational wave (centered at the Laser Interferometer Gravitational Observatories, or Laser Interferometer Gravitational Wave Observatories [LIGO], in the United States; GEO600 in Germany [involving the British]; VIRGO in Italy [with the French], and TAMA in Japan), want damping in some of their components to be as small as possible. Frequency standards the world over require very small damping to insure high precision for timekeeping.

For the specific components of a system, a successful design frequently requires identification of the specific mechanisms primarily responsible for the dissipation of energy. Even after identifying the dominant sources, the theoretical difficulty of their treatment can also range from great to small, depending on the type of damping. For dashpot fluid damping, adequate models have existed for decades. For material damping, on the other hand, theories of internal friction are numerous and largely lacking in self-consistency.

The fundamental mechanisms responsible for damping are in most cases nonlinear; however, the oscillator's motion can itself be approximated in many cases by a linear second-order differential equation. If the potential energy is quadratic in the displacement, then the undamped linear equation of motion is that of the simple harmonic oscillator, because its solution is a combination of the sine and cosine (harmonic) functions. This undamped equation comprises the sum of two terms, one being a displacement and the other term an acceleration. The constant parameter multiplying each term of the pair depends on the nature of the system. For example, in the case of a mass–spring oscillator, the acceleration is multiplied by the mass, and the displacement by the spring constant. Thus, the equation corresponds to Newton's Second Law applied to a Hooke's Law (idealized) spring. In an electronic L–C oscillator, the "displacement" corresponds to the charge on the capacitor (divided by C) and the "acceleration" corresponds to the second time derivative of the capacitor's charge (multiplied by inductance L).

The usual means to describe damping, which is always present with oscillation, is to add a velocity term to the aforementioned displacement and acceleration. Although the damping could derive from several causes, there is usually a single dominant process. For example, the damping of current in a series-connected resistor, inductor, capacitor (RLC) circuit may depend mostly on Joule heating in the resistor R, in spite of the fact that there must also be energy loss in the form of radiation. Thus, the equation of motion includes a first-time derivative of the capacitor's charge (current) multiplied by R, in accord with Ohm's law.

Whether radiation is important for damping of the RLC circuit depends on the amount of coupling to the environment. If the circuit communicates with a final amplifier connected to an antenna, then radiation may become more important than Joule losses. The frequency of oscillation is a key parameter in this case, and also for damping problems in general. Unfortunately for some common systems, theoretical efforts to account properly for the effects of frequency have proven largely unsuccessful — except for models of phenomenological type developed by empiricism.

2.2.2 Specific Considerations

The mass–spring oscillator is the textbook example of harmonic motion, for which one of the most sophisticated mechanical oscillators ever built is the LaCoste version of vertical seismometer. Significant portions of the experimental data presented in this chapter were generated with an instrument designed around the LaCoste zero-length spring (LaCoste, 1934). The instrument used for this data collection was part of the World Wide Standardized Seismograph Network (WWSSN) during the 1960s. The spring of this seismometer is responsible for hysteretic damping of the instrument, rather than viscous damping as commonly assumed. Contrary to popular belief, air damping is not important for this seismograph at its nominal operating period, which is typically greater than 15 sec. Since every long-period pendulum apparently exhibits similar behavior, we thus find strong synergetic evidence in support of an old (mostly unheeded) claim that hysteretic damping (friction force independent of frequency) is universal (Kimball and Lovell, 1927). Their claim in 1927 to have discovered a universal form of internal friction (damping) is strengthened since the same behavior is seen in three distinctly different systems: (i) a mass–spring oscillator (as demonstrated by Gunar Streckeisen, details given later); (ii) a pendulum whose restoration depends on the Earth's gravitational field (demonstrated by several independent groups); and (iii) a rotating rod strained by a transverse deflection (1927 experiments of Kimball and Lovell).

The assumption of universality for hysteretic damping is a key point of this chapter. It will be shown that the damping of even a vibrating gas column (Ruchhardt's experiment to measure the ratio of heat capacities) is likely also hysteretic. The models that are described represent a departure from common theories of damping. Interestingly, the author's model has similarities to ordinary sliding friction, as given to us by Charles Augustin Coulomb. It effectively modifies the Coulomb coefficient of kinetic friction to yield an effective energy-dependent internal friction coefficient. The energy dependence is necessary to obtain exponential decay, as opposed to the linear decay of Coulomb damping. Just as with conventional Coulomb damping, its form is nonlinear, involving the algebraic sign of the velocity. We will see that the damping capacity predicted by the model permits an equivalent viscous form. Yet the underlying physics is related to creep of secondary type as opposed to the primary creep of viscoelasticity.

It is this author's opinion that much of the existing theory of damping is not the best means for modeling dissipation. The difficulties arise from approximating oscillator decay with linear mathematics. Although most individuals recognize the oft-stated caveat that viscous damping is an approximation to the actual physics of dissipation, they do not recognize some of the many serious limitations of the approximation. The situation is similar to the place in which we found ourselves at the beginning of the era labeled "deterministic chaos." The "butterfly effect" (Lorenz, 1972) has radically altered the thinking of many, but only in relationship to large-amplitude motions of a pendulum, where the instrument is no longer iscochronous because of nonlinearity. As an archetype of chaos, the pendulum must be rigid and capable of "winding" (displacement greater than π) before chaos is possible. Nonlinearity is a prerequisite for the chaos, but it is not sufficient, since there are many examples of highly nonlinear but nonchaotic motions. For example, amplitude jumps of nonlinear oscillators, during a frequency sweep of an external drive, have been known for many years. They were observed before chaos was recognized, in systems like the Duffing and Van der Pol oscillators. Yet chaos, with its sensitive dependence on initial conditions (responsible for the butterfly effect), was not contemplated at the time. As with most significant advances, Lorenz's discovery was by accident, as he modeled convection in the atmosphere. The author's confrontation with complexity that derives from mesoscale structures in metals was likewise unexpected. "Strange phenomena" (as Richard Feynman would probably have labeled them) were encountered while using his patented fully differential capacitive sensor to study various mechanical systems, mainly oscillators.

As with chaos, the pendulum may ultimately serve as an archetype of complexity. When operated at low energy, especially through a combination of long period and small amplitude, the free decay of the physical pendulum departs radically from the predictions based on linear equations of motion. Such complexity can be easily demonstrated when the pendulum is fabricated from soft alloy metals. For example, Figure 2.1 illustrates the decay of a rod pendulum constructed with ordinary (heavy-gauge lead–tin) solder of the type used for joining electrical conductors (Peters, 2002a, 2002b, 2002c).

The "jerkiness" (discontinuities) in the record of Figure 2.1 is in no way related to amplitude jumps of the type previously mentioned; rather, these are jumps of the Portevin–Le Chatelier (PLC) type (Portevin and Le Chatelier, 1923). They are a fundamental, yet "dirty" phenomenon that physics has chosen for decades to try and ignore (even though materials science and engineering took early note of the PLC effect). The most obvious and profound thing that can be said about Figure 2.1 is the following: the presence of PLC jerkiness means that the concept of a potential energy function is not really valid, since the requirement for its definition is that a closed integral of the force with respect to displacement must vanish.

No matter the form of hysteresis, which is the cause for damping, it disallows the curl of the force to be zero, so that potential energy is never formally meaningful for a macroscopic oscillator (since there is always damping). In those cases where the damping is essentially continuous (not true for the example of Figure 2.1), the assumption of a potential energy function retains some computational meaning. For oscillators influenced by the PLC effect, this is no longer true. The resulting properties are important to a variety of technology issues, such as sensor performance, since noise is no longer the simple thermal form predicted by the fluctuation–dissipation theorem (used to characterize white, i.e., Johnson, noise).

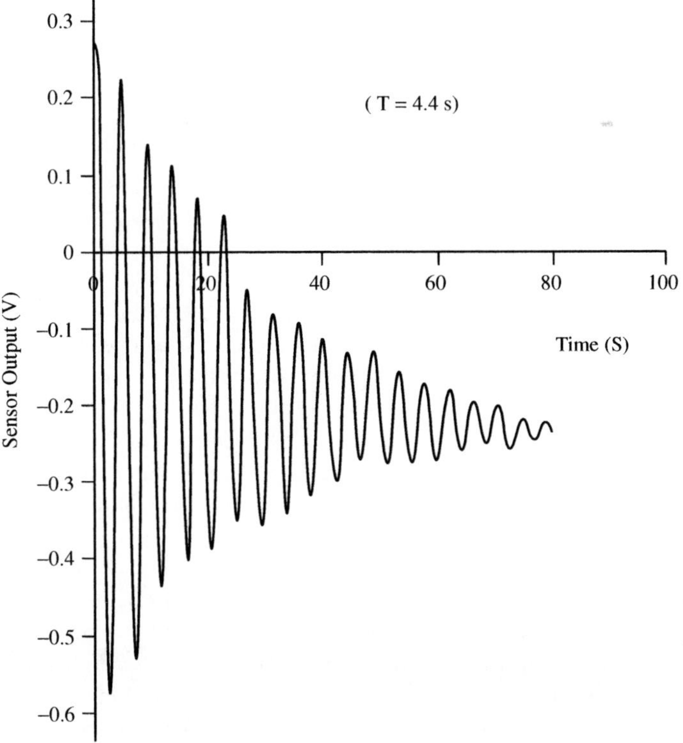

FIGURE 2.1 Free-decay of a rod pendulum fabricated from solder.

Practical means for dealing with systems influenced by "stiction" have been known by engineers for decades. Because of metastabilities in the assumed potential function, the system is prone to latching (stuck in a localized potential well). One means for mastering the metastabilities (unstick the part designed to move) is to "dither" the system. The process has become more sophisticated in the last decade, which saw a major growth of interest in stochastic resonance. In the definition by Bulsara and Gammaitoni (1996):

> A stochastic resonance is a phenomenon in which a nonlinear system is subjected to a periodic modulated signal so weak as to be normally undetectable, but it becomes detectable due to resonance between the weak deterministic signal and stochastic noise.

The phenomenon is related to dithering (Gammaitoni, 1995). It is a case where the signal-to-noise ratio (SNR) can be increased by the counterintuitive act of raising the level of noise. Such a gain in SNR is not possible with a harmonic potential.

In recent studies of granular materials, "tapping" has become a popular means to study behavior that violates the fundamental theorem of calculus. Years ago, this author used tapping as a means to accelerate creep in wires under tension. Evidently, hammering the table on which the extensometer rested caused vibrational excitations of the wire that stimulated length changes of discontinuous PLC type. Because of the broad spectral character of an impulse, various eigenmodes of the wire (de Silva, 2006) could be thus readily excited. After "hammering-down" under load, a silver wire could by this same means be stimulated, after partial load removal, to exhibit length contractions. Since the total number of atoms is fixed, the process must involve exchange of atoms between the surface of the sample and its volume.

Extensometer studies of wires at elevated temperature have also displayed strange behavior. A polycrystalline silver wire of diameter 0.1 mm and approximate length 30 cm was found to exhibit large fluctuations when heated in air to within 100 K of its melting point, using a vertical furnace (Peters, 1993a, 1993b). The large fluctuation in length at these temperatures (reminiscent of critical phenomena and visible to the naked eye) may be associated with oxide states of the metal, since the experiments were not performed in vacuum. When cycled in temperature, fluctuations in the length of a gold wire were found to exhibit dramatic hysteresis. With influence from Prof. Tom Erber of Illinois Tech University, it was postulated in (Peters, 1993a) that there may be some mesoscale quantization of fundamental type responsible for the thermal hysteresis (hysteron).

Mechanical hysteresis resulting from mesoanelastic defect structures is evidently ubiquitous. Piezotranslators, which are used as actuators in atomic force microscopes and other nanotechnology applications, are afflicted with high levels of hysteresis when operating open loop. This behavior is consistent with the anomalously large damping that was observed with a pendulum (reported elsewhere in this document) in which there was a steel/PZT interface for the knife-edge.

Even the common strain gauge exhibits complex hysteresis behavior. The normally large hysteresis that is observed in preliminary cycling of a gauge is typically reduced by significant amounts after repeated cycling, a type of work hardening (if strained well below failure limits). It should be noted that the hysteresis of all the discussions in this chapter is not to be confused with backlash (as in a gear train).

All of these experiments are in keeping with the premise that *mesoanelastic complexity* determines the nature of hysteretic damping. It is seen that there are a plethora of examples where strain (and thus damping) of a sample is not simple, is not smooth, but more like the complex behavior of granular materials. In the case of polycrystalline metals, the same grains that are made visible by methods of acid etching decoration are evidently responsible for mesoscale (nonsmooth) internal friction damping. To assume that damping is quantal at the atomic scale, rather than the mesoscale, is without experimental justification. Nevertheless, this is a popular assumption with which estimates of the noise floor of an instrument such as a seismometer is estimated, to calculate SNR.

2.2.3 The Pendulum as an Instrument for the Study of Material Damping

Because of its early contributions to physics, which in those days was called natural philosophy, one might be tempted to believe that the pendulum is only important to (i) the history of science or (ii) teaching of fundamental principles. A single observation should be sufficient to resist this temptation — (as already noted) the pendulum has in the last 30 years become the primary archetype for the new science of chaos. Additionally, many of the data sets of this document, which show significant and previously unpublished results, were generated with a pendulum.

To a student of elementary physics, the choice of a pendulum may seem unsophisticated. Yet, to the author, who has spent 15 intense years trying to understand harmonic oscillators, the pendulum is the most versatile instrument with which to understand damping. It has been central to the development of science in general. It was studied by Galileo, Huygens, Newton, Hooke, and all the best-known scientists of the Renaissance period. It served to establish collision laws, conservation laws, the nature of Earth's gravitational field and, most of all, it was the basis for Newton's two-body central force theory. This theory was foundational to the development of classical mechanics, which is central to all of physics and engineering. Historian Richard Westfall has remarked: "Without the pendulum, there would be no Principia" (Westfall, 1990).

In 1850, Sir George Gabriel Stokes published a foundational paper (Stokes, 1850). His treatment of pendulum damping permitted the understanding, decades later, of a number of important phenomena in physics and engineering. For example, his studies were foundational to the Navier–Stokes equations of fluid mechanics. Moreover, viscous flow known as Stokes' Law was the basis for Millikan's famous oil drop experiment that determined the charge of the electron.

Stokes noted in his paper that, "… pendulum observations may justly be ranked among those most distinguished by modern exactness." He also noted

> The present paper contains one or two applications of the theory of internal friction to problems which are of some interest, but which do not relate to pendulums. … the resistance thus determined proves to be proportional, for a given fluid and a given velocity, not to the surface, but to the radius of the sphere. … Since the index of friction of air is known from pendulum experiments, we may easily calculate the terminal velocity of a globule [water] of given size. … The pendulum thus, in addition to its other uses, affords us some interesting information relating to the department of meteorology.

The last statement of this quotation speaks to some of the errors in the "common theory" of his day. In similar manner, some of the common-to-physics damping models of today are erroneously applied. Those who hold the viscous damping linear model in unwarranted regard, fail to recognize the limitations under which it is valid. There are frequent misapplications for reason of experimental deficiencies. We can all profit by taking seriously the following well-known words of Kelvin:

> When you can measure what you are speaking about, and express it in numbers, you know something about it. But when you cannot measure it, when you cannot express it in numbers, your knowledge is of a meager and unsatisfactory kind. It may be the beginning of knowledge but you have scarcely in your thoughts advanced to the state of science. William Thomson, Lord Kelvin (1824 to 1907)

Simple (viscous) flow of the Stokes' Law type is possible only according to the restrictive conditions that Stokes spelled out in his paper. We now specify those conditions for viscous flow according to the nondimensional parameter given us late in the 19th century by Osborne Reynolds. Specifically, Stokes' Law is valid only for $\text{Re} = \rho v L/\eta < 60$ (approximately, for spheres); where ρ is the density of the retarding fluid, v is the speed of the object relative to the fluid, L is a characteristic dimension of the object, and η is the viscosity of the fluid. The requirement is not generally met for oscillators, and recent experiments have shown that contributions to the damping from air drag proportional to the square of the velocity cannot generally be ignored (Nelson and Olssen, 1986). This is just one example of how two or more damping types must sometimes be folded into an adequate model of dissipation. A novel method for combining all the common forms of damping in one mathematical expression is provided in this document. Additionally, it is shown how to calculate analytically the history of the amplitude of free-decay for such cases.

Considering the importance of Stokes' work, it is surprising that some of his requests for further experiments were apparently never seriously considered. On page 75 of his paper, one reads the following: "Moreover, experiments on the decrement of the arc of vibration are almost wholly wanting." Having noted this, Stokes appealed to experimentalists to generate such data. In the 19th century, collecting the data he requested would have been labor-intensive and therefore the experiments were probably never attempted. Sensors and data processing of the modern age now make them straightforward, but the pendulum has by now been viewed by too many as a relic rather than the important instrument described by Stokes. Much of the author's efforts have been directed at showing that the pendulum is still an important research instrument. For example, one physical pendulum of simple design was the basis for the generalized model of damping (modified Coulomb) that is here presented. Another has been used to illustrate surprisingly rich complexities of the motion that results from the ubiquitous defects of its structure (Peters 2002a, 2002b, 2002c). Thus studying the complex motions of "low and slow" physical pendula could yield significant new insight into the defect properties of materials — a field where relatively little first-principles progress has been made.

2.2.4 "Plenty of Room at the Bottom"

Richard Feynman gave a now-famous talk in 1959 titled, "There's plenty of room at the bottom" (presented at the American Physical Society's annual meeting at CalTech). Drawing on observations from biology, he spoke of a solid-state physics world involving "... strange phenomena that occur in complex situations." In the 44 years since Feynman's prophetic comments, there have been spectacular achievements in very large-scale integrated (VLSI) electronics, microelectromechanical systems (MEMS), and even nanotechnology. Progress in the mechanical (including sensor) realm has been much slower than in electronics; consequently, our present processing power far exceeds our acquisition (and actuator) capabilities.

One of the major obstacles to miniaturization involves dramatic change to physical properties that can occur as the size of a system shrinks below the mesoscale toward the atomic. For example, VLSI electronics is already beginning to be impacted by quantum properties of the atom, as component size continues to decrease in accord with Moore's law (Moore, 1965). Among other things, Feynman predicted that lubrication would no longer be "classical" at such a scale. On a related note, a paper by Nobel Laureate Edward M. Purcell (Purcell, 1977) draws a striking contrast between our macroscopic world and that of micro-organisms. At low Reynolds number, inertia becomes unimportant, and mechanics is dominated by viscous effects. The adoption of a new paradigm will be necessary for engineers to deal with these differences.

In the article "Plenty of room indeed" (Roukes, 2001), it is noted that there is an anticipated "dark side" of efforts to build truly useful micro- and nano-sized devices. Gaseous atoms and molecules constantly adsorb and desorb from device surfaces. This process is known to exchange momentum with the surface, even permitting scientific study of the gas–solid interface (Peters, 1990). The smaller the device, the less stable it will be because of adsorption/desorption. As Roukes has noted, this instability may pose a real disadvantage in various futuristic electromechanical signal-processing applications (Cleland and Roukes, 2002).

There is direct evidence, provided in the present chapter, that we need to be more concerned with noise: (i) the evacuated pendulum where it is speculated that outgassing influenced its free-decay, and (ii) the seismometer free-decay that showed both amplitude and phase noise and evidence for nonlinear damping. Concerning (i), when the vacuum chamber pressure is reduced, the preexisting steady state (normal rate balance between adsorption and desorption) becomes disturbed, so that there is a complex emission of gases from the surface of the pendulum. The emission is not likely to be spatially uniform, but more like the jets seen on Halley's comet when photographed by the Giotto spacecraft in 1986. In case (ii), the noise is seen to derive from mesoanelastic complexity of the structure of the pendulum itself rather than involving gases.

Miniaturized devices have the potential to serve as on-chip clocks, and the importance of phase noise to clocks is well documented. There is another, more subtle issue that points out the importance of phase noise. One of the best means for improving SNR is the technique of phase-sensitive detection, first employed by Robert Dicke at Princeton to improve solar experiments. The performance of miniaturized electromechanical sensors using "lock-in" amplifier methods may be influenced significantly by mechanical phase noise.

Phase noise of miniaturized devices is still mostly speculative. In addition to the mechanism just mentioned (adsorption/desorption), there is the matter that constitutes the theme of this chapter, defect organization. It is not possible to grow materials without dislocations and/or other disturbances to crystalline order, such as vacancies, interstitials, or substitutional impurities. Thus, "when mother nature fills the vacuum she abhors, she rarely does so with perfection."

Long before defects organize to the point of incipient failure (at much larger strains), they still influence vibration. They may even be a primary source of $1/f$ noise. Electronic noise of $1/f$ type is known to involve defects by means of trapping states, and these states derive from crystalline defects sometimes involving the surface. The interaction of the surface and the volume of a solid are important. For example, consider pure copper single crystals of the type used by the author in his doctoral work. A practical joke suggested by Vic Pare (that we never conducted because of the cost of these samples) would be to have a 98-lb weakling bend one by hand, then ask an NFL linebacker to straighten it back out! The striking irreversibility is the result of work hardening as dislocations develop at the surface and propagate into the bulk where they entangle.

In the case of polycrystalline materials, the memory features of hysteresis may be important according to the method of their fabrication. Wires are typically produced by pulling through successively smaller dies. This "swaging" may be conducive to the exchange of monolayer groups of atoms between the volume and the surface during fluctuation length changes. The fluctuation–dissipation theorem does not hold or, if it does, only in terms of larger entities than the atom. Thus, there are many yet-to-be-quantified elements of noise in the vibration of miniaturized devices. Feynman was right when he spoke of strange phenomena of the solid state.

Technology of the future is expected to be confronted increasingly with damping problems that must address issues of scaling — to deal with some factors discussed in this chapter, which, to the author's knowledge, have not been previously published. Until small (MEMS) oscillators become more common to the engineering world, we must study the mechanisms responsible for their damping by other than traditional means. One approach is similar to experimental techniques for the verification of the kinetic theory of gases. As noted by Present in his textbook (Present, 1958), there are two ways that Brownian motion can be studied: either (i) with small objects and an unsophisticated detector, or (ii) with larger objects and a very sensitive detector. It is the latter that provided some of our present knowledge of damping at the mesoscale. The fully differential capacitive transducer, whose patent label is "symmetric differential capacitive" (SDC), is a robust new technology that is sensitive, linear, and user-friendly (Peters, 1993a, 1993b). As with other sensitive detectors that have been used to predict the properties of small objects by studying larger ones, small-energy studies of various macroscopic pendula are demonstrating some of the "strange phenomena" of complex type predicted by Feynman.

From the author's perspective, we of the physics community have been guilty of two significant errors: (i) oversimplification of many problems by assuming a linear equation of motion based on viscous damping, and (ii) losing sight of fundamental issues by working with inappropriate, overly complicated damping models. The goal of this chapter is to assist progress toward a healthier balance between these extremes. It is hoped that readers will be thus better equipped to identify, and then dismantle, some of the impediments to the development of future technologies.

2.3 Background

2.3.1 Terminology

The large number of mechanisms capable of energy dissipation has resulted in a host of adjectives to describe damping phenomena in mechanical systems. They include (nonexhaustive list): viscous, eddy current, Coulomb, sliding, friction, structural, fluid, thermoelastic, internal friction, viscoelastic, material, solid, phonon–phonon, phonon–electron, and hysteretic. For present purposes, damping types will be grouped according to one of the following three categories: (i) fluid (including viscous), (ii) Coulomb, and (iii) hysteretic. Although hysteretic damping has come to be associated in engineering circles with a particular form of material damping in solids, it should be noted that all forms of damping involve hysteresis (for which the Greek meaning of the word is "to come late"). In a plot of periodic stress vs. strain, which is a straight line for displacements of a nondissipative, idealized substance, hysteresis causes the line to open into a loop. The size of the loop — more specifically the area inside this hysteresis loop — is a measure of the amount of nonrecoverable work done per cycle because of the damping. An actual force of friction is not readily recognizable in those cases that are labeled "internal friction." The word friction is used in a generic sense, meaning any process responsible for conversion of the oscillator's coherent motion into incoherent thermal activity.

With each of viscous, eddy current, and Coulomb damping, a force external to the oscillator is responsible for the dissipation of energy. The external force is associated respectively with (i) laminar fluid flow, (ii) induced currents, and (iii) surface friction. The surface friction case is not necessarily the trivial textbook presentation involving a coefficient of kinetic friction and a normal force. The cases just mentioned, along with thermoelastic damping; which is of internal rather than external origin, are much easier to treat theoretically than other cases. Viscous damping and eddy current damping (over the full range of the motion) are adequately described by a velocity term, which yields a linear equation of motion. Coulomb damping, however, is not proportional to velocity, but rather depends only on the algebraic sign of the velocity. The equation of motion is consequently nonlinear. Additionally, and unlike most other forms, Coulomb damping is not exponential. The turning points lie along a straight line when the motion is plotted vs. time. Similarly, if eddy current damping exists only over a small part of the motion, the decay is linear rather than exponential (Singh et al., 2002).

2.3.2 General Technical Features

Historically, viscous damping has been the model of choice because the resulting equation of motion is mathematically attractive and, for the RLC circuit, the form is appropriate. For mechanical oscillators, it is not generally appropriate, since viscous damping amounts to some part of the system moving in an external Newtonian fluid that removes energy because of a friction force that is proportional to velocity.

The defects responsible for material damping, such as dislocations, are also responsible for creep, so that high strength and high damping tend to be incompatible attributes. Magnesium alloys tend to be better than many other metals in this regard. Hardness of a material is neither a prerequisite for toughness nor for small damping, as recognized by those familiar with the mechanical properties of cast iron.

On a different scale, defects determine "how things break"; concerning which Marder and Fineberg have stated the following:

> the strength of solids calculated from an excessively idealized starting point comes out completely wrong; it is not determined by performance under ideal conditions, but instead by the survival of the most vulnerable spot under the most adverse of conditions. (Marder and Fineberg, 1996)

Three famous scientists are primarily responsible for the highly popular viscous damping model of the simple harmonic oscillator; they are Lord Kelvin (Thomson and Tait, 1873), G.G. Stokes and

H.A. Lorentz. Stokes is best known for his equations of fluid dynamics that also include the name Navier. Stokes' Law, which describes the terminal velocity of a raindrop, was developed through his treatment of the damping of a pendulum. Not only does his law provide a basis for the simplest approximation for damping of a macroscopic oscillator, it was used by Robert Millikan to determine the charge of the electron. It should be noted that harmonic oscillation in a fluid (even at low Reynolds number) is much more complicated than steady-flow viscous friction. This topic is treated in Chapter 3, Section 3.9.

The first individual to use the term "simple harmonic oscillator" was probably Lord Kelvin. Such an oscillator is a key tool of experimental physics and also the foundation for much of theoretical physics. It is the basis for communication via electromagnetic waves and even esoteric theories of superfluids and superconductors.

Much of the underpinnings of theory involving harmonic oscillation derive from the work of Hendrik Anton Lorentz (1853–1928). Lorentz is well known for a variety of classical physics contributions, such as (i) the transformation of special relativity associated with Einstein and (ii) the force law for the acceleration of charged particles, both of which bear his name. Before the existence of electrons was proved, Lorentz proposed that light waves were due to oscillations of an electric charge in the atom. For his development of a mathematical theory of the electron, he received the Nobel Prize in 1902. The importance of his contributions is further realized by noting that it is common practice to describe the lineshape of atomic spectra by the term Lorentzian. The Lorentzian is equivalent to the resonance response of the driven viscous-damped simple harmonic oscillator.

It is easy to show how resistance in an electric network is responsible for damping; however, it is a challenge to understand anelastic processes of mechanical damping in terms of viscosity. From comments of his Ph.D. dissertation, it has been said that even Lorentz was never apparently satisfied with the velocity damping term in his equation — not knowing just how to relate it to the underlying physics. It is also clear from Stokes' paper that he recognized the need for caution in the use of his law of viscous friction. It appears that both Lorentz and Stokes were very careful compared with the carelessness with which the viscous model has been employed by many individuals in recent years.

The failure of solids influenced by "hysteretic" damping to be adequately described by the methods of viscoelasticity is not widely appreciated. It is unfortunate that too few people have expanded their view of damping to include other important types, such as derive from the anelasticity of solids. It is important in this work to recognize some subtle differences, for example, inelastic (not elastic) is not to be equated with anelastic (other than elastic).

2.3.3 Active vs. Passive Damping

With improvements in cost/performance of electronics, active damping is increasingly popular. Using force-feedback with integration/differentiation circuitry (opamps), a mechanical oscillator can sometimes be tailored for a specific purpose. A sophisticated example of this technology is the broadband seismometer that began to replace earlier version (passive) instruments roughly 35 years ago. The Sprengnether–LaCoste spring instrument that was used for some of the experiments reported in this document has been superceded by force-feedback units such as the Streckeisen STS-1 and STS-2.

In lieu of feedback, another way actively to influence the damping of a mechanical oscillator is to connect the sensor to an amplifier having a negative input resistance. The seismometers marketed by Lennartz Electronics in Germany use this in a patented technique to improve the performance of ordinary, off-the-shelf electrodynamic geophones.

Active damping depends on the nature of the transfer function of the composite system (electronics plus mechanical). The characteristics of the transfer function are determined by the location of its poles and zeros in the complex plane. Seismometers operate nominally near 0.707 of critical damping. This is done for two reasons: (i) the instrument is easier to adjust and (ii) the interpretation of earthquake records is simpler. Of course, to increase damping is to decrease sensitivity because of the fluctuation–dissipation theorem.

The force-feedback technique is not practical for some situations, regardless of cost. Additionally, it must be recognized that the method is not the answer to all problems, since electronics cannot compensate for a poor mechanical design. The description of commercial products is in some cases highly exaggerated, giving the impression that almost any sensor can perform flawlessly in this manner. Some accelerometers have employed dithering to offset the effects of "stiction" in bearings. The dithering was necessary because the potential energy function is not truly harmonic, being afflicted with the consequences of nonlinear damping. Even with sensing schemes that do not use a bearing, the effects of nonlinearity persist. In "Seismic Sensors and their Calibration" (Bormann and Bergmann, 2002), Erhard Wielandt, in talking about transient disturbances in the spring of a seismometer, says the following:

> Most new seismometers produce spontaneous transient disturbances, quasi-miniature earthquakes caused by stress in the mechanical components.

In other words, internal friction from defects at the mesoscale cause behavior that is in some ways similar to ordinary sliding friction, where the static coefficient is greater than the kinetic coefficient. The postulate of Bantel and Newman is consistent with this idea (Bantel and Newman, 2000) when they refer to their observations as being consistent with a "stick–slip" model of internal friction.

It is seen then that one must use a detector that responds faithfully to the signal around which the servo-network functions. The linearity and sensitivity of that sensor are of paramount importance, since the basis for force-feedback design is linear system theory. For some less-challenging cases, the design approach is straightforward, since software packages like MATLAB® (see de Silva, 2006) have built in functions to describe behavior.

2.3.4 Magnetorheological Damping

A recent approach to damping control, that is quite different from the servo-networks mentioned above, is one that uses an magnetorheological (MR) fluid. It takes advantage of the large variation in viscosity of certain compound fluids according to the size of an applied magnetic field. J. David Carlson (Carlson, 2002) describes how an MR sponge damper is activated during the spin cycle of a washing machine to keep it from "walking out of the room." The peak in the Lorentzian (resonance response) of the machine is shown in his article to be substantially lowered by supplying current to the electromagnet of the damper.

2.3.5 Portevin–LeChatelier Effect

Physics, engineering, geoscience, and mathematics have all contributed greatly to a better understanding of damping phenomena; however, there has been little cross-discipline exchange of ideas and lessons learned. Some of the impediments to strong interdisciplinary programs derive from (i) the complexity of damping problems in general and (ii) the tendency for physics and mathematics research to be, on the one hand, less pragmatic and, on the other hand, highly specialized — focused on specific energy dissipation mechanisms. A good example of (i) involves the PLC effect, discovered in 1923. Why physics mostly ignored this early example of "dirty science" by two of their own number is not easily understood, although the birthing of quantum mechanics around this time may have been a factor. Had history turned in a different direction, perhaps we would already be able to explain from first principles the most important, but still barely understood, form of noise known as $1/f$, or flicker, or pink noise. Even though R.B. Johnson (well known for his discovery of white electronics noise in a resistor) was one of the first to see this form of noise, it still is not explained from first principles — although recent discoveries suggest an intimate connection with fractal geometry involving self-similarity. Such geometry is associated with the mesoscale of materials where the grain, rather than the atom, is the basic element of statistical mechanics.

For alloys, the PLC effect appears to be, in some ways, what the Barkhausen effect is to magnetic systems. In the case of ferrous materials, the noise which derives from the mesoscale has long been recognized; however, similar noise of mechanical type has not been seriously studied. This oversight is even more puzzling when one considers the admonition by G. Venkataraman, as recorded in the proceedings of a Fermi conference, for scientists to get involved in what he felt should become an important new field (Venkataraman, 1982).

2.3.6 Noise

Noise is purposely discussed in this chapter (also see the chapters in Section IX of this handbook) because it has been a, largely, missing component of efforts to understand the physics of damping. A feel for the importance of noise to damping research is to be gleaned from a comment by Kip Thorne in his foreword to the English translation of a book by V.B. Braginsky et al. (1985). Mainly because of instrumental needs of the Laser Interferometer Gravitational Wave Observatories (LIGO), Thorne writes,

> The central problem of such experiments is to construct an oscillator that is as perfectly simple harmonic as possible, and the largest obstacle to such construction is the oscillator's dissipation. If dissipation were perfectly smooth, it would not be much of an obstacle, but the fluctuation–dissipation theorem of statistical mechanics guarantees that any dissipation is accompanied by fluctuating forces. The stronger the dissipation, the larger the fluctuating forces, and the more seriously they mask the signals that the experimenter seeks to detect.

This comment by Thorne suggests a frequently important impediment to dialogue between engineering and physics — concern for different issues. LIGO is trying to minimize damping, whereas many engineering problems are concerned with just the opposite — making the damping as large as possible without compromising strength. More detailed discussions of noise are provided later.

2.3.7 Viscoelasticity

Within the world of polymers, damping is frequently described by the expression "viscoelasticity." This word, around which handbooks have been written (e.g., Lakes, 1998), is a combination of the two words, viscous and elastic. We like to think of ideal fluids as being viscous in the manner described by Newton. Likewise, ideal solids that obey Hooke's Law (stress proportional to strain) are described as elastic. Unfortunately, nature contains neither ideal solids nor ideal fluids. Real springs do not obey Hooke's Law, but rather are influenced by "anelasticity" (other than elastic) which gives rise to hysteresis in the stress–strain relationship. Real fluids usually have some (if not near total) degree of non-Newtonian character. Thus an envisioned "mixing" of fluid-like and solid-like character has dominated the thinking of those who, through the decades, attempted to develop theoretical models of damping.

It should be noted that the springs and dashpots used in models of viscoelasticity do not actually exist. They serve as a phenomenological means for (hopefully) understanding the elementary processes which their arrangement is designed to mimic. Consider, for example, high polymers, in which the interwoven structure of the long-chain molecules is one of extensive mechanical interference. (One popular visualization is that of an entanglement of a huge number of long, writhing snakes.) An increase of temperature is met with overall length reduction (negative temperature coefficient of expansion for the so-called entropy spring), which stands in stark contrast with metals. Such behavior is clearly important to damping since, as noted by Gross years ago, "...thermal movement interferes with the orientation and disorientation of the molecules and ultimately causes delay in the expansion and contraction of the specimen" (Gross, 1952).

2.3.8 Memory Effects

In this same article, Gross is one of the first to mention "memory" properties of creep. He describes a by-then old demonstration in which a "firmly suspended metal or plastic wire is twisted first in one direction for a long time and then in the other direction for a short time. Immediately after release, the deflection will be in the direction of the last twisting, but it decreases rapidly. Presently, a reversal occurs, and the wire begins to turn in the other direction, corresponding to the first twisting — the memory of the recent short-term handling has been obliterated by that of the more remote but longer lasting and therefore more impressive one!" Perhaps this old demonstration (sometimes today called the anelastic after-effect) is not so startling to those familiar with more modern shape-memory-alloys, which are expected by many to play increasingly important roles in the applied science of damping.

2.3.9 Early History of Viscoelasticity

Those who provided seminal influence in the development of the theories of viscoelasticity during the 19th century were some of the most famous names in physics, like Maxwell and Kelvin. Maxwell is best known for the electromagnetic equations associated with his name. He is far less known for two other significant contributions: (i) kinetic theory of gases and (ii) viscoelasticity — both of which are important to theories of oscillator damping. Maxwell's interest in the problem of viscoelasticity is first documented in a paper during his teen years, titled "The Equilibrium of Elastic Solids." Through his development with Boltzmann of the kinetic theory of gases, Maxwell showed a counterintuitive property of the viscosity of a gas. The viscosity does not decrease significantly as the pressure is reduced, until the mean free path between collisions of the molecules begins to approach dimensions of the chamber holding the gas. Important even to modern innovations such as MEMS oscillators, his surprising prediction was quickly verified by experiment. Maxwell's model of viscoelasticity combines a purely elastic spring with a purely viscous dashpot (fluid damper in which the friction force is proportional to the velocity).

Kelvin, probably the first to include a viscous damping term in the equation of motion of the simple harmonic oscillator, developed a similar model of viscoelasticity. Each of the two models is usually represented in literature (without original references) as containing a single spring and a single dashpot. They differ in that one connects the pair in series (Maxwell), while the other connects them in parallel (Kelvin–Voigt).

Both the Maxwell model and the Kelvin–Voigt model have been found by engineers to be less useful than the standard linear model (SLM) of anelasticity, largely advanced in the 20th century by Clarence Zener (Zener, 1948). In the three-component Zener model, a spring is connected in series with a parallel combination of spring and dashpot. Curiously, Zener is widely associated with electronics because of the common diode named after him, but fewer people know of his work in anelasticity. No doubt, his understanding of anelasticity helped him to better understand the complex processes at work in his diode.

2.3.10 Creep

The prevailing models of anelasticity appear frequently in the literature, but mostly in relationship to primary creep. Some of the papers exceptional to this rule are those by Berdichevsky (Berdichevsky et al., 1997). Recent work of a more heuristic type has shown that the equations of viscoelasticity are also able to accommodate secondary creep, in which the decay of strain rate with time has disappeared (Peters, 2001a, 2001b).

The importance of creep (and relaxation) physics to damping warrants some discussion. When a sample is subjected to a constant stress, the strain evolves through three phases of creep: (i) primary, (ii) secondary, and (iii) tertiary. An example of the first two of these phases is shown in Figure 2.2.

In the primary stage, the sample is deformed by anelastic processes involving defects of the crystalline structure. Influence of the disordering mechanisms is progressively reduced as the sample undergoes work hardening (such as pinning of dislocations). Work hardening would result in a purely exponential creep, in the absence of thermal effects which strive to undo the hardening (via diffusion processes). (At zero Kelvin, the creep would eventually cease, if described by a single time constant.) In the secondary stage, a balance between work hardening and thermal softening is attained, in which the strain vs. time has converted from exponential to linear. This balance cannot continue forever, if the stress is larger than

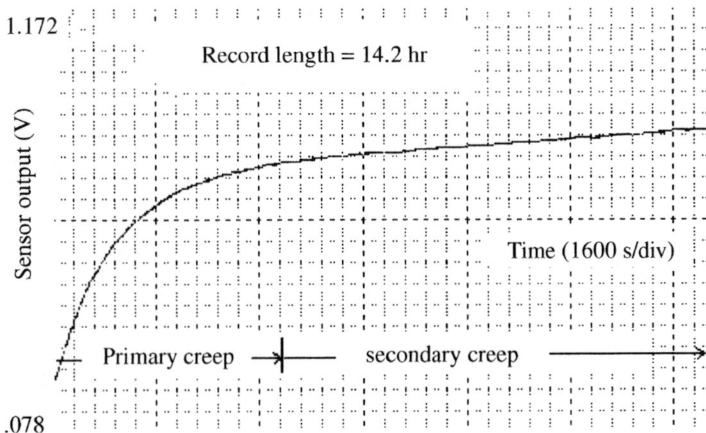

FIGURE 2.2 Example of creep in the spring of a vertical seismometer.

a threshold associated with failure (the elastic limit), and thus a final complex fracture of the sample finally occurs as the sample passes through the tertiary stage. Although one might want to divorce the issues of tertiary creep from considerations of damping, there is clearly a link between damping and failure, involving defects. We will return to this point later.

In Figure 2.2, creep resulted from the instrument having been severely disturbed (relocated accidentally by plumbers working in the building). As with any long-period mechanical oscillator, it is necessary for this instrument to stabilize after a major rebalancing. Primary creep is seen to have endured for about 5 h and what is labeled secondary creep in the figure does not continue indefinitely with the implied constant rate. Thus, the instrument will typically stabilize after one or several days, when the period of oscillation is of the order of 20 sec.

The total amount of creep in Figure 2.2 deserves mention. In the indicated 14.2 h, the mass of the seismometer moved a vertical distance of only 0.25 mm, which can be ascertained from the ordinate axis using the sensor calibration constant of 2000 V/m.

2.3.11 Stretched Exponentials

Systems typically demonstrate a more complex behavior than can be simply described by the SLM of viscoelasticity. In 1847, Kohlrausch (1847) discovered that the decay of the residual charge on a glass Leyden jar followed a *stretched exponential* law. The functional form that he discovered is often associated with a broad distribution of relaxation times, and has been found to describe a remarkably wide range of physical processes. To describe damping that is of the stretched exponential type, Kelvin chains or Maxwell elements in parallel have been used. Although an improved fit to the data can be realized by this means, the technique results in a high number of material parameters which have to be identified.

2.3.12 Fractional Calculus

A promising alternative to multiexponential decay models is to replace classical rheological dashpots by "fractional" elements. It is claimed that with only a few parameters, material behavior of many viscoelastic media can be described over large ranges of time and/or frequency (Hilfer, 2000). It may also be possible with fractional derivatives to treat the discontinuities that are sometimes present in decay (Asa, 1996). The disadvantages of fractional calculus are (i) the increasing computational/storage requirements and (ii) the esoteric mathematics, which is alien to the training of most.

2.3.13 Modified Coulomb Damping Model

Published here formally for the first time, with details described later, the heuristic "modified Coulomb" model is an alternative to all of the aforementioned damping models. It is thought to be closely related to secondary creep (Peters, 2001a, 2001b) and (like fractional calculus) accomplishes good fits with a small number of parameters. Developed from energy considerations, its equations are expressed in canonical form involving the quality factor Q.

2.3.14 Relaxation

Formally, relaxation is defined by the behavior of a sample subjected to a constant strain. Because of the mechanisms just discussed in relationship to creep, the stress relaxes exponentially toward zero (in the simplest approximation). In practice, the definition just given can be misleading since the word relaxation is used to describe a host of processes in which some quantity decays exponentially in time — for example, the relaxation of strain at constant stress in the Kelvin–Voigt model of viscoelasticity.

Some of the viscoelastic models using dashpots and springs have been quite successful in the limited regime of their applicability. For example, the Zener (Debye) model, which will be discussed again later, has been used for years to describe a particular form of damping in solids, which derives from relaxations associated with dislocations. Seminal experimental work of this type was conducted by Berry and Nowick in the 1950s (Berry and Norwich, 1958). A well-known theoretical model to describe dislocation damping was developed by Granato and Lucke (Granato and Lucke, 1956). The Granato model is that of a vibrating string (bowed Frank–Read source), where the end points of the "string" are points on the dislocation line that have been pinned. Recent theory shows that the Granato model is not always adequate; that "dislocation interactions may alter substantially the dislocation component of the spectrum observed during internal friction experiments." (Greaney et al., 2002) (excellent introductory material on this subject is to be found online at http://mid-ohio.mse.berkeley.edu/alex/rachel/rachel/rachel.html).

Bordoni (1954) performed experiments that led to his observation of relaxation-type internal friction processes where the acoustic attenuation is seen to peak at certain temperatures. The so-called Bordoni peaks occur at low temperatures or at ultrasonic frequencies. These losses, which are maximum when dislocation relaxations can take place in step with the driving frequency, were first observed in the FCC metals: lead, copper, aluminum, and silver.

Dislocation damping as just described is characterized by a temperature-dependent relaxation that exhibits Arrhenius behavior. By plotting the internal friction vs. reciprocal temperature, one may estimate the activation energy of the process responsible for the damping. The following quotation from the introduction of the Berry paper assists in defining some of the many expressions used historically to describe damping:

> Internal friction is often loosely described as the ability of a solid to damp out vibrations. More strictly, it is a measure of the vibrational energy dissipated by the operation of specific mechanisms within the solid. Internal friction arises even at the smallest stress levels if Hooke's Law does not properly describe the static stress–strain curve of the material. The nonelastic behavior which Zener has called anelasticity arises when the strain in the material is dependent on variables other than stress.

In a recent private communication, Prof. Granato has indicated the following:

> Dislocations do follow the Zener (or Debye) form fairly well for the damping, but not for the elastic modulus. This is because the response to a stress is given as a Fourier series. The higher order terms in the series have little effect on the damping, but a strong effect on the modulus at high frequencies. This makes the modulus fall off more slowly than with the reciprocal frequency.

2.4 Hysteresis — More Details

Hysteresis and creep are common to many systems, such as electromechanical actuators, especially when used at high drive levels. Their transfer function is influenced by "rate-independent memory effects." The state of the actuator depends not only on the present value of the input signal but also on the nature of their past amplitudes, especially the extremum values, but not on rates of the past (Visintin, 1996). This statement is in support of the author's secondary creep model of hysteretic damping, where the amplitude of the previous turning point determines the magnitude of the internal friction force for the half-cycle that follows. One of the most dramatic examples of a memory effect is the demonstration mentioned above, by Gross in the 1950s, concerning a twisted wire.

Damping complexities derive from the defect structures that are found in real materials and which give rise to hysteresis, which in the Greek language means to "come late." Although, almost everybody seems to appreciate magnetic hysteresis at some level, too few individuals (at least in physics) have been trained in the mechanisms of mechanical hysteresis responsible for damping. Dislocations, for example, are usually an add-on chapter to a solid-state physics text — even though they are known to be indispensable with regard to actual, as opposed to idealized, properties of materials.

In the case of ferrous materials, the magnetization of a specimen lags behind the field generated by an electric current, to which the specimen responds. In the case of real springs that do not obey Hooke's Law $F = -kx$, the displacement x lags behind the spring's restoring force F. It is convenient to express the resulting hysteresis in terms of "intrinsic" variables instead of x and F. Thus, the strain ε (fractional change in the spring's length if it were a wire in tension) lags the stress σ (force per unit area). Usually in engineering practice, the stress is reckoned with respect to the external force (negative of the spring F), so that the equivalent to Hooke's Law is $\sigma = E\varepsilon$, where E is an elastic modulus descriptive of the material from which the spring is fabricated. In the case of a straight wire, E would be Young's modulus but, for coil springs, E is determined primarily by the shear modulus. Some of the ways in which hysteresis can be represented for a freely decaying oscillator are shown in Figure 2.3. The generalized coordinate q would be spring elongation for the force case shown, or it would be strain when the ordinate quantity is stress. The graph of velocity vs. displacement is referred to as a phase-space plot. It is commonly used in describing chaotic systems and, if "strobed" at the frequency of the oscillator, becomes the Poincaré section. Notice that the circulation is of opposite sign when using external force as opposed to spring force, in addition to the curves occupying different quadrants. It is important to recognize this difference, particularly when discussing negative damping where the oscillation amplitude builds in time, as illustrated in the right hand part of the figure.

Although not very common in mechanical oscillators, it is possible to realize negative damping. One example is that of an optically driven pendulum, because of the LiF crystals that were placed in its support structure (containing a high density of color centers produced by radiation) (Coy and Molnar, 1997). An interesting feature of this pendulum was its unwillingness to entrain to the driving laser.

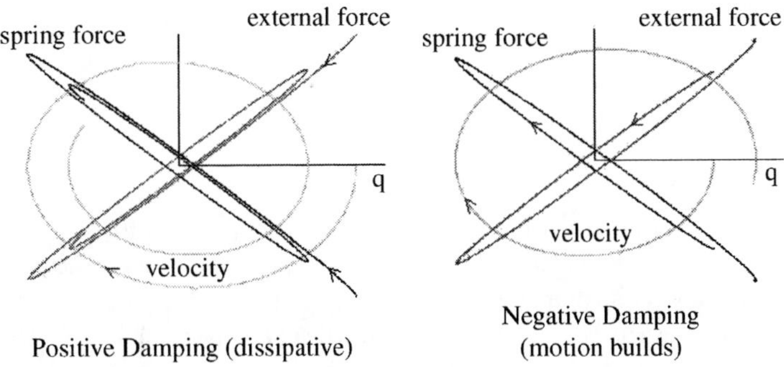

FIGURE 2.3 Three different ways to represent hysteresis damping for an oscillator in free-decay. Cases of both positive and negative damping are illustrated.

There are also examples of negative damping from aerodynamics, such as flutter. Since buildings and bridges can experience negative damping in catastrophic manner (Tacoma Narrows bridge as an example), it is not a subject to be ignored.

Another example of hysteresis that is very much like negative damping (though not usually labeled as such) is to be found in a heat engine (Peters, 2001a, 2001b). The motion is not simple harmonic; rather, the speed with which the hysteresis curve is traversed (in pressure vs. volume) increases as the size of the hysteresis loop increases. A larger loop (greater work done by the gas) results in higher revolutions per minute (r/min) of the engine as opposed to a larger amplitude of the motion at constant period. The gas pressure provides a force similar to the Hooke's Law force of the spring in a mass/spring oscillator.

It is usually assumed that hysteresis loops are "smooth," which is not necessarily true. For example, in the case of magnetic hysteresis, the "jerky" parts known as the Barkhausen effect (Barkhausen, 1919) are well known. The equivalent jerky behavior in metallic alloys is known as the Portevin–LeChatelier effect (Portevin and LeChatelier, 1923). Although we have historically avoided these cases that appear to be intractable in a mathematics sense (not obeying the fundamental theorem of calculus), their presence is undeniable testimony of the complex nature of hysteresis.

2.5 Damping Models

2.5.1 Viscous Damped Harmonic Oscillator

As first seen by students in a textbook, the equation of motion for a damped, driven harmonic oscillator is likely as follows:

$$m\ddot{x} + c\dot{x} + kx = F(t) \tag{2.1}$$

where m is mass, k is spring constant, c is a "constant" of viscous damping, and $F(t)$ is the external force driving the oscillator. It is convenient to work with a coefficient of performance, or quality factor Q, and rewrite Equation 2.1 in canonical form as

$$\ddot{x} + \frac{\omega_0}{Q}\dot{x} + \omega_0^2 x = \omega_0^2 \frac{F(t)}{k}, \quad \text{with } \omega_0^2 = \frac{k}{m} \tag{2.2}$$

For $F(t) = 0$ and an assumed solution, $x = A\exp(p\theta)$ with $\theta = \omega_0 t$, the differential equation becomes algebraic (quadratic) in $x(\theta)$, with the roots given by

$$p = -\frac{1}{2Q} \pm \sqrt{\frac{1}{4Q^2} - 1} \tag{2.3}$$

Depending on the value of Q, the motion is either overdamped (nonoscillatory), critically damped, or underdamped. Here, we restrict our attention to the last case corresponding to $Q > 1/2$, in which the square root term of Equation 2.3 is imaginary. Moreover, we are mostly concerned with systems in which $Q >> 1$.

2.5.2 Definition of *Q*

The quality factor Q is in general defined as 2π times the ratio of the energy of the oscillator to the energy lost to friction per cycle. For viscous damping (and hysteretic damping, later discussed), the Q is independent of the amplitude of oscillation. For other types of damping, we will see that the Q is not constant. In the case of the viscous damped oscillator, $Q = \omega_0/2\beta$ where β appears in the solution as an amplitude decay "constant." The parameter β is not really constant, as discussed in Chapter 3, Section 3.9.

$$x = A\,e^{-\beta t} e^{\pm j\omega_0 t\sqrt{1-1/4Q^2}} \tag{2.4}$$

Since x is real, we use the real part of Equation 2.4 and employ Euler's identity to obtain

$$x(t) = A\,e^{-\beta t}\cos(\omega_1 t - \phi), \text{ with } \omega_1 = \sqrt{\omega_0^2 - \beta^2} \tag{2.5}$$

where ϕ is a constant determined by the initial conditions.

2.5.3 Damping "Redshift"

It is seen that the frequency of oscillation depends on the damping constant, β; however, the fractional change $\Delta\omega/\omega_0$ is almost always negligibly small. For example, the reduction in frequency is only 1.4% for $Q = 3$, which is close to critical damping of $Q = 0.5$. At these small values of Q, the lifetime of a freely decaying oscillator is so short that the frequency is ill-defined because of the Heisenberg uncertainty principle. At larger Qs, where the frequency is well-defined, the shift is negligible; i.e., at $Q = 100$, the fractional shift is only 1.3×10^{-5}. In the case of internal friction (hysteretic) damping, there is no redshift anyway because the oscillator is isochronous.

2.5.4 Driven System

When $F(t)$ is not zero, but rather corresponds to harmonic drive at angular frequency ω and amplitude A, the response involves the sum of Equation 2.5 (transient) and a particular solution (steady state).

$$x_p(t) = A_p \cos(\omega t - \delta), \text{ with } \delta = \tan^{-1}\left(\frac{2\omega\beta}{\omega_0^2 - \omega^2}\right). \tag{2.6}$$

The system resonates (amplitude a maximum) at $\omega \rightarrow \omega_R = \sqrt{\omega_0^2 - 2\beta^2}$, and the variation of the amplitude with ω at steady state at any drive frequency ω is given by

$$A_p = \frac{A\omega^2}{\sqrt{(\omega_0^2 - \omega^2)^2 + 4\omega^2\beta^2}} \tag{2.7}$$

The resonance response curve described by Equation 2.7 is called the Lorentzian. More frequently in physics, the term is used to describe pressure-broadened line widths (Milonni and Eberly, 1988). As noted previously, Lorentz was never apparently content with the damping term, $2\beta\, dx/dt$. In his Ph.D. dissertation concerned with the damping of electron oscillators through electromagnetic radiation, he was not able to satisfactorily describe the damping from first principles. Although we might be tempted to say that this failure derived from his classical (prequantum mechanics) description of the problem, such a viewpoint is an oversimplification.

2.5.5 Damping Capacity

2.5.5.1 Viscous Damping

The loss per cycle, called the damping capacity, is computed for the viscous damping case as follows (per unit mass):

$$d_v = 2\beta \oint \dot{x}\, dx = 2\beta\omega A^2 \int_0^{2\pi} \sin^2\theta\, d\theta = 2\pi\beta\omega A^2 \tag{2.8}$$

where A is the amplitude of the oscillation. Because the total energy per unit mass is $\omega^2 A^2/2$, we see that $Q = \omega/(2\beta)$.

2.5.5.2 Hysteretic Damping, Linear Approximation

The equation of motion in this case is given by $m\ddot{x} + h/\omega\dot{x} + kx = 0$ where h is a constant. The energy loss in one cycle is given by

$$-\Delta E = md_h = \frac{h}{\omega}\int_0^T \dot{x}^2\,\mathrm{d}t = \frac{h}{\omega}\omega A^2 \int_0^{2\pi} \cos^2\theta\,\mathrm{d}\theta = \pi h A^2 \rightarrow d_h = \frac{\pi}{m}hA^2 \tag{2.9}$$

so that $Q = m\omega^2/h$.

2.5.5.3 Hysteretic Damping, Modified Coulomb Model

The nonlinear equation of motion introduced in this chapter to describe hysteretic damping is as follows:

$$\ddot{x} + c\sqrt{\frac{2E}{k}}\mathrm{sgn}(\dot{x}) + \omega^2 x = 0 = \ddot{x} + \frac{\pi\omega}{4Q_h}\sqrt{\omega^2x^2 + \dot{x}^2}\mathrm{sgn}(\dot{x}) + \omega^2 x = \ddot{x} + cA_{\mathrm{prev}}\mathrm{sgn}(\dot{x}) + \omega^2 x \tag{2.10}$$

where, in the last expression, the subscript "prev" implies amplitude at the last (previous) turning point of the motion. This particular form for the damping term (Peters, 2002a, 2002b, 2002c), thought to result from secondary as opposed to primary creep (Peters, 2001a, 2001b), is not as computationally useful as the middle expression involving the Q. The damping capacity is given by

$$-\Delta E = md_h = 4cmA\int_0^{\pi/2} A\cos\theta\,\mathrm{d}\theta \rightarrow d_h = 4cA^2 \tag{2.11}$$

yielding $Q = \pi\omega^2/(4c)$, so that the constant in the nonlinear model is related to the linear approximation constant through

$$c = \pi h/(4m)$$

2.5.6 Coulomb Damping

One of the simplest friction models is that in which a Hooke's Law spring is connected on one end to a mass that slides on a level table. The other end of the spring is connected to a stationary wall. The friction force of the mass against the table is of the type first described quantitatively by Charles Augustin Coulomb (1736–1806), although Leonardo da Vinci is probably the first to consider it scientifically. The equation of motion and its solution, for the free-decay of an oscillator damped by Coulomb friction, is given by

$$\begin{aligned} &m\ddot{x} + f\,\mathrm{sgn}(\dot{x}) + kx = 0 \\ &\text{Solution} \\ &x(t) = [x_0 - (2n+1)\Delta_x]\cos\omega t + (-1)^n\Delta_x \end{aligned} \tag{2.12}$$

The equation is nonlinear because of the sign of the velocity term, but it is easily integrated numerically; additionally, it is one of the few nonlinear equations for which an analytic solution is known and is given above (for more details, the reader is referred to Peters and Pritchett, 1997). The integer, n, specifies the number of half-cycle turning points from $t = 0$, and Δ_x is the decrement (linear, not logarithmic, having units of m) per half-cycle. There are occasions to use Equation 2.12; for example, problems in civil engineering where relative motion of members (slipping) occurs at a structural joint. The work against friction in one cycle can be obtained from energy considerations and is given by

$$f(4x_0 - 8\Delta_x) = \frac{1}{2}kx_0^2 - \frac{1}{2}k(x_0 - 4\Delta_x)^2 \tag{2.13}$$

which, for small decrement, yields

$$\Delta_x = \frac{f}{k} = \frac{f}{m\omega^2} \tag{2.14}$$

Damping characteristics for the models presently treated are summarized in Box 2.1.

Box 2.1

DAMPING CHARACTERISTICS

Type	Equation of Motion	Damping Capacity	Q
Viscous	$\ddot{x} + 2\beta\dot{x} + \omega_0^2 x = 0$	$2\pi\beta\omega A^2 m$	$\dfrac{\omega}{2\beta}$
Hysteretic (linear approximation)	$\ddot{x} + \dfrac{h}{m\omega}\dot{x} + \omega^2 x = 0$	$\pi h A^2$	$\dfrac{m\omega^2}{h}$
Hysteretic (modified Coulomb)	$\ddot{x} + c_h A \operatorname{sgn}(\dot{x}) + \omega^2 x = 0$	$4c_h A^2 m$	$\dfrac{\pi\omega^2}{4c_h}$
Coulomb	$\ddot{x} + \dfrac{f}{m}\operatorname{sgn}(\dot{x}) + \omega^2 x = 0$	$4fA$	$\dfrac{\pi m\omega^2 A}{4f}$
Amplitude dependent	$\ddot{x} + c_f A^2 \operatorname{sgn}(\dot{x}) + \omega^2 x = 0$	$4c_f A^3 m$	$\dfrac{\pi\omega^2}{4c_f A}$

2.5.7 Thermoelastic Damping

A microphone with Labview was used to analyze vibratory data of an aluminum rod. A rod of 1 m length can be excited to ear-piercing intensities by holding it at its center between thumb and finger of one hand, and stroking along the length with the other hand that is coated with violin-bow rosin. The decay of this "singing rod", which is a common part of physics demonstration equipment, was found to be in agreement with the following theoretical expression for thermoeleastic damping (Landau and Lifshitz, 1965):

$$\frac{1}{Q_{\text{Th.d}}} = \frac{\kappa T \alpha^2 \rho \omega}{9C^2} \tag{2.15}$$

where ω is the vibrational angular frequency, T is the temperature, ρ is the density of the bar, C is the heat capacity per unit volume, α is its thermal expansion coefficient, and κ is the thermal conductivity. The expression assumes adiabatic vibrations and there is no thermoelastic dissipation in pure shear oscillations (e.g., torsional oscillations of a bar) because the volume does not change and hence there is no local oscillation of the temperature. Notice, in particular, that the Q is inversely proportional to frequency, unlike viscous damping that is proportional to the frequency, or hysteretic damping that is proportional to the square of the frequency. Thermoelastic damping is important for high-frequency compressional oscillations in materials with significant thermal coefficients, and especially for metals because of their large thermal conductivity.

The demonstration of comparable behavior in polymers (entropy spring, but opposite sign compared with metals) is quite easy. Stretch a rubber band between the hands and immediately touch it to the forehead. The increase in temperature is easily sensed. Conversely, releasing the tension in the band cools it enough to be sensed by placing the band to a part of the face that is sensitive to temperature change. Equation 2.15 does not apply to polymers.

2.6 Measurements of Damping

2.6.1 Sensor Considerations

The challenge to any measurement is to accomplish the task without significantly altering the system under study (see de Silva, 2006). For measurements on mechanical oscillators of the type described in this document, two types of sensor are generally superior to every other kind: (i) optical and (ii) capacitive. Optical sensors are probably the least perturbative but they do not readily yield themselves to large dynamic range with good linearity (and small quantization errors for digital type). Inductive sensors, such as the linear variable differential transformer (LVDT), are known from seismology to be inherently more noisy (up to 100 times) because of ferromagnetic granularity. Additionally, transformers are not amenable to miniaturization, and the components are inherently less stable. It is therefore a mystery why the widespread use of the fully differential inductive sensor (LVDT) continues when we have available the superior fully differential capacitive sensor, which is electrically equivalent (apart from its reactance type) and capable of miniaturization to the MEMS level. The challenge with really small capacitive sensors is the increase in output reactance of the device as they approach femtoFarad levels of individual capacitors.

All measurements reported in this document were taken with the fully differential unit whose patent name is "symmetric differential capacitive" (see Peters, 1993a, 1993b). It is especially useful for studying mechanical oscillators of macroscopic size and, morphed to various forms, it recently has found application in MEMS. It is capable of great sensitivity when configured in the form of an array, as shown in Figure 2.4.

Various lines in Figure 2.4 correspond to narrow insulator strips, such as the single vertical line in the set that connects to the amplifier. In the cross-connected static set, the plates labeled "1" are electrically distinct from the others labeled "2". The total-plate arrangement constitutes a symmetric AC bridge, and the central position of the moving set ($x = 0$ as shown in the figure) corresponds to bridge balance with $V_0 = 0$. Displacement away from balance gives a voltage output that is linear between $-w/4$ and $w/4$, as illustrated in the graph at the bottom of Figure 2.4.

The oscillator frequency is typically tens of kHz, and the amplifier is of instrumentation type (Horowitz and Hill, 1989). Unlike a bridge null detector, the linear response through $x = 0$ is realized

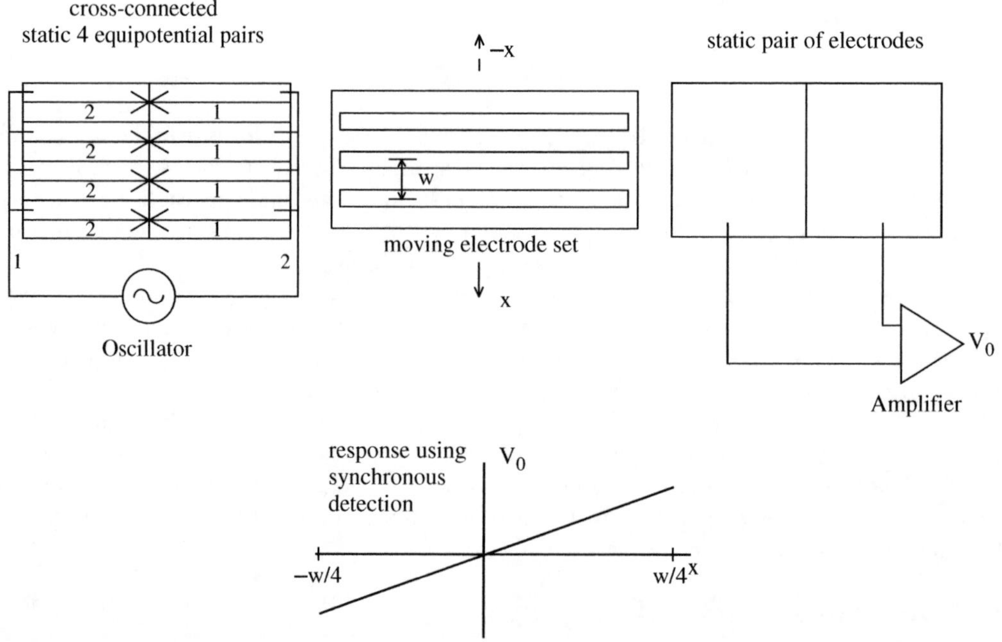

FIGURE 2.4 Illustration of a fully differential capacitive transducer array. For clarity, the three electrode-sets are shown separated from their operating positions (parallel with a small separation gap, with the moving electrodes in the middle).

when synchronous detection is employed. This can be accomplished with a lock-in amplifier, but the most recent Cavendish balance to employ the sensor uses diodes (Tel-Atomic Inc., online at http://www.telatomic.com/sdct1.html).

A tutorial ("detailed explanation") of the SDC sensor using diodes can also be found at this website.

In Figure 2.4, four individual SDC units have been shown connected in parallel. The total number, N, of individual units in an array depends on the characteristic width, w, for which the total range of detectable motion is $w/2$. If the requirement on range is small, then N can in principle be made very large, which is desirable for the following reason. The sensitivity of this position sensor is inversely proportional to w if output capacitance of the device is not a factor. As w is reduced, however, the degrading influence of increased output reactance (capacitive) is more significant than the improved sensitivity that would result if the sensor could be connected to an amplifier with infinite input impedance. Since the instrumentation amplifier's input capacitance is not negligible, shrinking w is beneficial only if the output reactance can by some means be kept low. This is accomplished with the array of individual units. In principle, the output capacitance could be held reasonably constant as N approaches 100, by using photolithographic techniques and small spacing between the parallel electrodes. The concept has been deemed feasible because of existing technologies as well as the following: although not in the form of an array, Auburn University has fabricated a mesoscale accelerometer around the SDC sensor. The prototype was built on printed wire board (PWB) under U.S. Army contract (Dean, 2002).

No doubt the popular silicon-based MEMS accelerometers marketed by Analog Devices utilize the impedance advantages of an array, employing a large number of "fingers" in a force-feedback arrangement. Although employed mostly otherwise, the first case of a fully differential capacitive transducer using force-feedback was one based on simultaneous action of actuator and sensor functions in a single unit of nonlinear type (Peters et al., 1991).

2.6.2 Common-Mode Rejection

In attempts to measure damping, one can be confronted with difficulties of mode mixing. For example, the historical Cavendish experiment, using optical detection, has been traditionally difficult unless the instrument is placed in a very quiet location to avoid pendulous swinging of the boom. The high-frequency pendulous motion (of the order of 1 Hz) as a "noise" becomes superposed on the low-frequency torsional signal. The computerized Cavendish balance sold by Tel-Atomic overcomes this problem by means of a mechanical common-mode rejection feature. An SDC sensor placed near one boom end is connected in electrical phase opposition to a second SDC sensor placed near the other end of the boom. The boom itself serves as the moving electrode for both sensors. Neither sensor has a first-order response to boom motion parallel to its long axis. Pendulous motion perpendicular to the boom orientation is largely canceled.

2.6.3 Example of Viscous Damping

The aforementioned Sprengnether–LaCoste spring seismometer is well-suited to the demonstration of viscous damping, when damping is imposed in the following manner: the instrument was built with a Faraday Law (velocity) detector; i.e., a coil that moves with the mass of the instrument, in the field of a stationary magnet. As originally employed, the coil was connected to the amplifier of a recorder. In the present configuration, however, the velocity detector is not employed, since its sensitivity is severely limited at low frequencies. Instead, an SDC array of the type shown in Figure 2.4 is used to measure the position of the mass (a pair of lead weights, total mass 11 kg). If the instrument is operated with the coil open-circuit, there is no induced current. By connecting a resistor across the coil (through very fine copper wires that go to terminals on the case), mass motion induces a current. The induced current opposes the motion through Lenz's Law, resulting in damping. The damping depends on the size of the

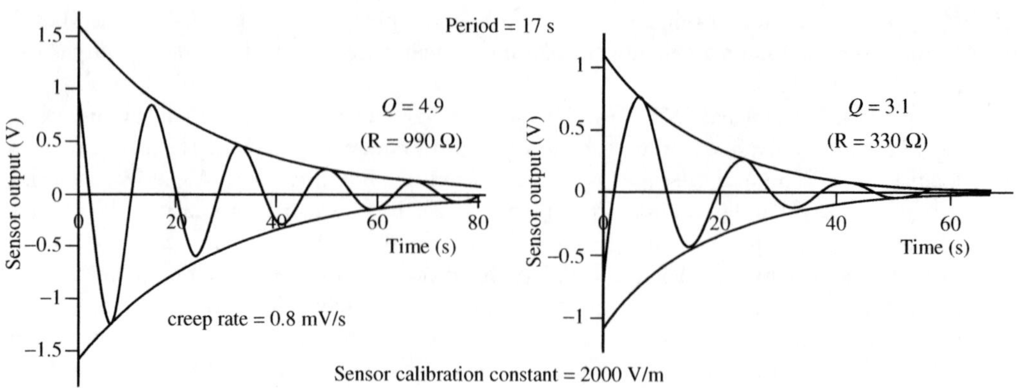

FIGURE 2.5 Examples of induced current damping of a vertical seismometer using two different resistors.

current and is thus an inverse function of the resistor's magnitude through Ohm's Law. The phenomenon is illustrated in Figure 2.5.

As compared with the "undamped" instrument, whose Q is approximately 80 at a period of 17 sec, it is seen that the addition of a 990-ohm resistor lowered the Q by more than an order of magnitude to 4.9. A 330-ohm resistor reduced it even further to 3.1. The amount of damping is also governed by the resistance of the coil winding, which is 480 ohm.

The envelopes that have been fitted to the decay curves were the basis for estimating the Q. The decay data were imported to Excel by first outputting the Dataq DI-154RS A/D generated record as a *.dat (CSV) file. The fits were produced by trial and error using the drag and autofill functions. Notice that the 990-ohm resistor (first) case is not as pure an exponential decay as the other case because of creep. The rate of creep is greater at large initial amplitudes of the motion.

2.6.4 Another Way to Measure Damping

Curve fitting (full nonlinear, in general) is the best way to estimate damping parameters, especially if the decay is not exponential. For more routine cases, simpler methods can be used. Among the host of ways that have been defined to specify the damping of an oscillator, one of the most common uses the logarithmic decrement. The solution to Equation 2.1 with zero right-hand side is given by

$$x(t) = x_0\, \mathrm{e}^{-\beta t}\cos(\omega t + \phi). \tag{2.16}$$

The full-cycle turning points, $x_N = x_0\, \mathrm{e}^{-\beta NT}$, with $N = 0, 1, 2, \ldots$ can be used to compute the logarithmic decrement through

$$\beta T = \frac{1}{N}\ln\frac{x_0}{x_N} \tag{2.17}$$

Unfortunately, an estimate based on Equation 2.17 can be difficult due to the presence of either or both of two problems: (i) mean position offset in the decay record or (ii) asymmetry of the decay, where the turning points on one side of equilibrium decay at a different rate than those on the other side. Case (ii) occurs more often than one might expect; it is frequently a consequence of material complexity and not the result of nonlinearity in the electronics of the detector. It is important, however, to be sure that the detector is either linear or that corrections for the nonlinearity be utilized before estimating the damping.

A method to provide partial compensation uses half-cycle turning points $n = 2N$, and works with a minimum of three such points.

$$\beta T = -2\ln[1 - (x_{n-1} - x_{n+1})/(x_{n-1} - x_n)] \tag{2.18}$$

Advantage is taken of random error reduction by using Equation 2.18 on a set of turning points (optimal number sometimes being about a dozen). The calculations are straightforward in a spreadsheet such as Excel by means of the autofill function.

2.7 Hysteretic Damping

2.7.1 Equivalent Viscous (Linear) Model

The few mechanical oscillators governed by Equation 2.1 tend to be those in which there is an external control, such as eddy current damping. For oscillators in which the damping derives from internal friction of its members, the following linear approximate form of the hysteretic damping model has been used:

$$m\ddot{x} + \frac{h}{\omega}\dot{x} + kx = F \tag{2.19}$$

It should be noted that hysteresis is the cause for all damping; however, engineers have come to use the term "hysteretic damping" for systems described by Equation 2.19. This equation differs in two important ways from Equation 2.1. For the viscous damped oscillator, Q is proportional to the frequency, but for the hysteretic damped oscillator, Q is proportional to the square of the frequency. Also, viscous damping changes the frequency of the oscillator, since $\omega_1 < \omega_0$ and, for resonance, the frequency is even lower. However, the hysteretic oscillator is isochronous, requiring only a single frequency $\omega = \sqrt{k/m} \rightarrow \omega_r$ to describe all features of the motion. For example, it is easy to show that the oscillator resonates at this frequency. Off resonance, the response is not the standard Lorentzian. To show this, assume steady state and use the phasor method given to us by Steinmetz, 1893 (complex exponential form for the variables); i.e., $F = F_0\,e^{j\omega t}$ and $x = x_0\,e^{j\omega t}$ to get the frequency transfer function

$$\frac{kx}{F} = \frac{1}{1 - \omega^2\dfrac{m}{k} + j\dfrac{h}{k}} = \frac{1}{1 - r^2 + j\alpha} = Z, \text{ with } r = \frac{\omega}{\omega_r} \text{ and } \alpha = \frac{h}{k} = \frac{1}{Q} \tag{2.20}$$

for which the real and imaginary parts are given by

$$\text{Re}\,Z = \frac{1 - r^2}{(1 - r^2)^2 + \alpha^2}, \qquad \text{Im}\,Z = \frac{-\alpha}{(1 - r^2)^2 + \alpha^2} \tag{2.21}$$

which is expressible in polar form as

$$Z = |Z|\,e^{j\delta}, \text{ where } |Z| = \frac{1}{\sqrt{(1 - r^2)^2 + \alpha^2}} \text{ and } \delta = -\tan^{-1}\frac{\alpha}{1 - r^2} \tag{2.22}$$

It is interesting to compare the steady-state response of the driven, hysteretic damped oscillator with that of the driven, viscous damped oscillator; i.e., Equation 2.22 compared with normalized Equation 2.7. A Bode plot comparison (log–log, for the amplitude case) is provided in Figure 2.6. At small values of the damping parameter α (large Q), there is insignificant difference between the two cases. At large values, however, the difference is significant.

2.7.2 Examples from Experiment of Hysteretic Damping

The vertical seismometer that was used for several of the present studies is known to decay according to hysteretic damping. In Section 2.16.4 titled "Failure of Viscoelasticity", details are provided of the work by Gunar Streckeisen (1974) that showed this to be true. Decay curves of the instrument are

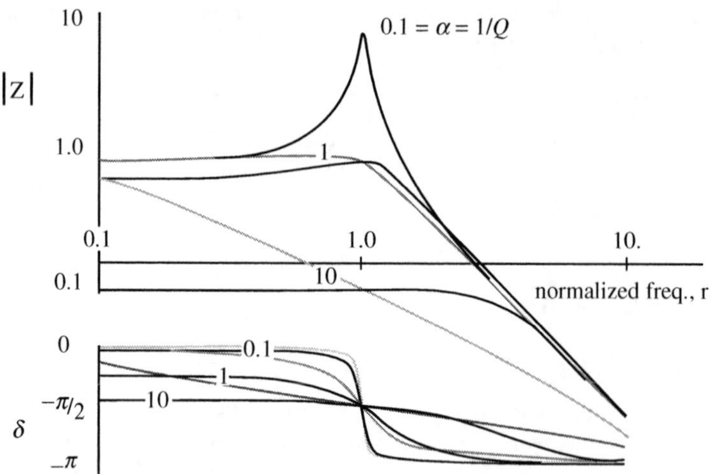

FIGURE 2.6 Bode plot comparison of steady-state driven system with (i) hysteretic damping (dark curves) and (ii) viscous damping (light curves).

frequently a near-perfect exponential, once corrected for secular drift of the record. Sometimes, this drift is the result of creep in the spring of the instrument, but it may also be the result of other factors, such as (i) temperature change, or (ii) barometric pressure variation, or even (iii) tidal influence. The temperature sensitivity is due to the difference of thermal coefficients of the materials from which the instrument is constructed, and the pressure variation is a buoyancy effect. Tidal influence is the smallest of the three, which causes minute accelerations of the crust of the Earth with a period of about 12 h.

In the discussions which follow, two different decay records are provided. In both cases, the initial amplitude of oscillation is quite large, being a significant fraction of 1 mm, and the period for the two cases is different — the first case being 17 sec and the second one 21 sec. The first case time record, shown in Figure 2.7, contains 9800 points. Once a 12 μV/s (upward) drift was removed, the decay (left curve) is seen to be "nearly textbook" exponential.

The adjective "nearly" is appropriate because there is a 12% difference in the decay constants defining the upper and lower turning points (0.0022 top, 0.0025 bottom), which were determined by trial and error "eyeball" exponential fits using Excel. In this author's experience, such is the norm for virtually all mechanical oscillations; perfectly symmetric exponential decays have rarely been seen in the hundreds of cases studied.

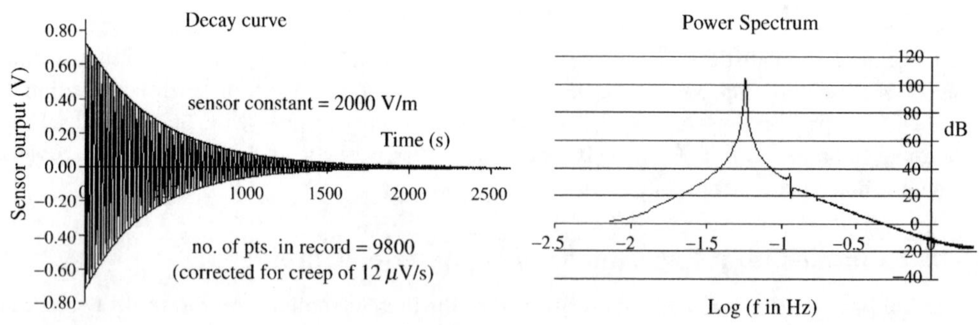

FIGURE 2.7 Free-decay of a vertical seismometer due to hysteretic damping. The period of oscillation is 17 sec.

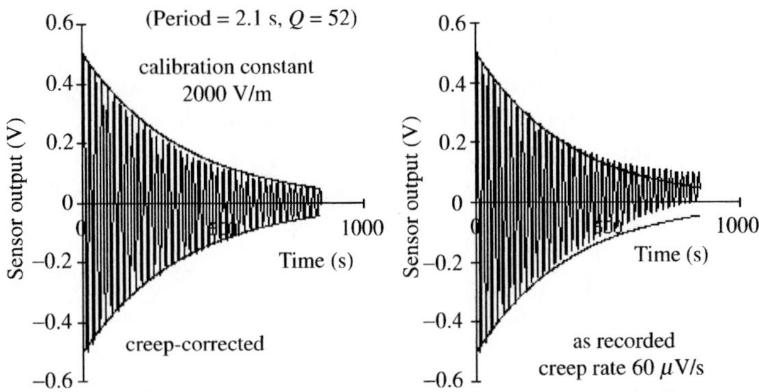

FIGURE 2.8 Free-decay of a seismometer due to hysteretic damping.

Because there are roughly 150 oscillations in the time record of Figure 2.7, it is not possible to resolve individual turning points of the motion, but the oscillations are very nearly that of a pure damped sinusoid, as noted from the right-side graph of the figure. This spectrum was generated with a 4096-point FFT, comprising the first 1090 sec of the time record. The second harmonic is the only distortion observed, and it is about 65 dB below the fundamental. For the case presented in Figure 2.7, $Q = 80$.

Another example of free-decay hysteretic damping is provided in Figure 2.8. As usual, the record was afflicted by drift, possibly from creep in the spring, in this case a constant rate of 60 μV/s, as observed in the graph on the right. All of these graphs were produced with Excel, as noted earlier in the discussion of induced current damping. As with the decay curve of Figure 2.7, the creep-corrected graph on the left was generated by adding a secular term to the raw data. Once corrected, the decay is a near-perfect exponential of hysteretic type. We will see other examples (from pendulum studies) in which two damping mechanisms are simultaneously active in a decay.

The Q values corresponding to Figure 2.7 and Figure 2.8 are consistent with hysteretic damping; i.e., 80 for the 17-sec oscillation and 52 for the 21-sec oscillation. As noted elsewhere in this document, $Q\alpha\omega^2$ for hysteretic damping as opposed to an exponent of 1 for viscous damping. Of course, one must collect data over a very much larger range of frequencies to verify this, as was done by Streckeisen (1974).

2.8 Failure of the Common Theory

Many mechanical oscillator studies in decades past, mainly by engineers, have shown that the so-called decay constant β is proportional to ω^{-1} instead of being constant (e.g., Bert, 1973). The damping for these cases came to be called "material", "structural", or "hysteretic." A common way to obtain the correct frequency dependence was to divide the velocity by frequency and call the result an "equivalent viscous" form of damping. The adjective "equivalent" draws attention to the fact that internal friction in a solid cannot really result from fluid effects. Moreover, elsewhere in this document, there is plenty of support for the position that the linear equations of viscous damping type cannot produce truly meaningful (predictive) models when doing modal analysis on multibody systems.

An important early work by Kimball and Lovell (1927) is evidently the first experiment to show that internal friction ("force") of many solids is virtually independent of frequency. In other words, their elegant technique, in which a rotating rod is deflected by a transverse force, was the first to demonstrate the "universality" of hysteretic damping. Although both researchers were physicists at General Electric in the time of Steinmetz, few physicists of the 21st century know of this important work. As with the important contributions of Portevin and LeChatelier, their study of systems influenced by "dirty physics" was evidently ignored in favor of the "clean" new quantum mechanics of that era.

It is interesting that a bell made of lead does not tinkle at room temperature, but it can be made to do so at 77 K, by immersion in liquid nitrogen. This demonstration, which is often employed in physics "circuses," shows clearly that the internal friction of lead at audio frequencies can be reduced substantially by lowering the temperature. An important lesson to be learned from these observations is that damping, in general, is a complex function of temperature, frequency, conductivity, ...(who knows where to terminate this list). Not only is a multitude of state variables necessary for a complete description of dissipation, but the previous history of stress–strain cycling may also be critical. Such is the nature of defect structures responsible for damping.

2.9 Air Influence

Even when operating an oscillator in high vacuum, there is a significant remanent damping that derives from internal friction. This fact is illustrated in Figure 2.9, which provides data for two different "simple" pendula. They are simple in the sense that the bob mass is concentrated near the bottom of the pendulum structure. In the figure, decay time (reciprocal of the decay constant, β) has been plotted against the natural log of the pressure in mtorr. Pressure reduction was done with a high-quality roughing pump, and the pressure was measured with (i) a mechanical gauge in the range 8 torr $< P <$ 760 torr and (ii) a thermocouple vacuum gauge for $0 < P <$ 100 mtorr. In the range from 100 mtorr to 8 torr, the pressure could not be accurately measured with either of these gauges. Similarly, pressures below 1 mtorr could not be presently measured, but in similar other experiments with this pump, and using an ion gauge, it was easy to pump below 0.01 mtorr.

The period of each pendulum was very close to 1 sec, and the starting amplitude of the motion for every case was about 25 mrad. The heavier pendulum used a pair of pointed steel supports resting on single-crystalline silicon wafers to provide the axis of rotation. At the bottom of the pendulum was attached a solid lead ball whose mass was approximately 1 kg. The lighter pendulum was supported by a steel knife-edge resting on hard ceramic flats, and a large (10.3 cm dia.) lightweight (143 g) hollow metal sphere was attached at the bottom to provide as much air drag as possible. The motion was measured with an SDC sensor feeding the computer through a Dataq DI-154RS A/D converter.

Although air damping is evident in Figure 2.9, it is not as influential as one might expect, at least for the heavy pendulum. Moreover, at atmospheric pressure, it was easy to demonstrate the importance of nonlinear drag. As also noted in Nelson and Olssen (1986), this form of fluid friction caused a significant amplitude-dependent damping.

The remanent damping, once air influence is eliminated (pressure below 1 mtorr), is substantial relative to atmospheric damping, for both pendula. Removing the air increased the Q from 7500 to

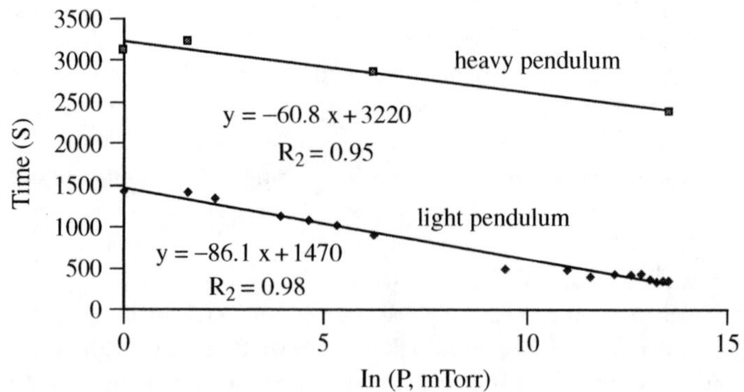

FIGURE 2.9 Pendulum damping as a function of pressure in a vacuum chamber.

10,100 for the heavy pendulum and, for the light pendulum, the increase was from 1000 to 4600. We thus see that even a pendulum designed to be heavily influenced by air drag also has significant damping that depends on the material from which the pendulum is fabricated or on the material upon which it rests.

The difference in internal friction damping between the heavy and light instruments was not expected to be so great. Although this might be due to the difference in axis-type (points for the heavy instrument and knife-edge for the light), no systematic effort was made to determine the primary source of the damping difference. In addition to different axis designs, the means for holding the instruments together was different. The light pendulum used a large-diameter solid brass wire between the axis and the lower mass, and the heavy pendulum used an aluminum tube.

Both of the pendula used to generate Figure 2.9 were relatively high-frequency instruments (period of 1 sec). The pivot was located, in each case, near the top end of the instrument. As such, they stand in stark contrast with the instruments that motivated this paper, where long-period pendula were used. A simple instrument to demonstrate some of the complexities of long-period instruments is a rod-pendulum of adjustable period (refer to Figure 2.1 above). The closer the axis to the center, the longer the period and the greater the influence of internal friction. It is easy to show that the sensitivity of a pendulum to external forces is proportional to the square of the period. Similarly, the ability to detect influence of internal configurational change is quadratic in the period.

2.10 Noise and Damping

2.10.1 General Considerations

Damping is inseparable from noise issues having nothing to do with undesirable sounds that might be produced by oscillation. In the simplest cases, the noise associated with damping can be described by the fluctuation–dissipation theorem. The viscous damped, thermally driven oscillator is a classic example of thermodynamic equilibrium, for which this theorem is applicable. The classic electronics analogous case is the Johnson noise of a resistor, described by the Nyquist (white) noise formula.

The largest obstacle to constructing a perfectly simple harmonic oscillator is the oscillator's dissipation. If damping were perfectly smooth, this would not be so great a challenge. However, the fluctuation–dissipation theorem of statistical mechanics guarantees that damping is accompanied by fluctuating forces. The larger the damping, the larger the fluctuating forces, i.e., the larger the noise. It is a standard problem in statistical mechanics to show that the magnitude of relative fluctuation is inversely proportional to the square root of the number of particles involved. In the case of internal friction noise, defects associated with mesoscale structures cause the effective number of particles responsible for the noise to be much smaller than the total number of atoms in a sample. Unfortunately, the fluctuation–dissipation theorem probably does not apply. It has been long known that it does not apply to the Barkhausen effect (Barkhausen, 1919). It has been recently demonstrated that it does not apply to structural glass (Grigera and Israeloff, 1999). The close relationship postulated by the author between the PLC effect and the Barkhausen effect implies that the fluctuation–dissipation theorem should also not generally apply to internal friction damping.

Internal friction noise is not white but rather more like $1/f$ (or flicker = pink) noise, a ubiquitous form that has not yet been explained from first principles. A frequently cited paper on self-organized criticality states the following:

> We shall see that the dynamics of a critical state has a specific temporal fingerprint, namely "flicker noise," in which the power spectrum $S(f)$ scales as $1/f$ at low frequencies. Flicker noise is characterized by correlations extended over a wide range of timescales, a clear indication of some sort of cooperative effect. Flicker noise has been observed, for example, in the light from quasars, the intensity of sunspots, the current through resistors, the sand flow in an hourglass, the flow of rivers such as the Nile, and even stock exchange price indices. Despite the ubiquity of flicker noise, its origin is not well understood. Indeed, one may say

> that because of its ubiquity, no proposed mechanism to date can lay claim as the single general underlying root of 1/*f* noise. We shall argue that flicker noise is in fact not noise but reflects the intrinsic dynamics of self-organized critical systems. Another signature of criticality is spatial self-similarity. It has been pointed out that nature is full of self-similar "fractal" structures, though the physical reason for this is not understood. (Bak, 1988)

It should be noted that controversy exists concerning this self-organized criticality paper, summarized in the following excerpt from Bak's book on 1/*f* noise called *How Nature Works: The Science of Self-Organized Criticality* (page 95):

> In an earlier work (CFJ), performed while an undergraduate student in Aarhus, Denmark, (Kim Christensen) showed that our analysis of 1/*f* noise in the original sandpile article was not fully correct. Fortunately, we have since been able to recover from that fiasco in a joint project by showing that for a large class of models, 1/*f* noise does indeed emerge in the SOC state.

In the last few years, mathematicians have been drawn to "... an analogy, in which three areas of mathematics and physics, usually regarded as separate, are intimately connected. The analogy is tentative and tantalizing, but nevertheless fruitful. The three areas are eigenvalue asymptotics in wave physics, dynamical chaos, and prime number theory" (Berry and Keating, 1999). Some mathematicians speculate that a dynamical system (perhaps some form of a mesoanelastic pendulum, in the thinking of this author) could become a "machine" to generate prime numbers.

2.10.2 Example of Mechanical 1/*f* Noise

Shown in Figure 2.10 is an example of mechanical flicker noise made worse by creep that originates in the spring (LaCoste type) of a Sprengnether vertical seismometer. The data are from two separate time records, the first run preceding the second run by about a half-hour. Just before collecting the data of the first run, a clamping pin was removed from the seismometer. Used to constrain the mass from moving, this pin had been left in place overnight to determine the amount of electronics noise, including drift. The measured electronics noise (white = $1/f^0$) was more than an order of magnitude smaller than the smallest (high frequency) noise components of mechanical (seismometer) type. The peak-to-peak amplitude of the oscillation in both cases was 0.5 mm (calibration constant for the sensor being 2000 V/m). The peak-to-peak amplitude for SNR = 1 for this system is of the order of 1 μm.

Although the spring force was not unloaded with the pin in place overnight, nevertheless, its removal caused a significant change to defect structures in the spring, as noted by the residuals between the data and their harmonic fits (magnified by a factor of ten).

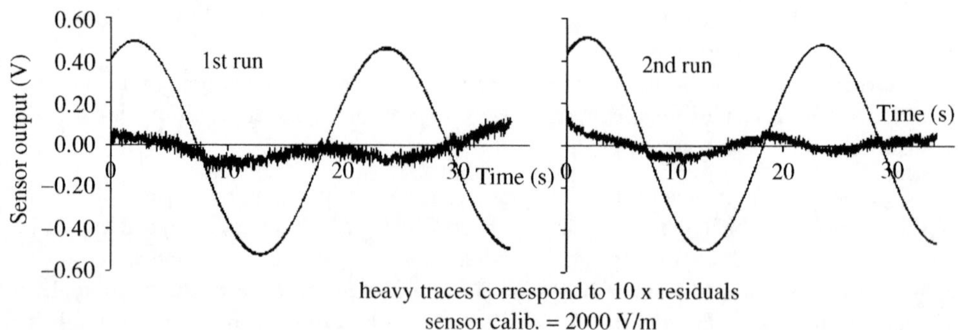

FIGURE 2.10 Evidence in support of 1/*f* mechanical noise in a seismometer.

The flicker character of the noise was demonstrated by computing power spectra of the residuals (not shown). The log–log plot, generated with the FFT (Cooley–Tukey FFT), showed $1/f$ frequency dependence for the second run. The larger noise of the first run was concentrated in the upper frequencies. The relaxation toward $1/f$ character suggests that flicker noise is a remanent of "work hardening." To demonstrate that the flicker noise was not due to the electronics, sensor output was recorded with the mass of the instrument locked. The electronics noise proved to be more than an order of magnitude smaller and "white" in character, probably mainly the result of A/D quantization (see Chapter 16).

Because the spring was not in equilibrium at the time the pin was removed (perhaps because of temperature change while the system was clamped), a great deal of initial molecular rearrangement occurred, involving atoms at grain boundaries. It is seen that the amount of fluctuations has noticeably decreased during the half-hour separating the two runs. Although the creep-noise would be undoubtedly much greater if the spring were relaxed altogether, it is not easy in such a case to quickly rebalance the seismometer to oscillate with a period in excess of 20 sec. These observations are in keeping with known properties of sensitive seismometers, as noted by Erhardt Wieland in "Instrumental self noise — transient disturbances" (ed. Borman and Bergmann, 2002):

> Most new seismometers produce spontaneous transient disturbances, quasi miniature earthquakes caused by stresses in the mechanical components. Although they do not necessarily originate in the spring, their waveform at the output seems to indicate a sudden and permanent (step-like) change in the spring force. Long-period seismic records are sometimes severely degraded by such disturbances. The transients often die out within some months or years; if not, and especially when their frequency increases, corrosion must be suspected. Manufacturers try to mitigate the problem with a low-stress design and by aging the components or the finished seismometer (by extended storage, vibrations, or alternate heating and cooling cycles). It is sometimes possible to relieve internal stresses by hitting the pier around the seismometer with a hammer, a procedure that is recommended in each new installation. (Wielandt, 2001)

Material damping noise appears to have features that are similar to Barkhausen noise — a magnetic phenomenon involving a system far-from-equilibrium. Such noise is associated with the granular nature of ferromagnetic domains and has consequence in the design of electronic instruments using iron alloys. For example, it is known (though not widely) that the popular LVDT is inherently less sensitive than a capacitive sensor of equivalent electrical type, because of its ferrous component (the rod-component that moves). As noted by Wielandt (2001), the capacitive sensor "…can be a hundred times better than that of the inductive type." Fully differential capacitive sensors, being electrically equivalent to the LVDT have still greater advantages borne of the higher symmetry. Additionally, by configuring the capacitive device as an array, it is possible for the sensitivity to also be greater.

Barkhausen noise and hysteretic damping noise may be much more similar than has been realized — involving granularity at the mesoscale, intermediate between micro- and macrophenomena. For such systems, first principle methods are very difficult to employ due to complexities that originate from a host of nonlinear interactions. For example, in the case of internal friction of solids, damping derives from stress–strain hysteresis determined by defect structures in the solid. Involving roughly 10^{12} atoms per "grain" in metal specimens, flicker noise evidently derives from self-similar structures of fractal geometry with a higher degree of spatial correlation than is true of white noise. The ubiquitousness of $1/f$ noise is consistent with the labeling of hysteretic damping as "universal," as first suggested by Kimball and Lovell (1927).

2.10.3 Phase Noise

The previous example was concerned with amplitude noise. It is also possible to see phase noise of mechanical type, as illustrated in Figure 2.11.

For the sensor calibration constant of 2000 V/m, it is seen that the initial amplitude of oscillation is 1.3 mrad, which is much too large to observe the discontinuities of mechanical Barkhausen type.

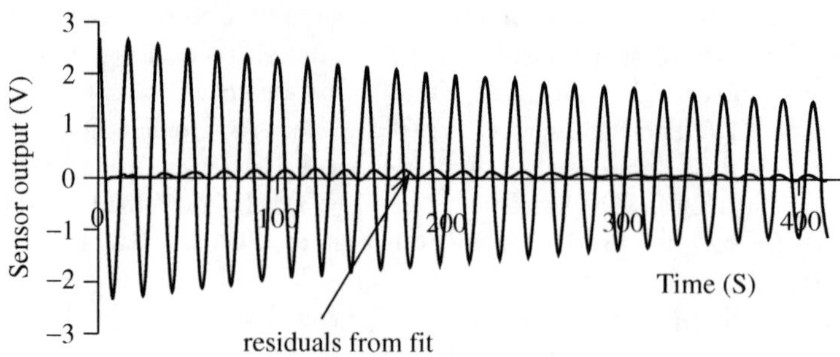

FIGURE 2.11 Illustration of phase noise in the free-decay of a vertical seismometer.

The phase noise is made obvious by comparing the decay to a "reference," i.e., by looking at the residuals from an Excel-generated fit to the data. By adjusting in a computer-generated damped sinusoid, (i) the initial values of amplitude and phase, (ii) the decay parameter, and (iii) the frequency, one can visually by trial and error come close to an optimum fit to the data. Having done so with Figure 2.11, the striking feature of the residuals (difference between data and fit) is the structure that looks something like "beats" but is not. The phase noise responsible for this behavior is thought to be consistent with $1/f$ mechanical noise. To visualize the noise, it is convenient to think in terms of a small randomizing (noise) vector whose tail is positioned at the head of the phasor used to generate the record. The component of the noise vector that is in the direction of the phasor generates amplitude noise as in Figure 2.10, whereas the perpendicular component is responsible for the phase noise of Figure 2.11.

Vibration phase noise imposes a serious limit on the performance of precision quartz crystal oscillators, since they are sensitive to acceleration. The phase noise in these oscillators can be observed by beating against a reference oscillator of known character; i.e., the reference serves the same purpose as the computer "fit" of Figure 2.11. To reduce the phase noise, crystals are isolated with low natural frequency vibration isolators (as described in the marketing literature of Wenzel Assoc., Austin, X).

2.11 Transform Methods

2.11.1 General Considerations

For linear systems, the Laplace and Fourier transforms (Laplace being more general) have been pre-eminent tools with which to study equations of motion (see Appendix 2A and Chapter 10). The author's transform experience (like most physicists) is mainly with Fourier transforms (FT). The discrete FT can be understood in terms of phasors (Peters, 1992). For linear differential equations, transforms are the means to convert differential forms to an equivalent algebraic form. Unfortunately, they cannot be directly employed on nonlinear equations due to the failure of superposition. Nevertheless, the linear approximations continue to be very valuable, so a chapter on damping deserves to mention some of their properties.

Ideas concerning the FFT were evidently originally treated by Gauss in the early 1800s, but the digital signal processing (DSP) "explosion" of the 1960s was largely due to the work of Cooley and Tukey (1965). For an interesting historical account about an "accident" in the publication of their paper, the reader is referred to Cipra (1993), who says the following about the FFT:

> The Fourier transform stands at the center of signal processing, which encompasses everything from satellite communications to medical imaging, from acoustics to spectroscopy. Fourier analysis, in the guise of x-ray crystallography, was essential to Watson and Crick's discovery of the double helix, and it continues to be important for the study of protein and viral structures. The Fourier transform is a fundamental tool, both theoretically and

computationally, in the solution of partial differential equations. As such, it is at the heart of mathematical physics, from Fourier's analytic theory of heat to the most modern treatments of quantum mechanics. Any kind of wave phenomenon — be it seismic, tidal, or electromagnetic — is a candidate for Fourier analysis. Many statistical processes, such as the removal of "noise" from data and computing correlations, are also based on working with Fourier transforms.

Concerning the last statement about noise, this author has used autocorrelation as a powerful means for identifying short-lived, low-frequency periodic signals in time records that do not readily show up in power spectra (FFTs). For example, they are the means for studying free-earth oscillations — eigenmodes excited by rapid relaxations of the Earth under tidal stressing (12 h periodic) (Peters, 2000). The FFT is used to generate the autocorrelation by means of the Wiener–Khintchine theorem (Press et al., 1986).

The great advantage of the FFT compared with the DFT has to do with degeneracy. The DFT proceeds to calculate the components of every "vector" in the reciprocal space (frequency reciprocal to time, units of "second", or wave number (spatial frequency) reciprocal to displacement, units of "meter") with disregard for the fact that many components have the same value, apart from a change of sign.

2.11.2 Bit Reversal

The key to the power of the FFT (central processor unit [CPU] time proportional to $n \log n$) compared with the discrete Fourier transform (DFT) (CPU time proportional to n^2) is the bit reversal scheme of the Cooley–Tukey algorithm. It is illustrated very simply by the following. Instead of a practically sized number of samples in the record to be transformed (minimum of $n = 1024$, typically), consider (for pedagogy) $n = 8$, distributed on the unit circle as shown in Figure 2.12.

Observe that the roots of unity in the complex plane, which have been numbered 0 to 7, divide the "pie" into eight equal pieces. (The algorithm requires that n be expressible as a power of 2). The usual decimal counting scheme for the eight "vectors" is as indicated, traversing the phasor diagram (circle on left) sequentially. In the Cooley–Tukey algorithm, a choice is made to reverse the bits of the binary representation of the vector. Usually, the least significant bit is on the right and the most significant bit on the left, so that decimal counting is as shown on the right in the table, from 0 to 7. With bit reversal, "lsb" becomes the leftmost binary digit and the "msb" is the rightmost digit. Thus, for example, binary 110 (usually 6) becomes 3.

Using bit reversal, the phasor diagram is not traversed in the usual phasor (circulatory) sense, but rather in a "flip-flop" back and forth across the circle. By this means, there is no needless repetition in the calculation of "vector" components (real and imaginary values of a given term in the transform). For example, 5 is the simple negative of 1. It is much faster to reverse the sign on 1 to get 5 than to

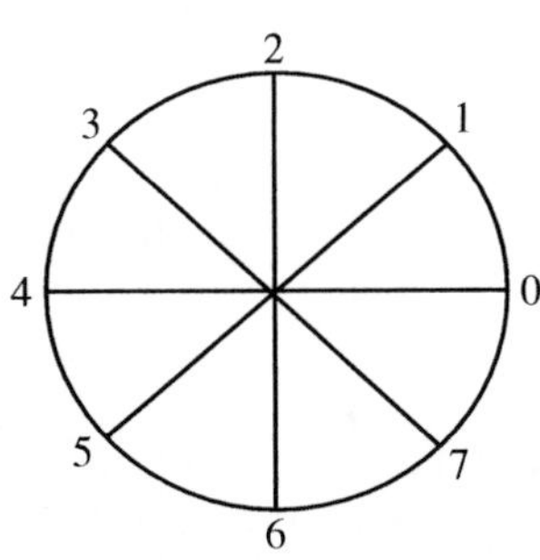

decimal (bit-reversed)	binary number	decimal (usual)
0	000	0
4	001	1
2	010	2
6	011	3
1	100	4
5	101	5
3	110	6
7	111	7

FIGURE 2.12 Graphical illustration of why the Cooley–Tukey FFT algorithm is significantly faster than the original DFT.

needlessly calculate values for sine and cosine terms a second time. The saving in time is substantial as n gets large, since there are then a great number of circulations of the phasor circle. For a 1K record, the FFT computes the transform 102.4 times faster than does the DFT. Additional details are provided in Peters (2003a, 2003b, 2003c).

2.11.3 Wavelet Transform

Recent work suggests that the wavelet transform (WT) may in the future replace the FT in some applications. It uses the Haar function, which is orthogonal on $[0,1]$, as opposed to the orthogonality of the harmonic functions (sine and cosine) corresponding to $[0,2\pi]$ (Strang, 1993). It is claimed that the WT is better able to address features of the Heisenberg uncertainty principle than the FFT.

2.11.4 Heisenberg's Famous Principle

The heart and soul of quantum mechanics is the Heisenberg uncertainty principle. As noted elsewhere in this chapter, it has things to say about damping models. According to well-known physicist Hans Bethe (1992), the principle has received "bad press":

> Many people believe that the uncertainty principle has made everything uncertain. It is quite the opposite. Without the uncertainty principle there could not exist any atoms, there could not be any certainty in the behavior of matter. So it is in fact a certainty principle.

Curiously, a failure figured in Heisenberg's discovery of the principle. During his thesis defense, in front of great theoretical physicist Arnold Sommerfeld (his director) and the famous experimentalist Wilhelm Wien, he proved unable to derive the magnifying power of a simple microscope. The scandal culminated with Professor Wien asking him to explain how a battery works, and he could not answer that question either. Knowing his extraordinary theoretical giftings, Sommerfeld gave him the highest possible grade to compensate for Wien's choice of an F. Thus, Heisenberg was awarded his doctorate.

Later, in an ironic turn of events, Heisenberg chose a microscope to illustrate features of the matrix quantum mechanics that he originated, and which corrected problems with the Bohr wave mechanics theory. His greatest source of embarrassment served to make Heisenberg famous!

2.12 Hysteretic Damping

2.12.1 Physical Basis

The model of simple harmonic oscillation with viscous damping assumes dissipation from an externally acting force. It is not suited to a conceptual understanding of hysteretic damping. To accommodate internal friction requires more than a single mass connected to the elastic component responsible for restoration. Two systems are pedagogically useful in this regard, one being a long-period physical pendulum (mechanical), and the other being the oscillator used by Ruchhardt to measure the ratio of heat capacities of a gas (mainly thermodynamic). Because of widespread confusion concerning the difference between viscous and hysteretic damping, both cases are presented here. The treatments are provided as evidence for the premise that hysteretic damping is the more important case for applied physics and engineering.

It is common knowledge that the damping of a mechanical oscillator results from the conversion of mechanical energy into thermal energy. One might expect, then, that a direct consideration of thermodynamics could yield conceptual understanding of the underlying physics. Although an ideal gas is rarely considered in this context, there is a classic experiment which speaks to its relevance. It is the ingenious technique used first in 1929 by Ruchhardt to measure γ, the ratio of heat capacity at constant pressure to that at constant volume (Zemansky, 1957).

2.12.2 Ruchhardt's Experiment

Consider a piston of mass m moving in a cylinder of cross-sectional area A, alternately compressing and expanding a volume of ideal gas V_0 about the residual pressure P_0. Assume that there is no sliding friction between the piston and the cylinder. A small displacement x of the mass results in volume change $\Delta V = V - V_0 = Ax$. There is a restoring force $F = A\Delta P$, where the pressure difference ΔP relates to ΔV through an assumed adiabatic process; i.e., the period of the motion is assumed too short for appreciable heat transfer into and out of the gas. Using $PV^\gamma =$ constant, one obtains

$$\gamma P_0 V_0^{\gamma-1}\Delta V + V_0^\gamma \Delta P = 0 \tag{2.23}$$

from which one obtains

$$m\ddot{x} + \frac{\gamma P_0 A^2}{V_0}x = 0 \tag{2.24}$$

This is the equation of motion of a simple harmonic oscillator. There is no damping because of the assumed adiabatic process. By measuring the period $T = 2\pi/\omega = 2\pi\sqrt{V_0 m/\gamma P_0 A^2}$, one can estimate γ.

Historically, it appears that such measurements slightly underestimate γ, which can be understood as follows.

The ideal gas equation of state $PV = NkT$ yields, through differentiation

$$P_0 xA + V_0\frac{F}{A} = Nk\Delta T$$

$$m\ddot{x} + \frac{P_0 A^2}{V_0}x = \frac{NkA}{V_0}\Delta T(t) = F_d(t) \tag{2.25}$$

Notice the difference between Equation 2.24 and Equation 2.25. In Equation 2.25, damping is possible (a type of "negative drive" term) from temperature variations associated with heat transfer during traversal of the cycle. If it were possible for the oscillation to be isothermal ($\Delta T = 0$ at very low frequency, essentially quasistatic), then the frequency would be lower than that of the adiabatic case, since $\gamma > 1$ is missing from Equation 2.25. In the isothermal case, there would also be no damping, since the heat into the gas during compression would be balanced by that which leaves during expansion. The only way to get damping is for the paths of compression and expansion in a plot of pressure vs. volume to separate, i.e., for there to be hysteresis. Reality must correspond to something between the two extremes of adiabatic and isothermal, with experiment obviously favoring adiabatic. The process must depart somewhat from adiabatic, however, since there is damping, which Equation 2.25 shows to derive from temperature variations yielding hysteresis. It is interesting to look at the temperature variations relative to a "driving force," $F'_d(t)$. In the Ruchhardt experiment, there must be small variations $\Delta T'(t)$ that lag behind $x(t)$. (These are not the reversible temperature variations of the adiabat, onto which the $\Delta T'(t)$ are superposed.) By comparing with Equation 2.25, the right-hand zero of Equation 2.24 may be replaced with a damping force that can be written in terms of the velocity as

$$F'\mathrm{d}(t) \propto \Delta T'(t) \rightarrow -\frac{c}{\omega}\dot{x} \tag{2.26}$$

where $c =$ constant. Notice that the multiplier on the velocity is not simply a constant, but rather a constant divided by the angular frequency. The use of velocity is mathematically convenient, but the magnitude of the velocity (speed) is not expected to be a first order influence on the temperature changes of hysteresis type. The derivative of x with respect to time not only shifts the phase by 90°, which accommodates the lag with which heat is transferred, but it also introduces a frequency multiplier through the chain rule. Thus, to make damping proportional to the velocity would cause increased dissipative heat flow and thus increased damping as the frequency is increased. Since this does not happen, and lest we introduce a nonphysical term into the equation, it is necessary to divide by the frequency. Replacing the right-hand-side zero of Equation 2.24 with Equation 2.26, we obtain the modified equation of motion, with damping

$$m\ddot{x} + \frac{c}{\omega}\dot{x} + \frac{\gamma P_0 A^2}{V_0}x = 0 \tag{2.27}$$

Additional justification for the form of the damping term in Equation 2.27 can be realized by looking at cases where there is negative damping, i.e., $c < 0$. Such is true when the gas is caused to cycle as an engine. An illustrative case study was that of a low temperature Stirling engine (Peters, 2002a, 2002b, 2002c), in which reasonable agreement between theory and experiment was realized through the use of an equation based on the same arguments used to derive Equation 2.27.

It is seen that a straightforward modeling of Ruchhardt's experiment to include damping yields an equation of motion that is in the form of hysteretic damping. It appears that, for many systems in which the dissipation is dominated by internal friction, hysteretic damping is a near universal form.

2.12.3 Physical Pendulum

In the paper by Speake et al. (1999), one finds the following statement:

> the logarithmic decrement (Q^{-1}) varied as the inverse of the square of the frequency. We interpreted this as evidence that, in Cu–Be over this frequency range, the imaginary component of Young's modulus was independent of frequency, contrary to that which was predicted by the Maxwell model.

To fit their theory with experiment, they used a "modified" Maxwell model with a distribution of time constants that ranged from 30 sec to more than 4000 sec. Motivation for their continued modeling efforts derived partly from the observation by Kuroda (1995) that anelasticity was cause for some of the huge errors that have been present in estimates of the Newtonian gravitational constant, G, by the time-of-swing method.

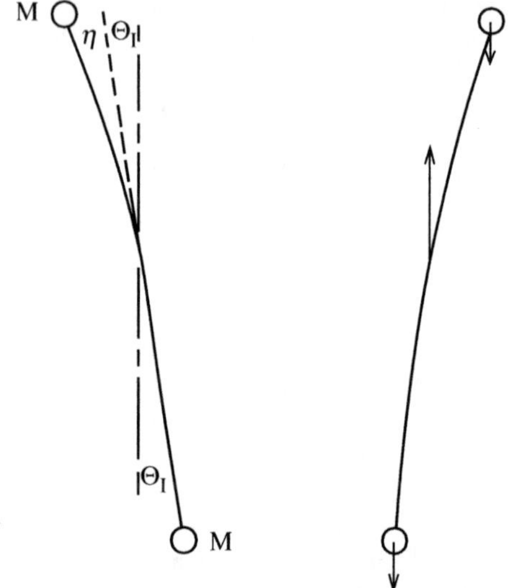

FIGURE 2.13 Idealized physical pendulum used to develop the modified Coulomb damping model.

Although it gives agreement with their particular experiment, the model of Speake et al. (1999) does not have the blessing of Occam's razor. Moreover, their claim that damping derived primarily from their flex pivot of Cu–Be may not be true. Other studies suggest that the material defining the axis of a long-period pendulum is for many cases no more important (and sometimes much less important) to the damping than the material from which the pendulum proper is constructed. A model which also agrees with experiment of the type they conducted, but which is simpler, is now presented.

Illustrated in Figure 2.13 is an idealized long-period mechanical oscillator which could be labeled a "physical pendulum." The top and bottom masses are the same, M, assumed to be much greater than the mass of the connecting structure, which is represented by the curved line.

A primary mechanism for internal friction damping can be understood by looking at the external forces acting on the pendulum, which are pictured in the "negative displacement" (b) case. The upward "normal" force that acts through the pivot (usually a knife-edge) is opposed by the pair of bob-weights situated left and right, respectively, of the axis of rotation. As the pendulum swings alternately between positive and negative displacements, the structure undergoes periodic flexure. It should be pointed out that internal friction could still be operative throughout the structure even without net bending; i.e., there could be complementary pieces of the structure undergoing compression and tension. Even if the oscillator were in the weightlessness of space, a drive torque would result in dynamic reactionary forces that give rise to damping by this means.

Assume that the masses are separated a distance $2L$ and the axis of rotation is ΔL above the geometric center. Applying Newton's Second Law, with the lower mass causing a restoring torque and the upper mass a "destoring" torque, yields

$$\ddot{\theta}_1 + \frac{g}{2L}\left(1 + \frac{\Delta L}{L}\right)\theta_1 - \frac{g}{2L}\left(1 - \frac{\Delta L}{L}\right)\theta_2 = 0, \quad \theta_2 = \theta_1 + \eta \tag{2.28}$$

(*Note:* Equation 2.28 can be rewritten to accommodate larger displacements, where elastic nonlinearity gives rise to unusual behavior. The amplitude trend of the period is opposite to that of the gravitational nonlinearity, thus providing for improved isochronism. For details refer to Peters, 2003a, 2003b, 2003c).

The difference in displacement of the masses involves an elastic term proportional to θ_1 and a dissipative term that depends on its time rate of change, i.e.

$$\eta = c\left(\theta_1 \cos\delta - \frac{\dot{\theta}_1}{\omega}\sin\delta\right), \quad \omega = \sqrt{g\frac{\Delta L}{L^2}} \tag{2.29}$$

where c is a dimensionless constant. This result can be obtained by the complex exponential Steinmetz (phasor) method. The equation is consistent with the common assumption that stress and strain are related through a complex constant. The angle δ is the phase angle with which η strain) lags behind θ_1 (stress). To describe the motion of the lower mass, we can ignore the elastic part of η, since it does not contribute to the damping (or if the rod does not bend, assuming there still is damping as noted previously). We thus remove the subscript, and after some algebra obtain the result

$$\ddot{\theta} + \frac{\alpha}{\omega}\dot{\theta} + \omega^2\theta = 0, \quad \alpha = \frac{gc}{2L}\sin\delta, \text{ for } \Delta L \ll L \tag{2.30}$$

which can also, in terms of $Q = 2\pi E/(-\Delta E)$, be expressed as

$$\ddot{\theta} + \frac{\omega}{Q}\dot{\theta} + \omega^2\theta = 0, \quad Q = \frac{2L}{gc\delta}\omega^2, \quad \delta \ll 1 \tag{2.31}$$

If, as a material property, δ is independent of frequency, then Q is quadratic in the frequency; i.e., the damping of the pendulum due to internal friction is inversely proportional to the square of the frequency — even though the internal friction (determined by δ) is itself frequency-independent. It is important to note that the frequency dependence of internal friction is not to be equated with the frequency dependence of the Q of the oscillator, even though internal friction is frequently stated as simply $1/Q$. This will be discussed in greater detail in Section 2.13.2.

2.12.3.1 Test of *Q* Dependencies

The dependence of Q on frequency and length in Equation 2.31 was tested experimentally with a physical pendulum. Two Pb spheres, each of mass approximately 1 kg, were each drilled through a diameter to allow the insertion of the shaft of an aluminum alloy arrow (length approx. 70 cm) of the type commonly used with compound hunting bows. A second hole was drilled perpendicular to the first and tapped for a set screw. The shaft was sawed into two pieces, which were rigidly rejoined around a carbon–steel knife-edge using force fit and epoxy to machined protuberances above and below the knife-edges. The knife-edges extend perpendicularly outward on opposite sides of the arrow at its center.

2.12.3.2 Simple Method to Measure Damping

Although an SDC sensor could have been employed instead, the experiments to be described were performed with a measurement technique that warrants description because of its novel simplicity — yet it is reasonably accurate. To measure both period and damping, a small "flag" was super-glued to the top of the upper shaft. This flag was a small, thin, U-shaped piece of plastic in which the upper legs of the U were about 1 mm wide, with a spacing between centers of about 0.5 cm. An infra-red photogate of the type used in general education laboratories was mounted so the flag would trip the photogate during pendulum oscillation. Two different timing measurements were then performed, using a Pasco Smart Timer. In every run, the pendulum was displaced initially about 10° by hand and then released. There was no need for precision initialization.

In the pendulum mode of the timer, the period was directly measured. For this case, the photogate was positioned, relative to the U-shaped flag (for which one vertical arm is slightly longer than the other), so as to be interrupted only once by the pendulum per pass. In the time-interval mode, the flag was positioned so that both arms interrupted the photogate beam. The reciprocal of this time of interruption proved to be a reasonable measure of the instantaneous speed of the pendulum at the position of the photogate, which was that of maximum kinetic energy. The time intervals were recorded manually for traversals separated by one period, through five cycles of oscillation. These numbers were then typed into Excel and their reciprocals graphed. A trendline (using the option to print the slope) was applied to the near linearly declining graph. The decrement of this line (fractional decrease per cycle) proved to be a good approximation to the logarithmic decrement of the motion, which could have been estimated with exceptional precision by means of the other techniques mentioned in this chapter.

In the first set of experiments, the sphere on the lower shaft was maintained at a constant distance from the knife-edge, while the mass on the upper shaft was positioned at increasingly greater distances from the knife-edge to lengthen the period. Over the full range of periods considered, the distance between the two masses changed by a small amount around its nominal value of 67 cm. The results of this first study are shown in the left graph of Figure 2.14, where the log-decrement has been plotted vs. the square of the period. The Q of the pendulum (π/Δ) may be calculated for any value of the period using the indicated slope of 0.0004. For example, the Q at a period of 10 sec was 76, this being near the shortest period considered. Near the other extreme of $T = 35$ sec, $Q = 6$. At the shortest possible periods, damping due to air drag would begin to become important.

The reasonable fit of the linear regression vs. period squared is consistent with the prediction by Equation 2.31 that Q should be quadratic in the frequency.

The Equation 2.31 also indicates that the log-decrement should be proportional to the reciprocal of the distance, L, between the masses. To test this prediction, the period of the pendulum was measured as a function of mass separation, also using the smart timer. In generating the data for the right graph of Figure 2.14, the period was maintained constant at 20 sec. For every datum, the top sphere was always only slightly closer to the knife-edge than the lower sphere. At 0.049, the intercept of the trendline differs enough from zero, relative to the size of the error bars, to imply a systematic error. Possible sources of the error include: (i) the masses are of finite size, rather than being points as assumed by the model, and (ii) a nonnegligible mass from parts other than the spheres. Nevertheless, the data show a clear size dependence of the Q.

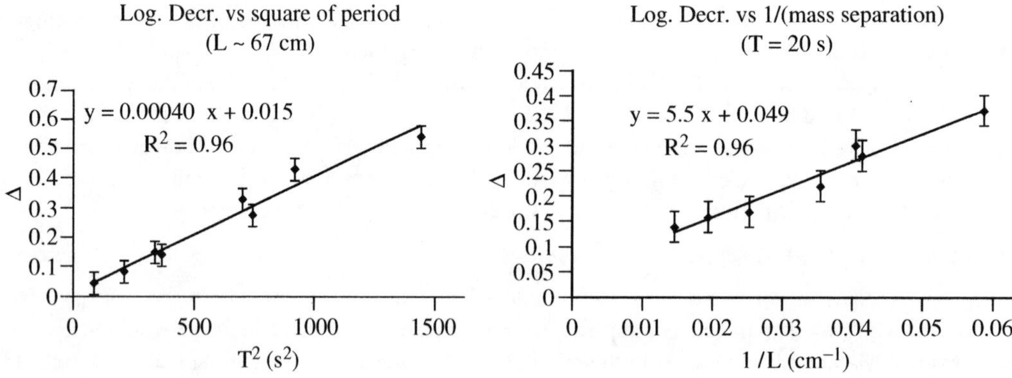

FIGURE 2.14 Results of experiments to test the dependencies of Q on (i) frequency and (ii) length of pendulum.

The experiments just described do not permit one readily to isolate the source of the damping, which, for the cases in Figure 2.14, had the knife-edge resting on silicon wafers (integrated circuit stock material). It is not known to what extent the dissipation was dominated by strain in the knife-edge–silicon interface or by flexure of the aluminum arrow. Although the model that generated equation 2.31 assumed only the latter, there is nevertheless theoretical and experimental basis for model acceptance, regardless of the details of the damping.

2.12.3.3 Highly Dissipative, yet Hard Materials

The same pendulum was used to demonstrate some counterintuitive features of internal friction damping by replacing the Si wafers with various materials. When very soft material, such as lead, was the support for the knife-edges, there was a significant increase of the damping, as expected. It was also found, however, that cast iron increased the log-decrement (10-sec period) by more than 40%. The same was also true of ceramic PZT wafers of the type used to ignite gas grills by striking the wafer impulsively. Both the cast iron and the lead–zirconate–titanate samples are very hard, so the internal friction must derive from large defect densities in which atomic disorder is a sensitive function of stress. Some other hard interfaces, such as steel on glass, or steel on sapphire, did not show a difference from steel on Si, which suggests that the dominant source of damping for the pendulum in all these cases was flexure of the aluminum shaft.

The observation involving cast iron is consistent with its known excellent damping properties at higher frequencies — important, for example, to engine blocks. Some magnesium alloys have also been developed that have excellent damping characteristics without seriously sacrificing strength.

2.13 Internal Friction

2.13.1 Measurement and Specification of Internal Friction

Mechanical spectroscopy is a popular means for measuring internal friction of materials (Fantozzi, 1982). Typically, a torsion pendulum is used to stress harmonically a sample and the lag of the response (strain), relative to the stress, provides the loss tangent and thus the internal friction. In such experiments, it is widespread practice to report internal friction as Q^{-1}. There can be confusion because of this practice, depending on the nature of the measurement technique, i.e., whether one actually measures Q as opposed to measuring something proportional to the stress–strain lag angle. If Q is obtained from an oscillatory free decay, using the logarithmic decrement defined as follows, then there is no problem.

$$\Delta = \ln\frac{x_n}{x_{n+1}} = \beta T = \frac{\pi}{Q} \tag{2.32}$$

Here, x_n and x_{n+1} are adjacent turning point amplitudes separated by one period of the motion, T. In practice, it is very difficult to adjust a mechanical system to oscillate over a wide frequency range. The widest range known to the author, for a mass–spring system, involved the work of Gunar Streckeisen (1974), in which a vertical seismometer using the LaCoste spring was adjusted to have periods in the range between 7 and 140 sec. Because of the difficulties in attaining a wide range of eigenmodes, internal friction is typically determined with a specimen that does not oscillate. We now consider that case.

2.13.2 Nonoscillatory Sample

In the typical torsional pendulum used to measure internal friction, the sample is of very small mass. Such a pendulum was built, for example, around the original version of the fully differential capacitive sensor, to study magnetoelastic wires (Atalay and Squire, 1992). As with many delicate instruments, the Atalay and Squire instrument was of the type labeled "inverted." A silk fiber at the top of the specimen was used to provide minimal tension in the sample. They used one linear rotary differential capacitance transducer (LRDCT) (Peters, 1989) in the drive mode to provide a known stress to the delicate magnetoelastic sample and a second LRDCT to measure the strain magnitude and the angle by which it lags behind the stress because of an elasticity. As such, they were measuring the lag angle and not Q, as will now be shown.

Without an inertial term, the sample response x to a periodic external force F is governed by

$$F = Kx = (k + j\zeta)x = F_0 e^{j\omega t} \tag{2.33}$$

so that the transfer function is given by

$$\frac{x}{F} = k^{-1} - j\frac{\zeta}{k^2} \tag{2.34}$$

from which it is seen that the measurement does not yield Q^{-1} but rather the lag angle ζ/k, where k is constant. Perhaps the measured angle, which is an indicator of the internal friction, has been called Q^{-1} because $k = m\omega_0^2$ for an oscillator of frequency ω_0, and $Q = m\omega_0^2/\zeta$ for the freely decaying oscillator. Bear in mind, however, that this expression for k does not apply to the nonoscillatory measurement just described. There is a frequency square difference between such a measurement and what would be measured if an adjustable oscillator were being considered.

An example of the importance of this issue is found in the article by Lakes and Quakenbush (1996), in which one reads from the abstract the following statement:

> The damping, tan δ, followed a ν^{-n} dependence, with $n \approx 0.2$, over many decades of frequency ν. This dependence corresponds to a stretched exponential relaxation function, and is attributed to a dislocation-point defect mechanism. It is not consistent with a self-organized criticality dislocation model which predicts tan $\delta \propto A^{-2}$. Dislocation damping in metals is relevant to development of high damping metals, the behavior of solders and of support wires in Cavendish balances.

The present arguments suggest that the experiment by Lakes and Quackenbush is (1996) not in strong disagreement with the SOC model; that the magnitude of the exponent difference between theory and experiment is really 0.2 and not 1.8 as they have indicated.

2.13.3 Isochronism of Internal Friction Damping

It is well known that, in the viscous damping free-decay case, the frequency of oscillation is lowered by damping according to

$$\omega_1 = \sqrt{\omega_0^2 - \beta^2} = \omega_0\sqrt{1 - (2Q_v)^{-2}} \tag{2.35}$$

and the resonance frequency of the driven oscillator is lowered even further (Marion and Thornton, 1998). It is not well known how difficult it is to measure this damping "red-shift," which brings in features of the Heisenberg uncertainty principle. Additionally, it is not well known that extensive damping experiments suggest that the frequency may not, for some systems, depend on the damping at all; i.e., the oscillator is isochronous. Isochronism cannot be realized with a linear homogeneous differential equation, but it can be realized with a nonlinear form that is obtained by modifying the damping term as follows:

$$\frac{\omega}{Q}\dot{x} \rightarrow \frac{\pi}{4}\frac{\omega}{Q}\sqrt{\omega^2[x(t)]^2 + [\dot{x}(t)]^2}\,\mathrm{sgn}(\dot{x}) \tag{2.36}$$

where sgn(dx/dt) is the algebraic sign of the velocity — it causes the equation of motion to be nonlinear even if the square root term were not present. For small damping, the square root term can be shown to be equal to the time-dependent amplitude of the motion multiplied by the angular frequency.

Other damping types are possible and are indicated in Peters (2002a, 2002b, 2002c) (...universal...) where evidence is also provided for harmonic distortion in the waveform because of the nonlinearity. It is shown in Peters and Pritchett (1997) that the oscillation is isochronous.

For large values of Q, the lag angle (radian measure) is given by $\delta = 1/Q$. Researchers usually measure δ and specify the magnitude of the internal friction as Q^{-1}. As noted previously, Q is proportional to frequency for the viscous damped oscillator. Thus, for viscous damping, the internal friction is inversely proportional to the frequency.

For hysteretic damping we obtain the result

$$\tan\delta = \alpha = \frac{h}{k} \tag{2.37}$$

where the variables are defined in Equation 2.19. For small damping in which $\tan\delta = \delta = Q^{-1}$, we find that the internal friction for hysteretic damping is inversely proportional to the square of the frequency, since h is constant and $k = m\omega^2$.

2.14 Mathematical Tricks — Linear Damping Approximations

2.14.1 Viscous Damping

In the Hooke's Law expression, $F = -kx$, it is common practice to approximate hysteresis of oscillatory motion by letting k become a complex coefficient. This is also standard practice in a variety of fields, such as the description of lossy electromagnetic media. No doubt the practice has been further popularized by the standard approach of solving electrical engineering ac circuit problems by means of phasors, the technique developed by Steinmetz (1893).

We recognize in the expression $x(t) = x_0\, e^{j\omega t} = x_0 \cos\omega t + jx_0 \sin\omega t$ that harmonic variation is contained in the real part (or alternatively the imaginary part) of the complex exponential form. Using Newton's Second Law, and representing the spring constant by $k\, e^{j\delta}$ with $\delta \ll 1$ (small damping), we obtain the damped harmonic oscillator equation

$$m\ddot{x} + kx + (jk\delta)x = 0 \tag{2.38}$$

where the approximations $\cos\delta \to 1$ and $\sin\delta \to \delta$ have been employed.

However, since $\dot{x} = j\omega x$, and $\dfrac{k}{m} = \omega^2$, Equation 2.38 can be rewritten as

$$\ddot{x} + \omega\delta\dot{x} + \omega^2 x = 0 \tag{2.39}$$

We thus see that the damping constant $\omega\delta = \omega/Q = 2\beta$ permits us to express the logarithmic decrement Δ in terms of the angle δ with which x lags F; i.e., $\Delta = \beta T = \pi\delta$. (Note that we are making no distinction here between the periods with and without damping, since the difference is small and hard to measure.) If β were independent of frequency, then δ would be inversely proportional to the frequency, which is rarely realized in practice.

2.14.2 Hysteretic Damping

Equation 2.39 does not properly represent some of the most important engineering systems. For those labeled "hysteretic," we must use a different form for the complex spring constant. We assume that $F = -(k_{\text{complex}})x = (k + jh)x$ where h is a real constant. Since $dx/dt = j\omega x$, this yields the equation of motion

$$\ddot{x} + \frac{h}{m\omega}\dot{x} + \omega^2 x = 0 \tag{2.40}$$

Since h is assumed to be a true constant (independent of frequency), the lag angle between displacement and force is given by

$$\delta = \frac{h}{k} = \frac{1}{Q} = \frac{h}{m\omega^2} \tag{2.41}$$

which is seen to be inversely proportional to the square of the frequency. (Note that δ here is the same as α in Figure 2.6.) It should be noted that the complex form for the spring constant is not simply obtained using the common theory of viscoelasticity. Such theory requires a multitude of relaxation times (stretched exponentials) (Speake et al., 1999).

2.15 Internal Friction Physics

2.15.1 Basic Concepts

All damping derives from varying degrees of complexity because of the myriad interactions that are present, either internal of nonconservative type or external involving the environment. This is true even for systems that come closest to being governed by the textbook equations. For example, the author has attempted to produce ideal harmonic oscillators using viscous liquids for damping. Even they are complicated and do not strictly obey Stokes' Law of drag force proportional to the velocity. The nonlinear Navier–Stokes equation may be capable of describing them, but not in a simple form except to a first approximation that is not really very good relative to the precision that is possible with modern sensors.

Perhaps the closest to being an ideal viscous damped oscillator is that in which the damping force derives from eddy currents through Faraday's Law. A magnet is attached to the oscillator and, as it moves in proximity to a conductor, the time rate of change of magnetic flux gives rise to a retarding force that is proportional to velocity. Because there really is a force involved, and because of Lenz's Law, the damping term makes sense physically. This case might be completely ideal except for one factor — the magnet is part of a mechanical system that must possess structural integrity if it is to oscillate. Because of loads present in the structure (reactionary normal forces to the various weights), there will always be some creep. The creep is ultimately unavoidable, since there is apparently no stress threshold below which plastic deformation ceases to exist. It is important to realize that forces associated with inertial mass (Newton's Second Law) are just as important as the weights. Systems designed around an elastic member (such as a spring, in contrast to a simple pendulum) will experience damping in the weightlessness and the airlessness of space.

2.15.2 Dislocations and Defects

The extent to which mechanical defects, such as dislocations, have been ignored by large segments of the scientific community is surprising. The surprise is even greater when one considers the importance of defects in another field — that of electronics. Our present information age (world of computing) came into existence only after widespread recognition of the importance of the defects called impurities. The n-type and p-type semiconductor materials necessary to our modern age result from the substitution of silicon atoms with others of pentavalent and trivalent type in surprisingly small concentrations.

The strength of solids is very much less than as predicted by theories of an ideal (perfect) crystal. Dislocations are the primary culprits. Their influence on materials used in engineering has prompted the statement: "when mother nature fills the vacuum she abhors, she rarely does so with perfection." Unfortunately, few students exposed to fundamental science receive training in defect physics. Moreover, it is difficult to provide a self-consistent fundamental description of their properties, so very few scientists have more than a superficial knowledge of their importance.

"Viscoelasticity" is a misleading term. To combine the words viscous and elastic suggests that the state variables vary smoothly in time, i.e., as a fluid in the viscous part. Unfortunately, this is not true of hysteresis associated with either "domains" or with "grains." In the case of magnetic domains, it is quite easy to demonstrate nonsmooth (jerky) behavior that is called Barkhausen noise. Although the phenomenon was demonstrated by Barkhausen in 1919, only recent studies have begun to understand some of its complexities better (Urbach et al., 1995a,b).

A similar phenomenon, that must relate in some manner to the Barkhausen effect, is the PLC effect. Under applied stress, alloys frequently display discontinuous strain increase (jumps). The author has even demonstrated strain recovery of a similar type, catalyzed by "tapping." The polycrystalline metals that demonstrate these effects are obviously influenced by "granularity." They differ from the "granular materials" that have become a hot topic of recent research. Even pure polycrystalline metals exhibit these features. The German word to describe the deformation of tin under large stresses is *zinngeschrei* (=tin cries). Anyone who has ever bent large-diameter tungsten wire has experienced this phenomenon, since the nonsmooth strain can be both felt and heard.

There is still another type of material, thought to have great engineering potential in the future, that shows "granular" behavior — that of shape memory alloys (SMA). If an SMA specimen is cycled in temperature around the martensitic phase, it generates acoustic emissions (Amengual et al., 1987). For a figure taken from their work and other good pages about hysteresis, refer to the webpages of Prof. Sethna at http://www.lassp.cornell.edu/sethna/hysteresis/ReturnPointMemory.html. These emissions are probably related to the PLC effect and are characterized by surprising reproducibilities in spite of their complex behavior.

Thus, there is abundant experimental evidence against the overly simplistic view that hysteretic damping can be meaningfully described by simple, linear differential equations. The nonlinear terms necessary for a good mathematical treatment go beyond "chaos" to the world of "complexity." Chaos of deterministic type, though bewildering to many, is in many cases tractable (using equations that can be integrated numerically). Damping problems are much more complex than deterministic chaos. The challenges to our understanding derive in part from the long time that it has taken before there were any serious investigations of the mesoscale, the place where defect structures abide. If, as with Zener, we use the word anelasticity to describe systems that are "other than" elastic, then the term *mesoanelastic complexity* is an appropriate label for this poorly understood physics that is important and yet mostly unknown to many fields of both science and engineering.

2.16 Zener Model

2.16.1 Assumptions

The SLM of viscoelasticity provides a sound basis for some damping phenomena, yet it fails badly as an approximation for hysteretic damping. Its prominence in both the worlds of physics and engineering warrants the following detailed discussion so that the failure case may be properly documented.

Following the example of Zener, the following linear differential equation relates the stress, σ, the strain, ε, and their first time derivatives:

$$\sigma(t) + \tau_\varepsilon \dot{\sigma} = E_1(\varepsilon + \tau_\sigma \dot{\varepsilon}) \tag{2.42}$$

The τs are relaxation times (subscript ε meaning at constant strain and subscript σ at constant stress), and E_1 is the relaxed elastic modulus (ratio of stress to strain in a very slow process). Nominally, $\tau_\sigma > \tau_\varepsilon$, consistent with strain lagging stress. For periodic variations

$$\sigma(t) = \sigma_0 \, e^{j\omega t}, \quad \varepsilon(t) = \varepsilon_0 \, e^{j\omega t} \tag{2.43}$$

which, when substituted into Equation 2.42, yields

$$(1 + j\omega\tau_\varepsilon)\sigma_0 = E_1(1 + j\omega\tau_\sigma)\varepsilon_0 \tag{2.44}$$

The complex modulus of elasticity is defined by

$$E_C = \frac{1 + j\omega\tau_\sigma}{1 + j\omega\tau_\varepsilon} E_1 \tag{2.45}$$

and is seen to relate stress and strain according to

$$\sigma(t) = E_C \varepsilon(t) \tag{2.46}$$

From Equation 2.45, the real and imaginary parts of the modulus are found to be

$$\text{Real }(E_C) = \frac{1 + \omega^2 \tau_\varepsilon \tau_\sigma}{1 + \omega^2 \tau_\varepsilon^2} E_1 \tag{2.47}$$

$$\text{Imag }(E_C) = \frac{\omega(\tau_\sigma - \tau_\varepsilon)}{1 + \omega^2 \tau_\varepsilon^2} E_1 \tag{2.48}$$

The independent variable, or "frequency," for all cases is the convenient dimensionless parameter, $\omega\tau = \omega\sqrt{\tau_\varepsilon \tau_\sigma}$.

It is convenient to use polar form, so that

$$E_C = |E_C|\, e^{j\delta} \tag{2.49}$$

where $|E_C|$ is obtained by computing the square root of the sum of the squares of the real and imaginary parts. In this form, it is apparent that δ is a lag angle which determines the damping loss for the system. Moreover, from Equation 2.47 and Equation 2.48, it is seen to obey

$$\tan \delta = \frac{\omega(\tau_\sigma - \tau_\varepsilon)}{1 + \omega^2 \tau_\sigma \tau_\varepsilon} \tag{2.50}$$

2.16.2 Frequency Dependence of Modulus and Loss

The essential features of the Zener model are illustrated in Figure 2.15, where the "unrelaxed" high-frequency modulus obeys the relation $(E_1E_2)/(E_1 + E_2) = E_1(\tau_\sigma/\tau_\varepsilon)$.

In viscous damping models, the damping is quantified by the product βT, which is equal to the logarithmic decrement. The logarithmic decrement is directly proportional to the period when the damping "constant" β is truly constant. The graph in Figure 2.16 compares the logarithmic decrement computed by the standard model against a case where β = constant. Also shown in the figure is a set of hysteresis curves for $\omega\tau = 10, 1$, and 0.1, respectively. Notice that the damping is large only for $\omega\tau$ near 1, in accord with the bottom plot of Figure 2.15. For that case, points (a) to (f) and back to (a) are shown, labels to illustrate work done by the stress in traversing the hysteresis loop. The algebraic sign of the work changes around the loop and the net work done in one cycle is just the area enclosed by the loop.

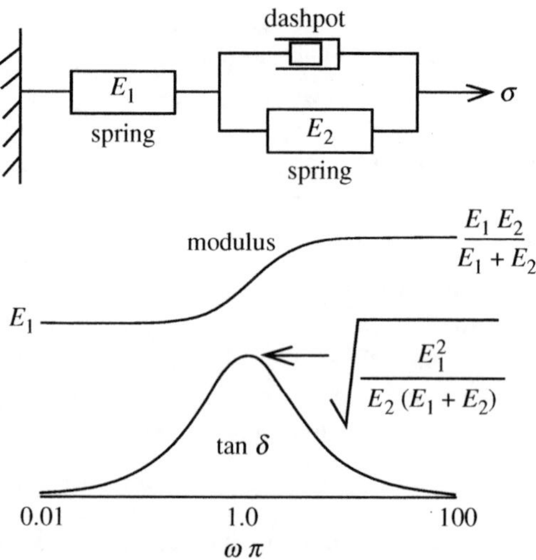

FIGURE 2.15 Zener Model of anelasticity. Bottom curves are "frequency" variation of modulus and loss respectively.

For damping based on the Zener (standard linear) model to agree with the simple viscous approximation, it is necessary that $\omega\tau >> 1$; i.e., the period of the oscillator must be significantly shorter than the smaller of the relaxation times, as illustrated in the bottom graph of Figure 2.16.

2.16.3 Successes — Models of Viscoelasticity

Viscoelasticity, as an approximation for damping, is evidently quite adequate for some materials. The assumption of fluid character as a basis for hysteresis is expected to be closest to correct when

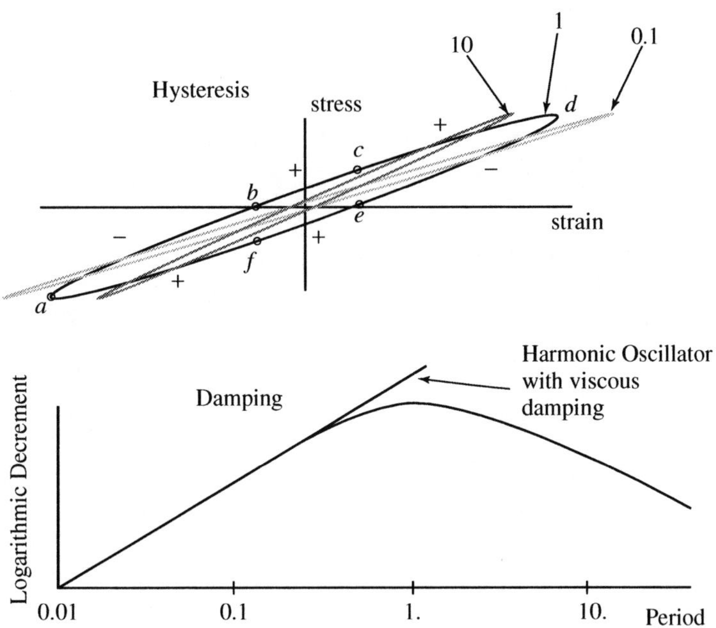

FIGURE 2.16 Characteristics of the Zener model.

applied to those cases in which variations in strain are almost continuous. The materials of rheological type for which this appears to be most true are solids built from long chain polymers, i.e., various plastics. Such materials can yield surprising results, however. Shown in Figure 2.17 are results from a study that used a nylon monofilament sample (8-lb fishing line). The pair of torsional free-decay records corresponds to two different temperatures — 290 K (room temperature) and 390 K (above the glass transition temperature of the nylon). Although a significant increase in the period was observed as the temperature was increased above the glass transition temperature (changing from 18.2 to 27.8 sec), the logarithmic decrement was found to be almost unchanged. This was not in keeping with the expectation that softening of the material at the higher temperature would result in significantly greater damping. The effect is just the opposite of what was mentioned concerning cast iron, which, though very hard, does not have small damping. Here, a softening does not result in significantly increased damping.

Although there was some creep observed for both the decays of Figure 2.17, the creep was more pronounced in the higher temperature case. This is illustrated by the lower curve of the bottom graph, which is a computer fit in which the secular term necessary for best fit was removed to illustrate the creep. In both decay cases, the log decrement was calculated by importing the A/D data (Dataq DI-154RS) to Excel and then using trial and error adjustment of parameters to achieve the best fit.

Although the damping of glasses is normally treated using the theory of viscoelasticity, Granato (2002) has recently modeled these materials via defects. In his paper, Granato states the following: "As dislocations carry the deformation in crystals, interstitials are the basic microscopic elements carrying the deformation in glasses near and above the glass temperature."

2.16.4 Failure of Viscoelasticity

Unfortunately for the elegant theory of the Zener model that has been presented, there are many mechanical systems for which the Q is not proportional to frequency, but rather proportional to the square of the frequency. The logarithmic decrement ($\Delta = \pi/Q$) has been measured for a host of

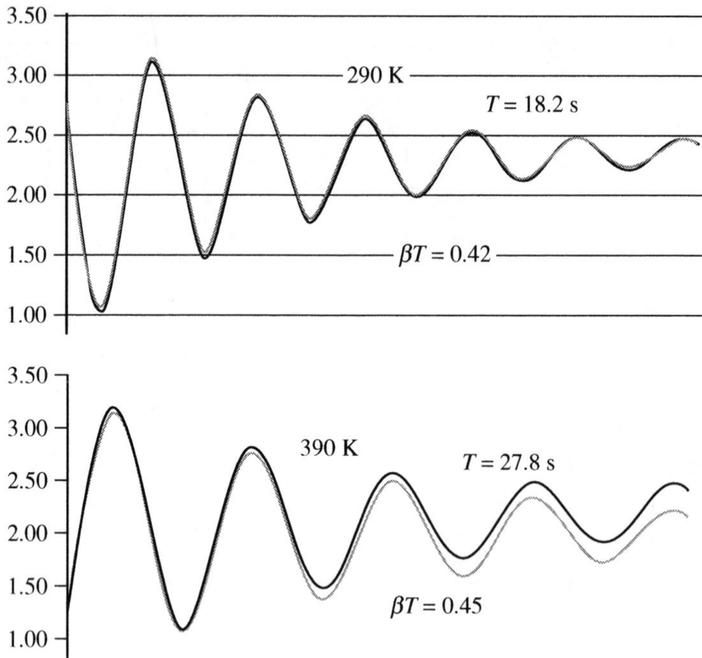

FIGURE 2.17 Torsional free-decay records of monofilament nylon at temperatures first below and then above the glass transition temperature. Although the modulus decreased dramatically at the higher temperature, the damping did not.

long-period mechanical oscillators, configured as some form of a pendulum. In all cases, these systems were described approximately by $\Delta = \beta T \alpha T^2$ rather than by $\beta T \alpha T$. Similar behavior has been noted in mechanical oscillators other than the pendulum — for example, in the geophysics research of Gunar Streckeisen and Erhard Wielandt, who are well known for the development of the widely employed STS-1 seismometer. During his pursuit of the Ph.D., Streckeisen (1974) measured the numerical damping (fraction of critical damping) of a vertical Sprengnether long-period seismometer 5100-V. After removing the magnet of the velocity transducer (to eliminate eddy currents and reduce viscous air damping, he found that the numerical damping was proportional to the square of the period between periods of 7 and 140 sec. He took about 30 measurements over this interval of periods, and showed that the damping increased from about 0.0008 to about 0.3 — a factor of roughly 400, not 20 as one would expect for viscous damping. To quote Wielandt (private communication), "the data are very clear."

2.17 Toward a Universal Model of Damping

2.17.1 Damping Capacity Quadratic in Frequency

The quadratic dependence on frequency of Q (log decrement proportional to period squared) is equivalent to friction force being frequency-independent. In support for the claim of universality, it was noted in the Introduction (Section 2.2.2) that three very different systems showed this characteristic: (i) the vertical seismometer just discussed, (ii) various pendula, and (iii) the rotating rod direct measurement of internal friction first done by Kimball and Lovell (1927), who measured the transverse deflection of the end of a rod when it was rotated about a horizontal axis.

2.17.2 Pendula and Universal Damping

An example of one of the author's experiments that illustrate universal (hysteretic) damping is provided in Figure 2.14. Other works that illustrate hysteretic damping include those by Peter Saulson of Syracuse University, who has been frequently cited in the literature (see Saulson et al., 1994).

The pioneering work of Braginsky (important to LIGO) has already been mentioned in the context of small force measurements and noise. He and his Moscow group members argue that the internal friction in fused silica may be roughly independent of frequency from 0.1 Hz to 10 kHz (Braginsky et al., 1993).

An oft-cited paper speaking to the issues of hysteretic damping is an article by Quinn et al. (1992) concerned with material problems in the construction of long-period pendula. (The type of pendulum on which they based their studies was first described in the scientific literature 2 years earlier (Peters, 1990).) In a follow-on paper, Speake et al. (1999) state the following: "The analogues of anelasticity and its resultant $1/f$ noise are seen in a wide range of other processes (for example, dielectric and magnetic ones) described in terms of frequency-dependent susceptibilities."

The jerkiness (discontinuous change) that is the hallmark of the Barkhausen effect may have been first seen mechanically in the experiments that generated the metastable states paper. From a consideration of the chapter by James Brophy (Brophy, 1965), it was postulated in this 1990 paper that the jerky behavior of a mesodynamic pendulum is a type of mechanical Barkhausen effect.

2.17.3 Modified Coulomb Model — Background

The results that follow grew naturally out of the application of fully differential capacitive sensors to the study of mechanical oscillators. Efforts to model internal friction influence on long-period pendula uncovered something surprising to most — that the foundation for physics laid by Charles Augustin Coulomb may be much broader than had been realized. Most individuals in the physics community do not associate Coulomb's name with contributions other than to the laws of electrostatics. Engineers, however, have long used his name in the context of sliding friction, since, in fact, Coulomb gave us the empirical description which involves static and kinetic coefficients. Because of his interest in the civil engineering of soils (Heyman, 1997), Coulomb also provided something else — a basis for understanding granular flows and even some types of fracture. Concerning the latter, the Mohr criterion, applied to the Coulomb failure envelope, defines a "coefficient of internal friction," which is used to predict brittle failure (Gere and Timoshenko, 1996).

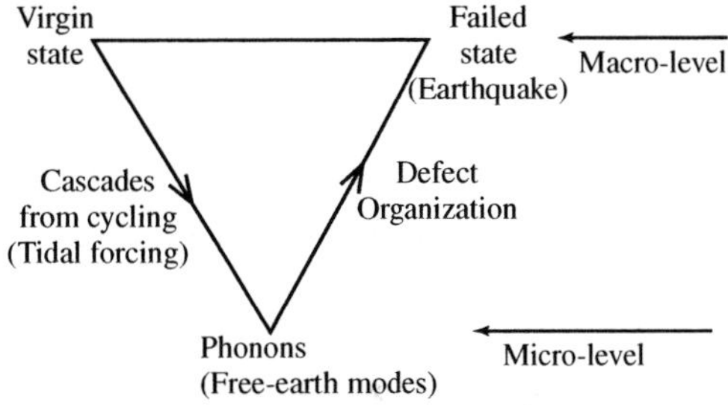

FIGURE 2.18 Heuristic description of how materials fail — processes connected with damping.

Coulomb friction is empirically simple, at least as a first approximation, since it depends only on the normal force between surfaces and the algebraic sign of the velocity when there is relative motion. Like so many problems of multibody type, a complete theory of sliding friction is far from being realized. Simplistic textbook efforts to explain energy loss, by picturing "hills and valleys" of the surface of two solids in contact, are useless. An example of such naivete can be realized by trying to understand the phenomenon of optical contacting. Two orthogonally oriented fused silica cylindrical fibers, allowed to touch, can experience atomic bonding forces that are surprisingly strong, being much greater than the weak attraction of the van der Waals type. Cleanliness of the surfaces is paramount for success in such a demonstration, which speaks to another issue — a connection between internal friction and surface physics.

The conversion of mechanical energy to thermal energy must involve nonlinear (avalanche or cascade) processes. A heuristic description of defect structural interactions that generate heat and eventual failure is the phonon triangle of Tom Erber (Illinois Tech University) shown in Figure 2.18. The author has extended Erber's triangle to include the larger-scale Earth in an attempt to explain earthquakes. Everybody recognizes that the bending of a wire does not take it from the virgin initial state to the failed state along the macroscopic upper leg.

There must first be a downward path to the microlevel, through cascading. These cascades can cause Barkhausen noise in the case of ferromagnetic samples, and acoustic emission in nonmagnetic metallic alloys (PLC effect). Failure requires the upward path of defect organization, the mechanisms of which are not yet understood. One of the first theories with possible implications to the organization leg is that of self-organized criticality. In the magnetic case, Erber has used a fluxgate magnetometer to improve failure predictions, since magnetic hysteresis is proving to be a sensitive indicator of mesoscale structure changes during cycling toward failure. Inferred from these studies is some yet-to-be discovered connectivity between noise, damping, and failure.

Surface friction is expected somehow to be connected with internal friction, the biggest difference being that the surface has many more defect states with which to redistribute energy. The larger density of states of the surface (reduced order) is probably an important factor in the difference between surface friction and the modified internal friction model which follows.

2.17.4 Modified Coulomb Damping Model — Equations of Motion

In the following damping model for internal friction, Coulomb's law of sliding friction is modified by assuming that the coefficient of friction is not constant, but rather involves the energy of oscillation E in a power law; i.e.

$$m\ddot{x} + cm\left[\frac{2E}{k}\right]^{\lambda} \operatorname{sgn}(\dot{x}) + kx = 0, \qquad E = \frac{1}{2}m\dot{x}^2 + \frac{1}{2}kx^2 \tag{2.51}$$

where $c =$ constant that is different for each λ. For Coulomb (sliding) friction $\lambda = 0$. For amplitude-independent damping of hysteretic type, $\lambda = \frac{1}{2}$. For amplitude-dependent (such as large Reynolds number fluid) damping, $\lambda = 1$. In all cases, if $c \ll 1$ (small damping), the damping capacity is quadratic in the frequency, so that the internal friction $Q^{-1} \sim \omega^{-2}$. Equation 2.51 is easily implemented, in spite of its nonlinearity, which we will see later to be a cause for harmonics in the decay.

It is convenient to rewrite Equation 2.51 in canonical form so as to involve the Q of the oscillator. For the case of hysteretic damping ($\lambda = \frac{1}{2}$), the equation becomes

$$\ddot{x} + \frac{\pi\omega}{4Q_h}\sqrt{\omega^2x^2 + \dot{x}^2}\,\operatorname{sgn}(\dot{x}) + \omega^2 x = 0 \tag{2.52}$$

Similarly, for amplitude-dependent damping ($\lambda = 1$)

$$\ddot{x} + \frac{\pi}{4y_0Q_{f0}}(\omega^2x^2 + \dot{x}^2)\operatorname{sgn}(\dot{x}) + \omega^2 x = 0 \tag{2.53}$$

where y_0 is the initial value of the amplitude of x (largest maximum of x), and Q_f is found not to be constant, as in the case of hysteretic damping. Rather, in this case, the Q increases as the amplitude decreases. On the other hand, the Q of an oscillator influenced only by Coulomb ($\lambda = 0$, sliding) friction decreases with the amplitude, and the equation of motion in canonical form is given by

$$\ddot{x} + \frac{\pi\omega^2 y_0}{4Q_{c0}}\operatorname{sgn}(\dot{x}) + \omega^2 x = 0 \tag{2.54}$$

In Equation 2.53 and Equation 2.54, the subscript 0 is used to identify the initial value of the time varying Q. Equation 2.54 is equivalent to equation 2.12 with $Q_{c0}/y_0 = \pi/(4\Delta_x)$.

As will be illustrated with some examples, it is possible for an oscillator to be influenced simultaneously by all three types of friction. One may treat such a system with the following equation of motion

$$\ddot{x} + \left[\frac{\pi\omega^2 y_0}{4Q_{c0}} + \frac{\pi\omega}{4Q_h}\sqrt{\omega^2 x^2 + \dot{x}^2} + \frac{\pi}{4y_0 Q_{f0}}(\omega^2 x^2 + \dot{x}^2) \right] \mathrm{sgn}(\dot{x}) + \omega^2 x = 0 \tag{2.55}$$

At any instant during the decay, the total (time-dependent Q) is given by

$$\frac{1}{Q(t)} = \frac{1}{Q_c} + \frac{1}{Q_h} + \frac{1}{Q_f} \tag{2.56}$$

in which it is seen that the smallest Q in the set (largest damping term) is dominant in a manner reminiscent of capacitors connected in series.

It is instructive to look at the analytical solution for the time dependence of the amplitude (turning points, $y(t) = |x_{max}|$), when all the Qs >> 1. Such a solution is obtained from energy considerations by noting first that the time rate of change of the energy is zero in the absence of friction, i.e.

$$\dot{E} = \frac{d}{dt}\left(\frac{1}{2}m\dot{x}^2 + \frac{1}{2}kx^2\right) = \dot{x}(m\ddot{x} + kx) = 0, \qquad \text{no friction} \tag{2.57}$$

With friction, dE/dt is determined by the rate of doing work against the friction force; i.e., dE/dt is proportional to $\omega y f$, where f is the friction force. In the case of Coulomb friction, f is constant (determined by y_0,) so dE/dt is proportional to $E^{1/2}$. For hysteretic damping, f is proportional to y, so dE/dt is proportional to E. For fluid damping, f is proportional to y^2, so dE/dt is proportional to $E^{3/2}$. Thus, the general case is described by

$$\dot{E} = -\left(c_1 + c_2\sqrt{E} + c_3 E\right)\sqrt{E} \tag{2.58}$$

Because the energy is proportional to y^2, we can write down the equation for the time varying amplitude as

$$\dot{y} = -c - by - ay^2 \tag{2.59}$$

where a, b, and c are constants. The solution to this first-order equation can be found in integral tables, and the result depends on the size of c relative to the product ab. For present purposes, we will restrict ourselves to the case where Coulomb damping is not dominant, in which the solution involves an exponential. (For large c, one may develop the corresponding general case in terms of the tangent or its inverse.) The present result is as follows, using $r = (b^2 - 4ac)^{1/2}$, where $4ac < b^2$

$$\text{with } \alpha = 2ay_0 + b - r, \qquad \beta = 2ay_0 + b + r, \qquad p = \frac{\alpha}{\beta}e^{-rt}$$

$$y = \frac{b(p-1) + r(p+1)}{2a(1-p)} \tag{2.60}$$

In the case where $c = 0$, Equation 2.60 can be simplified to the following form, which is useful for curve fitting:

$$\frac{1}{y} = \left(\frac{a}{b} + \frac{1}{y_0}\right)e^{bt} - \frac{a}{b} \tag{2.61}$$

For the case where $a = 0$, the better form for curve fitting is

$$y = \left(y_0 + \frac{c}{b}\right)e^{-bt} - \frac{c}{b} \qquad (\text{until } y = 0) \tag{2.62}$$

Curve-fits based on the modified Coulomb damping model are summarized in Box 2.2.

Box 2.2

Curve-Fit to the Turning Points

If no damping

$$\dot{E} = \frac{d}{dt}\left(\frac{1}{2}m\dot{x}^2 + \frac{1}{2}kx^2\right) = \dot{x}(m\ddot{x} + kx) = 0, \qquad \text{no friction}$$

with damping (E prop. to y^2, $\dot{E}$ prop. to $\omega y \cdot$ friction force)

$$\dot{E} = -\left(c_1 + c_2\sqrt{E} + c_3E\right)\sqrt{E}$$

equivalent to (c for Coulomb, b for hysteretic, a for fluid)

$$\dot{y} = -c - by - ay^2$$

general solution

$$\text{with } \alpha = 2ay_0 + b - r, \qquad \beta = 2ay_0 + b + r, \qquad p = \frac{\alpha}{\beta}e^{-rt}$$

$$y = \frac{b(p-1) + r(p+1)}{2a(1-p)}$$

special case, $c = 0$

$$\frac{1}{y} = \left(\frac{a}{b} + \frac{1}{y_0}\right)e^{bt} - \frac{a}{b}$$

special case, $a = 0$

$$y = \left(y_0 + \frac{c}{b}\right)e^{-bt} - \frac{c}{b} \qquad (\text{until } y = 0)$$

2.17.5 Model Output

Shown in Figure 2.19 is a case in which the decay is influenced by all three types of friction. Notice how the Q rises initially, peaks at a value less than what would be true for hysteretic damping alone (constant Q case), and then later declines. The initial rise is due to the amplitude-dependent damping term (size determined by coefficient a), and the later decline is due to the Coulomb damping term (determined by coefficient b).

The code in Table 2.1 that was used to generate Figure 2.19 has been reproduced here for two reasons: (i) to show the ease with which the modified Coulomb model may be numerically applied in general to a damping problem, and (ii) to illustrate an integration algorithm that has proven to be intuitive, simple, and powerful — the Cromer–Euler technique, which Alan Cromer first described as the "last point approximation (LPA)" (Cromer, 1981) in contrast to the unstable "first point approximation" given to us by Euler. Over the last 20 years, the author has employed the LPA in a host of applications that span from the generation of satellite ephemerides in the U.S. antisatellite program to both simple and several-body nonlinear problems of deterministic chaos type.

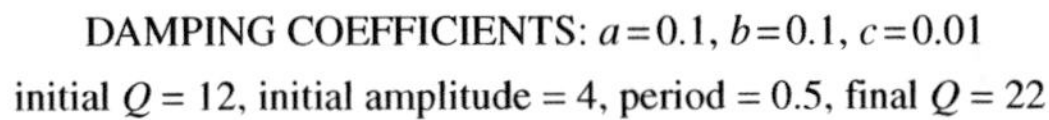

DAMPING COEFFICIENTS: a=0.1, b=0.1, c=0.01
initial Q = 12, initial amplitude = 4, period = 0.5, final Q = 22

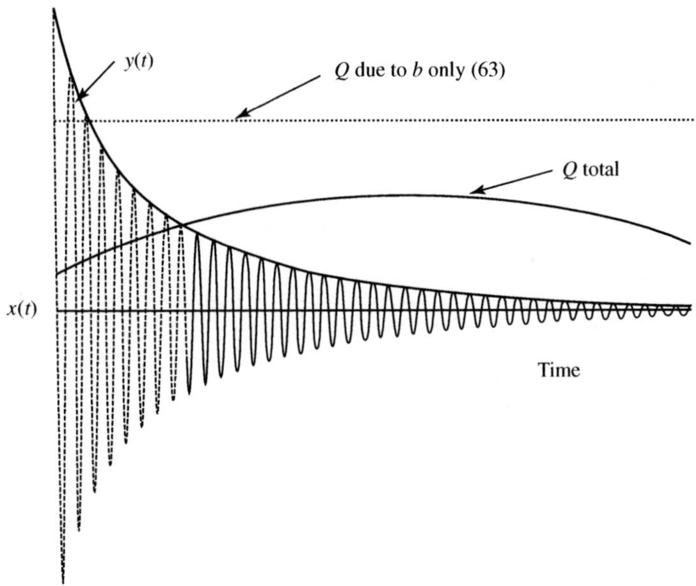

FIGURE 2.19 Model generated results based on Equation 20.55 and Equation 2.60.

2.17.6 Experimental Examples

The code of Table 2.1 is useful in determining the nature of a given experimental case. Too frequently in the past, it has been naively assumed that the entire decay record was exponential. Particularly when longer records are collected, it is found that most damping is nonlinear. In the two experimental examples that follow to backup this claim, one is a near-perfect (nonlinear) particular case of amplitude-dependent (fluid) damping, and the other is a mixture of amplitude-dependent and hysteretic damping, but devoid of any Coulombic influence. Coulomb friction frequently tends to be either "all or nothing," depending on whether there is an unwanted mechanical contact that involves slippage. A notable exception is found in the case where a pendulum is influenced by eddy current damping in a narrow region of its total motion (Singh et al., 2002).

The pendulum that was used to generate the data displayed in Figure 2.14 was also used as follows. A large flat piece of plastic was attached to the bottom of the pendulum, so that its movement (normal vector to the surface in the direction of motion) would disturb a great amount of air in turbulent manner. As expected, there was a dominant initial (large level) amplitude-dependent damping, as shown in Figure 2.20.

The speed (maximum) was measured with a photogate as previously discussed. The expressions shown in the figure are consistent with Equation 2.59 and Equation 2.61. The data, which were collected by hand and typed into Excel, produced the "jagged" curve, and the computerized fit according to Equation 2.61 is the smooth curve of the pair. It is noteworthy that the quadratic drag of the air (determined by $a = 0.036$) is 40% greater than the viscous drag at the start of the decay. By cycle 37, the quadratic part has become much less significant than the constant Q viscous part, having become roughly 60% smaller.

The fluid damping "soup-can" pendulum data of Figure 2.21 was generated with a can of Bush's black-eye peas. The container with enclosed unbroken contents, being a right circular cylinder of length 11 cm × diameter 7.4 cm, was suspended horizontally by a pair of knife edges under opposing end-lips

TABLE 2.1 QuickBasic Code to Calculate Amplitude History $y(t)$ and Integrate Equation of Motion to Obtain $x(t)$; Accommodates Three Common Forms of Friction

```
CLS
  REM: setup display
SCREEN 12: VIEW (0, 0) − (600, 470): WINDOW (−.2, − 5) − (1, 5)
  REM: assign constants and initialize variables
pi = 3.1416: dt = 0.002: t = 0
x0 = 4: x = x0: y0 = x0: xd = 0
Period = .5: omega = 2*pi/period: b = .1: a = .1: c = .01
  REM: print damping coefficients
PRINT "DAMPING COEFFICIENTS: a = "; a; ", b = "; b; ", c = "; c
r = SQR(b^2 − 4*a*c): alpha = 2*a*x0 + b − r
Beta = 2*a*x0 + b + r
  REM: Use a, b and c — set Q's to dampen (quadratic, linear, and constant resp.)
qf = omega/2/a/y0: qh = omega/2/b: qc = y0*omega/2/c
  REM: start integration loop
LOOP0:
t = t + dt
  REM: analytically compute amplitude (y = magnitude of x) at each time point
p = alpha*EXP(−r*t)/beta
y = (b*(p − 1) + r*(p + 1))/2/a/(1 − p)
  REM: integr. the eq. of motion to get x(t), using 3 fric. force/mass terms
  REM: The coeff.'s ff, fh & fc correspond to: quadratic in speed (fluid),
  REM: linear in speed (hysteretic), and independ. of speed (Coulomb) resp.
ff = (pi/4)*(1/y0)*(1/qf)*(omega^2*x^2 + xdot^2)
fh = (pi/4)*(omega/qh)*SQR(omega^2*x^2 + xdot^2)
fc = (pi/4)*omega^2*y0/qc
  REM: check algebraic sign – USE SIGN BUT NOT MAGNITUDE OF VELOCITY
IF xdot > 0 THEN GOTO SKIP
ff = −ff: fh = −fh: fc = −fc
SKIP: xdoubledot = −ff − fh −fc − omega^2*x
xdot = xdot + xdoubledot*dt: x = x + xdot*dt
  REM: calculate the energy and then the amplitude to evaluate Q
  REM: could instead use analytical result q = (pi/4)*omega2*x/abs(ff + fh + fc)
Energy = .5*xdot^2 + .5*omega^2*x^2
Amplitude = SQR(2 * energy)/omega
  REM: calculate loss per period due to friction
loss = ABS(ff + fh + fc)*4*amplitude
q = 2*pi*energy/loss
IF t < 1.2*dt THEN PRINT "initial Q = "; 10*INT(q)/10;
IF t < 20 THEN GOTO SKIP2
PRINT ", initial Amplitude = "; x0; ", Period = "; period;
PRINT ", final Q = "; 10*INT(q)/10
  REM: DO GRAPH
SKIP2: PSET(.04*t, .5*q/omega): PSET (.04*t, .5*qh/omega), 4
PSET (.04*t, 4*x/y0): PSET (.04*t, 0): PSET (.04*t, 48*y/y0)
IF t > 20 OR y < 0 THEN GOTO pause
GOTO LOOP0
Pause: GOTO pause
RETURN: END: STOP
```

(Peters, 2002a, 2002b, 2002c). The motion of the can was measured with an SDC sensor connected to a Dataq A/D converter. Whereas experiments of similar type, with homogeneous fluid contents, have produced viscous decay records, the present case involved only friction of so-called "fluid" type; i.e., quadratic in the "velocity." To generate the figure, the A/D record was exported to the Microsoft software package, Excel. Fits to the data were then obtained by adjusting, through trial and error, the a, b, and c

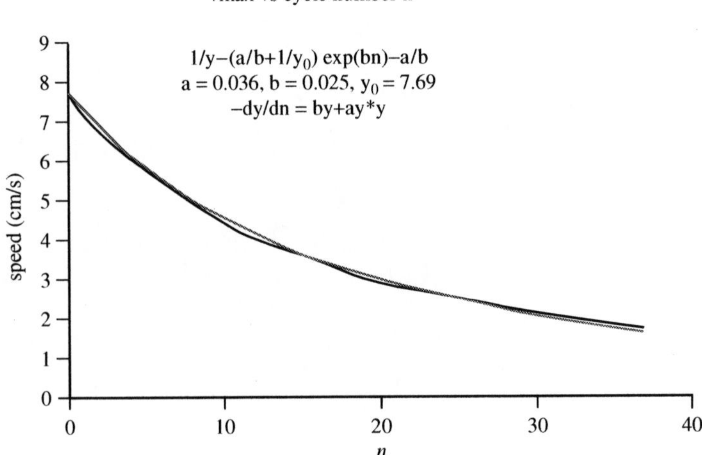

FIGURE 2.20 Decay of an air-damped pendulum as a function of cycle number n.

coefficients of a "fit" to the amplitude. For this case, the fit was easily accomplished because both b and c proved to be essentially zero.

The second case, involving an evacuated pendulum, was not a single pure type of damping, but can be seen in Figure 2.22 to have both hysteretic and amplitude-dependent contributions. Although fluid damping is amplitude-dependent in the same manner, with the damping term being proportional to the square of the amplitude, the word "fluid" is not used to describe this case since the system involved exclusively solid materials.

Not all decay records of this pendulum in vacuum yielded a mix of friction types as displayed in the figure. The effect was observed to be transient, and it is speculated that outgassing of components may have been a factor.

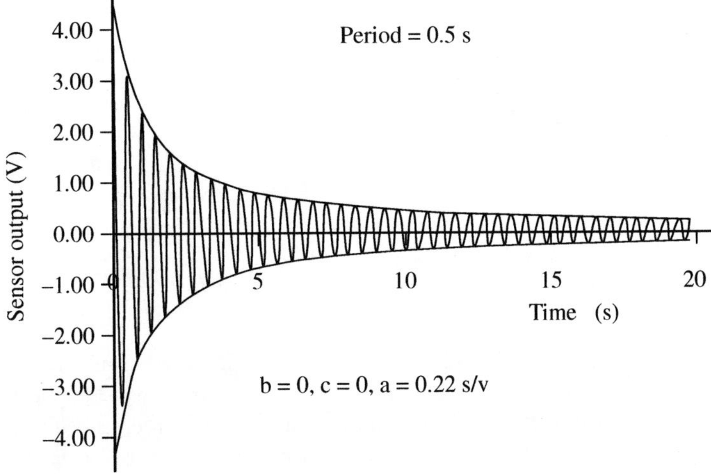

FIGURE 2.21 Example of fluid damping of a "soup-can" pendulum. The granular contents (black-eye peas and water) result in a friction force that is quadratic in the velocity.

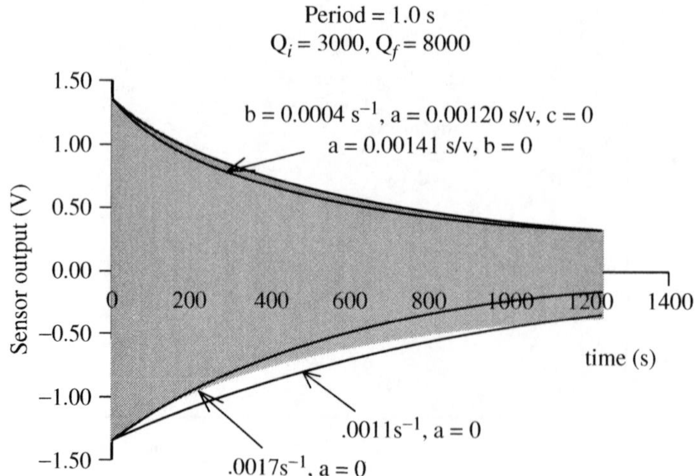

FIGURE 2.22 Example of a mix of two damping types, hysteretic and amplitude-dependent.

2.17.6.1 Numerical Integration

Instead of integrating the second-order equation of motion twice — first the acceleration, followed by the resulting velocity — more accurate results are obtained by integrating the equivalent pair of first-order equations.

For example, the equation of the simple harmonic oscillator with viscous damping is expressible as

$$\dot{p} = -q - kp \qquad \dot{q} = p \tag{2.63}$$

where the position variable has been represented by the generalized coordinate q (x elsewhere), and for the momentum $p = m\, dq/dt$, and here the mass, m, has been set to unity. Likewise, the spring constant has been set to unity. It is generally useful to distill a given problem to its most basic form when attempting to understand the physics. Constants that provide no useful information for trend analysis purposes are conveniently "normalized." Such is common practice, for example, in modeling chaotic systems.

The second-order set can always be reduced mathematically to a first-order pair; however, the pair results naturally from the use of the Hamiltonian as opposed to the Newtonian formulation of mechanics.

2.17.7 Damping and Harmonic Content

Equation 2.52 to Equation 2.54 are the nonlinear, modified Coulomb damping model forms that correspond, respectively, to (i) hysteretic, (ii) amplitude-dependent, and (iii) Coulomb damping. The damping term for each of the three cases can be expressed as follows:

$$\frac{f}{m} = \frac{\pi}{4}\frac{\omega}{Q}[v] \tag{2.64}$$

where f is the friction force, and $[v]$ is the square wave whose fundamental in a Fourier series expansion is equal to the velocity of the oscillator times $4/\pi$; i.e., for a square wave $\pm h$, the amplitude of the fundamental is $\pm(4/\pi)h$. We see that all the damping types that have been considered in this chapter, when expressed in canonical form, correspond to a fundamental friction force $f = m\omega v/Q$. The simplicity of this result is probably why viscous damping has been viewed by so many physicists as "inviolate." One must be careful, however, because (as noted in the previous section) only for the case of hysteretic damping is Q constant. For amplitude-dependent damping $Q_f = Q_{f0}(y_0/y)$ and for Coulomb damping

$Q_c = Q_{c0}(y/y_0)$. The time-dependent Q of nonexponential cases will have significant influence on mode development in many-body systems because of elastic nonlinearity (necessary for mode coupling).

There is another important subtlety of Equation 2.64. When only the fundamental of $[v]$ is retained, equivalent to viscous damping, Q is proportional to frequency. When all odd harmonics are included (full square wave), Q becomes proportional to frequency squared. This means that *harmonics in the friction force are responsible for the primary difference between hysteretic damping and viscous damping.* Something being presently considered is how, in an algorithmic sense, to modify Equation 2.52 to Equation 2.54 to provide for "dispersion," i.e., means for providing Q dependence other than frequency squared. We posit the following: that hysteretic (exponential) damping is the idealized universal form of damping due to secondary creep. When there is an activation process of Zener (Debye) type, such as dislocation relaxation, then additional terms must be added to the hysteretic "background." It may be that this can be accommodated by a suitable removal of harmonics from the square wave of the hysteretic case, and it may happen that Q is constant for systems that vary continuously. It is conjectured that the PLC effect, responsible for discontinuous changes, plays a role in those cases where Q is not constant. Equations of motion based on the modified Coulomb damping model are summarized in Box 2.3.

Box 2.3

Equations of Motion Based on Nonlinear Damping

Equation of motion in terms of energy

$$m\ddot{x} + cm\left[\frac{2E}{k}\right]^2 \mathrm{sgn}(\dot{x}) + kx = 0, \qquad E = \frac{1}{2}m\dot{x}^2 + \frac{1}{2}kx^2$$

Hysteretic-only damping (exponential)

$$\ddot{x} + \frac{\pi\omega}{4Q_k}\sqrt{\omega^2x^2 + \dot{x}^2}\,\mathrm{sgn}(\dot{x}) + \omega^2 x = 0$$

Velocity-square (fluid) damping

$$\ddot{x} + \frac{\pi}{4y_0 Q_{f0}}(\omega^2x^2 + \dot{x}^2)\,\mathrm{sgn}(\dot{x}) + \omega^2 x = 0$$

Coulomb damping

$$\ddot{x} + \frac{\pi\omega^2 y_0}{4Q_{c0}}\,\mathrm{sgn}(\dot{x}) + \omega^2 x = 0$$

All three damping types simultaneously active

$$\ddot{x} + \left[\frac{\pi\omega^2 y_0}{4Q_{c0}} + \frac{\pi\omega}{4Q_k}\sqrt{\omega^2x^2 + \dot{x}^2} + \frac{\pi}{4y_0 Q_{f0}}(\omega^2x^2 + \dot{x}^2)\right]\mathrm{sgn}(\dot{x}) + \omega^2 x = 0$$

Quality factor

$$\frac{1}{Q(t)} = \frac{1}{Q_c} + \frac{1}{Q_k} + \frac{1}{Q_f}$$

2.18 Nonlinearity

2.18.1 General Considerations

Electrical nonlinearity is the type with which most engineers are familiar. It is the very basis for common nondigital forms of communication, such as that of frequency modulation type. A popular form of radio amateur communication is one in which the carrier and one of the two normal sidebands of a signal are suppressed before going to the antenna. At the receiver, the carrier is "regenerated" before going to the demodulator. The demodulator required for ultimate transduction by speaker is also a nonlinear device.

Nonlinearity of mechanical type is encountered throughout nature. The human ear, for example, is not linear, but rather characterized by both quadratic and cubic nonlinearities. If an intense, pure low frequency (inaudible) sound of frequency f is present with a higher frequency audible one of frequency F, then one typically hears (in addition to F) tones at $F \pm f$ due to the quadratic nonlinearity and $F \pm 2f$ due to the cubic nonlinearity.

Very high frequency acoustics (ultrasound) is employed for studies of elasticity. The quasi-linear features of ultrasonic propagation have been the basis for measuring second-order elastic constants (determined by velocity of propagation) and internal friction (by attenuation of the beam, i.e., damping). A commonly employed ultrasonic technique that has been used to study both linear and nonlinear phenomena is the pulse-echo method. By using a thin specimen and extending the pulse width, the overlapped signal can add constructively or destructively and, in the former case, resonance is approached as the width gets very large (Peters, 1973). The pulse-echo method was the basis for this author's Ph.D. dissertation ("Temperature dependence of the nonlinearity parameters of copper single crystals," The University of Tennessee, 1968). The distinguished career of his professor, M.A. Breazeale, has focused on ultrasonic harmonic generation as a means to determine the shape of the interatomic potential of solids (Breazeale and Leroy, 1991). A longitudinal wave distorts because of the anharmonic potential (acoustic equivalence of optical frequency doubling with lasers in a KdP crystal). In like manner, phonon–phonon interactions are possible only because of nonzero elastic constants of order higher than second (second-order constants determining the harmonic potential). Because phonon–phonon interactions are part of damping, there must be consequences, at least for some cases, from nonlinear damping terms.

The unifying theme for this chapter is that damping is fundamentally nonlinear, in spite of the fact that linear approximations have prevailed in modeling and, for many purposes these linear models appear to be acceptable (Richardson and Potter, 1975). In their paper, Richardson and Potter state that "... an equivalent viscous damping component can always be derived, which will account for all of the energy loss from the system. Thus, in measuring the modal vibration parameters for the linear motion of a system, we don't care what the detailed damping mechanism really is."

Although their statement may be true for steady state, it is not expected to be true for the transient processes that lead to steady conditions of oscillation. As demonstrated elsewhere in this chapter, mixtures of different damping types are common among oscillators, and only with viscous or hysteretic damping is the Q independent of amplitude. Other cases may result, for example, from the decay being a combination of hysteretic damping and amplitude-dependent damping. An example used to illustrate this combination was an outgassing pendulum oscillating in vacuum. Similarly, a long, "simple" pendulum, oscillating in air, is found to require a pair of terms — viscous damping and "fluid" damping (Nelson and Olssen, 1986). In the Nelson and Olsson experiment, the drag was found, because of the size of the Reynolds number, to involve both first- and second-power velocity terms. Their case can, incidentally, be treated by the modified Coulomb, generalized damping model of this document.

The presence of either amplitude-dependent damping or Coulomb damping is expected to play a role in determining what modes of a multibody system are actually excited by external forcing. Concerning the latter, Coulomb friction is the basis for exciting chaotic vibrations in mechanical systems (Moon, 1987). Without the nonlinear friction, the excitation would be impossible. In similar manner (although chaotic motion may be present but not in an obvious way), friction from rosin on a violin bow is used to play the violin. Still another example of similar physics is the "singing rod" that was mentioned elsewhere as exhibiting thermoelastic damping.

Whatever combinations of normal modes are initially excited in a linear system are the only ones that can exist thereafter. Such is not the case, however, for many systems and, since nonlinearity is required for mode coupling, there must be nonlinearity in the equations of motion. There is no question about the existence and importance of elastic nonlinearity. Indeed, thermal expansion would be impossible in the absence of higher order elastic constants. The importance of nonlinear damping remains yet to be quantified, since models to include it have been few in number. For those who have found it advantageous to include the oldest and simplest type of nonlinearity in a damping model — Coulomb damping (sliding friction) — the improvements realized by their choice are unlikely to cause them to revisit the problem and try to solve it in terms of a viscous equivalent linear approximation.

There are many examples of damping of a single type other than viscous. In their efforts to improve the knowledge of the Newtonian gravitational constant $G = 6.67 \times 10^{-11}$ Nm^2/kg^2 (approx.), Bantel and Newman (2000) discovered a pure form of amplitude-dependent damping of internal friction type. They did their experiments at liquid helium temperature (4.2 K) and noted the following: "A striking feature noted in our data is the linearity of the amplitude dependence of Q^{-1} for the three metal fiber materials," and also "Linearity implies that Q may depend on frequency but not on amplitude, while in fact Fig.1 displays a significant amplitude dependence (and hence nonlinearity) of internal friction in all fibers tested." They also considered the temperature dependence of damping and note that there are two independent contributions in Cu–Be. One is linear and temperature-independent and the other amplitude-dependent and independent of temperature. Finally, it is worth noting their statement, "...our results are strongly suggestive of some kind of 'stick–slip' mechanism ...," which lends strong support to the modified Coulomb internal friction damping model of the present document.

Repetition is felt to be warranted — such systems cannot always be reasonably described by an equivalent viscous form! For a case of amplitude-dependent Q, the equivalent form has no meaning unless the amplitude is fixed, i.e., it oscillates at steady state. Unfortunately, the evolution of the system to steady state is expected to depend on the damping form(s). Surely a model (not yet realized) that predicts what modes survive is worth much more than one which only characterizes the modes after they have reached steady state. The author and Prof. Dewey Hodges of Georgia Tech's Aerospace School are planning projects to try to develop such predictive capability. The present state of the art applied to structures suggests that a truly predictive model cannot ignore damping nonlinearity.

As demonstrated by Bantel and Newman (2000), the mixture of damping types that can co-exist in a system may change with temperature. Early experiments by Berry and Nowick (1958) also showed, as have many investigators subsequently, that damping generally depends on aging. It is naive to believe that aging would not also change the mix of damping types, when there is more than one type. Thus, an adequate damping model must be able to easily accommodate several damping types that are simultaneously active. A variety of engineering techniques have evolved to treat such problems. The most "successful" ones suffer from the fact that an excessive number of parameters or coupled equations must be adjusted by trial and error to yield decent agreement with experiment. This is reminiscent of the state of high-energy (nuclear) physics before the standard model. The hallmark of physics success has always been *simplification*. As noted by Albert Einstein: "All physics is either impossible or trivial. It is impossible until you understand it. Then it becomes trivial." It is hard to imagine, however, that certain damping physics could ever become trivial. Nevertheless, the simplifying nature of better conceptual understanding is a goal to strive for.

One of the remarkable things about the majority of damping models has been the absence of a direct consideration of energy in describing the dissipation process. After all, the most important quantity transformed by the damping is energy, so its inclusion is natural.

2.18.2 Harmonic Content

When the damping is nonlinear, the waveform of the oscillator in free-decay contains harmonics. The harmonic content is most obvious in the residuals (difference) after fitting a damped sinusoid to the record, as shown in Figure 2.23.

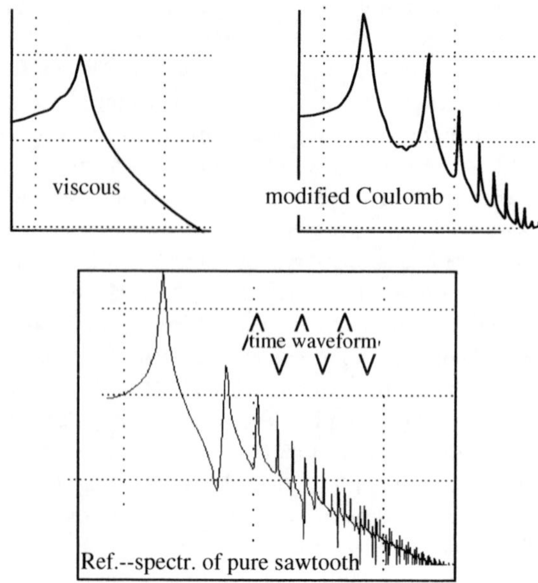

FIGURE 2.23 Harmonic differences between the residuals of the modified Coulomb damping model and the classic viscous damping model. For reference purposes, a pure sawtooth is included in the figure.

Residuals are still present for the viscous case because the equation of motion was integrated numerically and compared against the classic exponentially decaying sinusoid (solution to the equation) that was used for fitting in all cases. There is always some degree of mismatch with the fit because of rounding errors in the computer. In Figure 2.23, the fundamental is smaller for the viscous case because the fit is inherently more perfect by about an order of magnitude in most of the "eye-ball" fits that were performed by Excel after importation of the data.

A test for harmonic content was performed on the seismometer (17-sec period) data displayed in Figure 2.11 illustrating phase noise. The power spectrum of the residuals for that case is shown in Figure 2.24.

The third harmonic is especially noticeable in this case. That the other harmonics are not so "cleanly" displayed may result from the significant phase noise of the record.

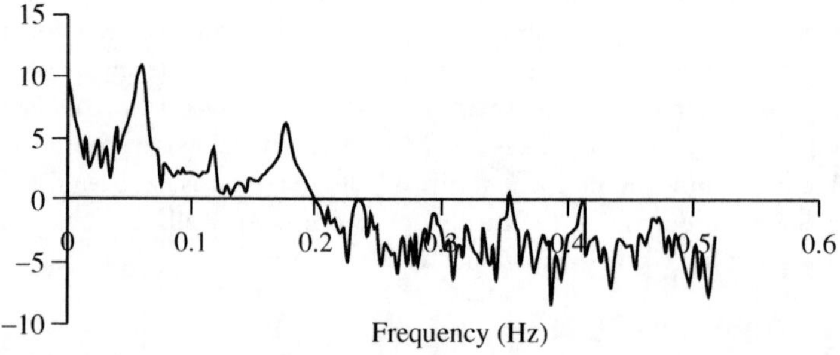

FIGURE 2.24 Power spectrum of residuals, Sprengnether vertical seismometer free-decay, showing harmonic content.

By looking at the FFT of residuals, rather than the experimental record itself, one finds evidence for a combination of both mechanical and electronic noise. At lower frequencies, the noise (largely mechanical) is approximately $1/f$, while at higher frequencies the noise (largely electronic) begins to be more nearly "white" (frequency-independent) because of discretization errors of the resolution-limited 12-bit A to D converter.

In general, more spectral information can be gleaned from a consideration of the residuals than from the experimental data alone, particularly as one looks for harmonic distortion of mechanical type. Spectral "fingerprints" may prove ultimately useful in determining to what extent damping models of engineering type need to be implemented in full nonlinear form as opposed to an "equivalent viscous" form that is more convenient mathematically.

The importance of the harmonics observed in Figure 2.24 in determining system evolution is not completely known. It was noted earlier that they are expected to influence the evolution of a multibody system to steady state. Presently, it appears that they may serve to validate damping models. From one model type to another, there can be significant differences in the spectral character of the residuals, as shown in Figure 2.25. As compared with Figure 2.23, the fit with the modified Coulomb (hysteretic case) model has been tweaked to reduce the fundamental somewhat, but the odd harmonics remain significant. Observe that the spectrum of the residuals is almost the same for this model and the simplified structural model (see de Silva, 2000, p. 354). This is true even though the temporal variation of the friction force is dramatically different for the two, as seen from the lower time traces that were used to obtain the residuals (which are too small to be seen in the graphs).

From this author's perspective, the simplified structural model is unrealistic, since the friction force, given by $f = c|x|\ \mathrm{sgn}(\dot{x})$, vanishes for zero displacement (the absolute value of the displacement being used to get the hysteretic form of frequency dependence). This is seen in Figure 2.26, which compares hysteresis curves for several models. The modified Coulomb case shown is slightly different from Equation 2.52 that was used to generate Figure 2.25; Figure 2.26 was generated with the A_{prev} shown in Equation 2.10.

More studies of this type are obviously called for. The spectrum of residuals is a powerful means for the study of damping physics, and it needs to be more widely employed.

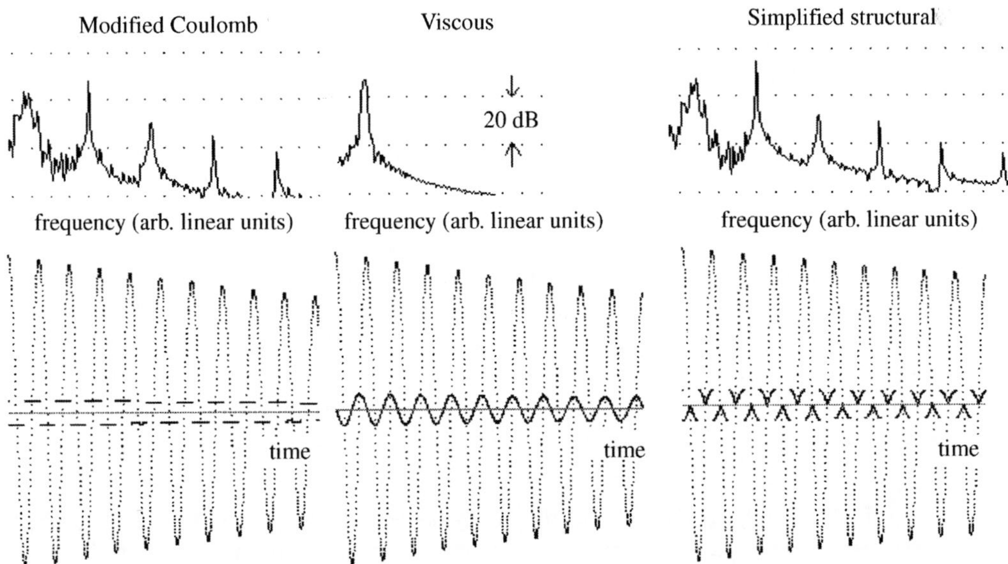

FIGURE 2.25 Illustration of the spectral difference of the residuals for three different damping models. The corresponding temporal records used to generate the spectra are also shown underneath each case.

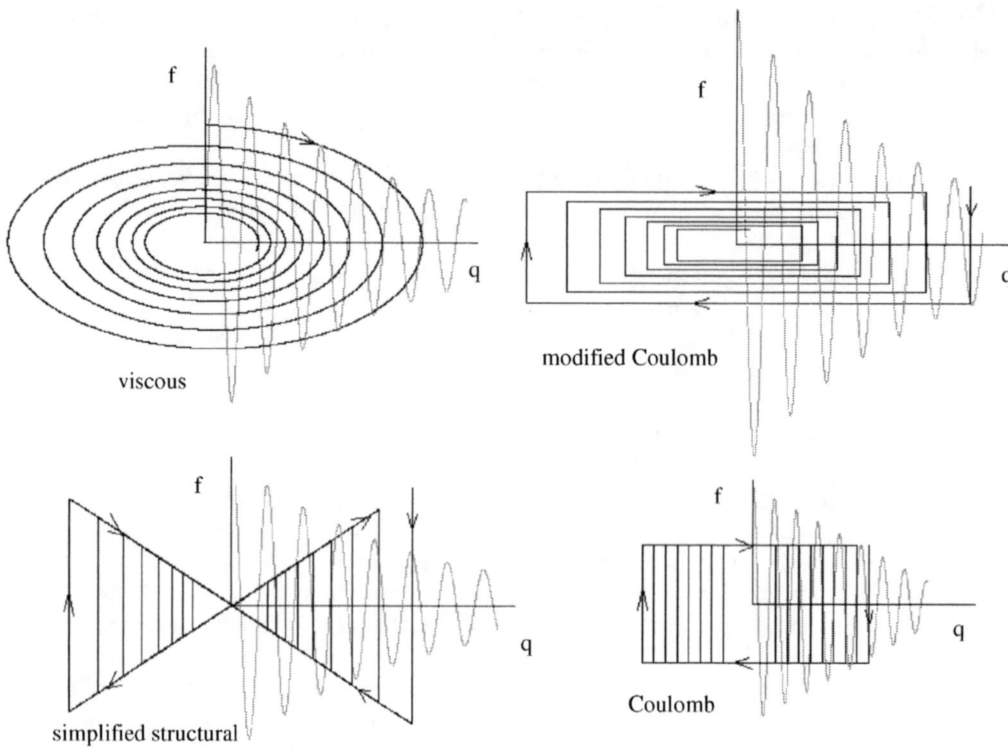

FIGURE 2.26 Comparison of hysteresis curves for some damping models.

2.18.3 Nonlinearity/Complexity and Future Technologies

Nonlinear damping models must improve if we are to overcome various technological barriers. One barrier is in the area of civil engineering. One of the pioneers of finite element modeling (FEM) is Prof. Emeritus Edward L. Wilson, of the University of California Berkeley. In Technical Note 19 (pertaining to "structural analysis programs") — a document published by his company Computers and Structures Inc — Dr. Wilson says the following:

> Linear viscous damping is a property of the computer model and is not a property of a real structure.

Expanding upon the statement, he notes:

> the use of linear modal damping, as a percentage of critical damping, has been used to approximate the nonlinear behavior of structures. The energy dissipation in real structures is far more complicated and tends to be proportional to displacements rather than proportional to the velocity. The use of approximate "equivalent viscous damping" has little theoretical or experimental justification...the standard "state of the art" assumption of modal damping needs to be re-examined and an alternative approach must be developed [in reference to Rayleigh damping].

One of the hi-tech areas where modeling improvements are also sorely needed is that involving miniaturized mechanical systems. For example, MEMS devices have already encountered some of the "strange phenomena" of solid-state physics mentioned by Richard Feynman in his famous 1959 talk. To master or compensate for these phenomena, better understanding of the physics will be necessary.

2.18.4 Microdynamics, Mesomechanics, and Mesodynamics

At least three different broad fields of research have focused on problems associated with the structural defects that cause hysteresis. These are as follows.

2.18.4.1 Microdynamics

In the microdynamics world, the emphasis appears to have been primarily on "contact" friction. The 6th Microdynamics Workshop held at the Jet Propulsion Laboratory in 1999 produced the following statements (quoting Marie Levine's Program Overview): (1) "We have demonstrated that microdynamics exist. The next step is to qualify and quantify microdynamics through rigorous testing and analysis techniques." (2) Microdynamics is "defined as sub-micron nonlinear dynamics of materials, mechanisms (latches, joints, etc.) and other interface discontinuities."

In this workshop, it was noted that frequency-based computational methods *cannot* be used to model quasi-static, transient, and nonstationary disturbances. One of the flight operations they have recommended to minimize adverse effects of microdynamics is dithering.

2.18.4.2 Mesomechanics

Ostermeyer and Popov (1999) have the following to say about mesomechanics: "Real physical objects inherently possess discrete internal structures. Great efforts are needed to formulate continuum models of really granular bodies. The history of the last two centuries in a multitude of ways has been marked by highly successful attempts at formulating and analyzing the continuum models of the discrete world. In spite of great advances of continuum mechanics, a number of physical processes are amenable to simulation within the framework of continuum approaches only to a very limited extent. Among these are primarily all the processes whereby the medium continuity is impaired; i.e., those of nucleation and accumulation of damages and cracks and failure of materials and constructions."

Their paper speaks to one of the difficulties concerning granular materials that was mentioned earlier in this chapter — that the potential energy cannot be defined in the common manner. They introduce a temperature-dependent nonequilibrium interaction potential that is not constant in time due to the relaxation processes occurring in the system.

2.18.4.3 Mesodynamics

The author of this chapter is singlehandedly responsible for the use of the term "mesodynamics" in the context of mechanical oscillators. His research has been conducted independently of those doing mesomechanics; he came only recently to know of the latter. Whereas mesomechanics seems to have been largely concerned with failure, mesodynamics has been concerned with low-level hysteresis. It is probably closely related to the aforementioned microdynamics, except that the latter seems to have focused on surfaces (sliding friction), whereas mesodynamics is concerned with internal friction.

A group of individuals using "mesodynamics" to describe some of their computational physics is part of the Materials Science Division of Argonne National Laboratory. Their description of computational theory includes: (i) atomic-level simulation (using molecular dynamics); (ii) mesoscale simulation, i.e., "mesodynamics" (using FEM); and (iii) macroscale (continuum) simulation (FEM). Like the author of this chapter, they recognize that the mesoscale is not a continuum (meaning, for example, that the foundation of viscoelasticity is, for many cases, on shaky ground). They employ "dynamical simulation methods in which the microstructural elements (grain boundaries and grain junctions) are considered as the fundamental entities whose dynamical behavior determines microstructural evolution in space and time."

At the Theoretical Division of Los Alamos National Laboratory, Brad Lee Holian has been modeling mesodynamics via nonequilibrium molecular-dynamics (NEMD). In his paper, "Mesodynamics from Atomistics: A New Route to Hall-Petch," he notes that (i) the mesoscopic nonlinear elastic behavior must agree with the atomistic in compression; and (ii) the mesoscale cold curve in tension represents surface, rather than bulk cohesion, thereby decreasing inversely with grain size (Holian, 2003).

The complexity of mesodynamics, which this author has labeled "mesoanelastic complexity," is responsible for much of the aforementioned "strange phenomena." To those familiar with the Barkhausen effect and the PLC effect, they are less strange. It is thought that Richard Feynman, if he were still alive, would identify with mesodynamics because of material in his three-volume series (Feynman, 1970). For example, we have already noted his discussion of the Barkhausen effect, and he included in its entirety a reprint of the Bragg–Nye paper on bubbles which show two-dimensional defect structures such as dislocations, "grains," and "recrystallization" boundaries after stirring (Bragg et al., 1947).

Another famous individual, whose work related in an unexpected way to the material of this chapter, was Enrico Fermi. In one of the first dynamics calculations carried out on a computer, he and colleagues treated a chain of harmonic oscillators coupled together by a nonlinear term (Fermi, 1940). The continuum limit of their model is the remarkable nonlinear partial differential equation known as the Korteweg–deVries equation, whose solution is a soliton, used to advantage in optical fibers. Damping of solitons, whether of the KdV type or the Sine Gordon (kink/antikink) type, is not to be described by linear mathematics. Incidentally, the Sine Gordon soliton is used in modeling dislocations (Nabarro, 1987). The earliest theory to describe dislocation damping using kink/anti-kink pairs was that of Seeger (1956).

2.18.5 Example of the Importance of Mesoanelastic Complexity

As noted earlier in this chapter, once hysteretic damping was finally recognized to be important to the Cavendish experiment, better agreement with theory and experiment was possible. Curiously, Henry Cavendish may have been the first person to encounter a "strange" phenomenon (which he did not discuss) (Cavendish, 1798). In his first mass swing to perturb the balance, which used a "fiber" made of copper (silvered), there was an anomalously small period of oscillation that was only 55 sec. The period reported for subsequent trials was about 421 sec.

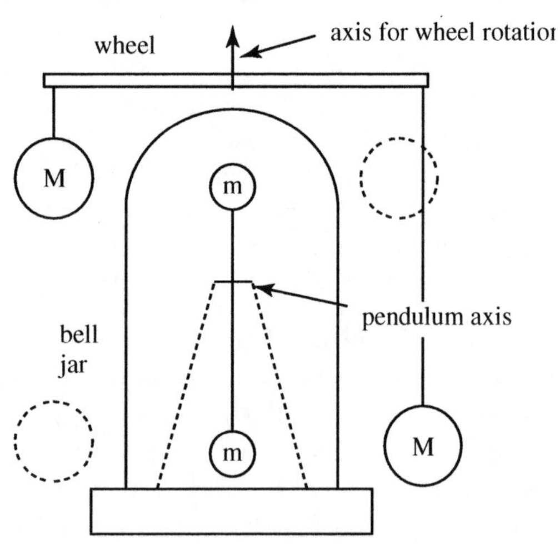

FIGURE 2.27 Physical pendulum used in the late 1980s to try and measure the Newtonian gravitational constant.

Whereas the Michell–Cavendish apparatus was a torsion balance, the instrument of Figure 2.27 is a physical pendulum. The perturbing masses, M, were hung from a bicycle wheel whose axle was suspended from the ceiling. The long-period pendulum was placed under a bell jar so that the instrument would not be driven by air currents. By rotating the wheel at constant angular velocity, the driving force on the pendulum was harmonic. (In the figure, the position of each M one-half period later are shown by the dashed circles.) Knowing the amount of damping, as determined from large amplitude free-decay, it was easy to estimate the number of orbits of the bell jar, at the resonance frequency of the pendulum, required to excite motion to a level above noise in the sensor. Surprisingly, if it were initially at rest, no amount of drive by this means was able to get the pendulum oscillating! The reason involves metastabilities of the defect structures. The potential well is not harmonic (parabolic), but is rather modulated by "fine structure." When located in a deep metastability, the small gravitational force of the drive (in nanoNewtons) is not able to "unlatch" the system. If the pendulum had been dithered (a practice used in engineering) this problem could have been, at least partly, avoided. As it was, the pendulum rested on an isolation table of the type used in optics experiments.

More recently, a Hungarian research team has used a similar apparatus and postulated that the anomalies of their experiment derive from gravity being other than prescribed by Newton (Sarkadi and Badonyi, 2001). Although they claim that there is a "strong dependence of gravitational attraction on the mass ratio of interacting bodies," this author believes that additional experiments must be performed before such a claim has merit. It may be that the anomalous behavior of their pendulum is instead the result of mesoanelastic complexity, i.e., phenomena related to nonlinear damping.

The author's most recent research on damping complexity is based on the premise that the most important scale for the treatment of internal friction is the mesoscale, and not the atomic scale (Peters 2004). Experiments to support this position center around a study of the SMA NiTinol.

2.19 Concluding Remark

Much of the material of this part of the chapter on damping is clearly not appropriate to direct engineering application. It was deemed important to present some of the extensive background information responsible for birthing the practical equations of Section 2.17. In Chapter 3, the reader will find practical aids to the measurement of damping.

Bibliography

Amengual, A., Manosa, L.L., Marco, F., Picornell, C., Segui, C.,, and Torra, V., Systematic study of the martensitic transformation in a Cu–Zn–Al alloy, reversibility versus irreversibility via acoustic emission, *Thermochim. Acta*, 116, 195–308, 1987.

Asa, F., Modal synthesis when modeling damping by use of fractional derivatives, *AIAA J.*, 34(5), 1051–1058, 1996.

Atalay, S. and Squire, P., Torsional pendulum system for measuring the shear modulus and internal friction of magnetoelastic amorphous wires, *Meas. Sci. Technol.*, 3, 735–739, 1992.

Bak, P., Tang, and Wiesenfeld, Self-organized criticality, *Phys. Rev. A*, 38, 364–374, 1988.

Bantel, M.K. and Newman, R.D., High precision measurement of torsion fiber internal friction at cryogenic temperatures, *J. Alloys Compd.*, 310, 233–242, 2000.

Barkhausen, H., *Phys. Z.*, 20, 401, 1919.

Berdichevsky, V., Hazzledine, P., and Shoykhet, B., Micromechanics of diffusional creep, *Int. J. Eng. Sci.*, 35, 10/11, 1003–1032, 1997.

Berry, M.V. and Keating, J.P., The Riemann zeros and eigenvalue asymptotics, *Siam Rev.*, 41, 2, 236–266, 1999.

Berry, B. and Nowick, A., Internal friction study of aluminum alloys containing 4 weight percent copper, *National Advisory Committee for Aeronautics, Technical Note 4225*, online at http://naca.larc.nasa.gov/reports/1958/naca-tn-4225, 1958.

Bert, C.W., Material damping: an introductory review of mathematical models. measures and experimental techniques, *J. Sound Vib.*, 29, 129–153, 1973.

Bethe, H., Lecture at Cornell University online info. at http://www.nd.edu/~bjanko/Copenhagen/Cop3.pdf, 1992.

Bordoni, P., Elastic and anelastic behavior of some metals at very low temperatures, *J. Acoust. Soc. Am.*, 26, 495, 1954.

Bormann P. and Bergmann E., eds. 2002. *New Manual of Observatory Practice*, Institute of Geophysics, University of Stuttgart, online at http://www.seismo.com/msop/nmsop/nmsop.html.

Bragg, Sir Lawrence, F.R.S., and Nye, J.F., A dynamical model of a crystal structure, *Proc. R. Soc. Lond., Ser. A, Math. Phys. Sci.*, 190, 1023, 474–481, 1947.

Braginsky, V.B., Mitrofanov, V.P., and Panov, V.I. 1985. *Systems with Small Dissipation*, The University of Chicago Press, Chicago.

Braginsky, V.B., Mitrofanov, V.P., and Okhrimenko, O.A., Isolation of test masses for gravitational wave antennae, *Phys. Lett. A*, 175, 82–84, 1993.

Breazeale, M.A. and Leroy, O. 1991. *Physical Acoustics: Fundamentals and Applications*, Kluwer Academic/Plenum Publishers, New York.

Brophy, J. 1965. Fluctuations in magnetic and dielectric solids. In *Fluctuation Phenomena in Solids*, R.E. Burgess, Ed., Academic Press, New York.

Bulsara, A.R. and Gammaitoni, L., Tuning in to noise, *Phys. Today*, 49, 39–45, 1996.

Carlson, J.D., Controlling vibration with MR fluid damping, *Sensor Technol. Design*, 19, 2, 2002, online at http://www.sensorsmag.com/articles/0202/30/main.shtml.

Cavendish, H., Experiments to determine the density of the Earth, *Philos. Trans. R. Soc. Lond.*, 469–526, 1798.

Cipra, B., The FFT: making technology fly, *SIAM News*, 26, 3, 1993, online at http://www.siam.org/siamnews/mtc/mtc593.htm.

Cleland, A. and Roukes, M., Noise processes in nanomechanical resonators, *J. Appl. Phys.*, 92, 5, 2758–2770, 2002.

Cooley, J. and Tukey, J., An algorithm for the machine calculation of complex Fourier Series, *Math. Comput.*, 19, 90, 297–301, 1965.

Coy, D. and Molnar, M., Optically driven pendulum, *Proc. NCUR* XI, 1621–1626, 1997.

Cromer, A., Stable solutions using the Euler approximation, *Am. J. Phys.*, 49, 5, 455–459, 1981.

Dean, R., Low-cost, high precision MEMS accelerometer fabricated in laminate online at http://www.eng.auburn.edu/ee/leap/MEMSFabricateTable.htm, 2002.

de Silva, C.W. 2006. *Vibration — Fundamentals and Practice*, 2nd ed., Taylor & Francis, CRC Press, Boca Raton, FL.

Fantozzi, G., Esnouf Benoit, W., and Richie, I., *Prog. Mater. Sci.*, 27, 311, 1982.

Fermi E., Pasta J., and Ulam S., Studies in nonlinear problems I, Los Alamos report (reproduced in R. Feynman, 1970: *Lectures on Physics*, Addison Wesley, Boston, 1940).

Fraden, J. 1996. *Handbook of Modern Sensors, Physics, Designs, & Applications*, 2nd ed., AIP Press (Springer), Secaucus, NJ.

Gammaitoni, L., Stochastic resonance and the dithering effect in threshold physical systems, *Phys. Rev. E*, 52, 469, 1995.

Gere, J., Timoshenko, S. 1996. *Mechanics of Materials*, Chapman & Hall, London.

Granato, A. 2002. High damping and the mechanical response of amorphous materials, Submitted to the Proceedings of the International Symposium on High Damping Materials in Tokyo, August 22, 2002 for publication in the Journal of Alloys and Compounds, private communication preprint.

Granato, A. and Lucke, K., Theory of mechanical damping due to dislocations, *J. Appl. Phys.*, 27, 583, 1956.

Greaney, P., Friedman, L.,, and Chrzan, D., Continuum simulation of dislocation dynamics: predictions for internal friction response, *Comput. Mater. Sci.*, 25, 387–403, 2002.

Grigera, T. and Israeloff, N., Observation of fluctuation–dissipation theorem violations in a structural glass, *Phys. Rev. Lett.*, 83, 24, 5038–5041, 1999.

Gross, B., The flow of solids, *Phys. Today*, 5, 8, 6–10, 1952.

Heyman, J. 1997. *Coulomb's Memoir on Statics: An Essay in the History of Civil Engineering*. Imperial College Press, London, ISBN: 1860940560.

Hilfer, P., ed. 2000. *Applications of Fractional Calculus in Physics*, World Scientific, London.

Hodges, D. 2003. Private communication (Georgia Tech School of Aerospace).

Holian, B. 2003. Mesodynamics from atomistics: a new route to Hall–Petch, private communication preprint.

Horowitz, H. and Hill, W. 1989. *Art of Electronics*, 2nd ed., Cambridge University Press.

Kimball, A. and Lovell, D., Internal friction in solids, *Phys. Rev.*, 30, 948–959, 1927.

Kohlrausch, R., *Ann. Phys.*, 12, 392, 1847, online information at http://www.ill.fr/AR-99/page/74magnetism.htm.

Kuroda, K., Does the time of swing method give a correct value of the Newtonian gravitational constant?, *Phys. Rev. Lett.*, 75, 2796–2798, 1995.

LaCoste, L., A new type long period vertical seismograph, *Physics*, 5, 178–180, 1934.

Lakes, R. 1998. *Viscoelastic Solids*, CRC Press, Boca Raton, FL.

Lakes, R. and Quackenbush, J., Viscoelastic behavior in indium tin alloys over a wide range of frequency and time, *Philos. Mag Lett.*, 74, 227–232, 1996.

Landau, L. and Lifshitz, E. 1965. *Theory of Elasticity*, Nauka, Moscow.

Lorenz, E. 1972. Predictability: Does the flap of a butterfly's wings in Brazil set off a tornado in Texas, presented to AAAS, Washington, DC.

Marder, M. and Fineberg, J., How things break, *Phys. Today*, 49, 24–29, 1996.

Marion, J. and Thornton, S. 1988. *Classical Mechanics of Particles and Systems*, 3rd ed., HBJ, Academic Press, New York, p. 114.

Milonni, P. and Eberly, J. 1988. *Lasers*. Wiley Interscience, Hoboken, NJ, p. 93.

Moon, F. 1987. *Chaotic Vibrations, An Introduction for Applied Scientists and Engineers*, Wiley Interscience, Hoboken, NJ.

Moore, G., Moore's law is described online at http://www.intel.com/research/silicon/mooreslaw.htm, 1965.

Nabarro, F. 1987. *Theory of Crystal Dislocations*, Dover, New York.

Nelson, R. and Olssen, M., The pendulum—rich physics from a simple system, *Am. J. Phys.*, 54, 112–121, 1986.

Ostermeyer, G. and Popov, V., Many-particle non-equilibrium interaction potentials in the mesoparticle method, *Phys. Mesomech.*, 2, 31–36, 1999.

Peters, R., Resonance generation of ultrasonic second harmonic in elastic solids, *J. Acoust. Soc. Am.*, 53, 6, 1673, 1973.

Peters, R., Linear rotary differential capacitance transducer, *Rev. Sci. Instrum.*, 60, 2789, 1989.

Peters, R., Metastable states of a low-frequency mesodynamic pendulum, *Appl. Phys. Lett.*, 57, 1825, 1990.

Peters, R., Fourier transform construction by vector graphics, *Am. J. Phys.*, 60, 439, 1992.

Peters, R., Fluctuations in the length of wires, *Phys. Lett. A*, 174, 3, 216, 1993a.

Peters, R., Full-bridge capacitive extensometer, *Rev. Sci. Instrum.*, 64, 8, 2250–2255, 1993b. This paper describes an SDC sensor with cylindrical geometry. Other geometries, including the more common planar one, are described online at http://physics.mercer.edu/petepag/sens.htm.

Peters, R. 2000. Autocorrelation Analysis of Data from a Novel Tiltmeter, abstract, *Amer. Geo. Union annual mtg.*, San Francisco.

Peters, R. 2001a. The Stirling engine refrigerator — rich pedagogy from applied physics, online at http://xxx.lanl.gov/html/physics/0112061.

Peters, R. 2001b. Creep and Mechanical Oscillator Damping, http://arXiv.org/html/physics/0109067/.

Peters, R. 2002a. The pendulum in the 21st century—relic or trendsetter, *Proc. The Int'l Pendulum Project*, University of New South Wales, Australia, Proceedings, October, 2002.

Peters, R. 2002b. The soup-can pendulum, *Proc. The Int'l Pendulum Project*, University of New South Wales, Australia, Proceedings, October. 2002.

Peters, R. 2002c. Toward a universal model of damping—modified Coulomb friction online at http://arxiv.org/html/physics/0208025.

Peters, R. 2003a. Graphical explanation of the speed of the Fast Fourier Transform, online at http://arxiv.org/html/math.HO/0302212.

Peters, R. 2003b. Nonlinear damping of the 'linear' Pendulum, online at http://arxiv.org/pdf/physics/03006081.

Peters, R. 2003c. Flex-Pendulum—basis for an improved timepiece, online at http://arxiv.org/pdf/physics/0306088.

Peters, 2004. Friction at the Mesoscale. In *Contemporary Physics*, P. Knight, Ed., Vol. 45, no. 6, 475–490, Imperial College, London 2004.

Peters, R. and Kwon, M., Desorption studies using Langmuir recoil force measurements, *J. Appl. Phys.*, 68, 1616, 1990.

Peters, R. and Pritchett, T., The not-so-simple harmonic oscillator, *Am. J. Phys.*, 65, 1067–1073, 1997.

Peters, R., Breazeale, M., and Pare, V., Temperature dependence of the nonlinearity parameters of Copper, *Phys. Rev.*, B1, 3245, 1970.

Peters, R., Cardenas-Garcia, J., and Parten, M., Capacitive servo-device for microrobotic applications, *J. Micromech. Microeng.*, 1, 103, 1991.

Portevin, A. and Le Chatelier, M., Tensile tests of alloys undergoing transformation, *C. R. Acad. Sci.*, 176, 507, 1923.

Present, R. 1958. *The Kinetic Theory of Gases*. McGraw-Hill, New York.

Press, W., Flannery, B., Teukolsky, S., and Vetterling, W. 1986. *Numerical Recipes—the Art of Scientific Computing*. Cambridge University Press.

Purcell, E., Life at low Reynolds number, *Am. J. Phys.*, 45, 3–11, 1977.

Richardson, M., and Potter, R. 1975. Viscous vs structural damping in modal analysis, *46th Shock and Vibration Symposium*.

Roukes, M., Plenty of room indeed, *Scientific American*, 285, 48–57, 2001.

Sarkadi, D. and Badonyi, L., A gravity experiment between commensurable masses, *J. Theor.*, 3–6, 2001.

Saulson, P., Stebbins, R., Dumont, F., and Mock, S., The inverted pendulum as a probe of anelasticity, *Rev. Sci. Instrum.*, 65, 182–191, 1994.

Seeger, A., On the theory of the low-temperature internal friction peak observed in metals, *Philos. Mag.*, 1, 1956.

Singh, A., Mohapatra, Y., and Kumar, S., Electromagnetic induction and damping, quantitative experiments using a PC interface, *Am. J. Phys.*, 70, 424–427, 2002.

Speake, C., Quinn, T., Davis, R., and Richman, S., Experiment and theory in anelasticity, *Meas. Sci. Technol.*, 10, 430–434, 1999. See also Quinn, Speake and Brown, 1992: Materials problems in the construction of long-period pendulums, *Philos. Mag.* A 65, 261–276, 1999.

Steinmetz, C.P., *Complex Number Technique, paper given at the International Electrical Congress*, Chicago, 1893.

Stokes, G., On the effect of the internal friction of fluids on the motion of pendulums, *Trans. Cambridge Philos. Soc.*, IX, 8, 1850, read December 9, 1850.

Strang, G., Wavelet transforms versus Fourier Transforms, *Bull. Am. Math. Soc.*, 28, 288–305, 1993.

Streckeisen, G. 1974. *Untersuchungen zur Messgenauigkeit langperiodischer Seismometer*, Diplomarbeit, Institut für Geophysik der ETH Zürich (communicated privately by E. Wielandt).

Tabor, M. 1989. The FUP Experiment. In *Chaos and Integrability in Nonlinear Dynamics: Introduction*. Wiley, New York.

Thomson, W., Tait, G. 1873. *Elements of Natural Philosophy, Part I*. The Clarendon Press, Oxford, (Thomson was later known as Lord Kelvin).

Urbach, J., Madison, R., and Markert, J., Reproducibility of magnetic avalanches in an Fe–Ni–Co alloy, *Phys. Rev. Lett.*, 75, 4694, 1995a.

Urbach, J., Madison, R., and Markert, J., Interface depinning, self-organized criticality, and the Barkhausen effect, *Phys. Rev. Lett.*, 75, 276, 1995b.

Venkataraman, G. 1982. Fluctuations and mechanical relaxation. In *Mechanical and Thermal Behavior of Metallic Materials*, Caglioti, G. and Milone, A., eds., pp. 278–414. North-Holland, Amsterdam.

Visintin, A. 1996. *Differential Models of Hysteresis*. Springer, Berlin.

Westfall, R. 1990. Making a world of precision: Newton and the construction of a quantitative physics. In *Some Truer Method. Reflections on the Heritage of Newton*, F. Durham and R.D. Purrington, eds., pp. 59–87. Columbia University Press, New York.

Wielandt, E. 2001. *Seismometry*, section Electronic Displacement Sensing, online at http://www.geophys.uni-stuttgart.de/seismometry/hbk_html/node1.html.

Zemansky, M.W. 1957. *Heat and Thermodynamics*, 4th ed., McGraw-Hill, New York, p. 127.

Zener, C. 1948. *Elasticity and Anelasticity of Metals*, Chicago Press, Chicago.

3

Experimental Techniques in Damping

Randall D. Peters
Mercer University

Summary

This chapter is a continuation of Chapter 1 and Chapter 2, and is concerned with practical experimental techniques for measuring damping. It begins with a discussion of the requirements placed on electronics. After demonstrating the importance of sensor linearity using a computer simulation, the issues of data acquisition and processing are addressed. The power of the Fast Fourier Transform is illustrated, not just for spectral analysis, but also (in "short time" form) for measuring the damping of each component when a system oscillates with multiple modes. Various sensor types are discussed in relation to their advantages and disadvantages for specific applications through the treatment of seven different systems studied in free decay. These seven cases differ with respect to factors such as (i) eigenfrequency, (ii) material type, and (iii) method of estimating the logarithmic decrement, and thus the Q of the decay. In the case of some solids, damping is shown to result largely from defects in the structures. A powerful test for nonlinear damping is demonstrated: simply looking at a graph of Q to see whether it changes with time. Two examples of driven oscillators are given. The first being very nearly linear, and the second being highly nonlinear, due to an anharmonic restoring force involving magnets plus several simultaneously acting damping mechanisms. The nonlinear system is used to illustrate difficulties in interpretation that can arise in driven systems due to phenomena such as frequency and/or amplitude jumps involving hysteresis. Then an illustration is given of how elastic-type nonlinearities may couple with damping-type nonlinearities in order to determine which modes of a complex system survive during the transient approach to steady state. Mechanical noise, another important feature of nonlinear damping, is also examined. The magnitude of the 1/f character in an evacuated pendulum is shown to decrease with time, as the oscillator is allowed to stabilize against creep. The final sections address the common misconception that viscous air friction is the most important form of mechanical oscillator damping. Cases are chosen to demonstrate that (i) internal friction is nearly always also important, if not the most important, and moreover, that (ii) fluid damping is rarely simple — involving the density as well as the viscosity of the fluid. It is shown that damping has a complicated frequency dependence, as opposed to the simple (overly idealized) form predicted by common theory.

3.1 Electronic Considerations

3.1.1 Sensor Linearity

The importance of sensor linearity is often overlooked. It is naively assumed that one can simply employ a lookup table to provide calibration corrections. This assumption can result in serious misinterpretations of spectral data, especially in a multimode system. A classic example of artifacts (nonreal signals) that result from a nonlinear sensor is to be found in the ear. The phenomenon, known as aural harmonics, is well known to musicians and figures in the use of "fortissimo" and "pianissimo" in orchestral music. In this chapter we describe how the artifacts mentioned in Chapter 2 are generated. Figure 3.1 illustrates differences according to the nature of the nonlinearity.

The only "real" signals in Figure 3.1 are at frequencies f_1 and f_2. The number and type of other "unreal" (artifact) signals depends on the type of nonlinearity. The sensor response for the left graph (quadratic) is of the form $V = ax + bx^2$, whereas for the right graph $V = ax + bx^2 + cx^3$. The influence of terms other than $V = ax$ (ideal, linear output voltage) was generated by (i) simulating the pair of harmonic signals, (ii) inputting these signals to each simulated sensor, respectively, and (iii) performing a Fast Fourier Transform (FFT) on the output.

Although it is possible to understand mathematically the various artifacts using trigonometric identities, the phenomenon is much easier to demonstrate with a computer. For Figure 3.1, all

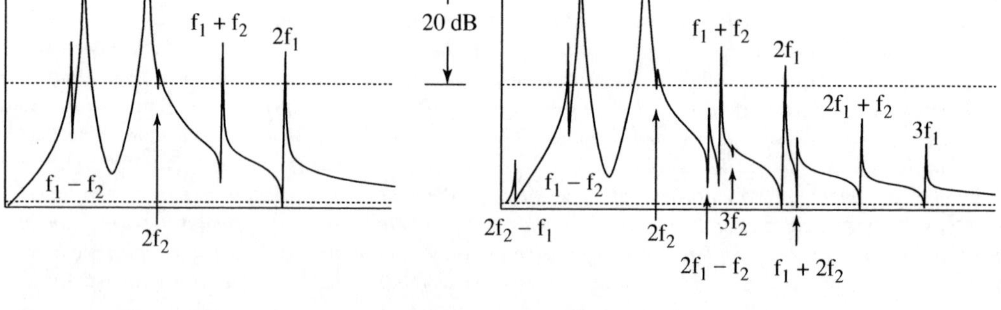

FIGURE 3.1 Spectral illustration of sum and difference artifact frequencies according to nonlinear sensor type.

numerical operations were performed with code written by the author using QuickBasic. It was used to (i) simulate the harmonic signal that was written to a data file, after which it was (ii) read by the FFT algorithm based on the details supplied in *Numerical Recipes* (Press et al., 1986).

3.1.2 Frequency Issues

The choice of a sensor depends largely on the frequencies to be measured. For higher frequencies an excellent instrument for data collection is a digital (storage) oscilloscope, where a microphone can often be directly connected to the instrument. At lower frequencies, a serial-port analog-to-digital converter (ADC) is generally adequate and user-friendly. Examples of each will be provided. The majority of examples considered in this chapter involve low frequencies, where the eigenmode is typically described not in terms of frequency but rather the period (reciprocal of frequency).

3.1.3 Data Acquisition

In the absence of sophisticated data collection and analysis tools, the true character of damping is not readily discovered. Proper characterization is important, since a crude estimate of the damping, based on a single parameter (such as the viscous linear model), may be inappropriate if the oscillator is driven at places (either frequency or amplitude) other than where the parameter was measured. Some of the examples from the experiment that follows were selected to demonstrate the importance of nonlinearity. The probability that an oscillator, selected at random, might have a Q that varies in time is proving to be more significant than anticipated. Were it not for dramatic improvements in numerical-type technology, this improved understanding of damping would not have been possible.

As with computer technology in general over the last decade, ADCs have become much more powerful. The Dataq model 700, for example, is superior (at lower frequencies) to many of the "plug-in" boards of the previous generation that were several times more expensive. The Dataq ADC operates through the USB port (Windows 98 and later), has 16-bit resolution and the software support is excellent. Especially useful for the present purposes are its ability to (i) easily perform data compression with which to view long records, (ii) quickly compute an FFT according to different, useful options, and (iii) easily output files to a spreadsheet.

3.2 Data Processing

3.2.1 Language Type

The author's experience with software began with early computers and even included the loading of the Fortran compiler of a PDP-11 using punch-tape. He has programmed computers (or hardware-specific processors) with (i) machine code, (ii) assembly language, (iii) Fortran, and (iv) Basic, and he has acquired a rudimentary knowledge of Pascal and C++. The drudgery of machine coding was a factor in his quest to better understand the Fourier transform (Peters, 1992, 2003a, 2003b, 2003c, 2003d). His philosophy with regard to numerical methods is similar to his view of hardware: choose the simplest package (lowest level of sophistication) consistent with the desired results for the problem at hand. The reader may be surprised to learn that QuickBasic (which some have modernized to Visual Basic for Windows) is his favorite language. Nearly all simulation results presented in this chapter were generated with the DOS version of QuickBasic.

3.2.2 Integration Technique

Too few have discovered the powerful integration scheme in which Cromer (1981) modified the unstable Euler algorithm. The difference between the two methods involves the sequencing (order) of updates to the state vectors in the discrete approximation of the integrals. The method was called the "last point approximation" (LPA) by Cromer, whereas the Euler technique would be called the "first point approximation" according to this nomenclature. The LPA was discovered by a high school student working for Cromer. She was attempting to simulate planetary motion with the Euler method and accidentally coded the LPA. The author first used the LPA to do intercept analyses for the U.S. antisatellite program — computing, among other things, orbital ephemerides. More recently he has used it in place of Runge Kutta techniques to do all kinds of mechanical system simulations, including nonlinear types with several DoF. A physics theorist at Texas Technical University, Professor Thomas Gibson, now regularly uses the LPA as part of the graduate-level course which he teaches in numerical methods.

3.2.3 Fourier Transform

With the Cooley–Tukey improvement to make it fast, the Fourier transform has become a tool of major software importance. Just as the integrated circuit dramatically changed hardware development, the FFT has had a profound influence on the evolution of scientific code.

Whereas many recognize the value of the FFT for viewing "raw" spectral data, few have discovered other powerful tools based in the FFT. For example, autocorrelation is unrivaled in its ability to uncover low-frequency signals of fairly short duration that are corrupted by noise. The number of cycles is not great enough (Heisenberg effect) for a well-defined line to be observed in the FFT by itself. Through the Wiener–Khintchin theorem, the autocorrelation overcomes this limitation. It is computed by multiplying the transform by its conjugate and then taking the inverse transform. The author has used this technique to study free oscillations of the earth (Peters, 2004).

3.2.3.1 Short Time Fourier Transform

A powerful software tool is one in which the Fourier transform is not computed over the entire length of a record. Instead, the record is subdivided (usually with some degree of overlap between adjacent subsections), and the FFT is computed for each subsection. Because the data are generally of the temporal (rather than spatial) type, the technique is called the short time Fourier transform (STFT). For equivalent processing, where the independent variable has units of meters rather than seconds (as in optics applications), the technique could just as well be called the short space Fourier transform.

The STFT is especially useful when waveforms are not pure harmonic, as from a single-degree-of-freedom (single-DoF) oscillator. For systems with multiple modes, whether they derive from eigenmodes as recognized by most, or from mechanical noise as recognized by a few (generated as part of the internal friction of load bearing members); the STFT is a powerful means for isolating and thus determining the temporal history of individual spectral components.

The most common form of the STFT is the canned programs that are a part of software packages such as LabVIEW. With the Dataq software it is easy to accomplish the same thing manually, since one can readily step in time from place to place of a stored record, computing the FFT at any position. The intensity at a given position is obtained by clicking on the displayed spectral line of interest, which provides the value either in dB (Dataq version) or in volts. The amplitude history in dB of the line is thus obtained (equally spaced-in-time values) with a simple click of the mouse. In this way, the free decay of a single component of the system can be readily extracted from the total system response. For the present purposes, the individual intensities were copied by hand to paper and later typed into a spreadsheet for plotting. The process is not laborious, since the number of necessary points is typically less than two dozen.

An example of a manually generated STFT is provided in Figure 3.2. Unlike the methodology described above (operating on experimental data residing in a Dataq folder), the record of Figure 3.2 was generated by computer and written to an output file. The data correspond to three superposed, exponentially damped sinusoids.

From the upper graph (time record), it is not clear how the individual components are changing with time. The triplet of components becomes obvious in the frequency domain (lower left), and when the

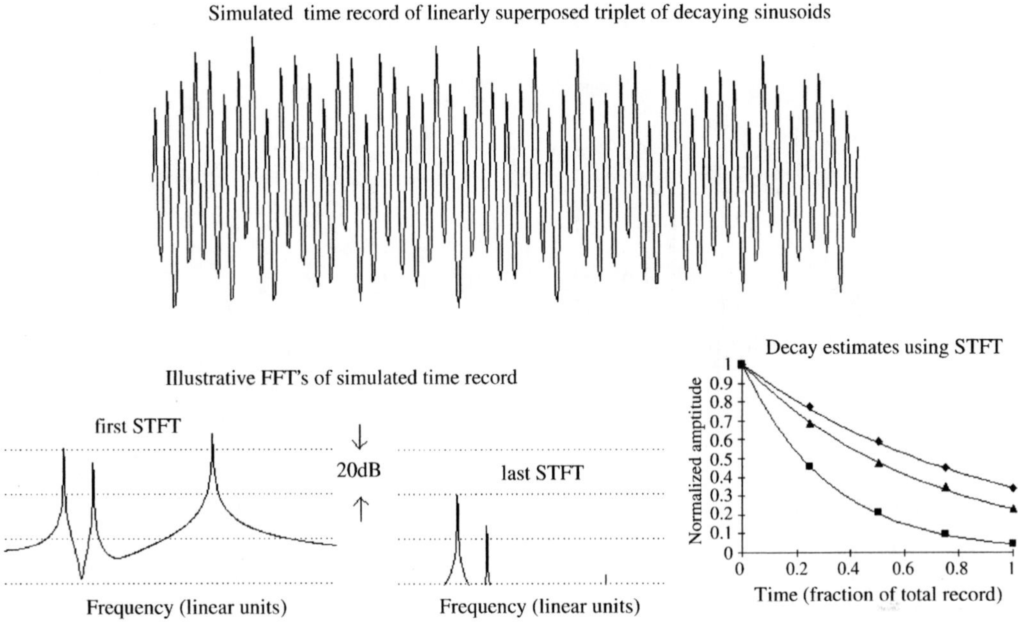

FIGURE 3.2 Example of separating the time decay of superposed components using the STFT.

intensities of each line are plotted *versus* time (lower right) the exponential character of each becomes visible. To test the viability of this method, the individual damping parameters were evaluated from a given STFT graph and found to be in excellent agreement with the values supplied to the simulation.

3.2.3.2 Example Use of the STFT

The majority of the examples given in this chapter avoid low-energy oscillations as space does not allow mesoanelastic regimes to be treated at length. The following case was chosen to illustrate (i) the importance of the Portevin–LeChatelier (PLC) effect on damping (Portevin and Le Chatelier, 1923), and (ii) the power of the STFT in eliminating the influence of clutter. The STFT benefit was expected, since the difference between a noisy record and a multimode case like the previous example is in the number of modes. The PLC effect is significant for high-energy internal friction dissipation even though the jumps for which the effect is known are not obvious at these energies. Evidently, this is a consequence of the large number of events, for which the average effect is a fairly smooth decay.

The scales for the two graphs of Figure 3.3 are different by nearly three orders of magnitude, with the level of the sensor output for each case being indicated (mid-range values). Two features are evident from a direct visual inspection: (i) the change from a smooth to a jerky decay in going from high to low levels, and (ii) a 4% increase in the frequency of oscillation at the lower energy. The latter is recognizable from the vertical lines that have been added (every fifth peak). One of the more interesting (and surprising to most) features of the lower graph is that phase of the oscillation is not significantly altered as the result of mean position jumps.

The following information is provided for those skeptical of the comments concerning the lower trace of Figure 3.3. From the study of hundreds of low-level decays in different mechanical oscillators, the author has become confident that the jumps shown are not sensor (or other electronic) artifacts. Prejudice against this conclusion has been considerable over the last 14 years, in spite of the fact that a similar phenomenon was noted (and accepted) in magnetic materials many years ago, i.e., the Barkhausen effect (Barkhausen, 1919). Unfortunately, the related phenomenon in mechanical systems (PLC effect) is hardly known among physicists.

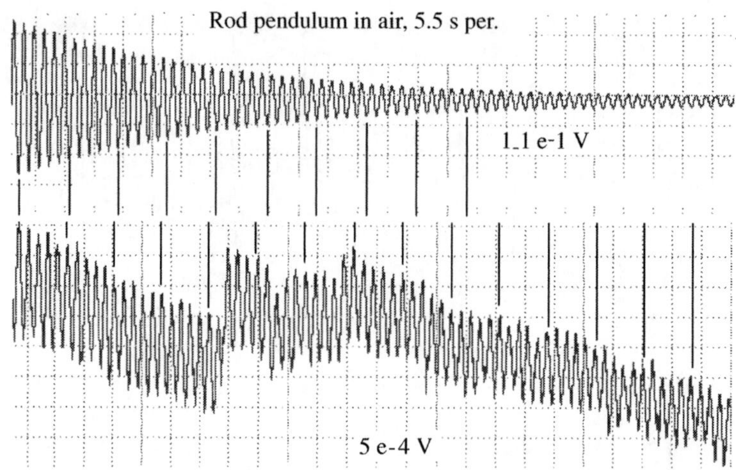

FIGURE 3.3 Illustration of the character changes in decay in going from high to low energies of a pendulum.

The value of the STFT for the study of data such as that of Figure 3.3 is illustrated in Figure 3.4. Whereas Figure 3.3 showed only the first and the last portions of a long record, here the STFT was applied to the entire record using 27 different FFTs.

As seen in the upper left curve, there is a sharp decrease in the damping in STFT No. 5. Thus, the intensity was replotted in each of the intervals from 0 to 5 and 5 to 27. Both intervals show a near perfect linear trendline fit, indicating exponential decay. Thus the Q was found to quickly change from 78 to 340 at an energy level in the region of 10^{-10} J. Although air damping was a factor in the early part of the record, it is not thought to be capable of causing the rapid change in Q that was observed. A similar sharp

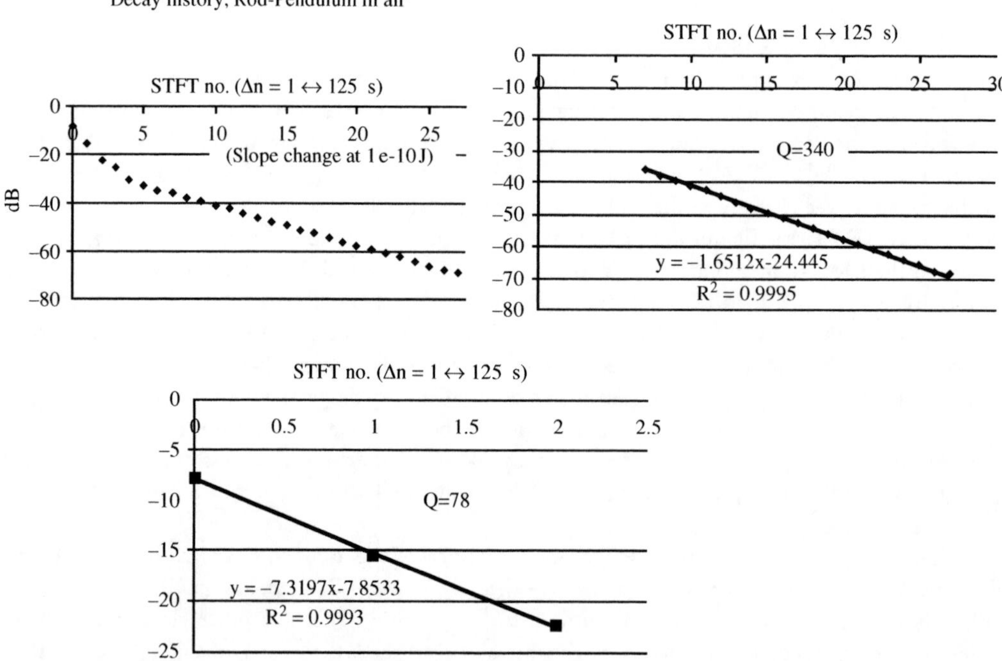

FIGURE 3.4 Example of the use of STFT analysis applied to a dataset, the first and last portions of which are shown in Figure 3.3.

change in the damping of this same pendulum was seen at roughly 10^{-11} J with the pendulum swinging in a high vacuum. The slope change was equally rapid for the vacuum case, but the change in Q was from 120 to 210. It is not known why the damping at low-level energy in a vacuum would have a lower Q than in air. Perhaps the difference derives from a different placement of the knife-edges on the silicon flats. Some of the damping of this pendulum is the result of the knife-edges being fabricated from brass rather than a harder metal such as carbon steel.

3.3 Sensor Choices

Box 3.1 shows some representative sensors for damping measurements. The list is far from exhaustive; for a detailed description of each (plus discussion of other types), the reader is referred to Fraden (1996). Of the transducer types indicated, position sensors are generally the most versatile; but the present chapter also provides examples of the use of (i) velocity, (ii) microphone, and (iii) photogate measurements.

In addition to the need for linearity (discussed in Section 3.1 above), the ideal sensor will be noninvasive. In reality, it is not possible to perform a measurement that does not at some level perturb the system under study. The least perturbative types of direct measurement are optical and electrical — capacitive, followed by inductive.

Some of the advantages and disadvantages of the devices indicated in Box 3.1 are provided below.

Box 3.1

Some Sensor Types

Representative Sensors for Damping Measurements					
Position	Velocity	Pressure	Time Interval	Acceleration	Force/Strain
Capacitive	Faraday law (electromagnetic)	Microphone	Photogate	Accelerometer	Strain gauge
LVDT		Pressure gauge			
Optical		Capacitive			
Encoder		Optoelectronic			
Shadow		Piezoresistive			
Potentiometric					

3.3.1 Direct Measurement

3.3.1.1 Position Sensors

The inductive linear variable differential transformer (LVDT) is a sensor that is commonly used in engineering applications. Thus, it has been a natural choice for many position-sensing purposes; but it is both more invasive and noisier than capacitive sensors. Wielandt (2001) notes the following concerning the advantage of capacitive over inductive sensors: "Their sensitivity is ... typically a hundred times better than that of the inductive type."

Optical encoders are also readily available and have been used extensively. Because of their digital nature, based in a finite number of elements, their low-level resolution is poor compared to capacitive devices.

Optical sensing by shadow means is easy to employ — for example, using a solar cell of the type discussed later. The method is afflicted, however, by (i) an offset voltage, and (ii) the degrading influence of background light.

Potentiometers are very easy to use, but compared to other position sensors they are extremely invasive because of Coulomb friction in the slider and also the bearings that support it.

3.3.1.2 Velocity Sensor

The most important velocity sensor is that which functions on the basis of Faraday's law. Using a magnet and a coil, an electromagnetic force is generated in the wire of the coil when it experiences a changing magnetic flux. Prevalent in seismometers before the advent of broadband (feedback) instruments, its primary shortcoming is poor sensitivity at low frequencies.

3.3.1.3 Time Interval

Photogates have become the primary means for kinematic studies in introductory physics laboratories. Combined with compact, user-friendly timers, it is possible to measure both period and velocity. As illustrated later, they can be easily used to measure damping in slowly oscillating systems, but only in a limited amplitude range.

3.3.2 Indirect Measurement

In the cases of (i) pressure, (ii) acceleration, and (iii) force/stress sensing, the measurement is an indirect one. Consider, for example, the Ruchhardt experiment to measure the ratio of heat capacities of a gas (discussed in Chapter 2). The oscillation of the piston could be measured in several different ways. For instance, direct position sensing could be accomplished by attaching a small electrode to the piston and allowing it to move between stationary capacitor plates. Alternatively, a "flag" on the piston could be used to interrupt the light beam of a photogate. Depending on constraints, however, the easiest method might be an indirect measurement in which a pressure sensor monitors the gas through a catheter communicating with the cylinder of the apparatus.

Accelerometers can sometimes be connected directly to an oscillator, but only if the mass of the instrument is very small compared to the system being studied. As with the measurement of velocity, their sensitivity at low frequency is very poor (the second derivative of position yielding a response that is proportional to frequency-squared).

Strain gauges are easy to employ but also lack sensitivity (compared to position measurement), since they communicate with a very small portion of the oscillating sample (if noninvasive).

3.4 Damping Examples

3.4.1 Case 1: Vibrating Bar — Linear with Significant Noise

The simplest means to measure the Q of an oscillator whose frequency is in the range of the human ear is to use a microphone connected to a digital oscilloscope. In this case, the microphone was an inexpensive dynamic type and the oscilloscope was a Tektronix TDS 3054. A better choice, had it been available, would be an electret microphone. The ring-down of a xylophone bar, following a strong (sharp) hammer strike, is shown in Figure 3.5.

The voltage *versus* time of the microphone output was saved to memory in the oscilloscope, from which the digital record was output to a floppy disk, using the CSV format. Data from the disk were read into columns A and B of an Excel spreadsheet using "Open file." An envelope fit was then performed on the turning points by placement of trial and error data into column C, using "autofill." A separate graph was generated for each value of the constant b in the expression "$= 0.04 * \exp(-b * \mathrm{A1})$" typed into Cell C1. (The lower turning points were obtained by typing "$= -b1$" into Cell D1 and using autofill.

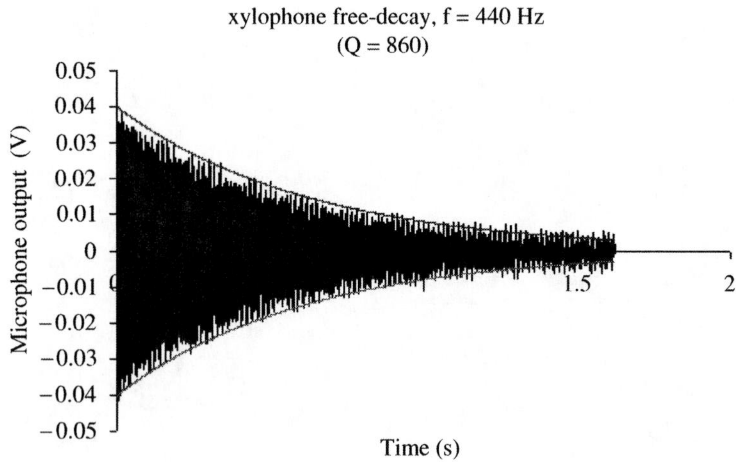

FIGURE 3.5 Free-decay record of a vibrating bar.

(Additional details concerning the use of Excel in this manner will be provided in the discussion of seismometer damping that follows.)

Although optimizing algorithms could be generated to perform such a fit (with a probable slight increase in accuracy), this visual technique is preferred here as it is more understandable, user-friendly, and its performance is proven. The total time required in Excel to generate Figure 3.5 using a 2K record (2048 points) is typically only a few minutes with a modern Pentium computer.*

Once a satisfactory fit was obtained ($b = 1.6$ in Figure 3.5), the Q was estimated using

$$Q = \frac{\pi}{\Delta} = \frac{\pi f}{b} \tag{3.1}$$

There is a fair amount of electronic noise in Figure 3.5 because the microphone was connected directly to the oscilloscope. The smallest bandwidth of the oscilloscope, at 20 MHz, causes a large amount of Johnson (white) noise. Narrowing the bandpass by means of a preamplifier would improve the quality of the data dramatically. Such is typically true of signal to noise ratio (SNR) improvement by tailoring the electronics to the need.

3.4.2 Case 2: Vibrating Reed — Example of Nonlinear Damping

To illustrate another sensing technique, the system shown in Figure 3.6 was used.

A 90° twist was given to a hacksaw blade after heating with a torch and quenching. One end was clamped to the vertical post shown and a piece of cardboard was taped to the other end. An incandescent lamp is placed above the cardboard, which vibrates horizontally, and a solar panel below the cardboard is used as a sensor. The solar panel in this case is a commercial unit that comes with a cigarette lighter plug for charging automobile batteries. The output from the panel goes to the Tektronix digital scope also in the picture.

Unlike other sensing schemes described in this chapter, the solar panel output is not bipolar but instead has a constant voltage offset corresponding to the equilibrium position of the reed.

The frequency of oscillation is too low to operate the oscilloscope with a.c. coupling. Thus, it is important to make the d.c. offset as small as possible. This was accomplished by shielding nonactive parts

*A disclaimer is in order at this point. Present comments by the author should not be interpreted as an endorsement of Microsoft products in general. Although QuickBasic and Excel have both proven unusually beneficial to the work described in this chapter, they are the only software packages marketed by the company to have received a strong endorsement from the author.

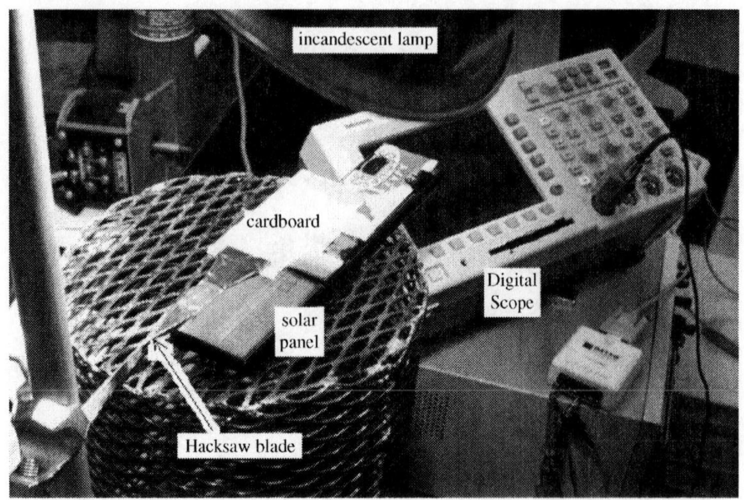

FIGURE 3.6 Setup for measurement of vibrating reed free-decay.

of the solar cell from the lamp. Although the d.c. offset could be removed with a voltage bucking battery, this was found to introduce unacceptable noise spikes. With the offset, the gain of the electronics was limited by the amount of vertical position shift allowed by the scope. The results of this study are illustrated in Figure 3.7.

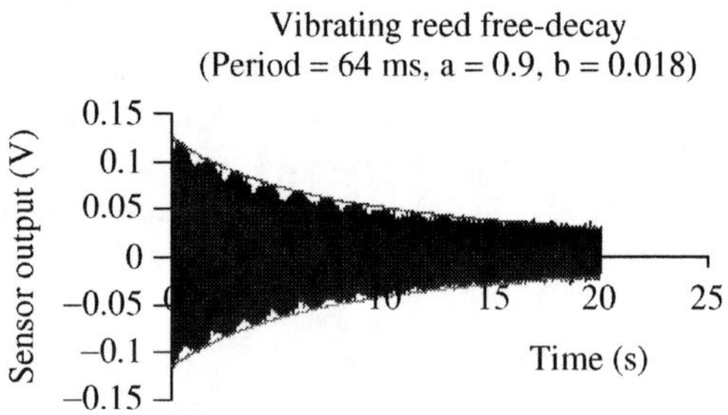

FIGURE 3.7 Vibrating reed decay with amplitude-dependent damping.

The nonzero value of b (0.018) in Figure 3.7 indicates the presence of amplitude-dependent damping (refer to Equation 2.61, Chapter 2). The nonlinear damping in this case probably derives from the air, rather than internal friction of the hacksaw blade. Its presence causes the Q of the system to increase with time.

The Q of the system is calculated from the expression

$$Q = \frac{\pi}{(b + ay)\tau} \tag{3.2}$$

where τ is the period of oscillation. At the start of the record ($y = 0.12$) the Q is 390 and it approaches 2700 as the amplitude approaches zero.

3.4.3 Case 3: Seismometer

Since the ability of a seismometer to detect tremors is proportional to the square of the period of the instrument, they require a good low-frequency sensor. The most common sensor for the latest generation commercial instruments is a half-bridge (differential) capacitive type. Because of the greater sensitivity and linearity of the full-bridge symmetric differential capacitive (SDC) sensor mentioned in Chapter 2, it is well suited to these applications, being easy to employ. (The full-bridge character is described in a TEL-Atomic tutorial (Peters, 2002).) A significant advantage of the SDC symmetry (equivalent electrically to the inductive LVDT) is its relative insensitivity to construction imperfections, such as roughness of surface and nonparallelism of electrodes. Thus, construction can be done crudely without serious degradation of performance. For example, electrodes of the first prototype of the SDC sensor were fabricated from sheet

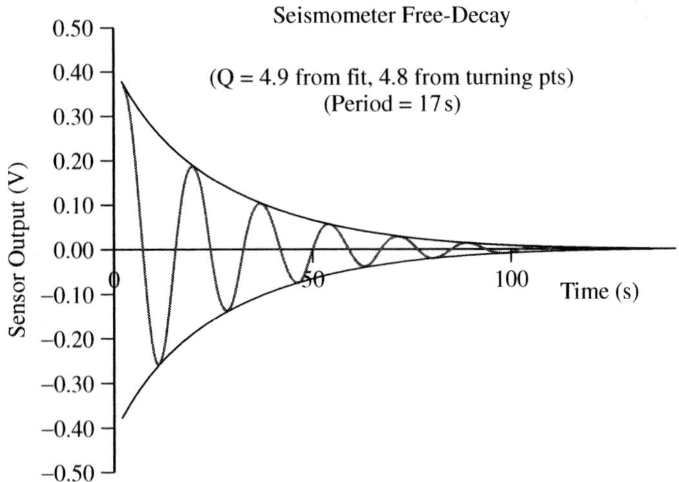

FIGURE 3.8 Free-decay record of the Sprengnether vertical seismometer, period 17 sec.

copper that was cut with shears and subsequently flattened by hammer on a hard plane surface. This stands in stark contrast to the optically polished surfaces necessary for the use of some sensors.

The Sprengnether vertical seismometer discussed at several points in Chapter 2 was studied in a configuration for which $Q = 4.9$, as determined by an external 990 Ω resistor to provide induced-current damping. The resistor was connected across the coil that is part of the original equipment and which moves (with the mass of the instrument) in the field of a stationary magnet, i.e., a Faraday's law (velocity) detector. Excitation to initiate the free-decay study was accomplished by applying an alternating (square wave) current to the coil, reversing the direction of the current at each turning point of the motion of the mass. The fundamental (Fourier series) of a square-drive generated this way is shifted 90° from the mass motion, corresponding therefore to resonance. After cessation of the drive, data as shown in Figure 3.8 were collected with a Dataq DI-700 ADC (16-bit).

The graph in Figure 3.8 was generated with Excel after the Dataq record was saved to floppy disc as an ∗.dat (CSV) file. It was imported to Excel using "open file" with "comma delimiter." Once in Excel, these data were shifted one place to the right (from the default A column to the B column) to accommodate computer generation of a time-data column. The column of time values was generated according to the sample rate, the value of which is by default saved to the data file. To generate the time column, a 0 was placed in the first row, n corresponding to the start of data. Dropping down one row in the A column, "$= A_n + 1/(\text{sample rate})$" was typed, to increment the time. Then the lower right hand corner "small solid square" of the box containing this time was grabbed and held with the left button of the mouse to autofill all the way to the last time point of the data. The computer-generated exponentials, which correspond to the turning points, were obtained by generating two additional columns. These were obtained by placing the cursor at a row corresponding to the time A_n (in column C) and then typing "$= A_0 * \exp(-(omega/2/Q) * A_n)$." The value of A_0 is obvious from the data and a first estimate for Q can be quickly obtained from about a dozen turning points (read with the Dataq software before the data are ever saved).

The technique is illustrated in Table 3.1. For example, in the case of Figure 3.8, $Q = 4.8$ from the 13 turning points. Thus the argument of the exponential was set to 0.0385. Using autofill, the columnar (upper) exponential was then quickly produced. Then a second (adjacent) column D was generated in similar manner, by taking the negative of the last point and then autofilling to the top row. (When one autofills downward, the rate with which Excel traverses the rows increases exponentially after the last row of data has been passed; thus, it is much easier to fill upwards rather than downwards.)

Once the pair of exponentials being fitted to the data have been graphed, along with the data, it is simple to adjust the curves by varying the argument (in this case, small changes around 0.0385) until a

TABLE 3.1 Estimation of Q from the Turning Points

Use of Excel to estimate logarithmic decrement from turning points of the motion

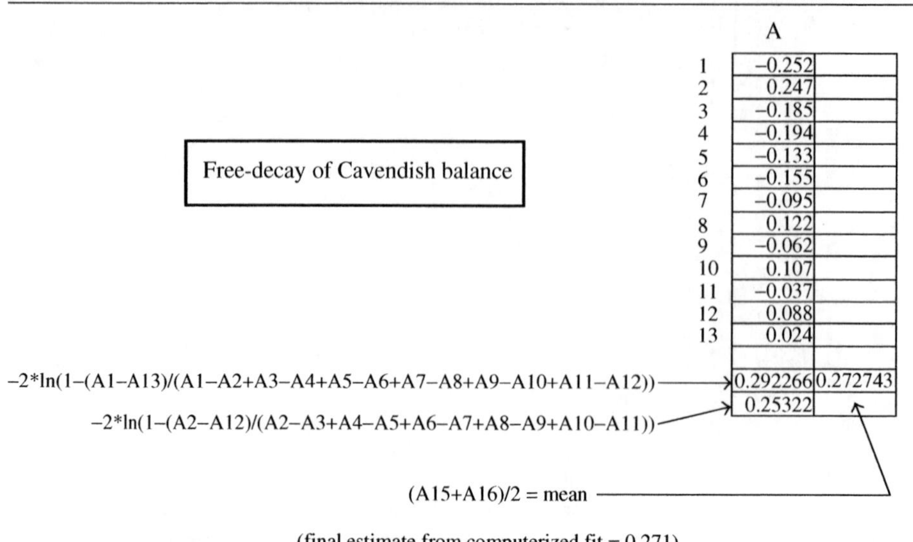

Free-decay of Cavendish balance

	A	
1	−0.252	
2	0.247	
3	−0.185	
4	−0.194	
5	−0.133	
6	−0.155	
7	−0.095	
8	0.122	
9	−0.062	
10	0.107	
11	−0.037	
12	0.088	
13	0.024	
	0.292266	0.272743
	0.25322	

−2*ln(1−(A1−A13)/(A1−A2+A3−A4+A5−A6+A7−A8+A9−A10+A11−A12)) → 0.292266

−2*ln(1−(A2−A12)/(A2−A3+A4−A5+A6−A7+A8−A9+A10−A11)) → 0.25322

(A15+A16)/2 = mean → 0.272743

(final estimate from computerized fit = 0.271)

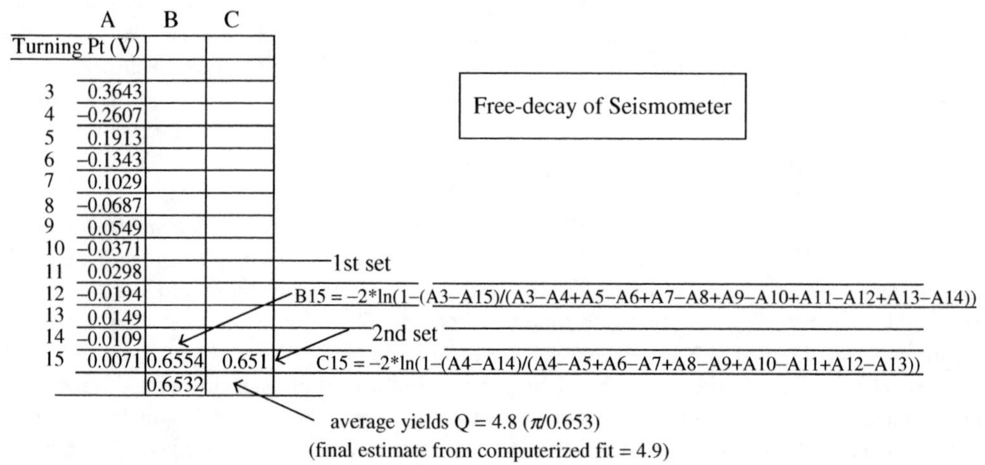

Free-decay of Seismometer

	A	B	C
	Turning Pt (V)		
3	0.3643		
4	−0.2607		
5	0.1913		
6	−0.1343		
7	0.1029		
8	−0.0687		
9	0.0549		
10	−0.0371		
11	0.0298		
12	−0.0194		
13	0.0149		
14	−0.0109		
15	0.0071	0.6554	0.651
		0.6532	

1st set

B15 = −2*ln(1−(A3−A15)/(A3−A4+A5−A6+A7−A8+A9−A10+A11−A12+A13−A14))

2nd set

C15 = −2*ln(1−(A4−A14)/(A4−A5+A6−A7+A8−A9+A10−A11+A12−A13))

average yields Q = 4.8 (π/0.653)

(final estimate from computerized fit = 4.9)

good fit is obtained (using autofill each time). The fit is rapid and accurate when there are not too many parameters to vary, since the eye is well suited to this operation. Upon obtaining the best-fit by this means, the preliminary value of $Q = 4.8$ was altered to the final value of $Q = 4.9$.

Note that in Table 3.1 a new Excel worksheet was employed (column A no longer the time as in the discussion above). The voltages corresponding to the turning points (maximum and minimum) were each read by placing the cursor at an extremum and manually recording the value displayed in turn by the Dataq software. These values were then typed into Excel, as opposed to the "file open" method for importing large datasets, i.e., as used to generate Figure 3.7.

3.4.4 Case 4: Rod Pendulum with Photogate Sensor

One of the simplest ways to measure damping at larger levels is to use a photogate of the type common to general education physics laboratories. The infrared beam of the photogate is tripped by a "flag" attached to the oscillator. The rod pendulum pictured in Figure 3.9 was studied by this means. Figure 3.10 shows a closeup of the flag which is attached to the top of the pendulum and which passes through the photogate during oscillation.

As seen in the figures, various parts of the pendulum are clamped to a vertical steel rod. Both upper and lower masses are made of lead, each with a pair of holes drilled in it — one to pass the brass rod of the pendulum through and the other (after tapping) to hold a thumbscrew for securing the mass at different vertical positions on the rod.

This mechanical oscillator is a compound pendulum; the period can be made long as the knife-edge (also clamped to the rod) approaches the center of mass. At long periods, the instrument is not very responsive to external accelerations of the supporting frame; but it is sensitive to internal structural changes. Low-frequency instability is encountered as the upper parts of the instrument experience creep, particularly in the materials just above the knife-edge. The upper pendulum has similarities to an inverted pendulum, except that it is rigidly connected to the lower pendulum, and causes the oscillator to eventually exhibit double-well (Duffing) characteristics. This happens at larger amplitudes as the period is increased toward really long times. The tendency toward mesoanelastic complexity depends on the dimensions of the rod. As expected because of the well-known engineering properties of rods and tubes, a large diameter, thin-wall tube will behave differently from a solid rod made from the same amount of material (same total mass). This will be true if the tube does not experience localized (sharp) deformation prone to creasing.

Damping measurements with a photogate require that the time required for the flag to pass through the beam be fairly small — thus larger amplitudes of motion are required than with other sensors. Of course really large motion would result in a period increase, consistent with long-understood pendulum dynamics. For amplitudes within the acceptable range (which in practice is not overly restrictive), the velocity of the pendulum as it passes through the equilibrium position is inversely proportional to the time

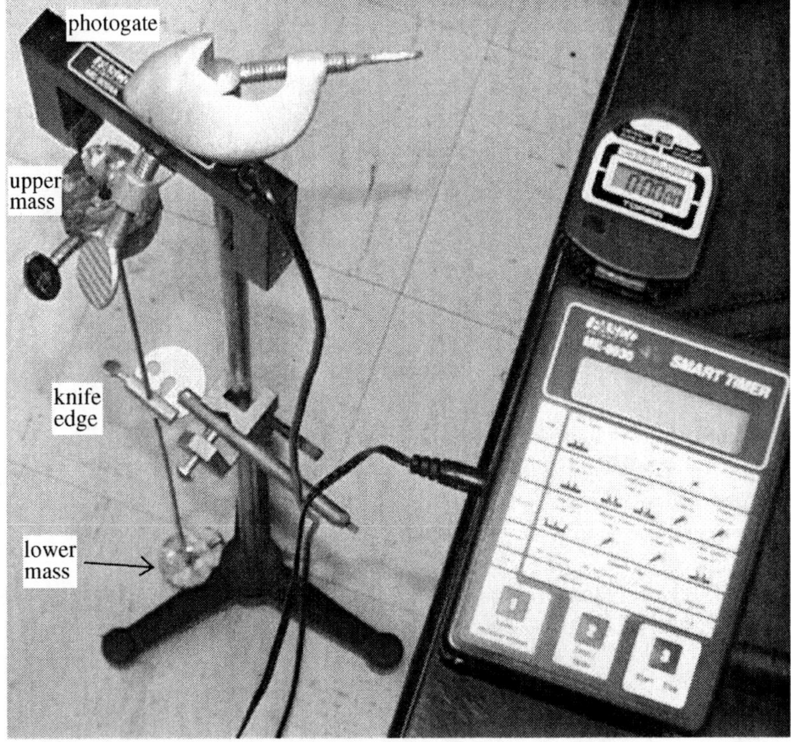

FIGURE 3.9 Rod pendulum in which damping measurements are made with a photogate.

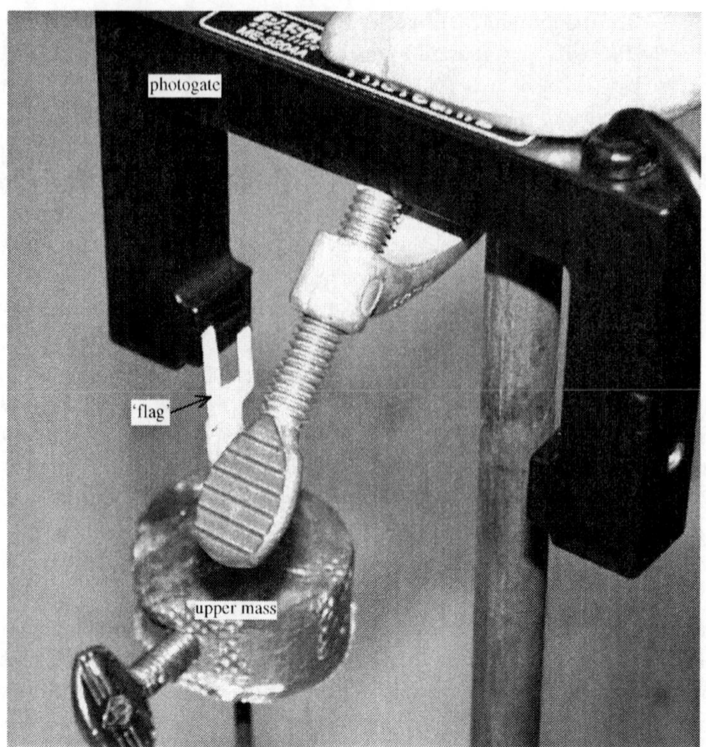

FIGURE 3.10 Top of the pendulum showing the upper mass and "flag" for tripping the photogate.

interval between interrupts of the photogate beam by the two vertical arms of the flag. If the period does not change with amplitude, then there is also an inverse relationship between the gate time and the amplitude. A plot of the inverse of these times *versus* cycle number is, for the constraints indicated, a reasonable approximation of the turning points of the free-decay.

In this case the single gate time interval measurements were made by a Pasco Smart Timer. It is a user-friendly instrument that also permits the period of the pendulum to be accurately measured by the flag (by using two different lengths of the flag arms). For experiments of this type, it may prove more convenient to measure the period with a stopwatch (infrequently as compared to the velocity). The sequentially increasing time intervals are read manually from the Smart Timer and recorded by hand, once per cycle. Of course, to do so requires that the period be long enough to permit these operations. The recorded values are conveniently analyzed by typing to a spreadsheet, which is then used to graph damping curves such as shown in Figure 3.11.

A pure exponential fit is not appropriate to the decay of Figure 3.11, in which the upper mass had been removed and a business card taped to the bottom of the pendulum to cause turbulent air damping (period near 1 sec). The fit shown, however, involving both linear and quadratic dampings, is seen to be quite reasonable. As in the case of the vibrating reed discussed earlier, this system is adequately described by the nonlinear damping equation 2.61 given in Chapter 2. For the data of Figure 3.11, $Q = 25$ initially and increases to 70 at the end of the record.

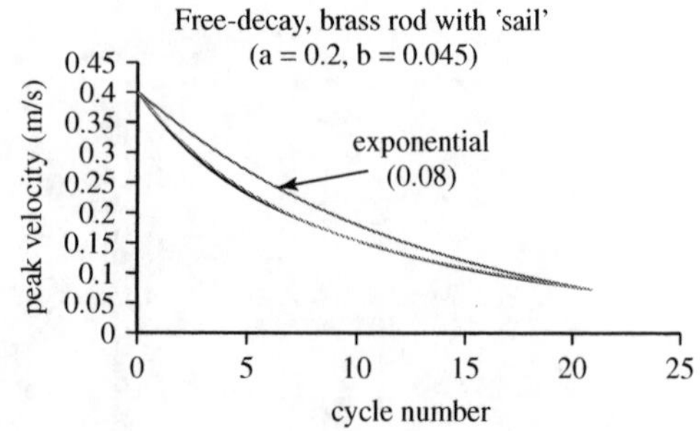

FIGURE 3.11 Free-decay of a pendulum as determined by photogate measurements.

Without the business card, and with the upper mass in place, the decay of this pendulum was found with the photogate measurement technique to be exponential, as expected for viscous damping with periods of about 5 sec. At periods in excess of about 10 sec, however, internal friction of the rod becomes more important than air damping. Although the decay is then still exponential at larger levels, the frequency dependence is not the same as required by linear air damping.

3.4.5 Case 5: Rod Pendulum Influenced by Material under the Knife-Edge

The data in Figure 3.11 were collected with the knife-edges resting on hard ceramic alumina flats. When supported by other materials, the damping of a rod pendulum can be influenced by anelastic flexure other than that of the rod. Hardness of the material does not guarantee low damping, as will be seen in the following examples. The data that follows were collected with a different pendulum, depicted in Figure 3.12.

The sensor in this case was an SDC unit, connected to the computer through the Dataq DI-700 A/D converter. The upper and lower masses are each approximately 1 kg and their separation distance on the aluminum hunting arrow from which the pendulum was fabricated was about 70 cm.

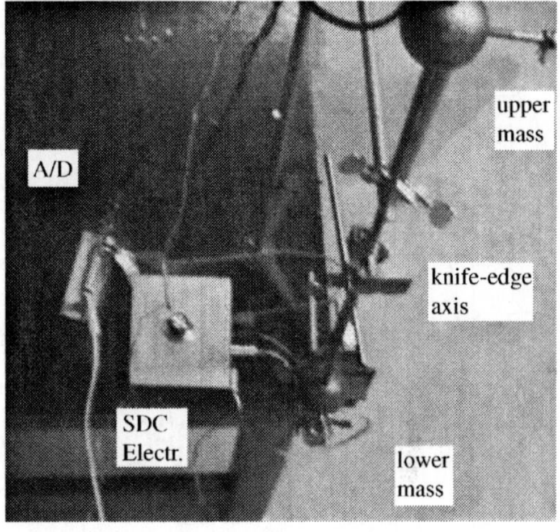

FIGURE 3.12 Long period rod pendulum used to study the influence of different materials under the knife-edge.

3.4.5.1 Lithium Fluoride Samples

The samples used to collect the data in Figure 3.13 were identical pairs, except that one pair had been irradiated with a huge dose of gamma rays. The resulting changes to the structure of the crystal are responsible not only for color centers as noted in the photograph in Figure 3.14, but also a dramatic change in the internal friction. It is clear from Figure 3.13 that internal friction in the LiF is the dominant source of damping of the rod pendulum that was used (Peters, 2003a, 2003b, 2003c, 2003d).

Lithium fluoride is used in thermoluminescent film badges (radiation monitors). When exposed to energetic radiation, atoms are "knocked" from their crystal lattice sites into metastable states corresponding to interstitial positions of the lattice. Upon ramping the temperature of the sample in an oven fitted with a photomultiplier tube, jumps from the metastable state are accompanied by the release of photons. The amount of light so generated is a measure of the dose that was received by the crystal. Because light flashes are observed with rather small changes in the temperature, it is reasonable to expect that mechanical strains might also cause a significant change to the defect state of such crystals. This postulate is confirmed by the data in Figure 3.13, which show a dramatic difference in the decay character of the pure (clear) crystals (bottom figure) and those which were extensively damaged by gammas (top figure).

In both of the decays in Figure 3.13 there is significant nonlinear damping, as evidenced in the early portions of each of the two records. The top case is nearly pure Coulombic, and the bottom case is partially amplitude dependent. This is revealed from estimates of the Q, shown in Figure 3.15.

The Q values in Figure 3.15 were computed from successive triplet-values of the turning points of the motion, read directly from the decay pattern displayed on the monitor by the Dataq software. The equation used is

$$Q = \frac{\pi}{-2\ln[1-(\theta_n-\theta_{n+2})/(\theta_n-\theta_{n+1})]}, \quad n = 0, 1, 2, \ldots \tag{3.3}$$

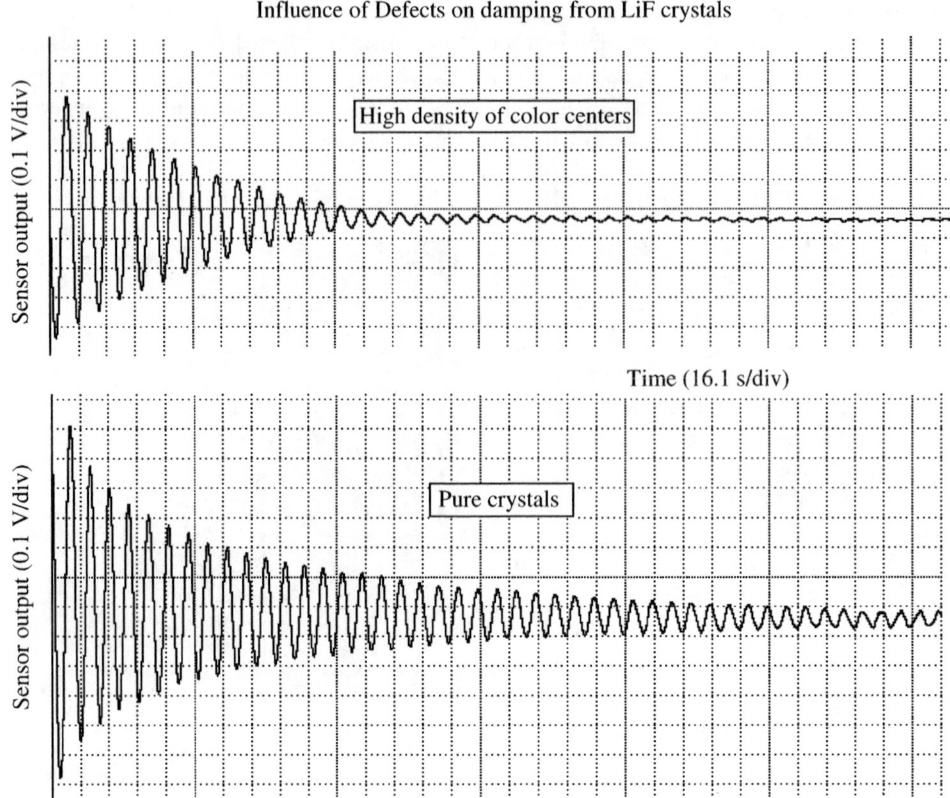

FIGURE 3.13 Illustration of damping difference according to specimen type under the knife-edge.

3.4.6 Hard Materials with Low Q

It is commonly (and mistakenly) thought that hard materials must necessarily also have low damping. The following two examples show that this is not necessarily so. Even though cast iron is very hard, it is also quite dissipative, which makes it an ideal material for engine blocks. Figure 3.16 shows a decay curve for the steel knife-edges of the pendulum resting on cast iron samples.

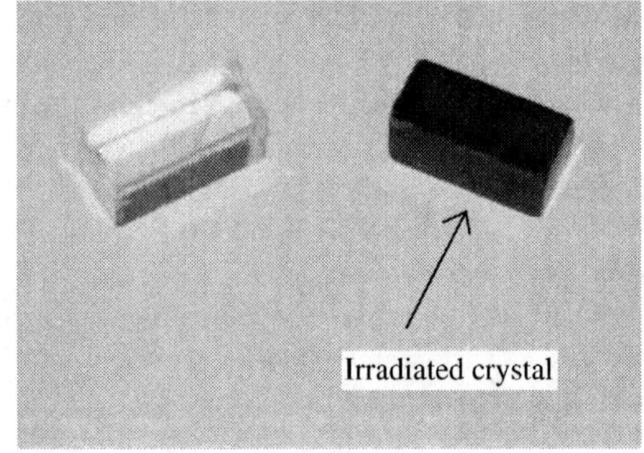

FIGURE 3.14 Photograph of LiF single crystals used to obtain the data in Figure 3.13.

At the start of the record the damping with cast iron is nearly twice as great as that of steel-on-sapphire or steel-on-silicon, where the Q was found to be of the order of 80. This large damping measurement is consistent with the known excellent properties of cast iron for use in engine blocks, although the frequencies for such applications are much higher.

Figure 3.17 is another very hard material which has large damping — the ceramic piezoelectric wafer formed from lead, zirconium, and titanium (PZT), which by means of a mechanical impulse is commonly used to generate an electric spark to ignite a gas grill. The secular decline of Q based on the short temporal record indicates Coulomb damping. It is consistent with the nearly straight-line turning points for the early part of the long-term record, also shown. The long-term record

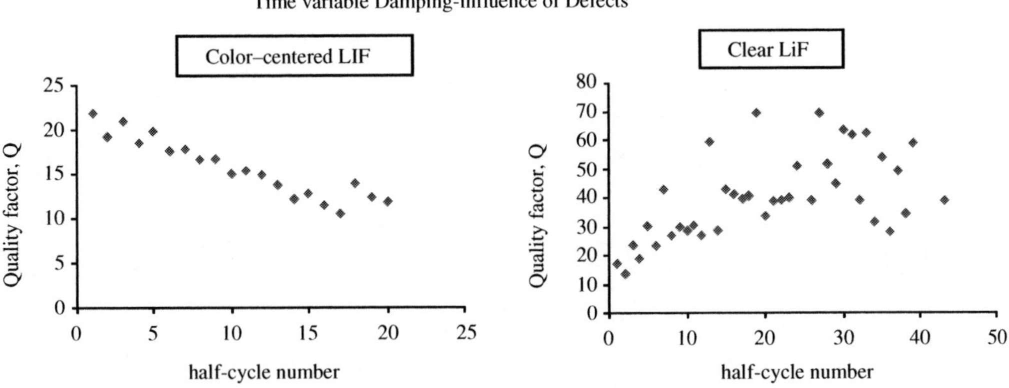

FIGURE 3.15 Temporal dependence of the Q, LiF crystal experiments.

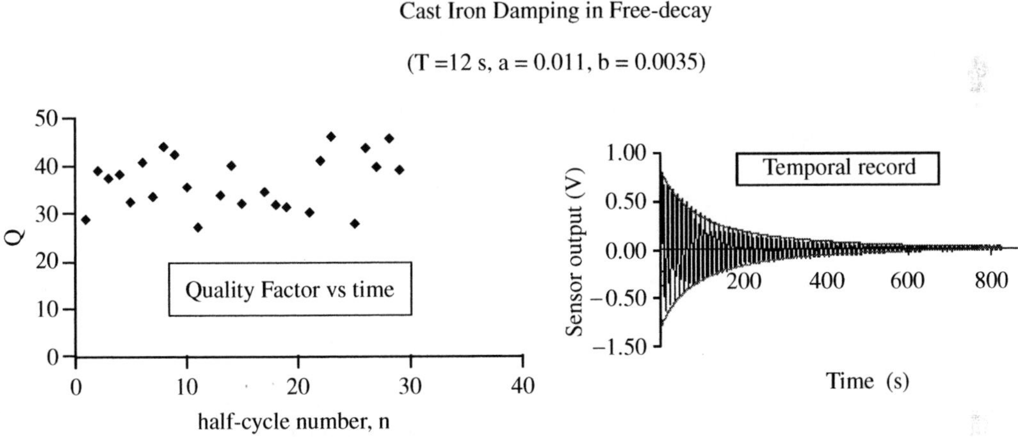

FIGURE 3.16 Data collected using cast iron samples.

is labeled as anomalous because it does not appear to be consistent with several simultaneously acting dissipation mechanisms. Instead, the strong Coulomb damping seen early on seems to disappear later, once the amplitude has dropped below a particular level. This suggests activation processes of a quantal type. It would be interesting to study the PZT wafers in a different pendulum configuration, and not operating "open-circuit" as in the present case, but rather with different resistors connected between the top and bottom of the wafers.

3.4.7 Anisotropic Internal Friction

With Polaroid material (H sheet) placed under the knife-edges it was found that the damping depends on the direction of the long-chain polymeric molecules. The direction of the molecules in a sample is readily determined by looking through the Polaroid at reflected light from a polished floor. When the reflection occurs close to the Brewster angle, only the horizontal component of the electric field is significant in the reflected light for unpolarized incident light. The direction of the molecules is thus determined by rotating the sample until the minimum of level of light is found. When this occurs the molecular chains are situated horizontally.

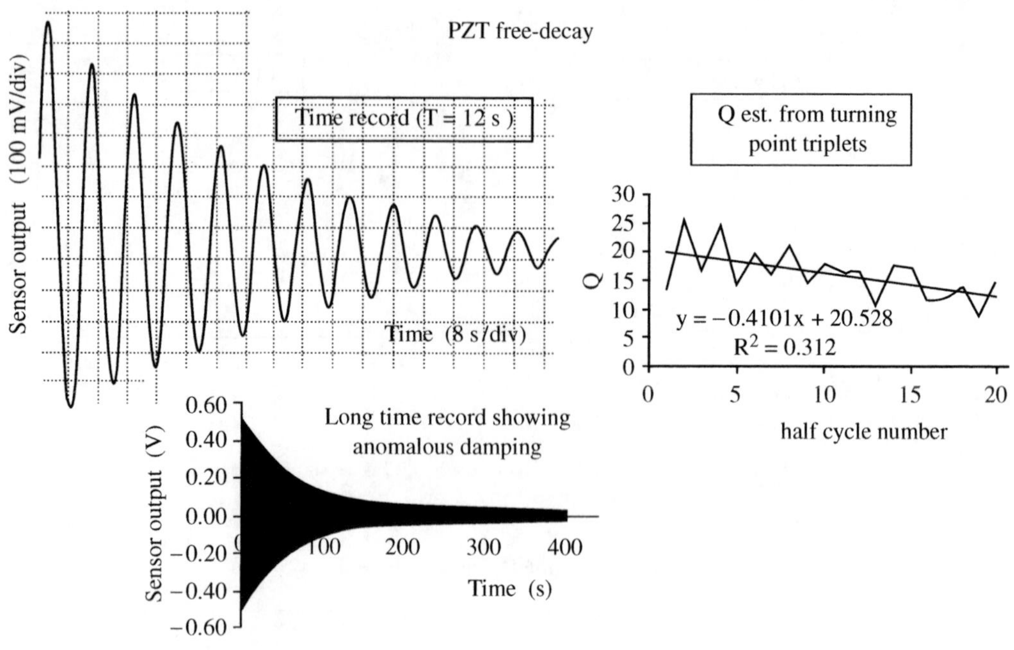

FIGURE 3.17 Data from an experiment involving PZT ceramic wafers.

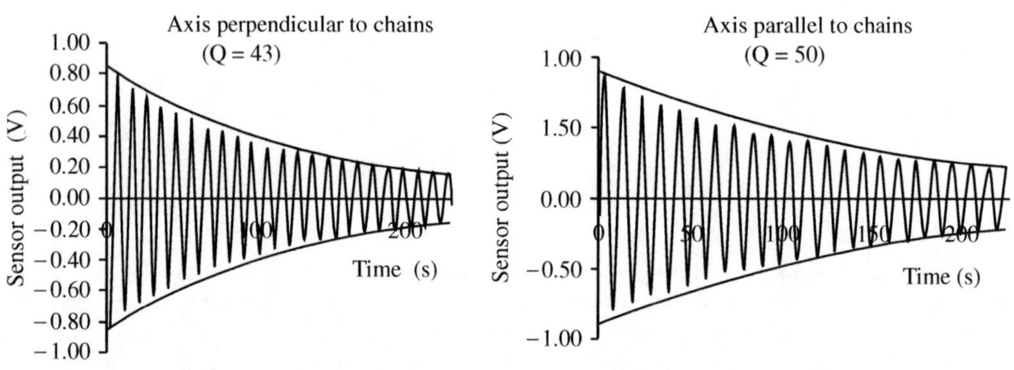

FIGURE 3.18 Free-decay curves showing anisotropy of the internal friction in polaroid material.

It was reasoned that the molecular properties of Polaroid might result in mechanical as well as optical anisotropies. This postulate proved to be true, as shown in Figure 3.18.

When oscillating on silicon at a period of 10 sec, previous studies have found that the instrument decays consistently with a Q of 80 (uncertainty 3%). In the present study, half a dozen free-decay records were obtained for (i) edges parallel to the chains and (ii) edges perpendicular to the chains. The average Q of oscillation was estimated at 50 for the parallel case and 43 for the perpendicular case. Reproducibility proved slightly better for the parallel case (4%) as compared to the perpendicular case (5%). Additional details are documented elsewhere (Peters, 2003a, 2003b, 2003c, 2003d).

3.4.7.1 Summary, Free-Decay Q Estimation

All of the techniques so far described are methods based on free-decay, which is especially important for nonlinear systems. With linear systems it is also possible to use steady-state methods, as noted in Box 3.2 (last column; de Silva, 2006). Box 3.2 summarizes the techniques used in the present chapter to estimate the logarithmic decrement, βT, from which $Q = \pi/(\beta T)$.

Box 3.2

Methods for Quantifying Damping

Damping (Q Estimation) Techniques ($Q = \pi/\beta T$, T = Period)

Logarithmic Decrement (Full-Cycle, N)	Turning Points (Half-Cycle, n)	Nonlinear Fit to Envelope	Time τ to l/e ($0.3679x_0$)	Short Time Fourier Transform	Bandwidth, Magnification Factor, Hysteresis Loop, Step-Response
$\beta T = \frac{1}{N}\ln\frac{x_0}{x_N}$	$\beta T = -2\ln\left[\frac{1-(x_{n-1}-x_{n+1})}{(x_{n-1}-x_{n+1})}\right]$	$-\dot{y} = ay^2 + by + c$	$\beta T = T/\tau$	$-\beta T = \frac{T\ln 10}{20}$ [slope(dB/s)]	de Silva (2006), Chapter 7

The best method is to use a full nonlinear fit; the worst is to measure the time to $1/e$. The expression in Box 3.2 for the logarithmic decrement, using the STFT, is equivalent to

$$Q = 27.29\frac{f}{\left|\frac{\mathrm{dB}}{\mathrm{s}}\right|} \tag{3.4}$$

where f is the frequency in Hz and the STFT slope is specified in dB per s.

3.5 Driven Oscillators with Damping

This chapter has been mainly concerned with oscillators in free-decay. It is also possible to make quantitative predictions from measurements at steady state. Confidence in predictions, however, depends on the nature of the damping. Such data are of limited value for most nonlinear systems, unless supplemented with free-decay data.

3.5.1 MUL Apparatus

Some of the techniques applicable to driven systems are illustrated by the multipurpose undergraduate laboratory (MUL) apparatus shown in Figure 3.19, that has been used by students in the physics department at Mercer University.

For the purpose of measuring the Lorenz force (basis for defining the current unit, the ampere) a constant current is supplied through the posts to the pivoted-on-points brass wire on which a weight, W, is shown hanging on one of the horizontal arms of the wire. Current enters the wire through one post *via* the banana plug inserted into a drilled hole. It thereafter travels through the lower (invisible) shorter straight segment of the wire located between the poles of the drive magnet; and it finally exits through the banana plug on the opposite post. When carrying a current, the force on the wire from the part inside the magnet causes vertical deflection, the direction up or down being determined by the direction of the current. This results in a rotation about the pivot points (indented tops of the posts). The position is measured by the capacitive sensor, *S* (one of several variants of the SDC patent).

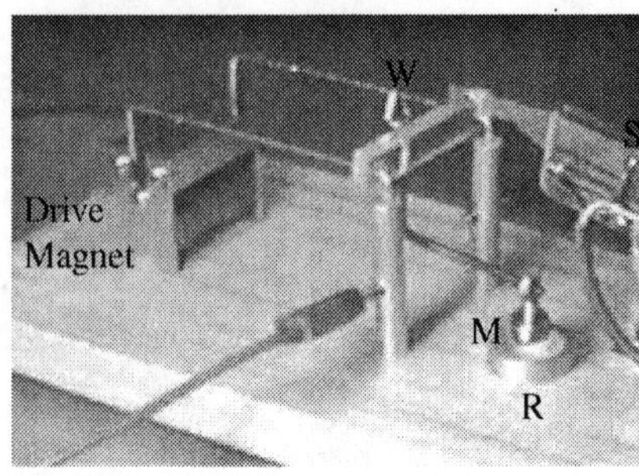

FIGURE 3.19 Apparatus for studying resonance and the Lorenz force law.

The sensitivity of this current balance depends on the location of the center of mass of the oscillatory wire, which is determined in part by the position of the rare earth magnet, M, which hangs from a steel nut on the threaded part of the heavier brass rod having a 90° bend. The upper end of this threaded rod is held by a plexiglass member that also holds the ends of the oscillatory wire.

3.5.2 Driven Harmonic Oscillator

The MUL becomes a driven harmonic oscillator when the excitation current is a.c. rather than the d.c. used for the Lorenz force study. The damping is determined primarily by eddy currents in the aluminum ring, R, that lies on the wooden base underneath and in close proximity to magnet M.

The apparatus is useful for studying both free-decay and driven oscillation. Engineering students Brandon R. Bowden and James D. Sipe have programmed LabVIEW to generate both free-decay curves and resonances.

An example Lorentzian (resonance response) is given in Figure 3.20. (Additional information is found in a laboratory writeup (Peters, 1998).)

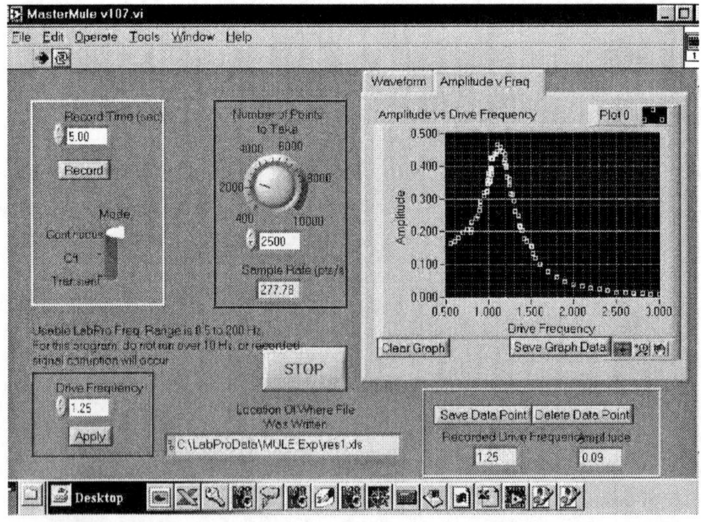

FIGURE 3.20 Screens from the LabVIEW program used with the MUL to study both transient and resonance phenomena.

3.6 Oscillator with Multiple Nonlinearities

An oscillator can have significant nonlinearities of both the elastic and damping types. An example is the mechanical system pictured in Figure 3.21.

The instrument is a modified extensometer that was sold by TEL-Atomic and which was designed around the SDC sensor to measure Young's modulus and thermal expansion coefficients. The wire sample normally used with the instrument (along with a hollow power resistor that fits in the black clamp) has been removed, and two rare earth magnets have been employed. One magnet is superglued to the bottom of the pan where the weights are normally placed, as shown in Figure 3.22; and the other magnet is attached to the bottom of the inductor that is sitting on the top of the oscillator (cased instrument, Pasco) used for drive. The pair of magnets are positioned in close proximity so as to repel each other, thus supporting the mass of the moveable arm of the extensometer.

The study of nonlinear systems requires a linear sensor; i.e., any nonlinear contributions from the detector must be negligible. Figure 3.23 shows the calibration results for the instrument and its linear response for the range of amplitudes used in the study.

The potential energy of this oscillator was assumed to have the following form

$$U(x) = \frac{b}{x^n} + cx \tag{3.5}$$

and the parameters were estimated by measuring x as small masses were placed on the pan. A linear regression fit to a log–log plot (using the sensor calibration constant of 550 V/m) yielded $b = 1.02 \times 10^{-5}$, $n = 1.526$, and $c = 0.304$ (system international units). Anharmonicity of the

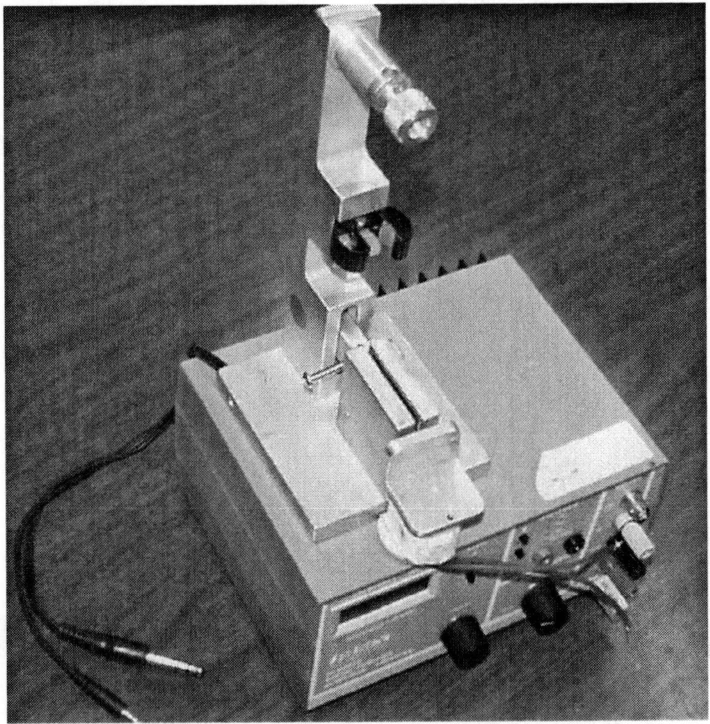

FIGURE 3.21 Mechanical oscillator with multiple nonlinearities.

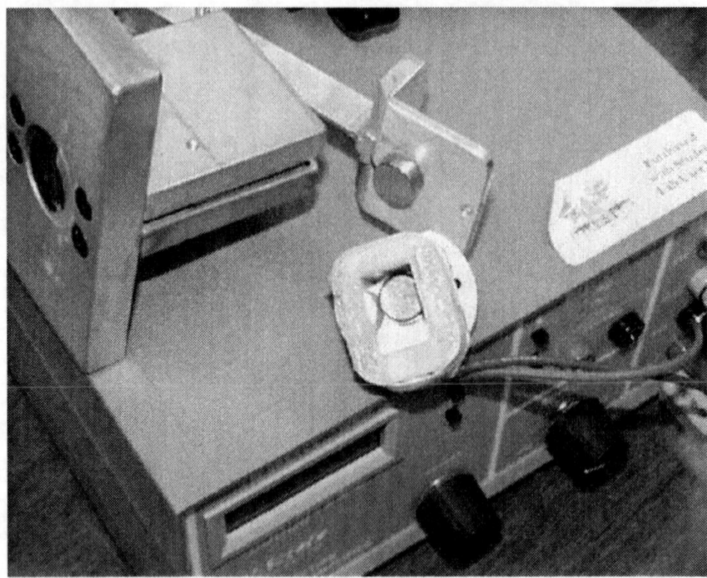

FIGURE 3.22 Closeup picture of the oscillator in Figure 3.21 (nonoperational configuration), showing placement of the rare earth magnets.

potential is readily apparent in the plot shown in Figure 3.24, with the force of restoration being greater in compression (x decreasing) than it is in extension. This feature is reminiscent of interatomic potentials, with anharmonicity being responsible for thermal expansion.

Because of the elastic nonlinearity, the mean position depends on the amplitude of the oscillation, as is evident in the free-decay curve in Figure 3.25.

The damping of this oscillator was also found to be nonlinear, as seen in Figure 3.26.

The oscillator exhibits hysteresis when driven at larger amplitudes, as shown in Figure 3.27, where it can be seen that the location of an amplitude jump depends on which way the oscillator is adjusted, either up or down in frequency. Such jumps (well known with oscillators with nonlinear elasticity) stand in stark contrast with the behavior of a linear oscillator, as can be seen by comparing Figure 3.27 with the screen picture in Figure 3.20.

A surprise from this study involves the frequency of oscillation. In general, the oscillator did not entrain to the drive. Moreover, the preferred frequencies were not necessarily the same as the free-decay frequency of 6.01 Hz. Some of the frequencies (measured with power spectra) are indicated in

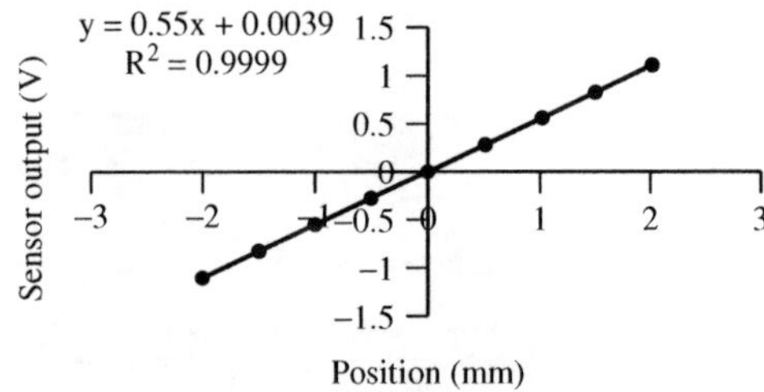

FIGURE 3.23 Calibration data for the sensor used with the oscillator having multiple nonlinearities.

Potential Energy, oscillator with multiple nonlinearities

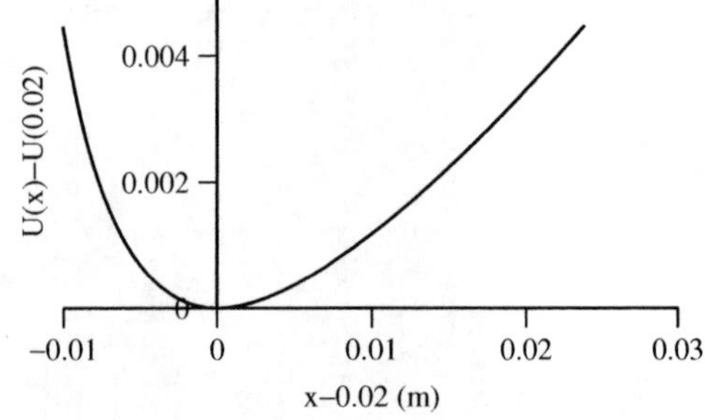

FIGURE 3.24 Plot of the potential energy function of the oscillator.

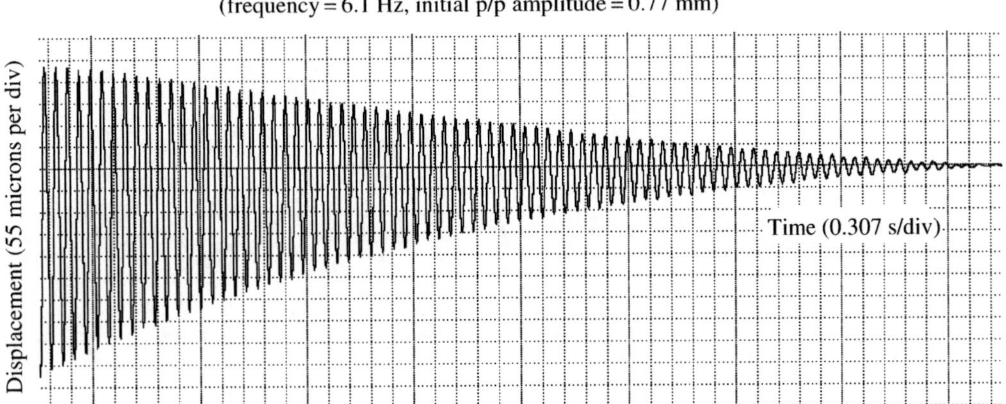

FIGURE 3.25 Free-decay curve showing the mean position shift as a function of oscillator amplitude. (Decreasing sensor voltage corresponds to increasing x.) The frequency of oscillation is 6.01 Hz.

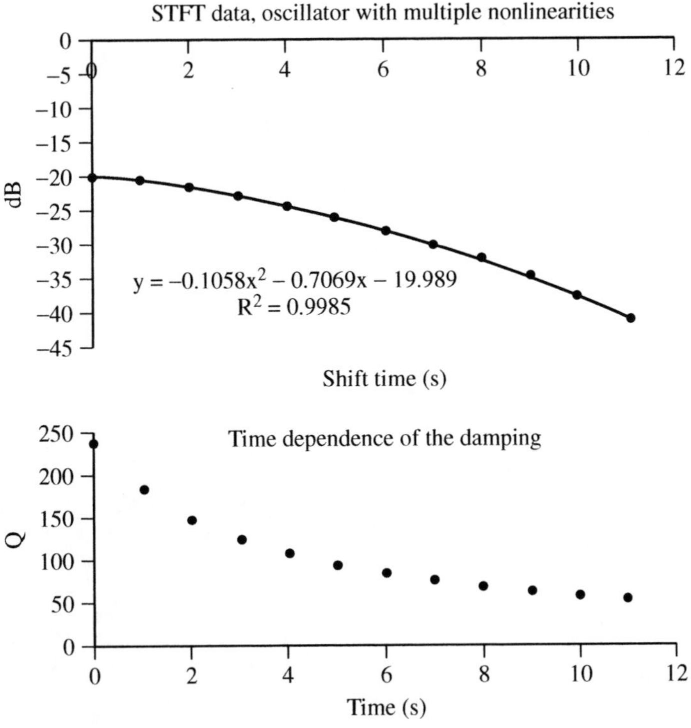

FIGURE 3.26 Free-decay character as determined using the short time Fourier transform.

Figure 3.27. The 3% frequency jumps observed in Figure 3.27 (in going from 5.46 Hz to 6.20 Hz) are not real but rather artifacts of the finite resolution of the 1024 point transforms that were employed.

Figure 3.27 demonstrates why nonlinear damping measurements should be done in free-decay. Figure 3.25 demonstrates how exponential fits make no sense for some oscillators. Fortunately, the STFT can be used to determine the amplitude dependence of the Q.

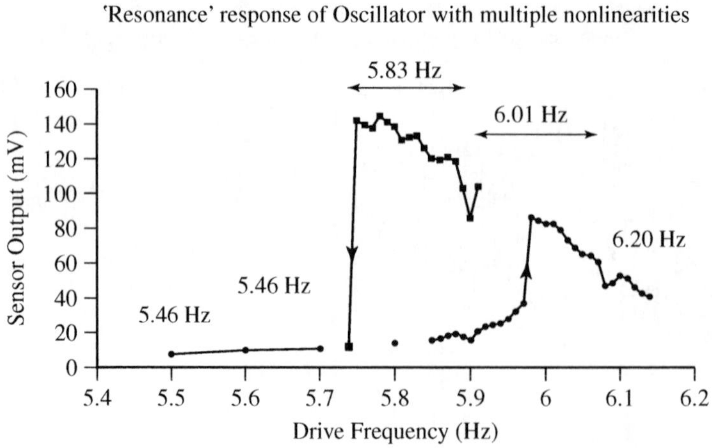

'Resonance' response of Oscillator with multiple nonlinearities

5.83 Hz
6.01 Hz
6.20 Hz
5.46 Hz
5.46 Hz
Sensor Output (mV)
160
140
120
100
80
60
40
20
0
5.4 5.5 5.6 5.7 5.8 5.9 6 6.1 6.2
Drive Frequency (Hz)

FIGURE 3.27 Resonance response (steady state) of the driven oscillator.

3.7 Multiple Modes of Vibration

3.7.1 The System

In engineering, multimode oscillations are common. Many, if not most, cases have mode mixing features even though they may in some cases be too small to be readily observed. The importance of nonlinearity to these problems is not widely appreciated, so a case to illustrate salient features is provided here. Free-decay records were obtained with an oscillator in the form of a vertically oriented (hanging) tungsten wire, of length 24 cm and diameter 0.31 mm. It was clamped at the top end, and at the bottom a rectangular plate was attached that was 11.3 cm long, 1.3 cm wide, and 0.8 mm thick. The plate was cut from double-sided copper circuit board. The board was positioned between the stationary plates of a capacitive sensor, as shown in Figure 3.28.

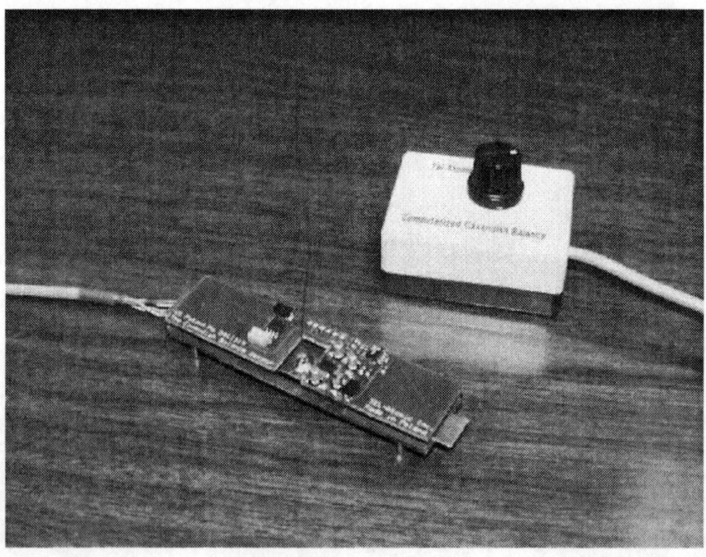

FIGURE 3.28 Photograph of the detector used to monitor the multimode oscillator.

For the picture, the apparatus was disassembled and the plate allowed to rest on the top of the bottom electrode set. Operationally, the plate was positioned midway between the upper and lower static electrode sets (separation distance of 4 mm); and there was no mechanical contact during oscillation. As can be seen, the top of the circuit board containing the upper electrode set contains more than a dozen electronic components; these are of the surface mount technology type. The detector is of the SDC type and this particular embodiment is manufactured in Poland for TEL-Atomic Inc., Jackson, MI, for use in the Computerized Cavendish Balance.

As can be seen in the picture, the wire was rather kinked instead of straight, which is expected to be a significant source of nonlinearity. For this reason, not to mention that it is very difficult to make larger diameter tungsten wires reasonably straight, no serious attempt was undertaken to remove the kinks.

3.7.2 Some Experimental Results

An example decay record generated with this apparatus is illustrated in Figure 3.29.

3.7.3 Short Time Fourier Transform

When multiple modes are present in a decay, as in Figure 3.29, it is not possible to readily estimate Q for all of the various modes using time data. The decays can be estimated using the FFT, in a technique called the short time Fourier transform, which is built-in to various software packages related to acquisition systems, such as LabVIEW (see Appendix 15A). With the versatile software supplied with the Dataq A/D converter, it is straightforward to employ an equivalent manual technique. Using the number of points to define the FFT a value (always a power of 2 total) that is substantially smaller than the number of points in the record, a manual scan is performed in which one simply increments from start to finish, calculating a separate FFT at each position in time along the way. As an illustration of this powerful tool, Figure 3.30 shows spectra corresponding to the start and the finish of the data in Figure 3.29.

All the modes decay in time, and the rate of decay is especially large for those modes that correspond to sum and difference frequencies of the primary modes at 1.19 and 2.19 Hz. Table 3.2 gives the spectral intensities in dB for the two times considered. Where the rows are blank for the end of record case, the values were insignificantly small.

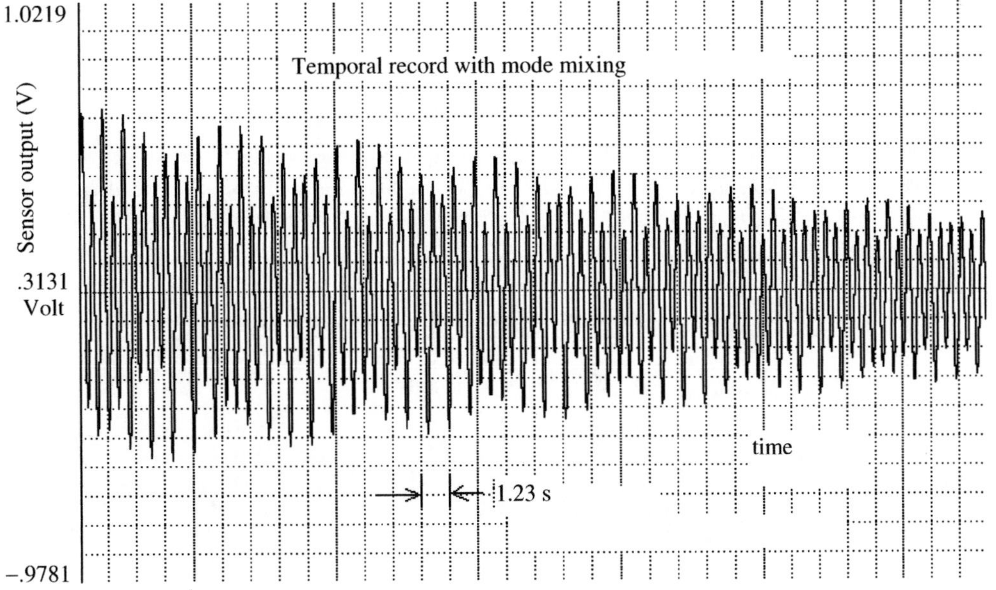

FIGURE 3.29 Example free-decay of a multimode wire oscillator.

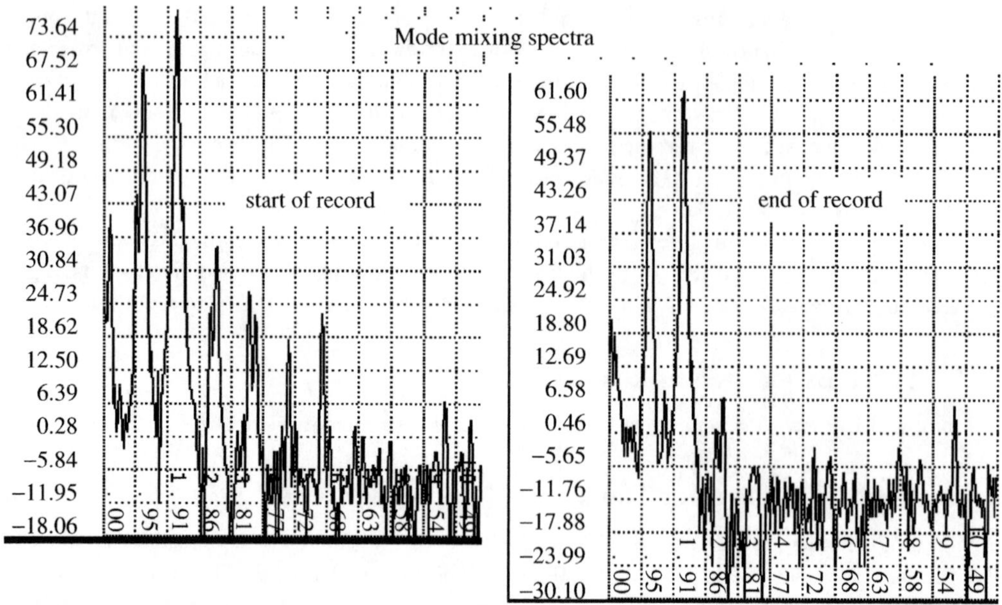

FIGURE 3.30 Beginning and end spectra corresponding to the temporal data from Figure 3.29. Ordinate values are spectral intensity in dB, abscissa values are frequency in Hz (linear scale).

The decibel values in the table are referenced to the bit-size (16 corresponding to 65536) of the ADC. In terms of the sensor output voltage, V, it is defined by Dataq as:

$$\mathrm{dB} = 20\,\log_{10}(32,768 \times V/\mathrm{FS}) \tag{3.6}$$

where FS is the full-scale voltage as determined by the gain setting.

Elsewhere in this chapter, the decibel is calculated with a different reference. For example, for an FFT spectral line having real and imaginary components R and I, respectively (voltage based), the intensity in dB is calculated using

$$\mathrm{dB} = 20\,\log_{10}\left[\sqrt{R^2 + I^2}\Big/\left(\frac{n}{2}\right)\right] \tag{3.7}$$

where n is the number of points in the FFT. This is convenient for determining noise levels. For example, from later graphs showing electronics noise, the floor of the SDC sensor is found to be of the order of -120 dB, corresponding to a microvolt. The position resolution defined by this noise level is about 500 nm, i.e., the wavelength of visible light.

TABLE 3.2 Spectral Intensities for Some of the Lines Shown in Figure 3.30

Frequency (Hz)	Start of Record (dB)	End of Record (dB)
2.19	78.3	63.0
1.19	68.1	55.6
1.00	44.6	
0.19	40.8	
3.38	35.0	6.7
4.34	26.7	
4.53	22.4	
6.53	22.9	
5.53	17.8	

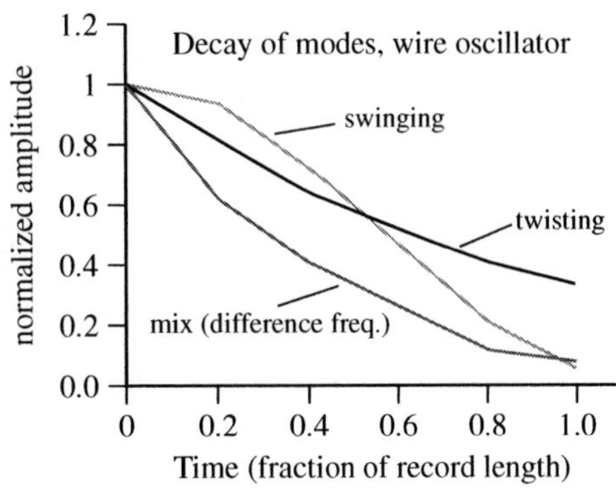

FIGURE 3.31 Decay of modes of the wire oscillator, determined by the manual STFT.

Of the two primary modes of this kinked-wire case study, the higher frequency (2.19 Hz) is the twisting mode and the lower frequency (1.19 Hz) is the swinging mode. The swinging mode is a little higher frequency than that which would result if the wire were completely flexible, yielding a near simple pendulum (1.02 Hz for 24 cm length). The swinging mode is two dimensional (pendulum equivalent called conical), but the sensor only responds (first order) to motion perpendicular to the long axis of the electrodes. It should also be noted that this motion is attenuated, relative to the twisting response, because of the mechanical common-mode rejection feature discussed in Chapter 2.

The manual STFT was used on the data that generated Figure 3.29 to estimate the decay history of three different modes — both of the primary ones (twist and swing) and also the mode whose frequency is the difference between the frequencies of the primaries, i.e., 1 Hz. Figure 3.31 shows the results, where a Hanning window was used, and the total number of points in the record permitted five equally time-spaced FFTs, when working with a 1024 point transform.

Although the decay of the twisting mode is seen to be reasonably exponential, there was large beating between the modes (readily observed in Figure 3.29). Beating alone would not yield a mix signal whose frequency is 1.0 Hz. However, beating in a linear system can cause amplitude variations in the weaker swinging mode.

3.7.4 Nonlinear Effects — Mode Mixing

At least two signals in the spectra are the result of nonlinearity, i.e., the lines corresponding to the sum and difference of the frequencies of the primary pair — at 3.38 and 1.00 Hz, respectively. If the system of oscillator and detector were completely linear, then no such sum and difference cases would be possible. It is also to be noted that these mixtures are not the result of sensor nonlinearity, which as noted previously one must be careful to avoid.

The amplitude of a mix signal was expected to approximately obey the following relation:

$$A_{\mathrm{m}} \propto A_1 A_2 \tag{3.8}$$

To test this premise, the STFT was used to estimate the amplitudes of each of the three components indicated in Equation 3.8. The amplitudes were all normalized, relative to the starting value for each case, and the results used to generate the graphs in Figure 3.32.

The amplitude of oscillation for a given mode, at the time of the transform, is found by using the peak value in dB of the intensity of the spectral line for that mode, according to

$$A \propto 10^{\mathrm{dB}/20} \tag{3.9}$$

where the factor of 20 is used since the spectral intensities were calculated in terms of voltages. Although calibration constants (in V/m and V/rad) could be used to express the amplitude in meters or in radians, corresponding to the mode, nothing is gained by doing so for the present purposes.

The mixing index for these cases is defined by the expression

$$\text{index} = \frac{A_{\mathrm{m}}}{\sqrt{A_1 A_2}} \tag{3.10}$$

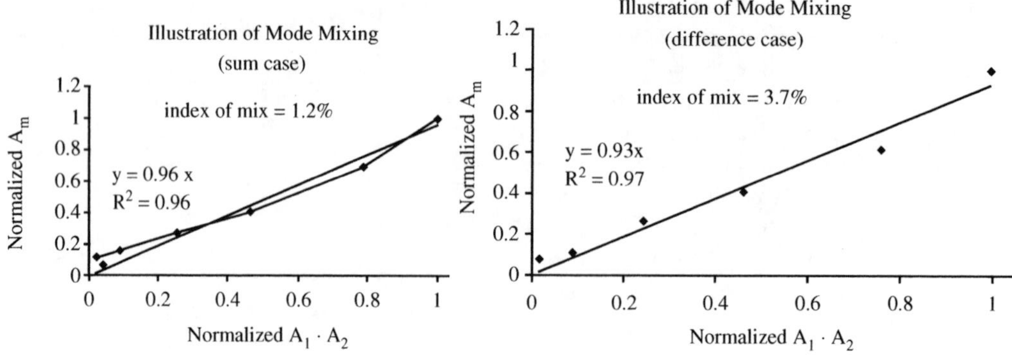

FIGURE 3.32 Evidence in support of nonlinear mixing according to Equation 3.8.

which is similar to expressions encountered in optics. It can be seen that the sum and difference frequencies are approximated reasonably well by theoretical expectation.

3.8 Internal Friction as Source of Mechanical Noise

Chapter 2 claims that internal friction is responsible not only for damping but also for significant mechanical noise of $1/f$ type. Figure 3.33 is provided in support of that claim.

The pendulum in these experiments (lead spheres near the ends of an aluminum tube with a pair of steel-points for the axis) was operated in a high vacuum to eliminate the influence of air. The electronics

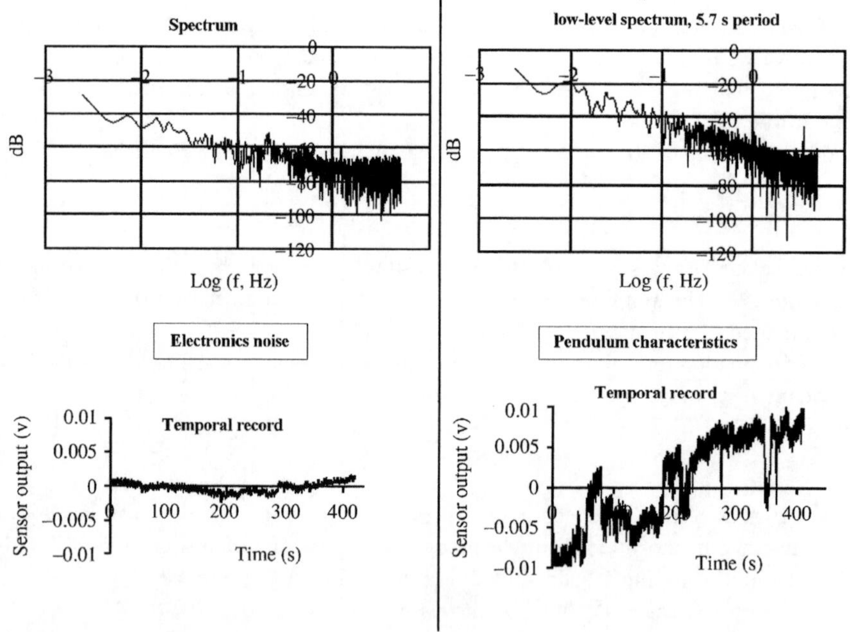

FIGURE 3.33 Power spectrum and associated temporal record showing mechanical $1/f$ noise (right pair). For reference, electronics noise is also provided (left pair).

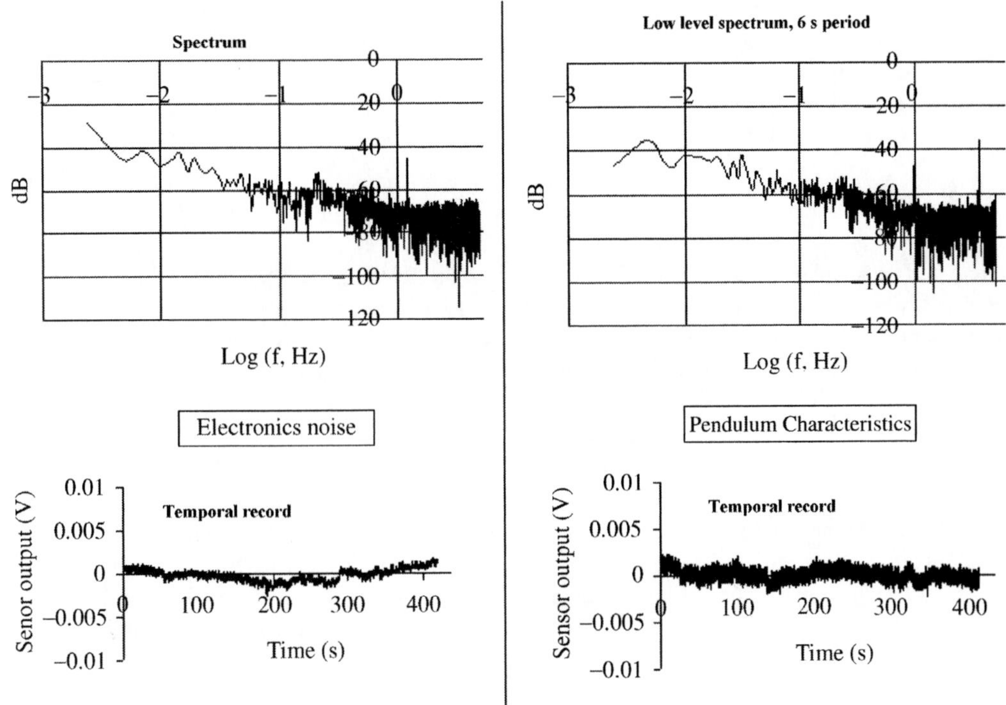

FIGURE 3.34 Same as Figure 3.33, but after the pendulum had stabilized.

noise was obtained by removing the top mass and measuring the motion after oscillation reached a minimum (frequency approximately 1 Hz); some pendulous mode remains because of pump noise transmitted through the vacuum hose. Similarly, the pump vibrations excite a vibratory mode in the long-period pendulum (sharp spectral line at 3.5 Hz, right spectrum). In this vibratory mode the lead masses move in the same direction relative to the stationary axis (similar to the bending mode of the carbon dioxide molecule). It is interesting to note that there is no coherent oscillation to be seen above noise corresponding to the period of 5.7 sec. The mechanical noise is seen to include bistability, which is not uncommon for this type of system before hardening, following a significant force disturbance. The data in Figure 3.33 were collected after replacing the upper mass, which had been removed to measure the electronics noise, and after pumping to the operating pressure (below 5 μm Hg).

The mechanical noise is seen to be $1/f$ for $f < 1.5$ Hz, which is where electronics begins to contribute noticeably. Everywhere below 1 Hz, the electronics noise is an order of magnitude smaller than the mechanical noise.

After the pendulum had stabilized overnight and been allowed to oscillate through a number of free-decays (initialization by tilting the chamber), the data shown in Figure 3.34 was collected.

It can be seen that the mechanical noise has mostly settled out, leaving the remnant electronics noise.

3.9 Viscous Damping — Need for Caution

Recent experiments have shown important subtleties of viscous damping (Peters, 2003a, 2003b, 2003c, 2003d). It is true that the dissipation at a specified frequency can be adequately modeled by simply multiplying the velocity term in the differential equation by a coefficient. It is not proper, however, to call this coefficient a constant, since the damping coefficient is frequency dependent and also involves the density as well as the viscosity of the fluid in which oscillation takes place.

Some engineers have known about the history term, which is most simply treated in the case of a sphere executing simple harmonic motion. The friction force acting on the sphere in this case can be reasonably approximated by

$$f_{\text{harmonic}} = 6\pi\eta a\left(1 + C_{\text{H}}\frac{a}{\delta}\right)v, \quad \delta = \sqrt{\frac{2\eta}{\omega\rho}}, \quad (C_{\text{H}} \to 1 \text{ as } v \to 0) \tag{3.11}$$

where ω is the angular frequency of oscillation, a is the radius of the sphere; and for the fluid, η and ρ are its viscosity and density, respectively. Only in the limit of zero frequency does the damping reduce to the form that one expects on the basis of Stokes' law of viscous friction (steady flow).

Using Equation 3.11 in the equation of motion for a pendulum yields for the Q

$$Q_{\text{v}} = \frac{I\omega}{6\pi\eta a\left(1 + \frac{a}{\delta}\right)L^2}, \quad \delta = \sqrt{\frac{2\eta}{\omega\rho}} \tag{3.12}$$

where I is the moment of inertia, and L is the distance from the axis to the center of the sphere. Typically, the ratio a/δ is significantly greater than unity so that the damping is governed by the surface area of the sphere rather than by its radius. These complexities of viscous damping are summarized in Box 3.3.

Reasonable experimental validation of the estimate for Q was provided, as demonstrated in Figure 3.35.

The instrument in this case was a compound pendulum in which a mass was located above the axis of rotation, as well as the usual situation of mass below the axis. The water damping was provided through a small sphere at the bottom of the pendulum, immersed in water held by a rectangular container.

If the history term in Equation 3.12 is ignored there can be huge errors. For example, in the case of water damping, the damping can be underestimated by 1000 to 3000%, as shown in Figure 3.36.

At low frequencies, it is also important to correct for the influence of hysteretic damping of the pendulum. Figure 3.37 shows the large errors that occur when one fails to do so.

For some cases, buoyancy and added mass of the fluid are also quite significant to the frequency of oscillation, as shown in Figure 3.38.

Box 3.3

Complexities of Viscous Damping

Friction force is not a function only of viscosity η; it also depends on density ρ and angular frequency ω

$$f_{\text{harmonic}} = 6\pi\eta a\left(1 + C_{\text{H}}\frac{a}{\delta}\right)\text{v}, \quad \delta = \sqrt{\frac{2\eta}{\omega\rho}}, \quad (C_{\text{H}} \to 1 \text{ as v} \to 0)$$

resulting in a complicated frequency dependence for the Q of viscous damping

$$Q_{\text{v}} = \frac{I\omega}{6\pi\eta a\left(1 + \frac{a}{\delta}\right)L^2}, \quad \delta = \sqrt{\frac{2\eta}{\omega\rho}}$$

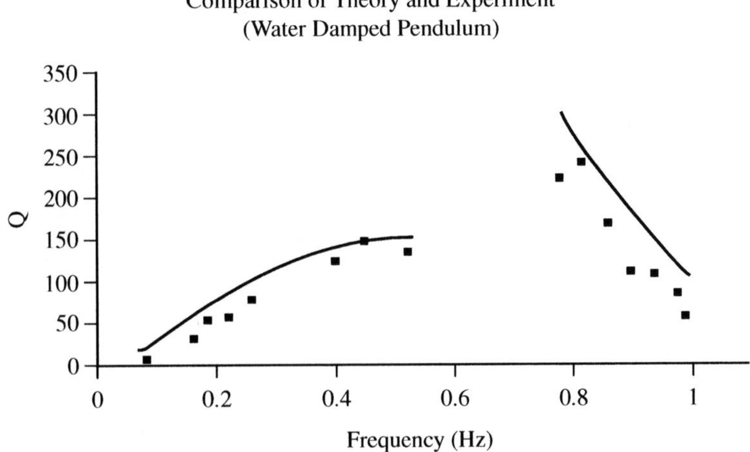

FIGURE 3.35 Comparison of theory and experiment for a pendulum damped by water.

3.10 Air Influence

As seen from Figure 3.37, low-frequency motions are likely to be influenced more by internal friction than by any fluids that interact with the oscillator. The most important fluid is of course air, and a true delineation between external and internal effects requires that the oscillator be studied in a high vacuum. It is not enough to just remove most of the air, since the viscosity of gases is surprisingly constant until the mean free path between collisions becomes a significant fraction of chamber dimensions.

Theoretically, it is possible to roughly estimate air influence, although only in the simplest of geometries, such as a sphere. In such cases, Equation 3.11 could be used (with accounts for the history term, using appropriate values for the viscosity and density). It is also possible in some cases to estimate air influence experimentally, as in the example that follows.

3.10.1 Brass and Solder Rod Pendula

Because of its malleability, the internal friction of solder (lead–tin alloy) is large, compared to that of much harder brass. A pendulum of each material was studied, both having a length of about 50 cm and a

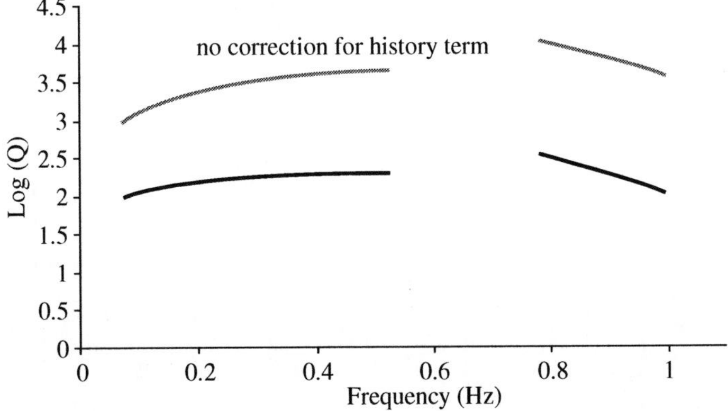

FIGURE 3.36 Illustration of how huge errors can occur in damping estimates if one ignores the history term.

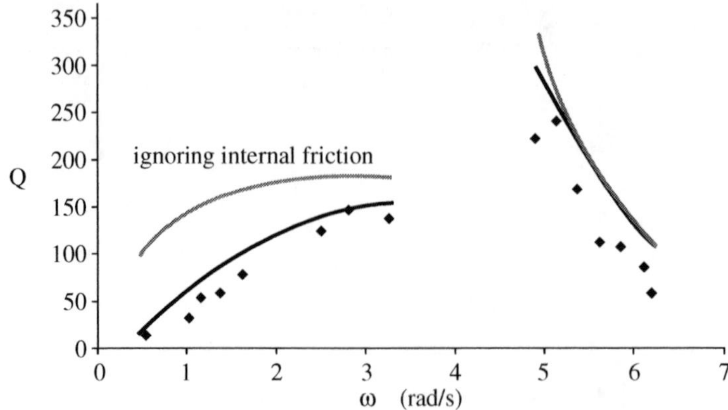

FIGURE 3.37 Illustration of significant low-frequency errors that result from a failure to recognize the hysteretic damping component of the pendulum.

diameter of about 3 mm. The technique used was the photogate method described in Section 3.4.4 (Case 4 above). Unlike the previous study, no lead masses were clamped on the rod — but it used the same adjustable knife-edge.

Figure 3.39 clearly shows that the internal friction for the solder pendulum is much greater than that of the brass pendulum.

A nonlinear fit was generated for each decay curve, from which the history of the quality factor was graphed as a function of velocity amplitude, as shown in Figure 3.40.

Consider the pair of brass curves in Figure 3.40. The large difference in Q at 10 cm/sec (387 compared to 266) is in stark contrast with their near equality at 50 cm/sec. This is primarily a consequence of air drag that is quadratic in the velocity at the larger amplitude. It is more important to brass than to solder because of the small internal friction of the brass.

From the large difference in internal friction of the two materials, a first order correction for air influence on the solder pendulum is to simply subtract $1/Q$ of the brass from $1/Q$ (raw data) of the solder, to yield the reciprocal Q (corrected) due to internal friction of the solder. This has been done in Figure 3.41.

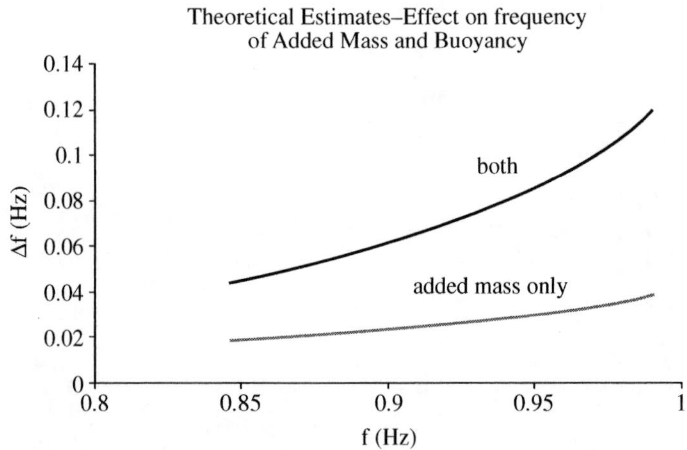

FIGURE 3.38 Example of how fluid properties influence the frequency as well as damping of an oscillator.

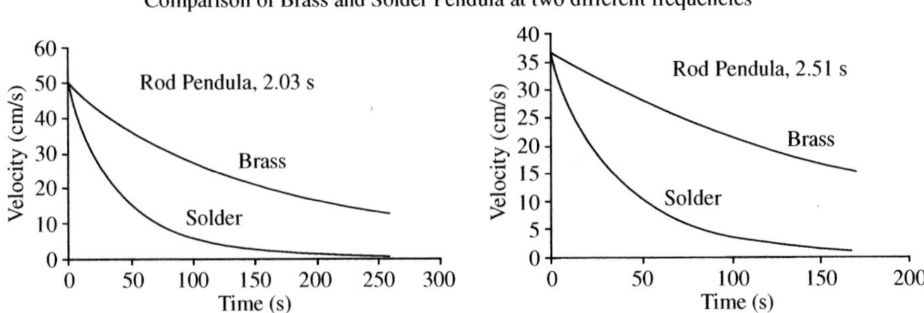

FIGURE 3.39 Free-decay curves for brass and solder pendula at two different frequencies, showing the larger internal friction of solder. The velocity is that of the peak value (amplitude) at the top of the pendulum, approx. 22 cm above the axis.

From Figure 3.41 it can be seen that the internal friction damping is not simply hysteretic (constant Q); rather it is a function of amplitude. It can also be seen, from the close proximity of the solid and dashed curves, that the air influence on the solder pendulum is much less than that of the internal friction. By contrast, air influence is of comparable magnitude to the internal friction in the case of the brass pendulum (or even larger, at large amplitude).

A minimum of two frequencies was considered for the study, since the frequency variation of the damping is different for external and internal frictions. (Note: although the period is a function of amplitude, the amount of nonisochronism is small compared to the damping changes and is ignored here.) The periods were matched for the two pendula at each of 2.03 and 2.51 sec. For hysteretic-only (internal friction) damping, the Q at the shorter period should in theory be 1.53 times that of the longer period, for both brass and solder. If the damping were viscous only, the factor should be 1.24. In the case of solder at 10 cm/sec (corrected), the ratio is $1.66 = 131/71$, and for brass it is $1.46 = 387/266$. Although the ratio for solder is greater than the expected 1.53, the difference is within experimental

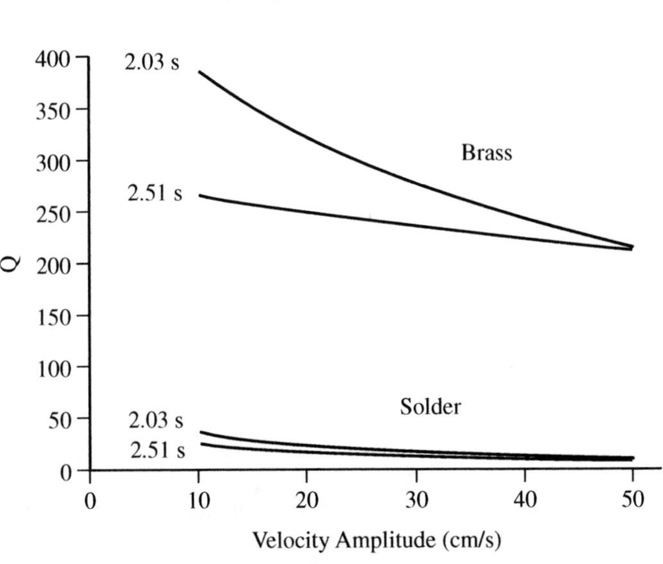

FIGURE 3.40 Illustration of amplitude-dependent damping in a rod pendulum made of (i) brass and (ii) solder. The two different matched periods of oscillation are indicated in seconds.

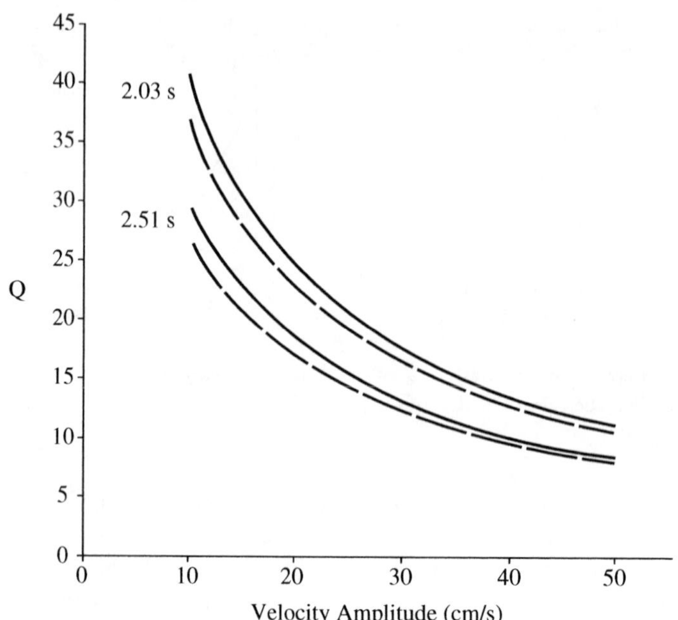

FIGURE 3.41 Amplitude dependence of the estimated quality factor due only to internal friction in the solder pendulum. Dashed lines show the Q before correction for air damping.

uncertainty for individual Q values, which from other, more detailed experiments were in the neighborhood of 5 to 10%.

The ratio for brass (1.46) is between 1.24 and 1.53, as expected, because of the comparable influence of air and internal friction.

References

Barkhausen, H., *Z. Phys.*, 20, 401, 1919.

Cromer, A., Stable solutions using the Euler approximation, *Am. J. Phys.*, 49(5), 455–459, 1981.

de Silva, C.W. 2006. *Vibration, Fundamentals and Practice*, 2nd ed., Taylor & Francis, CRC Press, Boca Raton, FL.

Fraden, J. 1996. *Handbook of Modern Sensors, Physics, Designs, & Applications*, 2nd ed., AIP Press (Springer), Secaucus, NJ.

Peters, R., Fourier transform construction by vector graphics, *Am. J. Phys.*, 60, 439, 1992.

Peters, R., Mercer physics laboratory experiment, described online at http://physics.mercer.edu/labs/sho.htm, 1998.

Peters, R., Tutorial description of the symmetric differential capacitive (SDC) sensor, online at http://www.telatomic.com/sdct2.html, 2002.

Peters, R., Graphical explanation of the speed of the Fast Fourier Transform, online at http://arxiv.org/html/math.HO/0302212, 2003a.

Peters, R., Nonlinear damping of the "linear" pendulum, online at http://arxiv.org/pdf/physics/03006081, 2003b.

Peters, R., Oscillator damping with more than one mechanism of internal friction dissipation, online at http://arxiv.org/html/physics/0302003/, 2003c.

Peters, R., Anisotropic internal friction damping, online at http://arxiv.org/html/physics/0302055, 2003d

Peters, R., Folded pendulum measurements of the Earth's free oscillations, online at http://physics.mercer.edu/petepag/eigen.html, 2004.

Portevin, A. and Le Chatelier, M., Tensile tests of alloys undergoing transformation, *C. R. Acad. Sci.*, 176, 507, 1923.

Press, W., Flannery, B., Teukolsky, S., and Vetterling, W. 1986. *Numerical Recipes — The Art of Scientific Computing*, Cambridge University Press, Cambridge.

Wielandt, E., *Seismometry*, section: Electronic Displacement Sensing, online at http://www.geophys.uni-stuttgart.de/seismometry/hbk_html/node1.html, 2001.

4

Structure and Equipment Isolation

Y.B. Yang
National Taiwan University

L.Y. Lu
National Kaohsiung First University of Science and Technology

J.D. Yau
Tamkang University

Summary

In this chapter, a brief review will be given of the concept of isolation for suppressing the vibrations in structures and equipment subjected either to harmonic or seismic ground excitations. The mechanism of various isolation devices, including the elastomeric bearing, sliding bearing, resilient-friction base isolator, and Electricite de France system, will first be described in Section 4.2, together with their mathematical models. In Section 4.3, a closed form solution will be derived for the dynamic response of a structure–equipment system isolated by bearings of the elastomeric type, subjected to harmonic motions. Such a solution enables us to interpret the various behaviors of the structure and equipment under excitation. The elastomeric bearings can help increase the fundamental period of the structure, thereby, reducing the accelerations transmitted to the superstructure. In Section 4.4 and Section 4.5, the seismic behavior of a structure–equipment system isolated by a sliding support, with and without resilient force, will be studied using a state-space incremental-integration approach. With the introduction of a frictional sliding interface, the motion of the structure–equipment system will be uncoupled from the ground excitation, and the influence of the latter will be mitigated. The residual base displacement caused by the sliding isolator can be reduced through inclusion of a resilient mechanism in the isolator. Nevertheless, the resilient mechanism can make the system more sensitive to the low-frequency components of excitation. In Section 4.6, issues related to design of base isolators will be discussed, along with the concepts underlying some design codes and guidelines. The notation used is listed at the end of the chapter.

4.1 Introduction

Conventionally, structural designers are concerned about the safety of buildings, bridges, and other civil engineering structures that are subjected to earthquakes. The recent history of earthquakes reveals that strong earthquakes, such as the 1994 Northridge earthquake (U.S.A.), 1995 Kobe earthquake (Japan), and 1999 Chi-Chi earthquake (Taiwan), can cause some badly designed structures or buildings to fail or collapse, and also cause some well-designed structures to malfunction due to the damage or failure of the equipment housed in the structure or building. Both the failures of structures and equipment, also known as *structural* and *nonstructural* failures, respectively, can cause serious harm to the residents or personnel working in a building. For the case where the equipment is part of a key service system, such as in hospitals, power stations, telecommunication centers, high-precision factories, and the like, the lives and economic losses resulting from the malfunctioning of the equipment can be tremendous. Thus, the maintenance of the safety of structures and attached equipment during a strong earthquake is a subject of high interest in earthquake engineering. In this regard, *base isolation* has been proved to be an effective means for protecting the structures and attached equipment, which is made possible through reduction of the seismic forces transmitted from the ground to the superstructure (Yang et al., 2002).

For light secondary systems mounted on heavier primary systems, it was concluded that the response of the light secondary system, that is, the equipment, is affected by four major dynamic characteristics in earthquakes (Igusa and Der Kiureghian, 1985a, 1985b, 1985c; Yang and Huang, 1993). The first issue is *tuning*, which means that the natural frequency of the equipment is coincident with that of the structure. Such an effect may amplify the response of the equipment due to the fact that the light secondary system behaves as if it were a vibration absorber of the heavier primary system. The second issue is *interaction*, which is related to the feedback effect between the motions of the primary and secondary systems. Ignoring the feedback effect of interaction may result in an overestimation of the true response of the combined system. The third issue is *non-classical damping*, which may occur when the damping properties of the two systems are drastically different, such that the natural frequencies and mode shapes of the combined system can only be expressed in terms of complex numbers. Under such a circumstance, the conventional response spectrum analysis, based on modal superposition, becomes inapplicable. The last issue is *spatial coupling*, which relates to the effect of multiple support motions when the secondary system of interest is mounted at multiple locations. By considering the inelastic effect, Igusa (1990) proposed an equivalent linearization technique for investigating the response characteristics of an inelastic primary–secondary system with two degrees of freedom (DoF) under random vibrations. His results indicated that the existence of small nonlinearity is helpful for reducing the coupling system responses. With the concept of equivalent linearization, Huang et al. (1994) explored the response and reliability of a linear secondary system mounted on a yielding primary structure under white-noise excitations. It was concluded that the response of the secondary system could be reduced by increasing the equipment damping or by locating equipment at higher levels of the primary structure.

Owing to the fact that the mass and stiffness of a secondary system are much smaller than those of the primary structure, the interaction effect of the combined system, as well as the ill-conditioning in system matrices, may take place when one performs the dynamic analysis. To deal with this problem, some researchers chose to evaluate the response of the secondary systems from the floor motions. To avoid solving large eigenvalue problems and to account for the interaction between the building and equipment components, Villaverde (1986) applied the response spectrum technique to the analysis of a combined building–equipment system, by which the maximum response of light equipment mounted on the building under the earthquake is expressed in terms of the natural frequencies and mode shapes of the building and equipment. To take into account the equipment–structure interactions, Suarez and Singh (1989) proposed an analytical scheme for computing the dynamic characteristics of the combined system, using the modal properties to compute the floor spectra. Lai and Soong (1990) presented a statistical energy analysis technique for evaluating the response of coupling primary–secondary structural systems, based on the concept of power-balance equation, that is, the power input to the primary system is equal to the dissipated energy of the primary system plus the transferred energy to the secondary system. Using a mean-square condensation procedure, Chen and Soong (1994) considered the effect of interaction by calculating the multi-DoF response of a primary–secondary system under random excitations. Later on, Chen and Soong (1996) derived an exact solution for the mean-square response of a structure–equipment system under dynamic loads, indicating that there exists an optimal damping ratio for reducing the vibration of equipment attached to the primary structure. Gupta and coworkers investigated the response of a secondary system with multiple supports on a primary structure subjected to earthquakes, taking into account the interaction effect between the equipment and structure (Dey and Gupta, 1998, 1999; Chaudhuri and Gupta, 2002). Their results indicated that when the soil–structure interaction (SSI) is taken into account, the response of the equipment–structure system will be affected by the SSI, unless a very stiff soil condition is considered.

On the other hand, a number of research works have been conducted by implementing isolation systems at the base of the equipment–structure system, aiming to reduce the earthquake forces transmitted from the ground. Based on a theoretical and experimental investigation, Kelly and Tsai (1985) showed that seismic protection can be achieved effectively for lightweight equipment mounted on an isolated structure installed with elastic bearings at the base. A hybrid isolation system with base-isolated floors was proposed by Inaudi and Kelly (1993), for the protection of highly sensitive devices mounted on a structure subjected to support motions. Considering the effects of torsion and translation, Yang and Huang (1998) studied the seismic response of light equipment items mounted on torsional buildings supported by elastic bearings. Their results indicated that the response of an equipment–structure system can be effectively reduced through installation of base isolators, and that there exists an optimal location for mounting the equipment. Juhn et al. (1992) presented a series of experimental results for the secondary systems mounted on a sliding base-isolated structure. They concluded that the acceleration response of the secondary system may be amplified when the input motions are composed of low-frequency vibrations. In this case, the sliding bearings are not considered to be an effective isolation device, which implies that the base-isolated structure is not suitable for a construction site with soft soil.

Concerning the use of sliding bearings (supports) as base isolators, Lu and Yang (1997) investigated the response of an equipment item attached to a sliding primary structure under earthquake excitations. Their results showed that the response of the equipment can be effectively reduced through the installation of a sliding support at the structural base, in comparison with that of a structure with fixed base. To overcome the discontinuous nature of the sliding and nonsliding phases of a structural system with sliding base, a fictitious spring model was proposed by Yang and coworkers for simulating the mechanism of sliding and nonsliding (Yang et al., 1990, 2000; Yang and Chen, 1999). Such a model will be described in a later section of this chapter. Agrawal (2000) adopted the same fictitious spring model in studying the response of an equipment item mounted on a torsionally coupled structure with sliding support. His results indicated that sliding supports could effectively reduce the equipment response, compared to that of a fixed-base structure. However, in the tuning region, where the natural frequency of the equipment coincides with the fundamental frequency of the structure, the equipment response may be adversely amplified due to the increase in eccentricity of the torsionally coupled structure.

The problem of building isolation has recently received more attention than ever from researchers and engineers, due to the construction of high-precision factories worldwide. More and more stringent requirements have been employed in this regard for removing the ambient or man-made vibrations (Rivin, 1995; Steinberg, 2000). To allow sensitive electronic equipment to operate in a harsh environment, Veprik and Babitsky (2000) proposed an optimization procedure for the design of vibration isolators aimed at minimizing the response of the internal components of electronic equipment. As for the protection of high-tech equipment from micro- or ambient vibrations, Yang and Agrawal (2000) showed that passive hybrid floor isolation systems are more effective in mitigating the equipment response than passive or hybrid base isolation systems. Xu and coworkers studied the response of a batch of high-tech equipment mounted on a hybrid platform, which in turn is mounted on a building floor (Xu et al., 2003; Yang et al., 2003). Both their theoretical and experimental studies showed that the hybrid platform, which is composed of leaf springs, oil dampers, and an electron-magnetic actuator with velocity feedback control, is more effective in mitigating the velocity response of the high-tech equipment than the passive platform.

The objective of this chapter is to give an overview on the seismic behavior of various base isolators. The organization of this chapter can be summarized as follows. In Section 4.2, the mechanisms of various seismic isolators that are currently in use are introduced and explained. In Section 4.3, a structure–equipment system isolated by bearings of the elastomeric type is modeled by a three-DoF system composed of a spring and dashpot unit, for which a closed-form solution is obtained for the dynamic response of the isolated system subjected to harmonic earthquakes; remarks on the dynamic response of the system components are also made. In Section 4.4 and Section 4.5, the seismic behaviors of a structure–equipment system isolated by a sliding support, with and without resilient capability, will be investigated. Also presented are numerical methods based on the incremental-integration procedure for the analysis of structural systems with sliding-type isolators.

4.2 Mechanisms of Base-Isolated Systems

Figure 4.1 shows a simplified model for a structural system subjected to a support motion.

For this single-DoF system, the equation of motion can be written as

$$m\ddot{x} + c\dot{x} + kx = -m\ddot{x}_g \tag{4.1}$$

where m denotes the mass, c the damping, k the stiffness, x the displacement of the system, and $\ddot{x}_g$ the ground acceleration. By assuming the system to be linearly elastic, the response $x(t)$ can be obtained using Duhamel's integral, as

$$x(t) = \frac{1}{\Omega_d}\int_0^t \ddot{x}_g(\tau)e^{-\zeta\Omega(t-\tau)} \sin \Omega_d(t-\tau)d\tau \tag{4.2}$$

where the natural angular frequency, Ω, damped natural frequency, Ω_d, and damping ratio, ζ, of the system are defined as follows:

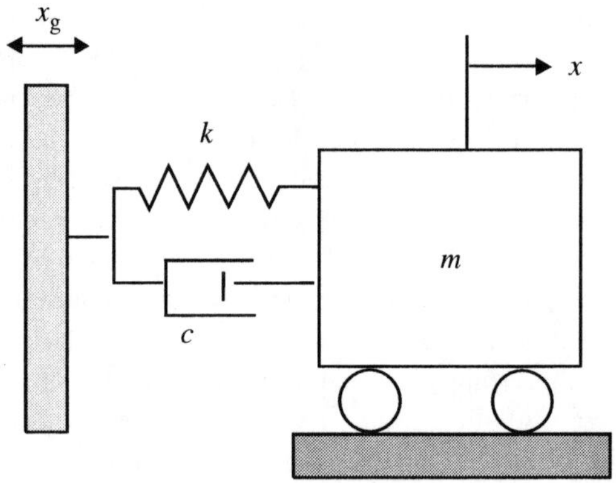

FIGURE 4.1 Model of a single-DoF system.

$$\Omega = \sqrt{\frac{k}{m}} \tag{4.3a}$$

$$\Omega_d = \Omega\sqrt{1-\zeta^2} \tag{4.3b}$$

$$\zeta = \frac{c}{2m\Omega} \tag{4.3c}$$

Correspondingly, the natural period, T, and damped period, T_d, of the structure are

$$T = \frac{2\pi}{\Omega} = 2\pi\sqrt{\frac{m}{k}} \tag{4.4a}$$

$$T_d = \frac{2\pi}{\Omega_d} = \frac{T}{\sqrt{1-\zeta^2}} \tag{4.4b}$$

For a given support acceleration, $\ddot{x}_g$, the displacement, x, and acceleration, $\ddot{x}$, of the single-DoF system can be related to the natural period, T, and damping ratio, ζ, of the system. Thus, for a specific earthquake, by first selecting a damping ratio, ζ, and using Equation 4.2, one can compute the peak displacement x, for a structure with a period of vibration, T, with given values of m, c, and k. Repeating the above procedure for a wide range of periods, T, while keeping the damping ratio, ζ, constant, one can obtain response curves similar to those shown in Figure 4.2. By varying the damping ratio, ζ, one can construct the *displacement response spectra* and *pseudo-acceleration response spectra* for all single-DoF structures under a given earthquake, as schematically shown in Figures 4.2 and 4.3, respectively.

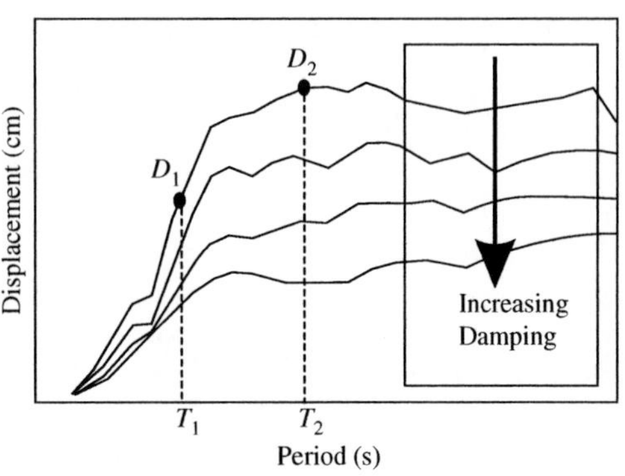

FIGURE 4.2 Schematic of displacement response spectra.

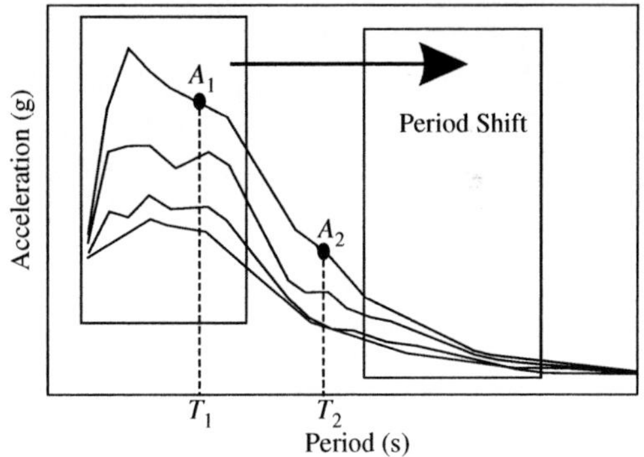

FIGURE 4.3 Schematic of pseudo-acceleration response spectra.

A general impression that is gained from Figure 4.2 and Figure 4.3 is that a structure with a shorter natural period has less displacement response when subjected to an earthquake, but it also has a larger acceleration response. Specifically, let us consider a structure of a constant damping ratio, ζ, with its period increased from T_1 to T_2. As can be observed from the figures, the displacement of the structure increases from D_1 to D_2, while the acceleration decreases from A_1 to A_2. Such a feature is known as the *period shift* effect. On the other hand, by increasing damping ratio of the structure, the displacement of the structure decreases significantly, as can be seen from Figure 4.2. The same is also true with the acceleration response, as can be seen from Figure 4.3. Moreover, a larger damping ratio also makes the structure less sensitive to the variation in ground vibration characteristics, as indicated by the smoother response curves for structures having higher damping ratios, in both figures. From the aforementioned two response spectra, one observes that the philosophy of base isolation is to lengthen the vibration period of the structure to be protected, using base isolators of some kind, by which the earthquake force transmitted to the structure can be greatly reduced. In the meantime, some additional damping must be introduced on the base isolators in order to control the relative displacements across the base isolators with tolerable limits.

To fulfill the function of lengthening the period of vibration of the structure to be protected, the base isolators that are inserted between the structure and its foundation must be flexible in the horizontal direction, but stiff enough in the vertical direction so as to carry the heavy loads transmitted from the superstructure. With such devices, the natural period of vibration of the structure will be significantly lengthened and shifted away from the dominant frequency range of the expected earthquakes. The following is a summary of the fundamental features of four types of isolators frequently used in engineering practice.

4.2.1 Elastomeric Isolation System

Elastomeric bearing is the type of base isolator most commonly known to researchers and engineers working on base isolation. It is usually composed of alternating layers of steel and hard rubber and, for this reason, it is also known as the *laminated rubber bearing*. This type of bearing is stiff enough to sustain the vertical loads, yet flexible under the lateral forces. The ability to deform horizontally enables the bearing to reduce significantly the structural base shear transmitted from the ground. While the major function of elastomeric bearings is to reduce the transmission of shear forces to the superstructure by lengthening the vibration period of the entire system, they must also provide sufficient rigidity under vertical loads. Let us consider a structure installed with elastomeric bearings, which is subjected to a support acceleration, $\ddot{x}_g$, as in Figure 4.4. By representing the isolated structure as a single-DoF system, based on the assumption that the superstructure is rigid in comparison with the stiffness of the elastic bearings, the equation of motion for the entire system can be written as

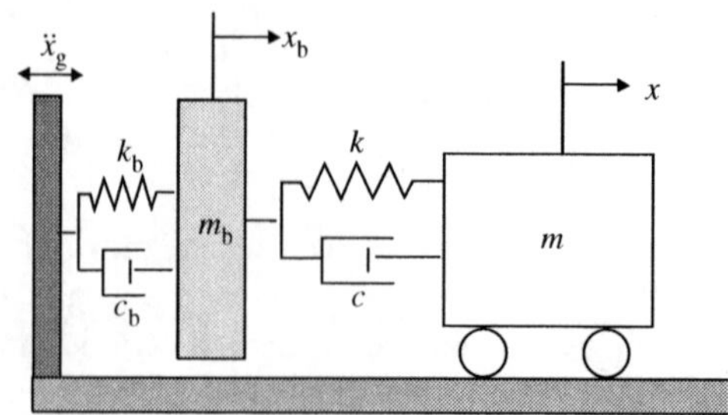

FIGURE 4.4 Model for base-isolation systems with elastic bearing.

$$\begin{bmatrix} m & 0 \\ 0 & m_b \end{bmatrix}\begin{Bmatrix} \ddot{x} \\ \ddot{x}_b \end{Bmatrix} + \begin{bmatrix} c & -c \\ -c & c+c_b \end{bmatrix}\begin{Bmatrix} \dot{x} \\ \dot{x}_b \end{Bmatrix} + \begin{bmatrix} k & -k \\ -k & k+k_b \end{bmatrix}\begin{Bmatrix} x \\ x_b \end{Bmatrix} = -\begin{Bmatrix} m\ddot{x}_g \\ m_b\ddot{x}_g \end{Bmatrix} \tag{4.5}$$

where m, c, and k denote the mass, damping, and stiffness of the superstructure, respectively, m_b, c_b, and k_b denote the mass, damping, and stiffness of the base raft, respectively, and x and x_b denote the displacements of the superstructure and the base, respectively.

In reality, the reduction in the seismic forces transmitted to a superstructure through the installation of laminated rubber bearings is achieved at the expense of large relative displacements across the bearings. If substantial damping can be introduced into the bearings or the isolation system, then the problem of large displacements can be alleviated. It is for this reason that the laminated rubber bearing with a central lead plug inserted has been devised (Yang et al., 2002). To simulate the dynamic properties of the lead–rubber bearing (LRB) system, an equivalent linearized system has been proposed, for which the equation of motion is

$$\begin{bmatrix} m & 0 \\ 0 & m_b \end{bmatrix}\begin{Bmatrix} \ddot{x} \\ \ddot{x}_b \end{Bmatrix} + \begin{bmatrix} c & -c \\ -c & c+c_{eq} \end{bmatrix}\begin{Bmatrix} \dot{x} \\ \dot{x}_b \end{Bmatrix} + \begin{bmatrix} k & -k \\ -k & k+k_{eq} \end{bmatrix}\begin{Bmatrix} x \\ x_b \end{Bmatrix} = -\begin{Bmatrix} m\ddot{x}_g \\ m_b\ddot{x}_g \end{Bmatrix} \tag{4.6}$$

where c_{eq} and k_{eq} respectively represent the equivalent linearized damping and stiffness coefficients of the LRB system. The dynamic behavior of a structure–equipment system isolated by elastomeric bearings with linearized damping and stiffness coefficients, when subjected to harmonic and earthquake excitations, will be investigated analytically and numerically, respectively, in Section 4.3.

4.2.2 Sliding Isolation System

Another means for increasing the horizontal flexibility of a base-isolated structure is to insert a *sliding* or *friction surface* between the foundation and the base of the structure. The shear force transmitted to the superstructure through the sliding interface is limited by the static frictional force, which equals the product of the coefficient of friction and the weight of the superstructure. The coefficient of friction is usually kept as low as is practical. However, it must be high enough to provide a frictional force that can sustain strong winds and minor earthquakes without sliding. Since the sliding system has no dominant natural period, it is generally frequency-independent when the structure is subjected to earthquakes with a wideband frequency content. As mentioned previously, when a sliding structure is subjected to a ground motion, transitions may occur repeatedly between the sliding and nonsliding phases. To take into account such a phase transition, Yang et al. (1990) proposed the use of a fictitious spring between the structural base raft and the underlying ground to simulate the static–dynamic frictional force of the sliding device. With reference to Figure 4.5, the equation of motion for the structure with sliding base can be written as follows:

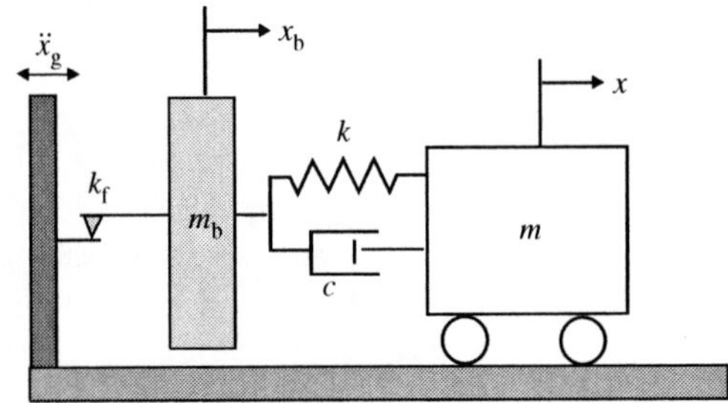

FIGURE 4.5 Model for base-isolation systems with sliding support.

$$\begin{bmatrix} m & 0 \\ 0 & m_b \end{bmatrix}\begin{Bmatrix} \ddot{x} \\ \ddot{x}_b \end{Bmatrix}+\begin{bmatrix} c & -c \\ -c & c \end{bmatrix}\begin{Bmatrix} \dot{x} \\ \dot{x}_b \end{Bmatrix}+\begin{bmatrix} k & -k \\ -k & k \end{bmatrix}\begin{Bmatrix} x \\ x_b \end{Bmatrix}+\begin{Bmatrix} 0 \\ f_r \end{Bmatrix}=-\begin{Bmatrix} m\ddot{x}_g \\ m_b\ddot{x}_g \end{Bmatrix} \tag{4.7}$$

where k_f is the stiffness of the fictitious spring and the frictional force, f_r, can be represented as

$$f_r=\begin{cases} k_f(x_b-x_{b0}) & \text{for non-sliding phase,} \\ \pm\mu(m+m_b)g & \text{for sliding phase} \end{cases} \tag{4.8}$$

with x_{b0} indicating the initial elongation of the fictitious spring in the current nonsliding phase, μ the coefficient of friction, and g the acceleration of gravity. The fictitious spring concept will be incorporated in the analysis of sliding structures in Section 4.4 of this chapter, when considering both harmonic and seismic excitations.

4.2.3 Sliding Isolation System with Resilient Mechanism

One particular problem with a sliding structure is the occurrence of residual displacements after earthquakes. To remedy such a drawback, the sliding surface is often made concave, so as to provide a recentering mechanism for the isolated structures. This is the idea behind the *friction pendulum system* (FPS), shown in Figure 4.6, which utilizes a spherical concave surface to produce a recentering force for the superstructure under excitations. To guarantee that a sliding structure can return to its original position, other mechanisms, such as high-tension springs and elastomeric bearings, can be used as an auxiliary system for providing the restoring forces. Previously, the sliding isolation systems have been successfully applied in the protection of important structures, such as nuclear power plants, emergency fire water tanks, large chemical storage tanks, and so on, from the damaging actions of severe earthquakes.

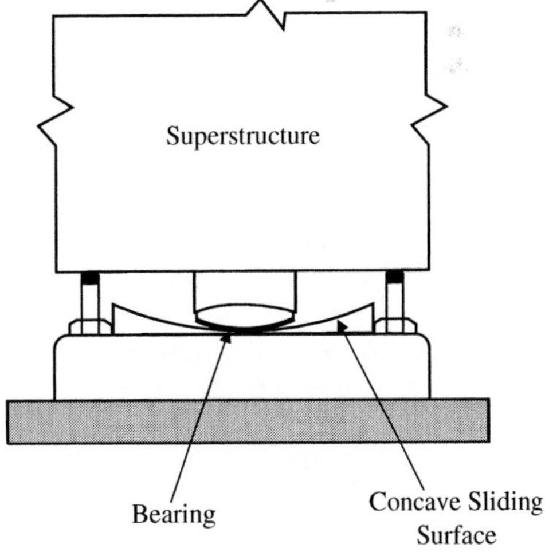

FIGURE 4.6 Friction pendulum system.

To improve the performance of sliding isolators under strong earthquakes, Mostaghel (1984) and Mostaghel and Khodaverdian (1987) proposed the *resilient-friction base isolator* (RFBI) for controlling the transmission of shear force to the superstructures, while keeping the residual displacements within an allowable level. The RFBI device is basically made of a central rubber core and Teflon-coated steel plates, and offers a friction resistance for keeping the system in the nonsliding mode under wind excitations and small earthquakes, and a restoring force by the rubber ingredient for limiting the maximum sliding displacements. The equation of motion for a structure installed with RFBI, as shown in Figure 4.7, can be written as

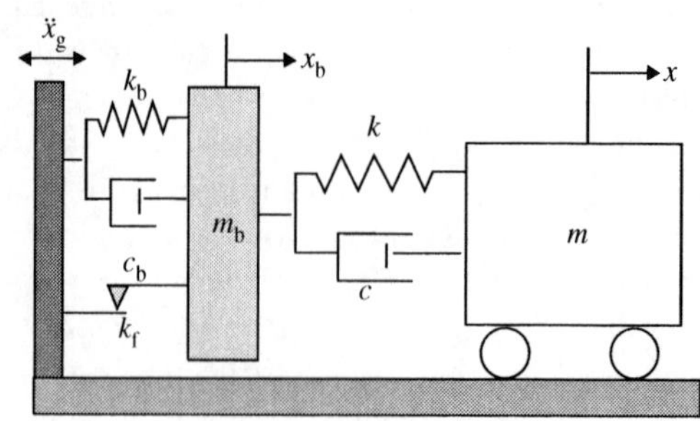

FIGURE 4.7 Model for base-isolation systems with RFBI device.

$$\begin{bmatrix} m & 0 \\ 0 & m_b \end{bmatrix}\begin{Bmatrix} \ddot{x} \\ \ddot{x}_b \end{Bmatrix} + \begin{bmatrix} c & -c \\ -c & c+c_b \end{bmatrix}\begin{Bmatrix} \dot{x} \\ \dot{x}_b \end{Bmatrix} + \begin{bmatrix} k & -k \\ -k & k+k_b \end{bmatrix}\begin{Bmatrix} x \\ x_b \end{Bmatrix} + \begin{Bmatrix} 0 \\ f_r \end{Bmatrix} = -\begin{Bmatrix} m\ddot{x}_g \\ m_b\ddot{x}_g \end{Bmatrix} \tag{4.9}$$

The interfacial frictional force, f_r, existing in the RFBI and appearing in Equation 4.9 serves as the outlet for energy dissipation. The behavior of a structure–equipment system supported by sliding isolators with resilient mechanism subjected to both harmonic and earthquake excitations will be investigated in Section 4.5.

4.2.4 Electricite de France System

To limit effectively the acceleration of base-isolated structures and internal secondary systems, such as those of nuclear power plants, when subjected to strong earthquakes, the *Electricite de France* (EDF) system was proposed by Gueraud et al. (1985). The design concept of an EDF system is to arrange the elastomeric bearing and sliding device at the base of a structure in series. For low-level ground motions, the EDF system will behave as an elastomeric bearing and return to the original position after support motions, while for strong earthquakes, the EDF system will behave as a sliding device. The EDF system may have a residual displacement after some major earthquakes. Because of the sliding mechanism of the EDF system, the maximum horizontal acceleration of the superstructure is kept within a certain range (Gueraud et al., 1985; Park et al., 2002), while the shear force transmitted to the superstructure through the frictional interface is smaller than the static frictional force. For the mathematical model shown for the EDF system in Figure 4.8, the equations of motion for the nonsliding and sliding phases can be written as

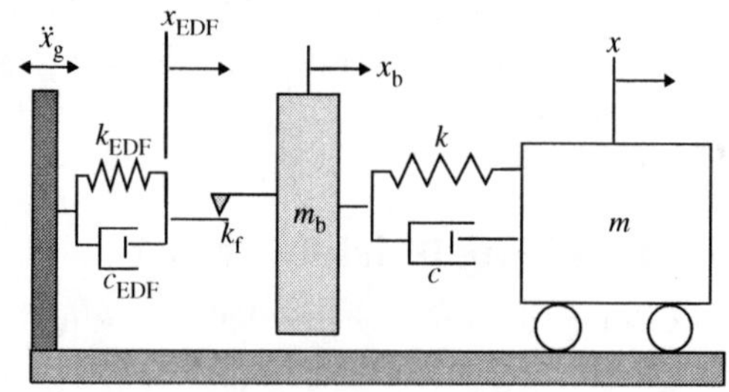

FIGURE 4.8 Model for base-isolation systems with EDF device.

(a) nonsliding phase:

$$\begin{Bmatrix} m\ddot{x} \\ m_b\ddot{x}_b \\ 0 \end{Bmatrix} + \begin{bmatrix} c & -c & 0 \\ -c & c & 0 \\ 0 & 0 & c_{EDF} \end{bmatrix}\begin{Bmatrix} \dot{x} \\ \dot{x}_b \\ \dot{x}_{EDF} \end{Bmatrix} + \begin{bmatrix} k & -k & 0 \\ -k & k+k_f & -k_f \\ 0 & -k_f & k_f+k_{EDF} \end{bmatrix}\begin{Bmatrix} x \\ x_b \\ x_{EDF} \end{Bmatrix}$$

$$= \begin{Bmatrix} -m\ddot{x}_g \\ -m_b\ddot{x}_g \\ 0 \end{Bmatrix} \tag{4.10a}$$

(b) sliding phase:

$$\begin{Bmatrix} m\ddot{x} \\ m_b\ddot{x}_b \\ 0 \end{Bmatrix} + \begin{bmatrix} c & -c & 0 \\ -c & c & 0 \\ 0 & 0 & c_{EDF} \end{bmatrix} \begin{Bmatrix} \dot{x} \\ \dot{x}_b \\ \dot{x}_{EDF} \end{Bmatrix} + \begin{bmatrix} k & -k & 0 \\ -k & k & 0 \\ 0 & 0 & k_{EDF} \end{bmatrix} \begin{Bmatrix} x \\ x_b \\ x_{EDF} \end{Bmatrix}$$

$$= \begin{Bmatrix} -m\ddot{x}_g \\ -m_b\ddot{x}_g \mp \mu(m+m_b)g \\ \pm\mu(m+m_b)g \end{Bmatrix} \tag{4.10b}$$

where c_{EDF} and k_{EDF}, respectively, denote the damping and stiffness of the EDF system, and x_{EDF} denotes the displacement of the system.

4.2.5 Concluding Remarks

To mitigate the transmission of earthquake forces to a structure, and the potentially earthquake-induced damage to the equipment attached to the structure, base isolation is an effective structural design philosophy. With the installation of base isolators, the natural period of vibration of the structure will be significantly lengthened and shifted away from the dominant frequency range of the expected earthquakes. In accordance, the earthquake force transmitted to the structure can be significantly reduced. In this section, the mechanisms of four types of base isolator frequently used in engineering practice are introduced. Since the base isolators, such as the elastomeric bearings or sliding isolations, have relatively flexible stiffness in the horizontal direction, the occurrence of residual displacements after earthquakes may cause certain problems on the structure to be protected. To remedy such a drawback and to further guarantee that a base-isolated structure can return to its original position, the RFBI is implemented for controlling the transmission of shear force to the superstructure, while keeping the residual displacement within an allowable level. On the other hand, to limit the acceleration level of internal secondary systems housed in a base-isolated structure under strong earthquakes, such as those of the nuclear power plants, the EDF system can be used as an alternative device for base isolation, even though some residual displacements may be induced after the earthquakes.

4.3 Structure–Equipment Systems with Elastomeric Bearings

Owing to the stringent requirements for normal functioning of high-tech facilities, such as printed circuit boards, semiconductor factories, and sensitive medical devices, the need to suppress excessive vibrations in sensitive structure–equipment systems has become an issue of great concern to structural designers. Besides, these high-tech facilities may suffer significant damages during a major earthquake. Using elastomeric isolation systems to reduce the earthquake forces transmitted from the ground is one of the most popular ways adopted by structural designers. In this section, the performance of elastomeric bearings in protecting structure–equipment systems against horizontal ground motions will be investigated.

4.3.1 Formulation of Base Isolation Systems with Elastic Bearing

By modeling the structure, internal equipment and the base of an isolated structure–equipment system as a lumped mass system, one can construct the mathematical model for the structure–equipment isolation system supported by an elastic bearing in Figure 4.9. The following is the

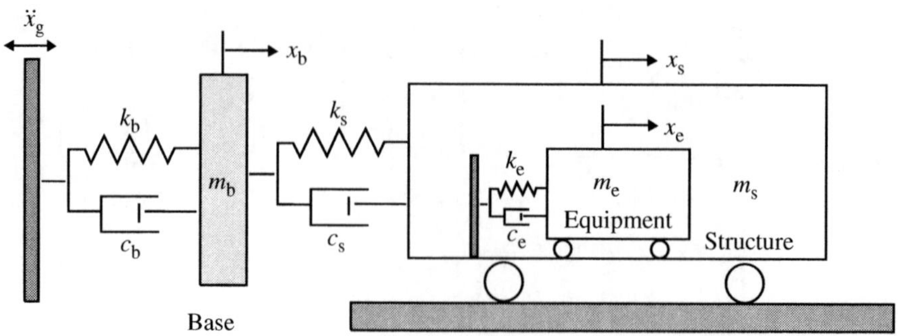

FIGURE 4.9 Model of a structure–equipment isolation system with elastic bearing.

equation of motion for the base-isolated structure–equipment system when it is subjected to a ground acceleration, $\ddot{x}_g$:

$$\begin{Bmatrix} m_e\ddot{x}_e \\ m_s\ddot{x}_s \\ m_b\ddot{x}_b \end{Bmatrix} + \begin{bmatrix} c_e & -c_e & 0 \\ -c_e & c_s + c_e & -c_s \\ 0 & -c_s & c_s + c_b \end{bmatrix} \begin{Bmatrix} \dot{x}_e \\ \dot{x}_s \\ \dot{x}_b \end{Bmatrix} + \begin{bmatrix} k_e & -k_e & 0 \\ -k_e & k_s + k_e & -k_s \\ 0 & -k_s & k_s + k_b \end{bmatrix} \begin{Bmatrix} x_e \\ x_s \\ x_b \end{Bmatrix} = -\begin{Bmatrix} m_e \\ m_s \\ m_b \end{Bmatrix} \ddot{x}_g \tag{4.11}$$

where m represents the mass, c the damping coefficient, and k the stiffness of the system. Also, the subscripts 'e', 's', and 'b' are associated with the DoF of the equipment, structure, and base, respectively. The notations employed in Figure 4.9 have been defined in Table 4.1. It should be mentioned that the elastic bearing stiffness, k_b, appearing in Equation 4.11, is a parameter relating to the boundary conditions of the system considered here. A small value of k_b relative to the structural stiffness, k_s, means that the system is isolated by a set of soft bearings. In contrast, a large value of k_b means that the structure is rigidly supported.

TABLE 4.1 Definition of Symbols

Symbol	Definition
c_e, c_s	Damping coefficients of equipment and superstructure
k_e, k_s	Stiffness of equipment and superstructure
k_b	Stiffness of elastic bearing or resilient stiffness of isolation system
m_e, m_s, m_b	Masses of equipment, superstructure and base mat
$x_e(t)$, $x_s(t)$, $x_b(t)$	Relative-to-the-ground displacements of equipment, superstructure and base mat
$\ddot{x}_g(t)$	Ground acceleration
μ	Frictional coefficient of sliding isolation system
ω_b	Frequency of isolation system
ω_e, ω_s	Frequencies of equipment and superstructure
ω_g	Frequency of ground excitation
ζ_e, ζ_s	Damping ratios of equipment and structure

4.3.2 Free Vibration Analysis

By neglecting the damping and forcing terms in Equation 4.11, the equation of motion for free vibration can be written as

$$\begin{Bmatrix} m_e\ddot{x}_e \\ m_s\ddot{x}_s \\ m_b\ddot{x}_b \end{Bmatrix} + \begin{bmatrix} k_e & -k_e & 0 \\ -k_e & k_s + k_e & -k_s \\ 0 & -k_s & k_s + k_b \end{bmatrix} \begin{Bmatrix} x_e \\ x_s \\ x_b \end{Bmatrix} = \begin{Bmatrix} 0 \\ 0 \\ 0 \end{Bmatrix} \tag{4.12}$$

By solving the preceding equation, one can obtain the natural frequencies and vibration modes of the structure–equipment system with elastic bearings. As for the present problem, the horizontal stiffness of the elastic bearing is designed to be quite low compared with that of the superstructure. It follows that the superstructure in its entirety behaves essentially as a *rigid body* for the fundamental vibration mode shape of the combined system, which implies that the displacement responses for the equipment, structure, and base under free vibration can be approximately taken as the same, that is, $x_e = x_s = x_b = x$. By introducing such a condition into Equation 4.12, the equation of motion for the equivalent single-DoF base-isolated system can be written as

$$(m_e + m_s + m_b)\ddot{x} + k_b x = 0 \tag{4.13}$$

Equation 4.13 indicates that the fundamental frequency, ω_1, of the base-isolated system can be approximated by $\omega_1 \approx \sqrt{k_b/(m_e + m_s + m_b)}$. Further, if the condition of fixed base is considered, that is, by letting the responses of the base be equal to zero, $x_b = \ddot{x}_b = 0$, the structure–equipment isolation system will be reduced to the case of a fixed-base system, such that the equation of motion becomes

$$\begin{Bmatrix} m_e\ddot{x}_e \\ m_s\ddot{x}_s \end{Bmatrix} + \begin{bmatrix} k_e & -k_e \\ -k_e & k_s + k_e \end{bmatrix} \begin{Bmatrix} x_e \\ x_s \end{Bmatrix} = \begin{Bmatrix} 0 \\ 0 \end{Bmatrix} \tag{4.14}$$

as is well known.

4.3.3 Dynamics of Structure–Equipment Isolation Systems to Harmonic Excitations

The advantage of a closed-form solution is that it allows us to examine the key parameters involved in the problem considered. This is what will be sought herein. For the case of a harmonic ground excitation, x_g, with amplitude, X_g, that is, with $x_g = X_g\, e^{i\omega t}$, one may derive from Equation 4.11 the following:

$$\begin{Bmatrix} m_e\ddot{x}_e \\ m_s\ddot{x}_s \\ m_b\ddot{x}_b \end{Bmatrix} + \begin{bmatrix} k_e & -k_e & 0 \\ -k_e & k_s + k_e & -k_s \\ 0 & -k_s & k_s + k_b \end{bmatrix} \begin{Bmatrix} x_e \\ x_s \\ x_b \end{Bmatrix} = \begin{Bmatrix} m_e \\ m_s \\ m_b \end{Bmatrix} X_g\omega^2\, e^{i\omega t} \tag{4.15}$$

Correspondingly, the steady-state responses of the system can be expressed as

$$\begin{Bmatrix} x_e \\ x_s \\ x_b \end{Bmatrix} = \begin{Bmatrix} X_e \\ X_s \\ X_b \end{Bmatrix} e^{i\omega t} \tag{4.16}$$

where (X_e, X_s, and X_b) represent the amplitudes of the equipment, structure, and base, respectively.

Substituting Equation 4.16 into Equation 4.15 yields

$$\begin{bmatrix} k_e - m_e\omega^2 & -k_e & 0 \\ -k_e & k_s + k_e - m_s\omega^2 & -k_s \\ 0 & -k_s & k_s + k_b - m_b\omega^2 \end{bmatrix} \begin{Bmatrix} X_e \\ X_s \\ X_b \end{Bmatrix} = \begin{Bmatrix} m_e \\ m_s \\ m_b \end{Bmatrix} X_g\omega^2 \tag{4.17}$$

from which the amplitudes (X_e, X_s, and X_b) for the system can be solved as follows:

$$X_e = \frac{X_s + X_g f_e^2}{1 - f_e^2} \tag{4.18a}$$

$$X_b = \frac{X_s + \varepsilon_b f_s^2 X_g}{1 + k_b/k_s - \varepsilon_b f_s^2} \tag{4.18b}$$

$$X_s = \frac{\left[m_s + \dfrac{m_e}{1 - f_e^2} + \dfrac{m_b}{1 + k_b/k_s - \varepsilon_b f_s^2}\right] X_g \omega^2}{\dfrac{k_b - m_b\omega^2}{1 + k_b/k_s - \varepsilon_b f_s^2} - \left(m_s + \dfrac{m_e}{1 - f_e^2}\right)\omega^2} \tag{4.18c}$$

where the amplitudes of the equipment and base, that is, X_e and X_b, have been expressed in terms of the amplitude of the base, X_s. The parameters in Equation 4.18 are defined as

$$f_e = \omega/\omega_e \tag{4.19a}$$

$$f_s = \omega/\omega_s \tag{4.19b}$$

$$\varepsilon_b = m_b/m_s \tag{4.19c}$$

$$\omega_e = \sqrt{k_e/m_e} \tag{4.19d}$$

$$\omega_s = \sqrt{k_s/m_s} \tag{4.19e}$$

Finally, the state-steady absolute acceleration responses of the structure, equipment, and base can be expressed in terms of the ground acceleration $\ddot{x}_g$ as

$$a_s = \ddot{x}_s + \ddot{x}_g = -(X_s + X_g)\omega^2\, e^{i\omega t} = \frac{k_b \ddot{x}_g}{D(\omega)} \tag{4.20a}$$

$$a_e = \ddot{x}_e + \ddot{x}_g = -\frac{(X_s + X_g)\omega^2\, e^{i\omega t}}{1 - f_e^2} = \frac{k_b \ddot{x}_g}{(1 - f_e^2)D(\omega)} \tag{4.20b}$$

$$a_b = \ddot{x}_b + \ddot{x}_g = \frac{-[X_s + X_g + (k_b/k_s)X_g]\omega^2\, e^{i\omega t}}{1 + k_b/k_s - \varepsilon_b f_s^2} = \frac{(D^{-1}(\omega) + k_s^{-1})k_b \ddot{x}_g}{1 + k_b/k_s - \varepsilon_b f_s^2} \tag{4.20c}$$

$$D(\omega) = (k_b - m_b\omega^2) - (1 + k_b/k_s - \varepsilon_b f_s^2)\left(m_s + \frac{m_e}{1 - f_e^2}\right)\omega^2 \tag{4.20d}$$

As can be seen, the acceleration response of each component in the structure–equipment system depends mainly on the stiffness of the elastic bearing, k_b. In particular, the use of a smaller bearing stiffness, k_b, can result in significant reduction of the shear forces transmitted to the superstructure, as indicated by Equation 4.20a. This explains why an elastic bearing can be effectively used as an isolator for reducing the base shear of the structure–equipment system. In contrast, if the bearing stiffness, k_b, is made to be infinitely large, that is, by letting $k_b \to \infty$, the acceleration responses in Equation 4.20, reduce to

$$a_s = \frac{(1 - f_e^2)\ddot{x}_g}{(1 - f_e^2)(1 - f_s^2) - \varepsilon_e f_s^2} \tag{4.21a}$$

$$a_e = \frac{\ddot{x}_g}{(1 - f_e^2)(1 - f_s^2) - \varepsilon_e f_s^2} \tag{4.21b}$$

$$a_b = \ddot{x}_g \tag{4.21c}$$

with the use of L'Hospital's Rule, where $\varepsilon_e = m_e/m_s$. As can be seen from Equation 4.21c, the acceleration of the structural base is equal to the ground acceleration. Clearly, the present problem has been reduced to a two-DoF system with a rigid base, for which the solutions have been given in Equation 4.21a and Equation 4.21b.

Some important high-tech facilities, such as semiconductor factories and medical devices, are very sensitive to vibrations, especially to those caused by resonance. To consider the effect of resonance, we shall let the ground excitation frequency, ω, coincide with the equipment frequency, ω_e, that is, by letting $f_e = 1$ or $\omega_e = \omega$. For this case, the acceleration responses of the system in Equation 4.20 reduce to

$$a_s = 0 \tag{4.22a}$$

$$a_e = \frac{-k_b \ddot{x}_g}{k_e[1 + (k_b - m_b\omega_e^2)/k_s]} \tag{4.22b}$$

$$a_b = \frac{k_b \ddot{x}_g}{k_s + k_b - m_b\omega_e^2} \tag{4.22c}$$

Because of the coincidence of the ground excitation frequency with the equipment frequency, the equipment behaves like a vibration absorber of the structure. For this reason, the response of the equipment is greatly amplified, as implied by Equation 4.22b, while the response of structure is completely suppressed, as indicated by Equation 4.22a. Moreover, if the frequency of the equipment is equal to the fundamental frequency of the structure–equipment isolation system, that is, $\omega_e(\approx \omega_1) = \sqrt{k_b/(m_e + m_s + m_b)}$, then the responses of the system in Equation 4.22 can further be expressed as follows:

$$a_s = 0 \tag{4.23a}$$

$$a_e = \frac{-k_b \ddot{x}_g}{k_e[1 + (m_s + m_e)\omega_e^2/k_s]} \tag{4.23b}$$

$$a_b = \frac{k_b \ddot{x}_g}{k_s + (m_s + m_e)\omega_e^2} \tag{4.23c}$$

Since the equipment mass is generally much smaller than the structural mass, the preceding equation can be further reduced to

$$a_s = 0 \tag{4.24a}$$

$$a_e = \frac{-k_b \ddot{x}_g}{k_e(1 + \omega_e^2/\omega_s^2)} \tag{4.24b}$$

$$a_b = \frac{k_b \ddot{x}_g}{k_s(1 + \omega_e^2/\omega_s^2)} \tag{4.24c}$$

As indicated by Equation 4.24b, the acceleration response of the equipment depends on the stiffness ratio, k_b/k_e, of the base to the equipment.

For the resonance condition of $\omega_e = \omega$, mentioned previously, let us consider the case when the structural frequency is equal to the equipment frequency, that is, $\omega_e = \omega_s$. For this case, the responses of the system in Equation 4.22 reduce to

$$a_s \approx 0 \tag{4.25a}$$

$$a_e \approx \frac{-k_b \ddot{x}_g}{\varepsilon_e[k_s(1 - \varepsilon_b) + k_b]} \tag{4.25b}$$

$$a_b = \frac{k_b \ddot{x}_g}{k_s(1 - \varepsilon_b) + k_b} \tag{4.25c}$$

which indicates that the acceleration response of the equipment may be greatly amplified, as implied by the relatively small mass ratio ε_e $(= m_e/m_s)$ and large stiffness ratio, k_b/k_s, in Equation 4.25b. Such a phenomenon has been referred to as the *tuning of equipment*.

On the other hand, when the excitation frequency, ω, coincides with the fundamental frequency, ω_1, of the isolated system, that is, $\omega(\approx \omega_1) = \sqrt{k_b/(m_e + m_s + m_b)}$, resonant response may be induced on the structure–equipment isolation system. Considering that the first priority in design of high-tech equipment is to reduce the vibrations of the equipment, rather than the structure, by comparing the denominators in Equation 4.20a and Equation 4.20b, one may assume that the condition $|1 - f_e^2| \geq 1$ or $f_e = \omega/\omega_e \geq \sqrt{2}$ remains satisfied for a good design, which is equivalent to

$$\frac{\omega_e}{\omega_s} \leq \sqrt{\frac{k_b/k_s}{2[1 + (m_b + m_e)/m_s]}} \tag{4.26}$$

Since the fundamental frequency, ω_s, of a base-isolated structure is generally low in practice, the horizontal stiffness of the equipment attached to the structure must be designed to be soft enough such that Equation 4.26 can be satisfied. Certainly, this is one of the guidelines to be obeyed in the design of equipment for the sake of vibration reduction.

4.3.4 Illustrative Example

The forgoing formulations have been made by neglecting the damping of the structural system and by assuming the ground motion to be of the harmonic type. In practice, there is always some damping with the structural system, while the ground motion may be random in nature. To deal with such problems, the only recourse is to use numerical methods that are readily available. In this section, the Newmark β method, proposed by Newmark (1959), with $\gamma = 1/2$ and $\beta = 1/4$, will be adopted for solving the second-order differential equation presented in Equation 4.11, which has the advantage of being numerically stable.

The example considered is the structure–equipment system isolated by elastomeric bearings, shown in Figure 4.9, with the data given in Table 4.2. As can be seen, the equipment has a frequency equal to five times the structural frequency, that is, $\omega_e = 5\omega_s$ (= 8.34 Hz). The 1940 El Centro earthquake (NS component) with a peak ground acceleration (PGA) of 341.55 gal is adopted as the ground excitation, as given in Figure 4.10. By an eigenvalue analysis, the natural frequencies solved for the base-isolated system are 2.46, 21.41, and 52.74 rad/sec. Because of the installation of elastic bearings on the structure–equipment system, the fundamental frequency of the system decreases significantly and is approximately equal to $\omega_1 \approx \sqrt{k_b(m_e + m_s + m_b)} = 2.51$ rad/sec, according to Section 4.3.2. From this example, one observes that the use of a single-DoF system to model a base-isolated system can give a generally good result for the first frequency of vibration.

As can be seen from Figure 4.11, for the structural acceleration of the system, the main-shock response of the fixed-base structure has been effectively eliminated due to installation of the elastic bearings. However, as indicated by Figure 4.12, because of the installation of soft bearings, the base displacement response of the isolated system is much larger and decays much slowly, even after the main shocks.

In Figure 4.13 the acceleration response of the equipment for the isolated and fixed-base cases are compared. As can be seen, the main-shock response of the fixed-base structure has been effectively suppressed through the installation of the elastic bearings. Furthermore, the equipment response appears to be almost identical to the structure response shown in Figure 4.11, due to the fact that the equipment

TABLE 4.2 System Parameters Used in Simulation (Section 4.3.4)

Equipment		Superstructure		Isolation System	
Parameter	Value	Parameter	Value	Parameter	Value
Mass m_e	3 t (= $m_s/100$)	Mass m_s	300 t	Mass m_b	100 t (= $m_s/3$)
Horizontal stiffness k_e	8258 kN/m	Horizontal stiffness k_s	33,030 kN/m	Horizontal stiffness k_b	2546 kN/m
Damping	15.74 kN m/s	Damping	314.79 kN m/s	Damping	50.46 kN m/s
Frequency $\omega_e = \sqrt{k_e/m_e}$	52.47 rad/sec	Frequency $\omega_s = \sqrt{k_s/m_s}$	10.46 rad/sec	Frequency $\omega_b = \sqrt{k_b/m_b}$	5.05 rad/sec

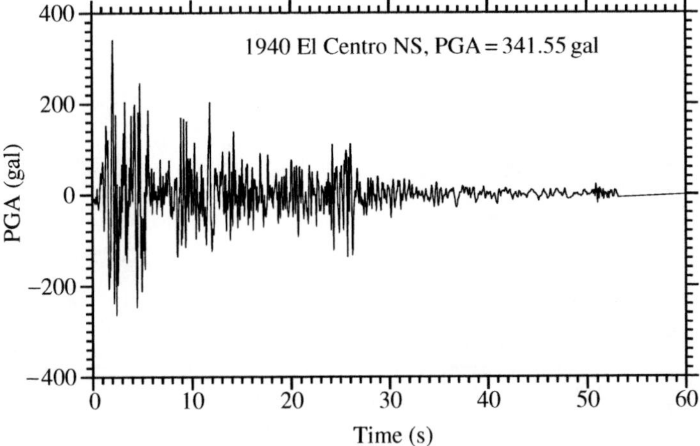

FIGURE 4.10 Waveform of 1940 El Centro earthquake (NS component).

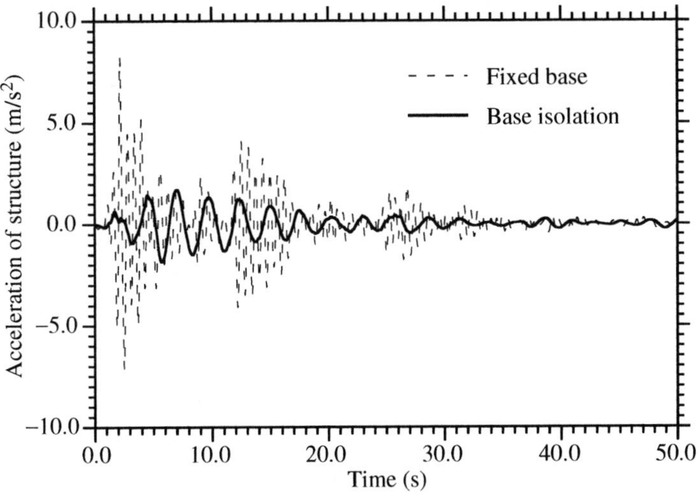

FIGURE 4.11 Comparison of structural accelerations.

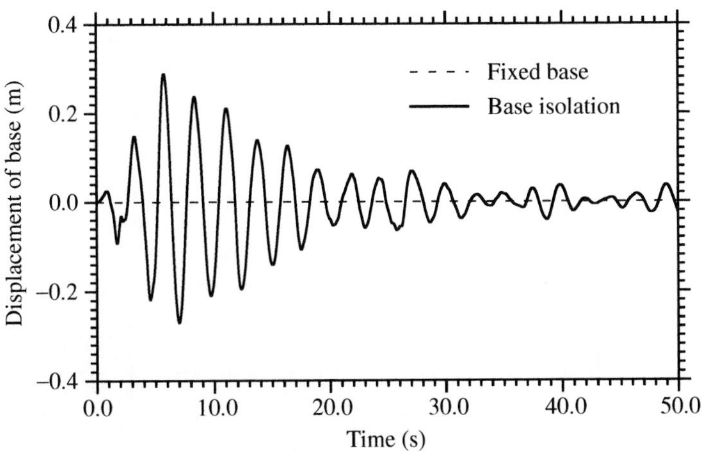

FIGURE 4.12 Comparison of base displacements.

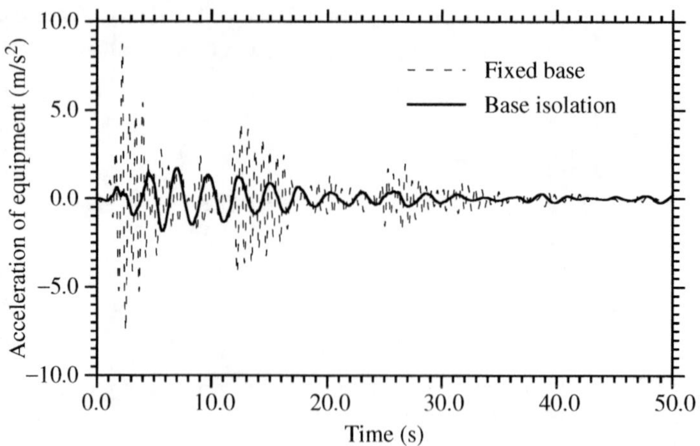

FIGURE 4.13 Comparison of equipment accelerations.

is rigidly attached to the structure, as implied by the relatively higher frequency of the equipment. As for the present example, the effectiveness of the elastic bearings in reducing the equipment response is ascertained.

For high-tech equipment, engineers may be concerned about the effect on equipment tuning induced by external excitations, such as earthquakes and traffic-induced vibrations. To investigate this effect on the structure–equipment isolation system considered, the maximum equipment acceleration was plotted as a function of the frequency ratio, ω_e/ω_s, in Figure 4.14. As can be seen, the response of the equipment is greatly amplified when it has a frequency close to the fundamental frequency of the structure–equipment isolation system, that is, when $\omega_e = 2.51$ rad/sec or $\omega_e/\omega_s = 0.24$. To avoid such a situation, it is suggested that isolators be mounted at both the structure base and equipment base. From Figure 4.14, one observes that the use of a small horizontal stiffness for the equipment will generally lead to greater equipment response due to tuning effect. However, as the stiffness of the equipment is further reduced, the equipment will reach another isolation state, in which the equipment response will be substantially suppressed, as indicated by the region with relatively small values of ω_e/ω_s. The margin for such a frequency ratio of ω_e/ω_s can be obtained by substituting the parameters in Table 4.2 into

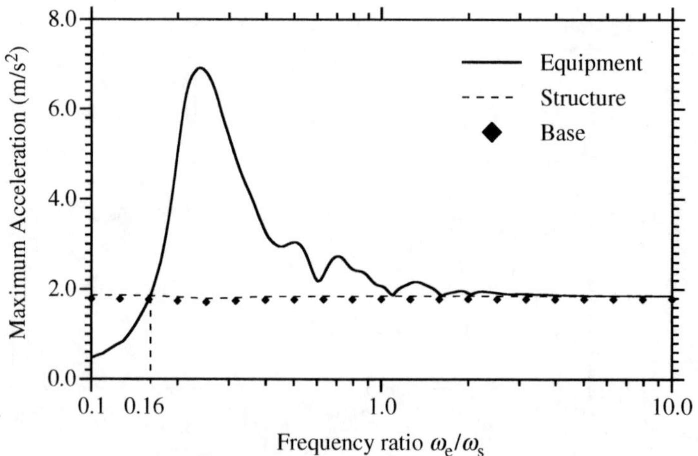

FIGURE 4.14 Maximum accelerations of structure–equipment isolation systems.

Equation 4.26, which yields a critical ratio of $\omega_e/\omega_s = 0.168$, very close to the value of 0.16 marked in Figure 4.14.

4.3.5 Concluding Remarks

This section investigates the dynamic response of a mathematical model of a structure–equipment system isolated by elastomeric bearings and subjected to ground excitations. Based on the closed-form solution of a structure–equipment isolation system subjected to harmonic support motions, one observes that the coincidence of the ground excitation frequency with the equipment frequency will make the equipment behave like a vibration absorber of the structure, of which the acceleration response will be greatly amplified. For the case that the first priority in design is to reduce the vibration of the equipment rather than that of the structure, Equation 4.26 provides a guideline for the design of equipment, which has been verified in the numerical example.

4.4 Sliding Isolation Systems

Sliding isolation can be an effective means for the seismic protection of structural systems. By implementing sliding isolators under the base mat of a structure, the transmission of ground excitation to the structure can be greatly reduced. Currently, applications of sliding isolation systems can be found elsewhere (Naeim and Kelly, 1999). A *sliding isolator* usually consists of a slider with frictional surfaces. For this reason, it is also referred to as a *friction isolator.* When subjected to an earthquake, the slider will slide along the frictional contact surfaces whenever the horizontal seismic force exceeds the maximum frictional force of the support, which, by Coulomb's theory, is equal to the normal contact force multiplied by the static (or dynamic) coefficient of friction of the sliding surfaces. Because of this, the seismic force transmitted to the superstructure is generally less than the maximum frictional force of the isolator. Obviously, the maximum frictional force is an important parameter for the design of a sliding isolation system, because it decides when the system starts to slide and how large the shear force is to be transmitted to the superstructure.

The motion of a sliding structure consists of two different states, namely, the sliding state and the stick (or nonsliding) state. At any instant of motion, the structure can only belong to one of the two states. Although in each state the sliding structure can be modeled as a linear system, the governing equations of motion for the two states are different. As a result, the overall behavior of the sliding structure is nonlinear. Such nonlinearity has resulted in the occurrence of subharmonic resonance in the frequency response of a sliding structure (Mostaghel et al., 1983; Westermo and Udwadia, 1983), making the dynamic response much more complicated. In some applications, a sliding isolation system has been designed with an automatic recentering mechanism (or resilient mechanism), so that the structure can slide back to its original position after the earthquake (Mokha et al., 1991). This type of sliding systems has been called the *resilient sliding isolation system*, which will be investigated in Section 4.5. The implementation of a recentering mechanism offers some advantages, but will inevitably introduce some disadvantages, as will be discussed in Section 4.5.

The purpose of this section is to investigate the seismic behavior of a sliding isolated structure and also the behavior of an equipment item mounted on the structure. No consideration will be made for the recentering mechanism. The nonlinear dynamic equation for a structure with an underneath friction element is first formulated. Next, two numerical approaches will be presented for solving the nonlinear equation, the *shear balance method* and *fictitious spring method.* Finally, using some assumed data, the harmonic response and seismic behavior of a sliding structure, together with the equipment mounted on it, will be presented. In this section, the frictional coefficient of the sliding system is assumed to be of the Coulomb type, that is, a time-invariant constant. For simplification, no distinction will be made between the static and dynamic frictional coefficients, or between the dynamic and maximum static frictional force.

4.4.1 Mathematical Modeling and Formulation

4.4.1.1 Equation of Motion

A sliding isolated structure with an attached equipment item, as schematically shown in Figure 4.15, can be represented as a mass–spring–dashpot system of three DoF, as shown in Figure 4.16, for which the notations employed have been defined in Table 4.1. When the structural system is excited by an earthquake, the equation of motion can be written as

$$\mathbf{M}\ddot{\mathbf{x}}(t) + \mathbf{C}\dot{\mathbf{x}}(t) + \mathbf{K}\mathbf{x}(t) = -\mathbf{M}\mathbf{L}_1\ddot{x}_g(t) + \mathbf{L}_2 f(t) \tag{4.27}$$

where the vector $\mathbf{x}$ denotes the dynamic responses of the whole structural system

$$\mathbf{x}(t) = \begin{Bmatrix} x_e(t) \\ x_s(t) \\ x_b(t) \end{Bmatrix} \tag{4.28}$$

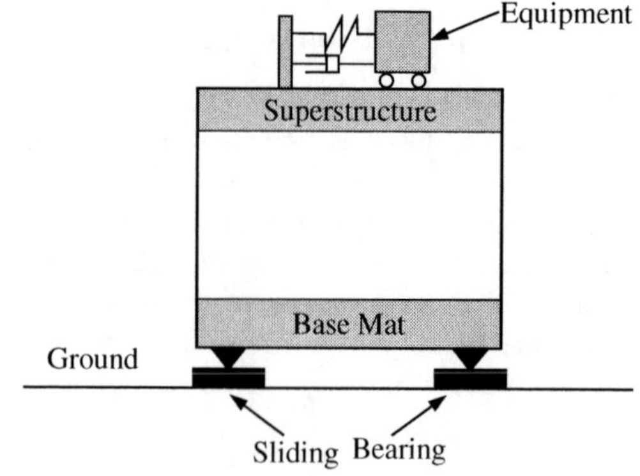

FIGURE 4.15 Schematic for an isolated structure–equipment system with sliding bearing.

The mass, damping, and stiffness matrices in Equation 4.27 are defined as

$$\mathbf{M} = \begin{bmatrix} m_e & 0 & 0 \\ 0 & m_s & 0 \\ 0 & 0 & m_b \end{bmatrix},$$

$$\mathbf{C} = \begin{bmatrix} c_e & -c_e & 0 \\ -c_e & c_e + c_s & -c_s \\ 0 & -c_s & c_s \end{bmatrix}, \tag{4.29}$$

$$\mathbf{K} = \begin{bmatrix} k_e & -k_e & 0 \\ -k_e & k_e + k_s & -k_s \\ 0 & -k_s & k_s \end{bmatrix}$$

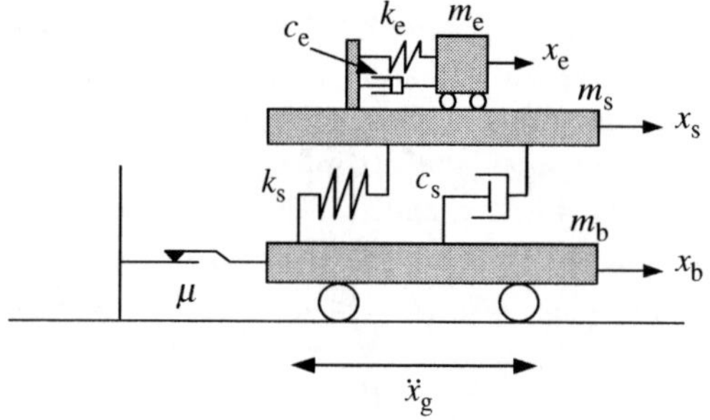

FIGURE 4.16 Model for an isolated structure–equipment system with sliding support.

and the force distribution vectors as

$$\mathbf{L}_1 = \begin{Bmatrix} 1 \\ 1 \\ 1 \end{Bmatrix}, \quad \mathbf{L}_2 = \begin{Bmatrix} 0 \\ 0 \\ 1 \end{Bmatrix} \tag{4.30}$$

Note that in Equation 4.27, the isolator frictional force, $f(t)$, which is not a constant, is moved to the right-hand side of the equation. This nonlinear force requires a special treatment in an analysis procedure, as will be explained later on.

For a systematic treatment, the above equation of motion can be further written in a state space form as shown below (Meirovitch, 1990):

$$\dot{\mathbf{z}}(t) = \mathbf{A}\mathbf{z}(t) + \mathbf{E}\ddot{x}_g(t) + \mathbf{B}f(t) \tag{4.31}$$

where the state vector $\mathbf{z}(t)$ and the system matrix $\mathbf{A}$ are defined as

$$\mathbf{z}(t) = \begin{bmatrix} \dot{\mathbf{x}}(t) \\ \mathbf{x}(t) \end{bmatrix}, \quad \mathbf{A} = \begin{bmatrix} -\mathbf{M}^{-1}\mathbf{C} & -\mathbf{M}^{-1}\mathbf{K} \\ \mathbf{I} & \mathbf{0} \end{bmatrix} \tag{4.32}$$

and the excitation and friction distribution vectors as

$$\mathbf{E} = \begin{bmatrix} -\mathbf{L}_1 \\ \mathbf{0} \end{bmatrix}, \quad \mathbf{B} = \begin{bmatrix} \mathbf{M}^{-1}\mathbf{L}_2 \\ \mathbf{0} \end{bmatrix} \tag{4.33}$$

4.4.1.2 Conditions for Stick and Sliding States

As mentioned above, the motion of a sliding structure at any instant has two possible states, namely, the stick (or nonsliding) and sliding states. The following are the conditions that must be satisfied by the sliding structure:

(1) In stick state

$$|f(t)| < f_{\max} = \mu W \tag{4.34a}$$

$$\dot{x}_b(t) = 0 \tag{4.34b}$$

where μ is the coefficient of friction and W is the total weight of the structure. According to the preceding equations, the frictional force in the stick state is an unknown with a magnitude less than the maximum frictional force, $f_{\max}$, which equals the product of μ and W, while the sliding velocity of the structure is simply zero. Whenever the frictional force satisfies Equation 4.34a, the sliding system remains in the stick state, otherwise it changes into the sliding state.

(2) In sliding state

$$f(t) = -\text{sgn}(\dot{x}_b(t))f_{\max} = -\text{sgn}(\dot{x}_b(t))\mu W \tag{4.35a}$$

$$\dot{x}_b(t) \neq 0 \tag{4.35b}$$

where the function $\text{sgn}(x)$ denotes the sign of the variable x. According to Equation 4.35a and Equation 4.35b, the frictional force in the sliding state has a magnitude equal to the maximum frictional force, but directed in a sense opposite to that of the sliding velocity. On the other hand, the sliding velocity of the isolator remains as an unknown.

4.4.2 Methods for Numerical Analysis

Two numerical methods commonly used for the analysis of sliding isolated structural systems, the shear balance method and fictitious spring method, will be introduced in this section. By employing the discrete-time state-space formula, both methods can be cast in an incremental form that is suitable for the analysis of sliding systems with multiple DoF.

4.4.2.1 Shear Balance Method

Consider the state-space equation, Equation 4.31, and assume that both the ground acceleration and frictional force vary linearly within each time interval, as shown in Figure 4.17. Equation 4.31 may be written in the following incremental form (Meirovitch, 1990)

$$\mathbf{z}[k+1] = \mathbf{A}_d\mathbf{z}[k] + \mathbf{E}_0\ddot{x}_g[k] + \mathbf{E}_1\ddot{x}_g[k+1] + \mathbf{B}_0 f[k] + \mathbf{B}_1 f[k+1] \tag{4.36}$$

where the symbol $x[k]$ denotes that the variable x is evaluated at the kth time step. The other coefficient matrices in equation 4.36 are defined as

$$\mathbf{A}_\mathrm{d} = e^{\mathbf{A}\Delta t} = \sum_{i=0}^{\infty} \frac{\Delta t^i}{i!}\mathbf{A}^i \tag{4.37}$$

$$\mathbf{B}_0 = \left[(\mathbf{A})^{-1}\mathbf{A}_\mathrm{d} + \frac{1}{\Delta t}(\mathbf{A})^{-2}(\mathbf{I} - \mathbf{A}_\mathrm{d})\right]\mathbf{B} \tag{4.38a}$$

$$\mathbf{B}_1 = \left[-(\mathbf{A})^{-1} + \frac{1}{\Delta t}(\mathbf{A})^{-2}(\mathbf{A}_\mathrm{d} - \mathbf{I})\right]\mathbf{B} \tag{4.38b}$$

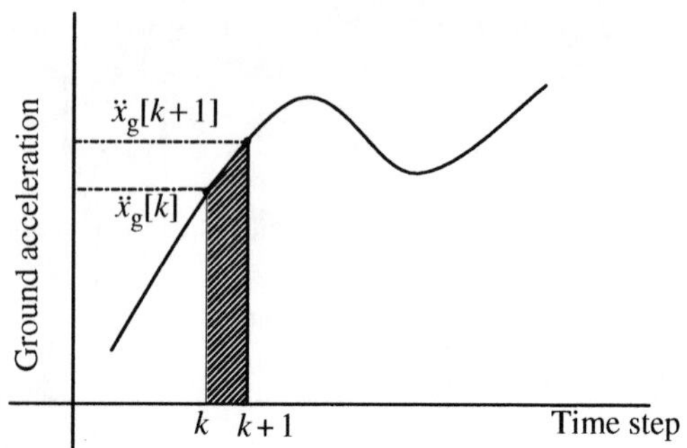

FIGURE 4.17 Force integration scheme with linear interpolation.

where Δt denotes the size of the time step considered for analysis. The matrices $\mathbf{E}_0$ and $\mathbf{E}_1$ can be computed in a way similar to $\mathbf{B}_0$ and $\mathbf{B}_1$, except that the matrix $\mathbf{B}$ in Equation 4.38a and Equation 4.38b should be replaced by the matrix $\mathbf{E}$. In some applications, the system matrix $\mathbf{A}$ may not be invertible, that is, $\mathbf{A}^{-1}$ does not always exist. If this is the case, one may compute $\mathbf{B}_0$ and $\mathbf{B}_1$ (and similarly $\mathbf{E}_0$ and $\mathbf{E}_1$) by the following formulas:

$$\mathbf{B}_0 = \mathbf{A}_\mathrm{d}^*\mathbf{B}, \quad \mathbf{B}_1 = (\hat{\mathbf{A}}_\mathrm{d} - \mathbf{A}_\mathrm{d}^*)\mathbf{B} \tag{4.39}$$

where

$$\mathbf{A}_\mathrm{d}^* = \sum_{i=0}^{\infty} \frac{(\Delta t)^{i+1}}{i!(i+2)}\mathbf{A}^i, \quad \hat{\mathbf{A}}_\mathrm{d} = \sum_{i=0}^{\infty} \frac{(\Delta t)^{i+1}}{(i+1)!}\mathbf{A}^i \tag{4.40}$$

Note that, on the right-hand side of Equation 4.36, the only unknown at the kth time step is the frictional force $f[k+1]$; therefore, $f[k+1]$ must be determined before the next time step response $\mathbf{z}[k+1]$ is computed. Wang et al. (1998) proposed the shear balance method for computing the frictional force $f[k+1]$. By this method, the sliding structure is first assumed to be in the stick state at the $(k+1)$th step, for which the condition given in Equation 4.34b must be satisfied

$$\dot{x}_\mathrm{b}[k+1] = \mathbf{D}\mathbf{z}[k+1] = 0 \tag{4.41}$$

where $\mathbf{D}$ is a relation matrix, equal to $\mathbf{D} = [0\ 0\ 1\ 0\ 0\ 0]$ for the model shown in Figure 4.16. Substituting $\mathbf{z}[k+1]$ in Equation 4.36 into Equation 4.41, one may solve for the *estimated frictional force* at the $(k+1)$th time step as

$$\bar{f}[k+1] = -(\mathbf{D}\mathbf{B}_1)^{-1}\mathbf{D}(\mathbf{A}_\mathrm{d}\mathbf{z}[k] + \mathbf{B}_0 f[k] + \mathbf{E}_0\ddot{x}_\mathrm{g}[k] + \mathbf{E}_1\ddot{x}_\mathrm{g}[k+1]) \tag{4.42}$$

where $\bar{f}[k+1]$ with an overbar signifies that the frictional force is an estimate obtained by assuming the sliding structure to be at the stick state. Such a value may not be the actual one if the system is not in the stick state. The physical meaning for $\bar{f}[k+1]$ is that it represents the *balanced shear force* required at the $(k+1)$th time step for the structure to remain in the stick state. Therefore, the sign of $\bar{f}[k+1]$ indicates the direction of the resistant force provided by the isolation system. In spite of the fact that $\bar{f}[k+1]$ may not be the actual frictional force, it plays an important role for determining the actual state (stick or sliding) and the actual frictional force of the sliding isolated structure, as will be described below based on Equation 4.34a and Equation 4.35a.

(1) The system is in the "stick state" if $|\bar{f}[k+1]| < f_{\max}$ and the frictional force is

$$f[k+1] = \bar{f}[k+1] \tag{4.43}$$

(2) The system is in the "sliding state" if $|\bar{f}[k+1]| \geq f_{\max}$ and the frictional force is

$$f[k+1] = \mathrm{sgn}(\bar{f}[k+1])f_{\max} \tag{4.44}$$

As can be seen, the term $-\mathrm{sgn}(\dot{x}_b[k+1])$ in Equation 4.35a is replaced by $\mathrm{sgn}(\bar{f}[k+1])$ in Equation 4.44. Such a replacement is justified since the sign of $\bar{f}[k+1]$ indicates the direction of the resistant force at the $(k+1)$th time step. Once the correct frictional force, $f[k+1]$, is determined by using either Equation 4.43 or Equation 4.44, it can be substituted into Equation 4.36 to obtain the response $\mathbf{z}[k+1]$ for the next time step. The computational flow-chart for the shear balance method has been given in Figure 4.18.

4.4.2.2 Fictitious Spring Method

The fictitious spring method was first proposed by Yang et al. (1990) for the analysis of a sliding structure. Later, Lu and Yang (1997) reformulated the method into a state-space form for the analysis of equipment

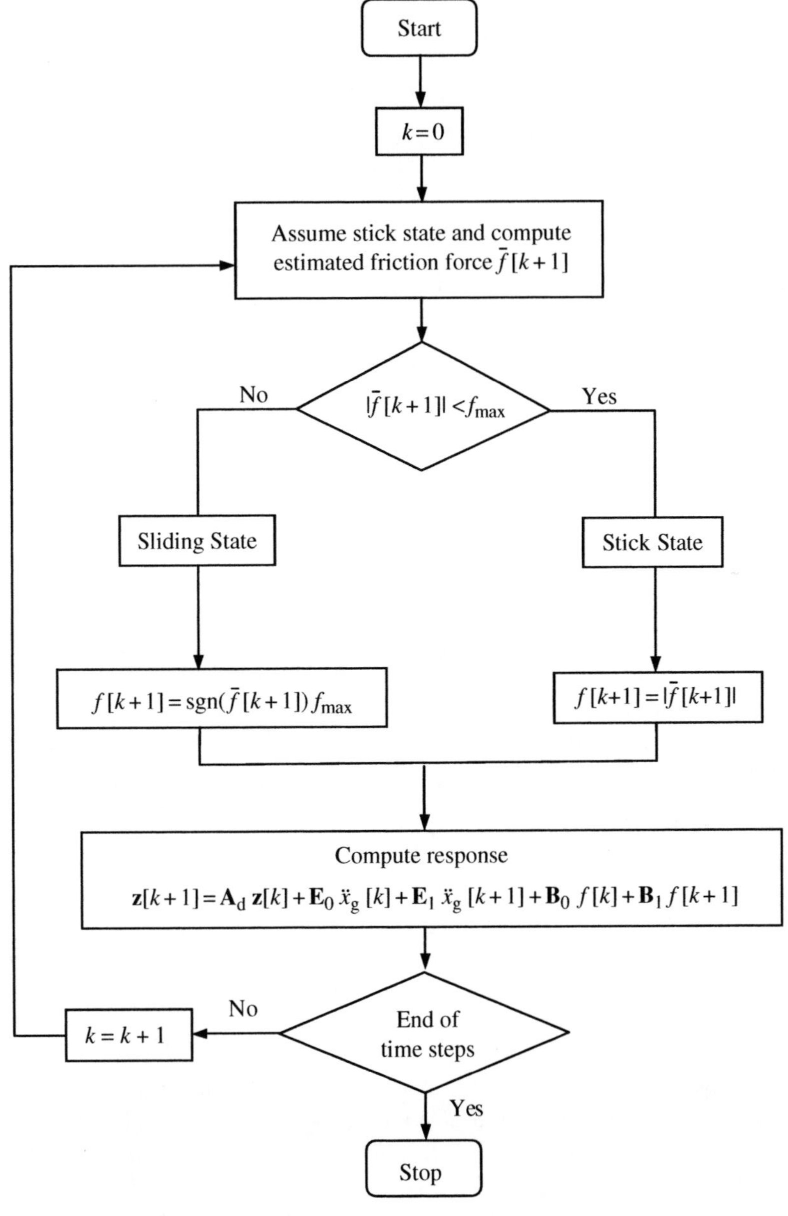

FIGURE 4.18 Computational flow-chart for shear balance method.

mounted on a sliding structure. By this method, a fictitious spring, k_f, is introduced between the base mat and the ground, as in Figure 4.19, to represent the mechanism of sliding or friction. The stiffness, k_f, of the fictitious spring is taken as zero for the sliding state and as a very large value for the stick state. With the introduction of the fictitious spring, the stiffness matrix, **K**, in Equation 4.29 should be modified as follows:

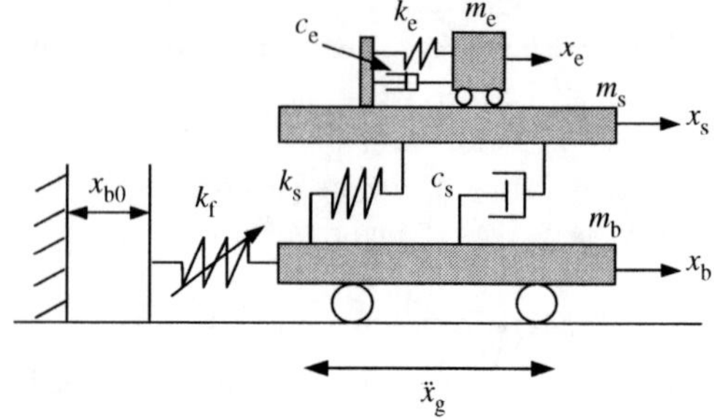

FIGURE 4.19 Model for isolated structures with fictitious spring.

$$\mathbf{K} = \mathbf{K}(k_f) = \begin{bmatrix} k_e & -k_e & 0 \\ -k_e & k_e + k_s & -k_s \\ 0 & -k_s & k_s + k_f \end{bmatrix} \tag{4.45}$$

Accordingly, the state-space dynamic equation, Equation 4.31, should be modified as

$$\dot{\mathbf{z}}(t) = \mathbf{A}(k_f)\mathbf{z}(t) + \mathbf{E}\ddot{x}_g(t) + \mathbf{B}\tilde{f}(t) \tag{4.46}$$

where

$$\mathbf{A} = \mathbf{A}(k_f) = \begin{bmatrix} -\mathbf{M}^{-1}\mathbf{C} & -\mathbf{M}^{-1}\mathbf{K}(k_f) \\ \mathbf{I} & \mathbf{0} \end{bmatrix} \tag{4.47}$$

Depending on the current state of the sliding system, the fictitious stiffness, k_f, and the modified friction term, $\tilde{f}(t)$, in Equation 4.46 may take one of the following two sets of values:

(1) In the stick state

$$k_f = \alpha k_s, \quad \tilde{f}(t) = k_f x_{b0} \tag{4.48}$$

(2) In the sliding state

$$k_f = 0, \quad \tilde{f}(t) = -\mathrm{sgn}(\dot{x}_b(t))\mu W \tag{4.49}$$

In Equation 4.48, the symbol α represents a constant of very large value, and x_{b0} the initial elongation of the fictitious spring in the current stick state (computation of x_{b0} will be explained later). Note that the modified friction term, $\tilde{f}(t)$, may not be the actual frictional force. The actual frictional force can be determined as follows:

(1) In the stick state

$$f(t) = k_f(x_b(t) - x_{b0}) \tag{4.50}$$

(2) In the sliding state

$$f(t) = \tilde{f}(t) \tag{4.51}$$

According to Equation 4.50 and Equation 4.51, the actual frictional force of the isolation system in the stick state is equal to the internal force of the fictitious spring, while in the sliding state it is equal to the modified frictional force, $\tilde{f}(t)$. The frictional force computed from the preceding two equations should obey the conditions given in Equation 4.34a and Equation 4.35a as well.

With the conditions imposed for the stick and sliding states in Equation 4.48 and Equation 4.49, respectively, the equation of motion in Equation 4.46 actually represents two different sets of equations.

Specifically, Equation 4.46 and Equation 4.48 collectively describe the motion of the structure in the stick state, while Equation 4.46 and Equation 4.49 represent the motion of the structure in the sliding state. Owing to the fact that a sliding system may switch between the two states at certain instants, the behavior of the entire system should undoubtedly be regarded as a nonlinear one. Nevertheless, within each particular state, the behavior of the system as represented either by Equation 4.46 and Equation 4.48 or Equation 4.46 and Equation 4.49 is a linear one.

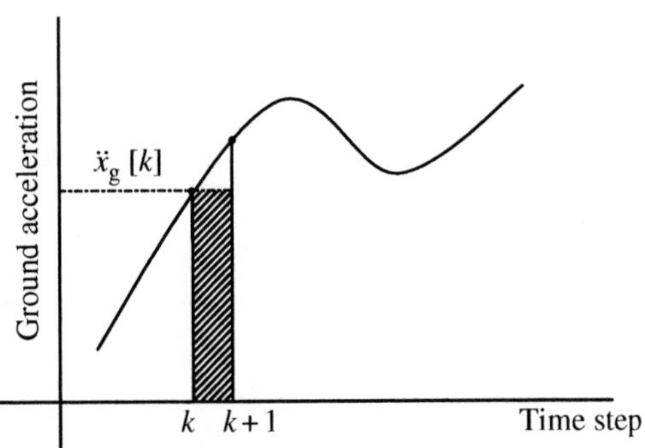

FIGURE 4.20 Constant force integration scheme.

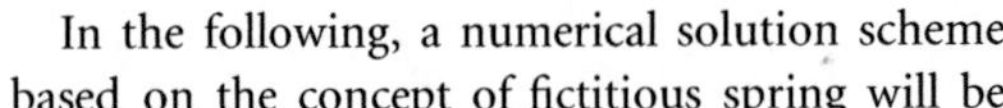

In the following, a numerical solution scheme based on the concept of fictitious spring will be introduced. Let Δt denote a time increment, which is usually taken as a very small value, and assume that the ground excitation and frictional force are constant within each time increment, Δt (see Figure 4.20). Accordingly, the discrete-time solution of Equation 4.46 can be rewritten in an incremental form (Meirovitch, 1990) as

$$\mathbf{z}[k+1] = \mathbf{A}_\mathrm{d}\mathbf{z}[k] + \mathbf{E}_\mathrm{d}\ddot{x}_\mathrm{g}[k] + \mathbf{B}_\mathrm{d}\tilde{f}[k] \tag{4.52}$$

where

$$\mathbf{A}_\mathrm{d} = \mathbf{A}_\mathrm{d}(k_\mathrm{f}) = \mathrm{e}^{\mathbf{A}(k_\mathrm{f})\Delta t} = \sum_{i=0}^{\infty}\frac{\Delta t^i}{i!}\mathbf{A}(k_\mathrm{f})^i \tag{4.53}$$

$$\mathbf{E}_\mathrm{d} = \mathbf{E}_\mathrm{d}(k_\mathrm{f}) = \mathbf{A}(k_\mathrm{f})^{-1}(\mathbf{A}_\mathrm{d}(k_\mathrm{f}) - \mathbf{I})\mathbf{E} \tag{4.54a}$$

$$\mathbf{B}_\mathrm{d} = \mathbf{B}_\mathrm{d}(k_\mathrm{f}) = \mathbf{A}(k_\mathrm{f})^{-1}(\mathbf{A}_\mathrm{d}(k_\mathrm{f}) - \mathbf{I})\mathbf{B} \tag{4.54b}$$

For the case where the system matrix $\mathbf{A}$ is invertible, $\mathbf{B}_\mathrm{d}$ and $\mathbf{E}_\mathrm{d}$ may be computed instead using the following formulas:

$$\mathbf{E}_\mathrm{d} = \mathbf{E}_\mathrm{d}(k_\mathrm{f}) = \left[\sum_{i=0}^{\infty}\frac{\Delta t^i}{i!}\mathbf{A}(k_\mathrm{f})^{i-1}\right]\mathbf{E} \tag{4.55}$$

$$\mathbf{B}_\mathrm{d} = \mathbf{B}_\mathrm{d}(k_\mathrm{f}) = \left[\sum_{i=0}^{\infty}\frac{\Delta t^i}{i!}\mathbf{A}(k_\mathrm{f})^{i-1}\right]\mathbf{B} \tag{4.56}$$

Equation 4.52 is the solution of the sliding system given in incremental form, because the response, $\mathbf{z}[k+1]$, can be computed from the solution of the previous step, $\mathbf{z}[k]$. Note that, in Equation 4.53 and Equation 4.54, the coefficient matrices $\mathbf{A}_\mathrm{d}$, $\mathbf{E}_\mathrm{d}$, and $\mathbf{B}_\mathrm{d}$ have two possible sets of values, as the fictitious spring constant, k_f, may take different values for the sliding and stick states. Nevertheless, once the time step size, Δt, is chosen, the coefficient matrices $\mathbf{A}_\mathrm{d}$, $\mathbf{E}_\mathrm{d}$, and $\mathbf{B}_\mathrm{d}$, remain constant for each state. As such, they need only be calculated once at the beginning of the incremental procedure. The computational flow-chart for the fictitious spring method described above has been given in Figure 4.21.

The dynamic equation and its discrete-time solution for the sliding structure in the two states have been presented above. In the following, we shall describe how to determine the *transition time* for the sliding structure to switch from one state to the other. Once the transition time is determined, the original step size should be scaled down accordingly to reflect the transition point (Yang et al., 1990).

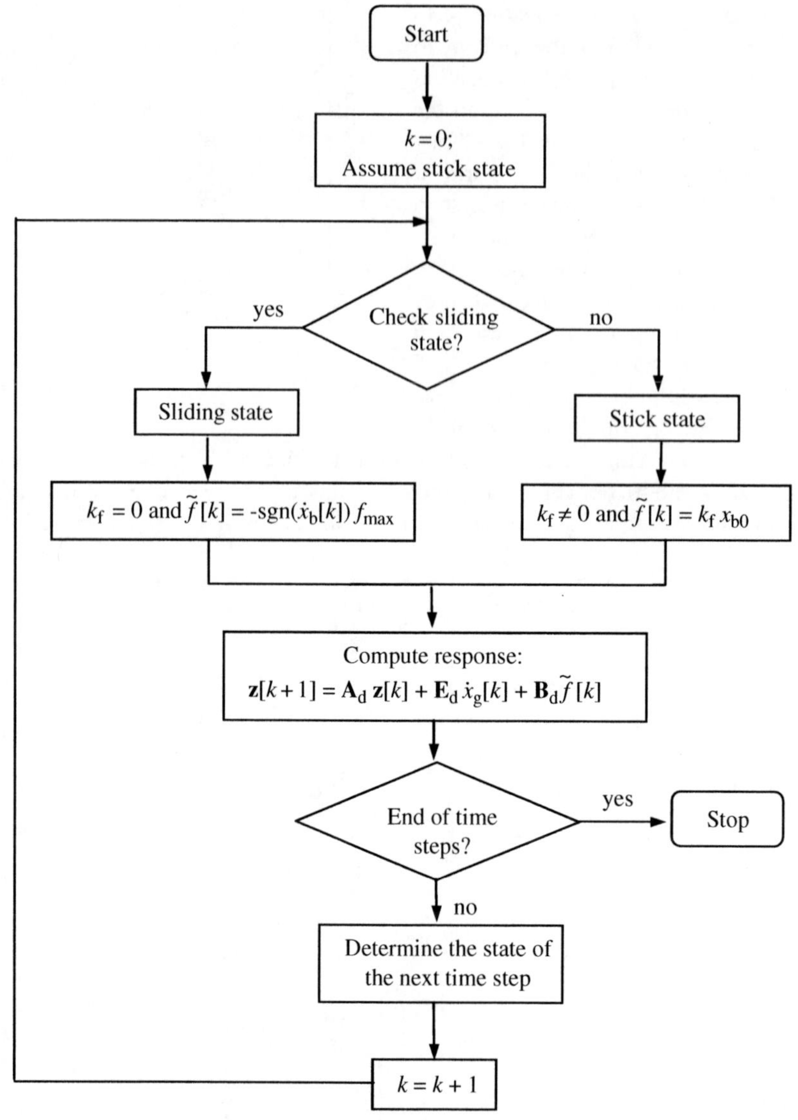

FIGURE 4.21 Computational flow-chart for fictitious spring method.

Transition from stick to sliding state. As stated in Equation 4.34a, the condition for a sliding isolated structure to remain in the stick state is that the static frictional force under the base mat be less than its maximum value, f_{max}. Once the frictional force exceeds this maximum value, the system starts to slide. With the fictitious spring method, the static frictional force is computed as the internal force of the fictitious spring, as in Equation 4.50. Based on the above considerations and Equation 4.50, the condition for the sliding system to transfer from the stick to the sliding state is

$$|f(t_0)| = |k_f(x_b(t_0) - x_{b0})| = f_{max} = \mu W \tag{4.57}$$

where t_0 denotes the transition time at which the structure starts to slide and k_f is the spring constant given in Equation 4.48 for the stick state. Because a very large value has been used for the fictitious spring constant, k_f, in the stick state, the deviation in base displacement due to spring elongation is very small in this state, which can just be neglected. In practice, the transition time, t_0 may not occur precisely at the discrete points of time considered. It is likely that the spring force is less than f_{max} at the current time step, say, the kth step, but exceeds f_{max} at the following time step. If this is the case, numerical methods such as the bisection method should be employed to locate the transition time, t_0 within the time interval $(k\Delta t, (k+1)\Delta t)$ considered based on Equation 4.57.

Transition from sliding state to stick state. The structure in the sliding state may return to the stick state whenever the following two conditions are satisfied. (1) The relative velocity of the base mat to the ground reaches zero, that is, $\dot{\xi}_b(t_0) = 0$ where t_0 is the transition time; (2) the estimated static frictional force, denoted by $\bar{f}(t_0)$, is less than the maximum static frictional force, that is, $\bar{f}(t_0) < f_{max}$. Here, the estimated static frictional force, $\bar{f}(t_0)$, is defined as the shear force required to balance the motion of the superstructure if the system is assumed to be in the stick state, similar to the one given in Equation 4.42 for $\bar{f}[k+1]$. By letting the relative velocity and relative acceleration between the base mat and the ground be equal to zero, that is, $\ddot{\xi}_b(t_0) = \dot{\xi}_b(t_0) = 0$, the estimated frictional force can be calculated from the free-body diagram of the base mat as

$$\bar{f}(t_0) = m_b\ddot{x}_g(t_0) - k_s(x_s(t_0) - x_b(t_0)) - c_s\dot{x}_s(t_0) \tag{4.58}$$

For the sliding structure to transfer from the sliding to the stick state, both the aforementioned conditions must be satisfied simultaneously. Once the structure enters the stick state, the term $f(t_0)$ should be set equal to $\bar{f}(t_0)$ and used as the initial frictional force. For the sake of equilibrium, the initial base mat displacement, x_{b0}, should be computed as

$$x_{b0} = x_b(t_0) - (\bar{f}(t_0)/k_f) \tag{4.59}$$

where the value of k_f is the one given for the stick state in Equation 4.48.

Concerning the two conditions mentioned above, it may happen that only the first condition, $\dot{\xi}_b = 0$ is satisfied, while the computed $\bar{f}(t_0)$ is still larger than f_{max}. If this is the case, the sliding system should not be regarded as a transition to the stick state. Rather, the situation should be regarded as an indication for reversing the direction of sliding in the next time step. Correspondingly, the frictional force, $f(t_0)$, should be set equal to the sliding frictional force, rather than the estimated one, $\bar{f}(t_0)$.

4.4.3 Simulation Results for Sliding Isolated Systems

4.4.3.1 Numerical Model and Ground Excitations

In this section, the dynamic behavior of the sliding isolated structure–equipment system shown in Figure 4.16 will be analyzed using the shear balance method. Although the sliding structure and equipment considered are both of single-DoF, there exists no difficulty for use of the method to solve problems with multi-DoF systems. In Table 4.3, the material properties adopted for the present model have been listed, which are intended to simulate a small, five-story, reinforced concrete frame. For the present purposes, two types of ground excitation are considered, namely, harmonic and earthquake excitations. The harmonic excitation is considered primarily for studying the frequency response of the sliding system, while the earthquake excitation is considered for the effect of earthquake intensity. For the

TABLE 4.3 System Parameters Used in Simulation (Section 4.4.3)

Equipment		Superstructure		Isolation System	
Parameter	Value	Parameter	Value	Parameter	Value
Mass m_e	3 t (= m_s/100)	Mass m_s	300 t	Mass m_b	100 t (= m_s/3)
Frequency ω_e	5ω_s or a variable	Frequency ω_s	1.67 Hz	Frictional coefficient μ	0.05, 0.1, 0.25
Damping ratio ζ_e	5%	Damping ratio ζ_s	5%	—	—

harmonic excitation, a sinusoidal ground acceleration of the following form is adopted:

$$\ddot{x}_g(t) = 0.5g \sin \omega_g t \tag{4.60}$$

where ω_g denotes the excitation frequency and g is the gravitational acceleration. For the earthquake excitation, the 1940 El Centro earthquake (NS component) is considered, for which the waveform has been given in Figure 4.10. The PGA level of the earthquake will be adjusted for reasons of research. Concerning the effectiveness of response reduction, three quantities are chosen as the indices, namely, the base displacement (base drift), structural acceleration, and equipment acceleration. For all the figures shown in Section 4.4.3 below, the symbol μ in the legend is used to denote the frictional coefficient of the sliding isolation system and the word "fixed" denotes the response of the corresponding fixed-base structure.

4.4.3.2 Harmonic Response of Structure

Time history. For the isolated system subjected to a harmonic excitation of $\omega_g = 1$ Hz, the base displacements computed for different coefficients of friction were plotted in Figure 4.22 and Figure 4.23. As can be seen, the base displacements quickly reach the steady-state response within the first few cycles. Meanwhile, the use of a smaller frictional coefficient results in a larger permanent displacement before the steady state is reached. From the structural accelerations plotted in Figure 4.24, one observes that, for a sliding structure with a smaller coefficient of friction, the steady-state response is achieved in a faster way, accompanied by a larger reduction on structural acceleration. Of interest in Figure 4.24 is that the response of the fixed-base case shows a clear period of 1 sec, while in the sliding case, the response is contaminated by high-frequency signals caused by the sliding-stick transitions.

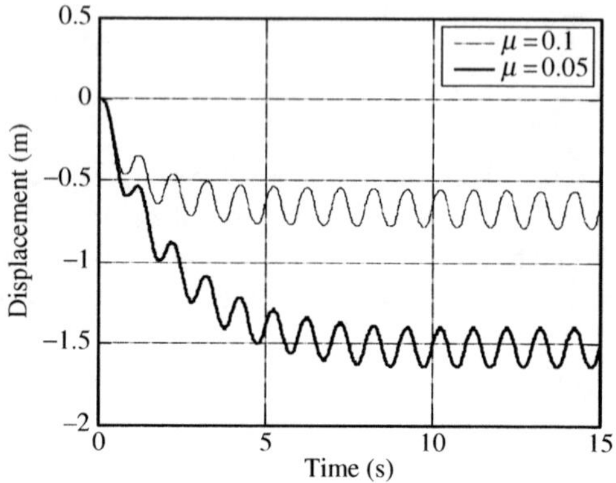

FIGURE 4.22 Comparison of base displacements ($\omega_g = 1$ Hz; $\omega_e = 5\omega_s$).

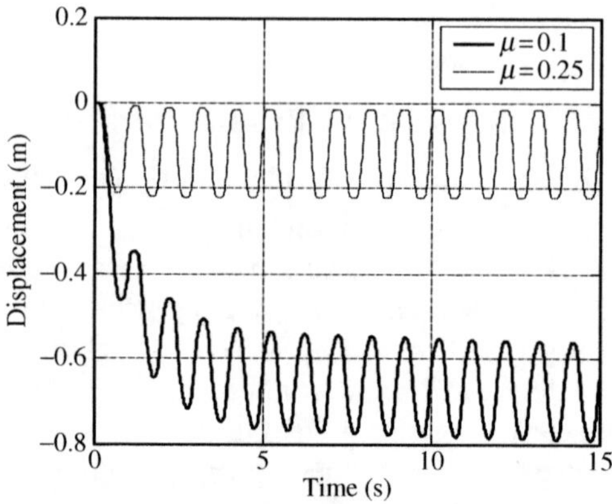

FIGURE 4.23 Comparison of base displacements ($\omega_g = 1$ Hz; $\omega_e = 5\omega_s$).

Hysteretic behavior. In order to understand the mechanical characteristics of a nonlinear device used for vibration control, it is common to present a diagram showing the force–deformation relation of the device, also referred to as the *hysteretic diagram* (Soong and Dargush, 1997). Figure 4.25 shows the hysteresis loops of the sliding isolation system (the sliding layer) for $\mu = 0.1$ and 0.25, when the system is subjected to a harmonic excitation of $\omega_g = 1$ Hz. In the figure, the horizontal and vertical axes, respectively, represent the base displacement and shear force, that is, the frictional force, under the mat. Just like many other frictional elements or devices, the shape of the hysteresis loop of a sliding bearing is rectangular. The height of the rectangle is equal to the maximum frictional force that depends on the coefficient of friction, while the width of the hysteresis loop is determined by the base-sliding displacement. As the coefficient of friction decreases, the height of the loop decreases, while the width increases. The total area of the hysteresis loop is equivalent to the portion of the energy dissipated by the sliding bearing.

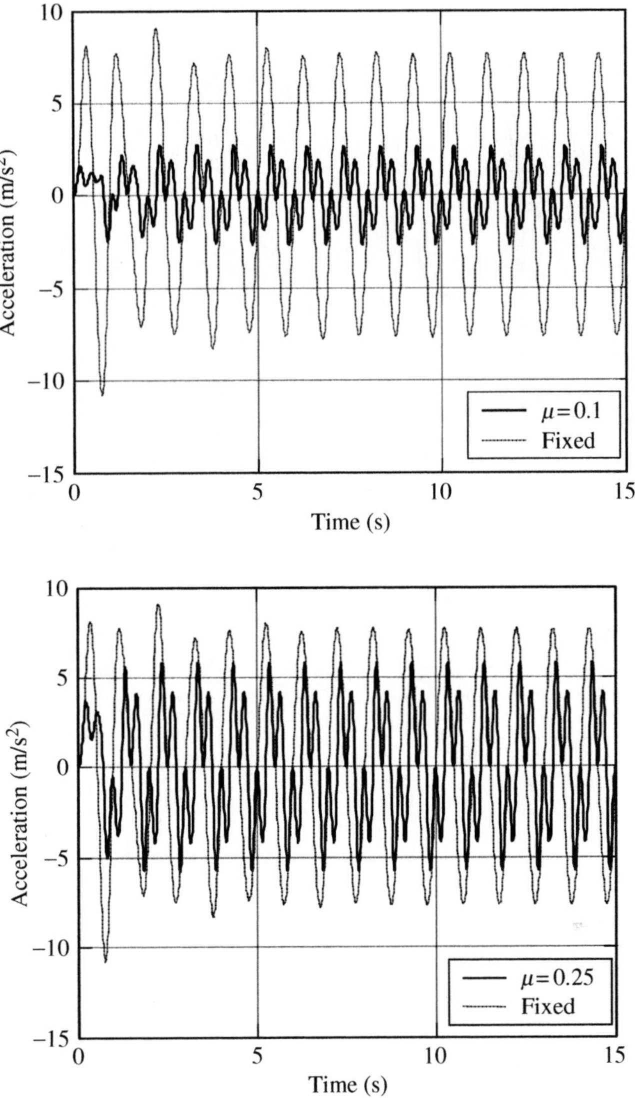

FIGURE 4.24 Comparison of structural accelerations ($\omega_g = 1$ Hz; $\omega_e = 5\omega_s$).

Frequency response. Figure 4.26 shows the maximum structural acceleration with respect to the excitation frequency for four different frictional coefficients, $\mu = 0.05, 0.1, 0.25$ and ∞ (for the fixed-base case). Here, the maximum acceleration means the steady-state acceleration response. The following observations can be made from Figure 4.26: (1) Compared with the fixed-base case, the use of a smaller frictional coefficient can reduce the structural acceleration for the frequency range considered. (2) The sliding mechanism can effectively suppress the main resonant response, associated with the natural frequency of 1.67 Hz of the superstructure system. (3) As the coefficient of friction, μ, decreases from ∞ to 0.05, the main resonant frequency associated with the structural natural frequency drifts from the fixed-base frequency of 1.67 Hz toward a higher value. (4) For the sliding cases of $\mu = 0.05$, 0.1, 0.25, there exist some minor peaks in the range of lower excitation frequencies, besides the main resonant peak. Such a phenomenon is called the *subharmonic resonance.* For a large frictional coefficient, say, with $\mu = 0.25$, the subharmonic resonant response may be even larger than that of the main resonance.

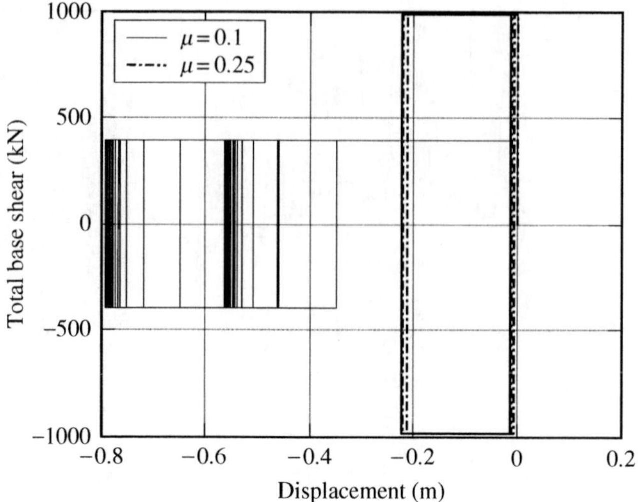

FIGURE 4.25 Hysteresis loops of a sliding bearing ($\omega_g = 1$ Hz; $\omega_e = 5\omega_s$).

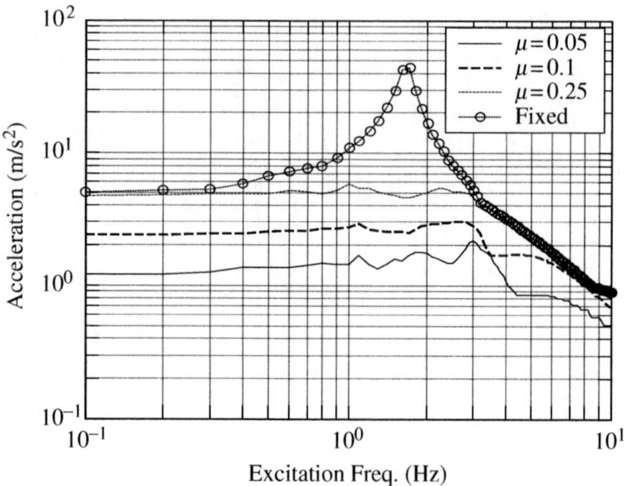

FIGURE 4.26 Maximum structural acceleration vs. ground excitation frequency ($\omega_e = 5\omega_s$).

Figure 4.27 shows the frequency responses of the maximum base displacement for various frictional coefficients. The following are observed: (1) The larger the frictional coefficient, the smaller the base drift is. (2) The base displacement has very large magnitudes in the lower excitation frequencies and decreases monotonically as the excitation frequency increases. (3) The extremely large base drift exhibited in the lower excitation frequency range is due to the initial permanent displacement observed in Figure 4.22.

4.4.3.3 Harmonic Response of Equipment

Time history. Consider an equipment item of a natural frequency equal to five times of the structural frequency, that is, $\omega_e = 5\omega_s$. For the case of a harmonic excitation of $\omega_g = 1$ Hz, the accelerations solved for the equipment mounted on the structure with $\mu = 0.1$ and 0.25, along with the fixed-base case, have

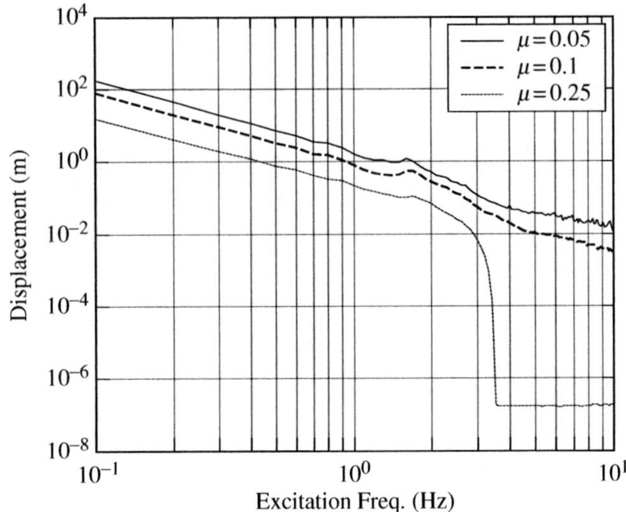

FIGURE 4.27 Maximum base displacement vs. ground excitation frequency ($\omega_e = 5\omega_s$).

been plotted in Figure 4.28. As can be seen, the equipment quickly reaches the steady state within a few cycles of oscillation. The equipment response is effectively suppressed for the case with a smaller frictional coefficient. Additionally, the waveforms shown in Figure 4.28 for the equipment appear to be marginally higher than those of the primary structure shown in Figure 4.24, which can be attributed to the use of a relatively stiff equipment, that is, with $\omega_e = 5\omega_s$.

Frequency responses. For an equipment item of the frequency $\omega_e = 5\omega_s$ (= 8.34 Hz), the maximum acceleration response has been plotted as a function of the excitation frequency in Figure 4.29. By comparing Figure 4.29 with Figure 4.26, one observes that the frequency response curves of the equipment and primary structure are generally similar, except that a secondary resonant peak occurs around the equipment natural frequency of 8.34 Hz in Figure 4.29. The other observations from Figure 4.29 are as follows: (1) In comparison with the fixed-base case, the sliding isolation alleviates both the structural and equipment resonant peaks around the frequencies of 1.67 and 8.34 Hz, respectively. However, the level of alleviation is more apparent for the former than for the latter. (2) The equipment also exhibits the same subharmonic resonance behavior as that of the primary structure, in terms of the resonance peaks and frequencies. (3) As the frictional coefficient μ decreases from ∞ (for the fixed-base case) to 0.05, the main resonant frequency associated with the structure drifts toward a higher value. However, the resonant frequency associated with the equipment remains the same.

Effect of equipment tuning. The effect of equipment tuning refers to the case when the equipment frequency is tuned to the structural frequency, that is, $\omega_e = \omega_s$. Figure 4.30 shows the frequency response of the equipment when the equipment tuning occurs. Compared with Figure 4.29, this figure shows the following: (1) When equipment tuning occurs, the sliding isolation system can still mitigate the main resonant peak of the equipment, but the effectiveness of mitigation is drastically reduced. (2) Although the subharmonic resonance can still be observed, the relevant frequencies of the equipment are different from those of the primary structure. (3) The frequency of the maximum resonant response remains equal to the tuned equipment's natural frequency of 1.67 Hz, regardless of the change in the frictional coefficient, μ, from ∞ to 0.05.

4.4.3.4 Earthquake Response of Structure

Time history. For the isolated system subjected to the El Centro earthquake with a PGA of 0.5*g*, the structural acceleration and base displacement of the sliding system have been plotted in Figure 4.31 and Figure 4.32, respectively, together with the response for the fixed-base case in Figure 4.31. As can be

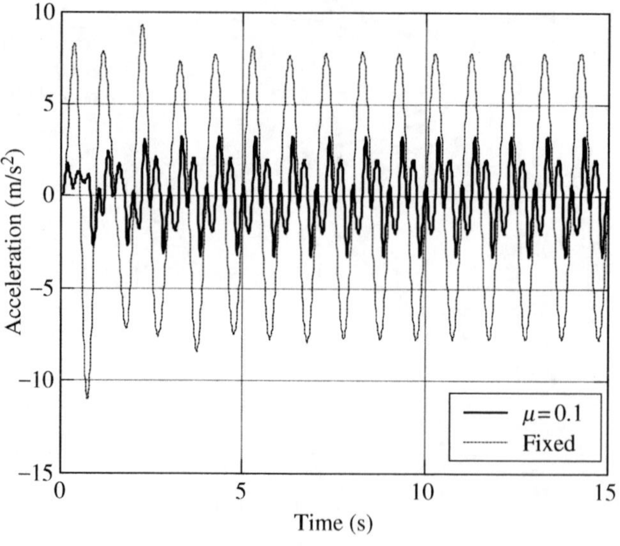

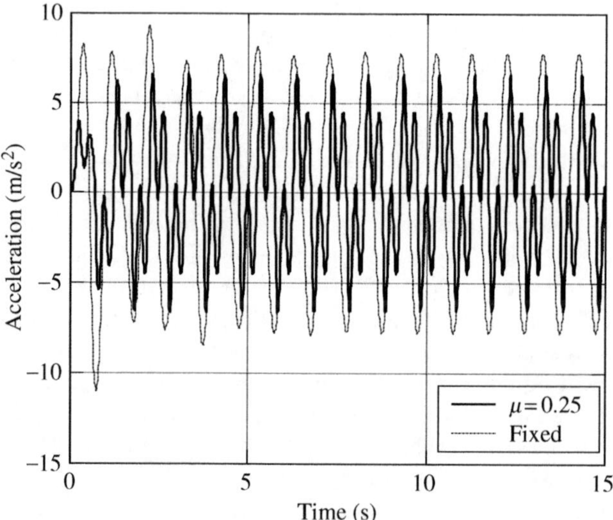

FIGURE 4.28 Comparison of equipment accelerations ($\omega_g = 1$ Hz; $\omega_e = 5\omega_s$).

seen from Figure 4.31, the main-shock response occurring between 0 and 10 sec for the fixed-base structure has been effectively suppressed by the sliding isolators with $\mu = 0.1$ and 0.25. About 80 and 60% of the maximum structural acceleration have been suppressed by the isolators with $\mu = 0.1$ and 0.25, respectively. On the other hand, Figure 4.32 demonstrates that the better suppression effect for the case with $\mu = 0.1$ is achieved at the expense of a larger base displacement. It is interesting to note that the horizontal segments in the curves of Figure 4.32 actually represent the stick state of the sliding system, which is useful for unveiling the sliding-stick mechanism involved.

Effect of earthquake intensity. The maximum structural acceleration and base displacement vs. the PGA have been plotted in Figure 4.33 and Figure 4.34, respectively. As can be seen from Figure 4.33, the maximum structural acceleration for the fixed-base case is proportional to the earthquake intensity, while in all the sliding cases it remains essentially as a constant after the PGA reaches a certain level. In other words, the reduction in structural maximum response and the efficiency of isolation have increased

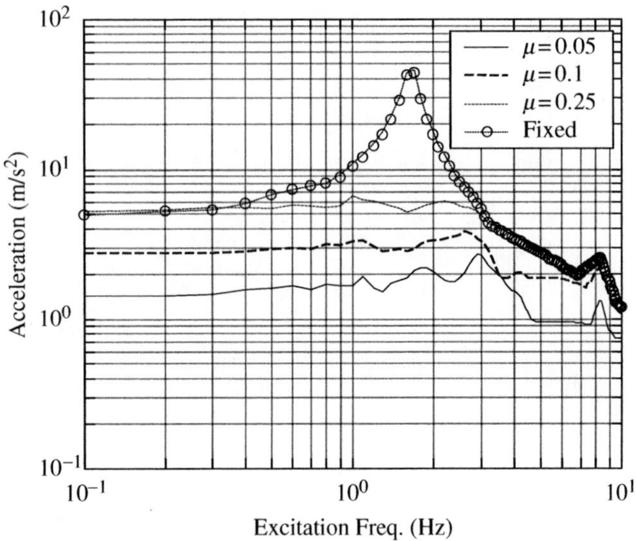

FIGURE 4.29 Maximum equipment acceleration vs. ground excitation frequency ($\omega_e = 5\omega_s$).

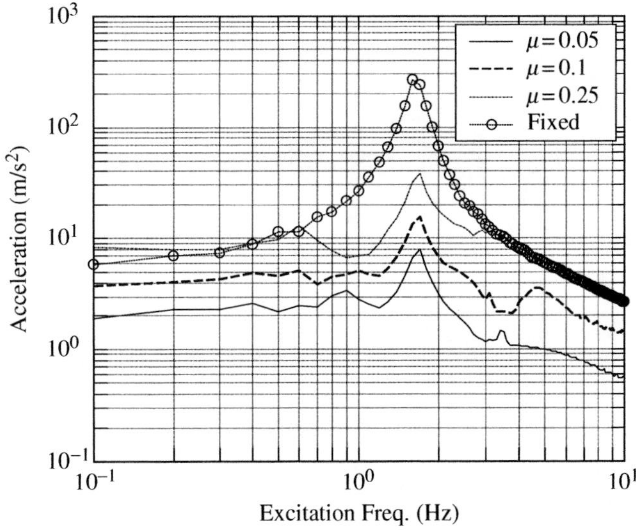

FIGURE 4.30 Maximum equipment acceleration vs. ground excitation frequency for equipment tuning ($\omega_e = \omega_s$).

with the increase in earthquake intensity. For example, for the case of $\mu = 0.1$, Figure 4.33 shows that the structural acceleration is reduced by around 60% at PGA $= 0.2g$, while it is reduced by more than 90% at PGA $= 1.0g$. However, as indicated by Figure 4.34, the above reduction in structural response has been achieved at the expense of increased base displacements. For the same PGA level, a sliding system with a smaller frictional coefficient has a better effect of vibration reduction, but this is accompanied by a larger base displacement.

Residual base displacement. The residual base displacement is defined as the permanent base displacement of the structure after it stops vibrating. This quantity is important in the study of sliding structures. Figure 4.35 shows the residual base displacement as a function of the earthquake PGA level. A first look at the figure reveals that no clear relation exists between the earthquake intensity and residual

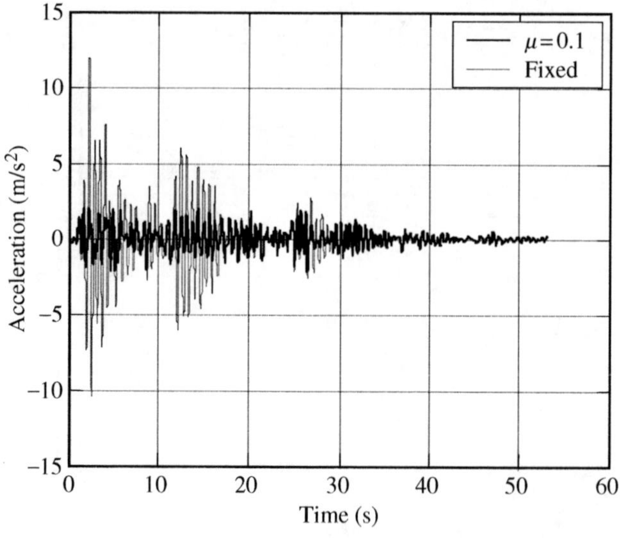

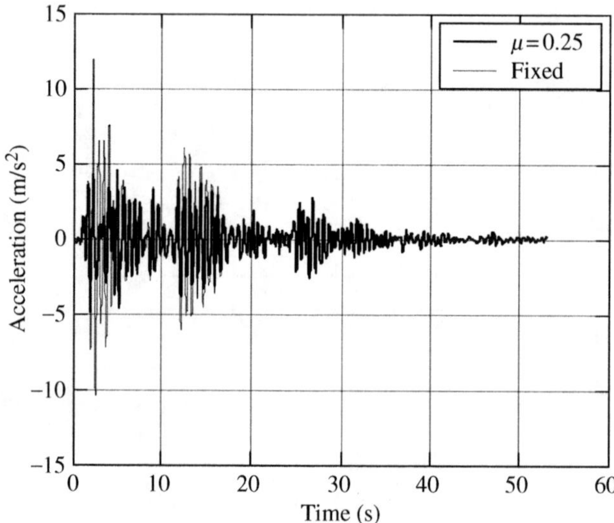

FIGURE 4.31 Comparison of structural accelerations ($\omega_e = 5\omega_s$, PGA $= 0.5g$).

displacement, because a larger PGA may lead to a smaller residual base displacement. Nevertheless, after taking the average of the residual displacements over the PGA range of 0.1 to 1g, we obtain $x_{res} = 0.083$, 0.084, 0.084 m for the case of $\mu = 0.05$, 0.1, 0.25, respectively. These values indicate that a smaller frictional coefficient leads to a larger residual base displacement in general. However, when the frictional coefficient, μ, approaches zero, the residual displacement approaches a constant equal to the permanent ground displacement.

4.4.3.5 Earthquake Response of Equipment

Time history. Consider an equipment item with a natural frequency equal to five times the structural frequency, that is, $\omega_e = 5\omega_s$ (= 8.34 Hz). For the isolated system subjected to the El Centro earthquake

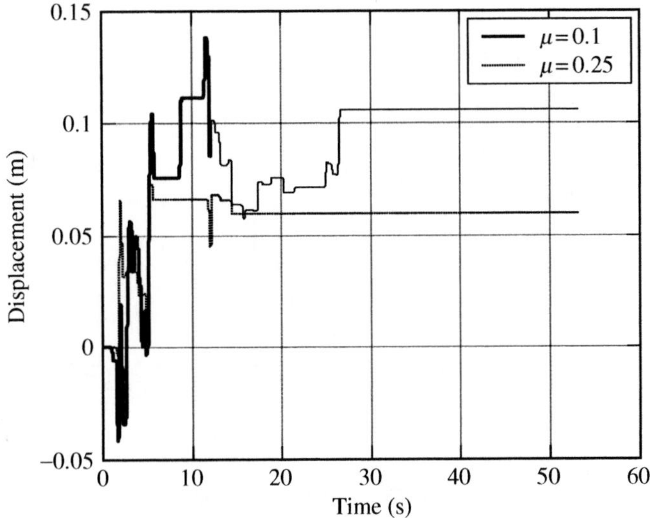

FIGURE 4.32 Comparison of base displacements ($\omega_e = 5\omega_s$, PGA $= 0.5g$).

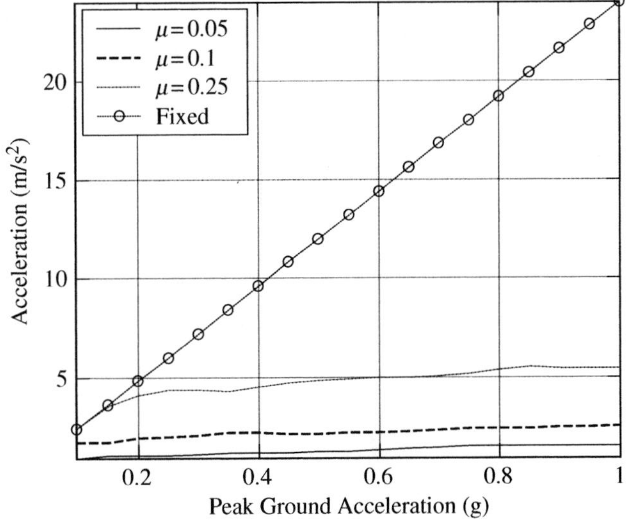

FIGURE 4.33 Maximum structural acceleration vs. PGA ($\omega_e = 5\omega_s$).

with PGA $= 0.5g$, the time histories computed for the equipment acceleration for the cases with $\mu = 0.1$ and 0.25, along with the fixed-base case, have been plotted in Figure 4.36a and b. As can be seen, the main-shock response of the fixed-base structure occurring for the first 10 sec has been effectively suppressed through installation of the sliding isolator with $\mu = 0.1$ and 0.25. A higher level of reduction can be achieved if a smaller frictional coefficient is chosen.

Effect of earthquake intensity. Figure 4.37 shows the maximum equipment acceleration as a function of the PGA level. Because of the use of a relatively stiff equipment ($\omega_e = 5\omega_s$), the curves shown in Figure 4.37 are similar to those for the primary structure in Figure 4.33, but with slightly higher values. Therefore, the observations made previously for Figure 4.33 are applicable to Figure 4.37. The maximum response of equipment items with other frequencies will be discussed below.

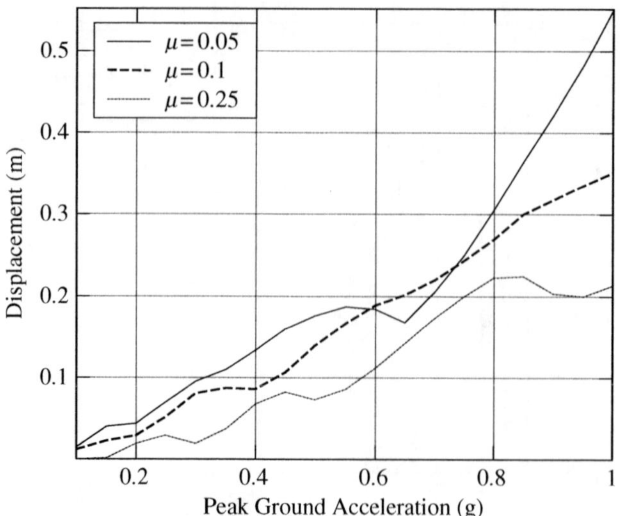

FIGURE 4.34 Maximum base displacement vs. PGA ($\omega_e = 5\omega_s$).

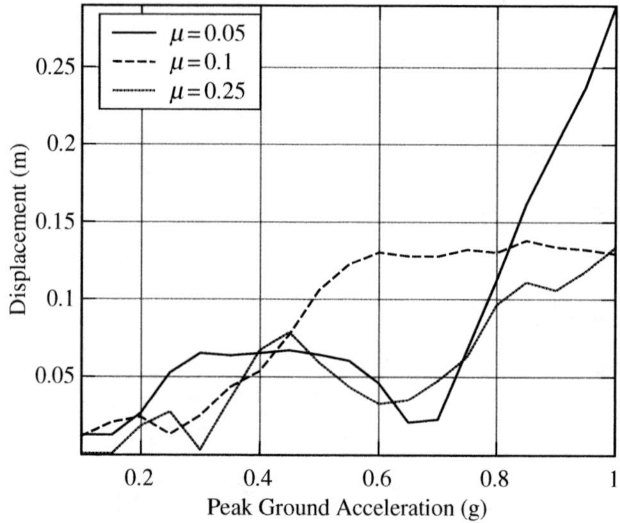

FIGURE 4.35 Residual base displacement vs. PGA ($\omega_e = 5\omega_s$).

Effect of equipment tuning. In order to study the equipment tuning effect, the maximum equipment acceleration has been plotted as a function of the equipment frequency in Figure 4.38. As can be seen, for all the values of μ considered, the equipment response is amplified when the equipment frequency moves close to the structural frequency of 1.67 Hz for the fixed-base case. Note that, since the resonant frequency of a sliding structure shifts to a higher value as the frictional coefficient decreases (see Figure 4.26), the frequency for which the most severe tuning effect occurs in Figure 4.38 also shifts from 1.67 Hz to a higher value as μ decreases. Nevertheless, it is concluded that, by choosing a smaller μ, the amplification of the equipment response due to tuning effect can be effectively suppressed.

4.4.4 Concluding Remarks

The dynamic behavior of a sliding isolated structural system with an attached equipment item was investigated in this section. A sliding isolated structure is classified as a nonlinear dynamic system, as the

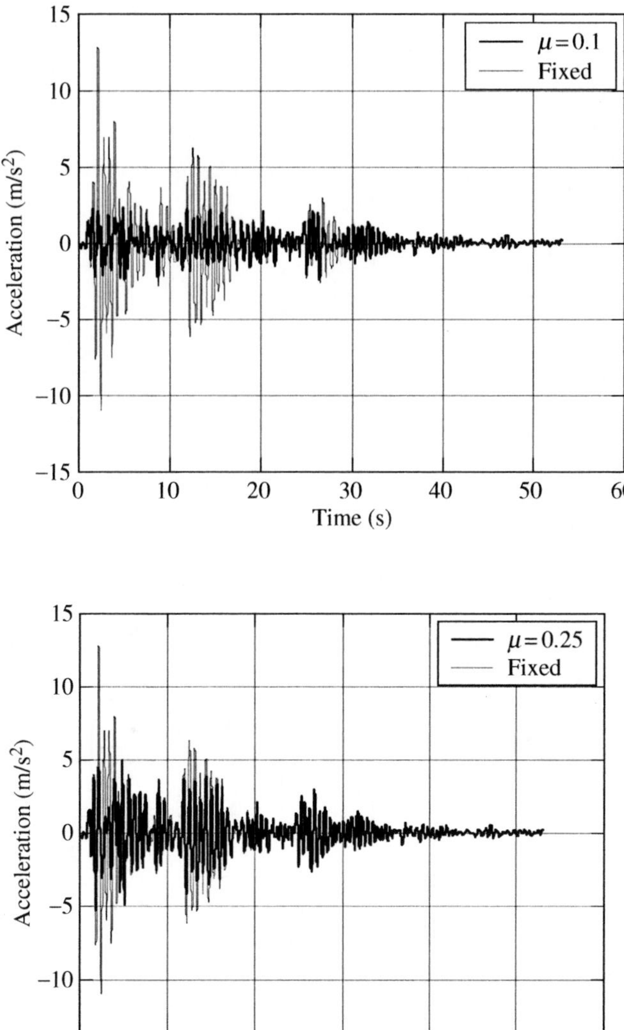

FIGURE 4.36 Comparison of equipment accelerations ($\omega_e = 5\omega_s$, PGA $= 0.5g$).

frictional forces induced on the sliding surface do not remain constant. To deal with such nonlinear systems, two analysis methods were formulated, the shear balance method and the fictitious spring method, both of which were presented in an incremental form that is suitable for direct implementation. Through the selection of a sliding isolated structure–equipment model, the responses of the structure and equipment subjected to both harmonic and earthquake excitations were analyzed. For the case of harmonic excitation, the results showed that the resonant responses of both the structure and attached equipment can be effectively suppressed, which remains good even when the equipment frequency is tuned to the structural frequency. For the case of seismic excitation, the results indicated that the level of reduction on the structural and equipment responses increases as the PGA level of the earthquake increases. Moreover, a sliding system with a smaller frictional coefficient has a higher isolation efficiency, at the expense of a larger base displacement.

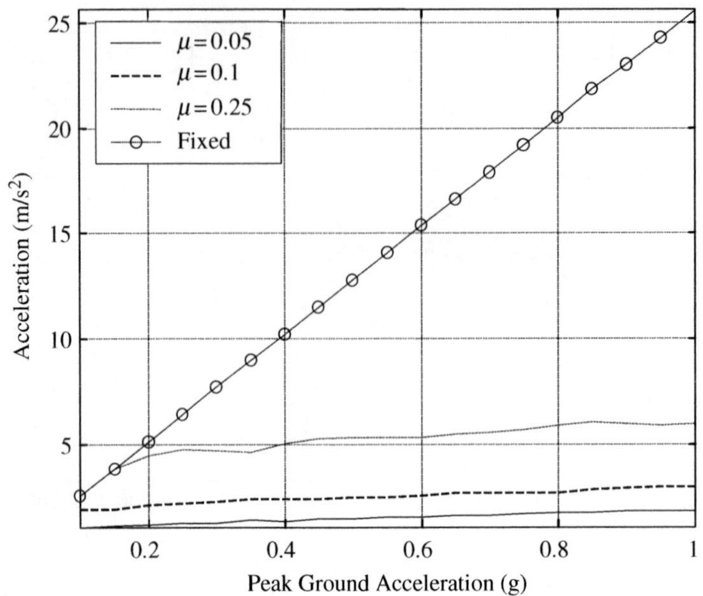

FIGURE 4.37 Maximum equipment acceleration vs. PGA ($\omega_e = 5\omega_s$).

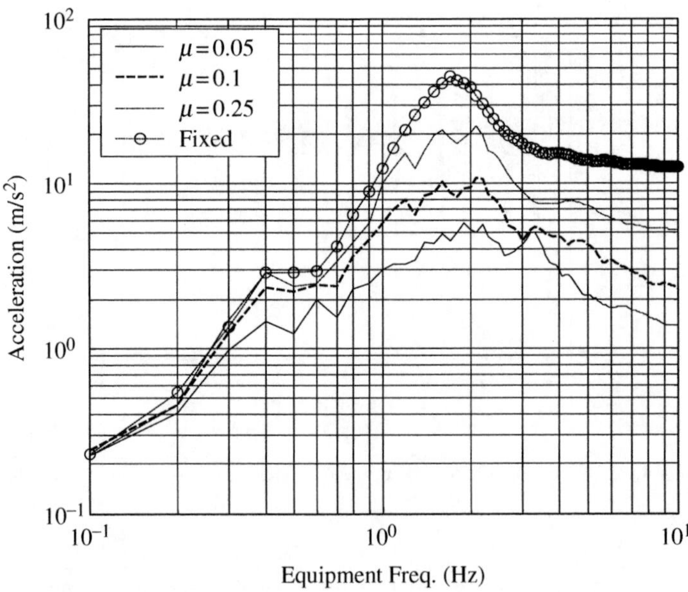

FIGURE 4.38 Maximum equipment acceleration vs. equipment frequency (PGA = 0.5g).

4.5 Sliding Isolation Systems with Resilient Mechanism

In Section 4.4, the relevant equations of analysis have been presented for a sliding isolated structural system, with no consideration made for the resilient (or recentering) mechanism. Because of this, rather large residual displacements may occur on the sliding isolation system, as have been numerically illustrated. If the concept of sliding isolators is to be applied to a real structure, it is important that the residual base displacements be controlled within certain limits, since they are not tolerable for some

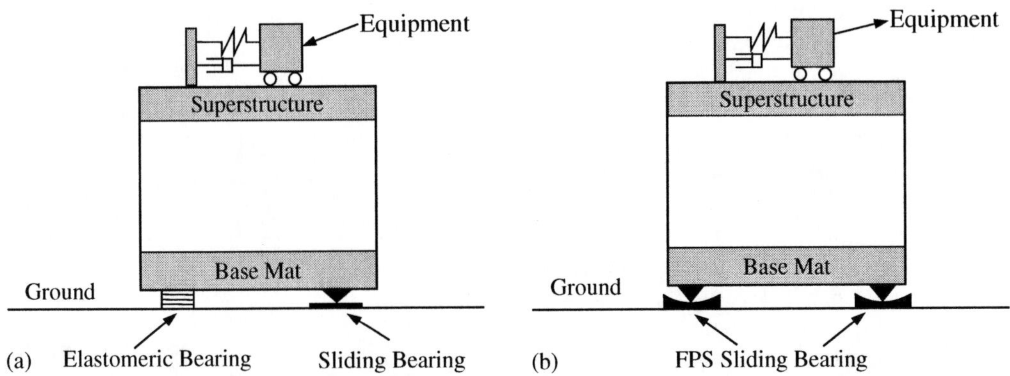

FIGURE 4.39 Schematic for a structure–equipment system isolated by a sliding bearing with resilience capability: (a) combined isolation system; (b) friction pendulum system (FPS).

engineering applications. For example, residual base displacements may distort the networking of water and power lines, change the space between the isolated structure and adjacent buildings, and widen the gaps at building entrances. Therefore, in practice, a sliding isolation system is usually enhanced through inclusion of mechanisms that can provide some resilient force. However, for certain ground motions, the added resilient force may also present some negative effects not readily transparent to structural designers, as will be illustrated in the numerical studies later on.

As shown in Figure 4.39, there are at least two ways of implementing the resilient mechanism in a sliding isolation system. Figure 4.39a shows an isolation system that combines the elastomeric bearings with sliding bearings (Chalhoub and Kelly, 1990), in which the elastomeric bearings are used to provide the resilient force, and the sliding bearings to uncouple the structural system from the ground motion. On the other hand, the resilient mechanism can also be incorporated into each single sliding bearing, in a way similar to that in the RFBI described in Section 4.2.3 (Mostaghel and Khodaverdian, 1987) or the FPS shown in Figure 4.39b (Mokha et al., 1991). The FPS isolation system has been implemented in many existing buildings and bridges. A typical FPS bearing consists a spherical sliding surface and a slider, which usually has a smooth coating of very low friction. When an FPS device is implemented under a structure, the slider will slide on the spherical surface during an earthquake, and the gravitational load of the structure, together with the curved sliding surface, will provide the resilient force for the system to return to its original position. The resilient stiffness of an FPS bearing depends on the radius of curvature of the sliding surface and the structural weight carried by the bearing.

This section is aimed at investigating the behavior of a structure–equipment system isolated by a sliding system with resilient device. For convenience of discussion, a system with resilient device will be referred to as the *resilient sliding isolation* (RSI), and a sliding system without resilient device, as the one studied previously, as the *pure sliding isolation* (PSI).

4.5.1 Mathematical Modeling and Formulation

Both the RSI systems shown in Figure 4.39 can be represented by the mathematical model given in Figure 4.40, for which the symbols used have been defined in Table 4.1. The RSI model shown in Figure 4.40 differs from the PSI model shown Figure 4.16 in that a linear spring of stiffness, k_b, is added to simulate the resilient force of the isolator. Obviously, an RSI model can be considered as the composition of a friction element and a spring element in parallel. Owing to addition of resilient stiffness, the number of vibration frequencies of the system is increased by one. The newly introduced frequency, which depends on the resilient stiffness, is called the *isolation frequency*, which can be approximated by

$$\omega_b = \sqrt{k_b/(W/g)} \tag{4.61}$$

where W is the total weight of the isolated structure–equipment system. The isolation frequency commonly used in design is between 0.33 and 0.5 Hz, which implies a period of 2 to 3 sec (Naeim and Kelly, 1999).

For the type of combined sliding system shown in Figure 4.39(a), the actual value of resilient stiffness, k_b, is decided by the total horizontal stiffness of the elastomeric bearings implemented. On the other hand, for the FPS shown in Figure 4.39(b), the resilient stiffness, k_b, is approximated by the following equation for small isolator displacements:

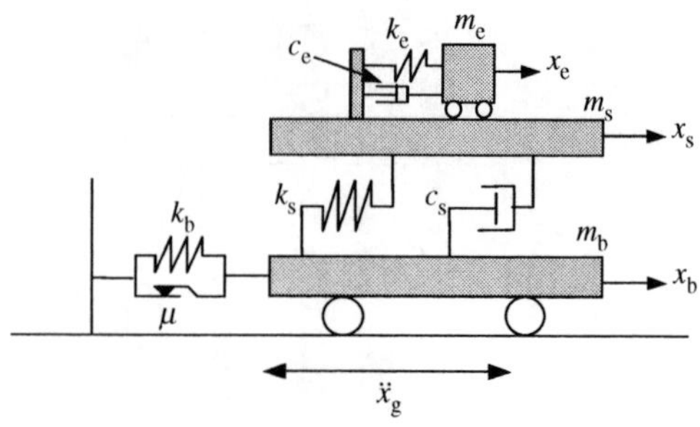

FIGURE 4.40 Model for a structure–equipment system isolated by a sliding bearing with resilience capability.

$$k_b = \frac{W}{R} \tag{4.62}$$

where R denotes the radius of curvature of the sliding surface. By substituting Equation 4.62 into Equation 4.61, one can verify that the isolation period of an FPS is equal to the oscillation period of a pendulum; that is

$$T_b = \frac{2\pi}{\omega_b} = \frac{2\pi}{\sqrt{k_b/(W/g)}} = 2\pi\sqrt{R/g} \tag{4.63}$$

When the sliding isolated system of Figure 4.40 is excited by a ground motion, its equation of motion may be written in exactly the same form as that of Equation 4.27; that is

$$\mathbf{M}\ddot{\mathbf{x}}(t) + \mathbf{C}\dot{\mathbf{x}}(t) + \mathbf{K}\mathbf{x}(t) = -\mathbf{M}\mathbf{L}_1\ddot{x}_g(t) + \mathbf{L}_2 f(t) \tag{4.64}$$

All the variables used in the preceding equation are the same as those defined in Equation 4.28 to Equation 4.30, except that the stiffness matrix, **K**, should be modified as

$$\mathbf{K} = \begin{bmatrix} k_e & -k_e & 0 \\ -k_e & k_e + k_s & -k_s \\ 0 & -k_s & k_s + k_b \end{bmatrix} \tag{4.65}$$

Note that, in Equation 4.64, the frictional force, $f(t)$, which does not remain constant, is placed on the right-hand side, while the resilient stiffness, k_b, which remains constant, is absorbed by the stiffness matrix, **K**, as in Equation 4.65. However, if the total shear force, $s(t)$ of the isolation system is of interest, it should be computed as the summation of the frictional force and resilient force (see Figure 4.40); that is

$$s(t) = k_b x_b(t) + f(t) \tag{4.66}$$

The equation of motion as given in Equation 4.64 can be recast in the following form of the first-order state-space equation:

$$\dot{\mathbf{z}}(t) = \mathbf{A}\mathbf{z}(t) + \mathbf{E}\ddot{x}_g(t) + \mathbf{B}f(t) \tag{4.67}$$

The definitions of the matrices $\mathbf{z}(t)$, **E**, **A**, and **B** are the same as those defined in Equation 4.32 and Equation 4.33, except that the system matrix, **A**, should be modified to account for the addition of the resilient stiffness k_b in the stiffness matrix, **K**.

4.5.2 Methods for Numerical Analysis

If one compares the equation of motion for the RSI system in Equation 4.67 with that for the PSI system in Equation 4.31, one will conclude that the only source of nonlinearity in both equations comes from the same term, namely, the frictional force, $f(t)$. As a result, the two methods of solution mentioned in Section 4.4.2, the shear balance method (Wang et al., 1998) and fictitious spring method (Yang et al., 1990), remain valid for the analysis of the RSI systems, with no modification required. Moreover, owing to inclusion of the resilient stiffness, k_b, in the structural stiffness matrix, $\mathbf{K}$, the system matrix, $\mathbf{A}$, becomes nonsingular and invertible. This introduces some advantage in computation of relevant coefficient matrices, including the $\mathbf{B}_0$ and $\mathbf{B}_1$ matrices in Equation 4.38a and Equation 4.38b. In Section 4.5.3, the shear balance method will be employed to simulate the response of an RSI system.

4.5.3 Simulation Results for Sliding Isolation with Resilient Mechanism

4.5.3.1 Numerical Model and Ground Excitations

In this section, the dynamic behavior of a sliding system represented by the model shown in Figure 4.40 will be investigated. The data adopted in the analysis for the equipment, structure, and the isolator have been listed in Table 4.4. To facilitate comparison, some of the data are selected to be the same as those in Table 4.3. In particular, the isolation frequency chosen is $\omega_b = 0.4$ Hz, falling in the common range of 0.33 to 0.5 Hz. Again, two types of ground excitations are considered, namely, the harmonic and earthquake excitations. For the harmonic excitation, a waveform of ground acceleration identical to the one given in Equation 4.60 is used. And for the earthquake excitation, the 1940 El Centro earthquake with different levels of PGA will be used, of which the acceleration waveform has been given in Figure 4.10. The harmonic excitation is adopted mainly for studying the frequency response of the sliding isolated system, while the earthquake excitation is for studying the effect of earthquake intensity. The dynamic responses computed for the RSI system, including the structure and equipment, will be presented, with emphasis placed on comparison with the PSI system of the same parameters. Similar to what was done in Section 4.4.3, the symbol μ will be used to denote the frictional coefficient of the sliding isolation system in all figures, and the word "fixed" denotes the fixed-base structure.

4.5.3.2 Harmonic Response of Structure

Time history. Consider an RSI system subjected to a harmonic excitation of $\omega_g = 1$ Hz. The base displacement and structural acceleration of the RSI system have been plotted in Figure 4.41 and Figure 4.42, respectively. Clearly, both the base displacement and structural acceleration of the RSI system reach their steady-state harmonic responses in the first few cycles. Moreover, a smaller sliding frictional coefficient ($\mu = 0.1$) is more effective for suppressing the structural acceleration, as indicated by Figure 4.42. However, this is achieved only at the expense of a larger base displacement, as indicated by Figure 4.41. By comparing the result for the RSI system in Figure 4.41 with those for the PSI system in Figure 4.23, the effect of resilient mechanism in eliminating the permanent base displacement for the case with a small frictional coefficient of $\mu = 0.1$ can be clearly appreciated. In spite of the large difference in base displacement, the structural accelerations for the RSI and PSI systems shown in Figure 4.42 and Figure 4.24, respectively, appear to be quite similar, when interpreted in terms of the waveform and response amplitude. This implies that, for the harmonic excitation considered, the resilient mechanism in RSI has little influence on the isolation effectiveness.

Hysteretic behavior. In Figure 4.43, the hysteresis loops for RSI systems with $\mu = 0.1$ and 0.25 subjected to a harmonic excitation of $\omega_g = 1$ Hz have been plotted, in which the vertical axis represents

TABLE 4.4 System Parameters Used in Simulation (Section 4.5.3)

Equipment		Superstructure		Isolation System	
Parameter	Value	Parameter	Value	Parameter	Value
Mass m_e	3 t (= m_s/100)	Mass m_s	300 t	Mass m_b	100 t (= m_s/3)
Frequency ω_e	$5\omega_s$ or a variable	Frequency ω_s	1.67 Hz	Frictional coefficient μ	0.05, 0.1, 0.25
Damping ratio ζ_e	5%	Damping ratio ζ_s	5%	Isolation frequency ω_b	0.4 Hz

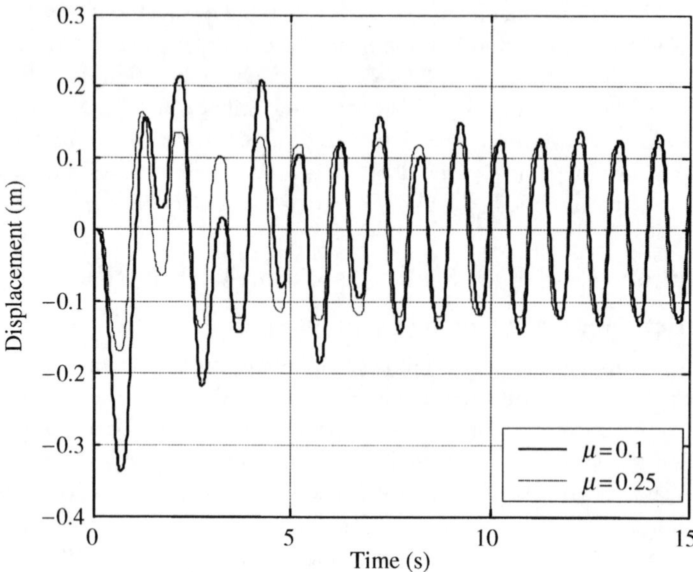

FIGURE 4.41 Comparison of base displacements ($\omega_g = 1$ Hz, $\omega_e = 5\omega_s$).

the total shear, $s(t)$, of the bearing computed from Equation 4.66. It is interesting to note that, for an RSI system, the hysteresis loop is a parallelogram, of which the slope of the inclined upper and lower sides exactly represents the total resilient stiffness, k_b, and the height and width, respectively, are decided by the maximum frictional force and maximum base displacement. As the frictional coefficient decreases, the height of the parallelogram decreases, but the width increases. The total area of the hysteresis loop represents the portion of energy dissipated by the RSI system. Noteworthy is the fact that, when the resilient stiffness, k_b, reduces to zero, the hysteresis parallelogram reduces to a square as well, identical to the one shown in Figure 4.25 for the PSI system.

Frequency responses. The maximum accelerations of the steady-state response of the structure for four different frictional coefficients, that is, $\mu = 0.05$, 0.1, 0.25 and ∞ (fixed-base), have been plotted in Figure 4.44. From this figure, the following observations can be made: (1) Compared with the fixed-base case, the resonant peak occurring around the structural frequency, ω_s, of 1.67 Hz was effectively suppressed by the RSI system, but a resonance of higher amplitude was induced in the lower frequency range (with frequencies lower than 0.6 Hz for the case studied). A further investigation reveals that the newly induced resonance is associated with the isolation frequency, ω_b, of 0.4 Hz. Such an observation remains valid for all values of frictional coefficients, μ. (2) The use of a lower frictional coefficient, μ, will result in a smaller response in the high-frequency range for the RSI system, for example, with frequencies higher than 0.6 Hz, but a larger response for the low-frequency range. (3) Although both the RSI and PSI systems can effectively remove the resonant peak around the structural frequency of 1.67 Hz, the RSI system has the side effect of creating a low-frequency resonant peak at the isolation frequency, ω_b. This implies that the RSI system is more sensitive to the excitation frequency.

The frequency responses of the maximum base displacement for the RSI system with various frictional coefficients have been plotted in Figure 4.45. When compared with the results for the PSI system in Figure 4.27, it is clear that the resilient mechanism of the RSI system considerably reduces the base displacement in the nonresonant excitation range, but it also amplifies the base displacement in the region when the excitation frequency is close to the isolation frequency, ω_b. From Figure 4.44 and Figure 4.45, we observe that both the structural acceleration and base displacement of an RSI system may resonate at the isolation frequency, which is usually designed to be less than 0.5 Hz. This implies that an RSI system may be ineffective or unsafe for a ground motion with enriched low-frequency

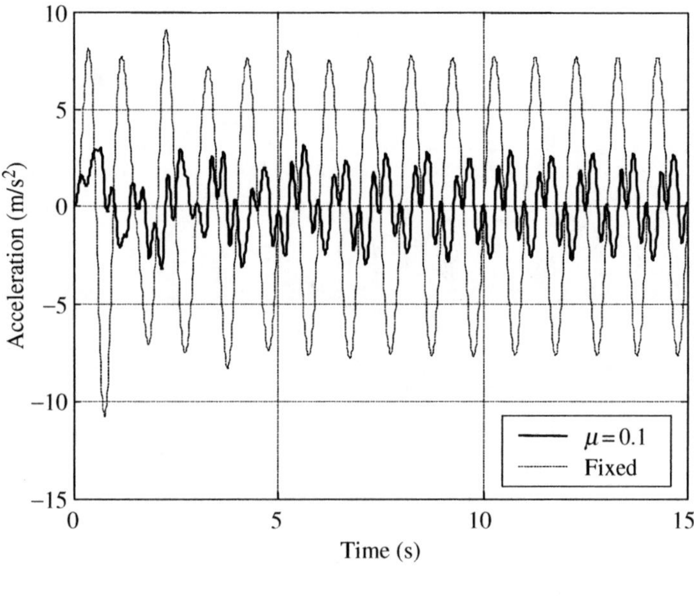

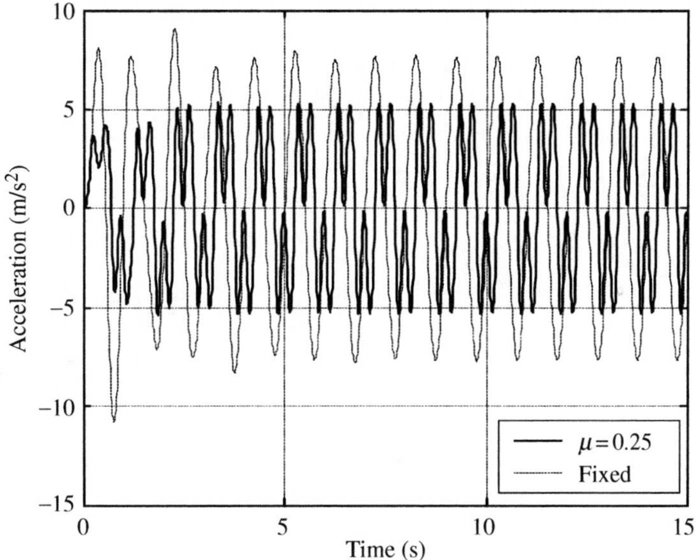

FIGURE 4.42 Comparison of structural accelerations ($\omega_g = 1$ Hz, $\omega_e = 5\omega_s$).

(long-period) vibrations, such as the case with a near-fault earthquake containing a long-period, pulse-like waveform (Jangid and Kelly, 2001; Lu et al., 2003). Structural designers should be aware of such a side effect when designing an RSI system.

4.5.3.3 Harmonic Response of Equipment

Time history. Consider an equipment item with a frequency of $\omega_e = 5\omega_s$ (= 8.34 Hz), attached to the RSI system. The harmonic acceleration responses of the equipment for the case with $\mu = 0.1$ and 0.25 have been plotted in Figure 4.46a and b, respectively, together with those for the fixed-base case. As can be seen, the equipment acceleration has been effectively suppressed by the RSI with the smaller frictional coefficient ($\mu = 0.1$). Moreover, the acceleration waveforms shown in Figure 4.46 are similar to those of

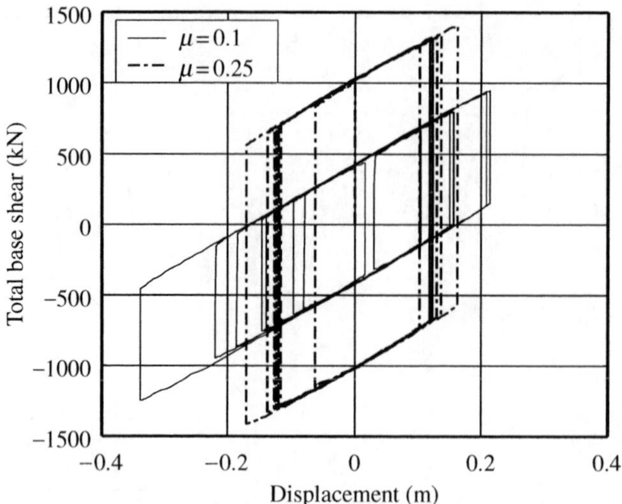

FIGURE 4.43 Hysteresis loops of a sliding bearing with resilience capability ($\omega_g = 1$ Hz, $\omega_e = 5\omega_s$).

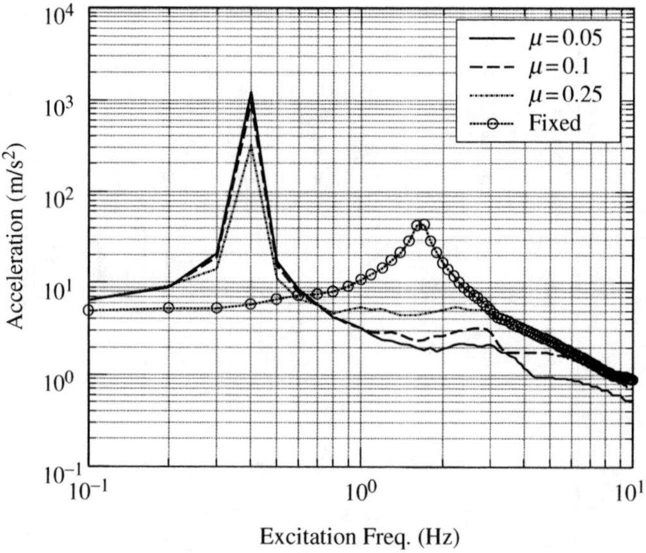

FIGURE 4.44 Maximum structural acceleration vs. ground excitation frequency ($\omega_e = 5\omega_s$).

the PSI system shown in Figure 4.28. This implies that for the given excitation, the behavior of the equipment was not altered by introduction of the resilient mechanism in the RSI system.

Frequency responses. Figure 4.47 shows the acceleration frequency response curve of the attached equipment with a frequency of $\omega_e = 5\omega_s$ (8.34 Hz). A comparison of Figure 4.47 with Figure 4.44 indicates that the frequency responses of the equipment and primary structure are generally similar, except that a resonant peak associated with the equipment frequency around 8.34 Hz appears in Figure 4.47. Owing to such a similarity, the observations made previously for Figure 4.44 are applicable to Figure 4.47 for the attached equipment.

Effect of equipment tuning. Figure 4.48 shows the frequency response of the attached equipment for the case when the equipment frequency is tuned to the structural frequency, that is, with $\omega_e = \omega_s$. Similar to

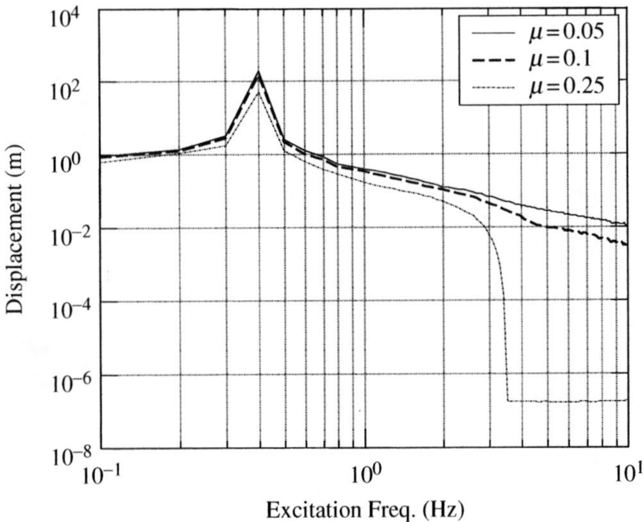

FIGURE 4.45 Maximum base displacement vs. ground excitation frequency ($\omega_e = 5\omega_s$).

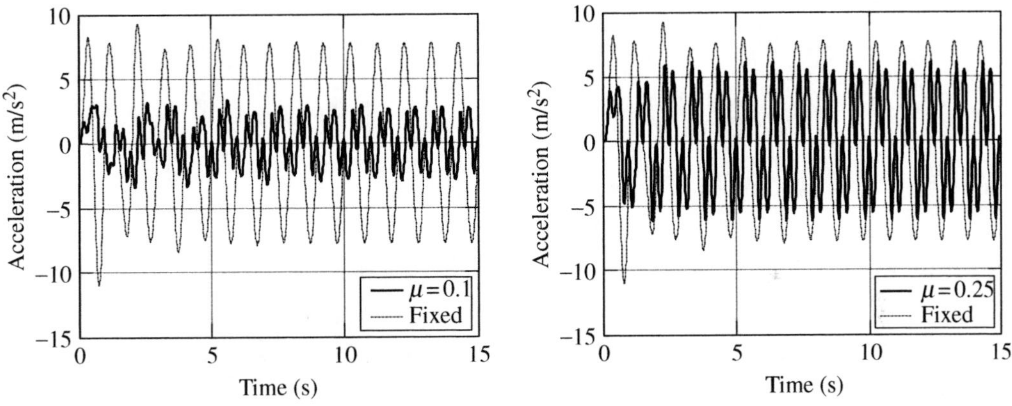

FIGURE 4.46 Comparison of equipment accelerations ($\omega_g = 1$ Hz, $\omega_e = 5\omega_s$).

the structural frequency response shown in Figure 4.44, the equipment attached to the RSI system also resonates at the isolation frequency ω_b of 0.4 Hz. Such a resonance does not occur for the equipment attached to the PSI system (see Figure 4.30). Through comparison of the tuned case in Figure 4.48 with the detuned case in Figure 4.47, the following observations can be made: (1) Even when the equipment tuning occurs, an RSI system mitigates the equipment's resonant peak associated with the structural frequency at 1.67 Hz, although the effectiveness of isolation has been reduced. (2) The tuning effect has no influence on the resonant response associated with the isolation frequency of 0.4 Hz.

4.5.3.4 Earthquake Response of Structure

Time history. For an RSI system subjected to the El Centro earthquake with PGA = 0.5g, the structural acceleration and base displacement have been shown in Figure 4.49 and Equation 4.50, respectively. By comparing Figure 4.49 with Figure 4.31 for the corresponding PSI system, one observes that the structural accelerations of the RSI and PSI systems are generally similar, in terms of the response waveform and the response magnitude. Both systems reduce the maximum structural acceleration quite effectively, for example, by about 80% for $\mu = 0.1$. However, significant difference does exist between the

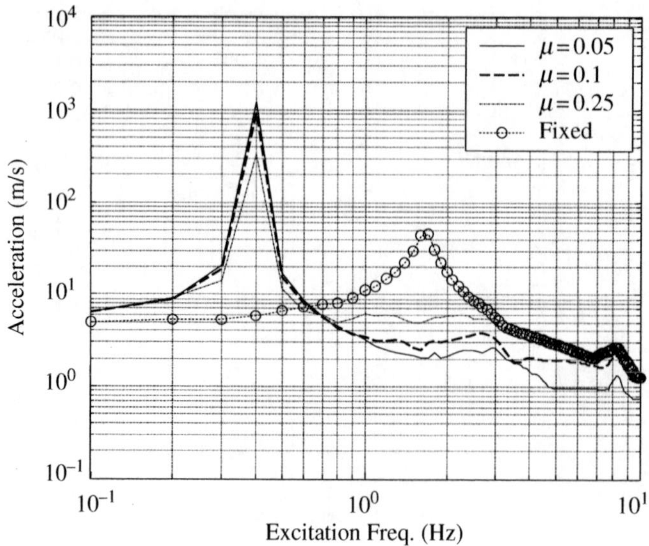

FIGURE 4.47 Maximum equipment acceleration vs. ground excitation frequency ($\omega_e = 5\omega_s$).

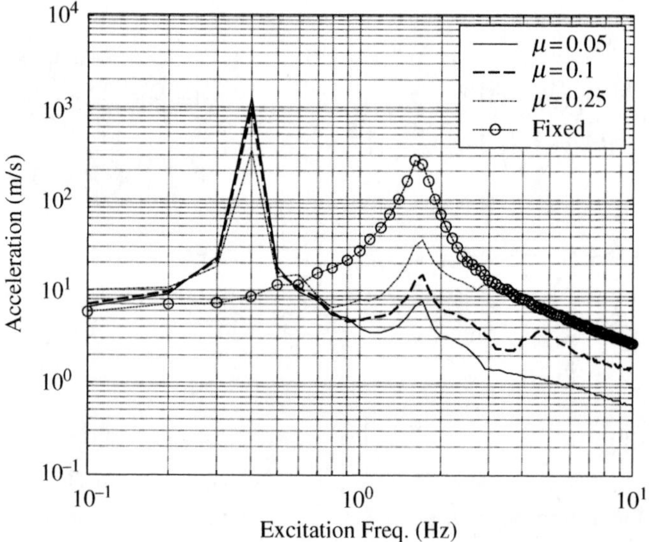

FIGURE 4.48 Maximum equipment acceleration vs. ground excitation frequency under tuning condition ($\omega_e = 5\omega_s$).

base displacements for the RSI system in Figure 4.50 and those for the PSI system in Figure 4.32. For example, for $\mu = 0.1$, the maximum base displacement experienced by the RSI system has been reduced by about 30%, while the residual base displacement has been reduced by about 70%, as can be seen by comparing Figure 4.50 with Figure 4.32. This implies that the resilient mechanism of the RSI system plays an important role in reducing the maximum and residual base displacements, especially the latter.

In spite of the observations made above, one should not forget that the frequency content of one earthquake may be different from an other. As was demonstrated in Figure 4.44 and Figure 4.45, an RSI system is generally sensitive to low-frequency excitations and may resonate at the isolation frequency.

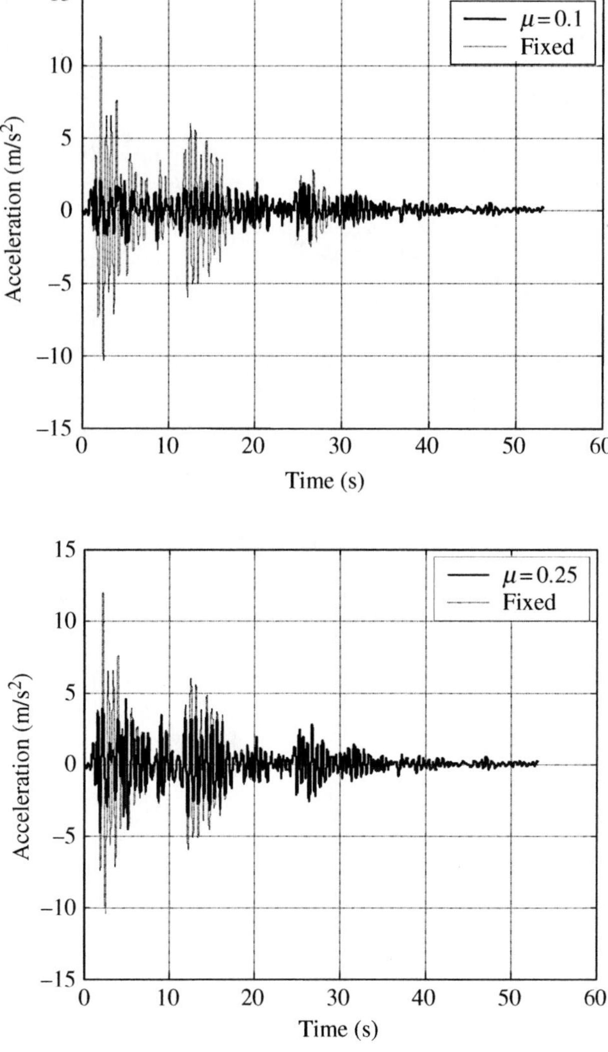

FIGURE 4.49 Comparison of structural accelerations ($\omega_e = 5\omega_s$, PGA $= 0.5g$).

Therefore, if the RSI system is subjected to an earthquake containing more low-frequency components, unlike the El Centro earthquake, it is likely that the maximum structural responses induced exceed those of the PSI system.

Effect of earthquake intensity. The maximum structural acceleration and base displacement of the RSI system have been plotted with respect to the PGA in Figure 4.51 and Figure 4.52, respectively. These figures indicate that as the earthquake intensity increases from 0.1 to 1g, the structural acceleration is reduced by an increasing amount by the RSI system, while the maximum base displacement also increases. By comparing Figure 4.51 and Figure 4.52 with Figure 4.33 and Figure 4.34 for the PSI system, one observes that both the RSI and PSI systems perform equally well for the El Centro earthquake, although the PSI system induces a slightly larger base displacement. On the other hand, unlike the response for the PSI system, the use of a smaller frictional coefficient for the RSI system does not always lead to a lower structural acceleration, as can be verified by comparing the responses for $\mu = 0.1$ and 0.05 with a PGA greater than 0.8g in Figure 4.33 and Figure 4.51. This can be attributed to the large resilient force induced by the large base displacement under higher PGA levels.

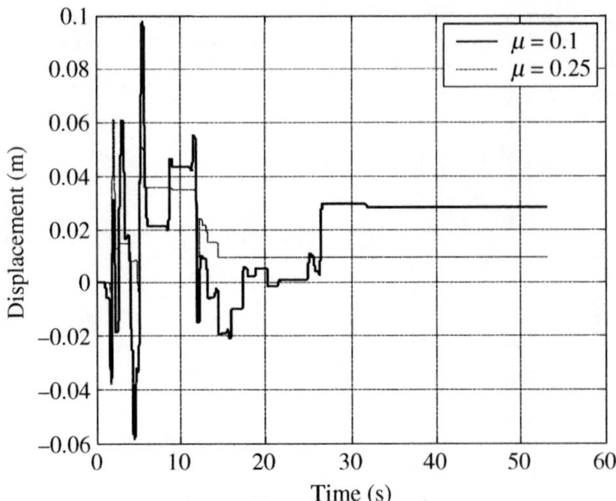

FIGURE 4.50 Comparison of base displacements ($\omega_e = 5\omega_s$, PGA $= 0.5g$).

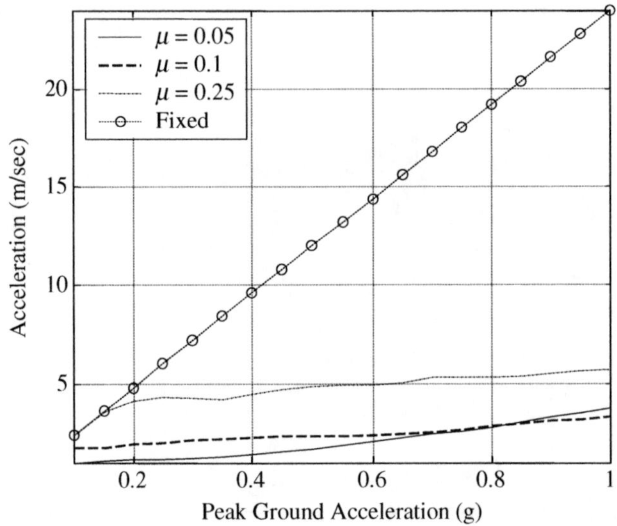

FIGURE 4.51 Maximum structural acceleration vs. PGA ($\omega_e = 5\omega_s$).

Residual base displacement. Figure 4.53 shows the residual base displacement of the RSI system vs. the PGA of the earthquake. For a given μ, it is difficult to establish a relation between the earthquake intensity and residual displacement, because a larger PGA may result in a smaller residual base displacement in some cases. However, if one takes the average of residual displacements over the PGA range from 0.1 to 1g, the following can be computed: $x_{res} = 0.0065, 0.011$, and 0.014 m for $\mu = 0.05, 0.1$, and 0.25, respectively. These values indicate that a smaller frictional coefficient leads to a smaller residual base displacement, which can be attributed to the fact that for a SRI system with a smaller coefficient of friction, it is easier for the resilient mechanism to return the structure to its initial position after an earthquake. On the other hand, a comparison of Figure 4.53 with Figure 4.35 for the PSI system indicates that for the same value of μ, the residual displacement was reduced substantially by the RSI system. This is certainly an advantage offered by the resilient mechanism of the RSI system.

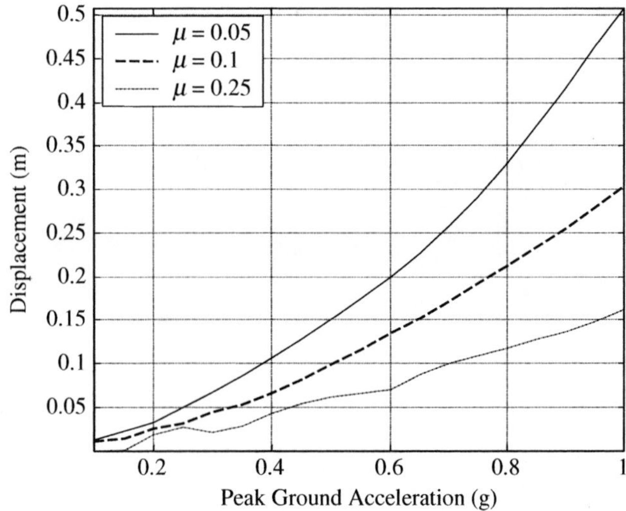

FIGURE 4.52 Maximum base displacement vs. PGA ($\omega_e = 5\omega_s$).

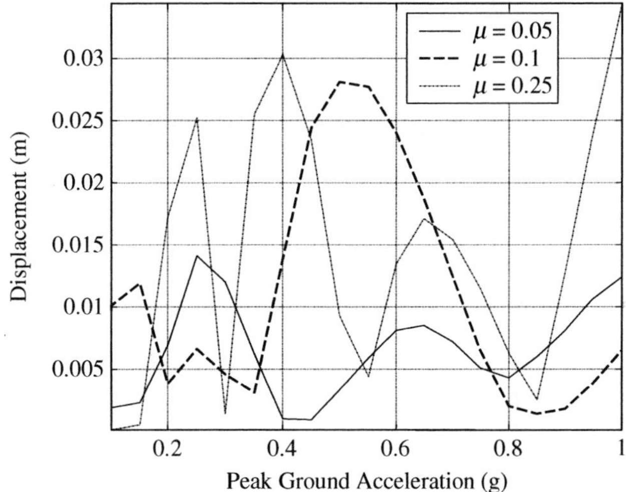

FIGURE 4.53 Residual base displacement vs. PGA ($\omega_e = 5\omega_s$).

4.5.3.5 Earthquake Response of Equipment

Time history. Let us consider an equipment item of natural frequency equal to five times the structural frequency, that is, $\omega_e = 5\omega_s$ (= 8.34 Hz). The acceleration responses of the equipment mounted on the RSI system that were subjected to the El Centro earthquake with a PGA of 0.5g for $\mu = 0.1$ and 0.25 have been plotted in Figure 4.54a and b, respectively, along with those for the fixed-base cases. As can be seen, the main-shock response of the equipment appearing during the first 10 sec for the fixed-base system was effectively suppressed by the RSI system with $\mu = 0.1$ or 0.25. The level of reduction is more pronounced for the case with a smaller frictional coefficient, that is, with $\mu = 0.1$. By comparing Figure 4.54 with Figure 4.36 for the PSI system, one concludes that the effect of the resilient mechanism of the RSI system on the equipment response is insignificant for the earthquake and equipment frequency considered.

Effect of earthquake intensity. Figure 4.55 shows the maximum equipment acceleration vs. the PGA of the earthquake. This figure illustrates that for all values of the frictional coefficient, μ, considered, an

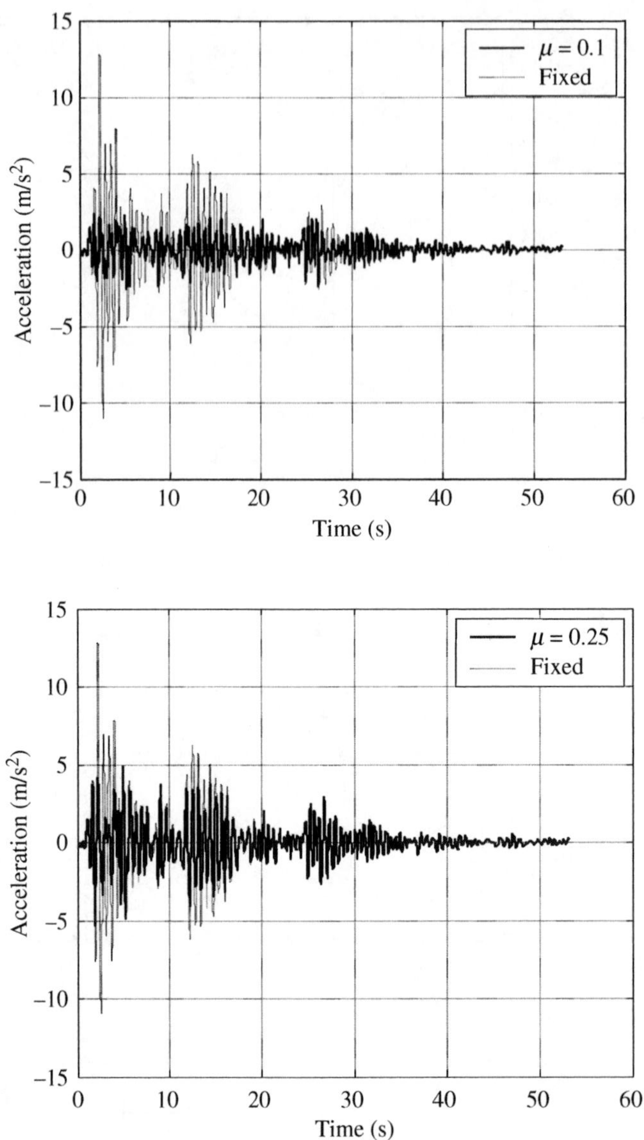

FIGURE 4.54 Comparison of equipment accelerations ($\omega_e = 5\omega_s$, PGA $= 0.5g$).

increasing amount of reduction can be achieved by the RSI system as the earthquake intensity increases from 0.1 to 1g. Because relatively stiff equipment (i.e., with $\omega_e = 5\omega_s$) was assumed in the simulation, the curves shown in Figure 4.55 are similar to those of Figure 4.51 for the primary structure. Therefore, the observations made previously for Figure 4.51 apply here. The maximum response of equipment items with other natural frequencies will be discussed below. Moreover, a comparison of Figure 4.55 with Figure 4.37 (for the PSI system) reveals that the resilient mechanism can have some minor effect on the equipment response, but only when a smaller frictional coefficient (i.e., $\mu = 0.05$ or 0.1) is used and when the PGA level is high.

Effect of equipment tuning. In order to study the effect of equipment tuning, the maximum acceleration of the equipment has been plotted in Figure 4.56 for equipment frequencies ranging from 0.1 to 10 Hz.

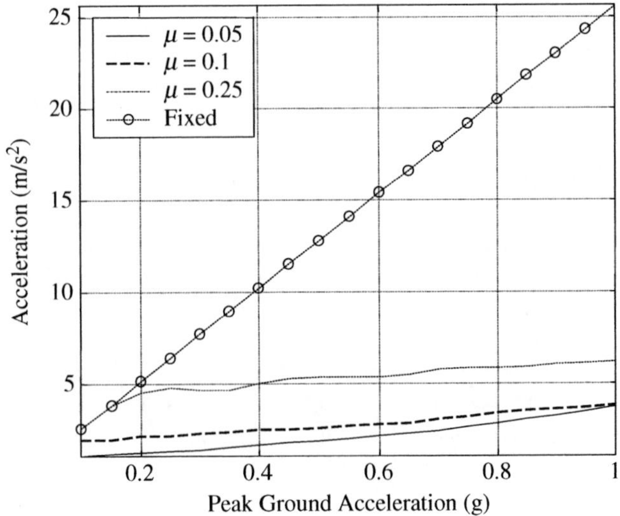

FIGURE 4.55 Maximum equipment acceleration vs. PGA ($\omega_e = 5\omega_s$).

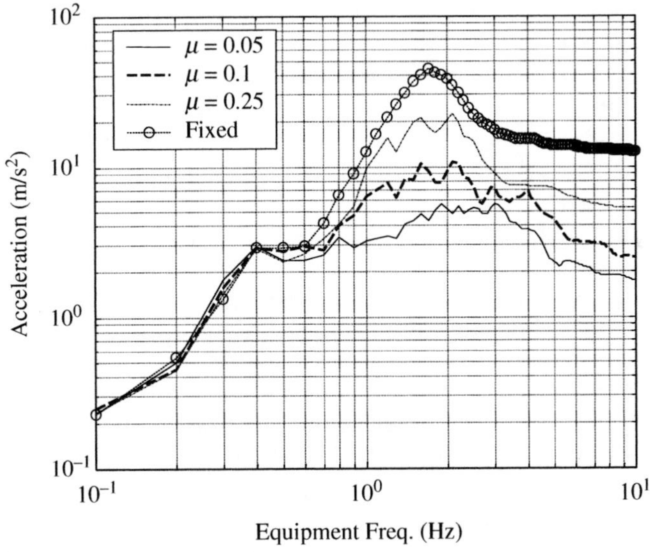

FIGURE 4.56 Maximum equipment acceleration vs. equipment frequency (PGA $= 0.5g$).

As can be seen, for all the values of μ considered, the equipment response is amplified when the equipment frequency is close to the structural frequency, ω_s, of 1.67 Hz, which means that the tuning effect tends to enlarge the equipment response. However, the use of a smaller μ can help in reducing the amplification of the equipment response resulting from the tuning effect. Finally, a comparison between Figure 4.56 and Figure 4.38 (for the PSI system) shows that the two diagrams are quite similar for an equipment item with a frequency higher than 1 Hz, but are different for that with a lower frequency. This implies that for the earthquake considered, the resilient mechanism of the RSI system has little effect on the response of the equipment with a higher stiffness.

4.5.4 Concluding Remarks

In this section, the behavior of a structure–equipment system isolated by an RSI system under both the harmonic and earthquake excitations has been investigated. Both the responses of the structure and equipment were studied, with special attention given to the effect of the resilient mechanism that characterizes an RSI system. The numerical results demonstrated that when subjected to a harmonic excitation, an RSI system is able to effectively suppress the resonant peaks associated with the structural frequency for both the structure and equipment, but it may also induce some resonant response near the isolation frequency due to the presence of resilient stiffness. Therefore, an RSI system is more sensitive to the frequency content of the ground excitation than a PSI system, especially to excitations of low-frequency components. As for the earthquake responses, the numerical results showed that the resilient mechanism of an RSI system can considerably reduce the residual base displacement. The resilient mechanism has a minor effect on the acceleration response of the structure and equipment, as long as no resonance is induced by the RSI system at the isolation frequency. By and large, both the RSI and PSI systems can be used as effective devices for reducing the acceleration responses of a structure and equipment.

4.6 Issues Related to Seismic Isolation Design

4.6.1 Design Methods

Having been developing for over 30 years, the technology of seismic isolation has matured. Many earthquake-prone countries, including the U.S., Japan, New Zealand, Taiwan, China, and European countries, have developed their own design codes, regulations, or guidelines (Fujita, 1998; Kelly, 1998; Martelli and Forni, 1998). Although most of the codes were developed based on the theory of structural dynamics, the design details outlined in the codes vary from one country to another. While a comprehensive explanation of the various design codes is not the purpose of this section, a brief overview of the concept underlying the design codes will be given. For more details, interested readers should refer to each code or to the books by Naeim and Kelly (1999) or Skinner et al. (1993). The design concept introduced herein is based on the series of Uniform Building Code (UBC, 1994, 1997).

Given the fact that base isolation devices are diverse, most design codes or regulations have been written in such a way as not to be specific with respect to the isolation systems. For instance, in the UBC (1997), no particular isolation system is identified as being acceptable; rather, it requires that every isolation system is stable for required displacement, has properties that do not degrade under repeated cyclic loadings, and provides increasing resistance with increasing displacement.

The design methods for base isolation can be classified as static analysis and dynamic analysis. The static analysis is applicable for stiff and regular buildings (in vertical and horizontal directions) that are constructed on soil of a relatively stiff condition. On the other hand, dynamic analysis is usually required for isolation systems with an irregular or long-period superstructure, or constructed on relatively soft soils. For a sophisticated design case, static analysis may be used in the preliminary design phase in order to draft or initiate the isolation design parameters, while dynamic analysis is employed in the final design phase for tuning or finalizing the design details of the isolation system. For simple design cases, static analysis alone is considered sufficient.

4.6.2 Static Analysis

For static analysis, a number of formulas have been specified in the design codes, so that engineers can easily calculate the following design parameters (shown in the design sequence): maximum isolator displacement, D; isolator total shear, V_b; total base shear, V_s, of superstructure; and seismic load, F_i, applied on each floor. These formulas were usually derived based on a simplified isolation model, assuming the isolation system can be linearized (even though most isolation systems are nonlinear) and the superstructure can be modeled as a rigid block. Such a simplified model is considered reasonable, since the displacements of an isolated structure are concentrated at the isolation level, which implies that the superstructure behaves as a rigid block. Based on such a model, only the first vibration mode with the superstructure treated as a rigid body has been considered in deriving the formulas. This explains why static analysis is suitable only for rigid and regular structures.

4.6.2.1 Computation of Maximum Isolator Displacement

An isolation design by static analysis usually starts with the calculation of the maximum isolator displacement, D, which depends on several factors:

$$D = D(Z, N, S, T_{ef}, \zeta_{ef}) \qquad (4.68)$$

where Z denotes the earthquake zone factor, N the near-fault factor, S the soil condition factor, T_{ef} the effective isolation period, and ζ_{ef} the effective isolation damping. For example, in the UBC (1994), the formula derived from the constant-velocity spectra over the period range of 1.0 to 3.0 sec has been given in the following form:

$$D = \frac{0.25ZNST_{ef}}{B} \qquad (4.69)$$

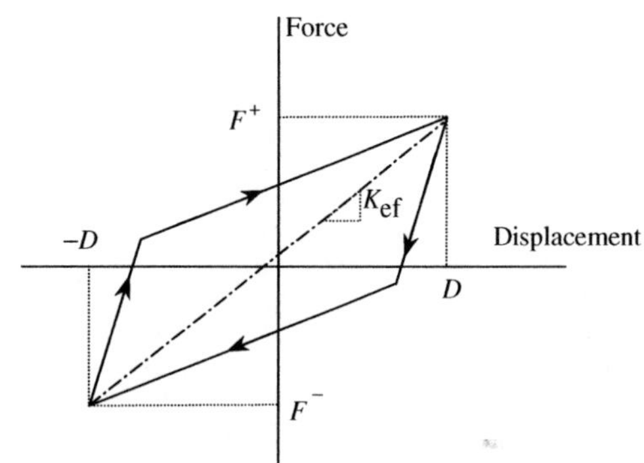

FIGURE 4.57 Typical force–displacement diagram for an isolation system.

where B is the damping factor, given as

$$B = B(\zeta_{ef}) \approx 0.25(1 - \ln \zeta_{ef}) \qquad (4.70)$$

In the above equations, the factors Z, N, and S depend on conditions of the construction site of the isolated structure; however, the factors T_{ef} and ζ_{ef} depend solely on the properties of the chosen isolation system. The factors T_{ef} and ζ_{ef} are called the "effective" period and damping of the isolation system, because they are frequently obtained by linearizing a nonlinear isolation system. The way to linearize an isolation system will be explained below, along with the formulas for computing T_{ef} and ζ_{ef}. Suppose that for a nonlinear isolation system, the force-displacement relation (hysteresis loop) obtained from a component test is shown in Figure 4.57. The effective stiffness of this isolation system can be computed by

$$K_{ef} = \frac{F^+ - F^-}{2D} \qquad (4.71)$$

where F^+ and F^-, respectively, denote the largest positive and negative forces in the test. After the linearized stiffness is obtained from Equation 4.71, the corresponding effective quantities T_{ef} and ζ_{ef} can be computed from the dynamic theory for a single DoF oscillation system; that is

$$T_{ef} = 2\pi\sqrt{\frac{W}{K_{ef}g}} \qquad (4.72)$$

$$\zeta_{ef} = \frac{1}{2\pi}\left(\frac{A}{K_{ef}D^2}\right) \qquad (4.73)$$

where W is the structural weight, g the gravitational acceleration, and A the total area enclosed by the hysteresis loop in Figure 4.57.

4.6.2.2 Computation of Maximum Isolator Shear

After the maximum isolator displacement, D, is obtained, the maximum isolator shear, V_b, can be estimated by the following formula:

$$V_b = K_{ef} D \tag{4.74}$$

Obviously, the above equation represents an equivalent static force exerted on the isolation system, when the system is displaced by an amount, D. In some design codes, V_b has also been referred to as the design force beneath the isolation system.

4.6.2.3 Computation of Total Base Shear

The total base shear, V_s, of the superstructure can be given as

$$V_s = \frac{K_{ef} D}{R_I} \tag{4.75}$$

where R_I is a reduction factor (ductility factor) to account for structural ductility, which will be developed when the structure is subjected to an earthquake with intensity above the design level. In some codes, V_s has also been referred to as the design force above the isolation system.

4.6.2.4 Computation of Shear Force for Each Floor

Having computed the above total base shear, V_s, a formula is employed to distribute this total shear to each floor of the isolated structure. For instance, in the, UBC (1997), the shear force, F_i, exerted on each floor is computed by

$$F_i = V_s \frac{h_i w_i}{\sum_{j=1}^{n} h_j w_j} \tag{4.76}$$

where n denotes the number of floors, w_i the weight of the ith floor, and h_i the height of the ith floor above the isolation level. Note that the sum of F_i ($i = 1$ to n) must be equal to V_s.

The general procedure for static analysis was illustrated in Figure 4.58. Once the design parameters, D, V_b, V_s, and F_i, are all determined according to the code, they can be used in the detailed design of structural elements as well as of isolator elements. Nevertheless, in most applications, because the test data of the isolation system may not be available in the beginning of design, the values of K_{ef}, T_{ef}, and ζ_{ef}, which are required in computing D, are not known to the designer. If this is the case, the design can begin with assumed values of K_{ef}, T_{ef}, and ζ_{ef}, which may be obtained from experience or previous test data on similar isolators. After the preliminary design is completed, prototype isolators will be fabricated and tested. The actual values of K_{ef}, T_{ef}, and ζ_{ef}, obtained from the tests will be used in the aforementioned code formulas to update the design parameters D, V_b, V_s. Moreover, one observes from Equation 4.71 that the linearized isolator stiffness, K_{ef}, is a function of the design parameter, D, itself, and so are T_{ef}, and ζ_{ef}, obtained from Equation 4.72 and 4.73. In order to obtain K_{ef}, as well as T_{ef}, and ζ_{ef}, an initial guess of D is required at the beginning of design. As a result, the design procedure may have to be repeated iteratively until the difference between the final value of D and the value D_0 computed in the last iteration is less than a preset tolerance. Such an iterative process is illustrated in Figure 4.58.

4.6.3 Dynamic Analysis

The dynamic analysis may be carried out in one of the two forms: response spectrum analysis and time-history analysis. Response spectrum analysis usually involves application of the concepts of response spectrum and modal superposition, and so on. Since these concepts primarily come from the dynamics of linear systems, the response spectrum analysis is only suitable for isolation systems with linear

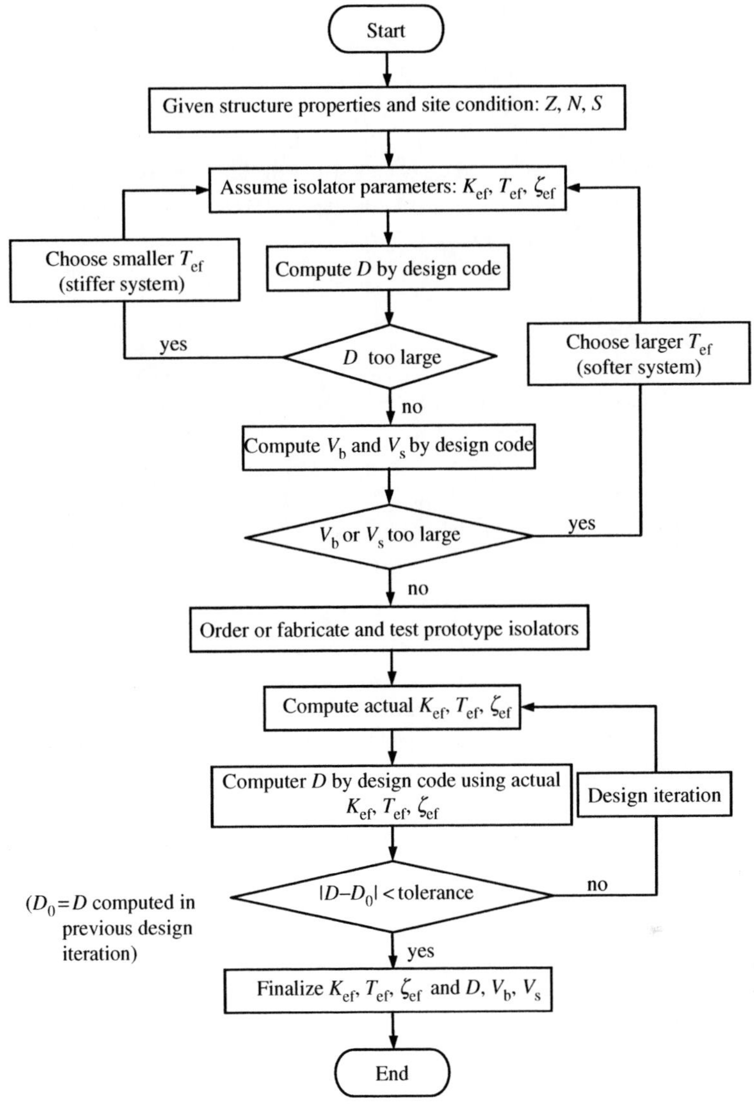

FIGURE 4.58 Flow chart of static analysis.

properties. For the case when the isolation system or the superstructure appears to be highly nonlinear, a time-history analysis is generally required.

Because dynamic analysis depends generally on the usage of computer programs, relatively few formulas have been given in the dynamic analysis sections of design codes. Nevertheless, for a successful time-history analysis, the designer must prepare the following three basic elements: (1) a set of representative input ground motions, (2) accurate mathematic models for isolators and superstructures, and (3) a computer program that is capable of performing the nonlinear time-history analysis. These three elements are explained below.

4.6.3.1 Input Ground Motions

The response of an isolated system depends greatly on the chosen input ground motions, which are usually expressed in the form of ground accelerations. Each ground motion is called one record event.

The chosen events must be representative of the site conditions and soil characteristics. Design codes usually specify the minimum number of events required for analysis. Each ground motion event must be scaled so that all events are compatible with each other and also with the code specified target spectrum. In the UBC (1997), the scaling factor for each event is obtained in response spectra, and then applied to the time domain of the record data. In particular, site specific ground motions are required in the UBC for the following cases: (1) an isolated structure located on a soft soil, (2) an isolated structure located within certain distance (e.g., 10 km) of an active fault, (3) an isolated structure with very long period of vibration (e.g., greater than 3 sec).

4.6.3.2 Mathematic Models

Before any time-history analysis can be carried out, a mathematic model that can accurately reflect the mechanical behavior of the isolation system and the superstructure must be constructed. If the isolation system is nonlinear, the nonlinear parameters must be identified so that the constructed mathematic model can correctly describe the hysteretic behavior of the isolation system. In many cases, the isolation system is assumed to be nonlinear, but the superstructure linear. Establishing an accurate mathematic model is curial for obtaining reliable results in a time-history analysis.

4.6.3.3 Computer Programs

In practice, the task of time-history analysis is executed through the use of a computer program. The mathematic model properties mentioned above will be input to the program for analysis. The computer program selected should be capable of simulating the three-dimensional behavior of structures with selected nonlinear elements. To serve the purpose of isolation design and analysis, several structural analysis programs running on the platform of personal computers have been developed for easy access. Some of the widely used programs include (but are not limited to): ETABS (ETABS, 2004), SAP-2000 Nonlinear (SAP, 2000), and 3D-BASIS (Nagarajaiah et al., 1993). Most of these programs provide a set of imbedded mathematic models for the widely used isolator elements with linear or nonlinear parameters. The designers using these programs can easily build up the mathematic model for the isolated structure considered, specify the parameters of the isolator elements selected, and execute a nonlinear time-history analysis on a personal computer.

4.6.4 Concluding Remarks

In this section, the design concept of seismic isolation for structures was briefly reviewed. The design methods can be based either on static or dynamic analysis. The fundamental issues that should be considered in each design method were highlighted, along with some relevant formulas for computing the relevant parameters. It is believed that, with the concepts and procedures presented in this section, the readers should have a general knowledge of the procedure for base isolation design of structures and equipment.

Acknowledgments

The authors are indebted to the graduate student, Cheng-Yan Wu, at the Department of Construction Engineering, National Kaohsiung First University of Science and Technology, for preparing some of the graphs presented in this chapter.

Nomenclature

Symbol	Quantity	Symbol	Quantity
c, c_s	damping coefficients of superstructure	V_b	total shear force of isolation system
c_e	damping coefficients of equipment	V_s	total base shear of superstructure
c_{EDF}	damping of the EDF system	w_i	weight of the ith floor
D	maximum isolator displacement	W	total weight of superstructure
f_r	interfacial frictional force	x, $x_s(t)$	relative-to-the-ground displacements of superstructure
F_i	seismic load applied on the ith floor		
k, k_s	stiffness of superstructure	$x_b(t)$	relative-to-the-ground displacements of base mat
k_b	stiffness of isolation system		
k_e	stiffness of equipment	$x_e(t)$	relative-to-the-ground displacements of equipment
k_{EDF}	stiffness of the EDF system		
k_f	stiffness of the fictitious spring in sliding isolation	$\ddot{x}_g(t)$	ground acceleration
		Z	zone factor
K_{ef}	effective stiffness of isolation system	μ	frictional coefficient of sliding isolation system
m, m_s	mass of superstructure		
m_b	mass of base mat	ω_b	frequency of isolation system
m_e	mass of equipment	ω_e	frequencies of equipment
n	number of building stories	ω_g	frequency of ground excitation
N	near fault factor	ζ_e	damping ratios of equipment
R_I	ductility factor	ζ_{ef}	effective damping ratios of isolation system
S	soil factor		
T	natural period of superstructure	ζ, ζ_s	damping ratios of superstructure
T_d	damped period of superstructure	Ω_d	damped natural frequency
T_{ef}	effective isolation period	Ω, ω_s	frequencies of superstructure

References

Agrawal, A.K., Behaviour of equipment mounted over a torsionally coupled structure with sliding support, *Eng. Struct.*, 22, 72–84, 2000.

Chalhoub, M.S. and Kelly, J.M. 1990. Earthquake simulator test of a combined sliding bearing and rubber bearing isolation system, Report No. UCB/EERC-87/04, Earthquake Engineering Research Center, University of California, Berkeley, CA.

Chaudhuri, S.R. and Gupta, V.K., A response-based decoupling criterion for multiply-supported secondary systems, *Earthquake Eng. Struct. Dyn.*, 31, 1541–1562, 2002.

Chen, G. and Soong, T.T., Energy-based dynamic analysis of secondary systems, *J. Eng. Mech., ASCE*, 120, 514–534, 1994.

Chen, G. and Soong, T.T., Exact solutions to a class of structure–equipment systems, *J. Eng. Mech., ASCE*, 122, 1093–1100, 1996.

Dey, A. and Gupta, V.K., Response of multiply supported systems to earthquakes in frequency domain, *Earthquake Eng. Struct. Dyn.*, 27, 187–201, 1998.

Dey, A. and Gupta, V.K., Stochastic seismic response of multiply-supported secondary systems in flexible-base structures, *Earthquake Eng. Struct. Dyn.*, 28, 351–369, 1999.

ETABS, Integrated Analysis, Design and Drafting of Building Systems, Version 8, Software by Computers and Structures Inc., Berkeley CA, 2004.

Fujita, T., Seismic isolation of civil buildings in Japan, *Prog. Struct. Eng. Mater.*, 1, 295–300, 1998.

Gueraud, R., Noel-Leroux, J.P., Livolant, M., and Michalopoulos, A.P., Seismic isolation using sliding-elastomer bearing pads, *Nucl. Eng. Des.*, 84, 363–377, 1985.

Huang, C.D., Zhu, W.Q., and Soong, T.T., Nonlinear stochastic response and reliability of secondary systems, *J. Eng. Mech., ASCE*, 120, 177–196, 1994.

Igusa, T., Response characteristic of inelastic 2-DOF primary–secondary system, *J. Eng. Mech., ASCE*, 116, 1160–1174, 1990.

Igusa, T. and Der Kiureghian, A.D., Dynamic characteristic of two-degree-of-freedom equipment–structure systems, *J. Eng. Mech., ASCE*, 111, 1–19, 1985a.

Igusa, T. and Der Kiureghian, A.D., Dynamic response of multiply supported secondary systems, *J. Eng. Mech., ASCE*, 111, 20–41, 1985b.

Igusa, T. and Der Kiureghian, A.D., Generation of floor response spectra including oscillator–structure interaction, *Earthquake Eng. Struct. Dyn.*, 13, 661–676, 1985c.

Inaudi, J.A. and Kelly, J.M., Minimum variance control of base-isolation floors, *J. Struct. Eng., ASCE*, 119, 438–453, 1993.

Jangid, R.S. and Kelly, J.M., Base isolation for near-fault motion, *Earthquake Eng. Struct. Dyn.*, 30, 691–707, 2001.

Juhn, G., Manolis, G.D., Constantinou, M.C., and Reinhorn, A.M., Experimental study of secondary systems in base-isolated structure, *J. Struct. Eng., ASCE*, 118, 2204–2221, 1992.

Kelly, J.M., Seismic isolation of civil buildings in USA, *Prog. Struct. Eng. Mater.*, 1, 279–285, 1998.

Kelly, J.M. and Tsai, H.C., Seismic response of light internal equipment in base-isolated structures, *Earthquake Eng. Struct. Dyn.*, 13, 711–732, 1985.

Lai, M.L. and Soong, T.T., Statistical energy analysis of primary–secondary structural systems, *J. Eng. Mech., ASCE*, 116, 2400–2413, 1990.

Lu, L.Y. and Yang, Y.B., Dynamic response of equipment in structures with sliding support, *Earthquake Eng. Struct. Dyn.*, 26, 61–77, 1997.

Lu, L.Y., Shih, M.H., Tzeng, S.W., and Chang, C.S. 2003. Experiment of a sliding isolated structure subjected to near-fault ground motion, In *Proceedings of the Seventh Pacific Conference on Earthquake Engineering*, February 13–15, Christchurch.

Martelli, A. and Forni, M., Seismic isolation of civil buildings in Europe, *Prog. Struct. Eng. Mater.*, 1, 286–294, 1998.

Meirovitch, L. 1990. *Dynamics and Control of Structures*, Wiley, New York.

Mokha, A.S., Constantinous, M.C., Reinhorn, A.M., and Zayas, V.A., Experimental study of friction-pendulum isolation system, *J. Struct. Eng., ASCE*, 117, 1201–1217, 1991.

Mostaghel, N., 1984. *Resilient-friction Base Isolator*, Report No. UTEC 84-097, University of Utah, Salt Lake City, UT.

Mostaghel, N. and Khodaverdian, M., Dynamics of resilient-friction base isolator (R-FBI), *Earthquake Eng. Struct. Dyn.*, 15, 379–390, 1987.

Mostaghel, N., Hejazi, M., and Tanbakuchi, J., Response of sliding structures to harmonic support motion, *Earthquake Eng. Struct. Dyn.*, 11, 355–366, 1983.

Naeim, F. and Kelly, J.M. 1999. *Design of Seismic Isolated Structures: From Theory to Practice*, Wiley, New York.

Nagarajaiah, S., Li, C., Reinhorn, A.M., and Constantinou, M.C. 1993. 3D-BASIS-TABS: Computer program for nonlinear dynamic analysis of three dimensional base isolated structures, Technical report NCEER-93-0011, National Center for Earthquake Engineering Research, Buffalo, NY.

Newmark, N.M., A method of computation for structural dynamics, *J. Eng. Mech. Div., ASCE*, 85, 67–94, 1959.

Park, K.S., Jung, H.J., and Lee, I.W., A comparative study on aseismic performances of base isolation systems for multi-span continuous bridge, *Eng. Struct.*, 24, 1001–1013, 2002.

Rivin, E.I., Vibration isolation of precision equipment, *Precision Eng.*, 17, 41–56, 1995.

SAP 2000. Integrated Structural Analysis and Design Software, Software by Computers and Structures, Inc., Berkeley, CA.

Skinner, R.I., Robinson, W.H., and Mcverry, G.H. 1993. *An Introduction to Seismic Isolation*, Wiley, New York.

Soong, T.T. and Dargush, G.F. 1997. *Passive Energy Dissipation Systems in Structural Engineering*, Wiley, New York.

Steinberg, D.S. 2000. *Vibration Analysis for Electronic Equipment*, 3rd ed., Wiley, New York.

Suarez, L.E. and Singh, M.P., Floor spectra with equipment–structure–equipment interaction effects, *J. Eng. Mech., ASCE*, 115, 247–264, 1989.

UBC 1994. Uniform building code, *International Conference of Building Officials*, Whittier, CA.

UBC 1997. Uniform building code, *International Conference of Building Officials*, Whittier, CA.

Veprik, A.M. and Babitsky, V.I., Vibration protection of sensitive electronic equipment from harsh harmonic vibration, *J. Sound Vib.*, 238, 19–30, 2000.

Villaverde, R., Simplified seismic analysis of secondary systems, *J. Eng. Mech., ASCE*, 112, 588–604, 1986.

Wang, Y.P., Chung, L.L., and Liao, W.H., Seismic response analysis of bridges isolated with friction pendulum bearings, *Earthquake Eng. Struct. Dyn.*, 27, 1069–1093, 1998.

Westermo, B. and Udwadia, F., Period response of a sliding oscillator system to harmonic excitation, *Earthquake Eng. Struct. Dyn.*, 11, 135–146, 1983.

Xu, Y.L., Liu, H.J., and Yang, Z.C., Hybrid platform for vibration control of high-tech equipment in buildings subject to ground motion. Part 1. Experiment, *Earthquake Eng. Struct. Dyn.*, 32, 1185–1200, 2003.

Yang, J.N. and Agrawal, A.K., Protective systems for high-technology facilities against microvibration and earthquake, *Struct. Eng. Mech.*, 10, 561–575, 2000.

Yang, Y.B. and Chen, Y.C., Design of sliding-type base isolators by the concept of equivalent damping, *Struct. Eng. Mech.*, 8, 299–310, 1999.

Yang, Y.B. and Huang, W.H., Seismic response of light equipment in torsional buildings, *Earthquake Eng. Struct. Dyn.*, 22, 113–128, 1993.

Yang, Y.B. and Huang, W.H., Equipment–structure interaction considering the effect of torsion and base isolation, *Earthquake Eng. Struct. Dyn.*, 27, 155–171, 1998.

Yang, Y.B., Lee, T.Y., and Tsai, I.C., Response of multi-degree-of-freedom structures with sliding supports, *Earthquake Eng. Struct. Dyn.*, 19, 739–752, 1990.

Yang, Y.B., Hung, H.H., and He, M.J., Sliding and rocking response of rigid blocks due to horizontal excitations, *Struct. Eng. Mech.*, 9, 1–16, 2000.

Yang, Y.B., Chang, K.C., and Yau, J.D. 2002. Base isolation. In *Earthquake Engineering Handbook*, W.F. Chen and C. Scawthorn, Eds., CRC Press, Boca Raton, FL, chap. 17.

Yang, Z.C., Liu, H.J., and Xu, Y.L., Hybrid platform for vibration control of high-tech equipment in buildings subject to ground motion. Part 2. Analysis, *Earthquake Eng. Struct. Dyn.*, 32, 1201–1215, 2003.

5

Vibration Control

Nader Jalili
Clemson University

Ebrahim Esmailzadeh
University of Ontario

Summary

The fundamental principles of vibration-control systems are formulated in this chapter (also see Chapter 7). There are many important areas directly or indirectly related to the main theme of the chapter. These include practical implementation of vibration-control systems, nonlinear control schemes, actual hardware implementation, actuator bandwidth requirements, reliability, and cost. Furthermore, in the process of designing a vibration-control system, in practice, several critical criteria must be considered. These include weight, size, shape, center-of-gravity, types of dynamic disturbances, allowable system response, ambient environment, and service life. Keeping these in mind, general design steps and procedures for vibration-control systems are provided.

5.1 Introduction

The problem of reducing the level of vibration in constructions and structures arises in various branches of engineering, technology, and industry. In most of today's mechatronic systems, a number of possible devices such as reaction or momentum wheels, rotating devices, and electric motors are essential to the system's operation and performance. These devices, however, can also be sources of detrimental vibrations that may significantly influence the mission performance, effectiveness, and accuracy of operation. Therefore, there is a need for vibration control. Several techniques are utilized either to limit or alter the vibration response characteristics of such systems. During recent years, there has been considerable interest in the practical implementation of these vibration-control systems. This chapter presents the basic theoretical concepts for vibration-control systems design and implementation, followed by an overview of recent developments and control techniques in this subject. Some related practical developments in variable structure control (VSC), as well as piezoelectric vibration control of flexible structures, are also provided, followed by a summary of design steps and procedures for vibration-control systems. A further treatment of the subject is found in Chapter 7.

5.1.1 Vibration Isolation vs. Vibration Absorption

In vibration isolation, either the source of vibration is isolated from the system of concern (also called "force transmissibility"; see Figure 5.1a), or the device is protected from vibration of its point of attachment (also called "displacement transmissibility", see Figure 5.1b). Unlike the isolator, a vibration absorber consists of a secondary system (usually mass–spring–damper trio) added to the primary device to protect it from vibrating (see Figure 5.1c). By properly selecting absorber mass, stiffness, and damping, the vibration of the primary system can be minimized (Inman, 1994).

5.1.2 Vibration Absorption vs. Vibration Control

In vibration-control schemes, the driving forces or torques applied to the system are altered in order to regulate or track a desired trajectory while simultaneously suppressing the vibrational transients in the system. This control problem is rather challenging since it must achieve the motion tracking objectives while stabilizing the transient vibrations in the system. Several control methods have been developed for such applications: optimal control (Sinha, 1998); finite element approach (Bayo, 1987); model reference adaptive control (Ge et al., 1997); adaptive nonlinear boundary control (Yuh, 1987); and several other techniques including VSC methods (Chalhoub and Ulsoy, 1987; de Querioz et al., 1999; de Querioz et al., 2000).

As discussed before, in vibration-absorber systems, a secondary system is added in order to mimic the vibratory energy from the point of interest (attachment) and transfer it into other components or dissipate it into heat. Figure 5.2 demonstrates a comparative schematic of vibration control (both single-input control and multi-input configurations) on translating and rotating flexible beams, which could represent many industrial robot manipulators as well as vibration absorber applications for automotive suspension systems.

5.1.3 Classifications of Vibration-Control Systems

Passive, active, and semiactive (SA) are referred to, in the literature, as the three most commonly used classifications of vibration-control systems, either as isolators or absorbers (see Figure 5.3; Sun et al., 1995). A vibration-control system is said to be active, passive, or SA depending on the amount of

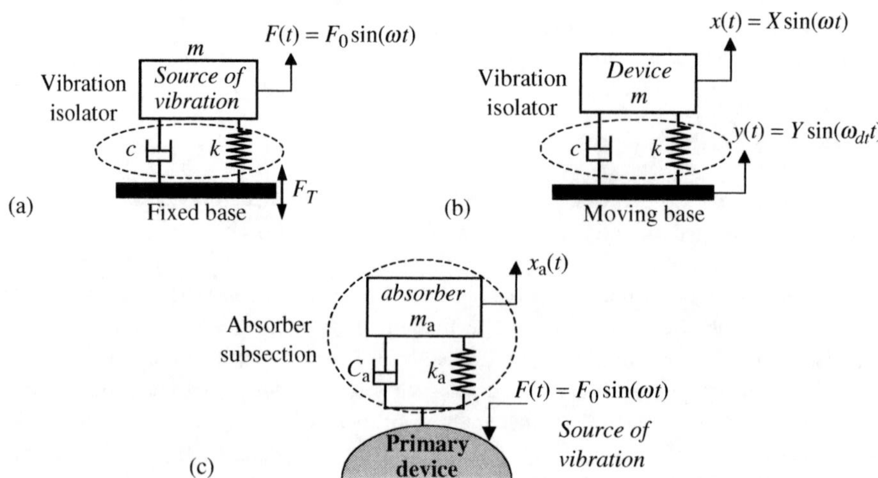

FIGURE 5.1 Schematic of (a) force transmissibility for foundation isolation; (b) displacement transmissibility for protecting device from vibration of the base and (c) application of vibration absorber for suppressing primary system vibration.

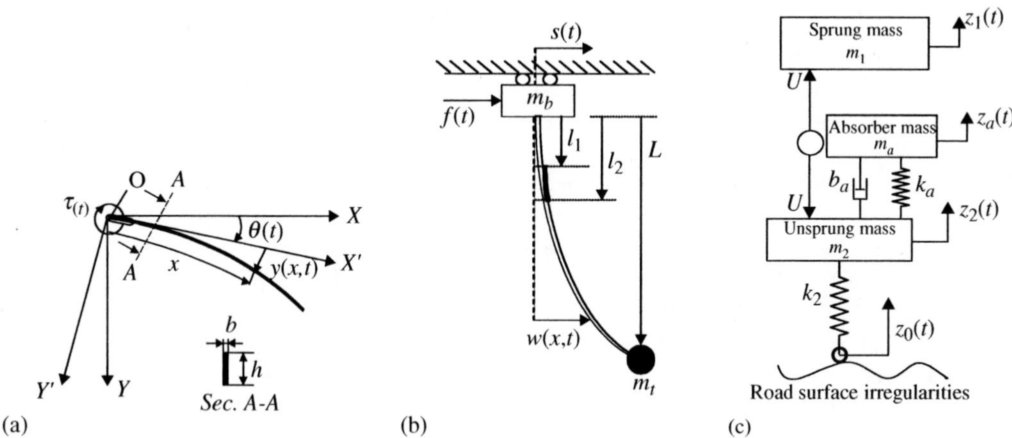

FIGURE 5.2 A comparative schematic of vibration-control systems: (a) single-input simultaneous tracking and vibration control; (b) multi-input tracking and vibration control and (c) a two-DoF vehicle model with dynamic vibration absorber.

external power required for the vibration-control system to perform its function. A passive vibration control consists of a resilient member (stiffness) and an energy dissipater (damper) either to absorb vibratory energy or to load the transmission path of the disturbing vibration (Korenev and Reznikov, 1993; Figure 5.3a). This type of vibration-control system performs best within the frequency region of its highest sensitivity. For wideband excitation frequency, its performance can be improved considerably by optimizing the system parameters (Puksand, 1975; Warburton and Ayorinde, 1980; Esmailzadeh and Jalili, 1998a). However, this improvement is achieved at the cost of lowering narrowband suppression characteristics.

The passive vibration control has significant limitations in structural applications where broadband disturbances of highly uncertain nature are encountered. In order to compensate for these limitations, active vibration-control systems are utilized. With an additional active force introduced as a part of absorber subsection, $u(t)$ (Figure 5.3b), the system is controlled using different algorithms to make it more responsive to source of disturbances (Soong and Constantinou, 1994; Olgac and Holm-Hansen, 1995; Sun et al., 1995; Margolis, 1998). The SA vibration-control system, a combination of active and passive treatment, is intended to reduce the amount of external power necessary to achieve the desired performance characteristics (Lee-Glauser et al., 1997; Jalili, 2000; Jalili and Esmailzadeh, 2002), see Figure 5.3c.

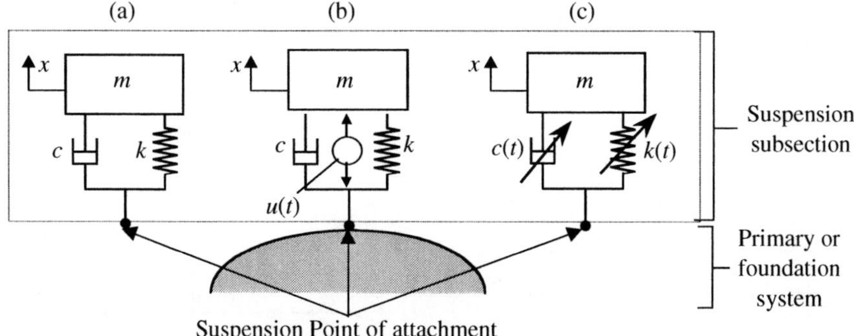

FIGURE 5.3 A typical primary structure equipped with three versions of suspension systems: (a) passive; (b) active and (c) SA configurations.

5.1.4 Performance Characteristics of Vibration-Control Systems

In the design of a vibration-control system, it often occurs that the system is required to operate over a wideband load and frequency range that is impossible to meet with a single choice of required stiffness and damping. If the desired response characteristics cannot be obtained, an active vibration-control system may provide an attractive alternative vibration control for such broadband disturbances. However, active vibration-control systems suffer from control-induced instability in addition to the large control effort requirement. This is a serious concern, which prevents them from the common usage in most industrial applications. On the other hand, passive systems are often hampered by a phenomenon known as "detuning." Detuning implies that the passive system is no longer effective in suppressing the vibration it was designed for. This occurs due to one of the following reasons: (1) the vibration-control system may deteriorate and its structural parameters can be far from the original nominal design, (2) the structural parameters of the primary device itself may alter, or (3) the excitation frequency or the nature of disturbance may change over time.

A semiactive (also known as adaptive-passive) vibration-control system addresses these limitations by effectively integrating a tuning control scheme with tunable passive devices. For this, active force generators are replaced by modulated variable compartments such as variable rate damper and stiffness (see Figure 5.3c; Hrovat et al., 1988; Nemir et al., 1994; Franchek et al., 1995). These variable components are referred to as "tunable parameters" of the suspension system, which are retailored *via* a tuning control, thus resulting in semiactively inducing optimal operation. Much attention is being paid to these systems because of their low energy requirement and cost. Recent advances in smart materials, and adjustable dampers and absorbers have significantly contributed to applicability of these systems (Garcia et al., 1992; Wang et al., 1996; Shaw, 1998).

5.2 Vibration-Control Systems Concept

5.2.1 Introduction

With a history of almost a century (Frahm, 1911), the dynamic vibration absorber has proven to be a useful vibration-suppression device, widely used in hundreds of diverse applications. It is elastically attached to the vibrating body to alleviate detrimental oscillations from its point of attachment (see Figure 5.3). This section overviews the conceptual design and theoretical background of three types of vibration-control systems, namely the passive, active and SA configurations, along with some related practical implementations.

5.2.2 Passive Vibration Control

The underlying proposition in all vibration control or absorber systems is to adjust properly the absorber parameters such that the system becomes absorbent of the vibratory energy within the frequency interval of interest. In order to explain the underlying concept, a single-degree-of-freedom (single-DoF) primary system with a single-DoF absorber attachment is considered (Figure 5.4). The governing dynamics is expressed as

$$m_a\ddot{x}_a(t) + c_a\dot{x}_a(t) + k_a x_a(t) = c_a\dot{x}_p(t) + k_a x_p(t) \quad (5.1)$$

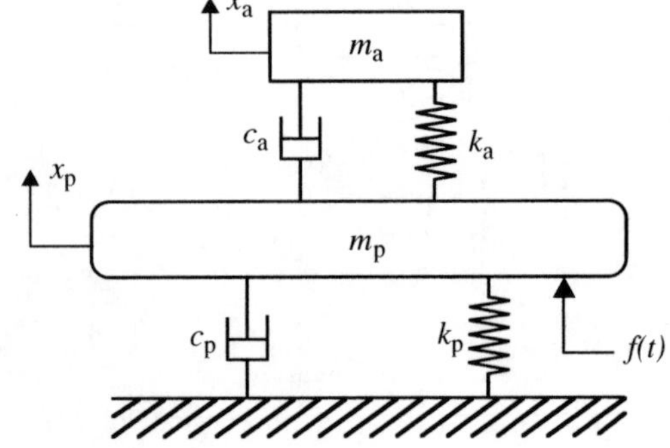

FIGURE 5.4 Application of a passive absorber to single-DoF primary system.

$$m_p\ddot{x}_p(t) + (c_p + c_a)\dot{x}_p(t) + (k_p + k_a)x_p(t) - c_a\dot{x}_a(t) - k_a x_a(t) = f(t) \tag{5.2}$$

where $x_p(t)$ and $x_a(t)$ are the respective primary and absorber displacements, $f(t)$ is the external force, and the rest of the parameters including absorber stiffness, k_a, and damping, c_a, are defined as per Figure 5.4. The transfer function between the excitation force and primary system displacement in the Laplace domain is then written as

$$\text{TF}(s) = \frac{X_p(s)}{F(s)} = \left\{ \frac{m_a s^2 + c_a s + k_a}{H(s)} \right\} \tag{5.3}$$

where

$$H(s) = \{m_p s^2 + (c_p + c_a)s + k_p + k_a\}(m_a s^2 + c_a s + k_a) - (c_a s + k_a)^2 \tag{5.4}$$

and $X_a(s)$, $X_p(s)$, and $F(s)$ are the Laplace transformations of $x_a(t)$, $x_p(t)$, and $f(t)$, respectively.

5.2.2.1 Harmonic Excitation

When excitation is tonal, the absorber is generally tuned at the disturbance frequency. For this case, the steady-state displacement of the system due to harmonic excitation can be expressed as

$$\left| \frac{X_p(j\omega)}{F(j\omega)} \right| = \left| \frac{k_a - m_a\omega^2 + jc_a\omega}{H(j\omega)} \right| \tag{5.5}$$

where ω is the disturbance frequency and $j = \sqrt{-1}$. An appropriate parameter tuning scheme can then be selected to minimize the vibration of primary system subject to external disturbance, $f(t)$.

For complete vibration attenuation, the steady state, $|X_p(j\omega)|$, must equal zero. Consequently, from Equation 5.5, the ideal stiffness and damping of absorber are selected as

$$k_a = m_a\omega^2, \quad c_a = 0 \tag{5.6}$$

Notice that this tuned condition is only a function of absorber elements (m_a, k_a, and c_a). That is, the absorber tuning does not need information from the primary system and hence its design is stand alone. For tonal application, theoretically, zero damping in the absorber subsection results in improved performance. In practice, however, the damping is incorporated in order to maintain a reasonable trade-off between the absorber mass and its displacement. Hence, the design effort for this class of application is focused on having precise tuning of the absorber to the disturbance frequency and controlling the damping to an appropriate level. Referring to Snowdon (1968), it can be proven that the absorber, in the presence of damping, can be most favorably tuned and damped if adjustable stiffness and damping are selected as

$$k_{opt} = \frac{m_a m_p^2 \omega^2}{(m_a + m_p)^2}, \quad c_{opt} = m_a\sqrt{\frac{3k_{opt}}{2(m_a + m_p)}} \tag{5.7}$$

5.2.2.2 Broadband Excitation

In broadband vibration control, the absorber subsection is generally designed to add damping to and change the resonant characteristics of the primary structure in order to dissipate vibrational energy maximally over a range of frequencies. The objective of the absorber design is, therefore, to adjust the *absorber parameters* to minimize the peak magnitude of the frequency transfer function ($\text{FTF}(\omega) = |\text{TF}(s)|_{s=j\omega}$) over the absorber parameters vector $\mathbf{p} = \{c_a\ k_a\}^T$. That is, we seek $\mathbf{p}$ to

$$\min_{\mathbf{p}} \{ \max_{\omega_{min} \le \omega \le \omega_{max}} \{|\text{FTF}(\omega)|\}\} \tag{5.8}$$

Alternatively, one may select the mean square displacement response (MSDR) of the primary system for vibration-suppression performance. That is, the absorber parameters vector, $\mathbf{p}$, is selected such that the MSDR

$$E\{(\bar{x}_p)^2\} = \int_0^\infty \{\mathrm{FTF}(\omega)\}^2 S(\omega)\mathrm{d}\omega \tag{5.9}$$

is minimized over a desired wideband frequency range. $S(\omega)$ is the power spectral density of the excitation force, $f(t)$, and FTF was defined earlier.

This optimization is subjected to some constraints in $\mathbf{p}$ space, where only positive elements are acceptable. Once the optimal absorber suspension properties, c_a and k_a, are determined, they can be implemented using adjustment mechanisms on the spring and the damper elements. This is viewed as a SA adjustment procedure as it adds no energy to the dynamic structure. The conceptual devices for such adjustable suspension elements and SA treatment will be discussed later in Section 5.2.5.

5.2.2.3 Example Case Study

To better recognize the effectiveness of the dynamic vibration absorber over the passive and optimum passive absorber settings, a simple example case is presented. For the simple system shown in Figure 5.4, the following nominal structural parameters (marked by an overscore) are taken:

$$\begin{aligned} \bar{m}_p &= 5.77 \text{ kg}, \quad \bar{k}_p = 251.132 \times 10^6 \text{ N/m}, \quad \bar{c}_p = 197.92 \text{ kg/sec} \\ \bar{m}_a &= 0.227 \text{ kg}, \quad \bar{k}_a = 9.81 \times 10^6 \text{ N/m}, \quad \bar{c}_a = 355.6 \text{ kg/sec} \end{aligned} \tag{5.10}$$

These are from an actual test setting, which is optimal by design (Olgac and Jalili, 1999). That is, the peak of the FTF is minimized (see thin lines in Figure 5.5). When the primary stiffness and damping increase 5% (for instance during the operation), the FTF of the primary system deteriorates considerably (the dashed line in Figure 5.5), and the absorber is no longer an optimum one for the present primary. When the absorber is optimized based on optimization problem 8, the retuned setting is reached as

$$k_a = 10.29 \times 10^6 \text{ N/m}, \quad c_a = 364.2 \text{ kg/sec} \tag{5.11}$$

which yields a much better frequency response (see dark line in Figure 5.5).

The vibration absorber effectiveness is better demonstrated at different frequencies by frequency sweep test. For this, the excitation amplitude is kept fixed at unity and its frequency changes every 0.15 sec from

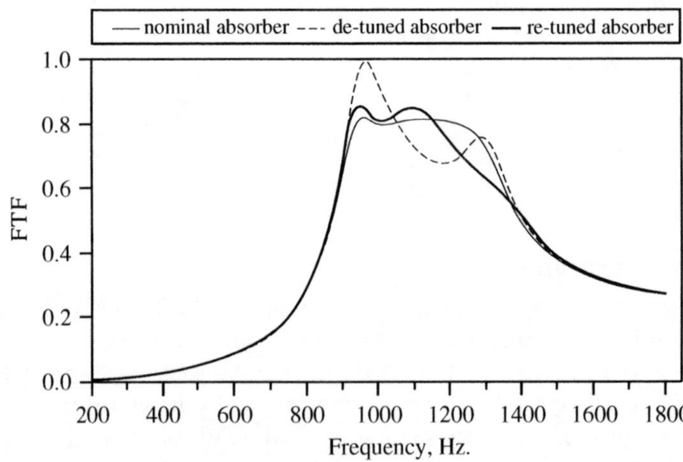

FIGURE 5.5 Frequency transfer functions (FTFs) for nominal absorber (thin-solid line), detuned absorber (thin-dotted line), and retuned absorber (thick-solid line) settings. (*Source:* From Jalili, N. and Olgac, N., *AIAA J. Guidance Control Dyn.*, 23, 961–970, 2000a. With permission.)

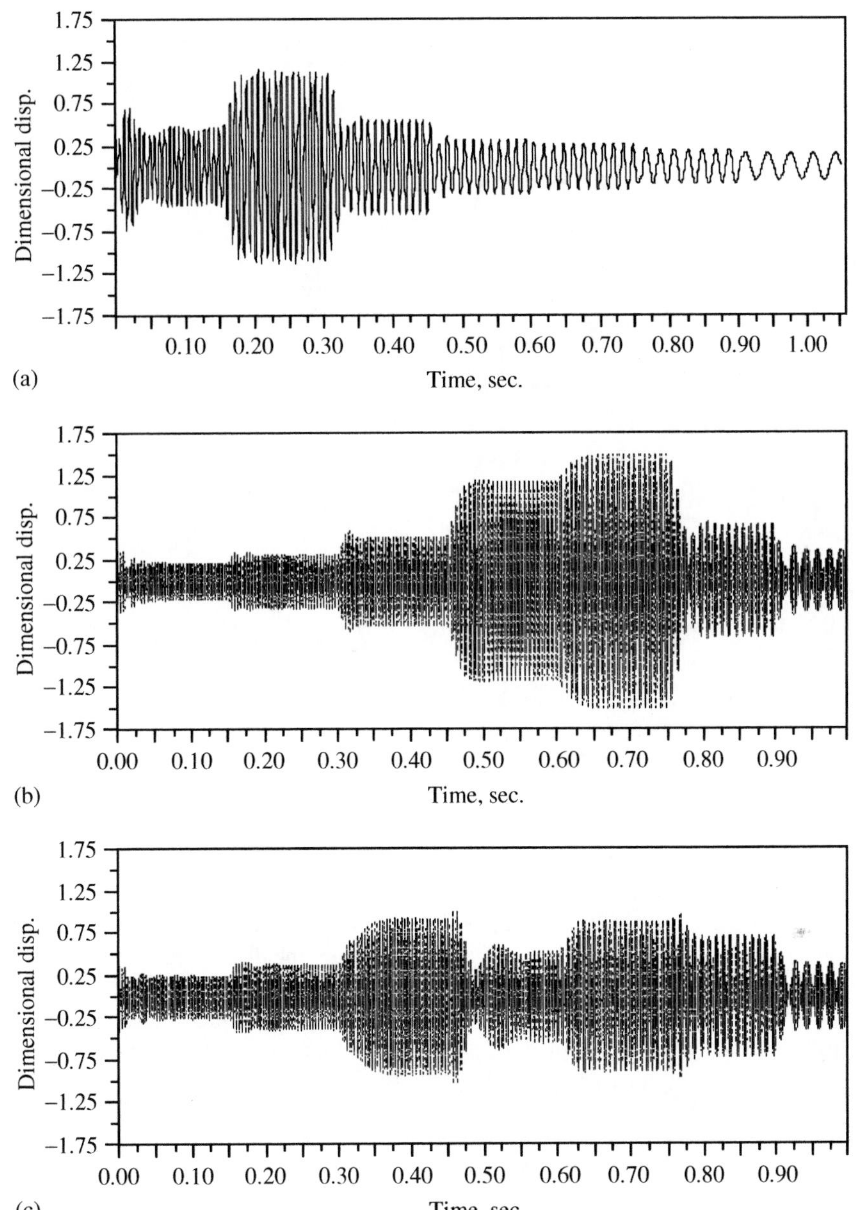

FIGURE 5.6 Frequency sweep each 0.15 with frequency change of 1860, 1880, 1900, 1920, 1930, 1950, and 1970 Hz: (a) nominally tuned absorber settings; (b) detuned absorber settings and (c) retuned absorber settings. (*Source:* From Jalili, N. and Olgac, N., *AIAA J. Guidance Control Dyn.*, 23, 961–970, 2000a. With permission.)

1860 to 1970 Hz. The primary responses with nominally tuned, with detuned, and with retuned absorber settings are given in Figure 5.6a–c, respectively.

5.2.3 Active Vibration Control

As discussed, passive absorption utilizes resistive or reactive devices either to absorb vibrational energy or load the transmission path of the disturbing vibration (Korenev and Reznikov, 1993; see Figure 5.7, top). Even with optimum absorber parameters (Warburton and Ayorinde, 1980;

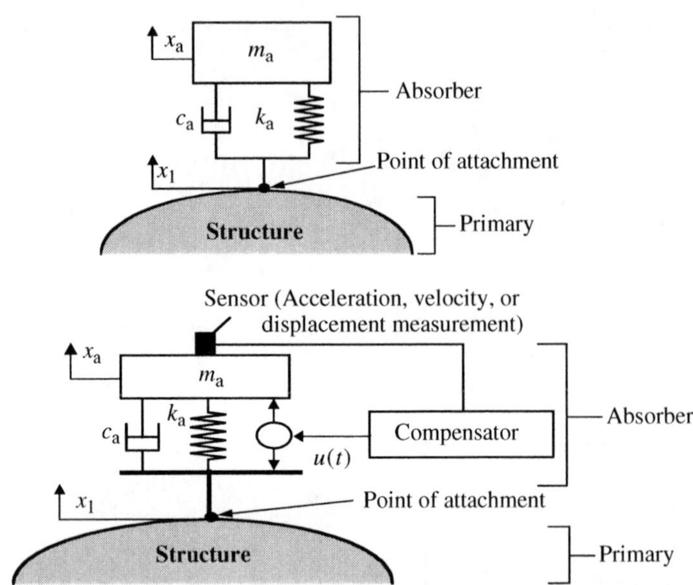

FIGURE 5.7 A general primary structure with passive (top) and active (bottom) absorber settings.

Esmailzadeh and Jalili, 1998a), the passive absorption has significant limitations in structural applications where broadband disturbances of highly uncertain nature are encountered.

In order to compensate for these limitations, active vibration-suppression schemes are utilized. With an additional active force, $u(t)$ (Figure 5.7, bottom), the absorber is controlled using different algorithms to make it more responsive to primary disturbances (Sun et al., 1995; Margolis, 1998; Jalili and Olgac, 1999). One novel implementation of the tuned vibration absorbers is the active resonator absorber (ARA) (Knowles et al., 2001b). The concept of the ARA is closely related to the concept of the delayed resonator (Olgac and Holm-Hansen, 1994; Olgac, 1995). Using a simple position (or velocity or acceleration) feedback control within the absorber subsection, the delayed resonator enforces that the dominant characteristic roots of the absorber subsection be on the imaginary axis, hence leading to resonance. Once the ARA becomes resonant, it creates perfect vibration absorption at this frequency. The conceptual design and implementation issues of such active vibration-control systems, along with their practical applications, are discussed in Section 5.3.

5.2.4 Semiactive Vibration Control

Semiactive (SA) vibration-control systems can achieve the majority of the performance characteristics of fully active systems, thus allowing for a wide class of applications. The idea of SA suspension is very simple: to replace active force generators with continually adjustable elements which can vary and/or shift the rate of the energy dissipation in response to instantaneous condition of motion (Jalili, 2002).

5.2.5 Adjustable Vibration-Control Elements

Adjustable vibration-control elements are typically comprised of variable rate damper and stiffness. Significant efforts have been devoted to the development and implementation of such devices for a variety of applications. Examples of such devices include electro-rheological (ER) (Petek, 1992; Wang et al., 1994; Choi, 1999), magneto-rheological (MR) (Spencer et al., 1998; Kim and Jeon, 2000) fluid dampers, and variable orifice dampers (Sun and Parker, 1993), controllable friction braces (Dowell and Cherry, 1994), and variable stiffness and inertia devices (Walsh and Lamnacusa, 1992;

Nemir et al., 1994; Franchek et al., 1995; Abe and Igusa, 1996). The conceptual devices for such adjustable properties are briefly reviewed in this section.

5.2.5.1 Variable Rate Dampers

A common and very effective way to reduce transient and steady-state vibration is to change the amount of damping in the SA vibration-control system. Considerable design work on SA damping was done in the 1960s to the 1980s (Crosby and Karnopp, 1973; Karnopp et al., 1974) for vibration control of civil structures such as buildings and bridges (Hrovat et al., 1983) and for reducing machine tool oscillations (Tanaka and Kikushima, 1992). Since then, SA dampers have been utilized in diverse applications ranging from trains (Stribersky et al., 1998) and other off-road vehicles (Horton and Crolla, 1986) to military tanks (Miller and Nobles, 1988). During recent years, there has been considerable interest in the SA concept in the industry for improvement and refinements of the concept (Karnopp, 1990; Emura et al., 1994). Recent advances in smart materials have led to the development of new SA dampers, which are widely used in different applications.

In view of these SA dampers, ER and MR fluids probably serve as the best potential hardware alternatives for the more conventional variable-orifice hydraulic dampers (Sturk et al., 1995). From a practical standpoint, the MR concept appears more promising for suspension, since it can operate, for instance, on vehicle battery voltage, whereas the ER damper is based on high-voltage electric fields. Owing to their importance in today's SA damper technology, we briefly review the operation and fundamental principles of SA dampers here.

5.2.5.1.1 Electro-Rheological Fluid Dampers

ER fluids are materials that undergo significant instantaneous reversible changes in material characteristics when subjected to electric potentials (Figure 5.8). The most significant change is associated with complex shear moduli of the material, and hence ER fluids can be usefully exploited in SA absorbers where variable-rate dampers are utilized. Originally, the idea of applying an ER damper to vibration control was initiated in automobile suspensions, followed by other applications (Austin, 1993; Petek et al., 1995).

The flow motions of an ER fluid-based damper can be classified by shear mode, flow mode, and squeeze mode. However, the rheological property of ER fluid is evaluated in the shear mode (Choi, 1999). As a result, the ER fluid damper provides an adaptive viscous and frictional damping for use in SA system (Dimarogonas-Andrew and Kollias, 1993; Wang et al., 1994).

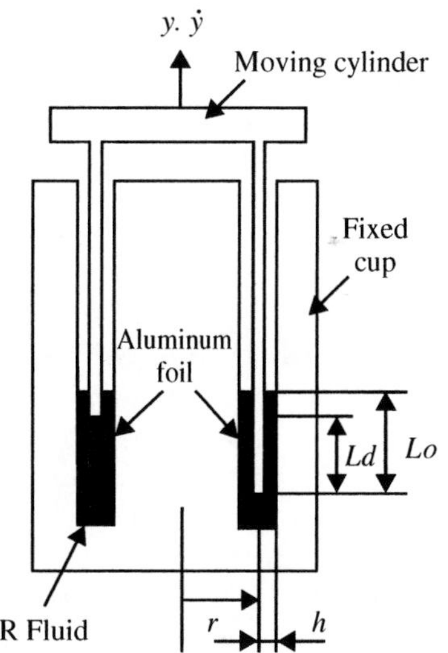

FIGURE 5.8 A schematic configuration of an ER damper. (*Source:* From Choi, S.B., *ASME J. Dyn. Syst. Meas. Control*, 121, 134–138, 1999. With permission.)

5.2.5.1.2 Magneto-Rheological Fluid Dampers

MR fluids are the magnetic analogies of ER fluids and typically consist of micron-sized, magnetically polarizable particles dispersed in a carrier medium such as mineral or silicon oil. When a magnetic field is applied, particle chains form and the fluid becomes a semisolid, exhibiting plastic behavior similar to that of ER fluids (Figure 5.9). Transition to rheological equilibrium can be achieved in a few milliseconds, providing devices with high bandwidth (Spencer et al., 1998; Kim and Jeon, 2000).

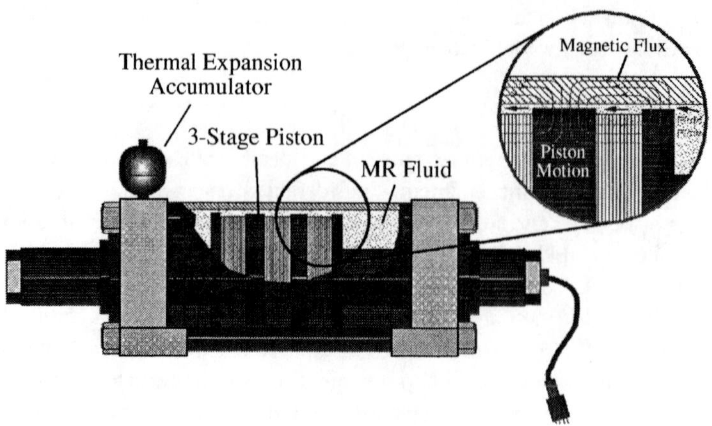

FIGURE 5.9 A schematic configuration of an MR damper. (*Source:* From Spencer, B.F. et al., *Proc. 2nd World Conf. on Structural Control*, 1998. With permission.)

5.2.5.2 Variable-Rate Spring Elements

In contrast to variable dampers, studies of SA springs or time-varying stiffness have also been geared for vibration-isolation applications (Hubard and Marolis, 1976), for structural controls and for vibration attenuation (Sun et al., 1995 and references therein). The variable stiffness is a promising practical complement to SA damping, since, based on the discussion in Section 5.2, both the absorber damping and stiffness should change to adapt optimally to different conditions. Clearly, the absorber stiffness has a significant influence on optimum operation (and even more compared to the damping element; Jalili and Olgac, 2000b).

Unlike the variable rate damper, changing the effective stiffness requires high energy (Walsh and Lamnacusa, 1992). Semiactive or low-power implementation of variable stiffness techniques suffers from limited frequency range, complex implementation, high cost, and so on. (Nemir et al., 1994; Franchek et al., 1995). Therefore, in practice, both absorber damping and stiffness are concurrently adjusted to reduce the required energy.

5.2.5.2.1 Variable-Rate Stiffness (Direct Methods)

The primary objective is to directly change the spring stiffness to optimize a vibration-suppression characteristic such as the one given in Equation 5.8 or Equation 5.9. Different techniques can be utilized ranging from traditional variable leaf spring to smart spring utilizing magnetostrictive materials. A tunable stiffness vibration absorber was utilized for a four-DoF building (Figure 5.10), where a spring is threaded through a collar plate and attached to the absorber mass from one side and to the driving gear from the other side (Franchek et al., 1995). Thus, the effective number of coils, N, can be changed resulting in a variable spring stiffness, k_a:

$$k_a = \frac{d^4 G}{8D^3 N} \tag{5.12}$$

where d is the spring wire diameter, D is the spring diameter, and G is modulus of shear rigidity.

5.2.5.2.2 Variable-Rate Effective Stiffness (Indirect Methods)

In most SA applications, directly changing the stiffness might not be always possible or may require large amount of control effort. For such cases, alternatives methods are utilized to change the effective tuning ratio ($\tau = \sqrt{k_a/m_a}/\omega_{primary}$), thus resulting in a tunable resonant frequency.

In Liu et al. (2000), a SA flutter-suppression scheme was proposed using differential changes of external store stiffness. As shown in Figure 5.11, the motor drives the guide screw to rotate with slide block, G, moving along it, thus changing the restoring moment and resulting in a change of store

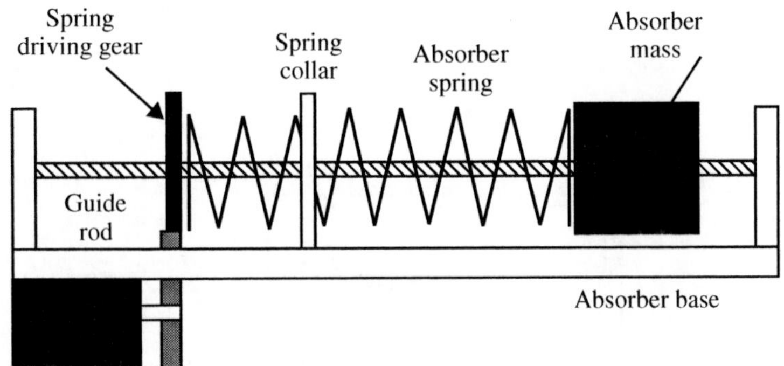

FIGURE 5.10 The application of a variable-stiffness vibration absorber to a four-DoF building. (*Source:* From Franchek, M.A. et al., *J. Sound Vib.*, 189, 565–585, 1995. With permission.)

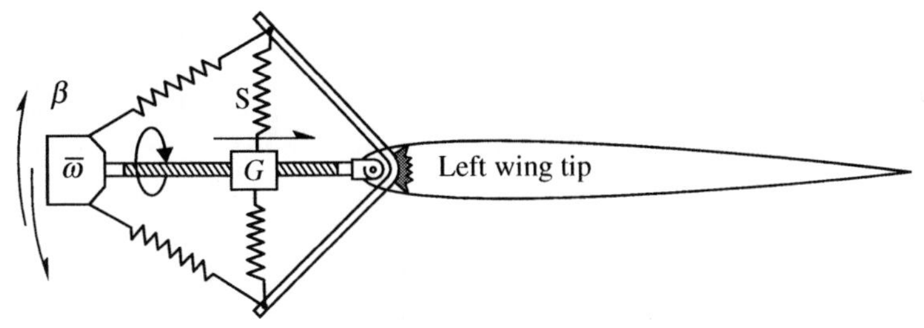

FIGURE 5.11 A SA flutter control using adjustable pitching stiffness. (*Source:* From Liu, H.J. et al., *J. Sound Vib.*, 229, 199–205, 2000. With permission.)

pitching stiffness. Using a double-ended cantilever beam carrying intermediate lumped masses, a SA vibration absorber was recently introduced (Jalili and Esmailzadeh, 2002), where positions of moving masses are adjustable (see Figure 5.12). Figure 5.13 shows an SA absorber with an adjustable effective inertia mechanism (Jalili et al., 2001; Jalili and Fallahi, 2002). The SA absorber consists of a rod carrying a moving block and a spring and damper, which are mounted on a casing. The position of the moving block, r_v, on the rod is adjustable which provides a tunable resonant frequency.

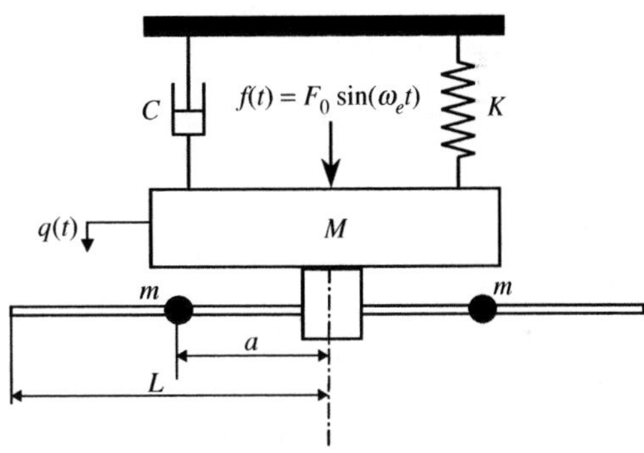

FIGURE 5.12 A typical primary system equipped with the double-ended cantilever absorber with adjustable tuning ration through moving masses, *m*. (*Source:* From Jalili, N. and Esmailzadeh, E., *J. Multi-Body Dyn.*, 216, 223–235, 2002. With permission.)

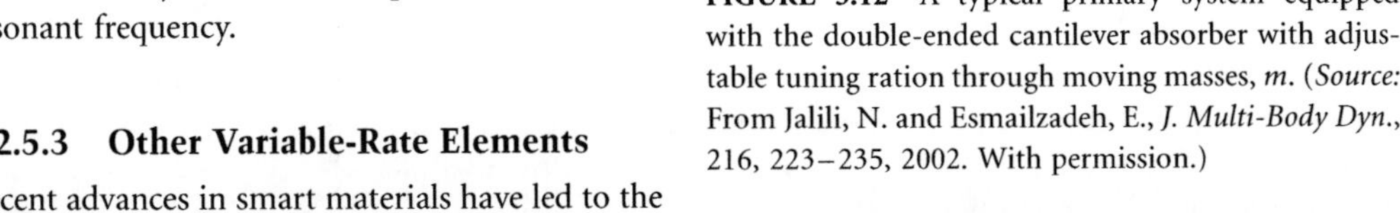

5.2.5.3 Other Variable-Rate Elements

Recent advances in smart materials have led to the development of new SA vibration-control systems using indirect influence on the suspension elements. Wang et al. used a SA piezoelectric network (1996) for structural-vibration control. The variable resistance and inductance in an external RL circuit are used as real-time adaptable control parameters.

Another class of adjustable suspensions is the so-called hybrid treatment (Fujita et al., 1991). The hybrid design has two modes, an active mode and a passive mode. With its aim of lowering the

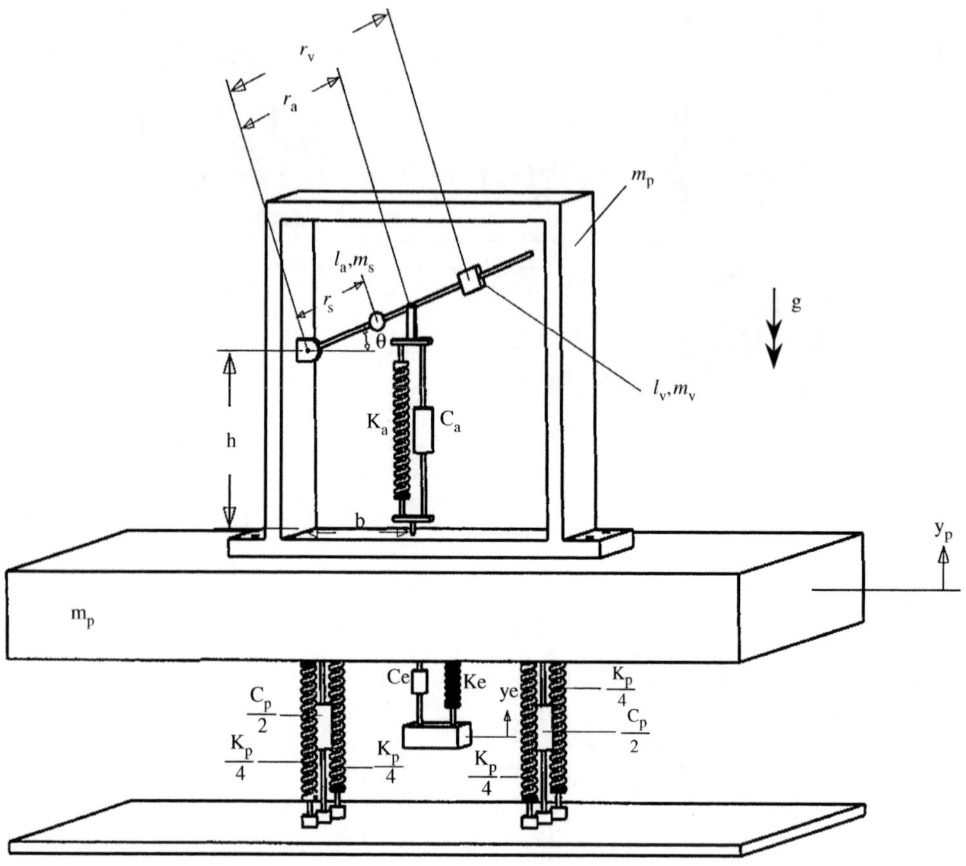

FIGURE 5.13 Schematic of the adjustable effective inertia vibration absorber. (*Source:* From Jalili, N. et al., *Int. J. Model. Simulat.*, 21, 148–154, 2001. With permission.)

control effort, relatively small vibrations are reduced in active mode, while passive mode is used for large oscillations. Analogous to hybrid treatment, the semiautomated approach combines SA and active suspensions to benefit from the advantages of individual schemes while eliminating their shortfalls (Jalili, 2000). By altering the adjustable structural properties (in the SA unit) and control parameters (in the active unit), a search is conducted to minimize an objective function subject to certain constraints, which may reflect performance characteristics.

5.3 Vibration-Control Systems Design and Implementation

5.3.1 Introduction

This section provides the basic fundamental concepts for vibration-control systems design and implementation. These systems are classified into two categories: vibration absorbers and vibration-control systems. Some related practical developments in ARAs and piezoelectric vibration control of flexible structures are also provided.

5.3.2 Vibration Absorbers

Undesirable vibrations of flexible structures have been effectively reduced using a variety of dynamic vibration absorbers. The active absorption concept offers a wideband of vibration-attenuation frequencies as well as real-time tunability as two major advantages. It is clear that the active control could be a destabilizing factor for the combined system, and therefore, the stability of the combined system (i.e., the primary and the absorber subsystems) must be assessed.

An actively tuned vibration-absorber scheme utilizing a resonator generation mechanism forms the underlying concept here. For this, a stable primary system (see Figure 5.7, top, for instance) is forced into a marginally stable one through the addition of a controlled force in the active unit (see Figure 5.7, bottom). The conceptual design for generating such resonance condition is demonstrated in Figure 5.14, where the system's dominant characteristic roots (poles) are moved and placed on the imaginary axis. The absorber then becomes a resonator capable of mimicking the vibratory energy from the primary system at the point of attachment. Although there seem to be many ways to generate such resonance, only two widely accepted practical vibration-absorber resonators are discussed next.

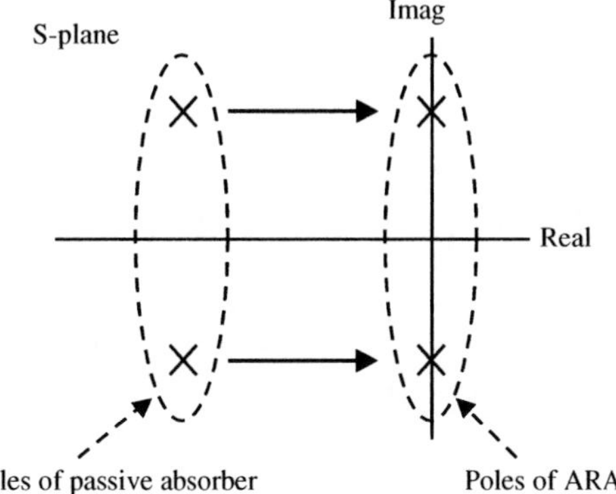

FIGURE 5.14 Schematic of the active resonator absorber concept through placing the poles of the characteristic equation on the imaginary axis.

5.3.2.1 Delayed-Resonator Vibration Absorbers

A recent active vibration-absorption strategy, the delayed resonator (DR), is considered to be the first type of active vibration absorber when operated on a flexible beam (Olgac and Jalili, 1998; Olgac and Jalili, 1999). The DR vibration absorber offers some attractive features in eliminating tonal vibrations from the objects to which it is attached (Olgac and Holm-Hansen, 1994; Olgac, 1995; Renzulli et al., 1999), some of which are real-time tunability, the stand-alone nature of the actively controlled absorber, and the simplicity of the implementation. Additionally, this single-DoF absorber can also be tuned to handle multiple frequencies of vibration (Olgac et al., 1996). It is particularly important that the combined system, that is, the primary structure and the absorber together, is asymptotically stable when the DR is implemented on a flexible beam.

5.3.2.1.1 *An Overview of the Delayed-Resonator Concept*

An overview of the DR is presented here to help the reader. The equation of motion governing the absorber dynamics alone is

$$m_a\ddot{x}_a(t) + c_a\dot{x}_a(t) + k_a x_a(t) - u(t) = 0, \quad u(t) = g\ddot{x}_a(t-\tau) \tag{5.13}$$

where $u(t)$ represents the delayed acceleration feedback. The Laplace domain transformation of this equation yields the characteristics equation

$$m_a s^2 + c_a s + k_a - g s^2 \mathrm{e}^{-\tau s} = 0 \tag{5.14}$$

Without feedback ($g = 0$), this structure is dissipative with two characteristic roots (poles) on the left half of the complex plane. For g and $\tau > 0$, however, these two finite stable roots are supplemented by infinitely many additional finite roots. Note that these characteristic roots (poles) of Equation 5.14 are discretely located (say at $s = a + j\omega$), and the following relation holds:

$$g = \frac{|m_a s^2 + c_a s + k_a|}{|s^2|}\mathrm{e}^{\tau a} \tag{5.15}$$

where $|\cdot|$ denotes the magnitude of the argument.

Using Equation 5.15, the following observation can be made:

- For $g = 0$: there are two finite stable poles and all the remaining poles are at $a = -\infty$.
- For $g = +\infty$: there are two poles at $s = 0$, and the rest are at $a = +\infty$.

Considering these and taking into account the continuity of the root loci for a given time delay, τ, and as g varies from 0 to ∞, it is obvious that the roots of Equation 5.14 move from the stable left half to the unstable right half of the complex plane. For a certain critical gain, g_c, one pair of poles reaches the imaginary axis. At this operating point, the DR becomes a perfect resonator and the imaginary characteristic roots are $s = \pm j\omega_c$, where ω_c is the resonant frequency and $j = \sqrt{-1}$. The subscript "c" implies the crossing of the root loci on the imaginary axis. The control parameters of concern, g_c and τ_c, can be found by substituting the desired $s = \pm j\omega_c$ into Equation 5.14 as

$$g_c = \frac{1}{\omega_c^2}\sqrt{(c_a\omega_c)^2 + (m_a\omega_c - k_a)^2}, \quad \tau_c = \frac{1}{\omega_c}\left\{\tan^{-1}\left[\frac{c_a\omega_c}{m_a\omega_c^2 - k_a}\right] + 2(\ell - 1)\pi\right\}, \quad \ell = 1, 2, \ldots \tag{5.16}$$

When these g_c and τ_c are used, the DR structure mimics a resonator at frequency ω_c. In turn, this resonator forms an *ideal absorber* of the tonal vibration at ω_c. The objective of the control, therefore, is to maintain the DR absorber at this marginally stable point. On the DR stability, further discussions can be found in Olgac and Holm-Hansen (1994) and Olgac et al. (1997).

5.3.2.1.2 *Vibration-Absorber Application on Flexible Beams*

We consider a general beam as the primary system with absorber attached to it and subjected to a harmonic force excitation, as shown in Figure 5.15. The point excitation is located at b, and the absorber is placed at a. A uniform cross section is considered for the beam and Euler–Bernoulli assumptions are made. The beam parameters are all assumed to be constant and uniform. The elastic deformation from the undeformed natural axis of the beam is denoted by $y(x, t)$ and, in the derivations that follow, the dot ($\cdot$) and prime ($'$) symbols indicate a partial derivative with respect to the time variable, t, and position variable x, respectively.

Under these assumptions, the kinetic energy of the system can be written as

$$T = \frac{1}{2}\rho\int_0^L \left(\frac{\partial y}{\partial t}\right)^2 \mathrm{d}x + \frac{1}{2}m_a\dot{q}_a^2 + \frac{1}{2}m_e\dot{q}_e^2 \tag{5.17}$$

The potential energy of this system using linear strain is given by

$$U = \frac{1}{2}\mathrm{EI}\int_0^L \left(\frac{\partial^2 y}{\partial x^2}\right)^2 \mathrm{d}x + \frac{1}{2}k_a\{y(a,t) - q_a\}^2 + \frac{1}{2}k_e\{y(b,t) - q_e\}^2 \tag{5.18}$$

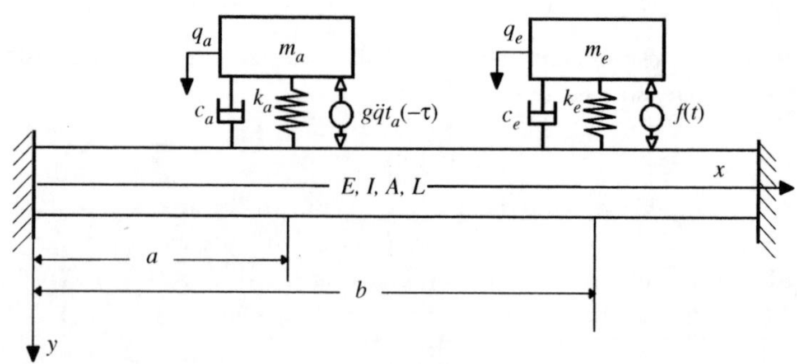

FIGURE 5.15 Beam–absorber–exciter system configuration. (*Source:* From Olgac, N. and Jalili, N., *J. Sound Vib.*, 218, 307–331, 1998. With permission.)

The equations of motion may now be derived by applying Hamilton's Principle. However, to facilitate the stability analysis, we resort to an assumed-mode expansion and Lagrange's equations. Specifically, y is written as a finite sum "Galerkin approximation":

$$y(x,t) = \sum_{i=1}^{n} \Phi_i(x) q_{bi}(t) \tag{5.19}$$

The orthogonality conditions between these mode shapes can also be derived as (Meirovitch, 1986)

$$\int_0^L \rho \Phi_i(x)\Phi_j(x)\mathrm{d}x = N_i\delta_{ij}, \quad \int_0^L \mathrm{EI}\Phi_i''(x)\Phi_j''(x)\mathrm{d}x = S_i\delta_{ij} \tag{5.20}$$

where $i,\ j = 1,\ 2, \ldots, n$, δ_{ij} is the Kronecker delta, and N_i and S_i are defined by setting $i = j$ in Equation 5.20.

The feedback of the absorber, the actuator excitation force, and the damping dissipating forces in both the absorber and the exciter are considered as non-conservative forces in Lagrange's formulation. Consequently, the equations of motion are derived.

Absorber dynamics is governed by

$$m_a\ddot{q}_a(t) + c_a\left\{\dot{q}_a(t) - \sum_{i=1}^{n} \Phi_i(a)\dot{q}_{bi}(t)\right\} + k_a\left\{q_a(t) - \sum_{i=1}^{n} \Phi_i(a)q_{bi}(t)\right\} - g\ddot{q}_a(t-\tau) = 0 \tag{5.21}$$

The exciter is given by

$$m_e\ddot{q}_e(t) + c_e\left\{\dot{q}_e(t) - \sum_{i=1}^{n} \Phi_i(b)\dot{q}_{bi}(t)\right\} + k_e\left\{q_e(t) - \sum_{i=1}^{n} \Phi_i(b)q_{bi}(t)\right\} = -f(t) \tag{5.22}$$

Finally, the beam is represented by

$$\begin{aligned} &N_i\ddot{q}_{bi}(t) + S_i q_{bi}(t) + c_a\left\{\sum_{i=1}^{n} \Phi_i(a)\dot{q}_{bi}(t) - \dot{q}_a(t)\right\}\Phi_i(a) + c_e\left\{\sum_{i=1}^{n} \Phi_i(b)\dot{q}_{bi}(t) - \dot{q}_e(t)\right\}\Phi_i(b) \\ &+ k_a\left\{\sum_{i=1}^{n} \Phi_i(a)q_{bi}(t) - q_a(t)\right\}\Phi_i(a) + k_e\left\{\sum_{i=1}^{n} \Phi_i(b)q_{bi}(t) - q_e(t)\right\}\Phi_i(b) + g\Phi_i(a)\ddot{q}_a(t-\tau) \\ &= f(t)\Phi_i(b), \quad i = 1, 2, \ldots, n \end{aligned} \tag{5.23}$$

Equation 5.21 to Equation 5.23 form a system of $n+2$ second-order coupled differential equations.

By proper selection of the feedback gain, the absorber can be tuned to the desired resonant frequency, ω_c. This condition, in turn, forces the beam to be motionless at a, when the beam is excited by a tonal force at frequency ω_c. This conclusion is reached by taking the Laplace transform of Equation 5.21 and using feedback control law for the absorber. In short,

$$Y(a,s) = \sum_{i=1}^{n} \Phi_i(a)Q_{bi}(s) = 0 \tag{5.24}$$

where $Y(a,s) = \Im\{y(a,t)\}$, $Q_a(s) = \Im\{q_a(t)\}$ and $Q_{bi}(s) = \Im\{q_{bi}(t)\}$. Equation 5.24 can be rewritten in time domain as

$$y(a,t) = \sum_{i=1}^{n} \Phi_i(a)q_{bi}(t) = 0 \tag{5.25}$$

which indicates that the steady-state vibration of the point of attachment of the absorber is eliminated. Hence, the absorber mimics a resonator at the frequency of excitation and absorbs all the vibratory energy at the point of attachment.

5.3.2.1.3 *Stability of the Combined System*

In the preceding section, we have derived the equations of motion for the beam–exciter–absorber system in its most general form. As stated before, inclusion of the feedback control for active absorption is, indeed, an invitation to instability. This topic is treated next.

The Laplace domain representation of the combined system takes the form (Olgac and Jalili, 1998)

$$\mathbf{A}(s)\mathbf{Q}(s) = \mathbf{F}(s) \tag{5.26}$$

where

$$\mathbf{Q}(s) = \begin{Bmatrix} Q_a(s) \\ Q_e(s) \\ Q_{b1}(s) \\ \vdots \\ Q_{bn}(s) \end{Bmatrix}_{(n+2)\times 1}, \quad \mathbf{F}(s) = \begin{Bmatrix} 0 \\ -F(s) \\ 0 \\ \vdots \\ 0 \end{Bmatrix}_{(n+2)\times 1},$$

$$\mathbf{A}(s) = \begin{pmatrix} m_a s^2 + c_a s + k_a - g s^2 e^{-\tau s} & 0 & -\Phi_1(a)(c_a s + k_a) & \cdots & -\Phi_n(a)(c_a s + k_a) \\ 0 & m_e s^2 + c_e s + k_e & -\Phi_1(b)(c_e s + k_e) & \cdots & -\Phi_n(b)(c_e s + k_e) \\ m_a \Phi_1(a) s^2 & m_e \Phi_1(b) s^2 & N_1 s^2 + cs + S_1(1 + j\delta) & \cdots & 0 \\ \vdots & \vdots & \vdots & \ddots & \vdots \\ m_a \Phi_n(a) s^2 & m_e \Phi_n(b) s^2 & 0 & \cdots & N_n s^2 + cs + S_n(1 + j\delta) \end{pmatrix} \tag{5.27}$$

In order to assess the combined system stability, the roots of the characteristic equation, $\det(\mathbf{A}(s)) = 0$ are analyzed. The presence of feedback (transcendental delay term for this absorber) in the characteristic equations complicates this effort. The root locus plot observation can be applied to the entire system. It is typical that increasing feedback gain causes instability as the roots move from left to right in the complex plane. This picture also yields the frequency range for stable operation of the combined system (Olgac and Jalili, 1998).

5.3.2.1.4 *Experimental Setting and Results*

The experimental setup used to verify the findings is shown in Figure 5.16. The primary structure is a 3/8′ in. × 1′ in. × 12′ in. steel beam (2) clamped at both ends to a granite bed (1). A piezoelectric actuator with a reaction mass (3 and 4) is used to generate the periodic disturbance on the beam. A similar actuator-mass setup constitutes the DR absorber (5 and 6). The two setups are located symmetrically at one quarter of the length along the beam from the center. The feedback signal used to implement the DR is obtained from the accelerometer (7) mounted on the reaction mass of the absorber structure. The other accelerometer (8) attached to the beam is present only to monitor the vibrations of the beam and to evaluate the performance of the DR absorber in suppressing them. The control is applied *via* a fast data acquisition card using a sampling of 10 kHz.

The numerical values for this beam–absorber–exciter setup are taken as

- *Beam*: $E = 210$ GPa, $\rho = 1.8895$ kg/m
- *Absorber*: $m_a = 0.183$ kg, $k_a = 10{,}130$ kN/m, $c_a = 62.25$ N sec/m, $a = L/4$
- *Exciter:* $m_e = 0.173$ kg, $k_e = 6426$ kN/m, $c_e = 3.2$ N sec/m, $b = 3L/4$

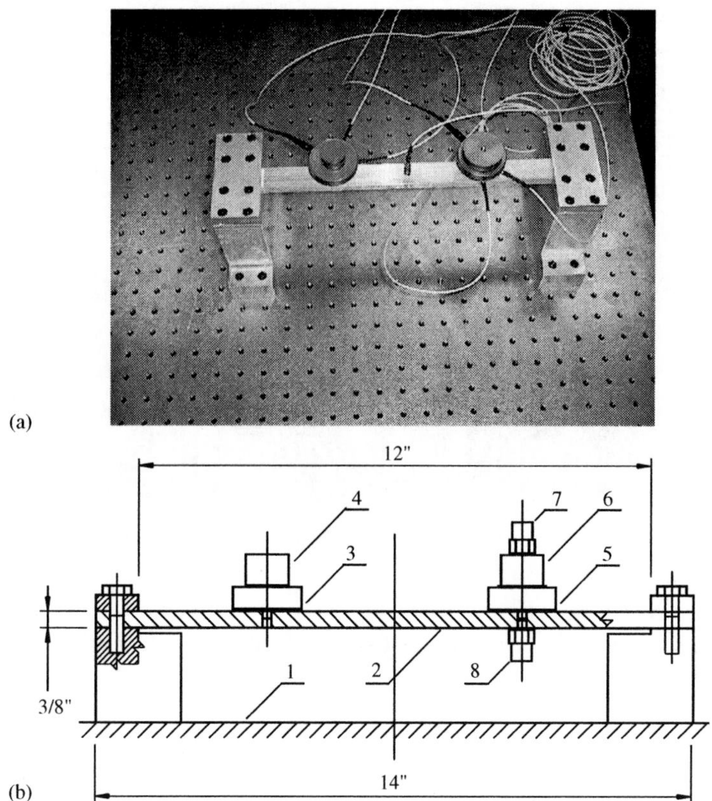

FIGURE 5.16 (a) Experimental structure and (b) schematic depiction of the setup. (*Source:* From Olgac, N. and Jalili, N., *J. Sound Vib.*, 218, 307–331, 1998. With permission.)

5.3.2.1.5 Dynamic Simulation and Comparison with Experiments

For the experimental set up at hand, the natural frequencies are measured for the first two natural modes, ω_1 and ω_2. These frequencies are obtained much more precisely than those of higher-order natural modes. Table 5.1 offers a comparison between the experimental (*real*) and analytical (*ideal*) clamped–clamped beam natural frequencies.

The discrepancies arrive from two sources: first, the experimental frequencies are structurally damped natural frequencies, and second, they reflect the effect of partially clamped BCs. The theoretical frequencies, on the other hand, are evaluated for an undamped ideal clamped–clamped beam.

After observing the effect of the number of modes used on the beam deformation, a minimum of three natural modes are taken into account. We then compare the simulated time response vs. the experimental results of vibration suppression. Figure 5.17 shows a test with the excitation frequency $\omega_c = 1270$ Hz. The corresponding theoretical control parameters are $g_{c\ \text{theory}} = 0.0252$ kg and $\tau_{c\ \text{theory}} = 0.8269$ msec. The experimental control parameters for this frequency are found to be $g_{c\ \text{exp.}} = 0.0273$ kg and $\tau_{c\ \text{exp.}} = 0.82$ msec. The exciter disturbs the beam for 5 msec, then the DR tuning is triggered. The acceleration of the beam at the point of attachment decays exponentially. For all intents and purposes, the suppression takes effect in approximately 200 ms. These results match very closely with the experimental data, Figure 5.18. The only noticeable difference is in the frequency content of the exponential decay. This property is dictated by the dominant poles of the combined system. The imaginary part, however, is smaller in the analytical study. This difference is a nuance that does not affect the earlier observations.

TABLE 5.1 Comparison between Experimental and Theoretical Beam Natural Frequencies (Hz)

Natural Modes	Peak Frequencies (Experimental)	Natural Frequencies (Clamped–Clamped)
First mode	466.4	545.5
Second mode	1269.2	1506.3

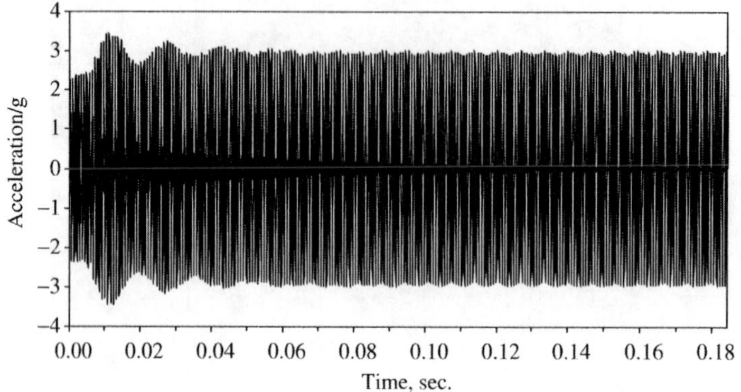

FIGURE 5.17 Beam and absorber response to 1270 Hz disturbance, analytical. (*Source:* From Olgac, N. and Jalili, N., *J. Sound Vib.*, 218, 307–331, 1998. With permission.)

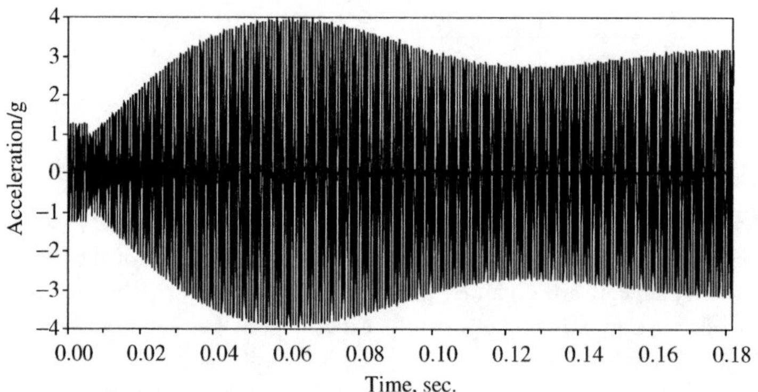

FIGURE 5.18 Beam and absorber response to 1270 Hz disturbance, experimental. (*Source:* From Olgac, N. and Jalili, N., *J. Sound Vib.*, 218, 307–331, 1998. With permission.)

5.3.2.2 Active Resonator Vibration Absorbers

One novel implementation of the tuned vibration absorbers is the ARA (Knowles et al., 2001b). The concept of the ARA is closely related to the concept of the delayed resonator (Olgac and Holm-Hansen, 1994; Olgac and Jalili, 1999). Using a simple position (or velocity or acceleration) feedback control within the absorber section, it enforces the dominant characteristic roots of the absorber subsection to be on the imaginary axis, and hence leading to resonance. Once the ARA becomes resonant, it creates perfect vibration absorption at this frequency.

A very important component of any active vibration absorber is the actuator unit. Recent advances in smart materials have led to the development of advanced actuators using piezoelectric ceramics, shape memory alloys, and magnetostrictive materials (Garcia et al., 1992; Shaw, 1998). Over the past two decades, piezoelectric ceramics have been utilized as potential replacements for conventional transducers.

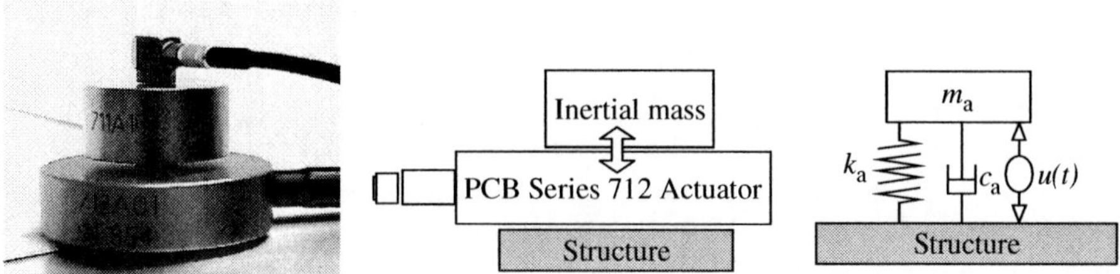

FIGURE 5.19 A PCB series 712 PZT inertial actuator (left), schematic of operation (middle), and a simple single-DoF mathematical model (right). (Active Vibration Control Instrumentation, A Division of PCB Piezotronics, Inc., www.pcb.com.)

These materials are compounds of lead zirconate–titanate (PZT). The PZT properties can be optimized to suit specific applications by appropriate adjustment of the zirconate–titanate ratio. Specifically, a piezoelectric inertial actuator is an efficient and inexpensive solution for active structural vibration control. As shown in Figure 5.19, it applies a point force to the structure to which it is attached.

5.3.2.2.1 *An Overview of PZT Inertial Actuators*

PZT inertial actuators are most commonly made out of two parallel piezoelectric plates. If voltage is applied, one of the plates expands as the other one contracts, hence producing displacement that is proportional to the input voltage. The resonance of such an actuator can be adjusted by the size of the inertial mass (see Figure 5.19). Increasing the size of the inertial mass will lower the resonant frequency and decreasing the mass will increase it. The resonant frequency, f_r, can be expressed as

$$f_r = \frac{1}{2\pi}\sqrt{\frac{k_a}{m_a}} \tag{5.28}$$

where k_a is the effective stiffness of the actuator, m_a, is defined as

$$m_a = m_{e_{PZT}} + m_{inertial} + m_{acc} \tag{5.29}$$

The PZT effective mass is $m_{e_{PZT}}$, $m_{inertial}$ is the inertial mass, and m_{acc} is the accelerometer mass. Using a simple single-DoF system (see Figure 5.19), the parameters of the PZT inertial actuators can be experimentally determined (Knowles et al., 2001a). This "parameter identification" problem is an inverse problem. We refer interested readers to Banks and Ito (1988) and Banks and Kunisch (1989) for a general introduction to parameter estimation or inverse problems governed by differential equations.

5.3.2.2.2 *Active Resonator Absorber Concept*

The concept of ARA is closely related to that of the DR (Olgac, 1995; Olgac and Holm-Hansen, 1995; Olgac and Jalili, 1998). Instead of a compensator, the DR uses a simple delayed position (or velocity, or acceleration) feedback control within the absorber subsection for the mentioned "sensitization."

In contrast to that of the DR absorber, the characteristic equation of the proposed control scheme is rational in nature and is hence easier to implement when closed-loop stability of the system is concerned. Similar to the DR absorber, the proposed ARA requires only one signal from the absorber mass, absolute or relative to the point of attachment (see Figure 5.7 bottom). After the signal is processed through a compensator, an additional force is produced, for instance, by a PZT inertial actuator. If the compensator parameters are properly set, the absorber should behave as an ideal resonator at one or even more frequencies. As a result, the resonator will absorb vibratory energy from the primary mass at given frequencies. The frequency to be absorbed can be tuned in real time. Moreover, if the controller or the actuator fails, the ARA will still function as a passive absorber, and thus it is inherently fail-safe. A similar vibration absorption methodology is given by Filipović and Schröder (1999) for linear systems. The ARA, however, is not confined to the linear regime.

For the case of linear assumption for the PZT actuator, the dynamics of the ARA (Figure 5.7, bottom) can be expressed as

$$m_a\ddot{x}_a(t) + c_a\dot{x}_a(t) + k_a x_a(t) - u(t) = c_a\dot{x}_1(t) + k_a x_1(t) \tag{5.30}$$

where $x_1(t)$ and $x_a(t)$ are the respective primary (at the absorber point of attachment) and absorber mass displacements. The mass, m_a, is given by Equation 5.29 and the control, $u(t)$, is designed to produce designated resonance frequencies within the ARA.

The objective of the feedback control, $u(t)$, is to convert the dissipative structure (Figure 5.7, top) into a conservative or marginally stable one (Figure 5.7, bottom) with a designated resonant frequency, ω_c. In other words, the control aims the placement of dominant poles at $\pm j\omega_c$ for the combined system, where $j = \sqrt{-1}$ (see Figure 5.14). As a result, the ARA becomes marginally stable at particular frequencies in the determined frequency range. Using simple position (or velocity or acceleration) feedback within the absorber section (i.e., $U(s) = \bar{U}(s)X_a(s)$), the corresponding dynamics of the ARA, given by Equation 5.30, in the Laplace domain become

$$(m_a s^2 + c_a s + k_a)X_a(s) - \bar{U}(s)X_a(s) = C(s)X_a(s) = (c_a s + k_a)X_1(s) \tag{5.31}$$

The compensator transfer function, $\bar{U}(s)$, is then selected such that the primary system displacement at the absorber point of attachment is forced to be zero; that is

$$C(s) = (m_a s^2 + c_a s + k_a) - \bar{U}(s) = 0 \tag{5.32}$$

The parameters of the compensator are determined through introducing resonance conditions to the absorber characteristic equation, $C(s)$; that is, the equations $\mathrm{Re}\{C(j\omega_i)\} = 0$ and $\mathrm{Im}\{C(j\omega_i)\} = 0$ are simultaneously solved, where $i = 1, 2, \ldots, l$ and l is the number of frequencies to be absorbed. Using additional compensator parameters, the stable frequency range or other properties can be adjusted in real time.

Consider the case where $U(s)$ is taken as a proportional compensator with a single time constant based on the acceleration of the ARA, given by

$$U(s) = \bar{U}(s)X_a(s), \quad \text{where } \bar{U}(s) = \frac{gs^2}{1 + Ts} \tag{5.33}$$

Then, in the time domain, the control force, $u(t)$, can be obtained from

$$u(t) = \frac{g}{T}\int_0^t e^{-(t-\tau)/T}\ddot{x}_a(\tau)d\tau \tag{5.34}$$

To achieve ideal resonator behavior, two dominant roots of Equation 5.32 are placed on the imaginary axis at the desired crossing frequency, ω_c. Substituting $s = \pm j\omega_c$ into Equation 5.32 and solving for the control parameters, g_c and T_c, one can obtain

$$g_c = m_a\left(\frac{c_a^2}{m_a^2\left(\omega^2 - \dfrac{k_a}{m_a}\right)} - \frac{k_a}{m_a\omega^2} + 1\right), \quad T_c = \frac{c_a\sqrt{k_a/m_a}}{\sqrt{m_a k_a}\left(\omega^2 - \dfrac{k_a}{m_a}\right)}, \quad \text{for } \omega = \omega_c \tag{5.35}$$

The control parameters, g_c and T_c, are based on the physical properties of the ARA (i.e., c_a, k_a, and m_a) as well as the frequency of the disturbance, ω, illustrating that the ARA does not require any information from the primary system to which it is attached. However, when the physical properties of the ARA are not known within a high degree of certainty, a method to autotune the control parameters must be considered. The stability assurance of such autotuning proposition will bring primary system parameters into the derivations, and hence the primary system cannot be totally decoupled. This issue will be discussed later in the chapter.

5.3.2.2.3 Application of ARA to Structural-Vibration Control

In order to demonstrate the effectiveness of the proposed ARA, a simple single-DoF primary system subjected to tonal force excitations is considered. As shown in Figure 5.20, two PZT inertial actuators are used for both the primary (model 712-A01) and the absorber (model 712-A02) subsections. Each system consists of passive elements (spring stiffness and damping properties of the PZT materials) and active compartment with the physical parameters listed in Table 5.2. The top actuator acts as the ARA with the controlled force, $u(t)$, while the bottom one represents the primary system subjected to the force excitation, $f(t)$.

The governing dynamics for the combined system can be expressed as

$$m_a\ddot{x}_a(t) + c_a\dot{x}_a(t) + k_a x_a(t) - u(t) = c_a\dot{x}_1(t) + k_a x_1(t) \tag{5.36}$$

$$m_1\ddot{x}_1(t) + (c_1 + c_a)\dot{x}_1(t) + (k_1 + k_a)x_1(t) - \{c_a\dot{x}_a(t) + k_a x_a(t) - u(t)\} = f(t) \tag{5.37}$$

where $x_1(t)$ and $x_a(t)$ are the respective primary and absorber displacements.

5.3.2.2.4 Stability Analysis and Parameter Sensitivity

The sufficient and necessary condition for asymptotic stability is that all roots of the characteristic equation have negative real parts. For the linear system, Equation 5.36 and Equation 5.37, when utilizing controller (Equation 5.34), the characteristic equation of the combined system (Figure 5.20, right) can be determined and the stability region for compensator parameters, g and T, can be obtained using the Routh–Hurwitz method.

5.3.2.2.5 Autotuning Proposition

When using the proposed ARA configuration in real applications where the physical properties are not known or vary over time, the compensator parameters, g and T, only provide partial vibration suppression. In order to remedy this, a need exists for an autotuning method to adjust the compensator

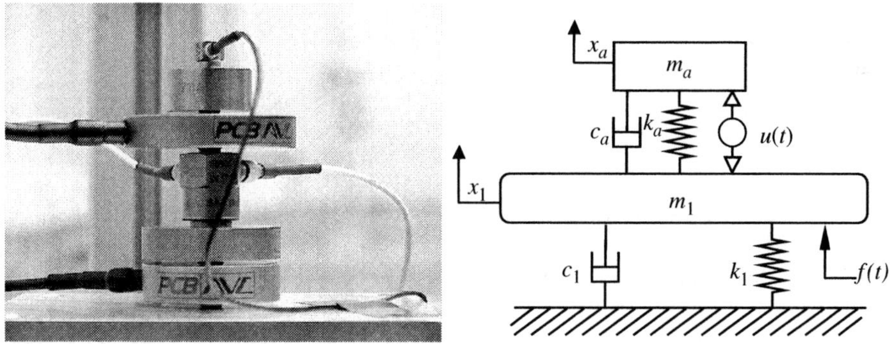

FIGURE 5.20 Implementation of the ARA concept using two PZT actuators (left) and its mathematical model (right).

TABLE 5.2 Experimentally Determined Parameters of PCB Series 712 PZT Inertial Actuators

PZT System Parameters	PCB Model 712-A01	PCB Model 712-A02
Effective mass, m_{ePZT} (gr)	7.20	12.14
Inertial mass, $m_{inertial}$ (gr)	100.00	200.00
Stiffness, k_a (kN/m)	3814.9	401.5
Damping, c_a (Ns/m)	79.49	11.48

parameters, g and T, by some quantities, Δg and ΔT, respectively (Jalili and Olgac, 1998b; Jalili and Olgac, 2000a). For the case of the linear compensator with a single time constant, given by Equation 5.33, the transfer function between primary displacement, $X_1(s)$, and absorber displacement, $X_a(s)$, can be obtained as

$$G(s) = \frac{X_1(s)}{X_a(s)} = \frac{m_a s^2 + c_a s + k_a - \dfrac{gs^2}{1+Ts}}{c_a s + k_a} \tag{5.38}$$

The transfer function can be rewritten in the frequency domain for $s = \mathrm{j}\omega$ as

$$G(\mathrm{j}\omega) = \frac{X_1(\mathrm{j}\omega)}{X_a(\mathrm{j}\omega)} = \frac{-m_a\omega^2 + c_a\omega\mathrm{j} + k_a + \dfrac{g\omega^2}{1+T\omega\mathrm{j}}}{c_a\omega\mathrm{j} + k_a} \tag{5.39}$$

where $G(\mathrm{j}\omega)$ can be obtained in real time by convolution of accelerometer readings (Renzulli et al., 1999) or other methods (Jalili and Olgac, 2000a). Following a similar procedure as is utilized in Renzulli et al. (1999), the numerator of the transfer function (Equation 5.39) must approach zero in order to suppress primary system vibration. This is accomplished by setting

$$G(\mathrm{j}\omega) + \Delta G(\mathrm{j}\omega) = 0 \tag{5.40}$$

where $G(\mathrm{j}\omega)$ is the real-time transfer function and $\Delta G(\mathrm{j}\omega)$ can be written as a variational form of Equation 5.39 as

$$\Delta G(\omega \mathrm{i}) = \frac{\partial G}{\partial g}\Delta g + \frac{\partial G}{\partial T}\Delta T + \text{higher order terms} \tag{5.41}$$

Since the estimated physical parameters of the absorber (i.e., c_a, k_a, and m_a) are within the vicinity of the actual parameters, Δg and ΔT should be small quantities and the higher-order terms of Equation 5.41 can be neglected. Using Equation 5.40 and Equation 5.41 and neglecting higher-order terms, we have

$$\begin{aligned} \Delta g &= \mathrm{Re}[G(\mathrm{j}\omega)]\left[\frac{2Tc_a\omega^2 - k_a + k_aT^2\omega^2}{\omega^2}\right] + \mathrm{Im}[G(\mathrm{j}\omega)]\left[\frac{c_a - T^2\omega^2 c_a + 2k_aT}{\omega^2}\right], \\ \Delta T &= \mathrm{Re}[G(\mathrm{j}\omega)]\left[\frac{c_a - T^2\omega^2 c_a + 2k_aT}{g\omega^2}\right] + \mathrm{Im}[G(\mathrm{j}\omega)]\left[\frac{k_a - 2Tc_a\omega^2 - k_aT^2\omega^2}{g\omega^3}\right] + \frac{T}{g}\Delta g \end{aligned} \tag{5.42}$$

In the above expressions, g and $T$are the current compensator parameters given by Equation 5.35, c_a, k_a, and m_a are the estimated absorber parameters, ω is the absorber base excitation frequency, and $G(\mathrm{j}\omega)$ is the transfer function obtained in real time. That is, the retuned control parameters, g and T, are determined as follows

$$g_{\text{new}} = g_{\text{current}} + \Delta g \quad \text{and} \quad T_{\text{new}} = T_{\text{current}} + \Delta T \tag{5.43}$$

where Δg and $\Delta T$are those given by Equation 5.42. After compensator parameters, g and T, are adjusted by Equation 5.43, the process can be repeated until $|G(\mathrm{j}\omega)|$ falls within the desired level of tolerance. $G(\mathrm{j}\omega)$ can be determined in real time as shown in Liu et al. (1997) by

$$G(\mathrm{j}\omega) = |G(\mathrm{j}\omega)|e^{(\mathrm{j}\phi(\mathrm{j}\omega))} \tag{5.44}$$

where the magnitude and phase are determined assuming that the absorber and primary displacements are harmonic functions of time given by

$$x_a(t) = X_a \sin(\omega t + \phi_a), \quad x_1(t) = X_1 \sin(\omega t + \phi_1) \tag{5.45}$$

With the magnitudes and phase angles of Equation 5.44, the transfer function can be determined from Equation 5.44 and the following:

$$|G(j\omega)| = \frac{X_1}{X_a}, \quad \phi(j\omega) = \phi_1 - \phi_a$$

5.3.2.2.6 Numerical Simulations and Discussions

To illustrate the feasibility of the proposed absorption methodology, an example case study is presented. The ARA control law is the proportional compensator with a single time constant as given in Equation 5.34. The primary system is subjected to a harmonic excitation with unit amplitude and a frequency of 800 Hz. The ARA and primary system parameters are taken as those given in Table 5.2. The simulation was done using Matlab/Simulink® and the results for the primary system and the absorber displacements are given in Figure 5.21.

As seen, vibrations are completely suppressed in the primary subsection after approximately 0.05 sec, at which the absorber acts as a marginally stable resonator. For this case, all physical parameters are assumed to be known exactly. However, in practice these parameters are not known exactly and may vary with time, so the case with estimated system parameters must be considered.

To demonstrate the feasibility of the proposed autotuning method, the nominal system parameters (m_a, m_1, k_a, k_1, c_a, c_1) were fictitiously perturbed by 10% (i.e., representing the actual values) in the simulation. However, the nominal values of m_a, m_1, k_a, k_1, c_a, and c_1 were used for calculation of the compensator parameters, g and T. The results of the simulation using nominal parameters are given in Figure 5.22. From Figure 5.22, top, the effect of parameter variation is shown as steady-state oscillations of the primary structure. This undesirable response will undoubtedly be encountered when the experiment is implemented. Thus, an autotuning procedure is needed.

The result of the first autotuning iteration is given in Figure 5.22, middle, where the control parameters, g and T, are adjusted based on Equation 5.43. One can see tremendous improvement in the primary system response with only one iteration (see Figure 5.22, middle). A second iteration is

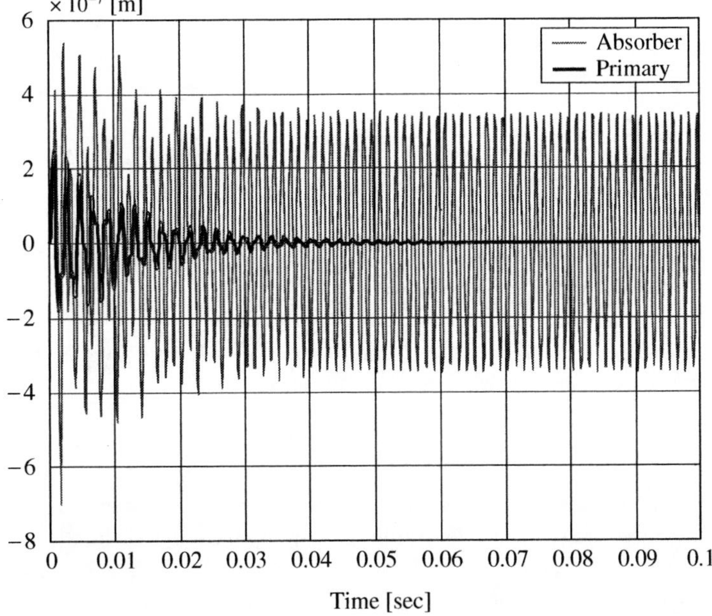

FIGURE 5.21 Primary system and absorber displacements subjected to 800 Hz harmonic disturbance.

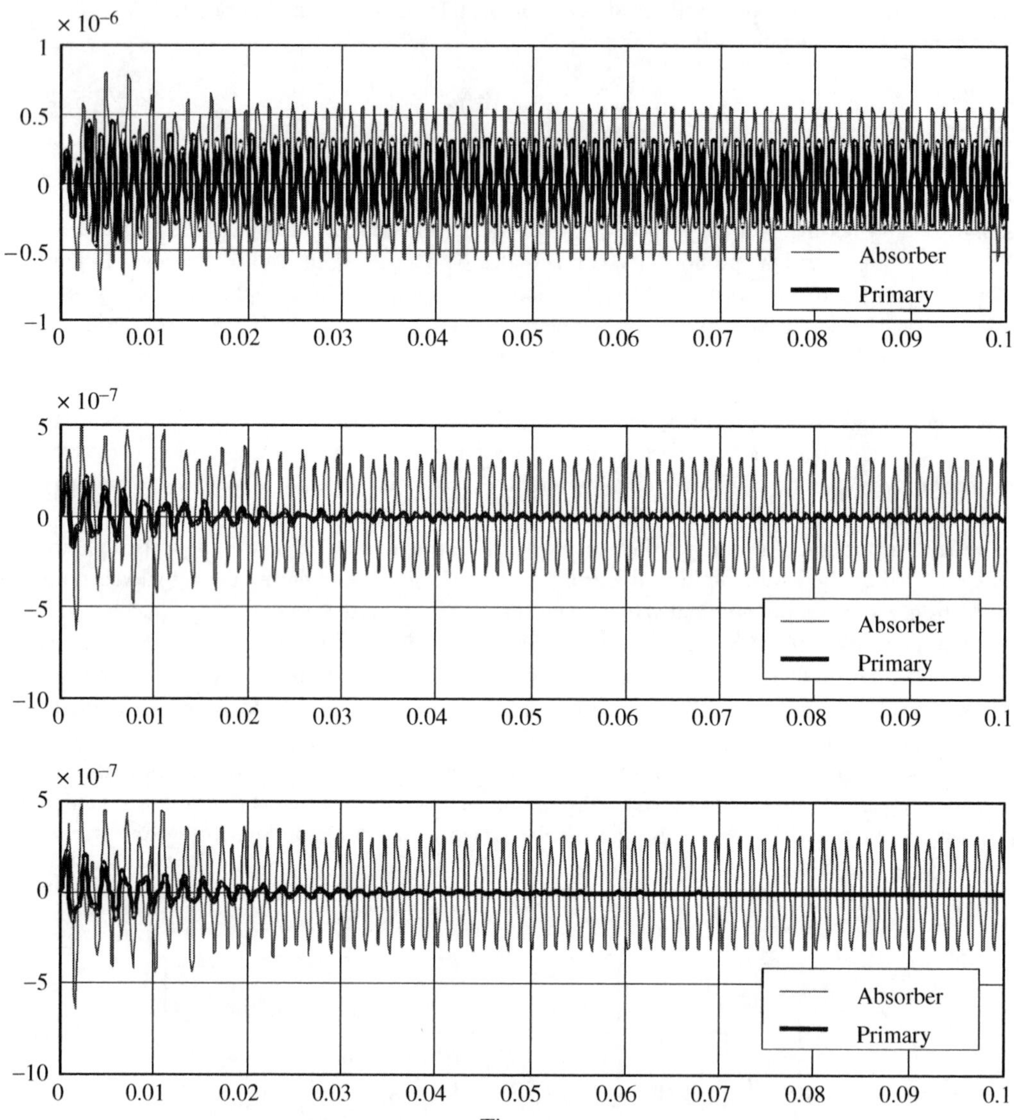

FIGURE 5.22 System responses (displacement, m) for (a) nominal absorber parameters; (b) after first autotuning procedure and (c) after second autotuning procedure.

performed, as shown in Figure 5.22, bottom. The response closely resembles that from Figure 5.21, where all system parameters are assumed to be known exactly.

5.3.3 Vibration-Control Systems

As discussed, in vibration-control schemes, the control inputs to the systems are altered in order to regulate or track a desired trajectory while simultaneously suppressing the vibrational transients in the system. This control problem is rather challenging since it must achieve the motion-tracking objectives while stabilizing the transient vibrations in the system. This section provides two recent control methods developed for the regulation and tracking of flexible beams. The experimental implementations are also discussed. The first control method is a single-input vibration-control system discussed in

Section 5.3.3.1, while the second application utilizes a secondary input control in addition to the primary control input to improve the vibrational characteristics of the system (see Section 5.3.3.2).

5.3.3.1 Variable Structure Control of Rotating Flexible Beams

The vibration-control problem of a flexible manipulator consists in achieving the motion-tracking objectives while stabilizing the transient vibrations in the arm. Several control methods have been developed for flexible arms (Skaar and Tucker, 1986; Bayo, 1987; Yuh, 1987; Sinha, 1988; Lou, 1993; Ge et al., 1997; de Querioz et al., 1999). Most of these methods concentrate on a model-based controller design, and some of these may not be easy to implement due to the uncertainties in the design model, large variations of the loads, ignored high frequency dynamics, and high order of the designed controllers. In view of these methods, VSC is particularly attractive due to its simplicity of implementation and its robustness to parameter uncertainties (Yeung and Chen, 1989; Singh et al., 1994; Jalili et al., 1997).

5.3.3.1.1 Mathematical Modeling

As shown in Figure 5.23, one end of the arm is free and the other end is rigidly attached to a vertical gear shaft, driven by a DC motor. A uniform cross section is considered for the arm, and we make the Euler–Bernoulli assumptions. The control torque, τ, acting on the output shaft, is normal to the plane of motion. Viscous frictions and the ever-present unmodeled dynamics of the motor compartment are to be compensated *via* a perturbation estimation process, as explained later in the text. Since the dynamic system considered here has been utilized in literature quite often, we present only the resulting partial differential equation (PDE) of the system and refer interested readers to Junkins and Kim (1993) and Luo et al. (1999) for detailed derivations.

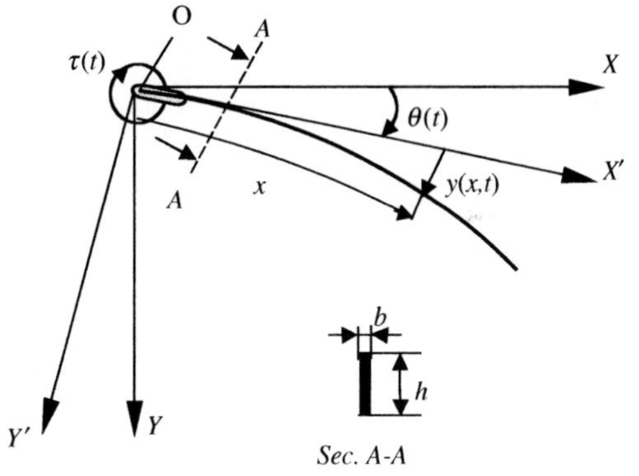

FIGURE 5.23 Flexible arm in the horizontal plane and kinematics of deformation. (*Source:* From Jalili, N., *ASME J. Dyn. Syst. Meas. Control*, 123, 712–719, 2001. With permission.)

The system is governed by

$$I_h\ddot{\theta}(t) + \rho\int_0^L x\ddot{z}(x,t)\mathrm{d}x = \tau \tag{5.46}$$

$$\rho\ddot{z}(x,t) + EIz''''(x,t) = 0 \tag{5.47}$$

with the corresponding boundary conditions

$$z(0,t) = 0, \quad z'(0,t) = \theta(t), \quad z''(L,t) = 0, \quad z'''(L,t) = 0 \tag{5.48}$$

where ρ is the arm's linear mass density, L is the arm length, E is Young's modulus of elasticity, I is the cross-sectional moment of inertia, I_h is the equivalent mass moment of inertia at the root end of the arm, $I_t = I_h + \rho L^3/3$ is the total inertia, and the global variable z is defined as

$$z(x,t) = x\theta(t) + y(x,t) \tag{5.49}$$

Clearly, the arm vibration equation (Equation 5.47) is a homogeneous PDE but the boundary conditions (Equation 5.48) are nonhomogeneous. Therefore, the closed form solution is very tedious to obtain, if not impossible. Using the application of VSC, these equations and their associated boundary conditions can be converted to a homogeneous boundary value problem, as discussed next.

5.3.3.1.2 Variable Structure Controller

The controller objective is to track the arm angular displacement from an initial angle, $\theta_d = \theta(0)$, to zero position, $\theta(t \to \infty) = 0$, while minimizing the flexible arm oscillations. To achieve the control insensitivity against modeling uncertainties, the nonlinear control routine of sliding mode control with an additional perturbation estimation (SMCPE) compartment is adopted here (Elmali and Olgac, 1992; Jalili and Olgac, 1998a). The SMCPE method, presented in Elmali and Olgac (1992), has many attractive features, but it suffers from the disadvantages associated with the truncated-model-base controllers. On the other hand, the infinite-dimensional distributed (IDD)-base controller design, proposed in Zhu et al. (1997), has practical limitations due to its measurement requirements in addition to the complex control law.

Initiating from the idea of the IDD-base controller, we present a new controller design approach in which an online perturbation estimation mechanism is introduced and integrated with the controller to relax the measurement requirements and simplify the control implementation. As utilized in Zhu et al. (1997), for the tip-vibration suppression, it is further required that the sliding surface enable the transformation of nonhomogeneous boundary conditions (Equation 5.48) to homogeneous ones. To satisfy vibration suppression and robustness requirements simultaneously, the sliding hyperplane is selected as a combination of tracking (regulation) error and arm flexible vibration as

$$s = \dot{w} + \sigma w \tag{5.50}$$

where $\sigma > 0$ is a control parameter and

$$w = \theta(t) + \frac{\mu}{L} z(L, t) \tag{5.51}$$

with the scalar, μ, being selected later. When $\mu = 0$, controller (Equation 5.50) reduces to a sliding variable for rigid-link manipulators (Jalili and Olgac, 1998a; Yeung and Chen, 1988). The motivation for such sliding a variable is to provide a suitable boundary condition for solving the beam Equation 5.47, as will be discussed next and is detailed in Jalili (2001).

For the system described by Equation 5.46 to Equation 5.48, if the variable structure controller is given by

$$\tau = \psi_{\text{est}} + \frac{I_t}{1+\mu}\left(-k\,\text{sgn}(s) - Ps - \frac{\mu}{L}\ddot{y}(L,t) - \sigma(1+\mu)\dot{\theta} - \frac{\sigma\mu}{L}\dot{y}(L,t)\right) \tag{5.52}$$

where ψ_{est} is an estimate of the beam flexibility effect

$$\psi = \rho \int_0^L x\ddot{y}(x,t)\text{d}x \tag{5.53}$$

k and P are positive scalars $k \geq 1 + \mu/I_t|\psi - \psi_{\text{est}}|$, $-1.2 < \mu < -0.45$, $\mu \neq -1$ and sgn( ) represents the standard signum function, then, the system's motion will first reach the sliding mode $s = 0$ in a finite time, and consequently converge to the equilibrium position $w(x, t) = 0$ exponentially with a time-constant $1/\sigma$ (Jalili, 2001).

5.3.3.1.3 *Controller Implementation*

In the preceding section, it was shown that by properly selecting control variable, μ, the motion exponentially converges to $w = 0$ with a time-constant $1/\sigma$, while the arm stops in a finite time. Although the discontinuous nature of the controller introduces a robustifying mechanism, we have made the scheme more insensitive to parametric variations and unmodeled dynamics by reducing the required measurements and hence easier control implementation. The remaining measurements and ever-present modeling imperfection effects have all been estimated through an online estimation process. As stated before, in order to simplify the control implementation and reduce the measurement effort, the effect of all uncertainties, including flexibility effect ($\int_0^L x\ddot{y}(x,t)\text{d}x$) and the ever-present unmodeled dynamics, is gathered into a single quantity named perturbation, ψ, as given by Equation 5.53. Noting Equation 5.46, the perturbation term can be expressed as

$$\psi = \tau - I_t\ddot{\theta}(t) \tag{5.54}$$

which requires the yet-unknown control feedback τ. In order to resolve this dilemma of causality, the current value of control torque, τ, is replaced by the most recent control, $\tau(t-\delta)$, where δ is the small time-step used for the loop closure. This replacement is justifiable in practice since such an algorithm is implemented on a digital computer and the sampling speed is high enough to claim this. Also, in the absence of measurement noise, $\ddot{\theta}(t) \cong \ddot{\theta}_{cal}(t) = [\dot{\theta}(t) - \dot{\theta}(t-\delta)]/\delta$.

In practice and in the presence of measurement noise, appropriate filtering may be considered and combined with these approximate derivatives. This technique is referred to as "switched derivatives". This backward differences are shown to be effective when δ is selected to be small enough and the controller is run on a fast DSP (Cannon and Schmitz, 1984). Also, $\ddot{y}(L,t)$ can be obtained by attaching an accelerometer at arm-tip position. All the required signals are, therefore, measurable by currently available sensor facilities and the controller is thus realizable in practice. Although these signals may be quite inaccurate, it should be pointed out that the signals, either measurements or estimations, need not be known very accurately, since robust sliding control can be achieved if k is chosen to be large enough to cover the error existing in the measurement/signal estimation (Yeung and Chen, 1989).

5.3.3.1.4 Numerical Simulations

In order to show the effectiveness of the proposed controller, a lightweight flexible arm is considered ($h >> b$ in Figure 5.23). For numerical results, we consider $\theta_d = \theta(0) = \pi/2$ for the initial arm base angle, with zero initial conditions for the rest of the state variables. The system parameters are listed in Table 5.3. Utilizing assumed mode model (AMM), the arm vibration, Equation 5.47, is truncated to three modes and used in the simulations. It should be noted that the controller law, Equation 5.52, is based on the original infinite dimensional equation, and this truncation is utilized only for simulation purposes.

We take the controller parameter $\mu = -0.66$, $P = 7.0$, $k = 5$, $\varepsilon = 0.01$ and $\sigma = 0.8$. In practice, σ is selected for maximum tracking accuracy taking into account unmodeled dynamics and actuator hardware limitations (Moura et al., 1997). Although such restrictions do not exist in simulations (i.e., with ideal actuators, high sampling frequencies and perfect measurements), this selection of σ was decided based on actual experiment conditions.

The sampling rate for the simulations is $\delta = 0.0005$ sec, while data are recorded at the rate of only 0.002 sec for plotting purposes. The system responses to the proposed control scheme are shown in Figure 5.24. The arm-base angular position reaches the desired position, $\theta = 0$, in approximately 4 to 5 sec, which is in agreement with the approximate settling time of $t_s = 4/\sigma$ (Figure 5.24a). As soon as the system reaches the sliding mode layer, $|s| < \varepsilon$ (Figure 5.24d), the tip vibrations stop (Figure 5.24b), which demonstrates the feasibility of the proposed control technique. The control torque exhibits some residual vibration, as shown in Figure 5.24c. This residual oscillation is expected since the system

TABLE 5.3 System Parameters Used in Numerical Simulations and Experimental Setup for Rotating Arm

Properties	Symbol	Value	Unit
Arm Young's modulus	E	207×10^9	N/m^2
Arm thickness	b	0.0008	m
Arm height	h	0.02	m
Arm length	L	0.45	m
Arm linear mass density	ρ	$0.06/L$	kg/m
Total arm base inertia	I_h	0.002	kg m^2
Gearbox ratio	N	14:1	—
Light source mass	—	0.05	kg
Position sensor sensitivity	—	0.39	V/cm
Motor back EMF constant	K_b	0.0077	V/rad/sec
Motor torque constant	K_t	0.0077	N m/A
Armature resistance	R_a	2.6	Ω
Armature inductance	L_a	0.18	mH
Encoder resolution	—	0.087	Deg/count

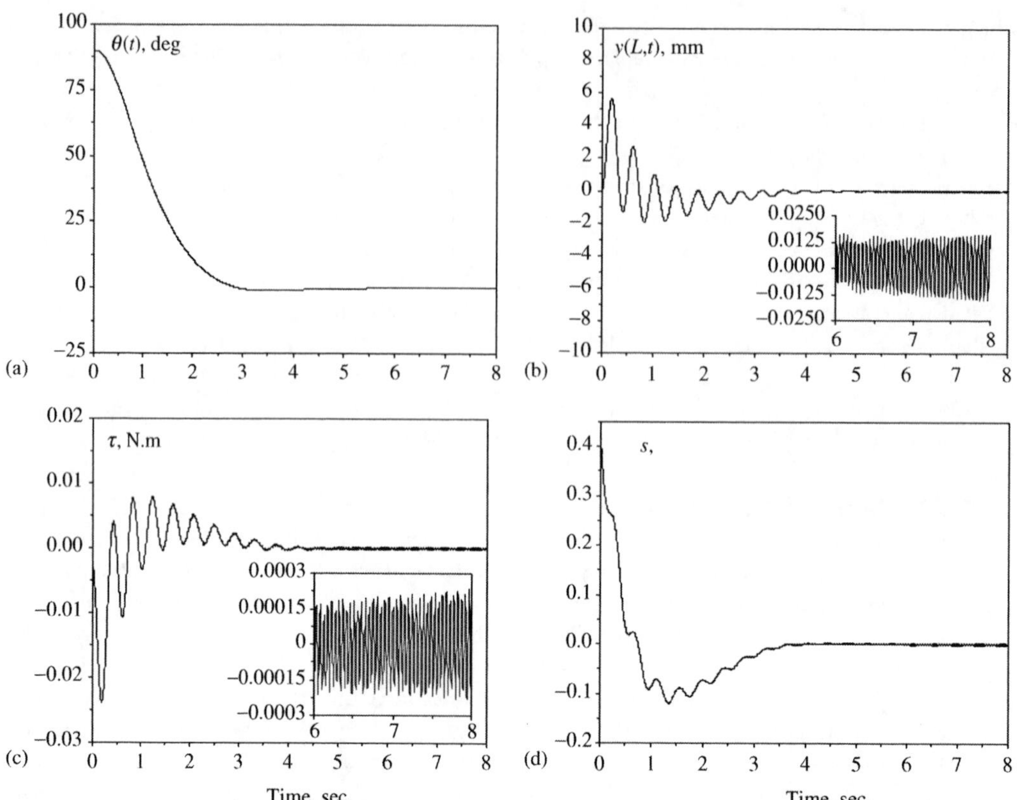

FIGURE 5.24 Analytical system responses to controller with inclusion of arm flexibility, that is, $\mu = -0.66$: (a) arm angular position; (b) arm-tip deflection; (c) control torque and (d) sliding variable sec. (*Source:* From Jalili, N., *ASME J. Dyn. Syst. Meas. Control*, 123, 712–719, 2001. With permission.)

motiono is not forced to stay on $s = 0$ surface (instead it is forced to stay on $|s| < \varepsilon$) when saturation function is used. The sliding variable s is also depicted in Figure 5.24d. To demonstrate better the feature of the controller, the system responses are displayed when $\mu = 0$ (Figure 5.25). As discussed, $\mu = 0$ corresponds to the sliding variable for the rigid link. The undesirable oscillations at the arm tip are evident (see Figure 5.25b and c).

5.3.3.1.5 Control Experiments

In order to demonstrate better the effectiveness of the controller, an experimental setup is constructed and used to verify the numerical results and concepts discussed in the preceding sections. The experimental setup is shown in Figure 5.26. The arm is a slender beam made of stainless steel, with the same dimensions as used in the simulations. The experimental setup parameters are listed in Table 5.3. One end of the arm is clamped to a solid clamping fixture, which is driven by a high-quality DC servomotor. The motor drives a built-in gearbox ($N = 14{:}1$) whose output drives an antibacklash gear. The antibacklash gear, which is equipped with a precision encoder, is utilized to measure the arm base angle as well as to eliminate the backlash. For tip deflection, a light source is attached to the tip of the arm, which is detected by a camera mounted on the rotating base.

The DC motor can be modeled as a standard armature circuit; that is, the applied voltage, v, to the DC motor is

$$v = R_a i_a + L_a \, \mathrm{d}i_a/\mathrm{d}t + K_b \dot{\theta}_m \tag{5.55}$$

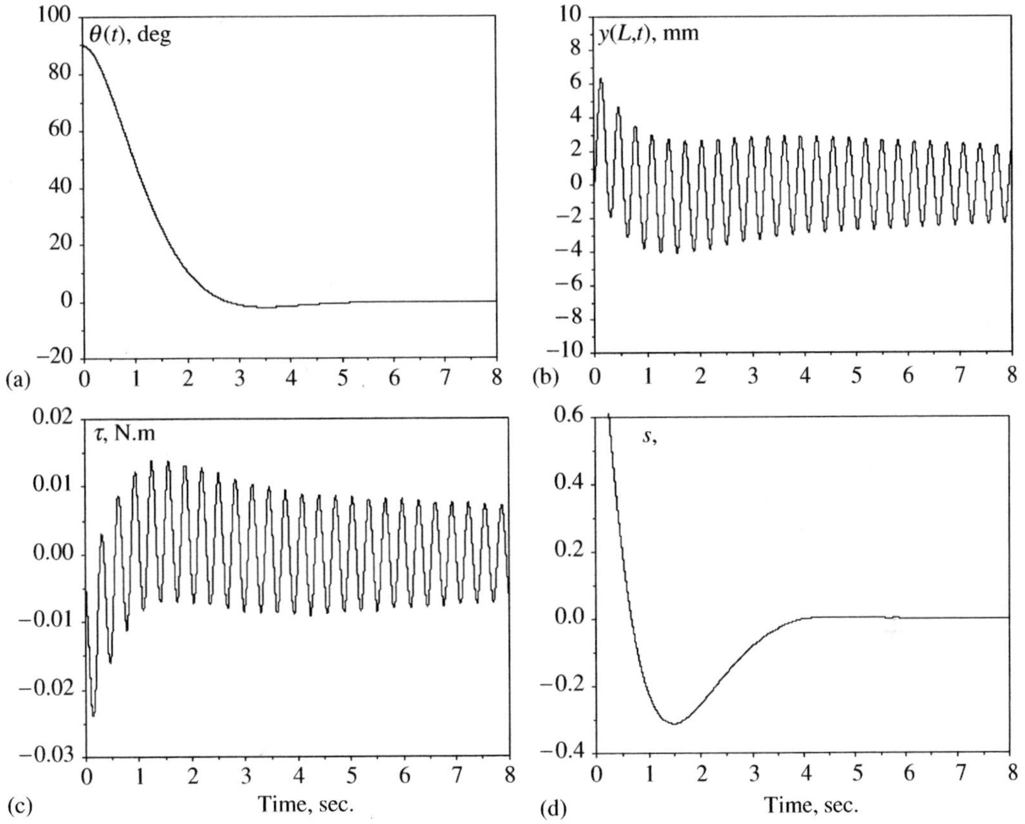

FIGURE 5.25 Analytical system responses to controller without inclusion of arm flexibility, that is, $\mu = 0$: (a) arm angular position; (b) arm-tip deflection; (c) control torque and (d) sliding variable sec. (*Source:* From Jalili, N., *ASME J. Dyn. Syst. Meas. Control*, 123, 712–719, 2001. With permission.)

FIGURE 5.26 The experimental device and setup configuration. (*Source:* From Jalili, N., *ASME J. Dyn. Syst. Meas. Control*, 123, 712–719, 2001. With permission.)

where R_a is the armature resistance, L_a is the armature inductance, i_a is the armature current, K_b is the back-EMF (electro-motive-force) constant, and θ_m is the motor shaft position. The motor torque, τ_m from the motor shaft with the torque constant, K_t, can be written as

$$\tau_m = K_t i_a \tag{5.56}$$

The motor dynamics thus become

$$I_e \ddot{\theta}_m + C_v \dot{\theta}_m + \tau_a = \tau_m = K_t i_a \tag{5.57}$$

where C_v is the equivalent damping constant of the motor, and $I_e = I_m + I_L/N^2$ is the equivalent inertia load including motor inertia, I_m, and gearbox, clamping frame and camera inertia, I_L. The available torque from the motor shaft for the arm is τ_a.

Utilizing the gearbox from the motor shaft to the output shaft and ignoring the motor electric time constant, (L_a/R_a), one can relate the servomotor input voltage to the applied torque (acting on the arm) as

$$\tau = \frac{NK_t}{R_a} v - \left(C_v + \frac{K_t K_b}{R_a}\right) N^2 \dot{\theta} - I_h \ddot{\theta} \tag{5.58}$$

where $I_h = N^2 I_e$ is the equivalent inertia of the arm base used in the derivation of governing equations. By substituting this torque into the control law, the reference input voltage, V, is obtained for experiment.

The control torque is applied *via* a digital signal processor (DSP) with sampling rate of 10 kHz, while data are recorded at the rate 500 Hz (for plotting purposes only). The DSP runs the control routine in a single-input–single-output mode as a free standing CPU. Most of the computations and hardware commands are done on the DSP card. For this setup, a dedicated 500 MHz Pentium III serves as the host PC, and a state-of-the-art dSPACE® DS1103 PPC controller board equipped with a Motorola Power PC 604e at 333 MHz, 16 channels ADC, 12 channels DAC, as microprocessor.

The experimental system responses are shown in Figure 5.27 and Figure 5.28 for similar cases discussed in the numerical simulation section. Figure 5.27 represents the system responses when controller (Equation 5.52) utilizes the flexible arm (i.e., $\mu = -0.66$). As seen, the arm base reaches the desired position (Figure 5.27a), while tip deflection is simultaneously stopped (Figure 5.27b). The good correspondence between analytical results (Figure 5.24) and experimental findings (Figure 5.27) is noticeable from a vibration suppression characteristics point of view. It should be noted that the controller is based on the original governing equations, with arm-base angular position and tip deflection measurements only. The unmodeled dynamics, such as payload effect (owing to the light source at the tip, see Table 5.3) and viscous friction (at the root end of the arm), are being compensated through the proposed online perturbation estimation routine. This, in turn, demonstrates the capability of the proposed control scheme when considerable deviations between model and plant are encountered. The only noticeable difference is the fast decaying response as shown in Figure 5.27b and c. This clearly indicates the high friction at the motor, which was not considered in the simulations (Figure 24b and c). Similar responses are obtained when the controller is designed based on the rigid link only, that is, $\mu = 0$. The system responses are displayed in Figure 5.28. Similarly, the undesirable arm-tip oscillations are obvious. The overall agreement between simulations (Figure 24 and Figure 25) and the experiment (Figure 27 and Figure 28) is one of the critical contributions of this work.

5.3.3.2 Observer-Based Piezoelectric Vibration Control of Translating Flexible Beams

Many industrial robots, especially those widely used in automatic manufacturing assembly lines, are Cartesian types (Ge et al., 1998). A flexible Cartesian robot can be modeled as a flexible cantilever beam with a translational base support. Traditionally, a PD control strategy is used to regulate the movement of the robot arm. In lightweight robots, the base movement will cause undesirable vibrations at the arm tip because of the flexibility distributed along the arm. In order to eliminate

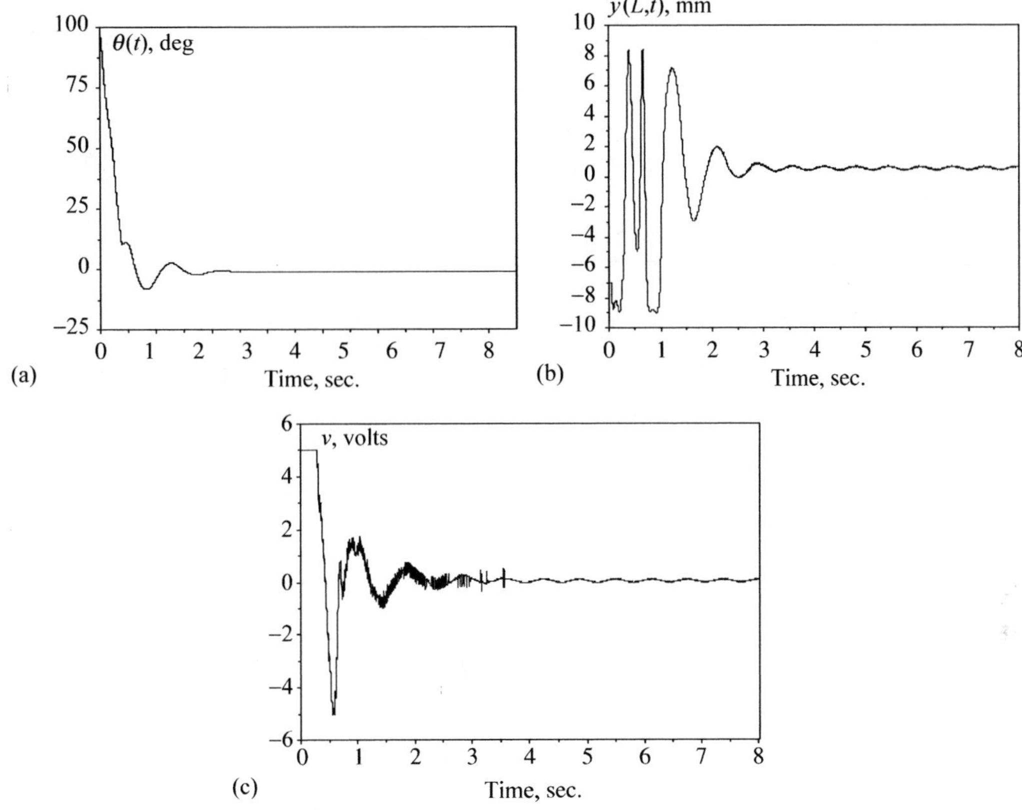

FIGURE 5.27 Experimental system responses to controller with inclusion of arm flexibility, that is, $\mu = -0.66$: (a) arm angular position; (b) arm-tip deflection; (c) control voltage applied to DC servomotor. (*Source:* From Jalili, N., *ASME J. Dyn. Syst. Meas. Control*, 123, 712–719, 2001. With permission.)

such vibrations, the PD controller must be upgraded with additional compensating terms. In order to improve further the vibration suppression performance, which is a requirement for the high-precision manufacturing market, a second controller, such as a piezoelectric (PZT) patch actuator attached on the surface of the arm, can be utilized (Oueini et al., 1998; Ge et al., 1999; Jalili et al., 2002).

In this section, an observer-based control strategy is presented for regulating the arm motion (Liu et al., 2002). The base motion is controlled utilizing an electrodynamic shaker, while a piezoelectric (PZT) patch actuator is bonded on the surface of the flexible beam for suppressing residual arm vibrations. The control objective here is to regulate the arm base movement, while simultaneously suppressing the vibration transients in the arm. To achieve this, a simple PD control strategy is selected for the regulation of the movement of the base, and a Lyapunov-based controller is selected for the PZT voltage signal. The selection of the proposed energy-based Lyapunov function naturally results in velocity-related signals, which are not physically measurable (Dadfarnia et al., 2003). To remedy this, a reduced-order observer is designed to estimate the velocity related signals. For this, the control structure is designed based on the truncated two-mode beam model.

5.3.3.2.1 *Mathematical Modeling*

For the purpose of model development, we consider a uniform flexible cantilever beam with a PZT actuator bonded on its top surface. As shown in Figure 5.29, one end of the beam is clamped into

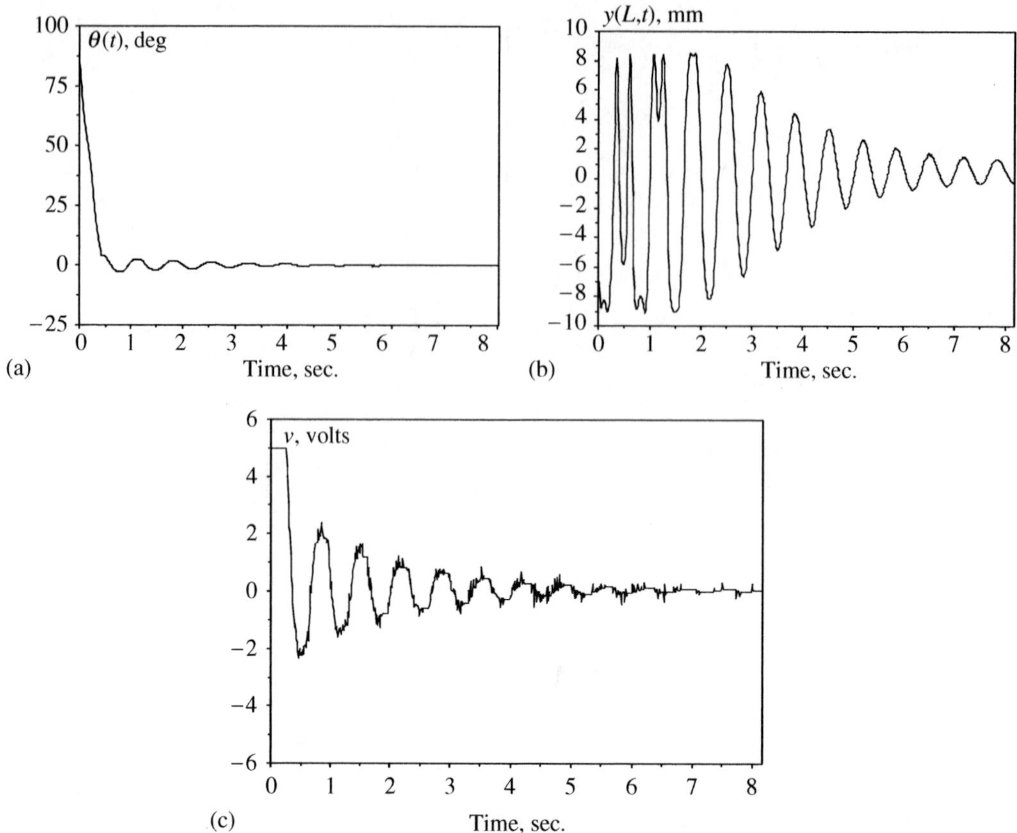

FIGURE 5.28 Experimental system responses to controller without inclusion of arm flexibility, that is, $\mu = 0$: (a) arm angular position; (b) arm-tip deflection; (c) control voltage applied to DC servomotor. (*Source:* From Jalili, N., *ASME J. Dyn. Syst. Meas. Control*, 123, 712–719, 2001. With permission.)

a moving base with the mass of m_b, and a tip mass, m_t, is attached to the free end of the beam. The beam has total thickness t_b, and length L, while the piezoelectric film possesses thickness and length t_b and $(l_2 - l_1)$, respectively. We assume that the PZT and the beam have the same width, b. The PZT actuator is perfectly bonded on the beam at distance l_1 measured from the beam support. The force, $f(t)$, acting on the base and the input voltage, $v(t)$, applied to the PZT actuator are the only external effects.

To establish a coordinate system for the beam, the x-axis is taken in the longitudinal direction and the z-axis is specified in the transverse direction of the beam with midplane of the beam to be $z = 0$, as shown in Figure 5.30. This coordinate is fixed to the base.

The fundamental relations for the piezoelectric materials are given as (Ikeda, 1990)

$$\mathbf{F} = \mathbf{cS} - \mathbf{hD} \tag{5.59}$$

$$\mathbf{E} = -\mathbf{h}^{\mathrm{T}}\mathbf{S} + \boldsymbol{\beta}\mathbf{D} \tag{5.60}$$

where $\mathbf{F} \in \Re^6$ is the stress vector, $\mathbf{S} \in \Re^6$ is the strain vector, $\mathbf{c} \in \Re^{6\times 6}$ is the symmetric matrix of elastic stiffness coefficients, $\mathbf{h} \in \Re^{6\times 3}$ is the coupling coefficients matrix, $\mathbf{D} \in \Re^3$ is the electrical displacement vector, $\mathbf{E} \in \Re^3$ is the electrical field vector, and $\boldsymbol{\beta} \in \Re^{3\times 3}$ is the symmetric matrix of impermittivity coefficients.

An energy method is used to derive the equations of motion. Neglecting the electrical kinetic energy, the total kinetic energy of the system is expressed as (Liu et al., 2002; Dadfarnia et al., 2004)

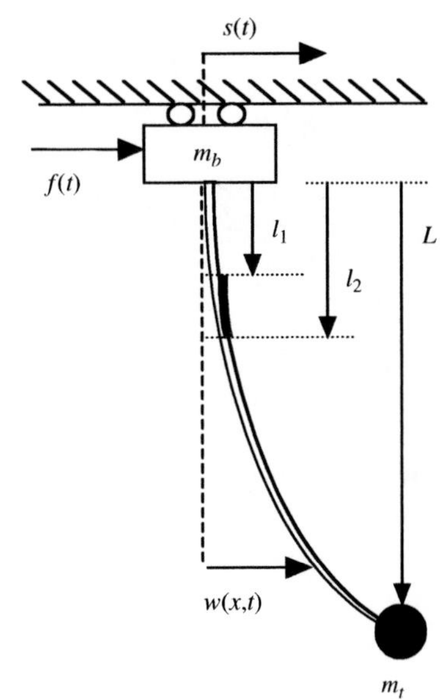

FIGURE 5.29 Schematic of the SCARA/Cartesian robot (last link).

$$\begin{aligned}E_k &= \frac{1}{2}m_b\dot{s}(t)^2 + \frac{1}{2}b\int_0^{l_1}\rho_b t_b(\dot{s}(t) + \dot{w}(x,t))^2 dx \\ &+ \frac{1}{2}b\int_{l_1}^{l_2}(\rho_b t_b + \rho_p t_p)(\dot{s}(t) + \dot{w}(x,t))^2 dx \\ &+ \frac{1}{2}b\int_{l_2}^{L}\rho_b t_b(\dot{s}(t) + \dot{w}(x,t))^2 dx \\ &+ \frac{1}{2}m_t(\dot{s}(t) + \dot{w}(L,t))^2 \\ &= \frac{1}{2}m_b\dot{s}(t)^2 + \frac{1}{2}\int_0^{L}\rho(x)(\dot{s}(t) + \dot{w}(x,t))^2 dx \\ &+ \frac{1}{2}m_t(\dot{s}(t) + \dot{w}(L,t))^2 \qquad (5.61)\end{aligned}$$

where

$$\begin{aligned}\rho(x) &= [\rho_b t_b + G(x)\rho_p t_p]b \\ G(x) &= H(x - l_1) - H(x - l_2)\end{aligned} \qquad (5.62)$$

and $H(x)$ is the Heaviside function, ρ_b and ρ_p are the respective beam and PZT volumetric densities. Neglecting the effect of gravity due to planar motion and the higher-order terms of quadratic in w' (Esmailzadeh and Jalili, 1998b), the total potential energy of the system can be expressed as

$$\begin{aligned}E_p &= \frac{1}{2}b\int_0^{l_1}\int_{-t_b/2}^{t_b/2}\mathbf{F}^T\mathbf{S}\,dy\,dx + \frac{1}{2}b\int_{l_1}^{l_2}\int_{-t_b/2}^{t_b/2}\mathbf{F}^T\mathbf{S}\,dy\,dx + \frac{1}{2}b\int_{l_1}^{l_2}\int_{t_b/2}^{(t_b/2)+t_p}[\mathbf{F}^T\mathbf{S} + \mathbf{E}^T\mathbf{D}]dy\,dx \\ &+ \frac{1}{2}b\int_{l_2}^{L}\int_{-t_b/2}^{t_b/2}\mathbf{F}^T\mathbf{S}\,dy\,dx \\ &= \frac{1}{2}\int_0^L c(x)\left[\frac{\partial^2 w(x,t)}{\partial x^2}\right]^2 dx + h_l D_y(t)\int_{l_1}^{l_2}\frac{\partial^2 w(x,t)}{\partial x^2}dx + \frac{1}{2}\beta_l(l_2 - l_1)D_y(t)^2 \qquad (5.63)\end{aligned}$$

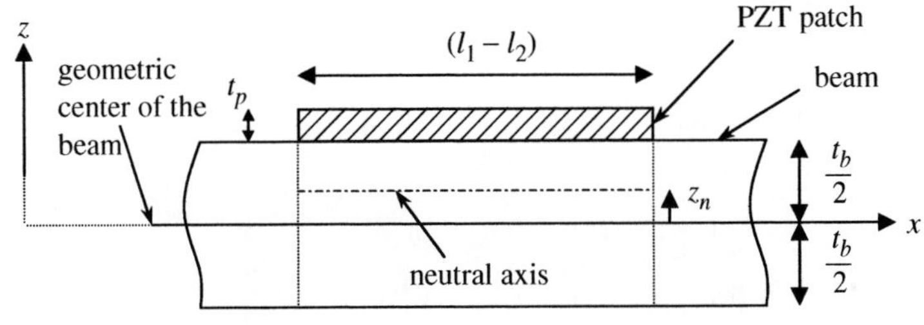

FIGURE 5.30 Coordinate system.

where

$$c(x) = \frac{b}{3}\left\{\left(\frac{c_{11}^{b}t_{b}^{3}}{4}\right) + G(x)\left\{3c_{11}^{b}t_{b}z_{n}^{2} + c_{11}^{p}\left(t_{p}^{3} + 3t_{p}\left(\frac{t_{b}}{2} - z_{n}\right)^{2} + 3t_{p}^{2}\left(\frac{t_{b}}{2} - z_{n}\right)\right)\right\}\right\} \tag{5.64}$$

$$h_{l} = h_{12}t_{p}b(t_{p} + t_{b} - 2z_{n})/2, \quad \beta_{l} = \beta_{22}bt_{p}$$

and

$$z_{n} = \frac{c_{11}^{p}t_{p}(t_{p} + t_{b})}{c_{11}^{b}t_{b} + c_{11}^{p}t_{p}}$$

The beam and PZT stiffnesses are c_{11}^{b} and c_{11}^{p}, respectively.

Using the AMM for the beam vibration analysis, the beam deflection can be written as

$$w(x,t) = \sum_{i=1}^{\infty} \phi_{i}(x)q_{i}(t), \quad P(x,t) = s(t) + w(x,t) \tag{5.65}$$

The equations of motion can now be obtained using the Lagrangian approach

$$\left[m_{b} + m_{t} + \int_{0}^{L} \rho(x)dx\right]\ddot{s}(t) + \sum_{j=1}^{\infty} m_{j}\ddot{q}_{j}(t) = f(t) \tag{5.66a}$$

$$m_{i}\ddot{s}(t) + m_{di}\ddot{q}_{i}(t) + \omega_{i}^{2}m_{di}q_{i}(t) + h_{l}(\phi_{i}'(l_{2}) - \phi_{i}'(l_{1}))D_{y}(t) = 0 \tag{5.66b}$$

$$h_{l}\sum_{j=1}^{\infty}\{(\phi_{j}'(l_{2}) - \phi_{j}'(l_{1}))q_{j}(t)\} + \beta_{l}(l_{2} - l_{1})D_{y}(t) = b(l_{2} - l_{1})v(t) \tag{5.66c}$$

where

$$m_{dj} = \int_{0}^{L} \rho(x)\phi_{j}^{2}(x)dx + m_{t}\phi_{j}^{2}(L), \quad m_{j} = \int_{0}^{L} \rho(x)\phi_{j}(x)dx + m_{t}\phi_{j}(L) \tag{5.67}$$

Calculating $D_{y}(t)$ from Equation 5.66b and substituting into Equation 5.66c results in

$$m_{i}\ddot{s}(t) + m_{di}\ddot{q}_{i}(t) + \omega_{i}^{2}m_{di}q_{i}(t) - \frac{h_{l}^{2}(\phi_{i}'(l_{2}) - \phi_{i}'(l_{1}))}{\beta_{l}(l_{2} - l_{1})}\sum_{j=1}^{\infty}\{(\phi_{j}'(l_{2}) - \phi_{j}'(l_{1}))q_{j}(t)\}$$

$$= -\frac{h_{l}b(\phi_{i}'(l_{2}) - \phi_{i}'(l_{1}))}{\beta_{l}}v(t), \quad i = 1, 2, \ldots \tag{5.68}$$

which will be used to derive the controller, as discussed next.

5.3.3.2.2 *Derivation of the Controller*

Utilizing Equation 5.66a and Equation 5.68, the truncated two-mode beam with PZT model reduces to

$$\left[m_{b} + m_{t} + \int_{0}^{L} \rho(x)dx\right]\ddot{s}(t) + m_{1}\ddot{q}_{1}(t) + m_{2}\ddot{q}_{2}(t) = f(t) \tag{5.69a}$$

$$\begin{aligned} &m_{1}\ddot{s}(t) + m_{d1}\ddot{q}_{1}(t) + \omega_{1}^{2}m_{d1}q_{1}(t) - \frac{h_{l}^{2}(\phi_{1}'(l_{2}) - \phi_{1}'(l_{1}))}{\beta_{l}(l_{2} - l_{1})} \\ &\times\{(\phi_{1}'(l_{2}) - \phi_{1}'(l_{1}))q_{1}(t) + (\phi_{2}'(l_{2}) - \phi_{2}'(l_{1}))q_{2}(t)\} \\ &= -\frac{h_{l}b(\phi_{1}'(l_{2}) - \phi_{1}'(l_{1}))}{\beta_{l}}v(t) \end{aligned} \tag{5.69b}$$

$$m_2\ddot{s}(t) + m_{d2}\ddot{q}_2(t) + \omega_2^2 m_{d2}q_2(t) - \frac{h_l^2(\phi_2'(l_2) - \phi_2'(l_1))}{\beta_l(l_2 - l_1)} \times \{(\phi_1'(l_2) - \phi_1'(l_1))q_1(t) + (\phi_2'(l_2) - \phi_2'(l_1))q_2(t)\} = -\frac{h_l b(\phi_2'(l_2) - \phi_2'(l_1))}{\beta_l} v(t) \tag{5.69c}$$

The equations in Equation 5.69 can be written in the following more compact form

$$\mathbf{M}\ddot{\boldsymbol{\Delta}} + \mathbf{K}\boldsymbol{\Delta} = \mathbf{F}_e \tag{5.70}$$

where

$$\mathbf{M} = \begin{bmatrix} \psi & m_1 & m_2 \\ m_1 & m_{d1} & 0 \\ m_2 & 0 & m_{d2} \end{bmatrix}, \quad \mathbf{K} = \begin{bmatrix} 0 & 0 & 0 \\ 0 & k_{11} & k_{12} \\ 0 & k_{12} & k_{22} \end{bmatrix}, \quad \mathbf{F}_e = \begin{Bmatrix} f(t) \\ \epsilon_1 v(t) \\ \epsilon_2 v(t) \end{Bmatrix}, \quad \Delta = \begin{Bmatrix} s(t) \\ q_1(t) \\ q_2(t) \end{Bmatrix} \tag{5.71}$$

and

$$\psi = m_b + m_t + \int_0^L \rho(x)\mathrm{d}x, \quad \epsilon_1 = -\frac{h_l b}{\beta_l}(\phi_1'(l_2) - \phi_1'(l_1)), \quad \epsilon_2 = -\frac{h_l b}{\beta_l}(\phi_2'(l_2) - \phi_2'(l_1)),$$

$$k_{11} = \omega_1^2 m_{d1} - \frac{h_l^2}{\beta_l(l_2 - l_1)}(\phi_1'(l_2) - \phi_1'(l_1))^2,$$

$$k_{12} = -\frac{h_l^2}{\beta_l(l_2 - l_1)}(\phi_1'(l_2) - \phi_1'(l_1))(\phi_2'(l_2) - \phi_2'(l_1)), \tag{5.72}$$

$$k_{22} = \omega_2^2 m_{d2} - \frac{h_l^2}{\beta_l(l_2 - l_1)}(\phi_2'(l_2) - \phi_2'(l_1))^2$$

For the system described by Equation 5.70, if the control laws for the arm base force and PZT voltage generated moment are selected as

$$f(t) = -k_p\Delta s - k_d\dot{s}(t) \tag{5.73}$$

$$v(t) = -k_v(\epsilon_1\dot{q}_1(t) + \epsilon_2\dot{q}_2(t)) \tag{5.74}$$

where k_p and k_d are positive control gains, $\Delta s = s(t) - s_d$, s_d is the desired set-point position, and $k_v > 0$ is the voltage control gain, then the closed-loop system will be stable, and in addition

$$\lim_{t\to\infty}\{q_1(t), q_2(t), \Delta s\} = 0$$

See Dadfarnia et al. (2004) for a detailed proof.

5.3.3.2.3 *Controller Implementation*

The control input, $v(t)$, requires the information from the velocity-related signals, $\dot{q}_1(t)$ and $\dot{q}_2(t)$, which are usually not measurable. Sun and Mills (1999) solved the problem by integrating the acceleration signals measured by the accelerometers. However, such controller structure may result in unstable closed-loop system in some cases. In this paper, a reduced-order observer is designed to estimate the velocity signals, $\dot{q}_1$ and $\dot{q}_2$. For this, we utilize three available signals: base displacement, $s(t)$, arm-tip deflection, $P(L, t)$, and beam root strain, $\epsilon(0, t)$; that is

$$y_1 = s(t) = x_1 \tag{5.75a}$$

$$y_2 = P(L, t) = x_1 + \phi_1(L)x_2 + \phi_2(L)x_3 \tag{5.75b}$$

$$y_3 = \epsilon(0, t) = \frac{t_b}{2}(\phi_1''(0)x_2 + \phi_2''(0)x_3) \tag{5.75c}$$

It can be seen that the first three states can be obtained by

$$\begin{Bmatrix} x_1 \\ x_2 \\ x_3 \end{Bmatrix} = \mathbf{C}_1^{-1}\mathbf{y} \tag{5.76}$$

Since this system is observable, we can design a reduced-order observer to estimate the velocity-related state signals. Defining $\mathbf{X}_1 = [\,x_1 \quad x_2 \quad x_3\,]^{\mathrm{T}}$ and $\mathbf{X}_2 = [\,x_4 \quad x_5 \quad x_6\,]^{\mathrm{T}}$, the estimated value for $\mathbf{X}_2$ can be designed as

$$\hat{\mathbf{X}}_2 = \mathbf{L}_r\mathbf{y} + \hat{\mathbf{z}} \tag{5.77}$$

$$\dot{\hat{\mathbf{z}}} = \mathbf{F}\hat{\mathbf{z}} + \mathbf{G}\mathbf{y} + \mathbf{H}\mathbf{u} \tag{5.78}$$

where $\mathbf{L}_r \in R^{3\times3}$, $\mathbf{F} \in R^{3\times3}$, $\mathbf{G} \in R^{3\times3}$, and $\mathbf{H} \in R^{3\times2}$ will be determined by the observer pole placement. Defining the estimation error as

$$\mathbf{e}_2 = \mathbf{X}_2 - \hat{\mathbf{X}}_2 \tag{5.79}$$

the derivative of the estimation error becomes

$$\dot{\mathbf{e}}_2 = \dot{\mathbf{X}}_2 - \dot{\hat{\mathbf{X}}}_2 \tag{5.80}$$

Substituting the state-space equations of the system (Equation 5.77 and Equation 5.78) into Equation 5.80 and simplifying, we obtain

$$\dot{\mathbf{e}}_2 = \mathbf{F}\mathbf{e}_2 + (\mathbf{A}_{21} - \mathbf{L}_r\mathbf{C}_1\mathbf{A}_{11} - \mathbf{G}\mathbf{C}_1 + \mathbf{F}\mathbf{L}_r\mathbf{C}_1)\mathbf{X}_1 + (\mathbf{A}_{22} - \mathbf{L}_r\mathbf{C}_1\mathbf{A}_{12} - \mathbf{F})\mathbf{X}_2 + (\mathbf{B}_2 - \mathbf{L}_r\mathbf{C}_1\mathbf{B}_1 - \mathbf{H})\mathbf{u} \tag{5.81}$$

In order to force the estimation error, $\mathbf{e}_2$, to go to zero, matrix **F** should be selected to be Hurwitz and the following relations must be satisfied (Liu et al., 2002):

$$\mathbf{F} = \mathbf{A}_{22} - \mathbf{L}_r\mathbf{C}_1\mathbf{A}_{12} \tag{5.82}$$

$$\mathbf{H} = \mathbf{B}_2 - \mathbf{L}_r\mathbf{C}_1\mathbf{B}_1 \tag{5.83}$$

$$\mathbf{G} = (\mathbf{A}_{21} - \mathbf{L}_r\mathbf{C}_1\mathbf{A}_{11} + \mathbf{F}\mathbf{L}_r\mathbf{C}_1)\mathbf{C}_1^{-1} \tag{5.84}$$

The matrix **F** can be chosen by the desired observer pole placement requirement. Once **F** is known, $\mathbf{L}_r$, **H**, and **G** can be determined utilizing Equation 5.82, to Equation 5.84, respectively. The velocity variables, $\hat{\mathbf{X}}_2$, can now be estimated by Equation 5.77 and Equation 5.78.

5.3.3.2.4 *Numerical Simulations*

In order to show the effectiveness of the controller, the flexible beam structure in Figure 5.29 is considered with the PZT actuator attached on the beam surface. The system parameters are listed in Table 5.4.

First, we consider the beam without PZT control. We take the PD control gains to be $k_p = 120$ and $k_d = 20$. Figure 5.31 shows the results for the beam without PZT control (i.e., with only PD force control for the base movement). To investigate the effect of PZT controller on the beam vibration, we consider the voltage control gain to be $k_v = 2 \times 10^7$. The system responses to the proposed controller with a piezoelectric actuator based on the two-mode model are shown in Figure 5.32. The comparison between the tip displacement, from Figure 5.31 and Figure 5.32, shows that the beam vibration can be suppressed significantly utilizing the PZT actuator.

5.3.3.2.5 *Control Experiments*

In order to demonstrate better the effectiveness of the controller, an experimental setup is constructed and used to verify the numerical results. The experimental apparatus consists of a flexible beam with a PZT actuator and strain sensor attachments, as well as data acquisition,

TABLE 5.4 System Parameters Used in Numerical Simulations and Experimental Setup for Translational Beam

Properties	Symbol	Value	Unit
Beam Young's modulus	c_{11}^{b}	69×10^9	N/m^2
Beam thickness	t_b	0.8125	mm
Beam and PZT width	b	20	mm
Beam length	L	300	mm
Beam volumetric density	ρ_b	3960.0	kg/m^3
PZT Young's modulus	c_{11}^{p}	66.47×10^9	N/m^2
PZT coupling parameter	h_{12}	5×10^8	V/m
PZT impermittivity	β_{22}	4.55×10^7	m/F
PZT thickness	t_p	0.2032	mm
PZT length	$l_2 - l_1$	33.655	mm
PZT position on beam	l_1	44.64	mm
PZT volumetric density	ρ_p	7750.0	kg/m^3
Base mass	m_b	0.455	kg
Tip mass	m_t	0	kg

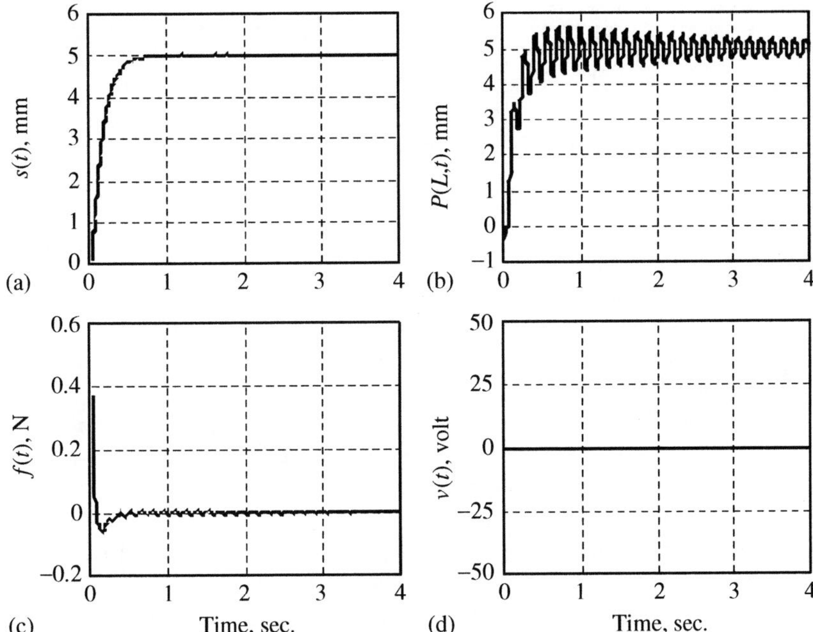

FIGURE 5.31 Numerical simulations for the case without PZT control: (a) base motion; (b) tip displacement; (c) control force and (d) PZT voltage.

amplifier, signal conditioner and the control software. As shown in Figure 5.33, the plant consists of a flexible aluminum beam with a strain sensor and a PZT patch actuator bound on each side of the beam surface. One end of the beam is clamped to the base with a solid clamping fixture, which is driven by a shaker. The shaker is connected to the arm base by a connecting rod. The experimental setup parameters are listed in Table 5.4.

Figure 5.34 shows the high-level control block diagram of the experiment, where the shaker provides the input control force to the base and the PZT applies a controlled moment on the beam. Two laser sensors measure the position of the base and the beam-tip displacement. A strain-gauge sensor, which is attached near the base of the beam, is utilized for the dynamic strain measurement. These three signals

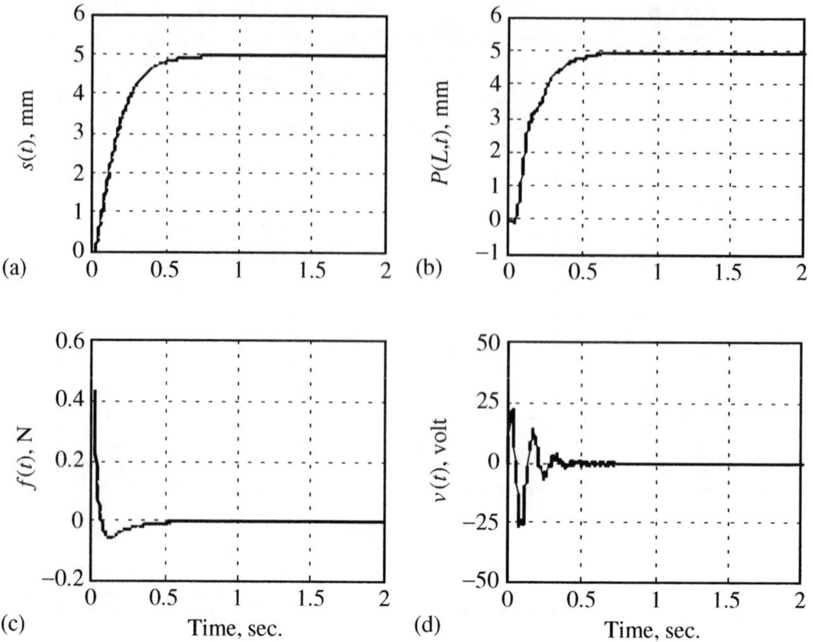

FIGURE 5.32 Numerical simulations for the case with PZT control: (a) base motion; (b) tip displacement; (c) control force and (d) PZT voltage.

are fed back to the computer through the ISA MultiQ data acquisition card. The remaining required signals for the controller (Equation 5.66) are determined as explained in the preceding section. The data acquisition and control algorithms are implemented on an AMD Athlon 1100 MHz PC running under the RT-Linux operating system. The Matlab/Simulink environment and Real Time Linux Target are used to implement the controller.

The experimental results for both cases (i.e., without PZT and with PZT control) are depicted in Figure 5.35 and Figure 5.36, respectively. The results demonstrate that with PZT control, the arm vibration is eliminated in less than 1 sec, while the arm vibration lasts for more than 6 sec when PZT control is not used. The experimental results are in agreement with the simulation results except for some differences at the beginning of the motion. The slight overshoot and discrepancies at the beginning of the motion are due to the limitations of the experiment (e.g., the shaker saturation limitation) and unmodeled dynamics in the modeling (e.g., the friction modeling). However, it is still apparent that the PZT voltage control can substantially suppress the arm vibration despite such limitations and modeling imperfections.

5.4 Practical Considerations and Related Topics

5.4.1 Summary of Vibration-Control Design Steps and Procedures

In order to select a suitable vibration-control system, especially a vibration isolator, a number of factors must be considered.

5.4.1.1 Static Deflection

The static deflection of the vibration-control system under the deadweight of the load determines to a considerable extent the type of the material to be used in the isolator. Organic materials, such as rubber

FIGURE 5.33 The experimental setup: (a) the whole system; (b) PZT actuator, ACX model No. QP21B; (c) dynamic strain sensor (attached on the other side of the beam), model No. PCB 740B02.

and cork, are capable of sustaining very large strains provided they are applied momentarily. However, if large strains remain for an appreciable period of time, they tend to drift or creep. On the other hand, metal springs undergo permanent deformation if the stress exceeds the yield stress of the material, but show minimal drift or creep when the stress is maintained below the yield stress.

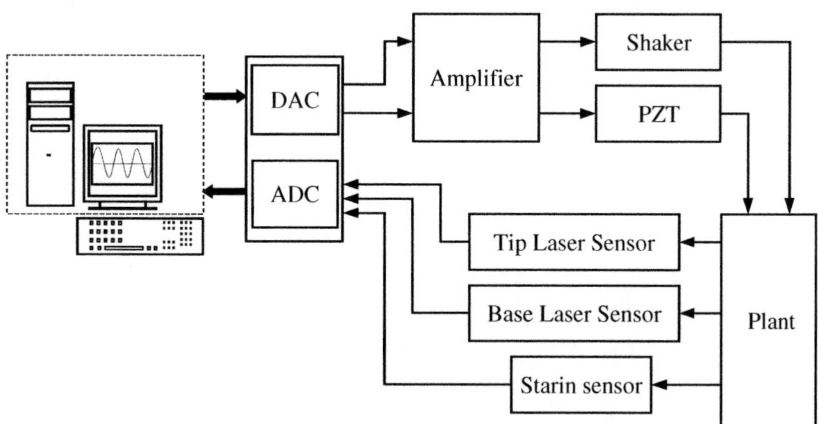

FIGURE 5.34 High-level control-block diagram.

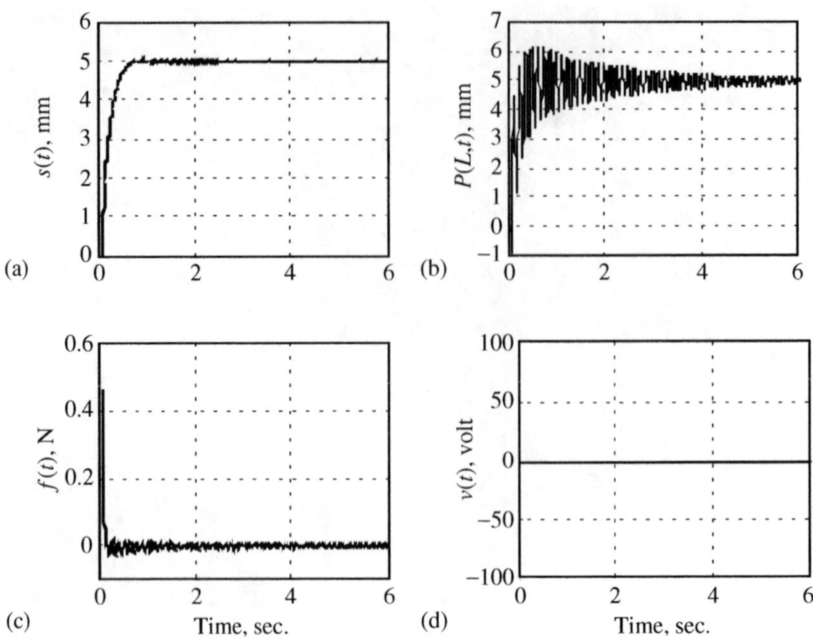

FIGURE 5.35 Experimental results for the case without PZT control: (a) base motion; (b) tip displacement; (c) control force and (d) PZT voltage.

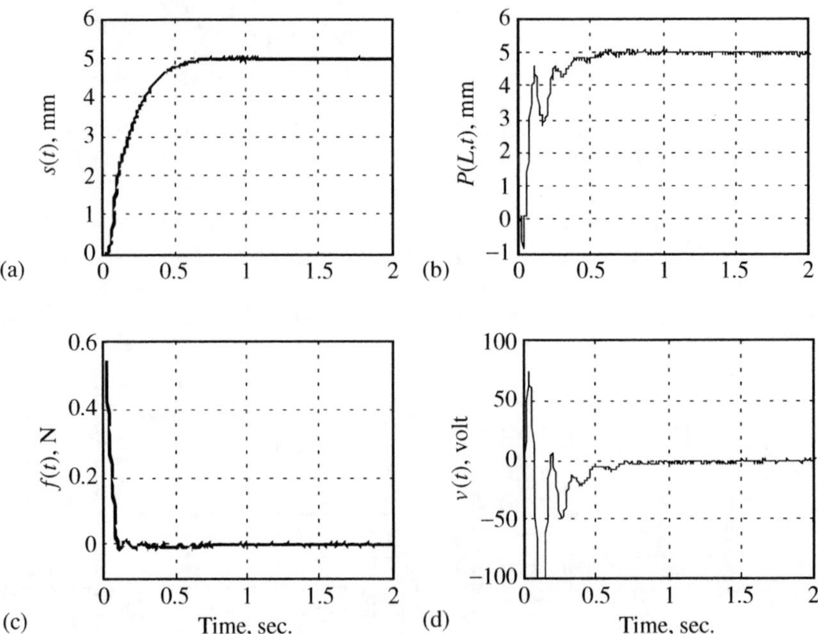

FIGURE 5.36 Experimental results for the case with PZT control: (a) base motion; (b) tip displacement; (c) control force and (d) PZT voltage.

5.4.1.2 Stiffness in Lateral Directions

Resilient materials strained in compression are most useful when the load is relatively large and the static deflection is small. Such applications are difficult to design for a small load, unless the required static deflection is small. Otherwise, the small area and great thickness tend to cause a condition of instability. To a considerable extent, this limitation can be overcome by using sponge rubber, a material of lower modulus. In general, when the load is small, it is preferable to use rubber springs that carry the load in shear.

5.4.1.3 Environmental Conditions

It is highly common for vibration-control systems to be subjected to harsh environmental conditions. Especially in military applications, extreme ambient temperatures are encountered in addition to exposure to substances like ozone, rocket fuels, and so on. Organic materials are usually more susceptible to these environmental conditions than metal materials. However, owing to the superior mechanical properties of organic materials, such as lighter weight, smaller size, greater damping, and the ability to store large amounts of energy under shock, organic materials that are capable of withstanding the harsh conditions are being developed.

5.4.1.4 Damping Characteristics

In most of the vibration-control applications, the excitations cover a wide range of frequencies and may have random properties requiring the vibration-control systems to possess adequate damping. Elastomers possess very good damping properties when compared with metal springs, and they also eliminate the trouble of standing waves that occurs at high frequencies. If a metal spring is used in vibration-control applications requiring isolation of vibration at high frequencies, it is common to employ rubber pads in series with the metal spring, which also results in the damping of vibrations due to the addition of damping material.

5.4.1.5 Weight and Space Limitations

The amount of load-carrying resilient material is determined by the quantity of energy to be stored. In most of the cases, the vibration amplitude tends to be small relative to the static deflection, and the amount of material may be calculated by equating the energy stored in the material to the work done on the vibration control system.

5.4.1.6 Dynamic Stiffness

In the case of organic materials like rubber, the natural frequency calculated using the stiffness determined from a static-force deflection test of the spring is almost invariably lower than that experienced during vibration; that is, the dynamic modulus is greater than static modulus. The ratio between the dynamic and static modulus is generally between one and two. In many vibration-control applications, it is not feasible to mount the equipment directly upon the vibration-control system (isolator). Instead, a heavy, rigid block, usually made of concrete or heavy steel, supported by the isolator is employed.

5.4.2 Future Trends and Developments

During recent years, there has been considerable interest in the design and implementation of a variety of vibration-control systems. Recent developments in multivariable control design methodology and microprocessor implementation of modern control algorithms have opened a new era for the design of externally controlled passive systems for use in such systems: fuzzy reasoning (Yoshimura, 1998); adaptive algorithms (Venhovens, 1994); observer design (Hedrick et al., 1994); and many others.

Observing these developments combined with the substantial ongoing theoretical advances in the areas of adaptive and nonlinear controls (Astrom and Wittenmark, 1989; Alleyne and Hedrick, 1995), it is expected that the future will bring applications of these techniques in advanced vibration-control system design. For practical implementation, however, it is preferable to simplify these strategies, thus leading to simpler software implementations. Suboptimal policy neglecting some performance requirements can serve as an example of such simplifications.

References

Abe, M. and Igusa, T., Semiactive dynamic vibration absorbers for controlling transient response, *J. Sound Vib.*, 198, 5, 547–569, 1996.

Alleyne, A. and Hedrick, J.K., Nonlinear adaptive control of active suspensions, *IEEE Trans. Control Syst. Technol.*, 3, 94–101, 1995.

Astrom, J.J. and Wittenmark, B. 1989. *Adaptive Control*, Addison-Wesley, Reading, MA.

Austin, S.A., The vibration damping effect of an electrorheological fluid, *ASME J. Vib. Acoust.*, 115, 136–140, 1993.

Banks, H.T. and Ito, K., A unified framework for approximation in inverse problems for distributed parameter systems, *Control Theor. Adv. Technol.*, 4, 73–90, 1988.

Banks, H.T. and Kunisch, K. 1989. *Estimation Techniques for Distributed Parameter Systems*, Birkhauser, Boston, MA.

Bayo, E., A finite-element approach to control the end-point motion of a single-link flexible robot, *J. Robotic Syst.*, 4, 63–75, 1987.

Cannon, R.H. Jr. and Schmitz, E., Initial experiments on the end-point control of a flexible one-link robot, *Int. J. Robotics Res.*, 3, 62–75, 1984.

Chalhoub, N.G. and Ulsoy, A.G., Control of flexible robot arm: experimental and theoretical results, *ASME J. Dyn. Syst. Meas. Control*, 109, 299–309, 1987.

Choi, S.B., Vibration control of flexible structures using ER dampers, *ASME J. Dyn. Syst. Meas. Control*, 121, 134–138, 1999.

Crosby, M. and Karnopp, D.C. 1973. The active damper—a new concept for shock and vibration control, *Shock Vibration Bulletin*, Part H, Washington, DC.

Dadfarnia, M., Jalili, N., Xian, B. and Dawson, D.M. 2003. Lyapunov-based piezoelectric control of flexible cartesian robot manipulators, In *Proceedings of the 22nd American Control Conference (ACC'03)*, Denver, CO.

Dadfarnia, M., Jalili, N., Liu, Z., and Dawson, D.M., An observer-based piezoelectric control of flexible cartesian robot manipulators: theory and experiment, *J. Control Eng. Practice*, 12, 1041–1053, 2004.

de Querioz, M.S., Dawson, D.M., Agrawal, M., and Zhang, F., Adaptive nonlinear boundary control of a flexible link robot arm, *IEEE Trans. Robotics Automat.*, 15, 779–787, 1999.

de Querioz, M.S., Dawson, D.M., Nagarkatti, S.P., and Zhang, F. 2000. *Lyapunov-Based Control of Mechanical Systems*, Birkhauser, Boston, MA.

Dimarogonas-Andrew, D. and Kollias, A., Smart electrorheological fluid dynamic vibration absorber, *Intell. Struct. Mater. Vib. ASME Des. Div.*, 58, 7–15, 1993.

Dowell, D.J., Cherry, S. 1994. Semiactive friction dampers for seismic response control of structures, Vol. 1, pp. 819–828. In *Proceedings of the Fifth US National Conference on Earthquake Engineering*, Chicago, IL.

Elmali, H. and Olgac, N., Sliding mode control with perturbation estimation (SMCPE): a new approach, *Int. J. Control*, 56, 923–941, 1992.

Emura, J., Kakizaki, S., Yamaoka, F., and Nakamura, M. 1994. Development of the SA suspension system based on the sky-hook damper theory, *SAE* Paper No. 940863.

Esmailzadeh, E. and Jalili, N., Optimal design of vibration absorbers for structurally damped Timoshenko beams, *ASME J. Vib. Acoust.*, 120, 833–841, 1998a.

Esmailzadeh, E. and Jalili, N., Parametric response of cantilever timoshenko beams with tip mass under harmonic support motion, *Int. J. Non-Linear Mech.*, 33, 765–781, 1998b.

Filipović, D. and Schröder, D., Vibration absorption with linear active resonators: continuous and discrete time design and analysis, *J. Vib. Control*, 5, 685–708, 1999.

Frahm, H., Devices for damping vibrations of bodies, US Patent #989958, 1911.

Franchek, M.A., Ryan, M.W., and Bernhard, R.J., Adaptive-passive vibration control, *J. Sound Vib.*, 189, 565–585, 1995.

Fujita, T., Katsu, M., Miyano, H., and Takanashi, S., Fundamental study of active-passive mass damper using *XY*-motion mechanism and hydraulic actuator for vibration control of tall building, *Trans. Jpn Soc. Mech. Engrs*, Part C, 57, 3532–3539, 1991.

Garcia, E., Dosch, J., and Inman, D.J., The application of smart structures to the vibration suppression problem, *J. Intell. Mater. Syst. Struct.*, 3, 659–667, 1992.

Ge, S.S., Lee, T.H., and Zhu, G., A nonlinear feedback controller for a single-link flexible manipulator based on a finite element method, *J. Robotic Syst.*, 14, 165–178, 1997.

Ge, S.S., Lee, T.H., and Zhu, G., Asymptotically stable end-point regulation of a flexible SCARA/cartesian robot, *IEEE/ASME Trans. Mechatron.*, 3, 138–144, 1998.

Ge, S.S., Lee, T.H., and Gong, J.Q., A robust distributed controller of a single-link SCARA/cartesian smart materials robot, *Mechatronics*, 9, 65–93, 1999.

Hedrick, J.K., Rajamani, R., and Yi, K., Observer design for electronic suspension applications, *Vehicle Syst. Dyn.*, 23, 413–440, 1994.

Horton, D.N. and Crolla, D.A., Theoretical analysis of a SA suspension fitted to an off-road vehicle, *Vehicle Syst. Dyn.*, 15, 351–372, 1986.

Hrovat, D., Barker, P., and Rabins, M., Semiactive versus passive or active tuned mass dampers for structural control, *J. Eng. Mech.*, 109, 691–705, 1983.

Hrovat, D., Margolis, D.L., and Hubbard, M., An approach toward the optimal SA suspension, *ASME J. Dyn. Syst. Meas. Control*, 110, 288–296, 1988.

Hubard, M. and Marolis, D. 1976. The SA spring: is it a viable suspension concept?, pp. 1–6. In *Proceedings of the Fourth Intersociety Conference on Transportation*, Los Angeles, CA.

Ikeda, T. 1990. *Fundamental of Piezoelectricity*, Oxford University Press, Oxford, New York.

Inman, D.J. 1994. *Engineering Vibration*, Prentice Hall, Englewood Cliffs, NJ.

Jalili, N., A new perspective for semi-automated structural vibration control, *J. Sound Vib.*, 238, 481–494, 2000.

Jalili, N., An infinite dimensional distributed base controller for regulation of flexible robot arms, *ASME J. Dyn. Syst. Meas. Control*, 123, 712–719, 2001.

Jalili, N., A comparative study and analysis of SA vibration-control systems, *ASME J. Vib. Acoust.*, 124, 593–605, 2002.

Jalili, N., Dadfarnia, M., Hong, F., and Ge, S.S. 2002. An adaptive non model-based piezoelectric control of flexible beams with translational base, pp. 3802–3807. In *Proceedings of the American Control Conference (ACC'02)*, Anchorage, AK.

Jalili, N., Elmali, H., Moura, J., and Olgac, N. 1997. Tracking control of a rotating flexible beam using frequency-shaped sliding mode control, pp. 2552–2556. In *Proceedings of the 16th American Control Conference (ACC'97)*, Albuquerque, NM.

Jalili, N. and Esmailzadeh, E., Adaptive-passive structural vibration attenuation using distributed absorbers, *J. Multi-Body Dyn.*, 216, 223–235, 2002.

Jalili, N. and Fallahi, B., Design and dynamics analysis of an adjustable inertia absorber for SA structural vibration attenuation, *ASCE J. Eng. Mech.*, 128, 1342–1348, 2002.

Jalili, N., Fallahi, B., and Kusculuoglu, Z.K., A new approach to SA vibration suppression using adjustable inertia absorbers, *Int. J. Model. Simulat.*, 21, 148–154, 2001.

Jalili, N. and Olgac, N., Time-optimal/sliding mode control implementation for robust tracking of uncertain flexible structures, *Mechatronics*, 8, 121–142, 1998a.

Jalili, N. and Olgac, N. 1998b. Optimum delayed feedback vibration absorber for MDOF mechanical structures, In *Proceedings of the 37th IEEE Conference on Decision Control* (CDC'98), Tampa, FL.

Jalili, N. and Olgac, N., Multiple identical delayed-resonator vibration absorbers for multi-DoF mechanical structures, *J. Sound Vib.*, 223, 567–585, 1999.

Jalili, N. and Olgac, N., Identification and re-tuning of optimum delayed feedback vibration absorber, *AIAA J. Guidance Control Dyn.*, 23, 961–970, 2000a.

Jalili, N. and Olgac, N., A sensitivity study of optimum delayed feedback vibration absorber, *ASME J. Dyn. Syst. Meas. Control*, 121, 314–321, 2000b.

Junkins, J.L. and Kim, Y. 1993. *Introduction to Dynamics and Control of Flexible Structures*. AIAA Educational Series, Washington, DC.

Karnopp, D., Design principles for vibration-control systems using SA dampers, *ASME J. Dyn. Syst. Meas. Control*, 112, 448–455, 1990.

Karnopp, D.C., Crodby, M.J., and Harwood, R.A., Vibration control using SA force generators, *J. Eng. Ind.*, 96, 619–626, 1974.

Kim, K. and Jeon, D., Vibration suppression in an MR fluid damper suspension system, *J. Intell. Mater. Syst. Struct.*, 10, 779–786, 2000.

Knowles, D., Jalili, N., and Khan, T. 2001a. On the nonlinear modeling and identification of piezoelectric inertial actuators, In *Proceedings of the 2001 International Mechanical Engineering Congress and Exposition* (IMECE'01), New York.

Knowles, D., Jalili, N., and Ramadurai, S. 2001b. Piezoelectric structural vibration control using active resonator absorber, In *Proceedings of the 2001 International Mechanical Engineering Congress and Exposition* (IMECE'01), New York.

Korenev, B.G. and Reznikov, L.M. 1993. *Dynamic Vibration Absorbers: Theory and Technical Applications*, Wiley, Chichester, England.

Lee-Glauser, G.J., Ahmadi, G., and Horta, L.G., Integrated passive/active vibration absorber for multistory buildings, *ASCE J. Struct. Eng.*, 123, 499–504, 1997.

Liu, Z., Jalili, N., Dadfarnia, M., and Dawson, D.M. 2002. Reduced-order observer based piezoelectric control of flexible beams with translational base, In *Proceedings of the 2002 International Mechanical Engineering Congress and Exposition* (IMECE'02), New Orleans, LA.

Liu, J., Schönecker, A., and Frühauf, U. 1997. Application of discrete Fourier transform to electronic measurements, pp. 1257–1261. In *International Conference on Information, Communications and Signal Processing*, Singapore.

Liu, H.J., Yang, Z.C., and Zhao, L.C., Semiactive flutter control by structural asymmetry, *J. Sound Vib.*, 229, 199–205, 2000.

Luo, Z.H., Direct strain feedback control of flexible robot arms: new theoretical and experimental results, *IEEE Trans. Automat. Control*, 38, 1610–1622, 1993.

Luo, Z.H., Guo, B.Z., and Morgul, O. 1999. *Stability and Stabilization of Finite Dimensional Systems with Applications*, Springer, London.

Margolis, D., Retrofitting active control into passive vibration isolation systems, *ASME J. Vib. Acoust.*, 120, 104–110, 1998.

Meirovitch, L. 1986. *Elements of Vibration Analysis*, McGraw-Hill, New York.

Miller, L.R. and Nobles, C.M. 1988. The design and development of a SA suspension for military tank, *SAE* Paper No. 881133.

Moura, J.T., Roy, R.G., and Olgac, N., Frequency-shaped sliding modes: analysis and experiments, *IEEE Trans. Control Syst. Technol.*, 5, 394–401, 1997.

Nemir, D., Lin, Y., and Osegueda, R.A., Semiactive motion control using variable stiffness, *ASCE J. Struct. Eng.*, 120, 1291–1306, 1994.

Olgac, N., Delayed resonators as active dynamic absorbers, US Patent #5431261, 1995.

Olgac, N., Elmali, H., Hosek, M., and Renzulli, M., Active vibration control of distributed systems using delayed resonator with acceleration feedback, *ASME J. Dyn. Syst., Meas. Control*, 119, 380–389, 1997.

Olgac, N., Elmali, H., and Vijayan, S., Introduction to dual frequency fixed delayed resonator (DFFDR), *J. Sound Vib.*, 189, 355–367, 1996.

Olgac, N. and Holm-Hansen, B., Novel active vibration absorption technique: delayed resonator, *J. Sound Vib.*, 176, 93–104, 1994.

Olgac, N. and Holm-Hansen, B., Tunable active vibration absorber: the delayed resonator, *ASME J. Dyn. Syst., Meas. Control*, 117, 513–519, 1995.

Olgac, N. and Jalili, N., Modal analysis of flexible beams with delayed-resonator vibration absorber: theory and experiments, *J. Sound Vib.*, 218, 307–331, 1998.

Olgac, N. and Jalili, N., Optimal delayed feedback vibration absorber for flexible beams, *Smart Struct.*, 65, 237–246, 1999.

Oueini, S.S., Nayfeh, A.H., and Pratt, J.R., A nonlinear vibration absorber for flexible structures, *Nonlinear Dyn.*, 15, 259–282, 1998.

Petek, N.K., Shock absorbers uses electrorheological fluid, *Automot. Eng.*, 100, 27–30, 1992.

Petek, N.K., Romstadt, D.L., Lizell, M.B., and Weyenberg, T.R. 1995. Demonstration of an automotive SA suspension using electro-rheological fluid, *SAE* Paper No. 950586.

Puksand, H., Optimum conditions for dynamic vibration absorbers for variable speed systems with rotating and reciprocating unbalance, *Int. J. Mech. Eng. Educ.*, 3, 145–152, 1975.

Renzulli, M., Ghosh-Roy, R., and Olgac, N., Robust control of the delayed resonator vibration absorber, *IEEE Trans. Control Syst. Technol.*, 7, 683–691, 1999.

Shaw, J., Adaptive vibration control by using magnetostrictive actuators, *J. Intell. Mater. Syst. Struct.*, 9, 87–94, 1998.

Singh, T., Golnaraghi, M.F., and Dubly, R.N., Sliding-mode/shaped-input control of flexible/rigid link robots, *J. Sound Vib.*, 171, 185–200, 1994.

Sinha, A., Optimum vibration control of flexible structures for specified modal decay rates, *J. Sound Vib.*, 123, 185–188, 1988.

Skaar, S.B. and Tucker, D., Point Control of a one-link flexible manipulator, *J. Appl. Mech.*, 53, 23–27, 1986.

Snowdon, J.C. 1968. *Vibration and Shock in Damped Mechanical Systems*, Wiley, New York.

Soong, T.T. and Constantinou, M.C. 1994. *Passive and Active Structural Control in Civil Engineering*, Springer, Wien.

Spencer, B.F., Yang, G., Carlson, J.D., and Sain, M.K. 1998. Smart dampers for seismic protection of structures: a full-scale study, In *Proceedings of the Second World Conference on Structural Control*, Kyoto, Japan.

Stribersky, A., Muller, H., and Rath, B., The development of an integrated suspension control technology for passenger trains, *Proc. Inst. Mech. Engrs*, 212, 33–41, 1998.

Sturk, M., Wu, M., and Wong, J.Y., Development and evaluation of a high voltage supply unit for electrorheological fluid dampers, *Vehicle Syst. Dyn.*, 24, 101–121, 1995.

Sun, J.Q., Jolly, M.R., and Norris, M.A., Passive, adaptive, and active tuned vibration absorbers — a survey, *ASME Trans.*, 117, 234–242, 1995 (Special 50th Anniversary, Design issue).

Sun, D. and Mills, J.K. 1999. PZT actuator placement for structural vibration damping of high speed manufacturing equipment, pp. 1107–1111. In *Proceedings of the American Control Conference* (ACC'99), San Diego, CA.

Sun, Y. and Parker, G.A., A position controlled disc valve in vehicle SA suspension systems, *Control Eng. Practice*, 1, 927–935, 1993.

Tanaka, N. and Kikushima, Y., Impact vibration control using a SA damper, *J. Sound Vib.*, 158, 277–292, 1992.

Venhovens, P.J., The development and implementation of adaptive SA suspension control, *Vehicle Syst. Dyn.*, 23, 211–235, 1994.

Walsh, P.L. and Lamnacusa, J.S., A variable stiffness vibration absorber for minimization of transient vibrations, *J. Sound Vib.*, 158, 195–211, 1992.

Wang, K.W., Kim, Y.S., and Shea, D.B., Structural vibration control via electrorheological-fluid-based actuators with adaptive viscous and frictional damping, *J. Sound Vib.*, 177, 227–237, 1994.

Wang, K.W., Lai, J.S., and Yu, W.K., An energy-based parametric control approach for structural vibration suppression via SA piezoelectric networks, *ASME J. Vib. Acoust.*, 118, 505–509, 1996.

Warburton, G.B. and Ayorinde, E.O., Optimum absorber parameters for simple systems, *Earthquake Eng. Struct. Dyn.*, 8, 197–217, 1980.

Yeung, K.S. and Chen, Y.P., A new controller design for manipulators using the theory of variable structure systems, *IEEE Trans. Automat. Control*, 33, 200–206, 1988.

Yeung, K.S. and Chen, Y.P., Regulation of a one-link flexible robot arm using sliding mode control technique, *Int. J. Control*, 49, 1965–1978, 1989.

Yoshimura, T., A SA suspension of passenger cars using fuzzy reasoning and the filed testing, *Int. J. Vehicle Des.*, 19, 150–166, 1998.

Yuh, J., Application of discrete-time model reference adaptive control to a flexible single-link robot, *J. Robotic Syst.*, 4, 621–630, 1987.

Zhu, G., Ge, S.S., Lee, T.H. 1997. Variable structure regulation of a flexible arm with translational base, pp. 1361–1366. In *Proceedings of the 36th IEEE Conference on Decision and Control*, San Diego, CA.

6

Helicopter Rotor Tuning

Kourosh Danai

University of Massachusetts

Summary

Before a helicopter leaves the plant, its rotors need to be tuned so that the helicopter vibration meets the required specifications during different flight regimes. For this, three different adjustments can be made to each rotor blade in response to the magnitude and phase of vibration. In this chapter (also see Chapter 7), the basic concepts for determining the blade adjustments are discussed, and three methods with fundamentally different approaches are described. A neural network-based method is described, which trains a feedforward network as the inverse model of the effect of the blade adjustments on helicopter vibrations, and uses the inverse model to determine the blade adjustments. Another is a probability-based method that maximizes the likelihood of success of the selected blade adjustments based on a stochastic model of the probability densities of the vibration components. The third method is an adaptive method that uses an interval model to represent the range of effect of blade adjustments on helicopter vibration, so as to cope with the nonlinear and stochastic nature of aircraft vibration. This method includes the a priori *knowledge of the process by defining the initial coefficients of the interval model according to sensitivity coefficients between the blade adjustments and helicopter vibration, but then transforms these coefficients into intervals and updates them after each tuning iteration, to improve the model estimation accuracy. The details of rotor tuning are described through a case study, which demonstrates the application of the adaptive method.*

6.1 Introduction

Helicopter rotor tuning (track and balance) is the process of adjusting the rotor blades so as to reduce the aircraft vibration and the spread of rotors. Rotor tuning as applied to Sikorsky's Black Hawk (H-60) helicopters is performed as follows. For initial measurements, the aircraft is flown through six different regimes, during which measurements of rotor track and vibration (balance) are recorded. Rotor track is measured by optical sensors, which detect the vertical position of the blades. Vibration is measured at the frequency of once per blade revolution (per rev) by two accelerometers, A and B, attached to the sides of the cockpit (see Figure 6.1, detail B). The vibration data are vectorially combined into two components: A + B, representing the vertical vibration of the aircraft, and A − B, representing its roll vibration. A sample of peak vibration levels for the six flight regimes, as well as the peak angular positions relative to a reference blade, are given in Table 6.1, along with a sample of track data.

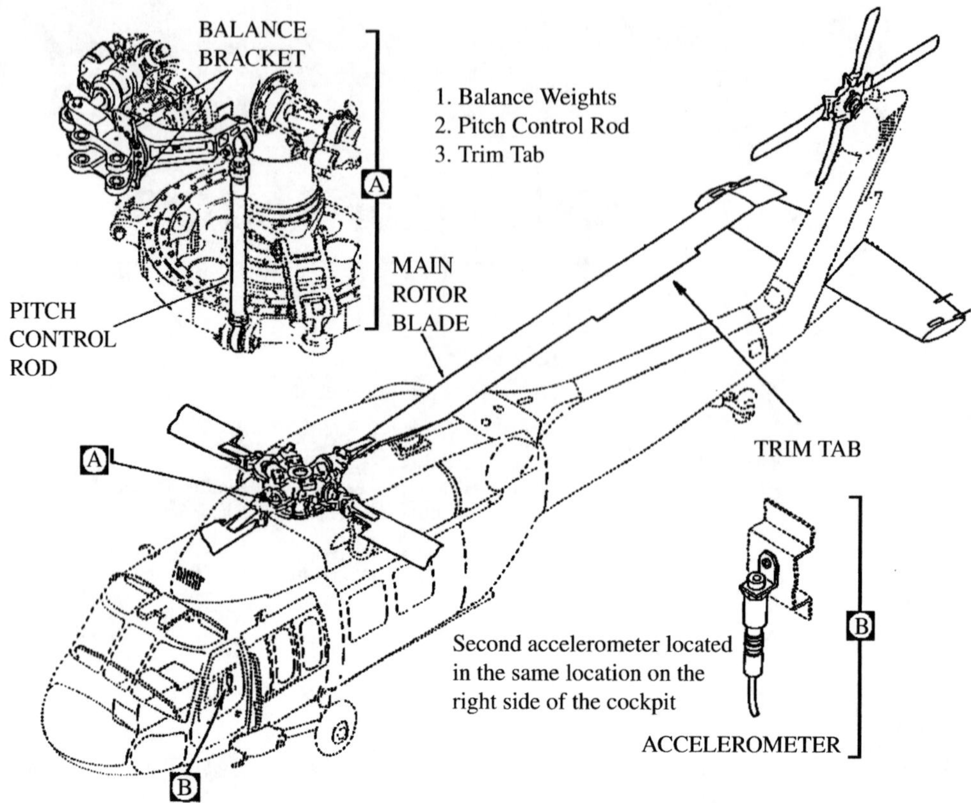

FIGURE 6.1 Illustration of the position of accelerometers A and B on the aircraft, and the rotor blade adjustments (push rod, trim tab, and hub weights).

TABLE 6.1 Typical Track and Balance Data Recorded during a Flight

Flight Regime	Vibration			
	A + B		A − B	
	Magnitude (ips)	Phase (°)	Magnitude (ips)	Phase (°)
fpm	0.19	332	0.38	272
hov	0.07	247	0.10	217
80	0.02	86	0.04	236
120	0.04	28	0.04	333
145	0.02	104	0.07	162
vh	0.10	312	0.12	211
	Track (mm)			
	Blade #			
	1	2	3	4
fpm	−2	3	1	−2
hov	−1	3	0	−2
80	1	11	1	−13
120	2	13	−1	−14
145	5	18	−3	−20
vh	2	13	−1	−14

The six flight regimes in Table 6.1 are: ground (fpm), hover (hov), 80 knots (80), 120 knots (120), 145 knots (145), and maximum horizontal speed (vh). The track data indicate the vertical position of each blade relative to a mean position.

In order to bring track and one per rev vibration within specification, three types of adjustments can be made to the rotor system: pitch control rod adjustments, trim tab adjustments, and balance weight adjustments (see Figure 6.1). Pitch control rods can be extended or contracted by a certain number of notches to alter the pitch of the rotor blades. Positive push rod adjustments indicate extension. Trim tabs, which are adjustable surfaces on the trailing edge of the rotor blades, affect the aerodynamic pitch moment of the air foils and consequently their vibration characteristics. Tab adjustments are measured in thousandths of an inch, with positive and negative changes representing upward and downward tabbing, respectively. Finally, balance weights can be either added to or removed from the rotor hub to tune vibrations through changes in the blade mass. Balance weights are measured in ounces, with positive adjustments representing the addition of weight. In the case of the Sikorsky H-60 helicopter, which has four main rotor blades, a total of 12 adjustments can be made to tune the rotors (i.e., three adjustments per blade). Among them, balance weights primarily affect the ground vibration, so they are not commonly used for in-flight tuning. Furthermore, since the symmetry of rotor blades in four-bladed aircraft produces identical effects for adjustments to opposite blades, the combined form of blade adjustments to opposite blade pairs can be used as inputs. Accordingly, the input vector can be defined as

$$\mathbf{\Delta x} = [\Delta x_1, \Delta x_2, \Delta x_3, \Delta x_4]^{\mathrm{T}} \tag{6.1}$$

where Δx_1 and Δx_3 denote the combined (condensed) trim tab adjustments (ΔT) to blade combinations one/three and two/four, respectively, and Δx_2 and Δx_4 represent the combined pitch control rod adjustments (ΔP) to blade combinations one/three and two/four, respectively. The relationships between the combined and individual adjustments are in the form:

$$\Delta x_1 = \Delta T_3 - \Delta T_1 \tag{6.2}$$

$$\Delta x_2 = \Delta P_3 - \Delta P_1 \tag{6.3}$$

$$\Delta x_3 = \Delta T_4 - \Delta T_2 \tag{6.4}$$

$$\Delta x_4 = \Delta P_4 - \Delta P_2 \tag{6.5}$$

Ideally, identical adjustments made to any two aircraft with different tail numbers should result in identical changes in vibration. In reality, however, significant inconsistencies in vibration changes may be present for identical adjustments to different tail numbers. This is perhaps due more to nonuniformity of flight conditions from weather or error in implementing the blade adjustments than factors such as dissimilarities between aircraft and rotor blades.

Virtually all of the current systems of rotor track and balance rely on the strategy shown in Figure 6.2, whereby the measurements of the flight just completed are used as the basis of search for the new blade adjustments. The search for blade adjustments is guided by the "process model" (see Figure 6.2), which represents the relationship between vibration changes and blade adjustments. A difficulty of rotor tuning is the excess of equations compared to degrees of freedom (four inputs to control 24 outputs), which translates into one-to-many mapping. Another difficulty is caused by the high level of noise present in the vibration measurements.

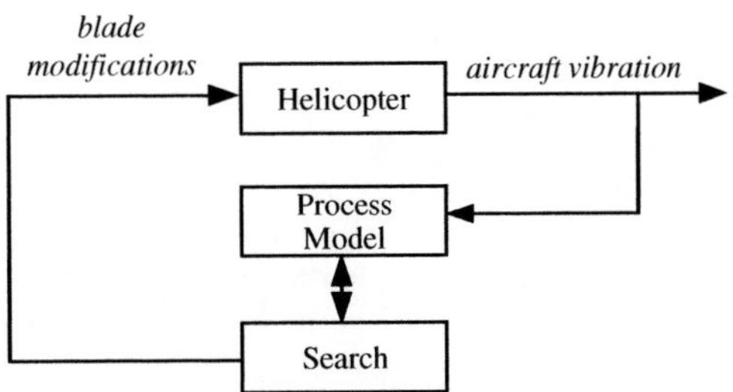

FIGURE 6.2 Tuning strategy of the current methods.

The traditional approach to rotor tuning uses linear relationships to define the process model and uses model inversion to streamline the search. The drawback of the traditional approach, therefore, is its neglect of the potential nonlinearity of track and balance, and the vibration noise, as well as its limited capacity to produce comprehensive solutions to facilitate model inversion due to its consideration of the most extreme vibration components. In an attempt to include the potential nonlinearity of the process, Taitel et al. (1995) trained a set of neural networks with actual track and balance data to map vibration measurements to blade adjustments as well as to evaluate the goodness of the solution. In effect, they developed an inverse model based on the solutions available in the historical track and balance data, and provided a forward model to evaluate the solution. The potential advantage of this method is that it can interpolate among the historical solutions to address potential nonlinearity and vibration noise. Its disadvantages are that it is only applicable to helicopters with extensive track and balance history, and that its solutions are constrained by those contained in the historical data.

Another deviation from the traditional approach is introduced by Ventres and Hayden (2000), who define the relationships between blade adjustments and vibration in frequency domain, and provide an extension of these relationships to higher order vibrations. They use an optimization method to search for the adjustments to reduce per rev vibration as well as higher-order vibrations. Accordingly, this approach has the capacity to provide a comprehensive solution, but it too neglects the potential nonlinearity between the blade adjustments and aircraft vibration as well as the noise in the measurements.

The most recent solutions to rotor tuning are those by Wang et al. (2005a, 2005b), which are designed to address both the stochastics of vibration and the potential nonlinearity of the tuning process. In the first solution, which is a probability-based method, the underlying model comprises two components: a deterministic component and a probability component. The method relies on the probability model to estimate the likelihood of the measured vibration satisfying the specifications and to search for blade adjustments that will maximize this likelihood. The likelihood measures in the probability model are computed according to the probability distribution of vibration derived from historical track and balance data. The second solution is an adaptive method that uses an interval model to cope with the potential nonlinearity of the process and to account for vibration noise. This method, which also incorporates learning to provide adaptation to the rotor tuning process, initializes the coefficients of the interval model according to the sensitivity coefficients between the blade adjustments and helicopter vibration. However, it modifies these coefficients after the first iteration to better represent the vibration measurements acquired. This method takes into account vibration data from all of the flight regimes during the search for the appropriate blade adjustments; therefore, it has the capacity to provide comprehensive solutions. The remainder of this chapter describes three of the methods discussed above to provide a representation of various solutions proposed for rotor tuning, followed by a case study to demonstrate the application of the adaptive method.

6.2 Neural Network-Based Tuning

As mentioned earlier, rotor tuning in four-bladed aircraft is performed by first specifying a condensed set of adjustments to reduce vibrations, and then expanding these adjustments into a detailed set to satisfy the track requirements. This same strategy is implemented in the system of neural networks shown in Figure 6.3 (Taitel et al., 1995). The first network in this system, called the selection net, determines the condensed blade adjustments (output) that will bring about a given change in vibration (input). To eliminate vibration, the negatives of the vibration measurements from the flight are utilized as inputs to this network. The validity of the condensed adjustments is then checked by predicting their effect on vibration via the condensed simulation net. Theoretically, these simulated vibration changes should be the negative of the vibration measurements from the aircraft so that their summation will be zero. However, owing to the inexactness of the neural network models and noise, the resultant vibration will most likely not equal zero. In cases where the resultant vibration is not within specifications (usually less than 0.20 inches per second [ips]), the condensed adjustments may be refined by feeding the resultant

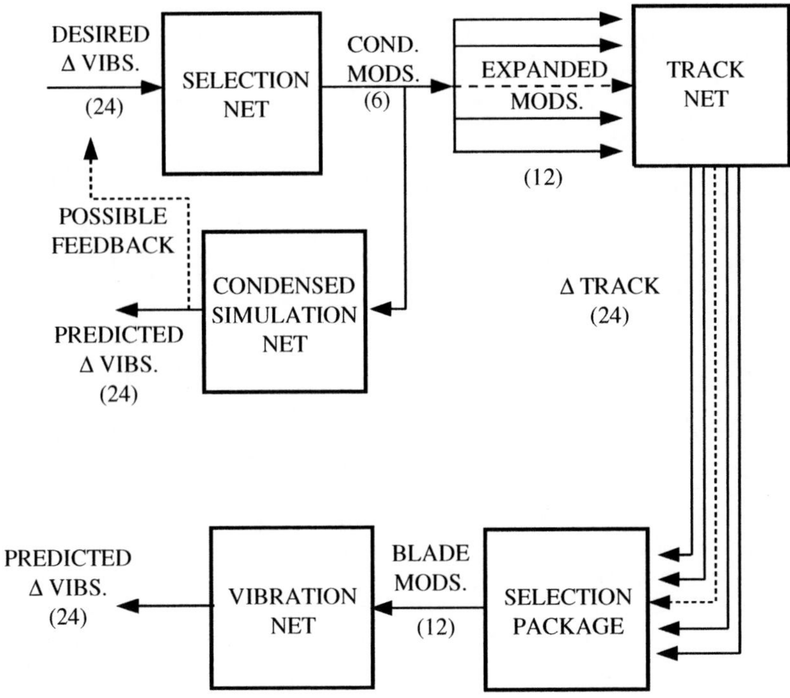

FIGURE 6.3 Schematic of the rotor tuning system. The numbers inside parentheses represent the number of inputs or outputs of individual nets.

vibration back into the selection net. This feedback is depicted by the dashed feedback line in Figure 6.3. It should be noted that the condensed simulation net may also serve as a diagnostic tool by indicating behavior out of the norm. For example, an aircraft with vibrations significantly different from those predicted by this network may suffer from defective components.

Just as with the traditional approach, once the condensed solution has been specified, it needs to be expanded into a detailed form to satisfy the rotor track requirements. As previously mentioned, the condensed set of adjustments may be viewed as the constraint on detailed adjustments so as to ensure that the vibration solution is not compromised for track. Each one of these detailed sets of adjustments is a candidate for the final rotor tuning solution, and it is left to the track net and the selection package to determine which set of detailed adjustments provides the best tracking performance. For selection purposes, the track net simulates the changes in track due to a candidate set of detailed adjustments, and then adds these changes to the initial track measurements from the flight to estimate the resultant track. The set of detailed adjustments that yields the smallest estimated track (i.e., smallest maximum blade spread) is selected as the solution to the rotor tuning problem. The selected set of detailed adjustments is then checked via the vibration net, which, similar to the condensed simulation net, serves as an independent evaluator of the selected adjustments.

6.3 Probability-Based Tuning

The noted contribution of this method is its introduction of the likelihood of success as a criterion in the search for the blade adjustments (Wang et al., 2005). This method speculates the effectiveness of various adjustment sets in reducing the vibration and selects the set with the maximum probability of producing acceptable vibration. The concept of this method is explained in the context of a simple example. If the measured vibration from the current flight is denoted by $V_j(k-1)$ and the estimated vibration change according to the model is represented by $\Delta\hat{V}_j(k) = f(\Delta x)$ as a function of the blade adjustments, Δx, then the predicted vibration of the next flight, $\hat{V}_j(k)$, can be defined as

$$\hat{V}_j(k) = V_j(k-1) + \Delta\hat{V}_j(k) \tag{6.6}$$

$$V_j(k) = \hat{V}_j(k) + \hat{e}_j(k) \tag{6.7}$$

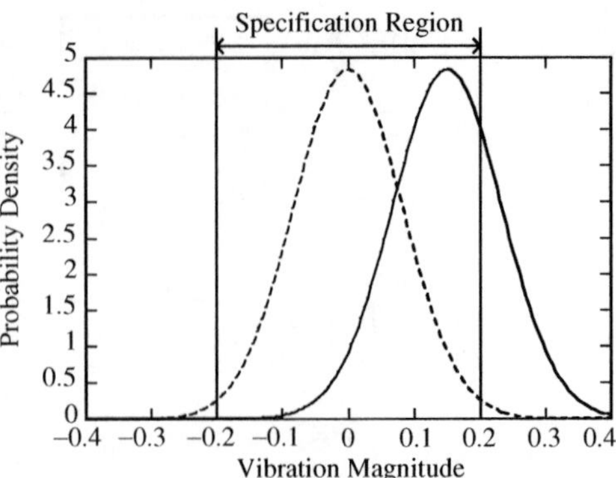

FIGURE 6.4 Illustration of improved placement of the predicted vibration within the specification range.

where $V_j(k)$ denotes the measured vibration for the next flight. In rotor tuning, the adjustments are selected according to the predicted vibration, $\hat{V}_j(k)$, whereas the objective is defined in terms of the measured vibration. The inclusion of the probability model here is to account for the inevitable uncertainty in the actual position of the measured vibration. According to Equation 6.7, the mean value of the measured vibration is equal to the value of the predicted vibration plus the mean value of the prediction error. However, since the predicted vibration is a deterministic entity, the probability distribution of the measured vibration is the same as that of the prediction error. Accordingly, whereas the nominal value of the measured vibration can be controlled by the blade adjustment, its optimal position within the specification region should be determined according to its probability distribution. For a case where the prediction error, $\hat{e}_j(k)$, is zero-mean and normally distributed, as illustrated in Figure 6.4, placing the predicted vibration at the center of the specification range will be synonymous with maximizing the probability that the measured vibration will be within the range. The likelihood of success of blade adjustments can therefore be measured by the area under the probability density function of prediction error located within the specification region. The blade adjustment set that produces the highest likelihood will be the preferred adjustment.

The main difficulty with rotor tuning, however, is the limited number of DoFs, which precludes perfect positioning of the predicted vibration. This point is illustrated in Figure 6.5 for a case where two vibration components are to be positioned at the center of the specification region with only one adjustment. If one assumes that the effect of adjustment, Δx, on the change in the two vibration components, $\Delta\hat{V}_j(k)$, can be represented by a linear model, as

$$\Delta\hat{V}_j(k) = a_{ij}\Delta x$$

then the position of the predicted vibration components will be constrained to the line L in Figure 6.5. As illustrated in this figure, since it will be impossible to place the predicted vibration components at the center, a compromised position needs to be selected. In this method, the best compromised position for the predicted vibration is that which renders the largest probability of satisfying the specifications for the measured vibration. This position, for the two-component vibration example, is one that maximizes $P_r[(V_1, V_2) \in S] = \int_{(V_1,V_2)\in S} p(V_1, V_2)\mathrm{d}V_1\,\mathrm{d}V_2$. The above formulation indicates that the placement of the predicted vibration requires knowledge of

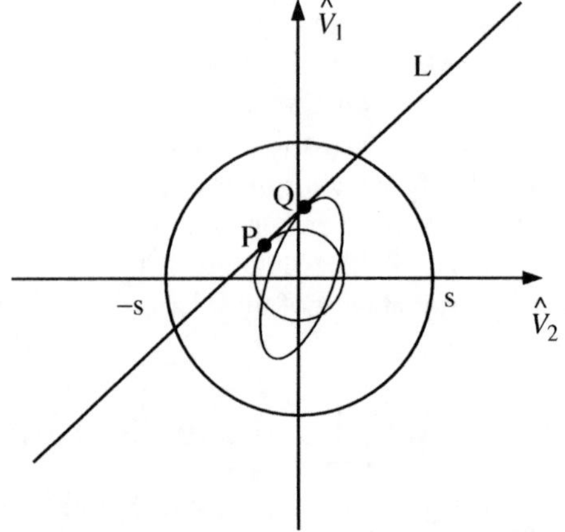

FIGURE 6.5 Restricted placement of vibration components within the specification region for a two-dimensional case.

the joint probability density function, $p(V_1, V_2)$, of the vibration components. In the ideal case of independent vibration components with equal probability distributions, the loci of the points with equal probabilities $P_r[(V_1,\ldots,V_n)\in S]$ are surfaces of hyperspheres. Such ideal loci for the two-component vibration example of Figure 6.5 are circles centered at the origin (see Figure 6.5), which lead to point P as the best compromised position closest on line L to the center of the specification circle. Point P, however, does not represent the best position if the two vibration components are dependent or have unequal distributions. The loci of equal probabilities for this more general case are elliptical, as also shown in Figure 6.5, indicating point Q as the best position on line L for placing the predicted vibration. The inadequacy of the DoFs illustrated here is exacerbated in rotor tuning, where 24 correlated vibration components need to be positioned within the specification region using only four condensed blade adjustments. For the 24-component vector of measured vibration $\mathbf{V}(k)=[V_{c1}(k), V_{s1}(k),\ldots, V_{c12}(k), V_{s12}(k)]^{\mathrm{T}}$, where V_c and V_s represent the cosine and sine components of each vibration measurement, respectively, the joint probability density function of measured vibration for the kth flight, $\mathbf{V}(k)$, can be characterized as an N-dimensional Gaussian function:

$$p(\mathbf{V}(k))=\frac{1}{(2\pi)^{N/2}|\Phi|^{1/2}}\exp\left[-\frac{1}{2}\hat{\mathbf{e}}(k)^{\mathrm{T}}\Phi^{-1}\hat{\mathbf{e}}(k)\right] \tag{6.8}$$

$$\hat{\mathbf{e}}(k)=\mathbf{V}(k)-\mathbf{V}(k-1)-C\Delta\mathbf{x}(k) \tag{6.9}$$

where Φ represents the covariance matrix of the prediction error. Now, if $\Gamma=\{|\mathbf{V}_j|=\sqrt{V_{cj}^2+V_{sj}^2}\leq\alpha, j=1,\ldots,12\}$ denotes the specification region in 24-dimensional Euclidean space, the blade adjustments, $\Delta\mathbf{x}^*$, can be selected such that the probability that the measured vibration is within the acceptable range is maximized (see also Table 6.2). Formally,

$$\Delta\mathbf{x}^*=\arg_{\Delta\mathbf{x}}\max\left[\Pr(\mathbf{V}(k)\in\Gamma)=\int_\Gamma p(\mathbf{V}(k))\mathrm{d}\mathbf{V}(k)\right] \tag{6.10}$$

TABLE 6.2 Summary of Probability-Based Tuning

For the input vector:

$$\Delta\mathbf{x}=[\Delta x_1, \Delta x_2, \Delta x_3, \Delta x_4]^{\mathrm{T}}$$

where Δx_1 and Δx_3 denote the combined trim tab adjustments to blade combinations one to three and two to four, respectively, and Δx_2 and Δx_4 represent the combined pitch control rod adjustments to blade combinations one to three and two to four, respectively, the blade adjustments, $\Delta\mathbf{x}$, can be selected such that the probability that the measured vibration is within the acceptable range is maximized. Formally,

$$\Delta\mathbf{x}^*=\arg_{\Delta\mathbf{x}}\max[\Pr(\mathbf{V}(k)\in\Gamma)=\textstyle\int_\Gamma p(\mathbf{V}(k))\mathrm{d}\mathbf{V}(k)]$$

where $\Pr(\mathbf{V}(k))$ denotes the probability of the measured vibration, Γ denotes the specification region in 24-dimensional Euclidean space, and $p(\mathbf{V}(k))$ represents the joint probability density of the measured vibration for the kth flight characterized as an N-dimensional Gaussian function:

$$p(\mathbf{V}(k))=\frac{1}{(2\pi)^{N/2}|\Phi|^{1/2}}\exp[-\frac{1}{2}\hat{\mathbf{e}}(k)^{\mathrm{T}}\Phi^{-1}\hat{\mathbf{e}}(k)]$$

with

$$\hat{\mathbf{e}}(k)=\mathbf{V}(k)-\mathbf{V}(k-1)-C\Delta\mathbf{x}(k)$$

representing the predicted error in vibration.

6.4 Adaptive Tuning

The schematic of this method is shown in Figure 6.6 (Wang et al., 2005). As in the other methods, it uses a process model as the basis of search for the appropriate blade adjustments, but instead of using a linear model, it uses an interval model to accommodate process nonlinearity and measurement noise. According to this model, the feasible region of the process is estimated first, to include the adjustments that will result in acceptable vibration estimates. This feasible region is then used to search for the blade adjustments that will minimize the modeled vibration. If the application of these adjustments does not result in satisfactory vibration, the interval model will be updated to better estimate the feasible region and improve the choice of blade adjustments for the next flight. Important parameters of adaptive tuning are summarized in Table 6.3.

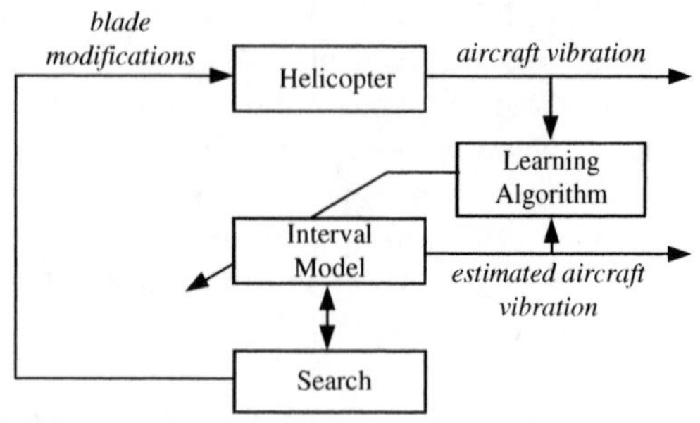

FIGURE 6.6 The strategy of the proposed tuning method.

6.4.1 The Interval Model

In order to account for the stochastics and nonlinearity of vibration, an interval model (Moore, 1979) is defined to represent the range of aircraft vibration caused by blade adjustments. The interval model used here has the form:

$$\Delta\overleftrightarrow{y}_j = \sum_{i=1}^{n} \overleftrightarrow{C}_{ji}\Delta x_i, \quad j = 1, \ldots, m \tag{6.11}$$

where each coefficient is defined as an interval:

$$\overleftrightarrow{C}_{ji} = [C_{\mathrm{L}ji}, C_{\mathrm{U}ji}]$$

In the above model, the variables with the two-sided arrow, $\leftrightarrow$, denote intervalled variables, $C_{\mathrm{L}ji}$ and $C_{\mathrm{U}ji}$ represent, respectively, the current values of the lower and upper bounds of the sensitivity coefficients between each input, Δx_i, and output, $\Delta\overleftrightarrow{y}_j$. The interval $\Delta\overleftrightarrow{y}_j$ denotes the estimated range of change of the jth output caused by the change to the current inputs, $\Delta x_1, \ldots, \Delta x_n$.

TABLE 6.3 Summary of Adaptive Tuning

In adaptive tuning, each vibration component is defined as

$$\Delta\overleftrightarrow{y}_j = \textstyle\sum_{i=1}^{n} \overleftrightarrow{C}_{ji}\Delta x_i, \quad j = 1, \ldots, m$$

where each coefficient is defined as an interval:

$$\overleftrightarrow{C}_{ji} = [C_{\mathrm{L}ji}, C_{\mathrm{U}ji}]$$

with $C_{\mathrm{L}ji}$ and $C_{\mathrm{U}ji}$ representing, respectively, the current values of the lower and upper bounds of the sensitivity coefficients between each input, Δx_i, and output, $\Delta\overleftrightarrow{y}_j$. The blade adjustments are then sought by minimizing the objective function:

$$S = \frac{\sum_{e=1}^{N_e} \mathrm{Distance}(x_c, x_e)}{\left(\prod_{s=1}^{N_s} \mathrm{Distance}(x_c, x_s)\right)^{1/N_s}}$$

where x_c represents a candidate set of blade adjustments within the feasible region, x_e represents any set of blade adjustments within the selection region, x_s denotes each of the previously selected blade adjustments, and N_e and N_s represent the number of the estimated feasible blade adjustments and the previously selected blade adjustments, respectively.

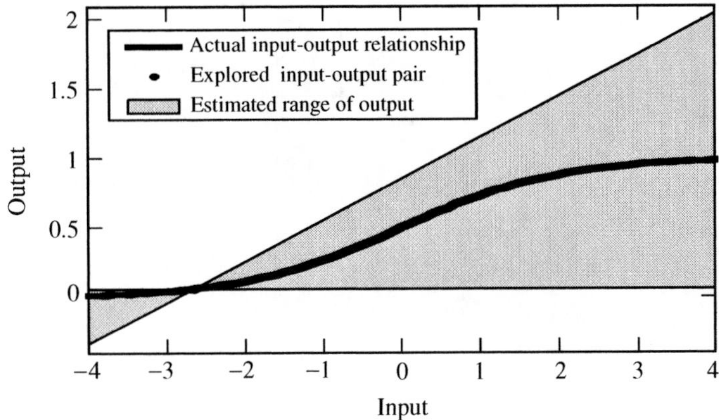

FIGURE 6.7 Estimated range of output by the interval model using one reference input.

The fit provided by the interval model for a mildly nonlinear input/output relationship is illustrated in Figure 6.7, where the output range is estimated relative to one explored input.[1] According to Equation 6.11, the estimated range of the output becomes larger, and therefore less accurate, as the potential input is selected farther from the current input (producing a large Δx_i). This potential drawback of the interval model is considerably reduced when multiple inputs have been explored so that the interval model can take advantage of several inputs for estimating the output range. The estimated output, $\vec{y}_j$, at a potential input, x_i, may be computed relative to any set of previously explored inputs, yielding different estimates of $\vec{y}_j$ (due to different values of Δx_i). In order to cope with the multiplicity of estimates, $\vec{y}_j$ is defined as the common range among all of the $\vec{y}_j$ estimates (Yang, 2000). The estimation of $\vec{y}_j$ using this commonality rule is illustrated in Figure 6.8, which indicates that using this estimation approach enables representation of the system nonlinearities in a piecewise fashion. It can be shown that the lack of commonality between the estimated ranges of output will cause a part of the input–output relationship to not be represented by the interval model. In such cases, however, the lack of compliance between the interval model and the *input–output* relationship can be corrected by adaptation of the coefficient intervals through learning.

6.4.2 Estimation of Feasible Region

The feasible region comprises all sets of blade adjustments that will reduce the aircraft vibration within specifications. The feasible region is estimated here by comparing the individually estimated $\vec{y}_j$ values with their corresponding constraints, so as to decide whether the corresponding blade adjustments belong to the feasible region. In this method, even when the interval $\vec{y}_j$ partly overlaps the vibration constraint, the corresponding blade adjustments are included in the estimated feasible region. The above procedure of estimating the feasible region based on individual outputs is then extended to multiple outputs by forming the conjunction of the estimated feasible regions from each output.

6.4.3 Selection of Blade Adjustments

The blade adjustments provide the coordinates of the feasible region, therefore, they need to provide a balanced coverage of the input space. As such, blade adjustment selection becomes synonymous with maximizing the distance of the selected blade adjustments from the previous blade adjustments, as well as

[1]An explored input represents an input for which the exact value of the output is available. In rotor tuning, an explored input would denote a blade adjustment that has been applied to the helicopter, and for which the corresponding vibration changes have been measured.

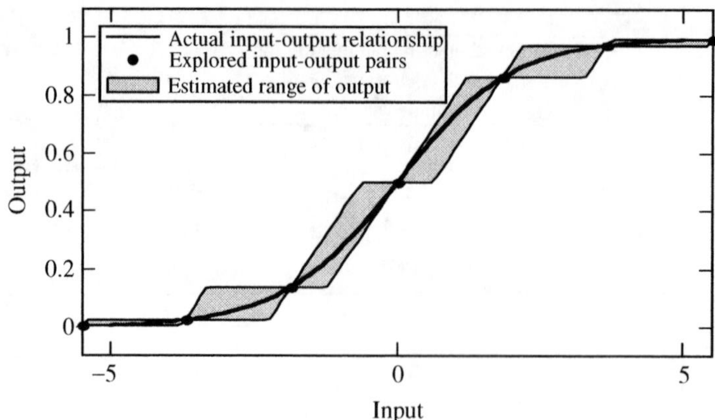

FIGURE 6.8 Estimated range of output by the interval model using seven reference inputs.

bringing them closer to the center of the feasible region. This objective can be pursued by minimizing the following objective function:

$$S = \frac{\sum_{e=1}^{N_e} \text{Distance}(x_c, x_e)}{\left(\prod_{s=1}^{N_s} \text{Distance}(x_c, x_s)\right)^{1/N_s}} \tag{6.12}$$

where x_c represents a candidate set of blade adjustments within the feasible region, x_e represents any set of blade adjustments within the selection region, x_s denotes each of the previously selected blade adjustments, and N_e and N_s represent, respectively, the numbers of the estimated feasible blade adjustments and the previously selected blade adjustments. Note that when the candidate set, x_c, is close to the previously selected blade adjustments, $\left(\prod_{s=1}^{N_s} \text{Distance}(x_c, x_s)\right)^{1/N_s}$ becomes small, and when the candidate set of blade adjustments, x_c, is far from the center of the feasible region, the value of $\sum_{e=1}^{N_e}$ Distance(x_c, x_e) becomes large. By minimizing S, the candidate blade adjustments are selected such that the above extremes are avoided.

6.4.4 Learning

Although an interval model defined according to the sensitivity coefficients may provide a suitable initial basis for tuning, it may not be the most representative of the rotor tuning process. As such, it may not be able to carry the search process to the end. A noted feature of the proposed method is its learning capability, which enables it to refine its knowledge base. To this end, the coefficients of the model are updated by considering new values for each of the upper and lower limits of individual coefficients. The objective is to make the range of the coefficients as small as possible while making sure that the interval model envelopes the acquired *input–output* data. The learning problem can be defined as

$$\text{Minimize } E = \sum_{m=1}^{K-1} \sum_{k>m}^{K} \{[y_L(m,k) - y(k)]^2 + [y_U(m,k) - y(k)]^2\} \tag{6.13}$$

subject to

$$y_U(m,k) \geq y(k) \tag{6.14}$$

$$y_L(m,k) \leq y(k) \tag{6.15}$$

$$C_{Ui} - \gamma \geq C_{Li} \tag{6.16}$$

where K represents the total number of sample points collected so far, $y_L(m,k)$ and $y_U(m,k)$ represent, respectively, the lower and upper limits of the estimated output range at the kth sample point relative to the mth sample point, $y(k)$ denotes the actual output value at the kth sample point, and C_{Ui} and C_{Li} represent the upper and lower limits of the ith coefficient interval, respectively. The parameter γ is a small positive number to control the range of the coefficients.

Most of the approaches that can be potentially used for adapting the coefficient intervals, such as gradient descent (Ishibuchi et al., 1993) or nonlinear programming, cannot be applied to rotor tuning due to their demand for rich training data and their impartiality to the initial value of the coefficients representing the *a priori* knowledge of the process. As an alternative, a learning algorithm is devised here to cope with the scarcity of track and balance data while staying true to the initial values of the coefficients. In this algorithm, the coefficients of the interval model, initially set pointwise at the sensitivity coefficients, are adapted after each flight in two steps: enlargement and shrinkage. First, the vibration measurements from all of the flights completed for the present tail number are matched against the estimated output ranges from the current interval model. If any of the measurements do not fit the upper or lower limits of the estimates, the coefficient intervals are enlarged in small steps, iteratively, and the output ranges are re-estimated at each iteration using the updated interval model. The enlargement of the coefficient intervals stops when the estimated output ranges include all of the measurements. At this point, even though the updated interval model provides a fit for the *input–output* data, it may be overcompensated. In order to rectify this situation, the coefficient intervals are shrunk individually by selecting new candidates for their upper and lower limits.

The shrinkage–enlargement learning algorithm has the form:

$$\Delta C_{Li} = -\eta\delta_L \Delta x_i(m,k) \tag{6.17}$$

$$\Delta C_{Ui} = -\eta\delta_U \Delta x_i(m,k) \tag{6.18}$$

where, during the enlargement phase, δ_L and δ_U are defined as

$$\delta_L = \begin{cases} \Delta y_L & \text{If } \Delta x_i(m,k) > 0 \text{ and } \Delta y_L > 0 \\ \Delta y_U & \text{If } \Delta x_i(m,k) < 0 \text{ and } \Delta y_U < 0 \\ 0 & \text{otherwise} \end{cases} \tag{6.19}$$

$$\delta_U = \begin{cases} \Delta y_U & \text{If } \Delta x_i(m,k) > 0 \text{ and } \Delta y_U < 0 \\ \Delta y_L & \text{If } \Delta x_i(m,k) < 0 \text{ and } \Delta y_L > 0 \\ 0 & \text{otherwise} \end{cases} \tag{6.20}$$

and during the shrinkage phase, they are defined as

$$\delta_L = \begin{cases} \Delta y_L & \text{If } \Delta x_i(m,k) > 0 \text{ and } \Delta y_L < 0 \\ \Delta y_U & \text{If } \Delta x_i(m,k) < 0 \text{ and } \Delta y_U > 0 \\ 0 & \text{otherwise} \end{cases} \tag{6.21}$$

$$\delta_U = \begin{cases} \Delta y_U & \text{If } \Delta x_i(m,k) > 0 \text{ and } \Delta y_U > 0 \\ \Delta y_L & \text{If } \Delta x_i(m,k) < 0 \text{ and } \Delta y_L < 0 \\ 0 & \text{otherwise} \end{cases} \tag{6.22}$$

with

$$\Delta x_i(m,k) = x_i(k) - x_i(m) \tag{6.23}$$

$$\Delta y_L = y_L(m,k) - y(k) \tag{6.24}$$

$$\Delta y_U = y_U(m,k) - y(k) \tag{6.25}$$

This procedure is repeated for each coefficient interval in an iterative fashion until the objective function E (Equation 6.13) is minimized. The minimization of E ensures limited adaptation of the coefficient intervals within the smallest possible range.

At the beginning of tuning, the limited number of *input–output* data available for learning will not provide a comprehensive representation of the process. Therefore, the coefficient intervals should not be shrunk drastically until enough *input–output* data have become available. For this, the length of each coefficient interval $[C_{Li}, C_{Ui}]$ is constrained by the minimal interval length for each tuning iteration as

$$\min L = \{C_{Ui}(0) - C_{Li}(0)\}(1 - \beta)^n \tag{6.26}$$

where $\beta \in [0, 1]$ controls the shrinkage rate of the coefficient interval, and n denotes the number of tuning iterations. The coefficient interval cannot be shrunk when $\beta = 0$ and can be shrunk without limit when $\beta = 1$. Usually, β is selected closer to 0.

6.5 Case Study

The utility of the Interval Model (IM) method is demonstrated in application to Black Hawks. Ideally, the performance of the proposed method should be evaluated side by side against that of the traditional method. However, such an evaluation would require tuning the aircraft with one method, undoing changes, and tuning the aircraft with another. Since such testing is prohibitively costly and infeasible, a compromised approach of evaluating the method in simulation is utilized. A process simulation model is therefore used to represent the block "helicopter" in Figure 6.6, with the block "forward model" represented by an interval model.

6.5.1 Simulation Model

Considering the potential nonlinearity of the effect of blade adjustments on the helicopter vibration and the high level of noise present in vibration measurements, multilayer neural networks offer the most suitable framework for modeling. A series of neural networks were trained with historical balance data to represent the relationships between vibration changes and blade adjustments, and the stochastic aspects of vibration were represented by the addition of random numbers to the outputs of the networks.

A total of 102 sets of vibration data were used to train and test the neural networks. The inputs to these networks were the combined blade adjustments of push rods and trim tabs to opposite blade pairs, and their outputs were the resulting vibration changes between two consecutive flights. Since the vibration data are vector quantities that are represented by both magnitude and phase components (see Table 6.1), the vibration data were transformed into Cartesian coordinates, so that each vector element would denote the change in the cosine or sine component of the A + B or A − B vibration of each of the six flight regimes (see Table 6.1). In this study, each neural network model consisted of four inputs and one output, so a total of 24 networks were trained to represent all of the vibration components. Alternatively, all of the vibration measurements may be represented by one neural network, but such a network is more difficult to train. Formally, the outputs of the neural networks, which represent the cosine and sine components of the vibration at different regimes, $v_{cj}(k)$ and $v_{sj}(k)$, respectively, are defined as

$$\hat{V}_{sj}(k) = V_{sj}(k-1) + \Delta V_{sj}(k) + R_{sj}(k) \tag{6.27}$$

$$\hat{V}_{cj}(k) = V_{cj}(k-1) + \Delta V_{cj}(k) + R_{cj}(k) \tag{6.28}$$

$$\Delta V_{sj}(k) = F_{sj}(\Delta x) \tag{6.29}$$

$$\Delta V_{cj}(k) = F_{cj}(\Delta x) \tag{6.30}$$

$$\hat{V}_j(k) = \sqrt{\hat{V}_{sj}(k)^2 + \hat{V}_{cj}(k)^2} \tag{6.31}$$

where the input vector $\mathbf{\Delta x} = \{\Delta x_1, \Delta x_2, \Delta x_3, \Delta x_4\}$ denotes the set of combined blade adjustments, each of the functionals, F_{sj} and F_{cj}, represent the change in vibration between two consecutive flights as represented by a neural network, and $R_{sj}(k)$ and $R_{cj}(k)$ denote random numbers added to the outputs of the networks to account for measurement noise. Each of the networks consisted of two hidden layers, with four and eight processing elements in the first and second layers, respectively. To avoid overtraining, the 102 sets of data were divided into two equal subsets, one set to train the network and the other to test its performance. The random numbers, R_{cj} and R_{sj}, were generated according to the Gaussian distribution $N(\mu, \sigma^2)$, with the mean μ and variance σ^2 defined as

$$\hat{\mu} = \frac{1}{M}\sum_{i=1}^{M} e_i \tag{6.32}$$

$$\hat{\sigma}^2 = \frac{1}{M-1}\sum_{i=1}^{M} (e_i - \hat{\mu})^2 \tag{6.33}$$

In the above formulation, M represents the total number of data sets and e_i denotes the difference between the measured and expected value of vibration, defined as

$$e_j(k) = V_j(k) - V_j(k-1) - \Delta V_j(k) \tag{6.34}$$

A sample of estimated vibration changes generated by the neural network model is compared side by side with the actual vibration changes in Figure 6.9. The results indicate close agreement between the predicted and actual vibration changes.

6.5.2 Interval Modeling

In application to the Black Hawks, a total of 24 interval models need to be constructed to approximate the changes in the cosine and sine components of the A + B and A − B vibrations at each of the six flight regimes. The interval models have the form

$$\overleftrightarrow{V}_{cj}(k) = V_{cj}(k-1) + \sum_{i=1}^{4} \overleftrightarrow{C}_{cji}(k-1)\Delta x_i(k) \tag{6.35}$$

$$\overleftrightarrow{V}_{sj}(k) = V_{sj}(k-1) + \sum_{i=1}^{4} \overleftrightarrow{C}_{sji}(k-1)\Delta x_i(k) \tag{6.36}$$

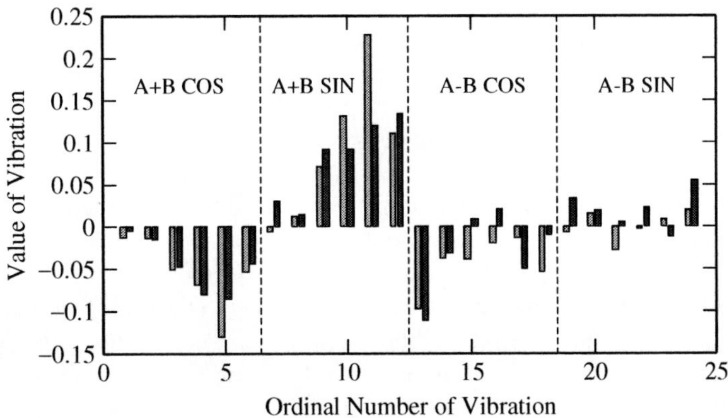

FIGURE 6.9 A sample set of simulated vibration changes shown side by side with the actual vibration changes.

$$\vec{V}_j(k) = \sqrt{\vec{V}_{sj}^2(k) + \vec{V}_{cj}^2(k)}, \quad i = 1,\ldots,4 \text{ and } j = 1,\ldots,12 \tag{6.37}$$

where the $\vec{V}_{cj}(k)$ and $\vec{V}_{sj}(k)$ represent, respectively, the estimated cosine and sine components of A + B or A − B vibration at each of the six flight regimes, $\vec{V}_j(k)$ denotes the magnitude of the vibration, and Δx_i are the same as those in Equation 6.2 to Equation 6.5. For this study, the feasible region was defined to include all of the blade adjustments associated with vibration estimates that satisfied the specification: $\max\{\min(\vec{V}_1),\ldots,\min(\vec{V}_{12})\} \leq 0.2$. The above specification ensures that the lower limit of the estimated vibration range of the largest vibration component will be less than 0.2 ips (an industry standard). The selection of the lower limit here is to ensure that the feasible region is as large as possible, so as not to eliminate any potentially good candidate blade adjustments. The computation of the feasible region was based on the range $[-0.015, 0.015]$ for push rods and $[-0.035, 0.035]$ for trim tabs, within which 20,000 random sets of blade adjustments were evaluated for their feasibility. The blade adjustments associated with vibration ranges satisfying the specification were included in the feasible region.

As noted earlier, the proposed method uses the feasible region as the basis of search for the blade adjustments. For this study, the blade adjustments set that produced the smallest value for the objective function S (Equation 6.12) was selected to be applied to the helicopter. It should be noted that, given the stringent constraints on the vibration components, there were cases where the search algorithm could not find any feasible blade adjustments that would satisfy all of the constraints. In such cases, the set of blade adjustments that produced the smallest lower limit of the maximum estimated vibration was used as a compromised solution.

The interval model was updated after each tuning iteration. For shrinkage–enlargement learning, the parameter β in Equation 6.26 was set to 1 and γ to 0, so that the coefficient intervals could be shrunk without limits. Learning was performed separately for each tail number to customize the interval model to individual tail numbers; that is, the interval model was set to the sensitivity coefficients for each tail number and was adapted after the first tuning iteration. Accordingly, the interval model was actually a pointwise model for the first iteration and took the form of an interval model thereafter.

6.5.3 Performance Evaluation

The interval model (IM) method was tested on 39 tail numbers, for which actual track and balance data were available from the field. For each tail number, the IM method was applied iteratively until either the simulated vibrations were within their specifications, or an upper limit of five process iterations had been reached.

Since the stochastic aspects of vibration measurements impose randomness on the rotor tuning process, rotor tuning solutions cannot be evaluated by deterministic measures. This calls for the creation of performance measures that account for uncertainty. One such measure that assesses tuning efficiency is the *average tuning iteration number* (ATIN) which represents the average number of iterations taken for tuning each tail number. The number of flights used by the IM method for the 39 tail numbers is included in Table 6.4 along with those actually performed in the field. The results indicate that the IM method requires a smaller ATIN relative to that actually performed.

Another potentially significant aspect of the IM method is its adaptation capability, which enables it to transform a pointwise model into an IM, and to subsequently update it after the first iteration. Adaptation capability, however, may not be as significant in rotor tuning, which offers limited possibility for training. In order to evaluate the significance of learning in the performance of the IM method, the results in Table 6.4 were reproduced in Table 6.5 with the learning feature turned off. The ATINs indicate that with learning, the IM method requires fewer iterations for tuning each tail number, despite the small number of iterations taken to tune each tail number. This, in turn, indicates that the interval model enhances the performance of the IM method, since without learning, the model remains pointwise at the sensitivity coefficients. However, perhaps an equally interesting set of results in Table 6.5 are those indicating that even without learning, the IM method requires fewer iterations than actually

TABLE 6.4 The Number of Tuning Iterations Required by the Interval Model Method and Those Applied in the Field

Tail # (39)	Number of Tuning Iterations	
	Actual	IM Method
176	1	1
178	1	1
179	3	2
180	1	1
184	2	1
⋮	⋮	⋮
260	4	1
⋮	⋮	⋮
861	1	1
Total	71	48
ATIN	1.82	1.23

TABLE 6.5 The Number of Tuning Iterations Required by the IM Method (with and without Learning) along with Those Actually Applied in the Field

Tail # (39)	Tuning Iteration Number		
	Actual	IM Method	
		With Learning	Without Learning
185	3	2	3
186	3	2	3
208	2	2	3
245	3	2	3
260	4	2	3
802	3	2	3
822	3	2	3
⋮	⋮	⋮	⋮
Total	71	48	62
ATIN	1.82	1.23	1.59

performed in the field. Given that the adjustments associated with both sets of results were selected from the same model (i.e., sensitivity coefficients), the better performance of the IM method can only be attributed to its more effective search strategy that leads to more comprehensive solutions.

A preferred aspect of a system of rotor tuning is its ability to tune the aircraft within one iteration. This aspect of the method was evaluated by checking the number of tail numbers tuned within one iteration. For these results, in order to eliminate the difference between the simulation model and the helicopter, only the vibration estimates from simulation were used to evaluate the suitability of the adjustments. The results of this study are shown in Table 6.6, where the tail numbers tuned within one iteration are shown by a $\surd$ and those requiring more than one iteration are denoted by $\times$. The results indicate that the IM method satisfies this more stringent criterion better than the actual adjustments, further validating the claim that the IM method benefits from a more effective search engine.

Owing to the randomness of the vibration measurements, repeated applications of an adjustment set may lead to slightly different vibration measurements. This, in turn, may cause a variance in the number of iterations produced by adjustments when the resulting vibration is close to the specified threshold. It would be beneficial, therefore, to devise a measure for the probability of success of adjustments.

TABLE 6.6 Tally of the Tail Numbers Tuned within One Iteration According to Simulated Vibration

Tail # (39)	Tuned within One Iteration	
	Actual	IM Method
176	√	√
178	×	√
179	√	√
⋮	⋮	⋮
822	×	√
858	×	√
859	√	√
861	√	√
Total	19	30

The empirical measure, the acceptability index (AI), is defined here as

$$\mathrm{AI} = \frac{1}{N}\sum_{l=1}^{N} s_l \tag{6.38}$$

to denote the percentage of times an adjustment set will result in the vibration satisfying the specification. In the above equation, N represents the total number of flights simulated to represent the repeated applications of the same adjustment set, and

$$s_l = \begin{cases} 1 & \text{if vibration of the } l\text{th simulation flight is acceptable} \\ 0 & \text{if vibration of the } l\text{th simulation flight is unacceptable} \end{cases}$$

The AIs computed for both the actual and selected adjustments at the first iteration are included in Table 6.7. The results indicate that the IM method provides adjustments with a higher probability of success as judged by the acceptability of vibration estimates from the simulation model. These results, which indicate that the selected adjustments from the IM method can more consistently tune the rotors within one iteration, imply the better positioning of the adjustments within the feasible region.

TABLE 6.7 The Values of Acceptability Index (Trial Mode) Computed for Both the Actual and Selected Adjustments at the First Flight

Tail # (39)	Acceptability Index	
	Actual	IM Method
176	0.92	0.87
178	0	0.54
179	0.61	0.52
⋮	⋮	⋮
260	0.40	0.89
261	0.09	0.95
263	0.18	0.12
⋮	⋮	⋮
822	0.00	0.64
857	0.62	0.55
858	0.64	0.93
859	0.94	0.67
861	0.96	0.74
Average	0.581	0.724

TABLE 6.8 Comparison of the First Iteration Solutions of IM Method and Actual Solutions from Sikorsky's Production Line with the Cumulative Acceptable Adjustments

Tail #	Modifications		
	Actual Iteration 1 Modifications	CAM	IM Iteration 1 Modifications
801	6, −4, −10, 11	2, −4, −4, 14	3, −5, −6, 12
802	5, 2, 0, 0	9, 0, −10, 10	8, −2, −5, 10
822	6, 0, −20, 0	10, −4, −23, 13	8, −4, −22, 6
858	7, 0, −14, 0	9, −2, −10, 3	9, −2, −15, 4

Another evaluation basis for the adjustments can be established by comparing them to the actual cumulative adjustments performed in the field. The cumulative adjustment set, $\sum x$, can be defined as

$$\sum x = \sum_{k=1}^{N} \Delta x_k \tag{6.39}$$

where N represents the total number of tuning iterations performed in the field for the tail number and Δx_k denotes the adjustments applied at the kth iteration. A sample of actual first iteration adjustments, actual cumulative adjustments, and first iteration adjustments from the IM method is shown in Table 6.8. The results indicate that the adjustments from the IM method are closer to the actual cumulative adjustments than are the actual first iteration adjustments. Although the cumulative adjustments may not be the most desirable ones for the aircraft, they represent an acceptable set that has been proven in the field. The closeness of the IM method's solutions to the actual cumulative adjustments further validates its effectiveness.

6.6 Conclusion

A logical feature for future rotor tuning systems will be the capability to adjust the blades during the flight. For this, these systems will need to have the capability to learn from their mistakes. They will also need to be able to monitor the condition of the rotor system in-flight, so they will stop modifying the blade parameters when more drastic actions are necessary for saving the aircraft. As such, these systems will need to be used with strong operator interaction to prevent implementation of inappropriate adjustments, and must have the ability to explain the recommended adjustments to the operator.

References

Ishibuchi, H., Tanaka, H., and Okada, H., An architecture of neural networks with interval weights and its applications to fuzzy regression analysis, *Fuzzy Sets Syst.*, 57, 27–59, 1993.

Moore, R.E. 1979. *Methods and Applications of Interval Analysis*, Society for Industrial and Applied Mathematics, Philadelphia.

Taitel, H., Danai, K., and Gauthier, D.G., Helicopter track and balance with artificial neural nets, *ASME J. Dyn. Syst. Meas. Contr.*, 117, 226–231, 1995.

Ventres, S. and Hayden, R.E. 2000. *Rotor tuning using vibration data only*, American Helicopter Society 56th Annual Forum, Virginia Beach, VA, May 2–4, 2000.

Wang, S., Danai, K., and Wilson, M., A probability-based approach to helicopter track and balance, *J. Am. Helicopter Soc.*, 50, 1, 56–64, 2005a.

Wang, S., Danai, K., and Wilson, M., An adaptive method of helicopter track and balance, *ASME J. Dyn. Syst. Meas. Contr.*, 2005b, March, in press.

Yang, D.Z. 2000. Knowledge-based interval modeling method for efficient global optimization and process tuning, Ph.D. Thesis. Department of Mechanical and Industrial Engineering, University of Massachusetts, Amherst.

7

Vibration Design and Control

Clarence W. de Silva
The University of British Columbia

Summary

There are desirable and undesirable types and situations of mechanical vibration. Undesirable vibrations are those that cause human discomfort and hazards, structural degradation and failure, performance deterioration and malfunction of machinery and processes, and various other problems. This chapter discusses ways of either eliminating or reducing the undesirable effects of vibration. Specifically, some useful topics on design for vibration suppression and the control of vibration are addressed. General approaches to vibration mitigation may be identified from the dynamic systems point of view. Typically, a set of vibration specifications is given as simple threshold values (bounds) or frequency spectra, and the goal is to either design or control the system to meet these specifications. Frequency-domain techniques based on transfer functions such as transmissibility, and time-domain techniques using the state-space representation, optimal control and modal control are presented. Applications considered here include vibration isolation, balancing of rotating and reciprocating machinery, whirling suppression, and passive and active control of vibration.

7.1 Introduction

Consider the schematic diagram of a vibratory system shown in Figure 7.1. Forcing excitations $\mathbf{f}(t)$ to the mechanical system S cause the vibration responses $\mathbf{y}$. Our objective is to suppress $\mathbf{y}$ to a level that is acceptable. Clearly, there are three general ways of doing this:

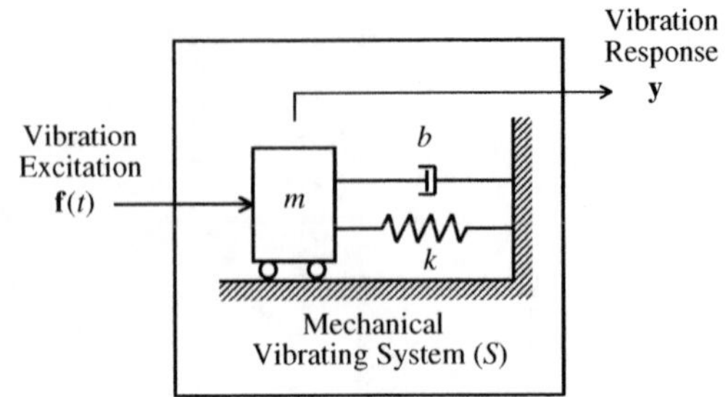

FIGURE 7.1 A vibrating mechanical system.

1. *Isolation.* Suppress the excitations of vibration. This method deals with **f**.
2. *Design modification.* Modify or redesign the mechanical system so that for the same levels of excitation, the resulting vibrations are acceptable. This method deals with S.
3. *Control.* Absorb or dissipate the vibrations using external devices, through implicit or explicit sensing and control. This method deals with **y**.

Within each of these three categories, several approaches can be used to achieve the objective of vibration mitigation. Essentially, each of these approaches involves designing (either complete through redesign or incremental design modification) of the system on the one hand, and controlling the vibration through external means (passive or active devices) on the other. Note that removal of faults (e.g., misalignments and malfunctions by repair or parts replacement) can also remove vibrations. This approach may fall into any of the three categories listed above.

The category of vibration isolation involves "isolating" a mechanical system (S) from vibration excitations (**f**) so that the excitation signals are "filtered" out or dissipated prior to reaching the system. The use of properly designed suspension systems, mounts, and damping layers falls within this category. The category of design modification will involve making changes to the components and the structure of a mechanical system according to a set of specifications and design guidelines. Balancing of rotating machinery and structural modification through modal analysis and design techniques fall into this category. The category of control will involve either passive devices (which do not use external power), such as dynamic absorbers and dampers, or active control devices (which need external power for operation). In the passive case, the control device implicitly senses the vibration response and dissipates it (as in the case of a damper) or absorbs and stores its energy where it is slowly dissipated (as in the case of a dynamic absorber). In the active case, the vibrations **y** are explicitly sensed through *sensors* and *transducers*. The forces that should be acted on the system to counteract and suppress vibrations are determined by a controller, and the corresponding forces or torques are applied to the system through one or more *actuators*.

Note that there may be some overlap in the three general categories of vibration mitigation that were mentioned above. For example, the addition of a mount (category 1) may also be interpreted as a design modification (category 2) or as incorporating a passive damper (category 3). It should also be noted that the general approach, commonly known as *source alteration*, may fall into either category 1 or category 2. In this case, the purpose is to alter or remove the source of vibration. The source could either be external (e.g., road irregularities that result in vehicle vibrations) — a category 1 problem, or internal (imbalance or misalignment in rotating devices that results in periodic forces, moments, and vibrations) — a category 2 problem. It can be more difficult to alter external vibration sources (e.g., resurfacing the roadways) than to modify the internal sources (e.g., balancing of rotating machinery).

Furthermore, the external source of vibration may be quite random and may not be accessible for alteration at all (e.g., aerodynamic forces on an aircraft).

7.1.1 Shock and Vibration

Sometimes, response to shock loads is considered separately from response to vibration excitations for the purpose of design and control of mechanical systems. For example, shock isolation and vibration isolation are treated under different headings in some literature. This is actually unnecessary. Even though vibration analysis predominantly involves periodic excitations and responses, transient and random oscillations (vibrations) are also commonly found in practice. The frequency band of the latter two types of signals is much broader than that of a simple periodic signal. A shock signal is transient by definition, and has a very short duration (in comparison to the predominant time constants of the mechanical system to which the shock load is applied). Hence, it will possess a wide band of frequencies. Consequently, frequency-domain techniques are still applicable. Furthermore, time-domain techniques are particularly suited to dealing with transient signals in general and shock signals in particular. In that context, a shock excitation may be treated as an impulse whose effect is to instantaneously change the velocity of an inertia element. Then, in the time domain, a shock load may also be treated as an initial-velocity excitation of an otherwise free (unforced) system.

7.2 Specification of Vibration Limits

Design and control procedures of vibration have the primary objective of ensuring that under normal operating conditions, the system of interest does not encounter vibration levels that exceed the specified values. In this context, then, the ways of specifying vibration limits become important. This section will present some common ways of vibration specification.

7.2.1 Peak Level Specification

Vibration limits for a mechanical system may be specified in either the time domain or the frequency domain. In the time domain, the simplest specification is the peak level of vibration (typically, acceleration in units of g — the acceleration due to gravity). Here, the techniques of isolation, design, or control should ensure that the peak vibration response of the system do not exceed the specified level. In this case, the entire time interval of operation of the system is monitored, and the peak values are checked against the specifications. Note that, in this case, it is the instantaneous peak value at a particular time instant that is of interest, and what is used in representing vibration is an instantaneous amplitude measure rather than an average amplitude or an energy measure.

7.2.2 Root-Mean-Square Value Specification

The root-mean-square (RMS) value of a vibration signal $y(t)$ is given by the square root of the average of the squared signal as

$$y_{\mathrm{RMS}} = \left[\frac{1}{T} \int_0^T y^2 \mathrm{d}t \right]^{1/2} \tag{7.1}$$

Note that, by squaring the signal, its sign is eliminated and, essentially, the energy level of the signal is used. The period T, over which the squared signal is averaged, will depend on the problem and the nature of the signal. For a periodic signal, one period is adequate for averaging. For transient signals, several time constants (typically four times the larger time constant) of the vibrating system would be sufficient. For random signals, a value that is as large as feasible should be used.

In the method of RMS value specification, the RMS value of the acceleration response (typically, acceleration in *g*s) is computed using Equation 7.1 and is compared with the specified value. In this method, instantaneous bursts of vibration do not have a significant effect as they are filtered out because of the integration. It is the average energy or power of the response signal that is considered. The duration of exposure enters into the picture indirectly and in an undesirable manner. For instance, a highly transient vibration signal can initially have a damaging effect. However, the larger the T that is used in Equation 7.1, the smaller the computed RMS value. Hence, in this case, the use of a large value for T would lead to diluting or masking the damage potential. In practice, the longer the exposure to a vibration signal, the greater the harm caused. Hence, when using specifications such as peak and RMS values, they have to be adjusted according to the period of exposure. Specifically, a larger specification should be used for longer periods of exposure.

7.2.3 Frequency-Domain Specification

It is not realistic to specify the limitation to vibration exposure of a complex dynamic system by just a single threshold value. Usually, the effect of vibration on a system depends on at least the following three parameters:

1. Level of vibration (peak, RMS, power, etc.).
2. Frequency content (range) of excitation.
3. Duration of exposure to vibration.

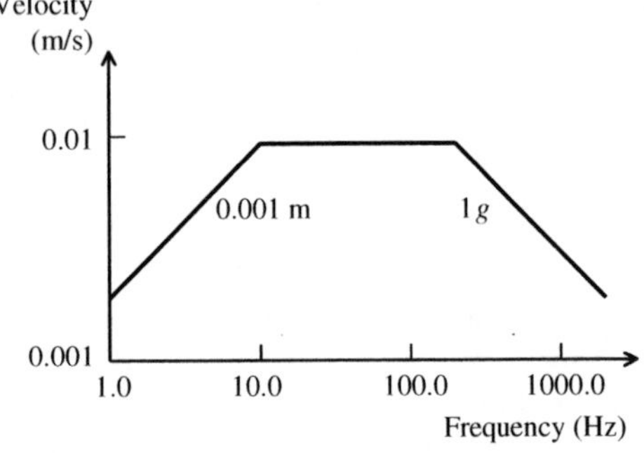

FIGURE 7.2 Operating vibration specification (nomograph) for a machine.

This is particularly true because the excitations that generate the vibration environment may not necessarily be a single-frequency (sinusoidal) signal and may be broadband and random. Furthermore, the response of the system to the vibration excitations will depend on its frequency transfer function, which determines its resonances and damping characteristics. Under these circumstances, it is desirable to provide specifications in a *nomograph* where the horizontal axis gives frequency (Hz) and the vertical axis could represent a motion variable such as displacement (m), velocity (m/s), or acceleration (m/s^2 or g). It is not important which of these motion variables represents the vertical axis of the nomograph. This is true because in the frequency domain

$$\text{Velocity} = j\omega \times \text{displacement}$$

$$\text{Acceleration} = j\omega \times \text{velocity}$$

and one form of motion may be easily converted into one of the remaining two motion representations. In each of the forms, assuming that the two axes of the nomograph are graduated in a logarithmic scale, the constant displacement, constant velocity, and constant acceleration lines are straight lines.

Consider a simple specification of machinery vibration limits as given by the following values:

$$\text{Displacement limit (peak)} = 0.001\ \text{m}$$

$$\text{Velocity limit} = 0.01\ \text{m/s}$$

$$\text{Acceleration limit} = 1.0g$$

This specification may be represented in a velocity vs. frequency nomograph (log–log) as in Figure 7.2.

Usually, such simple specifications in the frequency domain are not adequate. As noted above, the system behavior will vary depending on the excitation frequency range. For example, motion sickness in humans may be predominant in low frequencies in the range of 0.1 to 0.6 Hz and passenger discomfort in

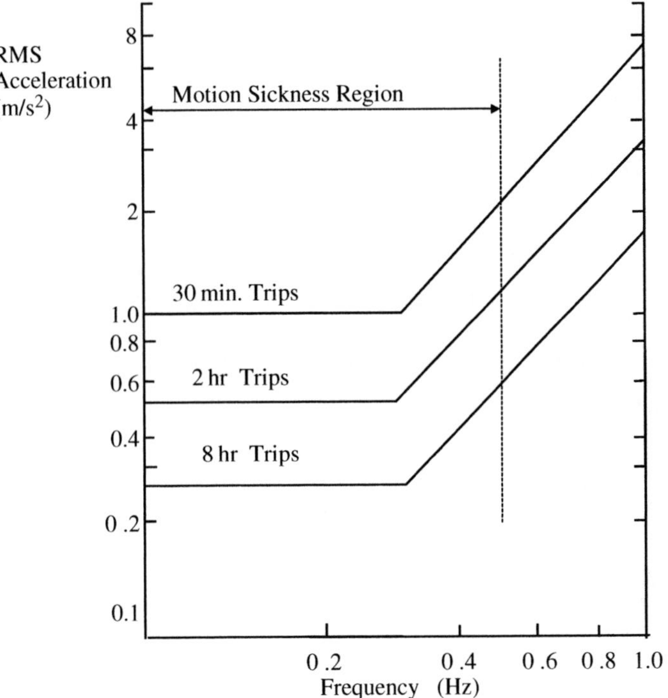

FIGURE 7.3 A severe-discomfort vibration specification for ground transit vehicles.

ground transit vehicles may be most serious in the frequency range of 4 to 8 Hz for vertical motion and 1 to 2 Hz for lateral motion. In addition, for any dynamic system, particularly at low damping levels, the neighborhoods of resonant frequencies should be avoided and, hence, should be specified by low vibration limits in the resonant regions. Furthermore, the duration of vibration exposure should be explicitly accounted for in specifications. For example, Figure 7.3 presents a ride comfort specification for a ground transit vehicle, where lower vibration levels are specified for longer trips.

Finally, it should be noted that the specifications we are concerned with in the present context of design and control are upper bounds of vibration. The system should perform below (within) these specifications under normal operating conditions. Test specifications are lower bounds. The test should be conducted at or above these vibration levels so that the system would meet the test specifications. Some considerations of vibration engineering are summarized in Box 7.1.

7.3 Vibration Isolation

The purpose of vibration isolation is to "isolate" the system of interest from vibration excitations by introducing an *isolator* in between them. Examples of isolators are machine mounts and vehicle suspension systems. Two general types of isolation can be identified:

1. Force isolation (related to force transmissibility)
2. Motion isolation (related to motion transmissibility)

In force isolation, vibration forces that would be ordinarily transmitted directly from a source to a supporting structure (isolated system) are filtered out by an isolator through its flexibility (spring) and dissipation (damping) so that part of the force is routed through an inertial path. Clearly, the concepts of *force transmissibility* are applicable here. In motion isolation, vibration motions that are applied at a moving platform of a mechanical system (isolated system) are absorbed by an isolator through its flexibility and dissipation so that the motion that is transmitted to the system of interest is weakened.

Box 7.1

VIBRATION ENGINEERING

Vibration mitigation approaches:

- Isolation (buffers system from excitation)
- Design modification (modifies the system)
- Control (senses vibration and applies a counteracting force: passive/active)

Vibration specification:

- Peak and RMS values
- Frequency-domain specs on a nomograph
  - *Vibration levels
  - *Frequency content
  - *Exposure duration

Note:

$|\text{Velocity}| = \omega \times |\text{Displacement}|$

$|\text{Acceleration}| = \omega \times |\text{Velocity}|$

Limiting specifications:

Operation (design) specifications: specify upper bounds
Testing specifications: specify lower bounds

The concepts of motion transmissibility are applicable in this case. The design problem in both cases is to select applicable parameters for the isolator so that the vibrations entering the system are below specified values within a frequency band of interest (the operating frequency range).

Let us revisit the main concepts of force transmissibility and motion transmissibility. Figure 7.4(a) gives a schematic model of force transmissibility through an isolator. Vibration force at the source is $f(t)$. In view of the isolator, the source system (with impedance Z_m) is made to move at the same speed as the isolator (with impedance Z_s). This is a parallel connection of impedances. Hence, the force $f(t)$ is split so that part of it is taken up by the inertial path (broken line) of Z_m. Only the remainder (f_s) is transmitted through Z_s to the supporting structure, which is the isolated system. Force transmissibility is

$$T_f = \frac{f_s}{f} = \frac{Z_s}{Z_m + Z_s} \tag{7.2}$$

Figure 7.4(b) gives a schematic model of motion transmissibility through an isolator. Vibration motion $v(t)$ of the source is applied through an isolator (with impedance Z_s and mobility M_s) to the isolated system (with impedance Z_m and mobility M_m). The resulting force is assumed to transmit directly from the isolator to the isolated system and hence, these two units are connected in series. Consequently, we have the motion transmissibility:

$$T_m = \frac{v_m}{v} = \frac{M_m}{M_m + M_s} = \frac{Z_s}{Z_s + Z_m} \tag{7.3}$$

It can be seen that, according to these two models, we have

$$T_f = T_m \tag{7.4}$$

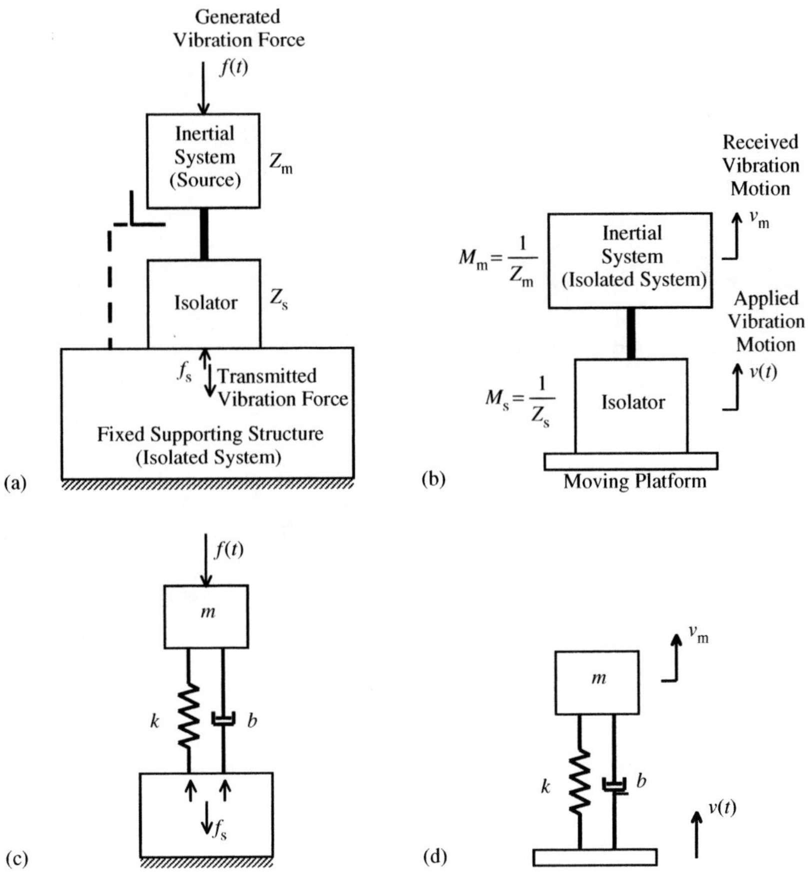

FIGURE 7.4 (a) Force isolation; (b) motion isolation; (c) force isolation example; (d) motion isolation example.

As a result, the concepts of force transmissibility and motion transmissibility may be studied using just one common transmissibility function T.

Simple examples of force isolation and motion isolation are shown in Figure 7.4(c) and (d). For both cases, the transmissibility function is given by

$$T = \frac{k + bj\omega}{(k - m\omega^2 + bj\omega)} \tag{7.5}$$

where ω is the frequency of vibration excitation. Note that the model (Equation 7.5) is not restricted to sinusoidal vibrations. Any general vibration excitation may be represented by a Fourier spectrum, which is a function of frequency ω. Then, the response vibration spectrum is obtained by multiplying the excitation spectrum by the transmissibility function T. The associated design problem is to select the isolator parameters k and b to meet the specifications of isolation.

Equation 7.5 may be expressed as

$$T = \frac{\omega_n^2 + 2\zeta\omega_n\omega j}{(\omega_n^2 - \omega^2 + 2\zeta\omega_n\omega j)} \tag{7.6}$$

where

$\omega_n = \sqrt{k/m}$ = undamped natural frequency of the system

$\zeta = \dfrac{b}{2\sqrt{km}}$ = damping ratio of the system

Equation 7.6 may be written in the nondimensional form:

$$T = \frac{1 + 2\zeta rj}{1 - r^2 + 2\zeta rj} \tag{7.7}$$

where the nondimensional excitation frequency is defined as

$$r = \omega/\omega_n$$

The transmissibility function has a phase angle as well as magnitude. In practical applications, the level of attenuation of the vibration excitation (rather than the phase difference between the vibration excitation and the response) is of primary importance. Accordingly, the transmissibility magnitude

$$|T| = \sqrt{\frac{1 + 4\zeta^2 r^2}{(1 - r^2)^2 + 4\zeta^2 r^2}} \tag{7.8}$$

is of interest. It can be shown that $|T| < 1$ for $r > \sqrt{2}$, which corresponds to the isolation region. Hence, the isolator should be designed such that the operative frequencies ω are greater than $\sqrt{2}\omega_n$. Furthermore, a threshold value for $|T|$ would be specified, and the parameters k and b of the isolator should be chosen so that $|T|$ is less than the specified threshold in the operating frequency range (which should be given). This procedure may be illustrated using an example.

Example 7.1

A machine tool and its supporting structure are modeled as the simple mass–spring–damper system shown in Figure 7.5.

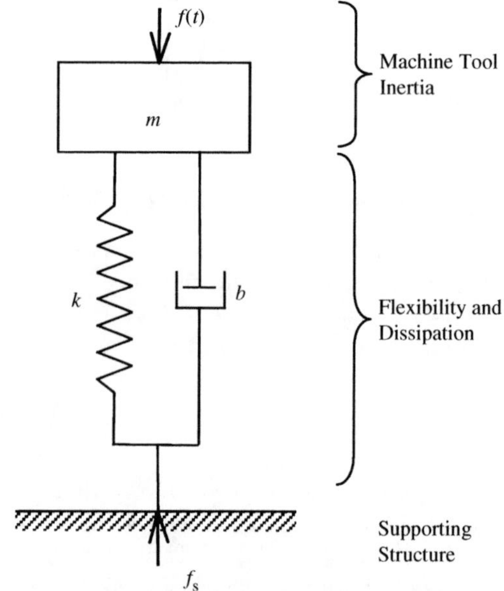

FIGURE 7.5 A simplified model of a machine tool and its supporting structure.

1. Draw a mechanical-impedance circuit for this system in terms of the impedances of the three elements: mass (m), spring (k), and viscous damper (b).
2. Determine the exact value of the frequency ratio r in terms of the damping ratio ζ, at which the force transmissibility magnitude will peak. Show that for small ζ, this value is $r = 1$.
3. Plot $|T_f|$ vs. r for the interval $r = [0, 5]$, with one curve for each of the five ζ values 0.0, 0.3, 0.7, 1.0, and 2.0, on the same plane. Discuss the behavior of these transmissibility curves.
4. From part (3), determine for each of the five ζ values and the excitation frequency range with respect to ω_n, for which the transmissibility magnitude is:
   - Less than 1.05
   - Less than 0.5
5. Suppose that the device in Figure 7.5 has a primary, undamped natural frequency of 6 Hz and a damping ratio of 0.2. It is necessary that the system has a force transmissibility magnitude of less than 0.5 for operating frequency values greater than 12 Hz. Does the existing system meet this requirement? If not, explain how you should modify the system to meet the requirement.

Solution

1. Here, the elements m, b, and f are in parallel with a common velocity v across them, as shown in Figure 7.6. In the circuit, $Z_m = mj\omega$, $Z_b = b$, and $Z_k = k/j\omega$.
Force transmissibility

$$T_f = \frac{F_s}{F} = \frac{F_s/V}{F/V} = \frac{Z_s}{Z_s + Z_0}$$
$$= \frac{Z_b + Z_k}{Z_m + Z_b + Z_k} \quad \text{(i)}$$

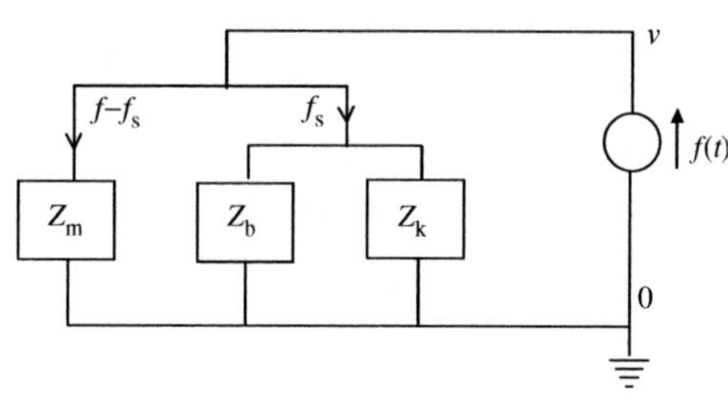

FIGURE 7.6 The mechanical impedance circuit of the force isolation problem.

Substitute the element impedances. We obtain

$$T_f = \frac{b + \dfrac{k}{j\omega}}{mj\omega + b + \dfrac{k}{j\omega}} = \frac{bj\omega + k}{-\omega^2 m + bj\omega + k} = \frac{j\omega b/m + k/m}{-\omega^2 + j\omega b/m + k/m} \quad \text{(ii)}$$

The last expression is obtained by dividing the numerator and the denominator by m. Now, use the fact that

$$\frac{k}{m} = \omega_n^2 \quad \text{and} \quad \frac{b}{m} = 2\zeta\omega_n$$

and divide Equation ii throughout by ω_n^2. We obtain

$$T_f = \frac{\omega_n^2 + 2\zeta\omega_n j\omega}{\omega_n^2 - \omega^2 + 2\zeta\omega_n j\omega} = \frac{1 + 2\zeta rj}{1 - r^2 + 2\zeta rj} \quad \text{(iii)}$$

The transmissibility magnitude is

$$|T_f| = \sqrt{\frac{1 + 4\zeta^2 r^2}{(1 - r^2)^2 + 4\zeta^2 r^2}} \quad (7.9)$$

where $r = \omega/\omega_n$ is the normalized frequency.

2. To determine the peak point of $|T_f|$, differentiate the expression within the square-root sign in Equation iv and equate to zero:

$$\frac{[(1 - r^2)^2 + 2\zeta^2 r^2]8\zeta^2 r - [1 + 4\zeta^2 r^2][2(1 - r^2)(-2r) + 8\zeta^2 r]}{[(1 - r^2)^2 + 4\zeta^2 r^2]^2} = 0$$

Hence,

$$4r\{[(1 - r^2)^2 + 2\zeta^2 r^2]2\zeta^2 + [1 + 4\zeta^2 r^2][(1 - r^2) - 2\zeta^2]\} = 0$$

which simplifies to

$$r(2\zeta^2 r^4 + r^2 - 1) = 0$$

The roots are

$$r = 0 \quad \text{and} \quad r^2 = \frac{-1 \pm \sqrt{1 + 8\zeta^2}}{4\zeta^2}$$

The root $r = 0$ corresponds to the initial stationary point at zero frequency. That does not represent a peak. Taking only the positive root for r^2 and then its positive square root, the peak point of the transmissibility magnitude is given by

$$r = \frac{\left[\sqrt{1 + 8\zeta^2} - 1\right]^{1/2}}{2\zeta} \quad \text{(iv)}$$

For small ζ, Taylor series expansion gives

$$\sqrt{1+8\zeta^2} \approx 1 + \frac{1}{2} \times 8\zeta^2 = 1 + 4\zeta^2$$

With this approximation, Equation iv equates to 1. Hence, for small damping, the transmissibility magnitude will have a peak at $r = 1$, and from Equation 7.9, its value is

$$|T_f| \approx \frac{\sqrt{1+4\zeta^2}}{2\zeta} \approx \frac{1 + \frac{1}{2} \times 4\zeta^2}{2\zeta}$$

or

$$|T_f| \approx \frac{1}{2\zeta} + \zeta \approx \frac{1}{2\zeta} \tag{v}$$

3. The five curves of $|T_f|$ vs. r for $\zeta = 0$, 0.3, 0.7, 1.0, and 2.0 are shown in Figure 7.7. Note that these curves use the exact expression (see Equation 7.9).
From the curves, we observe the following:
    1. There is always a nonzero frequency value at which the transmissibility magnitude will peak. This is the resonance.
    2. For small ζ, the peak transmissibility magnitude is obtained at approximately $r = 1$. As ζ increases, this peak point shifts to the left (i.e., a lower value for peak frequency).
    3. The peak magnitude decreases with increasing ζ.
    4. All the transmissibility curves pass through the magnitude value 1.0 at the same frequency $r = \sqrt{2}$.
    5. The isolation (i.e., $|T_f| < 1$) is given by $r > \sqrt{2}$. In this region, $|T_f|$ increases with ζ.
    6. The transmissibility magnitude decreases for large r.
4. From the curves in Figure 7.7, we obtain:
    - For $|T_f| < 1.05$; $r > \sqrt{2}$ for all ζ.
    - For $|T_f| < 0.5$; $r > 1.73$, 1.964, 2.871, 3.77, 7.075 for $\zeta = 0.0$, 0.3, 0.7, 1.0, and 2.0, respectively.

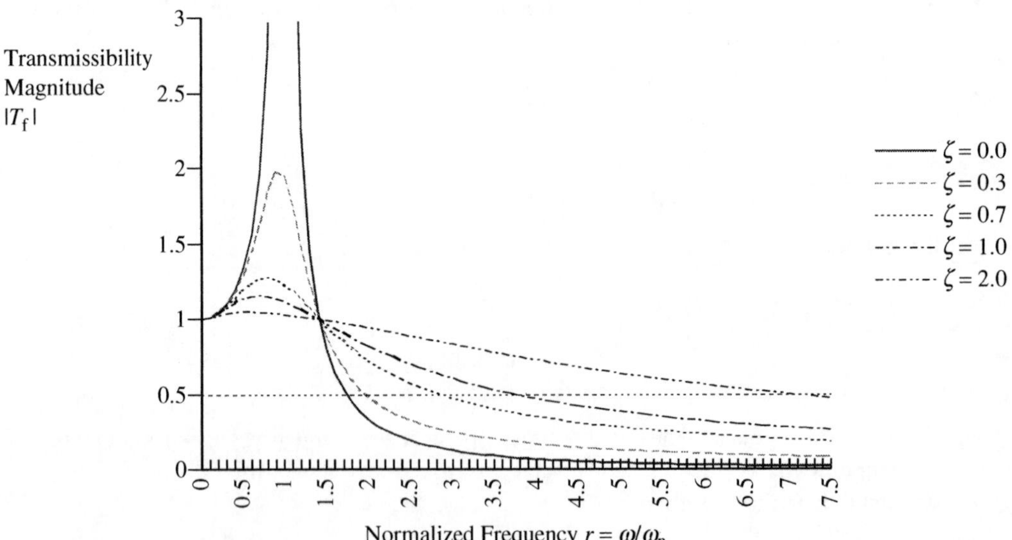

FIGURE 7.7 Transmissibility curves for a simple oscillator model.

5. We need

$$\sqrt{\frac{1+4\zeta^2r^2}{(1-r^2)^2+4\zeta^2r^2}} < \frac{1}{2}$$

or

$$\frac{1+4\zeta^2r^2}{(1-r^2)^2+4\zeta^2r^2} < \frac{1}{4}$$

or

$$4+16\zeta^2r^2 < (1-r^2)^2+4\zeta^2r^2$$

or

$$r^4-2r^2-12\zeta^2r^2-3>0$$

For $\zeta = 0.2$ and $r = 12/6 = 2$, the left-hand-side expression computes to

$$2^4-2\times2^2-12\times(0.2)^2\times2^2-3=3.08>0$$

Hence, the requirement is met. In fact, since, for $r = 2$

$$\text{LHS}=2^4-2\times2^2-12\times2^2\zeta^2-3=5-48\zeta^2$$

it follows that the requirement would be met for

$$5-48\zeta^2>0$$

or

$$\zeta<\sqrt{\frac{5}{48}}=0.32$$

If the requirement was not met (e.g., if $\zeta = 0.4$), the option would be to reduce damping.

7.3.1 Design Considerations

The *level of isolation* is defined as $1 - T$. It was noted that in the isolation region ($r > \sqrt{2}$) the transmissibility decreases (hence, the level of isolation increases) as the damping ratio ζ decreases. Thus, the best conditions of isolation are given by $\zeta = 0$. This is not feasible in practice, but we should maintain ζ as small as possible. For small ζ in the isolation region, Equation 7.8 may be approximated by

$$T=\frac{1}{(r^2-1)} \tag{7.10}$$

Note that T is real, in this case, of $\zeta \cong 0$, and also is positive since $r > \sqrt{2}$. However, in general, T may denote the magnitude of the transmissibility function. Substitute

$$r^2=\omega^2/\omega_n^2=\omega^2m/k$$

We get

$$k=\frac{\omega^2mT}{(1+T)} \tag{7.11}$$

This equation may be used to determine the design stiffness of the isolator for a specified level of isolation $(1 - T)$ in the operating frequency range $\omega > \omega_0$ for a system of known mass (including the isolator mass). Often, the static deflection δ_s of spring is used in design procedures and is given by

$$\delta_s = \frac{mg}{k} \tag{7.12}$$

Substituting Equation 7.12 into Equation 7.11, we obtain

$$\delta_s = (1 + T)\frac{g}{\omega^2 T} \tag{7.13}$$

Since the isolation region is $\omega > \sqrt{2}\omega_n$ it is desirable to make ω_n as small as possible in order to obtain the widest frequency range of operation. This is achieved by making the isolator as soft as possible (k as low as possible). However, there are limits to this in terms of structural strength, stability, and availability of springs. Then, m may be increased by adding an inertia block as the base of the system, which is then mounted on the isolator spring (with a damping layer) or an air-filled pneumatic mount. The inertia block will also lower the centroid of the system, thereby providing added desirable effects of stability and reducing rocking motions and noise transmission. For improved load distribution, instead of just one spring of design stiffness k, a set of n springs, each with stiffness k/n and uniformly distributed under the inertia block, should be used.

Another requirement for good vibration isolation is low damping. Usually, metal springs have very low damping (typically ζ less than 0.01). On the other hand, higher damping is needed to reduce resonant vibrations that will be encountered during start-up and shutdown conditions when the excitation frequency will vary and pass through the resonances. In addition, vibration energy has to be effectively dissipated, even under steady operating conditions. Isolation pads made of damping material such as cork, natural rubber, and neoprene may be used for this purpose. They can provide damping ratios of the order of 0.01.

The basic design steps for a vibration isolator in force isolation are as follows:

1. The required level of isolation (1 to T) and the lowest frequency of operation (ω_0) are specified. The mass of the vibration source (m) is known.
2. Use Equation 7.11 with $\omega = \omega_0$ to compute the required stiffness k of the isolator.
3. If the component k is not satisfactory, then increase m by introducing an inertia block and recompute k.
4. Distribute k over several springs.
5. Introduce a mounting pad of known stiffness and damping. Modify k and b accordingly and compute T using Equation 7.8. If the specified T is exceeded, then modify the isolator parameters as appropriate and repeat the design cycle.

Box 7.2 gives some relations that are useful in a design for vibration isolation.

Example 7.2

Consider a motor and fan unit of a building ventilation system weighing 50 kg and operating in the speed range of 600 to 3600 rpm. Since offices are located directly underneath the motor room, a 90% vibration isolation is desired. A set of mounting springs, each having a stiffness of 100 N/cm, is available. Design an isolation system to mount the motor-fan unit on the room floor.

Solution

For an isolation level of 90%, the required force transmissibility is $T = 0.1$. The lowest frequency of operation is $\omega = (600/60)2\pi$ rad/s. First, we try four mounting points. The overall spring stiffness is $k = 4 \times 100 \times 10^2$ N/m. Substitute in Equation 7.11.

$$4 \times 100 \times 100 = \frac{(10 \times 2\pi)^2 \times m \times 0.1}{1.1}$$

Box 7.2

Vibration Isolation

Transmissibility (force/force or motion/motion):

$$|T| = \sqrt{\frac{1 + 4\zeta^2 r^2}{(1 - r^2)^2 + 4\zeta^2 r^2}} \cong \frac{1}{(r^2 - 1)} \text{ for } r > 1 \text{ and small } \zeta$$

Properties:

1. $T_{\text{peak}} \cong \dfrac{\sqrt{1 + 4\zeta^2}}{2\zeta} \cong \dfrac{1}{2\zeta}$ for small ζ
2. T_{peak} occurs at $r_{\text{peak}} = \dfrac{[\sqrt{1 + 8\zeta^2} - 1]^{1/2}}{2\zeta} \cong 1$ for small ζ
3. All $|T|$ curves coincide at $r = \sqrt{2}$ for all ζ
4. Isolation region: $r > \sqrt{2}$
5. In isolation region:
   $|T|$ decreases with r (i.e., better isolation at higher frequencies)
   $|T|$ increases with ζ (i.e., better isolation at lower damping)

Design formulas:

Level of isolation $= 1 - T$

Isolator stiffness:

$$k = \frac{\omega^2 mT}{(1 + T)}$$

where

m = system mass
ω = operating frequency

Static deflection:

$$\delta_s = \frac{mg}{k} = (1 + T)\frac{g}{\omega^2 T}$$

which gives $m = 111.5$ kg. Since the mass of the unit is 50 kg, we should use an inertia block of mass 61.5 kg or more.

7.3.2 Vibration Isolation of Flexible Systems

The simple model shown in Figure 7.4(c) and (d) may not be adequate in the design of vibration isolators for sufficiently flexible systems. A more appropriate model for this situation is shown in Figure 7.8. Note that the vibration isolator has an inertia block of mass m in addition to damped flexible mounts of stiffness k and damping constant b. The vibrating system itself has a stiffness K and damping constant B in addition to its mass M.

In the absence of K, B, and the inertia block (m) as in Figure 7.4(c), the vibrating system becomes a simple inertia (M). Then, y_a and y are the same and the equation of motion is

$$M\ddot{y} + b\dot{y} + ky = f(t) \tag{7.14}$$

with the force transmitted to the support structure, f_s, given by

$$f_s = b\dot{y} + ky \tag{7.15}$$

The force transmissibility in this case is

$$T_{\text{inertial}} = \frac{f_s}{f} = \frac{bs + k}{Ms^2 + bs + k} \quad \text{with } s = j\omega \tag{7.16}$$

For the flexible system and isolator shown in Figure 7.8, the equation of motion:

$$M\ddot{y}_a + B(\dot{y}_a - \dot{y}) + K(y_a - y) = f(t) \tag{7.17}$$

$$m\ddot{y} + B(\dot{y} - \dot{y}_a) + K(y - y_a) + b\dot{y} + ky = 0 \tag{7.18}$$

Hence, in the frequency domain, we have

$$(Ms^2 + Bs + K)y_a - (Bs + K)y = f \tag{7.19}$$

$$[ms^2 + (B + b)s + K + k]y = (Bs + K)y_a \quad \text{with } s = j\omega \tag{7.20}$$

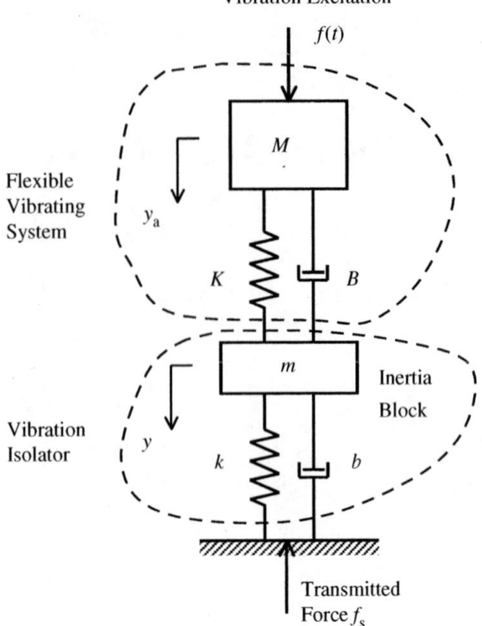

FIGURE 7.8 A model for vibration isolation of a flexible system.

Substitute Equation 7.20 into Equation 7.19 for eliminating y_a. We obtain

$$\left\{(Ms^2 + Bs + K)\frac{[ms^2 + (B + b)s + K + k]}{(Bs + K)} - (Bs + K)\right\}y = f$$

which simplifies

$$\left\{\frac{Ms^2[ms^2 + (B + b)s + K + k] + (Bs + K)(ms^2 + bs + k)}{(Bs + K)}\right\}y = f \tag{7.21}$$

The force transmitted to the supporting structure is still given by Equation 7.15. Hence, the transmissibility with the flexible system is

$$T_{\text{flexible}} = \frac{(Bs + K)(bs + k)}{\{Ms^2[ms^2 + (B + b)s + K + k] + (Bs + K)(ms^2 + bs + k)\}} \quad \text{with } s = j\omega \tag{7.22}$$

From Equation 7.16 and Equation 7.22, the transmissibility magnitude ratio is

$$\frac{T_{\text{flexible}}}{T_{\text{inertial}}} = \left|\frac{(Bs + K)(Ms^2 + bs + k)}{Ms^2[ms^2 + (B + b)s + K + k] + (Bs + K)(ms^2 + bs = k)}\right| \quad \text{with } s = j\omega \tag{7.23}$$

or

$$\frac{T_{\text{flexible}}}{T_{\text{inertial}}} = \left|\frac{(Ms^2 + bs + k)}{Ms^2(ms^2 + bs + k)/(Bs + K) + Ms^2 + ms^2 + bs + k}\right| \quad s = j\omega \tag{7.24}$$

In the nondimensional form, we have

$$\frac{T_{\text{flexible}}}{T_{\text{inertial}}} = \left|\frac{1 - r^2 + 2j\zeta_b r}{-r^2(1 - r_m r^2 + 2j\zeta_b r)/(r_\omega^2 + 2j\zeta_a r_\omega r) + 1 - (1 + r_m)r^2 + 2j\zeta_b r}\right| \tag{7.25}$$

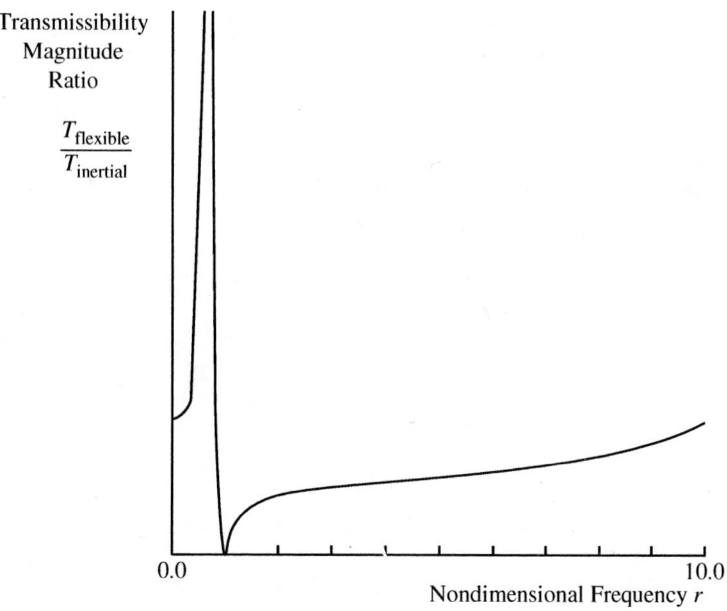

FIGURE 7.9 The effect of system flexibility on the transmissibility magnitude in the undamped case (mass ratio = 1.0; natural frequency ratio = 10.0).

where

$$r = \frac{\omega}{\sqrt{k/M}}$$

$$r_\text{m} = \frac{m}{M}$$

$$r_\omega = \frac{\sqrt{K/M}}{\sqrt{k/M}} = \sqrt{\frac{K}{k}}$$

$$\zeta_\text{a} = \frac{B}{2\sqrt{KM}}$$

$$\zeta_\text{b} = \frac{b}{2\sqrt{kM}}$$

Again, the design problem of vibration isolation is to select the parameters r_m, r_ω, ζ_a, and ζ_b so that the required level of vibration isolation is realized for an operating frequency range of r.

A plot of Equation 7.25 for the undamped case with $r_\text{m} = 1.0$ and $r_\omega = 10.0$ is given in Figure 7.9. Generally, the transmissibility ratio will be zero at $r = 1$ (the resonance of the inertial system) and there will be two values of r (the resonances of the flexible system), for which the ratio will become infinity in the undamped case. The latter two neighborhoods should be avoided under steady operating conditions.

7.4 Balancing of Rotating Machinery

Many practical devices that move contain rotating components. Examples are wheels of vehicles, shafts, gear transmissions of machinery, belt drives, motors, turbines, compressors, fans, and rollers. An unbalance (imbalance) is created in a rotating part when its center of mass does not coincide with the axis of rotation. The reasons for this *eccentricity* include the following:

1. Inaccurate production procedures (machining, casting, forging, assembly, etc.)
2. Wear and tear
3. Loading conditions (mechanical)
4. Environmental conditions (thermal loads and deformation)
5. Use of inhomogeneous and anisotropic material (which does not have a uniform density distribution)
6. Component failure
7. Addition of new components to a rotating device

For a component of mass m, eccentricity e, and rotating at angular speed ω, the centrifugal force that is generated is $me\omega^2$. Note the quadratic variation with ω. This rotating force may be resolved into two orthogonal components, which will be sinusoidal with frequency ω. It follows that harmonic forcing excitations are generated due to the unbalance, which can generate undesirable vibrations and associated problems.

Problems caused by unbalance include wear and tear, malfunction and failure of components, poor quality of products, and undesirable noise. The problem becomes increasingly important given the present trend of developing high-speed machinery. It is estimated that the speed of operation of machinery has doubled during the past 50 years. This means that the level of unbalance forces may have quadrupled during the same period, causing more serious vibration problems.

An unbalanced rotating component may be balanced by adding or removing material to or from the component. We need to know both the magnitude and location of the balancing masses to be added to, or removed. The present section will address the problem of component balancing for vibration suppression.

Note that the goal to remove the source of vibration (namely, the mass eccentricity) typically by adding one or more balancing mass elements. Two methods are available:

1. Static (single-plane) balancing
2. Dynamic (two-plane) balancing

The first method concerns balancing of planar objects (e.g., pancake motors, disks) whose longitudinal dimension about the axis of rotation is not significant. The second method concerns balancing of objects that have a significant longitudinal dimension. We will discuss both methods.

7.4.1 Static Balancing

Consider a disk rotating at angular velocity ω about a fixed axis. Suppose that the mass center of the disk has an eccentricity e from the axis of rotation, as shown in Figure 7.10(a). Place a fixed coordinate frame

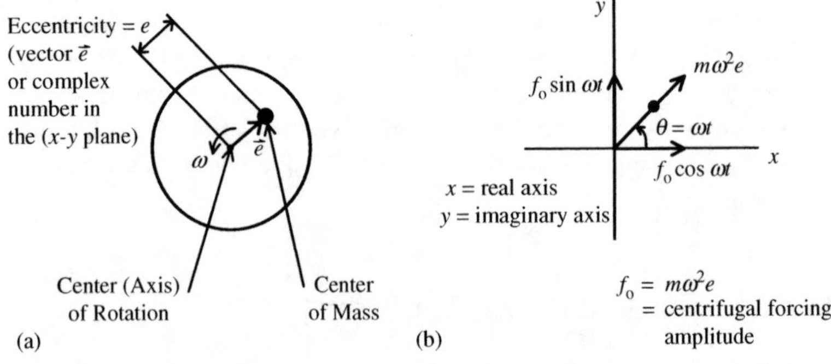

FIGURE 7.10 (a) Unbalance in a rotating disk due to mass eccentricity; (b) rotating vector (phasor) of centrifugal force due to unbalance.

x–y at the center of rotation. The position $\vec{e}$ of the mass center in this coordinate frame may be represented as:

1. A position vector rotating at angular speed ω,.
2. A complex number, with x-coordinate denoting the real part and y-coordinate denoting the imaginary part.

The centrifugal force due to the mass eccentricity is also a vector in the direction of $\vec{e}$, but with a magnitude $f_o = m\omega^2 e$, as shown in Figure 7.10(b). It is seen that harmonic excitations result in both x and y directions, given by $f_o \cos \omega t$ and $f_o \sin \omega t$, respectively, where $\theta = \omega t$ = orientation of the rotating vector with respect to the x-axis. To balance the disk, we should add a mass m at $-\vec{e}$. However, we do not know the value of m and the location of $\vec{e}$.

7.4.1.1 Balancing Approach

1. Measure the amplitude V_u and the phase angle ϕ_1 (e.g., by the signal from an accelerometer mounted on the bearing of the disk) of the unbalance centrifugal force with respect to some reference.
2. Mount a known mass (trial mass) M_t at a known location on the disk. Suppose that its own centrifugal force is given by the rotating vector $\vec{V}_w$, and the resultant centrifugal force due to both the original unbalance and the final mass is $\vec{V}_r$.
3. Measure the amplitude V_r and the phase angle ϕ_2 of the resultant centrifugal force as in step 1, with respect to the same phase reference.

A vector diagram showing the centrifugal forces $\vec{V}_u$ and $\vec{V}_w$ due to the original unbalance and the trial-mass unbalance, respectively, is shown in Figure 7.11. The resultant unbalance is $\vec{V}_r = \vec{V}_u + \vec{V}_w$. Note that $-\vec{V}_u$ represents the centrifugal force due to the balancing mass. Therefore, if we determine the angle ϕ_b in Figure 7.11, it will give the orientation of the balancing mass. Suppose also that the balancing mass is M_b and it is mounted at an eccentricity equal to that of the trial mass M_t. Then,

$$\frac{M_b}{M_t} = \frac{V_u}{V_w}$$

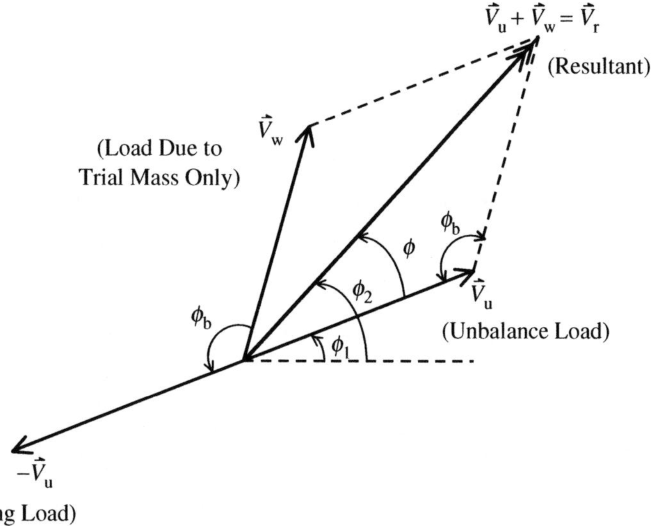

FIGURE 7.11 A vector diagram of the single-plane (static) balancing problem.

We need to determine the ratio V_u/V_w and the angle ϕ_b. These values can be derived as follows:

$$\phi = \phi_2 - \phi_1 \tag{7.26}$$

The cosine rule gives

$$V_w^2 = V_u^2 + V_r^2 - 2V_u V_r \cos\phi \tag{7.27}$$

This will provide V_w since V_u, V_r, and ϕ are known. Apply the cosine rule again:

$$V_r^2 = V_u^2 + V_w^2 - 2V_u V_w \cos\phi_b$$

Hence,

$$\phi_b = \cos^{-1}\left[\frac{V_u^2 + V_w^2 - V_r^2}{2V_u V_w}\right] \tag{7.28}$$

Note: One may think that since we measure ϕ_1, we know exactly where $\vec{V}_u$ is. This is not the case because we do not know the reference line from which ϕ_1 is measured. We only know that this reference is kept fixed (through strobe synchronization of the body rotation) during measurements. Hence, we need to know ϕ_b, which gives the location of $-\vec{V}_u$ with respect to the known location of $\vec{V}_w$ on the disk.

7.4.2 Complex Number/Vector Approach

Again, suppose that the imbalance is equivalent to a mass of M_b that is located at the same eccentricity (radius) r as the trial mass M_t. Define complex numbers (mass location vectors in a body frame)

$$\vec{M}_b = M_b \angle \theta_b \tag{7.29}$$

$$\vec{M}_t = M_t \angle \theta_t \tag{7.30}$$

as shown in Figure 7.12.

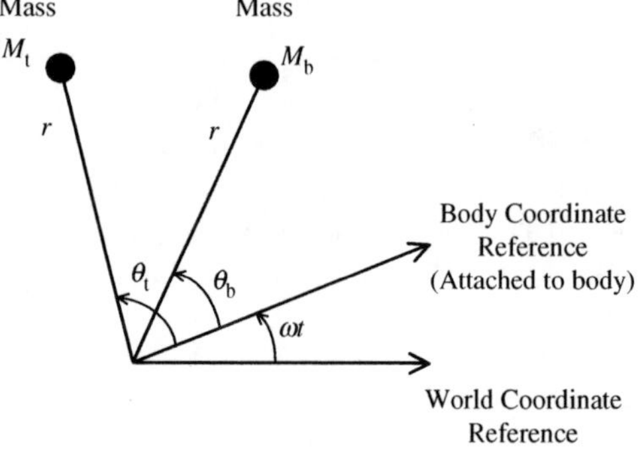

FIGURE 7.12 Rotating vectors of mass location.

Associated force vectors are

$$\vec{V}_u = \omega^2 r e^{j\omega t} M_b \angle \theta_b \tag{7.31}$$

$$\vec{V}_w = \omega^2 r e^{j\omega t} M_t \angle \theta_t \tag{7.32}$$

or

$$\vec{V}_u = \vec{A}\vec{M}_b \tag{7.33}$$

$$\vec{V}_w = \vec{A}\vec{M}_t \tag{7.34}$$

where $\vec{A} = \omega^2 r e^{j\omega t}$ is the conversion factor (complex) from the mass to the resulting dynamic force (rotating). This factor is the same for both cases since r is the same. We need to determine $\vec{M}_b$.

From Equation 7.33

$$\vec{M}_b = \frac{\vec{V}_u}{\vec{A}} \tag{7.35}$$

Substitute Equation 7.34:

$$\vec{M}_b = \frac{\vec{V}_u}{\vec{V}_w} \cdot \vec{M}_t \tag{7.36}$$

However, since

$$\vec{V}_r = \vec{V}_u + \vec{V}_w \tag{7.37}$$

we have

$$\vec{M}_b = \frac{\vec{V}_u}{(\vec{V}_r - \vec{V}_u)} \cdot \vec{M}_t \tag{7.38}$$

Since we know $\vec{M}_t$ and we measure $\vec{V}_u$ and $\vec{V}_r$ to the same scaling factor, we can compute $\vec{M}_b$. Locate the balancing mass at $-\vec{M}_b$ (with respect to the body frame).

Example 7.3

Consider the following experimental steps:

Measured: Accelerometer amplitude (oscilloscope reading) of 6.0 with a phase lead (with respect to strobe signal reference, which is synchronized with the rotating body frame) of 50°.
Added: Trial mass $M_t = 20$ g at angle 180° with respect to a body reference radius.
Measured: Accelerometer amplitude of 8.0 with a phase lead of 60° (with respect synchronized strobe signal).

Determine the magnitude and location of the balancing mass.

Solution

Method 1:

We have the data

$$\phi = 60 - 50° = 10°$$

$$V_u = 6.0; \quad V_r = 8.0$$

Hence, from Equation 7.27:

$$V_w = \sqrt{6^2 + 8^2 - 2 \times 6 \times 8 \cos 10°} = 2.37$$

Balancing mass:

$$M_b = \frac{6.0}{2.37} \times 20 = 50.63 \text{ g}$$

Equation 7.28 gives

$$\phi_b = \cos^{-1}\left[\frac{6^2 + 2.37^2 - 8^2}{2 \times 6 \times 2.37}\right] = \cos^{-1}(-0.787) = 142° \text{ or } 218°$$

Pick the result $0° \leq \phi_b \leq 180°$, as clear from the vector diagram shown in Figure 7.11.

Hence,

$$\phi_b = 142°$$

However,

$$\vec{M}_t = 20\angle 180° \text{ g}$$

It follows that

$$-\vec{M}_b = 50.63\angle(180° + 142°) \text{ g} = 50.63\angle 322° \text{ g}$$

Method 2:

We have

$$\vec{M}_t = 20\angle 180° \text{ g}$$

$$\vec{V}_u = 6.0\angle 50°$$

$$\vec{V}_r = 8.0\angle 60°$$

Then, from Equation 7.38 we obtain

$$\vec{M}_b = \frac{6.0\angle 50^\circ}{(8.0\angle 60^\circ - 6.0\angle 50^\circ)} 20\angle 180^\circ \text{ g}$$

First, we compute

$$8.0\angle 60^\circ - 6.0\angle 50^\circ = (8.0 \cos 60^\circ + j8.0 \sin 6.0^\circ) - (6.0 \cos 50^\circ + j6.0 \sin 50^\circ)$$

$$= (8.0 \cos 60^\circ - 6.0 \cos 50^\circ) + j(8.0 \sin 60^\circ - 6.0 \sin 50^\circ) = 0.1433 + j2.332$$

$$= 2.336\angle 86.48^\circ$$

Hence,

$$\vec{M}_b = \frac{6.0\angle 50^\circ}{2.336\angle 86.48^\circ} 20\angle 180^\circ = \frac{6.0 \times 20}{2.336}\angle(50^\circ + 180^\circ - 86.48^\circ) = 51\angle 143.5^\circ \text{ g}$$

The balancing mass should be located at

$$-\vec{M}_b = 51\angle 323.5^\circ \text{ g}$$

Note: This angle is measured from the same body reference as for the trial mass.

7.4.3 Dynamic (Two-Plane) Balancing

Instead of an unbalanced disk, consider an elongated rotating object supported at two bearings, as shown in Figure 7.13. In this case, in general, there may not be an equivalent single unbalance force at a single plane normal to the shaft axis. To show this, recall that a system of forces may be represented by a single force at a specified location and a couple (two parallel forces that are equal and opposite). If this single force (resultant force) is zero, we are left with only a couple. The couple cannot be balanced by a single force.

All the unbalance forces at all the planes along the shaft axis can be represented by an equivalent single unbalance force at a specified plane and a couple. If this equivalent force is zero, then to balance the couple we will need two equal and opposite forces at two different planes.

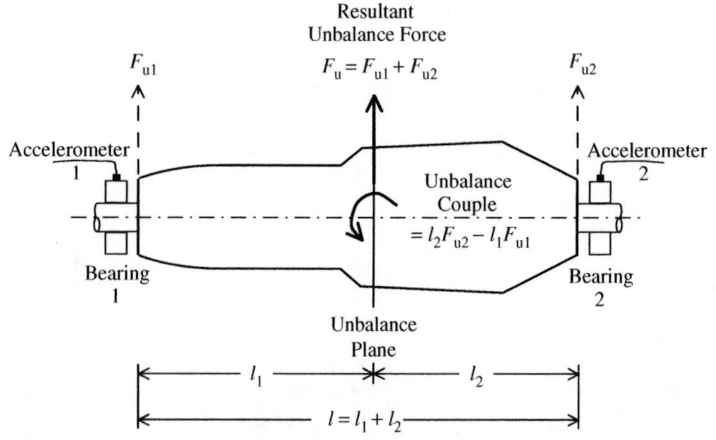

FIGURE 7.13 A dynamic (two-plane) balancing problem.

On the other hand, if the couple is zero, then a single force in the opposite direction at the same plane of the resultant unbalance force will result in complete balancing. However, this unbalance plane may not be reachable, even if it is known, for the purpose of adding the balancing mass.

In the present (two-plane) balancing problem, the balancing masses are added at the two bearing planes so that both the resultant unbalance force and couple are balanced, in general. It is clear from Figure 7.13 that even a sole unbalance mass $\vec{M}_b$ at a single unbalance plane may be represented by two unbalance masses $\vec{M}_{b1}$ and $\vec{M}_{b2}$ at the bearing planes 1 and 2. Likewise, in the presence of an unbalance couple, we can simply add two equal and opposite forces at the planes 1 and 2 so that its couple is equal to the unbalance couple. Hence, a general unbalance can be represented by the two unbalance masses $\vec{M}_{b1}$ and $\vec{M}_{b2}$ at planes 1 and 2, as shown in Figure 7.13. As for the single-plane balancing problem, the resultant unbalance forces at the two bearings (which would be measured by the accelerometers at 1 and 2) are

$$\vec{V}_{u1} = \vec{A}_{11}\vec{M}_{b1} + \vec{A}_{12}\vec{M}_{b2} \tag{7.39}$$

$$\vec{V}_{u2} = \vec{A}_{21}\vec{M}_{b1} + \vec{A}_{22}\vec{M}_{b2} \tag{7.40}$$

Suppose that a trial mass of $\vec{M}_{t1}$ (at a known location with respect to the body reference line) was added at plane 1. The resulting unbalance forces at the two bearings are

$$\vec{V}_{r11} = \vec{A}_{11}(\vec{M}_{b1} + \vec{M}_{t1}) + \vec{A}_{12}\vec{M}_{b2} \tag{7.41}$$

$$\vec{V}_{r21} = \vec{A}_{21}(\vec{M}_{b1} + \vec{M}_{t1}) + \vec{A}_{22}\vec{M}_{b2} \tag{7.42}$$

Next, suppose that a trial mass of $\vec{M}_{t2}$ (at a known location with respect to the body reference line) was added at plane 2, after removing $\vec{M}_{t1}$. The resulting unbalance forces at the two bearings are

$$\vec{V}_{r12} = \vec{A}_{11}\vec{M}_{b1} + \vec{A}_{12}(\vec{M}_{b2} + \vec{M}_{t2}) \tag{7.43}$$

$$\vec{V}_{r22} = \vec{A}_{21}\vec{M}_{b1} + \vec{A}_{22}(\vec{M}_{b2} + \vec{M}_{t2}) \tag{7.44}$$

The following subtractions of equations are now made.

Equation 7.41 minus Equation 7.39:

$$\vec{V}_{r11} - \vec{V}_{u1} = \vec{A}_{11}\vec{M}_{t1} \quad \text{or} \quad \vec{A}_{11} = \frac{\vec{V}_{r11} - \vec{V}_{u1}}{\vec{M}_{t1}} \tag{7.45}$$

Equation 7.42 minus Equation 7.40:

$$\vec{V}_{r21} - \vec{V}_{u2} = \vec{A}_{21}\vec{M}_{t1} \quad \text{or} \quad \vec{A}_{21} = \frac{\vec{V}_{r21} - \vec{V}_{u2}}{\vec{M}_{t1}} \tag{7.46}$$

Equation 7.43 minus Equation 7.39:

$$\vec{V}_{r12} - \vec{V}_{u1} = \vec{A}_{12}\vec{M}_{t2} \quad \text{or} \quad \vec{A}_{12} = \frac{\vec{V}_{r12} - \vec{V}_{u1}}{\vec{M}_{t2}} \tag{7.47}$$

Equation 7.44 minus Equation 7.40:

$$\vec{V}_{r22} - \vec{V}_{u2} = \vec{A}_{22}\vec{M}_{t2} \quad \text{or} \quad \vec{A}_{22} = \frac{\vec{V}_{r22} - \vec{V}_{u2}}{\vec{M}_{t2}} \tag{7.48}$$

Hence, generally

$$\vec{A}_{ij} = \frac{\vec{V}_{rij} - \vec{V}_{ui}}{\vec{M}_{tj}} \tag{7.49}$$

These parameters A_{ij} are called *influence coefficients*.

Next, in Equation 7.39 and Equation 7.40 eliminate $\vec{M}_{b2}$ and $\vec{M}_{b1}$ separately to determine the other. Thus,

$$\vec{A}_{22}\vec{V}_{u1} - \vec{A}_{12}\vec{V}_{u2} = (\vec{A}_{22}\vec{A}_{11} - \vec{A}_{12}\vec{A}_{21})\vec{M}_{b1}$$

$$\vec{A}_{21}\vec{V}_{u1} - \vec{A}_{11}\vec{V}_{u2} = (\vec{A}_{21}\vec{A}_{12} - \vec{A}_{11}\vec{A}_{22})\vec{M}_{b2}$$

or

$$\vec{M}_{b1} = \frac{\vec{A}_{22}\vec{V}_{u1} - \vec{A}_{12}\vec{V}_{u2}}{(\vec{A}_{22}\vec{A}_{11} - \vec{A}_{12}\vec{A}_{21})} \tag{7.50}$$

$$\vec{M}_{b2} = \frac{\vec{A}_{21}\vec{V}_{u1} - \vec{A}_{11}\vec{V}_{u2}}{(\vec{A}_{21}\vec{A}_{12} - \vec{A}_{11}\vec{A}_{22})} \tag{7.51}$$

Substitute Equation 7.45 to Equation 7.48 into Equation 7.50 and Equation 7.51 to determine $\vec{M}_{b1}$ and $\vec{M}_{b2}$. Balancing masses that should be added are $-\vec{M}_{b1}$ and $-\vec{M}_{b2}$ in planes 1 and 2, respectively.

The single-plane and two-plane balancing approaches are summarized in Box 7.3.

Example 7.4

Suppose that the following measurements are obtained.

Without trial mass:
Accelerometer at 1: amplitude = 10.0; phase lead = 55°.
Accelerometer at 2: amplitude = 7.0; phase lead = 120°

With trial mass 20 g at location 270° of plane 1:
Accelerometer at 1: amplitude = 7.0; phase lead = 120°
Accelerometer at 2: amplitude = 5.0; phase lead = 225°

With trial mass 25 g at location 180° of plane 2:
Accelerometer at 1: amplitude = 6.0; phase lead = 120°
Accelerometer at 2: amplitude = 12.0; phase lead = 170°

Determine the magnitude and orientation of the necessary balancing masses in planes 1 and 2 in order to completely balance (dynamic) the system.

Solution

In the phasor notation, we can represent the given data as follows:

$$\vec{V}_{u1} = 10.0\angle 55°; \quad \vec{V}_{u2} = 7.0\angle 120°$$

$$\vec{V}_{r11} = 7.0\angle 120°; \quad \vec{V}_{r21} = 5.0\angle 225°$$

$$\vec{V}_{r12} = 6.0\angle 120°; \quad \vec{V}_{r22} = 12.0\angle 170°$$

$$\vec{M}_{t1} = 20\angle 270° \text{ g}; \quad \vec{M}_{t2} = 25\angle 180° \text{ g}$$

From Equation 7.45 to Equation 7.48, we have

$$\vec{A}_{11} = \frac{7.0\angle 120° - 10.0\angle 55°}{20\angle 270°}; \quad \vec{A}_{21} = \frac{5.0\angle 225° - 7.0\angle 120°}{20\angle 270°}$$

$$\vec{A}_{12} = \frac{6.0\angle 120° - 10.0\angle 55°}{25\angle 180°}; \quad \vec{A}_{22} = \frac{12.0\angle 170° - 7.0\angle 120°}{25\angle 180°}$$

These phasors are computed as below:

$$\vec{A}_{11} = \frac{(7.0\cos 120° - 10\cos 55°) + j(7\sin 120° - 10\sin 55°)}{20\angle 270°} = \frac{-9.235 - j2.129}{20\angle 270°} = \frac{9.477\angle 193°}{20\angle 270°}$$

$$= 0.474\angle -77°$$

Box 7.3

Balancing of Rotating Components

Static or single-plane balancing (balances a single equivalent dynamic force)

Experimental approach:

1. With respect to a body reference line of accelerometer signal at bearing, measure magnitude (V) and phase (ϕ):
   (a) Without trial mass: (V_u, ϕ_1) or $\vec{V}_u = V_u\angle\phi_1$
   (b) With trial mass M_t: (V_r, ϕ_2) or $\vec{V}_r = V_r\angle\phi_2$.
2. Compute balancing mass M_b and its location with respect to M_t.
3. Remove M_t and add M_b at determined location.

Computation approach 1:

$$V_w = [V_u^2 + V_r^2 - 2V_uV_w\cos(\phi_2 - \phi_1)]^{1/2}$$

$$\phi_b = \cos^{-1}\left[\frac{V_u^2 + V_w^2 - V_r^2}{2V_uV_w}\right] \text{ and } M_b = \frac{V_u}{V_w}M_t$$

Locate M_b at ϕ_b from M_t.

Computation approach 2:

Unbalance mass phasor

$$\vec{M}_b = \frac{\vec{V}_u}{(\vec{V}_r - \vec{V}_u)}\vec{M}_t$$

where $\vec{M}_t = M_t\angle\theta_t$ (trial mass phasor).

Locate balancing mass at $-\vec{M}_b$.

Dynamic or two-plane balancing (balances an equivalent dynamic force and a couple)

Experimental approach:

1. Measure $\vec{V}_{ui}$ at bearings $i = 1, 2$, with a trial mass.
2. Measure $\vec{V}_{rij}$ at bearings $i = 1, 2$, with only one trial mass $\vec{M}_{tj}$ at $j = 1, 2$.
3. Compute unbalance mass phasor $\vec{M}_{bi}$ in planes $i = 1, 2$.
4. Remove trial mass and place balancing masses $-\vec{M}_{bi}$ in planes $i = 1, 2$.

Computations:

Influence coefficients: $\vec{A}_{ij} = (\vec{V}_{rij} - \vec{V}_{ui})/\vec{M}_{tj}$

Unbalance mass phasors:

$$\vec{M}_{b1} = \frac{\vec{A}_{22}\vec{V}_{u1} - \vec{A}_{12}\vec{V}_{u2}}{(\vec{A}_{22}\vec{A}_{11} - \vec{A}_{12}\vec{A}_{21})} \quad \text{and} \quad \vec{M}_{b2} = \frac{\vec{A}_{21}\vec{V}_{u1} - \vec{A}_{11}\vec{V}_{u2}}{(\vec{A}_{21}\vec{A}_{12} - \vec{A}_{11}\vec{A}_{22})}$$

$$\vec{A}_{21} = \frac{(5\cos 225° - 7\cos 120°) + j(5\sin 225° - 7\sin 120°)}{20\angle 270°} = \frac{-7.036 - j9.6}{20\angle 270°} = \frac{11.9\angle 234°}{20\angle 270°}$$

$$= 0.595\angle -36°$$

$$\vec{A}_{12} = \frac{(6\cos 120° - 10\cos 55°) + j(6\sin 120° - 10\sin 55°)}{25\angle 180°} = \frac{-8.736 - j3.0}{25\angle 180°} = \frac{9.237\angle 199°}{25\angle 180°}$$

$$= 0.369\angle 19°$$

$$\vec{A}_{22} = \frac{(12\cos 170° - 7\cos 120°) + j(12\sin 170° - 7\sin 120°)}{25\angle 180°} = \frac{-8.318 - j4.0}{25\angle 180°} = \frac{9.23\angle 205.7°}{25\angle 180°}$$

$$= 0.369\angle 25.7°$$

Next, the denominators of the balancing mass phasors (in Equation 7.50 and Equation 7.51) are computed as

$$\begin{aligned}\vec{A}_{22}\vec{A}_{11} - \vec{A}_{12}\vec{A}_{21} &= (0.369\angle 25.7° \times 0.474\angle -77°) - (0.369\angle 19° \times 0.595\angle -36°)\\ &= 0.1749\angle -51.3° - 0.2196\angle -17°\\ &= (0.1749\cos 51.3° - 0.2196\cos 17°) - j(0.1749\sin 51.3° - 0.2196\sin 17°)\\ &= -0.1 - j0.0723 = 0.1234\angle 216°\end{aligned}$$

and, hence

$$-(\vec{A}_{22}\vec{A}_{11} - \vec{A}_{12}\vec{A}_{21}) = 0.1234\angle 36°$$

Finally, the balancing mass phasors are computed using Equation 7.50 and Equation 7.51 as

$$\begin{aligned}\vec{M}_{b1} &= \frac{0.369\angle 25.7° \times 10\angle 55° - 0.369\angle 19° \times 7.0\angle 120°}{0.1234\angle 216°} = \frac{3.69\angle 80.7° - 2.583\angle 139°}{0.1234\angle 216°}\\ &= \frac{(3.69\cos 80.7° - 2.583\cos 139°) + j(3.69\sin 80.7° - 2.5838\sin 139°)}{0.1234\angle 216°} = \frac{2.546 + j1.947}{0.1234\angle 216°}\\ &= \frac{3.205\angle 37.4°}{0.1234\angle 216°} = 26\angle -178.6°\end{aligned}$$

$$\begin{aligned}\vec{M}_{b2} &= \frac{0.595\angle -36° \times 10\angle 55° - 0.474\angle -77° \times 7.0\angle 120°}{0.1234\angle 36°} = \frac{5.95\angle 19° - 3.318\angle 43°}{0.1234\angle 36°}\\ &= \frac{(5.95\cos 19° - 3.318\cos 43°) - j(5.95\sin 19° - 3.318\sin 43°)}{0.1234\angle 36°} = \frac{3.2 + j0.326}{0.1234\angle 36°}\\ &= \frac{1.043\angle 5.8°}{0.1234\angle 36°} = 8.45\angle -30.0°\end{aligned}$$

Finally, we have

$$-\vec{M}_{b1} = 26\angle 1.4°\ \text{g}; \qquad -\vec{M}_{b2} = 8.45\angle 150°\ \text{g}$$

7.4.4 Experimental Procedure of Balancing

The experimental procedure for determining the balancing masses and locations for a rotating system should be clear from the analytical developments and examples given above. The basic steps are:

1. Determine the magnitude and the phase angle of accelerometer signals at the bearings with and without trial masses at the bearing planes.
2. Using this data, compute the necessary balancing masses (magnitude and location) at the bearing planes.
3. Place the balancing masses.
4. Check whether the system is balanced. If not, repeat the balancing cycle.

A laboratory experimental setup for two-plane balancing is shown schematically in Figure 7.14. A view of the system is shown in Figure 7.15. The two disks rigidly mounted on the shaft are driven by a DC motor. The drive speed of the motor is adjusted by the manual speed controller. The (two) shaft

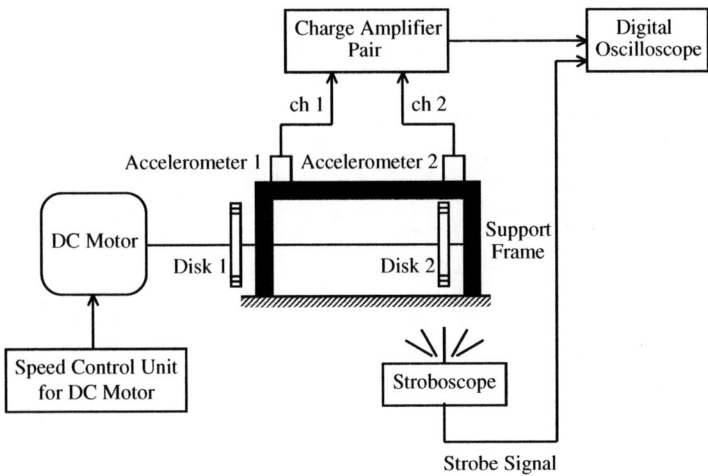

FIGURE 7.14 Schematic arrangement of a rotor balancing experiment.

bearings are located very close to the disks, as shown in Figure 7.14. Two accelerometers are mounted on the top of the bearing housing so that the resulting vertical accelerations can be measured. The accelerometer signals are conditioned using the two-channel charge amplifier and read and displayed through two channels of the digital oscilloscope. The output of the stroboscope (tachometer) is used as the reference signal with respect to the phase angles of the accelerometer signals that are measured.

In Figure 7.15, the items of equipment are seen from left to right. The first item is the two-channel digital oscilloscope. The manual speed controller with control knob for the DC motor follows. Next is the pair of charge amplifiers for the accelerometers. The strobe-light unit (strobe-tacho) is placed on top of the common housing of the charge amplifier pair. The two-disk rotor system with the drive motor is shown as the last item to the right. Also, note the two accelerometers (seen as small vertical projections) mounted on the bearing frame of the shaft directly above the two bearings.

Because this reference always has to be fixed prior to reading the oscilloscope data, the strobe-tacho is synchronized with the disk rotation. This is achieved as follows (note that all the readings are taken with

FIGURE 7.15 A view of the experimental setup for two-plane balancing at the University of British Columbia.

the same rotating speed, which is adjusted by the manual speed controller): First, make a physical mark (e.g., a black spot in a white background) on one of the disks. Aim the strobe flash at this disk. As the motor speed is adjusted to the required fixed value, the strobe flash is synchronized such that the mark on the disk "appears" stationary at the same location (e.g., at the uppermost location of the circle of rotation). This ensures not only that the strobe frequency is equal to the rotating speed of the disk, but also that the same phase angle reference is used for all readings of accelerometer signals.

The two disks have slots at locations whose radius is known, and whose angular positions in relation to a body reference line (a radius representing the 0° reference line) are clearly marked. Known masses (typically, bolts and nuts of known mass) can be securely mounted in these slots. Readings obtained through the oscilloscope are:

1. Amplitude of each accelerometer signal
2. Phase lead of the accelerometer signal with respect to the synchronized and reference-fixed strobe signal (note: a phase lag should be represented by a negative sign in the data)

The measurements taken and the computations made in the experimental procedure should be clear from Example 7.4.

7.5 Balancing of Reciprocating Machines

A reciprocating mechanism has a slider that moves rectilinearly back and forth along some guideway. A piston-cylinder device is a good example. Often, reciprocating machines contain rotatory components in addition to the reciprocating mechanisms. The purpose is to either covert a reciprocating motion to a rotary motions (as in the case of an automobile engine), or to convert a rotary motion to a reciprocating motions (as in the opto-slider mechanism of a photocopier). Irrespective of the reciprocating machine employed, it is important to remove the vibratory excitations that arise in order to realize the standard design goals of smooth operation, accuracy, low noise, reliability, mechanical integrity, and extended service life. Naturally, in view of their rotational asymmetry, reciprocating mechanisms with rotary components are more prone to unbalance than purely rotary components. Removing the "source of vibration" by proper balancing of the machine would be especially applicable in this situation.

7.5.1 Single-Cylinder Engine

A practical example of a reciprocating machine with integral rotary motion is the internal combustion (IC) engine of an automobile. A single-cylinder engine is sketched in Figure 7.16. Observe the nomenclature of the components. The reciprocating motion of the piston is transmitted through the connecting rod and crank into a rotatory motion of the crankshaft. The crank, as sketched in Figure 7.16, has a counterbalance mass, the purpose of which is to balance the rotary force (centrifugal).

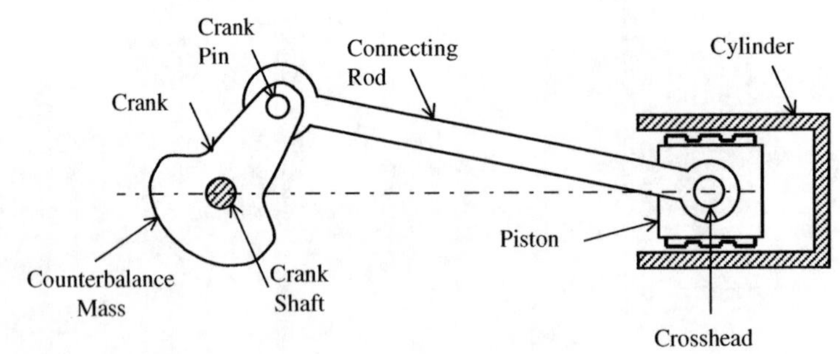

FIGURE 7.16 A single-cylinder reciprocating engine.

We will ignore this in our analysis because the goal is to determine the unbalance forces and ways to balance them.

Clearly, both the connecting rod and the crank have distributed mass and moment of inertia. To simplify the analysis, we approximate as follows:

1. Represent the crank mass by an equivalent lumped mass at the crank pin (equivalence may be based on either centrifugal force or kinetic energy).
2. Represent the mass of the connecting rod by two lumped masses, one at the crank pin and the other at the cross head (piston pin).

The piston itself has a significant mass, which is also lumped at the crosshead. Hence, the equivalent system has a crank and a connecting rod, both of which are considered massless, with a lumped mass m_c at the crank pin and another lumped mass m_p at the piston pin (crosshead).

Furthermore, under normal operation, the crankshaft rotates at a constant angular speed (ω). Note that this steady speed is realized not by natural dynamics of the system, but rather by proper speed control (a topic which is beyond the scope of the present discussion).

It is a simple matter to balance the lumped mass m_c at the crank pin. Simply place a countermass m_c at the same radius in the radially opposite location (or a mass in inverse proportion to the radial distance form the crankshaft, but remaining in the radially opposite direction). This explains the presence of the countermass in the crank shown in Figure 7.16. Once complete balancing of the rotating inertia (m_c) is thus achieved, we still need to completely eliminate the effect of the vibration source on the crankshaft. To achieve this, we must compensate for the forces and moments on the crankshaft that result from:

1. The reciprocating motion of the lumped mass m_p
2. Time-varying combustion (gas) pressure in the cylinder

Both types of forces act on the piston in the direction of its reciprocating (rectilinear) motion. Hence, their influence on the crankshaft can be analyzed in the same way, except that the combustion pressure is much more difficult to determine.

The above discussion justifies the use of the simplified model shown in Figure 7.17 for analyzing the balancing of a reciprocating machine. The characteristics of this model are as follows:

1. A light crank OC of radius r rotates at constant angular speed ω about O, which is the origin of the x–y coordinate frame.
2. A light connecting rod CP of length l is connected to the connecting rod at C and to the piston at P with frictionless pins. Since the rod is light and the joints are frictionless, the force f_c supported by it will act along its length. (Assume that the force f_c in the connecting rod is compressive, for the purpose of the sign convention). Connecting rod makes an angle ϕ with OP (the negative x axis).
3. A lumped mass m_p is present at the piston. A force f acts at P in the negative x direction. This may be interpreted as either the force due to the gas pressure in the cylinder or the inertia force $m_p a$ where a is the acceleration m_p in the positive x direction. These two cases of forcing are considered separately.
4. A lateral force f_l acts on the piston by the cylinder wall, in the positive y direction.

Again, note that the lumped mass m_c at C is not included in the model of Figure 7.17 because it is assumed to be completely balanced by a countermass in the crank. Furthermore, the lumped mass m_p includes both the mass of the piston and also part of the inertia of the connecting rod.

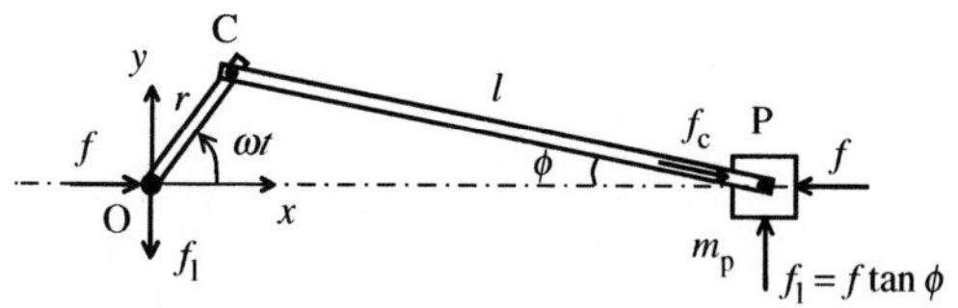

FIGURE 7.17 The model used to analyze balancing of a reciprocating engine.

There are no external forces at C. Furthermore, the only external forces at P are f and f_l, where f is interpreted as either the inertia force in m_p or the gas force on the piston. Hence, there should be equal and opposite forces at the crankshaft O, as shown in Figure 7.17, to support the forces acting at P. Now, let us determine f_l.

Equilibrium at P gives

$$f = f_c \cos \phi$$

$$f_l = f_c \sin \phi$$

Hence,

$$f_l = f \tan \phi \tag{7.52}$$

This lateral force f_l acting at both O and P, albeit in the opposite directions, forms a couple $\tau = xf_l$ or, in view of Equation 7.52:

$$\tau = xf \tan \phi \tag{7.53}$$

This couple acts as a torque on the crankshaft. It follows that, once the rotating inertia m_c at the crank is completely balanced by a countermass, the load at the crankshaft is due only to the piston load f and it consists of:

1. A force f in the direction of the piston motion (x)
2. A torque $\tau = xf \tan \phi$ in the direction of rotation of the crankshaft (z)

As discussed below, the means of removing f at the crankshaft will also remove τ to some extent. Hence, we will discuss only the approach of balancing f.

7.5.2 Balancing the Inertia Load of the Piston

First, consider the inertia force f due to m_p. Here,

$$f = m_p a \tag{7.54}$$

where a is the acceleration $\ddot{x}$, with the coordinate x locating the position P of the piston (in other words OP $= x$). We notice from Figure 7.17 that

$$x = r \cos \omega t + l \cos \phi \tag{7.55}$$

However,

$$r \sin \omega t = l \sin \phi \tag{7.56}$$

Hence,

$$\cos \phi = \left[1 - \left(\frac{r}{l}\right)^2 \sin^2 \omega t\right]^{1/2} \tag{7.57}$$

which can be expanded up to the first term of Taylor series as

$$\cos \phi \cong 1 - \frac{1}{2}\left(\frac{r}{l}\right)^2 \sin^2 \omega t \tag{7.58}$$

This approximation is valid because l is usually several times larger than r and, hence, $(r/l)^2$ is much small than unity. Next, in view of

$$\sin^2 \omega t = \frac{1}{2}[1 - \cos 2\omega t] \tag{7.59}$$

we have

$$\cos \phi \cong 1 - \frac{1}{4}\left(\frac{r}{l}\right)^2 [1 - \cos 2\omega t] \tag{7.60}$$

Substitute Equation 7.60 into Equation 7.55. We get, approximately

$$x = r\cos\omega t + \frac{l}{4}\left(\frac{r}{l}\right)^2\cos 2\omega t + l - \frac{l}{4}\left(\frac{r}{l}\right)^2 \tag{7.61}$$

Differentiate Equation 7.61 twice with respect to t to get the acceleration

$$a = \ddot{x} = -r\omega^2\cos\omega t - l\left(\frac{r}{l}\right)^2\omega^2\cos 2\omega t \tag{7.62}$$

Hence, from Equation 7.54, the inertia force at the piston (and its reaction at the crankshaft) is

$$f = -m_p r\omega^2\cos\omega t - m_p l\left(\frac{r}{l}\right)^2\omega^2\cos 2\omega t \tag{7.63}$$

It follows that the inertia load of the reciprocating piston exerts a vibratory force on the crankshaft which has a *primary component* of frequency ω and a smaller *secondary component* of frequency 2ω, where ω is the angular speed of the crank. The primary component has the same form as that created by a rotating lumped mass at the crank pin. However, unlike the case of a rotating mass, this vibrating force acts only in the x direction (there is no $\sin\omega t$ component in the y direction) and, hence, cannot be balanced by a rotating countermass. Similarly, the secondary component cannot be balanced by a countermass rotating at double the speed. To eliminate f, we use multiple cylinders whose connecting rods and cranks are connected to the crankshaft with their rotations properly phased (delayed), thus canceling out the effects of f.

7.5.3 Multicylinder Engines

A single-cylinder engine generates a primary component and a secondary component of vibration load at the crankshaft, and they act in the direction of piston motion (x). Because there is no complementary orthogonal component (y), it is inherently unbalanced and cannot be balanced using a rotating mass. It can be balanced, however, by using several piston-cylinder units with their cranks properly phased along the crankshaft. This method of balancing multicylinder reciprocating engines is addressed now.

Consider a single cylinder whose piston inertia generates a force f at the crankshaft in the x direction given by

$$f = f_p\cos\omega t + f_s\cos 2\omega t \tag{7.64}$$

Note that the primary and secondary forcing amplitudes f_p and f_s, respectively, are given by Equation 7.63. Suppose that there is a series of cylinders in parallel, arranged along the crankshaft, and the crank of cylinder i makes an angle α_i with the crank of cylinder 1 in the direction of rotation, as schematically shown in Figure 7.18(a). Hence, force f_i on the crankshaft (in the x direction, shown as vertical in Figure 7.18) due to cylinder i is

$$f_i = f_p\cos(\omega t + \alpha_i) + f_s\cos(2\omega t + 2\alpha_i) \quad \text{for } i = 1, 2, \ldots, \text{ with } \alpha_1 = 0 \tag{7.65}$$

Not only do the cranks need to be properly phased, but the cylinders should also be properly spaced along the crankshaft to obtain the necessary balance. Consider two examples.

7.5.3.1 Two-Cylinder Engine

Consider the two-cylinder case, as shown schematically in Figure 7.18(b) where the two cranks are in radially opposite orientations (i.e., 180° out of phase). In this case, $\alpha_2 = \pi$. Hence,

$$f_1 = f_p\cos\omega t + f_s\cos 2\omega t \tag{7.66}$$

$$f_2 = f_p\cos(\omega t + \pi) + f_s\cos(2\omega t + 2\pi) = -f_p\cos\omega t + f_s\cos 2\omega t \tag{7.67}$$

It follows that the primary force components cancel out. However, they form a couple $z_0 f_p\cos\omega t$ where z_0 is the spacing of the cylinders. This causes a bending moment on the crankshaft, and it will not vanish

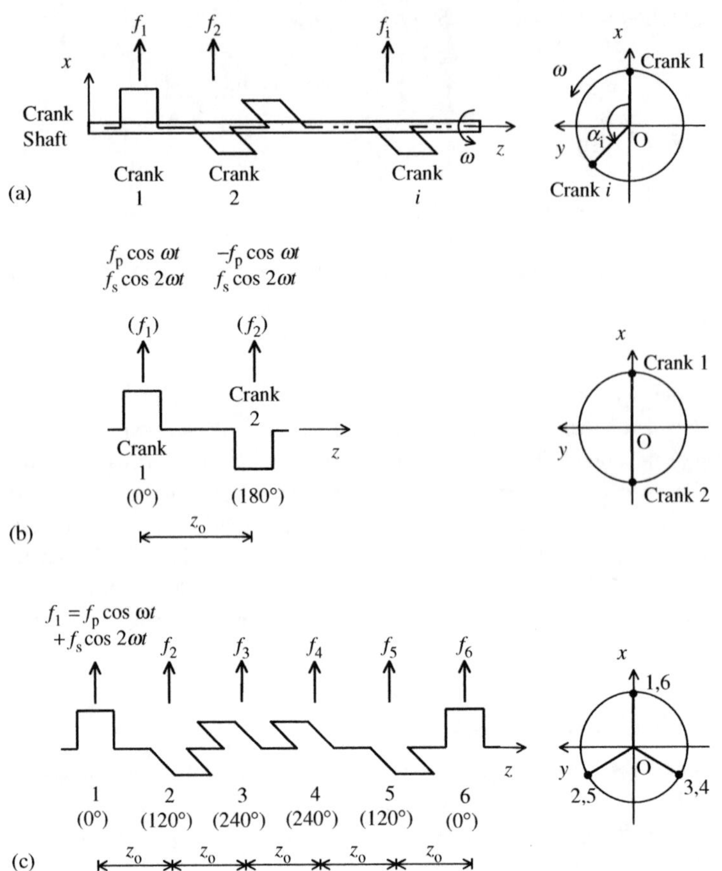

FIGURE 7.18 (a) Crank arrangement of a multicylinder engine; (b) two-cylinder engine; (c) six-cylinder engine (balanced).

unless the two cylinders are located at the same point along the crankshaft. Furthermore, the secondary components are equal and additive to $2f_s \cos 2\omega t$. This resultant component acts at the midpoint of the crankshaft segment between the two cylinders. There is no couple due to the secondary components.

7.5.3.2 Six-Cylinder Engine

Consider the six-cylinder arrangement shown schematically in Figure 7.18(c). Here, the cranks are arranged such that $\alpha_2 = \alpha_5 = 2\pi/3$, $\alpha_3 = \alpha_4 = 4\pi/3$, and $\alpha_1 = \alpha_6 = 0$. Furthermore, the cylinders are equally spaced, with spacing z_0. In this case, we have

$$f_1 = f_6 = f_p \cos \omega t + f_s \cos 2\omega t \tag{i}$$

$$f_2 = f_5 = f_p \cos(\omega t + 2\pi/3) + f_s \cos(2\omega t + 4\pi/3) \tag{ii}$$

$$f_3 = f_4 = f_p \cos(\omega t + 4\pi/3) + f_s \cos(2\omega t + 8\pi/3) \tag{iii}$$

Now, we use the fact that

$$\cos \theta + \cos\left(\theta + \frac{2\pi}{3}\right) + \cos\left(\theta + \frac{4\pi}{3}\right) = 0 \tag{iv}$$

which may be proved either by straightforward trigonometric expansion or by using geometric interpretation (i.e., three sides of an equilateral triangle, the sum of whose components in any direction vanishes). The relation iv holds for any θ, including $\theta = \omega t$ and $\theta = 2\omega t$. Furthermore,

$$\cos(2\omega t + 8\pi/3) = \cos\left(2\omega t + \frac{2\pi}{3}\right)$$

Thus, from Equation i to Equation iii, we can conclude that

$$f_1 + f_2 + f_3 + f_4 + f_5 + f_6 = 0 \tag{7.68}$$

This means that the lateral forces on the crankshaft that are exerted by the six cylinders will completely balance. Furthermore, by taking moments about the location of crank 1 of the crankshaft, we have

$$\begin{aligned}&(z_0 + 4z_0)\left[f_p \cos\left(\omega t + \frac{2\pi}{3}\right) + f_s \cos\left(2\omega t + \frac{4\pi}{3}\right)\right]\\&+ (2z_0 + 3z_0)\left[f_p \cos\left(\omega t + \frac{4\pi}{3}\right) + f_s \cos\left(2\omega t + \frac{8\pi}{3}\right)\right]\\&+ 5z_0[f_p \cos \omega t + f_s \cos 2\omega t]\end{aligned} \tag{v}$$

which also vanishes in view of relation iv. Hence, the set of six forces is in complete equilibrium and, as a result, there will be neither a reaction force nor a bending moment on the bearings of the crankshaft from these forces.

In addition, it can be shown that the torques $x_i f_i \tan \phi_i$ on the crankshaft due to this set of inertial forces f_i will add to zero, where x_i is the distance from the crankshaft to the piston of the ith cylinder and ϕ_i is the angle ϕ of the connecting rod of the ith cylinder. Hence, this six-cylinder configuration is in complete balance with respect to the inertial load.

Example 7.5

An eight-cylinder in-line engine (with identical cylinders that are placed in parallel along a line) has its cranks arranged according to the phasing angles 0, 180, 90, 270, 270, 90, 180, and 0° on the crankshaft. The cranks (cylinders) are equally spaced, with spacing z_0. Show that this engine is balanced with respect to primary and secondary components of reaction forces and bending moments of inertial loading on the bearings of the crankshaft.

Solution

The sum of the reaction forces on the crankshaft are

$$\begin{aligned}&2\left[f_p \cos \omega t + f_s \cos 2\omega t + f_p \cos(\omega t + \pi) + f_s \cos(2\omega t + 2\pi) + f_p \cos\left(\omega t + \frac{\pi}{2}\right) + f_s(2\omega t + \pi)\right.\\&\left.+ f_p \cos\left(\omega t + \frac{3\pi}{2}\right) + f_s \cos(2\omega t + 3\pi)\right] = 2\left[f_p \cos \omega t - f_p \cos \omega t - f_p \sin \omega t + f_p \sin \omega t\right.\\&\left.+ f_s \cos 2\omega t + f_s \cos 2\omega t - f_s \cos 2\omega t - f_s \cos 2\omega t\right] = 0\end{aligned}$$

Hence, both primary forces and secondary forces are balanced. The moment of the reaction forces about the crank 1 location of the crankshaft is

$$\begin{aligned}&(z_0 + 6z_0)[f_p \cos(\omega t + \pi) + f_s \cos(2\omega t + 2\pi)] + (2z_0 + 5z_0)\left[f_p \cos\left(\omega t + \frac{\pi}{2}\right) + f_s \cos(2\omega t + \pi)\right]\\&+ (3z_0 + 4z_0)\left[f_p \cos\left(\omega t + \frac{3\pi}{2}\right) + f_s \cos(2\omega t + 3\pi)\right] + 7z_0[f_p \cos \omega t + f_s \cos 2\omega t]\\&= 7z_0[-f_p \cos \omega t + f_s \cos 2\omega t - f_p \sin \omega t - f_s \cos 2\omega t + f_p \sin \omega t - f_s \cos 2\omega t + f_p \cos \omega t + f_s \cos 2\omega t] = 0\end{aligned}$$

Hence, both primary bending moments and secondary bending moments are balanced. Therefore, the engine is completely balanced.

The formulas applicable for balancing reciprocating machines are summarized in Box 7.4.

Box 7.4

Balancing of Reciprocating Machines

Single cylinder engine:

Inertia force at piston (and its reaction on crankshaft)

$$f = -m_p r\omega^2 \cos \omega t - m_p l\left(\frac{r}{l}\right)^2 \omega^2 \cos 2\omega t = f_p \cos \omega t + f_s \cos 2\omega t$$

where

ω = rotating speed of crank
m_p = equivalent lumped mass at piston
r = crank radius
l = length of connecting rod
f_p = amplitude of the primary unbalance force (frequency ω)
f_s = amplitude of the secondary unbalance force (frequency 2ω)

Multicylinder engine:

Net unbalance reaction force on crankshaft $= \sum_{i=1}^{n} f_i$
Net unbalance moment on crankshaft $= \sum_{i=1}^{n} z_i f_i$
where

$f_i = f_p \cos(\omega t + \alpha_i) + f_s \cos(2\omega t + 2\alpha_i)$
α_i = angular position of the crank of ith cylinder, with respect to a body (rotating) reference (i.e., crank phasing angle)
z_i = position of the ith crank along the crankshaft, measured from a reference point on the shaft
n = number of cylinders (assumed identical)

Note: For a completely balanced engine, both the net unbalance force and the net unbalance moment should vanish.

Finally, it should be noted that, in the configuration considered above, the cylinders are placed in parallel along the crankshaft. These are termed in-line engines. Their resulting forces f_i act in parallel along the shaft. In other configurations such as V6 and V8, the cylinders are placed symmetrically around the shaft. In this case, the cylinders (and their inertial forces, which act on the crankshaft) are not parallel. Here, a complete force balance may be achieved without having to phase the cranks. Furthermore, the bending moments of the forces can be reduced by placing the cylinders at nearly the same location along the crankshaft. Complete balancing of the combustion/pressure forces is also possible by such an arrangement.

7.5.4 Combustion/Pressure Load

In the balancing approach presented above, the force f on the piston represents the inertia force due to the equivalent reciprocating mass. Its effect on the crankshaft is an equal reaction force f in the lateral direction (x) and a torque $\tau = xf \tan \phi$ about the shaft axis (z). The balancing approach is to use a series of cylinders so that their reaction forces f_i on the crankshaft from an equilibrium set so that no net reaction or bending moment is transmitted to the bearings of the shaft. The torques τ_i also can be balanced by the same approach, which is the case, for example, in the six-cylinder engine.

Another important force that acts along the direction of piston reciprocation is the drive force due to gas pressure in the cylinder (e.g., created by combustion of the fuel–air mixture of an internal combustion engine). As above, this force may be analyzed by denoting it as f. However, several important observations should be made first:

1. The combustion force f is not sinusoidal of frequency ω. It is reasonably periodic but the shape is complex and depends on the firing/fuel-injection cycle and the associated combustion process.
2. The reaction forces f_i on the crankshaft, which are generated from cylinders i, should be balanced to avoid the transmission of reaction forces and bending moments to the shaft bearings (and hence, to the supporting frame — the vehicle). However, the torques τ_i in this case are in fact the drive torques. Obviously, they are the desired output of the engine and should not be balanced, unlike the inertia torques.

Therefore, although the analysis completed for balancing the inertia forces cannot be directly used here, we can employ similar approaches to the use of multiple cylinders for reducing the gas-force reactions. This is a rather difficult problem given the complexity of the combustion process itself. In practice, much of the leftover effects of the ignition cycle are suppressed by properly designed engine mounts. Experimental investigations have indicated that in a properly balanced engine unit, much of the vibration transmitted through the engine mounts is caused by the engine firing cycle (internal combustion) rather than the reciprocating inertia (sinusoidal components of frequency ω and 2ω). Hence, active mounts, where stiffness can be varied according to the frequency of excitation, are being considered to reduce engine vibrations in the entire range of operating speeds (e.g. 500 to 2500 rpm).

7.6 Whirling of Shafts

In the previous two sections, we studied the vibration excitations caused on rotating shafts and their bearings due to some form of mass eccentricity. Methods of balancing these systems to eliminate the undesirable effects were also presented. One limitation of the given analysis is the assumptions that the rotating shaft is rigid and, thus, does not deflect from its axis of rotation due to the unbalanced excitations. In practice, however, rotating shafts are made lighter than the components they carry (rotors, disks, gears, etc.) and will undergo some deflection due to the unbalanced loading. As a result, the shaft will bow out and this will further increase the mass eccentricity and associated unbalanced excitations and gyroscopic forces of the rotating elements (disks, rotors, etc.). The nature of damping of rotating machinery (which is rather complex and incorporates effects of rotation at bearings, structural deflections, and lateral speeds) will further affect the dynamic behavior of the shaft under these conditions. In this context, the topic of whirling of rotating shafts becomes relevant.

Consider a shaft that is driven at a constant angular speed ω (e.g., by using a motor or some other actuator). The central axis of the shaft (passing through its bearings) will bow out. The deflected axis itself will rotate, and this rotation is termed *whirling* or *whipping*. The whirling speed is not necessarily equal to the drive speed ω (at which the shaft rotates about its axis with respect to a fixed frame). However, when the whirling speed is equal to ω, the condition is called *synchronous whirl*, and the associated deflection of the shaft can be quite excessive and damaging.

To develop an analytical basis for whirling, consider a light shaft supported on two bearings carrying a disk of mass m in between the bearings, as shown in Figure 7.19(a). Note that C is the point on the disk at which it is mounted on the shaft. Originally, in the neutral configuration when the shaft is not driven ($\omega = 0$), the point C coincides with point O on the axis joining the two bearings. If the shaft were rigid, then the points C and O would continue to coincide during motion. The mass center (centroid or center of gravity for constant g) of the disk is denoted as G in Figure 7.19. During motion, C will move away from O due to the shaft deflection. The whirling speed (speed of rotation of the shaft axis) is the speed of rotation of the radial line OC with respect to a fixed reference. Denoting the angle of OC with respect to a fixed reference as θ, the whirling speed is $\dot{\theta}$. This is explained in Figure 7.19(b) where an end view of the

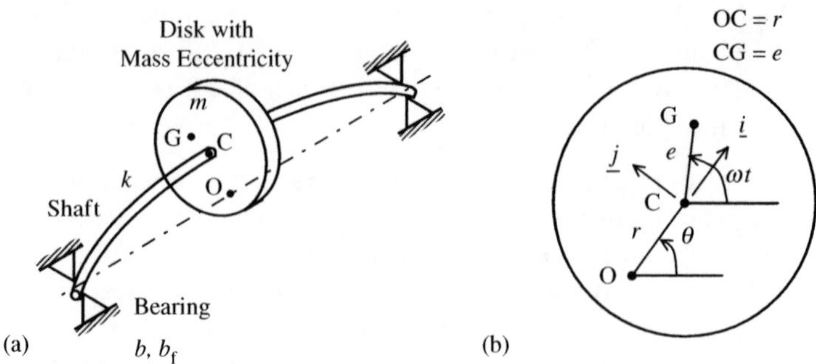

FIGURE 7.19 (a) A whirling shaft carrying a disk with mass eccentricity; (b) end view of the disk and whirling shaft.

disk is given under deflected conditions. The constant drive speed ω of the shaft is the speed of the shaft spin with respect to a fixed reference, and is the speed of rotation of the radial line CG with respect to the fixed horizontal line shown in Figure 7.19(b). Hence, the angle of shaft spin is ωt, as measured with respect to this line. The angle of whirl, θ, is also measured from the direction of this fixed line, as shown.

7.6.1 Equations of Motion

Under practical conditions, the disk moves entirely in a single plane. Hence, its complete set of equations of motion consists of two equations for translatory (planar) motion of the centroid (with lumped mass m), and one equation for rotational motion about the fixed bearing axis. The latter equation depends on the motor torque that drives the shaft at constant speed ω, and is not of interest in the present context. Thus, we will limit our development to the two translatory equations of motion. The equations may be written either in a Cartesian coordinate system (x, y) or a polar coordinate system (r, θ). Here, we will use the polar coordinate system.

Consider a coordinate frame $(\mathbf{i}, \mathbf{j})$ that is fixed to the disk with its $\mathbf{i}$ axis lying along OC as shown in Figure 7.19(b). Note that the angular speed of this frame is $\dot{\theta}$ (about the $\mathbf{k}$ axis that is orthogonal to $\mathbf{i}$ and $\mathbf{j}$). Hence, as is well known, we have

$$\frac{\mathrm{d}\mathbf{i}}{\mathrm{d}t} = \dot{\theta}\mathbf{j} \quad \text{and} \quad \frac{\mathrm{d}\mathbf{j}}{\mathrm{d}t} = -\dot{\theta}\mathbf{i} \tag{7.69}$$

The position vector of the mass point G from O is

$$\vec{OG} = \mathbf{r}_G = \vec{OC} + \vec{CG} = r\mathbf{i} + e\cos(\omega t - \theta)\mathbf{i} + e\sin(\omega t - \theta)\mathbf{j} \tag{7.70}$$

The velocity vector $\mathbf{v}_G$ of the mass point G can be obtained simply by differentiating Equation 7.70 with the use of Equation 7.69. However, this can be simplified because ω is constant. Here, line CG has a velocity $e\omega$ that is perpendicular to it about C. This can be resolved along the axes $\mathbf{i}$ and $\mathbf{j}$. Hence, the velocity of G relative to C is

$$\mathbf{v}_{G/C} = -e\omega\sin(\omega t - \theta)\mathbf{i} + e\omega\cos(\omega t - \theta)\mathbf{j}$$

However, the velocity of point C is

$$\mathbf{v}_C = \frac{\mathrm{d}}{\mathrm{d}t} r\mathbf{i} = \dot{r}\mathbf{i} + r\frac{\mathrm{d}\mathbf{i}}{\mathrm{d}t} = \dot{r}\mathbf{i} + r\dot{\theta}\mathbf{j}$$

Hence, the velocity of G, which is given by $\mathbf{v}_G = \mathbf{v}_C + \mathbf{v}_{G/C}$, can be expressed as

$$\mathbf{v}_G = \dot{r}\mathbf{i} + r\dot{\theta}\mathbf{j} - e\omega\sin(\omega t - \theta)\mathbf{i} + e\omega\cos(\omega t - \theta)\mathbf{j} \tag{7.71}$$

Similarly, the acceleration of C is

$$\mathbf{a}_C = \frac{d}{dt}\mathbf{v}_C = \frac{d}{dt}[\dot{r}\mathbf{i} + r\dot{\theta}\mathbf{j}] = \ddot{r}\mathbf{i} + \dot{r}\dot{\theta}\mathbf{j} + \dot{r}\dot{\theta}\mathbf{j} + r\ddot{\theta}\mathbf{j} - r\dot{\theta}^2\mathbf{i} = (\ddot{r} - r\dot{\theta}^2)\mathbf{i} + (r\ddot{\theta} + 2\dot{r}\dot{\theta})\mathbf{j}$$

Also, since line CG rotates at constant angular speed ω about C, the point G has only a radial (centrifugal) acceleration $e\omega^2$ along GC. This can be resolved along $\mathbf{i}$ and $\mathbf{j}$ as before. Hence, the acceleration of G relative to C is

$$\mathbf{a}_{G/C} = -e\omega^2\cos(\omega t - \theta)\mathbf{i} - e\omega^2\sin(\omega t - \theta)\mathbf{j}$$

It follows that the acceleration of point G, given by $\mathbf{a}_G = \mathbf{a}_C + \mathbf{a}_{G/C}$, may be expressed as

$$\mathbf{a}_G = (\ddot{r} - r\dot{\theta}^2)\mathbf{i} + (r\ddot{\theta} + 2\dot{r}\dot{\theta})\mathbf{j} - e\omega^2\cos(\omega t - \theta)\mathbf{i} - e\omega^2\sin(\omega t - \theta)\mathbf{j} \tag{7.72}$$

The forces acting on the disk are as follows:

Restraining elastic force due to lateral deflection of the shaft $= -kr\mathbf{i}$
Viscous damping force (proportional to the velocity of C) $= -b\dot{r}\mathbf{i} - br\dot{\theta}\mathbf{j}$

In addition, there is a frictional resistance at the bearing, which is proportional to the reaction and, hence, the shaft deflection is r and also depends on the spin speed ω. The following approximate model may be used:

$$\text{Bearing friction force} = -b_f r\omega\mathbf{j}$$

Here,

k = lateral deflection stiffness of the shaft at the location of the disk
b = viscous damping constant for lateral motion of the shaft
b_f = bearing frictional coefficient

The overall force acting on the disk is

$$\mathbf{f} = -b\dot{r}\mathbf{i} - (br\dot{\theta} + b_f r\omega)\mathbf{j} \tag{7.73}$$

The equation of rectilinear motion

$$\mathbf{f} = m\mathbf{a}_G \tag{7.74}$$

on using Equation 7.72 and Equation 7.73, reduces to the following pair in the $\mathbf{i}$ and $\mathbf{j}$ directions:

$$-kr - b\dot{r} = m[\ddot{r} - r\dot{\theta}^2 - e\omega^2\cos(\omega t - \theta)] \tag{7.75}$$

$$-br\dot{\theta} - b_f r\omega = m[r\ddot{\theta} + 2\dot{r}\dot{\theta} - e\omega^2\sin(\omega t - \theta)] \tag{7.76}$$

These equations may be expressed as

$$\ddot{r} + 2\zeta_v\omega_n\dot{r} + (\omega_n^2 - \dot{\theta}^2)r = e\omega^2\cos(\omega t - \theta) \tag{7.77}$$

$$r\ddot{\theta} + 2(\zeta_v\omega_n r + \dot{r})\dot{\theta} + 2\zeta_f\omega_n\omega r = e\omega^2\sin(\omega t - \theta) \tag{7.78}$$

where the undamped natural frequency of lateral vibration is

$$\omega_n = \sqrt{\frac{k}{m}} \tag{7.79}$$

and

ζ_v = viscous damping ratio of lateral motion
ζ_f = frictional damping ratio of the bearings

Equation 7.77 and Equation 7.78, which govern the whirling motion of the shaft-disk system, are a pair of coupled nonlinear equations, with excitations (depending on ω) that are coupled with a motion variable (θ). Hence, a general solution would be rather complex. A relatively simple solution is possible, however, under steady-state whirling.

7.6.2 Steady-State Whirling

Under steady-state conditions, the whirling speed $\dot{\theta}$ is constant at $\dot{\theta} = \omega_w$, hence, $\ddot{\theta} = 0$. Also, the lateral deflection of the shaft is constant, hence, $\dot{r} = \ddot{r} = 0$. Therefore, Equation 7.77 and Equation 7.78 become

$$(\omega_n^2 - \omega_w^2)r = e\omega^2 \cos(\omega t - \theta) \tag{7.80}$$

$$2\zeta_v \omega_n \omega_w r + 2\zeta_f \omega_n \omega r = e\omega^2 \sin(\omega t - \theta) \tag{7.81}$$

In Equation 7.80 and Equation 7.81, the left-hand side is independent of t. Hence, the right-hand side should also be independent of t. For this, we must have

$$\theta = \omega t - \phi \tag{7.82}$$

where ϕ is interpreted as the phase lag of whirl with respect to the shaft spin (ω), and should be clear from Figure 7.19(b). It follows from Equation 7.82 that, for steady-state whirl, the whirling speed $\dot{\theta} = \omega_w$ is

$$\omega_w = \omega \tag{7.83}$$

This condition is called synchronous whirl because the whirl speed (ω_w) is equal to the shaft spin speed (ω). It follows that under steady-state conditions, we should have the state of synchronous whirl. The equations governing steady-state whirl are

$$(\omega_n^2 - \omega^2)r = e\omega^2 \cos \phi \tag{7.84}$$

$$2\zeta\omega_n \omega r = e\omega^2 \sin \phi \tag{7.85}$$

along with Equation 7.82 and, hence, Equation 7.83. Here, $\zeta = \zeta_v + \zeta_f$ is the overall damping ratio of the system. Note that the phase angle ϕ and the shaft deflection r are determined from Equation 7.84 and Equation 7.85. In particular, squaring these two equations and adding to eliminate ϕ, we obtain

$$r = \frac{e\omega^2}{\sqrt{(\omega_n^2 - \omega^2)^2 + (2\zeta\omega_n\omega)^2}} \tag{7.86}$$

which is of the form of magnitude of the frequency transfer function of a simple oscillator with an acceleration excitation. Divide Equation 7.85 by Equation 7.84 to get the phase angle:

$$\phi = \tan^{-1}\frac{2\zeta\omega_n\omega}{(\omega_n^2 - \omega^2)} \tag{7.87}$$

Using simple calculus (differentiate the square and equate to zero), we can show that the maximum deflection occurs at the critical spin speed ω_c given by

$$\omega_c = \frac{\omega_n}{\sqrt{1 - 2\zeta^2}} \tag{7.88}$$

This *critical speed* corresponds to a resonance. For light damping, we have approximately $\omega_c = \omega_n$. Hence, critical speed for low damping is equal to the undamped natural frequency of bending vibration of the shaft-rotor unit. The corresponding shaft deflection is (see Equation 7.86)

$$r_c = \frac{e}{2\zeta} \tag{7.89}$$

which is also a good approximation of r at critical speed, with light damping. From Equation 7.84 and Equation 7.85, we can see that, at critical speed (with low damping), $\sin \phi = 1$ and $\cos \phi = 0$, which gives $\phi = \pi/2$. Also, note from Equation 7.86 that the steady-state shaft deflection is almost zero at low speeds and approaches e at very high speeds. However, Equation 7.87 shows that, for small ω, $\tan \phi$ is positive and small. We can see from Equation 7.85 that $\sin \phi$ is positive. This means ϕ itself is small for small ω. For large ω, we can see from Equation 7.86 that r approaches e. Thus, we can see from

Equation 7.87 that tan ϕ is small and negative, whereas Equation 7.85 shows that sin ϕ is positive. Hence, ϕ approaches π for large ω.

It is seen from Equation 7.89 that, at critical speed, the shaft deflection increases with mass eccentricity and decreases with damping. This indicates that the approaches for reducing the damaging effects of whirling are:

1. Eliminate or reduce the mass eccentricity through proper construction practices and balancing.
2. Increase damping.
3. Increase shaft stiffness.
4. Avoid operation near critical speed.

There will be limitations to the use of these approaches, particularly making the shaft stiffer. Note also that our analysis did not include the mass distribution of the shaft. A Bernoulli–Euler type beam analysis has to be incorporated for a more accurate analysis of whirling for shafts whose mass cannot be accurately represented by a single parameter that is lumped at the location of the rotor. Formulas related to whirling of shafts are summarized in Box 7.5.

Example 7.6

The fan of a ventilation system has a normal operating speed of 3600 rpm. The blade set of the fan weighs 20 kg and is mounted in the mid-span of a relatively light shaft that is supported on lubricated bearings at its two ends. The bending stiffness of the shaft at the location of the fan is 4.0×10^6 N/m. Equivalent damping ratio that acts on the possible whirling motion of the shaft is 0.05. Owing to fabrication error, the centroid of the fan has an eccentricity of 1.0 cm from the neutral axis of rotation of the shaft:

1. Determine the critical speed of the fan system and the corresponding shaft deflection at the location of the fan at steady state.
2. What is the steady-state shaft deflection at the fan during normal operation?

The fan was subsequently balanced using a mass of 5 kg. The centroid eccentricity was reduced to 2 mm by this means. What is the shaft deflection at the fan during normal operation now? Comment on the improvement that has been realized.

Solution

1. The system is lightly damped. Hence, the critical speed is given by the undamped natural frequency; thus

$$\omega_c \cong \omega_n = \sqrt{\frac{k}{m}} = \sqrt{\frac{4 \times 10^6}{20}} \text{ rad/s} = 447.2 \text{ rad/s}$$

The corresponding shaft deflection is

$$r_c = \frac{e}{2\zeta} = \frac{1.0}{2 \times 0.05} \text{ cm} = 10.0 \text{ cm}$$

2. Operating speed $\omega = (3600/60) \times 2\pi$ rad/s $= 377$ rad/s. Using Equation 7.86, the corresponding shaft deflection, at steady state, is

$$r = \frac{1.0 \times (377)^2}{[(447.2^2 - 377^2)^2 + (2 \times 0.05 \times 447.2 \times 377)^2]^{1/2}} \text{ cm} = 2.36 \text{ cm}$$

After balancing, the new eccentricity $e = 0.2$ cm.
The new natural frequency (undamped) is

$$\omega_n = \sqrt{\frac{4 \times 10^6}{25}} \text{ rad/s} = 400 \text{ rad/s}$$

Box 7.5
Whirling of Shafts

Whirling: A shaft spinning at speed ω about its axis, may bend due to flexure. The bent (bowed out) axis will rotate at speed ω_w. This is called whirling.

Equations of motion:

$$\ddot{r} + 2\zeta_v \omega_n \dot{r} + (\omega_n^2 - \dot{\theta}^2)r = e\omega^2 \cos(\omega t - \theta)$$

$$r\ddot{\theta} + 2(\zeta_v \omega_n r + \dot{r})\dot{\theta} + 2\zeta_f \omega_n \omega r = e\omega^2 \sin(\omega t - \theta)$$

where (r, θ) are polar coordinates of shaft deflection at the mounting point of lumped mass.

e = eccentricity of the lumped mass from the spin axis of shaft
$\dot{\theta} = \omega_w$ = whirling speed
ω = spin speed of shaft
$\omega_n = \sqrt{k/m}$ = natural frequency of bending vibration of shaft
k = bending stiffness of shaft at lumped mass
m = lumped mass
ζ_v = damping ratio of bending motion of shaft
ζ_f = damping ratio of shaft bearings

Steady-state whirling (synchronous whirl):
Here, whirling speed ($\dot{\theta}$ or ω_w) is constant and equals the shaft spin speed ω (i.e., $\omega_w = \omega$ for steady-state whirling).

Shaft deflection at lumped mass

$$r = \frac{e\omega^2}{[(\omega_n^2 - \omega^2)^2 + (2\zeta\omega_n\omega)^2]^{1/2}}$$

Phase angle between shaft deflection (r) and mass eccentricity (e)

$$\phi = \tan^{-1}\frac{2\zeta\omega_n\omega}{(\omega_n^2 - \omega^2)}$$

where $\zeta = \zeta_v + \zeta_f$

Note: For small spin speeds ω, we have small r and ϕ. For large ω, we have $r \cong e$ and $\phi \cong \pi$

Critical speed:

$$\text{Spin speed } \omega = \frac{\omega_n}{\sqrt{1 - 2\zeta^2}} \omega_n \text{ for small } \zeta$$

$$\phi = \pi/2$$

The corresponding shaft deflection during steady-state operation is

$$r = \frac{0.2 \times (377)^2}{[(400^2 - 377^2)^2 + (2 \times 0.05 \times 400 \times 377)^2]^{1/2}} \text{ cm} = 1.216 \text{ cm}$$

Note that, even though the eccentricity has been reduced by a factor of five by balancing, the operating deflection of the shaft has been reduced only by a factor of less than two. The main reason for this is that the operating speed is close to the critical speed. Methods of improving the performance include changing the operating speed, using a smaller mass to balance the fan, using more damping, and making the shaft stiffer. However, some of these methods may not be feasible. Operating speed is determined by the task requirements. A location may not be available that is sufficiently distant to place a balancing mass that is appropriately small. Increased damping will increase heat generating, cause bearing problems, and will also reduce the operating speed. Replacement or stiffening of the shaft may require too much modification to the system and add cost. A preferable alternative would be to balance the fan by removing some mass. This will move the critical frequency (natural frequency) away from the operating speed rather than closer to it, while reducing the mass eccentricity at the same time. For example, suppose that a mass of 3 kg is removed from the fan, which results in an eccentricity of 2.0 mm. The new natural/critical frequency is

$$\sqrt{\frac{4\times 10^6}{17}}\ \text{rad/s} = 485.1\ \text{rad/s}$$

The corresponding shaft deflection during steady operation is

$$r = \frac{0.2\times(377)^2}{[(485.1^2 - 377^2)^2 + (2\times 0.05\times 485.1\times 377)^2]^{1/2}}\ \text{cm} = 0.3\ \text{cm}$$

In this case, the deflection has been reduced by a factor of eight.

7.6.3 Self-Excited Vibrations

Equation 7.77 and Equation 7.78, which represent the general whirling motion of a shaft, are nonlinear and coupled. In these equations, the motion variables (r and θ) occur as (nonlinear) products of the excitation (ω). Such systems are termed self-excited. Note that, in general (before reaching the steady state) the response variables r and θ will exhibit vibratory characteristics in view of the presence of the excitation functions $\cos(\omega t - \theta)$ and $\sin(\omega t - \theta)$. Hence, a whirling shaft may exhibit self-excited vibrations. Because the excitation forces directly depend on the motion itself, it is possible that a continuous energy flow into the system could occur. This will result in a steady growth of the motion amplitudes and represents an *unstable* behavior.

A simple example of self-excited vibration is provided by a pendulum whose length is time variable. Although the system is stable when the length is fixed, it can become unstable under conditions of variable length. Practical examples of self-excited vibrations with possible exhibition of instability include the flutter of aircraft wings due to coupled aerodynamic forces, wind-induced vibrations of bridges and tall structures, galloping of ice-covered transmission lines due to air flow-induced vibrations, and chattering of machine tools due to friction-related excitation forces. Proper design and control methods, as discussed in this chapter, are important in suppressing self-excited vibrations.

7.7 Design through Modal Testing

Experimental modal analysis (EMA) involves extracting modal parameters (natural frequencies, modal damping ratios, mode shapes) of a mechanical system through testing (notably, through excitation-response data) and then developing a dynamic model of the system (mass, stiffness, and damping matrices) on that basis. The techniques of EMA are useful in modeling and model validation (i.e., verification of the accuracy of an existing model that was obtained, for example, through analytical modeling). In addition to these uses, EMA is also a versatile tool for design development. In the context of "design for vibration," EMA may be employed in the design and design modification of mechanical systems with the goal of achieving desired performance under vibrating conditions. This section will introduce this approach.

In applying EMA for design development of a mechanical system, three general approaches are employed:

1. Component modification
2. Modal response specification
3. Substructuring

The method of component modification allows us to modify (i.e., add, remove, or vary) physical parameters (inertia, stiffness, damping) in a mechanical system, and to determine the resulting effect on the modal response (natural frequencies, damping ratios, and mode shapes) of the system. The method of modal response specification provides the capability to establish the best changes, from the design viewpoint, in system parameters (inertia, stiffness, damping values, and associated directions) in order to realize a specified change in the modal response. In the techniques of substructuring, two or more subsystem models are combined using proper components of interfacing (interconnection), and the overall model of the integrated system is determined. Some of the subsystems used in this approach could be of analytical or computational origin (e.g., finite element models). It should be clear how these methods could be used in the design development of a mechanical system for proper vibration performance. The first method is essentially a trial and error technique of incremental design. Here, some appropriate parameters are changed and the resulting modal behavior is determined. If the resulting performance is not satisfactory, further changes are made in discrete steps until an acceptable performance (with regard to natural frequencies, response magnification factors, etc.) is achieved. The second method is clearly a direct design approach, where the design specifications are first developed in terms of modal characteristics, and then the design procedure will generate the size and type of the physical parameters to meet the specifications. In the third method, a suitable set of subsystems is first designed to meet performance characteristics of each subsystem. Then, these subsystems are linked through suitable mechanical interfacing components, and the performance of the overall system is determined to verify acceptance. In this manner, a complex system may be designed through the systematic design of its subsystems.

7.7.1 Component Modification

The method of component modification involves changing a mass, stiffness, or damping element in the system and determining the corresponding dynamic response, particularly the natural frequencies, modal damping ratios, and mode shapes. This is relatively straightforward because a single modal analysis or modal test (EMA) will give the required information. Because single step of component modification might not be acceptable as an appropriate design (e.g., a natural frequency might be too close to a significant frequency component of a vibration excitation), a number of modifications may be necessary. For such incremental procedures, modal analysis would be more convenient and cost effective than EMA because, in the latter case, physical modification and retesting would be needed, whereas the former involves the same computational steps as before, but with a new set of parameter values.

For example, consider an aluminum I beam that has a number of important modes of vibration, including bending and torsional modes. Figure 7.20(a) shows the fourth mode shape of vibration at natural frequency 678.4 Hz. The dotted line in Figure 7.20(b) shows the transfer function magnitude when the beam is excited at some location in the vertical direction and the response is measured in the vertical direction, at some other location, where neither of the locations are node points. The curve shows the first six natural frequencies.

Next, a lumped mass is added to the top flange at the shown location. The corresponding transfer function magnitude is shown by the solid curve in Figure 7.20(b). Note that all the natural frequencies have decreased due to the added mass, but the effect is larger for higher modes. Similarly, mode shapes also will change. If the new modes are not satisfactory (e.g., a particular natural frequency has not shifted enough) further modification and evaluation will be required.

Consider a mechanical vibrating system whose free response $\mathbf{y}$ is described by

$$\mathbf{M}\ddot{\mathbf{y}} + \mathbf{K}\mathbf{y} = 0 \tag{7.90}$$

Damping has been ignored for simplicity, but the following discussion can also be extended to a damped system (quite directly, for the case of proportional damping). If the mass matrix $\mathbf{M}$ and the stiffness matrix $\mathbf{K}$ are modified by $\delta\mathbf{M}$ and $\delta\mathbf{K}$, respectively, the corresponding response (as well as the natural frequencies and mode shapes) will be different from that of the original system. To illustrate, let the

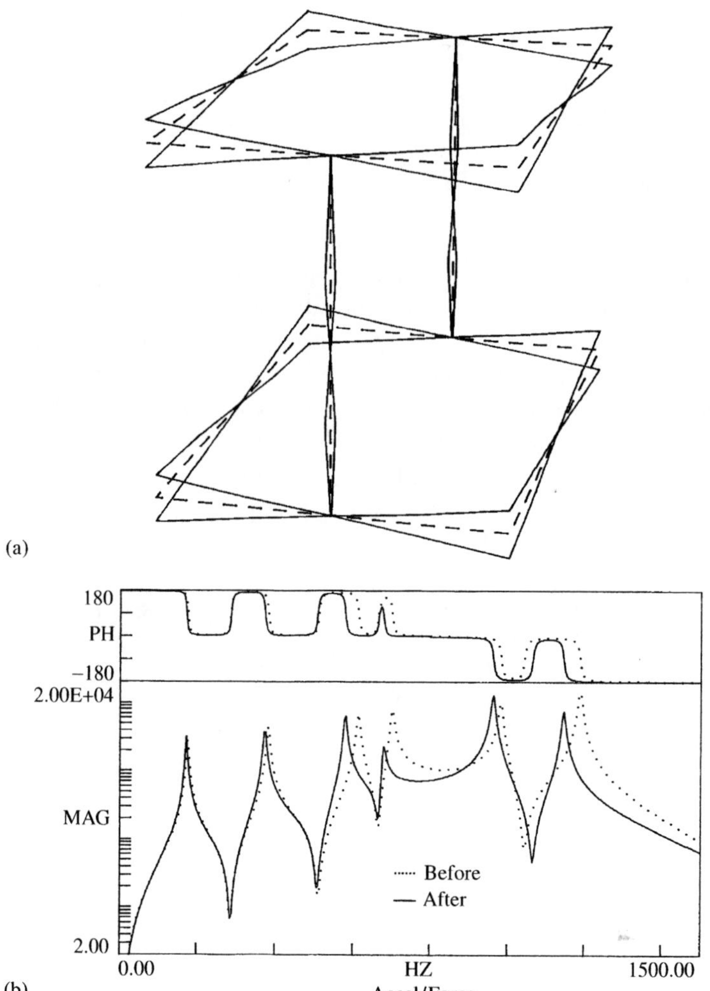

FIGURE 7.20 An example of component modification: (a) the shape of Mode 4 prior to modification; (b) transfer function magnitude before and after modification.

modal matrix (the matrix whose columns are the independent mode shape vectors of the original system) be $\boldsymbol{\Psi}$. Then, using the modal transformation

$$\mathbf{y} = \boldsymbol{\Psi}\mathbf{q} \tag{7.91}$$

Equation 7.90 can be expressed in the canonical form, with modal generalized coordinates $\mathbf{q}$, as

$$\bar{\mathbf{M}}\ddot{\mathbf{q}} + \bar{\mathbf{K}}\mathbf{q} = 0 \tag{7.92}$$

where

$$\boldsymbol{\Psi}^{\mathrm{T}}\mathbf{M}\boldsymbol{\Psi} = \bar{\mathbf{M}} = \mathrm{diag}[M_1, M_2, \ldots, M_n] \tag{7.93}$$

$$\boldsymbol{\Psi}^{\mathrm{T}}\mathbf{K}\boldsymbol{\Psi} = \bar{\mathbf{K}} = \mathrm{diag}[K_1, K_2, \ldots, K_n] \tag{7.94}$$

If the same transformation (Equation 7.91) is used for the modified system

$$(\mathbf{M} + \delta\mathbf{M})\ddot{\mathbf{y}} + (\mathbf{K} + \delta\mathbf{K})\mathbf{y} = 0 \tag{7.95}$$

we obtain

$$(\bar{\mathbf{M}} + \boldsymbol{\Psi}^{\mathrm{T}}\delta\mathbf{M}\boldsymbol{\Psi})\ddot{\mathbf{q}} + (\bar{\mathbf{K}} + \boldsymbol{\Psi}^{\mathrm{T}}\delta\mathbf{K}\boldsymbol{\Psi})\mathbf{q} = 0 \tag{7.96}$$

Since both $\boldsymbol{\Psi}^{\mathrm{T}}\delta\mathbf{M}\boldsymbol{\Psi}$ and $\boldsymbol{\Psi}^{\mathrm{T}}\delta\mathbf{K}\boldsymbol{\Psi}$ are not diagonal matrices in general, $\boldsymbol{\Psi}$ would not remain the modal matrix for the modified system. Furthermore, the original natural frequencies $\omega_i = \sqrt{K_i/M_i}$ will change due to the component modification. For the special case of proportional modifications ($\delta\mathbf{M}$ proportional to $\mathbf{M}$ and $\delta\mathbf{K}$ proportional to $\mathbf{K}$), the mode shapes will not change. However, in general, the natural frequencies will change.

The reverse problem is the modal response specification. Here, a required set of modal parameters (ω_{ir} and $\boldsymbol{\psi}_{ir}$) is specified and the necessary changes $\delta\mathbf{M}$ and $\delta\mathbf{K}$ to meet the specifications must be determined. Note that the solution is not unique and is more difficult than the direct problem. In this case, a sensitivity analysis may initially be performed to determine the directions and magnitudes of the modal shift for a particular physical parameter shift. Then, the necessary magnitudes of physical shift to achieve the specified modal shift are estimated on that basis. The corresponding modifications are made and the modified system is analyzed/tested to check whether it is within specification. If not, further cycles of modification should be performed.

Example 7.7

As an example of component modification, consider the familiar problem of a two-DoF system, as shown in Figure 7.21. The squared nondimensional natural frequencies $r_i^2 = (\omega_i/\omega_0)^2$ of the systems are given by

$$r_1^2, r_2^2 = \frac{1}{2\alpha} \times \{\alpha + \beta + \alpha\beta\}\left\{1 \pm \sqrt{1 - \frac{4\alpha\beta}{(\alpha + \beta + \alpha\beta)^2}}\right\}$$

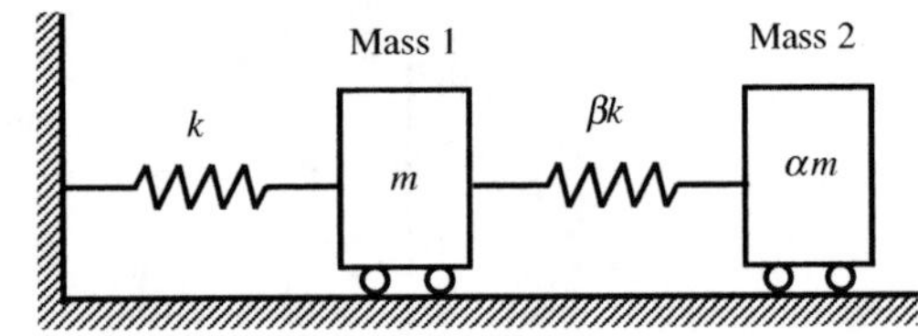

FIGURE 7.21 A two-degree-of-freedom example.

where $\omega_0 = \sqrt{k/m}$. We also showed that the mode shapes, as given by the ratio of the displacement of mass 2 to that of mass 1 at a natural frequency, are

$$\left(\frac{\psi_2}{\psi_1}\right)_i = \frac{1 + \beta - r_i^2}{\beta} \quad \text{for mode } i$$

Consider a system with $\alpha = 0.5$ and $\beta = 0.5$. By direct computation, we can show that $r_1 = 0.71$ and $r_2 = 1.41$. Estimate the modification of β (the relative stiffness of the second spring) that would be necessary to shift the system natural frequencies to approximately $r_1 = 0.8$ and $r_2 = 2.0$. Check the corresponding shift in mode shapes.

Solution

For $\alpha = 0.5$ and $\beta = 0.5$, direct substitution yields $r_1 = 0.71$ and $r_2 = 1.41$ with $(\psi_2/\psi_1)_1 = 2.0$ and $(\psi_2/\psi_1)_2 = -1.0$. Now, consider an incremental change in β by 0.1. Then, $\beta = 0.6$. The corresponding natural frequencies are computed as

$$r_i^2 = \frac{1}{2 \times 0.5}\{0.5 + 0.6 + 0.5 \times 0.6\}\left\{1 \mp \left[1 - \frac{4 \times 0.5 \times 0.6}{(0.5 + 0.6 + 0.5 \times 0.6)^2}\right]^{1/2}\right\} = 0.528, 2.272$$

Hence,

$$r_1, r_2 = 0.727, 1.507$$

This step may be interpreted as a way of establishing the sensitivity of the system to the particular component modification. Clearly, the problem of modification is not linear. However, as a first approximation, assume a linear variation of r_i^2 with β, and make modifications according to

$$\frac{\delta\beta}{\delta\beta_0} = \frac{\delta r_i^2}{\delta r_{i0}^2} \tag{7.97}$$

where the subscript 0 refers to the initial trial variation ($\delta\beta_0 = 0.1$). Equation 7.97 is intuitively satisfying given the nature of the physical problem and the fact that, for a single-DoF problem, squared frequency varies with k_0. Then, we have

For mode 1:

$$\frac{\delta\beta}{0.1} = \frac{0.8^2 - 0.71^2}{0.727^2 - 0.71^2} = 5.634$$

or

$$\delta\beta = 0.56$$

For mode 2:

$$\frac{\delta\beta}{0.1} = \frac{2^2 - 1.41^2}{1.507^2 - 1.41^2} = 7.09$$

or

$$\delta\beta = 0.709$$

Therefore, we use $\delta\beta = 0.71$, which is the larger of the two. This corresponds to

$$\beta = 0.5 + 0.71 = 1.21$$

The natural frequencies are computed as usual:

$$r_1^2, r_2^2 = \frac{1}{2\times 0.5}\{0.5 + 1.21 + 0.5\times 1.21\}\left\{1 \mp \left[1 - \frac{4\times 0.5\times 1.21}{(0.5 + 1.21 + 0.5\times 1.21)^2}\right]^{1/2}\right\} = 0.60, 4.03$$

or

$$r_1, r_2 = 0.78, 2.01$$

In view of the nonlinearity of the problem, this shift in frequencies is satisfactory. The corresponding mode shapes are

$$\left(\frac{\psi_2}{\psi_1}\right)_1 = \frac{(1 + 1.21) - 0.6}{1.21} = 1.33$$

$$\left(\frac{\psi_2}{\psi_1}\right)_2 = \frac{(1 + 1.21) - 4.03}{1.21} = -1.50$$

It follows that, as the stiffness of the second spring is increased, the motions of the two masses become closer in mode 1. Furthermore, in mode 2, the node point becomes closer to mass 1. Note the limitation of this particular component modification. As $\beta \to \infty$, the two masses become rigidly linked giving a frequency ratio of $r_1 = \sqrt{k/(m + \alpha m)}/\sqrt{k/m} = 1/\sqrt{1+\alpha} = 1/\sqrt{1.5} = 0.816$, with $r_2 \to \infty$. Hence, it is unreasonable to expect a frequency ratio that is closer to this value of r_1 by a change in β alone.

7.7.2 Substructuring

For large and complex mechanical systems with many components, the approach of substructuring can make the process of "design for vibration" more convenient and systematic. In this approach, the system is first divided into a convenient set of subsystems that are more amenable to testing and analysis. The subsystems are separately modeled and designed through the approaches of modal analysis and testing, along with any other convenient approaches (e.g., finite element technique). Note that the performance of the overall system depends on the interface conditions that link the subsystems, as well as the characteristics of the individual subsystems. Hence, it is not possible to translate the design specifications for the overall system into those for the subsystems without taking the interface conditions into account. The overall system is *assembled* from the designed subsystems by using *compatibility* requirements at the assembly locations together with dynamic equations of the interconnecting components such as spring–mass–damper units or rigid linkages. If the assembled system does not meet the design specifications, then modifications should be made to one or more of the subsystems and interfacing (assembly) linkages, and the procedure should be repeated. Thus, the main steps of using the approach of substructuring for vibration design of a complex system are as follows:

1. Divide the mechanical system into convenient subsystems (substructuring) and represent the interconnection points of subsystems by forces/moments.
2. Develop models for the subsystems through analysis, modal testing, and other standard procedures.
3. Design the subsystems so that their performance is well within the performance specifications provided for the overall system.
4. Establish the interconnecting (assembling) linkages for the subsystems, and obtain dynamic equations for them in terms of the linking forces/moments and motions (displacements/rotations).
5. Establish continuity (force balancing) and compatibility (motion consistency) conditions at the assembly locations.
6. Using matrix methods, eliminate the unknown variables and assemble the overall system.
7. Analyze (or test) the overall system to determine its vibration performance. If satisfactory, stop. If not, make modifications to the systems or assembly conditions and repeat step 4 to step 7.

As a simple example, consider two single-DoF systems that are interconnected by a spring linkage, as shown in Figure 7.22. The two subsystems may be represented by

$$\begin{bmatrix} m_1 & 0 \\ 0 & m_2 \end{bmatrix}\ddot{\mathbf{y}} + \begin{bmatrix} k_1 & 0 \\ 0 & k_2 \end{bmatrix}\mathbf{y} = 0$$

and the corresponding natural frequencies are

$$\omega_{s1} = \sqrt{k_1/m_1} \quad \text{and} \quad \omega_{s2} = \sqrt{k_2/m_2}$$

The overall interconnected system is given by

$$\begin{bmatrix} m_1 & 0 \\ 0 & m_2 \end{bmatrix}\ddot{\mathbf{y}} + \begin{bmatrix} k_1 + k_c & -k_c \\ -k_c & k_2 + k_c \end{bmatrix}\mathbf{y} = 0$$

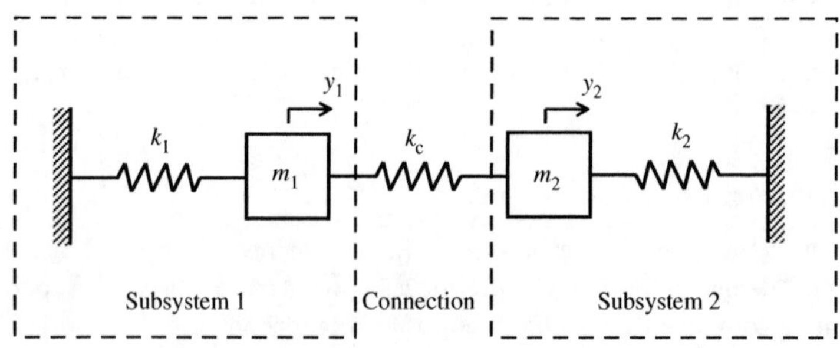

FIGURE 7.22 An example of substructuring.

Its natural frequencies are obtained by solving the equation

$$\det\begin{bmatrix} k_1 + k_c - \omega^2 m_1 & -k_c \\ -k_c & k_2 + k_c - \omega^2 m_2 \end{bmatrix} = 0$$

or

$$(k_1 + k_c - \omega^2 m_1)(k_2 + k_c - \omega^2 m_2) - k_c^2 = 0$$

which simplifies to

$$\omega^4 - \left[\frac{k_1 + k_c}{m_1} + \frac{k_2 + k_c}{m_2}\right]\omega^2 + \frac{k_1 k_2 + k_c(k_1 + k_2)}{m_1 m_2} = 0$$

The sum of the roots is

$$\omega_1^2 + \omega_2^2 = \frac{k_1 + k_c}{m_1} + \frac{k_2 + k_c}{m_2} > \omega_{s1}^2 + \omega_{s2}^2$$

The product of the roots is

$$\omega_1^2 \omega_2^2 = \frac{k_1 k_2 + k_c(k_1 + k_2)}{m_1 m_2} > \omega_{s1}^2 \omega_{s2}^2$$

This does not mean that both frequencies will increase due to the interconnection. Note that the limit on the lower frequency, as $k_c \to \infty$, is given by that of a single-DoF system with mass $m_1 + m_2$ and stiffness $k_1 + k_2$, which is, $\sqrt{(k_1 + k_2)/(m_1 + m_2)}$. This value can be larger or smaller than the natural frequency of a subsystem depending on the relative values of the parameters. Hence, even for this system, exact satisfaction of a set of design natural frequencies would be somewhat challenging because these frequencies depend on the interconnection as well as the subsystems.

Substructuring is a design development technique where complex designs can be accomplished through parallel and separate development of several subsystems and interconnections. Through this procedure, dynamic interactions among subsystems can be estimated and potential problems can be detected, which will allow redesigning of the subsystems or interfacing linkages prior to building the over prototype. Design approaches using EMA that may be used in vibration problems are summarized in Box 7.6.

7.8 Passive Control of Vibration

The techniques discussed in this chapter for reduction of the effects of mechanical vibration fall into the categories of vibration isolation and design for vibration. The third category, vibration control, is addressed now. Characteristic of vibration control is the use of a sensing device to detect the level of vibration in a system, and an actuation (forcing) device to apply a forcing function to the system to counteract the effects of vibration. In some such devices, the sensing and forcing functions are implicit and integrated together.

Vibration control may be subdivided into the following two broad categories:

1. Passive control
2. Active control

Passive control of vibration employs passive controllers. By definition, passive devices do not require external power for their operation. The two passive controllers of vibration discussed in the present section are vibration absorbers (or dynamic absorbers or Frahm absorbers, named after H. Frahm, who first employed the technique for controlling ship oscillations) and dampers. In both types of devices, sensing is implicit and control is achieved through a force generated by the device from its response to the vibration excitation. A dynamic absorber is a mass–spring-type mechanism with little or no damping,

Box 7.6
Test-Based Design Approaches for Vibration

1. **Component modification**:
Modify a component (mass, spring, damper) and determine modal parameters (natural frequencies, damping ratios, mode shapes).
    - Can determine sensitivity to component changes.
    - Can check whether a particular change is satisfactory.

2. **Modal response specification**:
Specify a desired modal response (natural frequencies, damping ratios, mode shapes) and determine the "best" component changes (mass, spring, damper) that will realize the modal specs.
    - Can be accomplished by first performing a sensitivity study (as in item 1).

3. **Substructuring**:
    (i) Design subsystems to meet specs (analytically, experimentally, or by a mixed approach).
    (ii) Establish interconnections between subsystems, and obtain continuity (force balance) and compatibility (motion consistency) at assembly locations.
    (iii) Assemble the overall system by eliminating unknown variables at interconnections.
    (iv) Analyze or test the overall system. If satisfactory, stop. Otherwise, make changes to the subsystems or interconnections, and repeat the above steps.

which can "absorb" the vibration excitation through energy transfer into it, thereby reducing the vibrations of the primary system. The energy received by the absorber will be slowly dissipated due to its own damping. A damper is a purely dissipative device which, unlike a dynamic absorber, directly dissipates the energy received from the system rather than storing it. Hence, it is a more wasteful device, which also may exhibit problems related to wear and thermal effects. However, it has advantages over an absorber, having, for example, a wider frequency of operation.

7.8.1 Undamped Vibration Absorber

A dynamic vibration absorber (or a dynamic absorber, vibration absorber, or Frahm absorber) is a simple mass–spring oscillator with very low damping. An absorber that is tuned to a frequency of vibration of a mechanical system and is able to receive a significant portion of the vibration energy from the primary system at that frequency. In effect, the resulting vibration of the absorber applies an oscillatory force opposing the vibration excitation of the primary system and thereby virtually cancels the effect. In theory, the vibration of the system can be completely removed while the absorber itself undergoes vibratory motion. Since damping is quite low in practical vibration absorbers, we will first consider the case of an undamped absorber.

A vibration absorber may be used for vibration control in two common types of situations, as shown in Figure 7.23. Here, the primary system whose vibration needs to be controlled is modeled as an

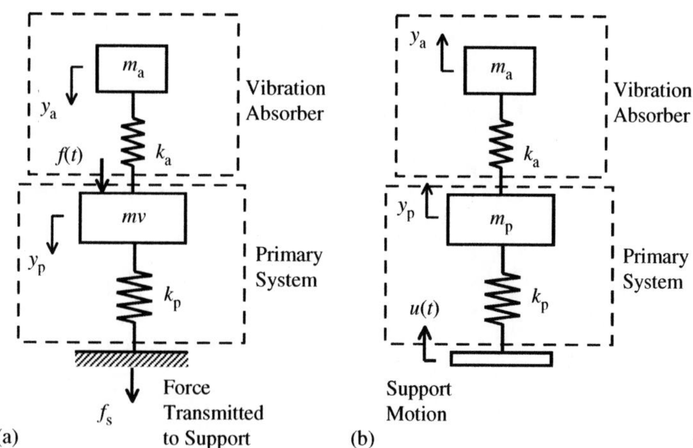

FIGURE 7.23 Two types of applications of a vibration absorber: (a) reduction of the response to forcing excitation (or reducing the force transmitted to support structure); (b) reduction of the response to support motion.

undamped, single-DoF mass–spring system (denoted by the subscript p). An undamped vibration absorber is also a single-DoF mass–spring system (denoted by the subscript a). In the application shown in Figure 7.23(a), the objective of the absorber is to reduce the vibratory response y_p of the primary system as a result of a vibration excitation $f(t)$. The force f_s that is transmitted to the support structure due to the vibratory response of the system is given by

$$f_s = k_p y_p \tag{7.98}$$

Therefore, the objective of reducing y_p may also be interpreted as one of reducing this transmitted vibratory force (a goal of vibration isolation). In the second type of application, represented in Figure 7.23(b), the primary system is excited by a vibratory support motion and the objective of the absorber is to reduce the resulting vibratory motions y_p of the primary system. Note that, in both classes of application, the purpose is to reduce the vibratory responses. Hence, static loads (e.g., gravity) are not considered in the analysis.

Table 7.1 shows the development of the equations of motion for the two systems shown in Figure 7.23. Because we are interested mainly in the control of oscillatory responses to oscillatory excitations, the frequency-domain model is particularly useful. Note from Table 7.1 that the transfer function f_s/f of System a is simply k times the transfer function y_p/f, and is in fact identical to the transfer function y_p/u of System b. The two problems are essentially identical and, thus, we need only address only one of them.

Before investigating the common transfer function for the two types of problems, let us look closely at the frequency-domain equations for the system shown in Figure 7.23(a). We have

$$(k_p + k_a - \omega^2 m_p)y_p - k_a y_a = f \tag{7.99}$$

$$(k_a - \omega^2 m_a)y_a = k_a y_p \tag{7.100}$$

along with Equation 7.98. Here, m_p and k_p are the mass and the stiffness of the primary system, m_a and k_a are the mass and the stiffness of the absorber, f is the excitation amplitude, ω is the excitation frequency, y_p is the primary mass response, and y_a is the absorber response. Now note from Equation 7.100 that if $\omega = \sqrt{k_a/m_a}$ then $y_p = 0$. This means that if the absorber is tuned so that its natural frequency is equal to the excitation frequency (drive frequency), the primary system will not (ideally) undergo any vibratory motion, and is perfectly controlled. The reason for this should be clear from Equation 7.99 which, when $y_p = 0$ is substituted, gives $k_a y_a = -f$. In other words, a tuned absorber applies to the primary system a spring force that is exactly equal and opposite to the excitation force, thereby neutralizing the effect. The absorber mass moves, albeit 180° out of phase with the excitation.

TABLE 7.1 Equations for the Two Types of Absorber Applications

	Absorber Application for the Reduction of Response to a:	
	Forcing Excitation	Support Motion
Time-domain equations	$m_p\ddot{y}_p = -k_p y_p - k_a(y_p - y_a) + f(t)$ $m_a\ddot{y}_a = k_a(y_p - y_a)$	$m_p\ddot{y}_p = k_p(u(t) - y_p) - k_a(y_p - y_a)$ $m_a\ddot{y}_a = k_a(y_p - y_a)$
Frequency-domain equations	$(-\omega^2 m_p + k_p + k_a)y_p = k_a y_a + f$ $(-\omega^2 m_a + k_a)y_a = k_a y_p$	$(-\omega^2 m_p + k_p + k_a)y_p = k_a y_a + k_p u$ $(-\omega^2 m_a + k_a)y_a = k_a y_p$
Matrix form	$\begin{bmatrix} k_p + k_a - \omega^2 m_p & -k_a \\ -k_a & k_a - \omega^2 m_a \end{bmatrix}\begin{bmatrix} y_p \\ y_a \end{bmatrix} = \begin{bmatrix} f \\ 0 \end{bmatrix}$	$\begin{bmatrix} k_p + k_a - \omega^2 m_p & -k_a \\ -k_a & k_a - \omega^2 m_a \end{bmatrix}\begin{bmatrix} y_p \\ y_a \end{bmatrix} = k_p\begin{bmatrix} u \\ 0 \end{bmatrix}$
Transfer-function matrix form	$\begin{bmatrix} y_p \\ y_a \end{bmatrix} = \frac{1}{\Delta}\begin{bmatrix} k_a - \omega^2 m_a & k_a \\ k_a & k_p k_a - \omega^2 m_p \end{bmatrix}\begin{bmatrix} f \\ 0 \end{bmatrix}$	$\begin{bmatrix} y_p \\ y_a \end{bmatrix} = \frac{k}{\Delta}\begin{bmatrix} k_a - \omega^2 m_a & k_a \\ k_a & k_p k_a - \omega^2 m_p \end{bmatrix}\begin{bmatrix} u \\ 0 \end{bmatrix}$
Vibration-control transfer function	$\frac{f_s}{f} = \frac{k_p y_p}{f} = \frac{k_p}{\Delta}(k_a - \omega^2 m_a)$	$\frac{y_p}{u} = \frac{k_p}{\Delta}(k_a - \omega^2 m_a)$
Characteristic polynomial	$\Delta = (k_p + k_a - \omega^2 m_p)(k_a - \omega^2 m_a) - k_a^2$ $= m_p m_a \omega^4 - [k_a(m_p + m_a) + k_p m_a]\omega^2 + k_p k_a$	

The frequency of these motions will be ω (the same as that of the excitation) and the amplitude is proportional to that of the excitation (f) and inversely proportional to the stiffness of the absorber spring. It follows that a vibration absorber "absorbs" vibration energy from the primary system. Furthermore, note from Equation 7.98 that with a tuned absorber the vibration force transmitted to the support structure is (ideally) zero as well. All this information is observed without any mathematical manipulation of the equations of motion.

Note that we are dealing with vibratory excitations and responses. Therefore, static loading (such as gravity and spring preloads) is not considered (we investigate responses with respect to the static equilibrium configuration of the system). In summary, we are now able to state the characteristics of a vibration absorber (undamped) as follows:

1. It is effective only for a single excitation frequency (i.e., a sinusoidal excitation).
2. For the best effect, it should be "tuned" such that its natural frequency $\sqrt{k_a/m_a}$ is equal to the excitation frequency.
3. In the case of forcing vibration excitation, a tuned absorber can (ideally) make the vibratory response of the primary system and the vibratory force transmitted to the support structure zero.
4. In the case of a vibratory support motion, a tuned absorber can make the resulting response of the primary system zero.
5. It functions by acquiring vibration energy from the primary system and storing it (as kinetic energy of the mass or potential energy of the spring) rather than by directly dissipating the energy.
6. It functions by applying a vibration force to the primary system that is equal and opposite to the excitation force, thereby neutralizing the excitation.
7. The amplitude of motion of the vibration absorber is proportional to the excitation amplitude and is inversely proportional to the absorber stiffness. The frequency of the absorber motion is the same as the excitation frequency.

Now, consider the transfer function (f_s/f or y_p/u) of an undamped vibration absorber, as given in Table 7.1. We have

$$G(\omega) = \frac{k_p(k_a - \omega^2 m_a)}{m_p m_a \omega^4 - [k_a(m_p + m_a) + k_p m_a]\omega^2 + k_p k_a} \tag{7.101}$$

It is convenient to use a nondimensional form in analyzing this frequency-transfer function. To that end, we define the following nondimensional parameters and frequency variable:

Fractional mass of the absorber $\mu = m_a/m_p$
Nondimensional natural frequency of the absorber $\alpha = \omega_a/\omega_p$
Nondimensional excitation (drive) frequency $r = \omega/\omega_p$

where

$\omega_a = \sqrt{k_a/m_a}$ = natural frequency of the absorber

$\omega_p = \sqrt{k_p/m_p}$ = natural frequency of the primary system

It is straightforward to divide the numerator and the denominator by $k_p k_a$ and then carry out simple algebraic manipulations to express the transfer function of Equation 7.101 in the nondimensional form as

$$G(r) = \frac{\alpha^2 - r^2}{r^4 - [\alpha^2(1+\mu)+1]r^2 + \alpha^2} \tag{7.102}$$

For this undamped system, there is no difference between the resonant frequencies (where the magnitude of the transfer function peaks) and the natural frequencies (roots of the characteristic equation that correspond to the "natural" or free time response oscillations). These are obtained by solving the equation

$$r^4 - [\alpha^2(1+\mu)+1]r^2 + \alpha^2 = 0 \tag{7.103}$$

which gives

$$r_1^2, r_2^2 = \frac{1}{2}[\alpha^2(1+\mu)+1] \mp \frac{1}{2}\sqrt{[\alpha^2(1+\mu)+1]^2 - 4\alpha^2} \tag{7.104}$$

These are squared frequencies, both of which are positive as clear from Equation 7.104. The actual, nondimensional natural frequencies are their square roots. The magnitude of the transfer function becomes infinite at either of these two natural/resonant frequencies. Furthermore, it is clear from Equation 7.102 that the transfer function magnitude becomes zero at $r = \alpha$, where the excitation frequency (ω) is equal to the natural frequency of the absorber (ω_a) as noted above. In the present undamped case, the transfer function $G(r)$ is real but it can be either positive or negative. The magnitude is thus the absolute value of $G(r)$, which is positive. The magnitude plot given in Figure 7.24 shows the resonant and control characteristics of a system with an undamped vibration absorber. Originally, the primary system had a resonance at $r = 1$ (i.e., $\omega = \omega_p$). When the absorber, which also has a resonance at $r = 1$, is added, the original resonance becomes an *antiresonance* with a zero response. Here, however, two new resonances are created, one at $r = 0.854$ and the other at $r = 1.171$, which are on either side of the tuned frequency ($r = 1$) of the absorber.

Owing to these two resonances, the effective region of the absorber is limited to a narrow frequency band centering its tuned frequency. Specifically, the absorber is not effective unless $|G| < 1$. The effective frequency band of a vibration absorber may be determined using this condition.

Example 7.8

A high-precision, yet high-power positioning system uses a hydraulic actuator and a valve. The pressurized oil to this hydraulic servo system is provided by a gear-type rotary pump. The pump and the positioning system are mounted on the same workbench. The mass of the pump is 25 kg. The normal operating speed of the pump is 3600 rpm. During operation, it was observed that the pump exhibits a vertical resonance at this speed and it affects the accuracy of the position servo system. To control the vibrations of the pump at its operating speed, a vibration absorber of mass 1.25 kg tuned to the normal operating speed of the pump is attached, as shown schematically in Figure 7.25. Because the speed of the pump normally fluctuates during operation, we must determine the speed range within which the vibration absorber is effective. What are the new resonant frequencies of the system? (Neglect damping.)

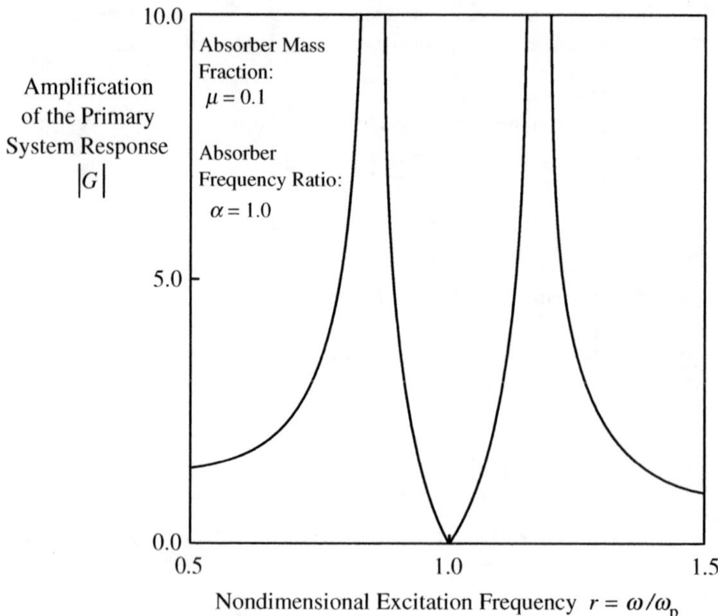

FIGURE 7.24 The effect of an undamped vibration absorber on the vibration response of a primary system.

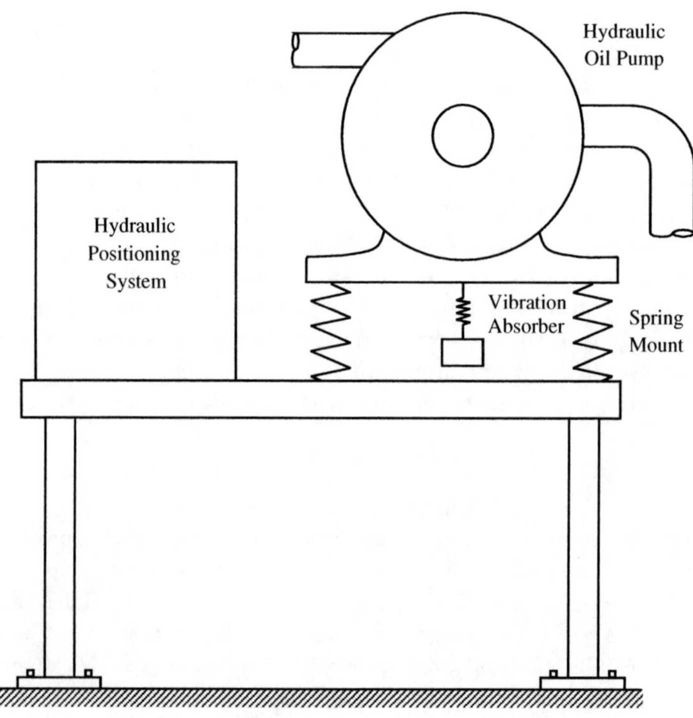

FIGURE 7.25 A hydraulic positioning system with a gear pump.

Solution

For this problem, the fractional mass $\mu = 1.25/25.0 = 0.05$. Since the absorber is tuned to the resonant frequency of the pump, $\alpha = 1.0$. From Equation 7.103, the characteristic equation of the modified system becomes

$$r^4 - 2.05r^2 + 1 = 0$$

which has roots $r_1 = 0.854$ and $r_2 = 1.171$. It follows that the new resonances are at 0.854×3600 and 1.171×3600 rpm. These are 3074.4 and 4215.6 rpm, which should be avoided. From Equation 7.102, the system transfer function is

$$G(r) = \frac{1 - r^2}{(r^4 - 2.05r^2 + 1)}$$

The effective frequency band of the absorber corresponds to $|G(r)| < 1.0$. Since a sign reversal of $G(r)$ occurs at $r = 1$, we need to solve both

$$\frac{1 - r^2}{(r^4 - 2.05r^2 + 1)} = 1 \text{ and } -1$$

The first equation gives the roots $r = 0$ and 1.025. The second equation gives the roots $r = 0.977$ and 1.45.

Hence, the effective frequency band corresponds to $\Delta r = [0.977, 1.025]$. In terms of the operating speed of the pump, we have an effective band of 3517.2 to 3690 rpm. Thus, a speed fluctuation of about ± 80 rpm is acceptable.

Finally, recall that the presence of the absorber generates two new resonances on either side of the resonance of the original system (to which the absorber is normally tuned). It is also clear from Equation 7.104 that these two resonances become farther and farther apart as the fractional mass μ of the vibration absorber is increased.

7.8.2 Damped Vibration Absorber

Damping is not the primary means by which vibration control is achieved in a vibration absorber. As noted above, the absorber acquires vibration energy from the primary system (and in turn, exerts a force on the system that is equal and opposite to the vibration excitation), thereby suppressing the vibratory motion. The energy received by the absorber has to be dissipated gradually and, hence, some damping should be present in the absorber. Furthermore, the two resonances that are created by adding the absorber have an infinite magnitude in the absence of damping. Hence, damping has the added benefit of lowering these resonant peaks.

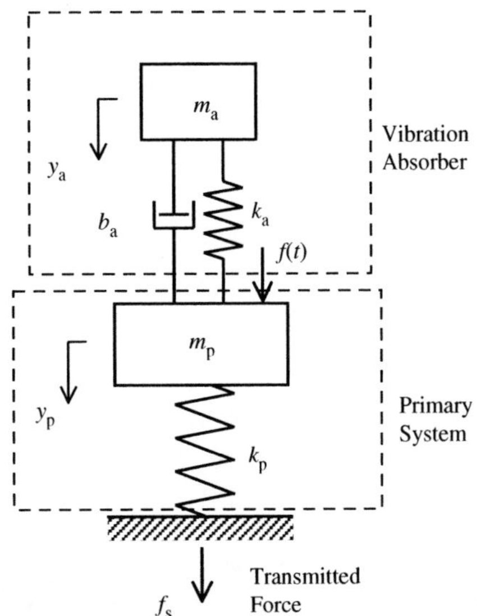

FIGURE 7.26 Primary system with a damped vibration absorber.

The analysis of a vibratory system with a damped absorber is similar to but somewhat more complex than, that involving an undamped absorber. Furthermore, an extra design parameter — the damping ratio of the absorber — enters into the scene. Consider the model shown in Figure 7.26. Another version of application of a damped absorber, which corresponds to Figure 7.23(b), may also be presented. However, because the two types of application have the same transfer function, it is sufficient to consider Figure 7.26 alone.

Again, the transfer function of vibration control may be taken as either y_a/f or f_s/f, the latter being simply k_p times the former. Although we will consider the dimensionless case of f_s/f, the results are equally valid for y_p/f, except that the responses must be converted from force to displacement by dividing by k_p.

There is no need to derive the transfer function anew for the damped system. Simply replace k_a in Equation 7.101 by the complex stiffness $k_a + j\omega b_a$, which incorporates the viscous damping constant b_a and the excitation frequency ω. Hence, the transfer function of the damped system is

$$G(\omega) = \frac{k_p(k_a + j\omega b_a - \omega^2 m_a)}{m_p m_a \omega^4 - [(k_a + j\omega b_a)(m_p + m_a) + k_p m_a]\omega^2 + k_p(k_a + j\omega b_a)} \tag{7.105}$$

With the parameters defined as before, the nondimensional form of this transfer function is obtained by dividing throughout by $k_p k_a$ and then substituting the appropriate parameters. In particular, we use the fact that

$$\frac{b_a}{k_a} = \frac{2b_a}{2\sqrt{k_a m_a}}\sqrt{\frac{m_a}{k_a}} = \frac{2\zeta_a}{\omega_a} = \frac{2\zeta_a}{\omega_p}\frac{\omega_p}{\omega_a} = \frac{2\zeta_a}{\alpha\omega_p} \tag{7.106}$$

where the damping ratio ζ_a of the absorber is given by

$$\zeta_a = \frac{b_a}{2\sqrt{k_a m_a}} \tag{7.107}$$

as usual. Then, we follow the same procedure that used to derive Equation 7.102 from Equation 7.101 to get

$$G(r) = \frac{\alpha^2 - r^2 + 2j\zeta_a \alpha r}{r^4 - [(\alpha^2 + 2j\zeta_a \alpha r)(1 + \mu) + 1]r^2 + (\alpha^2 + 2j\zeta_a \alpha r)} \tag{7.108}$$

Note that this result is equivalent to simply replacing α^2 by $\alpha^2 + 2j\zeta_a \alpha r$ in Equation 7.102.

It is important to note that the undamped natural frequencies are obtained by solving the characteristics equation with $\zeta_a = 0$. These are the same as before and given by the square roots of Equation 7.104. The damped natural frequencies are obtained by first setting $jr = \lambda$ (hence, $r^2 = -\lambda^2$ and $r^4 = \lambda^4$) and then solving the resulting characteristics equation (see the denominator of Equation 7.108).

$$\lambda^4 + 2\zeta_a \alpha(1 + \mu)\lambda^3 + (\alpha^2 + \alpha^2\mu + 1)\lambda^2 + 2\zeta_a \alpha\lambda + \alpha^2 = 0 \tag{7.109}$$

and then taking the imaginary parts of the roots of λ. These depend on ζ_a and are different from those obtained from Equation 7.104. The resonant frequencies correspond to the r values where the magnitude of $G(r)$ will peak. Generally, these are not the same as the undamped or damped natural frequencies. However, for low damping (small ζ_a compared with 1), these three types of system characteristics frequencies are almost identical.

The magnitude of the transfer function (Equation 7.108) is plotted in Figure 7.27 for the case $\mu = 1.0$ and $\alpha = 1.0$, as in Figure 7.24, but for damping ratios $\zeta_a = 0.01$, 0.1, and 0.5. Note that the curve for $\zeta_a = 0.01$ is very close to that in Figure 7.24 for the undamped case. When ζ_a is large, as shown in the case of $\zeta_a = 0.5$, the two masses m_p and m_a tend to become locked together and appear to behave like a single mass. Then, the system tends to act like a single-DoF one, and the primary system is modified only in its mass (which increases). Consequently, only one resonant frequency is produced, which is smaller than that of the original primary system. Furthermore, as expected in this high-damping case, the effect of a vibration absorber is no longer present.

All three curves in Figure 7.27 pass through the two common points A and B, as shown. This is true for all curves corresponding to all values of ζ_a, and particularly for the extreme cases of $\zeta_a = 0$ and $\zeta_a \to \infty$. Hence, these points can be determined as the points of intersection of the transfer function magnitude curves for the limiting cases $\zeta_a = 0$ and $\zeta_a \to \infty$.

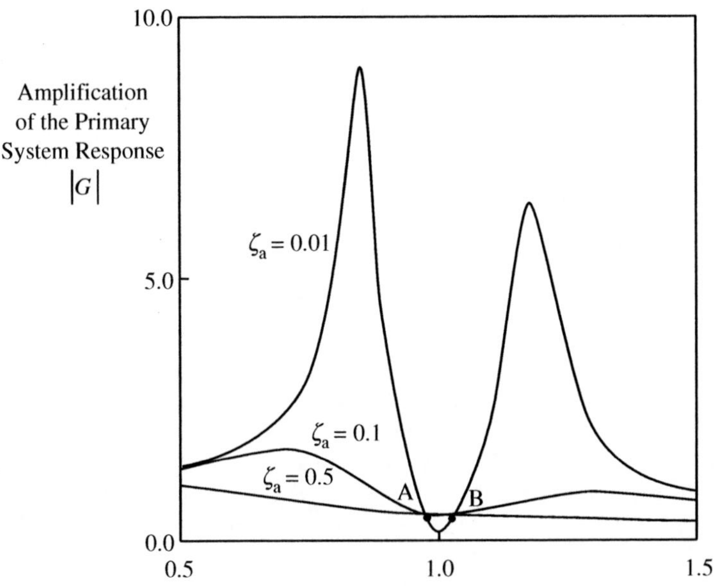

FIGURE 7.27 Vibration amplification (transfer function magnitude) curves for damped vibration absorbers (absorber mass $\mu = 0.1$, absorber resonant frequency $\alpha = 1.0$).

Equation 7.102 gives $G(r)$ for $\zeta_a = 0$. Next, from Equation 7.108, we note that, as $\zeta_a \to \infty$, all the terms not containing ζ_a can be neglected. Hence,

$$G(r) = \frac{2j\zeta_a \alpha r}{-2j\zeta_a \alpha r(1+\mu)r^2 + 2j\zeta_a \alpha r}$$

Cancel the common term and we get (for $r \neq 0$)

$$G(r) = \frac{1}{1-(1+\mu)r^2} \quad \text{for } \zeta_a \to \infty \tag{7.110}$$

Note that this is the normalized transfer function of a single-DoF system of natural frequency $1/\sqrt{1+\mu}$. This result confirms the fact that as $\zeta_a \to \infty$, the two masses m_p and m_a become locked together and act as a single mass $(m_p + m_a)$ supported on a spring of stiffness k_p. Its natural frequency is $\sqrt{k_p/(m_p+m_a)}$ which, when normalized with respect to $\sqrt{k_p/m_p}$, becomes

$$\sqrt{\frac{k_p}{(m_p+m_a)}\frac{m_p}{k_p}} = \sqrt{\frac{m_p}{(m_p+m_a)}} = \frac{1}{\sqrt{1+\mu}}$$

In determining the points of intersection between the functions (see Equation 7.102 and Equation 7.110), we should first note that at the first point of intersection (A), the function in Equation 7.102 is negative and positive in Equation 7.110, while the reverse is true for the second point of intersection (B). For either point, this means that the sign of one of the functions should be reversed before equating them. Thus,

$$\frac{\alpha^2 - r^2}{r^4 - [\alpha^2(1+\mu)+1]r^2 + \alpha^2} = -\frac{1}{1-(1+\mu)r^2}$$

which gives

$$(2+\mu)r^4 - 2[\alpha^2(1+\mu)+1]r^2 + 2\alpha^2 = 0 \tag{7.111}$$

This is the equation whose roots (e.g., r_1 and r_2) give the points A and B. Then, we have the sum of the squared roots equal to the negative coefficient of r^2 in the quadratic (in r^2) Equation 7.111. Thus,

$$r_1^2 + r_2^2 = \frac{2[\alpha^2(1+\mu)+1]}{(2+\mu)} \tag{7.112}$$

In addition, the product of the squared roots is equal to the constant term in the quadratic (r^2) in Equation 7.111. Hence,

$$r_1^2 r_2^2 = \frac{2\alpha^2}{(2+\mu)} \tag{7.113}$$

7.8.2.1 Optimal Absorber Design

It has been pointed out (primarily by J.P. Den Hartog) that an optimal absorber design should not only have equal response magnitudes at the common points of intersection (i.e., equal ordinates of points A and B in Figure 7.27), but also that the resonances should occur at these points to achieve some balance and uniformity in the response amplification in the region surrounding the tuned frequency of the absorber. It is expected that these (intuitive) design conditions would give relations between the parameters α, μ, and ζ_a, corresponding to an optimal absorber.

Consider the first requirement of equal transfer function magnitudes at A and B. As noted earlier, because these two points do not depend on ζ_a, we use Equation 7.110 to satisfy the requirement. Thus, keeping in mind the sign reversal of the transfer function between A and B (i.e., as the transfer function passes through the resonance), we have

$$\frac{1}{1-(1+\mu)r_1^2} = -\frac{1}{1-(1+\mu)r_2^2}$$

which gives

$$r_1^2 + r_2^2 = \frac{2}{1+\mu} \tag{7.114}$$

Substituting this result (for equal ordinates) in the intersection-point condition (see Equation 7.112), we have

$$\frac{2}{1+\mu} = \frac{2[\alpha^2(1+\mu)+1]}{(2+\mu)}$$

On simplification, we get the simple result

$$\alpha = \frac{1}{1+\mu} \tag{7.115}$$

Next, we turn to the task of achieving peak magnitudes of the transfer function at the points of intersection (A and B). Generally, when one point peaks the other does not. As reported by Den Hartog, with straightforward but lengthy analysis, we obtain

$$\zeta_a^2 = \frac{\mu[3-\sqrt{\mu/(\mu+2)}]}{8(1+\mu)^3} \tag{7.116}$$

for peak at the first intersection point, and

$$\zeta_a^2 = \frac{\mu[3+\sqrt{\mu/(\mu+2)}]}{8(1+\mu)^3} \tag{7.117}$$

for peak at the second intersection point.

So, for design purposes, a balance is obtained by taking the average value of the results of Equation 7.116 and Equation 7.117 as

$$\zeta_a^2 = \frac{3\mu}{8(1+\mu)^3} \tag{7.118}$$

Thus, Equation 7.115 and Equation 7.118 correspond to an optimal vibration absorber. In addition, practical requirements and limitations need to be addressed in any design procedure. In particular, since μ is considerably less than unity (i.e., absorber mass is a small fraction of the primary mass), the absorber mass should undergo relatively large amplitudes at the operating frequency in order to receive the energy of the primary system. The absorber spring must be designed accordingly, while meeting the tuning frequency conditions that determine the ratio m_a/k_a.

Example 7.9

The air compressor of a wind tunnel weighs 48 kg and normally operates at 2400 rpm. The first major resonance of the compressor unit occurs at 2640 rpm, with severe vibration amplitudes that are quite dangerous. Design a vibration absorber (damped) for installation on the mounting base of the compressor. What are the vibration amplifications of the compressor unit at the new resonances of the modified system? Compare these with the vibration amplitude of the original system in normal operation.

Solution

As usual, we will tune the absorber to the normal operating speed (2400 rpm). Then, we have the nondimensional resonant frequency of the absorber:

$$\alpha = \frac{\omega_a}{\omega_p} = \frac{2400}{2640} = \frac{12}{13}$$

Now, for an optimal absorber, from Equation 7.115

$$\mu = \frac{1}{\alpha} - 1 = \frac{13}{12} - 1 = \frac{1}{12}$$

Hence, the absorber mass

$$m_a = 48 \times \frac{1}{12}\ \text{kg} = 4.0\ \text{kg}$$

Then, from Equation 7.118, the damping ratio of the absorber is

$$\zeta_a = \left[\frac{3/12}{8(1+1/12)^3}\right]^{1/2} = 0.157$$

Now,

$$\omega_a = \sqrt{\frac{k_a}{m_a}} = \sqrt{\frac{k_a}{4.0}} = \frac{2400}{60} \times 2\pi\ \text{rad/s} = 88\pi\ \text{rad/s}$$

Hence,

$$k_a = (88\pi)^2 \times 4.0\ \text{N/m} = 2.527 \times 10^5\ \text{N/m}$$

Also,

$$\zeta_a = \frac{1}{2}\frac{b_a}{\sqrt{m_a k_a}}$$

Then, we have

$$b_a = 2 \times 0.157\sqrt{4.0 \times 2.527 \times 10^5}\ \text{N s/m} = 315.7\ \text{N s/m}$$

This gives us the damped absorber. Now, let us check its performance. We know that, in theory, the vibration amplitude at the operating speed should be almost zero now. However, two resonances are created around the operating point. Since damping is small, we use the undamped characteristic Equation 7.103 to compute these resonances:

$$r^4 - \left[\frac{12^2}{13^2}\left(1 + \frac{1}{12}\right) + 1\right]r^2 + \frac{12^2}{13^2} = 0$$

which gives

$$r^4 - \frac{25}{13}r^2 + \frac{12^2}{13^2} = 0$$

The roots of r^2 are 0.692 and 1.231. The (positive) roots of r are 0.832 and 1.109.

These correspond to compressor speeds of (multiply r by 2640 rpm) 2196 and 2929 rpm. Although they are approximately at -10% and $+20\%$ of the operating speed, the first resonance will be encountered during startup and shutdown conditions. To determine the corresponding vibration amplifications (force/force), use Equation 7.108 which, when the undamped characteristic equation is substituted into the denominator, becomes

$$G(r) = \frac{\alpha^2 - r^2 + 2j\zeta_a \alpha r}{[-2j\zeta_a \alpha r(1+\mu)r^2 + 2j\zeta_a \alpha r]} = \frac{1 - j(\alpha^2 - r^2)/(2\zeta_a \alpha r)}{1 - (1+\mu)r^2} \tag{7.119}$$

Substitute the resonant frequencies $r_1 = 0.832$ and $r_2 = 1.109$. We get $|G(r_1)| = 4.223$ and $|G(r_2)| = 4.634$.

Without the absorber, we approximate the system by a simple undamped oscillator with transfer function

$$G_p(r) = \frac{1}{1 - r^2}$$

The corresponding vibration amplification at the operating speed is

$$|G_p(r_0)| = \frac{1}{|1 - 12^2/13^2|} = 6.76$$

It is observed that after adding the absorber, the resonant vibrations are smaller than even the operating vibrations of the original system. Hence, the design is satisfactory. Note that we used the force/force transfer functions. To get the displacement/force transfer functions we divide by k_p. However, we have

$$\sqrt{\frac{k_p}{m_p}} = \frac{2640}{60} \times 2\pi\ \text{rad/s} = 88\pi\ \text{rad/s}$$

Hence,

$$k_p = (88\pi)^2 \times 48\ \text{N/m} = 3.67 \times 10^6\ \text{N/m} = 3.67 \times 10^3\ \text{N/mm}$$

Thus, the amplitude of operating vibrations of the original system is

$$\frac{6.76}{3.67 \times 10^3}\ \text{mm/N} = 1.84 \times 10^{-3}\ \text{mm/N}$$

The amplitudes of the resonant vibrations of the modified system are

$$\frac{4.223}{3.67 \times 10^3} \text{ and } \frac{4.634}{3.67 \times 10^3}\ \text{mm/N or } 1.15 \times 10^{-3} \text{ and } 1.26 \times 10^{-3}\ \text{mm/N}$$

Vibration absorbers are simple and passive devices, which are commonly used in the control of narrowband vibrations (limited to a very small interval of frequencies). Applications are found in vibration suppression of transmission wires (e.g., a stockbridge damper, which simply consists of a piece of cable carrying two masses at its ends), consumer appliances, automobile engines, and industrial machinery. It should be noted that the concepts presented for a rectilinear vibration absorber may be directly extended to a rotary vibration absorber. Figure 7.28 provides a schematic representation of a rotary vibration absorber. This model corresponds to vibration force excitations (compare with Figure 7.23(a)). The case of rotational support-motion excitations (see Figure 7.23(b)), which has essentially the same transfer function, may also be addressed. Approaches of vibration control are summarized in Box 7.7.

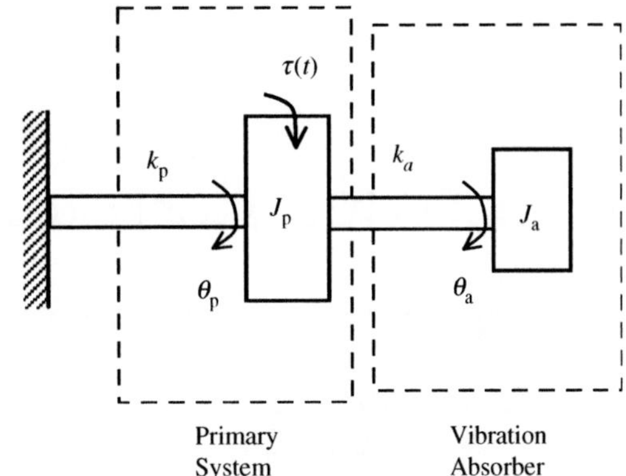

FIGURE 7.28 The application of a rotary vibration absorber.

7.8.3 Vibration Dampers

As discussed above, vibration absorbers are simple and effective passive devices, which are used in vibration control. They have the added advantage of being primarily nondissipative. The main disadvantage of a vibration absorber is that it is only effective over a very narrow band of frequencies enclosing its resonant frequency (tuned frequency). When passive vibration control over a wide band of frequencies is required, a damper would be a preferable choice.

Vibration dampers are dissipative devices. They control vibration through direct dissipation of the vibration energy of the primary (vibrating) system. As a result, however, there will be substantial heat generation, and associated thermal problems and component wear. Consequently, methods of cooling (e.g., use of a fan, coolant circulation, and thermal conduction blocks) may be required in some special situations.

Consider a vibrating system modeled as an undamped single-DoF mass–spring system (simple oscillator). The magnitude of the excitation-response transfer function will have a resonance with a theoretically infinite magnitude in this case. Operation in the immediate neighborhood of such a resonance would be destructive. Adding a simple viscous damper, as shown in Figure 7.29(a), will correct the situation. The equation of motion (about the static equilibrium position) is

$$m\ddot{y} + b\dot{y} + ky = f(t) \tag{7.120}$$

with the dynamic force that is transmitted through the support base (f_s) given by

$$f_s = ky + b\dot{y} \tag{7.121}$$

Hence, the transfer function between the forcing excitation f and the vibration response y is

$$\frac{y}{f} = \frac{1}{k - \omega^2 m + j\omega b} \tag{7.122}$$

and that between the forcing excitation and the force transmitted to the support structure is

$$\frac{f_s}{f} = \frac{k + j\omega b}{k - \omega^2 m + j\omega b} \tag{7.123}$$

Using the nondimensional frequency variable $r = \omega/\omega_n$ where $\omega_n = \sqrt{k/m}$ is the undamped natural frequency of the system and the damping ratio $\zeta = b/(2\sqrt{km})$, we can express Equation 7.122 and

Box 7.7
VIBRATION CONTROL

Passive control (no external power):

1. Dampers
   - A dissipative approach (thermal problems, degradation)
   - Useful over a wide frequency band
2. Vibration absorbers (dynamic absorbers, Frahm absorbers)
   - Absorbs energy from vibrating system and applies counteracting force
   - Useful over a very narrow frequency band (near the tuned frequency)
   - Absorber executes large motions

Undamped absorber design:

$$\text{Transfer function of system with absorber} = \frac{\alpha^2 - r^2}{r^4 - [\alpha^2(1+\mu)+1]r^2 + \alpha^2}$$

where

μ = absorber mass/primary system mass
α = absorber natural frequency/primary system natural frequency
r = excitation frequency/primary system natural frequency

The most effective operating frequency $r_{op} = \alpha$.
Avoid the two resonances.

Optimal damped absorber design:
Mass ratio

$$\mu = \frac{1}{\alpha} - 1$$

Damping ratio

$$\zeta_a = \frac{3\mu}{8(1+\mu)^3}$$

Active control (needs external power):

1. Measure vibration response using sensors/transducers.
2. Apply control forces to vibrating system through actuators, according to a suitable control algorithm.

Equation 7.123 in the form

$$\frac{y}{f} = \frac{1}{k(1 - r^2 + 2j\zeta r)} \tag{7.124}$$

$$\frac{f_s}{f} = \frac{1 + 2j\zeta r}{(1 - r^2 + 2j\zeta r)} \tag{7.125}$$

When vibration control of the primary system is desired, we use the transfer function in Equation 7.124. However, when force transmissibility is the primary consideration, we use Equation 7.125.

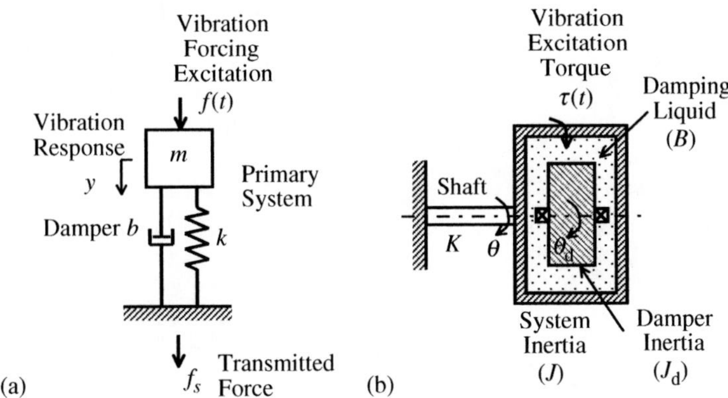

FIGURE 7.29 (a) A system with a linear viscous damper; (b) a rotary system with a Houdaille damper.

Furthermore, it is convenient to use the transfer function in Equation 7.124 in the nondimensional form:

$$\frac{ky}{f} = G(r) = \frac{1}{(1 - r^2 + 2j\zeta r)} \tag{7.126}$$

The magnitude of this transfer function is plotted in Figure 7.30 for several values of damping ratio. Note that the addition of significant levels of damping considerably lowers the resonant peak and flattens the overall response. This example illustrates the broadband nature of the effect of a damper. However, unlike a vibration absorber, it is not possible with a simple damper to bring the vibration levels to a theoretical zero. However, a damper is able to bring the response uniformly close to the static value (unity in Figure 7.30).

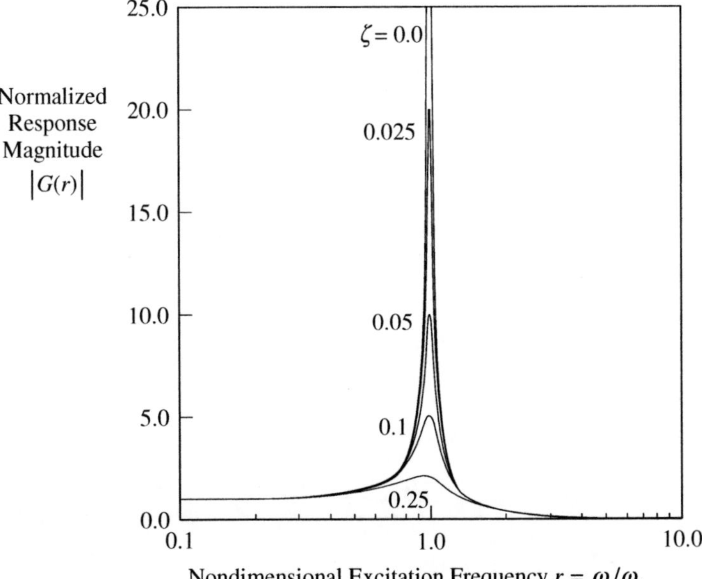

FIGURE 7.30 Frequency response of a system containing a linear damper.

Another common application of damper is connecting it through a free inertia element. For a rotational system, such an arrangement is know as the Houdaille damper, and is modeled as in Figure 7.29(b). The equations of motion are

$$J\ddot{\theta} + B(\dot{\theta} - \dot{\theta}_d) + K\theta = \tau(t) \tag{7.127}$$

$$J_d\ddot{\theta}_d + B(\dot{\theta}_d - \theta) = 0 \tag{7.128}$$

In this case, the transfer function between the vibratory excitation torque τ and the response angle θ is given by

$$\frac{\theta}{\tau} = \frac{B + J_d j\omega}{KB - B(J + J_d)\omega^2 - J_d J j\omega^3 + KJ_d j\omega} \tag{7.129}$$

Again, we use the normalized form of $K\theta/\tau$. Then, we obtain

$$\frac{K\theta}{\tau} = G(r) = \frac{2\zeta + jr\mu}{2\zeta[1 - (1+\mu)r^2] + jr\mu(1 - r^2)} \tag{7.130}$$

where $r = \omega/\omega_n$, $\zeta = B/(2\sqrt{KJ})$, $\mu = J_d/J$, and $\omega_n = \sqrt{K/J}$.

Note the two extreme cases. When $\zeta = 0$, the system becomes the original undamped system, as expected. When $\zeta \rightarrow \infty$, the system becomes an undamped simple oscillator, but with a lower natural frequency of $r = 1/\sqrt{1+\mu}$, instead of $r = 1$ that was present in the original system. This is to be expected because as $\zeta \rightarrow \infty$, the two inertia elements become locked together and act as a single combined inertia $J + J_d$. Clearly, in these two extreme systems, the effect of damping is not present. Optimal damping occurs somewhere in between, as is clear from the curves of response magnitude shown in Figure 7.31 for the case of $\mu = 0.2$.

Proper selection of the nature and values of damping is crucial for the effective use of a damper in vibration control. Damping in physical systems is known to be nonlinear and frequency dependent, as well as time-variant and dependent on the environment (e.g., temperature). Various models are available for different types of damping, but these are only approximate representations. In practice, such considerations as the type of damper used, the nature of the system, the specific application, and the

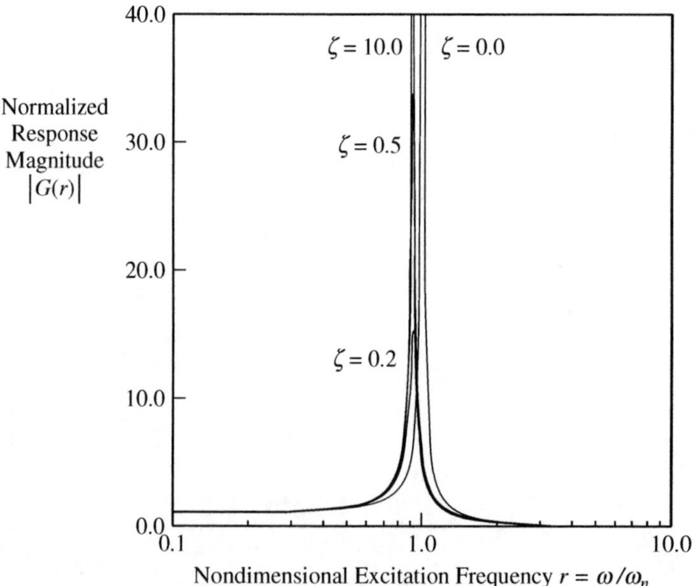

FIGURE 7.31 Response curves for a rotary system with Houdaille damper of inertia ratio $\mu = 0.2$.

speed of operation, determine which particular model (linear viscous, hysteric, Coulomb, Stribeck, quadratic aerodynamic, etc.) is suitable. In addition to the simple linear theory of viscous damper, specific properties of physical damping should be taken into consideration in practical designs.

7.9 Active Control of Vibration

Passive control of vibration is relatively simple and straightforward. Although it is robust, reliable, and economical, it has its limitations. Note that the control force that is generated in a passive device depends entirely on the natural dynamics. Once the device is designed (i.e., after the parameter values for mass, damping constant, stiffness, location, etc. are chosen), it is not possible to adjust the control forces that are naturally generated in real time. Furthermore, in a passive device there is no supply of power from an external source. Hence, even the magnitude of the control forces cannot be changed from their natural values. Since a passive device senses the response of the system as an integral process of the overall dynamics of the system, it is not always possible to directly target the control action at particular responses (e.g., particular modes). This can result in incomplete control, particularly in complex and high-order (e.g., distributed-parameter) systems. These shortcomings of passive control can be overcome using active control. Here, the system responses are directly sensed using sensor-transducer devices, and control actions of specific desired values are applied to desired locations/modes of the system.

7.9.1 Active Control System

Figure 7.32 presents a schematic diagram of an active control system. The mechanical dynamic system whose vibrations need to be controlled is the *plant* or *process*. The controller is the device that generates the signal (or command) according to some scheme (or control law) and controls vibrations of the plant. The plant and the controller are the two essential components of a *control system*. Usually, the plant must

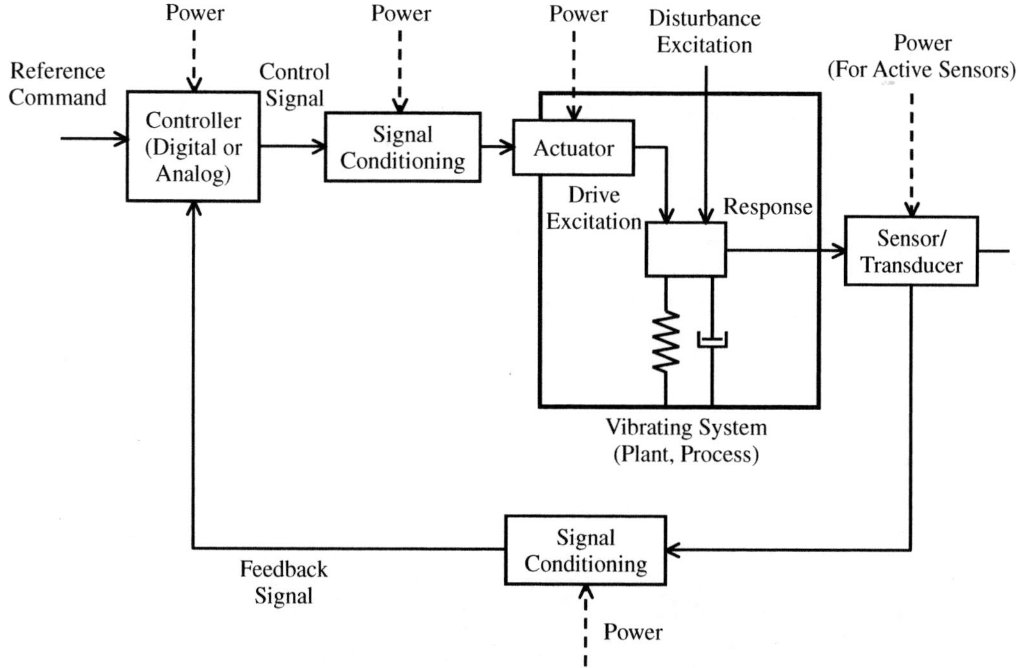

FIGURE 7.32 A system for active control of vibration.

be monitored and its response must be measured using sensors providing feedback into the controller. Then, the controller compares the sensed signal with a desired response specified externally, and uses the error to generate a proper control signal. In this manner, we have a feedback control system. In the absence of a sensor and feedback, we have an open-loop control system. In *feed-forward control*, the excitation (i.e., input signal), not the response (i.e., output signal), is measured and used (i.e., fed forward into the controller) for generating the control signal. Both feedback and feedforward schemes may be used in the same control system.

The actuator that receives a control signal and drives the plant may be an integral part of the plant (e.g., the motor that drives the blade of a saw). Alternatively, it may have to be added specifically as an external component for the control actuation (e.g., a piezoelectric or electromagnetic actuator for controlling blade vibrations of a saw). In the former case, in particular, proper signal conditioning is needed to convert the control signal to a form that is compatible with the existing actuator. In the latter case, both the controller and the actuator must be developed in parallel for integration into the plant. In digital control, the controller is a digital processor. The control signal is in digital form and, typically, it has to be converted into the analog form prior to using in the actuator. Hence, digital-to-analog conversion (DAC) is a form of signal conditioning that is useful here. Furthermore, the analog signal that is generated may have to be filtered and amplified to an appropriate level for use in the actuator. It follows that filters and amplifiers are signal conditioning devices, which are useful in vibration control. In *software control*, the control signal is generated by a computer, which functions as the digital controller. In *hardware control*, the control signal is rapidly generated by digital hardware without using software programs. Alternatively, *analog control* may be used where the control signal is generated directly using analog circuitry. In this case, the controller is quite fast and it does not require DAC. Note that the actuator may need high levels of power. Furthermore, the controller and associated signal conditioning will require some power. The need for an external power source for control distinguishes active control from passive control.

In a feedback control system, sensors are used to measure the plant response, which enables the controller to determine whether the plant operates properly. A sensor unit that "senses" the response may automatically convert (transduce) this "measurement" into a suitable form. A piezoelectric accelerometer senses acceleration and converts it into an electric charge, an electromagnetic tachometer senses velocity and converts it into a voltage, and a shaft encoder senses a rotation and converts it into a sequence of voltage pulses. Hence, the terms *sensor* and *transducer* are used interchangeably to denote a sensor-transducer unit. The signal that is generated in this manner may need conditioning before feeding into the controller. For example, the charge signal from a piezoelectric accelerometer has to be converted to a voltage signal of appropriate level using a charge amplifier, and then it has to be digitized using an analog-to-digital converter (ADC) for use in a digital controller. Furthermore, filtering may be needed to remove measurement noise. Hence, signal conditioning is usually needed between the sensor and the controller as well as between the controller and the actuator. External power is required to operate active sensors (e.g., potentiometer) whereas passive sensors (e.g., electromagnetic tachometer) employ self-generation and do not need an external power source. External power may be needed for conditioning the sensor signals. Finally, as indicated in Figure 7.32, a vibrating system may have unknown disturbance excitations, which can make the control problem particularly difficult. Removing such excitations at the source level through proper design or vibration isolation is desirable, as discussed above. However, in the context of control, if these disturbances can be measured or some information about them is available, then they can be compensated for within the controller itself. This is, in fact, the approach of feedforward control.

7.9.2 Control Techniques

The purpose of a vibration controller is to excite (activate) a vibrating system in order to control its vibration response in a desired manner. In the present context of active feedback control, the controller uses measured response signals and compares them with their desired values in its task of determining an appropriate action. The relationship that generates the control action from a measured desired value for the response) is called a *control law*. Sometimes, a *compensator* (a hardware or software) is employed to improve the system performance or to enhance the ... that the task of control is easier. However, for our purpose, we may consider a compensator as an integral part of the controller and thus a distinction between the two is not made.

Various control laws, both linear and nonlinear, have been developed for practical applications. Many of them are suitable in vibration control. A comprehensive presentation of all such control laws is outside the scope of this book. We will give several linear control laws that are common and representative of what is available. These techniques are based on a linear representation (linear model) of the vibrating system (plant). Even when the overall operating range of a plant (e.g., robotic manipulator) is nonlinear, it is often possible to linearize the vibration response (e.g., link vibrations and joint vibrations of a robot) about a reference configuration (e.g., robot trajectory). These linear control techniques would be still suitable even though the overall dynamics of the system is nonlinear.

7.9.2.1 State-Space Models

In applying many types of control techniques, it is convenient to represent the vibrating system (plant) by a state-space model. This is simply a set of ordinary first-order differential equations, which could be coupled or nonlinear, and could have time-varying parameters (time-variant models). Here, we limit our discussion to linear and time-invariant state-space models. Such a model is expressed as

$$\dot{\mathbf{x}} = \mathbf{A}\mathbf{x} + \mathbf{B}\mathbf{u} \tag{7.131}$$

$$\mathbf{y} = \mathbf{C}\mathbf{x} + \mathbf{D}\mathbf{u} \tag{7.132}$$

where

$\mathbf{x} = [x_1, x_2, \ldots, x_n]^{\mathrm{T}}$ = state vector (nth order column)
$\mathbf{u} = [u_1, u_2, \ldots, u_r]^{\mathrm{T}}$ = input vector (rth order column)
$\mathbf{y} = [y_1, y_2, \ldots, y_m]^{\mathrm{T}}$ = output vector (mth order column)
$\mathbf{A}$ = system matrix ($n \times n$ square)
$\mathbf{B}$ = input gain matrix ($n \times r$)
$\mathbf{C}$ = measurement gain matrix ($m \times n$)
$\mathbf{D}$ = feedforward gain matrix ($m \times r$)

Usually, for vibrating systems, it is possible to make $\mathbf{D} = 0$, and hence we will drop this matrix in the sequel. Furthermore, although a state variable x_i need not have a direct physical meaning, an output variable y_j should have some physical meaning and, in typical situations, should be measurable as well. The input variables are the "control variables" and are used for controlling the system (plant). The output variables are the "controlled variables," which correspond to the system response and are measured for feedback control.

It can be verified that the eigenvalues of the system matrix $\mathbf{A}$ occur in complex conjugates of the form $-\zeta_i\omega_i \pm j\sqrt{1-\zeta_i^2}\,\omega_i$ in the damped oscillatory case, or as $\pm j\omega_i$ in the undamped case, where ω_i is the ith natural frequency of the system and ζ_i is the corresponding damping ratio (of the ith mode). The mathematical verification requires some linear algebra. An intuitive verification can be made since Equation 7.131 is an equivalent model for a system having the traditional mass–spring–damper model

$$\mathbf{M}\ddot{\mathbf{y}} + \mathbf{C}\dot{\mathbf{y}} + \mathbf{K}\mathbf{y} = \mathbf{f}(t) \tag{7.133}$$

where $\mathbf{M}$ = mass matrix, $\mathbf{C}$ = damping matrix, $\mathbf{K}$ = stiffness matrix, $\mathbf{f}(t)$ = forcing input vector, and $\mathbf{y}$ = displacement response vector. Where both models (Equation 7.131 and Equation 7.133), are equivalent they should have the same characteristic equation, which by its roots determines the natural frequencies and modal damping ratios. This is the case because we are simply looking at two different mathematical representations of the same system. Hence, the parameters of its dynamics, such as ω_i and ζ_i, should remain unchanged. In fact, the state-space mode (Equation 7.131) is not unique, and different versions of state vectors and corresponding models are possible. Of course, all of them should have the

same characteristics polynomial (and hence, the same ω_i and ζ_i). One such state-space model may be derived from Equation 7.133 as follows:
Define the state vector as

$$x = \begin{bmatrix} y \\ \dot{y} \end{bmatrix} \text{ and } \mathbf{u} = \mathbf{f}(t) \tag{7.134}$$

Since (for nonsingular **M**, as required), Equation 7.133 may be written as

$$\ddot{\mathbf{y}} = -\mathbf{M}^{-1}\mathbf{K}\mathbf{y} - \mathbf{M}^{-1}\mathbf{C}\dot{\mathbf{y}} + \mathbf{M}^{-1}\mathbf{f}(t) \tag{7.135}$$

we have

$$\dot{\mathbf{x}} = \begin{bmatrix} \mathbf{0} & \mathbf{I} \\ -\mathbf{M}^{-1}\mathbf{K} & -\mathbf{M}^{-1}\mathbf{C} \end{bmatrix}\mathbf{x} + \begin{bmatrix} \mathbf{0} \\ \mathbf{M}^{-1} \end{bmatrix}\mathbf{u} \tag{7.136}$$

This is a state-space model that is equivalent to the conventional model (Equation 7.133), and can be shown to have the same characteristic equation. The development of a state-space model for a vibrating system can be illustrated using an example.

Example 7.10

Consider a machine mounted on a support structure, modeled as in Figure 7.33. Using the excitation forces $f_1(t)$ and $f_2(t)$ as the inputs and the displacements y_1 and y_2 of the masses m_1 and m_2 as the outputs, develop a state-space model for this system.

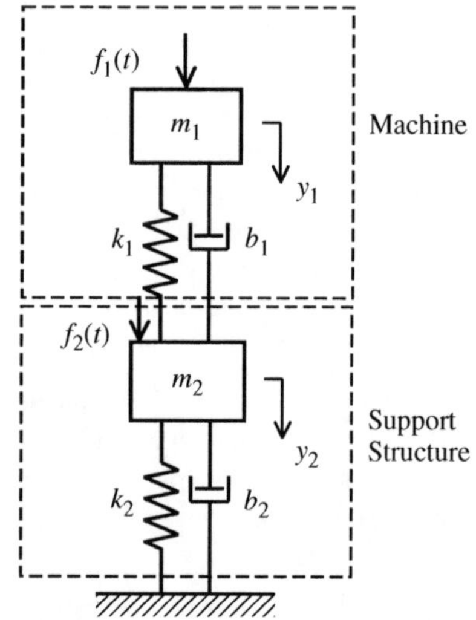

FIGURE 7.33 A model of a machine mounted on support structure.

Solution

Assume that the displacements are measured from the static equilibrium positions of the masses. Hence, the gravity forces do not enter into the formulation. Newton's Second law is applied to the two masses; thus

$$m_1\ddot{y}_1 = f_1 - k_1(y_1 - y_2) - b_1(\dot{y}_1 - \dot{y}_2)$$

$$m_2\ddot{y}_2 = f_2 - k_1(y_2 - y_1) - b_1(\dot{y}_2 - \dot{y}_1) - k_2y_2 - b_2\dot{y}_2$$

The following state variables are defined:

$$x_1 = y_1; \quad x_2 = \dot{y}_1; \quad x_3 = y_2; \quad x_4 = \dot{y}_2$$

Also, the input vector is $\mathbf{u} = [u_1\ u_2]^T$ and the output vector is $\mathbf{y} = [y_1\ y_2]^T$. Then, we have

$$\dot{x}_1 = x_2$$

$$m_1\dot{x}_2 = u_1 - k_1(x_1 - x_3) - b_1(x_2 - x_4)$$

$$\dot{x}_3 = x_4$$

$$m_2\dot{x}_4 = u_2 - k_1(x_3 - x_1) - b_1(x_4 - x_2) - k_2x_3 - b_2x_4$$

Accordingly, the state-space model is given by Equation 7.131 and Equation 7.132 with

$$\mathbf{A}=\begin{bmatrix} 0 & 1 & 0 & 0 \\ -k_1/m_1 & -b_1/m_1 & k_1/m_1 & b_1/m_1 \\ 0 & 0 & 0 & 1 \\ k_1/m_2 & b_1/m_2 & (k_1+k_2)/m_2 & (b_1+b_2)/m_2 \end{bmatrix},\quad \mathbf{B}=\begin{bmatrix} 0 & 0 \\ 1/m_1 & 0 \\ 0 & 0 \\ 0 & 1/m_2 \end{bmatrix},$$

$$\mathbf{C}=\begin{bmatrix} 1 & 0 & 0 & 0 \\ 0 & 0 & 1 & 0 \end{bmatrix},\quad \text{and } \mathbf{D}=0$$

Also, note that the system can be expressed as

$$\begin{bmatrix} m_1 & 0 \\ 0 & m_2 \end{bmatrix}\ddot{\mathbf{y}}+\begin{bmatrix} b_1 & -b_1 \\ -b_1 & (b_1+b_2) \end{bmatrix}\dot{\mathbf{y}}+\begin{bmatrix} k_1 & -k_1 \\ -k_1 & (k_1+k_2) \end{bmatrix}\mathbf{y}=\mathbf{f}(t)$$

Its characteristic equation may be expressed as the determinant equation:

$$\det\begin{bmatrix} m_1s^2+b_1s+k & -b_1s-k_1 \\ -b_1s-k_1 & m_2s^2+(b_1+b_2)s+(k_1+k_2) \end{bmatrix}=0$$

It can be verified through direct expansion of the determinants that this equation is equivalent to the characteristic equation of the matrix **A**, as given by $\det(\lambda\mathbf{I}-\mathbf{A})=0$, or

$$\det\begin{bmatrix} \lambda & -1 & 0 & 0 \\ k_1/m_1 & \lambda+b_1/m_1 & -k_1/m_1 & -b_1/m_1 \\ 0 & 0 & \lambda & -1 \\ k_1/m_2 & -b_1/m_2 & -(k_1+k_2)/m_2 & \lambda-(b_1+b_2)/m_2 \end{bmatrix}=0$$

Note that, in the present context, x and y represent the vibration response of the plant and the control objective is to reduce these to zero. We will give some common control techniques that can achieve this goal.

7.9.2.2 Position and Velocity Feedback

In this technique, the position and velocity of each DoF is measured and fed into the system with sign reversal (negative feedback) and amplification by a constant gain. Because velocity is the derivative of position and since the gains are constant (i.e., proportional), this method falls into the general category of proportional-plus-derivative (PD or PPD) control. In this approach, it is tacitly assumed that the degrees of freedom are uncoupled. Then, control gains are chosen so that the DoF in the controlled system are nearly uncoupled, thereby justifying the original assumption. To explain this control method, suppose that a DoF of a vibrating system is represented by

$$m\ddot{y}+b\dot{y}+ky=u(t) \tag{7.137}$$

where y is the displacement (position) of the DoF and u is the excitation input that is applied. Now suppose that u is generated according to the (active) control law

$$u=-k_cy-b_c\dot{y}+u_r \tag{7.138}$$

where k_c is the position feedback gain and b_c is the velocity feedback gain. The implication here is that the position y and the velocity $\dot{y}$ are measured and fed into the controller which in turn generates u according to Equation 7.138. Also, u_r is some reference input that is provided externally to the controller.

Then, substituting Equation 7.138 into Equation 7.137, we obtain

$$m\ddot{y} + (b + b_c)\dot{y} + (k + k_c)y = u_r \tag{7.139}$$

The closed-loop system (the controlled system) now behaves according to Equation 7.139. The control gains b_c and k_c can be chosen arbitrarily (subject to the limitations of the physical controller, signal conditioning circuitry, the actuator, etc.) and may even be negative. In particular, by increasing b_c, the damping of the system can be increased. Similarly, by increasing k_c the stiffness (and the natural frequency) of the system can be increased. Even though a passive spring and damper with stiffness k_c and damping constant b_c can accomplish the same task, once the devices are chosen it is not possible to conveniently change their parameters. Furthermore, it will not be possible to make k_c or b_c negative in this case of passive physical devices. The method of PPD control is simple and straightforward, but the assumptions of linear uncoupled DoF place a limitation on its general use.

7.9.2.3 Linear Quadratic Regulator Control

This is an optimal control technique. Consider a vibrating system that is represented by the linear state-space model:

$$\dot{\mathbf{x}} = \mathbf{A}\mathbf{x} + \mathbf{B}\mathbf{u} \tag{7.131}$$

Assume that all the states $\mathbf{x}$ are measurable and all the system modes are controllable. Then, we use the constant-gain feedback control law:

$$\mathbf{u} = \mathbf{K}\mathbf{x} \tag{7.140}$$

The choice of parameter values for the feedback gain matrix $\mathbf{K}$ is infinite. Therefore, we can use this freedom to minimize the cost function:

$$J = \frac{1}{2}\int_t^{\infty} [\mathbf{x}^{\mathrm{T}}\mathbf{Q}\mathbf{x} + \mathbf{u}^{\mathrm{T}}\mathbf{R}\mathbf{u}]\mathrm{d}\tau \tag{7.141}$$

This is the time integral of a quadratic function in both state and input variables, and the optimization goal may be interpreted as bringing $\mathbf{x}$ down to zero (regulating $\mathbf{x}$ to 0), but without spending a rather high control effort. Hence, the name linear quadratic regulation (LQR). In addition, $\mathbf{Q}$ and $\mathbf{R}$ are weighting matrices, with the former being at least positive semidefinite and the latter positive definite. Typically, $\mathbf{Q}$ and $\mathbf{R}$ chosen as diagonal matrices with positive diagonal elements whose magnitudes are determined by the degree of relative emphasis that should be given to various elements of $\mathbf{x}$ and $\mathbf{u}$. It is well known that $\mathbf{K}$ that minimizes the cost function (Equation 7.141) is given by

$$\mathbf{K} = -\mathbf{R}^{-1}\mathbf{B}^{\mathrm{T}}\mathbf{K}_{\mathrm{r}} \tag{7.142}$$

where $\mathbf{K}_{\mathrm{r}}$ is the positive-definite solution of the matrix Riccati algebraic equation

$$\mathbf{K}_{\mathrm{r}}\mathbf{A} + \mathbf{A}^{\mathrm{T}}\mathbf{K}_{\mathrm{r}} - \mathbf{K}_{\mathrm{r}}\mathbf{B}\mathbf{R}^{-1}\mathbf{B}^{\mathrm{T}}\mathbf{K}_{\mathrm{r}} + \mathbf{Q} = 0 \tag{7.143}$$

It is also known that the resulting closed-loop control system is stable. Furthermore, the minimum (optimal) value of the cost function (Equation 7.141) is given by

$$J_{\mathrm{m}} = \frac{1}{2}\mathbf{x}^{\mathrm{T}}\mathbf{K}_{\mathrm{r}}\mathbf{x} \tag{7.144}$$

where $\mathbf{x}$ is the present value of the state vector. Major computational burden of the LQR method is in the solution (Equation 7.143). Other limitations of the technique arise due to the need to measure all the state variables (which may be relaxed to some extent). Although stability of the controlled system is guaranteed, the level of stability that is achieved (i.e., stability margin or the level of modal damping) cannot be directly specified. Further, robustness of the control system in the presence of model errors, unknown disturbances and so on, may be questionable. Besides, the cost function incorporates an integral over an infinite time duration, which does not typically reflect the practical requirement of rapid vibration control.

7.9.2.4 Modal Control

The LQR control technique has the serious limitation of not being able to directly achieve specified levels of modal damping, which may be an important goal in vibration control. The method of modal control that accomplishes this objective is *pole placement*, where poles (eignevalues) of the controlled system are placed at specified values. Specifically, consider the plant (Equation 7.131) and the feedback control law (Equation 7.140). Then, the closed-loop system is given by

$$\dot{\mathbf{x}} = (\mathbf{A} + \mathbf{BK})\mathbf{x} \tag{7.145}$$

It is well known that if the plant (**A**, **B**) is controllable, then a control gain matrix **K** can be chosen that will arbitrarily place the eigenvalues of the closed-loop system matrix **A** + **BK**. Based on the given assumptions, the modal control technique assigns not only the modal damping but also the damped natural frequencies at specified values. The assumptions given above are quite stringent but they can be relaxed to some degree. However, a shortcoming of this method is the fact that it does not place a restriction on the control effort, for example, as the LQR technique does, in achieving a specified level of modal control.

7.10 Control of Beam Vibrations

Beam is a distributed-parameter system, which in theory has an infinite number of modes of vibration with associated mode shapes and natural frequencies. In this sense, it is an "infinite order" system with infinite DoF. Hence, the computation of modal quantities and associated control inputs can be quite complex. Fortunately, however, just a few modes may be retained in a dynamic model without sacrificing a great deal of accuracy, thereby facilitating simpler control. Some concepts of controlling vibrations in a beam are considered in this section. The present treatment is intended as an illustration of the relevant techniques and is not meant to be exhaustive. These techniques may be extended to other types of continuous system such as beams with different boundary conditions and plates. Because the control techniques that were outlined previously depend on a model, we will first illustrate the procedure of obtaining a state-space model for a beam.

7.10.1 State-Space Model of Beam Dynamics

Consider a Bernoulli–Euler-type beam with Kelvin–Voigt-type internal (material) damping. The beam equation may be expressed as

$$ELv(x,t) + E^{*}L\frac{\partial v(x,t)}{\partial t} + \rho A(x)\frac{\partial^2 v(x,t)}{\partial t^2} = f(x,t) \tag{7.146}$$

in which L is the partial differential operator given by

$$L = \frac{\partial^2 I(x)}{\partial x^2}\frac{\partial^2}{\partial x^2} \tag{7.147}$$

and

$f(x,t)$ = distributed force excitation per unit length of the beam
$v(x,t)$ = displacement response at location x along the beam at time t
$I(x)$ = second moment of area of the beam cross section about the neutral axis
E = Young's modulus of the beam material
E^{*} = Kelvin–Voigt material damping parameter

Note that a general beam with nonuniform characteristics is assumed and, hence, the variations of $I(x)$ and $\rho A(x)$ with x are retained in the formulation.

Using the approach of modal expansion, the response of the beam may be expressed by

$$v(x, t) = \sum_{i=1}^{\infty} Y_i(x) q_i(t) \tag{7.148}$$

where $Y_i(x)$ is the ith mode shape of the beam, which satisfies

$$LY_i(x) = \frac{\rho A(x)}{E} \omega_i^2 Y_i(x) \tag{7.149}$$

and ω_i is the ith undamped natural frequency. The orthogonality condition for this general example of a nonuniform beam is

$$\int_{x=0}^{l} \rho A Y_i Y_j \mathrm{d}x = \begin{cases} 0 & \text{for } i \neq j \\ \alpha_j & \text{for } i = j \end{cases} \tag{7.150}$$

Suppose that the forcing excitation on the beam is a set of r point forces $u_k(t)$ located at $x = l_k$, $k = 1, 2, \ldots, r$. Then, we have

$$f(x, t) = \sum_{k=1}^{r} u_k \delta(x - l_k) \tag{7.151}$$

where $\delta(x - l_i)$ is the *Dirac delta function.* Now, substitute Equation 7.148 and Equation 7.151 into Equation 7.146, use Equation 7.149, multiply throughout by $Y_j(x)$, and integrate over $x[0, l]$, using Equation 7.150. This gives

$$\ddot{q}_j + \gamma_j \dot{q}_j(t) + \omega_j^2 q_j = \frac{1}{\alpha_j} \sum_{k=1}^{r} u_k Y_j(l_k) \quad \text{for } j = 1, 2, \ldots \tag{7.152}$$

where

$$\gamma_j = \frac{E^*}{E} \omega_j^2 \tag{7.153}$$

Now, define the state variables x_j according to

$$x_{2j-1} = \omega_j q_j, \quad x_{2j} = \dot{q}_j \quad \text{for } j = 1, 2, \ldots \tag{7.154}$$

Assuming that only the first m modes are retained in the expansion, we then have the state equations

$$\dot{x}_{2j-1} = \omega_j x_{2j}, \quad \dot{x}_{2j} = -\omega_j x_{2j-1} - \gamma_j x_{2j} + \frac{1}{\alpha_j} \sum_{k=1}^{r} u_k Y_j(l_k) \quad \text{for } j = 1, 2, \ldots, m \tag{7.155}$$

This can be put in the matrix–vector form of a state-space model

$$\dot{\mathbf{x}} = \mathbf{A}\mathbf{x} + \mathbf{B}\mathbf{u} \tag{7.131}$$

where

$$\mathbf{A} = \begin{bmatrix} 0 & \omega_1 & & & \\ -\omega_1 & -\gamma_1 & & 0 & \\ & & \ddots & & \\ & 0 & & 0 & \omega_m \\ & & & -\omega_m & -\gamma_m \end{bmatrix}_{n \times n} \tag{7.156}$$

and

$$\mathbf{B} = \begin{bmatrix} 0 & & 0 \\ Y_1(l_1)/\alpha_1 & \cdots & Y_1(l_r)/\alpha_1 \\ \vdots & & \vdots \\ 0 & & 0 \\ Y_\mathrm{m}(l_1)/\alpha_\mathrm{m} & \cdots & Y_\mathrm{m}(l_r)/\alpha_\mathrm{m} \end{bmatrix}_{n\times r} \tag{7.157}$$

with $n = 2m$, where m is the number of modes retained in the modal expansion. Note that, as the number of modes used in this model increases, both the accuracy and the computational effort that is needed for the control problem increase because of the proportional increase of the system order. At some point, the potential improvement in accuracy by further increasing the model size would be insignificant in comparison with added computational burden. Hence, a balance must be struck in this tradeoff.

7.10.2 Control Problem

The state-space model (Equation 7.131) for the beam dynamics, with matrices (Equation 7.156 and Equation 7.157), is known to be *controllable*. Hence, it is possible to determine a constant-gain feedback controller $\mathbf{u} = \mathbf{Kx}$ that minimizes a quadratic-integral cost function of the form in Equation 7.141. Also, a similar controller can be determined that places the eigenvalues of the system at specified locations thereby achieving not only specified levels of modal damping but also a specified set of natural frequencies. However, there is a practical obstacle to achieving such an active controller. Note that, in the model given in Equation 7.156 and Equation 7.157, the state variables are proportional to the modal variables q_i and their time derivatives $\dot{q}_i$. They are not directly measurable. However, the displacements and velocities at a set of discrete locations along the beam can usually be measured. Let these locations (s) be denoted by $p_1, p_2, \ldots, p_s$. Thus, in view of the modal expansion (Equation 7.148), the measurements can be expressed as

$$v(p_j, t) = \sum_{i=1}^{m} Y_i(p_j)q_i(t), \quad \dot{v}(p_j, t) = \sum_{i=1}^{m} Y_i(p_j)\dot{q}_i(t) \quad \text{for } j = 1, 2, \ldots, s \tag{7.158}$$

Now, define the output (measurement) vector $\mathbf{y}$ according to

$$y = [v(p_1, t), \dot{v}(p_1, t), \ldots, v(p_s, t), \dot{v}(p_s, t)]^\mathrm{T} \tag{7.159}$$

In view of Equation 7.158 and the definitions of the state variable in Equation 7.154, we can write

$$\mathbf{y} = \mathbf{Cx} \tag{7.160}$$

with

$$C = \begin{bmatrix} Y_1(p_1)/\omega_1 & 0 & \cdots & Y_\mathrm{m}(p_1)/\omega_\mathrm{m} & 0 \\ 0 & Y_1(p_1) & \cdots & 0 & Y_\mathrm{m}(p_1) \\ \vdots & \vdots & \cdots & \vdots & \vdots \\ Y_1(p_s)/\omega_1 & 0 & \cdots & Y_\mathrm{m}(p_s)/\omega_\mathrm{m} & 0 \\ 0 & Y_1(p_s) & \cdots & 0 & Y_\mathrm{m}(p_s) \end{bmatrix}_{2s\times n} \tag{7.161}$$

Hence, an active controller is possible of the form:

$$\mathbf{u} = \mathbf{Hy} \tag{7.162}$$

which is an output feedback controller. Therefore, in view of Equation 7.160, we have

$$\mathbf{u} = \mathbf{HCx} \tag{7.163}$$

This is not the same as complete state feedback $\mathbf{u} = \mathbf{Kx}$ where $\mathbf{K}$ can take any real value (and, hence, the LQR solution in Equation 7.142 and the complete pole placement solution cannot be applied directly). In Equation 7.163, only $\mathbf{H}$ can be arbitrarily chosen, and $\mathbf{C}$ is completely determined according to Equation 7.161. The resulting product $\mathbf{HC}$ will not usually correspond to either the LQR solution or the complete pole assignment solution. Still, the output feedback controller in Equation 7.162 can provide a satisfactory performance. However, a sufficient number of displacement and velocity sensors (s) have to be used in conjunction with a sufficient number of actuators (r) for active control. This will increase the system complexity and cost. Furthermore, due to added components and their active nature, the reliability of fault-free operation may degrade somewhat. A satisfactory alternative would be to use passive control devices such as dampers and dynamic absorbers, which is illustrated below. Note that in the matrices $\mathbf{B}$ and $\mathbf{C}$ given by Equation 7.157 and Equation 7.161, both the actuator locations l_i and the sensor locations p_j are variable. Hence, there exists an additional design freedom (or optimization parameters) in selecting the sensor and actuator locations in achieving satisfactory control.

7.10.3 Use of Linear Dampers

Now, consider the use of a discrete set of linear dampers for controlling beam vibration. Suppose that r linear dampers with damping constants b_j are placed at locations l_j, $j = 1, 2, \ldots, r$ along the beam, as schematically shown in Figure 7.34. The damping forces are given by

$$u_j = -b_j\dot{v}(l_j, t) \quad \text{for } j = 1, 2, \ldots, r \tag{7.164}$$

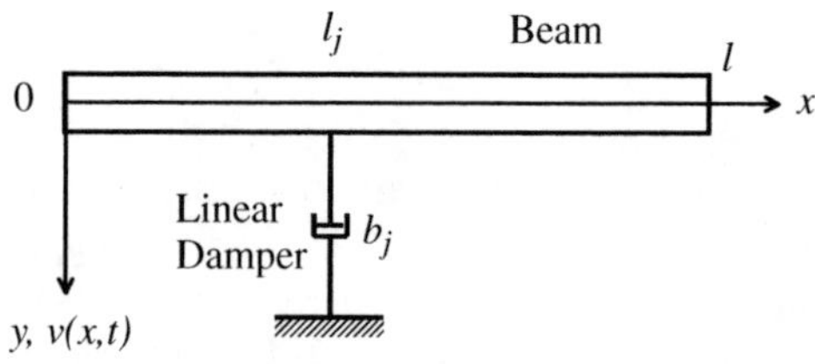

FIGURE 7.34 Use of linear dampers in beam vibration control.

Substituting the truncated modal expansion (m modes)

$$\dot{v}(l_j, t) = \sum_{i=1}^{m} Y_i(l_j)\dot{q}_i(t) \tag{7.165}$$

we get, in view of Equation 7.154, the passive feedback control action

$$\mathbf{u} = -\mathbf{Kx} \tag{7.166}$$

with

$$\mathbf{K} = \begin{bmatrix} 0 & b_1 Y_1(l_1) & \cdots & 0 & b_1 Y_m(l_1) \\ \vdots & \vdots & \cdots & \vdots & \vdots \\ 0 & b_r Y_1(l_r) & \cdots & 0 & b_r Y_m(l_r) \end{bmatrix}_{r\times n} \tag{7.167}$$

By substituting Equation 7.166 into Equation 7.131, we have the closed-loop system equation

$$\dot{\mathbf{x}} = (\mathbf{A} - \mathbf{F})\mathbf{x} = \mathbf{A}_c\mathbf{x} \tag{7.168}$$

where $\mathbf{F} = \mathbf{BK}$ and is given by

$$\mathbf{F} = \begin{bmatrix} 0 & 0 & \cdots & 0 & 0 \\ 0 & \sum b_i Y_{11}(l_i)/\alpha_1 & \cdots & 0 & \sum b_i Y_{1m}(l_i)/\alpha_1 \\ \vdots & \vdots & \cdots & \vdots & \vdots \\ 0 & 0 & \cdots & 0 & 0 \\ 0 & \sum b_i Y_{m1}(l_i)/\alpha_m & \cdots & 0 & \sum b_i Y_{mm}(l_i)/\alpha_m \end{bmatrix}_{n \times n} \tag{7.169}$$

with

$$Y_{ij}(x) = Y_i(x) Y_j(x) \tag{7.170}$$

In this case, the controller design involves the selection of the damping constants b_i and the damper locations l_j to achieve the required performance. This may be achieved, for example, by seeking to make the eigenvalues of the closed-loop system matrix $\mathbf{A}_c$ reach a set of desired values. This achieves the desired modal damping and natural frequency characteristics. However, given that the structure of the **F** matrix is fixed, as seen in Equation 7.169, this is not equivalent to complete state feedback (or complete output feedback). Hence, generally, it is not possible to place the poles of the system at the exact desired locations.

7.10.3.1 Design Example

In realizing a desirable modal response of a beam using a set of linear dampers, one may seek to minimize a cost function of the form

$$J = \mathrm{Re}(\boldsymbol{\lambda} - \boldsymbol{\lambda}_d)^T \mathbf{Q}\, \mathrm{Re}(\boldsymbol{\lambda} - \boldsymbol{\lambda}_d) + \mathrm{Im}(\boldsymbol{\lambda} - \boldsymbol{\lambda}_d)^T \mathbf{R}(\boldsymbol{\lambda} - \boldsymbol{\lambda}_d) \tag{7.171}$$

where $\boldsymbol{\lambda}$ are the actual eigenvalues of the closed-loop system matrix ($\mathbf{A}_c$), and $\boldsymbol{\lambda}_d$ are the desired eigenvalues that will give the required modal performance (damping ratios and natural frequencies). "Re" denotes the real part and "Im" denotes the imaginary part. Weighting matrices **Q** and **R**, which are real and diagonal with positive diagonal elements, should be chosen to relatively weight various eigenvalues. This allows the emphasis of some eigenvalues over others, with real parts and the imaginary parts weighting separately.

Various computational algorithms are available for minimizing the cost function (Equation 7.171). Although the precise details are beyond the scope of this book, we will present an example result. Consider a uniform simply supported 12×5 American Standard beam, with the following pertinent specifications: $E = 2 \times 10^8$ kPa (29×10^6 psi), $\rho A = 47$ kg/m (2.6 lb/in.), length $l = 15.2$ m (600 in.), $I = 9 \times 10^{-5}$ m^4 (215.8 in.^{4}). The internal damping parameter for the jth mode of vibration is given by

$$E^*(\omega_j) = (g_1/\omega_j) + g_2 \tag{7.172}$$

in which ω_j is the jth undamped natural frequency given by

$$\omega_j = (j\pi/l)^2 \sqrt{EI/\rho A} \tag{7.173}$$

The numerical values used for the damping parameters are $g_1 = 88 \times 10^4$ kPa (12.5×10^4 psi) and $g_2 = 3.4 \times 10^4$ kPa s (5×10^3 psi s). For the present problem, $Y_i(x) = \sqrt{2}\sin(j\pi x/l)$ and $\alpha_j = \rho A l$ for all j.

First, ω_j and γ_j are computed using Equation 7.173 and Equation 7.153, respectively, along with Equation 7.172. Next, the open-loop system matrix **A** is formed according to Equation 7.156 and its eigenvalues are computed. These are listed in Table 7.2, scaled to the first undamped natural frequency (ω_1). Note that in view of the very low levels of internal material damping of the beam, the actual natural frequencies, as given by the imaginary parts of the eigenvalues, are almost identical to the undamped natural frequencies.

Next, we attempt to place the real parts of the (scaled) eigenvalues all at -0.20 while exercising no constraint on the imaginary parts (i.e., damped natural frequencies) by using: (a) single damper, and

TABLE 7.2 Eigenvalues of the Open-Loop (Uncontrolled) Beam

Mode	Eigenvalue (rad/sec) (Multiply by 26.27)
1	− 0.000126 ± j1.0
2	− 0.000776 ± j4.0
3	− 0.002765 ± j9.0
4	− 0.007453 ± j16.0
5	− 0.016741 ± j25.0
6	− 0.033.75 ± j36.0

(b) two dampers. In the cost function (Equation 7.171), the first three modes are more heavily weighted than the remaining three. Initial values of the damper parameters are $b_1 = b_2 = 0.1$ lbf s/in. (17.6 N s/m) and the initial locations $l_1/l = 0.0$ and $l_2/l = 0.5$. At the end of the numerical optimization, using a modified gradient algorithm, the following optimized values were obtained:

1. Single-damper control

$$b_1 = 36.4 \text{ lbf s/in. } (6.4 \times 10^3 \text{ N s/m})$$

$$l_1/l = 0.3$$

The corresponding normalized eigenvalues (of the closed-loop system) are given in Table 7.3.

2. Two-damper control

$$b_1 = 22.8 \text{ lbf s/in. } (4.0 \times 10^3 \text{ N s/m})$$

$$b_2 = 12.1 \text{ lbf s/in. } (2.1 \times 10^3 \text{ N s/m})$$

$$l_1/l = 0.25, \quad l_2/l = 0.43$$

The corresponding normalized eigenvalues are given in Table 7.4.

It would be overly optimistic to expect perfect assignment all real parts at − 0.2. However, note that good levels of damping have been achieved for all modes except for Mode 3 in the single-damper control and Mode 4 in the two-damper control. In any event, because the contribution of the higher modes towards the overall response, is relatively smaller, it is found that the total response (e.g., at point $x = l/12$) is well damped in both cases of control.

TABLE 7.3 Eigenvalues of the Beam with an Optimal Single Damper

Mode	Eigenvalue (rad/s) (Multiply by 26.27)
1	− 0.225 ± j0.985
2	− 0.307 ± j3.955
3	− 0.037 ± j8.996
4	− 0.119 ± j15.995
5	− 0.355 ± j24.980
6	− 0.158 ± j35.990

TABLE 7.4 Eigenvalues of the Beam with Optimized Two Dampers

Mode	Eigenvalue (rad/s) (Multiply by 26.27)
1	− 0.216 ± j0.982
2	− 0.233 ± j3.974
3	− 0.174 ± j8.997
4	− 0.079 ± j15.998
5	− 0.145 ± j24.999
6	− 0.354 ± j35.989

Bibliography

Beards, C.F. 1996. *Engineering Vibration Analysis with Application to Control Systems*, Halsted Press, New York.

Cao, Y., Modi, V.J., de Silva, C.W., and Misra, A.K., On the control of a novel manipulator with slewing and deployable links, *Acta Astronaut.*, 49, 645–658, 2001.

Caron, M., Modi, V.J., Pradhan, S., de Silva, C.W., and Misra, A.K., Planar dynamics of flexible manipulators with slewing deployable links, *J. Guid. Control Dyn.*, 21, 572–580, 1998.

Chen, Y., Wang, X.G., Sun, C., Devine, F., and de Silva, C.W., Active vibration control with state feedback in woodcutting, *J. Vibr. Control*, 9, 645–664, 2003.

den Hartog, J.P., 1956. *Mechanical Vibrations*, Mc-Graw-Hill, NewYork.

de Silva, C.W., Optimal estimation of the response of internally damped beams to random loads in the presence of measurement noise, *J. Sound Vibr.*, 47, 485–493, 1976.

de Silva, C.W., An algorithm for the optimal design of passive vibration controllers for flexible systems, *J. Sound Vibr.*, 74, 495–502, 1982.

de Silva, C.W. and Wormley, D.N. 1983. *Automated Transit Guideways: Analysis and Design*, D.C. Heath & Co., Lexington, KY.

de Silva, C.W. 1989. *Control Sensors and Actuators*, Prentice Hall, Englewood Cliffs, NJ.

de Silva, C.W. 1995. *Intelligent Control: Fuzzy Logic Applications*, CRC Press, Boca Raton, FL.

de Silva, C.W. 2005. *Mechatronics—an Integrated Approach*, Taylor & Francis, CRC Press, Boca Raton, FL.

de Silva, C.W. 2006. *Vibration—Fundamentals and Practice*, 2nd ed., Taylor & Francis, CRC Press, Boca Raton, FL.

Goulet, J.F., de Silva, C.W., Modi, V.J., and Misra, A.K., Hierarchical control of a space-based deployable manipulator using fuzzy logic, *AIAA J. Guid. Control Dyn.*, 24, 395–405, 2001.

Irwin, J.D. and Graf, E.R. 1979. *Industrial Noise and Vibration Control*, Prentice Hall, Englewood Cliffs, NJ.

Karray, F. and de Silva, C.W. 2004. *Soft Computing and Intelligent System Design: Theory, Tools, and Applications*, Pearson, England.

MATLAB *Control Systems Toolbox*, The MathWorks, Inc., Natick, MA, 2004.

Van de Vegte, J. and de Silva, C.W., Design of passive vibration controls for internally damped beams by modal control techniques, *J. Sound Vibr.*, 45, 417–425, 1976.

Appendix 7A

MATLAB® Control Systems Toolbox

7A.1 Introduction

Modeling, analysis, design, data acquisition, and control are important activities within the field of vibration. Computer software tools and environments are available for effectively carrying out, both at the learning level and at the professional application level. Several such environments and tools are commercially available.

MATLAB[1] is an interactive computer environment with a high-level language and tools for scientific and technical computation, modeling and simulation, design, and control of dynamic systems. SIMULINK[1] is a graphical environment for modeling, simulation, and analysis of dynamic systems, and is available as an extension to MATLAB. The Control Systems Toolbox of MATLAB is suitable in the analysis, design, and control of mechanical vibrating systems.

[1]MATLAB and SIMULINK are registered trademarks and products of The MathWorks, Inc. LabVIEW is a product of National Instruments, Inc.

7A.2 MATLAB

MATLAB interactive computer environment is very useful in computational activities in Mechatronics. Computations involving scalars, vectors, and matrices can be carried out and the results can be graphically displayed and printed. MATLAB toolboxes are available for performing specific tasks in a particular area of study such as control systems, fuzzy logic, neural network, data acquisition, image processing, signal processing, system identification, optimization, model predictive control, robust control, and statistics. User guides, Web-based help, and on-line help from the parent company, MathWorks, Inc., and various other sources. What is given here is a brief introduction to get started in MATLAB for tasks that are particularly related to Control Systems and Mechatronics.

7A.2.1 Computations

Mathematical computations can be done by using the MATLAB command window. Simply type in the computations against the MATLAB prompt ">>" as illustrated next.

7A.2.2 Arithmetic

An example of a simple computation using MATLAB is given below:

```
>>x = 2; y = -3;
>>z = x^2 - x * y + 4
z =
   14
```

In the first line, we have assigned values 2 and 3 to two variables x and y. In the next line, the value of an algebraic function of these two variables is indicated. Then, MATLAB provides the answer as 14. Note that if you place a ";" at the end of the line, the answer will not be printed/displayed.

Table 7A.1 gives the symbols for common arithmetic operations used in MATLAB.

Following example shows the solution of the quadratic equation $ax^2 + bx + c = 0$:

```
>>a = 2; b = 3; c = 4;
>>x = (-b + sqrt(b^2 - 4*a*c))/(2*a)
x =
 - 0.7500 + 1.1990i
```

The answer is complex, where i denotes $\sqrt{-1}$. Note that the function sqrt( ) is used, which provides the positive root only. Some useful mathematical functions are given in Table 7A.2.

7A.2.3 Arrays

An array may be specified by giving the start value, increment, and the end value limit. An example is given below.

```
>>x = (0.9: - 0.1:0.42)
x =

0.9000   0.8000   0.7000   0.6000   0.5000
```

TABLE 7A.1 MATLAB Arithmetic Operations

Symbol	Operation
+	Addition
−	Subtraction
*	Multiplication
/	Division
^	Power

TABLE 7A.2 Useful Mathematical Functions in MATLAB

Function	Description
abs( )	Absolute value/magnitude
acos( )	Arc-cosine (inverse cosine)
acosh( )	Arc-hyperbolic-cosine
asin( )	Arc-sine
atan( )	Arc-tan
cos( )	Cosine
cosh( )	Hyperbolic cosine
exp( )	Exponential function
imag( )	Imaginary part of a complex number
log( )	Natural logarithm
log10( )	Log to base 10 (common log)
real( )	Real part of a complex number
sign( )	Signum function
sin( )	Sine
sqrt( )	Positive square root
tan( )	Tan function

Note that MATLAB is case sensitive.

The entire array may be manipulated. For example, all the elements are multiplied by π as below:

```
>>x = x*pi
x =
2.8274    2.5133    2.1991    1.8850    1.5708
```

The second and the fifth elements are obtained by

```
>>x([2 5])
ans =
   2.5133    1.5708
```

Next, we form a new array y using x, and then plot the two arrays, as shown in Figure 7A.1:

```
>>y = sin(x);
>>plot(x,y)
```

A polynomial may be represented as an array of its coefficients. For example, the quadratic equation $ax^2 + bx + c = 0$ as given before, with $a = 2$, $b = 3$, and $c = 4$, may be solved using the function "roots" as below:

```
>> p = [2   3   4];
>>roots(p)
ans =
  -0.7500 + 1.1990i
  -0.7500 - 1.1990i
```

The answer is the same as we obtained before.

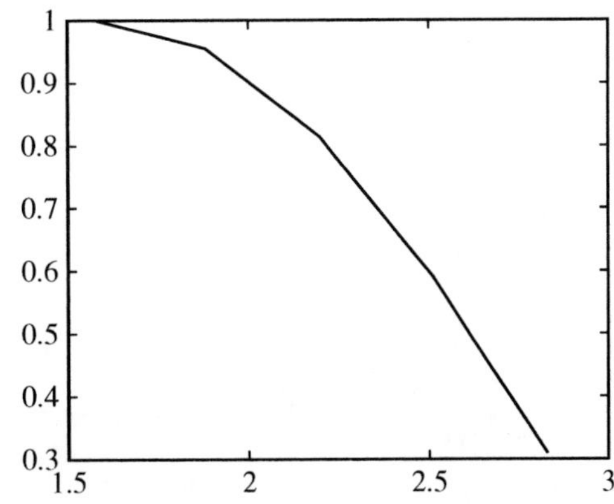

FIGURE 7A.1 A plot using MATLAB.

7A.2.4 Relational and Logical Operations

Useful relational operations in MATLAB are given in Table 7A.3. Basic logical operations are given in Table 7A.4.

TABLE 7A.3 Some Relational Operations

Operator	Description
<	Less than
<=	Less than or equal to
>	Greater than
>=	Greater than or equal to
==	Equal to
~=	Not equal to

TABLE 7A.4 Basic Logical Operations

Operator	Description
&	AND
|	OR
~	NOT

Consider the following example:

```
>> x = (0:0.25:1) * pi
x =
    0   0.7854   1.5708   2.3562   3.1416
>> cos(x) > 0
ans =
    1   1   1   0   0
>> (cos(x) > 0)&(sin(x) > 0)
ans =
    0   1   1   0   0
```

In this example, first an array is computed. Then the cosine of each element is computed. Next it is checked whether the elements are positive. (A truth value of 1 is sent out if true and a truth value of 0 if false.) Finally, the "AND" operation is used to check whether both corresponding elements of two arrays are positive.

7A.2.5 Linear Algebra

MATLAB can perform various computations with vectors and matrices (see Appendix 3A and Appendix 6A). Some basic illustrations are given here.

A vector or a matrix may be specified by assigning values to its elements. Consider the following example:

```
>> b = [1.5  -2];
>> A = [2  1;  -1  1];
>> b = b'
b =
      1.5000
     -2.0000
>> x = inv(A) * b
x =
      1.1667
     -0.8333
```

In this example, first a second-order row vector and 2×2 matrix are defined. The row vector is transposed to get a column vector. Finally the matrix–vector equation $\mathbf{Ax} = \mathbf{b}$ is solved according to $\mathbf{x} = \mathbf{A}^{-1}\mathbf{b}$. The determinant and the eigenvalues of $\mathbf{A}$ are determined by

TABLE 7A.5 Some Matrix Operations in MATLAB

Operation	Description
+	Addition
−	Subtraction
*	Multiplication
/	Division
^	Power
'	Transpose

TABLE 7A.6 Useful Matrix Functions in MATLAB

Function	Description
det( )	Determinant
inv( )	Inverse
eig( )	Eigenvalues
[,] = eig( )	Eigenvectors and eigenvalues

```
>> det(A)
ans =
   3
>> eig(A)
ans =
    1.5000 + 0.8660i
    1.5000 - 0.8660i
```

Both eigenvectors and eigenvalues of **A** computed as

```
>>[V,P] = eig(A)
V =
          0.7071                0.7071
    -0.3536 + 0.6124i    -0.3536 - 0.6124i
P =
    1.5000 + 0.8660i          0
          0             1.5000 - 0.8660i
```

Here, the symbol **V** is used to denote the matrix of eigenvectors. The symbol **P** is used to denote the diagonal matrix whose diagonal elements are the eigenvalues.

Useful matrix operations in MATLAB are given in Table 7A.5 and several matrix functions are given in Table 7A.6.

7A.2.6 M-Files

The MATLAB commands have to be keyed in on the command window, one by one. When several commands are needed to carry out a task, the required effort can be tedious. Instead, the necessary commands can be placed in a text file, edited as appropriate (using text editor), which MATLAB can use to execute the complete task. Such a file is called an M-file. The file name must have the extension "m" in the form *filename.m*. A toolbox is a collection of such files, for use in a particular application area (e.g., control systems, fuzzy logic). Then, by keying in the M-file name at the MATLAB command prompt, the file will be executed. The necessary data values for executing the file have to be assigned beforehand.

7A.3 Control Systems Toolbox

There are several toolboxes with MATLAB, which can be used to analyze, compute, simulate, and design control problems. Both time-domain representations and frequency-domain representations can be used. Also, both classical and modern control problems can be handled. The application is illustrated here through several control problems.

7A.3.1 MATLAB Modern Control Examples

Several examples in modern control engineering are given now to illustrate the use of MATLAB in control.

7A.3.1.1 *Pole Placement of a Third-Order Plant*

A mechanical plant is given by the input–output differential equation $\dddot{x} + \ddot{x} = u$, where u is the input and x is the output. Determine a feedback law that will yield approximately a simple oscillator with a damped natural frequency of 1 unit and a damping ratio of $1/\sqrt{2}$.

To solve this problem, first we define the state variables as $x_1 = x$, $x_2 = \dot{x}_1$, and $x_3 = \dot{x}_2$. The corresponding state-space model is

$$\dot{\mathbf{x}} = \begin{bmatrix} \dot{x}_1 \\ \dot{x}_2 \\ \dot{x}_3 \end{bmatrix} = \underbrace{\begin{bmatrix} 0 & 1 & 0 \\ 0 & 0 & 1 \\ 0 & 0 & -1 \end{bmatrix}}_{\mathbf{A}} \begin{bmatrix} x_1 \\ x_2 \\ x_3 \end{bmatrix} + \underbrace{\begin{bmatrix} 0 \\ 0 \\ 1 \end{bmatrix}}_{\mathbf{B}} u$$

$$y = \underbrace{[1 \quad 0 \quad 0]}_{\mathbf{C}} \mathbf{x}$$

The open-loop poles and zeros are obtained using the following MATLAB commands:

```
>> A = [0 1 0; 0 0 1; 0 0 − 1];
>> B = [0; 0; 1];
>> C = [1 0 0];
>> D = [0];
>> sys_open = ss(A,B,C,D);
>> [nat_freq_open,damping_open,poles_open] = damp(sys_open)
>> pzmap(sys_open)
```

The open-loop poles are: $[0\ 0\ -1]^T$.
The step response of the open-loop system is obtained using the command:

```
>> step(sys_open)
```

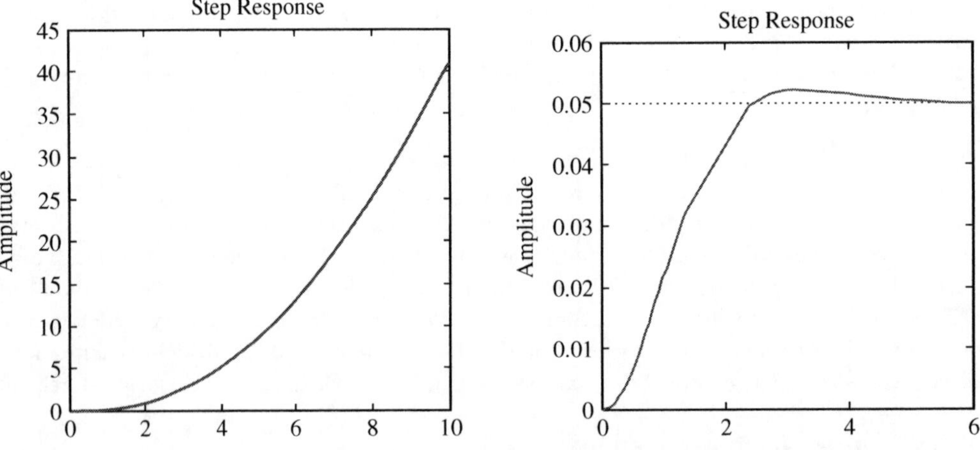

FIGURE 7A.2 (a) Step response of the open-loop system; (b) step response of the third-order system with pole-placement control.

The result is shown in Figure 7A.2(a). Clearly, the system is unstable.

With the desired damped natural frequency $\omega_d = 1$ and damping ratio $\zeta = 1/\sqrt{2}$, we get the undamped natural frequency $\omega_n = \sqrt{2}$ and, hence, $\zeta\omega_n = 1$. It follows that we need to place two poles at $-1 \pm j$. Also the third pole has to be far from these two on the left half plane (LHP); say, at -10. The corresponding control gain K can be computed using the "place" command in MATLAB:

```
>>p = [ − 1 + j − 1 − j − 10];
>>K = place(A,B,p)
place:ndigits = 15
K =
     20.0000    22.0000    11.0000
```

The corresponding step response of the closed-loop system is shown in Figure 7A.2(b).

7A.3.1.2 Linear Quadratic Regulator for a Third-Order Plant

For the third-order plant in the previous example, we design a linear quadratic regulator (LQR), which has a state feedback controller, using MATLAB Control Systems Toolbox. The MATLAB command K = lqr(A,B,Q,R) computes the optimal gain matrix **K** such that the state-feedback law $\mathbf{u} = -\mathbf{Kx}$ minimizes the quadratic cost function

$$J = \int_0^\infty (\mathbf{x}^\mathrm{T}\mathbf{Q}\mathbf{x} + \mathbf{u}^\mathrm{T}\mathbf{R}\mathbf{u})\mathrm{d}t$$

The weighting matrices **Q** and **R** are chosen to apply the desired weights to the various states and inputs. The MATLAB commands for designing the controller are

```
>> A = [0 1 0;  0 0 1; 0 0  − 1];
>> B = [ 0;  0;  1 ];
>> C = [ 1  0  0 ];
>> D = [0];
>> Q = [ 2 0 0;  0 2 0;  0 0 2 ];
>> R = 2;
>> Klqr = lqr(A,B,Q,R)
>> lqr_closed = ss(A−B*Klqr,B,C,D);
>> step(lqr_closed)
```

The step response of the system with the designed LQR controller is shown in Figure 7A.3.

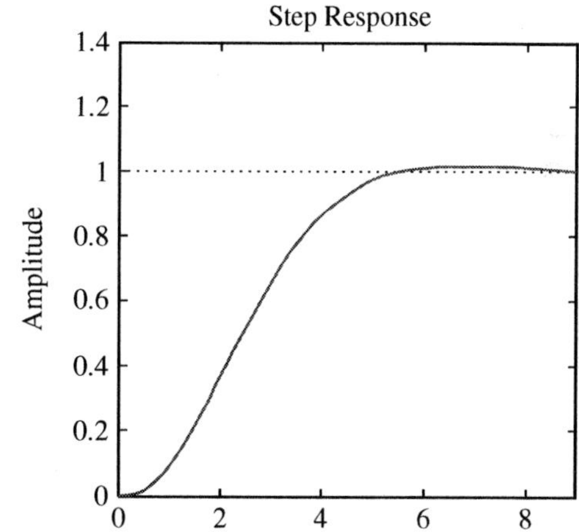

FIGURE 7A.3 Step response of the third-order system with LQR control.

7A.3.1.3 Modal Analysis Example

Consider the two-DoF mechanical system shown in Figure 7A.4. We now solve the modal analysis problem using MATLAB, for the numerical values

$$\alpha = 0.5,\ \ \beta = 0.5,\ \ m = 1\ \text{kg};\ \ k = 1\ \text{N/m}$$

For the given mass matrix **M** and the stiffness matrix **K**, the solution steps for the alternative approach of modal analysis are

1. Determine $\mathbf{M}^{1/2}$ and $\mathbf{M}^{-1/2}$.
2. Solve for the eigenvalues λ and the eigenvectors $\boldsymbol{\phi}$ of $\mathbf{M}^{-1/2}\mathbf{K}\mathbf{M}^{-1/2}$. These eigenvalues are the squares of the natural frequencies of the original system.

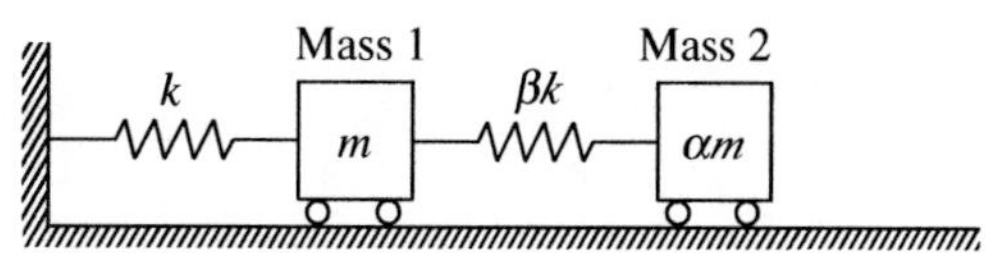

FIGURE 7A.4 A two-DoF system.

3. Determine the modal vectors $\boldsymbol{\psi}$ of the original system using the transformation $\boldsymbol{\psi} = \mathbf{M}^{-1/2}\boldsymbol{\phi}$.

The program code is given below:

```
%Modal Analysis Example
clear;
m = 1.0;
k = 1.0;
M = [ m 0;   0 m/2 ];
K = [3/2 * k  - k/2;
       - k/2 k/2];
M_sqrt = M^0.5;
M_s_inv = inv(M_sqrt);
lemda = eig(M_s_inv * K * M_s_inv);
[U,D] = eig(M_s_inv * K * M_s_inv);
V = M_s_inv * U;
disp('Natural frequencies')
fprintf('omega1 = %10.3f %14.3f \n',sqrt(lemda(2,1)));
fprintf('\n omega2 = %10.3f%14.3f \n',sqrt(lemda(1,1)));
fprintf('\n')
fprintf('\nMode shapes \n')
fprintf('First mode   Second Mode \n')
for i = 1:2
   fprintf('%10.3f %14.3f \n',V(i,2)/V(1,2),V(i,1)/V(1,1));
end
```

The necessary results are obtained as shown below:

```
>> Natural frequencies
   omega1 = 1.414
   omega2 = 0.707
```

Mode Shapes

First Mode	Second Mode
1.000	1.000
− 1.000	2.000

8

Structural Dynamic Modification and Sensitivity Analysis

Su Huan Chen
Jilin University

Summary

The matrix perturbation theory for structural dynamic modification and sensitivity analysis is presented in this chapter. The theory covers a broad spectrum of subjects, specifically, matrix perturbation of real modes of complex structures and matrix perturbation of complex modes. The contents include nine sections. Section 8.2 provides the preliminaries to matrix perturbation and sensitivity analysis. Section 8.3 presents the matrix perturbation method including first-order and second-order perturbation. Section 8.4 presents methods for design sensitivity analysis. In Section 8.5, high-accuracy modal superposition for sensitivity analysis of modes is given. Section 8.6 presents the sensitivity analysis of eigenvectors for free–free structures. In Section 8.7 and Section 8.8, matrix perturbations for repeated modes and closely spaced modes are discussed. In Section 8.9, the matrix perturbation approach for complex modes is presented.

8.1 Introduction

In modern engineering problems, the dynamic design of structures is becoming increasingly important. In order to achieve an optimal design, we repeatedly have to modify the structural parameters and solve the generalized eigenvalue problem. The iterative vibration analysis can be very tedious for large and complex structures. Therefore, it is necessary to seek a fast computation method for sensitivity analysis and reanalysis. The matrix perturbation method is an extremely useful tool for this purpose.

The matrix perturbation method is concerned with how the natural frequencies and modal vectors change if small modifications are imposed on the parameters of structures. Engineering problems often involve many small modifications in the structural parameters, such as material property variations, manufacturing errors, iterative design of structural parameters, design sensitivity analysis, random eigenvalue analysis, robustness analysis of control systems, and so on.

In this chapter, it is assumed that the reader has an undergraduate knowledge in vibration theory and a working knowledge in the finite element method.

The contents of the chapter include the basic preliminaries: vibration equations of the finite element model, eigenvalue problem, modal vectors, orthogonality conditions, modal expansion theorem, and the power series expansion of eigensolutions. The chapter also covers such topics as: the perturbation method for distinct eigenvalues and corresponding eigenvectors; sensitivities of eigenvalues and eigenvectors; the high-accuracy modal superposition method for eigenvector derivatives; eigenvector derivatives for free–free structures; perturbation method for systems with repeated eigenvalues and close eigenvalues; and perturbation method of the complex modes of systems with real unsymmetric matrices.

8.2 Structural Dynamic Modification of Finite Element Model

The finite element method is an important tool to obtain numerical and computational solutions to problems in structural vibration analysis. By applying the finite element method to a structure, a discrete analysis model to idealize the continuum can be obtained. The finite equation of vibrations of a structure in the global coordinate system is

$$\mathbf{M}\ddot{\mathbf{q}} + \mathbf{C}\dot{\mathbf{q}} + \mathbf{K}\mathbf{q} = \mathbf{Q} \tag{8.1}$$

where $\mathbf{M}$, $\mathbf{K}$, and $\mathbf{C}$ are the mass, stiffness, and damping matrices, respectively, $\ddot{\mathbf{q}}$, $\dot{\mathbf{q}}$, and $\mathbf{q}$ are the acceleration, velocity, and displacement vectors, respectively, and $\mathbf{Q}$ is the external load vector.

Neglecting the damping force and external load vector, Equation 8.1 becomes

$$\mathbf{M}\ddot{\mathbf{q}} + \mathbf{K}\mathbf{q} = \mathbf{0} \tag{8.2}$$

This is the natural vibration equation for the structure. Its solution (the natural vibration) is harmonic, and is given by

$$\mathbf{q} = \mathbf{u}\cos(\omega t - \varphi) \tag{8.3}$$

where $\mathbf{u}$ is modal vector, and ω the natural frequency of the system. Substituting Equation 8.3 into Equation 8.2, the eigenproblem of structural vibration can be obtained as

$$\mathbf{K}\mathbf{u} = \lambda\mathbf{M}\mathbf{u} \tag{8.4}$$

where λ ($\lambda = \omega^2$) denote the eigenvalues of the system.

In structural vibration analysis, the natural frequencies and the corresponding modal vectors can be obtained by solving the eigenproblem (Equation 8.4). The solutions for n eigenvalues and corresponding eigenvectors satisfy

$$\mathbf{KU} = \mathbf{MU\Lambda} \tag{8.5}$$

where $\mathbf{U}$, which is called the *modal matrix*, is an $(n \times n)$ matrix with its columns equal to the n eigenvectors, and $\mathbf{\Lambda}$ is an $(n \times n)$ diagonal matrix consisting of the corresponding eigenvalues as the diagonal elements; specifically

$$\mathbf{U} = [\mathbf{u}_1, \mathbf{u}_2, \ldots, \mathbf{u}_n] \tag{8.6}$$

$$\mathbf{\Lambda} = \mathrm{diag}(\lambda_i), \quad i = 1, 2, \ldots, n \tag{8.7}$$

An important relation for eigenvectors is that of $\mathbf{M}$ orthogonality and $\mathbf{K}$ orthogonality; that is, we have

$$\mathbf{u}_i^{\mathrm{T}} \mathbf{M} \mathbf{u}_j = \delta_{ij} \tag{8.8}$$

$$\mathbf{u}_i^{\mathrm{T}} \mathbf{K} \mathbf{u}_j = \lambda_i \delta_{ij} \tag{8.9}$$

where δ_{ij} is the Kronecker delta. For n eigenpairs, Equation 8.8 and Equation 8.9 can be written as

$$\mathbf{U}^{\mathrm{T}} \mathbf{M} \mathbf{U} = \mathbf{I} \tag{8.10}$$

$$\mathbf{U}^{\mathrm{T}} \mathbf{K} \mathbf{U} = \mathbf{\Lambda} \tag{8.11}$$

Since the modal vectors are independent, an arbitrary displacement vector, $\mathbf{u}$, can be expressed as a linear combination of $\mathbf{u}_i$, $i = 1, 2, \ldots, n$; that is

$$\mathbf{u} = \sum_{r=1}^{n} c_r \mathbf{u}_r = \mathbf{UC} \tag{8.12}$$

where c_r is a constant. Each constant c_r can be determined by

$$c_r = \mathbf{u}_r^{\mathrm{T}} \mathbf{M} \mathbf{u}, \quad r = 1, 2, \ldots, n \tag{8.13}$$

This is known as the *expansion theorem.*

Suppose the physical parameter of a given structure is given a small modification. This will cause a small change in the matrices $\mathbf{K}_0$ and $\mathbf{M}_0$; that is

$$\mathbf{M} = \mathbf{M}_0 + \varepsilon \mathbf{M}_1, \quad \mathbf{K} = \mathbf{K}_0 + \varepsilon \mathbf{K}_1 \tag{8.14}$$

where ε is a small parameter, $\mathbf{K}_0$ and $\mathbf{M}_0$ are the original mass and stiffness matrices, respectively, and $\varepsilon \mathbf{M}_1$ and $\varepsilon \mathbf{K}_1$ are the corresponding modifications. It is obvious that if $\mathbf{M}_0$ and $\mathbf{K}_0$ are symmetric, the matrices $\mathbf{M}_1$ and $\mathbf{K}_1$ are also symmetric.

If $\varepsilon \mathbf{M}_1$ and $\varepsilon \mathbf{K}_1$ are small, the changes of eigenvalues and eigenvectors of the structure are also small. According to the matrix perturbation theory, the eigensolutions of Equation 8.4 can be expressed in the form of a power series in ε; thus

$$\mathbf{u}_i = \mathbf{u}_{0i} + \varepsilon \mathbf{u}_{1i} + \varepsilon^2 \mathbf{u}_{2i} + \cdots \tag{8.15}$$

$$\lambda_i = \lambda_{0i} + \varepsilon \lambda_{1i} + \varepsilon^2 \lambda_{2i} + \cdots \tag{8.16}$$

where $\mathbf{u}_{0i}$ and λ_{0i} are the eigensolutions of the original structure, λ_{1i} and λ_{2i} are the first- and the second-order perturbations of the eigenvalues, and $\mathbf{u}_{1i}$ and $\mathbf{u}_{2i}$ are the first- and the second-order perturbation of the eigenvectors.

Since the eigensolutions of the original structure, $\mathbf{u}_{0i}$ and λ_{0i}, are known, only the first- and the second-order perturbations of the eigensolutions are required without solving Equation 8.4.

8.3 Perturbation Method of Vibration Modes

The perturbation methods of vibration modes are well developed (Fox and Kapoor, 1968; Rogers, 1977; Chen and Wada, 1979; Hu, 1987; Chen, 1993). In this section, it is assumed that all eigenvalues of the original structure are distinct.

8.3.1 First-Order Perturbation of Distinct Modes

According to the expansion theorem, the first-order perturbation, $\mathbf{u}_{1i}$, can be expanded by the modal vectors, $\mathbf{u}_{0s}$, of the original structure as

$$\mathbf{u}_{1i} = \sum_{s=1}^{n} c_{1s}\mathbf{u}_{0s} \tag{8.17}$$

where

$$c_{1s} = \frac{1}{\lambda_{0i} - \lambda_{0s}}(\mathbf{u}_{0s}^{\mathrm{T}} K_1 \mathbf{u}_{0i} - \lambda_{0i}\mathbf{u}_{0s}^{\mathrm{T}}\mathbf{M}_1\mathbf{u}_{0i}), \quad s \neq i \tag{8.18}$$

$$c_{1i} = -\tfrac{1}{2}\mathbf{u}_{0i}^{\mathrm{T}}\mathbf{M}_1\mathbf{u}_{0i} \tag{8.19}$$

The first-order perturbation of the eigenvalues is

$$\lambda_{1i} = \mathbf{u}_{0i}^{\mathrm{T}}\mathbf{K}_1\mathbf{u}_{0i} - \lambda_{0i}\mathbf{u}_{0i}^{\mathrm{T}}\mathbf{M}_1\mathbf{u}_{0i} \tag{8.20}$$

8.3.2 Second-Order Perturbation of Distinct Modes

If the parameter modification is fairly large, in order to obtain high computing accuracy, the second-order perturbation must be used. According to the expansion theorem, the second-order perturbation, $\mathbf{u}_{2i}$, can be expanded by the modal vectors, $\mathbf{u}_{0s}$, of the original structure as

$$\mathbf{u}_{2i} = \sum_{s=1}^{n} c_{2j}\mathbf{u}_{0s} \tag{8.21}$$

where

$$c_{2s} = \frac{1}{\lambda_{0i} - \lambda_{0s}}(\mathbf{u}_{0s}^{\mathrm{T}}\mathbf{K}_1\mathbf{u}_{1i} - \lambda_{0i}\mathbf{u}_{0s}^{\mathrm{T}}\mathbf{M}_1\mathbf{u}_{1i} - \lambda_{1i}\mathbf{u}_{0s}^{\mathrm{T}}\mathbf{M}_0\mathbf{u}_{1i} - \lambda_{1i}\mathbf{u}_{0s}^{\mathrm{T}}\mathbf{M}_1\mathbf{u}_{0i}), \quad s \neq i \tag{8.22}$$

$$c_{2s} = -\frac{1}{2}(\mathbf{u}_{1i}^{\mathrm{T}}\mathbf{M}_0\mathbf{u}_{1i} + \mathbf{u}_{0i}^{\mathrm{T}}\mathbf{M}_1\mathbf{u}_{1i} + \mathbf{u}_{1i}^{\mathrm{T}}\mathbf{M}_1\mathbf{u}_{0i}) \tag{8.23}$$

The second perturbation of the eigenvalues is

$$\lambda_{2i} = \mathbf{u}_{0i}^{\mathrm{T}}\mathbf{K}_1\mathbf{u}_{1i} - \lambda_{0i}\mathbf{u}_{0i}^{\mathrm{T}}\mathbf{M}_1\mathbf{u}_{1i} - \lambda_{1i}\mathbf{u}_{0i}^{\mathrm{T}}\mathbf{M}_0\mathbf{u}_{1i} - \lambda_{1i}\mathbf{u}_{0i}^{\mathrm{T}}\mathbf{M}_1\mathbf{u}_{0i} \tag{8.24}$$

8.3.3 Numerical Examples

As illustrations of the matrix perturbation method, several numerical examples are given now.

Example 8.1

Consider the five-degree-of-freedom (five-DoF) system shown in Figure 8.1. The physical parameters are given as

$$m_1 = m_2 = m_3 = m_4 = 1.0, \quad m_5 = 0.5, \quad k_1 = k_2 = k_3 = k_4 = k_5 = 1.0$$

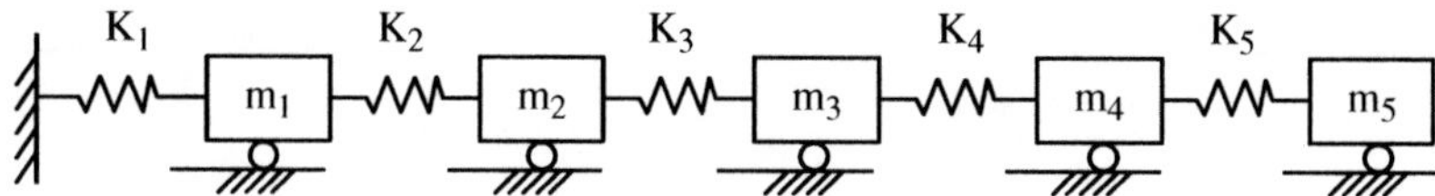

FIGURE 8.1 Mass–spring system for Example 8.1.

In order to study the computing accuracy of first- and second-order perturbations, let us assume that the fifth mass undergoes a decrement of 5 to 30%, and the stiffness of the first spring undergoes a decrement of 5 to 30%.

The computed results for the natural frequencies are presented in Table 8.1, in which the initial solutions mean the eigensolutions of the original structure.

TABLE 8.1 Comparison of Natural Frequencies

Mode Number		Changes of Structural Parameter (%)					
		5	10	15	20	25	30
1	A	0.3022	0.2922	0.2821	0.2724	0.2629	0.2536
	B	0.3128	0.3128	0.3128	0.3128	0.312	0.3128
	C	3.52	7.14	10.9	14.38	18.97	23.3
	D	0.3017	0.2903	0.2783	0.2658	0.2527	0.2388
	C	0.14	0.58	1.34	2.43	3.9	5.81
	E	0.3022	0.2922	0.2827	0.2740	0.2660	0.2588
	C	0.0033	0.068	0.24	0.58	0.93	2.08
2	A	0.8788	0.8512	0.8249	0.7998	0.7756	0.7523
	B	0.9079	0.98079	0.9079	0.98079	0.9079	0.98079
	C	3.31	6.66	10.1	13.52	17.06	20.68
	D	0.8775	0.8460	0.8133	0.7792	0.7435	0.7060
	C	0.15	0.61	1.41	2.58	4.14	6.15
	E	0.3789	0.8518	0.8267	0.8089	0.7835	0.7659
	C	0.0076	0.062	0.21	0.52	0.91	1.80
3	A	1.3732	1.3348	1.2989	1.2650	1.2332	1.2031
	B	1.1421	1.1421	1.1421	1.1421	1.1421	1.1421
	C	2.99	5.94	8.88	11.79	14.7	17.5
	D	1.371	1.3266	1.2806	1.2328	1.1832	1.1313
	C	0.15	0.62	1.41	2.54	4.05	5.96
	E	1.3733	1.3356	1.3015	1.2712	1.2449	1.2231
	C	0.0074	0.060	0.20	0.49	0.74	1.66
4	A	1.7355	1.6923	1.6520	1.6143	1.5790	1.5457
	B	1.7820	1.7820	1.7820	1.7820	1.7820	
	C	2.68	5.30	7.78	10.38	12.85	15.28
	D	1.7331	1.6828	1.6319	1.5773	1.5209	1.5209
	C	0.14	0.56	1.28	2.29	3.62	5.27
	E	1.7356	1.6933	1.6552	1.6216	1.5830	1.5695
	C	0.0070	0.057	0.19	0.45	0.68	1.53
5	A	1.9273	1.8825	1.8408	1.8017	1.7649	1.7304
	B	1.9753	1.9753	1.9753	1.9753	1.9753	1.9753
	C	2.49	4.93	7.31	9.63	11.92	14.16
	D	1.9248	1.8929	1.896	1.7696	1.7079	1.6492
	C	0.1300	0.51	1.15	2.06	3.23	4.69
	E	1.9274	1.8835	1.8439	1.8090	1.7790	1.7541
	C	0.0064	0.051	0.17	0.41	0.83	1.37

TABLE 8.2 Comparison of Natural Frequencies

Mode No.	A (Hz)	B (Hz)	F (%)	D (Hz)	G (%)	E (Hz)	H (%)
1	27.78	25.45	8.39	29.27	5.36	28.14	1.29
2	109.1	107.7	1.28	110.39	1.18	109.35	0.28
3	157.4	153.2	2.67	159.53	1.35	158.79	0.88
4	230.5	233.2	1.17	231.24	0.32	231.02	0.022
5	320.7	325.6	1.53	320.58	0.04	320.89	0.04
6	391.1	393.7	0.66	392.06	0.24	319.35	0.06

As can be seen from the results, if the change of the structural parameter is 15%, the average change of the natural frequencies is 9.0%. Using the first-order perturbation, the average error of the frequencies is reduced to 1.32%.

If the change of the structural parameter is 30%, the average error of the natural frequencies is 18%. Using the first-order perturbation, the average error of natural frequencies is reduced to 1.6%.

The notation used in Table 8.1 and Table 8.2 is as follows:

- A: the exact solutions of the modified structure
- B: the initial solutions of the original structure
- C: percent error
- D: the first-order perturbation solutions
- E: the second-order perturbation solutions
- F: the percent errors of the initial solutions
- G: the percent errors of the first-order perturbation
- H: the percent errors of the second perturbation

Example 8.2

Consider a truss structure (as shown in Figure 8.2) with 20 rods. The cross section area of the second rod is changed from 1.0 to 2.0 cm^2. The results calculated are listed in Table 8.2.

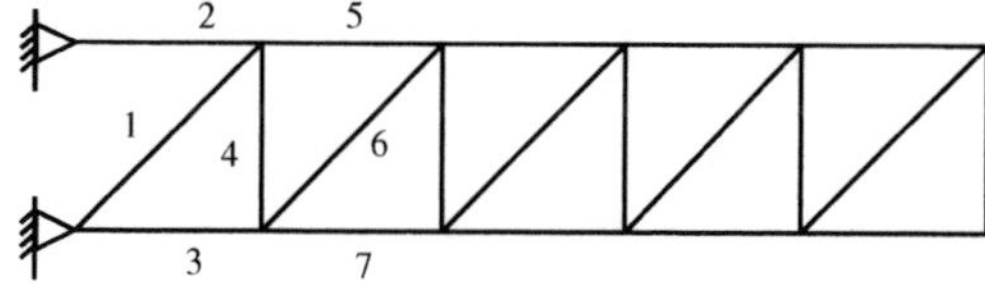

FIGURE 8.2 Truss structure for Example 8.2.

Example 8.3

Consider a torsional vibration system with five disks, as shown in Figure 8.3. The physical parameters of the system are as follows:

- $I_1 = 10.78$ kg cm sec^2
- $I_2 = 82.82$ kg cm sec^2
- $I_3 = 14.27$ kg cm sec^2
- $I_4 = 29.56$ kg cm sec^2
- $I_5 = 21.66$ kg cm sec^2
- $K_1 = 10.48 \times 10^4$ kg cm/rad
- $K_2 = 34.30 \times 10^4$ kg cm/rad
- $K_3 = 24.40 \times 10^4$ kg cm/rad
- $K_4 = 40.60 \times 10^4$ kg cm/rad

The corresponding constrained system is shown in Figure 8.3b, in which the hung stiffness is $K_s = 4060$ kg cm/rad.

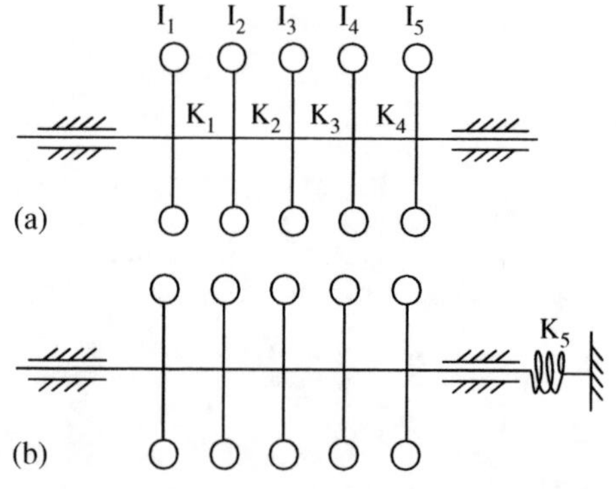

FIGURE 8.3 Torsional vibration system for Example 8.3.

The exact eigensolutions of the constrained system are taken as the initial results. Using matrix

TABLE 8.3 Comparison for Natural Frequencies

	No.				
	1	2	3	4	5
ω_0	0.000000	62.554934	105.668169	177.680405	224.361444
ω_x	1.595950	62.604815	105.669360	177.704224	224.386235
δ		0.079700	0.001070	0.013406	0.011000
$\varepsilon\omega_1$	−1.525707	−0.049895	−0.001132	−0.023834	−0.002508
ω'_0	0.080243	62.554920	105.448168	177.680390	224.383727
δ'		2.24×10^{-5}	9.46×10^{-7}	8.22×10^{-8}	9.93×10^{-5}

TABLE 8.4 Comparison for Eigenvectors

No.	u_0^i	u_x^i	εu_1^i	u'^i_0
1	0.079283	0.097363	−0.000080	0.079823
	0.079283	0.079342	−0.000059	0.079283
	0.079283	0.079287	−0.000004	0.079283
	0.079283	0.079198	0.000085	0.079283
	0.079283	0.079189	0.000153	0.079342
2	−0.095647	−0.095611	0.000036	−0.095647
	−0.057148	−0.057064	−0.000084	−0.057148
	0.008612	0.008719	−0.000107	0.008612
	0.099081	0.099192	−0.000111	0.099081
	0.125223	0.125259	0.000036	0.125223
3	0.277894	0.277884	0.000010	0.277894
	−0.041278	−0.041284	0.000005	−0.41279
	−0.027509	−0.027497	−0.000013	−0.027510
	0.009810	0.009840	−0.000030	0.009810
	0.024263	0.024279	−0.000016	0.024263
4	−0.009038	−0.009033	−0.000005	−0.009038
	0.020312	−0.020309	0.000002	0.020312
	−0.125560	−0.125580	0.000024	−0.125556
	−0.098791	0.098735	−0.000053	−0.098788
	0.144375	0.144413	−0.000038	0.144375
5	0.004823	0.004823	0.000001	0.004824
	−0.020156	−0.020154	−0.0000002	−0.020156
	0.217244	0.217227	0.000018	0.217245
	−0.088717	−0.088725	0.000008	−0.088717
	0.052618	0.052652	−0.000034	0.052618

perturbation, the eigensolutions of the free–free system can be obtained. The results are listed in Table 8.3 and Table 8.4.

The notation used in Table 8.3 is as follows:

- ω_0: exact solution of the natural frequency of the free–free system (l/sec)
- ω'_0: perturbation solutions of the natural frequency of the free–free system (l/sec)
- ω_x: natural frequency of the constrained system (l/sec)

$\varepsilon\omega_1 = \omega'_0 - \omega_x$ the perturbation of the natural frequency

$$\delta = \frac{|\omega_0 - \omega_x|}{\omega_0} \quad (\%)$$

$$\delta' = \frac{|\omega_0 - \omega'_0|}{\omega_0} \quad (\%)$$

The notation used in Table 8.4 is as follows:

- $\mathbf{u}_0^i$: exact solution of eigenvectors of the free–free system
- $\mathbf{u}'^i_0$: perturbation solution of eigenvectors of the free–free system
- $\mathbf{u}_x^i$: eigenvector of the constrained system
- $\varepsilon\mathbf{u}_1^i$: first-order perturbation of eigenvectors

As can be seen from Table 8.3, the natural frequencies of the free–free system are increased by the hung elastic elements. For example, the frequency of the rigid mode is increased to 1.595950 (l/sec), and the frequency of the first elastic mode is increased by 0.8124%. By modifying the eigensolutions with the perturbation method, the frequency of the rigid mode is reduced to 0.079700 (l/sec), which is nearly equal to zero, and all the frequencies of the elastic modes become almost exact solutions. The results in Table 8.4 show that the mode shapes of the free–free system, $\mathbf{u}'^i_0$, are close to the exact solution, u_0^i.

8.4 Design Sensitivity Analysis of Structural Vibration Modes

In the optimization of structural analysis, the design sensitivity analysis of eigenvalues and eigenvectors plays an essential role. The designer can use this information directly in an interactive computer-aided design procedure as a valuable guide. Significant work has been done in this area (Haug et al., 1985; Adelmen and Haftka, 1986; Chen and Pan, 1986; Wang, 1991).

8.4.1 Direct Differential Method for Sensitivity Analysis

Design sensitivity analysis of eigenvalues and eigenvectors will reveal how the changes in some design parameters in the system affect the dynamic characteristics of the structure.

Let $\lambda_{i,j}$ and $\mathbf{u}_{i,j}$ denote the sensitivity of the eigenvalue, λ_i, and the eigenvector, $\mathbf{u}_i$, respectively, with respect to the design variables b_j $(j = 1, 2, \ldots, L)$, and let $\mathbf{K}_{,j}$ and $\mathbf{M}_{,j}$ denote the derivative of the stiffness and mass matrices, respectively, with respect to b_j. The design sensitivity of the eigenvalue is

$$\lambda_{i,j} = \mathbf{u}_i^{\mathrm{T}}(\mathbf{K}_{,j} - \lambda_i \mathbf{M}_{,j})\mathbf{u}_i \tag{8.25}$$

The sensitivity of the eigenvector, $\mathbf{u}_{i,j}$, can be expressed as the following series:

$$\mathbf{u}_{i,j} = \sum_{s=1}^{n} c_{ijs}\mathbf{u}_s \tag{8.26}$$

where

$$c_{ijs} = \frac{1}{\lambda_i - \lambda_s}\mathbf{u}_s^{\mathrm{T}}(\mathbf{K}_{,j} - \lambda_i \mathbf{M}_{,j})\mathbf{u}_i, \quad i \neq s, \quad i, s = 1, 2, \ldots, n, \quad i \neq s \tag{8.27}$$

$$c_{iji} = -\frac{1}{2}\mathbf{u}_i^{\mathrm{T}}\mathbf{M}_{,j}\mathbf{u}_i \tag{8.28}$$

8.4.2 Perturbation Sensitivity Analysis

Let $\Delta\mathbf{K}$ and $\Delta\mathbf{M}$ denote the increments of the stiffness and the mass matrices resulting from an incremental change of the design variable, Δb_j, and let $\Delta\lambda_i$ and $\Delta\mathbf{u}_i$ denote the corresponding perturbations of the eigenvalue and eigenvector, respectively. The direct differential method of design sensitivity analysis of vibration modes can now be put into perturbation form, approximately as

$$\lambda_{i,j} = \frac{\Delta\lambda_i}{\Delta b_j} \tag{8.29}$$

$$\mathbf{u}_{i,j} = \frac{\Delta\mathbf{u}_i}{\Delta b_j} \tag{8.30}$$

where $\Delta\lambda_i$ and $\Delta\mathbf{u}_i$ can be evaluated by the perturbation formulas presented in this chapter. In practical analysis, the design variables could be the cross-sectional area of the truss members, bending moment of inertia, equivalent torsional moment of inertia of a beam, the thickness of a plate, or other variable. In some complex structures, a mass, m_r, may be placed at a node point and moving in the direction of the rth DoF, or an elastic support with spring stiffness, K_r, may be placed at a certain node point. It is also possible that an elastic connector of stiffness, K_j, might exist between two components. They can also be considered as design variables. In finite element analysis, $\Delta\mathbf{K}$ and $\Delta\mathbf{M}$ are known to be the sum of the element increments, $\Delta\mathbf{K}^e$ and $\Delta\mathbf{M}^e$; thus

$$\Delta\mathbf{K} = \sum_e \Delta\mathbf{K}^e \tag{8.31}$$

$$\Delta\mathbf{M} = \sum_e \Delta\mathbf{M}^e \tag{8.32}$$

Hence, the sensitivity formulas of vibration modes as given above can be transformed into the finite element perturbation form (Chen and Pan, 1986)

$$\lambda_{i,j} = \frac{1}{\Delta b_j}\sum_e \bar{\mathbf{u}}_i^{\mathrm{T}}(\Delta\mathbf{K}^e - \lambda_i\Delta\mathbf{M}^e)\bar{\mathbf{u}}_i \tag{8.33}$$

and

$$\mathbf{u}_{i,j} = \frac{1}{\Delta b_j}\sum_e\left(\sum_{\substack{s=1\\ s\neq i}}^{n}\frac{1}{\lambda_i-\lambda_s}\bar{\mathbf{u}}_s^{\mathrm{T}}(\Delta\mathbf{K}^e - \lambda_i\Delta\mathbf{M}^e)\bar{\mathbf{u}}_i\mathbf{u}_s - \frac{1}{2}\bar{\mathbf{u}}_i^{\mathrm{T}}\Delta\mathbf{M}^e\bar{\mathbf{u}}_i\mathbf{u}_i\right) \tag{8.34}$$

In these formulas, the overbar signifies that the eigenvector concerned contains only the components needed for the eth finite element. It is important to observe that, in Equation 8.33 and Equation 8.34, calculations are done on the element basis, and as a result, the calculations are greatly simplified.

Using the shorthand notations

$$\lambda_{i,j}^e = \frac{1}{\Delta b_j}\bar{\mathbf{u}}_i^{\mathrm{T}}\Delta\mathbf{K}^e - \lambda_i\Delta\mathbf{M}^e\bar{\mathbf{u}}_i \tag{8.35}$$

$$\mathbf{u}_{i,j}^e = \frac{1}{\Delta b_j}\left(\sum_{\substack{s=1\\ s\neq i}}^{n}\frac{1}{\lambda_i-\lambda_s}\bar{\mathbf{u}}_s^{\mathrm{T}}\Delta\mathbf{K}^e - \lambda_i\Delta\mathbf{M}^e\bar{\mathbf{u}}_i\mathbf{u}_s - \frac{1}{2}\bar{\mathbf{u}}_i^{\mathrm{T}}\Delta\mathbf{M}^e\bar{\mathbf{u}}_i\mathbf{u}_i\right) \tag{8.36}$$

Equation 8.33 and Equation 8.34 can be written as

$$\lambda_{i,j} = \sum_e \lambda_{i,j}^e \tag{8.37}$$

and

$$\mathbf{u}_{i,j} = \sum_e \mathbf{u}_{i,j}^e \tag{8.38}$$

where $\lambda_{i,j}^e$ and $\mathbf{u}_{i,j}^e$ are the design sensitivity of the eth element for the eigenvalue λ_i and the eigenvector $\mathbf{u}_i$, respectively. Let us consider the following important cases.

For a concentrated mass, m_r, placed at a node point and moved in the direction of the rth DoF, Equation 8.33 and Equation 8.34 become

$$\lambda_{i,r} = \frac{\Delta\lambda_i}{\Delta m_r} = -\lambda_i u_{ir}^2 \tag{8.39}$$

and

$$\mathbf{u}_{i,r} = \frac{\Delta \mathbf{u}_i}{\Delta m_r} = \sum_{\substack{s=1 \\ s \neq i}}^{n} \frac{-\lambda_i}{\lambda_i - \kappa_s} u_{sr} u_{ir} \mathbf{u}_s - \frac{1}{2} u_{ir}^2 \mathbf{u}_i \tag{8.40}$$

where u_{ir} is the rth element of the ith eigenvector $\mathbf{u}_i$.

For an elastic connector with stiffness K_j between the rth and the lth DoF of two components, Equation 8.33 and Equation 8.34 become

$$\lambda_{i,j} = \frac{\Delta \lambda_i}{\Delta k_j} = (u_{ir} - u_{il})^2 \tag{8.41}$$

and

$$\mathbf{u}_{i,j} = \frac{\Delta \mathbf{u}_i}{\Delta k_j} = \sum_{\substack{s=1 \\ s \neq i}}^{n} \frac{1}{\lambda_i - \lambda_s} (u_{sr} u_{ir} - u_{sl} u_{ir} - u_{sr} u_{il} + u_{sl} u_{il}) \mathbf{u}_s \tag{8.42}$$

For an elastic support with spring stiffness K_r placed in the direction of the rth DoF, Equation 8.33 and Equation 8.34 become

$$\lambda_{i,r} = \frac{\Delta \lambda_i}{\Delta k_r} = u_{ir}^2 \tag{8.43}$$

and

$$\mathbf{u}_{i,r} = \frac{\Delta \mathbf{u}_i}{\Delta k_r} = \sum_{\substack{s=1 \\ s \neq i}}^{n} \frac{1}{\lambda_i - \lambda_s} u_{sr} u_{ir} \mathbf{u}_s \tag{8.44}$$

8.4.3 Numerical Example

The design sensitivity analysis of an automotive chassis is presented here as an illustration of the method.

Example 8.4

The finite element model of an automobile chassis consists of 39 beam elements involving 30 nodal points and 180 DoF (Figure 8.4).

The design variables for the sensitivity analysis of eigenvalues in this example are the equivalent torsional moment of inertia, J, and the bending moment of inertia, I_y, of the beam element of

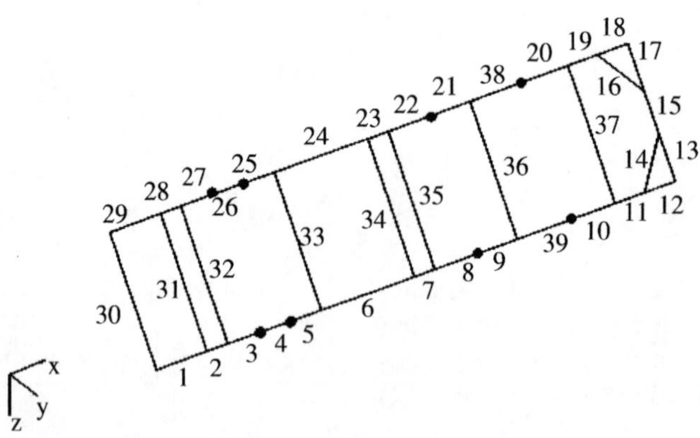

FIGURE 8.4 Finite element model of the automotive chassis for Example 8.4.

TABLE 8.5 Sensitivities of the First Four Chassis Frequencies

1	NE	15	11	19	24	6	37	34	36
	λ_{1J}^e	9.97	5.23	5.23	3.64	3.64	3.34	3.32	3.31
	NE	24	6	23	7	22	8	33	21
	λ_{1Iy}^e	0.015	0.15×10^{-2}	0.33×10^{-3}	0.33×10^{-3}	0.28×10^{-3}	0.19×10^{-3}	0.11×10^{-3}	0.11×10^{-3}
2	NE	16	14	11	19	18	12	13	17
	λ_{2J}^e	6.67×10^{-2}	0.67×10^{-2}	0.45×10^{-2}	0.45×10^{-2}	0.11×10^{-2}	0.11×10^{-2}	0.53×10^{-3}	0.53×10^{-3}
	NE	38	10	21	9	22	8	24	6
	λ_{2Iy}^e	2.92	2.92	1.78	1.78	1.49	1.49	1.44	1.44
3	NE	15	19	11	37	31	30	32	1
	λ_{3J}^e	95.9	47.2	47.2	26.5	25.4	25.2	23.8	12.5
	NE	10	38	9	21	8	22	6	24
	λ_{3Iy}^e	2.96	2.96	1.97	1.97	1.74	1.74	1.70	1.70
4	NE	14	16	19	11	12	18	17	13
	λ_{4J}^e	0.19	0.19	0.12	0.12	0.03	0.03	0.015	0.015
	NE	10	38	39	20	5	25	6	24
	λ_{4Iy}^e	22.4	22.4	15.7	15.7	11.8	11.8	10.9	10.9

the structure. Results for the sensitivity of eigenvalues with respect to J and I_y are given in Table 8.5, in which NE denotes the number of the element. Only the highest eight values are given, and they are listed in descending order.

From Table 8.5, it is seen that for this particular chassis, the sensitivities of the first natural frequency, λ_{1Iy}^e, are much smaller than λ_{1J}^e. This indicates that there is very little effect of the change of bending moment of inertia, I_y, of the beams on the vibration of the chassis at its first natural frequency. Thus, we can conclude that the first mode is a torsional mode. Similarly, the results indicate that the third mode is also a torsional mode. On the other hand, the second and the fourth modes are recognized to be bending modes. This information is very useful to the designer when deciding on a change in the design. For example, if he wants to increase the first torsional frequency, the efficient way is for him to increase the equivalent torsional moments of inertia of beam elements 15, 11, 19, 24, 6, and so on.

It should be noted that only the first low-frequency modes are available and can be used as basis vectors of eigenvector derivatives in Equation 8.26. However, modal truncation induces errors, and the errors become significant if more high-frequency modes are truncated. An improvement to truncated modal superposition representation of eigenvector derivatives is presented in the next section.

8.4.4 Concluding Remarks

As can be seen from the numerical examples given above, the matrix perturbation method is an extremely useful tool for fast reanalysis of a modified structure. It is widely used in a range of structural modifications, such as the modification of various types of elements, local modification of structures, sensitivity analysis of vibration modes, and so on. Therefore, matrix perturbation plays an important role in dynamic analysis and optimization of structures.

8.5 High-Accuracy Modal Superposition for Sensitivity Analysis of Modes

The modal superposition method is often used to compute the derivatives of modal vectors. Because of the cost of generating computer solutions for a dynamic analysis, it is impractical to obtain all modes.

Therefore, only the first L low-frequency modes are computed and are used as basis vectors of eigenvector derivatives. However, as noted above, modal truncation induces errors, which can be significant if more high-frequency modes are truncated. An explicit method to improve the truncated modal superposition representation of eigenvector derivatives is presented (Wang, 1991), in which a residual static mode is used to approximate the contribution due to unavailable high-frequency modes (method one).

In this section a more accurate modal superposition method (method two; Chen, 1993a; Liu and Chen, 1994a) than method one is given. In this method, the contribution of the truncated modes to the eigenvector derivatives is expressed exactly, as a convergent series that can be evaluated by a simple iterative procedure.

8.5.1 Method One

The modal sensitivity can be expressed as

$$u_{i,j} = \sum_{s=1}^{N} c_s u_s = \sum_{j=1}^{L} c_j \mathbf{u}_j + \mathbf{S}_R \tag{8.45}$$

where

$$\mathbf{S}_R = \sum_{j=L+1}^{N} c_j \mathbf{u}_j \tag{8.46}$$

Since $\lambda_i << \lambda_{L+1}$, Equation 8.46 can be approximated as

$$\mathbf{S}_R \approx \mathbf{S}_{RA} = \bar{\mathbf{H}}_0 - \bar{\mathbf{W}}_0 \tag{8.47}$$

where

$$\bar{\mathbf{H}}_0 = \mathbf{K}^{-1}(-\mathbf{K}_{,j} + \lambda_{i,j}\mathbf{M} + \lambda_i \mathbf{M}_{,j}) \tag{8.48}$$

$$\bar{\mathbf{W}}_0 = \sum_{j=1}^{L} \frac{1}{\lambda_j} \mathbf{u}_j^{\mathrm{T}}(-\mathbf{K}_{,j} + \lambda_{i,j}\mathbf{M} + \lambda_i \mathbf{M}_{,j})\mathbf{u}_j \tag{8.49}$$

8.5.2 Method Two

The contribution of $u_{i,j}$, S_R due to truncated high-frequencies modes is as follows:

$$\mathbf{S}_R = \sum_{j=0}^{\infty} \lambda_i^j (\mathbf{H}_j - \mathbf{W}_j) \tag{8.50}$$

where

$$\mathbf{W}_j = \mathbf{U}_L \boldsymbol{\Lambda}_L^{-j-1} \mathbf{U}_L^{\mathrm{T}}(-\mathbf{K}_{,j} + \lambda_{i,j}\mathbf{M} + \lambda_i \mathbf{M}_{,j}) \tag{8.51}$$

$$\mathbf{U}_L = [\mathbf{u}_1, \mathbf{u}_2, \ldots, \mathbf{u}_L] \tag{8.52}$$

$\mathbf{H}_j$ can be obtained with the following iterative procedure:

$$\left.\begin{aligned} &\mathbf{H}_0 = \mathbf{K}^{-1}(-\mathbf{K}_{,j} + \mathbf{K}_{i,j}\mathbf{M} + \lambda_i \mathbf{M}_{,j}) \\ &\mathbf{F}'_{j-1} = \mathbf{M}\mathbf{H}_{j-1}, \quad j \geq 1 \\ &\mathbf{H}_j = \mathbf{K}^{-1}\mathbf{F}'_{j-1} \end{aligned}\right\} \tag{8.53}$$

Define $\mathbf{S}_R(k)$ as

$$\mathbf{S}_R(k) = \sum_{j=0}^{k} \lambda_i^j(\mathbf{H}_j - \mathbf{W}_j) \tag{8.54}$$

Using this definition, the given iterative process can be terminated if the following inequality

$$\|\mathbf{S}_R(k) - \mathbf{S}_R(k-1)\|_2 \leq \varepsilon \tag{8.55}$$

is satisfied, where ε is a specified accuracy requirement.

It should be noted that, if only the first term in the series (Equation 8.50) is retained with all the other terms neglected, then method two is reduced to method one. In addition, the series (Equation 8.50) can be used to estimate the errors induced by the modal truncation.

8.6 Sensitivity of Eigenvectors for Free–Free Structures

As can be seen from Equation 8.48 and Equation 8.53, both method one and method two fail to deal with the free–free structures with rigid-body modes because they involve the inversion of the stiffness matrix. However, we can transform the eigenproblem with a singular stiffness matrix into its equivalent eigenproblem with a nonsingular stiffness matrix, in the sense that these two eigenproblems have the same derivatives of eigenvalues and eigenvectors (Liu and Chen, 1994b).

Consider the eigenvalue problem

$$\bar{\mathbf{K}}\bar{\mathbf{u}}_i = \bar{\lambda}_i\mathbf{M}\bar{\mathbf{u}}_i \tag{8.56}$$

where

$$\bar{\mathbf{K}} \equiv \mathbf{K} - \mu\mathbf{M} \tag{8.57}$$

Here, μ is a nonzero scalar parameter and $\bar{\mathbf{K}}$ is nonsingular if $\mu \neq \lambda_i$ $(i = 1, 2, \ldots, n)$

It can be shown that

$$\bar{\lambda}_i = \lambda_i - \mu \tag{8.58}$$

$$\bar{\mathbf{u}}_i = \mathbf{u}_i, \quad i = 1, 2, \ldots, N \tag{8.59}$$

and

$$\frac{d\bar{\lambda}_i}{db} = \frac{d\lambda_i}{db} \tag{8.60}$$

$$\frac{d\bar{\mathbf{u}}_i}{db} = \frac{d\mathbf{u}_i}{db} \tag{8.61}$$

The derivatives $d\mathbf{u}_i/db$ can be obtained from the derivatives $d\bar{\mathbf{u}}_i/db$ of the eigenproblem of Equation 8.56, in which $\bar{\mathbf{K}}$ is nonsingular. In this context, both method one and method two, discussed in Section 8.5.1 and Section 8.5.2, can be applied to deal with the free–free structures with rigid-body modes.

To achieve a faster average convergent speed for all the first m eigenvector derivatives, μ can be determined as

$$\begin{cases} \mu = \dfrac{\left(\sum_{j=1}^{m} \lambda_j\right)}{m}, & j = 1, 2, \ldots, m \\ \mu \neq \lambda_j, & \end{cases} \tag{8.62}$$

8.7 Matrix Perturbation Theory for Repeated Modes

8.7.1 Basic Equations

In this section, let us consider the case of repeated eigenvalues, namely, $\lambda_{0i} = \lambda_{0i+1} = \cdots = \lambda_{0i+m-1}$. The system is known as a *degenerate system*. In engineering, many complex and large structures, such as airplanes, rockets, tall towers, bridges, and ocean platforms, often have multiple or cluster eigenvalues. The matrix perturbation for the repeated modes is presented in Haug, et al. (1980), Chen and Pan (1986), Hu (1987), Mills-Curran (1988), Ojalvo (1988), Dailey (1989), Lim et al. (1989) and Shaw and Jayasuriya (1992).

Assume that $\lambda_0 = \lambda_{01} = \lambda_{02} = \cdots = \lambda_{0m}$; that is, λ_0 is a repeated eigenvalue with multiplicity equal to m, and $\mathbf{u}_{01}, \mathbf{u}_{02}, \cdots, \mathbf{u}_{0m}$ are the eigenvectors associated with λ_0. Then, a linear combination of $\mathbf{u}_{0j}$ $(j = 1, 2, \ldots, m)$, denoted as $\mathbf{U}_0$, will also be the eigenvector associated with λ_0:

$$\mathbf{U}_0 = \mathbf{U}_{0m}\alpha \tag{8.63}$$

where

$$\mathbf{U}_{0m} = [\mathbf{u}_{01}, \mathbf{u}_{02}, \ldots, \mathbf{u}_{0m}] \tag{8.64}$$

$$\boldsymbol{\alpha}^{\mathrm{T}}\boldsymbol{\alpha} = \mathbf{I} \tag{8.65}$$

and

$$\boldsymbol{\alpha} = [\alpha_1, \alpha_2, \ldots, \alpha_m]^{\mathrm{T}} \tag{8.66}$$

Note that $\boldsymbol{\alpha}$ is a constant matrix to be determined.

According to the matrix perturbation method, the eigenvalues and eigenvectors of the structure with repeated eigenvalues for the perturbed structure can be expressed as

$$\boldsymbol{\Lambda}_m = \boldsymbol{\Lambda}_0 + \varepsilon\boldsymbol{\Lambda}_1 \tag{8.67}$$

$$\mathbf{U}_m = \mathbf{U}_{0m}\alpha + \varepsilon(\mathbf{U}_0\mathbf{C}_m + \mathbf{U}_A\mathbf{C}_A) = \mathbf{U}_{0m}\alpha + \varepsilon(\mathbf{U}_{0m}\alpha\mathbf{C}_m + \mathbf{U}_A\mathbf{C}_A) \tag{8.68}$$

where $\mathbf{U}_A$ is the $n \times (n-m)$ modal matrix containing all the eigenvectors except $\mathbf{U}_{0m}$, $\boldsymbol{\Lambda}_m$ is the $m \times m$ eigenvalue diagonal matrix of the perturbed structure, $\boldsymbol{\Lambda}_1$ is the $m \times m$ diagonal matrix with its diagonal elements equal to the first-order perturbations of eigenvalues, $\mathbf{C}_m$ is an $m \times m$ matrix to be determined, and $\mathbf{C}_A$ is an $(n-m) \times (n-m)$ matrix to be determined.

8.7.2 The First-Order Perturbation of Eigensolutions

$\boldsymbol{\Lambda}_1$ and α can be computed from the following $(m \times m)$ eigenproblem:

$$\mathbf{W}\boldsymbol{\alpha} = \boldsymbol{\alpha}\boldsymbol{\Lambda}_1, \quad \boldsymbol{\alpha}^{\mathrm{T}}\boldsymbol{\alpha} = \mathbf{I} \tag{8.69}$$

where

$$\mathbf{W} = \mathbf{U}_{0m}^{\mathrm{T}}(\mathbf{K}_1 - \lambda_0\mathbf{M}_1)\mathbf{U}_{0m} \tag{8.70}$$

Solving the $m \times m$ eigenproblem of Equation 8.69 can produce $\boldsymbol{\Lambda}_1$ and $\boldsymbol{\alpha}$.

If matrix $\mathbf{W}$ has no repeated eigenvalues, $\boldsymbol{\alpha}$ can be uniquely determined; if matrix $\mathbf{W}$ has repeated eigenvalues, $\boldsymbol{\alpha}$ can be determined using the higher order perturbation equations. Here, we assume that matrix $\mathbf{W}$ has no repeated eigenvalues; that is, $\lambda_{1i} \neq \lambda_{1j}$, $(i \neq j)$, where λ_{1k} $(0 < k \leq m)$ are the elements of the diagonal matrix $\boldsymbol{\Lambda}_1$.

The matrix $\mathbf{C}_A$ is

$$\mathbf{C}_A = (\boldsymbol{\Lambda}_A - \lambda_0\mathbf{I})^{-1}\mathbf{U}_A^{\mathrm{T}}(\lambda_0\mathbf{M}_1 - \mathbf{K}_1)\mathbf{U}_{0m}\boldsymbol{\alpha} \tag{8.71}$$

The elements of C_m are

$$C_{ij}^m = \frac{R_{ij}}{\lambda_{jm}^{(1)} - \lambda_{im}^{(1)}}, \quad i \neq j \quad i,j = 1,2,\ldots,m \tag{8.72}$$

where R_{ij} are the elements of $\mathbf{R}$ given by

$$\mathbf{R} = -\boldsymbol{\alpha}^{\mathrm{T}}\mathbf{U}_{0m}^{\mathrm{T}}\mathbf{M}_1\mathbf{U}_{0m}\boldsymbol{\alpha}\boldsymbol{\Lambda}_1 + \boldsymbol{\alpha}^{\mathrm{T}}\mathbf{U}_{0m}\mathbf{K}_1\mathbf{U}_A\mathbf{C}_A - \lambda_0\boldsymbol{\alpha}^{\mathrm{T}}\mathbf{U}_{0m}\mathbf{M}_1\mathbf{U}_A\mathbf{C}_A - \boldsymbol{\alpha}^{\mathrm{T}}\mathbf{U}_{0m}^{\mathrm{T}}\mathbf{M}_0\mathbf{U}_A\mathbf{C}_A\boldsymbol{\Lambda}_1 \tag{8.73}$$

and

$$C_{ii}^m = \frac{1}{2}Q_{ii} \tag{8.74}$$

where Q_{ii} is the diagonal elements of $\mathbf{Q}$, given by

$$\mathbf{Q} = -\alpha^{\mathrm{T}}\mathbf{U}_{0m}^{\mathrm{T}}\mathbf{M}_1\mathbf{U}_{0m}\alpha \tag{8.75}$$

8.7.3 High-Accuracy Modal Superposition for the First-Order Perturbation of Repeated Modes

In Section 8.5, the high-accuracy modal superposition for the first-order perturbation of eigenvectors of distinct eigenvalues is given. In this section, we extend these methods to the situation with repeated modes.

8.7.3.1 Method One for Computing $\mathbf{U}_1$

Assuming $\mathbf{U}_{AL}$ and $\boldsymbol{\Lambda}_{AL}$ are the first L modes and eigenvalues excluding the repeated modes, the first-order perturbation of eigenvectors is

$$\mathbf{U}_1 = \mathbf{U}_{0m}\alpha\mathbf{C}_m + \mathbf{U}_{AL}\mathbf{C}_{AL} + \mathbf{S}_R \tag{8.76}$$

$$\mathbf{S}_R = \mathbf{U}_{\mathrm{S}} - [\mathbf{U}_{0m}\vdots\mathbf{U}_{AL}]\mathrm{diag}(\lambda_0^{-1}, \boldsymbol{\Lambda}_{AL}^{-1})[\mathbf{U}_{0m}\vdots\mathbf{U}_{AL}]^{\mathrm{T}}\mathbf{T} \tag{8.77}$$

where $\mathbf{U}_{\mathrm{S}}$ is the static displacement obtained by

$$\mathbf{K}\mathbf{U}_{\mathrm{S}} = \mathbf{T} \tag{8.78}$$

and

$$\mathbf{T} = \mathbf{M}_0\mathbf{U}_{0m}\boldsymbol{\alpha}\boldsymbol{\Lambda}_1 + \lambda_0\mathbf{M}_1\mathbf{U}_{0m}\boldsymbol{\alpha} - \mathbf{K}_1\mathbf{U}_{0m}\boldsymbol{\alpha} \tag{8.79}$$

In Equation 8.79, $\boldsymbol{\Lambda}_1$ and $\boldsymbol{\alpha}$ can be obtained from Equation 8.69.

The matrix $\mathbf{C}_{AL}$ is given by

$$\mathbf{C}_{AL} = (\boldsymbol{\Lambda}_{AL} - \lambda_0\mathbf{I})^{-1}\mathbf{U}_{AL}^{\mathrm{T}}(\lambda_0\mathbf{M}_1 - \mathbf{K}_1)\mathbf{U}_{0m}\boldsymbol{\alpha} \tag{8.80}$$

and the elements of matrix $\mathbf{C}_m$ are

$$C_{ij}^m = \frac{R_{ij}}{\lambda_{jm}^{(1)} - \lambda_{im}^{(1)}}, \quad i \neq j, \quad i,j = 1,2,\ldots,m \tag{8.81}$$

where $\mathbf{R}$ is given by

$$\mathbf{R} = \boldsymbol{\alpha}^{\mathrm{T}}\mathbf{U}_{0m}^{\mathrm{T}}\mathbf{M}_1\mathbf{U}_{0m}\boldsymbol{\Lambda}_1 - \boldsymbol{\alpha}^{\mathrm{T}}\mathbf{U}_{0m}^{\mathrm{T}}(\lambda_0\mathbf{M}_1 - \mathbf{K}_1)(\mathbf{U}_{AL}\mathbf{C}_{AL} + \mathbf{S}_R) - \boldsymbol{\alpha}^{\mathrm{T}}\mathbf{U}_{0m}^{\mathrm{T}}\mathbf{M}_0\mathbf{S}_R\boldsymbol{\Lambda}_1 \tag{8.82}$$

The diagonal elements of $\mathbf{C}_m$ are

$$C_{ii}^m = \frac{1}{2}Q_{ii} \tag{8.83}$$

where

$$\mathbf{Q} = -\boldsymbol{\alpha}^{\mathrm{T}}\mathbf{U}_{0m}^{\mathrm{T}}\mathbf{M}_1\mathbf{U}_{0m}\boldsymbol{\alpha} - \boldsymbol{\alpha}^{\mathrm{T}}\mathbf{U}_{0m}^{\mathrm{T}}\mathbf{M}_0\mathbf{S}_R - \mathbf{S}_R^{\mathrm{T}}\mathbf{M}_0\mathbf{U}_{0m}\boldsymbol{\alpha} \tag{8.84}$$

8.7.3.2 Method Two for Computing $\mathbf{U}_1$

The first-order perturbation of eigenvectors can be expressed as

$$\mathbf{U}_1 = \mathbf{U}_{0m}\boldsymbol{\alpha}\mathbf{C}_m + \mathbf{U}_{AL}\mathbf{C}_{AL} + \mathbf{S}_R \tag{8.85}$$

where $\mathbf{C}_{AL}$ can also be calculated using Equation 8.80; that is

$$\mathbf{C}_{AL} = (\boldsymbol{\Lambda}_{AL} - \lambda_0\mathbf{I})^{-1}\mathbf{U}_{AL}^{\mathrm{T}}(\lambda_0\mathbf{M}_1 - \mathbf{K}_1)\mathbf{U}_{0m}\boldsymbol{\alpha}$$

and $\mathbf{S}_R$ is given by

$$\mathbf{S}_R = \sum_{j=0}^{\infty}\lambda_0^j(\mathbf{H}_j - \mathbf{W}_j) \tag{8.86}$$

where

$$\mathbf{W}_j = [\mathbf{U}_{0m}\vdots\mathbf{U}_{AL}]\boldsymbol{\Lambda}_0^{-j-1}[\mathbf{U}_{0m}\vdots\mathbf{U}_{AL}]^{\mathrm{T}}\mathbf{T}, \quad j \geq 0 \tag{8.87}$$

$$\mathbf{T} = \mathbf{M}_0\mathbf{U}_{0m}\boldsymbol{\alpha}\boldsymbol{\Lambda}_1 + \lambda_0\mathbf{M}_1\mathbf{U}_{0m}\boldsymbol{\alpha} - \mathbf{K}_1\mathbf{U}_{0m}\boldsymbol{\alpha} \tag{8.88}$$

The iterative method for computing $\mathbf{H}_j$ is as follows:

$$\begin{aligned} &\mathbf{H}_0 = \mathbf{K}^{-1}\mathbf{T}, \\ &\mathbf{F}'_{j-1} = \mathbf{M}\mathbf{H}_{j-1}, \quad j \geq 1 \\ &\mathbf{H}_j = \mathbf{K}^{-1}\mathbf{F}'_{j-1}, \end{aligned} \tag{8.89}$$

This iterative process can be terminated according to the accuracy requirement. If we define $\mathbf{S}_R(k)$ as

$$\mathbf{S}_R(k) = \sum_{j=0}^{k}\lambda_0^j(\mathbf{H}_j - \mathbf{W}_j) \tag{8.90}$$

the termination condition can be stated as

$$\|S_R(k) - S_R(k-1)\|_2 \leq \varepsilon, \quad j = 1, 2, \ldots, m \tag{8.91}$$

where ε is a specified accuracy requirement.

The computation method for $\mathbf{C}_m$ in Equation 8.85 is similar to that of Equation 8.81 to Equation 8.84. The only difference is that $\mathbf{S}_R$ in Equation 8.82 and Equation 8.84 can be replaced with $\mathbf{S}_R(k)$ in Equation 8.90.

8.8 Matrix Perturbation Method for Closely Spaced Eigenvalues

The vibration modes with close frequencies, that is, with clusters of frequencies, often occur in certain structural systems including large space structures, multispan beams, and in some nearly periodic structures and symmetric structures. Therefore, it is important here to present the perturbation method for vibration modes with close eigenvalues (Liu, 2000).

The perturbation analysis of close eigenvalues can be transformed into a problem with a repeated eigenvalue, which is equal to the average value of the close eigenvalues (Chen, 1993).

8.8.1 Method One of Perturbation Analysis for Close Eigenvalues

Consider vibration eigenproblem

$$\mathbf{K}_0[\mathbf{U}_0 \vdots \mathbf{U}_A] = \mathbf{M}_0[\mathbf{U}_0 \vdots \mathbf{U}_A]\text{diag}(\mathbf{\Lambda}_0, \mathbf{\Lambda}_A) \tag{8.92}$$

$$[\mathbf{U}_0 \vdots \mathbf{U}_A]^{\mathrm{T}}\mathbf{M}_0[\mathbf{U}_0 \vdots \mathbf{U}_A] = \mathbf{I} \tag{8.93}$$

where $\mathbf{K}_0$ and $\mathbf{M}_0$ are $n \times n$ real symmetric matrices, and $\mathbf{\Lambda}_0$ and $\mathbf{U}_0$ are the $m \times m$ diagonal matrix of close eigenvalues and the corresponding $n \times m$ modal matrix.

Using the spectral decomposition of $\mathbf{K}_0$, the problem can be expressed as

$$\mathbf{K}_0 = \bar{\mathbf{K}}_0 + \varepsilon\,\delta\mathbf{K}_0 \tag{8.94}$$

where

$$\bar{\mathbf{K}}_0 = \mathbf{M}_0(\lambda_0\mathbf{U}_0\mathbf{U}_0^{\mathrm{T}})\mathbf{M}_0 + \mathbf{M}_0(\mathbf{U}_A\mathbf{\Lambda}_A\mathbf{U}_A^{\mathrm{T}})\mathbf{M}_0 \tag{8.95}$$

$$\varepsilon\,\delta\mathbf{K}_0 = \mathbf{M}_0(\mathbf{U}_0(\varepsilon\,\delta\mathbf{\Lambda}_0)\mathbf{U}_0^{\mathrm{T}})\mathbf{M}_0 \tag{8.96}$$

$$\varepsilon\,\delta\mathbf{\Lambda}_0 = \mathbf{\Lambda}_0 - \lambda_0\mathbf{I} = \mathbf{\Lambda}_0 - \left(\frac{\sum_{i=1}^{m}\lambda_{0i}}{m}\right)\mathbf{I} \tag{8.97}$$

It can be seen that $\bar{\mathbf{K}}_0$ given by Equation 8.95 satisfies

$$\bar{\mathbf{K}}_0[\mathbf{U}_0 \vdots \mathbf{U}_A] = \mathbf{M}_0[\mathbf{U}_0 \vdots \mathbf{U}_A]\text{diag}(\lambda_0\mathbf{I}, \mathbf{\Lambda}_A) \tag{8.98}$$

$$[\mathbf{U}_0 \vdots \mathbf{U}_A]^{\mathrm{T}}\mathbf{M}_0[\mathbf{U}_0 \vdots \mathbf{U}_A] = \mathbf{I} \tag{8.99}$$

This indicates that λ_0 and $\mathbf{U}_0$ are the repeated eigenvalues and the corresponding eigenvector subspace with multiplicity m of the eigenproblem (Equation 8.92), and $\mathbf{\Lambda}_A$ and $\mathbf{U}_A$ are also the eigensolution of eigenproblem (Equation 8.92).

If $\mathbf{\Lambda}_0 \to \lambda_0\mathbf{I}$, $\varepsilon\,\delta\mathbf{\Lambda}_0 \to \mathbf{0}$, and $\bar{\mathbf{K}}_0 \to \mathbf{K}_0$, and if the small parameter modifications $\varepsilon\mathbf{K}_1$ and $\varepsilon\mathbf{M}_1$ are introduced to the matrices $\mathbf{K}_0$ and $\mathbf{M}_0$, the eigenproblem with close eigenvalues becomes

$$(\mathbf{K}_0 + \varepsilon\mathbf{K}_1)\mathbf{U} = (\mathbf{M}_0 + \varepsilon\mathbf{M}_1)\mathbf{U}\mathbf{\Lambda} \tag{8.100}$$

$$\mathbf{U}(\mathbf{M}_0 + \varepsilon\mathbf{M}_1)\mathbf{U}^{\mathrm{T}} = \mathbf{I} \tag{8.101}$$

Substituting Equation 8.94 into Equation 8.100, we obtain

$$(\bar{\mathbf{K}}_0 + \varepsilon\bar{\mathbf{K}}_1)\mathbf{U} = (\mathbf{M}_0 + \varepsilon\mathbf{M}_1)\mathbf{U}\mathbf{\Lambda} \tag{8.102}$$

$$\mathbf{U}(\mathbf{M}_0 + \varepsilon\mathbf{M}_1)\mathbf{U}^{\mathrm{T}} = \mathbf{I} \tag{8.103}$$

where

$$\varepsilon\bar{\mathbf{K}}_1 = \varepsilon\,\delta\mathbf{K}_0 + \varepsilon\mathbf{K}_1 \tag{8.104}$$

$$\mathbf{\Lambda} = \lambda_0\mathbf{I} + \varepsilon\mathbf{\Lambda}_1 + \varepsilon^2\mathbf{\Lambda}_2 + \cdots \tag{8.105}$$

$$\mathbf{U} = \mathbf{U}_0\alpha + \varepsilon\mathbf{U}_1 + \varepsilon^2\mathbf{U}_2 + \cdots \tag{8.106}$$

Therefore, the eigenproblem of Equation 8.102 and Equation 8.103 can be considered to be a perturbed eigenproblem with the perturbation matrices equal to $(\delta\mathbf{K}_0 + \mathbf{K}_1)$ and $\mathbf{M}_1$, respectively. The eigensolutions, $\mathbf{\Lambda}$ and $\mathbf{U}$, can be obtained from Equation 8.102 and Equation 8.103 by using the perturbation method for repeated eigenvalues as discussed in Section 8.7. Accordingly, the perturbation problem of modes with close eigenvalues is transformed into one of the repeated eigenvalues.

The complete algorithm for $\mathbf{\Lambda}$ and $\mathbf{U}$ is given below.

(1) Compute

$$\lambda_0 = \frac{\sum_{i=1}^{m} \lambda_{0i}}{m}$$

(2) Compute

$$\mathbf{W} = \mathbf{U}_0^{\mathrm{T}}(\delta\mathbf{K}_0 + \mathbf{K}_1 - \lambda_0\mathbf{M}_1)\mathbf{U}_0$$

(3) Solve the eigenvalue problem

$$\mathbf{W}\boldsymbol{\alpha} = \boldsymbol{\alpha}\mathbf{\Lambda}_1$$

$$\boldsymbol{\alpha}^{\mathrm{T}}\boldsymbol{\alpha} = \mathbf{I}$$

(4) Compute the perturbed eigenvalues of the close eigenvalues

$$\mathbf{\Lambda} = \lambda_0\mathbf{I} + \mathbf{\Lambda}_1$$

(5) Compute the new eigenvectors $\mathbf{U}_0\boldsymbol{\alpha}$ corresponding to λ_0.

(6) Compute the matrix $\mathbf{C}_A$

$$\mathbf{C}_A = (\mathbf{\Lambda}_A - \lambda_0\mathbf{I})^{-1}\mathbf{U}_A^{T}(\lambda_0\mathbf{M}_1 - \mathbf{K}_1 - \delta\mathbf{K}_0)\mathbf{U}_0\boldsymbol{\alpha}$$

(7) Compute

$$\mathbf{R} = [R_{ij}]$$

$$\mathbf{R} = -\boldsymbol{\alpha}^{\mathrm{T}}\mathbf{U}_0^{\mathrm{T}}\mathbf{M}_1\mathbf{U}_0\boldsymbol{\alpha}\mathbf{\Lambda}_1 - \lambda_0\boldsymbol{\alpha}^{\mathrm{T}}\mathbf{U}_0^{\mathrm{T}}\mathbf{M}_1\mathbf{U}_A\mathbf{C}_A - \boldsymbol{\alpha}^{\mathrm{T}}\mathbf{U}_0^{\mathrm{T}}(\delta\mathbf{K}_0 + \mathbf{K}_1)\mathbf{U}_A\mathbf{C}_A - \boldsymbol{\alpha}\mathbf{U}_0^{\mathrm{T}}\mathbf{M}_0\mathbf{U}_A\mathbf{C}_A\mathbf{\Lambda}_1$$

(8) Compute

$$\mathbf{C}_m = [C_{ij}^m]$$

$$C_{ij}^m = \frac{R_{ij}}{\lambda_{1j} - \lambda_{1i}}, \quad i \neq j, \quad i,j = 1,2,\ldots,m$$

$$C_{ii}^m = \frac{1}{2}Q_{ii}$$

$$\mathbf{Q} = -\alpha^{\mathrm{T}}\mathbf{U}_0^{\mathrm{T}}\mathbf{M}_1\mathbf{U}_0\alpha$$

(9) Compute the perturbed eigenvectors $\mathbf{U}$

$$\mathbf{U} = \mathbf{U}_0\boldsymbol{\alpha} + \mathbf{U}_0\boldsymbol{\alpha}\mathbf{C}_m + \mathbf{U}_A\mathbf{C}_A$$

8.8.2 Method Two of Perturbation Analysis for Close Eigenvalues

Because of the importance of the problem in both theory and practice, we now present method two of perturbation analysis for close eigenvalues, which is equivalent to method one given above.

Using the spectral decomposition of $\mathbf{M}_0$, the problem can be expressed as

$$\mathbf{M}_0 = \bar{\mathbf{M}}_0 + \varepsilon\,\delta\mathbf{M}_0 \tag{8.107}$$

Then, the following equations hold:

$$\mathbf{K}_0[\mathbf{U}_0 \vdots \mathbf{U}_A] = \bar{\mathbf{M}}_0[\mathbf{U}_0 \vdots \mathbf{U}_A]\mathrm{diag}(\lambda_0\mathbf{I}, \boldsymbol{\Lambda}_A) \tag{8.108}$$

$$[\mathbf{U}_0 \vdots \mathbf{U}_A]^{\mathrm{T}}\mathbf{M}_0[\mathbf{U}_0 \vdots \mathbf{U}_A] = \mathbf{I} \tag{8.109}$$

where

$$\bar{\mathbf{M}} = \lambda_0^{-2}\mathbf{K}_0\mathbf{U}_0\mathbf{U}_0^{\mathrm{T}}\mathbf{K}_0 + \mathbf{K}_0[\mathbf{U}_A(\boldsymbol{\Lambda}_A^{-1})^2\mathbf{U}_A^{\mathrm{T}}]\mathbf{K}_0 \tag{8.110}$$

$$\varepsilon\,\delta\mathbf{M}_0 = \mathbf{K}_0[\mathbf{U}_0\varepsilon\,\delta(\boldsymbol{\Lambda}_0^{-1})^2\mathbf{U}_0^{\mathrm{T}}]\mathbf{K}_0 \tag{8.111}$$

$$\varepsilon\,\delta\boldsymbol{\Lambda}_0^{-2} = \boldsymbol{\Lambda}_0^{-2} - \lambda_0^{-2}\mathbf{I} \tag{8.112}$$

and

$$\lambda_0 = \frac{\sum_{i=1}^{m}\lambda_{0i}}{m} \tag{8.113}$$

It can be seen that $\bar{\mathbf{M}}_0$ and $\varepsilon\,\delta\mathbf{M}_0$ given by Equation 8.110 and Equation 8.111 satisfy Equation 8.108 and Equation 8.109; that is, λ_0 and $\mathbf{U}_0$ are the repeated eigenvalues and corresponding modal matrix of Equation 8.108 and Equation 8.109. $\boldsymbol{\Lambda}_A$ and $\mathbf{U}_A$ are the eigenvalue diagonal matrix and the corresponding modal matrix excluding $\boldsymbol{\Lambda}_0$ and $\mathbf{U}_0$, respectively.

If $\mathbf{K}_0$ and $\mathbf{M}_0$ are modified to $\mathbf{K}_0 + \varepsilon\mathbf{K}_1$ and $\mathbf{M}_0 + \varepsilon\mathbf{M}_1$, the eigenvalue problem becomes

$$(\mathbf{K}_0 + \varepsilon\mathbf{K}_1)\mathbf{U} = (\mathbf{M}_0 + \varepsilon\mathbf{M}_1)\mathbf{U}\boldsymbol{\Lambda} \tag{8.114}$$

$$\mathbf{U}(\mathbf{M}_0 + \varepsilon\mathbf{M}_1)\mathbf{U}^{\mathrm{T}} = \mathbf{I} \tag{8.115}$$

Substituting Equation 8.107 into Equation 8.114 yields

$$(\mathbf{K}_0 + \varepsilon\mathbf{K}_1)\mathbf{U} = (\bar{\mathbf{M}}_0 + \varepsilon\bar{\mathbf{M}}_1)\mathbf{U}\boldsymbol{\Lambda} \tag{8.116}$$

$$\mathbf{U}(\bar{\mathbf{M}}_0 + \varepsilon\bar{\mathbf{M}}_1)\mathbf{U}^{\mathrm{T}} = \mathbf{I} \tag{8.117}$$

where

$$\varepsilon\bar{\mathbf{M}}_1 = \varepsilon\,\delta\mathbf{M}_0 + \varepsilon\mathbf{M}_1 \tag{8.118}$$

Thus, Equation 8.116 and Equation 8.117 can be considered to be a perturbed eigenproblem with repeated eigenvalues, and the perturbation method for repeated eigenvalues can be used to obtain the perturbed eigensolutions of Equation 8.116 and Equation 8.117:

$$\boldsymbol{\Lambda} = \lambda_0\mathbf{I} + \varepsilon\boldsymbol{\Lambda}_1 + \cdots \tag{8.119}$$

$$\mathbf{U} = \mathbf{U}_0\alpha + \varepsilon\mathbf{U}_1 + \cdots \tag{8.120}$$

Example 8.5

For the six-DoF mass–spring system shown in Figure 8.5, the stiffness and mass matrices $\mathbf{K}_0$ and $\mathbf{M}_0$ are given by

$$\mathbf{K}_0 = \begin{bmatrix} 1500 & -1000 & & & & \\ -1000 & 1200 & -200 & & & \\ & -200 & 15,200 & -5000 & -5000 & -5000 \\ & & -5000 & 5000 & & \\ & & -5000 & & 5000 & \\ & & -5000 & & & 5000 \end{bmatrix} \text{(N/m)}$$

$$\mathbf{M}_0 = \text{diag}(200, 300, 50, 20, 20, 20.004) \text{ (kg)}$$

The perturbation eigensolutions are computed for the following three cases:

Case 1

$$\varepsilon\mathbf{K}_1 = \begin{bmatrix} 0 & 0 & 0 & 0 & 0 & 0 \\ 0 & 0 & 0 & 0 & 0 & 0 \\ 0 & 0 & 5 & 0 & -5 & 0 \\ 0 & 0 & 0 & 0 & 0 & 0 \\ 0 & 0 & -5 & 0 & 5 & 0 \\ 0 & 0 & 0 & 0 & 0 & 0 \end{bmatrix} \text{(N/m)}$$

$$\varepsilon\mathbf{M}_1 = \text{diag}(0, \ldots, 0) \text{ (kg)}$$

Case 2

$$\varepsilon\mathbf{K}_1 = \begin{bmatrix} 5.00 & 0 & 0 & 0 & 0 & 0 \\ 0 & 0 & 0 & 0 & 0 & 0 \\ 0 & 0 & 0 & 0 & 0 & 0 \\ 0 & 0 & 0 & 0 & 0 & 0 \\ 0 & 0 & 0 & 0 & 0 & 0 \\ 0 & 0 & 0 & 0 & 0 & 0 \end{bmatrix} \text{(N/m)}$$

$$\varepsilon\mathbf{M}_1 = \text{diag}(0, 0, 0, 0, 0, 0.5) \text{ (kg)}$$

Case 3

$$\varepsilon\mathbf{K}_1 = \mathbf{0},\ \varepsilon\boldsymbol{M}_1 = \text{diag}(0, 0, 0, 0, 2.0, 0) \text{ (kg)}$$

The unperturbed eigensolutions have a single pair of close eigenvalues given by

$$\boldsymbol{\Lambda}_0 = \text{diag}(249.966642, 250.000000)$$

$$\mathbf{U}_0^{\mathrm{T}} = \begin{bmatrix} 0.00000 & 0.00000 & -0.00012 & -0.091283 & -0.091283 & 0.182560 \\ 0.00000 & 0.00000 & 0.00000 & -0.158114 & 0.158114 & 0.00000 \end{bmatrix}$$

The other unperturbed eigensolutions are as follows:

$$\boldsymbol{\Lambda}_A = \text{diag}(0.594885, 2.478725, 10.234656, 552.175102)$$

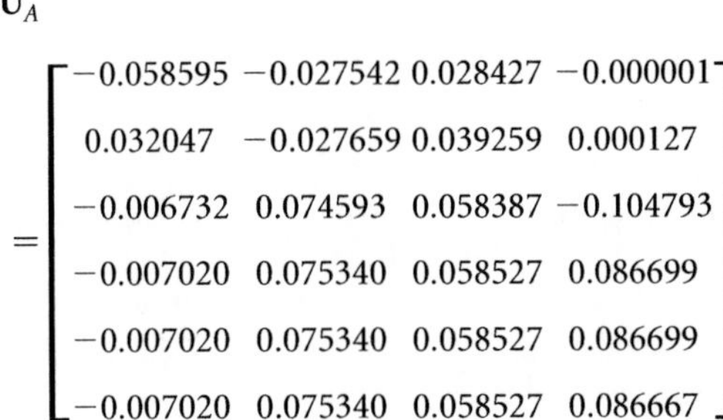

$$
\mathbf{U}_A = \begin{bmatrix} -0.058595 & -0.027542 & 0.028427 & -0.000001 \\ 0.032047 & -0.027659 & 0.039259 & 0.000127 \\ -0.006732 & 0.074593 & 0.058387 & -0.104793 \\ -0.007020 & 0.075340 & 0.058527 & 0.086699 \\ -0.007020 & 0.075340 & 0.058527 & 0.086699 \\ -0.007020 & 0.075340 & 0.058527 & 0.086667 \end{bmatrix}
$$

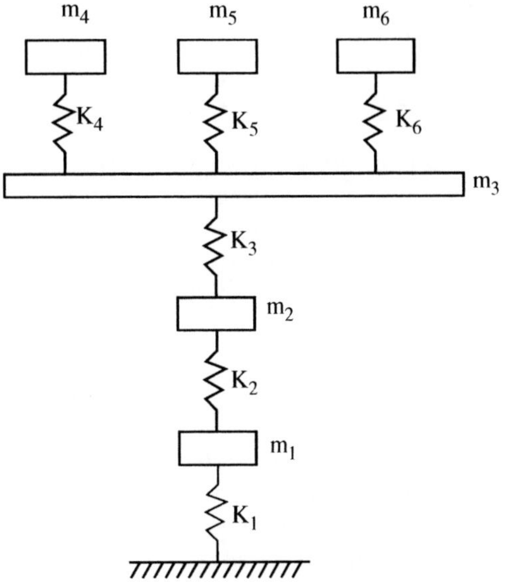

FIGURE 8.5 Six-DoF mass–spring system for Example 8.5.

The perturbed eigensolutions associated with the single pair of close eigenvalues for the three cases are summarized in Table 8.6. These results show that the perturbation analysis of distinct eigenvalues is not only inaccurate but also misleading when applied to close eigenvalues, and that the perturbed eigensolutions given by the present method are in good agreement with the exact solutions.

For example, in Case 3, the eigenvalue errors induced by the present method are reduced to 1.047950 and 0.000000, while the errors induced by the perturbation of distinct eigenvalues are 4.174100 and 3.025595. The eigenvectors obtained by the perturbation method of distinct eigenvalues are not only

TABLE 8.6 Comparison of Eigensolutions with Close Eigenvalues

	Exact		Perturbation Method of Distinct Eigenvalues		Perturbation Method of Close Eigenvalues	
			Case 1			
Eigenvalues	249.973872	250.159351	250.008324	250.125000	249.973872	250.159451
Eigenvectors	0.000000	0.000000	0.000000	0.000000	0.000000	0.000000
	0.000000	0.000000	0.000000	0.000000	0.000000	0.000000
	0.000018	−0.000066	−0.000043	0.000079	0.000011	0.000005
	0.150492	0.103411	−0.433574	0.039471	0.150501	−0.103354
	0.014254	−0.181965	0.251058	0.355699	0.014251	0.182015
	−0.164753	0.078701	0.182585	−0.395285	−0.164745	−0.078658
			Case 2			
Eigenvalues	243.878599	247.932109	244.759777	246.875000	243.725997	247.908461
Eigenvectors	0.000000	0.000000	0.000000	0.000000	0.000000	0.000000
	0.000000	−0.000004	0.000002	0.000005	0.000000	0.000000
	−0.000015	0.001506	−0.000772	−0.001976	0.000011	0.000006
	−0.000622	0.182090	8.467433	−5.098718	0.001103	0.183060
	−0.155859	−0.091163	−8.648272	−4.784467	−0.157555	−0.092069
	0.156472	−0.090082	0.180521	9.882118	0.158052	−0.090988
			Case 3			
Eigenvalues	234.474405	249.975015	245.800915	237.500000	233.326455	249.975015
Eigenvectors	0.000000	0.000000	0.000000	0.000000	0.000000	0.000000
	0.000016	0.000000	−0.000008	0.000021	0.000000	−0.000000
	−0.005558	−0.000015	0.003030	−0.007905	0.000006	0.000010
	−0.089507	−0.158189	34.141638	−19.920531	−0.091153	0.158189
	0.175420	0.000158	−34.321586	−19.612210	0.182572	−0.000158
	−0.089778	0.158022	0.181586	39.528471	−0.091415	−0.158025

inaccurate but also misleading, while good agreement with the exact eigenvectors has been obtained by the present method.

8.8.3 Concluding Remarks

Perturbation analysis of vibration modes with close frequencies is presented in this section. It can be regarded as a general treatment of perturbation analysis, because the perturbation analysis of both distinct eigenvalues and repeated eigenvalues is contained in the present method. The results obtained by this method allow one to analyze the influence of parameter changes in a system on the dynamic characteristics of the system, which is very important for effective structural design.

8.9 Matrix Perturbation Theory for Complex Modes

In Section 8.2 to Section 8.8, the matrix perturbation for real modes of systems with real symmetric mass and stiffness matrices, **M** and **K**, was given. However, in many engineering problems such as systems with nonproportional damping (see Chapter 1), dynamic systems under nonconservative forces, analysis of aero-elastic flutter, and structural vibration control systems, the system matrices are not symmetric and may not be diagonalizable. In this case, the matrix perturbation for real modes cannot be used, and we must use the matrix perturbation for complex modes (Murthy and Haftka, 1988; Chen, 1993; Liu, 1999; Adhikari and Friswell, 2001). In the following, we assume that the system is not defective; that is, the system has a complete eigenvector set to span the eigenspace. The discussion in this chapter is limited to the nondefective systems.

8.9.1 Basic Equations

The vibration equation of a linear system with n-DoFs is given by

$$\mathbf{M}\ddot{\mathbf{q}} + \mathbf{C}\dot{\mathbf{q}} + \mathbf{K}\mathbf{q} = \mathbf{Q}(t) \tag{8.121}$$

where the matrices **M**, **C**, and **K**, are assumed to be real and unsymmetric. The free vibration equation of the system is

$$\mathbf{M}\ddot{\mathbf{q}} + \mathbf{C}\dot{\mathbf{q}} + \mathbf{K}\mathbf{q} = \mathbf{0} \tag{8.122}$$

The corresponding right eigenvalues problem is

$$(\mathbf{M}s^2 + \mathbf{C}s + \mathbf{K})\mathbf{x} = \mathbf{0} \tag{8.123}$$

and its adjoint eigenvalue problem is

$$\begin{gathered}(\mathbf{M}s^2 + \mathbf{C}s + \mathbf{K})^{\mathrm{T}}\mathbf{y} = 0 \\ \mathbf{y}^{\mathrm{T}}(\mathbf{M}s^2 + \mathbf{C}s + \mathbf{K}) = 0\end{gathered} \tag{8.124}$$

It is common in literature to call **y** the *left eigenvector*, while **x** in the original system, a column vector, is called the *right eigenvector.*

Let us introduce a state vector

$$\mathbf{u} = \begin{Bmatrix} s\mathbf{x} \\ \mathbf{x} \end{Bmatrix} = \mathbf{T}\mathbf{x} \tag{8.125}$$

where **T** is the state transformation matrix

$$\mathbf{T} = \begin{Bmatrix} s\mathbf{I} \\ \mathbf{I} \end{Bmatrix} \tag{8.126}$$

Similarly, we introduce the state vector

$$\mathbf{v} = \begin{Bmatrix} s\mathbf{y} \\ \mathbf{y} \end{Bmatrix} = \mathbf{Ty} \tag{8.127}$$

Hence, Equation 8.123 and Equation 8.124 become

$$(\mathbf{A}s + \mathbf{B})\mathbf{u} = \mathbf{0} \tag{8.128}$$

$$(\mathbf{A}s + \mathbf{B})^{\mathrm{T}}\mathbf{v} = \mathbf{0} \tag{8.129}$$

or

$$\mathbf{v}^{\mathrm{T}}(\mathbf{A}s + \mathbf{B}) = \mathbf{0}$$

where

$$\mathbf{A} = \begin{bmatrix} -\mathbf{C} & -\mathbf{K} \\ \mathbf{I} & \mathbf{0} \end{bmatrix}$$

$$\mathbf{B} = \begin{bmatrix} \mathbf{M} & \mathbf{0} \\ \mathbf{0} & \mathbf{I} \end{bmatrix}$$

It is well known that the eigenvalues of the adjoint eigenproblem (Equation 8.129) are identical to that of the original eigenproblem (Equation 8.128). The *characteristic equation* is

$$\det(\mathbf{A} + s\mathbf{B}) = 0$$

This characteristic determinant is a polynomial of $2n$ order in s, and $2n$ eigenvalues s_i $(i = 1, 2, \ldots, 2n)$ can be found in the complex domain. The left and right modal vectors, $\mathbf{v}_i$ and $\mathbf{u}_i$, corresponding to s_i satisfy

$$\mathbf{A}\mathbf{u}_i = s_i\mathbf{B}\mathbf{u}_i \tag{8.130}$$

and

$$\mathbf{A}^{\mathrm{T}}\mathbf{v}_i = s_i\mathbf{B}^{\mathrm{T}}\mathbf{v}_i \tag{8.131}$$

The *orthogonality conditions* are

$$\mathbf{v}_j^{\mathrm{T}}\mathbf{B}\mathbf{u}_i = \mathbf{0} \tag{8.132}$$

$$\mathbf{v}_j^{\mathrm{T}}\mathbf{A}\mathbf{u}_i = \mathbf{0} \tag{8.133}$$

The *normalization conditions* are

$$\begin{aligned} \mathbf{v}_i^{\mathrm{T}}\mathbf{B}\mathbf{u}_i &= 1 \\ \mathbf{u}_i^{\mathrm{T}}\mathbf{B}\mathbf{u}_i &= 1 \end{aligned} \tag{8.134}$$

Therefore, the orthogonality conditions can be written as

$$\begin{aligned} \mathbf{v}_j^{\mathrm{T}}\mathbf{B}\mathbf{u}_i &= \delta_{ij} \\ \mathbf{v}_j^{\mathrm{T}}\mathbf{A}\mathbf{u}_i &= s_i\delta_{ij} \end{aligned} \tag{8.135}$$

8.9.2 Matrix Perturbation Method for Distinct Modes

If small changes are made on the structural parameters, the mass, damping, and stiffness matrices of the system also have small changes given by

$$\begin{aligned} \mathbf{M} &= \mathbf{M}_0 + \varepsilon\mathbf{M}_1 \\ \mathbf{C} &= \mathbf{C}_0 + \varepsilon\mathbf{C}_1 \\ \mathbf{K} &= \mathbf{K}_0 + \varepsilon\mathbf{K}_1 \end{aligned} \tag{8.136}$$

and we have

$$\mathbf{A} = \mathbf{A}_0 + \varepsilon \mathbf{A}_1 \tag{8.137}$$

$$\mathbf{B} = \mathbf{B}_0 + \varepsilon \mathbf{B}_1 \tag{8.138}$$

where ε is a small parameter.

In the following, we first consider the case of distinct eigenvalues, s_{0i}, of the original system. According to the matrix perturbation theory, the eigenvalues and eigenvectors can be expressed as a power series in ε, that is

$$\mathbf{S} = \mathbf{S}_0 + \varepsilon \mathbf{S}_1 + \varepsilon^2 \mathbf{S}_2 + \cdots \tag{8.139}$$

$$\mathbf{U} = \mathbf{U}_0 + \varepsilon \mathbf{U}_1 + \varepsilon^2 \mathbf{U}_2 + \cdots \tag{8.140}$$

$$\mathbf{V} = \mathbf{V}_0 + \varepsilon \mathbf{V}_1 + \varepsilon^2 \mathbf{V}_2 + \cdots \tag{8.141}$$

where $\mathbf{S}_0$, $\mathbf{U}_0$, and $\mathbf{V}_0$ are the eigensolutions of the original system; $\mathbf{S}_1$, $\mathbf{U}_1$, and $\mathbf{V}_1$ are the first-order perturbations of eigensolutions; and $\mathbf{S}_2$, $\mathbf{U}_2$, and $\mathbf{V}_2$ the second-order perturbations.

$\mathbf{U}_1$ can be expressed as a linear combination of the right eigenvectors of the original system as

$$\mathbf{U}_1 = \mathbf{U}_0 \mathbf{C}^1 \tag{8.142}$$

where $\mathbf{C}^1$ is to be the determined matrix given by

$$C_{ij}^1 = \frac{1}{s_{0j} - s_{0i}} P_{ij}^1, \quad j \neq i, \quad i, j = 1, 2, \ldots \tag{8.143}$$

Also

$$\mathbf{S}_1 = \mathrm{diag}(P_{11}^1, P_{22}^1, \ldots) \tag{8.144}$$

where P_{ij}^1 are the elements of $\mathbf{P}^1$ given by

$$\mathbf{P}^1 = \mathbf{V}_0^{\mathrm{T}}(-\mathbf{A}_1 \mathbf{U}_0 + \mathbf{B}_1 \mathbf{U}_0 \mathbf{S}_0) \tag{8.145}$$

The $\mathbf{V}_1$ can be expressed as the expansion of $\mathbf{V}_0$

$$\mathbf{V}_1 = \mathbf{V}_0 \mathbf{D}^1 \tag{8.146}$$

where $\mathbf{D}^1$ is to be the determined coefficient matrix given by

$$D_{ij}^1 = \frac{1}{s_{0j} - s_{0i}} R_{ij}^1, \quad j \neq i, \quad i, j = 1, 2, \ldots \tag{8.147}$$

and R_{ij}^1 are the nondiagonal elements of $\mathbf{R}^1$

$$\mathbf{R}^1 = \mathbf{U}_0^{\mathrm{T}}(\mathbf{B}_1^{\mathrm{T}} \mathbf{V}_0 \mathbf{S}_0 - \mathbf{A}_1^{\mathrm{T}} \mathbf{V}_0) \tag{8.148}$$

If the modification of the parameter is fairly large, the second-order perturbation must be used to obtain high accuracy. According to the expansion theorem, the second-order perturbation of eigenvectors, $\mathbf{U}_2$, can be expressed as

$$\mathbf{U}_2 = \mathbf{U}_0 \mathbf{C}^2 \tag{8.149}$$

In a similar manner, $\mathbf{S}_2$ and the elements C_{ij}^2 can be obtained as

$$\mathbf{S}_2 = \mathrm{diag}(P_{11}^2, P_{22}^2, \ldots) \tag{8.150}$$

$$C_{ij}^2 = \frac{1}{s_{0j} - s_{0i}} P_{ij}^2, \quad j \neq i, i, \quad j = 1, 2, \ldots \tag{8.151}$$

where P_{ij}^2 are the nondiagonal elements of $\mathbf{P}^2$

$$\mathbf{P}^2 = \mathbf{V}_0^{\mathrm{T}} \mathbf{B}_0 \mathbf{U}_1 \mathbf{S}_1 + \mathbf{V}_0^{\mathrm{T}} \mathbf{B}_1 \mathbf{U}_0 \mathbf{S}_1 + \mathbf{V}_0^{\mathrm{T}} \mathbf{B}_1 \mathbf{U}_1 \mathbf{S}_0 - \mathbf{V}_0^{\mathrm{T}} \mathbf{A}_1 \mathbf{U}_1 \tag{8.152}$$

$\mathbf{V}_2$ can be expressed as

$$\mathbf{V}_2 = \mathbf{V}_0\mathbf{D}^2 \tag{8.153}$$

where $\mathbf{D}^2$ is to be the determined coefficient matrix given by

$$D_{ij}^2 = \frac{1}{S_{0j} - S_{0i}} R_{ij}^2, \quad j \neq i, \quad i, j = 1, 2, \ldots \tag{8.154}$$

and R_{ij}^2 are the nondiagonal elements of $\mathbf{R}^2$

$$\mathbf{R}^2 = \mathbf{U}_0^{\mathrm{T}}(\mathbf{B}_0^{\mathrm{T}}\mathbf{V}_1\mathbf{S}_1 + \mathbf{B}_1^{\mathrm{T}}\mathbf{V}_0\mathbf{S}_1 + \mathbf{B}_1^{\mathrm{T}}\mathbf{V}_1\mathbf{S}_0 - \mathbf{A}_1^{\mathrm{T}}\mathbf{V}_1) \tag{8.155}$$

If $j = i$, the coefficients C_{ii}^1, D_{ii}^1, C_{ii}^2, and D_{ii}^2 can be computed as

$$C_{ii}^1 = \frac{-1}{\mathbf{u}_{0i}^{\mathrm{T}}(\mathbf{B}_0 + \mathbf{B}_0^{\mathrm{T}})\mathbf{u}_{0i}} \left(\mathbf{u}_{0i}^{\mathrm{T}}\mathbf{B}_1\mathbf{u}_{0i} + \sum_{\substack{j=1 \\ j \neq i}}^{n} C_{ij}\mathbf{u}_{0j}^{\mathrm{T}}(\mathbf{B}_0 + \mathbf{B}_0^{\mathrm{T}})\mathbf{u}_{0i} \right) \tag{8.156}$$

$$D_{ii}^1 = Q_{ii}^1 - C_{ii}^1 \tag{8.157}$$

where Q_{ii}^1 is the diagonal element of $\mathbf{Q}_1$

$$\mathbf{Q}_1 = -\mathbf{V}_0^{\mathrm{T}}\mathbf{B}_1\mathbf{U}_0 \tag{8.158}$$

$$C_{ii}^2 = \frac{-\mathbf{u}_{1i}^{\mathrm{T}}\mathbf{B}_0\mathbf{u}_{1i} + \mathbf{u}_{0i}^{\mathrm{T}}\mathbf{B}_1\mathbf{u}_{1i} + \mathbf{u}_{1i}^{\mathrm{T}}\mathbf{B}_1\mathbf{u}_{0i} - \sum_{\substack{j=1 \\ j \neq i}}^{n} C_{ij}\mathbf{u}_{0j}^{\mathrm{T}}(\mathbf{B}_0 + \mathbf{B}_0^{\mathrm{T}})\mathbf{u}_{0i}}{\mathbf{u}_{0i}^{\mathrm{T}}(\mathbf{B}_0 + \mathbf{B}_0^{\mathrm{T}})\mathbf{u}_{0i}} \tag{8.159}$$

$$D_{ii}^2 = Q_{ii}^2 - C_{ii}^2 \tag{8.160}$$

and Q_{ii}^2 is the diagonal element of $\mathbf{Q}^2$

$$\mathbf{Q}^2 = -\mathbf{V}_0^{\mathrm{T}}\mathbf{B}_0\mathbf{U}_1 - \mathbf{V}_1^{\mathrm{T}}\mathbf{B}_0\mathbf{U}_1 - \mathbf{V}_1^{\mathrm{T}}\mathbf{B}_1\mathbf{U}_0 \tag{8.161}$$

8.9.3 High-Accuracy Modal Superposition for Eigenvector Derivatives

For a large-scale structure, only a small number of the first lower L modes are extracted, and the higher modes are truncated in order to reduce the computational cost. The modal superposition method may not only give inaccurate result, but also may be misleading if the truncation is considerable. In this section, we give a high-accuracy modal superposition method for derivatives of the complex mode of nonsymmetric matrices.

8.9.3.1 Improved Modal Superposition

An improved modal superposition (IMS) to reduce the computation errors by modal truncation was proposed (Lim et al., 1989). The derivatives of modes can be expressed as

$$\frac{\partial \mathbf{u}_i}{\partial \mathbf{b}} = \bar{\alpha}_{ii}\mathbf{u}_i + \bar{z}_i \tag{8.162}$$

$$\bar{z}_i = \sum_{\substack{j=1 \\ j \neq i}}^{L} \frac{\mathbf{v}_j^{\mathrm{T}}\mathbf{F}_i}{S_i - S_j}\mathbf{u}_j + \mathbf{A}^{-1}\mathbf{F}_i + \sum_{j=1}^{L} \frac{\mathbf{v}_j^{\mathrm{T}}\mathbf{F}_i}{S_j}\mathbf{u}_j \tag{8.163}$$

$$\mathbf{F}_i = \left(\frac{\partial \mathbf{A}}{\partial \mathbf{b}} - \frac{\partial S_i}{\partial \mathbf{b}}\mathbf{B} - S_i\frac{\partial \mathbf{B}}{\partial \mathbf{b}} \right)\mathbf{u}_i \tag{8.164}$$

where $\mathbf{A}^{-1}\mathbf{F}_i$ is the contribution of the truncated higher modes to the derivatives of modes, as given by

$$\bar{\alpha}_{ii} = -\frac{1}{2}\left(\mathbf{u}_i^{\mathrm{T}}\frac{\partial \mathbf{B}}{\partial \mathbf{b}}\mathbf{u}_i + \mathbf{u}_i^{\mathrm{T}}\mathbf{B}\bar{z}_i + \bar{z}_i^{\mathrm{T}}\mathbf{B}\mathbf{u}_i\right) \tag{8.165}$$

8.9.3.2 High-Accuracy Modal Superposition

Assume that the eigenvalues are ordered according to their modular magnitude, and satisfy the following condition:

$$|S_i| < |S_j|, \quad j > L \tag{8.166}$$

The derivatives of modes can be expressed as

$$\frac{\partial \mathbf{u}_i}{\partial \mathbf{b}} = \bar{\alpha}_{ii}\mathbf{u}_i + \bar{z}_i = \bar{\alpha}_{ii}\mathbf{u}_i + z_{iL} + z_{iH} \tag{8.167}$$

where

$$\bar{z}_i = z_{iL} + z_{iH} \tag{8.168}$$

$$z_{iL} = \sum_{\substack{j=1 \\ j\neq i}}^{L} \frac{\mathbf{v}_j^{\mathrm{T}}\mathbf{F}_i}{S_i - S_j}\mathbf{u}_j \tag{8.169}$$

$$z_{iH} = -\sum_{j=1}^{K}\left(\mathbf{A}^{-1}(\mathbf{B}\mathbf{A}^{-1})^{j-1} - \mathbf{U}_L((\mathbf{S}_L)^{-1})^j\mathbf{V}_L^{\mathrm{T}}\right)\mathbf{F}_i \tag{8.170}$$

Also, $\bar{\alpha}_{ii}$ can be obtained from Equation 8.165, and K denotes the number of terms used in series (Equation 8.170).

It can be shown that for $K = 1$, Equation 8.167 is equivalent to Equation 8.162.

8.9.3.3 Numerical Example

Example 8.6

Consider a 20-DoF system, as shown in Figure 8.6, with the parameters given by

$$m_1 = m_2 = \cdots = m_{19} = 2m, \quad m_{20} = m = 1.0 \text{ kg}$$

$$k_1 = k_2 = \cdots = k_{21} = 1.0 \times 10^3 \text{ N/m}$$

$$c_1 = c_2 = \cdots = c_7 = 3c, \quad c_8 = c_9 = \cdots = c_{14} = 2c$$

$$c_{15} = c_{16} = \cdots = c_{21} = c = 0.1 \text{ N sec/m}$$

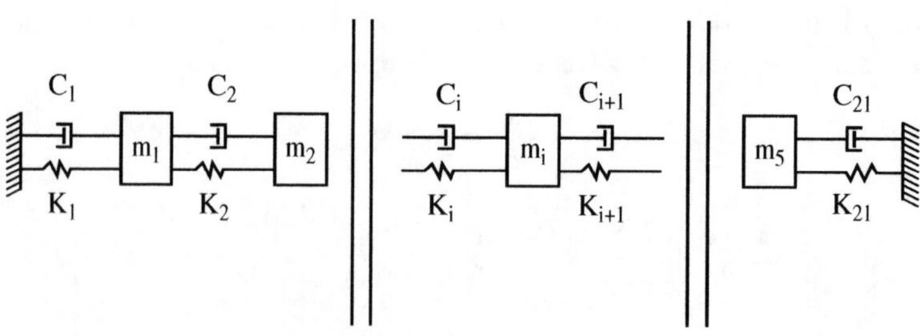

FIGURE 8.6 The 20-DoF system for Example 8.6.

TABLE 8.7 Errors of Eigenvector Derivatives (%)

	Modes used	TMS	IMS	HAMS		
				2	3	4
$\partial u_1/\partial b_1$	4	60.18	13.14	1.41	0.01	0.00
	8	26.66	1.01	0.03	0.00	0.00
	12	16.76	0.66	0.01	0.00	0.00
$\partial u_1/\partial b_1$	4	101.00	50.94	27.72	7.64	1.51
	8	48.65	10.32	1.29	0.38	0.03
	12	23.98	3.37	0.35	0.06	0.01
$\partial u_1/\partial b_1$	4	69.74	14.82	3.07	0.02	0.00
	8	26.60	1.39	0.23	0.00	0.00
	12	25.92	0.85	0.12	0.00	0.00
$\partial u_1/\partial b_1$	4	90.63	43.79	21.33	9.01	2.00
	8	44.86	6.65	1.56	0.35	0.04
	12	23.76	3.23	0.37	0.11	0.02
$\partial u_1/\partial b_1$	4	72.52	15.07	8.46	1.11	0.03
	8	37.95	1.22	0.57	0.03	0.00
	12	28.54	0.52	0.03	0.00	0.00
$\partial u_1/\partial b_1$	4	63.33	9.19	1.92	0.42	0.00
	8	27.56	0.75	0.49	0.21	0.00
	12	25.00	0.47	0.15	0.01	0.00

For the purpose of comparison, the errors of the truncation modal superposition (TMS), the IMS, and the high-accuracy superposition (HAMS) in computing the derivatives of eigenvectors are listed in Table 8.7.

For the sake of simplicity, the errors of eigenvector derivatives are represented by

$$\left|\left(\frac{\partial u_i}{\partial b_j}\right)_\varepsilon - \left(\frac{\partial u_i}{\partial b_j}\right)_\alpha\right|$$

where $(\partial u_i/\partial b_j)_\varepsilon$ denotes the exact solution and $(\partial u_i/\partial b_j)_\alpha$ denotes those obtained by the three methods presented above. In the computation of the eigenvector derivatives, the parameters m_1 and m_{10} are functions of the design variable b_1, the parameters c_8 and c_{15} are functions of design variable b_2, and K denotes the number of terms used in series (Equation 8.170).

The results in Table 8.7 confirm that the solution accuracy of the high-accuracy modal superposition is much higher than that of the TMS and the IMS. For example, if only the first four modes are used, the errors of $\partial u_2/\partial b_1$ are 101.00 and 50.94% for the truncated modal superposition and the IMS, and the errors are reduced to about 27.27, 7.64, and 1.51% for the high-accuracy modal superposition. If the first 12 modes are used, the error of $\partial u_1/\partial b_1$ is 16.76% for the truncated modal superposition, and the errors are reduced to 0.01, 0.00 and 0.00% for the case of $K = 2, 3, 4$ in the series (Equation 8.170), where the first four modes are used, respectively.

8.9.4 Matrix Perturbation for Repeated Eigenvalues of Nondefective Systems

8.9.4.1 Basic Equation

Consider a system having repeated eigenvalues, $S_0 = S_1 = \cdots = S_{0m}$, with multiplicity m, and the corresponding right and left modal matrices

$$\mathbf{U}_{0m} = [\,\mathbf{u}_{01} \quad \mathbf{u}_{02} \quad \cdots \quad \mathbf{u}_{0m}\,] \tag{8.171}$$

$$\mathbf{V}_{0m} = [\,\mathbf{v}_{01} \quad \mathbf{v}_{02} \quad \cdots \quad \mathbf{v}_{0m}\,] \tag{8.172}$$

the remaining eigenvalues being distinct.

The repeated eigenvalues satisfy the following equations:

$$\mathbf{A}_0\mathbf{U}_{0m} = \mathbf{B}_0\mathbf{U}_{0m}\mathbf{S}_0 \tag{8.173}$$

$$\mathbf{A}_0^{\mathrm{T}}\mathbf{V}_{0m} = \mathbf{B}_0^{\mathrm{T}}\mathbf{V}_{0m}\mathbf{S}_0 \tag{8.174}$$

$$\mathbf{V}_{0m}^{\mathrm{T}}\mathbf{B}_0\mathbf{U}_{0m} = \mathbf{I} \tag{8.175}$$

$$\mathbf{u}_{0i}^{\mathrm{T}}\mathbf{B}_0\mathbf{u}_{0i} = 1 \tag{8.176}$$

If small changes are made to the parameters, we have

$$\mathbf{A} = \mathbf{A}_0 + \varepsilon\mathbf{A}_1 \tag{8.177}$$

$$\mathbf{B} = \mathbf{B}_0 + \varepsilon\mathbf{B}_1 \tag{8.178}$$

The eigenvalues and eigenvectors of the perturbed system can be expressed as power series expansions in ε:

$$\mathbf{S}_m = \mathbf{S}_0 + \varepsilon\mathbf{S}_1 + \varepsilon^2\mathbf{S}_2 + \cdots \tag{8.179}$$

$$\mathbf{U}_m = \mathbf{U}_0 + \varepsilon\mathbf{U}_1 + \varepsilon^2\mathbf{U}_2 + \cdots \tag{8.180}$$

$$\mathbf{V}_m = \mathbf{V}_0 + \varepsilon\mathbf{V}_1 + \varepsilon^2\mathbf{V}_2 + \cdots \tag{8.181}$$

where

$$\mathbf{U}_0 = \mathbf{U}_{0m}\alpha \tag{8.182}$$

$$\mathbf{V}_0 = \mathbf{V}_{0m}\beta \tag{8.183}$$

and $\alpha_{m\times m}$ and $\beta_{m\times m}$ are to be determined coefficient matrices.

8.9.4.2 The First-Order Perturbation of Eigenvalues

The first-order perturbation diagonal matrix, $\mathbf{S}_1$, of the repeated eigenvalues and the coefficient matrix, $\boldsymbol{\alpha}$, can be obtained from the equations:

$$\mathbf{W}\boldsymbol{\alpha} = \boldsymbol{\alpha}\mathbf{S}_1 \tag{8.184}$$

$$\mathbf{W}^{\mathrm{T}}\beta = \beta\mathbf{S}_1 \tag{8.185}$$

$$\mathbf{W} = \mathbf{V}_{0m}^{\mathrm{T}}(\mathbf{A}_1 - \mathbf{S}_0\mathbf{B}_1)\mathbf{U}_{0m} \tag{8.186}$$

and the normalization conditions

$$\boldsymbol{\alpha}^{\mathrm{T}}\boldsymbol{\alpha} = \mathbf{I} \tag{8.187}$$

$$\boldsymbol{\beta}^{\mathrm{T}}\boldsymbol{\alpha} = \mathbf{I} \tag{8.188}$$

If matrix $\mathbf{W}$ has no repeated eigenvalues, $\boldsymbol{\alpha}$ and $\boldsymbol{\beta}$ can be uniquely determined. If $\mathbf{W}$ has repeated eigenvalues, we must consider the higher order perturbation equations for determining $\boldsymbol{\alpha}$ and $\boldsymbol{\beta}$. Here, we assume that S_{1i} are distinct eigenvalues, that is, $S_{1i} \neq S_{1j}$ $(i \neq j)$, where S_{1i} are the diagonal elements of $\mathbf{S}_1$.

8.9.4.3 The First-Order Perturbation of Eigenvectors

According to the modal expansion theorem, the first-order perturbation of the right and left eigenvectors, $\mathbf{U}_1$ and $\mathbf{V}_1$, can be expressed as

$$\mathbf{U}_1 = \mathbf{U}_{0m}\boldsymbol{\alpha}\mathbf{C}_m^1 + \mathbf{U}_A\mathbf{C}_A^1 \tag{8.189}$$

$$\mathbf{V}_1 = \mathbf{V}_{0m}\boldsymbol{\beta}\mathbf{D}_m^1 + \mathbf{V}_A\mathbf{D}_A^1 \tag{8.190}$$

where $\mathbf{C}_m^1$ and $\mathbf{C}_A^1$ are coefficient matrices which are to be determined, and $\mathbf{U}_A$ and $\mathbf{V}_A$ are the right and left modal matrices corresponding to the distinct eigenvalues:

$$\mathbf{C}_A^1 = (\mathbf{S}_A - \mathbf{S}_0)^{-1}\mathbf{V}_A^{\mathrm{T}}(\mathbf{S}_0\mathbf{B}_1 - \mathbf{A}_1)\mathbf{U}_{0m}\alpha \tag{8.191}$$

$$\mathbf{D}_A^1 = (\mathbf{S}_A - \mathbf{S}_0)^{-1}\mathbf{U}_A^{\mathrm{T}}(\mathbf{S}_0\mathbf{B}_1^{\mathrm{T}} - \mathbf{A}_1^{\mathrm{T}})\mathbf{V}_{0m}\boldsymbol{\beta} \tag{8.192}$$

The elements of matrix $\mathbf{C}_m^1$ can be computed by

$$C_{mij}^1 = \frac{R_{ij}^1}{\lambda_{1j} - \lambda_{1i}}, \quad i \neq j, \ \ i,j = 1,2,\ldots,m \tag{8.193}$$

where R_{ij}^1 are the nondiagonal elements of $\mathbf{R}^1$:

$$\mathbf{R}^1 = \boldsymbol{\beta}^{\mathrm{T}}\mathbf{V}_{0m}^{\mathrm{T}}\mathbf{B}_0\mathbf{U}_A\mathbf{C}_A^1\mathbf{S}_1 + \boldsymbol{\beta}^{\mathrm{T}}\mathbf{V}_{0m}^{\mathrm{T}}\mathbf{B}_1\mathbf{U}_{0m}\boldsymbol{\alpha}\mathbf{S}_1 + \boldsymbol{\beta}^{\mathrm{T}}\mathbf{V}_{0m}^{\mathrm{T}}\mathbf{B}_1\mathbf{U}_A\mathbf{C}_A^1\mathbf{S}_0 - \boldsymbol{\beta}^{\mathrm{T}}\mathbf{V}_{0m}^{\mathrm{T}}\mathbf{A}_1\mathbf{U}_A\mathbf{C}_A^1 \tag{8.194}$$

$$C_{mii}^1 = \frac{-1}{\mathbf{u}_{0i}^{\mathrm{T}}(\mathbf{B}_0 + \mathbf{B}_0^{\mathrm{T}})\mathbf{u}_{0i}}\left(\mathbf{u}_{0i}^{\mathrm{T}}\mathbf{B}_1\mathbf{u}_{0i} + \sum_{\substack{j=1\\ j\neq i}}^{m} c_{mij}^1\mathbf{u}_{0j}^{\mathrm{T}}(\mathbf{B}_0 + \mathbf{B}_0^{\mathrm{T}})\mathbf{u}_{0i} + \sum_{j=m+1}^{n} c_{Aij}^1\mathbf{u}_{0j}^{\mathrm{T}}(\mathbf{B}_0 + \mathbf{B}_0^{\mathrm{T}})\mathbf{u}_{0i}\right) \tag{8.195}$$

$$D_{mij}^1 = \frac{R_{ij}^2}{S_{1j} - S_{1i}}, \quad i \neq j, \ \ i,j = 1,2,\ldots,m \tag{8.196}$$

where R_{ij}^2 are the nondiagonal elements of $\mathbf{R}^2$:

$$\mathbf{R}^2 = \boldsymbol{\alpha}^{\mathrm{T}}\mathbf{U}_{0m}^{\mathrm{T}}\mathbf{B}_0\mathbf{V}_A\mathbf{D}_A^1\mathbf{S}_1 + \boldsymbol{\alpha}^{\mathrm{T}}\mathbf{U}_{0m}^{\mathrm{T}}\mathbf{B}_1^{\mathrm{T}}\mathbf{V}_{0m}\boldsymbol{\beta}\mathbf{S}_1 + \boldsymbol{\alpha}^{\mathrm{T}}\mathbf{U}_{0m}^{\mathrm{T}}\mathbf{B}_1^{\mathrm{T}}\mathbf{V}_A\mathbf{D}_A^1\mathbf{S}_0 - \boldsymbol{\alpha}^{\mathrm{T}}\mathbf{U}_{0m}^{\mathrm{T}}\mathbf{A}_1^{\mathrm{T}}\mathbf{V}_A\mathbf{D}_A^1 \tag{8.197}$$

$$D_{mii}^1 = Q_{ii}^2 - C_{mii}^1 \tag{8.198}$$

where Q_{ii}^2 are the diagonal elements of $\mathbf{Q}^2$:

$$\mathbf{Q}^2 = -\boldsymbol{\beta}^{\mathrm{T}}\mathbf{V}_{0m}^{\mathrm{T}}\mathbf{B}_1\mathbf{U}_{0m}\boldsymbol{\alpha} \tag{8.199}$$

8.9.5 Matrix Perturbation for Close Eigenvalues of Unsymmetric Matrices

Assume that $\mathbf{S}_0$ is a diagonal matrix with m close eigenvalues; $\mathbf{U}_{0n\times m}$ and $\mathbf{V}_{0n\times m}$ are the corresponding right and left eigenvectors matrices; $\mathbf{S}_A$ is the remaining distinct eigenvalue diagonal matrix; $\mathbf{U}_{An\times(n-m)}$ and $\mathbf{V}_{An\times(n-m)}$ are the corresponding right and left eigenvector matrices. They satisfy the following equations:

$$\mathbf{A}_0[\mathbf{U}_0 \vdots \mathbf{U}_A] = \mathbf{B}_0[\mathbf{U}_0 \vdots \mathbf{U}_A]\mathrm{diag}(\mathbf{S}_0, \mathbf{S}_A) \tag{8.200}$$

$$\mathbf{A}_0^{\mathrm{T}}[\mathbf{V}_0 \vdots \mathbf{V}_A] = \mathbf{B}_0^{\mathrm{T}}[\mathbf{V}_0 \vdots \mathbf{V}_A]\mathrm{diag}(\mathbf{S}_0, \mathbf{S}_A) \tag{8.201}$$

$$[\mathbf{V}_0 \vdots \mathbf{V}_A]^{\mathrm{T}}\mathbf{B}_0[\mathbf{U}_0 \vdots \mathbf{U}_A] = \mathbf{I} \tag{8.202}$$

$$\mathbf{u}_{0i}^{\mathrm{T}}\mathbf{B}_0\mathbf{u}_{0i} = 1 \tag{8.203}$$

Construct the matrix $\bar{\mathbf{A}}_0$ as

$$\mathbf{A}_0 = \bar{\mathbf{A}}_0 + \varepsilon\bar{\mathbf{A}}_0 \tag{8.204}$$

where

$$\bar{\mathbf{A}}_0 = \mathbf{B}_0[\mathbf{U}_0 \vdots \mathbf{U}_A]\mathrm{diag}(S_0\mathbf{I}, \mathbf{S}_A)[\mathbf{V}_0 \vdots \mathbf{V}_A]^{\mathrm{T}}\mathbf{B}_0 \tag{8.205}$$

$$\varepsilon\bar{\mathbf{A}}_0 = \mathbf{B}_0\mathbf{U}_0(\varepsilon[\delta\mathbf{S}_0])\mathbf{V}_0^{\mathrm{T}}\mathbf{B}_0 \tag{8.206}$$

$$\varepsilon[\delta\mathbf{S}_0] = \mathbf{S}_0 - S_0\mathbf{I} \tag{8.207}$$

$$S_0 = \frac{1}{m}\left(\sum_{k=1}^{n} S_{0i}\right) \tag{8.208}$$

Here, S_{0i} are the close eigenvalues, and S_0 is the average of S_{0i} $(i = 1, 2, \ldots, m)$.

It can be shown that the following equations hold:

$$\bar{\mathbf{A}}_0\mathbf{U}_0 = \mathbf{B}_0\mathbf{U}_0 S_0 \mathbf{I} \tag{8.209}$$

$$\bar{\mathbf{A}}_0^{\mathrm{T}}\mathbf{V}_0 = \mathbf{B}_0^{\mathrm{T}}\mathbf{V}_0 S_0 \mathbf{I} \tag{8.210}$$

These equations indicate that S_0 is the repeated eigenvalue with multiplicity, m, for the eigenproblem defined by Equation 8.209 and Equation 8.210, and $\mathbf{U}_0$ and $\mathbf{V}_0$ are the corresponding right and left modal matrices, respectively.

If small modifications $\varepsilon\mathbf{A}_1$ and $\varepsilon\mathbf{B}_1$ are imposed on the matrices $\mathbf{A}_0$ and $\mathbf{B}_0$, then the eigenproblems of the perturbed system become

$$(\bar{\mathbf{A}}_0 + \varepsilon\bar{\mathbf{A}}_1)\mathbf{U} = (\mathbf{B}_0 + \varepsilon\mathbf{B}_1)\mathbf{U}\mathbf{S} \tag{8.211}$$

$$(\bar{\mathbf{A}}_0 + \varepsilon\bar{\mathbf{A}}_1)^{\mathrm{T}}\mathbf{V} = (\mathbf{B}_0 + \varepsilon\mathbf{B}_1)^{\mathrm{T}}\mathbf{V}\mathbf{S} \tag{8.212}$$

where

$$\varepsilon\bar{\mathbf{A}}_1 = \varepsilon\bar{\mathbf{A}}_0 + \varepsilon\mathbf{A}_1 \tag{8.213}$$

The eigensolutions of Equation 8.211 and Equation 8.212 are given by

$$\mathbf{U} = \mathbf{U}_0\boldsymbol{\alpha} + \varepsilon\mathbf{U}_1 \tag{8.214}$$

$$\mathbf{V} = \mathbf{V}_0\boldsymbol{\alpha} + \varepsilon\mathbf{V}_1 \tag{8.215}$$

$$\mathbf{S} = \mathbf{S}_0 + \varepsilon\mathbf{S}_1 \tag{8.216}$$

It should be noted that Equation 8.211 and Equation 8.212 are the eigenproblem for repeated eigenvalues. That is, the perturbation analysis for close eigenvalues has been transferred into that of repeated eigenvalues. Hence, the methods given by Section 8.9.4 can be used to compute $\mathbf{S}_1$, $\boldsymbol{\alpha}$, $\boldsymbol{\beta}$, $\mathbf{U}_1$, and $\mathbf{V}_1$ in Equation 8.214 to Equation 8.216.

References

Adelmen, H.M. and Haftka, R.T., Sensitivity analysis of discrete structural systems, *AIAA J.*, 24, 823, 1986.

Adhikari, S. and Friswell, M.I., Eigenderivative analysis of asymmetric non-conservative systems, *Int. J. Numer. Methods Eng.*, 39, 1813, 2001.

Chen, S.H. 1993. *Matrix Perturbation Theory in Structural Dynamics*, International Academic Publishers, Beijing.

Chen, S.H. and Liu, Z.S., High accuracy modal superposition for eigenvector derivatives, *Chin. J. Mech.*, 25, 432, 1993a.

Chen, S.H. and Liu, Z.S., Perturbation analysis of vibration modes with close frequencies, *Commun. Numer. Methods Eng.*, 9, 427, 1993b.

Chen, S.H. and Pan, H.H., Design sensitivity analysis of vibration modes by finite element perturbation, *Proc. of the Fourth IMAC*, 38, 1986.

Chen, J.C. and Wada, B.K., Matrix perturbation for structural dynamics, *AIAA J.*, 15, 1095, 1979.

Dailey, R.L., Eigenvector derivatives with repeated eigenvalues, *AIAA J.*, 27, 486, 1989.

Fox, R.L. and Kapoor, M.P., Rates of change of eigenvalues and eigenvectors, *AIAA J.*, 12, 2426, 1968.

Haug, E.J., Komkov, V., and Choi, K.K. 1985. *Design Sensitivity Analysis of Structural Systems*, Academic Press, Orlando, FL.

Haug, E.J. and Rousselet, B., Design sensitivity analysis in structural dynamics, eigenvalue variations, *J. Struct. Mech.*, 8, 161, 1980.

Hu, H. 1987. *Natural Vibration Theory for Multi-DoF Structures*, Science Press, Beijing (in Chinese).

Lim, K.B., Juang, J.N., and Ghaemmaghani, P., Eigenvector derivatives of repeated eigenvalues using singular value decomposition, *J. Guidance Control Dyn.*, 12, 282, 1989.

Liu, J.K., Perturbation technique for non-self-adjoint systems with repeated eigenvalues, *AIAA J.*, 37, 222, 1999.

Liu, X.L., Derivation of formulas for perturbation analysis with modes of close eigenvalues, *Struct. Eng. Mech.*, 10, 427, 2000.

Liu, Z.S. and Chen, S.H., Contribution of the truncated modes to eigenvector derivatives, *AIAA J.*, 32, 1551, 1994a.

Liu, Z.S. and Chen, S.H., An accurate method for computing eigenvector derivatives for free–free structures, *Int. J. Comput. Struct.*, 52, 1135, 1994b.

Mills-Curran, W.C., Calculation of eigenvector derivatives for structures with repeated eigenvalues, *AIAA J.*, 26, 867, 1988.

Murthy, D.V. and Haftka, R.T., Derivatives of eigenvalues and eigenvectors of a general complex matrix, *Int. J. Numer. Methods Eng.*, 26, 293, 1988.

Ojalvo, E.U., Efficient computation of modal sensitivities for systems with repeated frequencies, *AIAA J.*, 26, 361, 1988.

Rogers, L.C., Derivatives of eigenvalues and eigenvectors, *AIAA J.*, 8, 943, 1977.

Shaw, J. and Jayasuriya, S., Modal sensitivities for repeated eigenvalues and eigenvalues derivatives, *AIAA J.*, 30, 850, 1992.

Wang, B.P., Improved approximate methods for computing eigenvector derivatives in structural dynamics, *AIAA J.*, 29, 1018, 1991.

9

Vibration in Rotating Machinery

H. Sam Samarasekera
Sulzer Pumps (Canada), Inc.

Summary

This chapter concerns vibration in rotating machinery. Although it is impractical to totally eliminate such vibrations, it is essential that they be controlled to within acceptable limits for safe and reliable operation of such machines. The two major categories of vibration phenomena that occur in rotating machinery are forced vibration and self-excited instability. Monitoring, diagnosis and control of these vibrations requires a sound understanding of rotor dynamics in machinery. Predicting the vibration behavior of a rotating machine by analytical means has become customary in many industries. With the advent of computer technology, several computer-based programs have been developed to accurately predict the behavior of rotating machinery. Significant strides in modeling techniques have also been made over the past century to accurately represent components such as shaft sections, disks, impellers, bearings, seals, rotor dampers, and rotor–stator interactions. This has enhanced the accuracy and reliability of both analytical and computational procedures. The chapter presents useful techniques of analysis, measurement, diagnosis, and control of vibration in rotating machinery.

9.1 Introduction

Vibrations are an inherent part of all rotating machinery. Residual mass imbalance and dynamic interaction forces between the stationary and rotating components, which are practically impossible to eliminate, cause these vibrations. The challenge is to identify the source of vibration and control it to within reasonable limits. Because of economic advantages, the trend in industry has been to move towards high speed, high power, lighter and more compact machinery. This has resulted in machines operating above their first critical speeds, which was unheard of in the past. The new operating parameters have required concurrent development of vibration technology without which it is not possible to safely and reliably operate such machinery. Industry has also come to realize that *vibration* is an essential phenomenon, which could be used to assess the performance, durability, and reliability of rotating machinery.

Engineers at different levels approach the subject of vibration in rotating machinery differently. The machinery designer has to recognize the potential sources of vibration and control them to within acceptable levels. In the past few decades, owing to the advancement in computers and modeling techniques, better understanding of the dynamics of rotating machinery, including the identification of potential sources of vibration, has been realized. This has enabled designers to accurately predict the rotordynamic behavior of machinery, allowing it to reach higher operating speeds and larger energy capacities safely and reliably.

Approaching vibration from a different perspective, the maintenance engineer uses vibration standards and guidelines to monitor the health of equipment for their timely repair and refurbishment. Reliable vibration monitoring and diagnostics techniques have moved industry into *predictive* rather than *preventive* maintenance practices, which considerably reduce plant downtimes that rely on key rotating machinery. Premature replacement of machinery components has also been minimized. The resulting financial and economic benefits provide an added incentive for the study and understanding of vibration in rotating machinery.

The vibration specialist or troubleshooter has to use his knowledge of rotordynamics and his diagnostic capabilities to solve vibration problems in rotating machinery. In most cases, it is also important to have an understanding of the interfacial dynamics of the rotating machinery with the surrounding system in order to solve a vibration problem.

From a safety and reliability standpoint, the public must be concerned with vibration in rotating machinery. Their concerns are addressed through vibration standards and guidelines. These procedures have been developed for rotating machinery by numerous organizations, both at the national and international levels. Some of these standards are industry specific and some are equipment type specific, while a number of them try to cover a wide range of rotating machinery. The objective of most of these standards is to establish and control quality, safety, durability and reliable performance of rotating machinery for the benefit of those who use or operate it.

9.1.1 History of Vibration in Rotating Machinery

Although various types of rotating machinery have been in use for many centuries, understanding of their rotordynamic behavior did not begin until 1869 (Rankine, 1869). Since that time, there has been steady growth in the development and understanding of the vibration behavior of rotating machinery. A tabulation of major historical events that have contributed to this growth is presented in Table 9.1.

- All rotating machinery vibrates to some degree. For public safety and machine reliability, the vibrations have to be controlled to within acceptable limits.
- Modern trends towards more sophisticated, higher speed compact rotating machinery have contributed to the rapid development in vibration technology through a better understanding of their rotordynamics.
- Vibration technology is integrated into the areas of design, maintenance, and troubleshooting of rotating machinery.
- From a safety and reliability standpoint, the public is protected by the implementation of vibration standards and guidelines.
- The first publicly reported rotordynamic study was made in 1869.

TABLE 9.1 A Chronological Listing of Major Contributions that Have Led to the Development and Understanding of Vibration in Rotating Machinery

Year	Contributor	Description
1869	Rankine, W.J.M.	He examined the equilibrium of a frictionless, uniform shaft disturbed from its initial position. The resulting recorded article is recognized to be the first on the subject of rotor dynamics. He proposed that motion is stable below the first critical speed, is neutral or indifferent at the critical speed, and unstable above the critical speed He also developed numerical formulae for critical speeds for the cases of a shaft resting freely on a bearing at each end and for an overhanging shaft fixed in direction at one end
1883	Greenhill, A.G.	He studied the effect of end thrust and torque on the stability of a long shaft and concluded that they were both unimportant. He also obtained formulae for the cases of an unloaded shaft resting on bearings at each end and fixed in direction at each end
Circa 1890	Reynolds, O.	He extended the theory developed by Rankine and Greenhill for the case of a shaft loaded with pulleys
1893	Dunkerley, S.	He developed formulae for critical speeds for loaded shafts in terms of the diameter of the shaft, weights of pulleys, the manner in which the shaft is supported, and so on, and verified them by experiment He postulated that any degree of unbalance will excite the shaft at the critical speed to very high amplitudes and that it is possible to operate above the first critical speed. The dependence of critical speed on the moment of inertia of the rotating pulley was identified
1894	Rayleigh, J.W.S.	He developed an approximate method to calculate the natural frequency of a continuous beam with distributed mass and flexibility using the energy method
1895	DeLaval, G.	He was responsible for the first experimental demonstration that a steam turbine is capable of sustained operation above the first critical speed
1916	Timoshenko, T.	He discovered the effects of transverse shear deflection on the natural frequency of a continuous beam and applied the principle to the case of the rotating shaft
1919	Jeffcott, H.H.	He examined the effect of unbalance on the whirl amplitudes and the forces transmitted to the bearings. The case of a light uniform shaft supported freely on bearings at its ends and carrying a thin pulley of mass m at the center of the span was studied. He assumed the moment of inertia of the pulley to be negligible. Using this model, later known as the *Jeffcott model*, a comprehensive theory was developed to explain the behavior of the rotor as it passed through the critical speed The effect of damping on the whirl amplitude, a phase change of angle π as it passes through the critical speed, and the concept of synchronous rotor whirling (precession) were introduced and explained. He also recognized that with a separation margin of 10% on either side of a critical speed, the amplitude of vibration would not be excessive. He demonstrated that it is better from the vibration point of view to design the shaft with its critical speed below the working speed rather than to have a critical speed the same proportion above the working speed. Accordingly, he explained the behavior of the De Laval steam turbine and the economic advantages of operation above the critical speed
1921	Southwell, R.V. and Gough, B.S.	They found that a torque and an end thrust of constant magnitude lowers the critical speed of a rotating shaft, disproving Greenhill's earlier (1883) conclusions

(*continued on next page*)

TABLE 9.1 (*continued*)

Year	Contributor	Description
1921	Holzer, H.	He developed a numerical method to calculate torsional critical speeds and mode shapes for a multidisk rotor system
1924	Newkirk, B.L.	He observed that a rotor operating at a speed above the first critical speed can enter into high, violent whirling and the center of the rotor will precess in the forward direction at a rate equal to that of the critical speed. Unlike in the case of synchronous whirling, if the speed is increased beyond the initial whirl speed, the whirl amplitude will continue to increase, eventually leading to failure. This was the first time that it was realized that nonsynchronous unstable motion can exist in a high-speed rotor
		Based on experiments, he made the following key observations on nonsynchronous whirling. The amplitude and the onset speed of whirling are independent of the rotor balance. Whirling always occurs at speeds above the critical speed, and the whirl speed is always constant at the critical speed, regardless of the rotor speed. The whirl threshold speed can vary even for machines of similar construction. Whirling occurs only in built up rotors, and not in single piece constructions. Increasing the foundation flexibility, distortion or misalignment of the bearing housings, or introducing damping to the foundation or increasing the axial thrust bearing load, increased the threshold speed of whirling
1924	Kimball, A.T.	Suggested that internal friction or viscous action due to bending may cause a shaft to whirl when rotating at any speed above the first critical speed. He postulated that the nonsynchronous whirling observed by Newkirk was due to this phenomenon
1924	Newkirk, B.L.	Based on Mr Kimball's theory, he concluded that similar frictional forces are generated at the mating face between the shrunk on disk and the shaft of a built-up rotor, and the nonsynchronous whirling observed by him was due to this effect. However, he was unable to explain some of his experimental findings, in particular, the effects of bearing or foundation flexibility, damping, and misalignment
1925	Newkirk, B.L.	He experienced another form of nonsynchronous whirling, similar but different to that caused by the frictional effects of a shrink-fit disk. It occurred at rotor speeds just exceeding twice the first critical speed on shafts mounted on journal bearings. He recognized that the oil in the journal bearing was responsible for the violent motion and called it *oil whip*. The whirl speed and direction of whirling were the same as that for friction induced whirling, that is, the first critical speed in the forward direction. A theory to explain how the oil film can produce the whirling motion of a journal and to account for why it took the same direction as rotation of the shaft was proposed. However, the theory does not explain why whirling does not commence until the rotor speed reaches twice the critical speed value. The influence of foundation flexibility on the rotor stability was also found to be confusing to Newkirk. In the case of friction-induced whirl, he was able to totally eliminate the rotor instability by means of a flexibly mounted bearing. When this was tried with the journal bearings, the whirl amplitudes magnified. External damping at the bearing was found to have a favorable influence on whirl amplitudes
1925	Stodola, A.	He developed an iterative procedure to calculate the fundamental frequency of a vibrating system based on an assumed mode shape
1927	Stodola, A.	He provided an explanation and formulae for the gyroscopic moment effect on the critical speed of a rotor. He also introduced the notion of synchronous and nonsynchronous reverse precession of a rotor under specific conditions

TABLE 9.1 *(continued)*

Year	Contributor	Description
1933	Robertson, D.	In order to understand oil whip, he studied the stability of the ideal 360° infinitely long journal bearing, and erroneously concluded that the rotor will be unstable at all speeds and not only at speeds above twice the critical speed value
1933	Smith, D.M.	He studied the case of unsymmetrical rotors on unsymmetrical supports and obtained four different critical speed values in comparison to the single value for a symmetrical system. He also discussed the presence of additional critical speeds due to gyroscopic effects of large disks
1944	Myklestad, N.	A lumped parameter transfer matrix method to calculate natural frequencies for airplane wings was developed by him
1945	Prohl, M.	He developed a lumped parameter transfer matrix method for calculating critical speeds of flexible rotors
1953	Poritsky, H.	Using the small displacement theory, he derived a radial stiffness coefficient for the journal bearings and analyzed the rotor behavior under oil whirl conditions. He concluded that the rotor was stable below twice the critical speed and indicated that increasing the rotor or bearing flexibility will reduce the threshold speed of instability. He also proposed a stability criterion for a rotor based on the bearing and rotor stiffness
1953	Miller, D.F.	He introduced a solution to the steady-state forced vibration problem, for a beam or rotating shaft on damped, flexible end supports. The response of the rotor to an unbalance force and the damped resonance frequencies are calculated by this method
1955	Pinkus, O.	He investigated oil whirl in various journal bearing types and made the following major conclusions. The unbalance of the rotor has minimal effect on stability. The threshold of instability occurs at approximately twice the first critical speed of the rotor. In the unstable region, the whirl frequency remained constant at the first critical speed, irrespective of the shaft rotating speed. At speeds nearly equal to three times the first critical speed, whipping motion stops with a heavy shaft rotor, whereas with a light shaft rotor it does not cease. High loads, high viscosity, flexible mountings, and bearing asymmetry favor stability
1958	Lomakin, A.	The influence of the dynamic characteristics of seals on the critical speeds and stability of pump rotors were introduced by him
1958	Thomas, H.	He proposed that an eccentric turbine rotor would generate a destabilizing force due to the circumferential variation in clearance
1966	Gunter, E.J. Jr.	He combined the different theories on whirling developed by the rotor dynamist and the bearing specialist, and elegantly explained some of the conflicting experimental evidence gathered thus far. He emphasized the importance of considering the combined effects of rotor parameters and the bearing and foundation characteristics on rotor stability, and developed more comprehensive criteria for self-excited whirl instability
1969	Black, H.F.	He provided a comprehensive analysis of annular pressure seals on the vibrations of pump rotors
1970	Ruhl, R.	He introduced finite element models for flexible rotors for calculating rotor critical speeds and mode shapes. These models did not take into account gyroscopic effects and axial loading
1974	Lund, J.	A transfer matrix method to calculate damped critical speeds of a rotor taking into account the cross coupling terms as well were introduced by him
1976	Nelson, H. and McVaugh, J.	They extended the finite element model of a rotor to account for rotary inertia, gyroscopic effect, and axial loads

(*continued on next page*)

TABLE 9.1 (*continued*)

Year	Contributor	Description
1978–1980	Benckert, H. and Wachter, J.	A method to calculate flow induced spring constants for labyrinth gas seals and the use of *swirl breaks* to reduce the destabilizing force caused by tangential velocity in labyrinth seals was introduced by them
1980	Nelson, H.	He further developed a finite element model of a rotor to include shear deflection and axial torque effects
1980	Brennen, C. et al.	They recognized the presence of substantial shroud forces, which influences the rotor dynamics of a pump
1986	Muszynska, A.	She demonstrated that oil whirl occurs at about one-half the running speed in a vertical rotor. With further increase in speed, oil whip will commence when the whirl frequency approaches the critical speed of the rotor

9.2 Vibration Basics

The vibration phenomena that manifest in rotating machinery can be divided into two major categories: *forced vibration* and *self-excited instability.* A stimulus or a source of excitation is required to initiate and sustain vibratory motion in a rotor. When the stimulus is a forcing phenomenon such as mass unbalance, it will produce forced flexural vibration in the rotor analogous to linear forced vibration response in a simple spring–mass system. On the other hand, self-excited vibration (instability) does not require a forcing phenomenon for its initiation or sustenance. A description of these phenomena is given next.

9.2.1 Forced Vibration

A rotating force vector (unbalance), a steady directional force (gravity), or a periodic force (pump impeller/diffuser interface action), will cause forced vibration in a rotating machine. The response of the rotor will depend on the nature of the forcing function and how it relates to rotor characteristics. The rotor responses to the most common excitation phenomena are examined below.

9.2.1.1 Unbalance Response — Synchronous Whirling

As an introduction to the theory on rotating machinery vibration and understanding unbalance response, it is most appropriate to examine Jeffcott's (1919) rotor, which is a simple model that has many of the basic characteristics of more complex rotating machinery. The Jeffcott rotor represents a massless elastic shaft supported freely in bearings at its ends and carrying a disk of mass m at the center of its span. The mass center of the disk is eccentric to its geometric center by a distance e. Refer to Figure 9.1.

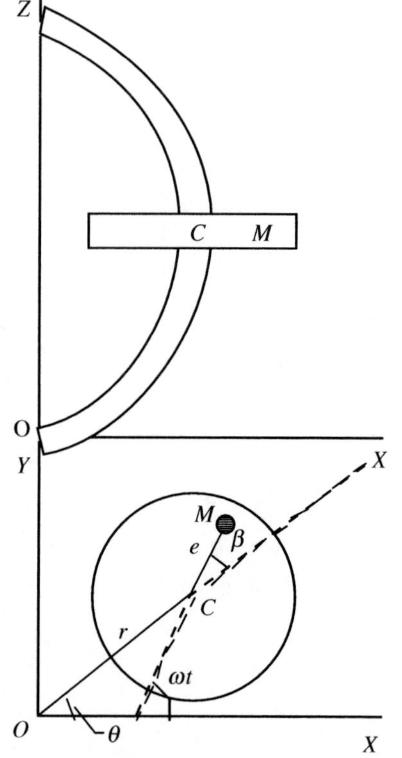

FIGURE 9.1 Jeffcott rotor.

$C =$ geometric center of the disk
$\beta =$ phase angle

$M =$ mass center of the disk
$c =$ viscous damping coefficient of on the rotor
$O =$ bearing center
$r =$ deflection of rotor from origin
$\theta =$ angle of precession
$k =$ shaft stiffness
$\omega =$ angular velocity of the rotor $= \dot{\theta} + \dot{\beta}$

Whirling is defined as the angular velocity of rotation of the rotor geometric center (c) or the time derivative ($\dot{\theta}$) of the angle of precession (θ) (also see Chapter 7). *Synchronous whirling* is when the rate of whirling, $\dot{\theta}$, is equal to the total angular velocity, ω, of the system.

Applying Newton's Laws of motion to the rotor, the differential equations of motion in polar coordinates (r, θ) are obtained as

$$\ddot{r} + \frac{c}{m}\dot{r} + \left(\frac{k}{m} - \dot{\theta}^2\right)r = e\omega^2 \cos\beta \tag{9.1}$$

$$r\ddot{\theta} + \left(\frac{c}{m}r + 2\dot{r}\right)\dot{\theta} = e\omega^2 \sin\beta \tag{9.2}$$

For a steady-state condition, the values of r, β, $\dot{\theta}$, and ω are constant. For synchronous whirling, Equation 9.1 and Equation 9.2 reduce to

$$\left(\frac{k}{m} - \omega^2\right)r = e\omega^2 \cos\beta \tag{9.3}$$

$$\frac{c}{m}\omega r = e\omega^2 \sin\beta \tag{9.4}$$

From Equation 9.3 and Equation 9.4

$$r = \frac{e\omega^2}{\sqrt{\left(\frac{k}{m} - \omega^2\right)^2 + \left(\frac{c\omega}{m}\right)^2}} \tag{9.5}$$

$$\beta = \tan^{-1}\frac{c\omega}{m\left(\frac{k}{m} - \omega^2\right)} \tag{9.6}$$

$$F = \frac{kr}{2} = \frac{ke\omega^2}{2\sqrt{\left(\frac{k}{m} - \omega^2\right)^2 + \left(\frac{c\omega}{m}\right)^2}} \tag{9.7}$$

Using the following relationships:

$\omega_N = \sqrt{\frac{k}{m}}$ — Natural frequency of rotor without damping

$c_{cr} = 2\sqrt{km}$ — Critical damping coefficient

$\zeta = \frac{c}{c_{cr}}$ — Damping ratio

Equation 9.6 and Equation 9.7 are reduced to the following nondimensional form:

$$\frac{r}{e} = \frac{2F}{ke} = \frac{(\omega/\omega_N)^2}{\sqrt{(1 - (\omega/\omega_N)^2)^2 + (2\zeta\omega/\omega_N)^2}} \tag{9.8}$$

$$\beta = \tan^{-1}\frac{2\zeta(\omega/\omega_N)}{(1 - (\omega/\omega_N)^2)} \tag{9.9}$$

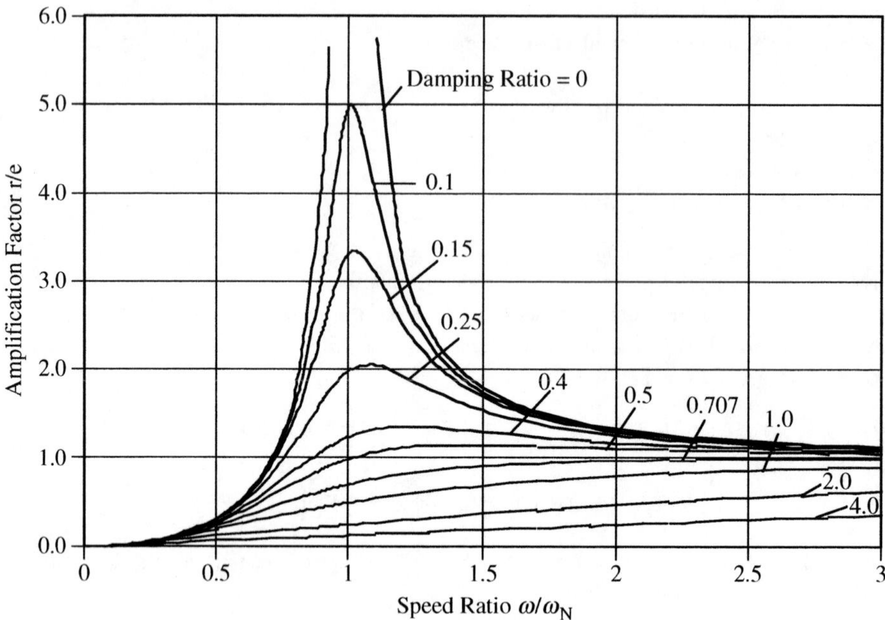

FIGURE 9.2 Jeffcott rotor response with mass eccentricity — amplification vs. speed.

Figure 9.2 is a graphical representation of the unbalance response of the rotor as a function of rotating speed, ω. Upon examination of the phase relationship, it is important to note that the phase angle, β, changes from approximately 0° at low speed to values approaching 180° at the higher speed. At ω_N, $\beta = 90°$. A pictorial illustration of this phenomenon is given in Figure 9.3.

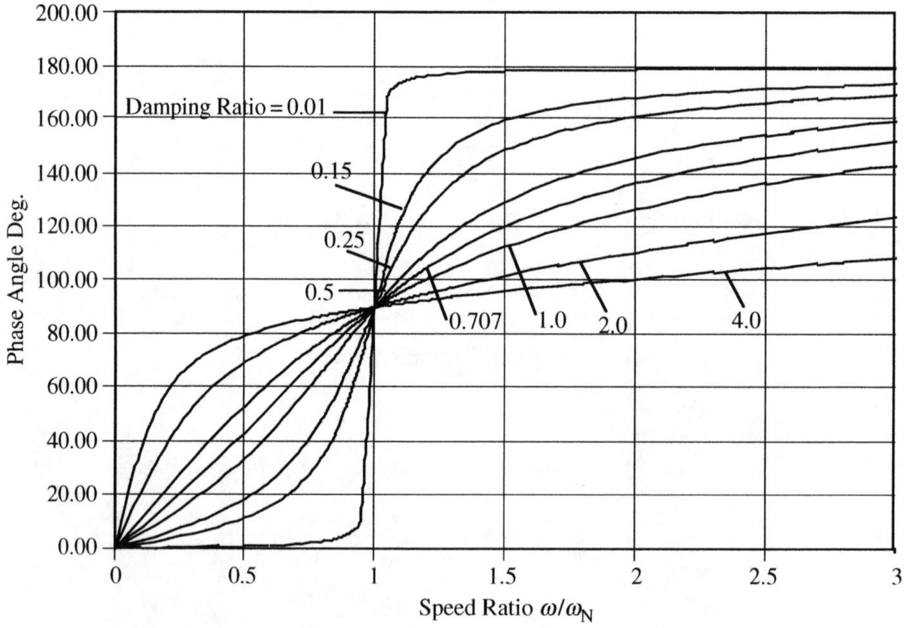

FIGURE 9.3 Jeffcott rotor response to unbalance — phase angle vs. speed.

For the case of zero damping, when $\omega = \omega_N$, the rotor deflection and the bearing forces are unbounded. For all other cases, the rotor deflection and the bearing forces are bounded, and their amplitude depends on the damping ratio. If a shaft is quickly accelerated through its critical speed to a higher working speed, then there may not be enough time for large rotor deflection to take place. At high speeds, $\omega >> \omega_N$, the amplitude of the rotor deflection decreases and approaches the value e, the eccentricity of the rotor.

The critical speed, ω_{cr}, of a rotor in the general case, is the speed at which the rotor deflection amplitude or the force amplitude transmitted to the bearings is a maximum.

This implies that, at $\omega = \omega_{cr}$

$$\frac{dr}{d\omega} = \frac{dF}{d\omega} = 0$$

Using Equation 9.8, the following relationship between the natural frequency of the rotor and its critical speed is derived:

$$\omega_{cr} = \frac{\omega_N}{\sqrt{1 - 2\zeta^2}} \tag{9.10}$$

From Equation 9.10, it is evident that the critical speed of a rotor is not a fixed value and is dependent on the degree of rotor damping. When $\zeta = 1/\sqrt{2}$, the system is said to be critically damped.

It is important to note that rotor response to unbalance (or imbalance) is recognizable and controllable. The amplitude of the force transmitted to the bearing can be reduced by operation at speeds above the critical speed, reducing unbalance, increasing viscous damping, and avoiding operation close to critical speeds.

9.2.1.2 Shaft Bow

A rotor with a bent shaft will behave in a similar manner to a rotor with an eccentric mass (Ehrich, 1999). At high rotor speeds ($\omega >> \omega_{cr}$), the shaft will tend to correct the bow as illustrated in Figure 9.4. When shaft bow is combined with mass eccentricity, unique behavior patterns are produced depending on the phase angle between the bow and the eccentric mass (Childs, 1993).

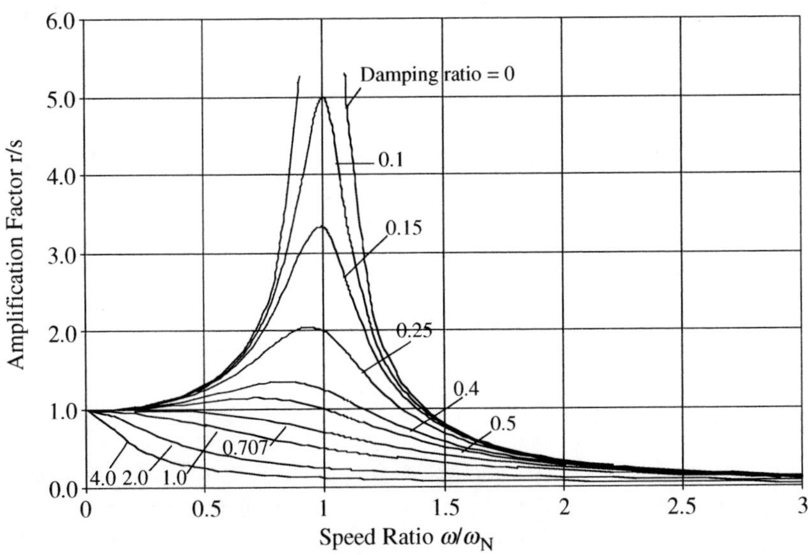

FIGURE 9.4 Jeffcott rotor response with shaft bow — amplification vs. speed.

9.2.1.3 Gravity Critical

A special case of synchronous whirling may occur in certain types of horizontal rotors due to the gravitational force. It is a secondary critical speed commonly called the *gravity critical*, which can occur in a very heavy lightly damped rotor. The critical speed will occur at approximately half the natural frequency of the rotor and its amplitudes of deflection at the critical speed are bounded and approximately twice the static deflection of the rotor (Gunter, 1966).

9.2.1.4 The Influence of Rotor Inertia and Gyroscopic Action

The effect of rotor inertia is ignored in the Jeffcott model. However, in practice, it is recognized that rotor inertia and gyroscopic action has an influence on the natural frequencies, critical speeds, and unbalance response of the rotor, including reverse whirling. In the case of the natural frequency of the rotor (zero speed), the diametral or rotary inertia provides an additional natural frequency associated with the rotational degree of freedom (DoF). Also, the inertia effect lowers the first natural frequency (Childs, 1993). In the rotating case, the effect of inertia generates both forward and reverse whirling critical speeds (Childs, 1993). These forward whirling critical speeds tend to be higher (stiffening effect) and the reverse whirling critical speed lower than the natural frequency of the rotor. At the forward critical speeds, large amplitude whirling motion due to imbalance occurs, whereas the reverse critical speeds are insensitive to imbalance of the rotor.

9.2.1.5 Rotor Housing Response across an Annular Clearance

If the rotor deflection due to imbalance exceeds the uniform annular gap, continuous contact would occur between the rotor and stator resulting in coupled motion between the rotor and stator (Childs, 1993). For low contact frictional forces, synchronous forward whirling driven by the imbalance forces will occur. If the contact friction force is large enough to prevent slipping between the rotor and stator, reverse whirling will take place. For the case of synchronous forward whirling in a certain range of running speeds, instability will occur due to engagement between the rotor and stator (Black, 1968). The zones of instability depend on the coupled natural frequency of the rotor and stator and the degree of rotor deflection with respect to the annular gap.

9.2.1.6 Effect of Nonlinearity and Asymmetry on Forced Vibration Response

The foregoing analysis has assumed that stiffness and damping are linear and symmetric and the resulting forces are proportional to the deflection and velocity of the rotor. However, in reality, rotating machinery components have inherent nonlinearities and asymmetries that can have a profound influence on their rotordynamic behavior. At large amplitudes of motion, stiffness and damping coefficients become nonlinear and result in modifying the response amplitude and critical speeds of the rotor. Nonlinearity in the support stiffness will introduce considerable distortion to the otherwise simple harmonic vibration behavior of a purely linear system. The stiffness and damping coefficients of the bearings and their supports are asymmetric in most cases, in particular in horizontal machines. As a result, the forced vibratory responses in the two principal directions are different and can behave independent of each other. Each principal direction will display a critical speed unique to itself. Ehrich (1999) has presented a discussion on how nonlinearity and asymmetry of stator systems influence forced vibration response.

The influence of rotor stiffness asymmetry and inertia asymmetry on rotor stability is discussed in Section 9.2.3.

9.2.2 Self-Excited Vibration

Instability (nonsynchronous whirling) is a self-induced excitation phenomenon, sometimes described as *sustained transient motion*, that can occur in rotating machinery. At the inception of instability, the rotor deflection will continue to build up with increase in speed, whereas in the case of a critical speed resonance, the amplitude of the deflection reaches a maximum value and then decreases. If the rotor speed is increased above the instability threshold speed, the large amplitudes of motion will normally

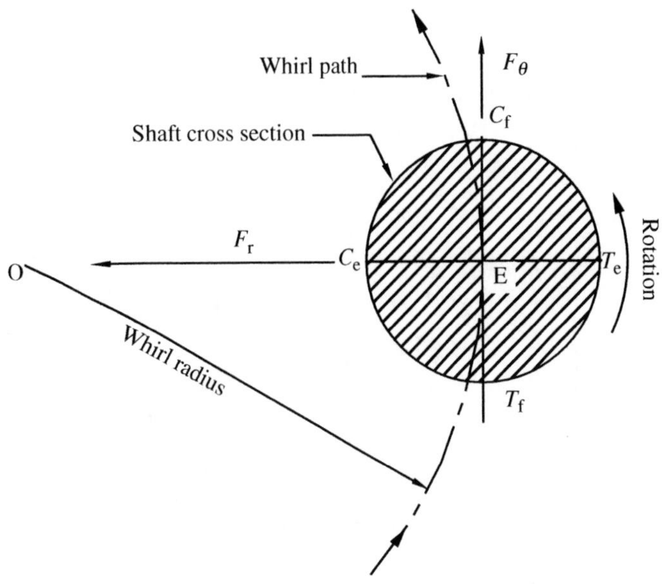

FIGURE 9.6 Internal friction damping forces acting on a rotor.

whirling motion increases. The force, F_θ, will oppose the external damping force, $c\dot{\theta}r$. At the eshold of instability, the two forces nullify each other. It is also know that the frequency of whirling, $\dot{\theta}$, he threshold of instability equals the natural frequency, ω_n, of the rotor. Mathematically it can be ressed as follows:

$$F_\theta = c_i r(\omega - \dot{\theta}) \quad (9.11)$$

ere c_i is the rotor internal damping coefficient:

$$c\dot{\theta}r = c_i r(\omega - \dot{\theta}) \quad (9.12)$$

$$\dot{\theta} = \omega_N \quad (9.13)$$

uation 9.12 and Equation 9.13 yield the following relationship between the threshold speed of tability, the first critical speed, and the damping factors (both internal and external):

$$\frac{\omega}{\omega_N} = 1 + \frac{c}{c_i} \quad (9.14)$$

.2.2 Tip Clearance Excitation (Alford's Force, Steam Whirl)

omas (1958) investigated the instability of steam turbines and suggested that nonsymmetric dial clearances caused by an eccentric rotor could result in destabilizing forces, and called them *arance excitation forces*. Subsequently Alford (1965) discovered a similar phenomenon in aircraft gas bines and, as a result, the destabilizing forces are sometimes referred to as *Alford forces* in orth America.

The destabilizing force is created as a result of the variation in the gap between the blade tip and the tor. When the gap decreases, the leakage decreases and consequently the efficiency increases, resulting a torque higher than the average torque produced by a uniform gap. When the gap increases, there is a rresponding decrease in the torque relative to the average. The variation in torque produced by the centricity results in a tangential force, which is normal to the radial deflection and is in the direction of e whirling motion as shown in Figure 9.7. Furthermore, it has been illustrated that the magnitude the resulting force increases proportionately with the increase in rotor deflection, that is, the decrease

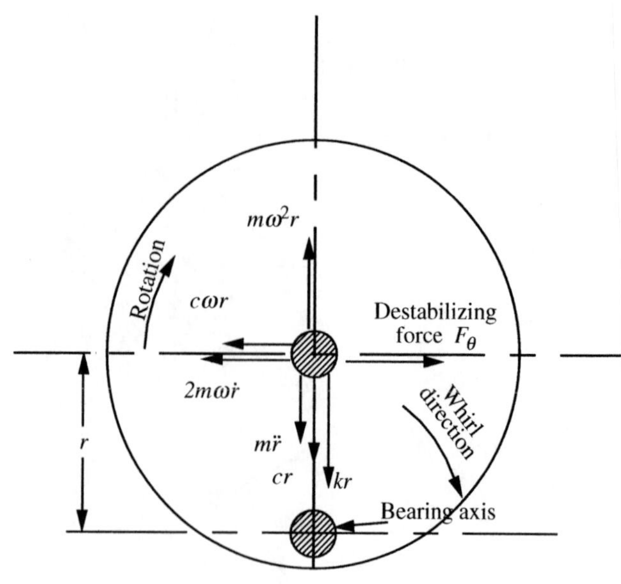

FIGURE 9.5 Rotor instability — general case.

result in damage to the machine. Unlike forced flexural vibration, rotor instability does not require a sustained forcing phenomenon to initiate or maintain the motion. only in machines operating at speeds well above the critical speeds of the rotor. Fur whirling speed is different to the rotor speed (nonsynchronous whirling), and it is ide speed irrespective of the rotor speed.

In general, the rotor instability is associated with the existence of a tangential force right angles to the deflection vector and directly opposing the damping force vect Figure 9.5. The nature of F_θ is such that its magnitude increases proportionat deflection. At the point where F_θ equals the external damping force, rotor instabil due to the nullification of the stabilizing force. This will produce a whirling motion amplitude. Several phenomena inherent in the rotor system that generates such tang have been identified and are discussed below. Rotordynamists believe that there still phenomena to be discovered.

9.2.2.1 Internal Friction Damping

This type of instability was first experienced in the early 1920s in blast furnace compre General Electric Company. These machines were subject to occasional fits of violent v (1924) carried out a series of experiments to understand the unusual behavior of these on the *internal friction theory* of Kimball (1924). Newkirk (1924) concluded that the i damping forces at the disk shaft interface caused the subsynchronous whirling.

In order to understand the internal friction damping phenomena let us examine th the whirling Jeffcott rotor (Figure 9.1). Figure 9.6 is a cross section of the shaft disk to its deflection, all of the fibers in the right half of the cross section are in tension, T_e, an half are in compression, C_e. These fiber stresses tend to straighten the shaft and produce F_r, which opposes the centrifugal force, $m\dot{\theta}r$. Furthermore, a set of frictional forces are shaft disk interface due to stretching and compression of the fibers. The fibers in the b will be stretched and are under frictional tension, T_f, and those in the upper half are b under frictional compression C_f. Similarly to the reaction force, F_r, produced from right C_e, a reaction force, F_θ, from bottom to top will be produced by the frictional stre The disturbing force, F_θ, is in the same direction as the whirling motion and as a result

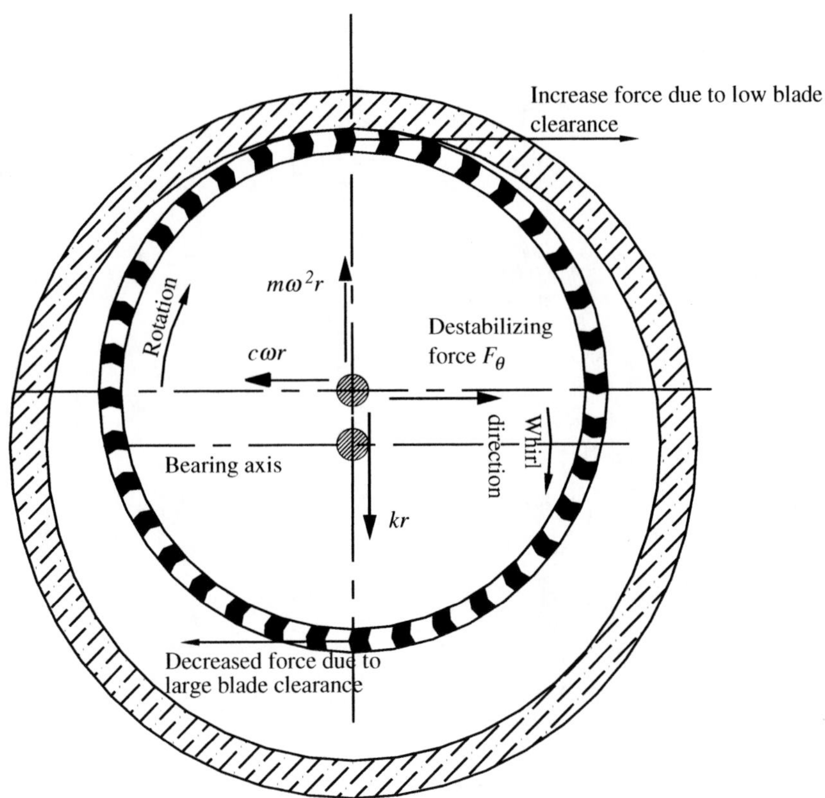

FIGURE 9.7 Tip clearance excitation (Alford forces).

in the gap. The resulting force is a destabilizing force that will oppose the external damping force, and at some point when they balance each other, rotor instability will occur. A detailed analysis of the tip clearance forces has been made by Urlichs (1977).

9.2.2.3 Impeller Diffuser Excitation Forces

Instability experienced in centrifugal compressors provides a strong suspicion that impeller–diffuser interaction phenomena are involved in the development of destabilizing forces. However, to date, no satisfactory destabilizing mechanism or source involving impellers has been identified. In the last two decades there have been several studies related to rotordynamic forces arising from shrouded centrifugal pump impellers, but very little work has been done on compressor impellers. The destabilizing force arising from the impeller–diffuser/volute interaction of a pump has been determined to be relatively small (Jery et al., 1984; Bolleter et al., 1985; Adkins and Brennen, 1986; Ohashi et al., 1986). The major portion of the destabilizing force is known to be generated in the narrow gap region between the casing and shroud of the impeller (Childs, 1986; Bolleter et al., 1989; Baskharone et al., 1994; Moore and Palazzolo, 2001). In the case of centrifugal compressors, an empirical method to determine stability of multistage machines has been proposed (Kirk and Donald, 1983). The stability maps proposed by them for flow through and back-to-back centrifugal compressors are shown in Figure 9.8 and Figure 9.9.

9.2.2.4 Propeller Whirl

Propeller whirl (Taylor and Browne, 1938; Houbolt and Reed, 1961) is another form of instability which occurs in aircraft rotors when there is a mismatch in the angular velocity vector of the propeller and the linear velocity vector of the aircraft. This angular mismatch results in the generation of a moment whose vector has a component of significant magnitude, which contributes to the instability of the

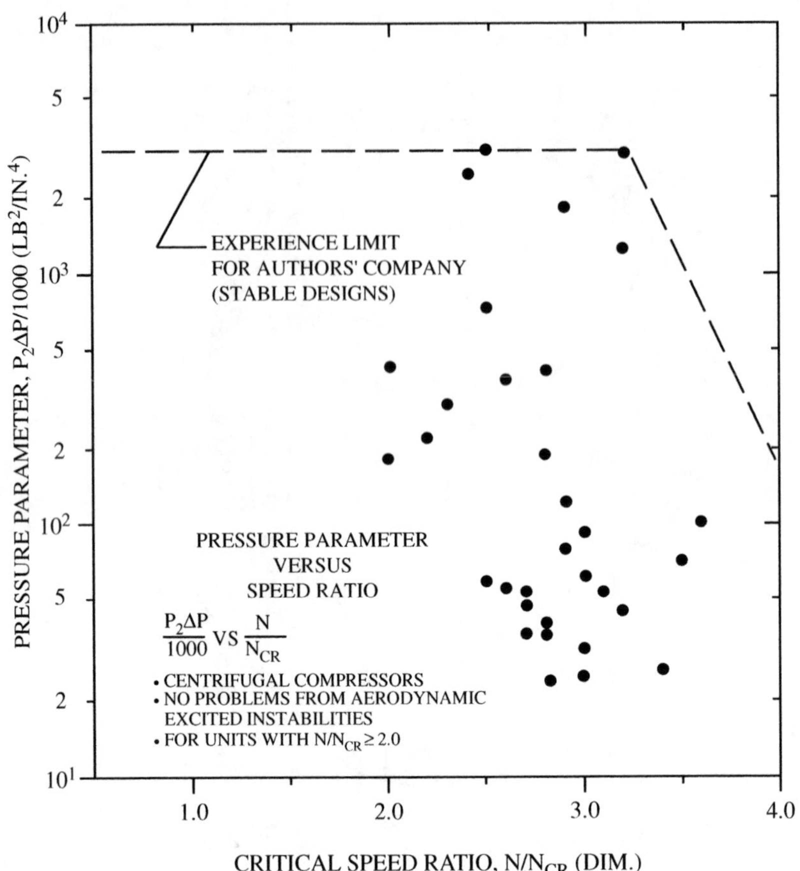

FIGURE 9.8 Proposed stability map for flow through centrifugal compressors. (*Source:* Rotor Dynamical Instability, 1983. With permission.)

propeller (Vance, 1988). Its magnitude is proportional to both the angular mismatch and the linear speed of the aircraft. With increasing speed, the magnitude of the destabilizing moment will exceed the rotor viscous damping moment and result in propeller instability (refer to Figure 9.5). Since the propeller is supported only from one end, the whirling motion is conical and is found to be in the reverse direction to propeller rotation. The instability is sensitive to the velocity and density of the air and not a function of the torque of the machine.

9.2.2.5 Fluid Trapped in a Hollow Rotor

Wolf (1968) has demonstrated that trapped fluid inside a hollow rotor can produce a force component tangential to the whirl orbit due to viscous drag forces. Under subsynchronous whirling speeds, this force component acts in the same direction of whirling motion and its magnitude is proportional to the rotor deflection. With reference to Figure 9.5, this force has all the markings of a destabilizing force, which can produce instability in the rotor. The threshold speed of instability is reached when the whirling speed equals the first critical speed of the rotor. It has been shown (Ehrich, 1999) that, at the threshold of instability, the rotor speed is less than twice the first critical speed. This results in a ratio of whirl speed to rotor speed in the range of 0.5 to 1.0.

9.2.2.6 Dry Friction Rubs

In Section 9.2.1.5, a dry friction rub situation was identified, where slipping was prevented between the rotor and stator under contact conditions. The contact was made possible by the deflection of the rotor

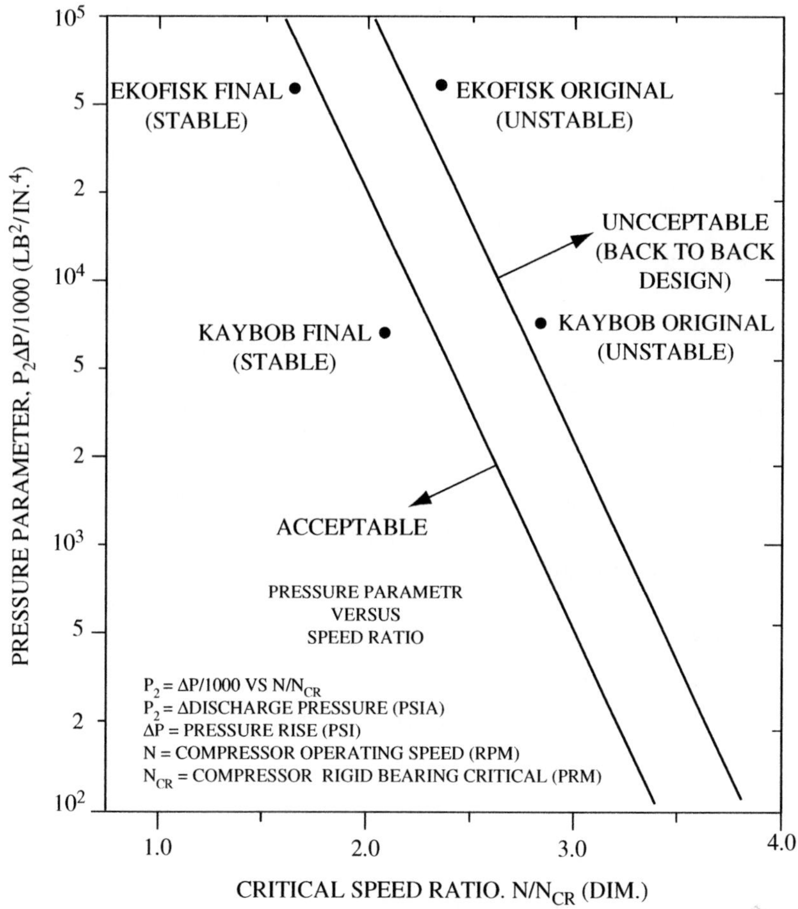

FIGURE 9.9 Proposed stability map for back-to-back centrifugal compressors. (*Source:* Rotor Dynamical Instability, 1983. With permission.)

due to unbalance forces. When contact is made between the rotor and stator, Coulomb friction produces a tangential force in the direction opposite to shaft rotation. Since the frictional force prevents slipping, whirling in the reverse direction to rotation occurs. The whirling speed is equal to $r\omega/C$, where r is the radius of the rotor, C the radial gap, and ω the speed of the rotor. Since the frictional force is in the same direction of whirling, it will cause the magnitude of whirling to increase, resulting in further increase of the frictional force. When the magnitude of this force exceeds the viscous damping force, rotor instability will occur. Another possibility is for the dry friction whirling speed to approach the coupled natural frequency of the rotor and stator, in which case unstable motion termed *dry friction whipping* takes place (Ehrich, 1999).

In addition to the case described above, dry friction rubs can occur in journal bearings, seals, wear rings, or any situation where a small clearance between a rotor and stator exists. The inadvertent closure of the clearance due to unbalance or lack of proper lubrication can initiate dry friction rub induced instability in these cases as well.

9.2.2.7 Torque Whirl/Load Torque

When the rotor disk axis is not aligned with the bearing axis, as in the case with an overhung rotor, Vance (1988) has shown that nonsynchronous whirling (torque whirl) can occur as a result of the misalignment between the load torque and the driving torque. His findings are based on the analysis of a simple rotor model. It appears that torque whirl instability can occur only in the case of long slender shafts with high-load torque values. The practical implications of this theory are still to be fully explored.

9.2.2.8 Oil Whirl/Whip

Newkirk and Taylor (1925) first experienced shaft whipping due to oil action in journal bearings during their investigation into internal friction induced whirling of rotors. They found that under certain conditions, a rotor mounted on journal bearings whipped when the rotor was running at any speed above double the critical speed; the whirling motion was in the forward direction and its speed matched the critical speed of the rotor. He provided a qualitative explanation of the phenomenon based on the fact that the oil film rotates at half the velocity of the shaft due to friction drag. Hence, for rotational speeds near twice the critical speed, the oil film provides the stimulus as its speed matches the critical speed value resulting in large displacements and whipping. Others have also drawn similar conclusions based on the suggestion of an oil wedge rotating at half speed, or rotating fluid force fields at half shaft speed. However, the foregoing fails to explain why oil whip persists at speeds greater than twice the critical speed. Ehrich (1999) has also provided a qualitative explanation for oil whirl based on the general theory of rotor instability.

Although a comprehensive explanation of the physical phenomena of oil whirl is still outstanding, numerous analytical models to identify where it could be encountered have been suggested. Gunter (1966) has analytically demonstrated that the instability in a rotor supported on journal bearings can be attributed to the cross coupling bearing coefficients. As a result, most of the research on oil whirl instability has narrowed to accurate estimation of bearing cross coupling coefficients.

9.2.2.9 Influence of Bearings and Supports on Rotor Instability

The results of the Jeffcott model can be easily adopted to include bearing stiffness and bearing support stiffness effects, provided they are both linear and circumferentially symmetric (isotropic). For this particular case, the rotor stiffness, k, is the equivalent stiffness resulting from the series connection of the shaft, bearings, and support stiffness. The resulting values for ω_N and ω_{cr} will be less than those for the simply supported Jeffcott model. This will result in lowering the threshold speed of instability.

If the bearing stiffness or the bearing-support stiffness is not symmetric (orthotropic) then it can be shown (Childs, 1993) that the threshold speed of instability is increased and the maximum amplitude of deflection of the rotor is reduced in comparison to the case with symmetric bearings.

The effect of damping at the bearings or at the bearing-support is very similar to the influence of stiffness. It reduces the amplitude of the synchronous rotor response at the critical speed, and elevates the threshold speed of instability. However, there is a limit to the amount of damping that can be applied. Excessive damping causes a reduction in stability (Childs, 1993).

The mass of the bearings plays a significant role on rotor stability. If the bearing mass is significantly larger than the rotor mass, the threshold speed of instability is lowered.

9.2.3 Parametric Instability

The instability phenomena described in Section 9.2.2 can be represented by linear differential equations where the system parameters such as mass, inertia, stiffness, damping, and natural frequency are assumed to be constants. There is another subcategory of self-excited motion, referred to as *parametric instability*, since it is induced by the periodic variation of the system parameters such as inertia, mass, and stiffness. A discussion of the more common forms of this phenomenon follows.

9.2.3.1 Shaft Stiffness Asymmetry

If the shaft of a rotor contains a sufficient level of stiffness asymmetry in the two principal axis of flexure, rotor instability could occur. Smith (1933) investigated the rotor behavior under unsymmetrical flexibility of the bearing supports and unsymmetrical transverse flexibility of the shaft, taking into consideration the damping effects as well. The following conclusions were derived based on his investigation:

In the presence of stiffness asymmetry, the onset speed of internal-friction induced instability is lowered.

When there is no external damping, the rotor becomes unstable at all speeds between the two undamped natural frequencies in the two orthogonal directions. However, if external damping is significant, parametric instability may be eliminated.

Within the unstable range, the whirling motion is in the forward direction and is synchronous with the shaft speed. Further, unlike in the case of internal-friction induced instability; it is theoretically possible to run through parametric instability. This makes parametric instability quite similar to the case of unbalance response.

When the asymmetric rotor is acted upon by a transverse disturbing steady force such as gravitational force, the rotor whirls at twice the speed of the shaft. This motion exhibits a resonant increase in amplitude at a speed that is approximately half the mean of the two natural frequencies.

9.2.3.2 Rotor Inertia Asymmetry

Crandall and Brosens (1961) analyzed the parametric excitation of a rotor with nonsymmetrical principal moments of inertia. Their results indicate that the rotor behavior is very similar to the case of rotors with stiffness asymmetry described in Section 9.2.3.1, and parametric instability over similar speed ranges occurs.

9.2.3.3 Pulsating Torque

Constant torque acting on a rotor is known to lower its critical speeds because they effectively reduce the rotor's lateral stiffness. A pulsating torque introduces lateral vibrations and instabilities into a rotor. When a combination of a pulsating and constant torque is applied to a rotor, it will induce unstable lateral vibrations in a specific range of rotor speeds and certain combinations of torque amplitudes. In the region of unstable lateral motion, the whirling speed of the rotor will coincide with the first critical speed of the rotor regardless of the rotor speed or the frequency of the pulsating torque. At rotor speeds outside the unstable region, the whirling speed of the rotor will be coincident with the pulsating torque frequency.

9.2.3.4 Pulsating Longitudinal Loads

Pulsating axial forces on a shaft that are in the order of magnitude of the buckling load will effectively cause a periodic variation in its lateral stiffness. This will result in a proportionate reduction of the lateral natural frequency of the shaft. Therefore, pulsating axial loads are capable of inducing parametric instability in a shaft for both the rotating and the stationary cases.

9.2.3.5 Nonsymmetric Clearance Effects

Bentley (1974) recognized that large subsynchronous whirling can occur in rotating machinery due to certain types of nonsymmetric clearance conditions. One such condition is when a rotor's whirling motion causes rubbing with a stationary surface over a portion of the rotor orbit. This effectively results in an increase in the rotor stiffness during the contact portion of the orbit, producing a periodic variation in rotor stiffness during each cycle. Another situation that produces cyclic variation in rotor stiffness can occur in the case of a rotor supported on antifriction bearings mounted with a clearance fit to the housing. The cyclic variation of the effective rotor stiffness produces a subsynchronous whirling motion at exactly half the rotational speed. Occasionally, whirling at one third and one fourth the running speed has also been observed by Childs (1993). Instability will occur when the whirling speed corresponds to a critical speed of the rotor.

9.2.4 Torsional Vibration

Rotating machinery with rotors that have relatively large moments of inertia are susceptible to torsional vibration problems. Torsional vibration is an oscillatory angular motion that is superimposed on the steady rotational motion of the rotor. In practice it can easily go undetected, as standard vibration monitoring equipment is not geared to measure torsional vibration. Special equipment must be used to detect torsional vibration in rotating machinery. Since the introduction of electric motor variable frequency drives (VFD), the incidence of field torsional vibration problems has increased. This has been attributed to the inherent torsional excitation forces present in current designs of VFDs. An added complication when using VFDs, as compared with using fixed speed drives, is the requirement of eliminating torsional natural frequencies over a wider speed range. Large synchronous electric motors are known to produce a large pulsating torque at a frequency that changes from twice the line frequency at the start, down to zero at the synchronous operating speed. In this case, any torsional natural frequency between zero and twice the line frequency may be subject to excitation. Most industries have come to recognize that torsional vibration is a potential hazard, and therefore, needs to be investigated at the design stage of rotating machinery. Several standard design specifications now require that torsional analysis is part of the design procedure.

The standard design practice in modeling the system for torsional analysis has been to calculate the undamped eigenvalues of the rotor as a free body in space. This practice is acceptable for most types of rotating machinery since the torsional stiffness and damping of bearings is insignificant. Also, in most cases, the torsional damping of the rotors itself is extremely low. Although the absence of damping is favorable from an analysis point of view, it makes it extremely difficult to eliminate a torsional vibration problem when encountered. This deficiency has been partly addressed with the introduction of several new lines of couplings that have a significant degree of torsional damping. Although not commonly used in rotating machinery, several torsional dampers such as the Lanchester damper have been developed for use on reciprocating machines. Dampers of similar design could be developed for use in rotating machinery to solve torsional vibration problems.

The torsional critical speeds of a simple rotor with one or two DoFs can be calculated by analytical methods. Numerical methods are used to calculate critical speeds and mode shapes of more complex systems with higher DoF. The Holzer numerical method (Holzer, 1921) described in Section 9.3.1.11 is one of the common methods used for these analyses.

- The two major categories of vibration phenomena that occur in rotating machinery are forced vibration and self-excited instability.
- Parametric instability is a special case of self-excited instability where some of the normally constant parameters vary, influencing rotor motion.
- Torsional vibrations are similar to lateral vibrations but occur in the planes perpendicular to the shaft axis.

9.3 Rotordynamic Analysis

Rotordynamic analysis is a part of current design procedures for rotating machinery that is carried out to predict the vibration behavior during operation of the machine. Potential problems are identified and eliminated by analytical means well before the manufacturing of components is begun. Furthermore, when a machine in operation displays unusual vibration behavior, analytical means are employed to study, identify, and help resolve the problem. In order for the analysis to be useful, it must be accurate and cost-effective.

During the last 100 years, several analytical procedures have been developed to understand the vibration behavior of rotating machinery. Some of these techniques are of historical interest only, and their usefulness in practical systems is very limited. With the advent of computer technology and advanced modeling techniques, several computer-based procedures have been developed to predict the vibration behavior of rotating machinery quite precisely. Of the most commonly used procedures, two are based on the lumped-parameter model where the distributed elastic and inertial properties are represented as a collection of rigid bodies connected by massless elastic beam elements. These two procedures are the transfer matrix formulations introduced by Myklestad (1944) and Prohl (1945), and the direct stiffness matrix formulations proposed by Biezeno and Grammel (1954). Ruhl and Booker (1972) introduced the third commonly used method, based on the finite element analysis (FEA) model in which the rotor is represented as an assemblage of elements with distributed elastic and inertial properties. Several of the well-known procedures are discussed next, some of them for historical significance.

9.3.1 Analysis Methods

9.3.1.1 Rankine's Numerical Method

Rankine (1869) proposed that, for a shaft of a given length, diameter, and material, there is a limit of speed, and for a shaft of a given diameter and material, turning at a given speed, there is a limit of length, below which centrifugal whirling is impossible. The limits of length and speed depend on the way the shaft is supported. The critical speed of the shaft is given by the following equation:

$$\omega = \frac{k(Hg)^{1/2}}{b^2} \tag{9.15}$$

where

ω = critical speed in rad/sec
k = radius of gyration of the cross section of the shaft
g = acceleration due to gravity
H = modulus of elasticity expressed in units of height of itself ($H = E/\rho$)
E = Young's modulus
ρ = density of the material
l = shaft length
$b = l/\pi$ for a simply supported shaft
$b = l/0.595\pi$ for an overhanging shaft

9.3.1.2 Greenhill's Formulae

Greenhill (1883) introduced the following differential equation of motion for a uniform shaft slightly deformed from straightness by centrifugal whirling:

$$\frac{d^4y}{dx^4} - \frac{m\omega^2}{gEak^2}y = 0 \tag{9.16}$$

The general solution to Equation 9.16 is given by

$$y = B \cosh \mu x + A \cos \mu x \tag{9.17}$$

where

$\mu^4 = m\omega^2/gEak^2$
m = weight of the shaft per unit length
ω = rotational speed
a = cross-sectional area of the shaft

The constants A and B depend on the boundary conditions at the support locations.

9.3.1.3 Reynolds' Equations

Reynolds extended the differential equation of motion for a uniform rotating shaft (Equation 9.16) to include shafts loaded with pulleys (disks) and for multispan rotors (Dunkerly, 1894).

At a bearing support, the difference in shear force must equal the bearing load P:

$$\frac{dM_r}{dx} - \frac{dM_l}{dx} = P \tag{9.18}$$

where M_r and M_l are bending moments in the right (r) and left (l) sides of the load.

At a load consisting of a revolving weight, W, the above equation becomes

$$\frac{dM_r}{dx} - \frac{dM_l}{dx} = \frac{W}{g}\omega^2 y \tag{9.19}$$

A further equation may be obtained by considering the *centrifugal couple* (gyroscopic moment) as given by

$$M_r - M_l = \omega^2 I' \frac{dy}{dx} \tag{9.20}$$

where I' = moment of inertia of the pulley.

The solution to Equation 9.20 is given by

$$y = A \cosh mx + B \sinh mx + C \cos mx + D \sin mx \tag{9.21}$$

The values of A, B, C, and D will depend on the boundary conditions between any two singular points.

9.3.1.4 Dunkerley Method

When considering the effects of the pulleys and the shaft together, the formulae derived by Reynolds were found to be limited for practical purposes. Dunkerly (1894) proposed an empirical method to consider the effects of the shaft and each of the pulleys separately, and then combine them using the following formula to obtain the critical speed of the rotor:

$$\frac{1}{\omega_c^2} = \frac{1}{\omega_s^2} + \sum_{i=1}^{n} \frac{1}{\omega_i^2} \tag{9.22}$$

where

ω_c = critical speed of the rotor
ω_s = critical speed of the shaft alone
ω_i = critical speed of the ith disk on a weightless shaft

In the case of the unloaded shaft, the critical speed, ω_s, is given by the following formula

$$\left(\frac{m\omega_s^2}{gEI}\right)^{1/4} l = a \tag{9.23}$$

where

I = sectional inertia of shaft
l = length of the span
a = a coefficient dependent on the manner of support of the shaft

The critical speed of the rotor ω_i with a single disk of weight W_i on a weightless shaft is given by

$$\omega_i = \theta\left(\frac{gEI}{W_i c^3}\right)^{1/2} \tag{9.24}$$

where

c = distance of disk from nearest support
θ = a coefficient dependant on the manner in which the shaft is supported, the position of the disk within the span and the dimensions of the disk

9.3.1.5 Rayleigh Method

The Rayleigh method is based on the premise that, when a system vibrates at its natural frequency, the maximum potential energy stored in the elastic components is equal to the maximum kinetic energy stored in the masses (Rayleigh, 1945).

The first natural frequency of a vibrating uniform beam is given by the following equation:

$$\omega^2 = \frac{EI \int_0^l \left(\frac{d^2 y}{dx^2} \right)^2 dx}{\mu \int_0^l y^2 \, dx} \tag{9.25}$$

The Rayleigh formula for a lumped mass system is

$$\omega^2 = \frac{\sum_{i=1}^{n} m_i y_i}{\sum_{i=1}^{n} m_i y_i^2} \tag{9.26}$$

where

ω = first natural frequency
y = deflection of the beam
x = distance along x-axis
l = length of the beam
m_i = ith lumped mass
y_i = static deflection of ith mass

The accuracy of the Rayleigh method depends upon the selection of a suitable deflection curve that approximates the fundamental mode shape. If the assumed curve represents the true mode shape, then the correct fundamental natural frequency will result. All deviations from the true mode shape will yield frequencies that are higher than the correct value.

9.3.1.6 Ritz Method

The Ritz method is an improvement on the Rayleigh method (Timoshenko et al., 1974) where the mode shape is represented by several orthogonal functions with unknown coefficients that satisfy the boundary conditions. The orthogonal functions are represented by a series of functions, $\Phi_i(x)$, where i varies from 1 to n. The mode shape is represented by the following expression:

$$y = \sum_{i=1}^{n} a_i \Phi_i(x) \tag{9.27}$$

In order for the coefficients a_i in the above equation to yield minimum values when substituted in the energy balance equation proposed by Rayleigh, the following expression needs to be satisfied:

$$\frac{\partial}{\partial a_i} \frac{\int_0^l \left(\frac{d^2 y}{dx^2} \right)^2 dx}{\int_0^l y^2 \, dx} = 0 \tag{9.28}$$

From Equation 9.25 and Equation 9.28, we find

$$\frac{\partial}{\partial a_i} \int_0^l \left[\left(\frac{d^2 y}{dx^2} \right)^2 - \frac{\omega^2 \mu}{EI} y^2 \right] dx = 0 \tag{9.29}$$

Substituting Equation 9.27 for y in Equation 9.29 and performing the mathematical operations, a system of linear equations in a_i is obtained. The number of such equations will be equal to n. These equations will yield solutions different from zero only if the determinant of the coefficients of a_i is equal to zero. This condition yields the frequency equation, from which the frequency of each mode can be derived.

9.3.1.7 Stodola–Vianello Method

The Stodola–Vianello method is a numerical iterative process (Timoshenko et al., 1974) that can be used to calculate the natural frequencies and mode shapes of vibrating systems. An approximate mode shape is first assumed and by successive iterations it is refined until convergence is obtained to the desired level of accuracy. This method can be used to refine the assumed mode shape when using the Rayleigh formulae or in the more general case of the matrix iteration process illustrated below.

Using Newton's Second law, the equations of motion for a multi-DoF system in matrix notation are

$$\{Y\} = \omega_i^2[A][m]\{Y\} \tag{9.30}$$

where

$$[A] = [K]^{-1} \tag{9.31}$$

$[A]$ is the flexibility matrix
$[m]$ is the mass matrix
$[K]$ is the stiffness matrix

To start the iterative process a trial vector, $\{Y\}_1$, representing the mode shape is substituted to both sides of Equation 9.30 and solve for the natural frequency, ω_i. For this reason, let the product of $[A]$, $[m]$, and $\{Y\}_1$ be $\{Y\}_2'$. The first approximation for ω_i may be obtained by dividing any one of the elements on $\{Y\}_1$ by $\{Y\}_2'$ (Note that, if $\{Y\}_1$ was the true mode shape, then the ratio for all such elements will be equal.) The vector $\{Y\}_2'$ is then normalized by dividing all the elements by the first element to produce $\{Y\}_2$. The vector $\{Y\}_2$ is premultiplied by $[m]$ and $[A]$ to produce $\{Y\}_3'$. Once again the ratio of corresponding elements of $\{Y\}_3'$ and $\{Y\}_2$ are compared for equality.

This procedure is repeated until the mode shape and the associated frequency is determined to the desired level of accuracy. In the above iteration procedure, the mode shape converges to the one corresponding to the lowest natural frequency. If the stiffness matrix had been used instead of the flexibility matrix, then convergence at the highest natural frequency is obtained. After the first mode of vibration is determined, it is removed from the system matrices by the use of a sweeping matrix so that higher modes can be obtained. This procedure is repeated until all the desired mode shapes and natural frequencies are determined.

9.3.1.8 Myklestad–Prohl Method (Transfer Matrix Method)

The Myklestad–Prohl transfer matrix formulation (Myklestad, 1944; Prohl, 1945) is commonly used to analyze lumped parameter models of rotating machinery. The distributed elastic and inertial properties of the rotor are represented as a collection of rigid bodies connected by massless elastic beam elements as illustrated in Figure 9.10. This method is best suited to calculate critical speeds and mode shapes of rotors neglecting the effects of viscous damping. The Myklestad–Prohl procedure can also be adopted to perform synchronous response and stability analysis, including for the effects of damping.

In order to demonstrate the transfer matrix procedure, an axisymmetric rotor is analyzed to determine its undamped critical speeds and mode shapes. Refering to Figure 9.10, the rotor is divided into n nodes, and each node is connected to the adjacent node by a massless elastic beam with uniform cross-sectional properties. The mass of components such as disks, impellers, and so on, together with the mass of the adjacent portion of the shaft, is lumped at the nodes. The Myklestad–Prohl method is based on the solution of the Bernoulli–Euler equation and the variables of interest are displacement (y), slope (θ), moment (M), and shear (V). The development of the following procedure follows Childs (1993).

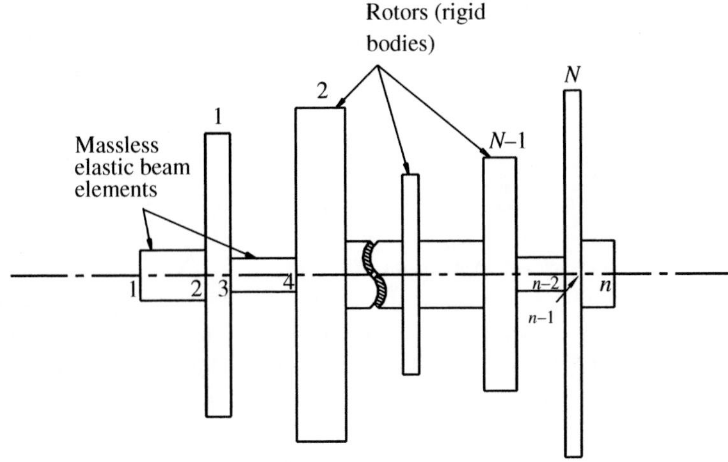

FIGURE 9.10 Lumped-parameter model of rotor.

At a typical nodal point (n), the variables on the left-hand side (l) are related to the variables on the right-hand side (r) by the following relationship:

$$\begin{Bmatrix} y_n^{\mathrm{r}} \\ \theta_n^{\mathrm{r}} \\ M_n^{\mathrm{r}} \\ V_n^{\mathrm{r}} \end{Bmatrix} = \begin{bmatrix} 1 & 0 & 0 & 0 \\ 0 & 1 & 0 & 0 \\ 0 & -J_n\omega^2 & 1 & 0 \\ -m_n\omega^2 & 0 & 0 & 1 \end{bmatrix} \begin{Bmatrix} y_n^{\mathrm{l}} \\ \theta_n^{\mathrm{l}} \\ M_n^{\mathrm{l}} \\ V_n^{\mathrm{l}} \end{Bmatrix} \tag{9.32}$$

For the purpose of abbreviation:

$$(Q)_n^{\mathrm{T}} = (y_n \quad \theta_n \quad M_n \quad V_n) \tag{9.33}$$

and Equation 9.32 can be written in a more compact form as follows:

$$(Q)_n^{\mathrm{r}} = [T_{\mathrm{m}n}](Q)_n^{\mathrm{l}} \tag{9.34}$$

where $[T_{\mathrm{m}n}]$ represents the transfer mass matrix at node n.

At a massless beam section, connecting node n to node $n+1$ the transfer matrix is given by

$$\begin{Bmatrix} y_{n+1}^{\mathrm{l}} \\ \theta_{n+1}^{\mathrm{l}} \\ M_{n+1}^{\mathrm{l}} \\ V_{n+1}^{\mathrm{l}} \end{Bmatrix} = \begin{bmatrix} 1 & l_n & \dfrac{l_n^2}{2EI_n} & \dfrac{-l_n^3}{6EI_n} \\ 0 & 1 & \dfrac{l_n}{EI_n} & \dfrac{l_n^2}{2EI_n} \\ 0 & 0 & 1 & -l_n \\ 0 & 0 & 0 & 1 \end{bmatrix} \begin{Bmatrix} y_n^{\mathrm{r}} \\ \theta_n^{\mathrm{r}} \\ M_n^{\mathrm{r}} \\ V_n^{\mathrm{r}} \end{Bmatrix} \tag{9.35}$$

Equation 9.35 may be written in a more abbreviated form as

$$(Q)_{n+1}^{\mathrm{l}} = [T_{\mathrm{b}n}](Q)_n^{\mathrm{r}} \tag{9.36}$$

where $[T_{\mathrm{b}n}]$ represents the beam element transfer matrix connecting node n to node $n+1$.

From Equation 9.34 and Equation 9.36, we obtain the combined transfer matrix for nodes n and $n+1$:

$$(Q)_{n+1}^{\mathrm{l}} = [T_{\mathrm{b}n}][T_{\mathrm{m}n}](Q)_n^{\mathrm{l}} = [T_n](Q)_n^{\mathrm{l}} \tag{9.37}$$

Starting with node one, successive matrix multiplications are carried out until node $n+1$ is reached. The last node $(n+1)$ is a dummy node with the beam length, l, equal to zero, and the mass and inertias also equal to zero. This makes the nodal parameters on the left-hand side of node $n+1$ equal to those on the right-hand side of node n. The result is as follows:

$$(Q)_n^r = [T_n][T_{n-1}]\cdots[T_1](Q)_1^l \quad \text{or} \quad (Q)_n^r = [T](Q)_1^l \tag{9.38}$$

The matrix $[T]$ is a function of the rotational speed, ω. The Myklestad–Prohl method uses a trial and error solution to determine the values of ω which satisfy the boundary conditions and Equation 9.38 simultaneously. It is not necessary to store and multiply all the matrices together. The transfer matrix procedure is used to proceed from one end to the other without having to store all the nodal matrices. In all cases, two boundary conditions each are known at the two ends of the shaft, and the frequencies that satisfy these boundary conditions are the critical speeds of the rotor. Once the critical speeds are calculated, the corresponding mode shapes can also be determined using the transfer matrix procedure. It should be noted that other types of elements, such as elastic supports, flexible couplings, and so on, could also be introduced very conveniently.

9.3.1.9 Direct Stiffness Method

The direct stiffness method uses a lumped-parameter formulation to evaluate the dynamic characteristics of a flexible rotor. The general differential equation of motion that characterizes its behavior (less the damping and gyroscopic forces) is as follows:

$$\begin{bmatrix} [m] & 0 \\ 0 & [J] \end{bmatrix} \begin{Bmatrix} (\ddot{Y}) \\ (\ddot{\theta}) \end{Bmatrix} + [K] \begin{Bmatrix} (Y) \\ (\theta) \end{Bmatrix} = \begin{Bmatrix} (F) \\ (T) \end{Bmatrix} \tag{9.39}$$

where $[m]$ and $[J]$ are diagonal matrices which contains the nodal masses, m_i, and nodal moments of inertia, J_i, respectively. The stiffness matrix, $[K]$, contains the internal stiffness terms of the beam elements as well as any external spring stiffness at the supports. The vectors (F) and (T) represent external forces and moments acting on the system, respectively.

The stiffness matrix for a typical beam element based on the Bernoulli–Euler equations is as follows (Childs, 1993):

$$[K^i] = \frac{2EI_i}{l_i^3} \begin{bmatrix} 6 & 3l_i & -6 & 3l_i \\ 3l_i & 2l_i^2 & -3l_i & l_i^2 \\ -6 & -3l_i & 6 & -3l_i \\ 3l_i & l_i^2 & -3l_i & 2l_i^2 \end{bmatrix} \tag{9.40}$$

The overall stiffness matrix, $[K]$, has to be assembled by combining the individual component matrices in a systematic manner. The following procedure illustrates the process.

The stiffness matrix of the ith beam element in matrix notation is

$$[K^i] = [k_{j,k}^i] \tag{9.41}$$

where j and k vary from $(2i-1)$ to $(2i+2)$.

To form the overall stiffness matrix, the elements with the same subscripts of adjacent beam elements are added over n beam elements as given by the following equation:

$$[K] = [K_{j,k}] = \sum_{i=1}^{n} \sum_{j=2i-1}^{2i+2} \sum_{k=2i-1}^{2i+2} k_{j,k}^i \tag{9.42}$$

Once the inertia matrix and the stiffness matrix for the entire system are assembled, the eigenvalues and eigenvectors can be evaluated by solving the following homogeneous equation derived from Equation 9.39:

$$[M](\ddot{Y}) + [K](Y) = 0 \tag{9.43}$$

There are numerous analysis procedures (Meirovitch, 1986) for the solution of Equation 9.43 that yield the eigenvalues and eigenvectors of the system. The method of choice will depend on the complexity and nature of the inertia and stiffness matrices. Perhaps the most widely known is the matrix iteration using the power method in conjunction with the sweeping technique. However, this method is not necessarily the most efficient, particularly for higher-order systems. The Jacobi's method, which uses matrix iteration to diagonalize a matrix by successive rotations, is more commonly used owing to its higher efficiency. Details of these techniques are given in the text by Meirovitch (1986).

When the damping matrix and the gyroscopic matrix is also included in Equation 9.39, the direct stiffness method can be used to calculate damped critical speeds, forced rotor response, and instability of the rotor in addition to the eigenvalues using similar methods of solution.

9.3.1.10 The Finite Element Analysis Method

The basis of the FEA method is to provide formulation for complex and irregular systems that can utilize the automation capabilities of computers (also see Chapter 9). The FEA method considers a rotordynamic system as an assemblage of discreet elements, where every such element has distributed and continuous properties, namely, the consistent representation of both mass and stiffness as distributed parameters. As illustrated in Section 9.3.1.9, the lumped-parameter method uses a consistent stiffness matrix equation (Equation 9.40) in its formulation, and therefore, the identical procedure can be adopted for the finite element method as well. For the distributed mass representation of an element, Archer (1963) procedure, which is based on the assumption that the mass distribution is proportional to the elastic distribution similar to the Rayleigh–Ritz formulation, is utilized. The resulting mass matrix is as follows:

$$[m^i] = \frac{m^i l_i}{420}\begin{bmatrix} 156 & 22l_i & 54 & -13l_i \\ 22l_i & 4l_i^2 & 13l_i & -3l_i^2 \\ 54 & 13l_i & 156 & -22l_i \\ -13l_i & -3l_i^2 & -22l_i & 4l_i^2 \end{bmatrix} \tag{9.44}$$

The overall stiffness matrix, $[K]$, for the entire system is assembled by combining the individual component matrices in a systematic manner according to Equation 9.41 and Equation 9.42. The overall mass matrix can also be assembled in precisely the same manner, as given by Equation 9.45 and Equation 9.46.

The mass matrix of the ith beam element in matrix notation can be represented as

$$[m^i] = [m^i_{j,k}] \tag{9.45}$$

where j and k varies from $(2i - 1)$ to $(2i + 2)$:

$$[M] = [M_{j,k}] = \sum_{i=1}^{n} \sum_{j=2i-1}^{2i+2} \sum_{k=2i-1}^{2i+2} m^i_{j,k} \tag{9.46}$$

Once the mass matrix and the stiffness matrix for the entire system are assembled, Equation 9.43 that describes the free vibration of the complete system can be solved. The solution methods of the eigenvalue problem, which can be utilized, are the same as those used for the direct stiffness method illustrated in Section 9.3.1.9 above. Details of the FEA methods are given in Ruhl and Booker (1972).

9.3.1.11 Torsional Analysis (Holzer Method)

The development of torsional analysis methods have gone through a similar evolutionary process to lateral vibration methods. Holzer (1921) first introduced the lumped-parameter numerical method to calculate torsional natural frequencies of a multi-DoF system. Even to-date, this is the most commonly used method because of its simplicity and reasonable degree of accuracy. The Holzer method is a transfer matrix formulation that uses a lumped parameter model similar to that used in the Myklestad–Phrol method described in Section 9.3.1.8. The only difference is that the transfer matrices represented by Equation 9.32 and Equation 9.35 are replaced by the equations

$$\begin{Bmatrix} \theta \\ T \end{Bmatrix}_n^{\mathrm{r}} = \begin{bmatrix} 1 & 0 \\ -\omega^2 J & 1 \end{bmatrix}_n \begin{Bmatrix} \theta \\ T \end{Bmatrix}_n^{\mathrm{l}} \tag{9.47}$$

$$\begin{Bmatrix} \theta \\ T \end{Bmatrix}_{n+1}^{\mathrm{l}} = \begin{bmatrix} 1 & \frac{1}{k} \\ 0 & 1 \end{bmatrix}_{n+1} \begin{Bmatrix} \theta \\ T \end{Bmatrix}_n^{\mathrm{r}} \tag{9.48}$$

Starting with node one, successive matrix multiplications are carried out until node $n+1$ is reached. The result can be represented by Equation 9.38. The matrix $[T]$ is a function of the rotational speed, ω. In all cases, one boundary condition at each end of the rotor is known. A trial-and-error solution to determine the values of ω which satisfy the boundary conditions and Equation 9.38 are simultaneously determined. These values are the torsional critical speeds of the rotor. Once the critical speeds are calculated, the corresponding torsional mode shapes can also be determined using the transfer matrix procedure.

In the case of branched systems and geared systems, particular attention has to be paid to the relative rotational speeds of the components. The rule is quite simple: multiply all stiffness and inertias of the geared shaft by N^2, where N is the speed ratio of the geared shaft to the reference shaft.

Other methods such as the distributed mass matrix method, direct stiffness method, and finite element method can also be used to determine torsional critical speeds of rotors. These procedures are very similar to those for lateral critical speed analysis.

9.3.2 Modeling

The design and analysis of rotordynamic systems require the development of models that simulate the behavior of the physical system. In the past, the critical speed of the rotor was considered to be the main criterion for stable operation. Today, stable, well-damped rotordynamic response to the exciting forces within a machine is considered to be a necessary condition for high reliability. The accuracy and reliability of the results greatly depends on the credibility of the system model and its adaptability to the analytical procedure. Even the most accurate and efficient analytical method cannot produce good results from a bad model. The methods that are commonly used to model shaft sections and disks and other such elements attached to shafts have been discussed in the previous sections. Useful formulae for calculating critical speeds of simple systems are given in Table 9.2. Models to represent bearings, rotor dampers, seals, and rotor–stator interactions are discussed in the following sections.

9.3.2.1 Journal Bearings

Journal bearings were used in rotating machinery for a long time before their dynamic characteristics were fully understood. Considerable effort has been expended in the last few decades to understand and develop techniques for their accurate representation in rotordynamic analysis. A variety of bearing types with improved characteristics have been developed over the years. Figure 9.11 shows the most commonly used types in rotating machinery. Hagg and Sankey (1958) were amongst the first to provide dynamic stiffness and damping coefficients for a number of these bearing types. However, these coefficients are considered incomplete as cross-coupling terms were not considered. Soon after, there was a flurry of activity related to the analysis of journal bearings; Sternlicht (1959), Warner (1963),

TABLE 9.2 Useful Formulas in Vibration Analysis and Design

Rankine formula	$\omega = \dfrac{k(Hg)^{1/2}}{b^2}$	*Note*: This formula is of historical interest only and has limited practical value
Greenhill formula	$\dfrac{d^4y}{dx^4} - \dfrac{m\omega^2}{gEak^2}y = 0$	
Dunkerly equation	$\dfrac{1}{\omega_c^2} = \dfrac{1}{\omega_s^2} + \sum_{i=1}^{n} \dfrac{1}{\omega_i^2}$ The above equation reduces to $\omega_c^2 = \dfrac{g}{\sum y_{stat}}$	
Formulas for natural frequency calculation (Blevins, 2001; Gorman, 1975)	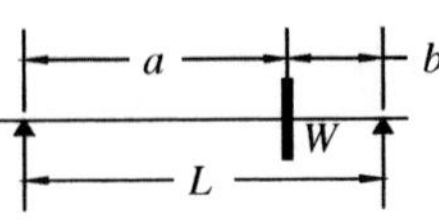	$\omega_c = \dfrac{1}{ab}\sqrt{\dfrac{3EIL}{W}}$
	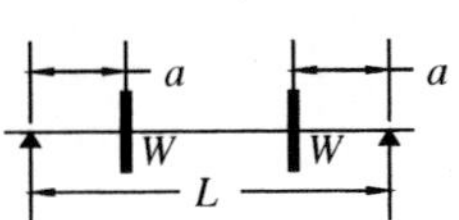	$\omega_c = \dfrac{1}{a}\sqrt{\dfrac{6EI}{W(3L-4a)}}$
	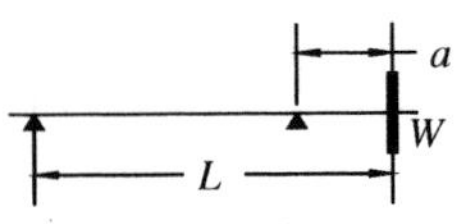	$\omega_c = \dfrac{1}{a}\sqrt{\dfrac{3EI}{WL}}$
	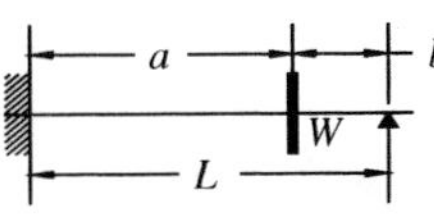	$\omega_c = \left(\dfrac{12EIL^3}{Wa^3b^2(3L+b)}\right)^{1/2}$
	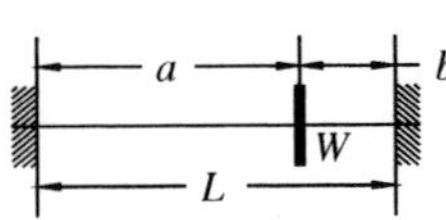	$\omega_c = \left(\dfrac{3L^3EI}{Wa^3b^3}\right)^{1/2}$
	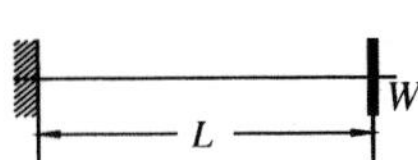	$\omega_c = \left(\dfrac{3EI}{WL^3}\right)^{1/2}$
	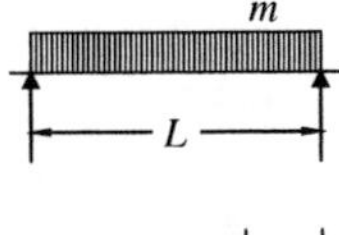	$\omega_c = \left(\dfrac{98EI}{mL^4}\right)^{1/2}$
		$\omega_c = \left(\left(\dfrac{64a}{9L}\right)^2 - \left(\dfrac{6a}{L}\right) + \dfrac{16}{3}\right)^2\left(\dfrac{EI}{mL^4}\right)^{1/2}$, $\dfrac{a}{L} \geq 0.25$

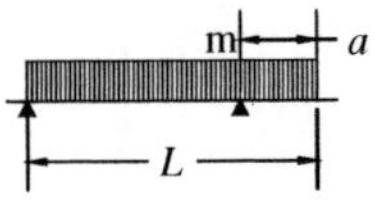

(*continued on next page*)

TABLE 9.2 *(continued)*

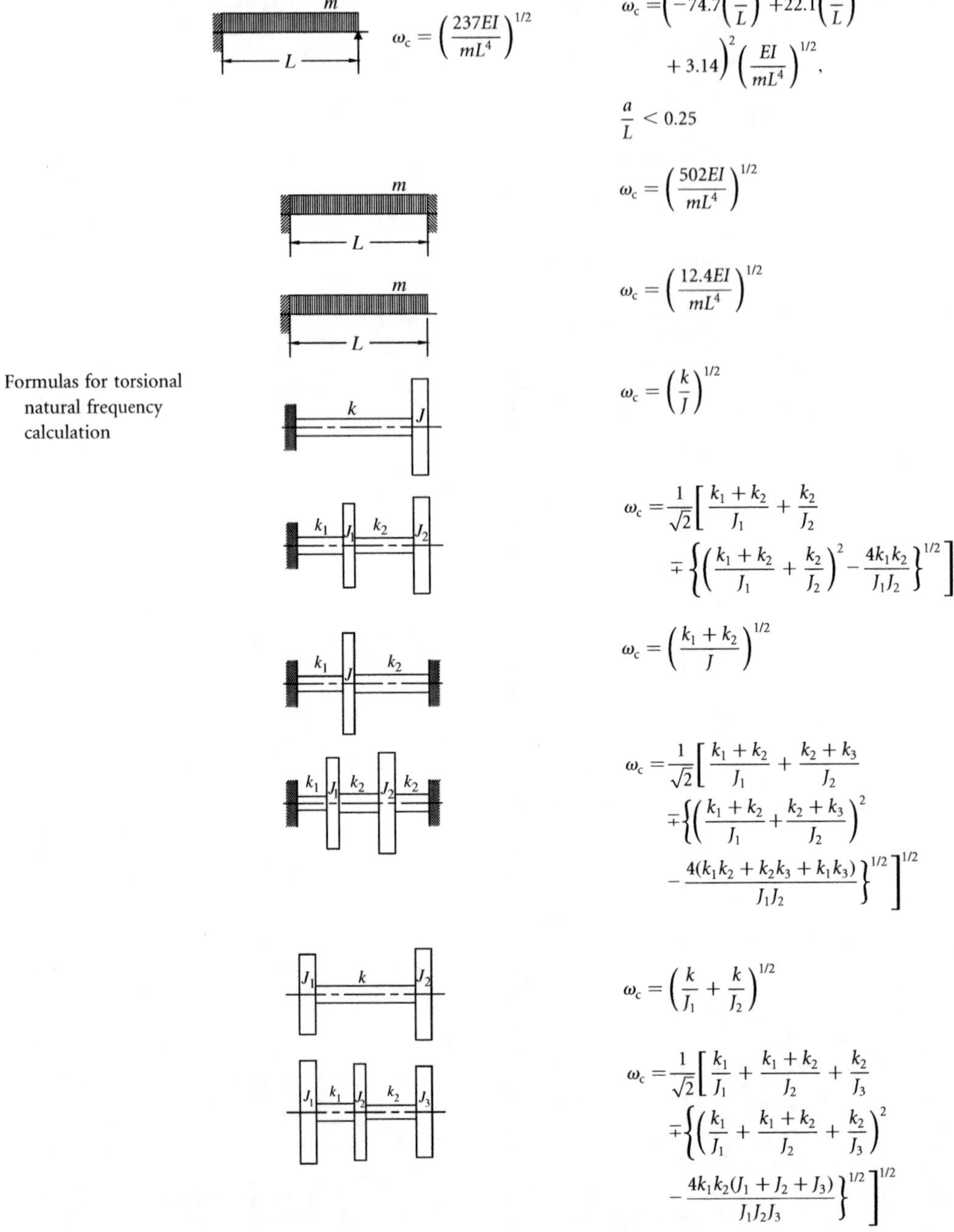

$$\omega_c = \left(\frac{237EI}{mL^4}\right)^{1/2}$$

$$\omega_c = \left(-74.7\left(\frac{a}{L}\right)^2 + 22.1\left(\frac{a}{L}\right) + 3.14\right)^2 \left(\frac{EI}{mL^4}\right)^{1/2}, \quad \frac{a}{L} < 0.25$$

$$\omega_c = \left(\frac{502EI}{mL^4}\right)^{1/2}$$

$$\omega_c = \left(\frac{12.4EI}{mL^4}\right)^{1/2}$$

Formulas for torsional natural frequency calculation

$$\omega_c = \left(\frac{k}{J}\right)^{1/2}$$

$$\omega_c = \frac{1}{\sqrt{2}}\left[\frac{k_1 + k_2}{J_1} + \frac{k_2}{J_2} \mp \left\{\left(\frac{k_1 + k_2}{J_1} + \frac{k_2}{J_2}\right)^2 - \frac{4k_1k_2}{J_1J_2}\right\}^{1/2}\right]^{1/2}$$

$$\omega_c = \left(\frac{k_1 + k_2}{J}\right)^{1/2}$$

$$\omega_c = \frac{1}{\sqrt{2}}\left[\frac{k_1 + k_2}{J_1} + \frac{k_2 + k_3}{J_2} \mp \left\{\left(\frac{k_1 + k_2}{J_1} + \frac{k_2 + k_3}{J_2}\right)^2 - \frac{4(k_1k_2 + k_2k_3 + k_1k_3)}{J_1J_2}\right\}^{1/2}\right]^{1/2}$$

$$\omega_c = \left(\frac{k}{J_1} + \frac{k}{J_2}\right)^{1/2}$$

$$\omega_c = \frac{1}{\sqrt{2}}\left[\frac{k_1}{J_1} + \frac{k_1 + k_2}{J_2} + \frac{k_2}{J_3} \mp \left\{\left(\frac{k_1}{J_1} + \frac{k_1 + k_2}{J_2} + \frac{k_2}{J_3}\right)^2 - \frac{4k_1k_2(J_1 + J_2 + J_3)}{J_1J_2J_3}\right\}^{1/2}\right]^{1/2}$$

Rayleigh equations

$$\omega^2 = \frac{EI\int_0^l \left(\frac{d^2y}{dx^2}\right)^2 dx}{\mu\int_0^l y^2\, dx}$$

TABLE 9.2 (*continued*)

	$\omega^2 = \dfrac{\sum_{i=1}^{n} m_i y_i}{\sum_{i=1}^{n} m_i y_i^2}$	
Ritz method	$y = \sum_{i=1}^{n} a_i \Phi_i(x)$ $\dfrac{\partial}{\partial a_i} \dfrac{\int_0^l \left(\frac{d^2 y}{dx^2}\right)^2 dx}{\int_0^l y^2\, dx} = 0$ $\dfrac{\partial}{\partial a_i} \int_0^l \left[\left(\frac{d^2 y}{dx^2}\right)^2 - \frac{\omega^2 \mu}{EI} y^2\right] dx = 0$	
Stodola–Vianello method	$\{Y\} = \omega_i^2 [A][m]\{Y\},\ [A] = [K]^{-1}$	
Transfer matrix — Myklestad–Phrol method	$\begin{Bmatrix} y_{n+1}^{l} \\ \theta_{n+1}^{l} \\ M_{n+1}^{l} \\ V_{n+1}^{l} \end{Bmatrix} = \begin{bmatrix} 1 & l_n & \frac{l_n^2}{2EI_n} & \frac{-l_n^3}{6EI_n} \\ 0 & 1 & \frac{l_n}{EI_n} & \frac{l_n^2}{2EI_n} \\ 0 & 0 & 1 & -l_n \\ 0 & 0 & 0 & 1 \end{bmatrix} \begin{Bmatrix} y_n^{r} \\ \theta_n^{r} \\ M_n^{r} \\ V_n^{r} \end{Bmatrix}$	
Stiffness matrix for a beam element	$[K^i] = \dfrac{2EI_i}{l_i^3} \begin{bmatrix} 6 & 3l_i & -6 & 3l_i \\ 3l_i & 2l_i^2 & -3l_i & l_i^2 \\ -6 & -3l_i & 6 & -3l_i \\ 3l_i & l_i^2 & -3l_i & 2l_i^2 \end{bmatrix}$	
Mass matrix for a beam element	$[m^i] = \dfrac{m^i l_i}{420} \begin{bmatrix} 156 & 22l_i & 54 & -13l_i \\ 22l_i & 4l_i^2 & 13l_i & -3l_i^2 \\ 54 & 13l_i & 156 & -22l_i \\ -13l_i & -3l_i^2 & -22l_i & 4l_i^2 \end{bmatrix}$	
Squeeze-film damper coefficients	$k = \dfrac{24R^3 L\mu\omega\varepsilon}{C_r^3(2+\varepsilon^2)(1-\varepsilon^2)}$ $c = \dfrac{12\pi R^3 L\mu}{C_r^3(2+\varepsilon^2)(1-\varepsilon^2)^{1/2}}$ $k = \dfrac{2RL^3\mu\omega\varepsilon}{C_r^3(1-\varepsilon^2)^2}$ $c = \dfrac{\pi R L^3 \mu}{2C_r^3(1-\varepsilon^2)^{\frac{3}{2}}}$	
Unbalance sensitivity	$\mathrm{SF} = \dfrac{a}{U} M$	
Rolling element bearing defect frequencies	$f_{\mathrm{bor}} = \dfrac{ND}{60d}\left[1 - \left(\dfrac{d}{D}\cos\theta\right)^2\right]$	N= rotational speed (rpm), D = rolling element pitch diameter

(*continued on next page*)

TABLE 9.2 (*continued*)

	$f_{\text{ir}} = \frac{Nn}{120}\left(1 + \frac{d}{D}\cos\theta\right)$	d = rolling element diameter, N = number of rolling elements
	$f_{\text{or}} = \frac{Nn}{120}\left(1 - \frac{d}{D}\cos\theta\right)$	θ = contact angle with respect to axis, bor = ball or roller defect
	$f_c = \frac{N}{120}\left(1 - \frac{d}{D}\cos\theta\right)$	ir = inner race defect, or = outer race defect, c = cage defect
Lomakin formula for radial stiffness for a close clearance bushing	$k = \frac{\pi}{8}(1 + \varsigma)\lambda\mu^4\left(\frac{l}{b_m}\right)^2 \Delta pD$	
	$\mu^2 = \frac{1}{1 + \varsigma + (\lambda l/2b_m)}$	
	b_m = radial clearance, l = length of bushing,	D = diameter, Δp = differential pressure across bushing,
	ζ = inlet loss coefficient,	λ = friction coefficient

Lund (1964), Lund (1965), Glienicke (1966), Orcutt (1967), Lund (1968), Someya et al. (1988), and several others provided complete bearing coefficients, including cross-coupling terms, for several bearing types. This information is considered to be a valuable resource for those engaged in rotordynamic analysis. The general form of the rotordynamic model for a journal bearing resulting from the above contributions is given by the following equation:

$$\begin{Bmatrix} F_X \\ F_Y \end{Bmatrix} = -\begin{bmatrix} k_{11} & k_{12} \\ k_{21} & k_{22} \end{bmatrix}\begin{Bmatrix} X \\ Y \end{Bmatrix} - \begin{bmatrix} c_{11} & c_{12} \\ c_{21} & c_{22} \end{bmatrix}\begin{Bmatrix} \dot{X} \\ \dot{Y} \end{Bmatrix} \tag{9.49}$$

Since the dawn of the digital computer era, several computer codes have been developed to analyze all aspects of journal bearings, including stiffness and damping coefficients. Many of these codes have been developed by equipment manufactures and research centers for their exclusive use. Several commercially available software codes popularized in North America are given in Table 9.3. Although bearing coefficients given in the form of charts and tables from the earlier studies are still in use, computer-based codes are growing in popularity.

9.3.2.2 Rolling Element Bearings

Rolling element bearings are used in numerous types of rotating machinery which are required to be compact, manage high loads, and have low heat rejection and simple lubrication systems. Unlike journal bearings, their load-carrying capacity is not speed-dependent and as a result is capable of full load capacity down to zero speed. Some of these salient features make rolling element bearings very attractive to many industries.

From a rotordynamic standpoint, rolling element bearings are modeled as linear spring elements with direct spring coefficients only. The damping terms are insignificant and as a result do not attenuate rotor deflections at critical speeds. A typical rolling element bearing is represented by the following equation:

$$\begin{Bmatrix} F_X \\ F_Y \end{Bmatrix} = -\begin{bmatrix} k & 0 \\ 0 & k \end{bmatrix}\begin{Bmatrix} X \\ Y \end{Bmatrix} \tag{9.50}$$

The absence of cross-coupling stiffness and damping terms signifies that bearing induced rotor instability will not occur. Although, for convenience of analysis, the spring stiffness is considered linear, its

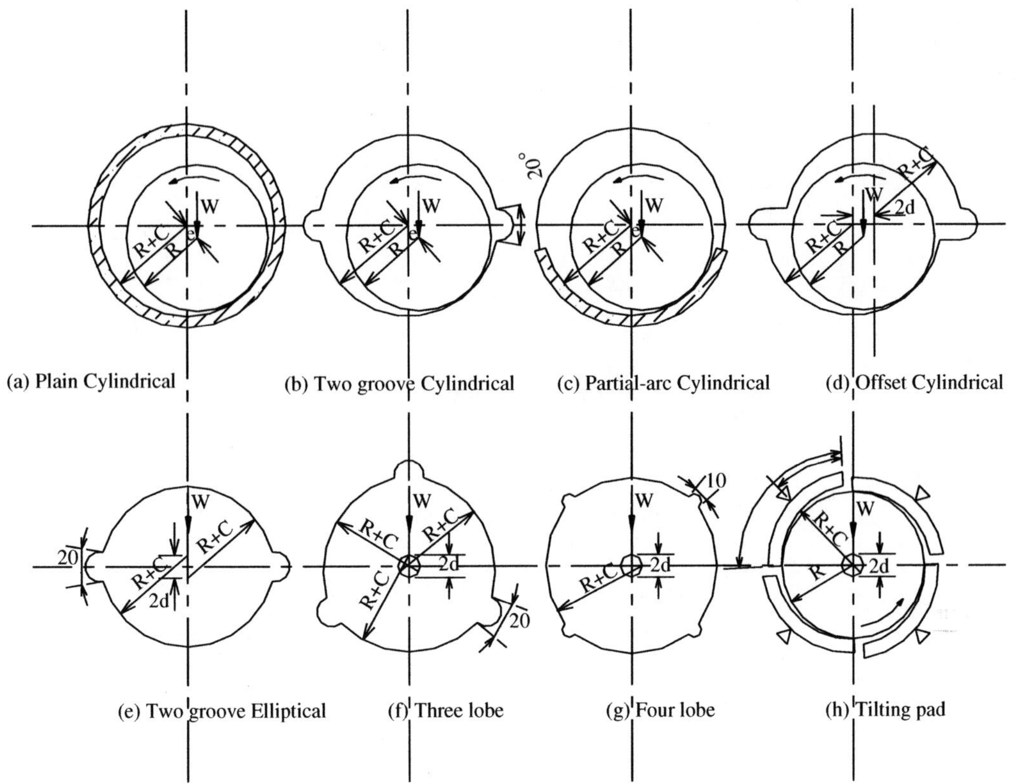

FIGURE 9.11 Common types of journal bearings.

true behavior can be quite the opposite, leading to some calculation inaccuracies. The nonlinearities are most significant where the bearings have no preload and some internal clearance in the bearings exists. Preloaded bearings with little or no internal clearance behave quite linearly. Jones (1960), Harris (1991), and Kramer (1993) have analyzed the bearing stiffness coefficients for the common types of rolling element bearings, and this data can be utilized for rotordynamic study of rotating machinery.

9.3.2.3 Squeeze-Film Dampers

Squeeze-film dampers are used to introduce damping capacity to a rolling element bearing or, in the case of journal bearings, to provide additional damping and stiffness to eliminate rotor instability problems. Squeeze-film dampers have come into prominence through the modern aircraft gas turbine industry where the bearing of choice is the rolling element bearing. In the mid-1970s, several designs were introduced to add damping capacity and predictable stiffness to the rolling element bearings. In its basic form, a squeeze-film damper is very similar to a nonrotating cylindrical journal bearing where the outer race of the rolling element bearing forms the journal as illustrated in Figure 9.12. The addition of end seals (to control leakage) and centering springs are modifications that have been introduced to enhance its performance. The interactive force at a bearing using a squeeze-film damper can be represented by the following equation:

$$\begin{Bmatrix} F_X \\ F_Y \end{Bmatrix} = -\begin{bmatrix} k & 0 \\ 0 & k \end{bmatrix}\begin{Bmatrix} X \\ Y \end{Bmatrix} - \begin{bmatrix} c & 0 \\ 0 & c \end{bmatrix}\begin{Bmatrix} \dot{X} \\ \dot{Y} \end{Bmatrix} \tag{9.51}$$

The stiffness and damping coefficients for the squeeze-film dampers have been derived (Ehrich, 1999), from the solution of the Reynolds' equation for the case of a nonrotating journal bearing. For dampers with end seals, the long journal bearing theory is used to generate the following stiffness and damping

TABLE 9.3 Rotordynamic Analysis Software

Name of Software	Type of Analysis	Supplier
CAD20	Lateral critical speeds of flexible rotors	CADENSE Programs, Foster Miller Technologies Inc., Albany, NY, USA
CAD21	Unbalance response of flexible rotors	
CAD21a	Response of flexible rotors to nonsynchronous sinusoidal excitation	
CAD22	Torsional critical speeds and response of geared systems	
CAD24	Transient torsional critical speeds of geared system	
CAD25	Dynamic stability of flexible rotors	
CAD25a	Transient response of flexible rotors	
CAD26	Lateral critical speeds of multilevel rotors	
CAD27	Unbalance response of multilevel rotors	
CAD30	Dynamic coefficients of liquid lubricated journal bearings	
CAD30a	Dynamic coefficients of ball bearings	
CAD31	Dynamic coefficients of liquid lubricated tilting pad journal bearings	
CAD32	Dynamic coefficients of liquid lubricated axial-groove and single pad journal bearings	
CAD34a	Performance of tilting pad thrust bearings	
CAD34b	Performance of tapered-land thrust bearings	
CAD36	Dynamic coefficients of liquid lubricated pressure dam journal bearings	
CAD38	Dynamic coefficients of liquid lubricated deep-pocket hydrostatic journal bearings	
CAD40	Dynamic coefficients of gas lubricated journal bearings	
CAD41	Dynamic coefficients of gas lubricated tilting pad journal bearings	
CAD42	Dynamic coefficients of gas lubricated spiral groove journal bearings	
CAD42i	Dynamic coefficients of liquid lubricated spiral groove journal bearings	
FEATURE	Rotor bearing system analysis	
COJOUR	Analysis of journal bearings	
DYNROT	A program designed to perform a complete study of the rotordynamic behavior of rotors. It is capable of linear, nonlinear and torsional analysis of rotors	Dipartimento di Meccanica, Politecnico di Torino, Torino, Italy
DyRoBeS	Comprehensive rotordynamic analysis software for lateral and torsional analysis, including bearing analysis of rotor-bearing systems	AGILE SOFTWARE CONCEPTS NREC White River Junction, VT, USA
RotorLab	A software package for agile modeling of rotor systems, bearings, and seals. It combines the tasks of design, modeling, analysis, post processing, and data management into a consistent user interface	
DAMBRG2	Coefficients and rigid rotor stability information for two-lobe isoviscous bearings with a pressure dam in only one pad	ROMAC—Rotating Machinery and Controls Laboratory, University of Virginia, Charlottesville, VA, USA
HYDROB	Predicts the steady state and dynamic operating characteristics of hybrid journal bearings	
PDAM2D	This program can analyze stiffness and damping coefficients, and the rigid rotor stability threshold of multipad pressure dam bearings	
SQFDAMP	Determines stiffness and damping coefficients for short and long squeeze-film bearings with and without fluid film cavitation	
THBRG	Dynamic coefficients of multilobe journal bearings with incompressible fluid	

TABLE 9.3 (*continued*)

Name of Software	Type of Analysis	Supplier
THPAD	Dynamic coefficients of tilting pad journal bearings with incompressible fluid	
THRUST	Predicts the steady-state operating characteristics of tilting-pad and fixed geometry fluid-film thrust bearings	
CRTSP2	Undamped lateral critical speeds of dual-level rotor systems	
MODFR2	Undamped lateral critical speeds of single or dual-level rotor systems	
TWIST2	Undamped torsional critical speeds and mode shapes of rotor systems	
FRESP2	Predicts the modal frequency forced response of dual rotor systems with a flexible substructure	
RESP2V3	Nonplanar synchronous unbalance response of dual-level multimass flexible rotors	
HCOMB	Dynamic coefficients of straight-through honeycomb seals with a compressible gas	
LABY3	Dynamic coefficients for straight-through and uniform interlocking type labyrinth seals with a compressible fluid	
SEAL2	Stiffness and damping coefficients for plain and grooved seals with incompressible turbulent axial flow	
SEAL3	Stiffness, damping and mass coefficients of both plain and circumferentially grooved seals	
TURSEAL	Stiffness and damping coefficients of turbulent flow annular seals or water lubricated bearings	
FSTB3	Stability, damped critical speeds, and whirl mode shapes of multispool rotor systems	
ROTSTB	Stability, damped critical speeds, and whirl mode shapes of single spool rotor systems	
COTRAN	Nonlinear time transient analysis of multilevel rotors with substructure	
TORTRAN3	Transient torsional rotor response	
hydrosealt	Stiffness and damping coefficients, and threshold speed of instability of cylindrical-pad journal bearings and pad-hydrostatic bearings of arbitrary arc lengths and preloads	Rotordynamics Laboratory, Texas A&M University, College Station, TX, USA
hydroflext	Stiffness and damping coefficients, and threshold speed of instability of a variety of bearing and seal types	
hydrotran	Predicts the transient force response of a rigid rotor supported on fluid film bearings	
hydrojet	Force coefficients for a variety of hybrid bearing and seal types handling process fluids	
hydroTRC	Stiffness and damping coefficients for a variety of bearing and seal types and for different types of fluids	
hseal2p	Stiffness and damping coefficients of seals that operate under two-phase flow conditions	
fembear	Stiffness and damping coefficients of cylindrical and fixed arc pad hydrostatic and hydrodynamic bearings for laminar and isothermal flow conditions	
sfdfem	Damping force coefficients of finite length squeeze-film dampers executing circular centered motion	
sfdflexs	Instantaneous fluid film forces for arbitrary journal motions and circular centered orbits in multiple pad integral squeeze-film dampers	
hsealm	Stiffness and damping coefficients of cylindrical annular pressure seals	

(*continued on next page*)

TABLE 9.3 *(continued)*

Name of Software	Type of Analysis	Supplier
lubsealn	Stiffness and damping coefficients of single-land and multiple-land high pressure oil seal rings and cylindrical journal bearings	
ROTECH	Lateral rotordynamic analysis for critical speeds; unbalance response, linear stability and nonlinear transient response of rotors. Also includes a torsional rotordynamic analysis program	ROTECH Engineering Services, Delmont, PA, USA
ROTOR-E	A comprehensive software package for lateral rotordynamic analysis of rotating equipment	Engineering Dynamics Inc., San Antonio, TX, USA
ROTORINSA	A software package devoted to the prediction of the steady-state lateral dynamic behavior of rotors	Laboratoire de Mecanique des Structures, LMST INSA Lyon, Lyon, France
TURBINE-PAK	A software package for rotordynamic analysis of nonlinear multibearing rotor-bearing-foundation systems	Scientific Engineering Research, Mt Best, Vic., Australia.
TURBINE-PAK NONLINEAR	Designed to study transient responses of rotor-bearing-foundation systems, including the loss of stability of the system	
XLrotor	A complete suite of analysis tools for rotating machinery dynamics. Handles both lateral and torsional analysis of rotors. Also includes codes for calculating coefficients for fluid film and antifriction bearings	Rotating Machinery Analysis Inc., Austin, TX, USA
XLTRC	A suite of codes for executing a complete lateral rotordynamic analysis of rotating machinery	The Turbomachinery Laboratory, Texas A&M University, College Station, TX, USA
XLAnSeal	Force and moment coefficients for annular turbulent seals in the laminar, turbulent, and transition flow regimes	
XLCGrv	Coefficients for centered grooved-stator, turbulent flow, annular pump seals	
XLLaby	Stiffness and damping coefficients for tooth-on-rotor or tooth-on-stator gas labyrinth seals	
XLIsotSL	Coefficients for smooth rotor/honeycomb stator annular seals	
XLLubGT	Coefficients for high-pressure oil bushing seals of compressors or smooth pump seals in the laminar flow regime	
XLJrnl	Stiffness and damping coefficients for fixed-arc and tilting-pad bearings	
HLHydPad	Stiffness and damping coefficients for hydrostatic and hybrid journal-pad bearings in the laminar flow regime	
XLTFPBrg	Stiffness and damping coefficients for fixed-arc, tilting-pad and flexure-pivot hydrostatic bearings	
XLPresDm	Stiffness and damping coefficients for multilobed, rigid-pad arc bearings with preload and pressure-dam bearings with relief tracks	
XLBalBrg	Stiffness coefficients for ball bearings	
XLLSFD	Damping and mass coefficients for locally sealed squeeze-film dampers	
XLOSFD	Damping and mass coefficients for open ended squeeze-film dampers	
XLSFDFEM	Damping coefficients for squeeze-film dampers with various types of end seals	
XLPIMPLR	Stiffness, damping, and mass matrices for centrifugal pump impellers	
XLWachel	Destabilizing cross-coupled force coefficients for impellers of centrifugal compressors	
XLClrEx	Destabilizing cross-coupled stiffness coefficients for unshrouded turbines	

coefficients:

$$k = \frac{24R^3 L\mu\omega\varepsilon}{C_r^3(2+\varepsilon^2)(1-\varepsilon^2)} \tag{9.52}$$

$$c = \frac{12\pi R^3 L\mu}{C_r^3(2+\varepsilon^2)(1-\varepsilon^2)^{1/2}} \tag{9.53}$$

where

R = the damper radius
ω = whirl speed
L = length of damper
μ = viscosity of oil
C_r = the radial clearance
ε = eccentricity ratio (orbit radius/C_r)

Similarly, for a damper without end seals, the short journal bearing theory yields the following stiffness and damping coefficients:

$$k = \frac{2RL^3\mu\omega\varepsilon}{C_r^3(1-\varepsilon^2)^2} \tag{9.54}$$

$$c = \frac{\pi RL^3\mu}{2C_r^3(1-\varepsilon^2)^{\frac{3}{2}}} \tag{9.55}$$

Although the above equations, based on the Reynolds' equation, have been proposed to predict damper characteristics, the experimental evidence does not validate these equations. Therefore, these equations should be used with caution for practical purposes.

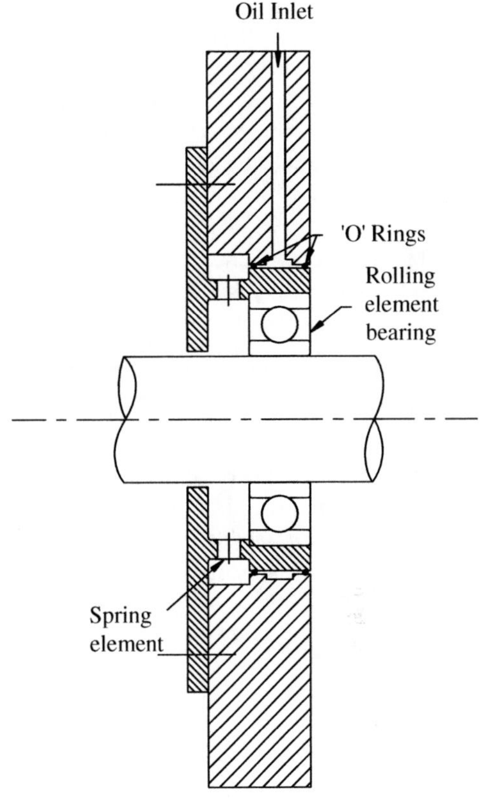

FIGURE 9.12 A squeeze-film damper.

9.3.2.4 Annular Seals

Annular seals are primarily used in pumps, compressors, gas turbines, and steam turbines to minimize leakage and thereby improve the volumetric efficiency of the machine. In addition to their basic function, they also play a vital role in the rotordynamics of the machine, especially in multistage machines, providing stiffness and damping and thereby enhancing high-speed operational capability. In fact, in the last few decades, most of the development work on seals has focused on understanding and improving their dynamic vibration characteristics rather than improving their efficiency in sealing.

Lomakin (1958) was the first to publish on the restoring forces in smooth annular clearances in pumps. However, it was more than a decade later that Black (1968) provided the major initial impetus for the understanding and development of seals. Childs (1993) provided an excellent compendium of the research work conducted in the area of seals. His book also provides the most comprehensive coverage of the subject of seal dynamics.

In the present context, seals are handled in the same manner as the stiffness and damping characteristics of journal bearings with some degree of modifications. In particular, fluid inertia effects are included, and it is assumed that the center of the shaft orbit is the same as the center of the stationary seal ring. Assuming rotational symmetry the reaction force-seal motion model can be represented by the following equation:

$$\begin{Bmatrix} F_X \\ F_Y \end{Bmatrix} = -\begin{bmatrix} k & k_c \\ -k_c & k \end{bmatrix}\begin{Bmatrix} X \\ Y \end{Bmatrix} - \begin{bmatrix} c & c_c \\ -c_c & c \end{bmatrix}\begin{Bmatrix} \dot{X} \\ \dot{Y} \end{Bmatrix} - \begin{bmatrix} m & 0 \\ 0 & m \end{bmatrix}\begin{Bmatrix} \ddot{X} \\ \ddot{Y} \end{Bmatrix} \tag{9.56}$$

An added complexity is the predominance of turbulent flow in annular seals. This invalidates the use of Reynolds' equation for the derivation of seal coefficients. The highest degree of accuracy can be obtained by the direct solution of the Navier–Stokes and continuity equations. However, at the present moment, such methods are considered to be excessively costly and impractical. As a result, two practical semiempirical methods have been developed to derive seal coefficients. In the first approach, the semiempirical turbulent model is directly substituted in the Navier–Stokes equation and a numerical technique is used for its solution. The second, most commonly used technique uses a bulk flow model together with control volume formulations, namely, the continuity equation and momentum equation, to obtain the desired results. For a detailed discussion of these methods, solution techniques, the influence of various physical parameters on the coefficients, and an excellent compilation of computational and experimental results, the publication by Childs (1993) is recommended.

9.3.2.5 Impeller–Diffuser/Volute Interface

It is widely known that the flow fields within certain types of rotating machinery can significantly influence its vibration behavior. Thomas (1958) recognized and explained the presence of destabilizing clearance excitation forces in axial flow steam turbines. Black (1974) was the first to suggest that centrifugal pump impellers could also develop destabilizing forces. The nature of these forces and their influence on rotor instability has been explained in Section 9.2.2.2 and Section 9.2.2.3 of this chapter. The impeller–diffuser/volute forces assuming rotational symmetry can generally be modeled by an equation of the following form:

$$\begin{Bmatrix} F_X \\ F_Y \end{Bmatrix} = -\begin{bmatrix} k & k_c \\ -k_c & k \end{bmatrix}\begin{Bmatrix} X \\ Y \end{Bmatrix} - \begin{bmatrix} c & c_c \\ -c_c & c \end{bmatrix}\begin{Bmatrix} \dot{X} \\ \dot{Y} \end{Bmatrix} - \begin{bmatrix} m & m_c \\ -m_c & m \end{bmatrix}\begin{Bmatrix} \ddot{X} \\ \ddot{Y} \end{Bmatrix} \tag{9.57}$$

For analytical procedures for the derivation of impeller interaction coefficients and a comparison of experimental data, the work by Childs (1993) is recommended. It is well recognized that a considerable amount of work still needs to be done towards understanding the complex nature of impeller–diffuser/volute interactive forces, especially at off-design conditions.

9.3.3 Design

Since the real machine is not available for tests, at the preliminary design stage it is a common practice to develop an accurate mathematical model of the machine to predict its dynamic behavior in operation. It is also prudent to understand and estimate how the machine will interact with its operating environment and how the environment could influence the operation of the machine. A suitable model of the rotor can be developed using the techniques described in Section 9.3.2, and the rotordynamic characteristics of the machine can be analyzed using one of the methods described in Section 9.3.1, above. Based on these methods, numerous computer-based rotordynamic analysis programs have been developed. A listing of the most widely known computer programs in North America is given in Table 9.3. The objectives of the analysis are to predict the critical speeds, excitation frequencies, the amplitudes of deflection, and the magnitude of the forces of the rotor within its full operating range. In certain situations, evaluation of the energy content of the excitation may also be required.

Once the mathematical model is developed, the eigenvalues of the rotor and the mode shapes can be determined. The results can then be presented in the form of a Campbell diagram, where the eigenvalues along with the excitation frequencies are plotted as a function of rotor speed. Critical speeds occur at the speeds corresponding to the points of intersection of the excitation frequency lines and the eigenvalue lines. The Campbell diagram presentation (Figure 9.13) of the results is very useful since the influence of key parameters such as stiffness, damping, clearances (new and worn conditions), and so on can all be shown on the same diagram. A critical speed, although present, may be of little consequence if it is associated with sufficient damping. As illustrated in Figure 9.4, when the damping ratio $\zeta \geq 0.707$, the system is critically damped and above this level of damping there is no amplification of the rotor deflection. At or near a critical speed the amplification factor is $\approx 1/2\zeta$. Using this estimated value,

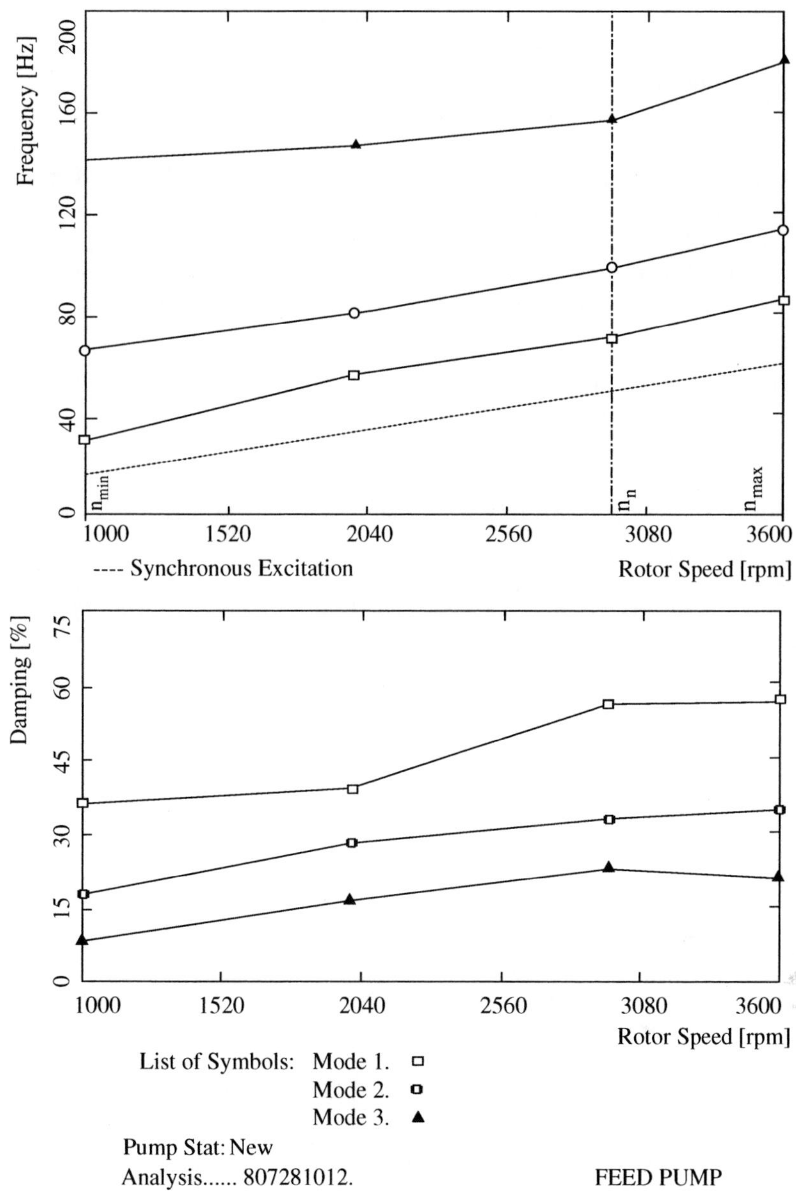

FIGURE 9.13 Campbell diagram for a multistage pump.

depending on the internal clearances of the machine, it is possible to assess if the rotor can pass through a critical speed without causing damage to the components. An amplification factor of 2.5 or below is a typical acceptance limit for centrifugal pumps, even for continuous operation at or near it. However, if the amplification factor exceeds the acceptable limits or the critical speeds are too close to the continuous operating speed, then design modifications have to be made to change the critical speed values. At the design stage, it is considered good practice to ensure that the critical speeds are not within ±10% of the continuous operating speed; these limits are sometimes referred to as *separation margins*. The mode shapes of the rotor are important from the standpoint of identifying where the maximum deflections occur. It also provides a good guide for assessing design modifications to improve damping or reduce sensitivity to unbalance forces.

The eigen analysis only provides relative deflections of the rotor. In order to estimate true deflections, a forced response analysis has to be made. Forcing functions of estimated magnitude are applied at selected locations to determine the resulting deflections at specific points on the rotor. This type of analysis is typically carried out for synchronous excitation forces only. The nature of the forcing function depends on the type of the machine; mechanical unbalance is common to all types of machines, whereas hydraulic unbalance is relevant to centrifugal pumps and electrical unbalance to electric motors. The challenge, of course, is to determine the magnitudes, directions, and locations of the forces to apply and how the resulting rotor response should be judged. Of course, these criteria are machine type-dependent and not necessarily applicable to all types of rotating machinery. An example of how forced response analysis on centrifugal pumps is evaluated is given below (Bolleter et al., 1992):

1. Maximum amplification factors and required separation margins are defined by specifications; example as shown in API 610, 8th edn., 1995.
2. Excitation forces are defined and the response is judged relative to admissible shaft vibration limits, and relative to clearances.
3. Apply unbalance forces of such a magnitude that maximum permissible vibration limits at the vibration probe locations are reached, and then evaluate if the deflections exceed the minimum clearances in the machine.
4. Apply an unbalance force of arbitrary magnitude and determine the resulting response at the same or another location, and calculate the sensitivity factor (SF) using the following formula:

$$\mathrm{SF} = \frac{a}{U} M \tag{9.58}$$

where
$a =$ rotor deflection
$U =$ unbalance force
$M =$ rotor mass

The sensitivity factor should then be compared with experimental base values of similar machines for acceptance. The rotor responses to the applied forces can be further analyzed to extract other parameters of interest, such as phase angles and force magnitudes at the bearings, in order to evaluate the design.

In order to optimize the rotating machine design in terms of placement of critical speeds and control of deflections and forces, a parameter sensitivity coefficients analysis (Lund, 1979; Rajan et al., 1986; Rajan et al. 1987) may be carried out. For speed and convenience of analysis, the optimization routine can be automated.

- Rotordynamic analysis is a part of the current rotating machinery design practice used to predict their vibration behavior.
- The most current rotordynamic analytical procedures are computer-based and are derived from the lump-parameter model or the transfer matrix method.
- In the lumped-parameter model method, the distributed elastic and inertial properties of the rotor are represented as a collection of rigid bodies connected by massless elastic beams.
- In the transfer matrix method, commonly called the FEA method, the rotor is represented as an assembly of elements with distributed elastic and inertial properties.
- Accurate modeling and representation of rotor components is vital to the accuracy and reliability of analysis results. As a result, significant advancement in modeling shaft sections, disks, impellers, bearings, seals, rotor dampers, and rotor–stator interactions have been made.

9.4 Vibration Measurement and Techniques

9.4.1 Units of Measurement

The most commonly used units of measurement of vibration are as follows:

- Displacement (peak to peak): millimeter (mm) or micron (10^{-6} m) in metric units, and *thou* or *mil* (0.001 in) in imperial units
- Velocity (peak, RMS or "true peak"): mm/sec or m/sec in metric units, and in./sec in imperial units
- Acceleration (RMS): g or m/sec^2 in metric units, and g or in./sec^2 in imperial units
- Frequency: hertz (Hz) or cycles/min (cpm) in both systems of units
- Phase angle: degrees in both systems of units

9.4.2 Measured Parameters and Methods

Under steady-state conditions, the vibration from a rotating machine is a periodic signal of a complex waveform. During unstable operation or upset conditions, the signal may become random in nature. In certain types of machines, transient signals that are nonperiodic could also be present due to internal impacts and damping. Based on the simple spring–mass system and the mathematical Fourier analysis procedure, all periodic complex waveforms can be reduced to the sum of a series of sinusoidal functions. In the case of random signals, averaging techniques are used to reduce them to periodic signals for convenience of analysis.

A quantitative assessment of the vibration can be made in terms of the amplitude, velocity, acceleration, or the magnitude of force of the motion. Other key parameters such as frequency, phase angle, and the time-varying nature of the signal are important in fully characterizing it. Because the true signal is not purely sinusoidal, it is important to identify its magnitude as, *peak*, *peak-to-peak* or *root mean square* (RMS). The preferred parameter of measurement varies throughout the industry and depends on the nature, complexity, and type of machine, and the purpose for which it is measured. A general classification of measured parameters and techniques based on industry and rotating machine type is given in Table 9.4.

- The displacement amplitude, velocity, acceleration, frequency, and phase angle of the vibration signal are some of the parameters that can be used to assess the condition of a rotating machine.
- The vibration signals generated by a typical rotating machine is complex in nature and, therefore, requires various mathematical analysis procedures and signal averaging techniques to reduce them to simple and interpretable forms.

9.5 Vibration Control and Diagnostics

9.5.1 Standards and Guidelines

Given the fact that some degree of vibration is always present in rotating machinery, some means of judging "how much is too much" has to be established so that vibrations can be controlled within reasonable limits. When such a judgment is not based on a scientific method, there is room for speculation and it will depend on the one making the decision, the manufacturer, end user, the governing authority, and so on. Since vibrations are a key indication parameter of the performance of rotating machinery, different interest groups monitor it for a variety of purposes. It can be used as a measure of

TABLE 9.4 Measurement Parameters and Techniques

Measurement/Technique	Description	When/Where Used
Acceleration, RMS	When high frequency or force of the vibration is of interest	Gear boxes, rolling element bearings, gas/steam turbines. Mainly used by defense and aerospace industry
Bode/Nyquist plot	Plot of displacement amplitude and phase angle vs. speed	Observe critical speeds and instability in machines using journal bearings
Cepstrum analysis	Inverse Fourier transform of logarithmic power spectrum	To detect families of harmonics and sidebands in gearboxes, rolling element bearings, and electric motors
Condition monitoring	Analysis of signals generated by the machine to determine its condition on a continuous or periodic basis	On critical equipment where predictive maintenance programs are used. Can reduce equipment redundancy
Displacement peak-to-peak		
(a) Absolute	Absolute displacement amplitude of rotor vibration	Where the rotor mass is very much larger than the stator mass. Large motors, generators, and fans
(b) Relative	Relative displacement amplitude of rotor vibration	Commonly used on machines with journal bearings or in close clearance seals
(c) External	Absolute displacement amplitude of stator component vibration	Low speed machines (less than 1000 rpm)
Modal analysis	To measure the vibration response of a structure to an applied force. The force can be periodic or an impact force	To determine the modal mass, stiffness and damping properties of a structure. Also used to measure structural natural frequencies
Orbit analysis	The path of the shaft centerline motion during rotation	Diagnostics of machines using journal bearings. Provides a picture of the motion of the journal in the bearing
Polar plots	A polar graph of amplitude versus phase at various machine speeds	Similar to Bode plots, can be used to detect critical speeds and instability. Modal properties can also be extracted from polar plots
Phase angle	Phase angle of vibration signal	Useful in balancing, diagnosing critical speeds, and misalignment problems
Rolling element bearing analysis		
(a) Acceleration	Measure the amplitude of all pass and discreet frequency accelerations	When damage has progressed to generate audible noise amplitudes increase. Useful from 5–5 kHz
(b) Shock pulse method	A high frequency resonance technique tuned to the detector natural frequency	Early detection of failure, measures ultrasonic noise. Proprietary technique
(c) Envelope technique	Bearing defects cause periodic impacts, which make bearing components resonate. Demodulation and enveloping techniques are used to detect the impact (fault) frequencies	Used for early failure detection (ultrasonic noise) as well as for advanced stages of damage (audible noise)
(d) Spike energy method	Measure the broadband acceleration over the 5–45 kHz range	Used for early failure detection (ultrasonic noise) as well as for advanced stages of damage (audible noise)
(e) Kurtosis method	The normalized fourth moment of the probability distribution of acceleration over the 2–80 kHz range	Used for early failure detection (ultrasonic noise) as well as for advanced stages of damage (audible noise)
Run-up, run-down analysis (waterfall/cascade plots)	Three-dimensional plot of frequency or time spectrum vs. time or speed	Used for diagnosing a variety of vibration problems. Helpful in analyzing transient signals

TABLE 9.4 *(continued)*

Measurement/Technique	Description	When/Where Used
Spectrum analysis	Plot of amplitude vs. frequency of vibration	Used for diagnostics, to determine frequency, harmonics, side bands, beats, transfer functions, etc., and to control of vibrations levels at discreet frequencies
Trend analysis	Vibration data collected periodically over an extended time domain	Useful in predictive maintenance programs in assessing machine conditions
Time averaging	Averaging of time records using triggering at the same point of the waveform of a repetitive signal	Used in the analysis of faulty gearboxes. Can reduce asynchronous components in the signal and improve signal-to-noise ratio
Time-domain analysis	Plot of amplitude versus time	To observe amplitude modulation, beats, impacts, transients, and phase angle. Very useful in diagnostics
Velocity-peak or RMS	Velocity amplitude of vibration signal	Parameter most commonly used in many industries to monitor vibrations. Peak readings relate to peak stress levels and rms to energy of vibrations

quality and workmanship or the common basis for acceptance between the user and manufacturer. From a safety point of view, the operator can establish *normal*, *alarm*, and *shutdown levels* based on vibration limits. Vibration levels are also used in making maintenance decisions in rotating machinery.

Rathbone (1939) was the first to publish guidelines for vibration limits for machinery, based on his experience as an insurance agent. Since that time, numerous individuals, organizations, and governing bodies have developed a variety of guidelines and standards for vibration levels in rotating machinery. A listing of the more commonly used guidelines and standards is given in Table 9.5. It should be recognized that these are experience-based standards, and therefore, will grow and develop with technology.

ISO 10816 Part 1 to Part 5 is a comprehensive set of standards that has been developed for the evaluation of vibration of rotating machinery by measuring the vibration response on nonrotating, structural components such as bearing housings. Vibration measuring points as specified by these standards are shown in Figure 9.14. In a similar vein, ISO 7919 Part 1 to Part 5 has been developed for the evaluation of vibration by measuring the vibration on rotating shafts. These standards cover the most widely used types of rotating machinery and they relate to both operational monitoring and acceptance testing of equipment. Table 9.6 and Table 9.7 are derived from these standards and are presented as a general guideline for vibration limits of rotating machinery. For specific details, including the limitations of the standards, the reader is advised to refer to the relevant sections of ISO 10816 and ISO 7919 Standards.

9.5.2 Vibration Cause Identification

Vibrations are an inherent part of all rotating machinery. Vibration can be due to many causes: improper design, practical manufacturing limits, poor installation, the effect of system environment, component deterioration, operation outside of design limits, or a combination of the above. At times, finding the exact cause of vibration can be quite a challenge, as several of the causes have similar symptoms. Table 9.8 is a list of the more commonly known causes of vibration in rotating machinery and their symptoms.

9.5.3 Vibration Analysis — Case Study

In the past, it was common practice to operate centrifugal pumps at a fixed speed and attain required flow changes by means of throttling. This forces the pump to operate at low efficiency conditions

TABLE 9.5 Vibration Guidelines and Standards

Year	Author/ Organization	Reference Number	Title/Description
2002	AGMA	ANSI/AGMA 6000-B96	Specification for Measurement of Linear Vibration on Gear Units
2003	API	ANSI/API std 541-2003	Form-Wound Squirrel-Cage Induction Motors 500 hp and Larger
1997	API	API STD 546, second edition	Brushless Synchronous Machines, 500 kVA and Larger
2004	API	API STD 610/ISO 13709, ninth edition	Centrifugal Pumps for Petroleum, Petrochemical and Natural Gas Industries
1997	API	API STD 611, fourth edition	General Purpose Steam Turbines for Petroleum, Chemical and Gas Industry Services
2005	API	API STD 612/ISO 10437, sixth edition	Petroleum, Petrochemical and Natural Gas Industries – Steam Turbines – Special-Purpose Applications
2003	API	API STD 613, fifth edition	Special Purpose Gear Units for Petroleum, Chemical and Gas Industry Services
1998	API	API STD 616, fourth edition	Gas Turbines for the Petroleum, Chemical, and Gas Industry Services
2002	API	API STD 617, seventh edition	Axial and Centrifugal Compressors and Expander-compressors for Petroleum, Chemical and Gas Industry Services
2000	API	API STD 670, fourth edition	Mechanical Protection Systems
2004	API	API STD 672, fourth edition	Packaged Integrally Geared, Centrifugal Air Compressors for Petroleum, Chemical, and Gas Industry Services
2001	API	API STD 673, second edition	Special Purpose Fans
1997	API	API STD 677, second edition	General Purpose Gear Units for Petroleum, Chemical, and Gas Industry Services
1996	API	API STD 681, first edition	Liquid Ring Vacuum Pumps for Petroleum, Chemical, and Gas Industry Services
2000	API	API STD 685, first edition	Sealless Centrifugal Pumps for Petroleum, Heavy-Duty Chemical, and Gas Industry Services
1965	BDS	BDS 5626-65	Measurement of Vibration on Electrical Rotating Machines
1964	Blake, M.P.	Hydrocarbon Processing, January 1964	New Vibration Standards for Maintenance
1963	CAGI		In-Service Standards for Centrifugal Compressors
1975	CAGI		Standard for Centrifugal Air Compressors
1971	CSN	CSN 011410	Permitted Limits for Unbalanced Solid Machine Elements
1968	Dresser Industrial		General Guidelines for Vibration on Clark Centrifugal Compressors
1966	Gosstandart	GOST 12379-66	Measurement of Vibration on Electrical Rotating Machines
2002	HI	ANSI/HI 9.6.4	Centrifugal and Vertical Pumps — Vibration Measurement and Allowable Values
1996	IEC	IEC 60034-14	Rotating Electrical Machines, Part 14: Mechanical Vibrations of Certain Machines with Shaft Heights 56 mm and Higher — Measurement, Evaluation and Limits of Vibration
1964	IRD	IRD #305D	General Machinery Vibration Severity Chart
1995–2001	ISO		Mechanical Vibration — Evaluation of Machine Vibration by Measurements on Nonrotating Parts:
		ISO 10816-1:1995	Part 1: General Guidelines
		ISO 10816-2:2001	Part 2: Land-Based Steam Turbines and Generators in Excess of 50 MW with Normal Operating Speeds of 1500, 1800, 3000 and 3600 rpm
		ISO 10816-3:1998	Part 3: Industrial Machines with Nominal Power above 15 kW and Nominal Speeds between 120 and 15,000 rpm when Measured *In Situ*
		ISO 10816-4:1998	Part 4: Gas Turbine Driven Sets Excluding Aircraft Derivations

TABLE 9.5 (*continued*)

Year	Author/ Organization	Reference Number	Title/Description
		ISO 10816-5: 2000	Part 5: Machine Sets in Hydraulic Power Generating and Pumping Plants
2002	ISO		Mechanical Vibration — Vibration of Active Magnetic Bearing Equipped Rotating Machinery
		ISO 14839-1: 2002	Part 1: Vocabulary
		ISO/CD 14839-2:2004	Part 2: Evaluation of Vibration
1996–2001	ISO		Mechanical Vibrations of Nonreciprocating Machines — Measurement on Rotating Shafts and Evaluation Criteria
		ISO 7919-1: 1996	Part 1 (1996): General Guidelines
		ISO 7919-2: 2001	Part 2 (2001): Land-Based Steam Turbines and Generators in Excess of 50 MW with Normal Operating Speeds of 1500, 1800, 3000 and 3600 rpm
		ISO 7919-3: 1996	Part 3 (1996): Coupled Industrial Machines
		ISO 7919-4: 1996	Part 4 (1996): Gas Turbine Sets
		ISO 7919-5: 1997	Part 5 (1997): Machine Sets in Hydraulic Power Generating and Pumping Plants
1993	ISO	ISO 8579-2	Acceptance Code for Gears, Part 2: Determination of Mechanical Vibration of Gear Units During Acceptance Testing
2004	ISO		Rolling Bearings — Measuring Methods for Vibration
		ISO 15242-1:2004	Part 1: Fundamentals
		ISO 15242-2:2004	Part 2: Radial Ball Bearings with Cylindrical Bore and Outside surface
		ISO/CD 15242-3	Part 3: Spherical and Taper Radial Roller Bearings with Cylindrical Bore and Outside Diameter
1959	Kruglov, N.V.	Teplonerg, 8 (85), 1959	Turbomachine Vibration Standards
1967	Maten, S	Hydrocarbon Processing, January 1967	New Vibration Velocity Standards
1983	McHugh, J.D.	J. Lub. Tech., Trans. ASME, 1983, 105	Estimating the Severity of Shaft Vibration within Fluid Film Journal Bearings
1974	MIL	MIL-STD-167-1	Mechanical Vibration of Shipboard Equipment, Type I: Environmental, Type II: Internally Excited
2003	NEMA	NEMA MG 1-2003	Motors and Generators, Part 7 — Mechanical Vibration — Measurement, Evaluation and Limits
1991	NEMA	NEMA SM 23-1991	Steam Turbines for Mechanical Drive Service
1991	NEMA	NEMA SM 24-1991	Land Based Steam Turbine Generator Sets 0 to 33,000 kW
1965	PKN	PN-65/E-04255	Measurement of Vibration of Electrical Rotating Machines
1939	Rathbone, T.C.	Power Plant Engineering, November 1939	Vibration Tolerances
1964	VDI	VDI 2056	Evaluation Criteria for Mechanical Vibrations in Machines
1982	VDI	VDI 2059 P1	Shaft Vibrations of Turbosets Principles for Measurement and Evaluation
1990	VDI	VDI 2059 P2	Shaft Vibrations of Steam Turbosets for Power Station Measurement and Evaluation
1985	VDI	VDI 2059 P3	Shaft Vibrations of Industrial Turbosets Measurement and Evaluation
1981	VDI	VDI 2059 P4	Shaft Vibrations of Gas Turbosets Measurement and Evaluation
1982	VDI	VDI 2059 P5	Shaft Vibrations of Hydraulic Machinesets Measurement and Evaluation
1949	Yates, H.G.	Trans. N.E. Coast Inst. Engrs Ship Builders, Vol. 65, 1949	Vibration Diagnosis of Marine Geared Turbines

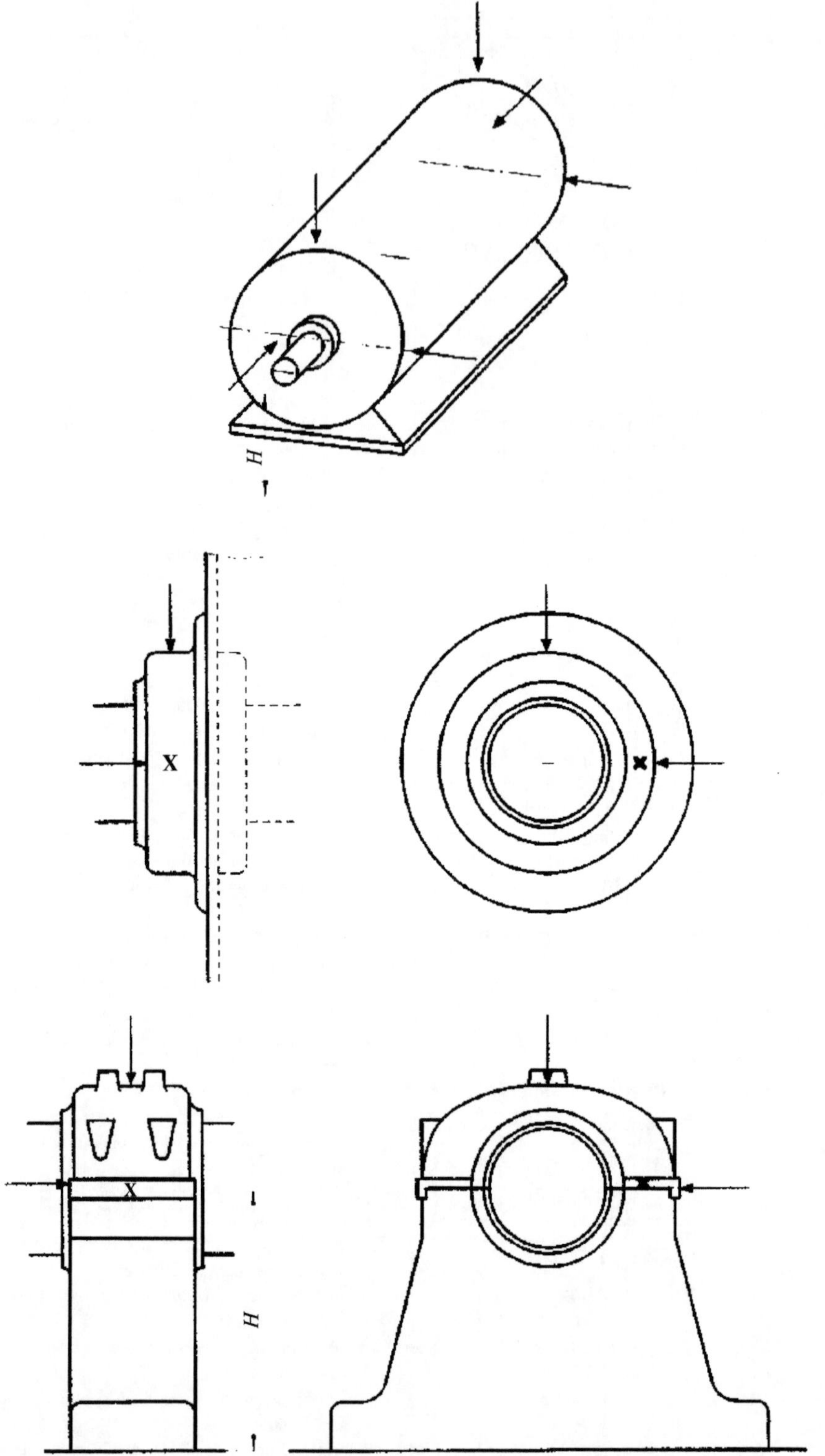

FIGURE 9.14 (a) Measuring points; (b) measuring points for vertical machine sets. (*Source:* ISO 10816-3, 1998-05-15. With permission.)

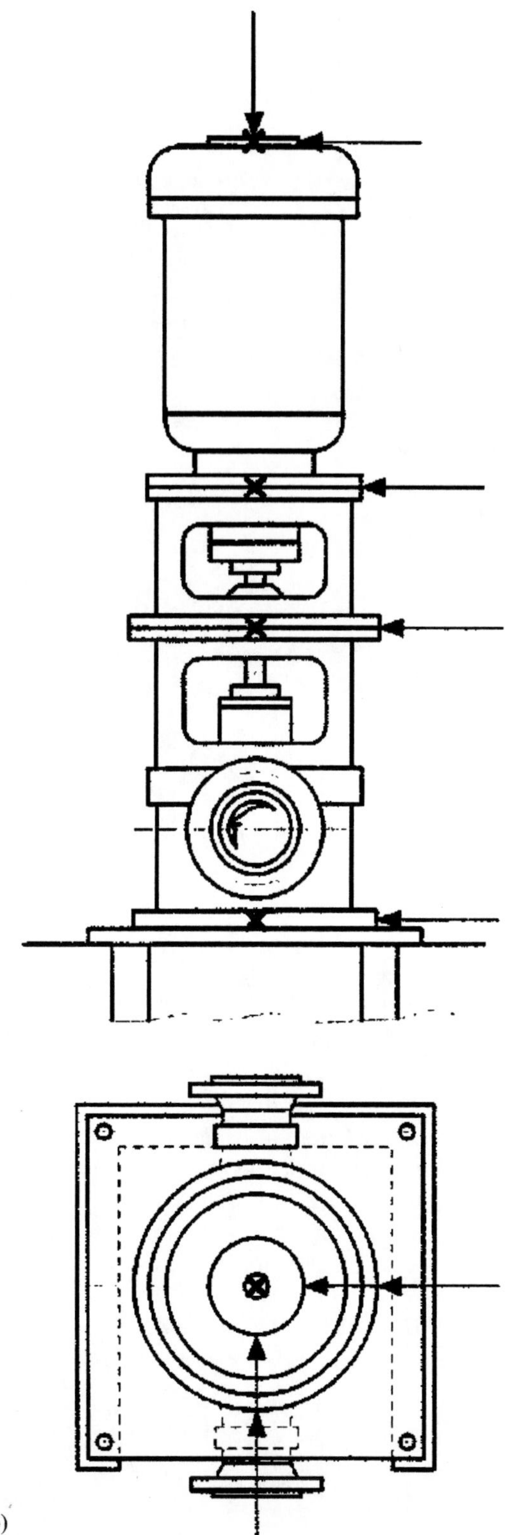

(b)

FIGURE 9.14 (*continued*)

TABLE 9.6A Acceptable Vibration Levels for Rotating Machinery Measured on Nonrotating Parts

Machinery Type	Power Level	Speed Range (rpm)	Applicable Vibration Level	
			Rigid Support	Flexible Support
Steam turbines	$15 \leq P \leq 300$ kW	$120 \leq N \leq 15{,}000$	V1 and D3	V3 and D7
	300 kW $\leq P \leq 50$ MW	$120 \leq N \leq 15{,}000$	V3 and D5	V6 and D8
	$P > 50$ MW	$N < 1{,}500$ or $N > 3{,}600$	V3 and D5	V6 and D8
	$P > 50$ MW	$N = 1{,}500$ or 1,800	V5	V5
	$P > 50$ MW	$N = 3{,}000$ or 3,600	V7	V7
Gas turbines	$15 \leq P \leq 300$ kW	$120 \leq N \leq 15{,}000$	V1 and D3	V3 and D7
	300 kW $\leq P \leq 3$ MW	$120 \leq N \leq 15{,}000$	V3 and D5	V6 and D8
	$P > 3$ MW	$3{,}000 \leq N \leq 20{,}000$	V8	V8
Hydraulic turbines and pump turbine	Horizontal machines			
	$P > 1$ MW	$60 \leq N \leq 300$	N/A	V4
	$P > 1$ MW	$300 < N \leq 1{,}800$	V2 and D6	N/A
Vertical machines	$P > 1$ MW	$60 < N \leq 1{,}800$	V2 and D6	N/A
	$P > 1$ MW	$60 < N \leq 1{,}000$	V2 and D6	V4 and D9
Centrifugal pumps				
Separate driver	$P > 15$ kW	$120 \leq N \leq 15{,}000$	V3 and D2	V6 and D4
Integral driver	$P > 15$ kW	$120 \leq N \leq 15{,}000$	V1 and D1	V3 and D2
Electric motors				
Shaft height $H \geq 315$ mm	$P > 15$ kW	$120 \leq N \leq 15{,}000$	V3 and D5	V6 and D8
Shaft height $160 \leq H < 315$ mm	$P > 15$ kW	$120 \leq N \leq 15{,}000$	V1 and D3	V3 and D7
Generators, excluding those used in hydraulic power generation	$15 \leq P \leq 300$ kW	$120 \leq N \leq 15{,}000$	V1 and D3	V3 and D7
	300 kW $\leq P \leq 50$ MW	$120 \leq N \leq 15{,}000$	V3 and D5	V6 and D8
	$P > 50$ MW	$N < 1{,}500$ or $N > 3{,}600$	V3 and D5	V6 and D8
	$P > 50$ MW	$N = 1{,}500$ or 1,800	V5	V5
	$P > 50$ MW	$N = 3{,}000$ or 3,600	V7	V7
Generators and motors used in hydraulic power generation	Horizontal machines			
	$P > 1$ MW	$60 \leq N \leq 300$	N/A	V4
	$P > 1$ MW	$300 < N \leq 1{,}800$	V2 and D6	N/A
Vertical machines	$P > 1$ MW	$60 < N \leq 1{,}800$	V2 and D6	N/A
	$P > 1$ MW	$60 < N \leq 1{,}000$	V2 and D6	V4 and D9
Compressors, rotary, blowers, and fans	$15 \leq P \leq 300$ kW	$120 \leq N \leq 15{,}000$	V1 and D3	V3 and D7
	300 kW $\leq P \leq 50$ MW	$120 \leq N \leq 15{,}000$	V3 and D5	V6 and D8

resulting in wasted energy and premature failure of components due to high vibration. The current practice to obtain flow changes in the pump is by means of speed change. This eliminates flow throttling and allows the pump to operate close to its best efficiency point, where energy is not wasted and vibrations are a minimum. However, as illustrated below, variable speed operation of a pump-motor set over a wide speed range could pose several challenging problems.

TABLE 9.6B Maximum Vibration Velocity Limits for Different Levels (mm/sec, RMS)

Vibration Level	Zone A	Zone B	Zone C	Alarm	Trip
V1	1.4	2.8	4.5	3.5	5.6
V2	1.6	2.5	4.0	3.1	5.0
V3	2.3	4.5	7.1	5.6	8.9
V4	2.5	4.0	6.4	5.0	8.0
V5	2.8	5.3	8.5	6.6	10.6
V6	3.5	7.1	11.0	8.9	13.8
V7	3.8	7.5	11.8	9.4	14.8
V8	4.5	9.3	14.7	11.6	18.4

TABLE 9.6C Maximum Vibration Displacement Limits for Different Levels (μm, RMS)

Vibration Level	Zone A	Zone B	Zone C	Alarm	Trip
D1	11	22	36	28	45
D2	18	36	56	45	70
D3	22	45	71	56	89
D4	28	56	90	70	113
D5	29	57	90	71	113
D6	30	50	80	63	100
D7	37	71	113	89	141
D8	45	90	140	113	175
D9	65	100	160	125	200

Zone A: Newly commissioned machines should fall within this zone. *Zone B*: Machines with vibrations within this zone are considered acceptable for long-term operation. *Zone C*: Machines with vibrations within this zone are normally considered unsatisfactory for long-term operation. Such a machine may be operated for a short period in this condition. *Alarm*: The values chosen will normally be set relative to a baseline value determined from experience. However, it is recommended that the alarm value shall not exceed those given herein. *Trip*: The values will generally relate to the mechanical integrity of the machine. They will generally be the same for all machines with similar design. It is recommended that the trip value shall not exceed those given herein. *Notes*: (1) The measured vibration is broadband, and the frequency range will depend on the type of machine being considered. A range from 2 to 1000 Hz is typical except for in high-speed machines, >10,000 rpm, where the upper limit should at least be six times the rotational frequency. (2) It is common practice to evaluate rotating machinery based on the broadband RMS vibration velocity, since it can be related to the vibration energy levels. However, other quantities such as vibration displacement or acceleration may be preferred. Especially low speed machines can have unacceptably large vibration displacements when the 1 × rpm component is dominant. Therefore, where specified, both the velocity and displacement criteria are met. (3) Since typical vibration waveforms measured on rotating machinery are complex in nature, there is no simple relationship between broadband velocity, displacement, and acceleration. (4) Vibration measurements shall be taken on bearing support housings, or other structural components, which adequately respond to the dynamic forces of the machine. Recommended locations for bearing housings are shown in Figure 9.14. (5) For certain types of machines, the axial vibration limits may differ from those for radial directions. Also, within the same machine set, in particular hydraulic power-generating sets, the applicable level may differ from bearing to bearing depending on its classification as a rigid or flexible support. (6) Above vibration limits apply to steady-state/normal operating conditions of the machine. If the vibration levels are sensitive to the operational conditions, then evaluation of the machine for operating conditions outside steady-state conditions will have to be based on different criteria. (7) The vibration limits specified herein should not be used to assess the condition of rolling element type bearings although it encompasses machines that may have these types of bearings. (8) It must be recognized that the vibration measurement on nonrotating parts alone does not form the only basis for judging the condition of a machine. In certain types of machines, it is common practice also to judge the vibration based on measurements taken on rotating shafts. (9) A support may be considered as rigid in a specific direction only if its natural frequency in that direction exceeds the main excitation frequency by at least 25%, otherwise it is considered to be flexible. In some cases, a support may be rigid in one direction and flexible in another. (10) In the case of hydraulic machine sets, major differences in radial bearing support arrangement can occur. For evaluation of the support type it is recommended that the reader refer to ISO 10816-5.

Description. The following case study is taken from a petroleum pipeline pump application where pump-motor sets with VFDs were installed in a new pipeline starting in Alberta, Canada and terminating in Minnesota, USA. VFDs are frequently used in the pipeline industry to power high horsepower pumps to eliminate power wasted by throttling, reduce inrush current at motor startup, and to provide greater operating flexibility. However, variable speed operation can cause vibration problems in the pump, motor, and the couplings that are not normally experienced with fixed speed pumps. Unexpected high torsional and lateral vibrations were experienced with these pumps and motors at certain operating speeds. A rotor torsional resonance, motor housing resonance, acoustic resonance in the internals of the pump, and discharge piping were identified to be the causes of the high vibration in the pump-motor set. Details on diagnosing the problems and the corrective measures taken to resolve them are given below:

Pump type: The pump was a centrifugal, two-stage, double volute horizontal pump, with six vane impellers, normally designed to operate at a fixed speed. Generation of pressure pulsations at the vane passing frequencies of 6 × and 12 × rotational speed is normally expected.

TABLE 9.7A Acceptable Vibration Levels for Rotating Machinery, Measured on Rotating Shafts

Machinery Type	Power Level	Speed Range (RPM)	Applicable Vibration Level	
			Relative Displacement	Absolute Displacement
Steam turbines	$P \leq 50$ MW	$1{,}000 \leq N \leq 30{,}000$	D8	—
	$P > 50$ MW	$N = 1{,}500$	D5	D7
	$P > 50$ MW	$N = 1{,}800$	D4	D6
	$P > 50$ MW	$N = 3{,}000$	D2	D5
	$P > 50$ MW	$N = 3{,}600$	D1	D3
Gas turbines	$P > 3$ MW	$3{,}000 \leq N \leq 30{,}000$	D8	—
	$P \leq 3$ MW	$1{,}000 \leq N \leq 30{,}000$	D8	—
Hydraulic turbines and pumps used in hydraulic power generation and pumping plants	$P > 1$ MW	$60 \leq N \leq 1{,}800$	D9	D9
Centrifugal pumps	All	$1{,}000 \leq N \leq 30{,}000$	D8	—
Electric motors	All	$1{,}000 \leq N \leq 30{,}000$	D8	—
Generators, excluding those used in hydraulic power generation	$P \leq 50$ MW	$1{,}000 \leq N \leq 30{,}000$	D8	—
	$P > 50$ MW	$N = 1{,}500$	D5	D7
	$P > 50$ MW	$N = 1{,}800$	D4	D6
	$P > 50$ MW	$N = 3{,}000$	D2	D5
	$P > 50$ MW	$N = 3{,}600$	D1	D3
Generators and motors used in hydraulic power generation	$P > 1$ MW	$60 \leq N \leq 1{,}000$	D9	D9
	$P > 1$ MW	$1{,}000 < N \leq 1{,}800$	D8	—
Compressors, rotary, blowers, and fans	All	$1{,}000 \leq N \leq 30{,}000$	D8	—

Motor: The motor was a 3000 hp, two pole horizontal induction motor, designed to operate at 3600 rpm.

Supply: The supply was a VFD of the current source inverter type. These drives are known to generate an oscillatory torque at 6 × and 12 × the operating frequency.

Coupling: Flexible disc type coupling with a spacer was used. These couplings have very little torsional damping capacity.

Speed range: The speed range was from 1440 rpm (24 Hz) to 3900 rpm (65 Hz).

Reference: Refer to Figure 9.15 to Figure 9.18.

As for the resonance at second and third torsional critical speeds (Figure 9.15) the second and third torsional modes are excited when the 6 × component of rotational speed corresponds to the critical speeds of 92 and 268 Hz, respectively. The 6 × rpm torsional excitation is caused by the pressure pulsations in the pump. The waterfall plot (Figure 9.15a) was taken during a run down of the set with the power to the motor turned off.

A similar plot taken during run up of the motor (Figure 9.15b) shows excitations at the same frequencies but having different amplitude. Since both the pump and motor generate 6 × excitation, it suggests a phase difference between the excitation torques.

It is important to note that the conventional vibration monitoring devices cannot detect the torsional resonance problem. The only indication of a problem was the unusual chattering noise emitted by the coupling. Special techniques to measure dynamic torque using strain gauges had to be used to detect the torsional vibrations.

As for the motor housing resonance (Figure 9.16), the 2 × rotational speed vibration of the motor is dominant and peaks at 118 Hz corresponding to a natural frequency of the motor frame. In the waterfall plot, the natural frequencies corresponds to excitations that are parallel to the axis. Excitation at harmonics, including the 6 × component, is present but is not dominant.

As for the pump vibrations at the vane passing frequency (Figure 9.17), the 6 × vane passing frequency is dominant at all operating speeds. It peaks at 238 Hz, possibly due to an acoustic resonance in

TABLE 9.7B Maximum Vibration Displacement S_{p-p} Limits (μm) Peak-to-Peak Limits for Different Levels

	Zone A	Zone B	Zone C
D1	75	150	240
D2	80	165	260
D3	90	180	290
D4	90	185	290
D5	100	200	320
D6	110	220	350
D7	120	240	385
D8	$4800/\sqrt{n}$	$9000/\sqrt{n}$	$13{,}200/\sqrt{n}$
D9	$10^{(2.3381-0.0704 \log n)}$	$10^{(2.5599-0.0704 \log n)}$	$10^{(2.8609-0.0704 \log n)}$

Zone A: Newly commissioned machines should fall within this zone. *Zone B*: Machines with vibrations within this zone are considered acceptable for long-term operation. *Zone C*: Machines with vibrations within this zone are normally considered unsatisfactory for long-term operation. Such a machine may be operated for a short period in this condition. *Alarm*: The values chosen will normally be set relative to a baseline value determined from experience. However, it is recommended that the alarm value shall not exceed those given herein. *Trip*: The values will generally relate to the mechanical integrity of the machine. They will generally be the same for all machines with similar design. It is recommended that the trip value shall not exceed those given herein. *Notes*: (1) The measured vibration is broadband and is shaft vibration displacement peak to peak. Where applicable, vibration limits for both absolute and relative radial shaft vibrations are given in certain cases. (2) Relative displacement is the vibratory displacement between the shaft and an appropriate structural component such as the bearing housing. Absolute displacement is the vibratory displacement of the shaft with reference to an inertial frame of reference. (3) Relative measurements are carried out with a noncontacting transducer. Absolute readings are obtained by one of the following methods: by a shaft riding probe on which a seismic transducer is mounted so that it measures absolute shaft displacement directly, or with the combination of a noncontacting transducer which measures relative shaft displacement and a seismic transducer which measures support vibration. Their conditioned outputs are vectorially added to provide a measure of the absolute shaft motion. (4) The vibration evaluation criteria are dependent upon a variety of factors and the criteria adopted will vary for different types of machines. Some of these factors are the bearing type, clearance, and diameter. The adopted criteria have to be compared with the bearing diametral clearance (C) and adjusted to suit. Typical values are: Zone A $\leq 0.4C$; Zone B $\leq 0.6C$; and Zone C $\leq 0.7C$. (5) Above vibration limits apply to steady state/normal operating conditions of the machine. If the vibration levels are sensitive to the operational conditions then evaluation of the machine for operating conditions outside steady-state conditions will have to be based on different criteria. (6) It is recommended that vibration readings at each location be made with a pair of transducers and that the transducers are mounted perpendicular to the shaft axis and they are at an angle of 90° to one another. The vibration limits apply to each measured direction. (7) The mechanical and electrical run-out at each measurement location must be assessed and should be $<25\%$ of the allowable limit or 6 μm, whichever is greater. (8) It must be recognized that the vibration measurement on rotating shafts does not form the only basis for judging the condition of a machine. In certain types of machine, it is common practice also to judge the vibration based on measurements taken on nonrotating parts. (9) ALARM levels should be set relative to a baseline value determined from experience for the measurement position, direction and type of machine. It must provide a warning that a defined value, which is significantly above the baseline value, has been reached. The maximum ALARM setting should be $\leq 0.75C$. (10) The TRIP values should be based on protecting the mechanical integrity of the machine. Consideration of damage to bearings is typical; therefore, maximum TRIP setting should be $\leq 0.9C$.

the discharge pipe. It does not seem to correspond to a structural natural frequency due to the absence of excitations at 238 Hz at all speeds.

As for the acoustic resonance in the pump cross-over pipe (Figure 9.18), dynamic pressure pulsation measurements made on the pump cross-over pipe from the first stage discharge to the second stage suction show an acoustic resonance at 540 Hz. The consistent presence of some excitation at 540 Hz at all speeds confirms that it is an acoustic natural frequency of the cross-over pipe. When the 6 × rpm pressure pulsation frequency coincides with the acoustic natural frequency, a resonance condition occurs and the magnitude of the pressure pulsation increases by almost a factor of 30.

As for corrective action, for a pump that has to operate over a wide speed range, totally eliminating the coincidence of all the frequencies of exciting forces with the system natural frequencies is impractical. Therefore, the system has to be designed such that the resulting magnitudes of the forces are controlled to within tolerable levels so that safe and reliable operation can take place. This can be accomplished by a

TABLE 9.8 Vibration Cause Identification

Cause	Dominant Frequency	Spectrum, Time Domain, Orbit Shape	Characteristics, Corrections, Comments
Mass unbalance	1 ×	High 1 × with much lower harmonics; circular or elliptic orbits	Corrected by shop or field balancing
Shaft bow	1 ×	Run down plot shows decrease of vibration at critical speed	The shaft has to be straightened using an acceptable method
Misalignment	1 × and 2 ×	Equally high 1 × and 2 × , figure 8 orbits	Realign at operating conditions; loads causing misalignment, such as nozzle loads, may have to be reduced
Worn journal bearings	1 × , 1/2 ×	Equally high 1 × and 1/2 ×	Difficult to balance
Gravity critical	2 ×	Run down plot will show excitation at 1/2 critical speed	Can be corrected by balancing
Asymmetric shaft	2 ×	Run down plot will show excitation at 1/2 critical speed	Typically occurs on multistage machines when all the keyways lie in the same plane; correct by staggering them
Shaft crack	1 × and 2 ×	High 1 × and run down plots may show excitation at 1/2 critical speed	Confirmation and detection of location of the crack may require NDE techniques
Loose components	1 × and higher orders plus fractional subharmonics	High 1 × with lower level orders and fractional subharmonics	Shimming and peening may be used as temporary methods to fix the problem
Coupling lockup	1 × and 2 ×	Equally high 1 × and 2 × , figure 8 orbits	Stop starts may change vibration pattern
Thermal instability	1 ×	High 1 × varies with temperature. Phase angle may change	Proper prewarming or compromise balancing can correct the problem
Oil whirl	< 1/2 × , typically 0.35 × to 0.47 ×	Run-up plot will show 1/2 × increasing and locking into fixed value < 1/2 ×	Temporary problem may be caused by excess clearances, oil viscosity, or unloading of the bearing; if it is a design problem, correct by changing to tilting pad bearings
Internal rubs	1/4 × , 1/3 × , 1/2 × , 2 × , 3 × , 4 × , etc.	Run down plots may show decreasing amplitudes and disappearance; loops in orbits	May get progressively worse; galling between contact surfaces or heat build-up may cause seizure and shaft failure
Trapped fluids in rotor	0.8 × to 0.9 ×	Time domain signal will show beating	Balancing the rotor may reduce the vibration
Defective rolling element bearings	At bearing defect frequency	Peaks at defect frequencies in spectrum	Shock pulse measurements can also be used to detect problem
Damaged gears	Gear mesh frequency	High peaks at gear mesh frequency with side bands. Time domain may also show pulses	To determine exact nature of damage further analysis may be required

TABLE 9.8 (*continued*)

Cause	Dominant Frequency	Spectrum, Time Domain, Orbit Shape	Characteristics, Corrections, Comments
Electric motor problems	1 × (line frequency), 2 × (line frequency)	High peaks at 1 × and 2 × line frequency with side bands; disappears when power to motor is turned off	In the case of two pole motors, it can be confused with mechanical causes as the rotational speed is the same as line frequency
Casing distortion	1 ×	High 1 ×, may change with time	Caused by high nozzle loads, casing not free to expand, soft foot or foundation distortion
Piping forces	1 ×, 2 ×	Equally high 1 × and 2 ×	Causes misalignment between bearings or between coupled equipment
Rotor and bearing critical	1 ×	High 1 ×, on rundown plot 1 × decreases rapidly, may also show a large phase angle change	More common in machines originally designed for fixed speed operation, later converted to variable speed operation
Structural resonance	1 ×, 2 ×	High 1 × and some 2 ×;can be easily identified on run down plot	Increase or decrease stiffness of structure or add or remove mass to change natural frequency
Rotor hysteresis	0.65 × to 0.85 ×	Spectrum will show high magnitudes at 0.65 × to 0.85 ×	Occurs in built up rotors with transitional fits
Hydraulic causes	1 × (vane pass frequency), 2 × (vane pass frequency)	High 1 × and 2 × vane pass frequency	Common in centrifugal pumps due to flow recirculation or inadequate gap between impeller and casing

direct reduction of the exciting force or by means of increased damping. Based on these guidelines, the following modifications were proposed to correct the problem:

1. Torsional resonance
   - Use an electrometric type coupling that has a high degree of torsional damping to reduce the magnitude of the torsional excitation forces such that the torsional stresses within the rotors are within acceptable limits.
   - Since both the pump and VFD generate excitation at 6 × rpm, their effects could be compounding one another. Introducing either five vane or seven vane impellers into the pump will eliminate this possibility.
   - Additional filters could be introduced into the VFD to reduce the 6 × and 12 × component periodic torsional excitation.
   - Consider not operating (lock out) the pump within ±10% of the frequency at which torsional resonance occurs.
2. Motor housing resonance at 2 ×
   - Although the 2 × vibration is dominant, its magnitude is within tolerable levels. The fact that some 2 × vibration is also present in the pump indicates that the 2 × vibration is perhaps caused by misalignment between the pump and the motor. This can be corrected by proper

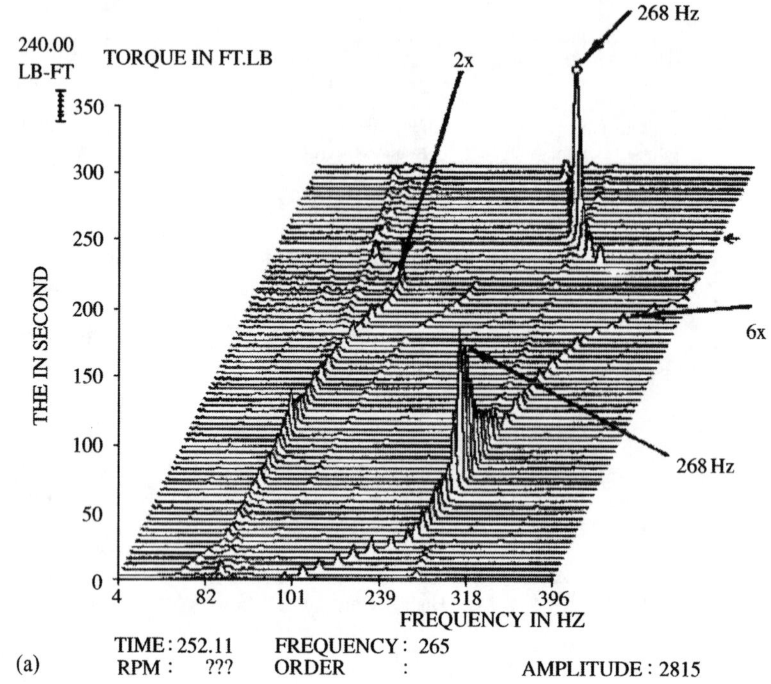

(a)

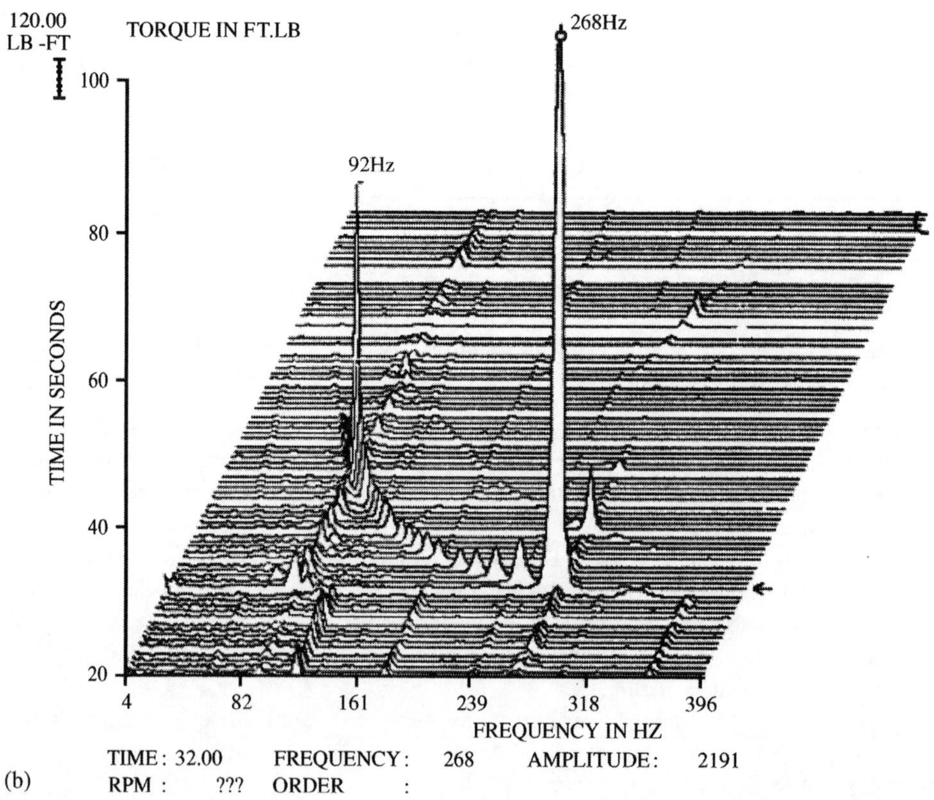

(b)

FIGURE 9.15 (a) Torsional resonance run-up and run-down plots; (b) torsional resonance run-down plot. (*Source:* Private communique, Insight Engineering Services Ltd. Alta., Canada. With permission.)

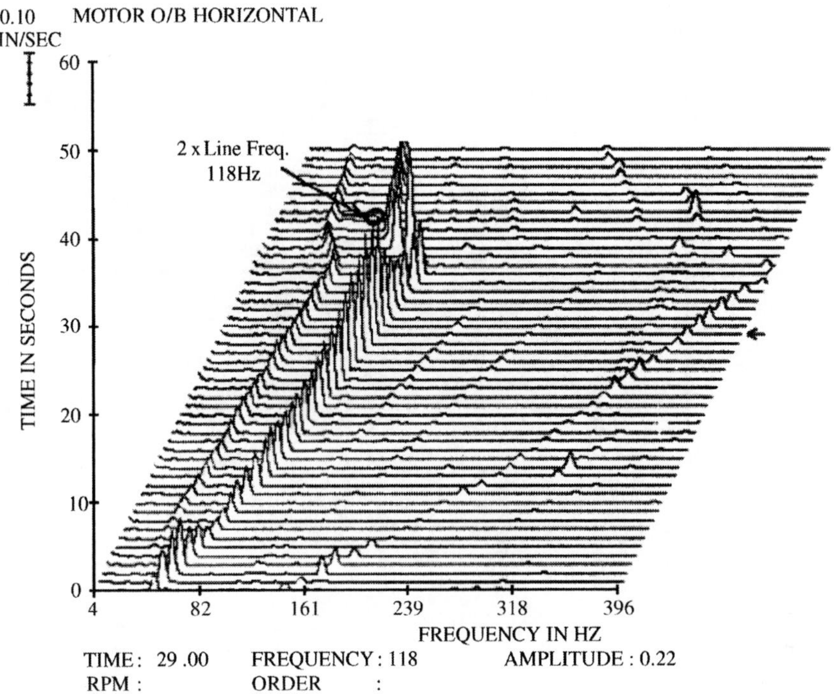

FIGURE 9.16 Motor frame resonance. (*Source:* Private communique, Insight Engineering Services Ltd., Alta., Canada. With permission.)

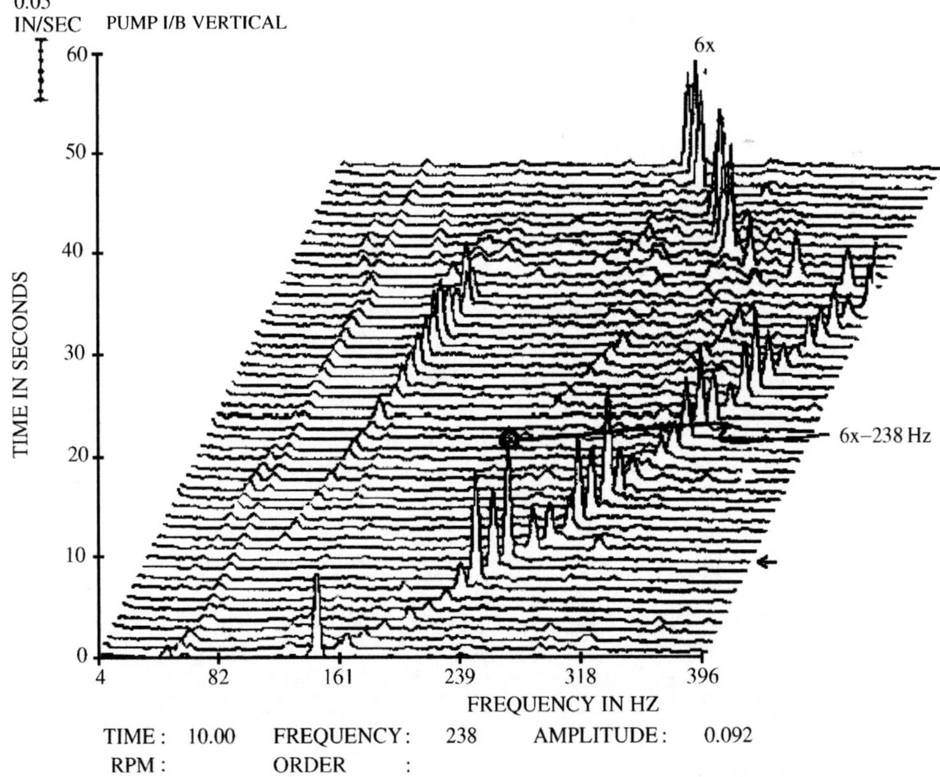

FIGURE 9.17 Pump bearing housing resonance. (*Source:* Private communique, Insight Engineering Services Ltd., Alta., Canada. With permission.)

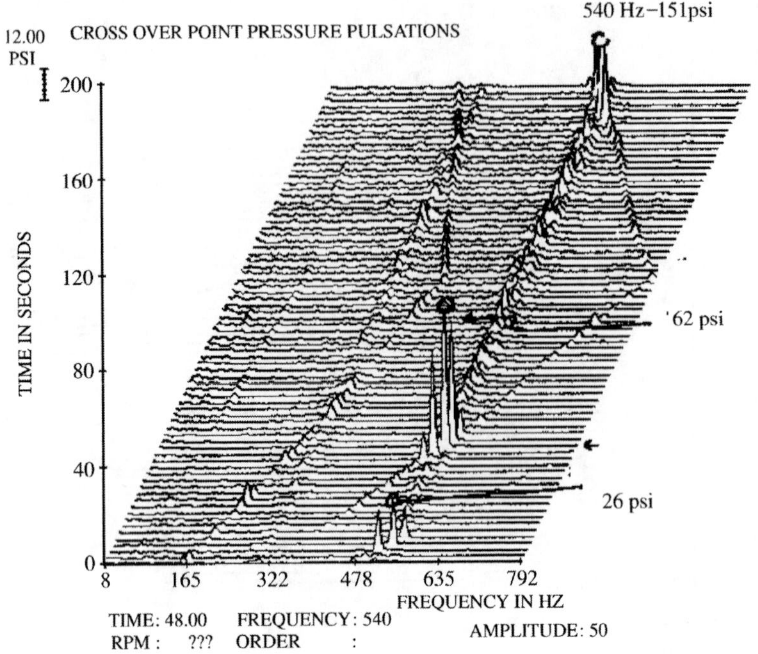

FIGURE 9.18 Pump cross-over pipe acoustic resonance. (*Source:* Private communique, Insight Engineering Services Ltd., Alta., Canada. With permission.)

alignment and thus reducing the 2 × excitation forces. In some cases, due to an unequal air gap between the rotor and stator of the motor, the motor could generate the 2 × vibration. Under such conditions, accurate centering of the motor bearings will generally correct the problem.

3. Pump vibrations at vane passing frequency
   - Generally, high vibrations at vane passing frequency are caused by pressure pulsations generated at the discharge of the impeller. There are several hydraulic modifications that can be made to the pump to reduce the amplitude of these pulsations that occur at vane passing frequency. The most common method is to increase the gap between the impeller discharge vanes and diffuser/volute. Also, changing the ratio of the number of impeller vanes to diffuser/volute vanes can help in reducing vane passing frequency pressure pulsations and the resulting vibration.
4. Acoustic resonance in the pump cross-over pipe
   - Once the pump is constructed, it is not possible to change the acoustic natural frequency of the cross-over pipe. However, the excitation force, pressure pulsations generated at the impeller discharge, can be reduced by the methods outlined above.

- The root cause of a vibration problem in a rotating machine can be determined by careful study and analysis of the vibration signals.
- Industrial and international vibration standards and guidelines have been developed to ensure safe and reliable operation of rotating machinery.
- Equipment manufacturers, users, insurance companies, and public interest groups use vibration standards to control vibration to within acceptable levels.

References

Adkins, D. and Brennen, C. 1986. Origins of hydrodynamic forces on centrifugal pump impellers, NASA CP No 2443, p. 467, In *Proceedings of a Workshop held at Texas A&M University*, Dallas, TX.

Alford, J., Protecting turbomachinery from self-excited rotor whirl, *Trans. ASME, J. Eng. Power*, 87, 333, 1965.

API Standard 610, 8th ed., Centrifugal Pumps for General Refinery Service, 1995.

Archer, J.S., Consistent mass matrix for distributed mass systems, *J. Struct. Div., Proc. ASCE*, 89, ST4, 161, 1963.

Baskharone, E.A., Daniel, A.S., and Hensel, S.J., Rotordynamic effects of the shroud-to-housing leakage flow in centrifugal pumps, *Trans. ASME, J. Fluid Eng.*, 116, 558, 1994.

Benckert, H., Wachter, J. 1980. Flow induced spring coefficients of labyrinth seals for application in turbomachinery, NASA CP No. 2133, p. 189, In *Proceedings of a Workshop held at Texas A&M University*, Dallas, TX.

Bentley, D. 1974. *Forced Subrotative Speed Dynamic Action of Rotating Machinery*, ASME, Dallas, TX, 74-PET-16.

Biezeno, C.B. and Grammel, R. 1954. Engineering Dynamics, Steam Turbines, Vol. III. D. Van Nostrand Co. Inc., New York (originally published in German in 1939 as Technische Dynamik by Julius Springer, Berlin, Germany).

Black, H.F., Interaction of a whirling rotor with a vibrating stator across a clearance annulus, *J. Mech. Eng. Sci.*, 10, 1, 1968.

Black, H.F. 1974. Lateral stability and vibrations of high speed centrifugal pump rotors, p. 56, In *Proceedings IUTAM Symposium on Dynamics of Rotors*, Lyngby, Denmark.

Blevins, Robert, D. 2001. Formulas for Natural Frequency and Mode Shapes, Krieger Publishing Co. Inc., Melbourne, FL.

Bolleter, U., Frei, A., Florjancic, S., Leibundgut, E., and Stürchler, R., *Rotordynamic Modeling and Testing of Boiler Feedpumps*, EPRI TR-100980, 1992.

Bolleter, U., Leibundgut, E., Stürchler, R., and McCloskey, T. 1989. Hydraulic interaction and excitation forces of high head pump impellers, p. 187, In *Proceedings of the Third ASCE/ASME Mechanical Conference*, La Jolla, CA.

Bolleter, U., Wyss, A., Welte, I., and Stürchler, R., Measurement of hydrodynamic interaction matrices of boiler feed pump impellers, *Trans. ASME, J. Vib. Stress Reliab. Des.*, 1985, SME, 85-DET-147, New York.

Brennen, C., Acosta, A., and Caughey, T. 1980. A test program to measure cross-coupling forces in centrifugal pumps and compressors, NASA CP No. 2133, p. 229, In *Proceedings of a Workshop held at Texas A&M University*, Dallas, TX.

Childs, D.W. 1986. Force and moment rotordynamic coefficients for pump-impeller shroud surfaces, NASA CP No. 2443, p. 467, In *Proceedings of a workshop held at Texas A&M University Dallas*, TX.

Childs, D. 1993. Turbomachinery Rotordynamics, Wiley, New York.

Chree, C., The whirling and transverse vibration of rotating shafts, *Phil. Mag.*, 7, 504, 1904.

COJOUR, User's Guide: Dynamic Coefficients for Fluid Film Journal Bearings, EPRI CS-4093.

Crandall, S.H. and Brosens, P.J., Whirling of unsymmetrical rotors, *J. Appl. Mech.*, 28, 567, 1961.

Dunkerly, S., On the whirling and vibration of shafts, *Phil. Trans. R. Soc., London A*, 185, 279, 1894.

Ehrich, F.F. 1999. Handbook of Rotordynamics, Revised ed., Krieger Publishing Co. Inc., Melbourne, FL.

Foppl, A., Das Problem der Laval'schen Turbinewelle, *Civilingenieur*, 41, 333, 1885.

Glienicke, J. 1966. Experimental investigation of the stiffness and damping coefficients of turbine bearings and their application to instability prediction, p. 122, In *Proceedings of the Journal Bearings for Reciprocating and Turbo Machinery Symposium*, Nottingham, UK.

Gorman and Daniel, J. 1975. *Free Vibration Analysis of Beams and Shafts*, Wiley, New York.

Greenhill, A.G., On the strength of shafting when exposed both to torsion and to end thrust, *Proc. I. Mech. Eng (London)*, 182, 1883.

Gunter, E.J. Jr., Dynamic stability of rotor-bearing systems, *NASA SP-113*, 1966.

Hagg, A.C. and Sankey, G.O., Elastic and damping properties of oil-film journal bearings for application to unbalance vibration calculations, *Trans. ASME, J. Appl. Mech.*, 25, 141, 1958.

Harris, T. 1991. Rolling Bearing Analysis, 3rd ed., Wiley, New York.

Holzer, H. 1921. Die Berechnung der Drehschwingungen, Springer, Berlin.

Houbolt, J.C. and Reed, W.H., Propeller nacelle whirl flutter, *Inst. Aerospace Sci.*, 1, 61, 1961.

ISO 10816-1. *Mechanical vibration—evaluation of machine vibration by measurements on non-rotating parts.* Part 1. General Guidelines, ISO, Geneva, Switzerland, 1995.

ISO 10816-2. *Mechanical vibration—evaluation of machine vibration by measurements on non-rotating parts.* Part 2. Land-based Steam Turbines and Generators in excess of 50 MW with normal operating speeds of 1500 r/min, 1800 r/min, 3000 r/min and 3600 r/min, ISO, Geneva, Switzerland, 2001.

ISO 10816-3. *Mechanical vibration—evaluation of machine vibration by measurements on non-rotating parts.* Part 3. Industrial machines with nominal power above 15 kW and nominal speeds between 120 r/min and 15 000 r/min when measured in situ, ISO, Geneva, Switzerland, 1998.

ISO 10816-4. *Mechanical vibration—evaluation of machine vibration by measurements on non-rotating parts.* Part 4. Gas Turbine Driven Sets Excluding Aircraft Derivations, ISO, Geneva, Switzerland, 1998.

ISO 10816-5. *Mechanical vibration—evaluation of machine vibration by measurements on non-rotating parts.* Part 5. Machine Sets in Hydraulic Power Generating and Pumping Plants, ISO, Geneva, Switzerland, 2000.

ISO 7919-1. *Mechanical vibrations of non-reciprocating machines—measurement on Rotating Shafts and Evaluation Criteria.* Part 1. General Guidelines, ISO, Geneva, Switzerland, 1996.

ISO 7919-2. *Mechanical vibrations of non-reciprocating machines—Measurement on Rotating Shafts and Evaluation Criteria.* Part 2. Land-Based Steam Turbines and Generators in Excess of 50 MW with Normal Operating Speeds of 1500 r/min, 1800 r/min, 3000 r/min and 3600 r/min, ISO, Geneva, Switzerland, 2001.

ISO 7919-3. *Mechanical vibrations of non-reciprocating machines—Measurement on Rotating Shafts and Evaluation Criteria.* Part 3. Coupled Industrial Machines, ISO, Geneva, Switzerland, 1996.

ISO 7919-4. *Mechanical vibrations of non-reciprocating machines—Measurement on Rotating Shafts and Evaluation Criteria.* Part 4. Gas Turbine Sets, ISO, Geneva, Switzerland, 1996.

ISO 7919-5. *Mechanical vibrations of non-reciprocating machines—Measurement on Rotating Shafts and Evaluation Criteria.* Part 5 Machine Sets in Hydraulic Power Generating and Pumping Plants, ISO, Geneva, Switzerland, 1997.

Jeffcott, H.H., The lateral vibration of loaded shafts in the neighbourhood of a whirling speed—the effect of want of balance, *Phil. Mag.*, 37, 304, 1919.

Jery, B., Acosta, A., Brennen, C., and Caughey, T. 1984. Hydrodynamic impeller stiffness, damping, and inertia in the rotordynamics of centrifugal flow pumps, NASA CP No. 2338, p. 137, In *Proceedings of a workshop held at Texas A&M University*, Dallas, TX.

Jones, A., A general theory for elastically constrained ball and radial roller bearings under arbitrary load and speed conditions, *Trans. ASME J. Basic Eng.*, 82, 309, 1960.

Kimball, A.L. Jr., Internal friction theory of shaft whirling, *Gen. Electr. Rev.*, 27, 244, 1924.

Kirk, R.G., Donald, G.N. 1983. Design Criteria of Improved Stability of Centrifugal Compressors, AMD-Vol. 55, Rotor Dynamical Instability. ASME, New York, p. 59.

Kramer, E. 1993. Dynamics of Rotors and Foundations, Springer, Berlin.

Lomakin, A.A., Calculating the critical speed and the conditions to ensure dynamic stability of the rotors in high pressure hydraulic machines, taking account of the forces in the seals, *Energomashinostroenie*, 4, 1, 1958.

Lund, J.W., Spring and damping coefficients for the tilting-pad journal bearing, *Trans. ASLE*, 7, 342, 1964.

Lund, J.W. 1965. Rotor-bearing Dynamics Design Technology. Part III. Design Handbook for Fluid-film Bearings, AFAPL-TR-64-45. Wright-Patterson Air Force Base, Dayton, OH.

Lund, J.W. 1968. Rotor-bearing Dynamics Design Technology. Part VII. The Three Lobe Bearing and Floating Ring Bearing, AFAPL-TR-65-45. Wright-Patterson Air Force Base, Dayton, OH.

Lund, J.W., Sensitivity of the critical speeds of a rotor to changes in the design, *Trans. ASME, J. Mech. Des.*, 102, 115, 1979.

Lund, J.W. and Orcutt, F.K., Calculation and experiments on the unbalance response of a flexible rotor, *Trans. ASME, J. Eng. Ind.*, 89, 785, 1967.

Meirovitch, L. 1986. Elements of Vibration Analysis, 2nd ed., McGraw-Hill, New York.

Miller, D.F., Forced lateral vibration of beams on damped flexible end supports, *Trans. ASME, J. Appl. Mech.*, 20, 167, 1953.

Moore, J.J. and Palazzolo, A.B., Rotordynamic force prediction of Whirling Centrifugal Impeller Shroud passages using Computational Fluid Dynamic techniques, *Trans. ASME, J. Eng. Gas Turbine Power*, 123, 910, 2001.

Muszynska, A., Whirl and whip—rotor/bearing stability problems, *J. Sound Vib.*, 110, 443, 1986.

Myklestad, N.O., A new method for calculating natural modes of uncoupled bending vibrations of airplane wings and other types of beams, *J. Aeronaut. Sci.*, 11, 153, 1944.

Nelson, H., A Finite rotating shaft element using Timoshenko beam theory, *Trans. ASME, J. Mech. Des.*, 102, 793, 1980.

Nelson, H. and McVaugh, J., The dynamics of rotor-bearing systems using finite elements, *Trans. ASME, J. Eng. Ind.*, 98, 593, 1976.

Newkirk, B.L., *Shaft Whipping Gen. Electr. Rev.*, 27, 169, 1924.

Newkirk, B.L. and Taylor, H.D., Shaft whipping due to oil action in journal bearings, *Gen. Electr. Rev.*, 28, 559, 1925.

Ohashi, H., Hatanaka, R., and Sakurai, A. 1986. Fluid force testing machine for whirling centrifugal impeller, In *Proceedings of the International Federation for Theory of Machines and Mechanisms, International Conference on Rotordynamics*, JSME, Tokyo, Japan.

Orcutt, F.K., The steady-state and dynamic characteristics of the tilting-pad journal bearing in laminar and turbulent flow regimes, *Trans. ASME, J. Lubricat. Technol.*, 89, 392, 1967.

Perera, L. 2002. Private communiqué, Insight Engineering Services Ltd, Alta., Canada.

Pinkus, O. and Sternlicht, B. 1961. Theory of Hydrodynamic Lubrication, McGraw-Hill, New York.

Poritsky, H., Contribution to the theory of oil whip, *Trans. ASME*, 75, 1153, 1953.

Prohl, M.A., A general method for calculating Critical Speeds of flexible rotors, *Trans. ASME, J. Appl. Mech.*, 12, A-142, 1945.

Rajan, M., Nelson, H.D., and Chen, W.J., Parameter sensitivity in the dynamics of rotor-bearing systems, *Trans. ASME, J. Vib. Acoust. Stress Reliab. Des.*, 108, 197, 1986.

Rajan, M., Rajan, S.D., and Nelson, H.D., and Chen, W.J., Optimal placement of critical speeds in rotor-bearing systems, *Trans. ASME, J. Vib. Acoust. Stress Reliab. Des.*, 109, 152, 1987.

Rankine, W.J.M., On the centrifugal force of rotating shafts, *Engineer*, 249, 9, 1869.

Rathbone, T.C., Vibration tolerances, *Power Plant Eng.*, November, 1939.

Rayleigh, J.W.S. 1945. Theory of Sound, Dover Publications, New York.

Robertson, D., Whirling of a journal in a sleeve bearing, *Phil. Mag.*, 15, 96, 113, 1933.

Robertson, D., Transient whirling of a rotor, *Phil. Mag.*, 20, 793, 1935.

Ruhl, R.L. and Booker, J.F., A finite element model for distributed parameter turborotor systems, *Trans. ASME, J. Eng. Ind.*, 94, 126, 1972.

Smith, D.M., The motion of a rotor carried by a flexible shaft in flexible bearings, *Proc. R. Soc. London A*, 142, 92, 1933.

Someya, T. 1989. *Journal Bearing Databook*, Springer, New York.

Southwell, R.V. and Gough B.S., 1921, *Complex Stress Distributions in Engineering Materials*, British Association For Advancement of Science Reports, 345.

Sternlicht, B., Elastic and damping properties of cylindrical journal bearings, *Trans. ASME, J. Basic Eng.*, 81, 101, 1959.

Stodola, A. 1927. *Steam and Gas Turbines*, Vol. I, McGraw-Hill, New York.

Taylor, E.S. and Browne, K.A., Vibration isolation of aircraft power plants, *J. Aeronaut. Sci.*, 6, 43, 1938.

Thomas, H.J., Unstable natural vibration of turbine rotors excited by the axial flow in stuffing boxes and blading, *Bull. AIM*, 71, 1039, 1958.

Timoshenko, S., Young, D.H., and Weaver, W. Jr. 1974. Vibration Problems in Engineering, 4th ed., Wiley, New York.

Urlichs, K., Leakage flow in thermal turbo-machines as the origin of vibration exciting lateral forces, NASA, TT-17409, 1977.

Vance, J.M. 1988. Rotordynamics of Turbomachinery, Wiley, New York.

Warner, P.C., Static and dynamic properties of partial journal bearings, *Trans. ASME, J. Basic Eng.*, 85, 247, 1963.

Wolf, J.A. 1968. Whirl Dynamics of a Rotor Partially Filled with Liquids, ASME, New York, 68-WA/APM-25.

10

Regenerative Chatter in Machine Tools

Robert G. Landers
University of Missouri at Rolla

Summary

Regenerative chatter, a result of unstable interactions between machining forces and structural deflections, is a great limitation in machining operations. This chapter describes the modeling, analysis, simulation, detection, and control of regenerative chatter in machining operations, and, in particular, turning and face milling. An analytical method is applied to calculate the limiting stable depth-of-cut and corresponding spindle speeds to generate stability lobe diagrams. The method is applied to both turning and face-milling operations. Time-domain simulation is described and applied to turning and face-milling operations. Methods for chatter detection are presented and experimental results from a face-milling operation are given. Chatter suppression techniques, namely spindle-speed selection, feed selection, depth-of-cut selection, and spindle-speed variation, are presented and two simulations of a turning operation are used to illustrate the spindle-speed selection and spindle-speed variation techniques. Finally, a case study of a face-milling operation is presented. The nomenclature used in the presentation is listed at the end of the chapter.

10.1 Introduction

Regenerative chatter is a major limitation in machining operations. This phenomenon is a result of an unstable interaction between the machining forces and the structural deflections. The forces generated when the cutting tool and part come into contact produce significant structural deflections. These structural deflections modulate the chip thickness that, in turn, changes the machining forces. For certain cutting conditions, this closed-loop, self-excited system becomes unstable and regenerative chatter occurs. Regenerative chatter may result in excessive machining forces and tool wear, tool failure, and scrap parts due to unacceptable surface finish, thus severely decreasing operation productivity and part quality.

A typical chatter stability chart, the so-called stability lobe diagram, is shown in Figure 10.1. If the process parameters are above the stability borderline, chatter will occur, and if the process parameters are below the stability borderline, chatter will not occur. The asymptotic stability borderline is the depth-of-cut below which stable machining is guaranteed regardless of the spindle speed. The lobed nature of the stability borderline allows stable pockets to form; thus, at specific ranges of spindle speeds, the depth-of-cut may be substantially increased beyond the asymptotic stability limit. These pockets become smaller as the spindle speed decreases. The stability borderline is "pulled up" for low spindle speeds due to process damping (i.e., the back side of the tool rubbing on the part surface). If accurate models of the structural components and the cutting process are available, the stability lobe diagram may be used to plan chatter-free machining operations.

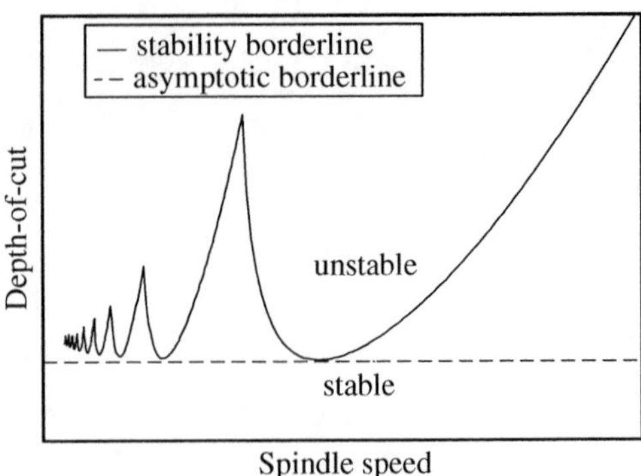

FIGURE 10.1 Stability lobe diagram.

The analysis of regenerative chatter as the interaction between the cutting forces and structural vibrations was established by Tobias (1965) and Koenigsberger and Tlusty (1971). Merritt (1965) used systems theory to determine stability and construct the stability lobe diagram by generating specialized plots from the harmonic solutions of the system's characteristic equation. Chatter analysis reveals a natural delay in the system leading many researchers to use Nyquist techniques to generate stability lobe diagrams (Minis et al., 1990a, 1990b; Lee and Liu, 1991a, 1991b; Minis and Yanushevsky, 1993). A set of process parameters is selected and the characteristic equation is formed. The Nyquist criterion is applied to determine if the system for this process parameter set is stable. The depth-of-cut is adjusted and the procedure is repeated until the critical depth-of-cut is determined. Another chatter analysis technique capable of generating stability lobe diagrams analytically for linear systems has recently been introduced (Altintas and Budak, 1995; Budak and Altintas, 1998a, 1998b). This technique is utilized in this chapter.

The theoretical analysis of regenerative chatter laid the foundation for developing techniques to automatically detect its occurrence and to automatically suppress it. Since there is a dominant chatter frequency, which is near a structural frequency, that occurs when chatter develops, most monitoring techniques analyze the frequency of a process variable, and chatter is detected when significant energy is present near a structural frequency. Most automatic chatter suppression routines either adjust the spindle speed to be in a pocket of the stability lobe diagram or vary the spindle speed to bring the current and previous tooth passes into phase. While automatic monitoring and control of regenerative chatter shows great promise, it has been mostly limited to laboratory applications. Therefore, commercial tools are not currently available.

While regenerative chatter in turning and face-milling operations is discussed in this chapter, this phenomenon is not limited to these specific manufacturing operations. Other machining operations for which chatter has been analyzed include end milling (Budak and Altintas, 1998a, 1998b), grinding (Inasaki et al., 2001), drilling (Tarng and Li, 1994), and so on. Also, the regenerative chatter phenomenon occurs in other manufacturing operations, most notably in rolling (Yun et al., 1998; Tlusty, 2000).

Section 10.2 and Section 10.3 present an analytical method to examine regenerative chatter in turning and face-milling operations, respectively. Section 10.4 discusses a numerical technique known as time domain simulation that may be used to analyze regenerative chatter for nonlinear systems. The subject of chatter detection is presented in Section 10.5, then methods to perform chatter suppression are discussed and illustrated in Section 10.6. Section 10.7 presents a case study of a face-milling operation.

10.2 Chatter in Turning Operations

A schematic of a turning operation is shown in Figure 10.2. The part structure is assumed to be perfectly rigid, while the cutting-tool structure is capable of vibrations in the longitudinal (i.e., the z) direction only. The machining force in the longitudinal direction is

$$F(t) = Pdf(t) \tag{10.1}$$

The depth-of-cut is assumed to be constant; however, the feed, and hence the machining force, is time-varying due to structural vibrations. It is assumed here that the machining force does not explicitly depend upon the cutting speed.

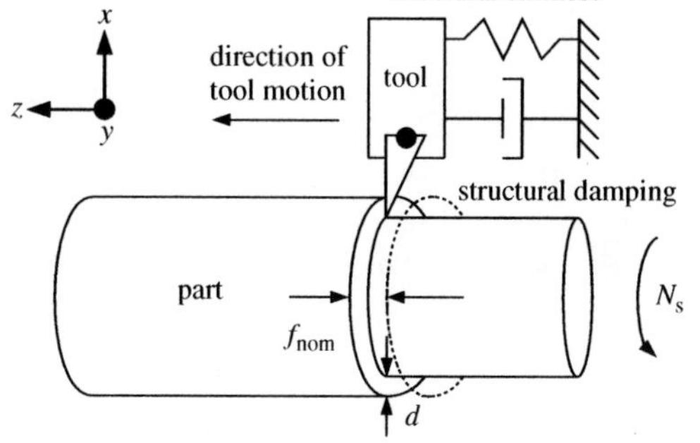

FIGURE 10.2 Turning operation schematic: current pass (solid line) and previous pass (dotted line).

The feed is the chip thickness in the longitudinal direction. The nominal feed is the distance the tool advances relative to the part each spindle revolution and is constant once the tool fully engages the part. However, the cutting tool vibrates, leaving an undulated surface on the part and, thus, modulates the feed. The instantaneous feed is

$$f(t) = f_{\text{nom}} + \Delta z(t) = f_{\text{nom}} + z(t) - z(t - T) \tag{10.2}$$

The term f_{nom} is the nominal feed, also known as the static feed. The term $\Delta z(t)$ is the feed due to the cutting-tool vibrations and is known as the dynamic feed. The parameter T is the spindle-rotation period. The structural vibration, $z(t)$, known as the inner modulation, is the cutting-tool vibration at the current time. The delayed structural vibration $z(t - T)$, known as the outer modulation, is the cutting-tool vibration as of when the part was at the current angular during the previous spindle rotation. The modulation in feed due to structural vibrations is illustrated in Figure 10.3.

Inserting Equation 10.2 into Equation 10.1:

$$F(t) = Pdf_{\text{nom}} + Pd\Delta z(t) = F_{\text{nom}} + \Delta F(t) \tag{10.3}$$

The force $F_{\text{nom}} = Pdf_{\text{nom}}$ is due to the nominal chip thickness and does not vary since the depth-of-cut and nominal feed are constant. The force $\Delta F(t) = Pd\Delta z(t)$ is due to changes in the nominal feed caused by structural vibrations.

The structural vibrations are related to the machining force by

$$z(s) = -g(s)F(s) \tag{10.4}$$

where $g(s)$ is the transfer function relating the structural vibrations to the machining forces. Since $F(t) - F(t - T) = \Delta F(t) - \Delta F(t - T)$, the structural vibrations are related to the machining

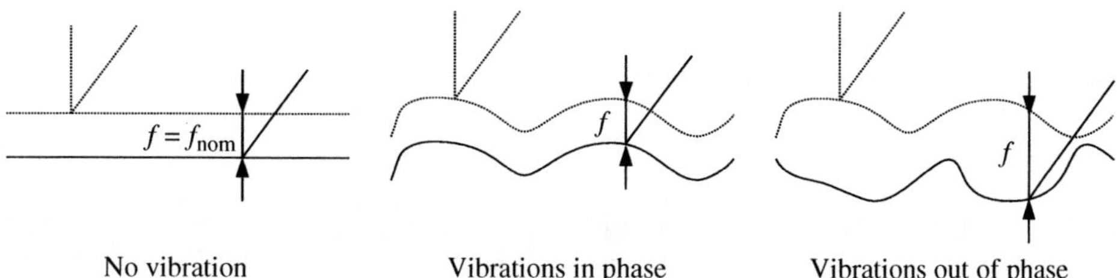

FIGURE 10.3 Modulation in feed due to structural vibrations in a turning operation: current pass (solid line) and previous pass (dotted line).

forces by

$$\Delta z(s) = -(1 - e^{-sT})g(s)\Delta F(s) \tag{10.5}$$

Substituting for Δz in Equation 10.5 and rearranging:

$$\Delta F(s)\left\{1 + Pd\left[1 - e^{-sT}\right]g(s)\right\} = 0 \tag{10.6}$$

Equation 10.6 is now solved, based on the method presented by Budak and Altintas (1998a, 1998b), to determine the stability lobe diagram. Assuming the steady-state solution is a harmonic function at a single chatter frequency ω_c, Equation 10.6 becomes

$$\Delta F(j\omega_c)e^{j\omega_c t}\left\{1 + Pd\left[1 - e^{-j\omega_c T}\right]g(j\omega_c)\right\} = 0 \tag{10.7}$$

where $j^2 = -1$. For nontrivial solutions of Equation 10.7, the following eigenvalue problem is derived:

$$\det\left\{1 + Pd\left[1 - e^{-j\omega_c T}\right]g(j\omega_c)\right\} = 0 \tag{10.8}$$

Since the structural dynamics are one-dimensional, Equation 10.8 reduces to

$$1 + Pd\left[1 - e^{-j\omega_c T}\right]g(j\omega_c) = 0 \tag{10.9}$$

The parameter Λ is defined as

$$\Lambda = Pd\left[1 - e^{-j\omega_c T}\right] = \Lambda_R + j\Lambda_I \tag{10.10}$$

Using the Euler identity $1 - e^{-j\omega_c T} = 1 - \cos(\omega_c T) + j\sin(\omega_c T)$, the limiting stable depth-of-cut is

$$d_{lim} = \left(\frac{1}{P}\right)\frac{\Lambda_R + j\Lambda_I}{1 - \cos(\omega_c T) + j\sin(\omega_c T)} \tag{10.11}$$

Equation 10.11 is rewritten as

$$d_{lim} = \left(\frac{1}{2P}\right)\left\{\frac{\Lambda_R[1 - \cos(\omega_c T)] + \Lambda_I\sin(\omega_c T)}{1 - \cos(\omega_c T)} + j\frac{\Lambda_R\sin(\omega_c T) + \Lambda_I[1 - \cos(\omega_c T)]}{1 - \cos(\omega_c T)}\right\} \tag{10.12}$$

Since the limiting depth-of-cut must be a real number:

$$\Lambda_R\sin(\omega_c T) + \Lambda_I[1 - \cos(\omega_c T)] = 0 \tag{10.13}$$

The parameter κ is defined as

$$\kappa = \frac{\Lambda_I}{\Lambda_R} = \frac{\sin(\omega_c T)}{1 - \cos(\omega_c T)} \tag{10.14}$$

The limiting stable depth-of-cut is solved explicitly as

$$d_{lim} = \frac{\Lambda_R}{2P}(1 + \kappa^2) \tag{10.15}$$

Note that Λ_R must be positive for d_{lim} to be positive. From Equation 10.9, the parameter Λ is

$$\Lambda = -\frac{1}{g(j\omega_c)} \tag{10.16}$$

Equation 10.16 is used to determine Λ_R and Λ_I, and these values are used to solve for d_{lim}. Next, the spindle speed at which the limiting depth-of-cut occurs is determined. The trivial solution to Equation 10.14 is

$$\omega_c T = 0 + 2l\pi, \quad l = 0, 1, 2, \ldots \tag{10.17}$$

The quantity $\omega_c T$ may be interpreted as the number of vibration cycles during a spindle rotation. The trivial solution indicates that the successive vibrations are in phase (i.e., there is no regeneration). The nontrivial solution to Equation 10.14 is

$$\cos(\omega_c T) = \frac{\kappa^2 - 1}{\kappa^2 + 1} \tag{10.18}$$

and may be rewritten as

$$\omega_c T = \varepsilon + 2l\pi, \quad l = 0, 1, 2, \ldots \tag{10.19}$$

where

$$\varepsilon = \cos^{-1}\left(\frac{\kappa^2 - 1}{\kappa^2 + 1}\right) \tag{10.20}$$

The parameter ε is the fraction of the vibration cycles during a spindle rotation. The angle of Λ in the complex plane is

$$\varphi = \tan^{-1}\left(\frac{\Lambda_I}{\Lambda_R}\right) = \tan^{-1}(\kappa) \tag{10.21}$$

Substituting $\kappa = \tan(\varphi)$ into Equation 10.18 yields

$$\cos(\omega_c T) = -\cos(2\varphi) \tag{10.22}$$

A solution to Equation 10.22 is

$$\omega_c T = \pi - 2\varphi + 2l\pi, \quad l = 0, 1, 2, \ldots \tag{10.23}$$

Comparing Equation 10.19 and Equation 10.23, it is seen that the fraction of vibration cycles is $\varepsilon = \pi - 2\varphi$. Since $0 \le \varepsilon \le 2\pi$, one must ensure that $-\pi/2 \le \varphi \le \pi/2$ when computing φ. For example, if Equation 10.21 is solved using a four-quadrant inverse tangent function whose solution is bounded between $-\pi$ and π, then $-\pi/2 \le \varphi \le \pi/2$ since Λ_R is positive. For milling applications, it will be seen that Λ_R must be negative; therefore, the following conditions must be enforced to ensure $0 \le \varepsilon \le 2\pi$:

$$\begin{aligned} &\text{if } \Lambda_I < 0 \ \text{ then } \varphi \rightarrow \varphi + \pi \\ &\text{if } \Lambda_I > 0 \ \text{ then } \varphi \rightarrow \varphi - \pi \end{aligned} \tag{10.24}$$

The spindle speed is

$$N_s = \frac{60}{T} = \frac{60\omega_c}{\varepsilon + 2l\pi}, \quad l = 0, 1, 2, \ldots \tag{10.25}$$

To construct a stability lobe diagram, the following steps are implemented:

1. Select a chatter frequency (ω_c) near a dominant structural frequency.
2. Calculate Λ_R and Λ_I using Equation 10.16.
3. Calculate d_{lim} using Equation 10.15.
4. Select a stability lobe number (l) and calculate N_s using Equation 10.25. The point (N_s, d_{lim}) is the point on the stability lobe diagram corresponding to the chatter frequency, ω_c, and the stability lobe number, l.
5. Repeat Step 4 for the desired number of stability lobes. The result is a vector of spindle speeds, $\vec{N}_s = \{N_{s_1} \quad N_{s_2} \quad \cdots \quad N_{s_n}\}$. Each point $\{(N_{s_1}, d_{lim}) \quad (N_{s_2}, d_{lim}) \quad \cdots \quad (N_{s_n}, d_{lim})\}$. corresponds to a different stability lobe, and all of the points correspond to the chatter frequency ω_c.
6. Select another chatter frequency and repeat Steps 2 to 5. In this manner, the stability lobe diagram is constructed. The smaller the difference between successive chatter frequencies, the greater the resolution of the stability lobe diagram. In general, the lobes will overlap. In this case, the

minimum limiting depth-of-cut is the smallest depth-of-cut. If the lobes do not overlap, then the range of chatter frequencies must be increased.

10.2.1 Example 1

The feed force for a turning operation is given by Equation 10.1 and the structural dynamics are given by Equation 10.26. An analytical expression for the limiting depth-of-cut and corresponding spindle speed for a given chatter frequency and stability lobe number is developed. The stability lobe diagram is plotted for $P = 0.6$ kN/mm^2, $\omega_n = 600$ Hz, $\zeta = 0.2$, and $k = 12$ kN/mm. The stability lobe diagram is compared to stability lobe diagrams for $\zeta = 0.1, 0.3$, and 0.4. The stability lobe diagram is then compared with stability lobe diagrams for $\omega_n = 500$, 700, and 800 Hz. The first ten lobes are included for all stability lobe diagrams:

$$\ddot{z}(t) + 2\zeta\omega_n\dot{z}(t) + \omega_n^2 z(t) = -\frac{\omega_n^2}{k}F(t) \tag{10.26}$$

The parameter Λ is

$$\Lambda = \frac{-1}{g(j\omega_c)} = \Lambda_R + j\Lambda_I = \frac{k}{\omega_n^2}(\omega_c^2 - \omega_n^2) - j\frac{k}{\omega_n^2}(2\zeta\omega_c\omega_n) \tag{10.27}$$

The limiting depth-of-cut is

$$d_{\text{lim}} = \frac{k(\omega_c^2 - \omega_n^2)}{2K\omega_n^2}\left[1 + \frac{4\zeta^2\omega_c^2\omega_n^2}{(\omega_c^2 - \omega_n^2)^2}\right] \tag{10.28}$$

The spindle speed is

$$N_s = \frac{60\omega_c}{\pi - 2\tan^{-1}\left(\dfrac{-2\zeta\omega_c\omega_n}{\omega_c^2 - \omega_n^2}\right) + 2l\pi}, \qquad l = 0, 1, 2, \ldots \tag{10.29}$$

where l is the stability lobe number. Note that the chatter frequency must be greater than the structural natural frequency for the limiting depth-of-cut to be positive. The first ten lobes of the

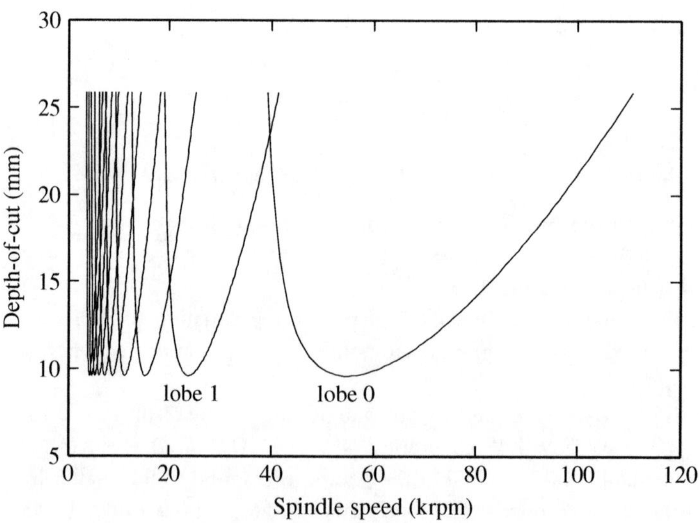

FIGURE 10.4 Unprocessed stability lobe diagram for Example 1.

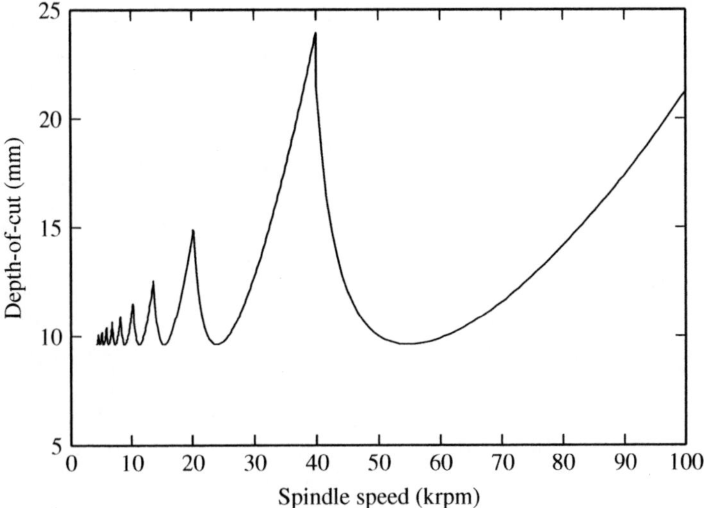

FIGURE 10.5 Processed stability lobe diagram for Example 1.

stability lobe diagram are plotted in Figure 10.4 and Figure 10.5. In Figure 10.4, the entire solution for each of the ten stability lobes is shown. The largest stability lobe is the zeroth lobe on the right. The lobe number increases from right to left on the stability lobe diagram and successive lobes become closer together. The stability lobe diagram is processed in Figure 10.5 such that the minimum depth-of-cut is selected at each spindle speed showing the true stability borderline.

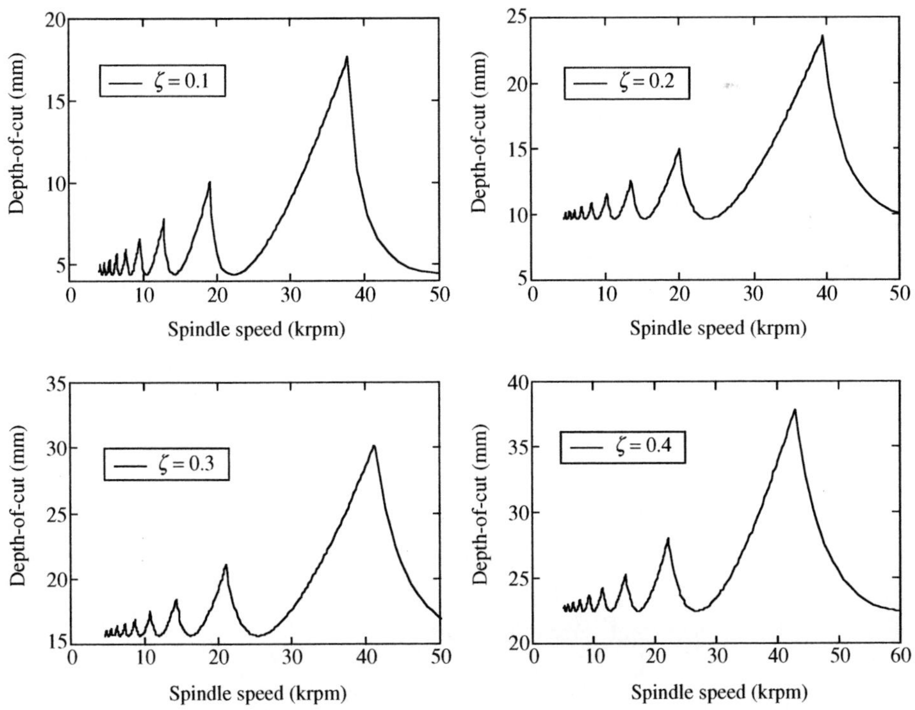

FIGURE 10.6 Stability lobe diagrams for Example 1 with $\zeta = 0.1$, 0.2, 0.3, and 0.4.

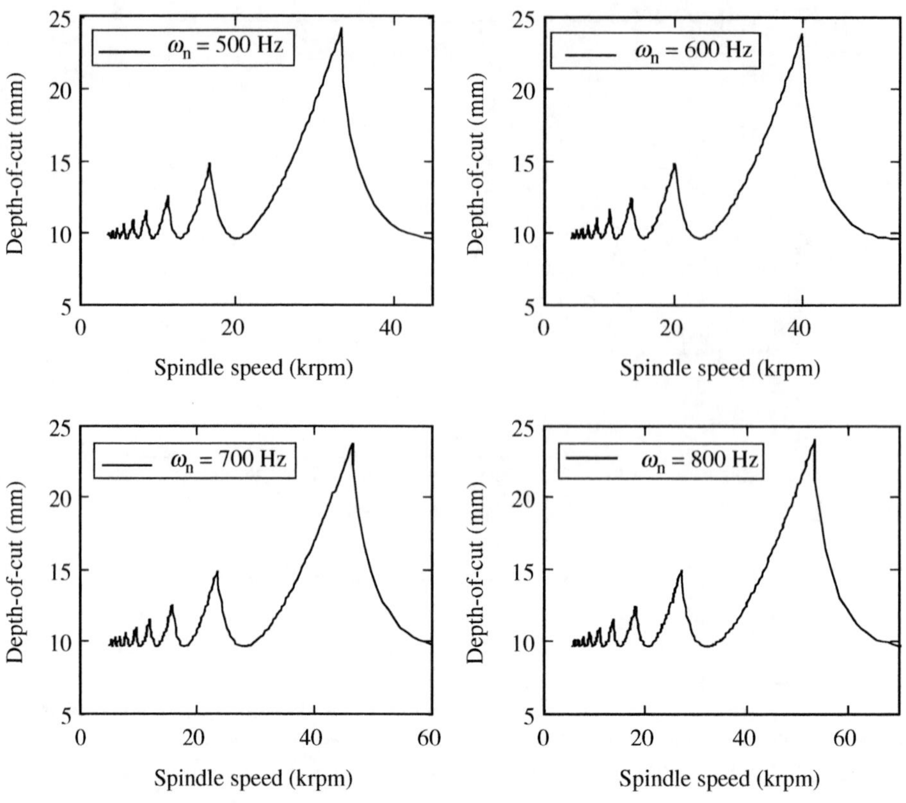

FIGURE 10.7 Stability lobe diagrams for Example 1 with $\omega_n = 500$, 600, 700, and 800 Hz.

In Figure 10.6, the effect of the structural damping ratio is illustrated: as the structural damping ratio increases, the lobes shift slightly to the left and the asymptotic stability boundary shifts up dramatically. The effect of the structural natural frequency is illustrated in Figure 10.7: as the structural natural frequency increases, the lobes shift to the right but the magnitude remains the same.

This section presented an analytical method to generate stability lobe diagrams for turning operations. The limiting depth-of-cut in a turning operation is given by

$$d_{\text{lim}} = \frac{\Lambda_R}{2P}(1 + \kappa^2)$$

where Λ_R is the real part of $-1/g(j\omega_c)$, $g(j\omega_c)$ is the structural transfer function evaluated at the chatter frequency, ω_c, $\kappa = \Lambda_I/\Lambda_R$, and Λ_I is the imaginary part of $-1/g(j\omega_c)$. The corresponding spindle speed is

$$N_s = \frac{60\omega_c}{\varepsilon + 2l\pi}$$

where $l = 0, 1, 2, \ldots$ is the stability lobe number and

$$\varepsilon = \cos^{-1}\left(\frac{\kappa^2 - 1}{\kappa^2 + 1}\right)$$

10.3 Chatter in Face-Milling Operations

A schematic of a face-milling operation is shown in Figure 10.8. In milling operations, multiple teeth may be in contact with the part simultaneously, the feed naturally varies as a function of the tooth angle even when structural vibrations are not present, and each tooth enters and leaves contact with the part every spindle revolution. The depth-of-cut is the chip thickness in the z direction and is assumed to be constant, since the machine tool and part structures are typically much stiffer in the z direction than in the x and y directions.

The instantaneous feed of the ith tooth, illustrated in Figure 10.9, is

$$f_i(t) = f_t \cos[\theta_i(t)] + \Delta x(t) \cos[\theta_i(t)] + \Delta y(t) \sin[\theta_i(t)] \quad (10.30)$$

where

$$\Delta x(t) = \{x_t(t) - x_t(t - T_t)\} - \{x_p(t) - x_p(t - T_t)\} \quad (10.31)$$

$$\Delta y(t) = \{y_t(t) - y_t(t - T_t)\} - \{y_p(t) - y_p(t - T_t)\} \quad (10.32)$$

The term $f_t \cos[\theta_i(t)]$ in Equation 10.30 represents the feed due to the distance the part advances relative to the cutting tool each tooth rotation and is known as the static feed. The terms $\Delta x(t)\cos[\theta_i(t)]$

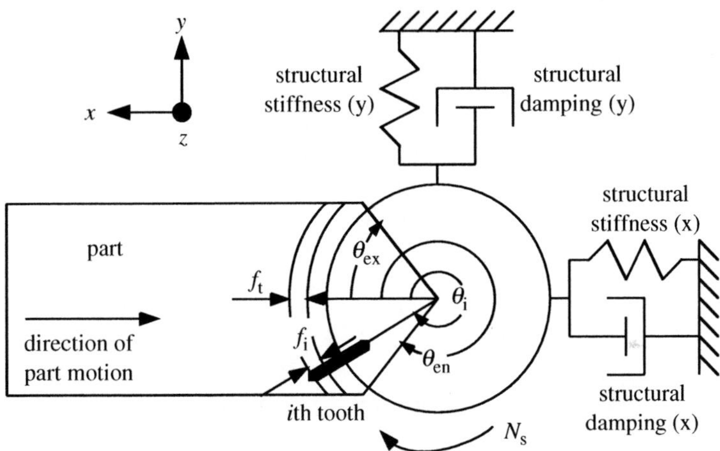

FIGURE 10.8 Face milling operation schematic: current pass (solid line), previous pass (dotted line), and depth-of-cut in z direction.

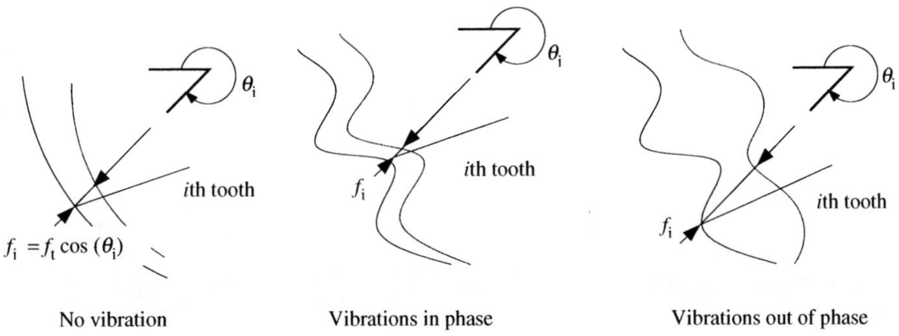

FIGURE 10.9 Modulation in feed due to structural vibrations in a face-milling operation: current pass (solid line) and previous pass (dotted line).

and $\Delta y(t)\sin[\theta_i(t)]$ in Equation 10.30 represent the feed due to tool and part vibrations in the x and y directions, respectively, at the tooth angle $\theta_i(t)$, and are known as the dynamic feed.

The machining forces in the x and y directions, respectively, are

$$\begin{aligned} F_x(t) = {} & df_t \sum_{i=1}^{N_t} \Big\{ -P_T \cos(\psi_r) \cos^2[\theta_i(t)] + P_C \cos[\theta_i(t)] \sin[\theta_i(t)] \Big\} \sigma[\theta_i(t)] \\ & + d\Delta x(t) \sum_{i=1}^{N_t} \Big\{ -P_T \cos(\psi_r) \cos^2[\theta_i(t)] + P_C \cos[\theta_i(t)] \sin[\theta_i(t)] \Big\} \sigma[\theta_i(t)] \\ & + d\Delta y(t) \sum_{i=1}^{N_t} \Big\{ -P_T \cos(\psi_r) \sin[\theta_i(t)] \cos[\theta_i(t)] + P_C \sin^2[\theta_i(t)] \Big\} \sigma[\theta_i(t)] \end{aligned} \tag{10.33}$$

$$\begin{aligned} F_y(t) = {} & df_t \sum_{i=1}^{N_t} \Big\{ -P_T \cos(\psi_r) \cos[\theta_i(t)] \sin[\theta_i(t)] - P_C \cos^2[\theta_i(t)] \Big\} \sigma[\theta_i(t)] \\ & + d\Delta x(t) \sum_{i=1}^{N_t} \Big\{ -P_T \cos(\psi_r) \cos[\theta_i(t)] \sin[\theta_i(t)] - P_C \cos^2[\theta_i(t)] \Big\} \sigma[\theta_i(t)] \\ & + d\Delta y(t) \sum_{i=1}^{N_t} \Big\{ -P_T \cos(\psi_r) \sin^2[\theta_i(t)] - P_C \cos[\theta_i(t)] \sin[\theta_i(t)] \Big\} \sigma[\theta_i(t)] \end{aligned} \tag{10.34}$$

where

$$\sigma[\theta_i(t)] = \begin{cases} 1 & \text{if } \theta_{en} \le \theta_i(t) \le \theta_{ex} \\ 0 & \text{if } \theta_{en} > \theta_i(t) > \theta_{ex} \end{cases} \tag{10.35}$$

The function $\sigma[\theta_i(t)]$ determines if the ith tooth is in contact with the part at the tooth angle, $\theta_i(t)$. The first terms in Equation 10.33 and Equation 10.34 are the machining forces acting on the tool in the x and y directions, respectively, due to the static feed. The second terms in Equation 10.33 and Equation 10.34 are the machining forces acting on the tool in the x and y directions, respectively, due to the dynamic feed resulting from structural vibrations in the x direction. The third terms in Equation 10.33 and Equation 10.34 are the machining forces acting on the tool in the x and y directions, respectively, due to the dynamic feed resulting from structural vibrations in the y direction.

The dynamic portion of the face milling force process model may be written compactly as

$$\begin{bmatrix} \Delta F_x(t) \\ \Delta F_y(t) \end{bmatrix} = dA(t) \begin{bmatrix} \Delta x(t) \\ \Delta y(t) \end{bmatrix} = d \begin{bmatrix} A_{11}(t) & A_{12}(t) \\ A_{21}(t) & A_{22}(t) \end{bmatrix} \begin{bmatrix} \Delta x(t) \\ \Delta y(t) \end{bmatrix} \tag{10.36}$$

where

$$A_{11}(t) = \sum_{i=1}^{N_t} \Big\{ -P_T \cos(\psi_r) \cos^2[\theta_i(t)] + P_C \cos[\theta_i(t)] \sin[\theta_i(t)] \Big\} \sigma[\theta_i(t)] \tag{10.37}$$

$$A_{12}(t) = \sum_{i=1}^{N_t} \Big\{ -P_T \cos(\psi_r) \sin[\theta_i(t)] \cos[\theta_i(t)] + P_C \sin^2[\theta_i(t)] \Big\} \sigma[\theta_i(t)] \tag{10.38}$$

$$A_{21}(t) = \sum_{i=1}^{N_t} \Big\{ -P_T \cos(\psi_r) \cos[\theta_i(t)] \sin[\theta_i(t)] - P_C \cos^2[\theta_i(t)] \Big\} \sigma[\theta_i(t)] \tag{10.39}$$

$$A_{22}(t) = \sum_{i=1}^{N_t} \Big\{ -P_T \cos(\psi_r) \sin^2[\theta_i(t)] - P_C \cos[\theta_i(t)] \sin[\theta_i(t)] \Big\} \sigma[\theta_i(t)] \tag{10.40}$$

These coefficients modulate the instantaneous feed as the tooth angular displacement changes. The summation from $i = 1$ to N_t represents the contribution to this modulation for each of the N_t teeth. Note the matrix $A(t)$ is time-varying and periodic with the tooth-passing period, T_t. For chatter analysis, the matrix $A(t)$ is typically expanded in a Fourier series using the zeroth term (Minis and Yanushevsky, 1993; Budak and Altintas, 1998a). The zeroth term of the Fourier expansion of the force process matrix $A(t)$ is

$$A^0 = \frac{N_t}{2\pi}\begin{bmatrix} A_{11}^0 & A_{12}^0 \\ A_{21}^0 & A_{22}^0 \end{bmatrix} \tag{10.41}$$

where

$$A_{11}^0 = \frac{1}{2}\left[-P_T\cos(\psi_r)\left\{\theta + \frac{1}{2}\sin(2\theta)\right\} + P_C\sin^2(\theta)\right]_{\theta=\theta_{en}}^{\theta=\theta_{ex}} \tag{10.42}$$

$$A_{12}^0 = \frac{1}{2}\left[-P_T\cos(\psi_r)\sin^2(\theta) + P_C\left\{\theta - \frac{1}{2}\sin(2\theta)\right\}\right]_{\theta=\theta_{en}}^{\theta=\theta_{ex}} \tag{10.43}$$

$$A_{21}^0 = \frac{1}{2}\left[-P_T\cos(\psi_r)\sin^2(\theta) - P_C\left\{\theta + \frac{1}{2}\sin(2\theta)\right\}\right]_{\theta=\theta_{en}}^{\theta=\theta_{ex}} \tag{10.44}$$

$$A_{22}^0 = \frac{1}{2}\left[-P_T\cos(\psi_r)\left\{\theta - \frac{1}{2}\sin(2\theta)\right\} - P_C\sin^2(\theta)\right]_{\theta=\theta_{en}}^{\theta=\theta_{ex}} \tag{10.45}$$

The dynamic force process is now approximated by the linear, time-invariant relationship:

$$\begin{bmatrix} \Delta F_x(t) \\ \Delta F_y(t) \end{bmatrix} = dA^0\begin{bmatrix} \Delta x(t) \\ \Delta y(t) \end{bmatrix} \tag{10.46}$$

The tool and part vibrations, respectively, are related to the machining forces by

$$\begin{bmatrix} x_t(s) \\ y_t(s) \end{bmatrix} = G_t(s)\begin{bmatrix} F_x(s) \\ F_y(s) \end{bmatrix} = \begin{bmatrix} G_{t_{11}}(s) & G_{t_{12}}(s) \\ G_{t_{21}}(s) & G_{t_{22}}(s) \end{bmatrix}\begin{bmatrix} F_x(s) \\ F_y(s) \end{bmatrix} \tag{10.47}$$

$$\begin{bmatrix} x_p(s) \\ y_p(s) \end{bmatrix} = -G_p(s)\begin{bmatrix} F_x(s) \\ F_y(s) \end{bmatrix} = -\begin{bmatrix} G_{p_{11}}(s) & G_{p_{12}}(s) \\ G_{p_{21}}(s) & G_{p_{22}}(s) \end{bmatrix}\begin{bmatrix} F_x(s) \\ F_y(s) \end{bmatrix} \tag{10.48}$$

where $G_t(s)$ and $G_p(s)$ are the transfer functions relating the tool structural and part structural vibrations, respectively, to the machining forces. The negative sign in Equation 10.48 is due to the fact that the forces acting on the part are equal in magnitude and opposite in direction to the machining forces given in Equation 10.33 and Equation 10.34. Since

$$\begin{bmatrix} F_x(t) \\ F_y(t) \end{bmatrix} - \begin{bmatrix} F_x(t-T_t) \\ F_y(t-T_t) \end{bmatrix} = \begin{bmatrix} \Delta F_x(t) \\ \Delta F_y(t) \end{bmatrix} - \begin{bmatrix} \Delta F_x(t-T_t) \\ \Delta F_y(t-T_t) \end{bmatrix}$$

the structural vibrations can be related to the machining forces by

$$\begin{bmatrix} \Delta x \\ \Delta y \end{bmatrix} = (1 - e^{-sT_t})[G_t(s) + G_p(s)]\begin{bmatrix} \Delta F_x(s) \\ \Delta F_y(s) \end{bmatrix} \tag{10.49}$$

The machine tool and part vibrations are assumed to occur at a chatter frequency, ω_c, when a marginally stable depth-of-cut is taken. Assuming the steady-state solution is a harmonic function at a chatter frequency, ω_c, and substituting for the structural vibrations, Equation 10.49 becomes

$$\begin{bmatrix} \Delta F_x \\ \Delta F_y \end{bmatrix} e^{j\omega_c t} = \frac{dN_t}{2\pi}(1 - e^{j\omega_c T_t})G^0(j\omega_c)\begin{bmatrix} \Delta F_x \\ \Delta F_y \end{bmatrix} e^{j\omega_c t} \tag{10.50}$$

where the matrix G^0 is

$$G^0(j\omega_c) = \frac{2\pi}{N_t}A^0[G_t(j\omega_c) + G_p(j\omega_c)] \tag{10.51}$$

Equation 10.50 is now solved based on the method presented by Budak and Altintas (1998a, 1998b) to determine the stability lobe diagram. The characteristic equation of Equation 10.50 is

$$\det\left[I_2 - \frac{dN_t}{2\pi}(1 - e^{j\omega_c T_t})G^0(j\omega_c)\right] = 0 \tag{10.52}$$

where I_2 is the 2×2 identity matrix. The solution of Equation 10.52 yields the limiting stable depth-of-cut. The inverse of the eigenvalue of G^0 is defined as

$$\Lambda(j\omega_c) = \Lambda_R(j\omega_c) + j\Lambda_I(j\omega_c) = -\frac{dN_t}{2\pi}(1 - e^{j\omega_c T_t}) \tag{10.53}$$

Expanding the exponential term in Equation 10.53 and noting that the depth-of-cut must be a real number, the limiting stable depth-of-cut may be written as

$$d_{lim} = -\frac{\pi\Lambda_R}{N_t}(1 + \kappa^2) \tag{10.54}$$

where the parameter κ is defined by the transcendental equation

$$\kappa = \frac{\Lambda_I}{\Lambda_R} = \frac{\sin(\omega_c T_t)}{1 - \cos(\omega_c T_t)} \tag{10.55}$$

Equation 10.55 is solved for the tooth-passing period of the lth stability lobe and the tooth-passing period is related to the spindle speed to yield

$$N_s = \frac{60\omega_c}{N_t[\pi - 2\varphi + 2l\pi]}, \quad l = 0, 1, 2, \ldots \tag{10.56}$$

where, again, $\varphi = \tan^{-1}(\kappa)$. A chatter frequency is selected and the limiting stable depth-of-cut is calculated from Equation 10.54 corresponding to the spindle speed on the lth lobe as given by Equation 10.56.

10.3.1 Example 2

The cutting and thrust pressures in a face-milling operation are given by $P_C = 2.0$ kN/mm^2 and $P_T = 0.8$ kN/mm^2, respectively, and the lead angle is 45°. The part is assumed to be perfectly rigid and the tool structural dynamics for the x and y directions are given by Equation 10.57 and Equation 10.58, respectively. The nominal parameters are $\theta_{en} = -45°$, $\theta_{ex} = 45°$, $N_t = 4$, $k_x = 14$ kN/mm, $k_y = 17$ kN/mm, $\zeta_x = 0.15$, $\zeta_y = 0.1$, $\omega_x = 3000$ rad/s, and $\omega_y = 4000$ rad/s. Stability lobe diagrams are generated for the nominal parameters and $N_t = 1$, 2, and 8 teeth. Next, stability lobe diagrams are generated for the nominal parameters and $\theta_{ex} = 30°$, 60°, and 75°. The first 15 lobes are included for all stability lobe diagrams.

$$\ddot{x}_t(t) + 2\zeta_x\omega_x\dot{x}_t(t) + \omega_x^2 x_t(t) = \frac{\omega_x^2}{k_x}F_x(t) \tag{10.57}$$

$$\ddot{y}_t(t) + 2\zeta_y\omega_y\dot{y}_t(t) + \omega_y^2 y_t(t) = \frac{\omega_y^2}{k_y}F_y(t) \tag{10.58}$$

The effect of the number of teeth is illustrated in Figure 10.10: as the number of teeth increases, the lobes shift to the left and the asymptotic stability borderline decreases. In Figure 10.11, the effect of the exit angle is illustrated: as the exit angle increases, the asymptotic stability borderline decreases.

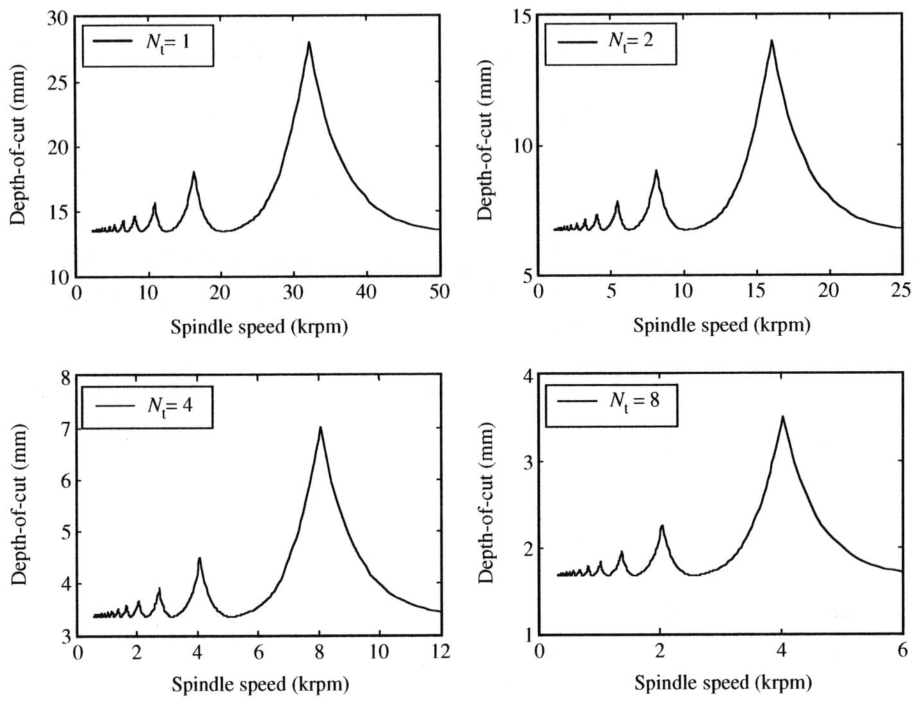

FIGURE 10.10 Stability lobe diagrams for Example 2, with $N_t = 1, 2, 4$, and 8.

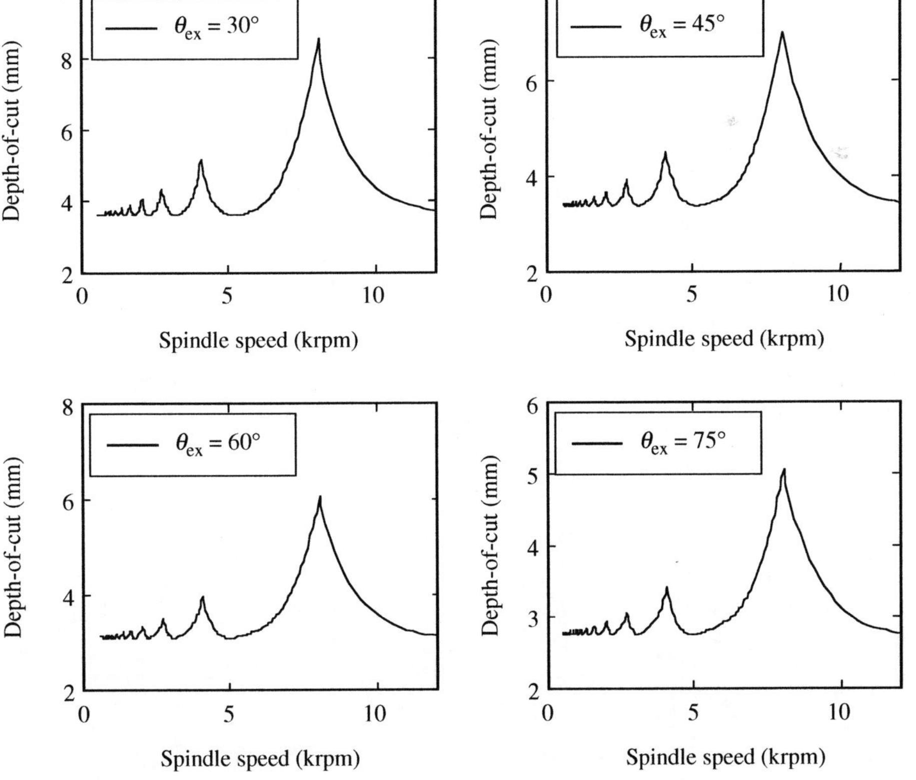

FIGURE 10.11 Stability lobe diagrams for Example 2, with $\theta_{ex} = 30°, 45°, 60°$, and 75°.

This section presented an analytical method to generate stability lobe diagrams for face-milling operations. The limiting depth-of-cut in a face-milling operation is given by

$$d_{\lim} = -\frac{\pi \Lambda_R}{N_t}(1+\kappa^2)$$

where N_t is the number of teeth, Λ_R is the inverse eigenvalue of $(2\pi/N_t)A^0[G_t(j\omega_c) + G_p(j\omega_c)]$, A^0 is the zeroth term of the Fourier expansion of the force process matrix, $G_t(j\omega_c)$ and $G_p(j\omega_c)$ are the transfer functions relating the tool structural and part structural vibrations, respectively, to the machining forces evaluated at the chatter frequency ω_c, $\kappa = \Lambda_I/\Lambda_R$, and Λ_I is the imaginary part of $(2\pi/N_t)A^0[G_t(j\omega_c) + G_p(j\omega_c)]$. The corresponding spindle speed is

$$N_s = \frac{60\omega_c}{N_t[\pi - 2\tan^{-1}(\kappa) + 2l\pi]}$$

where $l = 0, 1, 2, \ldots$ is the stability lobe number.

10.4 Time-Domain Simulation

Time-domain simulation (Tlusty and Ismail, 1981, 1983; Tlusty, 1986; Tsai et al., 1990; Lee and Liu, 1991a, 1991b; Smith and Tlusty, 1993; Elbestawi et al., 1994; Tarng and Li, 1994; Weck et al., 1994) is an alternative method for determining regenerative chatter. In a time-domain simulation, the machining forces and structural vibrations are simulated in the time domain for a specific set of process parameters and the resulting signals (i.e., forces and displacements) are examined to determine if chatter is present. The analyses presented above for the turning and face-milling operations assume that the tool always maintains contact with the part and that the cutting and thrust pressures are independent of the process parameters. Further, the face milling analysis approximated the time-varying force process matrix, $A(t)$, by the zeroth term of its Fourier expansion. With time domain simulations, nonlinear effects may be directly incorporated into the simulation; thus, more accurate stability prediction is possible. The disadvantage of time-domain simulations is the extreme computational cost that is required. For a specific spindle speed, several simulations must be conducted at different depths-of-cut; thus, the stability boundary for that spindle speed is determined iteratively. This procedure is repeated for a range of spindle speeds to construct a complete stability lobe diagram.

For turning operations, the machining force is calculated using Equation 10.1, the feed is calculated using Equation 10.2, and the tool displacement is calculated using Equation 10.4. For face-milling operations, the feed is calculated using Equation 10.30, the machining forces in the x and y directions are calculated using Equation 10.33 to Equation 10.35, and the tool and part displacements, respectively, are calculated using Equation 10.47 and Equation 10.48. To calculate the machining forces in the face-milling operation, the angular displacement of each tooth is required. The angular displacement of the ith tooth is

$$\theta_i(t) = \frac{2\pi}{60}N_s t + \frac{2\pi}{N_t}(i-1) \tag{10.59}$$

The feed and force equations are static, while the structural displacement equations are dynamic and must be solved via a numerical integration technique. A sufficiently small time step must be utilized in the numerical integrations to account for the small system time constants associated with the large structural frequencies.

10.4.1 Example 3

The feed force for a turning operation is given by $F(t) = 0.6df^{0.7}(t)$. The structural dynamics are given by Equation 10.26 with the following parameters: $\omega_n = 600$ Hz, $\zeta = 0.2$, and $k = 12$ kN/mm.

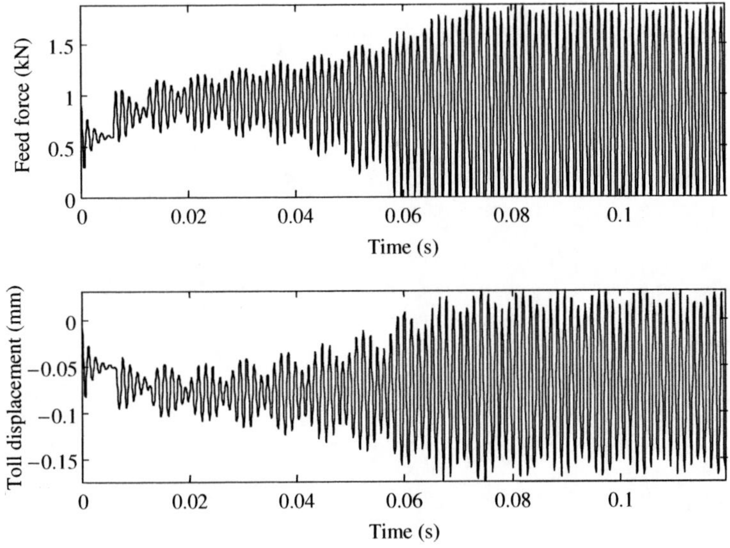

FIGURE 10.12 Time-domain simulations for Example 3 with $f_{\text{nom}} = 0.1$ mm.

A time-domain simulation of the system including the effect of the tool disengaging from the part is constructed, and simulations for $N_s = 10{,}000$ rpm, $d = 8$ mm, and $f_{\text{nom}} = 0.1$ mm are conducted. The simulation is repeated for $f_{\text{nom}} = 0.2$ mm. For both simulations, the time history of the feed force and the tool displacement are plotted.

To account for the phenomenon of the tool disengaging from the part, the feed in Equation 10.2 must be modified as follows:

$$f(t) = \begin{cases} f_{\text{nom}} + z(t) - f_p(t-T) & \text{if } f_{\text{nom}} + z(t) - f_p(t-T) \geq 0 \\ 0 & \text{if } f_{\text{nom}} + z(t) - f_p(t-T) < 0 \end{cases} \tag{10.60}$$

where

$$f_p(t) = \begin{cases} z(t) & \text{if } f_{\text{nom}} + z(t) - f_p(t-T) \geq 0 \\ -f_{\text{nom}} + f_p(t-T) & \text{if } f_{\text{nom}} + z(t) - f_p(t-T) < 0 \end{cases} \tag{10.61}$$

If the feed at the current time is calculated to be negative, then the cutting tool has disengaged from the part and the feed is zero. The term $f_p(t)$ accounts for feed due to structural vibrations at the previous spindle rotation, even when the cutting tool disengages from the part. The results for $f_{\text{nom}} = 0.1$ mm and $f_{\text{nom}} = 0.2$ mm are shown in Figure 10.12 and Figure 10.13, respectively. As the nominal feed is increased, chatter is suppressed.

10.4.2 Example 4

The cutting and thrust forces in a face-milling operation are given by $F_C(t) = 1.4df^{0.6}(t)$ and $F_T(t) = 0.4df^{0.8}(t)$, respectively. The lead angle is 45°, the entry angle is −60°; the exit angle is 60°, the number of teeth is four, and the feed per tooth is $f_t = 0.15$ mm. The part is assumed to be perfectly rigid, and tool structural dynamics for the x and y directions are given by Equation 10.58 and Equation 10.59, respectively. A time-domain simulation is developed to determine the limiting stable depth-of-cut for spindle speeds of 1000 and 32,000 rpm. For both spindle speeds, the system is simulated for a depth-of-cut 10% below the limiting stable depth-of-cut and for a depth-of-cut 10% above the limiting stable depth-of-cut. The cutting force, thrust force, x tool displacement, and y tool displacement are plotted.

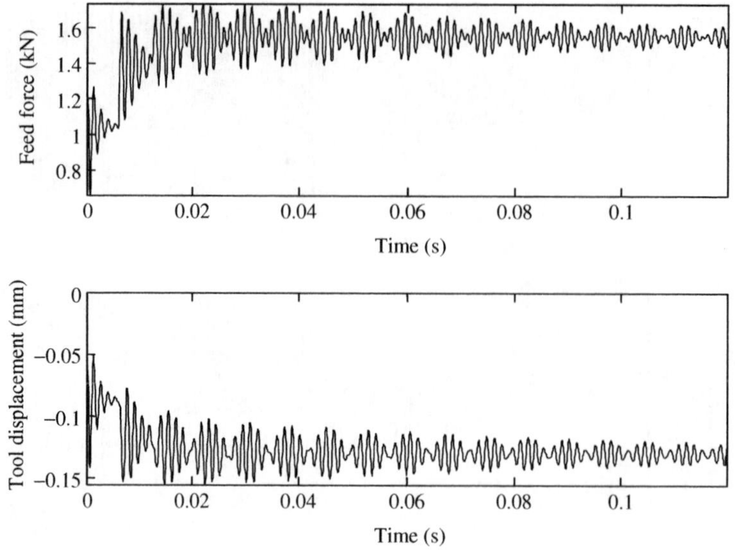

FIGURE 10.13 Time-domain simulations for Example 3 with $f_{nom} = 0.2$ mm.

The nonlinear effect of tooth disengagement is included:

$$\ddot{x}_t(t) + 2(0.15)(3000)\dot{x}_t(t) + 3000^2 x_t(t) = \frac{3000^2}{15} F_x(t) \tag{10.62}$$

$$\ddot{y}_t(t) + 2(0.1)(4000)\dot{y}_t(t) + 4000^2 y_t(t) = \frac{4000^2}{17} F_y(t) \tag{10.63}$$

To account for the phenomenon of the tool disengaging from the part, the feed in Equation 10.30 must be modified as follows:

$$f_i(t) = \begin{cases} f_{ci}(t) - f_{pi}(t - T_t) & \text{if } f_{ci}(t) \geq 0 \\ 0 & \text{if } f_{ci}(t) < 0 \end{cases} \tag{10.64}$$

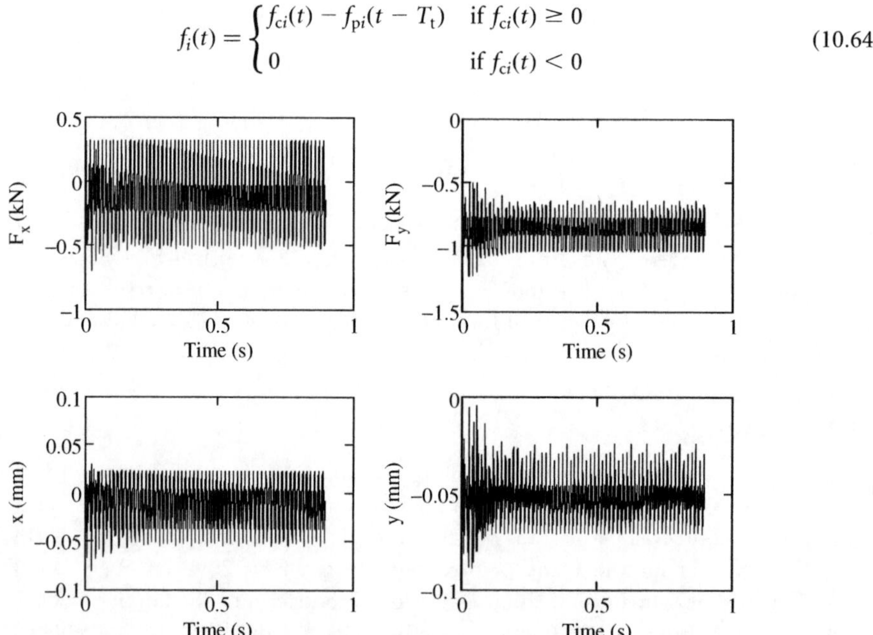

FIGURE 10.14 Time-domain simulation for Example 4 with $N_s = 1000$ rpm and $d = 2.025$ mm.

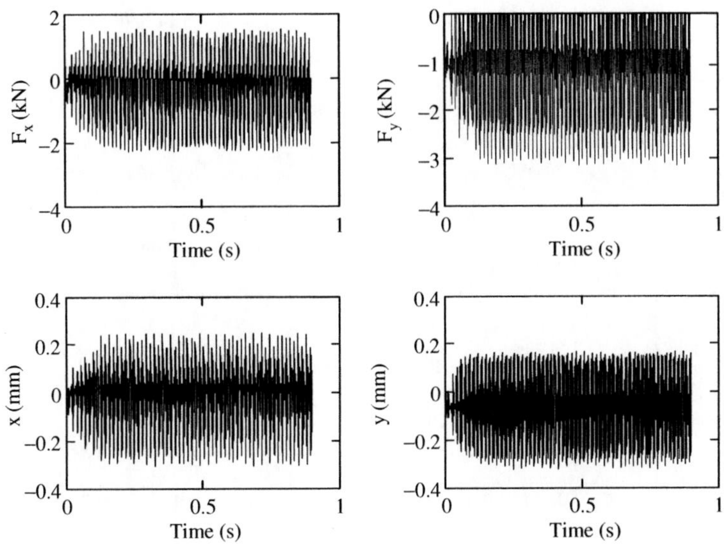

FIGURE 10.15 Time-domain simulation for Example 4 with $N_s = 1000$ rpm and $d = 2.475$ mm.

where

$$f_{ci}(t) = f_t \cos[\theta_i(t)] + \{x_t(t) - x_p(t)\} \cos[\theta_i(t)] + \{y_t(t) - y_p(t)\} \sin[\theta_i(t)] \tag{10.65}$$

$$f_{pi}(t) = \begin{cases} \{x_t(t-T_t) - x_p(t-T_t)\}\cos[\theta_i(t)] + \{y_t(t-T_t) - y_p(t-T_t)\}\sin[\theta_i(t)] & \text{if } f_{ci}(t) \geq 0 \\ -f_t \cos[\theta_i(t)] + f_{pi}(t-T_t) & \text{if } f_{ci}(t) < 0 \end{cases} \tag{10.66}$$

Note that $\theta_i(t) = \theta_{i+1}(t - T_t)$ and $i - 1 \rightarrow N_t$ if $i = 1$. The term $f_{pi}(t)$ represents the contribution to the instantaneous feed when the previous tooth was at the same angular location as the ith tooth. If the tooth and part are in contact, this contribution is due to the tool and part vibrations. If the tooth and part are not in contact, this contribution is the previous contribution added to the static portion and the instantaneous feed is set to zero. Through time-domain simulations, the limiting stable depth-of-cut for

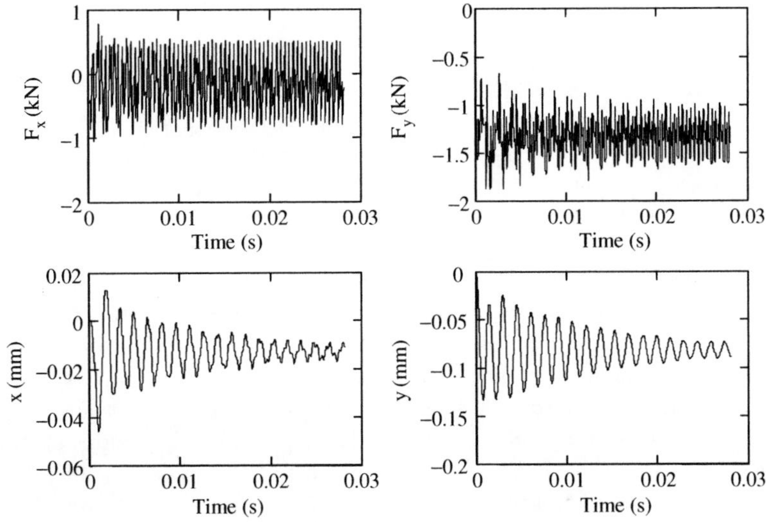

FIGURE 10.16 Time-domain simulation for Example 4 with $N_s = 32{,}000$ rpm and $d = 3.105$ mm.

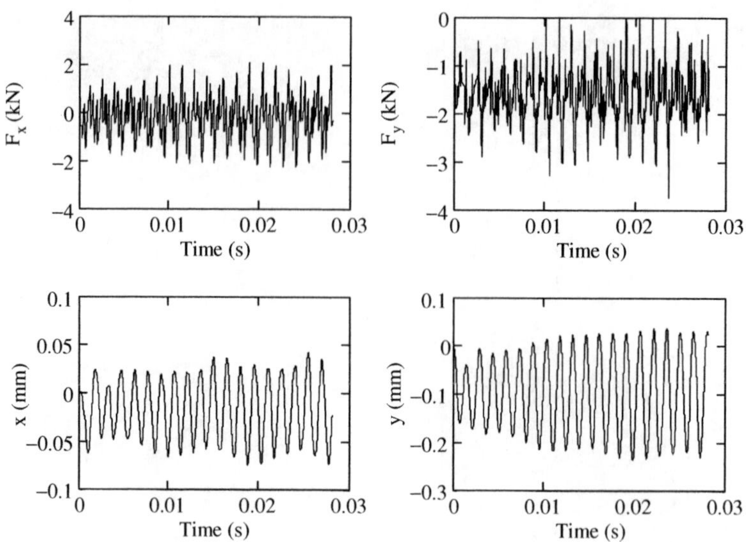

FIGURE 10.17 Time-domain simulation for Example 4 with $N_s = 32{,}000$ rpm and $d = 3.795$ mm.

$N_s = 1000$ rpm is found to be 2.25 mm and the limiting depth-of-cut for $N_s = 32{,}000$ rpm is found to be 3.45 mm. The results are shown in Figure 10.14 to Figure 10.17. The system is stable in Figure 10.14 and Figure 10.16, while instability is evidenced in Figure 10.15 and Figure 10.17 by the force in the y direction saturating at 0 kN.

> This section presented the technique of time-domain simulation as an alternative means to analyze regenerative chatter. Time-domain simulations are the direct numerical simulations of the force process and structural vibrations. A process parameter is changed iteratively from simulation to simulation to determine the critical value at which chatter occurs. A sufficiently small time step must be utilized in the numerical integrations to account for the small system time constants associated with the large structural frequencies.

10.5 Chatter Detection

Regenerative chatter is easily detected by an operator due to the loud, high-pitched noise it produces and the distinctive "chatter marks" it leaves on the part surface. However, automatic detection is required for intelligent manufacturing (Cho and Ehmann, 1988; Delio et al., 1992). At the onset of chatter, process signals (e.g., force, vibration) contain significant energy at the chatter frequency. It is a well-known fact that the chatter frequency will be close to a dominant structural frequency. The most common method to detect the presence of chatter is to threshold the frequency signal of a process signal. To analyze the frequency content of a signal, a Fourier transform, or fast Fourier transform, is performed. If the frequency content of the resulting signal near a dominant chatter frequency is above a threshold value, then chatter is determined to be present. It should be noted that machining process signals also contain significant energy at the tooth-passing frequency. If the dominant structural frequencies and tooth-passing frequency are sufficiently separated, then the tooth-passing frequency may be ignored when determining the presence of chatter. If the dominant structural frequencies and tooth-passing frequency are close, then the signal must be filtered at the tooth-passing frequency using a notch filter. Also, forced vibrations, such as those resulting from the impact between the cutting tool and part, must not be allowed to falsely trigger the chatter detection algorithm. These thresholding algorithms all suffer from the lack of an analytical method of selecting a threshold value. This value is typically selected empirically and will not be valid over a wide range of cutting conditions and machining operations.

10.5.1 Example 5

An experimental face-milling operation, a complete description of which is given in Landers (1997), is conducted with a spindle speed of 1500 rpm and a tool with four teeth. The dominant structural frequencies are 334, 414, 653, and 716 Hz. The machining force F_z is sampled at a frequency of 2000 Hz, and the time-domain signal is transformed into the frequency domain via a Fourier transform using 80 points (i.e., one spindle revolution). The power spectral density of the force signal is shown for depths-of-cut of 1.0 and 1.5 mm in Figure 10.18 and Figure 10.19, respectively. In Figure 10.18, there is significant energy at 100 Hz, which is the tooth-passing frequency. There is also significant energy at 750 Hz due to structural vibrations; however, the system did not chatter, as evidenced by the lack of chatter marks on the part and a high-pitched sound during machining. In Figure 10.19, there is significant energy at 665 Hz as well as 100 Hz. Chatter was evidenced by the chatter marks left on the part surface and the high-pitched sound during machining. The results demonstrate that the chatter frequency is 665 Hz, which is near the dominant structural frequency of 653 Hz. Note that the power spectral density at the frequency of 0 Hz is ignored in Figure 10.18 and Figure 10.19. This component is stronger than the components at all other frequencies since the machining force F_z fluctuates about a static, nonzero value. In this application, a thresholding algorithm may ignore the low frequencies where the tooth-passing frequency is strong; however, if the operation were to be performed at a higher spindle speed, say 7500 rpm, or the number of teeth were increased from 4 to 20, the tooth-passing frequency would be 500 Hz, close to the structural frequencies. In this case, the force signal would have to be filtered at the tooth-passing frequency.

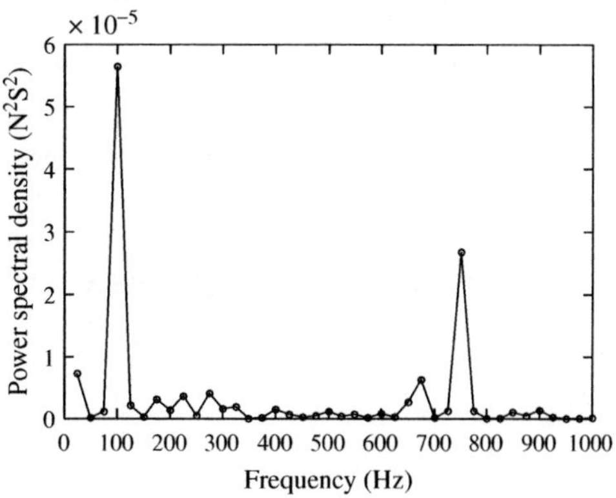

FIGURE 10.18 Power spectral density of F_z in a face-milling operation with $d = 1.0$ mm.

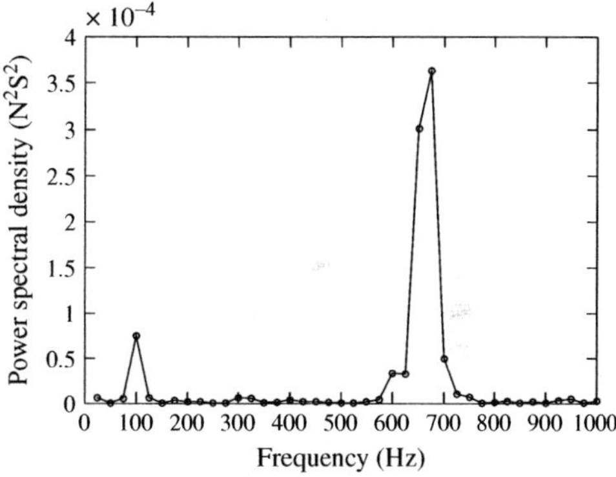

FIGURE 10.19 Power spectral density of F_z in a face-milling operation with $d = 1.5$ mm.

> This section presented techniques to detect the occurrence of regenerative chatter. The phenomenon of regenerative chatter is easily detected by an operator due to the loud, high-pitched noise it produces and the distinctive "chatter marks" it leaves on the part surface. The most common method to detect the presence of chatter is to threshold the frequency signal of a process signal. In this case, one must be careful to separate out the spindle rotation and tooth-passing frequencies.

10.6 Chatter Suppression

Most machining process plans are derived from handbooks or from a database. Since these plans do not consider the physical machine that will be used, chatter-free operations cannot be guaranteed. Thus, multiple iterations, where the feed or spindle speed are adjusted using the operator's experience, are typically required. The tool position may also be adjusted (e.g., the depth-of-cut may be decreased) to suppress chatter and, while this is guaranteed to be effective due to the presence of the asymptotic stability borderline, this approach is typically not employed since part program must be rewritten to add multiple passes, thereby drastically decreasing productivity. The stability lobe diagram can be used as a tool to plan chatter-free machining operations and productivity can be greatly increased by selecting the process parameters to lie in a pocket between two lobes. A cutting tool design methodology (Altintas et al., 1999) has also been proposed for milling tools where the pitch is slightly adjusted such that the teeth are not evenly spaced. The variable pitch has the effect of changing the phase difference between successive teeth vibrations and, if designed properly, will suppress chatter. These design techniques are very sensitive to parameter variations and model uncertainty, and may not be used reliably for a large range of operating conditions. This section will describe methods for automatic chatter suppression.

10.6.1 Spindle-Speed Selection

For the stability lobe diagram generated from a system modeled as having a one-dimensional structure, it is seen that the maximum depths-of-cut are located at the tooth-passing frequencies (i.e., the number of teeth multiplied by the spindle speed) corresponding to the dominant structural frequency and integer fractions thereof. If the dominant structural frequency is known, it may be used as an aid in selecting spindle speeds; however, the structural dynamics are often unknown and may be determined only through costly testing. Further, structural dynamics change drastically over time.

It is known, however, that during chatter, the dominant frequency seen in the cutting-process output is close to a dominant structural frequency. This fact is used in Smith and Delio (1992) to suppress chatter automatically. The following steps are taken:

1. Implement a chatter detection routine to determine the presence of chatter.
2. If chatter is detected, determine the chatter frequency, ω_c. This will be the frequency at which the process signal has the greatest energy.
3. Set the new spindle speed to be $N_s = \omega_c/[N_t(N+1)]$, where N is the smallest positive integer such that the new spindle speed does not violate the maximum spindle speed constraint.
4. Repeat Steps 1 to 3 until the chatter has been suppressed.

The equation $N_s = \omega_c/[N_t(N+1)]$ may be interpreted as selecting the tooth-passing frequency, or an integer fraction thereof, corresponding to the approximate dominant structural frequency. Note that if the depth-of-cut is too large and the maximum spindle speed is too small, this technique will not be effective and the feed or depth-of-cut must be adjusted, or the spindle speed must be continuously varied.

10.6.2 Example 6

The feed force for a turning operation is given by Equation 10.1, and the structural dynamics are given by Equation 10.26. The system parameters are $P = 0.75\ \text{kN/mm}^2$, $f_{\text{nom}} = 0.1$ mm, $\omega_n = 750$ Hz, $\zeta = 0.1$, and $k = 15$ kN/mm. The depth-of-cut is 5 mm. The spindle speed that should be selected to suppress chatter if the chatter frequency is 725 Hz, when the spindle speed is not constrained, is determined. The spindle speed that should be selected to suppress chatter if the maximum spindle speed is 15,000 rpm is also determined. The system is simulated for a spindle speed of 10,000 rpm for ten spindle revolutions and then for ten spindle revolutions for the spindle

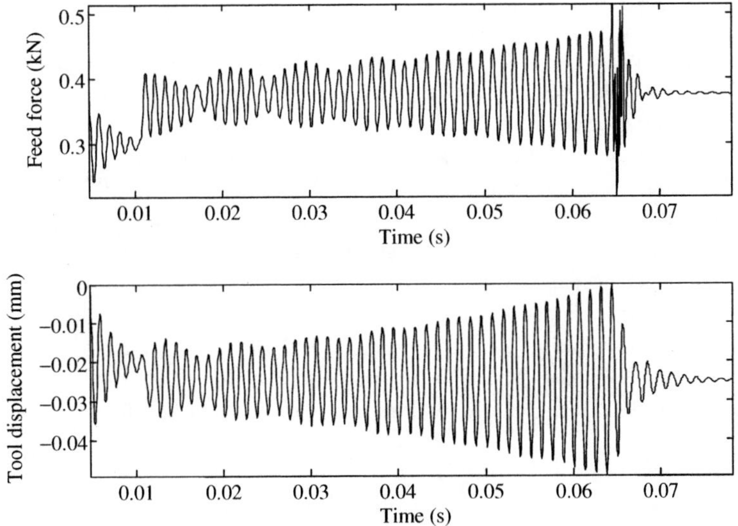

FIGURE 10.20 Time-domain simulations using spindle speed selection with $N_s = 43{,}500$ rpm.

speed calculated when the spindle speed is not constrained. The simulation is then repeated for the spindle speed calculated when the spindle speed is constrained. Feed force and tool displacement are plotted for both cases.

For a chatter frequency of 725 Hz, the optimal spindle speed is 60(725) = 43,500 rpm. Other possible spindle speeds are 43,500/2 = 21,750 rpm, 43,500/3 = 14,500 rpm, 43,500/4 = 10,875 rpm, and so on. Therefore, when the maximum spindle speed is 15,000 rpm, a spindle speed of 14,500 rpm is used. The time domain simulations are in Figure 10.20 and Figure 10.21. The results illustrate that a depth-of-cut of 5.3 mm is stable at 43,500 rpm, but not at 14,500 rpm. Therefore, if the spindle speed is limited to 15,000 rpm, spindle-speed selection may not be used to suppress the chatter present in the machining operation.

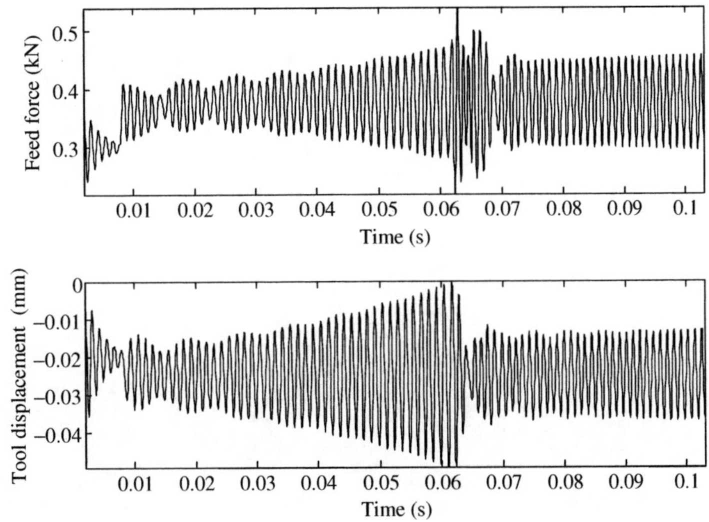

FIGURE 10.21 Time-domain simulations using spindle speed selection with $N_s = 14{,}500$ rpm.

10.6.3 Feed and Depth-of-Cut Selection

When chatter occurs, operators will sometimes increase the feedrate via the feedrate override button on the machine tool control panel. This has the effect of increasing the feed, assuming the spindle speed remains constant. When linear chatter analysis techniques are employed, the force process gains are linearized about the nominal feed, and stability does not appear to be affected by the nominal feed. However, the stability results are only valid for a small region about the nominal feed. It is well known that there is a nonlinear relationship between the machining forces and the feed of the form $F = P(f)df$. The pressure can be expressed in the form $P(f) = Kf^{\alpha}$ where $\alpha < 0$; thus, the pressure decreases as the feed increases. Since the stable depth-of-cut is inversely proportional to the pressure, the stability limit will increase as the feed increases, assuming the spindle speed remains constant. An illustration of this phenomenon was shown in Example 3: when the feed was increased from 0.1 to 0.2 mm, chatter was suppressed. While increasing the feed can suppress chatter, the sensitivity of chatter to feed is limited and other adverse phenomenon, such as tooth chippage, may occur.

Another method to suppress chatter is to decrease the depth-of-cut (Weck et al., 1975). This method is guaranteed to work as evidenced by stability lobe diagrams. However, this method is typically not preferred as it dramatically decreases operation productivity by increasing the total number of tool passes that are required to complete the operation.

10.6.4 Spindle-Speed Variation

Spindle speed variation (SSV) is another technique that has shown the ability to suppress chatter (Inamura and Sata, 1974; Lin et al., 1990). The spindle speed is varied about some nominal value, typically in a sinusoidal manner. Although SSV is a promising technique, the theory required to guide the designer in the selection of suitable amplitudes and frequencies is in its infancy (Radulescu et al., 1997a, 1997b; Sastry et al., 2002). Also, in some cases, SSV may create chatter that would not occur when using a constant spindle speed.

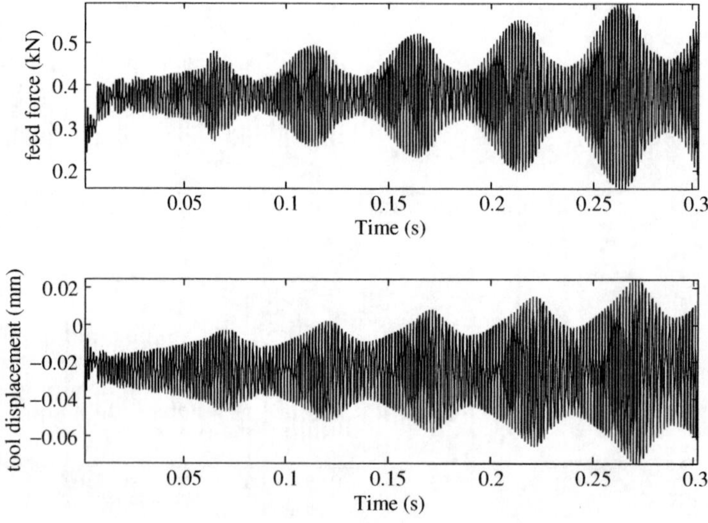

FIGURE 10.22 Time-domain simulations using spindle-speed variation with $A = 0.1$ and $\Omega = 20$ Hz.

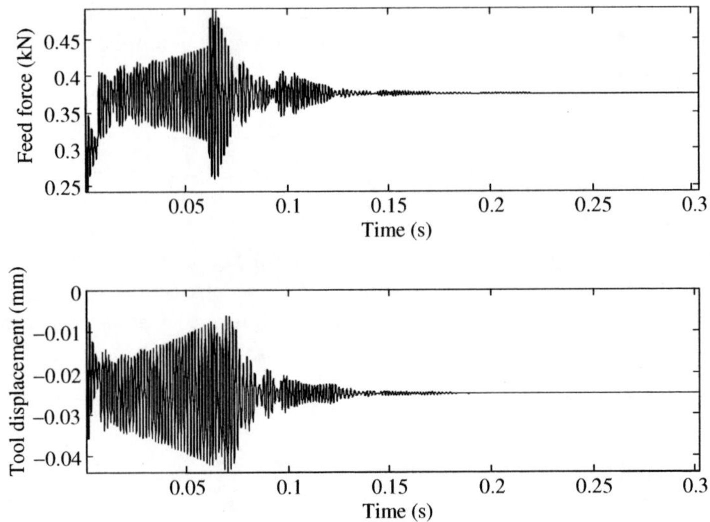

FIGURE 10.23 Time-domain simulations using spindle-speed variation with $A = 0.25$ and $\Omega = 20$ Hz.

10.6.5 Example 7

The feed force for a turning operation is given by Equation 10.1 and the structural dynamics are given by Equation 10.26. The system parameters are $P = 0.75$ kN/mm^2, $f_{\text{nom}} = 0.1$ mm, $\omega_{\text{n}} = 750$ Hz, $\zeta = 0.1$, and $k = 15$ kN/mm. The depth-of-cut is 5 mm. The system is simulated for a nominal spindle speed of $N_{\text{nom}} = 10{,}000$ rpm for 10 spindle revolutions and then for 30 spindle revolutions for the spindle speed calculated from Equation 10.67 for the following three cases: $A = 0.1$ and $\Omega = 20$ Hz, $A = 0.25$ and $\Omega = 20$ Hz, and $A = 0.25$ and $\Omega = 160$ Hz. Feed force and tool displacement are plotted for all three cases.

$$N_s(t) = N_{\text{nom}}[1 + A \sin(\Omega t)] \tag{10.67}$$

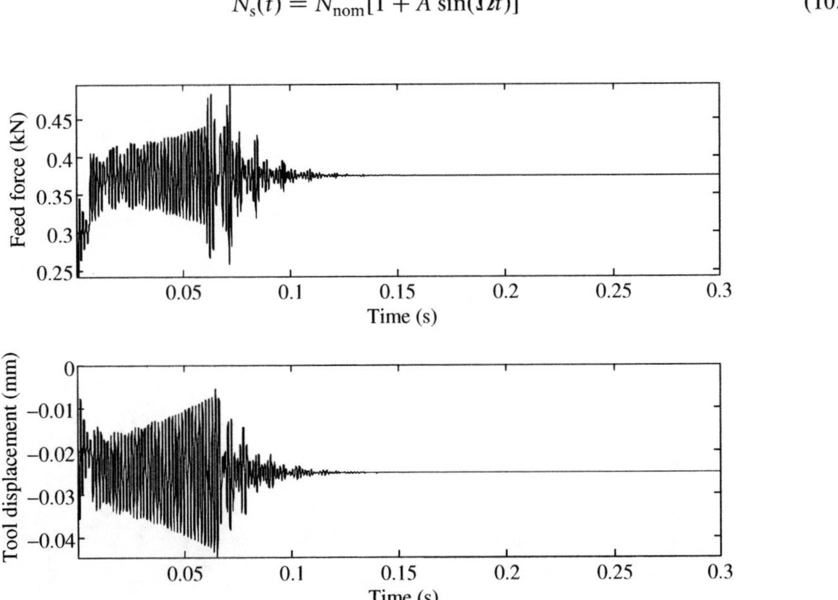

FIGURE 10.24 Time-domain simulations using spindle-speed variation with $A = 0.25$ and $\Omega = 160$ Hz.

The time-domain simulations are in Figure 10.22 to Figure 10.24 for the respective cases. The results illustrate that SSV may be utilized to suppress chatter; however, the amplitude and frequency of the spindle speed vibration must be carefully chosen.

> This section presented several techniques to suppress regenerative chatter. The three major techniques to suppress chatter are spindle-speed selection, feed selection, and SSV. In spindle-speed selection, the spindle speed is adjusted to be a multiple of the chatter frequency to place the spindle speed in a pocket of the stability lobe diagram. In feed selection, the feed is increased to suppress chatter. In SSV, the spindle speed is varied in a sinusoidal manner to decrease the phase difference between the current and previous tooth passes.

10.7 Case Study

A case study of regenerative chatter for a face-milling operation is now presented. Further details are presented in Landers (1997). The machine tool is a three-axis vertical milling machine (Figure 10.25). Each axis has a linear encoder with a resolution of 10 μm mounted on it. The axis motors (186 W) drive pulleys that rotate leadscrews and provide motion to the linear axes. The spindle (2240 W) drives the face mill (Carboloy R/L220.13-02.00-12, 50 mm diameter). The tool holds four carbide inserts (Carboloy SEAN 42AFTN-M14 HX, 45° lead angle). The part is 6061 aluminum. The spindle is run open-loop. A Kistler 9293 piezoelectric three-component dynamometer was utilized for force process modeling and chatter detection. The x and y channels have a natural frequency of 4.5 kHz, rigidity of 0.7 kN/μm, and range of -20 to 20 kN. The z channel has a natural frequency of 5 kHz, rigidity of 7 kN/μm, and range of -100 to 200 kN. A Bently Nevada 3000 Series Type 190 proximity transducer was utilized to measure the static stiffnesses of the structural components. The sensor gain is 8 V/mm, the response is flat to 10 kHz, and the range is 1.02 mm. A Kistler Quartz Model #802A accelerometer (resonant frequency 36.7 kHz) was utilized to measure the dynamic characteristics of the structural components.

The cutting and thrust pressures, respectively, are

$$P_C = 0.29f^{-0.25}d^{-0.13}\left(\frac{V}{1000}\right)^{-0.72} \tag{10.68}$$

$$P_T = 0.16f^{-0.40}d^{-0.41}\left(\frac{V}{1000}\right)^{-0.58} \tag{10.69}$$

The transfer function matrices, respectively, between the tool structure and machining forces, and the part structure and machining forces are modeled as

$$\begin{bmatrix} x_t(s) \\ y_t(s) \end{bmatrix} = \begin{bmatrix} \dfrac{(4500^2/14)}{s^2 + 2(0.07)(4500)s + 4500^2} & 0 \\ 0 & \dfrac{(4100^2/14)}{s^2 + 2(0.11)(4100)s + 4100^2} \end{bmatrix} \begin{bmatrix} F_x(s) \\ F_y(s) \end{bmatrix} \tag{10.70}$$

$$\begin{bmatrix} x_p(s) \\ y_p(s) \end{bmatrix} = -\begin{bmatrix} \dfrac{(2600^2/9.5)}{s^2 + 2(0.09)(2600)s + 2600^2} & 0 \\ 0 & \dfrac{(2100^2/9.5)}{s^2 + 2(0.22)(2100)s + 2100^2} \end{bmatrix} \begin{bmatrix} F_x(s) \\ F_y(s) \end{bmatrix} \tag{10.71}$$

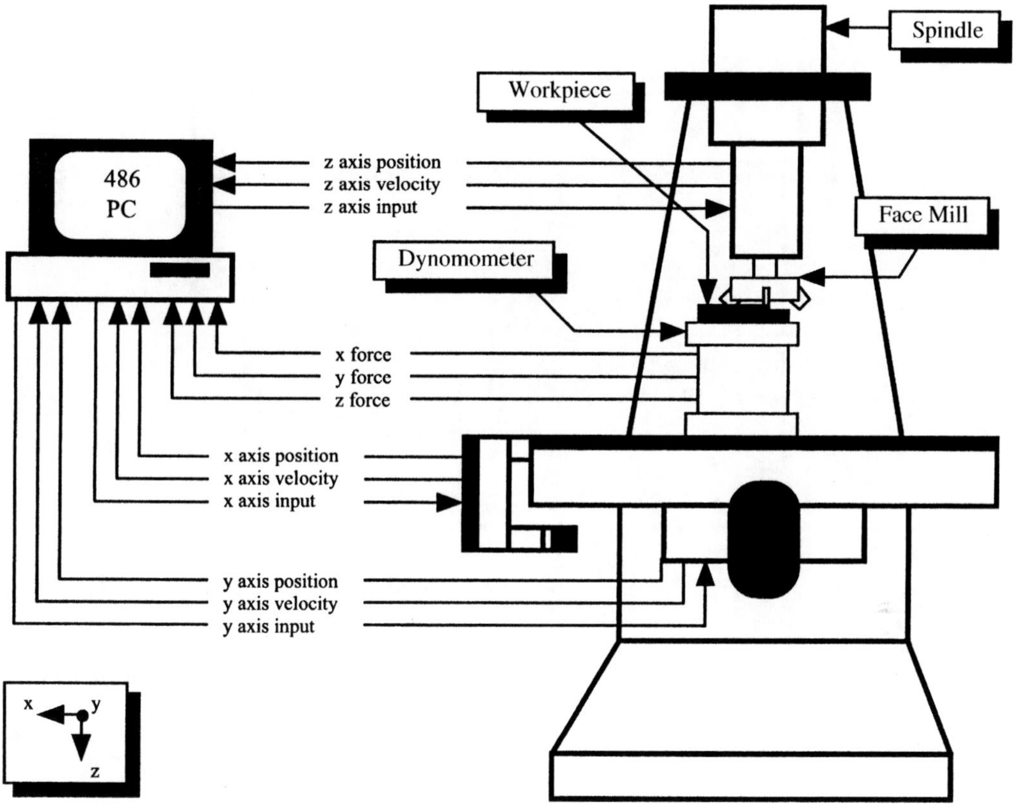

FIGURE 10.25 Three-axis vertical machine tool schematic.

Using the machining force and structural models, a stability lobe diagram was constructed using time domain simulations. Experimental data were collected by adjusting the depth-of-cut in increments of 0.1 mm until chatter occurred. The time-domain simulations and the experimental data are plotted in Figure 10.26. The cutting conditions were $f_t = 0.10$ mm/tooth, $N_t = 4$ teeth, $\theta_{en} = -90°$, and $\theta_{ex} = 90°$. The chatter detection methodology for this system was described in Example 5.

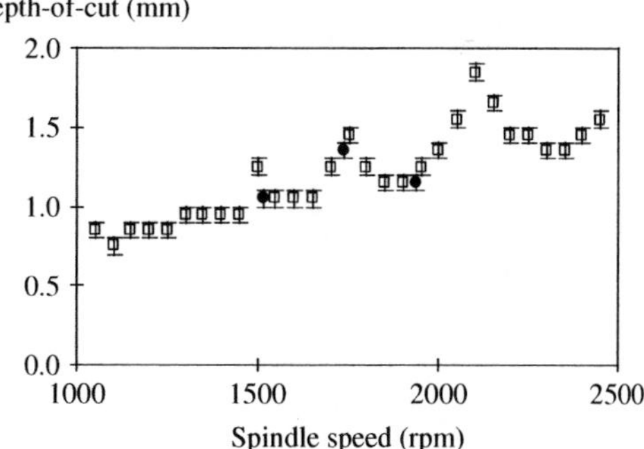

FIGURE 10.26 Stability lobe diagram for a face-milling operation — time-domain simulations (empty boxes) and experimental points (filled circles).

Spindle-speed adjustment is typically a more productive chatter suppression option. However, the machine tool in this case study is not equipped with automatic spindle speed control and, thus, the depth-of-cut is adjusted to suppress chatter. When chatter is detected, the chatter suppressor rewrites the part program to accommodate one additional tool pass (Figure 10.27). Therefore, the new operation depth-of-cut is

$$d_n = \frac{d_p}{1 + N_c} \tag{10.72}$$

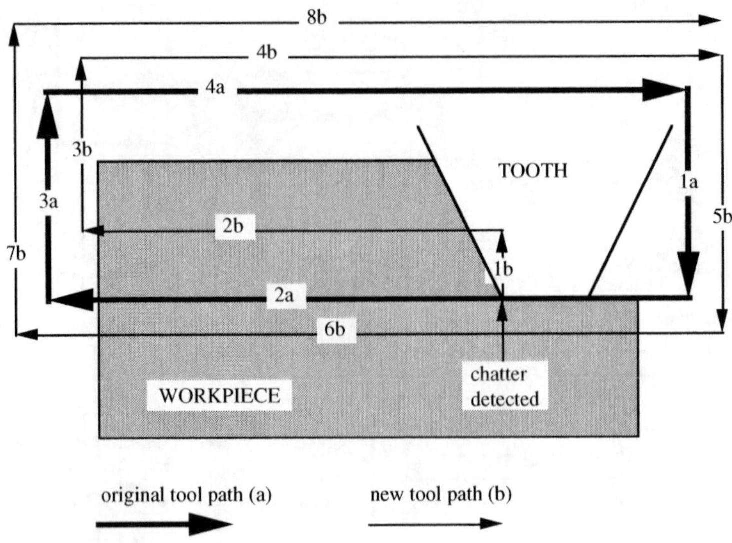

FIGURE 10.27 Original tool path (a) is rewritten when chatter occurs. New tool path (b) contains an additional tool pass.

where d_p is the previous operation depth-of-cut and N_c is the number of times the chatter suppression routine has been invoked. The new value may be well below the stability limit; however, making all passes an equal depth-of-cut provides a good balance between productivity and the search for a stable depth-of-cut. Results of this controller are presented in Landers and Ulsoy (1998, 2001).

Nomenclature

Symbol	Quantity
d	depth-of-cut (mm)
d_{lim}	limiting stable depth-of-cut (mm)
f	feed (mm)
f_t	feed per tooth (mm)
F	machining force (kN)
F_C	cutting force (kN)
F_T	thrust force (kN)
F_x	force acting on cutting tool in x direction (kN)
F_y	force acting on cutting tool in y direction (kN)
F_z	force acting on cutting tool in z direction (kN)
k	structural stiffness (kN/mm)
N_s	spindle speed (rpm)
N_t	number of teeth
P	machining pressure (kN/mm^2)
P_C	cutting pressure (kN/mm^2)
P_T	thrust pressure (kN/mm^2)
t	time (sec)
T	spindle rotational period (s)
T_t	tooth rotational period (s)
x_p	part structural displacement in x direction (mm)
x_t	cutting tool structural displacement in x direction (mm)
y_p	part structural displacement in y direction (mm)
y_t	cutting tool structural displacement in y direction (mm)
z	cutting tool structural displacement in z direction (mm)
θ	tooth angle (rad)
θ_{en}	tooth entry angle (rad)
θ_{ex}	tooth exit angle (rad)
ω_c	chatter frequency (rad/sec)
ω_n	structural natural frequency (rad/sec)
ψ_r	lead angle (rad)
ζ	structural damping ratio

References

Altintas, Y. and Budak, E., Analytical prediction of stability lobes in milling, *Ann. CIRP*, 44/1, 357–362, 1995.

Altintas, Y., Engin, S., and Budak, E., Analytical stability prediction and design of variable pitch cutters, *ASME J. Manuf. Sci. Eng.*, 121, 173–178, 1999.

Budak, E. and Altintas, Y., Analytical prediction of chatter stability in milling, part I: general formulation, *ASME J. Dyn. Syst. Meas. Control*, 120, 22–30, 1998a.

Budak, E. and Altintas, Y., Analytical prediction of chatter stability in milling, part II: application of the general formulation to common milling systems, *ASME J. Dyn. Syst. Meas. Control*, 120, 31–36, 1998b.

Cho, D.W. and Ehmann, K.F., Pattern recognition for on-line chatter detection, *Mech. Syst. Signal Process.*, 2, 279–290, 1988.

Delio, T., Tlusty, J., and Smith, S., Use of audio signals for chatter detection and control, *ASME J. Eng. Ind.*, 114, 146–157, 1992.

Elbestawi, M.A., Ismail, F., Du, R., and Ullagaddi, B.C., Modeling machining dynamics including damping in the tool–workpiece interface, *ASME J. Eng. Ind.*, 116, 435–439, 1994.

Inamura, T. and Sata, T., Stability analysis of cutting under varying spindle speed, *Ann. CIRP*, 23/1, 119–120, 1974.

Inasaki, I., Karpuschewshi, B., and Lee, H.-S., Grinding chatter — origin and suppression, *Ann. CIRP*, 50/2, 515–534, 2001.

Koenigsberger, I., Tlusty, J. 1971. *Structures of Machine Tools*, Pergamon Press, New York.

Landers, R.G. 1997. Supervisory machining control: a design approach plus force control and chatter analysis components, Ph.D. Dissertation. Department of Mechanical Engineering and Applied Mechanics, University of Michigan, Ann Arbor.

Landers, R.G. and Ulsoy, A.G., Supervisory machining control: design and experiments, *Ann. CIRP*, 47/1, 301–306, 1998.

Landers, R.G. and Ulsoy, A.G., Supervisory control of a face-milling operation in different manufacturing environments, *Trans. Control Automat. Syst. Eng.*, 3, 1–9, 2001.

Lee, A.-C. and Liu, C.-S., Analysis of chatter vibration in a cutter–workpiece system, *Int. J. Mach. Tools Manuf.*, 31, 221–234, 1991a.

Lee, A.-C. and Liu, C.-S., Analysis of chatter vibration in the end milling process, *Int. J. Mach. Tools Manuf.*, 31, 471–479, 1991b.

Lin, S.C., DeVor, R.E., and Kapoor, S.G., The effects of variable speed cutting on vibration control in face milling, *ASME J. Eng. Ind.*, 112, 1–11, 1990.

Merritt, H.E., Theory of self-excited machine tool chatter: contribution to machine-tool chatter research — 1, *ASME J. Eng. Ind.*, 87, 447–454, 1965.

Minis, I., Magrab, E., and Pandelidis, I., Improved methods for the prediction of chatter in turning, Part III: a generalized linear theory, *ASME J. Eng. Ind.*, 112, 28–35, 1990a.

Minis, I. and Yanushevsky, R., A new theoretical approach for the prediction of machine tool chatter in milling, *ASME J. Eng. Ind.*, 115, 1–8, 1993.

Minis, I., Yanushevsky, R., and Tembo, A., Analysis of linear and nonlinear chatter in milling, *Ann. CIRP*, 39/1, 459–462, 1990b.

Radulescu, R., Kapor, S.G., and DeVor, R.E., An investigation of variable spindle speed face milling for tool work structures with complex dynamics, part 1: simulation results, *ASME J. Manuf. Sci. Eng.*, 119, 266–272, 1997a.

Radulescu, R., Kapor, S.G., and DeVor, R.E., An investigation of variable spindle speed face milling for tool work structures with complex dynamics, part 2: physical explanation, *ASME J. Manuf. Sci. Eng.*, 119, 273–280, 1997b.

Sastry, S., Kapor, S.G., and DeVor, R.E., Floquet theory based approach for stability analysis of the variable speed face-milling process, *ASME J. Manuf. Sci. Eng.*, 124, 10–17, 2002.

Smith, S. and Delio, T., Sensor-based chatter detection and avoidance by spindle speed selection, *ASME J. Dyn. Syst. Meas. Control*, 114, 486–492, 1992.

Smith, S. and Tlusty, J., Efficient simulation programs for chatter in milling, *Ann. CIRP*, 42/1, 463–466, 1993.

Tarng, Y.S. and Li, T.C., Detection and suppression of drilling chatter, *ASME J. Dyn. Syst. Meas. Control*, 116, 729–734, 1994.

Tlusty, J., Dynamics of high-speed milling, *ASME J. Eng. Ind.*, 108, 59–67, 1986.

Tlusty, J. 2000. *Manufacturing Processes and Equipment*, Prentice Hall, Upper Saddle River, NJ.

Tlusty, J. and Ismail, F., Basic nonlinearity in machining chatter, *Ann. CIRP*, 30/1, 299–304, 1981.

Tlusty, J. and Ismail, F., Special aspects of chatter in milling, *ASME J. Vib. Acoust. Stress Reliab. Des.*, 105, 24–32, 1983.

Tobias, S.A. 1965. *Machine Tool Vibration*, Wiley, New York.

Tsai, M.D., Takata, S., Inui, M., Kimura, F., and Sata, T., Prediction of chatter vibration by means of a model-based cutting simulation system, *Ann. CIRP*, 39/1, 447–450, 1990.

Weck, M., Altintas, Y., and Beer, C., CAD assisted chatter-free NC tool path generation in milling, *Int. J. Mach. Tools Manuf.*, 34, 879–891, 1994.

Weck, M., Verhagg, E., and Gather, M., Adaptive control of face-milling operations with strategies for avoiding chatter-vibrations and for automatic cut distribution, *Ann. CIRP*, 24/1, 405–409, 1975.

Yun, I.S., Wilson, W.R.D., and Ehmann, K.F., Review of chatter studies in cold rolling, *Int. J. Mach. Tools Manuf.*, 38, 1499–1530, 1998.

11

Fluid-Induced Vibration

Seon M. Han
Texas Tech University

Summary

This chapter gives an overview on the subject of fluid-induced vibration in an ocean environment. The main objective is to show how the fluid forces on an offshore structure due to current and random waves are modeled. The chapter is divided into three sections. The first section describes the ocean environment, especially the currents and random waves. The second section is dedicated to obtaining fluid forces utilizing the results from the first section and the third section gives some examples to show how the results from the first two sections can be used in practice. In the first section, the concept of spectral density is introduced. For a given spectrum, methods to obtain a sample time series are given. In the second section, the forces that the fluid can exert on a body are discussed. The regimes in which inertia, drag, or diffraction forces are dominant are shown in terms of the ratio of the wave height to the structural diameter and the ratio of the structural diameter to the wavelength. The Morison equation is extended to the case of a moving inclined cylinder. The Morison equation requires the use of experimentally determined fluid coefficients such as added mass, inertia, and drag coefficients. Plots of these fluid coefficients for various values of the fluid parameters are reproduced here. The vortex shedding force is discussed briefly. In the third section, four examples are given to show how fluid forces affect the static and dynamics of ocean structures, how the significant wave height can be chosen to represent the condition in a certain area for a long time, and how the time series can be constructed from a given spectrum. Finally, the available numerical codes for modeling slender flexible bodies in fluids are listed.

11.1 Description of the Ocean Environment

In modeling offshore structures, one needs to account for the forces exerted by the surrounding fluid. In-depth studies are given in Kinsman (1965), Sarpkaya and Isaacson (1981), Wilson (1984), Chakrabarti (1987), and Faltinsen (1993). The vibration characteristics of a structure can be significantly altered when it is surrounded by water. For example, damping by the fluid (or the added mass) lowers the natural frequency of vibration. When considering the dynamics of an offshore structure, one must also consider the forces due to the surrounding fluid. The two important sources of fluid motion are ocean waves and ocean currents.

Most steady large currents are generated by the drag of the wind passing over the surface of the water, and they are confined to a region near the ocean surface. Tidal currents are generated by the gravitational attraction of the sun and the moon, and they are most significant near coasts. The ultimate source of the ocean circulation is the uneven radiation heating of the Earth by the Sun.

Isaacson (1988) suggested an empirical formula for the current velocity in the horizontal direction as a function of depth:

$$U_c(x) = (U_{\mathrm{tide}}\,(d) + U_{\mathrm{circulation}}\,(d))\left(\frac{x}{d}\right)^{1/7} + U_{\mathrm{drift}}\,(d)\left(\frac{x-d+d_0}{d_0}\right) \tag{11.1}$$

where U_{drift} is the wind-induced drift current, U_{tide} is the tidal current, $U_{\mathrm{circulation}}$ is the low-frequency long-term circulation, x is the vertical distance measured from the ocean bottom, d is the depth of the water, and d_0 is the smaller of the depth of the thermocline and 50 m. The value of U_{tide} is obtained from tide tables, and U_{drift} is about 3% of the 10 min mean wind velocity at 10 m above the sea level.

It should be noted that these currents evolve slowly compared with the time scales of engineering interests. Therefore, they can be treated as a quasisteady phenomenon. Waves, on the other hand, cannot be treated as a steady phenomenon. The underlying physics that govern wave dynamics are too complex and, therefore, waves must be modeled stochastically. The subsequent section discusses the concept of the spectral density, available ocean wave spectral densities, a method to obtain the spectral density from wave time histories, methods to obtain a sample time history from a spectral density, the short-term and long-term statistics, and a method to obtain fluid velocities and accelerations from wave elevation using linear wave theory.

11.1.1 Spectral Density

Here, we will consider only surface gravity waves. Let us first consider a regular wave in order to familiarize ourselves with the terms that are used to describe a wave. The wave surface elevation is denoted as $\eta(x,t)$ and can be written as $\eta(x,t) = A\cos(kx - \omega t)$, where k is the wave number, and ω is the angular frequency. Figure 11.1 shows the surface elevation at two time instances ($t = 0$ and $t = \tau$) and the surface elevation at a fixed location ($x = 0$). A is the amplitude, H is the wave height or the distance between the maximum and minimum wave elevation or twice the amplitude, and T is the period given by $T = 2\pi/\omega$.

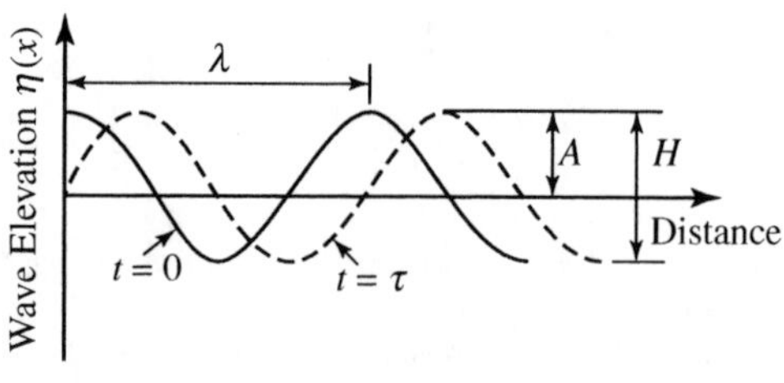

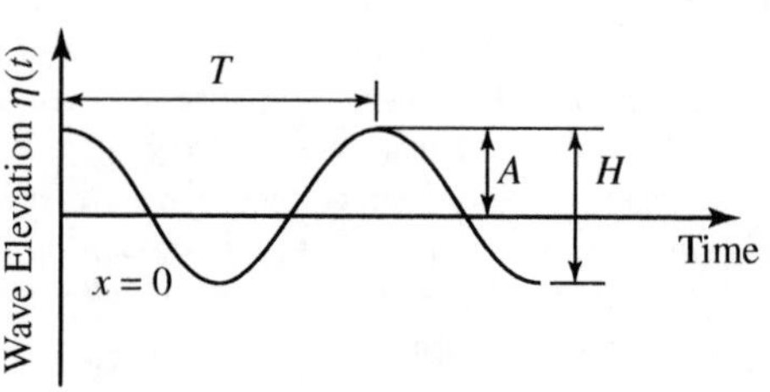

FIGURE 11.1 Regular wave.

In practice, waves are not regular. Figure 11.2 shows a schematic time history of an irregular wave surface elevation. The wave height and frequency are not easy to find. Therefore, we rely on a statistical description for the wave elevation such as the wave spectral density. The spectral density tells us how the energy of the system is distributed among frequencies. The random surface elevation $\eta(t)$ can be thought of as a summation of regular waves with different frequencies. The surface elevation $\eta(t)$ is related to its Fourier transform $X(\omega)$ by

$$\eta(t) = \frac{1}{2\pi}\int_{-\infty}^{\infty} X(\omega)\exp(-\mathrm{i}\omega t)\mathrm{d}\omega$$

Suppose that the energy of the system is proportional to $\eta^2(t)$ so that we can write the energy as

$$E = \frac{1}{2}C\eta^2(t)$$

where C is the proportionality constant.

Let us assume that the expected value of the energy is given by

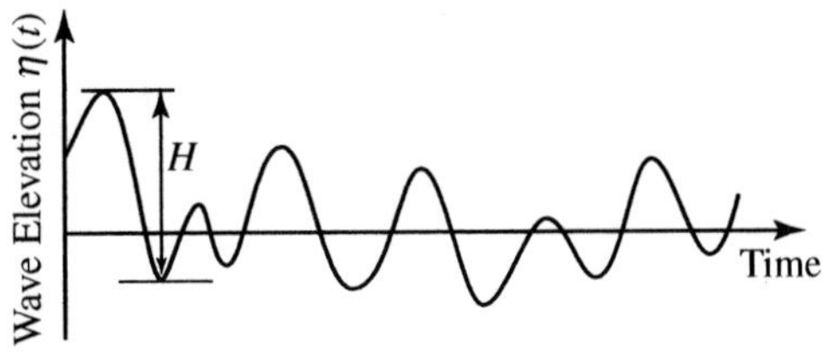

FIGURE 11.2 Time history of random wave.

$$E\{E\} = \frac{1}{2}CE\{\eta^2(t)\}$$

where $E\{\eta^2(t)\}$ is the mean square of $\eta(t)$. If $\eta(t)$ is an *ergodic process*, then the mean square of $\eta(t)$ can be approximated by the time average over a long period of time:

$$E\{\eta^2(t)\} = \lim_{T_s \to +\infty} \frac{1}{T_s}\int_{-T_s/2}^{T_s/2} \eta^2(t)\mathrm{d}t = \lim_{T_s \to +\infty} \frac{1}{T_s}\frac{1}{2\pi}\int_{-\infty}^{\infty} |X(\omega)|^2\,\mathrm{d}\omega \tag{11.2}$$

where we have used Parseval's theorem

$$\int_{-\infty}^{\infty} \eta^2(t)\mathrm{d}t = \frac{1}{2\pi}\int_{-\infty}^{\infty} |X(\omega)|^2\,\mathrm{d}\omega \tag{11.3}$$

where

$$|X(\omega)|^2 = X(\omega)X^*(\omega), \quad X(\omega) = \int_{-\infty}^{\infty} \eta(t)\exp(-\mathrm{i}\omega t)\mathrm{d}t, \quad X^*(\omega) = \int_{-\infty}^{\infty} \eta(t)\exp(\mathrm{i}\omega t)\mathrm{d}t$$

We define the power spectral density (or simply the spectrum) as

$$S_{\eta\eta}(\omega) \equiv \frac{1}{2\pi T_s}|X(\omega)|^2 \tag{11.4}$$

so that $E\{\eta^2(t)\}$ is given by

$$E\{\eta^2(t)\} = \int_{-\infty}^{\infty} S_{\eta\eta}(\omega)\mathrm{d}\omega \tag{11.5}$$

For a zero-mean process, $E\{\eta^2(t)\}$ is also the variance σ_η^2. The spectral density has units of $\eta^2 t$. Where η is the wave elevation, the spectral density has a unit of m^2 sec.

It can also be shown that $S_{\eta\eta}(\omega)$ is related to the autocorrelation function, $R(\tau)$, by the Wiener–Khinchine relations (Wiener, 1930; Khinchine, 1934):

$$S_{\eta\eta}(\omega) = \frac{1}{2\pi}\int_{-\infty}^{\infty} R_{\eta\eta}(\tau)\exp(-\mathrm{i}\omega\tau)\mathrm{d}\tau, \quad R_{\eta\eta}(\tau) = \int_{-\infty}^{\infty} S_{\eta\eta}(\omega)\exp(\mathrm{i}\omega\tau)\mathrm{d}\omega \tag{11.6}$$

It should be noted that, in some textbooks, the factor $1/2\pi$ appears in the second equation instead of the first. Figure 11.3 shows some important pairs of $S_{\eta\eta}(\omega)$ and $R_{\eta\eta}(\tau)$.

There are a few properties of the spectral density that readers should become familiar with. The first property is that the spectral density function of a real-valued stationary process is both real and symmetric. That is, $S_{\eta\eta}(\omega) = S_{\eta\eta}(-\omega)$ (Equation 11.4). Secondly, the area under the spectral density is equal to $E\{\eta^2(t)\}$ (Equation 11.5) and is also equal to $R_{\eta\eta}(0) = \sigma_\eta^2 - \mu_\eta^2$, where σ_η^2 is the variance and μ_η^2 is the mean of $\eta(t)$. In most cases, we only consider a zero-mean process so that the area under the spectral density is just σ_η^2. If the process does not have a zero mean, the mean can be subtracted from it so that the process has a zero mean.

For ocean applications, a one-sided spectrum in terms of cycles per second (cps) or hertz is often used. We will denote the one-sided spectrum with a superscript "o". The one-sided spectrum can be obtained from the two-sided spectrum by

$$S_{\eta\eta}^{\mathrm{o}}(\omega) = 2S_{\eta\eta}(\omega), \quad \omega \geq 0$$

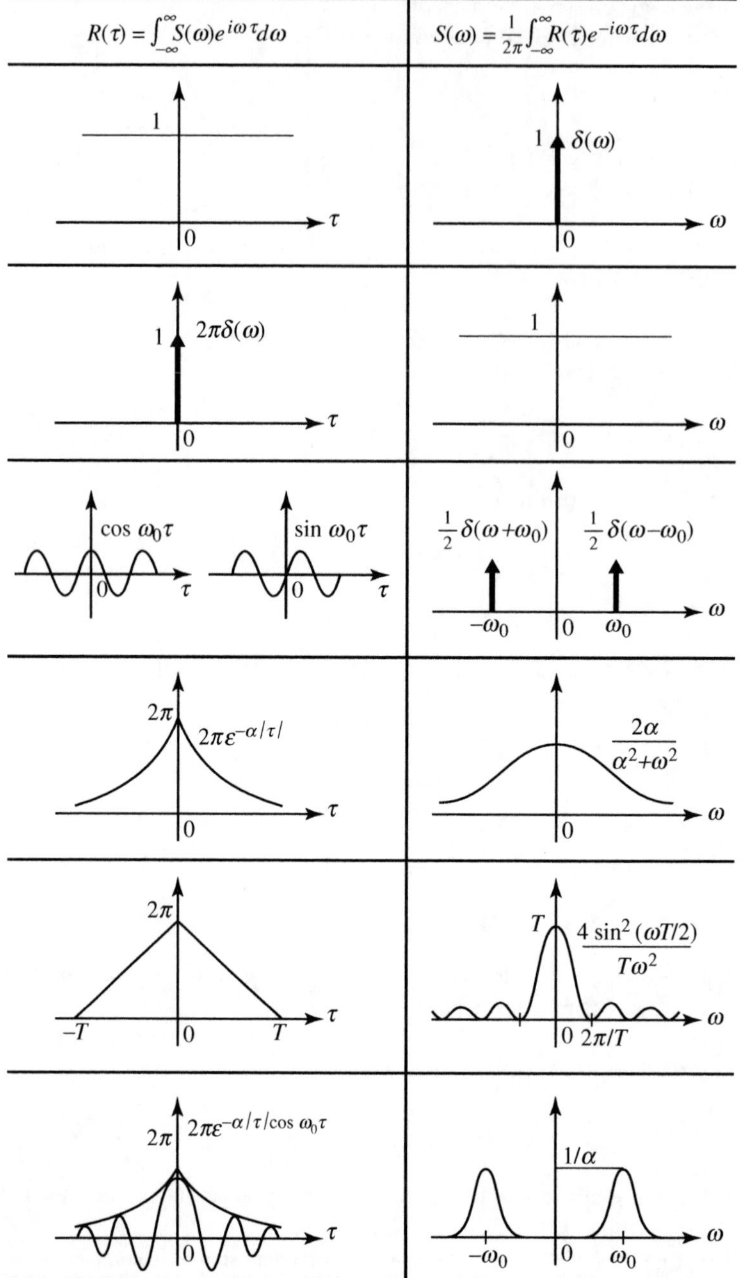

FIGURE 11.3 Relationship between the autocorrelation function and the power spectral density.

The two-sided spectrum in terms of ω can be transformed to the spectrum in terms of f (where $\omega = 2\pi f$) by

$$S_{\eta\eta}(f) = 2\pi S_{\eta\eta}(\omega), \quad f, \omega \geq 0$$

Then, the two-sided spectrum in terms of ω can be transformed to the one-sided spectrum in terms of cps (or hertz) by

$$S^{\circ}_{\eta\eta}(f) = 4\pi S_{\eta\eta}(\omega), \quad f, \omega \geq 0$$

It should be noted that the spectral density that we have defined here is the *amplitude half-spectrum*. The amplitude, height, and height double spectra are related to the amplitude half-spectrum by

$$S^A(\omega) = 2S(\omega), \quad S^H(\omega) = 8S(\omega), \quad S^{2H}(\omega) = 16S(\omega)$$

11.1.2 Ocean Wave Spectral Densities

In this section, we will discuss spectral density models to describe a random sea. An excellent review of existing spectral density models is given in Chapter 4 of Chakrabarti (1987).

The ocean wave spectrum models are semiempirical formulas. That is, they are derived mathematically but the formulation requires one or more experimentally determined parameters. The accuracy of the spectrum depends significantly on the choice of these parameters.

In formulating spectral densities, the parameters that influence the spectrum are *fetch limitations*, *decaying* vs. *developing seas*, *water depth*, *current*, and *swell*. The fetch is the distance over which a wind blows in a wave-generating phase. Fetch limitation refers to the limitation on the distance due to some physical boundaries so that full wave development is prohibited. In a developing sea, the sea has not yet reached its stationary state under a stationary wind. In contrast, a wind has blown for a sufficient time in a fully developed sea, and the sea has reached its stationary state. In a decaying sea, the wind has dropped off from its stationary value. Swell is the wave motion caused by a distant storm and persists even after the storm has died down or moved away.

The Pierson–Moskowitz (P–M) spectrum (Pierson and Moskowitz, 1964) is the most extensively used spectrum for representing a fully developed sea. It is a one-parameter model in which the sea severity can be specified in terms of the wind velocity. The P–M spectrum is given by

$$S^{\mathrm{o}}_{\eta\eta}(f) = \frac{8.1\times 10^{-3}g^2}{\omega^5}\exp\left(-0.74\left(\frac{g}{U_{\mathrm{w},19.5\ \mathrm{m}}}\right)^4\omega^{-4}\right)$$

where g is the gravitational constant and $U_{\mathrm{w},19.5\ \mathrm{m}}$ is the wind speed at a height of 19.5 m above the still water. The P–M spectrum is also called the wind-speed spectrum because it requires wind data. It can also be written in terms of the modal frequency ω_{m} as

$$S^{\mathrm{o}}_{\eta\eta}(f) = \frac{8.1\times 10^{-3}g^2}{\omega^5}\exp\left(-1.25\left(\frac{\omega_{\mathrm{m}}}{\omega}\right)^4\right) \tag{11.7}$$

Note that the modal frequency is the frequency at which the spectrum is the maximum.

In some cases, it may be more convenient to express the spectrum in terms of significant wave height rather than the wind speed or modal frequency. For a narrowband Gaussian process[1], the significant wave height is related to the standard deviation by $H_{\mathrm{s}} = 4\sigma_\eta$. The standard deviation is the square root of the area under the spectral density, $\int_{-\infty}^{\infty} S_{\eta\eta}(\omega)\mathrm{d}\omega = \sigma_\eta^2$. Then, the spectrum can be written as

$$S^{\mathrm{o}}_{\eta\eta}(f) = \frac{8.1\times 10^{-3}g^2}{\omega^5}\exp\left(-\frac{0.0324g^2}{H_{\mathrm{s}}^2}\omega^{-4}\right) \tag{11.8}$$

and the peak frequency and the significant wave height are related by

$$\omega_{\mathrm{m}} = 0.4\sqrt{g/H_{\mathrm{s}}} \tag{11.9}$$

The P–M spectrum is applicable for deep water, unidirectional seas, fully developed and local-wind-generated seas with unlimited fetch, and was developed for the North Atlantic. The effect of swell is not accounted for. Although it was developed for the North Atlantic, the spectrum is valid for other locations. However, the limitation that the sea is fully developed may be too restrictive because it cannot model the

[1]See Section 11.1.5 for details.

effect of waves generated at a distance. Therefore, we consider a two-parameter spectrum, such as the Bretschneider spectrum, in order to model a sea that is not fully developed as well as a fully developed sea.

The Bretschneider spectrum (Bretschneider, 1959, 1969) is a two-parameter spectrum in which both the sea severity and the state of development can be specified. The Bretschneider spectrum is given by

$$S^{o}_{\eta\eta}(f) = 0.169\frac{\omega_s^4}{\omega^5}H_s^2 \exp\left(-0.675\left(\frac{\omega_s}{\omega}\right)^4\right)$$

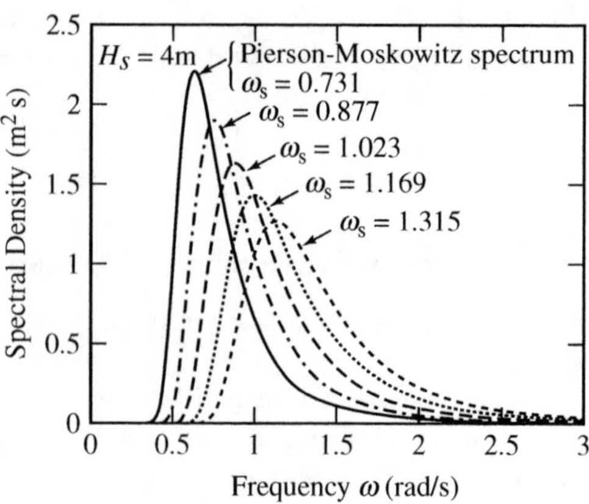

FIGURE 11.4 Bretschneider spectrum with various values of ω_s.

where $\omega_s = 2\pi/T_s$ and T_s is the significant period. The sea severity can be specified by H_s and the state of development can be specified by ω_s. It can be shown that the relationship $\omega_s = 1.167\omega_m$ (equivalent to $\omega_s = 1.46/\sqrt{H_s}$) renders the Bretschneider spectrum and the P–M spectrum equivalent. Figure 11.4 shows the Bretschneider spectra for $H_s = 4$ m. When $\omega_s = 0.731$ rad/sec, the P–M and the Bretschneider spectra are identical. It should be noted that the developing sea will have a slightly higher modal frequency than the fully developed sea, and can be described by ω_s greater than $1.46/\sqrt{H_s}$.

Other two-parameter spectral densities that are often used are the International Ship Structures Congress (ISSC) and the International Towing Tank Conference (ITTC) spectra. The ISSC spectrum is written in terms of the significant wave height and the mean frequency, where the mean frequency is given by

$$\bar{\omega} = \sqrt{\frac{\int_0^\infty \omega S(\omega)\mathrm{d}\omega}{\int_0^\infty S(\omega)\mathrm{d}\omega}} = 1.30\omega_m$$

Thus, the ISSC spectrum is given by

$$S^{o}_{\eta\eta}(f) = 0.111\frac{\bar{\omega}^4}{\omega^5}H_s^2 \exp\left(-0.444\left(\frac{\bar{\omega}}{\omega}\right)^4\right)$$

The ITTC spectrum is based on the significant wave height and the zero crossing frequency and is given by

$$S^{o}_{\eta\eta}(f) = 0.0795\frac{\omega_z^4}{\omega^5}H_s^2 \exp\left(-0.318\left(\frac{\omega_z}{\omega}\right)^4\right)$$

where the zero crossing frequency, ω_z, is given by

$$\omega_z = \sqrt{\frac{\int_0^\infty \omega^2 S(\omega)\mathrm{d}\omega}{\int_0^\infty S(\omega)\mathrm{d}\omega}} = 1.41\omega_m$$

The Bretschneider, ITTC, and ISSC spectra are called two-parameter spectra, and they can be written as

$$S^{o}_{\eta\eta}(f) = \frac{A}{4}\frac{\tilde{\omega}^4}{\omega^5}H_s^2 \exp\left(-A\left(\frac{\tilde{\omega}}{\omega}\right)^4\right)$$

with A and ω given in Table 11.1.

TABLE 11.1 Two-Parameter Spectrum Models
$S^{o}_{\eta\eta}(\omega) = (A/4)H_s^2\bar{\omega}^4/\omega^5 \exp(-A(\omega/\bar{\omega})^{-4})$

Model	A	$\bar{\omega}$
Bretschneider	0.675	ω_s
ITTC	0.318	ω_z
ISSC	0.4427	$\bar{\omega}$

The spectra that we have discussed so far do not allow us to generate spectra with two peaks to represent local or distant storms or to specify the sharpness of the peaks. The Ochi–Hubble (O–H) spectrum (Ochi and Hubble, 1976) is a six-parameter spectrum with the form:

$$S^{o}_{\eta\eta}(\omega) = \frac{1}{4}\sum_{i=1}^{2} \frac{((4\lambda_i+1)\omega_{mi}^4/4)^{\lambda_i}}{\Gamma(\lambda_i)} \frac{H_{si}^2}{\omega^{4\lambda_i+1}} \exp\left(-\left(\frac{4\lambda_i+1}{4}\right)\left(\frac{\omega_{mi}}{\omega}\right)^4\right)$$

where $\Gamma(\lambda_i)$ is the Gamma function, H_{s1}, ω_{m1}, and λ_1 are the significant wave height, modal frequency, and shape factor for the lower frequency components, respectively, and H_{s2}, ω_{m2}, and λ_2 are those for the higher frequency component. Assuming that the entire spectrum is that of a narrow band, the equivalent significant wave height is given by

$$H_s = \sqrt{H_{s1}^2 + H_{s2}^2}$$

For $\lambda_1 = 1$ and $\lambda_2 = 0$, the spectrum reduces to the P–M spectrum. With the assumption that the entire spectrum is narrowband, the value of λ_1 is much higher than λ_2. The O–H spectrum represents unidirectional seas with unlimited fetch. The sea severity and the state of development can be specified by H_{si} and ω_{mi}, respectively. In addition, λ_i can be selected to control the frequency width of the spectrum. For example, a small λ_i (wider frequency range) describes a developing sea, and a large λ_i (narrower frequency range) describes a swell condition. Figure 11.5 shows the O–H spectrum with $\lambda_1 = 2.72$, $\omega_{m1} = 0.626$ rad/sec, $H_{s1} = 3.35$ m, $\lambda_2 = 2.72$, $\omega_{m2} = 1.25$ rad/sec, and $H_{s2} = 2.19$ m.

Finally, another spectrum that is commonly used is the Joint North Sea Wave Project (JONSWAP) spectrum developed by Hasselmann et al. (1973). It is a fetch-limited spectrum because the growth over a limited fetch is taken into account. The attenuation in shallow water is also taken into account. The JONSWAP spectrum is written as

$$S^{o}_{\eta\eta}(\omega) = \frac{\alpha g^2}{\omega^5}\exp\left(-1.25\left(\frac{\omega_m}{\omega}\right)^4\right)\gamma^{\exp(-(\omega-\omega_m)/2\tau^2\omega_m^2)}$$

where γ is the peakedness parameter and τ is the shape parameter. The peakedness parameter γ is the ratio of the maximum spectral energy to the maximum spectral energy of the corresponding P–M spectrum. That is, when $\gamma = 7$, the peak spectral energy is seven times that of the P–M spectrum.

$$\gamma = \begin{cases} 7.0 & \text{for very peaked data} \\ 3.3 & \text{for mean of selected JONSWAP data} \\ 1.0 & \text{for P–M spectrum} \end{cases}$$

$$\tau = \begin{cases} 0.07 & \text{for } \omega \le \omega_m \\ 0.09 & \text{for } \omega > \omega_m \end{cases}$$

$\alpha = 0.076(\bar{X})^{-0.22}$ or 0.0081 if fetch independent

$\bar{X} = gX/U_w^2$

X = fetch length (nautical miles)

U_w = wind speed (knots)

$$\omega_m = 2\pi \times 3.5(g/U_w)\bar{X}^{-0.33}$$

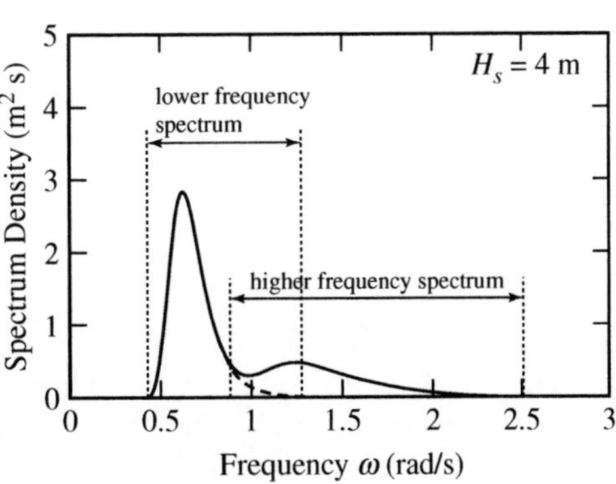

FIGURE 11.5 Ochi–Hubble spectrum.

Figure 11.6 shows the JONSWAP spectrum when $\alpha = 0.0081$ and $\omega_m = 0.626$ rad/sec for three peakedness parameters.

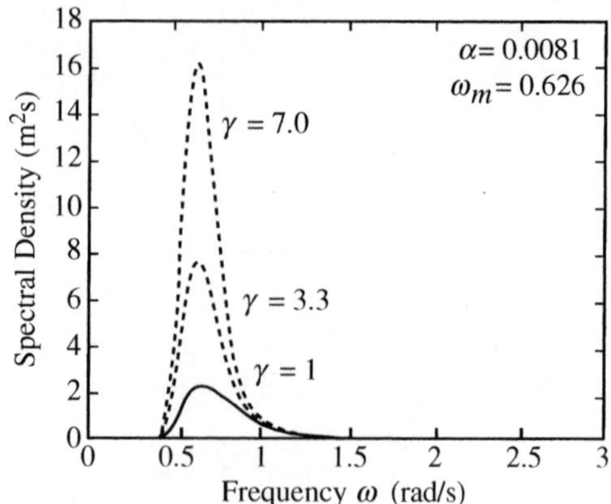

FIGURE 11.6 JONSWAP spectrum for $\gamma = 1.0$, 3.3, and 7.0.

11.1.3 Approximation of Spectral Density from Time Series

From the time history of the wave elevation, the spectral density function can be obtained by two methods.

The first method is to use the autocorrelation function $R_{\eta\eta}(\tau)$, which is related to the spectral density function $S_{\eta\eta}(\omega)$ by the Wiener–Khinchine relations (Equation 11.6).

The autocorrelation $R_{\eta\eta}(\tau)$ is the expected value of $\eta(t)\eta(t+\tau)$ or $R_{\eta\eta}(\tau) = E\{\eta(t)\eta(t+\tau)\}$, where t is an arbitrary time and τ is the time lag. For a weakly stationary process, the autocorrelation is a function of the time lag only.

Assuming that the process is ergodic, the autocorrelation function for a given time history of length T_s can be approximated as

$$\hat{R}_{\eta\eta}(\tau) = \lim_{T_s \to \infty} \frac{1}{T_s - \tau} \int_0^{T_s - \tau} \eta(t)\eta(t+\tau)\mathrm{d}t \quad \text{for } 0 < \tau < T_s$$

Note that the superscript ^ is used to emphasize that the variable is an approximation based on a sample time history of length T_s. The spectral density is then obtained by taking the Fourier cosine transform of $\hat{R}_{\eta\eta}(\tau)$,

$$\hat{S}_{\eta\eta}(\omega) = \frac{1}{\pi} \int_0^{T_s} \hat{R}_{\eta\eta}(\tau) \cos \omega\tau \, \mathrm{d}\tau \tag{11.10}$$

The second method for obtaining the spectral density function is to use the relationship between the spectral density and the Fourier transform of the time series. They are related by

$$\hat{S}_{\eta\eta}(\omega) = \lim_{T_s \to \infty} \frac{1}{2\pi T_s} |\hat{X}(\omega)\hat{X}^*(\omega)| \tag{11.11}$$

where $\hat{X}(\omega)$ is given by

$$\hat{X}(\omega) = \int_0^{T_s} \eta(t) \exp(-\mathrm{i}\omega t)\mathrm{d}t$$

and $\hat{X}^*(\omega)$ is the complex conjugate given by

$$\hat{X}^*(\omega) = \int_0^{T_s} \eta(t) \exp(\mathrm{i}\omega t)\mathrm{d}t$$

In order to obtain the Fourier transforms of the time series, the discrete Fourier transform (DFT) or the fast Fourier transform (FFT) procedure can be used. For detailed descriptions of how this is done, see Appendix 1 in Tucker (1991). Nowadays, spectral analysis is almost always carried out via FFTs because it is easier to use and faster than the formal method via correlation function.

It should be noted that the length of the sample time history only needs to be long enough so that the limits converge. Taking a longer sample will not improve the accuracy of the estimate. Instead, one should take many samples or break one long sample into many parts. For n samples, the spectral densities are obtained for each sample time history using either Equation 11.10 or Equation 11.11, and they are averaged to give the estimate.

The determination of the spectral density from wave records depends on the details of the procedure such as the length of the record, sampling interval, degree and type of filtering and smoothing, and time discretization.

11.1.4 Generation of Time Series from a Spectral Density

In a nonlinear analysis, the structural response is found by a numerical integration in time. Therefore, one needs to convert the wave elevation spectrum into an equivalent time history. The wave elevation can be represented as a sum of many sinusoidal functions with different angular frequencies and random phase angles. That is, we write $\eta(t)$ as

$$\eta(t) = \sum_{i=1}^{N} \cos(\omega_i t - \varphi_i)\sqrt{2S_{\eta\eta}(\omega_i)\Delta\omega_i} \tag{11.12}$$

where φ_i is a uniform random number between 0 and 2π, ω_i are discrete sampling frequencies, $\Delta\omega_i = \omega_i - \omega_{i-1}$, and N is the number of partitions. Recall that the area under the spectrum is equal to the variance, σ_η^2. The incremental area under the spectrum, $S_{\eta\eta}(\omega_i)\Delta\omega_i$, can be denoted as σ_i^2 such that the sum of all the incremental area equals the variance of the wave elevation or $\sigma_\eta^2 = \sum_{i=1}^{N} \sigma_i^2$. The time history can be written as

$$\eta(t) = \sum_{i=1}^{N} \cos(\omega_i t - \varphi_i)\sqrt{2}\sigma_i$$

The sampling frequencies, ω_i, can be chosen at equal intervals such that $\omega_i = i\omega_1$. However, the time history will then have the lowest frequency of ω_1 and will have a period of $T = 2\pi/\omega_1$. In order to avoid this unwanted periodicity, Borgman (1969) suggested that the frequencies are chosen so that the area under the spectrum curve for each interval is equal or $\sigma_i^2 = \sigma^2 = \sigma_\eta^2/N$. The time history is written as

$$\eta(t) = \sqrt{\frac{2}{N}}\sigma_\eta \sum_{i=1}^{N} \cos(\bar{\omega}_i t - \varphi_i) \tag{11.13}$$

where $\bar{\omega}_i = (\omega_i + \omega_{i-1})/2$. The discrete frequencies, ω_i, are chosen such that the area between the interval $0 < \omega < \omega_i$ is equal to i/N of the total area under the curve between the interval $0 < \omega < \omega_N$ or

$$\int_0^{\omega_i} S_{\eta\eta}(\omega)\mathrm{d}\omega = \frac{i}{N}\int_0^{\omega_N} S_{\eta\eta}(\omega)\mathrm{d}\omega \quad \text{for } i = 1, \ldots, N$$

where it is assumed that the area under the spectrum beyond ω_N is negligible. If $\eta(t)$ is a narrowband Gaussian process, the standard deviation can be replaced by $\sigma_\eta = H_s/4$, and the time history can be written as

$$\eta(t) = \frac{H_s}{4}\sqrt{\frac{2}{N}} \sum_{i=1}^{N} \cos(\bar{\omega}_i t - \varphi_i)$$

Shinozuka (1972) proposed that the sampling frequencies, $\bar{\omega}_i$, in Equation 11.13 should be randomly chosen according to the density function, $f(\omega) \equiv S_{\eta\eta}^{o}(\omega)/\sigma_\eta^2$. This is equivalent to performing an integration using the Monte Carlo method. The random frequencies ω distributed according to $f(\omega)$ can be obtained from uniformly distributed random numbers, x, by $\omega = F^{-1}(x)$, where $F(\omega)$ is the cumulative distribution of $f(\omega)$.

The random frequencies obtained this way are used in Equation 11.13 to generate a sample time series. It should be noted that many sample time histories should be obtained and averaged to synthesize a time history for use in numerical simulations.

11.1.5 Short-Term Statistics

In discussing wave statistics, we often use the term *significant wave* to describe an irregular sea surface. The significant wave is not a physical wave that can be seen but rather a statistical description of random waves. The concept of significant wave height was first introduced by Sverdrup and Munk (1947) as the average height of the highest one third of all waves. Usually, ships co-operate in programs to find sea statistics by reporting a rough estimate of the storm severity in terms of an observed wave height. This observed wave height is consistently very close to the significant wave height.

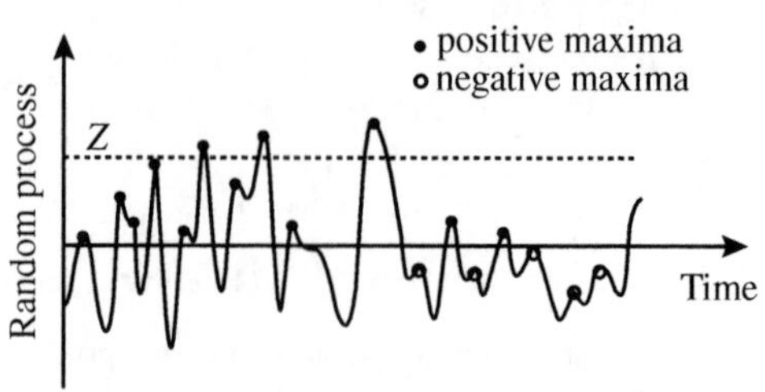

FIGURE 11.7 A sample time history.

Stationarity and ergodicity are two assumptions that are made in describing short-term waves statistics. These assumptions are valid only for "short" time intervals — approximately two hours or the duration of a storm — but not for weeks or years. The wave elevation is assumed to be weakly stationary so that its autocorrelation is a function of time lag only. As a result, the mean and the variance are constant, and the spectral density is invariant with time. Therefore, the significant wave height and the significant wave period are constant when we consider short-term statistics. In this case, the individual wave height and wave period are the stochastic variables. We then need to determine certain statistics for the analysis and design of offshore structures when we consider short time intervals.

Consider a sample time history of a zero-mean random process, as shown in Figure 11.7. The questions that we ask are how often is a certain level (e.g., z in the figure) exceeded, and how are the maxima distributed? Likewise, we can ask when we can expect to see that a certain level is exceeded for the first time, and what are the values of the peaks of a random process? The first question is important when a structure may fail due to a one-time excessive load, and the second question is important when a structure may fail due to cyclic loads.

It is found that the rate at which a random process $X(t)$ crosses Z with a positive slope (zero up-crossing) may be calculated from

$$\nu_{z^+} = \int_0^\infty \nu f_{X\dot{X}}(z, \nu)\mathrm{d}\nu$$

where $f_{X\dot{X}}(x, \dot{x})$ is the joint probability density function of X and $\dot{X}(t)$. The expected time of the first up-crossing is then the inverse of the crossing rate or

$$E\{T\} = 1/\nu_{z^+}$$

The probability density function of the maxima, A, can be calculated from

$$f_A(a) = \frac{\displaystyle\int_{-\infty}^{0} -\omega f_{X\dot{X}\ddot{X}}(a, 0, \omega)\mathrm{d}\omega}{\displaystyle\int_{-\infty}^{0} -\omega f_{\dot{X}\ddot{X}}(0, \omega)\mathrm{d}\omega}$$

where $f_{X\dot{X}\ddot{X}}(x, \dot{x}, \ddot{x})$ is the joint probability density function of X, $\dot{X}$, and $\ddot{X}$.

If $X(t)$ is a Gaussian process, then we can write the joint probability density functions as

$$f_{X\dot{X}}(x, \dot{x}) = \frac{1}{2\pi\sigma_X\sigma_{\dot{X}}} \exp\left[-\frac{1}{2}\left(\frac{x}{\sigma_X}\right)^2 - \frac{1}{2}\left(\frac{\dot{x}}{\sigma_{\dot{X}}}\right)^2\right], \quad -\infty < x < \infty, \quad -\infty < \dot{x} < \infty$$

and

$$f_{X\dot{X}\ddot{X}}(x, \dot{x}, \ddot{x}) = \frac{1}{(2\pi)^{3/2}|M|^{1/2}} \exp\left[-\frac{1}{2}(\{x\} - \{\mu_X\})^{\mathrm{T}}[M]^{-1}(\{x\} - \{\mu_X\})\right]$$

where

$$[M] = \begin{bmatrix} \sigma_X^2 & 0 & \sigma_{\dot{X}}^2 \\ 0 & \sigma_{\dot{X}}^2 & 0 \\ \sigma_{\dot{X}}^2 & 0 & \sigma_{\ddot{X}}^2 \end{bmatrix} \quad \text{and} \quad \{x\} - \{\mu_X\} = \begin{bmatrix} x - \mu_X \\ \dot{x} - \mu_{\dot{X}} \\ \ddot{x} - \mu_{\ddot{X}} \end{bmatrix}$$

Then, for a stationary Gaussian process, the up-crossing rate is given by

$$\nu_z^+ = \int_0^\infty f_{X\dot{X}}(Z, \dot{x})\dot{x}\,\mathrm{d}\dot{x} = \frac{1}{2\pi\sigma_X\sigma_{\dot{X}}}\exp\left[-\frac{1}{2}\left(\frac{Z}{\sigma_X}\right)^2\right]\int_0^\infty \exp\left[-\frac{1}{2}\left(\frac{\dot{x}}{\sigma_{\dot{X}}}\right)^2\right]\dot{x}\,\mathrm{d}\dot{x}$$

$$= \frac{\sigma_{\dot{X}}}{2\pi\sigma_X}\exp\left[-\frac{1}{2}\left(\frac{Z}{\sigma_X}\right)^2\right] \qquad (11.14)$$

and the probability density function of maxima is given by the Rice density function (Rice, 1954)

$$f_A(a) = \frac{\sqrt{1-\alpha^2}}{\sqrt{2\pi}\sigma_\eta}\exp\left[-\frac{1}{2}\frac{\alpha^2}{\sigma_\eta^2(1-\alpha^2)}\right] + a\frac{\alpha}{\sigma_\eta^2}\Phi\left(\frac{a\alpha}{\sigma_\eta\sqrt{\alpha^2-1}}\right)\exp\left(-\frac{1}{2}\frac{a^2}{\sigma_\eta^2}\right) \quad \text{for } -\infty < a < \infty$$

where $\Phi(x)$ is the cumulative distribution function of standard normal random variable

$$\Phi(x) = \frac{1}{\sqrt{2\pi}}\int_{-\infty}^{x}\exp(-z^2/2)\mathrm{d}z$$

and α is the irregularity factor equivalent to the ratio of the number of zero up-crossings (number of times that $\eta[t]$ crosses zero with a positive slope) to the number of peaks. α ranges from 0 to 1, and it is also equal to

$$\alpha = \frac{\sigma_{\dot{\eta}}^2}{\sigma_\eta^2\sigma_{\ddot{\eta}}^2}$$

If $X(t)$ is a broadband process, $\alpha = 0$ and the Rice distribution is reduced to the Gaussian probability density function given by

$$f_A(a) = \frac{1}{\sqrt{2\pi}\sigma_\eta}\exp\left[-\frac{1}{2}\frac{\alpha^2}{\sigma_\eta^2}\right] \quad \text{for } -\infty < a < \infty$$

If $X(t)$ is a narrowband process, it is guaranteed that it will have a peak whenever $\eta(t)$ crosses its mean. In this case, the irregularity factor is close to unity, and the Rice distribution is reduced to the Rayleigh probability density function given by

$$f_A(a) = \frac{a}{\sigma_\eta^2}\exp\left[-\frac{1}{2}\frac{\alpha^2}{\sigma_\eta^2}\right] \quad \text{for } 0 < a < \infty$$

In other words, the amplitudes of a narrowband stationary Gaussian process are distributed according to the Rayleigh distribution.

Figure 11.8 shows the Rice distribution for various values of α. Note that the Rice distribution includes both positive and negative maxima except when $\alpha = 1$, in which case all the maxima are positive. The positive maxima are the local maxima that occur above the mean of $X(t)$, and the negative maxima are the local maxima that occur below the mean, as shown in Figure 11.7. In some cases, the negative maxima may not mean

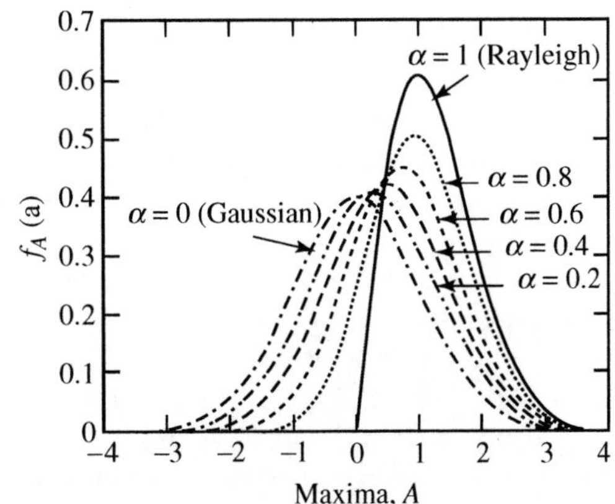

FIGURE 11.8 Rice distribution for maxima.

much physically. In those cases, we can use the truncated Rice distribution, where only the positive portion of $f_A(a)$ is used. $f_A(a)$ is normalized by the area under the probability density for positive maxima (Longuet-Higgins, 1952; Ochi, 1973):

$$f_A^{\text{trunc}}(a) = \frac{f_A(a)}{\int_0^\infty f_A(a)\mathrm{d}a}, \quad a \geq 0$$

The truncated Rice distribution is shown in Figure 11.9.

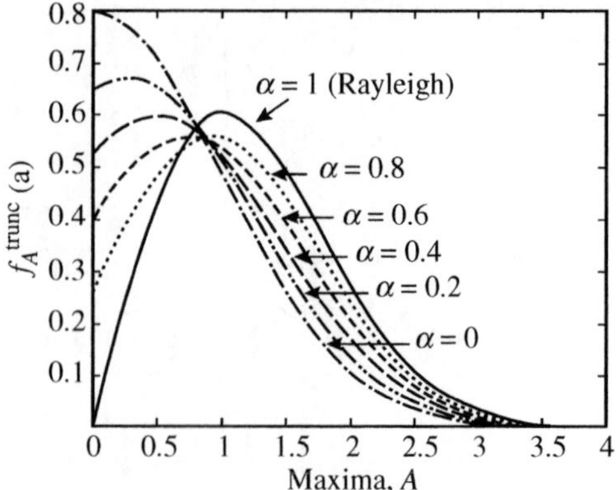

FIGURE 11.9 Truncated Rice distribution.

If $X(t)$ is the wave elevation, its maxima, A, are the amplitudes of the wave elevation. The wave height, $H = 2A$, is then distributed according to

$$f_H(h) = f_A(H/2)\frac{\mathrm{d}A}{\mathrm{d}H} = \frac{h}{4\sigma_\eta^2}\exp\left[-\frac{1}{2}\frac{h^2}{4\sigma_\eta^2}\right] \quad \text{for } 0 < h < \infty$$

For any given wave, the probability that the height is less than h (the cumulative distribution) is

$$F_H(h) = 1 - \exp\left[-\frac{1}{2}\frac{h^2}{4\sigma_\eta^2}\right] \quad \text{for } 0 < h < \infty$$

If $\eta(t)$ is a stationary narrowband process so that the peaks are distributed according to the Rayleigh distribution, we find that the root-mean-square wave height, $\sqrt{E\{H^2\}}$, is given by

$$\sqrt{E\{H^2\}} = \int_0^\infty h^2 f_H(h)\mathrm{d}h = 2\sqrt{2}\sigma_\eta$$

In addition, it can be shown that the average and the significant wave heights are given by

$$H_0 \equiv E\{H\} = \sqrt{2\pi}\sigma_\eta, \quad H_s \equiv E\{H_{1/3}\} = 4\sigma_\eta \qquad (11.15)$$

where $E\{H_{1/3}\}$ means that it is the expectation of the highest one third of the waves.

11.1.6 Long-Term Statistics

Because offshore structures are designed for long life spans, we must also consider long-term wave statistics. Previously, when we considered the short-term statistics, the significant wave height and spectrum were assumed to be invariant with time. This assumption is valid only over time periods of days at most. For longer time periods, the significant wave height has its own statistics and is a random variable.

When one uses short-term statistics to describe long-term events, improbable events seem unjustifiably probable. For example, let us consider the probability that the wave height, distributed according to the Rayleigh distribution, exceeds a certain extreme value. Let us assume that the mean period of this wave is 10 sec and the probability that the height of any given wave is greater than 300 ft is 10^{-10}. The value is small and the occurrence of a 300 ft wave seems improbable. However, the probability that the height will exceed 300 ft at least once in 10 years (3×10^8 sec) is given by

$$1 - (1 - 10^{-10})^{3\times10^8/10} = 0.997$$

Thus, the statistical description states that it is almost certain that the wave height will exceed 300 ft at least once in 10 years. This prediction is a shortcoming of the short-term statistics since waves of this magnitude do not arrive at this probability.

In order to compute the probability that a wave height will exceed a certain extreme value, we require statistics for these extreme events. The actual maximum amplitude in a sequence of random amplitudes is a random variable itself. It has a probability distribution with mean value, standard deviation and other statistical properties. In fact, the distributions of these maximum values are called the extreme value distributions (EVDs). Gumbel (1958) obtained three methods of extrapolation known as three asymptotes. They are the Gumbel, Fretchet, and Weibull distributions. We will discuss the Gumbel and Weibull distributions in the next section. For the moment, we will discuss the concept of the N-year storm.

In long-term statistics, we often speak of an N-year storm. It means that, for any given year, the probability that we will have an N-year storm is

$$p = \frac{1}{N}$$

It follows that the probability that we will have m storms in n years is given by

$$\Pr\{mN\text{-year storms in } n \text{ years}\} = {}_n P_m \left(\frac{1}{N}\right)^m \left(1 - \frac{1}{N}\right)^{n-m}$$

where ${}_n P_m$ is the permutation given by

$${}_n P_m = \frac{n!}{(n-m)!}$$

The probability that we will have at least one N-year storm in n years is

$$\Pr\{\text{at least one } N\text{-year storms in } n \text{ years}\} = 1 - \left(1 - \frac{1}{N}\right)^n$$

For a large N, the probability can be approximated as $1 - \exp(n/N)$. It should be noted that the probability that we will have exactly one N-year storm in N years is not one, but

$$\Pr\{\text{one } N\text{-year storm in } N \text{ years}\} = \left(1 - \frac{1}{N}\right)^{M-1}$$

As $N \rightarrow \infty$, we find that

$$P = 1/e \approx 0.3679$$

The probability that we will have at least one N-year storm in N years is

$$\Pr\{\text{at least one } N\text{-year storms in } n \text{ years}\} = 1 - \left(1 - \frac{1}{N}\right)^N$$

As $N \rightarrow \infty$, we find that

$$P = 1 - 1/e \approx 0.6321 \tag{11.16}$$

11.1.6.1 Weibull Distribution

The Weibull distribution fits probabilities of extremes quite satisfactorily. In long-term statistics, the significant wave height follows the Weibull distribution closely. The probability density and the cumulative distribution are given by

$$f(h) = \frac{m}{\beta}\left(\frac{h-\gamma}{\beta}\right)^{m-1} \exp\left(-\left(\frac{h-\gamma}{\beta}\right)^m\right), \quad F(h) = 1 - \exp\left(-\left(\frac{h-\gamma}{\beta}\right)^m\right) \text{ for } \gamma < h \tag{11.17}$$

where m is called the shape parameter. Manipulating the cumulative distribution, we can write

$$\ln(-\ln\{1-F(h)\}) = m\{\ln(h-\gamma) - \ln(\beta)\}$$

where the left-hand side is known from data. If we let $y = \ln(-\ln\{1-F(h)\})$ and $x = \ln(h-\gamma)$, y is a straight line with slope of m and a y-intercept of $-m\ln\beta$:

$$y = mx - m\ln\beta$$

Suppose we have significant wave height data over a long period of time, and our goal is to find the Weibull parameters, γ, β, and m that best fit the distribution of the significant wave heights. These parameters can be determined by the least-squares method or using the Weibull paper. Using the latter method, we first guess γ so that the discrete points (x,y) or $(\ln(h-\gamma), \ln(-\ln\{1-F(h)\}))$ form a straight line. The slope of this line is m, and the value of y when the line intersects the y axis is $-m\ln\beta$. This method will be illustrated in Section 11.3.3.

The Gumbel distribution is given by

$$\begin{aligned} f(h) &= \alpha\exp(-\alpha(h-\beta))\exp\{\exp[-\alpha(h-\beta)]\} \\ F(h) &= \exp\{-\exp[-\alpha(h-\beta)]\} \quad \text{for } -\infty < h < \infty \end{aligned} \tag{11.18}$$

When $\ln[-\ln F(h)]$ is plotted against h, the result is a line with a slope of $-\alpha$ and y intercept of $\alpha\beta$.

Another distribution that may be used is the lognormal distribution given by (Jasper, 1954)

$$f(h) = \frac{1}{\sqrt{2\pi}\sigma h}\exp\left[-\frac{1}{2}\frac{(\ln h-\mu)^2}{\sigma^2}\right], \quad F(h) = \Phi\left(\frac{\ln h-\mu}{\sigma}\right) \quad \text{for } 0 \le h \tag{11.19}$$

11.1.6.2 Wave Velocities *via* Linear Wave Theory

The wave velocities that correspond to the wave elevation given in Equation 11.12 can be obtained by *linear wave theory*. Linear wave theory, also called airy wave theory, sinusoidal wave theory, and small-amplitude theory, is the simplest wave theory. It is also the most important wave theory because it forms the basis for the probabilistic spectral description of waves.

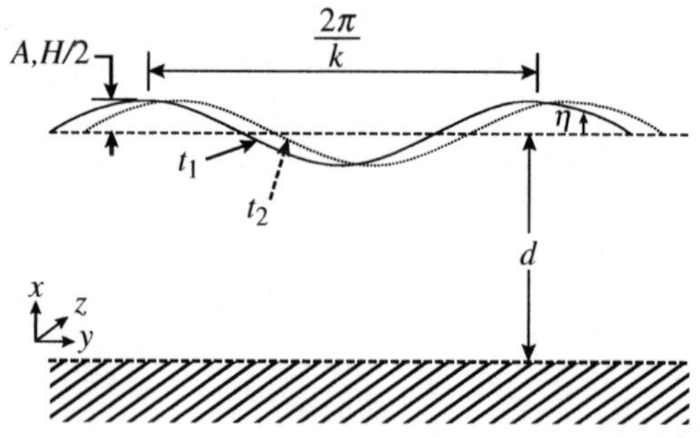

FIGURE 11.10 A schematic of a simple sinusoidal wave shown at two different times.

Linear wave theory assumes that the wave height is small compared with the wavelength and wave depth. In addition, fluid particles are assumed to follow a circular orbit. The readers should refer to Kinsman (1965) and LeMehaute (1976) for detailed descriptions.

In linear wave theory, the surface elevation is given by

$$\eta(y,t) = A\cos(\omega t - ky) \tag{11.20}$$

which is a plane wave traveling to the right in Figure 11.10. Linear wave theory relates this sinusoidal surface elevation to the wave velocities given by

$$w_y(x,y,t) = A\omega\frac{\cosh kx}{\sinh kd}\cos(\omega t - ky), \quad w_y(x,y,t) = A\omega\frac{\sinh kx}{\sinh kd}\sin(\omega t - ky) \tag{11.21}$$

where k, ω, and A are wave number, angular frequency, and amplitude of a surface wave, respectively. The velocities vary with time, horizontal coordinate y, and depth x measured from the ocean floor. The wave velocities are sinusoidal in y and t, but exponentially decrease with the distance from the surface.

The frequency ω is related to the wave number k by the dispersion relation given by

$$\omega^2 = gk \tanh kd$$

where d is the water depth. For deep water, tanh kd approaches unity and the frequency is given by

$$\lim_{d\to\infty} \omega^2 = gk$$

For the surface elevation given in Equation 11.12, the surface elevation and the wave velocities are given by

$$\begin{aligned}
\eta(y,t) &= \sum_{i=1}^{N} \cos(\omega_i t - k_i y - \varphi_i)\sqrt{2S_{\eta\eta}(\omega_i)\Delta\omega_i} \\
w_y(x,y,t) &= \sum_{i=1}^{N} \omega_i \frac{\cosh k_i x}{\sinh k_i d}\cos(\omega_i t - k_i y - \varphi_i)\sqrt{2S_{\eta\eta}(\omega_i)\Delta\omega_i} \\
w_y(x,y,t) &= \sum_{i=1}^{N} \omega_i \frac{\sinh k_i x}{\sinh k_i d}\sin(\omega_i t - k_i y - \varphi_i)\sqrt{2S_{\eta\eta}(\omega_i)\Delta\omega_i}
\end{aligned} \tag{11.22}$$

The wave accelerations can be obtained by differentiating the wave velocities with respect to time. Sample time histories of the wave velocity and acceleration can be obtained using either Borgman's or Shinozuka's method.

11.1.7 Summary

In this section, the concept of spectral density is introduced. It is then shown how the concept is used to describe the ocean wave heights. The spectral density or spectrum is related to the autocorrelation function by the Wiener–Khinchine relations:

$$\begin{aligned}
S_{\eta\eta}(\omega) &= \frac{1}{2\pi}\int_{-\infty}^{\infty} R_{\eta\eta}(\tau)\exp(-\mathrm{i}\omega\tau)\mathrm{d}\tau \\
R_{\eta\eta}(\tau) &= \int_{-\infty}^{\infty} S_{\eta\eta}(\omega)\exp(-\mathrm{i}\omega\tau)\mathrm{d}\omega
\end{aligned}$$

In addition, the spectral density function of a real-valued stationary process is also real and symmetric, or

$$S_{\eta\eta}(\omega) = S_{\eta\eta}(-\omega) \tag{11.23}$$

and the area under the spectral density is given by

$$\int_{-\infty}^{\infty} S_{\eta\eta}(\omega)\mathrm{d}\omega = R_{\eta\eta}(0) = \sigma_\eta^2 - \mu_\eta^2 \tag{11.24}$$

If the spectral density is given only for $\omega \geq 0$, then this one-sided spectrum is related to the two-sided spectrum by

$$S^{\mathrm{o}}_{\eta\eta}(\omega) = 2S_{\eta\eta}(\omega), \quad \omega \geq 0$$

When the frequency is given in Hertz instead of in rad/sec, the spectra are related by

$$S_{\eta\eta}(f) = 2\pi S_{\eta\eta}(\omega), \quad \omega \geq 0$$

The spectra that are often used to describe wave heights are the P–M, Bretschneider, ITTC, ISSC, O–H, and JONSWAP spectra. The most widely used spectrum is the P–M spectrum, which is a single parameter spectrum. The P–M spectrum is applicable for deep water, unidirectional seas, fully developed and local-wind-generated sea with unlimited fetch, and was originally developed for the North Atlantic. The single parameter for this spectrum can be expressed as the wind velocity at 19.5 m above sea level or the significant wave height that specifies the sea severity. When it is written in terms of the significant wave height, it is given by

$$S^{o}_{\eta\eta}(f) = \frac{8.1 \times 10^{-3} g^2}{\omega^5} \exp\left(-\frac{0.0324 g^2}{H_s^2} \omega^{-4} \right)$$

Bretschneider, ITTC, and ISSC spectra are two-parameter spectra in which the state of development as well as the sea severity can be specified. The O–H spectrum is a six-parameter spectrum that allows us to represent local and distant storm effects and to specify sharpness of the peaks as well as to specify the sea severity and the state of development. The JONSWAP spectrum allows us to account for growth over a limited fetch.

For a given ocean spectrum, a sample time history can be obtained by

$$\eta(t) = \sqrt{\frac{2}{N}} \sigma_\eta \sum_{i=1}^{N} \cos(\bar{\omega}_i t - \varphi_i)$$

In Borgman's method, the sampling frequencies $\bar{\omega}_i = (\omega_i + \omega_{i-1})/2$ are chosen so that the area between ω_{i-1} and ω_i are equal. In Shinozuka's method, the sampling frequencies are chosen randomly. The traditional method of choosing the sampling frequency at even intervals is not recommended.

When a relatively short interval of time is considered, for example, about two hours or the duration of a storm, it can be assumed that the spectrum and its statistics are invariant with time. In this case, the distribution of local maxima or peaks of a stationary Gaussian process is described by the Rice distribution. The Rice distribution can be reduced to the Rayleigh distribution when the process is narrowband and to the Gaussian distribution when the process is broadband. In the long-term statistics, the spectrum and its statistics may vary with time. In this case, the term "*N*-year storm" is often used to indicate the sea severity, and the significant wave heights closely follow the Weibull distribution.

Finally, the wave velocities and accelerations are related to the wave velocities using linear wave theory.

11.2 Fluid Forces

The following is a list of several types of forces that the fluid can exert on a body:

1. *Drag force.* This is due to the pressure difference between the downstream and upstream flow region. It can be thought of as the force required to hold a body stationary in a fluid of constant velocity. The drag force is proportional to the square of the velocity of the fluid relative to the structure.
2. *Inertia force.* This is the force exerted by the fluid while it accelerates and decelerates as it passes the structure. It is also the force required to hold a rigid structure in a uniformly accelerating flow, and it is proportional to the fluid acceleration. The concept of the inertia force in an inviscid flow was first formulated by Lamb (1945).
3. *Added mass.* As the body accelerates or decelerates in a stationary fluid, the body carries a certain amount of the surrounding fluid along with it. This entrained fluid is called the *added, apparent,* or *virtual mass.* In order to accelerate the body, additional force is required to accelerate or decelerate the added mass.
4. *Diffraction force.* This is due to the scattering of an incident wave on the surface of the structure. It is important when the body is large compared with the wavelength of the incident wave.
5. *Froude–Kryloff force.* This is the pressure force on the structure due to the incident wave, assuming that the structure does not exist and does not interfere with the incident wave.
6. *Lift force.* This is due to nonsymmetrical separation of the fluid or due to vortices that are shed in a nonsymmetrical way. The component of the force perpendicular to the flow direction is the lift force.

7. *Wave slamming force.* This is due to a single occasional wave with a particularly high amplitude and energy, and it may be important at the free surface. Sarpkaya and Isaacson (1981) reviewed the research on slamming of water against circular cylinders. Miller (1977, 1980) found that the peak wave slamming force on a rigidly held horizontal circular cylinder is proportional to the square of the horizontal water particle velocity.

11.2.1 Wave Force Regime

Previously, we discussed various types of forces caused by waves and currents. In some cases, one type of force may be dominant. Hogben (1976) gave a literature review of the fluid force in various regimes. The load regime of importance can be demonstrated for the case of a vertical cylinder in Figure 11.11 in terms of H/D and $\pi D/\lambda$, where H is the wave height, D is the cylinder diameter, and λ is the wavelength. When linear wave theory is used, H/D is related to the Keulegan–Carpenter number by

$$K = \pi H/D$$

The Keulegan–Carpenter number gives a measure of the importance of drag force relative to the inertia force. The term $\pi D/\lambda$ is called the diffraction parameter, and it determines the importance of the diffraction effect. As H/D increases, the drag force becomes more important and the inertia force becomes less important. As $\pi D/\lambda$ increases, the diffraction force becomes important.

Using linear wave theory, the maximum drag force to the maximum inertia force can be written as

$$\frac{f_{\text{drag}}}{f_{\text{inertia}}} = \frac{1}{2\pi}\frac{H}{D} = \frac{K}{2\pi^2}$$

From the last relation, we find that the drag force is 5% of the inertial force when $H/D = 0.314$. The Morison equation may be used for $D/\lambda < 0.2$ and $f_{\text{drag}}/f_{\text{inertia}} > 0.1$ or thereabouts. It should be noted that Figure 11.11 is valid only near the surface. The drag force is predominant for a cylinder that extends from the bottom to the near surface, so that the Morison equation may be used.

For example, consider a fixed jacket platform with legs with a diameter of 10 m and bracings with a diameter of 0.8 m. For a 10-year storm with $\lambda = 100$ m and $H = 8$ m, the ratios H/D and D/λ for the leg are 0.8 and 0.1, respectively. Similarly, the ratios H/D and D/λ for the bracings are 10 and 0.08, respectively. Figure 11.11 shows that the inertia force is dominant for the legs, and both inertia and the drag forces are important for the bracings.

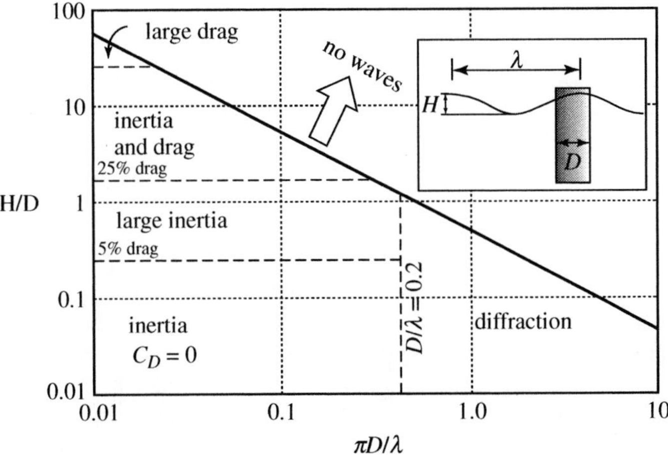

FIGURE 11.11 Load regimes near surface.

11.2.2 Wave Forces on Small Structures — Morison Equation

The added mass, M_A, can be written as

$$M_A = C_A M_{disp}$$

where C_A is called the *added mass coefficient* and M_{disp} is the mass of the fluid displaced by the structure. For a cylinder with a diameter, D, and height, h, the displaced fluid mass is $\pi D^2 h/4$. It should be noted that the added mass is a tensor quantity. That is, we can speak of the added mass force in the x_i direction due to the acceleration of the body in the x_j direction, denoted as M_{ij}^A. M_{ij}^A is symmetric so that the added mass force in the x_i direction due to the acceleration in the x_j direction is equal to the added mass force in the x_j direction due to the acceleration in the x_i direction. The off-diagonal terms are not zero if the cross-section is not symmetric.

Similarly, the inertia force can be written as

$$F_M = C_M M_{disp} \dot{w} \tag{11.25}$$

where the proportionality constant, C_M, is called the *inertia coefficient.*

It should be noted that the added mass and the inertia effects are often neglected for a body vibrating in air since the displaced air mass is negligible.

The drag force is proportional to the square of the fluid velocity, w, the density of the fluid, ρ, and the area of the body projected onto the plane perpendicular to the flow direction, A_f,

$$F_D = \frac{1}{2} C_D \rho A_f w|w|$$

where C_D is the *drag coefficient.* The absolute value sign is used to ensure that the drag force always acts in the direction of the flow. For a cylinder with a diameter D and height h, the projected area A_f is Dh.

For a body with nonzero velocity, the drag force is given by

$$F_D = \frac{1}{2} C_D \rho A_f (w - v)|w - v| \tag{11.26}$$

where $w - v$ is the velocity of the fluid relative to the body.

Morison et al. (1950) combined the inertia and drag terms (Equation 11.25 and Equation 11.26) so that the fluid force on a body is given by

$$f = \frac{1}{2} C_D \rho A_f w|w| + C_M M_{disp} \dot{w}$$

For a cylinder, the fluid force per unit length can be written as

$$f = \frac{1}{2} C_D \rho D w|w| + C_M \rho \pi \frac{D^2}{4} \dot{w}$$

For a moving cylinder with velocity v, the Morison force is given by

$$f = \frac{1}{2} C_D \rho D (w - v)|w - v| + C_M \rho \pi \frac{D^2}{4} \dot{w}$$

11.2.2.1 Inclined Cylinder

Let us now consider the inclined cylinder shown in Figure 11.13. The direction of the flow makes an angle of θ with the cylinder. Often, only the fluid force in the normal direction is considered. The normal component is given by

$$f^n = \frac{1}{2} C_D \rho D (w^n - v^n)|w^n - v^n| + C_M \rho \pi \frac{D^2}{4} \dot{w}^n \tag{11.27}$$

where the superscript is used for the normal component. The term, $w^n - v^n$, is the normal component of the relative velocity of the fluid with respect to the structure. Suppose that fluid is flowing to the right, and the cylinder is also moving to the right, as shown in Figure 11.12. The normal components of the fluid and cylinder velocities are

$$w^{\mathrm{n}} = |w| \cos \theta, \quad v^{\mathrm{n}} = |v| \cos \theta$$

In three dimensions, it may be difficult to picture what the normal component should be. Here, we can find the normal component using the formula

$$(w^{\mathrm{n}} - v^{\mathrm{n}})\vec{n} = \vec{t} \times (\vec{w} - \vec{v}) \times \vec{t} \qquad (11.28)$$

where $\vec{t}$ is the unit vector tangent to the cylinder and $\vec{n}$ is the unit vector normal to the cylinder.

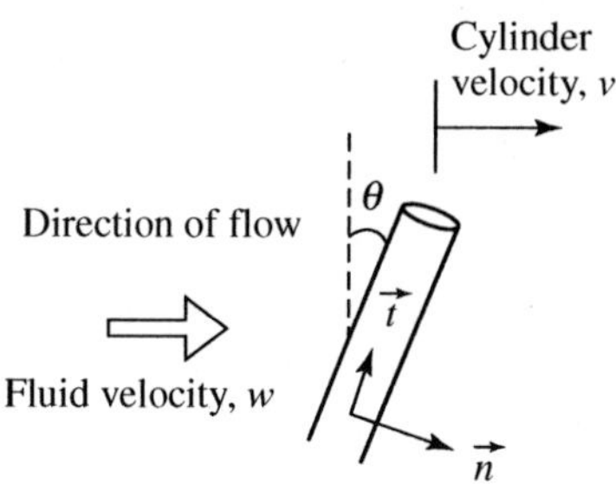

FIGURE 11.12 Inclined cylinder.

Note that the normal direction depends on the direction of the flow as well as the inclination of the cylinder.

In some cases, the tangential drag force may be included, and it can be written as

$$f^{\mathrm{t}} = \frac{1}{2} C_{\mathrm{T}} \rho D (w^{\mathrm{t}} - v^{\mathrm{t}})|w^{\mathrm{t}} - v^{\mathrm{t}}| \qquad (11.29)$$

where C_{T} is the tangential drag coefficient. Note that C_{T} is usually a very small number.

The normal component of the fluid force is more dominant than the tangential component. It may seem strange that the fluid force does not act in the direction of the fluid motion. Instead, the force is predominantly in the normal direction defined by Equation 11.28. In Section 11.3.1, we will demonstrate what this means by considering a towing cable.

11.2.2.1.1 Determination of Fluid Coefficients

The drag, inertia, and added mass coefficients must be obtained by experiment. However, for a long cylinder, C_{M} approaches its theoretical limiting value (uniformly accelerated inviscid flow) of 2, and C_{A} approaches unity (Lamb, 1945; Wilson, 1984). In reality, the inertia and drag coefficients are functions of at least three parameters (Wilson, 1984):

$$C_{\mathrm{M}} = C_{\mathrm{M}}(\mathrm{Re}, K, \text{cylinder roughness})$$
$$C_{\mathrm{D}} = C_{\mathrm{D}}(\mathrm{Re}, K, \text{cylinder roughness})$$

where Re is the Reynolds number and K is the Keulegan–Carpenter number given by

$$Re \equiv \frac{\rho_{\mathrm{f}} U D}{\mu}, \quad K \equiv \frac{UT}{D} \qquad (11.30)$$

where ρ_{f} is the density of the fluid, U is the free stream velocity, D is the diameter of the structure, μ is the dynamic or absolute viscosity, and T is the wave period.

Sarpkaya looked at the variation of these hydrodynamic coefficients extensively and obtained the plots shown in Figure 11.13 to Figure 11.15 (Sarpkaya, 1976; Sarpkaya et al., 1977). Figure 11.13 shows the inertia and drag coefficients for a smooth cylinder as a function of K for various values of Re and the reduced frequency β, defined by $\beta = \mathrm{Re}/K$. From this figure, we find that for low Re and β, the inertial coefficient decreases and the drag coefficient increases at about $10 < K < 15$. It is found that the drop and the increase in these coefficients are due to shedding vortices, which also exert forces perpendicular to the structure and the flow.

Figure 11.14 and Figure 11.15 show the inertia and drag coefficients for a rough cylinder, whose roughness is measured by k/D. Figure 11.14a shows a drop in the drag coefficient for Re between 10^4 and 10^5, and this is called the "drag crisis." For a larger Re, the drag coefficient stays constant. As the surface becomes rougher, the drop occurs at lower Re and the drag coefficients for the larger Re increases.

Figure 11.14 to Figure 11.16 can be used to obtain proper values of the drag and inertia coefficients for fluid with known Re, Keulegan–Carpenter number, and cylinder roughness.

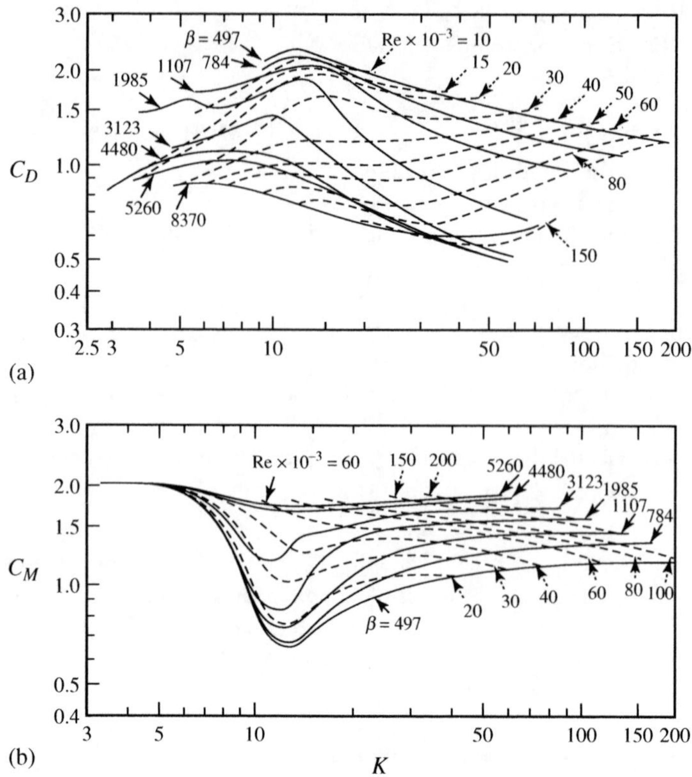

FIGURE 11.13 Drag and inertia coefficients as functions of K for various values of Re and β. (*Source*: Sarpkaya, 1976, *Proceedings of the Eighth Offshore Technology Conference.* With permission.)

11.2.3 Vortex-Induced Vibration

When the flow passes around a fixed cylinder, for a very low Re ($0 < \mathrm{Re} < 4$), the flow separates and reunites smoothly. When the Re is between 4 and 40, eddies are formed and are attached to the downstream side of cylinder. They are stable and there is no oscillation in the flow. For a flow with a Reynolds number greater than about 40, the fluid near the cylinder starts to oscillate due to shedding vortices. These shedding vortices exert an oscillatory force on the cylinder in the direction perpendicular to both the flow and the structure. The frequency of oscillation is related to the nondimensionalized parameter, the Strouhal number, defined by

$$St = \frac{f_v D}{U} \tag{11.31}$$

where f_v is the frequency of oscillation, U is the steady velocity of the flow, and D is the diameter of the cylinder. For circular cylinders, the Strouhal number stays roughly at 0.22 for laminar flow ($10^3 < \mathrm{Re} < 2 \times 10^5$) and 0.3 for turbulent flow (Patel, 1989).

The lift force due to these shedding vortices can be written as

$$f_L = \frac{1}{2} C_L \rho A_f U^2 \cos 2\pi f_v t \tag{11.32}$$

where C_L is the lift coefficient, which is also a function of Re, K, and the surface roughness. The experimental data of the lift coefficients show considerable scatter with typical values ranging from 0.25 to 1. For smooth cylinders, the lift coefficient approaches about 0.25 as Re and K increase.

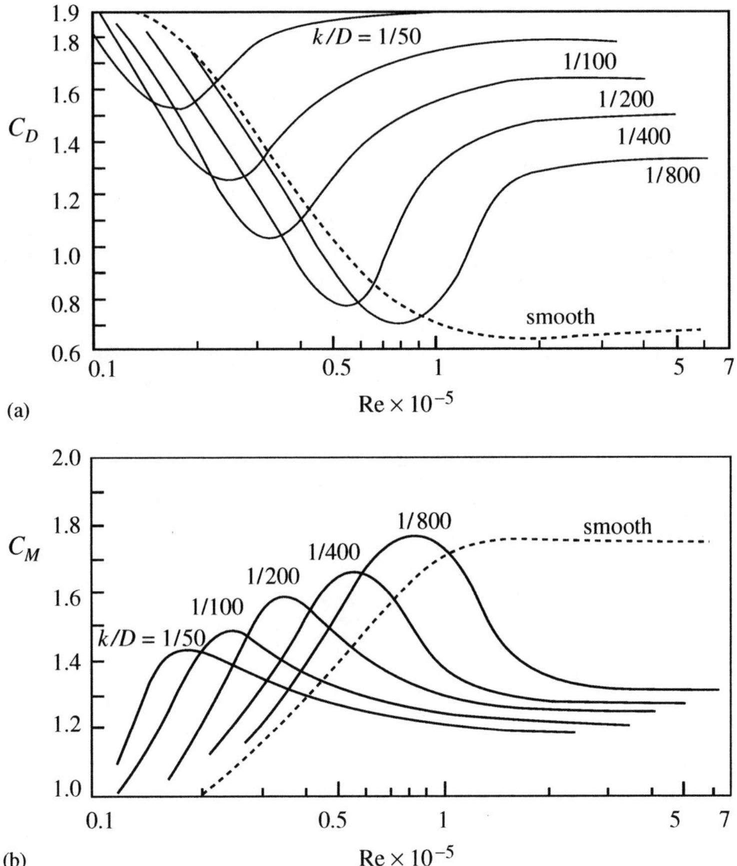

FIGURE 11.14 Drag and inertia coefficients for a rough cylinder as functions of Re for various values of cylinder roughness (as measured by *k/D*) for $K = 20$. (*Source*: Sarpkaya et al., 1977, *Proceedings of the Ninth Offshore Technology Conference*. With permission.)

It should be noted that the vortex forces are not generally correlated on the entire cylinder length. That is, the phase of the vortex shedding forces varies over the length. The correlation length — the length over which vortex shedding is synchronized — for a stationary cylinder is about three to seven diameters for laminar flow. If sectional forces are randomly phased, the net effect will be small. The total force on a cylinder of length L will be only a fraction of Lf_L. This fraction is called the joint acceptance and depends on the ratio of the correlation length to the total length.

When the flow passes by a cylinder that is free to vibrate, the shedding frequency is also controlled by the movement of the cylinder. When the shedding frequency is close to the first natural frequency of the cylinder (± 25 to 30% of the natural frequency [Sarpkaya and Isaacson, 1981]), the cylinder takes control of the vortex shedding. The vortices will shed at the natural frequency instead of at the frequency determined by the Strouhal number. This is called lock-in or synchronization, which is a result of nonlinear interaction between the oscillation of the body and the action of the fluid. Figure 11.16 shows the shedding frequency, as a function of flow velocity in the presence of a structure. f_1 and f_2 are the natural frequencies of the structure.

The amplitude of the structural response and the range of the fluid velocity over which the lock-in phenomenon persists are functions of a reduced damping parameter — the ratio of the damping force to the exciting force (Vandiver, 1985, 1993). If the reduced damping parameter is small, the lock-in can persist over a greater range of flow velocity.

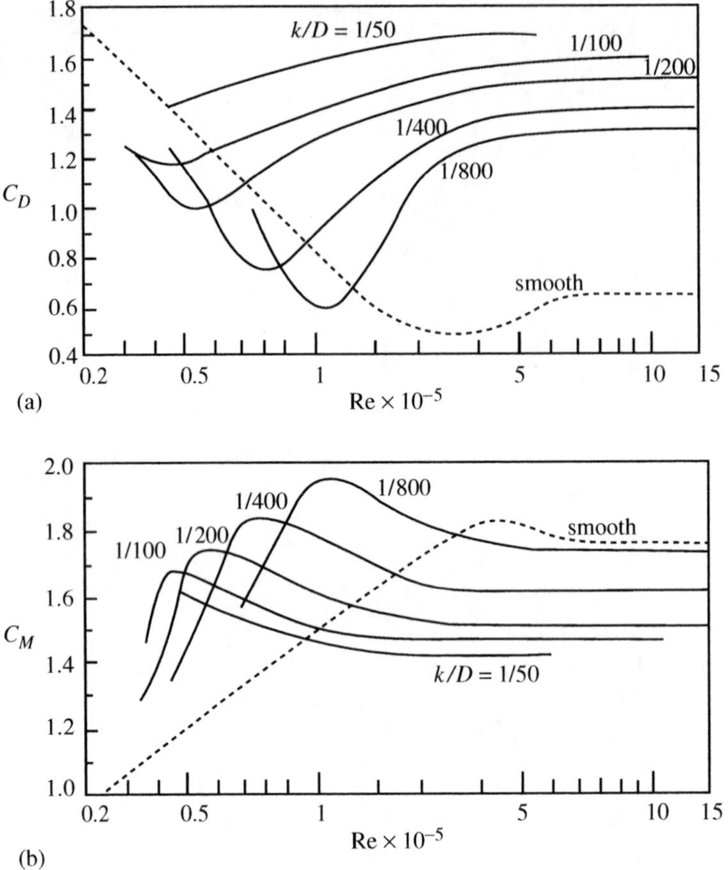

FIGURE 11.15 Drag and inertia coefficients for a rough cylinder as functions of *Re* for various values of cylinder roughness (as measured by *k/D*) for $K = 60$. (*Source*: Sarpkaya et al., 1977, *Proceedings of the Ninth Offshore Technology Conference*. With permission.)

The existing models for vortex-induced oscillation for a rigid cylinder include single-degree-of-freedom models and coupled models. The single-DoF models assume that the effect of vortex shedding is an external forcing function, which is not affected by the motion of the body. The coupled models assume that the equations that govern the motion of the structure and the lift coefficients are coupled so that the fluid and the structure affect each other (Billah, 1989).

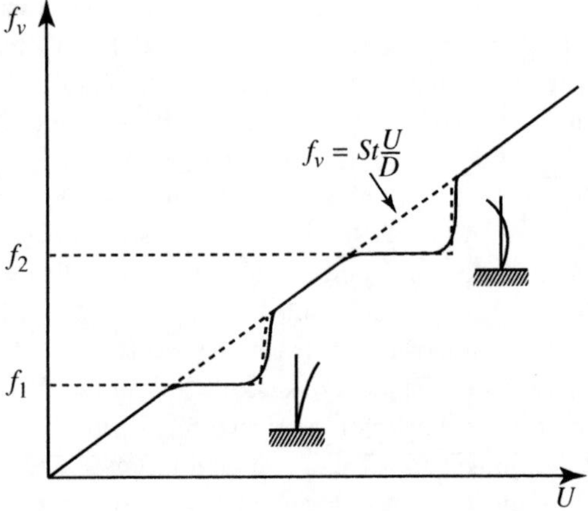

FIGURE 11.16 An example of fluid elastic resonance.

11.2.4 Summary

Some of the fluid forces are discussed briefly, and the regimes where inertia, drag, and diffraction forces are important are shown as functions of the ratio of the structural diameter to the wave length, D/λ, and the ratio of the wave height to the structural diameter, H/D. The wave forces on small structures are modeled by the Morison equation, and it is valid for $D/\lambda < 0.2$ and $H/D > 0.63$ or thereabouts. The Morison equation includes the effects of added mass, inertia, and drag. The added mass is simply

$$M_A = C_A M_{\text{disp}}$$

For a cylinder with transverse velocity, v, the normal and the tangential components of the drag and the inertia forces are given by

$$f^n = \frac{1}{2} C_D \rho D (w^n - v^n)|w^n - v^n| + C_M \rho \pi \frac{D^2}{4} \dot{w}^n$$

$$f^t = \frac{1}{2} C_T \rho D (w^t - v^t)|w^t - v^t|$$

The fluid coefficients are at least functions of three parameters: the Reynolds number, the Keulegan–Carpenter number, and the cylinder roughness. The plots of these coefficients are reproduced in Figure 11.13 to Figure 11.15.

The frequency of the lift force that is exerted by shedding vortices is closely related to the Strouhal number given by

$$St = \frac{f_v D}{U}$$

The lift force due to these shedding vortices can be written as

$$f_L = \frac{1}{2} C_L \rho A_f U^2 \cos 2\pi f_v t$$

If the structure is free to vibrate, then lock-in or synchronization may occur when the shedding frequency is close to the structure's natural frequency. The structure takes control of the vortex shedding. Many nonlinear models are available to capture this phenomenon.

11.3 Examples

Four examples are given in this section. The first example illustrates the roles of the normal and the tangential components of the drag force in the static configuration of a towing cable. The second example shows how the equation of motion of an articulated tower can be formulated in the presence of surrounding fluid. The third example shows how to choose a single significant wave height to represent a certain condition from significant wave height data over a long period of time. The final example shows how to reconstruct time series data from a given spectrum.

11.3.1 Static Configuration of a Towing Cable

For the purpose of ocean surveillance, oceanographic or geographic measurements, or ocean exploration, marine cables with instrument packages or Remotely Operated Vehicles are often towed behind ships or submarines. For example, the goal of the VENTS program by the National Oceanic and Atmospheric Administration (NOAA) is to conduct research on the impacts and consequences of submarine volcanoes and hydrothermal venting on the global ocean. In attempts to locate and map the distributions of hydrothermal plumes in the Mid-Ocean Ridge system, an instrument package called a CTD (Conductivity, Temperature and Depth Sensors) is towed behind a ship.

Let us consider a cable and a body towed behind a ship at a constant velocity with no current as shown in Figure 11.17. What kind of shape will the cable take? What will be the distance between the ship and the towed body?

We immediately recognize that this is equivalent to having a stationary ship with a steady current in the opposite direction. The equation of motion is given by

$$\sum \vec{F} = m\vec{a}(s,t) = \vec{0} = \frac{\partial}{\partial s}(T\vec{t}) + f^{n}\vec{n} + f^{t}\vec{t} + mg\vec{k}$$

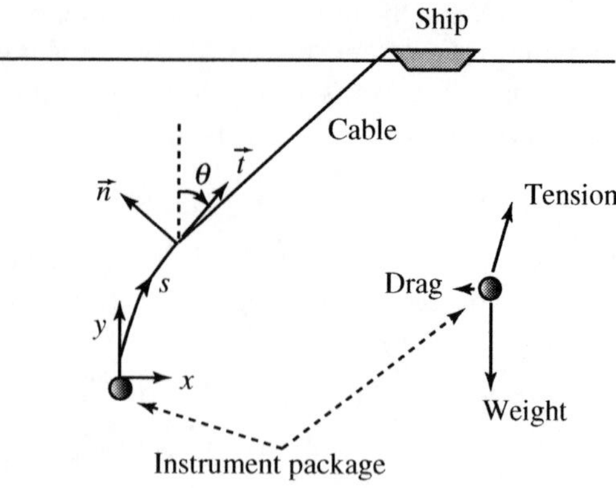

FIGURE 11.17 Towed system in equilibrium and the forces acting on the towed body.

where m is the mass of the cable per unit length, $\vec{a}(s,t)$ is the acceleration of the cable, s is the coordinate along the cable, T is the tension which is a function of s, $(\vec{t}, \vec{n}, \vec{b})$ is the set of unit vectors of the curvilinear coordinate system, $\vec{k}$ is the unit vector downward in the direction of gravity, g is the gravitational acceleration, f^{n} is the normal drag force, and f^{t} is the tangential drag force. The added mass and the inertial terms are zero because the fluid acceleration and the cable acceleration are zero. The normal and tangential drag forces are given in Equation 11.27 and Equation 11.29. In our case, they are given by

$$f^{n} = C_{D}\rho\frac{D}{2}U^{2}\cos^{2}\theta, \quad f^{n} = -C_{T}\rho\frac{D}{2}U^{2}\sin^{2}\theta$$

The corresponding scalar equations are given by

$$\begin{aligned} &\frac{dT}{ds} - C_{T}\rho\frac{D}{2}U^{2}\sin^{2}\theta - mg\cos\theta = 0 \\ &-T\frac{d\theta}{ds} + C_{D}\rho\frac{D}{2}U^{2}\cos^{2}\theta - mg\sin\theta = 0 \end{aligned} \tag{11.33}$$

where θ is the angle that the tangential vector makes with the vertical and measured positive clockwise. Note that we have used $\partial\vec{t}/\partial s = (-\partial\theta/\partial s)\vec{n}$ and $\vec{k} = -\cos\theta\vec{t} - \sin\theta\vec{n}$. Equation 11.33 shows that the tangential components of the external forces act to increase the tension, while the normal components cause the towline to bend. Because the normal component of the drag force is much larger than the tangential component, most of the fluid force is used to turn the cable.

From the force diagram (in Figure 11.17), the angle that the cable makes with the vertical where it is connected to the towed body is given by

$$T(0)\cos\theta(0) = W, \quad T(0)\sin\theta(0) = \text{Drag}$$

Once we know the weight and the drag force on the towed body, the tension and the angle at $s = 0$ can be found. If the drag is negligible compared with the weight, then the cable must be near vertical and the tension must be equal to the weight of the towed body at $s = 0$:

$$T(0) \approx W \quad \text{and} \quad \theta(0) \approx 0$$

For now, let us assume that this is the case. Then, with these initial conditions, the system of ordinary differential equations (Equation 11.33) can be solved numerically for $T(s)$ and $\theta(s)$. For example, even very simple finite difference equations will work. A set of equations

$$T_{i+1} = T_i + \left(mg \cos\theta_i - C_{\mathrm{T}}\rho \frac{D}{2} U^2 \sin^2\theta_i\right)\Delta s \tag{11.34}$$

$$\theta_{i+1} = \theta_i - \left(mg \sin\theta_i + C_{\mathrm{D}}\rho \frac{D}{2} U^2 \cos^2\theta_i\right)\Delta s/T_i \tag{11.35}$$

where $T_i = T(i\Delta s)$, are used here, and it works very well for $\Delta s = 0.05$.

The Cartesian coordinates, x and y, are related to θ by

$$\frac{\mathrm{d}x}{\mathrm{d}s} = \sin\theta \quad \text{and} \quad \frac{\mathrm{d}y}{\mathrm{d}s} = \cos\theta$$

and can also be obtained by integrating them numerically.

Figure 11.18 shows the results when $mg = 1.5$ N/m, $C_{\mathrm{D}}\rho DU^2/2 = 10$ N/m, $C_{\mathrm{T}}\rho DU^2/2 = 0.1$ N/m, $W = 100$ N, and the cable is 100 m long. Care is taken so that the ship is located at $x = 0$ and $y = 0$.

It is interesting to note that θ approaches a critical value, and the shape gradually becomes linear toward the ship. Mathematically, $\mathrm{d}\theta/\mathrm{d}s$ becomes zero. This is when the drag force is completely balanced by the normal component of the cable weight. The angle at which this occurs, θ_{cr}, can be obtained from the second governing equation and

$$mg \sin\theta_{\mathrm{cr}} = -f^{\mathrm{n}}, \quad \frac{\sin\theta_{\mathrm{cr}}}{\cos\theta_{\mathrm{cr}}} = C_{\mathrm{D}}\rho \frac{D}{2} U^2 \frac{1}{mg}$$

In our case, $\theta_{\mathrm{cr}} = 1.184$ rad, and this value agrees with Figure 11.18.

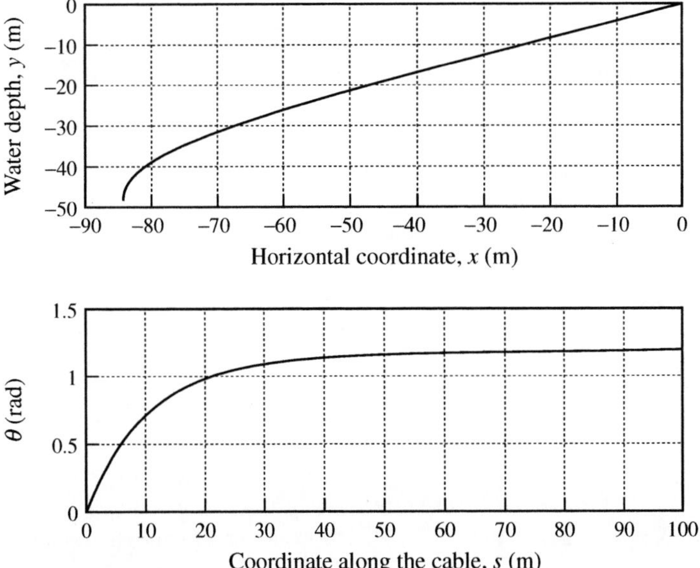

FIGURE 11.18 The equilibrium configuration of a towed cable and the angle that the cable makes with the vertical when $mg = 1.5$ N/m, $C_{\mathrm{D}}\rho DU^2/2 = 10$ N/m, $C_{\mathrm{T}}\rho DU^2/2 = 0.1$ N/m, and $W = 100$ N.

11.3.2 Fluid Forces on an Articulated Tower

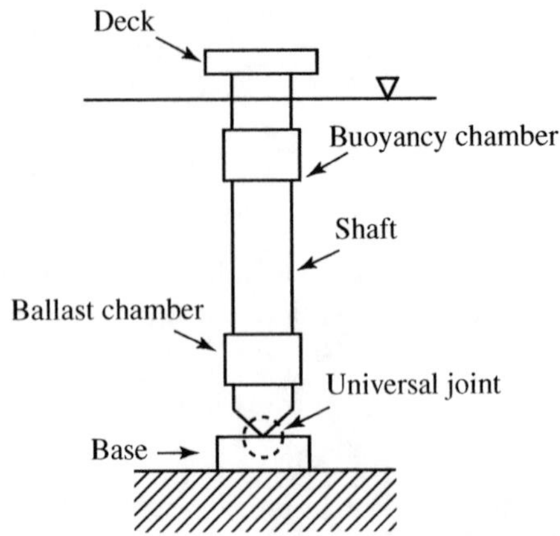

FIGURE 11.19 Schematic of an articulated tower.

Offshore structures are used in the oil industry as exploratory, production, oil storage, and oil landing facilities. They are designed to be self-supporting and sufficiently stable for offshore activities such as drilling and production of oil. An articulated tower as seen in Figure 11.19 is an example of an offshore platform that consists of a base, shaft, universal joint that connects the base and the shaft, ballast chamber, buoyancy chamber, and deck. The ballast chambers provide the extra weight so that the tower's bottom stays on the ocean floor, and the buoyancy chamber adds the necessary buoyancy so that the tower does not fall.

An articulated tower can be effectively modeled as a rigid inverted pendulum, where the deck is modeled as a point mass, the shaft as a uniform rigid bar, and the buoyancy chamber by a point buoyancy. In two dimensions, motion of the tower can be described with a single DoF (Chakrabarti and Cottor, 1979; Bar-Avi, 1996). The equation of motion in terms of the tower's deflection angle is obtained by summing the moment about the point O in Figure 11.20 and is given by

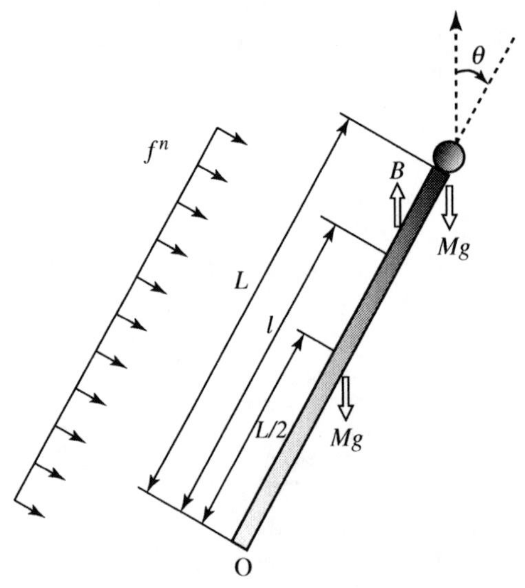

FIGURE 11.20 Free-body diagram.

$$I\frac{d^2\theta}{dt^2} = \sum M_O = mg\frac{L}{2}\sin\theta + MgL\sin\theta - Bl\sin\theta + \int_0^L f^n x\,dx$$

where I is the mass moment of inertia about the point O given by $I = mL^2/3 + ML^2$, m is the mass of the shaft, g is the gravitational acceleration, L is the length of the shaft, M is the point mass at the top, B is the buoyancy provided by the buoyancy chamber, l is its moment arm, f^n is the normal fluid force per unit length, and x is the coordinate along the shaft from O.

The fluid force per unit length in the normal direction is given by

$$f^n = C_D\rho\frac{D}{2}(w^n - v^n)|w^n - v^n| + C_M\rho\pi\frac{D^2}{4}\dot{w}^n - C_A\rho\pi\frac{D^2}{4}a^n$$

where the last term is the force in the normal direction due to the added mass. v^n and a^n are the velocity and the acceleration of the body in the normal direction and are given by

$$v^n = x\frac{d\theta}{dt} \quad \text{and} \quad a^n = x\frac{d^2\theta}{dt^2}$$

If we assume that the surrounding fluid is stationary, then the normal velocity and the acceleration of the fluid (w and $\dot{w}$) are zeros. Thus, the moment due to the fluid force is given by

$$\int_0^L f^{\mathrm{n}} x\,\mathrm{d}x = \int_0^L \left(-C_{\mathrm{D}}\rho \frac{D}{2} x^2 \left(\frac{\mathrm{d}\theta}{dt}\right)^2 \mathrm{sign}\left(\frac{\mathrm{d}\theta}{dt}\right) + C_{\mathrm{A}}\rho\pi \frac{D^2}{4} x \frac{\mathrm{d}^2\theta}{dt^2} \right) x\,\mathrm{d}x$$

$$= -C_{\mathrm{D}}\rho \frac{D}{2}\frac{L^4}{4}\left(\frac{\mathrm{d}\theta}{dt}\right)^2 \mathrm{sign}\left(\frac{\mathrm{d}\theta}{dt}\right) + C_{\mathrm{A}}\rho\pi \frac{D^2}{4}\frac{L^3}{3}\frac{\mathrm{d}^2\theta}{dt^2}$$

and the equation of motion is given by

$$\left(m\frac{L^2}{3} + ML^2 + C_{\mathrm{A}}\rho\pi \frac{D^2}{4}\frac{L^3}{3} \right) \frac{\mathrm{d}^2\theta}{dt^2} = \left(mg\frac{L}{2} + MgL - Bl_{\mathrm{b}} \right) \sin\theta - C_{\mathrm{D}}\rho \frac{D}{2}\frac{L^4}{4} \mathrm{sign}\left(\frac{\mathrm{d}\theta}{dt}\right)\left(\frac{\mathrm{d}\theta}{dt}\right)^2$$

Note that the normal fluid drag force adds directly to the restoring moment in the case of a rigid bar. The equation of motion can be solved numerically once the initial conditions ($\theta[0]$ and $\mathrm{d}\theta/dt[0]$) are given.

The equation of motion can be simplified if we assume that the angle of rotation θ is small. More specifically, if we assume that θ^2 is negligible when compared with 1, then we find that[2]

$$\sin\theta \approx \theta$$

The equation of motion can be simplified to

$$\left(m\frac{L^2}{3} + ML^2 + C_{\mathrm{A}}\rho\pi \frac{D^2}{4}\frac{L^3}{3} \right) \frac{\mathrm{d}^2\theta}{dt^2} - \left(mg\frac{L}{2} + MgL - Bl_{\mathrm{b}} \right) \theta + C_{\mathrm{D}}\rho \frac{D}{2}\frac{L^4}{4} \mathrm{sign}\left(\frac{\mathrm{d}\theta}{dt}\right)\left(\frac{\mathrm{d}\theta}{dt}\right)^2 = 0$$

which resembles the equation for a linear oscillator with a nonlinear damping term. Note that the system becomes unstable when the stiffness term (the coefficient of θ) becomes negative. This occurs when the buoyancy is not sufficient or

$$B < \frac{1}{l_{\mathrm{b}}}\left(mg\frac{L}{2} + MgL \right)$$

11.3.3 Distribution of Significant Wave Heights — Weibull and Gumbel Distributions

The National Buoy Data Center (NBDC) run by NOAA collects ocean data such as wind, current, wave, pressure, and temperature data in various locations and the records are made public. Let us say that we are to design an articulated tower (in Section 11.3.2) in one of these locations where the data are available. The first task is to characterize the environment. Using all of the information that is collected is inefficient and impractical. Instead, we are interested in choosing a single number that can represent typical and extreme situations such as 10- and 50-year storms. For now, let us only consider random waves. We are then interested in finding the significant wave heights representing 10- and 50-year storms.

From NBDC data for a buoy outside Monterey Bay, the number of occurrences for ranges of significant wave heights is constructed in Table 11.2. The measurements were taken every hour for about 12 years. We first construct the corresponding Weibull distribution using the method described in Section 11.1.6. We first guess γ so that a pair of $\ln(-\ln\{1 - F(h)\})$ and $\ln(h - \gamma)$ form a

[2]This is called the small angle assumption.

TABLE 11.2 Number of Occurrences of Various Sea States

Significant Wave Height, h (m)	Number of Occurrences	Sum
<1	2,367	2,367
1–2	46,353	48,720
2–3	3,4285	83,005
3–4	1,3181	96,186
4–5	3,813	99,999
5–6	716	100,715
6–7	145	100,860
7–8	32	100,892
8–9	8	100,900
9–10	2	100,902
Total	100,902	

straight line. Figure 11.21 shows that the pair yields nearly a straight line when $\gamma \approx 0.84$. The slope and the y intercept of this line are 1.6 and -0.78, respectively. The Weibull parameters are then $m = 1.6$ and $\beta = 1.6$.

Similarly, we can find the corresponding Gumbel probability density function by plotting pairs of $(h, \ln(-\ln\{F(h)\}))$ to form a line. For the data shown in Table 11.2, the line has a slope of -1.52 and y intercept of 2.84 so that $\alpha = -1.52$ and $\beta = 1.87$.

Figure 11.22 shows the Weibull probability density and the cumulative distribution (Equation 11.17) in solid lines, the Gumbel probability density and the cumulative distribution in dotted lines (Equation 11.18), and the discrete probability density and the cumulative distribution derived from Table 11.2 in symbols.

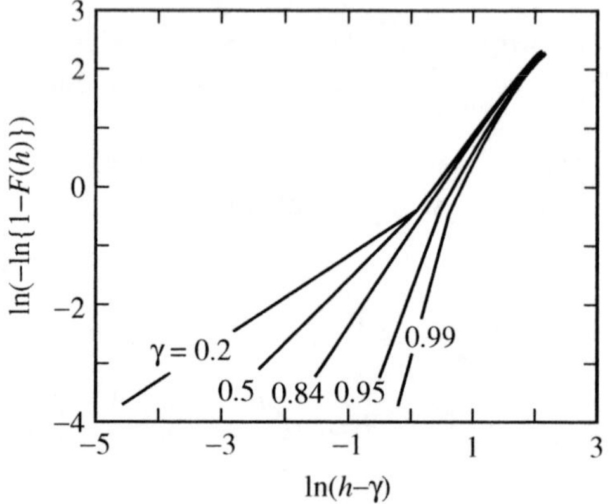

FIGURE 11.21 Plots of $(\ln(h - \gamma), \ln[-\ln\{1 - F(h)\}])$ for various values of γ.

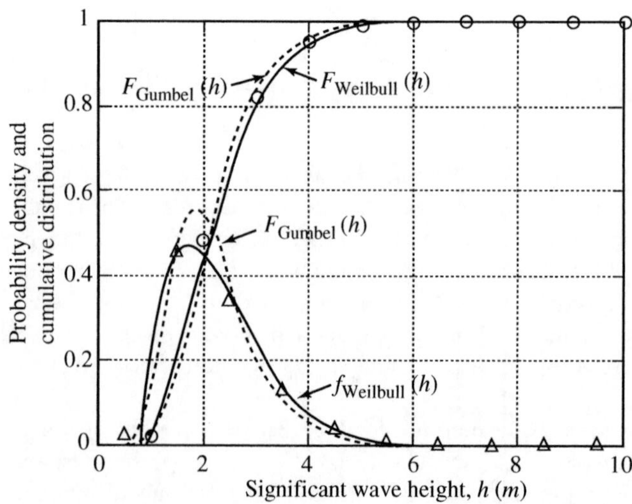

FIGURE 11.22 Weibull approximations of the probability density and cumulative distribution of significant wave heights measured in the outer Monterey Bay area. The symbols are the values given in Table 11.2.

TABLE 11.3 Comparison of Representative Significant Wave Heights for Long-Term Predictions from Gumbel and Weibull Distributions

	5-Year (m)	10-Year (m)	50-Year (m)
Weibull	7.84	8.15	8.79
Gumbel	8.83	9.33	10.4

The next step is to find a significant wave height that can represent an N-year storm, h_N. The probability that we will not have an N-year storm in any given year is $1 - 1/N$ and is equivalent to the probability that the significant wave height will not exceed h_N in the same year. The probability that $h < h_N$ in a single measurement is $F(h_N)$, and the probability that $h < h_N$ in every measurement taken in a year is $F(h_N)^{24\times365}$. Then, we have

$$1 - \frac{1}{N} = F(h_N)^{24\times365}$$

Table 11.3 shows significant wave heights that represent 5-, 10-, and 50-year storms obtained using the Weibull and Gumbel distributions.

The Gumbel probability distribution gives higher significant wave heights. For this particular set of data, the Weibull distribution seems to fit the data better (Figure 11.22), and the Weibull distribution is the most often used distribution in the offshore industry.

11.3.4 Reconstructing Time Series for a Given Significant Wave Height

Previously, we found significant wave heights that could represent 5-, 10-, and 25-year storms for a given site. Recall that the significant wave height can entirely characterize the Pierson–Moskowitz spectra. Once the spectral density is determined, a sample time history of the wave profile, $\eta(t)$, can be determined using either Borgman's or Shinozuka's method (Section 11.1.4). Here, Shinozuka's method is used to generate the random wave elevations.

Let us first find the random frequencies distributed according to $S^{\text{o}}_{\eta\eta}(\omega)/\sigma^2_\eta$. The P–M spectrum in terms of the significant wave height is given by Equation 11.8.

$$S^{\text{o}}_{\eta\eta}(\omega) = 0.7795\omega^{-5}\exp\left(-\frac{3.118}{H_s^2}\omega^{-4}\right)$$

The variance is given by

$$\sigma^2_\eta = \int_0^\infty S^{\text{o}}_{\eta\eta}(\omega)\mathrm{d}\omega = \frac{H_s^2}{16}$$

The probability density and the cumulative distribution functions are given by

$$f(\omega) = S^{\text{o}}_{\eta\eta}(\omega)/\sigma^2_\eta = \frac{12.472}{H_s^2}\omega^{-5}\exp\left(-\frac{3.118}{H_s^2}\omega^{-4}\right), \quad F(\omega) = 1 - \exp\left(-\frac{3.118}{H_s^2}\omega^{-4}\right)$$

The inverse of the cumulative distribution function is given by

$$F^{-1}(x) = \left(-\frac{H_s^2}{3.118}\ln(1-x)\right)^{-1/4}$$

The random frequencies distributed according to $f(\omega)$ can be obtained from uniformly distributed random numbers x from 0 and 1. Table 11.4 shows uniform random numbers between

TABLE 11.4 Generation of Random Frequencies Distributed According to $f(\omega)$ from Uniform Random Numbers

Uniform Random Numbers $0 < x < 1$	Random Frequencies ω Distributed According to $f(\omega)$
0.950	$(-19.713\ln[1-0.950])^{-1/4} = 0.360$
0.231	$(-19.713\ln[1-0.231])^{-1/4} = 0.662$
0.606	$(-19.713\ln[1-0.606])^{-1/4} = 0.483$
⋮	⋮

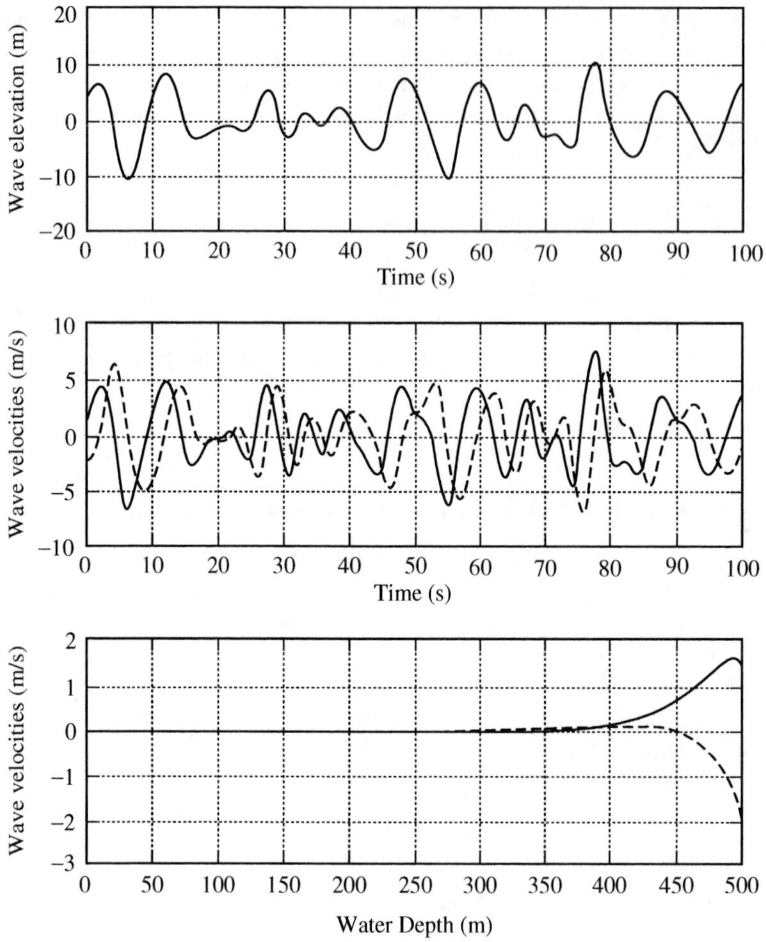

FIGURE 11.23 Wave elevation and velocities.

0 and 1 and the random frequencies distributed according to $f(\omega)^3$. The significant wave height of 7.84 m is used.

We can obtain 100 in this way, and the wave elevation is also obtained using Equation 11.13. The random phase φ_i is obtained by multiplying uniform random numbers (different from the ones used to generate the random frequencies) by 2π.

Figure 11.23 shows the surface elevation as a function of time, the corresponding wave velocities at the water surface (Section 11.1.7) as functions of time, and the wave velocities at $t = 0$ as functions of the water depth. Note that the wave velocities decay with depth.

[3]The uniform random numbers can be generated by the MATLAB® rand function.

11.3.5 Available Numerical Codes

Many numerical codes are available for modeling the dynamics of slender structures such as risers, tether, umbilicals, and mooring lines. The first example in this section was solved by a numerical code, WHOI Cable, developed at Woods Hole Oceanographic Institution. WHOI Cable is a time-domain program that can be used for analyzing the dynamics of towed and moored cable systems in both two and three dimensions. It takes into account bending and torsion as well as extension.

Comparative studies investigating flexible risers were carried out by ISSC Committee V7 from computer programs developed by 11 different institutions in the period between 1988 and 1991, and the results were reported by Larsen (1992). More recently, Brown and Mavrakos (1999) conducted a comparative study on the dynamic analysis of suspended wire and chain mooring lines and reported results from 15 different numerical codes. The participants included engineering consultancies, and academic and research institutions involved in marine technology. Some of the time-domain programs that were included in the comparative study are MODEX by Chalmers University of Technology, FLEXAN-C by Institute Francais du Petrole, DYWFLX95 by MARIN, R.FLEX by MARINTEK, CABLEDYN by National Technical University of Athens, DMOOR by Noble Denton Consultancy Services Ltd, V.ORCAFLEX by Orcina Ltd Consulting Engineers, ANFLEX by Petrobras SA, TDMOOR-DYN by University College London, FLEXRISER by Zentech International. Some of these programs are available to academic institutions and government laboratories at no cost.

Acknowledgments

The author wishes to express gratitude for the funding from the Woods Hole Oceanographic Institution and the Department of Mechanical Engineering at Texas Tech University.

References

Bar-Avi, P. 1996. Dynamic response of an offshore articulated tower, Ph.D. thesis, The State University of New Jersey, Rutgers, May 1996.

Billah, K. 1989. A study of vortex induced vibration, Ph.D. thesis, Princeton University, May 1989.

Borgman, L., Ocean wave simulation for engineering design, *J. Waterway Harbors Div.*, 95, 557–583, 1969.

Bretschneider, C. 1959. *Wave variability and wave spectra for wind-generated gravity waves.* Technical Memorandum No.118, Beach Erosion Board, U.S. Army Corps of Engineers, Washington D.C.

Bretschneider, C. 1969. Wave forecasting. In *Handbook of Ocean and Underwater Engineering*, Ed. J.J. Myers, McGraw-Hill, New York.

Brown, D. and Mavrakos, S. Comparative study on mooring line dynamic loading, *Marine Structures*, 12, 131–151, 1999.

Chakrabarti, S.K. 1987. *Hydrodynamics of Offshore Structures*, Computational Mechanics Publications, Southampton, U.K.

Chakrabarti, S.K. and Cottor, D., Motion analysis of articulated tower, *J. Waterway Port Coast. Ocean Div.*, 105, 281–292, 1979.

Faltinsen, O.M. 1993. *Sea Loads on Ships and Offshore Structures*, Cambridge University Press, Cambridge.

Gumbel, E. 1958. *Statistics of Extremes*, Columbia University Press, New York.

Hasselmann, K., Barnett, T., Bouws, E., Carlson, H., Cartwright, D., Enke, K., Ewing, J., Gienapp, H., Hasselmann, D., Kruseman, P., Meerburg, A., Muller, P., Olbers, D., Richter, K., Sell, W., and Walden, H. 1973. Measurement of wind-wave growth and swell decay during the joint North Sea wave project (JONSWAP), *Technical Report 13 A*. Deutschen Hydrographischen Zeitschrift.

Hogben, N. 1976. Wave loads on structures, *Behavior of Offshore Structures (BOSS)*,Oslo, Norway.

Isaacson, M., Wave and current forces on fixed offshore structures, *Can. J. Civil Eng.*, 15, 937–947, 1988.

Jasper, N., Statistical distribution patterns of ocean waves and of wave induced ship stresses and motions with engineering applications, *Trans. Soc. Nav. Arch. Mar. Engrs*, 64, 375–432, 1954.

Khinchine, A., Korrelations theorie der stationaren stochastischen prozesse, *Math. Ann.*, 109, 604–615, 1934.

Kinsman, B. 1965. *Wind Waves*, Prentice Hall, Englewood Cliffs, NJ.

Lamb, H. 1945. *Hydrodynamics*, 6th Ed., Cambridge University Press, New York.

Larsen, C. Flexible riser analysis — comparison of results from computer programs, *Marine Structure*, 5, 103–119, 1992.

LeMehaute, B. 1976. *Introduction to Hydrodynamics and Water Waves*, Springer-Verlag, New York.

Longuet-Higgins, M., On the statistical distribution of the height of sea waves, *J. Mar. Res.*, 11, 3, 245–266, 1952.

Miller, B., Wave slamming loads on horizontal circular elements of offshore structures, *Nav. Arch.*, 3, 81–98, 1977.

Miller, B. 1980. Wave slamming on offshore structures, *Technical Report No. NMI-R81*. National Maritime Institute.

Morison, J., O'Brien, M., Johnson, J., and Schaaf, S., The force exerted by surface waves on piles, *Pet. Trans., AIME*, 189, 149–157, 1950.

Ochi, M., On prediction of extreme values, *J. Ship Res.*, 17, 29–37, 1973.

Ochi, M. and Hubble, E. 1976. Six parameter wave spectra. ASCE pp. 301–328. *Proceedings of the Fifteenth Coastal Engineering Conference*, Honolulu, HI.

Patel, M. 1989. *Dynamics of Offshore Structures*, Butterworths, London.

Pierson, W. and Moskowitz, L., A proposed spectral form for fully developed wind seas based on the similarity theory of S.A. Kitaigorodskii, *J. Geophys. Res.*, 69, 24, 5181–5203, 1964.

Rice, S.O. 1954. Mathematical analysis of random noise. In *Selected Papers on Noise and Stochastic Processes*, N. Wax, Ed., Dover Publications, New York.

Sarpkaya, T. 1976. In-line and transverse forces on cylinders in oscillating flow at high Reynolds numbers, OTC 2533, pp. 95–108. *Proceedings of the Eighth Offshore Technology Conference*, Houston, TX.

Sarpkaya, T., Collins, and N., Evans, S. 1977. Wave forces on rough-walled cylinders at high Reynolds numbers, OTC 2901, pp. 175–184. In *Proceedings of the Ninth Offshore Technology Conference*, Houston, TX.

Sarpkaya, T., and Isaacson, M. 1981. *Mechanics of Wave Forces on Offshore Structures*, Van Nostrand Reihold, New York.

Shinozuka, M., Monte Carlo Solution of structural dynamics, *Comput. Struct.*, 2, 855–874, 1972.

Sverdrup, H., and Munk, W. 1947. Wind, sea, and swell: theory of relations for forecasting, *Technical Report 601*. U.S. Navy Hydrographic Office.

Tucker, M. 1991. *Waves in Ocean Engineering: Measurements, Analysis, and Interpretation*, Ellis Horwood, Chichester, U.K.

Vandiver, J. 1985. Prediction of lockin vibration on flexible cylinders in sheared flow, May 1985. In *Proceedings of the 1985 Offshore Technology Conference*, Paper No. 5006, Houston, TX.

Vandiver, J., Dimensionless parameters important to the prediction of vortex-induced vibration of long, flexible cylinders in ocean currents, *J. Fluids Struct.*, 7, 5, 423–455, 1993.

Wiener, N., Generalized harmonic analysis, *Acta Math.*, 55, 117–258, 1930.

Wilson, J. 1984. *Dynamics of Offshore Structures*, Wiley, New York.

12

Sound Levels and Decibels

S. Akishita
Ritsumeikan University

Summary

In this chapter, the basic characteristics of sound and sound propagation are described. Levels and decibels, which represent the magnitude of sound waves, are defined and explained.

12.1 Introduction

Sound is related to vibration, and is described as a propagating perturbation through a fluid, which is air or water in most cases. A very wide variety of noise sources exists. Each source is peculiar to its generation mechanism, which may cover a wide range of phenomena including fluid mechanics and the vibration of structures. Sound is perceived by the ear of the listener as a pressure wave superimposed upon the ambient air pressure. The *sound pressure* is the incremental variation about the ambient atmospheric pressure. Generally, it is detected by a microphone and expressed as oscillatory electric signal output from an audio measurement instrument. We shall present a mathematical description of these pressure waves that are known as sound. The field of acoustics concerns sound and vibration, and is treated in Chapter 12 to Chapter 20 of this book.

12.2 Sound Wave Characteristics

The characteristics of a sound wave are described by a pressure oscillation of a pure tone. A "pure tone" is a sinusoidal pressure wave of a specific frequency and amplitude, propagating at a velocity determined by the temperature and pressure of the medium (air).

Let us consider a hypothetical sound field in a duct with constant cross-sectional area, as shown in Figure 12.1a. A reciprocating piston at the left end emits the sound wave and it propagates toward the right-side end along the indicated axis. It is detected by a microphone at the right end. Figure 12.1b shows the instantaneous pressure distribution in a duct at time $t = t_0$. Figure 12.1c shows the pressure variation of the time history detected by the microphone at $x = x_0$.

The wavelength, λ, is the distance between successive two peaks in the waveform in Figure 12.1b. Wavelength is related to the frequency, f, and the velocity of wave propagation, c, by

$$\lambda = \frac{c}{f} \quad \text{(ft or m)} \tag{12.1}$$

The period, T, of the sinusoidal wave is the time interval required for one complete cycle, as depicted in Figure 12.1b. The period, T, is related to the frequency, f, by

$$T = \frac{1}{f} \quad \text{(s)} \tag{12.2}$$

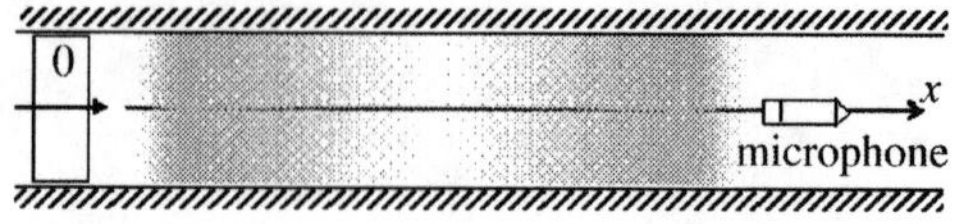

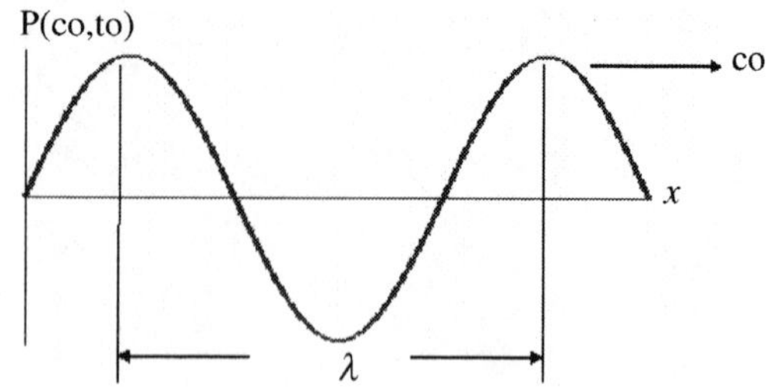

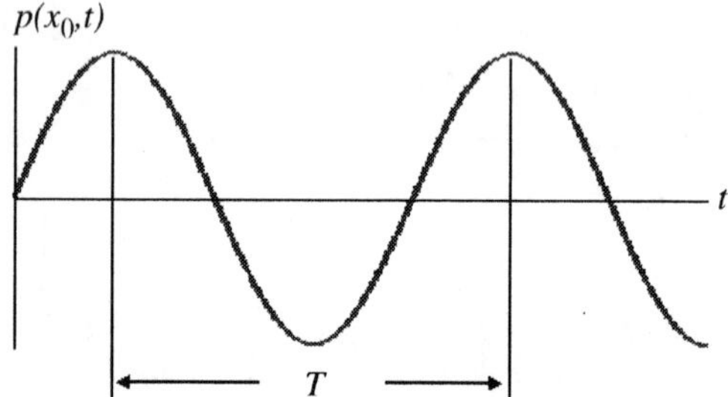

FIGURE 12.1 (a) Propagating sound wave in a duct; (b) instantaneous pressure distribution; (c) pressure variation in time history detected by a microphone at $x = x_0$.

12.2.1 Velocity of Sound

The velocity of sound is identical to the velocity of wave propagation, c, and in air it is given by

$$c = \sqrt{\frac{\gamma p_0}{\rho}} \quad \text{(ft/s or m/s)} \tag{12.3}$$

where γ denotes the ratio of specific heat, p_0 denotes the ambient or equilibrium pressure, and ρ denotes the ambient or equilibrium density. For air, γ is taken as 1.4. Equation 12.3 then becomes

$$c = \sqrt{\frac{1.4 p_0}{\rho}} \quad \text{(ft/s or m/s)} \tag{12.4}$$

which can be further simplified by the fact that the ratio p_0/ρ is related to the temperature of the gas. On assuming that the air behaves virtually as an ideal gas, the velocity, c, is related to the absolute temperature in degrees Kelvin (K) by

$$c = 20.05\sqrt{T} \quad \text{(m/s)} \tag{12.5}$$

where T, the temperature in degrees Kelvin, is

$$T = 273.2^\circ + (^\circ\text{C})\ \text{K} \tag{12.6}$$

Example 12.1

Calculate the velocity of sound, c, giving the temperature of 15°C.

Solution

$T = 273.2^\circ + 15^\circ = 288.2$ K, then
$c = 20.05\sqrt{288.2} = 340.4$ m/s

is obtained. This value means a typical velocity of sound in the air.

12.3 Levels and Decibels

Sound pressure and power are commonly expressed in terms of *decibel levels*. This allows us to use a logarithmic rather than a linear scale. It provides the distinct advantage of allowing accurate computations using small numerical values, while accommodating a wide range of numerical values.

12.3.1 Sound Power Level

Sound power level describes the acoustical power radiated by a given source with respect to the international reference of 10^{-12} W. The sound power level, L_W, is defined as

$$L_W = 10 \log\left(\frac{W}{W_{re}}\right) \quad \text{(dB)} \tag{12.7}$$

where W denotes sound power in question and $W_{re} = 10^{-12}$ W (reference).

Example 12.2

Determine the sound power level of a small ventilation fan that generates 10 W of sound power.

Solution

$$L_W = 10 \log\left(\frac{W}{W_{re}}\right) = 10 \log\left(\frac{10}{10^{-12}}\right) = 130 \text{ dB}$$

12.3.2 Sound Pressure Level

Sound pressure levels are expressed in decibels, as are sound power levels. The sound pressure level, L_p, is defined as

$$L_p = 10 \log\left(\frac{\bar{p}^2}{p_{re}^2}\right) = 20 \log\left(\frac{\bar{p}}{p_{re}}\right) \quad \text{(dB)} \tag{12.8}$$

where $\bar{p}$ denotes root-mean-square (RMS) sound pressure in question Pa or N/m^2 and $p_{re} = 20 \times 10^{-6}$ Pa $= 0.0002$ μbar. The pressure of 20×10^{-6} Pa has been chosen as a reference because it has been found that the average young adult can perceive a 10^3 Hz tone at this pressure. This reference is often referred to as the threshold of hearing at 10^3 Hz.

Example 12.3

Giving $L_p = 50$ dB for the Aeolian tone of 200 Hz, determine the RMS pressure of the tone.

Solution

Given L_p as 50 dB, then $\bar{p}$ is determined by using Equation 12.8.

$$50 = 20 \log\left(\frac{\bar{p}}{p_{re}}\right)$$

then

$$\bar{p} = 10^{50/20}\, p_{re} = 316.2 p_{re}$$

$$\bar{p} = 6.32 \times 10^{-3} \text{ Pa} = 0.0632 \ \mu\text{bar}$$

Note that this value is very small, contradicting the magnitude of the sensory impression of the human ear.

12.3.3 Overall Sound Pressure Level

The sound pressure level is defined assuming "pure tone" sound. However, practically any real sound contains various components of pure tone sound. Let us consider a set of n components of pure tone, denoted by

$$\begin{cases} p_1(t) = a_1 \sin(2\pi f_1 t + \phi_1) \\ \vdots \\ p_n(t) = a_n \sin(2\pi f_n t + \phi_n) \end{cases} \tag{12.9}$$

$$p(t) = p_1(t) + \cdots + p_n(t) \tag{12.10}$$

If L_p of $p(t)$ is evaluated in RMS pressure, $\bar{p}$, we have

$$\bar{p} = \left[\lim_{T\to\infty} \frac{1}{T}\int_0^T p^2(t)\mathrm{d}t\right]^{1/2} = \left[\lim_{T\to\infty} \frac{1}{T}\int_0^T (p_1(t) + \cdots + p_n(t))^2\mathrm{d}t\right]^{1/2} \tag{12.11}$$

Since

$$\lim_{T\to\infty} \frac{1}{T}\int_0^T p_i(t)p_j(t)\mathrm{d}t = 0, \quad i \neq j$$

is valid, $\bar{p}$ is obtained as

$$\bar{p} = \left[\overline{p_1(t)^2} + \cdots + \overline{p_n(t)^2}\right]^{1/2} = \left[\bar{p}_1^2 + \cdots + \bar{p}_n^2\right]^{1/2} \tag{12.12}$$

where

$$\bar{p}_i^2 \equiv \overline{p_i(t)^2} \equiv \lim_{T\to\infty} \frac{1}{T}\int_0^T p_i^2(t)\mathrm{d}t = \frac{1}{2}a_i^2 \tag{12.13}$$

Let us define $L_{pi} \equiv 10\log(\bar{p}_i^2/p_{re}^2)$ $(i = 1, 2, \ldots, n)$. Then the overall sound pressure level, L_p, of $p(t)$ is expressed by

$$L_p \equiv 20\log\frac{\bar{p}}{p_{re}} = 10\log\frac{1}{p_{re}^2}(\bar{p}_1^2 + \cdots + \bar{p}_n^2)$$

or L_p is expressed by L_{pi} $(i = 1, 2, \ldots, n)$ as follows:

$$L_p = 10\log(10^{L_{p1}/10} + 10^{L_{p2}/10} + \cdots + 10^{L_{pn}/10}) \tag{12.14}$$

Example 12.4

Determine the overall sound pressure level of the combination of three pure tones, the sound pressure levels of which are expressed by

$L_{p1} = 60$ dB ($f_1 = 250$ Hz), $L_{p2} = 65$ dB ($f_2 = 500$ Hz), $L_{p3} = 55$ dB ($f_3 = 1000$ Hz)

Solution

We have $10^{L_{p1}/10} = 10^6$, $10^{L_{p2}/10} = 10^{6.5}$, and $10^{L_{p3}/10} = 10^{5.5}$. Then the overall level, L_p, is determined by using Equation 12.14 as follows:

$$L_p = 10\log(10^6 + 10^{0.5}\times 10^6 + 10^{-0.5}\times 10^6)$$

$$= 10\log 10^6(1 + 10^{0.5} + 10^{-0.5}) = 60 + 10\log 4.479 = 66.5 \text{ dB}$$

Note that the sum of 65, 60 and 55 dB is just 66.5 dB.

13

Hearing and Psychological Effects

S. Akishita
Ritsumeikan University

Summary

In this chapter, first the characteristics of human hearing are discussed, including a brief description of the anatomy and function of the hearing mechanisms. Next, the frequency and loudness responses of the human hearing are explained, and then hearing loss causing permanent damage is described. Finally, the psychological response to noise is discussed by defining the indices, loudness (sones), noise-criteria curves, and sound level.

13.1 Introduction

This chapter considers the characteristics of human hearing. After a brief description of the anatomy and function of the hearing mechanism, those aspects of hearing that are important in noise control are discussed. The perception of sound by the human ear is a complicated process, dependent both on the frequency and pressure amplitude of the sound. We shall consider the structure of the ear and hearing mechanism. We will also briefly discuss various means of measuring the psychological effects of noise.

13.2 Structure and Function of the Ear [1]

The main components of the human ear are depicted in Figure 13.1(a). The ear is commonly divided into three main components: (1) the outer ear, (2) the middle ear, and (3) the inner ear.

The visible portion of the ear is called the *pinna*. Because of its small size compared with the primary wavelengths that we hear, the pinna serves only to produce a small enhancement of the sounds that arrive from the front of the listener as compared to those which arrive from behind; that is, the human sound reception system has a small frontal directivity. The remainder of the outer ear, which consists of the ear canal terminated in the ear drum, forms a resonant cavity at about 3 kHz. This resonant or near-resonant condition allows for a nearly reflection-free termination of the *ear canal* and thus a good impedance match of the *ear drum* to the air in which the sound wave was propagated.

The middle ear consists of three small *ear bones*, the hammer, anvil, and stirrup. The middle ear serves as an impedance transformer, which matches the low impedance of the air in which sound travels and in

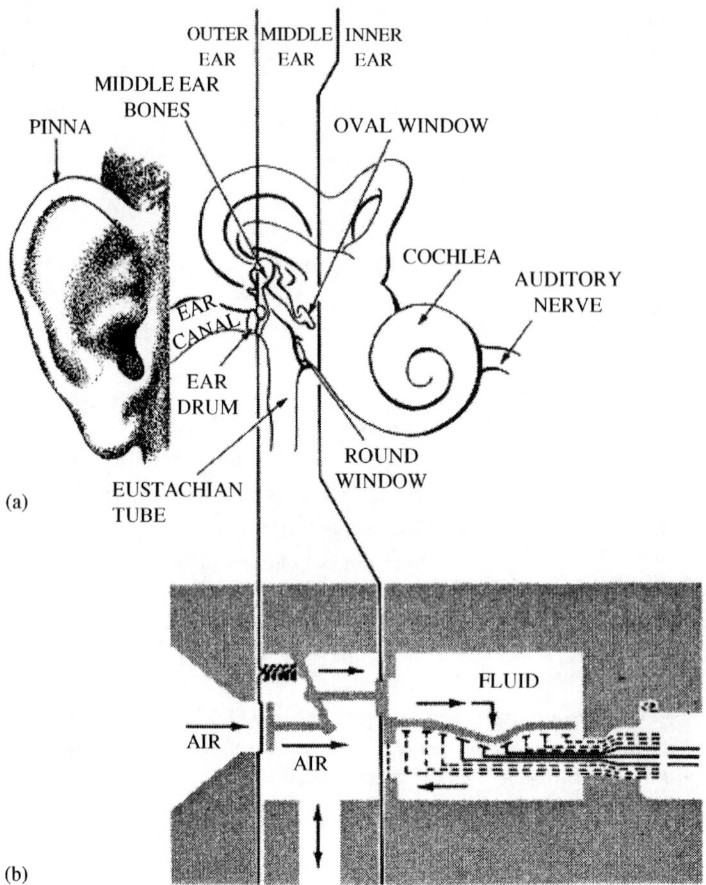

FIGURE 13.1 (a) Main components of the human ear; (b) functional diagram of the ear.

which the ear drum is located to the high impedance of the lymphatic fluid of the *cochlea* beyond the *oval window.* Without this impedance-matching transformation, a mismatch would occur, resulting in a loss of approximately 30 dB.

In the inner ear, the cochlea is the main component where the actual reception of sound takes place. The schematic extended structure of the cochlea is depicted in Figure 13.1(b). The cochlea, which is located in extremely hard temporal bone, is divided almost its entire length by the *basilar membrane.* At the end of the cochlea, the two canals are connected by the *helicotrema*, which allows for the flow of the *lymphatic fluid* between the two sections. The basilar membrane, which is about 3 cm long and 0.02 cm wide, has about 24,000 nerve ends terminated in *hair cells* located on the membrane. The motion of the oval window is transmitted to the basilar membrane and its associated sensing cells. This motion is sensed as sound.

13.3 Frequency and Loudness Response

The threshold of hearing, defined for binaural listening, is that sound pressure in the free field which one can just still hear as the signal is decreased. The threshold of hearing, for what is considered normal hearing, is shown in Figure 13.2. As seen from the curve, human hearing is most sensitive in the range of 2000 to 5000 Hz; furthermore, we note that the response in this range is very close to 0 dB, or 20×10^{-6} Pa. At the other end of the scale, there is the threshold of pain, which is usually taken as about 135 to 140 dB. Thus, there is a dynamic range of normal hearing of approximately 140 dB.

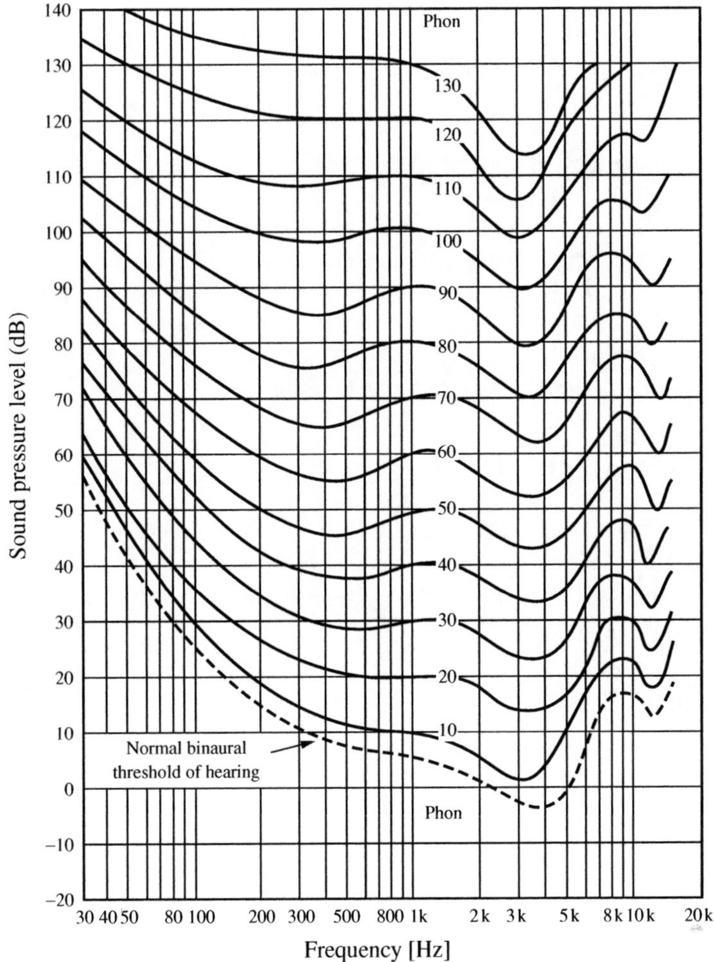

FIGURE 13.2 Equal-loudness contour for free-field binaural listening.

One also readily notes from the curve of Figure 13.2 that the threshold of hearing is a function of frequency. For example, with normal hearing, one would just be able to hear a 2000 Hz tone at a 0 dB level. However, one would require a pressure level of about 15 dB to be able to barely hear a 200 Hz tone. Thus, in describing the subjective loudness of sound, it is necessary to consider the characteristics of the human ear. This concept of loudness is quantized by the loudness level.

The *loudness* level of a particular sound is determined by the subjective comparison of the loudness of the sound to that of a 1000 Hz pure tone. The level, measured in *phons*, is equal numerically to the sound pressure level, in dB, of the 1000 Hz tone, which was regarded to be of equal loudness. A set of internationally standardized equal loudness contours is plotted in Figure 13.2. In keeping with the definition of loudness level, note that at 1000 Hz, all the equal loudness contours are equal in phons to the sound pressure level in dB.

Example 13.1

Determine the sound pressure level of a 100 Hz tone with a loudness level of 30 phon.

Solution

From Figure 13.2, we find the sound pressure level to be 44 dB.

13.4 Hearing Loss

Excessive and prolonged noise exposure causes permanent hearing loss. Various theories have been put forth in an effort to characterize and predict the possible damage that might be caused by a given exposure. Absolute proof of any theory concerning such a complex biological phenomenon is virtually impossible to achieve. However, reliable data have been collected, which deal with situations where workers have been continuously exposed to more or less the same noise environment for many years.

It is well established that excessive noise exposure causes permanent hearing damage by destroying the auditory sensor cells. These cells are hair cells located on the basilar membrane. Furthermore, other types of inner ear damage include harm to the auditory neurons, as well as damage to the structure of the organ of Corti. In all, the various theories and data have been taken advantage of establishing the noise-exposure criteria set forth in the noise exposure regulations.

13.5 Psychological Effects of Noise

In this section, certain generally accepted aspects of the psychological effects of noise will be discussed and quantified. Various indexes have been proposed that quantify the psychological effects of noise. However, only a few of indices, *loudness* (*sones*), *noise-criteria* (NC) *curve*, and *sound level*, are introduced in the following presentation.

13.5.1 Loudness Interpretation

As was discussed relating to Figure 13.2, loudness level is measured in phons, and the related quantity, loudness, is measured in *sones*. A sone is defined as the loudness of a 1000 Hz pure tone with a sound pressure level of 40 dB. On recalling the definition of loudness level, or by referring to Figure 13.2, one notes that 40 phon have a loudness equal to 1 sone. This relationship may be simply expressed as

$$S = 2^{(L_L - 40)/10} \text{ sone} \tag{13.1}$$

where S = loudness (sones), L_L = loudness level (phons), or conversely

$$L_L = 33.2 \log S + 40 \text{ phon} \tag{13.2}$$

Example 13.2

Make the following two conversions using the appropriate equation (Equation 13.1 or Equation 13.2): (1) convert 80 phon to sone, (2) convert 100 sone to phon.

Solution

1. To convert phons to sones, use Equation 13.2:

$$S = 2^{(L_L - 40)/10} = 2^{(80-40)/10} = 2^4 = 16 \text{ sone}$$

2. To convert sones to phons, use Equation 13.2:

$$L_L = 33.2 \log S + 40 = 33.2 \log 100 + 40 = 66.4 + 40 = 106.4 \text{ phon}$$

How should we determine the "total loudness" (sones), when the sound is composed of multiple frequency components? Probably the most widely used method for establishing the loudness of a complex noise is that developed by Stevens [2]. The method is based on the measurement of the 1-octave, 1/3-octave, or 1/2-octave band pressure levels. The measured band pressure levels are used

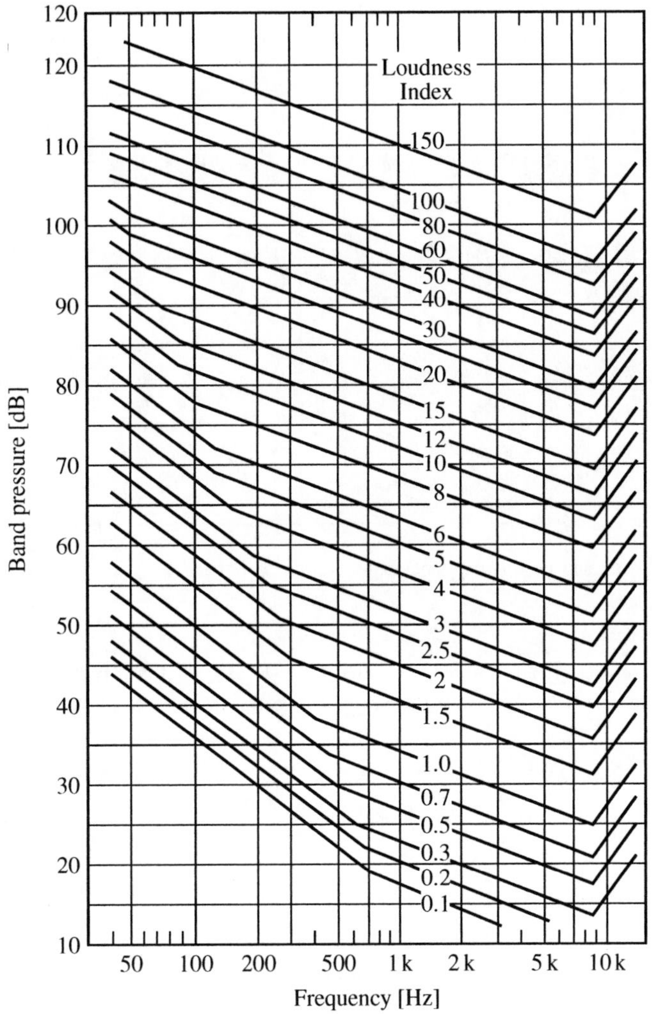

FIGURE 13.3 Equal loudness index contour [2].

in conjunction with the equal loudness index contours shown in Figure 13.3 to determine the loudness or loudness level by means of a simple calculation.

A step-by-step outline of the procedure is as follows:

1. Measure the band pressure levels (1-octave, 1/2-octave, or 1/3-octave) over the frequency range of interest. Usually, the range chosen is from about 50 to 10,000 Hz.
2. Enter the center frequency and band pressure level for each band in the contour of Figure 13.3, and determine the loudness index for each band.
3. Calculate the total loudness, S_t, in sones, by using

$$S_t = I_m(1 - K) + K\sum_{i=1}^{n} I_i \text{ (sone)} \tag{13.3}$$

where S_t = the total loudness (sones), I_m = the largest of the loudness indices, I_i = the loudness indices, including I_m, K = weighting factor for the bands chosen. $K = 0.3$ for 1-octave bands, $K = 0.2$ for 1/2-octave bands, $K = 0.15$ for 1/3-octave bands.

4. If so desired, one may calculate the loudness level in phons using Equation 13.2, or one may convert to loudness level by means of the conversion curve of Figure 13.3.

Example 13.3

A particular complex noise was measured to yield the one-octave band pressure given in the following table

Center Frequency (Hz)	Band Pressure Level (dB)	Loudness Index (sone)
63	66	2.5
125	63	3.2
250	65	4.8
500	70	7.5
1000	73	10.6
2000	76	15.2
4000	81	25.1
8000	79	29.0

Compute the loudness level using the procedure described before.

Solution

As a first step, the loudness indices are determined from Figure 13.3 and recorded in tabular form with the band pressure levels. Next, we note that one-octave bands have been used. Therefore $K = 0.3$ in Equation 13.3, and

$$S_t = I_m(1 - 0.3) + 0.3\sum_{i=1}^{8} I_i$$

From the table above, we find that $I_m = 29.0$ sone and, summing up, find $\sum I_i = 97.9$ sone. Therefore,

$$S_t = 29(1 - 0.3) + 0.3(97.9) = 49.67 \text{ sone}$$

We find that the loudness, $S_t \cong 50$ sone. The loudness level may now be calculated by means of Equation 13.2 as

$$L_L = 33.2 \log S_t + 40 = 33.2 \log 50 + 40 = 96.4 \text{ phon}$$

Therefore, L_L, the loudness level, is 96 phon.

13.5.2 Noise-Criteria Curves

Noise-criteria curves, which are neglected here, were established in 1957 for rating indoor noise. The curves have been utilized as one method of rating background noise level in a room. Each curve specifies the maximum octave-band sound pressure level for a given NC rating. If the octave band levels for a given noise spectrum are known, the rating of that noise in terms of the NC curves is given by plotting the noise spectrum on the set of NC curves to determine the point of highest penetration.

In 1971, some objections to the NC curves led to their modification. The new curves, which are shown in Figure 13.4, are called the *preferred noise-criteria* (PNC) *curves*. Although these curves differ from the NC curves, they are used in exactly the same manner.

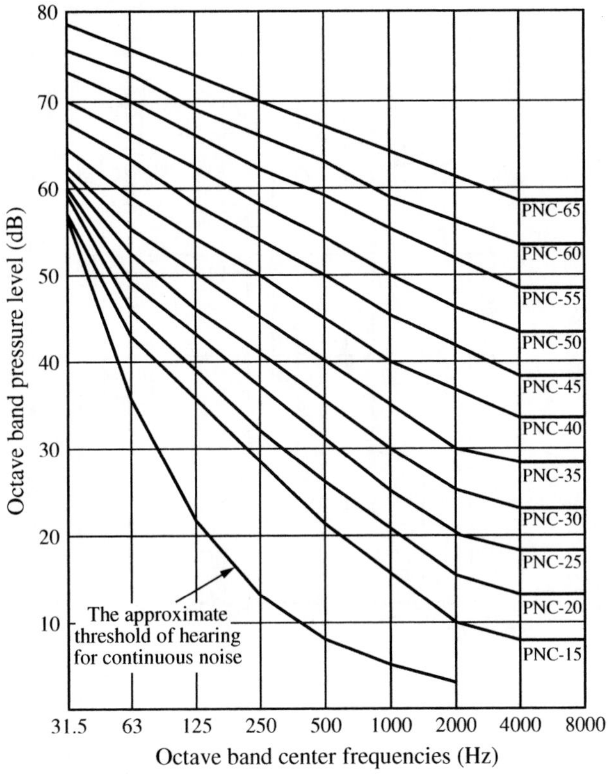

FIGURE 13.4 1971 preferred noise-criteria curves.

Example 13.4

Determine the PNC rating for the octave-band noise spectrum tabulated below.

Center frequency (Hz)	63	125	250	500	1000	2000	4000	8000
Band pressure level (dB)	65	60	60	63	55	50	45	40

Solution

The highest penetration is found at 500 Hz on PNC-60. Hence, the answer is PNC-60.

13.5.3 Sound Level

Sound levels are sound pressure levels that have been weighted according to a particular weighting curve. Three weightings, A, B, and C, and associated sound levels, have been developed as a method to subjectively evaluate the impact of noise upon the human ear, in a proper manner. The frequency response and decibel conversions from a flat response for each of these weightings are given in Figure 13.5 and Table 13.1, respectively.

The A-weighting network is now used almost exclusively in measurements that relate directly to the human response to noise, both from the viewpoint of hearing damage and of annoyance. Such measurements are referred to as *sound level measurements*. Sound level is designated by **L** and the designated unit is the dBA. Similarly, dBB and dBC are used to designate sound level weighted by B weighting and C weighting networks, respectively.

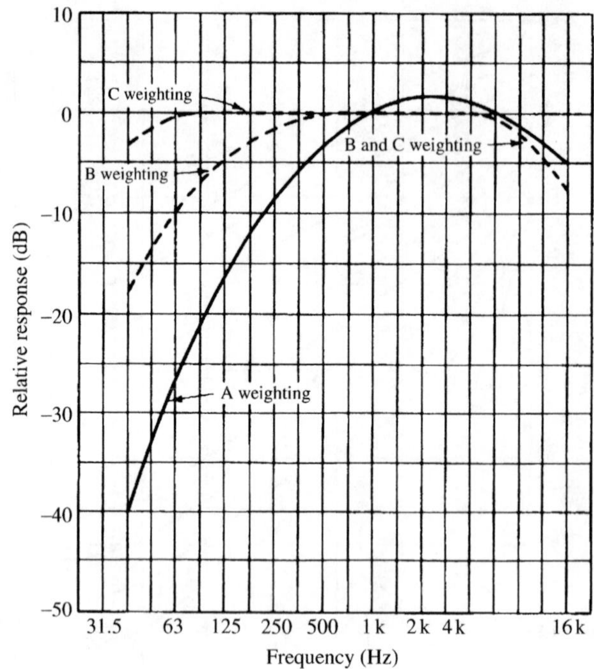

FIGURE 13.5 Frequency response for the A, B, and C weighting networks.

TABLE 13.1 Sound Level Conversion Chart from Flat Response to A Weighting

Frequency (Hz)	A Weighting (dB)
50	−30.2
63	−26.2
80	−22.5
100	−19.1
125	−16.1
160	−13.4
200	−10.9
250	−8.6
315	−6.6
400	−4.8
500	−3.2
630	−1.9
800	−0.8
1,000	0
1,250	+0.6
1,600	+1.0
2,000	+1.2
2,500	+1.3
3,150	+1.2
4,000	+1.0
5,000	+0.5
6,300	−0.1
8,000	−1.1
10,000	−2.5
12,500	−4.3
16,000	−6.6

TABLE 13.2 Octave-Band Sound Pressure Levels

f_c (Hz)	L_{flat} (dB)	ΔL_A (dB)	$L_A = L_{fla}t + \Delta L$ (dB)	I_{iA}
63	74	− 26.2	47.8	0.60×10^5
125	71	− 16.1	54.9	3.09
250	61	− 8.6	52.4	1.74
500	60	− 3.2	56.8	6.31
1000	62	0	62.0	1.585×10^6
2000	60	1.2	61.2	1.318
4000	62	1.0	63.0	1.995
8000	69	− 1.1	67.9	6.166
Sum				12.238×10^6

Note:

- f_c: band center frequency
- L_{flat}: sound pressure level with flat weighting
- ΔL_A: A-weighting level
- I_{iA}: sound pressure intensity with A weighting.

In noise-abatement problems, it is often necessary to convert calculated a 1-octave-band or 1/3-octave-band sound pressure level to a total sound level in dBA. Table 13.1 gives sound level conversion by A weighting from flat response pressure.

Example 13.5

Determine the total A weight sound level, L, of the set of octave-band sound pressure levels given in Table 13.2.

Solution

Refer Table 13.1 for the dB conversion from a flat response level, L_{flat}, for each of the octave bands to a sound pressure intensity with A weighting I_{iA}, and then the sum of I_{iA}. Finally, the total sound level with A weighting $L_{total,A}$ is given by

$$L_{total,A} = 10 \log \sum_{i=1}^{n} I_{iA} = 10 \log 12.238 \times 10^6 = 70.9 \text{ (dB)}$$

References

1. Irwin, J.D. and Graf, E.R. 1979. *Industrial Noise and Vibration Control*, Prentice Hall, Englewood Cliffs, NJ.
2. American National Standard USAS S3.4-1968. 1968. *Procedure for the Computation of Loudness of Noise*, America National Standards Institute, New York, NY.

14

Noise Control Criteria and Regulations

S. Akishita
Ritsumeikan University

Summary

In this chapter, the basic ideas behind the development of noise control criteria and regulations are discussed, taking into consideration that the standards and criteria vary from country to country and depend on governments in power. Legislations in the European Union and regulations in Japan are introduced as typical examples. Some indexes as measures of noise evaluation are described.

14.1 Introduction

In order to protect people from being exposed to excessive noise, different communities have implemented different types of legislative control. While the controls vary in scope, control mechanisms, and technical requirements, and are based on different control philosophies, they are intended to achieve a balance between the demand for a tranquil environment and the need for maintaining economic and social activities. In general, the noise standards vary according to the time of day and the use of the land concerned, with the more stringent standards applied to rest periods and areas where the noise sensitivity is high, such as those with schools and hospitals, and exclusive residential areas. Different countries have adopted different noise standards and regulations to meet their local situations and requirements. This chapter cannot describe all major control criteria and regulations in the world, or even in the major industrialized countries. Only the main issues of legislation on noise emission and reception are briefly introduced in the chapter. More details in the on-going noise control issues are found in Refs. [2,3].

14.2 Basic Ideas behind Noise Policy

Every noise policy originates from the idea of protecting the quality of life from noise pollution of all kinds. When establishing a noise policy, it is useful to consider the distinction between noise *emission* and *immission* (or *reception*) [6]. The former means literally emitting or radiating sound energy or power from a noise source, whereas the latter means receiving, perceiving, or observing radiated noise, which leads to the extent of the noise exposure at a position near the noise source. Therefore, noise emission is controlled with noise regulation law by the government, whereas noise immission is legislated with environmental quality standards. The measure of the extent describing the former is the "sound power level," and that describing the latter is the sound pressure level.

The global professional organization on noise control, the International Institute of Noise Control Engineering (I-INCE) recently started its activities to develop a global noise policy [5]. In response to the question "is noise policy a global issue, or is it a local issue?", I-INCE had a common theme presented in special session. It was felt that noise is primarily a global policy issue, but many noise problems can only be solved with the active participation of local authorities. The task of the technical study group is to take a global approach to noise in order to define the requirements for an international noise control policy to be effective, stated as follows:

> *All vehicles, devices, machinery,* and *equipment* that emit audible sound are manufactured products; most are entered into world trade and many are produced in two or more different countries by companies with worldwide operations. The *noise emission* of these products is an appropriate subject of international agreements and regulations. The *noise immisions* resulting from the operation of these products are growing in severity as traffic flow and the pace of industrialization continues to increase in many parts of the globe.

The technical study group reports the classification of noise areas as follows:

1. OCCUPATION NOISE—noise received at the workplace, indoors and outdoors, caused by all noise sources in the vicinity of the workplace.
2. ENVIRONMENTAL NOISE—noise perceived by individuals in the domestic environment, indoors and outdoors, caused by sources controlled by others.
3. CONSUMER PRODUCT NOISE—noise perceived by users and bystanders of noise generating products over which the individual has some control, including noise in the passenger compartment of vehicles, excluding occupational and environmental noise.

14.3 Legislation

The World Health Organization (WHO) published the historic "Guidelines for Community Noise" in 2000, which has been accepted as the most significant recommendations for noise exposure criteria. The bodies that are responsible for enacting the regulations as law include the Federal Government in the USA, the European Union (EU) in Europe, and the Japanese Government in Japan. In the following, the EU's legislation on noise immission is shown, as an example of the flow of legislation process [8].

On July 18th, 2002, a European Directive on the assessment and management of environmental noise was published in the *Official Journal of European Communities.* It was required to be implemented in the national legislation of the EU Member States no later than July 18th, 2004. From then on, a program was to start, containing periodic noise mapping, the making of action plans, and information of the public. The directive also has strengthened the position of the European Commission regarding the reduction of noise emission.

In 2002, the development of the European Directive on environmental noise resulted in an approved directive relating to the assessment and management of environmental noise, for which the acronym "DAMEN" is used. According to Article 1 of the DAMEN its objective is to "define a common approach to avoid, prevent or reduce harmful effects, including annoyance, due to exposure to environmental noise." A rough description of actions in the DAMEN is shown in Figure 14.1. Brief notes are given next to supplement Figure 14.1.

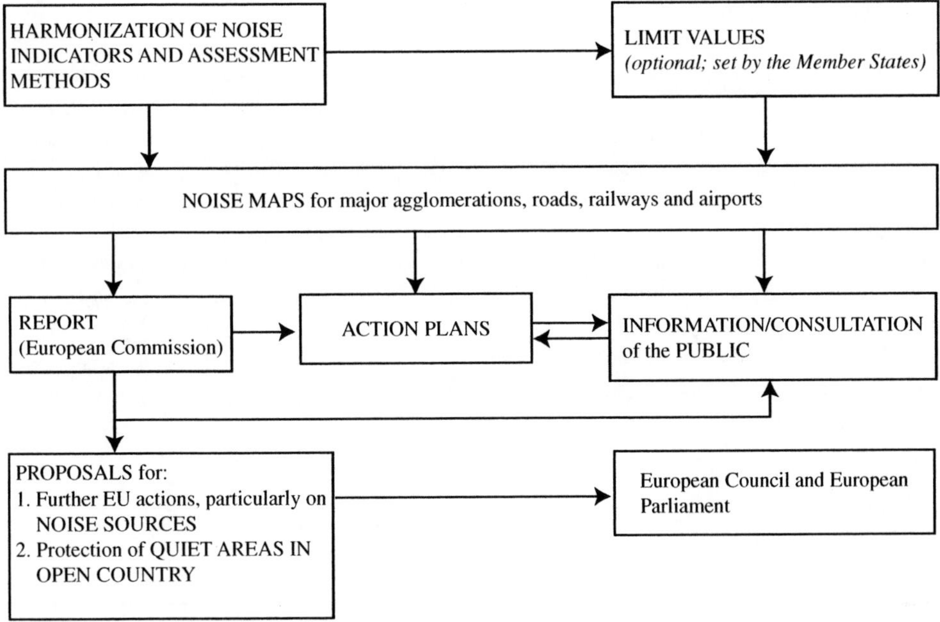

FIGURE 14.1 Overview of the DAMEN. (*Source*: Wolde, T.T. 2003. The European Union's legislation on noise immission, pp. 4367–4371. In *Proceedings of Inter-noise 2003* (N832). With permission.)

14.3.1 Action Plans

In a case where the mapping results are such that they violate the local or national limit value, or are found unsatisfactory for other reasons, action plans shall be developed for the improvement of the situation. These action plans shall be discussed with the citizens involved. A summary of the action plans shall be sent to the European Commission.

14.3.2 Publication of Data by the European Commission

Every five years, starting in 2009, the Commission shall publish a summary report from the noise maps and the action plans.

14.3.3 Proposal for Further European Union Action

In 2004, the European Commission was to submit a report to the European Parliament and the Council containing a review of existing EU measures relating to sources of environmental noise and present proposals for improvement, if appropriate. In 2009, the European Commission will submit to the European Parliament and the Council a report on the implementation of the directive. That report will in particular assess the need for further EU action and, if appropriate, propose implementing strategies on aspects such as:

- Long-term and medium-term goals for the reduction of the number of persons harmfully affected by environmental noise
- Additional measures on noise emission by specific sources
- The protection of quiet areas on the open country

14.4 Regulation

Noise regulation is executed by local governments once the central government enacts a noise regulation law. The law is considered the "national minimum." For example, factory noise, construction work noise, and road traffic noise are under the purview of the Noise Control Act, which means the central government is responsible of regulating these kinds of noise. On the other hand, community noise and factory noise are under the purview of the original regulation of local governments. It can be said that local governments are responsible for a great part of the noise policy, although they may not always fully understand the situations concerned. In what follows, an outline of the legal system for environmental noise problem in Japan is given as an example of a typical legal system for noise regulation [6].

In Japan, the "Environmental Quality Standards for Noise" was revised in 1999 after 27 years with the old law. Figure 14.2 outlines of the legal system in Japan.

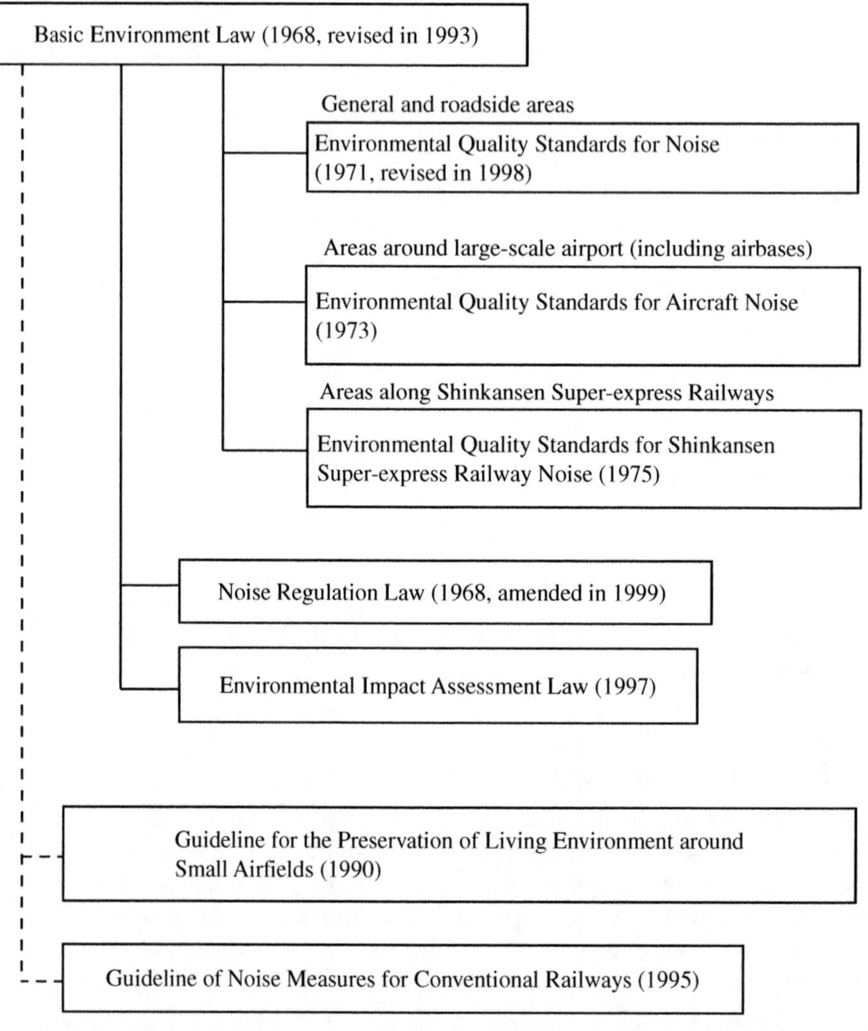

FIGURE 14.2 Legal system for environmental noise in Japan. (*Source:* Tachibana, H. and Kaku, J. 2003. Acoustic measures for the environmental noise assessment in Japan, pp. 3317–3322. In *Proceedings of Inter-noise 2003* (N1007). With permission.)

Each of these laws and standards is legislated for a specific noise problem (a noise source), and therefore different noise indices are specified according to the respective noise problems. To review this situation from a historical viewpoint, it can be said that each law or standard responds to a specific noise problem promptly, through the use of available measurement technology at that time. However, some laws and standards have become outdated since their establishment, when considering the current situation, international dynamics, and the current acoustic measurement technology.

14.5 Measures of Noise Evaluation

Basically, the A weighting networks are applied to obtain a measure of noise evaluation. As the measures are legislated by governments, they are dependent on the legislative regulations and standards. In what follows, the measure of noise evaluation legislated in Japan is given to show the concepts behind the legislation [6].

In the regulations and standards for environmental noise problems, a variety of noise measures are used. In order to improve these legislative regulations and standards in the future, the present measures shall be reviewed considering the difference between noise emission and immission and the difference between noise measurement and monitoring, and impact assessment and prediction. These measures legislated in Japan are listed and classified in Table 14.1 by considering the difference between noise characteristics.

TABLE 14.1 Assessment Methods Specified in Laws and Standards for Environmental Noises in Japan

Noise Sources	Law and Standards	Noise Indices	Assessment Time
Roads	Environmental Quality Standards for Noise	$L_{Aeq,T}$[a]	Daytime (6:00–22:00); nighttime (22:00–6:00)
Shinkansen superexpress railways	Environmental Quality Standards for Shinkansen Superexpress Railways	$L_{A,Smax}$[b]	Every event
Conventional railways	Guideline of Noise Measures for Conventional Railways	$L_{Aeq,T}$	Daytime (7:00–22:00); nighttime (22:00–7:00)
Aircrafts	Environmental Quality Standards for Aircrafts Noise	WECPNL[c]	Time weighting
	Guideline for the Preservation of Living Environment around Small Airfields	L_{den}[d]	
Construction works Factories Large-scale retail stores	Noise Regulation Law (specific noise sources) Law concerning the measures by large scale retail stores for preservation of living environment	According to time variation: L_A[e]; $L_{A,Fmax}$[f]; L_{A5}[g]; $L_{A,Fmax,5}$[h]	Not specified; every event

[a] $L_{Aeq,T}$, equivalent continuous A-weighted sound pressure level.
[b] $L_{A,Smax}$, SLOW maximum value of A-weighted sound pressure level.
[c] WECPNL, weighted equivalent continuous perceived noise level(calculated from $L_{A,Smax}$).
[d] L_{den}, day/evening/night equivalent continuous A-weighted sound pressure level.
[e] L_A, FAST maximum value of A-weighted sound pressure level.
[f] $L_{A,Fmax}$, A-weighted sound pressure level.
[g] L_{A5}, upper value of the 90% range of A-weighted sound pressure level.
[h] $L_{A,Fmax,5}$, upper value of the 90% range of the FAST maximum Aweighted sound pressure level.

Source: Tachibana, H. and Kaku, J. 2003. Acoustic measures for the environmental noise assessment in Japan, pp. 3317–3322. In *Proceedings of Inter-noise 2003* (N1007). With permission.

When considering the consistency between noise measurement and monitoring, and noise prediction for impact assessment, it is most reasonable to use energy based indices such as L_{Aeq}. Of course, L_{Aeq} is not a panacea and some secondary adjustment may be needed for the exact assessment of environmental noise with different characteristics. Nevertheless, the possibility of unification by L_{Aeq} should be considered in the near future in Japan. Although L_{Aeq} is now being widely used for the assessment of aircraft noise in almost all countries, WECPNL is still being used in Japan. WECPNL is very close to L_{Aeq} in concept and it is not difficult to change the assessment index from WECPNL to L_{Aeq}.

The aim of the laws and standards shown in Figure 14.2 is to measure and assess the environmental noise for prevention or maintenance of the present situation. Therefore, any noise index should be appropriately used for each of noise problems, as shown in Table 14.1, which presents assessment methods specified in laws and standards for environmental noise in Japan. In particular, when predicting the future noise situation in environmental impact assessments, the indices should be suitable for theoretical calculation. The statistical noise indices such as the percentile level (L_{A5}) and maximum level ($L_{A,Fmax}$ or $L_{A,Smax}$) specified in the laws and standards have to be predicted statistically. It is difficult to predict these quantities by a simple physical calculation model, in principle. In this respect, the energy-based noise indices such as L_{Aeq} can be easily treated in energy based calculation, and the prediction model becomes simple and clear in physical meaning. In an environmental impact assessment, the predicted results are to be compared with the related laws or standards. In the case of road traffic noise, L_{Aeq} has been adopted in the new environmental quality standards, and therefore prediction has become very simple in theory, founded on energy-based indices.

In the prediction of road traffic noise, a motor vehicle as the noise source can be treated as a stationary sound sources of a constant sound power for a limited path. On the other hand, in the case of predicting construction noise, there are many complicated problems because various kinds of machines and equipment with various temporal variations of characteristics must be treated. Therefore, in the construction noise prediction method given in the "Acoustic Society of Japan CN-model 2002" [7], various noise indices for describing the acoustic output of various types of noise sources are specified as given in Table 14.2, which presents classification of noise sources and indices for expressing their acoustic output. Finally, Table 14.3 presents definitions and indices of measurement for acoustical output of noise sources.

TABLE 14.2 Classification of Noise Sources and Indices for Expressing Their Acoustic Output

Temporal Variation	Indices for Expressing Acoustic Output Sign	Terms
Stationary	L_{WA}	A-weighted sound power level
	$L_A(r_0)$	A-weighted sound power level at the reference distance ($r_0 = 1$ m)
Fluctuating randomly and widely	L_{WAeff}	Effective A-weighted sound power level
	$L_{Aeff}(r_0)$	Effective A-weighted sound pressure level at the reference distance ($r_0 = 1$ m)
	$L_{A,Fmax,5}(r_0)$	5% value of A-weighted sound pressure level at the reference distance ($r_0 = 1$ m)
Intermittent impulsive	L_{JA}	A-weighted sound energy level
	L_{WAeff}	Effective A-weighted sound power level
	$L_{AE}(r_0)$	Single event sound exposure level at the reference distance ($r_0 = 1$ m)
	$L_{A,Fmax}(r_0)$	FAST max. of A-weighted sound pressure level at the reference distance ($r_0 = 1$ m)

Source: Tachibana, H. and Kaku, J. 2003. Acoustic measures for the environmental noise assessment in Japan, pp. 3317–3322. In *Proceedings of Inter-noise 2003* (N1007). With permission.

TABLE 14.3 Definitions and Measurements of Indices for Acoustical Output of Noise Sources

Indices	Definition	Measurement Method
L_{WA}	$L_{WA} = 10\log\frac{P_A}{P_0}$ (1) Here, $P_0 = 1\ pW$	$L_{WA} = L_A(r) + 20\log\frac{r}{r_0} + 8$ (2) Here, $L_A(r)$ is the A-weighted sound pressure level measured at a distance of r, $r_0 = 1$ m
L_{WAeff}	Effective A-weighted sound power level applied to fluctuating, intermittent and impulsive sounds	$L_{WAeff} = L_{Aeff}(r) + 20\log\frac{r}{r_0} + 8$ (3) Here, L_{Aeff} is the A-weighted sound pressure level measured at a distance of r $L_{Aeff} = 10\log\left[\frac{1}{T}\int_1^2 \frac{p_A^2(t)}{p_0^2}\mathrm{d}t\right]$ (4) Here, $T(t_1 - t_2)$ is averaging time (s), $p_0 = 20\ \mu$Pa
L_{JA}	$L_{JA} = 10\log\frac{E_A}{E_0}$ (5) Here, $E_0 = 1$ pJ	$L_{JA} = L_{AE}(r) + 20\log\frac{r}{r_0} + 8$ (6) Here, L_{AE} is the single event sound exposure level measured at a distance of r $L_{AE} = 10\log\left[\frac{1}{T}\int_1^2 \frac{p_A^2(t)}{p_0^2}\mathrm{d}t\right]$ (7) Here, $T_0 = 1$ s, $t_1 - t_2$ is the time including the event (s)
$L_A(r_0)$ $L_{A,Fmax}(r_0)$	A-weighted sound pressure level converted to the value at the reference distance $(r_0) = 1$ m	$L_A(r_0) = L_A(r) + 20\log\frac{r}{r_0} + 8$ (8) Here, $L_A(r)$ is the A-weighted sound pressure level measured at a distance of r

Source: Tachibana, H. and Kaku, J. 2003. Acoustic measures for the environmental noise assessment in Japan, pp. 3317–3322. In *Proceedings of Inter-noise 2003* (N1007). With permission.

References

1. Fahy, F. 1985. *Sound and Structural Vibration, Radiation, Transmission and Response*, Academic Press, New York, chap. 2.
2. Fields, J.M. and de Jong, R.G., Standardized general-purpose noise reaction questions for community noise survey: research and a recommendation, *J. Sound Vib.*, 242, 641–679, 2001.
3. Harris, C.M., Ed. 1979. *Handbook of Noise Control*, 2nd ed., McGraw-Hill, New York, chap. 37.
4. Irwin, J.D. and Graf, E.R. 1979. *Industrial Noise and Vibration Control*, Prentice Hall, New York, chap. 5.
5. Lang, W.W. and Wolde, T.T. 2003. Progress report for TSG#5 'Global Noise Policy', pp. 98–101. In *Proceedings of Inter-noise 2003* (N872).

6. Tachibana, H. and Kaku, J. 2003. Acoustic measures for the environmental noise assessment in Japan, pp. 3317–3322. In *Proceedings of Inter-noise 2003* (N1007).
7. Tachibana, H. and Yamamoto, K. 2003. *Construction Noise Prediction Model*, ASJ CN-Model 2002, proposed by the Acoustical Society of Japan, EURONOISE, in Naples (2003.5).
8. Wolde, T.T. 2003. The European Union's legislation on noise immission, pp. 4367–4371. In *Proceedings of Inter-noise 2003* (N832).

15

Instrumentation

Kiyoshi Nagakura

Railway Technical Research Institute

Summary

This chapter describes some measuring methods for the identification and ranking of noise source that are of benefit in noise control projects. Sound intensity measurement and directional measuring devices such as the mirror–microphone system and microphone array are introduced and their principles and applications are described.

15.1 Sound Intensity Measurement

Every noise control project starts with the identification and ranking of the noise sources. Several methods have been proposed for the purpose and have proved to be useful and widely utilized. In this chapter, sound intensity measurement and directional measuring devices such as the mirror–microphone system and microphone array are introduced and their principles and applications are described. Other useful measurements, such as acoustic holography method [1,2] and spatial transformation of sound fields [3], are described in the literature.

15.1.1 Theoretical Background

Sound intensity is a measure of the magnitude and direction of the flow of sound energy. The instantaneous intensity vector, $\mathbf{I}(t)$, is given by the product of the instantaneous sound pressure, $p(t)$, and the corresponding particle velocity, $\mathbf{u}(t)$, that is, $\mathbf{I}(t) = p(t)\mathbf{u}(t)$.

In practice, the time-averaged intensity, $\bar{\mathbf{I}}$, is more important, and is given by the equation:

$$\bar{\mathbf{I}} = \lim_{T\to\infty} \frac{1}{T} \int_{-T/2}^{T/2} p(t)\mathbf{u}(t)\mathrm{d}t \tag{15.1}$$

The intensity vector denotes the net rate of flow of energy per unit area (watts/m^2). Thus, the acoustic power, W, of the source located in a closed surface, S, is given by the integral of the intensity passing through the surface, S, as

$$W = \iint_s \bar{\mathbf{I}}\cdot \mathrm{d}S \tag{15.2}$$

Equation 15.2 indicates that the measurement of sound intensity over a surface enclosing a source enables the estimation of its sound power, which shows the usefulness of the sound intensity concept.

15.1.2 Measurement Method

The principle of intensity measurement systems in commercial production employs two closely spaced pressure microphones [4,5], as shown in Figure 15.1.

The particle velocity, $u_r(t)$, in a particular direction, r, can be approximated by integrating over time the difference of sound pressures at two points separated by a distance Δr in that direction:

$$u_r(t) = -\frac{1}{\rho_0}\int_{-\infty}^{t}\frac{p_2(\tau) - p_1(\tau)}{\Delta r}\mathrm{d}\tau \qquad (15.3)$$

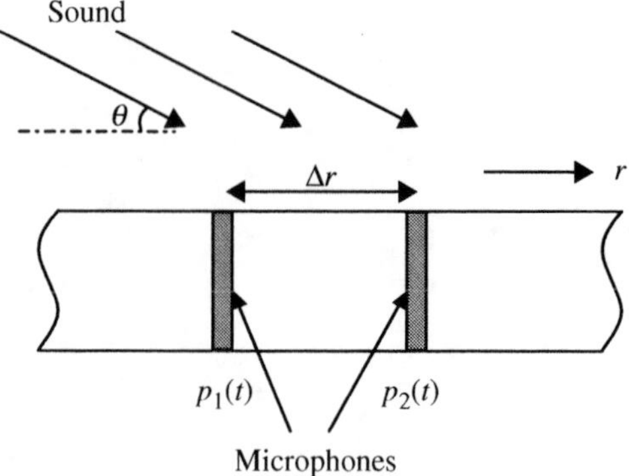

FIGURE 15.1 Microphone arrangement used to measure sound intensity.

where p_1 and p_2 are the sound pressure signals from the two microphones. The sound pressure at the center of two microphones is approximated by

$$p(t) = \frac{p_1(t) + p_2(t)}{2} \qquad (15.4)$$

Thus, the intensity in the direction r can be calculated as

$$I_r(t) = -\frac{1}{2\rho_0\Delta r}[p_1(t) + p_2(t)]\int_{-\infty}^{t}[p_2(\tau) - p_1(\tau)]\mathrm{d}\tau \qquad (15.5)$$

Some commercial intensity analyzers use Equation 15.5 to measure the intensity. Another type of analyzer uses the equation in the frequency domain:

$$I_r(\omega) = -\frac{\mathrm{Im}[G_{12}]}{\omega\rho_0\Delta r} \qquad (15.6)$$

where G_{12} is the cross spectrum between the two microphone signals. Equation 15.6 makes it possible to calculate sound intensity with a dual-channel fast fourier transform (FFT) analyzer.

15.1.3 Errors in Measurement of Sound Intensity

The principal systematic error of the two-microphone method is due to the approximation of the pressure gradient by a finite pressure difference. When the incident sound is a plane wave, the ratio of the measured intensity, $\hat{I}_r$, and the true intensity, I_r, is given by

$$\hat{I}_r/I_r = \frac{\sin(k\Delta r\cos\theta)}{k\Delta r\cos\theta} \qquad (15.7)$$

where the angle θ is as defined in Figure 15.1 and k is the wave number. Equation 15.7 indicates that the upper frequency limit is inversely proportional to the distance between the microphones.

Another serious error is caused by the phase mismatch between the two measurement channels. In the calculation of intensity from Equation 15.5, the phase difference, φ, between the two microphone signals, p_1 and p_2, is very important. Hence, the phase mismatch between the two measurement channels, $\Delta\varphi$, must be much smaller than φ. Since φ increases with frequency, this error is serious in lower frequencies. Other possible errors, such as in the sensitivity of microphones and random errors associated with a given finite averaging time, are usually less serious.

15.1.4 Applications

One important application of sound intensity measurement is the determination of the sound power level using Equation 15.2. Furthermore, measurement of the intensity in the very near field of a source surface makes it possible to identify and rank the noise-sources. Plots of the sound intensity measured on a surface near a sound source are useful for investigating noise source distributions. Figure 15.2 shows sound intensity of noise from a wheel of a railway car. An intensity probe is located in the vicinity of the wheel and the normal component of sound intensity is measured by traversing the probe on a plane 100 mm away from the side surface of the wheel. These figures show a free vibration behavior of the wheel at each frequency; the wheel vibrates with one nodal diameter at 700 Hz and with three nodal diameters at 1150 Hz. Visualization by intensity vectors also gives valuable information about a noise source. Figure 15.3 shows the sound intensity vectors at each octave band measured in the vicinity of a railway car running at 120 km/h. These results suggest that the main radiator of rolling noise is the rail at the 500 Hz to 1 kHz band and the wheels at the 2 to 4 kHz band.

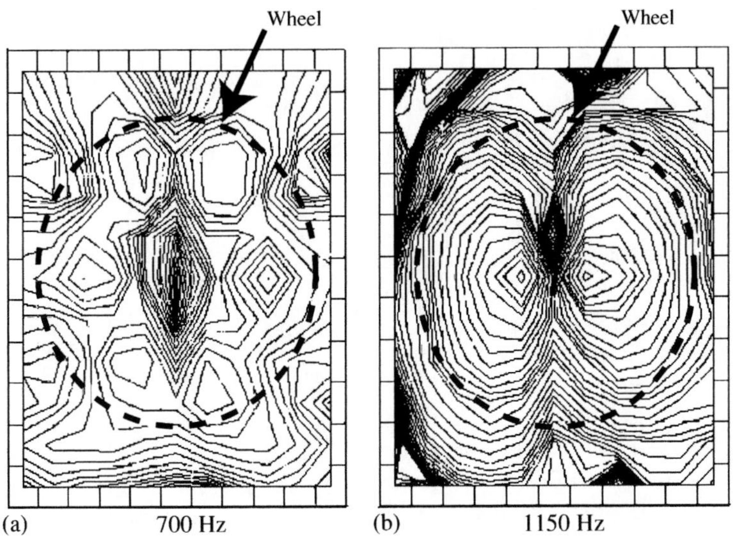

FIGURE 15.2 Measurements of the sound intensity radiated by a wheel of a railway car (1 dB contour).

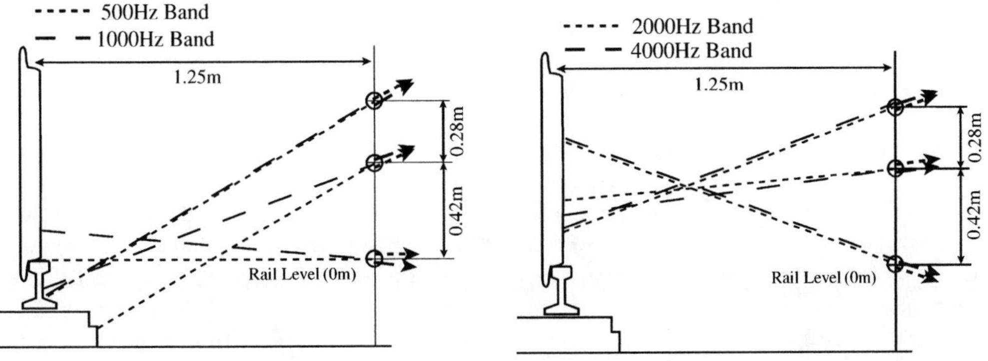

FIGURE 15.3 Sound intensity vectors measured in the vicinity of a railway car running at 120 km/h.

15.2 Mirror–Microphone System

15.2.1 Principle of Measurement

A mirror–microphone system consists of a reflector of elliptic or parabolic shape and an omnidirective microphone located at its focus [6,7]. Figure 15.4 shows the layout of a reflector of elliptic shape, an omnidirective microphone, and a noise source. Here, S and S' denote the front and back surfaces of the mirror, respectively; $P(\mathbf{r})$ denotes the pressure field on this configuration; $P_i(\mathbf{r})$ denotes the pressure field of free space; $\mathbf{r}_m$ is the position of the microphone; $\mathbf{r}$ is a point on the mirror surface. The normal, $\mathbf{n}_0$, directs toward the medium.

Using Green's theorem, the pressure at the microphone position $P(\mathbf{r}_m)$ is obtained by

$$P(\mathbf{r}_m) = P_i(\mathbf{r}_m) + \iint_{(s+s')} P(\mathbf{r}) \frac{\partial}{\partial \mathbf{n}_0} \times \left[\frac{e^{ikR_m}}{4\pi R_m} \right] d^2\mathbf{r} \tag{15.8}$$

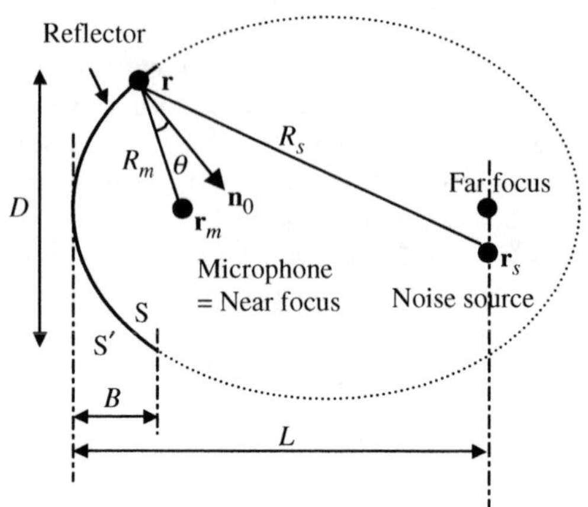

FIGURE 15.4 Layout of a reflector, microphone, and noise source.

where $k = 2\pi f / c_0$ is the wave number, f is the frequency of sound, c_0 is the speed of sound, and $R_m = |\mathbf{r} - \mathbf{r}_m|$ is the distance between the microphone and the mirror surface. If the wavelength is sufficiently smaller than the diameter of the reflector, the pressure field $P(\mathbf{r})$ is approximated by $2P_i(\mathbf{r})$ on the front surface, S, and by zero on the back surface, S'. In such a frequency range, the incident field term $P_i(\mathbf{r}_m)$ can be ignored. With these approximations, assuming that the noise source is a monopole type point source located at a position, $\mathbf{r}_s$, Equation 15.8 reduces to

$$P(\mathbf{r}_m) = -\frac{m(f)}{8\pi^2} \iint_s \frac{e^{ik(R_m + R_s)}}{R_m R_s} \left(ik - \frac{1}{R_m} \right) \times \cos\theta(\mathbf{r}) d^2\mathbf{r} \tag{15.9}$$

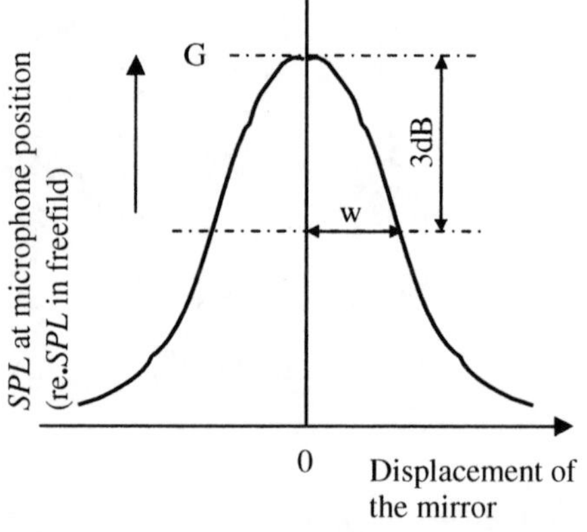

FIGURE 15.5 Directivity pattern of a mirror–microphone system.

Here, $m(f)$ is the amplitude of the mass-flux rate of the source, $R_s = |\mathbf{r} - \mathbf{r}_s|$ is the distance between the sound source and the mirror surface and the angle $\theta(\mathbf{r})$ is defined in Figure 15.4. When the noise source is located at the far focus of the mirror, the sound pass length $R_m + R_s$ is constant with respect to $\mathbf{r}$, and a strong signal is obtained. As the noise source is moved away in the direction perpendicular to the mirror axis, the variance of the sound pass length, $R_m + R_s$, due to the position $\mathbf{r}$ increases, and thus the microphone signal drops off due to interference (see Figure 15.5, which we call the "directivity pattern"). The ratio of the peak level to the free field level at the microphone, G, is referred to as the "gain factor." The spatial resolution of the mirror is characterized by the displacement of the mirror position, w, at which the microphone signal drops off by a given relative amount, such as 3 dB. The quantities G and w

can be related to the mirror geometries in Figure 15.4 by

$$G \approx 10\log(CD^4/\lambda^2 B^2) \quad (C = \text{const.}) \tag{15.10}$$

$$w \propto \lambda L/D \tag{15.11}$$

The gain factor, G, increases with frequency at the rate of 6 dB per octave, and the spatial resolution, w, is inversely proportional to the frequency. The lower frequency limit is decided by the size of the mirror. On the other hand, there is no higher frequency limit, except for the capacity of an omnidirectional microphone itself. Thus, measurements with the mirror–microphone system are more suited to a scaled model test.

15.2.2 Applications

The mirror–microphone system has proved useful for identification of a noise source because of its directional property [8–10]. A scan of the source region produces a noise source map. It has an advantage in that the measurement is possible at a far field and it needs only one sensor, but has a disadvantage in that the measuring process is a time-consuming task.

Figure 15.6 shows an example of source maps of aerodynamic noise generated by a one-fifth scale high-speed train model, obtained from measurements by a mirror–microphone system, in a wind tunnel test. The surface of the car model is divided into several noise-source areas and the noise-source distribution in each area is measured by traversing the mirror–microphone system over the surface. The diameter and focal distance of the reflector are 1.7 and 3 m, respectively. Detailed maps of noise-source strength are obtained, which show that aerodynamic noise from high-speed trains is generated in relatively localized areas, namely, the local surface structures. The mirror–microphone system can be used for the measurement of the source distribution of a moving noise source. Figure 15.7 gives a time

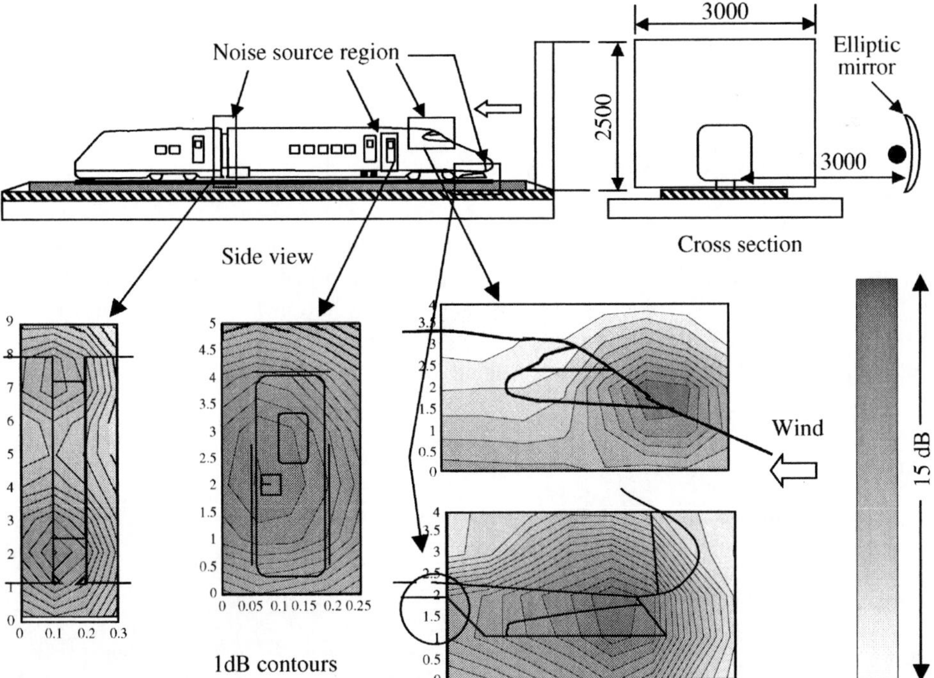

FIGURE 15.6 Noise-source distribution of a one-fifth scale Shinkansen car model in a wind tunnel test measured with an elliptic mirror–microphone system.

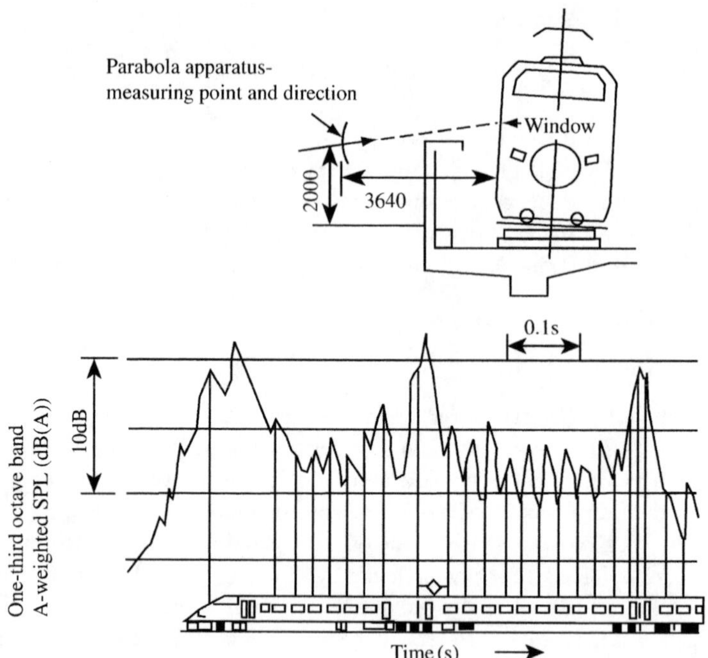

FIGURE 15.7 Time history of the A-weighted one-third octave band ($f_0 = 8$ kHz) sound pressure level measured with a parabolic mirror–microphone system ($D = 1$ m, train speed = 274 km/h).

history of noise from a high-speed train measured with a parabolic mirror–microphone system, the diameter of which is 1 m. Peaks of the time history correspond to pantographs, doors, gaps between cars and the step-up of windows, which shows that they are main noise sources.

15.3 Microphone Array

15.3.1 Principle of Microphone Array

A microphone array [11] consists of several microphones distributed spatially to measure an acoustic field. The time signals from each microphone are added, accounting for the time delay between sound sources and microphones, and a directional output signal can be obtained as a result. The algorithm is called "beamforming." Now, consider M omnidirectional microphones distributed in a far field of noise sources. The output signal of the array focused to a particular location in the source region, $\mathbf{r}$, and $z(\mathbf{r}, t)$, is calculated as a sum of delayed and weighted signals of each microphone:

$$z(\mathbf{r}, t) = \sum_{m=1}^{M} w_m p_m(t - \Delta_m) \tag{15.12}$$

Here, $p_m(t)$ is the signal from the mth microphone, w_m is a weighting factor, and Δ_m is a time delay applied to signal of the mth microphone, as given by

$$\Delta_m = \frac{r_o - r_m}{c_0} \tag{15.13}$$

where r_o and r_m are the distances from the focus point to the reference point o and the mth microphone, respectively. When the focus location coincides with the source location, a strong signal is obtained (see Figure 15.8). If this process is repeated for various focus locations, $\mathbf{r}$, on the source surface, then a noise-source map can be obtained.

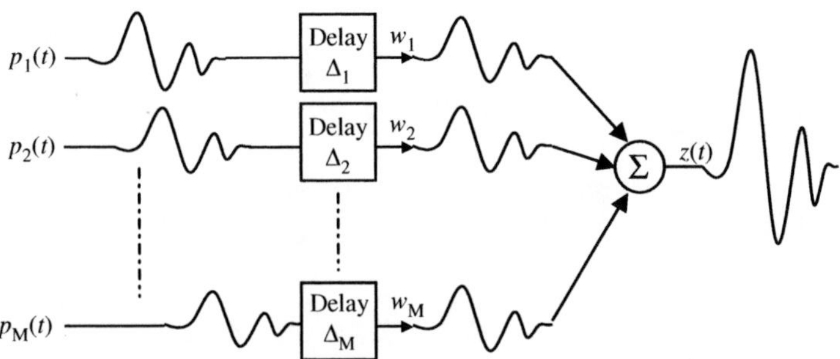

FIGURE 15.8 Principle of a microphone array. Individual time delays are chosen such that signals arriving from a given point will be added up coherently.

15.3.2 Array's Directivity Pattern

The performance of a microphone array is characterized by the spatial resolution and signal-to-noise ratio. For simplicity, consider a linear array of $M = 2N + 1$ microphones spaced equally by d. When a harmonic plane wave is propagating with an incident angle θ, and weighting factors all equal $1/M$, the ratio of the output signal of the array to that of the center microphone is computed using

$$W(\theta) = \frac{1}{M}\frac{\sin((M/2)kd\sin\theta)}{\sin((1/2)kd\sin\theta)} \tag{15.14}$$

where k is the wave number. Figure 15.9 shows the directivity patterns for different values of the product kd based on Equation 15.14. The highest peak appears at $\theta = 0$, which we call a "main lobe," and lower peaks also appear at some locations that are separate from a true source direction, which we call "side lobes." The width of the main lobe decides the performance of the array to separate two closely lying sources (which we call spatial resolution), and the ratio of main lobe to side lobe decides the signal-to-noise ratio of the array. The spatial resolution improves as kd increases, that is, in proportion to the ratio of the array length to the wavelength. However, when $kd = 2\pi$, a peak of the same strength as the true source appears due to a spatial aliasing at $\theta = 90°$, which occurs when $d > \lambda/2$, where λ is the wavelength. Thus, the acoustic frequency, f, is restricted by $f < c_0/2d$, to avoid aliasing.

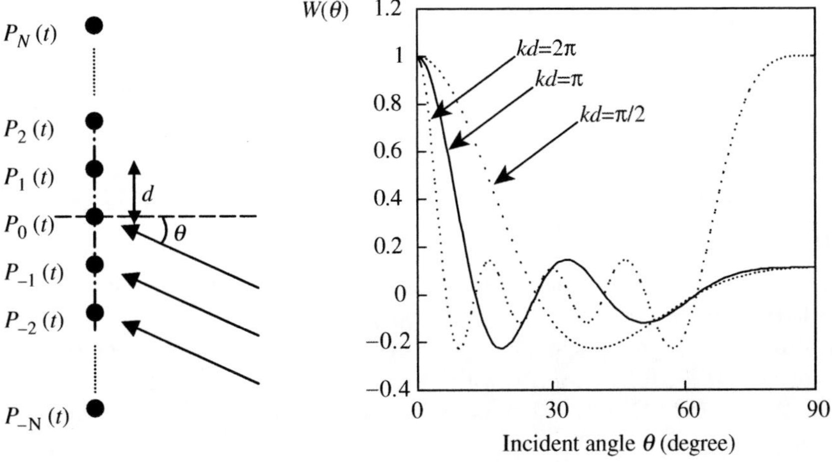

FIGURE 15.9 Directivity patterns of a linear array for different values of the product kd ($M = 2N + 1 = 9$).

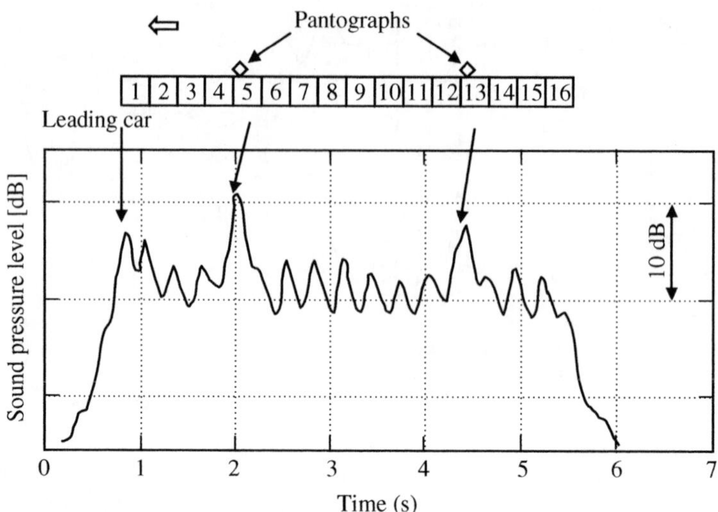

FIGURE 15.10 Time history of sound pressure level for a passing train measured with a linear microphone array located at a point 25 m away from a track (train velocity = 285 km/h).

In the above case of the linear array, the directivity exists only in the direction of the array (one-dimensional). If microphones are arranged in a two-dimensional plane, a two-dimensional directivity can be obtained. Recently, many microphone arrangements have been proposed that obtain better spatial resolution and to reduce side lobes [12–15].

15.3.3 Applications

Microphone arrays have been used for identification of the noise source in various situations, for example, in wind tunnel tests. Many actual examples can be found in published literature [10,16–18]. The measurement with a microphone array has the advantage of much shorter measuring time than that of a mirror–microphone system. Furthermore, the lower frequency limit is not so serious because the size of the apparatus can be easily extended.

Another fundamental example is given now. Nine microphones are arranged equally spaced by $d = \lambda/2$ for each one-third octave band, and their signals are summed without time delay. In this case, the array is focused to a fixed direction, perpendicular to the array axis. Figure 15.10 shows a time history of noise generated by a high-speed train, measured with a linear microphone array located at a point 25 m away from the track. It is found that pantographs, the leading car, and gaps between cars are the main noise sources in this example.

References

1. Ferris, H.G., Computation of far field radiation patterns by use of a general integral solution to the time independent scalar wave equation, *J. Acoust. Soc. Am.*, 41, 1967.
2. Maynard, J.D., Nearfield acoustic holography: theory of generalized holography and the development of NAH, *J. Acoust. Soc. Am.*, 78, 1985.
3. Ginn, K. and Hald, J., The effect of bandwidth on spatial transformation of sound field measurements, *Inter-Noise*, 87, 1987.
4. Fahy, F.J., Measurement of acoustic intensity using the cross-spectral density of two microphone signals, *J. Acoust. Soc. Am.*, 62, 1977.

5. Chung, J.Y., Cross-spectral method of measuring acoustic intensity without error caused by instrument phase mismatch, *J. Acoust. Soc. Am.*, 64, 1978.
6. Grosche, F.R., Stiewitt, H., and Binder, B., On aero-acoustic measurements in wind tunnels by means of a highly directional microphone system, *Paper AIAA-76-535*, 1976.
7. Sen, R., Interpretation of acoustic source maps made with an elliptic-mirror directional microphone system, *Paper AIAA-96-1712*, 1996.
8. Blackner, A.M. and Davis, C.M., Airframe noise source identification using elliptical mirror measurement techniques, *Inter-Noise 95*, 1995.
9. Dobrzynski, W., Airframe noise studies on wings with deployed high-lift devices, *Paper AIAA-98-2337*, 1998.
10. Dobrzynski, W., Research into landing gear airframe noise reduction, *Paper AIAA-2002-2409*, 2002.
11. Johnson, D.H. and Dudgeon, D.E. 1993. *Array Signal Processing*, Prentice Hall, Englewood Cliffs, NJ.
12. Elias, G., Source localization with a two-dimensional focused array: optimal signal processing for a cross-shaped array, *Inter-Noise 95*, 1995.
13. Dougherty, R.P. and Stoker, R.W., Sidelobe suppression for phased array aeroacoustic measurements, *Paper AIAA-98-2242*, 1998.
14. Nordborg, A., Optimum array microphone configuration, *Inter-Noise 2000*, 2000.
15. Hald, J. and Christensen, J.J., A class of optimal broad band phased array geometries designed for easy construction, *Inter-Noise 2002*, 2002.
16. Piet, J.F. and Elias, G., Airframe noise source localization using a microphone array, *Paper AIAA-97-1643*, 1997.
17. Hayes, J.A., Airframe noise characteristics of a 4.7% scale DC-10 model, *Paper AIAA-97-1594*, 1997.
18. Stoker, R.W., Underbrink, J.R., and Neubert, G.R., Investigation of airframe noise in pressurized wind tunnels, *Paper AIAA-2001-2107*, 2001.

16

Source of Noise

S. Akishita
Ritsumeikan University

Summary

In this chapter, a mathematical description of sound radiation is briefly presented accompanied by an introduction to sound sources, monopole, and dipole. The modeling of a simple source of noise is discussed, introducing Green's function. As an example of simple sound radiation, the sound field generated by a source embedded in a plane surface is described. Finally, an estimation of noise-source sound power is presented by introducing the power conversion factor of actual machinery.

16.1 Introduction

A careful examination of noise measurement data reveals that there exists a very wide variety of noise sources. Each source is peculiar to its generation mechanism, which can be any of a wide range of phenomena including fluid mechanics and the vibration of structures. However, in analysis, sources are normally simplified to rather simple and typical models in their generation mechanism.

The vibration of a solid body, which may be in contact with the fluid medium, generates sound waves or vibratory forces acting directly on a fluid, will result in the emission of acoustic energy in the medium. In the next section, an expression for an idealized sound source is introduced. We will assume that the fluid medium outside the source region is initially uniform and at rest. Also, we will concentrate on wave propagation in an infinite medium.

Generally, acoustic waves sensed as a sound represent a very small energy density in the medium. Only a very small fraction of the mechanical energy of a source body is converted into acoustic energy. The conversion factor, defined as the ratio of sound power to the mechanical power of the source, is in the order of 10^{-7} to 10^{-5}. Some examples of estimated sound power conversion factor are given for typical common noise sources.

16.2 Radiation of Sound

16.2.1 Point Source

16.2.1.1 Simple Source: Spherical Wave by a Monopole

Propagation of sound pressure wave $p(x, y, z, t)$ is described by the following partial differential equation for a medium where a field point is expressed by orthogonal coordinate system O-xyz, as shown in Figure 16.1(a):

$$\nabla^2 p - \frac{1}{c^2}\frac{\partial^2 p}{\partial t^2} = 0,$$
$$\nabla^2 \equiv \left(\frac{\partial^2}{\partial x^2} + \frac{\partial^2}{\partial y^2} + \frac{\partial^2}{\partial z^2}\right) \tag{16.1}$$

where c denotes the velocity of sound propagation. If the source region is compact and the generating motion has no preferred direction, it will produce a wave, which spreads spherically outwards. As the medium is assumed infinite in extent, the waveform will depend on the distance, r, from the center of the source. The wave equation in this case is

$$\frac{1}{r^2}\frac{\partial}{\partial r}\left(r^2\frac{\partial p}{\partial r}\right) - \frac{1}{c^2}\frac{\partial^2 p}{\partial t^2} = 0 \tag{16.2}$$

When a monopole source of angular frequency ω is assumed, the simplest solution for the outward propagating waveform is expressed as

$$p(r,t) = p_\omega(\mathbf{r})\mathrm{e}^{-\mathrm{i}\omega t},$$
$$p_\omega(\mathbf{r}) = \frac{-\omega\rho}{4\pi r}S_\omega\,\mathrm{e}^{ikr}, \quad k = \frac{\omega}{c} = \frac{2\pi}{\lambda} \tag{16.3}$$

where ρ denotes the density of the medium and k denotes the wave number. Here, p_ω is used to denote the sinusoidal component of the sound pressure with angular frequency, ω. The subscript ω on a variable typically indicates the sinusoidal component of a variable, but the variables related to sound energy and sound power, such as w, I, W, do not have the subscript ω even if they mean the sinusoidal component. In this case, the source of sound is taken as a "pulsating globe" of radius a and radial velocity U_ω on the surface. Therefore, the flow outward from the origin, S_ω, is related to U_ω as follows, and as shown in Figure 16.1(a):

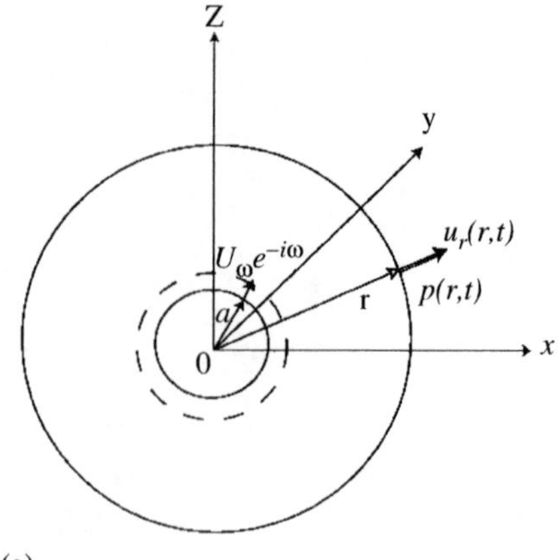

(a)

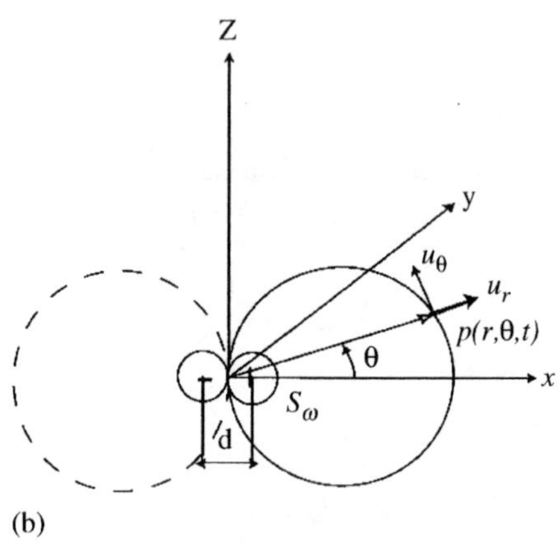

(b)

FIGURE 16.1 Directivity of monopole and dipole, (a) Spherical sound field by a monopole; (b) Axi-symmetric sound field by a dipole.

$$S_\omega = 4\pi a^2 U_\omega \tag{16.4}$$

We should note that, while a pulsating globe with a finite radius a is assumed as the physical sound source, the sound field by a monopole with infinitesimal small size and finite magnitude, as expressed mathematically in Equation 16.3, is used.

The other quantities related to the spherical wave are described next [1]:

$$\begin{cases} u_{r\omega} = \dfrac{-1}{4\pi r^2}(ikr - 1)S_\omega\,\mathrm{e}^{ik(r-ct)}; & \text{radial velocity} \\ w = \rho\left(\dfrac{1}{4\pi r^2}\right)^2 |S_\omega|^2\left[(kr)^2 + \dfrac{1}{2}\right]; & \text{energy density} \\ I = \rho c\left(\dfrac{k}{4\pi r}\right)^2 |S_\omega|^2 = \dfrac{|p|^2}{\rho c}; & \text{energy flux intensity} \\ W = (4\pi r^2)I = \rho c\dfrac{\pi}{\lambda^2}|S_\omega|^2 = \dfrac{\rho\omega^2}{4\pi c}|S_\omega|^2; & \text{total power} \end{cases} \tag{16.5}$$

We should note that the first term in the parentheses, kr, is negligible in the region where $|kr| << 1$ is valid. Hence,

$$w = \frac{\rho}{2}(|S_\omega|/4\pi r^2)^2 = \frac{\rho}{2}u_{r\omega}^2 = \left(\frac{\lambda}{\omega}\right)\frac{I}{4\pi r^2}$$

is reduced. Conversely, when $|kr| >> 1$ is valid, then $w = \rho(k/4\pi r)^2|S_\omega|^2 = I/c$ is valid.

16.2.1.2 Simple Source: Plane Wave by an Alternating Piston

Another example of simple sound wave is generated in the one-dimensional field of fluid medium, as shown in Figure 16.1. Let us set the coordinate x along the axis of wave propagation, for example, the axis of duct with a constant cross-sectional area. Then the wave equation is

$$\frac{\partial^2 p}{\partial x^2} - \frac{1}{c^2}\frac{\partial^2 p}{\partial t^2} = 0 \tag{16.6}$$

The solution for a periodic source is given by

$$p(x,t) = p_\omega(x)\mathrm{e}^{-i\omega t}, \quad p_\omega(x) = \rho c u_\omega \, \mathrm{e}^{ikx} \tag{16.7}$$

This is known as a *plane wave*, which is generated by the piston motion at the origin, the velocity of which is expressed by

$$u(t) = u_\omega(x)\mathrm{e}^{-i\omega t} \tag{16.8}$$

The other quantities related with the plane wave are given below:

$$\begin{cases} u_\omega(x) = u_\omega \, \mathrm{e}^{ikx}; & \text{particle velocity} \\ w = I = \rho c|u_\omega|^2 = |p_\omega|^2/\rho c^2; & \text{energy density, sound intensity} \\ W = SI \ (S;\ \text{cross-sectional area}); & \text{total power} \end{cases} \tag{16.9}$$

A plane wave is generated in very limited situations, but its utility is rather wide since the sound wave propagating through a duct or duct-like space with a gradually varying cross section is approximated as the plane wave. Network theory is applied to the sound wave propagating through a branch and junction by using the description of a plane wave.

16.2.1.3 Dipole and Multipoles and Their Sound Field

Let us return to the three-dimensional sound field. The second simple solution to Equation 16.1 is the "dipole" sound field. Suppose that a pair of monopoles, close together, opposite in sign, and equal in magnitude, S_ω, are located along the x-axis as shown in Figure 16.1(b). Since only a preferred direction is assigned along the x-axis, the sound field is axisymmetric as represented by

$$p(r,\theta,t) = -k^2 D_\omega \frac{\rho c \cos\theta}{4\pi r}\left(1 + \frac{i}{kr}\right)\mathrm{e}^{-i(\omega t - kr)} \tag{16.10}$$

D_ω is defined by

$$D_\omega = S_\omega d \tag{16.11}$$

where d denotes the separation of the monopoles as shown in Figure 16.1(b). Mathematically, d tends to zero, keeping D_ω finite, and the preferred axis is the x-axis. Physically, a sound field is commonly realized by a pair of monopoles with a finite separation that is short compared with the wavelength, λ, as illustrated next with realistic examples.

The characteristic quantities relating with dipole sound field are described below [1]:

$$\begin{cases} u_{r\omega} = -\dfrac{k^2 D_\omega \cos\theta}{4\pi r}\left(1 + \dfrac{2\mathrm{i}}{kr} - \dfrac{2}{k^2r^2}\right)\mathrm{e}^{-\mathrm{i}\omega t + kr}; & \text{radial velocity} \\ u_{\theta\omega} = \mathrm{i}\dfrac{k^2 D_\omega \sin\theta}{4\pi r^2}\left(1 + \dfrac{\mathrm{i}}{kr}\right)\mathrm{e}^{-\mathrm{i}\omega t + kr;} & \text{peripheral velocity} \\ w = \rho\left(\dfrac{k^2|D_\omega|}{4\pi r}\right)^2\left[\cos^2\theta + \dfrac{1}{2}\left(\dfrac{1}{kr}\right)^2 + \dfrac{1}{2}\left(\dfrac{1}{kr}\right)^4(1 + 3\cos^2\theta)\right]; & \text{energy density} \\ I_r = \rho c\left(\dfrac{k^2|D_\omega|}{4\pi r}\right)^2\cos^2\theta, \quad I_\theta = 0; & \text{sound intensity} \\ W = \dfrac{\rho\omega^4}{12\pi c^3}|D_\omega|^2; & \text{total power} \end{cases} \tag{16.12}$$

We should add the following notes on the dipole sound field.

When $|kr| >> 1$ is assumed, the second term in parentheses in Equation 16.10 is negligible. Then the directivity for $p(r, \theta, t)$ is expressed by $\cos\theta$. A similar directivity is found on I_r and on $u_{r\omega}$, $u_{\theta\omega}$, and w with the assumption $|kr| >> 1$.

A pair of dipoles produces a *quadrupole*, a pair of quadrupoles produce an *octopole*, and so on. These are called *multipole* in general. Out of multipoles, the quadrupole is common in representing a sound field generated by mixing fluid flow, especially jet flow. More details on multipoles are found in Ref. [1].

16.2.2 Sources of Finite Volume

16.2.2.1 Description of Sound Field by Green's Function

In order to describe the sound field from distributed sources, source terms are introduced to the right side of Equation 16.1. The partial differential equation with source term is derived from the equation system representing the dynamics of fluid flow with periodic motion at angular frequency ω:

$$\nabla^2 p_\omega + k^2 p_\omega = -m_\omega + \operatorname{div}\mathbf{F}_\omega \tag{16.13}$$

where p_ω denotes the acoustic pressure amplitude according to $p(x, y, z, t) = p_\omega(x, y, z)\mathrm{e}^{-\mathrm{i}\omega t}$, m_ω denotes the effective monopole source density expressed by $-\mathrm{i}\omega\rho s_\omega$, s_ω denotes the generalization of the point source strength, S_ω, of Equation 16.4 for a distributed source, and $\mathbf{F}_\omega$ denotes the vector representation of point force-density in the fluid. Introduction of Green's function, g_ω, of angular frequency, ω, satisfying the following equation is useful in general:

$$\nabla^2 g_\omega + k^2 g_\omega = -\delta(x - x_0)\delta(y - y_0)\delta(z - z_0)\mathrm{e}^{\mathrm{i}\omega t_0} \tag{16.14}$$

Here, $\delta(z)$ denotes the Dirac impulse (delta) function of the variable z; the coordinate (x_0, y_0, z_0) denotes the position of unit source, r_0, with periodic angular velocity, ω, and t_0 denotes the time pertaining to the source. The solution of Equation 16.14 is

$$g_\omega(\mathbf{r}, \mathbf{r}_0) = \frac{1}{4\pi R}\mathrm{e}^{\mathrm{i}kr}, \quad |\mathbf{r} - \mathbf{r}_0| = \sqrt{(x - x_0)^2 + (y - y_0)^2 + (z - z_0)^2} \tag{16.15}$$

where $\mathbf{r}$ denotes the position vector of sound field with coordinates (x, y, z). The solution of Equation 16.13 is described by using $g_\omega(\mathbf{r}, \mathbf{r}_0)$ as follows:

$$p_\omega(\mathbf{r}) = \iiint [m_\omega(\mathbf{r}_0) - \operatorname{div}\mathbf{F}_\omega(\mathbf{r}_0)]g_\omega(\mathbf{r}, \mathbf{r}_0)\mathrm{d}x_0\mathrm{d}y_0\mathrm{d}z_0 \tag{16.16}$$

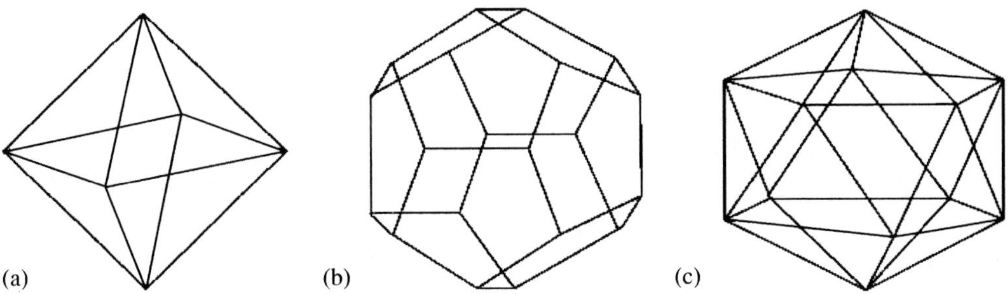

FIGURE 16.2 Practical monopole sources.

16.2.2.2 Radiation from Vibrating Small Body

A pulsating globe, as the simplest sound source, is rarely realized in the real world. The approximated monopole source required in most measurements is an eight-sided polyhedron, as depicted in Figure 16.2, where a loud speaker is installed on each of the surfaces. The sound field radiated by a thus approximated source is almost the same as that by a pulsating globe at the far field, where $kr >> 1$ is valid.

In this case, the sound field is represented by Equation 16.16, assuming that $\mathbf{F}_\omega(\mathbf{r}_0) = 0$, $m_\omega(\mathbf{r}_0) = -\mathrm{i}\omega\rho U_\omega \delta(n)$, and $\mathrm{d}x_0\mathrm{d}y_0\mathrm{d}z_0 = \mathrm{d}n\mathrm{d}S$, where the source $m_\omega(\mathbf{r}_0)$ is distributed on a thin layer of thickness, dn, on the spherical surface element, dS. Substitution of these relationships into Equation 16.16 yields the formula

$$\begin{aligned} p_\omega(\mathbf{r}) &= -\iint_S \frac{\mathrm{i}\omega\rho U_\omega}{4\pi R}\mathrm{e}^{\mathrm{i}kR}\mathrm{d}S \\ &= -\frac{\mathrm{i}\omega\rho}{4\pi R_0}(4\pi a^2)U_\omega\,\mathrm{e}^{\mathrm{i}kR_0} \end{aligned} \tag{16.17}$$

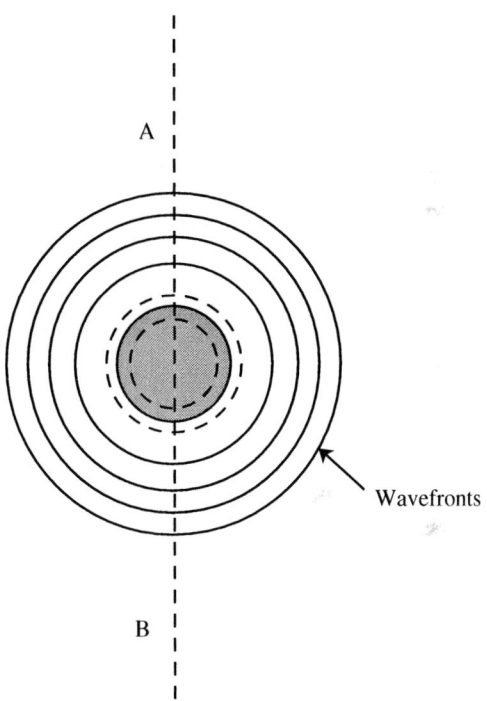

FIGURE 16.3 Field of a pulsating spherical source.

where $\iint_S \mathrm{d}S$ means integration on the approximately spherical surface, and then $R \equiv |\mathbf{r} - \mathbf{r}_0| \cong R_0 = \sqrt{x^2 + y^2 + z^2}$ is applied. The final reduction of Equation 16.17 gives the same expression as is deduced from Equation 16.3 and Equation 16.4.

The sound field generated by a monopole source is illustrated in Figure 16.1. This field is unchanged with the presence of a rigid plane AB in Figure 16.3. It is clear, on account of symmetry, that the presence of the rigid plane does not alter the sound field in any way, because only the tangential component of particle velocity is induced on the plane. This utilization of the symmetry and construction of the semi-infinite field is applied to the expression of the sound field generated by the baffled structure, as shown in Figure 16.4.

Let us imagine a sound field generated by an oscillating small body in an infinite medium, as shown in Figure 16.4(a). Since the oscillation is caused by an external force, the sound source is modeled by the force $-\mathbf{F}_\omega(\mathbf{r}_0)$. Therefore, the sound field is described by Equation 16.16, assuming $m_\omega(\mathbf{r}_0) = 0$. After applying a law of vector analysis, such as $g_\omega(\mathbf{r}, \mathbf{r}_0)\mathrm{div}\,\mathbf{F}_\omega(\mathbf{r}_0)\cdot\mathrm{grad}_0\, g_\omega + \mathrm{div}_0(g_\omega\mathbf{F}_\omega)$, the sound field is

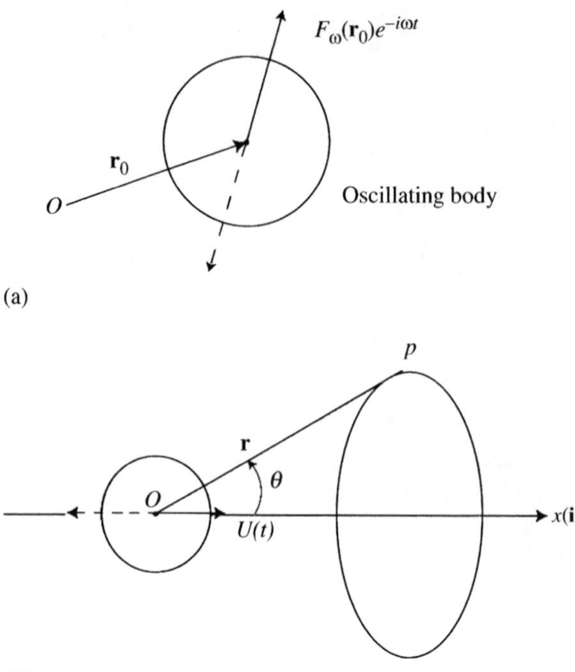

FIGURE 16.4 Dipole field generated by an oscillating small body: (a) oscillating small body; (b) oscillating sphere along x-axis.

represented by

$$p_\omega(\mathbf{r}) = \iiint \mathbf{F}_\omega(\mathbf{r}_0)\cdot\boldsymbol{\nabla}_0 g_\omega(\mathbf{r},\mathbf{r}_0)\mathrm{d}x_0\mathrm{d}y_0\mathrm{d}z_0 \tag{16.18}$$

where $\iiint \mathrm{div}_0(g_\omega\mathbf{F}_\omega)\mathrm{d}x_0\mathrm{d}y_0\mathrm{d}z_0 = \iint_S g_\omega\mathbf{F}_\omega\mathrm{d}S_n(x_0,y_0,z_0) = 0$ is applied. Assume that a small sphere of radius a_0 is oscillating along the x-axis with angular frequency ω. Instead of an oscillating sphere with velocity $U_t = U_\omega\,\mathrm{e}^{-\mathrm{i}\omega t}$, the sound field is generated by the concentrated body force, $F(t) = -m(\mathrm{d}/\mathrm{d}t)U(t) = \mathrm{i}\omega m U_\omega\,\mathrm{e}^{-\mathrm{i}\omega t}$, ($m$ is the mass of the sphere), at the origin, $\mathbf{r}_0 = 0$. Then $\mathbf{F}_\omega(\mathbf{r}_0)$ is expressed by $F_\omega(\mathbf{r}_0) = (F_{x\omega}\mathbf{i})\delta(x_0)\delta(y_0)\delta(z_0)$; $F_{x\omega} = im\omega U_\omega$. This approximate reduction is appropriate when $ka_0 << 1$ is valid. Substituting the approximation, Equation 16.18 is rewritten as

$$p_\omega(\mathbf{r}) \cong F_{x\omega}\frac{\partial}{\partial x_0}g_\omega(\mathbf{r},\mathbf{r}_0,\theta) = -k^2D_\omega\frac{\rho c}{4\pi r}\cos\theta\left(1+\frac{\mathrm{i}}{kr}\right)\mathrm{e}^{ikr} \tag{16.19a}$$

$$D_\omega = \frac{\mathrm{i}}{k\rho c}F_{\omega x} \tag{16.19b}$$

As the sound field is axisymmetric about x-axis, the sound pressure, $p_\omega(\mathbf{r},\theta)$, depends only on $\mathbf{r}$ and θ, where $\mathbf{r}$ and θ are defined in Figure 16.4(b). The expression is applicable to the sound field generated by an oscillating small body in a free space.

16.2.3 Radiation from a Plane Surface

16.2.3.1 Radiation from a Small Body in Infinite Plane Surface

The introduction of an infinite rigid plane surface to the sound field, as shown in Figure 16.3, simplifies the formulation of the sound field generated by an oscillating body adjacent to a large plane.

The configuration discussed in this section relates to a source in the presence of an infinite plane barrier, so that the medium is confined to one side of the plane.

By taking the effect of the image caused by the rigid plane surface, the Green's function, $g_\omega(\mathbf{r}, \mathbf{r}_0)$, in this case simplifies as given below, to what is called Rayleigh's formula [2]:

$$g_\omega(\mathbf{r}, \mathbf{r}_0) = \frac{1}{2\pi R} \mathrm{e}^{ikr}, \quad R = |\mathbf{r} - \mathbf{r}_0| \tag{16.20}$$

where $\mathbf{r}_0$ denotes the projection position of the source on the surface. Therefore, Equation 16.16 is rewritten as follows:

$$p_\omega(\mathbf{r}) = \iiint \frac{\mathrm{e}^{ikR}}{2\pi R} [m_\omega(\mathbf{r}_0) - \operatorname{div} \mathbf{F}_\omega(\mathbf{r}_0)] \mathrm{d}x_0 \mathrm{d}y_0 \mathrm{d}z_0 \tag{16.21}$$

16.2.3.2 Radiation from a Circular Piston

Let us consider the sound field generated by a rigid circular piston of radius a mounted flush with the surface of an infinite baffle and vibrating with simple harmonic motion of angular frequency, ω. The solution of this example is applicable to a number of related problems, including the radiation from the open end of a flanged organ pipe.

The coordinate system is shown in Figure 16.5, where the infinite baffle and the piston are placed in the Oxy-plane. Note that the observation point, P, is denoted by $\mathbf{r} = \overrightarrow{OP}$, while the source point, Q, is denoted by $\mathbf{r}_0 = \overrightarrow{OQ}$.

Since sound is observed only in the semi-infinite plane where $z > 0$ is valid, only $m_\omega(\mathbf{r}_0)$ remains nonzero in Equation 16.21. As the velocity of the piston is denoted by $U_\omega \, \mathrm{e}^{-\mathrm{i}\omega t}$ along the z-axis, $m_\omega(\mathbf{r}_0) = -\mathrm{i}\omega\rho U_\omega \delta(z_0)$, is distributed only on the circular piston. Finally, Equation 16.21 is rewritten as

$$p_\omega(\mathbf{r}) = \frac{-\mathrm{i}\omega\rho U_\omega}{2\pi} \iint_S \frac{\mathrm{e}^{ikR}}{R} \mathrm{d}x_0 \mathrm{d}y_0 \tag{16.22}$$

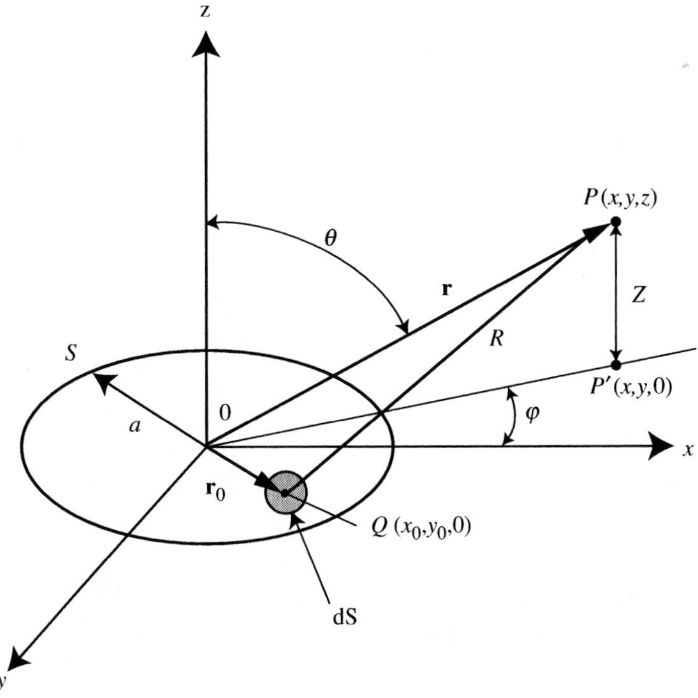

FIGURE 16.5 The rigid circular piston and the coordinate system.

Assuming $a << r$, the following approximation can be made:

$$R \cong r - (x_0 \cos\varphi + y_0 \sin\varphi)\sin\theta, \qquad \frac{1}{R} \cong \frac{1}{r} \tag{16.23}$$

where φ and θ denote the angles defining the observation point, P, as shown in Figure 16.5. By changing from Cartesian coordinates, x_0, y_0, to polar coordinates, ρ, φ_0, such that $x_0 = \rho\cos\varphi_0$, $y_0 = \rho\sin\varphi_0$ in the integral above, we rewrite Equation 16.22 as

$$p_\omega(\mathbf{r}) = -\frac{\mathrm{i}\omega\rho U_\omega}{2\pi R}\mathrm{e}^{\mathrm{i}kr}\int_0^{2\pi}\mathrm{d}\varphi_0 \times \int_0^a \exp[-k\rho\cos(\varphi_0 - \varphi)\sin\theta]\rho\mathrm{d}\rho \tag{16.24}$$

The integration is performed by introducing the Bessel function of the first order, $J_1(z)$ as follows:

$$p_\omega(\mathbf{r}) = -\frac{\mathrm{i}\omega\rho\,\mathrm{e}^{\mathrm{i}kr}}{2\pi r}f_\omega(\theta), \qquad f_\omega(\theta) = \pi a^2 U_\omega\left[\frac{2J_1(ka\sin\theta)}{ka\sin\theta}\right] \tag{16.25}$$

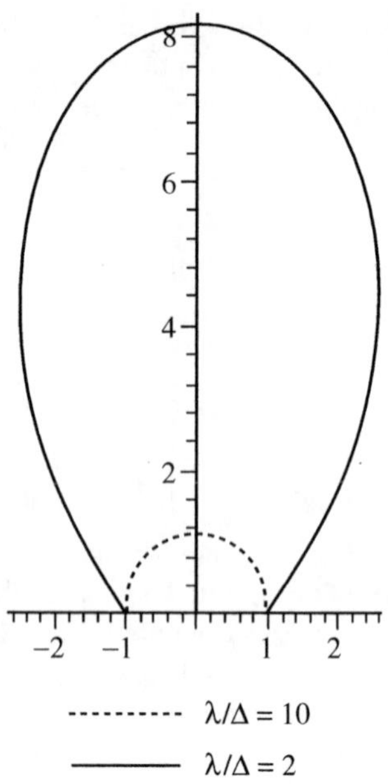

FIGURE 16.6 Directivity of sound intensity generated by an oscillating piston.

Note that dependency on φ has disappeared in the integration process, which follows from the axisymmetry of the sound field. The corresponding intensity at $\mathbf{r}$ is given by

$$I_r(\theta) \cong \frac{|p_\omega|}{\rho c} = \frac{\rho c U_\omega^2 a^2}{4r^2}(ka)^2\left[\frac{2J_1(ka\sin\theta)}{ka\sin\theta}\right]^2 \tag{16.26}$$

Figure 16.6 illustrates the dependency of I_r on θ for two cases of ka. Note that, for the smaller $ka = 2\pi a/\lambda$, I_r is almost independent on θ, which is similar to the dependence of the monopole.

16.2.3.3 Radiation from a Rectangular Plate

An normal velocity distribution, $u_\omega(x_0, y_0)$, is prescribed over a baffled planar radiator located in the plane $z_0 = 0$ in the region $-L_x \le x_0 \le L_x$, $-L_y \le y_0 \le L_y$, as shown in Figure 16.7.

In this case, the sound pressure field is represented by the following equation, similar to the previous section:

$$p_\omega(\mathbf{r}) = \frac{-\mathrm{i}\omega\rho}{2\pi}\int_{-L_y}^{L_y}\mathrm{d}y_0\int_{-L_x}^{L_x}\frac{U_\omega(x_0, y_0)\mathrm{e}^{\mathrm{i}kR}}{R}\mathrm{d}x_0 \tag{16.27}$$

Assuming $2L_x$, $2L_y << r$, the same approximation in as Equation 16.23 is acceptable. Therefore, $p_\omega(\mathbf{r}) = p_\omega(r, \theta, \phi)$ takes the form [2]:

$$p_\omega(r, \theta, \phi) = \frac{-\mathrm{i}\omega\rho\,\mathrm{e}^{\mathrm{i}kr}}{2\pi r}\int_{-L_y}^{L_y}\mathrm{d}y_0\int_{-L_x}^{L_x}U_\omega(x_0, y_0)\exp[-\mathrm{i}k\sin\theta(x_0\cos\phi + y_0\sin\phi)]\mathrm{d}x_0 \tag{16.28}$$

The result of the above integration, assuming $U_\omega(x_0, y_0) = U_\omega = \text{const}$, which means the rigid rectangular piston oscillates with amplitude U_ω along the z-axis, is

$$p_\omega(r, \theta, \phi) = \frac{\mathrm{i}\omega\rho U_\omega}{2\pi r}(4L_xL_y)\mathrm{e}^{\mathrm{i}kr}S(kL_x\sin\theta\cos\phi)S(kL_y\sin\theta\sin\phi) \tag{16.29a}$$

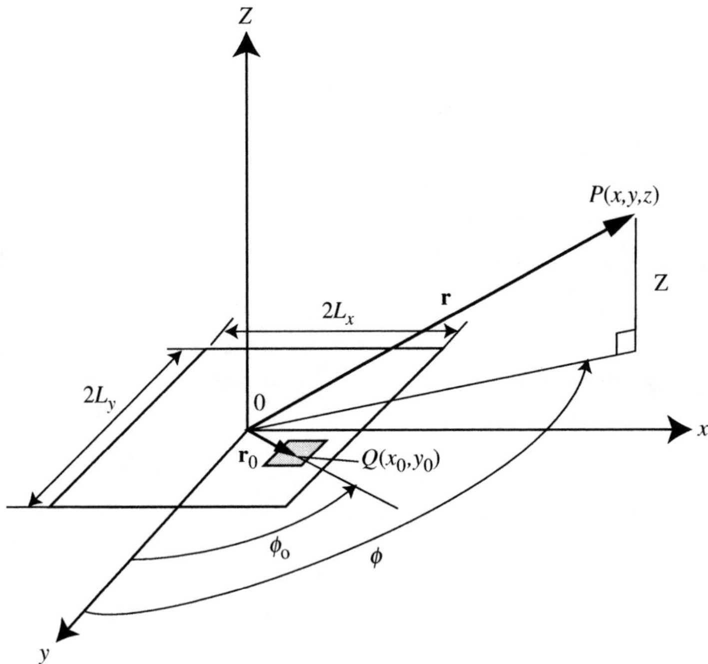

FIGURE 16.7 A flexural rectangular plate and the coordinate system.

where the function $S(z)$ is defined by

$$S(z) = \frac{\sin z}{z} \tag{16.29b}$$

Now, let us discuss the case of flexural plate, which has the same lengths, $2L_x$ and $2L_y$, but $U_\omega(x_0, y_0)$ represents the flexural vibration mode. Flexural mode patterns of rectangular panels take the general form of contiguous regions of roughly equal area and shape, which vary alternately in vibration phase, and are separated by nodal lines of zero vibration.

For simply supported edges, the normal vibration velocity distribution is

$$U_\omega(x_0, y_0) = U_{mn} \sin\left(\frac{m\pi x}{2L_x}\right) \sin\left(\frac{n\pi y}{2L_y}\right), \quad (0 \leq x \leq 2L_x,\ 0 \leq y \leq 2L_y) \tag{16.30}$$

Note that the coordinate system $O\text{-}x_0y_0z_0$ shifts its origin from the center point of the panel to the edge at the leftmost and frontmost point in Figure 16.7.

The approximation represented by Equation 16.23 is valid as well in this case. The result of integration is rather simple as given below [3]:

$$p_\omega(r, \theta, \phi) = \frac{-\mathrm{i}\omega\rho U_{mn}\,\mathrm{e}^{kr}}{2\pi r}\,\frac{4L_xL_y}{mn\pi^2}\left[\frac{(-1)^m\,\mathrm{e}^{\mathrm{i}\alpha} - 1}{(\alpha/m\pi)^2 - 1}\right]\left[\frac{(-1)^n\,\mathrm{e}^{\mathrm{i}\beta} - 1}{(\beta/n\pi)^2 - 1}\right] \tag{16.31}$$

where α and β are defined by

$$\alpha = 2kL_x \sin\theta\cos\phi, \quad \beta = 2kL_y \sin\theta\sin\phi \tag{16.32}$$

The corresponding intensity, $I_r(\theta, \varphi)$, at **r** is expressed by

$$I_r(r, \theta, \phi) = \frac{|p_\omega|^2}{\rho c} = 4\rho c|U_{mn}|^2\left(\frac{4kL_xL_y}{\pi^3 rmn}\right)^2\left\{\frac{\begin{matrix}\cos\\ \sin\end{matrix}\left(\frac{\alpha}{2}\right)\begin{matrix}\cos\\ \sin\end{matrix}\left(\frac{\beta}{2}\right)}{[(\alpha/m\pi)^2 - 1][(\beta/n\pi)^2 - 1]}\right\} \tag{16.33}$$

where cos(α/2) is used when m is an odd integer, and sin(β/2) is used when m is an even integer; cos(β/2) is used for even n and sin(β/2) for odd n.

16.2.4 Estimation of Noise-Source Sound Power

16.2.4.1 Power Conversion Factor of Machinery

It is often necessary to estimate the expected sound power that a particular machine might introduce into an environment [4]. One way where such an estimate may be approached for a particular class of machine is by means of the sound power conversion factor, η_n. This factor is defined as

$$\eta_n = \frac{P}{P_{\mathrm{m}}} \tag{16.34}$$

where P = sound power of the machine (W), and P_{m} = power of the machine (W). This relationship is valid for both mechanical and electrical machinery. The conversion factors for some common noise sources are given in Table 16.1.

Example 16.1

Estimate the sound power level of a typical 1-kW electric motor that operates at 1200 rpm.

Solution

From Table 16.1, we find that for typical electric motors, $\eta_n = 1 \times 10^{-7}$. Thus, using Equation 16.34, we obtain

$$P = \eta_n P_{\mathrm{m}} = (1 \times 10^{-7}) \times 1000 = 10^{-4} (\mathrm{W})$$

as the total sound power of motor. Then using Equation 37.7, L_{W} is given by

$$L_{\mathrm{W}} = 10 \log\left(\frac{10^{-4}}{10^{-12}}\right) = 80 (\mathrm{dB})$$

16.2.4.2 Fan Noise

We are familiar with noise nuisance caused by a domestic ventilating fan. The mechanical power of the fan is expressed by

$$P_{\mathrm{m}} = p_{\mathrm{T}} Q \tag{16.35}$$

where p_{T} denotes the total pressure rise through the fan and Q denotes the volumetric flow rate. According to the law of sixth power of flow velocity deduced by the aeroacoustics theory, sound power of the fan is proportional to $p_{\mathrm{T}^{2.5}} Q$. Therefore, the specific ratio k_{T} defined below is more useful than η_n for the fan:

$$k_{\mathrm{T}} = \frac{P_{\mathrm{m}}}{p_{\mathrm{T}^{2.5}} Q} \tag{16.36}$$

TABLE 16.1 Estimated Sound Power Conversion Factors for Common Noise Sources[a]

Noise Source	Conversion Factor		
	Low	Midrange	High
Compressor, air (1–100 hp)	3×10^{-7}	5.3×10^{-7}	1×10^{-6}
Gear trains	1.5×10^{-8}	5×10^{-7}	1.5×10^{-6}
Loud speakers	3×10^{-2}	5×10^{-2}	1×10^{-1}
Motors, diesel	2×10^{-7}	5×10^{-7}	2.5×10^{-6}
Motors, electric (1200 rpm)	1×10^{-8}	1×10^{-7}	3×10^{-7}
Pumps, over 1600 rpm	3.5×10^{-6}	1.4×10^{-5}	5×10^{-5}
Pumps, under 1600 rpm	1.1×10^{-6}	4.4×10^{-6}	1.6×10^{-5}
Turbines, gas	2×10^{-6}	5×10^{-6}	5×10^{-5}

[a] Total sound power for the four octave bands from 500 to 4000 Hz.

Source: Irwin, J. D. and Graf, E. R. 1979. *Industrial Noise and Vibration Control*, Prentice Hall, Englewood Cliffs, NJ.

TABLE 16.2 Specific Fan Noise Level for Low-Pressure Fans

Type	K_T (dB)
Axial	−87.5 to −70.7
Centrifugal/cirroco	−87.5 to −85.7
Centrifugal/radial	−98.7 to −84.7
Centrifugal/turbo	−104.7 to −89.7
Cross-flow	−76.7 to −68.7

Furthermore, usually the specific fan noise level, K_T, as given below is more useful than k_T.

$$K_T = L_W - 10\log(p_{T^{2.5}}Q)\text{dB} \tag{16.37}$$

where p_T denotes the total pressure rise in mmAg, Q denotes the flow rate in m^3/min, L_W denotes the total sound power level in dB, and K_T represents the radiation efficiency of the fan noise. This efficiency varies with the type of the fan and with the flow rate when the model is assigned. In particular, K_T will be the lowest at the flow rate at which the aerodynamic power efficiency is the highest. Therefore, we will have the best advantage on the sound environment when the fan is operated at the highest efficiency. For convenience of design, the L_W is often evaluated by A-weighted total sound level (dB-A) not by linear total level (dB). Table 16.2 gives the specific fan noise level evaluated with dB-A for five types of the low-pressure fans.

Example 16.2

Consider a ventilating axial-flow fan, the specifications of which are: $p_T = 10$ mmAq, $Q = 30\ m^3$/min, $D = 30$ cm (diameter of the duct containing fan rotor). Estimate the directional distribution of intensity $I_r(\theta)$, assuming $r = 3$ m and $f = 150$ Hz for the main component of the fan noise, and $K_T = -79.0$ dB-A from Table 16.2. The fan noise is assumed to be radiated as a plane sound wave at the mouth of the duct.

Solution

First, we modify the K_T in dB-A to that of linear scale. From the frequency response for A-weighting network shown in Figure 2.5, for $f = 150$ Hz, the modification is found as $\Delta K_T = 15$ dB. Then, the modified specific fan noise level is $K_T = -79.0 + 15.0 = -64.0$ dB. By using Equation 16.37, $L_W = K_T + 25\log P_T + 10\log Q = -64 + 25 + 15 = -24$ dB is obtained. This means the emitted total sound power is $W = 10^{-24/10} = 3.98 \times 10^{-3}$ W. Since we assume a plane sound wave at the mouth of the duct for the fan noise, we can use the sound radiation model of a circular piston with $a = 15$ cm for the radiated sound wave. In our case $k_a = 2\pi a/(c/f) = 0.415$, or $\lambda/a = 15.1$. A monopole model will be valid for the directional distribution of intensity from Figure 16.6. Therefore, $I_r = W/2\pi r^2 = 0.704 \times 10^{-4}\ W/m^2$ or $|p|^2 = I_r\rho c = 2.92 \times 10^{-2}(Pa^2)$. This is the same as the sound pressure level $L_p = 10\log|p|^2/p_{ref} = 78.6$ dB or $L_p(A) = L_p - 15 = 63.6$ dB-A.

References

1. Morse, P.M. and Ingard, K.U. 1986. *Theoretical Acoustics*, Princeton University Press, Princeton, NJ, chap. 7.
2. Junger, M.C. and Feit, D. 1986. *Sound, Structures, and Their Interaction*, 2nd ed., MIT Press, Cambridge, chap. 4.
3. Fahy, F. 1985. *Sound and Structural Vibration. Radiation, Transmission and Response*, Academic Press, New York, chap. 2.
4. Irwin, J.D. and Graf, E.R. 1979. *Industrial Noise and Vibration Control*, Prentice Hall, Englewood Cliffs, NJ, chap. 5.

17

Design of Absorption

Teruo Obata
Teikyo University

Summary

This chapter presents the basics of designing devices for sound absorption. The absorption coefficient and acoustic impedance are introduced. Characteristic properties and parameters of sound absorption material and basic elements are presented. Acoustic modeling, analysis, and design considerations of components, such as ducts, and noise attenuation devices, such as mufflers, are presented. A practical design example is presented for illustration of the concepts presented in the chapter.

17.1 Introduction

Sound-absorption equipment is used for multiple purposes in architectural acoustics, mechanical noise countermeasures, and so on. In this context, the necessity for designing sound-absorption equipment from the viewpoint of noise control is explained. Architectural acoustics is an important area of study, which involves architecture, sound-absorption design, and sound-measurement facility.

In the area of noise reduction, the characteristics of sound are important, and proper sound-absorbing material should be selected based on how much attenuation is necessary for each frequency of sound. In particular, acoustic characteristics of sound-absorbing material such as the type of material and the sound-absorption mechanism are important. In this chapter, the basics of sound absorption are given, and the prediction and calculation methods for attenuation of lined or dissipative mufflers are outlined.

17.2 Fundamentals of Sound Absorption

17.2.1 Attenuation of Sound

When an acoustic wave propagates in a medium, the sound energy attenuates due to such reasons as viscosity, heat conduction, and the effects of molecular absorption. In a medium of small volume surrounded by a boundary surface, the attenuation is particularly considerable, for example, when the medium is a thin tube. This is because there is the dissipation of the energy controlled by the viscosity of the medium and heat conduction between the material and the medium of tube wall. A sound-absorbing material may be utilized to adjust such dissipation of acoustic energy.

17.2.1.1 Absorption Coefficient and Normal Acoustic Impedance

Some amount of energy is lost when an acoustic wave hits the surface of a sound-absorbing material. Figure 17.1 illustrates an infinite medium of absorbing material separated by air and the reflected wave (sound pressure p_r) from the boundary surface with the air where a plane wave of sound pressure p_i is emitted in the direction indicated by an arrow, at an angle θ. When $\theta = 0$, sound pressure p in air is given by

$$p = p_i + p_r = (Ae^{-jkx} + Be^{jkx})e^{j\omega t} \tag{17.1}$$

where

A, B = the amplitude of sound pressure of incident and reflected waves (in Pa),
$j = \sqrt{-1}$,
$k = 2\pi f/c$; wave number (1/m),
ω = angular frequency (rad/sec).

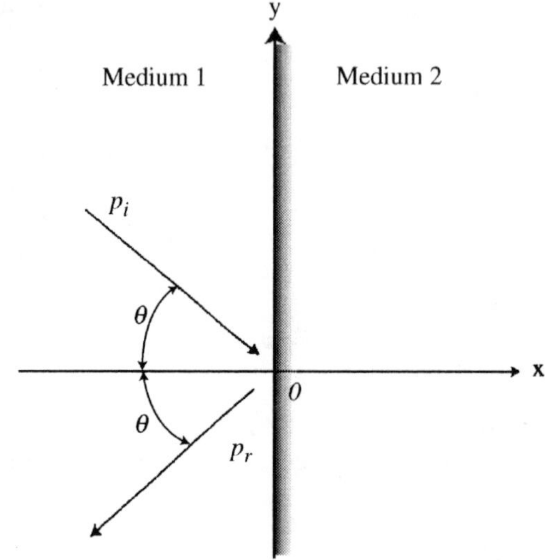

FIGURE 17.1 Plane wave incidence on an infinite absorbing material.

The sound pressure, p_m, in the absorbing material may be expressed using a complex propagation constant, by the equation:

$$p_m = p_{x=0}e^{-\gamma x}e^{j\omega t} \tag{17.2}$$

where

γ = the propagation constant in the absorbing material (m^{-1}). *Note*: $\gamma = \delta + j\beta$. γ is a property of the material itself and is not dependent on the mounting conditions when large areas of material are considered.

δ = attenuation constant. *Note*: δ tells us how much of the sound wave will be reduced as it travels through the material.

β = phase constant. *Note*: β is a measure of the velocity of propagation of the sound wave through the material.

The relation for determining the velocity of sound in the material is given by

$$c_m = \omega/\beta \tag{17.3}$$

Boundary conditions must be satisfied on the boundary surface. The acoustic impedance of a unit area of air and of absorbing material are, respectively, denoted by z and z_a. The pressure and the particle velocity on both sides of the boundary are equal. We have

$$\left.\begin{aligned} p_i + p_r &= p_{x=0} \\ \frac{p_i - p_r}{z_a} &= \frac{p_{x-0}}{z} \end{aligned}\right\} \tag{17.4}$$

The amplitude of reflectance of sound pressure, r, is obtained from Equation 17.4, and is given by

$$r = \frac{p_r}{p_i} = \frac{z_a - z}{z_a + z} \tag{17.5}$$

The reflectivity is the energy reflection rate. The absorption coefficient, α, of an absorbing material is defined as

$$\alpha = 1 - |r^2| \tag{17.6}$$

The impedance, z_n, through a surface is the quantity that represents the dissipation of energy of sound as well as the absorption coefficient. It is given as a ratio between sound pressure and particle velocity on boundary surface in the reflecting acoustic wave:

$$z_n = \left(\frac{p}{u}\right)_{x=0} = \left(\frac{\rho c}{\cos\theta}\right)\frac{p_i + p_r}{p_i - p_r} \tag{17.7}$$

Note that z_n is a complex quantity and involves both amplitude and phase, both of which depend on the sound pressure at the boundary surface in the reflecting acoustic wave.

In the case of oblique incidence, the surface impedance can be expressed by following equation:

$$z_n = Z\gamma z/q \tag{17.8}$$

where $z =$ the acoustic impedance (Pa sec/m^3). Here,

$$Z = \frac{z_1 \cosh(ql) + (\gamma z/q)\sinh(ql)}{z_1 \sinh(ql) + (\gamma z/q)\cosh(ql)}$$

$$q = \sqrt{\gamma^2 + k^2 \sin^2\theta}$$

The absorption coefficient, $\alpha(\theta)$, for an oblique incidence with angle θ may be expressed by

$$\alpha(\theta) = 1 - \left|\frac{z_n \cos\theta - \rho c}{z_n \cos\theta + \rho c}\right|^2 \tag{17.9}$$

17.3 Sound-Absorbing Materials

17.3.1 Porous Material

Porous acoustical materials are a special category of a more general class of gas–solid mixtures. They range from porous solids, for example, porous rocks, fibrous granular solids, expanded plastics, and form materials, to porous or turbid gases, for example, suspensions and emulsions. Sound is attenuated in a gas-saturated porous solid due to the restriction on the gas movement within it. A convenient microstructure model for such materials is one of a rigid solid matrix through which run cylindrical, capillary pores (tubing) with constant radius, normal to its surface. This model enables the use of Kirchoff's theory of sound propagation in narrow tubes with rigid walls. Accordingly, this mechanism of dissipation may be identified as (1) a viscous loss in the boundary layer at the wall of each capillary tube associated with the relative motion between the viscous gas and the solid wall, or (2) heat conduction between compressions and rarefactions of the gas and the conducting solid walls.

17.3.2 Tubular Material

Consider the absorption of low-frequency sound using the tubular absorbing material. By itself, sound absorption is not satisfactory with the tubular absorbing material. The material produces bending vibration due to an acoustic wave through it, and sound absorption occurs by the internal friction of the material. For hard plywood and gypsum boards, there is a natural frequency in the range 100 to 200 Hz, and the absorption coefficient ranges from 0.3 to 0.5. It is possible to increase the absorption coefficient by coating the board surface with fibrous absorbing material.

17.3.3 Membrane Material

For membrane material, the sound-absorption mechanism makes use of resonant vibration. Hence, resonant frequency is a governing parameter. The imaginary part (the reactance term) of the acoustic impedance of a membrane gives rise to a resonance. The associated natural frequency is given by

$$f_r = \frac{1}{2}\left\{\frac{1}{m}\left(\frac{1.4\times 10^5}{L}\right) + K_m\right\} \tag{17.10}$$

where

f_r = natural frequency (Hz)
m = surface density (kg/m^2)
L = thickness of air space (m)
K_m = board rigidity (kg/m^2 sec^2)

The K_m values of some boards are shown in Figure 17.2. The absorption coefficient is approximately 0.3 to 0.4 in the frequency range of 300 to 1000 Hz, when the thickness of the air space between the membrane and the rigid wall behind it is 50 to 100 mm.

17.3.4 Perforated Plate

A perforated board of sound absorbing material (i.e., a board with holes) is placed over a rigid wall at a fixed clearance, as shown in Figure 17.3. The sound-absorption characteristics depend on the board thickness, t, the hole diameter of the perforations, d, the clearance, L, between the perforated board and the rigid wall, and so on. The absorption coefficient becomes a maximum at resonant frequency. In the present case, the resonant frequency is given by

$$f_r = \frac{c}{2\pi}\sqrt{\frac{\varepsilon}{(t+0.8d)L}} \tag{17.11}$$

where sound speed is c, the airspace thickness is L (typically, 300 mm or less), and the ratio of the total area of holes to the total area of the board is ε. The absorption coefficient is approximately 0.3 to 0.4.

17.3.5 Acoustic Resonator

Yet another method of achieving sound absorption is using an acoustic resonator of Helmholtz type, which consists of a vessel of any shape containing a volume air, as shown in Figure 17.4. The air volume is in direct communication with the ambient air in the room through an interconnecting tube, which may be long or short and of any cross-sectional shape. An example of a resonator of Helmholtz type may be a 1 gal jar. When a sound wave impinges on the aperture of neck of the jar, the air in the neck will be set in oscillation, periodically expanding and compressing the air in the vessel.

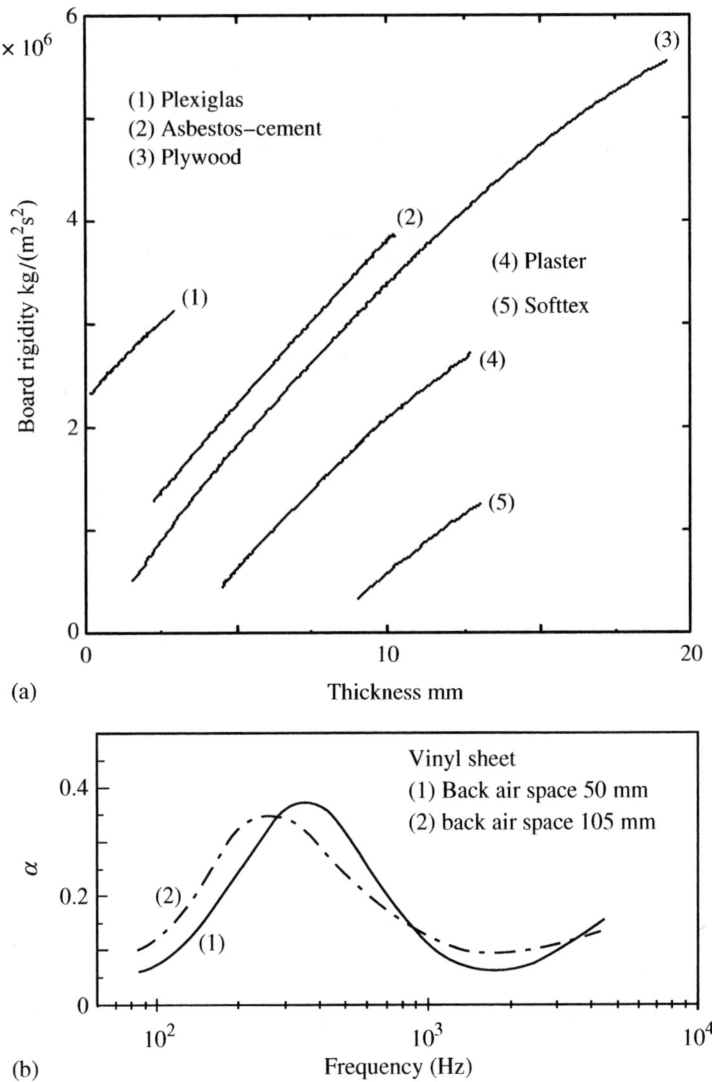

FIGURE 17.2 Some K_m values for membrane absorbing materials.

The resulting amplified motion of the air particles in the neck of the jar, due to phase cancellation between the air plug in the neck and the air volume in the vessel, causes energy dissipation due to friction in and around the neck. This type of absorber can be designed to produce maximum absorption over a very narrow frequency band or even a wide frequency band. The resonant frequency of a Helmholtz resonator may be expressed as

$$f_r = \frac{c}{2m}\sqrt{\frac{\varepsilon}{(t + 0.8d)L}} \qquad (17.12)$$

where

c = speed of sound (m/s)
S_n = cross-sectional area of neck of jar (m^2)
d_n = diameter of neck of jar (m)
V = volume of vessel (m^3)

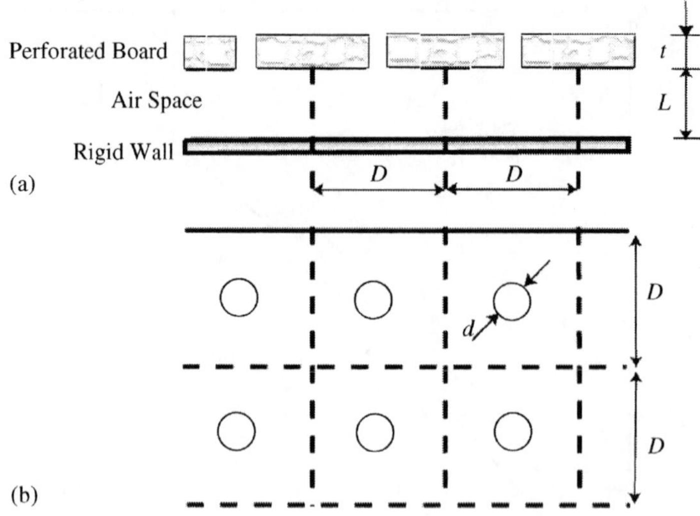

FIGURE 17.3 Sound-absorption characteristics of a perforated plate structure: (a) cross sectional view; (b) plan view.

17.4 Acoustic Characteristic Computation of Compound Wall

17.4.1 Absorption Coefficient of Combined Plate with Porous Blanket

A common form of problem in noise control is the need to reduce the sound radiated from a duct or some other object. A way to achieve this is by lining the duct with several centimeters of porous acoustic material, and covering it with a solid plate of some type, as indicated in Figure 17.5.

Consider the case of normal incidence with the sound-absorbing structure of Figure 17.5. Assume that the boundary conditions for the sound pressure and the volume flow-rate are identical. For plane wave incidence on the hard wall, the magnitude of reflection coefficient is -1 [1]. The following equation is obtained:

$$\begin{bmatrix} 1 & -1 & -1 & 0 \\ -1 & -m_1 & m_1 & 0 \\ 0 & e^{-\gamma l_1} & e^{\gamma l_1} & -(1+e^{-2jkl_2}) \\ 0 & m_2 e^{-\gamma l_1} & m_2 e^{\gamma l_1} & -(1-e^{-2jkl_2}) \end{bmatrix} \begin{bmatrix} B_1 \\ A_1 \\ B_1 \\ B_2 \end{bmatrix} = \begin{bmatrix} -1 \\ -1 \\ 0 \\ 0 \end{bmatrix} \tag{17.13}$$

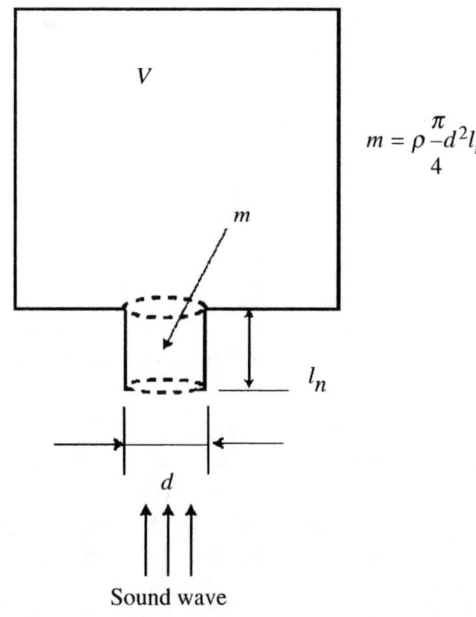

FIGURE 17.4 Geometry of a Helmholtz resonator. Volume, V, is connected to an infinitely open area by a neck tube of diameter d and length l_n.

where

$j = \sqrt{-1}$
$m_1 = z_0/z_1$, $m_2 = z_1/z_2$
z_0, z_1, z_2: acoustic impedance of each medium (Pa s/m^3)
γ = complex propagation constant (1/m)

The absorption coefficient for normal incidence is given by the following equation:

$$\alpha_0 = 1 - \left|\frac{B_0}{A_0}\right|^2 = 1 - |B_0|^2 \quad (17.14)$$

The absorption coefficient for random incidence may be approximated by

$$\alpha = \frac{1}{n}\sum_{i=1}^{n} \alpha(\theta)_i \quad (17.15)$$

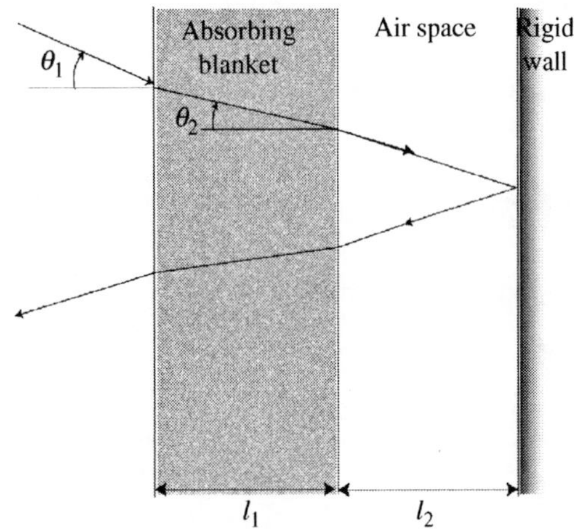

FIGURE 17.5 Structure for sound absorption using a blanket and an air space showing angles θ_1 in the air and θ_2 in the blanket.

where θ = the incident angle of sound, $0 < \theta < \pi/2$.

It is known that the propagation speed of the sound in fibrous materials changes with air, and the following equation holds on the boundary surface:

$$\sin\theta/\sin\theta' = c/c_m \quad (17.16)$$

Here, c_m is the sound speed in fibrous materials, which is calculated from the imaginary part of Equation 17.18, given later. The angle of reflection, θ', in the boundary surface of the back air space is obtained in a similar way. Hence, the following equation is substituted in Equation 17.13 instead of the thickness of the absorber, l_1, and the thickness of the air space, l_2, to obtain the absorption coefficient in oblique incidence:

$$l_1' = l_1/\cos\theta', \quad l_2' = l_2/\cos\theta'' \quad (17.17)$$

The complex propagation constant, γ, is an important physical quantity in absorbing material of propagated sound, which is given per unit length of acoustic attenuations, and phase changes. Between the aeroelasticity rate, K_a, of absorbing material and the bulk modulus, Q, of absorbing material, γ is given by the following equation, for $K_a > 20\,Q$ [2,3]:

$$\gamma = j\omega\sqrt{Y/K}\sqrt{\langle\rho_1\rangle - j\langle R_1\rangle/\omega} \quad (17.18)$$

$$\langle R_1\rangle = \frac{R_1[1 - \rho_0(1 - Y)/\rho_m]}{\left[1 + \dfrac{\rho_0(\kappa - 1)}{\rho_m}\right]^2\left[1 + \dfrac{R_1^2}{\rho_m^2\omega^2[1 + \rho_0(\kappa - 1)/\rho_m]^2}\right]}$$

$$\langle\rho_1\rangle = \rho_0\kappa - \frac{\dfrac{R_1^2(Y/\kappa + \rho_m/\rho_0\kappa)}{\rho_m^2\omega^2[1 + \rho_0(\kappa - 1)/\rho_m]^2} + \dfrac{1 + \rho_0 Y(\kappa - 1)/\rho_m\kappa}{1 + \rho_0(\kappa - 1)/\rho_m}}{1 + \dfrac{R_1^2}{\rho_m^2\omega^2[1 + \rho_0(\kappa - 1)/\rho_m]^2}}$$

where

ρ_m = density of acoustical material (kg/m^3)
ρ_0 = density of air (kg/m^3)
c_0 = speed of sound in air (m/s)

K = volume coefficient of elasticity of air (N/m^2)
R_1 = alternating flow resistance for unit thickness of material due to the difference between the velocity of the skeleton and the velocity of air in the interstices (Pa s/m^2). R_1 values are given in Table 17.1
Y = porosity = the ratio of the volume of the voids in the material to the total volume; porosity equals the total volume minus the fiber volume, all divided by total volume
$\kappa = 5.5 - 4.5Y$, the structure factor of the interstices in the skeleton
$\omega = 2\pi f$, the angular frequency (radians/s)

The acoustic impedance, z_1, of absorbing material is given by

$$z_1 = R + \mathrm{j}X = -\frac{\mathrm{j}K\gamma}{\omega Y} \tag{17.19}$$

in which

$$R = \rho_0 c_0 \left\{1 + 0.0571(\rho_0 f/R_f)^{-0.754}\right\}$$
$$X = -\rho_0 c_0 \left\{0.0870(\rho_0 f/R_f)^{-0.732}\right\}$$

17.4.2 Transmission Loss through a Single Porous Board

Assume that a sound wave impinges on the left side of a porous board at normal incidence and emerges with a reduced amplitude from the right side. The associated transmission loss of the porous board is obtained from

$$\left.\begin{aligned} \mathrm{TL}_0 &= 10\log_{10}(X+Y) \\ X &= \left\{1 + \frac{\omega^2 m^2 P R_f}{2\rho_0 c_0(\omega^2 m^2 P^2 + R_f^2)}\right\}^2 \\ Y &= \left\{\frac{\omega m R_f^2}{2\rho_0 c_0(\omega^2 m^2 P^2 + R_f^2}\right\}^2 \end{aligned}\right\} \tag{17.20}$$

where

m = surface density of the blanket (kg/m^2)
P = porosity of the blanket (porosity = the total volume minus the fiber volume, all divided by the total volume)
R_f = specific flow resistance of material (Pa s/m)

TABLE 17.1 Flow Resistance Values of Glass-Wool Board (Quality Regulation Range by JIS)

Board Type	K value	Gross Specific Gravity (kg/m^3)	Specific Flow Resistance ($\times 10^{-3}$ N s/m^4)	Standard of JIS for Glass Wool
#1 Glass-wool board	8	8 ± 2	1.5 ~ 7.0	JIS A 9505-A
	12	12 ± 2	2.5 ~ 12.0	
	16	16 ± 2	4.7 ~ 17.0	
	20	20 ± 3	5.0 ~ 22.0	
	24	24 ± 3	6.5 ~ 27.0	
#2 Glass-wool board	12	12 ± 2	1.5 ~ 7.0	JIS A 9505-B
	16	16 ± 2	2.5 ~ 10.0	
	20	20 ± 3	3.0 ~ 13.0	
	24	24 ± 3	4.0 ~ 16.0	
	32	32 ± 4	6.0 ~ 22.0	
	48	48 ± 5	11.0 ~ 38.0	
	64	64 ± 6	18.0 ~ 60.0	
	96	96 ± 10	27.0 ~ 95.0	
#3 Glass-wool board	96	96 ± 10	15.0 ~ 40.0	JIS A 9505-C

ρ_0 = density of air (kg/m^3)
c_0 = sound speed in air (m/s)

17.4.3 Transmission Loss through a Sandwich Board

Consider a wide wall formed by two panels (sheets) of infinite area separated with a homogeneous filling of fibrous acoustical material, as shown in Figure 17.6. Suppose that a plane wave impinges at an angle θ. As the pressure of both sides of the wall is equal with regard to the amplitude of the progressing wave and the reflected wave in each boundary surface, the following result may be established [4]:

$$\left.\begin{aligned} A_0 + B_0 &= A_1 + B_1 \\ (A_0 - B_0)/z_0 &= (A_1 - B_1)/z_1 \\ A_1 e^{-jkl'_1} + B_1 e^{jkl'_1} &= A_2 + B_2 \\ (A_1 e^{-jkl'_1} - B_1 e^{jkl'_1})/z_1 &= (A_2 - B_2)/z_2 \\ A_2 e^{-\gamma l'_2} + B_2 e^{\gamma l'_2} &= A_3 + B_3 \\ (A_2 e^{-\gamma l'_2} - B_2 e^{\gamma l'_2})/z_2 &= (A_3 - B_3)/z_3 \\ A_3 e^{-\gamma l'_3} + B_3 e^{\gamma l'_3} &= A_4 + B_4 \\ (A_3 e^{-\gamma l'_2} - B_3 e^{\gamma l'_2})/z_3 &= (A_4 - B_4)/z_0 \end{aligned}\right\} \quad (17.21)$$

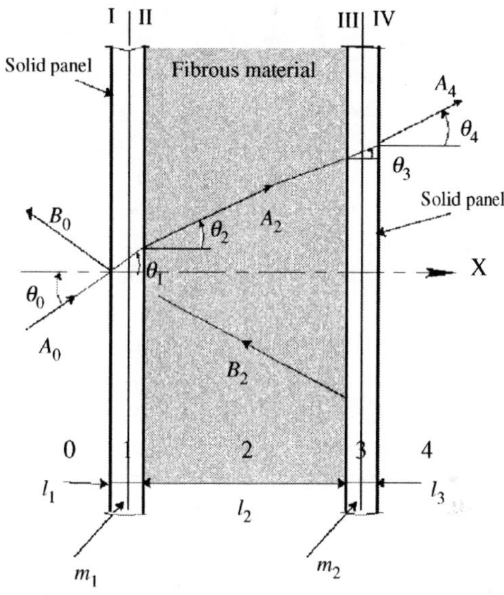

FIGURE 17.6 Cross-sectional view of a sandwich panel.

where A and B are the amplitude of sound pressures.

From Equation 17.17, $l'_1 = l_1/\cos\,\theta_1$, $l'_2 = l_2/\cos\,\theta'_2$, and l'_1 and l'_2 may be calculated.

The speed of sound in the walls is given by the following equation in terms of the modulus of longitudinal elasticity, E_i:

$$c_i = \sqrt{E_i/\rho_i} \quad (17.22)$$

The real part of acoustic impedance, z_i ($i = 1, 3$), is given by $R_i = r_i/\cos\,\theta$, and of the imaginary part is given at $X_i = m_i\omega$. The internal resistances, r_i, are functions of such factors as the material, frequency, temperature, and density. Some typical values are given in Table 17.2.

If the space of the transmission side is infinite, B_4 in Equation 17.21 becomes equal to zero. Then, the transmission loss is given is given by

$$\mathrm{TL}(\theta) = 10\,\log_{10}\left|\frac{A_4}{A_0}\right|^2 \quad (17.23)$$

TABLE 17.2 Internal Resistance Values of Several Useful Materials

Material	Thickness (mm)	Internal Resistance (Pa sec/m^3)
Aluminum	0.4	3.0
Plywood	3.0	7.5
Plaster board	7.0	15.0

17.5 Attenuation of Lined Ducts

17.5.1 Computation of Attenuation in a Lined Duct

A lined duct is an air passage with one or more of the interior surfaces covered with an acoustical material such as a glass or mineral fiber blanket. The parallel baffles are merely a series of side-by-side ducts that generally have a rectangular or round cross section. If the walls are covered with absorptive material, attenuation will occur because of the viscous motion of the air in and out of the porous of blanket.

Figure 17.7 shows an isometric illustration of a lined duct. The attenuation of sound for a lined duct is dependent primarily on the duct length, l_e, the thickness of the lining, b, the density of the lining, ρ, the width of the air passage, l, and the wavelength of sound, λ. At low frequencies ($l/\lambda < 0.1$), the attenuation of sound in a lined duct may be calculated from the following empirical formula:

$$\text{ATT} = K_l P/S \qquad (17.24)$$

where

K_l = the coefficient, which is determined from the random incidence absorption coefficient of lined material, given in the chart of Figure 17.8

P = acoustically lined perimeter of duct (m)

S = cross-sectional open area of duct (m^2)

If the absorbing material is lined in the rectangular cross section as shown in Figure 17.9 to Figure 17.11, the attenuation can be estimated using the formulas given in Table 17.3 [5].

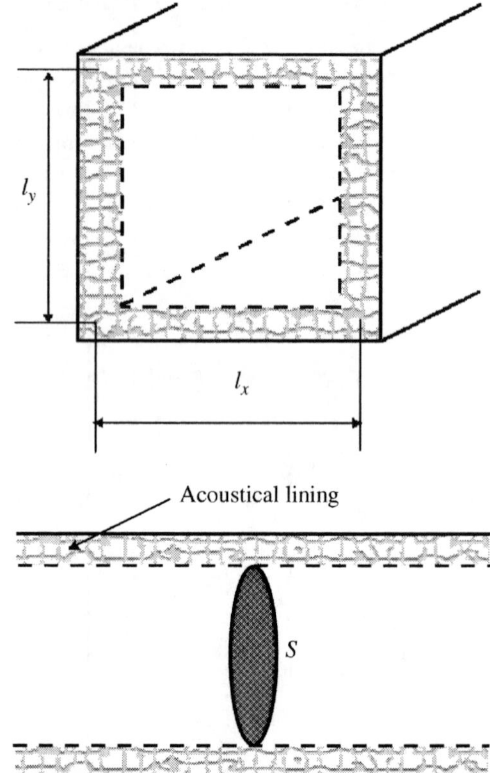

FIGURE 17.7 Illustration of a lined duct.

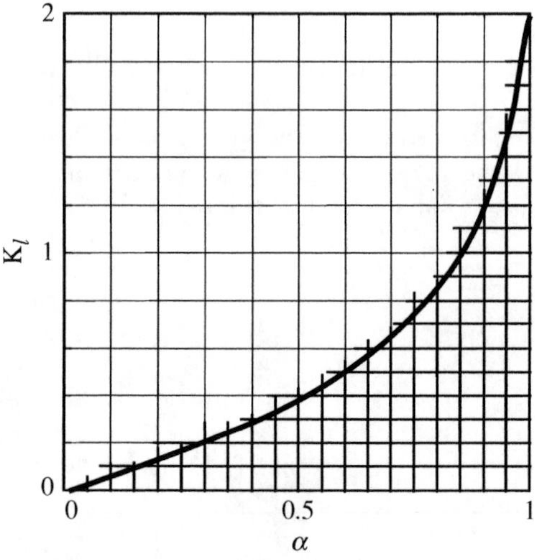

FIGURE 17.8 K_l value for sound-absorption coefficient by reverberation room method.

17.5.2 Attenuation in a Lined Bend

A lined bend duct is shown in Figure 17.12. The insertion loss, IL, of a lined bend results from two mechanisms: the reflection of sound back toward the source side, and the scattering of sound energy into the high-frequency region is rapidly attenuated by the lining beyond the bend. Higher-frequency modes will be attenuated by even an unlined duct for frequencies below the ratio of the air passage between the linings to the wavelength of sound equal to 0.5. At frequencies well above this ratio, the insertion loss of a lined bend is expected to be comparable to the reverberant-field

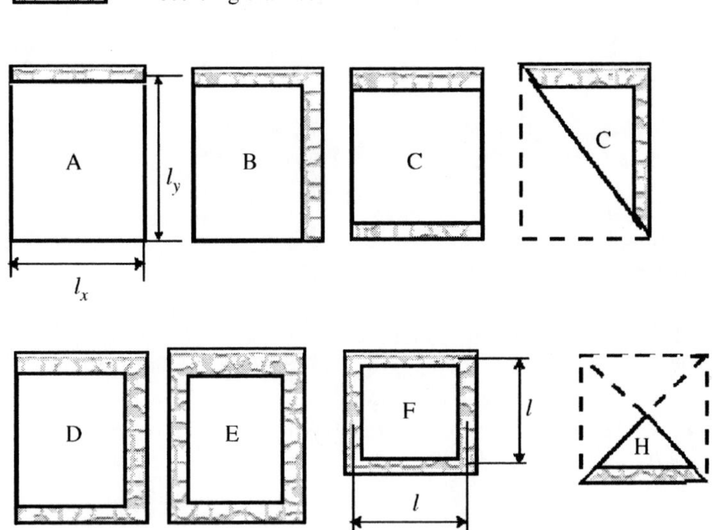

FIGURE 17.9 Duct-liner configurations corresponding to Table 42.3.

end correction derived for the duct. The insertion loss of a lined bend may be obtained as following equation [6]:

$$\mathrm{IL} = \frac{K_1 P}{S} + (l_1 + l_2) + \Phi \tag{17.25}$$

where Φ is obtained from Figure 17.13.

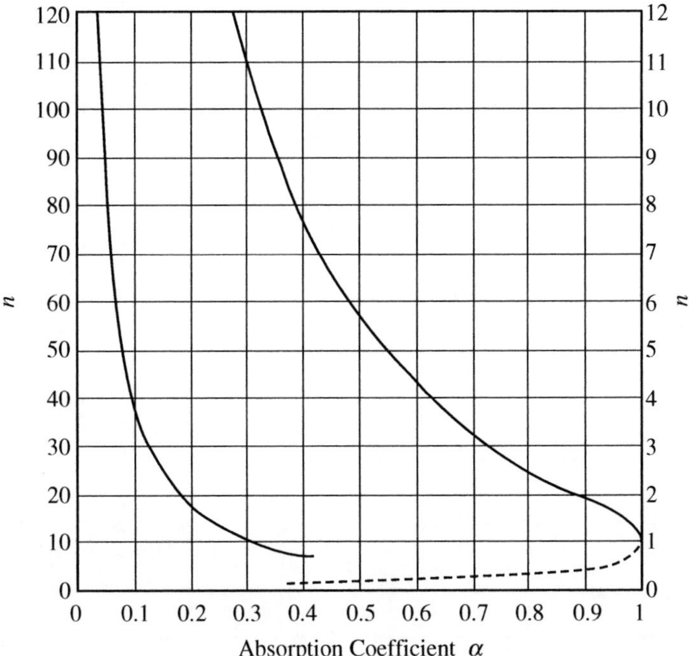

FIGURE 17.10 Relationship between absorption coefficient and stationary wave factor, n.

The total insertion loss for a lined bend is given in Figure 17.13 along with the attenuation of the lining beyond the bend.

17.5.3 Attenuation in Splitter Lined Duct

The use of parallel or zigzag baffle-type separators (splitters) to increase the perimeter–area ratio results in more compact attenuators. In rock-wool blankets, the attenuation of a parallel type splitter duct may be obtained directly from Figure 17.14. The peak value of the attenuation is related to wavelength of sound and the splitter interval. With the zigzag arrangement of acoustic blankets, the attenuation of high frequencies is improved over that of the parallel splitter [7].

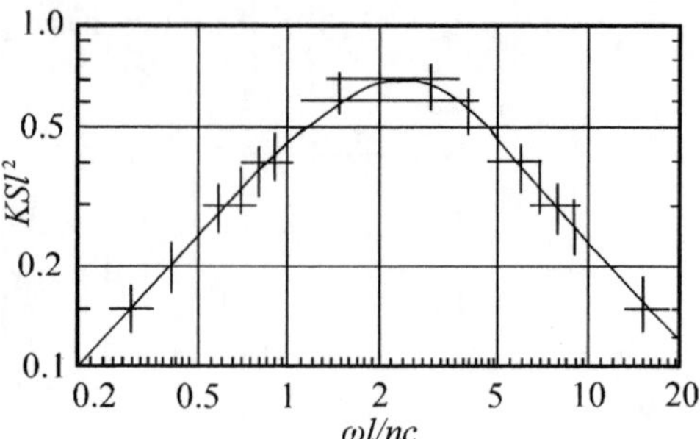

FIGURE 17.11 Damping function KSL^2 as a function of dimensionless frequency, $\omega l/nc$.

17.6 Attenuation of Dissipative Mufflers

17.6.1 Transmission Loss of Lined Expansion Chamber

The geometry and nomenclature for a dissipative muffler are given in Figure 17.15. For $f < 1.2c/D$, the assumption of plane wave is acceptable where $D =$ the diameter of the muffler.

The transmission loss for the light lining in the chamber may be obtained using [8,9]:

$$\mathrm{TL} = 10\log_{10}\left[\left\{\cosh(\delta_e l_e/2) + \frac{m+1}{2m}\sinh(\delta_e l_e/2)\right\}^2 \cos^2 kl_e + \left\{\sinh(\delta_e l_e/2) + \frac{m+1}{2m}\cosh(\delta_e l_e/2)\right\}^2 \sin^2 kl_e\right] \tag{17.26}$$

TABLE 17.3 Formulas for Attenuation of Several Lined Ducts

See Figure 17.7	Low-Frequencies Range: $\frac{\omega l}{nc} < 1$	Middle-Frequencies Range ($K_yS_yl_y$; see Figure 17.7)	High-Frequencies Range: $\frac{\omega l}{nc} > 5$
(A)	$\beta = \frac{4.34}{nl_y}$	$\beta = \frac{8.7c}{l_y^2\omega}(K_yS_yl_y^2)$	$\beta = 21.4\frac{c^2n}{\omega^2 l_y^3}$
(B)	$\beta = 4.34\left(\frac{1}{n_yl_y} + \frac{1}{n_xl_x}\right)$	$\beta = \frac{8.7c}{\omega}\left(\frac{K_yS_yl_y^2}{l_y^2} + \frac{K_xS_xl_x^2}{l_x^2}\right)$	$\beta = 21.4\frac{c^2}{\omega^2}\left(\frac{n_y}{l_y^3} + \frac{n_x}{l_x^3}\right)$
(C)	$\beta = \frac{8.7}{nl_y}$	$\beta = \frac{34.7c}{l_y^2\omega}\left(\frac{K_yS_yl_y^2}{4}\right)$	$\beta = 171\frac{c^2}{\omega^2}\frac{n_y}{l_y^3}$
(D)	$\beta = 4.34\left(\frac{2}{n_yl_y} + \frac{1}{n_xl_x}\right)$	$\beta = \frac{8.7c}{\omega}\left(\frac{K_yS_yl_y^2}{l_y^2} + \frac{K_xS_xl_x^2}{l_x^2}\right)$	$\beta = 21.4\frac{c^2}{\omega^2}\left(\frac{8n_y}{l_y^3} + \frac{n_x}{l_x^3}\right)$
(E)	$\beta = 8.7\left(\frac{1}{n_yl_y} + \frac{1}{n_xl_x}\right)$	$\beta = \frac{34.7c}{\omega}\left(\frac{K_yS_yl_y^2}{4l_y^2} + \frac{K_xS_xl_x^2}{4l_x^2}\right)$	$\beta = \frac{171c^2}{\omega^2}\left(\frac{n_y}{l_y^3} + \frac{n_x}{l_x^3}\right)$
(F)	$\beta = \frac{17.4}{nl}$	$\beta = \frac{69.5c}{4l^2\omega}KSl^2$	$\beta = \frac{341c^2n}{\omega^2l^3}$

β is attenuation (dB/m), n is absorbing factor plotted in Figure 17.7, $K_yS_yl_y$ is damping function, plotted in Figure 17.8, c is sound speed, l is the width of the duct, $\omega = 2\pi f$: angular frequency, x, y: coordinates, see Figure 17.7.
Source: Brüel, P.V. 1951. *Sound Insulation and Room Acoustics*, Chapman & Hall, London, p.159. With permission.

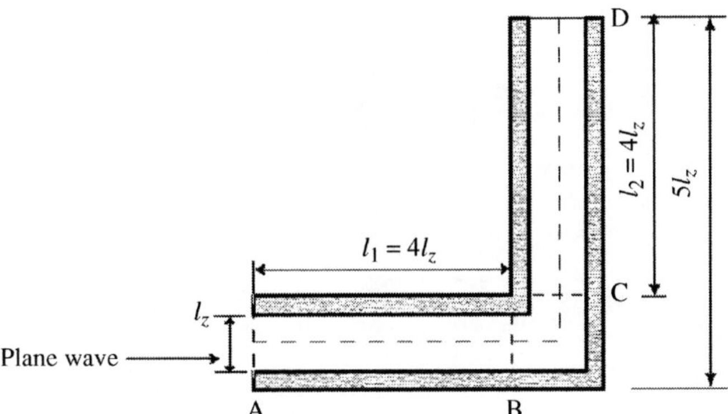

FIGURE 17.12 Sketch of a typical lined bend with plane wave incidence. (*Source*: Beranek, L.L. *Noise Reduction*, McGraw-Hill, 1960. With permission.)

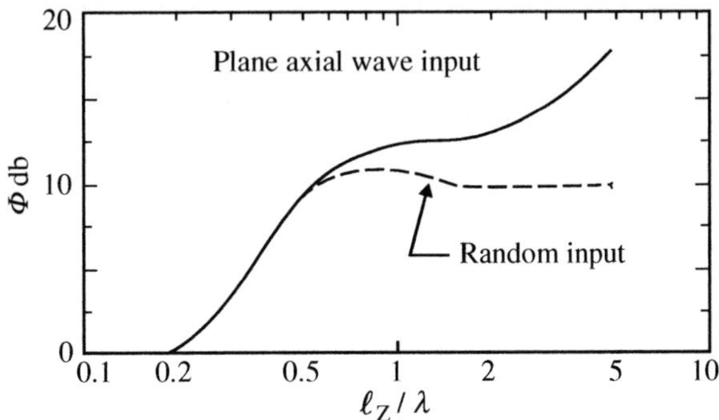

FIGURE 17.13 Insertion loss for lined bend. (The lining must extend two to four duct widths beyond the bend for this data to be valid.) (*Source*: Beranek, L.L. *Noise Reduction*, McGraw-Hill, 1960. With permission.)

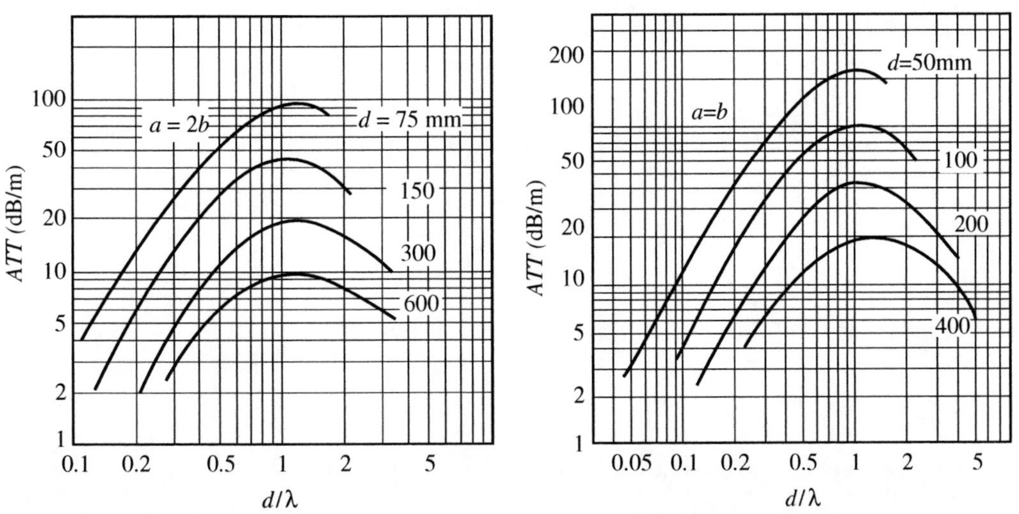

FIGURE 17.14 Sound attenuation for a splitter duct. Each baffle is constructed with two sheets of perforated metal filled with mineral wool, with about 100 to 140 kg/m^3 gross density; a = the width of the open space, b = the width of the baffle, d = the center-to-center distance of baffles, λ = the wavelength of the sound.

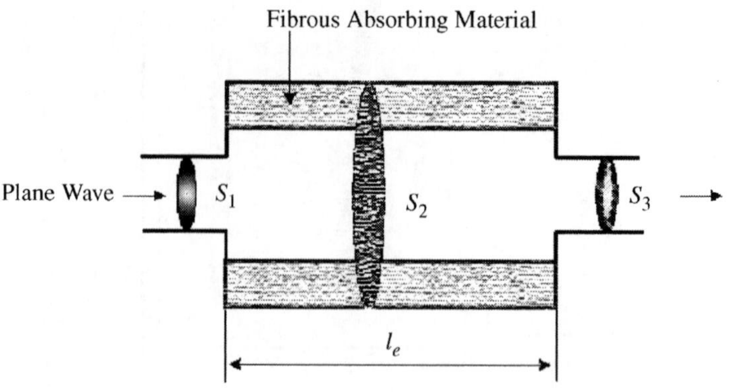

FIGURE 17.15 A dissipative muffler.

TABLE 17.4 Filled up Factor of Glass Wool, α_g

V_g/V	0.05	0.10	0.15	0.20	0.30	0.40	0.50	0.60	0.70	0.80	0.90	1.00
α_g	0.106	0.124	0.288	0.365	0.529	0.677	0.794	0.885	0.935	0.960	0.987	1.0

V_g = Filled up volume (factors of 100 kg/m^3), V = Volume of chamber

in which δ_e = the attenuation per unit length for the lined duct, which is given by the following equation:

$$20\log_{10}(\delta_e l_e) = \frac{K_l P l_e}{S} \tag{17.27}$$

The K_l values are obtained from the absorption coefficient, as shown earlier (see Figure 17.8). In particular, δ_e is given by

$$\delta_e = \frac{1}{l_e} 10^{0.05 K_l P l_e / S} \tag{17.28}$$

where m = the ratio of the area of expanded or lined sections to the area of inlet or outlet sections of muffler; $k = 2\pi f/c$, and l_e = the length of the muffler.

The transmission loss for the case of a thick lining of glass wool in the chamber is obtained using the empirical formula [10]

$$\mathrm{TL} = 10\log_{10}\left[1 + \left\{\frac{1}{2}\alpha_g m k l_e\right\}^2\right] \tag{17.29}$$

where

α_g = the coefficient, which is obtained from Table 17.4, using the filling volume and the density of glass wool

m = the ratio of the area of expanded or lined sections to the area of inlet or outlet sections of muffler

$k = 2\pi f/c$

l_e = the length of muffler

17.6.2 Transmission Loss of a Plenum Chamber

The geometry and nomenclature for a plenum chamber are given in Figure 17.16. A plenum chamber is similar in many ways to a lined expansion chamber. The main difference is that the inlet and outlet of a plenum chamber are not located in line. Generally, there is an offset to direct transmission of sound. Sound is reflected at the square-cornered bend as the cross section dimension of the duct is sufficiently large. Particularly at high frequencies, almost all of the sound energy may reflect many times off the lined sides when propagating from the inlet to the outlet. The transmission loss of a single plenum chamber can be obtained approximately from [11]:

$$\mathrm{TL} = 10\log_{10}\left\{S_w\left(\frac{\cos\theta}{2\pi d^2} + \frac{1}{R}\right)\right\} \quad (17.30)$$

where

$S_w = lW =$ area of the inlet and outlet
$d = \{(L - l)^2 + H^2\}^{1/2} =$ the slant distance from inlet to outlet
$\cos\theta = H/d$
$R = a/(1 - \alpha_m)$
$a =$ the total lined area in chamber times absorption coefficient
$\alpha_m =$ the statistical absorption coefficient of the lining

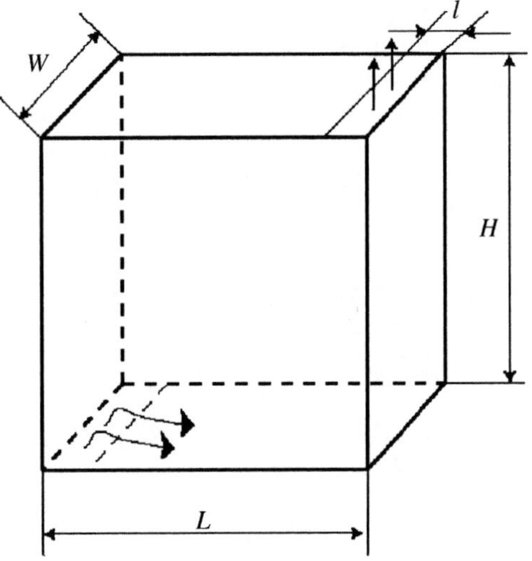

FIGURE 17.16 A single-plenum chamber showing the nomenclature used in Equation 17.26.

17.7 General Considerations

In order to carry out the design of noise-control measures for a particular problem, we must consider not only the fundamental acoustical properties of the material as discussed before, but also such practical aspects of the problem as (1) gas flow velocities, (2) temperature of gas, (3) moisture exposure, and (4) head losses for gas-flow. The client depends heavily on the expertise of the designer to realize adequate protection of the noise-control equipment under operating conditions.

17.7.1 Surface Treatment with Lining of Acoustic Material

Fibrous material in the market has some form of resin binder. Comparatively long fiber flocculent and comparatively short fiber are available. The packaging density of flocculent is about 60 to 100 kg/m^3. It is necessary to cover with perforated thin metal or wire netting so that an arbitrary shape may be maintained in the absorbing material. The perforated metal does not take into account the numerical aperture, hole shape, hole diameter, and metal thickness. From the acoustic viewpoint, a suitable numerical aperture is given in Table 17.5.

17.7.2 Gas Flow Velocity

Noise control problems often involve the use of an acoustical material in high-velocity gas-flow such as those found in the exhaust of engines or ventilating systems. Deterioration of the acoustical

TABLE 17.5 Perforated Metal for Treatment of Absorbing Material (Gas Flow Velocity is 25 m/sec or Less)

Perforation rate: 30 to 50%
Hole diameter: 5 to 10 mm
Hole shape: round, plus, slit and interminglement
Metal: iron, stainless steel (used in case of the corrosive gas)

material due to high-velocity gas flowing past it can be a serious problem. In addition, turbulence in the gas flow subjects the materials to vibration and can cause further deterioration. One solution to this problem is to install the acoustical material behind some form of protective facing, which will vary in complexity depending on the gas velocity.

A limited amount of information on this subject is available through field experience, as shown in Figure 17.17 [12]. However, the parameters of the treatment structure are not well established, for example, those concerning perforated metal, wire net, absorbing material, and gas flow. Multiple layers are used under conditions of flow velocity exceeding 25 m/sec, and the associated performance analysis can become rather complex.

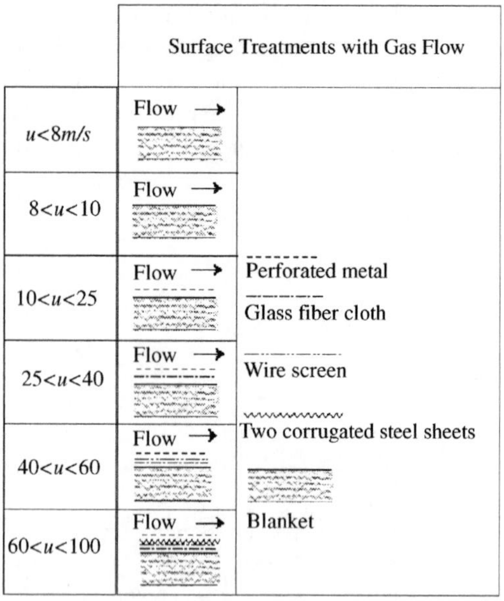

	Surface Treatments with Gas Flow	
$u<8m/s$	Flow →	
$8<u<10$	Flow →	
$10<u<25$	Flow →	Perforated metal Glass fiber cloth
$25<u<40$	Flow →	Wire screen
$40<u<60$	Flow →	Two corrugated steel sheets
$60<u<100$	Flow →	Blanket

FIGURE 17.17 Protective surface for absorbing material subjected to high-velocity gas flow.

17.7.3 Gas Temperature

In many noise-control problems, temperature is a very important factor. Sometimes high-temperature ducts that are radiating noise, for example, in diesel engines, and induced draft fans, must be wrapped. With a proper choice, it is possible to combine thermal and acoustical insulations using one single material. Under extremely high temperatures, the tensile strength of materials tends to decrease, and the material may be subjected to thermal shock.

Examples of absorbing materials that are currently available for use where temperature is an important consideration are given in Table 17.6.

17.7.4 Dust and Water Exposure

The holes of perforated metal can be blocked if a dust treatment is not carried out, and the sound absorption performance will deteriorate with adhesion to the surface of the absorbing material. Methods of dust accumulation and removal may be designed into cavity type mufflers used on the sound absorption equipment.

A fan of a cooling tower, for example, experiences a considerable amount of moisture. Precautions must be taken so that water droplets are not deposited on the sound absorbing material. The underside of the equipment should be treated with rust prevention material. Figure 17.18 shows the degradation

TABLE 17.6 Fibrous Materials of Use in Hot Gas Flows

Materials	Maximum Allowable Temperature (°C)
Glass fibers with binder	320 ~ 360
Glass wool	960 ~ 1060
Mineral wool felts	1160
Mineral wool	1660
Asbestos fibers	760
Alumina-silica	1900

of the acoustic characteristic of absorbing material due to moisture, using the normal incidence absorption coefficient [13].

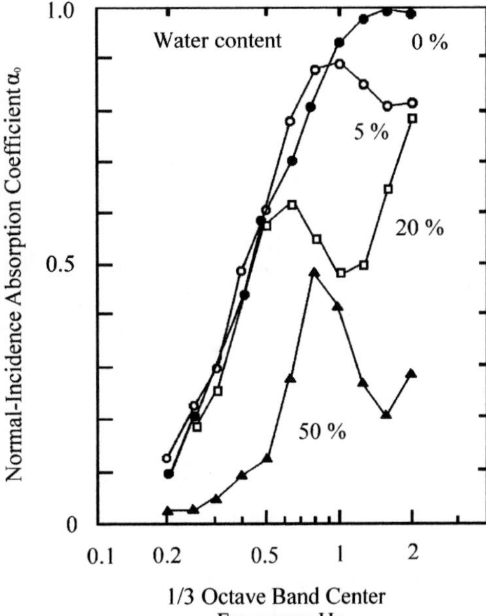

FIGURE 17.18 Degradation of the absorption coefficient by water content.

17.8 Practical Example of Dissipative Muffler

An example is given on the design of a dissipative muffler for noise reduction in an axial-flow fan for a ventilation system.

1. Specification of the axial-flow fan
   - Volume flow rate: $Q = 125$ m^3/min
   - Wind pressure: $p = 80$ mm Aq
   - Rotor blade number: $Z = 10$
   - Stationary blade number: $Z_s = 5$
   - Rotational speed: $N = 2580$ rpm
   - Shaft horsepower: $P = 3.75$ kW
2. The desired values of attenuation and head loss with the muffler installation follow. The noise of the fan propagates both intake and discharge sides. A performance level (noise reduction) of about 37.5 dB is required, when specific sound level K_s is obtained on the basis of the axial-flow fan specification given by

$$K_s = L_A - 10\log_{10}(p^2 Q) \tag{17.31}$$

3. Figure 17.19 gives the noise spectrum for the axial-flow fan. The blade passing frequency, BPF, is a fundamental component of the velocity fluctuation as the flow passes the blades. It is seen in the noise spectrum in Figure 17.19 at 430 Hz ($Q \times Z = (2580/60) \times 10$). By adding the background noise spectrum to this spectrum, it is seen that a muffler that provides an attenuation over 20 dB near 430 Hz, and about 15 dB in the frequency range of 800 to 1000 Hz is necessary.
   The head loss value of the muffler is to be maintained within 4 mm Aq.

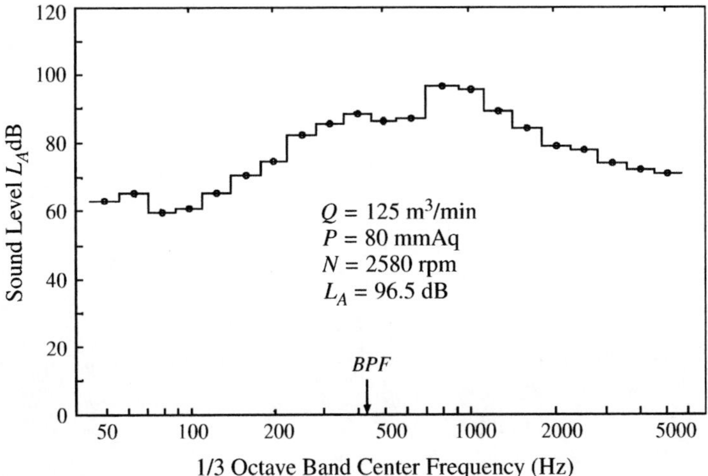

FIGURE 17.19 Noise spectrum of the axial flow fan of a factory ventilation system.

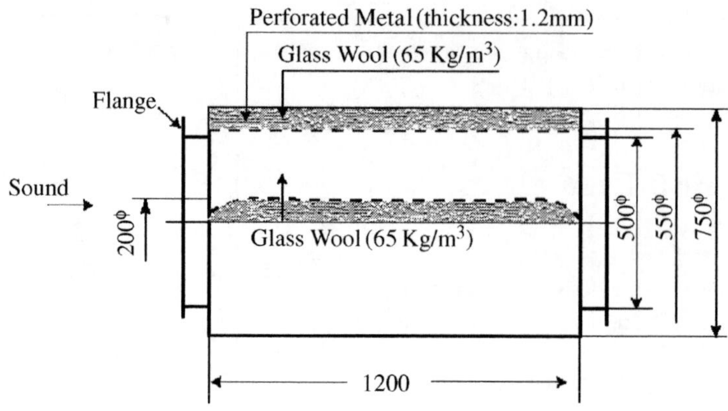

FIGURE 17.20 Half cross-sectional view of dissipative muffler for the axial flow fan.

4. The structure of the dissipative muffler is shown in Figure 17.20. The maximum value of outer diameter of the muffler is 750 mm, and the length chosen to optimize the performance.
5. The packing density of glass wool is chosen as 65 kg/m^3. The surface treatment of glass wool uses perforated metal with 1 mm thickness, 36% open area with 6 mm hole diameter. A sound absorption body of 200 mm diameter is supported in the center part, and it is welded to the outside cylinder by three props in the flow direction, and two in the circumferential direction.
6. The attenuation characteristics of the dissipative muffler may be calculated using Equation 17.27

$$\mathrm{TL} = 10 \log_{10}\left[1 + \left\{\frac{1}{2}\alpha_g mkl_e\right\}^2\right]$$

The proportion of the volume of glass wool filled into the muffler is approximately $(0.75^2 - 0.55^2 + 0.2^2)/0.75^2 = 0.53$. For a packing density of 100 kg/m^3, we have $0.53 \times 65/100 = 0.347$. The value of $\alpha_g \approx 0.6$ is obtained from Table 17.2.
The required expansion ratio, m, length, l_e, and wave number, k, are given by

$$m = (750/500)^2 = 2.25, \quad l_e = 1.2 \text{ m}, \quad k = 2\pi f/c$$

The speed of sound c depends on the environmental temperature. For a temperature of 25°C, we get 346.5 m/sec (= 331.5 + 0.6 × 25). The TL values at 100 to 1000 Hz are calculated. We have

$$f = 100 \text{ Hz}; \quad \mathrm{TL} = 5.0 \text{ dB}$$

$$f = 430 \text{ Hz}; \quad \mathrm{TL} = 16.1 \text{ dB}$$

$$f = 1000 \text{ Hz}; \quad \mathrm{TL} = 23.4 \text{ dB}$$

The flow velocity satisfies desired value of head loss p_{loss} (in mm Aq), and is calculated by following empirical equation [12]:

$$p_{\mathrm{loss}} = \left\{0.142 m_f^{-0.1}\left(\frac{l_e}{d_1}\right)^{3/4}\left(\frac{d_m}{d_1}\right)^{-1/3}\right\}\frac{u^2}{g} \tag{17.32}$$

Use the numerical values as follows:

- $m_f = (550/500)^2 = 1.21$, ratio of cross-sectional area between air passage and muffler.
- $d_1 = 500$ mm, diameter of inlet.
- $l_e = 1.2$ m, length of muffler.
- $d_m = 200$ mm, diameter of absorption body.
- $u = 10$ m/sec or less, flow velocity at inlet.
- $g = 9.8$ m/sec^2, acceleration of gravity force.

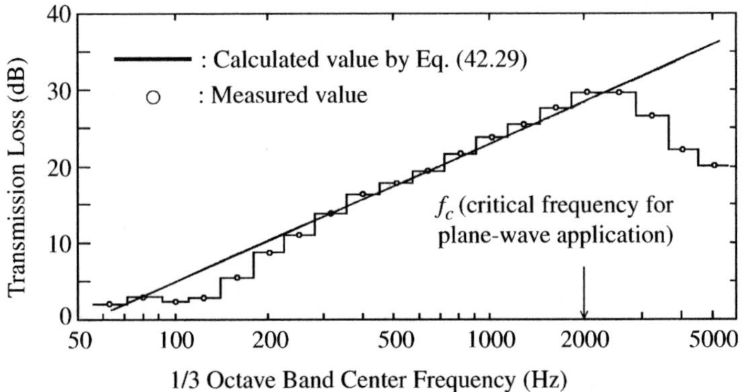

FIGURE 17.21 Noise-reduction characteristics of the designed dissipative muffler.

The corresponding head loss is 3.25 mm Aq, which corresponds to JIS B 833, and nearly agrees with the predicted value. Specifically, the condition of 4 mm Aq or less of the designed value is satisfied. The connection of axial-flow fan and the muffler uses vibration isolation, using the thick synthetic rubber.

7. The result of the attenuation realized from the spectrum after the muffler installation is shown in Figure 17.21. It is proven that the attenuation characteristics almost parallel the designed value. The frequency range where the approximation is valid is given by

$$f < c/d = 346.5/0.175 = 1980 \text{ Hz}$$

For frequencies below 250 Hz, the estimated result of the attenuation becomes slightly overestimated.

References

1. Obata, T., Hirata, M., Nishiwaki, N., Ohnaka, I., and Kato, K., Noise reduction characteristics of dissipative mufflers, 1st report, acoustical characteristics of fibrous materials, *Trans. Japan Soc. Mech. Eng.*, 42, 363, 3500, 1976.
2. Zwikker, C. and Kosten, C.W. 1949. *Sound Absorbing Materials*, Elsevier, New York.
3. Beranek, L.L., Acoustical properties of homogeneous isotropic rigid tiles and flexible blankets, *J. Acoust. Soc. Am.*, 19, 4, 556, 1947.
4. Obata, T. and Hirata, M., Estimation of acoustical transmission loss for combined walls, *Proc. Japan Soc. Mech. Eng. Annu. Meet.*, 780, 1, 42, 1978.
5. Brüel, P.V. 1951. *Sound Insulation and Room Acoustics*, Chapman & Hall, London, p. 159.
6. Beranek, L.L. 1971. *Noise and Vibration Control*, McGraw-Hill, New York, chap. 17, p. 390.
7. King, A.J., Attenuation of lined ducts, *J. Acoust. Soc. Am.*, 30, 6, 505, 1958.
8. Davis, D.D. Jr. and Stokes, G.M., 1954. *Natl. Advisory Comm. Aeronaut. Ann. Rept.*, 1192.
9. Davis, D.D. Jr. 1957. *Acoustical Filters and Mufflers. Handbook of Noise Control*, C.M. Harris, Ed., McGraw-Hill, New York, chap. 21.
10. Hagi, S., Studies on Silencer for Ventilating System, A Doctoral Thesis of University of Tokyo, 1961.
11. Wells, R.J., Acoustical plenum chambers, *Noise Control*, 4, 4, 9, 1958.
12. Obata, T. and Hirata, M., Estimation of acoustic power of flow-generated noise within silencer and head losses, *J. Acoust. Soc. Japan*, 34, 9, 532, 1978.
13. Koyasu, M., Acoustical properties of fibrous materials, personal letter, RC-SC35, *Japan Soc. Mech. Eng. Div. Meet.*, 15, 1975.

18

Design of Reactive Mufflers

Teruo Obata
Teikyo University

Summary

This chapter concerns the design of noise suppression devices such as mufflers. In particular, reactive mufflers that are inserted into long ducts are considered in detail. Analytical and empirical equations and information that are useful in the modeling and analysis of mufflers are presented, with an indication of their application ranges and limitations. A design procedure, complete with the necessary computations, is given. Methodologies and parameters of the performance analysis of acoustic systems with mufflers are indicated. An illustrative example of a muffler for a double-acting reciprocating compressor is presented.

18.1 Introduction

In noise-reduction applications, the need for a reactive muffler usually arises when transporting gas through a duct. For sound transmission through the duct to be minimized, an acoustic suppression device must be incorporated into the duct system. For example, in internal-combustion engines, it is required to reduce the intake and exhaust noise to acceptable levels. This may be accomplished by inserting a muffler in the intake and exhaust ducting to attenuate the pressure pulsations before they reach the environment.

A successful muffler design must satisfy at least the following three criteria: (1) muffler performance as a function of frequency (the maximum permissible noise generated by the gas flow through the muffler may have to be specified as well); (2) the maximum permissible average pressure drop through the muffler at a given temperature and mass flow; (3) the maximum allowable volume and restrictions on space utilization.

The customer may ask for a muffler with unrealistically high noise attenuation, virtually no backpressure, and very small size. In addition, it is important to the customer that the muffler is inexpensive and durable, and presents no maintenance problems. Needless to say, in practice, these criteria for muffler design are unrealistic, and have to be modified to practical levels. In this chapter, we will present some of the analytical and empirical tools that are helpful in muffler design.

18.2 Fundamental Equations

18.2.1 Analytical Model

The physical behavior of a reactive muffler may be adequately modeled by linear differential equations. The law of conservation of mass must hold, while three simultaneous equations, Newton's, Boyle–Charles, and that of conservation, must be satisfied. When these equations are combined, we obtain the wave equation for the plane, one-dimensional sound-pressure wave:

$$\frac{\partial^2 \xi}{\partial t^2} = c^2 \frac{\partial^2 \xi}{\partial x^2} \tag{18.1}$$

$$p = \rho c^2 \frac{\partial \xi}{\partial x} \tag{18.2}$$

where

$\xi =$ displacement of particle motion (m)
$c =$ velocity of sound (m/s)
$p =$ sound pressure (Pa)
$\rho =$ density of air (kg/m^3)
$t =$ time (s)
$x =$ coordinate system along which wave travels (m)

The stationary solutions for angular frequency ω of Equation 18.1 and Equation 18.2 are given by

$$\xi = (A\,\mathrm{e}^{-jkx} - B\,\mathrm{e}^{jkx})\mathrm{e}^{j\omega t} \tag{18.3}$$

$$p = -\rho c^2 k(A\,\mathrm{e}^{-jkx} + B\,\mathrm{e}^{jkx})\mathrm{e}^{j\omega t} \tag{18.4}$$

where A, $B =$ amplitudes of sound pressure or particle motion for traveling and reflecting waves, $k = 2\pi f/c$, wave number, and $j = \sqrt{-1}$.

18.2.2 Boundary Conditions

The boundary conditions are given below.

(1) *Sound source.* The sound source is assumed to be independent of the existence of the muffler, and the volume rate of the particles is assumed constant, as given by

$$S\dot{\xi} = \text{const} \tag{18.5}$$

in which $\cdot$ denotes the time derivative.

(2) *Open end of duct.* The reflection coefficient, R, at the open end of an unflanged circular pipe is available, and is given by

$$R = \frac{B\,\mathrm{e}^{jkx}}{A\,\mathrm{e}^{-jkx}} \tag{18.6}$$

The magnitude of the reflection coefficient, $|R|$, is shown in Figure 18.1a as a function of ka, where a is the pipe radius. The phase shift can be determined from Figure 18.1b, which is a plot of α/a as a function of ka. Also, the reflection coefficient is [1]:

$$R = -|R|\mathrm{e}^{-2jk\alpha} \tag{18.7}$$

For the small values of ka that are most often encountered in reactive muffler design, $|R| \approx 1$ and $\alpha/a = 0.613$.

(3) *Closed end.* The displacement of particle motion is zero at a rigid wall. Hence, we have

$$\xi = 0 \tag{18.8}$$

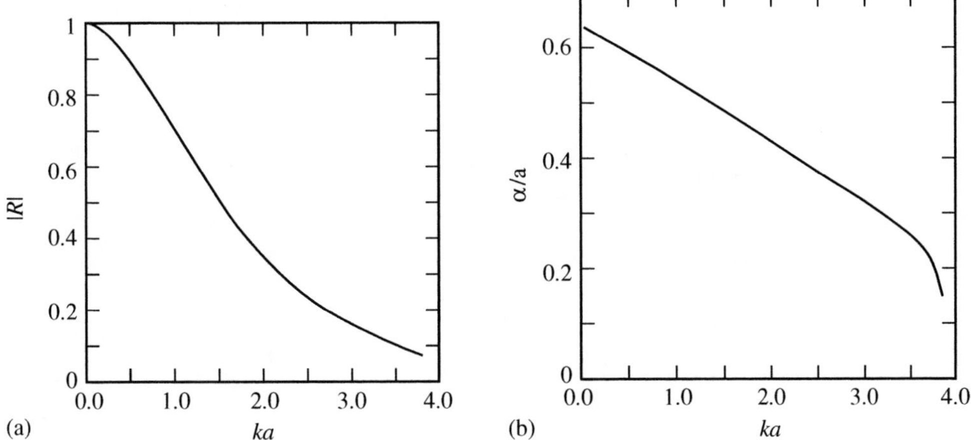

FIGURE 18.1 (a) Magnitude of the reflection coefficient at open end of an unflanged circular pipe; (b) End correction for an unflanged circular pipe. (*Source:* Levine, H. and Schwinger, J., On the radiation of sound from an unflanged circular pipe, *J. Phys. Rev.*, 73, 383, 1948. With permission.)

(4) *Junction conditions.* The following equations correspond to the continuity of volume flow rate of the particles and the continuity of pressure, even if the cross section changes suddenly:

$$S_i\dot{\xi}_i = S_{i+1}\dot{\xi}_{i+1} \tag{18.9}$$

$$p_i = p_{i+1} \tag{18.10}$$

18.3 Effects of Reactive Mufflers

The acoustic behavior of a reactive muffler may be expressed in term of the insertion loss, the difference in the noise levels measured at some external point with and without the muffler in the system. The transmission loss is defined as the insertion loss for a nonreflecting source and the end of exhaust duct.

18.3.1 Insertion Loss

A single expansion-type muffler installation is shown schematically in Figure 18.2. At the open end of a pipe, as in Figure 18.2, the traveled and reflected waves of the source become $A_0\,\mathrm{e}^{-jkl}$, $B_0\,\mathrm{e}^{jkl}$, over a length l, where the amplitudes are denoted by A_0, B_0. The reflective coefficient for length l_0 is given by

$$R_0 = \frac{B_0\,\mathrm{e}^{jkl_0}}{A_0\,\mathrm{e}^{-jkl_0}} \tag{18.11}$$

This is obtained from Equation 18.6 with $x = l_0$.

The energy, W_0, of the acoustic wave escaping from the open end of the pipe is given by

$$W_0 \propto \frac{S_0A_0^2(1 - R_0^2)}{\rho_0 c_0} \tag{18.12}$$

in which ρ_0 = density of air, and c_0 = speed of sound in air. The equation of the sound-pressure level measured at an open point at some distance is given by

$$p_0 = 10\,\log_{10}\left(\frac{Q_\mathrm{d}}{4\pi r_0^2} + \frac{4}{R_\mathrm{r}}\right) + PWL_{r_0} \tag{18.13}$$

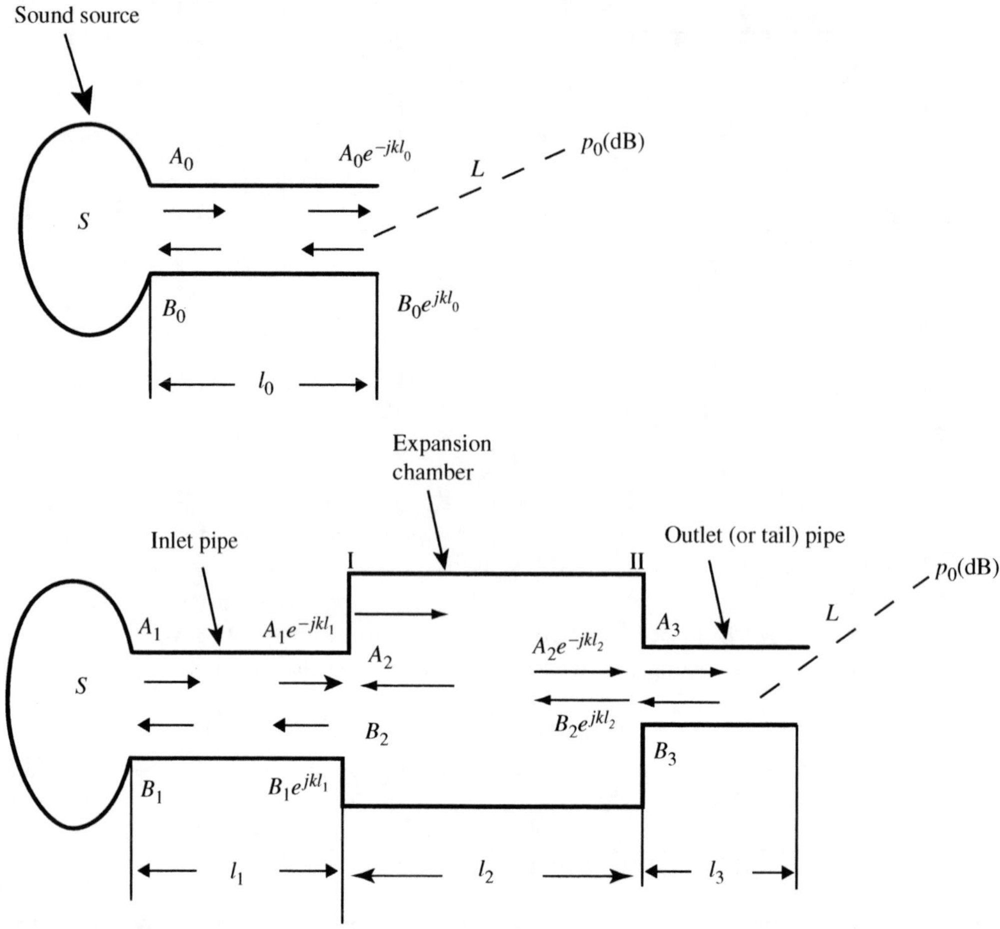

FIGURE 18.2 Measurement of insertion loss.

where

$r_{0,}$ = distance
$R_r = A/(1 - \alpha)$, room constant
A = the indoor sound absorbing power (indoor surface area times indoor average absorption coefficient)
α = the indoor average absorption coefficient
Q_d = the directivity factor from the open end

Therefore, the measured value of insertion loss can be obtained from Equation 18.14, when Q_d values are equal. Power level is defined as

$$\text{PWL} = 10\log_{10}\left(\frac{W}{10^{-12}}\right)$$

Now,

$$\text{IL} = \text{PWL}_{r_0} - \text{PWL}_r \tag{18.14}$$

Using Equation 18.14, it can be shown that IL can be expressed by

$$\text{IL} = 10\log_{10}\left|\frac{S_0}{S_3}\right|\left|\frac{A_0}{A_3}\right|^2\left|\frac{1-R_0^2}{1-R_3^2}\right| \tag{18.15}$$

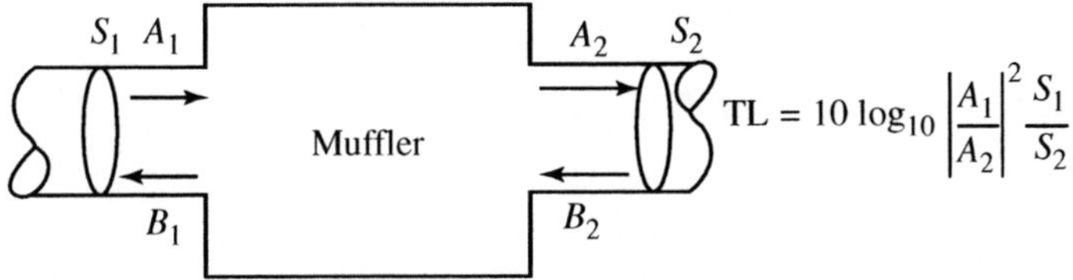

FIGURE 18.3 Definition of transmission loss for a muffler.

3.2 Transmission Loss

It is desirable to eliminate the source and radiation characteristics from the system in Figure 18.3, and to look only at some property of the muffler itself. This may be accomplished by defining a quantity called "transmission loss" (TL) as follows:

$$\mathrm{TL} = 10\,\log_{10}\left|\frac{A_1}{A_2}\right|^2 \frac{S_1}{S_2} \tag{18.16}$$

In the measurement of TL, it is difficult to separate the reflected wave. The theoretical calculation is easy and useful.

18.4 Calculation Procedure

For the reactive muffler shown in Figure 18.2, the following equations are obtained from Equation 18.5 to Equation 18.7 and Equation 18.9. The inlet pipe, cavity, and tail pipe are denoted by suffix in the figure [2].

(1) The sound source:

$$S_0(A_0 - B_0) = S_1(A_1 - B_1) \tag{18.17}$$

(2) The open end of pipes:

$$\left.\begin{aligned} B_0\,\mathrm{e}^{jkl_0} &= A_0\,\mathrm{e}^{-jkl_0} R_0 \\ B_3\,\mathrm{e}^{jkl_3} &= A_3\,\mathrm{e}^{-jkl_3} R_3 \end{aligned}\right\} \tag{18.18}$$

(3) The sudden expansion (junction I):

$$\left.\begin{aligned} A_1\,\mathrm{e}^{-jkl_1} + B_1\,\mathrm{e}^{jkl_1} &= A_2 + B_2 \\ S_1(A_1\,\mathrm{e}^{-jkl_1} - B_1\,\mathrm{e}^{jkl_1}) &= S_2(A_2 - B_2) \end{aligned}\right\} \tag{18.19}$$

(4) The sudden contraction (junction II):

$$\left.\begin{aligned} A_2\,\mathrm{e}^{-jkl_2} + B_2\,\mathrm{e}^{jkl_2} &= A_3 + B_3 \\ S_2(A_2\,\mathrm{e}^{-jkl_2} - B_2\,\mathrm{e}^{jkl_2}) &= S_3(A_3 - B_3) \end{aligned}\right\} \tag{18.20}$$

From these linear equations, the following equation can be obtained:

$$\begin{aligned} \frac{A_0}{B_0} = \frac{m_{10}}{1 - R_0\,\mathrm{e}^{-2jkl_0}} &\Big\{\mathrm{j}\left(1 + R_3\,\mathrm{e}^{-2jkl_3}\right)(\sin kl_1 \cos kl_2 + m_{21}kl_1 \sin kl_2) \\ &+ m_{32}\left(1 - R_3\,\mathrm{e}^{-2jkl_3}\right)(m_{21} \cos kl_1 \cos kl_2 - \sin kl_1 \sin kl_2)\Big\} \end{aligned} \tag{18.21}$$

When the reflection coefficients are absent, $R_0 = R_3 = 0$, we have

$$IL = 10\log_{10} m_{03}\{m_{10}^2(\sin kl_1 \cos kl_2 + m_{21}\cos kl_1 \sin kl_2)^2 + m_{10}^2 m_{32}^2 \times (m_{21}\cos kl_1 \cos kl_2 - \sin kl_1 \sin kl_2)^2\} \tag{18.22}$$

where $m_{10} = S_0/S_1$, $m_{21} = S_2/S_1$, $m_{32} = S_3/S_2$, $m_{03} = S_0/S_3$, which are ratios of cross-sectional areas of the pipes.

For the reflection factors, $R_0 = R_3 = -1$, $S_0 = S_1 = S_3$, and $m_{21} = m$, the equation of insertion loss becomes

$$\mathrm{IL} = 20\log_{10}\left\|\left[\frac{\cos kl_1}{\cos kl_0}\left\{\cos kl_{21}\cos kl_3 - m\sin kl_2 \sin kl_3 - \tan kl_1\left(\cos kl_2 \sin kl_3 + \frac{1}{m}\sin kl_2 \cos kl_3\right)\right\}\right]\right\| \tag{18.23}$$

The ratio $|A_1/A_3|$ may also be obtained from Equation 18.17 to Equation 18.20. When the magnitude of reflection coefficients $R_0 = R_3 = 0$, $S_0 = S_1 = S_3$, and expansion ratio of the cross-sectional open area of the pipes $m_{21} = m$, the transmission loss is given by

$$\mathrm{TL} = 10\log_{10}\left\{1 + \frac{1}{4}\left(m - \frac{1}{m}\right)^2 \sin^2 kl_2\right\} \tag{18.24}$$

When $R_0 = R_3 = -1$, $S_0 = S_1 = S_3$, and $m_{21} = m$, TL is obtained from the following equation:

$$\mathrm{TL} = 10\log_{10}\left|1 + \frac{1}{m^2}(m^2 - 1)\{(m^2 + 1)\sin^2 kl_3 - 1\}\sin^2 kl_2 + \frac{1}{2}\left(-m + \frac{1}{m}\right)\sin 2kl_2 \sin 2kl_3\right| \tag{18.25}$$

Computation formulas of insertion loss and transmission loss for the case of an expansion chamber with insertion pipe and resonator are shown in Table 18.1.

The principal structures of several reactive mufflers are shown in Figure 18.4.

18.5 Application Range of Model

18.5.1 Condition for Approximation of Plane Wave

The frequency range where the approximation of a plane wave is valid is given by

$$f_c < 1.22\frac{c}{D} \tag{18.26}$$

where

f_c = critical frequency of plane wave (Hz)
c = speed of sound (m/s)
D = diameter of muffler (m)

It is seen that the expansion ratio of an open area of a pipe increases with IL or TL. However, the application range of the analytical model decreases with increasing diameter of chamber.

18.5.2 Effect of Temperature

Under conditions of high-temperature and high-speed gas flow, as in an engine exhaust system, the primary effect of a change in pipe temperature is the corresponding change in the speed of sound, which is proportional to the square root of the absolute temperature. In the design of a reactive muffler, it is necessary to use the actual speed of sound in the gas inside the pipe. The most accurate values available for density (ρ_0) and the speed of sound (c_0) at each element should be used in calculating the impedance

TABLE 18.1 Transmission Loss of Reactive Mufflers and Insertion Loss of Reactive Mufflers

Muffler (see Figure 18.4)	TL (dB)	Application Limits and Comments
Transmission loss of reactive mufflers		
(a)	$\mathrm{TL} = 10\log_{10}\left\{1 + \left(\frac{kS}{4a}\right)^2\right\}$ a: radius of orifice	$a/\lambda < 0.1$, $R = 0$
(b)	$\mathrm{TL} = 20\log_{10}\left\lvert \begin{matrix} (1 + R_3\,\mathrm{e}^{-2jkl3})(\cos kl_2 + jm_{21}\sin kl_2) \\ + m_{32}(1 - R_3\,\mathrm{e}^{-2jkl3})(\mathrm{j}\sin kl_2 + m_{21}\cos kl_2) \end{matrix} \right\rvert$ $m_{10} = S_1/S_0$, $m_{21} = S_2/S_1$, $m_{32} = S_3/S_2$, R_0, R_3: the reflection coefficient of the open end, l_3 = length of the tail pipe (1) when $R_0 = R_3 = 0$, $S_0 = S_1 = S_3$ $\mathrm{TL} = 10\log_{10}\left\{1 + \frac{1}{4}\left(m_{21} - \frac{1}{m_{21}}\right)^2 \sin kl_2\right\}$ (2) when $R_0 = R_3 = -1$, $S_0 = S_1 = S_3$ $\mathrm{TL} = 10\log_{10}\left[1 + \left(\frac{m_{21}^2 - 1}{m_{21}^2}\right)\{(m_{21}^2 + 1)\sin^2 kl_3 - 1\}\sin^2 kl_2 + \frac{1}{2}\left(\frac{1}{m_{21}} - m_{21}\right)\sin 2kl_2 \sin 2kl_3\right]$	$f < 1.22c/D$
(c)	$\mathrm{TL} = 10\log_{10}\left[\left\{2\cos k(l_1 - l_{11} - l_{22}) - \frac{m-1}{m}\sin k(l_1 - l_{11} - l_{22})(\tan kl_{11} + \tan kl_{22})\right\}^2 + \left\{\left(m + \frac{1}{m}\right)\sin k(l_1 - l_{11} - l_{22}) + (m-1)\cos(l_1 - l_{11} - l_{22})(\tan kl_{11} + \tan kl_{22}) - \frac{(m-1)^2}{m}\tan kl_{11}\tan kl_{22}\sin k(l_1 - l_{11} - l_{22})\right\}^2\right]$ $M = S_1/S_0$	$R \approx 0$
(d)	$\mathrm{TL} = 10\log_{10}\left\{1 + \frac{1}{4}\left(\frac{m}{\frac{kS_2}{C_0} - \cot kl}\right)^2\right\}$ $C_0 = NC_i$; N: number of holes, $C_i = 2\pi a_i^2/(l_b + \pi a_i)$, l_b, l_b: thickness of the pipe, a_i: radius of a hole, $m = S_{12}/S$	$R \approx 0$

(continued on next page)

TABLE 18.1 *(continued)*

Muffler (see Figure 18.4)	TL (dB)	Application Limits and Comments
(e)	$\mathrm{TL} = 10\log_{10}\left[1 + \frac{1}{4}\left\lvert \frac{\frac{\sqrt{C_0 V}}{S}}{\frac{f}{f_r} - \frac{f_r}{f}} \right\rvert^2\right]$ $f_r = \frac{c}{2\pi}\sqrt{\frac{C_0}{V}}$, $C_0 = \frac{2\pi a^2}{2l_b + \pi a}$ l_b: length of the neck or thickness of the pipe, a: radius of the neck or hole	$R \approx 0$, $l_b \ll \lambda$ Resonator size $\ll \lambda$
(f)	$\mathrm{TL} = 10\log_{10}\left\{1 + \frac{m^2}{4}\left(\frac{\tan kl_b - \frac{S_b}{kV}}{\frac{S_b}{kV}\tan kl_b + 1}\right)\right\}$ $m = S_b/S$	$R \approx 0$, $l_b \ll \lambda$ Resonator size $\ll \lambda$
(g)	$\mathrm{TL} = 20\log_{10}\frac{1}{16m^2}\big\lvert[4m(m+1)^2\cos 2k(l+l_c) - 4m(m-1)^2\cos 2k(l-l_c)]$ $+\mathrm{j}\big\{2(m^2+1)(m+1)^2\sin 2k(l+l_c) - 2(m^2+1)(m-1)^2\sin 2k(l-l_c)$ $-4(m^2-1)^2\sin 2kl_c\big\}\big\rvert$ $m = S_2/S_1$	$R \approx 0$
(h)	$\mathrm{TL} = 10\log_{10}\big\lvert\{\cos 2kl - (m-1)\sin 2kl_c \tan kl_c\}^2 + \big\{\frac{\mathrm{j}}{2}\left(m + \frac{1}{m}\right)\sin 2kl$ $+(m-1)\tan kl_c\left(\left(m + \frac{1}{m}\right)\cos 2kl - \left(m - \frac{1}{m}\right)\right)\big\}^2\big\rvert$ $m = S_2/S_1$	$R \approx 0$

(i)	$$\mathrm{TL} = 10\log_{10}\left\{\frac{1}{4}\left|\frac{A_1 + jB_1}{A_2 + jB_2}\right|^2\right\}$$ $$A_1 = Y_3X_1^2 + Z_0Y_3^2 + Z_0(X_1 + X_3)^2$$ $$B_1 = X_1Y_3^2 + X_1X_3(X_1 + X_3)$$ $$A_2 = Y_3X_1^2 \cos kl + Z_0X_1Y_3 \sin kl$$ $$B_2 = X_1Y_3^2 + X_1X_3(X_1 + X_3)\cos kl - Z_0X_1(X_1 + X_3)\sin kl$$ $$X_1 = \frac{\omega\rho}{C_0} - \frac{\rho c^2}{\omega V_1},$$ $$X_2 = \frac{\omega\rho}{C_0} - \frac{\rho c^2}{\omega V_2}$$ $$X_3 = \frac{Z_0^2(X_2 \cos 2kl + \frac{1}{2}Z_0 \sin^2 kl)}{(X_2 \sin kl - Z_0 \cos kl)^2 + X_2^2 \cos^2 kl}$$ $$Y_3 = \frac{Z_0X_2^2}{(X_2 \sin kl - Z_0 \cos kl)^2 + X_2^2 \cos^2 kl}$$ $$Z_0 = \frac{\rho c}{S_0},$$ $$C_0 = \frac{2\pi a^2}{2l_b + \pi a}$$ l_b: thickness of the pipe, a: radius of hole	$R \approx 0$ Resonator size $\ll \lambda$
(j)	$$\mathrm{TL} = 10\log_{10}\left\{\left(\cos kl + \frac{\rho c}{4S_0X}\left(m + \frac{1}{m}\right)\sin kl - \frac{\rho c}{4S_0X}\left(m - \frac{1}{m}\right)\cos 2kl_b \sin kl\right)^2 + \left(\frac{1}{2}\left(m + \frac{1}{m}\right)\sin kl + \frac{\rho c}{4S_0X}\left(m - \frac{1}{m}\right)\sin 2kl_b \sin kl - \frac{\rho c}{2S_0X}\cos kl\right)^2\right\}$$ $$X = \frac{\omega\rho}{C_0} - \frac{\rho c^2}{\omega V},$$ $$m = \frac{S}{S_0}$$	$R \approx 0$

(continued on next page)

TABLE 18.1 *(continued)*

Muffler (see Figure 18.4)	IL (dB)	Application Limits and Comments
Insertion loss of reactive mufflers		
(b)	$\mathrm{IL} = 10\log_{10}\frac{1}{m_{30}} \times \left\lvert \frac{m_{10}}{1 - R_0\,\mathrm{e}^{-2jkl_0}}\{\mathrm{j}(1 + R_3\,\mathrm{e}^{-2jkl_3}(\sin kl_1 \cos kl_2 + m_{21}\cos kl_1 \sin kl_2). + m_{32}(1 - R_3\,\mathrm{e}^{-2jkl_3})(m_{21}\cos kl_1 \cos kl_2 - \sin kl_1 \sin kl_2)\} \right\rvert^2$	$f < 1.22c/D$, R is plotted in Fig.18.1
	(1) $\mathrm{IL} = 10\log_{10}\left\lvert \{1 + (m_{21}^2 - 1)\}\left(1 - \frac{m_{21}^2 + 1}{m_{21}^2}\sin^2 kl_1\right)\sin^2 kl_2 + \frac{1}{2}\left(m_{21} - \frac{1}{m_{21}}\right)\sin 2kl_1 \sin 2kl_2 \right\rvert$	$R_0 = R_3 = 0$, $S_0 = S_1 = S_3$
	(2) $\mathrm{IL} = 10\log_{10}\left\lvert \left(\frac{\cos kl_1}{\cos kl_0}\{\cos kl_2 \cos kl_3 - m_{21}\sin kl_2 \sin kl_3 - \tan kl_1(\cos kl_2 \sin kl_3 + \frac{1}{m_{21}}\sin kl_2 \cos kl_3)\}\right)\right\rvert^2$	$R_0 = R_3 = -1$, $S_0 = S_1 = S_3$
(c)	$\mathrm{IL} = 20\log_{10}\left\lvert \frac{\cos kl_1 \cos kl_2 - m \sin kl_1 \sin kl_2}{\cos kl_{11} \cos kl_{22}} \right\rvert$ $m = S_1/S_0$	$R = -1$
(d)	$\mathrm{IL} = 20\log_{10}\left[\cos^2 kl_2 + \frac{m}{\frac{kS_2}{C_0} - \cot kl}\sin 2kl_2 + \left(\frac{m}{\frac{kS_2}{C_0} - \cot kl}\right)^2 \sin^2 kl_2\right]$ $C_0 = NC_i$; $C_i = 2\pi a_i^2/(l_b + \pi a_i)$, N: number of holes, l_b: thickness of the pipe, a_i: radius of a hole, $m = S_{12}/S$	$R = -1$, $kl_0 << 1$

(e)

$$IL = 10 \log_{10} \left| \frac{\frac{\sqrt{C_0 V}}{S}}{\frac{f}{f_r} - \frac{f_r}{f}} \sin 2kl_2 + \frac{\frac{C_0 V}{S^2}}{\left(\frac{f}{f_r} - \frac{f_r}{f}\right)^2} \sin^2 kl_2 + \cos^2 kl_2 \right|$$

$$f_r = \frac{c}{2\pi}\sqrt{\frac{C_0}{V}},$$

$$C_0 = \frac{2\pi a^2}{2l_b + \pi a}$$

l_b: length of the neck or thickness of pipe, a: radius of the neck or hole

$R = -1,\ kl_0 << 1$

(f)

$$IL = 10 \log_{10} \left[\cos^2 kl_2 + m \sin 2kl_2 \frac{\frac{S_b}{kV}\tan kl_b + 1}{\tan kl_b - \frac{S_b}{kV}} + \left(m \frac{\frac{S_b}{kV}\tan kl_b + 1}{\tan kl_b - \frac{S_b}{kV}} \right)^2 \sin^2 kl_2 \right]$$

$$f_r = \frac{c}{2\pi}\sqrt{\frac{C_0}{V}},$$

$$C_0 = \frac{2\pi a^2}{2l_b + \pi a}$$

l_b: length of the neck or thickness of the pipe, a: radius of the neck or hole

$R = -1,\ kl_0 << 1$

(k)

$$IL = 20 \log_{10} |(\cos kl_1 \cos kl_{11} - m_1 \sin kl_1 \sin kl_{11}) + (\cos kl_2 \cos kl_{22} - m_2 \sin kl_2 \sin kl_{22}) + \cdots + (\cos kl_i \cos kl_{ii} - m_i \sin kl_i \sin kl_{ii})|$$

$R = -1,\ kl_0 << 1$

(l)

$$IL = 20 \log_{10} \left| \left\{ \frac{\cos kL_1 \cos kl_1 - m \sin kL_1 \sin kl_1}{\cos kl_{11}} + \frac{\cos kL_2 \cos kl_2 - m \sin kL_2 \sin kl_2}{\cos kl_{12} \cos kl_{21}} + \cdots + \frac{\cos kL_n \cos kl_n - m \sin kL_n \sin kl_n}{\cos kl_{(n-1)2}} \right\} \right|$$

$R = -1,\ kl_0 << 1$

A is the radius of tube in orifice hole or diameter of side branch, c is the sound speed, C_0 is the conductivity, D is the diameter of chamber, f is the frequency, f_r is the resonant frequency of the resonator, $k = 2\pi f/c$ is the wave number, L is the length, $m = S_i/S_{i+1}$ is the ratio of the cross section, IL is the insertion loss, N is the number of holes, R is the reflection coefficient, S is the cross section, TL is the transmission loss, V is the volume of chamber, Z is the acoustic impedance, ρ is the density, λ is the wavelength, $\omega = 2\pi f$, angular frequency.

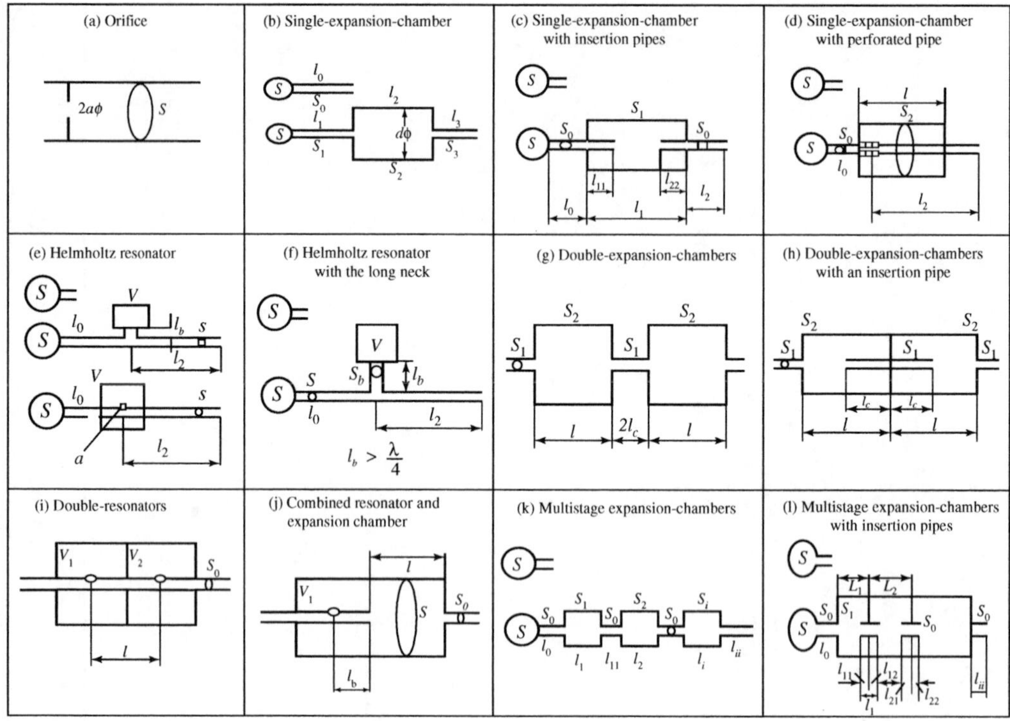

FIGURE 18.4 Sketches of the 12 principal structures of reactive mufflers.

of the elements. The impedance, z, is given by

$$z = -j\frac{\rho_0 c_0}{S}\frac{1}{kl} \tag{18.27}$$

where

$S =$ the cross-sectional open area of pipe
$k = 2\pi f / c_0$, wave number
$l =$ the length of the pipe element

Note that the impedance of the resonator chamber is proportional to $\rho_0 c_0^2$. However, c_0^2 is proportional to the absolute temperature of gas (T) and ρ_0 is proportional to $1/T$. Hence, the chamber impedance is independent of temperature. The connector impedance is a function of T, but in most cases the connector will be at the pipe temperature. For a resonator-type muffler, a temperature difference between the pipe and chamber is expected to have little effect on the performance of the muffler.

18.5.3 Effect of Gas Flow in Pipe

Under conditions of high-temperature and high-speed gas flow in a pipe, the pressure amplitude in the pipe is large, and is larger than what is predicted by theory. Analysis by the characteristic curve method is desirable under such conditions.

In a reactive muffler where the pipe flow passes through a sudden pipe expansion or an orifice, the computed transmission loss or insertion loss tends to be an overestimate because of new noise that is generated due to the resulting irregular air-flow within the muffler.

18.5.4 Effect of Friction Loss in Pipe

When an acoustic wave propagates in a pipe, it will attenuate due to viscous friction. The effect is large for long pipes of small diameter. Friction damping in a pipe may be incorporated into the propagation constant, γ, such that

$$\gamma = \delta + jk \tag{18.28}$$

where δ is the attenuation constant per unit length of pipe. By substituting Equation 18.28 into Equation 18.3 and Equation 18.4, we obtain

$$\xi = (A\,\mathrm{e}^{-\gamma l} - B\,\mathrm{e}^{\gamma l})\mathrm{e}^{j\omega t} \tag{18.29}$$

$$p = -\rho c^2 k(A\,\mathrm{e}^{-\gamma l} + B\,\mathrm{e}^{\gamma l})\mathrm{e}^{j\omega t} \tag{18.30}$$

Empirical formulas are given below for two cases of the attenuation coefficient δ [3].

(1) The formula for seamless steel or chloride-ethylene pipes (regression formula when the inside roughness is 4 to 8 μm and length under 3 m) is

$$\delta = 26,100\lambda^{-0.5}\frac{\mu}{\rho c d} \tag{18.31}$$

where

λ = wavelength of sound (m)
μ = viscosity of gas in the pipe (Pa s)
ρ = density of gas (kg/m^3)
d = diameter of the pipe (m)

(2) The equations for lining with glass wool are

$$\left.\begin{aligned} \delta_2 &= 2491\lambda^{-0.476}\left(\frac{\rho c d}{\mu}\right)^{-1.068} \\ \delta_6 &= 5175\lambda^{-0.476}\left(\frac{\rho c d}{\mu}\right)^{-1.303} \\ \delta_6 &= 11596\lambda^{-0.476}\left(\frac{\rho c d}{\mu}\right)^{-1.270} \end{aligned}\right\} \tag{18.32}$$

The suffix of δ gives the thickness of absorbing material in mm.

18.6 Practical Example

18.6.1 Expansion-Type Muffler for Reciprocating Compressor

Consider a double-acting (i.e., fluid on both sides of the piston in the cylinder) reciprocating compressor for supplying high-pressure air to a machine shop of a factory, for example.

The specifications of the reciprocating compressor follow:

- Delivery pressure: 6.9×10^5 Pa
- Rotational speed of driving shaft: 600 rpm
- Power of driving shaft: 450 kW
- Diameter of inlet pipe: 380 mm

Pressure pulsations of 10 and 20 Hz are produced by the compressor due to the rotational speed, as seen in Figure 18.5. The pressure wave from the inlet propagates the free space and can the damage nearby private houses. Wooden doors with glass paneling, wooden sliding-doors, and leaves of plants and foliage, and have been found to vibrate due to low-frequency audible sound. An attenuation of 15 to 20 dB was

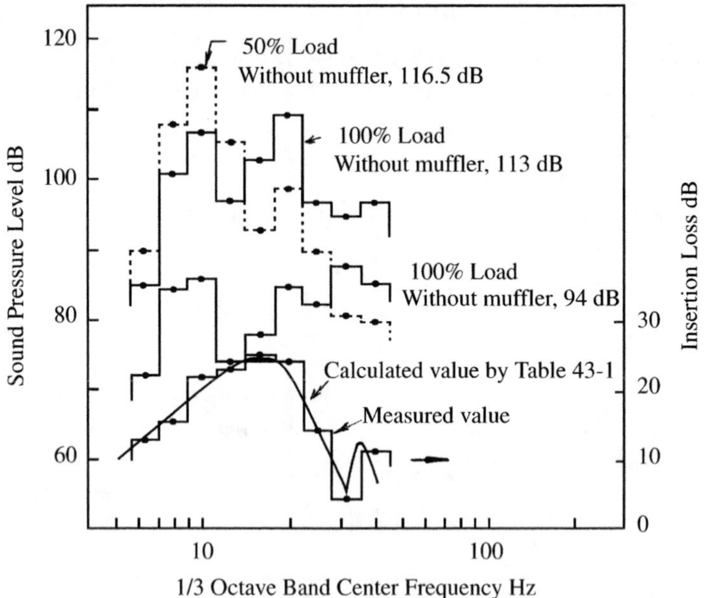

FIGURE 18.5 Noise spectrum at inlet of reciprocating compressor and insertion loss of designed muffler.

necessary at the frequencies 10 and 20 Hz. A muffler using an expansion and tail pipe type was suggested to handle the problem. The reflection coefficient of the tail pipe is approximately $R = -1$ and $ka = 2\pi fa/c = 2\pi \times 20 \times 0.19/345 = 0.0692$.

$$\mathrm{IL} = 20 \log_{10}|\cos kl_2 \cos kl_3 - m \sin kl_2 \sin kl_3|$$

where

$k = 2\pi f/c$; wave number (1/m)
l_2 = the length of the chamber (m)
l_3 = the length of the tail pipe (m)
m = the expansion ratio of the cross section between the chamber and inlet

With $kl_2 = kl_3 = \pi/2$, we have

$$\mathrm{IL} = 20 \log_{10}|m|$$

We need $m > 10$ in order to satisfy the desired value of IL. For 20 Hz, we use $kl_2 = kl_3 = \pi$. When $kl_2 = kl_3 = \pi/2$ at $f = 10$ Hz, we have IL = 0. Then, using $kl_2 = kl_3 = \pi/2$ at frequency 15 Hz, we can satisfy the IL condition of 20 dB at both frequencies. Hence, $l_2 = l_3 = 345/(4 \times 15) = 5.75$ (m) is chosen at a speed of sound $c = 345$ m/s.

The noise spectrum at the inlet of the reciprocating compressor under study and insertion loss of the muffler design in this example are shown in Figure 18.5.

The diameters or the lengths of the chamber and the tail pipe are properly selected, as shown and in Figure 18.6. At 10 Hz, IL is determined as

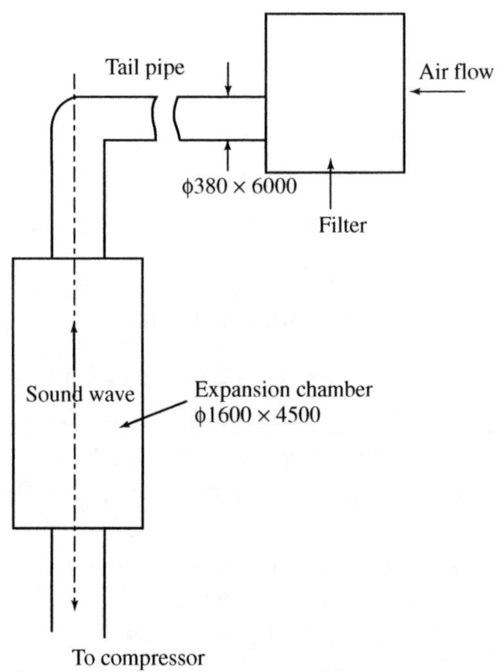

FIGURE 18.6 The muffler designed for the noise control of a reciprocating compressor.

indicated below:

$$\begin{aligned}\text{IL} &= 20\log_{10}|\cos(2\pi\times 10\times 4.5/345)\cos(2\pi\times 10\times 6/345) - (1.6/0.38)^2\sin(2\pi\times 10\times 4.5/345)\\ &\quad\times\sin(2\pi\times 10\times 6/345)|\\ &= 20\log_{10}|0.6825\times 0.4600 - 17.728\times 0.7308\times 0.8879| = 21.0\ (\text{dB})\end{aligned}$$

Similarly, at 20 Hz, we have

$$\begin{aligned}\text{IL} &= 20\log_{10}|\cos(2\pi\times 20\times 4.5/345)\cos(2\pi\times 20\times 6/345) - (1.6/0.38)^2\sin(2\pi\times 20\times 4.5/345)\\ &\quad\times\sin(2\pi\times 20\times 6/345)|\\ &= 20\log_{10}|(-0.0682)\times(-0.5767) - 17.728\times 0.9977\times 0.8170| = 23.2\ (\text{dB})\end{aligned}$$

Clearly, the attenuation at both frequencies satisfies the desired lower limit of 20 dB. Calculated values of IL at low frequencies are shown by a curved continuous line in Figure 18.5.

References

1. Levine, H. and Schwinger, J., On the radiation of sound from an unflanged circular pipe, *J. Phys. Rev.*, 73, 383, 1948.
2. Ohnaka, I., Lecture for noise reduction of machines, no. 2, *J. Marine Eng. Soc. Jpn.*, 4, 179, 1969.
3. Suyama, E. and Hirata, M., Attenuation constant of plane wave in a tube, *J. Acoust. Soc. Jpn.*, 35, 152, 1979.

19

Design of Sound Insulation

Kiyoshi Okura
Mitsuboshi Belting Ltd.

Summary

This chapter presents useful theory and design procedures for sound insulation. Related concepts and representations of transmission loss, the transmission coefficient, and impedance are given. Analysis and design procedures for sound insulation structures such as single and multiple panels and walls with sound absorption material are presented. Practical applications for the design of sound insulation components and systems are described.

19.1 Theory of Sound Insulation

19.1.1 Expressions of Sound Insulation [1]

19.1.1.1 Transmission Coefficient

Let us denote by I_i the acoustic energy incident on a wall per unit area and unit time. Some energy is dissipated in the wall, and, apart from the energy that is reflected by the wall, the rest is transmitted through the wall. Using I_t to denote the transmitted acoustic energy, the transmission coefficient of the wall is defined as

$$\tau = \frac{I_t}{I_i} \tag{19.1}$$

19.1.1.2 Transmission Loss

As an expression for sound insulation performance, we may use transmission loss (TL), which is defined as (also see Chapter 17 and Chapter 18)

$$\text{TL} = 10 \log\left(\frac{1}{\tau}\right) = 10 \log\left(\frac{I_i}{I_t}\right) \tag{19.2}$$

19.1.2 Transmission Loss of a Single Wall

Consider a plane sound wave incident on a impermeable infinite plate at angle θ, which is placed in a uniform air space as shown in Figure 19.1. The sound pressure of the incident, reflected, and transmitted waves, denoted by p_i, p_r, and p_t, respectively, are given by

$$
\begin{aligned}
p_i &= P_i e^{j\omega t - jk(x\cos\theta + y\sin\theta)} \\
p_r &= P_r e^{j\omega t - jk(-x\cos\theta + y\sin\theta)} \\
p_t &= P_t e^{j\omega t - jk(x\cos\theta + y\sin\theta)}
\end{aligned}
\tag{19.3}
$$

where P_i, P_r, and P_t are the sound pressure amplitudes of incident, reflected, and transmitted waves, respectively; ω is angular frequency; k is the wave number of the sound wave; c is the speed of sound, respectively in the air. Assuming that the plate is sufficiently thin compared with the wavelength of the incident sound wave, the vibration velocities on the incident and transmitted surfaces of the plate are equal. Then vibration velocity, u, of the plate in the x direction is equal to the particle velocity of the incident and transmitted sound waves, and we obtain relations

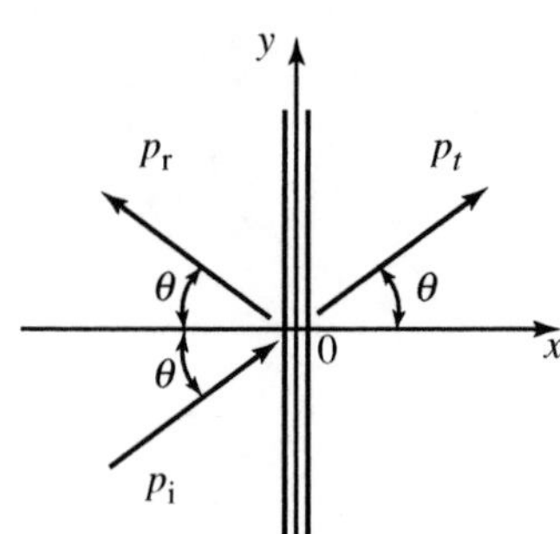

FIGURE 19.1 Plane sound wave incidence on an infinite plate.

$$u = -\frac{1}{j\omega\rho}\frac{\partial(p_i + p_r)}{\partial x} = -\frac{1}{j\omega\rho}\frac{\partial p_t}{\partial x} \tag{19.4}$$

$$\frac{p_i + p_r - p_t}{u} = Z_m \tag{19.5}$$

where ρ is the air density and Z_m is the *mechanical impedance* of the plate per unit area. From these equations, the transmission coefficient, τ_θ, and then the transmission loss, TL_θ, at the incident angle, θ, are obtained according to

$$TL_\theta = 10\log\frac{1}{\tau_\theta} = 10\log\left|\frac{p_i}{p_t}\right|^2 = 10\log\left|1 + \frac{Z_m\cos\theta}{2\rho c}\right|^2 \tag{19.6}$$

19.1.2.1 Coincidence Effect

Consider the vibration of the plate in the x–y plane shown in Figure 19.1. Denoting by m the surface density, and by B the bending stiffness per unit length of the plate, the equation of motion of the plate is given by

$$m\frac{\partial^2\xi}{\partial t^2} + B(1 + j\eta)\frac{\partial^4\xi}{\partial y^4} = p_i + p_r - p_t, \quad B = \frac{Eh^3}{12(1-\nu^2)} \cong \frac{Eh^3}{12} \tag{19.7}$$

where

ξ = displacement in the x direction
E = Young's modulus of the plate
h = thickness of the plate
η = loss factor of the plate
ν = Poisson's ratio of the plate

The plane sound wave of angular frequency, ω, and of incidence angle, θ, causes a bending wave in the plate where displacement is assumed to be $\xi = \xi_0 e^{j(\omega t - k_1 y)}$, as a solution of Equation 19.7. Hence, the mechanical impedance per unit area is obtained:

$$Z_m = \frac{p_i + p_r - p_t}{\partial\xi/\partial t} = \eta\frac{Bk_1^4}{\omega} + j\left(\omega m - \frac{Bk_1^4}{\omega}\right) \tag{19.8}$$

where $k_1 = k\sin\theta(k = \omega/c)$ is the wave number of the bending wave in y direction caused by the incident sound wave. Propagation speed of the forced bending wave, c_1, and a free bending wave of the plate, c_B, are given by

$$c_1 = \omega/k_1, \quad c_B = \left(\frac{\omega^2 B}{m}\right)^{1/4} \tag{19.9}$$

Equation 19.9 reduces Equation 19.8 to

$$Z_m = \eta\omega m\left(\frac{c_B}{c_1}\right)^4 + j\omega m\left[1 - \left(\frac{c_B}{c_1}\right)^4\right] \tag{19.10}$$

When the speed of forced bending wave, c_1, and the speed of free bending wave, c_B, are equal in Equation 19.10, the imaginary part of Z_m becomes 0, and a form of "resonance" occurs. Then the transmission loss decreases rapidly. This phenomenon is called the *coincidence effect*, and the resonant frequency dependent on the incident angle is given by

$$f = \frac{c^2}{2\pi \sin^2\theta}\sqrt{\frac{m}{B}} \tag{19.11}$$

The minimum of the resonant frequency is called coincidence critical frequency, or critical frequency for short, and it reduces to

$$f_c = \frac{c^2}{2\pi}\sqrt{\frac{m}{B}} \cong \frac{c^2}{1.8 c_L h} \tag{19.12}$$

where $c_L = \sqrt{E/\rho_P}$ is the speed of longitudinal wave in the plate, and ρ_P denotes the density of the plate.

Let us show the relations of the critical frequency and the plate thickness of typical material of sound insulation shown in Figure 19.2. Using the relation $c_B/c_1 = \sqrt{f/f_c}\sin\theta$, Equation 19.10

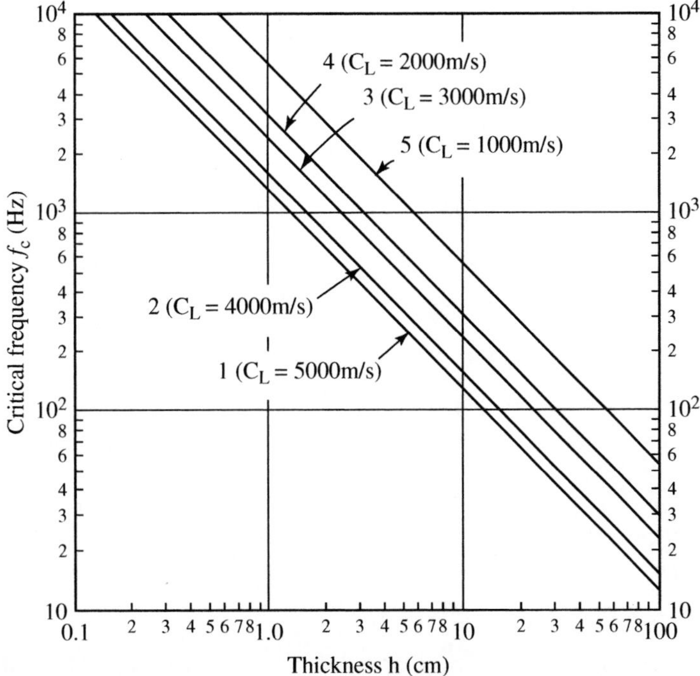

FIGURE 19.2 Critical frequency vs. plate thickness of typical sound insulation materials: (1) aluminum, steel, or glass; (2) hardboard or copper; (3) dense concrete, plywood, or brick; (4) gypsum board; (5) lead or light weight concrete. (*Source*: Beranek, L.L. 1988. *Noise and Vibration Control*, INCE/USA. With permission.)

becomes

$$Z_m = \eta\omega m\left(\frac{f}{f_c}\right)^2 \sin^4\theta + j\omega m\left[1 - \left(\frac{f}{f_c}\right)^2 \sin^4\theta\right] \tag{19.13}$$

19.1.2.2 Mass Law of Transmission Loss

When $f << f_c$, Equation 19.13 becomes $Z_m \cong j\omega m$. Then, the transmission loss depends on the incident angle, the frequency and the surface density of the plate. This is called the mass law of transmission loss.

Mass law of normal incidence represents the transmission loss at the incident angle $\theta = 0$, as given by

$$\mathrm{TL}_0 = 10 \log\left[1 + \left(\frac{\omega m}{2\rho c}\right)^2\right] \tag{19.14}$$

For $\omega m >> 2\rho c$, it becomes

$$\mathrm{TL}_0 \cong 10 \log\left(\frac{\omega m}{2\rho c}\right)^2 = 20 \log mf - 42.5; \quad \text{for air} \tag{19.15}$$

Mass law of random incidence represents the transmission loss at the angle averaged over a range of θ from 0 to 90°, which is realized for perfectly diffused sound field. We have

$$\mathrm{TL}_r = 10 \log(1/\tau_r) \cong \mathrm{TL}_0 - 10 \log(0.23\mathrm{TL}_0) \tag{19.16}$$

where the random incident transmission coefficient, τ_r, is defined as

$$\tau_r = \int_0^{\pi/2} \tau_\theta \cos\theta \sin\theta \, d\theta \Big/ \int_0^{\pi/2} \cos\theta \sin\theta \, d\theta \tag{19.17}$$

An approximation for Equation 19.16, as given below, is generally used for a practical use and this is often useful.

$$\mathrm{TL}_r = 18 \log mf - 44 \tag{19.18}$$

Mass law of field incidence represents the transmission loss at the angle averaged over a range of θ from 0 to about 78°, which is said to agree with actual sound field. We have

$$\mathrm{TL}_f = \mathrm{TL}_0 - 5 \tag{19.19}$$

The three types of transmission loss presented above are compared in Figure 19.3.

19.1.2.3 Stiffness Law of Transmission Loss [2]

The plate described above is assumed to be infinite. However, an actual plate is always supported by some structures at its boundaries and the plate size is finite. Transmission loss of a finite plate is considered to be related to the nature of excitation of vibration in the plate, for example, sound wave incidence, modes of vibration and characteristics of sound radiation. Therefore, the governing relationships become very complex. However, in the following frequency range, it is known that the transmission loss conforms to the mass law

$$\frac{c}{2a} < f << f_c \tag{19.20}$$

where a is length of shorter edge for rectangular plate.

When $f < c/2a$, the whole plate is excited in phase, and stiffness effects from the supports of its edges will appear. If we denote the equivalent stiffness of the plate as K and assume a loss factor of 0,

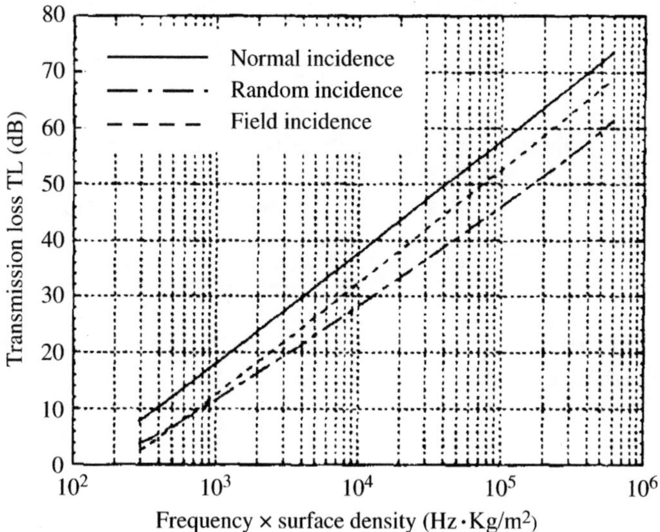

FIGURE 19.3 Theoretical transmission loss based on mass law.

the mechanical impedance of the plate is obtained by using Equation 19.8; thus

$$Z_{\mathrm{m}} = \mathrm{j}\left(\omega m - \frac{K}{\omega}\right) \tag{19.21}$$

The frequency at $Z_{\mathrm{m}} = 0$ corresponds to the first mode natural frequency, f_{11}, of the plate and consequently, the equivalent stiffness of rectangular plate with simple edge-support is given by

$$f_{11} = \frac{1}{2\pi}\sqrt{\frac{K}{m}} \equiv \frac{\pi}{2}\sqrt{\frac{B}{m}}\left[\left(\frac{1}{a}\right)^2 + \left(\frac{1}{b}\right)^2\right] \tag{19.22}$$

Then,

$$K = B\pi^4\left(\frac{1}{a^2} + \frac{1}{b^2}\right)^2$$

where a and b are the length of the short and long edges for the rectangular plate, respectively. When $f << f_{11}$ is assumed in Equation 19.21, the mass term can be neglected, and from Equation 19.6 the normal incidence transmission loss, $\mathrm{TL}_{\mathrm{S0}}$, is given by

$$\begin{aligned} \mathrm{TL}_{\mathrm{S0}} &= 10 \log\left|1 - \mathrm{j}\frac{K}{2\omega\rho c}\right|^2 \\ &= 10 \log\left[1 + \left(\frac{K}{2\omega\rho c}\right)^2\right] \\ &\cong 20 \log(K/f) - 74.5 \end{aligned} \tag{19.23}$$

This is called the stiffness law of the transmission loss, and it shows a 6 dB decay per octave.

The characteristics mentioned above for single wall transmission loss are shown in Figure 19.4 and summarized below.

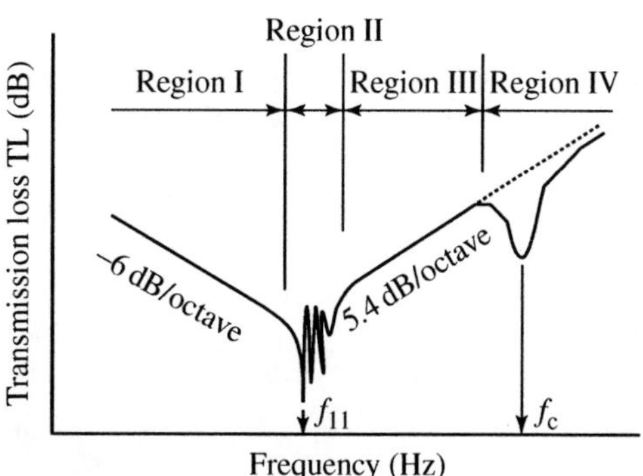

FIGURE 19.4 Transmission loss characteristics of a single wall.

1. Region I ($f << f_{11}$). Transmission loss is controlled by the stiffness of the panel:

$$\mathrm{TL} = \mathrm{TL}_0 - 40\log\left(\frac{f}{f_{11}}\right) \tag{19.24}$$

2. Region II ($f \approx f_{11}$). Transmission loss is controlled by the lower-mode natural frequencies of the panel, and the estimation becomes very complex.
3. Region III ($f_{11} << f \leq f_c/2$). Transmission loss is controlled by the mass (surface density) of the panel:

$$\mathrm{TL} = 18\log mf - 44 \tag{19.25}$$

4. Region IV ($f > f_c/2$). Transmission loss is controlled by the mass and the damping of the panel, and it is reduced by coincidence effects.

For $f_c/2 < f \leq f_c$:

TL is represented by a straight line connecting the value at $f = f_c/2$ of Equation 19.25 and the value at $f = f_c$ of Equation 19.26.

For $f > f_c$:

$$\mathrm{TL} = \mathrm{TL}_0 + 10\log\left(\frac{2\eta}{\pi}\frac{f}{f_c}\right) \tag{19.26}$$

19.1.3 Transmission Loss of Multiple Panels

To realize sound insulation of high performance, we often use a double wall or a multiple panel composed of insulation materials like steel plates and absorbing materials like fiber-glass. In this subsection, transmission loss of a multiple panel is described [3].

19.1.3.1 Calculation Method

Consider a multiple panel of infinite lateral extent as shown in Figure 19.5, which is composed of n acoustic elements, each element consisting of three basic materials, an impermeable plate, air space, and an absorption layer. Furthermore, consider a plane wave incident on the left-hand side surface of the nth element at angle θ. Let the sound pressure of the incident wave be p_i, and of the reflected wave be p_r, and the wave be propagating through the structure, and then radiating from the right-hand side of the first element as a plane wave of pressure p_t into a free field at transmission angle θ.

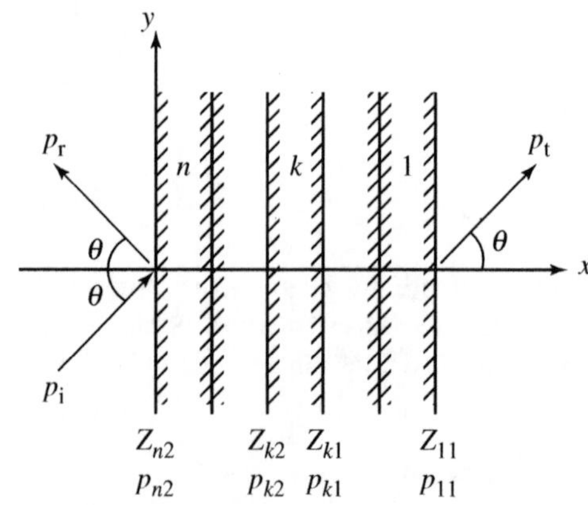

FIGURE 19.5 Calculation model of n-element multiple panel.

In the analysis, we append the subscript $k(= 1, 2, \ldots, n)$ to the physical parameters of the kth element, and "2" and "1" to the left- and right-hand side values of these parameters, respectively, as shown in Figure 19.5. The ratio of the sound pressure at the incident surface, p_{n2}, to the incident wave, p_i, is given by

$$\frac{p_{n2}}{p_i} = \frac{p_i + p_r}{p_i} = \frac{2Z_{n2}}{Z_{n2} + \rho c/\cos\theta} \tag{19.27}$$

where Z_{n2} is the acoustic impedance of the left-hand side normal to the surface of the nth element and $\rho c/\cos\theta$ is the acoustic impedance normal to the surface, which is equal to the radiation impedance of the first element, Z_{11}, shown in Figure 19.5. Using the usual condition of pressure matching at each interface, we can write the expression for the oblique incidence transmission coefficient as

$$\tau(\theta) = \left|\frac{p_t}{p_i}\right|^2 \equiv \left|\frac{p_{11}}{p_i}\right|^2 = \left|\frac{p_{n2}}{p_i}\right|^2 \cdot \left|\frac{p_{n1}}{p_{n2}} \cdots \frac{p_{k1}}{p_{k2}} \cdots \frac{p_{11}}{p_{12}}\right|^2 \tag{19.28}$$

Hence, we obtain the following expression for the random incidence transmission loss:

$$\text{TL} = 10 \log\left(\frac{\displaystyle\int_0^{\theta_l} \cos\theta \sin\theta \, d\theta}{\displaystyle\int_0^{\theta_l} \tau(\theta) \cos\theta \sin\theta \, d\theta}\right) \tag{19.29}$$

where θ_l is the limiting angle above which no sound is assumed to be received, and it varies between 78° and 85°.

If we know Z_{n2} in Equation 19.27 and the pressure ratio across each of the single elements in Equation 19.28, we can calculate the TL using Equation 19.29. We can obtain Z_{n2} by using the conditions of impedance matching at each interface from the rightmost to the leftmost element in order, if we know the impedance relations across each of the single elements.

Now, we present the pressure ratios and the acoustic impedance relations across three basic elements.

19.1.3.2 Impermeable Plate

Consider the vibration of an infinite impermeable plate of thickness, h, induced by the sound pressure difference on each side of the plate, as illustrated in Figure 19.6. In this case, the particle velocity on both sides of the plate must be the same as the plate vibration velocity. Then, from Equation 19.8, the following expressions are obtained:

$$Z_2 = Z_1 + Z_m \tag{19.30}$$

$$\frac{p_2}{p_1} = \frac{Z_2}{Z_1} \tag{19.31}$$

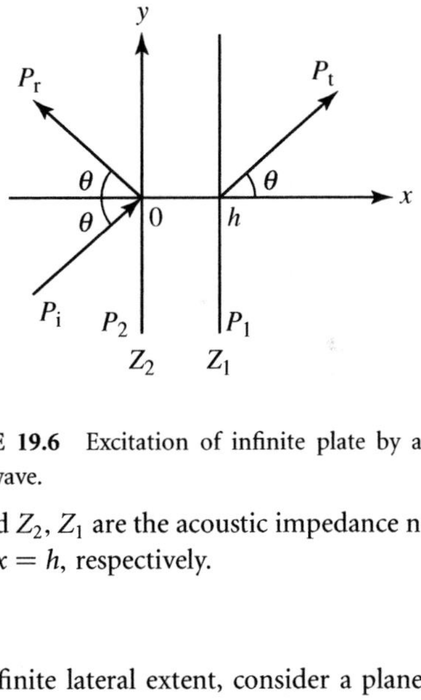

FIGURE 19.6 Excitation of infinite plate by a plane sound wave.

where p_2, p_1 are the sound pressure at the incident surface $x = 0$ and at the transmitted surface $x = h$, respectively, Z_m is the mechanical impedance of the plate, and Z_2, Z_1 are the acoustic impedance normal to the incident surface at $x = 0$ and the transmitted one at $x = h$, respectively.

19.1.3.3 Sound Absorbing Material

For a sound absorbing material layer of thickness d and infinite lateral extent, consider a plane wave incident at an angle θ to the normal, as shown in Figure 19.7. Deriving the wave equation in the sound absorbing material and applying the continuity conditions of the sound pressure across the surface at $x = 0$ and $x = d$, with some mathematical manipulation we get the following results:

$$Z_2 = \frac{\gamma Z_0}{q} \coth(qd + \varphi) \tag{19.32}$$

$$\frac{p_2}{p_1} = \frac{\cosh(qd + \varphi)}{\cosh\varphi} \tag{19.33}$$

$$q = \gamma\sqrt{1 + \left(\frac{k}{\gamma}\right)^2 \sin^2\theta}, \quad \varphi = \coth^{-1}\left(\frac{qZ_1}{\gamma Z_0}\right) \tag{19.34}$$

where γ is a propagation constant and Z_0 is a characteristic impedance of a homogeneous, isotropic absorbing material.

If porous material is used as the absorbing material, the following relations are applicable for γ and Z_0 [4]:

$$Z_0 = R + jX$$

$$R/\rho c = 1 + 0.0571(\rho f/R_l)^{-0.754} \quad (19.35)$$

$$X/\rho c = -0.0870(\rho f/R_l)^{-0.732}$$

$$\gamma = \alpha + j\beta$$

$$\alpha/k = 0.189(\rho f/R_l)^{-0.595} \quad (19.36)$$

$$\beta/k = 1 + 0.0978(\rho f/R_l)^{-0.700}$$

$$(0.01 \leq \rho f/R_l \leq 1)$$

where ρ is the air density, f is the frequency, and R_l is the flow resistivity, respectively. Specifically, note that R_l is defined as the flow resistance of the porous absorbing material per unit thickness. With data measured with a measuring tube of flow resistance, we can write

$$R_l = \frac{\Delta p}{l \cdot u} \quad (19.37)$$

where Δp is pressure difference between the inlet and the outlet of the absorbing material in the tube, u is the mean flow velocity in the tube, and l is the thickness of the absorbing material. It is known that the flow resistivity of porous absorbing material such as fiber-glass or rock wool is related to the bulk density, as shown in Figure 19.8.

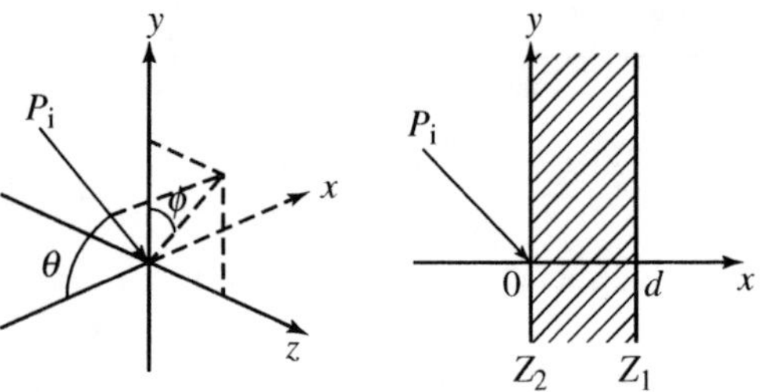

FIGURE 19.7 Schematic relation of sound wave directions.

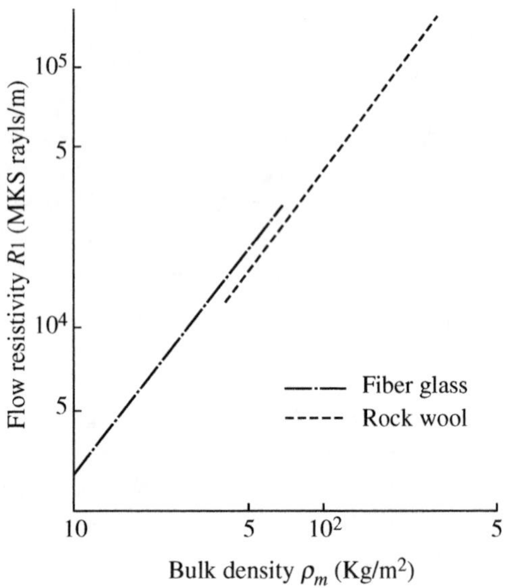

FIGURE 19.8 Flow resistivity vs. bulk density for porous, sound absorbing materials.

19.1.3.4 Air Space

For an air space, $Z_0 = \rho c$ and $\gamma = jk$. Hence, Equation 19.32 to Equation 19.34 reduce to

$$Z_2 = \frac{\rho c}{\cos\theta}\coth(jkd\cos\theta + \delta) \quad (19.38)$$

$$\frac{p_2}{p_1} = \frac{\cosh(jkd\cos\theta + \delta)}{\cosh\delta} \quad (19.39)$$

$$\delta = \coth^{-1}\left(\frac{Z_1\cos\theta}{\rho c}\right) \quad (19.40)$$

19.1.3.5 Double Wall [2]

Applying the theory formulated above, we can easily obtain the transmission loss of a double wall composed of the three elements: impermeable plate, air space, and impermeable plate, as shown in Figure 19.9. Assume that the two impermeable plates have the same surface density, m, and the mechanical impedance of the plates is $j\omega m$. Then, we can obtain following equations for element one and element three:

$$Z_{12} = Z_{11} + Z_{\mathrm{m}} = \frac{\rho c}{\cos\theta} + \mathrm{j}\omega m,$$
$$\frac{p_{12}}{p_{11}} = \frac{Z_{12}}{Z_{11}} = 1 + \mathrm{j}\frac{\omega m \cos\theta}{\rho c} \tag{19.41}$$

$$Z_{32} = Z_{31} + Z_m = Z_{22} + \mathrm{j}\omega m,$$
$$\frac{p_{32}}{p_{31}} = \frac{Z_{32}}{Z_{22}} = 1 + \mathrm{j}\frac{\omega m}{Z_{22}} \tag{19.42}$$

For element two:

$$Z_{22} = \frac{\rho c}{\cos\theta}\coth(jkd\cos\theta + \delta'),$$
$$\frac{p_{22}}{p_{21}} = \frac{\cosh(jkd\cos\theta + \delta')}{\cosh\delta'} \tag{19.43}$$

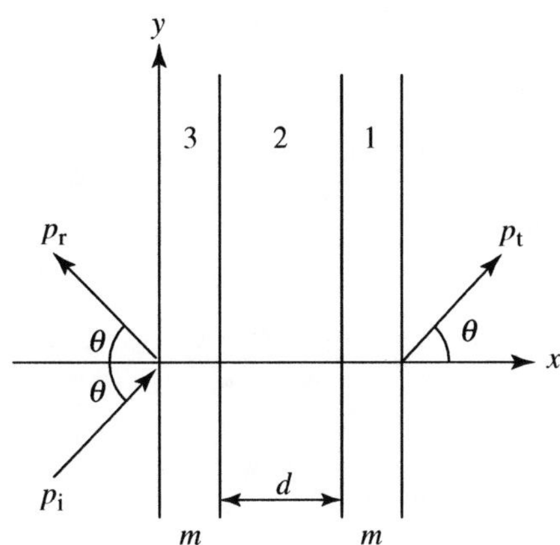

FIGURE 19.9 Calculation model of a double wall with air space.

where, by applying impedance matching conditions at the interface of element one and element two, the following definition is introduced:

$$\delta' = \coth^{-1}\left(\frac{Z_{21}\cos\theta}{\rho c}\right) = \coth^{-1}\left(\frac{Z_{12}\cos\theta}{\rho c}\right) = \coth^{-1}\left(1 + \mathrm{j}\frac{\omega m\cos\theta}{\rho c}\right) \tag{19.44}$$

In this case, Equation 19.27 reduces to

$$\frac{p_{\mathrm{i}}}{p_{32}} = \frac{Z_{32} + \rho c/\cos\theta}{2Z_{32}} = \frac{Z_{22} + \mathrm{j}\omega m + \rho c/\cos\theta}{2(Z_{22} + \mathrm{j}\omega m)} \tag{19.45}$$

Substituting Equation 19.41 to Equation 19.45 into Equation 19.28, we obtain the transmission loss of the double wall:

$$\mathrm{TL}_\theta = 10\log[1/\tau(\theta)] = 10\log[1 + 4a^2\cos^2\theta(\cos\beta - a\cos\theta\sin\beta)^2] \tag{19.46}$$

$$a = \omega m/2\rho c, \quad \beta = kd\cos\theta$$

In Equation 19.46, the transmission loss is zero, and full passage (i.e., "all-pass" in the filter terminology) of sound occurs when the following equation holds:

$$\cos\beta - a\cos\theta\sin\beta = 0 \tag{19.47}$$

When $\beta << 1(kd << 1)$, the frequency of full passage for normal incidence is given by

$$f_{\mathrm{r}} = \frac{1}{2\pi}\sqrt{\frac{2\rho c^2}{md}} \tag{19.48}$$

This is the natural frequency of a vibrating system consisting of two masses, m, connected by a spring of spring constant, $\rho c^2/d$.

When $\beta >> 1(kd >> 1)$, the solution of Equation 19.47 for β is $\beta \cong n\pi$, and the frequency of all passage for normal incidence is given by

$$f_{\mathrm{n}} = \frac{nc}{2d}\ (n = 1, 2, 3, \ldots) \tag{19.49}$$

These are the acoustic resonant frequencies of the air space d.

Characteristics of the transmission loss given by Equation 19.46, in case of normal incidence ($\theta = 0$), are as follows:

1. $f < f_{\mathrm{r}}\left(\beta < \sqrt{2\rho d/m}\right)$

$$\mathrm{TL} \cong 10\log(4a^2) = \mathrm{TL}_0 + 6 \tag{19.50}$$

This is equal to the transmission loss of a single wall of surface density $2m$.

2. $f_r \le f < f_1/\pi\left(\sqrt{2\rho d/m} \le \beta < 1\right)$

$$\text{TL} \cong 10\log(4a^4\beta^2) = 2\text{TL}_0 + 20\log(2kd) \quad (19.51)$$

This transmission loss indicates an 18 dB increase per octave.

3. $f = (2n-1)c/4d(\beta = n\pi - \pi/2)$

$$\text{TL} \cong 10\log(4a^4) = 2\text{TL}_0 + 6 \quad (19.52)$$

A straight line connecting the transmission losses at these frequencies in Figure 19.10 indicates a 12 dB increase per octave. When the two impermeable plates have different surface densities, m_1 and m_2, Equation 19.41 to Equation 19.52 reduce to

1. $f < f_r\left(\beta < \sqrt{2\rho d/m}\right)$

$$\text{TL} = 20\log[\omega(m_1 + m_2)/2\rho c] \quad (19.53)$$

2. $f_r \le f < f_1/\pi\left(\sqrt{2\rho d/m} \le \beta < 1\right)$

$$\text{TL} = \text{TL}_1 + \text{TL}_2 + 20\log(2kd) \quad (19.54)$$

3. $f = (2n-1)c/4d(\beta = n\pi - \pi/2)$

$$\text{TL} = \text{TL}_1 + \text{TL}_2 + 6 \quad (19.55)$$

In these equations, TL_1 and TL_2 are the transmission losses of each plate, which are given by Equation 19.15.

The transmission loss of a double wall, as mentioned above, is shown schematically in Figure 19.10. An actual double wall, however, is finite in size and the air space forms a closed acoustic field, which

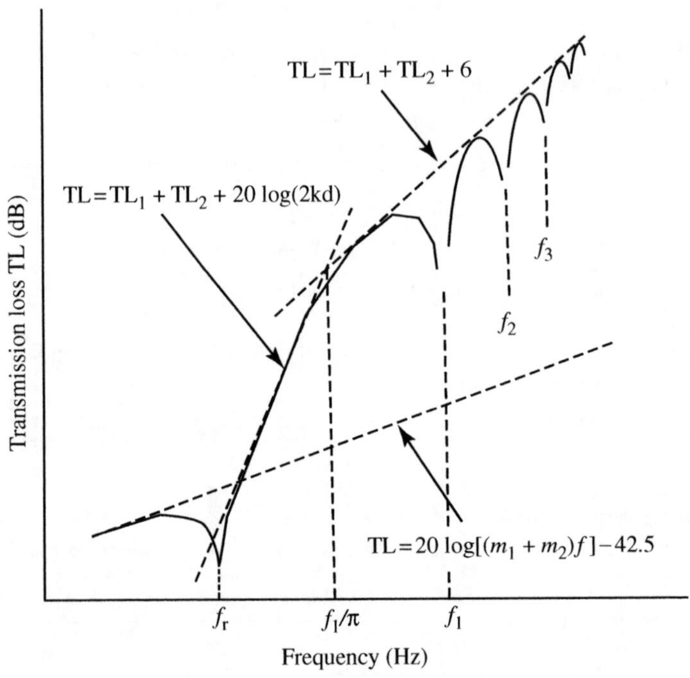

FIGURE 19.10 Transmission loss of a double wall with air space.

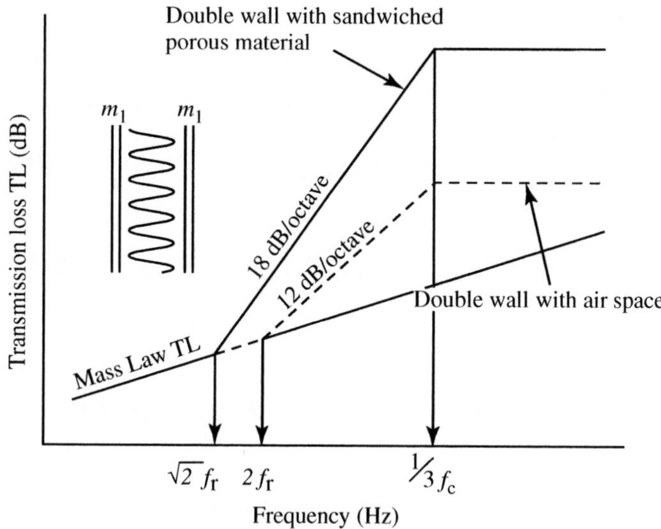

FIGURE 19.11 Design chart for estimating the transmission loss of a double wall with sandwiched porous material or air space.

makes the transmission loss deviate from the theoretical value. Figure 19.11 gives a design chart of an actual double wall, which is based on theory and experiments.

19.1.4 Transmission Loss of Double Wall with Sound Bridge [5]

In the previously presented theory, each plate of the multiple panel is considered to be structurally independent. In actual multiple panels, such as partitions of a building or sound insulation laggings of a duct, however, each plate is connected with steel sections, stud bolts, and the like, which are called sound bridges. This is illustrated in Figure 19.12.

The sound pressure of the transmitted wave through a double wall with sound bridges is given by the summation of radiated sound pressure from the vibration of the transmitted side plate excited by the sound in the air space and that mechanically excited by the sound bridges.

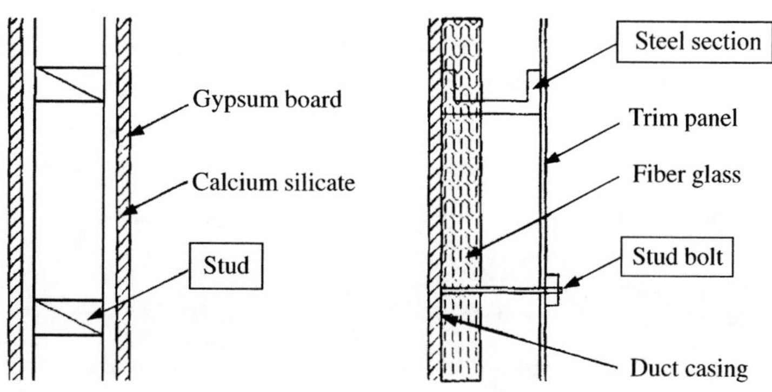

FIGURE 19.12 Examples of actual double wall with sound bridges.

The acoustic power radiated from the area, S, of an infinite plate excited by sound pressure is given by

$$W_P = \rho c S v_2^2 \tag{19.56}$$

where v_2^2 is the space averaged mean square vibration velocity over the plate. The acoustic power radiated from the plate mechanically excited by a point force or a line force is

$$W_B = \rho c \chi v^2 (f << f_c) \tag{19.57}$$

$$\chi = \begin{cases} \dfrac{8}{\pi^3}\lambda_c^2 & \text{(point force excitation)} \\ \dfrac{2}{\pi} l \lambda_c & \text{(line force excitation)} \end{cases} \tag{19.58}$$

where v^2 is the mean square vibration velocity of the plate at the excitation point, $\lambda_c = c/f_c$ is the wavelength of the bending wave at the critical frequency, and l is the length of the line force. By comparing Equation 19.56 and Equation 19.57, it is noted that χ is the effective area of the acoustic power radiated from the infinite plate excited by the point or line force. Acoustic power, W_B, is the power radiated from a small area near the excitation point, because a free bending wave propagating in an infinite plate can radiate little sound when $f < f_c$.

From the equations given above, the total acoustic power radiated from the transmitted side plate is obtained as

$$W_T = W_P + W_B = \rho c S v_2^2 \left[1 + \frac{n\chi}{S}\left(\frac{v}{v_2}\right)^2\right] \tag{19.59}$$

where n is the number of excitation forces applied to the area, S. Then, transmission loss TL_T of the double wall with sound bridges is given by

$$TL_T = 10 \log\left(\frac{W_I}{W_T}\right) = 10 \log\left(\frac{W_I}{W_P} \cdot \frac{W_P}{W_T}\right) = TL - TL_B \tag{19.60}$$

where W_I is the acoustic power incident on the double wall, TL is the transmission loss of the double wall without a sound bridge, and TL_B denotes the transmission loss reduction by the sound bridges, and is given by

$$TL_B = 10 \log\left(\frac{W_T}{W_P}\right) = 10 \log\left[1 + \frac{n\chi}{S}\left(\frac{v}{v_2}\right)^2\right] \tag{19.61}$$

We assume the following:

1. The vibration velocity of the incident side plate is not affected by the sound bridges.
2. The vibration velocity of the transmitted side plate at the excitation points (connecting points) is equal to the velocity v_1 of the incident side plate, and consequently, the next equation holds

$$\frac{v}{v_2} \cong \frac{v_1}{v_2}$$

With these assumptions, we apply the method presented in section 19.1.3, to determine v/v_2 as

$$\frac{v}{v_2} \cong \frac{v_1}{v_2} = \frac{\omega^2 m_2 d}{\rho c^2} \quad (f_r < f < f_l/\pi) = \frac{\omega m_2}{\rho c} \quad (f > f_l/\pi) \tag{19.62}$$

Using Equation 19.60 and Equation 19.53, we can obtain the increase in transmission loss, ΔTL, from the transmission loss of mass law based on the total mass of the double wall, as presented below.

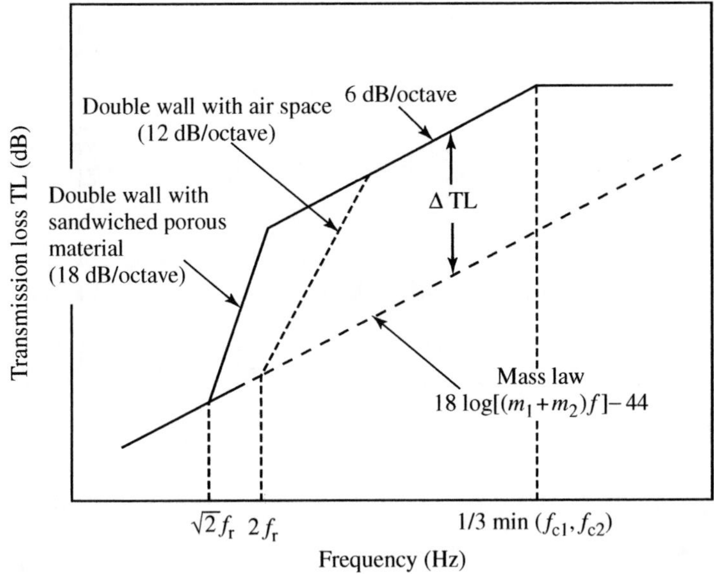

FIGURE 19.13 Design chart for estimating the transmission loss of a double wall with sound bridges.

1. Point connection

$$\Delta\mathrm{TL} = \mathrm{TL} - \mathrm{TL_B} - 20\log\left[\frac{\omega(m_1+m_2)}{2\rho c}\right]$$
$$= 20\log(ef_c) + 20\log\left(\frac{m_1}{m_1+m_2}\right) + 10\log\left(\frac{\pi^3}{8c^2}\right) \qquad (19.63)$$

2. Line connection

$$\Delta\mathrm{TL} = \mathrm{TL} - \mathrm{TL_B} - 20\log\left[\frac{\omega(m_1+m_2)}{2\rho c}\right]$$
$$= 10\log(bf_c) + 20\log\left(\frac{m_1}{m_1+m_2}\right) + 10\log\left(\frac{\pi}{2c}\right) \qquad (19.64)$$

where $e = \sqrt{S/n}$ is the distance between point forces and $b = S/nl$ is the distance between line forces.

Figure 19.13 presents a practical and useful design chart of the transmission loss for a double wall with sound bridges, which is based on Figure 19.11 and Equation 19.63 and 19.64.

19.2 Application of Sound Insulation

19.2.1 Acoustic Enclosure

Performance of an enclosure may be represented by the insertion loss (IL), which is the difference of acoustic power level before and after installation of the enclosure. When we assume a noise source and also an enclosure with one-dimensional model as shown in Figure 19.14a, the insertion loss through frequency is shown in Figure 19.14b. It is divided into the following four regions:

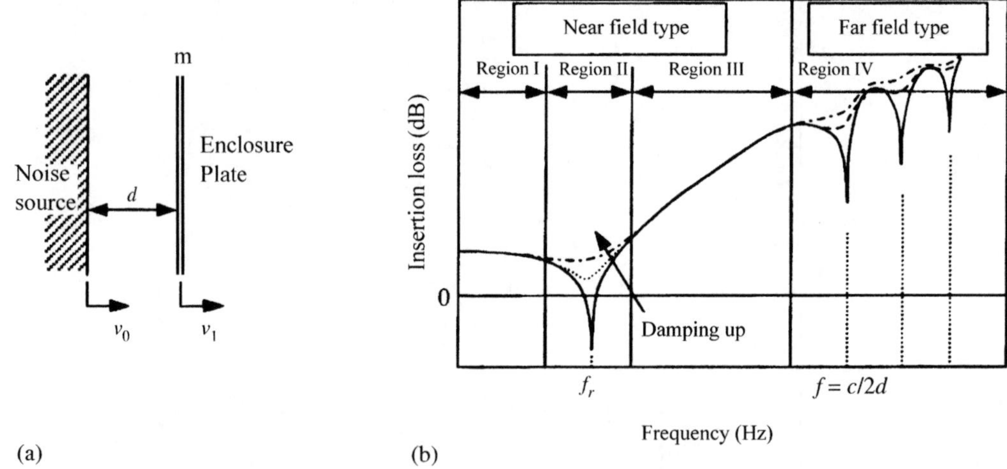

FIGURE 19.14 One-dimensional model for calculating the insertion loss characteristics of an acoustic enclosure: (a) One-dimensional model; (b) insertion loss of an enclosure.

1. Region I ($f < f_r$) is controlled by the stiffness of the enclosure plate and air space.
2. Region II ($f \cong f_r$) is the resonance region for a vibrating system consisting of the mass, the stiffness of the enclosure plate, and the capacitance of air space.
3. Region III ($f_r < f < c/2d$) is controlled by the mass of the enclosure plate.
4. Region IV ($f > c/2d$) is controlled by the diffused sound field. This region cannot be represented by a one-dimensional model.

19.2.1.1 Near Field Type [4]

When the distance between the noise source and the enclosure is less than half of a wavelength of the emitted sound from the source, insertion loss corresponds to the characteristics of the regions I to III, and we can represent them with a one-dimensional acoustic model, as shown in Figure 19.14a.

Consider an infinite flat enclosure plate with distance d from a noise source of plane sound wave, as shown in Figure 19.14a. The insertion loss of the plate is given by

$$\mathrm{IL} = 10\log\left(\frac{v_0}{v_1}\right)^2 = 10\log\left[1 - \frac{2\sin\theta(X\cos\theta - R\sin\theta)}{\rho c} + \frac{\sin^2\theta(X^2+R^2)}{\rho^2c^2}\right] \tag{19.65}$$

$$\theta = kd = \omega d/c, \quad R = \eta\omega m, \quad X = (\omega m - K/\omega) = \omega m[1 - (\omega_{11}/\omega)^2], \quad \omega_{11} = \sqrt{K/m}$$

where m and K are the density and equivalent stiffness of the plate per unit area, respectively, and η is the loss factor of the plate. If the enclosure is a rectangular plate of size $a \times b$ and is simply supported at its edges, the equivalent stiffness is given by Equation 19.22, and ω_{11} is the natural (angular) frequency of the first mode.

In Equation 19.65, the conditions in which the brackets of the right-hand side are equal to zero or $\mathrm{IL} = -\infty$ are satisfied by following frequencies:

(1) $\theta << \pi$

$$f_r = \frac{1}{2\pi}\sqrt{\frac{K + \rho c^2/d}{m}} = \frac{1}{2\pi}\sqrt{\omega_{11}^2 + \frac{\rho c^2}{md}} \tag{19.66}$$

This is the natural frequency of vibration of the one-degree-of-freedom (one-DoF) system determined by the stiffness of the plate, the spring constant of the air space, and the surface density of the plate, as shown in Figure 19.15.

(2) $\theta = n\pi$ $(n = 1, 2, \ldots)$

$$f_n = \frac{nc}{2d} \quad (n = 1, 2, \ldots) \qquad (19.67)$$

These are the resonant frequencies of the air space.

The frequency characteristics of the IL given by Equation 19.65 are shown in Figure 19.16, where the normal incidence transmission losses are shown by broken lines, as a reference. Equation 19.65 is approximated by

1. $f < f_r$

$$\text{IL} \cong 10 \log\left(1 + \frac{2Kd}{\rho c^2}\right) \qquad (19.68)$$

2. $f_r \leq f < f_1$

$$\text{IL} \cong 20 \log(mdf^2) + 20 \log\left(\frac{4\pi^2}{\rho c^2}\right)$$
$$= 20 \log(mdf^2) - 71 \qquad (19.69)$$

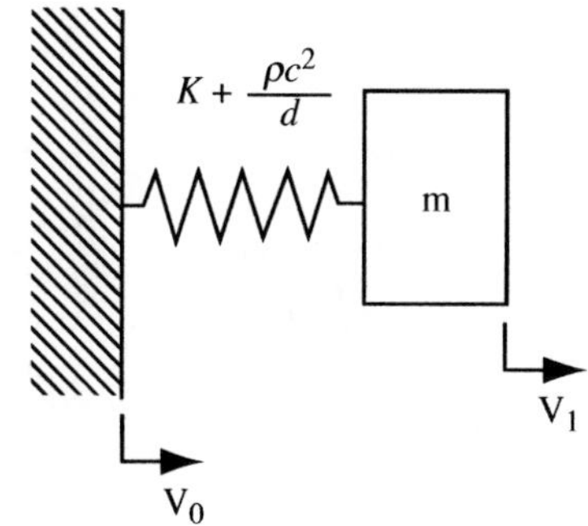

FIGURE 19.15 One-degree-of-freedom vibrating system.

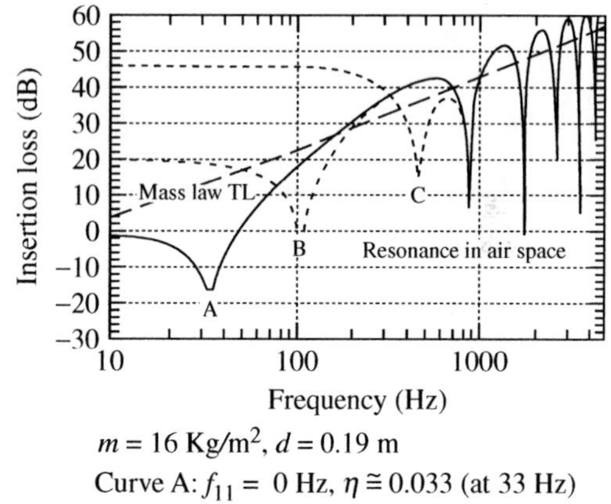

$m = 16$ Kg/m^2, $d = 0.19$ m

Curve A: $f_{11} = 0$ Hz, $\eta \cong 0.033$ (at 33 Hz)

Curve B: $f_{11} = 100$ Hz, $\eta \cong 0.033$ (at 105 Hz)

Curve C: $f_{11} = 475$ Hz, $\eta \cong 0.033$ (at 475 Hz)

FIGURE 19.16 Example of theoretical insertion loss of a near field-type enclosure.

19.2.1.2 Far Field-Type (Absorption Type) Enclosure [1]

When the distance between the noise source and the enclosure is larger than half of a wavelength of the emitted sound, insertion loss may be represented by the characteristics of Region IV, and it can be analyzed using the theory of room or hall acoustics.

Consider the enclosure shown in Figure 19.17, with a noise source of power level L_{WO}.

From the theory of room acoustics, the average sound pressure level, L_{PO}, on the inner surface of the enclosure plate is obtained as the sum of the direct and reverberant sound pressures:

$$L_{\text{PO}} = L_{\text{WO}} + 10 \log\left(\frac{1}{S} + \frac{4}{R}\right)$$
$$= L_{\text{WO}} - 10 \log S + 10 \log\left(1 + \frac{4S}{R}\right) \qquad (19.70)$$

$$R = \frac{\bar{\alpha} S}{(1 - \bar{\alpha})} \qquad (19.71)$$

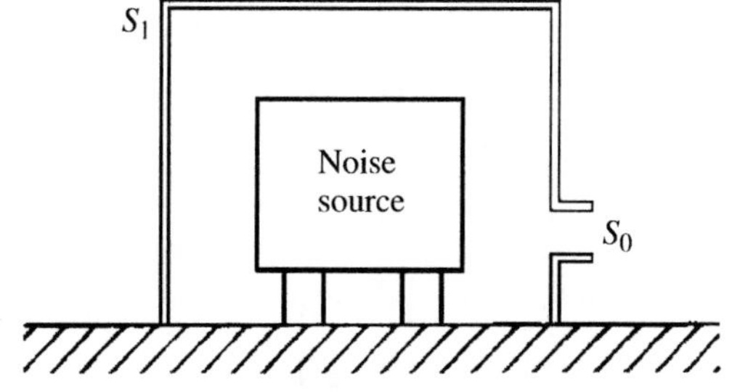

FIGURE 19.17 Calculation model of a far field type enclosure.

where S is the inner surface area of the enclosure and $\bar{\alpha}$ is the average absorption coefficient on the inner surface of the enclosure. In Equation 19.70, the first and the second terms of the right-hand side represent the influence of the direct sound field, and the third term represents the buildup caused by the covering.

When we use S_{P} for the area of the enclosure plate and $S_{\text{O}}(= S - S_{\text{P}})$ for the area of the enclosure opening, and assume diffusing condition for the sound field in the enclosure, acoustic power levels radiated from the plate and the opening are given, respectively, by

$$L_{WP} = L_{P0} - (\mathrm{TL} + 6) + 10 \log S_P,$$
$$L_{W0} = L_{P0} - 6 + 10 \log S_O \quad (19.72)$$

where TL is the random incident transmission loss of the enclosure plate. Then, the insertion loss of the enclosure is

$$\mathrm{IL} = L_{W0} - 10 \log\left(10^{L_{WP}/10} + 10^{L_{W0}/10}\right) \quad (19.73)$$

In the design of an acoustic enclosure, special attention should be paid to the following points:

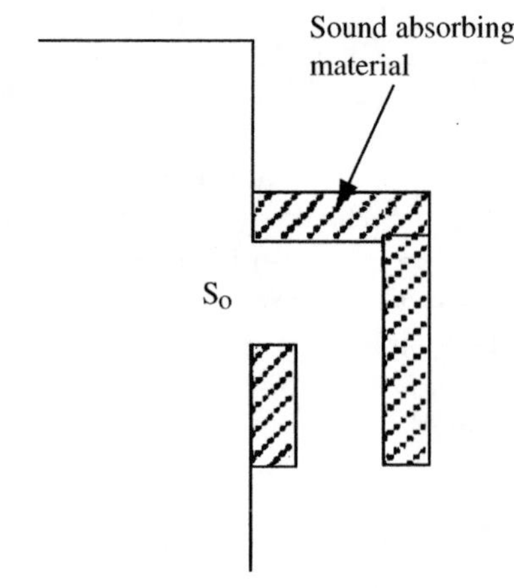

FIGURE 19.18 Example of a silencer at the opening.

1. *Buildup* increases the sound pressure in the enclosure and also the power level radiated from the enclosure. We must treat the inner surface of the enclosure with sound absorbing materials to reduce the absorption coefficient and to decrease the buildup.
2. *Opening* radiates more acoustic power than the enclosure plate by TL, as is clear from Equation 19.72. It is desirable to make the opening as small as possible within the range given by

$$\frac{S_O}{S_P} \leq 10^{-\mathrm{TL}/10} = \tau \quad (19.74)$$

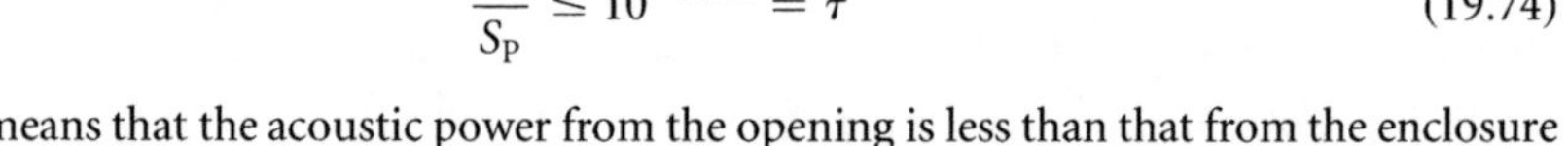

This equation means that the acoustic power from the opening is less than that from the enclosure plate. If the relation in Equation 19.74 cannot be satisfied because of ventilation requirements, and so on, some type of silencers should be provided at the opening, as shown in Figure 19.18.

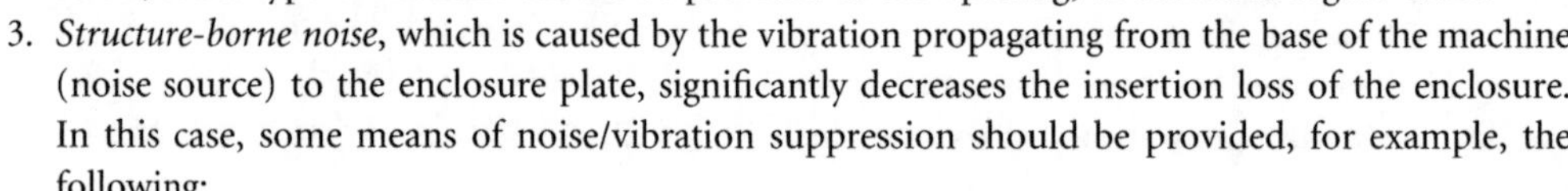

3. *Structure-borne noise*, which is caused by the vibration propagating from the base of the machine (noise source) to the enclosure plate, significantly decreases the insertion loss of the enclosure. In this case, some means of noise/vibration suppression should be provided, for example, the following:
   - Place supporting structures of the enclosure at the points of lowest vibration level, and the vibrations of the machine should be prevented from propagating to the enclosure plate, using vibration isolation materials.
   - Add damping materials to the enclosure plate so as to reduce the vibration level of the plate.

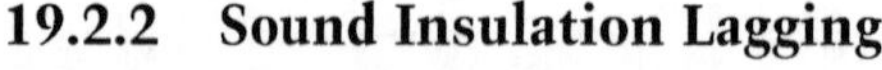

19.2.2 Sound Insulation Lagging

In electric power plants and chemical plants, for example, piping for high-pressure water or steam, and ducts for air or gas flow form major noise sources. For controlling these noise sources, we usually use sound insulation laggings, which cover the noise sources with heavy and impermeable plates or sheets with sound absorbing materials, as shown in Figure 19.19.

19.2.2.1 Pipe Lagging [2]

Approximate the cylindrical piping and pipe lagging with a one-dimensional model as shown in Figure 19.20. The insertion loss of one-layered lagging approximated by a one-DoF system is given by Equation 19.69 in the frequency region $f_r \leq f < f_1$, as mentioned before. It is not practical, however, to directly apply Equation 19.69 to actual laggings, and we approximate the insertion loss of actual laggings by

$$\mathrm{IL} = a \log(mdf^2) + b$$

where a and b are constants.

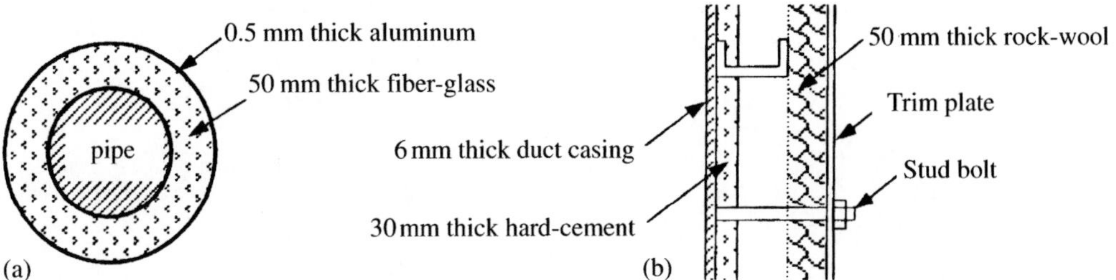

FIGURE 19.19 Examples of typical sound insulation laggings: (a) pipe lagging, (b) duct lagging.

By taking mdf^2 as the horizontal axis and plotting the insertion loss data from laboratory tests and field tests (the vertical coordinates), we obtain Figure 19.21. Apply regression analysis to the data in Figure 19.21 to obtain the insertion loss of one layered lagging as

$$\mathrm{IL} = 11.7 \log(mdf^2) - 43.3 \ (5 \times 10^3 \leq mdf^2 \leq 10^8) \tag{19.75}$$

Applying the same method to double layered lagging, approximated by a two-DoF vibrating system, we get Figure 19.22, and the insertion loss

$$\mathrm{IL} = 6.9 \log(m_1 m_2 d_1 d_2 f^4) - 40.3 \ (10^6 \leq m_1 m_2 d_1 d_2 f^4 \leq 10^{15}) \tag{19.76}$$

where the subscripts "1" and "2" denote the first layer and the second layer, respectively.

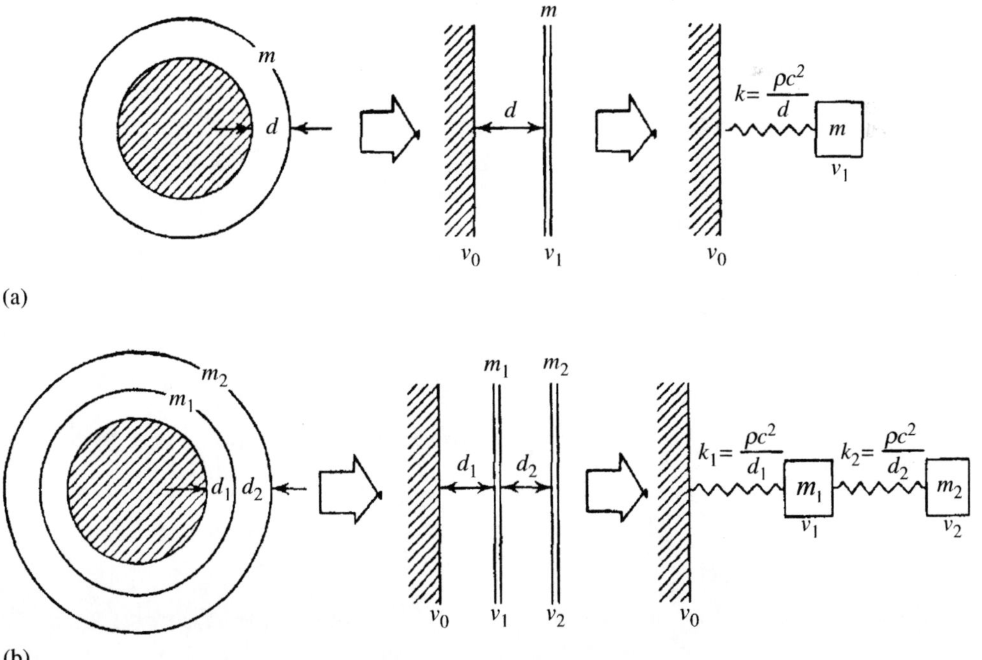

FIGURE 19.20 Examples of pipe laggings and calculation model: (a) one-layered lagging; (b) double-layered lagging.

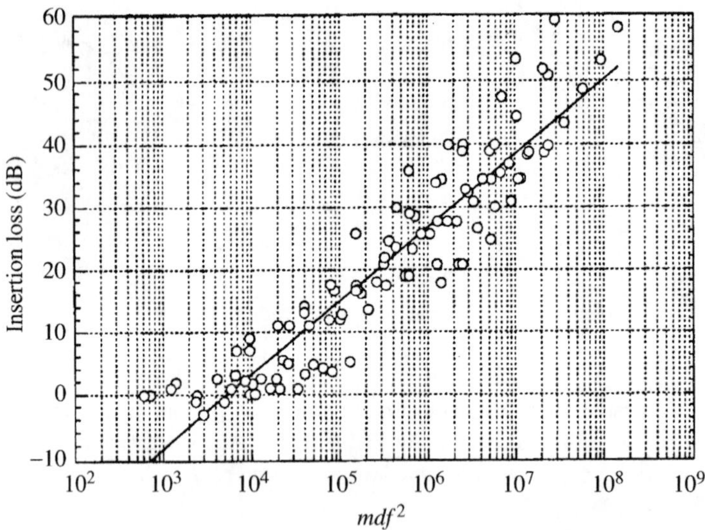

FIGURE 19.21 Measured insertion loss data obtained from laboratory and field tests for one-layered laggings and regression analysis.

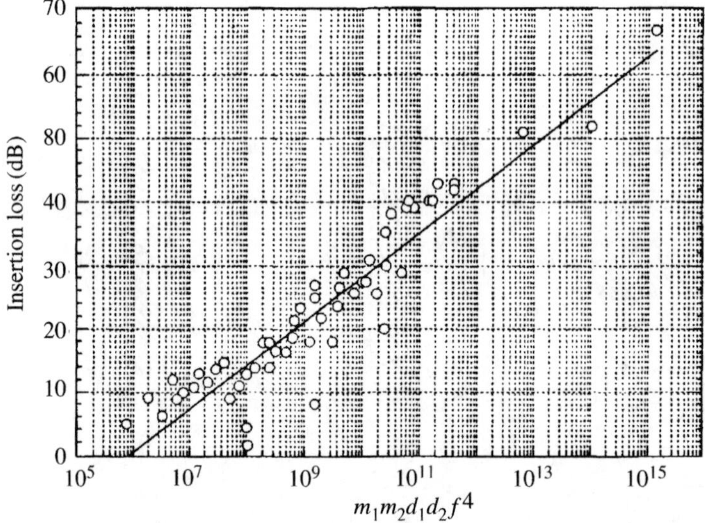

FIGURE 19.22 Measured data of insertion loss obtained from laboratory and field tests for double-layered laggings and regression analysis.

19.2.2.2 Duct Lagging [4]

Various types of the duct laggings are used according to the need, for example, as shown in Figure 19.19. A simpler and more practical approach is to place a thin plate on the duct casing through absorbing materials, as shown in Figure 19.23. In this case, assuming that vibration of the duct casing is not affected by the placed plate, the insertion loss of the duct lagging is obtained by following equations, which can be deduced from the method used in the transmission loss of double wall with sound bridges.

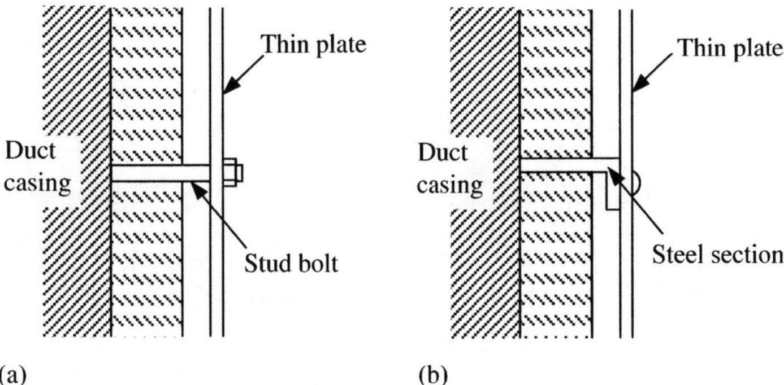

FIGURE 19.23 Examples of duct laggings and connection types of thin plate to the duct casing: (a) point connection; (b) line connection.

1. Point connection

$$\mathrm{IL} = -10\log\left[\beta^2 \frac{8}{\pi^3} n_P \frac{c^2}{f_c^2} + \left(\frac{f_r}{f}\right)^4\right] + 10\log\sigma \tag{19.77}$$

2. Line connection

$$\mathrm{IL} = -10\log\left[0.64 n_L \frac{c}{f_c} + \left(\frac{f_r}{f}\right)^4\right] + 10\log\sigma \tag{19.78}$$

where

β = vibration isolation factor of the flexible support ($\beta = 1$ for rigid support)
n_P = number of attachment points per unit area
n_L = number of studs per unit length

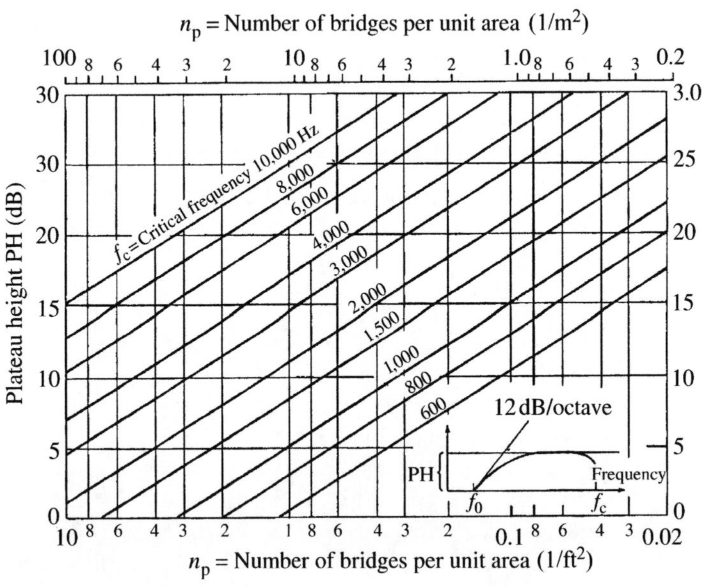

FIGURE 19.24 Plateau height for point connection as a function of the number of bridges per unit area [4]. (*Source*: Beranek, L. L. 1988. *Noise and Vibration Control*, INCE/USA. With permission.)

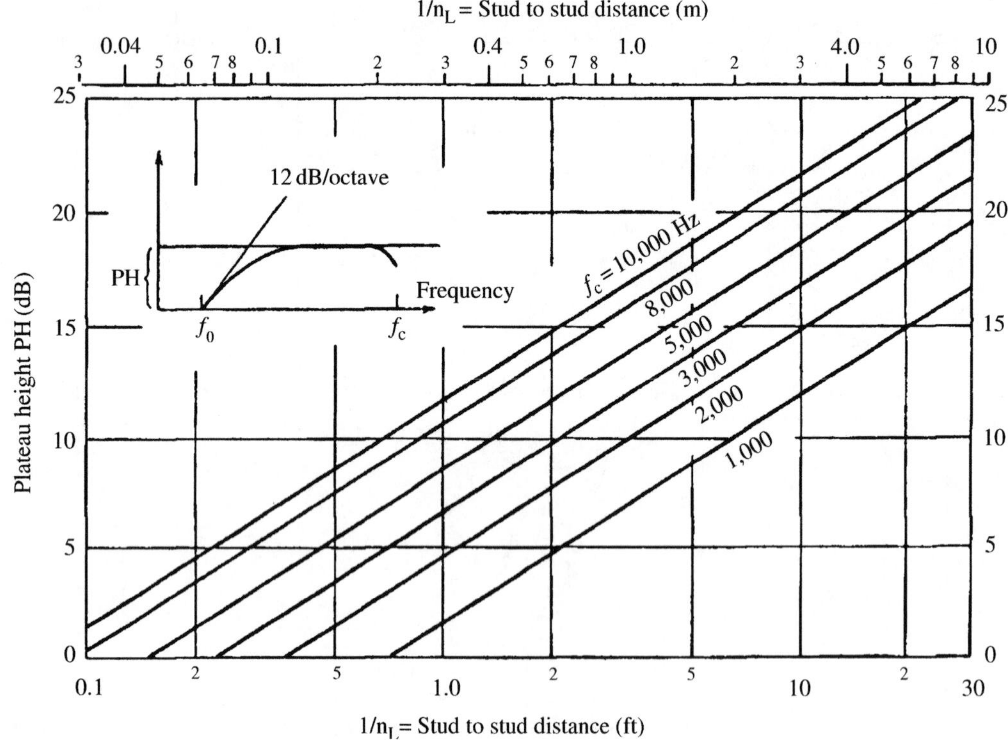

FIGURE 19.25 Plateau height for line connection as a function of the number of studs per unit length [4]. (*Source*: Beranek, L. L. 1988. *Noise and Vibration Control*, INCE/USA. With permission.)

f_c = critical frequency of the thin plate given by Equation 19.12
f_r = resonant frequency given by Equation 19.66
σ = sound radiation efficiency of the thin plate

Knowing the critical frequency, f_c, and the resonant frequency, f_r, we can obtain the insertion loss from the charts given in Figure 19.24 and Figure 19.25 instead of using Equation 19.77 and Equation 19.78. In Figure 19.24 and Figure 19.25, it is assumed that $\sigma = 1$. Note from Equation 19.77 and Equation 19.78 that we must consider the following measures to obtain a higher insertion loss.

1. Make the distance between attachment points or studs as large as possible (decrease n_P and n_L).
2. Make the air space as large as possible (decrease f_r).

References

1. Shiraki, K., ed. 1987. *From Designing of Noise Reduction to Simulation* (in Japanese), Ouyou-gijutsu Shuppan, Chiyoda-ku, Tokyo.
2. Tokita, Y., ed. 2000. *Sound Environment and Control Technology, Vol. I, Basic Engineering* (in Japanese), Fuji-techno-system, Bunkyo-ku, Tokyo.
3. Okura, K. and Saito, Y., Transmission loss of multiple panels containing sound absorbing materials in a random incidence field, *Inter-noise*, 78, 637, 1978.
4. Beranek, L.L., ed. 1988. *Noise and Vibration Control*, INCE/USA, Ames, IA.
5. Sharp, B.H. 1973. A study of techniques to increase the sound insulation of building elements, Wyle Laboratory Report WR 73-5, El Segundo, CA.

20

Statistical Energy Analysis

Takayuki Koizumi
Doshisha University

Summary

This chapter presents the basics of statistical energy analysis (SEA) as applied to acoustic problems in structural systems. Power flow equations for structures consisting of two or more subsystems are described. The modeling and analysis procedures for the structural subsystems and acoustic subsystems are given. An estimation procedure of the necessary SEA parameters is given. The practical application of the SEA procedure in structures is illustrated using a two-story building as an example.

20.1 Introduction

This chapter describes the basic concepts of the method of statistical energy analysis (SEA) and presents its application to structures. The analysis and computation techniques for vibration response and radiating sound in instruments and structures vary according to the characteristics of the physical object and the frequency range of interest. Here, we analyze vibration and noise in relation to a rather large-scale structure over a wide frequency band. Extensive computations are usually required, when, for example, the finite element method is used for the computations, with respect to a given oscillation mode. In particular, when the computations must be performed in the high-frequency range and when many modes are included in the frequency band, the level of computation becomes considerable, generally resulting in reduced computational accuracy. To supplement the weak point of the traditional approach, it is necessary to redistribute statistically the energy equally from all modes in the analytical frequency band. This allows computed results to be compared with experimental results for a structure across a wide frequency band. This is the SEA method [1]. Early in its development, the objective of this analytical method was to predict the vibration response of artificial satellites and rockets that receive sound excitation when the jet discharges, and to predict the response of vibration stress in the boundary layer noise of an aircraft's airframe. It also became a model that allows an exciting force to be statistically (randomly) diffused (distributed) over a wide frequency band. This technique considers energy of excitation of a diffused (distributed) sound field and its variables that represent the sound pressure, acceleration, and force. Thus, it can be applied to problems of solid-borne sound in which vibration propagates through each element [2] and problems of air-borne sound in which multiple barriers exist [3], even when more excitation points than one are present.

20.2 Power Flow Equations

With the SEA method, we do not deal with specific characteristic modes of the analyzed structure. Instead, we consider the structural components as a set of equivalent vibrating elements, and evaluate the vibration condition of the components as a macroscopic quantity averaged statistically over the frequency band and space (by describing the energy). We assume that the vibration modes within a given frequency band are distributed uniformly and are excited to the same degree.

Using the SEA method, we can formalize the relationships of power flows between subsystems, and by solving these relationships, we can compute the energy stored in each subsystem. Next, the equations of such basic power flow [4] are explained.

20.2.1 Power Flow Equations of a Two-Subsystem Structure

The power flow relationships of a structure consisting of a two-subsystem structure are shown in Figure 20.1. The equations for the power flows between subsystem 1 and subsystem 2 under typical conditions are expressed as

$$\text{Subsystem 1: } P_{i1} = P_{l1} + P_{12} \qquad (20.1)$$

$$\text{Subsystem 2: } P_{i2} = P_{l2} + P_{21} \qquad (20.2)$$

where P_{i1} is the input power to subsystem 1 from outside, P_{l1} is the internal power loss of subsystem 1, and P_{12} is the transmitted power from subsystem 1 to subsystem 2.

The internal power loss, P_{l1}, is written as

$$P_{l1} = \omega\eta_1 E_1 \qquad (20.3)$$

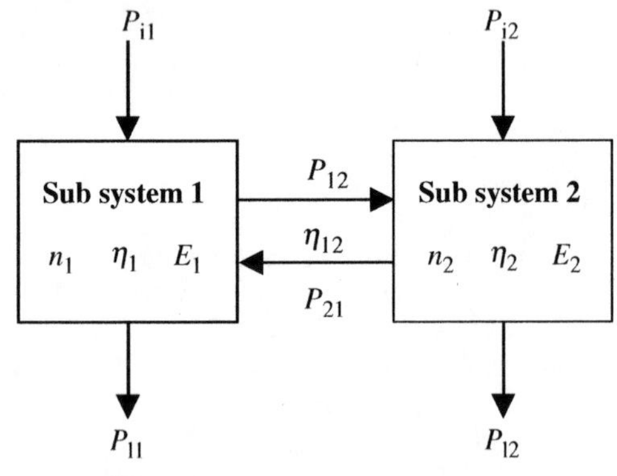

FIGURE 20.1 Power flow relationships between two subsystems.

where ω is the central angular frequency in the band, E_1 is the energy in the bandwidth $\Delta\omega$ of subsystem 1, and η_1 is the internal loss factor (ILF).

The average modal energy E_{m1} in subsystem 1, and E_{m2} in subsystem 2, are given by

$$E_{m1} = \frac{E_1}{N_1}, \quad E_{m2} = \frac{E_2}{N_2} \qquad (20.4)$$

where N_1 is number of modes in the bandwidth $\Delta\omega$ of subsystem 1, and N_2 is number of modes in the bandwidth $\Delta\omega$ of subsystem 2.

The transferred power, P_{12}, between subsystems 1 and 2 is expressed as

$$P_{12} = -P_{21} = P'_{12} - P'_{21} \qquad (20.5)$$

$$P'_{12} = \omega\eta_{12}E_1 = \omega\eta_{12}N_1E_{m1} \qquad (20.6)$$

$$P'_{21} = \omega\eta_{21}E_2 = \omega\eta_{21}N_2E_{m2} \qquad (20.7)$$

where η_{12} and η_{21} are the coupling loss factors (CLFs) from subsystem 1 to subsystem 2, and from subsystem 2 to subsystem 1. They satisfy the reciprocity relationship $\eta_{12}n_1 = \eta_{21}n_2$. Therefore, transferred power, P_{12}, becomes

$$P_{12} = \omega\eta_{12}N_1(E_{m1} - E_{m2}) = \omega\eta_{12}N_1\left(\frac{E_1}{N_1} - \frac{E_2}{N_2}\right) \tag{20.8}$$

Consequently, the power flow equations (Equation 20.1 and Equation 20.2) can be expressed as follows:

$$P_{i1} = \omega\eta_1 E_1 + \omega\eta_{12}N_1\left(\frac{E_1}{N_1} - \frac{E_2}{N_2}\right) \tag{20.9}$$

$$P_{i2} = \omega\eta_2 E_2 + \omega\eta_{21}N_2\left(\frac{E_2}{N_2} - \frac{E_1}{N_1}\right) \tag{20.10}$$

If the SEA parameters (i.e., the modal density, intrinsic loss factor, CLF, and input power) are given, then each subsystem's energy condition can be easily computed.

20.2.2 Power Flow Equations of a Multiple Subsystem Structure

By expanding the formulation in the previous section, it is possible to formalize the power flow relationships of a structure composed of multiple subsystems in the same way. The power flow equation for a structure composed of N subsystems is expressed by the following equation in the matrix form:

$$\omega\begin{bmatrix} \left(\eta_1 + \sum\limits_{i\neq 1}^{N}\eta_{1i}\right)n_1 & -\eta_{12}n_1 & \cdots & -\eta_{1N}n_1 \\ -\eta_{21}n_2 & \left(\eta_2 + \sum\limits_{i\neq 2}^{N}\eta_{2i}\right)n_2 & \cdots & -\eta_{2N}n_2 \\ \vdots & \vdots & \ddots & \vdots \\ -\eta_{N1}n_n & \cdots & \cdots & \left(\eta_N + \sum\limits_{i\neq N}^{N-1}\eta_{Ni}\right)n_N \end{bmatrix} \times \begin{bmatrix} E_1/n_1 \\ E_2/n_2 \\ \vdots \\ E_N/n_N \end{bmatrix} = \begin{bmatrix} P_{i1} \\ P_{i2} \\ \vdots \\ P_{iN} \end{bmatrix} \tag{20.11}$$

From Equation 20.11, if the SEA parameters are given in the same way as for the structure of two subsystems, then the energy equation of each subsystem can be obtained.

The average energy of a subsystem is expressed by the following equations by using the vibration velocity and sound pressure:

$$E = M\langle v^2\rangle \tag{20.12}$$

$$E = \frac{M\langle p^2\rangle}{Z_0^2} \tag{20.13}$$

where M is the mass of the subsystem, $\langle v^2\rangle$ is the average spatial square of the vibration velocity, $\langle p^2\rangle$ is the average spatial square of the sound pressure, and Z_0 is the specific acoustic impedance of air.

Accordingly, if each condition of component's energy is determined from Equation 20.11, it is possible to compute the vibration variable and the sound pressure with Equation 20.12 and Equation 20.13.

20.3 Estimation of SEA Parameters

To solve the power flow equations, it is necessary to determine the SEA parameters (i.e., the modal density, ILF, CLF, and input power). In this subsection, a method is given for computing the SEA parameters.

20.3.1 Modal Density

20.3.1.1 Structural Subsystem

Modal density is a key parameter for determining the dynamic characteristic of a structure. The number of modes, N, included in the frequency band (for estimation), is a factor denoting how easily energy, in transferring between subsystems, can be obtained. To determine N in the prescribed frequency band, it is first necessary to determine the modal density $n(f)$, that is, the gradient of N in the frequency band.

The modal density of a structural subsystem is computed by using the following equation [4,5]:

$$n(f) = \frac{\mathrm{d}N}{\mathrm{d}f} = \frac{1}{f_0} = \frac{A}{2t}\sqrt{\frac{12\rho(1-\nu^2)}{E}} \tag{20.14}$$

where A is the area of cross section, t is the thickness of the structural subsystem, ρ is the mass density, ν is the Poisson's ratio, E is the Young's modulus, and f_0 is the fundamental natural frequency of the structural subsystem.

20.3.1.2 Acoustic Subsystem

The modal density of an acoustic subsystem is determined by the following analytical equation [6]:

$$n(f) = \frac{\mathrm{d}N}{\mathrm{d}f} = \frac{4\pi V}{c^3}f^2 \tag{20.15}$$

where c is the speed of sound propagation within the acoustic subsystem, and V denotes the volume of the acoustic subsystem.

Modal density of the cavity in the low frequency band is deduced in a similar manner to that in the two-dimensional space. Define the depth of the cavity by d, and the frequency of the standing wave in the cavity by $f_d = c/2d$. If $f < f_d$, then the modal density is assumed to be uniformly distributed, and is estimated by

$$n(f) = \frac{2\pi S}{c^2}f \tag{20.16}$$

where S is the area of the cavity.

If $f > f_d$, modal density can be estimated using Equation 20.15, because the cavity is designated as acoustically three dimensional.

20.3.2 Internal Loss Factor

20.3.2.1 Structural Subsystem

The ILF, η_1, of a subsystem gives the loss ratio when the input power to the subsystem from the outside is converted to kinetic energy of the subsystem. An excitation test to measure the damping ratio is employed to estimate the ILF of the structural subsystem. There are several methods for estimating the internal loss factor. The ILF applied in the SEA method is estimated by measuring input energy and output energy simultaneously, or by measuring the attenuation ratio within a given period of time. Both methods require the same setup to conduct an excitation test, and one is able to improve the measurement precision by conducting both methods.

With the energy measuring methods mentioned above, the ILF can be estimated by

$$\eta = \frac{\int_{f_1}^{f_2} \mathrm{Re}(Y)F^2 \mathrm{d}f}{\omega_0 M \left\langle \int_{f_1}^{f_2} v^2 \mathrm{d}f \right\rangle} \tag{20.17}$$

where Y is the complex mobility at the driving point in the range of f_1 to f_2, F^2 is the power spectrum of the input vibration force, and v^2 is the power spectrum of the response speed. In addition, $\langle\ \rangle$ denotes the space average operator.

20.3.2.2 Acoustic Subsystem

The ILF of an acoustic subsystem is determined by [7]

$$\eta = \frac{cS\bar{\alpha}}{4V\omega} \tag{20.18}$$

where $\bar{\alpha}$ is the average acoustic absorption coefficient, V is the volume of the acoustic subsystem, and S denotes the surface area. The acoustic absorption coefficient can be estimated by measuring the reverberation time.

Both the ILF of the cavity in the low-frequency band and the modal density are deduced similar to that in two-dimensional spaces. For $f < f_d$, the ILF is estimated using

$$\eta = \frac{cS_p \alpha_p}{\pi \omega V_c} \tag{20.19}$$

where α_p is the acoustic absorption coefficient in the cavity, S_p is the peripheral area of the cavity, and V_c is the volume of the cavity.

For $f > f_d$, the modal density can be estimated using Equation 20.18 because the cavity is taken as an acoustic subsystem.

20.3.3 Coupling Loss Factor

20.3.3.1 Between Structural Subsystems

The CLF η_{ij} gives the loss ratio when power transmits between two subsystems [4]. For example, the CLF between two flat plates can be estimated using

$$\eta_{ij} = \frac{c_{gi} L_c \tau}{\pi \omega S_i} \tag{20.20}$$

where c_{gi} is the group velocity of the bending waves, L_c is the coupled length, S_i is the surface area, and τ_{ij} is the energy transmission factor from subsystem i to subsystem j. The transmission factor varies with the type of coupling, for example, I-type, L-type, or T-type shown in Figure 20.2. In this section, we use energy transmission efficiency of vertical incidence, reported by Cremer [7].

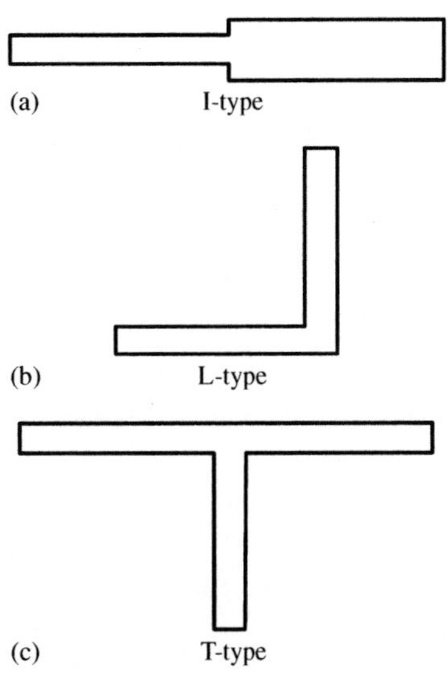

FIGURE 20.2 Coupled type.

20.3.3.2 Between a Structural Subsystem and an Acoustic Subsystem

Coupling power between a structural subsystem and an acoustic subsystem is the power flow based on resonance at transmission. The CLF between a structural subsystem and an acoustic subsystem is given by

$$\eta_{ij} = \frac{Z_0 S_c \sigma}{\omega M_i} \tag{20.21}$$

where Z_0 is the specific acoustic impedance of air, S_c is the surface area of coupling, σ is the acoustical radiation efficiency, and M_i is the mass of the structural subsystem.

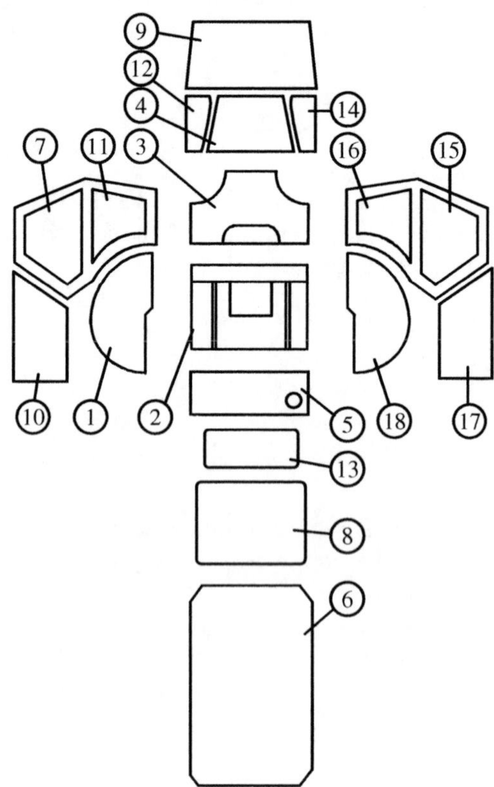

FIGURE 20.3 Modeling of the tractor cabin.

20.3.3.3 Between an Acoustic Subsystem and a Cavity

Coupling power between an acoustic subsystem and a cavity is power flow based on resonance in transmission. The CLF between an acoustic subsystem and a cavity is given by [8]

$$\eta_{sc} = \frac{c_s S_{cs} \tau_m}{4\omega V_s} \tag{20.22}$$

where c_s is the sound velocity in the acoustic space, V_s is the volume in the acoustic space, τ_m is the transmission factor at random incidence depending on mass flow through the partition, and S_{cs} is the coupling area.

20.3.4 Input Power

20.3.4.1 Vibration Input Power

The vibration input power P_{iN} is given by

$$P_{iN} = \omega M_i \langle v_i^2 \rangle \tag{20.23}$$

where M_i is the equivalent mass, $\langle v_i^2 \rangle$ is the spatial average of square of the vibration velocity, and ω is the central angular frequency.

20.3.4.2 Acoustical Input Power

The acoustical input power P_s is given by

$$P_s = \frac{\langle p^2 \rangle S^2 n(f)}{4M} \sigma_{rad} \frac{c^2}{2\pi S f^2} \tag{20.24}$$

where $\langle p^2 \rangle$ is the square average of the input sound pressure, S is the surface area of the component, M is the mass of the component, $n(f)$ is the modal density, and σ_{rad} is the sound radiation factor [2].

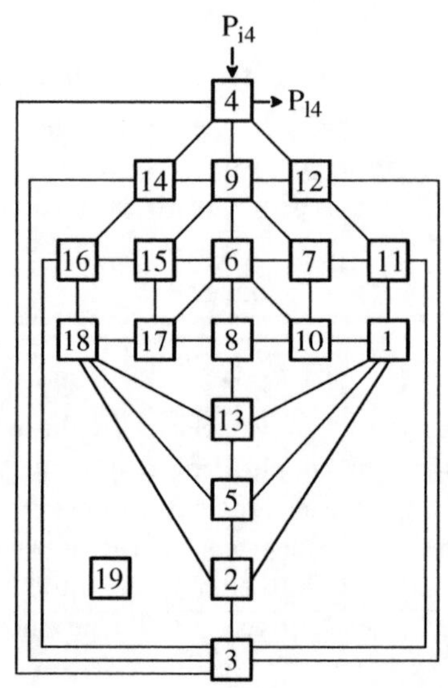

FIGURE 20.4 Power flows in the tractor cabin.

20.4 Application in Structures

In this section, we present an example of modeling with significant analytical accuracy, and discuss the application of the SEA method for structures.

20.4.1 Application for Prediction of Noise in a Tractor Cabin

Figure 20.3 shows a model of the tractor cabin. This figure shows that the cabin consists of a floor, a door, a ceiling, and other components. Figure 20.4 presents the power flow relationships within the cabin [9,10].

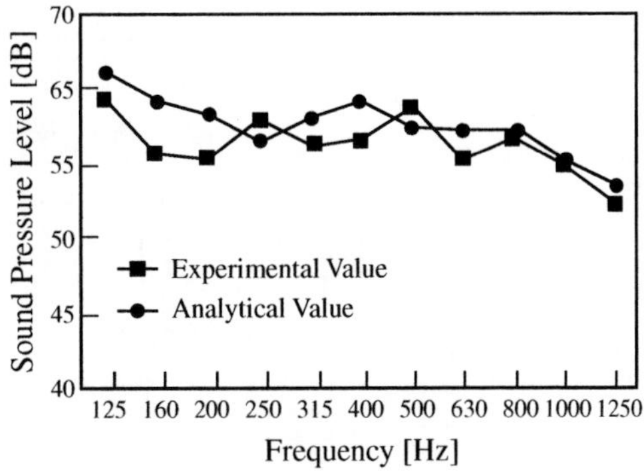

FIGURE 20.5 Results of estimating the sound pressure level in a cabin.

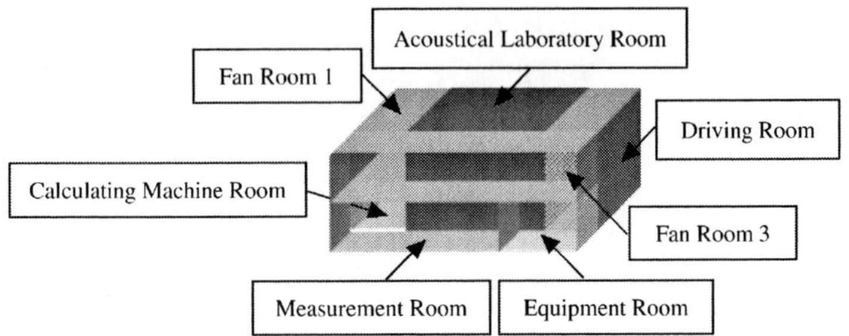

FIGURE 20.6 The configuration of the building.

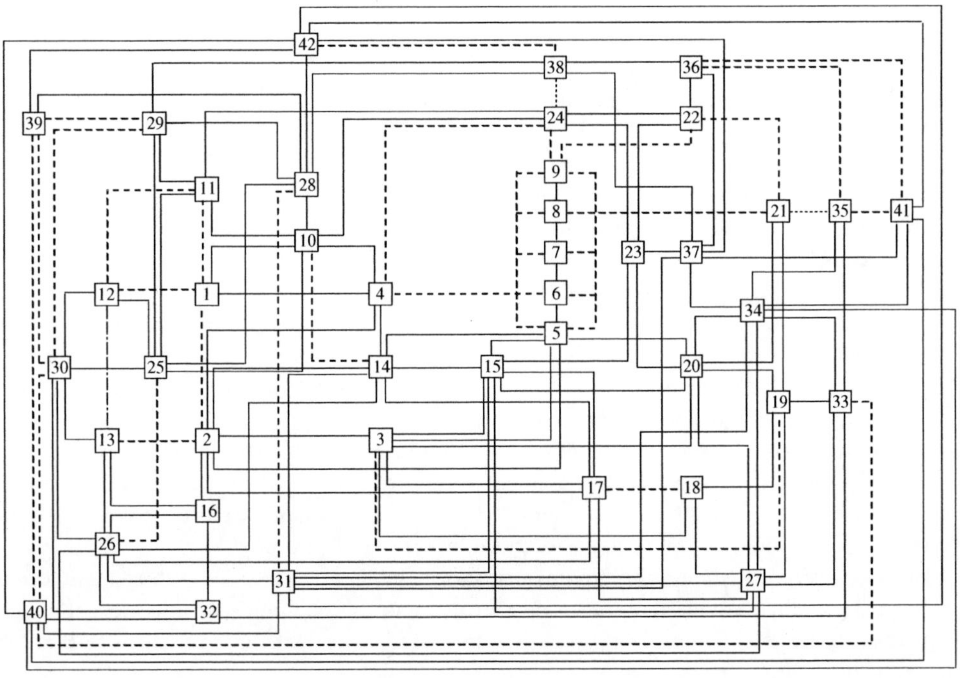

FIGURE 20.7 The power flow relationships between structural subsystems in the entire building.

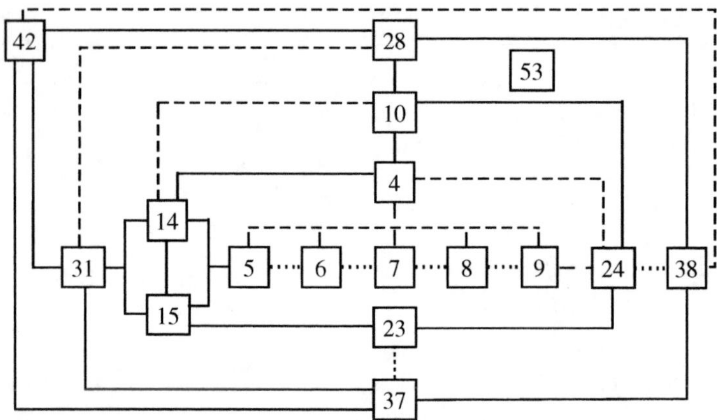

FIGURE 20.8 The power flow relationships in the acoustical laboratory room.

The results obtained for the cabin are shown in Figure 20.5. According to this figure, the disagreement between the computation and the measurement was found about 2 dB in the medium- to high-frequency band.

20.4.2 Application for Prediction of Noise and Vibration in a Building

Consider a two-story laboratory reinforced with concrete [11]. The building configuration is shown in Figure 20.6. This building comprises a driving room, an acoustical laboratory room, a computer room, a measurement room, an equipment room, and others.

We modeled the structural subsystem using I-, L-, or T-type connected points, and the acoustic subsystem as an element shown in Figure 20.6.

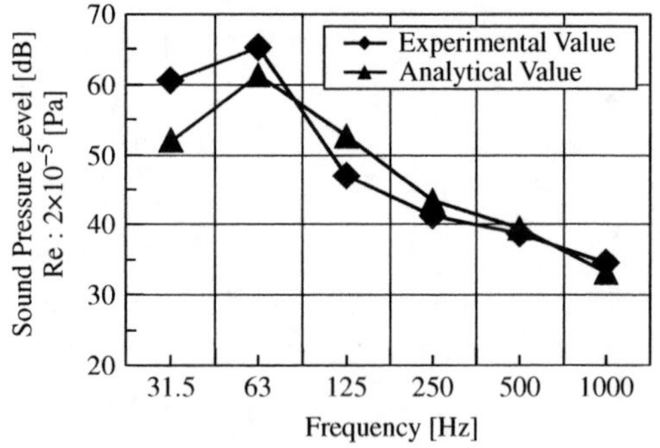

FIGURE 20.9 Estimated sound pressure level results for other rooms.

The SEA model constructed in this manner is composed of 61 elements, and has 244 connecting points. Subsystems 1 to 17 and subsystems 19 to 42 are concrete components. Subsystem 18 and subsystems 43 to 48 are plasterboard components; subsystems 49 to 55 are room components; and subsystems 56 to 61 are cavity components. For example, Figure 20.7 shows the power flow relationships between structural subsystems in the entire building, while Figure 20.8 shows them in the acoustical laboratory room. Here, the thin-dotted, dotted, and solid lines indicate the I-, L-, and T-type combinations, respectively. Subsystem 53 is the room component, and it is connected with all structural components shown in Figure 20.8. The plasterboards located between the computer room and the measurement room are considered as a partition; therefore, connections between subsystem 49 and subsystems 56 to 59 (cavity components), and subsystem 50 and subsystems 60 and 61 (cavity components) are derived from nonresonant modes.

The results obtained for some other rooms are shown in Figure 20.9. Computing accuracy in this building is worse than in the cabin because the structure of this building is complicated, although the differences between the computed values and the measured values were approximately 4 dB in the medium- to high-frequency band.

The computations take approximately 10 sec, so the workload on the personal computer is quite light.

References

1. Lyon, R.H. 1975. *Statistical Energy Analysis of Dynamical Systems: Theory and Applications*, MIT Press, Cambridge, MA.
2. Irie, Y., Solid propagation sound analysis by SEA, *Mitsubishi Heavy Ind. Tech. Rev.*, 21, 4, 571–578, 1984.
3. Crocker, M.J., Sound transmission using statistical energy method, *AIAA J.*, 15, 2, 75–83, 1977.
4. Irie, Y., Solid propagation sound analysis by SEA, *Nihon Onkyo Gakkai-shi*, 48, 6, 433–444, 1992.
5. Lyon, R.H. and Dejong, R.G. 1995. *Theory and Applications of Statistical Energy Analysis*, 2nd ed., Butterworth-Heinemann, Oxford.
6. Craik, R.J.M. 1996. *Sound Transmission Through the Buildings Using Statistical Energy Analysis*, Gower Press, England.
7. Cremer, L., Heckle, M., and Unger, E.E. 1973. *Structure Borne Sound*, Springer, New York, pp. 347–370.
8. *Mechanical Noise Handbook*, JSME, Japan Society of Mechanical Engineers, pp. 179–181, 1991 (in Japanese).
9. Koizumi, T., Tsujiuchi, N., Kubomoto, I., and Ishida, E., Estimation of the noise and vibration response in a tractor cabin using statistical energy analysis, *JSAE Paper*, 28, 4, 49–54, 1997.
10. Koizumi, T., Tujiuchi, N., Kubomoto, I., and Ishida, E., Estimation of the noise and vibration response in a tractor cabin using statistical energy analysis, *SAE Paper*, 1999-01-2821, 1999.
11. Koizumi, T., Tujiuchi, N., Tanaka, H., Okubo, M., and Shinomiya, M., Prediction of the vibration in building using statistical energy analysis, *IMAC Paper*, 7–13, 2002.

Index

A

D

E

F

G

H

J

K

L

M

N

O

P

T

Y

Z

Related Titles

Vibration Simulation Using MATLAB and ANSYS
Michael R. Hatch
ISBN: 1584882050

Vibration: Fundamentals and Practice, Second Edition
Clarence W. de Silva
ISBN: 0849319870